CHILTON®

ASIAN

MECHANICAL SERVICE
Mitsubishi - Toyota
2005 Edition

Australia • Canada • Mexico • Singapore • Spain • United Kingdom • United States

Chilton
Asian Mechanical Service
2005 Edition

Mitsubishi - Toyota

Vice President, Technology and Trades SBU:
Alar Elken

Executive Director, Professional Business Unit:
Gregory L. Clayton

Publisher, Professional Business Unit:
David Koontz

Marketing Director:
Beth A. Lutz

Marketing Specialist:
Brian McGrath

Marketing Coordinator:
Marissa Mariella

Production Director:
Mary Ellen Black

Production Editor:
Elizabeth Hough

Editorial Assistant:
Christine Wade

Publishing Coordinator
Paula Baillie

Editors:
Dennis Bailey
Terry Blomquist
Timothy A. Crain
Matthew Frederick
George Heinrich
Thomas A. Mellon
Richard J. Rivele
Christine L. Sheeky
Jon Wallace

Cover Design
Melinda Possinger

Printed in the United States of America
1 2 3 4 5 6 XX 07 06 05 04

For more information contact
Delmar Learning
Executive Woods
5 Maxwell Drive, PO Box 8007,
Clifton Park, NY 12065-8007
Or find us on the World Wide Web at :
www.trainingbay.com
or, www.chiltonsonline.com

ISBN: 1-4018-6717-0

ISSN: 1548-0887

NOTICE TO THE READER

Table of Contents

Sections

Model Index

EDITORIAL POLICY

Manufacturer and Model Coverage

This manual does not cover every Asian make and model that is currently available on the market. Rather, the Chilton editorial staff makes judicious decisions as to which models warrant coverage, based on which vehicles are serviced by most technicians.

Model Year Information

This manual is published toward the end of the year prior to the edition year. Every effort is made to gather current data from the Original Vehicle Manufacturers (OEMs) when they publish it. Different OEMs choose to release their new model information at different times of the year. Indeed, the same OEM can publish information early one season and late the next season. As a result, not all models are equally current when each edition of this manual is published.

Although information in this manual is based on industry sources and is as complete as possible at the time of publication, some vehicle manufacturers may make changes which cannot be included here. Information on very late models may not be available in some circumstances. While striving for total accuracy, the publisher cannot assume responsibility for any errors, changes, or omissions that may occur in the compilation of this data.

Safety Notice

Proper service and repair procedures are vital to the safe, reliable operation of all motor vehicles, as well as the personal safety of those performing the repairs. This manual outlines procedures for servicing and repairing vehicles using safe, effective methods. The procedures contain many NOTES, WARNINGS and CAUTIONS which should be followed along with standard safety procedures to reduce the possibility of personal injury or improper service which could damage the vehicle or compromise its safety.

Repair procedures, tools, parts, and technician skill and experience vary widely. It is not possible to anticipate all conceivable ways or conditions under which vehicles may be serviced, or to provide cautions for all possible hazards that may result. Standard and accepted safety precautions and equipment should be used when handling toxic or flammable fluids, and safety goggles or other protection should be used during cutting, grinding, chiseling, prying, or any other process that can cause material removal or projectiles.

Some procedures require the use of tools specially designed for a specific purpose. Before substituting another tool or procedure, you must be completely satisfied that neither your personal safety, nor the performance of the vehicle will be endangered.

LOCATING AND USING THE INFORMATION

Organization

To find where a particular model section or procedure is located, look in the Table of Contents. Main topics are listed with the page number on which they may be found. Following the main topics is an alphabetical listing of all of the procedures within the section and their page numbers.

Part Numbers

Part numbers listed in this book are not recommendations by the publisher for any product by brand name. They are references that can be used with interchanges manuals and aftermarket supplier catalogs to locate each brand supplier's discrete part number.

Special Tools

Special tools are recommended by the vehicle manufacturer to perform specific jobs. When necessary, special tools are referred to in the text by the part number of the tool manufacturer. These tools may be purchased, under the appropriate part number, from your local dealer or regional distributor, or an equivalent tool can be purchased locally from a tool supplier or parts outlet. Before substituting any tool for the one recommended, read the previous Safety Notice.

ACKNOWLEDGEMENT

The publisher would like to express appreciation to the following vehicle manufacturers for their assistance in producing this publication. No further reproduction or distribution of the material in this manual is allowed without the expressed written permission of the vehicle manufacturers and the publisher.

American Honda Motor Co., including Acura and Honda Division
Fuji Heavy Industries Ltd., including Subaru
Isuzu Motors Ltd.
Hyundai Group, including Hyundai and Kia Motor
Mazda Motor Corp.
Mitsubishi Motors North America, Inc.
Nissan North America, including Infiniti and Nissan Division
Suzuki Motor Corp.
Toyota Motor Sales USA, including Lexus and Toyota Division

MITSUBISHI

Diamante • Eclipse • Galant • Lancer • Mirage

1

SPECIFICATION AND MAINTENANCE CHARTS

ENGINE AND VEHICLE IDENTIFICATION

Engine							Model Year	
Code ①	Liters (cc)	Cu. In.	Cyl.	Fuel Sys.	Type	Eng. Mfg.	Code ②	Year
4G15/A	1.5 (1468)	87	4	MFI	SOHC	Mitsubishi	1	2001
4G93/C	1.8 (1834)	112	4	MFI	SOHC	Mitsubishi	2	2002
4G63/D	2.0 (1997)	122	4	MFI	DOHC	Mitsubishi	3	2003
4G94/E	2.0 (1999)	122	4	MFI	SOHC	Mitsubishi	4	2004
4G69/F	2.4 (2378)	145	4	MFI	SOHC	Mitsubishi	5	2005
4G64/G	2.4 (2351)	143	4	MFI	SOHC	Mitsubishi		
6G72/H	3.0 (2972)	181	6	MFI	SOHC	Mitsubishi		
6G72/L	3.0 (2972)	181	6	MFI	SOHC	Mitsubishi		
6G74/P	3.5 (3497)	213	6	MFI	SOHC	Mitsubishi		
6G75/S	3.8 (3828)	234	6	MFI	SOHC	Mitsubishi		

MFI: Multiport fuel injection

SOHC: Single overhead cam

DOHC: Double overhead camshafts

① Engine ID / 8th digit of the VIN

② 10th digit of the VIN

67170-GALA-C01

GENERAL ENGINE SPECIFICATIONS

Year	Model	Engine Displacement Liters	Engine ID/VIN	Net Horsepower @ rpm	Net Torque @ rpm (ft. lbs.)	Bore x Stroke (in.)	Compression Ratio	Oil Pressure @ rpm
2001	Mirage	1.5	4G15/A	92@6000	93@3000	2.97x3.23	9.2:1	54@2000
	Mirage	1.8	4G93/C	113@6000	116@4500	3.19x3.50	9.5:1	41@2000
	Eclipse	2.4	4G64/G	①	148@3000	3.41x3.94	9.5:1	41@2000
	Eclipse	3.0	6G72/L	175@5500	185@3000	3.59x2.99	8.9:1	30-80@2000
	Eclipse Spyder	2.4	4G64/G	①	148@3000	3.41x3.94	9.5:1	41@2000
	Eclipse Spyder	3.0	6G72/L	175@5500	185@3000	3.59x2.99	8.9:1	30-80@2000
	Galant	2.4	4G64/G	①	148@3000	3.41x3.94	9.5:1	41@2000
	Galant	3.0	6G72/L	175@5500	185@3000	3.59x2.99	8.9:1	30-80@2000
	Diamante	3.5	6G74/P	214@5000	228@3000	3.66x3.38	9.5:1	30-80@2000
2002	Mirage	1.5	4G15/A	92@6000	93@3000	2.97x3.23	9.2:1	54@2000
	Mirage	1.8	4G93/C	113@6000	116@4500	3.19x3.50	9.5:1	41@2000
	Lancer	2.0	4G94/E	120@5500	130@4250	3.21x3.77	9.5:1	43-100@3500
	Eclipse	2.4	4G64/G	①	148@3000	3.41x3.94	9.5:1	41@2000
	Eclipse	3.0	6G72/L	175@5500	185@3000	3.59x2.99	8.9:1	30-80@2000
	Eclipse Spyder	2.4	4G64/G	①	148@3000	3.41x3.94	9.5:1	41@2000
	Eclipse Spyder	3.0	6G72/L	175@5500	185@3000	3.59x2.99	8.9:1	30-80@2000
	Galant	2.4	4G64/G	①	148@3000	3.41x3.94	9.5:1	41@2000
	Galant	3.0	6G72/L	175@5500	185@3000	3.59x2.99	8.9:1	30-80@2000
	Diamante	3.5	6G74/P	214@5000	228@3000	3.66x3.38	9.5:1	30-80@2000
2003	Lancer	2.0	4G94/E	120@5500	130@4250	3.21x3.77	9.5:1	43-100@3500
	Eclipse	2.4	4G64/G	①	148@3000	3.41x3.94	9.5:1	41@2000
	Eclipse	3.0	6G72/H	175@5500	185@3000	3.59x2.99	8.9:1	30-80@2000
	Eclipse Spyder	2.4	4G64/G	①	148@3000	3.41x3.94	9.5:1	41@2000
	Eclipse Spyder	3.0	6G72/H	175@5500	185@3000	3.59x2.99	8.9:1	30-80@2000
	Galant	2.4	4G64/G	①	148@3000	3.41x3.94	9.5:1	41@2000
	Galant	3.0	6G72/H	175@5500	185@3000	3.59x2.99	8.9:1	30-80@2000
	Diamante	3.5	6G74/P	214@5000	228@3000	3.66x3.38	9.5:1	30-80@2000
2004	Lancer	2.0	4G94/E	120@5500	130@4250	3.21x3.77	9.5:1	43-100@3500
	Lancer	2.4	4G69/F	162@5750	162@4000	3.43x3.94	9.5:1	43-100@3500
	Lancer Evolution	2.0	4G63/D	271@6500	273@3500	3.35x3.46	8.8:1	43-100@3500
	Lancer Sportback	2.4	4G69/F	②	③	3.43x3.94	9.5:1	43-100@3500
	Eclipse	2.4	4G64/G	④	⑤	3.41x3.94	9.0:1	43-100@3500
	Eclipse	3.0	6G72/H	⑥	⑦	3.59x2.99	⑧	43-100@3500
	Eclipse Spyder	2.4	4G64/G	④	⑤	3.41x3.94	9.0:1	43-100@3500
	Eclipse Spyder	3.0	6G72/H	210@5750	205@3750	3.59x2.99	⑧	43-100@3500
	Galant	2.4	4G69/F	160@5500	157@4000	3.43x3.90	9.5:1	43-100@3500
	Galant	3.8	6G75/S	230@5250	250@4000	3.74x3.54	10:01	43-100@3500
	Diamante	3.5	6G74/P	214@5000	228@3000	3.66x3.38	9.5:1	30-80@2000

① California: 138@5500
Except California: 141@5500

② LS: 160@5750
Ralliart: 162@5750

③ LS: 161@4000
Ralliart: 162@4000

④ With M/T: 147@5500
With A/T: 142@5500

⑤ With M/T: 158@4000
With A/T: 155@4000

⑥ GT: 200@5500
GTS: 210@5750

⑦ GT: 205@4000
GTS: 205@3750

⑧ With IMT: 10.0:1
Without IMT: 9.0:1

67170-GALA-C02

ENGINE TUNE-UP SPECIFICATIONS

Year	Engine Displacement Liters	Engine ID/VIN	Spark Plugs Gap (in.)	Ignition Timing (deg.)		Fuel Pump (psi)	Idle Speed (rpm)		Valve Clearance	
				MT	AT		MT	AT	In.	Ex.
2001	1.5	4G15/A	0.039-0.043	2-8B	2-8B	38	600-800	600-800	HYD	HYD
	1.8	4G93/C	0.039-0.043	2-8B	2-8B	38	600-800	600-800	HYD	HYD
	2.4	4G64/G	0.039-0.043	2-8B	2-8B	38	650-850	650-850	HYD	HYD
	3.0	6G72/L	0.039-0.043	5B	5B	38	600-800	600-800	HYD	HYD
	3.5	6G74/P	0.039-0.043	—	2-8B	38	—	600-800	HYD	HYD
2002	1.5	4G15/A	0.039-0.043	2-8B	2-8B	38	600-800	600-800	HYD	HYD
	1.8	4G93/C	0.039-0.043	2-8B	2-8B	38	600-800	600-800	HYD	HYD
	2.0	4G94/E	0.039-0.043	2-8B	2-8B	38	600-800	600-800	HYD	HYD
	2.4	4G64/G	0.039-0.043	2-8B	2-8B	38	650-850	650-850	HYD	HYD
	3.0	6G72/L	0.039-0.043	5B	5B	38	600-800	600-800	HYD	HYD
	3.5	6G74/P	0.039-0.043	—	2-8B	38	—	600-800	HYD	HYD
2003	2.0	4G94/E	0.039-0.043	2-8B	2-8B	38	600-800	600-800	HYD	HYD
	2.4	4G64/G	0.039-0.043	2-8B	2-8B	38	600-800	650-850	HYD	HYD
	3.0	6G72/H	0.028-0.031	2-8B	2-8B	38	600-800	600-800	HYD	HYD
	3.5	6G74/P	0.039-0.043	—	2-8B	38	—	600-800	HYD	HYD
2004	2.0	4G94/E	0.039-0.043	2-8B	2-8B	38	600-800	600-800	HYD	HYD
	2.0	4G63/D	0.024-0.027	2-8B	—	33	800-900	—	HYD	HYD
	2.4	4G69/F	0.028-0.031	2-8B	2-8B	38	600-800	600-800	HYD	HYD
	2.4	4G64/G	0.039-0.043	2-8B	2-8B	38	600-800	650-850	HYD	HYD
	3.0	6G72/H	0.028-0.031	2-8B	2-8B	38	600-800	600-800	HYD	HYD
	3.5	6G74/P	0.039-0.043	—	2-8B	38	—	600-800	HYD	HYD
	3.8	6G75/S	0.028-0.031	—	2-8B	38	—	550-750	HYD	HYD

NOTE: The Vehicle Emission Control Information label often reflects specification changes made during production. The label figures must be used if they differ from those in this chart.

B: Before top dead center

HYD: Hydraulic

67170-GALA-C03

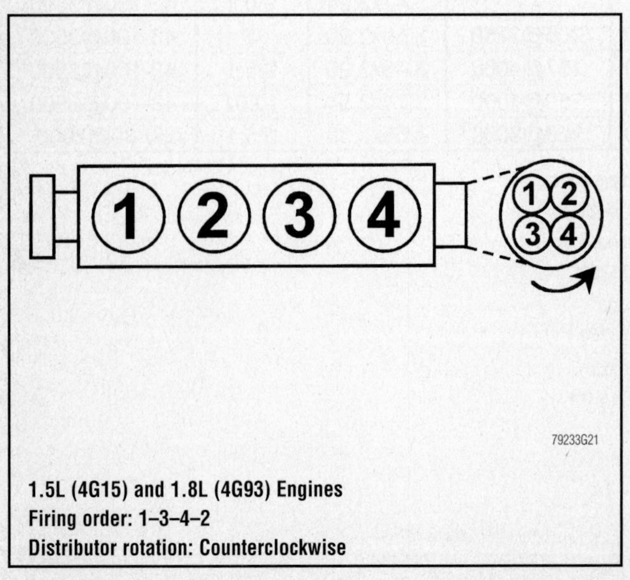

1.5L (4G15) and 1.8L (4G93) Engines
Firing order: 1–3–4–2
Distributor rotation: Counterclockwise

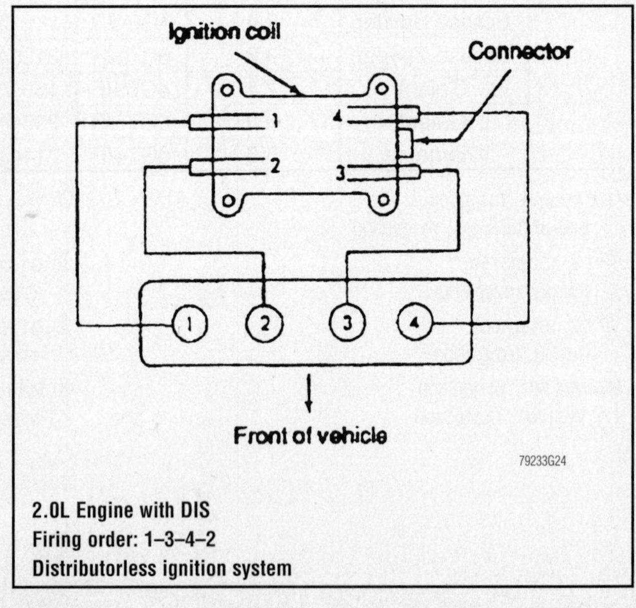

2.0L Engine with DIS
Firing order: 1–3–4–2
Distributorless ignition system

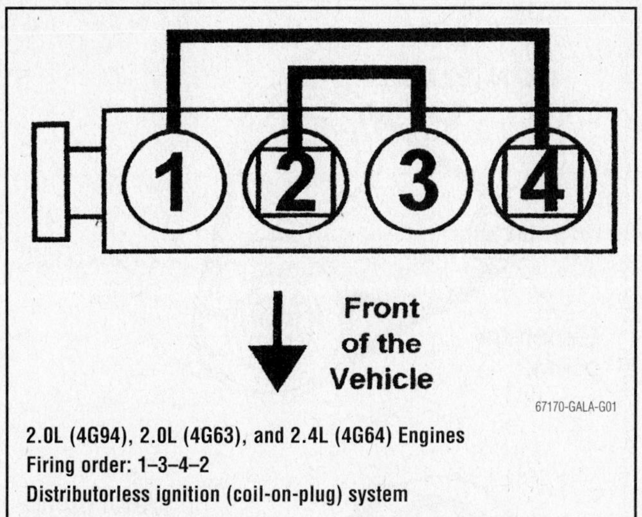

2.0L (4G94), 2.0L (4G63), and 2.4L (4G64) Engines
Firing order: 1–3–4–2
Distributorless ignition (coil-on-plug) system

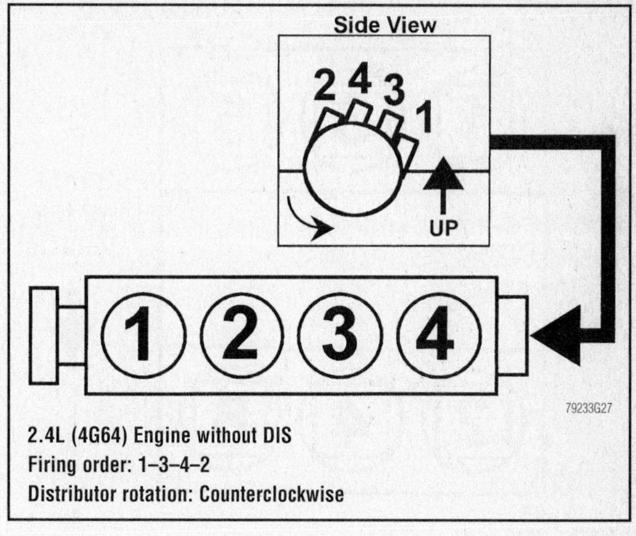

2.4L (4G64) Engine without DIS
Firing order: 1–3–4–2
Distributor rotation: Counterclockwise

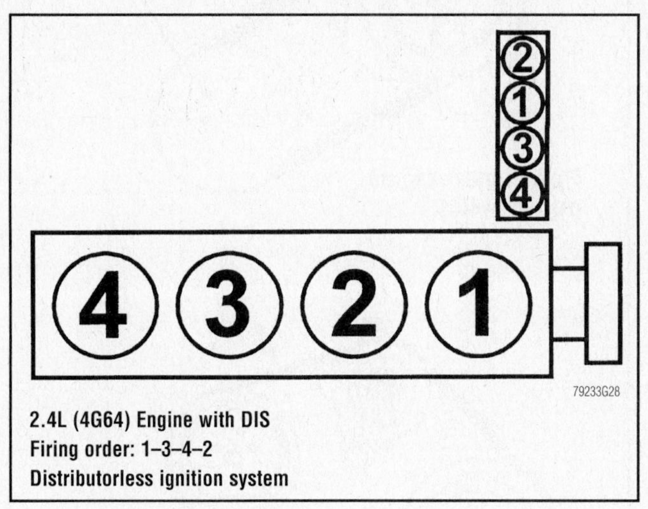

2.4L (4G64) Engine with DIS
Firing order: 1–3–4–2
Distributorless ignition system

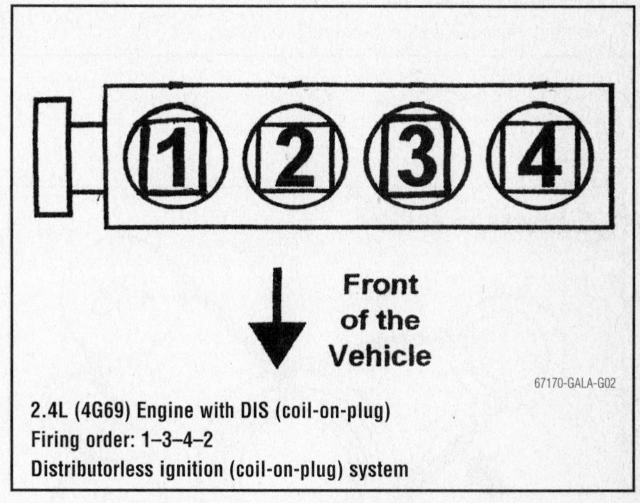

2.4L (4G69) Engine with DIS (coil-on-plug)
Firing order: 1–3–4–2
Distributorless ignition (coil-on-plug) system

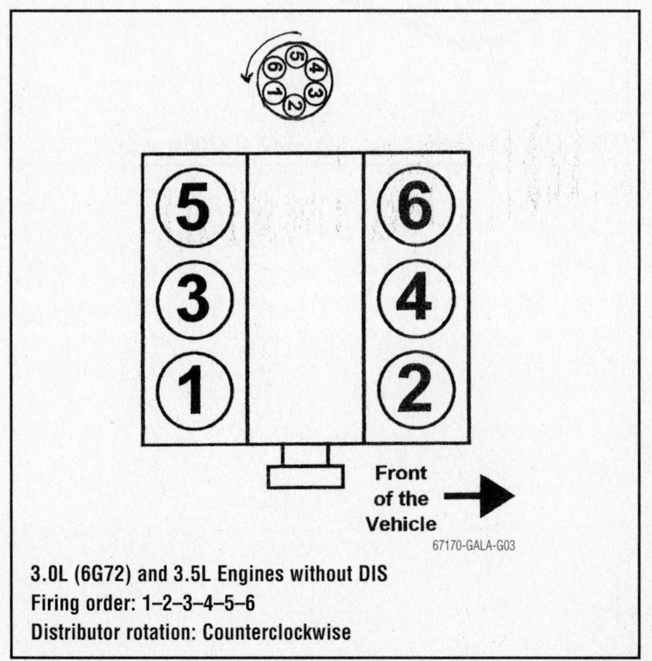

3.0L (6G72) and 3.5L Engines without DIS
Firing order: 1–2–3–4–5–6
Distributor rotation: Counterclockwise

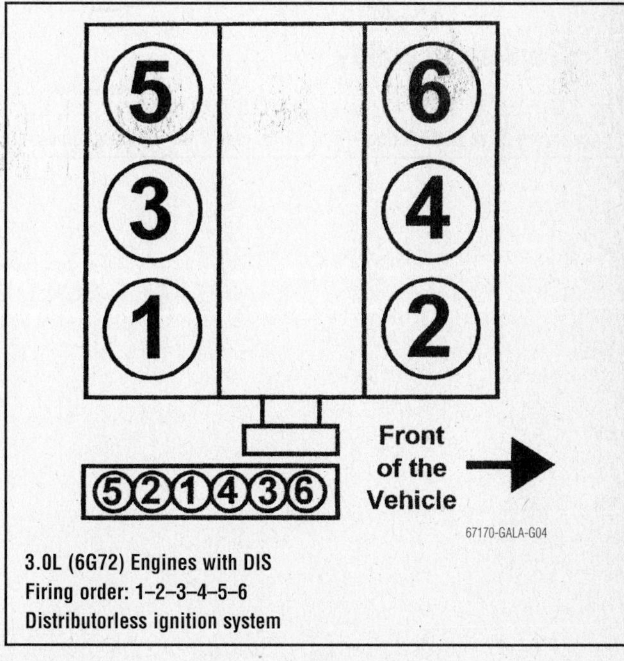

3.0L (6G72) Engines with DIS
Firing order: 1–2–3–4–5–6
Distributorless ignition system

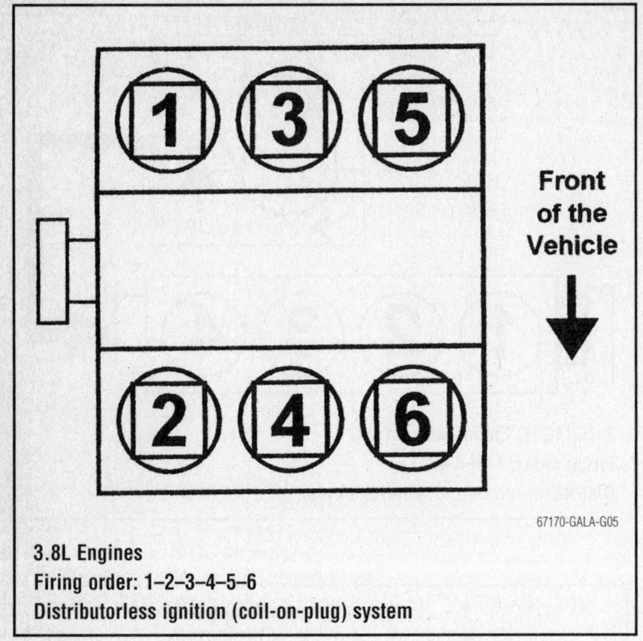

3.8L Engines
Firing order: 1–2–3–4–5–6
Distributorless ignition (coil-on-plug) system

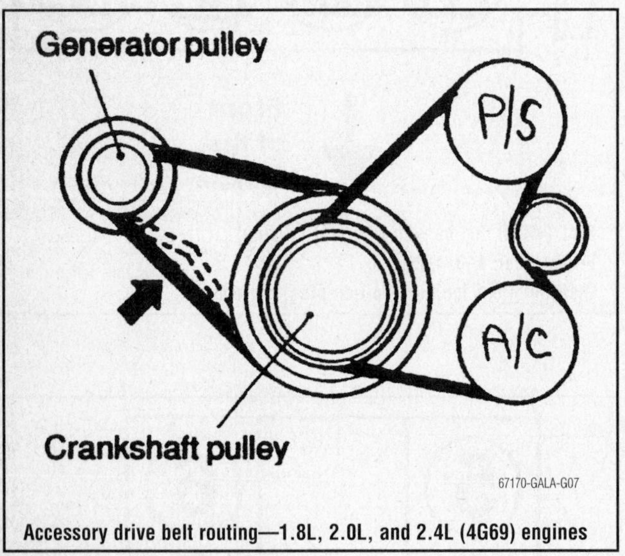

Accessory drive belt routing—1.8L, 2.0L, and 2.4L (4G69) engines

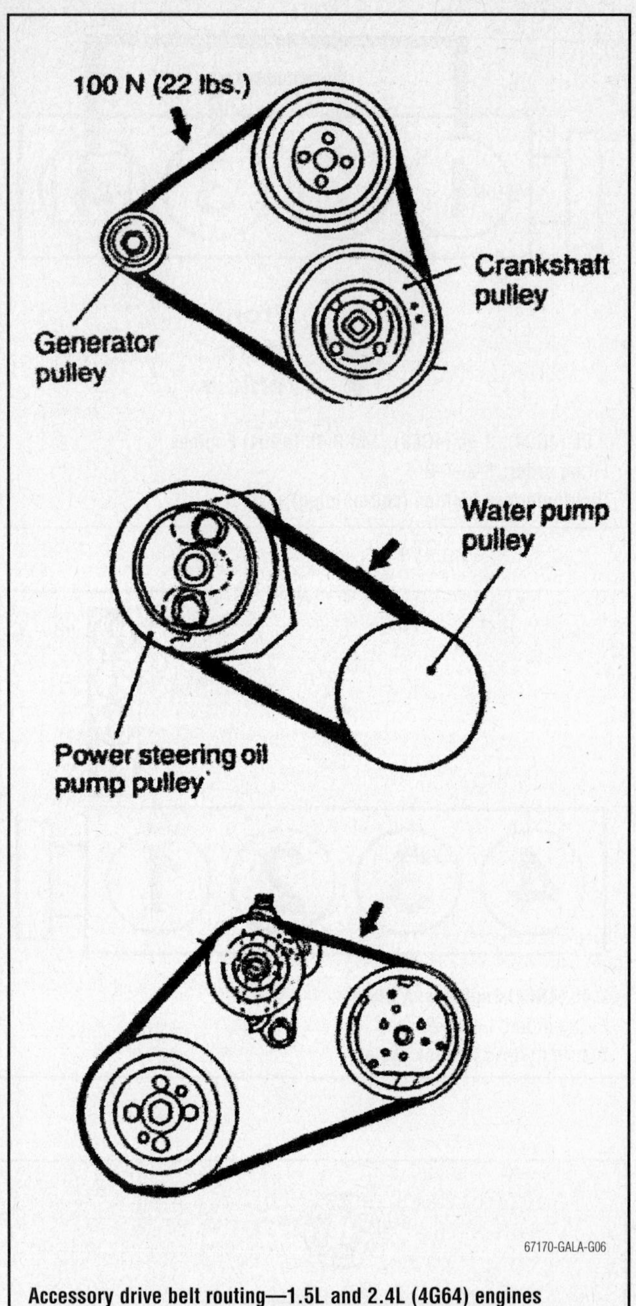

Accessory drive belt routing—1.5L and 2.4L (4G64) engines

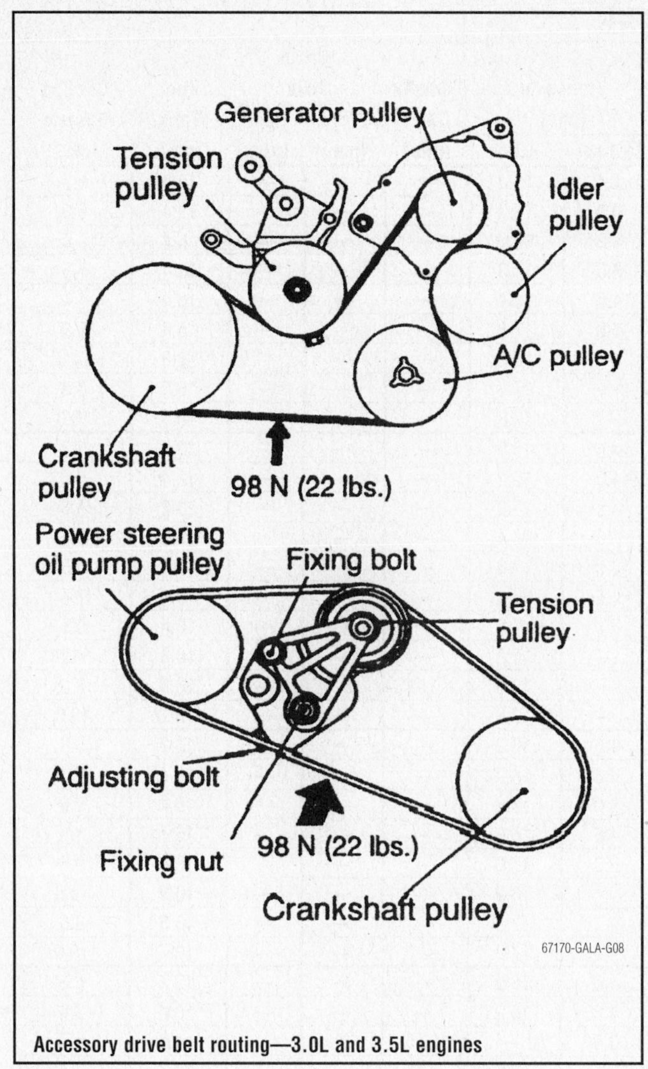

Accessory drive belt routing—3.0L and 3.5L engines

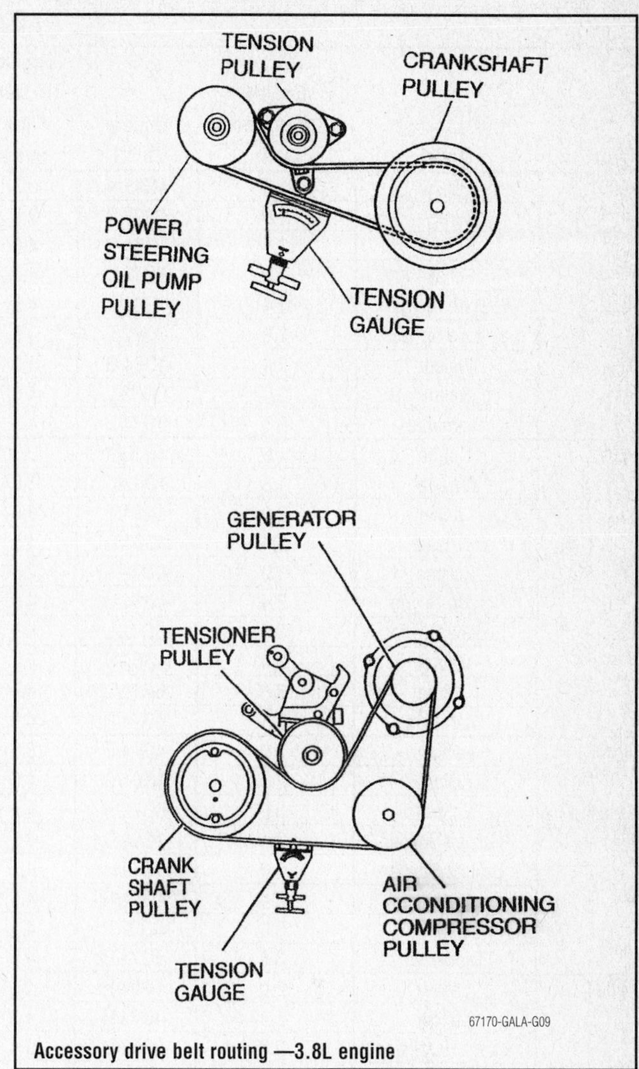

Accessory drive belt routing —3.8L engine

CAPACITIES

Year	Model	Engine Displacement Liters	Engine ID/VIN	Engine Oil with Filter (qts.)	Transaxle (pts.) Manual	Transaxle (pts.) Auto.	Transfer Case (pts.)	Drive Axle Front (pts.)	Drive Axle Rear (pts.)	Fuel Tank (gal.)	Cooling System (qts.)
2001	Mirage	1.5	4G15/A	3.3	4.4	16.2	—	—	—	12.4	5.3
	Mirage	1.8	4G93/C	3.9	4.6	16.2	—	—	—	12.4	6.3
	Eclipse	2.4	4G64/G	4.5	4.6	16.4	—	—	—	16.4	7.4
	Eclipse	3.0	6G72/L	4.5	6.0	17.8	—	—	—	16.4	8.5
	Eclipse Spyder	2.4	4G64/G	4.5	4.6	16.4	—	—	—	16.4	7.4
	Eclipse Spyder	3.0	6G72/L	4.5	6.0	17.8	—	—	—	16.4	8.5
	Galant	2.4	4G64/G	4.5	—	16.2	—	—	—	16.3	14.8
	Galant	3.0	6G72/L	4.5	—	17.8	—	—	—	16.3	8.5
	Diamante	3.5	6G74/P	4.5	—	17.8	—	—	—	18.7	10.0
2002	Mirage	1.5	4G15/A	3.3	4.4	16.2	—	—	—	12.4	5.3
	Mirage	1.8	4G93/C	3.9	4.6	16.2	—	—	—	12.4	6.3
	Lancer	2.0	4G94/E	3.8	4.6	16.2	—	—	—	13.2	5.3
	Eclipse	2.4	4G64/G	4.5	4.6	16.4	—	—	—	16.4	7.4
	Eclipse	3.0	6G72/L	4.5	6.0	17.8	—	—	—	16.4	8.5
	Eclipse Spyder	2.4	4G64/G	4.5	4.6	16.4	—	—	—	16.4	7.4
	Eclipse Spyder	3.0	6G72/L	4.5	6.0	17.8	—	—	—	16.4	8.5
	Galant	2.4	4G64/G	4.5	—	16.2	—	—	—	16.3	14.8
	Galant	3.0	6G72/L	4.5	—	17.8	—	—	—	16.3	8.5
	Diamante	3.5	6G74/P	4.5	—	17.8	—	—	—	18.7	10.0
2003	Lancer	2.0	4G94/E	3.8	4.6	16.2	—	—	—	13.2	5.3
	Eclipse	2.4	4G64/G	4.5	4.6	16.2	—	—	—	16.4	①
	Eclipse	3.0	6G72/H	4.5	6.0	17.6	—	—	—	16.4	9.6
	Eclipse Spyder	2.4	4G64/G	4.5	4.6	16.2	—	—	—	16.4	①
	Eclipse Spyder	3.0	6G72/H	4.5	6.0	17.6	—	—	—	16.4	9.6
	Galant	2.4	4G64/G	4.5	—	16.2	—	—	—	16.3	14.8
	Galant	3.0	6G72/L	4.5	—	17.8	—	—	—	16.3	8.5
	Diamante	3.5	6G74/P	4.5	—	17.8	—	—	—	18.7	10.0
2004	Lancer	2.0	4G94/E	4.0	4.6	16.2	—	—	—	13.2	6.3
	Lancer	2.4	4G69/F	4.5	4.6	16.2	—	—	—	13.2	7.4
	Lancer Evolution	2.0	4G63/D	5.4	6.0	—	1.16	—	1.16	14.0	6.3
	Lancer Sportback	2.4	4G69/F	4.5	—	16.2	—	—	—	13.2	7.4
	Eclipse	2.4	4G64/G	4.5	4.6	16.2	—	—	—	16.4	①
	Eclipse	3.0	6G72/H	4.5	6.0	17.6	—	—	—	16.4	9.6
	Eclipse Spyder	2.4	4G64/G	4.5	4.6	16.2	—	—	—	16.4	①
	Eclipse Spyder	3.0	6G72/H	4.5	6.0	17.6	—	—	—	16.4	9.6
	Galant	2.4	4G69/F	4.5	—	16.2	—	—	—	17.7	8.5
	Galant	3.8	6G75/S	4.8	—	17.8	—	—	—	17.7	9.2
	Diamante	3.5	6G74/P	4.5	—	17.8	—	—	—	18.7	10.0

NOTE: All capacities are approximate. Add fluid gradually and ensure a proper fluid level is obtained.

① M/T: 8.5 qts.
A/T: 8.3 qts.

67170-GALA-C04

VALVE SPECIFICATIONS

Year	Engine Displacement Liters	Engine ID/VIN	Seat Angle (deg.)	Face Angle (deg.)	Spring Test Pressure (lbs. @ in.)	Spring Installed Height (in.)	Stem-to-Guide Clearance (in.)		Stem Diameter (in.)	
							Intake	Exhaust	Intake	Exhaust
2001	1.5	4G15/A	44.5-45	45-45.5	①	1.570	0.0008-0.0020	0.0020-0.0035	0.260	0.260
	1.8	4G93/C	44.5-45	45-45.5	59@1.740	1.740	0.0008-0.0020	0.0020 0.0035	0.234	0.234
	2.4	4G64/G	44.5-45	45-45.5	60@1.740	1.740	0.0008-0.0020	0.0008-0.0028	0.236	0.232
	3.0	6G72/L	44.5-45	45-45.5	40.4@1.591	1.591	0.0012-0.0024	0.0020-0.0035	0.315	0.311
	3.5	6G74/P	44-44.5	45-45.5	60@1.740	1.740	0.0008-0.0020	0.0016-0.0028	0.236	0.236
2002	1.5	4G15/A	44.5-45	45-45.5	①	1.570	0.0008-0.0020	0.0020-0.0035	0.260	0.260
	1.8	4G93/C	44.5-45	45-45.5	59@1.740	1.740	0.0008-0.0020	0.0020 0.0035	0.234	0.234
	2.0	4G94/E	44.5-45	45-45.5	44.2@1.740	1.740	0.0008-0.0020	0.0016-0.0024	0.236	0.236
	2.4	4G64/G	44.5-45	45-45.5	60@1.740	1.740	0.0008-0.0020	0.0008-0.0028	0.236	0.232
	3.0	6G72/L	44.5-45	45-45.5	40.4@1.591	1.591	0.0012-0.0024	0.0020-0.0035	0.315	0.311
	3.5	6G74/P	44-44.5	45-45.5	60@1.740	1.740	0.0008-0.0020	0.0016-0.0028	0.236	0.236
2003	2.0	4G94/E	44.5-45	45-45.5	44.2@1.740	1.740	0.0008-0.0020	0.0016-0.0024	0.236	0.236
	2.4	4G64/G	44.5-45	45-45.5	60@1.740	1.740	0.0008-0.0020	0.0008-0.0028	0.236	0.232
	3.0	6G72/H	44.5-45	45-45.5	40.4@1.591	1.591	0.0012-0.0024	0.0020-0.0035	0.315	0.311
	3.5	6G74/P	44-44.5	45-45.5	60@1.740	1.740	0.0008-0.0020	0.0016-0.0028	0.236	0.236
2004	2.0	4G94/E	44.5-45	45-45.5	44.2@1.740	1.740	0.0008-0.0020	0.0016-0.0024	0.236	0.236
	2.0	4G63/D	44.5-45	45-45.5	63@1.570	1.570	0.0008-0.0020	0.0020 0.0035	0.236	0.236
	2.4	4G69/F	43-43.5	43.5-44	60@1.740	1.740	0.0008-0.0030	0.0016-0.0050	0.240	0.240
	2.4	4G64/G	44.5-45	45-45.5	60@1.740	1.740	0.0008-0.0020	0.0008-0.0028	0.236	0.232
	3.0	6G72/H	44.5-45	45-45.5	40.4@1.591	1.591	0.0012-0.0024	0.0020-0.0035	0.315	0.311
	3.5	6G74/P	44-44.5	45-45.5	60@1.740	1.740	0.0008-0.0020	0.0016-0.0028	0.236	0.236
	3.8	6G75/S	43-43.5	43.5-44	60@1.740	1.740	0.0008-0.0030	0.0016-0.0050	0.240	0.240

① Intake: 51@1.57
Exhaust: 64@1.57

67170-GALA-C05

CRANKSHAFT AND CONNECTING ROD SPECIFICATIONS

All measurements are given in inches.

Year	Engine Displacement Liters	Engine ID/VIN	Crankshaft Main Brg. Journal Dia.	Crankshaft Main Brg. Oil Clearance	Crankshaft Shaft End-play	Crankshaft Thrust on No.	Connecting Rod Journal Diameter	Connecting Rod Oil Clearance	Connecting Rod Side Clearance
2001	1.5	4G15/A	1.8900	0.0008-0.0040	0.0020-0.0120	3	1.6500	0.0008-0.0040	0.0039-0.0160
	1.8	4G93/C	1.9678-1.9685	0.0008-0.0040	0.0020-0.0098	3	1.7709-1.7717	0.0008-0.0040	0.0039-0.0160
	2.4	4G64/G	2.2436-2.2441	0.0008-0.0040	0.0020-0.0098	3	1.7709-1.7717	0.0008-0.0040	0.0039-0.0160
	3.0	6G72/L	2.3614-2.3622	0.0008-0.0040	0.0020-0.0120	3	2.1646-2.1654	0.0008-0.0040	0.0039-0.0160
	3.5	6G74/P	2.3614-2.3622	0.0008-0.0040	0.0020-0.0120	3	1.9700	0.0008-0.0040	0.0039-0.0160
2002	1.5	4G15/A	1.8900	0.0008-0.0040	0.0020-0.0120	3	1.6500	0.0008-0.0040	0.0039-0.0160
	1.8	4G93/C	1.9678-1.9685	0.0008-0.0040	0.0020-0.0098	3	1.7709-1.7717	0.0008-0.0040	0.0039-0.0160
	2.0	4G94/E	1.9700	0.0008-0.0012	0.0020-0.0098	3	NA	0.0008-0.0016	0.0039-0.0098
	2.4	4G64/G	2.2436-2.2441	0.0008-0.0040	0.0020-0.0098	3	1.7709-1.7717	0.0008-0.0040	0.0039-0.0160
	3.0	6G72/L	2.3614-2.3622	0.0008-0.0040	0.0020-0.0120	3	2.1646-2.1654	0.0008-0.0040	0.0039-0.0160
	3.5	6G74/P	2.3614-2.3622	0.0008-0.0040	0.0020-0.0120	3	1.9700	0.0008-0.0040	0.0039-0.0160
2003	2.0	4G94/E	1.9700	0.0008-0.0012	0.0020-0.0098	3	NA	0.0008-0.0016	0.0039-0.0098
	2.4	4G64/G	2.2436-2.2441	0.0008-0.0040	0.0020-0.0098	3	1.7709-1.7717	0.0008-0.0040	0.0039-0.0160
	3.0	6G72/H	2.3614-2.3622	0.0008-0.0040	0.0020-0.0120	3	2.1646-2.1654	0.0008-0.0040	0.0039-0.0160
	3.5	6G74/P	2.3614-2.3622	0.0008-0.0040	0.0020-0.0120	3	1.9700	0.0008-0.0040	0.0039-0.0160
2004	2.0	4G94/E	1.9700	0.0008-0.0012	0.0020-0.0098	3	NA	0.0008-0.0016	0.0039-0.0098
	2.0	4G63/D	2.2400	0.0008-0.0040	0.0020-0.0098	3	1.77	0.0008-0.0016	0.0039-0.0098
	2.4	4G69/F	2.2400	0.0008-0.0030	0.0020-0.009	3	1.77	0.0008-0.0030	0.0040-0.0150
	2.4	4G64/G	2.2436-2.2441	0.0008-0.0040	0.0020-0.0098	3	1.7709-1.7717	0.0008-0.0040	0.0039-0.0160
	3.0	6G72/H	2.3614-2.3622	0.0008-0.0040	0.0020-0.0120	3	2.1646-2.1654	0.0008-0.0040	0.0039-0.0160
	3.5	6G74/P	2.3614-2.3622	0.0008-0.0040	0.0020-0.0120	3	1.9700	0.0008-0.0040	0.0039-0.0160
	3.8	6G75/S	2.52	①	0.0020-0.0100	3	2.1650	0.0008-0.0019	0.0030-0.0090

NA - Not Available

① Number 1 & 4 mains: 0.0008-0.003
Number 2 & 3 mains: 0.0012-0.0016

67170-GALA-C06

PISTON AND RING SPECIFICATIONS

All measurements are given in inches.

Year	Engine Displacement Liters	Engine ID/VIN	Piston Clearance	Ring Gap			Ring Side Clearance		
				Top Compression	Bottom Compression	Oil Control	Top Compression	Bottom Compression	Oil Control
2001	1.5	4G15/A	0.0008-0.0016	0.0079-0.0310	0.0079-0.0310	0.0079-0.0390	0.0012-0.0040	0.0008-0.0040	NA
	1.8	4G93/C	0.0008-0.0016	0.0098-0.0310	0.0157-0.0310	0.0078-0.0390	0.0012-0.0039	0.0008-0.0039	NA
	2.4	4G64/G	0.0008-0.0016	0.0098-0.0310	0.0157-0.0310	0.0039-0.0390	0.0012-0.0040	0.0012-0.0040	NA
	3.0	6G72/L	0.0008-0.0020	0.0118-0.0310	0.0177-0.0310	0.0079-0.0390	0.0012-0.0040	0.0008-0.0040	NA
	3.5	6G74/P	0.0008-0.0020	0.0118-0.0310	0.0177-0.0310	0.0079-0.0390	0.0012-0.0040	0.0008-0.0040	NA
2002	1.5	4G15/A	0.0008-0.0016	0.0079-0.0310	0.0079-0.0310	0.0079-0.0390	0.0012-0.0040	0.0008-0.0040	NA
	1.8	4G93/C	0.0008-0.0016	0.0098-0.0310	0.0157-0.0310	0.0078-0.0390	0.0012-0.0039	0.0008-0.0039	NA
	2.0	4G94/E	0.0008-0.0016	0.0059-0.0118	0.0157-0.0217	0.0039-0.0138	0.0012-0.0028	0.0008-0.0024	NA
	2.4	4G64/G	0.0008-0.0016	0.0098-0.0310	0.0157-0.0310	0.0039-0.0390	0.0012-0.0040	0.0012-0.0040	NA
	3.0	6G72/L	0.0008-0.0020	0.0118-0.0310	0.0177-0.0310	0.0079-0.0390	0.0012-0.0040	0.0008-0.0040	NA
	3.5	6G74/P	0.0008-0.0020	0.0118-0.0310	0.0177-0.0310	0.0079-0.0390	0.0012-0.0040	0.0008-0.0040	NA
2003	2.0	4G94/E	0.0008-0.0016	0.0059-0.0118	0.0157-0.0217	0.0039-0.0138	0.0012-0.0028	0.0008-0.0024	NA
	2.4	4G64/G	0.0008-0.0016	0.0098-0.0310	0.0157-0.0310	0.0039-0.0390	0.0012-0.0040	0.0012-0.0040	NA
	3.0	6G72/H	0.0008-0.0020	0.0118-0.0310	0.0177-0.0310	0.0079-0.0390	0.0012-0.0040	0.0008-0.0040	NA
	3.5	6G74/P	0.0008-0.0020	0.0118-0.0310	0.0177-0.0310	0.0079-0.0390	0.0012-0.0040	0.0008-0.0040	NA
2004	2.0	4G94/E	0.0008-0.0016	0.0059-0.0118	0.0157-0.0217	0.0039-0.0138	0.0012-0.0028	0.0008-0.0024	NA
	2.0	4G63/D	0.0008-0.0016	0.0079-0.0118	0.0128-0.0197	0.0039-0.0157	0.0012-0.0028	0.0008-0.0024	NA
	2.4	4G69/F	0.0008-0.0015	0.0060-0.0300	0.0110-0.0300	0.0040-0.0300	0.0012-0.0030	0.0008-0.0030	NA
	2.4	4G64/G	0.0008-0.0016	0.0098-0.0310	0.0157-0.0310	0.0039-0.0390	0.0012-0.0040	0.0012-0.0040	NA
	3.0	6G72/H	0.0008-0.0020	0.0118-0.0310	0.0177-0.0310	0.0079-0.0390	0.0012-0.0040	0.0008-0.0040	NA
	3.5	6G74/P	0.0008-0.0020	0.0118-0.0310	0.0177-0.0310	0.0079-0.0390	0.0012-0.0040	0.0008-0.0040	NA
	3.8	6G75/S	0.0008-0.0016	0.0100-0.0300	0.0140-0.0300	0.0030-0.0300	0.0012-0.0030	0.0008-0.0030	NA

NA - Not Available

67170-GALA-C07

TORQUE SPECIFICATIONS

All readings in ft. lbs.

Year	Engine Displacement Liters	Engine ID/VIN	Cylinder Head Bolts	Main Bearing Bolts	Rod Bearing Bolts	Crankshaft Damper Bolts	Flywheel Bolts	Manifold Intake	Manifold Exhaust	Spark Plugs	Oil Pan Drain Plug
2001	1.5	4G15/A	①	37	12 ②	93	95	12	12	18	65-80
	1.8	4G93/C	③	18 ②	15 ②	131	71	15	④	18	65-80
	2.4	4G64/G	⑤	14.5 ②	14.5 ②	87	98	13	⑥	18	25-33
	3.0	6G72/L	⑦	57	38	136	54	13	14	18	25-33
	3.5	6G74/P	80	67	38	134	54	16	36	18	29
2002	1.5	4G15/A	①	37	12 ②	93	95	12	12	18	65-80
	1.8	4G93/C	③	18 ②	15 ②	131	71	15	④	18	65-80
	2.0	4G94/E	③	18 ②	15 ②	134	73	14	④	19	66-80
	2.4	4G64/G	⑤	14.5 ②	14.5 ②	87	98	13	⑥	18	25-33
	3.0	6G72/L	⑦	57	38	136	54	13	14	18	25-33
	3.5	6G74/P	80	67	38	134	54	16	36	18	29
2003	2.0	4G94/E	③	18 ②	15 ②	134	73	14	④	19	66-80
	2.4	4G64/G	⑤	14.5 ②	14.5 ②	87	98	13	⑥	18	25-33
	3.0	6G72/H	⑦	57	38	136	54	13	14	18	25-33
	3.5	6G74/P	80	67	38	134	54	16	36	18	29
2004	2.0	4G94/E	③	18 ②	15 ②	134	73	14	④	19	66-80
	2.0	4G63/D	⑩	18 ②	15 ②	17-21	95-101	⑪	⑫	16-22	26-32
	2.4	4G69/F	⑤	18 ②	15 ②	17-21	95-101	13	33-39	16-22	26-32
	2.4	4G64/G	⑤	14.5 ②	14.5 ②	87	98	13	⑥	18	25-33
	3.0	6G72/H	⑦	57	38	136	54	13	14	18	25-33
	3.5	6G74/P	80	67	38	134	54	16	36	18	29
	3.8	6G75/S	⑧	51-57	20 ②	134-140	54	15-17	⑨	14-22	25-33

① Step 1: Tighten all bolts to 35 ft. lbs.
Step 2: Loosen all bolts to 0 ft. lbs.
Step 3: Tighten all bolts to 15 ft. lbs.
Step 4: Tighten all bolts 90 degrees.
Step 5: Tighten all bolts an additional 90 degrees.

② Torque to specification plus an additional 90 degrees.

③ Step 1: Tighten all bolts to 54 ft. lbs.
Step 2: Loosen all bolts to 0 ft. lbs.
Step 3: Tighten all bolts to 15 ft. lbs.
Step 4: Tighten all bolts 90 degrees.
Step 5: Tighten all bolts an additional 90 degrees.

④ 8mm: 13 ft. lbs.
10mm: 21 ft. lbs

⑤ Step 1: Tighten all bolts to 58 ft. lbs.
Step 2: Loosen all bolts to 0 inch lbs.
Step 3: Tighten all bolts to 15 ft. lbs.
Step 4: Tighten all bolts 90 degrees.
Step 5: Tighten all bolts an additional 90 degrees.

⑥ 8mm: 20 ft. lbs.
10mm: 21 ft. lbs.

⑦ Step 1: Tighten all bolts in 3 steps to 80 ft. lbs.
Step 2: Loosen all bolts to 0 ft. lbs.
Step 3: Tighten all bolts in 3 steps to 80 ft. lbs.

⑧ Step 1: Tighten all bolts to 76-84 ft. lbs.
Step 2: Loosen all bolts to 0 inch lbs.
Step 3: Tighten all bolts to 76-84 ft. lbs.

⑨ 8mm: 12-16 ft. lbs.
10mm: 28-38 ft. lbs.
12mm bolt: 49-63 ft. lbs.
Mounting nut: 29-37 ft. lbs.

⑩ Step 1: Tighten all bolts to 58 ft. lbs.
Step 2: Loosen all bolts to 0 ft. lbs.
Step 3: Tighten all bolts to 15 ft. lbs.
Step 4: Tighten all bolts 90 degrees.
Step 5: Tighten all bolts an additional 90 degrees.

⑪ 8mm bolt: 15 ft. lbs.
10mm bolt: 22-30 ft. lbs.
Stay bolt: 21-25 ft. lbs.

⑫ 8mm: 20-28 ft. lbs.
10mm: 35-47 ft. lbs

67170-GALA-C08

WHEEL ALIGNMENT

Year	Model		Caster Range (+/-Deg.)	Caster Preferred Setting (Deg.)	Camber Range (+/-Deg.)	Camber Preferred Setting (Deg.)	Toe-in (in.)
2001	Diamante ①	F	0.50	+3.00	0.50	0	0 +/- 0.13
		R	—	—	0.50	-0.69	0.13 +/- 0.13
	Diamante ②	F	0.50	+3.00	0.50	0	0 +/- 0.13
		R	—	—	0.50	-0.81	0.13 +/- 0.13
	Eclipse ③	F	1.50	+4.69	0.50	-0.09	0 +/- 0.13
		R	—	—	0.50	-1.31	0.13 +/- 0.13
	Eclipse ②	F	1.50	+4.69	0.50	-0.31	0 +/- 0.13
		R	—	—	0.50	-1.69	0.13 +/- 0.13
	Galant	F	0.50	+3.00	0.50	0	0 +/- 0.13
		R	—	—	0.50	-1.00	0 +/- 0.13
	Mirage	F	0.50	+2.84	0.50	0	0 +/- 0.13
		R	—	—	0.50	-0.69	0.13 +/- 0.10
2002	Diamante	F	0.50	+3.00	0.50	0	0 +/- 0.13
		R	—	—	0.50	-0.81	0.13 +/- 0.13
	Eclipse ③	F	1.50	+4.69	0.50	-0.09	0 +/- 0.13
		R	—	—	0.50	-1.31	0.13 +/- 0.13
	Eclipse ②	F	1.50	+4.69	0.50	-0.31	0 +/- 0.13
		R	—	—	0.50	-1.69	0.13 +/- 0.13
	Lancer	F	0.30	+2.50	0.30	0	0.09 +/- 0.13
		R	—	—	0.30	-0.40	0.12 +/- 0.08
	Galant	F	0.50	+3.00	0.50	0	0 +/- 0.13
		R	—	—	0.50	-1.00	0 +/- 0.13
	Mirage	F	0.50	+2.84	0.50	0	0 +/- 0.13
		R	—	—	0.50	-0.69	0.13 +/- 0.10
2003	Diamante	F	0.50	+3.00	0.50	0	0 +/- 0.13
		R	—	—	0.50	-0.81	0.13 +/- 0.13
	Eclipse ③	F	+/- 30	+3.00	+/- 30	0	0 +/- 0.12
		R	—	—	+/- 30	④	0.12 +/- 0.12
	Eclipse ②	F	+/- 30	+3.00	+/- 30	0	0 +/- 0.12
		R	—	—	+/- 30	④	0.13 +/- 0.13
	Lancer	F	+/- 30	+2.50	0.30	0	0.09 +/- 0.13
		R	—	—	0.30	-0.40	0.12 +/- 0.08
	Galant	F	0.50	+3.00	0.50	0	0 +/- 0.13
		R	—	—	0.50	-1.00	0 +/- 0.13

67170-GALA-C09

WHEEL ALIGNMENT

Year	Model		Caster Range (+/-Deg.)	Caster Preferred Setting (Deg.)	Camber Range (+/-Deg.)	Camber Preferred Setting (Deg.)	Toe-in (in.)
2004	Diamante	F	0.50	+3.00	0.50	0	0 +/- 0.13
		R	—	—	0.50	-0.81	0.13 +/- 0.13
	Eclipse ③	F	+/- 30	+3.00	+/- 30	0	0 +/- 0.12
		R	—	—	+/- 30	④	0.12 +/- 0.12
	Eclipse ②	F	+/- 30	+3.00	+/- 30	0	0 +/- 0.12
		R	—	—	+/- 30	④	0.13 +/- 0.13
	Eclipse ⑤	F	+/- 30	+3.00	+/- 30	0	0 +/- 0.12
		R	—	—	+/- 30	④	0.13 +/- 0.13
	Lancer ③	F	+/- 30	+2.55	+/- 30	0	0.04 +/- 0.07
		R	—	—	+/- 30	-0.40	0.12 +/- 0.08
	Lancer ①	F	+/- 30	+2.45	+/- 30	0.05	0.04 +/- 0.07
		R	—	—	+/- 30	-0.40	0.12 +/- 0.08
	Lancer ②	F	+/- 30	+2.55	+/- 30	-0.05	0.04 +/- 0.07
		R	—	—	+/- 30	-0.40	0.12 +/- 0.08
	Lancer Evolution	F	+/- 30	+3.55	+/- 30	⑥	0 +/- 0.08
		R	—	—	+/- 30	-1.00	0.12 +/- 0.07
	Sportback ①	F	+/- 30	+2.35	+/- 30	0.05	0.04 +/- 0.07
		R	—	—	+/- 30	-0.40	0.12 +/- 0.08
	Sportback ②	F	+/- 30	+2.45	+/- 30	-0.05	0.04 +/- 0.07
		R	—	—	+/- 30	-0.40	0.12 +/- 0.08
	Galant	F	+/- 30	+3.00	+/- 30	0	0 +/- 0.12
		R	—	—	+/- 30	-0.50	0.12 +/- 0.12

① With 15 in. wheels

② With 16 in. wheels

③ With 14 in. wheels

④ Eclipse: -1.20
Eclipse Spyder: -1.10

⑤ With 17 in. wheels

⑥ -1.00
Optional: -2.00

67170-GALA-C10

TIRE, WHEEL AND BALL JOINT SPECIFICATIONS

Year	Model	OEM Tires Standard	OEM Tires Optional	Tire Pressures (psi) Front	Tire Pressures (psi) Rear	Wheel Size	Ball Joint Inspection	Wheel Lug Torque (ft. lbs.)
2001	Diamante	205/65HR15	215/60VR16	32	30	6-JJ	87-109 in. ①	65-80
	Galant LS/GT-Z/ES V6	205/55R16	None	32	29	6-JJ	3-13 in. ①	66-80
	Galant DE/ES 4-cyl	185/70HR14	None	29	26	5.5-JJ	3-13 in. ①	66-80
	Mirage DE	P175/70R13	None	31	31	5-J	9-56 in. ①	66-80
	Mirage LS	P185/65R14	None	31	31	5.5-JJ	9-56 in. ①	66-80
	Mirage GS Spyder	P215/50VR17	None	32	30	6.5-JJ	9-56 in. ①	66-80
	Eclipse RS	P195/70HR14	None	32	30	5.5-JJ	3-13 in. ①	66-80
	Eclipse GS	P195/70HR14	None	32	30	5.5-JJ	3-13 in. ①	66-80
	Eclipse GS-T	P205/55HR16	P215/50VR17	32	30	6-JJ	3-13 in. ①	66-80
2002	Diamante	215/60VR16	None	32	30	6-JJ	87-109 in. ①	65-80
	Galant LS/GT-Z/ES V6	205/55R16	None	32	29	6-JJ	3-13 in. ①	66-80
	Galant DE/ES 4-cyl	185/70HR14	None	29	26	5.5-JJ	3-13 in. ①	66-80
	Mirage DE	P175/70R13	None	31	31	5-J	9-56 in. ①	66-80
	Mirage LS	P185/65R14	None	31	31	5.5-JJ	9-56 in. ①	66-80
	Mirage GS Spyder	P215/50VR17	None	32	30	6.5-JJ	9-56 in. ①	66-80
	Lancer ES	P185/65R14	None	32	29	5-JJ	0-35 in. ①	66-80
	Lancer LS/OZ/Rally	P195/60R15	None	32	29	6-JJ	0-35 in. ①	66-80
	Eclipse RS	P195/70HR14	None	32	30	5.5-JJ	3-13 in. ①	66-80
	Eclipse GS	P195/70HR14	None	32	30	5.5-JJ	3-13 in. ①	66-80
	Eclipse GS-T	P205/55HR16	P215/50VR17	32	30	6-JJ	3-13 in. ①	66-80
2003	Diamante	215/60VR16	None	32	30	6-JJ	87-109 in. ①	65-80
	Galant LS/GT-Z/ES V6	205/55R16	None	32	29	6-JJ	3-13 in. ①	66-80
	Galant DE/ES 4-cyl	185/70HR14	None	29	26	5.5-JJ	3-13 in. ①	66-80
	Lancer ES	P185/65R14	None	32	29	5-JJ	0-35 in. ①	66-80
	Lancer LS/OZ/Rally	P195/60R15	None	32	29	6-JJ	0-35 in. ①	66-80
	Eclipse RS	P195/70HR14	None	32	29	5.5-JJ	3-13 in. ①	66-80
	Eclipse GS	P195/70HR14	None	32	29	5.5-JJ	3-13 in. ①	66-80
	Eclipse GS-T	P205/55HR16	P215/50VR17	32	29	6-JJ	3-13 in. ①	66-80

67170-GALA-C11

TIRE, WHEEL AND BALL JOINT SPECIFICATIONS

Year	Model	OEM Tires		Tire Pressures (psi)		Wheel	Ball Joint	Wheel Lug Torque
		Standard	Optional	Front	Rear	Size	Inspection	(ft. lbs.)
2004	Diamante	215/60VR16	None	32	30	6-JJ	87-109 in. ①	65-80
	Galant GTS	P215/55R17	None	32	32	7-JJ	31-61 ①	66-80
	Galant DE/ES/LS	P215/60R16	None	32	32	6.5-JJ	31-61 ①	66-80
	Lancer ES	P185/65R14	None	29	26	5.5-JJ	0-35 in. ①	66-80
	Lancer LS/OZ Rally	P195/60R15	None	29	26	6-JJ	0-35 in. ①	66-80
	Lancer Ralliart	P205/50R16	None	32	32	6-JJ	0-35 in. ①	66-80
	Lancer Evolution	P235/45R17	None	32	29	8-JJ	4.4-30 in. ①	66-80
	Lancer Sportback LS	P195/60R15	None	29	26	6-JJ	0-35 in. ①	66-80
	Lancer Sportback Ralliart	P205/50R16	None	32	32	6-JJ	0-35 in. ①	66-80
	Eclipse RS	P195/65HR15	None	32	29	6-JJ	3-13 in. ①	66-80
	Eclipse GS	P205/55HR16	None	32	29	6-JJ	3-13 in. ①	66-80
	Eclipse GT	P215/50VR17	None	32	29	6.5-JJ	3-13 in. ①	66-80
	Eclipse GTS	P215/50VR17	None	32	29	6.5-JJ	3-13 in. ①	66-80

OEM: Original Equipment Manufacturer

PSI: Pounds Per Square Inch

① Torque required in inch lbs. to rotate ball joint when removed from the knuckle.

67170-GALA-C12

BRAKE SPECIFICATIONS

All measurements in inches unless noted

			Brake Disc			Brake Drum Diameter			Minimum Lining Thickness		Brake Caliper	
Year	Model		Original Thickness	Minimum Thickness	Maximum Runout	Original Inside Diameter	Max. Wear Limit	Maximum Machine Diameter	Front	Rear	Bracket Bolts (ft. lbs.)	Mounting Bolts (ft. lbs.)
2001	Diamante	F	0.940	0.880	0.002	—	—	—	0.080	—	54	65
		R	0.410	0.330	0.0023	—	—	—	—	0.039	24	36-43
	Galant	F	①	②	0.002	—	—	—	0.080	—	③	65
		R	0.390	0.331	0.002	9.000	—	9.100	—	0.040	36-43	54
	Mirage	F	0.940	0.882	0.0012	—	—	—	0.080	—	36	67-81
		R	—	—	—	8.00	8.10	8.10	—	0.039	—	—
	Eclipse	F	0.940	0.882	0.002	—	—	—	0.080	—	58-72	46-62
		R	—	—	—	9.000	—	9.100	—	0.039	—	—
	Eclipse	F	0.940	0.882	0.003	—	—	—	0.080	—	③	46-62
	w/rear disc	R	0.390	0.331	0.003	—	—	—	—	0.080	36-43	54
2002	Diamante	F	0.940	0.880	0.002	—	—	—	0.080	—	54	65
		R	0.410	0.330	0.0023	—	—	—	—	0.039	24	36-43
	Galant	F	①	②	0.002	—	—	—	0.080	—	③	65
		R	0.390	0.331	0.002	9.000	—	9.100	—	0.040	36-43	54
	Mirage	F	0.940	0.882	0.0012	—	—	—	0.080	—	36	67-81
		R	—	—	—	8.00	8.10	8.10	—	0.039	—	—
	Lancer	F	0.900	0.880	0.002	—	—	—	0.080	—	37	62
		R	—	—	—	7.99	8.07	8.07	—	0.040	—	—
	Eclipse	F	0.940	0.882	0.002	—	—	—	0.080	—	58-72	46-62
		R	—	—	—	9.000	—	9.100	—	0.039	—	—
	Eclipse	F	0.940	0.882	0.003	—	—	—	0.080	—	③	46-62
	w/rear disc	R	0.390	0.331	0.003	—	—	—	—	0.080	36-43	54
2003	Diamante	F	0.940	0.880	0.002	—	—	—	0.080	—	54	65
		R	0.410	0.330	0.0023	—	—	—	—	0.039	24	36-43
	Galant	F	①	②	0.002	—	—	—	0.080	—	③	65
		R	0.390	0.331	0.002	9.000	—	9.100	—	0.040	36-43	54
	Lancer	F	0.900	0.880	0.002	—	—	—	0.080	—	37	62
		R	—	—	—	7.99	8.07	8.07	—	0.040	—	—
	Eclipse	F	0.940	0.882	0.002	—	—	—	0.080	—	58-72	46-62
		R	—	—	—	9.000	—	9.100	—	0.039	—	—
	Eclipse	F	0.940	0.882	0.003	—	—	—	0.080	—	③	46-62
	w/rear disc	R	0.390	0.331	0.003	—	—	—	—	0.080	36-43	54

67170-GALA-C13

BRAKE SPECIFICATIONS

All measurements in inches unless noted

			Brake Disc			Brake Drum Diameter			Minimum Lining Thickness		Brake Caliper	
Year	Model		Original Thickness	Minimum Thickness	Maximum Runout	Original Inside Diameter	Max. Wear Limit	Maximum Machine Diameter	Front	Rear	Bracket Bolts (ft. lbs.)	Mounting Bolts (ft. lbs.)
2004	Diamante	F	0.940	0.880	0.002	—	—	—	0.080	—	54	65
		R	0.410	0.330	0.0023	—	—	—	—	0.039	24	36-43
	Galant	F	1.020	0.960	0.0039	—	—	—	0.080	—	25-31	77-81
		R	0.390	0.330	0.0016	—	—	—	—	0.080	28-36	42-48
	Lancer	F	1.020	0.960	0.0006	—	—	—	0.080	—	④	67-81
		R	—	—	—	7.99	8.07	8.07	—	0.040	—	—
	Lancer w/rear disc	F	1.020	0.960	0.0006	—	—	—	0.080	—	④	67-81
		R	0.390	0.330	0.0015	—	—	—	—	0.080	28-36	42-48
	Lancer Evolution	F	1.260	1.170	0.0012	—	—	—	0.080	—	—	73-87
		R	0.870	0.80	0.0012	—	—	—	—	0.080	—	36-44
	Lancer Sportback	F	1.020	0.960	0.0015	—	—	—	0.080	—	25-31	67-81
		R	0.390	0.330	0.0015	—	—	—	—	0.080	28-36	42-48
	Eclipse	F	0.940	0.882	0.002	—	—	—	0.080	—	⑤	67-81
		R	—	—	—	9.000	—	9.100	—	0.039	—	—
	Eclipse w/rear disc	F	0.940	0.882	0.003	—	—	—	0.080	—	⑤	67-81
		R	0.390	0.331	0.003	—	—	—	—	0.080	36-43	41-44

F: Front

R: Rear

① 2.4L: 0.940
3.0L: 1.020

② 2.4L: 0.88
3.0L: 0.96

③ Lock pin (2.4L): 55 ft. lbs.
Lock bolt (3.0L): 28 ft. lbs.

④ Main slide pin (2.0L): 59-65 ft. lbs.
Sub slide pin (2.0L): 34-40 ft. lbs.
Lock pin bolt (2.4L): 25-31 ft. lbs.

⑤ Lock pin (2.4L): 46-62 ft. lbs.
Lock bolt (3.0L): 25-31 ft. lbs.

67170-GALA-C14

SCHEDULED MAINTENANCE INTERVALS

Mitsubishi—Diamante, Eclipse, Galant, Mirage, Lancer & Sportback

		VEHICLE MILEAGE INTERVAL (x1000)													
TO BE SERVICED	**TYPE OF SERVICE**	**7.5**	**15**	**22.5**	**30**	**37.5**	**45**	**52.5**	**60**	**67.5**	**75**	**82.5**	**90**	**97.5**	**105**
Engine oil & filter	R	✓	✓	✓	✓	✓	✓	✓	✓	✓	✓	✓	✓	✓	✓
Automatic transaxle fluid & filter	S/I		✓		✓		✓		✓		✓		✓		
Brake hoses	S/I		✓		✓		✓		✓		✓		✓		
Disc brake pads	S/I		✓		✓		✓		✓		✓		✓		
Driveshaft boots	S/I		✓		✓		✓		✓		✓		✓		
Valve clearance (Mirage)	S/I		✓		✓		✓		✓		✓		✓		
Air cleaner element	R				✓				✓				✓		
Engine coolant	R				✓				✓				✓		
Spark plugs (standard type)	R				✓				✓				✓		
Spark plugs (iridium type)	R														✓
Spark plugs (platinum type)	R								✓						
Ball joints & steering linkage seals	S/I				✓				✓				✓		
Drive belt(s)	S/I				✓				✓				✓		
Exhaust system	S/I				✓				✓				✓		
Fuel hoses	S/I				✓				✓				✓		
Manual transaxle fluid	S/I				✓				✓				✓		
Manual transaxle fluid (including transfer) (2004 Lancer Evolution)	S/I				✓				✓				✓		
Rear axle oil (2004 Lancer Evolution AWD)	S/I				✓				✓				✓		
Rear drum brake linings & rear wheel cylinders (Eclipse, Galant, Lancer & Mirage)	S/I				✓				✓				✓		
Ignition cables	R								✓						
Timing belt(s)	R								✓						
Distributor cap & rotor	S/I								✓						

67170-GALA-C15

SCHEDULED MAINTENANCE INTERVALS
Mitsubishi—Diamante, Eclipse, Galant, Mirage, Lancer & Sportback

		VEHICLE MILEAGE INTERVAL (x1000)													
TO BE SERVICED	TYPE OF SERVICE	7.5	15	22.5	30	37.5	45	52.5	60	67.5	75	82.5	90	97.5	105
EVAP system (except canister)	S/I								✓						
Fuel system (tank, pipe line, connection & fuel tank filler tube cap)	S/I								✓						

R: Replace S/I: Service or Inspect

FREQUENT OPERATION MAINTENANCE (SEVERE SERVICE)

If a vehicle is operated under any of the following conditions it is considered severe service:

- Extremely dusty or sandy areas.

- Extensive use of brakes while driving.

- 50% or more of the vehicle operation is in 32°C (90°F) or higher temperatures, or constant operation in temperatures below 0°C (32°F).

- Prolonged idling (vehicle operation in stop and go traffic).

- Frequent short running periods (engine does not warm to normal operating temperatures).

- Police, taxi, delivery usage or trailer towing usage.

Oil & oil filter: change every 3000 miles.

Disc brake pads: service or inspect ever 6000 miles.

Air filter element: service or inspect every 15,000 miles.

Automatic transaxle fluid & filter: replace every 15,000 miles.

Spark plugs: replace every 15,000 miles.

Manual transaxle oil (including transfer (Lancer Evolution): replace every 30,000 miles.

67170-GALA-C16

PRECAUTIONS

Before servicing any vehicle, please be sure to read all of the following precautions, which deal with personal safety, prevention of component damage, and important points to take into consideration when servicing a motor vehicle:

- Never open, service or drain the radiator or cooling system when the engine is hot; serious burns can occur from the steam and hot coolant.
- Observe all applicable safety precautions when working around fuel. Whenever servicing the fuel system, always work in a well-ventilated area. Do not allow fuel spray or vapors to come in contact with a spark, open flame, or excessive heat (a hot drop light, for example). Keep a dry chemical fire extinguisher near the work area. Always keep fuel in a container specifically designed for fuel storage; also, always properly seal fuel containers to avoid the possibility of fire or explosion. Refer to the additional fuel system precautions in this section.
- Fuel injection systems often remain pressurized, even after the engine has been turned **OFF**. The fuel system pressure must be relieved before disconnecting any fuel lines. Failure to do so may result in fire and/or personal injury.
- Brake fluid often contains polyglycol ethers and polyglycols. Avoid contact with the eyes and wash your hands thoroughly after handling brake fluid. If you do get brake fluid in your eyes, flush your eyes with clean, running water for 15 minutes. If eye irritation persists, or if you have taken brake fluid internally, IMMEDIATELY seek medical assistance.
- The EPA warns that prolonged contact with used engine oil may cause a number of skin disorders, including cancer. You should make every effort to minimize your exposure to used engine oil. Protective gloves should be worn when changing oil. Wash your hands and any other exposed skin areas as soon as possible after exposure to used engine oil. Soap and water, or waterless hand cleaner should be used.
- All new vehicles are now equipped with an air bag system. The system must be disabled before performing service on or around system components, steering column, instrument panel components, wiring and sensors. Failure to follow safety and disabling procedures could result in accidental air bag deployment, possible personal injury, and unnecessary system repairs.
- Always wear safety goggles when working with, or around, the air bag system. When carrying a non-deployed air bag, be sure the bag and trim cover are pointed away from your body. When placing a non-deployed air bag on a work surface, always face the bag and trim cover upward, away from the surface. This will reduce the motion of the module if it is accidentally deployed. Refer to the additional air bag system precautions later in this section.
- Clean, high quality brake fluid from a sealed container is essential to the safe and proper operation of the brake system. You should always buy the correct type of brake fluid for your vehicle. If the brake fluid becomes contaminated, completely flush the system with new fluid. Never reuse any brake fluid. Any brake fluid that is removed from the system should be discarded. Also, do not allow any brake fluid to come in contact with a painted surface; it will damage the paint.
- Never operate the engine without the proper amount and type of engine oil; doing so will result in severe engine damage.
- Timing belt maintenance is extremely important. Many models utilize an interference-type, non-freewheeling engine. If the timing belt breaks, the valves in the cylinder head may strike the pistons, causing potentially serious (also time-consuming and expensive) engine damage. Refer to the maintenance interval charts in the front of this section for the recommended replacement interval for the timing belt, and to the timing belt procedure for belt replacement and inspection.
- Disconnecting the negative battery cable on some vehicles may interfere with the functions of the on-board computer system(s) and may require the computer to undergo a relearning process once the negative battery cable is reconnected.
- When servicing drum brakes, only disassemble and assemble one side at a time, leaving the remaining side intact for reference.

ENGINE REPAIR

➡Disconnecting the negative battery cable on some vehicles may interfere with the functions of the on board computer systems and may require the computer to undergo a relearning process, once the negative battery cable is reconnected.

Distributor

REMOVAL

Before removing the distributor, position No. 1 cylinder at TDC on the compression stroke and align the timing marks.

1. Before servicing the vehicle, refer to the precautions at the beginning of this section.
2. Remove or disconnect the following:
   - Negative battery cable
   - Ignition wire cover, if equipped
   - Distributor harness connector
   - Distributor cap with all ignition wires still connected
   - Coil wire, if necessary
3. Matchmark the rotor to the distributor housing, and the distributor housing to the engine.
4. Remove or disconnect the following:
   - Hold-down nut
   - Distributor from the engine

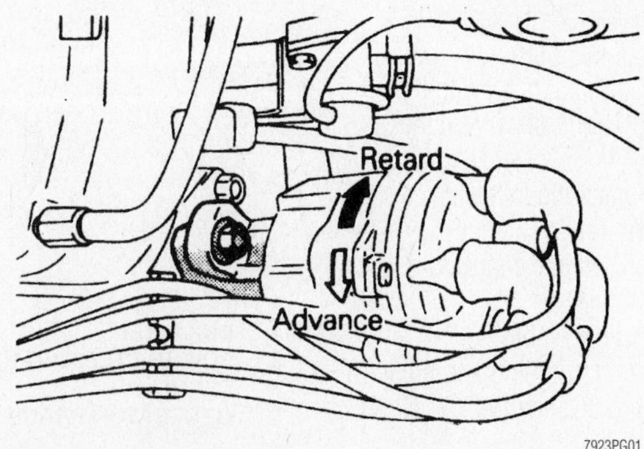

Adjusting the distributor—1.5L Mirage shown

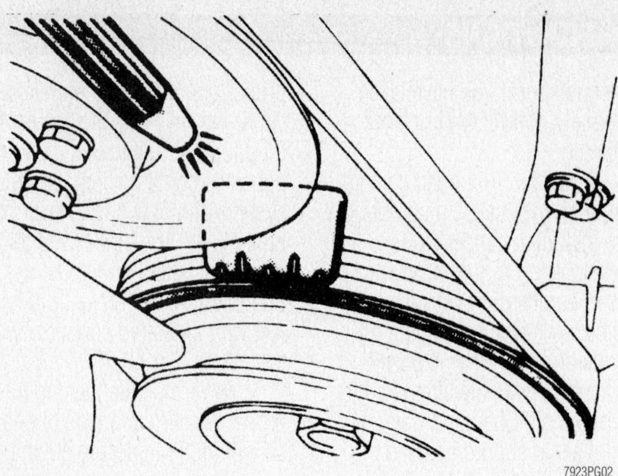

Checking the ignition timing—1.5L Mirage shown

INSTALLATION

Timing Not Disturbed

1. Install or connect the following:
   - New distributor housing O-ring and lubricate with clean oil
   - Distributor in the engine, matchmarks aligned
   - Hold-down nut
   - Distributor harness connectors
   - Distributor cap
   - Coil wire, if removed
   - Negative battery cable
2. Adjust the ignition timing and tighten the hold-down nut to 96 inch lbs. (11 Nm).

Timing Disturbed

1. Install a new distributor housing O-ring and lubricate with clean oil.
2. Position the engine so the No. 1 piston is at TDC of its compression stroke and the mark on the vibration damper is aligned with **0** on the timing indicator.
3. Align the distributor housing and gear mating marks. Install the distributor in the engine so the slot or groove of the distributor's installation flange aligns with the distributor installation stud in the engine block. Be sure the distributor is fully seated. Inspect alignment of the distributor rotor making sure the rotor is aligned with the position of the No. 1 ignition wire in the distributor cap.
4. Install or connect the following:
   - Hold-down nut
   - Distributor harness connectors
   - Distributor cap
   - Negative battery cable
5. Adjust the ignition timing and tighten the hold-down nut to 96 inch lbs. (11 Nm).

Alternator

REMOVAL

1.5L, 1.8L, and 2.4L Engines

1. Before servicing the vehicle, refer to the precautions at the beginning of this section.
2. Remove or disconnect the following:
   - Negative battery cable
   - Left side cover panel under the vehicle
   - Air intake hose, on turbocharged Galant models
   - Drive belts
   - Water pump pulleys
   - Alternator upper bracket/brace
   - Alternator wiring connectors
   - Alternator mounting bolts and remove the alternator

2.0L SOHC Engine

1. Before servicing the vehicle, refer to the precautions at the beginning of this section.
2. Remove or disconnect the following:
   - Negative battery cable
   - Drive belts
   - Power steering hose clamp
   - Alternator brace
   - Alternator electrical connector
   - Engine mount, on vehicles with Anti-lock Brake Systems (ABS)
   - Alternator, tilting the engine as necessary

2.0L DOHC Turbo Engine

1. Before servicing the vehicle, refer to the precautions at the beginning of this section.

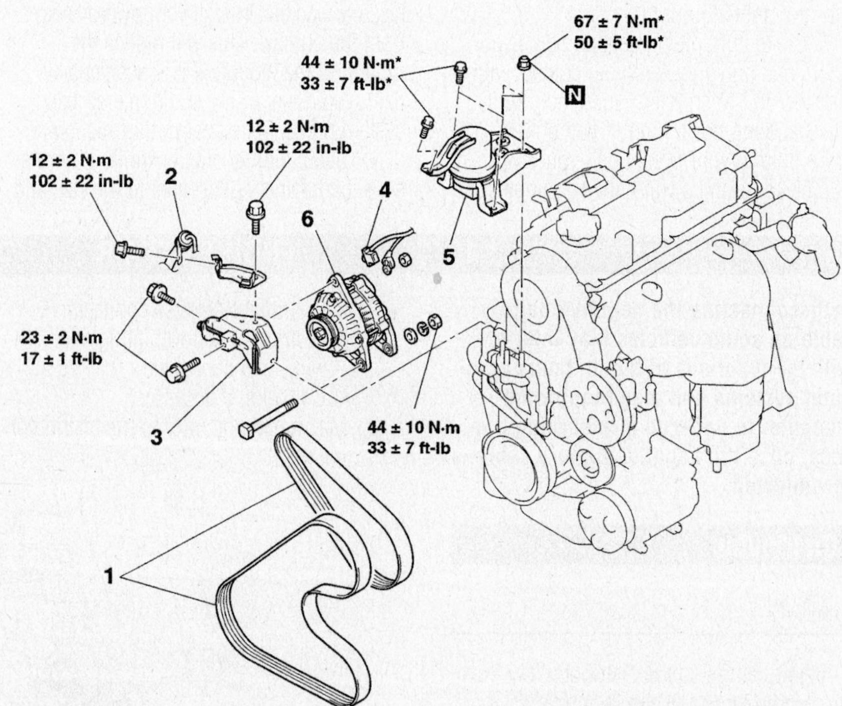

Exploded view of the alternator and related components—Lancer with 2.0L engine

2. Disconnect the negative battery cable.
3. Remove the oil dipstick and tube assembly.
4. Disconnect the fuel pressure solenoid connector.
5. Disconnect the knock sensor connector.
6. Disconnect the evaporative emissions purge solenoid connector.
7. Disconnect the ground cable connection.
8. Remove the fuel injectors.
9. Remove the accessory drive belt.
10. Disconnect the alternator wiring harness.
11. Remove engine mounting bolt.
12. Raise the engine slightly with a floor jack and remove the alternator from the vehicle.

3.0L and 3.5L Engines

1. Before servicing the vehicle, refer to the precautions at the beginning of this section.
2. Remove or disconnect the following:
   - Negative battery cable
   - Air intake hose
   - Alternator drive belt
3. On California models, remove the rear bank converter assembly.
4. Remove the engine roll stopper stay bracket assembly.
5. On the 3.0L engine, disconnect the Exhaust Gas Recirculation (EGR) temperature sensor wire and remove the EGR pipe assembly.
6. On the 3.0L engine, remove the intake plenum stay bracket assembly.
7. Remove or disconnect the following:
   - Alternator wiring harness connectors
   - Alternator upper and lower mounting bolts
8. From beneath the vehicle, remove the alternator.

3.8L Engine

1. Before servicing the vehicle, refer to the precautions at the beginning of this section.
2. Disconnect the negative battery cable.
3. Remove the accessory drive belt.
4. Disconnect the alternator wiring harness.
5. Disconnect the A/C compressor assembly connector.
6. Remove the A/C compressor assembly from its mounting bracket with the A/C hose still connected and move off to one side.

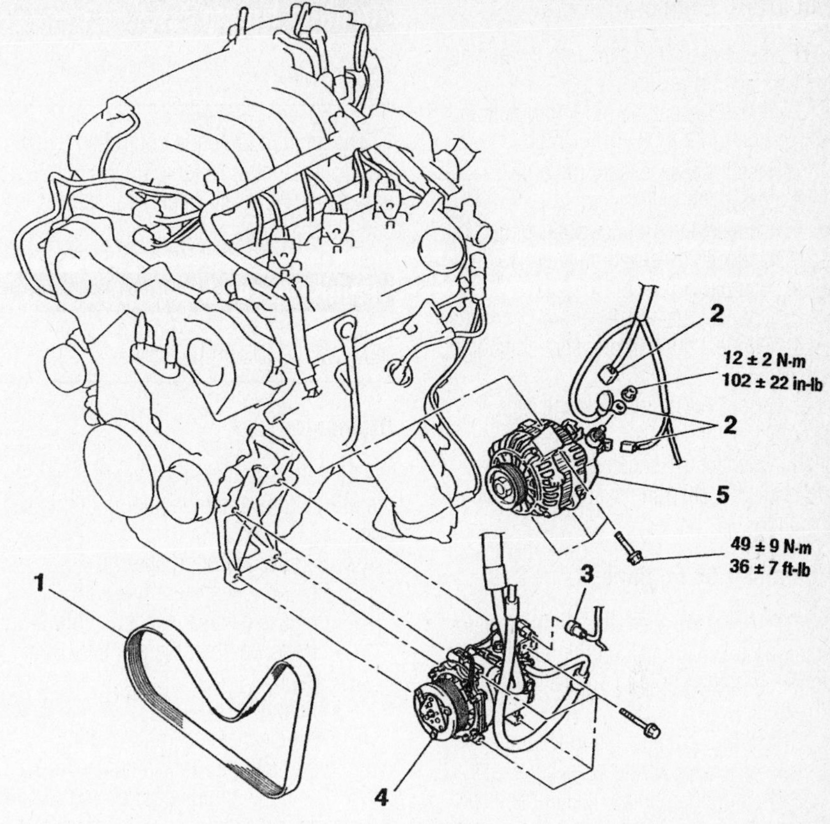

Exploded view of the alternator and related components—3.8L engine

7. Remove the alternator from under the vehicle.

INSTALLATION

1.5L, 1.8L, and 2.4L Engines

1. Position the alternator on the lower mounting fixture and install the lower mounting bolt and nut. Tighten nut just enough to allow for movement of the alternator.
2. Install or connect the following:
   - Alternator upper bracket/brace and connect the alternator electrical harness
   - Water pump pulleys
   - Drive belts and adjust to the proper tension
3. On turbocharged Galant models, install the air intake hose.
4. Install or connect the following:
   - Left side cover panel under the vehicle as required
   - Negative battery cable and check for proper operation

2.0L SOHC Engine

1. Install or connect the following:
   - Alternator
   - Engine mount, if removed
   - Alternator connector
   - Alternator brace
   - Power steering hose clamp
   - Drive belts
   - Negative battery cable
2. Adjust drive belts.

2.0L DOHC Engine

1. Install the alternator onto the engine and connect the wiring harness.
2. Install the engine mounting bolt and tighten bolt to 73 ft. lbs. (98 Nm).
3. Install the accessory drive belt.
4. Install the fuel injectors.
5. Connect the ground cable connection.
6. Connect the evaporative emissions purge solenoid connector.
7. Connect the knock sensor connector.
8. Connect the fuel pressure solenoid connector.
9. Install the oil dipstick and tube assembly.
10. Connect the negative battery cable and check the charging system for proper operation.

3.0L and 3.5L Engines

1. Position the alternator on the lower mounting fixture. Install and tighten the mounting bolt and nut to 14–18 ft. lbs. (20–25 Nm).
2. Install or connect the following:
   - Alternator wiring harness
3. On the 3.0L engine, install the intake plenum stay bracket and tighten the mounting bolt to 13 ft. lbs. (18 Nm).
4. On the 3.0L engine, install the EGR pipe and tighten the fitting connections to 43 ft. lbs. (60 Nm).
5. On the 3.0L engine, connect the EGR temperature sensor wire.
6. Install or connect the following:
   - Engine roll stopper stay and tighten the mounting bolt to 35 ft. lbs. (45 Nm) and the nut to 36–43 ft. lbs. (50–60 Nm)
   - Rear converter assembly, if removed
   - Drive belt and adjust the tensioner until the proper belt tension is achieved
   - Air intake hose
   - Negative battery cable and check the charging system for proper operation

3.8L Engine

1. Install the alternator assembly onto the engine.
2. Install the A/C compressor assembly onto its mounting bracket and reconnect the compressor wiring harness.
3. Connect the alternator wiring harness.
4. Install the accessory drive belt.
5. Connect the negative battery cable and check the charging system for proper operation.

Ignition Timing

ADJUSTMENT

The ignition timing is controlled by the Electronic Control Module (ECM) and is not adjustable. However it can be inspected using a scan tool.

Engine Assembly

REMOVAL & INSTALLATION

Diamante

1. Before servicing the vehicle, refer to the precautions at the beginning of this section.
2. Remove the hood assembly.
3. Relieve fuel system pressure.
4. Remove or disconnect the following:
   - Negative, then the positive battery cable
   - Battery
   - Air cleaner assembly and all adjoining air intake duct work
5. Drain the engine coolant and remove the radiator assembly and coolant reservoir (and bracket).
6. Remove or disconnect the following:
   - Engine undercover, if equipped
   - Front exhaust pipe
   - Transaxle assembly
   - Accelerator cable from the throttle body
   - Vacuum hoses from the intake manifold, label for installation
   - High pressure fuel line and the fuel return line
   - Vacuum hoses from the solenoid valves
   - Vacuum hoses from the purge canister
   - Heater hose connections from the engine
   - Harness for the Exhaust Gas Recirculation (EGR) temperature sensor connection, if equipped
   - Engine drive belts
   - Power steering pump oil pressure switch connection from the pump
   - Power steering pump and secure away from the work area
   - Air conditioning compressor. Wire the compressor aside. Do not discharge or disconnect the air conditioning lines.
   - Wiring to the alternator
   - Harness plugs for the Barometer (BARO) sensor, Idle Speed Control (ISC) motor, Throttle Position (TP) sensor, fuel injectors and Knock (KS) sensor
   - Harness plugs for the Engine Coolant Temperature (ECT) switch, sensor and gauge
   - Harness plugs for the ignition coil, condenser and ignition power transistor
   - Harness plugs for the variable induction control motor and the Manifold Absolute Pressure (MAP) sensor
   - Harness plugs for the Crankshaft Position (CKP) and Camshaft Position (CMP) sensors
   - Radiator overflow tank and remove the mounting bracket
   - Ground cable connections
7. Attach a hoist to the engine and take up the engine weight. Remove the engine mount bracket. Remove any torque control brackets (roll stoppers).
8. Lift the engine slowly and remove from the engine compartment.

To install:

9. Install or connect the following:
   - Engine and secure all control brackets
   - Transaxle assembly
   - Engine ground cable connections

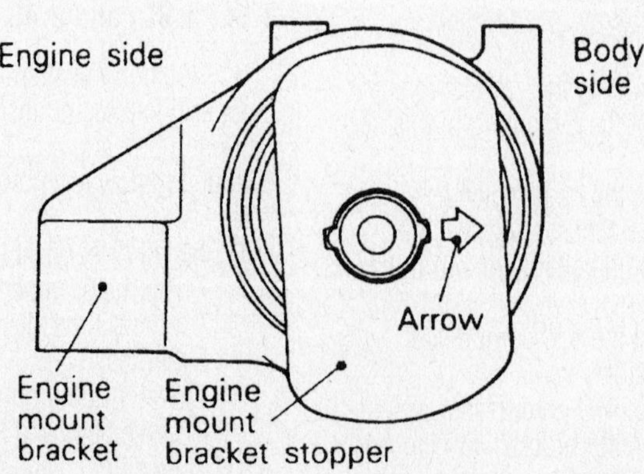

Alignment of the engine mount stopper bracket—Diamante shown

- Harness plugs for the CKP and CMP sensors
- Harness plugs for the variable induction control motor and the MAP sensor
- Harness plugs for the ignition coil, condenser and ignition power transistor
- Harness plugs for the ECT switch, sensor and gauge
- Harness plugs for the BARO sensor, ISC motor, TP sensor, fuel injectors and KS sensor
- Wiring to the alternator
- Air conditioning compressor assembly
- Power steering pump assembly
- Power steering pump oil pressure switch harness plug to the pump
- Engine drive belts, adjust
- Harness for the EGR temperature sensor
- Heater hose connections to the engine, using new hose clamps
- Vacuum hoses to the purge canister
- Vacuum hoses to the solenoid valves
- High pressure fuel line and the fuel return line, using new clamps or O-rings
- Vacuum hoses
- Accelerator cable to the throttle body
- Air cleaner assembly and all adjoining air intake duct work
- Radiator and coolant reservoir assembly
- Transaxle assembly
- Exhaust system to the engine, using new gaskets
- Battery to the vehicle
- Positive, then the negative battery cables
- Engine undercover, if equipped

10. Fill the engine with the proper amount of engine oil and coolant.
11. Install the hood.
12. Start the engine and check for leaks.

Eclipse

2.4L (4G64) ENGINES

1. Before servicing the vehicle, refer to the precautions at the beginning of this section.
2. Relieve the fuel system pressure.
3. Remove or disconnect the following:
   - Negative battery cable
   - Hood
   - Intake air duct
4. Drain the engine coolant.
5. Remove or disconnect the following:
   - Hoses and remove the radiator
   - Engine undercover
6. Attach an engine lifting fixture to the engine and remove the transaxle assembly.
7. Disconnect the following connectors:
   - Air conditioning compressor
   - Power steering pressure switch
   - Heated Oxygen (HO_2S) sensor
   - Engine Coolant Temperature (ECT) gauge sender
   - ECT sensor
   - Manifold Absolute Pressure (MAP) sensor
   - Intake Air Temperature (IAT) sensor
8. Remove or disconnect the following:
   - Power steering pump from the bracket and position the pump out of the way
   - Air conditioning compressor from the bracket and position it out of the way. Do not disconnect the hoses.
   - Accelerator cable from the throttle body and mounting bracket
9. Disconnect the following connectors:
   - Idle Air Control (IAC) motor
   - Knock (KS) sensor
   - Ignition module (power transistor)
   - Exhaust Gas Recirculation (EGR) solenoid
   - Oil pressure switch
   - Throttle Position (TP) sensor
   - Condenser
   - Injectors
   - Ignition coil
   - Camshaft Position (CMP) sensor
   - Crankshaft Position (CKP) sensor
   - Engine control wiring harness
10. Remove or disconnect the following:
    - Heater hoses from the engine
    - Fuel lines from the fuel supply rail
    - Purge air hose and the brake booster vacuum hose
    - Front exhaust pipe from the manifold
11. Place a floor jack against the oil pan with a piece of wood in between to protect the oil pan.
12. Raise the engine with the jack and remove the engine support fixture.
13. Install a chain hoist to the top of the engine.
14. Remove the engine mount bracket.
15. Lift the engine up slowly out of the engine compartment.

To install:

16. Slowly lower the engine assembly into the vehicle.
17. Position the floor jack under the oil pan with a piece of wood in between. Use the floor jack to adjust the height of the engine while installing the engine mount bracket.
18. Remove the chain hoist and install the engine support fixture.
19. Install or connect the following:
    - Front exhaust pipe to the manifold
    - Brake booster vacuum hose
    - New O-ring on the high pressure fuel line. Apply a small amount of clean engine oil to the O-ring and connect the fuel lines to the fuel supply rail.
20. Connect the following connectors:
    - IAC motor
    - KS sensor
    - Ignition module (power transistor)
    - EGR solenoid
    - Oil pressure switch
    - TP sensor
    - Condenser
    - Injectors
    - Ignition coil
    - CMP sensor
    - CKP sensor
    - Engine control wiring harness
21. Install or connect the following:
    - Accelerator cable, adjust
    - Air conditioning compressor and the power steering pump in their brackets
    - IAT sensor, MAP sensor, ECT sensor and gauge sender, HO_2S sensor, power steering pressure switch and the air conditioning compressor harness connectors
    - Radiator and hoses
    - Transaxle and remove the engine support fixture
    - Engine undercovers
    - Intake air duct
    - Negative battery cable
    - Hood
22. Refill the engine with the proper amount of coolant.

3.0L ENGINE

1. Before servicing the vehicle, refer to the precautions at the beginning of this section.
2. Disconnect the negative battery cable.
3. Drain the engine coolant.
4. Drain the engine oil and the transmission oil.
5. Relieve the fuel system pressure.
6. Remove or disconnect the following:
   - All wires, cables and hoses connected to the engine
   - Hood
   - Air intake and breather hoses
   - Radiator hoses and remove the radiator
   - Front exhaust pipe
   - Power steering pump and position it aside

- Air conditioning compressor drive belt
- Compressor from its mount and hang it out of the way. Do not disconnect the hoses and do not allow the compressor to hang by the hoses.

7. Install engine hoist equipment and make certain the attaching points on the engine are secure.
8. Raise the hoist enough to support the engine.
9. Remove or disconnect the following:
   - Front and rear engine roll stoppers
   - Left engine mount and support bracket
10. Slowly lift the engine and remove it from the vehicle.

Galant

1. Before servicing the vehicle, refer to the precautions at the beginning of this section.
2. Disconnect the negative battery cable.
3. Drain the engine coolant.
4. Drain the engine oil and the transmission oil.
5. Relieve the fuel system pressure.
6. Remove or disconnect the following:
   - Hood
   - Transaxle assembly
   - Radiator hoses and remove the radiator
   - Accelerator cable and remove the bracket
   - Air intake and breather hoses
   - Heater hoses
   - Brake booster vacuum hose at the engine
   - Vacuum hoses at the throttle body, label
   - Fuel feed and return hoses
7. Disconnect the following:
   - Power steering pressure switch
   - Alternator
   - Oil pressure switch
   - Air conditioning compressor
   - Each injector
   - Power transistor
   - Ignition coil
   - Throttle Position (TP) sensor
   - Idle Air Control (IAC) motor
   - Engine Coolant Temperature (ECT) switch
   - ECT sensor
   - Exhaust Gas Recirculation (EGR) temperature sensor
   - Engine control wiring harness
   - Heated Oxygen (HO_2S) sensor
   - Crankshaft Position (CKP) sensor
   - Camshaft Position (CMP) sensor
   - Refrigerant temperature switch
   - Condenser connection
8. Remove or disconnect the following:
   - Power steering pump and position it aside
   - Air conditioning compressor drive belt
   - Compressor from its mount and hang it out of the way. Do not disconnect the hoses and do not allow the compressor to hang by the hoses.
   - Front exhaust pipe
9. Install engine hoist equipment and make certain the attaching points on the engine are secure.
10. Raise the hoist enough to support the engine.
11. Remove or disconnect the following:
    - Front and rear engine roll stoppers
    - Left engine mount and support bracket
12. Slowly lift the engine and remove it from the vehicle.

To install:

13. Lower the engine into the vehicle.
14. Install the front and rear roll stoppers and the left engine mount. Do not torque the through-bolts at this time.
15. Remove the lifting apparatus from the engine.
16. Connect the exhaust system to the manifold, using a new gasket and new locking nuts. Tighten the nuts and the small bolt to 33 ft. lbs. (44 Nm)
17. Tighten the engine mount nuts and bolts. Correct torque values are:
    a. Upper mount to engine nuts: 42 ft. lbs. (57 Nm).
    b. Upper mount to engine bolt: 108 inch lbs. (12 Nm).
    c. Upper mount through-bolt: 72–87 ft. lbs. (98–118 Nm).
    d. Rear roll stopper through-bolt: 32 ft. lbs. (44 Nm).
    e. Front roll stopper through-bolt: 41 ft. lbs. (57 Nm).
18. Install or connect the following:
    - Air conditioning compressor, tightening the mounting bolts to 18 ft. lbs. (25 Nm)
    - Power steering pump, tightening the front bolts to 21 ft. lbs. (28 Nm) and the rear bolt to 16 ft. lbs. (22 Nm)
    - Accessory drive belts
19. Connect the following:
    - Power steering pressure switch
    - Alternator
    - Oil pressure switch
    - Air conditioning compressor
    - Each injector
    - Power transistor
    - Ignition coil
    - TP sensor
    - IAC motor
    - ECT switch
    - ECT sensor
    - EGR temperature sensor
    - Engine control wiring harness
    - HO_2S sensor
    - CKP sensor
    - CMP sensor
    - Refrigerant temperature switch
    - Condenser connection
20. Install or connect the following:
    - Fuel return hose and secure with the retaining clamp
    - New O-ring, connect the high pressure fuel line and tighten the bolts to 48 inch lbs. (6 Nm)
    - Vacuum lines running to the throttle body
    - Heater hoses
    - Accelerator cable bracket, tightening the bolts to 48 inch lbs. (6 Nm), and connect the accelerator cable
    - Radiator and connect the hoses
    - Transaxle
21. Fill the coolant system.
22. Connect the negative battery cable.
23. Start the engine and check for leaks.
24. Install the hood.

Lancer and Sportback

1. Before servicing the vehicle, refer to the precautions at the beginning of this section.
2. Relieve fuel system pressure.
3. Remove or disconnect the following:
   - Negative battery cable
   - Undercover, if equipped
   - Hood assembly
   - Air cleaner assembly and all adjoining air intake duct work
4. Drain the engine coolant and engine oil.
5. Remove or disconnect the following:
   - Radiator
   - Front exhaust pipe
   - Battery and battery tray
   - Accelerator cable
6. Detach the electrical connectors from the following components:
   - Air Conditioning (A/C) compressor
   - Power steering oil pressure switch
   - Crank angle sensor
   - Manifold differential pressure sensor
   - Evaporative emission (EVAP) purge solenoid

- Exhaust Gas Recirculation (EGR) solenoid valve
- Ignition coil
- Fuel injectors
- Throttle Position (TP) sensor
- Idle Air Control (IAC) motor
- Engine Coolant Temperature (ECT) sensor
- Camshaft Position (CMP) sensor
- Knock Sensor (KS)
- ECT gauge unit
- Heated Oxygen (HO_2S) sensor
- Starter and alternator
- Oil pressure switch

7. Remove or disconnect the following:
   - Brake booster vacuum hose
   - Power steering pump and A/C compressor drive belt
   - Power steering pump and brace. Position the assembly aside, but do NOT disconnect the fluid line.
   - A/C compressor, but do NOT disconnect the lines

➡Matchmark the installed position of the radiator hoses before disconnecting them.

   - Upper and lower radiator hoses
   - Heater and purge hoses
   - Fuel lines. Discard the O-rings
   - Transaxle assembly. Do NOT remove the flywheel bolt indicated by the arrow in the accompanying illustration. Removal of this bolt will cause the flywheel to be out of balance.

8. Remove the engine mount insulator and bracket as follows:
   a. Support the engine with a suitable floor jack.
   b. Remove the special tools that were installed for transaxle removal.
   c. Support the engine with special tool MB991453 attached to a chain block or engine hoist.
   d. Place a jack under the oil pan with a block of wood in between to protect the pan. Jack up the engine to take the weight off the engine mount insulator and bracket, then remove the insulator and bracket.

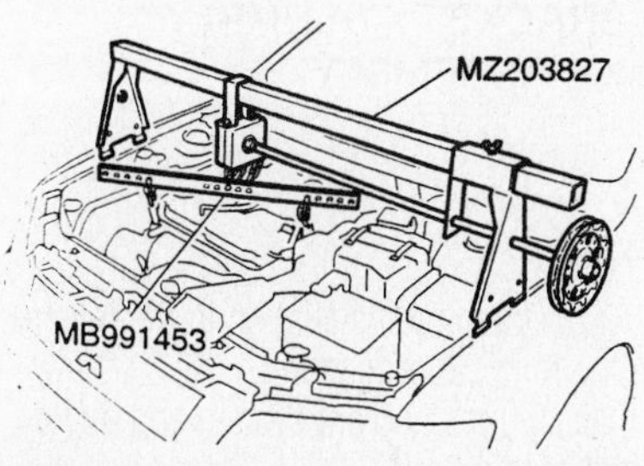

These special tools are needed to support the engine while the transaxle is removed—Lancer with 2.0L engine

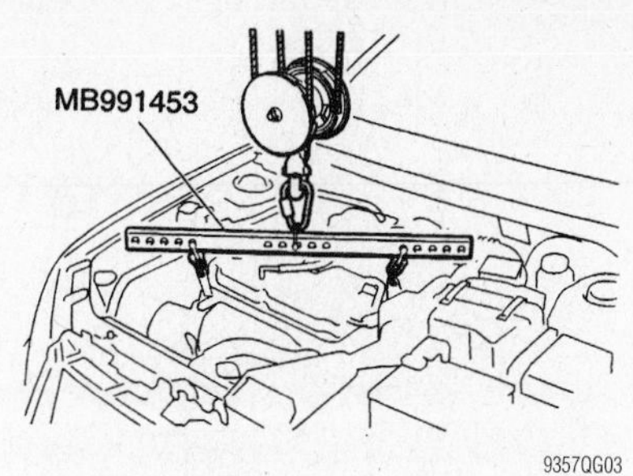

View of the special tool needed to support the engine during mount removal—Lancer with 2.0L engine

9. Make sure that all cables, hoses and harnesses are disconnected from the engine, then use the engine hose to slowly lift the engine up and out of the engine compartment

To install:

10. Installation is the reverse of the removal procedure. Torque the engine mounting fasteners to the specifications shown in the accompanying illustration.
11. Fill the coolant system and engine crankcase.
12. Connect the negative battery cable.
13. Start the engine and check for leaks.
14. Install the hood.

Lancer Evolution

1. Before servicing the vehicle, refer to the precautions at the beginning of this section.
2. Relieve fuel system pressure.
3. Disconnect the negative battery cable.
4. Disconnect the ignition coil connectors.
5. Disconnect the Heated Oxygen (HO_2S) sensor connector.
6. Disconnect the Crankshaft Position (CKP) sensor connector.
7. Disconnect the manifold differential pressure sensor connector.
8. Disconnect the fuel pressure solenoid connector.
9. Disconnect the Knock Sensor (KS) connector.
10. Disconnect the evaporative emission purge solenoid connector.
11. Disconnect the Throttle Position (TP) sensor connector.
12. Disconnect the Idle Air Control (IAC) motor connector.
13. Disconnect the fuel injector connectors.
14. Disconnect the Camshaft Position (CMP) sensor connector.
15. Disconnect the Engine Coolant Temperature (ECT) gauge unit connector.
16. Disconnect the ECT sensor connector.
17. Remove the control wiring harness and transaxle wiring harness combination.
18. Disconnect the ground cable connection.
19. Disconnect the alternator wiring connectors.
20. Disconnect the Exhaust Gas Recirculation (EGR) vacuum regulator solenoid valve connector.
21. Disconnect the engine oil pressure switch connector.
22. Remove the turbocharger wastegate actuator bolts.
23. Remove the accessory drive belt.
24. Disconnect the brake booster vacuum hose.
25. Disconnect the purge hose connection.
26. Disconnect the power steering oil pressure switch connector.
27. Remove the power steering oil pump heat protector.
28. Remove the power steering oil pump, mounting bracket and reservoir assembly.
29. Disconnect the A/C compressor connector.
30. Remove the A/C compressor and clutch assembly.
31. Disconnect the engine oil cooler feed and return hose connections.
32. Disconnect the heater water hose connections.
33. Disconnect the fuel return line and high-pressure hose connections. Discard the O-rings.
34. Remove the upper and lower radiator hoses.
35. Remove the transaxle and transfer case assemblies from the engine.
36. Remove the engine front roll stopper bracket and engine front mounting bracket through-bolts as follows:
   a. Support the engine with a suitable floor jack.
   b. Remove the special tools that were installed for transaxle removal.
   c. Support the engine with special tool MB991453 attached to a chain block or engine hoist.
   d. Place a jack under the oil pan with a block of wood in between to protect the pan. Jack up the engine to take the weight off the engine mount insulator and bracket, then remove the front roll stopper bracket and engine front mounting bracket through-bolts.

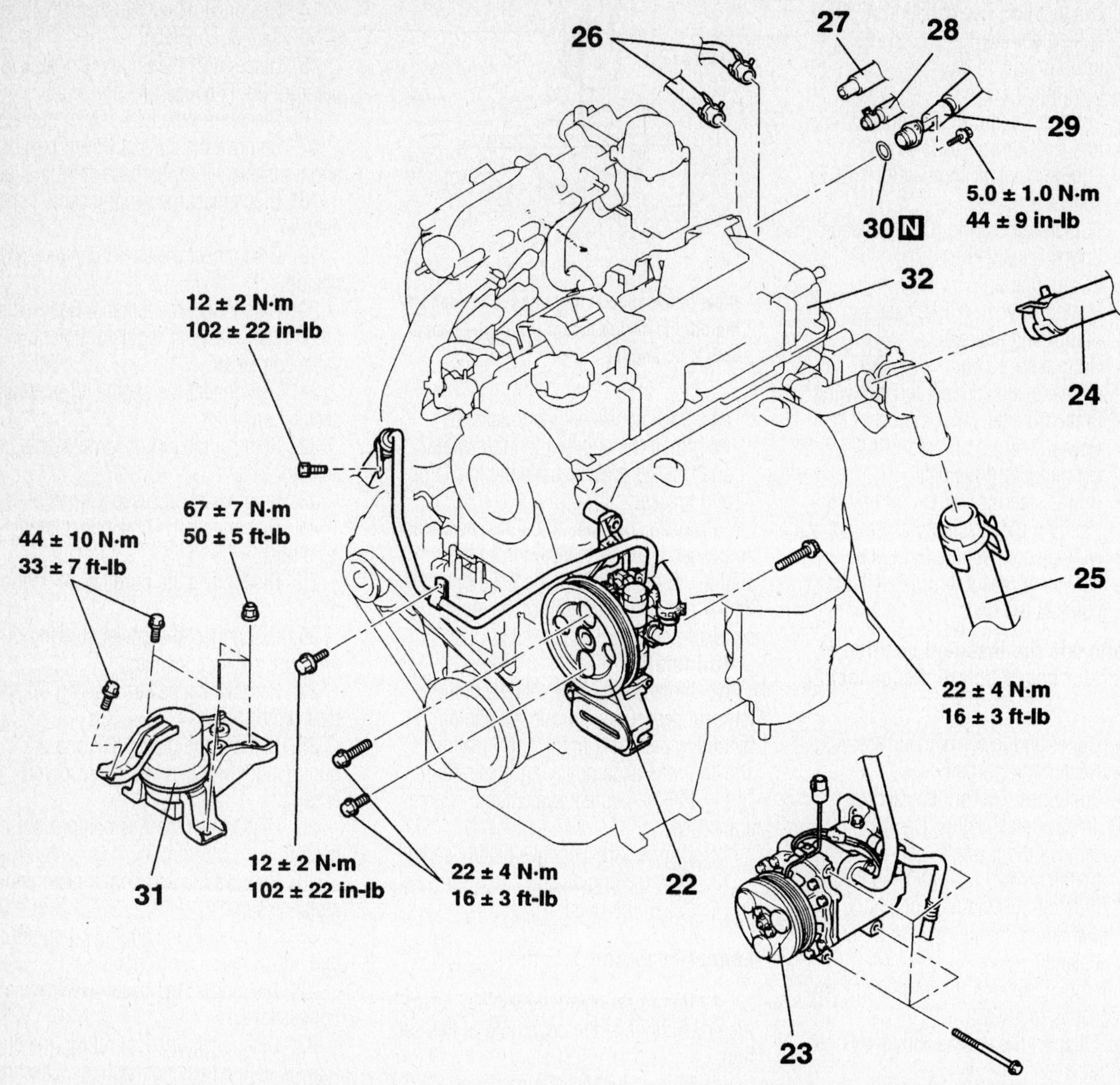

View of the engine mounts and related components—Lancer with 2.0L engine

37. Make sure that all cables, hoses and harnesses are disconnected from the engine, then use the engine hose to slowly lift the engine up and out of the engine compartment.

To install:

38. Install the engine and secure in position. The engine mount through-bolts should not be tightened until the full weight of the engine is on the mounts. Tighten the engine front roll stopper bracket through-bolt to 34–44 ft. lbs. (45–59 Nm). Tighten the engine front mounting bracket bolt to 66–80 ft. lbs. (88-108 Nm).

39. Continue installation in the reverse of the removal procedure.

40. Fill the coolant system and engine crankcase.

41. Connect the negative battery cable.

42. Start the engine and check for leaks.

43. Install the hood.

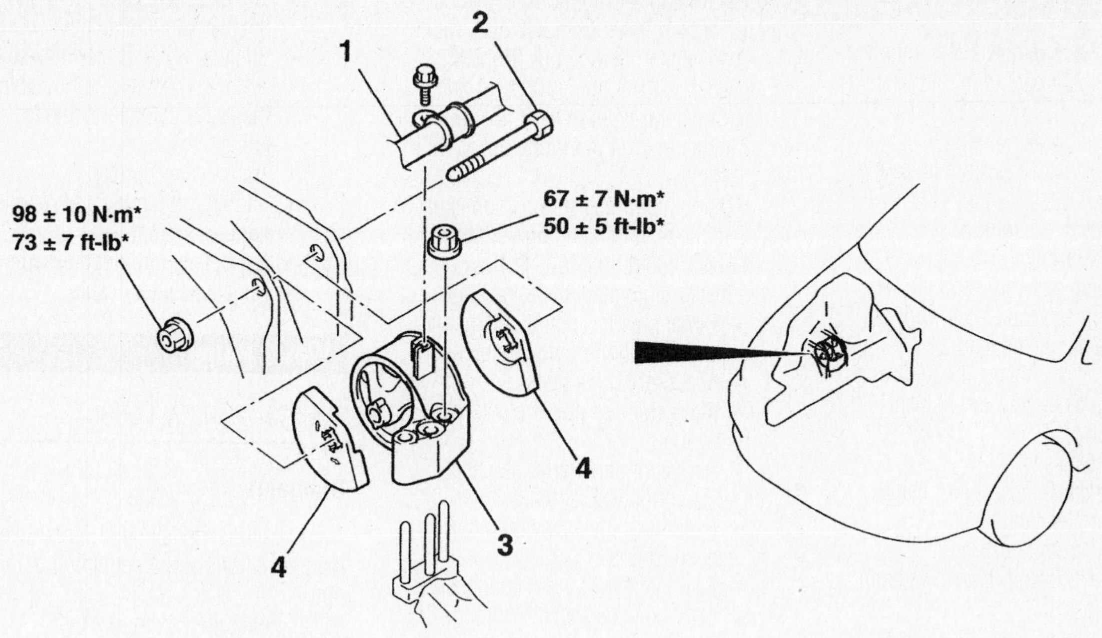

View of the engine front mounting bracket and related components—2.0L DOHC turbocharged engine

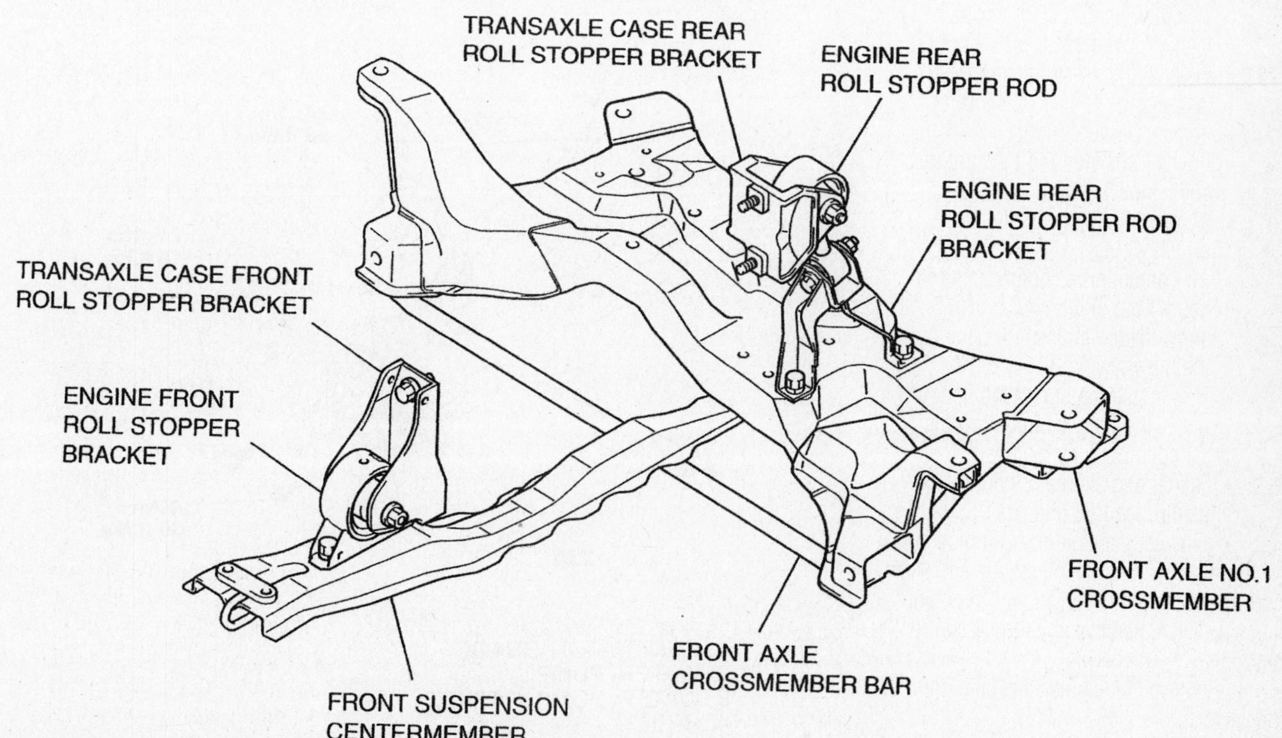

View of the engine front roll stopper bracket and related components—2.0L DOHC turbocharged engine

Mirage

1. Before servicing the vehicle, refer to the precautions at the beginning of this section.
2. Relieve fuel system pressure.
3. Remove or disconnect the following:
   - Negative battery cable
   - Undercover, if equipped
   - Hood assembly
   - Air cleaner assembly and all adjoining air intake duct work
4. Drain the engine coolant.
5. Remove or disconnect the following:
   - Radiator assembly and coolant reservoir
   - Transaxle assembly
   - Ground cable, accelerator cable, breather hose and heater hose connections from the engine
6. Note locations and remove vacuum hoses from engine.
7. Remove or disconnect the following:
   - Fuel feed and return hoses
   - Crankshaft Position (CKP) and Camshaft Position (CMP) sensor wiring
   - Heated Oxygen (HO_2S sensor), Engine Coolant (ECT) gauge and ECT sensor connections
   - Oil pressure switch
   - Thermo switch, with automatic transmissions
   - Harness connections for the Idle Speed Control (ISC) motor and Throttle Position (TP) sensor
   - Intake Air Temperature (IAT) sensor
   - Exhaust Gas Recirculation (EGR) temperature sensor (California)
   - Injector harness plugs
   - Power transistor and the ignition coil connections
   - Alternator and power steering switch wiring
   - Air conditioning compressor and hang it out of the way. Do NOT allow the compressor to hang by the hoses.
   - Power steering pump and hang it out of the way–Do not allow the pump to hang by the hoses.
   - Starter and alternator harness clamp, for 1.8L engines
   - Exhaust manifold to head pipe nuts
8. Attach a hoist to the engine and support the engine weight. Remove the engine mount bracket. Remove any torque control brackets (roll stoppers).
9. Remove the engine assembly from the vehicle.

To install:

10. Install the engine and secure in position. The front lower mount through-bolt nut should not be tightened until the full weight of the engine is on the mount. Tighten through-bolt to 72 ft. lbs. (100 Nm) and bracket mounting bolts to 42 ft. lbs. (58 Nm). Tighten bracket mounting nut to 38 ft. lbs. (53 Nm).
11. Using a new gasket, position exhaust pipe onto the manifold and tighten the flange nuts to 36 ft. lbs. (50 Nm).
12. Install or connect the following:
    - Power steering pump, alternator and air conditioning compressor
    - Accessory drive belts
    - Alternator and power steering wiring
    - Alternator and starter harness clamp for 1.8L engines
    - Ignition coil and power transistor connections
    - Fuel injector harness connections
    - EGR temperature sensor plug–California models
    - IAT sensor
    - IAC and TPS connectors
    - Thermo switch, automatic transmission
    - Oil pressure switch wiring
    - HO_2S sensor, ECT gauge and ECT sensor
    - CKP and CMP sensors
    - Fuel feed hose and tighten bolts to 44 inch lbs. (5 Nm), using new O-rings
    - Fuel return hose, using a new hose clamp
    - Vacuum hoses and the brake booster vacuum supply
    - Breather hose, heater hoses, accelerator cable and ground cables. Inspect accelerator cable for proper adjustment.
    - Transaxle assembly
    - Radiator assembly and refill the cooling system
    - Air cleaner and hood assembly
    - Negative battery cable

Water Pump

REMOVAL & INSTALLATION

Diamante

1. Before servicing the vehicle, refer to the precautions at the beginning of this section.
2. Drain the cooling system.
3. Disconnect the negative battery cable.
4. Remove the timing belt.
5. Remove the coolant hoses from the pump, if equipped.
6. Remove the alternator brace.

➡The water pump bolts are different in size. Note their locations for installation.

7. Remove the water pump, gasket and O-ring where the water inlet pipe joins the pump.

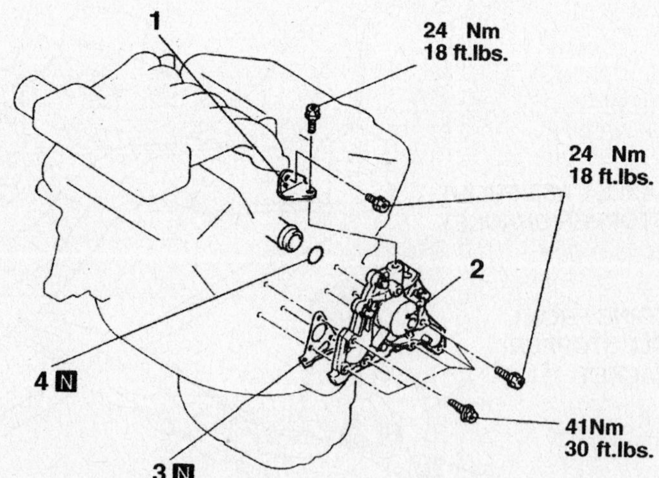

Water pump tightening torques—Diamante

To install:

8. Thoroughly clean both gasket surfaces of the water pump and block.
9. Install a new O-ring into the groove on the front end of the water inlet pipe. Do not apply oils or grease to the O-ring. Wet with water only.
10. Install the water pump assembly to the engine block, with new gasket. Torque the 8mm mounting bolts to 18 ft. lbs. (24 Nm), and the 10mm mounting bolt to 30 ft. lbs. (41 Nm).
11. Connect the hoses to the pump
12. Install the timing belt.
13. Install the engine drive belts.
14. Fill the system with coolant.
15. Connect the negative battery cable, run the vehicle until the thermostat opens and fill the radiator completely.
16. Once the vehicle has cooled, recheck the coolant level.

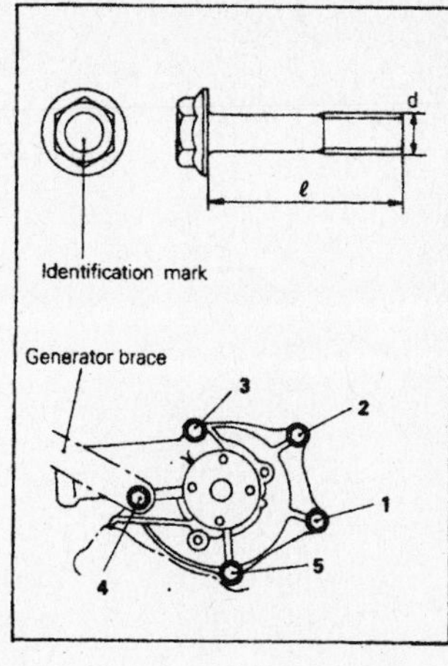

No.	Identification mark	Bolt diameter (d) x length (ℓ) mm (in.)	Torque Nm (ft.lbs.)
1	4	8 x 14 (.31 x .55)	12–15 (9–10)
2	4	8 x 22 (.31 x .87)	
3	4	8 x 30 (.31 x 1.18)	
4	7	8 x 65 (.31 x 2.56)	20–27 (15–19)
5	4	8 x 28 (.31 x 1.10)	12–15 (9–10)

7923PG11

Water pump bolt identification—Galant

Eclipse

1. Before servicing the vehicle, refer to the precautions at the beginning of this section.
2. Disconnect the negative battery cable.
3. Drain the engine coolant.
4. Remove or disconnect the following:
   - Timing belt
   - Alternator brace from the water pump
   - Timing belt rear cover
   - Water pump mounting bolts
   - Water pump, gasket and O-ring

To install:

5. Install or connect the following:
   - New O-ring on the water inlet pipe. Coat the O-ring with water or coolant. Do not allow oil or other grease to contact the O-ring.
   - Water pump to the engine block, with new gasket. Torque the mounting bolts to 10 ft. lbs. (13 Nm)
   - Alternator brace on the water pump. Torque the brace pivot bolt to 17 ft. lbs. (24 Nm).
   - Timing belt rear cover
   - Timing belt
   - Remaining components
6. Refill the engine with coolant.
7. Connect the negative battery cable, start the engine and check for leaks.

Galant

1. Before servicing the vehicle, refer to the precautions at the beginning of this section.
2. Disconnect the negative battery cable.
3. Drain the cooling system.
4. Remove or disconnect the following:
   - Engine undercover
   - Clamp bolt from the power steering hose
5. Support the engine with the appropriate equipment and remove the engine mount bracket.
6. Remove or disconnect the following:
   - Engine drive belts and the air conditioning tensioner bracket
   - Timing belt covers from the front of the engine
   - Camshaft and silent shaft timing belts
   - Alternator brace
   - Water pump, gasket and O-ring where the water inlet pipe(s) joins the pump

To install:

7. Thoroughly clean both gasket surfaces of the water pump and block.
8. Install a new O-ring into the groove on the front end of the water inlet pipe and wet with clean antifreeze only. Do not apply oils or grease to the O-ring.
9. Using a new gasket, install the water pump assembly. Tighten bolts with the head mark **4** to 10 ft. lbs. (14 Nm) and bolts with the head mark **7** to 18 ft. lbs. (24 Nm).
10. Install or connect the following:
    - Timing belts
    - Engine drive belts
    - Engine mount bracket
    - Engine undercover
11. Fill the system with coolant.
12. Connect the negative battery cable, run the vehicle until the thermostat opens and fill the radiator completely.
13. Once the vehicle has cooled, recheck the coolant level.

Lancer and Sportback

1. Before servicing the vehicle, refer to the precautions at the beginning of this section.
2. Disconnect the negative battery cable.
3. Drain the cooling system.
4. Remove the timing belt.
5. Remove the water pump bolts(s) and pump.
6. Remove the water pump gasket and O-ring.

To install:

7. Install the water pump along with a new gasket and O-ring. Tighten the mounting bolt(s) to 15–19 ft. lbs. (20–26 Nm).
8. Install the timing belt.
9. Refill the cooling system and connect the negative battery cable.

Mirage

1. Before servicing the vehicle, refer to the precautions at the beginning of this section.
2. Disconnect the negative battery cable.
3. Drain the cooling system.
4. Remove or disconnect the following:
   - Engine undercover
   - Clamp bolt from the power steering hose
   - Engine drive belts
5. Support the engine with the appropriate equipment and remove the engine mount bracket.
6. Remove or disconnect the following:
   - Timing belt

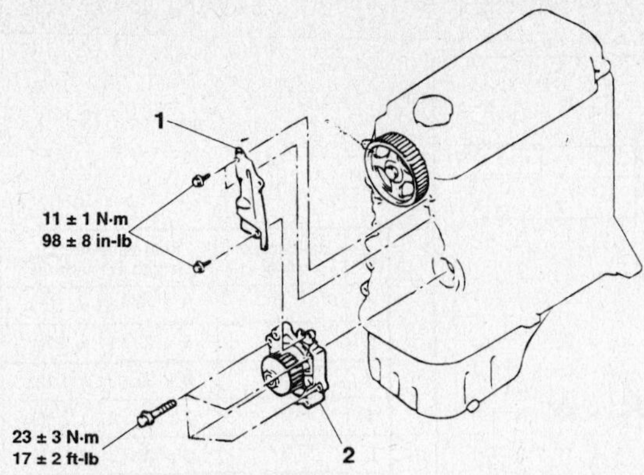

Exploded view of the water pump—Lancer with 2.0L engine

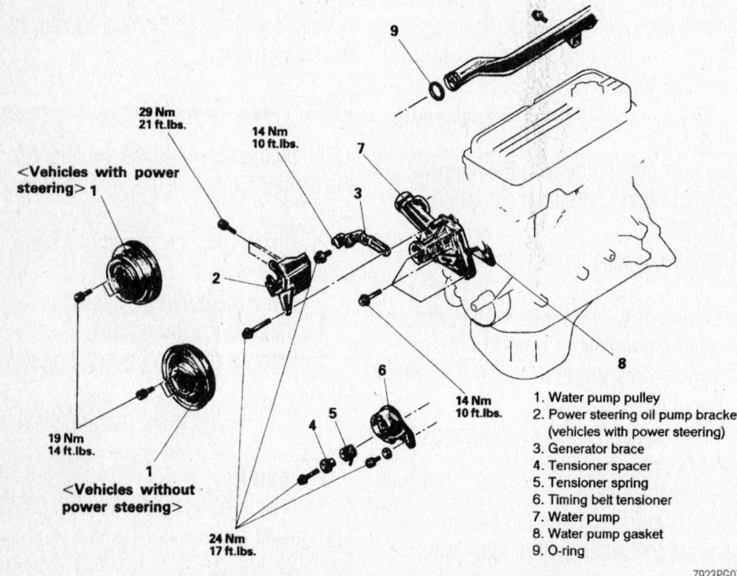

Water pump and related components—Mirage with 1.5L (4G15) engine

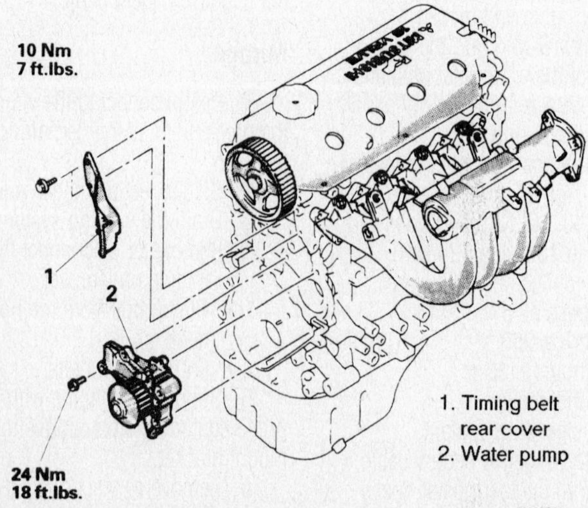

Water pump and related components—Mirage with 1.8L (4G93) engines

- Power steering pump bracket
- Alternator brace

➡The water pump mounting bolts are different in length, note their positioning for reassembly.

7. Remove the water pump, gasket and O-ring where the water inlet pipe(s) joins the pump.

To install:

8. Thoroughly clean both gasket surfaces of the water pump and block.
9. For 1.5L engines, install a new O-ring into the groove on the front end of the water inlet pipe. Do not apply oils or grease to the O-ring. Wet the O-ring with water only.
10. For 1.8L engines, apply a 0.09–0.12 in. (2.5–3.0mm) continuous bead of sealant to water pump and install the pump assembly. Install the water pump within 15 minutes of the application of the sealant. Wait 1 hour after installation of the water pump to refill the cooling system or starting the engine.
11. Install or connect the following:
    - Gasket and pump assembly and tighten the bolts to 17 ft. lbs. (24 Nm)
    - Remaining components in the reverse order of removal
12. Fill the system with coolant.
13. Connect the negative battery cable, run the vehicle until the thermostat opens and fill the radiator completely.
14. Once the vehicle has cooled, recheck the coolant level.

Heater Core

REMOVAL & INSTALLATION

Diamante

1. Before servicing the vehicle, refer to the precautions at the beginning of this section.
2. Disconnect the negative battery cable.
3. Drain the cooling system into a clean container for reuse.
4. Discharge and recover the air conditioning system refrigerant.
5. Remove or disconnect the following:
    - Heater hoses from the heater core
    - Refrigerant lines from the evaporator core and discard the O-rings

✻✻ CAUTION

After disconnecting the negative battery cable, wait at least 60 seconds before working on the SRS module or instrument panel.

6. Remove the passenger's side air bag by removing or disconnecting the following:

- Dash undercover
- Glove box assembly
- Glove box case
- Air bag-to-dash bolts and the air bag; then, disconnect the electrical connector

7. Remove or disconnect the following:
   - Floor console
   - Front pillar trim at both sides
8. Remove the instrument panel by removing or disconnecting the following:
   - Steering column covers
   - Hood lock release handle
   - Parking brake release handle
   - Lower left side instrument panel cover
   - Ignition key cylinder panel
   - Instrument panel Electronic Control Unit (ECU) and remove the ECU
   - Instrument panel meter bezel and the combination meter
   - Center air outlet assembly
   - Ashtray
   - Air conditioning control panel assembly and the audio unit
   - Console side cover assembly
   - Floor carpet rear reinforcement
   - Electrical harness connector and plug
   - Steering column mounting bolts and lower the steering column assembly
   - Instrument panel with the help of an assistant

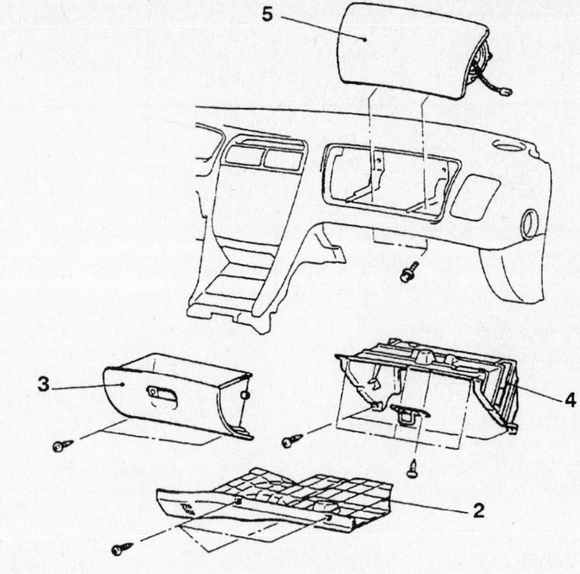

Exploded view of the passenger's air bag module—Diamante

9. Remove or disconnect the following:
   - ECU bracket
   - Center stay assembly
   - Heater hose connection and the center duct assembly
   - Foot distribution duct and the breather hose
   - Refrigerant lines from the evaporator and discard the O-rings
   - Air conditioning housing drain hose and remove the evaporator housing
   - Heater housing unit
   - Heater core support and the heater core

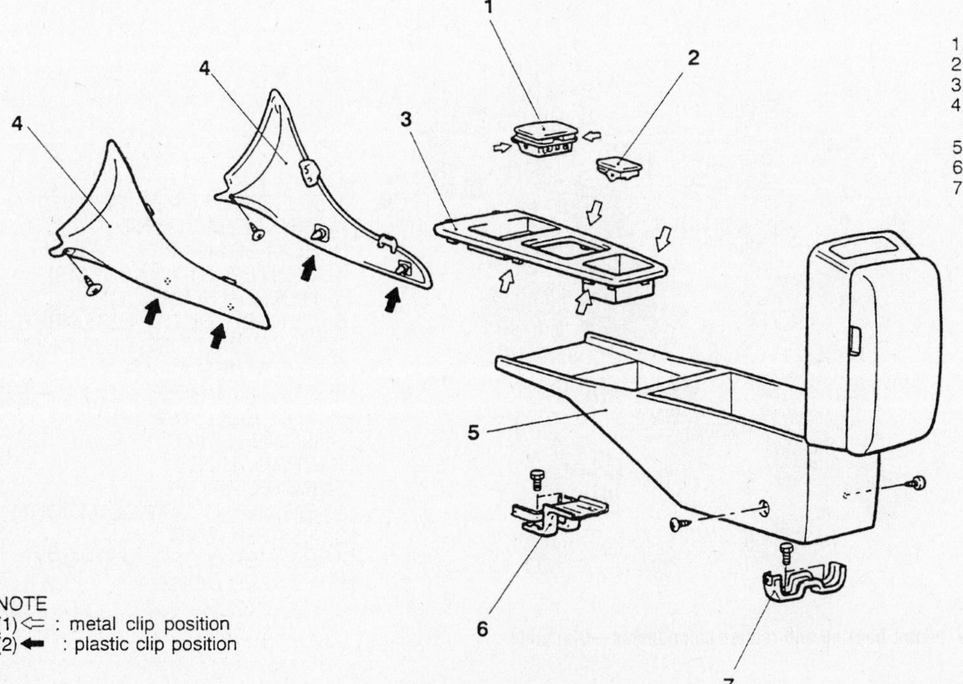

Exploded view of the floor console and related components—Diamante

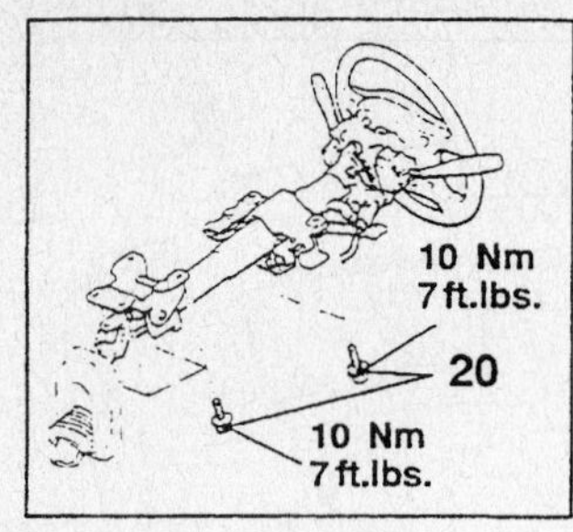

1. COLUMN COVER
2. HOOD LOCK RELEASE HANDLE
3. PARKING BRAKE RELEASE HANDLE
4. INSTRUMENT PANEL LOWER COVER ASSEMBLY (LH)
5. KEY CYLINDER PANEL
6. INSTRUMENT PANEL ECU
7. METER BEZEL
8. COMBINATION METER
9. CENTER AIR OUTLET ASSEMBLY
10. ASHTRAY
11. AIR CONTROL PANEL ASSEMBLY & AUDIO UNIT
12. UNDERCOVER ASSEMBLY
13. GLOVEBOX ASSEMBLY
14. GLOVEBOX OUTER CASE
15. PASSENGER SIDE AIRBAG MODULE
16. CONSOLE SIDE COVER ASSEMBLY
17. FLOOR CARPET REAR REINFORCEMENT
18. HARNESS CONNECTOR
19. PLUG
20. STEERING COLUMN MOUNTIN BOLT
21. INSTRUMENT PANEL

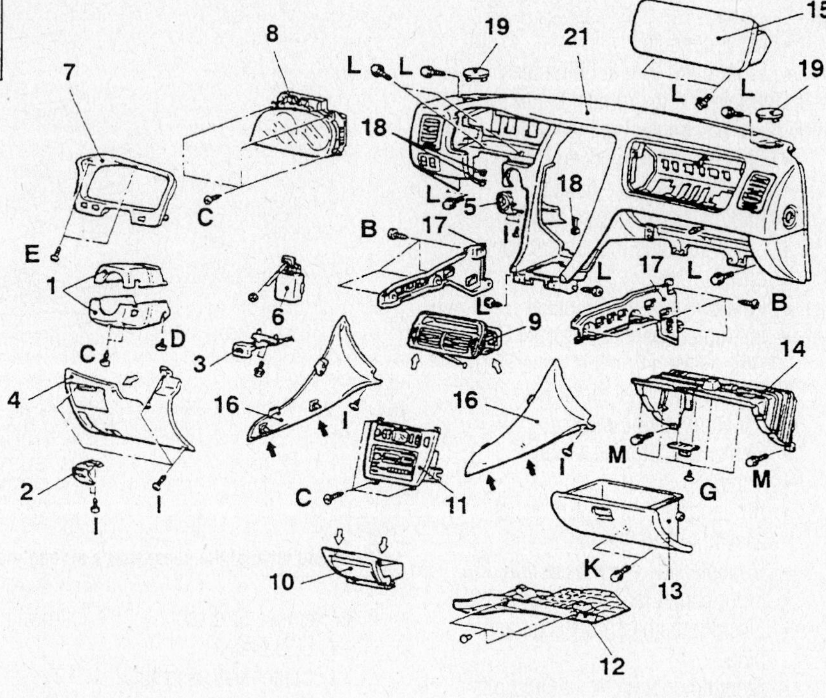

NOTE
(1) ⇦ : metal clip position
(2) ⬅ : plastic clip position

93112GG2

Exploded view of the instrument panel and steering column assembly—Diamante

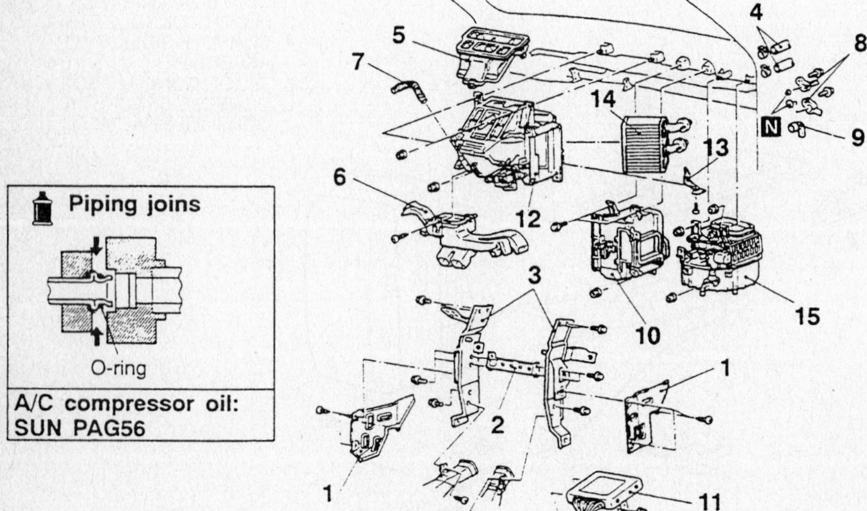

1. FLOOR CARPET FRONT REINFORCEMENT
3. ECU BRACKET
4. CENTER STAY ASSEMBLY
5. HEATER HOSE CONNECTION
6. CENTER DUCT ASSEMBLY
7. FOOT DISTRIBUTION DUCT
8. BREATHER HOSE
9. SUCTION PIPE, LIQUID PIPE B AND COOLING UNIT CONNECTION
10. DRAIN HOSE
11. EVAPORATOR
12. ENGINE CONTROL MODULE
13. HEATER UNIT
14. HEATER CORE SUPPORT
15. HEATER CORE

93112GG3

Exploded view of the heater core, heater housing and related components—Diamante

To install:

10. Install or connect the following:
    - Heater core support and the heater core
    - Heater housing unit
    - Air conditioning housing drain hose and Install the evaporator housing
    - Refrigerant lines to the evaporator using new O-rings
    - Foot distribution duct and the breather hose
    - Heater hose connection and the center duct assembly
    - Center stay assembly
    - ECU bracket
11. Install the instrument panel by installing or connecting the following:
    - Instrument panel with the help of an assistant
    - Steering column assembly and the steering column mounting bolts. Torque the bolts to 84 inch lbs. (10 Nm).
    - Electrical harness connector and plug
    - Floor carpet rear reinforcement
    - Console side cover assembly
    - Air conditioning control panel assembly and the audio unit
    - Ashtray
    - Center air outlet assembly
    - Instrument panel meter bezel and the combination meter
    - Instrument panel ECU and connect the ECU electrical connector
    - Ignition key cylinder panel
    - Lower left side instrument panel cover
    - Parking brake release handle
    - Hood lock release handle
    - Steering column covers
12. Front pillar trim at both sides
13. Floor console
14. Install the passenger's side air bag by installing or connecting the following:
    - Air bag-to-dash bolts and the air bag; then, connect the electrical connector
    - Glove box case
    - Glove box assembly
    - Dash undercover
    - Refrigerant lines to the evaporator core using new O-rings
    - Heater hoses to the heater core
15. Refill the cooling system.
16. Connect the negative battery cable.
17. Evacuate, charge and leak test the air conditioning system.
18. Operate the engine to normal operating temperatures; then, check the climate control operation and check for leaks.

Eclipse

✻✻ CAUTION

Wait for 1 minute after disconnecting the negative battery cable before working inside the vehicle. The air bag system is set to deploy for a short period of time after the battery is disconnected.

1. Before servicing the vehicle, refer to the precautions at the beginning of this section.
2. Disconnect the negative battery cable.
3. Drain the cooling system into a clean container for reuse.
4. Disconnect the heater hoses from the heater core tubes at the firewall. Do not allow the coolant to damage the vehicle speed sensor located below the heater hoses on the manual transmission vehicles.

✻✻ WARNING

To prevent damage to the air bag control unit during removal or installation of the floor console, avoid shocks or impact. Do not drop.

5. Remove the floor console by removing or disconnecting the following:
    - Center console trim panel
    - Ashtray and cup holder assembly
    - Shift lever knob on manual transmission
    - Retaining screws
    - Floor console assembly
6. Locate the rectangular plugs in the knee protector on either side of the steering column. Pry these plugs out and remove the screws.
7. Remove or disconnect the following:
    - Driver's side air bag assembly, the steering wheel and the passenger's side air bag assembly
    - Lap cooler duct and steering column covers
    - Instrument cluster bezel and then the instrument cluster
    - Radio
    - Glove box
    - Center air outlet assembly
    - Hood release handle and the lower cover
    - Heater control assembly
    - Front speakers and the instrument panel switch
    - Steering shaft support bolts and lower the steering column
    - Instrument panel mounting hardware and remove the instrument panel from the vehicle
    - Stamped steel center reinforcement
    - Lower ductwork from the heater box
    - Evaporator case mounting bolt and

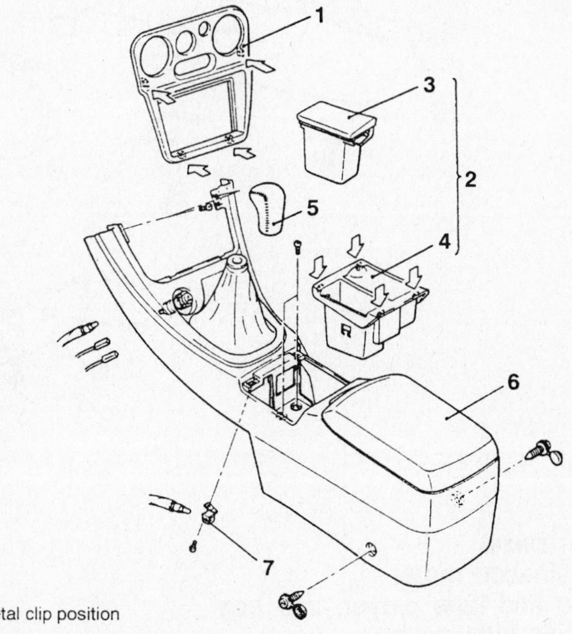

1. Center console panel
2. Ashtray and cupholder assembly
3. Ashtray
4. Cup holder
5. Shift lever knob <M/T>
6. Floor console assembly
7. Ashtray illumination light bracket

93112G56

Exploded view of the floor console and related components—Eclipse

nut to allow clearance for the heater unit removal
- Heater unit
- Heater core from the heater unit

To install:

8. Install or connect the following:
   - Heater core to the heater unit
   - Heater unit
   - Evaporator case mounting bolt and nut
   - Lower ductwork to the heater box
   - Stamped steel center reinforcement
   - Instrument panel and the instrument panel mounting hardware
   - Steering column and the steering shaft support bolts
   - Front speakers and the instrument panel switch
   - Heater control assembly
   - Hood release handle and the lower cover

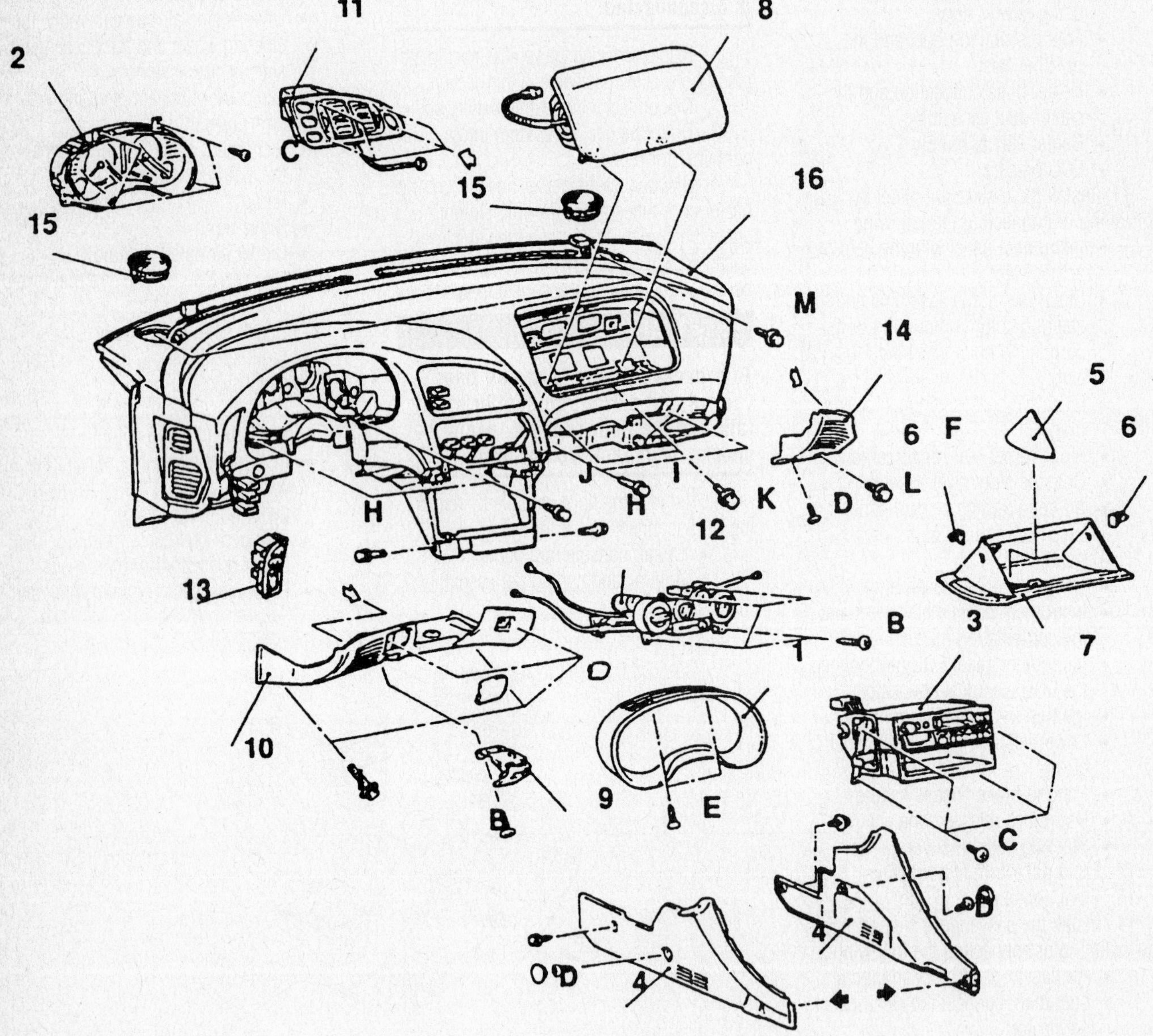

1. Meter bezel
2. Combination meter
3. Radio and tape player, and box
4. Console side cover
5. Sunglasses holder
6. Stopper
7. Glove box
8. Passenger's side air bag module assembly
9. Hood lock release handle
10. Instrument under cover L.H.
11. Center air outlet assembly
12. Heater control assembly
13. Instrument panel switch
14. Instrument under cover R.H.
15. Front speaker
16. Instrument panel assembly

93112G74

Exploded view of the instrument panel and related components—Eclipse

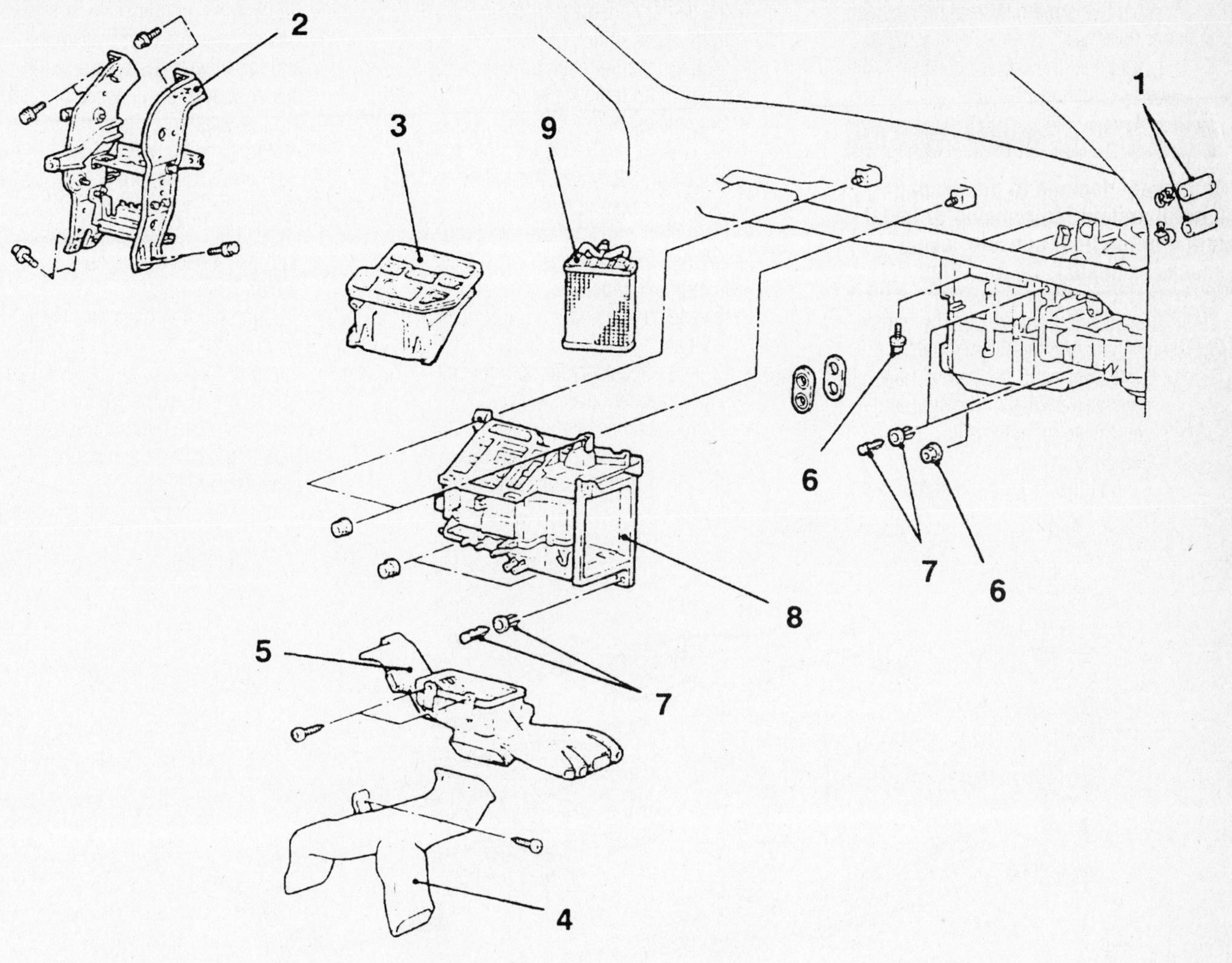

1. Heater hose connection
2. Center stay
3. Center duct
4. Semi rear heater duct
5. Foot distribution duct
6. Cooling unit installation bolt and nut
7. Clip
8. Heater unit
9. Heater core

93112GE0

Exploded view of the heater core, heater case and related components—Eclipse

- Center air outlet assembly
- Glove box
- Radio
- Instrument cluster and the instrument cluster bezel
- Steering column covers and the lap cooler duct
- Steering wheel, the driver's side air bag assembly and the passenger's side air bag assembly
- Screws and the rectangular plugs in the knee protector on either side of the steering column

9. Install the floor console by installing or connecting the following:

- Floor console assembly
- Retaining screws
- Shift lever knob on manual transmission
- Ashtray and cup holder assembly
- Center console trim panel
- Center console trim panel

10. Connect the heater hoses to the heater core tubes at the firewall.
11. Refill the cooling system.
12. Connect the negative battery cable.
13. Operate the engine to normal operating temperatures; then, check the climate control operation and check for leaks.

Galant

1. Before servicing the vehicle, refer to the precautions at the beginning of this section.
2. Disconnect the negative battery cable.

CAUTION

After disconnecting the negative battery cable, wait at least 60 seconds before working on the SRS module or instrument panel.

3. Drain the cooling system into a clean container for reuse.

4. Disconnect the heater hoses from the heater core at the firewall.

⁂ WARNING

To prevent damage to the air bag control unit during removal or installation of the floor console, avoid shocks or impact. Do not drop.

5. Remove the floor console by removing or disconnecting the following:
   - Shift lever knob on manual transmission vehicles or the shift indicator plate on automatic transmissions
   - Coin holder behind the shifter, then the center console trim cover in front of the shifter
   - Center console retaining bolt cover plugs, then remove the bolts
   - Console assembly, then the brackets

6. Remove or disconnect the following:
   - Steering column covers
   - Instrument cluster bezel and then the instrument cluster
   - Instrument panel switch, hood lock release handle and the lower duct work
   - Driver's knee protector and the left side air outlet cover
   - Center panel assembly
   - Glove box undercover, then the glove box and the right side panel cover
   - Radio and cup holder
   - Cables from the heater assembly and the blower, then pull out the heater control panel assembly, noting the location of the boss in the center reinforcement
   - Cool air bypass damper lever cable connection
   - Passenger's side air bag module and disconnect the harness connector, if equipped
   - Steering column bolts and lower the column
   - Harness connector at the lower left side of the instrument panel
   - Instrument panel mounting hardware and remove the instrument panel from the vehicle
   - Joint duct between the heater case

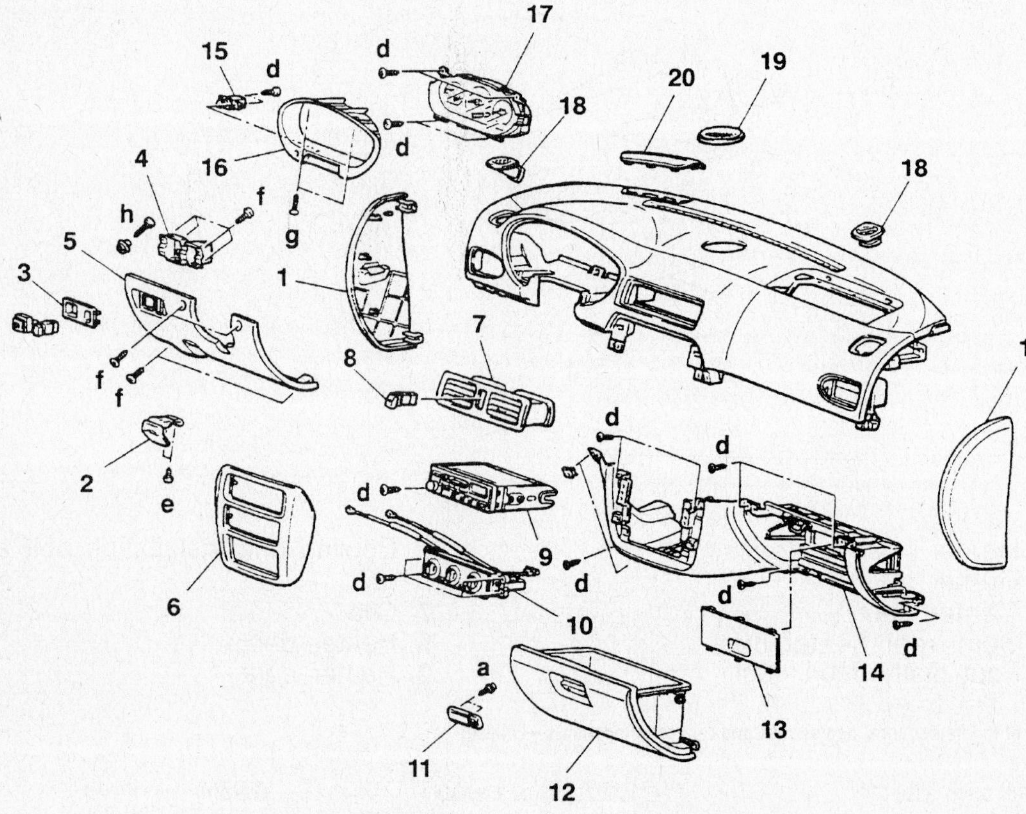

1. INSTRUMENT PANEL SIDE COVER
2. HOOD LOCK RELEASE HANDLE
3. SWITCH PANEL ASSEMBLY
4. CONNECTOR HOLDER
5. FRONT DRIVER'S SIDE UNDER COVER
6. CENTER PANEL ASSEMBLY
7. CENTER AIR OUTLET ASSEMBLY
8. HAZARD WARNING LIGHT SWITCH
9. RADIO AND TAPE PLAYER
10. HEATER CONTROL ASSEMBLY
11. GLOVE BOX STRIKER
12. GLOVE BOX
13. FRONT PASSENGER'S UNDER COVER PLUG
14. FRONT PASSENGER'S SIDE UNDER COVER
15. RHEOSTAT
16. METER BEZEL
17. COMBINATION METER
18. SIDE DEFROSTER GRILLE
19. SPEAKER GRILLE
20. INSTRUMENT PANEL UPPER PLUG

93112GF1

Exploded view of the instrument panel assembly—Galant

and the blower assembly (on models without air conditioning)
- Both stamped steel center reinforcement piece
- Electronic Control Module (ECM) bracket
- Evaporator retaining nut and remove the heater case assembly, if equipped with air conditioning
- Heater core from the case

To install:

7. Install or connect the following:
   - Heater core to the case
   - Evaporator retaining nut and Install the heater case assembly, if equipped with air conditioning
   - ECM bracket
   - Both stamped steel center reinforcement pieces
   - Joint duct between the heater case and the blower assembly (on models without air conditioning)
   - Instrument panel and install the instrument panel mounting hardware
   - Harness connector at the lower left side of the instrument panel
   - Steering column bolts
   - Passenger's side air bag module and connect the harness connector, if equipped
   - Cool air bypass damper lever cable connection
   - Cables to the heater assembly and the blower and install the heater control panel assembly
   - Radio and cup holder
   - Glove box undercover, then the glove box and the right side panel cover
   - Center panel assembly
   - Left side air outlet cover and the driver's knee protector
   - Lower duct work, the instrument panel switch and hood lock release handle
   - Instrument cluster and the instrument cluster bezel
   - Steering column covers
8. Install the floor console installing or connecting the following:
   - Console assembly brackets and the console
   - Center console retaining bolts, then the bolt cover plugs
   - Coin holder behind the shifter, then the center console trim cover in front of the shifter

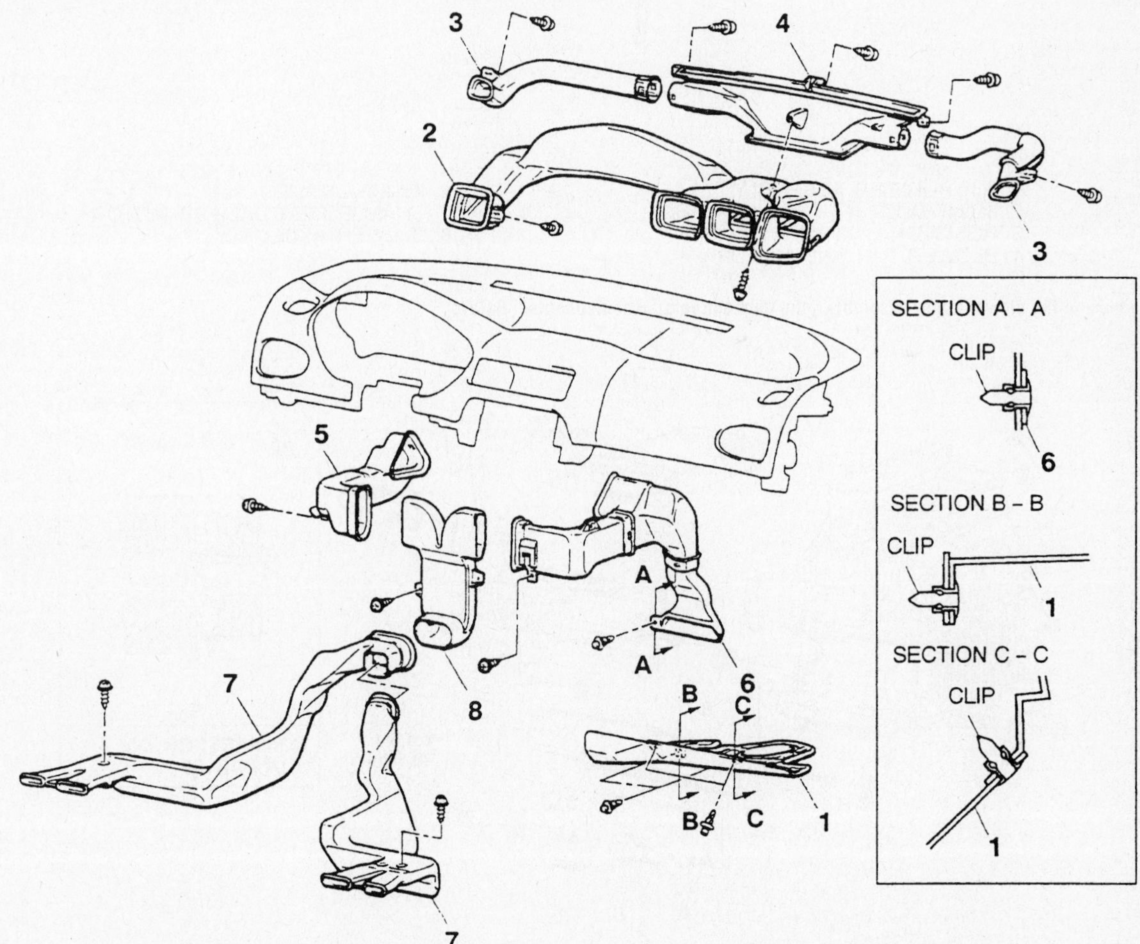

Exploded view of the ventilator assembly—Galant

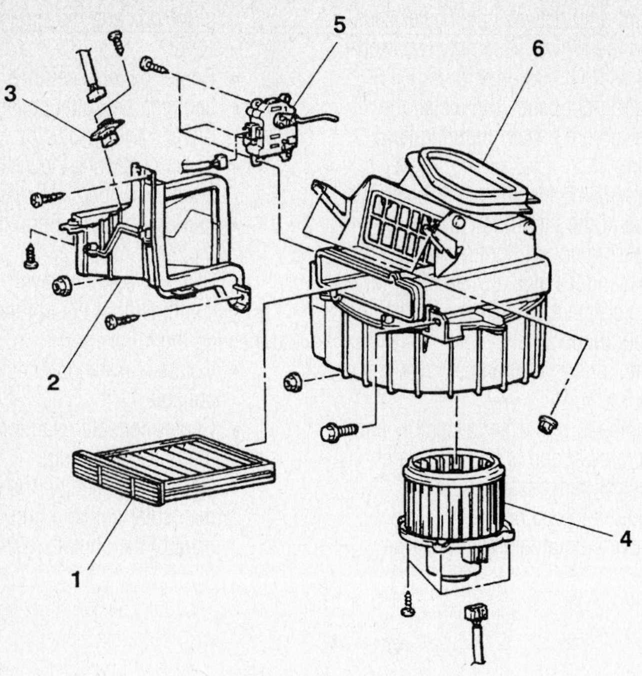

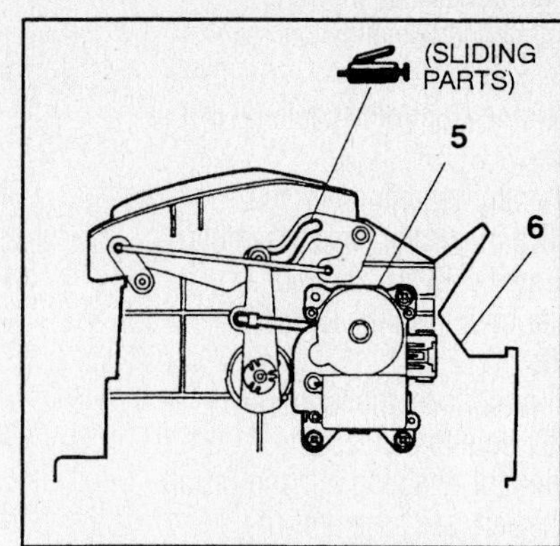

1. AIR PURIFIER ASSEMBLY
2. JOINT DUCT
3. RESISTOR
4. BLOWER FAN AND MOTOR
5. INSIDE/OUTSIDE AIR CHANGEOVER DAMPER MOTOR
6. BLOWER ASSEMBLY

93112GF3

Exploded view of the blower motor assembly, joint duct and related components—Galant

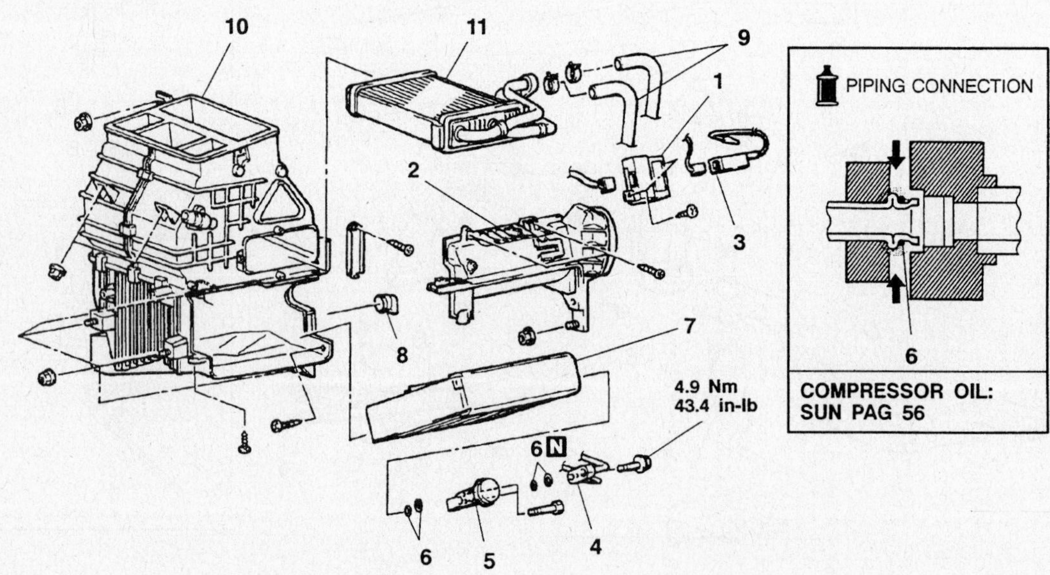

1. BELT LOCK CONTROLLER
2. COVER
3. AUTOMATIC COMPRESSOR CONTROLLER
4. A/C PIPE
5. EXPANSION VALVE
6. O-RING
7. EVAPORATOR
8. DRAIN HOSE
9. HEATER HOSE
10. HEATER/COOLER UNIT
11. HEATER CORE

93112GF4

Exploded view of the heater core, heater housing and related components—Galant

- Shift lever knob on manual transmission vehicles or the shift indicator plate on automatic transmissions
- Heater hoses to the heater core at the firewall

9. Refill the cooling system.
10. Connect the negative battery cable.
11. Operate the engine to normal operating temperatures; then, check the climate control operation and check for leaks.

Lancer and Sportback

1. Before servicing the vehicle, refer to the precautions at the beginning of this section.
2. Disconnect the negative battery cable.
3. Drain the cooling system into a clean container for reuse.
4. Discharge and recover the air conditioning system refrigerant.
5. Remove or disconnect the following:
   - Instrument panel
   - Front seat assembly
   - Front console assembly
   - Front floor carpet
   - Steering shaft attachment bolt
   - Front deck crossmember
   - Heater hoses
   - Flexible suction hose
   - Liquid pipe B connection
   - Center duct
   - Heater unit
   - Intake duct
   - Blower assembly
6. Disassemble the heater unit as necessary for access to components,

To install:

7. Assemble the heater unit as necessary.
8. Install or connect the following:
   - Blower assembly
   - Intake duct
   - Heater unit
   - Center duct
   - Liquid pipe B connection
   - Flexible suction hose
   - Heater hoses
   - Front deck crossmember
   - Steering shaft attachment bolt
   - Front floor carpet
   - Front console assembly
   - Front seat assembly
   - Instrument panel

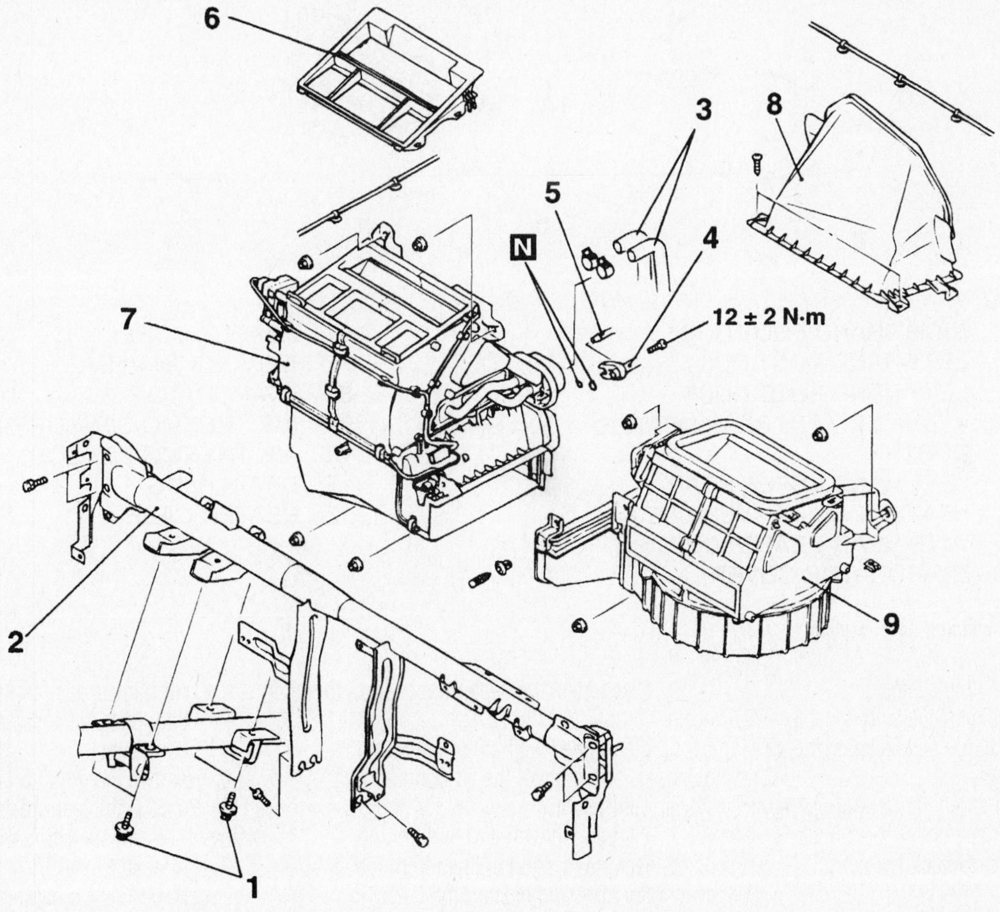

12 ± 2 N·m

1. STEERING SHAFT ATTACHMENT BOLT
2. FRONT DECK CROSSMEMBER
3. HEATER HOSE CONNECTION
4. FLEXIBLE SUCTION HOSE CONNECTION
5. LIQUID PIPE B CONNECTION
6. CENTER DUCT
7. HEATER UNIT
8. INTAKE DUCT
9. BLOWER ASSEMBLY

42356-GALA-G01

Exploded view of the heater core and related components—Lancer

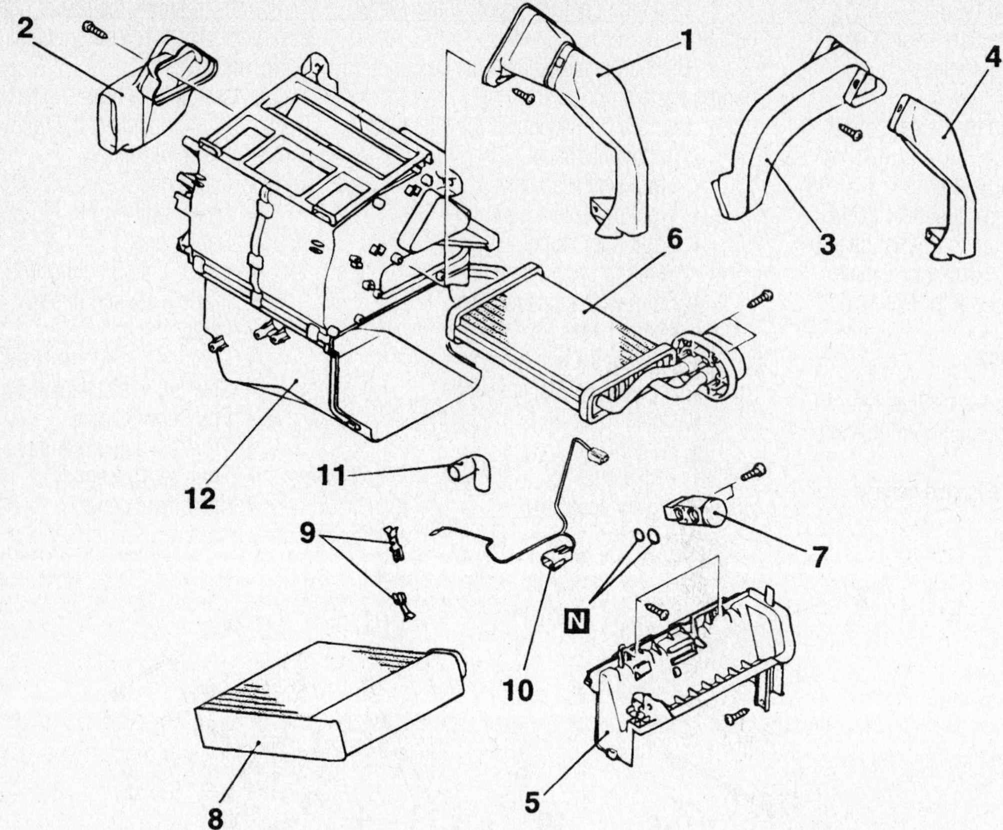

1. RIGHT-HAND FOOT DUCT
2. LEFT-HAND FOOT DUCT
3. LEFT-HAND FOOT DUCT <VEHICLE WITH REAR HEATER DUCT>
4. LEFT-HAND UPPER REAR HEATER DUCT A <VEHICLE WITH REAR HEATER DUCT>
5. EVAPORATOR COVER
6. HEATER CORE
7. EXPANSION VALVE
8. EVAPORATOR
9. AIR THERMO SENSOR CLIP
10. AIR THERMO SENSOR
11. DRAIN PLUG
12. HEATER CASE

42356-GALA-G02

Disassembly of the heater unit—Lancer

9. Refill the cooling system.
10. Connect the negative battery cable.
11. Evacuate, charge and leak test the air conditioning system.
12. Operate the engine to normal operating temperatures; then, check the climate control operation and check for leaks.

Mirage

1. Before servicing the vehicle, refer to the precautions at the beginning of this section.
2. Disconnect the negative battery cable.
3. Drain the cooling system into a clean container for reuse.
4. Remove the air cleaner cover and the air intake hose.
5. Disconnect the heater hoses from the heater core.
6. If equipped with air conditioning, discharge and recover the air conditioning system refrigerant.
7. If equipped with air conditioning, disconnect the refrigerant lines from the evaporator core and discard the O-rings.

⁂ CAUTION

After disconnecting the negative battery cable, wait at least 60 seconds before working on the SRS module or instrument panel.

8. Remove the passenger's side air bag by removing or disconnecting the following:
   - Glove box assembly
   - Air bag-to-dash bolts and the air bag; then, disconnect the electrical connector
9. Remove the floor console.
10. Remove the instrument panel by removing or disconnecting the following:
    - Rheostat
    - Hood release handle
    - Knee protector plug and the knee protector assembly
    - Steering column cover
    - Meter bezel and the combination meter
    - Mirror control switch or plug, if equipped
    - Auto-cruise main switch, fog light switch or plug, if equipped
    - Side air outlet assembly

- Radio and tape player
- Cup holder
- Heater control panel
- Heater control assembly
- Steering column bolts and lower the steering column
- Instrument panel assembly with the help of an assistant

11. If not equipped with air conditioning, remove the blower motor-to-heater housing joint duct.

12. If equipped with air conditioning, remove the evaporator housing-to-heater housing fasteners and remove the evaporator housing.

13. Remove or disconnect the following:
- Center reinforcement
- Center ventilation duct
- Foot distribution duct
- Heater housing
- Heater core from the heater housing

To install:

14. Install or connect the following:
- Heater core to the heater housing
- Heater housing
- Foot distribution duct
- Center ventilation duct
- Center reinforcement

15. If equipped with air conditioning, install the evaporator housing and the evaporator housing-to-heater housing fasteners.

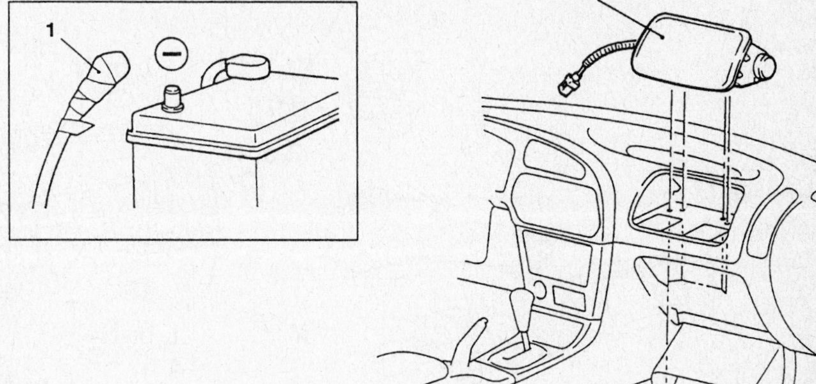

Exploded view of the passenger's air bag module—Mirage

16. If not equipped with air conditioning, install the blower motor-to-heater housing joint duct.

17. Install the instrument panel by installing or connecting the following:
- Instrument panel assembly with the help of an assistant
- Steering column and install the steering column bolts
- Heater control assembly
- Heater control panel
- Cup holder
- Radio and tape player
- Side air outlet assembly

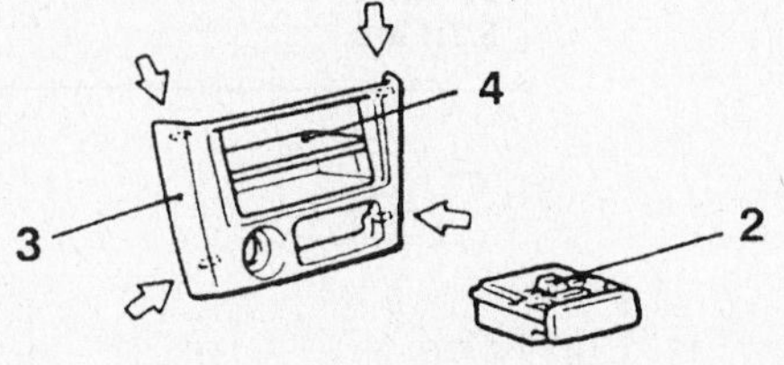

NOTE
⇦ : metal clip position

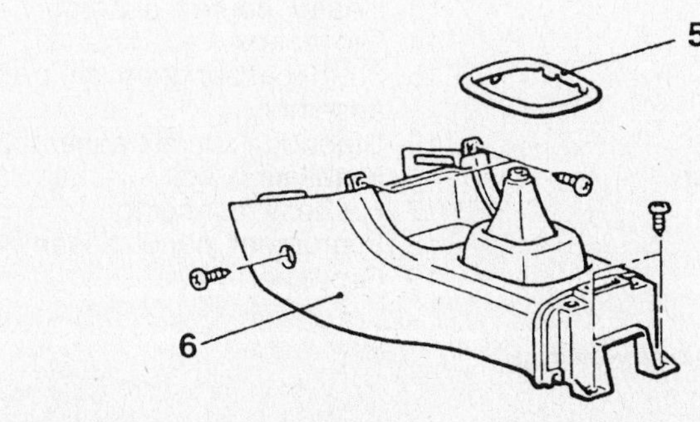

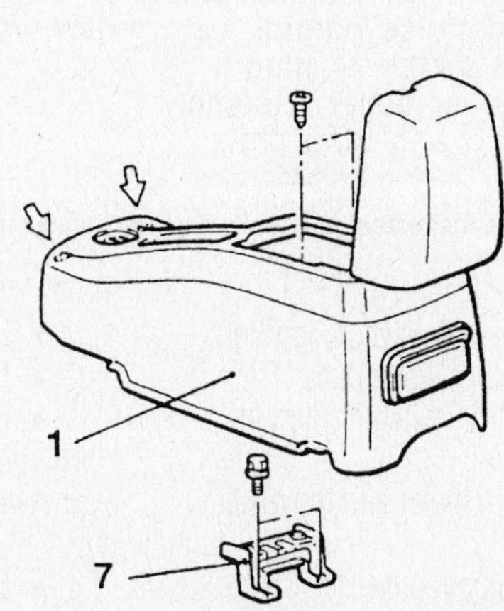

1. Rear floor console assembly
2. Ashtray
3. Audio panel
4. Box
- Shift lever knob
5. A/T panel
6. Front floor console assembly
7. Rear console bracket

Exploded view of the floor console and related components—Mirage

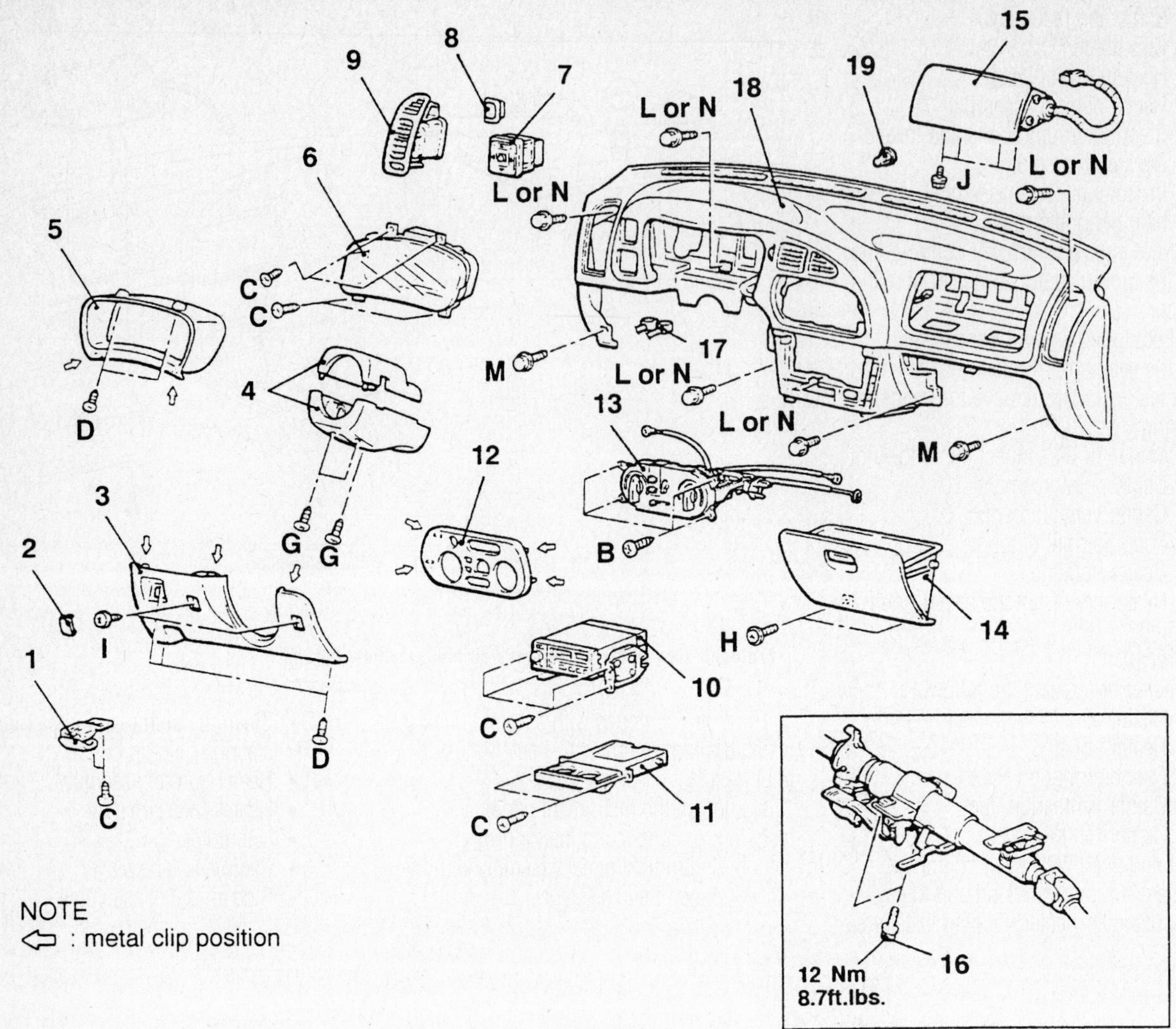

1. Hood lock release handle
2. Knee protector plug
3. Knee protector assembly
4. Column cover
5. Meter bezel
6. Combination meter
7. Door mirror control switch or plug
8. Auto-cruise control main switch, fog light switch or plug
9. Side air outlet assembly
10. Radio and tape player
11. Cup holder
12. Heater control panel
13. Heater control assembly
14. Glove box
15. Front passenger's air bag module assembly
16. Steering column assembly installation bolt
17. Harness connector
18. Instrument panel assembly
19. Grommet

93112GF7

Exploded view of the instrument panel and steering column assembly—Mirage

- Auto-cruise main switch, fog light switch or plug, if equipped
- Mirror control switch or plug, if equipped
- Combination meter and the meter bezel
- Steering column cover
- Knee protector assembly and the knee protector plug
- Hood release handle
- Rheostat
- Floor console

18. Install the passenger's side air bag by installing or connecting the following:

- Electrical connector; then, install the air bag and the air bag-to-dash bolts
- Glove box assembly

19. If equipped with air conditioning, use new O-rings and connect the refrigerant lines to the evaporator core.

20. Install or connect the following:

- Heater hoses to the heater core
- Air cleaner cover and the air intake hose

21. Refill the cooling system.

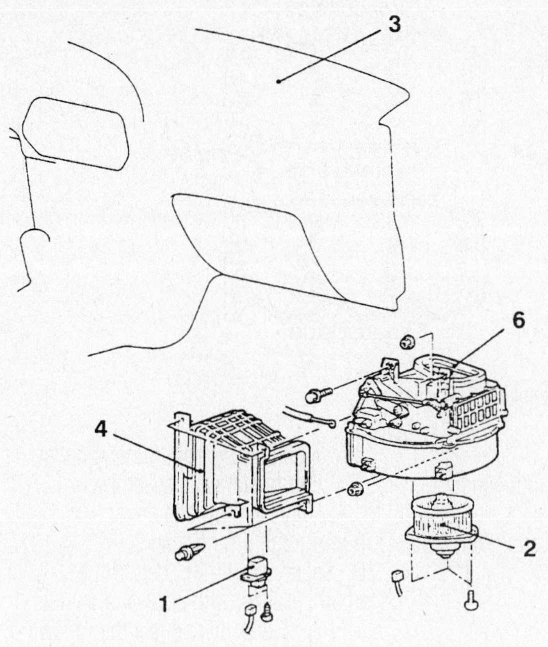

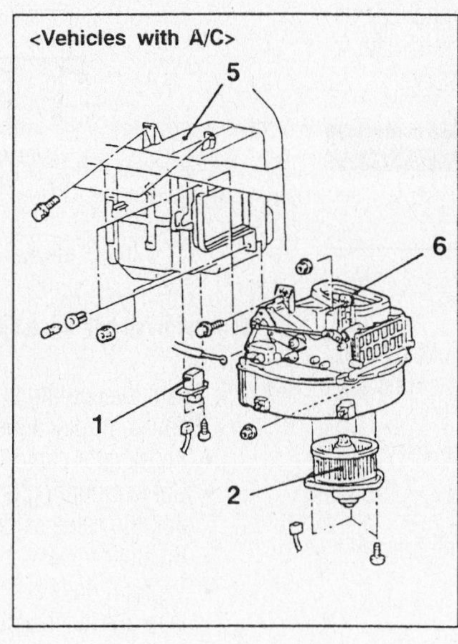

3
6
4
1
2
<Vehicles with A/C>
5
6
1
2
1. Resistor
2. Blower fan and motor
3. Instrument panel
4. Joint duct
5. Evaporator
6. Blower unit assembly
93112GF9

view of the joint duct and blower motor assembly—Mirage

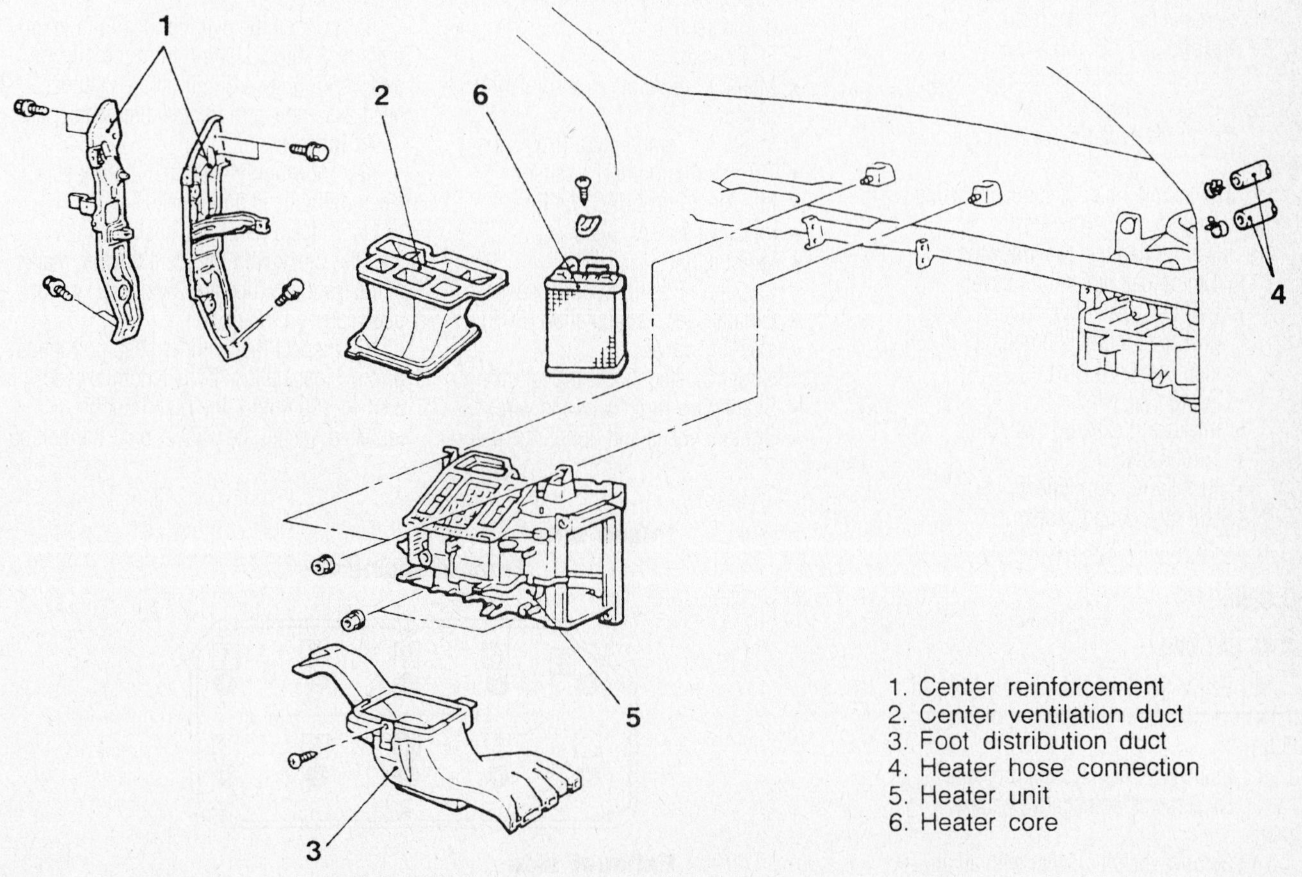

1
2
6
4
5
3
1. Center reinforcement
2. Center ventilation duct
3. Foot distribution duct
4. Heater hose connection
5. Heater unit
6. Heater core
93112GF8

Exploded view of the heater core, heater housing and related components—Mirage

22. Connect the negative battery cable.
23. If equipped with air conditioning, evacuate, charge and leak test the air conditioning system refrigerant.
24. Operate the engine to normal operating temperatures; then, check the climate control operation and check for leaks.

Cylinder Head

REMOVAL & INSTALLATION

Diamante

1. Before servicing the vehicle, refer to the precautions at the beginning of this section.
2. Disconnect the negative battery cable.
3. Drain the engine coolant
4. Remove or disconnect the following:
   - Timing belt
   - Intake and exhaust manifolds
   - Spark plug wires
   - Cylinder head covers
   - Timing belt rear center cover
5. Loosen the cylinder head bolts gradually in 3 stages, in the opposite of the installation sequence.
6. Remove the cylinder head.

To install:

7. Clean the cylinder head and mounting surface on the engine block.
8. Install the cylinder head using a new gasket.
9. Tighten the bolts in sequence using 3 stages to 80 ft. lbs. (105 Nm).
10. Install or connect the following:
    - Timing belt rear center cover
    - Cylinder head covers using new gaskets. Tighten the bolts to 24–36 inch lbs. (3–4 Nm).
    - Spark plug wires
    - Intake and exhaust manifolds
    - Timing belt
    - Remaining components
11. Refill the cooling system.
12. Connect the negative battery cable.

Eclipse

2.4L ENGINE

1. Before servicing the vehicle, refer to the precautions at the beginning of this section.
2. Relieve the fuel system pressure.
3. Disconnect the negative battery cable.
4. Remove the air cleaner with all air intake hoses.

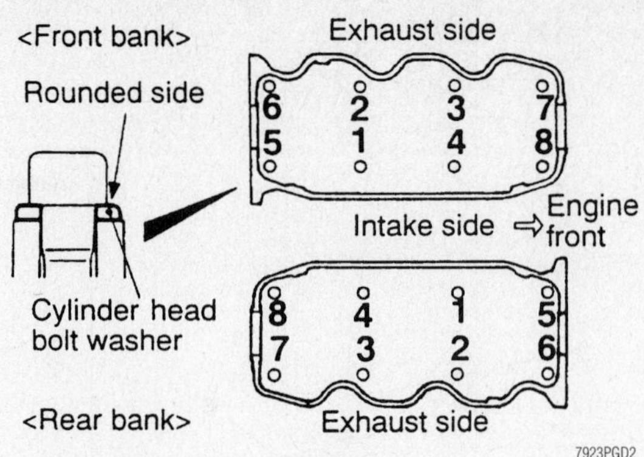

Cylinder head bolt tightening sequence—3.5L engine

5. Drain the cooling system.
6. Remove or disconnect the following:
   - Accelerator cable
   - Cable mounting brackets and position the cable aside
   - Breather hose
   - Vacuum lines at the throttle body, label for identification
   - High pressure fuel line, plug
   - Fuel return hose, plug
7. Disconnect the following connectors:
   - Air conditioning compressor
   - Power steering pressure switch
   - Heated Oxygen (HO_2S) sensor
   - Engine Coolant Temperature (ECT) gauge sender
   - ECT sensor
   - Manifold Absolute Pressure (MAP) sensor
   - Intake Air Temperature (IAT) sensor
   - Throttle Position (TP) sensor
   - Idle Air Control (IAC) motor
   - Injector harness
   - Ignition coil
   - Camshaft Position (CMP) sensor
   - Exhaust Gas Recirculation (EGR) solenoid valve
8. Remove or disconnect the following:
   - Spark plug wire cover and wires
   - Coolant hoses and unbolt the thermostat case from the engine, at the thermostat case assembly
   - Upper timing belt cover
9. Align all timing marks.
10. Secure the timing belt to the camshaft sprocket with cord or a wire tie.
11. Remove or disconnect the following:
    - Camshaft sprocket
    - Valve cover and the half-round seal
    - Intake manifold stay bracket from the intake manifold
    - Exhaust pipe self-locking nuts and separate the exhaust pipe from the exhaust manifold. Discard the gasket.
12. Loosen the cylinder head mounting bolts in 3 steps, starting from the outside and working inward. Lift off the cylinder head assembly and remove the head gasket.

To install:

13. Thoroughly clean the mating surfaces of the head and block.
14. Place a new head gasket on the cylinder block with the identification marks at the front top (upward) position. Do not use sealer on the gasket.
15. Inspect the cylinder head bolt length prior to installation. If the length exceeds 3.91 in. (99.4mm), the bolt must be replaced. Install the washer onto the bolt so

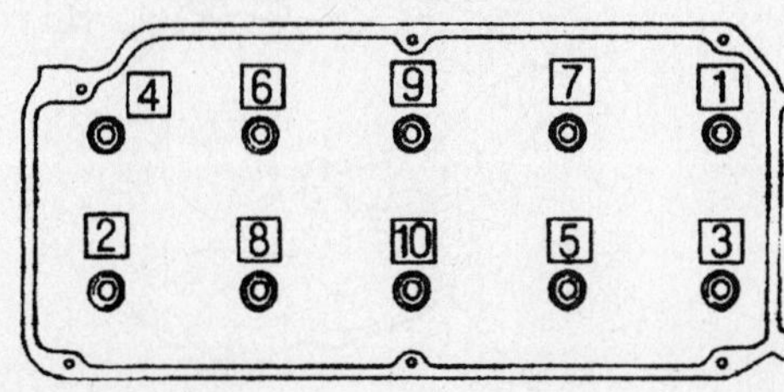

Cylinder head bolt removal sequence—2.4L engine

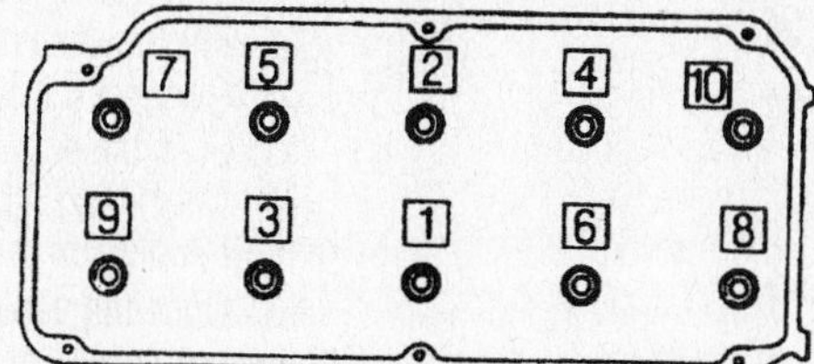

Cylinder head bolt installation sequence—2.4L engine

the chamfer on the washer faces towards the head of the bolt.

16. Carefully install the cylinder head on the block and tighten the cylinder head bolts as follows:
    a. Following the proper tightening sequence, tighten the cylinder head bolts to 58 ft. lbs. (78 Nm).
    b. Loosen all bolts completely.
    c. Torque bolts to 15 ft. lbs. (20 Nm).
    d. Tighten bolts an additional ¼ turn.
    e. Tighten bolts an additional ¼ turn.
17. Install or connect the following:
    - New exhaust pipe gasket and connect the exhaust pipe to the manifold. Tighten the bolts to 33 ft. lbs. (44 Nm).
    - Thermostat case and tighten the mounting bolts to 18 ft. lbs. (24 Nm)
    - Coolant hoses to the thermostat case
18. Apply sealer to the perimeter of the half-round seal and to the lower edges of the half-round portions of the belt-side of the new gasket. Install the valve cover.
19. Install or connect the following:
    - Camshaft sprocket with the timing belt attached. Remove the cord or wire tie.
    - Upper timing belt cover
    - Intake manifold stay and tighten the mounting bolts to 22 ft. lbs. (30 Nm)
20. Connect the following connectors:
    - Air conditioning compressor
    - Power steering pressure switch
    - HO_2S sensor
    - ECT gauge sender
    - ECT sensor
    - MAP sensor
    - IAT sensor
    - TP sensor
    - IAC motor
    - Injector harness
    - Ignition coil
    - CMP sensor
    - EGR solenoid valve
21. Remove or disconnect the following:
    - Spark plug wires and cover
    - Fuel lines using new O-rings
    - Air cleaner and intake hose
    - Breather hose
22. Fill the cooling system.
23. Connect the negative battery cable

3.0L ENGINE

1. Before servicing the vehicle, refer to the precautions at the beginning of this section.
2. Relieve the fuel system pressure. Disconnect the negative battery cable.
3. Drain the cooling system.
4. Remove or disconnect the following:
    - Air intake hose
    - Exhaust manifold
    - Air intake plenum and intake manifold
    - Timing belt
    - Camshaft sprockets and the rear timing belt cover
    - Power steering pump bracket. If removing the rear head, remove the alternator brace.
    - Water inlet pipe
    - Purge pipe assembly
    - Valve cover
5. Using the reverse sequence of the installation sequence, loosen the cylinder head mounting bolts in 3 steps. Lift off the cylinder head assembly and remove the head gasket.

To install:

6. Thoroughly clean the sealing surfaces of the head and block.
7. Place a new head gasket on the cylinder block making sure the identification mark on the cylinder head gasket is in the front top (upward) location. Do not use sealer on the gasket.
8. Carefully install the cylinder head on the block. Be sure the head bolt washers are installed with the chamfered edge upward. Using 3 even steps, torque the head bolts in sequence, to 76–83 ft. lbs. (105–115 Nm).
9. Apply sealer to the lower edges of the half-round portions and install the valve cover. Tighten valve cover bolts to 84 inch lbs. (9 Nm).
10. Install or connect the following:
    - Purge pipe assembly
    - Water inlet pipe
    - Power steering pump bracket and alternator brace
    - Rear timing belt cover and camshaft sprockets. Torque the retaining bolt to 65 ft. lbs. (90 Nm).
    - Timing belt

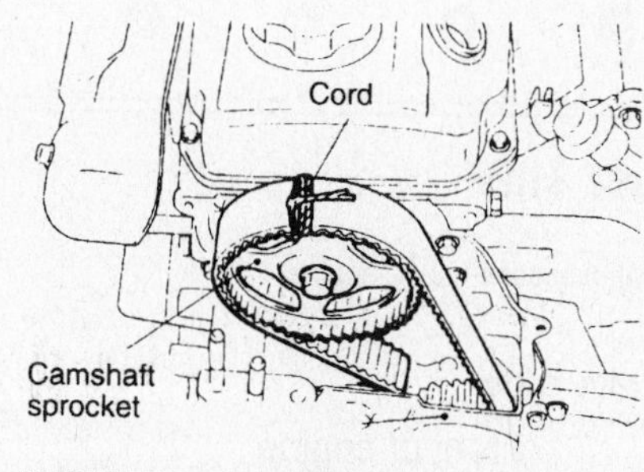

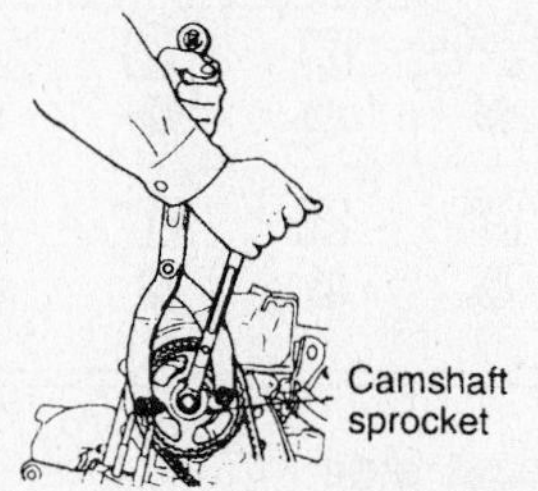

Secure the timing belt to the camshaft sprocket and remove the sprocket—2.4L (4G64) engine

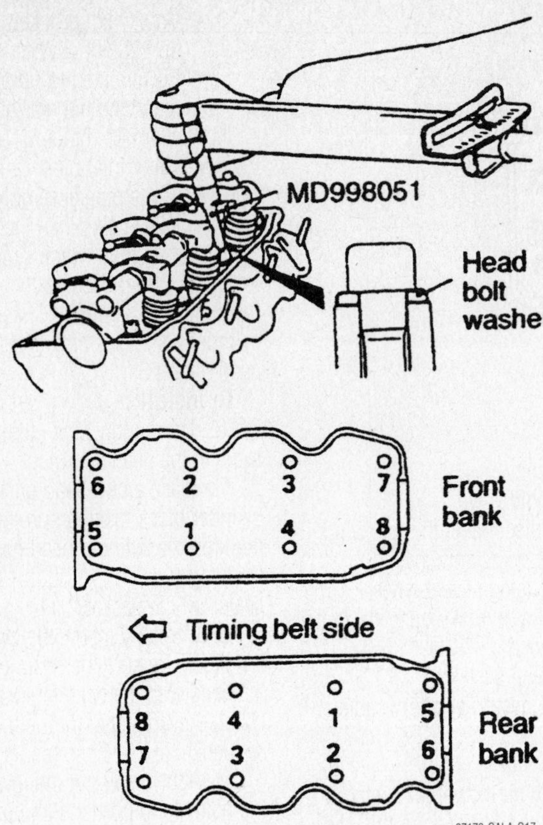

Tighten the cylinder head bolts according to the sequence shown—3.0L engines

Intake side

Front of engine

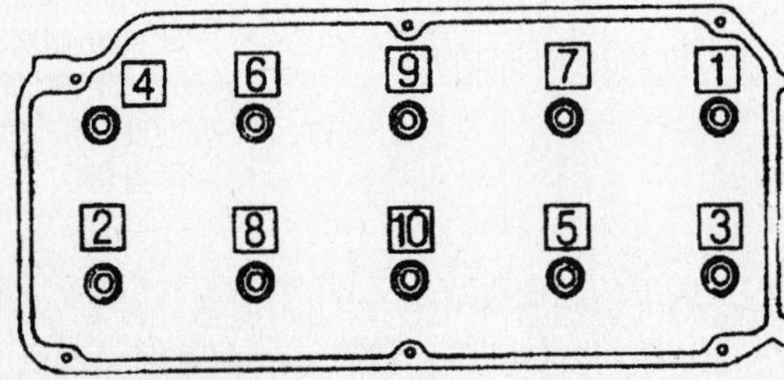

Exhaust side

67170-GALA-G15

Cylinder head bolt removal sequence—2.4L engine

Intake side

Front of engine

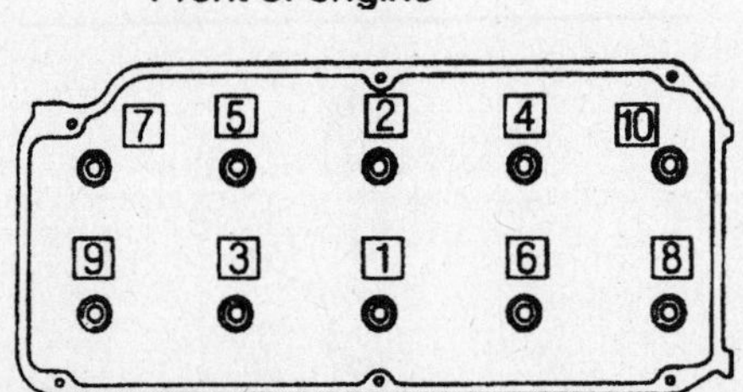

Exhaust side

67170-GALA-G16

Cylinder head bolt installation sequence—2.4L engine

- Intake manifold, air intake plenum and exhaust manifold, using new gaskets
- Air intake hose

11. Fill the system with coolant.
12. Connect the negative battery cable.
13. Start the engine.
14. Check and adjust the idle speed and ignition timing.
15. Once the vehicle has cooled, recheck the coolant level.

Galant

2.4L ENGINE

1. Before servicing the vehicle, refer to the precautions at the beginning of this section.
2. Relieve the fuel system pressure.
3. Disconnect the negative battery cable.
4. Remove the air cleaner with all air intake hoses.
5. Drain the cooling system.
6. Remove or disconnect the following:
   - Accelerator cable
   - Cable mounting brackets and position the cable aside
   - Breather hose
   - Vacuum lines at the throttle body, label for identification
   - High pressure fuel line and plug the line to avoid contamination
   - Fuel return hose and plug the hose to avoid contamination
7. Disconnect the following connectors:
   - Air conditioning compressor
   - Power steering pressure switch
   - Heated Oxygen (HO_2S) sensor
   - Engine Coolant Temperature (ECT) gauge sender
   - ECT sensor
   - Manifold Absolute Pressure (MAP) sensor
   - Intake Air Temperature (IAT) sensor
   - Throttle Position (TP) sensor
   - Idle Air Control (IAC) motor
   - Injector harness
   - Ignition coil
   - Camshaft Position (CMP) sensor
   - Exhaust Gas Recirculation (EGR) solenoid valve
8. Remove or disconnect the following:
   - Spark plug wire cover and wires
   - Coolant hoses and unbolt the thermostat case from the engine, at the thermostat case assembly
   - Upper timing belt cover
9. Align all timing marks.
10. Secure the timing belt to the camshaft sprocket with cord or a wire tie.
11. Remove or disconnect the following:

- Camshaft sprocket
- Valve cover and the half-round seal
- Intake manifold stay bracket from the intake manifold
- Exhaust pipe self-locking nuts and separate the exhaust pipe from the exhaust manifold. Discard the gasket.

12. Loosen the cylinder head mounting bolts in 3 steps, starting from the outside and working inward. Lift off the cylinder head assembly and remove the head gasket.

To install:

13. Thoroughly clean the mating surfaces of the head and block.
14. Place a new head gasket on the cylinder block with the identification marks at the front top (upward) position. Do not use sealer on the gasket.
15. Inspect the cylinder head bolt length prior to installation. If the length exceeds 3.91 in. (99.4mm), the bolt must be replaced. Install the washer onto the bolt so the chamfer on the washer faces towards the head of the bolt.
16. Carefully install the cylinder head on the block and tighten the cylinder head bolts as follows:
    a. Following the proper tightening sequence, tighten the cylinder head bolts to 58 ft. lbs. (78 Nm).
    b. Loosen all bolts completely.
    c. Torque bolts to 15 ft. lbs. (20 Nm).
    d. Tighten bolts an additional ¼ turn.
    e. Tighten bolts an additional ¼ turn.
17. Install or connect the following:
    - New exhaust pipe gasket and connect the exhaust pipe to the manifold. Tighten the bolts to 33 ft. lbs. (44 Nm).
    - Thermostat case and tighten the mounting bolts to 18 ft. lbs. (24 Nm)
    - Coolant hoses to the thermostat case
18. Apply sealer to the perimeter of the half-round seal and to the lower edges of the half-round portions of the belt-side of the new gasket. Install the valve cover.
19. Install or connect the following:
    - Camshaft sprocket with the timing belt attached. Remove the cord or wire tie.
    - Upper timing belt cover
    - Intake manifold stay and tighten the mounting bolts to 22 ft. lbs. (30 Nm)
20. Connect the following connectors:
    - Air conditioning compressor
    - Power steering pressure switch
    - HO_2S sensor
    - ECT gauge sender
    - ECT sensor
    - MAP sensor
    - IAT sensor
    - TP sensor
    - IAC motor
    - Injector harness
    - Ignition coil
    - CMP sensor
    - EGR solenoid valve
21. Remove or disconnect the following:
    - Spark plug wires and cover
    - Fuel lines using new O-rings
    - Air cleaner and intake hose
    - Breather hose
22. Fill the cooling system.
23. Connect the negative battery cable

3.0L ENGINE

1. Before servicing the vehicle, refer to the precautions at the beginning of this section.
2. Relieve the fuel system pressure. Disconnect the negative battery cable.
3. Drain the cooling system.
4. Remove or disconnect the following:
    - Air intake hose
    - Exhaust manifold
    - Air intake plenum and intake manifold
    - Timing belt
    - Camshaft sprockets and the rear timing belt cover
    - Power steering pump bracket. If removing the rear head, remove the alternator brace.
    - Water inlet pipe
    - Purge pipe assembly
    - Valve cover
5. Using the reverse sequence of the installation sequence, loosen the cylinder head mounting bolts in 3 steps. Lift off the cylinder head assembly and remove the head gasket.

To install:

6. Thoroughly clean the sealing surfaces of the head and block.
7. Place a new head gasket on the cylinder block making sure the identification mark on the cylinder head gasket is in the front top (upward) location. Do not use sealer on the gasket.
8. Carefully install the cylinder head on the block. Be sure the head bolt washers are installed with the chamfered edge upward. Using 3 even steps, torque the head bolts in sequence, to 76–83 ft. lbs. (105–115 Nm).
9. Apply sealer to the lower edges of the half-round portions and install the valve cover. Tighten valve cover bolts to 84 inch lbs. (9 Nm).
10. Install or connect the following:
    - Purge pipe assembly
    - Water inlet pipe
    - Power steering pump bracket and alternator brace
    - Rear timing belt cover and camshaft sprockets. Torque the retaining bolt to 65 ft. lbs. (90 Nm).
    - Timing belt
    - Intake manifold, air intake plenum and exhaust manifold, using new gaskets
    - Air intake hose
11. Fill the system with coolant.
12. Connect the negative battery cable.
13. Start the engine.
14. Check and adjust the idle speed and ignition timing.
15. Once the vehicle has cooled, recheck the coolant level.

3.8L ENGINE

1. Before servicing the vehicle, refer to the precautions at the beginning of this section.
2. Disconnect the negative battery cable.
3. Drain the engine coolant.
4. Remove the timing belt.
5. Remove the intake and exhaust manifolds.
6. Disconnect the spark plug wires.
7. Remove the cylinder head covers.
8. Remove the timing belt rear center cover.
9. Loosen the cylinder head bolts gradually in 3 stages, in the opposite of the installation sequence.
10. Remove the cylinder head.

To install:

11. Clean the cylinder head and mounting surface on the engine block.
12. Install the cylinder head using a new gasket.
13. Tighten the bolts in sequence using 3 stages to 76–84 ft. lbs. (103–113 Nm),

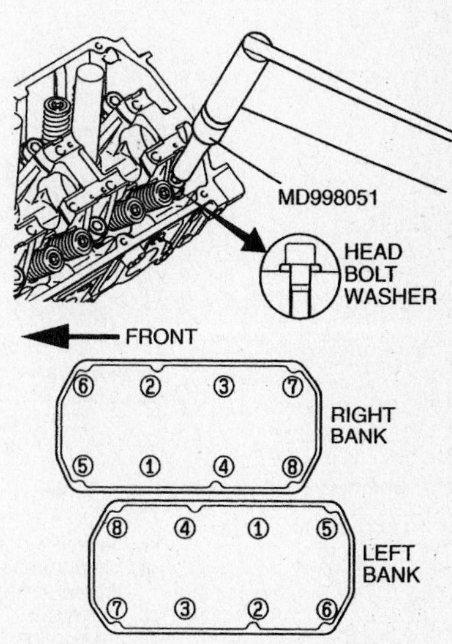

Cylinder head bolt tightening sequence—3.8L engine

then loosen all bolts to 0 inch lbs., then tighten them again to 76–84 ft. lbs. (103–113 Nm).

14. Install the timing belt rear center cover.
15. Install the cylinder head covers using new gaskets. Tighten the bolts to 24–36 inch lbs. (3–4 Nm).
16. Connect the spark plug wires.
17. Install the intake and exhaust manifolds.
18. Install the timing belt.
19. Install the remaining components.
20. Refill the cooling system.
21. Connect the negative battery cable.

Lancer, Evolution and Sportback

1. Before servicing the vehicle, refer to the precautions at the beginning of this section.
2. Relieve the fuel system pressure. Disconnect the negative battery cable.
3. Drain the cooling system.
4. If equipped, remove the turbocharger assembly.
5. Remove or disconnect the following:

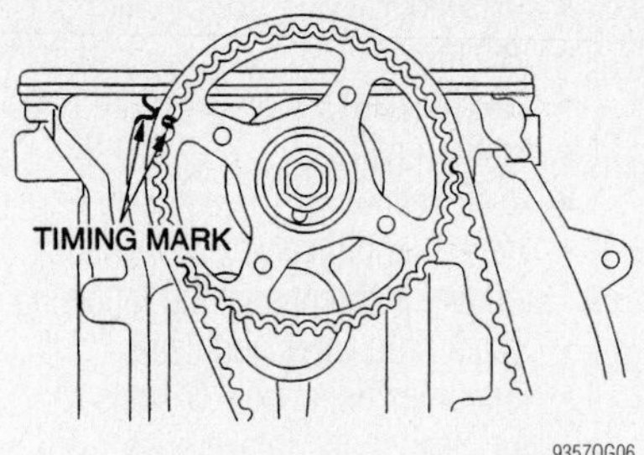

Proper timing mark alignment—Lancer with the 2.0L engine

- Engine undercover
- Air cleaner assembly
- Exhaust manifold
- Water hose and pipe

6. Detach the electrical connectors from the following components:

- Accelerator cable
- Air Conditioning (A/C) compressor
- Power steering oil pressure switch
- Crank angle sensor
- Manifold Differential Pressure (MDP) sensor
- Evaporative emission (EVAP) purge solenoid

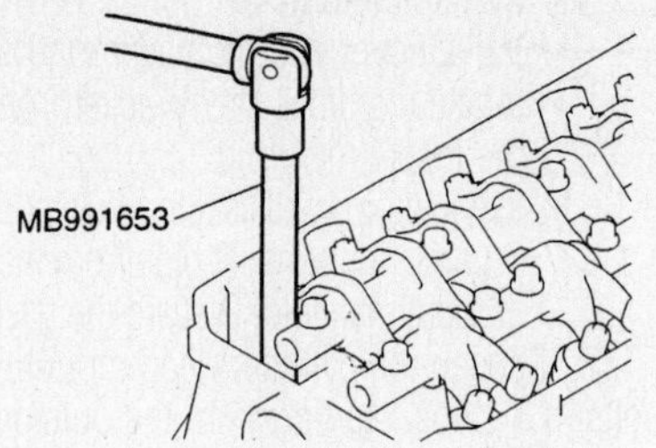

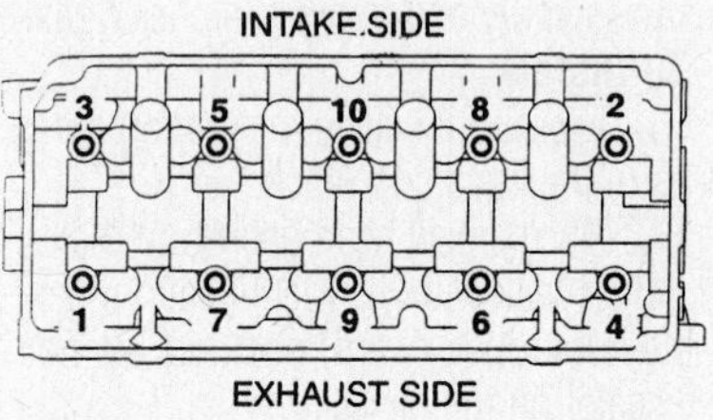

Cylinder head bolt removal sequence—Lancer with 2.0L engine

- Exhaust Gas Recirculation (EGR) solenoid valve
- Ignition coil
- Fuel injectors
- Throttle Position (TP) sensor

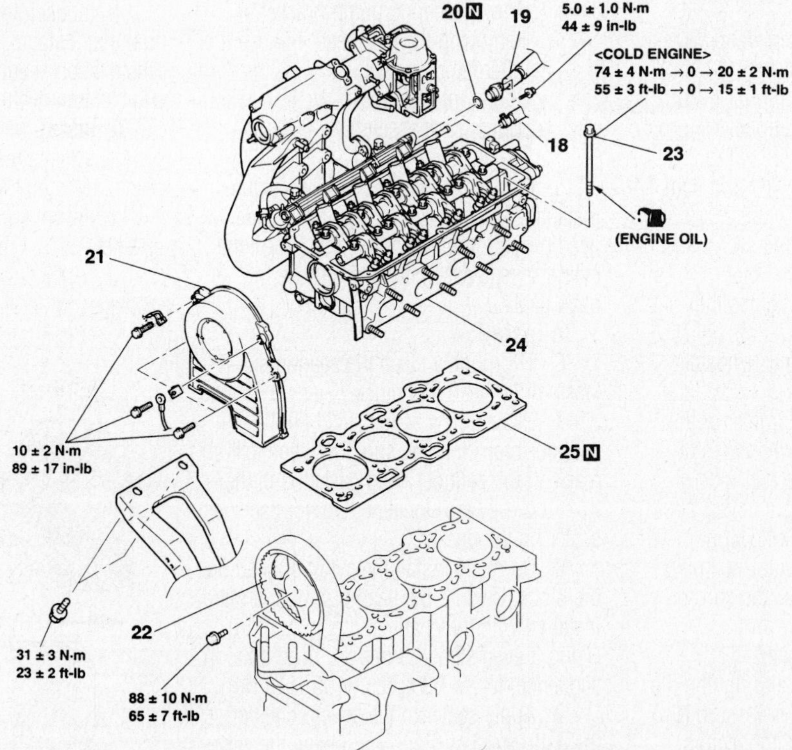

Exploded view of the cylinder head and related components—Lancer with 2.0L engine

- Idle Air Control (IAC) motor
- Engine Coolant Temperature (ECT) sensor
- Camshaft Position (CMP) sensor
- Knock Sensor (KS)
- ECT gauge unit
- Heated Oxygen (HO_2S) sensor
- Starter and alternator
- Oil pressure switch
- Brake booster vacuum hose

7. Remove or disconnect the following:
   - Purge hose connection
   - High pressure fuel line. Remove and discard the O-rings.
   - Timing belt front upper cover

⁂ WARNING

Always turn the crankshaft in the forward direction (clockwise) only!

8. Remove the camshaft sprocket, as follows:
   a. Turn the crankshaft clockwise to align the timing mark so that the No. 1 cylinder is at Top Dead Center (TDC) of the compression stroke.
   b. Secure the cam sprocket and timing belt with wire ties to prevent them from slipping out of place.
   c. While holding the sprocket from turning with special tools MB990767 and MD998719, remove the camshaft sprocket bolt and sprocket.
9. Remove or disconnect the following:
   - Cylinder head bolts, using the proper tool, in the proper sequence
   - Cylinder head gasket
   - Cylinder head assembly

To install:

10. Thoroughly clean the mating surfaces of the head and block.

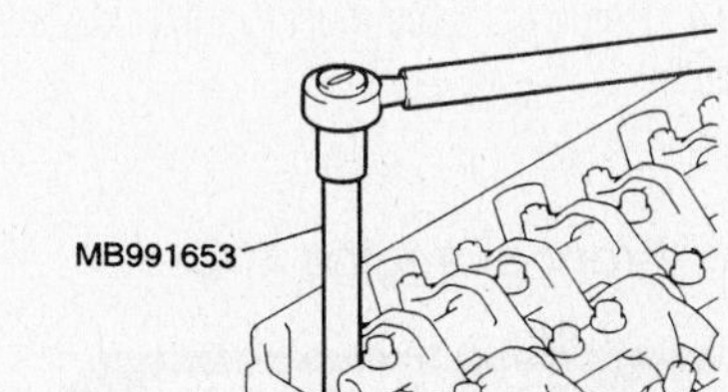

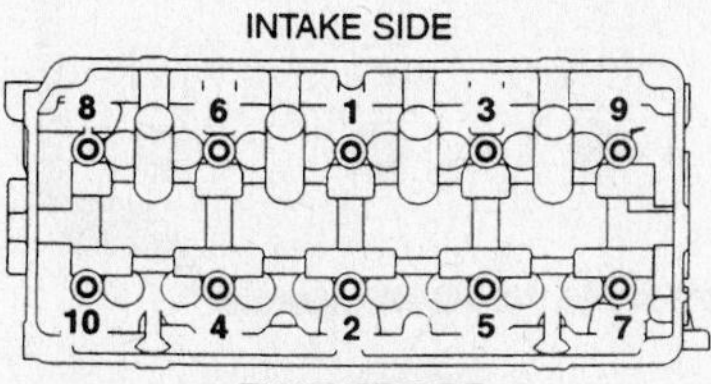

Cylinder head bolt installation sequence—2.0L engine

11. Place a new head gasket on the cylinder block with the identification marks at the front top (upward) position. Do not use sealer on the gasket.
12. Inspect the cylinder head bolt length prior to installation. If the length exceeds 3.8 in. (96.4mm), the bolt must be replaced. Install the washer onto the bolt so the chamfer on the washer faces towards the head of the bolt.
13. Carefully install the cylinder head on the block and tighten the cylinder head bolts as follows:
    a. Following the proper tightening sequence, tighten the cylinder head bolts to 52–58 ft. lbs. (70–78 Nm).
    b. Loosen all bolts completely.
    c. Torque bolts to 15 ft. lbs. (20 Nm).
    d. Tighten bolts an additional ¼ turn.
    e. Tighten bolts an additional ¼ turn.
14. The remainder of installation is the reverse of the removal procedure.
15. Fill the system with coolant.
16. Connect the negative battery cable.

Mirage

1.5L ENGINE

1. Before servicing the vehicle, refer to the precautions at the beginning of this section.
2. Relieve the fuel system pressure. Disconnect the negative battery cable.
3. Drain the cooling system.
4. Remove or disconnect the following:
   - Air intake hose and the air cleaner assembly
   - Ground cable connection and the accelerator cable
   - Positive Crankcase Ventilation (PCV) and the breather hose connection
   - Vacuum hoses from the intake and throttle body, label for reference
   - Vacuum line for the brake booster
   - Upper radiator hose, throttle body hoses, bypass hose and heater hose connections
   - Fuel feed and return lines
   - Spark plug wires
5. Disconnect the electrical harness plugs from the following:
   - Crankshaft Position (CKP) and Camshaft Position (CMP) sensors
   - Heated Oxygen (HO_2S) sensor
   - Engine Coolant Temperature (ECT) sensor and gauge sender
   - Idle Speed Control (ISC) motor
   - Throttle Position (TP) sensor
   - Intake Air Temperature (IAT) sensor
   - Exhaust Gas Recirculation (EGR) temperature sensor
6. Remove or disconnect the following:
   - Electrical harness plugs from the ignition distributor, fuel injectors, power transistor and ground cable
   - Engine control wiring harness
   - Clamp that holds the power steering pressure hose to the engine mounting bracket
7. Place a jack and wood block under the oil pan and carefully lift just enough to take the weight off the engine mounting bracket and remove the bracket.
   - Valve cover
   - Timing belt upper cover
8. Rotate the crankshaft clockwise and align the timing marks.
9. Attach the timing belt to the camshaft sprocket with cord or a wire tie.
10. Secure the camshaft from turning and remove the camshaft sprocket with the timing belt attached.
11. Remove the timing belt rear upper cover.
12. Remove the exhaust pipe from the exhaust manifold.
13. Loosen the cylinder head mounting bolts in sequence using 3 steps. Remove the cylinder head.

To install:

14. Thoroughly clean the mating surfaces of the head and block.
15. Place a new head gasket on the cylin-

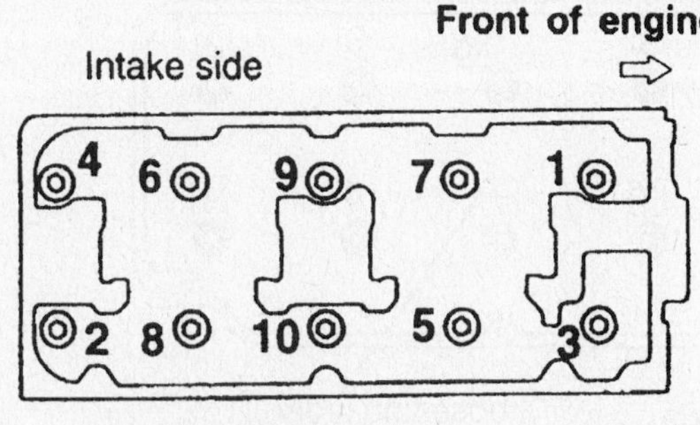

Cylinder head bolt loosening sequence—Mirage with 1.5L (4G15) engine

Intake side ⇦ Front of engine

8 6 1 3 9

10 4 2 5 7

Exhaust side

7923PG14

Cylinder head bolt tightening sequence—Mirage with 1.5L (4G15) engine

der block with the identification marks facing upward. Do not use sealer on the gasket.

16. Carefully install the cylinder head on the block. Tighten the cylinder head bolts as follows:
    a. 36 ft. lbs. (49 Nm) in the correct sequence.
    b. Loosen the bolts completely in the reverse order.
    c. Tighten the bolts in sequence to 14 ft. lbs. (20 Nm).
    d. Tighten each bolt in sequence 90 degrees.
    e. Tighten each bolt in sequence an additional 90 degrees.
17. Install or connect the following:
    - New exhaust pipe gasket and connect the exhaust pipe to the manifold
    - Upper rear timing cover
18. Align the timing marks and install the cam sprocket. Torque the retaining bolt to 51 ft. lbs. (70 Nm). Check the belt tension and adjust, if necessary. Install the outer timing cover.
19. Install or connect the following:
    - Valve cover and torque the retaining bolts to 16 inch lbs. (1.8 Nm)
    - Engine mount bracket and remove the support jack
    - Clamp that holds the power steering pressure hose to the engine mounting bracket
20. Connect the following:
    - CKP and CMP sensors
    - HO_2S sensor
    - ECT sensor and gauge sender
    - ISC motor
    - TP sensor
    - IAT
    - EGR temperature sensor
21. Install or connect the following:
    - Ignition distributor, fuel injectors, power transistor and ground cable
    - Engine control wiring harness
    - Fuel lines with new O-rings
    - Air cleaner assembly
    - Breather hose
22. Fill the system with coolant.
23. Connect the negative battery cable.

1.8L ENGINE

1. Before servicing the vehicle, refer to the precautions at the beginning of this section.
2. Relieve fuel system pressure. Disconnect the negative battery cable.
3. Remove the air cleaner assembly.
4. Drain the cooling system.
5. Disconnect the brake booster vacuum hose and PVC valve connection.
6. Note the locations and disconnect the vacuum hoses from the intake and throttle body.
7. Remove or disconnect the following:
    - Upper radiator hose, overflow tube and the water hose from the thermostat to the throttle body
    - Fuel feed and return lines
    - Accelerator cable connection from the throttle body
    - Oil pressure switch
8. Disconnect the following:
    - Heated Oxygen (HO_2S) sensor
    - Engine Coolant Temperature (ECT) sensor and gauge sender
    - IAC motor
    - Exhaust Gas Recirculation (EGR) temperature sensor
    - Throttle Position (TP) sensor
    - Knock (KS) sensor
    - Fuel injectors
    - Spark plug wires
    - Control harness assembly and position aside
    - Thermostat housing, thermostat and the thermostat case with O-ring from the engine
    - Rocker cover
    - Timing belt upper cover
9. Rotate the crankshaft clockwise and align the timing marks.
10. Attach the timing belt to the camshaft sprocket with cord or a wire tie.
11. Secure the camshaft from turning and remove the camshaft sprocket with the timing belt attached.
12. Remove the timing belt rear upper cover.
13. Loosen the cylinder head bolts in 2 or 3 steps in the proper sequence.
14. Remove the cylinder head from the engine.

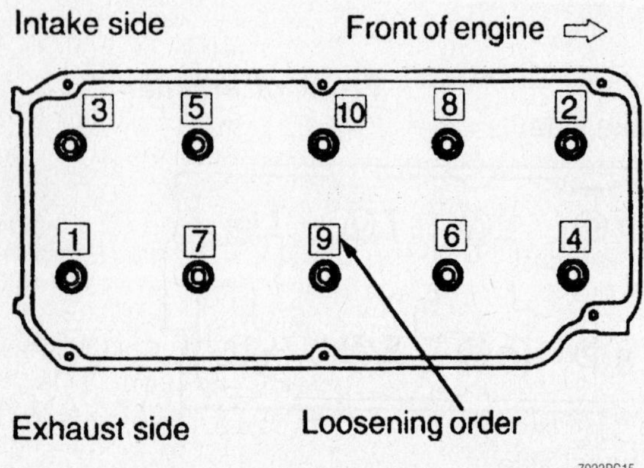

Cylinder head bolt loosening sequence—1.8L engine

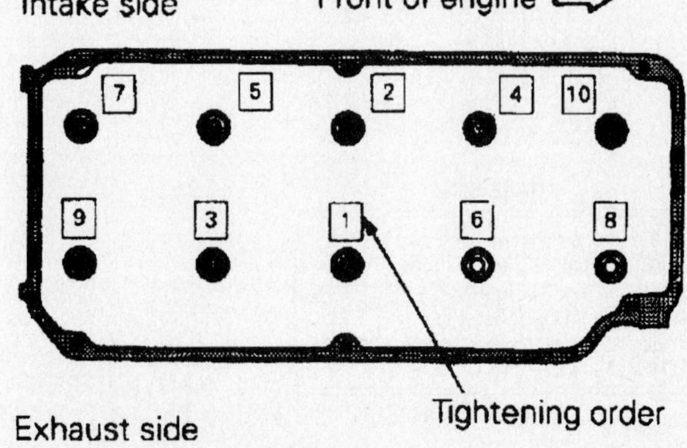

Cylinder head bolt torque sequence—1.8L engine

✻✻ CAUTION

When removing the cylinder head, take care not to bend or damage the plug guide. The plug guide can not be replaced.

To install:

15. Thoroughly clean the mating surfaces of the head and block.
16. Place a new head gasket on the cylinder block with the identification marks facing upward. Do not use sealer on the gasket.
17. Carefully install the cylinder head on the block.
18. Measure the cylinder head bolts prior to installation. Replace any that exceed 3.795 in. (96.4mm).
19. Apply a small amount of engine oil to the thread section of the bolt and install so the chamfer of the washer faces upward.
20. Tighten the cylinder head bolts as follows:
    a. In the proper tightening sequence, torque bolts to 54 ft. lbs. (75 Nm).
    b. In the reverse order of the tightening sequence, fully loosen all bolts.
    c. In the proper tightening sequence, torque bolts to 14 ft. lbs. (20 Nm).
    d. In the proper tightening sequence, tighten bolts ¼ turn (90 degrees).
    e. In the proper tightening sequence, tighten bolts an additional ¼ turn (90 degrees).
21. Install the camshaft sprocket and tighten the bolt to 65 ft. lbs. (90 Nm), while holding the sprocket in place using the appropriate wrench. Confirm proper timing mark alignment.
22. Install the upper timing belt cover and rocker cover. Torque the rocker cover bolts to 29 inch lbs. (3.3 Nm).
23. Loosen the water pipe mounting bolt for ease of thermostat housing installation.
24. Apply a thin bead of RTV sealant to the water tube connection on the thermostat case.
25. Apply a small amount of water to the O-ring of the water inlet pipe and press the thermostat case assembly onto the water inlet pipe. Install the thermostat case assembly mounting bolt tightening to 16 ft. lbs. (22 Nm).
26. Tighten the water pipe mounting bolt.
27. Install the thermostat into the housing so the jiggle valve is located at the top. Tighten the housing bolts to 10 ft. lbs. (14 Nm).
28. Connect the following:
    - HO_2S sensor
    - ECT sensor and gauge sender
    - IAC motor
    - EGR temperature sensor
    - TP sensor
    - KS sensor
    - Fuel injectors
29. Install or connect the following:
    - Upper radiator hose to the thermostat housing
    - Accelerator cable connection to the throttle body
    - Oil pressure switch
    - Spark plug wires
    - Control harness assembly
30. Replace the O-ring for the high pressure hose and install a new clamp on the return hose and reconnect the fuel lines.
    - Air intake hose
    - Breather hose and air cleaner case cover
    - Brake booster and the PCV vacuum hoses
31. Fill the system with coolant.
32. Connect the negative battery cable

Rocker Arms/Shafts

REMOVAL & INSTALLATION

Diamante

1. Before servicing the vehicle, refer to the precautions at the beginning of this section.
2. Disconnect the negative battery cable.
3. Remove the rocker arm cover.
4. Install the lash adjuster clips on the rocker arms, then loosen the bearing cap bolts. Do not remove the bolts from the bearing caps.

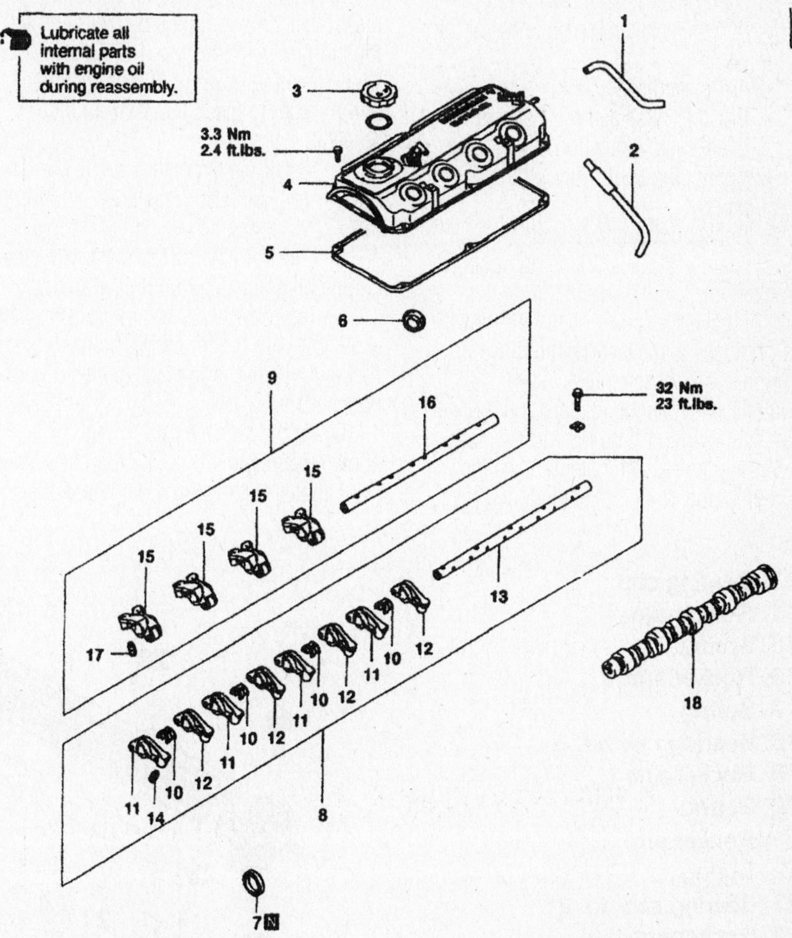

Rocker arm shafts and components—2.4L engines

5. Remove the rocker arms, shafts and bearing caps as an assembly.

To install:

6. Install the bearing caps/rocker arm assemblies. Tighten the bolts to 23 ft. lbs. (31 Nm).
7. Remove the lash adjuster clips.
8. Install the rocker arm cover using a new gasket.
9. Connect the negative battery cable.

Eclipse

2.4L ENGINE

1. Before servicing the vehicle, refer to the precautions at the beginning of this section.
2. Remove or disconnect the following:
   - Negative battery cable
   - Accelerator cable, remove the cable clamp mounting screws and position the accelerator cable out of the way.
   - Air intake hose
   - Breather hose and the Positive Crankcase Ventilation (PCV) hose
   - Spark plug cables from the spark plugs
   - Rocker cover and gasket
3. Install lash adjuster retainer tools to the rocker arm.
4. Remove the rocker shaft hold-down bolts gradually and evenly and remove the rocker shaft/arm assemblies.
5. Disassemble the rockers and the rocker shaft springs from the rocker shafts. If they are to be reused, note the location and positioning of all rocker shaft components. It is recommended that all lash adjusters and rockers be replaced as a complete set.

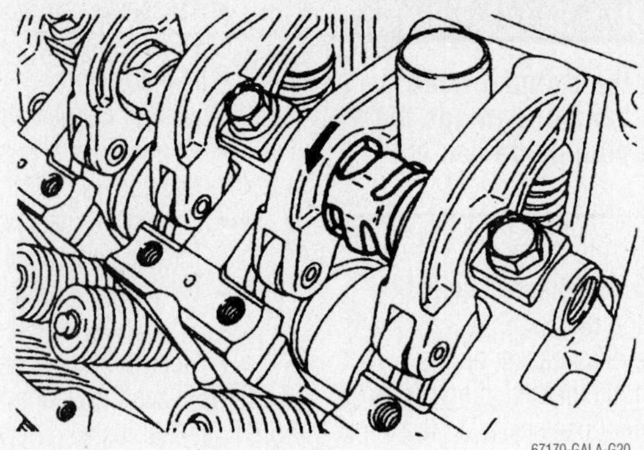

Installing the rocker shaft springs—2.4L engines

To install:

6. Immerse the lash adjusters in clean diesel fuel, and using a small wire, move the plunger up and down 4 or 5 times. While pushing down lightly on the check ball in order to bleed the air from the adjuster.
7. Install the lash adjusters to the rocker arms and attach the special holding tool.
8. Lubricate the rocker shaft with clean engine oil and install the rocker arms.
9. Temporarily tighten the rocker shaft assembly with the mounting bolts so that all rocker arms on the inlet valve side do not push on the valves.
10. Fit the rocker shaft springs from above and position them so that they are at right angles to the plug side. Install the rocker springs before installing the exhaust side rocker shaft and rocker arm assembly.
11. Install the exhaust side rocker shaft assembly in the engine. Tighten the rocker shaft mounting bolts gradually and evenly to 23 ft. lbs. (32 Nm).
12. Remove the lash adjuster retaining tools.
13. Install or connect the following:
    - Rocker cover and tighten the mounting bolts to 29 inch lbs. (3.3 Nm)
    - Spark plug wires to the spark plugs
    - PCV and breather hoses
    - Air intake hose
    - Accelerator cable brackets and reconnect the accelerator cable
    - Negative battery cable

3.0L ENGINE

On this engine, the hydraulic lash adjusters are built into the rocker arms.

13. Bearing cap
14. Rocker arm
15. Spring
16. Rocker arm
17. Spring
18. Bearing cap no. 3
19. Rocker arm
20. Spring
21. Rocker arm
22. Spring
23. Bearing cap no. 2
24. Rocker arm
25. Spring
26. Rocker arm
27. Spring
28. Rocker arm shaft
29. Rocker arm shaft
30. Bearing cap no. 1

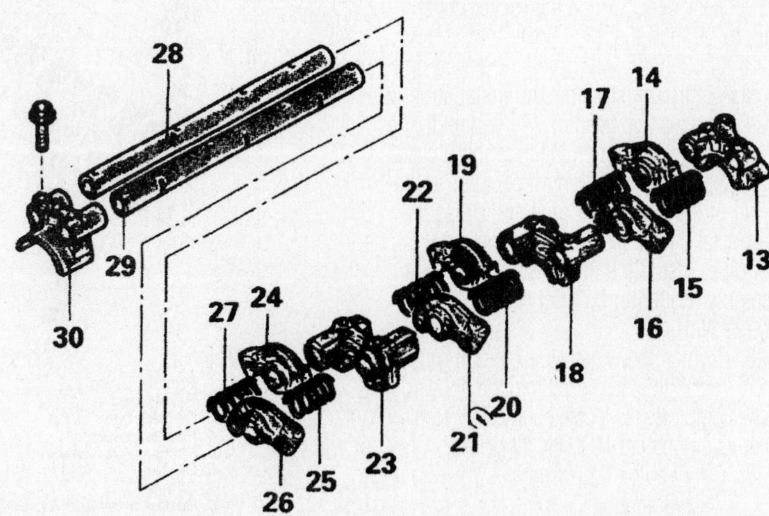

Rocker arm assembly—3.0L engines

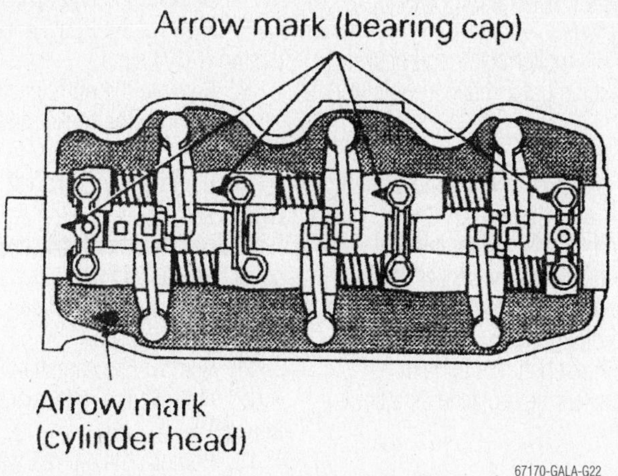

When installing the rocker arm/shaft assemblies, ensure that the arrow marks point in the same direction as the arrow stamped into the cylinder head—3.0L engines

1. Before servicing the vehicle, refer to the precautions at the beginning of this section.
2. Disconnect the negative battery cable.
3. Remove the valve cover. Install lash adjuster retainer tools to prevent the auto-lash adjuster from falling out of the rocker arm.
4. Loosen rocker arm and shaft assembly evenly in several steps. Remove the rocker arm and shaft assembly as a complete unit.
5. Remove the rear camshaft bearing cap and slide the rocker arms, springs and washers from the shaft. If they are to be reused, note the location and positioning of all rocker shaft components. It is recommended that all lash adjusters and rockers be replaced as a complete set.

To install:

6. Immerse the lash adjusters in clean diesel fuel. Using a small wire, move the plunger of the lash adjuster up and down 4 or 5 times while pushing down lightly on the check ball in order to bleed out the air. Install the lash adjusters in the rocker arms.
7. Using a light coat of engine oil, assemble the rocker arms to the shaft. Install the rear camshaft bearing cap.
8. Lubricate the camshaft and rocker shaft with clean engine oil and position on the cylinder head.
9. Apply a drop of sealant to the rear edges of the end caps.
10. Install or connect the following:
   - Assembly making sure the notches in the rocker shafts are facing up
   - Cap bolts and tighten evenly and gradually to 14 ft. lbs. (20 Nm). Remove the lash adjuster retainers.
   - Valve cover
   - Negative battery cable

Lancer and Sportback

2.0L SOHC ENGINE

1. Before servicing the vehicle, refer to the precautions at the beginning of this section.
2. Drain the engine oil.
3. Remove or disconnect the following:
   - Negative battery cable
   - Breather hose
   - Positive Crankcase Ventilation (PCV) hose
   - Oil filler cap
   - Rocker arm (valve) cover and gasket
   - Oil seals, if necessary
   - Rocker arm spring
4. Install special tool MD998443 to prevent the lash adjusters from falling out.

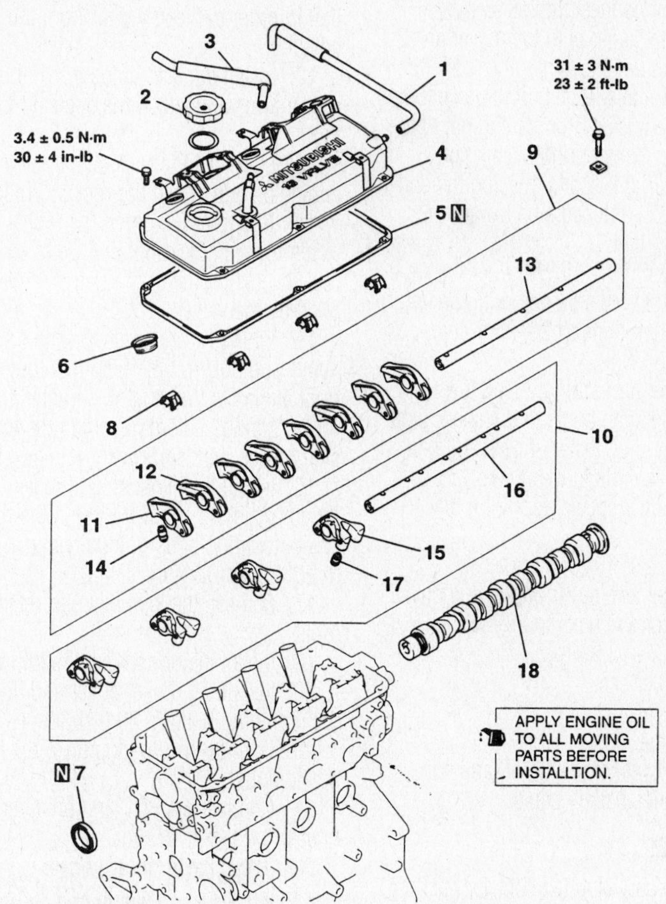

Exploded view of the rocker arms and related components—Lancer 2.0L engine

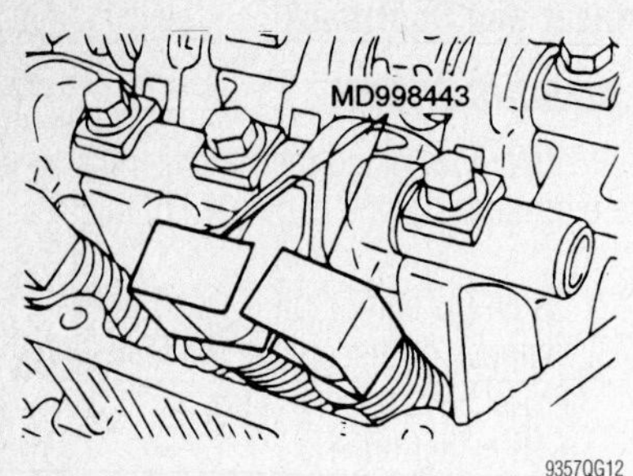

This special tool must be used to keep the lash adjusters in place

- Intake rocker arms and shafts
- Exhaust rocker arms and shafts
- Lash adjusters

To install:

5. Clean the lash adjusters if they are being reused.
6. Install or connect the following:
   - Lash adjuster onto the rocker arm, but do not allow the oil to spill out. Install the special tool to prevent the lash adjusters from falling out of the rocker arms during installation.
   - Rocker arm shafts, placing the end with the notched side toward the timing belt
   - Rocker arms. Move the rocker arms from side to side before tightening the shaft bolts to 21–25 ft. lbs. (28–34 Nm).
   - Rocker arm spring. Insert the spring at an angle to the spark plug guide, then install it so that it is at a right angle to the guide.
7. Remove the special tool from the rocker arms.
   - Oil seals, if necessary
   - Rocker arm cover gasket and cover. Tighten the bolts to 26–34 inch lbs. (2.9–3.9 Nm).
   - Oil filler cap
   - PCV hose
   - Breather hose
8. Fill the crankcase with oil and connect the negative battery cable.

2.4L ENGINE

1. Before servicing the vehicle, refer to the precautions at the beginning of this section.
2. Disconnect the negative battery cable.
3. Disconnect the accelerator cable, remove the cable clamp mounting screws and position the accelerator cable out of the way.
4. Remove the air intake hose.
5. Disconnect the breather hose and the Positive Crankcase Ventilation (PCV) hose.
6. Disconnect the spark plug cables from the spark plugs.
7. Remove the rocker cover and gasket.
8. Install lash adjuster retainer tools to the rocker arm.
9. Remove the rocker shaft hold-down bolts gradually and evenly and remove the rocker shaft/arm assemblies.
10. Disassemble the rockers and the rocker shaft springs from the rocker shafts. If they are to be reused, note the location and positioning of all rocker shaft components. It is recommended that all lash adjusters and rockers be replaced as a complete set.

To install:

11. Immerse the lash adjusters in clean diesel fuel, and using a small wire, move the plunger up and down 4 or 5 times. While pushing down lightly on the check ball in order to bleed the air from the adjuster.
12. Install the lash adjusters to the rocker arms and attach the special holding tool.
13. Lubricate the rocker shaft with clean engine oil and install the rocker arms.
14. Temporarily tighten the rocker shaft assembly with the mounting bolts so that all rocker arms on the inlet valve side do not push on the valves.
15. Fit the rocker shaft springs from above and position them so that they are at right angles to the plug side. Install the rocker springs before installing the exhaust side rocker shaft and rocker arm assembly.
16. Install the exhaust side rocker shaft assembly in the engine. Tighten the rocker shaft mounting bolts gradually and evenly to 23 ft. lbs. (32 Nm).
17. Remove the lash adjuster retaining tools.
18. Install the or connect the following:
19. Install the rocker cover and tighten the mounting bolts to 29 inch lbs. (3.3 Nm).
20. Connect the spark plug wires to the spark plugs.
21. Connect the PCV and breather hoses.
22. Install the air intake hose.
23. Install the accelerator cable brackets and reconnect the accelerator cable.
24. Connect the negative battery cable.

Lancer Evolution

1. Before servicing the vehicle, refer to the precautions at the beginning of this section.
2. Drain the engine oil.
3. Disconnect the negative battery cable.
4. Remove the breather hose.
5. Remove the Positive Crankcase Ventilation (PCV) hose.
6. Remove the oil filler cap.
7. Remove the rocker arm (valve) cover and gasket.
8. Remove the Camshaft Position (CMP) sensor.
9. Remove the CMP sensor support cover and gasket.
10. Remove the CMP sensing cylinder and sensor support.
11. Remove the camshaft oil seals.
12. Remove the right rear camshaft bearing cap.
13. Remove the left rear camshaft bearing cap.
14. Remove the right front camshaft bearing cap.
15. Remove the left front camshaft bearing cap.
16. Remove the No. 5 camshaft bearing cap(s).
17. Remove the No. 2 camshaft bearing cap(s).
18. Remove the No. 3 camshaft bearing cap(s).
19. Remove the No. 4 camshaft bearing cap(s). Be sure to clean the bearing caps and cylinder head surface of any old sealant.
20. Remove the camshaft(s).
21. Remove the rocker arms.
22. Remove the lash adjusters.
23. Remove the oil delivery body.

To install:

24. Clean the lash adjusters if they are being reused.
25. Install the oil delivery body and tighten the bolt to 90–106 inch lbs. (10–12 Nm).
26. Install the lash adjusters into the rocker arms.
27. Install the rocker arms.
28. Install the camshaft(s). The exhaust camshaft has a 0.16 inch (4mm) wide slit at the rear end, so not to confuse it with the intake camshaft. Make sure that the dowel pin on the end of each camshaft is in the top position.
29. Install the camshaft bearing caps. Intake and exhaust bearing caps for intake Nos. 2 through 5 are identical, except for an identification mark on each of them. (Example: E2=exhaust side No. 2; I3=Intake side No. 3.) Be sure to install the correct cap in the correct location.
30. Install the front and rear camshaft bearing caps. Apply fresh sealant to the mating surfaces between the caps and cylinder head.
31. Once the camshaft bearing caps are

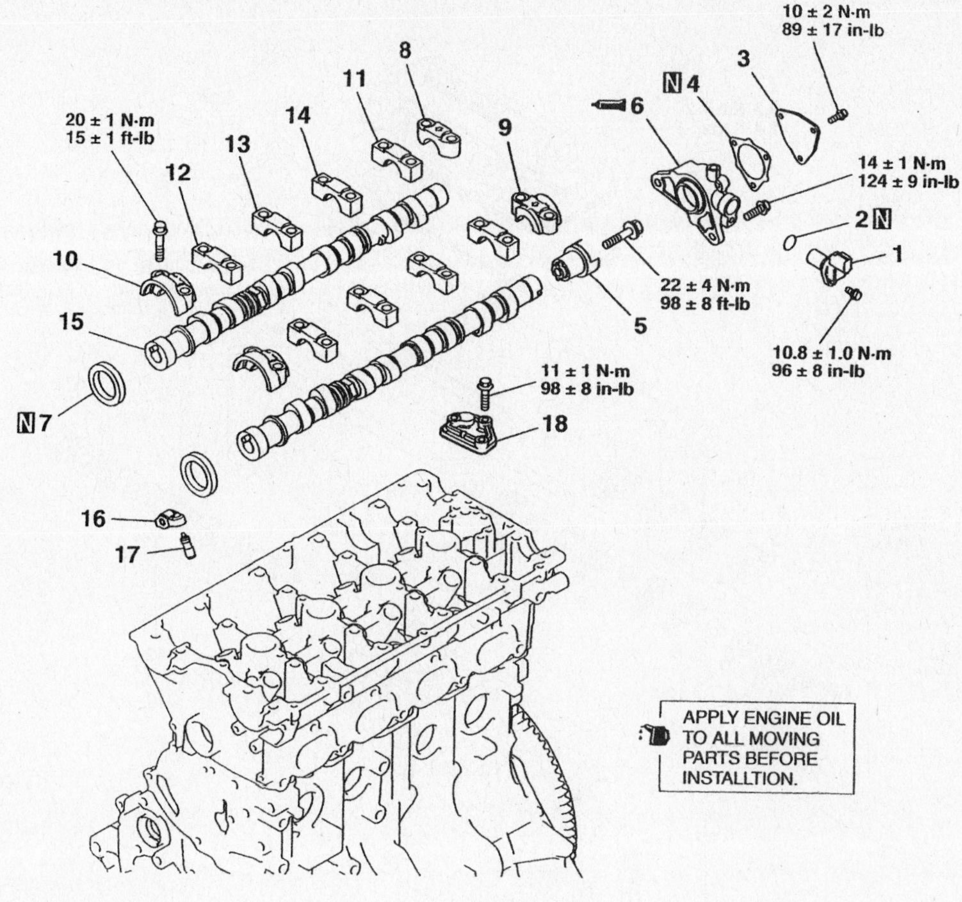

1. CAMSHAFT POSITION SENSOR
2. O-RING
3. COVER
4. GASKET
5. CAMSHAFT POSITION SENSING CYLINDER
6. CAMSHAFT POSITION SENSOR SUPPORT
7. CAMSHAFT OIL SEAL
8. BEARING CAP, REAR RIGHT
9. BEARING CAP, REAR LEFT
10. BEARING CAP, FRONT
11. BEARING CAP NO.5
12. BEARING CAP NO.2
13. BEARING CAP NO.3
14. BEARING CAP NO.4
15. CAMSHAFT
16. ROCKER ARM
17. LASH ADJUSTER
18. OIL DELIVERY BODY

67170-GALA-G23

Exploded view of the crankshafts, rocker arms and related components—2.0L DOHC engine

installed, tighten each bolt in 2 or 3 passes. Finally tighten the bolts to 15 ft. lbs. (20 Nm). Completely wipe off any squeezed out sealant.

32. Install the camshaft oil seals.
33. Apply a 0.12 inch (3mm) bead of sealant around the circumference of the sensor support that mates to the bearing cap/cylinder head side. Install the CMP sensor support.
34. Install the CMP sensing cylinder and tighten the bolt to 90–106 ft. lbs. (18–26 Nm).
35. Install the CMP sensor support cover and gasket.
36. Install CMP sensor.
37. Install or connect the rocker arm cover gasket and cover. Tighten the bolts to 26–34 inch lbs. (2.9–3.9 Nm).
38. Install or connect the oil filler cap.
39. Install or connect the PCV hose.
40. Install or connect the breather hose.
41. Fill the crankcase with oil and connect the negative battery cable.

Mirage

1. Before servicing the vehicle, refer to the precautions at the beginning of this section.
2. Remove or disconnect the following:
   - Negative battery cable
   - Spark plug cables—for 1.8L engine
   - Accelerator cable, breather hose and Positive Crankcase Ventilation (PCV) hose connections
   - Rocker cover
3. Loosen both rocker arm shaft assemblies gradually and evenly and remove the rocket shafts from the vehicle.
4. If disassembly is required, keep all parts in the exact order of removal.

To install:

5. Lubricate the rocker shaft with clean

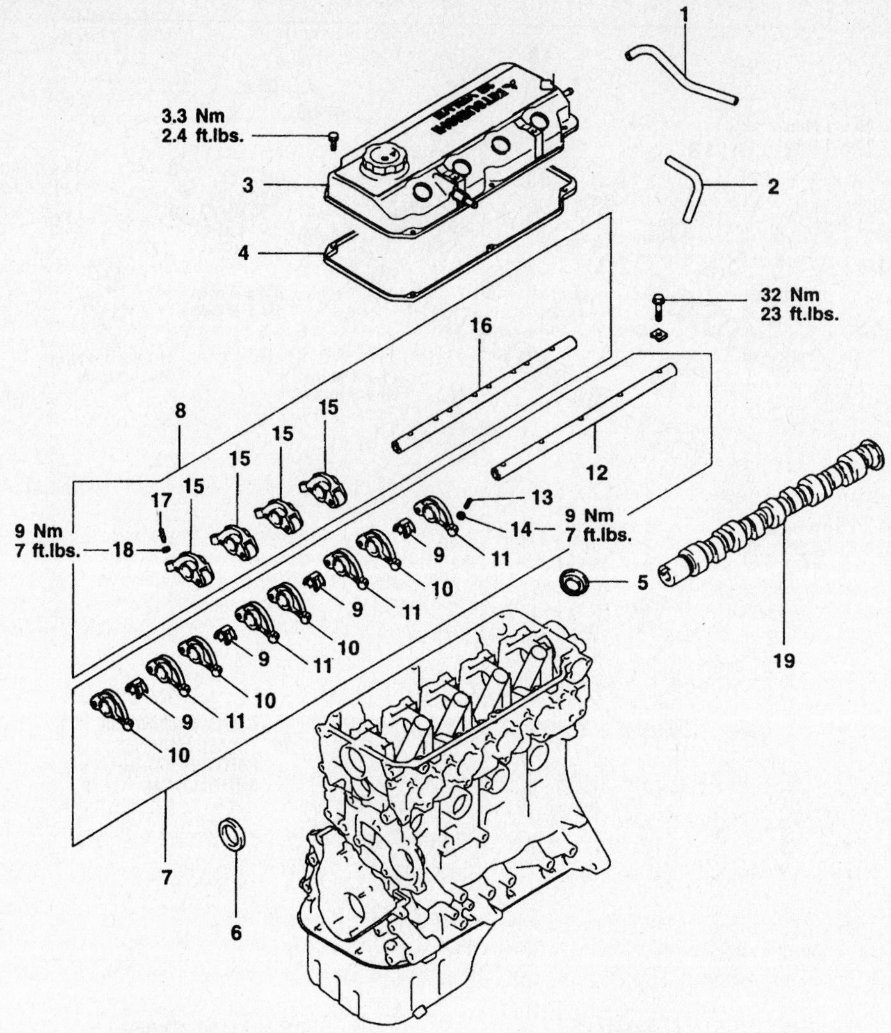

1. Breather hose
2. P.C.V. hose
3. Rocker cover
4. Rocker cover gasket
   Valve clearance pre-adjustment
5. Oil seal
6. Oil seal
7. Rocker arms and rocker arm shaft
8. Rocker arms and rocker arm shaft
9. Rocker shaft spring
10. Rocker arm A
11. Rocker arm B
12. Rocker arm shaft (Intake side)
13. Adjusting screw
14. Nut
15. Rocker arm C
16. Rocker arm shaft (Exhaust side)
17. Adjusting screw
18. Nut
19. Camshaft

7923PG27

Camshaft, rocker arm and shaft assemblies—Mirage 1.8L (4G93) engine

engine oil and install the rockers and springs.

6. Install the rocker arm and shaft assemblies. Tighten the rocker arm shaft retainer bolts to 23 ft. lbs. (32 Nm).

7. Check valve adjustment and install the valve cover. Tighten the valve cover bolts to 16 inch lbs. (1.8 Nm) for the 1.5L engine or to 29 inch lbs. (3.3 Nm) for the 1.8L engine.

8. Install or connect the following:
   - Spark plug cables, if detached
   - Accelerator cable, breather hose and PCV hose
   - Negative battery cable

Galant

2.4L ENGINE

1. Before servicing the vehicle, refer to the precautions at the beginning of this section.

2. Disconnect the negative battery cable.

3. Disconnect the accelerator cable, remove the cable clamp mounting screws and position the accelerator cable out of the way.

4. Remove the air intake hose.

5. Disconnect the breather hose and the Positive Crankcase Ventilation (PCV) hose.

6. Disconnect the spark plug cables from the spark plugs.

7. Remove the rocker cover and gasket.

8. Install lash adjuster retainer tools to the rocker arm.

9. Remove the rocker shaft hold-down bolts gradually and evenly and remove the rocker shaft/arm assemblies.

10. Disassemble the rockers and the rocker shaft springs from the rocker shafts. If they are to be reused, note the location and positioning of all rocker shaft components. It is recommended that all lash adjusters and rockers be replaced as a complete set.

To install:

11. Immerse the lash adjusters in clean diesel fuel, and using a small wire, move the plunger up and down 4 or 5 times. While pushing down lightly on the check ball in order to bleed the air from the adjuster.

12. Install the lash adjusters to the rocker arms and attach the special holding tool.

13. Lubricate the rocker shaft with clean engine oil and install the rocker arms.

14. Temporarily tighten the rocker shaft assembly with the mounting bolts so that all rocker arms on the inlet valve side do not push on the valves.

15. Fit the rocker shaft springs from above and position them so that they are at right angles to the plug side. Install the rocker springs before installing the exhaust side rocker shaft and rocker arm assembly.

16. Install the exhaust side rocker shaft assembly in the engine. Tighten the rocker shaft mounting bolts gradually and evenly to 23 ft. lbs. (32 Nm).

17. Remove the lash adjuster retaining tools.

18. Install the rocker cover and tighten the mounting bolts to 29 inch lbs. (3.3 Nm).

19. Connect the spark plug wires to the spark plugs.

20. Connect the PCV and breather hoses.

21. Install the air intake hose.

22. Install the accelerator cable brackets and reconnect the accelerator cable.

23. Connect the negative battery cable.

3.0L ENGINE

On this engine, the hydraulic lash adjusters are built into the rocker arms.

1. Before servicing the vehicle, refer to the precautions at the beginning of this section.

2. Disconnect the negative battery cable.

3. Remove the valve cover. Install lash adjuster retainer tools to prevent the auto-lash adjuster from falling out of the rocker arm.

4. Loosen rocker arm and shaft assembly evenly in several steps. Remove the rocker arm and shaft assembly as a complete unit.

5. Remove the rear camshaft bearing cap and slide the rocker arms, springs and washers from the shaft. If they are to be reused, note the location and positioning of all rocker shaft components. It is recommended that all lash adjusters and rockers be replaced as a complete set.

To install:

6. Immerse the lash adjusters in clean diesel fuel. Using a small wire, move the plunger of the lash adjuster up and down 4 or 5 times while pushing down lightly on the check ball in order to bleed out the air. Install the lash adjusters in the rocker arms.

7. Using a light coat of engine oil, assemble the rocker arms to the shaft. Install the rear camshaft bearing cap.

8. Lubricate the camshaft and rocker shaft with clean engine oil and position on the cylinder head.

9. Apply a drop of sealant to the rear edges of the end caps.

10. Install the assembly making sure the notches in the rocker shafts are facing up.

11. Install the cap bolts and tighten evenly and gradually to 14 ft. lbs. (20 Nm). Remove the lash adjuster retainers.

12. Install the valve cover.

13. Connect the negative battery cable.

3.8L ENGINE

1. Before servicing the vehicle, refer to the precautions at the beginning of this section.

2. Disconnect the negative battery cable.

3. Remove the rocker arm cover.

4. Install the lash adjuster clips on the rocker arms, then loosen the rocker shaft cap bolts. Do not remove the bolts from the rocker shaft caps.

5. Remove the rocker arms, shafts and rocker shaft caps as an assembly.

To install:

6. Install the rocker shaft caps/rocker arm assemblies. Tighten the bolts to 23 ft. lbs. (31 Nm).

7. Remove the lash adjuster clips.

8. Install the rocker arm cover using a new gasket.

9. Connect the negative battery cable.

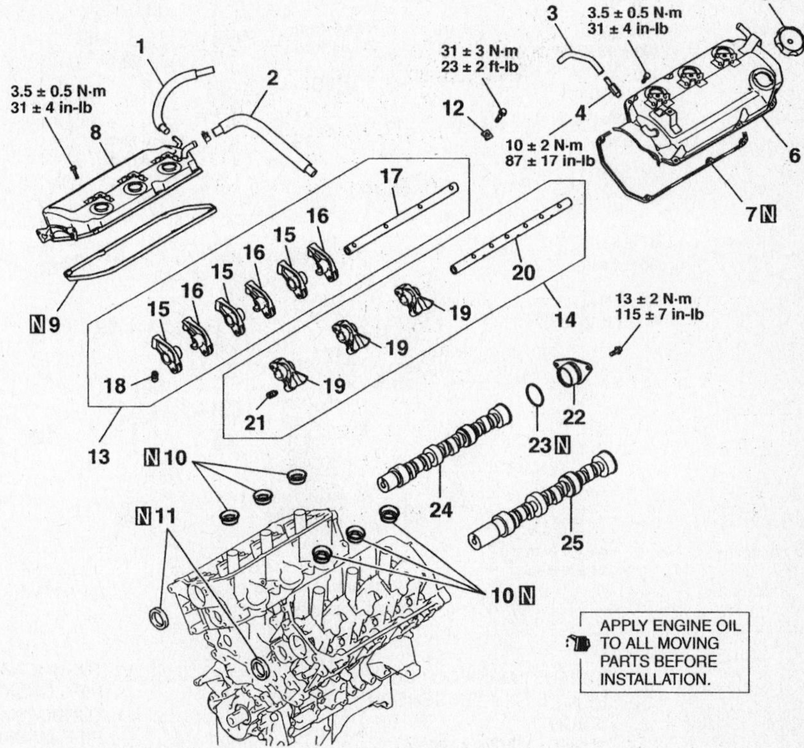

Rocker arm shafts and components—3.8L engines

Turbocharger

REMOVAL & INSTALLATION

Lancer Evolution

CAUTION

The air bag system must be disarmed before removing the turbocharger.

1. Before servicing the vehicle, refer to the precautions at the beginning of this section.
2. Disconnect the negative battery cable.
3. Drain the engine coolant.
4. Disengage the retaining clips and remove the front under cover, located under the front of the vehicle behind the front bumper face.
5. Remove the radiator assembly.
6. Remove the air intake hose from the air cleaner assembly.
7. Remove the air pipes & hoses from the charge air cooler assembly.
8. Remove the front axle crossmember bar from underneath the vehicle.
9. Remove the front exhaust pipe from the exhaust manifold.
10. Remove the exhaust manifold cover.
11. Remove the front Heated Oxygen (HO_2S) sensor.
12. Remove the turbocharger heat protector.
13. Disconnect the turbocharger water

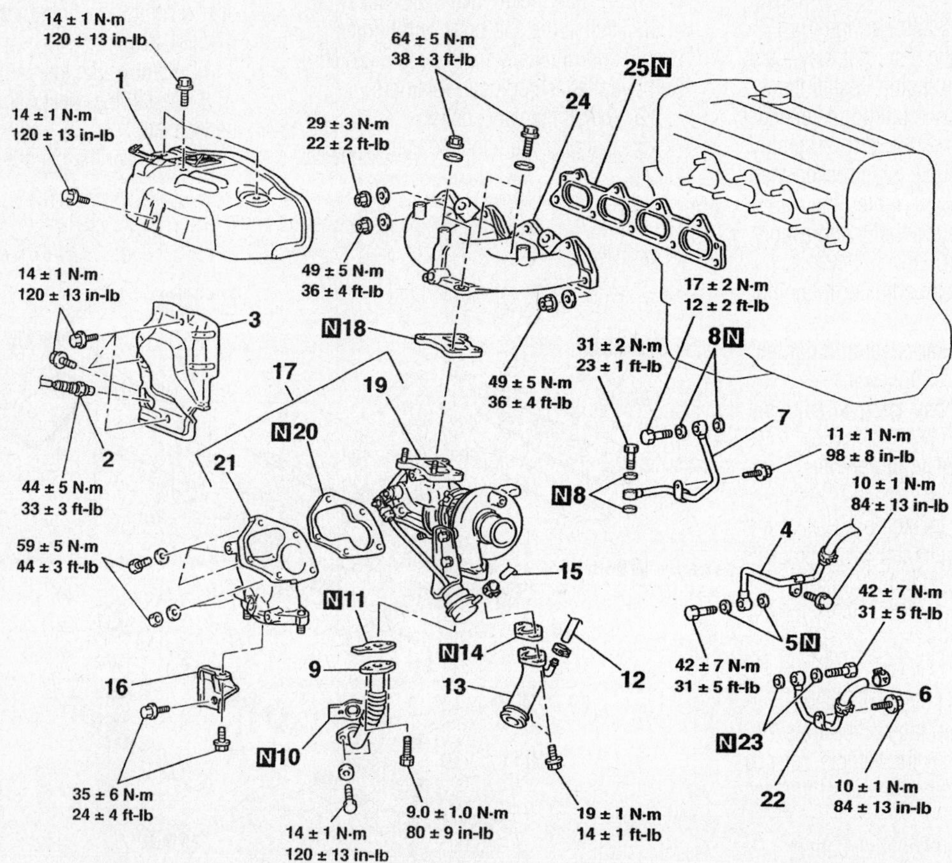

Exploded view of the turbocharger and exhaust manifold assembly—2.0L DOHC engine

feed pipe and return hose connections and remove gaskets.

14. Disconnect the turbocharger oil feed and oil return pipes and remove gaskets.
15. Remove the starter motor.
16. Disconnect the vacuum hose connections.
17. Remove the air outlet fitting and gasket.
18. Remove the exhaust fitting bracket.
19. Remove the turbocharger and exhaust fitting assembly. Remove the gasket.
20. Remove the exhaust manifold and gasket.

To install:

21. Clean all gasket material from the mating surfaces.
22. Install a new gasket and exhaust manifold. Tighten the retainers to 32–40 ft. lbs. (44–54 Nm).
23. Install the turbocharger assembly. Be sure to clean the oil and water pipe fittings, inside of eye bolts and individual pipes of any clogs. Clean or use compressed air to remove any carbon matter stuck in the oil passage of the turbocharger. Refill new engine oil at the oil feed pipe fitting hole.
24. Install the exhaust fitting assembly, along with a new gasket.
25. Install the exhaust fitting bracket.
26. Install the air outlet fitting and gasket.
27. Connect the vacuum hose connections.
28. Install the starter motor.
29. Connect the turbocharger oil feed and oil return pipes, along with new gaskets. Tighten the oil feed pipe fastener to 10–14 ft. lbs. (15–19 Nm). Tighten the oil return pipe fasteners to 71–89 inch lbs. (8–10 Nm).
30. Connect the turbocharger water feed pipe and return hoses, along with new gaskets. Tighten the pipe fasteners to 26–36 ft. lbs. (35–49 Nm).
31. Install the turbocharger heat protector.
32. Install the front Heated Oxygen (HO_2S) sensor.
33. Install the exhaust manifold cover.
34. Install the front exhaust pipe to the exhaust manifold.
35. Install the front axle crossmember bar from underneath the vehicle.
36. Install the air pipes & hoses to the charge air cooler assembly.
37. Install the air intake hose to the air cleaner assembly.
38. Install the radiator assembly.
39. Install the front under cover and secure with the retaining clips.
40. Refill the cooling system with the correct amount and type of coolant.
41. Connect the negative battery cable.

Intake Manifold

REMOVAL & INSTALLATION

Diamante

1. Before servicing the vehicle, refer to the precautions at the beginning of this section.
2. Relieve the fuel system pressure.
3. Remove or disconnect the following:
   - Negative cable and drain the cooling system
   - Air intake hose(s)
   - Accelerator control cables from the throttle body
   - Vacuum hoses including the brake booster hose
   - Wiring harness connectors
   - High pressure and return fuel hoses
   - Exhaust Gas Recirculation (EGR) pipe and remove the EGR valve and EGR temperature sensor from the intake plenum assembly
   - Manifold Absolute Pressure (MAP) sensor, if equipped
   - Plenum retaining bracket
   - Plenum retaining nuts and bolts and remove the air intake plenum from the intake manifold
   - Upper timing belt covers
   - Water pump stay bracket

➡It is not necessary to remove the fuel injectors from the intake unless the manifold assembly is being replaced.

   - Fuel rail with the injectors attached
   - Coolant hoses from the intake manifold

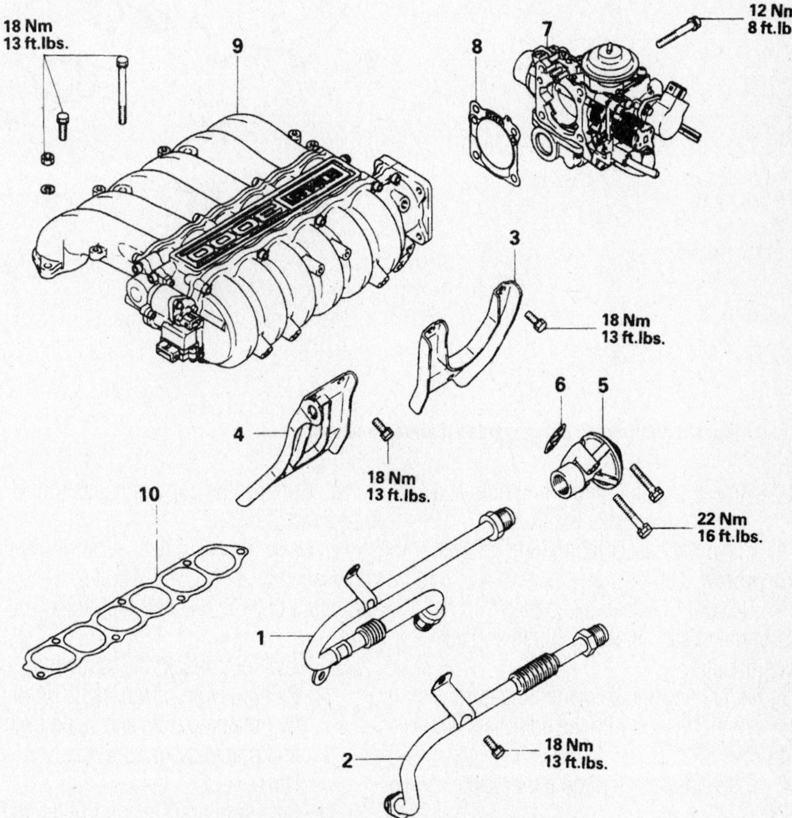

Exploded view of air intake plenum assembly—Diamante

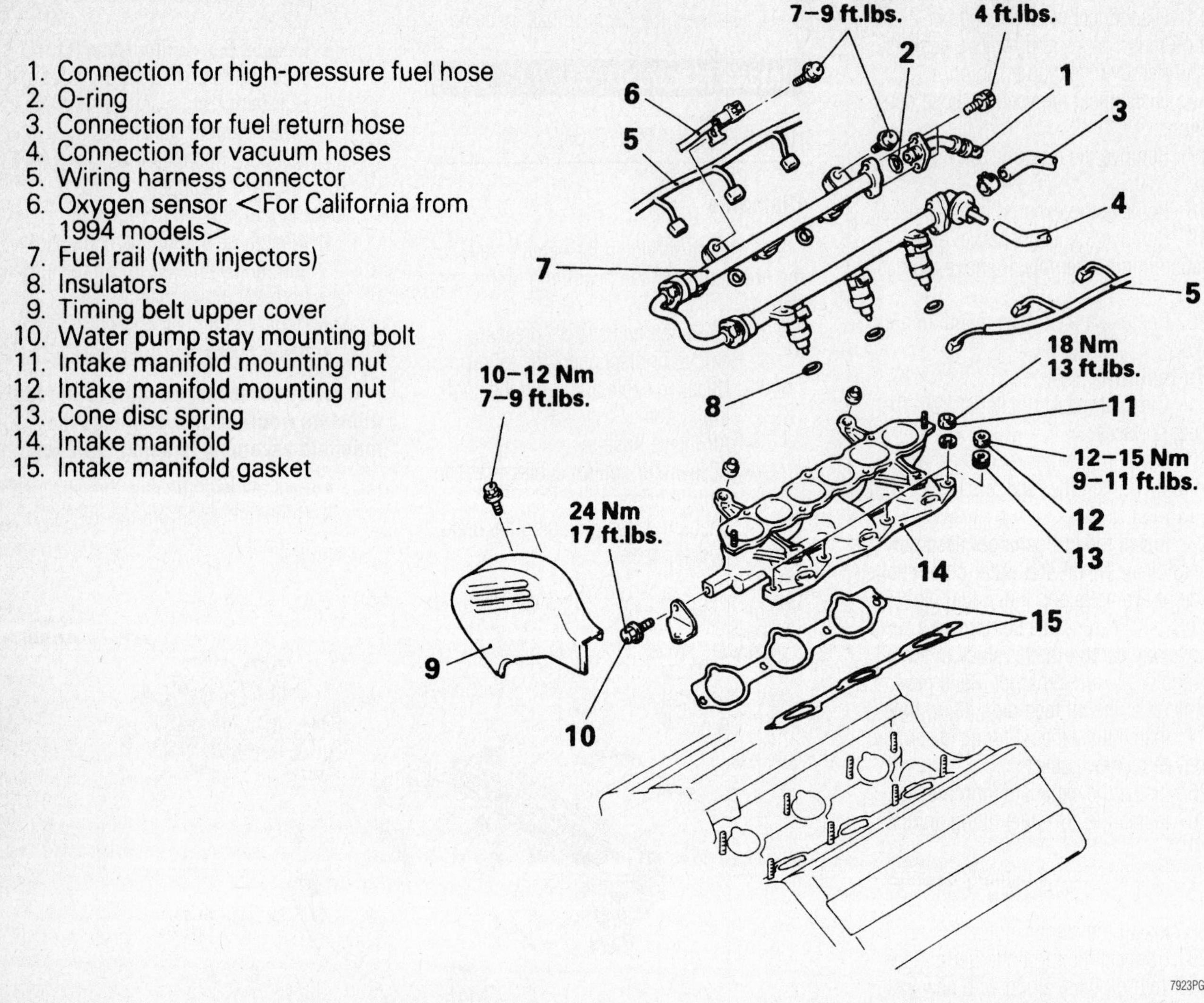

Intake manifold and related components—Diamante shown

- Intake manifold mounting nuts and remove the intake manifold

4. Clean the gasket mounting surfaces.

To install:

5. Thoroughly clean the mating surfaces of the heads, intake manifold and air intake plenum.
6. Install new intake manifold gaskets to the cylinder heads with the adhesive side facing up.
7. Place the manifold on the cylinder heads.
8. Lubricate the studs lightly with oil and install the nuts.
9. Tighten the mounting nuts as follows:
   a. Front bank nuts: 48–72 inch lbs. (5–8 Nm).
   b. Rear bank nuts: 14–17 ft. lbs. (20–23 Nm).
   c. Front bank nuts: 14–17 ft. lbs. (20–23 Nm).
10. Connect the coolant hoses to the intake manifold.
11. Using new O-rings, install the fuel rail assembly, if removed. Tighten the mounting bolts to 84–108 inch lbs. (10–13 Nm).
12. Install or connect the following:
    - Plenum, with new gasket. Tighten the retaining nuts and bolts evenly and gradually to 13 ft. lbs. (18 Nm).
    - Retaining bracket and tighten the retaining bolts to 13 ft. lbs. (18 Nm)
    - MAP sensor, if removed
    - EGR valve, using a new gasket. Tighten the bolts to 16 ft. lbs. (22 Nm).
    - EGR temperature sensor and tighten the fitting to 84–108 inch lbs. (10–12 Nm)
    - EGR pipe and tighten the fittings to 43 ft. lbs. (60 Nm)
    - High pressure fuel hose, use a new O-ring. Tighten the retaining bolts to 48 inch lbs. (5 Nm).
    - Fuel return hose, using a new clamp
    - Water pump stay bracket
    - Upper timing belt covers
    - Harness connectors and vacuum hoses
    - Accelerator cables, adjust
    - Air intake hose(s)
13. Fill the system with coolant.
14. Connect the negative battery cable.

Eclipse

2.4L ENGINE

1. Before servicing the vehicle, refer to the precautions at the beginning of this section.

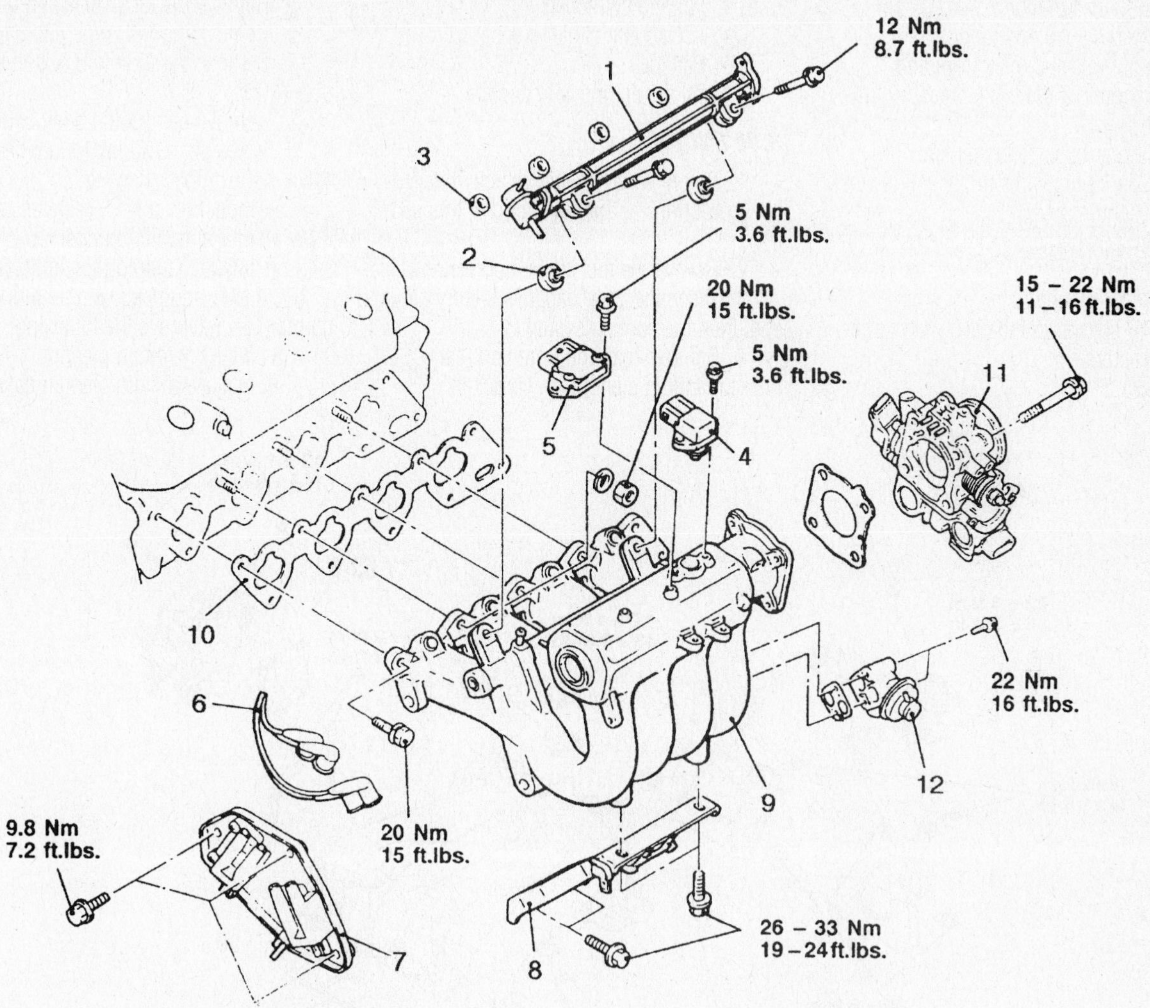

Intake manifold and related components—Eclipse and Galant 2.4L (4G64) engines

2. Relieve the fuel system pressure.
3. Remove the battery.
4. Drain the engine coolant.
5. Remove or disconnect the following:
   - Accelerator cable
   - Air intake hose
   - Ignition coil and the module wiring connectors
   - Manifold Absolute Pressure (MAP) sensor
   - Condenser
   - Throttle Position (TP) sensor and the Idle Air Control (IAC) motor connectors
   - Heated Oxygen (HO_2S) sensor connector
   - Crankshaft Position (CKP) sensor connector
   - Air conditioning compressor connector
   - Engine control wiring harness bracket and position the harness out of the way
   - Vacuum hoses, label for reference
   - Spark plug wires from the ignition coil
   - Fuel lines from the fuel rail
   - Heater hoses
   - Fuel rail assembly
   - Ignition coil and module
   - Exhaust Gas Recirculation (EGR) valve assembly
   - Intake manifold stay and the engine hanger
   - Intake manifold

To install:

6. Install or connect the following:
   - Intake manifold. Torque the intake manifold bolts to 15 ft. lbs. (20 Nm).
   - Intake manifold stay and the engine hanger. Torque the mounting bolts to 19–24 ft. lbs. (26–33 Nm).

- EGR assembly
- Ignition coil and module
- Fuel rail and insulators and reconnect the high-pressure fuel hose
- Heater hoses and fuel lines
- Spark plug wires to the coil towers
- Vacuum hoses
- Engine harness in the proper position
- MAP sensor
- TP sensor and the IAC motor connectors
- HO_2S
- Ignition condenser
- Accelerator cable, adjust
- Battery

7. Refill the engine with coolant.

3.0L ENGINE

1. Before servicing the vehicle, refer to the precautions at the beginning of this section.
2. Relieve the fuel system pressure.
3. Disconnect the negative battery cable and drain the cooling system.
4. Remove both the front and rear intake manifold plenum stay brackets.
5. Remove the EGR valve and gasket.
6. Remove the EGR pipe and gasket.
7. Remove the Manifold Absolute Pressure (MAP) sensor.
8. Remove the throttle body and gasket.
9. If equipped with an induction control valve, remove the following:
   - Induction control valve assembly
   - Intake manifold upper
   - Intake manifold plenum gasket
10. If NOT equipped with an induction control valve, remove the following:
   - Intake manifold plenum
   - Intake manifold plenum gasket

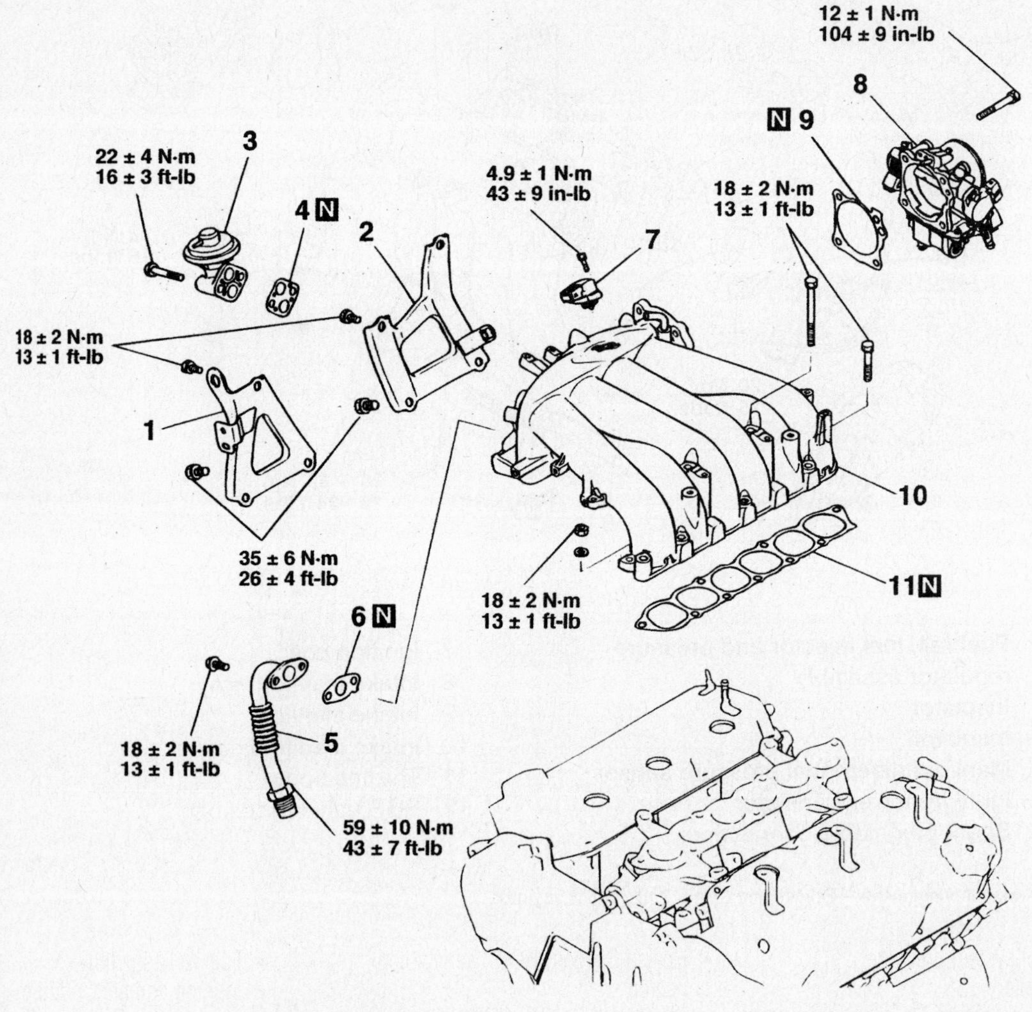

Exploded view of air intake plenum assembly—3.0L engine

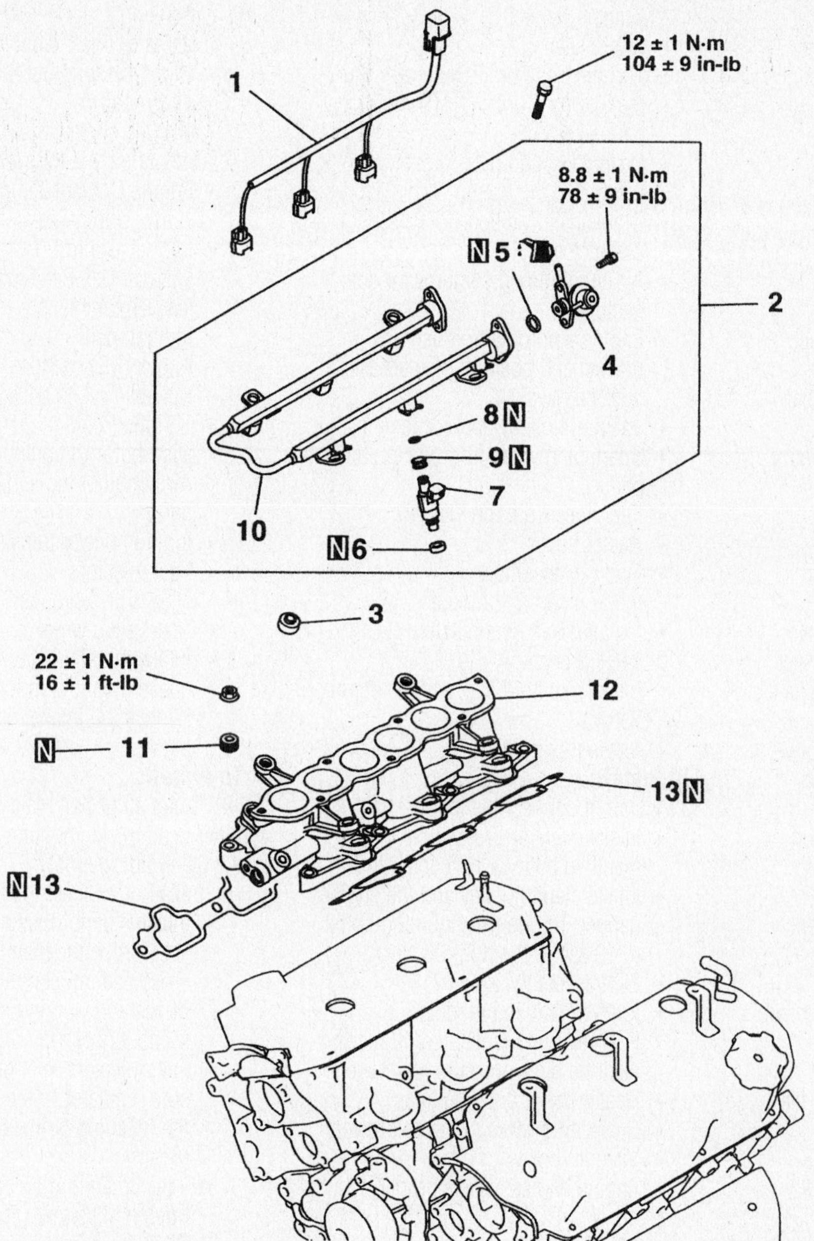

Intake manifold and related components—3.0L engine

11. Disconnect the fuel injector harness.
12. Remove the fuel rail and injectors assembly, along with the insulators.
13. Remove the intake manifold mounting nuts and remove the intake manifold and gaskets.
14. Clean the gasket mounting surfaces.

To install:

15. Thoroughly clean the mating surfaces of the heads, intake manifold and air intake plenum.
16. Install new intake manifold gaskets to the cylinder heads with the adhesive side facing up.
17. Place the manifold on the cylinder heads.
18. Lubricate the studs lightly with oil and install the nuts.

19. Tighten the mounting nuts as follows:
   a. Right bank nuts: 44–70 inch lbs. (5–8 Nm).
   b. Left bank nuts: 21–23 ft. lbs. (15–17 Nm).
   c. Right bank nuts: 21–23 ft. lbs. (15–17 Nm).
   d. Left bank nuts: 21–23 ft. lbs. (15–17 Nm).
   e. Right bank nuts: 21–23 ft. lbs. (15–17 Nm).
20. Using new O-rings, install the fuel rail and injectors assembly.
21. Connect the fuel injector wiring harness.
22. If equipped with an induction control valve, install the following:
   - Intake manifold plenum gasket
   - Intake manifold upper. Tighten retaining nuts to 12–14 ft. lbs. (16–20 Nm).
   - Install the induction control valve gasket. Apply sealant to the gasket surface and install while the sealant is still wet (within 15 minutes).
   - Induction control valve assembly. Tighten fasteners to 71–89 inch lbs. (8–10 Nm).
23. If NOT equipped with an induction control valve, install the following:
   - Intake manifold plenum gasket
   - Intake manifold plenum. Tighten retaining nuts to 12–14 ft. lbs. (16–20 Nm).
24. Install a new throttle body gasket and install the throttle body assembly.
25. Install the MAP sensor.
26. Install the EGR pipe and tighten the fittings to 12–14 ft. lbs. (16–20 Nm).
27. Install the EGR valve, using a new gasket. Tighten the bolts to 16 ft. lbs. (22 Nm).
28. Install the front and rear intake manifold plenum stay brackets.
29. Fill the system with coolant.
30. Connect the negative battery cable.

Galant

2.4L ENGINE

1. Before servicing the vehicle, refer to the precautions at the beginning of this section.
2. Relieve the fuel system pressure.
3. Remove the battery.
4. Drain the engine coolant.
5. Remove or disconnect the following:
   - Accelerator cable
   - Air intake hose
   - Ignition coil and the module wiring connectors
   - Manifold Absolute Pressure (MAP) sensor
   - Condenser
   - Throttle Position (TP) sensor and the Idle Air Control (IAC) motor connectors
   - Heated Oxygen (HO_2S) sensor connector
   - Crankshaft Position (CKP) sensor connector
   - Air conditioning compressor connector
   - Engine control wiring harness bracket and position the harness out of the way
   - Vacuum hoses, label for reference
   - Spark plug wires from the ignition coil
   - Fuel lines from the fuel rail
   - Heater hoses
   - Fuel rail assembly
   - Ignition coil and module
   - Exhaust Gas Recirculation (EGR) valve assembly
   - Intake manifold stay and the engine hanger
   - Intake manifold

To install:

6. Install or connect the following:
   - Intake manifold. Torque the intake manifold bolts to 15 ft. lbs. (20 Nm).
   - Intake manifold stay and the engine hanger. Torque the mounting bolts to 19–24 ft. lbs. (26–33 Nm).
   - EGR assembly
   - Ignition coil and module
   - Fuel rail and insulators and reconnect the high-pressure fuel hose
   - Heater hoses and fuel lines
   - Spark plug wires to the coil towers
   - Vacuum hoses
   - Engine harness in the proper position
   - MAP sensor
   - TP sensor and the IAC motor connectors
   - HO_2S
   - Ignition condenser
   - Accelerator cable, adjust
   - Battery
7. Refill the engine with coolant.

3.0L ENGINE

1. Before servicing the vehicle, refer to the precautions at the beginning of this section.
2. Relieve the fuel system pressure.
3. Disconnect battery negative cable.
4. Drain the cooling system.
5. Remove or disconnect the following:
   - Accelerator cable
   - Air intake hose
   - Coolant hose from the throttle housing
   - Vacuum lines, label for reference
   - High pressure fuel line and fuel return hose
   - Throttle control cable brackets
6. Disconnect the following:
   - Engine Coolant Temperature (ECT) sensor and gauge sender
   - Idle Air Control (IAC) motor
   - Exhaust Gas Recirculation (EGR) temperature sensor
   - Ignition coil
   - Knock (KS) sensor
   - Heated Oxygen (HO_2S) sensor
   - Throttle Position (TP) sensor
   - Distributor (if equipped)
   - Air conditioning temperature sensor
   - Ignition power transistor
   - Fuel injectors
7. Remove or disconnect the following:
   - Spark plug wires
   - Intake manifold stay bracket
   - Intake manifold mounting bolts and remove the intake manifold assembly

To install:

8. Clean all gasket material from the cylinder head intake mounting surface and intake manifold assembly.
9. Install or connect the following:
   - Intake manifold, using a new gasket. Torque the manifold in a crisscross pattern, starting from the inside and working outwards to 15 ft. lbs. (20 Nm).
   - Fuel injectors, fuel rail and pressure regulator to the engine. Torque the retaining bolts to 48 inch lbs. (6 Nm).
   - Intake manifold brace bracket and tighten bolts to 21 ft. lbs. (29 Nm)
   - Spark plug wires
10. Connect the following:
   - ECT sensor and gauge sender
   - IAC motor
   - EGR temperature sensor
   - Ignition coil
   - KS sensor
   - HO_2S sensor
   - TP sensor
   - Distributor (if equipped)
   - Air conditioning temperature sensor
   - Ignition power transistor
   - Fuel injectors
11. Install or connect the following:
   - Throttle control cable brackets
   - High pressure fuel line and fuel return hose
   - Vacuum lines
   - Coolant hoses

- Accelerator cable
- Air intake hose

12. Fill the system with coolant.
13. Connect the negative battery cable.

3.8L ENGINE

1. Before servicing the vehicle, refer to the precautions at the beginning of this section.
2. Relieve the fuel system pressure.
3. Disconnect the negative cable and drain the cooling system.
4. Remove the purge hose.
5. Remove the vacuum pipe and hose.
6. Remove the solenoid valve.
7. Remove the Exhaust Gas Recirculation (EGR) pipe and gasket.
8. Remove both the front and rear intake manifold plenum stay brackets.
9. Remove the EGR valve and gasket.
10. Remove the throttle body stay bracket, throttle body and gasket.
11. Remove the boost sensor.
12. Remove the intake manifold plenum and gasket.
13. Disconnect the fuel injector harness.
14. Remove the fuel rail and injectors assembly, along with the insulators.
15. Remove the intake manifold mounting nuts and remove the intake manifold and gaskets.
16. Clean the gasket mounting surfaces.

To install:

17. Thoroughly clean the mating surfaces of the heads, intake manifold and air intake plenum.
18. Install new intake manifold gaskets to the cylinder heads with the adhesive side facing up.
19. Place the manifold on the cylinder heads.
20. Lubricate the studs lightly with oil and install the nuts.
21. Tighten the mounting nuts as follows:
    a. Right bank nuts: 44–70 inch lbs. (5–8 Nm).
    b. Left bank nuts: 15–17 ft. lbs. (20–23 Nm).
    c. Right bank nuts: 15–17 ft. lbs. (20–23 Nm).
    d. Left bank nuts: 15–17 ft. lbs. (20–23 Nm).
    e. Right bank nuts: 15–17 ft. lbs. (20–23 Nm).
22. Using new O-rings, install the fuel rail and injectors assembly.
23. Connect the fuel injector wiring harness.
24. Install the intake manifold plenum gasket.
25. Install the intake manifold plenum. Tighten retaining bolts to 18–24 ft. lbs. (24–32 Nm).
26. Install the boost sensor. The sensor must be replaced if it gets dropped.
27. Install a new throttle body gasket and install the throttle body assembly. Tighten the throttle body fasteners to 18–24 ft. lbs. (24–32 Nm).
28. Install the throttle body stay bracket.
29. Install the EGR valve, using a new gasket. Tighten the bolts to 16–20 ft. lbs. (21–27 Nm).
30. Install the front and rear intake manifold plenum stay brackets.
31. Install the EGR pipe and tighten the fittings to 12–16 ft. lbs. (16–22 Nm).
32. Fill the system with coolant.
33. Connect the negative battery cable.

Lancer and Sportback

2.0L SOHC ENGINE

1. Before servicing the vehicle, refer to the precautions at the beginning of this section.
2. Relieve the fuel system pressure.
3. Disconnect the negative battery cable and drain the cooling system.
4. Remove the air cleaner assembly.
5. Remove the throttle body.
6. Remove the Exhaust Gas Recirculation (EGR) valve and gasket.
7. Remove the fuel rail and injector assembly.
8. Remove the bracket and the engine hanger.
9. Remove the Manifold Differential Pressure (MDP) sensor and O-ring.
10. Remove the intake manifold stay bracket.
11. Remove the intake manifold bolt and manifold.
12. Remove the intake manifold gasket and discard.

To install:

13. Clean all gasket material from the cylinder head intake mounting surface and intake manifold assembly.
14. Install the intake manifold, using a

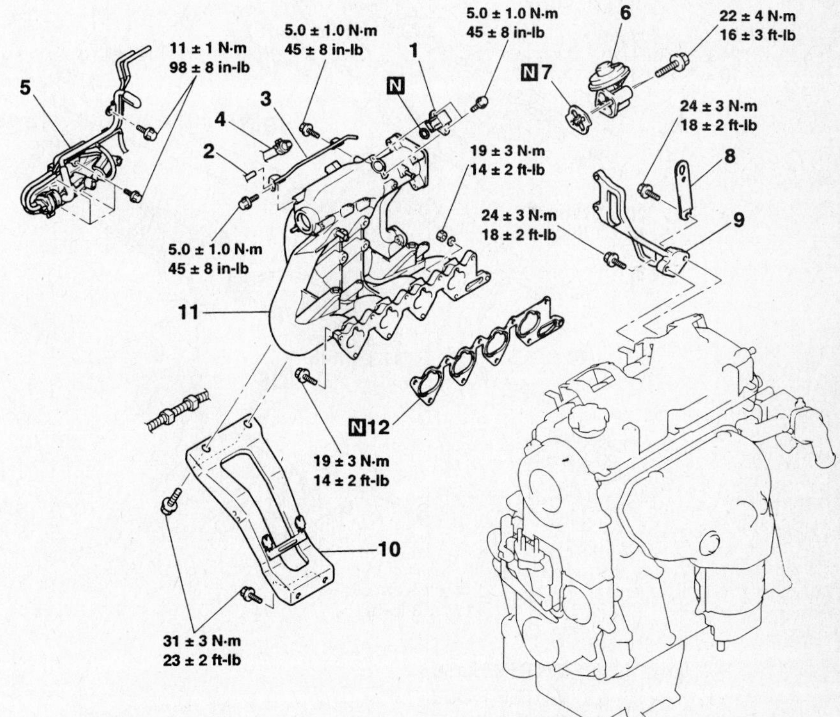

REMOVAL STEPS
1. MANIFOLD DIFFERENTIAL PRESSURE SENSOR
2. AUTO CRUISE VACUUM HOSE CONNECTION
3. VACUUM PIPE
4. BRAKE BOOSTER VACUUM HOSE CONNECTION
5. VACUUM HOSE AND PIPE ASSEMBLY

REMOVAL STEPS (Continued)
6. EGR VALVE
7. EGR VALVE GASKET
8. ENGINE HANGER
9. THROTTLE BODY STAY
10. INTAKE MANIFOLD STAY
11. INTAKE MANIFOLD
12. INTAKE MANIFOLD GASKET

9357QG13

Exploded view of the intake manifold and related components—Lancer with 2.0L engine

new gasket. Torque the manifold in a criss-cross pattern, starting from the inside and working outwards to 12–16 ft. lbs. (16–22 Nm).

15. Install the intake manifold stay.
16. Install the throttle body stay.
17. Install the engine hanger.
18. Install the EGR valve and gasket.
19. Install the vacuum hose and pipe assembly.
20. Connect the brake booster vacuum hose connection.
21. Connect the vacuum pipe.
22. Connect the auto cruise vacuum hose connection.
23. Install the MDP sensor.
24. Install the fuel rail assembly.
25. Install the throttle body.
26. Install the air cleaner assembly.
27. Fill the system with coolant.
28. Connect the negative battery cable.

2.4L ENGINE

1. Before servicing the vehicle, refer to the precautions at the beginning of this section.
2. Relieve the fuel system pressure.
3. Disconnect the negative battery cable and drain the cooling system.
4. Remove the air cleaner assembly.
5. Remove the throttle body.
6. Disconnect the vacuum pipes and hoses, label for reference.
7. Remove the solenoid valve and mounting bracket.
8. Remove the fuel rail and injector assembly.
9. Remove the Exhaust Gas Recirculation (EGR) valve and gasket.

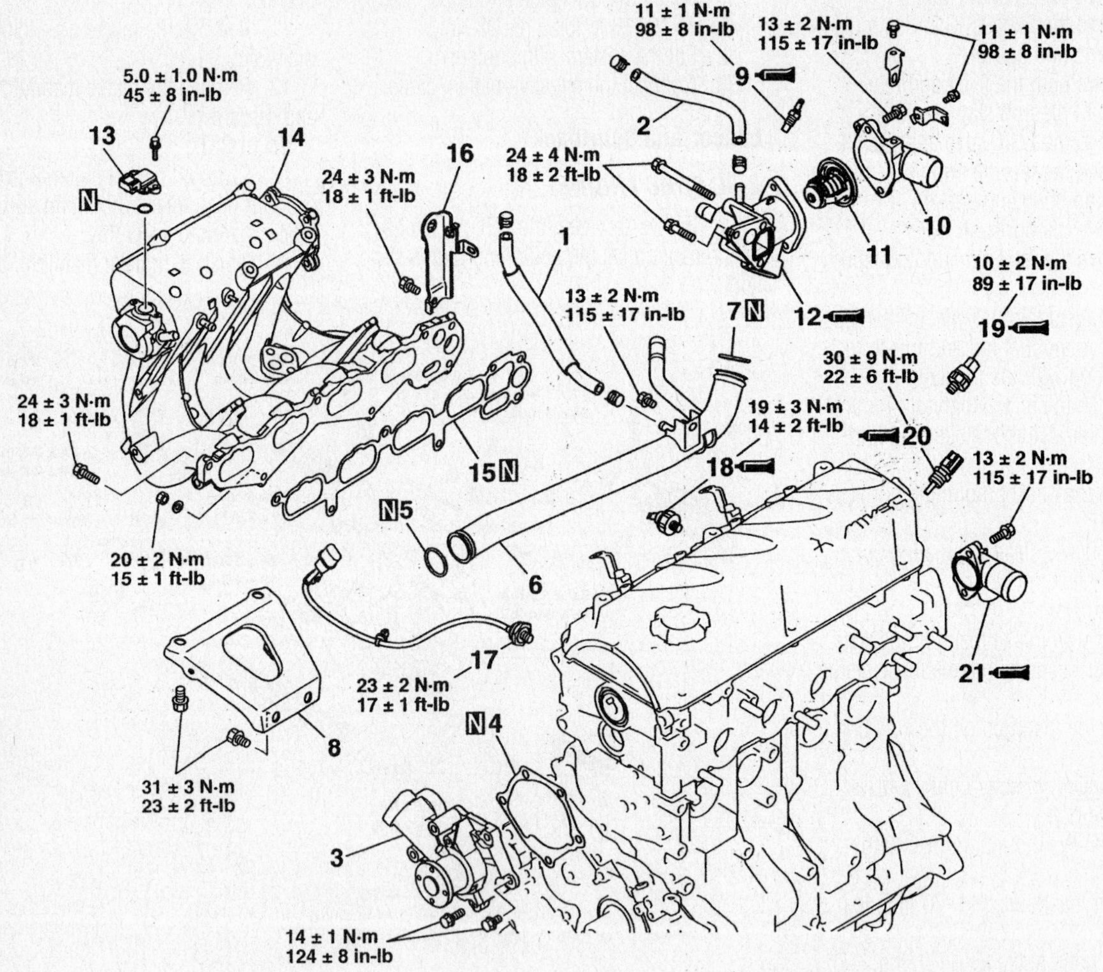

1. WATER HOSE
2. WATER HOSE
3. WATER PUMP
4. WATER PUMP GASKET
5. O-RING
6. WATER INLET PIPE
7. O-RING
8. INTAKE MANIFOLD STAY
9. ENGINE COOLANT TEMPERATURE GAUGE UNIT
10. WATER INLET FITTING
11. THERMOSTAT
12. THERMOSTAT HOUSING
13. MANIFOLD ABSOLUTE PRESSURE SENSOR
14. INTAKE MANIFOLD
15. INTAKE MANIFOLD GASKET
16. ENGINE HANGER
17. KNOCK SENSOR
18. ENGINE OIL PRESSURE SWITCH
19 ENGINE OIL PRESSURE SWITCH
20. ENGINE COOLANT TEMPERATURE SENSOR
21. WATER OUTLET FITTING

67170-GALA-G28

Intake manifold and related components—Lancer and Sportback 2.4L (4G69) engines

10. Disconnect the water hoses and remove the water inlet pipe.
11. Remove the intake manifold stay bracket.
12. Remove the Manifold Absolute Pressure (MAP) sensor.
13. Remove the engine hanger.
14. Remove the intake manifold bolt and manifold.
15. Remove the intake manifold gasket and discard.

To install:

16. Clean all gasket material from the cylinder head intake mounting surface and intake manifold assembly.
17. Install the intake manifold, using a new gasket. Torque the manifold in a criss-cross pattern, starting from the inside and working outwards to 17–19 ft. lbs. (21–27 Nm).
18. Install the engine hanger.
19. Install the MAP sensor.
20. Install the intake manifold stay.
21. Install the water inlet pipe and connect the water hoses.
22. Install the EGR valve and gasket.
23. Install the fuel rail and injectors assembly.
24. Install the solenoid valve and mounting bracket.
25. Connect the vacuum pipes and hoses.
26. Install the throttle body.
27. Install the air cleaner assembly.
28. Fill the system with coolant.
29. Connect the negative battery cable.

Lancer Evolution

2.0L DOHC ENGINE

1. Before servicing the vehicle, refer to the precautions at the beginning of this section.
2. Relieve the fuel system pressure.
3. Disconnect the negative battery cable and drain the cooling system.
4. Remove the air cleaner assembly.
5. Remove the throttle body and gasket.

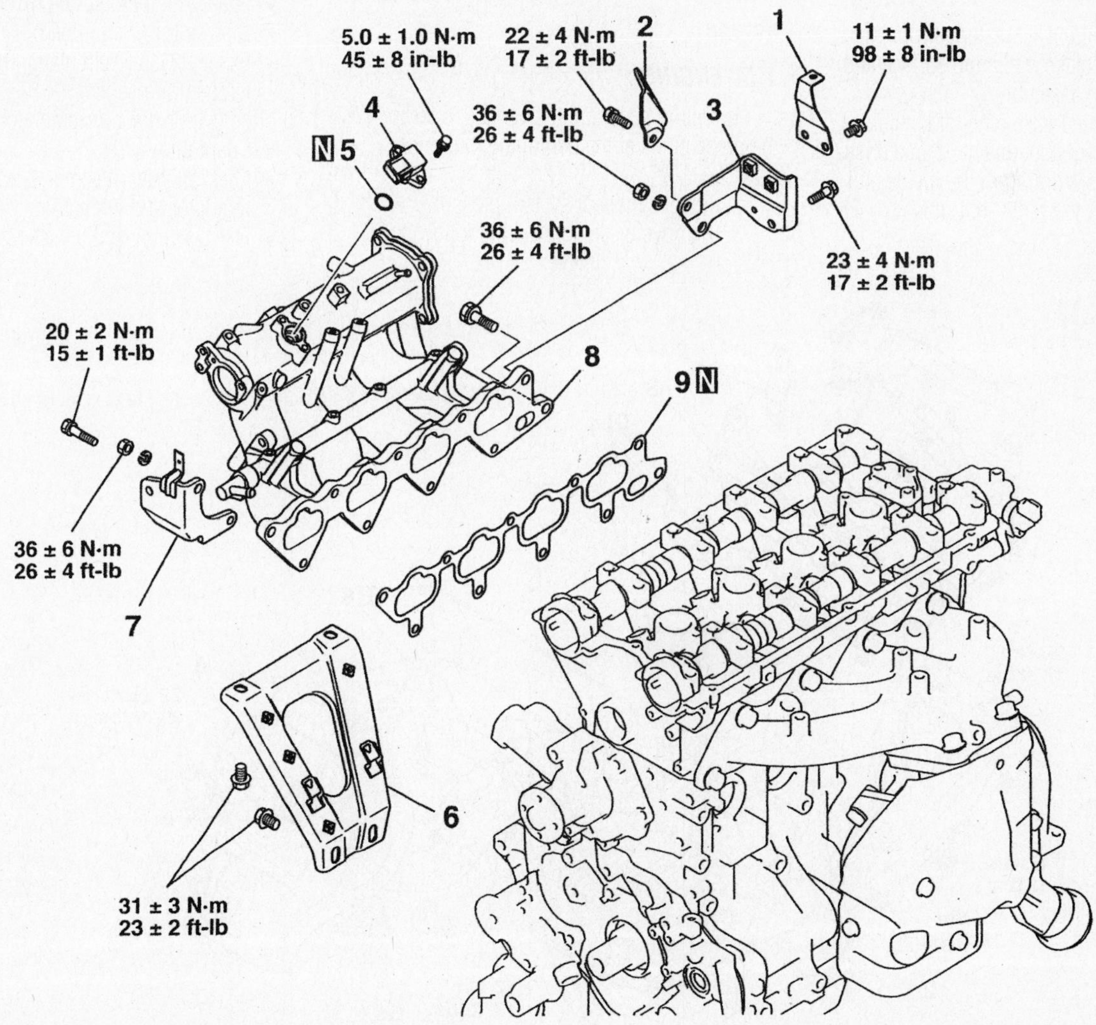

Exploded view of the intake manifold and related components—2.0L DOHC turbocharged engine

6. Remove the Exhaust Gas Recirculation (EGR) valve and gasket.
7. Remove the fuel rail and injector assembly.
8. Remove the bracket and the engine hanger.
9. Remove the air control valve bracket.
10. Remove the Manifold Differential Pressure (MDP) sensor and O-ring.
11. Remove the intake manifold stay bracket.
12. Remove the alternator brace stay bracket.
13. Remove the intake manifold bolt and manifold.
14. Remove the intake manifold gasket and discard.

To install:

15. Clean all gasket material from the cylinder head intake mounting surface and intake manifold assembly.
16. Install the intake manifold, using a new gasket. Torque the manifold in a criss-cross pattern, starting from the inside and working outwards to 22–30 ft. lbs. (30–42 Nm).
17. Install the alternator brace stay bracket.
18. Install the intake manifold stay. Tighten the bolts to 21–25 ft. lbs. (28–34 Nm).
19. Install the MDP sensor. Replace the MDP sensor if it has been dropped.
20. Install the air control valve bracket.
21. Install the engine hanger and bracket.
22. Install the fuel rail and injector assembly.
23. Install the EGR valve and gasket.
24. Install the throttle body.
25. Install the air cleaner assembly.
26. Fill the system with coolant.
27. Connect the negative battery cable.

Mirage

1.5L ENGINE

1. Before servicing the vehicle, refer to the precautions at the beginning of this section.
2. Relieve the fuel system pressure.
3. Remove or disconnect the following:
   - Battery negative cable and drain the cooling system
   - Upper radiator hose, heater hose and water bypass hose
   - Thermostat housing from intake manifold
   - Accelerator cable, breather hose and air intake hose
   - Vacuum hoses, label for reference
   - Throttle body assembly
   - High pressure fuel line and the fuel return hose
4. Disconnect the following:
   - Heated Oxygen (HO_2S) sensor
   - Engine Coolant Temperature (ECT) sensor
   - Idle Air Control (IAC) motor
   - Intake Air Temperature (IAT) sensor
   - Distributor (if equipped)
   - Exhaust Gas Recirculation (EGR) temperature sensor
5. Remove or disconnect the following:
   - Spark plug wires
   - Fuel rail, fuel injectors, pressure regulator and insulators
   - EGR valve from the intake manifold

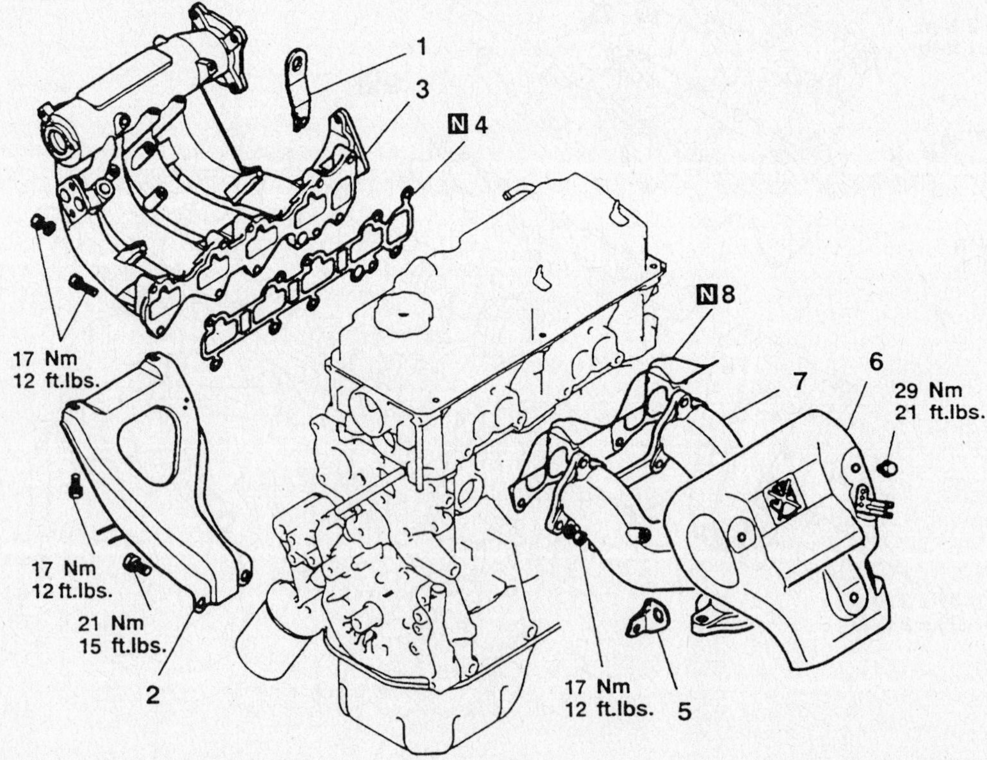

Exploded view of the intake and exhaust manifold mounting—Mirage 1.5L (4G15) engine

- Intake manifold support bracket and remove the engine mount support bracket
- Intake manifold mounting bolts and remove the intake manifold assembly

To install:

6. Clean all gasket material from the cylinder head intake mounting surface and intake manifold assembly.
7. Install or connect the following:
   - Intake manifold gasket, using a new gasket. Torque the manifold in a crisscross pattern, starting from the inside and working outwards to 13 ft. lbs. (18 Nm).
   - Intake manifold support bracket and tighten the mounting bolts to 16 ft. lbs. (22 Nm)
   - Engine mount support bracket and tighten the mounting bolts to 26 ft. lbs. (36 Nm)
   - EGR valve and tighten the mounting bolts to 15 ft. lbs. (21 Nm)
   - Install the fuel rail, fuel injectors and pressure regulator to the engine, using new insulators and O-rings. Torque the retaining bolts to 84–108 inch lbs. (10–13 Nm).
   - Spark plug wires
8. Connect the following:
   - HO_2S sensor
   - ECT sensor
   - IAC motor
   - IAT sensor
   - Distributor (if equipped)
   - EGR temperature sensor
9. Install or connect the following:
   - Fuel feed and return lines
   - Throttle body assembly
   - Vacuum hoses and pipes as necessary, including the brake booster vacuum line
   - Accelerator cable
   - Breather and air intake hose
   - Thermostat housing to the intake manifold and tighten the mounting bolts to 13 ft. lbs. (18 Nm)
   - Upper radiator hose, heater hose and water bypass hose
10. Fill the system with coolant.
11. Connect the negative battery cable.

1.8L ENGINE

1. Before servicing the vehicle, refer to the precautions at the beginning of this section.
2. Relieve the fuel system pressure.
3. Remove or disconnect the following:
   - Battery negative cable and drain the cooling system
   - Accelerator cable and the air intake hose
4. Disconnect the following:
   - Heated Oxygen (HO_2S) sensor
   - Engine Coolant Temperature (ECT) sensor
   - Idle Air Control (IAC) motor
   - Exhaust Gas Recirculation (EGR) temperature sensor
   - Throttle Position (TP) sensor
   - Oil pressure switch
   - Distributor (if equipped)
   - Fuel injectors
5. Label and remove all vacuum hoses.
6. Remove or disconnect the following:
   - Upper radiator hose, heater hose and water bypass hose
   - High pressure fuel line and the fuel return hose
   - Fuel rail, fuel injectors, pressure regulator and insulators
   - Intake manifold support bracket
   - Thermostat housing, if necessary for clearance
   - Intake manifold mounting bolts/nuts and remove the intake manifold assembly

To install:

7. Clean all gasket material from the cylinder head intake mounting surface and intake manifold assembly.
8. Install or connect the following:
   - Intake manifold, using a new gasket. Torque the manifold in a crisscross pattern, starting from the inside and working outwards to 14 ft. lbs. (20 Nm).
   - Thermostat housing
   - Intake manifold brace bracket
   - Fuel rail, fuel injectors and pressure regulator to the engine. Torque the retaining bolts to 108 inch lbs. (12 Nm).
   - Fuel feed and return lines
   - Upper radiator hose, heater hose and water bypass hoses
   - Vacuum hoses
9. Connect the following:
   - HO_2S sensor
   - ECT sensor
   - IAC motor
   - EGR temperature sensor
   - TP sensor
   - Oil pressure switch
   - Distributor (if equipped)
   - Fuel injectors
10. Connect and adjust the accelerator cable and install the air intake hose.
11. Fill the system with coolant.
12. Connect the negative battery cable.

Exhaust Manifold

REMOVAL & INSTALLATION

Diamante

1. Before servicing the vehicle, refer to the precautions at the beginning of this section.
2. Remove or disconnect the following:
   - Battery negative cable
   - Exhaust pipe from the exhaust manifold
   - Condenser electric cooling fan assembly
3. If removing the front manifold, remove the oil dipstick and tube from the engine.
4. If removing the rear manifold, disconnect the Exhaust Gas Recirculation (EGR) tube.
5. If removing the rear manifold, remove the intake plenum stay and the roll stopper bracket.
6. Remove or disconnect the following:
   - Electrical connector and remove the Heated Oxygen (HO_2S) sensor
   - Exhaust manifold mounting bolts the manifold

To install:

7. Clean all gasket material from the mating surfaces.
8. Install or connect the following:
   - New gasket and install the manifold. Tighten the nuts in a crisscross pattern to 21 ft. lbs. (30 Nm) for the J- engine or to 14 ft. lbs. (19 Nm) for the H- engine.
   - Heat shields
   - EGR tube and intake plenum stay and roll stopper bracket, if removed
   - HO_2S sensor
   - Electric cooling fan assembly, air conditioning compressor, dipstick tube and alternator, as required
   - New flange gasket and connect the exhaust pipe or converter assembly
   - Negative battery cable and check for exhaust leaks

Eclipse

2.4L ENGINE

⁂ CAUTION

The air bag system must be disarmed before removing the exhaust manifold.

1. Before servicing the vehicle, refer to the precautions at the beginning of this section.

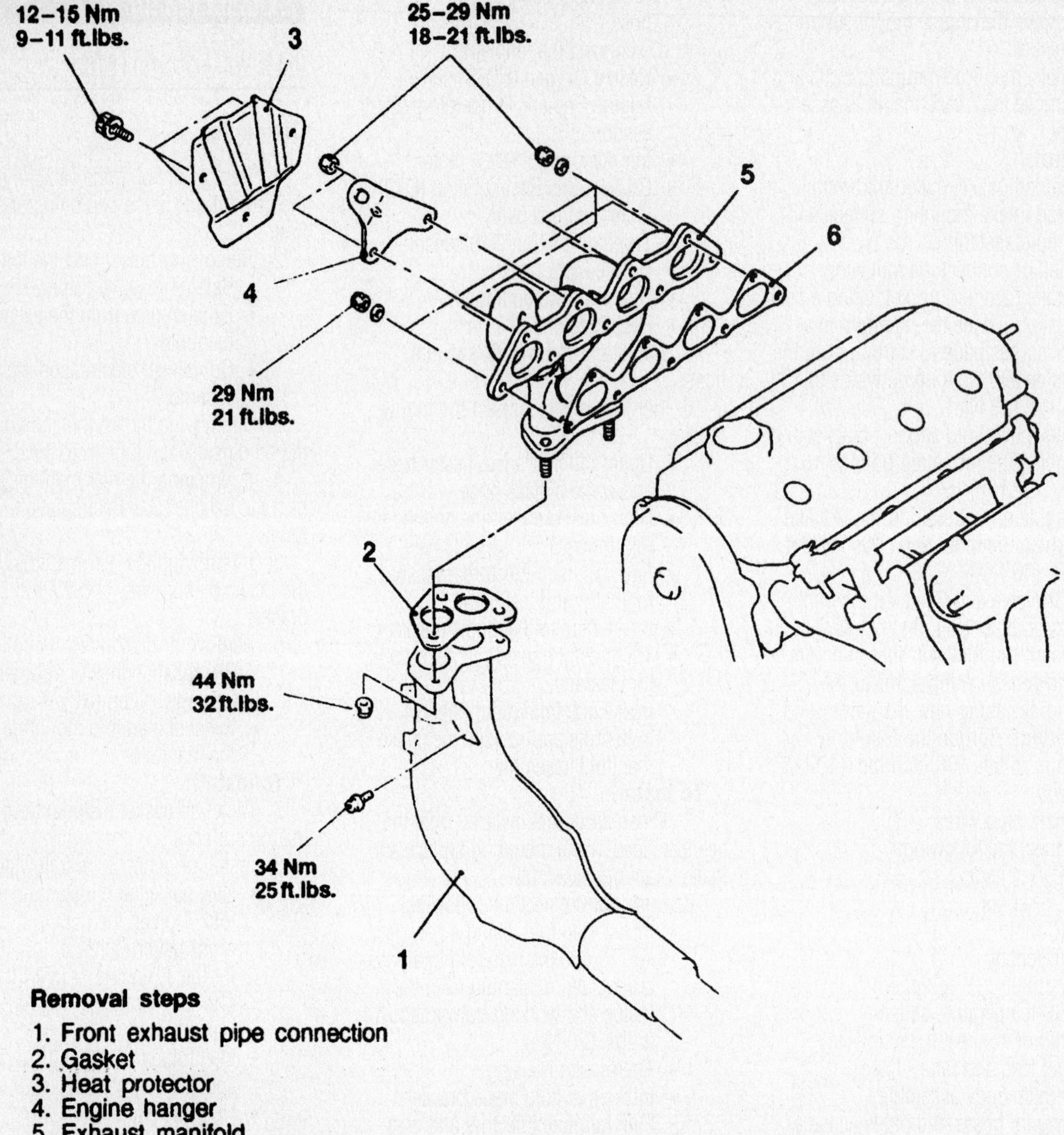

Exploded view of the exhaust manifold mounting—Eclipse 2.4L (4G64) engine shown, Galant similar

2. Remove or disconnect the following:
   - Negative battery cable
   - Front exhaust pipe from the exhaust manifold
   - Heat shield
   - Mounting nuts, the exhaust manifold, and the exhaust manifold gasket

To install:

3. Install or connect the following:
   - New exhaust manifold gasket to the cylinder head
   - Exhaust manifold. Torque the mounting nuts to 21 ft. lbs. (29 Nm).
   - Heat shield and tighten the bolts to 10 ft. lbs. (13 Nm)
   - New gasket between the exhaust manifold and the front exhaust pipe and reconnect the pipe. Torque the nuts to 32 ft. lbs. (34 Nm).
   - Negative battery cable
4. Start the engine and check for any exhaust leaks.

3.0L ENGINE

1. Before servicing the vehicle, refer to the precautions at the beginning of this section.
2. Disconnect the negative battery cable.
3. Remove the air cleaner assembly.
4. Remove the exhaust manifold heat protector shields.
5. Remove the Oxygen (O_2S) sensor.
6. Remove the exhaust manifold mounting nuts and exhaust manifold(s).
7. Remove the exhaust manifold gasket(s), and discard.
8. Remove the engine hanger.

To install:

9. Install the engine hanger.
10. Clean all gasket material from the mating surfaces.
11. Install a new gasket and the exhaust manifold. Torque the manifold nuts in a

crisscross pattern, starting from the inside and working outwards to 29–37 ft. lbs. (39–49 Nm).

12. Install the O_2S sensor. Tighten to 29–37 ft. lbs. (39–49 Nm).
13. Install the exhaust manifold heat protector shields.
14. Install the air cleaner assembly.
15. Connect the negative battery cable.

Galant

2.4L ENGINE

⁂ CAUTION

The air bag system must be disarmed before removing the exhaust manifold.

1. Before servicing the vehicle, refer to the precautions at the beginning of this section.
2. Remove or disconnect the following:
   - Negative battery cable
   - Front exhaust pipe from the exhaust manifold
   - Heat shield
   - Mounting nuts, the exhaust manifold, and the exhaust manifold gasket

To install:

3. Install or connect the following:
   - New exhaust manifold gasket to the cylinder head
   - Exhaust manifold. Torque the mounting nuts to 21 ft. lbs. (29 Nm).
   - Heat shield and tighten the bolts to 10 ft. lbs. (13 Nm)
   - New gasket between the exhaust manifold and the front exhaust pipe and reconnect the pipe. Torque the nuts to 32 ft. lbs. (34 Nm).
   - Negative battery cable
4. Start the engine and check for any exhaust leaks.

3.0L ENGINE

1. Before servicing the vehicle, refer to the precautions at the beginning of this section.
2. Disconnect the negative battery cable.
3. Remove the air cleaner assembly.
4. Remove the exhaust manifold heat protector shields.
5. Remove the Oxygen (O_2S) sensor.
6. Remove the exhaust manifold mounting nuts and exhaust manifold(s).
7. Remove the exhaust manifold gasket(s), and discard.
8. Remove the engine hanger.

To install:

9. Install the engine hanger.
10. Clean all gasket material from the mating surfaces.
11. Install a new gasket and the exhaust manifold. Torque the manifold nuts in a crisscross pattern, starting from the inside and working outwards to 29–37 ft. lbs. (39–49 Nm).
12. Install the oxygen sensor. Tighten to 29–37 ft. lbs. (39–49 Nm).
13. Install the exhaust manifold heat protector shields.
14. Install the air cleaner assembly.
15. Connect the negative battery cable.

3.8L ENGINE

1. Before servicing the vehicle, refer to the precautions at the beginning of this section.
2. Disconnect the negative battery cable.
3. Remove the air cleaner assembly.
4. Remove the front and rear Heated Oxygen (HO_2) sensors.
5. Remove the exhaust manifold heat protector shields.
6. Remove the exhaust manifold stay brackets.
7. Remove the exhaust manifold mounting nuts and exhaust manifold(s).
8. Remove the exhaust manifold gasket(s), and discard.
9. Remove the engine hanger.

To install:

10. Install the engine hanger.
11. Clean all gasket material from the mating surfaces.
12. Install a new gasket and the exhaust manifold. Torque the manifold nuts in a crisscross pattern, starting from the inside and working outwards to 29–37 ft. lbs. (39–49 Nm).
13. Install the exhaust manifold stay brackets.
14. Install the exhaust manifold heat protector shields.
15. Install the front and rear Heated Oxygen (HO_2) sensors. Tighten to 29–37 ft. lbs. (39–49 Nm).
16. Install the air cleaner assembly.
17. Connect the negative battery cable.

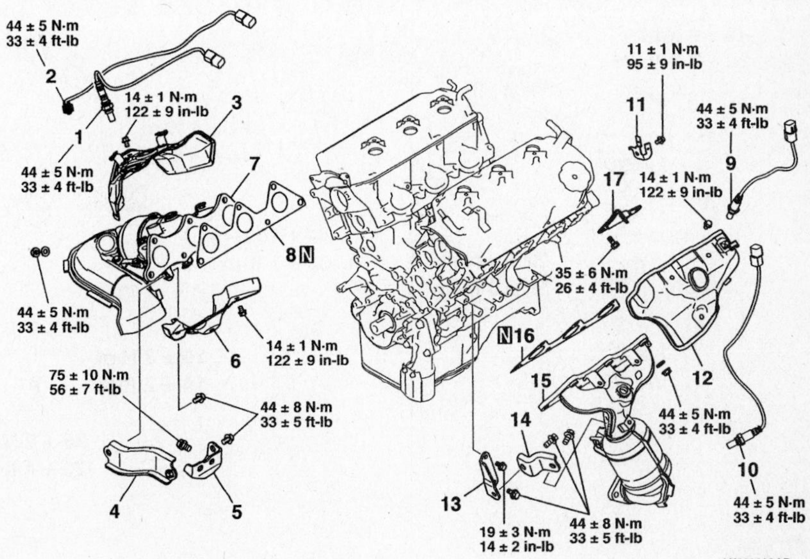

REMOVAL STEPS
1. RIGHT BANK HEATED OXYGEN SENSOR (FRONT)
2. RIGHT BANK HEATED OXYGEN SENSOR (REAR)
3. HEAT PROTECTOR, RIGHT
>>B<< 4. EXHAUST MANIFOLD STAY, RIGHT "B"
5. EXHAUST MANIFOLD STAY, RIGHT "A"
6. HEAT PROTECTOR, LOWER RIGHT
7. EXHAUST MANIFOLD, RIGHT
8. EXHAUST MANIFOLD GASKET

REMOVAL STEPS (Continued)
9. LEFT BANK HEATED OXYGEN SENSOR (FRONT)
10. LEFT BANK HEATED OXYGEN SENSOR (REAR)
11. CONNECTOR BRACKET
12. HEAT PROTECTOR, LEFT
>>A<< 13. EXHAUST MANIFOLD STAY, LEFT "B"
14. EXHAUST MANIFOLD STAY, LEFT "A"
15. EXHAUST MANIFOLD, LEFT
16. EXHAUST MANIFOLD GASKET
17. ENGINE HANGER

67170-GALA-G30

Exploded view of the exhaust manifold and related components—3.8L engine

Lancer and Sportback

2.0L SOHC ENGINE

1. Before servicing the vehicle, refer to the precautions at the beginning of this section.
2. Remove or disconnect the following:
   - Negative battery cable
   - Engine undercover
   - Pressure hose clamp bolt
   - Power steering pump bracket stay bolt
   - Drive belt
   - Power steering pump and bracket assembly and position it aside. Do NOT disconnect the fluid lines.
   - Front exhaust pipe and gasket from the manifold. Discard the gasket.
   - Exhaust manifold bracket "B" (see illustration)
   - Heated Oxygen (HO_2S) sensor
   - Heat shield
   - Exhaust manifold retainers and manifold
   - Exhaust manifold gasket, and discard
   - Exhaust manifold bracket "A" (see illustration)

To install:

3. Clean all gasket material from the mating surfaces.
4. Install or connect the following:
   - Exhaust manifold bracket "A"
   - New gasket and exhaust manifold. Tighten the upper retainers to 20–24 ft. lbs. (27–33 Nm) and the lower retainers to 10–14 ft. lbs. (15–19 Nm).
   - Heat shield

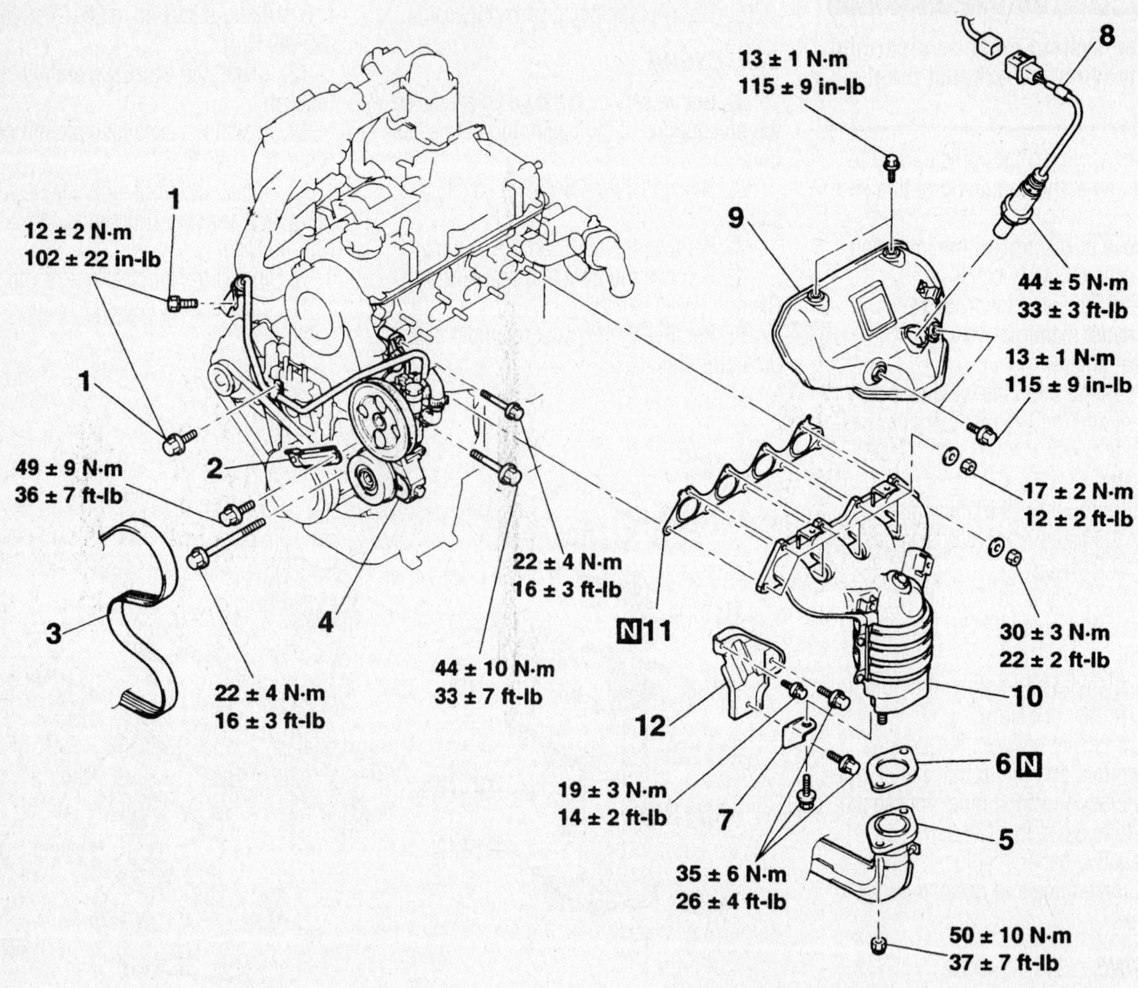

Exploded view of the exhaust manifold and related components—Lancer 2.0L engine

- HO_2S
- Exhaust manifold bracket "B"
- Front exhaust pipe to the exhaust manifold, using a new gasket
- Power steering pump assembly
- Drive belt
- Power steering pump bracket stay bolt
- Pressure hose clamp bolt
- Engine undercover
- Negative battery cable

2.4L ENGINE

1. Before servicing the vehicle, refer to the precautions at the beginning of this section.
2. Disconnect the negative battery cable.
3. Remove the air cleaner assembly.
4. Remove the engine hanger.
5. Remove the exhaust manifold cover.
6. Remove the exhaust manifold mounting nuts and exhaust manifold.
7. Remove the exhaust manifold gasket, and discard.

To install:

8. Clean all gasket material from the mating surfaces.
9. Install a new gasket and the exhaust manifold. Torque the manifold nuts in a crisscross pattern, starting from the inside and working outwards to 33–39 ft. lbs. (44–54 Nm).
10. Install the exhaust manifold cover.
11. Install the engine hanger.
12. Install the air cleaner assembly.
13. Connect the negative battery cable.

Lancer Evolution

The removal and installation procedure for the exhaust manifold is a part of the Turbocharger procedure, located earlier in this section.

Mirage

1. Before servicing the vehicle, refer to the precautions at the beginning of this section.
2. Remove or disconnect the following:
   - Battery negative cable
   - Exhaust pipe from the exhaust manifold
   - Electric cooling fan assembly
   - Heated Oxygen (HO_2S) sensor
   - Exhaust Gas Recirculation (EGR) pipe
   - Outer exhaust manifold heat shield and engine hanger
   - Exhaust manifold mounting bolts, the inner heat shield and the exhaust manifold

To install:

3. Clean all gasket material from the mating surfaces.
4. Using a new gasket and install the manifold. For 1.5L engines, tighten the nuts on a crisscross patter to 13 ft. lbs. (18 Nm). For 1.8L engines, tighten the inner nuts to in a crisscross pattern to 13 ft. lbs. (18 Nm) and tighten the 2 outer (larger) nuts to 22 ft. lbs. (30 Nm).
5. Install or connect the following:
   - Heat shields
   - EGR pipe
   - HO_2S sensor
   - Electric cooling fan assembly
   - New flange gasket and connect the exhaust pipe
   - Negative battery cable and check for exhaust leaks

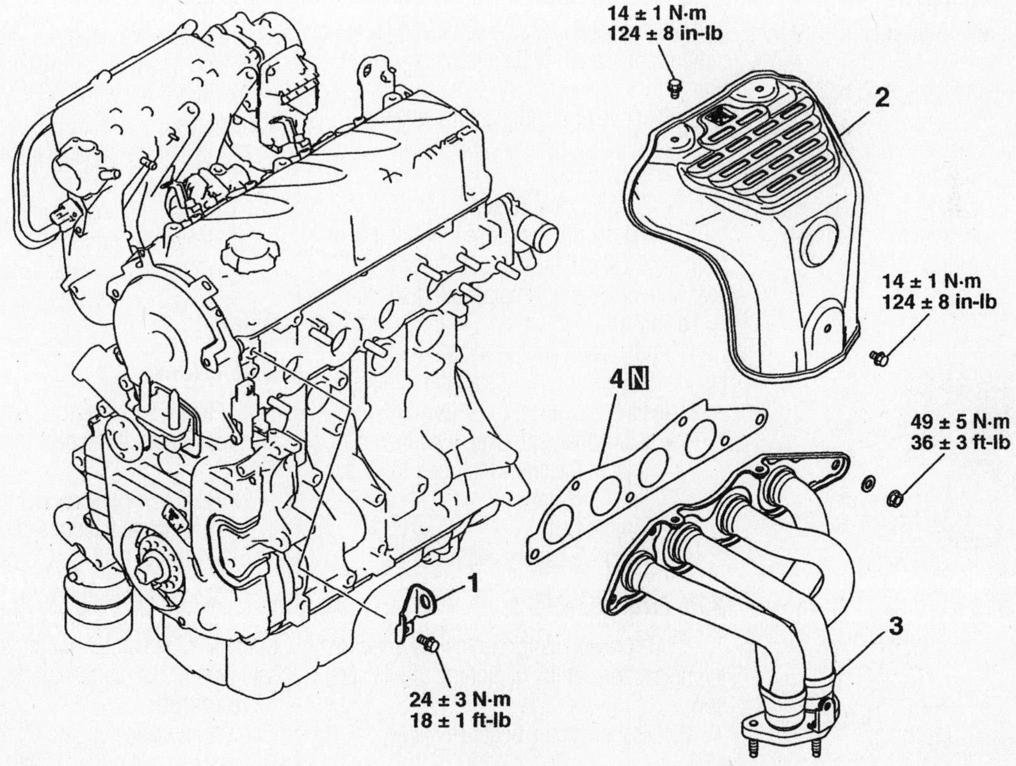

Exploded view of the exhaust manifold and related components—Lancer and Sportback 2.4L engine

Front Crankshaft Seal

REMOVAL & INSTALLATION

Diamante

1. Before servicing the vehicle, refer to the precautions at the beginning of this section.
2. Remove or disconnect the following:
   - Negative battery cable
   - Drive belts
   - Crankshaft pulley
   - Timing belt covers and the timing belt
   - Crankshaft Position (CKP) sensor
   - Crankshaft sprocket, the sensing blade, spacer and Woodruff® key
3. Pry the seal from the bore, using a suitable tool.

To install:

4. Using a seal driver, install the new crankshaft seal. Lubricate the lips of the seal with clean engine oil.
5. Install or connect the following:
   - Woodruff® key, spacer, sensing blade and the crankshaft sprocket
   - CKP sensor and tighten the retaining bolts to 84 inch lbs. (9 Nm)
   - Timing belt and the timing belt covers
   - Crankshaft pulley and retaining bolt. Torque the retaining bolt to 130–137 ft. lbs. (180–190 Nm) for the DOHC engine or to 108–116 ft. lbs. (150–160 Nm) for the SOHC engine.

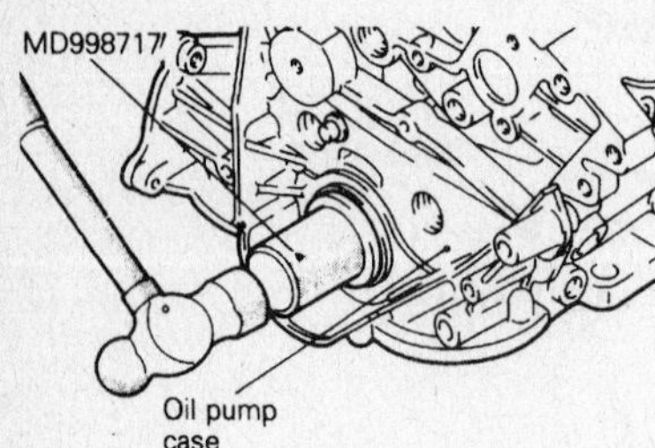

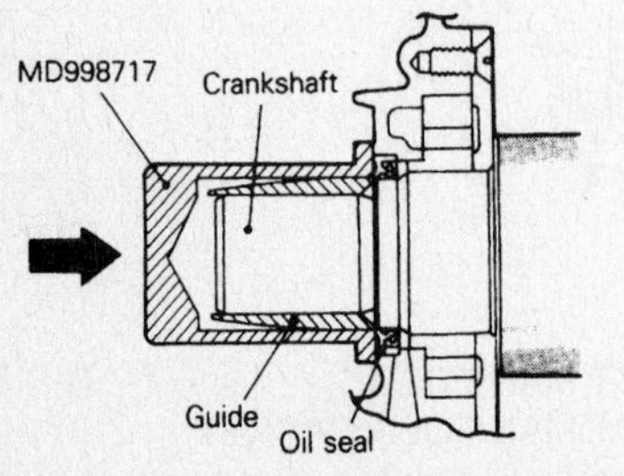

Crankshaft seal installation—Diamante

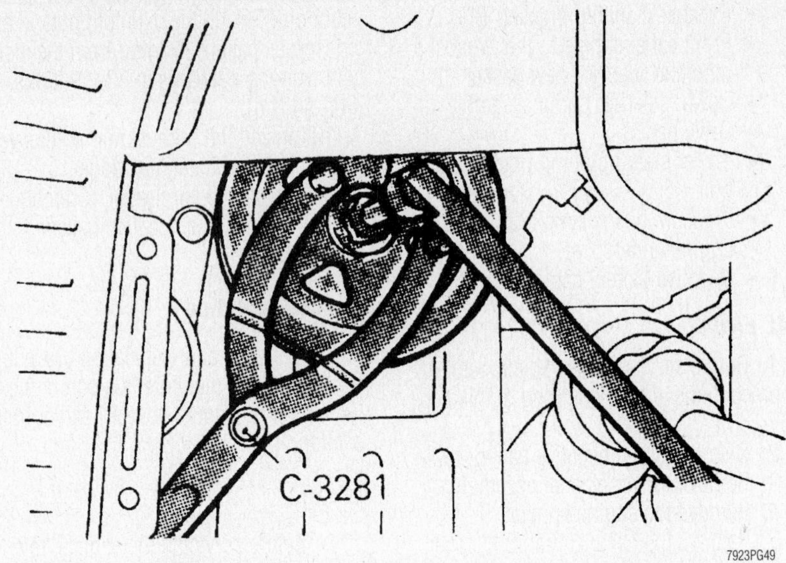

The crankshaft pulley pin spanner tool should be used to hold pulley while bolt is removed—Diamante

   - Drive belts, adjust
   - Negative battery cable

Eclipse

2.4L ENGINE

1. Before servicing the vehicle, refer to the precautions at the beginning of this section.
2. Remove or disconnect the following:
   - Negative battery cable
   - Timing belt
   - Crankshaft sprocket
3. Carefully pry the oil seal out of the front case. Be careful not to damage the oil seal bore or the crankshaft sealing surface.

To install:

4. Apply clean engine oil to the oil seal lip. Using a seal driver, install the oil seal.
5. Install or connect the following:
   - Crankshaft sprocket. If equipped, tighten the crankshaft bolt to 87 ft. lbs. (118 Nm).
   - Timing belt
   - Negative battery cable

3.0L ENGINE

1. Before servicing the vehicle, refer to the precautions at the beginning of this section.
2. Disconnect the negative battery cable.
3. Remove the timing belt.
4. Remove the crankshaft sprocket.
5. Remove the Crankshaft Position (CKP) sensor.
6. Remove the crankshaft sensing blade.
7. Remove the crankshaft spacer and woodruff key.
8. Carefully pry the oil seal out of the front case. Be careful not to damage the oil seal bore or the crankshaft sealing surface.

To install:

9. Apply clean engine oil to the oil seal lip. Using a seal driver, install the oil seal.
10. Install the woodruff key and crankshaft spacer.
11. Install the crankshaft sensing blade.
12. Install the CKP sensor.
13. Install the crankshaft sprocket.
14. Install the timing belt.
15. Connect the negative battery cable.

Galant

2.4L ENGINE

1. Before servicing the vehicle, refer to the precautions at the beginning of this section.
2. Remove or disconnect the following:
   - Negative battery cable
   - Timing belt
   - Crankshaft sprocket
3. Carefully pry the oil seal out of the front case. Be careful not to damage the oil seal bore or the crankshaft sealing surface.

To install:

4. Apply clean engine oil to the oil seal lip. Using a seal driver, install the oil seal.
5. Install or connect the following:
   - Crankshaft sprocket. If equipped, tighten the crankshaft bolt to 87 ft. lbs. (118 Nm).
   - Timing belt
   - Negative battery cable

3.0L AND 3.8L ENGINES

1. Before servicing the vehicle, refer to the precautions at the beginning of this section.
2. Disconnect the negative battery cable.
3. Remove the timing belt.
4. Remove the crankshaft sprocket.
5. Remove the Crankshaft Position (CKP) sensor.
6. Remove the crankshaft sensing blade.
7. Remove the crankshaft spacer and woodruff key.
8. Carefully pry the oil seal out of the front case. Be careful not to damage the oil seal bore or the crankshaft sealing surface.

To install:

9. Apply clean engine oil to the oil seal lip. Using a seal driver, install the oil seal.
10. Install the woodruff key and crankshaft spacer.
11. Install the crankshaft sensing blade.
12. Install the CKP sensor.
13. Install the crankshaft sprocket.
14. Install the timing belt.
15. Connect the negative battery cable.

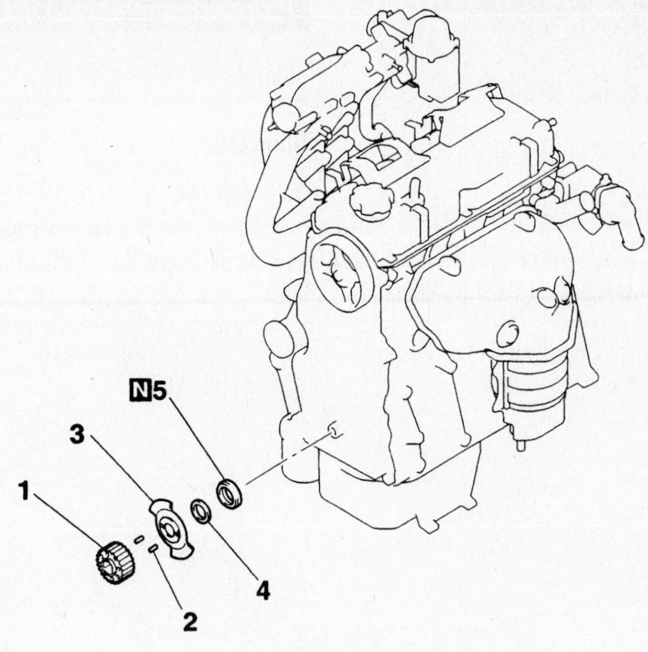

View of the front crankshaft seal—Lancer 2.0L engine

Lancer and Sportback

2.0L SOHC ENGINE

1. Before servicing the vehicle, refer to the precautions at the beginning of this section.
2. Remove or disconnect the following:
   - Negative battery cable
   - Timing belt
   - Crank angle sensor
   - Crankshaft sprocket
   - Spring pin
   - Crankshaft sending blade
   - Crankshaft spacer
   - Crankshaft front oil seal

To install:

3. Install or connect the following:
   - Engine oil to the oil seal lip
   - Crankshaft front using seal into the front case, using special driver tool MD998717
4. Remove all oil or other lubricants from the mounting surfaces on the crankshaft, spacer, sensing blade and crankshaft sprocket.
5. Assemble the spring pin, sensing blade, and crankshaft spacer together.
6. Install or connect the following:
   - Crankshaft sprocket assembly onto the crankshaft
   - Crank angle sensor
   - Timing belt
   - Negative battery cable

2.4L ENGINE

1. Before servicing the vehicle, refer to the precautions at the beginning of this section.
2. Disconnect the negative battery cable.
3. Remove the timing belt.
4. Remove the crankshaft balancer shaft drive sprocket.
5. Remove the crankshaft (woodruff) key.
6. Carefully pry the oil seal out of the front case. Be careful not to damage the oil seal bore or the crankshaft sealing surface.

To install:

7. Apply clean engine oil to the oil seal lip. Using a seal driver, install the oil seal.
8. Install the woodruff key.
9. Install the crankshaft balancer shaft drive sprocket.
10. Install the timing belt.
11. Connect the negative battery cable.

Lancer Evolution

2.0L DOHC ENGINE

1. Before servicing the vehicle, refer to the precautions at the beginning of this section.
2. Disconnect the negative battery cable.
3. Remove the timing belt.
4. Remove the crankshaft balancer shaft drive sprocket.
5. Remove the crankshaft (woodruff) key.
6. Carefully pry the oil seal out of the front case. Be careful not to damage the oil seal bore or the crankshaft sealing surface.

To install:

7. Apply clean engine oil to the oil seal lip. Using a seal driver, install the oil seal.
8. Install the woodruff key.
9. Install the crankshaft balancer shaft drive sprocket.
10. Install the timing belt.
11. Connect the negative battery cable.

Mirage

1. Before servicing the vehicle, refer to the precautions at the beginning of this section.
2. Remove or disconnect the following:
   - Negative battery cable
   - Crankshaft pulley retainer bolts and remove the pulley
   - Vibration damper retainer bolt and washer and remove damper
   - Timing belt
   - Crankshaft sprocket

3. Pry out the oil seal from front of engine.

To install:

4. Using proper size driver, install a new front seal.
5. Lubricate the lips of the new seal with clean engine oil.
6. Install or connect the following:
   - Timing belt, timing covers, valve cover and remaining components
   - Crankshaft sprocket and vibration damper
   - Engine undercover and connect the negative battery cable

Camshaft and Valve Lifters

REMOVAL & INSTALLATION

Diamante

3.5L ENGINE

1. Before servicing the vehicle, refer to the precautions at the beginning of this section.
2. Remove or disconnect the following:
   - Negative battery cable
   - Timing belt
   - Rocker arm cover
   - Lash adjuster clips on the rocker arms, then loosen the bearing cap bolts. Do not remove the bolts from the bearing caps.
   - Rocker arms, shafts and bearing caps as an assembly
   - Camshafts

To install:

3. Lubricate the camshafts with engine oil and position them on the cylinder heads.
4. Position the dowel pins as shown in the drawing.
5. Install or connect the following:

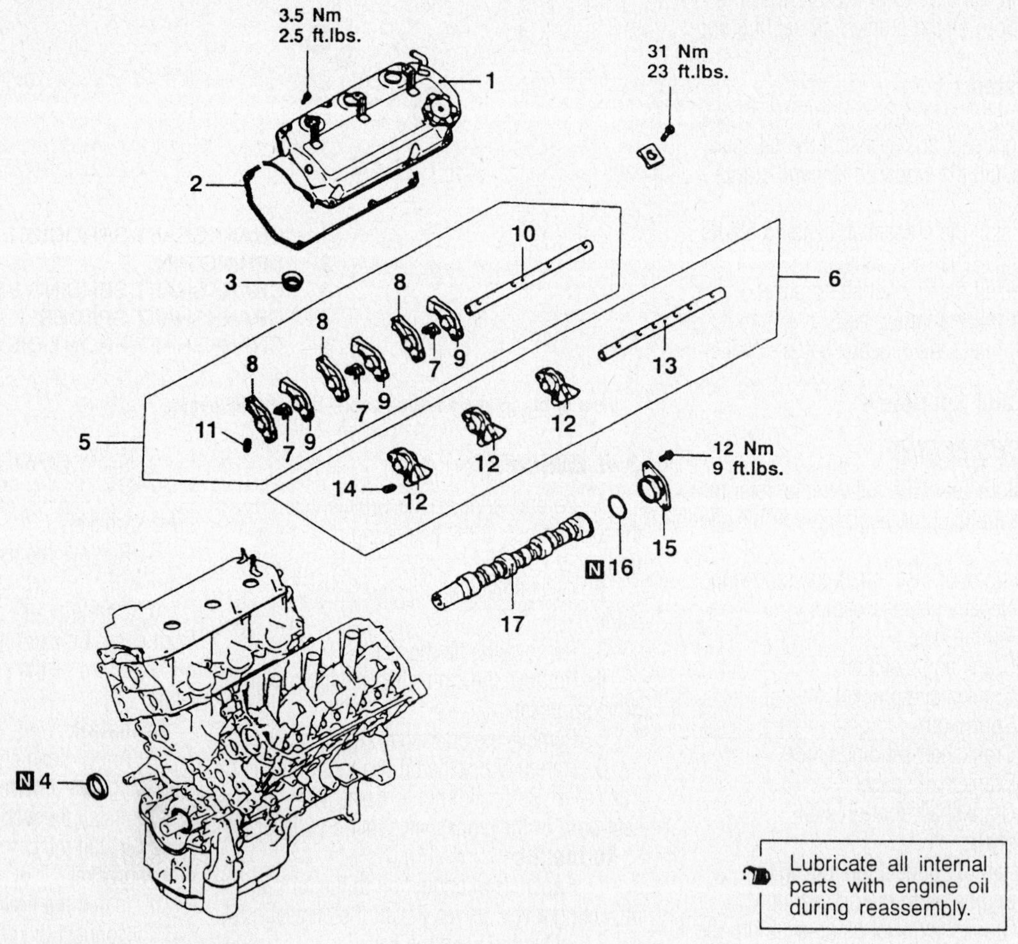

Removal steps

1. Rocker cover
2. Rocker cover gasket
3. Oil seal
4. Camshaft oil seal
5. Rocker arm, rocker arm shaft
6. Rocker arm, rocker arm shaft
7. Rocker shaft spring
8. Rocker arm A
9. Rocker arm B
10. Rocker arm shaft
11. Lash adjuster
12. Rocker arm C
13. Rocker arm shaft
14. Lash adjuster
15. Thrust case
16. O-ring
17. Camshaft

7923PGD3

Exploded view of the camshaft mounting—3.5L engine

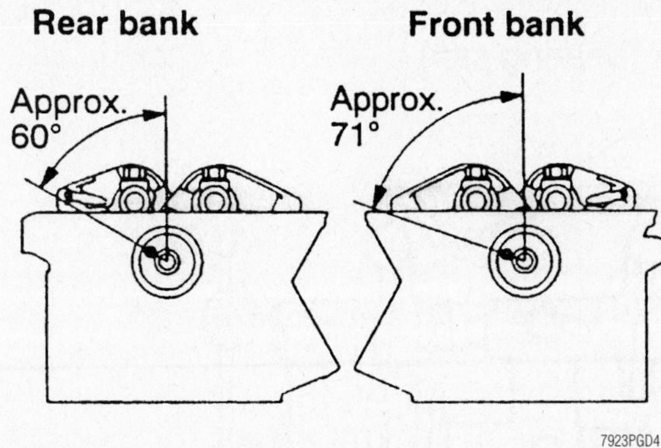

Camshaft dowel position during installation—3.5L engine

- Bearing caps/rocker arm assemblies. Tighten the bolts to 23 ft. lbs. (31 Nm).
- Rocker arm cover using a new gasket
- Timing belt and remaining components
- Negative battery cable

Eclipse

2.4L ENGINE

1. Before servicing the vehicle, refer to the precautions at the beginning of this section.
2. Remove or disconnect the following:
   - Remove the battery
   - Accelerator cable bracket and position the cable aside
   - Air intake hose
   - Breather hose and disconnect the Positive Crankcase Ventilation (PCV) hose
   - Spark plug cables
   - Rocker cover
3. Install lash adjuster retainer tools to the rocker arm.
   - Timing belt covers and the timing belt
   - Camshaft sprocket retainer bolt and remove the sprocket from the shaft
   - Camshaft oil seal
   - Both rocker arm shaft assemblies from the head
   - Camshaft from the cylinder head

To install:

4. Lubricate the camshaft journals and camshaft with clean engine oil and install the camshaft in the cylinder head.
5. Install the rocker arm and shaft assemblies. Tighten the rocker arm shaft retainer bolts to 21–25 ft. lbs. (29–35 Nm).
6. Apply a coating of engine oil to the oil seal. Using the proper size driver, press-fit the seal into the cylinder head.
7. Install or connect the following:
   - Camshaft sprocket and retainer bolt to 65 ft. lbs. (90 Nm)
   - Timing belt and belt covers
8. Remove the lash adjuster retaining tools.
9. Install or connect the following:
   - Rocker cover using new gasket material on mating surfaces
   - Spark plug cables
   - Air intake hose
   - Breather hose and connect the PCV hose
   - Battery
10. Run the engine at idle until normal operating temperature is reached. Check idle speed and ignition timing; adjust as required.

3.0L SOHC ENGINE

1. Before servicing the vehicle, refer to the precautions at the beginning of this section.
2. Remove or disconnect the following:
   - Negative battery cable
   - Intake manifold plenum stay bracket
   - Camshaft Position (CMP) sensor
   - Valve covers and the timing belt
3. Using a camshaft sprocket holding tool, hold the sprocket and loosen the bolt.
4. Remove the bolt and note the positioning of the knock pin at the end of the camshaft and remove the sprocket.
5. Install auto lash adjuster retainer tools on the rocker arms.

➡**Be sure to note the position of the rocker arms, rocker shafts and bearing caps for reinstallation purposes.**

6. Remove or disconnect the following:
   - Camshaft bearing caps but do not remove the bolts from the caps
   - Rocker arms, rocker shafts and bearing caps, as an assembly
   - Camshaft from the cylinder head
7. Inspect the bearing journals on the camshaft, cylinder head, and bearing caps.

To install:

➡**The right bank camshaft is identified by a 4mm slit at the rear end of the camshaft.**

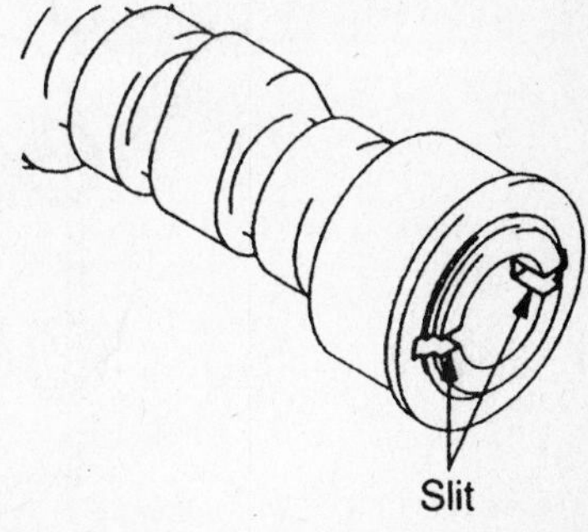

Right bank camshaft identification—Diamante 3.0L SOHC engine

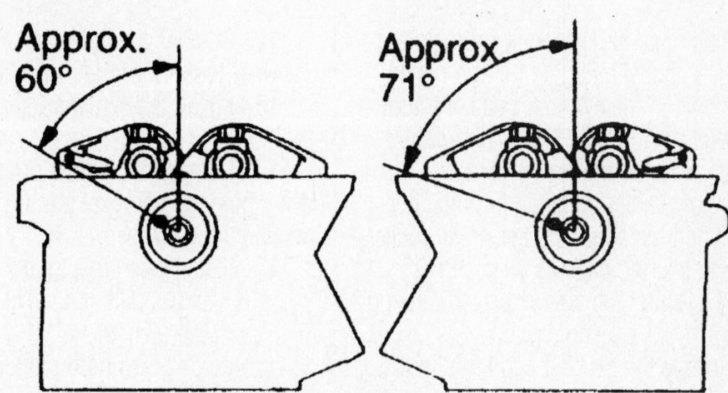

Proper positioning of the camshafts—Diamante 3.0L SOHC engine

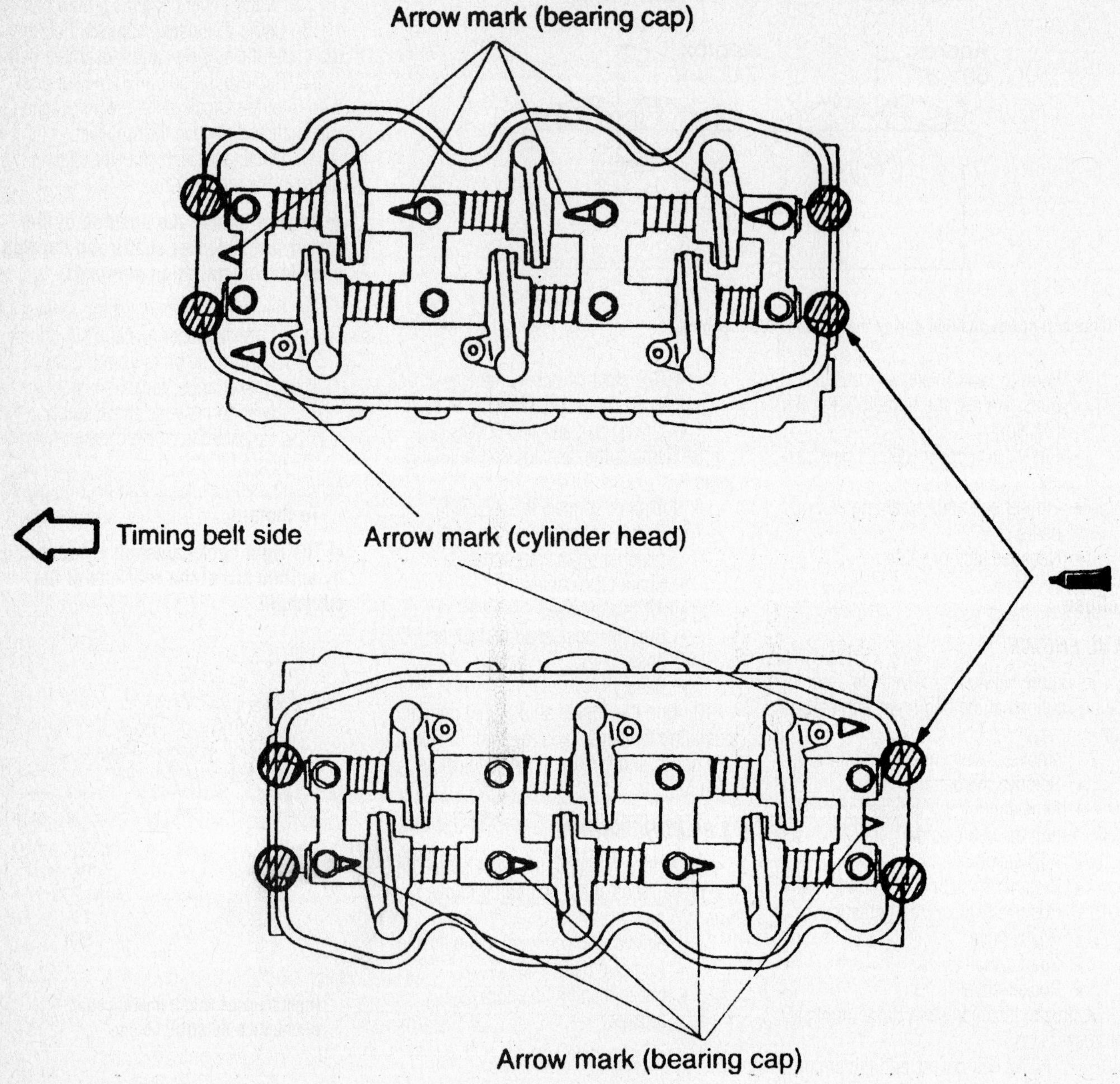

Alignment of the rocker shafts and application of sealant—Diamante 3.0L SOHC engine

8. Lubricate the camshaft journals and camshaft with clean engine oil and install the camshaft in the cylinder head. Be sure to properly position the knock pin of the camshaft as noted during removal.

9. Apply sealer at the ends of the bearing caps and install the rocker arms, rocker shafts and bearing caps as an assembly. Properly position the arrows on the bearing caps.

10. Torque the bearing cap bolts in the following sequence: No. 3, No. 2, No. 1 and No. 4 to 85 inch lbs. (10 Nm).

11. Repeat the sequence increasing the torque to 14 ft. lbs. (20 Nm).

12. Remove the auto lash adjuster retainer tools from the rocker arms.

13. Install the camshaft sprocket and bolt.

14. Using a camshaft sprocket holding tool, hold the sprocket and tighten the bolt to 65 ft. lbs. (90 Nm).

15. Install or connect the following:
- Timing belt and valve covers
- CMP sensor
- Intake manifold plenum stay bracket
- Negative battery cable and check for leaks

Galant

2.4L ENGINE

1. Before servicing the vehicle, refer to the precautions at the beginning of this section.
2. Relieve the fuel system pressure.
3. Remove or disconnect the following:
   - Negative battery cable
   - Accelerator cable, Positive Crankcase Ventilation (PCV) hoses, breather hoses, spark plug cables
   - Valve cover
   - Timing belt upper and lower covers

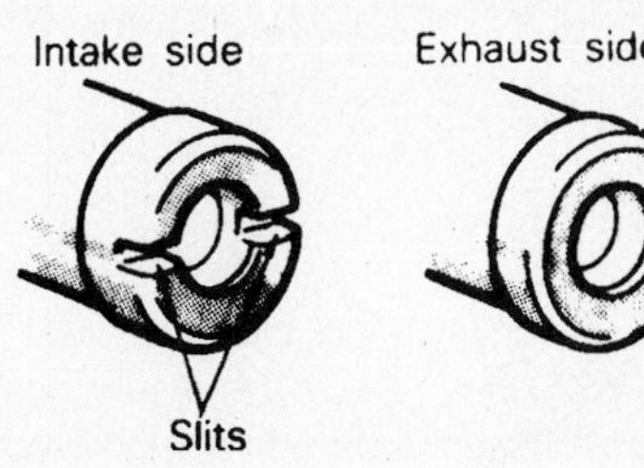

Camshaft identification—2.4L (4G64) engine

- Timing belt
- Camshaft sprockets

4. Loosen the bearing cap bolts in 2–3 steps. Label and remove all camshaft bearing caps.
   - Intake and exhaust camshafts
   - Rocker arms and lash adjusters

To install:

5. Install the lash adjusters and rocker arms into the cylinder head. Lubricate lightly with clean oil prior to installation.
6. Lubricate the camshafts with clean engine oil and position the camshafts on the cylinder head.
7. Be sure the dowel pin on both camshaft sprocket ends are located on the top.
8. Install the bearing caps. Tighten the caps in sequence and in 2 or 3 steps. No. 2 and 5 caps are of the same shape. Check the markings on the caps to identify the cap number and intake/exhaust symbol. Only **L** (intake) or **R** (exhaust) is stamped on No. 1 bearing cap. Also, be sure the rocker arm is correctly mounted on the lash adjuster and the valve stem end. Torque the retaining bolts to 15 ft. lbs. (20 Nm).
9. Apply a coating of engine oil to the oil seal. Using the proper size driver, press-fit the seal into the cylinder head.
10. Install or connect the following:
    - Camshaft sprockets and tighten the sprocket bolts to 58–72 ft. lbs. (80–100 Nm)
    - Timing belt, covers and related components
    - Valve cover and reconnect all related components
    - Negative battery cable

3.0L ENGINE

On this engine, the hydraulic lash adjusters are built into the rocker arms.

1. Before servicing the vehicle, refer to the precautions at the beginning of this section.
2. Remove or disconnect the following:
   - Negative battery cable
   - Intake manifold plenum stay bracket
   - Camshaft Position (CMP) sensor
   - Valve covers and the timing belt
3. Using a camshaft sprocket holding tool, hold the sprocket and loosen the bolt.
4. Remove the bolt and note the positioning of the knock pin at the end of the camshaft and remove the sprocket.
5. Install auto lash adjuster retainer tools on the rocker arms.

➡Be sure to note the position of the rocker arms, rocker shafts and bearing caps for reinstallation purposes.

6. Remove or disconnect the following:
   - Camshaft bearing caps but do not remove the bolts from the caps
   - Rocker arms, rocker shafts and bearing caps, as an assembly
   - Camshaft from the cylinder head
7. Inspect the bearing journals on the camshaft, cylinder head, and bearing caps.

To install:

➡The right bank camshaft is identified by a 4mm slit at the rear end of the camshaft.

8. Lubricate the camshaft journals and camshaft with clean engine oil and install the camshaft in the cylinder head. Be sure to properly position the knock pin of the camshaft as noted during removal.
9. Apply sealer at the ends of the bearing caps and install the rocker arms, rocker shafts and bearing caps as an assembly. Properly position the arrows on the bearing caps.
10. Torque the bearing cap bolts in the following sequence: No. 3, No. 2, No. 1 and No. 4 to 85 inch lbs. (10 Nm).
11. Repeat the sequence increasing the torque to 14 ft. lbs. (20 Nm).
12. Remove the auto lash adjuster retainer tools from the rocker arms.
13. Install the camshaft sprocket and bolt.
14. Using a camshaft sprocket holding tool, hold the sprocket and tighten the bolt to 65 ft. lbs. (90 Nm).
15. Install or connect the following:
    - Timing belt and valve covers
    - CMP sensor
    - Intake manifold plenum stay bracket
    - Negative battery cable and check for leaks

3.8L ENGINE

1. Before servicing the vehicle, refer to the precautions at the beginning of this section.
2. Disconnect the negative battery cable.
3. Remove the timing belt.
4. Remove the rocker arm cover.
5. Install the lash adjuster clips on the rocker arms, then loosen the rocker shaft cap bolts. Do not remove the bolts from the rocker shaft caps.
6. Remove the rocker arms, shafts and rocker shaft caps as an assembly.
7. Remove the Camshaft Position (CMP) sensor.
8. Remove the CMP sensing cylinder and sensor support (Left Bank) or thrust case (Right Bank).
9. Remove the camshaft sprocket.
10. Remove the camshaft.
11. Remove the camshaft oil seal.
12. Remove the hydraulic lash adjusters from the rocker arms.

To install:

13. Immerse the lash adjusters in clean diesel fuel, and using a small wire, move the plunger up and down 4 or 5 times. While pushing down lightly on the check ball in order to bleed the air from the adjuster.
14. Install the lash adjusters to the rocker arms.
15. Install the camshaft oil seal.
16. Install the camshaft.
17. Install the camshaft sprocket. Tighten the sprocket bolt to 58–72 ft. lbs. (78–98 Nm).
18. Install the CMP sensing cylinder and sensor support (Left Bank) or thrust case (Right Bank). Tighten the sensing cylinder bolt to 13–19 ft. lbs. (18–26 Nm).
19. Install the Camshaft Position (CMP) sensor.
20. Lubricate the rocker shaft with clean engine oil and install the rocker arms/shaft assembly. Tighten the rocker shaft mounting bolts gradually and evenly to 21–25 ft. lbs. (28–34 Nm).
21. Install or connect the rocker arm cover gasket and cover. Tighten the bolts to 27–35 inch lbs. (3–4 Nm).
22. Install the rocker arm cover using a new gasket.
23. Install the timing belt.
24. Fill the crankcase with oil and connect the negative battery cable.

Lancer and Sportback

2.0L ENGINE

1. Before servicing the vehicle, refer to the precautions at the beginning of this section.
2. Remove or disconnect the following:
   - Negative battery cable

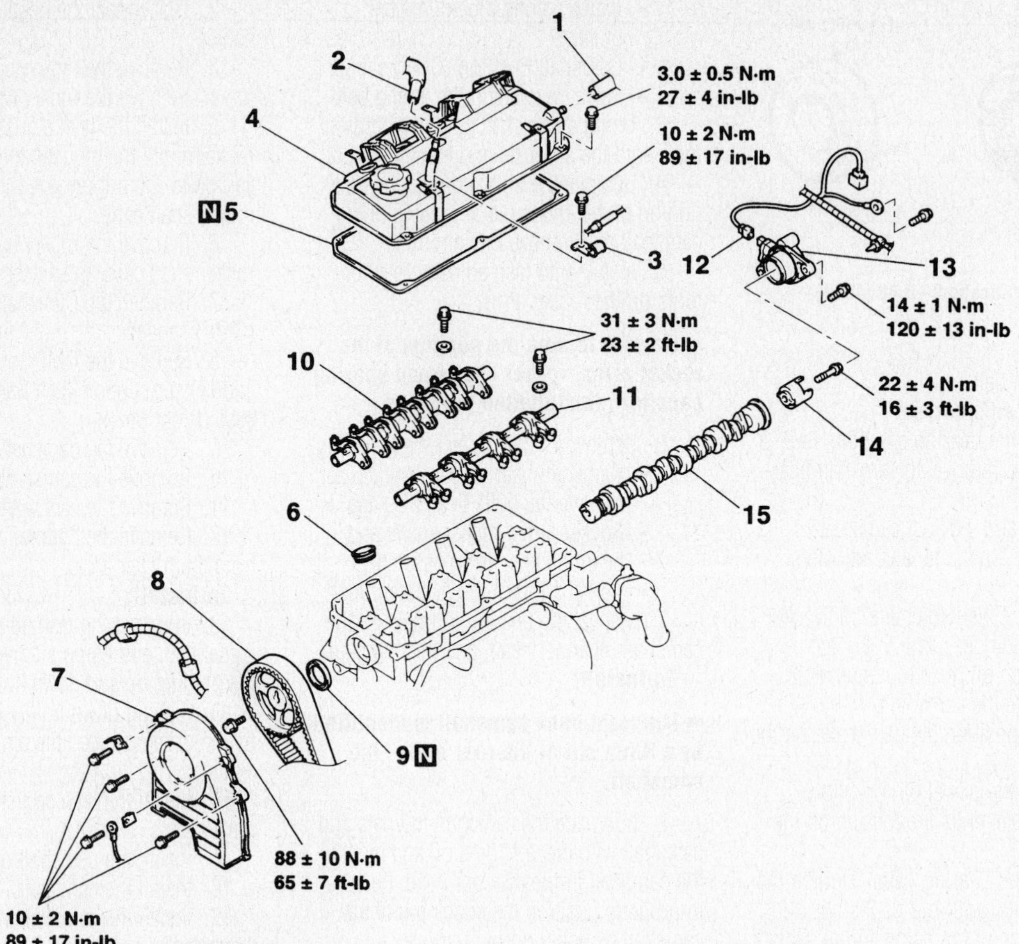

Exploded view of the camshaft and related components—Lancer

- Air cleaner assembly
- Ignition coil
- Breather hose
- Positive Crankcase Ventilation (PCV) hose
- Rocker arm (valve) cover and gasket
- Spark plug guide
- Timing belt front upper cover

3. Remove the camshaft sprocket, as follows:
   a. Secure the cam sprocket and timing belt with wire ties to prevent them from slipping out of place.
   b. While holding the sprocket from turning with special tools MB990767 and MD998719, remove the camshaft sprocket bolt and sprocket.
4. Remove or disconnect the following:
   - Intake and exhaust rocker arm and shaft assemblies. Loosen both rocker arm assemblies gradually and evenly and remove the rocker shafts from the vehicle. Do NOT disassemble!
   - Camshaft Position (CMP) sensor connector
   - CMP sensor support
   - CMP sensor sensing cylinder
   - Camshaft

To install:

5. Install or connect the following:
   - Camshaft
   - CMP sensor sensing cylinder, support and connector
   - Rocker arm and shaft assemblies
   - Camshaft oil seal
   - Camshaft sprocket. Tighten the bolt to 58–72 ft. lbs. (78–98 Nm).
   - Timing belt front upper cover
   - Spark plug guide
   - Rocker arm cover, with a new gasket. Tighten the bolts to 23–31 inch lbs. (2.5–3.5 Nm).
   - Accelerator cable clamp

- PCV and breather hoses
- Negative battery cable

2.4L ENGINE

1. Before servicing the vehicle, refer to the precautions at the beginning of this section.
2. Drain the engine oil.
3. Disconnect the negative battery cable.
4. Disconnect the accelerator cable, remove the cable clamp mounting screws and position the accelerator cable out of the way.
5. Remove the air intake hose.
6. Disconnect the breather hose and the Positive Crankcase Ventilation (PCV) hose.
7. Disconnect the spark plug cables from the spark plugs.
8. Remove the rocker cover and gasket.
9. Remove the Camshaft Position (CMP) sensor.
10. Remove the CMP sensing cylinder and sensor support.
11. Remove the camshaft sprocket
12. Remove the camshaft oil seal.
13. Remove the rocker shaft hold-down bolts gradually and evenly and remove the rocker shaft/arm assemblies.
14. Remove the hydraulic lash adjusters from the rocker arms.
15. Remove the camshaft.

To install:

16. Install the camshaft with the dowel pin at the end in the 12 o'clock position.
17. Immerse the lash adjusters in clean diesel fuel, and using a small wire, move the plunger up and down 4 or 5 times. While pushing down lightly on the check ball in order to bleed the air from the adjuster.
18. Install the lash adjusters to the rocker arms.
19. Lubricate the rocker shaft with clean engine oil and install the rocker arms/shaft assembly. Tighten the rocker shaft mounting bolts gradually and evenly to 23 ft. lbs. (32 Nm).
20. Install the camshaft oil seal.
21. Install the camshaft sprocket
22. Apply a 0.12 inch (3mm) bead of sealant around the circumference of the sensor support that mates to the cylinder head side. Install the CMP sensor support.
23. Install the CMP sensing cylinder and tighten the bolt to 13–19 ft. lbs. (18–26 Nm).
24. Install CMP sensor.
25. Install or connect the rocker arm cover gasket and cover. Tighten the bolts to 27–35 inch lbs. (3–4 Nm).
26. Connect the spark plug cables to the spark plugs.
27. Install or connect the PCV hose.
28. Install or connect the breather hose.
29. Install air intake hose.
30. Connect the accelerator cable.
31. Fill the crankcase with oil and connect the negative battery cable.

Lancer Evolution

2.0L DOHC ENGINE

1. Before servicing the vehicle, refer to the precautions at the beginning of this section.
2. Drain the engine oil.
3. Disconnect the negative battery cable.
4. Remove the breather hose.
5. Remove the Positive Crankcase Ventilation (PCV) hose.
6. Remove the oil filler cap.
7. Remove the rocker arm (valve) cover and gasket.
8. Remove the Camshaft Position (CMP) sensor.
9. Remove the CMP sensor support cover and gasket.
10. Remove the CMP sensing cylinder and sensor support.
11. Remove the camshaft oil seals.
12. Remove the right rear camshaft bearing cap.
13. Remove the left rear camshaft bearing cap.
14. Remove the right front camshaft bearing cap.
15. Remove the left front camshaft bearing cap.
16. Remove the No. 5 camshaft bearing cap(s).
17. Remove the No. 2 camshaft bearing cap(s).
18. Remove the No. 3 camshaft bearing cap(s).
19. Remove the No. 4 camshaft bearing cap(s). Be sure to clean the bearing caps and cylinder head surface of any old sealant.
20. Remove the camshaft(s).
21. Remove the rocker arms.
22. Remove the lash adjusters.

To install:

23. Clean the lash adjusters if they are being reused.
24. Install the lash adjusters into the rocker arms.
25. Install the rocker arms.
26. Install the camshaft(s). The exhaust camshaft has a 0.16 inch (4mm) wide slit at the rear end, so not to confuse it with the intake camshaft. Make sure that the dowel pin on the end of each camshaft is in the top position.
27. Install the camshaft bearing caps. Intake and exhaust bearing caps for intake Nos. 2 through 5 are identical, except for an identification mark on each of them. (Example: E2=exhaust side No. 2; I3=Intake side No. 3.) Be sure to install the correct cap in the correct location.
28. Install the front and rear camshaft bearing caps. Apply fresh sealant to the mating surfaces between the caps and cylinder head.
29. Once the camshaft bearing caps are installed, tighten each bolt in 2 or 3 passes. Finally tighten the bolts to 15 ft. lbs. (20 Nm). Completely wipe off any squeezed out sealant.
30. Install the camshaft oil seals.
31. Apply a 0.12 inch (3mm) bead of sealant around the circumference of the sensor support that mates to the bearing cap/cylinder head side. Install the CMP sensor support.
32. Install the CMP sensing cylinder and tighten the bolt to 90–106 ft. lbs. (18–26 Nm).
33. Install the CMP sensor support cover and gasket.
34. Install CMP sensor.
35. Install or connect the rocker arm cover gasket and cover. Tighten the bolts to 26–34 inch lbs. (2.9–3.9 Nm).
36. Install or connect the oil filler cap.
37. Install or connect the PCV hose.
38. Install or connect the breather hose.
39. Fill the crankcase with oil and connect the negative battery cable.

Mirage

1.5L ENGINE

1. Before servicing the vehicle, refer to the precautions at the beginning of this section.
2. Remove or disconnect the following:
    - Negative battery cable
    - Accelerator cable, breather hose and Positive Crankcase Ventilation (PCV) hose connections
    - Distributor, if equipped
    - Valve cover and discard the gasket
3. Loosen both rocker arm assemblies gradually and evenly and remove the rocker shafts from the vehicle.
4. Remove or disconnect the following:
    - Timing belt covers
    - Timing belt
    - Camshaft sprocket from the camshaft. Note the positioning of the dowel pin at the end of the camshaft.
    - Camshaft oil seal from the front of the cylinder head
    - Camshaft from the head

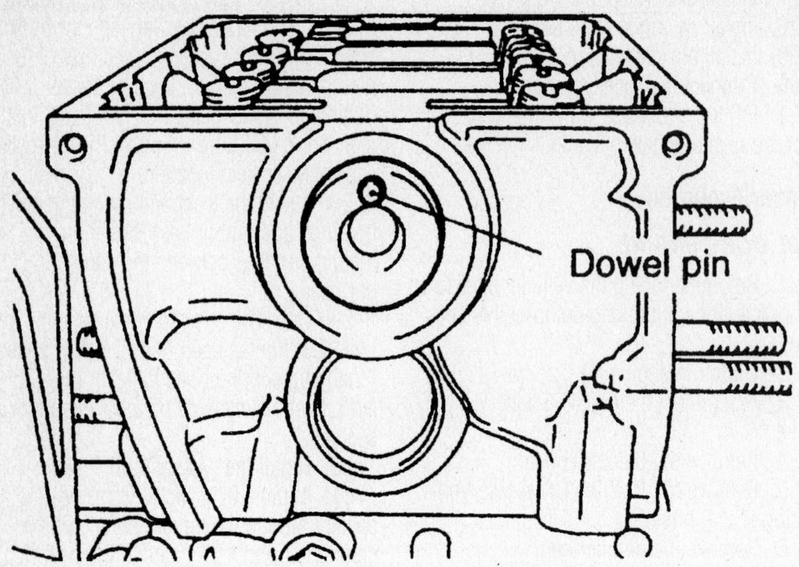

Positioning of the camshaft dowel pin—Mirage 1.5L (4G15) engine

To install:

5. Lubricate the camshaft with clean engine oil and slide it into the head. Be sure to position the dowel pin at the 12 o'clock position.
6. Install or connect the following:
   - New camshaft oil seal. Be sure to lubricate the lips of the seal with clean engine oil.
   - Camshaft sprocket and install the mounting bolt. Tighten the bolt to 51 ft. lbs. (70 Nm)
   - Timing belt
   - Timing belt covers
   - Rocker shaft assemblies. Torque the bolts gradually and evenly to 23 ft. lbs. (32 Nm).
7. Install the valve cover with a new gasket. Tighten the valve cover bolt to 16 inch lbs. (1.8 Nm).
8. Install or connect the following:
   - Distributor, if equipped
   - Accelerator cable, breather hose and PCV hose
   - Negative battery cable and check the ignition timing

1.8L ENGINE

1. Before servicing the vehicle, refer to the precautions at the beginning of this section.
2. Remove or disconnect the following:
   - Negative battery cable
   - Spark plug cables
   - Manifold Absolute Pressure (MAF) sensor connector and remove the air cleaner case cover
   - Accelerator cable, breather hose and Positive Crankcase Ventilation (PCV) hose connections
   - Rocker cover and discard the gasket
3. Loosen both rocker arm shaft assemblies gradually and evenly and remove the rocket shafts from the vehicle.
4. Remove or disconnect the following:
   - Timing belt covers
   - Timing belt
   - Camshaft sprocket from the camshaft. Note the positioning of the dowel pin at the end of the camshaft.
   - Camshaft oil seal from the front of the cylinder head
   - Camshaft from the head

To install:

5. Lubricate the camshaft journals and camshaft with clean engine oil and install the camshaft in the cylinder head. Be sure to position the dowel pin at the end of the camshaft as noted during the removal procedure.
6. Install or connect the following:
   - New camshaft oil seal.
   - Camshaft sprocket and tighten the retainer bolt to 65 ft. lbs. (90 Nm)
   - Timing belt
   - Timing belt covers
   - Rocker arm and shaft assemblies. Tighten the rocker arm shaft retainer bolts to 23 ft. lbs. (32 Nm).
   - Valve cover with a new gasket. Tighten the valve cover bolts to 29 inch lbs. (3.3 Nm).
   - Spark plug cables
   - Accelerator cable, breather hose and PCV hose
   - MAP sensor connector and install the air cleaner case cover
   - Negative battery cable

Valve Lash

ADJUSTMENT

Valve clearance is not adjustable on these vehicles.

Starter Motor

REMOVAL & INSTALLATION

1. Remove or disconnect the following:
   - Negative battery cable
   - Air-flow sensor assembly connector and remove the breather hose
   - Resonator retaining nuts and remove the air intake hose and resonator assembly as required

➡Use care when removing the air cleaner cover because the air-flow sensor is attached and is a sensitive component.

2. If equipped with Active-ECS suspension, remove the air compressor as follows:
   a. Disconnect the 2 electrical connectors, from the compressor.
   b. Disconnect the air line at the compressor.
   c. Remove the 3 mounting bolts, securing the compressor to the chassis.
3. Raise the vehicle and support safely.
4. Remove or disconnect the following:
   - Engine undercover
   - Heat shield from beneath the intake manifold on the 1.5L engine
   - Speedometer cable connector at the transaxle end, if necessary
   - Starter motor electrical connections
   - Starter motor mounting bolts and remove the starter
5. Install or connect the following:
   - Starter motor mounting bolts and remove the starter. Tighten the starter mounting bolts to 22 ft. lbs. (31 Nm).
   - Starter motor electrical connections
   - Speedometer cable connector at the transaxle end, if necessary
   - Hat shield from beneath the intake manifold on the 1.5L engine
   - Engine undercover
6. Lower the vehicle
7. Install or connect the following:

- Air compressor, if equipped with Active-ECS suspension
- Resonator retaining nuts and remove the air intake hose and resonator assembly as required
- Air-flow sensor assembly connector and remove the breather hose
- Negative battery cable and check the starter for proper operation

Oil Pan

REMOVAL & INSTALLATION

Diamante

3.5L ENGINE

1. Before servicing the vehicle, refer to the precautions at the beginning of this section.
2. Disconnect the negative battery cable.
3. Drain the engine oil.
4. Remove the mounting bolts from the lower oil pan.
5. Place a block of wood against the side of the pan and tap the block with a hammer to break the seal and remove the lower pan.
6. Remove or disconnect the following:
   - Starter
   - Dipstick tube
   - Upper oil pan

**** WARNING**

Do not pry or use seal breaker tool to remove the oil pan. Damage to the aluminum surface can result.

7. Screw a bolt into the threaded hole to force the oil pan from the engine block and remove the pan.
8. Remove the bolt used to remove the pan.

To install:

9. Clean and degrease the sealing surfaces of the upper oil pan and engine block.
10. Apply a bead of silicone sealant along the mounting surface of the upper oil pan.

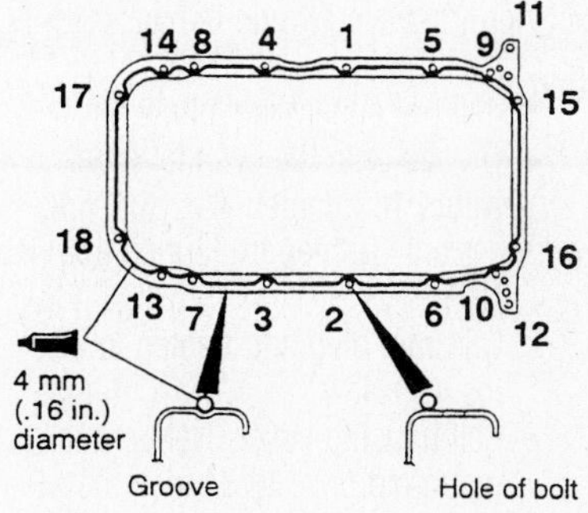

Apply sealant and tighten the bolts in the order shown—3.5L engine, upper oil pan

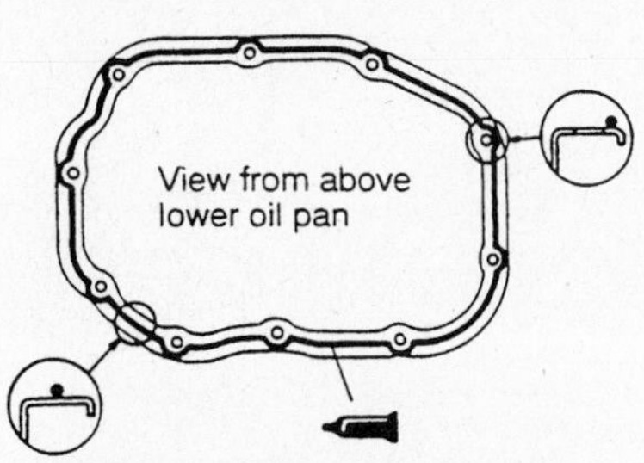

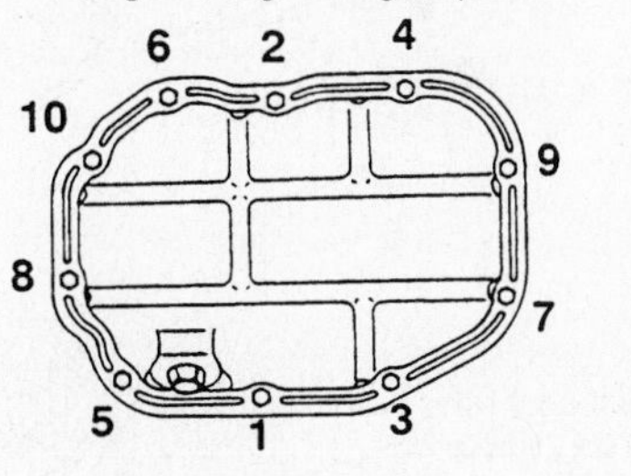

Apply sealant and tighten the bolts in the order shown—3.5L engine, lower oil pan

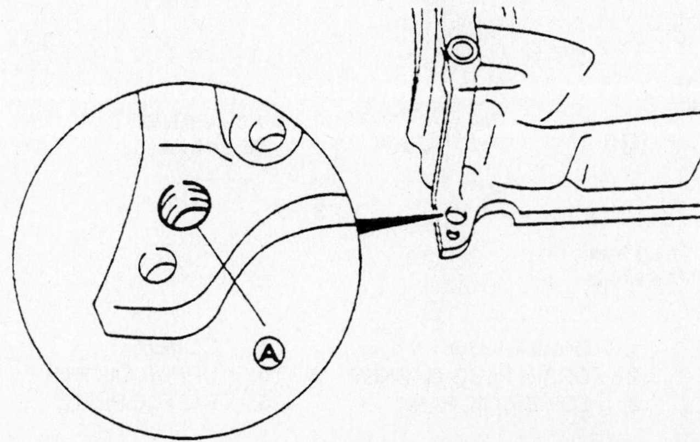

Install a bolt in the threaded hole to force the oil pan from the engine block—3.5L engine

11. Install or connect the following:
    - Upper oil pan. Tighten the bolts in sequence to 48 inch lbs. (6 Nm).
    - Dipstick tube using a new O-ring
    - Starter assembly
12. Clean and degrease the sealing surface of the lower oil pan.
13. Place a bead of sealant on the mounting surface of the lower oil pan. Install the lower pan. Tighten the bolts in sequence to 84–108 inch lbs. (10–12 Nm).
14. Install the drain plug using a new washer. Tighten the drain plug to 29 ft. lbs. (39 Nm).
15. Lower the vehicle and fill the crankcase to the correct level.
16. Connect the negative battery cable.
17. Start the engine and check for leaks.

Eclipse

2.4L (4G64) ENGINE

1. Before servicing the vehicle, refer to the precautions at the beginning of this section.
2. Remove the negative battery cable.
3. Drain the engine oil.
4. Remove or disconnect the following:
   - Engine dipstick and tube assembly
   - Front exhaust pipe
   - Bell housing inspection cover
   - Bolts attaching the oil pan to the cylinder block
5. Remove the oil pan assembly.

To install:

6. Clean the sealing surface on the oil pan and engine block. Apply a continuous bead of sealant to the oil pan.
7. Install or connect the following:
   - Oil pan to the cylinder block and tighten the bolts to 60 inch lbs. (7 Nm)
   - Bell housing inspection cover. Torque the bolts to 84 inch lbs. (9 Nm).
   - Front exhaust pipe
   - Engine dipstick and tube assembly using a new O-ring
8. Refill the engine with oil. Connect the negative battery cable. Start the engine and check for leaks.

3.0L ENGINE

1. Before servicing the vehicle, refer to the precautions at the beginning of this section.
2. Remove or disconnect the following:
   - Negative battery cable
   - Oil pan drain plug and drain the engine oil

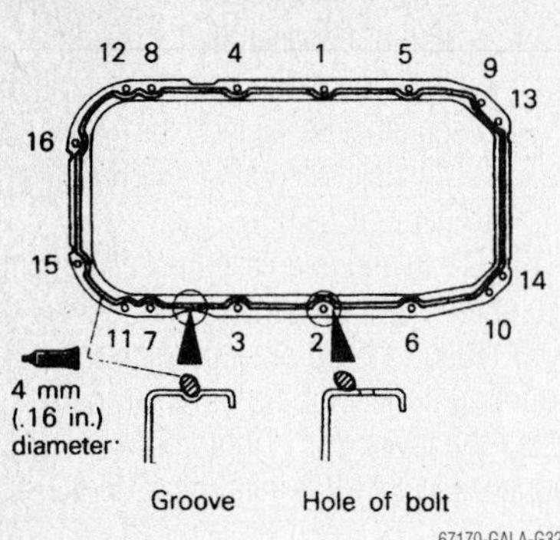

Oil pan bolt tightening sequence and application of sealant to the pan—3.0L

- Left side crossmember.
- Starter motor
- Roll stopper stay bracket, from the rear transaxle stay bracket
- Transaxle stay brackets
- Bell housing lower cover
- Oil pan mounting bolts
- The engine oil pan

To install:

3. Apply a 0.16 in. (4mm) continuous bead of sealer around the surface of the oil pan.

➡Assemble the oil pan to the cylinder block within 15 minutes after applying the sealant.

4. Install the oil pan mounting bolts. Following proper sequence, tighten mounting bolts to 48 inch lbs. (6 Nm).
5. Install or connect the following:
   - Lower bell housing cover and the starter motor
   - Transaxle stay brackets and connect the roll stopper bracket
   - Crossmember(s) and tighten the mounting bolts to 43–51 ft. lbs. (60–70 Nm)
6. Fill the engine with the proper amount of oil.
7. Connect the negative battery cable and check for leaks.

Galant

1. Before servicing the vehicle, refer to the precautions at the beginning of this section.
2. Remove or disconnect the following:
   - Negative battery cable
   - Oil pan drain plug and drain the engine oil
   - Oil dipstick and tube assembly
   - Heated Oxygen (HO_2S) sensor connector
   - Front exhaust pipe from the vehicle
   - Bell housing cover
   - Oil pan retainer bolts
3. Tap in between the engine block and the oil pan.

➡Do not use a pry tool when removing the oil pan. Damage to engine components may occur.

To install:

4. Apply sealant around the gasket surfaces of the oil pan.
5. Install or connect the following:
   - Oil pan onto the cylinder block within 15 minutes after applying sealant. Tighten to 72 inch lbs. (8 Nm).
   - Oil drain plug and tighten to 29 ft. lbs. (39 Nm)
   - Bell housing cover. Tighten the mounting bolts to 84 inch lbs. (9 Nm).
   - Front exhaust pipe and tighten the bolts at the catalytic converter to 36 ft. lbs. (49 Nm). Tighten the nuts at the exhaust manifold to 32 ft. lbs. (44 Nm).
   - HO_2S sensor connector
6. Fill the crankcase to the proper level.
7. Connect the negative battery cable. Start the engine and check for leaks.

Lancer and Sportback

2.0L SOHC ENGINE

1. Before servicing the vehicle, refer to the precautions at the beginning of this section.
2. Disconnect the negative battery cable.
3. Drain the engine oil.
4. Remove or disconnect the following:
   - Engine undercover
   - Front exhaust pipe
   - Lower oil pan bolts and lower pan
   - Cover
   - Upper oil pan bolt and upper pan
   - Baffle plate

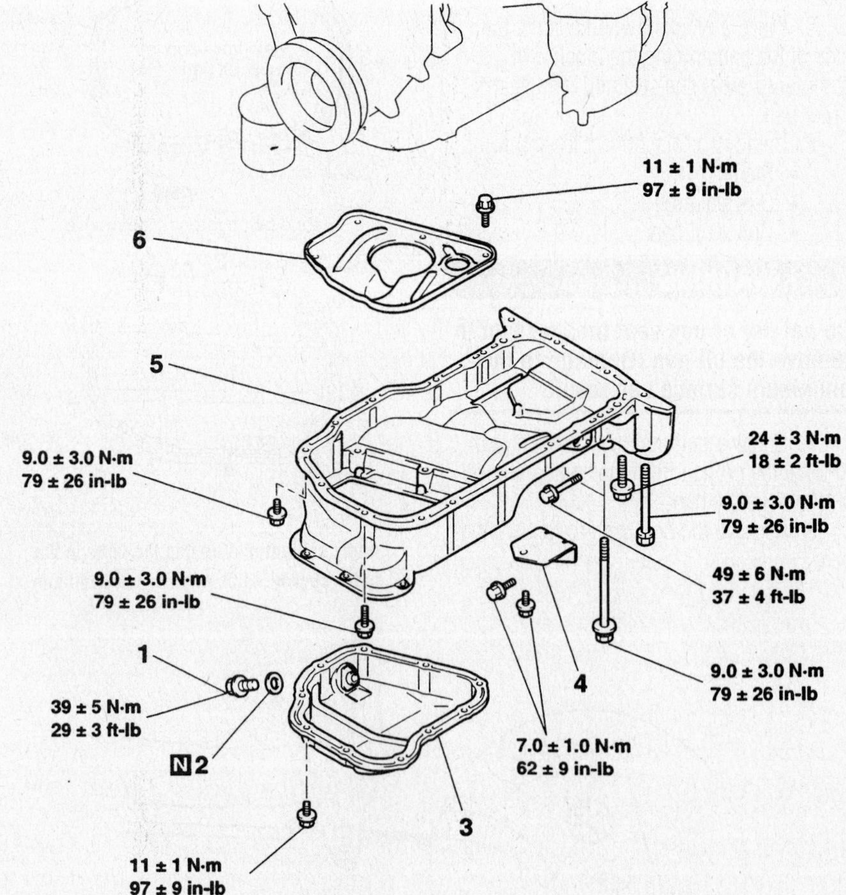

Exploded view of the oil pan mounting—Lancer with 2.0L engine

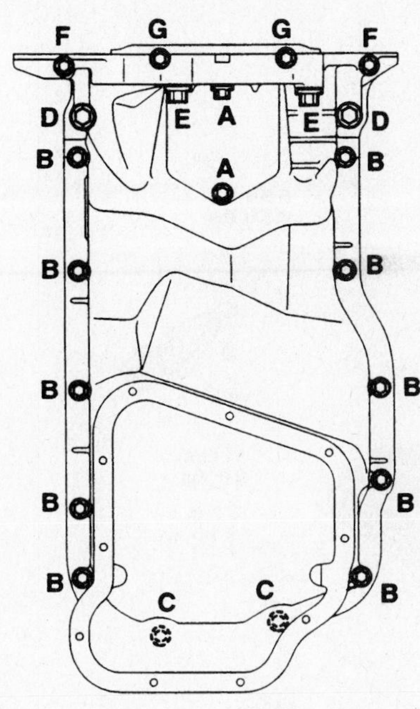

Designation	Symbol	Qty	Diameter × length mm (in)	Tightening torque
Flange Bolt	A	2	6 × 10 (0.2 × 0.4)	7.0 ± 1.0 N·m (62 ± 9 in-lb)
	B	10	6 × 18 (0.2 × 0.7)	9.0 ± 3.0 N·m (79 ± 26 in-lb)
	C	2	6 × 22 (0.2 × 0.9)	
	D	2	8 × 40 (0.3 × 1.6)	24 ± 3 N·m (18 ± 2 ft-lb)
	E	2	10 × 40 (0.4 × 1.6)	49 ± 6 N·m (37 ± 4 ft-lb)
Bolts with Washers	F	2	6 × 50 (0.2 × 2.0)	9.0 ± 3.0 N·m (79 ± 26 in-lb)
	G	2	6 × 127 (0.2 × 5.0)	

9357QG18

Upper oil pan bolt location and torque sequence—Lancer with 2.0L engine

To install:

5. Clean all gasket surfaces of the cylinder block and the upper and lower oil pan.
6. Install or connect the following:
   - Baffle plate
7. Apply a 0.16 in. (4mm) bead of sealant to the gasket surfaces of the upper oil pan.
   - Upper oil pan onto the cylinder block within 15 minutes after applying sealant. Tighten the bolts as shown in the accompanying figure.
8. Apply 0.16 in. (4mm) bead of sealant to the gasket surfaces of the lower oil pan.
   - Lower oil pan and tighten the bolts, in the sequence shown, to 88–106 inch lbs. (10–12 Nm)
   - Front exhaust pipe
   - Engine undercover
   - Oil drain plug with a new gasket and tighten to 29 ft. lbs. (40 Nm)
9. Lower the vehicle and fill the crankcase to the proper level with clean engine oil.
10. Connect the negative battery cable. Start the engine and check for leaks.

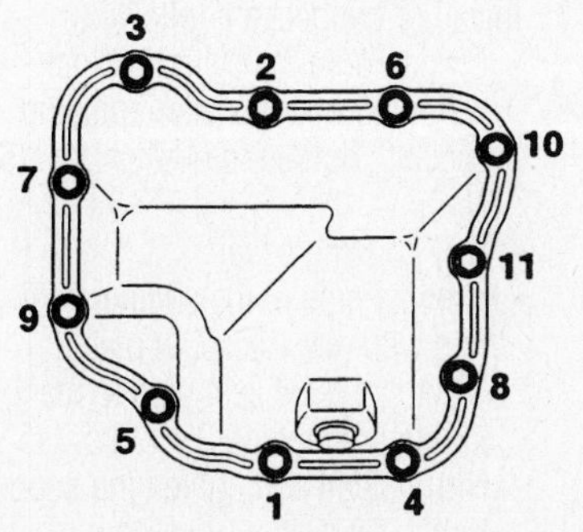

9357QG19

Lower oil pan bolt tightening sequence—Lancer with 2.0L engine

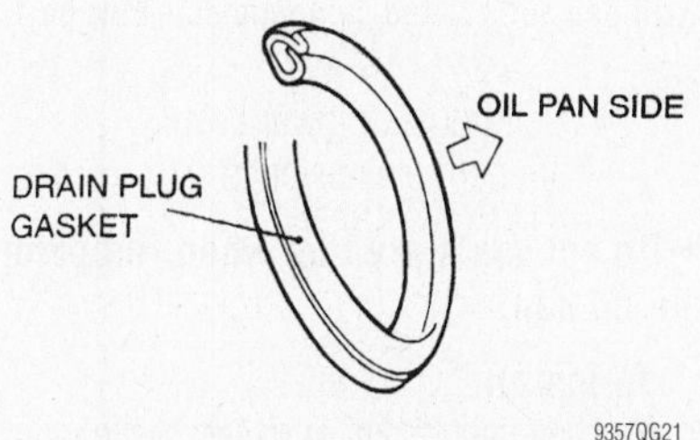

9357QG21

Make sure to the install the new drain plug gasket as shown, or leaks will occur

2.4L ENGINE

1. Before servicing the vehicle, refer to the precautions at the beginning of this section.
2. Disconnect the negative battery cable.
3. Drain the engine oil.
4. Remove the engine undercover.
5. Remove the front exhaust pipe.
6. Remove the oil filter.
7. Remove the oil pan bolts.
8. Remove the oil pan.

To install:

9. Clean all gasket surfaces of the cylinder block and the oil pan.
10. Apply a 0.15 in. (4mm) bead of sealant to the gasket surface of the oil pan.
11. Install the oil pan onto the cylinder block within 15 minutes after applying sealant.
12. Install a 0.30 in. (8mm) bolt in each of the two holes identified by the letter "S" in the accompanying illustration, and a 0.39 in. (10mm) bolt in each of the remaining 17 holes and tighten to 54–106 inch lbs. (6–12

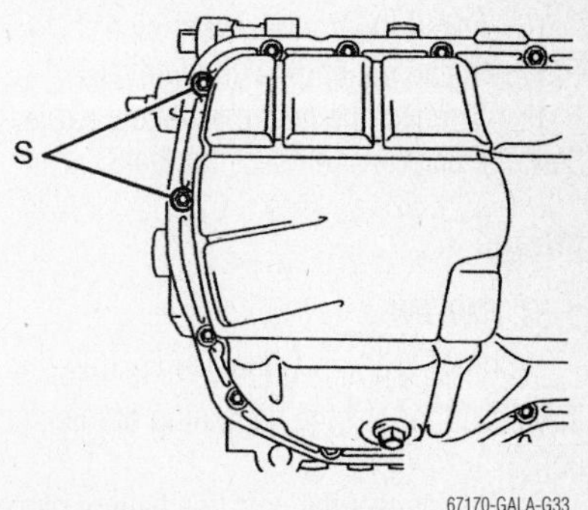

67170-GALA-G33

Install a 0.30 in. (8mm) bolt in each of the two holes identified by the letter "S"

Nm). Wait approximately 1 hour before filling the oil pan with engine oil.

13. Install the oil drain plug with a new gasket and tighten to 29 ft. lbs. (39 Nm).
14. Install a new oil filter.
15. Lower the vehicle and fill the crankcase to the proper level with clean engine oil.
16. Install the front exhaust pipe.
17. Install the engine undercover.
18. Connect the negative battery cable. Start the engine and check for leaks.

Lancer Evolution

2.0L DOHC ENGINE

1. Before servicing the vehicle, refer to the precautions at the beginning of this section.
2. Disconnect the negative battery cable.
3. Drain the engine oil.
4. Remove the engine undercover.
5. Remove the front exhaust pipe.
6. Remove the oil filter.
7. Remove the oil pan bolts.
8. Remove the oil pan.

To install:

9. Clean all gasket surfaces of the cylinder block and the oil pan.
10. Apply a 0.15 in. (4mm) bead of sealant to the gasket surface of the oil pan.
11. Install the oil pan onto the cylinder block within 15 minutes after applying sealant.
12. Install a 0.30 in. (8mm) bolt in each of the two holes identified by the letter "S" in the accompanying illustration, and a 0.39 in. (10mm) bolt in each of the remaining 17 holes and tighten to 54–106 inch lbs. (6–12 Nm). Wait approximately 1 hour before filling the oil pan with engine oil.
13. Install the oil drain plug with a new gasket and tighten to 29 ft. lbs. (39 Nm).
14. Install a new oil filter.
15. Lower the vehicle and fill the crankcase to the proper level with clean engine oil.
16. Install the front exhaust pipe.
17. Install the engine undercover.
18. Connect the negative battery cable. Start the engine and check for leaks.

Mirage

1.5L ENGINE

1. Before servicing the vehicle, refer to the precautions at the beginning of this section.
2. Disconnect the negative battery cable.
3. Drain the engine oil.
4. Remove or disconnect the following:

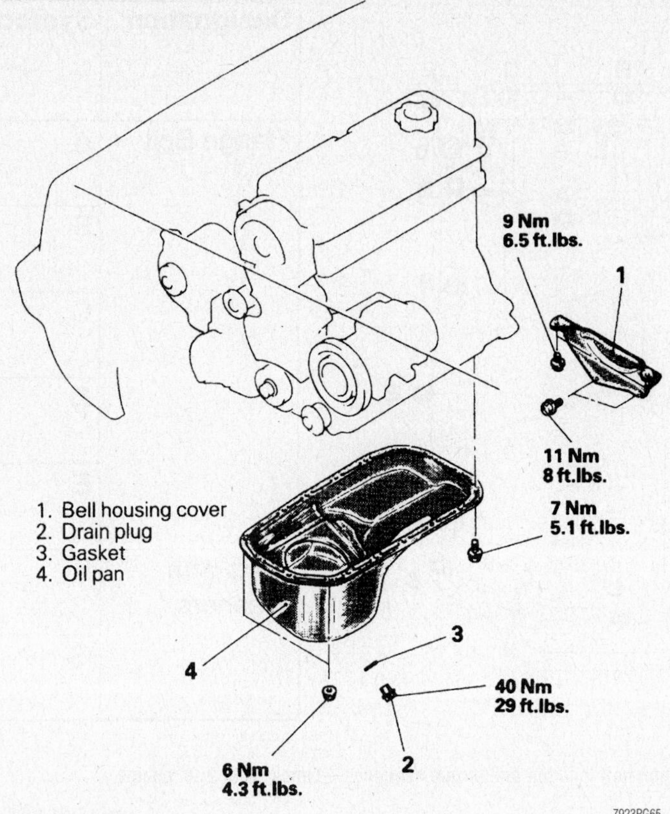

Oil pan and related components—Mirage 1.5L (4G15) engine

- Bell housing lower cover
- Oil pan retainer bolts

➡**Do not use a pry tool when removing the oil pan.**

To install:

5. Clean all gasket surfaces of the cylinder block and the oil pan.
6. Apply sealant to the gasket surfaces of the oil pan.
7. Install or connect the following:
   - Oil pan onto the cylinder block within 15 minutes after applying sealant. Tighten to 60 inch lbs. (7 Nm).
   - Bell housing cover
   - Oil drain plug with a new seal and tighten to 29 ft. lbs. (40 Nm)
8. Lower the vehicle and fill the crankcase to the proper level with clean engine oil.
9. Connect the negative battery cable. Start the engine and check for leaks.

1.8L ENGINE

1. Before servicing the vehicle, refer to the precautions at the beginning of this section.
2. Disconnect the negative battery cable.
3. Raise the vehicle and support safely.
4. Remove or disconnect the following:
   - Oil pan drain plug and drain the engine oil
   - Exhaust pipe from the engine manifold
   - Bell housing lower cover
   - Oil pan retainer bolts and remove the oil pan

➡**Do not use a pry tool when removing the oil pan.**

To install:

5. Clean all gasket surfaces of the cylinder block and the oil pan.
6. Apply sealant around the gasket surfaces of the oil pan.
7. Install or connect the following:
   - Oil pan onto the cylinder block within 15 minutes after applying sealant. Tighten to 60 inch lbs. (5 Nm).
   - Bell housing cover
   - Exhaust pipe to the engine manifold with new gasket in place. Tighten the exhaust pipe to manifold flange nuts to 33 ft. lbs. (45 Nm). Install and tighten the support bolt to 18 ft. lbs. (25 Nm).
   - Oil drain plug and tighten to 29 ft. lbs. (40 Nm)

8. Fill the crankcase to the proper level.
9. Connect the negative battery cable. Start the engine and check for leaks.

Oil Pump

REMOVAL & INSTALLATION

1.5L and 1.8L Engines

➡Whenever the oil pump is disassembled or the cover removed, the gear cavity must be filled with petroleum jelly to seal the pump and act as a prime. Do not use grease.

1. Before servicing the vehicle, refer to the precautions at the beginning of this section.
2. Remove or disconnect the following:
   - Negative battery cable
   - Front engine mount bracket and accessory drive belts
   - Timing belt upper and lower covers
   - Timing belt and crankshaft sprocket
   - Oil pan and remove the oil screen
   - Front cover mounting bolts. Note the lengths of the mounting bolts as they are removed for proper installation.
   - Front case assembly and oil pump assembly
   - Oil pump cover
   - Inner and outer gears from the front case

To install

3. Remove all gasket material from the mating surfaces and clean all parts.
4. Thoroughly coat both oil pump gears with clean engine oil and install them in the correct direction of rotation.
5. Install the pump cover and tighten the bolts to 84 inch lbs. (10 Nm).
6. Coat the relief valve and spring with clean engine oil. Install them and tighten the plug to 33 ft. lbs. (45 Nm).
7. Install or connect the following:
   - New front crankshaft seal and coat the lips of the seal with clean engine oil
   - Front case and oil pump assembly to the engine block using a new gasket. Tighten the bolts to 10 ft. lbs. (14 Nm).
   - Oil screen with new gasket. Torque the screen bolts to 14 ft. lbs. (19 Nm).
   - Oil pan
   - Crankshaft sprocket and timing belt
8. Fill the crankcase to the proper level.
9. Connect the negative battery cable.

2.0L SOHC Engine

1. Before servicing the vehicle, refer to the precautions at the beginning of this section.
2. Drain the engine oil.
3. Remove or disconnect the following:
   - Negative battery cable
   - Oil pressure switch
   - Oil filter
   - Drain plug and gasket. Discard the gasket.
   - Cover
   - Upper oil pan. Remove the 5 in. (127mm) bolt, which is closest to the flywheel/flexplate first, then, the other bolts.
   - Baffle plate
   - Lower oil pan. Remove the 5 in. (127mm) bolt, which is closest to the flywheel/flexplate first, then, the other bolts.
   - Oil screen and gasket
   - Relief plug and spring
   - Oil seal
   - Front case
   - O-ring
   - Oil pump case cover

➡Matchmark the installed position of the pump rotors before removing them.

   - Outer and inner oil pump rotors

To install:

4. Install or connect the following:
   - Inner and outer rotors, making sure the alignment marks are matched up
   - Oil pump case cover
   - O-ring

➡After installation or the front case, wait at least one hour before filling the crankcase with oil or starting the engine.

   - Front case. Apply a 0.12 inch (3mm) bead of sealant, then tighten the case bolts to 124 inch lbs. (14 Nm).
   - Oil seal
   - Relief plunger, spring and plug
   - Oil screen gasket and screen
   - Lower oil pan. Refer to the oil pan procedure for sealant application and torque specifications.
   - Baffle plate
   - Upper oil pan. Refer to the oil pan procedure for sealant application and torque specifications.
   - Cover
   - Drain plug with a new gasket. Tighten the plug to 29 ft. lbs. (35 Nm).
   - Oil filter
   - Oil pressure switch
5. Fill the engine with the correct amount of oil.

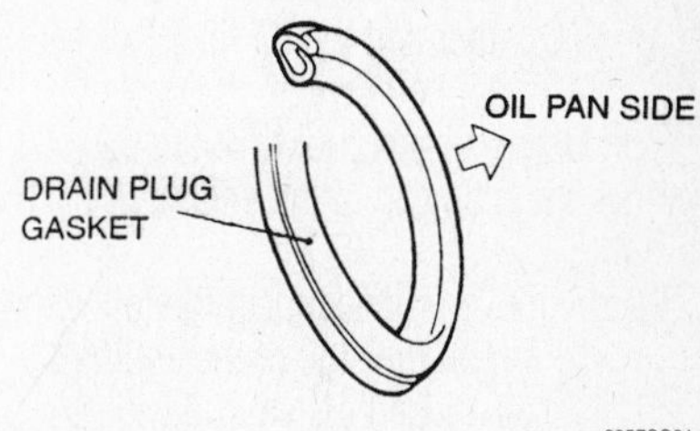

Make sure to the install the new drain plug gasket as shown, or leaks will occur

L = 20 (.79) L = 40 (1.57) *L = 30 (1.18) L = 20 (.79) L = 40 (1.57) L = 25 (.98) L = 75 (2.95) L = 55 (2.17)
Tighten together with belt tensioner
L = Bolt length below head [mm (in.)]
7923PG74

Exploded view of the oil pump and related components—2.0L (4G94) engine

6. Connect the negative battery cable.
7. Start the engine and check for leaks.

2.0L DOHC and 2.4L Engines

➡Whenever the oil pump is disassembled or the cover removed, the gear cavity must be filled with petroleum jelly to seal the pump and act as a prime. Do not use grease.

1. Before servicing the vehicle, refer to the precautions at the beginning of this section.
2. Disconnect the negative battery cable. Rotate the engine so No. 1 cylinder is on Top Dead Center (TDC) of its compression stroke.
3. Drain the engine oil.
4. Using the proper equipment, support the weight of the engine. Remove the front engine mount bracket and accessory drive belts.
5. Remove or disconnect the following:
   - Timing belt upper and lower covers
   - Timing belt and crankshaft sprocket
   - Electrical connector from the oil pressure sending unit
   - Oil pressure sensor
   - Oil filter and the oil filter bracket
   - Oil pan, oil screen and gasket
6. Using special tool MD998162, remove the plug cap in the engine front cover.
7. Remove or disconnect the following:
   - Plug on the side of the engine block. Insert a steel rod with a shank diameter of 0.32 in. (8mm) into the plug hole. This will hold the silent shaft.
   - Driven gear bolt that secures the oil pump driven gear to the silent shaft
   - Front cover mounting bolts. Note the lengths of the mounting bolts as they are removed for proper installation.
   - Front case cover and oil pump assembly. If necessary, the silent shaft can come out with the cover assembly.
   - Oil pump cover, located on the back of the engine front cover. Remove the oil pump drive and driven gears.
8. After disassembling the oil pump, clean all components and remove gasket material from mating surfaces.
9. Assemble the oil pump gears into the front case and rotate it to ensure smooth rotation and no looseness. Be sure there is no ridge wear on the contact surface between the front case and the gear surface of the oil pump front cover.

To install

10. Align the timing mark on the oil pump drive gear with that on the driven gear and install them into the engine front case. Apply engine oil to the gears.
11. Install the oil pump cover and tighten the retainer bolts to 13 ft. lbs. (18 Nm) on Eclipse models and 17 ft. lbs. (24 Nm) on Galant models.
12. Using the appropriate driver, install a new crankshaft seal into the front case.
13. Position new front case gasket in place. Set seal guide tool MD998285 on the front end of the crankshaft to protect the seal from damage. Apply a thin coat of oil to the outer circumference of the seal pilot tool.
14. Install the front case assembly through a new front case gasket and temporarily tighten the flange bolts.
15. Mount the oil filter on the bracket with new oil filter bracket gasket in place. Install the bolts with washers and tighten to 14 ft. lbs. (19 Nm).
16. Insert a Phillips screwdriver into the hole in the left side of the engine block to lock the silent shaft in place.
17. Install or connect the following:
   - Oil pump drive gear onto the left silent shaft. Tighten the driven gear bolt to 27 ft. lbs. (37 Nm).
   - New O-ring to the groove in the front case and install the plug cap. Tighten the cap to 17 ft. lbs. (24 Nm).
   - Oil screen in position with new gasket in place
18. Clean both mating surfaces of the oil pan and the cylinder block. Apply sealant in the groove in the oil pan flange.

➡After applying sealant to the oil pan, do not exceed 15 minutes before installing the oil pan.

19. Install or connect the following:
   - Oil pan to the engine and secure with the retainers. Tighten bolts to 60 inch lbs. (7 Nm).
   - Oil pressure gauge unit and the oil pressure switch
   - Electrical harness connector
   - Oil cooler. Oil cooler bolt to 31 ft. lbs. (43 Nm).
20. Refill the crankcase. Install new oil filter.
21. Connect the negative battery cable and start the engine. Verify correct oil pressure. Inspect for leaks.

3.0L Engines

➡Whenever the oil pump is disassembled or the cover removed, the gear cavity must be filled with petroleum jelly to seal the pump and act as a prime. Do not use grease.

1. Before servicing the vehicle, refer to the precautions at the beginning of this section.
2. Disconnect the negative battery cable.
3. Drain the engine oil.
4. Remove or disconnect the following:
   - Front engine mount bracket and accessory drive belts
   - Timing belt upper and lower covers
   - Timing belt and crankshaft sprocket

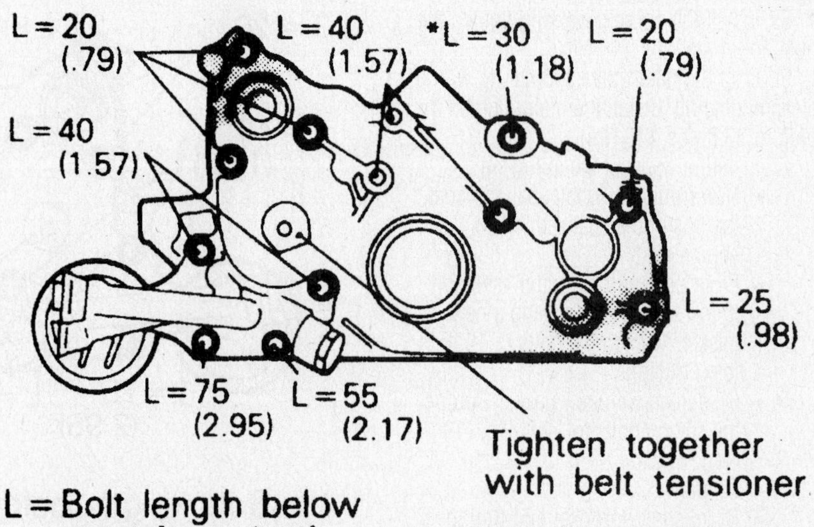

Front case bolt identification—Eclipse 2.4L (4G64) engines

- Oil pan
- Oil screen and gasket
- Front cover mounting bolts. Note the lengths of the mounting bolts as they are removed for proper installation.
- Front case cover and oil pump assembly

To install:

5. Thoroughly clean all gasket material from all mounting surfaces.
6. Apply engine oil to the entire surface of the gears or rotors.
7. Assemble the front case cover and oil pump assembly to the engine block.
8. Install or connect the following:
   - Oil screen with new gasket
   - Oil pan
   - Crankshaft sprocket and timing belt
   - Timing belt covers
   - Drive belts and the front engine mount bracket
   - Negative battery cable, refill the crankcase and check for adequate oil pressure

3.5L and 3.8L Engines

1. Before servicing the vehicle, refer to the precautions at the beginning of this section.
2. Remove or disconnect the following:
   - Negative battery cable
   - Timing belt
3. Drain the engine oil.
4. Remove or disconnect the following:
   - Splash shield from the wheel well
   - Oil filter adapter
   - Lower and upper oil pans
   - Lower baffle, oil pump pick-up and upper baffle
   - Oil pump case mounting bolts and the oil pump case
   - Oil pump gear cover

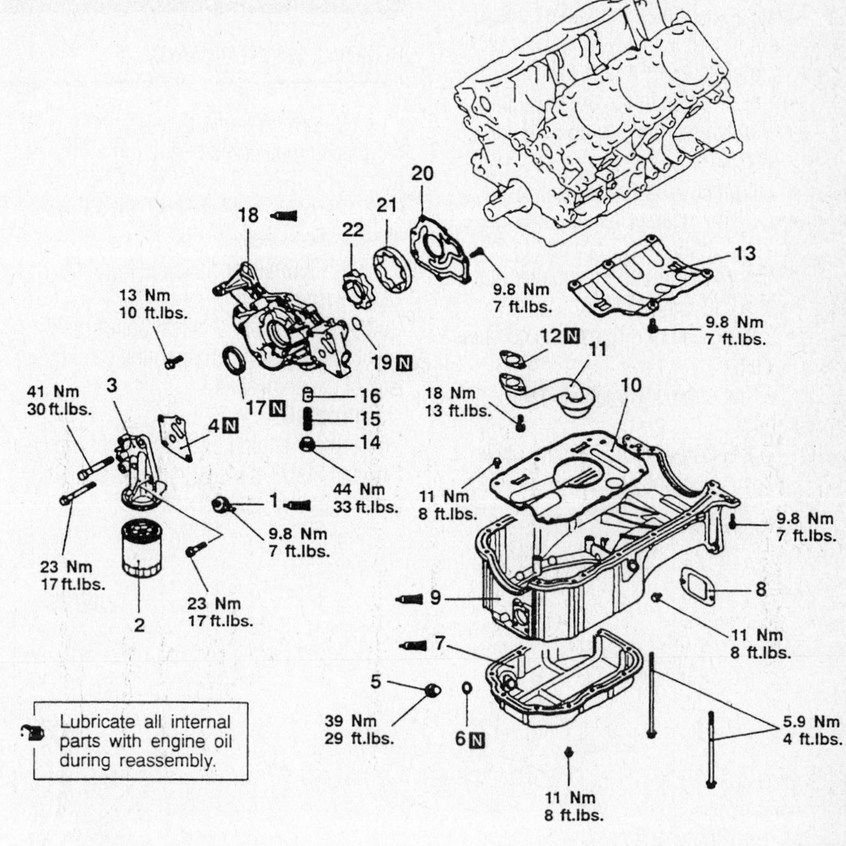

Exploded view of the oil pump mounting—3.5L engine

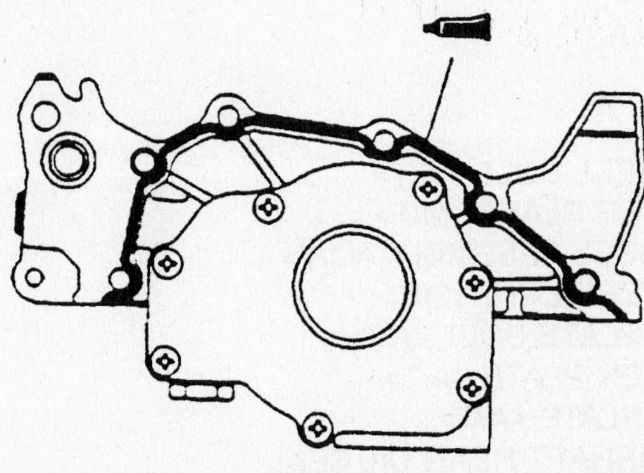

Apply sealant to the rear of the oil pump case—3.5L engine

5. Make matchmarks on the oil pump rotors before removing them.
   - Crankshaft seal from the oil pump case

To install:

6. Install or connect the following:
   - New crankshaft seal in the oil pump cover
7. Apply engine oil to the rotors, then align the matchmarks and install the rotors in the oil pump case.
   - Rotor cover and tighten the bolts to 84 inch lbs. (10 Nm)
8. Apply a 0.113 in. (3mm) bead of sealant to the back of the oil pump case. Install the case on the engine and tighten the bolts to 10 ft. lbs. (13 Nm).
   - Upper baffle plate and oil pump

pick-up using a new gasket–Tighten the baffle bolts to 84 inch lbs. (10 Nm) and the pick-up bolts to 13 ft. lbs. (18 Nm).
- Lower baffle in the upper oil pan. Tighten the bolts to 96 inch lbs. (11 Nm).
- Oil pans
- Oil filter adapter using a new gasket. Tighten the larger bolt to 30 ft. lbs. (41 Nm) and the smaller bolt to 17 ft. lbs. (23 Nm).
- Timing belt and remaining components

9. Fill the engine with the correct amount of oil.
10. Connect the negative battery cable.
11. Start the engine and check for leaks.

Rear Main Seal

REMOVAL & INSTALLATION

1. Before servicing the vehicle, refer to the precautions at the beginning of this section.
2. Remove or disconnect the following:
   - Transaxle
   - Flywheel or flexplate from the crankshaft
3. Carefully pry the seal out of the oil seal case without damaging the sealing surface of the crankshaft.

To install:

4. Apply engine oil to the lip of the new seal and install the seal in the case using the proper size seal driver.
5. Install or connect the following:
   - Flywheel or flexplate
   - Transaxle

Timing Belt

REMOVAL & INSTALLATION

1.5L and 1.8L Engines

1. Before servicing the vehicle, refer to the precautions at the beginning of this section.
2. Remove or disconnect the following:
   - Negative battery cable
   - Electrical connectors, tag before disconnecting
   - Timing belt upper and lower covers

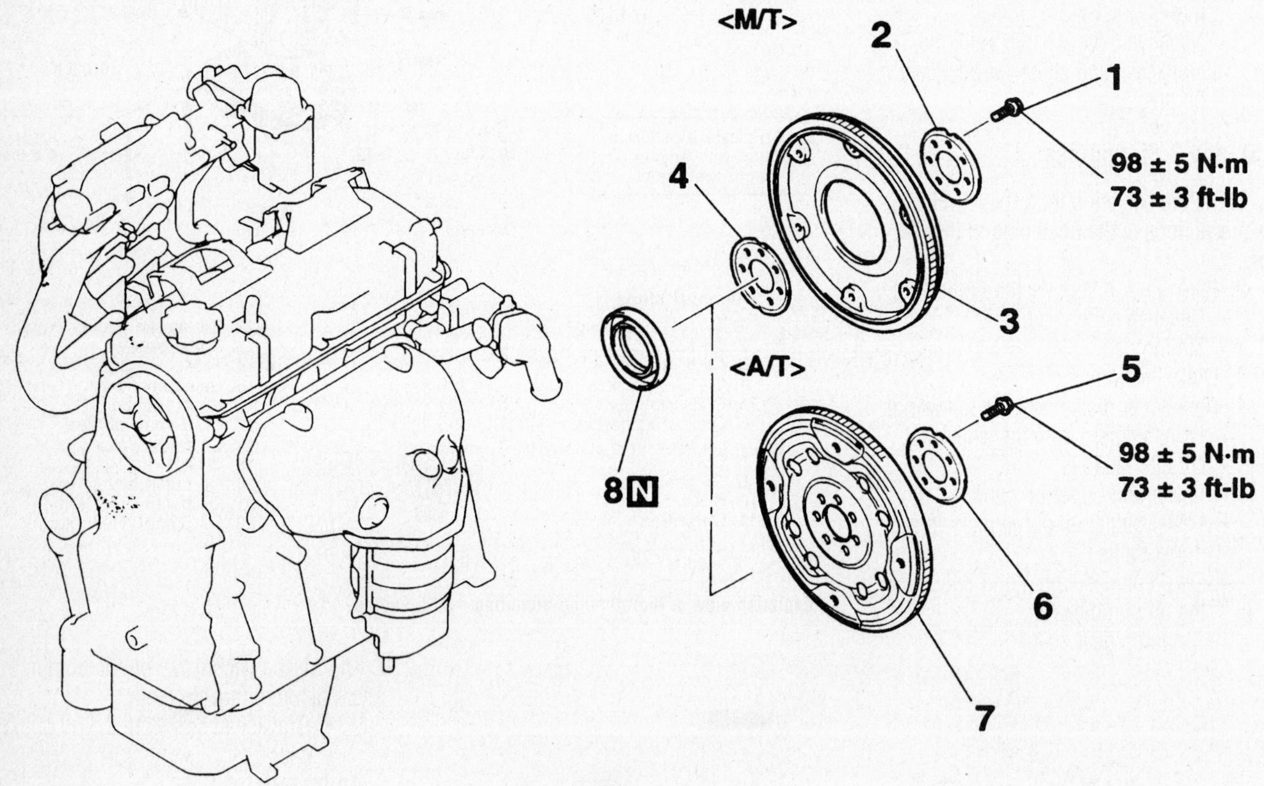

Exploded view of a typical rear main seal

3. Make a mark on the back of the timing belt indicating the direction of rotation so it may be reassembled in the same direction if it is to be reused. Loosen the timing belt tensioner and move the tensioner to provide slack to the timing belt. Tighten the tensioner in this position.

- Timing belt

WARNING

Coolant and engine oil will damage the rubber in the timing belt, drastically reducing its life. Do not allow engine oil or coolant to contact the timing belt, the sprockets or tensioner assembly.

4. If defective, replace the tensioner spacer, tensioner spring and tensioner assembly.

To install:

5. Position the tensioner, tensioner spring and tensioner spacer on engine block.
6. Align the timing marks on the camshaft sprocket and crankshaft sprocket. This will position No. 1 piston on Top Dead Center (TDC) on the compression stroke.
7. Position the timing belt on the crankshaft sprocket and keeping the tension side of the belt tight, set it on the camshaft sprocket, then the tensioner.
8. Apply slight counterclockwise force to the camshaft sprocket to give tension to the belt and be sure all timing marks are aligned.
9. Loosen the pivot side tensioner bolt and the slot side bolt. Allow the spring to remove the slack.
10. Tighten the slot side tensioner bolt, then the pivot side bolt. If the pivot side bolt is tightened first, the tensioner could turn with bolt, causing over tension.
11. For 1.5L engines, turn the crankshaft clockwise. Loosen the pivot side tensioner bolt, then the slot side bolt to allow the spring to take up any remaining slack. Tighten the slot bolt, then the pivot side bolt to 17 ft. lbs. (24 Nm).
12. For 1.8L engines, turn the crankshaft clockwise two rotations and tighten the adjuster bolt to 18 ft. lbs. (24 Nm) and the pivot (spring) bolt to 35 ft. lbs. (45 Nm).
13. Install the timing belt covers and tighten the cover bolts to 84–96 inch lbs. (10–11 Nm).
14. Install the remaining components in the reverse order of removal.

2.0L Engine

1. Before servicing the vehicle, refer to the precautions at the beginning of this section.

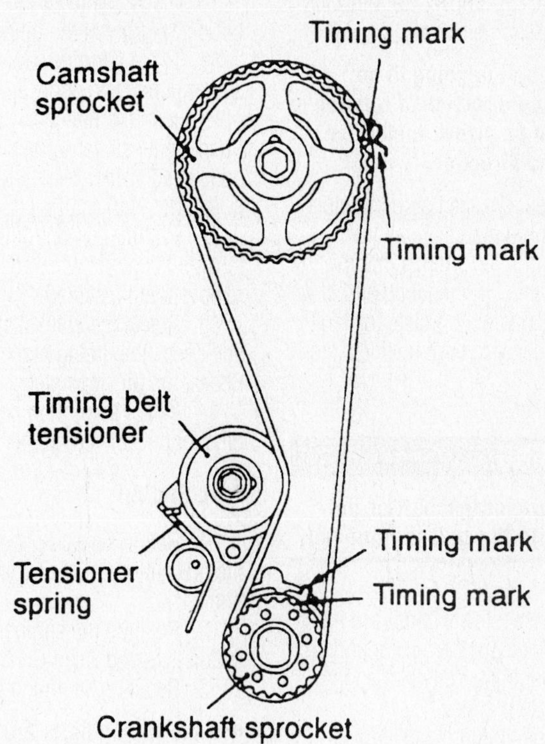

Timing mark locations with engine at Top Dead Center (TDC) of compression stroke—Mitsubishi 1.5L (4G15) engine

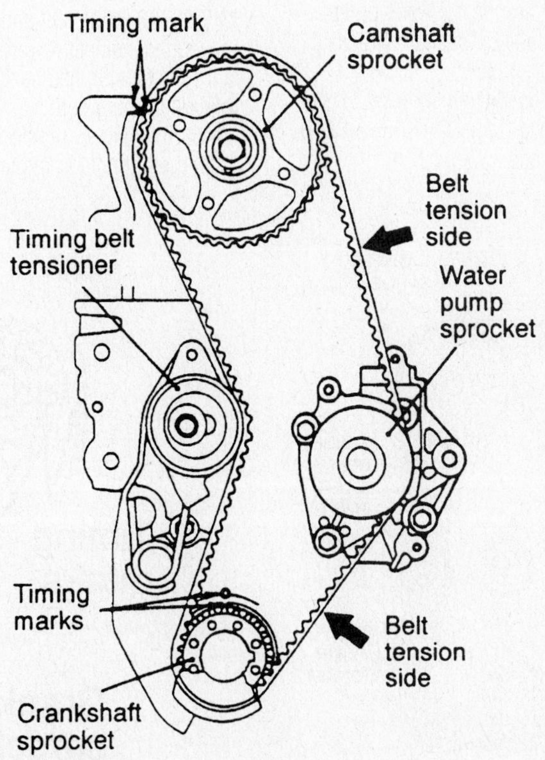

Timing mark locations with engine at Top Dead Center (TDC) of compression stroke—Mitsubishi 1.8L (4G93) engine

2. Remove or disconnect the following:
   - Front timing belt cover

➡If the timing belt is going to be reused, mark the direction of rotation on the belt with an arrow. Install the belt in the same direction.

3. Rotate the crankshaft sprocket clockwise until the timing marks are aligned.
4. Place 8mm Allen wrench into the belt tensioner, then using the long end of a ⅛ in. (3mm) Allen wrench, rotate the tensioner counterclockwise until it slides into the locking hole.
5. Remove the belt.

✲✲ WARNING

Do not rotate the crankshaft or the camshafts while the belt is removed.

To install:

6. Using a vise, slowly compress the plunger into the body of the tensioner and install a pin through the body of the tensioner to retain the plunger.
7. Be sure the timing marks are still aligned, if not, align the camshaft sprocket timing marks facing each other. Align the crankshaft sprocket timing mark with the mark on the oil pump housing, then turn the crankshaft sprocket backward ½ notch.
8. Install the timing belt, starting at the crankshaft, then around the water pump sprocket, idler pulley, camshaft sprockets and the tensioner pulley.
9. Turn the crankshaft sprocket ½ notch to Top Dead Center (TDC) to take up the slack in the belt.
10. Install the tensioner on the engine, but do not tighten the bolts.
11. Place a torque wrench on the tensioner pulley and apply 21 ft. lbs. (28 Nm) of torque in the direction of the water pump. Push the tensioner up against the tensioner pulley and tighten the mounting bolts to 23 ft. lbs. (31 Nm).
12. Pull the pin out of the tensioner. Belt tension is correct when the pin can be removed and installed.
13. Rotate the crankshaft two revolutions and check the timing marks for alignment. Repeat the previous steps, if necessary.
14. Install the timing belt cover and all other applicable components.

2.4L Engine

1. Before servicing the vehicle, refer to the precautions at the beginning of this section.
2. Position the engine so that the No. 1 piston is at Top Dead Center (TDC).
3. Remove the timing belt covers.

➡If the timing belts are going to be reused, mark the direction of rotation on the belt. This will ensure the belt is reinstalled in same direction, extending belt life.

4. To loosen the timing (outer) belt tensioner, install Mitsubishi Special tool MD998738 to the slot and screw inward to move the tensioner toward the water pump. Once the tension has been relieved, remove the outer timing belt.
5. If tensioner replacement is required, align the pin hole in the tensioner rod to the hole in the tensioner cylinder. Insert a 0.055 in. (1.4mm) wire in the hole and remove the special tool from the slot. With the cylinder tension relieved, remove the auto-tensioner cylinder assembly two mounting bolts.
6. Remove the outer crankshaft sprocket and flange.
7. Loosen the silent shaft (inner) belt tensioner and remove the belt.

To install:

✲✲ WARNING

Do not spray or immerse the sprockets or tensioners in cleaning solvent. The sprocket may absorb the solvent and transfer it to the belt. The tensioners are internally lubricated and the solvent will dilute or dissolve the lubricant.

8. Align the timing marks of the silent shaft sprockets and the crankshaft sprocket with the timing marks on the front case. Route the timing belt around the sprockets so there is no slack in the upper span of the belt and the timing marks are still aligned.
9. Install the tensioner pulley and move the pulley by hand so the long side of the belt deflects approximately ¼ in. (6mm).
10. Hold the pulley tightly so the pulley cannot rotate when the bolt is tightened. Tighten the bolt to 14 ft. lbs. (19 Nm) and recheck the deflection.
11. Align the timing marks of the camshaft, crankshaft and oil pump sprockets with their corresponding marks on the front case or rear cover.

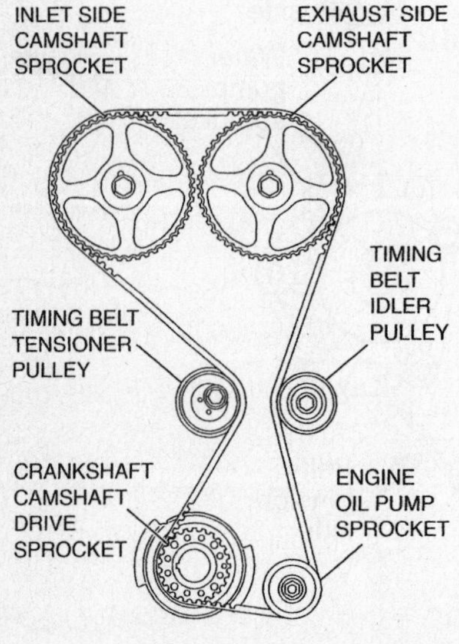

Timing belt sprocket mark alignment for belt service—Mitsubishi 2.0L turbo engine

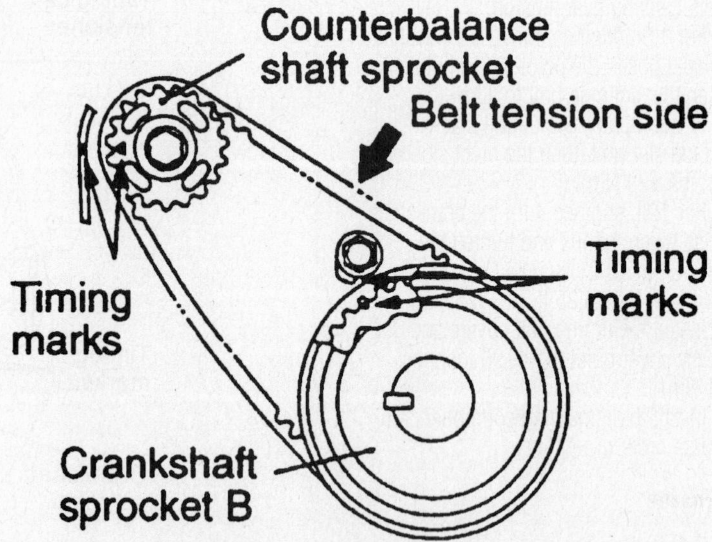

Timing belt "B" installation mark alignment—2.4L engines

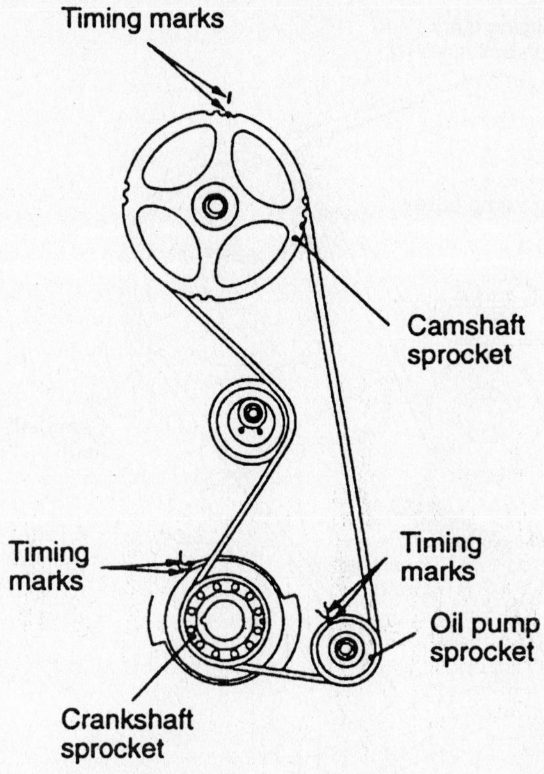

Proper alignment of the timing belt sprocket marks for belt service— 2.4L engines

➡There is a possibility to align all timing marks and have the oil pump sprocket and silent shaft out of time, causing an engine vibration during operation. If the following step is not followed exactly, there is a 50 percent chance that the silent shaft alignment will be 180 degrees (½ turn) off.

12. Before installing the timing belt, ensure that the left side (rear) silent shaft (oil pump sprocket) is in the correct position as follows:

a. Remove the plug from the rear side of the block and insert a tool with shaft diameter of 0.31 in. (8mm) into the hole.

b. With the timing marks still aligned, the shaft of the tool must be able to go in at least 2 ½ in. (63.5mm). If the tool can only go in approximately 1 in. (25mm), the shaft is not in the correct orientation and will cause a vibration during engine operation. Remove the tool from the hole and turn the oil pump sprocket one complete revolution. Realign the timing marks and insert the tool. The shaft of the tool must go in at least 2 ¼ in. (63.5mm).

c. Recheck and realign the timing marks.

d. Leave the tool in place to hold the silent shaft while continuing.

13. If the camshaft belt tensioner was removed, use a vise to carefully push the auto-tensioner rod in until the set hole in the rod is aligned with the hole in the cylinder. Place a wire into the hole to retain the rod. Mount the tensioner to the engine block and tighten the mounting bolt to 17 ft. lbs. (23 Nm).

14. Install the belt to the crankshaft sprocket, oil pump sprocket, then camshaft sprocket, in that order. While doing so, be sure there is no slack between the sprocket except where the tensioner is installed.

15. To adjust the timing (outer) belt perform the following steps:

a. Turn the crankshaft ¼ turn counterclockwise, then turn it clockwise to move No. 1 cylinder to TDC.

b. Loosen the center bolt. Using tool MD998752 and a torque wrench, apply a torque of 2.6 ft. lbs. (3.6 Nm) to the tensioner. Tighten the center bolt.

c. Screw the special tool into the engine left support bracket until its end makes contact with the tensioner arm. At this point, screw the special tool in some more and remove the set wire attached to the auto-tensioner, if the wire was not previously removed. Then, remove the special tool.

d. Rotate the crankshaft two complete turns clockwise and let it sit for approximately 15 minutes. Then, measure the auto-tensioner protrusion (the distance between the tensioner arm and auto-tensioner body) to ensure that it is within 0.15–0.18 in. (3.8–4.5mm). If out of specification, repeat substeps **a** through **d** until the specified value is obtained.

➡Do not manually overtighten the belt or it will howl.

16. Install the upper and lower timing belt covers.

3.0L Engine

1. Before servicing the vehicle, refer to the precautions at the beginning of this section.

2. Position the engine so the No. 1 cylinder is at Top Dead Center (TDC) of its compression stroke.

⁂ CAUTION

Wait at least 90 seconds after the negative battery cable is disconnected to prevent possible deployment of the air bag.

3. Remove all necessary components for access to the timing belt covers, then remove the covers from the engine.

4. If the same timing belt will be reused, mark the direction of the timing belt's rotation for installation in the same direction. Be sure the engine is positioned so the No. 1 cylinder is at the TDC of its compression stroke and the timing marks are aligned with the engine's timing mark indicators.

5. Loosen the timing belt tensioner bolt and remove the belt. If the tensioner is not being removed, position it as far away from the center of the engine as possible and tighten the bolt.

6. If the tensioner is being removed, mark the outside of the spring to ensure that it is not installed backwards. Unbolt the tensioner and remove it along with the spring.

⁂ WARNING

Do not rotate the camshafts when the timing belt is removed from the engine. Turning the camshaft when the timing belt is removed could cause the valves to interfere with the pistons thus causing severe internal engine damage.

To install:

7. Install the tensioner, if removed, and hook the upper end of the spring to the

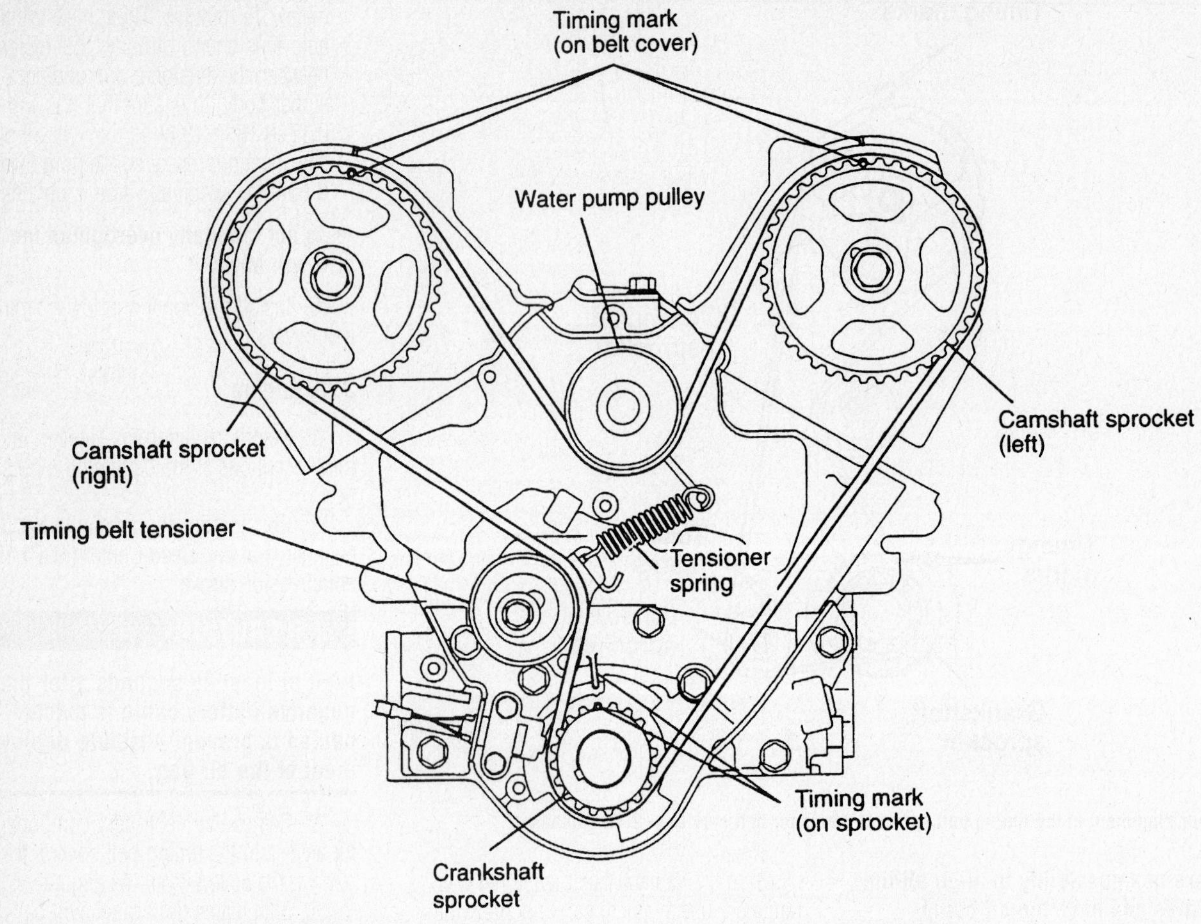

Align the sprockets properly before removing or installing the timing belt—Diamante with 3.0L (6G72) SOHC engine

water pump pin and the lower end to the tensioner in exactly the same position as originally installed.

8. Ensure both camshafts are still positioned so the timing marks align with those on the rear timing covers. Rotate the crankshaft so the timing mark aligns with the mark on the front cover.

9. Install the timing belt on the crankshaft sprocket and while keeping the belt tight on the tension side, install the belt on the front (left) camshaft sprocket.

10. Install the belt on the water pump pulley, then the rear (right) camshaft sprocket and the tensioner.

11. Loosen the bolt that secures the adjustment of the tensioner and lightly press the tensioner against the timing belt.

12. Check that the timing marks are in alignment.

13. Rotate the crankshaft 2 full turns in the clockwise direction only, then realign the timing marks.

14. Tighten the bolt that secures the tensioner to 19 ft. lbs. (26 Nm).

15. Install the lower and the upper timing belt covers, along with all other applicable components.

3.5L and 3.8L Engines

1. Before servicing the vehicle, refer to the precautions at the beginning of this section.

2. Position the engine so the No. 1 cylinder is at Top Dead Center (TDC) of its compression stroke.

3. Remove all necessary components for access to the timing belt covers, then remove the covers from the engine.

**** CAUTION**

Be sure to disconnect the negative battery cable. Wait at least 90 seconds after the negative battery cable is disconnected to prevent possible deployment of the air bag.

4. If the same timing belt will be reused, mark the direction of the timing belt's rotation for installation in the same direction. Be sure the engine is positioned so the No. 1 cylinder is at the TDC of its compression stroke and the timing marks are aligned with the engine's timing mark indicators on the rear timing covers.

5. Remove the timing belt.

**** WARNING**

Turning the camshaft sprocket when the timing belt is removed could cause the valves to contact with the pistons, resulting in severe engine damage.

6. Remove the bolts that secure the auto-tensioner to the engine block and remove the tensioner.

To install:

➡The auto-tensioner assembly must be reset to correctly adjust belt tension.

7. Loosen the center bolt of tensioner pulley to provide timing belt slack. Remove the timing belt tensioner assembly.

8. Position the auto-tensioner into a vise with soft jaws. The plug at the rear of

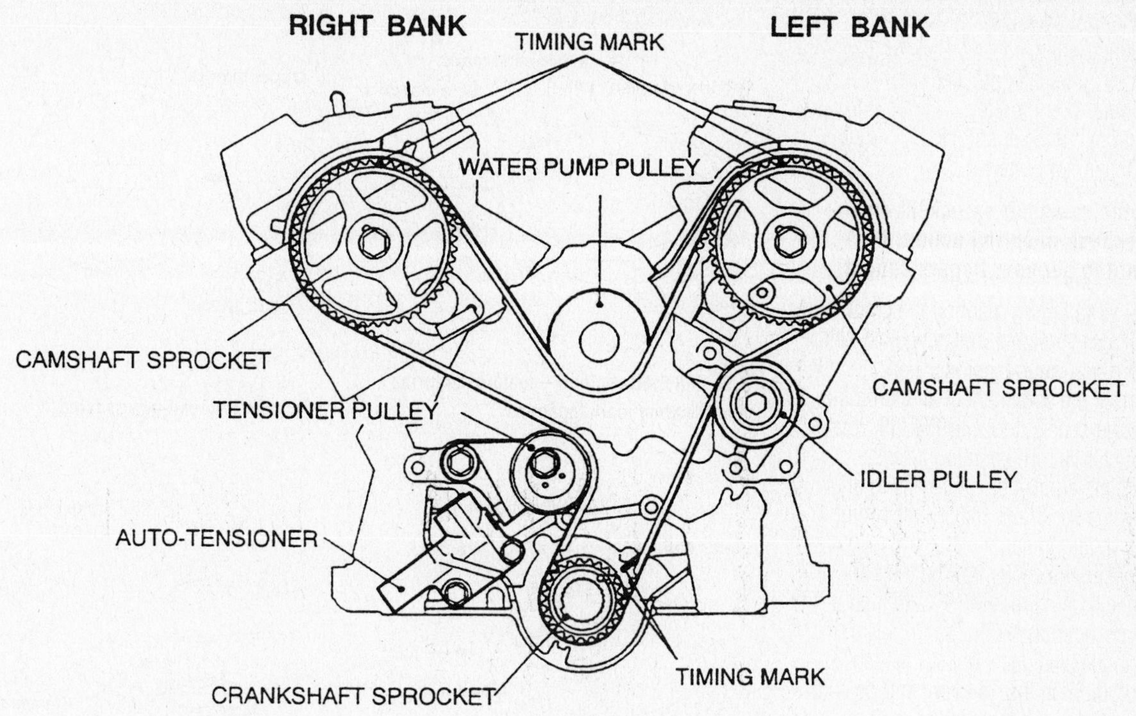

Sprocket alignment for timing belt installation—Mitsubishi Diamante with 3.5L (6G74) SOHC engine

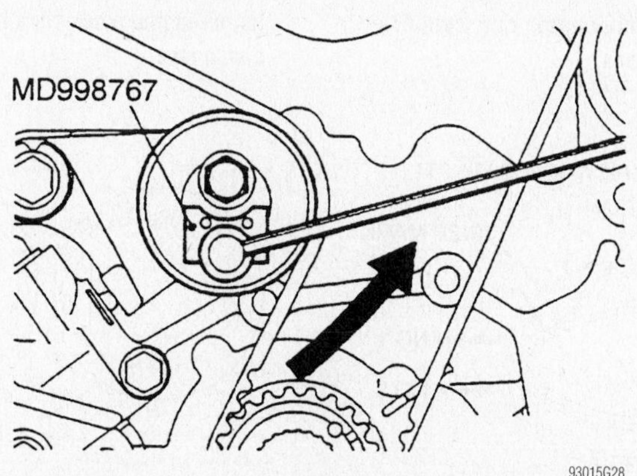

Special tool used for tightening timing belt—Mitsubishi Diamante with 3.5L (6G74) SOHC engine

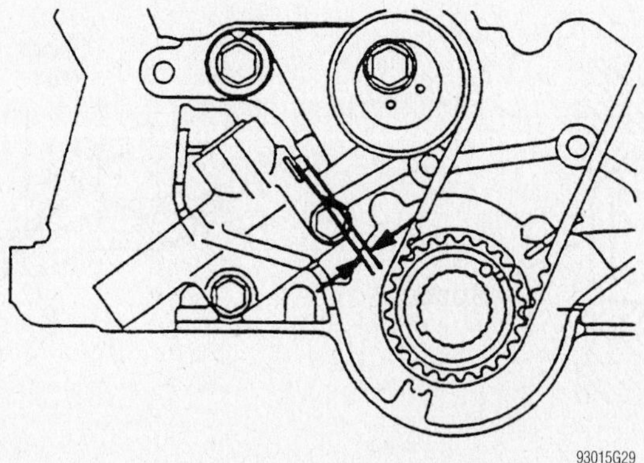

Measuring the standard value of the timing belt tensioner—Mitsubishi Diamante with 3.5L (6G74) SOHC engine

tensioner protrudes, be sure to use a washer as a spacer to protect the plug from contacting vise jaws.

9. Slowly push the rod into the tensioner until the set hole in rod is aligned with set hole in the auto-tensioner.

10. Insert a 0.055 in. (1.4mm) wire into the aligned set holes. Unclamp the tensioner from the vise and install it on the engine. Tighten tensioner mounting bolts to 17 ft. lbs. (24 Nm).

**** WARNING**

DO NOT rotate or turn the camshafts when removing the sprockets or severe engine damage will result from internal component interference.

11. Align the mark on the crankshaft sprocket with the mark on the front case. Then, move the crankshaft sprocket 3 teeth counterclockwise.

12. Align the timing marks of the camshafts with the marks on the rear covers.

13. Realign the crankshaft pulley with timing mark on the housing.

➡Be sure camshafts-to-cylinder heads and crankshaft-to-front cover timing marks are aligned.

14. Install the timing belt around the pulleys in the following order:

a. Crankshaft pulley.
b. Idler pulley.
c. Left camshaft sprocket.
d. Water pump pulley.
e. Right camshaft sprocket.
f. Tensioner pulley.

➡Since the camshaft sprockets turn easily because of spring action, be careful not to get your fingers caught.

15. Align all timing mark on the crankshaft and raise tensioner pulley against belt to remove slack, snug tensioner bolt.
16. Check the alignment of all the timing.
17. Using special tool MD998769, rotate the crankshaft ¼ turn counterclockwise, then rotate the crankshaft clockwise to align the timing marks. Check that all the timing marks are in alignment.
18. Loosen the center bolt on the tensioner pulley. Using tool MD998767 and a torque wrench, apply 3.3 ft. lbs. (4.4 Nm) to the tool on the tensioner. Tighten the tensioner bolt to 33 ft. lbs. (44 Nm) and be sure the tensioner does not rotate with the bolt.
19. Rotate the crankshaft two complete turns clockwise and let it sit for approximately five minutes. Then, check that the set pin can easily be inserted and removed from the hole in the auto-tensioner.
20. Remove the set wire attached to the auto-tensioner.
21. Measure the auto-tensioner protrusion (the distance between the tensioner arm and auto-tensioner body) to ensure that it is within 0.150–0.196 in. (3.8–5.0mm). If out of specification, repeat adjustment procedure until the specified value is obtained.
22. Check again that the timing marks on all sprockets are in proper alignment.
23. Install the timing belt covers and all other applicable components.

Piston and Ring

POSITIONING

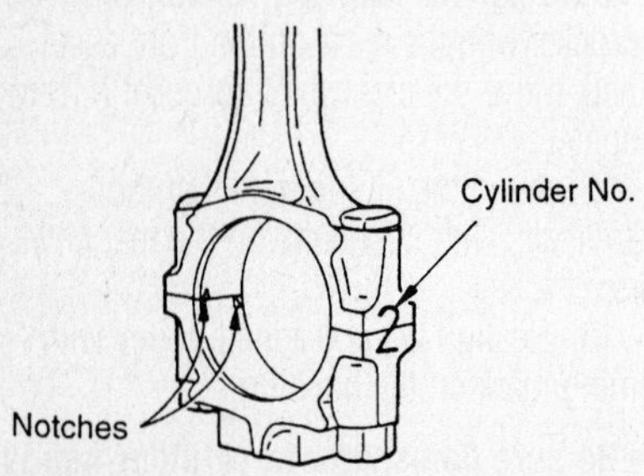

Before removing the caps from the connecting rods, be sure to matchmark them as shown

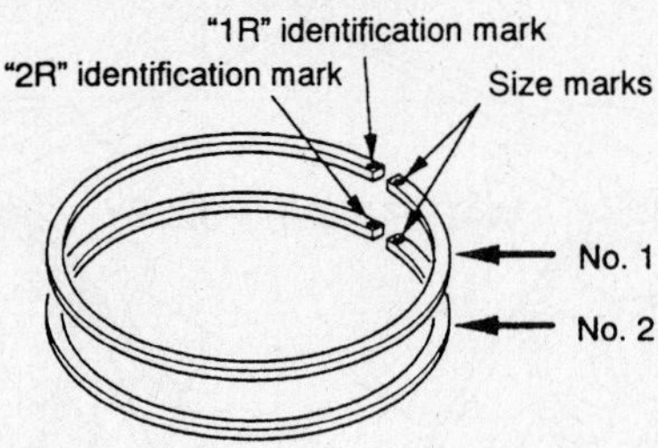

1.5L and 2.4L engines—compression ring identification mark locations

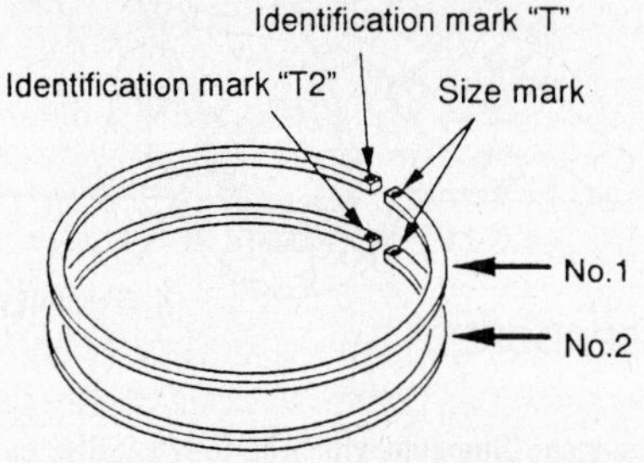

1.8L engine—compression ring identification mark locations

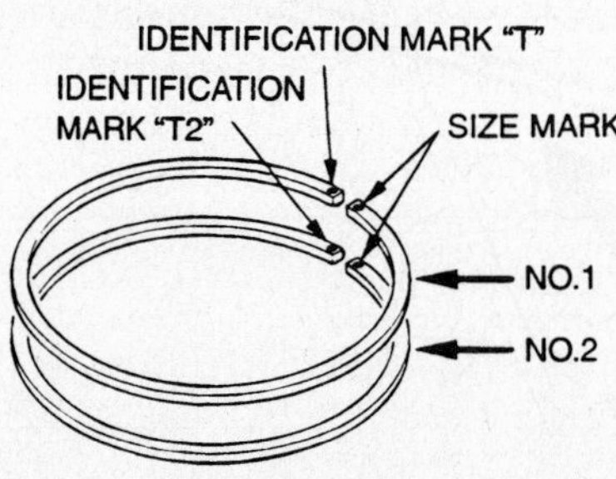

2.0L engines—compression ring identification mark locations

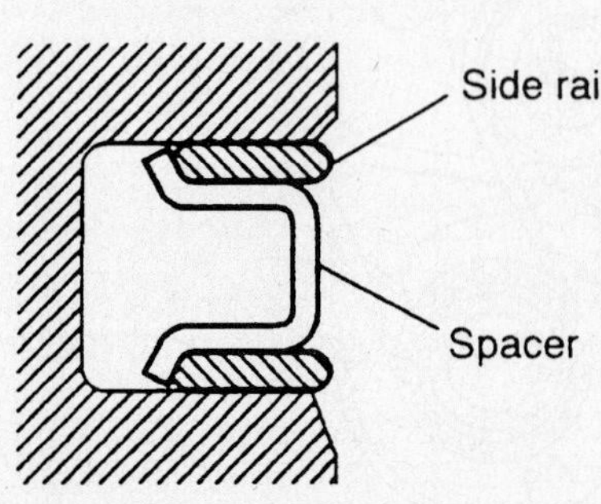

1.8L engine—oil side and spacer ring positioning

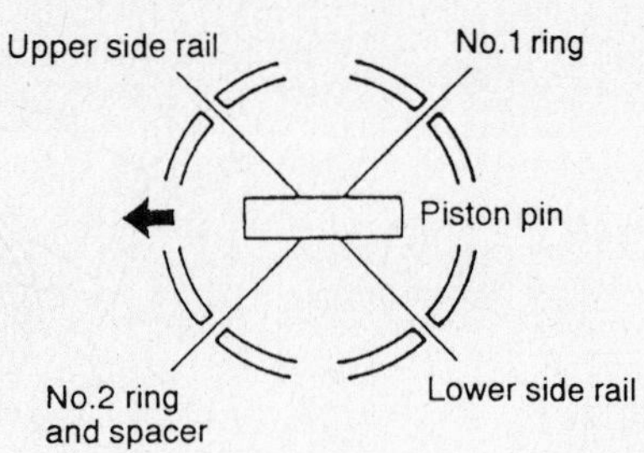

Piston ring end-gap spacing

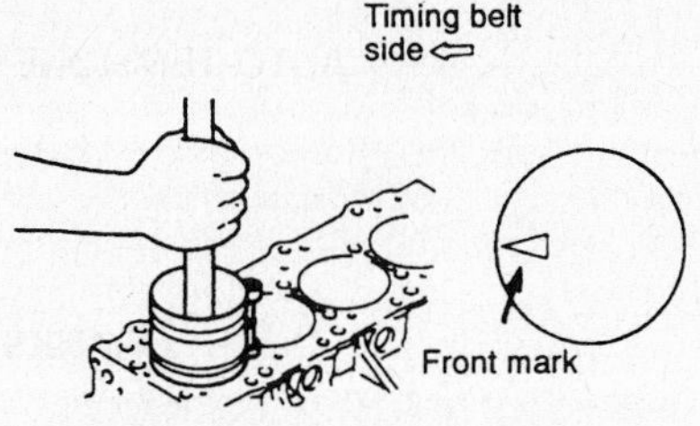

1.5L, 1.8L. 2.0L and 2.4L engines—piston-to-engine block mark location on the piston face

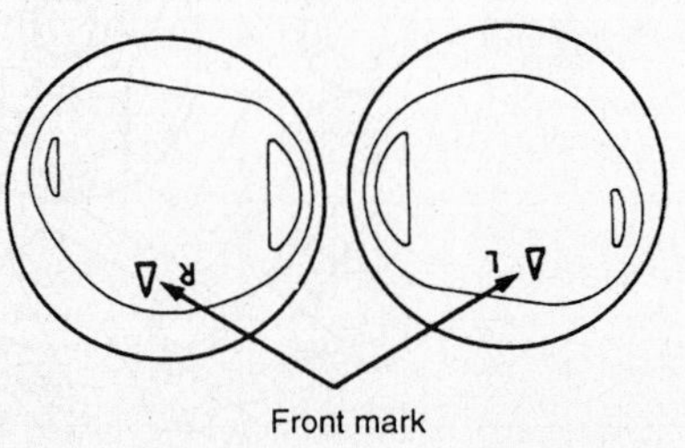

3.0L engine—piston-to-engine block mark locations

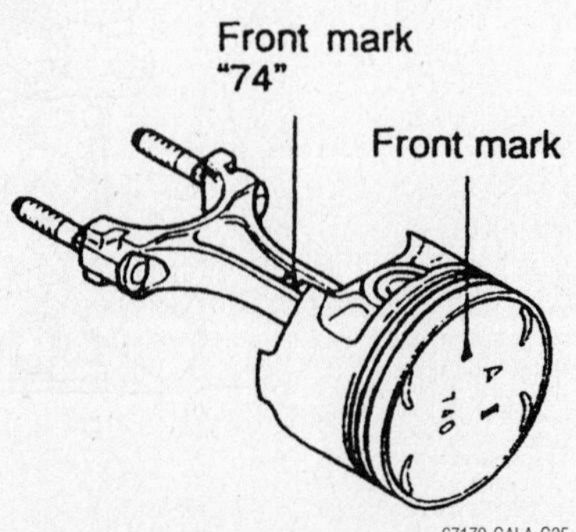

3.5L and 3.8L engines—piston and connecting rod assembly positioning

FUEL SYSTEM

Fuel System Service Precautions

Safety is the most important factor when performing not only fuel system maintenance but any type of maintenance. Failure to conduct maintenance and repairs in a safe manner may result in serious personal injury or death. Maintenance and testing of the vehicle's fuel system components can be accomplished safely and effectively by adhering to the following rules and guidelines.

- To avoid the possibility of fire and personal injury, always disconnect the negative battery cable unless the repair or test procedure requires that battery voltage be applied.
- Always relieve the fuel system pressure prior to disconnecting any fuel system component (injector, fuel rail, pressure regulator, etc.), fitting or fuel line connection. Exercise extreme caution whenever relieving fuel system pressure, to avoid exposing skin, face and eyes to fuel spray. Please be advised that fuel under pressure may penetrate the skin or any part of the body that it contacts.
- Always place a shop towel or cloth around the fitting or connection prior to loosening to absorb any excess fuel due to spillage. Ensure that all fuel spillage (should it occur) is quickly removed from engine surfaces. Ensure that all fuel soaked cloths or towels are deposited into a suitable waste container.
- Always keep a dry chemical (Class B) fire extinguisher near the work area.
- Do not allow fuel spray or fuel vapors to come into contact with a spark or open flame.
- Always use a back-up wrench when loosening and tightening fuel line connection fittings. This will prevent unnecessary stress and torsion to fuel line piping. Always follow the proper torque specifications.
- Always replace worn fuel fitting O-rings with new. Do not substitute fuel hose or equivalent, where fuel pipe is installed.

Fuel System Pressure

RELIEVING

Eclipse

1. Before servicing the vehicle, refer to the precautions at the beginning of this section.

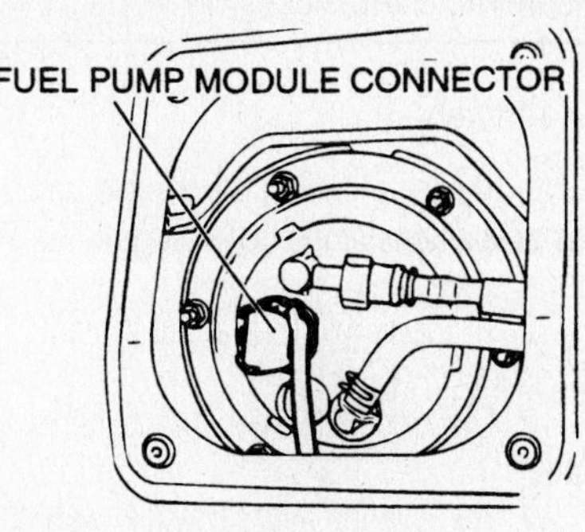

Location of the fuel pump connector—Lancer shown, other models similar

2. Turn the ignition to the **OFF** position.
3. Loosen the fuel filler cap to release fuel tank pressure.
4. Remove the driver side instrument panel side cover and disconnect the fuel pump relay.
5. Start the vehicle and allow it to run until it stalls from lack of fuel. Turn the key to the **OFF** position.
6. Disconnect the negative battery cable, then reconnect the fuel pump relay. Install the side cover.
7. Wrap shop towels around the fitting that is being disconnected to absorb residual fuel in the lines.
8. Place shop towels into proper safety container.

Except Eclipse

1. Before servicing the vehicle, refer to the precautions at the beginning of this section.
2. Turn the ignition to the **OFF** position.
3. Loosen the fuel filler cap to release fuel tank pressure.
4. Remove the rear seat cushion, then remove the service cover and disconnect the fuel pump harness connector.
5. Start the vehicle and allow it to run until it stalls from lack of fuel. Turn the key to the **OFF** position.
6. Disconnect the negative battery cable, then reconnect the fuel pump connector. Install the access cover and rear seat cushion.
7. Wrap shop towels around the fitting that is being disconnected to absorb residual fuel in the lines.
8. Place shop towels into proper safety container.

Fuel Filter

REMOVAL & INSTALLATION

Diamante

1. Before servicing the vehicle, refer to the precautions at the beginning of this section.
2. Properly relieve the fuel pressure.
3. Disconnect the negative battery cable.

➡The filter is located in the engine compartment, mounted on the inner fender panel.

4. Remove or disconnect the following:
   - Air cleaner assembly and intake hoses
   - Battery and battery tray
   - Fuel lines from the filter
   - Mounting bolts and remove the fuel filter from the vehicle

To install:

➡Install new gaskets or O-rings whenever fuel connections have been disassembled.

5. Install or connect the following:
   - Filter to its bracket finger-tight
   - New gaskets and connect the high pressure hose and eye bolt, then the main pipe and eye bolt. Tighten the eye bolts to 22 ft. lbs. (30 Nm). Tighten the flare nut to 25 ft. lbs. (35 Nm).
6. Tighten the mounting bolts fully.
7. Install or connect the following:
   - Air cleaner assembly
   - Battery and battery tray
   - Negative battery cable, install the fuel filler cap, turn the key to the **ON** position to pressurize the fuel system and check for leaks.

Lancer, Evolution, Sportback, Mirage and Galant

➡The fuel filter is located in the engine compartment.

1. Before servicing the vehicle, refer to the precautions at the beginning of this section.
2. Properly relieve the fuel system pressure.
3. Remove or disconnect the following:
   - Negative battery cable
   - Air intake hose and the battery
   - Fuel lines from the filter

- Mounting bolts and the fuel filter from the vehicle

To install:

4. If equipped with flare fitting, tighten the fitting by hand before installing the filter to the vehicle.
5. Install or connect the following:
   - Filter to its bracket finger-tight
   - New gaskets and connect the high pressure hose and eye bolt, then the main pipe. Tighten the eye bolts to 22 ft. lbs. (30 Nm). Tighten the flare nut to 27 ft. lbs. (37 Nm).
6. Tighten the filter mounting bolts fully.
7. Install or connect the following:
   - Air intake hose
   - Battery
   - Negative battery cable, install the fuel filler cap, turn the key to the **ON** position to pressurize the fuel system and check for leaks.

Eclipse

1. Before servicing the vehicle, refer to the precautions at the beginning of this section.
2. Properly relieve the fuel system pressure.
3. Disconnect the negative battery cable.
4. On models equipped with the 2.4L engine, remove the battery and the air intake hose.
5. Remove the fuel lines from the filter.
6. Remove clamp and the hose from the fuel pressure regulator.
7. Remove or disconnect the following:
   - Fuel filter mounting bracket bolts and remove the fuel filter
   - Bracket screw and remove the fuel filter from the mounting bracket

To install:

8. Install or connect the following:
   - Fuel filter to the mounting bracket with the screw
   - Fuel filter to the vehicle with the bracket mounting bolts
   - Main fuel pipe to the fuel filter connector or the filter itself. Torque the flare nut to 27 ft. lbs. (36 Nm).
9. Reconnect the high pressure fuel hose to the fuel filter. Torque the eye bolt to 22 ft. lbs. (29 Nm).
10. On 2.4L engine models, install the battery and the air intake hose.
11. Reconnect the negative battery cable, start the engine and check for fuel leaks.

Fuel Pump

REMOVAL & INSTALLATION

Diamante

1. Before servicing the vehicle, refer to the precautions at the beginning of this section.
2. Properly relieve the fuel system pressure.
3. Disconnect the negative battery cable.
4. Remove the left rear wheel well liner, if equipped.
5. Disconnect the center exhaust system from the main muffler. Disconnect the rear exhaust hangers, lower the exhaust and secure aside.
6. Remove the tank drain plug and drain the fuel into an approved container.
7. Remove or disconnect the following:
   - Fuel return hose, high pressure hose and vent hose from the sending unit
   - Electrical connector
   - Filler and vent hoses. Place a support under the tank and remove the retaining nuts.
8. Lower the tank from the vehicle.
9. Remove the fuel pump retaining nuts and remove the assembly from the tank.

To install:

10. Install or connect the following:
    - Pump assembly to the tank and tighten the retaining nuts to 24 inch lbs. (3 Nm)
    - Fuel tank and connect the filler and vent hoses. Tighten the tank retaining nuts and bolts to 19 ft. lbs. (26 Nm).
    - Return hose, high pressure hose and all other hoses and connectors connected to the pump/sending unit
    - Power cylinder unit and tighten the mounting bolts to 31 ft. lbs. (43 Nm), if equipped with 4WS
    - Exhaust pipe and secure the rear hangers
    - Left rear wheel well liner, if removed
11. Lower the vehicle and return fuel to the gas tank.
12. Connect the negative battery cable and check the entire system for proper operation and leaks.

Eclipse

1. Before servicing the vehicle, refer to the precautions at the beginning of this section.
2. Relieve the fuel system pressure.
3. Remove or disconnect the following:
   - Negative battery cable
   - Rear seat cushion by pulling the seat stopper near the floor and lifting the cushion up
   - Inspection cover on the right side of the vehicle
   - Harness connector and the fuel lines
   - Fuel pump assemble from the tank.

To install:

4. Install or connect the following:
   - Fuel pump in the tank
   - Hoses and the harness connector
   - Inspection cover
   - Rear seat
   - Negative battery cable

Galant

1. Before servicing the vehicle, refer to the precautions at the beginning of this section.
2. Properly relieve the fuel system pressure.
3. Remove or disconnect the following:
   - Negative battery cable
   - Rear seat cushion, by pulling the seat stopper outward and lifting the lower cushion upward
   - Access cover
   - Fuel pump wiring
   - Return hose and the high pressure fuel hose
   - Pump mounting nuts and remove the pump assembly

To install:

4. Install the fuel pump assembly to the tank and tighten the retaining nuts to 22 inch lbs. (2.5 Nm).

➡Tilt the float to the left of the vehicle, when installing the pump assembly.

5. Install or connect the following:
   - High pressure hose, return hose and the fuel tank wiring
   - Negative battery cable
6. Check the fuel pump for proper pressure and inspect the entire system for leaks.
7. Apply sealant to the access cover and install the cover.
8. Install the rear seat cushion.

Lancer, Evolution and Sportback

1. Before servicing the vehicle, refer to the precautions at the beginning of this section.
2. Properly relieve the fuel system pressure.
3. Disconnect the negative battery cable.

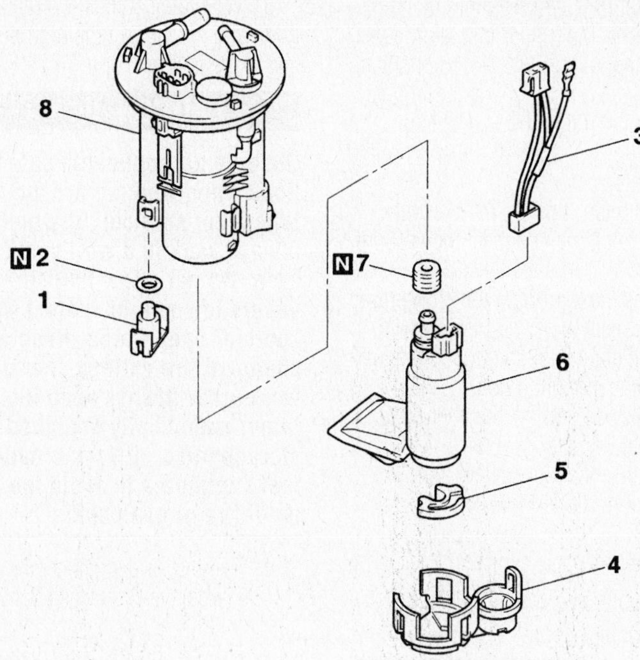

Exploded view of the fuel pump module—Lancer

4. Remove or disconnect the following:
   - Rear seat assembly
   - Retainer screws and service hole cover
   - Harness electrical connector
   - Fuel lines
   - Fuel pump module mounting nuts
   - Fuel pump module
5. Disassemble the fuel pump module as necessary. Refer to the accompanying figure.
6. Installation is the reverse of the removal procedure. Torque the fuel pump module nuts to 23 inch lbs. (2.5 Nm).

Mirage

1. Before servicing the vehicle, refer to the precautions at the beginning of this section.
2. Properly relieve the fuel system pressure.
3. Disconnect the negative battery cable.
4. Raise and safely support the vehicle.
5. Drain the fuel from the fuel tank into an approved container.
6. Disconnect the filler and vent hoses.
7. Support the tank with a transmission jack. Disconnect the retainer straps and lower the tank to gain access to the fitting on top of the tank.
8. Remove or disconnect the following:
   - Return hose, high pressure hose and vapor hoses from the pump/sending unit
   - Electrical connectors at the pump/sending unit
   - Fuel tank from the vehicle
   - Access plate to the fuel tank and remove the pump assembly

To install:

9. Install fuel pump into fuel tank, with new packing gasket, and tighten mounting nuts.
10. Raise the tank in position under the vehicle.
11. Attach all connections to the top of the tank.
12. Raise the tank completely and position the retainer straps around the fuel tank. Install new fuel tank self-locking nuts and tighten to 22 ft. lbs. (31 Nm).
13. Connect the return hose and high pressure hoses.
14. Install the vapor hose and the filler hose. Install the filler hose retainer screws to the fender, if removed.
15. Lower the vehicle and pour the drained fuel into the gas tank.
16. Connect the negative battery cable.

Check the fuel pump for proper pressure and inspect the entire system for leaks.

Fuel Injector

REMOVAL & INSTALLATION

1.5L and 1.8L Engines

1. Relieve the fuel system pressure as described in this section.
2. Remove or disconnect the following:
   - Positive Crankcase Ventilation (PCV) hose from the valve cover
   - Breather hose at the opposite end of the valve cover
   - High pressure fuel line

⁂ CAUTION

Observe all applicable safety precautions when working around fuel. Whenever servicing the fuel system, always work in a well ventilated area. Do not allow fuel spray or vapors to come in contact with a spark or open flame. Keep a dry chemical fire extinguisher near the work area. Always keep fuel in a container specifically designed for fuel storage; also, always properly seal fuel containers to avoid the possibility of fire or explosion.

3. Remove or disconnect the following:
   - Vacuum hose from the fuel pressure regulator
   - Fuel return hose from the pressure regulator
   - Electrical connector from each injector. Label for reference
   - Bolt(s) holding the fuel rail to the manifold. Carefully lift the rail up and remove it with the injectors attached. Take great care not to drop an injector. Place the rail and injectors in a safe location on the workbench; protect the tips of the injectors from dirt and/or impact.
   - Injector insulators from the intake manifold, discard. The insulators are not reusable.
   - Injectors from the fuel rail by pulling gently in a straight outward motion. Make certain the grommet and O-ring come off with the injector.

To install:

4. Install a new insulator in each injector port in the manifold.
5. Remove the old grommet and O-ring from each injector. Install a new grommet

and O-ring; coat the O-ring lightly with clean, thin oil.

6. If the fuel pressure regulator was removed, replace the O-ring with a new one and coat it lightly with clean, thin oil. Insert the regulator straight into the rail, then check that it can be rotated freely. If it does not rotate smoothly, remove it and inspect the O-ring for deformation or jamming. When properly installed, align the mounting holes and tighten the retaining bolts to 84 inch lbs. (9 Nm). This procedure must be followed even if the fuel rail was not removed.

7. Install or connect the following:
- Injector into the fuel rail, constantly turning the injector left and right during installation. When fully installed, the injector should still turn freely in the rail. If it does not, remove the injector and inspect the O-ring for deformation or damage.
- Delivery pipe and injectors to the engine. Make certain that each injector fits correctly into its port and that the rubber insulators for the fuel rail mounts are in position.
- Fuel rail retaining bolts and tighten them to 108 inch lbs. (12 Nm)
- Wiring harnesses to the appropriate injector
- Fuel return hose to the pressure regulator, then connect the vacuum hose

8. Replace the O-ring on the high pressure fuel line, coat the O-ring lightly with clean, thin oil and install the line to the fuel rail. Tighten the mounting bolts.
- PCV hose and the breather hose
- Negative battery cable

9. Pressurize the fuel system and inspect all connections for leaks.

2.0L Engine

1. Relieve the fuel system pressure as described in this section.
2. Disconnect the negative battery cable.
3. Wrap the connection with a shop towel and disconnect the high pressure fuel line at the fuel rail.

⁂ CAUTION

Observe all applicable safety precautions when working around fuel. Whenever servicing the fuel system, always work in a well ventilated area. Do not allow fuel spray or vapors to come in contact with a spark or open flame. Keep a dry chemical fire extinguisher near the work area. Always keep fuel in a container specifically designed for fuel storage; also, always properly seal fuel containers to avoid the possibility of fire or explosion.

4. Remove or disconnect the following:
- Positive Crankcase Ventilation (PCV) hose
- Exhaust Gas Recirculation (EGR) solenoid valve connector

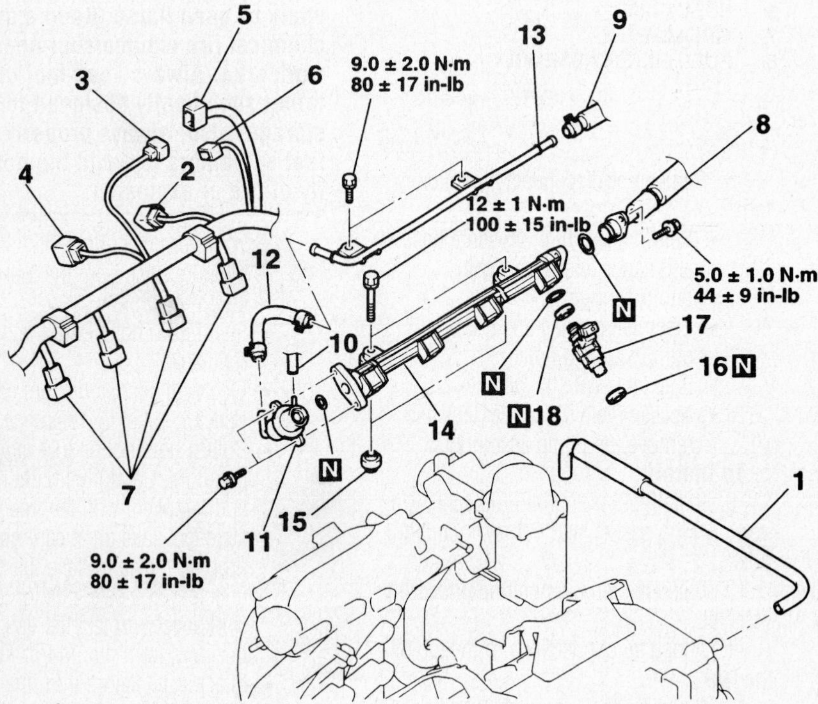

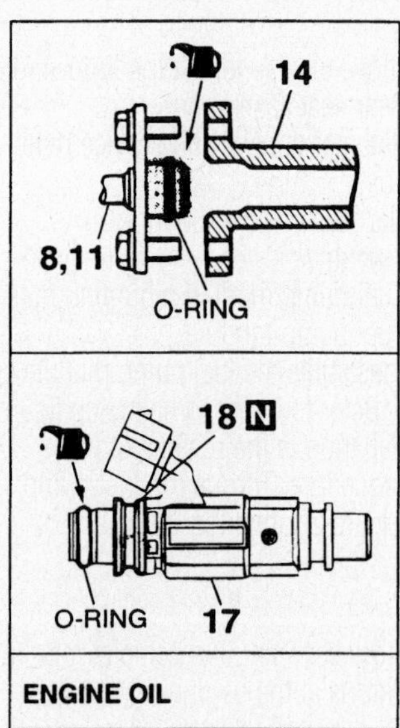

1. PCV HOSE CONNECTION
2. EGR SOLENOID VALVE CONNECTOR
3. MANIFOLD DIFFERENTIAL PRESSURESENSOR CONNECTOR
4. PURGE CONTROL SOLENOID VALVE CONNECTOR
5. THROTTLE POSITION SENSOR CONNECTOR
6. IDLE AIR CONTROL MOTOR CONNECTOR
7. INJECTOR CONNECTOR

93570G25

Exploded view of the fuel rail and injectors—2.0L engine

- Manifold Differential Pressure (MDP) sensor connector
- Purge control solenoid valve connector
- Throttle Position (TP) sensor connector
- Idle Air Control (IAC) motor connector
- Electrical connector from each injector, label for reference
- High pressure fuel hose
- Fuel return hose
- Vacuum hose(s)
- Fuel pressure regulator
- Fuel hose
- Fuel return pipe
- Bolt(s) holding the fuel rail to the manifold. Carefully lift the rail up and remove it with the injectors attached. Take great care not to drop an injector. Place the rail and injectors in a safe location on the workbench; protect the tips of the injectors from dirt and/or impact.
- Injector insulators from the intake manifold, discard. The insulators are not reusable.
- Injectors from the fuel rail by pulling gently in a straight outward motion. Make certain the grommet and O-ring come off with the injector.

To install:

5. Install a new insulator in each injector port in the manifold.
6. Remove the old grommet and O-ring from each injector. Install a new grommet and O-ring; coat the O-ring lightly with clean, thin oil.
7. If the fuel pressure regulator was removed, replace the O-ring with a new one and coat it lightly with clean, thin oil. Insert the regulator straight into the rail, then check that it can be rotated freely. If it does not rotate smoothly, remove it and inspect the O-ring for deformation or jamming. When properly installed, align the mounting holes and tighten the retaining bolts to 84 inch lbs. (9 Nm). This procedure must be followed even if the fuel rail was not removed.
8. Install or connect the following:
   - Injector into the fuel rail, constantly turning the injector left and right during installation. When fully installed, the injector should still turn freely in the rail. If it does not, remove the injector and inspect the O-ring for deformation or damage.
   - Fuel rail and injectors to the engine. Make certain that each injector fits correctly into its port and that the rubber insulators for the fuel rail mounts are in position.
   - Fuel return pipe and fuel hose
   - Fuel return hose to the fuel rail
   - High pressure fuel line to the fuel rail
   - Fuel injector connectors
   - IAC motor connector
   - TP sensor connector
   - Purge control solenoid connector
   - MDP sensor connector
   - EGR solenoid valve connector
   - PCV hose connector
   - Negative battery cable
9. Pressurize the fuel system and inspect all connections for leaks.

2.4L Engine

1. Relieve the fuel system pressure as described in this section.
2. Label and disconnect the spark plug wires. Position the wires aside.
3. Remove or disconnect the following:
   - Positive Crankcase Ventilation (PCV) hose from the valve cover
   - High pressure fuel line to the fuel rail and disconnect the line. Be prepared to contain fuel spillage; plug the line to keep out dirt and debris.

⁂ CAUTION

Observe all applicable safety precautions when working around fuel. Whenever servicing the fuel system, always work in a well ventilated area. Do not allow fuel spray or vapors to come in contact with a spark or open flame. Keep a dry chemical fire extinguisher near the work area. Always keep fuel in a container specifically designed for fuel storage; also, always properly seal fuel containers to avoid the possibility of fire or explosion.

4. Remove or disconnect the following:
   - Vacuum hose from the fuel pressure regulator
   - Fuel return hose from the pressure regulator
   - Electrical connector from each injector, label for reference
   - Bolt(s) holding the fuel rail to the manifold. Carefully lift the rail up and remove it with the injectors attached. Take great care not to drop an injector. Place the rail and injectors in a safe location on the workbench; protect the tips of the injectors from dirt and/or impact.
   - Injector insulators from the intake manifold, discard. The insulators are not reusable.
   - Injectors from the fuel rail by pulling gently in a straight outward motion. Make certain the grommet and O-ring come off with the injector.

To install:

5. Install a new insulator in each injector port in the manifold.
6. Remove the old grommet and O-ring from each injector. Install a new grommet and O-ring; coat the O-ring lightly with clean, thin oil.
7. If the fuel pressure regulator was removed, replace the O-ring with a new one and coat it lightly with clean, thin oil. Insert the regulator straight into the rail, then check that it can be rotated freely. If it does not rotate smoothly, remove it and inspect the O-ring for deformation or jamming. When properly installed, align the mounting holes and tighten the retaining bolts to 84 inch lbs. (9 Nm). This procedure must be followed even if the fuel rail was not removed.
8. Install or connect the following:
   - Injector into the fuel rail, constantly turning the injector left and right during installation. When fully installed, the injector should still turn freely in the rail. If it does not, remove the injector and inspect the O-ring for deformation or damage.
   - Delivery pipe and injectors to the engine. Make certain that each injector fits correctly into its port and that the rubber insulators for the fuel rail mounts are in position.
   - Fuel rail retaining bolts and tighten them to 108 inch lbs. (12 Nm)
   - Wiring harnesses to the appropriate injector
   - Fuel return hose to the pressure regulator, then connect the vacuum hose
   - O-ring on the high pressure fuel line, coat the O-ring lightly with clean, thin oil and install the line to the fuel rail. Tighten the mounting bolts to 48 inch lbs. (6 Nm).
   - PCV hose and spark plug wires
   - Negative battery cable
9. Pressurize the fuel system and inspect all connections for leaks.

3.0L, 3.5L and 3.8L Engines

1. Relieve the fuel system pressure.
2. Disconnect the negative battery cable.

✻✻ CAUTION

Work MUST NOT be started until at least 90 seconds after the ignition switch is turned to the LOCK position and the negative battery cable is disconnected from the battery. This will allow time for the air bag system backup power supply to deplete its stored energy preventing accidental air bag deployment which could result in unnecessary air bag system repairs and/or personal injury.

3. Drain the cooling system.
4. Disconnect all components from the air intake plenum and remove the plenum from the intake manifold.
5. Wrap the connection with a shop towel and disconnect the high pressure fuel line at the fuel rail.

✻✻ CAUTION

Observe all applicable safety precautions when working around fuel. Whenever servicing the fuel system, always work in a well ventilated area. Do not allow fuel spray or vapors to come in contact with a spark or open flame. Keep a dry chemical fire extinguisher near the work area. Always keep fuel in a container specifically designed for fuel storage; also, always properly seal fuel containers to avoid the possibility of fire or explosion.

6. Remove or disconnect the following:
   - Fuel return hose and remove the O-ring
   - Vacuum hose from the fuel pressure regulator
   - Electrical connectors from each injector
   - Fuel pipe connecting the fuel rails
   - Injector rail retaining bolts. Make sure the rubber mounting bushings do not get lost.
7. Lift the rail assemblies up and away from the engine.
8. Remove the injectors from the rail by pulling gently. Discard the lower insulator.

To install:

➡Some of the vehicles may have a clip that secures the injector to the fuel rail. Be sure to remove or install the injector clip where necessary.

9. Install or connect the following:
   - New grommet and O-ring to the injector. Coat the O-ring with light oil.
   - Injector to the fuel rail
10. Replace the seats in the intake manifold. Install the fuel rails and injectors to the manifold. Make sure the rubber bushings are in place before tightening the mounting bolts.
11. Tighten the retaining bolts to 84–108 inch lbs. (10–13 Nm). Install the fuel pipe with new gasket.
12. Install or connect the following:
   - Electrical connectors to the injectors
   - Fuel return hose
   - O-ring, lightly lubricate it and connect the high pressure fuel line
   - Intake plenum and all related items, using new gaskets
13. Fill the cooling system.
14. Connect the negative battery cable and check the entire system for proper operation and leaks.

DRIVE TRAIN

Transaxle Assembly

REMOVAL & INSTALLATION

Manual

ECLIPSE

1. Before servicing the vehicle, refer to the precautions at the beginning of this section.
2. Remove or disconnect the following:
   - Battery and the air intake hoses
   - Battery tray and support
   - Auto-cruise actuator and bracket, if equipped with cruise control
   - Charcoal canister and bracket
   - Shift and select cables from the transaxle
   - Back-up light switch and the Vehicle Speed Sensor (VSS) connectors
   - Starter assembly
   - Engine support fixture to the engine and remove the transaxle mounting bolts
3. Remove or disconnect the following:
   - Rear roll stopper bracket mounting bolts
   - Transaxle mounting bracket mounting nuts
   - Engine undercovers
   - Axle shafts
   - Slave cylinder from the bell housing without disconnecting the fluid line. Position it out of the way.
   - Bell housing cover and the right-hand center member stay (support)
   - Center member
4. Place a transmission jack under the transaxle and remove the transaxle mounting bolt.
5. Remove the transaxle mounting and lower the transaxle.

To install:

6. Raise the transaxle into position and install the transaxle mounting. Torque the through-bolt to 50 ft. lbs. (69 Nm).
7. Install or connect the following:

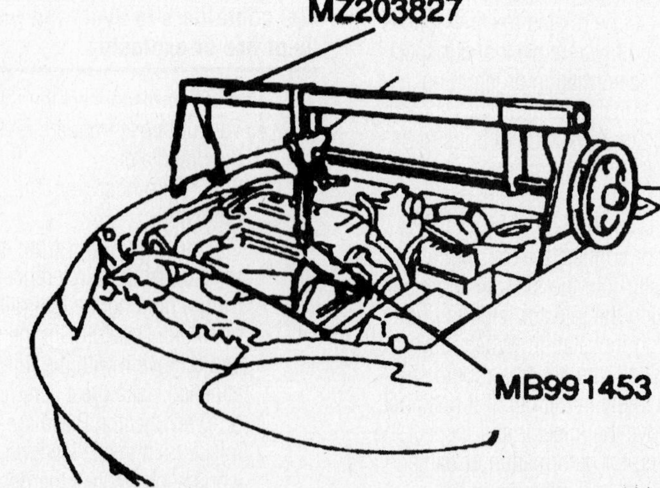

Proper method of supporting the engine assembly for transaxle removal

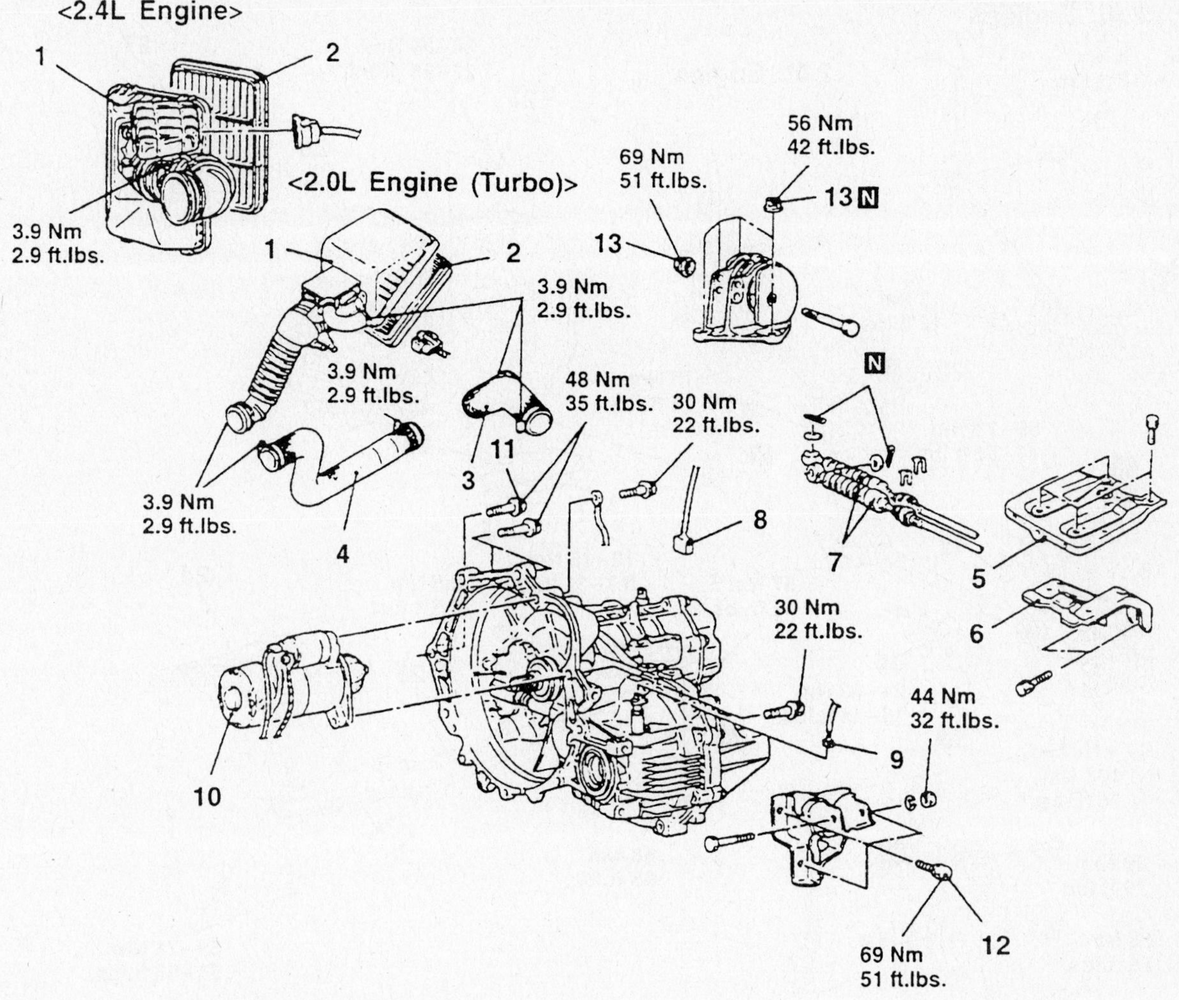

Removal steps

1. Air cleaner cover and air intake hose assembly
2. Air cleaner element
3. Air hose C <2.0L Engine (Turbo)>
4. Air hose A <2.0L Engine (Turbo)>
5. Battery tray
6. Battery tray stay
7. Shift cable and select cable connection
8. Backup light switch connector
9. Vehicle speed sensor connector
10. Starter motor
11. Transaxle assembly mounting bolts
12. Rear roll stopper bracket mounting bolts
13. Transaxle mounting bracket mounting nuts

- Supporting engine assembly

7923PG86

Exploded view of the manual transaxle mounting (1 of 2)—FWD Eclipse with 2.4L engine

- Transaxle assembly mounting bolt. Torque the bolt to 22–25 ft. lbs. (30–34 Nm).
- Center member assembly and the right-hand stay
- Bell housing cover and the slave cylinder
- Axle shafts. Be sure to install the washer in the proper direction.
- Engine undercovers and lower the vehicle
- Transaxle mounting bracket mounting nuts
- Rear roll stopper bracket mounting bolts
- Transaxle assembly mounting bolts. Torque the mounting bolts to 35 ft. lbs. (48 Nm).

8. Remove the engine support fixture.
9. Install or connect the following:
   - Starter assembly
   - VSS and the back-up light connectors
   - Cruise control actuator if removed
   - Battery tray support and the tray
   - Charcoal canister bracket and the canister

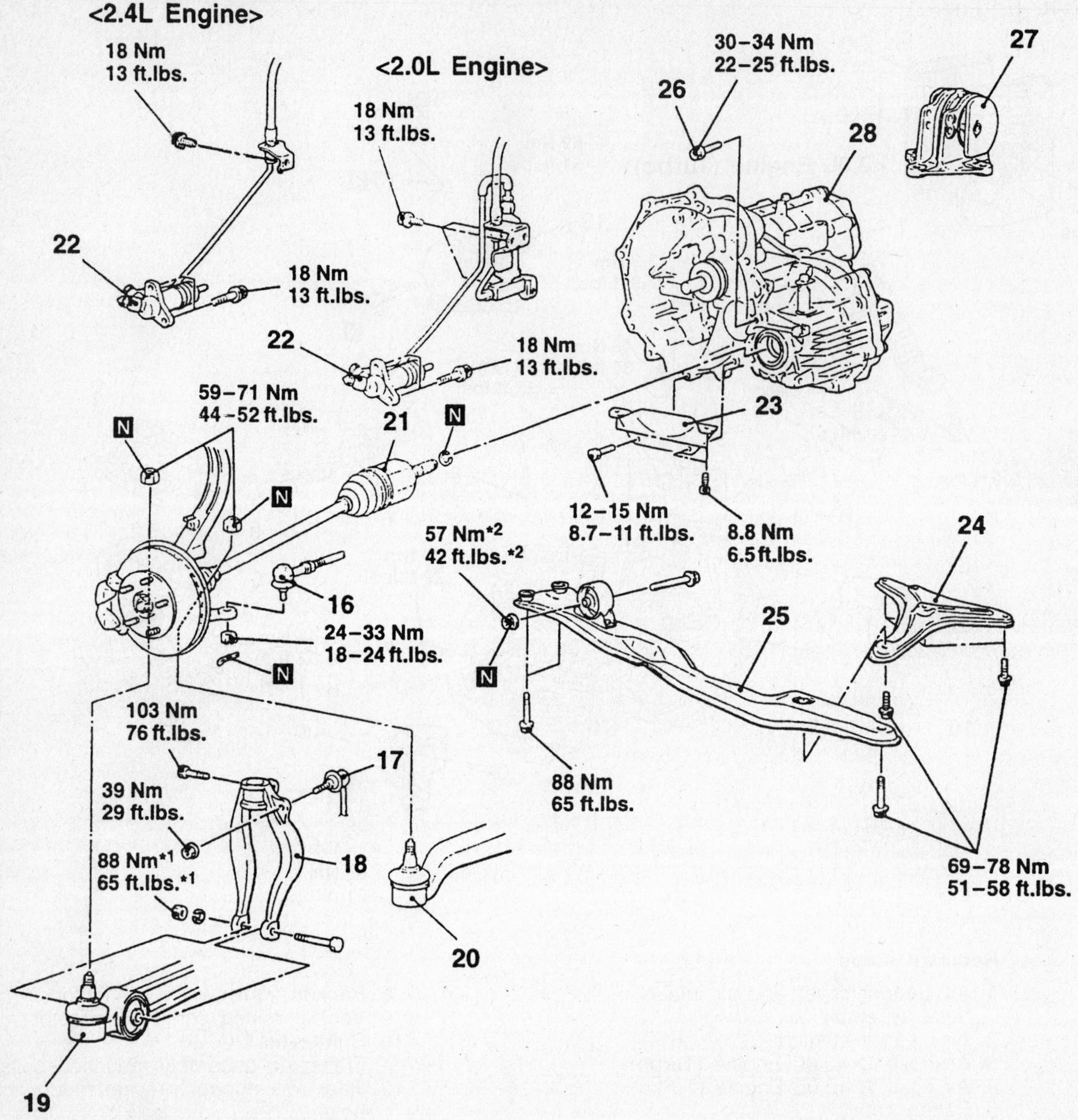

Lifiting up of the vehicle

16. Tie rod end ball joint and kunckle connection
17. Stabilizer link connection
18. Damper fork
19. Lateral lower arm ball joint and kunckle connection
20. Compression lower arm ball joint and kunckle connection
21. Drive shaft connection
22. Clutch release cylinder connection
23. Bell housing cover
24. Stay (R.H.)
25. Center member assembly
26. Transaxle assembly mounting bolt
27. Transaxle mounting
28. Transaxle assembly

Caution

*1: **Indicates parts which should be temporarily tightened, and then fully tightened with the vehicle on the ground in the unladen condition.**

*2: **For tightening locations indicated by the symbol, first tighten temporarily, and then make the final tightening with the entire weight of the engine applied to the vehicle body.**

7923PG87

Exploded view of the manual transaxle mounting (2 of 2)—FWD Eclipse with 2.4L engine

- Air duct and the air cleaner assembly

LANCER & EVOLUTION

1. Before servicing the vehicle, refer to the precautions at the beginning of this section.
2. Drain the transaxle fluid.
3. Remove or disconnect the following:
    - Negative battery cable
    - Engine undercover
    - Evaporative canister
    - Positive battery cable, battery and battery tray
    - Shifter cables
    - Back-up light switch and Vehicle Speed Sensor (VSS) connector
    - Starter motor
    - Clutch hose
    - Upper engine-to-transaxle bolts
    - Transaxle mount
    - Transaxle mount stopper
4. Install a suitable engine support assembly, then raise and safely support the vehicle.
    - Stabilizer bar
    - Wheel Speed Sensor (WSS) connector, if equipped with Anti-lock Brakes (ABS)
    - Brake hose clamp
    - Tie rod end
    - Lower control arm
    - Centermember
    - Halfshafts by inserting a prybar between the transaxle case and the driveshaft and prying the shaft from the transaxle. Do not pull on the driveshaft.
    - Bell housing lower cover
    - Transaxle to engine bolts and lower the transaxle from the vehicle

To install:

5. Install or connect the following:
    - Transaxle to the engine and install the lower mounting bolts
    - Bell housing cover

➡When installing the halfshafts, use new circlips on the axle ends.

6. Install or connect the following:
    - Halfshafts into the transaxle
    - Centermember
    - Lower control arm
    - Tie rod end
    - Brake hose clamp
    - WSS connector, if equipped
    - Stabilizer bar
7. Lower the vehicle, then remove the engine support.
    - Transaxle mount bracket. Tighten the nuts to 35 ft. lbs. (47 Nm).
    - Transaxle mount stopper. Tighten the nuts to 61 ft. lbs. (82 Nm).
    - Transaxle mount
    - Upper transaxle-to-engine bolts and torque to 36 ft. lbs. (48 Nm)
    - Clutch line
    - Starter motor
    - VSS connector
    - Back-up light switch connector
    - Shifter cables, adjust
    - Evaporative canister
    - Battery and battery tray
    - Engine undercover
    - Positive and negative battery cables
    - Transaxle with fluid
8. Bleed the clutch, check and adjust the front wheel alignment, then check the transaxle for proper operation.

MIRAGE

1. Before servicing the vehicle, refer to the precautions at the beginning of this section.
2. Remove or disconnect the following:
    - Negative battery cable
    - Front wheels and the inner wheel panels
    - Air cleaner assembly and vacuum hoses
3. Note the locations and disconnect the shifter cables.
    - Back-up lamp switch connector
    - Speedometer cable and remove the starter motor
    - Upper transaxle-to-engine mounting bolts
4. Remove the undercover and splash pan.
5. Drain the transaxle oil.
6. Support the engine and remove the crossmember.
7. Remove or disconnect the following:
    - Upper transaxle mounting bolt and bracket
    - Stabilizer bar, tie rod ends and the lower ball joint connections
    - Clutch release cylinder and clutch oil line bracket. Disconnect the clutch cable, if equipped with cable controlled clutch system.
    - Halfshafts by inserting a prybar between the transaxle case and the driveshaft and prying the shaft from the transaxle. Do not pull on the driveshaft.
    - Bell housing lower cover
    - Transaxle to engine bolts and lower the transaxle from the vehicle

To install:

8. Install or connect the following:
    - Transaxle to the engine and install the mounting bolts
    - Bell housing cover

➡When installing the halfshafts, use new circlips on the axle ends.

9. Install or connect the following:
    - Halfshafts into the transaxle
    - Slave cylinder or connect the clutch cable
    - Ball joints, tie rod ends and stabilizer bar connections
    - Upper transaxle mounting bracket and bolt
    - Crossmember
    - Undercover
    - Upper transaxle-to-engine mounting bolts
    - Starter motor
    - Back-up light switch connector and speedometer cable
    - Shifter cables, adjust
    - Air cleaner assembly
    - Front wheels
    - Negative battery cable and check the transaxle for proper operation

Automatic

DIAMANTE

1. Before servicing the vehicle, refer to the precautions at the beginning of this section.
2. Properly disarm the Supplemental Restraint System (SRS) system.
3. Raise and safely support the vehicle.
4. Remove or disconnect the following:
    - Front wheels
    - Engine side cover and undercovers
5. Drain the transaxle assembly.
6. If equipped, remove the front catalytic converter.
7. Remove or disconnect the following:
    - Exhaust pipe, main muffler and catalytic converter
    - Tie rod end and ball joint from the steering knuckle
    - Support bearing for the left side halfshaft
    - Halfshafts by inserting a prybar between the transaxle case and the driveshaft and prying the shaft from the transaxle
    - Air cleaner assembly and adjoining ductwork
    - Engine harness connection
    - Compressor assembly, if the vehicle is equipped with Active Electronic Controlled Suspension (Active-ECS)—suspend with wire—Do not allow the compressor to hang from the air hose.
    - Roll stopper stay bracket, if equipped

- Speedometer cable from the transaxle
- The clip that secures the shifter
- Shifter control cable from the transaxle
- Plug the oil cooler hoses from the transaxle

8. Disconnect the following:
   - Park/neutral switch electrical harness
   - Kickdown servo switch
   - Pulse generator
   - Oil temperature sensor electrical harness
   - Shift control solenoid valve harness.

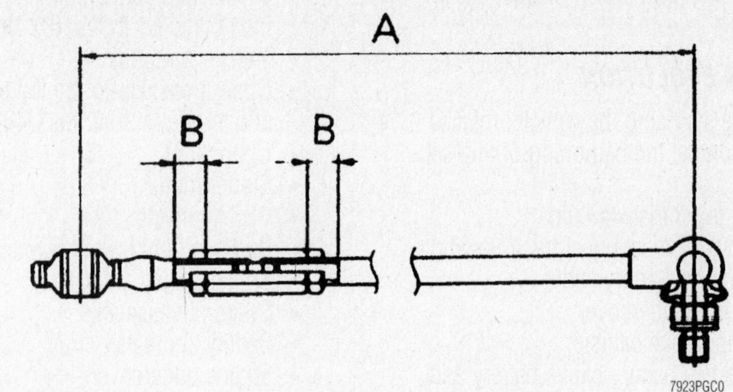

Height sensor rod adjustment—Diamante with a F4A33 automatic transaxle

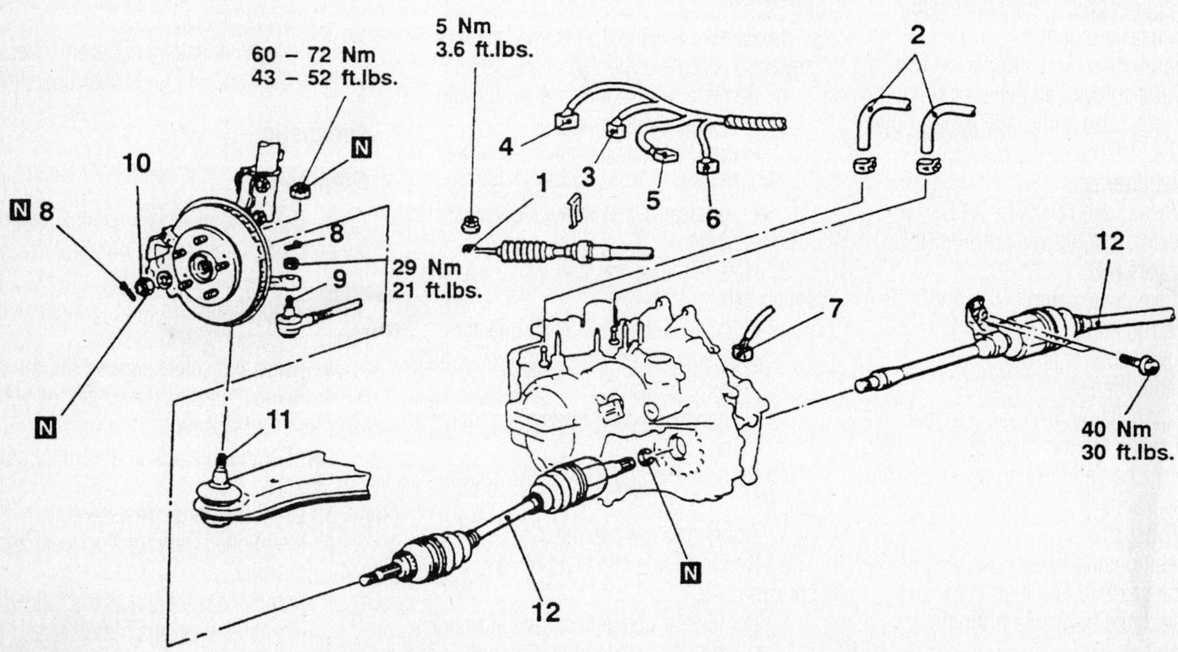

Removal steps

1. Transaxle control cable connection
2. Transaxle oil cooler hoses connection
3. PNP switch connector
4. A/T control solenoid valve connector
5. Input shaft speed sensor connector
6. Output shaft speed sensor connector
7. Vehicle speed sensor connector
8. Split pin
9. Connection of the tie rod end
10. Drive shaft nut
11. Connection for the lower arm ball joint
12. Drive shaft and inner shaft assembly (RH) and the drive shaft (LH)

Caution
Mounting locations marked by * should be provisionally tightened, and then fully tightened when the body is supporting the full weight of the engine.

7923PG84

Transaxle removal—Diamante with F4A51 transaxle 1 of 2

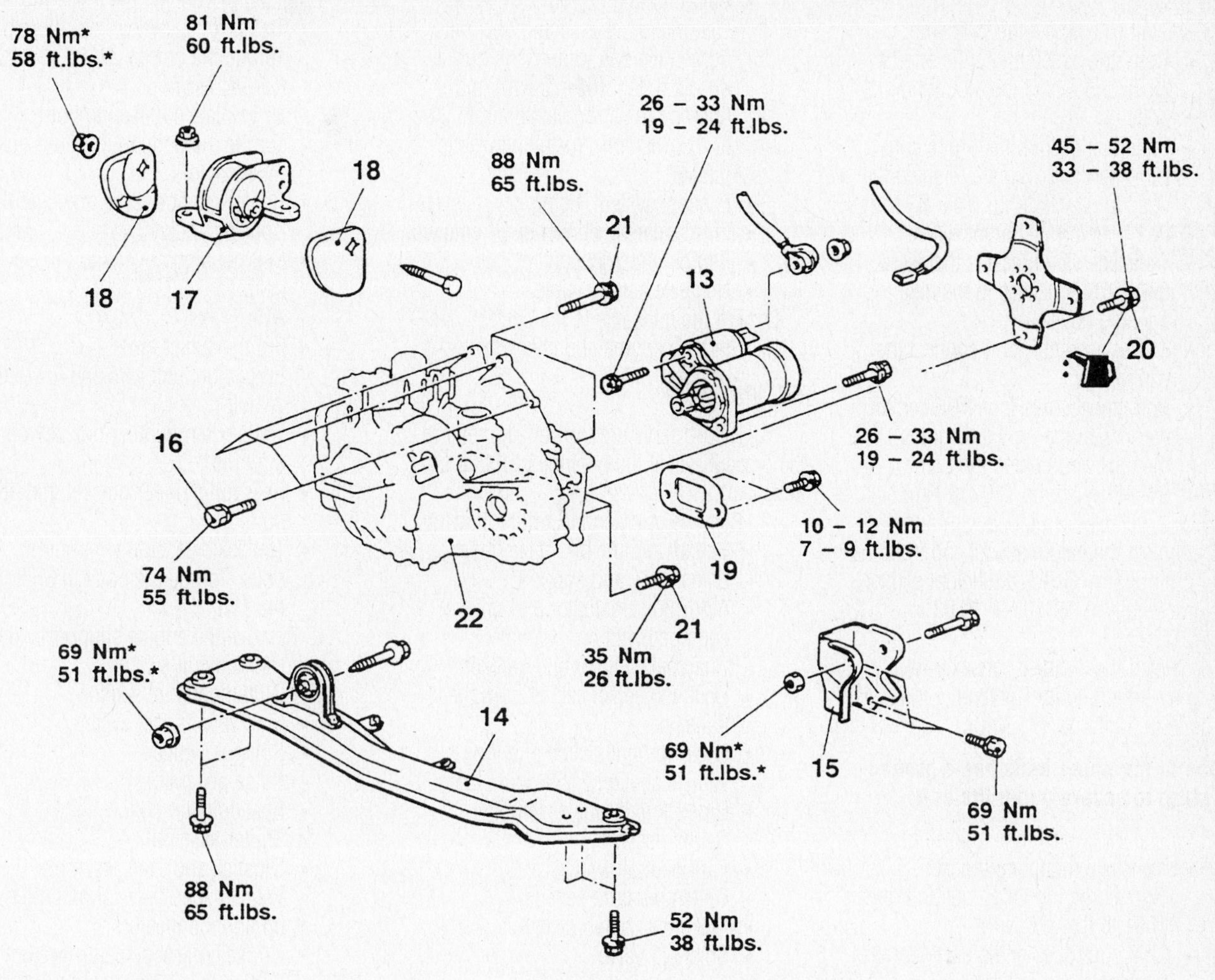

Lifting up of the vehicle

13. Starter motor
14. Center member assembly
15. Rear roll stopper bracket
16. Transaxle upper portion fixing bolt
17. Transaxle mounting bracket
18. Transaxle mount stopper
    - Support the engine and transaxle assembly
19. Bell housing cover
20. Drive plate attaching bolt
21. Transaxle lower portion fixing bolt
22. Transaxle assembly

Caution
Mounting locations marked by * should be provisionally tightened, and then fully tightened when the body is supporting the full weight of the engine.

7923PG85

Transaxle removal—Diamante with F4A51 transaxle 2 of 2

9. Support the transaxle and remove the transaxle mounting bracket.

10. Remove the 3 upper transaxle-to-engine mounting bolts.

11. For vehicles equipped with Active-ECS, disconnect the height sensor rod from the lower control arm.

12. Remove or disconnect the following:
    - Bolt that secures the Heated Oxygen (HO_2S) sensor harness to the right side crossmember
    - Starter assembly
    - Mounting brackets for access to the bell housing cover
    - Bell housing/oil pan covers assembly
    - Bolts holding the flexplate to the torque converter
    - Lower transaxle to engine bolts and remove the transaxle assembly

To install:

13. Install or connect the following:
   - Transaxle assembly to the engine block and install the mounting bolts
   - Bolts that secure the torque converter to the driveplate. Tighten the bolts to 34–38 ft. lbs. (46–53 Nm).
   - Bell housing/oil pan covers
   - Transaxle stay brackets that were removed for access to the bell housing cover
   - Starter assembly and connect the wiring
   - Bolt that secures the HO_2S sensor harness to the right side crossmember and tighten the bolt to 84–108 inch lbs. (10–12 Nm)

14. For vehicles equipped with Active-ECS, connect the height sensor rod from the lower control arm. Check the height sensor rod for a length (A) of 10.59–10.63 in. (269–270mm).

15. Install the 3 upper transaxle-to-engine mounting bolts. Tighten the mounting bolts to 54 ft. lbs. (75 Nm).

➡One of the upper bolts has a grounding strap to secure under the bolt.

16. Install or connect the following:
   - Transaxle mounting bracket. Tighten the mounting nut and bolts to 51 ft. lbs. (70 Nm).
   - Shift control solenoid valve harness
   - Kickdown servo switch, pulse generator and oil temperature sensor electrical harness
   - Park/neutral switch electrical harness
   - Oil cooler hoses to the transaxle, using new hose clamps
   - Shifter control cable to the transaxle and secure the cable with clip
   - Speedometer cable to the transaxle
   - Roll stopper stay bracket and tighten the one through nut and bolt to 36–43 ft. lbs. (50–60 Nm), if removed. Tighten the 2 mounting bolts to 16 ft. lbs. (22 Nm).
   - Active-ECS compressor assembly, if removed. Tighten the mounting bolts to 48 inch lbs. (5 Nm) and connect the electrical harness.
   - Engine harness connection
   - Air cleaner assembly and adjoining ductwork
   - Halfshafts and seat halfshafts into the transaxle, using new circlips
   - Bolt that secure the left side support bearing and tighten the bolts to 33 ft. lbs. (45 Nm)
   - Ball joint and tie rod end to the steering knuckle. Using new nuts, tighten the ball joint castle nut to 43–52 ft. lbs. (60–72 Nm) and tighten the tie rod castle nut to 22 ft. lbs. (30 Nm). Install new cotter pins.
   - Exhaust system, using new gaskets
   - Front catalytic converter, if removed
   - Engine undercovers
   - Negative battery cable

17. Fill the transaxle to the correct level.
18. Start the engine and check for leaks.

ECLIPSE

1. Before servicing the vehicle, refer to the precautions at the beginning of this section.
2. Remove or disconnect the following:
   - Battery and the air intake hoses
   - Battery tray and support
   - Auto-cruise actuator and bracket, if equipped with cruise control
   - Charcoal canister and bracket
   - Shift and select cables from the transaxle
   - Back-up light switch and the vehicle speed sensor connectors
   - Dipstick and tube assembly
   - Starter assembly
   - Park/neutral switch
   - Oil temperature sensor
   - Kick down servo switch
   - Solenoid valve
   - Pulse generator
   - Speedometer connections
3. Attach an engine support fixture to the engine and remove the transaxle mounting bolts.
4. Remove or disconnect the following:
   - Rear roll stopper bracket mounting bolts
   - Transaxle mounting bracket mounting nuts
5. Raise the vehicle and remove the engine undercovers.
6. Remove or disconnect the following:
   - Front exhaust pipe
   - Axle shafts
   - Bell housing cover and the right-hand center member stay (support)
   - Center member
   - Drive plate connecting bolts
7. Place a transmission jack under the transaxle and remove the transaxle mounting bolt.
8. Lower the transaxle.

To install:

9. Raise the transaxle into position and install the transaxle mounting bracket. Torque the through-bolt to 51 ft. lbs. (69 Nm).
10. Install or connect the following:
   - Transaxle assembly mounting bolt. Torque the bolt to 22–25 ft. lbs. (29–34 Nm).
   - Drive plate connecting bolts. Torque the bolts to 33–38 ft. lbs. (45–52 Nm).
   - Center member assembly and the right-hand stay
   - Bell housing cover and the slave cylinder
   - Axle shafts.
   - Front exhaust pipe
   - Engine undercovers and lower the vehicle
   - Transaxle mounting bracket mounting nuts
   - Rear roll stopper bracket mounting bolts
   - Transaxle assembly mounting bolts. Torque the bolts to 35 ft. lbs. (48 Nm).
11. Remove the engine support fixture.
   - Park/neutral switch
   - Oil temperature sensor
   - Kick down servo switch
   - Solenoid valve
   - Pulse generator
   - Speedometer connections
   - Starter assembly
   - Dipstick and tube assembly
   - Vehicle speed sensor and the back-up light connectors
   - Cruise control actuator if removed
   - Battery tray support and the tray
   - Charcoal canister bracket and the canister
   - Air duct and the air cleaner assembly
12. Refill the transaxle and the transfer case, if equipped, with the proper fluid.

GALANT

1. Before servicing the vehicle, refer to the precautions at the beginning of this section.
2. Remove or disconnect the following:
   - Negative battery cable
   - Air cleaner and intake hoses
3. Drain the transaxle into a suitable waste container.
   - Nut securing the shifter lever to the transaxle
   - Cable retaining clip and remove the cable from the transaxle
   - Shifter cable mounting bracket
   - Electrical connectors for the speedometer, neutral safety switch (inhibitor switch), the pulse generator, kickdown servo switch, and the oil temperature sensor
   - Oil cooler lines, at the transaxle

- Dipstick and tube from the transaxle
- Starter motor and position it aside

4. Support the engine assembly.
   - Rear roll stopper mounting bracket
   - Transaxle mount bracket
   - Upper transaxle mounting bolts
5. Raise and safely support the vehicle.
6. Remove or disconnect the following:
   - Front wheel assemblies
   - Right-hand undercover
   - Tie rod end from the steering knuckle
   - Stabilizer bar link from the damper fork
   - Damper fork from the lateral lower control arm
   - Later lower arm and the compression arm lower ball joints, from the steering knuckle
   - Halfshafts from the transaxle and secure aside
   - Cover from the transaxle bell housing
   - Engine front roll stopper through-bolt
   - Crossmember and the triangular right-hand stay
   - Bolts holding the flexplate to the torque converter
7. Support the transaxle using a transmission jack, and remove the transaxle lower coupling bolt.

➡The coupling bolt threads from the engine side into the transaxle and is located just above the halfshaft opening.

8. Slide the transaxle rearward and carefully lower it from the vehicle.

To install:

9. After the torque converter has been mounted on the transaxle, install the transaxle assembly to the engine. Install the mounting bolts and tighten to 35 ft. lbs. (48 Nm). Install the transaxle lower coupling bolt and tighten to 21–25 ft. lbs. (29–34 Nm).
10. Install or connect the following:
    - Torque converter to the flexplate and tighten the bolts to 33–38 ft. lbs. (45–52 Nm)
    - Cover to the transaxle bell housing and tighten the mounting bolts to 84 inch lbs. (9 Nm)
    - Crossmember and tighten the front mounting bolts to 65 ft. lbs. (88 Nm) and the rear bolt to 54 ft. lbs. (73 Nm)
    - Front engine roll stopper through-bolt and lightly tighten. Once the full weight of the engine is on the mounts, tighten the bolt to 42 ft. lbs. (57 Nm).
    - Triangular stay bracket and tighten the mounting bolts to 65 ft. lbs. (88 Nm)
    - Halfshafts, using new circlips on the axle ends
    - Tie rod and ball joints to the steering knuckle. Tighten the ball joint self-locking nuts to 48 ft. lbs. (65 Nm). Tighten the tie rod end nut to 21 ft. lbs. (28 Nm) and secure with a new cotter pin.
    - Damper fork to the lower control arm and tighten the through-bolt to 65 ft. lbs. (88 Nm)
    - Stabilizer link to the damper fork, and tighten the self-locking nut to 29 ft. lbs. (39 Nm)
    - Underpan
    - Wheels and lower the vehicle
    - Transaxle mount bracket to the transaxle, and tighten the mounting nuts to 32 ft. lbs. (43 Nm)
    - Rear roll stopper mounting bracket
    - Engine support. Tighten the transaxle mount through-bolt to 51 ft. lbs. (69 Nm) and tighten the front engine roll stopper through-bolt.
    - Upper transaxle mounting bolts and tighten to 35 ft. lbs. (48 Nm)
    - Starter motor
    - Dipstick tube and the dipstick
    - Shifter cable mounting bracket
    - Shifter lever and tighten the retaining nut to 14 ft. lbs. (19 Nm)
    - Oil cooler lines and secure with clamps
    - Electrical connectors for the speedometer, neutral safety switch (inhibitor switch), pulse generator, kickdown servo switch and oil temperature sensor
    - Air cleaner and the air intake hose
    - Negative battery cable
11. Fill the transaxle to the correct level.

LANCER & SPORTBACK

1. Before servicing the vehicle, refer to the precautions at the beginning of this section.
2. Install or connect the following:
   - Negative battery cable
   - Engine undercover
3. Drain the transaxle oil and engine coolant.
   - Front exhaust pipe
   - Battery and battery tray
   - Air cleaner assembly
   - Transaxle control cable
   - Vehicle Speed Sensor (VSS) connector
   - Input and output shaft speed sensor connectors
   - Inhibitor switch sensor connector
   - A/C control solenoid valve connector
   - Starter
   - Transaxle oil cooler line
   - Upper engine-to-transaxle bolts
   - Transaxle mount
   - Transaxle mount stopper
4. Install a suitable engine support assembly, then raise and safely support the vehicle.
   - Wheel Speed Sensor (WSS)
   - Brake hose clamp
   - Stabilizer bar
   - Lower control arm
   - Tie rod end
   - Lower control arm
   - Halfshafts by inserting a prybar between the transaxle case and the driveshaft and prying the shaft from the transaxle. Do not pull on the driveshaft.
   - Centermember
   - Bell housing lower cover
   - Transaxle to engine bolts and lower the transaxle from the vehicle
   - Front roll stopper installation bolt
   - Bell housing cover
5. Support the transaxle with a suitable jack.
   - Driveplate bolts
   - Lower transaxle-to-engine mounting bolts
   - Transaxle from the vehicle

To install:

6. Installation is the reverse of the removal procedure, noting the following:
   - Driveplate bolts: 36 ft. lbs. (49 Nm)
   - Centermember bolts: 51 ft. lbs. (69 Nm)
   - Hub nut: 181 ft. lbs. (245 Nm)
   - Transaxle mount bracket nuts: 36 ft. lbs. (49 Nm)
   - Transaxle mount stopper nuts: 60 ft. lbs. (81 Nm)
   - Transaxle-to-engine upper mounting bolts: 36 ft. lbs. (49 Nm)
7. Fill the transaxle and the engine cooling system to the correct level.
8. Check and adjust the front wheel alignment.
9. Check the speedometer and gear selector for proper operation.
10. Start the engine and check for leaks.

MIRAGE

1. Before servicing the vehicle, refer to the precautions at the beginning of this section.

2. Remove or disconnect the following:
   - Negative battery cable
   - Battery and battery tray
   - Air hose and air cleaner assembly
   - Under guard pan
3. Drain the transaxle oil.
   - Control cable and cooler lines
   - Shift control solenoid valve connector
   - Inhibitor switch, kickdown servo switch, the pulse generator and oil temperature sensor, if equipped
   - Speedometer cable and remove the starter
   - Transaxle mounting bolts and bracket
   - Stabilizer bar from the lower control arm
   - Steering tie rod end and the ball joint from the steering arm
   - Halfshafts at the inboard side from the transaxle. Tie the joint assembly aside.
4. Support the engine and remove the center member.
   - Bell housing cover and remove the driveplate bolts
   - Transaxle assembly lower connecting bolt, located just over the halfshaft opening
   - Transaxle

To install:

5. Install or connect the following:
   - Transaxle assembly on the engine. Tighten the driveplate bolts to 33–38 ft. lbs. (46–53 Nm).
   - Bell housing cover
   - Center member
   - Halfshafts to the transaxle, using new circlips
   - Tie rods, ball joints and stabilizer links to the steering arm
   - Transaxle mounting bracket and bolts
   - Starter
   - Speedometer cable
   - Inhibitor switch, kickdown servo switch, the pulse generator and oil temperature sensor, if disconnected
   - Shift control solenoid valve connector
   - Control cables and oil cooler lines
   - Air cleaner assembly
   - Battery tray and battery. Connect the positive, then the negative terminal.
6. Fill the transaxle to the correct level.
7. Start the engine and check for leaks.

Clutch

ADJUSTMENT

➡The following adjustment is for cable actuated clutch systems. Hydraulic systems are self-adjusting.

Mirage

1. Before servicing the vehicle, refer to the precautions at the beginning of this section.
2. Measure the clutch pedal height (measurement A). The specification is 6.38–6.50 in. (162–165mm).

➡The clutch pedal height is not adjustable. If not within specifications, part replacement is required.

3. Depress clutch pedal several times and check the pedal free-play (measurement B).
4. If measurement is not 0.67–0.87 in. (17–22mm), adjustment is required.
5. To adjust turn the outer cable adjusting nut, located at the firewall, until free-play is within range.
6. Depress clutch pedal several times and recheck measurement.

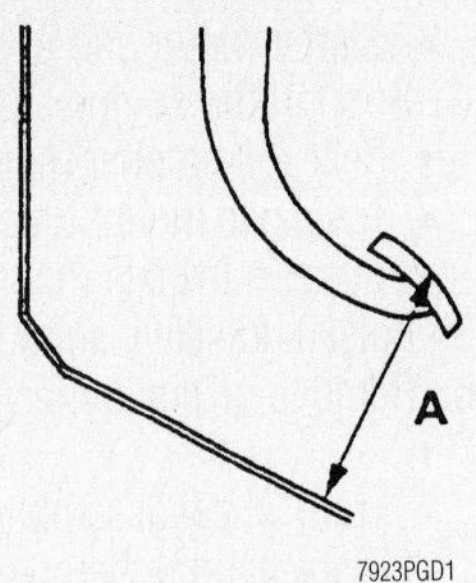

Clutch pedal height (A) measurement—Mirage

REMOVAL & INSTALLATION

Eclipse and Mirage

1. Before servicing the vehicle, refer to the precautions at the beginning of this section.
2. Remove or disconnect the following:
   - Negative battery cable
   - Transaxle assembly from the vehicle
   - Pressure plate attaching bolts, pressure plate and clutch disc. If the pressure plate is to be reused,

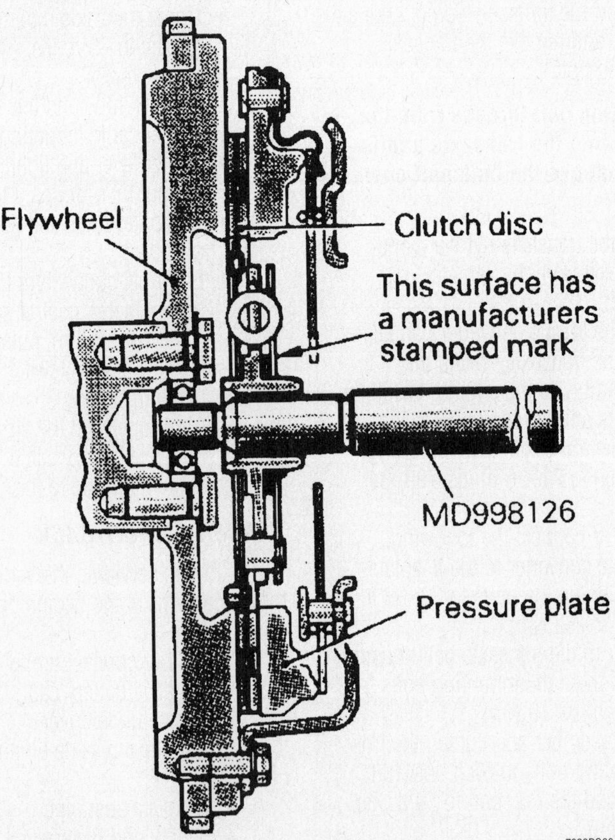

Use the alignment dowel to center the disc on the flywheel—Mirage

loosen the bolts in a diagonal pattern, 1 or 2 turns at a time. This will prevent warping the clutch cover assembly.
- Return clip and the pressure plate release bearing. Do not use solvent to clean the bearing.

3. Inspect the clutch release fork and fulcrum for damage or wear. If necessary, remove the release fork and unthread the fulcrum from the transaxle.

4. Carefully inspect the condition of the clutch components and replace any worn or damaged parts.

To install:

5. Inspect the flywheel for heat damage or cracks. Resurface or replace the flywheel as required.

6. Install the fulcrum and tighten to 25 ft. lbs. (35 Nm). Install the release fork. Apply a coating of multi-purpose grease to the point of contact with the fulcrum and the point of contact with the release bearing. Apply a coating of multi-purpose grease to the end of the release cylinder pushrod and the pushrod hole in the release fork.

7. Apply multi-purpose grease to the clutch release bearing. Pack the bearing inner surface and the groove with grease. Do not apply grease to the resin portion of the bearing. Place the bearing in position and install the return clip.

8. Using the proper alignment tool, install the clutch disc to the flywheel. Install the pressure plate assembly. Install the retainer bolts and tighten a little at a time, in a diagonal sequence. Tighten them to a final torque of 16 ft. lbs. (22 Nm). Remove the aligning tool.

9. Install the transaxle assembly.

10. Check for proper clutch operation.

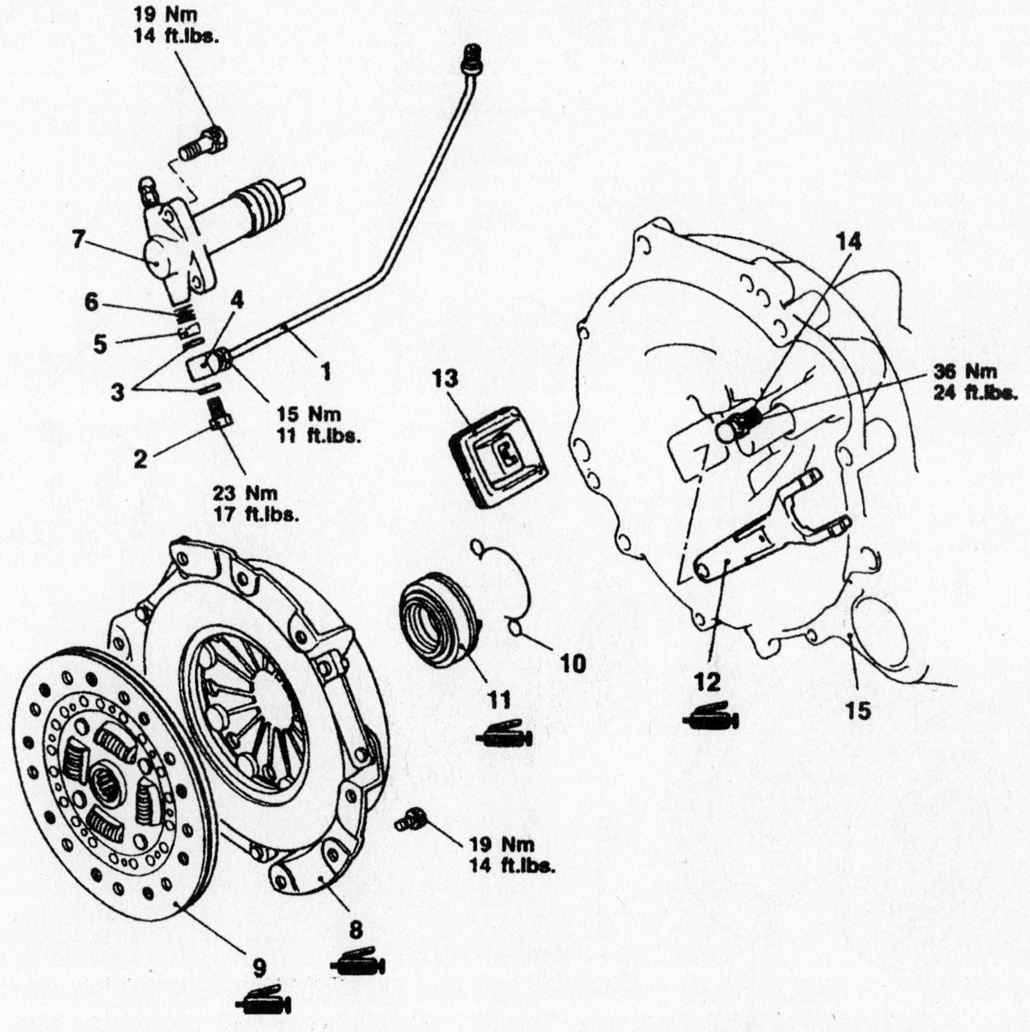

Exploded view of clutch assembly—Eclipse shown

Lancer and Evolution

1. Before servicing the vehicle, refer to the precautions at the beginning of this section.
2. Remove or disconnect the following:
   - Negative battery cable
   - Transaxle assembly from the vehicle
   - Clutch fluid line bracket, insulator and washer
   - Clutch fluid line
   - Clutch slave (release) cylinder
   - Boot
   - Clutch cover (pressure plate) attaching bolts, cover plate and clutch disc. If the pressure plate is to be reused, loosen the bolts in a diagonal pattern, 1 or 2 turns at a time. This will prevent warping the clutch cover assembly.
3. Carefully inspect the condition of the clutch components and replace any worn or damaged parts.

To install:

4. Inspect the flywheel for heat damage or cracks. Resurface or replace the flywheel as required.
5. Apply multi-purpose grease to the clutch release bearing. Pack the bearing inner surface and the groove with grease. Do not apply grease to the resin portion of the bearing. Place the bearing in position and install the return clip.
6. Using the proper alignment tool, install the clutch disc to the flywheel. Install the clutch cover (pressure plate) assembly. Install the retainer bolts and tighten a little at a time, in a diagonal sequence. Tighten them to a final torque of 14 ft. lbs. (19 Nm). Remove the aligning tool.

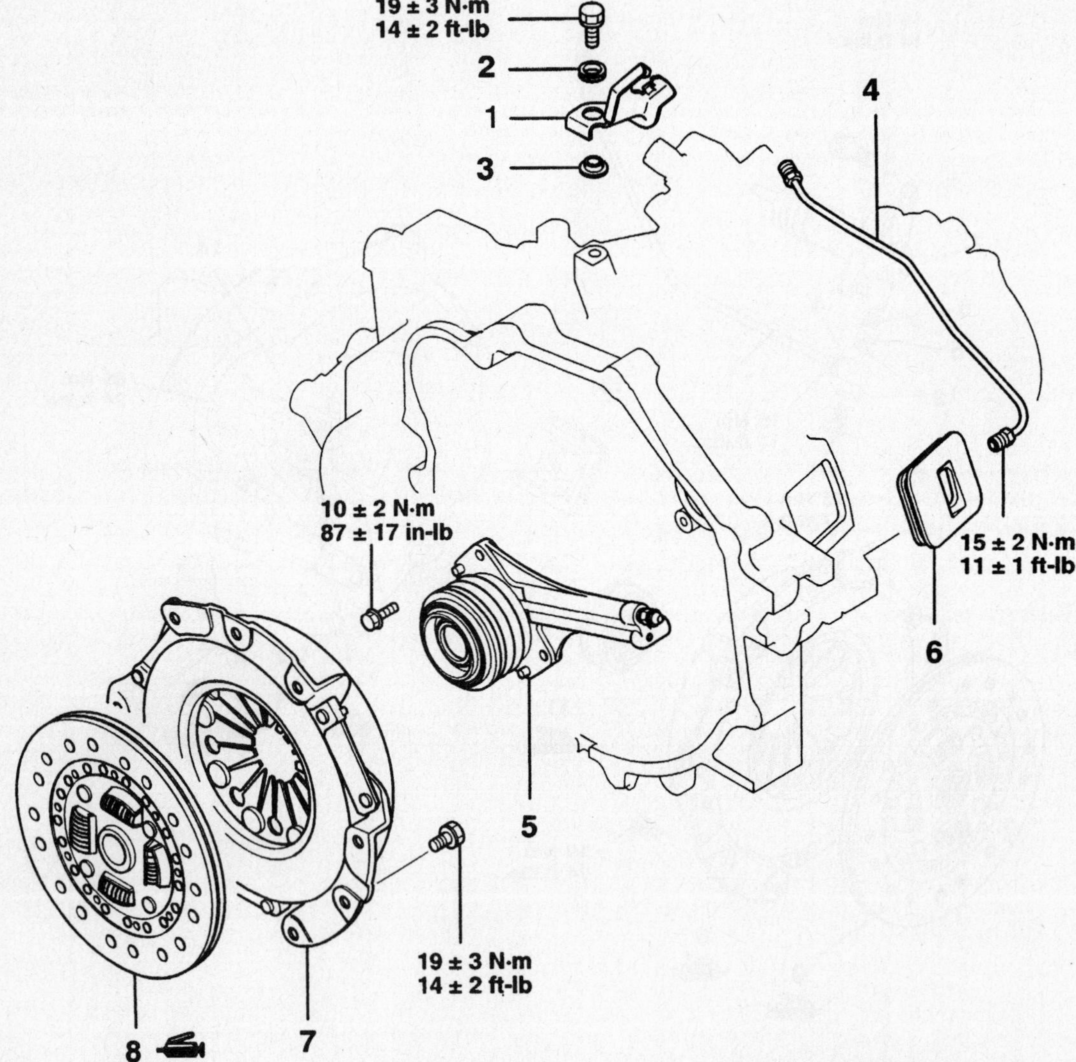

Exploded view of the clutch components—Lancer shown

7. Install or connect the following:
   - Boot
   - Clutch line, washer, insulator and bracket
   - Transaxle assembly
8. Check for proper clutch operation.

Hydraulic Clutch System

BLEEDING

1. Before servicing the vehicle, refer to the precautions at the beginning of this section.
2. Fill the reservoir with clean brake fluid meeting DOT 3 specifications.
3. Press the clutch pedal to the floor, then open the bleeder screw on the slave cylinder.
4. Tighten the bleed screw and release the clutch pedal.
5. Repeat the procedure until the fluid is free of air bubbles.

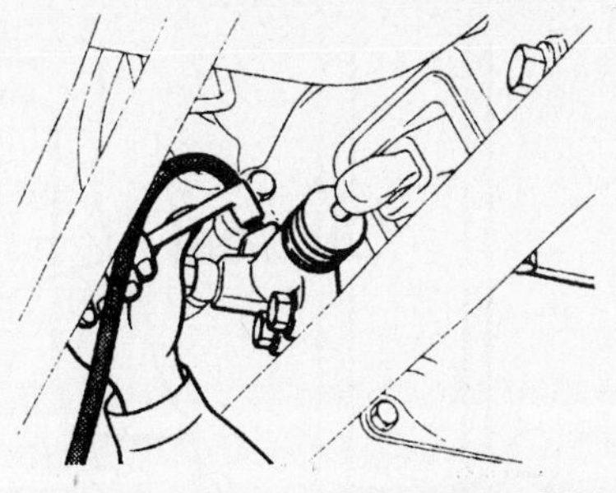

Bleeding a typical clutch hydraulic system

Transfer Case Assembly

REMOVAL & INSTALLATION

1. Before servicing the vehicle, refer to the precautions at the beginning of this section.
2. Remove or disconnect the following:
   - Engine undercovers
   - Front exhaust pipe
   - Transfer case mounting bolts
3. Support the driveshaft with wire or string and remove the transfer case from the transaxle.

To install:

4. Slide the driveshaft into the transfer case and install the transfer case to the transaxle. Torque the bolts to 40–44 ft. lbs. (54–59 Nm).
5. Install or connect the following:
   - Front exhaust pipe
   - Engine undercover

Halfshaft

REMOVAL & INSTALLATION

Front

DIAMANTE, GALANT, LANCER, EVOLUTION, SPORTBACK & MIRAGE

1. Before servicing the vehicle, refer to the precautions at the beginning of this section.

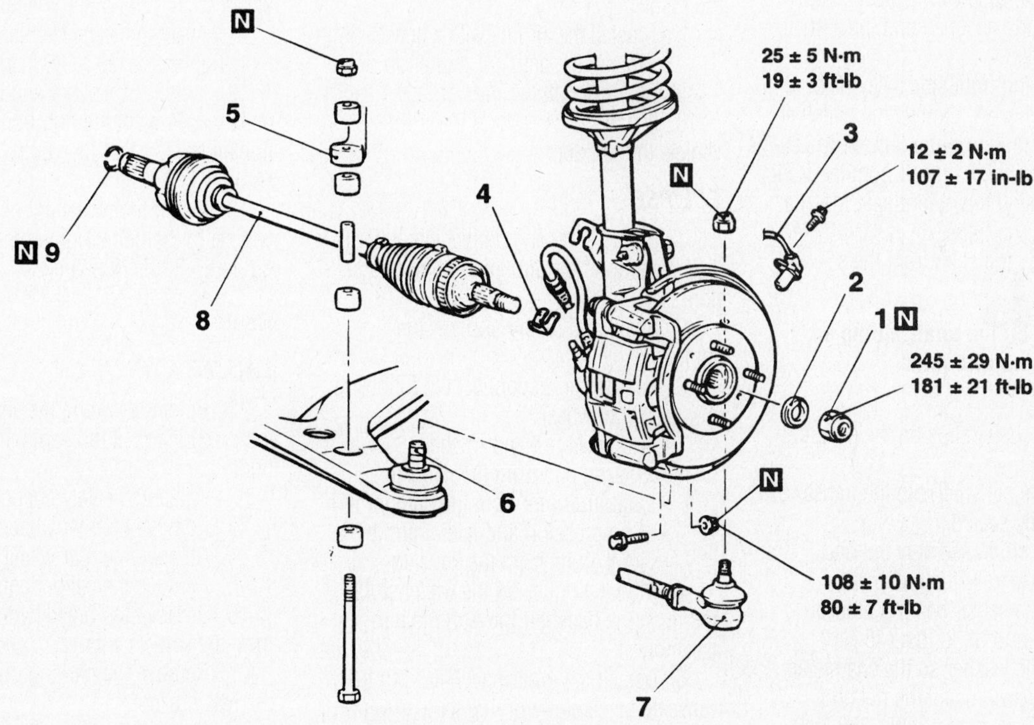

Exploded view of the halfshaft mounting—Lancer shown, others similar

2. Raise the vehicle and support it safely.

3. Remove the cotter pin, halfshaft nut and washer.

4. If equipped with Anti-Lock Brake (ABS), remove the front wheel speed sensor.

5. If equipped with Active Electronic Control Suspension (Active-ECS) perform the following:

a. Loosen the nut that secures the air line to the to the top of the strut and discard the O-ring.

b. Remove the bolts that secure the actuator to the top of the strut and remove the component. Disconnect the wiring harness.

6. Disconnect the lower ball joint and the tie rod end from the steering knuckle.

7. If removing the left side axle with an inner shaft, remove the center support bearing bracket bolts and washers. Then, remove the halfshaft by setting up a puller on the outside wheel hub and pushing the halfshaft from the front hub. Tap the shaft union at the joint case with a plastic hammer to remove the halfshaft and inner shaft from the transaxle.

8. If removing right side axle shafts without an inner shaft, remove the halfshaft by setting up a puller on the outside wheel hub and pushing the halfshaft from the front hub. After pressing the outer shaft, insert a prybar between the transaxle case and the halfshaft and pry the shaft from the transaxle.

➡Do not pull on the shaft; doing so damages the inboard joint.

To install:

9. Replace the circlips on the ends of the halfshafts.

10. Insert the halfshaft into the transaxle. Be sure it is fully seated.

11. Pull the strut assembly out and install the other end to the hub.

12. Install the center bearing bracket bolts and tighten to 33 ft. lbs. (45 Nm).

13. Install the washer so the chamfered edge faces outward. Install the nut and tighten to 145–188 ft. lbs. (200–260 Nm), for all models except Lancer. For Lancer, tighten the nut to 181 ft. lbs. (245 Nm). Secure with a new cotter pin.

14. Connect the ball joint to the steering knuckle. Torque the new retaining nut to 43–52 ft. lbs. (60–72 Nm) and secure with a new cotter pin.

15. Connect the tie rod end to the steering knuckle. Torque the retaining nut to 21 ft. lbs. (29 Nm) and secure with a new cotter pin.

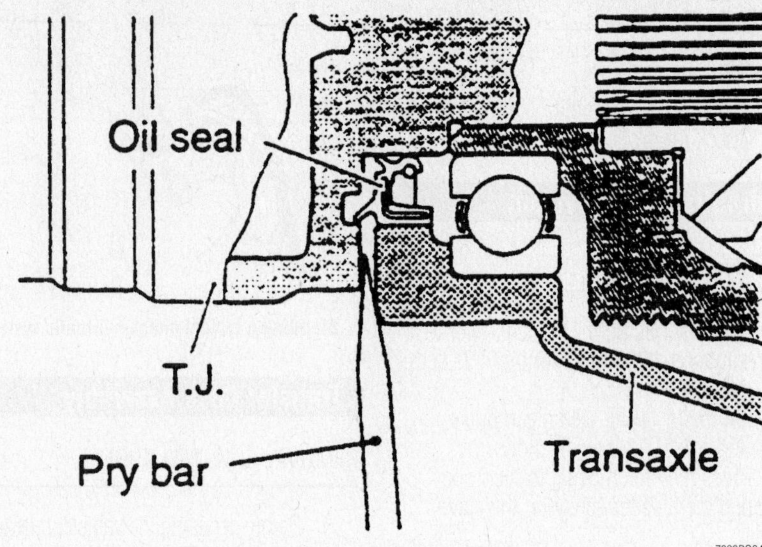

Proper method for removing the inner halfshaft from the transaxle or differential

16. If equipped with ABS, install the front wheel speed sensor.

17. If equipped with Active-ECS, perform the following:

a. Install the air line with a new O-ring.

b. Install the actuator to the top of the strut. Connect the wiring harness.

18. Install the wheel and lower the vehicle to the floor.

ECLIPSE

1. Before servicing the vehicle, refer to the precautions at the beginning of this section.

2. Raise and safely support the vehicle.

3. Remove or disconnect the following:

- Front wheel
- Halfshaft nut and washer
- Tie rod end from the knuckle
- Stabilizer link from the damper fork
- Compression and lateral arm ball joint studs from the knuckle

4. Mount a puller on the wheel studs and push the halfshaft through the hub assembly.

5. Detach the inner halfshaft from the transaxle by carefully prying the CV-joint housing out.

6. Pull the knuckle assembly outward and remove the halfshaft.

To install:

7. Place a new circlip on the inner halfshaft and install the halfshaft in the transaxle.

8. Push out on the knuckle assembly and install the halfshaft through the hub.

9. Using new nuts, install the lateral and compression arm ball joint studs in the knuckle. Tighten the nuts to 43–52 ft. lbs. (59–71 Nm). Install new cotter pins.

10. Install the damper fork on the knuckle. Do nut tighten the nut at this time.

11. Attach the stabilizer link to the damper fork. Tighten the nut to 29 ft. lbs. (39 Nm).

12. Install the washer and nut on the halfshaft. Prevent the hub from turning and tighten the nut to 145–188 ft. lbs. (196–255 Nm).

13. Install the wheel and lower the vehicle to the floor. Tighten the damper fork nut to 65 ft. lbs. (88 Nm).

Rear

LANCER EVOLUTION

1. Before servicing the vehicle, refer to the precautions at the beginning of this section.

2. Raise and safely support the vehicle.

3. Remove the rear wheel.

4. Remove the rear wheel speed sensor.

5. Remove the caliper and rotor.

6. Remove the brake hydraulic line from the wheel cylinder.

7. Remove the parking brake cable from the rear brakes.

8. Remove the lower end of the shock absorber from the knuckle.

9. Remove the trailing and lower arms from the knuckle.

10. Remove the toe control arm ball joint from the knuckle.

11. Prevent the hub assembly from turning by using a tool such as MB990767 and remove the halfshaft nut and washer.

12. Remove the differential mount support.

✲✲ WARNING

Do not pull on the halfshaft to remove it from the differential. Damage to the inner CV-joint will occur.

13. Push the lower part of the knuckle outward and pry the inner halfshaft out of the differential.
14. Push the outer end of the halfshaft through the hub/knuckle and remove it.

To install:

15. Install the outer end of the halfshaft through the hub/knuckle.
16. Place a new circlip on the inner halfshaft and install the halfshaft in the differential.
17. Install the differential mount support.
18. Install the washer and a new nut on the end of the halfshaft. Tighten the nut to 145–188 ft. lbs. (196–255 Nm).
19. Install the toe control arm to the knuckle. Tighten the new nut to 20 ft. lbs. (28 Nm).
20. Install the lower and trailing arms to the knuckle. Do not tighten the fasteners at this time.
21. Install the shock absorber. Tighten the bolt to 71 ft. lbs. (98 Nm).
22. Assemble the brake components.
23. Install the rear wheel speed sensor.
24. Install the rear wheel and lower the vehicle to the floor. Tighten the lower arm nut to 71 ft. lbs. (98 Nm) and the trailing arm nut to 85–99 ft. lbs. (118–137 Nm).

STEERING AND SUSPENSION

Air Bag

✲✲ CAUTION

All vehicles are equipped with an air bag system. The system must be disabled before performing service on or around system components, steering column, instrument panel components, wiring and sensors. Failure to follow safety and disabling procedures could result in accidental air bag deployment, possible personal injury and unnecessary system repairs.

PRECAUTIONS

Several precautions must be observed when handling the inflator module to avoid accidental deployment and possible personal injury.

- Never carry the inflator module by the wires or connector on the underside of the module.
- When carrying a live inflator module, hold securely with both hands, and ensure that the bag and trim cover are pointed away.
- Place the inflator module on a bench or other surface with the bag and trim cover facing up.
- With the inflator module on the bench, never place anything on or close to the module which may be thrown in the event of an accidental deployment.

DISARMING

1. Before servicing the vehicle, refer to the precautions at the beginning of this section.
2. Position the front wheels in the straight-ahead position and place the key in the **LOCK** position. Remove the key from the ignition lock cylinder.
3. Disconnect the negative battery cable and insulate the cable end with high-quality electrical tape or similar non-conductive wrapping.
4. Wait at least 1 minute before working on the vehicle. The air bag system is designed to retain enough voltage to deploy the air bag for a short period of time after the battery has been disconnected.

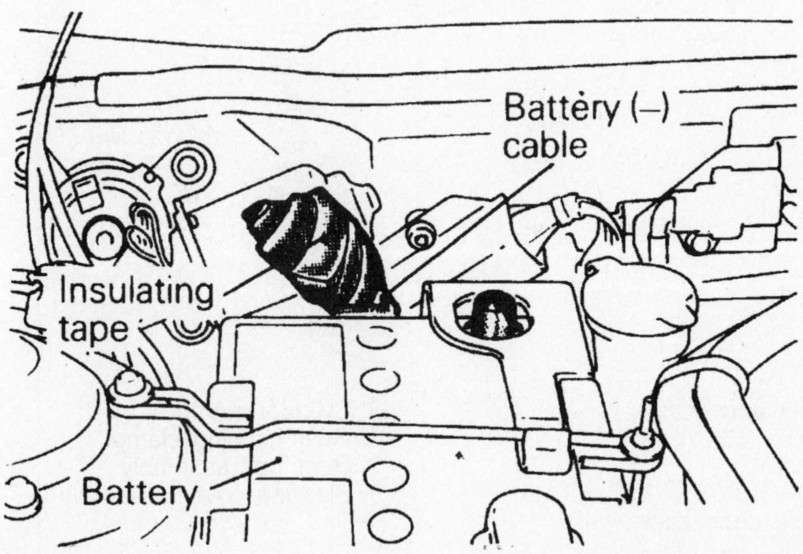

Insulate the negative battery cable to prevent accidental deployment of the air bag.

REARMING

1. Connect the negative battery cable, turn the ignition switch to the **ON** position and check the Supplemental Restraint (SRS) warning light for proper operation.

Rack and Pinion Steering Gear

REMOVAL & INSTALLATION

Manual

MIRAGE

➡Prior to removal of the steering rack, center the front wheels and remove the ignition key. Failure to do so may damage the SRS clockspring and render SRS system inoperative. Be sure to properly disarm the air bag system.

1. Before servicing the vehicle, refer to the precautions at the beginning of this section.
2. Remove or disconnect the following:
   - Battery negative cable
   - Wheels
   - Heated Oxygen (HO_2S) sensor and remove the front exhaust pipe
3. Properly support the engine. Remove both roll stopper mounting bolts and the 4 center member installation bolts.
   - Center member

➡Matchmark the pinion input shaft of the rack to the lower steering column joint for installation purposes.

4. Remove or disconnect the following:
   - Pinch bolt holding the lower steering column joint to the rack and pinion input shaft
   - Cotter pins and disconnect the tie rod ends from the steering knuckle
   - Rack and pinion steering assembly and its rubber mounts from the right side of the vehicle

To install:

5. Align the matchmarks of the input shaft and install the rack to the vehicle.
6. Secure the rack using the retainer clamps and bolts. Torque the bolts to 51 ft. lbs. (70 Nm).
7. Torque the steering column pinch bolt to 13 ft. lbs. (18 Nm).
8. Install or connect the following:
   - Center member
   - Front exhaust pipe
   - HO_2S sensor
   - Tie rod ends to the steering knuckles and tighten the castle nuts to 25 ft. lbs. (34 Nm). Install new cotter pins.
   - Wheels and connect the negative battery cable
9. Perform a front end alignment.

Power

DIAMANTE

➡Prior to removal of the steering gear box, center the front wheels and remove the ignition key. Failure to do so may damage the Supplemental Restraint System (SRS) clock spring and render SRS system inoperative.

1. Before servicing the vehicle, refer to the precautions at the beginning of this section.
2. Remove or disconnect the following:
   - Negative battery cable.
   - Front exhaust pipe
   - Transfer case assembly, if equipped with All Wheel Drive (AWD)
   - Bolt holding the lower steering column joint to the rack and pinion input shaft
   - Tie rod ends
   - Left and right frame members
   - Stabilizer bar bracket
   - Lines going to the rear pump, if equipped with Four Wheel Steering (4WS)
   - Rack and pinion steering assembly and its rubber mounts. Move the rack to the right to remove it from the crossmember.

To install:

3. Install or connect the following:
   - Rack and install the mounting bolts, tightening bolts to 51 ft. lbs. (70 Nm). When installing the rubber rack mounts, align the projection of the mounting rubber with

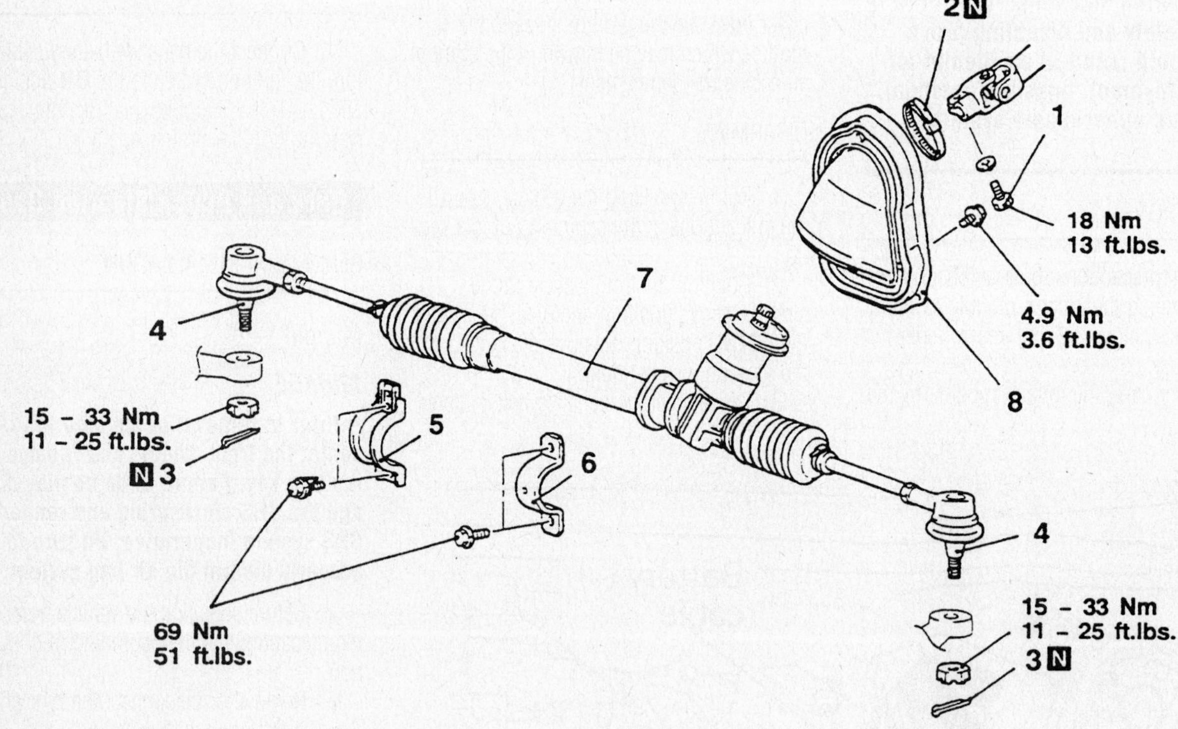

Exploded view of the manual steering gear mounting—Mirage

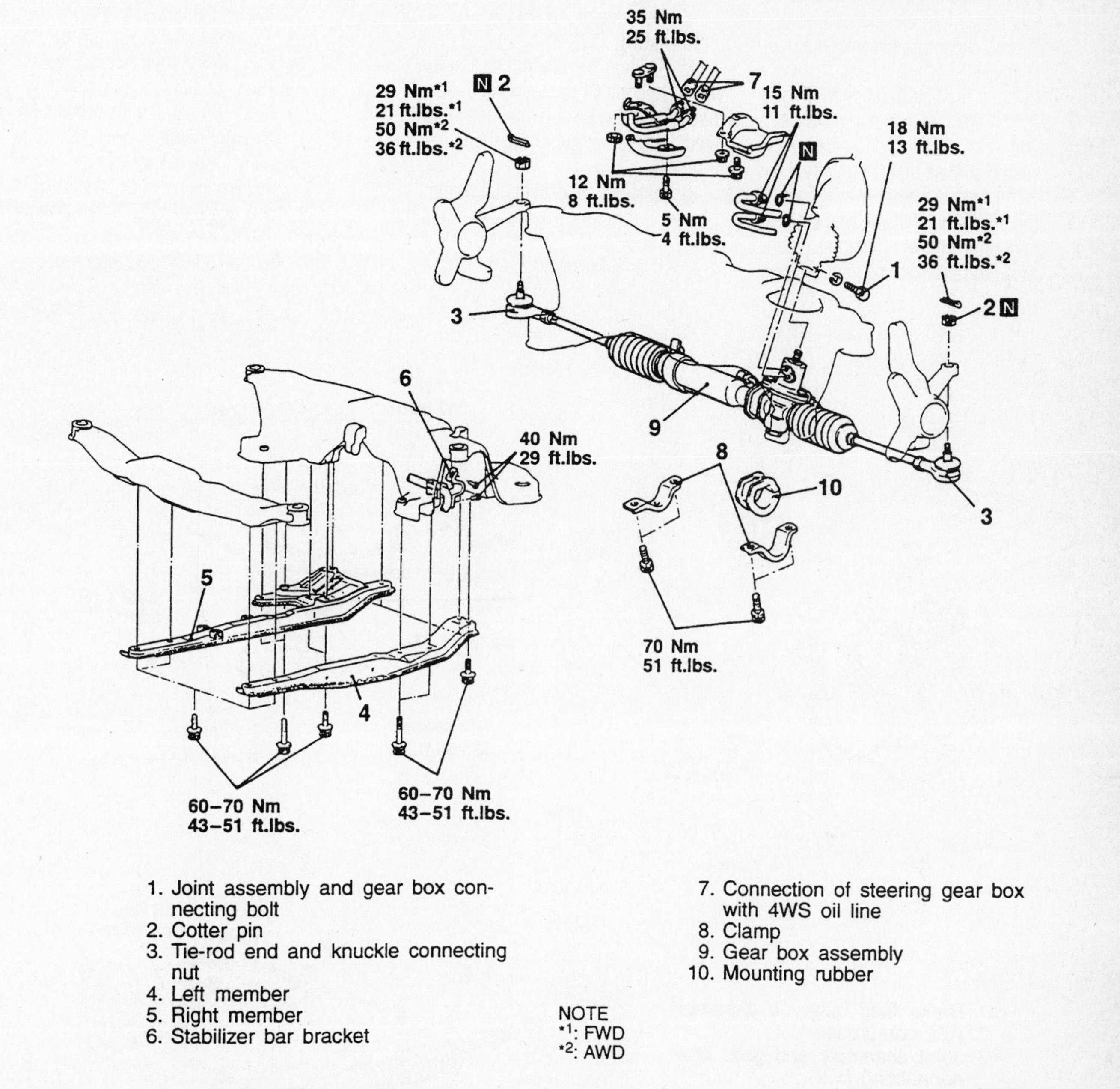

1. Joint assembly and gear box connecting bolt
2. Cotter pin
3. Tie-rod end and knuckle connecting nut
4. Left member
5. Right member
6. Stabilizer bar bracket
7. Connection of steering gear box with 4WS oil line
8. Clamp
9. Gear box assembly
10. Mounting rubber

NOTE
*1: FWD
*2: AWD

7923PGA5

Exploded view of the power steering gear removal—Diamante

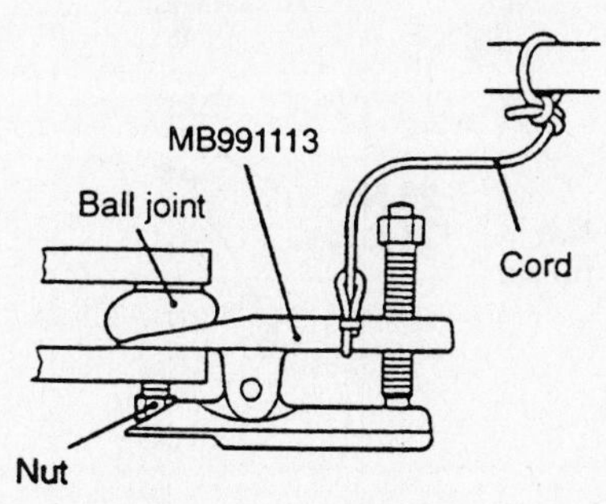

Proper tie-rod end removal method

the indentation in the crossmember. Install the pinch bolt.

- Pressure and return lines to the rack and to the rear pump, if equipped
- Frame members and tighten the bolts to 43–51 ft. lbs. (60–70 Nm)
- Tie rods and install new cotter pins
- Transfer case and front exhaust pipe

4. Refill the reservoir and bleed the system.
5. Perform a front end alignment.

ECLIPSE

➡Prior to removal of the steering gear box, center the front wheels and remove the ignition key. Failure to do so may damage the SRS clock spring and render SRS system inoperative.

1. Before servicing the vehicle, refer to the precautions at the beginning of this section.
2. Install or connect the following:
   - Negative battery cable
3. Drain the power steering fluid.
   - Stabilizer bar
   - Windshield washer reservoir

- Pinch bolt from the joint assembly
- Fluid lines from the steering rack
- Tie rod ends from the steering knuckles
- Left and right stays (supports)

4. Support the engine and remove the center member.

- Clamp and the mounting bolts
- Left lower compression arm from the body side of the vehicle and support it with wire or string
- Steering rack from the joint assembly and remove the rack from the left side of the vehicle

To install:

5. Position the steering rack in the vehicle and install the clamp and the mounting bolts. Be sure the rack is centered before connecting it to the joint assembly.
6. Install or connect the following:

- Left lower compression arm to the body
- Center member
- Left and right stays and remove the engine support fixture or jack
- Tie rods to the steering knuckles
- Fluid lines to the steering rack
- Pinch bolt in the joint assembly
- Stabilizer bar and the windshield washer reservoir

7. Safely lower the vehicle.
8. Connect the negative battery cable.
9. Refill and bleed the power steering system.
10. Perform a front end alignment.

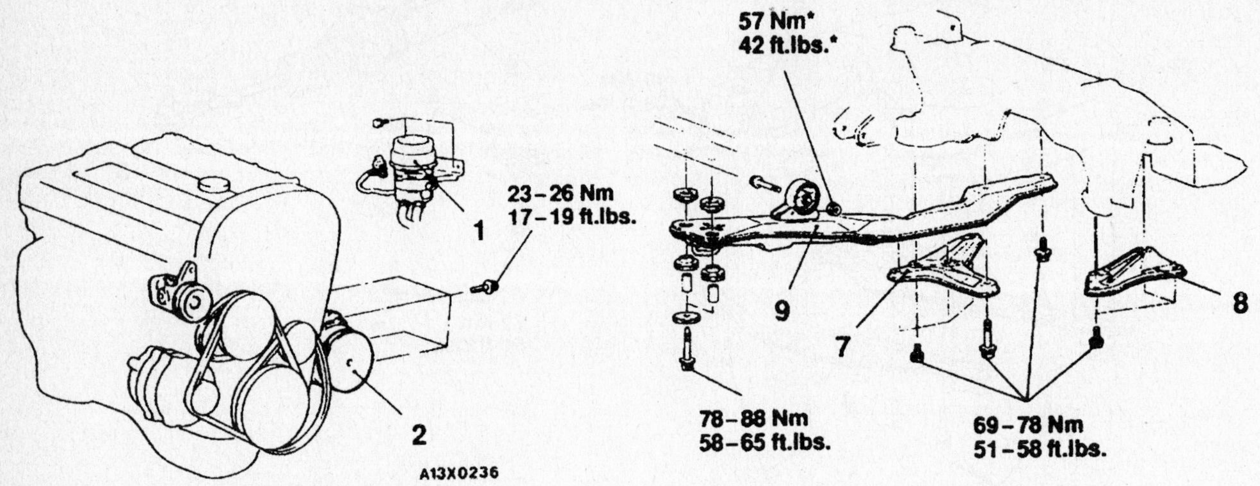

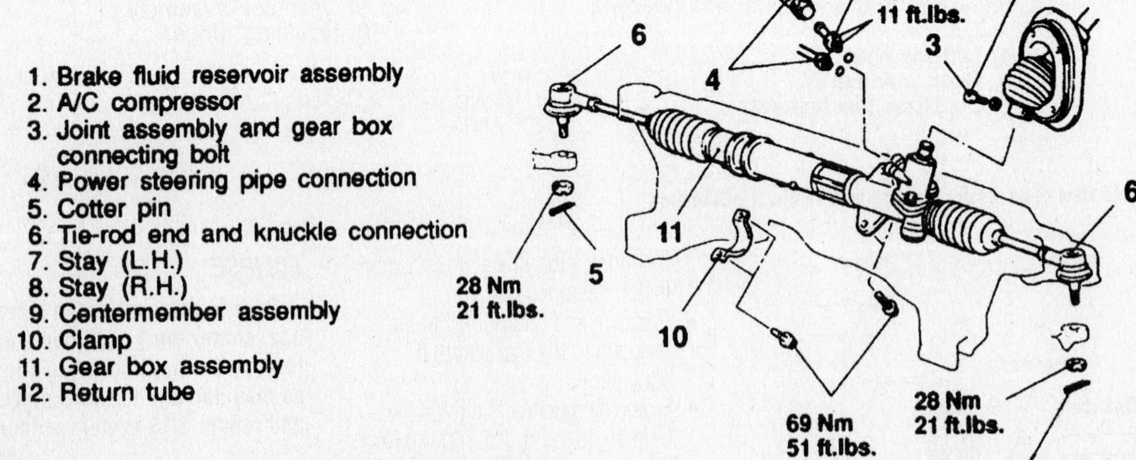

1. Brake fluid reservoir assembly
2. A/C compressor
3. Joint assembly and gear box connecting bolt
4. Power steering pipe connection
5. Cotter pin
6. Tie-rod end and knuckle connection
7. Stay (L.H.)
8. Stay (R.H.)
9. Centermember assembly
10. Clamp
11. Gear box assembly
12. Return tube

NOTE
The fasteners marked * should be temporarily tightened before they are finally tightened once the total weight of the engine has been placed on the vehicle body.

7923PGA2

Power steering rack assembly and related components—Eclipse

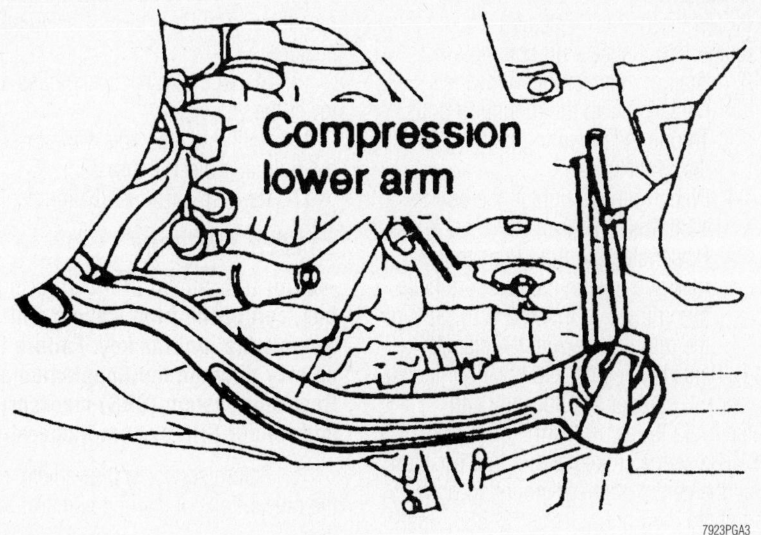

Disconnect the lower compression arm from the body—Eclipse

GALANT

⁂ WARNING

Prior to removal of the steering gear box, center the front wheels and remove the ignition key. Failure to do so may damage the Supplemental Restraint System (SRS) clock spring and render SRS system inoperative.

1. Before servicing the vehicle, refer to the precautions at the beginning of this section.
2. Install or connect the following:
   - Negative battery cable
   - Both front wheel assemblies
   - Bolt holding lower steering column joint to the rack and pinion input shaft
   - Stabilizer bar

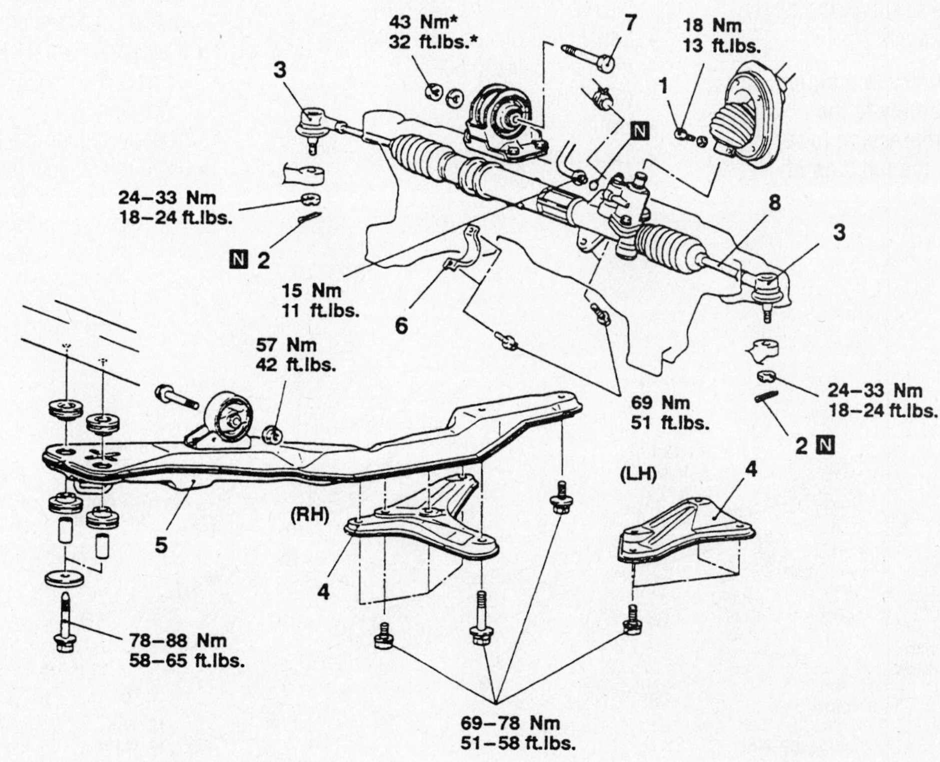

Exploded view of the power steering gear removal procedure—Galant

- Cotter pins and the tie rod ends from the steering knuckle

3. On vehicles equipped with Electronic Control Power steering (EPS), disconnect the wiring harness from the solenoid connector.

4. Locate the 2 triangular braces near the crossmember and remove both.

5. Support the center crossmember. Remove the through-bolt from the front round roll stopper and remove the bolts securing the center crossmember.

6. Remove the center crossmember.

7. Properly support the engine and remove the rear roll stopper through-bolt.

8. Remove or disconnect the following:
   - Power steering fluid pressure pipe and return hose from the rack fittings. Plug the fittings to prevent excessive fluid leakage.
   - Clamp bolts and the 2 bolts securing the rack assembly to the chassis
   - Rack and pinion steering assembly and its rubber mounts

➡When removing the rack and pinion assembly, tilt the assembly to the vehicle side of the compression lower arm and remove from the left side of the vehicle.

To install:

9. Center the rack assembly and insert the pinion into the steering column shaft.

10. Install or connect the following:
    - Rack and with the mounting bolts. Torque the mounting bolts to 51 ft. lbs. (69 Nm).
    - Pinch bolt and tighten the bolt to 13 ft. lbs. (18 Nm)
    - Power steering fluid lines to the rack and tighten the pressure hose fitting to 11 ft. lbs. (15 Nm). Secure the return hose with the clamp.

11. Raise the engine into position. Install the rear roll stopper through-bolt and tighten to 32 ft. lbs. (43 Nm).

12. Raise the crossmember into position. Install the center member mounting bolts and tighten the front bolts to 58–65 ft. lbs. (78–88 Nm) and the rear bolt to 51–58 ft. lbs. (69–78 Nm).

13. Install or connect the following:
    - Front roll stopper bolt and tighten the nut to 32 ft. lbs. (43 Nm)
    - 2 triangular braces and tighten the mounting bolts to 50–56 ft. lbs. (69–78 Nm)
    - Stabilizer bar
    - Tie rod ends and tighten nuts to 20 ft. lbs. (27 Nm)

14. On vehicles equipped with EPS, connect the wiring harness to the solenoid connector.

15. Install the wheel assemblies and lower the vehicle.

16. Refill the reservoir with power steering fluid and bleed the system.

17. Perform a front end alignment.

LANCER & MIRAGE

➡Prior to removal of the steering gear box, center the front wheels and remove the ignition key. Failure to do so may damage the Supplemental Restraint System (SRS) clockspring and render SRS system inoperative.

1. Before servicing the vehicle, refer to the precautions at the beginning of this section.

2. Drain the power steering system.

3. Remove or disconnect the following:
   - Battery negative cable. Raise the vehicle and support safely
   - Heated Oxygen (HO_2S) sensor and remove the front exhaust pipe, if necessary

4. Properly support the engine.
   - Both roll stopper mounting bolts and the 4 center member installa-

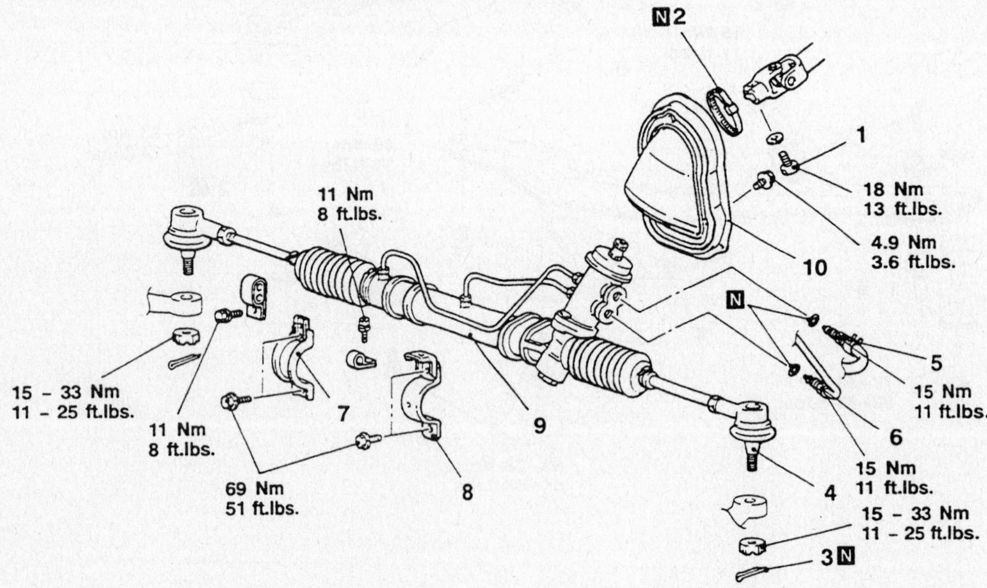

1. Steering shaft assembly and gear box connecting bolt
2. Band
3. Cotter pin
4. Tie-rod end and knuckle connection
5. Return tube connection
6. Pressure tube connection
7. Cylinder clamp
8. Gear housing clamp
9. Gear box assembly
10. Steering cover assembly

7923PGA1

Exploded view of the power steering gear assembly—Mirage

tion bolts. Remove the center member.

- Center member

➡**Matchmark the pinion input shaft of the rack to the lower steering column joint for installation purposes.**

5. Remove or disconnect the following:
   - Pinch bolt holding the lower steering column joint to the rack and pinion input shaft
   - Cotter pins and disconnect the tie rod ends from the steering knuckle
   - Power steering fluid pressure pipe and return hose from the rack fittings
   - Rack and pinion steering assembly and its rubber mounts from the right side of the vehicle

To install:

6. Align the matchmarks of the input shaft and install the rack to the vehicle.
7. Secure the rack using the retainer clamps and bolts. Torque the bolts to 52 ft. lbs. (70 Nm).
8. Torque the steering column pinch bolt to 13 ft. lbs. (18 Nm).
9. Using new O-rings, connect the power steering fluid lines to the rack fittings.
10. Install or connect the following:
    - Center member
    - Front exhaust pipe
    - HO_2S sensor
    - Tie rod ends to the steering knuckles and tighten the castle nuts to 25 ft. lbs. (34 Nm). Install new cotter pins.
    - Wheels and connect the negative battery cable
11. Refill the reservoir and bleed the system.
12. Perform a front end alignment.

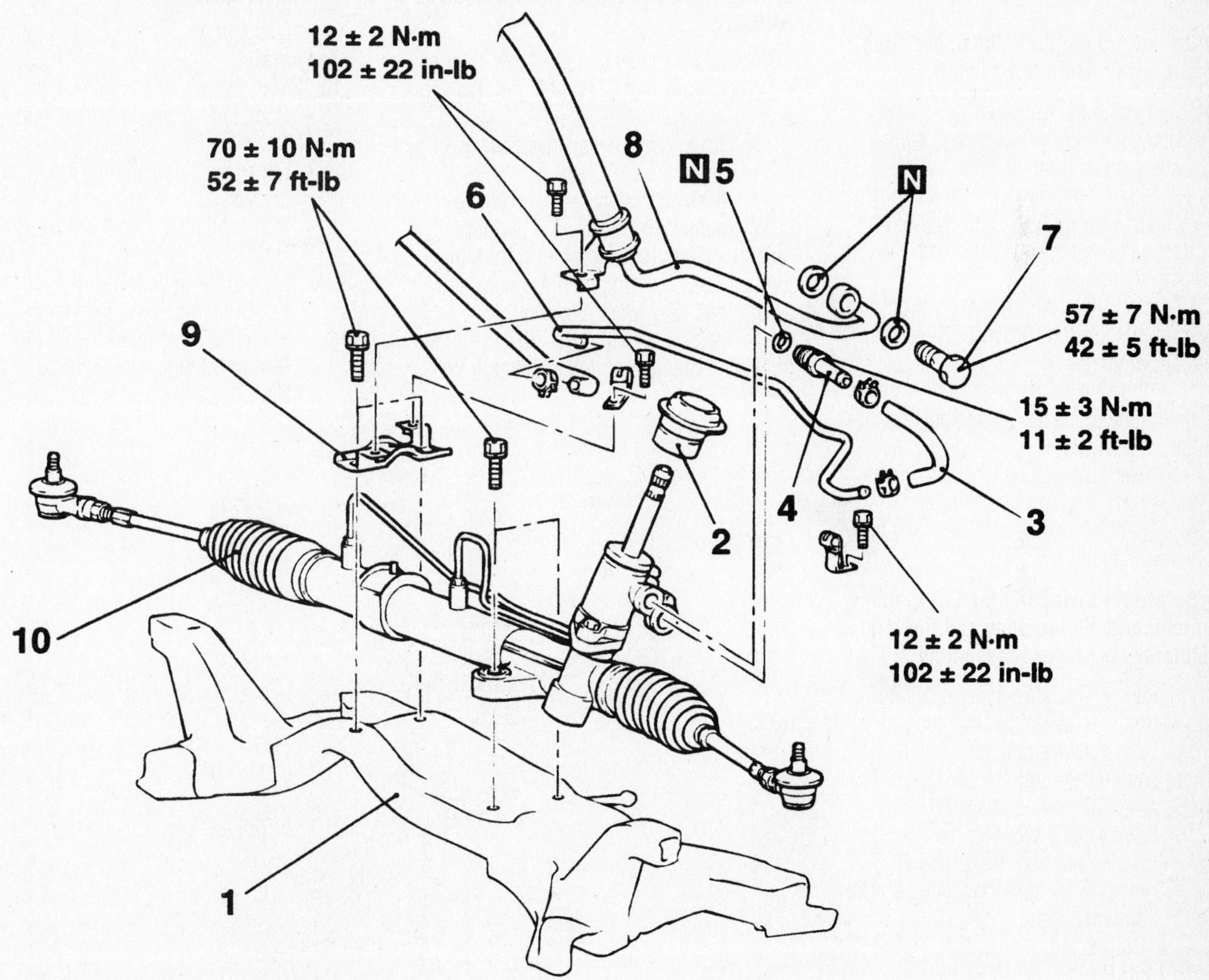

Exploded view of the power steering gear—Lancer

Strut and Coil Spring

REMOVAL & INSTALLATION

Front

DIAMANTE

1. Before servicing the vehicle, refer to the precautions at the beginning of this section.
2. Disconnect the negative battery cable.
3. Raise and safely support the vehicle.
4. Remove the brake hose and the tube bracket.

➡Do not pry the brake hose and tube clamp away when removing it.

5. If equipped with Anti-Lock Brake (ABS), disconnect the front speed sensor mounting clamp from the strut.
6. Support the lower arm and remove the strut to knuckle bolts. Use a piece of wire to suspend the knuckle to keep the weight off the brake hose.
7. If equipped with Active Electronic Control Suspension (Active-ECS) perform the following:
   a. Loosen the nut that secures the air line to the to the top of the strut and discard the O-ring.
   b. Remove the bolts that secure the actuator to the top of the strut and remove the component. Disconnect the wiring harness.

➡Before removing the top bolts, make matchmarks on the body and the strut insulator for proper reassembly.

8. Remove the strut upper nuts and remove the strut assembly from the vehicle.
9. Compress the coil spring using a spring compressor until the spring just comes away from one of the seats.
10. Remove or disconnect the following:
   - Center nut from the strut and remove the upper mounting bracket and bushings
   - Coil spring

To install:

11. Install or connect the following:
   - Compressed spring on the strut assembly
   - Upper bushings and the mounting bracket
   - Nut and tighten it to 43 ft. lbs. (59 Nm)
   - Strut to the vehicle and tighten the upper mounting nuts to 33 ft. lbs. (45 Nm)
12. Align the strut to the knuckle and connect with the mounting bolts. Torque the mounting bolts to 70–76 ft. lbs. (90–105 Nm).
13. If equipped with Active-ECS, perform the following:
   a. Install the air line with a new O-ring.
   b. Install the actuator to the top of the strut. Connect the wiring harness.
14. Install or connect the following:
   - Brake hose bracket and the ABS clamp, if equipped
   - Wheel and tire assembly
15. Perform a front end alignment.

LANCER, EVOLUTION, SPORTBACK & MIRAGE

1. Before servicing the vehicle, refer to the precautions at the beginning of this section.
2. Disconnect the negative battery cable.
3. Raise and safely support vehicle.
4. Remove the brake hose and tube bracket retainer bolt and bracket from the front strut. Do not pry the brake hose and tube clamp away when removing.
5. If equipped with ABS, disconnect the front speed sensor mounting clamp from the strut.
6. Support the lower arm using floor jack. Remove the lower strut to knuckle bolts.

➡Before removing the top bolts, make matchmarks on the body and the strut insulator for proper reassembly.

7. Remove the strut upper mounting bolts. Remove the strut assembly from the vehicle.
8. Compress the coil spring using a spring compressor until the spring just comes away from one of the seats.
9. Remove or disconnect the following:
   - Center nut from the strut and remove the upper mounting bracket and bushings
   - Coil spring

To install:

10. Install or connect the following:
   - Compressed spring on the strut assembly
   - Upper bushings and the mounting bracket
   - Nut and tighten it to 43 ft. lbs. (59 Nm)
   - Strut to the vehicle and install the top mounting bolts. Tighten the mounting bolts to 29 ft. lbs. (40 Nm) for Mirage and 33 ft. lbs. (44 Nm) for Lancer.

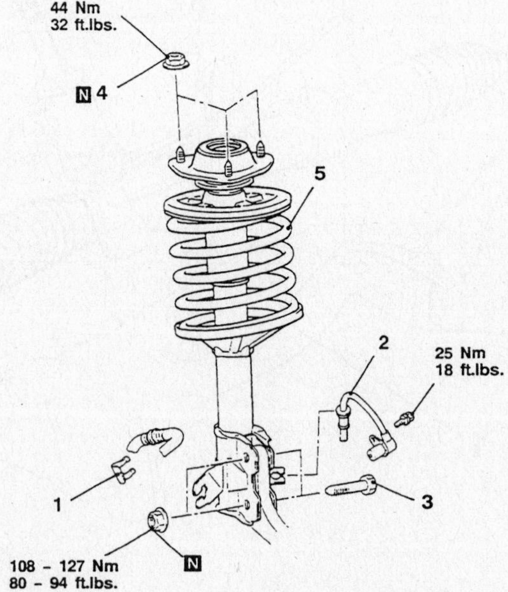

Removal steps
1. Brake hose clamp
2. Front speed sensor <Vehicles with ABS>
3. Bolts
4. Self-locking nut
5. Strut assembly

Caution
For vehicles with ABS, be careful when handling the pole piece at the tip of the speed sensor so as not to damage it by striking against other parts.

7923PGA6

Front strut assembly and related parts—Mirage shown, Lancer similar

11. Position the strut on the knuckle and install the mounting bolts. While holding the head of the lower mounting bolt, tighten the nuts to 80–94 ft. lbs. (110–130 Nm) for Mirage and 123 ft. lbs. (167 Nm) for Lancer.
12. Install or connect the following:
- Brake hose bracket and the ABS clamp, if equipped
- Wheel and tire assembly
13. Perform a front end alignment.

Shock Absorber and Coil Spring

REMOVAL & INSTALLATION

Front

ECLIPSE

1. Before servicing the vehicle, refer to the precautions at the beginning of this section.
2. Raise and safely support the vehicle.
3. Remove or disconnect the following:
- Front wheel
- 3 upper shock absorber mounting nuts. Do not remove the larger nut in the center of the strut at this time.
- Stabilizer link from the damper fork
- Damper fork mounting bolt

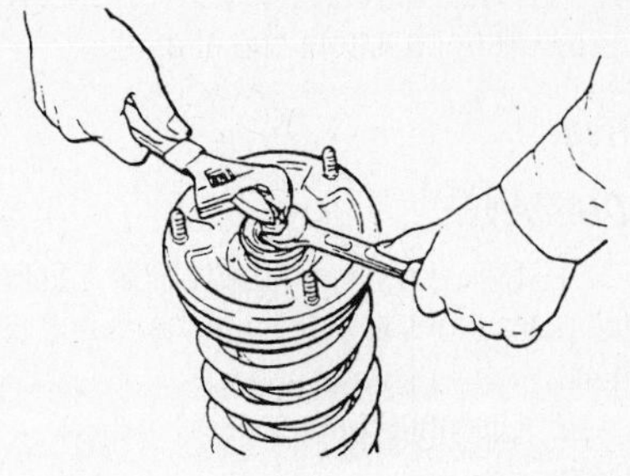

Removing the self-locking nut—Eclipse

- Shock absorber assembly from the vehicle
4. Use a coil spring compressor and compress the coil spring.
5. While holding the piston rod, remove the self-locking nut.
6. Remove or disconnect the following:
- Upper bracket assembly and spring pad
- Collar, upper bushing, cup assembly, bump rubber and dust cover
- Coil spring from the shock absorber

To install:

7. Align the end of the coil spring with the stepped part of the spring seat and install the compressed coil spring on the shock.

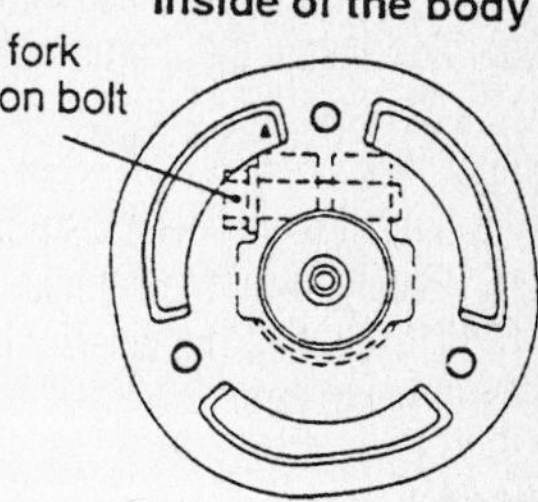

Upper bracket assembly alignment—Eclipse

8. Install the dust cover, bump rubber, cup assembly, upper bushing, collar, upper spring pad and bracket assembly on the strut.
9. Install or connect the following:
- Upper bushing and washer on the piston rod
- New self-locking nut on the piston rod. Temporarily tighten the nut.
10. Carefully remove the spring compressor from the spring. Torque the self-locking nut to 16 ft. lbs. (25 Nm).
11. Position the shock absorber assembly in the damper fork and install the mounting bolt.

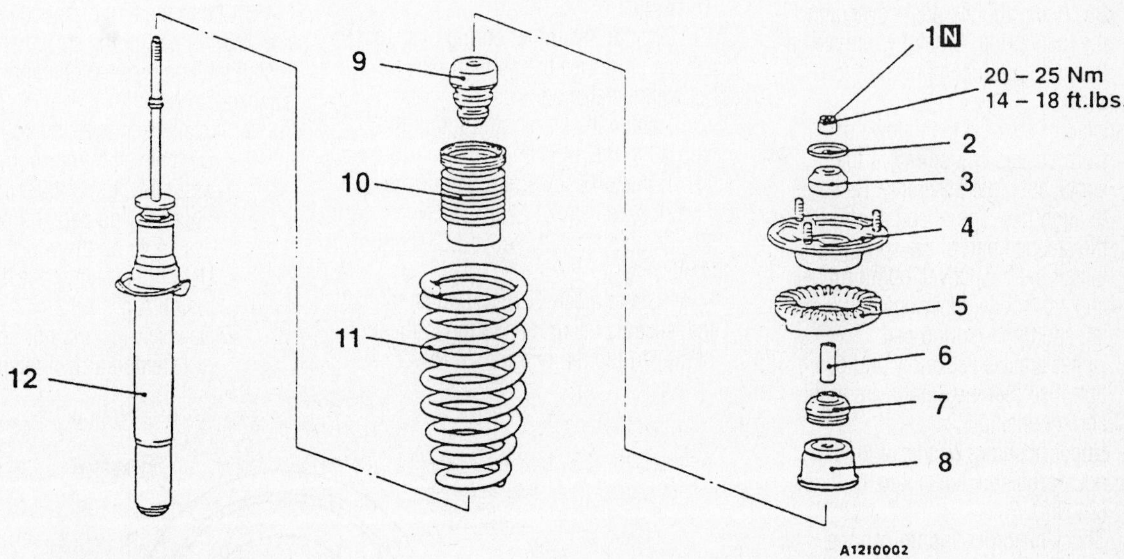

Exploded view of the coil spring removal procedure—Eclipse

12. Pass the studs in the upper bracket assembly through the holes in the inner fender and install the 3 mounting nuts.
13. Connect the stabilizer link to the damper fork.
14. Install the wheel assembly.
15. Safely lower the vehicle to the floor.
16. Check the front wheel alignment and adjust if necessary.

GALANT

1. Before servicing the vehicle, refer to the precautions at the beginning of this section.
2. Disconnect the negative battery cable.
3. Raise and safely support vehicle.
4. Remove or disconnect the following:
   - Appropriate wheel assembly
   - Sway bar link from the damper fork
   - Damper fork lower through-bolt and upper pinch bolt
   - Damper fork assembly
   - Shock absorber upper nuts and remove the strut assembly from the vehicle
5. Compress the coil spring with a special compression tool.
   - Self-locking nut and washer
   - Upper bushing, upper bracket assembly, the upper spring pad, and the collar
   - Other upper bushing, cup assembly, bump rubber, dust cover, and the coil spring. Carefully remove the coil spring compression tool

To install:

6. Install or connect the following:
   - Compressed coil spring to the shock absorber assembly. Be sure to align the edge of coil spring to the stepped part of the spring seat. Install the dust cover, bump rubber, cup assembly, upper bushing, collar, and upper spring pad.
   - Upper bracket assembly and position it so that the 3 bolts are in the correct position
   - Upper bushing, washer, and locknut. Torque the locknut to 18 ft. lbs. (24 Nm).
   - Shock absorber and tighten the upper mounting nuts to 32 ft. lbs. (44 Nm)
7. Align the shock to the damper fork and install the damper fork. Tighten the lower through-bolt/nut to 65 ft. lbs. (88 Nm) and the upper pinch bolt to 76 ft. lbs. (103 Nm).
8. Connect the sway bar link to the damper fork and tighten the link nut to 29 ft. lbs. (39 Nm).
9. Install the wheel and tire assembly.
10. Perform a front end alignment.

Rear

DIAMANTE

1. Before servicing the vehicle, refer to the precautions at the beginning of this section.
2. Disconnect the negative battery cable.
3. Raise and properly support vehicle. Remove both rear wheels.
4. Support the lower control arm with a jack.
5. Matchmark the positioning of the upper spring plate to the vehicle for reinstallation purposes.
6. If equipped with Active Electronic Control Suspension (Active-ECS), perform the following:
   a. Loosen the nut that secures the air line to the to the top of the strut and discard the O-ring.
   b. Remove the bolts that secure the actuator to the top of the strut and remove the component. Disconnect the wiring harness.
7. Remove the shock absorber lower mounting bolt and remove the 2 nuts that secure the shock upper plate to the vehicle.
8. Lower the support jack and remove the shock from the vehicle.

To install

9. Position the upper spring plate and install the strut. Use the support jack to assist with installation.
10. Tighten the upper strut mounting nuts to 33 ft. lbs. (45 Nm).
11. Tighten the lower strut mounting bolt to 71 ft. lbs. (98 Nm).
12. If equipped with Active-ECS perform the following:
   a. Using a new O-ring, tighten the nut that secures the air line to the to the top of the strut to 84 inch lbs. (9 Nm).
   b. Install the actuator to the top of the shock absorber and secure with mounting bolts. Connect the wiring harness.
13. Remove the support jack, install wheels and lower vehicle.
14. Connect the negative battery cable.

ECLIPSE & EVOLUTION

1. Before servicing the vehicle, refer to the precautions at the beginning of this section.
2. Remove or disconnect the following:
   - Service lid in the luggage compartment
   - Cap and flange nuts securing the upper mounting bracket to the body of the vehicle
3. Raise and safely support the vehicle.
4. Remove the bolt attaching the lower end of the shock to the knuckle and remove the shock absorber from the vehicle.
5. Use a coil spring compressor and compress the coil spring.
6. While holding the piston rod, remove the self-locking nut.
7. Remove or disconnect the following:
   - Upper bracket assembly and spring pad
   - Collar, upper bushing, cup assembly, bump rubber and dust cover
   - Coil spring from the shock absorber

To install:

8. Align the end of the coil spring with the stepped part of the spring seat and install the compressed coil spring on the shock absorber.
9. Install or connect the following:
   - Dust cover, bump rubber, cup assembly, upper bushing, collar, upper spring pad and bracket assembly on the shock absorber
   - Upper bushing and washer on the piston rod
   - New self-locking nut on the piston rod. Temporarily tighten the nut.

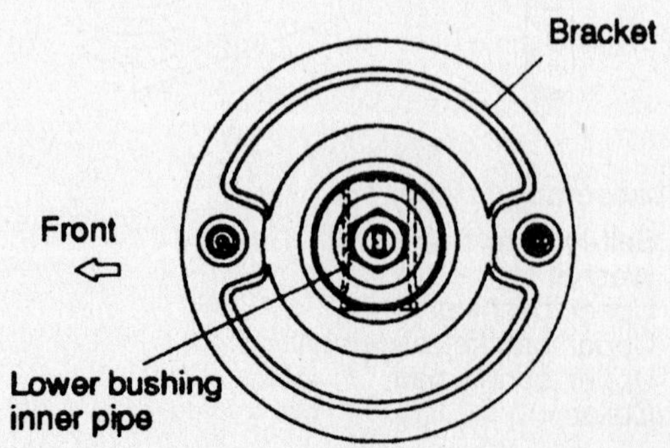

Correct upper bracket installed position—Eclipse

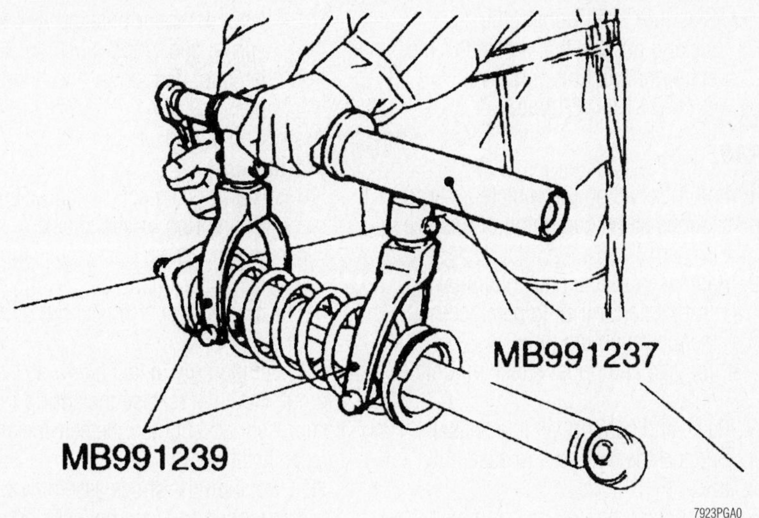

Correct method for compressing the coil spring

10. Remove the spring compressor from the spring. Torque the self-locking nut to 16 ft. lbs. (25 Nm).

11. Install the upper bracket of the shock to the vehicle. Torque the mounting nuts to 32 ft. lbs. (44 Nm).

12. Raise the suspension up with a jack or adjustable stand to align the shock absorber lower mounting holes.

13. Install the lower mounting bolt. Torque the bolt to 71 ft. lbs. (96 Nm).

14. Remove the jack or stand and safely lower the vehicle to the floor.

15. Install the cap and service lid.

GALANT

1. Before servicing the vehicle, refer to the precautions at the beginning of this section.

2. Raise and support the vehicle chassis.

3. Raise and support the lower control arm assembly slightly.

4. In order to gain access to the top mounting nuts, remove the rear seat as follows:

   a. While pulling the rear seat stopper outward, lift the lower cushion upward. Remove the lower cushion.

   b. Remove the seat back mounting bolts.

   c. Lift the seat back upward and remove the seat.

5. Remove or disconnect the following:
   - Shock upper mounting nuts
   - Shock lower mounting bolt and remove the assembly from the vehicle

6. Use a coil spring compressor and compress the coil spring.

7. Remove the shock cap.

8. While holding the piston rod, remove the self-locking nut.

9. Remove or disconnect the following:
   - Upper bracket assembly and spring pad
   - Collar, upper bushing, cup assembly, bump rubber and dust cover
   - Coil spring from the shock

To install

10. Align the end of the coil spring with the stepped part of the spring seat and install the compressed coil spring on the shock.

11. Install or connect the following:
    - Dust cover, bump rubber, cup assembly, upper bushing, collar, upper spring pad and bracket assembly on the shock
    - Upper bushing and washer on the piston rod
    - New self-locking nut on the piston rod. Temporarily tighten the nut.

12. Remove the spring compressor from the spring. Torque the self-locking nut to 16 ft. lbs. (25 Nm).

13. Install the shock cap.

14. Position the shock assembly so that the lower mounting bolt can be installed and lightly tightened.

15. Use a jack to raise or lower the lower control arm, so that the top shock plate studs align through the body. Raise the jack to hold the shock assembly in position.

16. Install the top plate nuts on the studs and tighten the mounting nuts to 32 ft. lbs. (44 Nm).

17. With the vehicle on the ground, tighten the lower mounting bolt to 71 ft. lbs. (98 Nm).

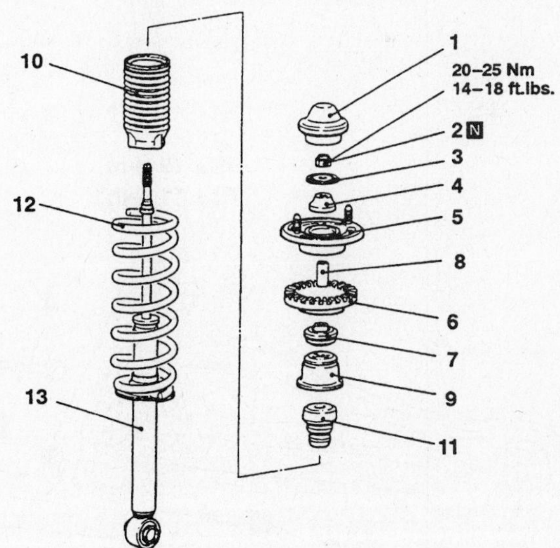

Exploded view of the rear shock absorber assembly—Galant and Mirage

18. Install the rear seat back and cushion.

LANCER & SPORTBACK

1. Before servicing the vehicle, refer to the precautions at the beginning of this section.
2. Remove or disconnect the following:
   - Stabilizer link connection
3. Support the lower control arm with a jack.
   - Lower control arm and trailing arm bolt
   - Upper shock absorber mounting nut
   - Shock absorber-to-lower control arm attaching bolt
   - Shock absorber assembly

To install:

4. Position the shock absorber into the vehicle. Install the spring seat stepped section so that it points toward the rear side of the vehicle.
5. Install or connect the following:
   - Shock absorber-to-lower control arm bolt and nut. Tighten the nut to 70 ft. lbs. (95 Nm).
   - Upper shock mounting nut and tighten to 32 ft. lbs. (44 Nm)
   - Lower control arm and trailing arm
   - Stabilizer link. Tighten the self-locking nuts so that the end of the stabilizer line bolt protrudes 0.24–0.31 in. (6–8mm).

MIRAGE

1. Before servicing the vehicle, refer to the precautions at the beginning of this section.
2. Remove or disconnect the following:
   - Trunk interior trim to gain access to the top mounting nuts
   - Top cap and upper shock mounting nuts
3. Raise and support vehicle chassis.
4. Support the trailing arm assembly with a jack.
5. Matchmark the upper spring plate to the vehicle chassis for reassembly and remove the upper spring plate mounting nuts.
6. Remove the shock lower mounting bolt and remove the assembly from the vehicle.
7. Compress the coil spring using the proper spring compressor.
8. Hold the piston rod with a wrench and remove the self-locking nut.
9. Remove or disconnect the following:
   - Washer, upper bushing A, bracket, spring pad, upper bushing B, collar, cup, dust cover and bump rubber
   - Coil spring

To install

10. Install or connect the following:
    - Coil spring on the shock
    - Bump rubber, dust cover, cup, collar, upper bushing A, spring pad, bracket, upper bushing B and the washer
11. Temporarily install a new self-locking nut, carefully release the spring from the compressor and tighten the self-locking nut to specifications.
12. Position the shock assembly so that lower mounting bolt can be installed and lightly tightened.
13. Use jack to raise or lower the axle assembly so that top shock plate studs aligns through body. Raise jack to hold the shock assembly in position.
14. Install the top plate nuts and tighten them to 20 ft. lbs. (28 Nm).
15. Lower the vehicle and tighten the lower mounting bolt to 65 ft. lbs. (90 Nm).
16. Install top cap and interior trim.

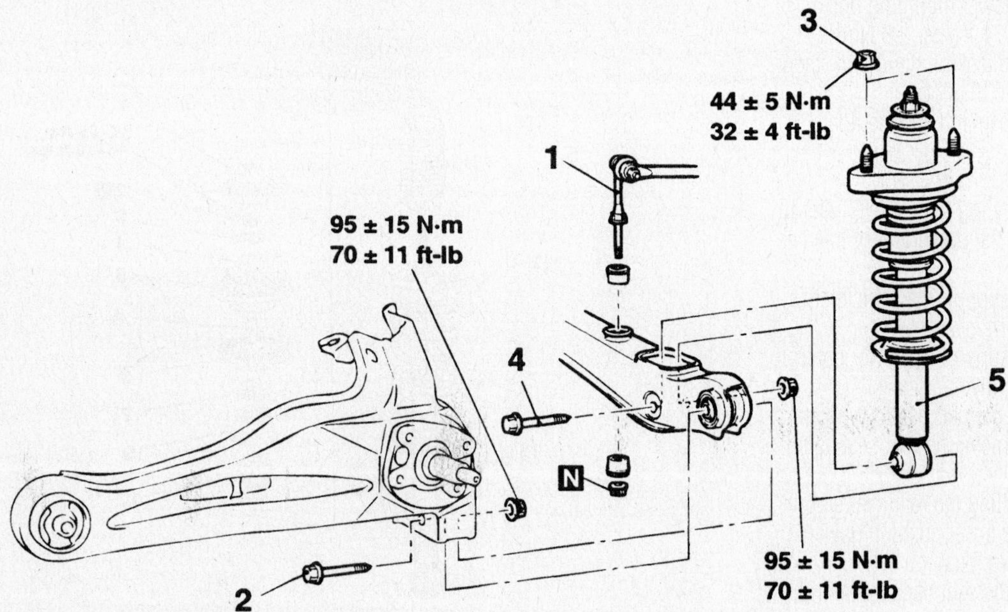

Exploded view of the rear shock absorber mounting—Lancer

Upper Ball Joint

REMOVAL & INSTALLATION

The upper ball joints are an integral part of the upper control arm. If the ball joint becomes worn or damaged, the control arm must be replaced.

Upper Control Arm

These vehicles use a strut type front suspension. No upper control arm is used.

Lower Ball Joint

REMOVAL & INSTALLATION

The lower ball joint is an integral part of the lower control arm assembly, and can not be serviced separately. A worn or damaged ball joint requires replacement of lower control arm assembly.

Lower Control Arm

REMOVAL & INSTALLATION

Front

DIAMANTE

1. Before servicing the vehicle, refer to the precautions at the beginning of this section.
2. Disconnect the negative battery cable.
3. Raise the vehicle and support safely allowing wheels and suspension to hang freely.
4. Remove or disconnect the following:
   - Sway bar links from the lower control arm
   - Ball joint stud from the steering knuckle
   - Inner mounting frame through-bolt and nut
   - Rear mount bolts. Remove the clamp if equipped
   - Rear rod bushing if servicing

To install:

5. Assemble the control arm and bushing.
6. Install or connect the following:
   - Control arm to the vehicle and install the through-bolt. Replace the nut and snug temporarily.
   - Rear mount clamp, bolts and replacement nuts. Torque the bolts to 72–87 ft. lbs. (100–120 Nm). Torque the nuts to 29 ft. lbs. (40 Nm).

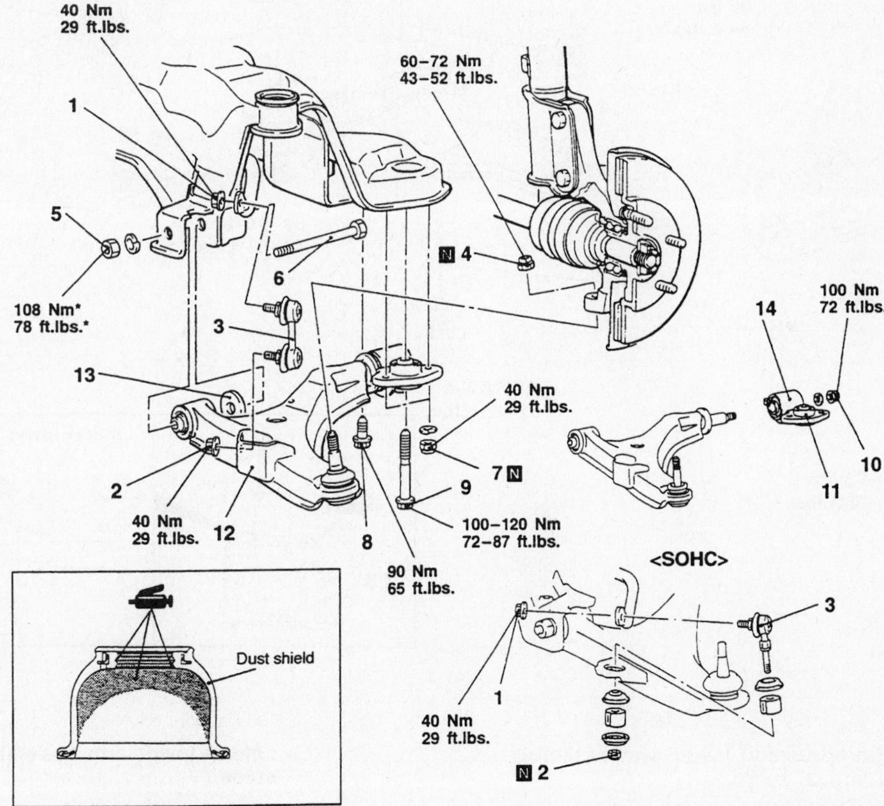

Removal steps

1. Stabilizer link mounting nut (stabilizer bar side)
2. Stabilizer link mounting nut (lower arm side)
3. Stabilizer link
4. Self-locking nut connecting lower arm ball joint to knuckle
5. Lower arm mounting nut
6. Lower arm mounting bolt
7. Clamp mounting self-locking nut
8. Clamp mounting bolt (small)
9. Clamp mounting bolt (large)
10. Lower arm clamp mounting self-locking nut
11. Lower arm mounting clamp
12. Lower arm
13. Stopper
14. Rod bushing

Caution
*: Indicates parts which should be temporarily tightened, and then fully tightened with the vehicle on the ground in an unladen condition.

7923PGB9

Exploded view of the lower control arm removal procedure—Diamante is similar

- Ball joint stud to the knuckle
- New nut and tighten to 43–52 ft. lbs. (60–72 Nm)
- Sway bar and links

7. Lower the vehicle to the floor for the final tightening of the frame mount through-bolt.

8. Once the full weight of the vehicle is on the floor, tighten the frame mount through-bolt nuts to 75–90 ft. lbs. (102–122 Nm).

9. Connect the negative battery cable.

10. Check the wheel alignment and adjust if necessary.

ECLIPSE & GALANT

The lower lateral arm ball joint and the compression arm ball joint are integral components of the lateral arm and the compression arm respectively. If the ball joints are to be serviced, the arms must be replaced.

1. Before servicing the vehicle, refer to the precautions at the beginning of this section.
2. Raise and support the vehicle safely.
3. Disconnect both ball joint studs from the steering knuckle.
4. To remove the lower lateral arm, remove the crossmember brackets.
5. Remove or disconnect the following:
   - Inner lateral arm mounting bolts and nut
   - Arm from the vehicle
   - 2 bolts holding the compression arm
   - Compression arm

To install:

6. Assemble the control arms and bushings.
7. Install or connect the following:
   - Lateral control arm to the vehicle and install the inner mounting bolts. Install a new nut and snug temporarily.

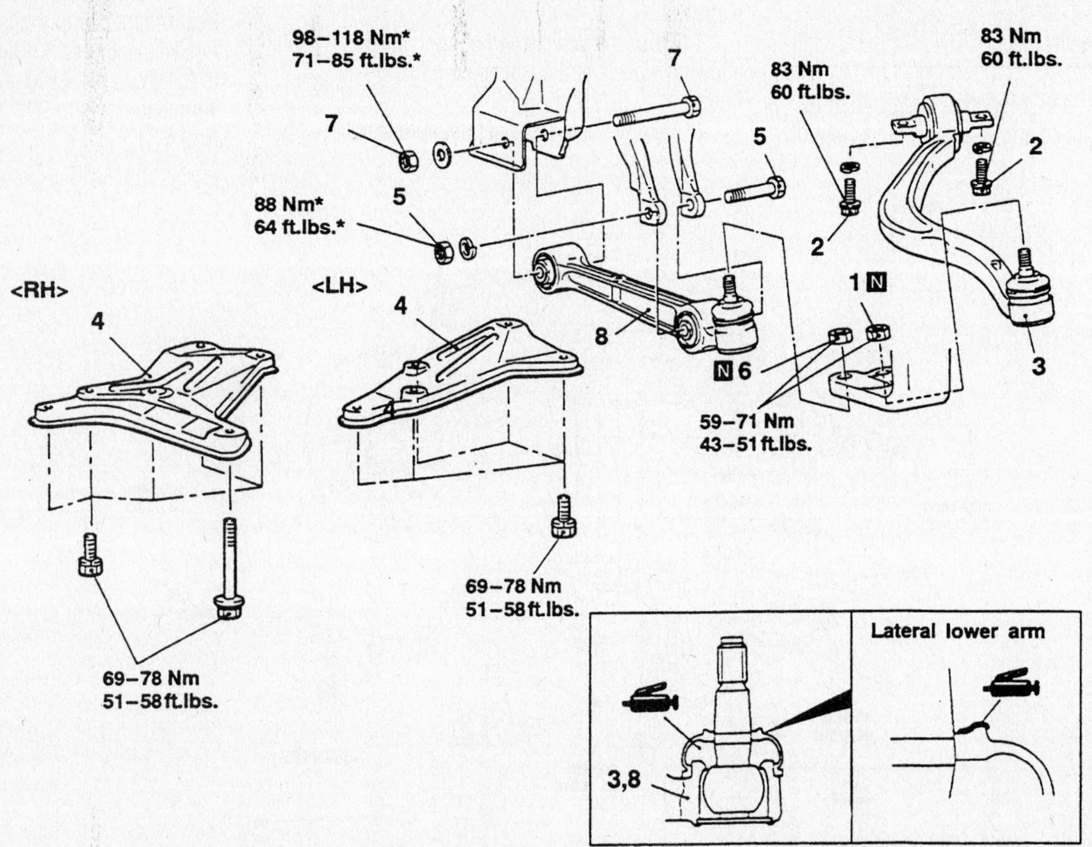

Compression lower arm assembly removal steps

1. Connection for compression lower arm ball joint and knuckle
2. Compression lower arm mounting bolt
3. Compression lower arm assembly

Lateral lower arm assembly removal steps

4. Stay
5. Shock absorber lower mounting bolt and nut
6. Connection for lateral lower arm ball joint and knuckle
7. Lateral lower arm mounting bolt and nut
8. Lateral lower arm assembly

Caution

***: Indicates parts which should be temporarily tightened, and then fully tightened with the vehicle on the ground in the unladen condition.**

7923PGB8

Exploded view of the lower control arms—Eclipse and Galant

- Compression arm to the vehicle
- Ball joint studs to the knuckle
- New nuts and tighten to 43–51 ft. lbs. (59–71 Nm)

8. Lower the vehicle to the floor for the final tightening.

9. Once the full weight of the vehicle is on the suspension, tighten the lateral arm rear bolt to 71–85 ft. lbs. (98–118 Nm) and the front bolt to the damper fork to 64 ft. lbs. (88 Nm).

10. Torque the bolts for the compression arm to 60 ft. lbs. (83 Nm).

11. Reinstall the crossmember brackets with their mounting bolts. Torque the mounting bolts to 51–58 ft. lbs. (69–78 Nm).

12. Perform an alignment on the vehicle.

LANCER, EVOLUTION & SPORTBACK

➡The suspension components should not be tightened until the vehicle's weight is resting on its wheels.

1. Before servicing the vehicle, refer to the precautions at the beginning of this section.
2. Raise the vehicle and support safely.
3. Remove or disconnect the following:
   - Wheel and tire assembly
   - Stabilizer bar self-locking nut, rubber bushings, stabilizer bar and collar. Discard the nut.
   - Lower control arm-to-knuckle bolt and nut
4. Lift the transaxle with a jack, then remove the front control arm-to-crossmember bolt.
   - Lower control arm

To install:

5. Install or connect the following:
   - Lower control arm into the vehicle
   - Lower control arm-to-crossmember bolts. Torque the bottom bolt to 123 ft. lbs. (167 Nm) and the side

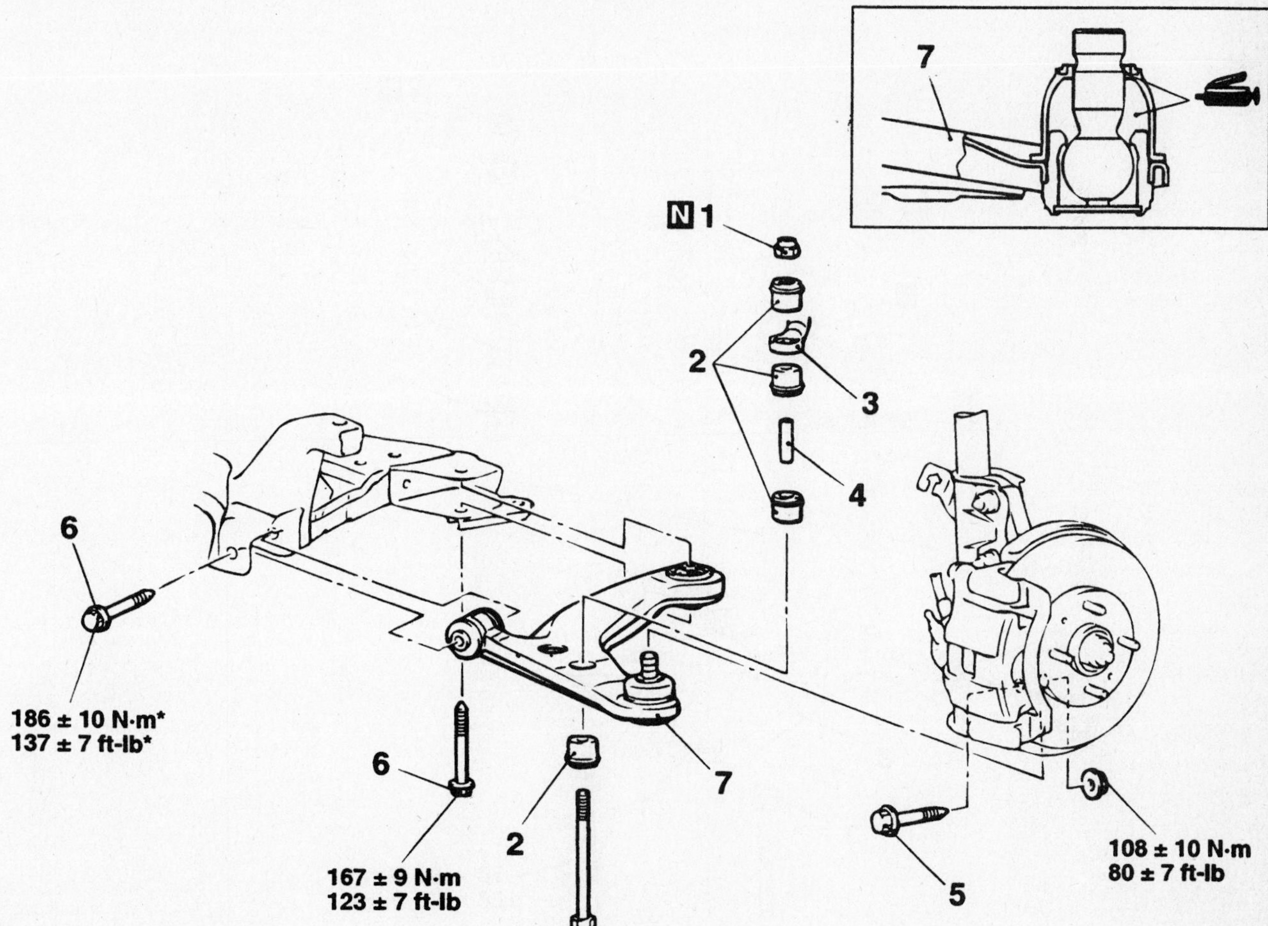

Exploded view of the lower control arm mounting—Lancer

bolt snug until the vehicle is lowered.

- Lower control arm-to-steering knuckle bolt. Torque to 80 ft. lbs. (108 Nm).
- Stabilizer bar collar, stabilizer bar, bushings and new self-locking nut.

6. Lower the vehicle, install the wheels, then with the weight of the vehicle on the wheels, torque the side control arm-to-crossmember bolt to 137 ft. lbs. (186 Nm).

MIRAGE

➡The suspension components should not be tightened until the vehicle's weight is resting on its wheels.

1. Before servicing the vehicle, refer to the precautions at the beginning of this section.
2. Raise the vehicle and support safely.
3. Remove or disconnect the following:
   - Wheel and tire assembly
   - Stabilizer bar links or mounting nuts and bolts from lower control arm. Remove the joint cups and bushings.
   - Ball joint stud from the steering knuckle
   - Inner lower arm mounting bolt and nut
   - Rear mount bolts from the retaining clamp. Remove the rear retainer clamp if equipped.
   - Arm from the vehicle

To install:

4. Install or connect the following:
   - Control arm to the vehicle and install the inner mounting bolt. Install new nut and tighten to 78 ft. lbs. (108 Nm).
   - Rear mount clamp and bolts. Torque the clamp mounting bolts to 65 ft. lbs. (90 Nm).
   - Ball joint stud to the knuckle. Install a new nut and tighten to 43–52 ft. lbs. (60–72 Nm).
   - Sway bar and links
5. Lower the vehicle to the floor for the final tightening of the inner frame mount bolt.
6. Install the wheel and tire assembly.

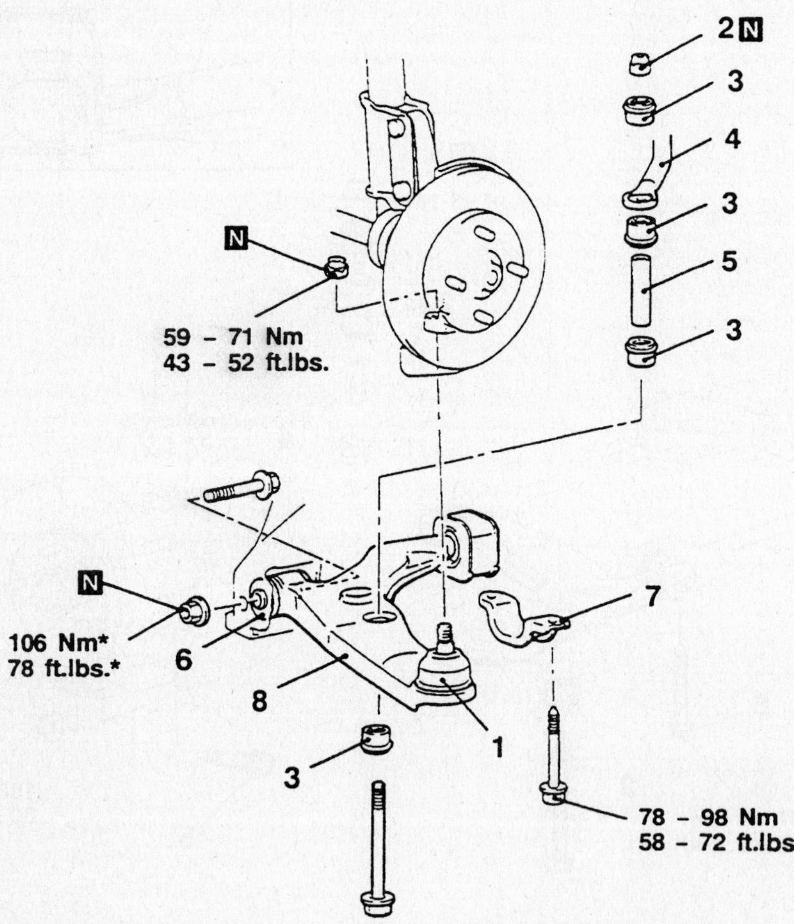

Removal steps

1. Lower arm ball joint connection
2. Self-locking nut
3. Stabilizer rubber
4. Stabilizer bar
5. Collar
6. Lower arm front bushing connection
7. Support bracket
8. Lower arm assembly

Caution

***: Indicates parts which should be temporarily tightened, and then fully tightened with the vehicle on the ground in the unladen condition.**

7923PGB0

Lower control arm assembly and related components—Mirage

Wheel Bearings

ADJUSTMENT

The front and rear wheel bearings on these vehicles are not adjustable. If the bearings are noisy or become loose, they must be replaced.

REMOVAL & INSTALLATION

Front

DIAMANTE & MIRAGE

1. Before servicing the vehicle, refer to the precautions at the beginning of this section.
2. Disconnect the negative battery cable.
3. Raise the vehicle and support safely. Remove the halfshaft nut.
4. If equipped with Anti-Lock Brake (ABS), remove the front wheel speed sensor.
5. If equipped with Active Electronic Control Suspension (Active-ECS), disconnect the height sensor from the lower control arm.
6. Remove the caliper assembly and brake pads. Suspend the caliper with a wire.
7. Ball joint and tie rod end from the steering knuckle.
8. Remove the halfshaft from the hub.
9. Unbolt the lower end of the strut and remove the hub and steering knuckle assembly from the vehicle.
10. Press the hub from the bearing and remove the bearing races from the knuckle.

To install:

11. Press the wheel bearing into the knuckle. Once the bearing is installed, install the inner race.

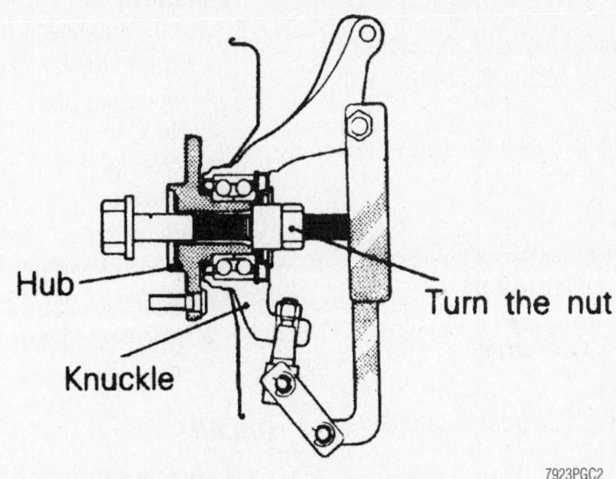

Use of press tool for hub removal—Mirage and Diamante

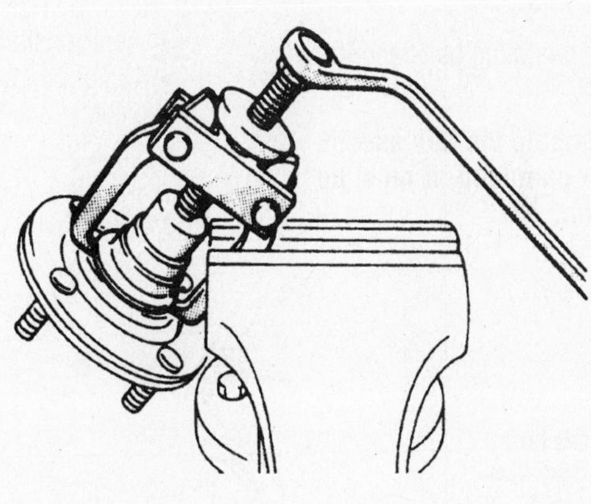

Removing inner race from hub—Mirage and Diamante

12. Install the grease seal.
13. Using a pressing, mount the front hub assembly into the knuckle. Tighten the nut of the pressing tool to 144–188 ft. lbs. (200–260 Nm). Rotate the hub to seat the bearing.
14. Install the hub and knuckle assembly onto the vehicle. Install the lower ball joint stud into the steering knuckle and install a new nut. Tighten to 52 ft. lbs. (72 Nm).
15. Install the halfshaft into the hub/knuckle assembly.
16. Install 2 front strut lower mounting bolts and tighten to 80–94 ft. lbs. (110–130 Nm) on Mirage or 65–76 ft. lbs. (90–105 Nm) on Diamante models.
17. Install the tie rod end and tighten the nut to 25 ft. lbs. (34 Nm) for Mirage and 21 ft. lbs. (29 Nm) on Diamante models.
18. Install the brake disc and caliper assembly.
19. If equipped with Active-ECS, connect the height sensor and tighten the mounting bolt to 15 ft. lbs. (20 Nm).
20. Install the front speed sensor, if removed.
21. Install the washer and new locknut to the end of the halfshaft. Tighten the locknut snugly to 144–188 ft. lbs. (200–260 Nm).
22. Install the tire and wheel assembly onto the vehicle. Lower the vehicle to the ground.

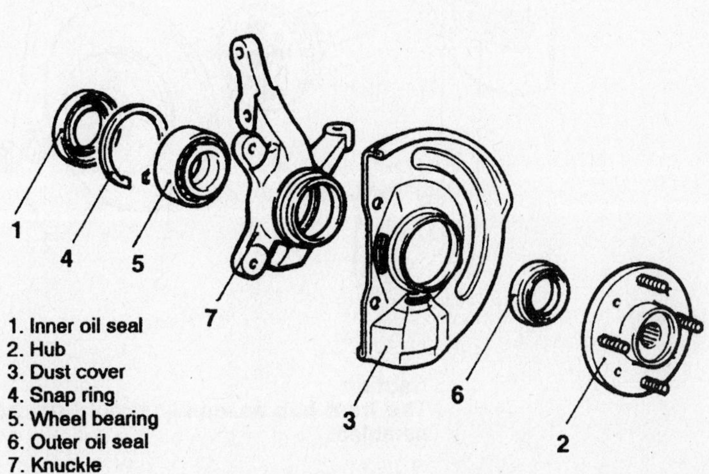

Front wheel bearing assembly exploded view—Mirage and Diamante

ECLIPSE

1. Before servicing the vehicle, refer to the precautions at the beginning of this section.
2. Remove or disconnect the following:
   - Front wheel
   - Axle nut
   - Wheel speed sensor, vehicles with Anti-Lock Brake (ABS)
   - Caliper and suspend it out of the way with wire or string
   - Brake rotor
   - Steering knuckle from the upper arm
3. Pull the knuckle away from the vehicle to access the hub mounting bolts on the inboard side of the hub. Be careful not to damage the ball joint boot or the ABS rotor if equipped.
4. Remove the mounting bolts and the front hub assembly.

➡Do not disassemble the hub assembly. If binding or damaged, it must be replaced as a unit.

To install:

5. Install or connect the following:
   - Hub to the knuckle. Torque the mounting bolts to 65 ft. lbs. (88 Nm).
   - Knuckle to the upper arm
   - Brake rotor and the caliper
   - Wheel speed sensor if removed
   - Axle nut and tighten to 145–188 ft. lbs. (196–255 Nm)
   - Wheel and lower the vehicle to the floor

GALANT

1. Before servicing the vehicle, refer to the precautions at the beginning of this section.
2. Raise the vehicle and support safely.
3. Remove or disconnect the following:
   - Appropriate wheel assembly
   - Cotter pin, halfshaft nut and washer
   - Vehicle Speed Sensor (VSS), if equipped with Anti-Lock Brake (ABS)
   - Caliper and brake pads. Support the caliper out of the way using wire.
   - Brake rotor from the hub assembly
   - Upper ball joint from the steering knuckle and pull the knuckle outward
4. From the back of the knuckle, remove the 4 bolts securing the hub to the knuckle.
5. Remove the hub and bearing assembly from the knuckle.

➡The hub assembly is not serviceable and should not be disassembled.

To install

6. Install or connect the following:
   - Hub to the steering knuckle and tighten the mounting bolts to 65 ft. lbs. (88 Nm)
   - Upper ball joint to the steering knuckle and tighten the self-locking nut to 21 ft. lbs. (28 Nm)
   - Axle washer and nut. Tighten the nut to 145–188 ft. lbs. (200–260 Nm).

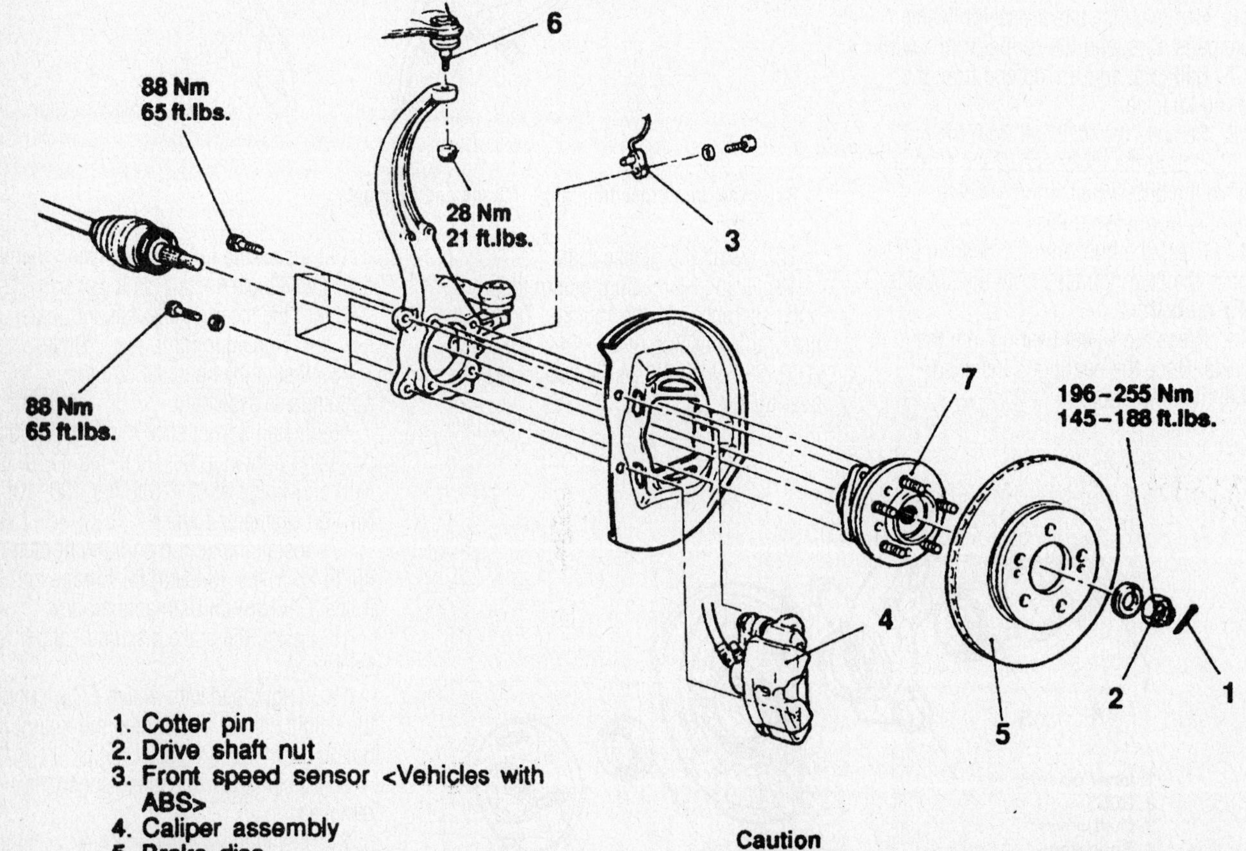

Front hub and related components—Eclipse

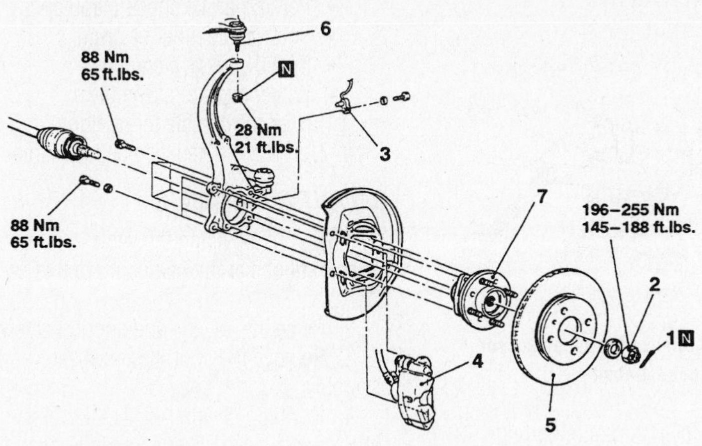

Removal steps

1. Cotter pin
2. Drive shaft nut
3. Front speed sensor <Vehicles with ABS>
4. Caliper assembly
5. Brake disc
6. Connection for upper arm
7. Front hub assembly

Caution
The front hub assembly should not be disassembled.

7923PGC5

Exploded view of the front hub removal—Galant

7. Position the rotor on the hub.
8. Install the caliper holder and the brake caliper.
9. If equipped with ABS, install the VSS.
10. Install the wheel assembly and lower the vehicle.

LANCER

1. Before servicing the vehicle, refer to the precautions at the beginning of this section.
2. Raise and safely support the vehicle.
3. Remove or disconnect the following:
   - Front wheel
   - Wheel Speed Sensor (WSS), vehicles with Anti-Lock Brake (ABS)
   - Caliper and suspend it out of the way with wire or string
   - Brake rotor
   - Axle nut, using special tool MB990767 to hold the hub secure while removing the nut. Discard the nut.

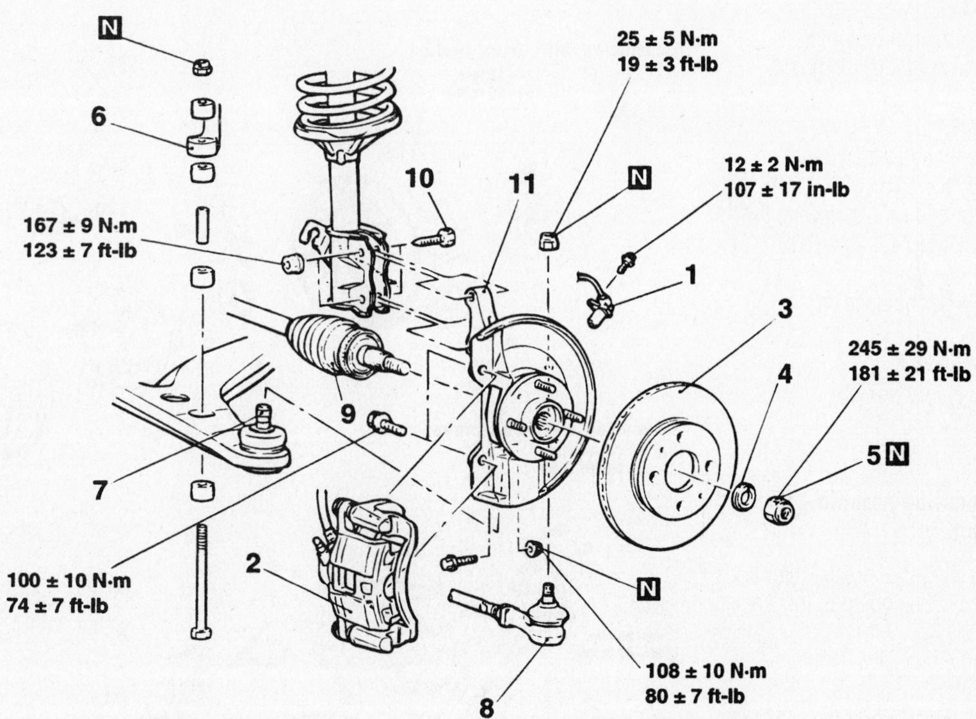

1. FRONT ABS SPEED SENSOR <VEHICLES WITH ABS>
2. CALIPER ASSEMBLY
3. BRAKE DISC
4. WASHER
5. DRIVESHAFT NUT
6. CONNECTION FOR STABILIZER BAR
7. CONNECTION FOR LOWER ARM BALL JOINT
8. CONNECTION FOR TIE ROD END
9. DRIVESHAFT
10. FRONT STRUT TO HUB AND KNUCKLE MOUNTING BOLT AND NUT
11. HUB AND KNUCKLE

93570G31

Exploded view of the front wheel bearing and related components

- Stabilizer bar from the lower control arm
- Lower ball joint from the steering knuckle
- Tie rod end from the steering knuckle. Do not remove the nut from the tie rod end. Loosen the nut and use special tool MB991113 or MB990635 to avoid damaging the threads.
- Halfshaft from the hub and knuckle using the proper puller
- Front strut-to-hub and knuckle mounting bolt and nut
- Hub and knuckle from the vehicle

➡Do not disassemble the hub assembly. If binding or damaged, it must be replaced as a unit.

To install:

4. Install or connect the following:
   - Hub and knuckle assembly
   - Front strut-to-hub bolt and nut. Torque to 123 ft. lbs. (167 Nm).
   - Halfshaft
   - Tie rod end. Torque the nut to 19 ft. lbs. (25 Nm).
   - Lower control arm ball joint. Tighten the nut to 19 ft. lbs. (25 Nm).
   - Stabilizer bar
   - New axle nut and washer and tighten to 181 ft. lbs. (245 Nm)
   - Brake rotor and caliper. Torque the caliper bolts to 74 ft. lbs. (100 Nm).
   - Front WSS, if equipped
   - Front wheel
5. Lower the vehicle

Rear

ECLIPSE

➡The hub and bearing assembly is serviced as a unit.

1. Before servicing the vehicle, refer to the precautions at the beginning of this section.
2. Disconnect the negative battery cable.
3. Raise and safely support the vehicle.
4. Remove or disconnect the following:
   - Wheel and tire assembly
   - Rear wheel speed sensor if equipped with Anti-Lock Brake (ABS)
   - Brake drum. Or, if equipped with disc brakes, remove the caliper assembly and rotor. Suspend the caliper out of the way with wire.
5. On vehicles with rear disc brakes, remove the parking brake shoes.

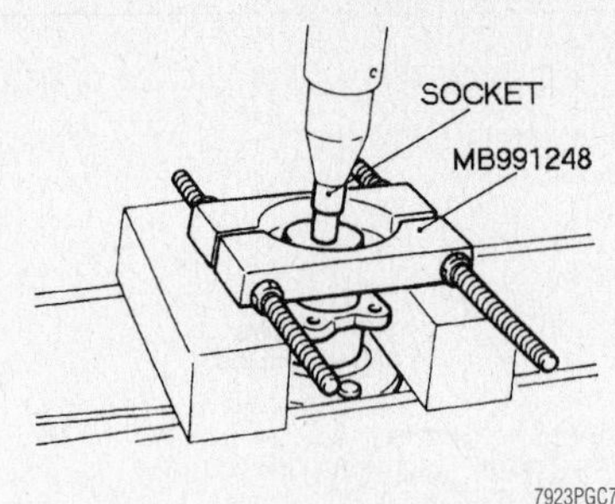

Use a press to remove the speed sensor rotor from the hub—Galant

6. Remove the hub mounting bolts from behind the backing plate and remove the hub.

➡The rotor for the ABS must be removed and installed using a press.

To install:

7. Press the rotor (ABS) to the hub.
8. Install or connect the following:
   - Hub and tighten the mounting bolts to 54–65 ft. lbs. (74–88 Nm)
   - Parking brake shoes if equipped
   - Rotor and caliper or drum
   - Speed sensor if equipped
   - Wheel and tire assembly
9. Lower the vehicle to the floor.
10. Connect the negative battery cable.

GALANT & DIAMANTE

1. Before servicing the vehicle, refer to the precautions at the beginning of this section.
2. Raise the vehicle and support safely.
3. Remove the appropriate wheel assembly.
4. If equipped with Anti-Lock Brake (ABS), remove the Vehicle Speed Sensor (VSS).
5. Remove the brake drum from the hub assembly.
6. From the back of the knuckle, remove the 4 bolts securing the hub to the knuckle.
7. Remove the hub and bearing assembly from the knuckle.

➡The hub assembly is not serviceable and should not be disassembled.

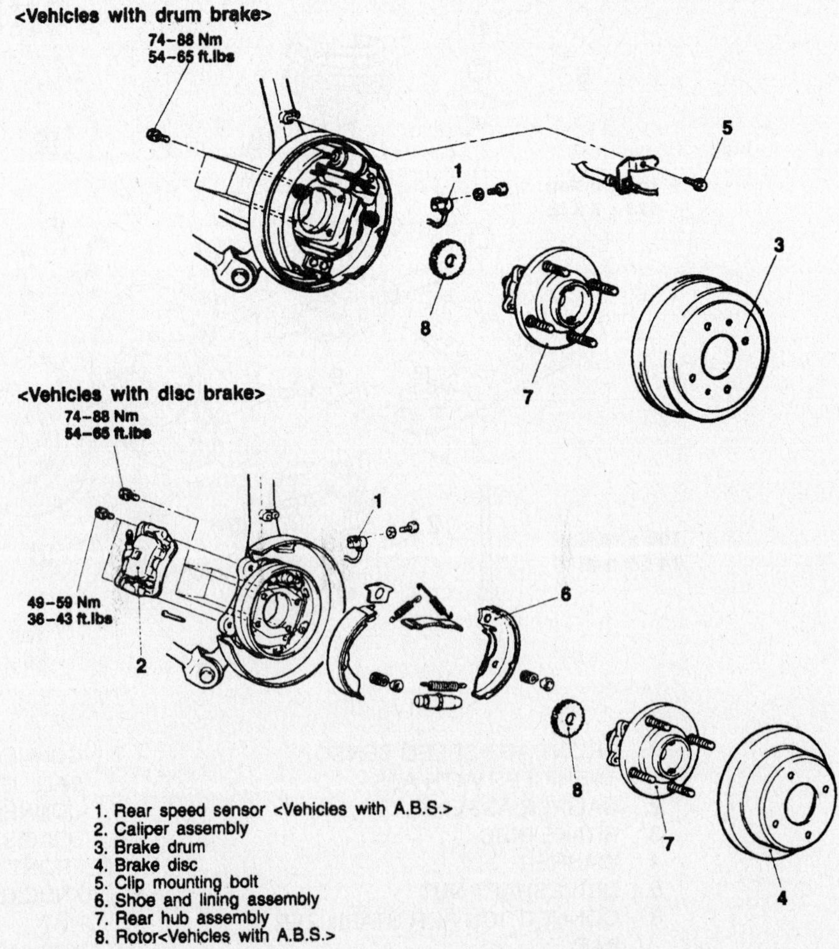

Exploded view of the rear hub/bearing assembly and related components—Galant

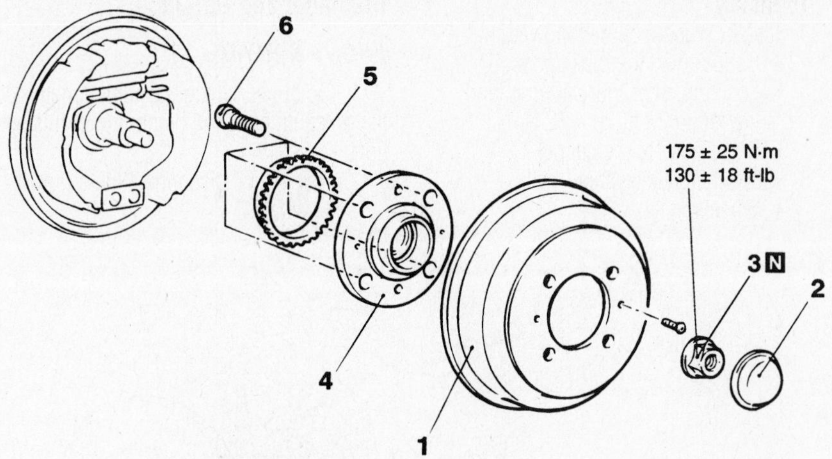

Exploded of the rear wheel hub and bearing—Lancer shown, Mirage similar

8. If replacing the hub, use special socket MB991248 and a press, to remove the wheel sensor rotor from the hub.

To install

9. Press the wheel sensor rotor onto the hub.
10. Install or connect the following:
    - Hub to the knuckle and tighten the mounting bolts to 54–65 ft. lbs. (74–88 Nm)
    - Brake drum on the hub
    - VSS, if equipped with ABS
    - Wheel assembly and lower the vehicle

LANCER & MIRAGE

➡The wheel bearing is serviced by replacement of the hub.

1. Before servicing the vehicle, refer to the precautions at the beginning of this section.
2. If equipped with Anti-Lock Brake (ABS), remove the wheel speed sensor.
3. Raise and safely support the vehicle.
4. Remove or disconnect the following:
    - Rear wheel
    - Caliper and brake disc or brake drum
    - Dust cap and flange nut
    - Rear hub assembly

To install:

5. Install or connect the following:
    - Rear hub assembly using a new flange nut. Torque the flange nut to 130 ft. lbs. (180 Nm).
    - Dust cap
    - Wheel speed sensor if removed. The air gap should be 0.012–0.035 in. (0.3–0.9mm).
    - Brake disc and caliper, or brake drum
    - Rear wheel assembly and lower the vehicle to the floor

BRAKES

Brake Caliper

REMOVAL & INSTALLATION

Mirage and Lancer

FRONT

1. Before servicing the vehicle, refer to the precautions at the beginning of this section.
2. Remove or disconnect the following:
    - Wheels
    - Brake hose
    - Caliper guide and lock pins
    - Caliper assembly from the caliper support

To install

3. Install or connect the following:
    - Brake caliper into position on the caliper support
    - Guide and lock pins
    - Brake hose. Bleed the brake system.
    - Wheels

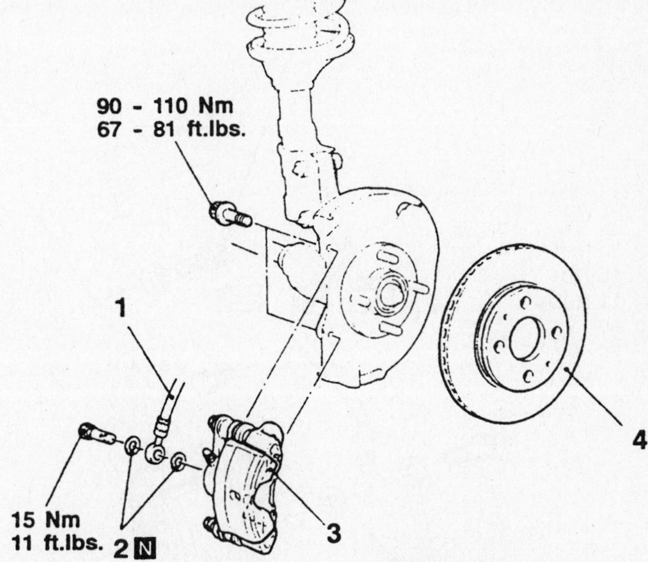

Front disc brakes—Mirage

Eclipse

FRONT

1. Before servicing the vehicle, refer to the precautions at the beginning of this section.
2. Remove or disconnect the following:
   - Wheels
   - Brake hose
   - Caliper guide and lock pins
   - Caliper assembly from the caliper support

To install

3. Install or connect the following:
   - Brake caliper into position on the caliper support
   - Guide and lock pins
   - Brake hose. Bleed the brake system.
   - Wheels

REAR

1. Before servicing the vehicle, refer to the precautions at the beginning of this section.
2. Loosen the parking brake cable adjustment from inside the vehicle.
3. Remove or disconnect the following:
   - Wheels
   - Brake hose
   - Caliper lock pin. Pivot the caliper upward, and slide the caliper assembly from the caliper support.

To install:

4. Install or connect the following:
   - Caliper over the brake pads
   - Lock pin, after lubricating it, and tighten to 23 ft. lbs. (32 Nm)
   - Brake hose to the caliper
5. Bleed the brake system.
   - Wheels

Diamante and Galant

FRONT & REAR

1. Before servicing the vehicle, refer to the precautions at the beginning of this section.
2. Remove or disconnect the following:
   - Wheels

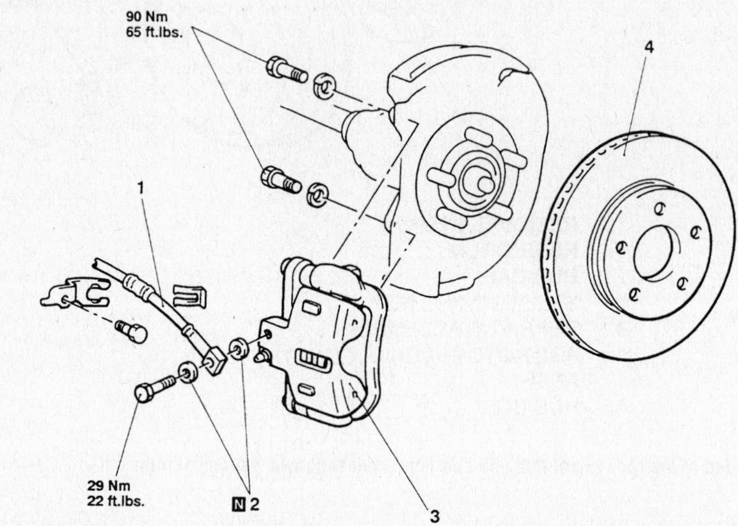

Front disc brakes—Diamante

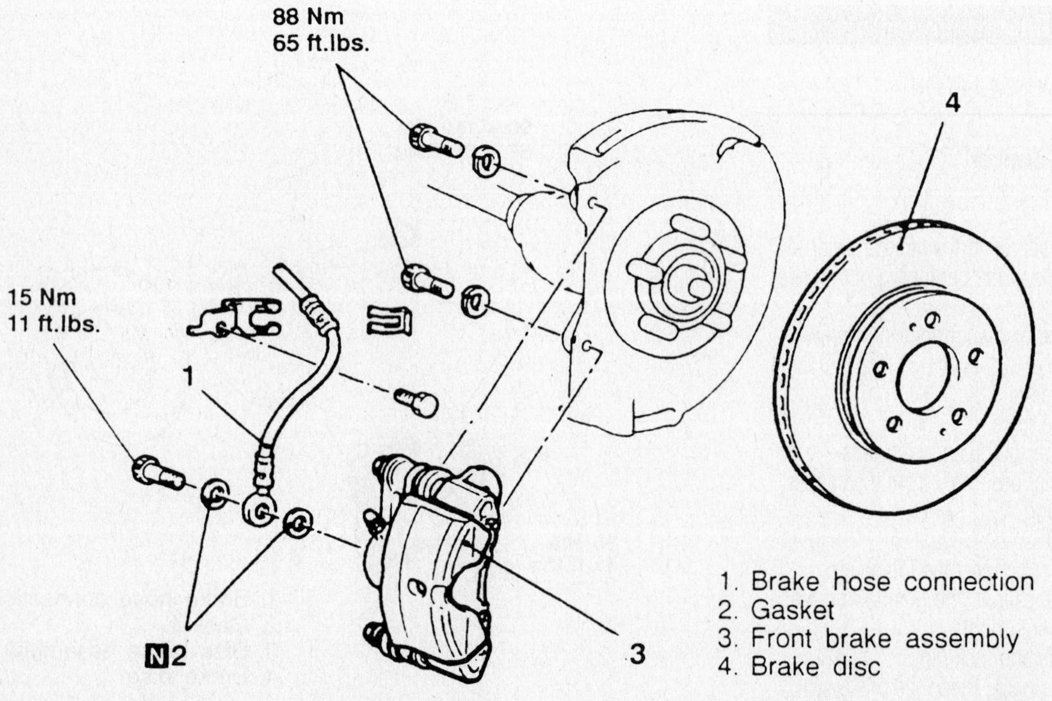

Front and rear disc brakes—Eclipse

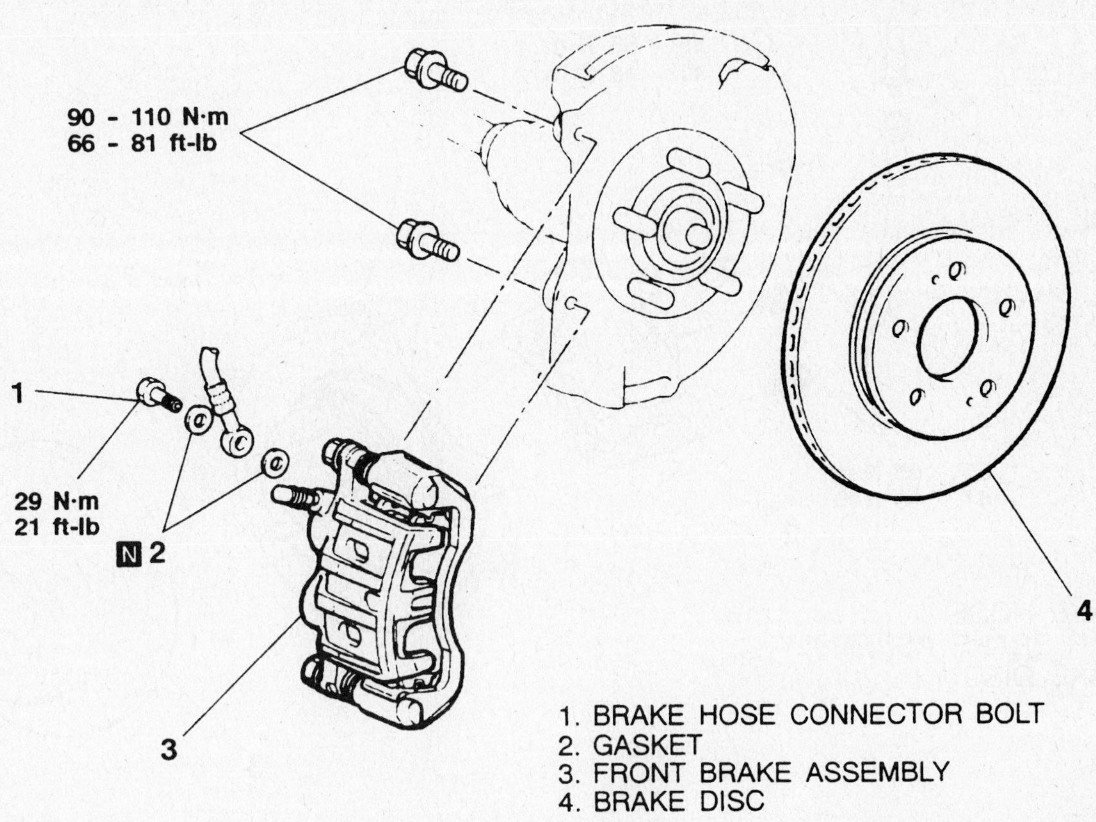

Front disc brakes—Galant

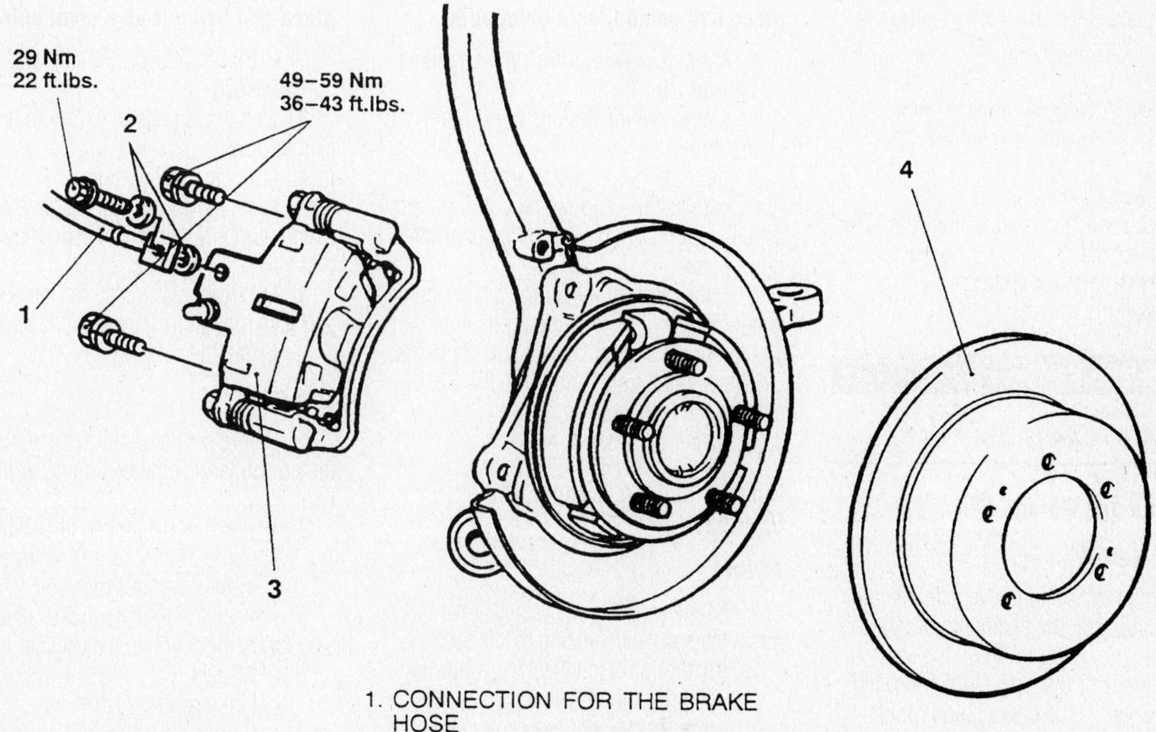

Rear disc brakes—Diamante

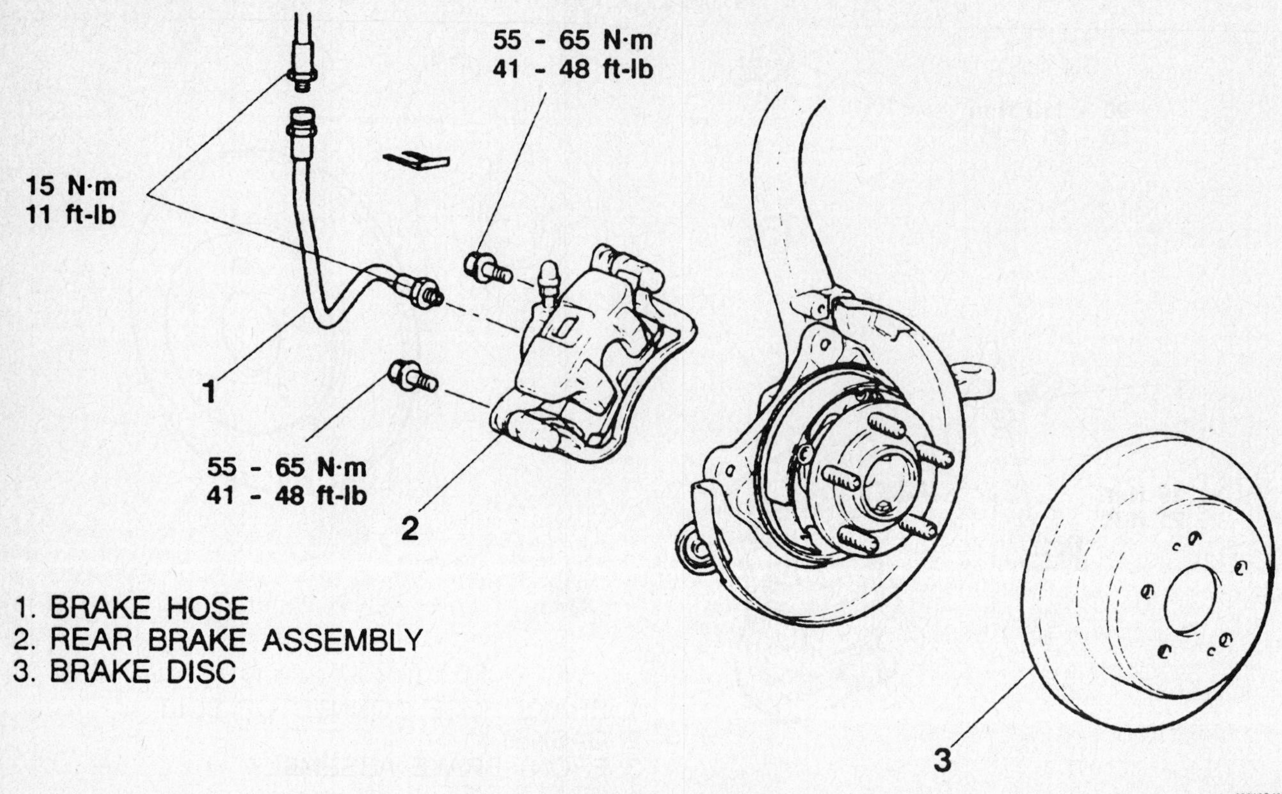

Rear disc brakes—Galant

- Brake hose from the caliper
- Caliper guide and lock pins and lift the caliper assembly from the caliper support

To install

3. Install or connect the following:
   - Caliper onto the caliper support
   - Guide pin and lock pin and tighten to specification
   - Brake hose or banjo bolt with new washers
4. Bleed the brake system.
5. Install the wheels.

Disc Brake Pads

REMOVAL & INSTALLATION

Diamante and Galant

FRONT & REAR

1. Before servicing the vehicle, refer to the precautions at the beginning of this section.
2. Remove or disconnect the following:
   - Some of the brake fluid from the master cylinder reservoir
   - Wheels
   - Caliper guide and lock pins and lift the caliper assembly from the caliper support.

➡On some vehicles, the caliper can be flipped up by leaving the upper pin in place and using it as a pivot point.

- Brake pads, spring clip and shims

To install:

3. Compress the pistons back into the caliper bore.
4. Lubricate slide points and install the brake pads, shims and spring clip onto the caliper support. Install the caliper over the brake pads.
5. Lubricate and install the caliper guide and lock pins in their original positions. Tighten guide and locking pins to 54 ft. lbs. (75 Nm) on the front, and 20 ft. lbs. (27 Nm) on the rear.
6. Install the wheels.

Mirage, Eclipse, Lancer and Sportback

FRONT

1. Before servicing the vehicle, refer to the precautions at the beginning of this section.
2. Remove or disconnect the following:
   - Some of the brake fluid from the master cylinder reservoir
   - Wheels
   - Caliper guide and lock pins and lift the caliper assembly from the caliper support.

➡On some vehicles, the caliper can be flipped up by leaving the upper pin in place and using it as a pivot point.

- Brake pads, spring clip and shims

To install:

3. Compress pistons back into the caliper bore.
4. Lubricate slide points and install the brake pads, shims and spring clips onto the caliper support. Install the caliper over the brake pads.
5. Lubricate and install the caliper guide and lock pins in their original positions.
6. Install the wheels.

REAR

1. Before servicing the vehicle, refer to the precautions at the beginning of this section.
2. Remove or disconnect the following:
   - Some of the brake fluid from the master cylinder reservoir
3. Loosen the parking brake cable adjustment from inside the vehicle.
   - Wheels
   - Parking brake cable
   - Caliper lower pin and swing the caliper assembly upwards
   - Outer shim, brake pads and spring clips from the caliper support.
4. On the Eclipse model, compress the

piston into the caliper bore. On the Mirage and Galant models, thread the piston into the caliper bore clockwise using disc brake driver tool MB990652.

To install:

5. Lubricate all sliding and pivot points. Install the brake pads, shims and spring clip to the caliper support. Install the caliper over the brake pads.
6. Lubricate, install and tighten the lower pin.
7. Install the wheels.

Brake Drums

REMOVAL & INSTALLATION

Mirage and Lancer

1. Before servicing the vehicle, refer to the precautions at the beginning of this section.
2. Remove or disconnect the following:
   - Wheels
   - Dust cap
   - Self-locking nut
   - Outer wheel bearing
   - Drum with the inner wheel bearing from the spindle
   - Grease seal

To install:

3. To determine if the self-locking nut is reusable:
   a. Screw in the self-locking nut until about ⅛ in. of the spindle is showing.
   b. Measure the torque required to turn the self-locking nut counterclockwise.
   c. The lowest allowable torque is 48 inch lbs. (5.5 Nm). If the measured torque is less than the specification, replace the nut.
4. Install or connect the following:
   - Inner wheel bearing, after lubricating it
   - New grease seal
   - Drum to the spindle
   - Outer wheel bearing, making sure to lubricate it first
   - Self-locking nut. Torque the nut to 108–145 ft. lbs. (150–200 Nm).
   - Grease cap
   - Wheels

Galant and Eclipse

1. Before servicing the vehicle, refer to the precautions at the beginning of this section.
2. Remove or disconnect the following:
   - Rear wheels
   - Drum from the rear hub assembly

To install:

3. Install or connect the following:
   - Drum on the rear hub assembly
   - Wheels

Brake Shoes

REMOVAL & INSTALLATION

Mirage and Lancer

1. Before servicing the vehicle, refer to the precautions at the beginning of this section.

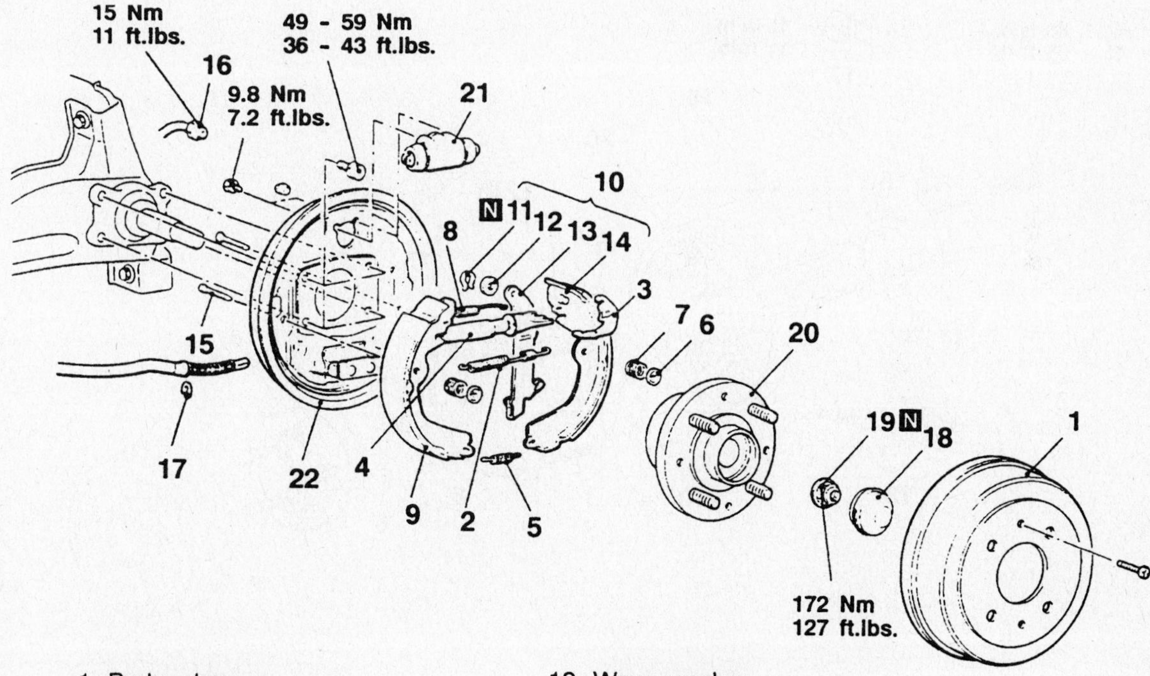

1. Brake drum
2. Shoe-to-lever spring
3. Adjuster lever
4. Auto adjuster assembly
5. Retainer spring
6. Shoe hold-down cup
7. Shoe hold-down spring
8. Shoe-to-shoe spring
9. Shoe and lining assembly
10. Shoe, lining and lever assembly
11. Retainer
12. Wave washer
13. Parking lever
14. Shoe and lining assembly
15. Shoe hold-down pin
16. Brake pipe connection
17. Snap ring
18. Hub cap
19. Flange nut
20. Rear hub assembly
21. Wheel cylinder
22. Backing plate

93016G46

Rear drum brakes—Mirage

2. Remove or disconnect the following:
   - Wheels
   - Brake drum
   - Shoe-to-shoe spring
   - Shoe-to-lever spring and adjuster assembly
   - Shoe hold-down clips and the brake shoes
   - Parking brake cable from the rear shoes by spreading the horseshoe clip apart.

To install

3. Lubricate the backing plate bosses, anchor pin, and parking brake actuating mechanism with lithium-based grease.
4. Install or connect the following:
   - Parking brake arm to the appropriate brake shoe
   - Brake shoes and the shoe hold-down clips
   - Adjuster assembly and the shoe-to-lever spring
   - Shoe-to-shoe spring
5. Pre-adjust the shoes so the drum slides on with a light drag and install brake drum.
   - New wheel bearing self-locking nut and torque to 130 ft. lbs. (180 Nm)
   - Wheel bearing dust cap and adjust the rear brake shoes

Galant and Eclipse

1. Before servicing the vehicle, refer to the precautions at the beginning of this section.
2. Remove or disconnect the following:
   - Rear wheels and drums
   - Lever return spring
   - Shoe-to-lever spring
   - Adjuster lever
   - Auto-adjuster assembly
   - Retainer spring
   - Brake shoe hold-down springs and spring cups
   - Shoe-to-shoe spring
   - Brake shoes
   - Parking brake cable from the lever on the rear shoe

To install:

3. Remove the parking brake lever from the used shoe and install it on the new brake shoe. Make sure the wave washer is installed in the proper direction.
4. Clean the backing plate and lightly apply brake grease to the 6 shoe support pads.
5. Clean the adjuster assembly and apply brake grease to the threads.
6. Install or connect the following:
   - Parking brake cable to the lever on the rear shoe
   - Rear shoe on the backing plate and the hold-down spring and pin
   - Front shoe on the backing plate and the hold-down spring and pin
   - Adjuster assembly between the 2 shoes
   - Shoe-to-shoe spring
   - Retainer spring
   - Adjuster lever
   - Shoe-to-lever spring
   - Lever return spring
7. Adjust the brake shoes
   - Drum
   - Wheels

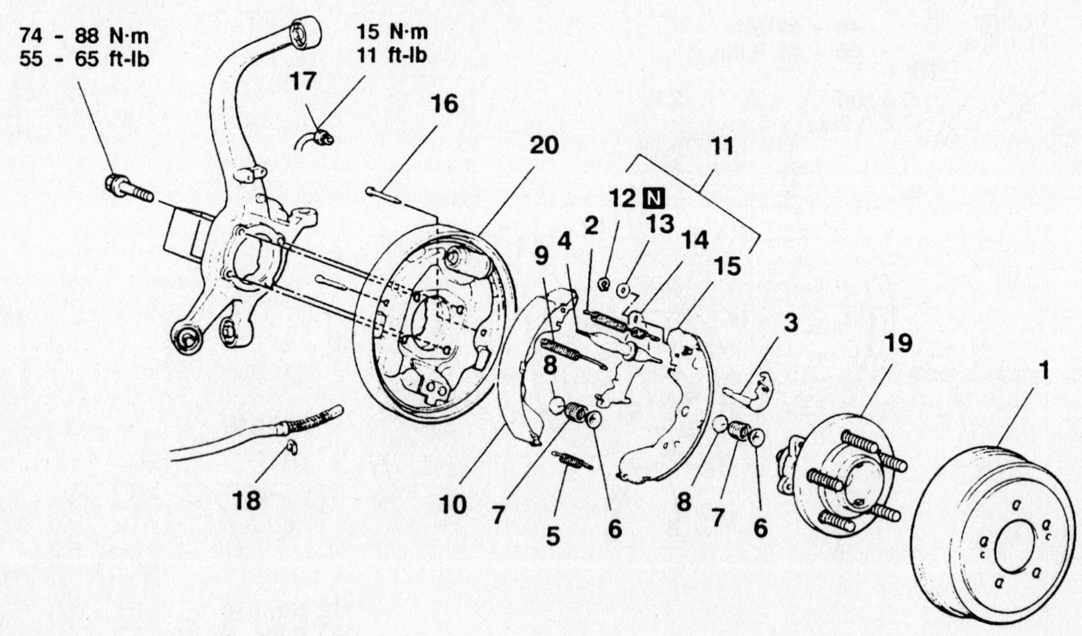

1. BRAKE DRUM
2. SHOE-TO-LEVER SPRING
3. ADJUSTER LEVER
4. AUTO ADJUSTER ASSEMBLY
5. RETAINER SPRING
6. SHOE HOLD-DOWN CUP
7. SHOE HOLD-DOWN SPRING
8. SHOE HOLD-DOWN CUP
9. SHOE-TO-SHOE SPRING
10. SHOE AND LINING ASSEMBLY
11. SHOE AND LEVER ASSEMBLY
12. RETAINER
13. WAVE WASHER
14. PARKING LEVER
15. SHOE AND LINING ASSEMBLY
16. SHOE HOLD-DOWN PIN
17. BRAKE TUBE CONNECTION
18. SNAP RING
19. REAR HUB ASSEMBLY
20. BACKING PLATE

93016G47

Rear drum brakes—Galant and Eclipse

MITSUBISHI

2

Endeavor

SPECIFICATION AND MAINTENANCE CHARTS

ENGINE AND VEHICLE IDENTIFICATION CHART

Engine						
Code	Liters (cc)	Cu. In.	Cyl.	Fuel Sys.	Engine Type	Eng. Mfg.
6G75/S	3.8 (3828)	233.6	6	MFI	SOHC	Mitsubishi

Model Year	
Code	Year
4	2004
5	2005

MFI: Multi-port Fuel Injection

67170-ENDE-C01

GENERAL ENGINE SPECIFICATIONS

Year	Engine Displacement Liters	Engine ID/VIN	Net Horsepower @ rpm	Net Torque @ rpm (ft. lbs.)	Bore x Stroke (in.)	Com-pression Ratio	Oil Pressure @ rpm
2004	3.8	6G75/S	225@5500	255@3750	3.74x3.54	10.0:1	①

① 4.2 or more psi @ idle

67170-ENDE-C02

ENGINE TUNE-UP SPECIFICATIONS

Year	Engine Displacement Liters	Engine ID/VIN	Spark Plugs Gap (in.)	Ignition Timing (deg.)		Fuel Pump (psi)	Idle Speed (rpm)		Valve Clearance	
				MT	AT		MT	AT	In.	Ex.
2004	3.8	6G75/S	0.028-0.031	—	①	38 ②	—	680	HYD	HYD

B: Before top dead center

HYD: Hydraulic

① Base ignition timing: 2-8 degrees BTDC
Actual ignition timing: 10 degrees BTDC

② With vacuum hose connected

67170-ENDE-C03

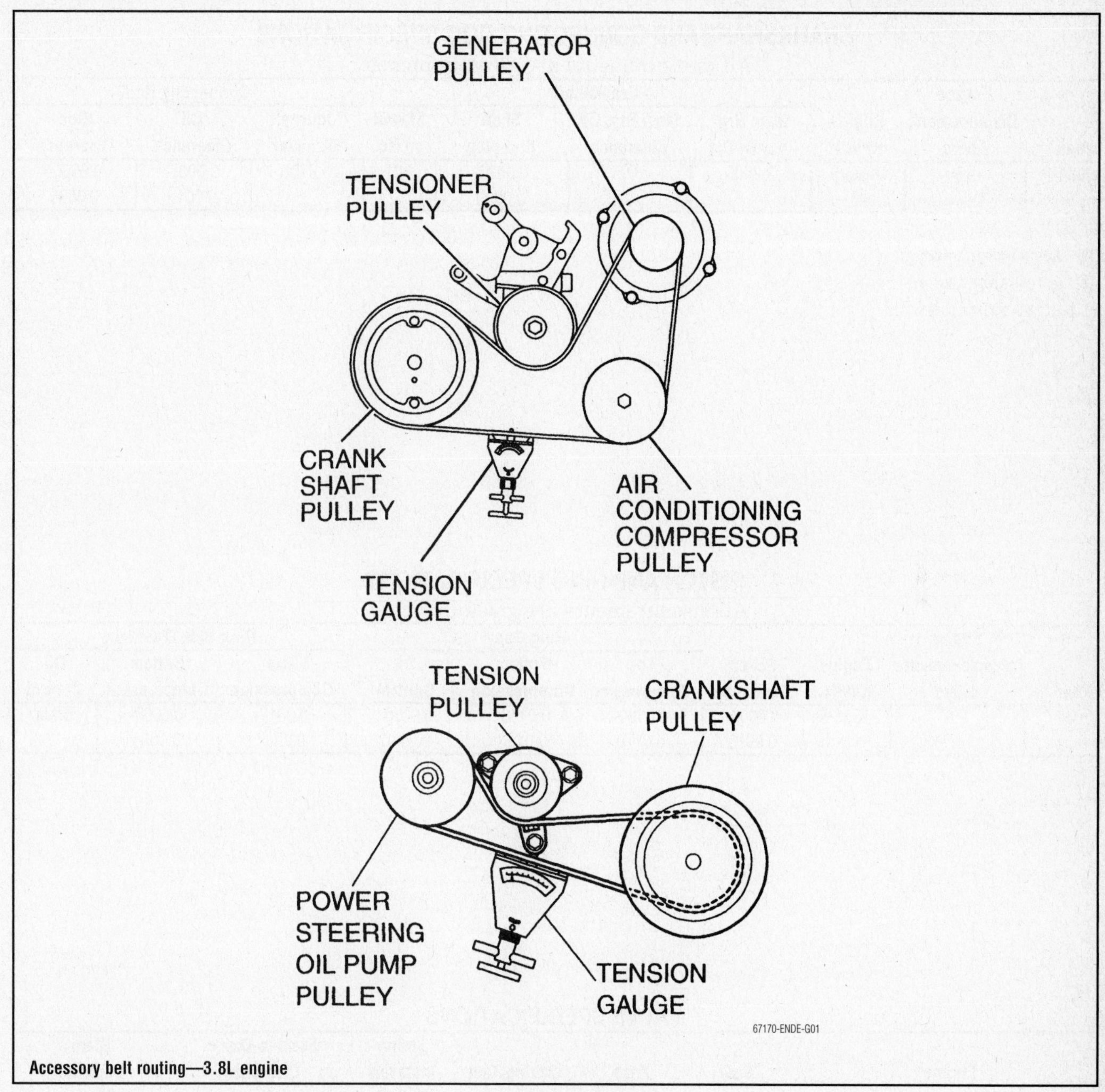

Accessory belt routing—3.8L engine

CAPACITIES

Year	Model	Engine Displacement Liters	Engine ID/VIN	Engine Oil with Filter (qts.)	Transmission (pts.) 5-Spd	Transmission (pts.) Auto.	Transfer Case (pts.)	Drive Axle Front (pts.)	Drive Axle Rear (pts.)	Fuel Tank (gal.)	Cooling System (qts.)
2004	Endeavor	3.8	6G75/S	4.5	—	①	1.12	—	2.4	21.4	②

① FWD: 17.8 pts.
AWD: 18.6 pts.

② FWD & AWD without tow kit: 9.5 qts.
AWD with tow kit: 10.1 qts.

67170-ENDE-C04

CRANKSHAFT AND CONNECTING ROD SPECIFICATIONS

All measurements are given in inches.

	Engine		Crankshaft				Connecting Rod		
Year	Displacement Liters	Engine ID/VIN	Main Brg. Journal Dia.	Main Brg. Oil Clearance	Shaft End-play	Thrust on No.	Journal Diameter	Oil Clearance	Side Clearance
2004	3.8	6G75/S	2.2500	①	0.0020-0.0090	NS	1.9700	0.0008-0.0015	0.0030-0.0090

NS - Not specified by manufacturer

① Nos. 1 & 4: 0.0008-0.0012 in

Nos. 2 & 3: 0.0012-0.0016 in.

67170-ENDE-C05

PISTON AND RING SPECIFICATIONS

All measurements are given in inches.

	Engine			Ring Gap			Ring Side Clearance		
Year	Displacement Liters	Engine ID/VIN	Piston Clearance	Top Compression	Bottom Compression	Oil Control	Top Compression	Bottom Compression	Oil Control
2004	3.8	6G75/S	0.0008-0.0015	0.0100-0.0170	0.0140-0.0190	0.0030-0.0140	0.0012-0.0027	0.0008-0.0023	Snug

67170-ENDE-C06

VALVE SPECIFICATIONS

Year	Engine Displacement Liters	Engine ID/VIN	Seat Angle (deg.)	Face Angle (deg.)	Spring Test Pressure (lbs. @ in.)	Spring Installed Height (in.)	Stem-to-Guide Clearance (in.) Intake	Stem-to-Guide Clearance (in.) Exhaust	Stem Diameter (in.) Intake	Stem Diameter (in.) Exhaust
2004	3.8	6G75/S	NS	43.5-44	60@1.74	1.740	0.0008-0.0019	0.0016-0.0023	0.240	0.240

NS - Not specified by manufacturer

67170-ENDE-C07

TORQUE SPECIFICATIONS

All readings in ft. lbs.

Year	Engine Displacement Liters	Engine ID/VIN	Cylinder Head Bolts	Main Bearing Bolts	Rod Bearing Bolts	Crankshaft Damper Bolts	Flywheel Bolts	Manifold Intake	Manifold Exhaust	Spark Plugs	Oil Pan Drain Plug
2004	3.8	6G75/S	①	54	②	136	55	16	③	18	29

① Step 1: 80 ft. lbs.
Step 2: Loosen completely
Step 3: 80 ft. lbs.

② 20 ft. lbs., plus an additional 90-94 degrees

② Exhaust manifold nut: 33 ft. lbs.
Exhaust manifold stay bolt (M8): 14 ft. lbs.
Exhaust manifold stay bolt (M10): 33 ft. lbs.
Exhaust manifold stay bolt (M12): 56 ft. lbs.

67170-ENDE-C08

BRAKE SPECIFICATIONS

All measurements in inches unless noted

			Brake Disc			Brake Drum Diameter			Minimum Lining Thickness		Brake Caliper	
Year	Model		Original Thickness	Minimum Thickness	Maximum Runout	Original Inside Diameter	Max. Wear Limit	Maximum Machine Diameter	Front	Rear	Bracket Bolts (ft. lbs.)	Mounting Bolts (ft. lbs.)
2004	Endeavor	F	1.020	0.960	0.0012	—	—	—	0.080	—	74	34
		R	0.390	0.330	0.0039	—	—	—	—	0.080	45	44

67170-ENDE-C09

WHEEL ALIGNMENT

			Caster		Camber		
Year	Model		Range (+/-Deg.)	Preferred Setting (Deg.)	Range (+/-Deg.)	Preferred Setting (Deg.)	Toe-in (in.)
2004	Endeavor	F	0.30	+3.00	0.30	0	0.0+/-0.12
		R	—	—	0.30	-0.50	0.12+/-0.12

67170-ENDE-C10

TIRE, WHEEL AND BALL JOINT SPECIFICATIONS

		OEM Tires		Tire Pressures (psi)		Wheel	Ball Joint	Lug Nut
Year	Model	Standard	Optional	Front	Rear	Size	Inspection	(ft. lbs.)
2004	Endeavor	P235/65R17	None	29	29	7-JJ	U: 7-30 in. ① L: 0.010 in.	100

OEM: Original Equipment Manufacturer

PSI: Pounds Per Square Inch

STD: Standard

OPT: Optional

① Torque required in inch lbs. to rotate ball joint when removed from the knuckle

67170-ENDE-C11

SCHEDULED MAINTENANCE INTERVALS
Mitsubishi—Endeavor

TO BE SERVICED	TYPE OF SERVICE	VEHICLE MILEAGE INTERVAL (x1000)												
		7.5	15	22.5	30	37.5	45	52.5	60	67.5	75	82.5	90	97.5
Engine oil & filter	R	✓	✓	✓	✓	✓	✓	✓	✓	✓	✓	✓	✓	✓
Automatic transmission & transfer oil	S/I		✓		✓		✓		✓		✓		✓	
Brake hoses	S/I		✓		✓		✓		✓		✓		✓	
Disc brake pads & rotors	S/I		✓		✓		✓		✓		✓		✓	
Drive shaft boots	S/I		✓		✓		✓		✓		✓		✓	
Air cleaner filter	R				✓				✓				✓	
Automatic transmission & transfer oil (4WD)	R				✓				✓				✓	
Engine coolant	R				✓				✓				✓	
Ball joints & steering linkage seals	S/I				✓				✓				✓	
Drive belt(s)	S/I				✓				✓				✓	
Exhaust system	S/I				✓				✓				✓	
Front & rear axle	S/I				✓				✓				✓	
Fuel hoses	S/I				✓				✓				✓	
Propeller shaft joint	S/I				✓				✓				✓	
Ignition cables	R								✓					
Timing belt	R								✓					
Distributor cap & rotor	S/I								✓					
EVAP system (except EVAP canister)	S/I								✓					
EGR valve ①	S/I													
EVAP canister ①	S/I													
PCV system ②	S/I													
Spark plugs ③	R													

R: Replace S/I: Service or Inspect

① Replace at 100,000 miles.

② PCV system (except EVAP canister): service or inspect at 100,000 miles.

③ Iron tips: 30,000 miles
Platinum tips: 60,000 miles
Irridium tips: 100,000 miles

FREQUENT OPERATION MAINTENANCE (SEVERE SERVICE)

If a vehicle is operated under any of the following conditions it is considered severe service:

- Extremely dusty areas.

- 50% or more of the vehicle operation is in 32°C (90°F) or higher temperatures, or constant operation in temperatures below 0°C (32°F).

- Prolonged idling (vehicle operation in stop and go traffic).

- Frequent short running periods (engine does not warm to normal operating temperatures).

- Police, taxi, delivery usage or trailer towing usage.

Oil & oil filter: replace every 3000 miles.

Front disc brake pads (dusty or salty conditions): service or inspect every 6000 miles.

Front disc brake pads: service or inspect every 7500 miles.

Air cleaner filter: service or inspect every 15,000 miles.

Rear drum brake linings & rear wheel cylinders: service or inspect every 15,000 miles.

Spark plugs (iron tip): replace every 15,000 miles.

PCV system: service or inspect every 60,000 miles.

67170-ENDE-C12

PRECAUTIONS

Before servicing any vehicle, please be sure to read all of the following precautions, which deal with personal safety, prevention of component damage, and important points to take into consideration when servicing a motor vehicle:

- Never open, service or drain the radiator or cooling system when the engine is hot; serious burns can occur from the steam and hot coolant.
- Observe all applicable safety precautions when working around fuel. Whenever servicing the fuel system, always work in a well-ventilated area. Do not allow fuel spray or vapors to come in contact with a spark, open flame, or excessive heat (a hot drop light, for example). Keep a dry chemical fire extinguisher near the work area. Always keep fuel in a container specifically designed for fuel storage; also, always properly seal fuel containers to avoid the possibility of fire or explosion. Refer to the additional fuel system precautions later in this section.
- Fuel injection systems often remain pressurized, even after the engine has been turned **OFF**. The fuel system pressure must be relieved before disconnecting any fuel lines. Failure to do so may result in fire and/or personal injury.
- Brake fluid often contains polyglycol ethers and polyglycols. Avoid contact with the eyes and wash your hands thoroughly after handling brake fluid. If you do get brake fluid in your eyes, flush your eyes with clean, running water for 15 minutes. If eye irritation persists, or if you have taken brake fluid internally, IMMEDIATELY seek medical assistance.
- The EPA warns that prolonged contact with used engine oil may cause a number of skin disorders, including cancer! You should make every effort to minimize your exposure to used engine oil. Protective gloves should be worn when changing oil. Wash your hands and any other exposed skin areas as soon as possible after exposure to used engine oil. Soap and water, or waterless hand cleaner should be used.
- All new vehicles are now equipped with an air bag system. The system must be disabled before performing service on or around system components, steering column, instrument panel components, wiring and sensors. Failure to follow safety and disabling procedures could result in accidental air bag deployment, possible personal injury and unnecessary system repairs.
- Always wear safety goggles when working with, or around, the air bag system. When carrying a non-deployed air bag, be sure the bag and trim cover are pointed away from your body. When placing a non-deployed air bag on a work surface, always face the bag and trim cover upward, away from the surface. This will reduce the motion of the module if it is accidentally deployed. Refer to the additional air bag system precautions later in this section.
- Clean, high quality brake fluid from a sealed container is essential to the safe and proper operation of the brake system. You should always buy the correct type of brake fluid for your vehicle. If the brake fluid becomes contaminated, completely flush the system with new fluid. Never reuse any brake fluid. Any brake fluid that is removed from the system should be discarded. Also, do not allow any brake fluid to come in contact with a painted surface; it will damage the paint.
- Never operate the engine without the proper amount and type of engine oil; doing so WILL result in severe engine damage.
- Timing belt maintenance is extremely important! Many models utilize an interference-type, non-freewheeling engine. If the timing belt breaks, the valves in the cylinder head may strike the pistons, causing potentially serious (also time-consuming and expensive) engine damage. Refer to the maintenance interval charts in the front of this section for the recommended replacement interval for the timing belt, and to the timing belt procedure for belt replacement and inspection.
- Disconnecting the negative battery cable on some vehicles may interfere with the functions of the on-board computer system(s) and may require the computer to undergo a relearning process once the negative battery cable is reconnected.
- When servicing drum brakes, only disassemble and assemble one side at a time, leaving the remaining side intact for reference.

ENGINE REPAIR

➡Disconnecting the negative battery cable on some vehicles may interfere with the functions of the on board computer systems and may require the computer to undergo a relearning process, once the negative battery cable is reconnected.

Distributor

The 3.8L engine is distributorless.

Alternator

REMOVAL

1. Before servicing the vehicle, refer to the precautions in the beginning of this section.
2. Remove or disconnect the following:
   - Negative battery cable
   - Under cover
   - Side under cover
   - Drive belt
   - Alternator connector
   - A/C compressor connector
   - A/C compressor and position aside. Do NOT disconnect the refrigerant lines.
   - Alternator

To install:

3. Install or connect the following:
   - Alternator. Torque the bolts to 36 ft. lbs. (49 Nm).
   - A/C compressor
   - A/C compressor connector
   - Alternator connector. Torque the nut to 102 inch lbs. (12 Nm).
   - Drive belt
   - Side under cover
   - Under cover
   - Negative battery cable

Ignition Timing

ADJUSTMENT

The ignition timing is controlled by the ECM and is not adjustable. The ECM determines the timing based on input from the crankshaft position sensor.

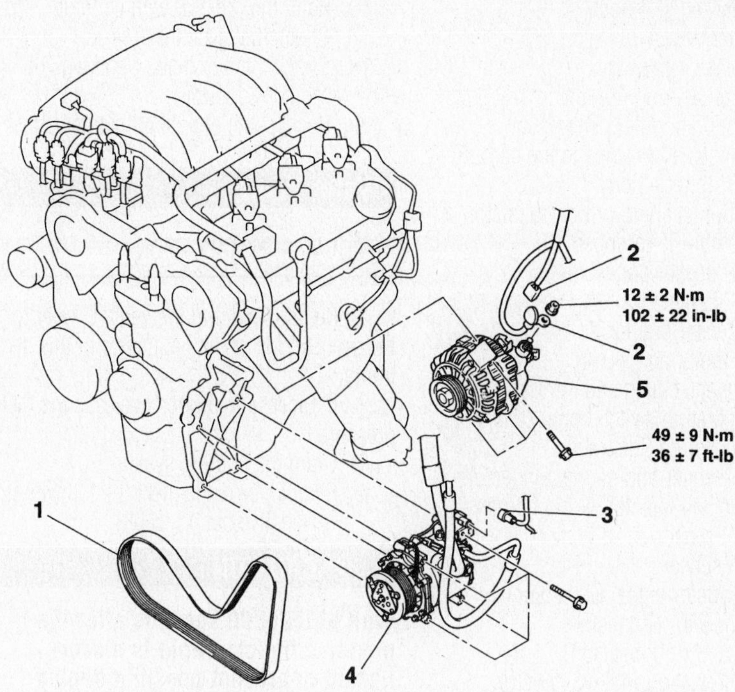

Alternator mounting and related components

TIMING CHECK

1. Before servicing the vehicle, refer to the precautions in the beginning of this section.

Before attempting to adjust the ignition timing, be sure of the following:

- The engine should be at normal operating temperature.
- The lights and all accessories should be OFF.
- The transmission should be in **P** or **N**.

2. Connect scan tool MB991958 to the data link connector
3. Set up the timing light.
4. Start the engine and run at idle.
5. Verify that the idle speed is about 680 rpm.
6. Select scan tool MB991958 actuator test "item number 17".
7. Check that basic timing is with in standard, it should be 2–8° BTDC.
8. If the base timing is out of specification:
   a. Check to see if the distributor is aligned properly
   b. Check to see if the timing belt cover and Crankshaft Position (CKP) sensor installation is conditions.
   c. Crankshaft sensing blade conditions.
9. Press the clear key on the scan tool, select forced drive stop mode and cancel the actuator test.

CAUTION

If the actuator test is not canceled, the forced drive will continue for 27 minutes. Driving in this state could lead to engine failure.

10. Check that the actual ignition timing is approximately 10° BTDC.

➡Keep in mind that the ignition timing may fluctuate by as much as +/-7° BTDC even under normal operation conditions. It is also further advanced by 5–10° BTDC at higher altitudes.

Engine Assembly

REMOVAL & INSTALLATION

1. Before servicing the vehicle, refer to the precautions in the beginning of this section.
2. Relieve the fuel system pressure.

CAUTION

The fuel injection system remains under pressure after the engine has beenOFF. Properly relieve fuel pressure before disconnecting any fuel lines. Failure to do so may result in fire or personal injury.

3. Drain the engine oil.
4. Drain the cooling system.
5. Remove or disconnect the following:
   - Battery
   - Hood, matchmark for reassembly
   - Oil dipstick
   - Engine undercover
   - Starter
   - Exhaust pipe from the exhaust manifolds
   - Transfer case, if equipped with 4WD
   - Transmission, if equipped with a manual transmission
6. If equipped with an automatic transmission and 2WD:
   a. Remove the inspection plate.
   b. Matchmark the flexplate to the converter; remove the torque converter bolts and move the torque converter back as far as it will go.
   c. Remove the lower bell housing bolts.
7. Remove or disconnect the following:
   - Air cleaner assembly, ducts and air intake hose
   - Linkages and cables from the throttle body
   - Fuel lines and plug the lines
   - Air conditioning compressor, if equipped and position it aside. It is not necessary to remove the lines from the compressor.
   - Radiator, shroud
   - Cooling fan
   - Heater hoses
   - Accessory belts
   - Power steering pump and wires from its brackets and position it to the side. Do not remove the hoses from the pump.
   - Alternator and wires
   - Ignition coil and power transistor assembly, if equipped
   - MDP sensor connector
   - EGR connector
   - TP sensor connector
   - IAC motor connector
   - Magnetic clutch and refrigerant temperature switch connector
   - EVAP Purge Solenoid
   - ECT sensor and gauge connectors
   - Front and injector wiring harness

- CMP sensor
- CKP sensor
- Distributor signal Generator
- Compactor connector
- Left and right heated O_2 sensor
- Oil pressure switch connector
- Vacuum hoses

8. Attach an engine removal device to the engine support eyes on the engine.

9. If equipped with an automatic transmission, support the transmission with a floor jack. Remove the remaining bell housing bolts.

10. Remove the engine mount nuts and remove the engine from the vehicle.

To install:

11. Lower the engine into position and install the engine mount nuts and bolts. Tighten the nuts to 18–20 ft. lbs. (25–27 Nm), and the bolts to tighten to 33 ft. lbs. (44 Nm).

12. Install or connect the following:
    - Bell housing bolts

13. Remove the engine removal device and the transmission support.

14. Install or connect the following:
    - Transfer case, if equipped
    - Manual transmission, if equipped
    - Automatic transmission, if equipped align the torque converter and flexplate and the bolts.
    - Inspection plate
    - Starter motor
    - Exhaust pipe to the exhaust manifolds using new gaskets
    - Lower radiator hose
    - Heater hoses
    - Alternator and wires
    - Power steering pump and brackets
    - Air conditioning compressor
    - Linkages and cables to the carburetor or throttle body
    - Ignition coil and power transistor assembly, if equipped
    - MDP sensor connector
    - EGR connector
    - TP sensor connector
    - IAC motor connector
    - Magnetic clutch and refrigerant temperature switch connector
    - EVAP Purge Solenoid
    - ECT sensor and gauge connectors
    - Front and injector wiring harness
    - CMP sensor
    - CKP sensor
    - Distributor signal Generator
    - Compactor connector
    - Left and right heated O_2 sensor
    - Oil pressure switch connector
    - Vacuum hoses
    - Air cleaner assembly, ducts and air intake hose
    - Accessory belts
    - Radiator, shroud and upper hose
    - Cooling fan
    - Battery
    - Oil dipstick
    - Hood

15. Refill the engine with the specified amount of oil.

16. Refill the radiator with coolant.

17. Check fuel system for leaks.

18. Check the automatic transmission fluid level, if equipped.

19. Recheck all engine adjustments.

Water Pump

REMOVAL & INSTALLATION

1. Before servicing the vehicle, refer to the precautions in the beginning of this section.

2. If necessary, properly release the fuel pressure.

3. Drain the cooling system.

4. Remove or disconnect the following:
    - Negative battery cable

⁂ CAUTION

Wait at least 90 seconds after the negative battery cable is disconnected to prevent possible deployment of the air bag.

- Timing belt
- Crankshaft Position (CKP) sensor
- Water pump bracket
- Retaining bolts, water pump assembly, gasket and O-ring

To install:

5. Clean and dry the mating surfaces of the block and water pump

6. Install or connect the following:

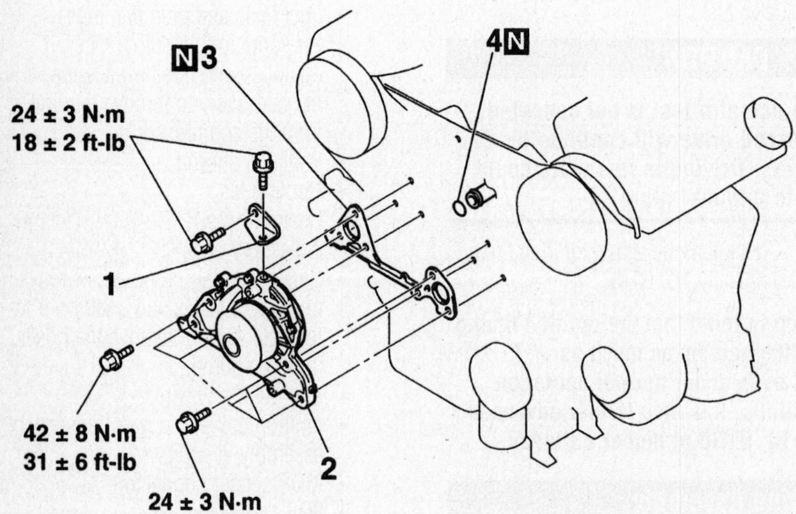

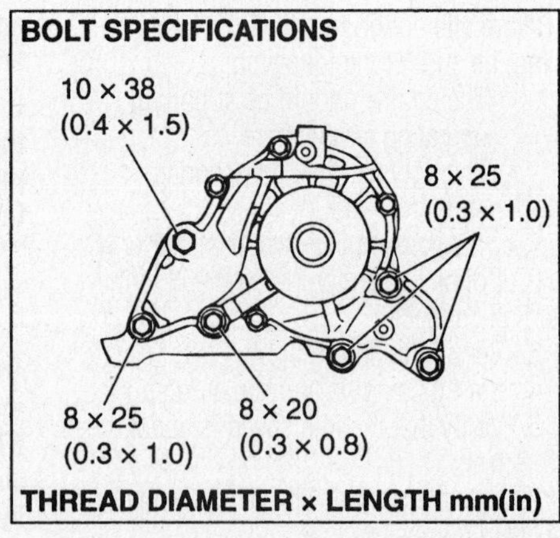

1. WATER PUMP BRACKET
2. WATER PUMP
3. WATER PUMP GASKET
4. O-RING

67170-ENDE-G03

Exploded view of the water pump and bolt tightening specifications—3.8L engine

- Water pump assembly, using a new gasket and O-ring. Torque the water pump bolts to the specifications shown in the accompanying figure.
- Water outlet fitting bracket
- CKP sensor
- Timing belt

7. Refill the radiator with coolant. This cooling system has a self-bleeding thermostat, so system bleeding is not required.
8. Run the vehicle until the thermostat opens and fill the overflow tank. Check for leaks.
9. Once the vehicle has cooled, recheck the coolant level.

Heater Core

REMOVAL & INSTALLATION

1. Before servicing the vehicle, refer to the precautions in the beginning of this section.
2. Place the wheels in the straight-ahead position.
3. Drain the cooling system into a clean container for reuse.
4. Discharge and recover the air conditioning system refrigerant.
5. Remove or disconnect the following:
   - Negative battery cable

✻✻ CAUTION

Wait at least 60 seconds after disconnecting the battery cable before performing any work on the air bag or instrument panel.

- Steering column-to-instrument panel cover screws and the cover
- Air bag module, carefully, from the steering wheel
- Electrical connectors from the air bag module

✻✻ CAUTION

Store the air bag module facing up.

- Steering wheel nut
- Press the steering wheel from the steering column, using a steering wheel puller
- Front scuff plate
- Cowl side trim
- Floor console assembly
- Front pillar trim
- Hood lock release handle
- Side cover
- Lower panel
- Plug
- Fog light switch or switch plug
- Rheostat
- Steering column lower cover
- Steering column cover protector
- Steering column upper cover
- Combination meter assembly
- Center panel assembly
- Air outlet
- Radio panel
- Heater/air conditioning control assembly
- Console box
- Heated seat switch, if equipped
- Hazard warning lamp switch
- Console meter hood
- Multi-center display unit
- Radio and CD player assembly
- Air bag caution label
- Glove box assembly, inner cover, lock assembly and damper
- Speaker panel
- Tweeter
- Clock spring connector
- Column switch connector
- Clock spring and column switch assembly
- Photo sensor and sensor harness clamp, if equipped
- Radio harness clamp
- Interior temperature sensor, if equipped
- Stay
- Passenger's air bag module nut and the air bag module
- Electrical connector from the air bag module
- Instrument panel assembly, carefully, with the help of an assistant
- Floor console side cover
- Front floor carpet
- Steering column shaft
- Blower motor assembly
- Both foot ducts
- Joint duct from the air conditioning evaporator housing assembly
- Foot distribution duct
- Center reinforcement
- Center ventilation duct
- Drain hose from the air conditioning evaporator housing assembly
- Heater hoses from the heater housing assembly
- Refrigerant lines from the air conditioning evaporator housing assembly and discard the O-rings
- Heater housing assembly
- Center duct assembly
- Heater core from the heater housing

To install:

6. Install or connect the following:
   - Heater core to the heater housing
   - Center duct assembly
   - Heater housing assembly
   - Refrigerant lines to the air conditioning evaporator housing assembly, using new O-rings
   - Heater hoses to the heater housing assembly
   - Drain hose to the air conditioning evaporator housing assembly
   - Center ventilation duct
   - Center reinforcement
   - Foot distribution duct
   - Joint duct to the air conditioning evaporator housing assembly
   - Both foot shower ducts
   - Blower motor assembly
   - Instrument panel, carefully with the help of an assistant
   - Side defroster grille
   - Multi-meter assembly and the multi-meter panel
   - Upper glove box frame and the glove box striker
   - Speaker and the combination meter
   - Heater/air conditioning control assembly
   - Radio and tape/CD player
   - Center under cover assembly
   - Glove box assembly
   - Under cover, the corner cover and the stopper
   - Meter bezel assembly
   - Knee protector assembly and bracket
   - Filler door lock release handle
   - Hood lock release handle
   - Electrical connector to the passenger's side air bag module
   - Passenger's air bag module and the air bag module nut
   - Glove box and the glove box stoppers
   - Foot shower duct
   - Steering wheel to the steering column
   - Steering wheel nut and torque the nut to 29 ft. lbs. (39 Nm)
   - Electrical connectors to the air bag module
   - Air bag module to the steering wheel
   - Steering column-to-instrument panel cover and the cover screws
   - Floor console assembly
7. Refill the cooling system.
8. Connect the negative battery.
9. Evacuate and charge the air conditioning system refrigerant.
10. Run the engine to normal operating temperatures; then, check the climate control operation and check for leaks.

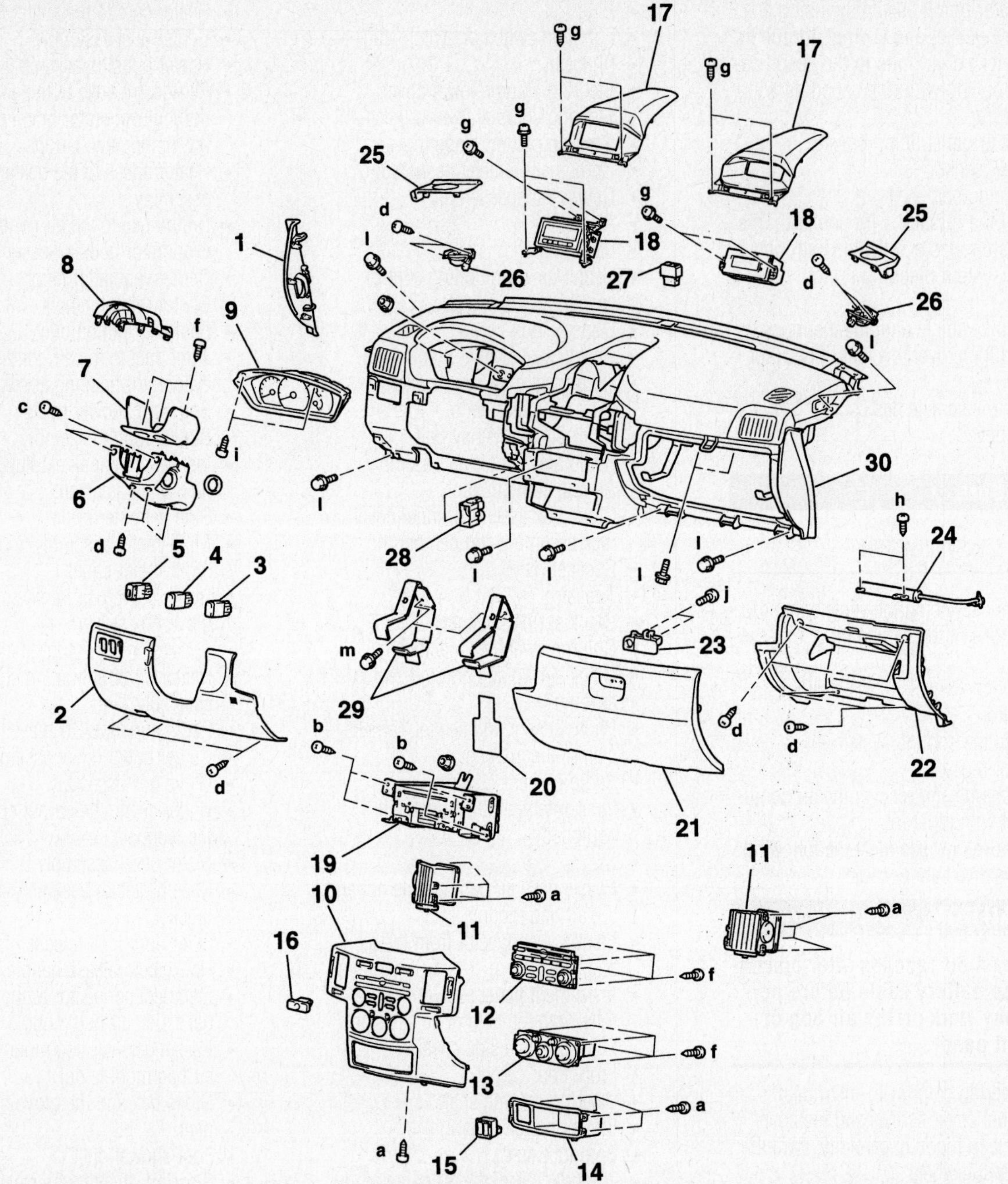

1. SIDE COVER
2. LOWER PANEL
3. PLUG
4. FOG LIGHT SWITCH OR FOG LIGHT SWITCH PLUG
5. RHEOSTAT
6. STEERING COLUMN LOWER COVER
7. STEERING COLUMN COVER PROTECTOR
8. STEERING COLUMN UPPER COVER
9. COMBINATION METER ASSEMBLY
10. CENTER PANEL ASSEMBLY
11. AIR OUTLET
12. RADIO PANEL
13. HEATER CONTROL (A/C-ECU)
14. CONSOLE BOX
15. HEATED SEAT SWITCH <VEHICLES WITH HEATED SEAT>
16. HAZARD WARNING LAMP SWITCH
17. CONSOLE METER HOOD
18. MULTI-CENTER DISPLAY UNIT
19. RADIO AND CD PLAYER OR RADIO, CD PLAYER AND CD CHANGER
20. AIR BAG CAUTION LABEL
21. GLOVE BOX ASSEMBLY
22. GLOVE BOX INNER COVER
23. GLOVE BOX LOCK ASSEMBLY
24. GLOVE BOX DAMPER
25. SPEAKER PANEL
26. TWEETER
27. PHOTO SENSOR <VEHICLES WITH AUTO A/C>
28. INTERIOR TEMPERATURE SENSOR <VEHICLES WITH AUTO A/C>
29. STAY
30. INSTRUMENT PANEL ASSEMB

67170-ENDE-G04

Exploded view of the instrument panel and related components

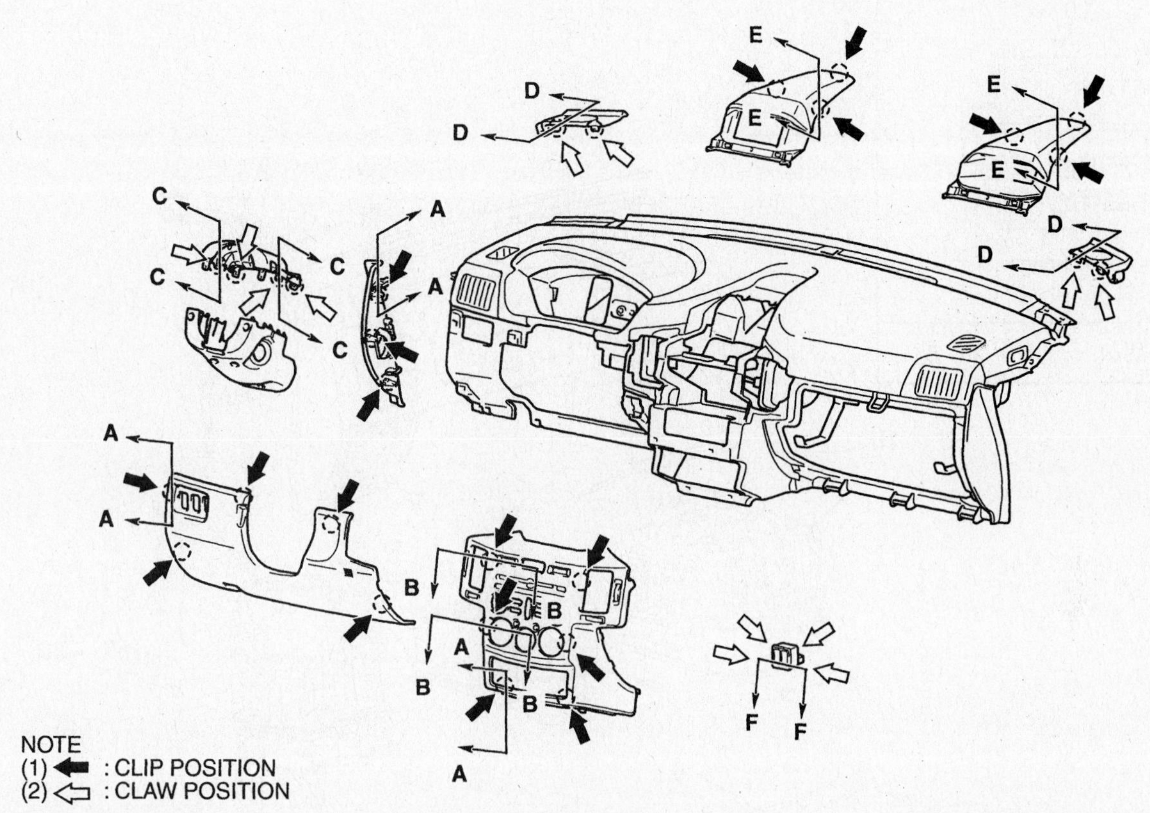

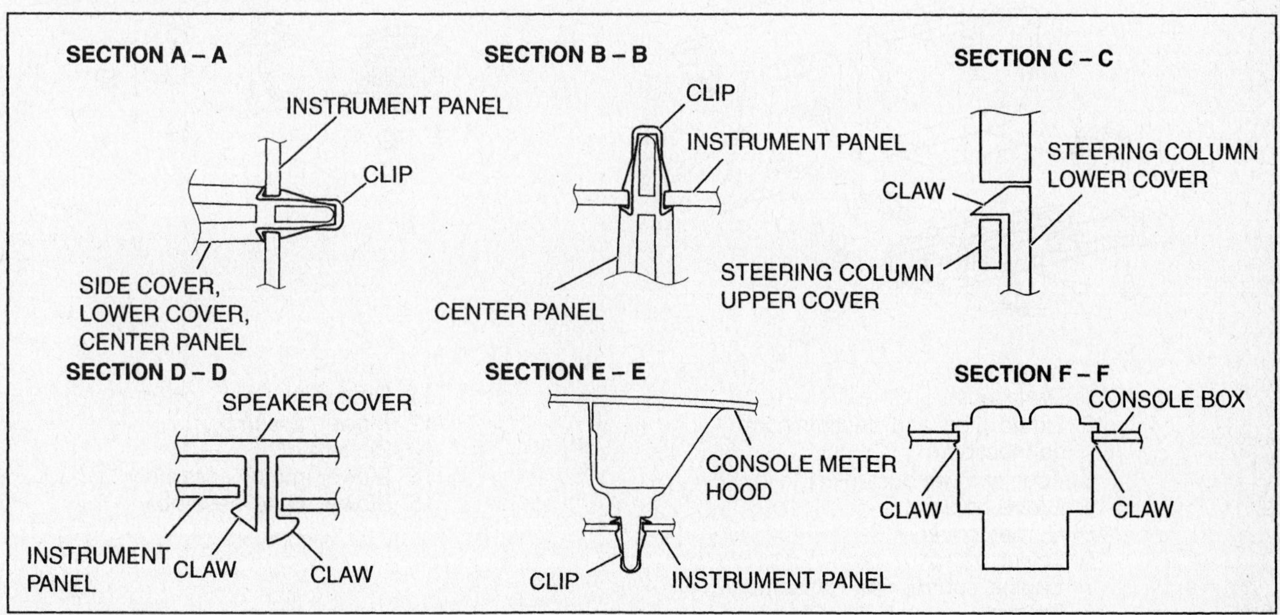

Location of the instrument panel retaining clips and claws

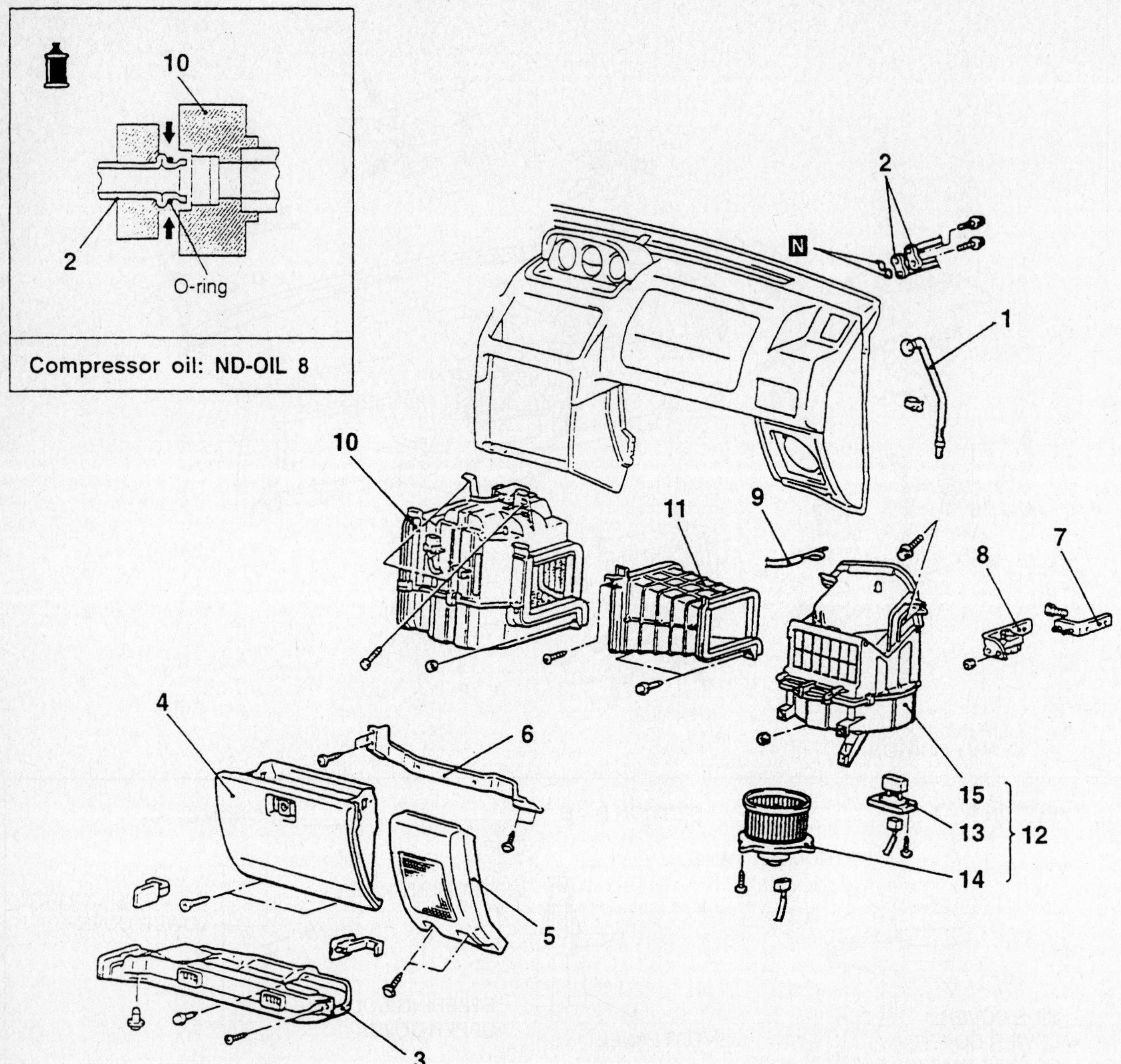

1. Drain hose
2. Liquid pipe and suction hose connection
3. Foot shower duct (R.H.)
4. Glove box
5. Corner cover
6. Lower frame
7. Engine control relay assembly
8. Bracket
9. Air selection control wire connection
10. Evaporator
11. Duct joint
12. Blower assembly
13. Resistor
14. Blower motor assembly
15. Blower case assembly

93113GE6

Exploded view of the air conditioning evaporator housing, blower motor assembly and related components

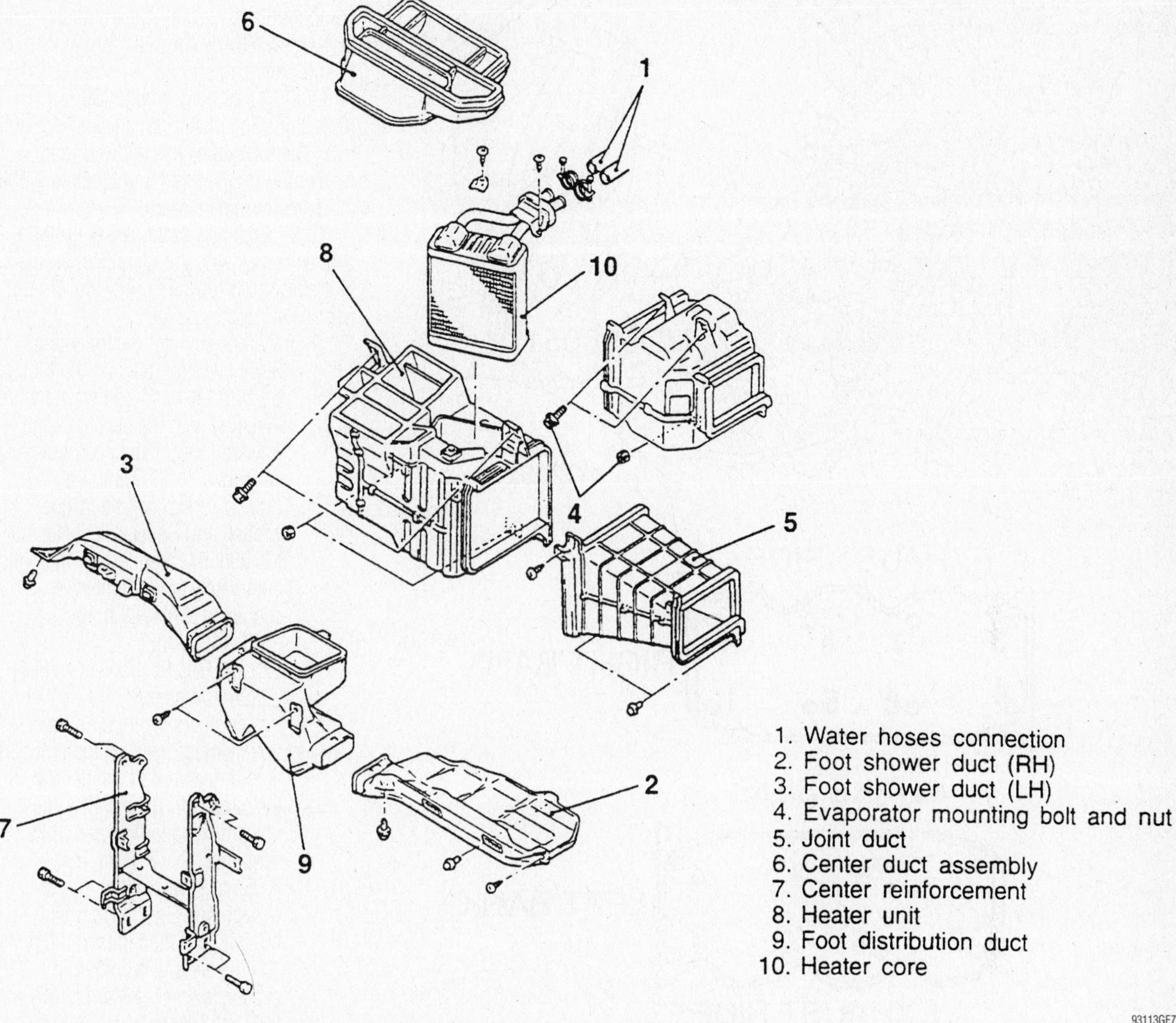

Exploded view of the heater housing, air conditioning evaporator housing and related components

Cylinder Head

REMOVAL & INSTALLATION

1. Before servicing the vehicle, refer to the precautions in the beginning of this section.

✲✲ CAUTION

The fuel injection system remains under pressure after the engine has been OFF. Properly relieve fuel pressure before disconnecting any fuel lines. Failure to do so may result in fire or personal injury.

2. Relieve fuel system pressure.
3. Drain the cooling system.
4. Remove or disconnect the following:
   - Negative battery cable

✲✲ CAUTION

Work must be started after 90 seconds from the time the ignition switch is turned to the LOCK position and the negative battery cable is disconnected.

   - Intake manifold
   - Exhaust manifold
   - Timing belt
   - Thermostat housing
   - Alternator
   - Blow-by hose from the left and right rocker arm covers
   - Positive Crankcase Ventilation (PCV) hose from the left and right rocker arm covers
   - Spark plug wires; tag before disconnecting
   - Ignition coil connectors and ignition coils
   - Engine control wiring harness clamp
   - Rocker covers and gaskets
   - Camshaft Position (CMP) sensor connector
   - Grounding
   - Timing belt rear center cover
5. Loosen the left cylinder head mounting bolts in 3 steps, in the sequence shown. Lift off the left cylinder head assembly and remove the head gasket.
   - Power steering oil pump bracket
   - Exhaust Gas Recirculation (EGR) pipe B and gasket
6. Loosen the right cylinder head mounting bolts in 3 steps, in the sequence shown. Lift off the right cylinder head assembly and remove the head gasket.

To install:

7. Thoroughly clean and dry the mating surfaces of the head and block. Check the cylinder head for cracks, damage or engine

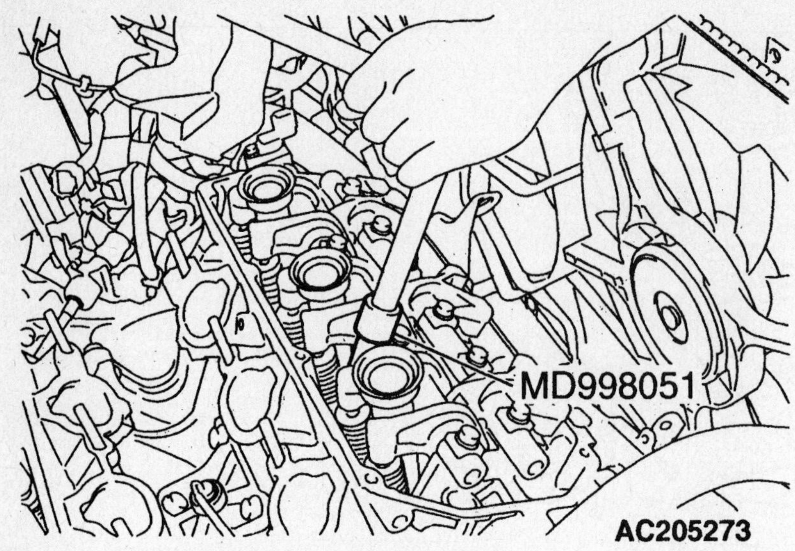

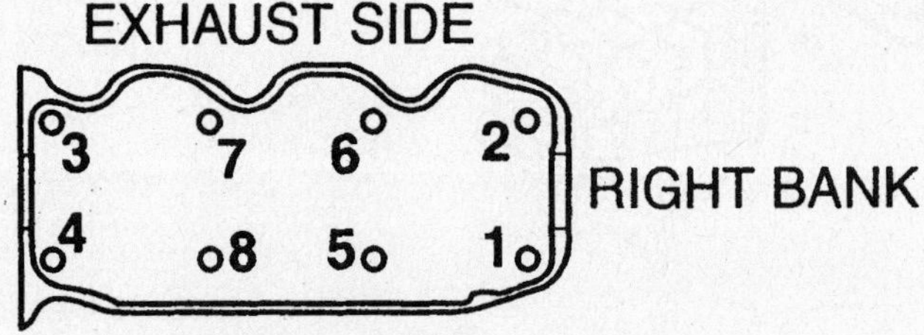

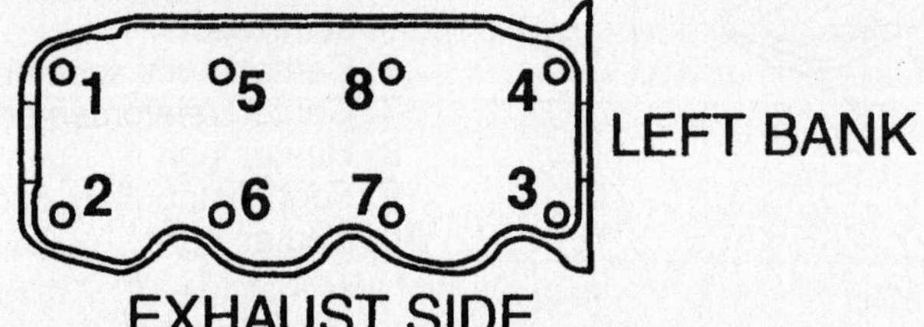

Loosen the cylinder head bolts in the proper sequence—3.8L engine

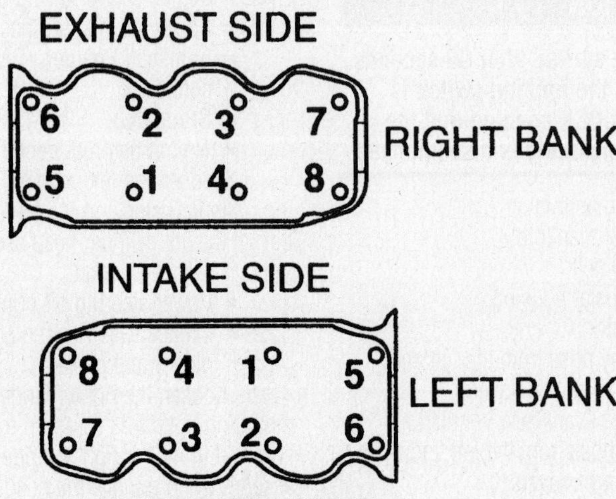

The cylinder head bolts must be installed with the beveled side up and tighten in the proper sequence—3.8L engine

coolant leakage. Remove scale, sealing compound and carbon. Clean oil passages thoroughly. Check the head for flatness. End to end, the head should be within 0.0012 in. (0.030mm), normally with 0.008 in. (0.203mm) the maximum allowed out of true. The total thickness allowed to be removed from the head and block is 0.008 in. (0.203mm) maximum.

8. Place a new head gasket on the cylinder block with the identification marks in the front top (upward) position. Do not use sealer on the gasket.
9. Install or connect the following:
   - Right cylinder head on the block. Be sure the head bolt washers are installed with the beveled edge upward. Torque the head bolts in sequence, to 77–83 ft. lbs. (105–113 Nm) with a torque wrench and Special Tool No. MD998501, then loosen the bolts completely and retighten in sequence to 77–83 ft. lbs. (105–113 Nm).
   - Exhaust Gas Recirculation (EGR) pipe B and gasket
   - Power steering oil pump bracket
   - Left cylinder head on the block. Be sure the head bolt washers are installed with the beveled edge upward. Torque the head bolts in sequence, to 77–83 ft. lbs. (105–113 Nm) with a torque wrench and Special Tool No. MD998501, then loosen the bolts completely and retighten in sequence to 77–83 ft. lbs. (105–113 Nm).
   - Timing belt rear center cover
   - Grounding
   - CMP sensor connector
   - Rocker covers with new gaskets
   - Engine control wiring harness clamp
   - Ignition coils and their connectors
   - Spark plug wires
   - Positive Crankcase Ventilation (PCV) hoses
   - Blow-by hoses
   - Alternator
   - Thermostat housing
   - Timing belt
   - Exhaust manifold
   - Intake manifold
   - Negative battery cable
10. Change the engine oil and oil filter.
11. Refill the system with coolant.
12. Run the vehicle until the thermostat opens.
13. Once the vehicle has cooled, recheck the coolant level.

Rocker Arms/Shafts

REMOVAL & INSTALLATION

1. Before servicing the vehicle, refer to the precautions in the beginning of this section.
2. Relieve the fuel system pressure.
3. Drain the engine oil and cooling system.
4. To remove the left bank rocker arms/shafts, remove or disconnect the following:

- Negative battery cable
- Intake plenum, if removing the right bank rocker arms
- Timing belt
- Thermostat housing
- Blow-by and Positive Crankcase Ventilation (PCV) hoses, if removing the left bank rocker arms
- PCV valve, if removing the left bank
- Breather and blow-by hoses, if removing the right bank rocker arms
- Ignition coil connectors and ignition coils

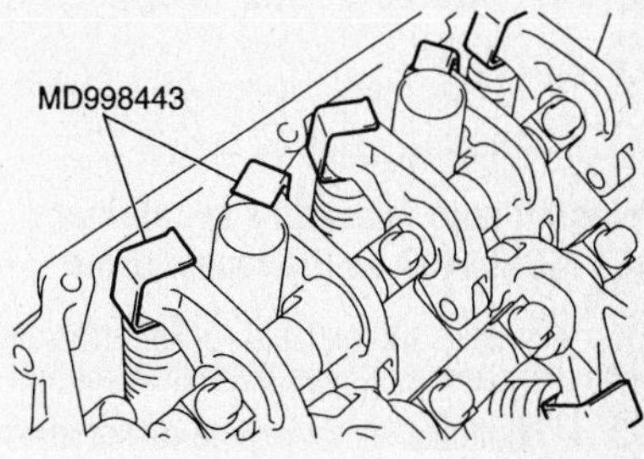

Install the special tool to prevent the lash adjusters from falling to the floor during rocker arm removal—3.8L engine

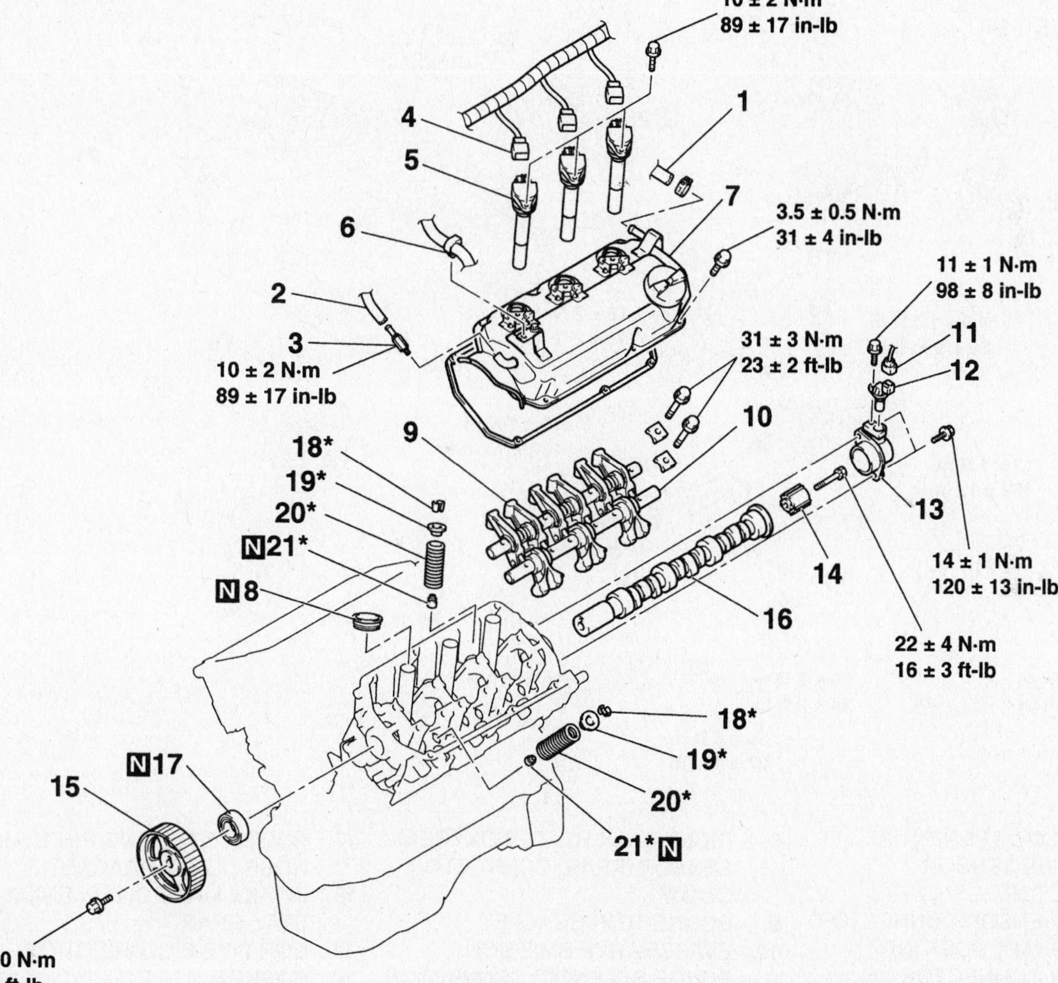

Exploded view of the rocker arms, camshafts and related components—Left bank shown, right bank similar

- Engine control wiring harness clamp
- Rocker arm (valve) cover, gasket and spark plug oil seal(s)

➡Install auto lash adjuster retainers SST MD998443 on the rocker arms

- Intake and exhaust rocker arms and shafts by loosening the mounting bolt and removing the rocker arm and shaft assembly with the bolt still attached
- Lash adjusters

To install:

➡Lubricate the valve train components with clean engine oil.

5. Bleed and install the lash adjusters to the to the original bores in the cylinder head.
6. Install or connect the following:
   - Rocker arms and shafts. Torque the bolts to 21–25 ft. lbs. (28–34 Nm).
   - Valve cover oil seals and gasket
   - Valve cover. Torque the bolts to 27–35 inch lbs. (3.0–4.0 Nm).
7. Install the remaining components in the reverse of the removal procedure.
8. Connect the negative. Fill the engine crankcase and cooling system.
9. Run vehicle and check for leaks.

Intake Manifold

REMOVAL & INSTALLATION

1. Before servicing the vehicle, refer to the precautions in the beginning of this section.

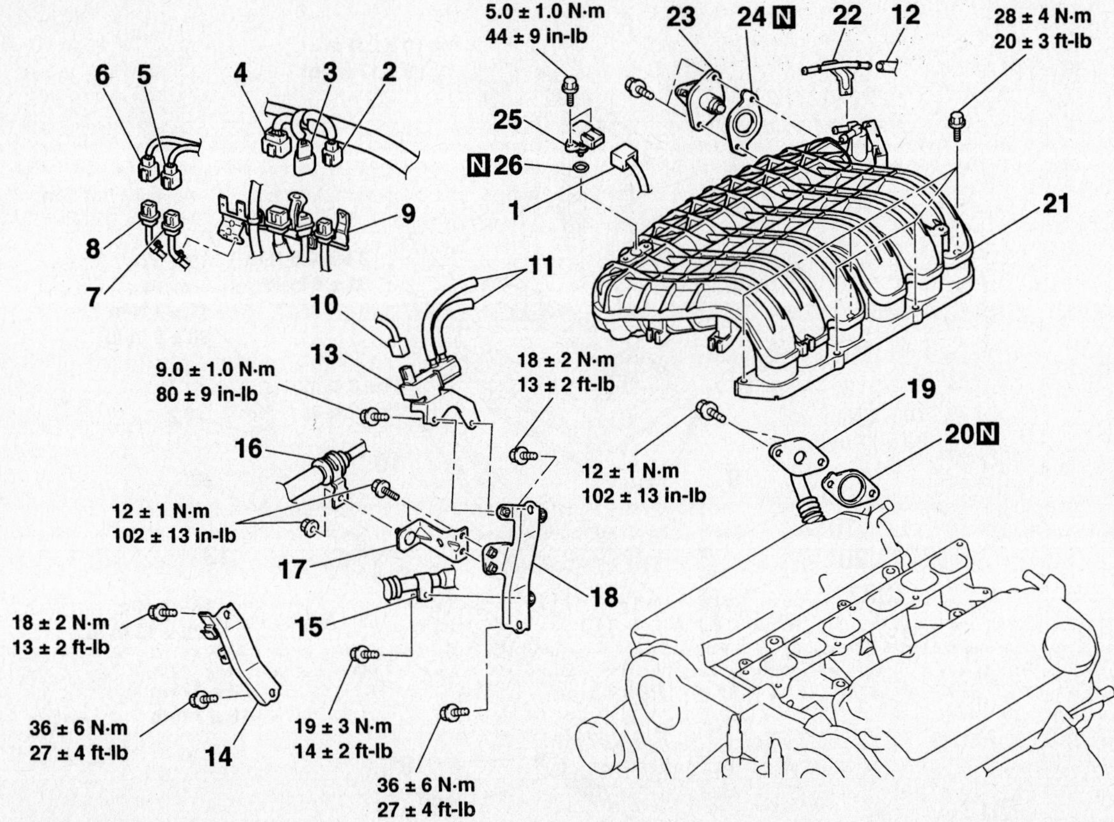

1. MANIFOLD DIFFERENTIAL PRESSURE SENSOR CONNECTOR
2. KNOCK SENSOR CONNECTOR
3. CRANKSHAFT POSITION SENSOR CONNECTOR
4. CONTROL WIRING HARNESS AND INJECTOR WIRING HARNESS COMBINATION CONNECTOR
5. RIGHT BANK HEATED OXYGEN SENSOR (FRONT) CONNECTOR
6. RIGHT BANK HEATED OXYGEN SENSOR (REAR) CONNECTOR
7. RIGHT BANK HEATED OXYGEN SENSOR (FRONT) CONNECTOR CLAMP
8. RIGHT BANK HEATED OXYGEN SENSOR (REAR) CONNECTOR CLAMP
9. CONNECTOR BRACKET
10. EVAPORATIVE EMISSION PURGE SOLENOID CONNECTOR
11. PURGE HOSE
12. PURGE HOSE CONNECTION
13. EVAPORATIVE EMISSION PURGE SOLENOID
14. INTAKE MANIFOLD PLENUM STAY, FRONT
15. EGR PIPE A CLAMP
16. POWER STEERING PRESSURE HOSE CLAMP
17. POWER STEERING PRESSURE HOSE CLAMP BRACKET
18. INTAKE MANIFOLD PLENUM STAY, REAR
19. EGR PIPE B CONNECTION
20. GASKET
21. INTAKE MANIFOLD PLENUM
22. VACUUM PIPE
23. EGR ADAPTER
24. EGR ADAPTER GASKET
25. MANIFOLD DIFFERENTIAL PRESSURE SENSOR
26. O-RING

67170-ENDE-G09

Exploded view of the intake manifold plenum—3.8L engine

CAUTION

The fuel injection system remains under pressure after the engine has been OFF. Properly relieve fuel pressure before disconnecting any fuel lines. Failure to do so may result in fire or personal injury.

2. Relieve the fuel pressure.
3. Partially drain the cooling system.
4. Remove or disconnect the following:

- Negative battery cable

CAUTION

Wait at least 90 seconds after the negative battery cable is disconnected to prevent possible deployment of the air bag.

- Air cleaner assembly
- Throttle body
- Manifold Differential Pressure (MDP) sensor connector
- Knock Sensor (KS) connector
- Crankshaft Position (CKP) sensor connector
- Control wiring harness and injector wiring harness combination connector
- Right bank Heated Oxygen Sensor (HO_2S) connector connections and clamps
- Connector bracket
- Evaporative emission (EVAP) purge solenoid connector
- Purge hose
- Purge hose connection
- EVAP purge solenoid
- Front intake manifold plenum stay

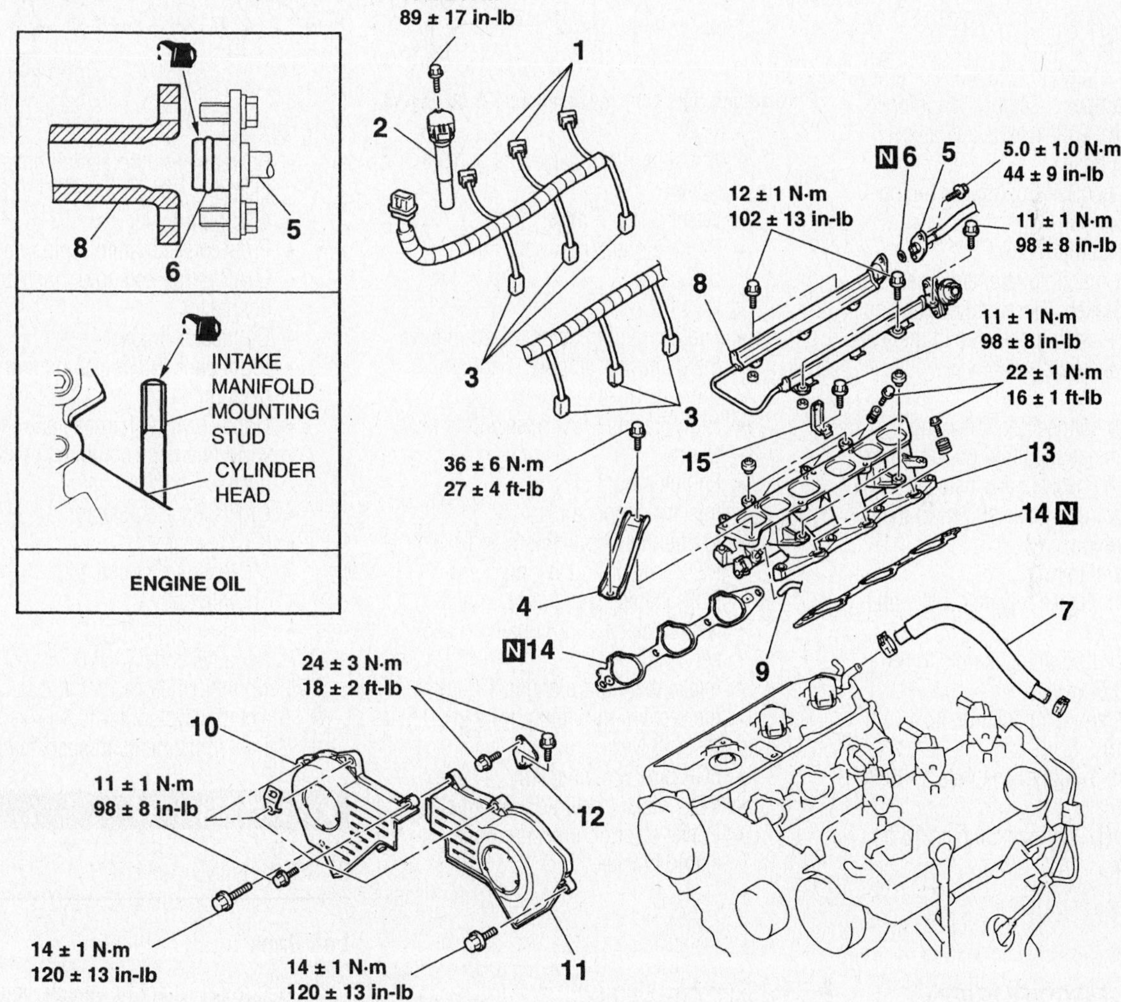

1. IGNITION COIL CONNECTOR
2. IGNITION COIL
3. INJECTOR CONNECTOR
4. ENGINE MOUNT STAY
5. FUEL HIGH-PRESSURE HOSE CONNECTION
6. O-RING
7. BLOW-BY HOSE
8. FUEL RAIL, INJECTOR AND FUEL DAMPER <UP TO DECEMBER 2003>, FUEL RAIL AND INJECTOR <FROM JANUARY 2004>
9. PCV HOSE CONNECTION
10. TIMING BELT FRONT UPPER COVER, RIGHT
11. TIMING BELT FRONT UPPER COVER, LEFT
12. WATER PUMP BRACKET
13. INTAKE MANIFOLD
14. INTAKE MANIFOLD GASKET
15. CONTROL HARNESS CLAMP

67170-ENDE-G10

Exploded view of the intake manifold—3.8L engine

- Exhaust Gas Recirculation (EGR) pipe A clamp
- Power steering pressure hose clamp
- Power steering pressure hose clamp bracket
- Rear intake manifold plenum stay
- EGR pipe B connection
- Retaining bolts and intake manifold plenum
- Intake manifold plenum gasket
- EGR adapter and gasket
- MDP sensor and O-ring
- Ignition coil connector and coil
- Injector connector
- Engine mount stay
- High pressure fuel hose and O-ring
- Blow-by hose
- Fuel rail and injectors. Also remove the damper assembly, if equipped.
- Positive Crankcase Ventilation (PCV) hose
- Right and left front upper timing belt cover
- Water pump bracket
- Intake manifold retainers, manifold and gasket. Thoroughly clean and dry the mating surfaces of the manifold and heads.

To install:

5. Install or connect the following:
   - New intake manifold gasket. Make sure the gaskets are installed with the protrusions as shown in the illustration.
   - Intake manifold

6. Install the intake manifold bolts and tighten as follows:

   a. 1st step: Right bank nuts to 58 inch lbs. (6.5 Nm).

   b. 2nd step: Left bank nuts to 16 ft. lbs. (22 Nm).

   c. 3rd step: Right bank nuts to 16 ft. lbs. (22 Nm).

   d. 4th step: Left bank nuts to 16 ft. lbs. (22 Nm).

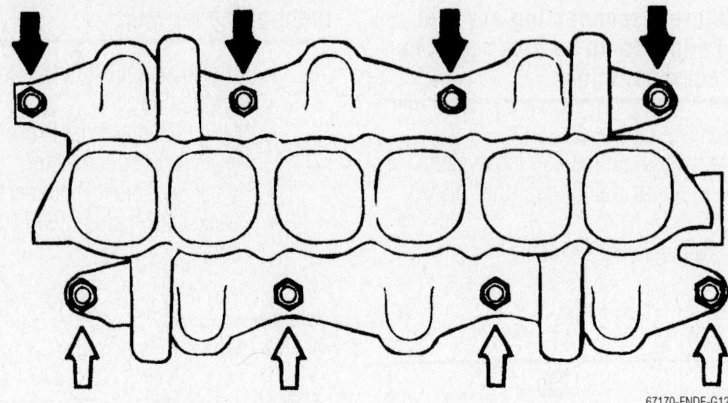

Left and right bank nut locations—3.8L engine

   e. 5th step: Right bank nuts to 16 ft. lbs. (22 Nm).

   - Water pump bracket
   - Right and left front upper timing belt cover
   - PCV hose
   - Fuel rail and injectors. Also remove the damper assembly, if equipped.
   - Blow-by hose
   - High pressure fuel hose with a new O-ring
   - Engine mount stay
   - Injector connectors
   - Ignition coil connectors and coils
   - MDP sensor and O-ring
   - EGR adapter and gasket
   - New intake manifold plenum gasket
   - Intake manifold plenum. Tighten the plenum mounting bolts to 20 ft. lbs. (28 Nm).
   - EGR pipe B connection
   - Rear intake manifold plenum stay
   - Power steering pressure hose clamp bracket
   - Power steering pressure hose clamp
   - EGR pipe A clamp
   - Front intake manifold plenum stay
   - EVAP purge solenoid, hose and connector
   - Connector bracket
   - Right bank HO_2S clamps and connectors
   - Control wiring harness and injector wiring harness combination connector
   - CKP sensor connector
   - KS connector
   - MDP sensor connector
   - Throttle body
   - Air cleaner assembly
   - Negative battery cable

7. Refill the radiator with coolant.
8. Check the system for fuel leaks.
9. Set all adjustments to specifications.

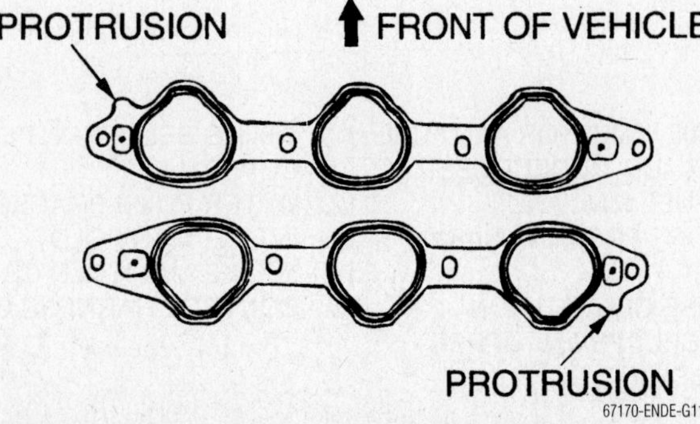

The intake manifold gaskets must be installed properly—3.8L engine

Exhaust Manifold

REMOVAL & INSTALLATION

Left Bank

1. Before servicing the vehicle, refer to the precautions in the beginning of this section.
2. Remove or disconnect the following:
   - Negative battery cable
   - Engine under cover
   - Air duct
   - Left Heated Oxygen Sensor (HO_2S) connectors and sensors
   - Engine oil dipstick, guide and O-ring
   - Heat shield

- Front exhaust pipe from the manifold. Remove and discard the gasket.
- Exhaust manifold stay (left B)
- Left side exhaust manifold and gasket
- Exhaust manifold stay (left A)

3. Clean the gasket mounting surfaces. Inspect the manifolds for cracks, flatness and/or damage.

To install:

4. Install or connect the following:
   - Exhaust manifold stay (left A)
   - New gasket and left side exhaust manifold. Torque the nuts to 30–36 ft. lbs. (39–49 Nm).
   - Exhaust manifold stay (left B)
   - Exhaust pipe to the manifold with a new gasket. Torque the bolts to 31–43 ft. lbs. (40–58 Nm).
   - Left side heat shield
   - Oil dipstick, guide and new O-ring
   - Left HO_2S and connectors
   - Air duct
   - Under cover
   - Negative battery cable
5. Start the engine and check for exhaust leaks.

Right Bank

1. Before servicing the vehicle, refer to the precautions in the beginning of this section.
2. Drain the transfer case fluid and the engine coolant.
3. Remove or disconnect the following:
   - Negative battery cable
   - Air cleaner assembly
   - Battery and battery tray
   - Under cover
   - Front exhaust pipe from the right exhaust manifold

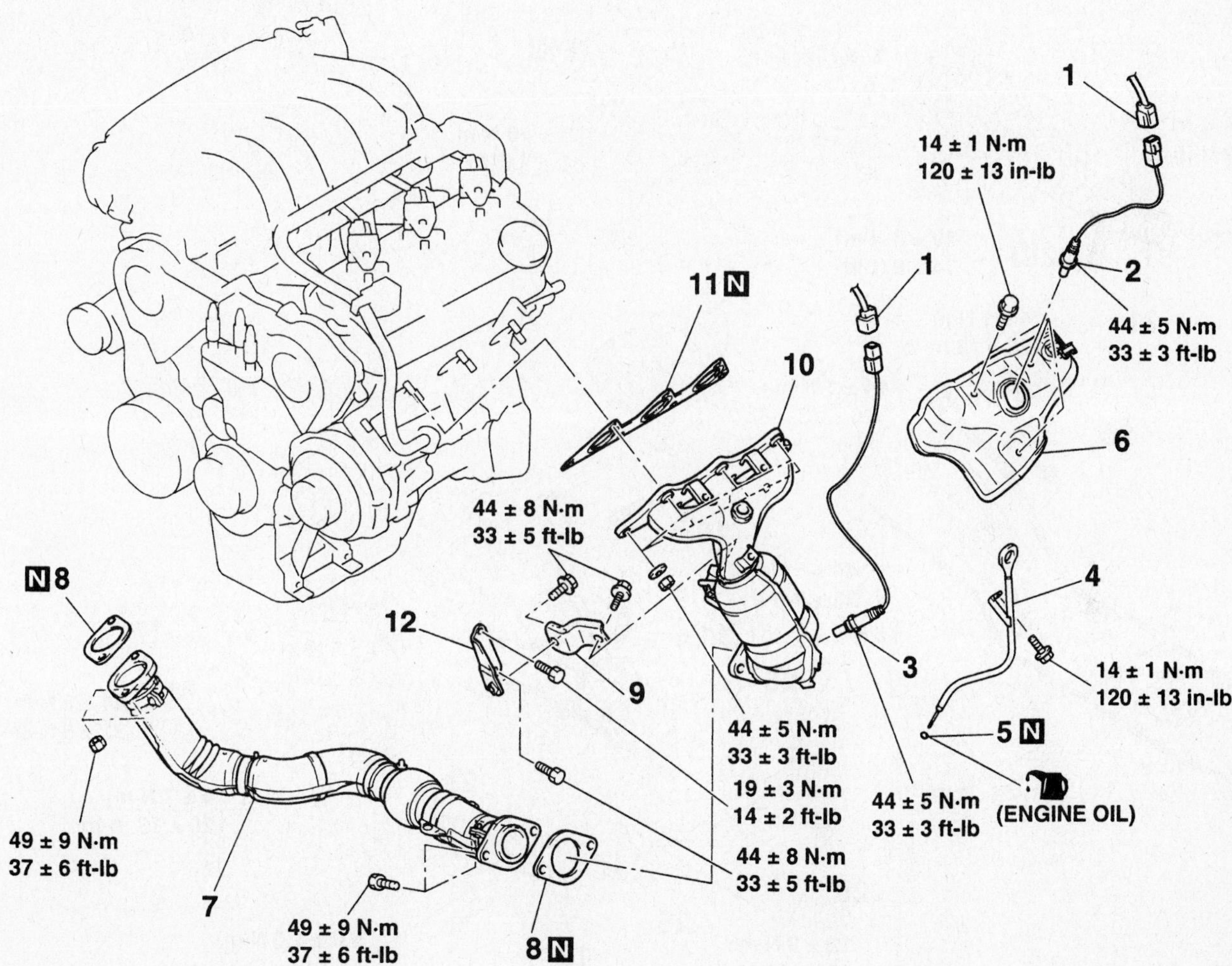

Exploded view of the left bank exhaust manifold—3.8L engine

- Propeller shaft, if AWD
- Manifold Differential Pressure (MDP) sensor connector
- Knock Sensor (KS) connector
- Crankshaft Position (CKP) sensor connector
- Control wiring harness and injector wiring harness combination connector
- Right Heated Oxygen Sensor (HO_2S) connectors
- Connector bracket
- Right HO_2S sensor connector clamps and sensors
- Exhaust Gas Recirculation (EGR) pipe A connection
- Water hoses
- EGR pipe A and gasket
- Exhaust manifold stay (right B)
- Steering gear and linkage protector
- Front floor backbone brace
- Front heat protector panel
- Transfer extension housing and O-ring, if AWD
- Lower and upper heat shields
- Power steering return pipe clamp connecting bolt and nut, if AWD
- Right side exhaust manifold and gasket

4. Clean the gasket mounting surfaces. Inspect the manifolds for cracks, flatness and/or damage.

To install:

5. Install or connect the following:
   - New gasket and right side exhaust manifold. Torque the

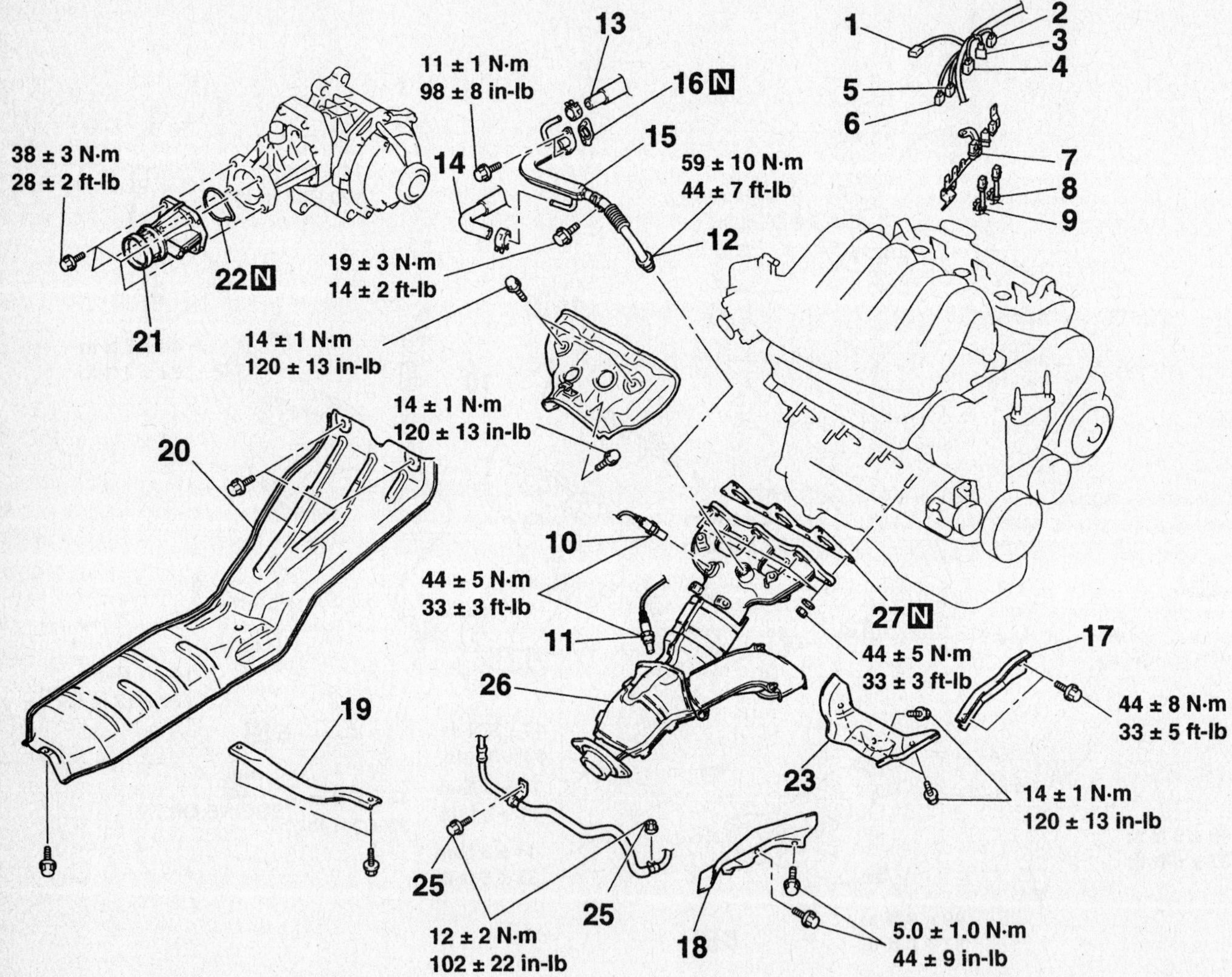

1. MANIFOLD DIFFERENTIAL PRESSURE SENSOR CONNECTOR
2. KNOCK SENSOR CONNECTOR
3. CRANKSHAFT POSITION SENSOR CONNECTOR
4. CONTROL WIRING HARNESS AND INJECTOR WIRING HARNESS COMBINATION CONNECTOR
5. RIGHT BANK HEATED OXYGEN SENSOR (FRONT) CONNECTOR
6. RIGHT BANK HEATED OXYGEN SENSOR (REAR) CONNECTOR

67170-ENDE-G14

Exploded view of the right exhaust manifold—3.8L engine

nuts to 30–36 ft. lbs. (39–49 Nm).

- Power steering return pipe clamp connecting bolt and nut, if AWD
- Lower and upper heat shields
- Transfer extension housing and O-ring, if AWD
- Front heat protector panel
- Front floor backbone brace
- Steering gear and linkage protector
- Exhaust manifold stay (right B)
- EGR pipe A and gasket
- Water hoses
- EGR pipe A connection
- Right HO_2S sensors and connector clamps
- Connector bracket
- Right HO_2S connectors
- Control wiring harness and injector wiring harness combination connector
- CKP sensor connector
- Knock Sensor (KS) connector
- Manifold Differential Pressure (MDP) sensor connector
- Propeller shaft, if AWD
- Front exhaust pipe to the exhaust manifold
- Under cover
- Battery and battery tray
- Air cleaner assembly
- Negative battery cable

6. Fill the transfer case with fluid. Fill the engine cooling system.
7. Start the engine and check for exhaust leaks.

Front Crankshaft Seal

REMOVAL & INSTALLATION

1. Before servicing the vehicle, refer to the precautions in the beginning of this section.
2. Drain the crankcase.
3. Drain and recycle the engine coolant.
4. Remove or disconnect the following:
   - Negative battery cable

CAUTION

Wait at least 90 seconds after the negative battery cable is disconnected to prevent possible deployment of the air bag.

- Timing belt
- Crankshaft Position (CKP) sensor
- Crankshaft sprocket
- Crankshaft sensing blade
- Crankshaft spacer and key
- Front oil seal

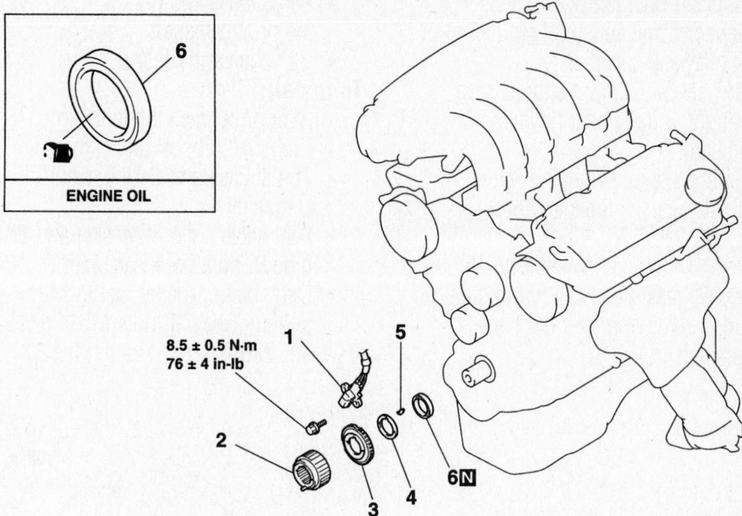

Exploded view of the crankshaft front oil seal—3.8L engine

To install:

- Front oil seal. Apply oil to the seal, and install using Crankshaft Front Oil Seal Installer tool no. MD998717.
- Crankshaft key and spacer.
- Crankshaft sensing blade
- CKP sensor

➡To be sure the crankshaft pulley bolt does not loosen, make sure the clean the mating areas of the crankshaft, spacer, sensing blade and sprocket.

- Crankshaft sprocket
- Timing belt

Camshaft and Valve Lifters

REMOVAL & INSTALLATION

1. Before servicing the vehicle, refer to the precautions in the beginning of this section.
2. Relieve the fuel system pressure.
3. Drain the engine oil and cooling system.
4. To remove the left bank rocker arms/shafts, remove or disconnect the following:
   - Negative battery cable
   - Intake plenum, if removing the right bank rocker arms
   - Timing belt
   - Thermostat housing
   - Blow-by and Positive Crankcase Ventilation (PCV) hoses, if removing the left bank rocker arms
   - PCV valve, if removing the left bank
   - Breather and blow-by hoses, if removing the right bank rocker arms
   - Ignition coil connectors and ignition coils
   - Engine control wiring harness clamp
   - Rocker arm (valve) cover, gasket and spark plug oil seal(s)

➡Install auto lash adjuster retainers SST MD998443 on the rocker arms

- Intake and exhaust rocker arms and shafts by loosening the mounting bolt and removing the rocker arm and shaft assembly with the bolt still attached
- Lash adjusters

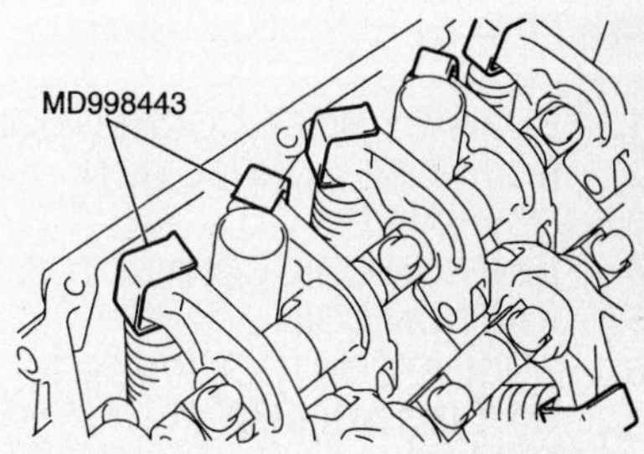

Install the special tool to prevent the lash adjusters from falling to the floor during rocker arm removal—3.8L engine

- Camshaft Position (CMP) sensor connector and sensor, if removing the left bank camshaft
- CMP sensor support and sensing cylinder, if removing the left camshaft
- Camshaft sprocket, using Mitsubishi Special tools MD998715 and MB990767
- Exhaust Gas Recirculation (EGR) pipe B, gasket and EGR valve support, if removing the right bank camshaft
- Thrust case and O-ring, if removing the right camshaft
- Camshaft and oil seal

To install:

5. Install or connect the following
   - Camshaft oil seal and camshaft
   - Thrust case and O-ring, right camshaft
   - EGR valve support, gasket and EGR pipe B, right bank camshaft
   - Camshaft sprocket, using Mitsubishi Special tools MD998715 and MB990767. Tighten the sprocket bolt to 65 ft. lbs. (88 Nm).
   - CMP sensing cylinder and sensor support, left camshaft
   - CMP sensor and connector and sensor, left bank camshaft

➡**Lubricate the valve train components with clean engine oil.**

6. Bleed and install the lash adjusters to the to the original bores in the cylinder head.
7. Install or connect the following:

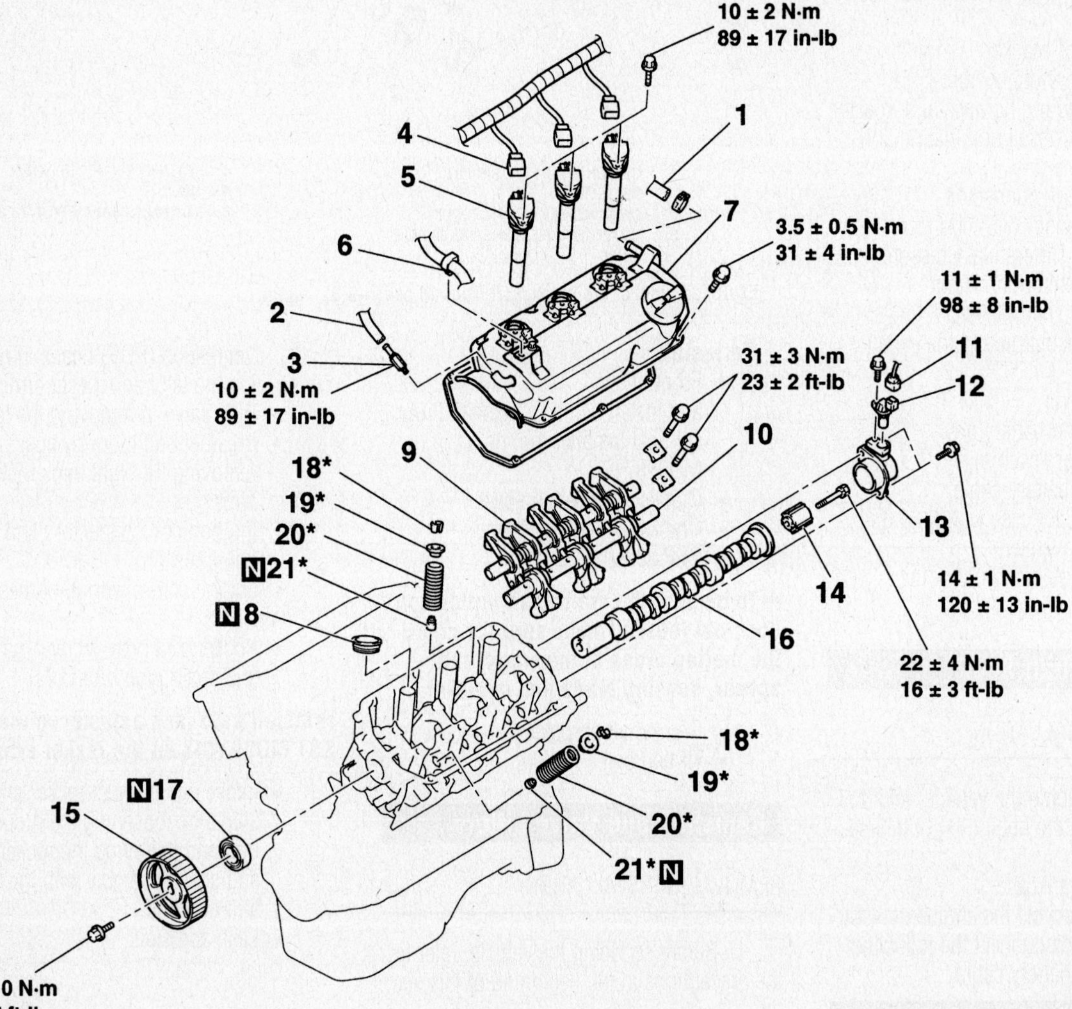

1. BLOW-BY HOSE CONNECTION
2. PCV HOSE CONNECTION
3. PCV VALVE
4. IGNITION COIL CONNECTOR
5. IGNITION COIL
6. ENGINE CONTROL WIRING HARNESS CLAMP
7. ROCKER COVER
8. SPARK PLUG GUIDE OIL SEAL
9. ROCKER ARM, SHAFT AND LASH ADJUSTER ASSEMBLY (INTAKE SIDE)
10. ROCKER ARM, SHAFT AND LASH ADJUSTER ASSEMBLY (EXHAUST SIDE)
11. CAMSHAFT POSITION SENSOR CONNECTOR
12. CAMSHAFT POSITION SENSOR
13. CAMSHAFT POSITION SENSOR SUPPORT
14. CAMSHAFT POSITION SENSING CYLINDER
15. CAMSHAFT SPROCKET
16. CAMSHAFT
17. CAMSHAFT OIL SEAL

671710-ENDE-G08

Exploded view of the camshafts and related components—Left bank shown, right bank similar

- Rocker arms and shafts. Torque the bolts to 21–25 ft. lbs. (28–34 Nm).
- Valve cover oil seals and gasket
- Valve cover. Torque the bolts to 27–35 inch lbs. (3.0–4.0 Nm).

8. Install the remaining components in the reverse of the removal procedure.
9. Connect the negative. Fill the engine crankcase and cooling system.
10. Run vehicle and check for leaks.

Starter Motor

REMOVAL & INSTALLATION

1. Before servicing the vehicle, refer to the precautions in the beginning of this section.
2. Remove or disconnect the following:
   - Negative battery cable
   - Engine under cover, if equipped
   - Wires
   - Starter cover
   - Starter motor

To install:

3. Install or connect the following:
   - Starter motor and cover. Torque the starter mounting bolts to 20–25 ft. lbs. (26–33 Nm) and the cover retainers to 36–42 inch lbs. (3.9–5.9 Nm).
   - Wires. Tighten the terminal nut to 80–124 inch lbs. (10–14 Nm).
   - Engine under cover, if equipped
   - Negative battery cable

Oil Pan

REMOVAL & INSTALLATION

1. Before servicing the vehicle, refer to the precautions in the beginning of this section.
2. Drain the engine oil.
3. Remove or disconnect the following:
   - Negative battery cable
   - Engine undercover
   - Front exhaust pipe
   - Oil pan drain plug and gasket
   - Starter motor
   - Engine oil dipstick and O-ring
   - Lower oil pan. If necessary, use a block of wood and hammer to carefully dislodge the lower oil pan.
   - Cover
   - 2 lower torque converter connecting bolts
   - Upper oil pan. Screw the M10 bolts holding the oil pan to the transmission assembly into the bolt holes shown in the illustration to remove the pan.

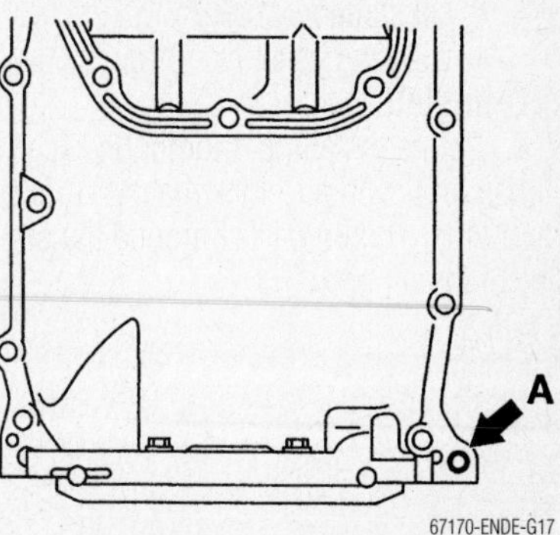

Screw the M10 bolts that hold the oil pan to the transmission assembly into the bolt holes shown by arrows—3.8L engine

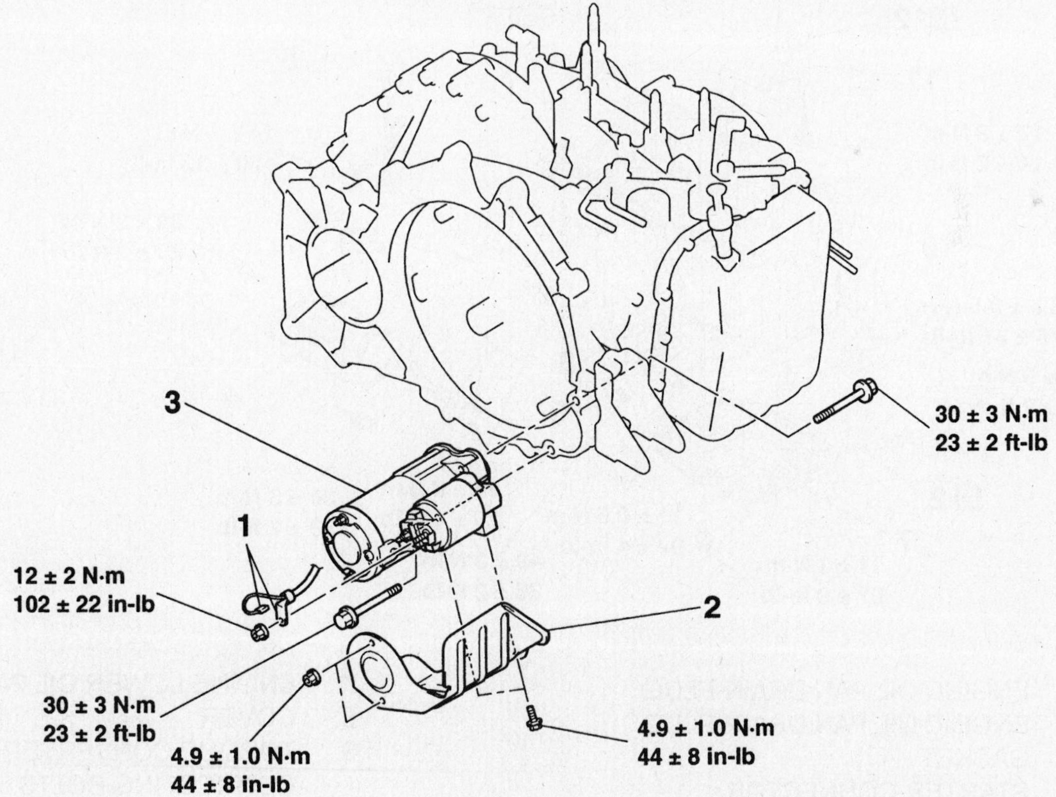

Starter motor mounting

- Oil screen
- Oil pan gasket

To install:

4. Before installing, thoroughly clean the oil pan and cylinder block mating surfaces.
5. Apply liquid gasket around the surface of the oil pan.

➡Assemble the oil pan to the cylinder block within 30 minutes after applying the liquid gasket.

6. Install or connect the following:
   - Oil screen. Torque the bolts to 12–16 ft. lbs. (16–22 Nm).
   - Upper oil pan. Torque the bolts to 76 inch lbs. (8.5 Nm), in the sequence shown the illustration. The bolt holes for bolts 13 and 14 are cut away on the transmission side. Be sure you do not insert the bolts at an angle.
   - 2 lower torque converter connect-

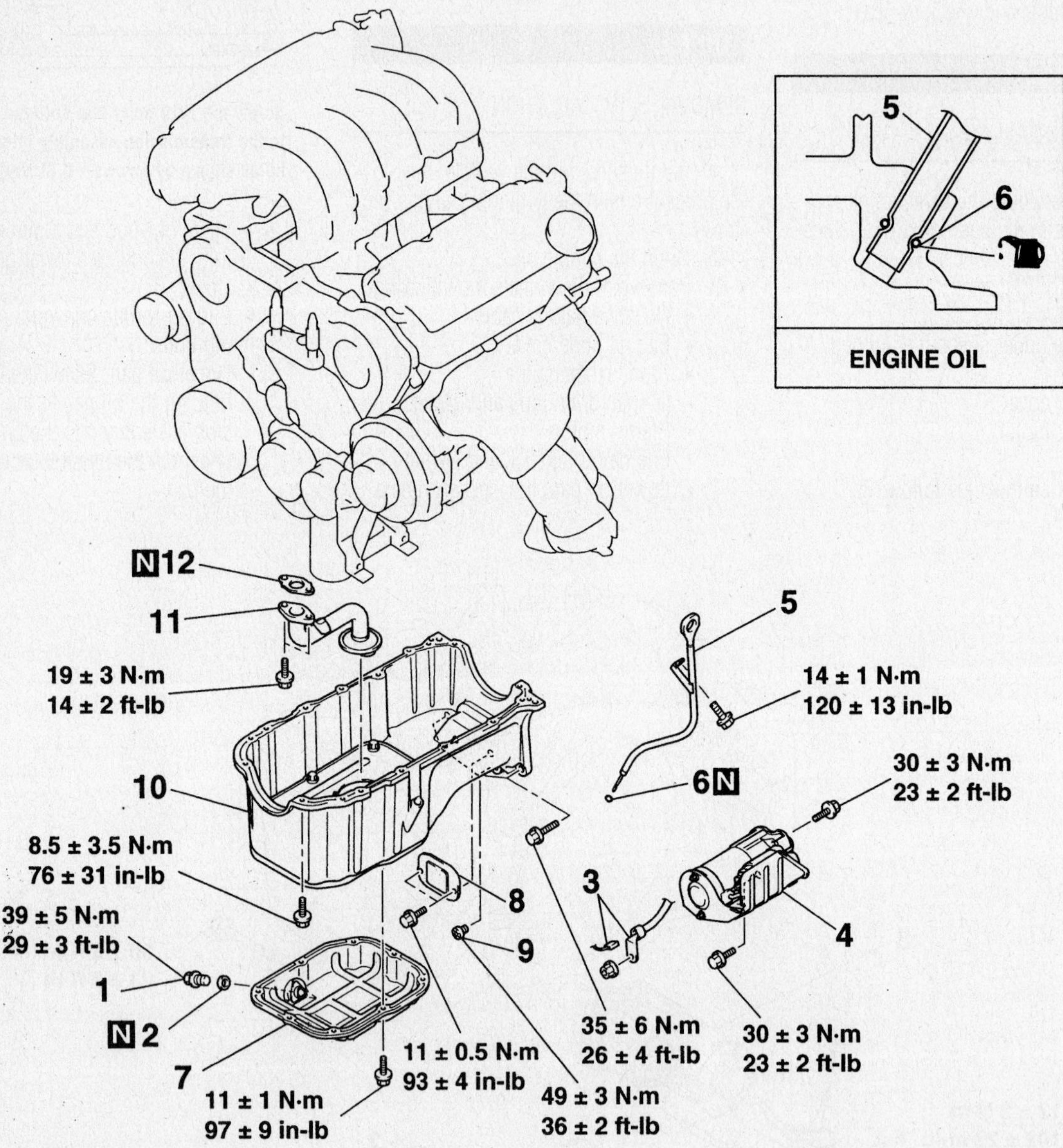

Exploded view of the oil pan and related components—3.8L engine

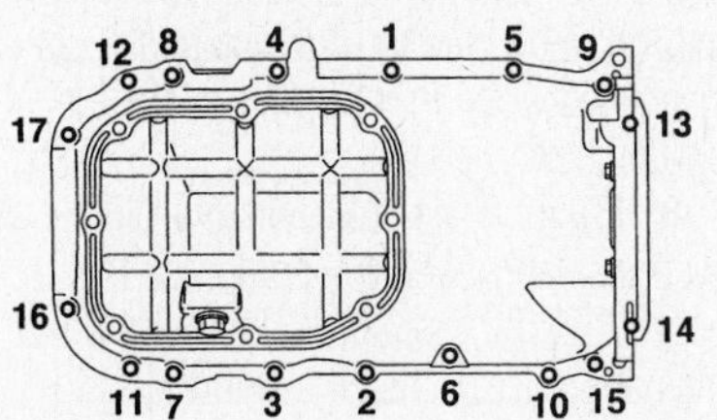

Upper oil pan bolt tightening sequence—3.8L engine

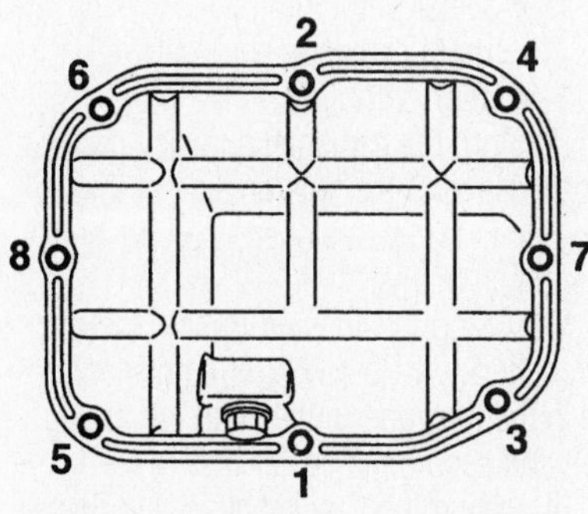

Lower oil pan bolt tightening sequence—3.8L engine

ing bolts. Tighten the bolts to 36 ft. lbs. (49 Nm).

- Cover
- Lower oil pan. Torque the bolts to 93 inch lbs. (11 Nm), in the sequence shown in the illustration.
- Engine oil dipstick and O-ring
- Starter motor
- Oil pan drain plug with a new gasket. Tighten to 29 ft. lbs. (39 Nm).
- Front exhaust pipe
- Engine undercover
- Negative battery cable

7. Fill the crankcase with oil. Start the engine and check for leaks.

Oil Pump

REMOVAL & INSTALLATION

1. Before servicing the vehicle, refer to the precautions in the beginning of this section.
2. Drain the engine oil.
3. Remove or disconnect the following:
   - Negative battery cable
   - Timing belt
   - Oil pressure switch
   - Oil dipstick
   - Oil pans from the engine
   - Oil baffle and screen
   - Oil pump mounting bolts and the pump from the front of the engine

➡Note the position of each oil pump case retaining bolts to facilitate installation. The bolts are of different length.

To install:

4. Clean the gasket mounting surfaces of the pump and engine block.
5. Prime the pump by pouring fresh oil into the inlet and turning the rotors or by packing pump with petroleum jelly. Using a new gasket, install the oil pump on the engine and tighten all bolts to 10 ft. lbs. (14 Nm).
6. Clean out the oil pick-up or replace as required. Replace the oil pick-up gasket ring and install the pick-up to the pump.
7. Install or connect the following:
   - Oil filter and the bracket. Torque the bolts to 17 ft. lbs. (23 Nm).
   - Oil baffle and screen. Torque the bolts to 13 ft. lbs. (18 Nm).
   - Oil pans
   - Oil pressure switch. Torque the switch to 87 inch lbs. (9.8 Nm).
   - Timing belt
   - Dipstick
   - Negative battery cable
8. Refill the engine with the proper amount of oil.
9. Start the engine and check for proper oil pressure. Check for leaks.

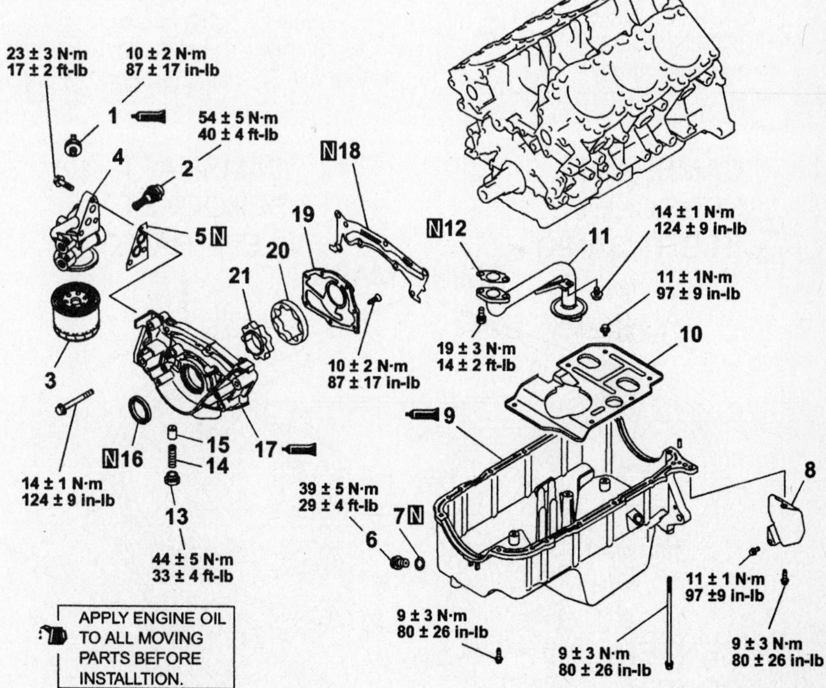

1. ENGINE OIL PRESSURE SWITCH
2. OIL COOLER BY-PASS VALVE
3. OIL FILTER
4. OIL FILTER BRACKET
5. OIL FILTER BRACKET GASKET
6. DRAIN PLUG
7. DRAIN PLUG GASKET
8. COVER
9. OIL PAN
10. BAFFLE PLATE
11. OIL SCREEN
12. OIL SCREEN GASKET
13. RELIEF PLUG
14. RELIEF SPRING
15. RELIEF PLUNGER
16. CRANKSHAFT OIL SEAL
17. OIL PUMP CASE
18. OIL PUMP CASE GASKET
19. OIL PUMP COVER
20. OIL PUMP OUTER ROTOR
21. OIL PUMP INNER ROTOR

42356-MONT-G16

Exploded view of the oil pump, oil pan and related components—3.8L engine

Timing Belt

REMOVAL & INSTALLATION

1. Before servicing the vehicle, refer to the precautions in the beginning of this section.
2. Drain the engine coolant.
3. Remove or disconnect the following:
   - Negative battery cable
   - Engine under cover and side under cover
   - Drive belts
   - Crankshaft pulley, using special tools MB991800 and MB991802
   - Manifold Differential Pressure (MDP) sensor connector
   - Knock Sensor (KS) connector
   - Crankshaft Position (CKP) sensor connector
   - Control wiring harness and injector wiring harness combination connector
   - Right bank Heated Oxygen Sensor (HO_2S) connectors
   - Connector bracket
   - Engine mount stay
   - Right and left timing belt upper covers
   - Tensioner pulley
   - Tensioner bracket
   - CKP sensor harness clamp
   - Lower timing belt cover
   - Engine mount
   - Engine support bracket
4. Turn the crankshaft clockwise to align the timing marks and set the No. 1 cylinder at Top Dead Center (TDC). If you are reusing the timing belt, mark the flat side of the belt with an arrow showing the clockwise direction.
5. Loosen the center bolt of the tension pulley, and then remove the timing belt.
   - Auto tensioner
   - Tensioner pulley
   - Tensioner arm
   - Shaft

To install:

6. Install or connect the following:
   - Idler pulley
   - Shaft
   - Tensioner arm assembly
   - Tension pulley
7. Press the end of the auto-tensioner inward with 72–145 ft. lbs. (98–196 Nm) of force and measure the distance that the pushrod is pushed in. If the standard distance is not 0.04 in. (1mm), replace the auto-tensioner.
8. Position the auto-tensioner in a soft-jawed vise and SLOWLY compress the pushrod until the pushrod and housing holes align; then, install a setting pin to secure the auto-tensioner in the retracted position.

➡If you are installing a new auto-tensioner, the pin will already be inserted into the pin holes of the new tensioner.

9. Install the auto-tensioner.
10. Align the camshaft and crankshaft TDC timing marks.
11. Install the timing belt (noting its rotational direction) so that there is no deflection between the sprockets and pulleys in the following manner:
   - Crankshaft sprocket
   - Idler pulley
   - Left camshaft sprocket
   - Water pump pulley
   - Right camshaft sprocket
   - Tension pulley
12. Turn the camshaft sprocket counterclockwise until the tension side of the timing belt is firmly stretched, then, recheck the timing marks.
13. Using the Tension Pulley Socket Wrench tool MD998767, or equivalent, push the tensioner pulley into the timing belt and secure the center bolt.
14. Using the Crankshaft Pulley Spacer tool MD998769, or equivalent, rotate the crankshaft ¼ turn counterclockwise, then, turn it again clockwise to align the timing marks.
15. Loosen the timing belt tensioner center bolt. Using the Tension Pulley Socket Wrench tool MD998767, or equivalent, and a torque wrench, apply 39 inch lbs. (4.4 Nm) pressure on the timing belt. Torque the tensioner pulley center bolt to 35 ft. lbs. (48 Nm).
16. Remove the setting pin from the auto-tensioner.
17. Rotate the crankshaft 2 complete revolutions and realign the timing marks. Then, wait for 5 minutes until the auto-tensioner pushrod extends to its standard value. If the standard value is not 0.19–0.24 in. (4.8–6.0mm), repeat the adjustment procedure. If the standard value is still not achieved, replace the auto-tensioner.
   - Install or connect the following:
   - Engine support bracket
   - Engine mount
   - Lower timing belt cover
   - CKP sensor harness clamp

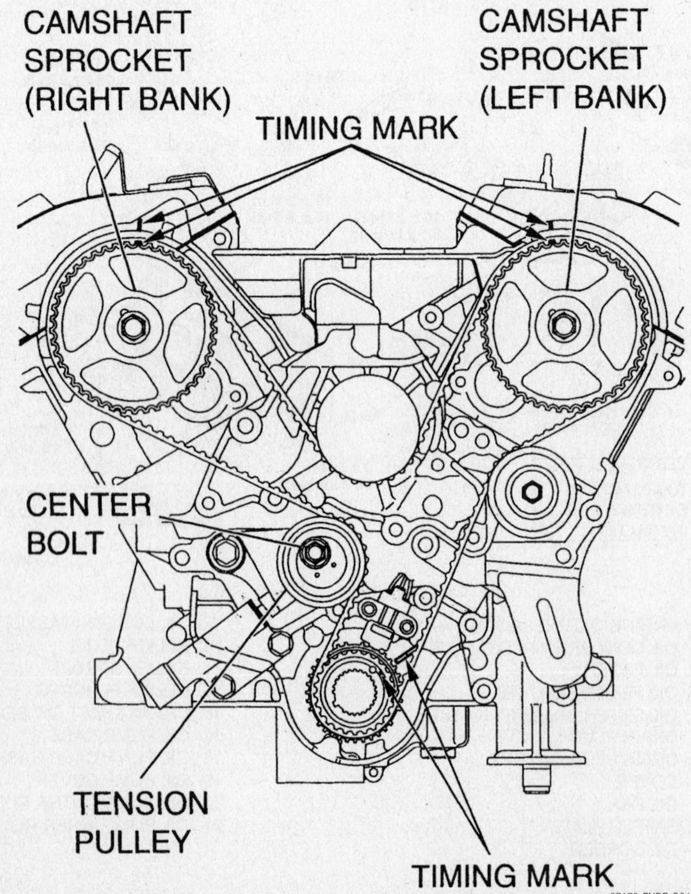

Installed view of the timing belt and alignment of the timing marks—3.8L engine

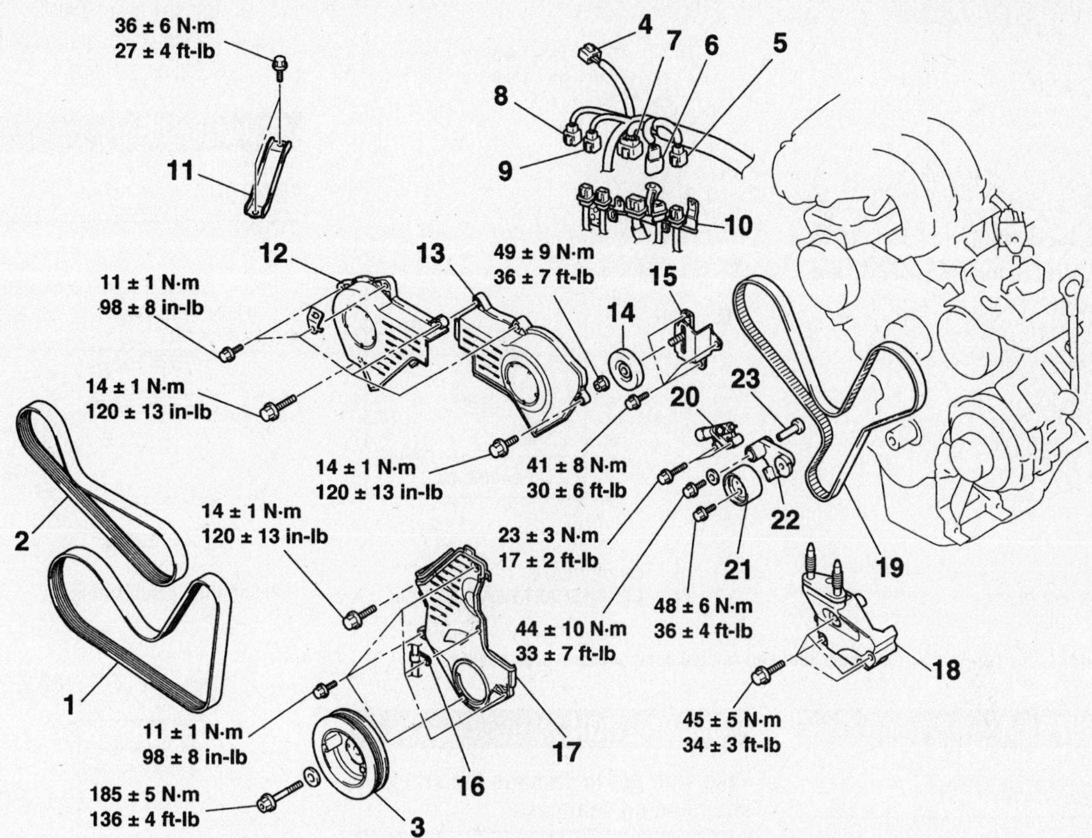

1. GENERATOR DRIVE BELT
2. POWER STEERING OIL PUMP DRIVE BELT
3. CRANKSHAFT PULLEY
4. MANIFOLD DIFFERENTIAL PRESSURE SENSOR CONNECTOR
5. KNOCK SENSOR CONNECTOR
6. CRANKSHAFT POSITION SENSOR CONNECTOR
7. CONTROL WIRING HARNESS AND INJECTOR WIRING HARNESS COMBINATION CONNECTOR
8. RIGHT BANK HEATED OXYGEN SENSOR (REAR) CONNECTOR
9. RIGHT BANK HEATED OXYGEN SENSOR (FRONT) CONNECTOR
10. CONNECTOR BRACKET
11. ENGINE MOUNT STAY
12. TIMING BELT FRONT UPPER COVER, RIGHT
13. TIMING BELT FRONT UPPER COVER, LEFT
14. TENSIONER PULLEY
15. TENSIONER BRACKET
16. CRANKSHAFT POSITION SENSOR HARNESS CLAMP
17. TIMING BELT LOWER COVER

67170-ENDE-G22

Exploded view of the timing belt and related components—3.8L engine

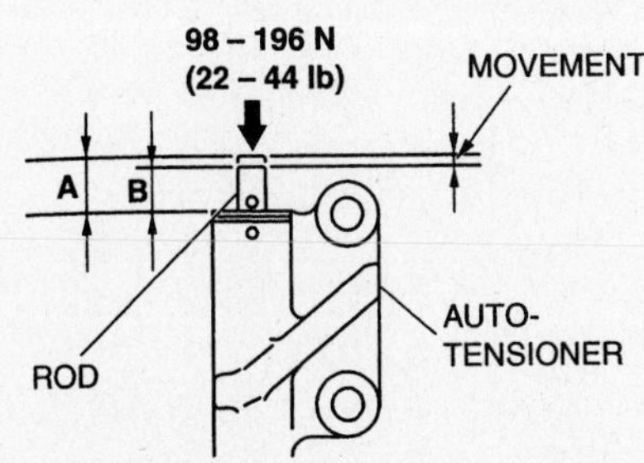

Inspecting the auto-tensioner movement—3.8L engines

- Tensioner bracket
- Tensioner pulley
- Left and right upper timing belt covers
- Engine mount stay
- Connector bracket
- HO_2S connectors
- Wiring harness and injector connector
- CKP sensor connector
- KS connector
- MDP sensor connector

18. Install the crankshaft pulley. Using Pulley Holder MB991800 and the 2 Crankshaft Pulley Holder Pin tools MD991802, or equivalent to hold the crankshaft pulley, and a socket torque wrench, torque the crankshaft pulley bolt to 136 ft. lbs. (185 Nm).
    - Drive belts
    - Side under cover
    - Under cover
    - Negative battery cable
19. Refill the cooling system.

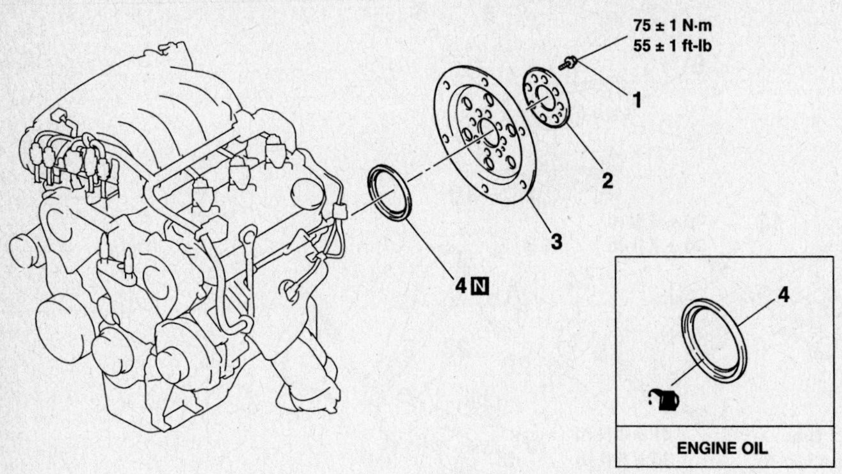

Exploded view of the driveplate, rear main seal and related components—3.8L engine

Rear Main Seal

REMOVAL & INSTALLATION

1. Before servicing the vehicle, refer to the precautions in the beginning of this section.
2. Remove or disconnect the following:
   - Transmission assembly
   - Transfer case, if AWD
   - Driveplate and adapter plate, matchmark for reassembly. Use Mitsubishi tool (MD998781) to hold the driveplate in position to remove the bolts.
3. Remove the rear oil seal as follows:
   a. Cut out a portion in the crankshaft oil seal lip.
   b. Cover the tip of a small prytool with a cloth and apply it to the cutout in the oil seal to pry the oil seal out.

CAUTION

Take care not to damage the crankshaft and oil seal case.

To install:

4. Inspect the sealing surface at the rear of the crankshaft. If a deep groove is worn into the surface, the crankshaft will have to be replaced. Coat the sealing lip of the seal with fresh, clean engine oil. Press the new seal into the case with a seal installing tool. The seal must be pressed in squarely until it bottoms in the case. It is necessary to use the proper tool (MD998718-01) to fit the seal into place.
5. Install or connect the following:
   - Drive plate and adapter. Use Mitsubishi tool (MD998781) to hold the driveplate in position while tightening the bolts to 55 ft. lbs. (75 Nm).
   - Transfer case, if AWD
   - Transmission and related components as necessary

Piston and Ring

POSITION

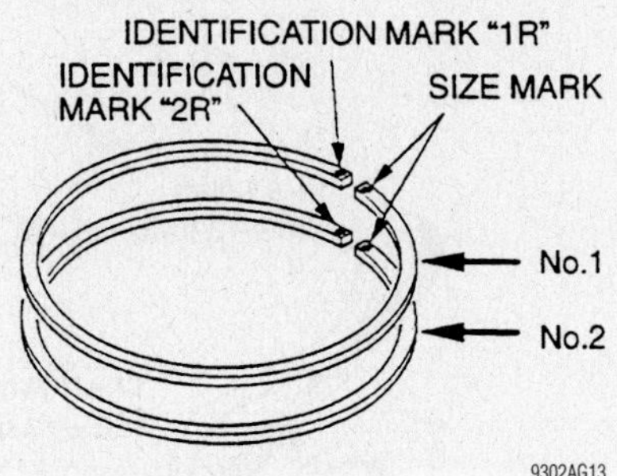

Piston ring identification

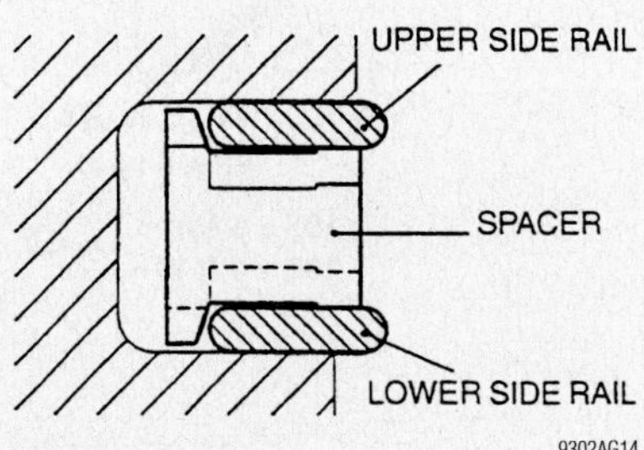

Oil ring identification

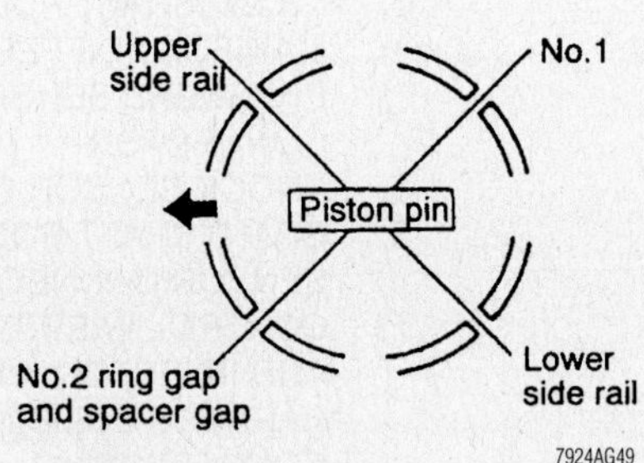

Piston ring end-gap spacing

FUEL SYSTEM

Fuel System Service Precautions

Safety is the most important factor when performing not only fuel system maintenance but any type of maintenance. Failure to conduct maintenance and repairs in a safe manner may result in serious personal injury or death. Maintenance and testing of the vehicle's fuel system components can be accomplished safely and effectively by adhering to the following rules and guidelines.

- To avoid the possibility of fire and personal injury, always disconnect the negative battery cable unless the repair or test procedure requires that battery voltage be applied.
- Always relieve the fuel system pressure prior to disconnecting any fuel system component (injector, fuel rail, pressure regulator, etc.), fitting or fuel line connection. Exercise extreme caution when relieving fuel system pressure, to avoid exposing your skin, face and eyes to fuel spray. Please be advised that fuel under pressure may penetrate the skin or any part of the body that it contacts.
- Always place a shop towel or cloth around the fitting or connection prior to loosening to absorb any excess fuel due to spillage. Ensure that all fuel spillage (should it occur) is quickly removed from engine surfaces. Ensure that all fuel soaked cloths or towels are deposited into a suitable waste container.
- Always keep a dry chemical (Class B) fire extinguisher near the work area.
- Do not allow fuel spray or fuel vapors to come into contact with a spark or open flame.
- Always use a back-up wrench when loosening and tightening fuel line connection fittings. This will prevent unnecessary stress and torsion to fuel line piping. Always follow the proper torque specifications.
- Always replace worn fuel fitting O-rings with new. Do not substitute fuel hose where fuel pipe is installed.

Fuel System Pressure

RELIEVING

⁂ CAUTION

The fuel system is under constant pressure, even with the engine off. This pressure must be relieved before disconnecting any fuel system component, fitting or fuel line connection. Failure to do so may result in personal injury.

1. Turn the ignition switch to **LOCK**.
2. Remove the left rear seat cushion mounting bolts, and lift the seat cushion to tie it to the head restraint.
3. Push on the floor mat to find the notches, then cut the carpet along the notches to access the service hole cover. Remove the service hole cover.
4. Detach the fuel pump module connector.
5. Start the engine and let it run out of fuel.
6. Attach the fuel pump module connector.
7. Install the service hole cover, floor mat and left rear seat cushion.

Fuel Filter

REMOVAL & INSTALLATION

The manufacturer does not provide a procedure or maintenance interval for replacing the fuel filter.

Fuel Pump

REMOVAL & INSTALLATION

1. Before servicing the vehicle, refer to the precautions in the beginning of this section.
2. Relieve the fuel system pressure.
3. Remove or disconnect the following:
   - Negative battery cable

⁂ CAUTION

The fuel injection system remains under pressure after the engine has been OFF. Properly relieve fuel pressure before disconnecting any fuel lines. Failure to do so may result in fire or personal injury. Do not allow fuel spray or fuel vapors to come in contact with a spark or open flame. Keep a dry chemical fire extinguisher nearby. Never store fuel in an open container due to risk of fire or explosion.

   - Left rear seat cushion and tie out of the way to the head restraint.

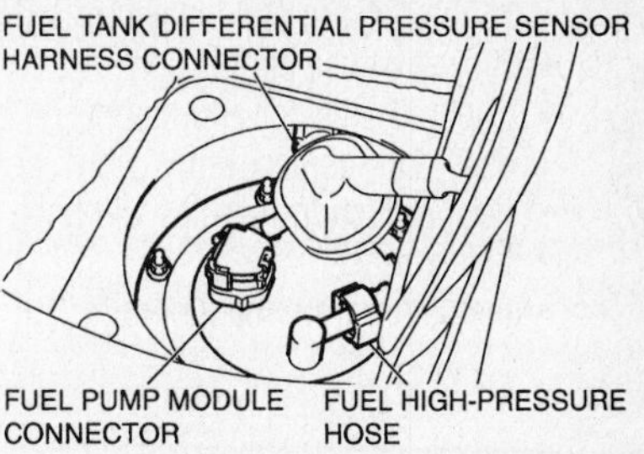

Location of the fuel pump connectors. The fuel pump on the Endeavor is removed through the rear floor pan

4. Push on the floor mat to find the notches, then cut the carpet along the notches to access the service hole cover. Remove the service hole cover.
   - Fuel pump module connector
   - Fuel tank differential pressure sensor connector
   - Fuel high-pressure hose
   - Mounting nuts and plate and fuel pump module assembly

⁂ WARNING

When removing the fuel pump module from the tank, be careful not to damage the module unit and float.

To install:

5. Install or connect the following:
   - Fuel pump assembly into the fuel tank. Torque the nuts to 24 inch lbs. (2.5 Nm).
   - Fuel lines
   - Fuel tank differential pressure sensor connector
   - Fuel pump module connector
   - Fuel pump cover. Torque the bolts to 14 inch lbs. (1.5 Nm).
   - Rear floor carpeting.
   - Negative battery cable
6. Start the vehicle; check for leaks and proper operation.

Fuel Injector

REMOVAL & INSTALLATION

1. Before servicing the vehicle, refer to the precautions in the beginning of this section.
2. Properly relieve the fuel system pressure.
3. Remove or disconnect the following:
   - Air cleaner assembly, ducts and air intake hose

- Intake manifold plenum
- Ignition coil connectors and coils
- Injector connectors
- High pressure fuel hose and O-ring
- Blow-by hose connection
- Fuel rail, injector and damper assembly, as necessary. Only earlier production vehicles will have a fuel damper.
- Insulators from the fuel rail
- Fuel injectors from the fuel rail
- O-rings from the injectors
- Fuel damper, if equipped

To install:

4. Install or connect the following:
   - Fuel damper, if equipped
   - New O-rings and grommets
   - Fuel injectors and insulators
   - Fuel rail and insulators. Torque the bolts to 102 inch lbs. (12 Nm).
   - Injector connectors
   - Blow-by hose connection
   - High pressure fuel hose and O-ring. Torque the bolts to 44 inch lbs. (5.0 Nm).
   - Ignition coils and connectors
   - Intake manifold plenum
   - Air cleaner assembly, ducts and air intake hose

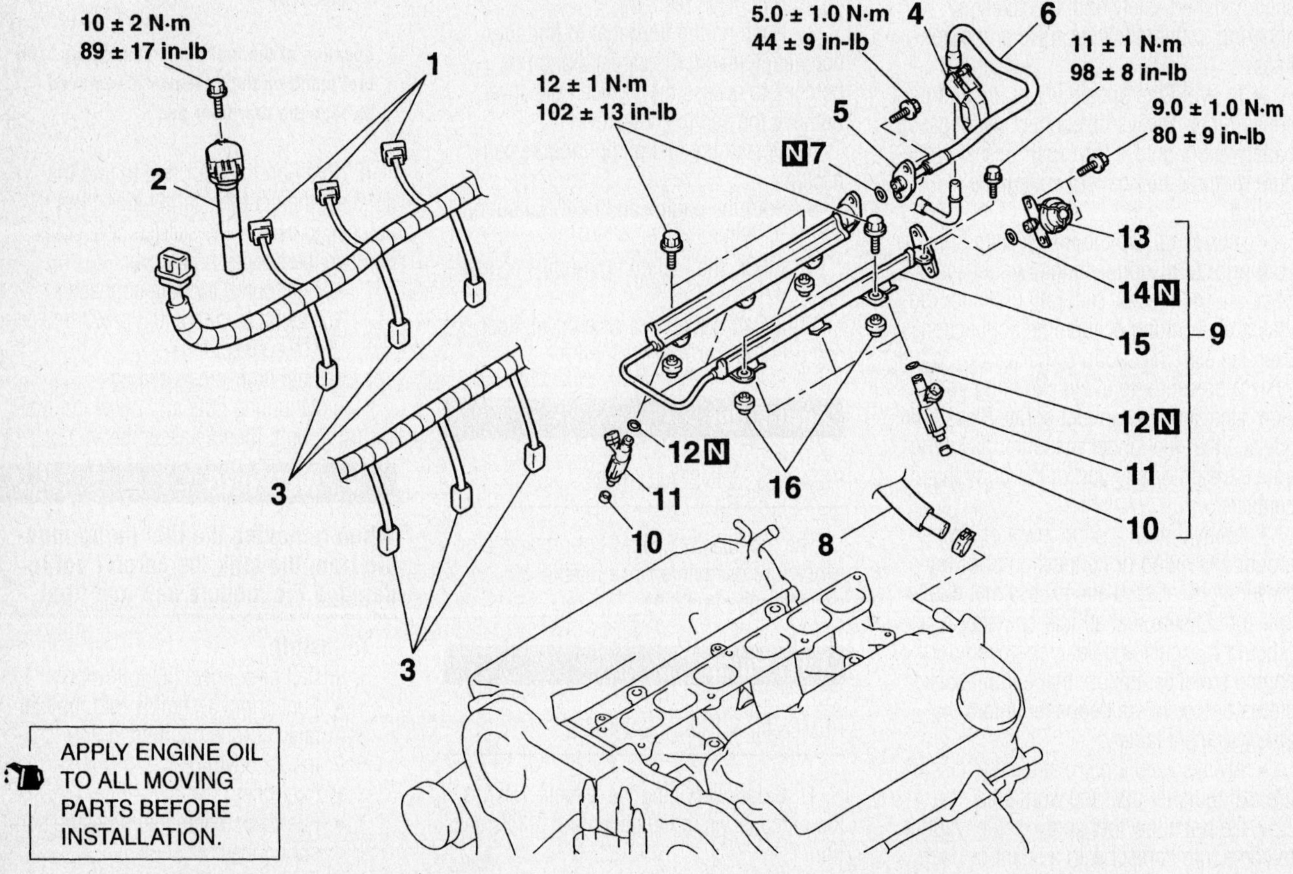

1. IGNITION COIL CONNECTOR
2. IGNITION COIL
3. INJECTOR CONNECTOR
4. FUEL HIGH-PRESSURE HOSE CONNECTION (FUEL LINE PIPE SIDE)
5. FUEL HIGH-PRESSURE HOSE CONNECTION (FUEL RAIL SIDE)
6. FUEL HIGH-PRESSURE HOSE
7. O-RING
8. BLOW-BY HOSE CONNECTION
9. FUEL RAIL, INJECTOR AND FUEL DAMPER <UP TO DECEMBER 2003>, FUEL RAIL AND INJECTOR <FROM JANUARY 2004>
10. INSULATORS
11. FUEL INJECTORS
12. O-RING
13. FUEL DAMPER <UP TO DECEMBER 2003>
14. O-RING
15. FUEL RAIL
16. INSULATORS

67170-ENDE-G26

Exploded view of the fuel rail and injectors

DRIVE TRAIN

Transmission Assembly

REMOVAL & INSTALLATION

1. Before servicing the vehicle, refer to the precautions in the beginning of this section.
2. Remove the engine under cover.
3. Drain the transmission fluid.
4. On AWD vehicles, drain the engine coolant.
5. Remove or disconnect the following:
   - Hood

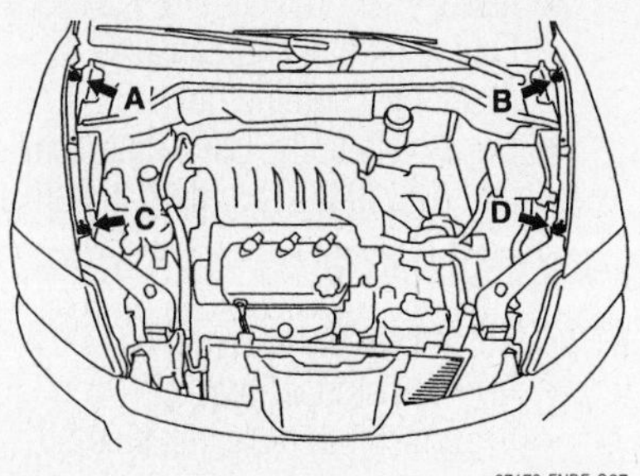

Front fender assembling bolts for special tool installation

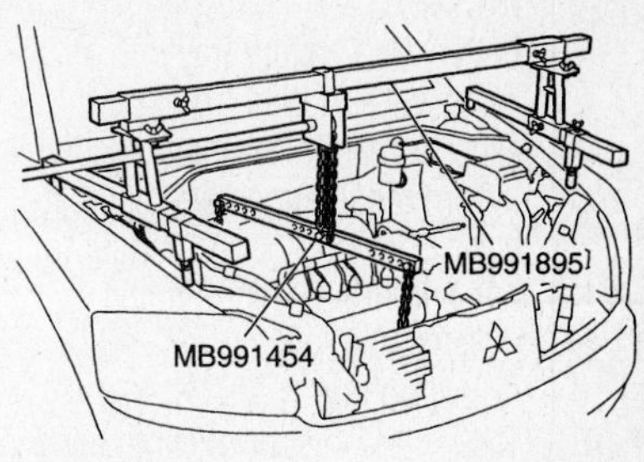

View of the engine hangers installed on the engine

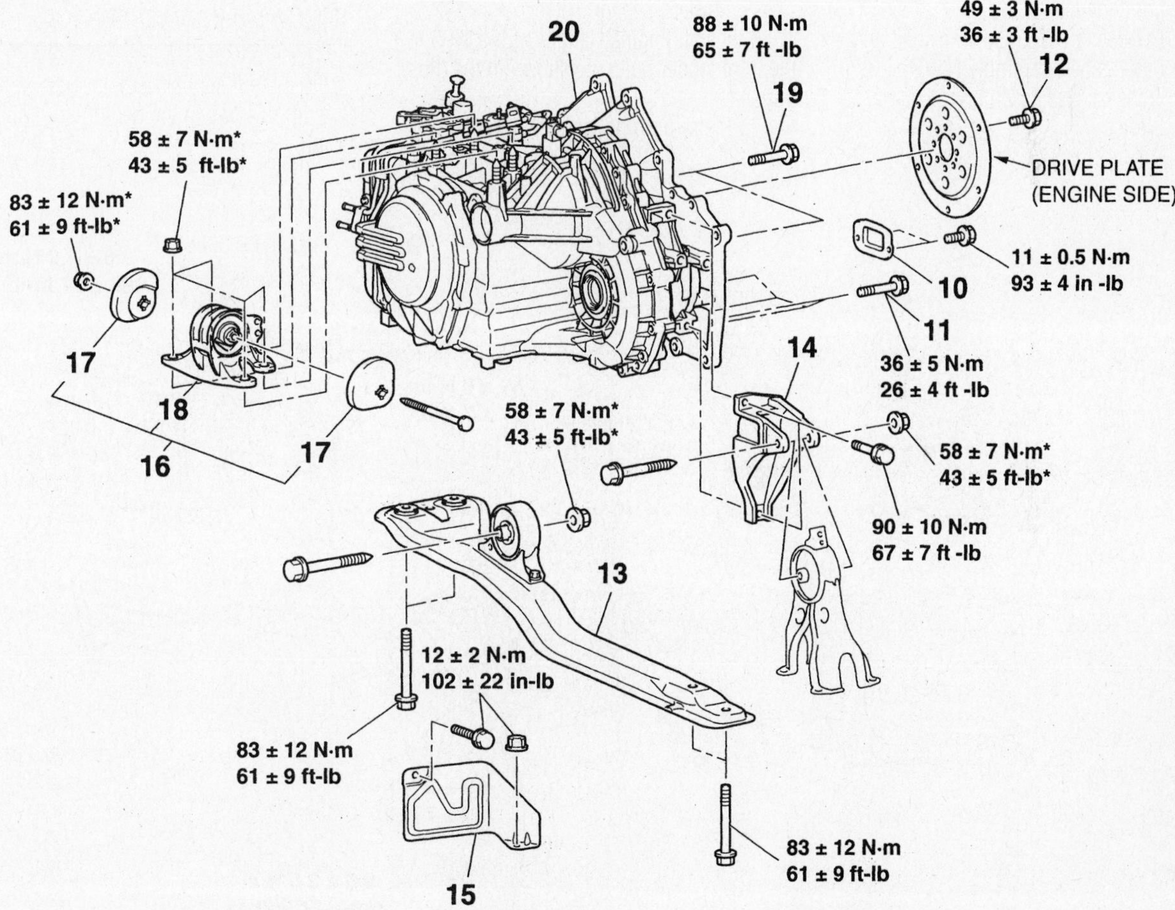

10. COVER
11. ENGINE OIL PAN AND TRANSAXLE ASSEMBLY COUPLING BOLTS
12. DRIVE PLATE BOLTS
13. CENTERMEMBER ASSEMBLY
14. REAR ROLL STOPPER BRACKET
15. CROSSMEMBER PLATE <LH SIDE>
16. TRANSAXLE MOUNT BRACKET ASSEMBLY
17. TRANSAXLE MOUNT STOPPER
18. TRANSAXLE MOUNT BRACKET
19. TRANSAXLE ASSEMBLY LOWER PART COUPLING BOLTS
20. TRANSAXLE ASSEMBLY

67170-ENDE-G28

Automatic transmission removal

- Negative battery cable
- Air cleaner assembly
- Powertrain Control Module (PCM)
- Battery and battery tray
- Front exhaust pipe from the 2 exhaust manifolds, then disconnect it from the intermediate pipe/catalytic converter (make certain to retain the bolts and nuts for reassembly).
- Adjusting nut
- Transmission control cable connection
- Inhibitor switch connector
- A/T control solenoid valve connector
- Input and output shaft speed sensor connectors
- Driveshaft (2WD) or driveshaft and output shaft assembly (AWD)
- Transfer case assembly
- Starter
- Transmission oil cooler hose
- Transmission assembly upper part coupling bolts
- Cover
- Engine oil pan and transaxle assembly coupling bolts
- Drive plate bolts. Use Mitsubishi tool (MD998781) to hold the driveplate in position to remove the bolts.
- Centermember assembly
- Rear roll stopper bracket
- Crossmember plate, left side
- Air cleaner bracket
- Transaxle mount bracket assembly
- Transaxle mount stopper
- Manifold Differential Pressure (MDP) sensor

6. Install Engine Hanger MB991895 to the front fender bolts, as shown in the illustration.

7. Install Engine Hanger Balancer MB991454 to the engine/transmission assembly. Place a thick towel on the end of the tool to prevent damaging the firewall.

8. Remove the transmission from the vehicle.

To install:

9. Installation is the reverse of the removal procedure. Note the tightening specifications on the accompanying illustrations.

10. Refill the transmission and transfer case with oil.

11. Start the vehicle and check for any leaks.

Transfer Case Assembly

REMOVAL & INSTALLATION

1. Before servicing the vehicle, refer to the precautions in the beginning of this section.

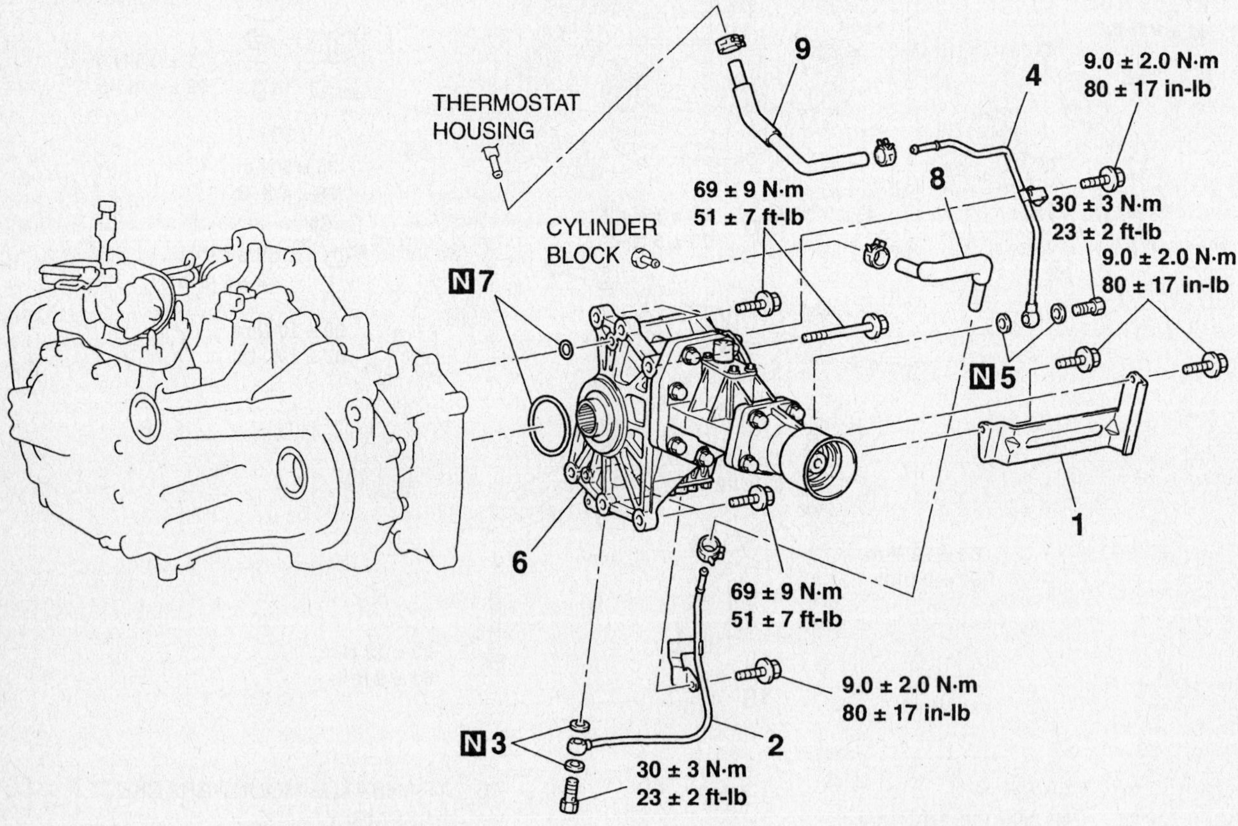

Exploded view of the transfer case assembly

2. Drain the transmission and transfer case fluid. Drain the engine coolant.

3. Remove or disconnect the following:

- Engine under cover
- Front exhaust pipe
- Front propeller shaft
- Driveshaft and output shaft
- Right exhaust manifold stay "B"
- Air guide
- Water feed tube assembly and gasket
- Water return hose and gasket
- Retaining bolts and detach the transfer case assembly from the transmission. Lower the transfer case between the engine and crossmember. Remove and discard the O-rings.
- Water feed hose
- Air cleaner assemby
- Battery and battery tray
- Water return hose

To install:

4. Install or connect the following:

- Water return hose
- Battery tray and battery
- Air cleaner assemby
- Water feed hose
- New O-rings onto the transfer case
- Transfer case, by maneuvering it up between the engine and crossmember. Tighten the retaining bolts to 51 ft. lbs. (69 Nm).
- Water return hose gasket and hose
- Water feed tube gasket and tube
- Air guide
- Right exhaust manifold stay "B"
- Driveshaft and output shaft
- Front propeller shaft
- Front exhaust pipe
- Engine under cover

5. Fill the engine cooling system. Fill the transmission and transfer case with fluid.

Halfshaft (Driveshaft)

REMOVAL & INSTALLATION

Front

1. Before servicing the vehicle, refer to the precautions in the beginning of this section.

2. Drain the transmission and transfer case fluid, as necessary.

3. Remove or disconnect the following:

- Negative battery cable
- Front and side under covers
- Front exhaust pipe
- Transfer case heat shield, if equipped
- Wheels
- Cotter pin
- Drive shaft nut and washer
- Speed sensor, if equipped with ABS
- Brake hose bracket
- Self locking nut and separate the lower ball joint
- Loosen the tie rod end nut and separate the tie rod end. Remove the nut.

4. Remove the driveshaft or driveshaft and inner shaft from the hub using Mitsubishi Special Tools MB990242, MB990244, MB991354 and MB990767 to push the driveshaft or driveshaft and inner shaft from the hub.

5. Remove the driveshaft from the hub by pulling the bottom of the brake rotor toward you.

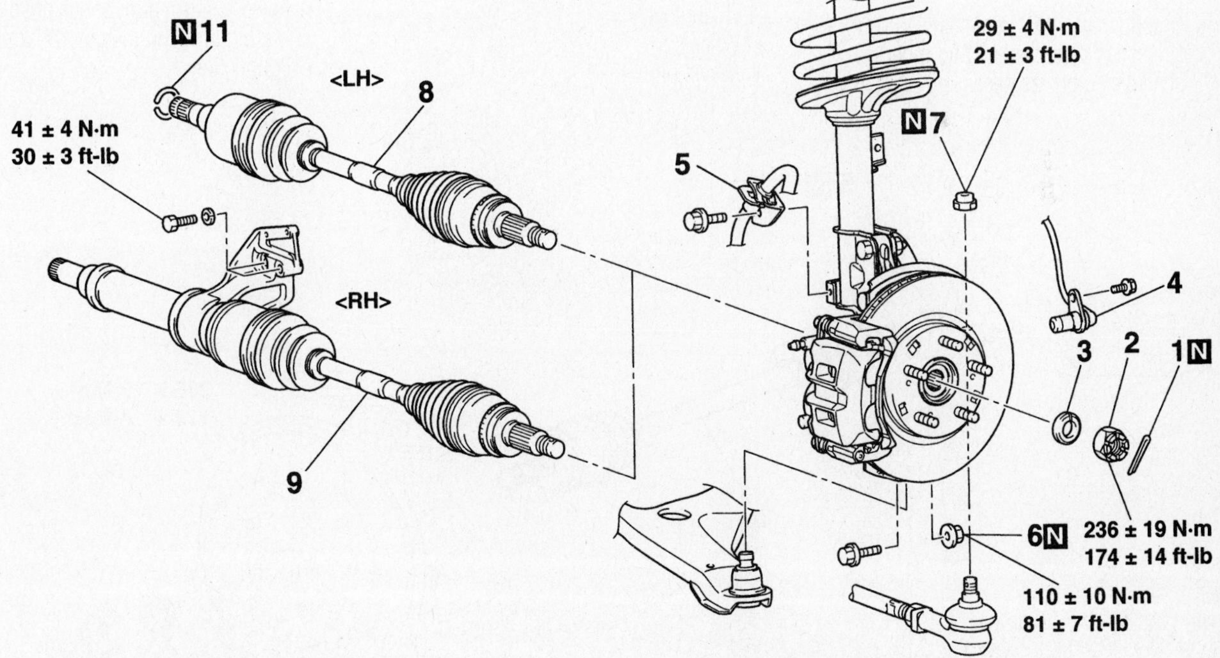

Exploded view of the front driveshaft

⁂ WARNING

Never pull on the driveshaft or you risk damaging the joints. Only use a prybar to remove the driveshaft from the transmission.

6. Insert a prybar between the transaxle case and driveshaft, then pry the driveshaft from the transmission. Make sure the spline of the driveshaft does not damage the oil seal.

7. If you have trouble removing the inner shaft from the transmission, hit the bracket assembly lightly with a plastic hammer and remove the inner shaft.

➡Do not apply pressure to the wheel bearing while the driveshafts are removed.

8. On AWD vehicles, use Mitsubishi Special tool MB991721 to remove the output shaft from the transmission. Make sure the splined part of the output shaft does not damage the oil seal.
   - Circlips from the ends of the shafts

To install:

9. Install or connect the following:
   - New circlips
   - Output shaft, AWD vehicles
   - Driveshaft or driveshaft and inner shaft assembly, as necessary
   - Tie rod end and tighten the nut to 21 ft. lbs. (29 Nm)
   - Lower ball joint and tighten the nuts to 81 ft. lbs. (110 Nm).
   - Brake hose bracket
   - Speed sensor, if equipped with ABS
   - New driveshaft washer, with the beveled edge facing out
   - Driveshaft nut. With no load on the wheel bearings, use Mitsubishi special tool MB990767 to tighten the nut to 174 ft. lbs. (236 Nm).
   - New cotter pin
   - Wheels
   - Transfer case heat shield, if equipped
   - Front exhaust pipe
   - Front and side under covers
   - Negative battery cable

10. Fill the transmission and transfer case (if equipped) with fluid

Rear

1. Before servicing the vehicle, refer to the precautions in the beginning of this section.

2. Remove or disconnect the following:
   - Negative battery cable
   - Undercover
   - Wheels
   - Clevis pin
   - Driveshaft nut, using and end yoke holder to hold the hub
   - Washer
   - Rear wheel speed sensor, if equipped with ABS
   - Lower control arm, shock absorber, trailing arm and toe control arm connection
   - Driveshaft. Use Mitsubishi Special Tools MB990242, MB990244, MB991354 and MB990767 to push the driveshaft from the hub.

⁂ WARNING

Never pull on the driveshaft or you risk damaging the joints. Only use a prybar to remove the driveshaft from the transmission.

3. Use a slide hammer to remove the driveshaft from the differential carrier. Make sure the spline of the driveshaft does not damage the oil seal.
   - Circlip

To install:

4. Install or connect the following:
   - New circlip on the driveshaft
   - Driveshaft into the differential carrier and hub
   - Lower control arm, shock absorber, trailing arm and toe control arm connection

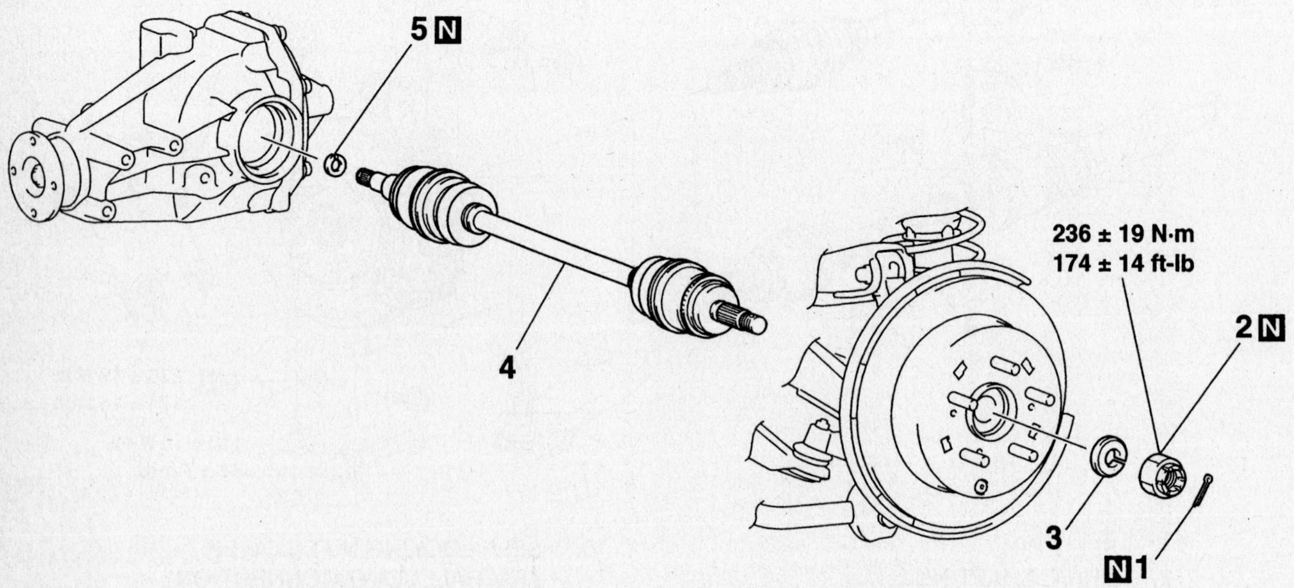

Exploded view of the rear driveshaft—AWD vehicles

- Rear wheel speed sensor, if equipped with ABS
- New driveshaft washer, with the beveled edge facing out
- Driveshaft nut. With no load on the wheel bearings, use Mitsubishi special tool MB990767 to tighten the nut to 174 ft. lbs. (236 Nm).
- New cotter pin
- Wheels

CV-Joints

OVERHAUL

1. Before servicing the vehicle, refer to the precautions in the beginning of this section.
2. Remove or disconnect the following:
   - Front wheel
   - Driveshaft from the car
   - Small and larger band
   - Circlip
   - Double Offset Joint (DOJ) outer race
   - Dust cover
   - Circlip
   - Balls from the cage
   - Cage from the inner race. Turn the cage so that the projections of the inner race align with the recesses of the cage.

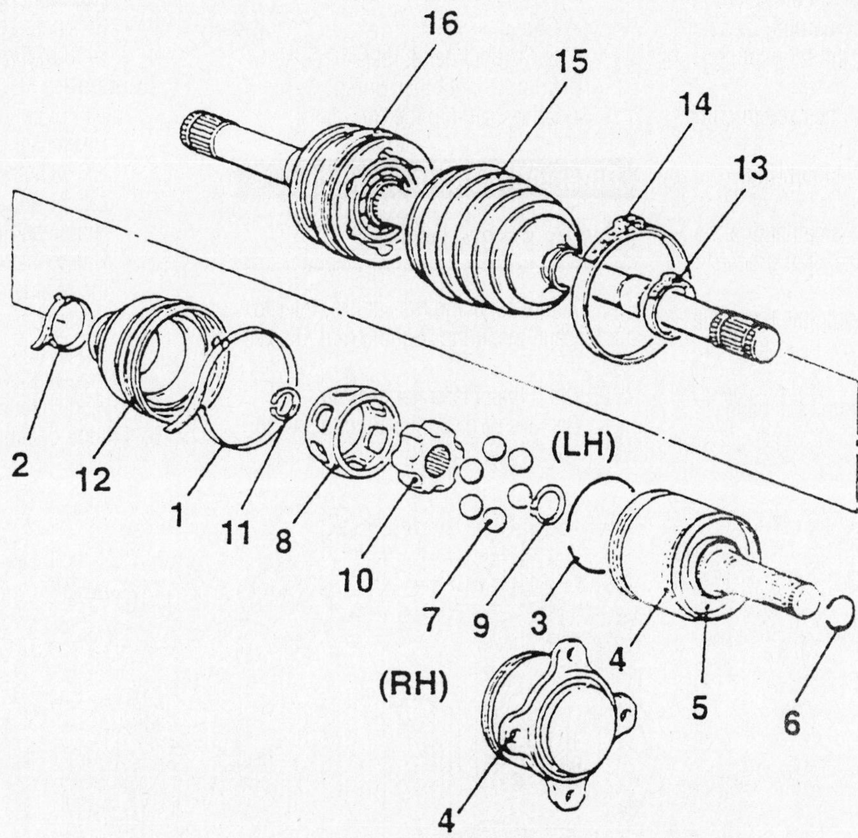

1. D.O.J. boot band (large)
2. D.O.J. boot band (small)
3. Circlip
4. D.O.J. outer race
5. Dust cover
6. Circlip
7. Ball
8. D.O.J. cage
9. Snap ring
10. D.O.J. inner race
11. Circlip
12. D.O.J. boot
13. B.J. boot band (small)
14. B.J. boot band (large)
15. B.J. boot
16. B.J. assembly

9308UG06

CV-Joint, exploded view

- Snapring from the shaft
- DOJ inner race
- Slide the boot off
- Birfield Joint (BJ) small and larger bands
- BJ boot
- BJ assembly

To install:

3. Check the shaft and splines for damage or wear. Inspect the cage, race and balls for any sign of corrosion, wear, cracking or damage. Clean all the parts thoroughly and air dry them completely before installation. Any remaining cleaning solvent can dissolve the lubricating grease.
4. Tool MB991561 can be used to crimp the bands in place.
5. Install or connect the following:
   - BJ assembly
   - BJ boot, slide the small end of the boot until only one shaft groove cone be seen.
   - BJ small band, crimp the band. Fill the BJ boot with 4.6 oz (130 g) of grease.
   - BJ larger band, crimp the band
   - DOJ small band and boot, fill the boot with grease
   - DOJ cage onto the driveshaft so that the smaller diameter side is installed first.
   - Circlip
   - DOJ inner race and new snap ring, apply grease to the inner race
   - Balls into the cage, grease to the ball areas of the cage and race
   - Outer race, fill the outer race about ⅓ full of grease.
   - Dust cover
   - Circlip
   - Large boot band, release the air from the boot then crimp
   - Driveshaft into the car

Differential Carrier Assemby

REMOVAL & INSTALLATION

1. Before servicing the vehicle, refer to the precautions in the beginning of this section.
2. Drain the differential gear oil.
3. Remove or disconnect the following:
   - Center exhaust pipe
   - Rear driveshaft
   - Stabilizer bar bushing

➡Matchmark the installed relation of the propeller shaft to the differential carrier before removal.

   - Propeller shaft from the differential carrier. Suspend the propeller shaft from the body to avoid damaging or bending the shaft.
   - Differential mount bracket
   - Hose and nipple
   - Retainers and differential carrier. Discard the washers.

To install:

4. Install or connect the following:
   - Differential carrier. Tighten the retainers, with new washers, to the specifications shown in the illustration.
   - Nipple and hose
   - Differential mount bracket. Tighten the retainers, with new washers, to the specifications shown in the illustration.
   - Propeller shaft, aligning the matchmarks made during removal.
5. Fill the differential with oil

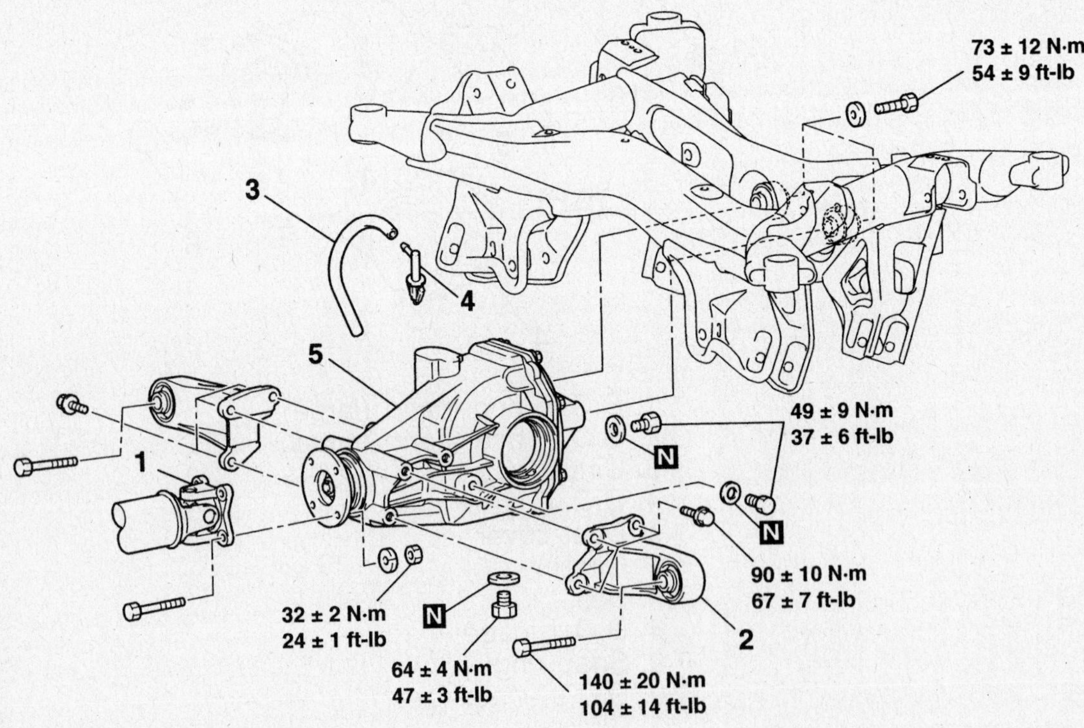

Exploded view of the differential carrier assembly

STEERING AND SUSPENSION

Air Bag

⁂ CAUTION

Some vehicles are equipped with an air bag system. The system must be disabled before performing service on or around system components, steering column, instrument panel components, wiring and sensors. Failure to follow safety and disabling procedures could result in accidental air bag deployment, possible personal injury and unnecessary system repairs.

PRECAUTIONS

Several precautions must be observed when handling the inflator module to avoid accidental deployment and possible personal injury.

- Never carry the inflator module by the wires or connector on the underside of the module.
- When carrying a live inflator module, hold securely with both hands, and ensure that the bag and trim cover are pointed away.
- Place the inflator module on a bench or other surface with the bag and trim cover facing up.
- With the inflator module on the bench, never place anything on or close to the module that may be thrown in the event of an accidental deployment.

DISARMING

To avoid personal injury when working on vehicles equipped with an air bag, the negative battery cable must be disconnected and at least 60 seconds must elapse before working on the system. Failure to do so may result in deployment of the air bag. You should also wrap or isolate the negative battery cable with electrical or other non-conductive tape.

Rack and Pinion Steering Gear

REMOVAL & INSTALLATION

1. Before servicing the vehicle, refer to the precautions in the beginning of this section.
2. On vehicles equipped with a supplemental restraint system (SRS), turn the front wheel to the straight ahead position and remove the ignition key to prevent the steering wheel from turning.

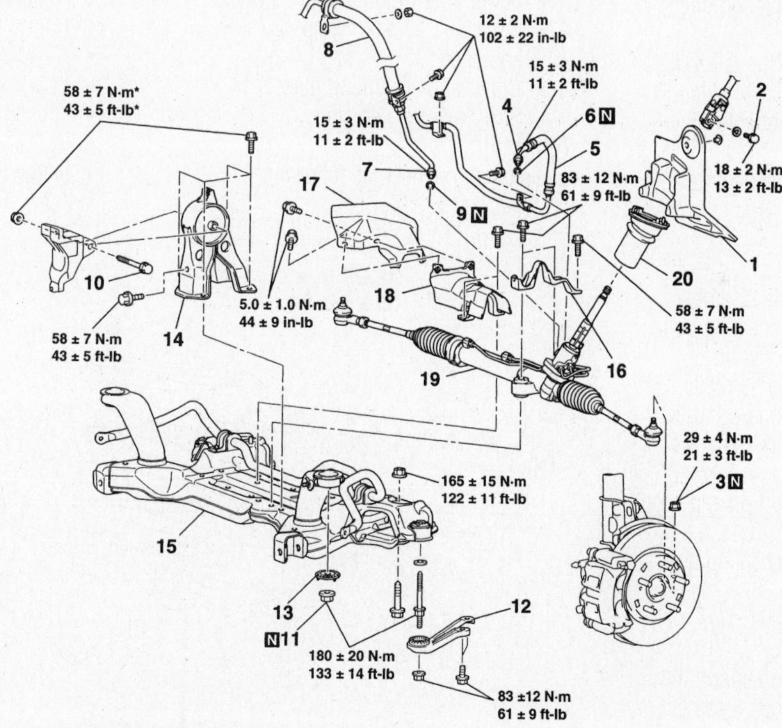

1. STEERING SHAFT PAD
2. STEERING COLUMN SHAFT ASSEMBLY AND STEERING GEAR CONNECTING BOLT
3. SELF LOCKING NUT (TIE ROD END AND KNUCKLE CONNECTION)
4. RETURN TUBE CONNECTION
5. RETURN TUBE
6. O-RING
7. PRESSURE TUBE CONNECTION
8. PRESSURE TUBE/PRESSURE HOSE
9. O-RING
10. REAR ROLL STOPPER CONNECTING BOLT
11. SELF LOCKING NUT
12. FRONT AXLE CROSSMEMBER STAY
13. PLATE STOPPER
14. REAR ROLL STOPPER
15. FRONT AXLE NO.1 CROSSMEMBER
16. POWER STEERING GEAR BRACKET
17. HEAT PROTECTOR
18. STEERING GEAR MOUNTING GEAR SIDE BRACKET
19. STEERING GEAR
20. STEERING JOINT COVER

67170-ENDE-G34

Exploded view of the power steering gear

➡Prior to removal of the steering gear box, center the front wheels and remove the ignition key. Failure to do so may damage the Supplemental Restraint System (SRS) clockspring and render SRS system inoperative.

3. Drain the power steering system.
4. Disconnect the negative battery cable. Insulate the cable end with high-quality electrical tape or similar non-conductive wrapping Wait at least 1 minute before proceeding.
5. Remove the steering wheel, as follows:
   a. Pry the cover from the steering wheel.
   b. Detach the connector, and retainers, as necessary then remove the air bag module assembly
   c. Remove the steering wheel nut, then use a puller to remove the steering wheel.
   d. Detach the sub-harness.
6. Remove the clock spring and put aside in a safe place.
7. Remove or disconnect the following:
   - Front scuff plate and cowl side trim
   - Front floor left side console cover
   - Front floor carpet
   - Stabilizer link and lower control arm assembly
   - Center member
   - Shaft pad
   - Steering column shaft assembly and steering gear connecting bolt
   - Tie rod end from the steering knuckle, using a suitable puller
   - Fluid return hose and return tube. Plug the hoses. Remove and discard the O-ring.
   - Pressure hose connection and hose gasket
   - Rear roll stopper connecting bolt
   - Self-locking nut
   - Front axle crossmember stay
   - Plate stopper
8. Use a transaxle jack to support the crossmember, then remove the crossmember mounting nuts and bolts. Lower the crossmember with the rear roll stopper, stabilizer bar, return tube and steering gear.
9. Remove the following from the crossmember.
   - Rear roll stopper
   - Front axle No. 1 crossemember
   - Power steering gear bracket
   - Heat protector
   - Steering gear mounting gear side bracket
   - Steering gear
   - Steering gear joint cover

To install:

10. Installation is the reverse of the removal procedure. Refer to the illustration for tightening specifications.
11. Refill the reservoir and bleed the system.
12. Perform a front end alignment.

Strut

REMOVAL & INSTALLATION

Front

1. Before servicing the vehicle, refer to the precautions in the beginning of this section.
2. Remove or disconnect the following:
   - Windshield wiper arm assemblies and front deck garnish
   - Tire and wheel
   - Stabilizer link
   - Brake hose bracket
   - Lower strut mounting bolt and nut
   - Front ABS sensor clamp, if equipped
   - Upper strut mounting nuts
   - Strut from the vehicle

To install:

3. Install or connect the following
   - Strut into the vehicle
   - Upper strut mounting nuts and tighten to 35 ft. lbs. (47 Nm)
   - Front ABS sensor clamp, if equipped
   - Lower strut mounting bolt and nut. Torque the nut to 225 ft. lbs. (305 Nm).
   - Brake hose bracket
   - Stabilizer link
   - Tire and wheel
   - Front deck garnish and wiper arms

Shock Absorber

REMOVAL & INSTALLATION

Rear

1. Before servicing the vehicle, refer to the precautions in the beginning of this section.
2. Remove or disconnect the following:
   - Luggage floor rear board
   - Tonneau cover
   - Luggage floor front board
   - Parcel strap hook
   - Luggage floor side board
   - Rear end trim
   - Luggage floor carpet bracket

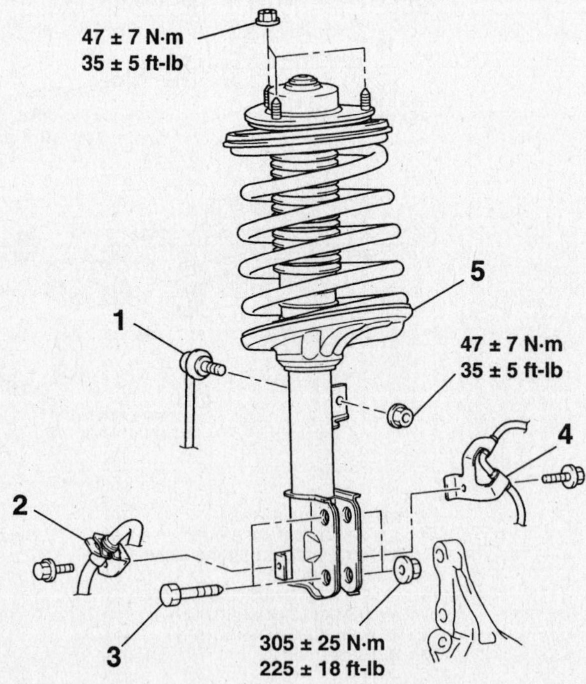

1. STABILIZER LINK
2. BRAKE HOSE BRACKET
3. STRUT BOLT
4. FRONT ABS SENSOR CLAMP <VEHICLES WITH ABS>
5. STRUT ASSEMBLY

67170-ENDE-G35

Front strut mounting

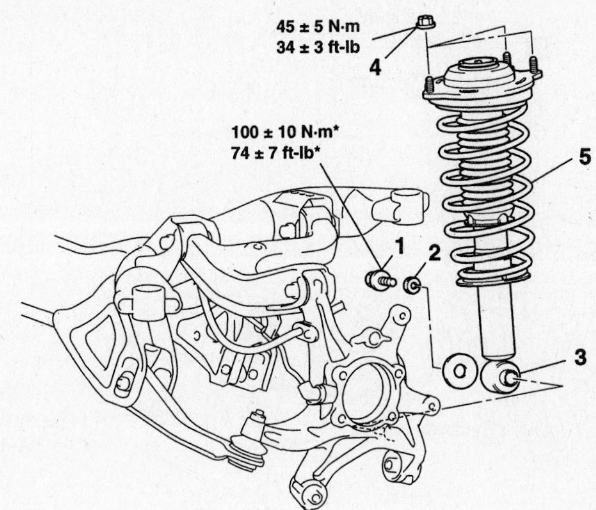

1. COIL SPRING BOLT
2. COIL SPRING WASHER
3. SHOCK ABSORBER ASSEMBLY AND KNUCKLE CONNECTION
4. COIL SPRING NUT
5. SHOCK ABSORBER ASSEMBLY

67170-ENDE-G37

Exploded view of the rear shock absorber

- Wheel and tire
- Lower shock absorber bolt and washer, and separate the shock from the knuckle
- Upper shock mounting nuts
- Shock absorber from the vehicle

To install:

3. Install or connect the following:
   - Shock absorber. Tighten the upper nuts to 34 ft. lbs. (45 Nm) and the lower shock-to-knuckle bolt hand-tight.
   - Wheel and tire
   - Luggage floor carpet bracket
   - Rear end trim
   - Luggage floor side board
   - Parcel strap hook
   - Luggage floor front board
   - Tonneau cover
   - Luggage floor rear board

4. With the weight of the vehicle resting on the suspension, tighten the lower shock-to-knuckle bolt to 74 ft. lbs. (100 Nm).

Coil Spring

REMOVAL & INSTALLATION

Front

1. Before servicing the vehicle, refer to the precautions in the beginning of this section.
2. Remove or disconnect the following:
   - Strut
   - Strut cover
3. Compress the coil spring until there is a clearance on both ends
   - Strut center nut
   - Insulator
   - Bearing
   - Spring upper seat
   - Spring upper pad
   - Strut cover
   - Strut damper
   - Coil spring
   - Spring lower pad
   - Strut
4. Installation is the reverse of removal. Torque the center nut to 48 ft. lbs. (65 Nm).

Rear

1. Before servicing the vehicle, refer to the precautions in the beginning of this section.
2. Remove or disconnect the following:
   - Shock absorber assembly
3. Install a suitable spring compressor on the spring.
   - Coil spring nut
   - Washer
   - Shock absorber insulator
   - Spring upper pad
   - Shock absorber cover
   - Shock absorber damper
   - Coil spring
   - Shock absorber

To install:

4. Assemble the following:
   - Shock absorber

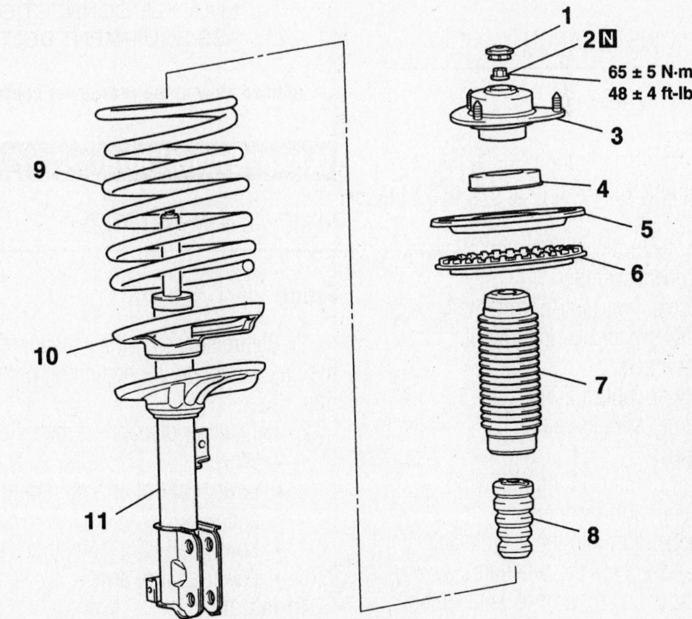

1. STRUT COVER
2. STRUT NUT
3. STRUT INSULATOR
4. STRUT BEARING
5. SPRING UPPER SEAT
6. SPRING UPPER PAD
7. STRUT COVER
8. STRUT DAMPER
9. COIL SPRING
10. SPRING LOWER PAD
11. FRONT SUSPENSION STRUT

67170-ENDE-G36

Exploded view of the front strut

- Coil spring
- Shock absorber damper
- Shock absorber cover
- Spring upper pad
- Shock absorber insulator
- Washer
- Coil spring nut and tighten to 17 ft. lbs. (23 Nm).

5. Remove the spring compressor.
   - Shock absorber assembly

Upper Control Arm

REMOVAL & INSTALLATION

Front

These vehicles use a strut type front suspension. No upper control arm is used.

Lower Ball Joint

REMOVAL & INSTALLATION

The lower ball joint is an integral part of the lower control arm assembly, and can not be serviced separately. A worn or damaged ball joint requires replacement of lower control arm assembly.

Upper Control Arm

REMOVAL & INSTALLATION

Rear

1. Before servicing the vehicle, refer to the precautions in the beginning of this section.
2. Remove or disconnect the following:
   - Wheel assembly
   - Upper control arm from the knuckle
   - ABS sensor clamp bolts, if equipped
   - Upper control arm
   - Upper arm stopper

To install:

3. Install or connect the following:
   - Upper arm stoppers
   - Upper arm into the vehicle and hand-tighten the retainers
   - Upper control arm-to-knuckle bolt and nut and tighten hand-tight
   - ABS sensor clamp bolts
   - Wheel assembly
4. Once the weight of the vehicle is on the suspension, tighten the control arm-to-knuckle nut and the upper control arm mounting nuts to 83 ft. lbs. (113 Nm).
5. Check and adjust the rear wheel alignment.

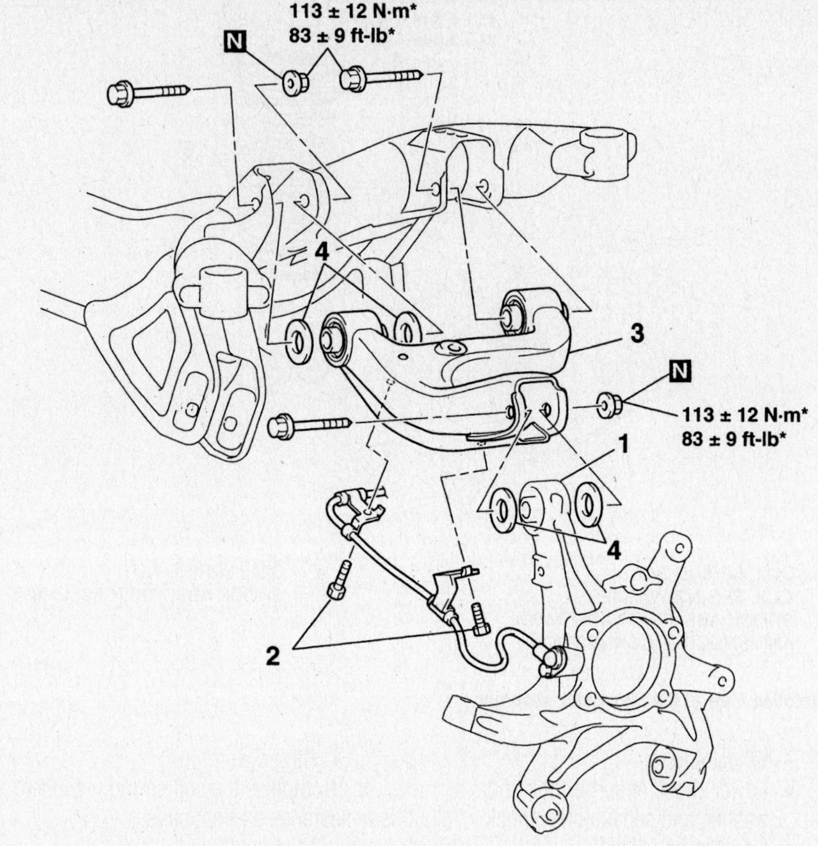

Exploded view of the rear upper control arm

Lower Control Arm

REMOVAL & INSTALLATION

Front

1. Before servicing the vehicle, refer to the precautions in the beginning of this section.
2. Remove or disconnect the following:
   - Wheel
   - Lower control arm mounting bolts and nuts
   - Lower ball joint from knuckle
   - Lower control arm

To install:

3. Install or connect the following:
   - Lower control arm
   - Lower control arm retainers. Tighten the knuckle nut and bolt to 81 ft. lbs. (110 Nm) and the other retainers hand-tight only at this time.
   - Wheel
4. Once the weight of the vehicle is resting on the suspension, torque the lower control arm mounting bolts and nuts 122 ft. lbs. (165 Nm).
5. Check and adjust the front wheel alignment.

Rear

1. Before servicing the vehicle, refer to the precautions in the beginning of this section.
2. Remove or disconnect the following:
   - Rear wheel
   - Lower control arm from the knuckle
   - Retaining nut, then separate the lower control arm from the stabilizer bar link
   - Lower control arm bolt, plate and lower control arm assembly

To install:

3. Install or connect the following:
   - Lower control arm
   - Lower control arm plate
   - Lower control arm bolt, hand-tight only at this time
   - Lower control arm to the stabilizer

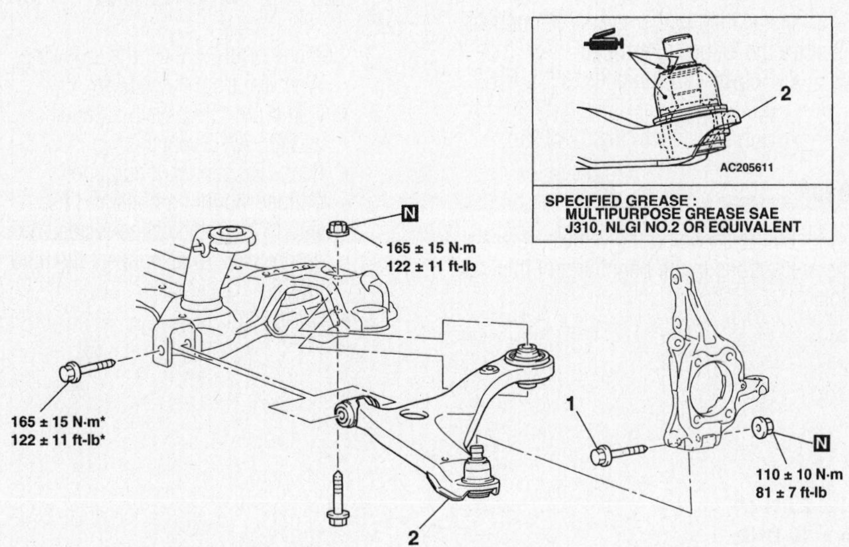

Exploded view of the front lower control arm

bar link. Tighten the retaining to 30 ft. lbs. (40 Nm).

- Lower control arm to the knuckle. Use a new nut and secure hand-tight.
- Rear wheel

4. Once the weight of the vehicle is resting on the suspension, torque the lower control arm mounting bolt to 57 ft. lbs. (78 Nm), and the lower control arm to knuckle nut to 83 ft. lbs. (113 Nm).

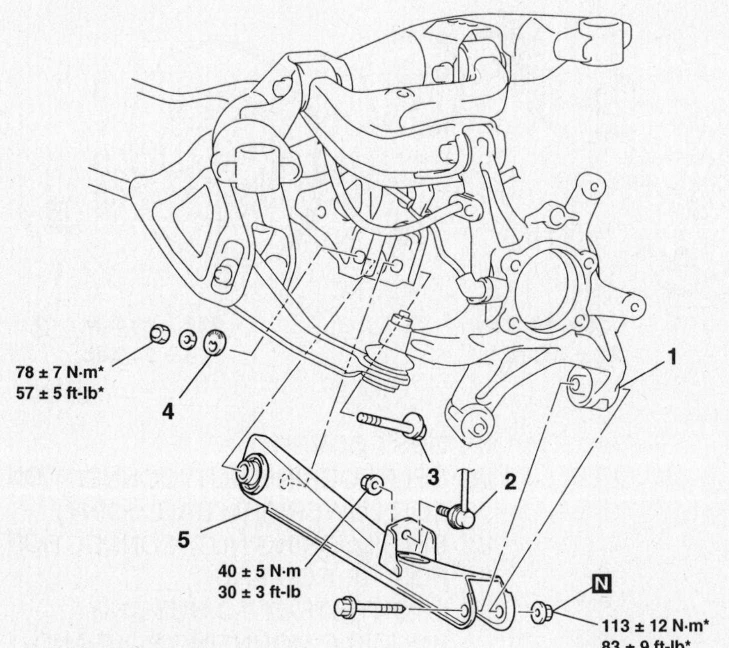

Exploded view of the rear lower control arm

5. Check and adjust the rear wheel alignment.

CONTROL ARM BUSHING REPLACEMENT

Front

1. Before servicing the vehicle, refer to the precautions in the beginning of this section.
2. Remove the wheel.
3. Remove the lower control arm and place in a vise.
4. Using tool MB990883 remove the bushing

To install:

5. Position the bushing with the larger end facing the front of the vehicle.

➡Coat the bushing with a soap solution and take care not to twist.

6. Using tool MB991522, press the bushing into the bracket.
7. Install the lower control arm.
8. Install the wheel.

Rear

1. Before servicing the vehicle, refer to the precautions in the beginning of this section.
2. Remove the wheel.
3. Remove the lower control arm.
4. Using tool MB991522 press out the bushing

To install:

5. Position the bushing with the larger end facing the front of the vehicle.
6. Using tool MB991522 press the bushing into the bracket.
7. Install the lower control arm.
8. Install the wheel.

Wheel Bearings

ADJUSTMENT

The wheel bearings are not adjustable. If the bearings are noisy or become loose, they must be replaced.

REMOVAL & INSTALLATION

Front

The wheel bearings are not replaceable. If defective, the hub/bearing assembly must be replaced.

1. Before servicing the vehicle, refer to the precautions in the beginning of this section.
2. Remove or disconnect the following:

- Tire and wheel assembly
- Cotter pin
- Driveshaft nut
- Washer
- Speed sensor, if equipped
- Brake hose bracket
- Caliper and rotor
- Front wheel hub assembly
- Dust cover
- Tie rod end
- Front strut-to-hub and knuckle mounting bolt and nut
- Hub/knuckle assembly

3. Installation is the reverse of removal. Observe the following torques:
   - Hub-to-knuckle bolts: 67 ft. lbs. (90 Nm)
   - Hub nut: 174 ft. lbs. (236 Nm)

Rear

1. Before servicing the vehicle, refer to the precautions in the beginning of this section.
2. Loosen the wheel lug nuts only ½ a turn.
3. Remove or disconnect the following:
   - Wheel(s) from the vehicle
   - Cotter pin, driveshaft nut and washer, AWD vehicles
   - Rear ABS sensor, if equipped
   - Caliper mounting bolts and use wire to hang the caliper aside. You do not have to disconnect the fluid line.
   - Brake rotor

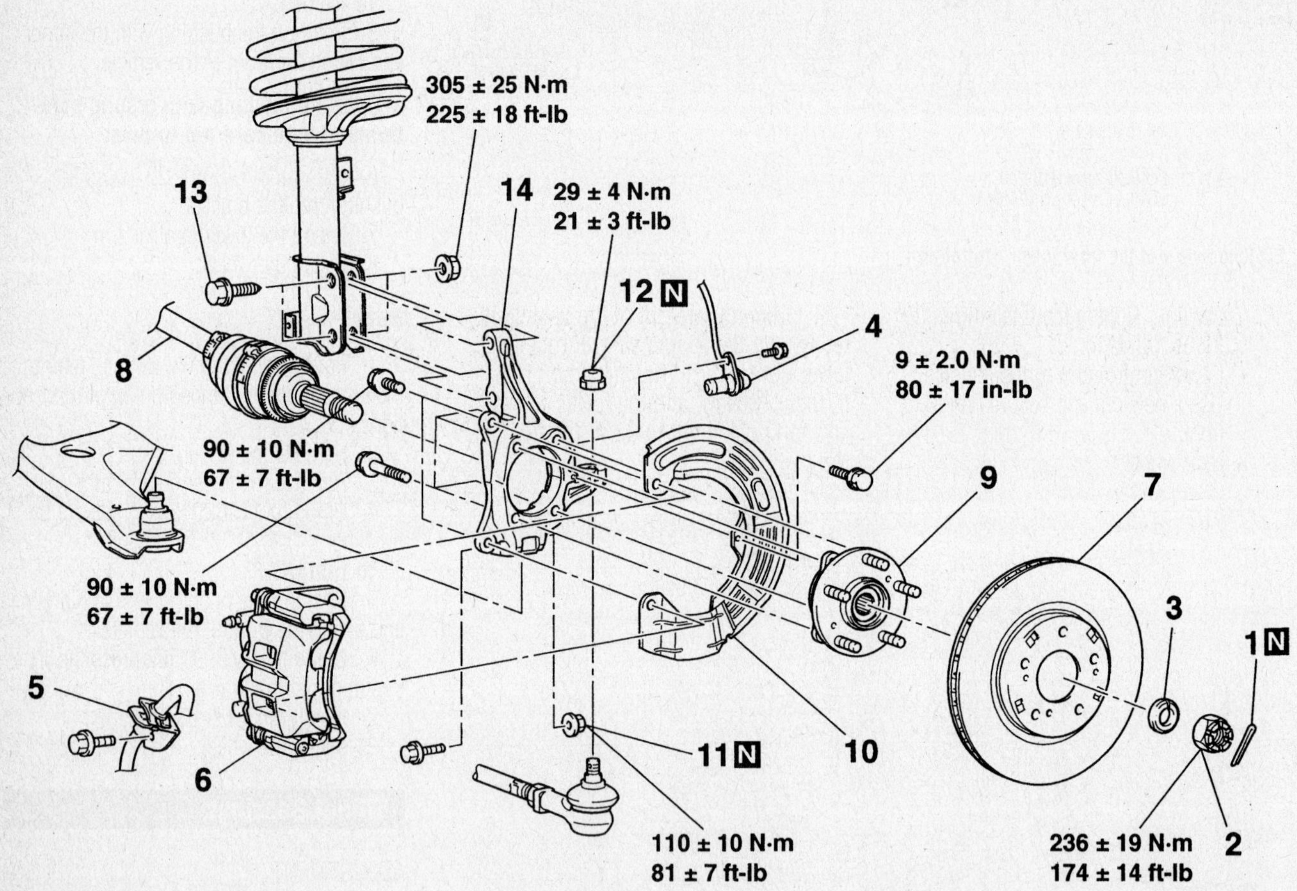

1. SPLIT PIN
2. DRIVE SHAFT NUT
3. WASHER
4. FRONT ABS SENSOR <VEHICLES WITH ABS>
5. BRAKE HOSE BRACKET
6. CALIPER ASSEMBLY
7. BRAKE DISC
8. DRIVE SHAFT
9. FRONT WHEEL HUB ASSEMBLY
10. DUST COVER
11. SELF LOCKING NUT (CONNECTION FOR LOWER ARM BALL JOINT)
12. SELF LOCKING NUT (CONNECTION FOR TIE ROD END)
13. FRONT STRUT TO HUB AND KNUCKLE MOUNTING BOLT AND NUT
14. KNUCKLE

67170-ENDE-G41

Exploded view of the front hub/bearing and knuckle assembly

- Rear hub assembly

To install:

- Rear hub assembly. Tighten mounting bolts to 67 ft. lbs. (90 Nm).
- Brake rotor
- Caliper and torque the mounting bolts to 45 ft. lbs. (60 Nm)
- Rear ABS sensor, if equipped
- Washer and driveshaft nut, AWD vehicles. Torque the nut to 174 ft. lbs. (236 Nm). Install a new cotter pin.
- Wheels

4. Road test the vehicle and check for leaks.

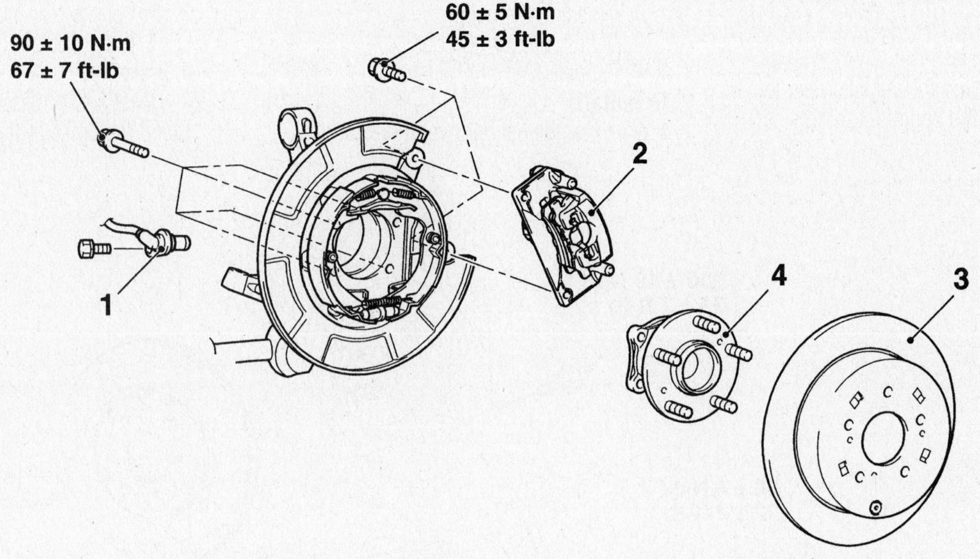

1. REAR ABS SENSOR<VEHICLES WITH ABS>
2. CALIPER ASSEMBLY
3. BRAKE DISC
4. REAR HUB ASSEMBLY

67170-ENDE-G42

Exploded view of the rear hub and bearing assembly—2WD vehicles

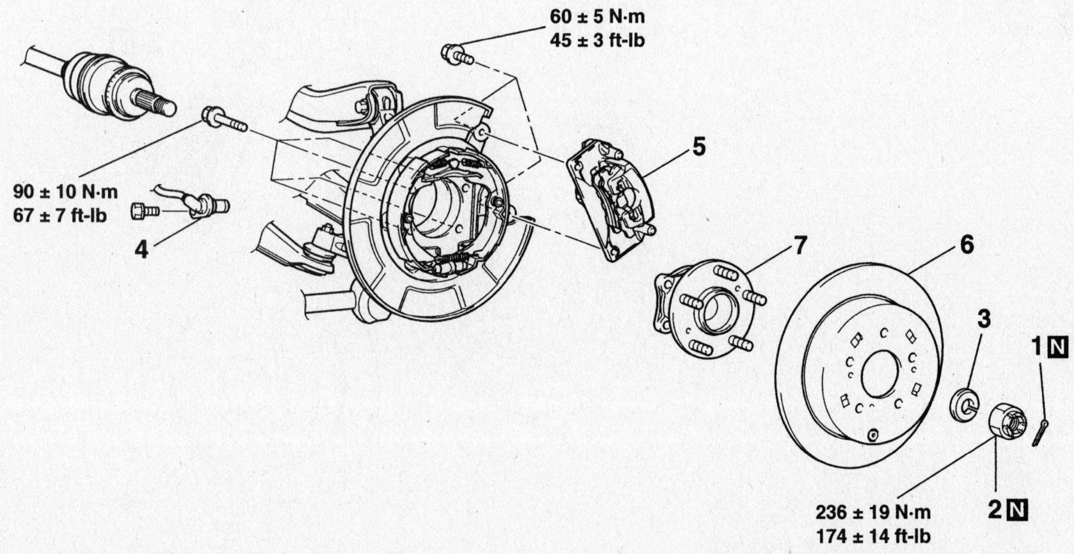

1. SPLIT PIN
2. DRIVE SHAFT NUT
3. WASHER
4. REAR ABS SENSOR<VEHICLES WITH ABS>
5. CALIPER ASSEMBLY
6. BRAKE DISC
7. REAR WHEEL HUB ASSEMBLY

67170-ENDE-G43

Exploded view of the rear hub and bearing assembly—AWD vehicles

BRAKES

Brake Caliper

REMOVAL & INSTALLATION

Front and Rear

1. Before servicing the vehicle, refer to the precautions in the beginning of this section.
2. Raise and safely support the vehicle.
3. Remove or disconnect the following:
   - Wheel and tire assembly
   - Brake hose connection, if you are replacing the caliper
   - Caliper mounting bolts
   - Caliper

To install:

4. Position the caliper over the rotor so the caliper engages the adapter correctly
5. Install or connect the following:
   - Caliper. Tighten the front caliper mounting bolts to 74 ft. lbs. (100 Nm) or the rear mounting bolts to 45 ft. lbs. (60 Nm).
   - Brake hose
   - Wheel and tire assembly
6. Bleed the brake system.

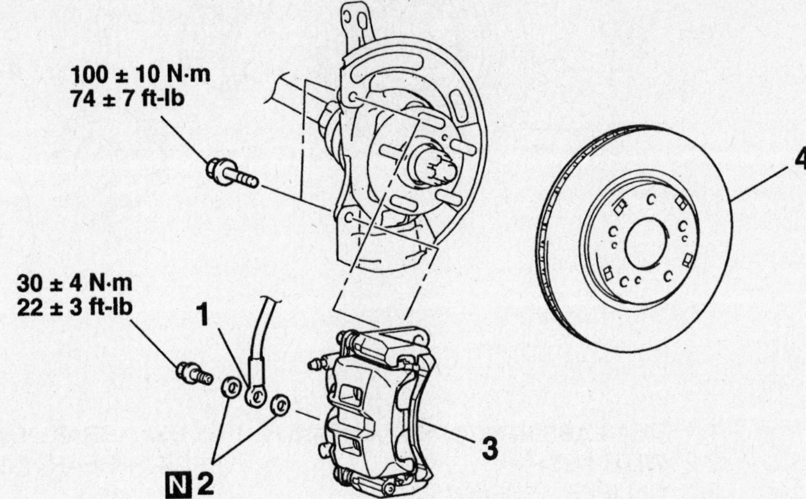

<REAR>

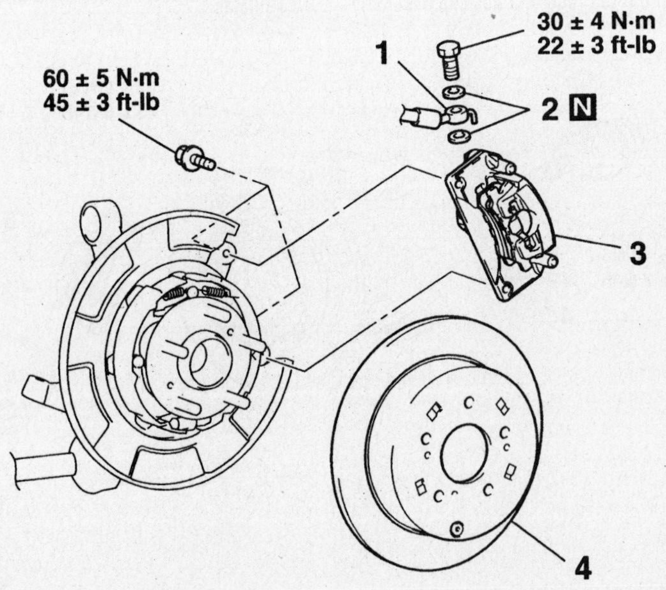

1. BRAKE HOSE CONNECTION
2. GASKET
3. BRAKE CALIPER ASSEMBLY
4. BRAKE DISC

67170-ENDE-G44

Exploded view of the front and rear caliper mounting

Disc Brake Pads

REMOVAL & INSTALLATION

Front and Rear

1. Before servicing the vehicle, refer to the precautions in the beginning of this section.
2. Remove ½ of the brake fluid from the master cylinder.
3. Raise and safely support the vehicle.
4. Remove or disconnect the following:
   - Wheel and tire assembly
   - Lower caliper guide pin bolt
   - Caliper from the caliper support
   - Disc brake pads, shims, and the clips from the caliper support

To install:

5. Clean the exposed portion of the caliper piston, then press the piston back into the caliper bore using the old inner brake pad and a C-clamp.
6. Install or connect the following:
   - Disc brake pads, shims, and the clips. Make sure the shims and clips are properly positioned.
   - Caliper over the rotor so the caliper engages the adapter correctly
   - Mounting pin(s)
   - Wheel and tire assembly and lower the vehicle
7. Apply the brake pedal several times until a firm pedal is obtained. Check the fluid level in the master cylinder and add fluid, as necessary.

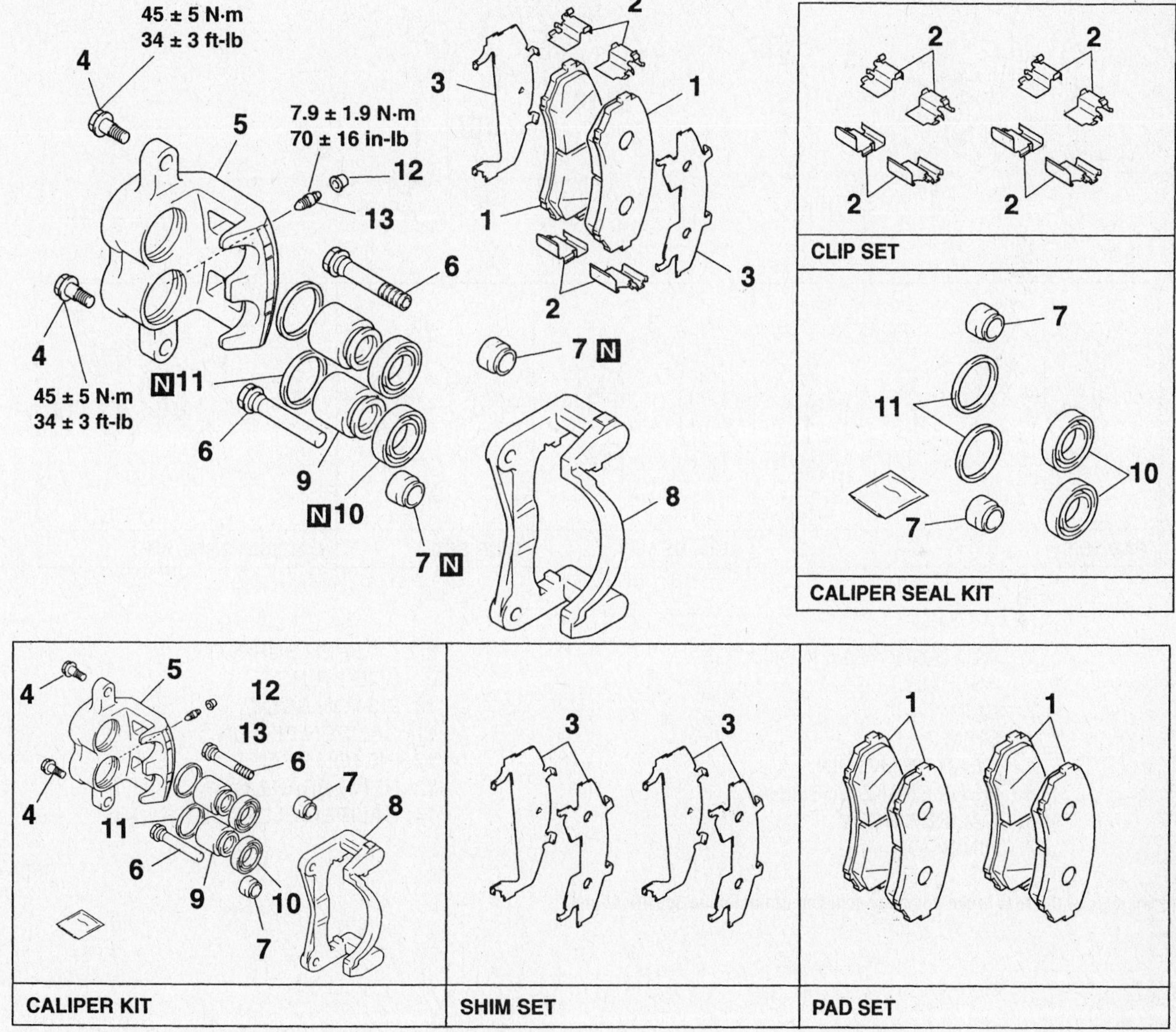

Exploded view of the front brake pads and component tightening specifications

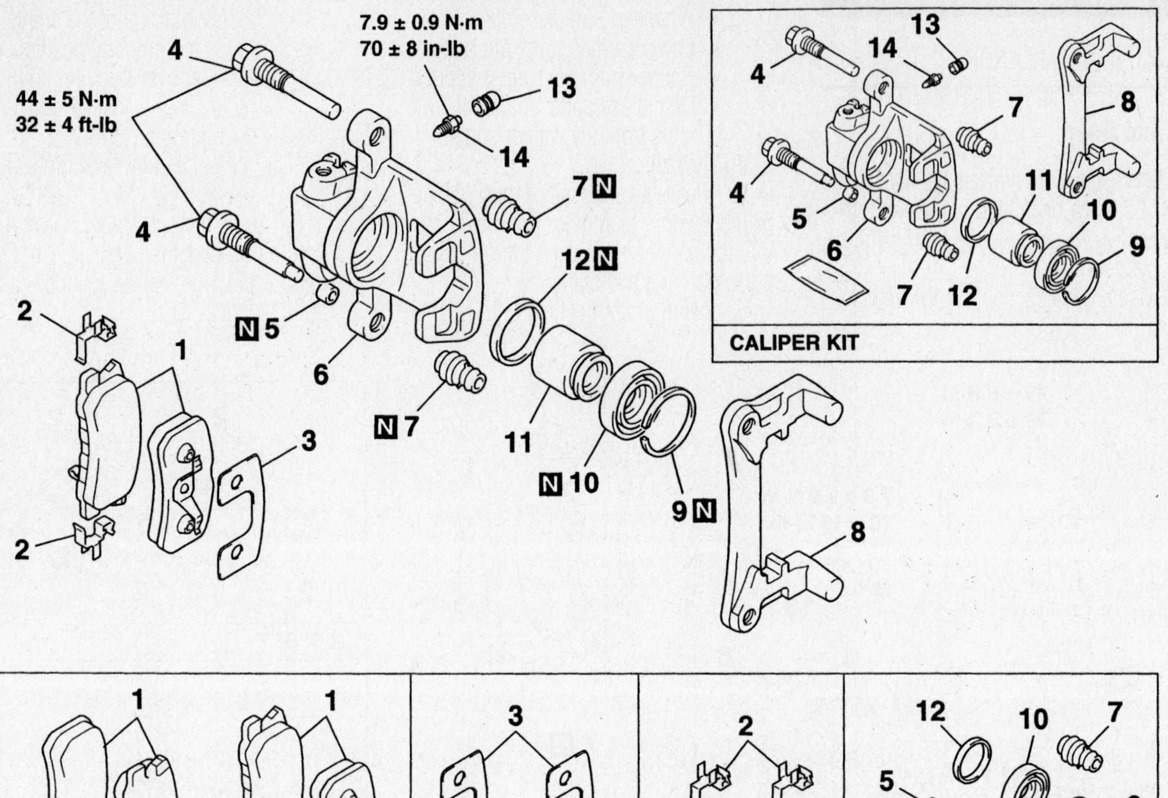

1. PAD (AND WEAR INDICATOR) ASSEMBLY
2. CLIP
3. SHIM
4. REAR BRAKE PIN
5. REAR BRAKE BUSHING
6. CALIPER BODY
7. PIN BOOT
8. CALIPER SUPPORT
9. BOOT RING
10. PISTON BOOT
11. CALIPER PISTON
12. PISTON SEAL
13. REAR BRAKE CAP
14. CALIPER BLEEDER

67170-ENDE-G46

Exploded view of the rear brake pads and component tightening specifications

MITSUBISHI

3

Montero • Montero Sport

SPECIFICATION AND MAINTENANCE CHARTS

ENGINE AND VEHICLE IDENTIFICATION CHART

Engine							Model Year	
Code	Liters (cc)	Cu. In.	Cyl.	Fuel Sys.	Engine Type	Eng. Mfg.	Code	Year
P	3.0 (2972)	181.4	6	MFI	SOHC	Mitsubishi	1	2001
H	3.0 (2972)	181.4	6	MFI	SOHC	Mitsubishi	2	2002
M	3.5 (3479)	213.4	6	MFI	SOHC	Mitsubishi	3	2003
R	3.5 (3479)	213.4	6	MFI	SOHC	Mitsubishi	4	2004
S	3.8 (3828)	233.6	6	MFI	SOHC	Mitsubishi	5	2005

MFI: Multi-port Fuel Injection

67170-MONT-C01

GENERAL ENGINE SPECIFICATIONS

Year	Engine Displacement Liters	Engine VIN	Net Horsepower @ rpm	Net Torque @ rpm (ft. lbs.)	Bore x Stroke (in.)	Com-pression Ratio	Oil Pressure @ rpm
2001	3.5	M	200@5000	228@3500	3.66x3.38	9.0:1	30-80@2000
	3.0	P	173@5500	188@4500	3.59x2.99	9.0:1	30-80@2000
2002	3.5	M	200@5000	228@3500	3.66x3.38	9.0:1	30-80@2000
	3.0	P	173@5500	188@4500	3.59x2.99	9.0:1	30-80@2000
2003	3.5	R	200@5000	228@3500	3.66x3.38	9.0:1	30-80@2000
	3.0	H	173@5500	188@4500	3.59x2.99	9.0:1	30-80@2000
	3.8	S	215@5500	248@3250	3.74x3.54	10.0:1	①
2004	3.5	R	200@5000	228@3500	3.66x3.38	9.0:1	30-80@2000
	3.8	S	215@5500	248@3250	3.74x3.54	10.0:1	①

① 11.6 psi @ idle

67170-MONT-C02

ENGINE TUNE-UP SPECIFICATIONS

Year	Engine Displacement Liters	Engine VIN	Spark Plugs Gap (in.)	Ignition Timing (deg.)		Fuel Pump (psi)	Idle Speed (rpm)		Valve Clearance	
				MT	AT		MT	AT	In.	Ex.
2001	3.5	M	0.039-0.043	—	5B	38 ①	—	700	HYD	HYD
	3.0	P	0.039-0.043	—	5B	38 ①	—	750	HYD	HYD
2002	3.5	M	0.039-0.043	—	5B	38 ①	—	700	HYD	HYD
	3.0	P	0.039-0.043	—	5B	38 ①	—	750	HYD	HYD
2003	3.5	R	0.039-0.043	—	5B	38 ①	—	700	HYD	HYD
	3.0	H	0.039-0.043	—	5B	38 ①	—	750	HYD	HYD
	3.8	S	0.028-0.031	—	5B	38 ①	—	750	HYD	HYD
2004	3.5	R	0.039-0.043	—	5B	38 ①	—	700	HYD	HYD
	3.8	S	0.028-0.031	—	5B	38 ①	—	750	HYD	HYD

B: Before top dead center

HYD: Hydraulic

① With vacuum hose connected

67170-MONT-C03

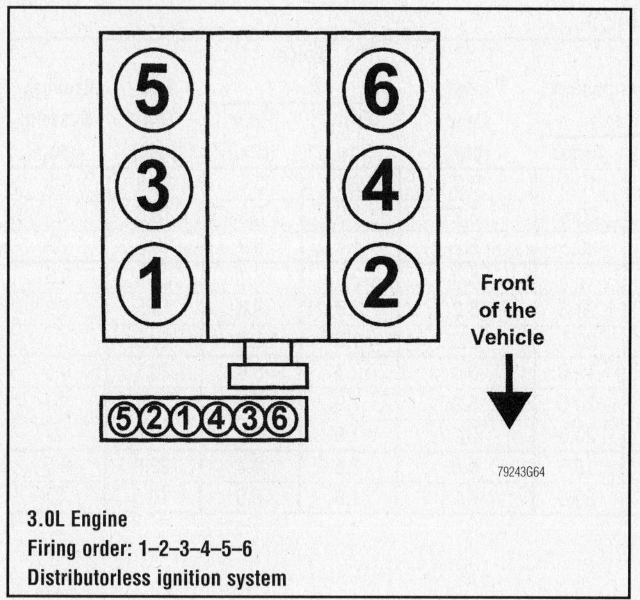

3.0L Engine
Firing order: 1–2–3–4–5–6
Distributorless ignition system

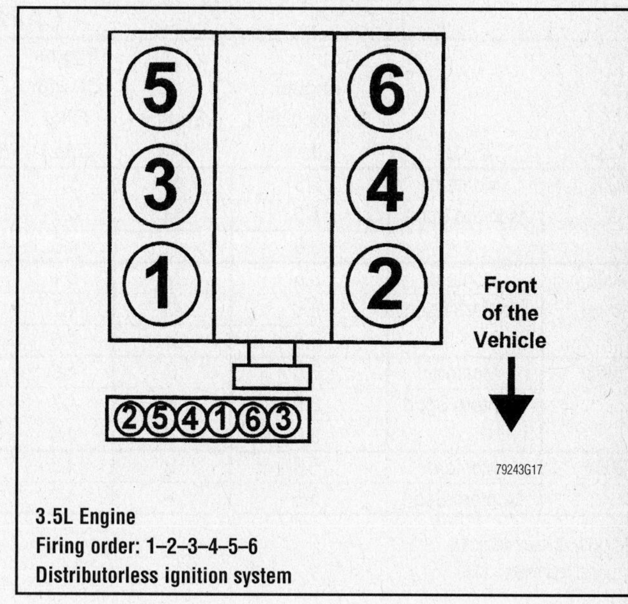

3.5L Engine
Firing order: 1–2–3–4–5–6
Distributorless ignition system

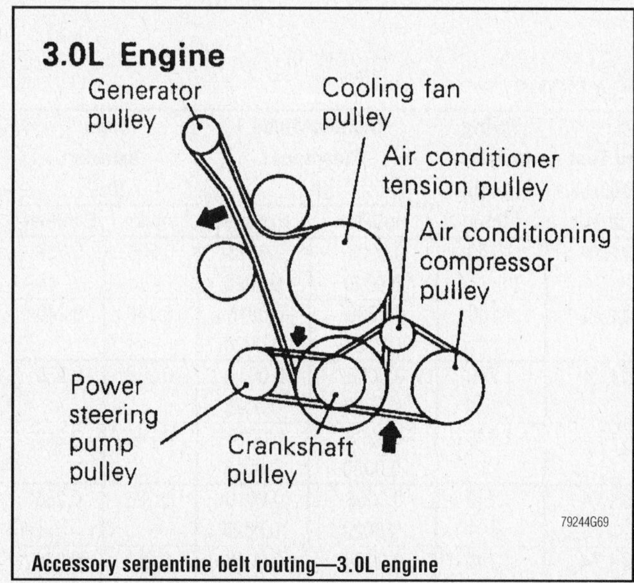

Accessory serpentine belt routing—3.0L engine

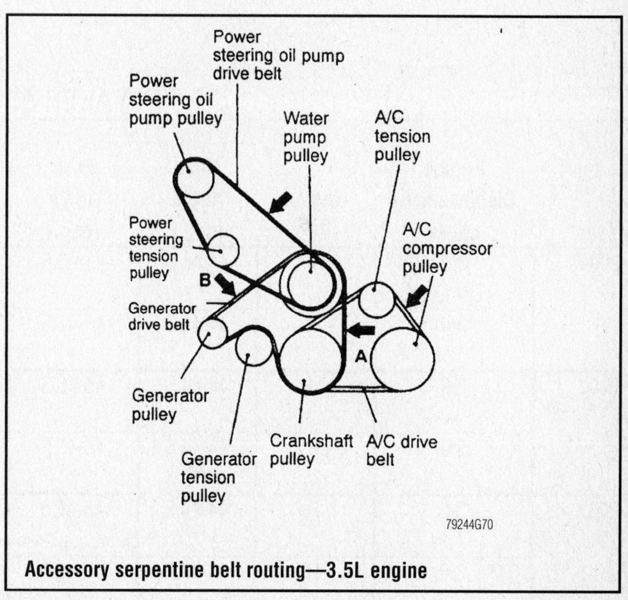

Accessory serpentine belt routing—3.5L engine

CAPACITIES

Year	Model	Engine Displacement Liters	Engine VIN	Engine Oil with Filter (qts.)	Transmission (pts.)		Transfer Case (pts.)	Drive Axle		Fuel Tank (gal.)	Cooling System (qts.)
					5-Spd	Auto.		Front (pts.)	Rear (pts.)		
2001	Montero	3.5	M	5.1	—	19.6	6.0	2.6	3.4	23.8	9.5
	Montero Sport	3.5	M	4.7	—	20.6	5.2	1.9	5.6	19.5	①
		3.0	P	4.6	—	20.6	5.2	1.9	6.8	19.5	①
2002	Montero	3.5	M	5.1	—	19.6	6.0	2.6	3.4	23.8	9.5
	Montero Sport	3.5	M	4.7	—	20.6	5.2	1.9	5.6	19.5	①
		3.0	P	4.6	—	20.6	5.2	1.9	6.8	19.5	①
2003	Montero	3.8	S	4.8	—	19.6	6.0	2.6	3.4	23.8	9.5
	Montero Sport	3.5	R	4.7	—	20.6	5.2	1.9	5.6	19.5	①
		3.0	H	4.6	—	20.6	5.2	1.9	6.8	19.5	①
2004	Montero	3.8	S	4.8	—	19.6	6.0	2.6	3.4	23.8	9.5
	Montero Sport	3.5	R	4.7	—	20.6	5.2	1.9	5.6	19.5	①

① without rear heater: 9.5
with rear heater: 10.6

67170-MONT-C04

VALVE SPECIFICATIONS

Year	Engine Displacement Liters	Engine VIN	Seat Angle (deg.)	Face Angle (deg.)	Spring Test Pressure (lbs. @ in.)	Spring Installed Height (in.)	Stem-to-Guide Clearance (in.)		Stem Diameter (in.)	
							Intake	Exhaust	Intake	Exhaust
2001	3.5	M	44-44.5	45-45.5	60@1.74	1.740	0.0008-0.0020	0.0016-0.0028	0.236	0.236
	3.0	P	44-44.5	45-45.5	60@1.74	1.740	0.0008-0.0020	0.0016-0.0028	0.240	0.240
2002	3.5	M	44-44.5	45-45.5	60@1.74	1.740	0.0008-0.0020	0.0016-0.0028	0.236	0.236
	3.0	P	44-44.5	45-45.5	60@1.74	1.740	0.0008-0.0020	0.0016-0.0028	0.240	0.240
2003	3.5	R	44-44.5	45-45.5	60@1.74	1.740	0.0008-0.0020	0.0016-0.0028	0.236	0.236
	3.0	H	44-44.5	45-45.5	60@1.74	1.740	0.0008-0.0020	0.0016-0.0028	0.240	0.240
	3.8	S	44-44.5	45-45.5	60@1.74	1.740	0.0008-0.0019	0.0016-0.0027	0.240	0.240
2004	3.5	M	44-44.5	45-45.5	60@1.74	1.740	0.0008-0.0020	0.0016-0.0028	0.236	0.236
	3.8	S	44-44.5	45-45.5	60@1.74	1.740	0.0008-0.0019	0.0016-0.0027	0.240	0.240

67170-MONT-C07

CRANKSHAFT AND CONNECTING ROD SPECIFICATIONS

All measurements are given in inches.

Year	Engine Displacement Liters	Engine VIN	Crankshaft Main Brg. Journal Dia.	Main Brg. Oil Clearance	Shaft End-play	Thrust on No.	Connecting Rod Journal Diameter	Oil Clearance	Side Clearance
2001	3.0	P	2.3614-2.3622	0.0008-0.0040	0.0020-0.0120	3	2.1646-2.1654	0.0008-0.0040	0.0039-0.0160
	3.5	M	2.3614-2.3622	0.0008-0.0040	0.0020-0.0120	3	1.9700	0.0008-0.0040	0.0039-0.0160
2002	3.0	P	2.3614-2.3622	0.0008-0.0040	0.0020-0.0120	3	2.1646-2.1654	0.0008-0.0040	0.0039-0.0160
	3.5	M	2.3614-2.3622	0.0008-0.0040	0.0020-0.0120	3	1.9700	0.0008-0.0040	0.0039-0.0160
2003	3.0	H	2.3614-2.3622	0.0008-0.0040	0.0020-0.0120	3	2.1646-2.1654	0.0008-0.0040	0.0039-0.0160
	3.5	R	2.3614-2.3622	0.0008-0.0040	0.0020-0.0120	3	1.9700	0.0008-0.0040	0.0039-0.0160
	3.8	S	2.3614-2.3622	0.0008-0.0040	0.0020-0.0120	3	1.9700	0.0008-0.0040	0.0039-0.0160
2004	3.5	R	2.3614-2.3622	0.0008-0.0040	0.0020-0.0120	3	1.9700	0.0008-0.0040	0.0039-0.0160
	3.8	S	2.3614-2.3622	0.0008-0.0040	0.0020-0.0120	3	1.9700	0.0008-0.0040	0.0039-0.0160

67170-MONT-C05

PISTON AND RING SPECIFICATIONS

All measurements are given in inches.

Year	Engine Displacement Liters	Engine VIN	Piston Clearance	Ring Gap Top Compression	Ring Gap Bottom Compression	Ring Gap Oil Control	Ring Side Clearance Top Compression	Ring Side Clearance Bottom Compression	Ring Side Clearance Oil Control
2001	3.0	P	0.0008-0.0020	0.0118-0.0310	0.0177-0.0310	0.0079-0.0390	0.0012-0.0040	0.0008-0.0040	Snug
	3.5	M	0.0008-0.0020	0.0118-0.0310	0.0177-0.0310	0.0079-0.0390	0.0012-0.0040	0.0008-0.0040	Snug
2002	3.0	P	0.0008-0.0020	0.0118-0.0310	0.0177-0.0310	0.0079-0.0390	0.0012-0.0040	0.0008-0.0040	Snug
	3.5	M	0.0008-0.0020	0.0118-0.0310	0.0177-0.0310	0.0079-0.0390	0.0012-0.0040	0.0008-0.0040	Snug
2003	3.0	H	0.0008-0.0020	0.0118-0.0310	0.0177-0.0310	0.0079-0.0390	0.0012-0.0040	0.0008-0.0040	Snug
	3.5	R	0.0008-0.0020	0.0118-0.0310	0.0177-0.0310	0.0079-0.0390	0.0012-0.0040	0.0008-0.0040	Snug
	3.8	S	0.0008-0.0019	0.0012-0.0170	0.0180-0.0230	0.0080-0.0230	0.0012-0.0027	0.0008-0.0023	Snug
2004	3.5	R	0.0008-0.0020	0.0118-0.0310	0.0177-0.0310	0.0079-0.0390	0.0012-0.0040	0.0008-0.0040	Snug
	3.8	S	0.0008-0.0019	0.0012-0.0170	0.0180-0.0230	0.0080-0.0230	0.0012-0.0027	0.0008-0.0023	Snug

67170-MONT-C06

TORQUE SPECIFICATIONS

All readings in ft. lbs.

Year	Engine Displacement Liters	Engine ID/VIN	Cylinder Head Bolts	Main Bearing Bolts	Rod Bearing Bolts	Crankshaft Damper Bolts	Flywheel Bolts	Manifold Intake *	Manifold Exhaust	Spark Plugs	Oil Pan Drain Plug
2001	3.5	M	80	54	38	134	54	16	22	18	29
	3.0	P	80	69	37	134	55	16	33	18	29
2002	3.5	M	80	54	38	134	54	16	22	18	29
	3.0	P	80	69	37	134	55	16	33	18	29
2003	3.5	R	80	54	38	134	54	16	22	18	29
	3.0	H	80	69	37	134	55	16	33	18	29
	3.8	S	80	74	①	137	54	②	33	18	29
2004	3.5	R	80	54	38	134	54	16	22	18	29
	3.8	S	80	74	①	137	54	②	33	18	29

① 20 ft. lbs., plus an additional 90 degrees

② 1st step (right bank nuts): 58 inch lbs.
2nd step (left bank nuts): 16 ft. lbs.
3rd step (right bank nuts): 16 ft. lbs.
4th step (left bank nuts): 16 ft. lbs.
5th step (right bank nuts): 16 ft. lbs.

67170-MONT-C08

WHEEL ALIGNMENT

Year	Model		Caster Range (+/-Deg.)	Caster Preferred Setting (Deg.)	Camber Range (+/-Deg.)	Camber Preferred Setting (Deg.)	Toe-in (in.)
2001	Montero	F	0.50	+3.50	0.50	0	0.10+/-0.10
		R	—	—	0.50	0	0.12+/-0.12
	Montero Sport	F	1.00	+2.66	0.50	①	0.14+/-0.14
		R	—	—	—	0	0
2002	Montero	F	0.50	+3.50	0.50	0	0.10+/-0.10
		R	—	—	0.50	0	0.12+/-0.12
	Montero Sport	F	1.00	+2.66	0.50	①	0.14+/-0.14
		R	—	—	—	0	0
2003	Montero	F	0.30	+3.50	0.30	0	0.10+/-0.10
		R	—	—	0.30	0	0.12+/-0.12
	Montero Sport	F	1.00	+2.66	0.50	①	0.14+/-0.14
		R	—	—	—	0	0
2004	Montero	F	0.30	+3.50	0.30	0	0.10+/-0.10
		R	—	—	0.30	0	0.12+/-0.12
	Montero Sport	F	1.00	+2.66	0.50	①	0.14+/-0.14
		R	—	—	—	0	0

① Right: 0.42
Left: 0.92

67170-MONT-C10

TIRE, WHEEL AND BALL JOINT SPECIFICATIONS

Year	Model	OEM Tires Standard	OEM Tires Optional	Tire Pressures (psi) Front	Tire Pressures (psi) Rear	Wheel Size	Ball Joint Inspection	Lug Nut (ft. lbs.)
2001	Montero	P265/70HR16	None	29	29	7-JJ	U: 7-30 in. ① L: 0.010 in.	100
	Montero Sport	P235/75R15	P255/70R16	26	26	Std: 6-JJ Opt: 7-JJ	U: 7-30 in. ① L: 0.010 in.	100
2002	Montero	P265/70HR16	None	29	29	7-JJ	U: 7-30 in. ① L: 0.010 in.	100
	Montero Sport	P235/75R15	P255/70R16	26	26	Std: 6-JJ Opt: 7-JJ	U: 7-30 in. ① L: 0.010 in.	100
2003	Montero	P265/70R16	None	29	29	7-JJ	U: 7-30 in. ① L: 0.010 in.	100
	Montero Sport	P235/75R15	P255/70R16	26	26	Std: 6-JJ Opt: 7-JJ	U: 7-30 in. ① L: 0.010 in.	100
2004	Montero	P265/70R16	None	29	29	7-JJ	U: 7-30 in. ① L: 0.010 in.	100
	Montero Sport	P235/75R15	P255/70R16	26	26	Std: 6-JJ Opt: 7-JJ	U: 7-30 in. ① L: 0.010 in.	100

OEM: Original Equipment Manufacturer

PSI: Pounds Per Square Inch

STD: Standard

OPT: Optional

① Torque required in inch lbs. to rotate ball joint when removed from the knuckle

67170-MONT-C11

BRAKE SPECIFICATIONS

All measurements in inches unless noted

Year	Model		Brake Disc Original Thickness	Brake Disc Minimum Thickness	Brake Disc Maximum Runout	Brake Drum Diameter Original Inside Diameter	Brake Drum Diameter Max. Wear Limit	Brake Drum Diameter Maximum Machine Diameter	Minimum Lining Thickness Front	Minimum Lining Thickness Rear	Brake Caliper Bracket Bolts (ft. lbs.)	Brake Caliper Mounting Bolts (ft. lbs.)
2001	Montero	F	1.023	0.960	0.002	—	—	—	0.079	—	65	54
		R	0.866	0.803	0.003	—	—	—	—	0.079	65	32
	Montero Sport	F	0.940	0.880	0.001	—	—	—	0.079	—	65	55
		R	0.710	0.650	0.003	10.63	—	10.71	—	①	94	32
2002	Montero	F	1.023	0.960	0.002	—	—	—	0.079	—	65	54
		R	0.866	0.803	0.003	—	—	—	—	0.079	65	32
	Montero Sport	F	0.940	0.880	0.001	—	—	—	0.079	—	65	55
		R	0.710	0.650	0.003	10.63	—	10.71	—	①	94	32
2003	Montero	F	1.023	0.960	0.002	—	—	—	0.080	—	83	66
		R	0.870	0.800	0.002	—	—	—	—	0.080	74	33
	Montero Sport	F	0.940	0.880	0.001	—	—	—	0.079	—	65	55
		R	0.710	0.650	0.003	10.63	—	10.71	—	①	94	32
2004	Montero	F	1.023	0.960	0.002	—	—	—	0.080	—	83	66
		R	0.870	0.800	0.002	—	—	—	—	0.080	74	33
	Montero Sport	F	0.940	0.880	0.001	—	—	—	0.079	—	65	55
		R	0.710	0.650	0.003	10.63	—	10.71	—	①	94	32

① disc pad: 0.79
brake shoe: 0.04

67170-MONT-C09

SCHEDULED MAINTENANCE INTERVALS

Mitsubishi—Montero, Montero Sport

TO BE SERVICED	TYPE OF SERVICE	VEHICLE MILEAGE INTERVAL (x1000)												
		7.5	15	22.5	30	37.5	45	52.5	60	67.5	75	82.5	90	97.5
Engine oil & filter	R	✓	✓	✓	✓	✓	✓	✓	✓	✓	✓	✓	✓	✓
Automatic transmission & transfer oil	S/I		✓		✓		✓		✓		✓		✓	
Brake hoses	S/I		✓		✓		✓		✓		✓		✓	
Disc brake pads & rotors	S/I		✓		✓		✓		✓		✓		✓	
Drive shaft boots	S/I		✓		✓		✓		✓		✓		✓	
Air cleaner filter	R				✓				✓				✓	
Automatic transmission & transfer oil (4WD)	R				✓				✓				✓	
Engine coolant	R				✓				✓				✓	
Ball joints & steering linkage seals	S/I				✓				✓				✓	
Drive belt(s)	S/I				✓				✓				✓	
Drum brake linings & wheel cylinders	S/I				✓				✓				✓	
Exhaust system	S/I				✓				✓				✓	
Front & rear axle	S/I				✓				✓				✓	
Fuel hoses	S/I				✓				✓				✓	
Manual transmission & transfer oil (4WD)	S/I				✓				✓				✓	
Propeller shaft joint	S/I				✓				✓				✓	
Ignition cables	R								✓					
Timing belt	R								✓					
Distributor cap & rotor	S/I								✓					
EVAP system (except EVAP canister)	S/I								✓					
EGR valve ①	S/I													
EVAP canister ①	S/I													
PCV system ②	S/I													
Spark plugs ③	R													

R: Replace S/I: Service or Inspect

① Replace at 100,000 miles.

② PCV system (except EVAP canister): service or inspect at 100,000 miles.

③ Iron tips: 30,000 miles
Platinum tips: 60,000 miles
Irridium tips: 100,000 miles

FREQUENT OPERATION MAINTENANCE (SEVERE SERVICE)

If a vehicle is operated under any of the following conditions it is considered severe service:

- **Extremely dusty areas.**
- **50% or more of the vehicle operation is in 32°C (90°F) or higher temperatures, or constant operation in temperatures below 0°C (32°F).**
- **Prolonged idling (vehicle operation in stop and go traffic).**
- **Frequent short running periods (engine does not warm to normal operating temperatures).**
- **Police, taxi, delivery usage or trailer towing usage.**

Oil & oil filter: replace every 3000 miles.

Front disc brake pads (dusty or salty conditions): service or inspect every 6000 miles.

Front disc brake pads: service or inspect every 7500 miles.

Air cleaner filter: service or inspect every 15,000 miles.

Rear drum brake linings & rear wheel cylinders: service or inspect every 15,000 miles.

Spark plugs (iron tip): replace every 15,000 miles.

PCV system: service or inspect every 60,000 miles.

67170-MONT-C12

PRECAUTIONS

Before servicing any vehicle, please be sure to read all of the following precautions, which deal with personal safety, prevention of component damage, and important points to take into consideration when servicing a motor vehicle:

- Never open, service or drain the radiator or cooling system when the engine is hot; serious burns can occur from the steam and hot coolant.
- Observe all applicable safety precautions when working around fuel. Whenever servicing the fuel system, always work in a well-ventilated area. Do not allow fuel spray or vapors to come in contact with a spark, open flame, or excessive heat (a hot drop light, for example). Keep a dry chemical fire extinguisher near the work area. Always keep fuel in a container specifically designed for fuel storage; also, always properly seal fuel containers to avoid the possibility of fire or explosion. Refer to the additional fuel system precautions later in this section.
- Fuel injection systems often remain pressurized, even after the engine has been turned **OFF**. The fuel system pressure must be relieved before disconnecting any fuel lines. Failure to do so may result in fire and/or personal injury.
- Brake fluid often contains polyglycol ethers and polyglycols. Avoid contact with the eyes and wash your hands thoroughly after handling brake fluid. If you do get brake fluid in your eyes, flush your eyes with clean, running water for 15 minutes. If eye irritation persists, or if you have taken brake fluid internally, IMMEDIATELY seek medical assistance.
- The EPA warns that prolonged contact with used engine oil may cause a number of skin disorders, including cancer! You should make every effort to minimize your exposure to used engine oil. Protective gloves should be worn when changing oil. Wash your hands and any other exposed skin areas as soon as possible after exposure to used engine oil. Soap and water, or waterless hand cleaner should be used.
- All new vehicles are now equipped with an air bag system. The system must be disabled before performing service on or around system components, steering column, instrument panel components, wiring and sensors. Failure to follow safety and disabling procedures could result in accidental air bag deployment, possible personal injury and unnecessary system repairs.
- Always wear safety goggles when working with, or around, the air bag system. When carrying a non-deployed air bag, be sure the bag and trim cover are pointed away from your body. When placing a non-deployed air bag on a work surface, always face the bag and trim cover upward, away from the surface. This will reduce the motion of the module if it is accidentally deployed. Refer to the additional air bag system precautions later in this section.
- Clean, high quality brake fluid from a sealed container is essential to the safe and proper operation of the brake system. You should always buy the correct type of brake fluid for your vehicle. If the brake fluid becomes contaminated, completely flush the system with new fluid. Never reuse any brake fluid. Any brake fluid that is removed from the system should be discarded. Also, do not allow any brake fluid to come in contact with a painted surface; it will damage the paint.
- Never operate the engine without the proper amount and type of engine oil; doing so WILL result in severe engine damage.
- Timing belt maintenance is extremely important! Many models utilize an interference-type, non-freewheeling engine. If the timing belt breaks, the valves in the cylinder head may strike the pistons, causing potentially serious (also time-consuming and expensive) engine damage. Refer to the maintenance interval charts in the front of this section for the recommended replacement interval for the timing belt, and to the timing belt procedure for belt replacement and inspection.
- Disconnecting the negative battery cable on some vehicles may interfere with the functions of the on-board computer system(s) and may require the computer to undergo a relearning process once the negative battery cable is reconnected.
- When servicing drum brakes, only disassemble and assemble one side at a time, leaving the remaining side intact for reference.

ENGINE REPAIR

➡Disconnecting the negative battery cable on some vehicles may interfere with the functions of the on board computer systems and may require the computer to undergo a relearning process, once the negative battery cable is reconnected.

Distributor

All of the engines covered in this section are distributorless.

Alternator

REMOVAL

1. Before servicing the vehicle, refer to the precautions in the beginning of this section.
2. Remove or disconnect the following:
   - Negative battery cable
   - Under cover
   - Air cleaner assembly, ducts and air intake hose
   - Drive belt(s)
   - Wires
   - Mounting bracket, if equipped
   - Alternator

To install:

3. Install or connect the following:
   - Alternator. Torque the through-bolt to 38 ft. lbs. (52 Nm) and the mounting bolt to 16 ft. lbs. (22 Nm).
   - Mounting bracket, if equipped. Torque the bolt to 14–18 ft. lbs. (20–25 Nm).
   - Wires. Torque the nut to 124 inch lbs. (14 Nm).
   - Drive belt(s)
   - Air cleaner assembly, ducts and air intake hose
   - Under cover
   - Negative battery cable

Ignition Timing

ADJUSTMENT

The ignition timing is controlled by the ECM and is not adjustable. The ECM determines the timing based on input from the crankshaft position sensor.

TIMING CHECK

1. Before servicing the vehicle, refer to the precautions in the beginning of this section.

Before attempting to adjust the ignition timing, be sure of the following:

- The engine should be at normal operating temperature.

<3.0L, 3.5L ENGINE>

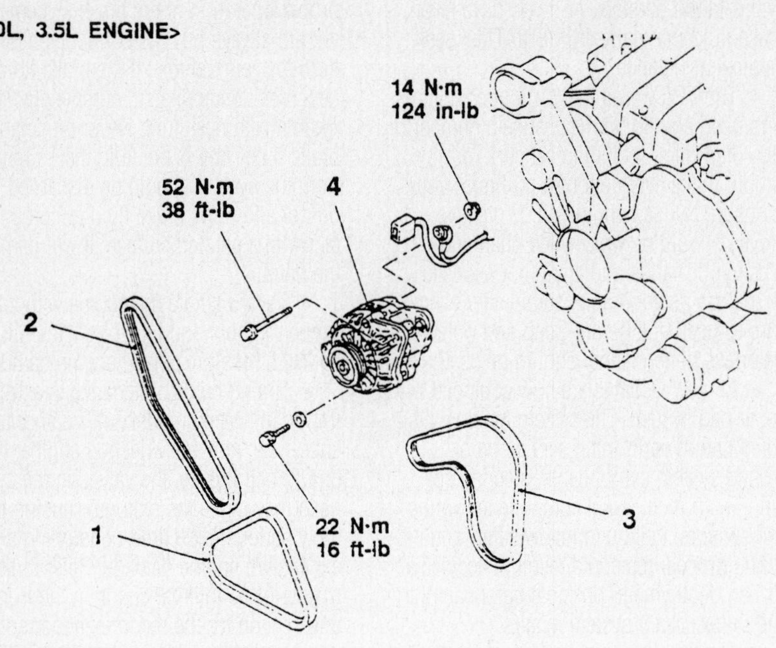

Alternator mounting and related components

- The lights and all accessories should be OFF.
- If equipped with an automatic transmission, the transmission should be in **P** or **N**.
- Connect scan tool MB991502 to the data link connector
- Set up the timing light.
- Start the engine and run at idle.
- Verify that the idle speed is 600–800 rpm.
- Select scan tool MB991502 actuator test "item number 17".
- Check that basic timing is with in standard, it should be 3–7° BTDC.
- Press the clear key on the scan tool, select forced drive stop mode and cancel the actuator test.

CAUTION

If the actuator test is not canceled, the forced drive will continue for 27 minutes. Driving in this state could lead to engine failure.

2. If the base timing is out of specification:
3. Check to see if the distributor is aligned properly
4. Check to see if the timing belt cover and Crankshaft Position (CKP) sensor installation is conditions.
5. Crankshaft sensing blade conditions.

Engine Assembly

REMOVAL & INSTALLATION

1. Before servicing the vehicle, refer to the precautions in the beginning of this section.
2. Relieve the fuel system pressure.

CAUTION

The fuel injection system remains under pressure after the engine has been OFF. Properly relieve fuel pressure before disconnecting any fuel lines. Failure to do so may result in fire or personal injury.

3. Drain the engine oil.
4. Drain the cooling system.
5. Remove or disconnect the following:
   - Battery
   - Hood, matchmark for reassembly
   - Oil dipstick
   - Engine undercover
   - Starter
   - Exhaust pipe from the exhaust manifolds
   - Transfer case, if equipped with 4WD
   - Transmission, if equipped with a manual transmission
6. If equipped with an automatic transmission and 2WD:
   a. Remove the inspection plate.
   b. Matchmark the flexplate to the converter; remove the torque converter bolts and move the torque converter back as far as it will go.
   c. Remove the lower bell housing bolts.
7. Remove or disconnect the following:
   - Air cleaner assembly, ducts and air intake hose
   - Linkages and cables from the throttle body
   - Fuel lines and plug the lines
   - Air conditioning compressor, if equipped and position it aside. It is not necessary to remove the lines from the compressor.
   - Radiator, shroud
   - Cooling fan
   - Heater hoses
   - Accessory belts
   - Power steering pump and wires from its brackets and position it to the side. Do not remove the hoses from the pump.
   - Alternator and wires
   - Ignition coil and power transistor assembly, if equipped
   - MDP sensor connector
   - EGR connector
   - TP sensor connector
   - IAC motor connector
   - Magnetic clutch and refrigerant temperature switch connector
   - EVAP Purge Solenoid
   - ECT sensor and gauge connectors
   - Front and injector wiring harness
   - CMP sensor
   - CKP sensor
   - Distributor signal Generator
   - Compactor connector
   - Left and right heated O_2 sensor

- Oil pressure switch connector
- Vacuum hoses

8. Attach an engine removal device to the engine support eyes on the engine.
9. If equipped with an automatic transmission, support the transmission with a floor jack. Remove the remaining bell housing bolts.
10. Remove the engine mount nuts and remove the engine from the vehicle.

To install:

11. Lower the engine into position and install the engine mount nuts and bolts. Tighten the nuts to 18–20 ft. lbs. (25–27 Nm), and the bolts to tighten to 33 ft. lbs. (44 Nm).
12. Install or connect the following:
    - Bell housing bolts
13. Remove the engine removal device and the transmission support.
14. Install or connect the following:
    - Transfer case, if equipped
    - Manual transmission, if equipped
    - Automatic transmission, if equipped align the torque converter and flexplate and the bolts.
    - Inspection plate
    - Starter motor
    - Exhaust pipe to the exhaust manifolds using new gaskets
    - Lower radiator hose
    - Heater hoses
    - Alternator and wires
    - Power steering pump and brackets
    - Air conditioning compressor
    - Linkages and cables to the carburetor or throttle body
    - Ignition coil and power transistor assembly, if equipped
    - MDP sensor connector
    - EGR connector
    - TP sensor connector
    - IAC motor connector
    - Magnetic clutch and refrigerant temperature switch connector
    - EVAP Purge Solenoid
    - ECT sensor and gauge connectors
    - Front and injector wiring harness
    - CMP sensor
    - CKP sensor
    - Distributor signal Generator
    - Compactor connector
    - Left and right heated O_{2} sensor
    - Oil pressure switch connector
    - Vacuum hoses
    - Air cleaner assembly, ducts and air intake hose
    - Accessory belts
    - Radiator, shroud and upper hose
    - Cooling fan
    - Battery
    - Oil dipstick
    - Hood
15. Refill the engine with the specified amount of oil.
16. Refill the radiator with coolant.
17. Check fuel system for leaks.
18. Check the automatic transmission fluid level, if equipped.
19. Recheck all engine adjustments.

Water Pump

REMOVAL & INSTALLATION

3.0L and 3.5L Engines

1. Before servicing the vehicle, refer to the precautions in the beginning of this section.
2. If necessary, properly release the fuel pressure.
3. Drain the cooling system.
4. Remove or disconnect the following:
    - Negative battery cable

✲✲ CAUTION

Wait at least 90 seconds after the negative battery cable is disconnected to prevent possible deployment of the air bag.

- Upper radiator shroud
- Accessory belts
- Air conditioning compressor tensioner pulley, if equipped
- Cooling fan and clutch assembly and the water pump pulley
- Thermostat and housing on 3.0L, 3.5L engines
- Water outlet, gasket and houses
- Radiator hoses from the water pump
- Crankshaft pulley(s)
- Timing belt covers. If the same timing belt will be reused, mark the direction of the timing belt's rotation, for installation in the same direction. Be sure the engine is positioned so the No. 1 cylinder is at the TDC of its compression stroke and the sprockets timing marks are aligned with the engine's timing mark indicators.
- Timing belt
- Water pump bolts are different lengths, note their positions before removing.
- Water pump from the block
- Water pipe connection and O-ring

To install:

5. Clean and dry the mating surfaces of the block and water pump
6. Install or connect the following:
    - New O-ring on the water pipe connection, wet the new O-ring with water to aid in installation
    - Water pump, with a new gasket, Torque the bolts to 17 ft. lbs. (23 Nm) on 3.0L and 3.5L engines
    - Alternator bracket bolt to 17 ft. lbs. (23 Nm)
    - Timing belt(s) and covers
    - Crankshaft pulley(s)
    - Thermostat and housing on 3.0L, 3.5L engines. Torque the bolts to 12–14 ft. lbs. (17–20 Nm).
    - Radiator hose to the water pump
    - Water outlet, new gasket and houses. Torque the bolts to 12–14 ft. lbs. (17–20 Nm).
    - Water pump pulley
    - Cooling fan and clutch assembly
    - Air conditioning compressor tensioner pulley, if equipped
    - Accessory belts
    - Upper radiator shroud
    - Thermostat and housing on 3.0L, 3.5L engines
    - Negative battery cable
7. Refill the radiator with coolant. This cooling system has a self-bleeding thermostat, so system bleeding is not required.
8. Run the vehicle until the thermostat opens and fill the overflow tank. Check for leaks.

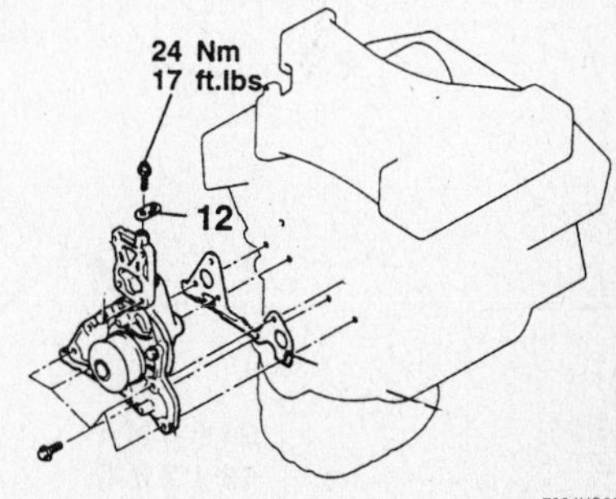

Water pump mounting—3.0L engine

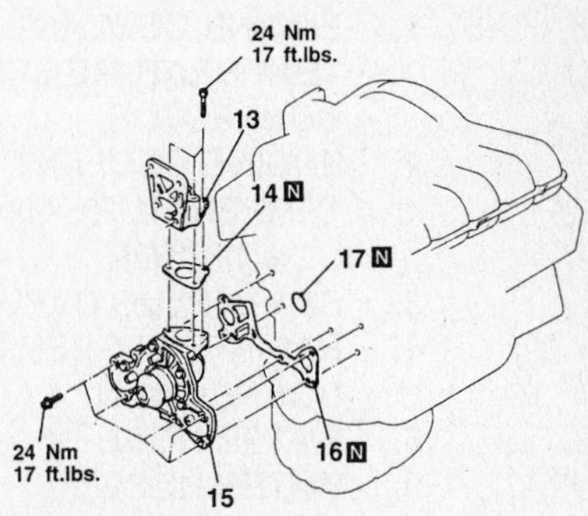

Water pump and related components—3.5L engine

9. Once the vehicle has cooled, recheck the coolant level.

3.8L Engine

1. Before servicing the vehicle, refer to the precautions in the beginning of this section.
2. If necessary, properly release the fuel pressure.
3. Drain the cooling system.
4. Remove or disconnect the following:
   - Negative battery cable

CAUTION

Wait at least 90 seconds after the negative battery cable is disconnected to prevent possible deployment of the air bag.

   - Timing belt
   - Camshaft sprocket
   - Engine Coolant Temperature (ECT) sensor connector
   - ECT gauge unit connector
   - Spark plug cable support
   - Upper radiator hose
   - Water hoses
   - Water outlet fitting bracket
   - Water outlet fitting, O-ring and gasket
   - Water pump assembly, gasket and O-ring
   - Fitting, gaskets and thermostat case

To install:

5. Clean and dry the mating surfaces of the block and water pump

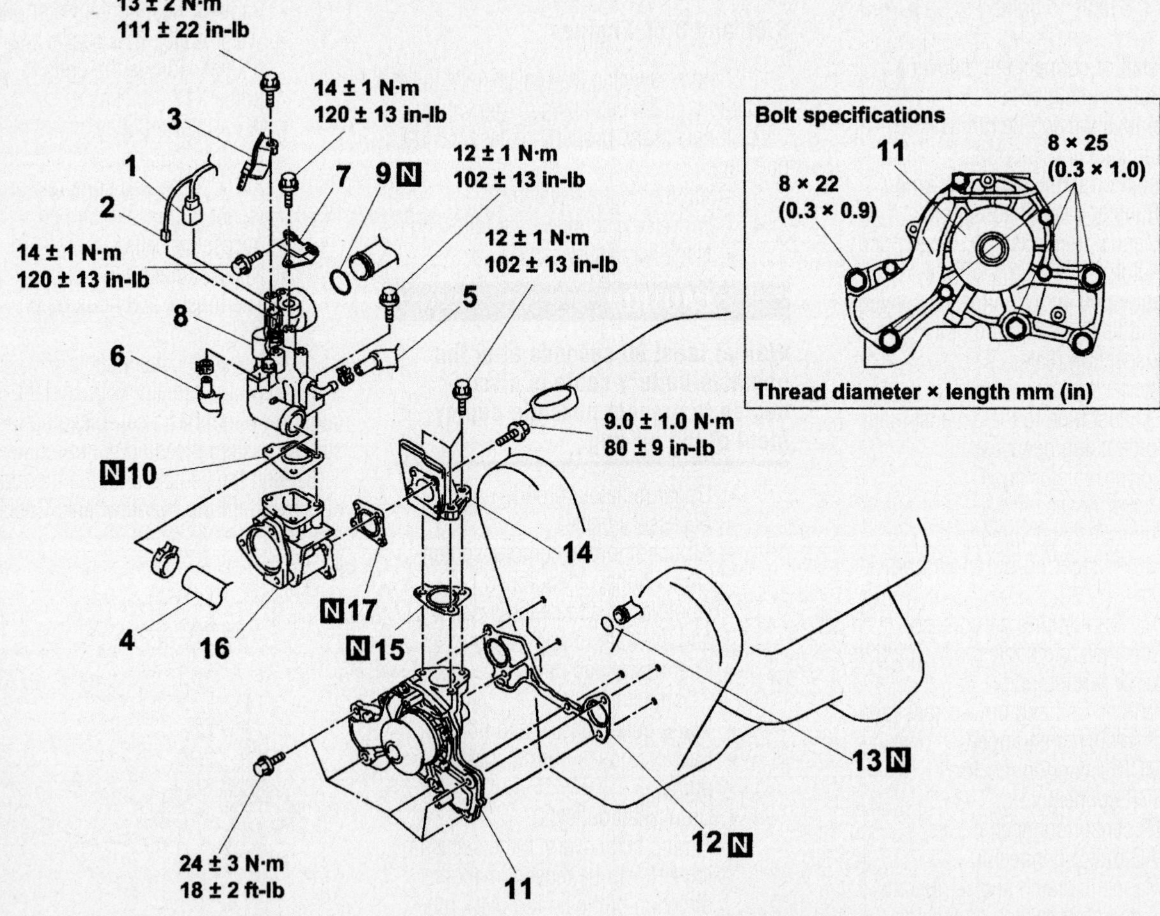

Exploded view of the water pump and related components—3.8L engine

6. Install or connect the following:
   - Thermostat case with new gaskets
   - Fitting
   - Water pump assembly, using a new gasket and O-ring. Torque the water pump bolts to 16–20 ft. lbs. (21–27 Nm).
   - Water outlet fitting, O-ring and gasket
   - Water outlet fitting bracket
   - Water hoses
   - Upper radiator hose
   - Spark plug cable support
   - ECT gauge unit connector
   - ECT sensor connector
   - Camshaft sprocket
   - Timing belt
7. Refill the radiator with coolant. This cooling system has a self-bleeding thermostat, so system bleeding is not required.
8. Run the vehicle until the thermostat opens and fill the overflow tank. Check for leaks.
9. Once the vehicle has cooled, recheck the coolant level.

Heater Core

REMOVAL & INSTALLATION

Montero

1. Before servicing the vehicle, refer to the precautions in the beginning of this section.
2. Place the wheels in the straight-ahead position.
3. Drain the cooling system into a clean container for reuse.
4. Discharge and recover the air conditioning system refrigerant.
5. Remove or disconnect the following:
   - Negative battery cable

✱✱ CAUTION

Wait at least 60 seconds after disconnecting the battery cable before performing any work on the air bag or instrument panel.

   - Floor console assembly
   - Steering column-to-instrument panel cover screws and the cover
   - Air bag module, carefully, from the steering wheel
   - Electrical connectors from the air bag module

✱✱ CAUTION

Store the air bag module facing up.

   - Steering wheel nut
   - Press the steering wheel from the steering column, using a steering wheel puller
   - Passenger's side foot shower duct
   - Glove box stoppers and the glove box
   - Passenger's air bag module nut and the air bag module
   - Electrical connector from the air bag module
   - Hood lock release handle
   - Fuel filler door lock release handle
   - Knee protector assembly and bracket

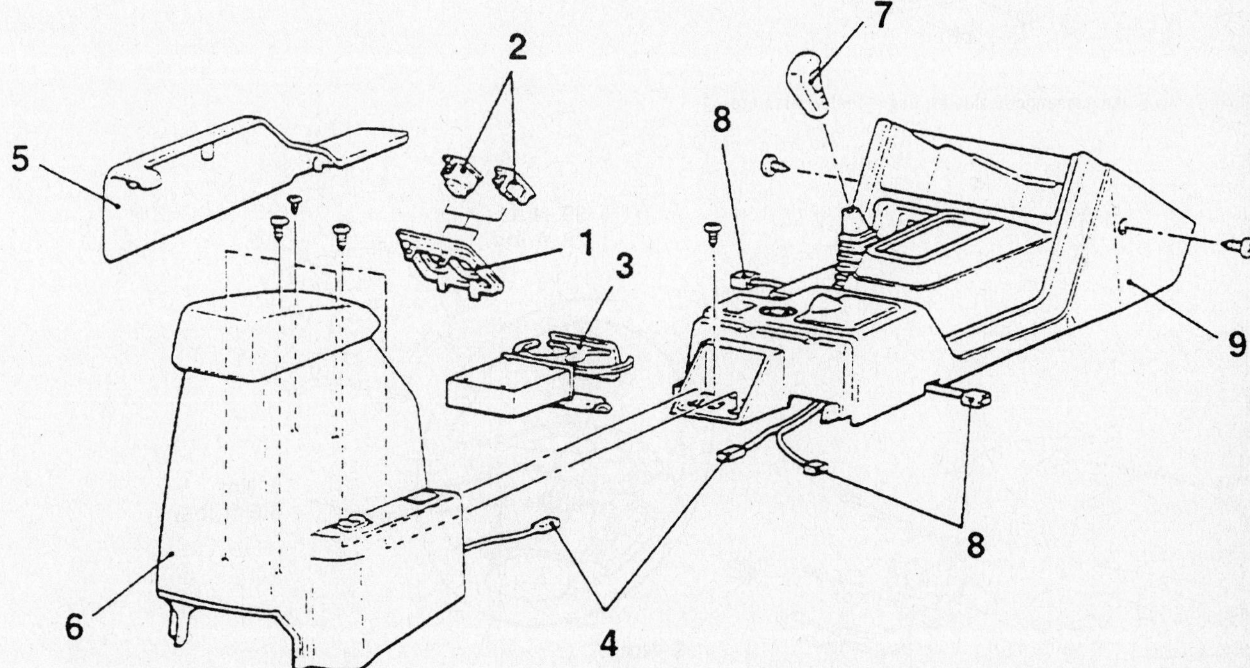

1. Switch panel
2. Suspension control switch or hole cover
3. Cup holder assembly
4. Rear console harness connector
5. Side panel A
6. Rear console assembly
7. Transfer shift lever knob
8. Floor console harness connector
9. Front console assembly

93113GE2

Exploded view of the floor console and related components—Montero

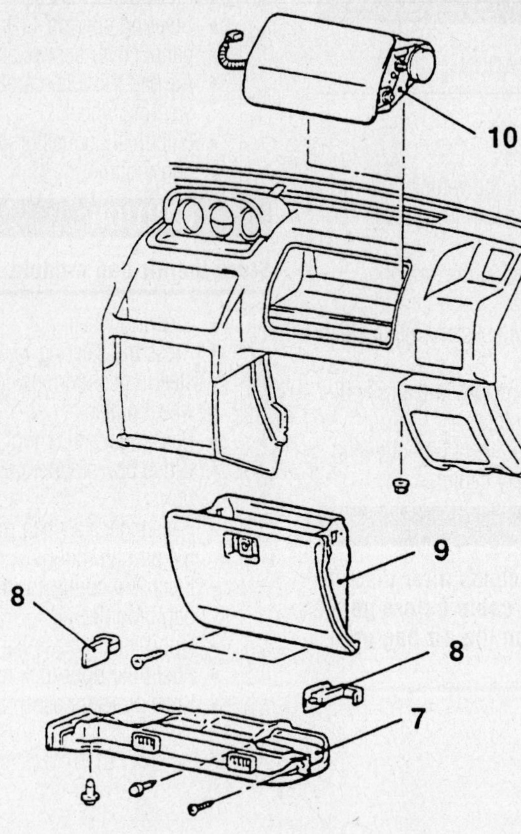

Exploded view of the passenger's side air bag module—Montero

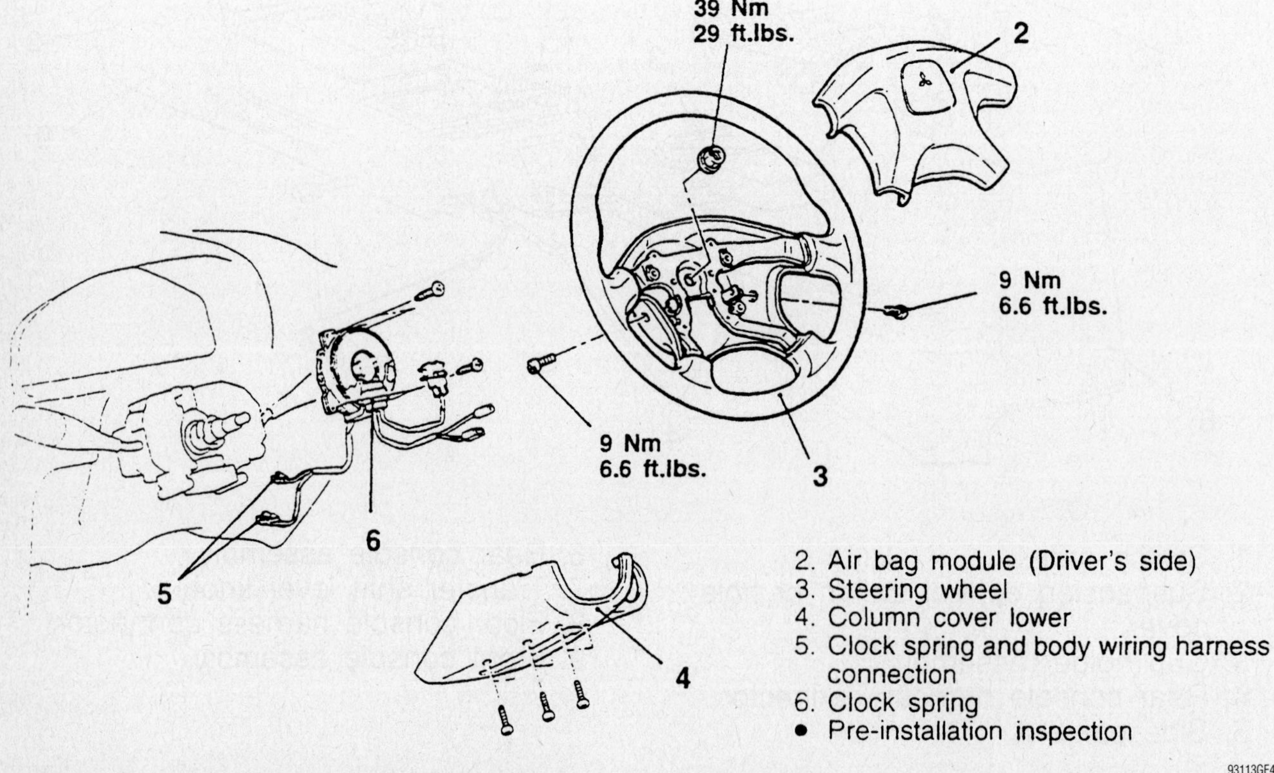

Exploded view of the steering wheel and air bag module—Montero

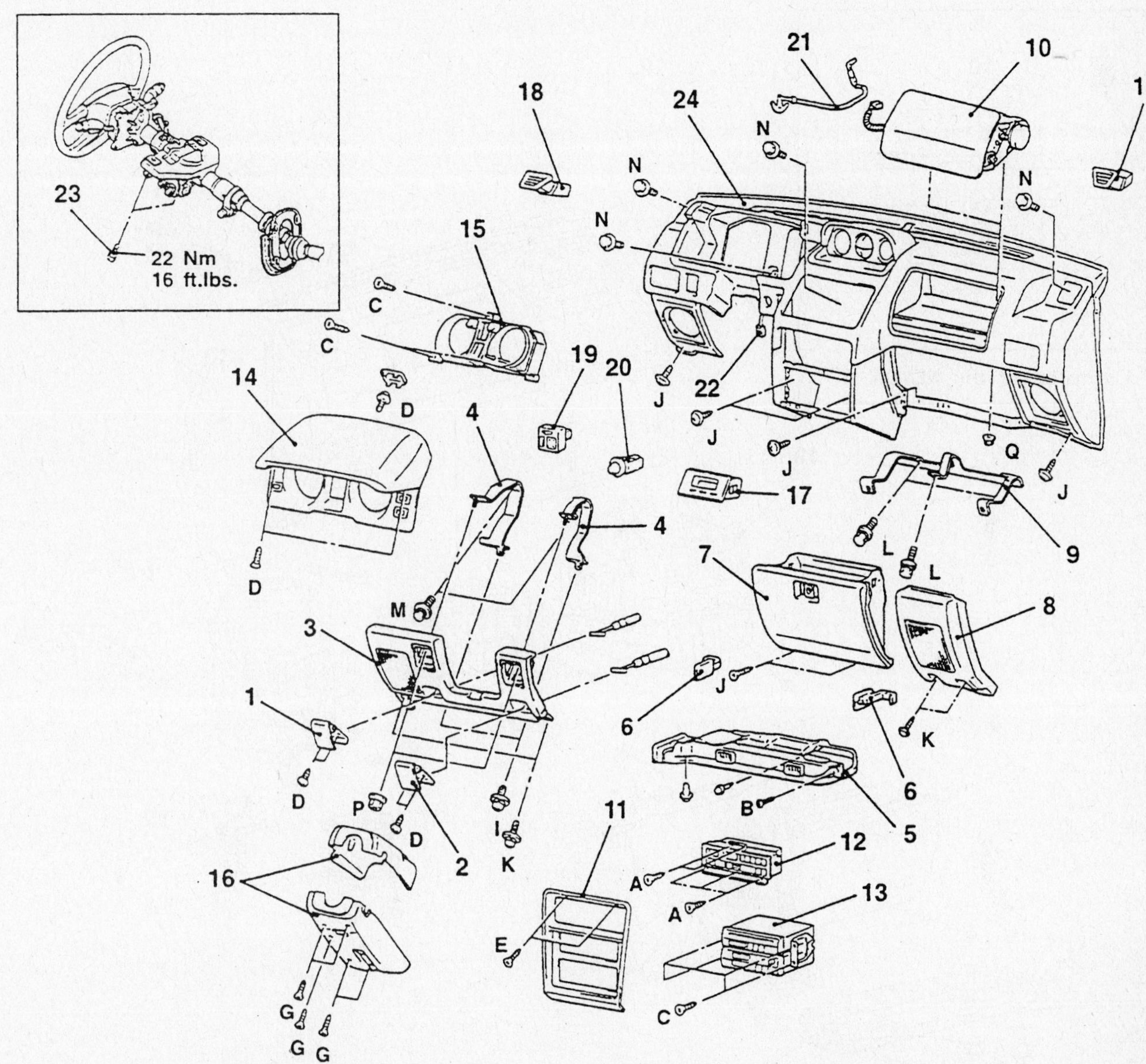

1. Hood lock release handle
2. Fuel filler door lock release handle
3. Knee protector
4. Stay
5. Foot shower duct (R.H.)
6. Glove box stopper
7. Glove box assembly
8. Corner cover
9. Stay
10. Passenger-side air bag module assembly
11. Center panel
12. Heater control assembly
13. Radio and tape player
14. Meter bezel assembly
15. Combination meter
16. Column cover
17. Clock
18. Side defroster garnish
19. Door mirror control switch
20. Rheostat
21. Ventilation control wire
22. Harness connector
23. Steering column installation bolts
24. Instrument panel assembly

93113GE5

Exploded view of the instrument panel and related components—Montero

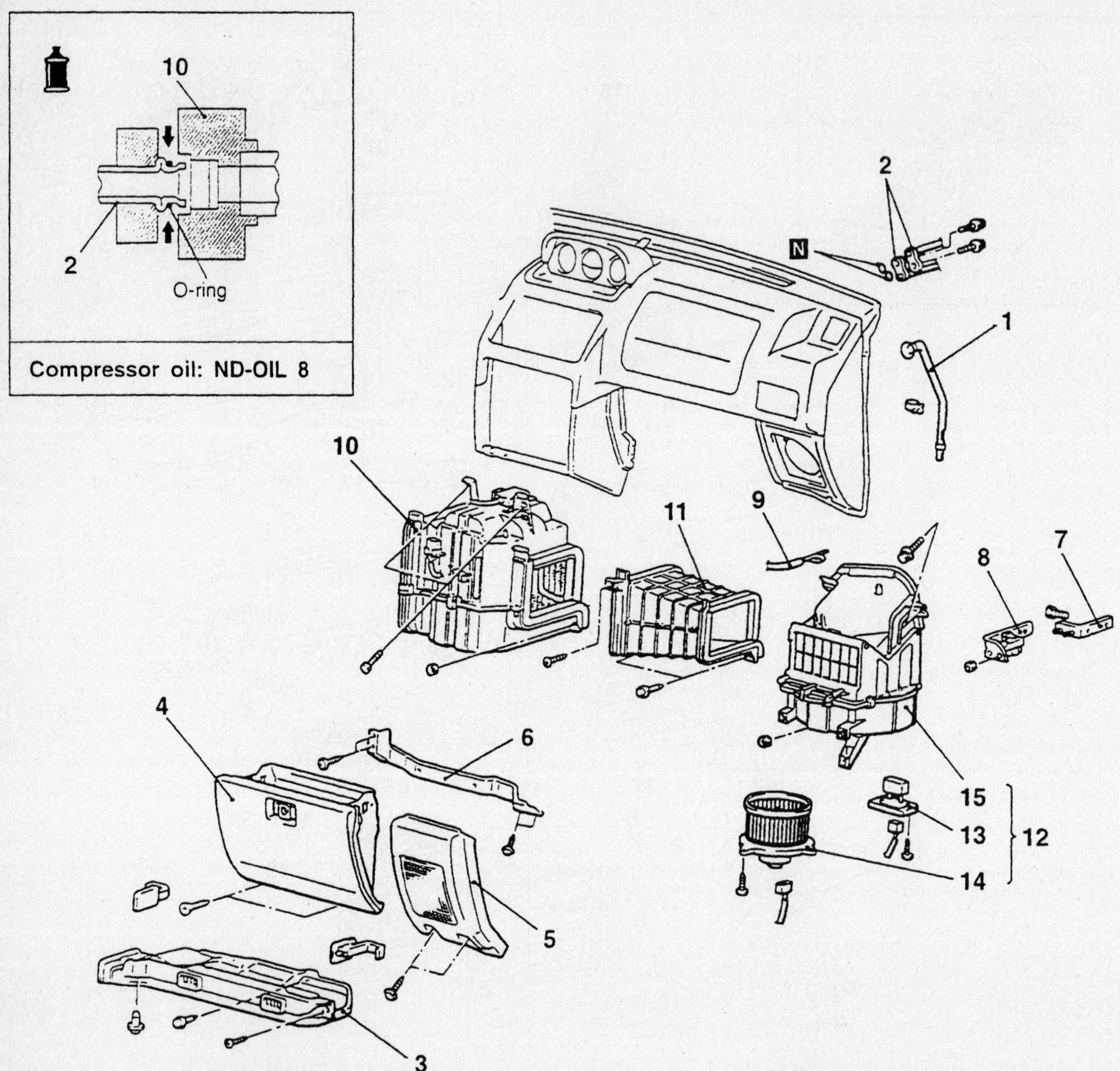

1. Drain hose
2. Liquid pipe and suction hose connection
3. Foot shower duct (R.H.)
4. Glove box
5. Corner cover
6. Lower frame
7. Engine control relay assembly
8. Bracket
9. Air selection control wire connection
10. Evaporator
11. Duct joint
12. Blower assembly
13. Resistor
14. Blower motor assembly
15. Blower case assembly

93113GE6

Exploded view of the air conditioning evaporator housing, blower motor assembly and related components—Montero

- Meter bezel assembly
- Under cover, the corner cover and the stopper
- Glove box assembly
- Center under cover assembly
- Radio and tape/CD player
- Heater/air conditioning control assembly
- Combination meter and the speaker
- Glove box striker and the upper glove box frame
- Multi-meter panel and the multi-meter assembly
- Side defroster grille
- Instrument panel assembly, carefully, with the help of an assistant
- Blower motor assembly
- Both foot ducts
- Joint duct from the air conditioning evaporator housing assembly
- Foot distribution duct
- Center reinforcement
- Center ventilation duct
- Drain hose from the air conditioning evaporator housing assembly
- Heater hoses from the heater housing assembly
- Refrigerant lines from the air conditioning evaporator housing assembly and discard the O-rings
- Heater housing assembly
- Center duct assembly
- Heater core from the heater housing

To install:

6. Install or connect the following:
   - Heater core to the heater housing
   - Center duct assembly
   - Heater housing assembly
   - Refrigerant lines to the air conditioning evaporator housing assembly, using new O-rings
   - Heater hoses to the heater housing assembly
   - Drain hose to the air conditioning evaporator housing assembly
   - Center ventilation duct
   - Center reinforcement
   - Foot distribution duct
   - Joint duct to the air conditioning evaporator housing assembly
   - Both foot shower ducts
   - Blower motor assembly
   - Instrument panel, carefully with the help of an assistant
   - Side defroster grille
   - Multi-meter assembly and the multi-meter panel
   - Upper glove box frame and the glove box striker
   - Speaker and the combination meter
   - Heater/air conditioning control assembly

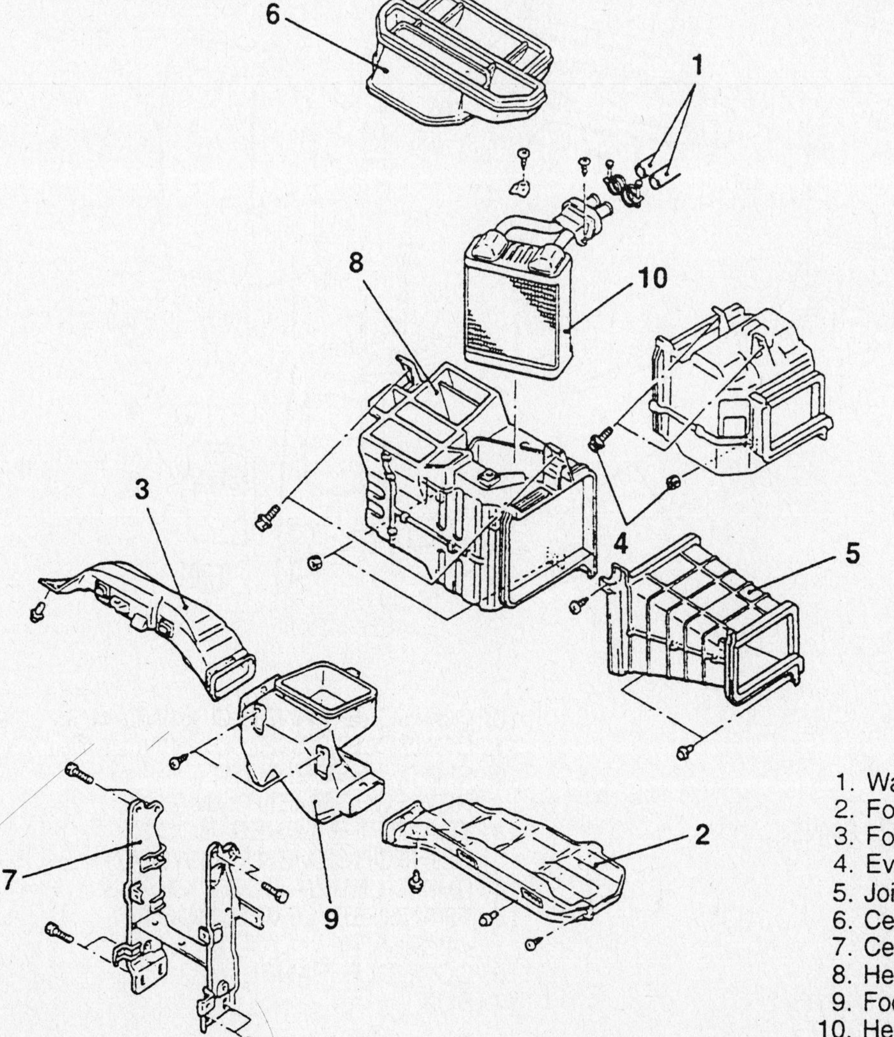

Exploded view of the heater housing, air conditioning evaporator housing and related components—Montero

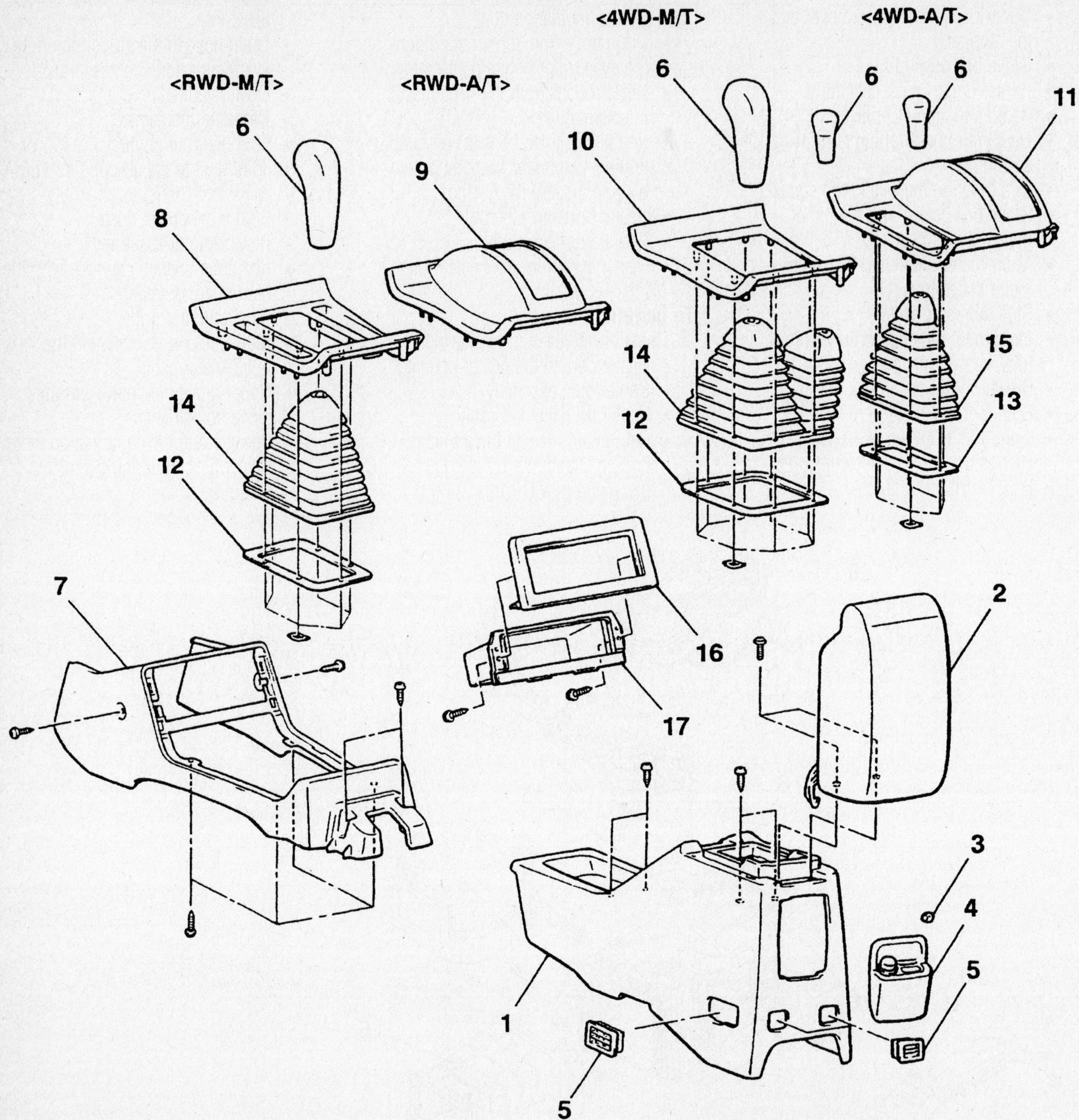

1. REAR FLOOR CONSOLE ASSEMBLY
2. CONSOLE LID ASSEMBLY
3. KNOB
4. REAR HEATER CONTROL PANEL ASSEMBLY
5. FOOT GRILL
6. SHIFT LEVER KNOB
7. FRONT FLOOR CONSOLE ASSEMBLY
8. CONSOLE PANEL A <RWD-M/T>
9. CONSOLE PANEL B <RWD-A/T>
10. CONSOLE PANEL C <4WD-M/T>
11. CONSOLE PANEL D <4WD-A/T>
12. SHIFT LEVER BOOT REINFORCEMENT <M/T>
13. TRANSFER LEVER BOOT REINFORCEMENT <4WD-A/T>
14. SHIFT LEVER BOOT <M/T>
15. TRANSFER LEVER BOOT <4WD-A/T>
16. CONSOLE PANEL
17. BOX

93113GD6

Exploded view of the floor console and related components—Montero Sport

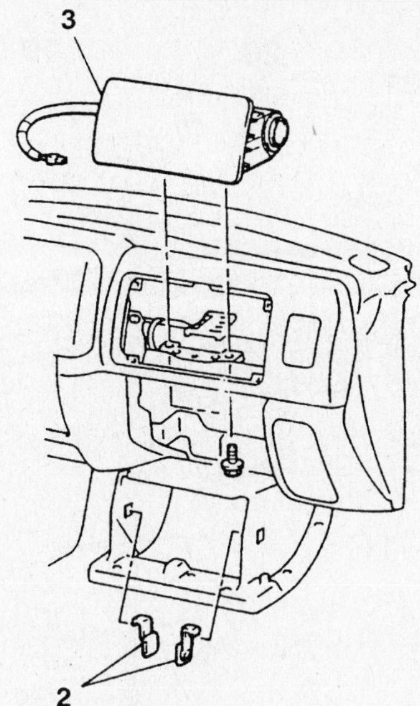

Exploded view of the passenger's side air bag module—Montero Sport

- Radio and tape/CD player
- Center under cover assembly
- Glove box assembly
- Under cover, the corner cover and the stopper
- Meter bezel assembly
- Knee protector assembly and bracket
- Filler door lock release handle
- Hood lock release handle
- Electrical connector to the passenger's side air bag module
- Passenger's air bag module and the air bag module nut
- Glove box and the glove box stoppers
- Foot shower duct
- Steering wheel to the steering column
- Steering wheel nut and torque the nut to 29 ft. lbs. (39 Nm)
- Electrical connectors to the air bag module
- Air bag module to the steering wheel
- Steering column-to-instrument panel cover and the cover screws
- Floor console assembly

7. Refill the cooling system.

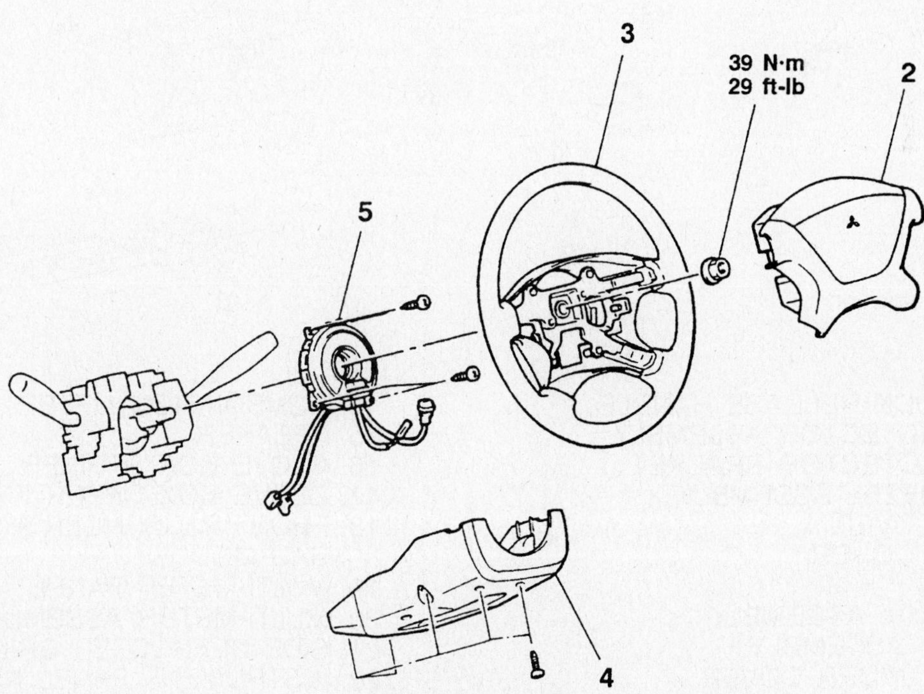

Exploded view of the steering wheel and air bag module—Montero Sport

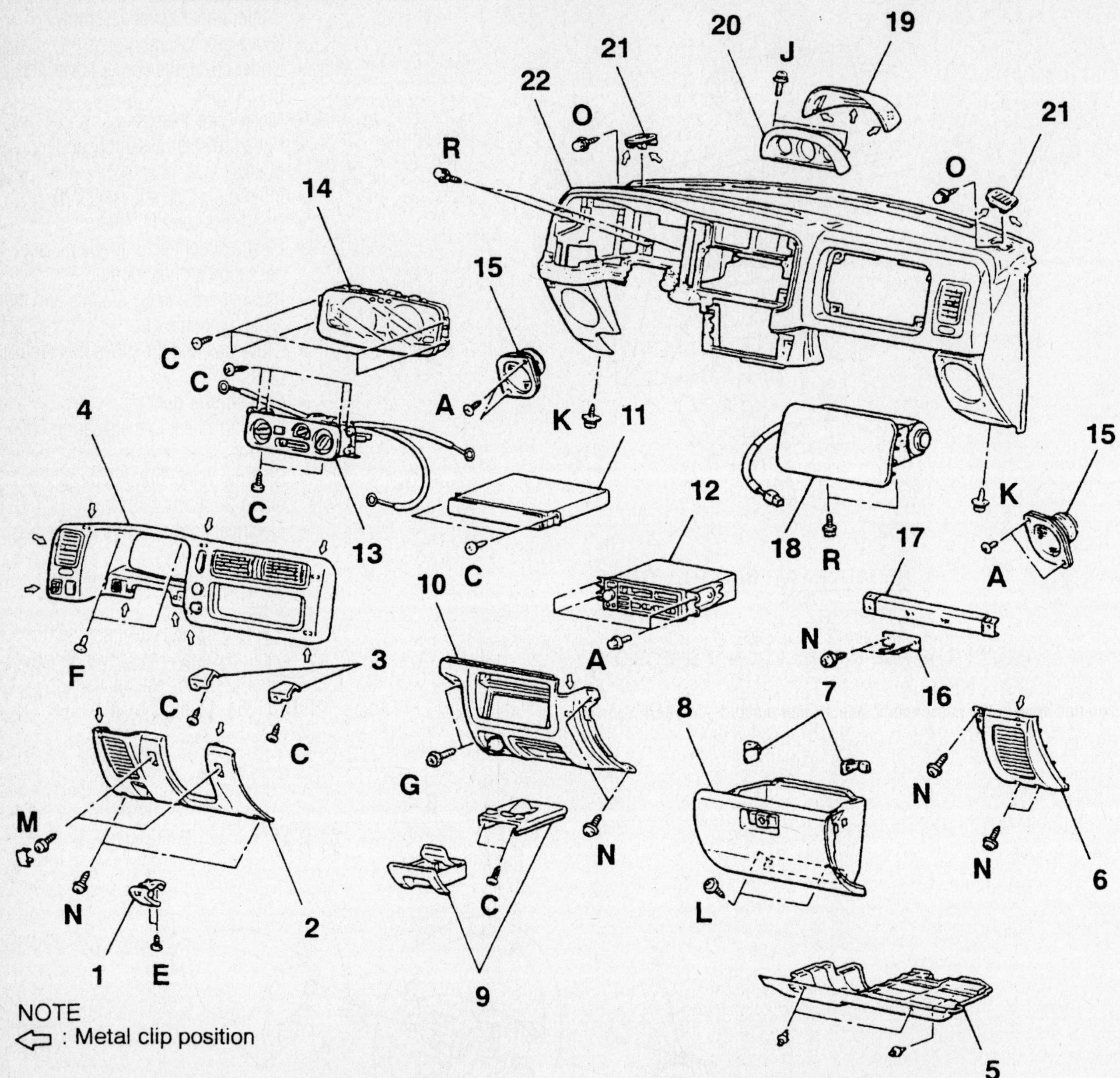

1. HOOD LOCK RELEASE HANDLE
2. KNEE PROTECTOR ASSEMBLY
3. KNEE PROTECTOR BRACKET
4. METER BEZEL ASSEMBLY
5. UNDER COVER
6. CORNER COVER
7. STOPPER
8. GLOVE BOX ASSEMBLY
9. ASHTRAY ASSEMBLY
10. CENTER UNDER COVER ASSEMBLY
11. CUP HOLDER ASSEMBLY
12. RADIO AND TAPE PLAYER
13. HEATER CONTROL ASSEMBLY
14. COMBINATION METER
15. SPEAKER
16. GLOVE BOX STRIKER
17. GLOVE BOX UPPER FRAME
18. FRONT PASSENGER'S SIDE AIR BAG MODULE
19. MULTI-METER PANEL
20. MULTI-METER ASSEMBLY
21. SIDE DEFROSTER GRILL
22. INSTRUMENT PANEL ASSEMBLY

93113GD9

Exploded view of the instrument panel and related components—Montero Sport

8. Connect the negative battery.
9. Evacuate and charge the air conditioning system refrigerant.
10. Run the engine to normal operating temperatures; then, check the climate control operation and check for leaks.

Montero Sport

FRONT HEATER SYSTEM

1. Before servicing the vehicle, refer to the precautions in the beginning of this section.
2. Place the wheels in the straight-ahead position.
3. Disconnect the negative battery.

✱✱ CAUTION

Wait at least 60 seconds after disconnecting the battery cable before performing any work on the air bag or instrument panel.

4. Drain the cooling system into a clean container for reuse.
5. Discharge and recover the air conditioning system refrigerant.
6. Remove or disconnect the following:
   - Floor console assembly
   - Steering column-to-instrument panel cover screws and the cover
   - Air bag module, carefully, from the steering wheel
   - Electrical connectors from the air bag module

✱✱ CAUTION

Store the air bag module facing up.

   - Steering wheel nut
   - Press the steering wheel from the steering column, using a steering wheel puller
   - Glove box stoppers and lower the glove box
   - Passenger's air bag module bolts and air bag module
   - Electrical connector from the air bag module

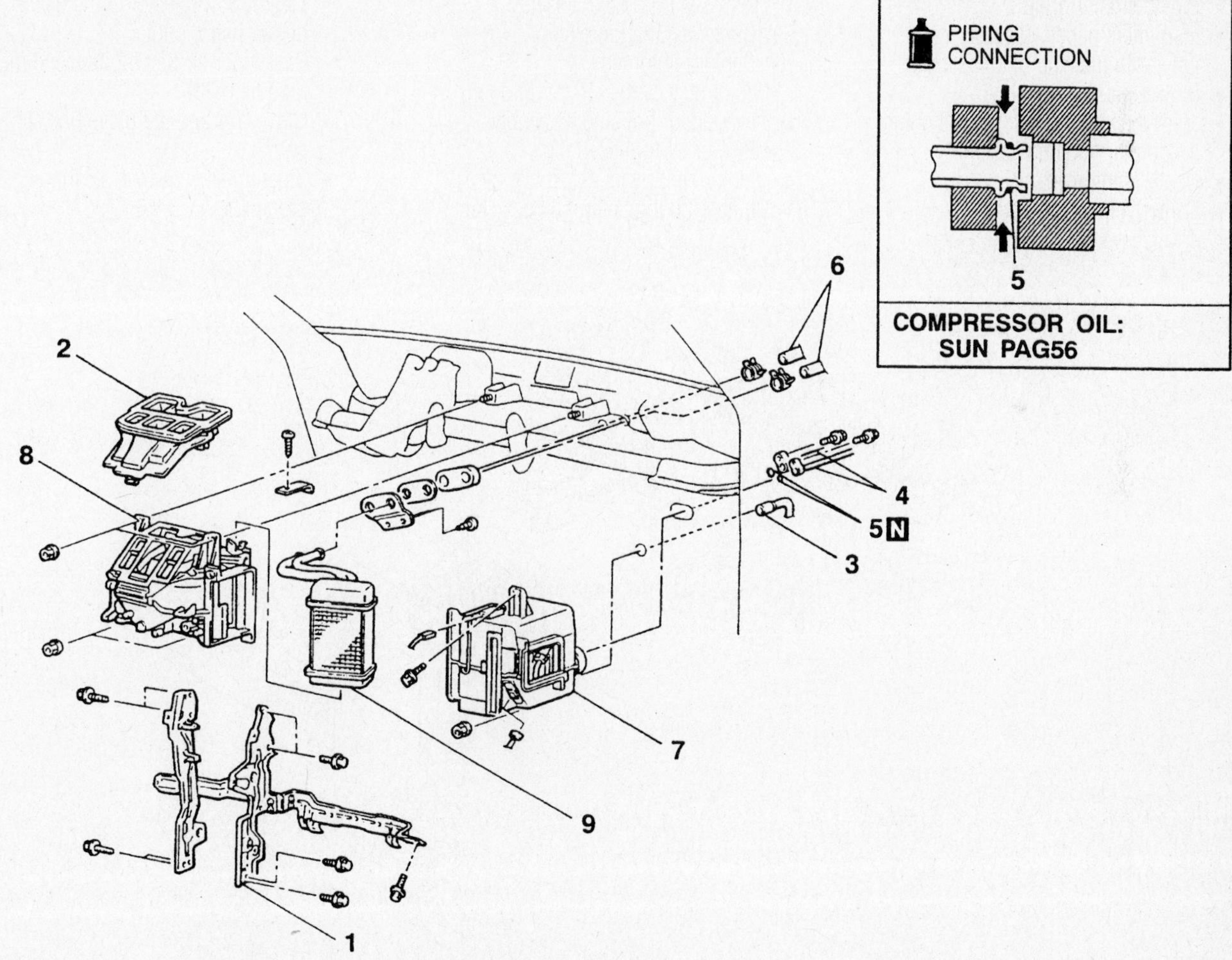

Exploded view of the heater housing, air conditioning evaporator housing and related components—Montero Sport

- Hood lock release handle
- Knee protector assembly and bracket
- Meter bezel assembly
- Under cover, the corner cover and the stopper
- Glove box assembly and the ashtray
- Center under cover assembly and the cup holder assembly
- Radio and tape/CD player
- Heater/air conditioning control assembly
- Combination meter and the speaker
- Glove box striker and the upper glove box frame
- Multi-meter panel and the multimeter assembly
- Side defroster grille
- Instrument panel assembly, carefully, with the help of an assistant
- Blower motor assembly
- Joint duct from the air conditioning evaporator housing assembly
- Center reinforcement
- Center ventilation duct
- Drain hose from the air conditioning evaporator housing assembly
- Heater hoses from the heater housing assembly
- Refrigerant lines from the air conditioning evaporator housing assembly and discard the O-rings
- Heater housing assembly
- Heater core from the heater housing

To install:

7. Install or connect the following:
   - Heater core to the heater housing
   - Heater housing assembly
   - Refrigerant lines to the air conditioning evaporator housing assembly, using new O-rings
   - Heater hoses to the heater housing assembly
   - Drain hose to the air conditioning evaporator housing assembly
   - Center ventilation duct
   - Center reinforcement
   - Joint duct to the air conditioning evaporator housing assembly
   - Blower motor assembly
   - Instrument panel assembly, carefully, with the help of an assistant
   - Side defroster grille
   - Multi-meter assembly and the multimeter panel
   - Upper glove box frame and the glove box striker
   - Speaker and the combination meter
   - Heater/air conditioning control assembly
   - Radio and tape/CD player
   - Center under cover assembly and the cup holder assembly
   - Glove box assembly and the ashtray
   - Under cover, the corner cover and the stopper
   - Meter bezel assembly
   - Knee protector assembly and bracket
   - Hood lock release handle
   - Electrical connector to the passenger air bag module
   - Passenger air bag module and the module bolts
   - Glove box and the glove box stoppers
   - Steering wheel to the steering column

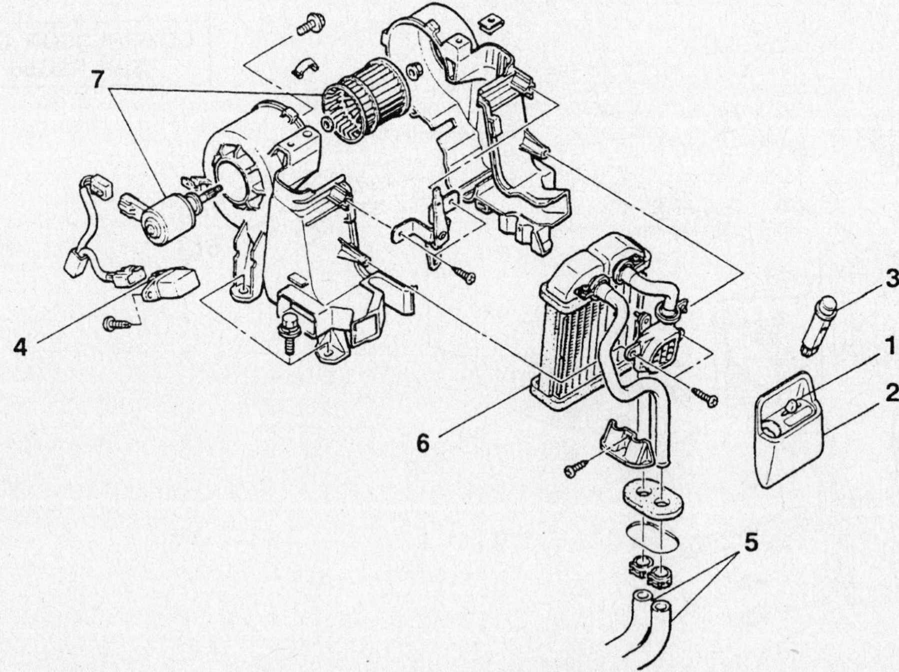

1. KNOB
2. REAR HEATER CONTROL PANEL ASSEMBLY
3. REAR HEATER SWITCH
4. RESISTOR
- DRAINING AND SUPPLYING OF COOLANT
5. REAR HEATER HOSE CONNECTION
6. REAR HEATER CORE ASSEMBLY
7. REAR BLOWER MOTOR ASSEMBLY

93113GE1

Exploded view of the rear heater core and related components—Montero Sport

- Steering wheel nut and torque the nut to 29 ft. lbs. (39 Nm)
- Electrical connectors to the air bag module
- Air bag module to the steering wheel
- Steering column-to-instrument panel cover and the cover screws
- Floor console assembly

8. Refill the cooling system.
9. Connect the negative battery.
10. Evacuate and charge the air conditioning system refrigerant.
11. Run the engine to normal operating temperatures; then, check the climate control operation and check for leaks.

REAR AUXILIARY SYSTEM

1. Before servicing the vehicle, refer to the precautions in the beginning of this section.
2. Drain the cooling system into a clean container for reuse.
3. Remove or disconnect the following:
   - Negative battery cable
   - Rear heater unit switch knob
   - Rear heater control panel assembly
   - Rear heater switch
   - Rear floor console
   - Resistor
   - Rear heater hoses from the rear heater core
   - Rear heater core from the rear heater housing

To install:

4. Install or connect the following:
   - Rear heater core to the rear heater housing
   - Rear heater hoses to the rear heater core
   - Resistor
   - Rear floor console
   - Rear heater switch
   - Rear heater control panel assembly
   - Rear heater unit switch knob
   - Negative battery cable
5. Refill the cooling system.
6. Run the engine to normal operating temperatures; then, check the climate control operation and check for leaks.

Cylinder Head

REMOVAL & INSTALLATION

3.0L Engines

1. Before servicing the vehicle, refer to the precautions in the beginning of this section.
2. Relieve the fuel system pressure.
3. Drain the cooling system.
4. Remove or disconnect the following:
   - Negative battery cable
   - Air cleaner assembly, ducts and air intake hose
   - Upper radiator hose
   - Accessory drive belts
   - Cooling fan and pulleys
   - Air conditioning compressor, if equipped
   - Power steering pump and mounting brackets and position them to the side, without disconnecting the lines.
   - Timing belt covers
5. Remove the timing belt as follows:
   a. Rotate the crankshaft and bring the No. 1 piston to Top Dead Center (TDC) on the compression stroke. Align the camshaft and crankshaft sprocket timing marks.
   b. Mark the timing belt in the direction of rotation for reinstallation purposes.
   c. Loosen the timing belt tensioner bolt and turn the tensioner counterclockwise.

⁂ WARNING

Do not rotate the crankshaft or camshaft sprockets after the timing belt has been removed.

6. Remove or disconnect the following:
   - Timing belt
   - Fuel lines and plug
   - Wiring connectors, vacuum lines and hoses from the air intake plenum, intake manifold and cylinder head.
   - Air intake plenum
   - Intake manifold
   - Exhaust manifold
   - Camshaft sprocket bolt and camshaft sprocket, if necessary
   - Alternator bracket and/or timing belt rear cover
   - Oil dipstick, on left side only
   - Crankshaft position (CKP) sensor on left side only
   - Spark plug wires from the spark plugs
   - Valve cover
   - Cylinder head bolts starting from the outside and working inward
   - Cylinder head from the engine

To install:

7. Clean the gasket mounting surfaces.
8. Install or connect the following:
   - New cylinder head gasket
   - Cylinder head on the engine.

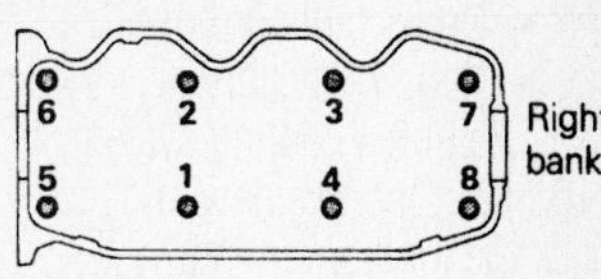

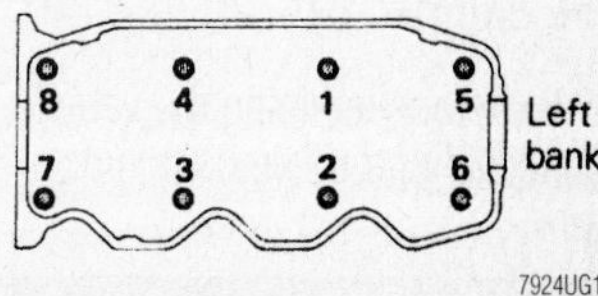

Cylinder head bolt tightening sequence—3.0L engines

Torque the cylinder head bolts in sequence using 3 even steps, to 80 ft. lbs. (108 Nm).

   - Exhaust manifold
   - Intake manifold and air intake plenum
   - Fuel lines
   - Wiring connectors, vacuum lines and hoses to the air intake plenum, intake manifold and cylinder head
   - Valve cover
   - Spark plug wires
   - CKP sensor, if removed
   - Oil dipstick, if removed
   - Alternator bracket and/or timing belt rear cover
   - Camshaft sprocket bolt and camshaft sprocket, if necessary

9. Be sure the camshaft and crankshaft sprocket timing marks are aligned.
10. Turn the timing belt tensioner to the extreme counter-clockwise position and temporarily tighten the bolt.
11. Install the timing belt in the original rotation direction. Loosen the timing belt tensioner bolt and allow the spring force of the tensioner to tension the belt.
12. Turn the crankshaft 2 turns in the normal direction of rotation and check the timing mark alignment.
13. If the timing is correct, tighten the tensioner bolt to 21 ft. lbs. (30 Nm). If the timing is incorrect, repeat the belt installation procedure.
14. Install or connect the following:
   - Timing belt covers
   - Alternator, alternator cover and alternator stay, if removed
   - Air conditioning compressor
   - Power steering pump with the brackets
   - Pulleys. Torque the crankshaft pulley bolt to 134 ft. lbs. (181 Nm).
   - Cooling fan
   - Accessory drive belts
   - Air cleaner assembly, ducts and air intake hose

- Upper radiator hose
- Negative battery

15. Refill the cooling system.
16. Start the engine and check for leaks. Check the ignition timing.

3.5L Engine

1. Before servicing the vehicle, refer to the precautions in the beginning of this section.

⁂ CAUTION

The fuel injection system remains under pressure after the engine has been OFF. Properly relieve fuel pressure before disconnecting any fuel lines. Failure to do so may result in fire or personal injury.

2. Relieve fuel system pressure.
3. Drain the cooling system.
4. Remove or disconnect the following:
   - Negative battery cable

⁂ CAUTION

Work must be started after 90 seconds from the time the ignition switch is turned to the LOCK position and the negative battery cable is disconnected.

- Air intake hoses
- Air intake plenum and intake manifold
- Exhaust manifold
- Engine under cover
- Radiator and shroud
- Alternator
- Cooling fan
- Timing belt
- Breather hose
- Oil dipstick
- Camshaft Position (CMP) sensor
- Spark plug cable center cover and remove the spark plug cables
- Valve cover
- Intake camshaft sprocket
- Rear timing belt cover
- Ignition coil
- Water hoses from the thermostat housing and the housing
- Water inlet from the front head and discard O-ring
- Water passage

5. Loosen the cylinder head mounting bolts in 3 steps, starting from the outside and working inward. Lift off the cylinder head assembly and remove the head gasket.

To install:

6. Thoroughly clean and dry the mating surfaces of the head and block. Check the cylinder head for cracks, damage or engine coolant leakage. Remove scale, sealing compound and carbon. Clean oil passages thoroughly. Check the head for flatness. End to end, the head should be within 0.0012 in. (0.030mm), normally with 0.008 in. (0.203mm) the maximum allowed out of true. The total thickness allowed to be removed from the head and block is 0.008 in. (0.203mm) maximum.

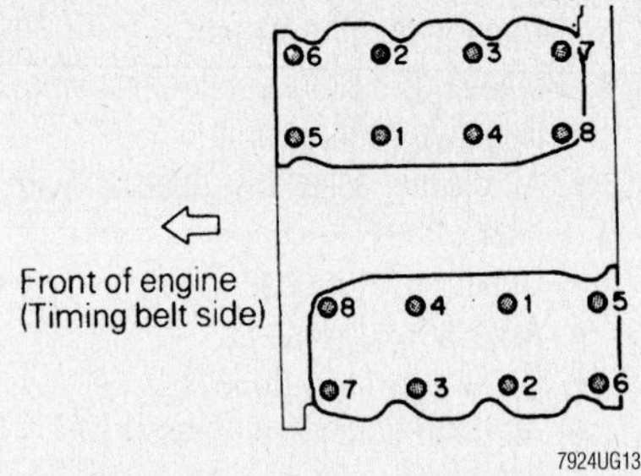

Cylinder head bolt tightening sequence—3.5L engines

7. Place a new head gasket on the cylinder block with the identification marks in the front top (upward) position. Do not use sealer on the gasket.
8. Install or connect the following:
   - Cylinder head on the block. Be sure the head bolt washers are installed with the chamfered edge upward. Using 3 even steps, torque the head bolts in sequence, to 76–83 ft. lbs. (105–115 Nm).
   - New O-ring and the water inlet to the front head
   - New gaskets, thermostat housing and connect the hoses
   - Water passage and new gaskets
   - Ignition coil and center rear timing belt cover
   - Intake camshaft sprocket. Use hex flange on camshaft to secure and tighten the retaining bolt to 65 ft. lbs. (90 Nm).
   - New gasket and the valve cover. Torque the bolts to 84 inch lbs. (10 Nm).
   - Spark plug cables and the center cover
   - Oil dipstick
   - CMP sensor
   - Breather hose
   - Radiator and shroud
   - Timing belt
   - Cooling fan
   - Alternator
   - Engine under cover
   - Intake manifold and new gasket. Torque the nuts to 16 ft. lbs. (21 Nm).
   - Air intake plenum and new gaskets. Torque the bolts to 13 ft. lbs. (18 Nm).
   - Exhaust manifold and new gaskets. Torque the nuts to 22 ft. lbs. (29 Nm).
   - Air intake hoses
   - Negative battery cable
9. Change the engine oil and oil filter.
10. Refill the system with coolant.
11. Run the vehicle until the thermostat opens.
12. Once the vehicle has cooled, recheck the coolant level.

3.8L Engine

1. Before servicing the vehicle, refer to the precautions in the beginning of this section.

⁂ CAUTION

The fuel injection system remains under pressure after the engine has been OFF. Properly relieve fuel pressure before disconnecting any fuel lines. Failure to do so may result in fire or personal injury.

2. Relieve fuel system pressure.
3. Drain the cooling system.
4. Remove or disconnect the following:
   - Negative battery cable

⁂ CAUTION

Work must be started after 90 seconds from the time the ignition switch is turned to the LOCK position and the negative battery cable is disconnected.

- Intake manifold
- Timing belt
- Front exhaust pipe
- Water outlet pipe assembly and O-ring
- Heater hose connections
- Water passage assembly and gasket
- Water pipe and O-ring
- Breather hose
- Spark plug wires; tag before disconnecting
- Ignition coil
- Oxygen (O_2) sensor connector
- Engine oil dipstick assembly and O-ring
- Intake manifold plenum stay
- Rocker cover
- Camshaft Position (CMP) sensor connector

5. Loosen the cylinder head mounting bolts in 3 steps, in the sequence shown. Lift

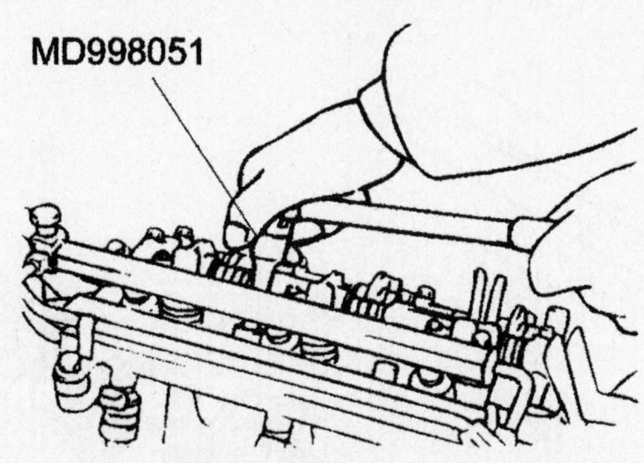

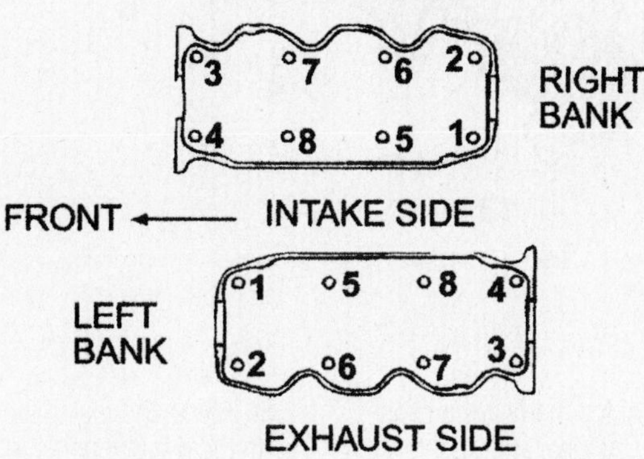

Loosen the cylinder head bolts in the proper sequence—3.8L engine

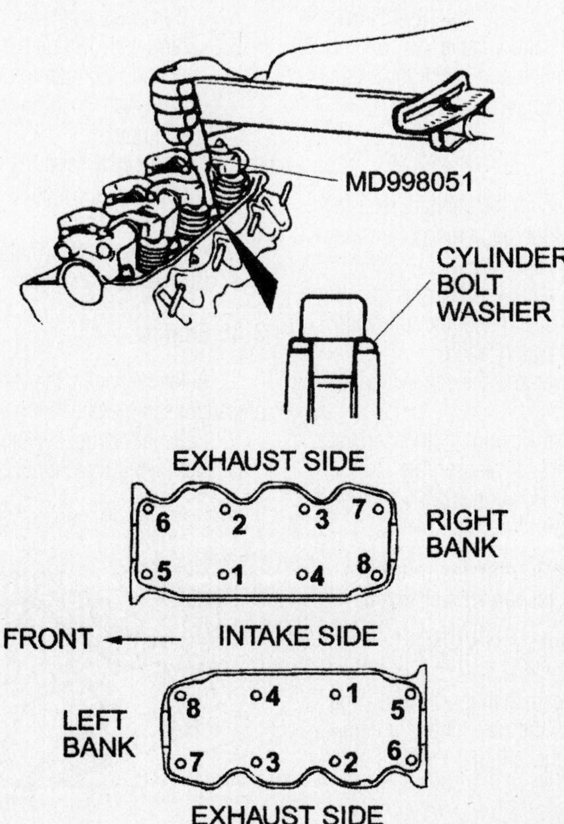

The cylinder head bolts must be installed with the chamfered edge up and tighten in the proper sequence—3.8L engine

off the cylinder head assembly and remove the head gasket.

To install:

6. Thoroughly clean and dry the mating surfaces of the head and block. Check the cylinder head for cracks, damage or engine coolant leakage. Remove scale, sealing compound and carbon. Clean oil passages thoroughly. Check the head for flatness. End to end, the head should be within 0.0012 in. (0.030mm), normally with 0.008 in. (0.203mm) the maximum allowed out of true. The total thickness allowed to be removed from the head and block is 0.008 in. (0.203mm) maximum.
7. Place a new head gasket on the cylinder block with the identification marks in the front top (upward) position. Do not use sealer on the gasket.
8. Install or connect the following:
   - Cylinder head on the block. Be sure the head bolt washers are installed with the chamfered edge upward. Using 3 even steps, torque the head bolts in sequence, to 77–83 ft. lbs. (105–113 Nm) with a torque wrench and Special Tool No. MD998501.
   - Rocker cover
   - Intake manifold plenum stay
   - Engine oil dipstick assembly and O-ring
   - O_2 sensor connector
   - Ignition coil
   - Spark plug wires
   - Breather hose
   - Water pipe and O-ring
   - Water passage assembly and gasket
   - Heater hose connections
   - Water outlet pipe assembly and O-ring
   - Front exhaust pipe
   - Timing belt
   - Intake manifold
   - Negative battery cable
9. Change the engine oil and oil filter.
10. Refill the system with coolant.
11. Run the vehicle until the thermostat opens.
12. Once the vehicle has cooled, recheck the coolant level.

Rocker Arms/Shafts

REMOVAL & INSTALLATION

3.0L Engines

1. Before servicing the vehicle, refer to the precautions in the beginning of this section.

1. Bearing cap No. 4
2. Rocker arm (B)
3. Spring
4. Rocker arm (A)
5. Spring
6. Bearing cap No. 3
7. Rocker arm (B)
8. Spring
9. Rocker arm (A)
10. Spring
11. Bearing cap No. 2
12. Rocker arm (B)
13. Spring
14. Rocker arm (A)
15. Spring
16. Rocker arm shaft (B)
17. Rocker arm shaft (A)
18. Bearing cap No. 1

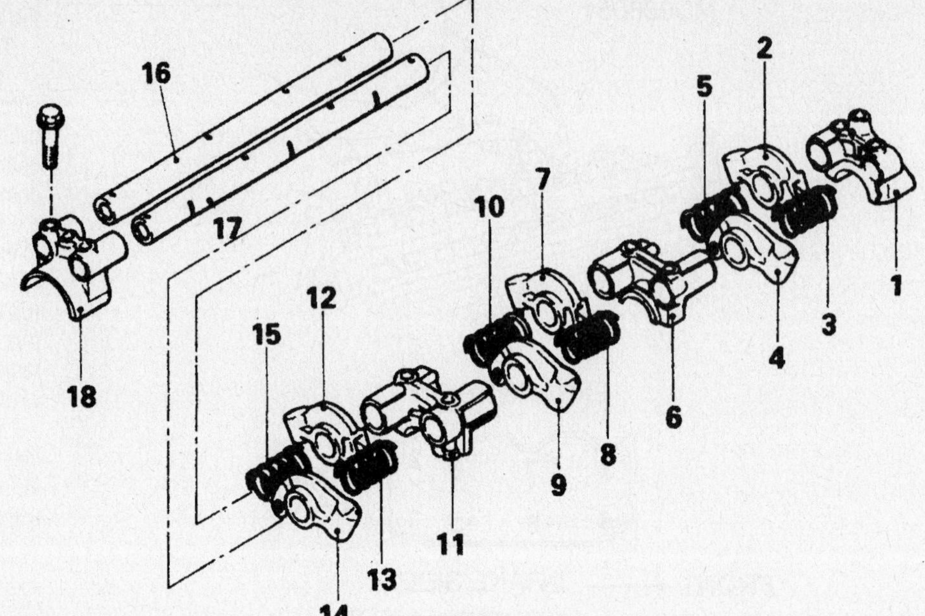

Exploded view of the rocker arms and shafts—3.0L engine

2. Remove or disconnect the following:
   - Negative battery cable

⁂ CAUTION

Work must be started after 90 seconds from the time the ignition switch is turned to the LOCK position and the negative battery cable is disconnected.

   - Valve cover
   - Auto lash adjuster retainers SST MD998443 on the rocker arms
   - Rocker arms, rocker shafts and bearing caps, as an assembly

To install:

3. Inspect the bearing journals on the camshaft and the cylinder head.
4. Lubricate the camshaft journals and camshaft with clean engine oil.
5. Install the rocker arms, rocker arm shaft and the rocker shaft spring as follows:
   a. Temporarily tighten the rocker shaft with the bolts so that the intake valve rocker arms do not push on the valves.
   b. Insert the rocker shaft spring from above and mount it at right angles to the plug guide.
   c. Before installing the exhaust rocker arms and the rocker arm shaft, mount the rocker shaft spring.
   d. Remove tool SST MD998443 used to hold the lash adjuster in position.
   e. Check to ensure that the flat side of the rocker shaft is perpendicular to the cylinder head, and facing the valves.
   f. Gradually tighten the bearing caps in 2 or 3 steps. In the final step, tighten to 23 ft. lbs. (31 Nm).
6. Install or connect the following:
   - Valve cover and new gasket. Torque the bolt to 2–3 ft. lbs. (3–4 Nm).
   - Negative battery cable
7. Start the engine and check for leaks and proper operation.

3.5L Engine

1. Before servicing the vehicle, refer to the precautions in the beginning of this section.
2. Relieve the fuel system pressure.
3. Remove or disconnect the following:
   - Negative battery cable
   - Valve cover and the semi-circular packing.
   - Crankshaft Position (CKP) sensor, matchmark for reassembly
   - Camshaft Position (CMP) sensor, if equipped

➡Install auto lash adjuster retainers SST MD998443 on the rocker arms

   - Rocker arms and shafts
   - Lash adjusters
4. Check the camshaft journals for wear or damage. Check the cam lobes for damage. Also, check the cylinder head oil holes for clogging.

To install:

➡Lubricate the valve train components with clean engine oil.

5. Bleed and install the lash adjusters to the to the original bores in the cylinder head.
6. Install or connect the following:
   - Rocker arms and shafts. Torque the bolts to 23 ft. lbs. (31 Nm).
   - Camshaft position sensor, if removed. Torque the mounting bolts to 78 inch lbs. (9 Nm).
   - Camshaft Position (CMP) sensor, if equipped
   - Valve cover and the semi-circular packing. Torque the bolts to 2.5 ft. lbs. (3.5 Nm).
   - Negative battery cable
7. Run vehicle and check for leaks.

3.8L Engine

1. Before servicing the vehicle, refer to the precautions in the beginning of this section.
2. Relieve the fuel system pressure.
3. Remove or disconnect the following:

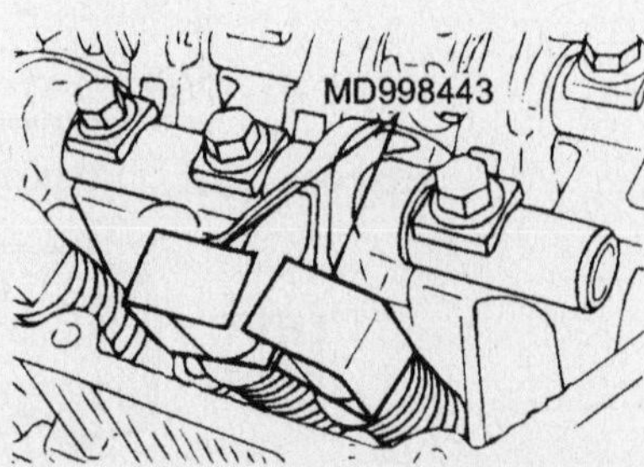

Install the special tool to prevent the lash adjusters from falling to the floor during rocker arm removal—3.8L engine

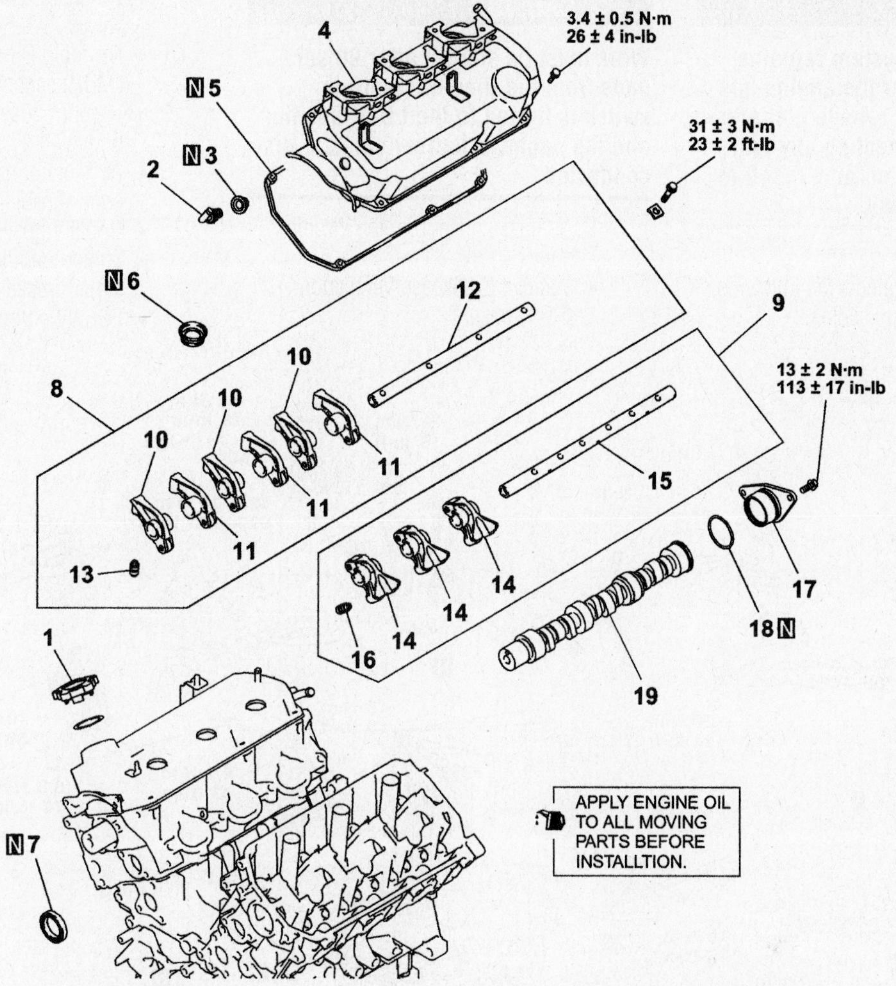

Exploded view of the rocker arms, camshafts and related components—3.8L engine

- Negative battery cable
- Oil filler cap
- Positive Crankcase Ventilation (PCV) valve and gasket
- Valve cover, gasket and oil seal(s)
- Camshaft and oil seal

➡Install auto lash adjuster retainers SST MD998443 on the rocker arms

- Rocker arms and shafts
- Lash adjusters

4. Check the camshaft journals for wear or damage. Check the cam lobes for damage. Also, check the cylinder head oil holes for clogging.

To install:

➡Lubricate the valve train components with clean engine oil.

5. Bleed and install the lash adjusters to the to the original bores in the cylinder head.
6. Install or connect the following:
   - Rocker arms and shafts. Torque the bolts to 21–25 ft. lbs. (28–34 Nm).
   - Valve cover oil seals and gasket
   - Valve cover. Torque the bolts to 22–30 inch lbs. (2.9–3.4 Nm).
   - Positive Crankcase Ventilation (PCV) valve and gasket
   - Oil filler cap
   - Negative battery cable
7. Run vehicle and check for leaks.

Intake Manifold

REMOVAL & INSTALLATION

3.0L Engine

1. Before servicing the vehicle, refer to the precautions in the beginning of this section.
2. Relieve the fuel pressure.

✱✱ CAUTION

The fuel injection system remains under pressure after the engine has been OFF. Properly relieve fuel pressure before disconnecting any fuel lines. Failure to do so may result in fire or personal injury.

3. Drain the engine coolant.
4. Remove or disconnect the following:
   - Negative battery cable

✱✱ CAUTION

Work must be started after 90 seconds from the time the ignition switch is turned to the LOCK position and the negative battery cable is disconnected.

   - Air intake hose from the throttle body
   - Positive Crankcase Ventilation (PCV) hose
   - Exhaust Gas Recirculation (EGR) valve
   - Manifold Differential Pressure (MDP) sensor
   - Vacuum hoses from the throttle body and air intake plenum
   - Accelerator cable and the throttle control cable
   - Coolant hoses
   - Engine oil filler neck bracket from the air intake plenum
   - EGR tube from the air intake plenum

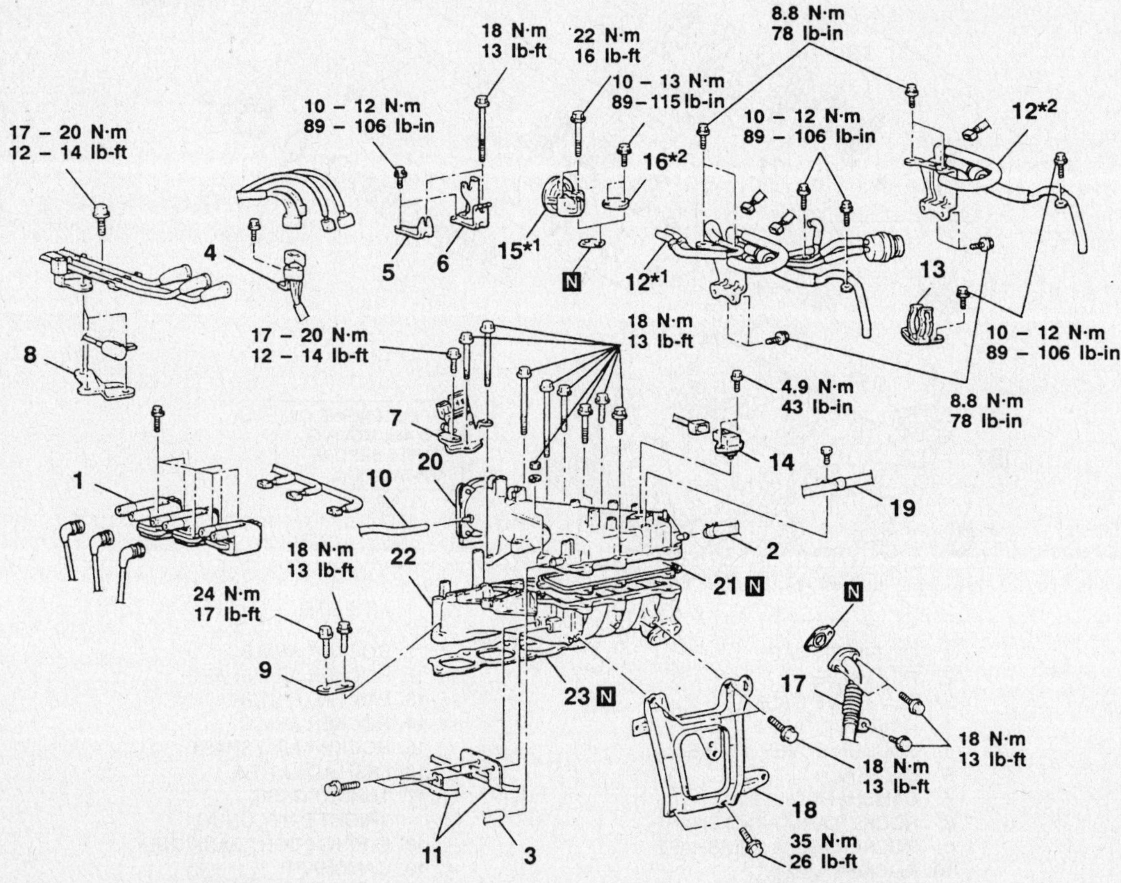

1. IGNITION COILS
2. BRAKE BOOSTER VACUUM HOSE CONNECTION
3. PCV HOSE CONNECTION
4. CRANKSHAFT POSITION SENSOR AND CAM POSITION SENSOR CONNECTOR
5. ACCELERATOR CABLE BRACKET <M/T>
6. THROTTLE CABLE BRACKET <A/T>
7. IGNITION POWER TRANSISTOR
8. WATER OUTLET FITTING BRACKET
9. WATER PUMP STAY
10. VACUUM HOSE CONNECTION
11. FUEL PIPE CONNECTION
12. SOLENOID VALVE AND VACUUM HOSE ASSEMBLY
13. VCV BRACKET
14. MDP SENSOR
15. EGR VALVE
16. COVER
17. EGR PIPE CONNECTION
18. INTAKE MANIFOLD PLENUM STAY
19. THROTTLE CABLE CONNECTION
20. AIR INTAKE FITTING
21. AIR INTAKE FITTING GASKET
22. UPPER INTAKE MANIFOLD
23. INTAKE MANIFOLD PLENUM GASKET

NOTE
*1: Vehicles for Federal
*2: Vehicles for California

7924UG41

Exploded view of the upper intake manifold and related components—3.0L 24-valve engine shown

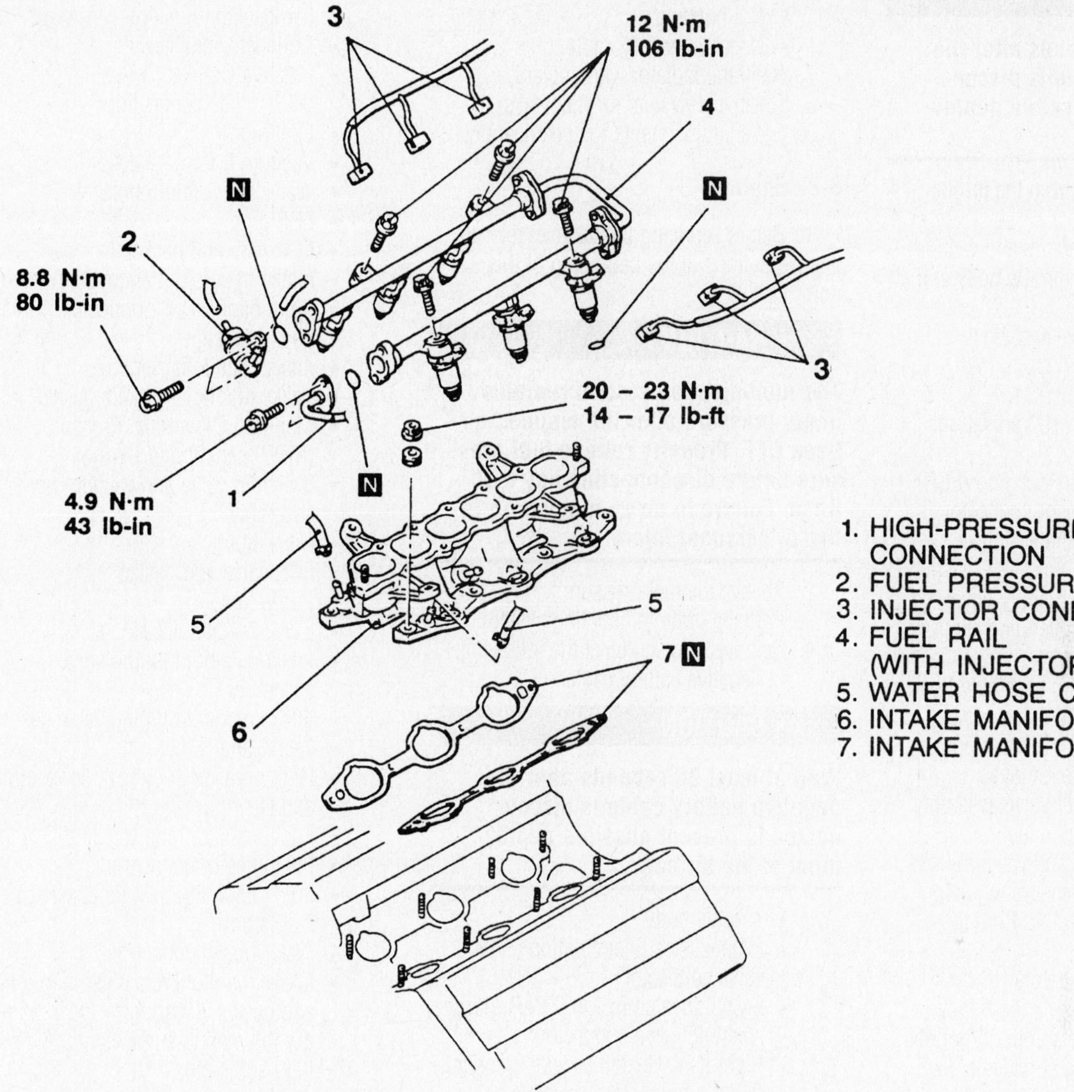

Exploded view of the lower intake manifold and related components—3.0L 24-valve engine shown

- Plenum brackets
- Air intake plenum assembly from the intake manifold and remove. Note the position of the mounting bolts as they are removed.
- Fuel hose from the fuel rail
- Fuel return line and vacuum hose from the fuel pressure regulator
- Electrical connectors from the injectors
- Fuel rail and injectors
- Intake manifold

5. Remove the gaskets and thoroughly clean and dry the mating surfaces of the manifold and heads.

To install:

6. Install or connect the following:
   - Intake manifold. Torque the nuts to 16 ft. lbs. (21 Nm) start from the center and working outward.
7. Connect the hoses and connect the wires to the coolant switches.
   - Fuel rail assembly and connect the fuel hoses
   - New gasket and the air intake plenum to the intake manifold. Torque the nuts/bolts to 13 ft. lbs. (17 Nm).
   - Plenum brackets
   - PCV hose and vacuum hose cluster to the plenum
   - EGR tube
   - EGR temperature sensor wire
   - Wires, hoses and linkages to the throttle body
   - Air intake hose to the throttle body
   - Upper radiator hose to the thermostat housing
   - Negative battery cable
8. Refill the radiator with coolant.
9. Check fuel system for leaks.

3.5L Engines

1. Before servicing the vehicle, refer to the precautions in the beginning of this section.

⁂ CAUTION

The fuel injection system remains under pressure after the engine has been OFF. Properly relieve fuel pressure before disconnecting any fuel lines. Failure to do so may result in fire or personal injury.

2. Relieve the fuel pressure.
3. Partially drain the cooling system.
4. Remove or disconnect the following:
   - Negative battery cable

⁂ CAUTION

Wait at least 90 seconds after the negative battery cable is disconnected to prevent possible deployment of the air bag.

- Air intake hose from the throttle body
- Electrical connectors and vacuum hoses from the throttle body and air intake plenum
- Accelerator cable and the throttle control cable
- Coolant hoses
- Positive Crankcase Ventilation (PCV) hose
- Exhaust Gas Recirculation (EGR) temperature sensor connector
- EGR tube from the air intake plenum
- Intake manifold plenum cover
- Intake manifold plenum stay brackets
- Air intake plenum assembly from the intake manifold and remove. Note the position of the mounting bolts as they are removed
- Induction control valve assembly
- Fuel hose from the fuel rail
- Fuel return line and vacuum hose from the fuel pressure regulator
- Electrical connectors from the injectors
- Fuel rail and injectors
- Intake manifold

5. Remove the gaskets and thoroughly clean and dry the mating surfaces of the manifold and heads.

To install:

6. Install or connect the following:
   - Intake manifold. Tighten the nuts to 16 ft. lbs. (21 Nm). Start from the center and work outward.
   - Fuel rail assembly and connect the fuel hoses
   - Induction control valve assembly and tighten to 72 inch lbs. (9 Nm).
   - New gasket and air intake plenum to the intake manifold. Torque the nuts/bolts to 13 ft. lbs. (18 Nm).
   - Plenum to engine brackets
   - Hoses and wires to the coolant switches
   - PCV hose and vacuum hose cluster to the plenum
   - EGR tube. Torque the bolts to 13 ft. lbs. (18 Nm).
   - EGR temperature sensor wire
   - Wires, hoses and linkages to the throttle body
   - Air intake hose to the throttle body
   - Upper radiator hose to the thermostat housing
   - Negative battery cable
7. Refill the radiator with coolant.
8. Check the system for fuel leaks.
9. Set all adjustments to specifications.

3.8L Engines

1. Before servicing the vehicle, refer to the precautions in the beginning of this section.

⁂ CAUTION

The fuel injection system remains under pressure after the engine has been OFF. Properly relieve fuel pressure before disconnecting any fuel lines. Failure to do so may result in fire or personal injury.

2. Relieve the fuel pressure.
3. Partially drain the cooling system.
4. Remove or disconnect the following:
   - Negative battery cable

⁂ CAUTION

Wait at least 90 seconds after the negative battery cable is disconnected to prevent possible deployment of the air bag.

- Throttle body
- Exhaust Gas Recirculation (EGR) valve connector
- Evaporative emission (EVAP) purge solenoid valve connector
- Right bank Heated Oxygen Sensor (HO_2S) connector connection
- Manifold Differential Pressure (MDP) sensor connector
- Capacitor connector
- Knock Sensor (KS) connector
- Control wiring harness and Camshaft Position (CMP) sensor wiring harness combination connector
- Ground cable
- Fuel injector connector
- Control wiring harness and injector wiring harness combination connector
- Intake manifold tuning solenoid connector
- Engine Coolant Temperature (ECT) sensor connector
- ECT gauge unit connector
- Crankshaft Position (CKP) sensor connector
- Ground cable
- Knock Sensor (KS) and Camshaft Position (CMP) sensor combination connector
- Control wiring harness and injector harness combination connector
- Connector bracket
- Positive Crankcase Ventilation (PCV) hose connection
- Fuel pipe
- Vacuum hose connection
- Water outlet fitting bracket
- EGR pipe
- EGR pipe and gasket
- Intake manifold plenum stay
- Right bank HO_2S connector
- Fuel pipe clip
- Intake manifold plenum
- Intake manifold plenum gasket
- Manifold Differential Pressure (MDP) sensor and O-ring
- Solenoid valve and vacuum hose assembly
- Capacitor
- EGR valve and gasket
- Purge hose
- EVAP purge solenoid valve
- Intake manifold tuning valve assembly
- Intake manifold tuning valve gaskets P and S
- High pressure fuel hose connection and O-ring
- Fuel pressure regulator and O-ring
- Fuel injector connectors
- Fuel rail (with injectors attached)
- Insulators
- Water hose connection
- Intake manifold retainers, manifold and gasket. Thoroughly clean and dry the mating surfaces of the manifold and heads.

To install:

5. Install or connect the following:
   - New intake manifold gasket. Make sure the gaskets are installed with the protrusions as shown in the illustration.
   - Intake manifold
6. Install the intake manifold bolts and tighten as follows:
   a. 1st step: Right bank nuts to 58 inch lbs. (6.5 Nm).
   b. 2nd step: Left bank nuts to 16 ft. lbs. (22 Nm).
   c. 3rd step: Right bank nuts to 16 ft. lbs. (22 Nm).
   d. 4th step: Left bank nuts to 16 ft. lbs. (22 Nm).
   e. 5th step: Right bank nuts to 16 ft. lbs. (22 Nm).
   - Water hose connection
   - Insulators
   - Fuel rail and injector assembly
   - Fuel injector connectors
   - Fuel pressure regulator and O-ring

- High pressure fuel hose connection and O-ring
- Intake manifold tuning valve gaskets P and S
- Intake manifold tuning valve assembly
- EVAP purge solenoid valve
- Purge hose
- EGR valve and gasket
- Capacitor
- Solenoid valve and vacuum hose assembly

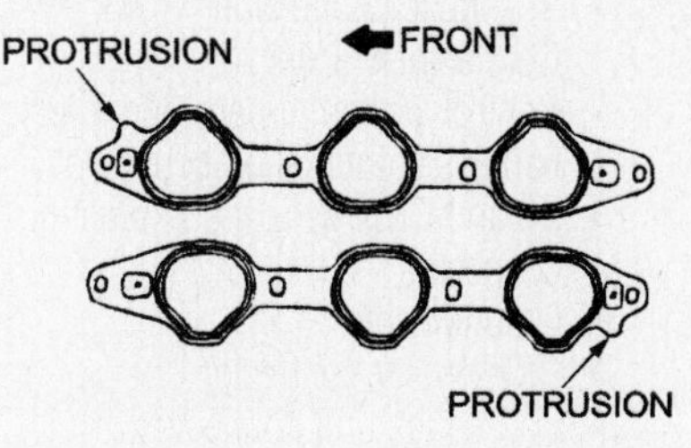

The intake manifold gaskets must be installed properly—3.8L engine

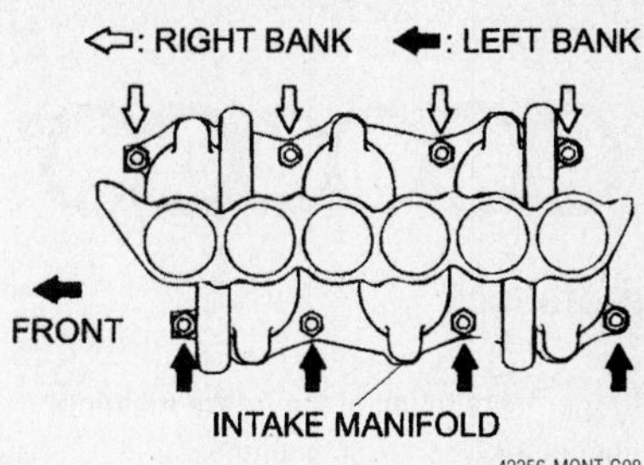

Left and right bank nut locations—3.8L engine

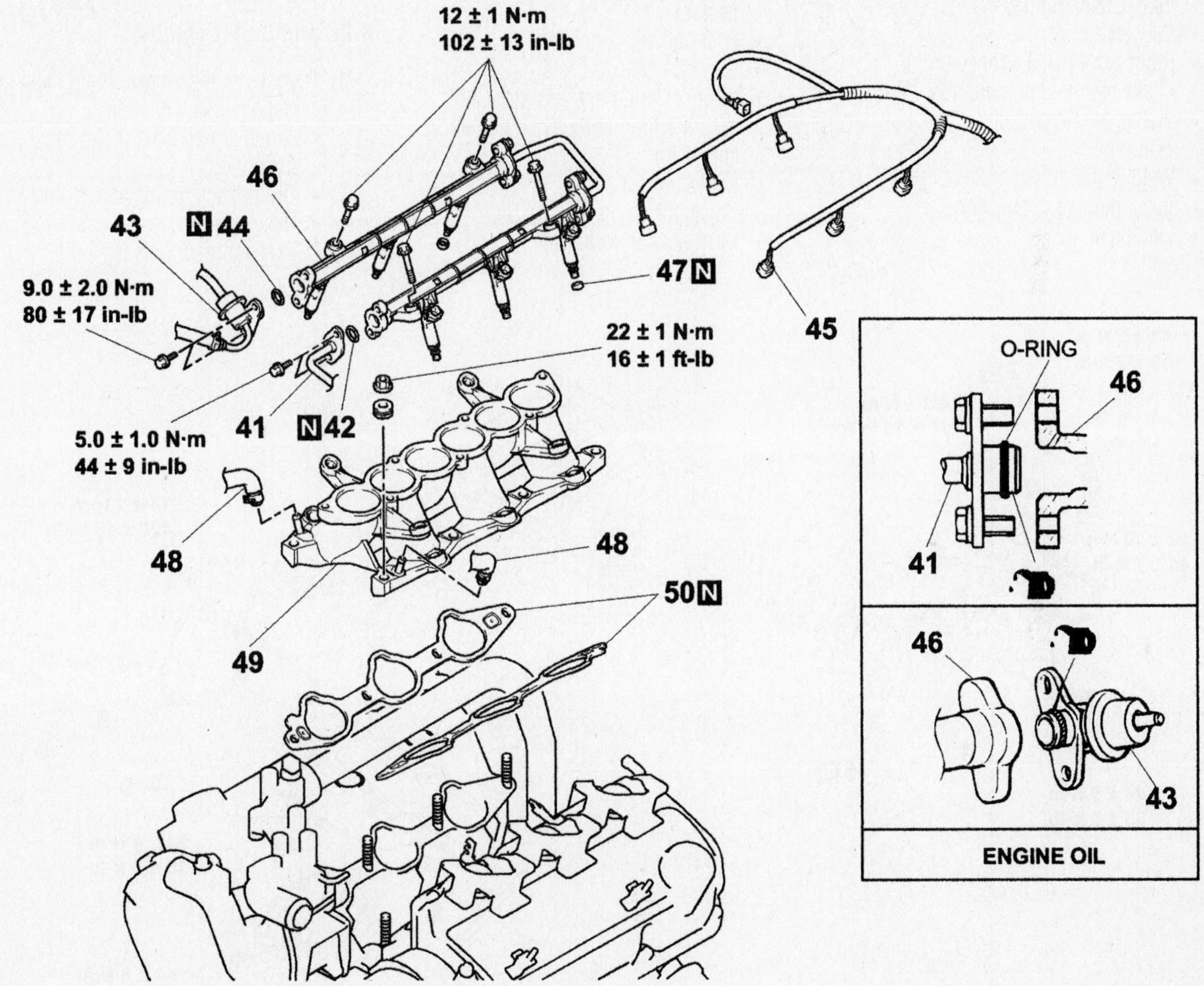

41. FUEL HIGH-PRESSURE HOSE CONNECTION
42. O-RING
43. FUEL PRESSURE REGULATOR
44. O-RING
45. INJECTOR CONNECTOR
46. FUEL RAIL (WITH INJECTORS)
47. INSULATORS
48. WATER HOSE CONNECTION
49. INTAKE MANIFOLD
50. INTAKE MANIFOLD GASKET

42356-MONT-G06

Exploded view of the intake manifold—3.8L engine

Proper installation of the intake manifold plenum gasket—3.8L engine

- Manifold Differential Pressure (MDP) sensor and O-ring
- Intake manifold plenum gasket
- Intake manifold plenum
- Fuel pipe clip
- Right bank HO_2S connector
- Intake manifold plenum stay
- EGR pipe and gasket
- EGR pipe
- Water outlet fitting bracket
- Vacuum hose connection
- Fuel pipe
- PCV hose connection
- Connector bracket
- Control wiring harness and injector harness combination connector
- KS and CMP sensor combination connector
- Ground cable
- CKP sensor connector
- ECT gauge unit connector
- ECT sensor connector
- Intake manifold tuning solenoid connector
- Control wiring harness and injector wiring harness combination connector
- Fuel injector connector
- Ground cable
- Control wiring harness and CMP sensor wiring harness combination connector
- KS connector
- Capacitor connector
- MDP sensor connector
- Right bank HO_2S connector
- EVAP purge solenoid valve connector
- EGR valve connector
- Throttle body
- Negative battery cable

7. Refill the radiator with coolant.
8. Check the system for fuel leaks.
9. Set all adjustments to specifications.

Exhaust Manifold

REMOVAL & INSTALLATION

3.0L and 3.5L Engines

1. Remove or disconnect the following:
   - Negative battery cable
   - Exhaust pipe from the exhaust manifolds
   - Oil dipstick, guide and O-ring
   - Heat shields
   - Exhaust manifolds

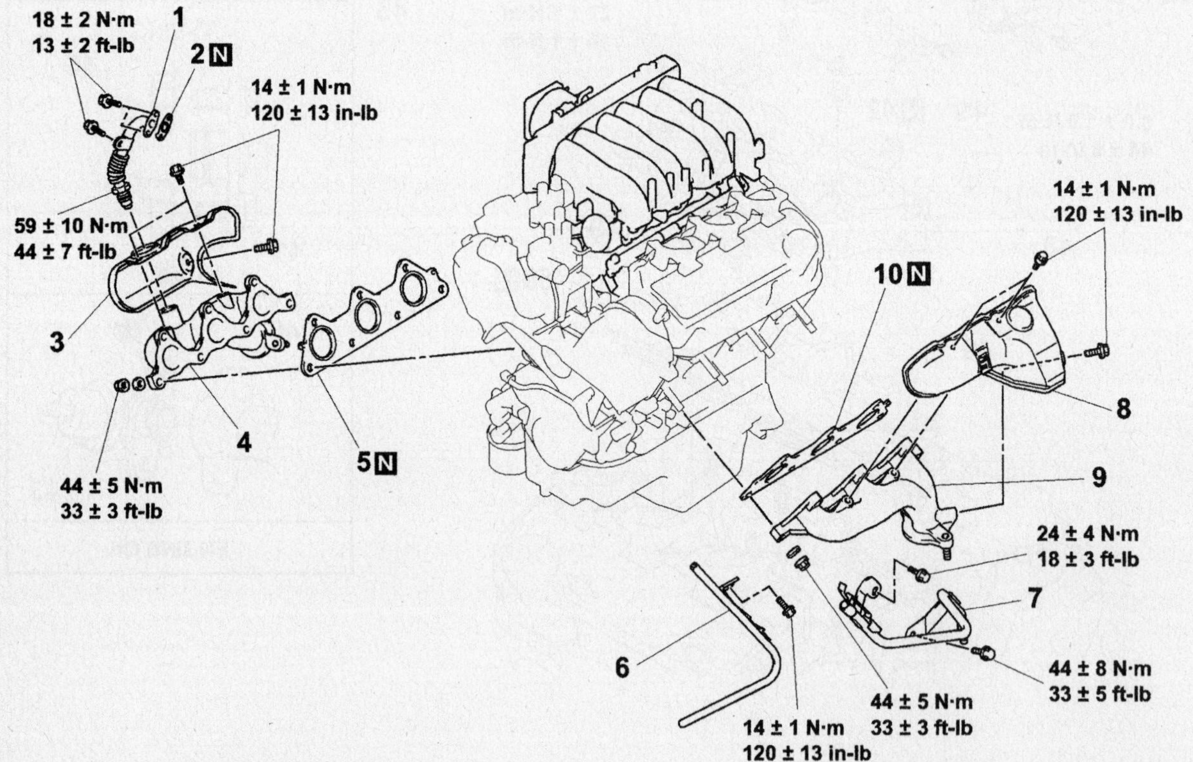

1. EGR PIPE
2. EGR PIPE GASKET
3. HEAT PROTECTOR <RH>
4. EXHAUST MANIFOLD <RH>
5. EXHAUST MANIFOLD GASKET <RH>
6. ENGINE OIL DIPSTICK GUIDE
7. TRANSMISSION FLUID DIPSTICK GUIDE
8. HEAT PROTECTOR <LH>
9. EXHAUST MANIFOLD <LH>
10. EXHAUST MANIFOLD GASKET <LH>

42356-MONT-G10

Exploded view of the exhaust manifolds—3.8L engine

2. Clean the gasket mounting surfaces. Inspect the manifolds for cracks, flatness and/or damage.

To install:

3. Install or connect the following:
   - New gasket and exhaust manifold. Torque the nuts to 33 ft. lbs. (44 Nm) on 3.0L engines and 22 ft. lbs. (29 Nm) on 3.5L engines.
   - Heat shield, Torque the bolts to 10 ft. lbs. (14 Nm).
   - Exhaust pipe to the exhaust manifolds. Torque the nuts to 35ft. (49 Nm).
   - Oil dipstick, guide and new O-ring
   - Negative battery cable
4. Start the engine and check for exhaust leaks.

3.8L Engines

1. Remove or disconnect the following:
   - Negative battery cable
   - Front exhaust pipe from the exhaust manifolds
   - Air cleaner assembly
   - Battery and battery tray
   - Exhaust Gas Recirculation (EGR) pipe and gasket
   - Right side heat shield
   - Right side exhaust manifold and gasket
   - Oil dipstick, guide and O-ring
   - Transmission fluid dipstick guide
   - Left side heat shield
   - Left side exhaust manifold and gasket
2. Clean the gasket mounting surfaces. Inspect the manifolds for cracks, flatness and/or damage.

To install:

3. Install or connect the following:
   - New gasket and left side exhaust manifold. Torque the nuts to 30–36 ft. lbs. (39–49 Nm).
   - Left side heat shield
   - Transmission fluid dipstick guide
   - Oil dipstick, guide and O-ring
   - New gasket and right side exhaust manifold. Torque the nuts to 30–36 ft. lbs. (39–49 Nm).
   - Right side heat shield
   - EGR gasket and pipe
   - Battery tray and battery
   - Air cleaner assembly
   - Front exhaust pipe to the exhaust manifolds. Torque the nuts to 35ft. (49 Nm).
   - Oil dipstick, guide and new O-ring
   - Negative battery cable
4. Start the engine and check for exhaust leaks.

Front Crankshaft Seal

REMOVAL & INSTALLATION

3.0L and 3.5L Engines

1. Before servicing the vehicle, refer to the precautions in the beginning of this section.
2. Drain the crankcase.
3. Drain and recycle the engine coolant.
4. Remove or disconnect the following:
   - Negative battery cable

⁂ CAUTION

Wait at least 90 seconds after the negative battery cable is disconnected to prevent possible deployment of the air bag.

   - Cooling fan
   - Accessory drive belts
   - Alternator
   - Engine undercover, if equipped
   - Power steering oil pump assembly
   - Air conditioner compressor and bracket, if equipped
   - Timing indicator bracket
   - Accessory mount assembly
   - Crankshaft pulley
   - Timing belt covers and the timing belt
   - Crankshaft sprocket
5. Cut out a portion in the crankshaft oil seal lip and pry out the oil seal with a flat prying tool, being careful not to damage the crankshaft.

To install:

6. Coat the lip of the new seal with oil and install the seal using the proper seal driver.
7. Install or connect the following:
   - Crankshaft sprocket and the timing belt
   - Timing belt covers
   - Crankshaft pulley. Torque the bolt to 134 ft. lbs. (181Nm).
   - Accessory mount Assembly. Torque the bolts to 33 ft. lbs. (44 Nm).
   - Timing indicator bracket. Torque the bolts to 97 inch lbs. (11 Nm).
   - Air conditioner compressor and bracket, if equipped
   - Power steering oil pump assembly
   - Engine undercover, if equipped
   - Alternator
   - Accessory drive belts
   - Cooling fan
   - Negative battery cable
8. Refill the crankcase.
9. Refill the cooling system.
10. Start the engine and check for proper operation.

3.8L Engine

1. Before servicing the vehicle, refer to the precautions in the beginning of this section.

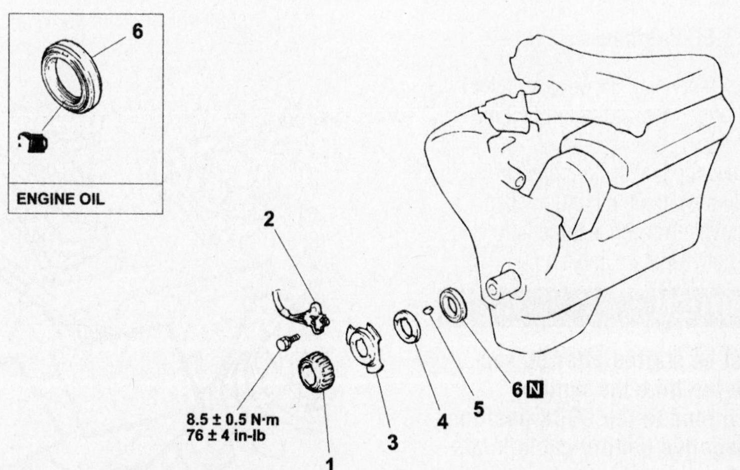

Exploded view of the crankshaft front oil seal—3.8L engine

2. Drain the crankcase.
3. Drain and recycle the engine coolant.
4. Remove or disconnect the following:
   - Negative battery cable

⁂ CAUTION

Wait at least 90 seconds after the negative battery cable is disconnected to prevent possible deployment of the air bag.

   - Timing belt
   - Crankshaft sprocket
   - Crankshaft Position (CKP) sensor
   - Crankshaft sensing blade
   - Crankshaft spacer and key
   - Front oil seal

To install:

   - Front oil seal. Apply oil to the seal, and install using Crankshaft Front Oil Seal Installer tool no. MD998717.
   - Crankshaft key and spacer.
   - Crankshaft sensing blade
   - CKP sensor

➡To be sure the crankshaft pulley bolt does not loosen, make sure the clean the mating areas of the crankshaft, spacer, sensing blade and sprocket.

   - Crankshaft sprocket
   - Timing belt

Camshaft and Valve Lifters

REMOVAL & INSTALLATION

3.0L and 3.5L Engines

1. Before servicing the vehicle, refer to the precautions in the beginning of this section.
2. Relieve the fuel system pressure.
3. Drain the engine oil and coolant.
4. Remove or disconnect the following:
   - Negative battery cable

⁂ CAUTION

Work must be started after 90 seconds from the time the ignition switch is turned to the LOCK position and the negative battery cable is disconnected.

   - Intake manifold plenum
   - Valve cover
   - Timing belt
   - Sprocket from the camshaft
5. Install auto lash adjuster retainers SST MD998443 on the rocker arms.

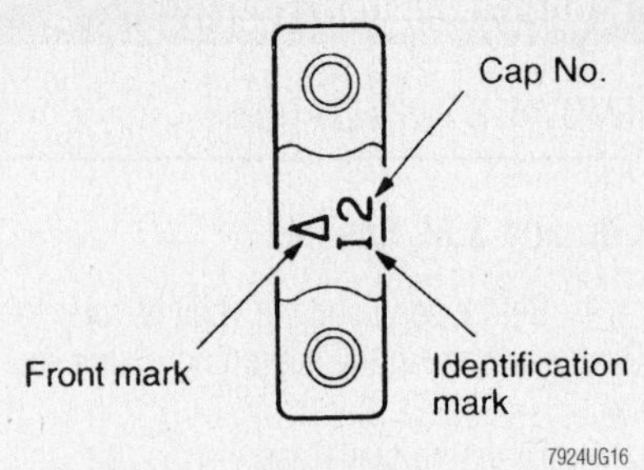

The camshaft bearing caps have identification marks on them—3.5L engine

6. Remove or disconnect the following:
   - Distributor and the distributor extension, if equipped
   - Rocker arms, rocker shafts and bearing caps, as an assembly
   - Thrust cage and O-ring
   - Camshaft from the cylinder head
7. Inspect the bearing journals on the camshaft and the cylinder head.

To install:

8. Lubricate the camshaft journals and camshaft with clean engine oil
9. Install or connect the following:
   - Camshaft in the cylinder head.
   - Thrust cage and O-ring. Torque the bolts to 109 inch lbs. (12 Nm).
   - Rocker arms, rocker arm shaft and the rocker shaft spring.
10. Temporarily tighten the rocker shaft with the bolts positioned so that the intake valve rocker arms do not push the valves.
11. Install or connect the following:
   - Rocker shaft spring from above and mount it at right angles to the plug guide.
12. Before installing the exhaust rocker arms and the rocker arm shaft, mount the rocker shaft spring.
13. Remove the SST used to hold the lash adjuster in position.
14. Check to ensure that the flat side of the rocker shaft is perpendicular to the cylinder head, and facing the valves.
15. Gradually tighten the bearing caps in 2 or 3 steps. In the final step tighten to 23 ft. lbs. (31 Nm).
16. Install or connect the following:
   - Distributor, if removed
   - Sprockets. Torque the bolts to 65 ft. lbs. (88 Nm).
   - Timing belt and timing belt cover
   - Valve cover. Torque the bolts to 26 inch lbs. (3.4 Nm).
   - Intake manifold plenum
   - Negative battery cable
17. Start the engine and check for leaks and proper operation.
18. Refill the coolant and crankcase.

3.8L Engine

1. Before servicing the vehicle, refer to the precautions in the beginning of this section.
2. Relieve the fuel system pressure.
3. Remove or disconnect the following:
   - Negative battery cable
   - Cylinder head assembly
   - Camshaft sprocket, using special tools M998715 and MB990767

Install the special tool to prevent the lash adjusters from falling to the floor during rocker arm removal—3.8L engine

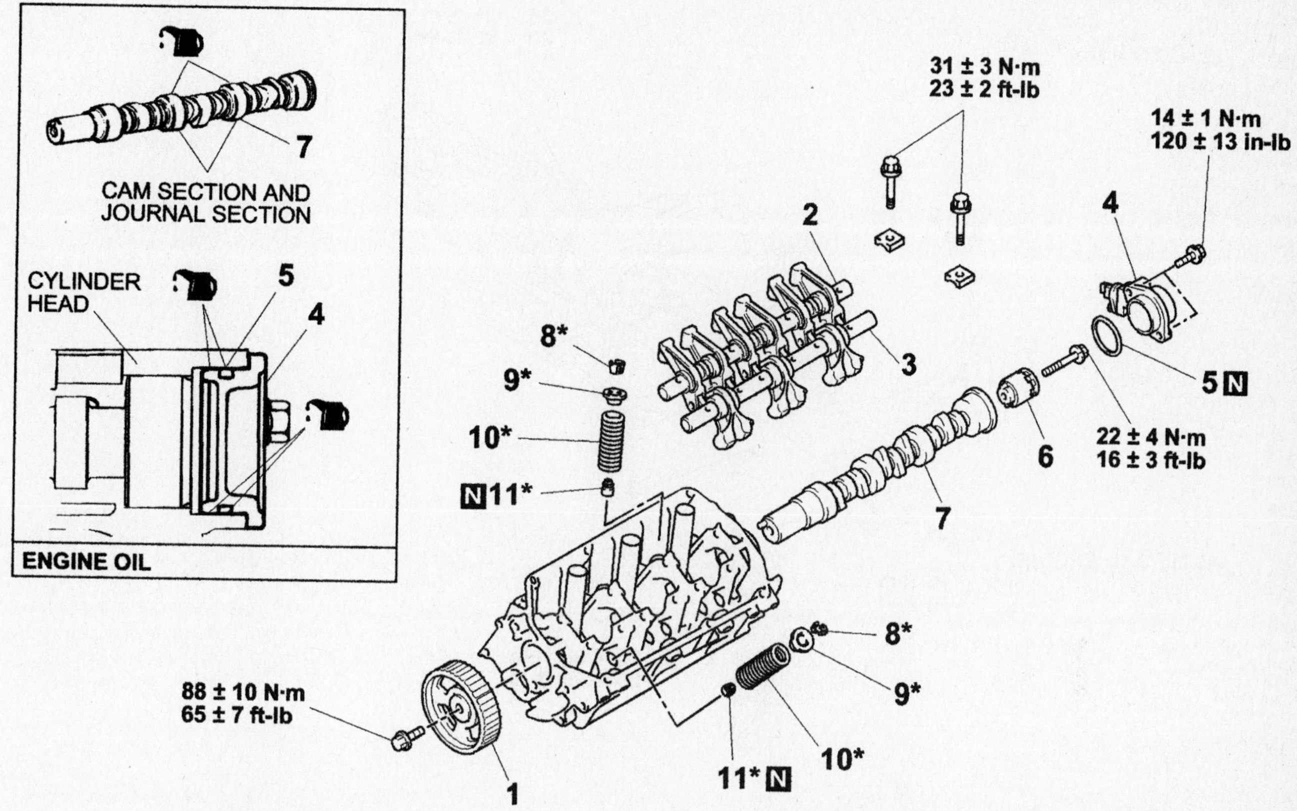

1. CAMSHAFT SPROCKET
2. ROCKER ARM, SHAFT AND LASH ADJUSTER ASSEMBLY (INTAKE SIDE)
3. ROCKER ARM, SHAFT AND LASH ADJUSTER ASSEMBLY (EXHAUST SIDE)
4. CAMSHAFT POSITION SENSOR SUPPORT
5. O-RING
6. SENSING CAMSHAFT POSITION CYLINDER
7. CAMSHAFT
8. VALVE SPRING RETAINER LOCKS
9. VALVE SPRING RETAINERS
10. VALVE SPRINGS
11. VALVE STEM SEALS

42356-MONT-G12

Exploded view of the camshaft and related components—3.8L engine

➡Install auto lash adjuster retainers SST MD998443 on the rocker arms

- Intake rocker arm, shaft and lash adjuster assembly. Loosen the rocker arm assembly mounting bolt and remove the assembly with the bolt still attached.
- Exhaust rocker arm, shaft and lash adjuster assembly. Loosen the rocker arm assembly mounting bolt and remove the assembly with the bolt still attached.
- Camshaft Position (CMP) sensor support and O-ring
- Sensing CMP cylinder
- Camshaft

To install:

- Camshaft
- Sensing CMP cylinder
- CMP sensor support and O-ring
- Exhaust rocker arm, shaft and lash adjuster assembly
- Intake rocker arm, shaft and lash adjuster assembly
- Camshaft sprocket
- Cylinder head assembly
- Negative battery cable

4. Run vehicle and check for leaks.

Starter Motor

REMOVAL & INSTALLATION

1. Before servicing the vehicle, refer to the precautions in the beginning of this section.
2. Remove or disconnect the following:
   - Negative battery cable
   - Engine under cover, if equipped
   - Front engine mount heat protector
   - Starter cover
   - Wires
   - Starter motor

To install:

3. Install or connect the following:
   - Starter motor and cover. Torque the bolts to 20–25 ft. lbs. (26–33 Nm).
   - Wires
   - Front engine mount heat protector. Torque the nut to 89 inch lbs. (10 Nm).
   - Engine under cover, if equipped
   - Negative battery cable

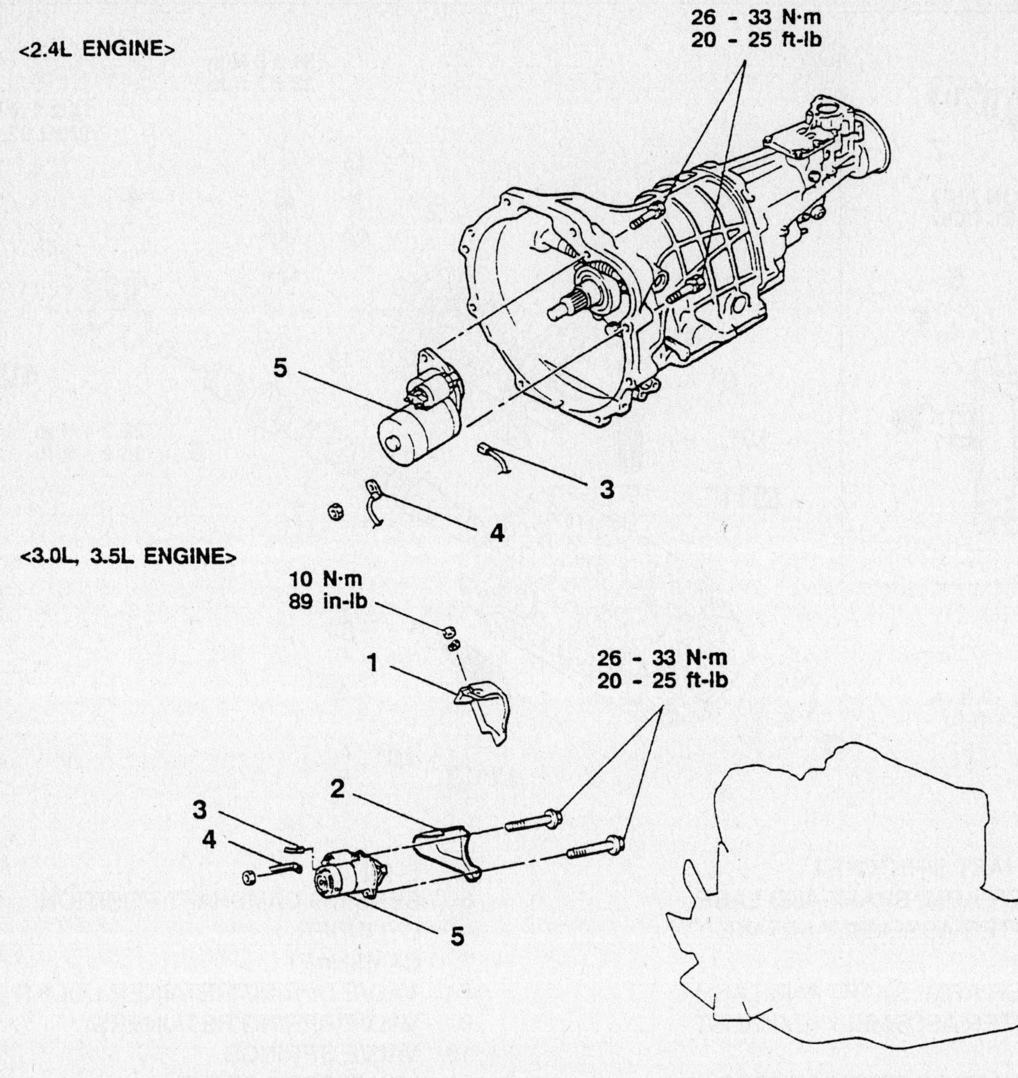

Starter motor mounting

Oil Pan

REMOVAL & INSTALLATION

3.0L Engines

1. Before servicing the vehicle, refer to the precautions in the beginning of this section.
2. Drain the engine oil.
3. Remove or disconnect the following:
   - Negative battery
   - Engine under cover
   - Alternator and belt
   - Stabilizer bar
   - Front exhaust pipe
   - Actuator assembly and heat protector
   - Oil dipstick
   - Crossmember assembly
   - Automatic transmission oil dipstick assembly
   - Exhaust pipe support bracket
   - Transmission stay
   - Oil pan, lower
   - Oil screen and baffle plate
   - Oil pan upper

To install:

4. Before installing, thoroughly clean the oil pan and cylinder block mating surfaces.
5. Apply liquid gasket around the surface of the oil pan.

➡Assemble the oil pan to the cylinder block within 15 minutes after applying the liquid gasket.

6. Install or connect the following:
   - Oil pan upper. Torque the bolts to 53 inch lbs. (6.0 Nm).

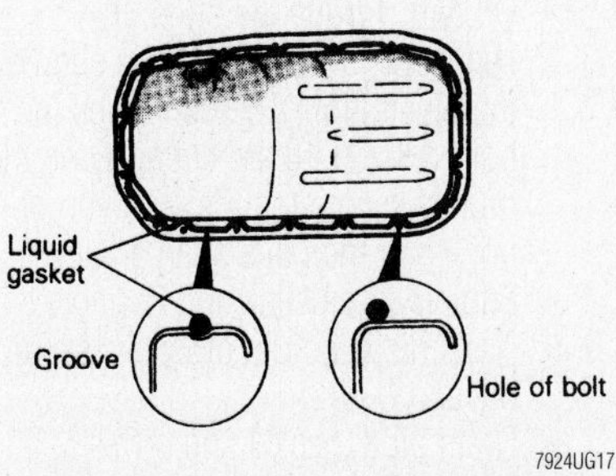

Apply a bead of sealant around the oil pan flange as shown—all engines are similar

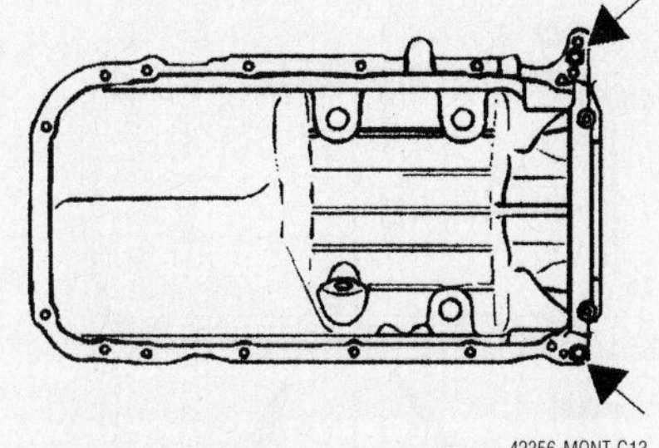

Screw the M10 bolts that hold the oil pan to the transmission assembly into the bolt holes shown by arrows—3.8L engine

- Oil screen and baffle plate. Torque the bolts to 14 ft. lbs. (19 Nm).
- Oil pan, lower. Torque the bolts to 53 inch lbs. (6.0 Nm).
- Transmission stay. Torque the bolts to 26 ft. lbs. (35 Nm).
- Exhaust pipe support bracket. Torque the bolts to 35 ft. lbs. (49 Nm).
- Automatic transmission oil dipstick assembly. Torque the bolts to 33 ft. lbs. (44 Nm).
- Crossmember assembly. Torque the bolts to 80 ft. lbs. (108 Nm).

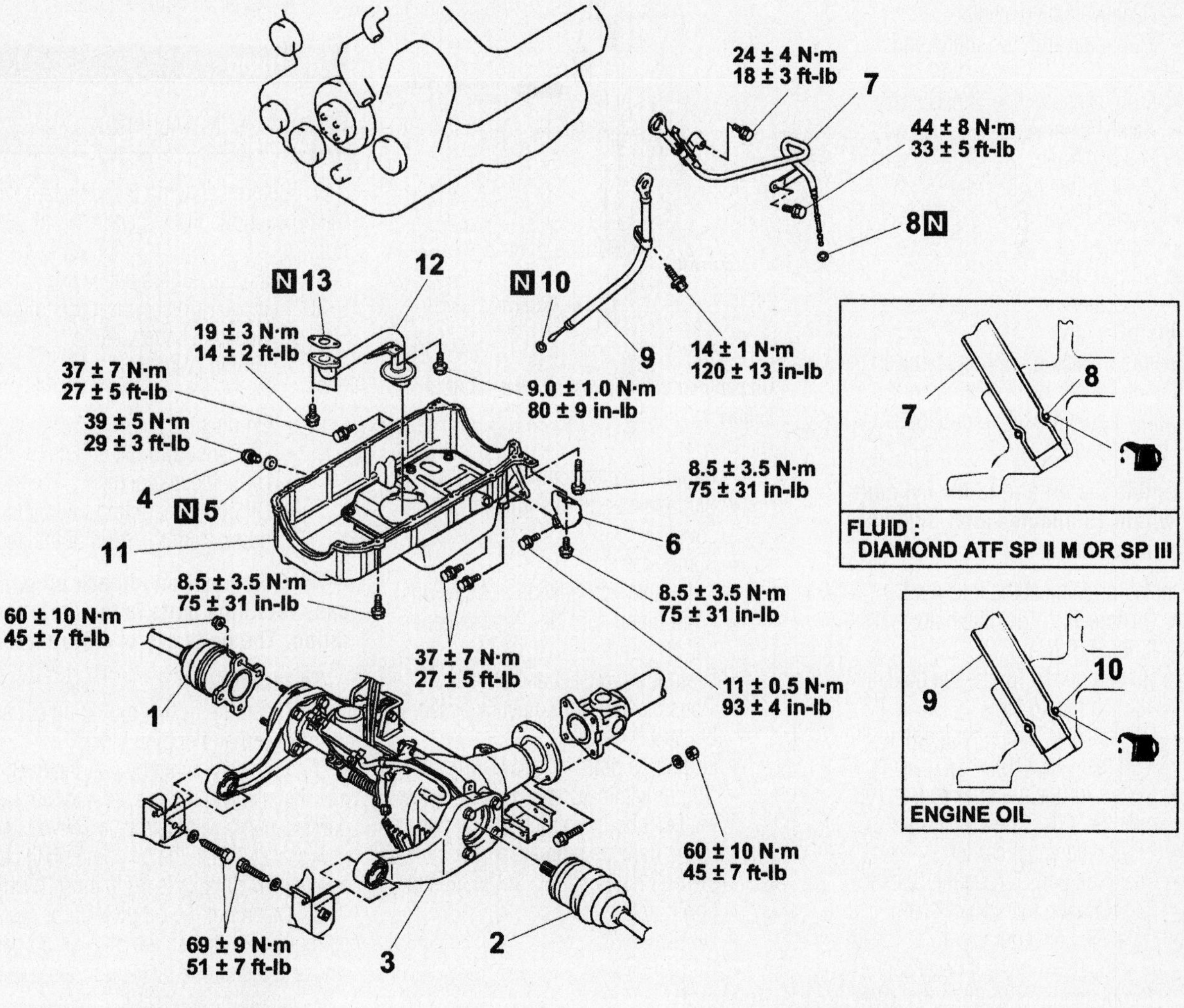

1. DRIVE SHAFT (RH) CONNECTION
2. DRIVE SHAFT (LH) CONNECTION
3. FRONT DIFFERENTIAL NUMBER 2 CROSSMEMBER ASSEMBLY
4. DRAIN PLUG
5. DRAIN PLUG GASKET
6. COVER
7. TRANSMISSION FLUID DIPSTICK ASSEMBLY
8. O-RING
9. ENGINE OIL DIPSTICK ASSEMBLY
10. O-RING
11. OIL PAN
12. OIL SCREEN
13. GASKET

42356-MONT-G14

Exploded view of the oil pan and related components—3.8L engine

- Oil dipstick. Torque the bolts to 35 ft. lbs. (48 Nm).
- Actuator assembly and heat protector
- Front exhaust pipe
- Stabilizer bar
- Alternator and belt
- Engine under cover
- Negative battery

3.5L Engines

1. Before servicing the vehicle, refer to the precautions in the beginning of this section.
2. Drain the engine oil.
3. Remove or disconnect the following:
   - Negative battery cable
   - Skid plate and the engine undercover
   - Front exhaust pipe, if necessary
   - Catalytic converter
   - Lower oil pan
   - Front differential carrier
   - Cover
   - Oil dipstick
   - Oil pan upper
   - Oil screen

To install:

4. Before installing, thoroughly clean the oil pan and cylinder block mating surfaces.
5. Apply liquid gasket around the surface of the oil pan.

➡Assemble the oil pan to the cylinder block within 15 minutes after applying the liquid gasket.

6. Install or connect the following:
   - Oil screen. Torque the bolts to 13 ft. lbs. (19 Nm).
   - Oil pan upper. Torque the bolts to 48 inch lbs. (6 Nm).
   - Oil dipstick. Torque the bolts to 39 inch lbs. (4.8 Nm).
   - Cover. Torque the bolts to 84–108 inch lbs. (10–12 Nm).
   - Front differential carrier
   - Lower oil pan. Torque the bolts to 84–108 inch lbs. (10–12 Nm).
   - Catalytic converter
   - Front exhaust pipe, if necessary. Torque the bolts to 35 ft. lbs. (49 Nm).
   - Skid plate and the engine undercover
   - Negative battery cable

3.8L Engines

1. Before servicing the vehicle, refer to the precautions in the beginning of this section.
2. Drain the engine oil.
3. Remove or disconnect the following:

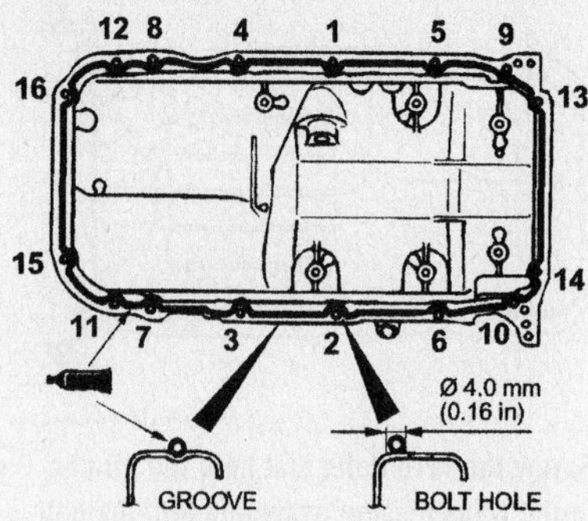

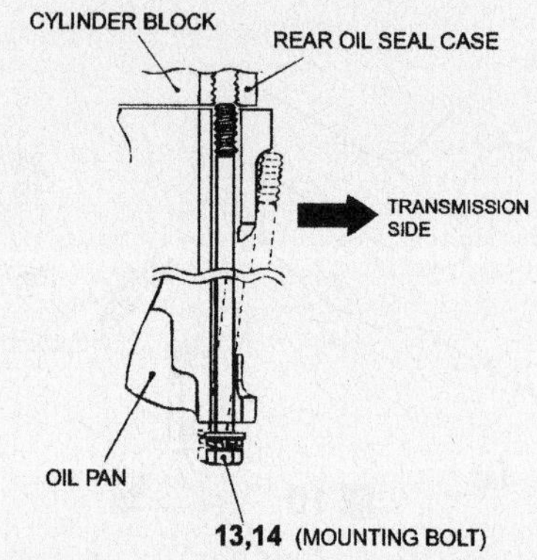

42356-MONT-G15

Oil pan bolt tightening sequence—3.8L engine

   - Negative battery cable
   - Skid plate and the engine undercover
   - Starter motor
   - Right and left halfshaft connections
   - Front differential No. 2 crossmember
   - Drain plug, gasket and cover
   - Transmission fluid dipstick assembly and O-ring
   - Engine oil dipstick and O-ring
   - Oil pan retainers and oil pan. Screw the M10 bolts holding the oil pan to the transmission assembly into the bolt holes shown in the illustration to remove the pan.
   - Oil screen
   - Oil pan gasket

To install:

4. Before installing, thoroughly clean the oil pan and cylinder block mating surfaces.
5. Apply liquid gasket around the surface of the oil pan.

➡Assemble the oil pan to the cylinder block within 30 minutes after applying the liquid gasket.

6. Install or connect the following:
   - Oil screen. Torque the bolts to 12–16 ft. lbs. (16–22 Nm).
   - Oil pan. Torque the bolts, in sequence, to specifications shown the illustration. The bolt holes for bolts 13 and 14 are cut away on the transmission side. Be sure you do not insert the bolts at an angle.
   - Engine oil dipstick and O-ring
   - Transmission fluid dipstick assembly and O-ring
   - Drain plug, gasket and cover
   - Front differential No. 2 crossmember
   - Right and left halfshaft connections
   - Skid plate and the engine undercover
   - Negative battery cable

Oil Pump

REMOVAL & INSTALLATION

1. Before servicing the vehicle, refer to the precautions in the beginning of this section.
2. Drain the engine oil.
3. Remove or disconnect the following:
   - Negative battery cable
   - Timing belt
   - Oil pressure switch
   - Oil dipstick
   - Oil pans from the engine
   - Oil baffle and screen
   - Oil pump mounting bolts and the pump from the front of the engine

➡Note the position of each oil pump case retaining bolts to facilitate installation. The bolts are of different length.

To install:

4. Clean the gasket mounting surfaces of the pump and engine block.
5. Prime the pump by pouring fresh oil into the inlet and turning the rotors or by packing pump with petroleum jelly. Using a new gasket, install the oil pump on the engine and tighten all bolts to 10 ft. lbs. (14 Nm).
6. Clean out the oil pick-up or replace as required. Replace the oil pick-up gasket ring and install the pick-up to the pump.
7. Install or connect the following:
   - Oil filter and the bracket. Torque the bolts to 17 ft. lbs. (23 Nm).
   - Oil baffle and screen. Torque the bolts to 13 ft. lbs. (18 Nm).
   - Oil pans. Torque the engines to 52 inch lbs. (5.9 Nm).
   - Oil pressure switch. Torque the switch to 87 inch lbs. (9.8 Nm).
   - Timing belt
   - Dipstick
   - Negative battery cable

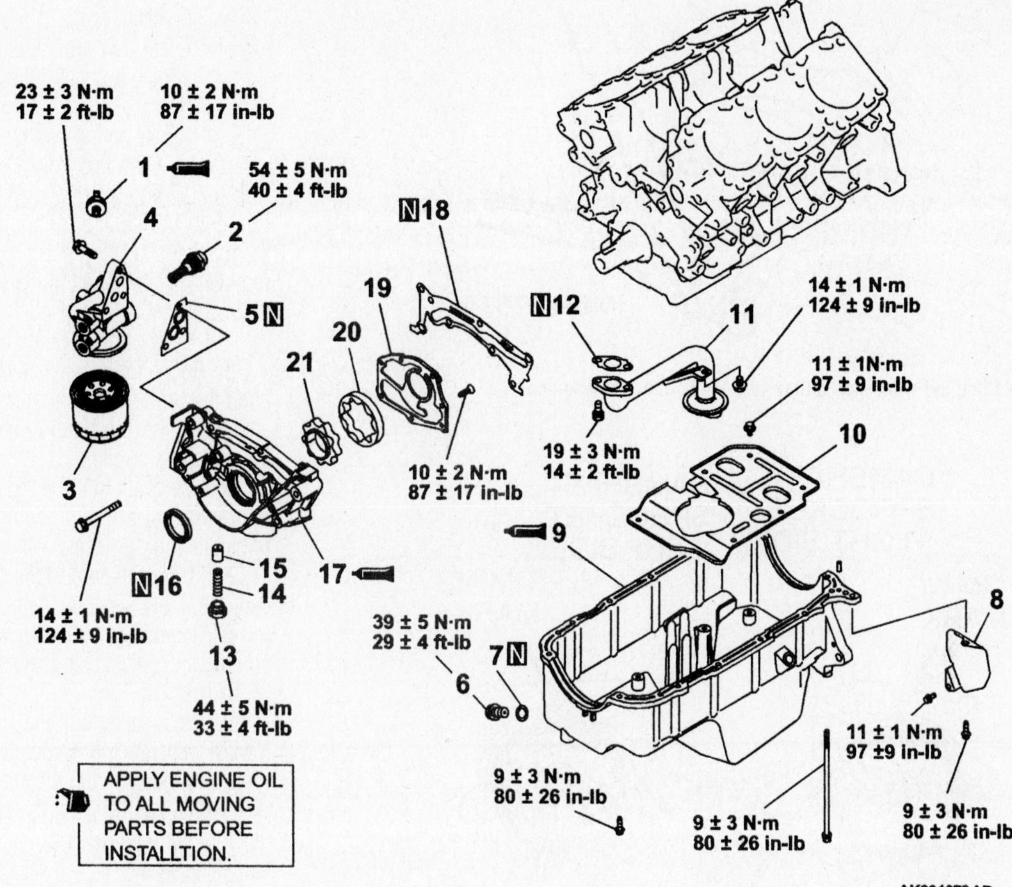

1. ENGINE OIL PRESSURE SWITCH
2. OIL COOLER BY-PASS VALVE
3. OIL FILTER
4. OIL FILTER BRACKET
5. OIL FILTER BRACKET GASKET
6. DRAIN PLUG
7. DRAIN PLUG GASKET
8. COVER
9. OIL PAN
10. BAFFLE PLATE
11. OIL SCREEN
12. OIL SCREEN GASKET
13. RELIEF PLUG
14. RELIEF SPRING
15. RELIEF PLUNGER
16. CRANKSHAFT OIL SEAL
17. OIL PUMP CASE
18. OIL PUMP CASE GASKET
19. OIL PUMP COVER
20 OIL PUMP OUTER ROTOR
21. OIL PUMP INNER ROTOR

42356-MONT-G16

Exploded view of the oil pump, oil pan and related components—3.8L engine shown, others similar

8. Refill the engine with the proper amount of oil.
9. Start the engine and check for proper oil pressure. Check for leaks.

Timing Belt

REMOVAL & INSTALLATION

3.0L and 3.5L Engines

1. Before servicing the vehicle, refer to the precautions in the beginning of this section.
2. Drain the engine coolant.
3. Remove or disconnect the following:
   - Negative battery cable
   - Upper radiator hose
   - Cooling fan shroud assembly
   - Cooling fan-to-clutch bolts and the fan
   - Cooling fan clutch-to-water pump nuts and the clutch assembly
   - Drive belts for the alternator, power steering pump and air conditioning compressor
   - Electrical connectors from the alternator
   - Alternator-to-engine bolts and the alternator bracket-to-engine bolts
   - Alternator and bracket from the engine
   - Power steering pump cover
   - Power steering pump-to-engine bolts and move the pump aside with the hoses and electrical connector attached
   - Air conditioning compressor-to-bracket bolts and move the compressor aside with the lines and electrical connector attached
   - Air conditioning compressor bracket-to-engine bolts and the bracket
   - Timing indicator bracket (near

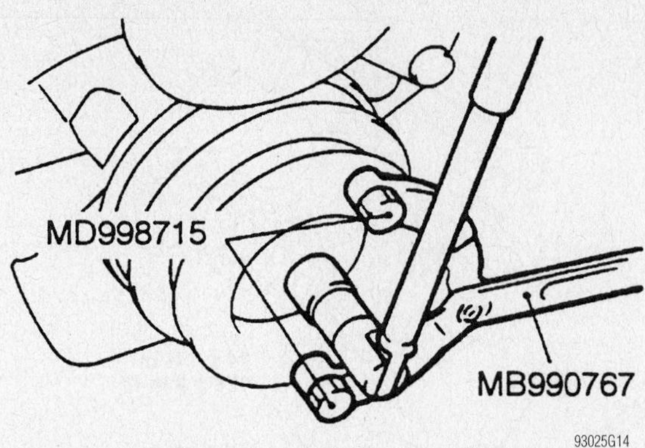

Removing or installing the crankshaft pulley bolt—3.0L and 3.5L engines

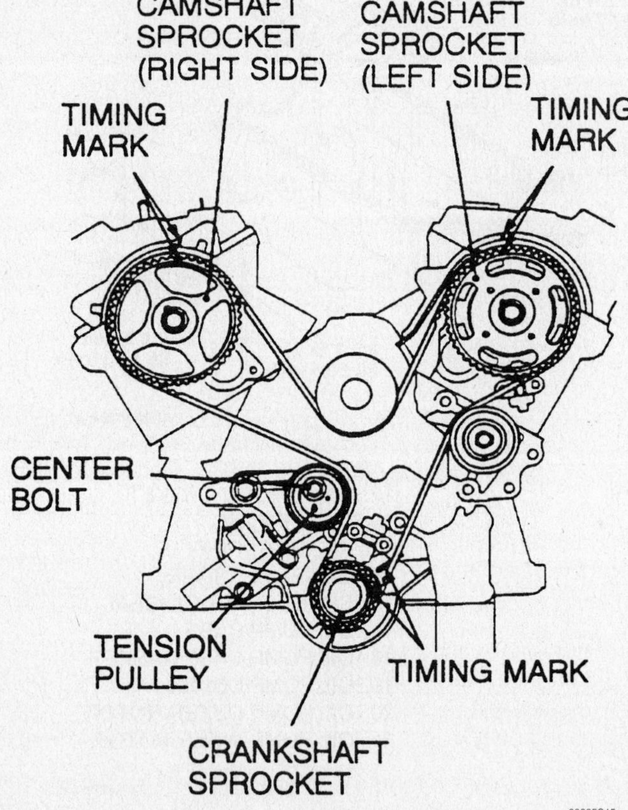

View of the timing belt alignment marks—3.0L and 3.5L engines

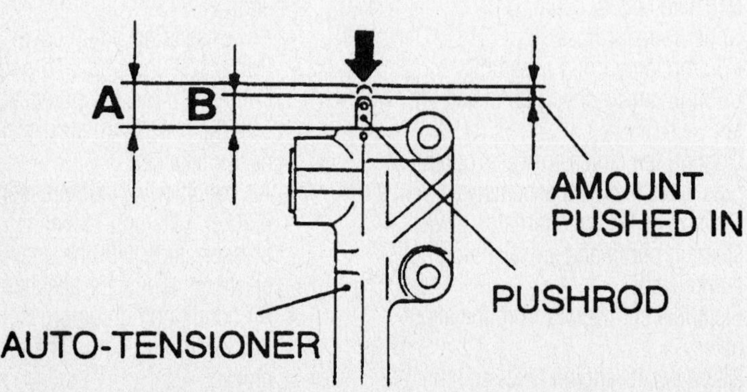

Inspecting the auto-tensioner movement—3.0L and 3.5L engines

crankshaft pulley) bolts and the bracket

- Accessory mount assembly-to-engine bolts and the mount assembly
- Upper timing belt cover assembly

4. Using the End Yoke Holder tool MD990767 and 2 Crankshaft Pulley Holder Pin tools MD998715, or equivalent to hold the crankshaft pulley, and a socket wrench, remove the crankshaft pulley bolt and the pulley.

- Lower timing belt cover

5. Rotate the crankshaft clockwise to align the timing marks to position the No. 1 cylinder at the Top Dead Center (TDC) of its compression stroke.
6. Use chalk to mark the rotating (clockwise) direction of the timing belt for reinstallation purposes.
7. Loosen the auto-tensioner pulley center bolt and remove the timing bolt.
8. Remove the auto-tensioner pulley and the auto-tensioner arm assembly.

To install:

9. Press the end of the auto-tensioner inward with 72–145 ft. lbs. (98–196 Nm) of force and measure the distance that the pushrod is pushed in. If the standard distance is not 0.04 in. (1mm), replace the auto-tensioner.
10. Position the auto-tensioner in a soft-jawed vise and SLOWLY compress the pushrod until the pushrod and housing holes align; then, install a setting pin to secure the auto-tensioner in the retracted position.
11. Align the camshaft and crankshaft TDC timing marks.
12. Install the timing belt (noting its rotational direction) so that there is no deflection between the sprockets and pulleys in the following manner:

- Crankshaft sprocket
- Idler pulley
- Left camshaft sprocket
- Water pump pulley
- Right camshaft sprocket
- Tension pulley

13. Turn the camshaft sprocket counterclockwise until the tension side of the timing belt is firmly stretched, then, recheck the timing marks.
14. Using the Tension Pulley Socket Wrench tool MD998767, or equivalent, push the tensioner pulley into the timing belt and secure the center bolt.
15. Using the Crankshaft Pulley Spacer tool MD998769, or equivalent, rotate the crankshaft ¼ turn counterclockwise, then, turn it again clockwise to align the timing marks.
16. Loosen the timing belt tensioner

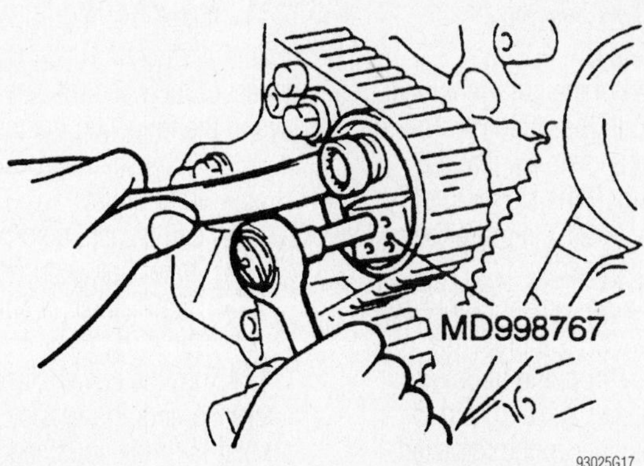

Adjusting the timing belt tensioner pulley—3.0L and 3.5L engines

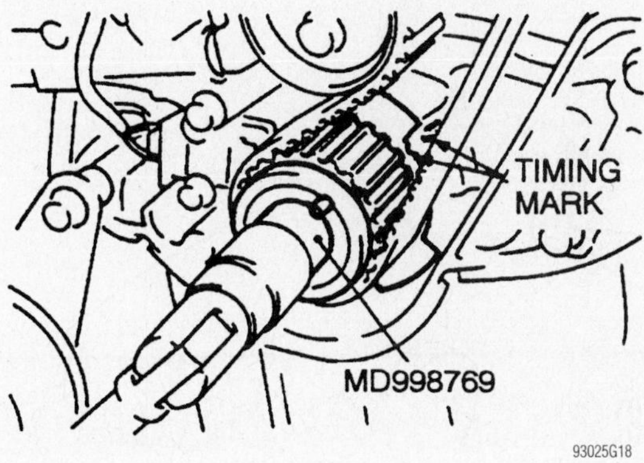

Using crankshaft spacer tool to rotate the crankshaft—3.0L and 3.5L engines

center bolt. Using the Tension Pulley Socket Wrench tool MD998767, or equivalent, and a torque wrench, apply 39 inch lbs. (4.4 Nm) pressure on the timing belt. Torque the tensioner pulley center bolt to 35 ft. lbs. (48 Nm).

17. Remove the setting pin from the auto-tensioner.
18. Rotate the crankshaft 2 complete revolutions and realign the timing marks. Then, wait for 5 minutes until the auto-tensioner pushrod extends to its standard value. If the standard value is not 0.15–0.20 in. (3.8–5.0mm), repeat the adjustment procedure. If the standard value is still not achieved, replace the auto-tensioner.
19. Install the lower timing belt cover and crankshaft pulley.
20. Using the End Yoke Holder tool MD990767 and 2 Crankshaft Pulley Holder Pin tools MD998715, or equivalent to hold the crankshaft pulley, and a socket torque wrench, torque the crankshaft pulley bolt to 134 ft. lbs. (181 Nm).
    - Install or connect the following:
    - Upper timing belt cover assembly
    - Remaining items by reversing the removal procedures
    - Negative battery cable
21. Refill the cooling system.

3.8L ENGINE

1. Before servicing the vehicle, refer to the precautions in the beginning of this section.
2. Drain the engine coolant.
3. Remove or disconnect the following:
    - Negative battery cable
    - Skid plate and under cover
    - Battery and battery tray
    - Air cleaner
    - Radiator shroud cover
    - Drive belt
    - Cooling fan and pulley
    - Drive belt auto tensioner
    - Accessory mount stay
    - Power steering pump. Unbolt and position aside; do not disconnect the fluid lines.
    - A/C compressor. Unbolt and position aside; do not disconnect the refrigerant lines.
    - Compressor bracket
    - Cooling fan bracket assembly
    - Accessory mount assembly
    - Timing belt upper cover
    - Crankshaft pulley, using special tools MB991800 and MB991802
    - Timing belt indicator bracket
    - Timing belt lower cover
    - Auto tensioner
4. Turn the crankshaft clockwise to

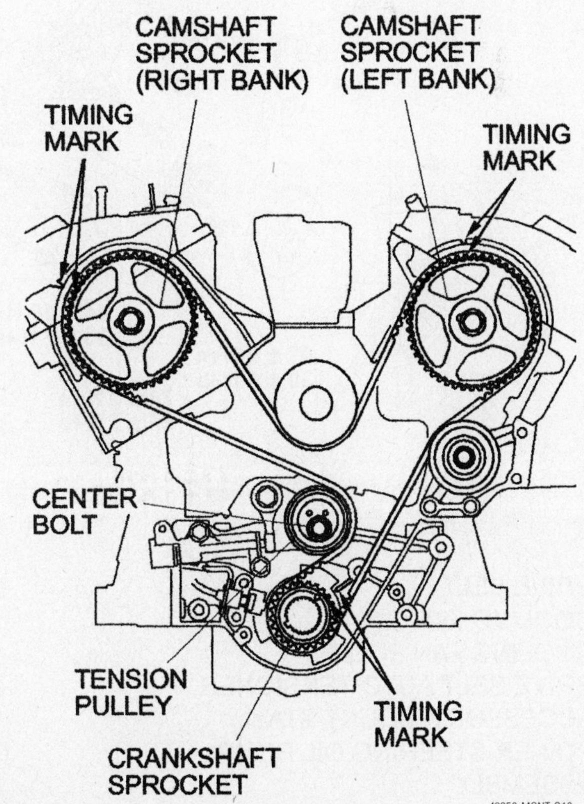

Installed view of the timing belt and alignment of the timing marks—3.8L engine

align the timing marks and set the No. 1 cylinder at Top Dead Center (TDC). If you are reusing the timing belt, mark the flat side of the belt with an arrow showing the clockwise direction.

5. Loosen the center bolt of the tension pulley, and then remove the timing belt.

6. Remove or disconnect the following:

- Tension pulley
- Tensioner arm assembly
- Shaft
- Idler pulley

To install:

7. Install or connect the following:

- Idler pulley
- Shaft
- Tensioner arm assembly
- Tension pulley

8. Press the end of the auto-tensioner inward with 72–145 ft. lbs. (98–196 Nm) of force and measure the distance that the pushrod is pushed in. If the standard distance is not 0.04 in. (1mm), replace the auto-tensioner.

9. Position the auto-tensioner in a soft-jawed vise and SLOWLY compress the pushrod until the pushrod and housing holes align; then, install a setting pin to secure the auto-tensioner in the retracted position.

10. Align the camshaft and crankshaft TDC timing marks.

11. Install the timing belt (noting its rotational direction) so that there is no deflection between the sprockets and pulleys in the following manner:

- Crankshaft sprocket
- Idler pulley
- Left camshaft sprocket
- Water pump pulley
- Right camshaft sprocket
- Tension pulley

12. Turn the camshaft sprocket counterclockwise until the tension side of the timing belt is firmly stretched, then, recheck the timing marks.

13. Using the Tension Pulley Socket Wrench tool MD998767, or equivalent,

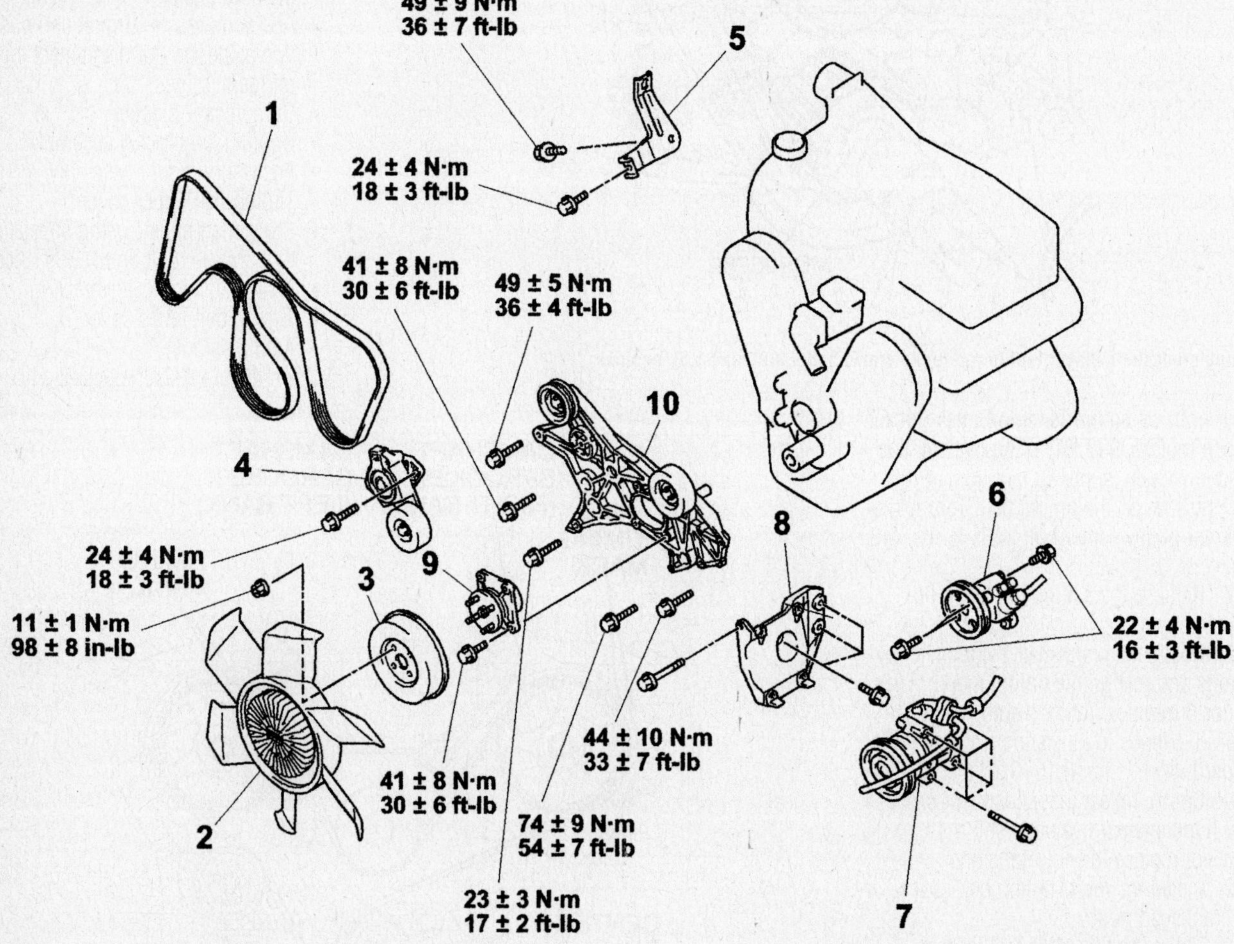

1. DRIVE BELT
2. COOLING FAN
3. COOLING FAN PULLEY
4. DRIVE BELT AUTO TENSIONER
5. ACCESSORY MOUNT STAY
6. POWER STEERING OIL PUMP ASSEMBLY
7. A/C COMPRESSOR ASSEMBLY
8. COMPRESSOR BRACKET
9. COOLING FAN BRACKET ASSEMBLY
10. ACCESSORY MOUNT ASSEMBLY

42356-MONT-G17

Exploded view of the components you need to remove for timing belt removal—3.8L engine

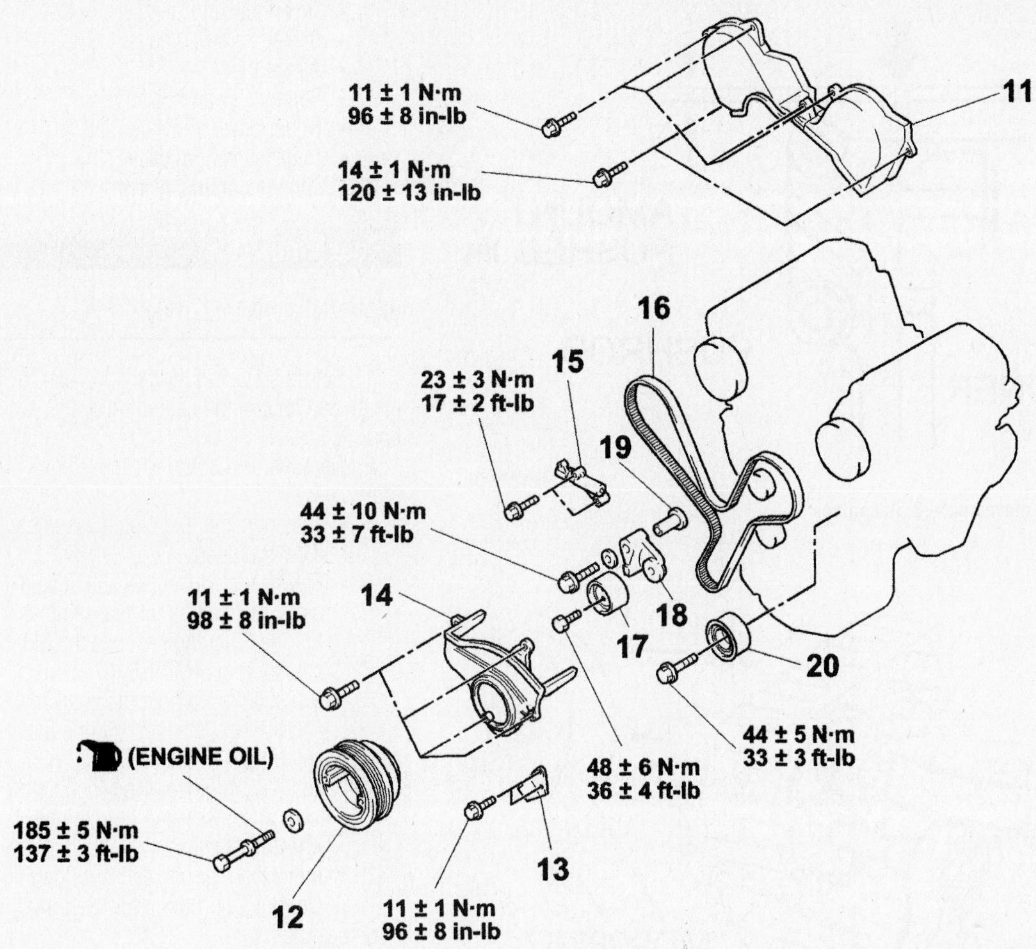

Exploded view of the timing belt and related components—3.8L engine

push the tensioner pulley into the timing belt and secure the center bolt.

14. Using the Crankshaft Pulley Spacer tool MD998769, or equivalent, rotate the crankshaft ¼ turn counterclockwise, then, turn it again clockwise to align the timing marks.

15. Loosen the timing belt tensioner center bolt. Using the Tension Pulley Socket Wrench tool MD998767, or equivalent, and a torque wrench, apply 39 inch lbs. (4.4 Nm) pressure on the timing belt. Torque the tensioner pulley center bolt to 35 ft. lbs. (48 Nm).

16. Remove the setting pin from the auto-tensioner.

17. Rotate the crankshaft 2 complete revolutions and realign the timing marks. Then, wait for 5 minutes until the auto-tensioner pushrod extends to its standard value. If the standard value is not 0.15–0.20 in. (3.8–5.0mm), repeat the adjustment procedure. If the standard value is still not achieved, replace the auto-tensioner.

18. Install the lower timing belt cover and crankshaft pulley.

19. Using the End Yoke Holder tool MD990767 and 2 Crankshaft Pulley Holder Pin tools MD998715, or equivalent to hold the crankshaft pulley, and a socket torque wrench, torque the crankshaft pulley bolt to 134 ft. lbs. (181 Nm).

- Install or connect the following:
- Timing belt upper cover
- Accessory mount assembly
- Cooling fan bracket assembly
- Compressor bracket
- A/C compressor
- Power steering pump
- Accessory mount stay
- Drive belt auto tensioner

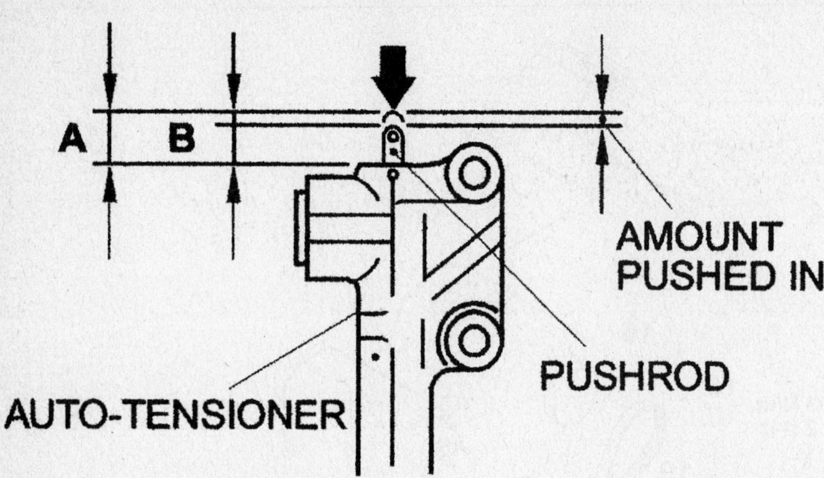

Inspecting the auto-tensioner movement—3.8L engines

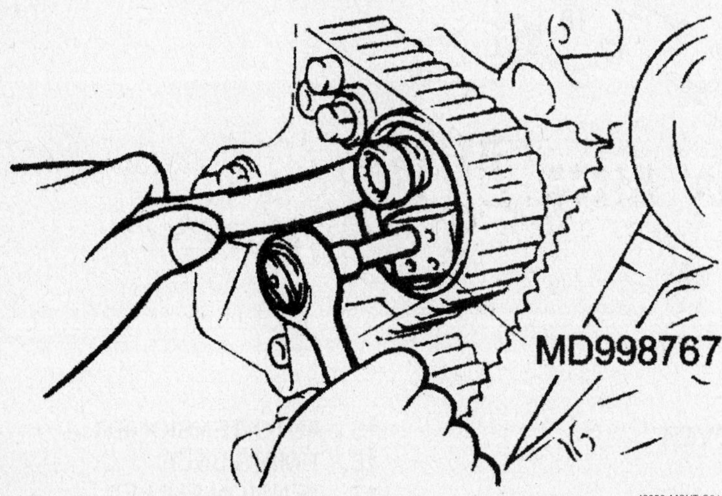

Adjusting the timing belt tensioner pulley—3.8L engines

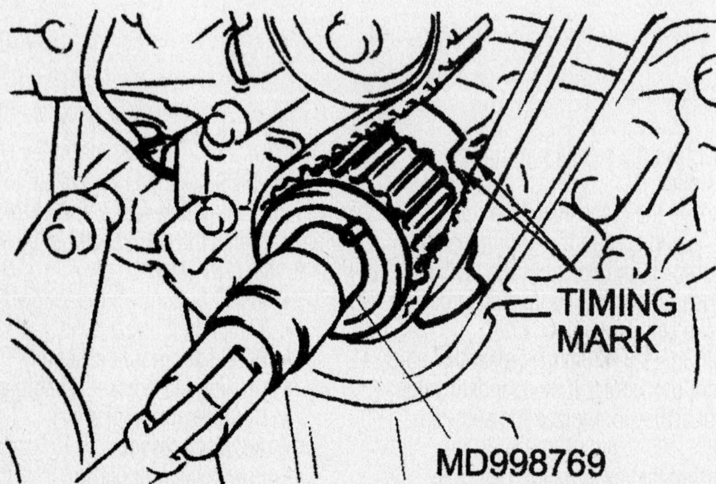

Using crankshaft spacer tool to rotate the crankshaft—3.0L and 3.5L engines

- Cooling fan pulley and fan
- Drive belt
- Radiator shroud cover
- Air cleaner
- Battery tray and battery
- Under cover and skid plate
- Negative battery cable

20. Refill the cooling system.

Rear Main Seal

REMOVAL & INSTALLATION

1. Before servicing the vehicle, refer to the precautions in the beginning of this section.
2. Remove or disconnect the following:
   - Transmission and clutch assembly, if so equipped.
   - Flywheel or driveplate and adapter plate, matchmark for reassembly. For the 3.0L engines, use the Mitsubishi tools (MB990767-01 and MIT308239) to hold the crankshaft and flywheel stationary while loosening the flywheel bolts. For 3.5L and 3.8L engines, use Mitsubishi tool (MD998781) to hold the flywheel in position.
3. Remove the rear oil seal as follows:
   a. Cut out a portion in the crankshaft oil seal lip.
   b. Cover the tip of a small prytool with a cloth and apply it to the cutout in the oil seal to pry the oil seal out.

⁂ CAUTION

Take care not to damage the crankshaft and oil seal case.

To install:

4. Inspect the sealing surface at the rear of the crankshaft. If a deep groove is worn into the surface, the crankshaft will have to be replaced. Coat the sealing lip of the seal with fresh, clean engine oil. Press the new seal into the case with a seal installing tool. The seal must be pressed in squarely until it bottoms in the case. It is necessary to use the proper tool (MD998718-01) to fit the seal into place.
5. Install or connect the following:
   - Rear plate
   - Transmission mounting plate
   - Flywheel or drive plate and adapter
   - Transmission and related components as necessary

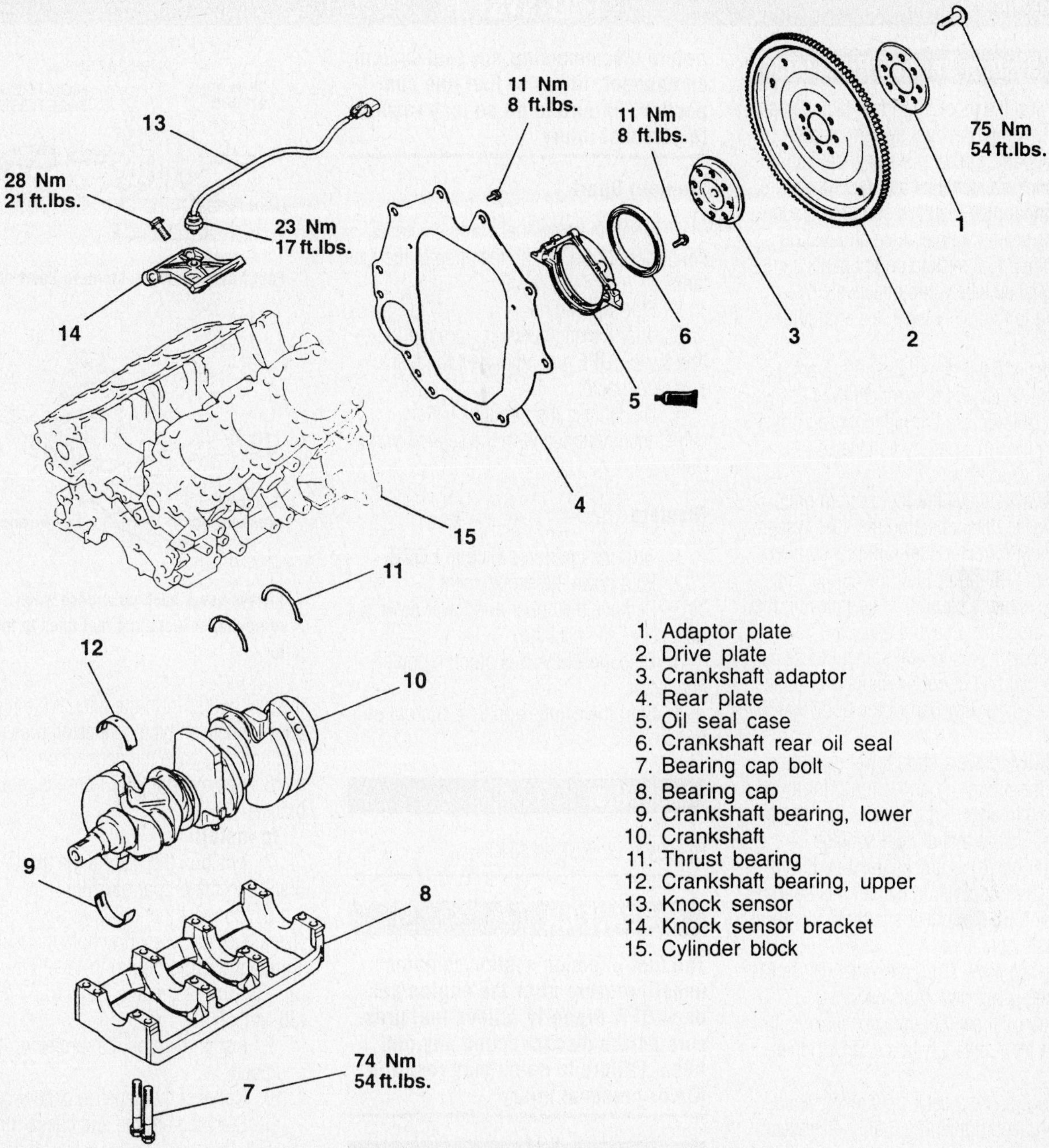

11 Nm
8 ft.lbs.
11 Nm
8 ft.lbs.
75 Nm
54 ft.lbs.
13
28 Nm
21 ft.lbs.
23 Nm
17 ft.lbs.
14
6
3
2
1
5
4
15
11
12
10
9
8
74 Nm
54 ft.lbs.
7
1. Adaptor plate
2. Drive plate
3. Crankshaft adaptor
4. Rear plate
5. Oil seal case
6. Crankshaft rear oil seal
7. Bearing cap bolt
8. Bearing cap
9. Crankshaft bearing, lower
10. Crankshaft
11. Thrust bearing
12. Crankshaft bearing, upper
13. Knock sensor
14. Knock sensor bracket
15. Cylinder block
7924UG22

Exploded view of the crankshaft, rear main seal and related components—3.5L engine shown, others similar

Piston and Ring

POSITION

IDENTIFICATION MARK "1R"
IDENTIFICATION MARK "2R"
SIZE MARK
No.1
No.2
9302AG13

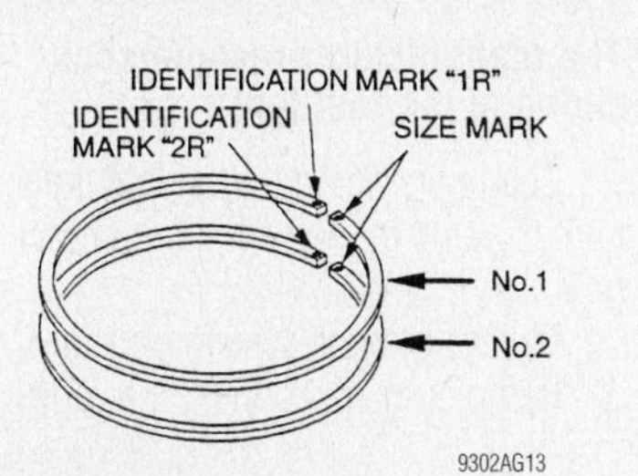

Piston ring identification

UPPER SIDE RAIL
SPACER
LOWER SIDE RAIL
9302AG14

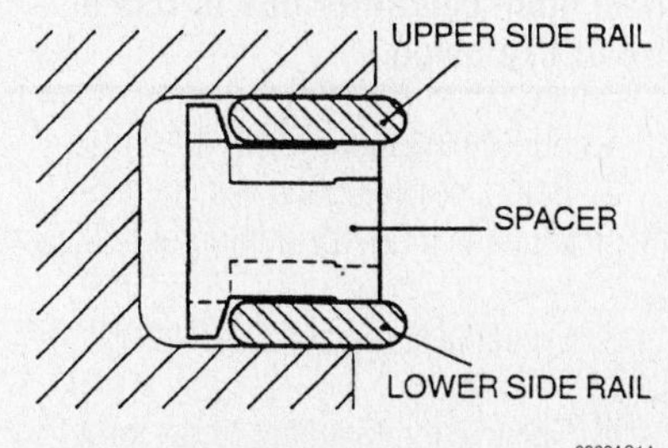

Oil ring identification

Upper side rail
No.1
Piston pin
No.2 ring gap and spacer gap
Lower side rail
7924AG49

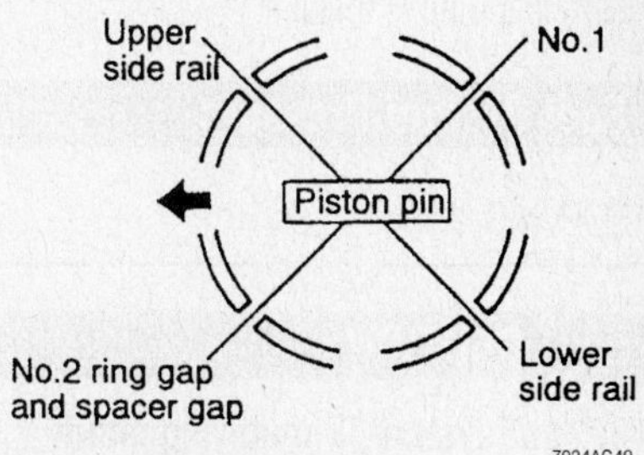

Piston ring end-gap spacing

FUEL SYSTEM

Fuel System Service Precautions

Safety is the most important factor when performing not only fuel system maintenance but any type of maintenance. Failure to conduct maintenance and repairs in a safe manner may result in serious personal injury or death. Maintenance and testing of the vehicle's fuel system components can be accomplished safely and effectively by adhering to the following rules and guidelines.

- To avoid the possibility of fire and personal injury, always disconnect the negative battery cable unless the repair or test procedure requires that battery voltage be applied.
- Always relieve the fuel system pressure prior to disconnecting any fuel system component (injector, fuel rail, pressure regulator, etc.), fitting or fuel line connection. Exercise extreme caution when relieving fuel system pressure, to avoid exposing your skin, face and eyes to fuel spray. Please be advised that fuel under pressure may penetrate the skin or any part of the body that it contacts.
- Always place a shop towel or cloth around the fitting or connection prior to loosening to absorb any excess fuel due to spillage. Ensure that all fuel spillage (should it occur) is quickly removed from engine surfaces. Ensure that all fuel soaked cloths or towels are deposited into a suitable waste container.
- Always keep a dry chemical (Class B) fire extinguisher near the work area.
- Do not allow fuel spray or fuel vapors to come into contact with a spark or open flame.
- Always use a back-up wrench when loosening and tightening fuel line connection fittings. This will prevent unnecessary stress and torsion to fuel line piping. Always follow the proper torque specifications.
- Always replace worn fuel fitting O-rings with new. Do not substitute fuel hose where fuel pipe is installed.

Fuel System Pressure

RELIEVING

⁂ CAUTION

The fuel system is under constant pressure, even with the engine off. This pressure must be relieved before disconnecting any fuel system component, fitting or fuel line connection. Failure to do so may result in personal injury.

Montero Sport

1. Disconnect the fuel pump electrical connector, located at the rear side of the fuel tank.
2. Start the engine.
3. After the engine stalls, turn the ignition switch **OFF** and reconnect the fuel pump connector.
4. Disconnect the negative battery cable, then continue with the service procedure.

Montero

1. Turn the ignition switch to **LOCK**.
2. Fold down the second seat.
3. Remove the upper and lower service hole cover and packing.
4. Disconnect the fuel pump module connector.
5. Start the engine and let it run out of fuel.

Fuel Filter

REMOVAL & INSTALLATION

⁂ CAUTION

The fuel injection system remains under pressure after the engine has been OFF. Properly relieve fuel pressure before disconnecting any fuel lines. Failure to do so may result in fire or personal injury.

⁂ CAUTION

Do not allow fuel spray or fuel vapors to come in contact with a spark or open flame. Keep a dry chemical fire extinguisher nearby. Never store fuel in an open container due to risk of fire or explosion.

1. Relieve the fuel system pressure.
2. Before servicing the vehicle, refer to the precautions in the beginning of this section.
3. Disconnect the negative battery cable.
4. Remove the fuel filter protector if equipped.
5. Using a back-up wrench disconnect the fuel line(s) from the filter. If the filter uses a push-on type connector, press the retainer to release the connection.
6. Remove the filter from the mounting bracket.

To install:

7. Position the filter to the mounting bracket in the proper direction.
8. Connect the fuel lines to the filter. Use a back-up wrench to hold the fuel filter. Torque the banjo bolt(s) to 18–25 ft. lbs. (25–35 Nm) or the line fitting to 27 ft. lbs. (36 Nm).
9. Install the fuel filter protector if equipped.
10. Connect the negative battery cable.
11. Start the engine and check for leaks.

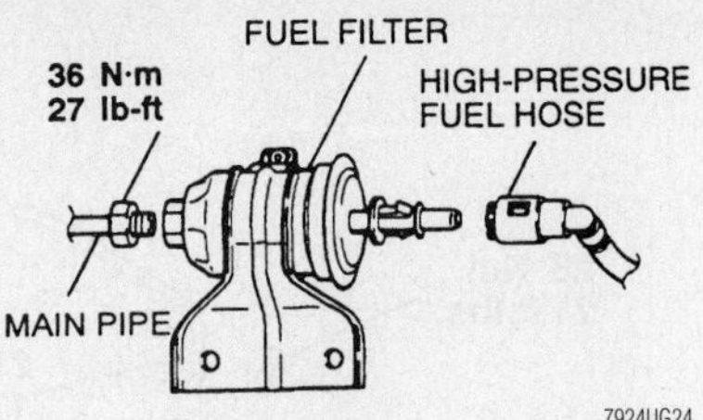

Fuel filter removal—Montero Sport shown

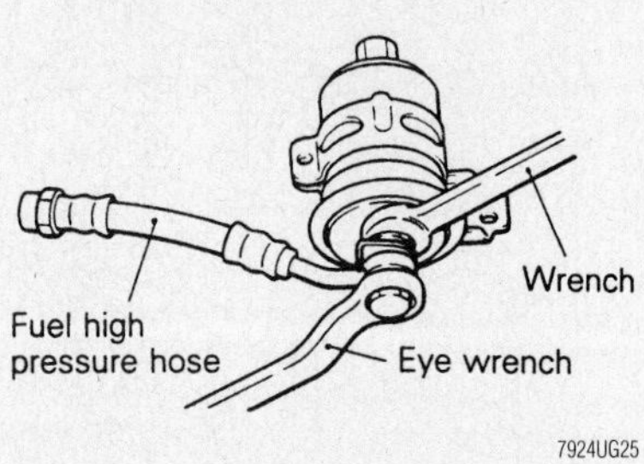

Always use a back-up wrench when removing or installing fuel lines to the filter

Fuel Pump

REMOVAL & INSTALLATION

Montero

➡The manufacturer recommends draining of the fuel tank.

1. Before servicing the vehicle, refer to the precautions in the beginning of this section.
2. Relieve the fuel system pressure.
3. Remove or disconnect the following:
   - Negative battery cable

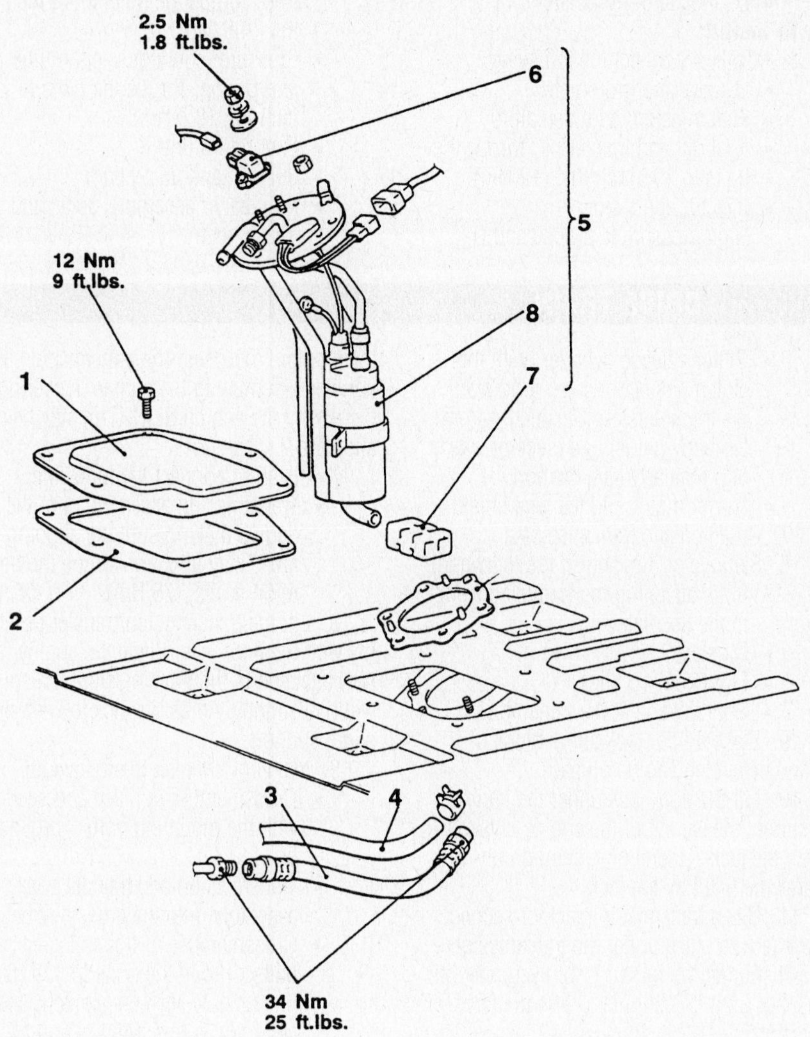

The fuel pump on the Montero is removed through the rear floor pan

⁂ CAUTION

The fuel injection system remains under pressure after the engine has been OFF. Properly relieve fuel pressure before disconnecting any fuel lines. Failure to do so may result in fire or personal injury. Do not allow fuel spray or fuel vapors to come in contact with a spark or open flame. Keep a dry chemical fire extinguisher nearby. Never store fuel in an open container due to risk of fire or explosion.

- Rear floor carpeting
- Fuel pump cover
- Fuel pump connector and the fuel hoses
- Fuel pump assembly

To install:

4. Install or connect the following:
   - Fuel pump assembly into the fuel tank. Torque the nuts to 24 inch lbs. (2.5 Nm).
   - Fuel lines and the fuel pump connector.
   - Fuel pump cover. Torque the bolts to 108 inch lbs. (12 Nm).
   - Rear floor carpeting.
   - Negative battery cable
5. Refill the fuel tank, if drained
6. Start the vehicle; check for leaks and proper operation.

Montero Sport

1. Before servicing the vehicle, refer to the precautions in the beginning of this section.
2. Properly relieve the fuel system pressure.
3. Remove the fuel tank drain plug and drain the fuel from the tank.

⁂ CAUTION

The fuel injection system remains under pressure after the engine has been OFF. Properly relieve fuel pressure before disconnecting any fuel lines. Failure to do so may result in fire or personal injury.

4. Remove or disconnect the following:
   - Negative battery cable

⁂ CAUTION

Wait at least 90 seconds after the negative battery cable is disconnected to prevent possible deployment of the air bag.

   - Fuel tank protector, if equipped
   - Fuel tank from the vehicle
   - Fuel pump retaining screws and the pump from the tank

To install:

5. Clean the seal area of the tank.
6. Install or connect the following:
   - New gasket
   - Fuel pump in the same position as originally installed.
   - Fuel pump retaining screws, Torque the nuts to 22 inch lbs. (2.5 Nm).
   - Fuel tank. Torque the bolts to 20 ft. lbs. (27 Nm).
   - Fuel tank drain plug and the fuel tank protector, if equipped
   - Negative battery cable
7. Refill the fuel tank and install the cap.
8. Check fuel system for leaks.

Fuel Injector

REMOVAL & INSTALLATION

3.0L and 3.5L Engines

1. Before servicing the vehicle, refer to the precautions in the beginning of this section.
2. Properly relieve the fuel system pressure.
3. Remove or disconnect the following:
   - Air cleaner assembly, ducts and air intake hose

- Intake manifold plenum
- Fuel return line
- Pressure regulator, vacuum line and O-ring
- High pressure fuel hose and O-ring
- Injector connector
- Fuel pipe and O-rings
- Fuel rails and insulators
- Fuel injectors and insulators
- O-rings and grommets

To install:

4. Install or connect the following:
   - O-rings and grommets
   - Fuel injectors and insulators
   - Fuel rail and insulators. Torque the bolts to 106 inch lbs. (12 Nm).
   - Injector connector
   - High pressure fuel hose and O-ring. Torque the bolts to 43 inch lbs. (4.9 Nm).
   - Pressure regulator, vacuum line and O-ring. Torque the bolts to 78 inch lbs. (8.8 Nm).
   - Fuel return line
   - Intake manifold plenum
   - Air cleaner assembly, ducts and air intake hose

DRIVE TRAIN

Transmission Assembly

REMOVAL & INSTALLATION

1. Before servicing the vehicle, refer to the precautions in the beginning of this section.
2. Drain the transmission fluid.
3. Remove or disconnect the following:
   - Negative battery cable
   - Transmission and transfer case shift lever assembly. On manual transmissions
   - Transfer case protector, if equipped
   - Front exhaust pipe from the 2 exhaust manifolds, then disconnect it from the intermediate pipe/catalytic converter (make certain to retain the bolts and nuts for reassembly).
   - Rear driveshaft at both the rear axle and the transfer case flanges. Matchmark for reassembly.
   - Front driveshaft from the front axle, by sliding it forward, plug the transfer case to prevent residual fluid leakage, on 4-wheel drive vehicles
   - Dust seal from the rear extension housing.
   - Ground cables
   - 4WD indicator light switch connector
   - Pulse generator connector
   - Speed Sensor Connector
   - Oxygen (O_2S) sensor connector
   - Back-up light switch connector
   - HI/LO detection switch connector
   - Center differential lock detection switch connection.

There may be others depending on the particular year, model and engine with which the vehicle came equipped.

4. Remove or disconnect the following:
   - Speedometer cable out of the transmission
5. On manual transmissions:
6. Remove or disconnect the following:
   - Clutch cylinder heat protector
   - Clutch release cylinder (with the clutch hose connected to it) from the transmission. Suspend it from the body by using a piece of wire or a similarly safe method.
   - Starter motor and the heat shield
7. On automatic transmissions:
8. Remove or disconnect the following:
   - Bolts attaching the torque converter to the flexplate
   - Dipstick
   - Fluid cooling lines
   - Shift linkage at the transmission
9. Place a floor jack and a block of wood below the engine oil pan.
10. Lift the floor jack under the engine just until the weight of the engine is taken onto the jack—the engine should only barely be lifted by the jack.
11. Use a transmission jack or second floor jack to place under the transmission. Don't support the transmission yet, only lift the jack until it is slightly below the transmission.
12. Remove or disconnect the following:
    - Left-hand and right-hand side transmission stays from the front of the transmission, if equipped
    - Bell housing lower cover
    - Transfer case mounting bracket, if equipped
13. Lift the floor jack up until the transmission is being slightly supported by it.
14. Remove or disconnect the following:
    - Transmission-to-crossmember bolts. Lift the jack about 1/4 in. (6mm) off of the crossmember support.
    - Crossmember from the vehicle
    - Transmission mounting blots. Pull the transmission away from the engine and lower it from the vehicle.

To install:

15. Lift the transmission and transfer assembly into position with the floor jack.
16. On the engine side, there are 2 centering locations. Be sure that the transmission mounting bolt holes are aligned with them before mounting the transmission and transfer assembly to the engine. Lowering the rear of the engine SLIGHTLY may help align the 2 assemblies.
17. Install or connect the following:
    - Transmission assembly onto the engine making sure the aligning areas stay aligned. Torque the bolts to 54 ft. lbs. (75 Nm).
18. Lift the transmission/transfer assembly with the floor jack. Since the engine is now attached to the transmission, it also will rise slightly. Adjust its jack to keep only slight support.
19. Install or connect the following:
    - Crossmember in place and secure with the mounting bolts. Torque the bolts to 47 ft. lbs. (65 Nm).
    - Transmission and transfer case assembly onto the crossmember
    - Crossmember-to-transmission bolts. Torque the bolts to 15–18 ft. lbs. (20–24 Nm) on Montero sport and to 36 ft. lbs. (49 Nm) on Montero.
    - Mounting bracket back onto the transfer case, if equipped
    - Bell housing lower cover
    - Left-hand and right-hand side transmission stays
    - Starter motor and heat shield
    - Clutch release cylinder
    - Speedometer cable into the transmission and secure it there with the retaining ring
    - Center differential lock detection switch connection
    - HI/LO detection switch connector
    - Back-up light switch connector
    - O_2 sensor connector
    - Speed Sensor Connector
    - Pulse generator connector
    - 4WD indicator light switch connector
    - Ground cables
    - Flexplate-to-torque converter bolts. Torque the bolts to 25–30 ft. lbs. (35–42 Nm). On automatic transmissions

- Dipstick tube
- Shift linkage
- Fluid cooler lines. Torque the line fittings to 32 ft. lbs. (44 Nm), on automatic transmissions

20. Tap the dust seal guard back onto the rear extension housing with a rubber or plastic mallet.
21. Install or connect the following:
    - Front driveshaft into the transfer case, then attach it to the front differential
    - Rear driveshaft, make certain that the matchmarks line up.
    - Front exhaust pipe to the catalytic converter and the exhaust manifolds
    - Transfer case protector, if equipped
    - Transmission and transfer case shift lever assembly, on manual transmissions
    - Negative battery cable to the battery
22. Refill the transmission and transfer case with oil.
23. Start the vehicle and check for any leaks.

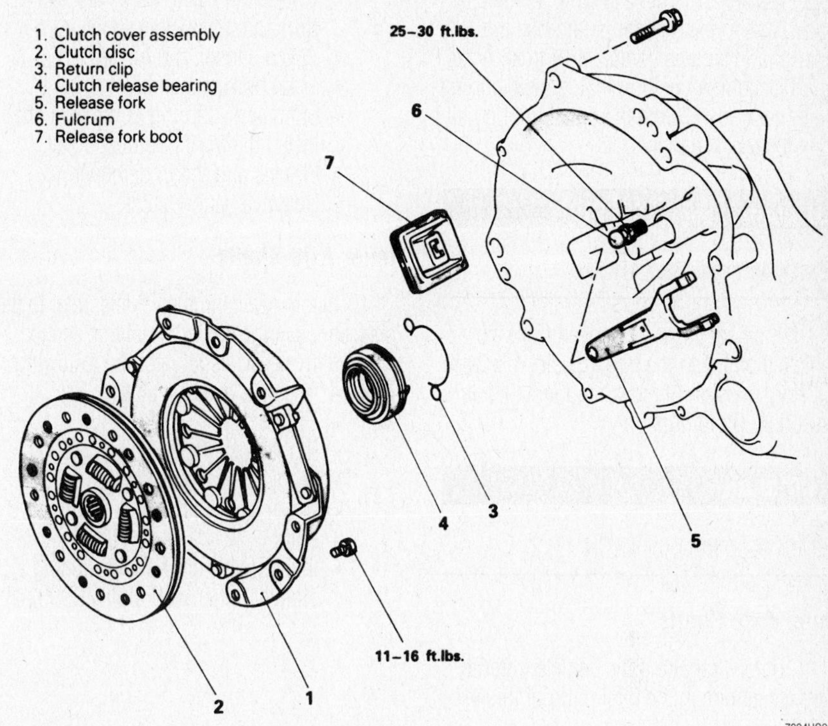

Exploded view of the typical clutch assembly components

Clutch

REMOVAL & INSTALLATION

1. Before servicing the vehicle, refer to the precautions in the beginning of this section.
2. Remove or disconnect the following:
    - Negative battery cable
    - Transmission assembly
3. Insert a suitable tool in the flywheel pilot bearing hole to keep the clutch disc from falling off. Loosen the clutch cover retainer bolts gradually in a crisscross fashion.
4. Remove or disconnect the following:
    - Clutch cover and disc
5. Check the release bearing for scorching, damage or strange noise. Replace, if necessary.
6. Inspect the flywheel surface for heat cracks or scoring. Reface or replace the flywheel as required.

To install:

7. Apply high temperature grease to the clutch disc splines, input shaft, contact points of the release fork and inside diameter of the release bearing.

➡Do not allow oil or grease to contact the clutch facing and pressure plate.

8. Install or connect the following:
    - Flywheel, align using a suitable tool

➡When installing the clutch disc, be sure that the surface having the manufacturer's stamped mark is on the pressure plate side.

- Clutch cover with the dowel pin holes in alignment with the dowel pins in the flywheel and tighten the bolt gradually in a crisscross fashion. Torque the bolts to 14 ft. lbs. (19 Nm).
- Transmission assembly
- Negative battery cable

9. Road test the vehicle for proper operation.

Hydraulic Clutch System

BLEEDING

✻✻ WARNING

When bleeding, keep the facial area well away from the slave cylinder and protect all painted surfaces from fluid contact. Brake fluid will damage painted surfaces and could cause physical injury.

1. Fill the clutch master cylinder with fresh DOT 3 brake fluid.
2. Have a helper sit in the vehicle.
3. Remove the bleeder screw cap.
4. If the system is empty, the most efficient way to get fluid down to the cylinder is:
    a. Loosen the bleeder about ½–¾ turn
    b. Place a finger firmly over the bleeder
    c. Have a helper pump the brakes slowly until fluid pressure is felt at the bleeder
    d. Once fluid is at the bleeder, close it before the pedal is released.

➡If the pedal is pumped rapidly, the fluid will churn and create small air bubbles, which are difficult and time consuming to remove from the system. These air bubbles will eventually congregate and will result in a spongy pedal.

5. Once fluid has been pumped to the slave cylinder:
    - Open the bleeder screw
    - Have a helper depress the clutch pedal
    - Lock the bleeder and have the helper release the pedal
    - Wait 15 seconds and repeat the procedure (including the 15 second wait) until no air bubbles flow from the bleeder.

Remember to close the bleeder before the pedal is released. If the bleeder is left open when the pedal is released, air will be induced into the system.

6. If a helper is not available, connect a small hose to the bleeder, submerge the other end in a clean container of fresh brake fluid placed in a position that is visible from the driver's seat. Pump the pedal until no air comes out of the tube.

Transfer Case Assembly

REMOVAL & INSTALLATION

The transfer case is removed from the vehicle along with the transmission. Refer to the Transmission Removal and Installation procedure for information.

Halfshaft

REMOVAL & INSTALLATION

Outer Axle Shafts

1. Before servicing the vehicle, refer to the precautions in the beginning of this section.
2. Remove or disconnect the following:
   - Negative battery cable
   - Undercover
   - Wheels
   - Hub cover dust cap
   - Snapring from the inside of the hub and the shim
   - Front brake caliper assembly and support with mechanics wire
   - Speed sensor, if equipped with ABS
   - Tie rod from the steering knuckle assembly
   - Upper and lower ball joints from the steering knuckle assembly
   - Front hub/knuckle assembly with the inner and outer bearings intact
3. On the left side, pull the halfshaft from the differential carrier.
4. For the right side, remove the fasteners and the halfshaft from the vehicle.

To install:

5. Install or connect the following:
   - New circlip, on the left side halfshaft
   - Inner shaft. Torque the nuts to 36–43 ft. lbs. (49–59 Nm), on the right side halfshaft.
   - Front hub/knuckle and bearing assembly
   - Upper ball joint to the knuckle. Torque the nut to 54 ft. lbs. (74 Nm).
   - Lower ball joint to knuckle. Torque the nut to 108 ft. lbs. (147 Nm).
   - New cotter pins
   - Tie rod end to the steering knuckle. Torque the nut to 33 ft. lbs. (44 Nm) and a new cotter pin.
   - Speed sensor, if removed
   - Front brake assembly
   - Shim and snapring to the axle shaft. Install the front hub dust cover
   - Wheels and the undercover
   - Negative battery cable

Inner Axle Shafts

1. Before servicing the vehicle, refer to the precautions in the beginning of this section.
2. Remove or disconnect the following:
   - Negative battery cable
   - Undercover
   - Right side wheel
   - Right outer halfshaft
   - Lower shock absorber mounting bolts
   - Inner shaft from housing, using a slide hammer with tool MB990241

To install:

3. Install or connect the following:
   - New circlip on the inner halfshaft
   - Inner shaft into the housing, drive the axle into position.
   - Lower shock absorber mounting bolts. Torque the bolts to 65–76 ft. lbs. (88–103 Nm).
   - Right halfshaft assembly
   - Undercover
   - Wheel
   - Negative battery cable

CV-Joints

OVERHAUL

1. Before servicing the vehicle, refer to the precautions in the beginning of this section.

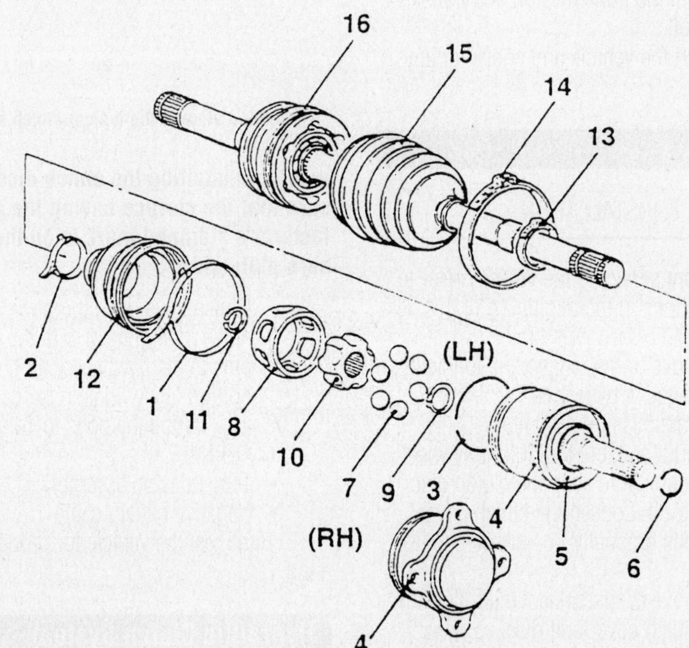

CV-Joint, exploded view

2. Remove or disconnect the following:
 - Front wheel
 - Driveshaft from the car
 - Small and larger band
 - Circlip
 - Double Offset Joint (DOJ) outer race
 - Dust cover
 - Circlip
 - Balls from the cage
 - Cage from the inner race. Turn the cage so that the projections of the inner race align with the recesses of the cage.
 - Snapring from the shaft
 - DOJ inner race
 - Slide the boot off
 - Birfield Joint (BJ) small and larger bands
 - BJ boot
 - BJ assembly

To install:

3. Check the shaft and splines for damage or wear. Inspect the cage, race and balls for any sign of corrosion, wear, cracking or damage. Clean all the parts thoroughly and air dry them completely before installation. Any remaining cleaning solvent can dissolve the lubricating grease.

4. Tool MB991561 can be used to crimp the bands in place.

5. Install or connect the following:
 - BJ assembly
 - BJ boot, slid the small end of the boot until only one shaft groove cone be seen.
 - BJ small band, crimp the band. Fill the BJ boot with 4.6 oz (130 g) of grease.
 - BJ larger band, crimp the band
 - DOJ small band and boot, fill the boot with grease
 - DOJ cage onto the driveshaft so that the smaller diameter side is installed first.
 - Circlip
 - DOJ inner race and new snap ring, apply grease to the inner race
 - Balls into the cage, grease to the ball areas of the cage and race
 - Outer race, fill the outer race about⅓ full of grease.
 - Dust cover
 - Circlip
 - Large boot band, release the air from the boot then crimp
 - Driveshaft into the car

Automatic Locking Hubs

REMOVAL & INSTALLATION

1. Place the locking hub in the free position. To do this, shift the transfer shift lever to the 2H position, then move the vehicle 4–7 ft. (1–2 m) backwards.

2. Before servicing the vehicle, refer to the precautions in the beginning of this section.

3. Remove or disconnect the following:
 - Front wheels of the vehicle
 - Hub cover
 - Snapring from the axle shaft
 - Shim
 - Drive flange
 - Front brake assembly
 - Speed sensor, if equipped
 - Lock washer
 - Lock nut
 - Front hub assembly

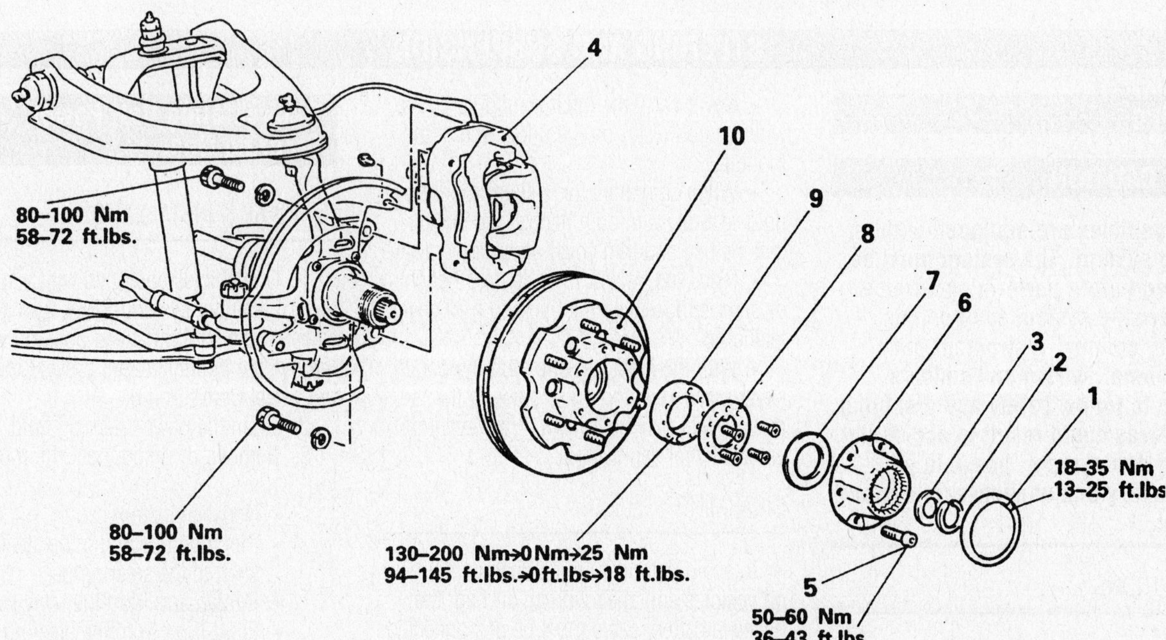

Axle hub and locking hub removal and installation—automatic hubs

To install:

4. Install or connect the following:
   - Front hub assembly
   - Lock nut. Torque the nut as follows:

a. Step 1: Torque the nut to 119 ft. lbs. (162 Nm).

b. Step 2: Loosen to 0 ft. lbs. (0 Nm).

c. Step 3: Torque the nut to 18 ft lbs. (25 Nm).

d. Step 4: Loosen the nut 30–40 degrees.

5. Install or connect the following:
   - Lock washer
   - Speed sensor, if equipped
   - Front brake assembly. Torque the bolts to 65 ft. lbs. (88 Nm).
   - Drive flange. Torque the bolts to 36–43 ft. lbs. (49–59 Nm).
   - Shim
   - Snapring to the axle shaft
   - Hub cover
   - Front wheels of the vehicle

Axle Shaft, Bearing and Seal

REMOVAL & INSTALLATION

Rear

1. Before servicing the vehicle, refer to the precautions in the beginning of this section.
2. Remove or disconnect the following:
   - Wheel assembly
   - Brake line
   - Rear brake assembly
   - Parking brake cable and assembly
   - Axle shaft
   - Snapring
   - Retainer
   - Bearing inner race, inner and outer
   - Oil seal
   - Bearing case
   - O-ring
   - Oil seal

To install:

3. Install or connect the following:
   - Bearing case
   - Bearing inner race, outer. Press the bearing into the bearing case.
   - Oil seal. Press the seal using tools MB990932 and MB990938.
   - Bearing inner race, inner. Press the bearing into the bearing case.
   - Axle shaft, Place into bearing case
   - Retainer, press onto shaft
   - Snapring
   - Oil seal and O-ring into axle shaft
4. Axle shaft assembly into axle.
   - Parking brake cable end
   - Parking brake cable attaching bolt
   - Rear brake assembly
   - Brake line
   - Wheel assembly

Pinion Seal

REMOVAL & INSTALLATION

1. Before servicing the vehicle, refer to the precautions in the beginning of this section.
2. Remove or disconnect the following:
   - Driveshaft, matchmark for reassembly
3. Check the turning torque of the pinion before proceeding. It should be 2.6–4.5 inch lbs. (0.4–0.5 Nm). This is the torque that must be reached during installation of the pinion nut.
   - Pinion nut and washer using a suitable pinion flange holding tool
   - Companion flange from the drive pinion
4. Pry the pinion seal out of the differential carrier.

To install:

5. Clean and inspect the sealing surface of the housing.
6. Install or connect the following:
   - New seal into the housing until the flange on the seal is flush with the carrier. Using a seal driver.
7. With the seal installed, the pinion bearing preload must be set.
   - Pinion nut (a new self-locking pinion nut must be used) while holding the flange, until the turning torque is the same as before removal. The final pinion nut torque must be between 137–181 ft. lbs. (190–250 Nm).
   - Driveshaft, align the matchmarks
8. Check the level of the differential lubricant when finished.

STEERING AND SUSPENSION

Air Bag

⁂ CAUTION

Some vehicles are equipped with an air bag system. The system must be disabled before performing service on or around system components, steering column, instrument panel components, wiring and sensors. Failure to follow safety and disabling procedures could result in accidental air bag deployment, possible personal injury and unnecessary system repairs.

PRECAUTIONS

Several precautions must be observed when handling the inflator module to avoid accidental deployment and possible personal injury.

- Never carry the inflator module by the wires or connector on the underside of the module.
- When carrying a live inflator module, hold securely with both hands, and ensure that the bag and trim cover are pointed away.
- Place the inflator module on a bench or other surface with the bag and trim cover facing up.
- With the inflator module on the bench, never place anything on or close to the module that may be thrown in the event of an accidental deployment.

DISARMING

To avoid personal injury when working on vehicles equipped with an air bag, the negative battery cable must be disconnected and at least 60 seconds must elapse before working on the system. Failure to do so may result in deployment of the air bag.

Recirculating Ball Power Steering Gear

REMOVAL & INSTALLATION

1. On vehicles equipped with a supplemental restraint system (SRS), turn the front wheel to the straight ahead position and remove the ignition key to prevent the steering wheel from turning.
2. Drain the power steering fluid.
3. Remove or disconnect the following:
   - Negative battery cable
   - Pinch bolt securing the steering shaft to the steering gear
   - Pitman arm from the relay rod
   - Fluid lines from the steering gear
   - Mounting bolts securing the gear to the frame rail and steering gear

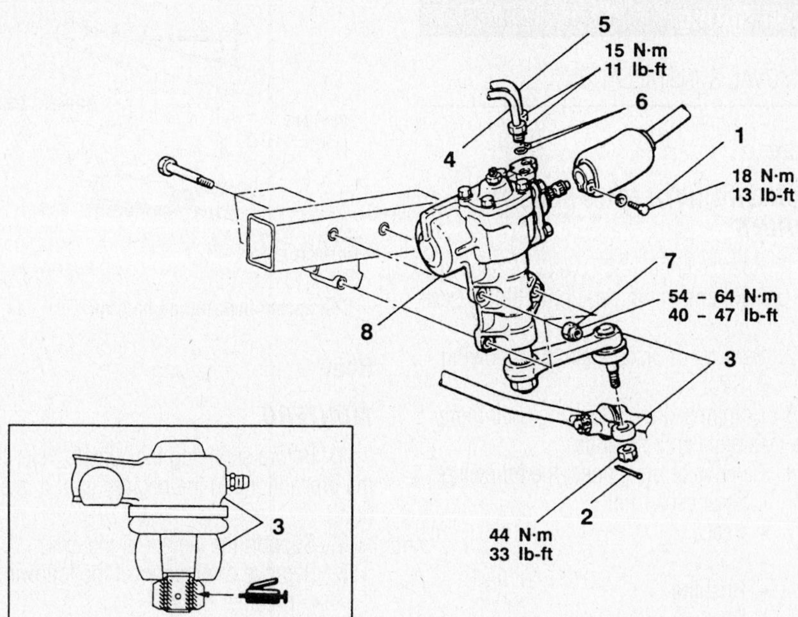

1. CONNECTING BOLT FOR STEERING GEAR BOX AND STEERING SHAFT
2. COTTER PIN
3. CONNECTION FOR PITMAN ARM AND RELAY ROD
4. PRESSURE TUBE
5. RETURN TUBE
6. O-RING
7. SELF-LOCKING NUT
8. POWER STEERING GEAR BOX

7924UG33

Exploded view of a typical power steering gear mounting

To install:

4. Install or connect the following:
   - Steering gear on the frame rail. Torque the nuts to 40–47 ft. lbs. (54–64 Nm).
   - Fluid lines to the steering gear use a new O-rings. Torque the fittings to 11 ft. lbs. (15 Nm).
   - Relay rod on the Pitman arm. Torque the nut to 33 ft. lbs. (44 Nm).
   - Steering shaft on the steering gear. Torque the bolt to 13 ft. lbs. (18 Nm).
   - Negative battery cable
5. Refill and bleed the power steering system.

Rack and Pinion Steering Gear

REMOVAL & INSTALLATION

1. On vehicles equipped with a supplemental restraint system (SRS), turn the front wheel to the straight ahead position and remove the ignition key to prevent the steering wheel from turning.
2. Drain the power steering fluid.
3. Remove or disconnect the following:
   - Negative battery cable
   - Engine under cover
   - Tie rod ends
   - Steering hoses
   - Left side differential mount bracket
   - Intermediate shaft-to-gear box bolt
   - Gear mounting clamps and gear
4. Installation is the reverse of removal. Observe the following torques:
   - Mounting clamp bolts: 51 ft. lbs. (69Nm)
   - Tie rod end ball stud nuts: 29 ft. lbs. (39Nm)
   - Intermediate shaft pinch bolt: 13 ft. lbs. (18Nm)

Shock Absorber

REMOVAL & INSTALLATION

Front

1. Before servicing the vehicle, refer to the precautions in the beginning of this section.
2. Remove the upper shock mounting nut, washer and bushing.
3. Remove the lower mounting bolts.
4. Remove the shock absorber.

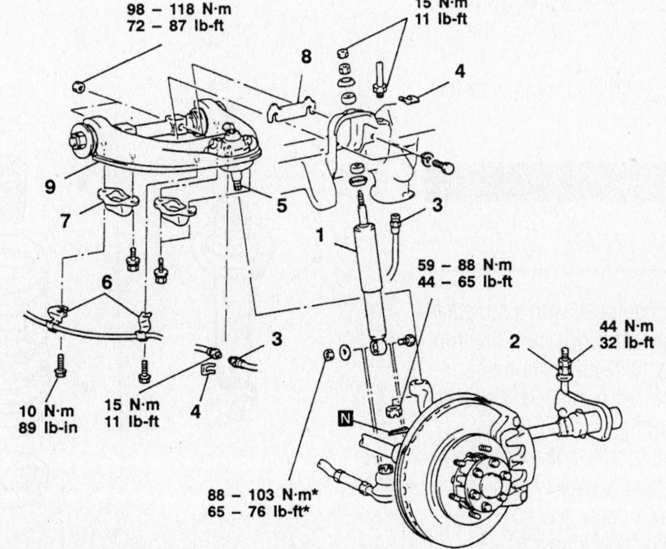

1. SHOCK ABSORBER
- BUMP STOPPER AND BUMP STOPPER BRACKET CLEARANCE ADJUSTMENT
2. REAR ANCHOR ARM ADJUSTING NUT
3. BRAKE HOSE CONNECTION
4. HOSE CLIP
5. UPPER ARM BALL JOINT CONNECTION
6. SPEED SENSOR BRACKET <VEHICLES WITH ABS>
7. REBOUND STOPPER
8. SHIMS
9. UPPER ARM
10. UPPER ARM BALL JOINT ASSEMBLY

Caution
***: Indicates parts which should be temporarily tightened, and then fully tightened with the vehicle on the ground in an unladen condition.**

7924UG34

Common shock absorber and upper control arm components

To install:

➡If the shock absorber has a white paint mark on the lower end, be sure the mark faces the outside of the vehicle when installed.

5. Install the shock absorber. Torque the lower nut to 65–76 ft. lbs. (88–103 Nm) and the upper nut to 11 ft. lbs. (15 Nm).
6. Test drive the vehicle and check the alignment.

Rear

1. Before servicing the vehicle, refer to the precautions in the beginning of this section.
2. Support the rear axle assembly with a hydraulic floor jack, so that the shock absorber may be removed.
3. Remove the upper and lower mounting nuts and bolts that attach the shock to the frame and bracket.
4. Remove the shock absorber from the vehicle.

To install:

5. Install the shock absorber
   - Lower bolt: 159–181 ft. lbs. (216–245 Nm) on the Montero Sport and 113 ft. lbs. (152 Nm) on the Montero
   - Upper mounting nut: 16 ft. lbs. (22 Nm) on the Montero Sport and 33 ft. lbs. (44 Nm) on the Montero
6. Remove the floor jack from under the axle assembly.

Strut

REMOVAL & INSTALLATION

1. On vehicles equipped with a supplemental restraint system (SRS), turn the front wheel to the straight ahead position and remove the ignition key to prevent the steering wheel from turning.
2. Remove or disconnect the following:
   - Negative battery cable
   - Upper control arm
   - Battery and battery tray
   - A/C condenser
   - Air cleaner
   - Wheel
   - Upper mounting nuts
   - Lower mounting bolt/nut
3. Installation is the reverse of removal. Torque the upper mounting nuts to 32 ft. lbs. (44 Nm); the lower mount nut to 119 ft. lbs. (162 Nm).

Coil Spring

REMOVAL & INSTALLATION

Front

MONTERO WITH MACPHERSON STRUTS

1. Before servicing the vehicle, refer to the precautions in the beginning of this section.
2. Remove or disconnect the following:
   - Strut
3. Compress the coil spring until there is a clearance on both ends
4. Remove or disconnect the following:
   - Strut center nut
   - Seat
   - Collar
   - Bushing
   - Bracket
   - Upper pad
   - Cup
   - Helper rubber
   - Spring
   - Lower pad
5. Installation is the reverse of removal. Torque the center nut to 17 ft. lbs. (22 Nm).

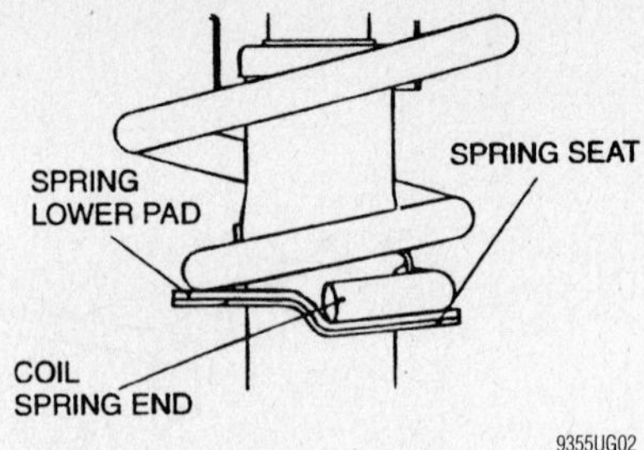

Coil spring installation on strut

Rear

MONTERO

1. Before servicing the vehicle, refer to the precautions in the beginning of this section.
2. Support the weight of the axle.
3. Remove or disconnect the following:
   - Breather hose
   - Parking brake cable attaching bolt
   - ABS speed sensor attaching bolt
   - Brake hose connection
   - Lower shock mounting bolt
   - Bolt that attaches the lateral rod to the body
   - Stabilizer bar
4. Lower the axle and remove the coil spring and seat

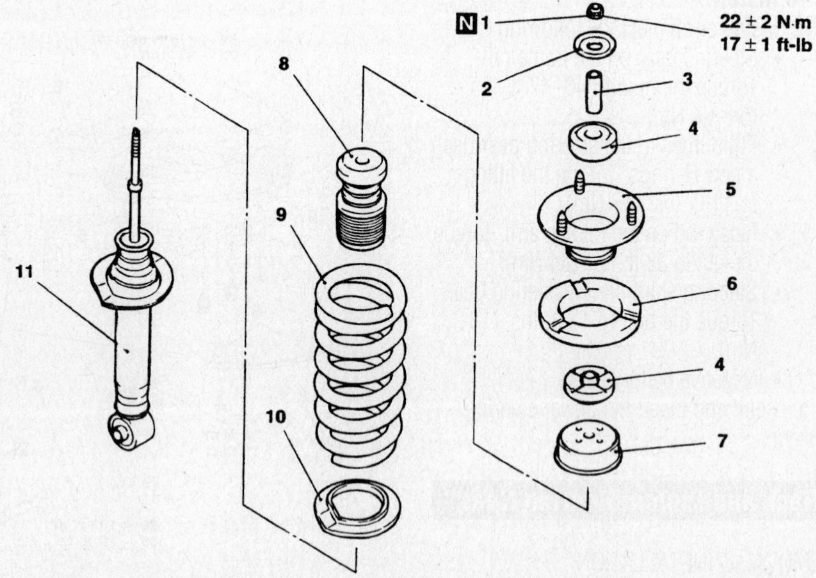

Strut exploded view

5. Installation is the reverse of removal. Torque the lateral rod-to-body bolt to 159–181 ft. lbs. (216–245 Nm).

MONTERO SPORT

1. Before servicing the vehicle, refer to the precautions in the beginning of this section.
2. Remove or disconnect the following:
   - Shock absorber lower bolt
   - Lower arm mounting bolt
   - Coil spring
3. Installation is the reverse of removal. Torque the lower arm bolt to 113 ft. lbs. (152 Nm).

Torsion Bars

REMOVAL & INSTALLATION

1. Before servicing the vehicle, refer to the precautions in the beginning of this section.
2. Support the lower arm with a jack.
3. Remove or disconnect the following:
   - Heat protector, right side only
   - Bump stopper
   - Anchor adjustment nut and arm assembly
   - Anchor collar
   - Torsion bar
   - Dust covers
   - Heat covers, right side only

To install:

4. Install or connect the following:
   - Heat covers, right side only
   - Dust covers
   - Torsion bar
   - Anchor collar
   - Anchor adjustment nut and arm assembly. Torque the nut to 32 ft. lbs. (44 Nm).
   - Heat protector, right side only

Upper Ball Joint

REMOVAL & INSTALLATION

1. Before servicing the vehicle, refer to the precautions in the beginning of this section.
2. Remove the front wheel.
3. Support the lower control arm.
4. Remove the upper ball joint from the steering knuckle.
5. Remove the ball joint from the upper control arm.

To install:

6. Install the ball joint in the upper control arm. Tighten the bolts to 22 ft. lbs. (30 Nm).

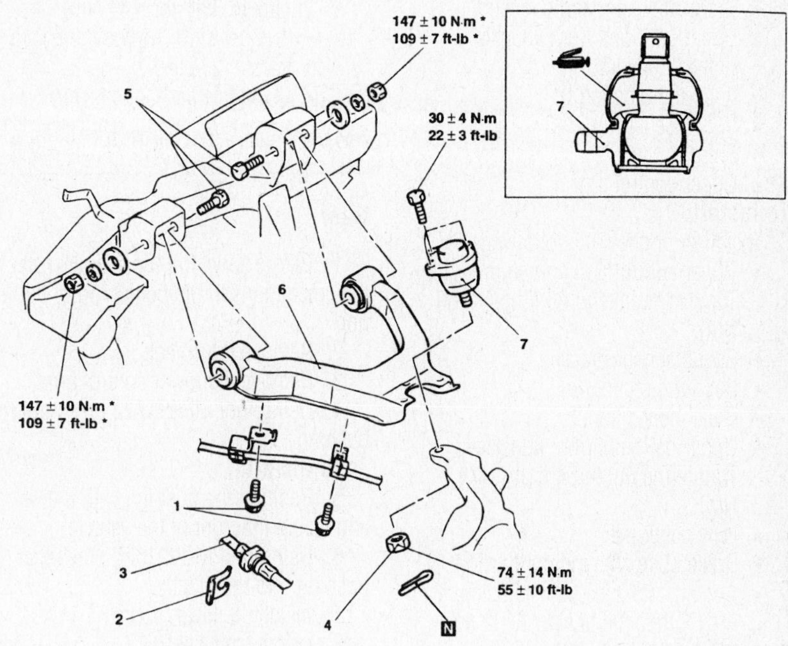

Exploded view of the upper ball joint and related components

7. Install the ball joint stud to the steering knuckle. Torque the nut to 54 ft. lbs. (74 Nm).
8. Install a new cotter pin.
9. Install the front wheel.
10. Grease the upper ball joint and all other suspension components with a grease fitting.

Lower Ball Joint

REMOVAL & INSTALLATION

1. Before servicing the vehicle, refer to the precautions in the beginning of this section.
2. Apply upward pressure to the lower control arm with a jack or an adjustable stand.

CAUTION

Do not disconnect the lower ball joint stud from the steering knuckle unless the lower control arm has a stand or a jack under it.

3. Remove the ball joint stud nut/stud from the steering knuckle.
4. Remove the ball joint retaining nuts/bolts and the ball joint from the arm

To install:

5. Install the lower ball joint on the control arm. Torque the ball joint retaining nuts/bolts to 60 ft. lbs. (81 Nm) for the Montero Sport, or to 70 ft. lbs. (95Nm) for Montero models.
6. Install the ball stud to the knuckle. Torque the nut to 108 ft. lbs. (147 Nm) and a new cotter pin.
7. Lubricate the ball joint with a grease gun.
8. Check and adjust the alignment if necessary.

Upper Control Arm

REMOVAL & INSTALLATION

1. Before servicing the vehicle, refer to the precautions in the beginning of this section.
2. Remove or disconnect the following:

- Wheel assembly
- Bumper stop
- Anchor arm assembly adjustment nut
- Brake hose clip and connection
- Upper ball joint from knuckle
- Brake hose clip
- Rebound stopper(s)
- Speed sensor bracket
- Shim
- Upper control arm

To install:

3. Install or connect the following:
   - Upper control arm and shim. Torque the nuts to 80 ft. lbs. (108 Nm).
   - Speed sensor bracket
   - Rebound stopper(s)
   - Brake hose clip
   - Upper ball joint from knuckle. Torque the nut to 54 ft. lbs. (74 Nm).
   - New cotter pin
   - Brake hose clip and connection
   - Anchor arm assembly adjustment nut. Torque the nuts to 33 ft. lbs. (44 Nm).
   - Bumper stop
   - Wheel assembly

Lower Control Arm

REMOVAL & INSTALLATION

1. Before servicing the vehicle, refer to the precautions in the beginning of this section.
2. Remove or disconnect the following:
   - Skid plate and undercover
   - Bumper stop
   - Rear anchor assembly
   - Torsion bar
   - Lower ball joint from knuckle
   - Stabilizer link assembly
   - Shock absorber mounting bolts
   - Lower arm shaft
   - Anchor arm
   - Lower control arm

To install:

3. Install or connect the following:
   - Lower control arm. Torque the mounting bolts to 108 ft. lbs. (147 Nm).
   - Anchor arm. Torque the bolt to 33 ft. Lbs. (44 Nm).
   - Lower arm shaft. Torque the bolt to 108 ft. lbs. (147 Nm).
   - Shock absorber mounting bolts. Torque the bolt to 65–76 ft. lbs. (88–103 Nm).
   - Stabilizer link assembly
   - Lower ball joint from knuckle. Torque the nut to 108 ft. lbs. (147 Nm).
   - Torsion bar
   - Rear anchor assembly. Torque the nuts to 33ft. lbs. (44 Nm).
   - Bumper stop. Torque the nut to 18 ft. lbs. (25 Nm).
   - Skid plate and undercover

CONTROL ARM BUSHING REPLACEMENT

Rear

1. Before servicing the vehicle, refer to the precautions in the beginning of this section.
2. Remove the wheel.
3. Remove the lower control arm.
4. Using tool MB991522 press out the bushing

To install:

5. Position the bushing with the larger end facing the front of the vehicle.
6. Using tool MB991522 press the bushing into the bracket.
7. Install the lower control arm.
8. Install the wheel.

Front

1. Before servicing the vehicle, refer to the precautions in the beginning of this section.
2. Remove the wheel.
3. Remove the lower control arm and place in a vise.
4. Using tool MB990883 remove the bushing

To install:

5. Position the bushing with the larger end facing the front of the vehicle.

➡Coat the bushing with a soap solution and take care not to twist.

6. Using tool MB991522, press the bushing into the bracket.
7. Install the lower control arm.
8. Install the wheel.

Wheel Bearings

ADJUSTMENT

Front

MONTERO

The bearings are integral with the hub. No adjustment is possible.

MONTERO SPORT

1. With the caliper removed, check the rotational starting torque. Rotational torque should be 2.7–11.5 inch lbs. (0.3–1.3 Nm). Rotational torque can be adjusted by tightening or loosening the adjusting nut.
2. Check the hub axial. Endplay should not exceed 0.002 inch (0.05mm). If adjusting nut tightening does not bring the axial play within specifications, the bearings must be replaced.
3. Check hub endplay. Endplay should be 0.02–0.03 inch (0.4–0.7mm). Shims are available to adjust endplay.
4. Install the hub assembly. Tighten the nut to 94–145 ft. lbs. (127–196 Nm). Loosen it completely. Tighten the nut to 18 ft. lbs. (25Nm), then back it off 30 degrees.

Rear

The rear wheel bearings are not adjustable. If the bearings are noisy or become loose, they must be replaced.

REMOVAL & INSTALLATION

Front

2001 MONTERO SPORT WITH 2WD

1. Before servicing the vehicle, refer to the precautions in the beginning of this section.
2. Remove or disconnect the following:
   - Tire and wheel assembly
   - Caliper assembly and suspend it from the upper arm
   - Dust cap
   - Cotter pin, castellated nut lock, wheel bearing nut and washer from the spindle
   - Outer wheel bearing
   - Hub and rotor as an assembly
   - Grease seal and inner wheel bearing
3. If required, press the inner and outer bearing outer races from the hub assembly.
4. If replacement of the hub is necessary, matchmark the brake disc with the hub, then separate the hub from the disc.

To install:

5. If removed, place the brake rotor on the hub, while aligning the matchmarks. Tighten the mounting bolts to 34–38 ft. lbs. (47–52 Nm).
6. If removed, press-fit the inner and outer bearing outer races into the hub assembly.
7. Lubricate the seal lip and inside surface of the front hub with MP grease.
8. Install or connect the following:
   - Inner wheel bearing and repack
   - New grease seal
   - Hub assembly on the spindle
   - Outer wheel bearing, washer and

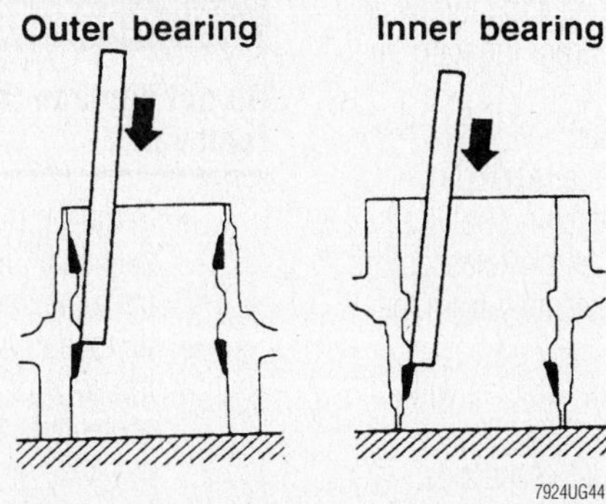

The bearing races can be removed from the hub using a drift and hammer

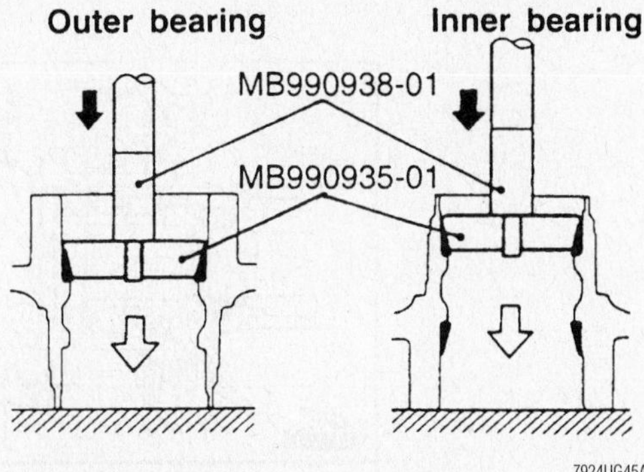

Install the new races into the hub using the proper size driver

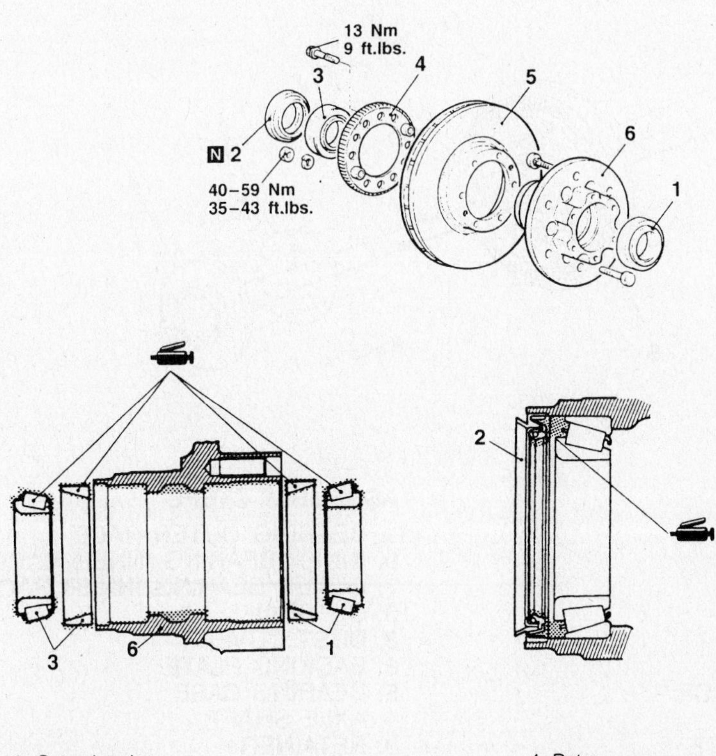

Exploded view of the hub and wheel bearing assembly—Montero

nut, lubricate. When the bearing preload is properly set, install the nut lock and a new cotter pin.

- Grease cap
- Caliper assembly
- Tire and wheel assembly

MONTERO

The wheel bearings are not replaceable. If defective, the hub/bearing assembly must be replaced.

1. Before servicing the vehicle, refer to the precautions in the beginning of this section.
2. Remove or disconnect the following:
   - Tire and wheel assembly
   - Hub cover
   - Nut
   - Washer
   - Brake hose
   - Speed sensor
   - Caliper
   - Rotor
   - Dust cover
   - Tie rod end
   - Upper and lower arms from the knuckle
   - Rotor shield
   - Hub/knuckle assembly
3. Mount the assembly in a vise. Install tool MB990998, or equivalent on the hub. Tighten the nut to 188 ft. lbs. (255 Nm). Check the rotation starting torque. Torque should be 15.48 inch lbs. (1.75 Nm). Wheel bearing backlash should be 0.
4. If the hub is to be replaced, remove the hub-to-knuckle bolts.
5. Installation is the reverse of removal. Observe the following torques:
   - Hub-to-knuckle bolts: 65 ft. lbs. (88 Nm)
   - Hub nut: 188 ft. lbs. (255 Nm)

2001–04 MONTERO SPORT

1. Before servicing the vehicle, refer to the precautions in the beginning of this section.
2. Remove or disconnect the following:
   - Wheel
   - Caliper
   - Hub cover
   - Snapring
   - Shim
   - Drive flange
   - Spring washer
   - Nut
   - Hub and outer bearing
   - Oil seal
   - Inner bearing
   - Races
3. Installation is the reverse of removal.
4. Install the hub assembly. Tighten the

nut to 94–145 ft. lbs. (127–196 Nm). Loosen it completely. Tighten the nut to 18 ft. lbs. (25 Nm), then back it off 30 degrees.

Rear

1. Before servicing the vehicle, refer to the precautions in the beginning of this section.
2. Loosen the wheel lug nuts only½ a turn.
3. Remove or disconnect the following:
   - Wheel(s) from the vehicle
4. Loosen the bleeder valve on the right rear caliper and drain the brake fluid into a container.
5. Remove or disconnect the following:
   - Rear brake hose from the hard line on the frame
   - Rear brake caliper
   - Rear disc off of the rear axle
   - Parking cable attaching bolt and cable end from the brake assembly
   - Parking brake assembly from the end of the axle
   - Speed sensor, on vehicles with Anti-lock Brakes (ABS)
   - Rear axle shaft out of the axle housing. If the rear axle shaft is difficult to remove, use a slide hammer (impact puller) to remove it.

**** WARNING**

Do not damage the oil seal during removal.

- Snapring from the inside end of the axle shaft. Remove 1 retainer bolt from the backing plate with a plastic mallet. Apply cloth tape around the edge of the bearing case for protection. Position the axle shaft in a vise or with a similar method. Using a grinder, grind down the retainer flat, on one side, until the thickness of the retainer is only

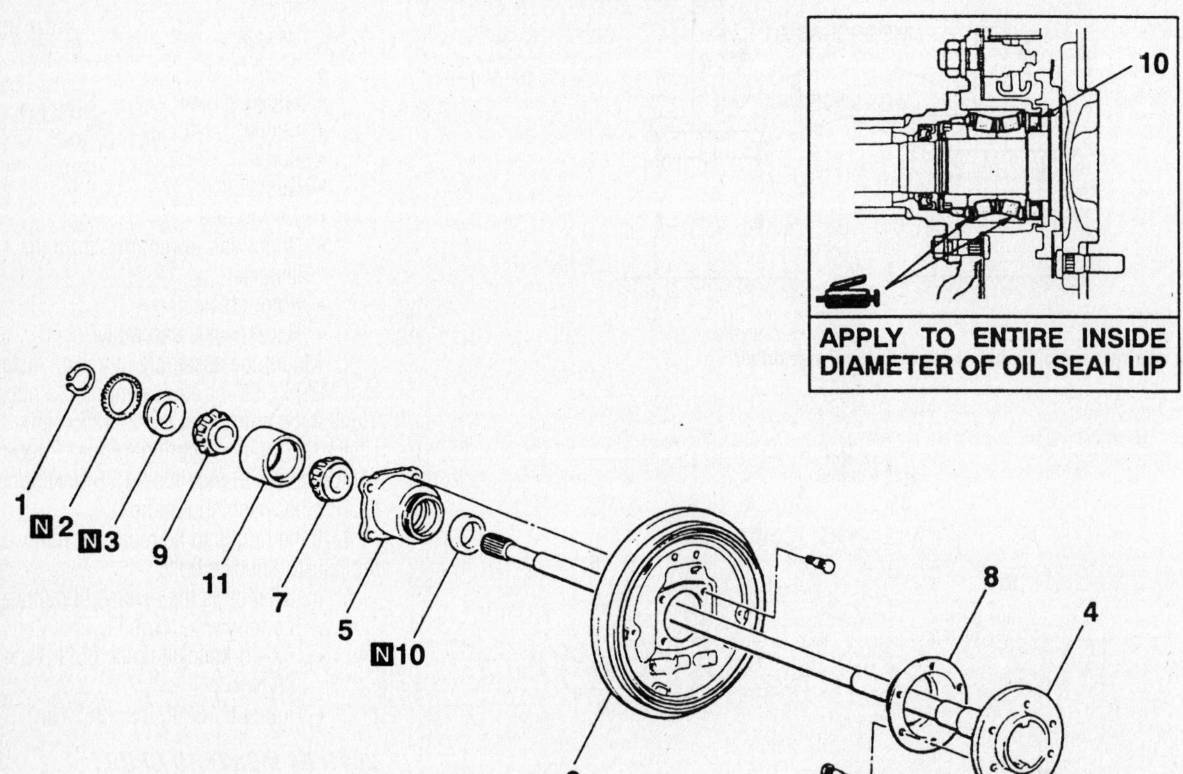

DISASSEMBLY STEPS

1. SNAP RING
2. ABS ROTOR
3. RETAINER
4. AXLE SHAFT
5. BEARING CASE
6. BACKING PLATE
7. OUTER BEARING INNER RACE
8. DUST COVER
9. INNER BEARING INNER RACE
10. OIL SEAL
11. BEARING OUTER RACE

ASSEMBLY STEPS

11. BEARING OUTER RACE
9. INNER BEARING INNER RACE
7. OUTER BEARING INNER RACE
10. OIL SEAL
8. DUST COVER
6. BACKING PLATE
5. BEARING CASE
4. AXLE SHAFT
3. RETAINER
2. ABS ROTOR
1. SNAP RING

7924UG48

Exploded view of the typical rear axle shaft, bearings and races

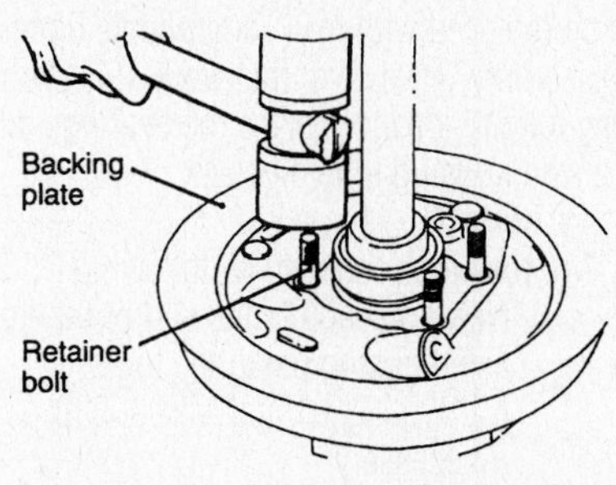

Remove one of the rear axle studs before attempting to grind down the retainer

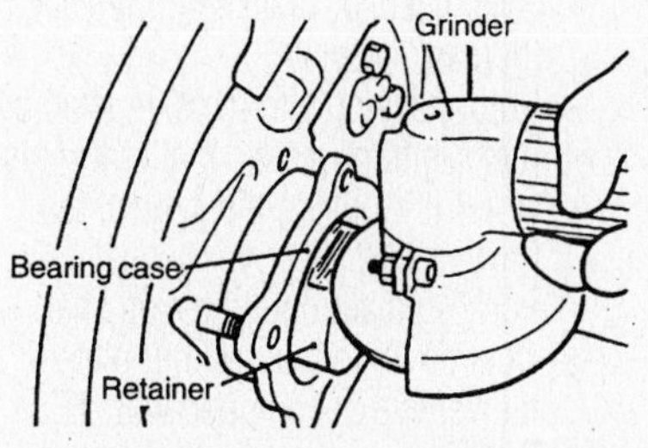

Using a grinder, grind the retainer, on one side, down to 1–2mm (0.04–0.08 in.) thickness

0.04–0.08 in. (1–2mm). That is that the retainer is ground down toward the axle shaft, not toward the flange. Cut, with a chisel, the place where the retainer ring has been shaven down and remove the retainer.

CAUTION

Be careful not to damage the bearing case and the axle shaft.

➡Only the retainer ring is to be ground down, NOT the axle shaft, the axle flange, the bearing or any other component.

6. Grind the plate of special tool MB990861 with a grinder (see illustration) so that there will be no interference between the plate and the bearing case. While adjusting the height of the hanger, secure the washers, plate and nuts in order so that the processed plate is as shown in the illustration.

➡The washers are used to eliminate the difference in height of the bearing case so that the plate and the bearing case are parallel.

Place the end of the bolt against the center of the axle shaft, then tighten the nuts to remove the axle shaft from the bearing case assembly.

➡The hanger and plate must be placed so that they are parallel.

7. Remove the bearing inner race and the bearing outer race. To remove the races, install the tool MB990560 and use a press to remove the bearing race from the axle shaft.

8. Remove the oil seal and the dust cover on vehicles without ABS.

9. On vehicles without ABS, insert an iron plate of approximately 0.04 in. (1mm) thickness between the rotor assembly and the axle shaft, then use a press to remove the rotor assembly.

WARNING

In order not to bend the rotor assembly plate, place the support in contact with the axle shaft when using the press.

10. Remove or disconnect the following:
- Axle shaft from the remaining bearings and components
- Backing plate

11. Reinstall the bearing inner race that was removed previously, then use the tool MB990799-01 and press to remove the bearing outer race.

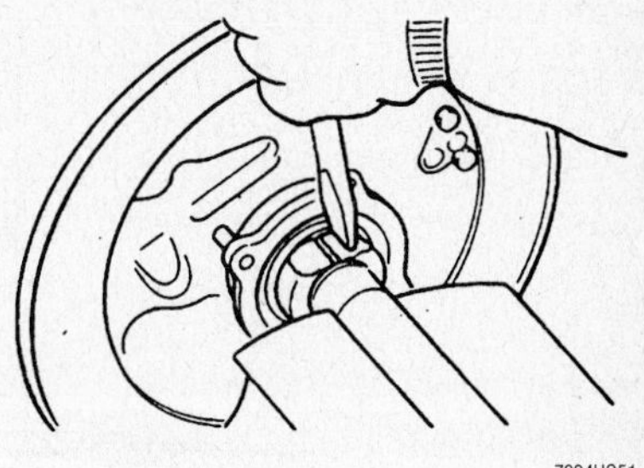

Use a chisel on the ground-down spot on the rear axle bearing retainer to split the retainer, then remove it

12. Remove or disconnect the following:
- Bearing case
- O-ring from the end of the axle housing tube
- Oil seal from the end of the rear axle housing using the tool MB990211-01 (slide hammer with a hooked end), if necessary

13. Check the dust cover for deformation or damage. Check the oil seal for damage. Check the inner and outer bearings for seizure, discoloration and rough raceway surface. Check the axle shaft for cracks, wear and damage. For there are any of these indications, replace the part with a new one. The retainer, the bearing inner (inner and

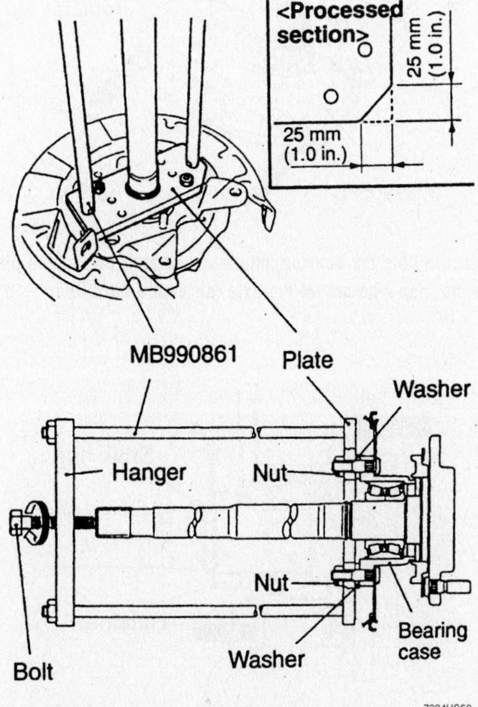

Use the special tool MB990861 to remove the rear axle shaft from the bearing case

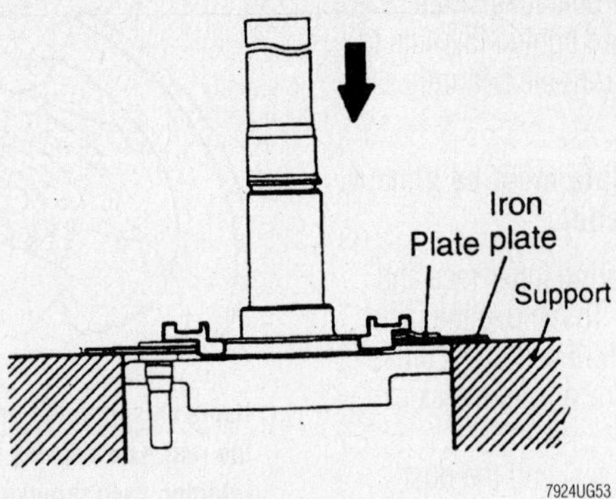

Use an iron plate and supports to remove the rotor assembly

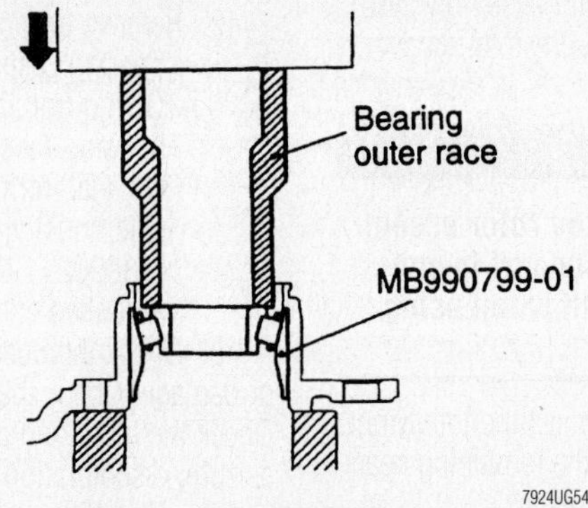

Use the tool MB990799-01 to install and remove the rear axle bearing races

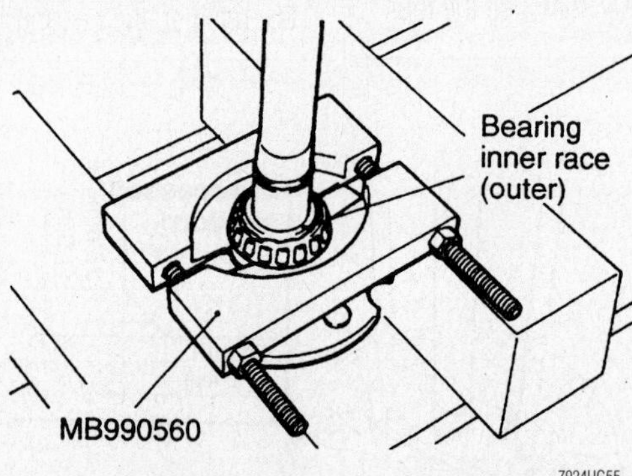

Use the MB990560 tool to hold the bearing inner race (outer), then use a plastic hammer to drive the axle out of the race—do not let the axle fall onto a hard floor

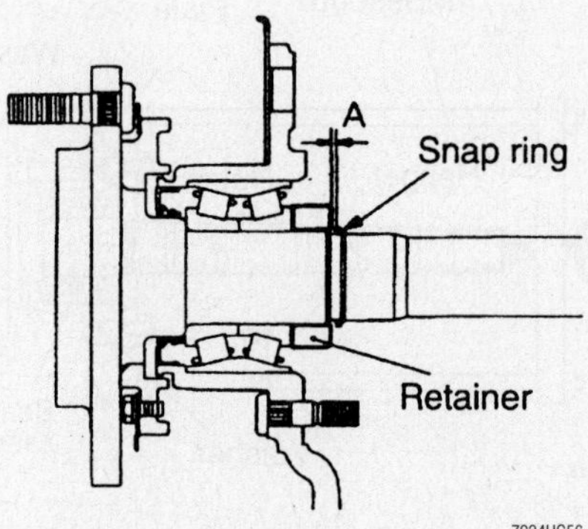

Measure the clearance (A) between the snapring and the retainer edge

outer) and outer races and the oil seal need to be replaced with new components upon reassembly. After all of this work, it is probably a good idea to replace the bearings and the axle housing tube oil seals.

To install:

14. Install or connect the following:
   - New oil seal into the end of the rear axle housing using the tools MB990932-01 and MB990938-01, if necessary.
   - New O-ring into the axle tube
15. Apply multi-purpose grease to the external surface of the bearing out race. Press-fit the bearing outer race into the bearing case by using the tool MB990890-01.
16. Install or connect the following:
   - Speed sensor bracket to the back of the backing plate
   - Rotor assembly to the axle shaft by press-fitting (plastic mallet will also work) it on using the special tool MB991388
   - Backing plate onto the axle shaft
   - Dust cover to the backing plate if the vehicle is equipped with ABS.
   - Bearing inner race (outer) to the bearing case
   - Oil seal to the front end of the bearing case. To do this, apply multi-purpose grease to the outside of the oil seal. Use the special tools MB990936-01 and MB990938-01 to press-fit the oil seal until it is flush with the end of the bearing case. Apply multi-purpose grease to the lip of the oil seal.
   - Axle shaft through the bearing inner race, the bearing case and the second bearing inner race in that order. Use the special tool MB990799 to press-fit the bearing inner race to the axle shaft.

⁂ WARNING

Both bearing inner race sets should be press-fitted together. The left and right lengths of the axle shaft are different in vehicles with rear differential locks. The right side is longer; be careful when installing it.

17. Use the tool MB990799-01 to press-fit the retainer onto the axle shaft, while checking that the press-fitting force is at the following values:
   - Initial press-fitting force: 11,016 lbs. (5000 kg) or more.
   - Final press-fitting force: 22,031–24,280 lbs. (98,000–108,000 N).

18. If the initial press-fitting force is less than the standard value, replace the axle shaft.

19. After installing the snapring, measure the clearance between the snapring and the retainer with a thickness gauge, and check that it is within the standard values. The standard value is 0.0065 in. (0.166mm) or less. If the clearance exceeds the standard value, change the snapring so that the clearance is at the standard value. Use the following adjusting snapring thicknesses:

- 0.0854 in. (2.17mm): no color.
- 0.0791 in. (2.01mm): yellow.
- 0.0728 in. (1.85mm): blue.
- 0.0665 in. (1.69mm): purple.
- 0.0602 in. (1.53mm): red.

20. Install or connect the following:
    - Axle assembly into the axle housing. Be sure that the grooves on the end of the axle shaft line up in the differential. Use a plastic or rubber mallet to help drive the axle shaft into the differential unit. Tighten the 4 retaining bolts for the axle shafts to 36–43 ft. lbs. (49–59 Nm).
    - Speed sensor
    - Parking brake assembly components to the axle flange.
    - Parking brake cable to the parking brake assembly, then secure it in place with the cable bracket.
    - Brake rotor onto the axle shaft, and the brake caliper. Torque the caliper bolts to 65 ft. lbs. (88 Nm).
    - Brake hose to the frame brake line. Torque the flare nut to 11 ft. lbs. (15 Nm).
    - Wheels. Torque the lug nuts as tight as possible with the vehicle not on the ground.
21. Bleed the brake system.
22. Lower the vehicle until the wheels are touching the ground, then finish tightening the lug nuts. Lower the vehicle the rest of the way to the ground.
23. Road test the vehicle and check for leaks.

BRAKES

Brake Caliper

REMOVAL & INSTALLATION

Front

1. Before servicing the vehicle, refer to the precautions in the beginning of this section.
2. Raise and safely support the vehicle.
3. Remove or disconnect the following:
    - Wheel and tire assembly
    - Brake hose from the caliper brake line and remove the retaining clip
    - Caliper guide pin bolts
    - Caliper, by lifting it from the caliper support

To install:

4. Make sure the disc brake pad shims and clips are properly positioned.
5. Position the caliper over the rotor so the caliper engages the adapter correctly
6. Install or connect the following:
    - Mounting pins and tighten to 54 ft. lbs. (74 Nm)
    - Brake hose to the caliper brake line
    - Retaining clip
7. Bleed the brake system.
    - Wheel and tire assembly

Rear

1. Before servicing the vehicle, refer to the precautions in the beginning of this section.
2. Raise and safely support the vehicle.
3. Remove or disconnect the following:
    - Wheel and tire assembly
    - Brake hose from the caliper brake line and remove the retaining clip
    - Caliper guide pin bolts
    - Caliper by lifting it from the caliper support

To install:

4. Make sure the disc brake pad shims and clips are properly positioned.
5. Position the caliper over the rotor so the caliper engages the adapter correctly.
6. Install or connect the following:
    - Mounting pins and tighten them to 32 ft. lbs. (44 Nm)
    - Brake hose to the caliper brake line

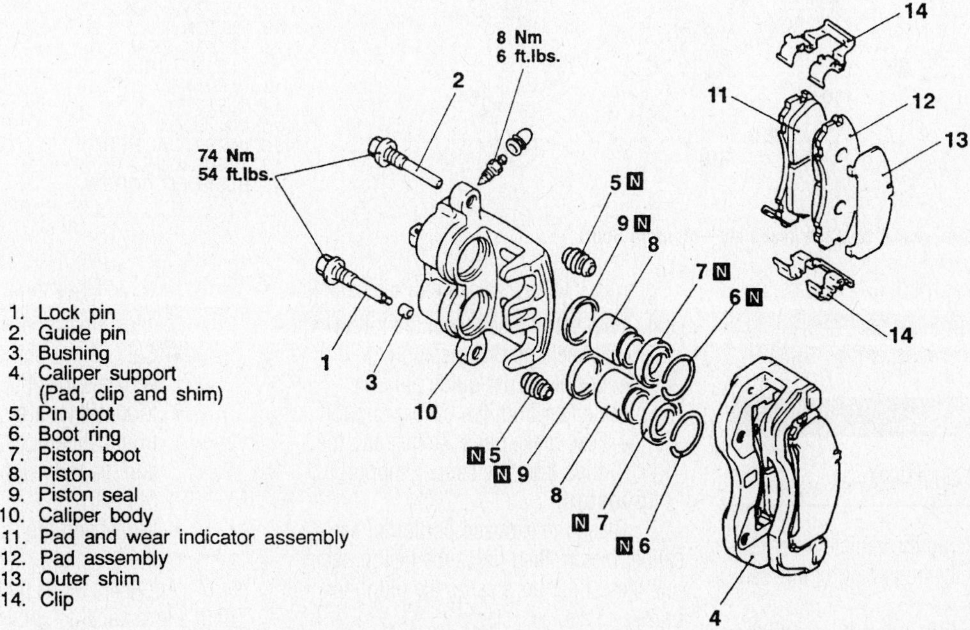

Exploded view of the front disc brake assembly—Montero and Montero Sport

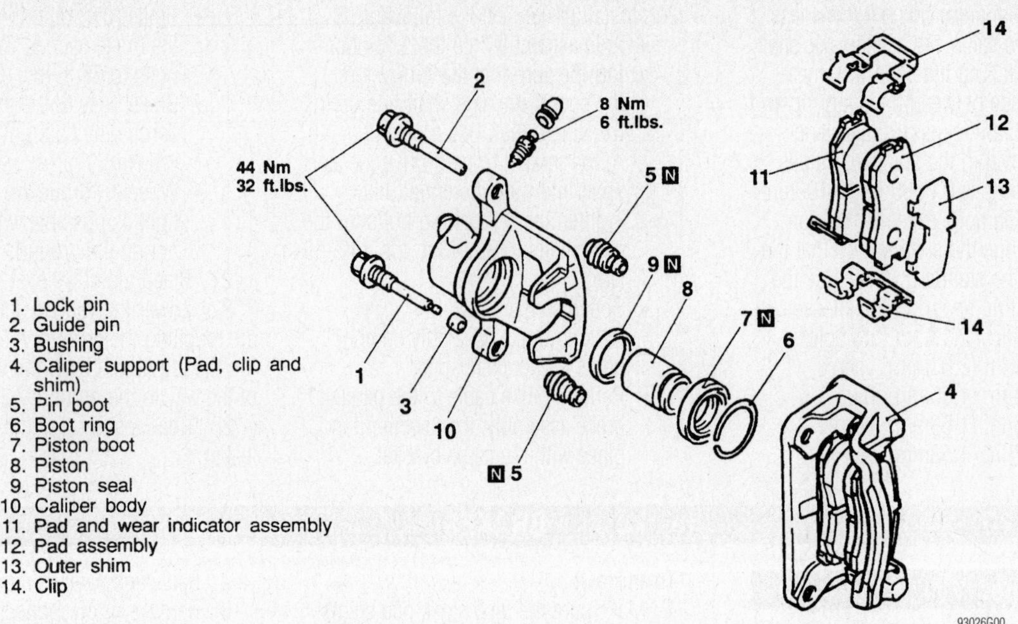

Exploded view of the rear disc brake assembly—Montero

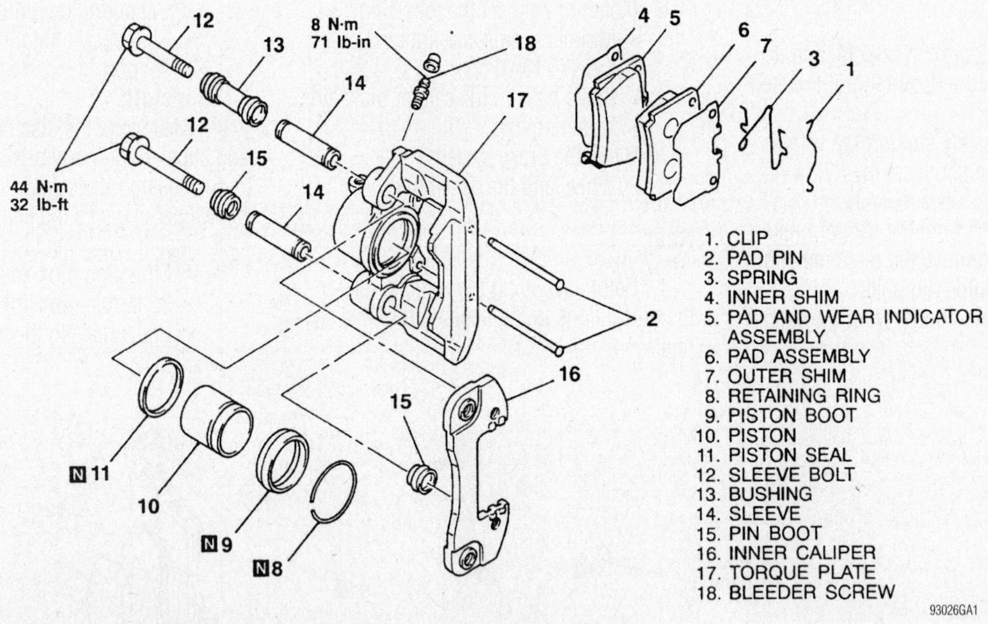

Exploded view of the rear disc brake assembly—Montero Sport

- Retaining clip

7. Bleed the brake system.
   - Wheel and tire assembly

Disc Brake Pads

REMOVAL & INSTALLATION

1. Before servicing the vehicle, refer to the precautions in the beginning of this section.
2. Remove ½ of the brake fluid from the master cylinder.
3. Raise and safely support the vehicle.
4. Remove or disconnect the following:
   - Wheel and tire assembly
   - Lower caliper guide pin bolt
   - Caliper from the caliper support
   - Disc brake pads, shims, and the clips from the caliper support

To install:

5. Clean the exposed portion of the caliper piston, then press the piston back into the caliper bore using the old inner brake pad and a C-clamp.
6. Install or connect the following:
   - Disc brake pads, shims, and the clips. Make sure the shims and clips are properly positioned.
   - Caliper over the rotor so the caliper engages the adapter correctly
   - Mounting pin(s) and tighten the front caliper to 54 ft. lbs. (74 Nm) and the rear caliper to 32 ft. lbs. (44 Nm).
   - Wheel and tire assembly and lower the vehicle
7. Apply the brake pedal several times until a firm pedal is obtained. Check the fluid level in the master cylinder and add fluid, as necessary.

MITSUBISHI

4

Outlander

SPECIFICATION AND MAINTENANCE CHARTS

ENGINE AND VEHICLE IDENTIFICATION

Engine						
Code ①	Liters (cc)	Cu. In.	Cyl.	Fuel Sys.	Type	Eng. Mfg.
4G69/G	2.4 (2378)	143	4	MFI	SOHC	Mitsubishi

Model Year	
Code ②	Year
3	2003
4	2004
5	2005

MFI: Multiport fuel injection

SOHC: Single overhead camsha

① Engine ID / 8th digit of the VIN

② 10th digit of the VIN

67170-OUTL-C01

GENERAL ENGINE SPECIFICATIONS

Year	Model	Engine Displacement Liters	Engine ID/VIN	Net Horsepower @ rpm	Net Torque @ rpm (ft. lbs.)	Bore x Stroke (in.)	Com-pression Ratio	Oil Pressure @ rpm
2003	Outlander	2.4	4G69/G	160@5750	162@4000	3.43x3.94	9.5:1	43-100@3500
2004	Outlander	2.4	4G69/G	160@5750	162@4000	3.43x3.94	9.5:1	43-100@3500

67170-OUTL-C02

ENGINE TUNE-UP SPECIFICATIONS

Year	Engine Displacement Liters	Engine ID/VIN	Spark Plugs Gap (in.)	Ignition Timing (deg.) MT	Ignition Timing (deg.) AT	Fuel Pump (psi)	Idle Speed (rpm) MT	Idle Speed (rpm) AT	Valve Clearance In.	Valve Clearance Ex.
2003	2.4	4G69/G	0.039	2-8B	2-8B	38	600-800	600-800	0.004	0.008
2004	2.4	4G69/G	0.039	2-8B	2-8B	38	600-800	600-800	0.004	0.008

NOTE: The Vehicle Emission Control Information label often reflects specification changes made during production. The label figures must be used if they differ from those in this chart.

B: Before top dead center

67170-OUTL-C03

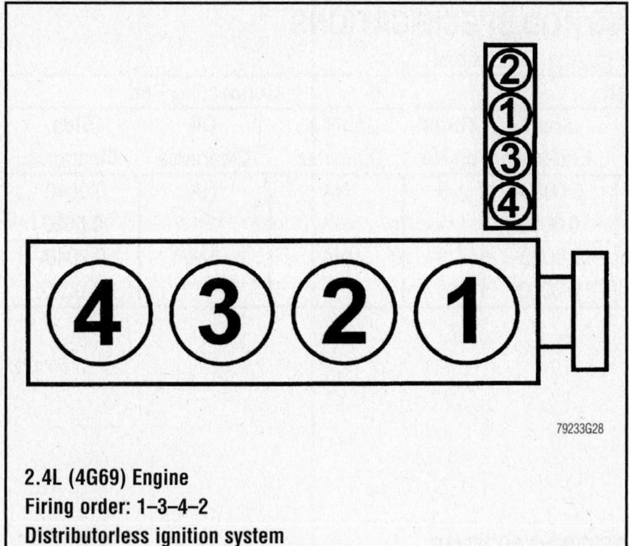

2.4L (4G69) Engine
Firing order: 1–3–4–2
Distributorless ignition system

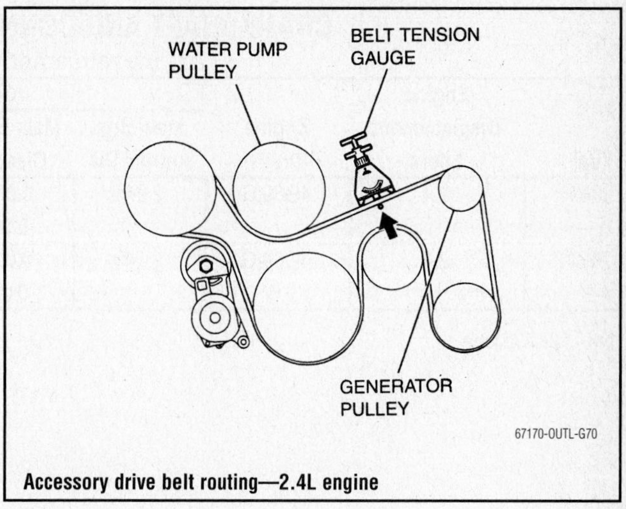

Accessory drive belt routing—2.4L engine

CAPACITIES

Year	Model	Engine Displacement Liters (cc)	Engine ID/VIN	Engine Oil with Filter	Transmission (pts.) 5-Spd	Transmission (pts.) Auto.	Transfer Case (pts.)	Drive Axle Front (pts.)	Drive Axle Rear (pts.)	Fuel Tank (gal.)	Cooling System (qts.)
2003	Outlander	2.4 (2350)	4G69/G	4.5	—	①	1.12	—	1.16	15.7	7.4
2004	Outlander	2.4 (2350)	4G69/G	4.5	—	①	1.12	—	1.16	15.7	7.4

NOTE: All capacities are approximate. Add fluid gradually and ensure a proper fluid level is obtained.

① F4A4B transaxle: 16.2 pts.
W4A4B tranaxle: 17.2 pts.

67170-OUTL-C04

VALVE SPECIFICATIONS

Year	Engine Displacement Liters	Engine ID/VIN	Seat Angle (deg.)	Face Angle (deg.)	Spring Test Pressure (lbs. @ in.)	Spring Installed Height (in.)	Stem-to-Guide Clearance (in.) Intake	Stem-to-Guide Clearance (in.) Exhaust	Stem Diameter (in.) Intake	Stem Diameter (in.) Exhaust
2003	2.4	4G69/G	NA	43.5-44	60@1.740	1.740	0.0008-0.0016	0.0016-0.0024	0.240	0.240
2004	2.4	4G69/G	NA	43.5-44	60@1.740	1.740	0.0008-0.0016	0.0016-0.0024	0.240	0.240

67170-OUTL-C07

CRANKSHAFT AND CONNECTING ROD SPECIFICATIONS

All measurements are given in inches.

Year	Engine Displacement Liters	Engine ID/VIN	Crankshaft Main Brg. Journal Dia.	Crankshaft Main Brg. Oil Clearance	Crankshaft Shaft End-play	Crankshaft Thrust on No.	Connecting Rod Journal Diameter	Connecting Rod Oil Clearance	Connecting Rod Side Clearance
2003	2.4	4G69/G	2.240	0.0008-0.0015	0.0020-0.0090	3	NA	NA	0.0040-0.0090
2004	2.4	4G69/G	2.240	0.0008-0.0015	0.0020-0.0090	3	NA	NA	0.0040-0.0090

NA - Not Available

67170-OUTL-C05

PISTON AND RING SPECIFICATIONS

All measurements are given in inches.

Year	Engine Displacement Liters	Engine ID/VIN	Piston Clearance	Ring Gap Top Compression	Ring Gap Bottom Compression	Ring Gap Oil Control	Ring Side Clearance Top Compression	Ring Side Clearance Bottom Compression	Ring Side Clearance Oil Control
2003	2.4	4G69/G	0.0008-0.0019	0.0080-0.0120	0.0120-0.0180	0.0040-0.0160	0.0012-0.0028	0.0008-0.0023	NA
2004	2.4	4G69/G	0.0008-0.0019	0.0080-0.0120	0.0120-0.0180	0.0040-0.0160	0.0012-0.0028	0.0008-0.0023	NA

NA - Not Available

67170-OUTL-C06

TORQUE SPECIFICATIONS

All readings in ft. lbs.

Year	Engine Displacement Liters	Engine ID/VIN	Cylinder Head Bolts	Main Bearing Bolts	Rod Bearing Bolts	Crankshaft Damper Bolts	Flywheel Bolts	Manifold Intake	Manifold Exhaust	Spark Plugs	Oil Pan Drain Plug
2003	2.4	4G69/G	①	18 ②	18 ②	123	98	③	④	18	29
2004	2.4	4G69/G	①	18 ②	18 ②	123	98	③	④	18	29

① Step 1: Tighten all bolts to 58 ft. lbs.
Step 2: Loosen all bolts to 0 ft. lbs.
Step 3: Tighten all bolts to 15 ft. lbs.
Step 4: Tighten all bolts 90 degrees.
Step 5: Tighten all bolts an additional 90 degrees.

② Torque to specification plus an additional 90 degrees.

③ Bolt: 18 ft. lbs.
Nut: 15 ft. lbs.

④ Bracket (bolt & washer assembly) bolt: 26 ft. lbs.
Bracket (flange) bolt: 27 ft. lbs.
Nut: 36 ft. lbs.

67170-OUTL-C08

WHEEL ALIGNMENT

Year	Model		Caster Range (+/-Deg.)	Caster Preferred Setting (Deg.)	Camber Range (+/-Deg.)	Camber Preferred Setting (Deg.)	Toe-in (in.)
2003	Outlander	F	0.30	+3.10	0.30	0	0.04 +/- 0.09
		R	—	—	0.30	-0.40	0.12 +/- 0.08
2004	Outlander	F	0.30	+3.10	0.30	0	0.04 +/- 0.09
		R	—	—	0.30	-0.40	0.12 +/- 0.08

67170-OUTL-C09

TIRE, WHEEL AND BALL JOINT SPECIFICATIONS

Year	Model	OEM Tires Standard	OEM Tires Optional	Tire Pressures (psi) Front	Tire Pressures (psi) Rear	Wheel Size	Ball Joint Inspection ①	Lug Nut (ft. lbs.)
2003	Outlander	P225/60R16	None	32	30	6-JJ	U: 4-26 in. L: 0-35 in.	73
2004	Outlander	P225/60R16	None	32	30	6-JJ	U: 4-26 in. L: 0-35 in.	73

OEM: Original Equipment Manufacturer

PSI: Pounds Per Square Inch

STD: Standard

OPT: Optional

L: Lower

U: Upper

① Torque required in inch lbs. to rotate ball joint when removed from the knuckle

67170-OUTL-C10

BRAKE SPECIFICATIONS

All measurements in inches unless noted

Year	Model		Brake Disc Original Thickness	Brake Disc Minimum Thickness	Brake Disc Maximum Runout	Brake Drum Diameter Original Inside Diameter	Brake Drum Diameter Max. Wear Limit	Brake Drum Diameter Maximum Machine Diameter	Minimum Lining Thickness Front	Minimum Lining Thickness Rear	Brake Caliper Bracket Bolts (ft. lbs.)	Brake Caliper Mounting Bolts (ft. lbs.)
2003	Outlander	F	1.020	0.960	0.0015	—	—	—	0.080	—	28	74
		R	—	—	—	9.000	—	9.080	—	0.040	—	—
2004	Outlander	F	1.020	0.960	0.0015	—	—	—	0.080	—	28	74
		R	—	—	—	9.000	—	9.080	—	0.040	—	—

NA: Not Available

F: Front

R: Rear

67170-OUTL-C11

SCHEDULED MAINTENANCE INTERVALS
Mitsubishi—Outlander

TO BE SERVICED	TYPE OF SERVICE	VEHICLE MILEAGE INTERVAL (x1000)													
		7.5	15	22.5	30	37.5	45	52.5	60	67.5	75	82.5	90	97.5	102.5
Engine oil & filter	R	✓	✓	✓	✓	✓	✓	✓	✓	✓	✓	✓	✓	✓	✓
Automatic transaxle fluid & filter	S/I		✓		✓		✓		✓		✓		✓		
Brake hoses	S/I		✓		✓		✓		✓		✓		✓		
Disc brake pads	S/I		✓		✓		✓		✓		✓		✓		
Driveshaft boots	S/I		✓		✓		✓		✓		✓		✓		
Valve clearance	S/I				✓				✓				✓		
Air cleaner element	R				✓				✓				✓		
Engine coolant	R				✓				✓				✓		
Spark plugs	R														✓
Ball joints & steering linkage seals	S/I				✓				✓				✓		
Drive belt(s)	S/I				✓				✓				✓		
Exhaust system	S/I				✓				✓				✓		
Fuel hoses	S/I				✓				✓				✓		
Transfer case fluid	S/I				✓				✓				✓		
Transfer case fluid	R								✓				✓		
Rear drum brake linings & rear wheel cylinders	S/I				✓				✓				✓		
Ignition cables	R								✓						
Timing belt(s)	R								✓						
EVAP system (except canister)	S/I								✓						
Fuel system (tank, pipe line, connection & fuel tank filler tube cap)	S/I								✓						

R: Replace S/I: Service or Inspect

FREQUENT OPERATION MAINTENANCE (SEVERE SERVICE)

If a vehicle is operated under any of the following conditions it is considered severe service:

- Extremely dusty areas.

- 50% or more of the vehicle operation is in 32°C (90°F) or higher temperatures, or constant operation in temperatures below 0°C (32°F).

- Prolonged idling (vehicle operation in stop and go traffic).

- Frequent short running periods (engine does not warm to normal operating temperatures).

- Police, taxi, delivery usage or trailer towing usage.

Oil & oil filter: change every 3750 miles.

Disc brake pads: service or inspect every 6000 miles.

Rear drum brake linings and rear wheel cylinders: service or inspect every 15,000 miles

Air filter element: service or inspect every 15,000 miles.

Automatic transaxle fluid & filter: replace every 15,000 miles.

67170-OUTL-C12

PRECAUTIONS

Before servicing any vehicle, please be sure to read all of the following precautions, which deal with personal safety, prevention of component damage, and important points to take into consideration when servicing a motor vehicle:

- Never open, service or drain the radiator or cooling system when the engine is hot; serious burns can occur from the steam and hot coolant.
- Observe all applicable safety precautions when working around fuel. Whenever servicing the fuel system, always work in a well-ventilated area. Do not allow fuel spray or vapors to come in contact with a spark, open flame, or excessive heat (a hot drop light, for example). Keep a dry chemical fire extinguisher near the work area. Always keep fuel in a container specifically designed for fuel storage; also, always properly seal fuel containers to avoid the possibility of fire or explosion. Refer to the additional fuel system precautions in this section.
- Fuel injection systems often remain pressurized, even after the engine has been turned **OFF**. The fuel system pressure must be relieved before disconnecting any fuel lines. Failure to do so may result in fire and/or personal injury.
- Brake fluid often contains polyglycol ethers and polyglycols. Avoid contact with the eyes and wash your hands thoroughly after handling brake fluid. If you do get brake fluid in your eyes, flush your eyes with clean, running water for 15 minutes. If eye irritation persists, or if you have taken brake fluid internally, IMMEDIATELY seek medical assistance.
- The EPA warns that prolonged contact with used engine oil may cause a number of skin disorders, including cancer. You should make every effort to minimize your exposure to used engine oil. Protective gloves should be worn when changing oil. Wash your hands and any other exposed skin areas as soon as possible after exposure to used engine oil. Soap and water, or waterless hand cleaner should be used.
- All new vehicles are now equipped with an air bag system. The system must be disabled before performing service on or around system components, steering column, instrument panel components, wiring and sensors. Failure to follow safety and disabling procedures could result in accidental air bag deployment, possible personal injury, and unnecessary system repairs.
- Always wear safety goggles when working with, or around, the air bag system. When carrying a non-deployed air bag, be sure the bag and trim cover are pointed away from your body. When placing a non-deployed air bag on a work surface, always face the bag and trim cover upward, away from the surface. This will reduce the motion of the module if it is accidentally deployed. Refer to the additional air bag system precautions later in this section.
- Clean, high quality brake fluid from a sealed container is essential to the safe and proper operation of the brake system. You should always buy the correct type of brake fluid for your vehicle. If the brake fluid becomes contaminated, completely flush the system with new fluid. Never reuse any brake fluid. Any brake fluid that is removed from the system should be discarded. Also, do not allow any brake fluid to come in contact with a painted surface; it will damage the paint.
- Never operate the engine without the proper amount and type of engine oil; doing so will result in severe engine damage.
- Timing belt maintenance is extremely important. Many models utilize an interference-type, non-freewheeling engine. If the timing belt breaks, the valves in the cylinder head may strike the pistons, causing potentially serious (also time-consuming and expensive) engine damage. Refer to the maintenance interval charts in the front of this section for the recommended replacement interval for the timing belt, and to the timing belt procedure for belt replacement and inspection.
- Disconnecting the negative battery cable on some vehicles may interfere with the functions of the on-board computer system(s) and may require the computer to undergo a relearning process once the negative battery cable is reconnected.
- When servicing drum brakes, only disassemble and assemble one side at a time, leaving the remaining side intact for reference.

ENGINE REPAIR

➡Disconnecting the negative battery cable on some vehicles may interfere with the functions of the on board computer systems and may require the computer to undergo a relearning process, once the negative battery cable is reconnected.

Alternator

REMOVAL & INSTALLATION

1. Before servicing the vehicle, refer to the precautions in the beginning of this section.
2. Remove or disconnect the following:
   - Negative battery cable
   - Engine undercover
   - Drive belt
   - A/C compressor connector and connector clamp
   - Alternator connector and terminal
   - Alternator connector bracket
   - Harness bracket
   - Alternator, lifting it up out of the vehicle
3. If necessary, remove the lower timing belt cover and alternator mounting bracket.

To install:

4. If removed, install the alternator mounting bracket and lower timing belt cover.
5. Install or connect the following:
   - Alternator, lifting it up out of the vehicle
   - Harness bracket
   - Alternator connector bracket
   - Alternator terminal and connector
   - A/C compressor connector clamp and connector
   - Drive belt
   - Engine undercover
   - Negative battery cable

Ignition Timing

ADJUSTMENT

The ignition timing is controlled by the Electronic Control Module (ECM) and is not adjustable. However it can be inspected using a scan tool.

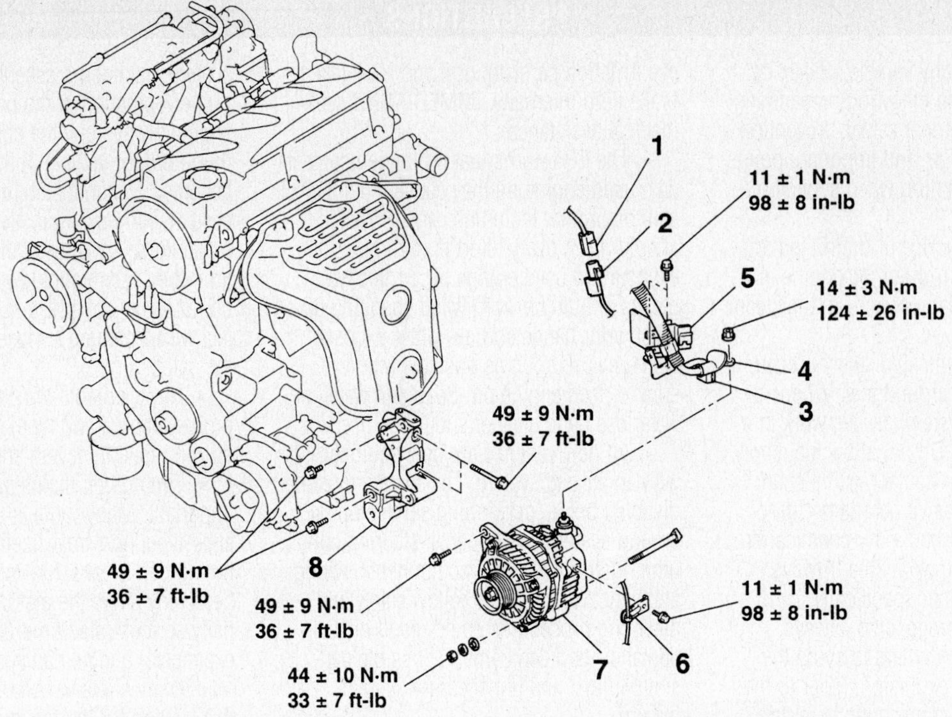

1. A/C COMPRESSOR ASSEMBLY CONNECTOR
2. A/C COMPRESSOR ASSEMBLY CONNECTOR CLAMP
3. GENERATOR CONNECTOR
4. GENERATOR TERMINAL
5. CONNECTOR BRACKET
6. HARNESS BRACKET
7. GENERATOR
8. GENERATOR MOUNTING BRACKET

67170-OUTL-G01

Exploded view of the alternator and related components

Engine Assembly

REMOVAL & INSTALLATION

1. Before servicing the vehicle, refer to the precautions in the beginning of this section.
2. Relieve fuel system pressure.
3. Remove or disconnect the following:
   - Negative battery cable
   - Undercover
   - Hood assembly
   - Air cleaner assembly and all adjoining air intake duct work
4. Drain the engine coolant, engine oil, transaxle and transfer case (if equipped).
5. Remove or disconnect the following:
   - Battery and battery tray
   - Accelerator cable
   - Radiator
   - Front exhaust pipe
   - Drive belt
6. Detach the electrical connectors and/or vacuum lines from the following components, as necessary:
   - Control wiring harness
   - Battery wiring harness
   - Evaporative emission (EVAP) vacuum connection
   - Brake booster vacuum hose connection
   - Air Conditioning (A/C) compressor
   - Power steering oil pressure switch
   - Crank angle sensor
   - Manifold differential pressure sensor
   - Exhaust Gas Recirculation (EGR) solenoid valve
   - Ignition coil
   - Fuel injectors
   - Throttle Position (TP) sensor
   - Idle Air Control (IAC) motor
   - Engine Coolant Temperature (ECT) sensor
   - Camshaft Position (CMP) sensor
   - Knock Sensor (KS)
   - ECT gauge unit
   - Heated Oxygen Sensor (HO_2S)
   - Starter and alternator
   - Oil pressure switch
7. Remove or disconnect the following:
   - Power steering pump and brace. Position the assembly aside, but do NOT disconnect the fluid line.
   - A/C compressor, but do NOT disconnect the lines

➡Matchmark the installed position of the radiator hoses before disconnecting them.

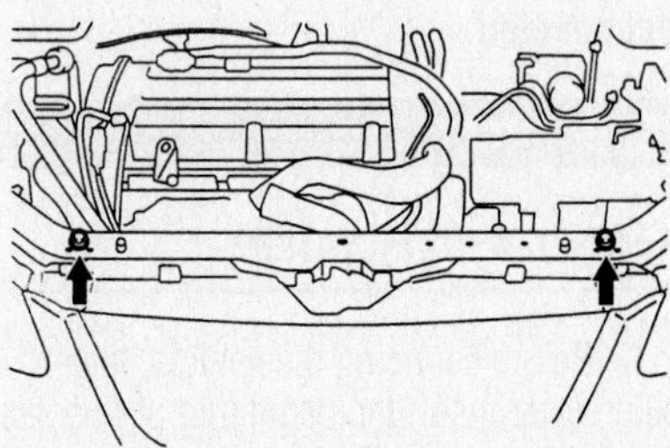

67170-OUTL-G02

Pre-tighten the two bolts on the vehicle to assemble the radiator support upper insulator for the engine hanger (MB991928 or MB991895)

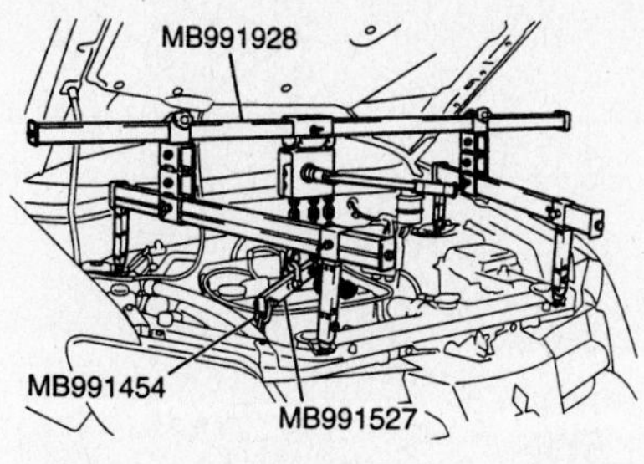

MB991928
MB991454
MB991527

67170-OUTL-G03

View of the tool MB991928 installed, necessary to support the engine when the transaxle is removed

MB991895
MB991527
MB991454

67170-OUTL-G04

View of the tool MB991895 installed, necessary to support the engine when the transaxle is removed

MB991454
MB991527

67170-OUTL-G05

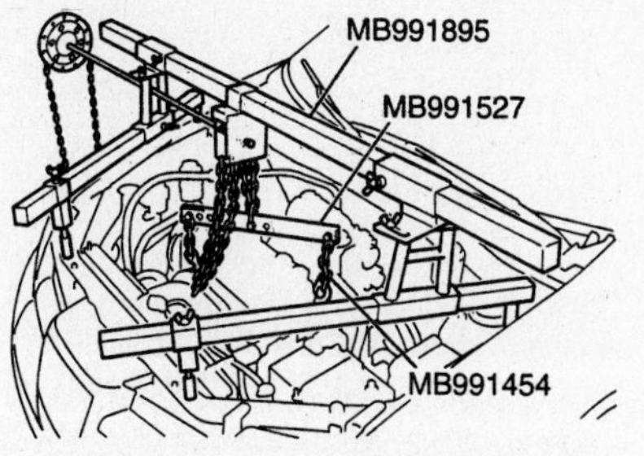

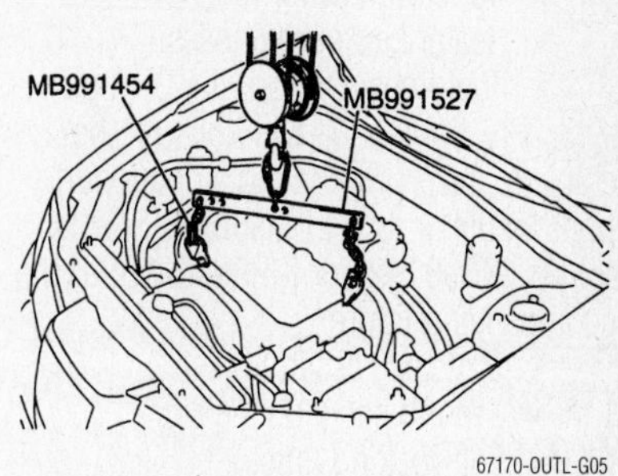

View of the special tools needed to support the engine during mount removal

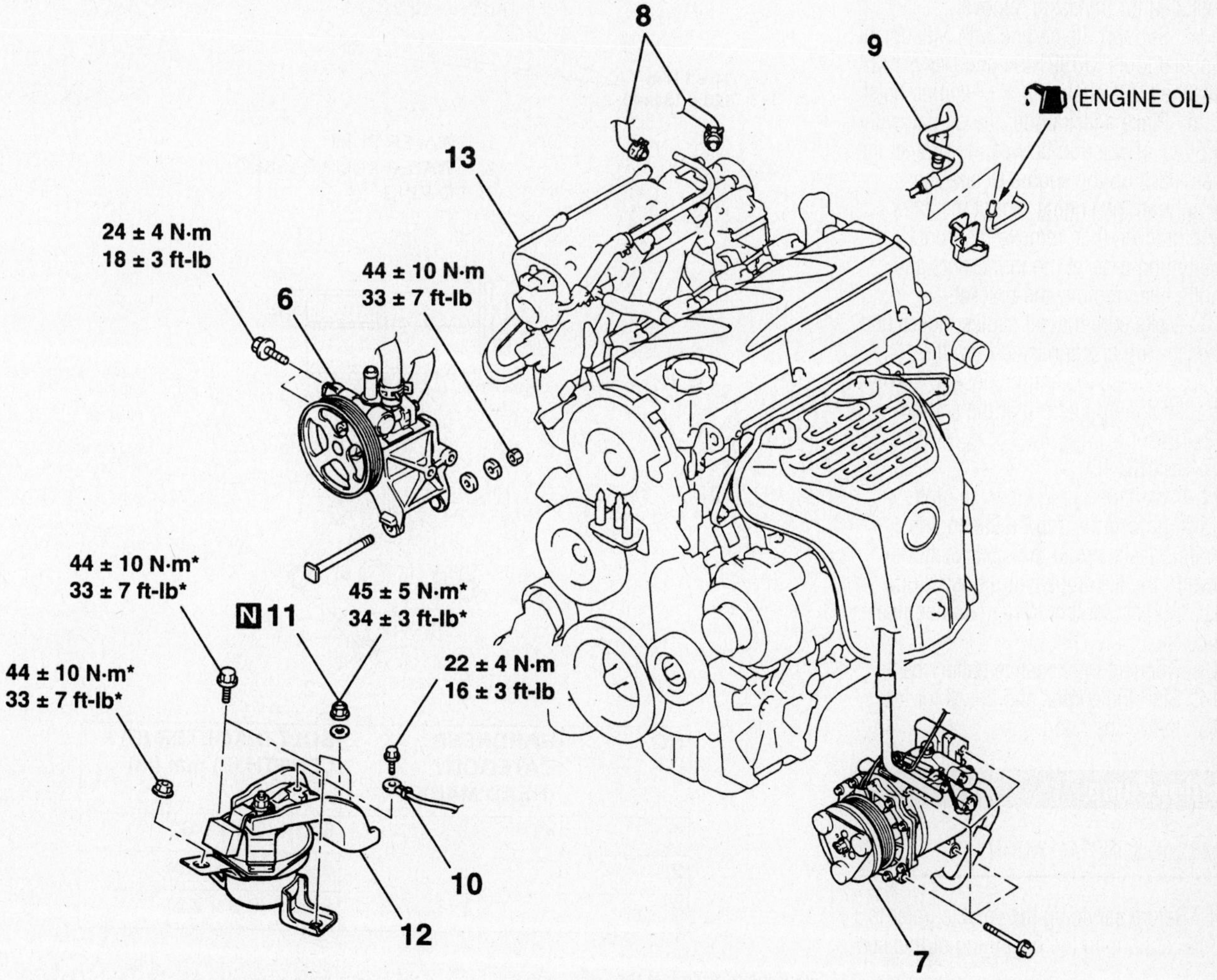

8
9
(ENGINE OIL)
13
24 ± 4 N·m
18 ± 3 ft-lb
44 ± 10 N·m
33 ± 7 ft-lb
6
44 ± 10 N·m*
33 ± 7 ft-lb*
N 11
45 ± 5 N·m*
34 ± 3 ft-lb*
44 ± 10 N·m*
33 ± 7 ft-lb*
22 ± 4 N·m
16 ± 3 ft-lb
10
12
7

6. POWER STEERING OIL PUMP AND BRACKET ASSEMBLY
7. A/C COMPRESSOR AND CLUTCH ASSEMBLY
8. HEATER WATER HOSES CONNECTION
9. FUEL HIGH-PRESSURE HOSE CONNECTION
10. GROUND CABLE CONNECTION
11. SELF LOCKING NUTS
12. ENGINE FRONT MOUNTING BRACKET
13. ENGINE ASSEMBLY

67170-OUTL-G06

View of the engine mounts and related components

- Upper and lower radiator hoses
- Heater and purge hoses
- Fuel lines. Discard the O-rings

8. Pre-tighten the two bolts on the vehicle to assemble the radiator support upper insulator to set Mitsubishi Special tool no. MB991928 or MB991895. See the accompanying figure.
    - Transaxle assembly
    - Ground cable connection
    - Self-locking nuts
9. Remove the engine front mounting bracket as follows:
    a. Support the engine with a suitable floor jack.
    b. Remove the special tools that were installed for transaxle removal.
    c. Support the engine with Mitsubishi Special tools MB991454 and MB991527 attached to a chain block or engine hoist.
    d. Place a jack under the oil pan with a block of wood in between to protect the pan. Jack up the engine to take the weight off the engine mount insulator and bracket, then remove the front mounting bracket mounting nuts and bolts, then remove the bracket.
10. Make sure that all cables, hoses and harnesses are disconnected from the engine, then use the engine hose to slowly lift the engine up and out of the engine compartment

To install:

11. Installation is the reverse of the removal procedure. Torque the engine mounting fasteners to the specifications shown in the accompanying illustration.
12. Fill the coolant system and engine crankcase.
13. Connect the negative battery cable.
14. Start the engine and check for leaks.
15. Install the hood.

Water Pump

REMOVAL & INSTALLATION

1. Before servicing the vehicle, refer to the precautions in the beginning of this section.
2. Disconnect the negative battery cable.
3. Drain the engine coolant.
4. Remove or disconnect the following:
    - Timing belt
    - Water pump mounting bolts
    - Water pump, gasket and O-ring

To install:

5. Install or connect the following:
    - New O-ring on the O-ring grooved at the tip of the water inlet pipe. Coat the O-ring with water or coolant. Do not allow oil or other grease to contact the O-ring. Insert the water inlet pipe.
    - Water pump to the engine block, with new gasket. Torque the mounting bolts to 10 ft. lbs. (14 Nm)
    - Timing belt
6. Refill the engine with coolant.
7. Connect the negative battery cable, start the engine and check for leaks.

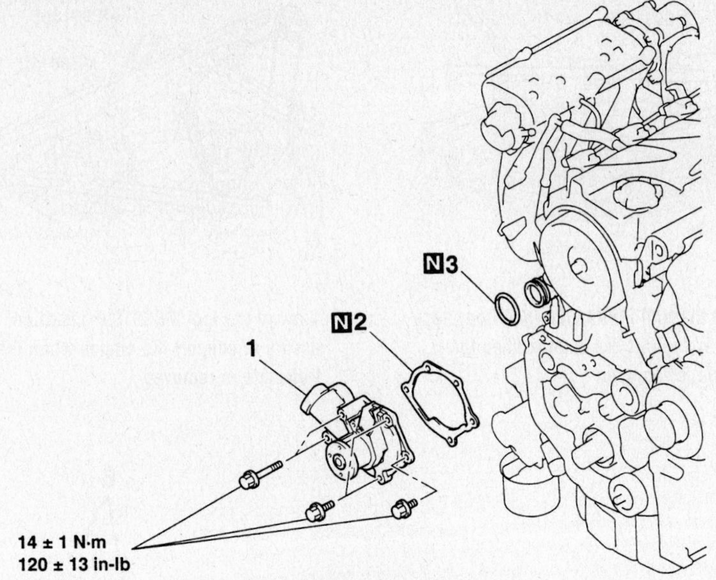

1. WATER PUMP
2. WATER PUMP GASKET
3. O-RING

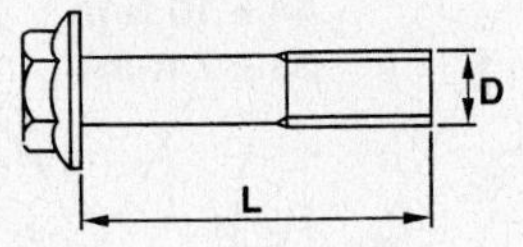

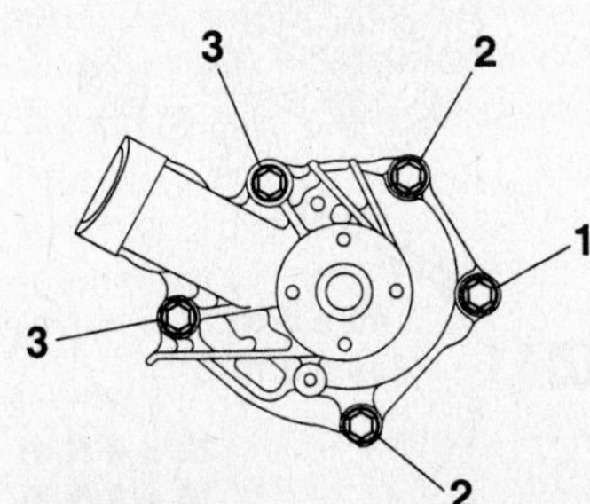

NO.	HARDNESS CATEGORY (HEAD MARK)	BOLT DIAMETER (D) × LENGTH (L) mm (in)
1	4T	8 ×14 (0.3 × 0.6)
2		8 × 22 (0.3 × 0.9)
3		8 × 55 (0.3 × 2.2)

67170-OUTL-G07

Exploded view of the water pump

Heater Core

REMOVAL & INSTALLATION

1. Before servicing the vehicle, refer to the precautions in the beginning of this section.
2. Disconnect the negative battery cable.

3. Drain the cooling system into a clean container for reuse.

4. Discharge and recover the air conditioning system refrigerant.

5. Remove or disconnect the following:

- Instrument panel
- Intake duct
- Right side foot duct
- Joint duct
- Center duct
- Heater hose connection
- Flexible suction hose connection
- Liquid pipe connection
- O-ring
- Left side foot duct
- Heater case

6. Disassemble the heater unit as necessary for access to components.

To install:

7. Assemble the heater unit as necessary.

8. Install or connect the following:

- Blower assembly
- Heater case
- Left side foot duct
- New O-ring
- Liquid pipe connection
- Flexible suction hose connection
- Heater hose connection
- Center duct
- Joint duct
- Right side foot duct
- Intake duct
- Instrument panel

9. Refill the cooling system.

10. Connect the negative battery cable.

11. Evacuate, charge and leak test the air conditioning system.

12. Operate the engine to normal operating temperatures; then, check the climate control operation and check for leaks.

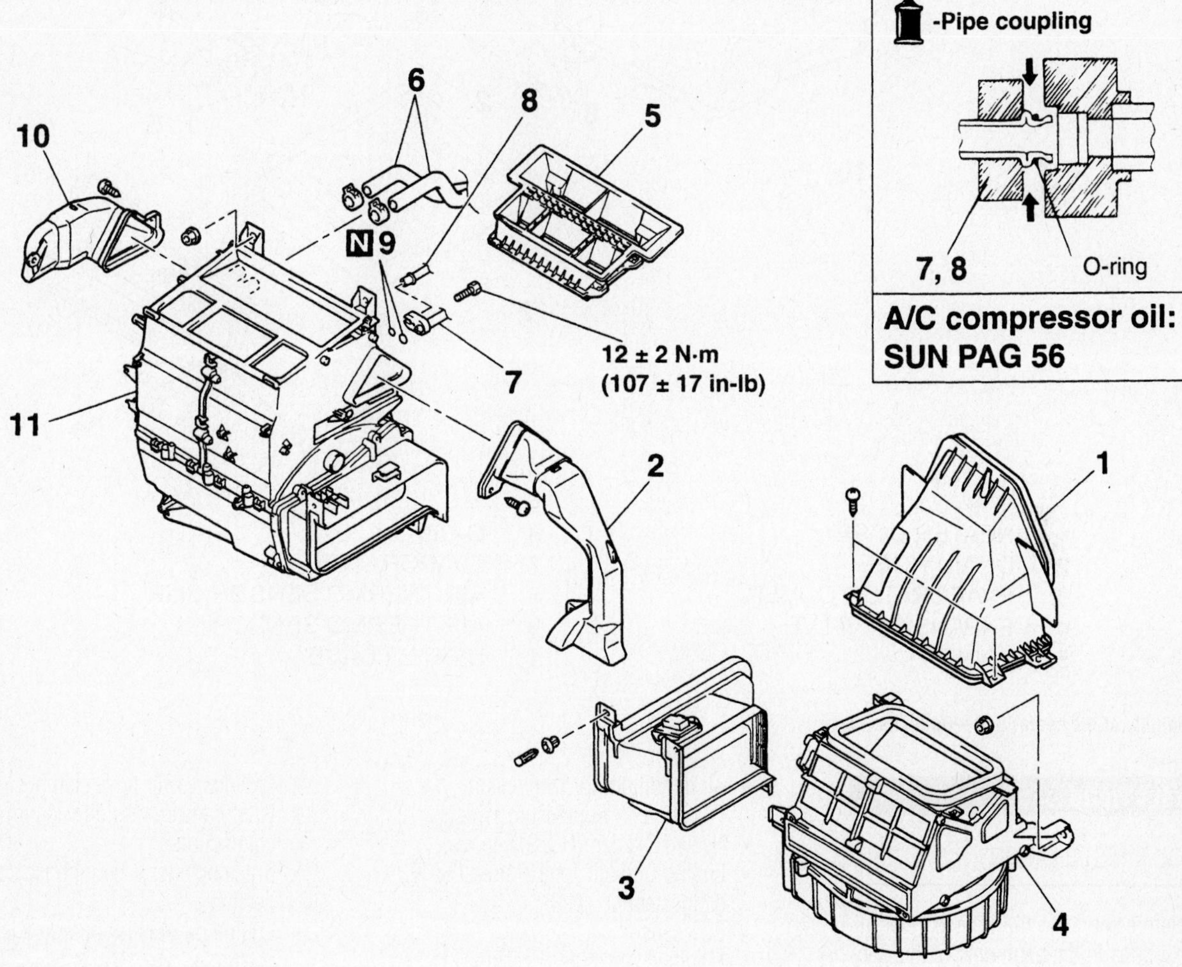

1. INTAKE DUCT
2. RIGHT-HAND FOOT DUCT
3. JOINT DUCT
4. BLOWER ASSEMBLY
5. CENTER DUCT
6. HEATER HOSE CONNECTION
7. FLEXIBLE SUCTION HOSE CONNECTION
8. LIQUID PIPE CONNECTION
9. O-RING
10. LEFT-HAND FOOT DUCT
11. HEATER CASE

67170-OUTL-G08

Exploded view of the heater unit —Outlander

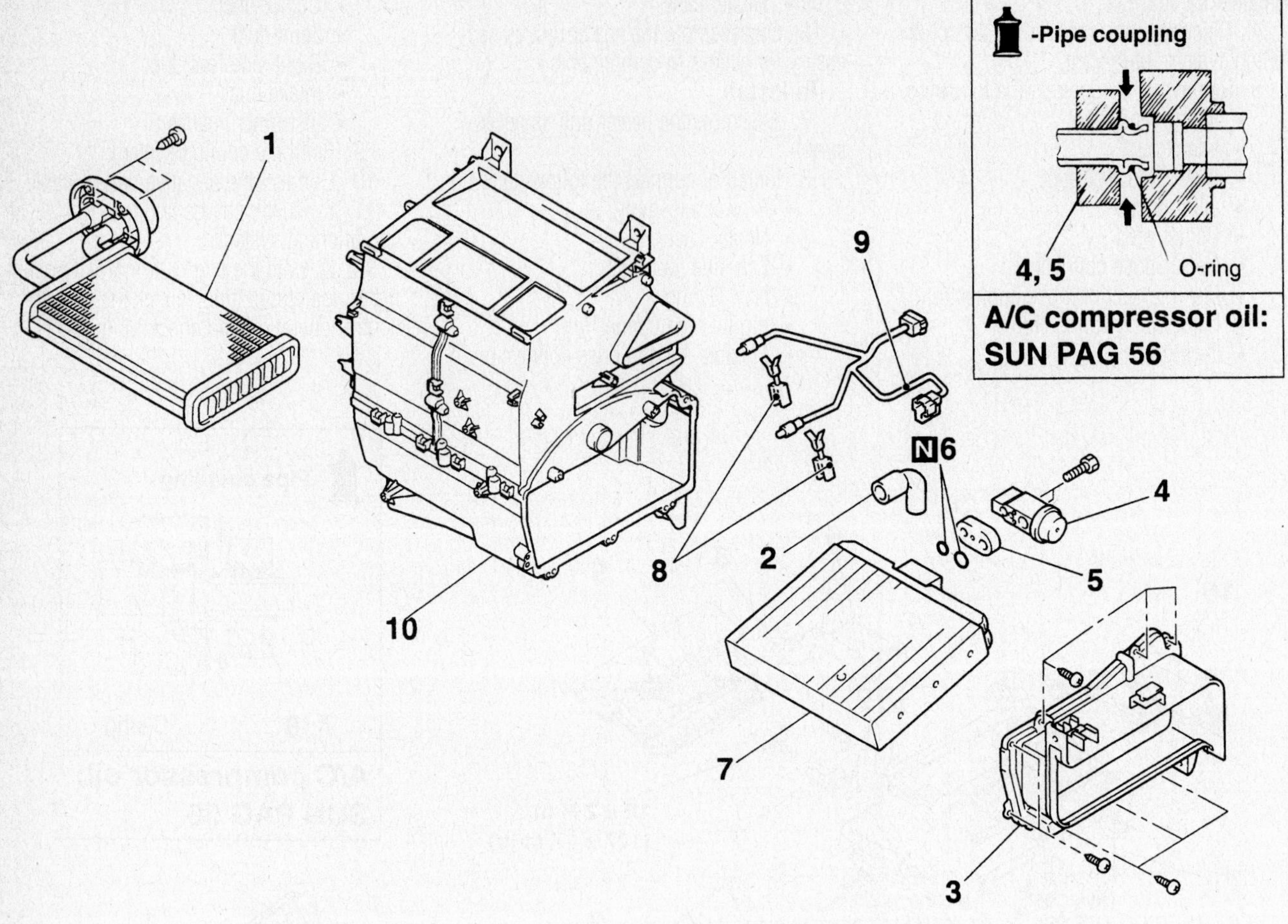

Disassembly of the heater unit—Outlander

Cylinder Head

REMOVAL & INSTALLATION

1. Before servicing the vehicle, refer to the precautions in the beginning of this section.
2. Relieve the fuel system pressure.
3. Disconnect the negative battery cable.
4. Remove the air cleaner with all air intake hoses.
5. Drain the cooling system.
6. Remove or disconnect the following:
   - Accelerator cable
7. Disconnect the following connectors necessary for control wire harness connection removal:
   - Air conditioning compressor
   - Power steering pressure switch
   - Heated Oxygen (HO_2S) sensor
   - Engine Coolant Temperature (ECT) gauge sender
   - ECT sensor
   - Manifold Absolute Pressure (MAP) sensor
   - Intake Air Temperature (IAT) sensor
   - Throttle Position (TP) sensor
   - Idle Air Control (IAC) motor
   - Injector harness
   - Ignition coil
   - Camshaft Position(CMP) sensor
   - Exhaust Gas Recirculation (EGR) solenoid valve
8. Remove or disconnect the following:
   - Battery wiring harness connection
   - Radiator lower hose clamp
   - Evaporative emission vacuum hose connection
   - Brake booster vacuum hose connection
   - Engine oil dipstick and dipstick guide
   - O-ring
   - Knock sensor connector
   - Battery wiring harness connection
   - Inlet manifold stay
   - Exhaust manifold
   - Spark plug wires
   - Coolant hoses and unbolt the thermostat case from the engine, at the thermostat case assembly
   - Upper timing belt cover
9. Align all timing marks.

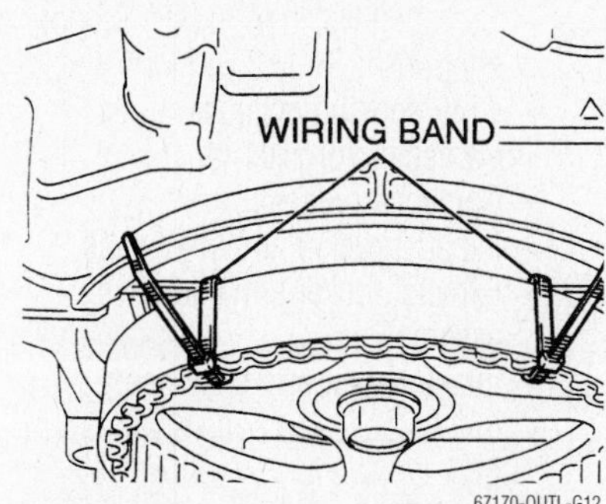

Secure the timing belt to the camshaft sprocket and remove the sprocket—2.4L (4G69) engine

10. Secure the timing belt to the camshaft sprocket with cord or a wire tie.
11. Remove or disconnect the following:
    - Camshaft sprocket
    - Valve cover and the half-round seal
    - Intake manifold stay bracket from the intake manifold
    - High pressure fuel line connection
    - Valve cover
12. Loosen the cylinder head mounting bolts in 3 steps, starting from the outside and working inward. Lift off the cylinder head assembly and remove the head gasket.

To install:

13. Thoroughly clean the mating surfaces of the head and block.
14. Place a new head gasket on the cylinder block with the identification marks at the front top (upward) position on the exhaust side. Do not use sealer on the gasket.
15. Inspect the cylinder head bolt length prior to installation. If the length exceeds 3.91 in. (99.4mm), the bolt must be replaced. Install the washer onto the bolt so the chamfer on the washer faces towards the head of the bolt.
16. Carefully install the cylinder head on the block and tighten the cylinder head bolts as follows:
    a. Following the proper tightening sequence, tighten the cylinder head bolts to 58 ft. lbs. (78 Nm).
    b. Loosen all bolts completely.
    c. Torque bolts to 15 ft. lbs. (20 Nm).
    d. Tighten bolts an additional ¼ turn.
    e. Tighten bolts an additional ¼ turn.
17. Apply sealer to the perimeter of the half-round seal and to the lower edges of the half-round portions of the belt-side of the new gasket. Install the valve cover.
18. Install or connect the following:
    - High pressure fuel line
    - Thermostat case and tighten the mounting bolts to 18 ft. lbs. (24 Nm)
    - Water and upper radiator hoses to the thermostat case

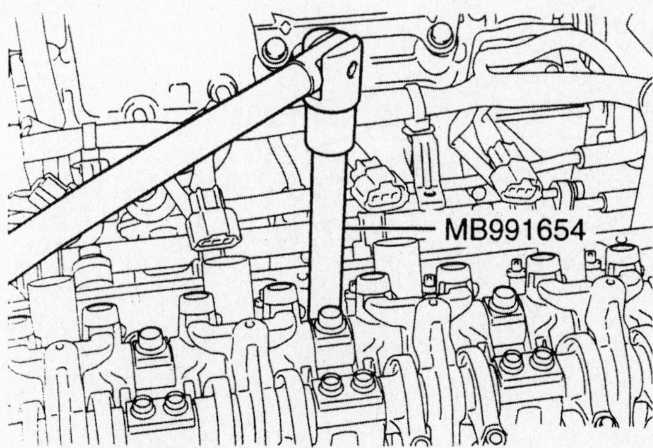

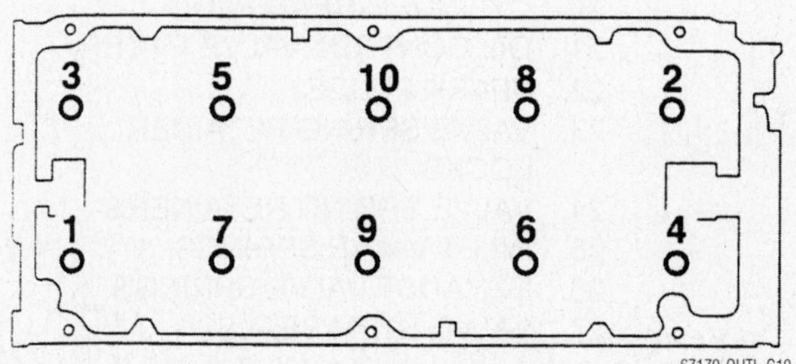

Cylinder head bolt removal sequence—2.4L (4G69) engine

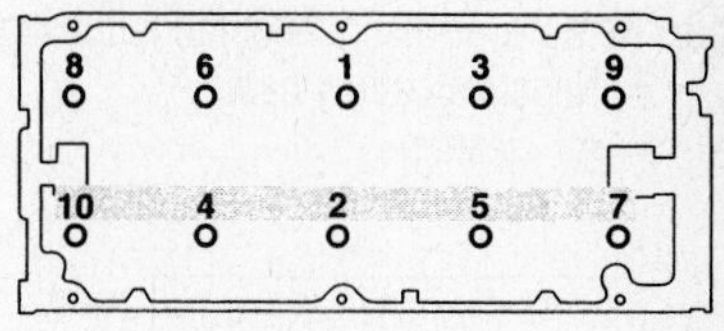

Cylinder head bolt installation sequence—2.4L (4G69) engine

- Camshaft sprocket with the timing belt attached. Remove the cord or wire tie.
- Upper timing belt cover
- Exhaust manifold
- Intake manifold stay and tighten the mounting bolts to 22 ft. lbs. (30 Nm)
- Battery wiring harness connector
- Knock Sensor connector

19. Install or connect the following:
    - Spark plug wires and cover
    - Fuel lines using new O-rings
    - Engine oil dipstick and guide, using new O-rings
    - Brake booster vacuum hose
    - Evaporative emission vacuum hose
    - Radiator lower hose clamp
    - Battery wiring harness connector
20. Connect the following connectors for the control wiring harness, as necessary:
    - Air conditioning compressor
    - Power steering pressure switch
    - HO_2S sensor
    - ECT gauge sender
    - ECT sensor
    - MAP sensor
    - IAT sensor
    - TP sensor
    - IAC motor
    - Injector harness
    - Ignition coil
    - CMP sensor
    - EGR solenoid valve
    - Air cleaner and intake hose
    - Breather hose
21. Install or connect the following:
    - Accelerator cable
    - Air cleaner assembly

22. Fill the cooling system.
23. Connect the negative battery cable

Rocker Arms/Shafts

REMOVAL & INSTALLATION

1. Before servicing the vehicle, refer to the precautions in the beginning of this section.
2. Drain the engine oil.
3. Remove or disconnect the following:
   - Negative battery cable
   - Air cleaner
   - Ignition coils
   - Upper timing belt cover
   - Positive Crankcase Ventilation (PCV) hose
   - Breather hose
   - Control wiring harness connection
   - Engine hanger
   - Oil control valve
   - Oil pressure switch
   - Rocker arm (valve) cover and gasket
   - Oil seals, if necessary
   - Accumulator assembly
   - Timing belt
   - Camshaft Position (CMP) sensor support
   - CMP sensing cylinder
   - Camshaft sprocket
   - Camshaft oil seal
   - Exhaust rocker arm shaft caps
   - Exhaust rocker arm and shaft assembly
   - Intake rocker arm shaft caps
   - Intake rocker arm and shaft assembly

To install:

4. Install the intake rocker arm and shaft assembly as follows:

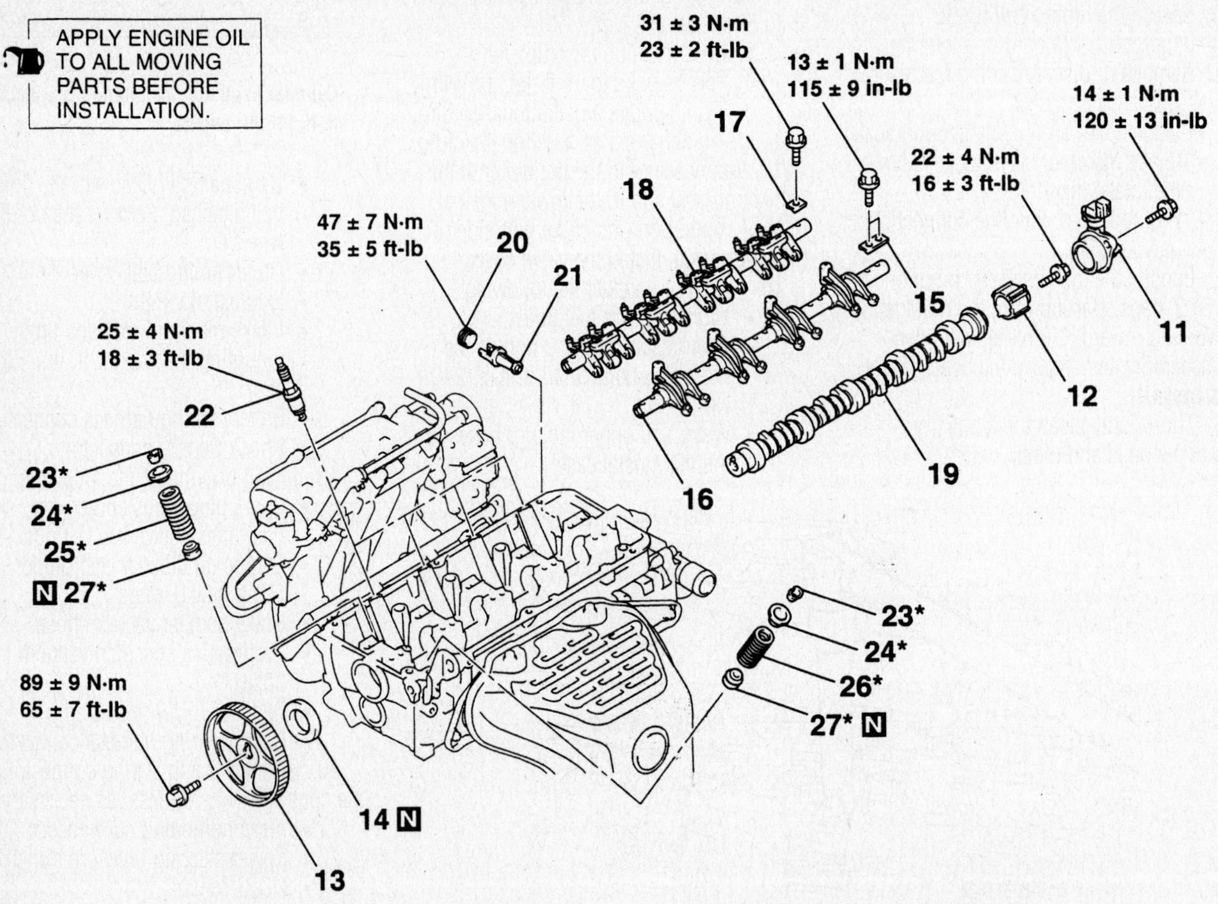

Exploded view of the rocker arms, camshaft and related components

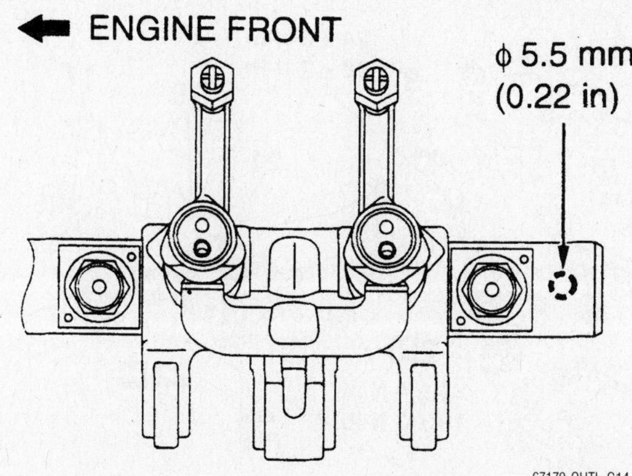

The hole on the intake rocker shaft must face the cylinder head

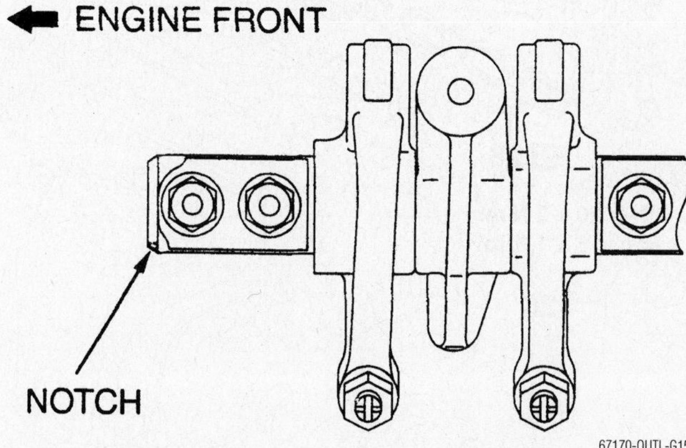

The exhaust rocker shaft must be properly positioned for installation

a. Place the intake rocker shaft so that the 0.22 in. (5.5mm) hole faces the cylinder head.

b. Install the intake rocker arm shaft caps.

c. Tighten the intake rocker shaft mounting bolts to 21–25 ft. lbs. (28–34 Nm).

- Rocker arm shafts, placing the end with the notched side toward the timing belt
- Rocker arms. Move the rocker arms from side to side before tightening the shaft bolts to 21

5. Install the exhaust rocker arm and shaft assembly as follows:

a. Place the exhaust rocker shaft so that its notch is positioned as shown in the accompanying figure.

b. Install the exhaust rocker arm shaft caps.

c. Tighten the exhaust rocker shaft mounting bolts to 106–124 inch lbs. (12–14 Nm).

6. The remainder of installation is the reverse of the removal procedure.

7. Fill the crankcase with oil and connect the negative battery cable.

Intake Manifold

REMOVAL & INSTALLATION

1. Before servicing the vehicle, refer to the precautions in the beginning of this section.
2. Relieve the fuel system pressure.
3. Remove or disconnect the following:
   - Battery negative cable and drain the cooling system
   - Air cleaner assembly
   - Throttle body
   - Fuel rail assembly
   - Exhaust Gas Recirculation (EGR) valve and gasket
   - Capacitor connector
   - Brake booster vacuum hose
   - Capacitor
   - Vacuum pipe
   - Brake booster vacuum hose connection
   - Evaporative emission (EVAP) canister vacuum hose
   - Manifold Absolute Pressure (MAP) sensor
   - Knock Sensor (KS) connector
   - EVAP purge solenoid valve connector
   - Harness clamp
   - KS connector clamp
   - Oil dipstick guide and O-ring
   - Harness clamp
   - Lower radiator hose. Matchmark the clamp to the hose before removal.
   - Heater hose
   - Engine Coolant Temperature (ECT) gauge unit connector
   - Thermostat housing and O-ring
   - KS harness clamp
   - MAP sensor and harness clamp
   - Intake manifold stay
   - Intake manifold stay
   - Intake manifold bolt and manifold.
   - Intake manifold gasket and discard

To install:

4. Clean all gasket material from the cylinder head intake mounting surface and intake manifold assembly.
5. Install the intake manifold, using a new gasket. Torque the manifold in a crisscross pattern, starting from the inside and working outwards as follows:
   a. Bolts: 16–20 ft. lbs. (21–27 Nm)
   b. Nuts: 14–16 ft. lbs. (18–22 Nm)
6. Install or connect the following:
   - Intake manifold stay. Torque the bolt to 21–25 ft. lbs. (29–33 Nm)
   - MAP sensor and harness clamp
   - KS harness clamp
   - Thermostat housing and O-ring. Clean the gasket mating surfaces, then apply a 3mm bead of 3M™ ADD part No. 8672, 3M™ ADD part No. 8679/8678 or equivalent.
   - ECT gauge unit connector
   - Heater hose
   - Lower radiator hose. Matchmark the clamp to the hose before removal.
   - Harness clamp
   - Oil dipstick guide and O-ring
   - KS connector clamp
   - Harness clamp
   - EVAP purge solenoid valve connector
   - KS connector
   - MAP sensor
   - EVAP canister vacuum hose
   - Brake booster vacuum hose connection
   - Vacuum pipe
   - Capacitor
   - Brake booster vacuum hose
   - Capacitor connector
   - EGR valve and gasket
   - Fuel rail assembly

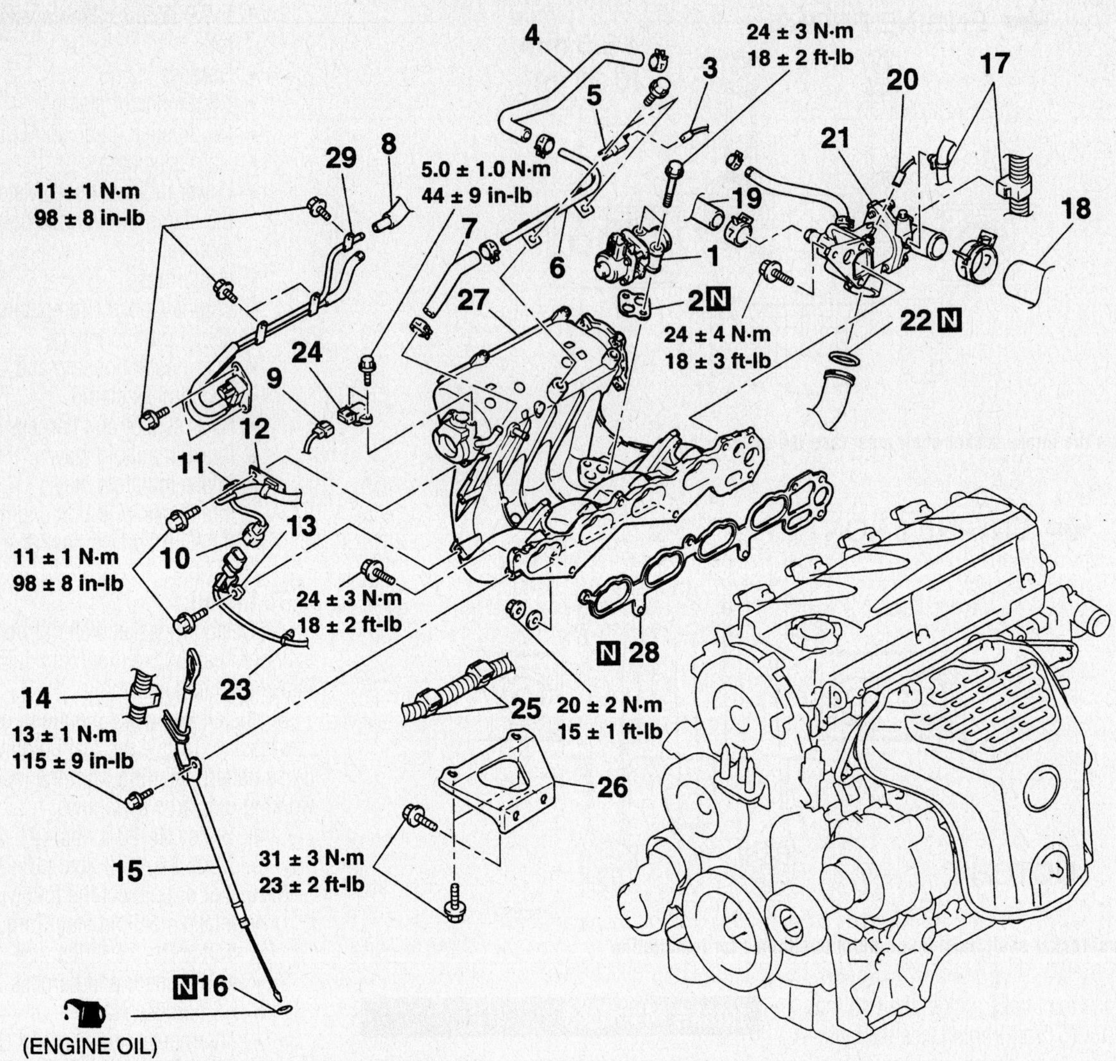

1. EXHAUST GAS RECIRCULATION VALVE
2. EXHAUST GAS RECIRCULATION VALVE GASKET
3. CAPACITOR CONNECTOR
4. BRAKE BOOSTER VACUUM HOSE
5. CAPACITOR
6. BRAKE BOOSTER VACUUM PIPE
7. BRAKE BOOSTER VACUUM HOSE
8. EVAPORATIVE EMISSION CANISTER VACUUM HOSE CONNECTION
9. MANIFOLD ABSOLUTE PRESSURE SENSOR CONNECTOR
10. KNOCK SENSOR CONNECTOR
11. EVAPORATIVE EMISSION PURGE SOLENOID VALVE CONNECTOR
12. HARNESS CRAMP
13. KNOCK SENSOR CONNECTOR CRAMP
14. HARNESS CRAMP
15. OIL DIPSTICK GUIDE
16. O-RING
17. HARNESS CRAMP
18. RADIATOR LOWER HOSE
19. HEATER HOSE CONNECTION
20. ENGINE COOLANT TEMPERATURE GAUGE UNIT CONNECTOR
21. THERMOSTAT HOUSING ASSEMBLY
22. O-RING
23. KNOCK SENSOR HARNESS CRAMP
24. MANIFOLD ABSOLUTE PRESSURE SENSOR
25. HARNESS CRAMP
26. INTAKE MANIFOLD STAY
27. INTAKE MANIFOLD
28. INTAKE MANIFOLD GASKET
29. EVAPORATIVE EMISSION PURGE SOLENOID VALVE, EVAPORATIVE EMISSION VACUUM HOSE AND PIPE ASSEMBLY

67170-OUTL-G16

Exploded view of the intake manifold and related components

- Throttle body
- Air cleaner assembly

7. Fill the system with coolant.
8. Connect the negative battery cable.

Exhaust Manifold

REMOVAL & INSTALLATION

1. Before servicing the vehicle, refer to the precautions in the beginning of this section.
2. Remove or disconnect the following:
   - Negative battery cable
   - Front exhaust pipe bracket
   - Front exhaust pipe from the exhaust manifold. Remove and discard the gasket.
   - Heat shield
   - Mounting nuts, the exhaust manifold, and the exhaust manifold gasket

To install:

3. Install or connect the following:
   - New exhaust manifold gasket to the cylinder head
   - Exhaust manifold. Torque the mounting nuts to 32–40 ft. lbs. (44–54 Nm).
   - Heat shield and tighten the bolts to 107–133 inch ft. lbs. (13–15 Nm)
   - New gasket between the exhaust manifold and the front exhaust pipe and reconnect the pipe. Torque the nuts to 30–44 ft. lbs. (39–59 Nm) and the bolts to 22–30 ft. lbs. (30–40 Nm).
   - Front exhaust pipe bracket and tighten the retainers to 22–30 ft. lbs. (30–40 Nm)
   - Negative battery cable
4. Start the engine and check for any exhaust leaks.

Front Crankshaft Seal

REMOVAL & INSTALLATION

1. Before servicing the vehicle, refer to the precautions in the beginning of this section.
2. Remove or disconnect the following:
   - Negative battery cable
   - Timing belt
   - Crankshaft sprocket
   - Crankshaft key
3. Carefully pry the oil seal out of the front case. Be careful not to damage the oil seal bore or the crankshaft sealing surface.

To install:

4. Apply clean engine oil to the oil seal lip. Using a seal driver, install the oil seal.
5. Install or connect the following:
   - Crankshaft key
   - Crankshaft sprocket. Clean and

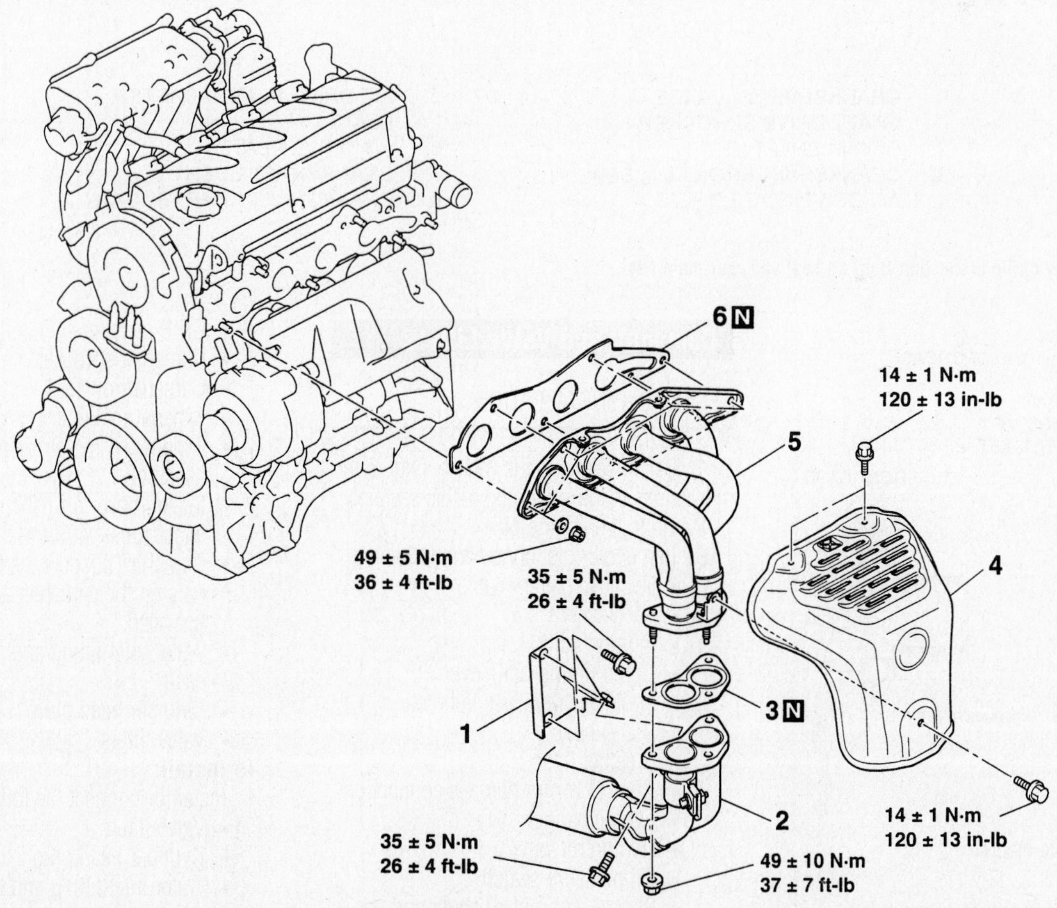

Exploded view of the exhaust manifold mounting

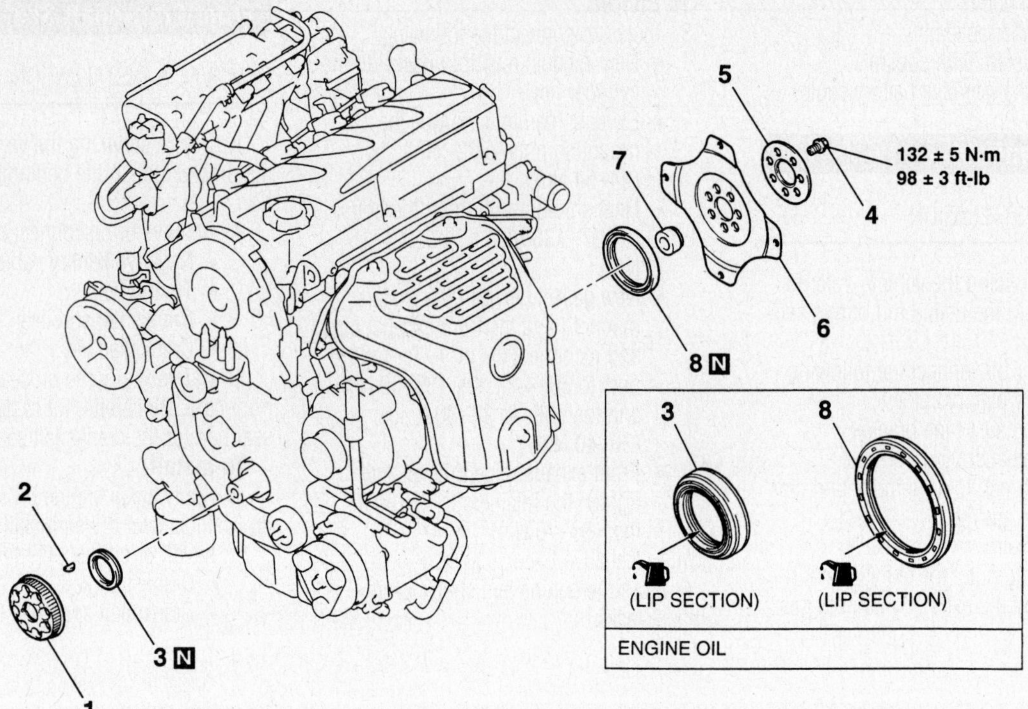

1. CRANKSHAFT BALANCER SHAFT DRIVE SPROCKET
2. CRANK SHAFT KEY
3. CRANKSHAFT FRONT OIL SEAL
4. A/T DRIVE PLATE BOLTS
5. A/T DRIVE PLATE ADAPTER PLATE
6. A/T DRIVE PLATE
7. CRANKSHAFT BUSH
8. CRANKSHAFT REAR OIL SEAL

67170-OUTL-G19

Exploded view of the crankshaft front oil seal and rear main seal

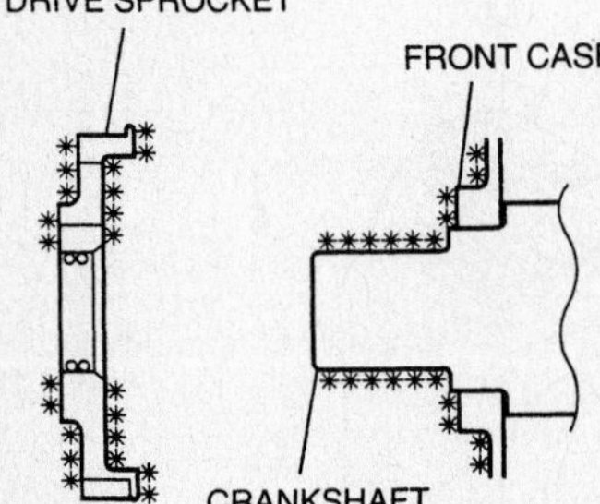

Clean and degrease, then install the crankshaft sprocket in the direction shown

degrease the sprocket and install as shown in the accompanying figure.

- Timing belt
- Negative battery cable

Camshaft and Valve Lifters

REMOVAL & INSTALLATION

1. Before servicing the vehicle, refer to the precautions in the beginning of this section.
2. Drain the engine oil.
3. Remove or disconnect the following:
   - Negative battery cable
   - Air cleaner
   - Ignition coils
   - Upper timing belt cover
   - Positive Crankcase Ventilation (PCV) hose
   - Breather hose
   - Control wiring harness connection
   - Engine hanger
   - Oil control valve
   - Oil pressure switch
   - Rocker arm (valve) cover and gasket
   - Oil seals, if necessary
   - Accumulator assembly
   - Timing belt
   - Camshaft Position (CMP) sensor support
   - CMP sensing cylinder
   - Camshaft sprocket
   - Camshaft oil seal
   - Exhaust rocker arm shaft caps
   - Exhaust rocker arm and shaft assembly
   - Intake rocker arm shaft caps
   - Intake rocker arm and shaft assembly
   - Camshaft. Inspect the bearing journals on the camshaft and the cylinder head.
   - Water inlet fitting and thermostat case
   - Cylinder head plug
   - Valve lifters

To install:

4. Install or connect the following:
   - Valve lifters
   - Cylinder head plug
   - Water inlet fitting and thermostat case assembly
5. Lubricate the camshaft journals and camshaft with clean engine oil
   - Camshaft. Make sure the dowel pin of the camshaft is in the proper position.

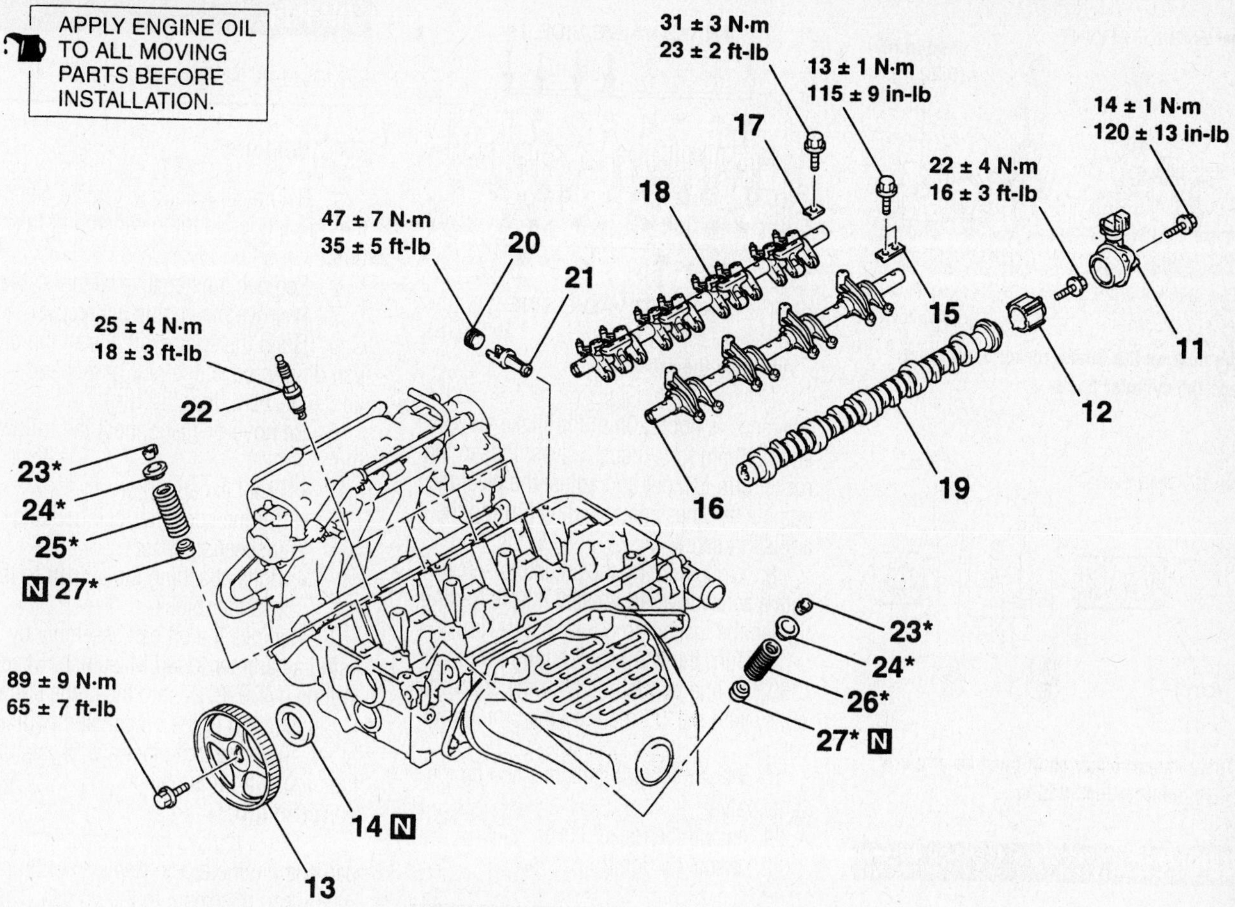

APPLY ENGINE OIL TO ALL MOVING PARTS BEFORE INSTALLATION.
31 ± 3 N·m
23 ± 2 ft-lb
13 ± 1 N·m
115 ± 9 in-lb
14 ± 1 N·m
120 ± 13 in-lb
17
22 ± 4 N·m
16 ± 3 ft-lb
18
47 ± 7 N·m
35 ± 5 ft-lb
20
21
15
11
25 ± 4 N·m
18 ± 3 ft-lb
12
22
23*
24*
19
16
25*
N 27*
23*
24*
26*
89 ± 9 N·m
65 ± 7 ft-lb
27* N
14 N
13

11. CAMSHAFT POSITION SENSOR SUPPORT
12. CAMSHAFT POSITION SENSING CYLINDER
13. CAMSHAFT SPROCKET
14. CAMSHAFT OIL SEAL
15. EXHAUST ROCKER ARM SHAFT CAPS
16. EXHAUST ROCKER ARM AND SHAFT ASSEMBLY
17. INLET ROCKER ARM SHAFT CAPS
18. INLET ROCKER ARM AND SHAFT ASSEMBLY
19. CAMSHAFT
20. CYLINDER HEAD PLUG
21. OIL CONTROL VALVE FILTER
22. SPARK PLUGS
23. VALVE SPRING RETAINER LOCKS
24. VALVE SPRING RETAINERS
25. INLET VALVE SPRINGS
26. EXHAUST VALVE SPRINGS
27. VALVE STEM SEALS

67170-OUTL-G13

Exploded view of the rocker arms, camshaft and related components

DOWEL PIN
67170-OUTL-G72

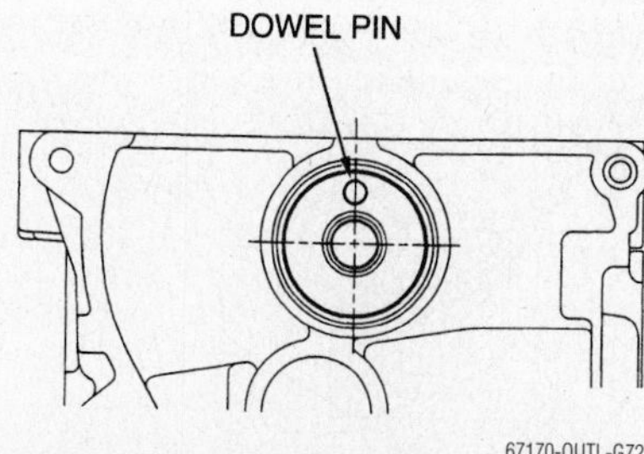

When installing the camshaft, make sure the dowel pin of the camshaft is in the proper position

6. Install the intake rocker arm and shaft assembly as follows:
   a. Place the intake rocker shaft so that the 0.22 in. (5.5mm) hole faces the cylinder head.
   b. Install the intake rocker arm shaft caps.
   c. Tighten the intake rocker shaft mounting bolts to 21–25 ft. lbs. (28–34 Nm).
7. Install the exhaust rocker arm and shaft assembly as follows:
   a. Place the exhaust rocker shaft so that its notch is positioned as shown in the accompanying figure.
   b. Install the exhaust rocker arm shaft caps.
   c. Tighten the exhaust rocker shaft mounting bolts to 106–124 inch lbs. (12–14 Nm).
8. The remainder of installation is the reverse of the removal procedure.
9. Fill the crankcase with oil and connect the negative battery cable.

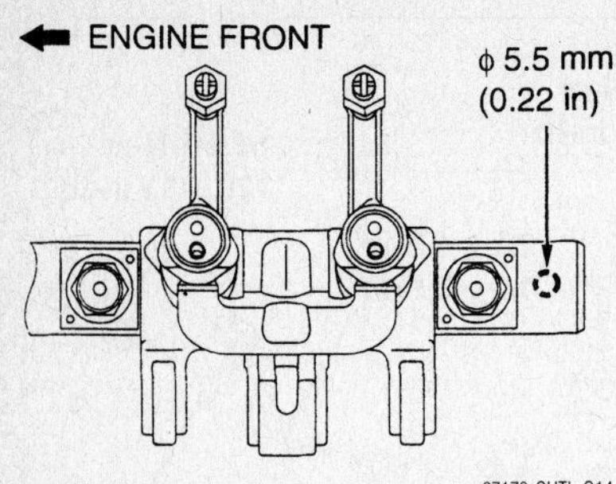

The hole on the intake rocker shaft must face the cylinder head

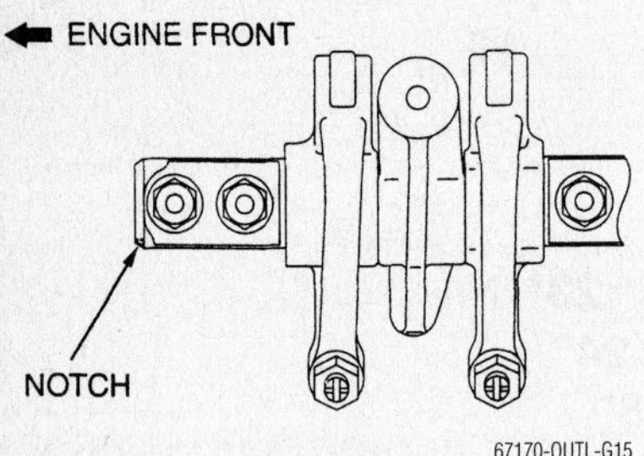

The exhaust rocker shaft must be properly positioned for installation

Valve Lash

ADJUSTMENT

➡Before inspection, check that the engine oil, starter and battery are operating at a normal range. Also make sure the engine coolant is 176–203°F (80–95°C), all lights and accessories are off and the transaxle is in Park.

1. Remove the ignition coils.
2. Remove the rocker arm (valve) cover.
3. Turn the crankshaft clockwise until the notch on the pulley is lines up with the "T" mark on the timing indicator.
4. Move the rocker arms on the No. 1 and 4 cylinders up and down by hand to determine which cylinder had its piston at Top Dead Center (TDC) of the compression stroke.
5. If both intake and exhaust rocker arms have a valve lash, the piston in the cylinder corresponding to these rocker arms is at TDC.
6. Valve clearance inspection and adjustment can be performed on the rocker arms denoted by the white arrows in the figure, when the No. 1 cylinder piston is at TDC and on rocker denoted by the black arrows in the figure when the No. 4 cylinder piston is at TDC.
7. Measure the valve clearance. If the clearance is not 0.20mm for intake valves and 0.30mm for exhaust valves, loosen the rocker arm locknut and adjust the clearance using a thickness gauge while turning the adjusting screw.

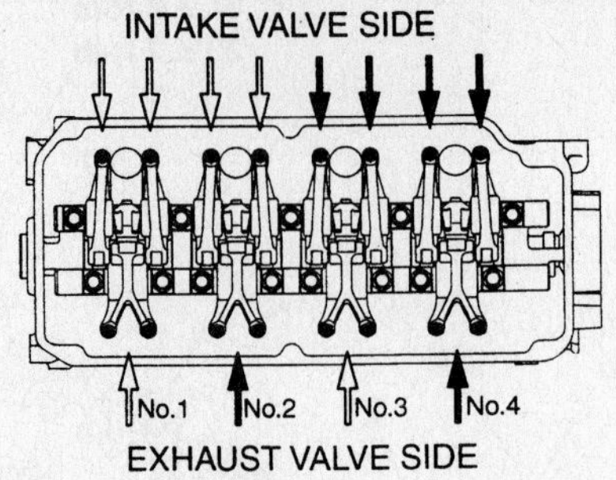

Adjusting the valve lash

8. While holding the adjusting screw with a screwdriver to prevent it from turning, tighten the locknut to 6.5 ft. lbs. (9 Nm).
9. Turn the crankshaft a full rotation (360°) to line up the notch on the crankshaft pulley with the "T" mark on the timing indicator.
10. Repeat the adjustment steps on the other valves.
11. Install the rocker cover.
12. Install the ignition coils.

Starter Motor

REMOVAL & INSTALLATION

1. Remove or disconnect the following:
   - Negative battery cable
   - Air duct
   - Exhaust manifold heat shield
   - Starter cover
   - Harness clamp
   - Starter electrical connector and terminal
   - Mounting bolts and starter
   - Starter cover bracket, if necessary

To install:

2. Install or connect the following:
   - Starter cover bracket, if removed. Tighten the bolts to 29–43 ft. lbs. (40–58 Nm).
   - Starter. Tighten the mounting bolts to 21–25 ft. lbs. (27–33 Nm)
   - Starter terminal and electrical connector
   - Harness clamp
   - Starter cover. Tighten the bolts to 63–97 inch lbs. (7–11 Nm).
   - Exhaust manifold heat shield
   - Air duct
   - Negative battery cable and check the starter for proper operation

Oil Pan

REMOVAL & INSTALLATION

FWD Vehicles

1. Before servicing the vehicle, refer to the precautions in the beginning of this section.
2. Remove the negative battery cable.
3. Remove the engine undercover.
4. Drain the engine oil. Install the oil pan drain plug with a new gasket and tighten to 29 ft. lbs. (39 Nm).
5. Remove or disconnect the following:
   - Front exhaust pipe
   - Torque converter/bell housing inspection cover
   - Bolts attaching the oil pan to the cylinder block
6. Remove the oil pan assembly by using a hammer to tap Mitsubishi Special tool no. MD998727 into the range shown in the illustration of the oil pan and engine block. Slide the tool sideways to separate the oil pan from the block.

To install:

7. Clean the sealing surface on the oil pan and engine block. Apply a continuous 4mm bead of sealant to the oil pan. The oil pan MUST be installed within 15 minutes of applying the sealant.
8. Install or connect the following:
   - Oil pan to the cylinder block and tighten the bolts to 80 inch lbs. (9 Nm)
   - Torque converter/bell housing inspection cover. Torque the bottom bolts to 19 ft. lbs. (26 Nm) and the flange bolts to 80 inch lbs. (9 Nm). Refer to accompanying illustration.
   - Front exhaust pipe
   - Engine undercover

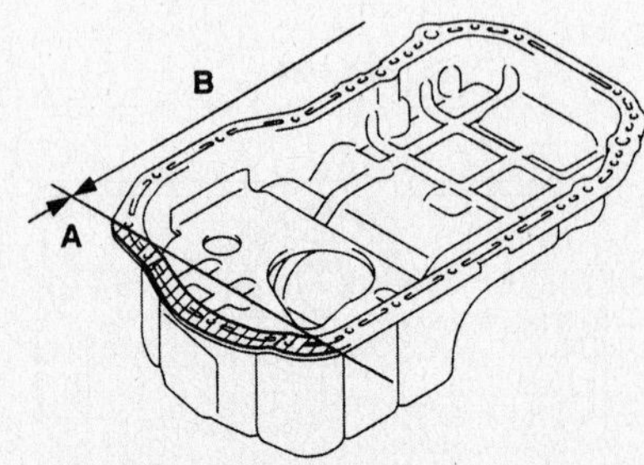

Use the special tool between the oil pan and block in area B only. Do NOT insert the pan separator tool into area A

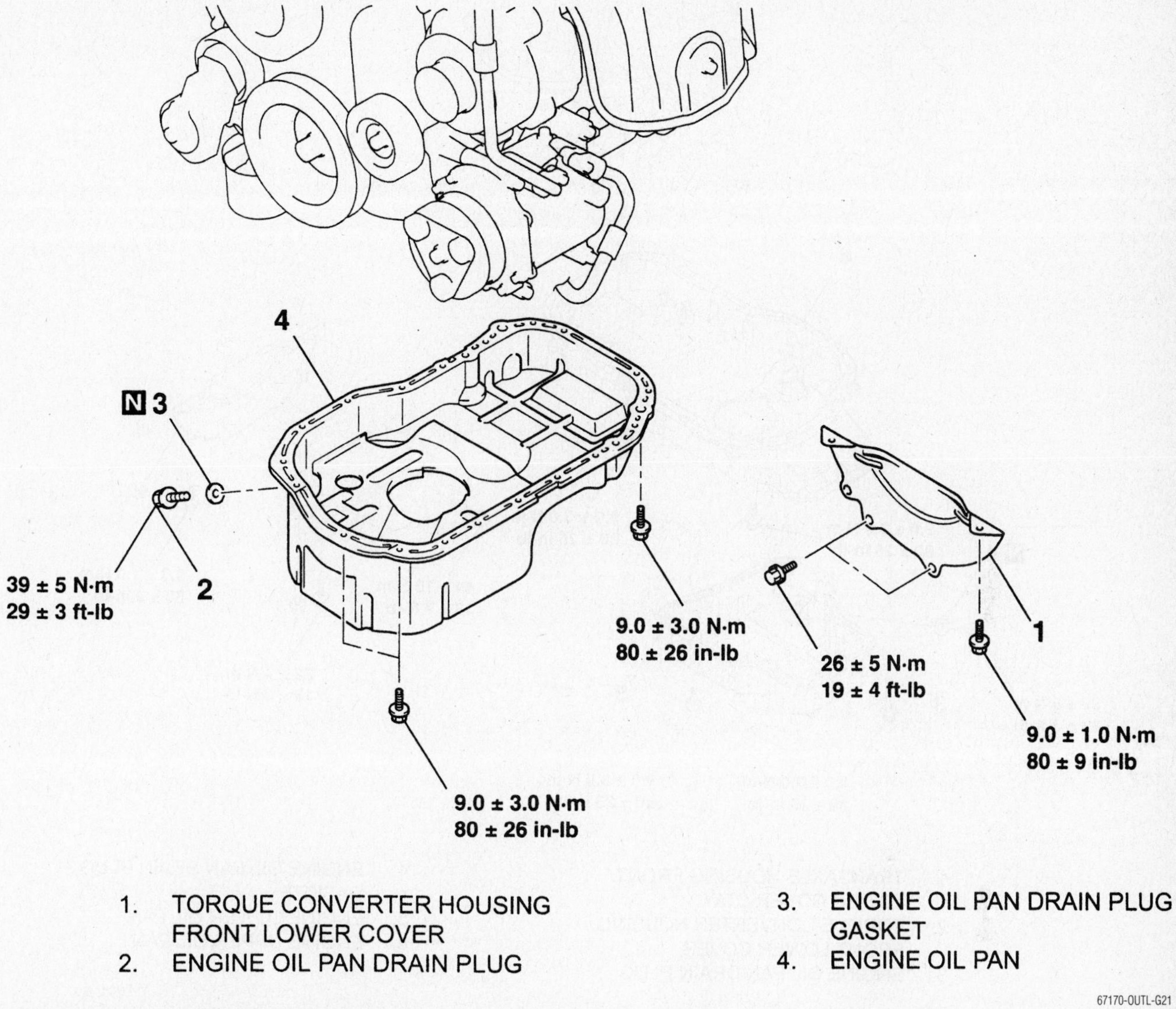

39 ± 5 N·m
29 ± 3 ft-lb

9.0 ± 3.0 N·m
80 ± 26 in-lb

26 ± 5 N·m
19 ± 4 ft-lb

9.0 ± 1.0 N·m
80 ± 9 in-lb

9.0 ± 3.0 N·m
80 ± 26 in-lb

1. TORQUE CONVERTER HOUSING FRONT LOWER COVER
2. ENGINE OIL PAN DRAIN PLUG
3. ENGINE OIL PAN DRAIN PLUG GASKET
4. ENGINE OIL PAN

67170-OUTL-G21

Exploded view of the oil pan and related components

⁂ WARNING

Wait at least 1 hour after tightening the oil pan bolts to refill the crankcase and start the engine.

9. Refill the engine with oil. Connect the negative battery cable. Start the engine and check for leaks.

AWD Vehicles

1. Before servicing the vehicle, refer to the precautions in the beginning of this section.
2. Remove the negative battery cable.
3. Remove the engine undercover.
4. Drain the engine oil. Install the oil pan drain plug with a new gasket and tighten to 29 ft. lbs. (39 Nm).
5. Remove or disconnect the following:
   - Front exhaust pipe
   - Center member
   - Transaxle housing front lower cover stay
   - Torque converter housing front lower (inspection) cover

➡There are 2 sizes of oil pan bolts used. Note the locations of the bolts as you remove them.

   - Lower oil pan mounting bolts. Place a piece of wood at the rear of the lower pan and strike it with a hammer to remove the lower pan.
   - Upper oil pan mounting bolts. Screw bolts into the bolt holes (A) shown in the illustration, then lift and remove the upper oil pan.

To install:

6. Clean the sealing surface on the oil pan and engine block. Apply a continuous 4mm bead of sealant to the oil pan. The oil pan MUST be installed within 15 minutes of applying the sealant.

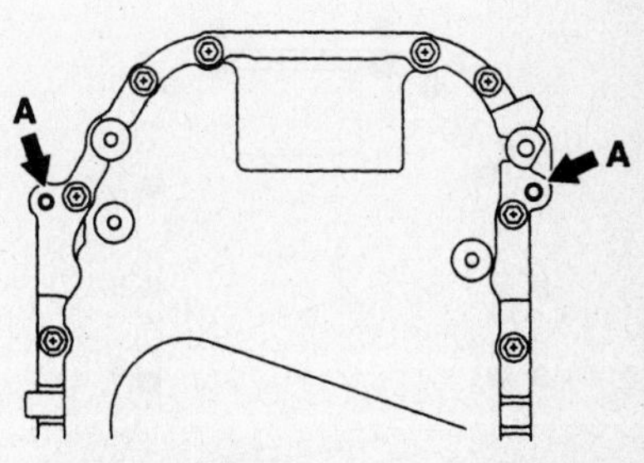

To remove the upper oil pan, screw bolts into the bolt holes A, then lift and remove the upper oil pan.

7. Install or connect the following:
   - Upper oil pan to the block and tighten the bolts to 80 inch lbs. (9 Nm)

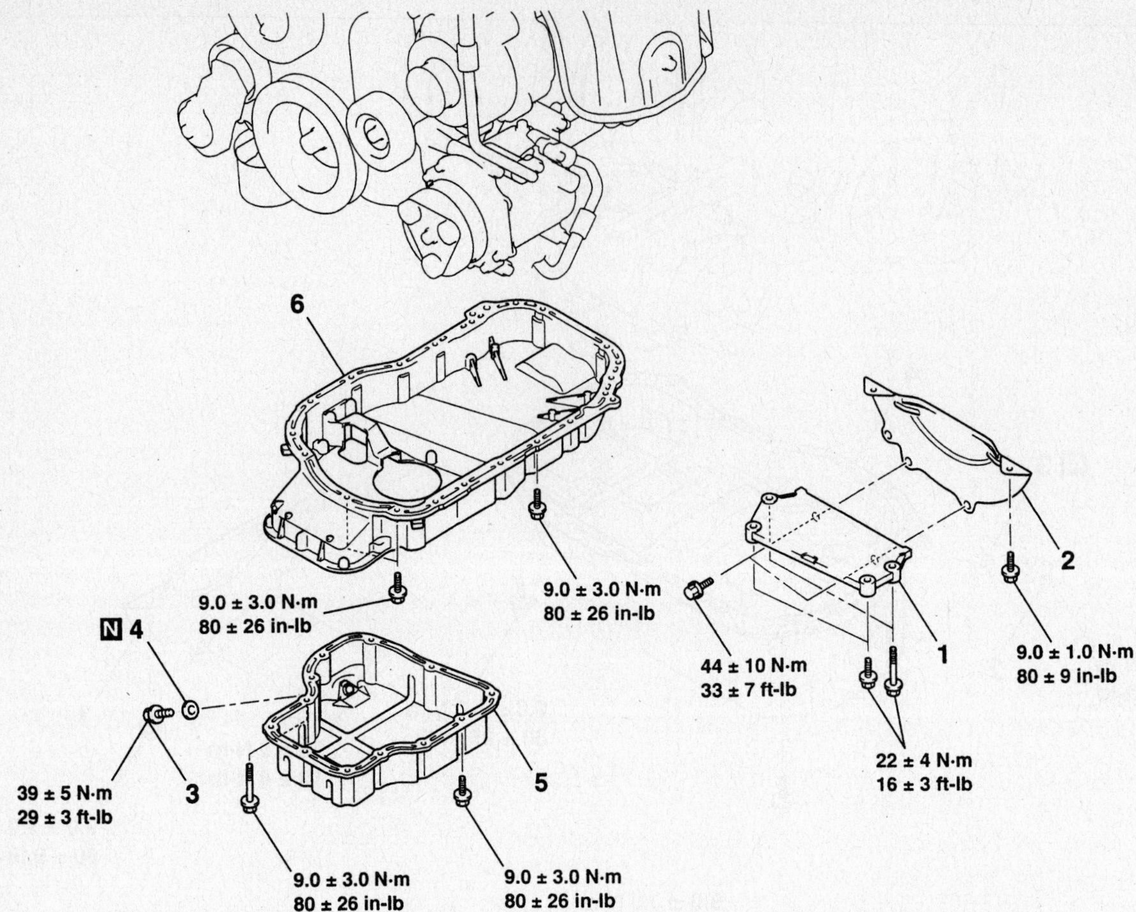

1. TRANSAXLE HOUSING FRONT LOWER COVER STAY
2. TORQUE CONVERTER HOUSING FRONT LOWER COVER
3. ENGINE OIL PAN DRAIN PLUG
4. ENGINE OIL PAN DRAIN PLUG GASKET
5. ENGINE LOWER OIL PAN
6. ENGINE UPPER OIL PAN

67170-OUTL-G23

Exploded view of the upper and lower oil pans used on AWD Outlanders

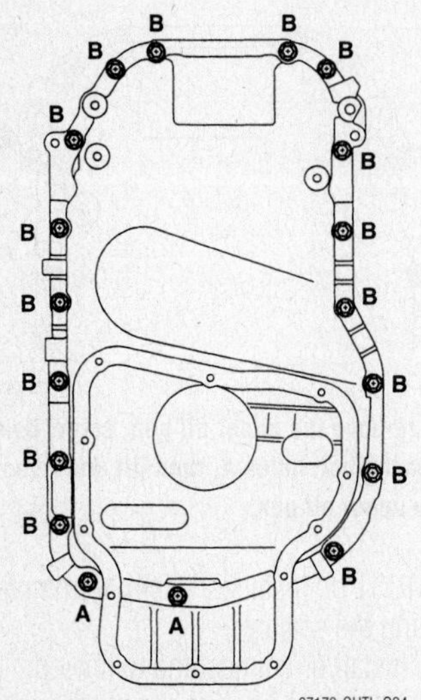

Upper oil pan bolt locations—AWD vehicles

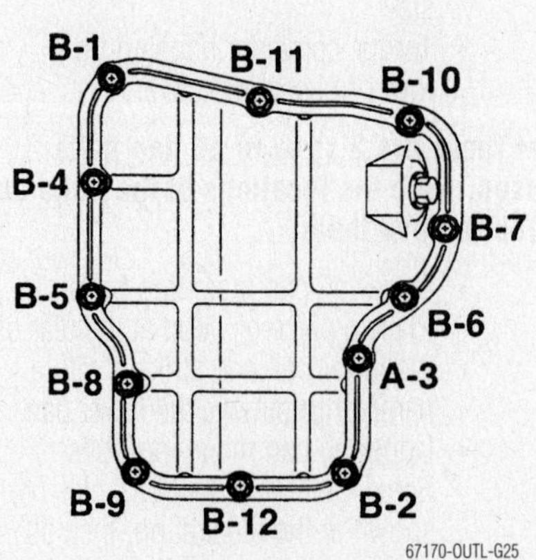

Lower oil pan bolt locations—AWD vehicles

- Lower oil pan to the block and tighten the bolts to 80 inch lbs. (9 Nm)
- Torque converter lower (inspection) cover. Torque the bolts to 80 inch lbs. (9 Nm).
- Transaxle housing front lower cover stay. Tighten the bolts as shown in the accompanying illustration.
- Center member
- Front exhaust pipe
- Engine undercover

WARNING

Wait at least 1 hour after tightening the oil pan bolts to refill the crankcase and start the engine.

8. Refill the engine with oil. Connect the negative battery cable. Start the engine and check for leaks.

Oil Pump

REMOVAL & INSTALLATION

1. Before servicing the vehicle, refer to the precautions in the beginning of this section.
2. Remove the negative battery cable.
3. Remove the engine undercover.
4. Drain the engine oil. Install the oil pan drain plug with a new gasket and tighten to 29 ft. lbs. (39 Nm).
5. Remove or disconnect the following:
   - Oil pan, FWD vehicles
   - Lower and upper oil pan, AWD vehicles
   - Oil screen and gasket
   - Relief spring
   - Relief plunger
   - Oil filter bracket and gasket
   - Plug. Fit Mitsubishi Special tool no. MD998162 on the plug, then hold it in position with special tool MD998783, loosen the plug, then remove the tools.
   - O-ring
   - Flange bolt by removing the plug on the side of the cylinder block. Insert a Phillips screwdriver into the hold to lock the counterbalance shaft, then loosen the flange bolt.
   - Front case and gasket
   - Oil pump cover
   - Oil pump driven gear
   - Oil pump drive gear
   - Crankshaft front oil seal
   - Oil pump oil seal
   - Left and right counterbalance shafts
   - Front counterbalance shaft bearing, using Mitsubishi Special tool no. MD998371
   - Rear left and right counterbalance shaft bearings using Mitsubishi Special tool nos. MD991603 and MD998372

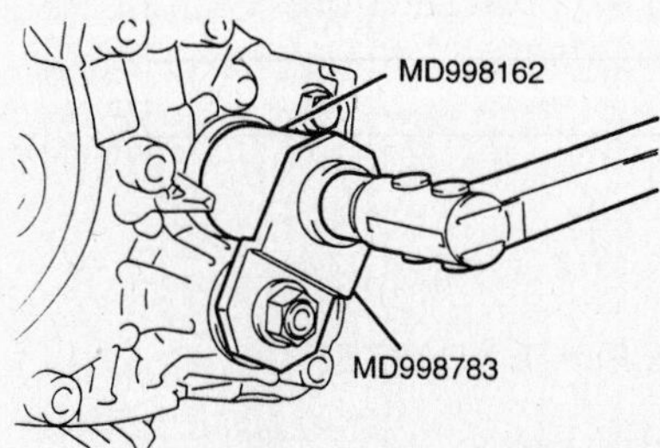

To remove or install the plug, fit special tool MD998162 on the plug and hold it in place with tool MD998783

To install:

6. Install the left counterbalance shaft rear bearing as follows:
   a. Install Mitsubishi Special tool no. MD991603 to the cylinder block.
   b. Apply engine oil to the rear bearing outer surface and bearing hole in the cylinder block.
   c. Use Mitsubishi Special tool MD998705 to install the rear bearing. The left rear bearing has no oil holes.
7. Install the right counterbalance shaft rear bearing as follows:
   a. Install the guide pin of special tool MD998705 in the threaded hole of the cylinder block.
   b. Align the ratchet ball of special tool MD998705 with the oil hold in the rear bearing to install the bearing of the special tool MD998705.
   c. Apply engine oil to the bearing outer surface and bearing hole in the cylinder block.
   d. Use special tool MD998705 to install the rear bearing. Make sure the oil hole of the bearing is aligned with the oil hole of the cylinder block.
8. Install the counterbalance shaft front bearing as follows:
   a. Remove the rear bearing installing part from special tool MD998705.
   b. Install the guide pin of special tool MD998705 into the threaded hole of the cylinder block.
   c. Align the ratchet ball of the tool with the oil hole in the rear bearing to install the bearing of the special tool.
   d. Using the tool, install the rear bearing. Make sure the oil hole of the bearing is aligned with the oil hole of the cylinder block.
9. Use a suitable socket wrench to install the counterbalance shaft oil seal into the front case.
10. Install the oil pump oil seal, using a suitable socket to press the seal into place.
11. Use special tool MD998375 to install the front oil seal into the front case.

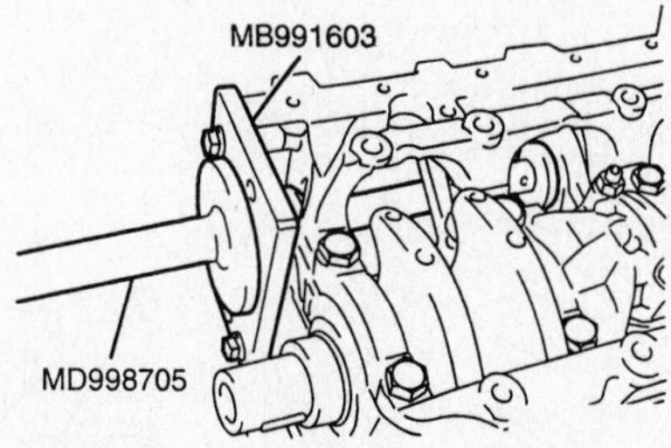

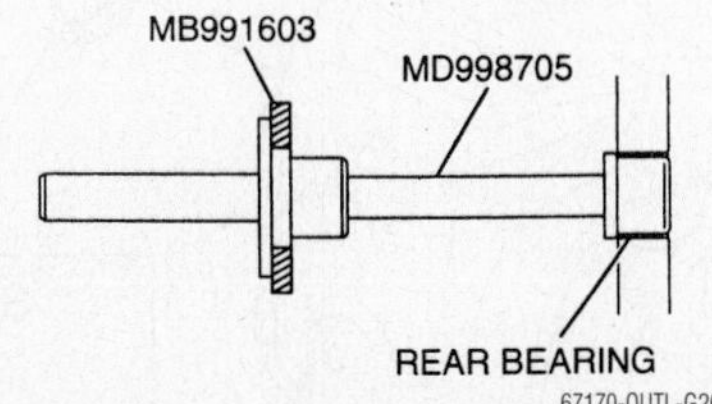

View of the special tools needed to install the left counterbalance shaft rear bearing

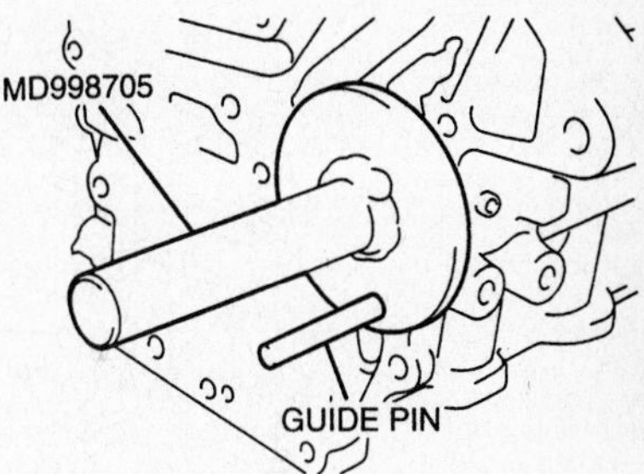

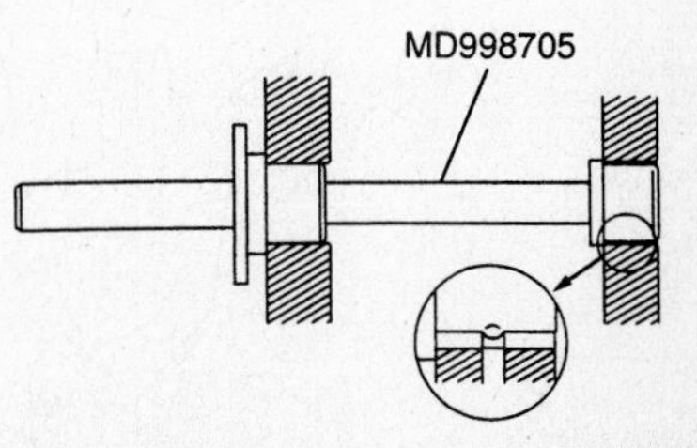

Use special tool MD998705 to install the rear bearing. Make sure the oil hole of the bearing is aligned with the oil hole of the cylinder block

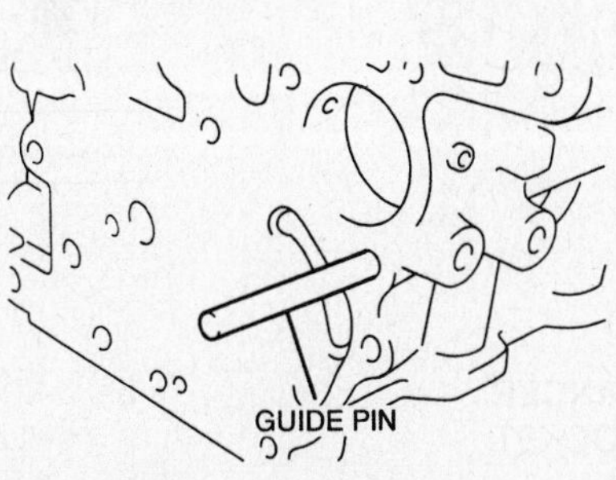

Install the guide pin of special tool MD998705 into the threaded hold of the cylinder block

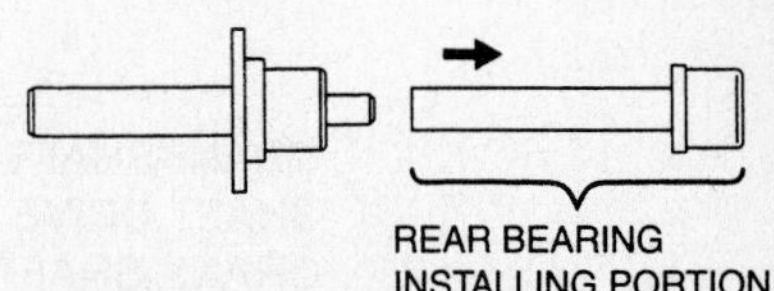

Remove the rear bearing installing part from special tool MD998705

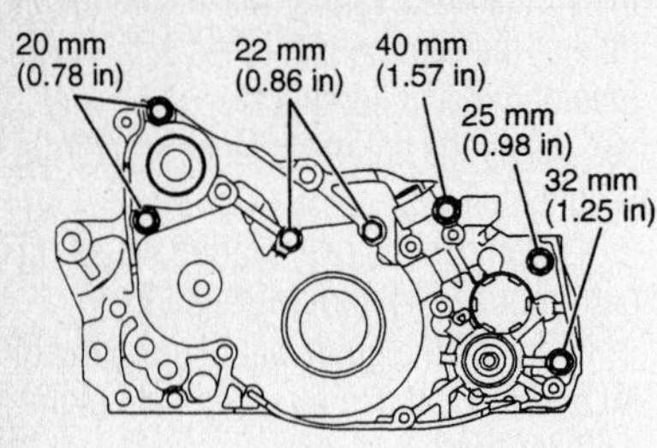

The front case bolts are all different lengths. Make sure to install the bolts in their proper locations

12. Install the oil pump gears into the front case and make sure the alignment marks line up.
13. Install the oil pump case as follows:
   a. Place special tool MD998285 on the front end of the crankshaft, then apply a thin coating of engine oil to the outer surface of the special tool.
   b. Apply engine oil to the lip of the crankshaft front oil seal.
   c. Install the front case. Be careful not to damage the oil seal.
   d. Install the bolts, noting their different lengths and tighten to 17 ft. lbs. (23 Nm).
14. Install the flange bolt, using a screwdriver to lock the counterbalance shaft and tighten to 27 ft. lbs. (36 Nm). Pull out the screwdriver and screw the plug back in.
15. Install the plug as follows:
   a. Install a new O-ring to the groove of the front case.
   b. Install the plug to the front case.
   c. Fit special tool MD998162 on the plug and hold it in place with tool MD998783.
   d. Tighten the plug to 17 ft. lbs. (23 Nm). Remove the special tools.
16. Install or connect the following:
   - Oil filter bracket and gasket
   - Relief plunger
   - Relief spring
   - Oil screen and gasket
   - Upper and lower oil pans, AWD vehicles
   - Oil pan, FWD vehicles

WARNING

Wait at least 1 hour after tightening the oil pan bolts to refill the crankcase and start the engine.

17. Refill the engine with oil. Connect the negative battery cable. Start the engine and check for leaks.

Rear Main Seal

REMOVAL & INSTALLATION

1. Before servicing the vehicle, refer to the precautions in the beginning of this section.
2. Remove or disconnect the following:

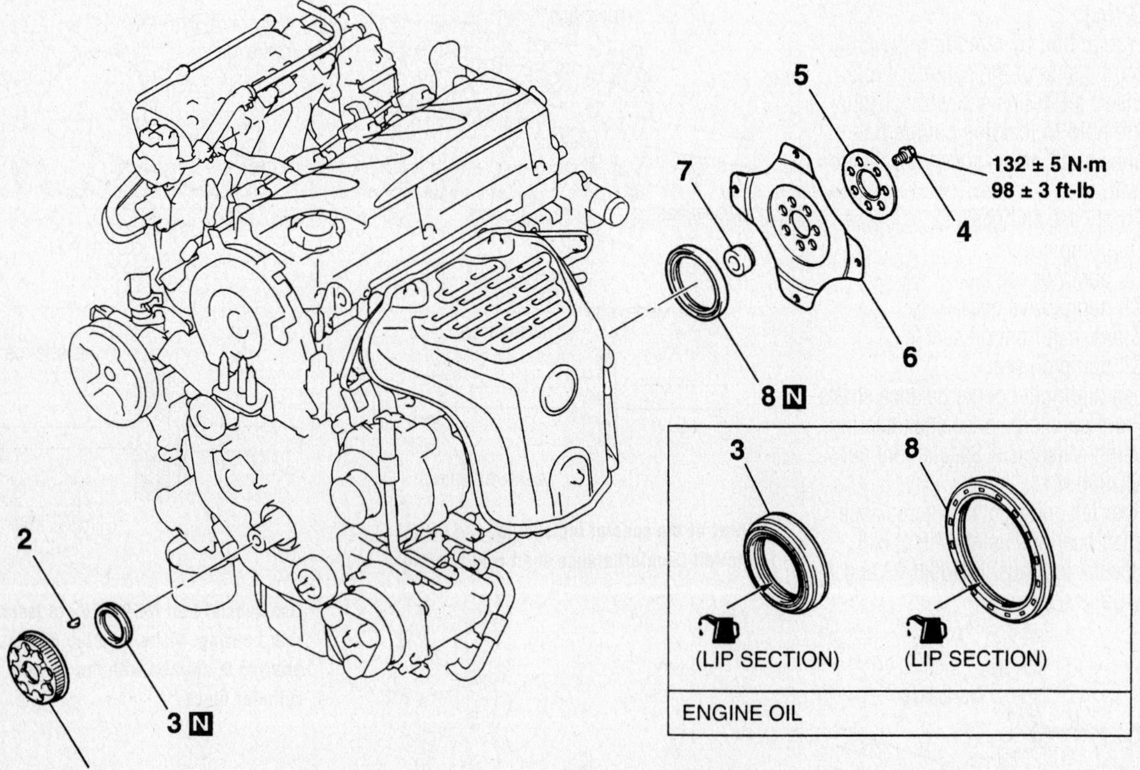

Exploded view of the crankshaft front oil seal and rear main seal

- Transaxle
- Driveplate bolts. Use special tool MD998781 to hold the driveplate while loosening the bolts.
- Driveplate adapter plate
- Driveplate
- Crankshaft bushing

3. Carefully pry the seal out of the oil seal case without damaging the sealing surface of the crankshaft.

To install:

4. Apply engine oil to the lip of the new seal and install the seal in the case using the proper size seal driver.
5. Install or connect the following:
   - Driveplate
   - Driveplate adapter plate
   - Driveplate bolts. Secure the driveplate with the special tool, as done during removal, then torque to 98 ft. lbs. (132 Nm).
   - Transaxle

Timing Belt

REMOVAL & INSTALLATION

1. Before servicing the vehicle, refer to the precautions in the beginning of this section.
2. Remove or disconnect the following:
   - Engine undercover
   - Drive belt
   - Crankshaft damper pulley
   - Control wiring harness connection
   - Battery wiring harness connection
   - Connector
   - Engine front mounting bracket
   - Harness bracket
   - Upper timing belt cover
   - Water pump pulley
   - Idler pulley
   - Auto-tensioner
   - Lower timing belt cover

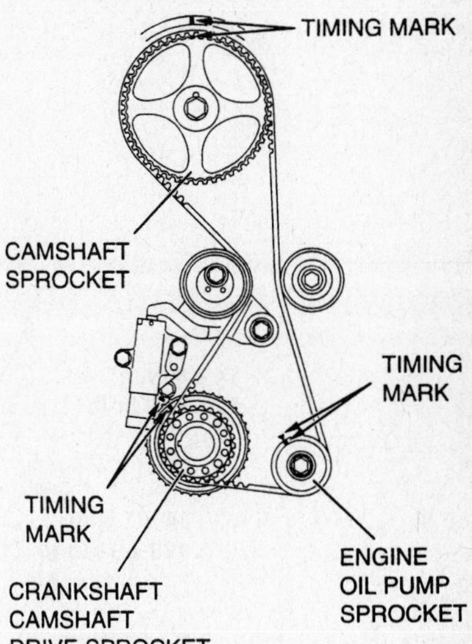

View of the alignment when the No. 1 piston is at TDC

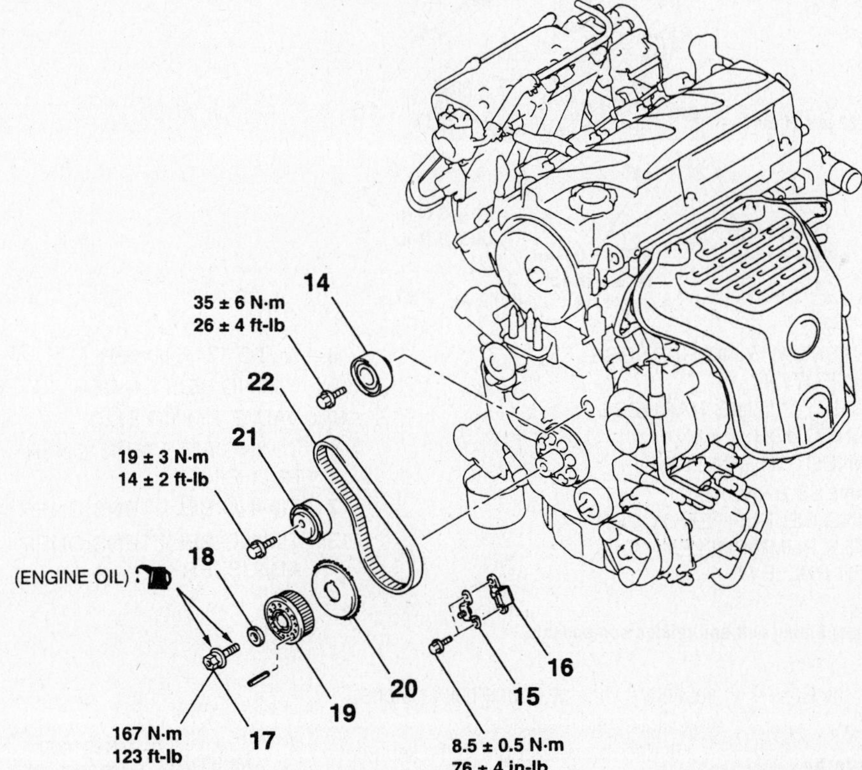

14. TIMING BELT IDLER PULLEY
15. TIMING BELT LOWER COVER BRACKET
16. CRANKSHAFT POSITION SENSOR
17. CRANKSHAFT PULLEY CENTER BOLT
18. CRANKSHAFT PULLEY WASHER
19. CRANKSHAFT CAMSHAFT DRIVE SPROCKET
20. CRANKSHAFT ANGLE SENSING BLADE
21. BALANCER TIMING BELT TENSIONER
22. BALANCER TIMING BELT

67170-OUTL-G40

Exploded view of the valve (outer) timing belt and related components

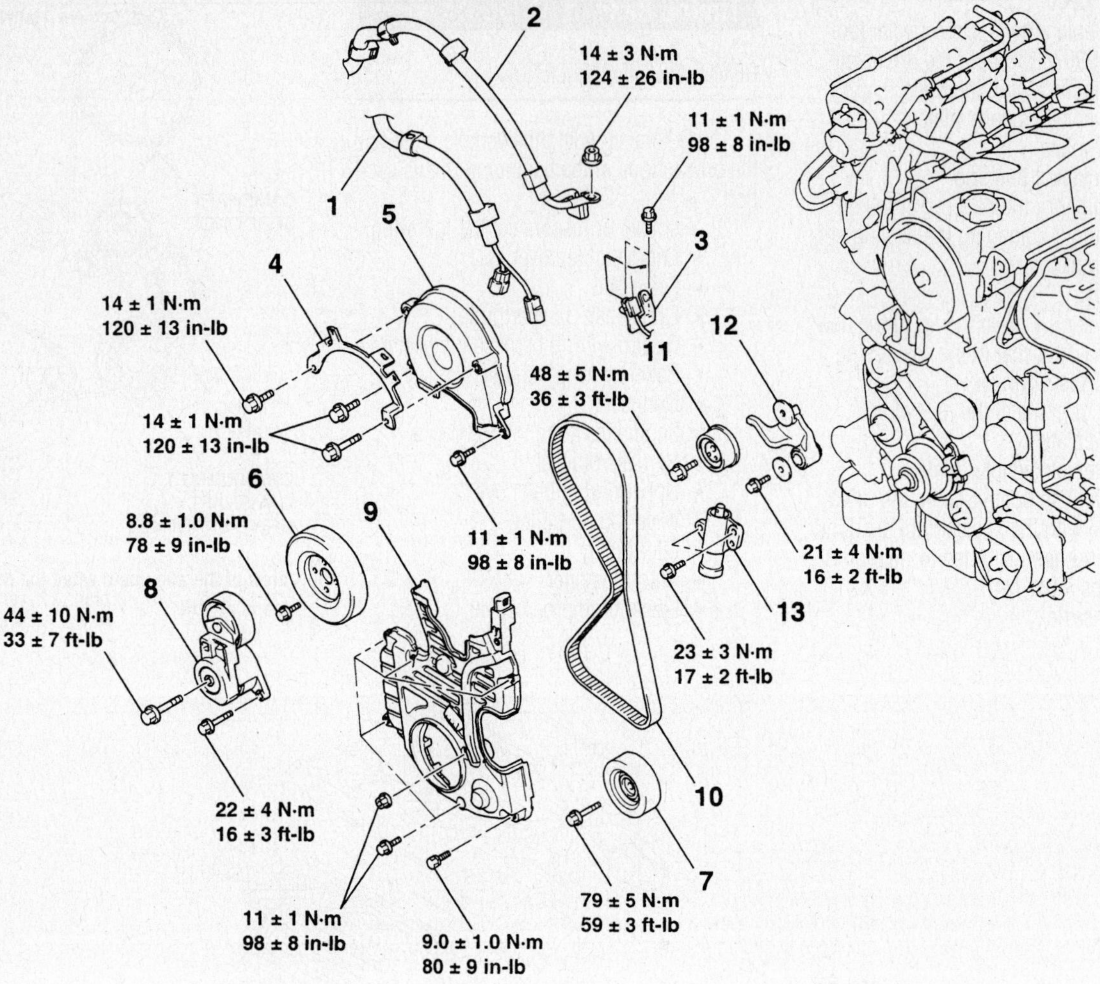

1. CONTROL WIRING HARNESS CONNECTION
2. BATTERY WIRING HARNESS CONNECTION
3. CONNECTOR BRACKET
4. HARNESS BRACKET
5. TIMING BELT UPPER COVER
6. WATER PUMP PULLEY
7. IDLER PULLEY
8. AUTO-TENSIONER
9. TIMING BELT LOWER COVER
10. VALVE TIMING BELT
11. TIMING BELT TENSIONER PULLEY
12. TIMING BELT TENSIONER ARM
13. TIMING BELT TENSIONER ADJUSTER

67170-OUTL-G41

Exploded view of the balancer (inner) timing belt and related components

3. Position the engine so that the No. 1 piston is at Top Dead Center (TDC).

➡If the timing belts are going to be reused, mark the direction of rotation on the belt. This will ensure the belt is reinstalled in same direction, extending belt life.

4. To loosen the timing (outer) belt tensioner, remove the rubber plug, then install Mitsubishi Special tool MD998738 to the slot and screw inward until the it contacts the timing belt tensioner arm.

5. Gradually screw in the tool, then align the timing belt tensioner adjuster rod set hold A with the timing belt tensioner adjusted cylinder set hole B.

6. Insert a wire or pin in the set hole aligned.

7. Loosen the timing belt tensioner pulley bolts and remove the outer timing belt.

8. Remove or disconnect the following:
   - Timing belt tensioner pulley
   - Timing belt tensioner arm
   - Timing belt tensioner adjuster
   - Timing belt idler pulley
   - Timing belt lower cover bracket
   - Crankshaft Position (CKP) sensor

9. Hold the crankshaft/camshaft drive

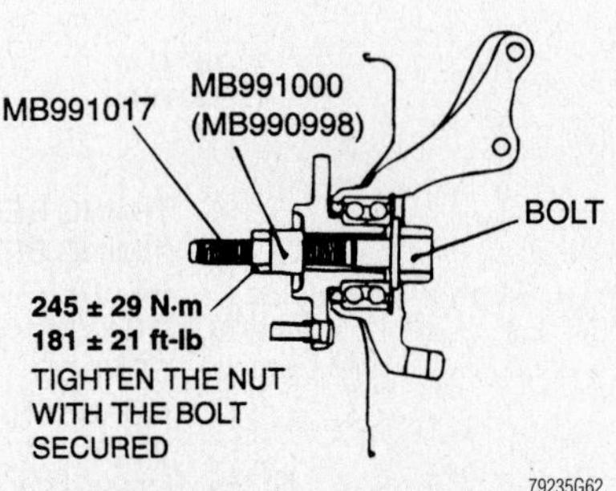

Proper alignment of the timing belt sprocket marks for belt service— 2.4L engines

sprocket with Mitsubishi Special tools MD991367 and MD991385. Loosen the crankshaft pulley center bolt, then remove the crankshaft pulley washer and sprocket.

- Crankshaft angle sensing blade
- Balancer timing belt tensioner and balancer timing (inner) belt.

To install:

WARNING

Do not spray or immerse the sprockets or tensioners in cleaning solvent. The sprocket may absorb the solvent and transfer it to the belt. The tensioners are internally lubricated and the solvent will dilute or dissolve the lubricant.

10. Align the timing marks of the balancer shaft sprockets and the crankshaft sprocket with the timing marks on the front case. Route the timing belt around the sprockets so there is no slack in the upper span of the belt and the timing marks are still aligned.

11. Assemble and temporarily hold the center of the balancer timing belt tensioner pulley so it is at the top left from the center of the assembling bolt, and the pulley flange is at the front side of the engine.

➡When tightening the mounting bolts, make sure the tensioner does not turn with the bolts.

12. Adjust the balancer belt tension as follows:

a. Use your fingers to lift the tensioner in the direction of the arrow shown in the accompanying figure. Apply

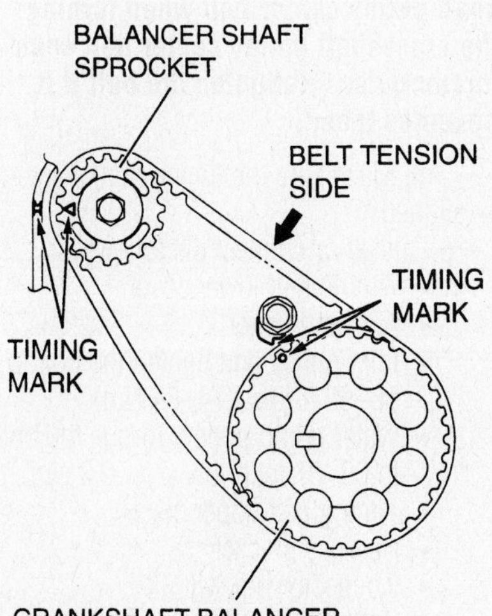

Balancer timing belt "B" installation mark alignment

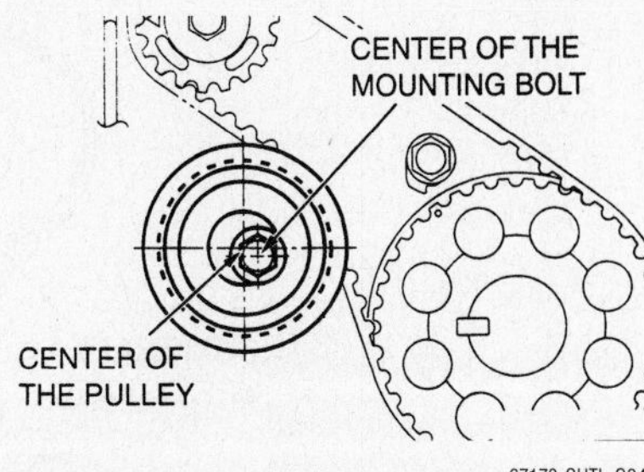

Assemble and temporarily hold the center of the balancer belt tensioner pulley so it's at the top left from the center of the assembling bolt, and the pulley flange is at the front side of the engine

22–30 inch lbs. (2.6–3.4 Nm) to the timing belt so the belt is taut without any looseness.

b. Tighten the assembling bolt to 12–16 ft. lbs. (16–22 Nm). Then, fix the timing belt tensioner.

c. Turn the crankshaft clockwise 2 revolutions to set the No. 1 cylinder to TDC of its compression stroke and check that the timing marks are still aligned.

d. Apply about 22 lbs. (100 N) of pressure to the center area between the sprockets, then check the belt deflection. It should measure 0.20–0.27 in. (5–7mm). If not within specifications, adjust the belt tension again.

13. Clean the crankshaft, crankshaft angle sensing blade, drive sprocket and crankshaft pulley washer, as shown.

14. Install the crankshaft angle sending blade and drive sprocket in the direction shown in the accompanying figure.

15. Place the larger chamfer side of the pulley washer in the direction shown, then assemble on the crankshaft pulley center bolt.

16. Apply a small amount of engine oil to the crankshaft pulley center bolt bearing surface and screw.

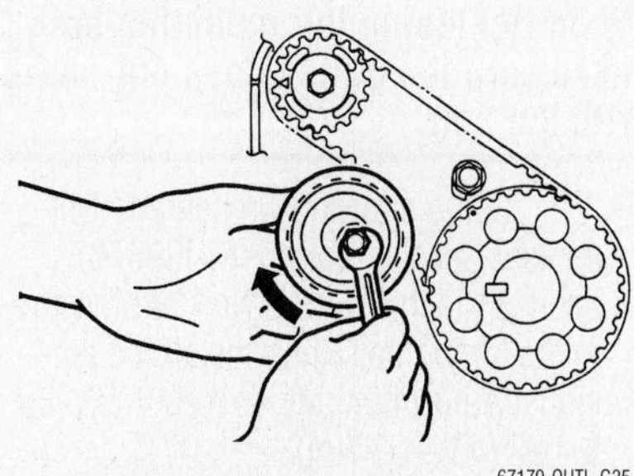

Lift the tensioner in the direction of the arrow, then apply 22–30 inch lbs. (2.6–3.4 Nm) of pressure to the timing belt so the it is taut without any looseness

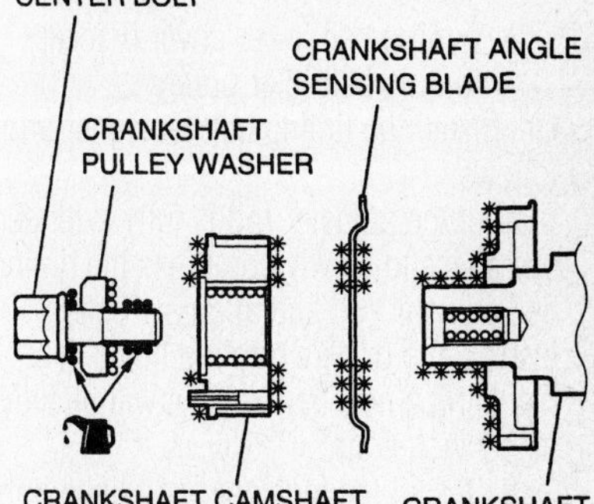

Crankshaft, sensing blade, drive sprocket and pulley washer cleaning and installation direction

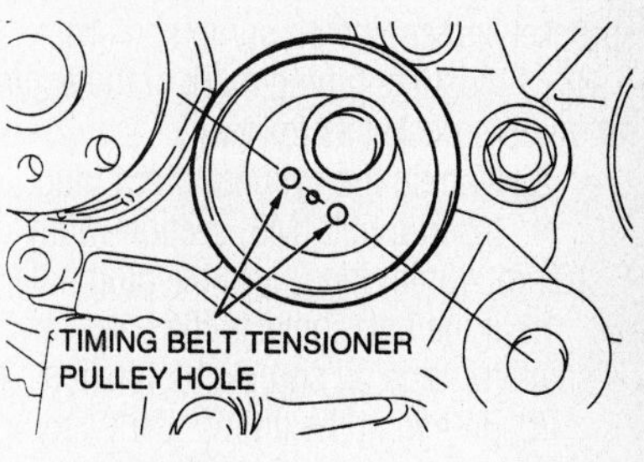

Temporarily tighten the timing belt tensioner pulley as shown

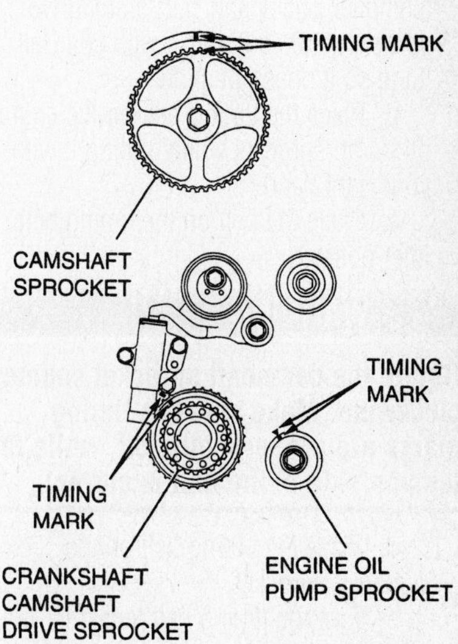

Properly align the timing marks on the sprockets

17. Hold the drive sprocket with Mitsubishi Special tool nos. MB991367 and MD991385, then torque the center bolt to 123 ft. lbs. (167 Nm).

18. Install or connect the following:
- CKP sensor
- Timing belt lower cover bracket
- Timing belt idler pulley

19. Install the timing belt tensioner arm as follows:

a. If the adjuster rod is fully extended, use a vise to slowly compress the timing belt adjuster rod and align set hole A . Insert a pin or wire into the holes. If installing a new tensioner, it will already be set with a pin.

b. Install the tensioner and tighten the bolts 15–19 ft. lbs. (20–26 Nm). Do not remove the pin until the timing belt is tensioned.

20. Install the timing belt tensioner arm.

21. Install the timing belt tensioner pulley. Temporarily tighten the pulley as shown in the accompanying figure.

22. Align the timing marks on the camshaft sprocket, crankshaft camshaft drive sprocket and engine oil pump sprocket.

23. Adjust the timing mark of the engine oil pump sprocket as follows:

a. Unplug the cylinder block plug,.

b. Insert a bolt (M6, section width 10mm, 45mm long) into the plug hole.

c. If the bolt contacts the balancer shaft, turn the oil sprocket 1 rotation.

d. Re-adjust the timing mark, then check to make sure the bolt fits. Do not remove the bolt until the timing belt is assembled.

24. Install the timing belt as follows:

a. Place the belt on the timing belt tensioner pulley and cam/crank drive sprocket, then support it with your left hand so it does not slide.

b. Place the timing belt on the engine oil pump sprocket while pulling it with your right hand.

c. Place the belt on the timing belt idler pulley.

⁂ WARNING

Rotate the camshaft sprocket counterclockwise. Make sure the timing marks are properly aligned, while the tension side of the belt is correct.

d. Place the timing belt on the camshaft sprocket.

e. Turn the timing belt tensioner pulley in the direction shown, using Mitsubishi special tool MD998767 to apply tension to the timing belt. Temporarily

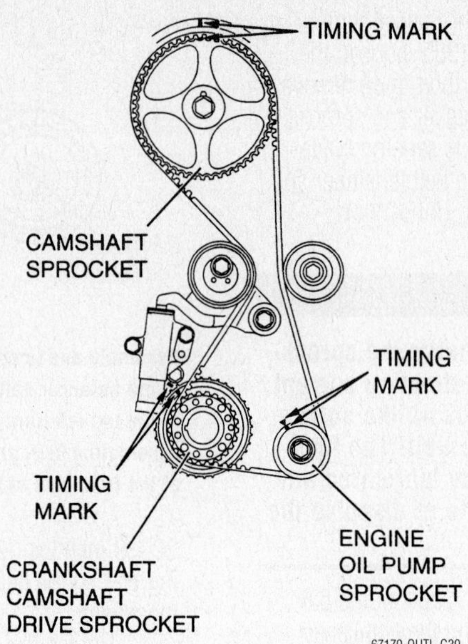

Throughout the procedure, you must make sure the timing marks are still aligned

tighten and fix the tensioner pulley mounting bolt.

f. Make sure the timing marks and still aligned.

g. Remove the bolt from the cylinder block plug hole.

h. Install the plug and tighten to 21–25 ft. lbs. (27–33 Nm).

25. Adjust the timing belt tension as follows:

a. Set Mitsubishi Special tool no. MD998738, used during belt removal.

b. Gradually screw the special tool to a position in which the wire or pin inserted in the tensioner adjuster moves slightly.

c. Rotate the crankshaft counterclockwise ¼-turn.

d. Rotate the crankshaft clockwise, aligning each timing mark to set the No. 1 cylinder to TDC.

e. Loosen the tensioner pulley mounting bolt.

⁂ WARNING

When tightening the mounting bolt, make sure the pulley does not rotate with the bolt.

f. Using a torque wrench and Mitsubishi Special tool no. MD998767, apply 31 inch lbs. (3.5 Nm) of pressure to the timing belt, then install the tensioner pulley bolt and tighten to 33–39 ft. lbs. (43—53 Nm).

g. Remove the wire or pin from the tensioner adjuster.

h. Remove Mitsubishi Special tool no. MD998738, and install the rubber plug to the timing belt undercover.

i. Rotate the crankshaft 2 full rotations in the clockwise direction, then wait 15 minutes to continue the procedure.

j. Insert the wire or pin removed in Step G, and make sure it can be pulled out with a light load. When it can be lightly removed, the timing belt is tensioned properly. If so, remove the pin.

k. If the projection of the timing belt tensioner adjuster rod is within 0.15–0.17 inch .(3.8–4.5mm), proper tension is applied.

l. If the wire or pin cannot be removed easily, the timing belt tension must be readjusted.

➡Always check the torque of the crankshaft pulley center bolt when turning the crankshaft pulley center bolt counterclockwise. Retighten the bolt if it becomes loose.

m. Make sure the timing marks are aligned.

26. Install or connect the following:
- Timing belt lower cover
- Auto-tensioner
- Idler pulley and tighten the bolt to 56–62 ft. lbs. (74–84 Nm)
- Water pump pulley. Torque the bolt to 69–87 inch lbs. (7.8–9.8 Nm).
- Timing belt upper cover
- Harness bracket
- Connector bracket
- Battery wiring harness connection
- Control wiring harness connection
- Crankshaft damper pulley. Torque the bolt to 18 ft. lbs. (25 Nm).
- Engine undercover

Piston and Ring

POSITIONING

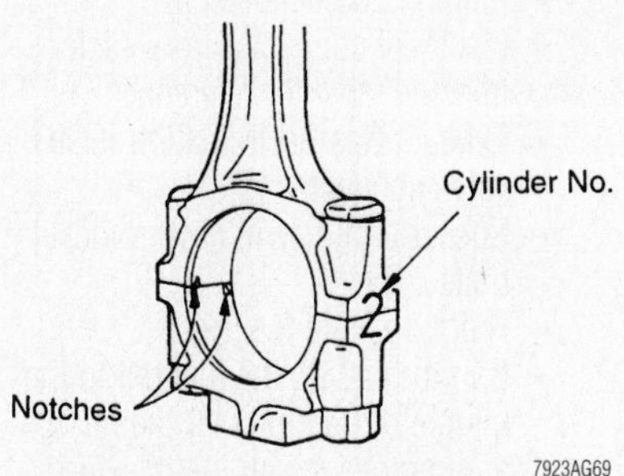

Before removing the caps from the connecting rods, be sure to matchmark them as shown

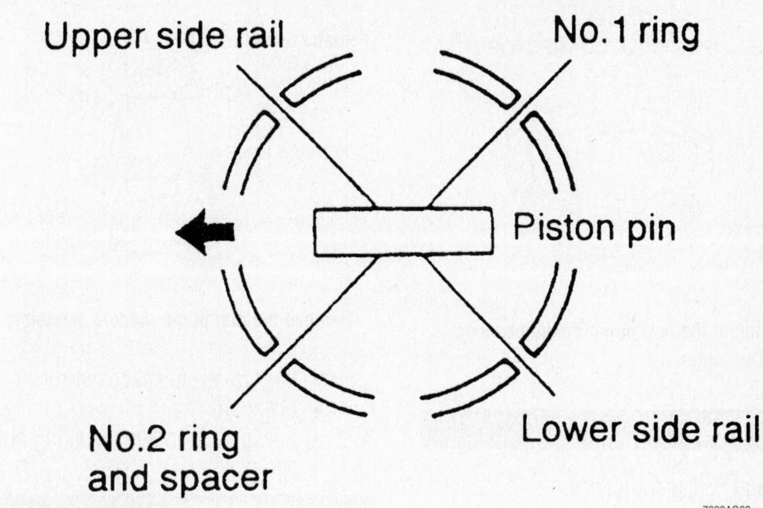

Piston ring end-gap spacing

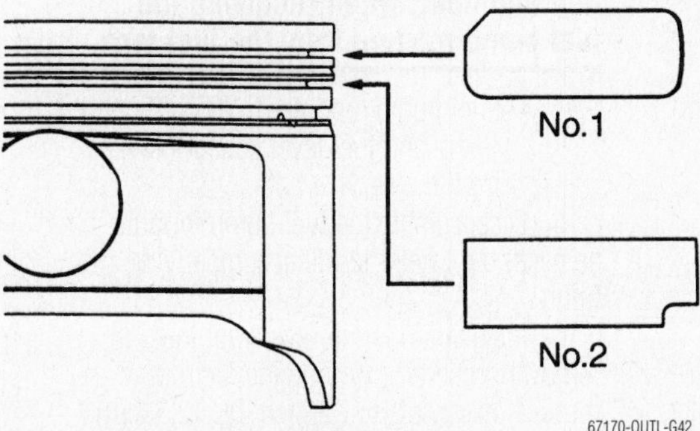

2.4L engines—compression ring identification marks. No. 1 ring: None, No. 2 ring: 2R

FUEL SYSTEM

Fuel System Service Precautions

Safety is the most important factor when performing not only fuel system maintenance but any type of maintenance. Failure to conduct maintenance and repairs in a safe manner may result in serious personal injury or death. Maintenance and testing of the vehicle's fuel system components can be accomplished safely and effectively by adhering to the following rules and guidelines.

- To avoid the possibility of fire and personal injury, always disconnect the negative battery cable unless the repair or test procedure requires that battery voltage be applied.
- Always relieve the fuel system pressure prior to disconnecting any fuel system component (injector, fuel rail, pressure regulator, etc.), fitting or fuel line connection. Exercise extreme caution whenever relieving fuel system pressure, to avoid exposing skin, face and eyes to fuel spray. Please be advised that fuel under pressure may penetrate the skin or any part of the body that it contacts.
- Always place a shop towel or cloth around the fitting or connection prior to loosening to absorb any excess fuel due to spillage. Ensure that all fuel spillage (should it occur) is quickly removed from engine surfaces. Ensure that all fuel soaked cloths or towels are deposited into a suitable waste container.
- Always keep a dry chemical (Class B) fire extinguisher near the work area.
- Do not allow fuel spray or fuel vapors to come into contact with a spark or open flame.
- Always use a back-up wrench when loosening and tightening fuel line connection fittings. This will prevent unnecessary stress and torsion to fuel line piping. Always follow the proper torque specifications.
- Always replace worn fuel fitting O-rings with new. Do not substitute fuel hose or equivalent, where fuel pipe is installed.

Fuel System Pressure

RELIEVING

1. Before servicing the vehicle, refer to the precautions in the beginning of this section.
2. Turn the ignition to the **OFF** position.
3. Loosen the fuel filler cap to release fuel tank pressure.
4. Remove the rear seat assembly, then remove the protector and disconnect the fuel pump module connector.
5. Start the vehicle and allow it to run until it stalls from lack of fuel. Turn the key to the **OFF** position.
6. Disconnect the negative battery cable, then reconnect the fuel pump connector. Install the rear seat assembly.
7. Wrap shop towels around the fitting that is being disconnected to absorb residual fuel in the lines.
8. Place shop towels into proper safety container.

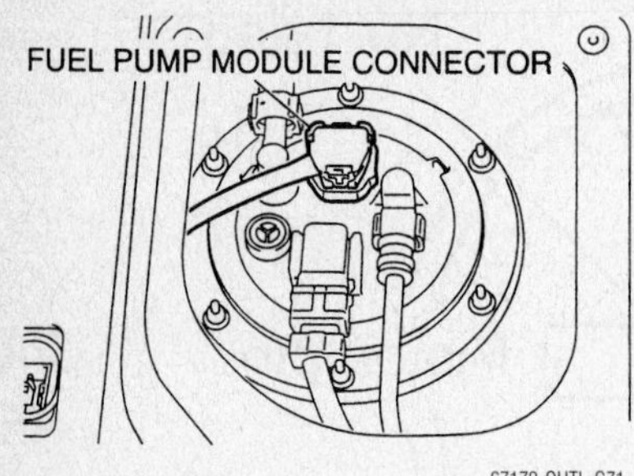

Location of the fuel pump module connector—Outlander

Fuel Filter

REMOVAL & INSTALLATION

1. Before servicing the vehicle, refer to the precautions in the beginning of this section.
2. Properly relieve the fuel system pressure.
3. Remove or disconnect the following:
   - Negative battery cable
   - Air intake hose and the battery
   - Fuel lines from the filter
   - Mounting bolts and the fuel filter from the vehicle

To install:

4. If equipped with flare fitting, tighten the fitting by hand before installing the filter to the vehicle.
5. Install or connect the following:
   - Filter to its bracket finger-tight
   - New gaskets and connect the high pressure hose and eye bolt, then the main pipe. Tighten the eye bolts to 22 ft. lbs. (30 Nm). Tighten the flare nut to 27 ft. lbs. (37 Nm).
6. Tighten the filter mounting bolts fully.
7. Install or connect the following:
   - Air intake hose
   - Battery
   - Negative battery cable, install the fuel filler cap, turn the key to the **ON** position to pressurize the fuel system and check for leaks.

Fuel Pump

REMOVAL & INSTALLATION

1. Before servicing the vehicle, refer to the precautions in the beginning of this section.
2. Properly relieve the fuel system pressure.
3. Disconnect the negative battery cable.
4. Remove or disconnect the following:
   - Rear seat cushion
   - Retainer screws and service hole cover

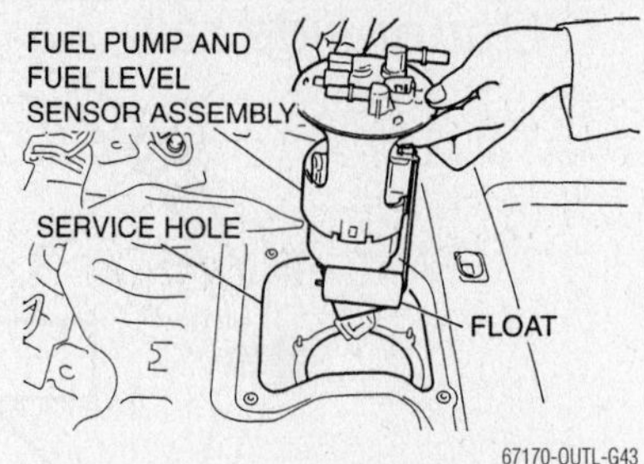

Remove the fuel pump module assembly

   - Harness electrical connector(s)
   - Fuel lines
   - Fuel pump module mounting nuts and plate

**** WARNING**

Be careful not to damage the module unit and float, when removing the fuel pump module from the fuel tank.

   - Fuel pump module (FWD) or Fuel pump and fuel level sensor (AWD) from the service hole.
5. Disassemble the fuel pump module as necessary. Refer to the accompanying figure.
6. Installation is the reverse of the removal procedure. Torque the fuel pump module place nuts to 22 inch lbs. (2.5 Nm).

Fuel Injector

REMOVAL & INSTALLATION

1. Before servicing the vehicle, refer to the precautions in the beginning of this section.
2. Relieve the fuel system pressure as described in this section.
3. Disconnect the negative battery cable.
4. Wrap the connection with a shop towel and disconnect the high pressure fuel line at the fuel rail.

**** CAUTION**

Observe all applicable safety precautions when working around fuel. Whenever servicing the fuel system, always work in a well ventilated area. Do not allow fuel spray or vapors to come in contact with a spark or open flame. Keep a dry chemical fire extinguisher near the work area. Always keep fuel in a container specifically designed for fuel storage; also, always properly seal fuel containers to avoid the possibility of fire or explosion.

5. Remove or disconnect the following:
   - Air cleaner
   - Positive Crankcase Ventilation (PCV) hose
   - Capacitor connector
   - Ignition coil connectors
   - Electrical connector from each injector, label for reference
   - Exhaust Gas Recirculation (EGR) valve connector
   - Rocker arm (valve) cover bracket bolts
   - High pressure fuel hose
   - Bolt(s) holding the fuel rail to the manifold. Carefully lift the rail up and remove it with the injectors attached. Take great care not to drop an injector. Place the rail and injectors in a safe location on the workbench; protect the tips of the injectors from dirt and/or impact.
   - Injector insulators from the intake manifold, discard. The insulators are not reusable.
   - Injectors from the fuel rail by pulling gently in a straight outward motion. Make certain the grommet and O-ring come off with the injector.

To install:

6. Install a new insulator in each injector port in the manifold.
7. Remove the old grommet and O-ring from each injector. Install a new grommet and O-ring; coat the O-ring lightly with clean, thin oil.
8. Install or connect the following:
   - Injector into the fuel rail, constantly turning the injector left and right during installation. When fully installed, the injector should still turn freely in the rail. If it does not, remove the injector and inspect the O-ring for deformation or damage.
   - Fuel rail and injectors to the engine. Make certain that each injector fits correctly into its port and that the rubber insulators for the fuel rail mounts are in position.
   - Engine oil to a new O-ring and install on the high pressure fuel line
   - High pressure fuel line to the fuel rail. Tighten to 44 inch lbs. (5 Nm).
   - Rocker arm (valve) cover bracket bolts
   - EGR valve connector
   - Fuel injector connectors
   - Ignition coil connectors
   - Capacitor connector
   - PCV hose connector
   - Negative battery cable
9. Pressurize the fuel system and inspect all connections for leaks.

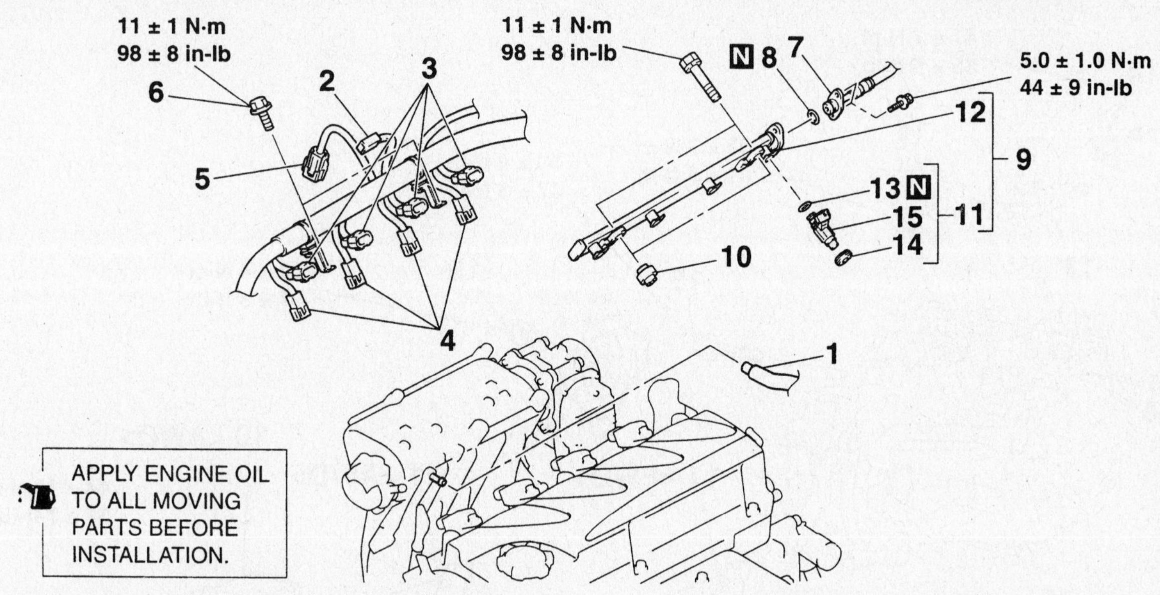

1. PCV HOSE CONNECTION
2. CAPACITOR CONNECTOR
3. IGNITION COIL CONNECTORS
4. INJECTOR CONNECTORS
5. EGR VALVE CONNECTOR
6. ROCKER COVER BRACKET INSTALLATION BOLTS
7. FUEL HIGH-PRESSURE HOSE CONNECTION
8. O-RING
9. FUEL RAIL AND INJECTOR ASSEMBLY
10. INSULATORS
11. INJECTOR ASSEMBLY
12. FUEL RAIL
13. O-RING
14. GROMMETS
15. INJECTORS

67170-OUTL-G44

Exploded view of the fuel rail and injectors

DRIVE TRAIN

Transaxle Assembly

REMOVAL & INSTALLATION

Automatic

1. Before servicing the vehicle, refer to the precautions in the beginning of this section.
2. Install or connect the following:
   - Negative battery cable
   - Front and side engine undercovers
3. Drain the transaxle oil and engine coolant. Drain the transfer fluid, if AWD.
   - Front exhaust pipe
   - Propeller shaft
   - Air cleaner assembly
   - Battery and battery tray
   - Hood
   - Transaxle control cable
   - A/T control solenoid valve assembly connector
   - Transmission range switch connector
   - Input and output shaft speed sensor connectors
   - A/T fluid cooler hose from the radiator. Plug the hose.
   - Starter mounting bolt
   - Loosen the upper engine-to-transaxle bolts. Do not remove the bolts yet.
   - A/T fluid cooler hose from the cooler. Plug the hose.
   - Halfshaft and inner shaft, FWD vehicles
   - Halfshaft and output shaft assembly, AWD vehicles
   - Transaxle stay
   - Bell housing cover
   - Drive plate bolts. Turn the crankshaft to access the bolts. Push the torque converter into the transaxle side and make sure the torque converter does not stay on the engine side.
   - Center member and front roll stopper assembly
   - Transfer case, AWD vehicles
   - Rear roll stopper, FWD vehicles
   - Roll rod bracket, AWD vehicles
4. Carefully support the transaxle with a jack, slightly raise the transaxle, then remove the transaxle mount.
   - Transaxle mount stopper
5. Install a suitable engine support assembly, then raise and safely support the vehicle.
   - Transaxle to engine bolts and lower the transaxle from the vehicle
6. Support the transaxle with a suitable jack.
   - Lower transaxle-to-engine mounting bolts
   - Transaxle from the vehicle

To install:

7. Installation is the reverse of the removal procedure, noting the following:
   - Transaxle-to-engine lower mounting bolts: 34–40 ft. lbs. (46–54 Nm)
   - Driveplate bolts: 34–40 ft. lbs. (46–54 Nm)

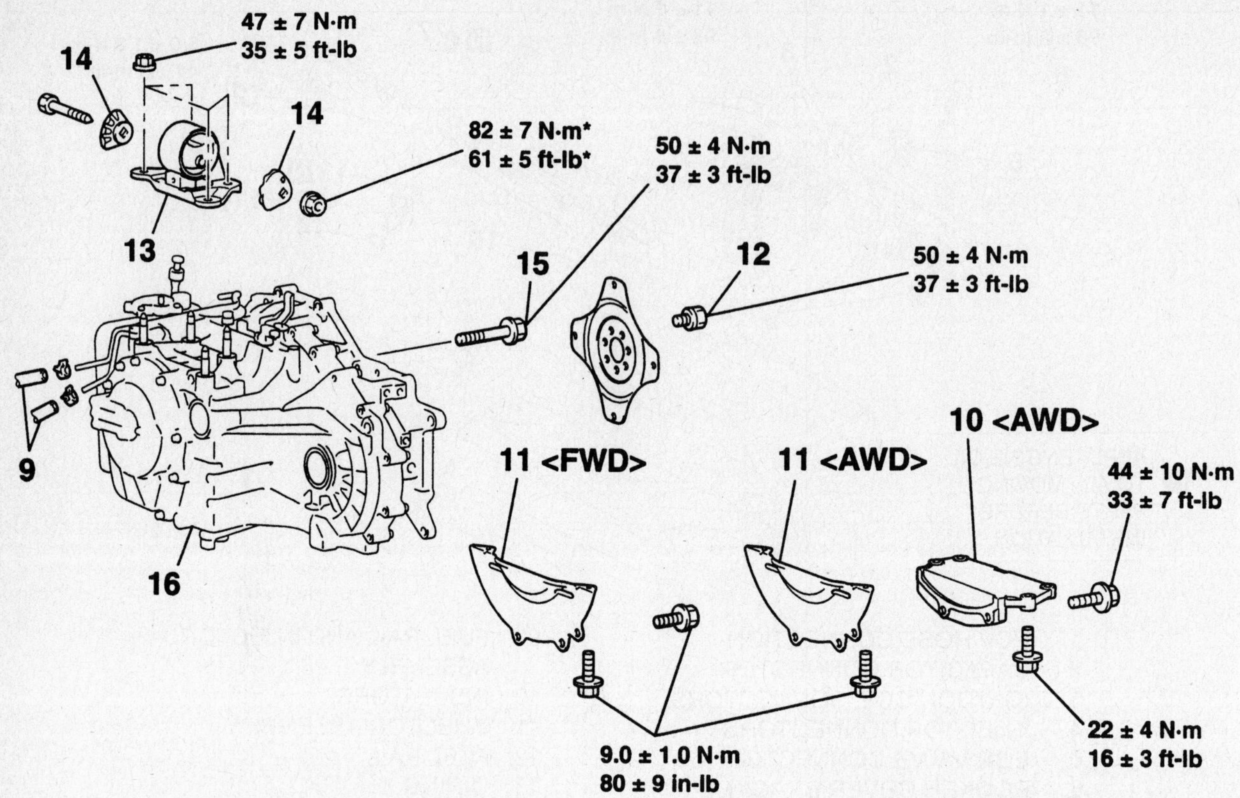

9. A/T FLUID COOLER HOSE (A/T FLUID COOLER SIDE)
10. TRANSAXLE STAY <AWD>
11. BELL HOUSING COVER
12. DRIVE PLATE BOLTS
- CENTERMEMBER AND FRONT ROLL STOPPER ASSEMBLY (REFER TO GROUP 32, ENGINE ROLL STOPPER AND CENTERMEMBER P.32-6).
- TRANSFER ASSEMBLY <AWD> (REFER TO P.23A-376).

13. TRANSAXLE MOUNT
14. TRANSAXLE MOUNT STOPPER
15. TRANSAXLE LOWER CONNECTING BOLTS
16. TRANSAXLE ASSEMBLY

67170-OUTL-G45

Automatic transaxle mounting

- Transaxle mount bracket nuts: 36 ft. lbs. (49 Nm)
- Transaxle mount stopper nuts: 56–66 ft. lbs. (75–89 Nm)
- Transaxle-to-engine upper mounting bolts: 34–40 ft. lbs. (46–54 Nm)

8. Fill the transaxle and the engine cooling system to the correct level.

9. If AWD, fill the transfer case to the correct level.

10. Check and adjust the front wheel alignment.

11. Check the speedometer and gear selector for proper operation.

12. Start the engine and check for leaks.

Transfer Case Assembly

REMOVAL & INSTALLATION

1. Before servicing the vehicle, refer to the precautions in the beginning of this section.

2. Drain the transaxle fluid and transfer case fluid.

3. Remove or disconnect the following:
   - Engine undercover(s)
   - Front exhaust pipe
   - Propeller shaft
   - Center member
   - Air guide
   - Halfshaft and output shaft
   - Rear roll stopper bolt and nut
   - Transfer case mounting bolts. Use a suitable tool to slide the transaxle to the front of the vehicle to make a suitable opening between the transaxle and crossmember. Pull the transfer case out of the opening.

To install:

4. Install or connect the following:
   - Transfer case to the transaxle. Torque the bolts to 44–58 ft. lbs. (60–78 Nm).
   - Rear roll stopper bolt and nut hand-tight. After the full weight of

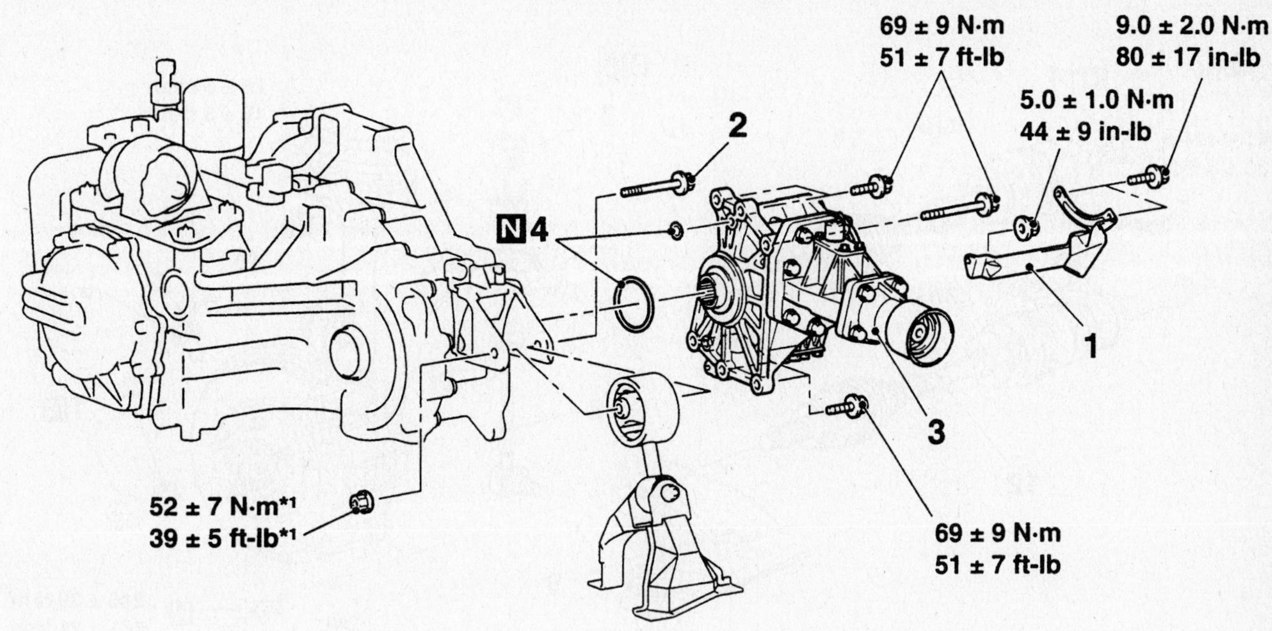

Exploded view of the transfer case

the vehicle is on the ground, torque the nut to 34–44 ft. lbs. (45–59 Nm).

- Halfshaft and output shaft
- Air guide and torque to 35–53 inch lbs. (4–6 Nm).
- Center member
- Propeller shaft
- Front exhaust pipe
- Engine undercover(s)

5. Fill the transaxle and the transfer case to the correct level.

Halfshaft

REMOVAL & INSTALLATION

Front

1. Before servicing the vehicle, refer to the precautions in the beginning of this section.
2. Raise the vehicle and support it safely.
3. Drain the transaxle fluid and the transfer case fluid, if AWD.
4. Remove the wheel and tire assembly.
5. Remove the cotter pin, halfshaft nut and washer.
6. If equipped with Anti-Lock Brake (ABS), remove the front wheel speed sensor and harness bracket.
7. Remove or disconnect the following:
   - Brake hose bracket
   - Self-locking nut
   - Stabilizer rubber insulator
   - Stabilizer link assembly
   - Lower control arm ball joint and tie rod end from the steering knuckle
8. To remove the halfshaft or halfshaft and inner shaft as follows:
   a. Use Mitsubishi Special tool Nos. MB990241, MB991354 and MB990767, or suitable puller, to push the halfshaft or halfshaft and inner shaft assembly from the hub.
   b. Remove the halfshaft from the hub by pulling the bottom of the rotor toward you.

⁂ WARNING

When pulling the halfshaft from the transaxle, be careful that the spline of the halfshaft does not damage the oil seal.

   c. Insert a prybar between the transaxle case and halfshaft, then pry the halfshaft out of the transaxle.
   d. If the inner shaft is difficult to remove, tap the bracket assembly lightly with a plastic hammer, then remove the inner shaft.
   e. Cover the halfshaft opening in the transaxle case to prevent foreign debris from entering.

➡Do not pull on the shaft; doing so damages the inboard joint.

9. For AWD vehicles, Use Mitsubishi Special tool No. MB991721 to remove the output shaft.

To install:

10. Replace the circlips on the ends of the halfshafts.
11. If AWD, install the output shaft.

➡When installing the output shaft, halfshaft or halfshaft and inner shaft assembly, make sure the splines do not damage the oil seal.

12. Insert the halfshaft or halfshaft and inner shaft into the transaxle. Be sure it is fully seated.
13. Push out on the knuckle assembly and install the halfshaft through the hub.
14. Install the self-locking nut. Tighten the nut so the protruding length of the stabilizer link is 0.35–0.39 inches (9.0–9.8mm).

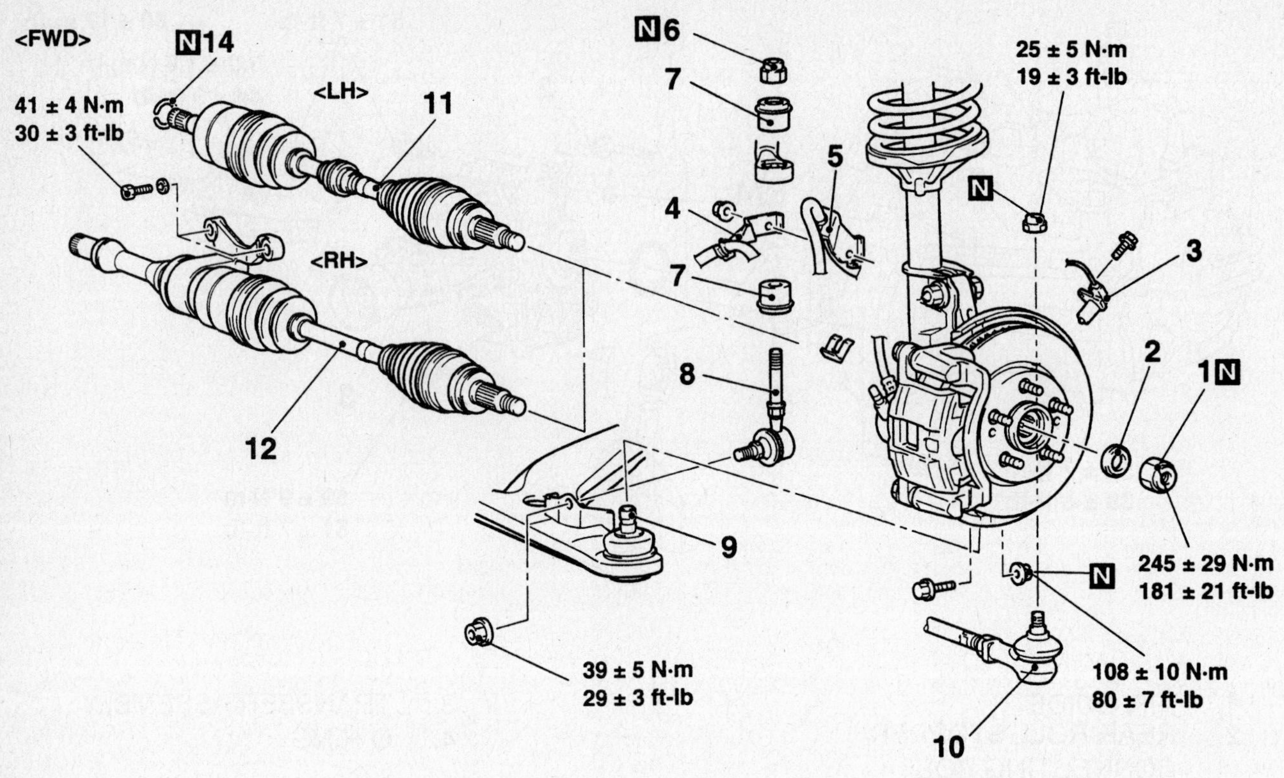

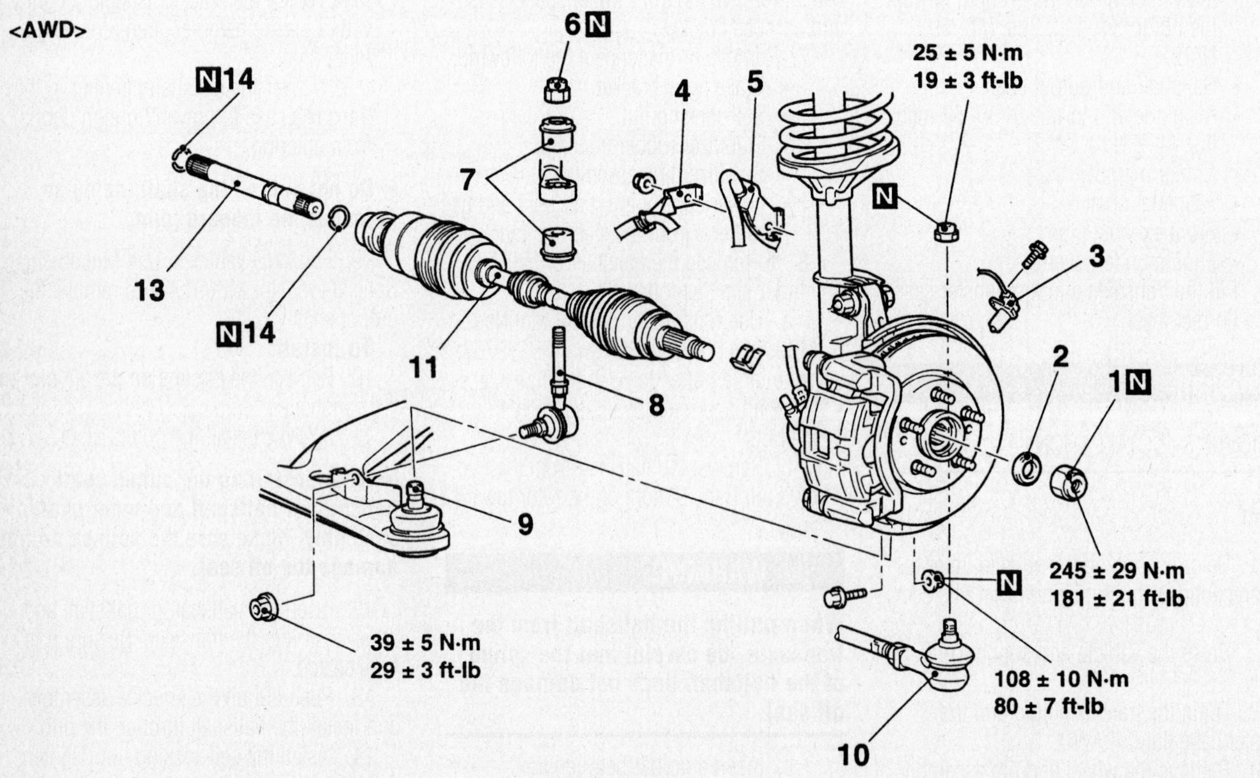

Exploded view of the halfshaft mounting

15. Connect the tie rod end to the steering knuckle. Torque the retaining nut to 21 ft. lbs. (29 Nm) and secure with a new cotter pin.

16. Connect the ball joint to the steering knuckle. Torque the new retaining nut to 43–52 ft. lbs. (60–72 Nm) and secure with a new cotter pin.

17. Install the stabilizer link and rubber insulator.

18. Install the brake hose bracket.

19. Install the ABS sensor harness bracket and ABS sensor, if equipped.

20. Install the washer so the chamfered edge faces outward. Install the halfshaft nut and tighten to 160–201 ft. lbs. (226–274 Nm). Secure with a new cotter pin.

21. Install the wheel and lower the vehicle to the floor.

22. Fill the transaxle and transfer case, if equipped, with fluid.

Rear

1. Before servicing the vehicle, refer to the precautions in the beginning of this section.
2. Raise and safely support the vehicle.
3. Remove or disconnect the following:
   - Rear wheel
4. Prevent the hub assembly from turning by using a tool such as MB990767 and remove the halfshaft nut and washer.
   - Anti-lock Brake System (ABS) sensor, if equipped
   - Brake drum and shoes, if equipped with drum brakes
   - Brake hydraulic line from the wheel cylinder
   - Parking brake cable from the rear brakes
   - Lower end of the shock absorber from the knuckle
   - Trailing and lower arms from the knuckle
   - Toe control arm ball joint from the knuckle
5. Remove the differential mount support.

✱✱ WARNING

Do not pull on the halfshaft to remove it from the differential. Damage to the inner CV-joint will occur.

6. Push the lower part of the knuckle outward and pry the inner halfshaft out of the differential.
7. Push the outer end of the halfshaft through the hub/knuckle and remove it.

To install:

8. Install the outer end of the halfshaft through the hub/knuckle.
9. Place a new circlip on the inner halfshaft and install the halfshaft in the differential.
10. Install or connect the following:
    - Differential mount support
    - Washer and a new nut on the end of the halfshaft. Tighten the nut to 160–201 ft. lbs. (226–274 Nm).
    - Toe control arm to the knuckle. Tighten the new nut to 20 ft. lbs. (28 Nm).
    - Lower and trailing arms to the knuckle. Do not tighten the fasteners at this time.
    - Shock absorber. Tighten the bolt to 70 ft. lbs. (95 Nm).
11. Assemble the brake components.
12. Install or connect the following:
    - Rear wheel speed sensor
    - Rear wheel and lower the vehicle to the floor. Tighten the lower arm nut to 71 ft. lbs. (98 Nm) and the trailing arm nut to 85–99 ft. lbs. (118–137 Nm).

CV-JOINT OVERHAUL

Front

1. Before servicing the vehicle, refer to the precautions in the beginning of this section.
2. Raise and safely support the vehicle.
3. Remove or disconnect the following:
   - Halfshaft assembly from the vehicle
   - Large Pillow Tripod Joint (PTJ) boot band
   - Small PTJ boot band
   - PTJ case and inner shaft assembly, right side FWD vehicles
   - PTJ case

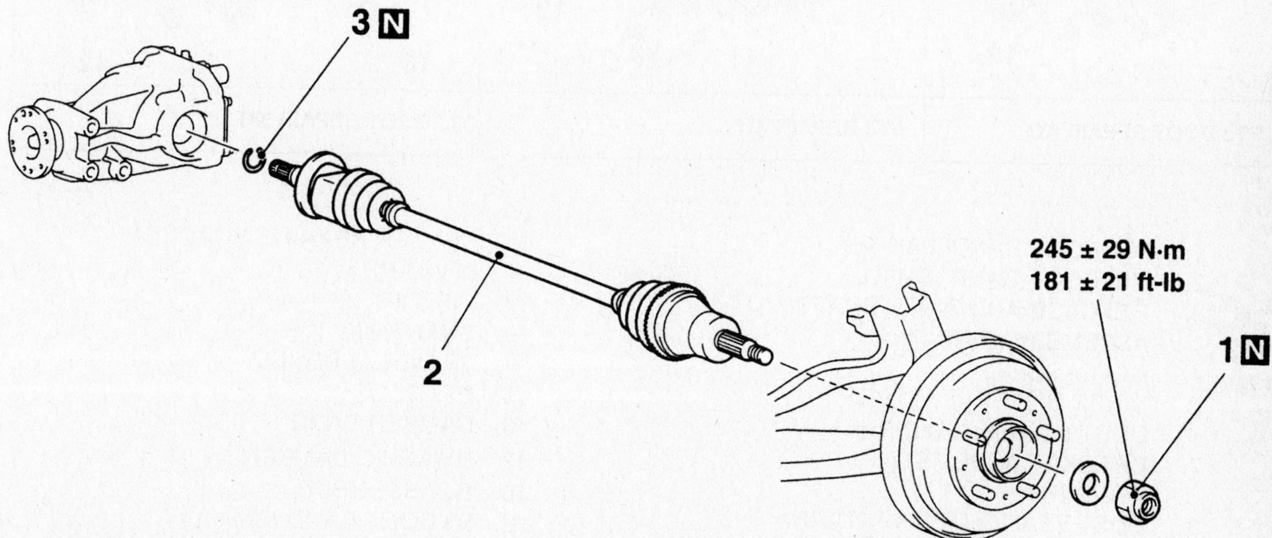

Exploded view of the rear halfshaft assembly—AWD vehicles

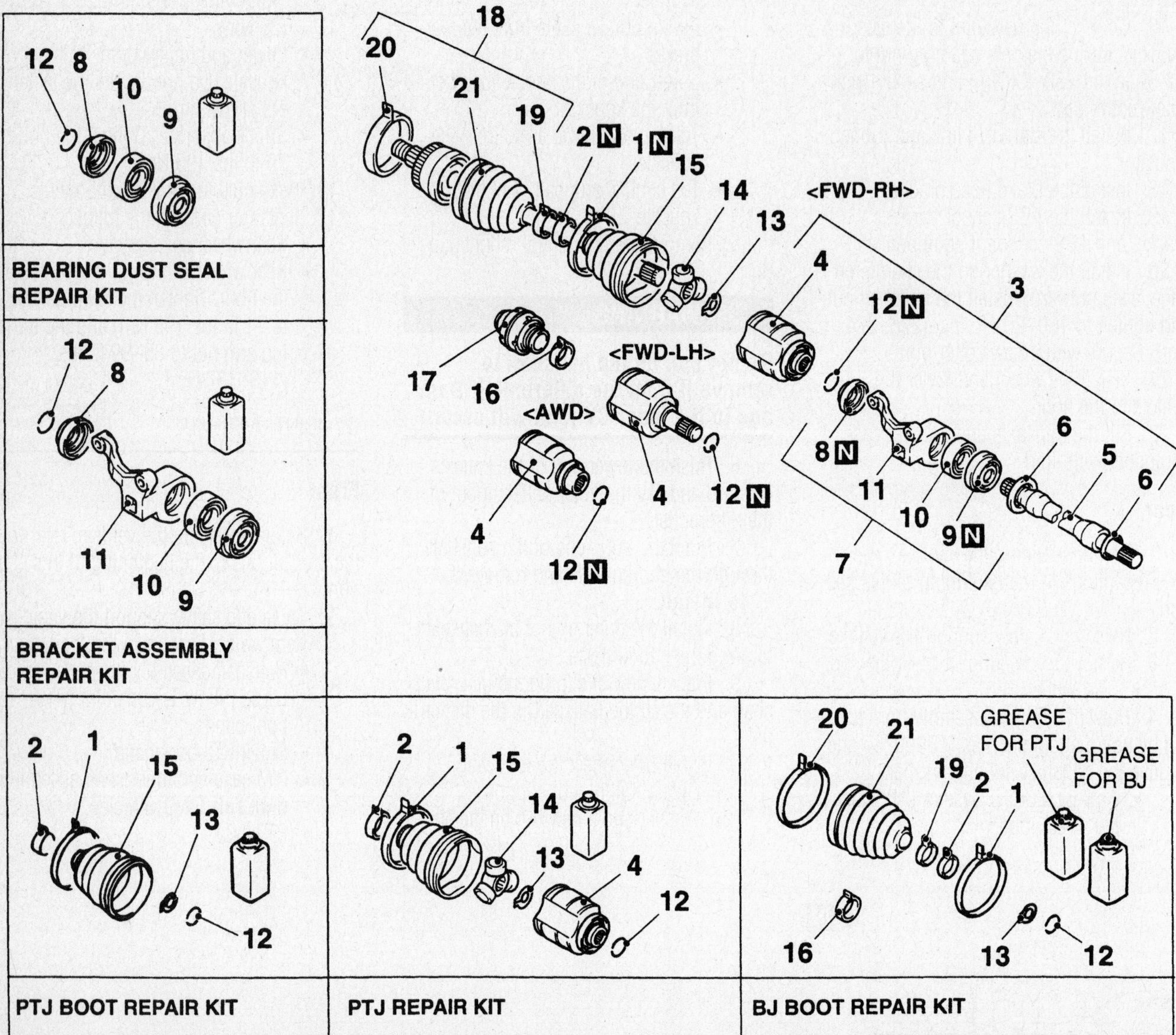

1. PTJ BOOT BAND (LARGE)
2. PTJ BOOT BAND (SMALL)
3. PTJ CASE AND INNER SHAFT ASSEMBLY <FWD-RH>
4. PTJ CASE
5. INNER SHAFT <FWD-RH>
6. DUST COVER <FWD-RH>
7. BRACKET ASSEMBLY <FWD-RH>
8. DUST SEAL OUTER <FWD-RH>
9. DUST SEAL INNER <FWD-RH>
10. CENTER BEARING <FWD-RH>
11. CENTER BEARING BRACKET <FWD-RH>
12. CIRCLIP
13. SNAP RING
14. SPIDER ASSEMBLY
15. PTJ BOOT
16. DAMPER BAND
17. DYNAMIC DAMPER
18. BJ ASSEMBLY
19. BJ BOOT BAND (SMALL)
20. BJ BOOT BAND (LARGE)
21. BJ BOOT

67170-OUTL-G49

Exploded view of the front halfshaft assembly

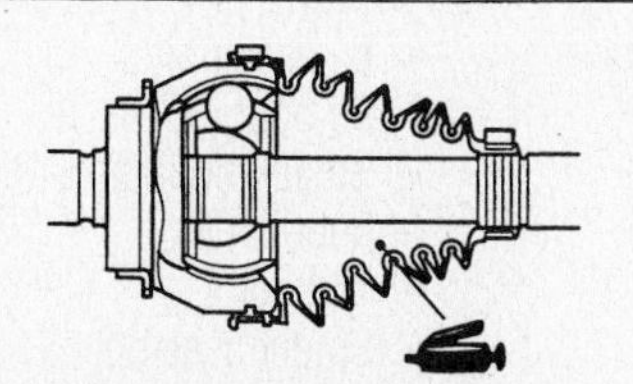
GREASE: REPAIR KIT GREASE
AMOUNT USED: <FWD>120 ± 10 g (4.2 ± 0.3 oz), <AWD> 85 ± 10 g (3.1 ± 0.3 oz)
CAUTION
THE DRIVE SHAFT JOINT USES SPECIAL GREASE. DO NOT MIX OLD AND NEW OR DIFFERENT TYPES OF GREASE.

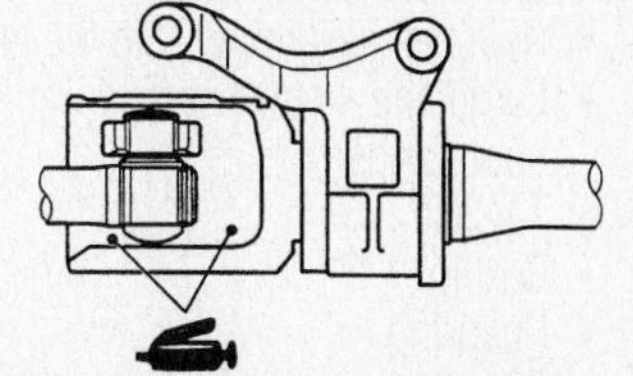
GREASE: REPAIR KIT GREASE
AMOUNT USED: <FWD> 210 ± 10 g (7.4 ± 0.3 oz), <AWD> 150 ± 10 g (5.3 ± 0.3 oz)
CAUTION
THE DRIVE SHAFT JOINT USES SPECIAL GREASE. DO NOT MIX OLD AND NEW OR DIFFERENT TYPES OF GREASE.

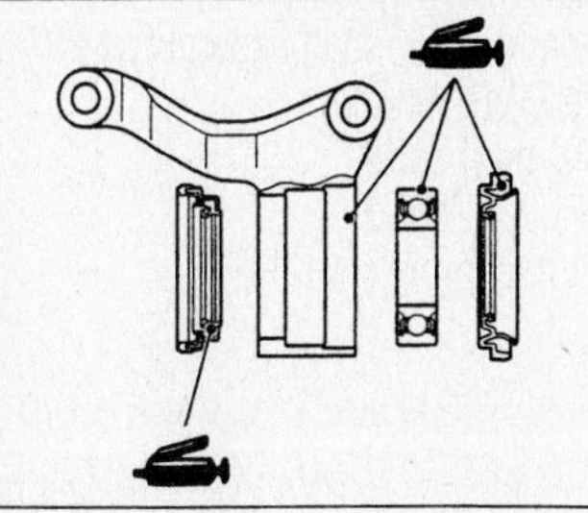
GREASE: REPAIR KIT GREASE
AMOUNT USED:
DUST SEAL INNER: 14 - 20 g (0.5 - 0.7 oz)
DUST SEAL OUTER: 8 - 12 g (0.3 - 0.4 oz)

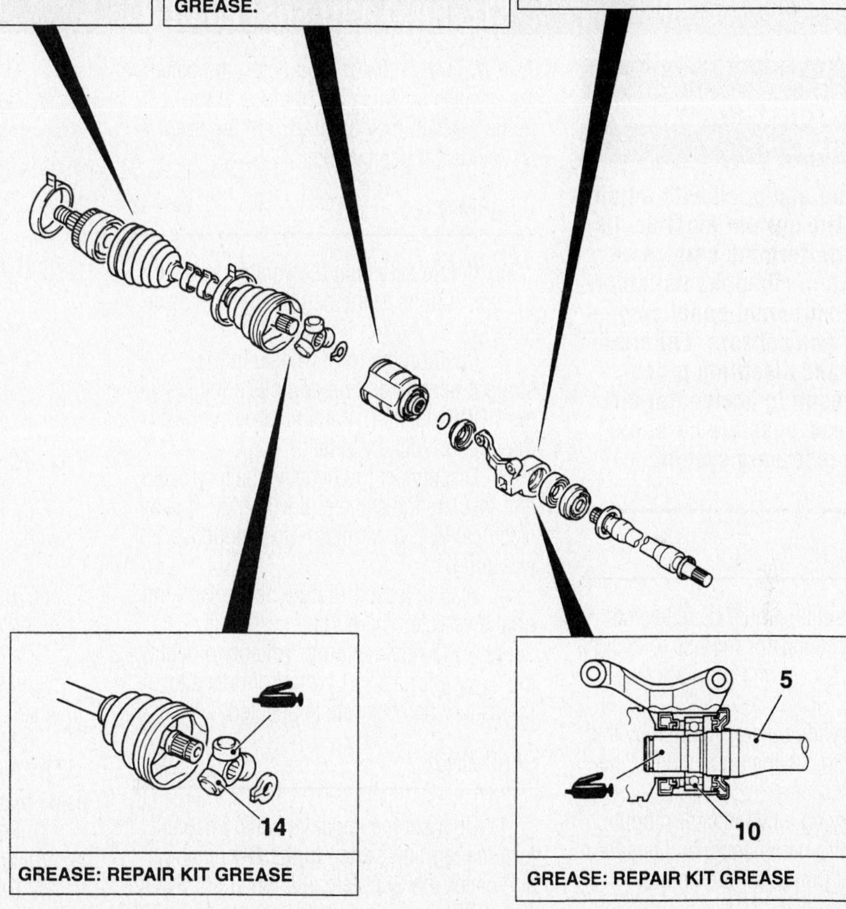
14
GREASE: REPAIR KIT GREASE
5
10
GREASE: REPAIR KIT GREASE
67170-OUTL-G74

Halfshaft assembly lubrication points and grease specifications

- Inner shaft and dust cover, right side FWD vehicles
- Bracket assembly, right side FWD vehicles
- Outer and inner dust seals, right side FWD vehicles
- Center bearing and bracket, right side FWD vehicles
- Circlip
- Snapring
- Spider assembly
- PTJ boot
- Damper band
- Dynamic Damper
- Birfield Joint (BJ) assembly
- Small and large BJ boot bands
- BJ boot

To install:

➡Refer to the illustration for the lubrication points and grease specifications.

- BJ boot
- Large and small BJ boot bands
- BJ assembly
- Dynamic Damper
- Damper band
- PTJ boot
- Spider assembly.
- Snapring
- Circlip
- Center bearing and bracket, right side FWD vehicles
- Outer and inner dust seals, right side FWD vehicles
- Bracket assembly, right side FWD vehicles
- Inner shaft and dust cover, right side FWD vehicles

- PTJ case
- PTJ case and inner shaft assembly, right side FWD vehicles
- Small PTJ boot band
- Large PTJ boot band
- Halfshaft assembly into the vehicle

Rear

1. Before servicing the vehicle, refer to the precautions in the beginning of this section.
2. Raise and safely support the vehicle.
3. Remove or disconnect the following:
   - Halfshaft assembly from the vehicle
   - Large and small Tripod Joint (TJ) boot bands
   - TJ case
   - Circlip
   - Snapring
   - Spider assembly
   - TJ boot
   - Large and small Birfield Joint (BJ) boot bands
   - BJ boot
   - BJ assembly

To install:

- BJ assembly
- BJ boot
- Small and large Birfield Joint (BJ) boot bands
- TJ boot
- Spider assembly
- Snapring
- Circlip
- TJ case
- Small and large TJ boot bands
- Halfshaft assembly to the vehicle

STEERING AND SUSPENSION

Air Bag

⁂ CAUTION

All vehicles are equipped with an air bag system. The system must be disabled before performing service on or around system components, steering column, instrument panel components, wiring and sensors. Failure to follow safety and disabling procedures could result in accidental air bag deployment, possible personal injury and unnecessary system repairs.

PRECAUTIONS

Several precautions must be observed when handling the inflator module to avoid accidental deployment and possible personal injury.

- Never carry the inflator module by the wires or connector on the underside of the module.
- When carrying a live inflator module, hold securely with both hands, and ensure that the bag and trim cover are pointed away.
- Place the inflator module on a bench or other surface with the bag and trim cover facing up.

INSULATING TAPE
BATTERY
BATTERY CABLE (–)
67170-OUTL-G50

Insulate the negative battery cable to prevent accidental deployment of the air bag

- With the inflator module on the bench, never place anything on or close to the module which may be thrown in the event of an accidental deployment.

DISARMING

1. Before servicing the vehicle, refer to the precautions in the beginning of this section.
2. Position the front wheels in the straight-ahead position and place the key in the **LOCK** position. Remove the key from the ignition lock cylinder.
3. Disconnect the negative battery cable and insulate the cable end with high-quality electrical tape or similar non-conductive wrapping.
4. Wait at least 1 minute before working on the vehicle. The air bag system is designed to retain enough voltage to deploy the air bag for a short period of time after the battery has been disconnected.

REARMING

1. Connect the negative battery cable, turn the ignition switch to the **ON** position and check the Supplemental Restraint System (SRS) warning light for proper operation.

Power Steering Gear

REMOVAL & INSTALLATION

➡Prior to removal of the steering gear box, center the front wheels and remove the ignition key. Failure to do so may damage the Supplemental Restraint System (SRS) clockspring and render SRS system inoperative.

1. Before servicing the vehicle, refer to the precautions in the beginning of this section.
2. Drain the power steering system.
3. Remove or disconnect the following:
   - Battery negative cable. Insulate the cable end with high-quality electrical tape or similar non-conductive wrapping Wait at least 1 minute before proceeding.
   - Heated Oxygen (HO_2) sensor and remove the front exhaust pipe
4. Properly support the engine.
   - Roll stopper mounting bolts and the 4 center member installation bolts. Remove the center member.
   - Center member
5. Remove the steering wheel, as follows:
   a. Pry the cover from the steering wheel.
   b. Detach the connector, and retainers, as necessary then remove the air bag module assembly
   c. Remove the steering wheel nut, then use a puller to remove the steering wheel.
   d. Detach the sub-harness.
6. Remove the clock spring and put aside in a safe place.
7. Remove the stabilizer link and lower control arm assembly.
8. Remove or disconnect the following:
   - Shaft cover
   - Steering column shaft assembly and steering gear
   - Tie rod end from the steering knuckle, using a suitable puller
   - Rear roll stopper bolt
   - Fluid return hose and return tube. Plug the hoses. Remove and discard the O-ring.
   - Pressure hose connection and hose gasket
9. Use a transaxle jack to support the crossmember, then remove the crossmember mounting nuts and bolts. Lower the crossmember with the rear roll stopper, stabilizer bar, return tube and steering gear.

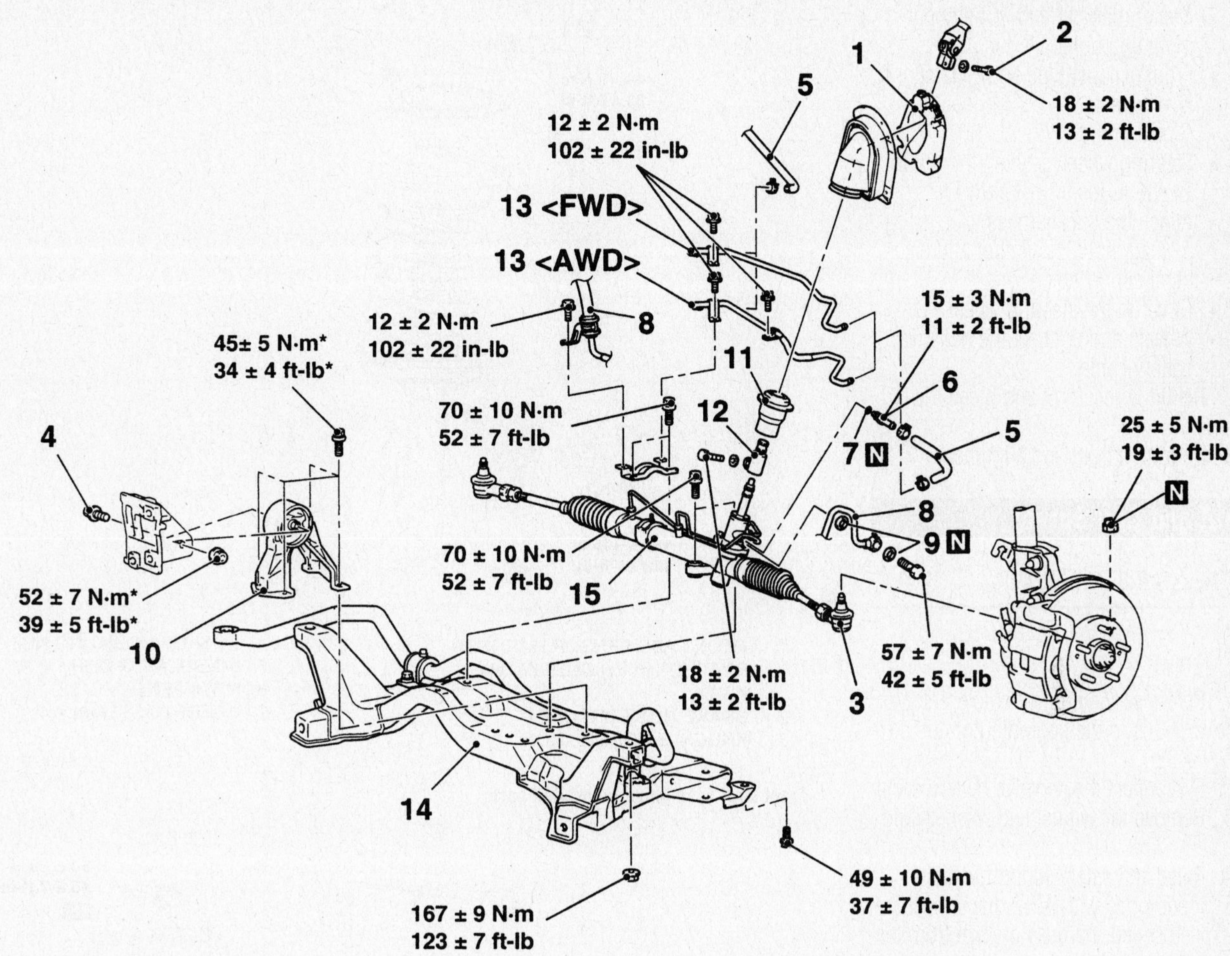

1. SHAFT COVER
2. STEERING COLUMN SHAFT ASSEMBLY AND STEERING GEAR CONNECTING BOLT
3. TIE ROD END AND KNUCKLE CONNECTION
4. REAR ROLL STOPPER CONNECTING BOLT
5. RETURN HOSE CONNECTION
6. RETURN TUBE
7. O-RING
8. PRESSURE HOSE CONNECTION
9. GASKET
10. REAR ROLL STOPPER
11. JOINT COVER GROMMET
12. STEERING EXTENSION
13. RETURN TUBE
14. CROSSMEMBER
15. STEERING GEAR

67170-OUTL-G51

Exploded view of the power steering gear

10. Remove the following from the crossmember.
    - Rear roll stopper
    - Joint cover grommet
    - Steering extension
    - Return tube
    - Crossmember
    - Steering gear

To install:

11. Install or connect the following:
    - Steering gear into the vehicle. Tighten the bolts to 45–59 ft. lbs. (60–80 Nm)
    - Crossmember to the vehicle with the return tube, joint cover grommet, steering extension and rear roll stopper. Tighten the crossmember assembly bolts to 30–44 ft. lbs. (39–59 Nm) and the nuts to 116–130 ft. lbs. (158–176 Nm).
    - Pressure hose with a new gasket. Tighten the bolt to 37–47 ft. lbs. (50–64 Nm).
    - Fluid return tube with a new O-ring
    - Return hose connection
    - Rear roll stopper connecting bolt
    - Tie rod end to the steering knuckle
    - Steering gear connecting bolt and steering column shaft assembly
    - Shaft cover
    - Stabilizer and lower control arm assembly

- Center member and roll stopper mounting bolts
- Front exhaust pipe and HO2 sensor, if removed
- Clock spring
- Steering wheel. Tighten the nut to 24–36 ft. lbs. (33–49 Nm).
- Attach the sub-harness
- Air bag module, retainer and electrical connector
- Cover to the steering wheel
- Wheels and connect the negative battery cable

12. Refill the reservoir and bleed the system.
13. Perform a front end alignment.

Strut and Coil Spring

REMOVAL & INSTALLATION

Front

1. Before servicing the vehicle, refer to the precautions in the beginning of this section.
2. Disconnect the negative battery cable.
3. Remove the windshield washer fluid tank.
4. Raise and safely support vehicle.
5. If equipped with ABS, disconnect the front speed sensor harness bracket from the strut.
6. Remove the brake hose bracket.
7. Support the lower arm using floor jack. Remove the lower strut to knuckle bolts.

➡Before removing the top bolts, make matchmarks on the body and the strut insulator for proper reassembly.

8. Remove the strut upper mounting nuts, insulator clip and stiffener plate.
9. Remove the strut assembly from the vehicle.

To install:

10. Install or connect the following:
    - Strut to the vehicle
    - Stiffener plate
    - Insulator clip
    - Top mounting nuts and tighten to 30–36 ft. lbs. (39–49 Nm)
11. Position the strut on the knuckle and install the mounting bolts. While holding the head of the lower mounting bolt, tighten the nuts to 116–130 ft. lbs. (158–175 Nm).
12. Install or connect the following:
    - Brake hose bracket and the ABS harness brake, if equipped
    - Wheel and tire assembly
    - Windshield washer fluid reservoir
13. Perform a front end alignment.

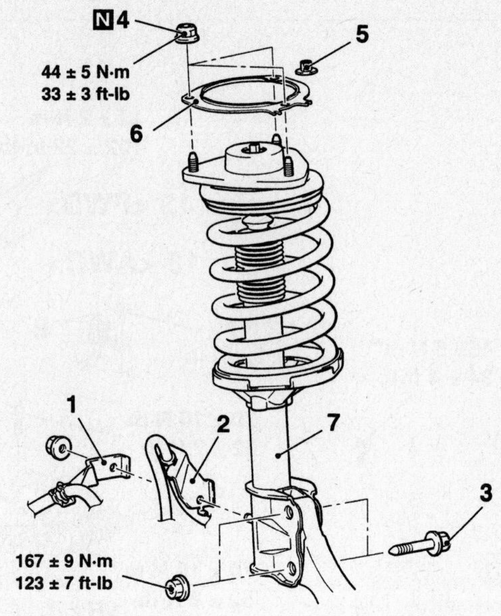

Front strut assembly and related parts

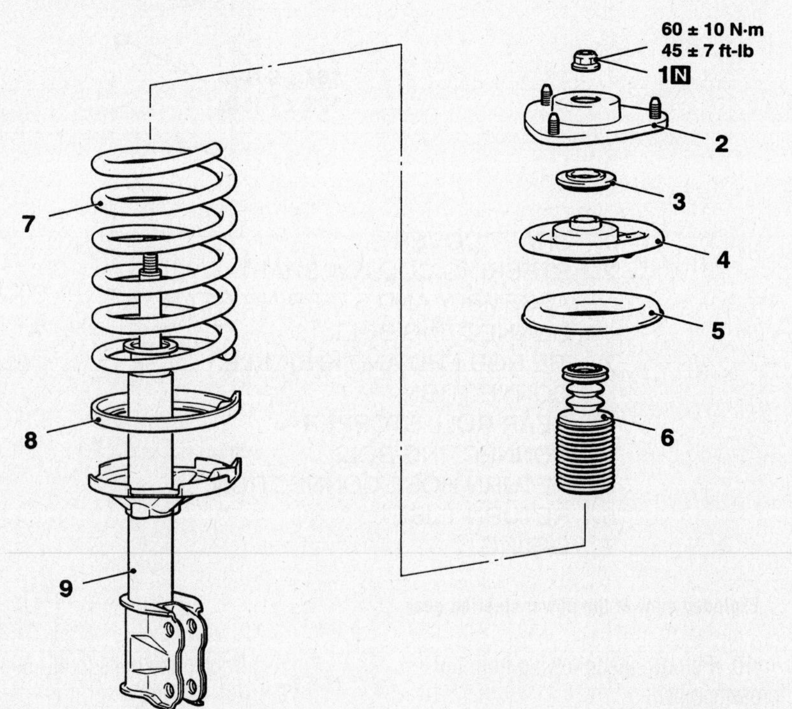

Exploded view of the strut overhaul

OVERHAUL

Front

1. Before servicing the vehicle, refer to the precautions in the beginning of this section.
2. Remove the strut assembly from the vehicle.
3. Compress the coil spring using a spring compressor until the spring just comes away from one of the seats.
4. Remove or disconnect the following:
    - Center nut from the strut. Do NOT use an impact wrench to remove the center nut.
    - Strut insulator assembly
    - Bearing
    - Upper spring seat and upper spring pad
    - Rubber bumper
    - Coil spring
    - Lower spring pad
    - Strut assembly

To install:

5. Install or connect the following:
    - Strut assembly
    - Lower spring pad
    - Compressed spring on the strut assembly
    - Rubber bumper
    - Upper spring pad and spring seat
    - Bearing
    - Strut insulator
    - Nut and tighten it to 38–52 ft. lbs. (50–70 Nm)
    - Strut to the vehicle

Shock Absorber and Coil Spring

REMOVAL & INSTALLATION

Rear

1. Before servicing the vehicle, refer to the precautions in the beginning of this section.
2. Remove or disconnect the following:
    - Stabilizer link connection
3. Support the lower control arm with a jack.
    - Lower control arm and trailing arm bolt
    - Upper shock absorber mounting nut
    - Shock absorber-to-lower control arm attaching bolt
    - Shock absorber assembly

To install:

4. Position the shock absorber into the vehicle. Install the spring seat stepped section so that it points toward the rear side of the vehicle.
5. Install or connect the following:
    - Shock absorber-to-lower control arm bolt and nut. Tighten the nut to 59–81 ft. lbs. (80–110 Nm).
    - Upper shock mounting nut and tighten to 30–36 ft. lbs. (39–49 Nm)
    - Lower control arm and trailing arm
    - Stabilizer link. Tighten the nuts to 26–32 ft. lbs. (34–44 Nm).

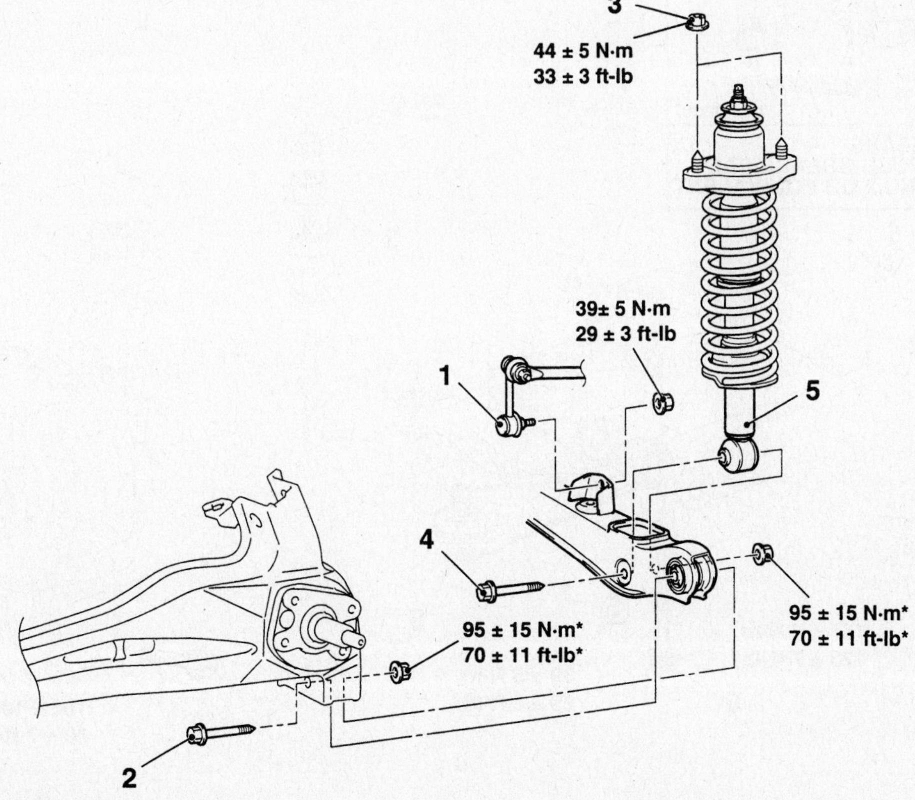

Exploded view of the rear shock absorber mounting

Upper Ball Joint

REMOVAL & INSTALLATION

The upper ball joints are an integral part of the upper control arm. If the ball joint becomes worn or damaged, the control arm must be replaced.

Upper Control Arm

REMOVAL & INSTALLATION

Front

These vehicles use a strut type front suspension. No upper control arm is used.

Lower Ball Joint

REMOVAL & INSTALLATION

The lower ball joint is an integral part of the lower control arm assembly, and can not be serviced separately. A worn or damaged ball joint requires replacement of lower control arm assembly.

Lower Control Arm

REMOVAL & INSTALLATION

Front

➡The suspension components should not be fully tightened until the vehicle's weight is resting on its wheels.

1. Before servicing the vehicle, refer to the precautions in the beginning of this section.
2. Raise the vehicle and support safely.
3. Remove or disconnect the following:
   - Lower control arm-to-knuckle bolt and nut
   - Wheel and tire assembly
   - Stabilizer bar self-locking nut, rubber bushings, and stabilizer link. Discard the nut.
4. Lift the transaxle with a jack, then remove the front control arm-to-crossmember bolt.
   - Lower control arm

To install:

5. Install or connect the following:
   - Lower control arm into the vehicle
   - Lower control arm-to-crossmember bolts. Torque the bottom bolt to

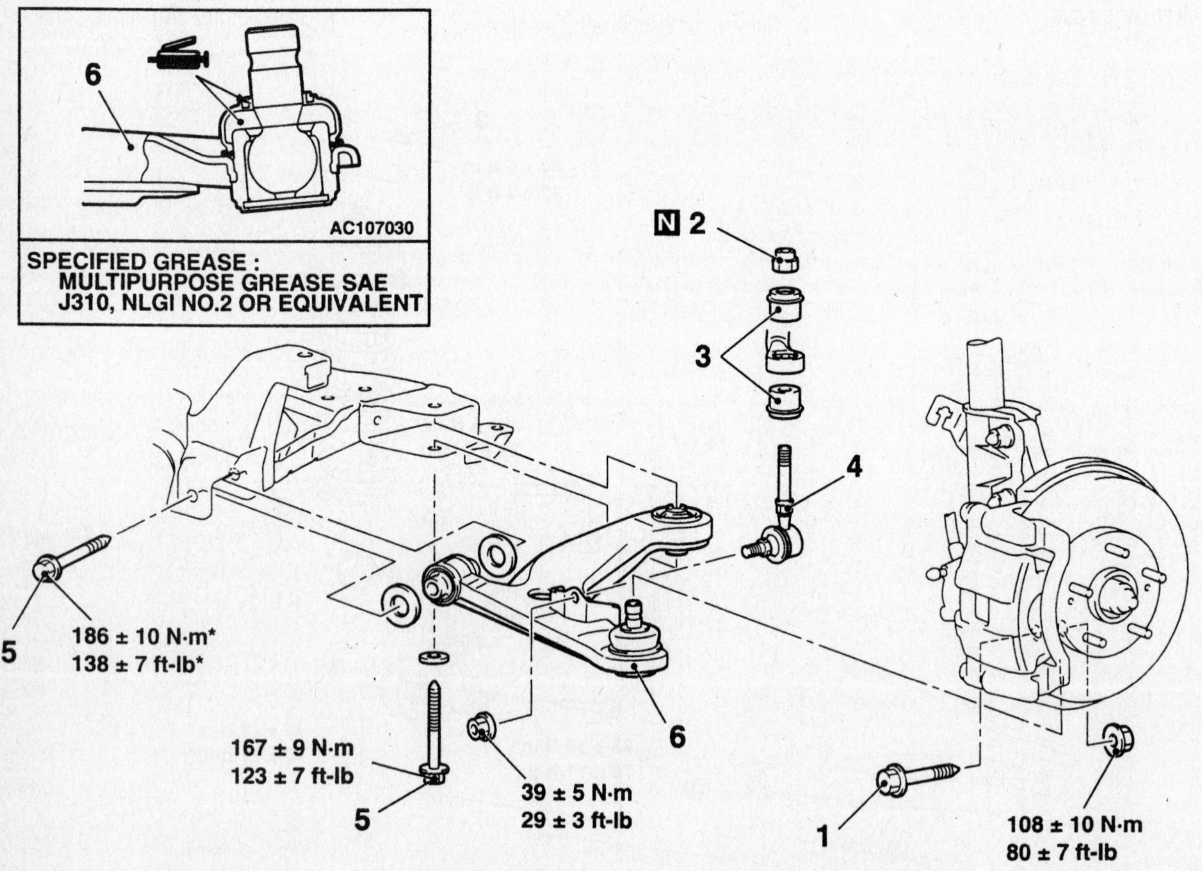

Exploded view of the lower control arm mounting

123 ft. lbs. (167 Nm) and the side bolt snug until the vehicle is lowered.

- Lower control arm-to-steering knuckle bolt. Torque to 80 ft. lbs. (108 Nm).
- Stabilizer link, bushings and new self-locking nut. Tighten the self-locking nut until the stabilizer link threads protrude 0.35–0.39 inch (9.0–9.8mm).

6. Lower the vehicle, install the wheels, then with the weight of the vehicle on the wheels, torque the side control arm-to-crossmember bolt to 137 ft. lbs. (186 Nm).
7. Check and adjust the front wheel alignment.

Control Link, Upper and Lower Control Arms

REMOVAL & INSTALLATION

Rear

➡The suspension components should not be fully tightened until the vehicle's weight is resting on its wheels.

1. Before servicing the vehicle, refer to the precautions in the beginning of this section.
2. Raise and safely support the vehicle.
3. Remove or disconnect the following:
   - Control link, after matchmarking its installed position
   - Mounting bolt and upper control arm
   - Stabilizer link nut and separate the link from the lower control arm
4. Support the lower control arm with a jack, remove the retaining bolts and separate the lower control arm from the trailing arm.
   - Lower shock absorber-to-lower control arm nut and bolt
   - Lower control arm

To install:

5. Install or connect the following:
   - Lower control arm and secure the retainers hand-tight only at this time
   - Shock absorber to the lower control arm. Secure with the nut and bolt and tighten until just snug
   - Lower control arm to the trailing arm. Secure the retainers hand-tight.
   - Stabilizer link and secure with the nut. Tighten to 26–35 ft. lbs. (34–44 Nm).
   - Upper arm. Install the arm so the "A" mark faces inside of the vehicle. Tighten the retainers snug. Final tightening will occur when the vehicle's weight is on the suspension.
   - Control link, so that its identification mark faces the front-outside of the vehicle
   - Rear wheel and tire assembly and lower the vehicle.
6. At this time, torque the components to the following specifications:
   - Lower control arm-to-trailing arm retainers: 59–81 ft. lbs. (80–110 Nm)
   - Lower shock absorber nut: 59–81 ft. lbs. (80–110 Nm)
   - Upper control arm retainers: 83–95 ft. lbs. (110–130 Nm)
   - Control link retainers: 83–95 ft. lbs. (110–130 Nm)
7. Check and adjust the rear wheel alignment.

Wheel Bearings

ADJUSTMENT

The front and rear wheel bearings on these vehicles are not adjustable. If the bearings are noisy or become loose, they must be replaced.

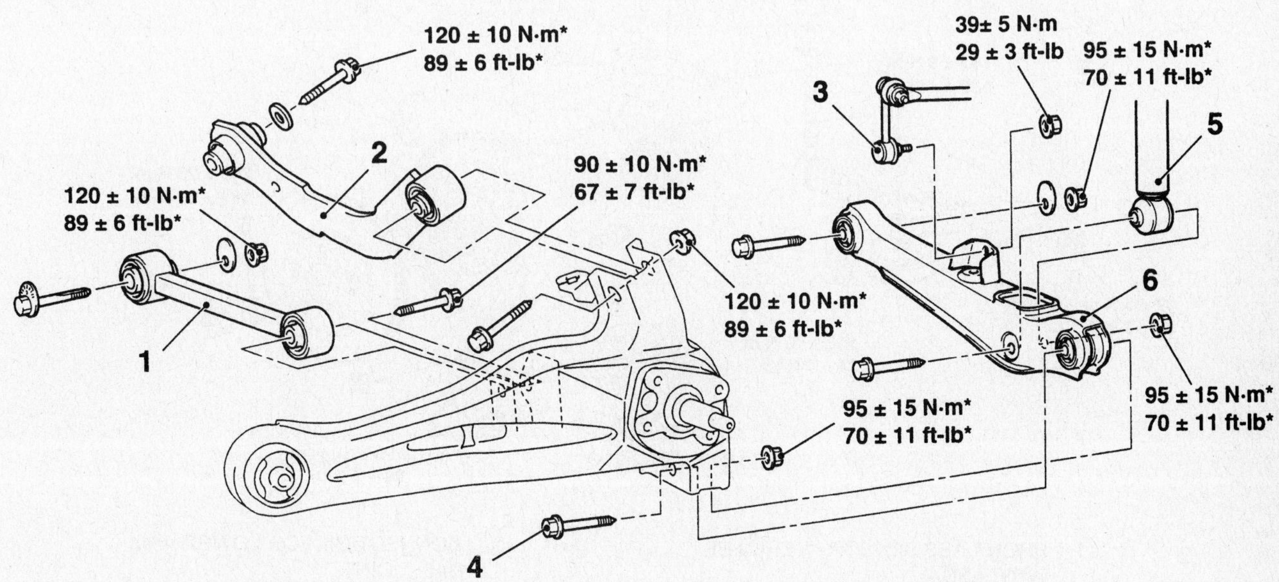

Exploded view of the rear control link, upper control arm and lower control arm

REMOVAL & INSTALLATION

Front

1. Before servicing the vehicle, refer to the precautions in the beginning of this section.
2. Raise and safely support the vehicle.
3. Remove or disconnect the following:
   - Front wheel
   - Wheel Speed Sensor (WSS), vehicles with Anti-Lock Brake (ABS)
   - Caliper and suspend it out of the way with wire or string
   - Brake rotor
   - Axle nut and washer, using special tool MB990767 to hold the hub secure while removing the nut. Discard the nut.
   - Stabilizer rubber insulator
   - Stabilizer link from the lower control arm
   - Lower ball joint from the steering knuckle
   - Tie rod end from the steering knuckle. Do not remove the nut from the tie rod end. Loosen the nut and use special tool MB991113 or MB990635 to avoid damaging the threads.
   - Halfshaft from the hub and knuckle using the proper puller
   - Front strut-to-hub and knuckle mounting bolt and nut
   - Hub and knuckle from the vehicle
4. If necessary, disassemble the hub and knuckle as follows:
   a. Remvoe the snap ring.
   b. Use Mitsbisishi Special tools MB991017, MB991056 or MB991355, MB991000 (MB990998) to pull the hub out of the knuckle.
   c. Remove the dust cover.
   d. Crush the oil seal on the wheel bearing in 2 places so the tabs of the special tool will catch on the inner race (outside).

➡Do not drop the hub when removing the inner race (outside).

   e. Use a suitable puller (Mitsubishi Special tool no MB990810 or equivalent) to remove the wheel bearing inner race (outside) from the hub.

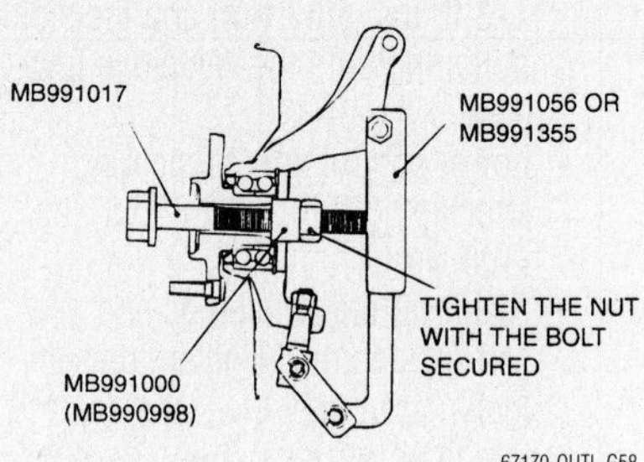

Use the special tools to pull the hub from the knuckle

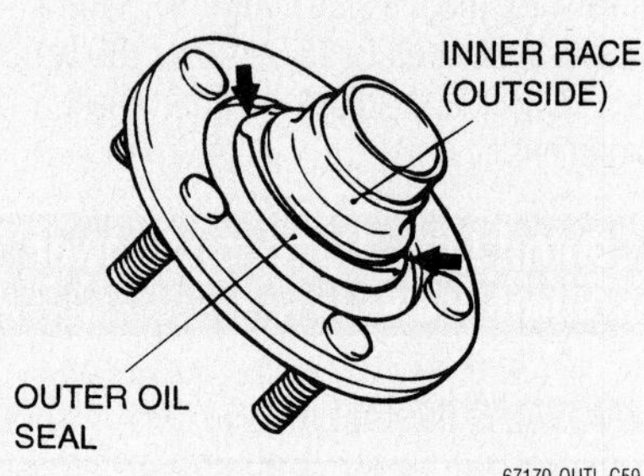

Crush the oil seal on the wheel bearing in the 2 places shown (arrows) so the tabs of the special tool will catch on the inner race

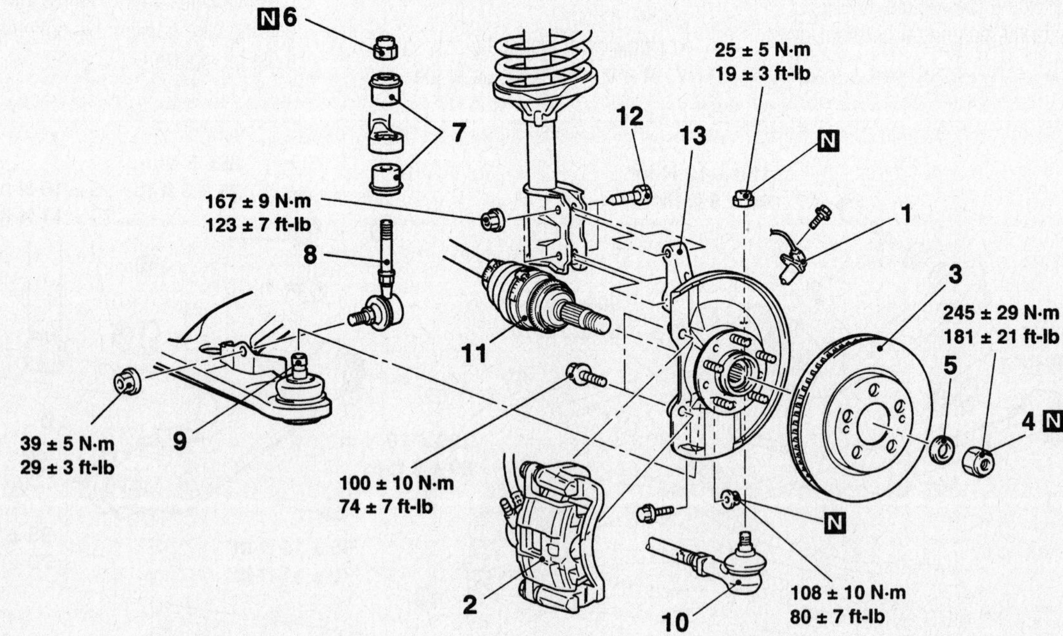

Exploded view of the front wheel bearing and related components

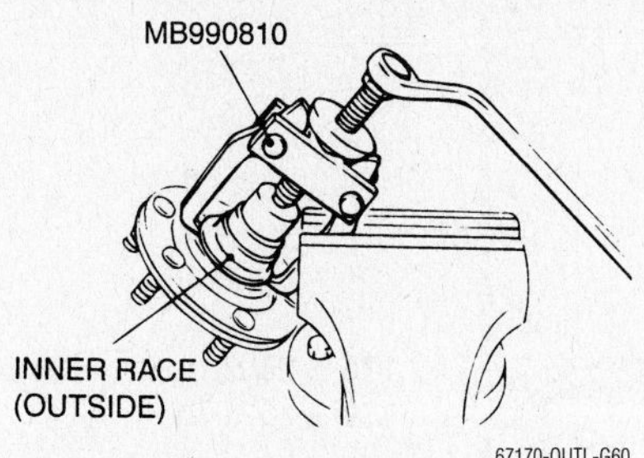

Use a puller as shown to separate the wheel bearing from the hub

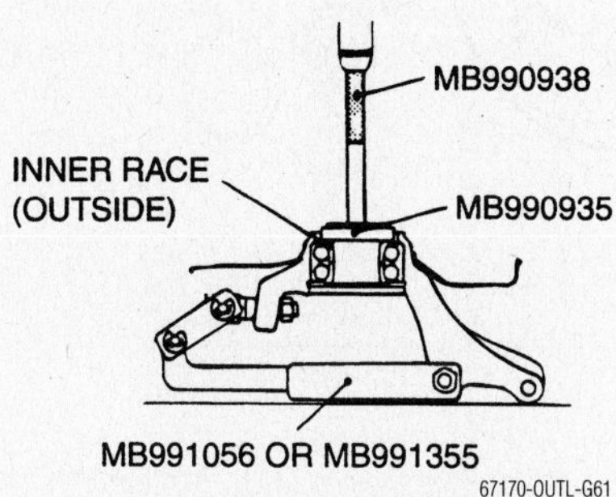

Place the inner race that was removed from the hub on the wheel bearing, then use the special tools to remove the wheel bearing

f. Install the inner race that was removed from the hub to the wheel bearing, then use Mitsubishi Special tools MB990935, 990938 and MB9911056 or MB991355 to remove the wheel bearing.

To install:

5. Assemble the hub and knuckle as follows:

a. Fill the wheel bearing with multipurpose grease. Apply a thin coat of grease fo the knuckle and bearing mating surfaces.

⁂ WARNING

To avoid damaging the wheel bearing, press the outer race when pressing in the wheel bearing.

b. Use Mitsubishi Special tools MB990883 and MB990890 to press in the bearing.

c. Install the snap ring.

d. Instlal the dust cover.

e. Install the hub. Tighten Mitsubishi Special tools MB991000 (MB990998) and MB991017 to 160–202 ft. lbs. (216–274 Nm), then press the hub into the knuckle.

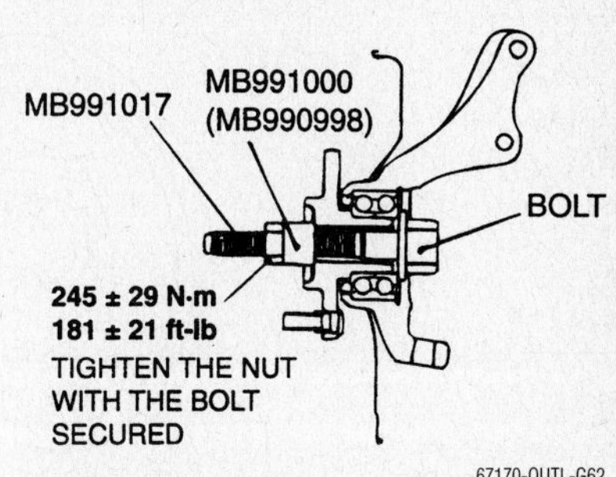

Tighten Mitsubishi Special tools to 160–202 ft. lbs. (216–274 Nm) to press the hub into the knuckle

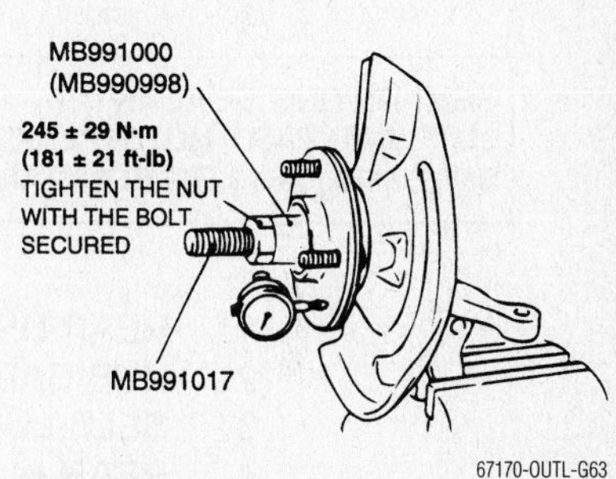

Check the wheel bearing play as shown

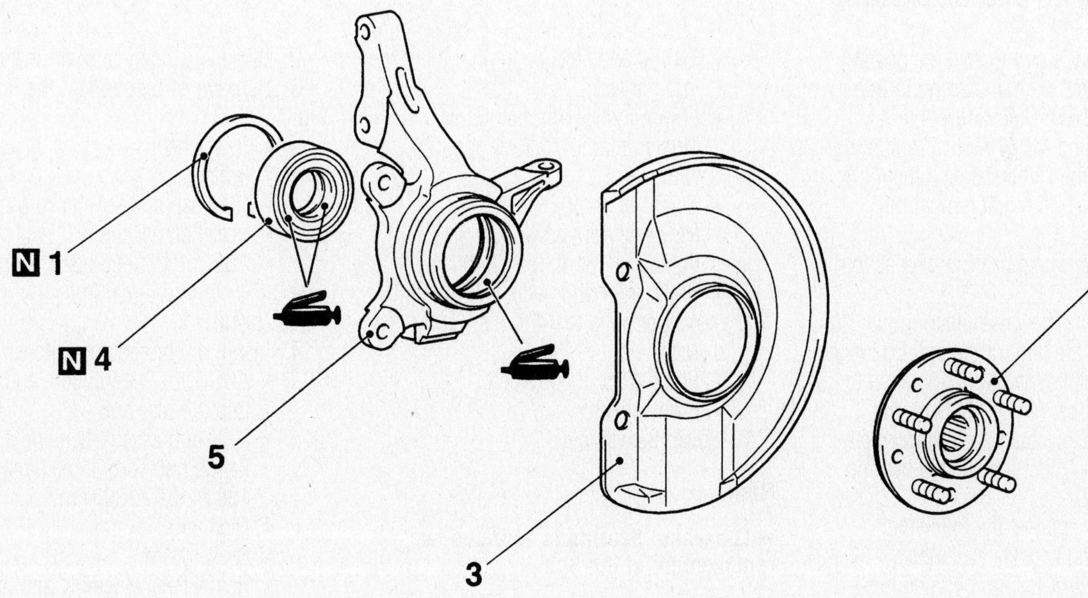

Disassembled view of the hub, bearing and knuckle

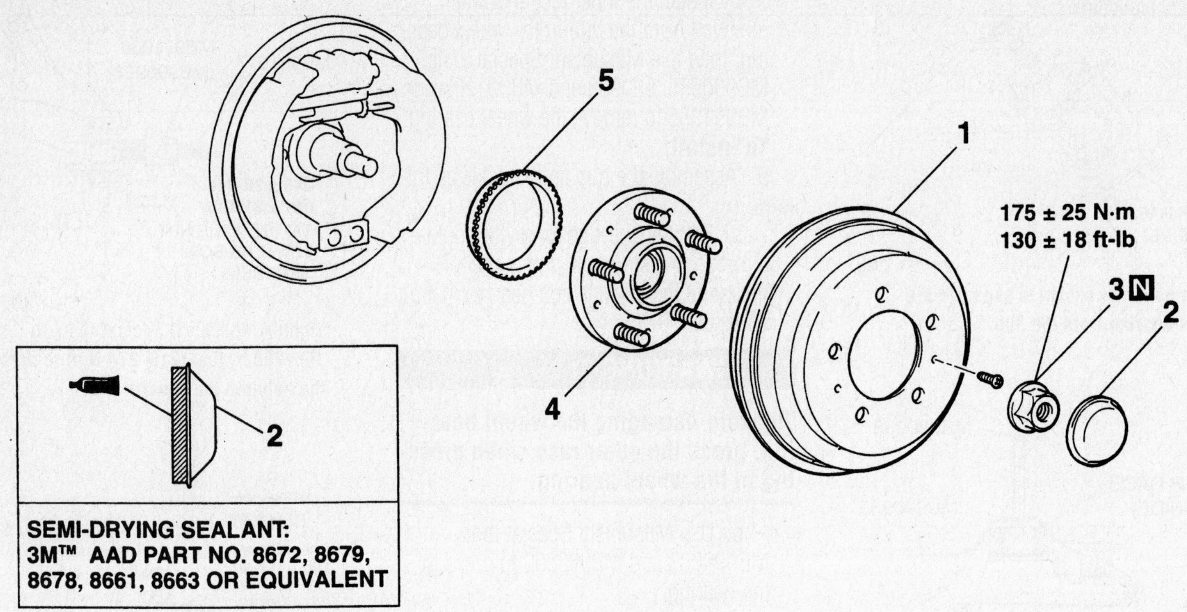

Exploded of the rear wheel hub and bearing

f. Rotate the hub to seat the bearing.

g. Measure the hub starting torque using Mitsubishi Special tools MB990326 and MB990685. The starting torque must be within 16 inch lbs. (1.8 Nm), and the hub rotation must be smooth.

6. Check the wheel bearing play as follows:

a. Measure the wheel bearing play. It it not within the limit range of 0.002 inch (0.05mm), while the nut is tightened to 160–202 ft. lbs. (216–274 Nm), the assembly has probably been incorrectly installed. You must replace the bearing and re-install.

7. Install or connect the following:
- Hub and knuckle assembly
- Front strut-to-hub bolt and nut. Torque to 123 ft. lbs. (167 Nm).
- Halfshaft
- Tie rod end. Torque the nut to 19 ft. lbs. (25 Nm).
- Lower control arm ball joint. Tighten the nut to 19 ft. lbs. (25 Nm).
- Stabilizer shaft
- New axle nut and washer and tighten to 181 ft. lbs. (245 Nm)
- Brake rotor and caliper. Torque the caliper bolts to 74 ft. lbs. (100 Nm).
- Front WSS, if equipped
- Front wheel

8. Lower the vehicle

Rear

➡The wheel bearing is serviced by replacement of the hub.

1. Before servicing the vehicle, refer to the precautions in the beginning of this section.
2. Raise and safely support the vehicle.
3. Remove or disconnect the following:
- Rear wheel
- Brake drum
- Dust cap and self-locking nut
- Rear hub assembly
- ABS rotor, if equipped, by pressing it out of the rear hub

To install:

4. Install or connect the following:
- ABS rotor, if equipped, and press it into the rear hub
- Rear hub assembly using a new flange nut. Torque the flange nut to 130 ft. lbs. (180 Nm).
- Dust cap
- Brake drum
- Rear wheel assembly and lower the vehicle to the floor

BRAKES

Brake Caliper

REMOVAL & INSTALLATION

Front

1. Before servicing the vehicle, refer to the precautions in the beginning of this section.
2. Remove or disconnect the following:
   - Wheels
   - Brake hose. Discard the gaskets.
   - Caliper mounting bolts
   - Caliper assembly

To install

3. Install or connect the following:
   - Brake caliper into position. Tighten the bolts to 74 ft. lbs. (100 Nm).
   - Brake hose with new gaskets. Tighten the bolt to 22 ft. lbs. (30 Nm). Bleed the brake system.
   - Wheels

Disc Brake Pads

REMOVAL & INSTALLATION

Front

1. Before servicing the vehicle, refer to the precautions in the beginning of this section.
2. Remove or disconnect the following:
   - Some of the brake fluid from the master cylinder reservoir
   - Wheels
   - Caliper guide and lock pins and lift the caliper assembly from the caliper support.

➡On some vehicles, the caliper can be flipped up by leaving the upper pin in place and using it as a pivot point.

   - Brake pads, spring clip and shims

To install:

3. Compress pistons back into the caliper bore.
4. Lubricate slide points and install the brake pads, shims and spring clips onto the caliper support. Install the caliper over the brake pads.
5. Lubricate and install the caliper guide and lock pins in their original positions. Tighten to 28 ft. lbs. (34 Nm).
6. Install the wheels.

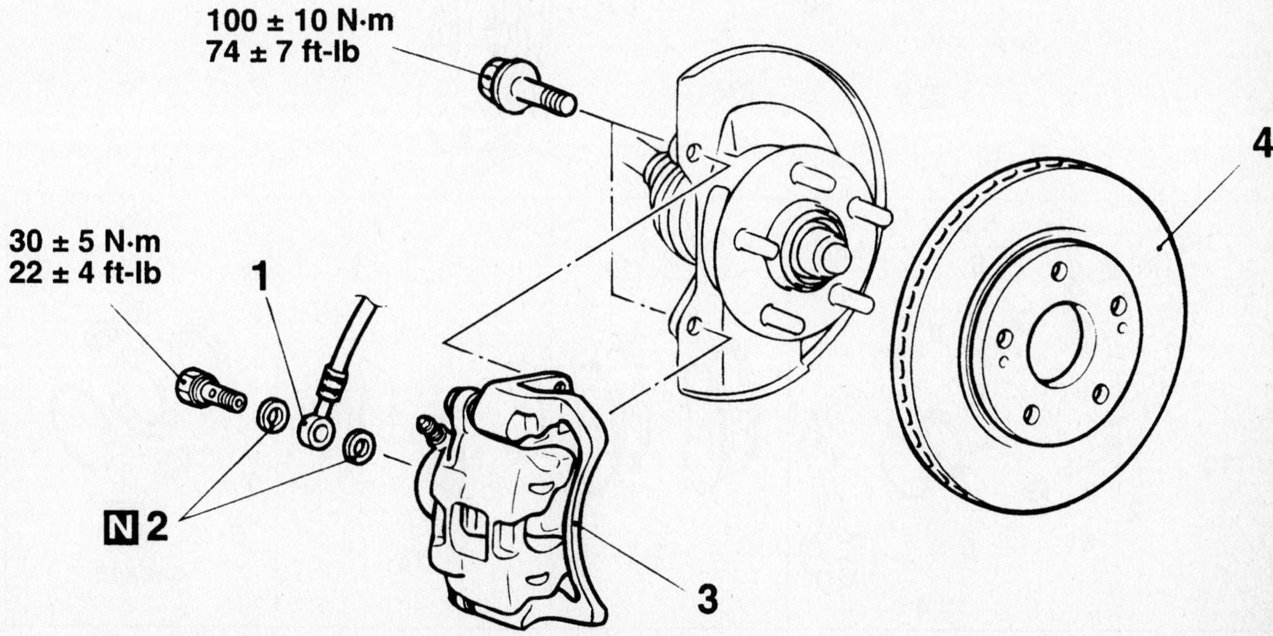

Front disc brake caliper mounting

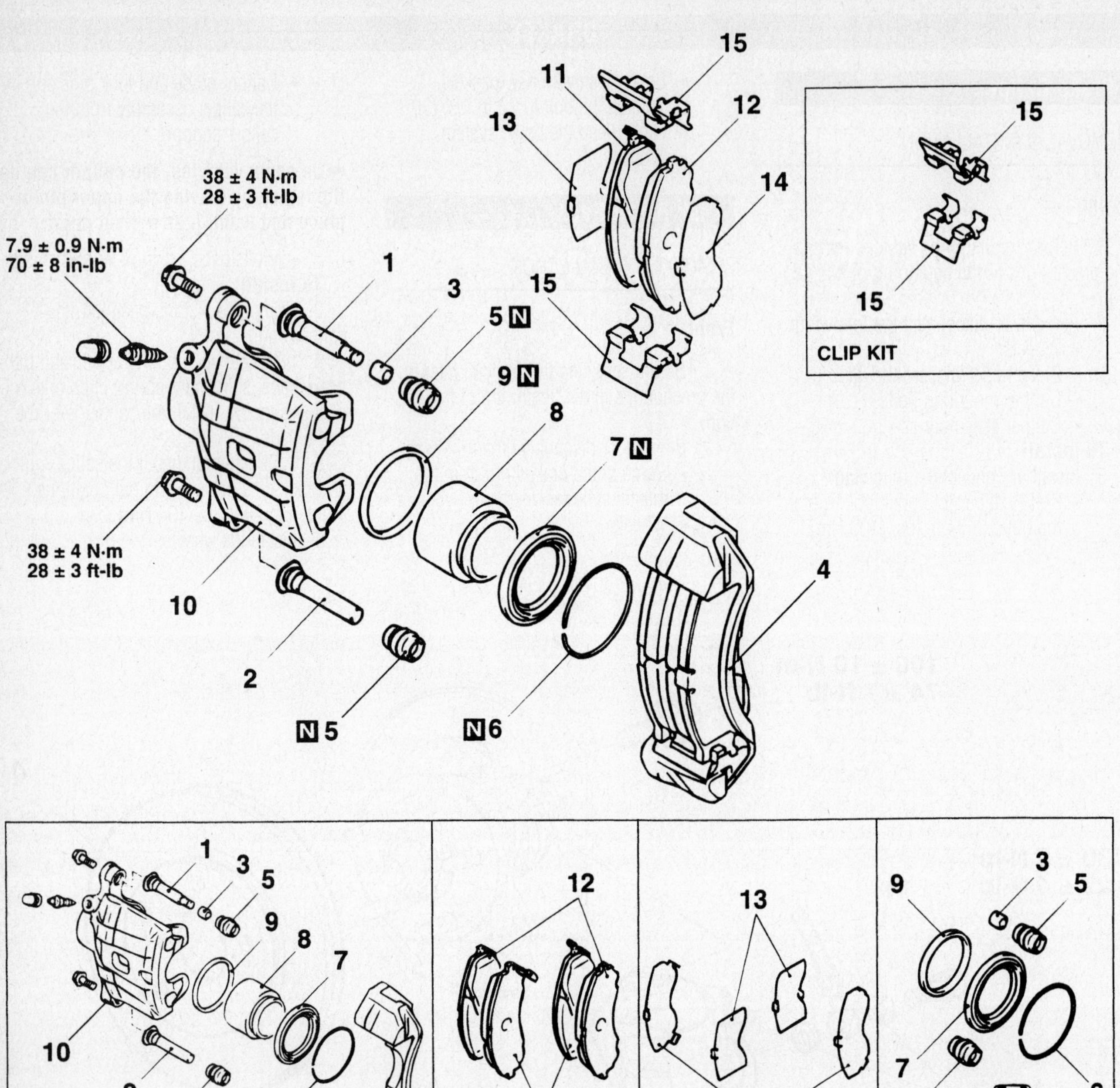

1. GUIDE PIN
2. LOCK PIN
3. BUSHING
4. CALIPER SUPPORT (INCLUDING PAD, CLIP, AND SHIM)
5. PIN BOOT
6. BOOT RING
7. PISTON BOOT
8. PISTON
9. PISTON SEAL
10. CALIPER BODY
11. PAD AND WEAR INDICATOR ASSEMBLY
12. PAD ASSEMBLY
13. INNER SHIM
14. OUTER SHIM
15. CLIP

67170-OUTL-G67

Exploded view of the caliper and brake pads

Brake Drums

REMOVAL & INSTALLATION

Rear

1. Before servicing the vehicle, refer to the precautions in the beginning of this section.
2. Remove or disconnect the following:
   - Wheels
   - Drum

To install:

- Drum
- Wheels

Brake Shoes

REMOVAL & INSTALLATION

Rear

1. Before servicing the vehicle, refer to the precautions in the beginning of this section.
2. Loosen the parking brake cable adjusting nut.
3. Remove or disconnect the following:
   - Wheels
   - Brake drum
   - Parking brake cable connection.

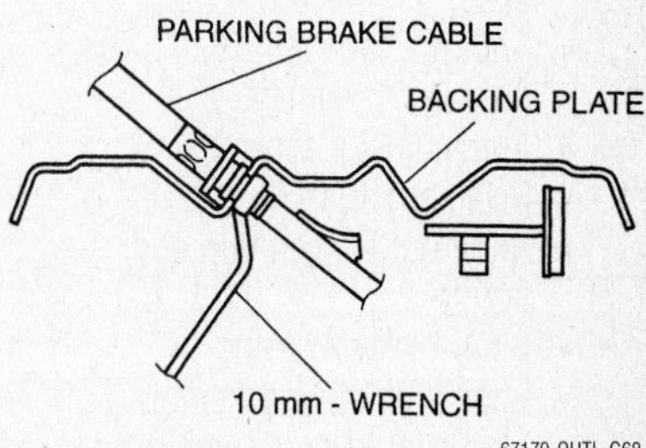

Disconnecting the parking brake cable

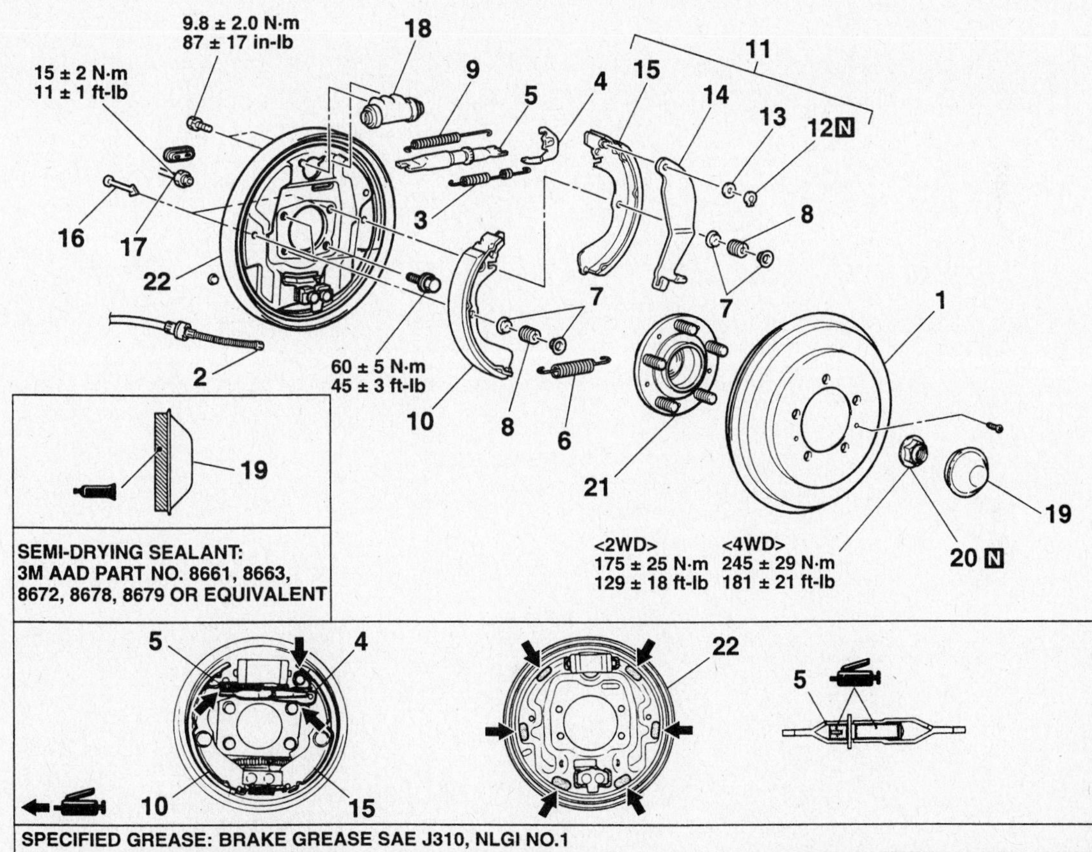

1. BRAKE DRUM
2. PARKING BRAKE CABLE CONNECTION
3. SHOE-TO-SHOE SPRING
4. ADJUSTER LEVER
5. AUTO ADJUSTER ASSEMBLY
6. RETAINER SPRING
7. SHOE HOLD-DOWN CUP
8. SHOE HOLD-DOWN SPRING
9. SHOE-TO-LEVER SPRING
10. SHOE AND LINING ASSEMBLY
11. SHOE AND LEVER ASSEMBLY
12. RETAINER
13. WAVE WASHER
14. PARKING LEVER
15. SHOE AND LINING ASSEMBLY
16. SHOE HOLD-DOWN PIN
17. BRAKE PIPE CONNECTION
18. WHEEL CYLINDER ASSEMBLY
19. HUB CAP (2WD)
20. LOCK NUT
21. REAR HUB ASSEMBLY
22. BACKING PLATE

67170-OUTL-G69

Exploded view of the rear drum brakes

Use a 10mm wrench to help remove the cable from the backing plate.

- Shoe-to-shoe spring
- Adjuster assembly
- Auto adjuster assembly
- Retainer spring
- Shoe hold-down cup and spring
- Shoe-to-lever spring
- Shoe and lining assembly
- Shoe and lever assembly
- Shoe from the lever by inserting a prytool to open the retainer joint, then remove the retainer

To install

4. Lubricate the backing plate bosses, anchor pin, and parking brake actuating mechanism with lithium-based grease.
5. Install or connect the following:
   - Lever to the shoe and secure with the retainer
   - Shoe and lever assembly
   - Shoe and lining assembly
   - Shoe-to-lever spring
   - Shoe hold-down spring
   - Shoe hold-down cup
   - Retainer spring
   - Auto adjuster assembly
   - Adjuster lever
   - Shoe-to-shoe spring
   - Parking brake cable
6. Pre-adjust the shoes so the drum slides on with a light drag and install brake drum.
   - Wheel bearing dust cap and adjust the rear brake shoes
7. Tighten the parking brake cable adjusting nut.

NISSAN

Altima • Maxima • Sentra • 350Z

5

SPECIFICATION AND MAINTENANCE CHARTS

ENGINE AND VEHICLE IDENTIFICATION

Engine						
Code ①	Liters (cc)	Cu. In.	Cyl.	Fuel Sys.	Engine Type	Eng. Mfg.
QG18DE	1.8 (1769)	108	4	MFI	DOHC	Nissan
SR20DE	2.0 (1998)	122	4	MFI	DOHC	Nissan
KA24DE	2.4 (2389)	146	4	MFI	DOHC	Nissan
QR25DE	2.5 (2488)	152	4	MFI	DOHC	Nissan
VQ30DE	3.0 (2988)	182	6	MFI	DOHC	Nissan
VQ35DE	3.5 (3498)	213	6	MFI	DOHC	Nissan

Model Year	
Code ②	Year
1	2001
2	2002
3	2003
4	2004
5	2005

MFI: Multi-port Fuel Injection

DOHC: Double Overhead Camshaft

① The Engine Code is stamped on the engine block near the starter.

② 10th position of the Vehicle Identification Number (VIN)

67170-NISS-C01

GENERAL ENGINE SPECIFICATIONS

Year	Model	Engine Displacement Liters	Engine Series (ID/VIN)	Net Horsepower @ rpm	Net Torque @ rpm (ft. lbs.)	Bore x Stroke (in.)	Com-pression Ratio	Oil Pressure @ rpm
2001	Altima	2.4	KA24DE	150@5600	154@5600	3.50x3.78	9.2:1	60@3000
	Maxima	3.0	VQ30DE	190@5600	205@4000	3.66x2.89	10.0:1	63@3000
	Sentra	1.8	QG18DE	126@6000	129@2400	3.15x3.46	9.5:1	50@3000
	Sentra	2.0	SR20DE	145@6400	136@4800	3.39x3.39	9.5:1	46@3200
2002	Altima	2.5	QR25DE	150@5600	154@5600	3.50X3.94	9.5:1	60@3000
	Altima	3.5	VQ35DE	260@6000	260@4800	3.76X3.20	10.3:1	43@2000
	Maxima	3.5	VQ35DE	260@6000	260@4800	3.76X3.20	10.3:1	43@2000
	Sentra	1.8	QG18DE	126@6000	129@2400	3.15x3.46	9.5:1	50@3000
	Sentra	2.5	QR25DE	150@5600	154@5600	3.50x3.94	9.5:1	60@3000
2003	Altima	2.5	QR25DE	150@5600	154@5600	3.50X3.94	9.5:1	60@3000
	Altima	3.5	VQ35DE	260@6000	260@4800	3.76X3.20	10.3:1	43@2000
	Maxima	3.5	VQ35DE	260@6000	260@4800	3.76X3.20	10.3:1	43@2000
	Sentra	1.8	QG18DE	126@6000	129@2400	3.15x3.46	9.5:1	50@3000
	Sentra	2.5	QR25DE	150@5600	154@5600	3.50x3.94	9.5:1	60@3000
	350Z	3.5	VQ35DE	260@6000	260@4800	3.76X3.20	10.3:1	43@2000
2004	Altima	2.5	QR25DE	150@5600	154@5600	3.50X3.94	9.5:1	60@3000
	Altima	3.5	VQ35DE	260@6000	260@4800	3.76X3.20	10.3:1	43@2000
	Maxima	3.5	VQ35DE	260@6000	260@4800	3.76X3.20	10.3:1	43@2000
	Sentra	1.8	QG18DE	126@6000	129@2400	3.15x3.46	9.5:1	50@3000
	Sentra	2.5	QR25DE	150@5600	154@5600	3.50x3.94	9.5:1	60@3000
	350Z	3.5	VQ35DE	260@6000	260@4800	3.76X3.20	10.3:1	43@2000

67170-NISS-C02

ENGINE TUNE-UP SPECIFICATIONS

Year	Engine Displacement Liters	Engine ID/VIN	Spark Plug Gap (in.)	Ignition Timing (deg.)		Fuel Pump (psi) ①	Idle Speed (rpm)		Valve Clearance	
				MT	AT		MT	AT ②	Intake ③	Exhaust ③
2001	1.8	QG18DE	0.043	9B	9B	36	625	725	0.015	0.016
	2.0	SR20DE	0.033	15B	15B	36	800	800	HYD	HYD
	2.4	KA24DE	0.041	20B	20B	33	650	650	0.015	0.016
	3.0	VQ30DE	0.041	15B	15B	34	650	700	0.014	0.015
2002	1.8	QG18DE	0.043	7B	18	51	600-700	750-850	0.015	0.017
	2.5	QR25DE	0.043	15B	20B	33	650	650	HYD	HYD
	3.5	VQ35DE	0.043	—	15B	34	—	600-700	HYD	HYD
2003	1.8	QG18DE	0.043	7B	18	51	600-700	750-850	0.015	0.017
	2.5	QR25DE	0.043	15B	20B	33	650	650	HYD	HYD
	3.5	VQ35DE	0.043	—	15B	51	600-700	600-700	HYD	HYD
2004	1.8	QG18DE	0.043	7B	18	51	600-700	750-850	0.015	0.017
	2.5	QR25DE	0.043	15B	15B	33	700	700	HYD	HYD
	3.5	VQ35DE	0.043	—	15B	51	600-700	600-700	HYD	HYD

NOTE: The Vehicle Emission Control Information label often reflects specification changes made during production.
The label figures must be used if they differ from those in this chart.

B: Before top dead center

HYD: Hydraulic

① System pressure at idle with vacuum hose connected; should increase to 43 psi when disconnected

② Automatic transmission in neutral

③ Engine warm

67170-NISS-C03

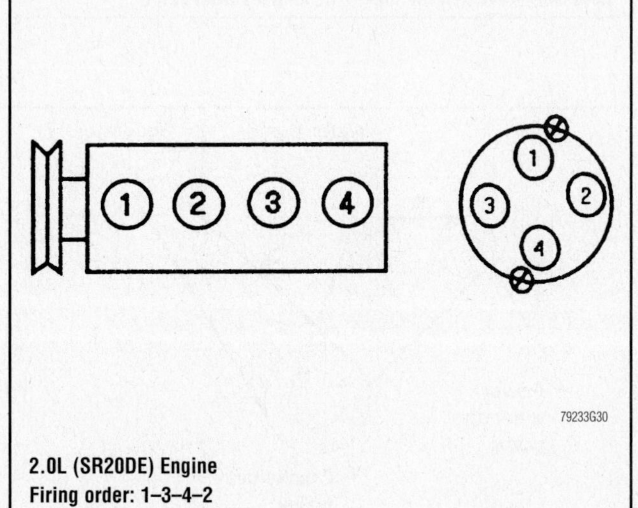

2.0L (SR20DE) Engine
Firing order: 1–3–4–2
Distributor rotation: Counterclockwise

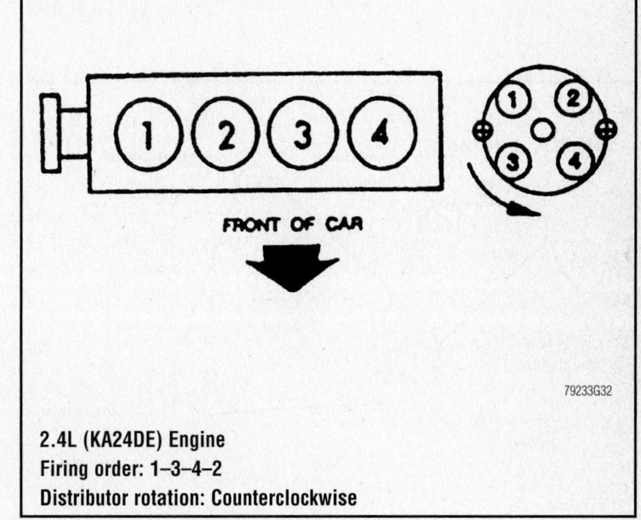

2.4L (KA24DE) Engine
Firing order: 1–3–4–2
Distributor rotation: Counterclockwise

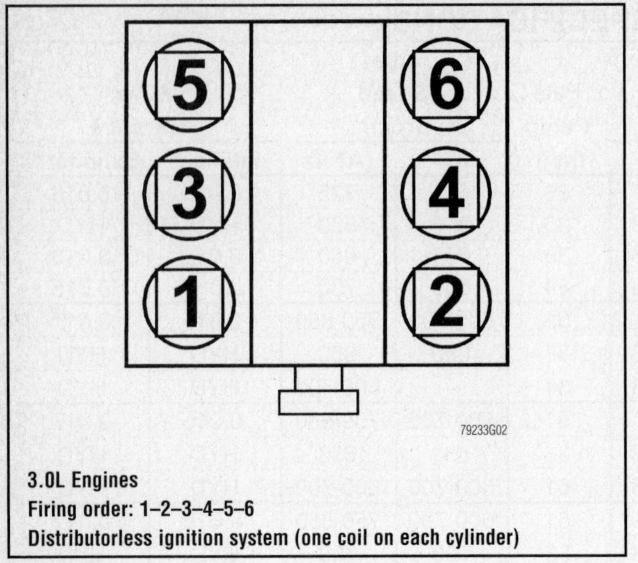

3.0L Engines
Firing order: 1–2–3–4–5–6
Distributorless ignition system (one coil on each cylinder)

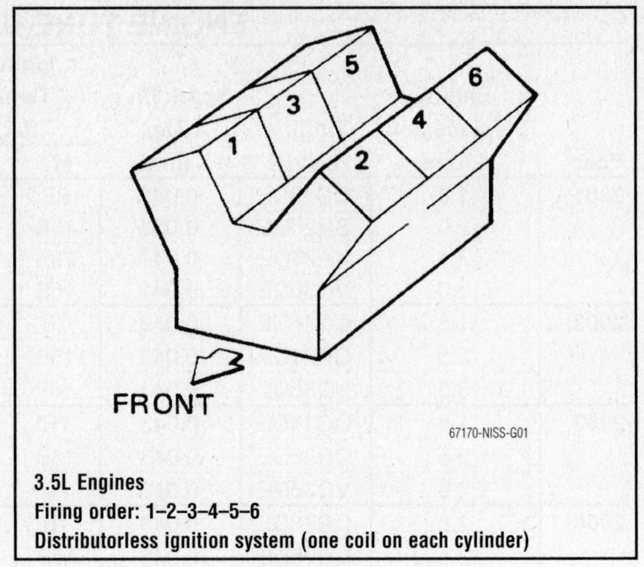

3.5L Engines
Firing order: 1–2–3–4–5–6
Distributorless ignition system (one coil on each cylinder)

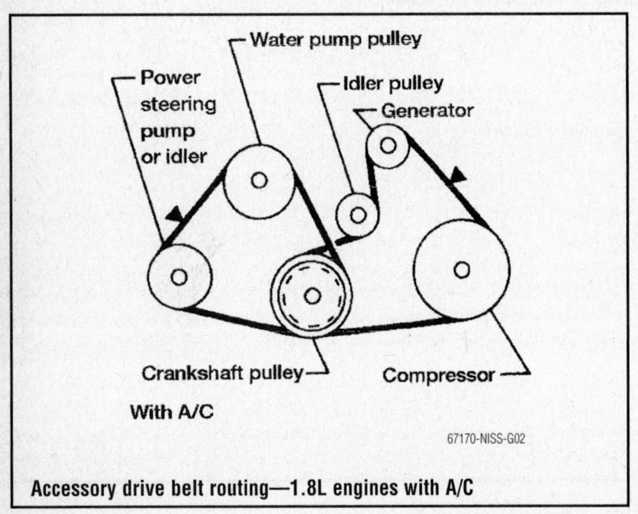

Accessory drive belt routing—1.8L engines with A/C

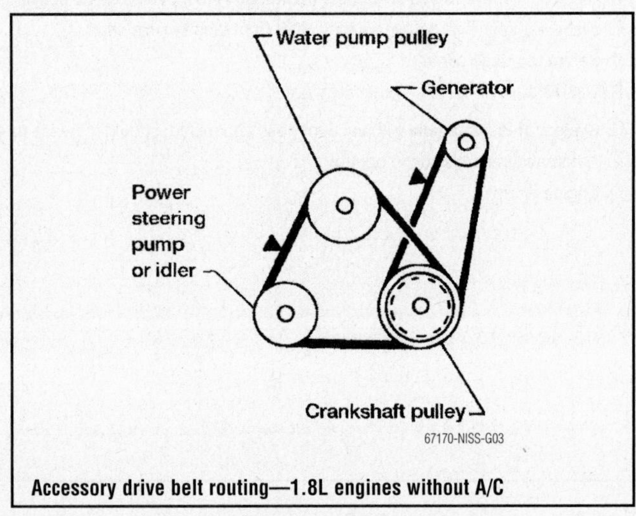

Accessory drive belt routing—1.8L engines without A/C

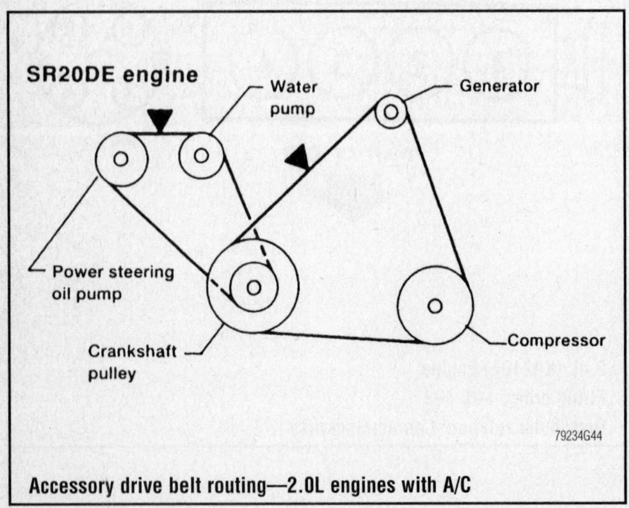

Accessory drive belt routing—2.0L engines with A/C

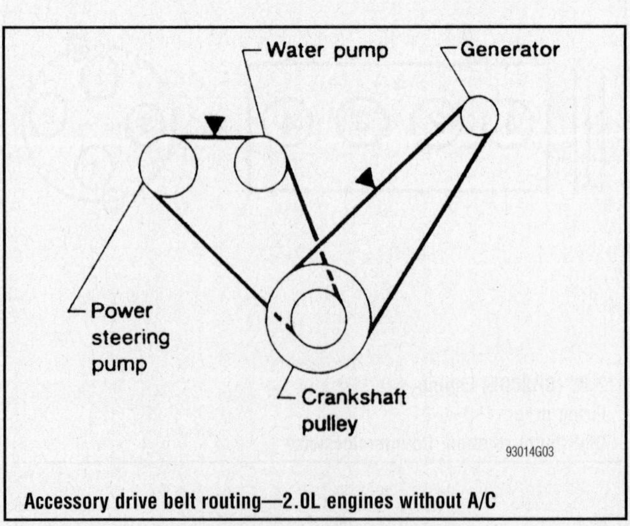

Accessory drive belt routing—2.0L engines without A/C

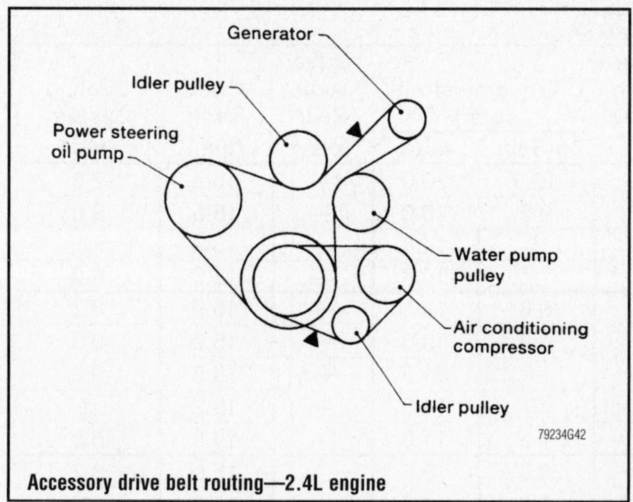

Accessory drive belt routing—2.4L engine

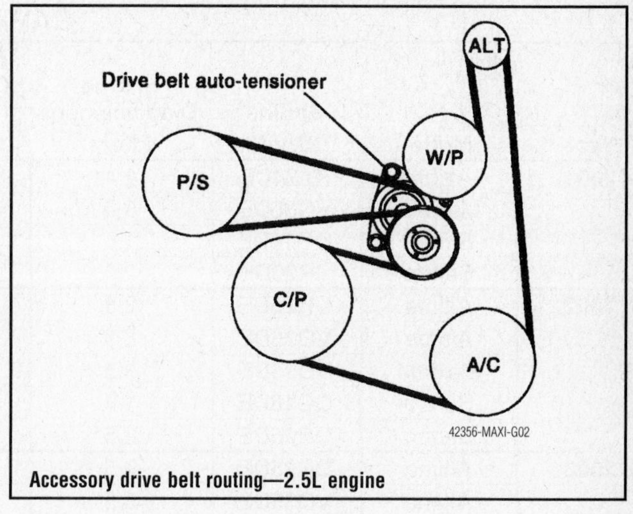

Accessory drive belt routing—2.5L engine

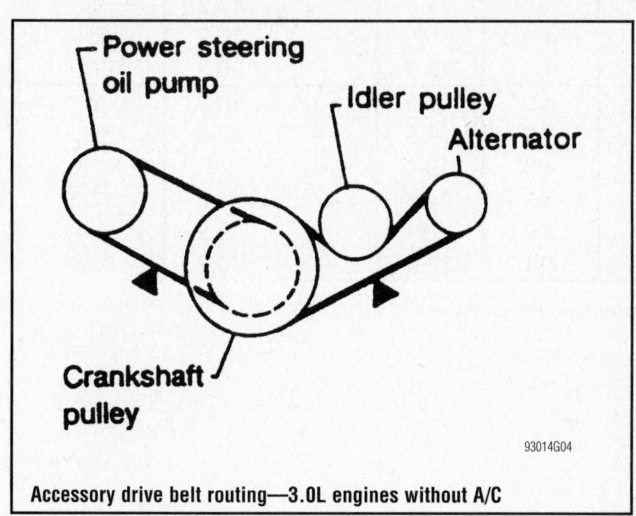

Accessory drive belt routing—3.0L engines without A/C

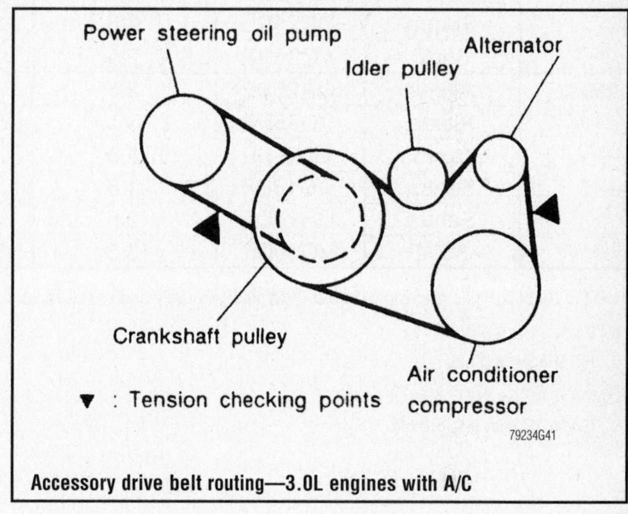

Accessory drive belt routing—3.0L engines with A/C

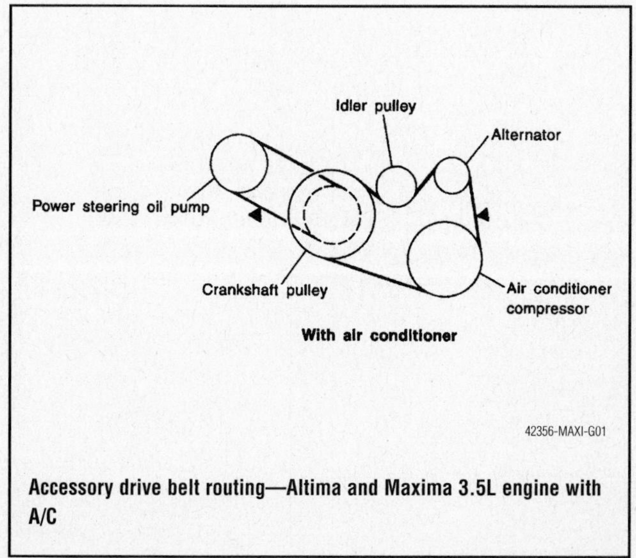

Accessory drive belt routing—Altima and Maxima 3.5L engine with A/C

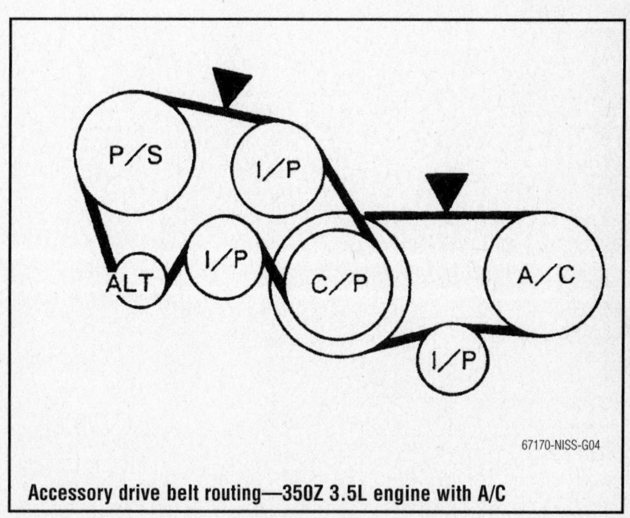

Accessory drive belt routing—350Z 3.5L engine with A/C

CAPACITIES

Year	Model	Engine ID/VIN	Engine Displacement Liters	Engine Oil with Filter (qts.)	Transmission (pts.) 5-Spd	Transmission (pts.) Auto.	Drive Axle Rear (pts.)	Fuel Tank (gal.)	Cooling System (qts.)
2001	Altima	KA24DE	2.4	4.1	10.0	20.0	—	15.9	7.8
	Maxima	VQ30DE	3.0	4.3	9.5	20.0	—	18.5	9.0
	Sentra	QG18DE	1.8	3.5	①	14.8	—	13.2	②
	Sentra	SR20DE	2.0	3.4	7.5	14.8	—	13.0	6.0
2002	Altima	QR25DE	2.5	4.5	6.8	19.5	—	15.9	8.2
	Altima	VQ35DE	3.5	4.3	6.8	19.0	—	15.9	9.3
	Maxima	VQ35DE	3.5	5.0	—	21.8	—	18.5	9.0
	Sentra	QG18DE	1.8	3.5	①	14.8	—	13.2	②
	Sentra	QR25DE	2.5	3.4	7.5	14.8	—	13.0	6.0
2003	Altima	QR25DE	2.5	4.5	6.8	19.5	—	15.9	8.2
	Altima	VQ35DE	3.5	4.3	6.8	19.0	—	15.9	9.3
	Maxima	VQ35DE	3.5	5.0	—	21.8	—	18.5	9.0
	Sentra	QG18DE	1.8	3.5	①	14.8	—	13.2	②
	Sentra	QR25DE	2.5	4.2	4.9	18	—	13.2	6.4
	350Z	VQ35DE	3.5	5	6.1	22.8	—	20.2	9.30
2004	Altima	QR25DE	2.5	4.5	4.9	19.5	—	20.0	8
	Altima	VQ35DE	3.5	4.3	4.9	19.5	—	20.0	8.5
	Maxima	VQ35DE	3.5	4.2	4.8	19.4	—	20	7.8
	Sentra	QG18DE	1.8	2.8	6.3	14.3	—	13.2	12.5
	Sentra	QR25DE	2.5	4.2	4.9	18	—	13.2	6.4
	350Z	VQ35DE	3.5	5	6.1	22.8	—	20.2	9.30

NOTE: All capacities are approximate. Add fluid gradually and check to be sure a proper fluid level is obtained.

① RS5F31A: 6.5 pts.
RS5F32V: 8.0 pts.

② GA16DE with MT: 5.5 qts.
GA16DE with AT: 6.0 qts.

67170-NISS-C04

VALVE SPECIFICATIONS

Year	Engine ID/VIN	Engine Displacement Liters	Seat Angle (deg.)	Face Angle (deg.)	Spring Test Pressure (lbs. @ in.)	Spring Installed Height (in.)	Stem-to-Guide Clearance (in.)		Stem Diameter (in.)	
							Intake	Exhaust	Intake	Exhaust
2001	QG18DE	1.8	45	45.25-45.75	83@0.931	NA	0.0008-0.0020	0.0016-0.0028	0.2152-0.2157	0.2144-0.2150
	SR20DE	2.0	45	45.25-45.75	137@1.181	NA	0.0008-0.0021	0.0016-0.0029	0.2348-0.2354	0.2341-0.2346
	KA24DE	2.4	45	45.25-45.75	123@1.024	NA	0.0008-0.0021	0.0016-0.0029	0.2742-0.2748	0.2734-0.2740
	VQ30DE	3.0	45	45.25-45.75	102@1.085	NA	0.0008-0.0021	0.0016-0.0029	0.2348-0.2354	0.2341-0.2346
2002	SR20DE	1.8	45	45.25-45.75	137@1.181	NA	0.0008-0.0021	0.0016-0.0029	0.2348-0.2354	0.2341-0.2346
	QR25DE	2.5	45.15-45.45	NA	34-39@1.39	NA	0.0009-0.0013	0.010-0.013	0.2348-0.2354	0.2344-0.2350
	VQ35DE	3.5	45.15-45.45	NA	91.5-103.2@1.094	1.796	0.010-0.013	0.011-0.015	0.2348-0.2354	0.2344-0.2350
2003	QG18DE	1.8	44.53-45.07	NA	83@0.931	NA	0.0008-0.0020	0.0016-0.0028	0.2152-0.2157	0.2144-0.2150
	QR25DE	2.5	45.15-45.45	NA	34-39@1.39	NA	0.0009-0.0013	0.010-0.013	0.2348-0.2354	0.2344-0.2350
	VQ35DE	3.5	45.15-45.45	NA	91.5-103.2@1.094	NA	0.010-0.013	0.011-0.015	0.2348-0.2354	0.2344-0.2350
2004	QG18DE	1.8	44.53-45.07	NA	83@0.931	NA	0.0008-0.0020	0.0016-0.0028	0.2152-0.2157	0.2144-0.2150
	QR25DE	2.5	45.15-45.45	NA	34-39@1.39	NA	0.0008-0.0021	0.0012-0.0025	0.2348-0.2354	0.2344-0.2350
	VQ35DE	3.5	45.15-45.45	NA	91.5-103.2@1.094	NA	0.010-0.013	0.011-0.015	0.2348-0.2354	0.2344-0.2350

NA: Not Available

67170-NISS-C05

CRANKSHAFT AND CONNECTING ROD SPECIFICATIONS

All measurements are given in inches.

	Engine		Crankshaft				Connecting Rod		
Year	Displacement Liters	Engine ID/VIN	Main Brg. Journal Dia.	Main Brg. Oil Clearance	Shaft End-play	Thrust on No.	Journal Diameter	Oil Clearance	Side Clearance
2001	1.8	QG18DE	1.9668-1.9671	0.0007-0.0017	0.0024-0.0071	3	1.6929-1.6934	0.0006-0.0015	0.0079-0.0185
	2.0	SR20DE	2.1643-2.1646	0.0002-0.0009	0.0039-0.0102	3	1.8885-1.8887	0.0008-0.0018	0.0079-0.0138
	2.4	KA24DE	2.3609-2.3612	0.0008-0.0019	0.0020-0.0070	3	1.9672-1.9675	0.0004-0.0014	0.0080-0.0160
	3.0	VQ30DE	2.3610-2.3612	0.0014-0.0021	0.0039-0.0098	3	1.7704-1.7706	0.0013-0.0023	0.0079-0.0138
2002	1.8	QG18DE	1.9668-1.9671	0.0007-0.0017	0.0024-0.0089	3	1.6929-1.6934	0.0006-0.0015	0.0079-0.0185
	2.5	QR25DE	2.1636-2.1645	①	0.0039-0.0102	3	1.8898-1.8903	0.0004-0.0014	0.0079-0.0138
	3.5	VQ35DE	2.3603-2.3612	0.0014-0.0021	0.0039-0.0098	3	1.7704-1.7706	0.0013-0.0023	0.0079-0.0138
2003	1.8	QG18DE	1.9668-1.9671	0.0007-0.0017	0.0024-0.0089	3	1.6929-1.6934	0.0006-0.0015	0.0079-0.0185
	2.5	QR25DE	2.1636-2.1645	①	0.0039-0.0102	3	1.8898-1.8903	0.0004-0.0014	0.0079-0.0138
	3.5	VQ35DE	2.3603-2.3612	0.0014-0.0021	0.0039-0.0098	3	1.7704-1.7706	0.0013-0.0023	0.0079-0.0138
2004	1.8	QG18DE	1.9668-1.9671	0.0007-0.0017	0.0024-0.0089	3	1.6929-1.6934	0.0006-0.0015	0.0079-0.0185
	2.5	QR25DE	2.1636-2.1645	①	0.0039-0.0102	3	1.8898-1.8903	0.0011-0.0018	0.0079-0.0138
	3.5	VQ35DE	2.3603-2.3612	0.0014-0.0018	0.0039-0.0098	3	1.7704-1.7706	0.0013-0.0023	0.0079-0.0138

① Nos. 1, 3 and 5: 0.0005-0.0009 in.
Nos. 2 and 4: 0.0007-0.0011

67170-NISS-C06

PISTON AND RING SPECIFICATIONS

All measurements are given in inches.

Year	Engine Displacement Liters	Engine ID/VIN	Piston Clearance	Ring Gap Top Compression	Ring Gap Bottom Compression	Ring Gap Oil Control	Ring Side Clearance Top Compression	Ring Side Clearance Bottom Compression	Ring Side Clearance Oil Control
2001	1.8	QG18DE	0.0010-0.0018	0.0079-0.0154	0.0126-0.0220	0.0079-0.0272	0.0018-0.0031	0.0012-0.0028	0.0026-0.0053
	2.0	SR20DE	0.0006-0.0014	0.0079-0.0118	0.0138-0.0197	0.0079-0.0236	0.0018-0.0031	0.0012-0.0026	SNUG
	2.4	KA24DE	0.0006-0.0014	0.0110-0.0205	0.0079-0.0272	0.0100-1.0000	0.0016-0.0031	0.0012-0.0028	SNUG
	3.0	VQ30DE	0.0006-0.0014	0.0087-0.0126	0.0126-0.0185	0.0079-0.0236	0.0016-0.0031	0.0012-0.0028	SNUG
2002	1.8	QG18DE	0.0010-0.0018	0.0079-0.0154	0.0126-0.0220	0.0079-0.0272	0.0018-0.0031	0.0012-0.0028	0.0026-0.0053
	2.5	QR25DE	0.0004-0.0012	0.0083-0.0122	0.0126-0.0185	0.0079-0.0236	0.0018-0.0031	0.0012-0.0028	0.0026-0.0053
	3.5	VQ35DE	0.0004-0.0012	0.0091-0.0130	0.0130-0.0189	0.0079-0.0197	0.0018-0.0031	0.0012-0.0028	0.0026-0.0053
2003	1.8	QG18DE	0.0010-0.0018	0.0079-0.0154	0.0126-0.0220	0.0079-0.0272	0.0018-0.0031	0.0012-0.0028	0.0026-0.0053
	2.5	QR25DE	0.0004-0.0012	0.0083-0.0122	0.0126-0.0185	0.0079-0.0236	0.0018-0.0031	0.0012-0.0028	0.0026-0.0053
	3.5	VQ35DE	0.0004-0.0012	0.0091-0.0130	0.0130-0.0189	0.0079-0.0197	0.0018-0.0031	0.0012-0.0028	0.0026-0.0053
2004	1.8	QG18DE	0.0010-0.0018	0.0079-0.0154	0.0126-0.0220	0.0079-0.0272	0.0018-0.0031	0.0012-0.0028	0.0026-0.0053
	2.5	QR25DE	0.0004-0.0012	0.0083-0.0122	0.0126-0.0185	0.0079-0.0236	0.0018-0.0031	0.0012-0.0028	0.0026-0.0053
	3.5	VQ35DE	0.0004-0.0012	0.0091-0.0130	0.0130-0.0189	0.0079-0.0197	0.0018-0.0031	0.0012-0.0028	0.0026-0.0053

67170-NISS-C07

TORQUE SPECIFICATIONS

All readings in ft. lbs.

Year	Engine Displacement Liters (cc)	Engine ID/VIN	Cylinder Head Bolts	Main Bearing Bolts	Rod Bearing Bolts	Crankshaft Damper Bolts	Flywheel Bolts	Manifold Intake	Manifold Exhaust	Spark Plugs	Oil Drain Plug
2001	1.8 (1769)	QG18DE	①	34-38	②	98-112	③	14	19	18	25
	2.0 (1998)	SR20DE	④	⑤	⑥	105-112	61-69	14	30	18	25
	2.4 (2389)	KA24DE	⑦	34-41	⑥	105-112	105-112	14	32	18	25
	3.0 (2988)	VQ30DE	⑧	⑨	⑩	⑪	61-69	⑫	23	18	25
2002	1.8 (1769)	QG18DE	①	34-38	②	98-112	③	14	19	18	25
	2.5 (2488)	QR25DE	⑬	⑭	⑮	⑪	76-83	13-15	29-32	18	25
	3.5 (3498)	VQ35DE	⑯	⑰	⑱	⑪	61-69	⑲	21-23	18	25
2003	1.8 (1769)	QG18DE	①	34-38	②	98-112	③	14	19	18	25
	2.5 (2488)	QR25DE	⑬	⑭	⑮	⑪	76-83	13-15	29-32	18	25
	3.5 (3498)	VQ35DE	⑯	⑰	⑱	⑪	61-69	⑲	21-23	18	25
2004	1.8 (1769)	QG18DE	①	34-38	②	98-112	③	14	19	18	25
	2.5 (2488)	QR25DE	⑬	⑭	⑮	⑪	76-83	13-15	29-32	18	25
	3.5 (3498)	VQ35DE	⑯	⑰	⑱	⑪	61-69	⑲	21-23	18	25

① Bolt Nos. 1-10:
Step 1: 22 ft. lbs.
Step 2: 43 ft. lbs.
Step 3: Loosen completely then retorque to 22 ft. lbs.
Step 4: 43 ft. lbs. or an additional 50-55 degrees
Bolt Nos. 11-14: Torque last, to 72 inch lbs.

② Step 1: 12 ft. lbs.
Step 2: 19 ft. lbs. or an additional 35-40 degrees

③ Manual transmission: 61-69 ft. lbs.
Automatic transmission: 69-76 ft. lbs.

④ Step 1: 29 ft. lbs.
Step 2: 58 ft. lbs.
Step 3: Loosen completely then retorque to 30 ft. lbs.
Step 4: Turn each bolt, in sequence,
an additional 90-100 degrees
Step 5: Repeat Step 4

⑤ Step 1: 20-24 ft. lbs.
Step 2: 75-80 degrees
Step 3: Loosen completely and retorque to 24-28 ft. lbs.
Step 4: 45-50 degree turn

⑥ 12 ft. lbs. plus an additional 60-65 degrees

⑦ Step 1: 22 ft. lbs.
Step 2: 58 ft. lbs.
Step 3: Loosen completely then retorque to 22 ft. lbs.
Step 4: 58 ft. lbs. or an additional 80-85 degrees

⑧ Step 1: 29-36 ft. lbs.
Step 2: Plus 60-65 degrees

⑨ Step 1: 3.6-7.2 ft. lbs.
Step 2: 20-23 ft. lbs.

⑩ Step 1: 29 ft. lbs.
Step 2: 90 ft. lbs.
Step 3: Loosen completely and retorque to 25-33 ft. lbs.
Step 4: Plus 90 ft. lbs. or 70 degrees
Step 5: Tighten two bolts marked with an "X" to 7-9 ft. lbs.

⑪ Step 1: 29-36 ft. lbs.
Step 2: 60-66 degrees

⑫ Step 1: 10-12 ft. lbs.
Step 2: 43-48 ft. lbs. or an additional 60-65 degrees

⑬ Step 1: 72 ft. lbs.
Step 2: Loosen completely, then retorque to 26-32 ft. lbs.
Step 3: Turn each bolt, in sequence, an additional 75-80 degrees
Step 4: Turn each bolt, in sequence, an additional 75-80 degrees

⑭ Bolt Nos. 1-10:
Step 1: 27-31 ft. lbs.
Step 2: Torque an additional 60-65 degrees
Bolt Nos. 11-14: Torque last, to 17-20 ft. lbs.

⑮ Step 1: 14-15 ft. lbs.
Step 2: 85-95 degrees

⑯ Step 1: 72 ft. lbs.
Step 2: Loosen bolts completely
Step 3: 25-33 ft. lbs.
Step 4: Tighten an additional 90-95 degrees
Step 5: Repeat Step 4

⑰ Step 1: Shift crankshaft to align the bearing beam
Step 2: Tighten all bolts to 24-28 ft. lbs.
Step 3: Tighten an additional 90-95 degrees

⑱ Step 1: Tighten to 15 ft. lbs.
Step 2: Tighten an additional 90-95 degrees

⑲ Step 1: Tighten to 4-7 ft. lbs.
Step 2: Tighten to 20-23 ft. lbs.
Step 3: Tighten, again, to 20-23 ft. lbs.

67170-NISS-C08

WHEEL ALIGNMENT

Year	Model		Caster Range (+/-Deg.)	Caster Preferred Setting (Deg.)	Camber Range (+/-Deg.)	Camber Preferred Setting (Deg.)	Toe-in (in.)
2001	Altima	F	0.75	+2.66	0.75	-0.10	0.04 +/- 0.04
		R	—	—	0.75	-1.25	—
	Maxima ①	F	0.75	+2.75	0.75	-0.25	0.04 +/- 0.04
		R	—	—	0.75	-1.00	0.04 +/- 0.04
	Maxima ②	F	0.75	+2.75	0.75	-0.33	0.04 +/- 0.04
		R	—	—	0.75	-1.00	0.04 +/- 0.04
	Sentra	F	0.75	+1.42	0.75	-0.58	—
		R	—	—	0.75	+1.00	—
2002	Altima	F	0.75	+2.66	0.75	-0.10	0.04 +/- 0.04
		R	—	—	0.75	-1.25	0.08 +/- 0.04
	Maxima ①	F	0.75	+2.75	0.75	-0.25	0.04 +/- 0.04
		R	—	—	0.75	-1.00	0.04 +/- 0.04
	Maxima ②	F	0.75	+2.75	0.75	-0.33	0.04 +/- 0.04
		R	—	—	0.75	-1.00	0.04 +/- 0.04
	Sentra	F	0.75	+1.42	0.75	-0.58	0.08 +/- 0.08
		R	—	—	0.75	+1.00	0.04 +/- 0.15
2003	Altima	F	0.75	+2.66	0.75	-0.10	0.04 +/- 0.04
		R	—	—	0.75	-1.25	0.08 +/- 0.04
	Maxima ①	F	0.75	+2.75	0.75	-0.25	0.04 +/- 0.04
		R	—	—	0.75	-1.00	0.04 +/- 0.04
	Maxima ②	F	0.75	+2.75	0.75	-0.33	0.04 +/- 0.04
		R	—	—	0.75	-1.00	0.04 +/- 0.04
	Sentra	F	0.75	+1.42	0.75	-0.58	0.08 +/- 0.08
		R	—	—	0.75	+1.00	0.04 +/- 0.15
	350Z	F	0.75	+8.17	0.75	-0.58	0.04 +/- 0.04
		R	—	—	0.50	+1.58	⑦
2004	Altima ③	F	0.75	+2.83	0.75	-0.25	0.04 +/- 0.04
		R	—	—	0.30	-1.25	0.15 +/- 0.06
	Altima ④	F	0.75	+2.83	0.75	-0.34	0.04 +/- 0.04
		R	—	—	0.30	-0.25	0.16 +/- 0.06
	Maxima	F	0.75	+2.83	0.75	-0.25	0.02 +/- 0.04
		R	—	—	0.50	-0.67	0.16 +/- 0.06
	Sentra ⑤	F	0.75	+1.60	0.75	-0.42	0.08 +/- 0.04
		R	—	—	0.75	-1.00	0.04 +/- 0.16
	Sentra ⑥	F	0.75	+1.72	0.75	-0.45	0.08 +/- 0.04
		R	—	—	0.75	-1.00	0.04 +/- 0.16
	350Z	F	0.75	+8.17	0.75	-0.58	0.04 +/- 0.04
		R	—	—	0.50	+1.58	⑦

① With P225/55R16, P215/55R16 tires
② With P205/65R15 tires
③ With 2.5L engine
④ With 3.5L engine
⑤ With 1.8L engine
⑥ With 2.5L engine
⑦ 0.04 +/- 0.03 with 17 inch tire
0.07 +/- 0.03 with 18 inch tire

67170-NISS-C09

TIRE, WHEEL AND BALL JOINT SPECIFICATIONS

		OEM Tires		Tire Pressures (psi)		Wheel	Lug Nut Torque
Year	Model	Standard	Optional	Front	Rear	Size	Ft. Lbs.
2001	Altima	P195/65R15	P205/60R15	30	30	6-JJ	80
	Maxima GLE	P215/55R16	None	29	29	6.5-JJ	80
	Maxima GXE	P205/65SR15	None	29	29	6JJ	80
	Maxima SE	P215/55R16	P225/50R17	29	29	6.5J/7J	80
	Sentra, base	P155/80R13	None	26	26	5-J	80
	Sentra XE	P175/70R13	None	26	26	5-J	80
	Sentra GXE	P175/65R14	None	26	26	5.5-JJ	80
	Sentra GLE	P175/65R14	None	26	26	5.5-JJ	80
	Sentra SE	P195/55R15	None	30	30	6-JJ	80
2002	Altima	P195/65R15	P205/60R15	30	30	6-JJ	80
	Maxima GLE	P215/55R16	None	29	29	6.5-JJ	80
	Maxima GXE	P205/65SR15	None	29	29	6JJ	80
	Maxima SE	P215/55R16	P225/50R17	29	29	6.5J/7J	80
	Sentra, base	P155/80R13	None	26	26	5-J	80
	Sentra XE	P175/70R13	None	26	26	5-J	80
	Sentra GXE	P175/65R14	None	26	26	5.5-JJ	80
	Sentra GLE	P175/65R14	None	26	26	5.5-JJ	80
	Sentra SE	P195/55R15	None	30	30	6-JJ	80
2003	Altima	P195/65R15	P205/60R15	30	30	6-JJ	80
	Maxima GLE	P215/55R16	None	29	29	6.5-JJ	80
	Maxima GXE	P205/65SR15	None	29	29	6JJ	80
	Maxima SE	P215/55R16	P225/50R17	29	29	6.5J/7J	80
	Sentra, base	P155/80R13	None	26	26	5-J	80
	Sentra XE	P175/70R13	None	26	26	5-J	80
	Sentra GXE	P175/65R14	None	26	26	5.5-JJ	80
	Sentra GLE	P175/65R14	None	26	26	5.5-JJ	80
	Sentra SE	P195/55R15	None	30	30	6-JJ	80
2004	Altima	P2055/65R16	P215/55R17	30	30	6.5-JJ/7-JJ	80
	Maxima	P225/55VR17	245/45VR18	29	29	7-JJ/7.5-JJ	80
	Sentra	P195/60HR15	P195/55HR16	26	26	6-JJ/7-JJ	80
	350Z Front	225/50R17	225/45R18	26	26	7.5-JJ/8-JJ	80
	350Z Rear	235/50R17	245/45R18	26	26	8-JJ/8.5-JJ	80

OEM: Original Equipment Manufacturer

PSI: Pounds Per Square Inch

STD: Standard

OPT: Optional

① Replace if any measurable movement is found.

67170-NISS-C10

BRAKE SPECIFICATIONS

All measurements in inches unless noted

			Brake Disc			Brake Drum Diameter			Minimum Lining Thickness		Brake Caliper	
Year	Model		Original Thickness	Minimum Thickness	Maximum Run-out	Original Inside Diameter	Max. Wear Limit	Maximum Machine Diameter	Front	Rear	Bracket Bolts (ft. lbs.)	Mounting Bolts (ft. lbs.)
2001	Altima	F	0.870	0.787	0.003	—	—	—	0.079	—	53-72	16-23
		R	0.390	0.315	0.003	9.00	NA	9.06	—	0.059	—	—
	Maxima	F	0.870	0.787	0.003	—	—	—	0.079	—	53-72	16-23
		R	0.350	0.315	0.003	—	—	—	—	0.059	—	—
	Sentra	F	0.710	0.630	0.003	—	—	—	0.079	—	40-47	12-14
		R	0.280	0.236	0.003	7.09	7.13	7.13	—	0.059	—	—
2002	Altima	F	1.020	0.866	0.003	—	—	—	0.079	—	53-72	16-23
		R	0.350	0.310	0.003	—	—	—	—	0.059	—	—
	Maxima	F	0.940	0.866	0.003	—	—	—	0.079	—	53-72	16-23
		R	0.350	0.315	0.003	—	—	—	—	0.059	—	—
	Sentra	F	0.710	0.630	0.003	—	—	—	0.079	—	40-47	12-14
		R	0.280	0.236	0.003	7.09	7.13	7.13	—	0.059	—	—
2003	Altima	F	1.020	0.866	0.003	—	—	—	0.079	—	53-72	16-23
		R	0.350	0.310	0.003	—	—	—	—	0.059	—	—
	Maxima	F	0.940	0.866	0.003	—	—	—	0.079	—	53-72	16-23
		R	0.350	0.315	0.003	—	—	—	—	0.059	—	—
	Sentra	F	0.710	0.630	0.003	—	—	—	0.079	—	40-47	12-14
		R	0.280	0.236	0.003	7.09	7.13	—	—	0.059	—	—
	350Z ①	F	0.945	0.866	0.002	—	—	—	0.079	—	114	17-22
		R	0.630	0.551	0.004	—	—	—	—	0.079	53-71	13-14
	350Z ②	F	1.181	1.118	0.002	—	—	—	0.079	—	112	—
		R	0.866	0.795	0.002	—	—	—	—	0.079	53-71	—
2004	Altima	F	1.020	0.866	0.003	—	—	—	0.079	—	53-72	16-23
		R	0.350	0.310	0.003	—	—	—	—	0.059	28-38	23-30
	Maxima	F	1.100	0.1.02	0.003	—	—	—	0.079	—	53-72	16-23
		R	0.350	0.315	0.002	—	—	—	—	0.059	101-129	53-71
	Sentra 1.8L	F	0.870	0.079	0.003	—	—	—	0.079	—	53-72	16-23
		R	—	—	—	8.05	—	—	—	0.059	—	—
	Sentra 2.5L	F	1.181	1.118	0.002	—	—	—	0.079	—	112	—
		R	0.350	0.315	0.003	—	—	—	—	0.059	28-38	16-23
	350Z ①	F	0.945	0.866	0.002	—	—	—	0.079	—	114	17-22
		R	0.630	0.551	0.004	—	—	—	—	0.079	53-71	13-14
	350Z ②	F	1.181	1.118	0.002	—	—	—	0.079	—	112	—
		R	0.866	0.795	0.002	—	—	—	—	0.079	53-71	—

NA: Not Available

① Front brake model CLZ25VD, rear brake model AD14VE.

② Front brake model OPB27VA, rear brake model OPB13VB.

67170-NISS-C11

SCHEDULED MAINTENANCE INTERVALS

Nissan—Altima, Sentra, Maxima & 350Z

TO BE SERVICED	TYPE OF SERVICE	VEHICLE MILEAGE INTERVAL (x1000)												
		7.5	15	22.5	30	37.5	45	52.5	60	67.5	75	82.5	90	97.5
Engine oil & filter	R	✓	✓	✓	✓	✓	✓	✓	✓	✓	✓	✓	✓	✓
Brake lines & cables	S/I		✓		✓		✓		✓		✓		✓	
Brake pads, discs, drums & linings	S/I		✓		✓		✓		✓		✓		✓	
Driveshaft boots	S/I		✓		✓		✓		✓		✓		✓	
Exhaust system	S/I				✓				✓				✓	
Transmission or transaxle fluid	S/I		✓		✓		✓		✓		✓		✓	
Air cleaner filter	R				✓				✓				✓	
Spark plugs (except platinum)	R				✓				✓				✓	
Spark plugs (platinum tip)	R								✓					
Idle RPM (Sentra)	S/I				✓				✓				✓	
Steering gear & linkage, axle & suspension parts	S/I				✓				✓				✓	
Engine coolant	R								✓					
Timing belt	R								✓					
Drive belts	S/I								✓					
Fuel lines	S/I								✓					
Vapor lines	S/I								✓					

R: Replace   S/I: Service or Inspect

FREQUENT OPERATION MAINTENANCE (SEVERE SERVICE)

If a vehicle is operated under any of the following conditions it is considered severe service:

- Extremely dusty areas.

- 50% or more of the vehicle operation is in 32°C (90°F) or higher temperatures, or constant operation in temperatures below 0°C (32°F).

- Prolonged idling (vehicle operation in stop and go traffic).

- Frequent short running periods (engine does not warm to normal operating temperatures).

- Police, taxi, delivery usage or trailer towing usage.

Oil & oil filter: change every 3750 miles.

Brake pads & discs: service or inspect every 7500 miles.

Driveshaft boots: service or inspect every 7500 miles.

Exhaust system: service or inspect every 7500 miles.

Steering gear & linkage, axle & suspension parts: service or inspect every 7500 miles.

Steering linkage ball joints & front suspension ball joints: service or inspect every 7500 miles.

Air cleaner filter: service or inspect every 15,000 miles.

67170-NISS-C12

PRECAUTIONS

Before servicing any vehicle, please be sure to read all of the following precautions, which deal with personal safety, prevention of component damage and important points to take into consideration when servicing a motor vehicle:

- Never open, service or drain the radiator or cooling system when the engine is hot; serious burns can occur from the steam and hot coolant.
- Observe all applicable safety precautions when working around fuel. Whenever servicing the fuel system, always work in a well-ventilated area. Do not allow fuel spray or vapors to come in contact with a spark, open flame, or excessive heat (a hot drop light, for example). Keep a dry chemical fire extinguisher near the work area. Always keep fuel in a container specifically designed for fuel storage; also, always properly seal fuel containers to avoid the possibility of fire or explosion. Refer to the additional fuel system precautions later in this section.
- Fuel injection systems often remain pressurized, even after the engine has been turned **OFF**. The fuel system pressure must be relieved before disconnecting any fuel lines. Failure to do so may result in fire and/or personal injury.
- Brake fluid often contains polyglycol ethers and polyglycols. Avoid contact with the eyes and wash your hands thoroughly after handling brake fluid. If you do get brake fluid in your eyes, flush your eyes with clean, running water for 15 minutes. If eye irritation persists, or if you have taken brake fluid internally, IMMEDIATELY seek medical assistance.
- The EPA warns that prolonged contact with used engine oil may cause a number of skin disorders, including cancer! You should make every effort to minimize your exposure to used engine oil. Protective gloves should be worn when changing oil. Wash your hands and any other exposed skin areas as soon as possible after exposure to used engine oil. Soap and water, or waterless hand cleaner should be used.
- All new vehicles are now equipped with an air bag system. The system must be disabled before performing service on or around system components, steering column, instrument panel components, wiring and sensors. Failure to follow safety and disabling procedures could result in accidental air bag deployment, possible personal injury and unnecessary system repairs.
- Always wear safety goggles when working with, or around, the air bag system. When carrying a non-deployed air bag, be sure the bag and trim cover are pointed away from your body. When placing a non-deployed air bag on a work surface, always face the bag and trim cover upward, away from the surface. This will reduce the motion of the module if it is accidentally deployed. Refer to the additional air bag system precautions later in this section.
- Clean, high quality brake fluid from a sealed container is essential to the safe and proper operation of the brake system. You should always buy the correct type of brake fluid for your vehicle. If the brake fluid becomes contaminated, completely flush the system with new fluid. Never reuse any brake fluid. Any brake fluid that is removed from the system should be discarded. Also, do not allow any brake fluid to come in contact with a painted surface; it will damage the paint.
- Never operate the engine without the proper amount and type of engine oil; doing so WILL result in severe engine damage.
- Timing belt maintenance is extremely important! Many models utilize an interference-type, non-freewheeling engine. If the timing belt breaks, the valves in the cylinder head may strike the pistons, causing potentially serious (also time-consuming and expensive) engine damage. Refer to the maintenance interval charts in the front of this section for the recommended replacement interval for the timing belt and to the timing belt procedure for belt replacement and inspection.
- Disconnecting the negative battery cable on some vehicles may interfere with the functions of the on-board computer system(s) and may require the computer to undergo a relearning process once the negative battery cable is reconnected.
- When servicing drum brakes, only disassemble and assemble one side at a time, leaving the remaining side intact for reference.
- The 350Z is equipped with an automatic window adjusting function. When a door is opened the window will automatically lower slightly to avoid contact with the roof opening. Always lower the driver and passenger windows before disconnecting the battery cables. This will prevent interference when the doors are closed.

ENGINE REPAIR

Distributor

REMOVAL

The Nissan 2.5L, 3.0L and 3.5L engines are equipped with a Distributorless Ignition System (DIS).

1.8L, 2.0L and 2.4L Engines

1. Before servicing the vehicle, refer to the precautions in the beginning of this section.
2. Set the engine to Top Dead Center (TDC) with the No. 1 piston on compression stroke.
3. Remove or disconnect the following:
   - Negative battery cable
   - Distributor spark plug wires from the distributor cap
   - Distributor cap. Scribe a mark on the engine block to show the rotor and distributor position prior to removal.
   - Wiring connections to the distributor
   - Bolt(s) holding distributor to engine
   - Distributor by pulling it upward from the cylinder block

➡Do not disturb the camshaft or crankshaft position after the distributor is removed from the engine. If any of these components are moved, TDC on cylinder No. 1 will have to be found again before reinstalling the distributor.

INSTALLATION

1.8L, 2.0L and 2.4L Engines

ENGINE NOT DISTURBED

1. Install or connect the following:
   - New distributor housing O-ring
   - Distributor so the rotor is aligned with the matchmark on the housing and the housing is aligned with the matchmark on the engine

➡Be sure the distributor is fully seated and the distributor gear is fully engaged.

   - Snug the hold-down bolt
   - Distributor pick-up lead wires

- Distributor cap and tighten the screws
- Splash shield
- Spark plug wires
- Negative battery cable

2. After the ignition timing has been adjusted, tighten the hold-down bolt(s) as follows:
   - 1.8L engines: 80–104 inch lbs. (9–11 Nm)
   - 2.0L and 2.4L engines: 108–144 inch lbs. (13–16 Nm)

ENGINE DISTURBED

1. Install the a new distributor housing O-ring
2. Position the engine so the No. 1 piston is at Top Dead Center (TDC) of its compression stroke and the mark on the vibration damper is aligned with **0** on the timing indicator.
   - Distributor in the engine so the rotor is aligned with the position of the No. 1 ignition wire on the distributor cap. Be sure the distributor is fully seated and that the distributor shaft is fully engaged.
   - Snug the hold-down bolt
   - Distributor pick-up lead wires
   - Distributor cap and tighten the screws. Install the splash shield, if equipped.
   - Spark plug wires
   - Negative battery cable.
3. After the ignition timing has been adjusted, tighten the hold-down bolt(s) as follows:
   - 1.8L engines: 80–104 inch lbs. (9–11 Nm)
   - 2.0L and 2.4L engines: 108–144 inch lbs. (13–16 Nm)

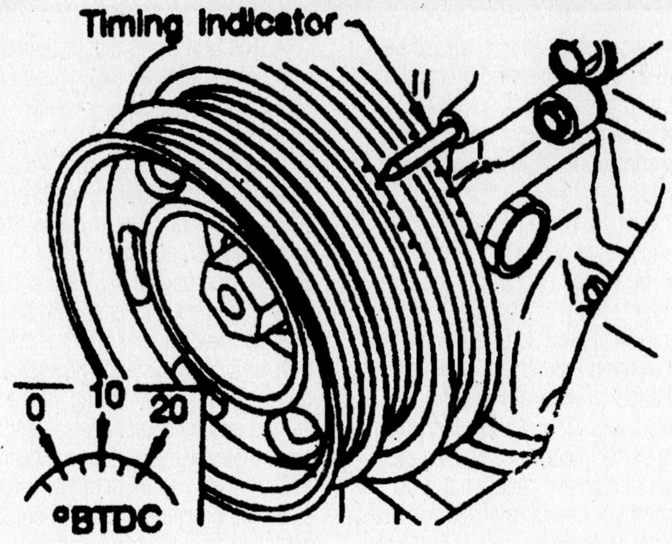

Point the timing light at the crankshaft pulley to see the timing marks—1.8L and 2.0L engines

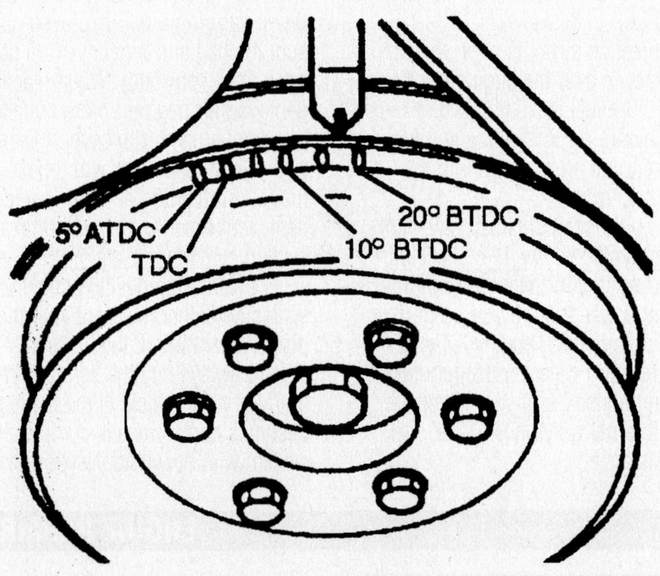

The timing marks are located on the crankshaft pulley—2.4L engine

Ignition Timing

ADJUSTMENT

1.8L, 2.0L and 2.4L Engines

Visually check the air cleaner, intake hoses, ducts, Exhaust Gas Recirculation (EGR) valve operation and electrical connections prior to the adjustment of the ignition timing. Correct or repair any problem as required. Be sure to inspect the throttle valve and the Throttle Position (TP) sensor for proper operation.

1. Before servicing the vehicle, refer to the precautions in the beginning of this section.
2. Locate the timing marks on the crankshaft pulley and the front of the engine.
3. Clean the timing marks.
4. Using chalk or white paint, color the mark on the crankshaft pulley and the mark on the scale which will indicate the correct timing when aligned with the notch on the crankshaft pulley.
5. Attach a tachometer to the engine.
6. Attach a timing light to the engine, to No. 1 cylinder's ignition wire.
7. Check to be sure all of the wires clear the fan; start the engine and allow it to reach normal operating temperatures.
8. Block the front wheels and set the parking brake. Shift the transmission into **NEUTRAL** for automatic and manual transaxles; do not stand in front of the vehicle when making adjustments.
9. Perform the following procedures:
   a. Race the engine at 2000 rpm for about 2 minutes under a no-load condition; be sure all of the accessories are turned off.
   b. Perform on board engine diagnostics and repair any fault code.
   c. Race the engine 2–3 times under no-load, then run the engine it for 1 minute at idle.
   d. Stop the engine and disconnect the Throttle Position (TP) sensor.
   e. Race the engine at 2000 rpm for

about 2 minutes under a no-load condition; be sure all of the accessories are turned **OFF**.

f. Run the engine at idle speed.

10. Aim the timing light at the timing marks. If the marks on the pulley and the engine are aligned when the light flashes, the timing is correct. The correct ignition timing is as follows:

a. 1.8L (QG18DE) engine: 6–10 degrees Before Top Dead Center (BTDC)

b. 2.0L (SR20DE) engine: 13–17 degrees BTDC

c. 2.4L (KA24DE) engine: 18–22 degrees BTDC

11. Turn the engine **OFF** and remove the tachometer and the timing light. If the marks are not in alignment, proceed with the following steps.

12. Turn the engine **OFF**.

13. Loosen the bolts that secure the distributor just enough so it can be turned.

14. Start the engine. Keep the wires of the timing light clear of the cooling fan.

15. With the timing light aimed at the pulley and the marks on the engine, turn the distributor for the proper adjustment.

16. Race the engine 2–3 times under no-load, then run the engine it for 1 minute at idle.

17. Aim the timing light at the timing marks. If the marks on the pulley and the engine are aligned when the light flashes, the timing is correct.

18. Tighten the bolt that secures the distributor and recheck the timing.

19. Turn the engine **OFF** and remove the tachometer and the timing light.

20. Connect the TP sensor.

2.5L, 3.0L and 3.5L Engines

➡The ignition timing is not adjustable. If not within specifications, further diagnostic inspection is required. The following procedure is for viewing the ignition timing setting.

Visually check the air cleaner, intake hoses, ducts, Exhaust Gas Recirculation (EGR) valve operation and electrical connections prior to the adjustment of the ignition timing. Correct or repair any problem as required. Be sure to inspect the throttle valve and Throttle Position (TP) sensor for proper operation.

1. Before servicing the vehicle, refer to the precautions in the beginning of this section.

2. Locate the timing marks on the crankshaft pulley and the front of the engine.

3. Clean the timing marks.

➡The ignition timing specification is 13–17 degrees Before Top Dead Center (BTDC).

4. Using chalk or white paint, color the mark on the crankshaft pulley and the mark on the scale, that will indicate the correct timing when aligned with the notch on the crankshaft pulley.

5. Attach a tachometer to the engine.

6. Attach a timing light to the engine to number 1 cylinder ignition wire.

7. Turn all electrical equipment and accessories **OFF**.

8. Check to be sure all of the wires clear the fan, then, start the engine and allow it to reach normal operating temperatures.

9. Block the front wheels and set the parking brake. Shift the transmission into **NEUTRAL** for manual transmission and automatic transmissions. Do not stand in front of the vehicle when making adjustments.

10. Perform the following procedures:

a. Race the engine at 2000 rpm for about 2 minutes under a no-load condition; be sure all of the accessories are turned **OFF**.

b. Perform on board engine diagnostics and repair any fault code.

c. Race the engine at 2000 rpm for about 2 minutes under a no-load condition.

d. Turn the engine **OFF** and disconnect the TP sensor.

e. Start and race the engine 2–3 times under no-load, then run the engine at idle speed.

➡The ignition timing specification is 13–17 degrees BTDC.

11. Aim the timing light at the timing marks. If the marks on the pulley and the engine are aligned when the light flashes, the timing is correct. Turn the engine **OFF** and remove the tachometer and the timing light. If the marks are not in alignment, proceed with the following steps.

12. Turn the engine **OFF**.

13. Check the Camshaft Position (CMP) sensor (PHASE), Crankshaft Position (CKP) sensor (REF) and CKP sensor (POS). Replace if necessary.

14. If the ignition timing is still not correct, substitute a known good Electronic Control Module (ECM).

➡The ECM may be the cause of the problem but this is rarely the case.

15. Turn the engine **OFF** and remove the tachometer and the timing light.

Alternator

REMOVAL

1.8L and 2.0L Engines

2.0L Engines

1. Before servicing the vehicle, refer to the precautions in the beginning of this section.

2. Remove or disconnect the following:
   - Negative battery cable
   - 2 lead wires and connector from the alternator
   - Drive belt adjusting bolt, loosen only
   - Drive belt
   - Alternator

2.4L Engine

1. Before servicing the vehicle, refer to the precautions in the beginning of this section.

2. Drain the cooling system below the upper radiator hose level.

3. Remove or disconnect the following:
   - Negative battery cable
   - Upper radiator hose
   - Alternator electrical harness, harness stay and the harness-to-A/C compressor
   - Throttle cable
   - Loosen the adjusting bolt
   - Accessory drive belt
   - Alternator mounting bolts
   - Alternator from the engine

2.5L Engine

1. Before servicing the vehicle, refer to the precautions in the beginning of this section.

2. Remove or disconnect the following:
   - Negative battery cable
   - Engine cover
   - Engine undercover
   - Alternator drive belt
   - Alternator mounting bolts
   - Alternator from the engine

3.0L Engines

1. Before servicing the vehicle, refer to the precautions in the beginning of this section.

2. Remove or disconnect the following:
   - Negative battery cable
   - Splash guard on the right side of the vehicle
   - Drive belt
   - Four A/C mounting bolts

- Radiator fan and shroud
- A/C compressor forward
- Alternator harness connector
- Alternator mounting bolts and lower the alternator from the vehicle

3.5L Engines

1. Before servicing the vehicle, refer to the precautions in the beginning of this section.
2. Remove or disconnect the following:
   - Negative battery cable
   - Right side engine undercover and side inspection cover
   - Radiator on Altima and Maxima
   - Radiator fan on 350Z
   - Drive belt
   - Alternator and A/C compressor harness connectors
   - Upper and lower alternator bolts
   - Alternator

INSTALLATION

1.8L and 2.0L Engines

1. Install or connect the following:
   - Alternator and retaining bolts loosely
   - Belt and connect the wiring
2. Adjust the drive belt.
3. Torque the retaining bolts to 25 ft. lbs. (34 Nm).
4. Connect the negative battery cable.

2.4L Engines

1. Install or connect the following:
   - Alternator and torque the bolts to 11–15 ft. lbs. (16–20 Nm)
   - Harness-to-A/C compressor, harness stay and the alternator electrical harness
   - Throttle cable
   - Drive belt. Properly tension the belt
   - Upper radiator hose
   - Negative battery cable
2. Top off the cooling system.
3. Start the vehicle, check for leaks and repair if necessary.

2.5L Engines

1. Install or connect the following:
   - Alternator and torque the upper bolt to 18–23 ft. lbs. (26–31 Nm) and the lower bolt to 33–38 ft. lbs. (44–52 Nm)
   - Drive belt. Properly tension the belt
   - Engine undercover
   - Engine cover
   - Negative battery cable

3.0L Engines

Install or connect the following:
- Alternator and torque the bolts to 38 ft. lbs. (52 Nm)
- Alternator harness connector
- A/C compressor back into location
- Radiator cooling fan and shroud
- A/C compressor mounting bolts
- Drive belt and tension the belt
- Splash guard
- Negative battery cable

➡Proper belt tension is important. A belt that is too tight may cause alternator bearing failure; one that is too loose will cause a gradual battery discharge and/or belt slippage, resulting in belt breakage from overheating.

3.5L Engine

Install or connect the following:
- Alternator
- Upper and lower alternator bolts. On Altima and Maxima tighten the upper bolt to 12–15 ft. lbs. (16 –20 Nm) and the lower bolt to 32–38 ft. lbs. (44–52 Nm). On 350Z, tighten the upper bolt to 48 ft. lbs. (65 Nm) and the lower bolts to 21 ft. lbs. (28 Nm).
- Alternator and A/C compressor harness connectors
- Drive belt
- Radiator on Altima and Maxima
- Radiator fan on 350Z
- Right side engine undercover and side inspection cover
- Negative battery cable

Engine Assembly

REMOVAL & INSTALLATION

Sentra

➡The engine and transaxle are removed as one unit from the underside of the vehicle.

1. Before servicing the vehicle, refer to the precautions in the beginning of this section.
2. Relieve the fuel system pressure.
3. Drain the coolant from the radiator and the engine block.
4. Drain the engine oil.
5. Remove or disconnect the following:
   - Negative and positive battery cables
   - Battery and tray from the vehicle
   - Both front wheels
   - Engine undercovers and the engine side covers
   - Air cleaner assembly and air duct
   - Vacuum hoses. Make sure to note the locations prior to disconnection them.
   - Heater hoses from the engine
   - Automatic transmission cooler hoses from the transaxle, if equipped
   - Power steering hoses
   - Fuel hoses from the engine
   - Harness and wiring connections. Make sure to note the locations prior to disconnecting them.
   - Throttle cable and the cruise control cable
   - Control cable, if equipped with an automatic transmission
   - Cooling fans, radiator and the recovery tank
   - Front halfshafts from the vehicle
   - Front exhaust pipe
   - Starter motor and intake manifold support brackets
   - Engine drive belts
   - Alternator and adjusting brackets
   - Power steering pump and A/C compressor. It is not necessary to disconnect the lines.
6. Position a transmission jack under the transaxle and support the engine with engine slinger.
   - Center crossmember
   - Front stabilizer bar, if necessary
   - Engine mounting bolts from both sides of the engine
7. Slowly lower the jacking devices and remove the engine and transaxle from the vehicle.

To install:

8. Install or connect the following:
   - Engine and transaxle assembly
   - Mounting bolts to both sides of the engine and torque the bolts to 44 ft. lbs. (60 Nm)
9. For vehicles with manual transaxles, adjust the height of the mounting bracket (buffer rod). The distance between the 2 through-bolts should be 2.13–2.20 in. (54–56mm).
10. Install or connect the following:
    - Center crossmember and torque the bolts to 40 ft. lbs. (54 Nm)
11. Remove the engine support jacks and engine slinger.
12. Install or connect the following:
    - A/C compressor and power steering pump
    - Alternator and brackets
    - Starter motor and intake manifold support bracket
    - Front exhaust pipe
    - Drive belts

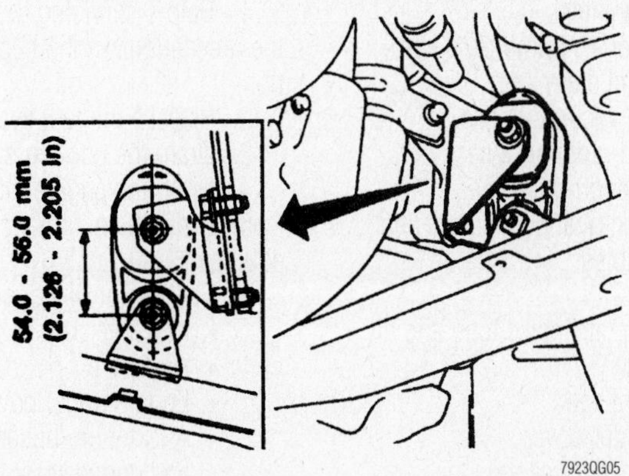

Be sure to adjust the height of the engine mount for manual transmission vehicles—2.0L

- Both front halfshafts
- Radiator, cooling fans and recovery tank
- Control cable, automatic transmissions only
- Throttle and cruise control cables, if equipped
- Wiring harness and electrical connections
- Power steering hoses and fuel line
- Transmission cooler lines, if equipped
- Vacuum hoses
- Air cleaner assembly
- Engine side and under covers
- Both front wheels
- Battery tray and battery
- Both battery cables

13. Fill the engine with clean oil.
14. Fill the cooling system.
15. Start the engine and check for leaks. Make all the necessary adjustments.

Altima

➡The engine and transaxle must be removed as a single unit. The engine and transaxle are removed from under the vehicle.

1. Before servicing the vehicle, refer to the precautions in the beginning of this section.
2. Release fuel system pressure.
3. Drain the cooling system.
4. Drain the engine oil.
5. Remove or disconnect the following:
   - Battery cables and the battery tray
   - Air cleaner assembly
   - Both front wheels
   - Engine under cover and engine hood
   - Cooler lines from the radiator, if equipped with an automatic transaxle
   - Upper and lower hoses from the radiator
   - Radiator assembly
   - Heater hoses from the engine
   - Throttle cable and cruise control cable, if equipped
   - Fuel feed and return hoses
   - All the necessary vacuum hoses and electrical connectors. Label all wires and hoses before disconnecting them.
   - Wiring from starter motor
   - Slave cylinder from the transaxle, if equipped. It is not necessary to disconnect the hydraulic hose.
   - Engine drive belts. Be sure to mark belts for reinstallation.
   - Alternator, A/C compressor and the power steering pump
   - Both halfshafts from the transaxle and support the engine with slinger and support the transaxle with proper jack
   - Left and right engine mounting through-bolts
   - Crossmember
   - Front and rear engine mounts
6. Lower the transaxle and engine assembly from the vehicle.

➡The engine and transaxle assembly should be removed through the bottom of the vehicle. Do not attempt to remove the assembly from above.

To install:

7. Raise the transaxle and engine assembly to the vehicle.
8. Install or connect the following:
   - Front and rear engine mounts. Torque the mounting bolts to 55 ft. lbs. (75 Nm).
   - Crossmember and torque the bolts to 57–72 ft. lbs. (77–98 Nm)
   - Left and right engine mounting through bolts. Torque the bolts to 72 ft. lbs. (98 Nm).
9. Remove the engine and transaxle support jacks.
   - Both halfshafts
   - Power steering pump, A/C compressor and alternator
   - Slave cylinder, if equipped
   - Starter motor
   - Drive belts
   - Vacuum hoses and electrical connectors
   - Fuel feed and return lines
   - Throttle and cruise control cables, if equipped
   - Radiator
   - Heater and radiator hoses
   - Cooler lines, if equipped
   - Engine side and under covers
   - Both front wheels
   - Air cleaner assembly
   - Battery tray and battery
   - Both battery cables
   - Hood
10. Fill the cooling system.
11. Fill the engine with clean oil.
12. Start the vehicle, check for leaks and repair if necessary.

Maxima

3.0L ENGINE

It is recommended the engine and transaxle be removed as a single unit. If need be, the units may be separated after removal.

➡The engine and transaxle assembly must be removed from the underside of the vehicle.

1. Before servicing the vehicle, refer to the precautions in the beginning of this section.
2. Release the fuel system pressure.
3. Drain the cooling system.
4. Drain the engine oil.
5. Drain the automatic transaxle, if equipped.
6. Remove or disconnect the following:
   - Negative battery cable
   - Hood
   - Engine under cover
   - Air cleaner, the air intake tube, the air flow meter and the throttle linkage
   - Drive belts
   - Engine ground cable
   - Electrical connector from the crank angle sensor
   - Engine electrical harness connectors
   - Fuel feed and fuel return hoses
   - Upper and lower radiator hoses
   - Heater inlet and outlet hoses
   - Engine vacuum hoses
   - Power steering pump, A/C compressor and alternator

- Carbon canister
- Auxiliary fan, washer tank and the radiator (with the fan assembly)
- Clutch release cylinder from the clutch housing, if equipped with a manual transaxle
- Shift control rod and the shift support rod, on some models with a manual transaxle
- Control cable from the transaxle, on models with an automatic transaxle

7. Install engine slingers to the block and connect a suitable lifting device to the slingers. Do not tension the lifting device at this point.
   - Exhaust pipe at both the manifold connections
   - Front exhaust pipe from the vehicle and support the engine and transaxle assembly with proper jack
   - Right and left side halfshafts from their side flanges
   - Bolt holding the radius link support
8. Lower the shifter and selector rods and remove the bolts from the motor mount brackets. Remove the nuts holding the front and rear motor mounts to the frame.
9. On some models it will be necessary to remove the center crossmember assembly from the vehicle.
10. Lower the engine/transaxle assembly onto an engine stand.

To install:

11. Raise the engine/transaxle assembly into the vehicle. When raising the engine onto the mounts, be sure to keep it as level as possible.
12. After installing the motor mounts, adjust and install the buffer rods; the front should be 3.50–5.58 in. (89–91mm) and the rear should be 3.90–3.98 in. (99–101mm).
13. Check the clearance between the frame and clutch housing and be sure the engine mount bolts are seated in the groove of the mounting bracket.
14. Remove the transaxle and engine jack assembly.
15. Install or connect the following:
    - Center crossmember, if removed. Torque the bolts to 72 ft. lbs. (98 Nm).
    - Halfshafts
    - Radius link support
    - Front exhaust pipe and remove the engine slingers and supports
    - Control cable, if equipped
    - Shift control and support rods, if equipped
    - Clutch release cylinder, if equipped
    - Radiator, auxiliary fan and washer tank
    - Carbon canister
    - Power steering pump, A/C compressor and alternator
    - All engine vacuum hoses
    - Heater and radiator hoses
    - Fuel feed and return lines
    - Engine electrical connectors and ground cables
    - Drive belts
    - Air cleaner, air intake tube and air flow meter
    - Throttle linkage
    - Engine under cover
    - Negative battery cable
    - Hood
16. Fill the transmission fluid.
17. Fill the cooling system.
18. Fill the engine with clean oil.
19. Start the vehicle, check for leaks and repair if necessary.

3.5L ENGINE

It is recommended the engine and transaxle be removed as a single unit. If need be, the units may be separated after removal.

➡The engine and transaxle assembly must be removed from the underside of the vehicle.

1. Before servicing the vehicle, refer to the precautions in the beginning of this section.
2. Release the fuel system pressure.
3. Drain the cooling system.
4. Drain the engine oil.
5. Drain the automatic transaxle, if equipped.
6. Remove or disconnect the following:
   - Negative battery cable
   - Hood
   - Engine under cover
   - All vacuum hoses, fuel lines, wires and connectors; tag before disconnecting
   - Front exhaust pipe from the manifold
   - Ball joints from the steering knuckle
   - Halfshafts
   - Radiator and fans
   - Drive belts
   - Alternator
   - A/C compressor. Position it aside with the lines attached. Do NOT disconnect the refrigerant lines.
   - Power steering pump and position aside with the lines attached. Do NOT disconnect the fluid lines.

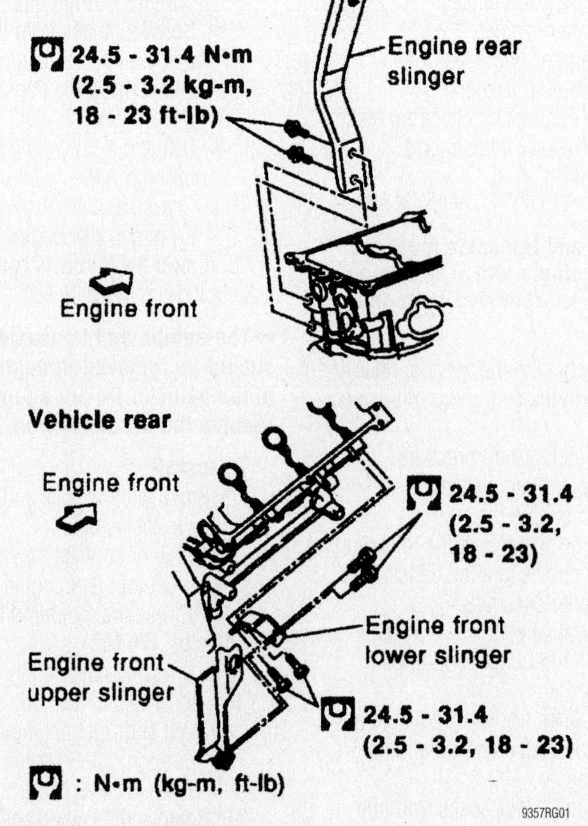

Installation of engine slingers to lift the engine

Vehicle front
A
Actuator harness
B
9357RG02

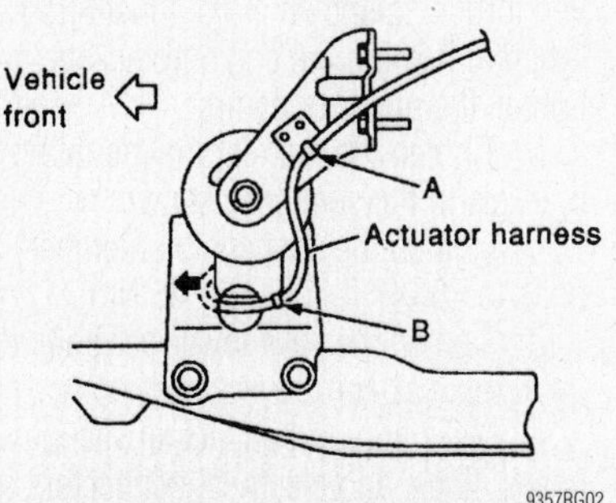

For electronically controlled engine mounts, the proper length from A to B is 6.69 in. (170mm)

7. Place a suitable jack under the transaxle. Install engine slingers and a suitable engine hoist. Raise the engine for access to the left side engine mount.
- Left side engine mount
- Control and support rods from the transaxle, manual transaxle only
- Control cable from the transaxle, automatic transaxle only
- Right side engine mount
- Center member, then carefully and slowly lower the transmission jack

8. Lower the engine/transaxle assembly onto an engine stand.

➡When lowering the engine out, guide it carefully to avoid hitting any other components.

To install:

9. Installation is the reverse of the removal procedure, noting the following points:
   a. If equipped with electronically controlled engine mounts, install them to the specifications shown in the accompanying figure
   b. Make sure to connect all vacuum hoses, lines, and electrical connectors as tagged during removal.

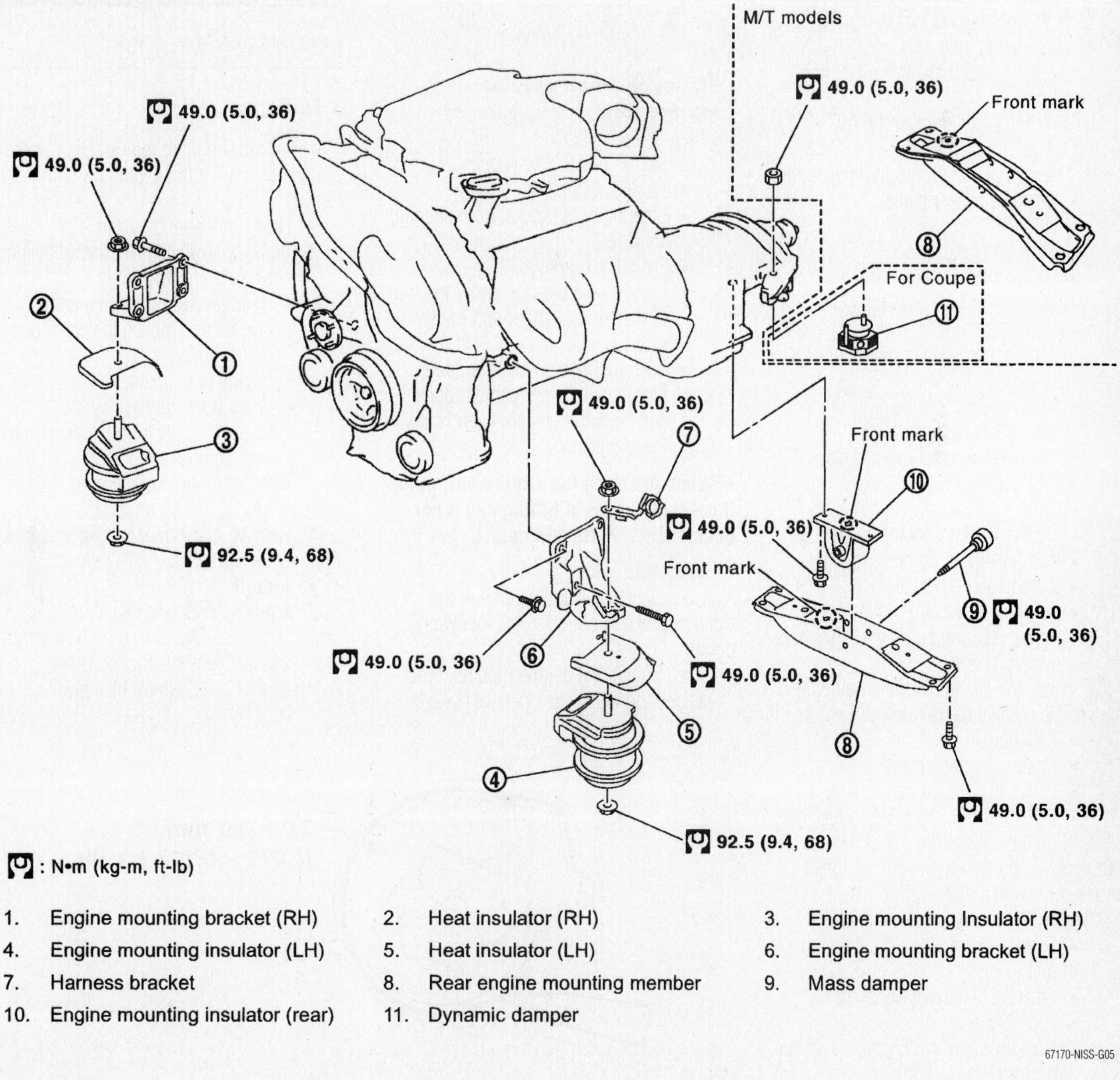

M/T models
49.0 (5.0, 36)
Front mark
For Coupe
49.0 (5.0, 36)
49.0 (5.0, 36)
92.5 (9.4, 68)
49.0 (5.0, 36)
Front mark
49.0 (5.0, 36)
Front mark
49.0 (5.0, 36)
49.0 (5.0, 36)
49.0 (5.0, 36)
92.5 (9.4, 68)
49.0 (5.0, 36)
: N•m (kg-m, ft-lb)
1. Engine mounting bracket (RH)
2. Heat insulator (RH)
3. Engine mounting Insulator (RH)
4. Engine mounting insulator (LH)
5. Heat insulator (LH)
6. Engine mounting bracket (LH)
7. Harness bracket
8. Rear engine mounting member
9. Mass damper
10. Engine mounting insulator (rear)
11. Dynamic damper
67170-NISS-G05

Exploded view of the engine mounting assemblies—350Z

c. Fill the cooling system.
d. Fill the engine with clean oil.
e. Start the vehicle, check for leaks and repair if necessary.

350Z

It is recommended the engine and transaxle be removed as a single unit. If need be, the units may be separated after removal.

➡The engine and transaxle assembly must be removed from the underside of the vehicle.

1. Before servicing the vehicle, refer to the precautions in the beginning of this section.
2. Release the fuel system pressure.
3. Drain the cooling system.
4. Drain the engine oil.
5. Drain the automatic transaxle, if equipped.
6. Discharge and recover the A/C refrigerant
7. Remove or disconnect the following:
   - Negative battery cable
   - Hood
   - Strut tower bar
   - Engine under cover
   - Wiper arms and cowl top
   - All vacuum hoses, fuel lines, wires and connectors; tag before disconnecting
   - Front wheels
   - Air cleaner case and duct
   - Cooling fan, reservoir and hoses
   - Heater hoses
   - Battery ground at cylinder head
   - Battery positive cable harness
   - A/C lines from compressor
   - 2 body ground cables
   - Fuel feed and EVAP hoses
   - Power steering pump and lines and wire to engine
8. From inside the passenger side of the vehicle remove the following:
   - Kick panel
   - Dash side finish panel
   - Lower instrument panel cover
   - ECM and TCM harness connectors
9. Pull the connectors out of the passenger side into the engine compartment and secure them to the engine.
10. Remove or disconnect the following:
    - Front exhaust pipe from the manifold
    - Steering column lower shaft
    - Propeller shaft
    - Shift lever and clutch slave cylinder on man. trans. models
    - Automatic transmission control rod

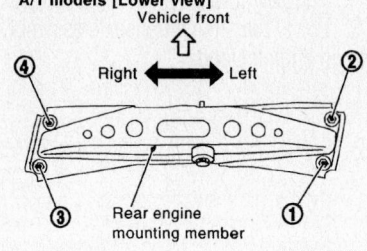

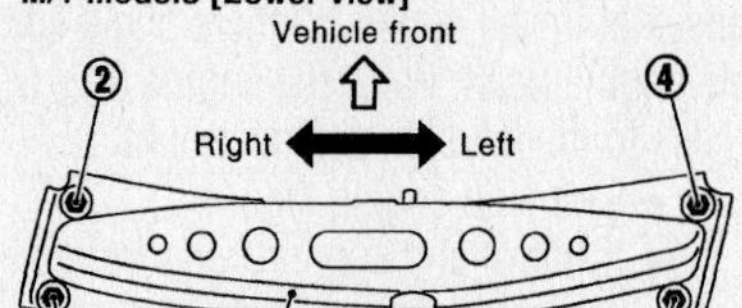

Rear engine member tightening sequence—350Z

- Upper rear oil pan plate
- Front stabilizer bar
- Steering outer socket from steering knuckle
- Front transverse link
- Place a suitable jack under the front suspension member and transmission
- Rear engine crossmember bolts
- Front suspension member bolts and nuts, then carefully and slowly lower the jack

➡When lowering the engine out, guide it carefully to avoid hitting any other components. bolts and nuts.

To install:

11. Installation is the reverse of the removal procedure, noting the following points:
    a. Tighten the engine mounts, brackets and mounting members to the specified torque as shown in the illustration. When tightening the engine brackets, tighten the upper bolts first.
    b. Tighten the rear engine mounting brackets in the sequence shown.
    c. Tighten the suspension member bolts to 45–47 ft. lbs. (60–65 Nm).
    d. Tighten the strut tower bar bolts to 24 ft. lbs. (32 Nm).
    e. Make sure to connect all vacuum hoses, lines, and electrical connectors as tagged during removal.
    f. Fill the cooling system.
    g. Fill the engine with clean oil.
    h. Start the vehicle, check for leaks and repair if necessary.

Water Pump

REMOVAL & INSTALLATION

1.8L Engine

1. Before servicing the vehicle, refer to the precautions in the beginning of this section.
2. Drain the cooling system.
3. Remove or disconnect the following:
   - Negative battery cable
   - Cylinder head front mounting bracket and loosen the water pump pulley bolts
   - Engine drive belts
   - Water pump pulley
   - Coolant hoses from the water inlet and thermostat housing
   - Water pump and thermostat housing
4. Remove all traces of gasket material from sealing surfaces.

To install:

5. Apply a continuous bead of liquid sealer to the sealing surface of the thermostat housing. The sealant should be 0.079–0.118 in. (2–3mm) diameter.

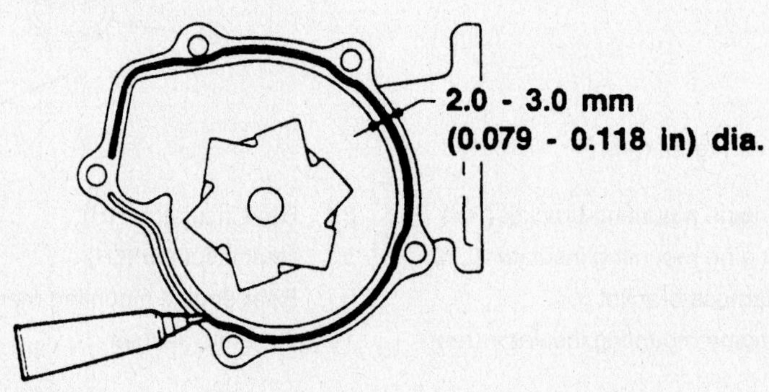

Apply RTV sealant to the water pump sealing surface as shown— 1.8L engines

6. Install or connect the following:
   - Water pump. Torque the bolts to 56–73 inch lbs. (7–8 Nm).
   - Pulley to the water pump and tighten the mounting bolts to 56–73 inch lbs. (7–8 Nm)
   - Coolant hoses to the thermostat housing
   - Drive belts and adjust as needed
   - Cylinder head front mounting bracket
   - Negative battery cable
7. Fill the cooling system
8. Start the engine, check for leaks and repair if necessary.

2.0L Engine

1. Before servicing the vehicle, refer to the precautions in the beginning of this section.
2. Drain the cooling system.
3. Remove or disconnect the following:
   - Negative battery cable
   - Right front wheel
   - Engine side and front covers
   - Loosen the water pump pulley bolts
   - Drive belts
   - 3 lower water pump bolts and position a jackstand under the engine
   - Front engine mount
   - Water pump
4. Remove all traces of liquid gasket material from sealing surfaces.

To install:

5. Apply a continuous bead of liquid sealer to the mating surface of the water pump. Sealer should be 0.079–0.118 in. (2–3mm) wide.
6. Install or connect the following:
   - Water pump and torque the bolts to 12–15 ft. lbs. (16–21 Nm)
   - Front engine mount and remove the engine support
   - Water pump pulley. Torque the mounting bolts to 55–73 inch lbs. (6–8 Nm).
   - Drive belts and adjust as needed
   - Engine front and side cover
   - Right front wheel
   - Negative battery cable
7. Fill the cooling system.
8. Start the vehicle, check for leaks and repair if necessary.

2.4L Engine

1. Before servicing the vehicle, refer to the precautions in the beginning of this section.
2. Drain the cooling system.
3. Remove or disconnect the following:
   - Negative battery cable
   - Right lower splash cover
   - Alternator and A/C compressor
   - Coolant tube
   - Water pump

➡Do not disconnect the air conditioning compressor lines. Unbolt the compressor and lay it off to the side.

➡The mounting bolts are different sizes and must be reinstalled in the correct location; therefore it is a good idea to arrange the bolts so that they can be easily identified during installation.

To install:

4. Be sure all gasket surfaces are clean and properly apply a continuous bead of silicone sealer to the pump.
5. Install or connect the following:
   - Water pump and torque the 6mm bolts to 57–66 inch lbs. (6–8 Nm) and the 8mm bolts to 12–14 ft. lbs. (16–19 Nm)
   - Coolant tube
   - Alternator and A/C compressor
   - Right side lower splash shield
   - Negative battery cable
6. Fill the cooling system.
7. Start the engine, check for leaks and repair if necessary.

2.5L Engine

1. Before servicing the vehicle, refer to the precautions in the beginning of this section.
2. Drain the cooling system.
3. Disconnect the negative battery cable
4. On Altima remove or disconnect the following:
   - Upper and lower engine covers
   - Coolant reservoir
   - Power Distribution Module (PDM) and move aside
   - Passenger front wheel
   - Engine ground
5. On all models remove or disconnect the following:
   - Drive belt
   - Radiator hose
   - Alternator
   - Water pump

To install:

6. Be sure all gasket surfaces are clean and properly apply a continuous bead of silicone sealer to the pump.
7. Install or connect the following:
   - Water pump and torque the bolts to 16–20 ft. lbs. (21–28 Nm)
8. Reverse the removal procedure to complete installation.

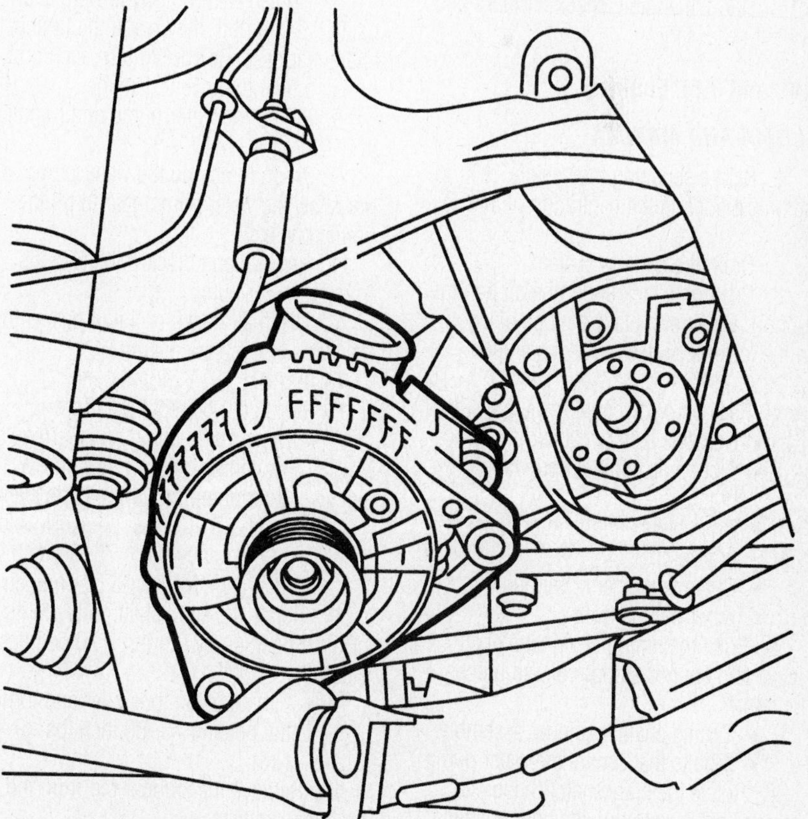

Exploded view of the water pump assembly–2.4L engine

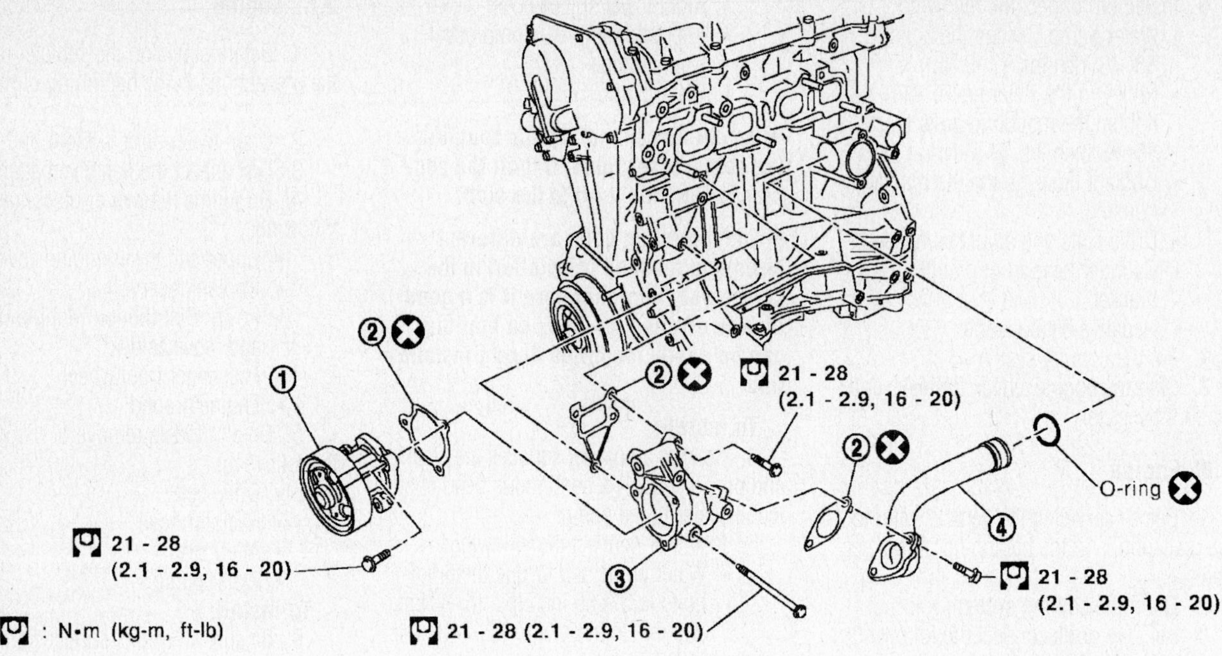

Exploded view of the water pump assembly–2.5L engine

9. Connect the negative battery cable.
10. Fill the cooling system.
11. Start the engine, check for leaks and repair if necessary.

3.0L and 3.5L Engines

ALTIMA AND MAXIMA

1. Before servicing the vehicle, refer to the precautions in the beginning of this section.
2. Drain the cooling system.
3. Position a jack under the oil pan for support. Be sure to place a block of wood on the jack for protection to the engine parts.
4. Remove or disconnect the following:
   - Negative battery cable
   - Right side engine mount and bracket
   - Drive belts and the idler pulley bracket
   - Chain tensioner cover and the water pump cover
5. Push the timing chain tensioner sleeve and apply a stopper pin so it does not return.
   - Timing chain tensioner assembly
   - 3 bolts that secure the water pump
6. Rotate the crankshaft 20 degrees counterclockwise to provide timing chain slack.
7. Put M8 bolts in 2 M8 threaded holes of the water pump.
8. Tighten each bolt by turning alternately ½ turn until they reach the timing chain rear case. Be sure to turn each bolt ½ turn at a time to prevent damage.
9. Lift up the water pump and remove it.
10. When removing the water pump, do not allow the water pump gear to hit the timing chain.
11. Remove and discard the O-rings from the water pump.
12. Clean all traces of liquid gasket from the water pump and covers.

To install:

13. Install or connect the following:
   - Water pump using new O-rings to the engine block. Torque the 3 water pump mounting bolts evenly to 62–89 inch lbs. (7–10 Nm) for 3.0L engines, or to 75–95 inch lbs. (8.5–10.7 Nm) for 3.5L engines.
14. Rotate the crankshaft pulley to its original position by turning it 20 degrees clockwise.
   - Timing chain tensioner and torque the bolts to 75–89 inch lbs. (9–10 Nm)
15. Remove the stopper pin from the timing chain tensioner.
16. Apply a continuous 0.091–0.130 in. (2.3–3.3mm) bead of liquid sealant to the mating surfaces of the timing chain tensioner and water pump covers.
   - Timing chain tensioner and water pump covers to the engine block. Torque the bolts to 89–108 inch lbs. (10–13 Nm).
   - Drive belts and the idler pulley bracket
   - Right side engine mounting bracket and the engine mount
   - Negative battery cable
17. Remove the jack from under the engine and install the drain plugs to the cylinder block.
18. Fill the cooling system.
19. Start the engine, check for leaks and repair if necessary.

350Z

1. Before servicing the vehicle, refer to the precautions in the beginning of this section.
2. Drain the cooling system.
3. Remove or disconnect the following:
   - Negative battery cable
   - Accessory drive belts
   - Radiator hoses
   - Cooling fan
   - Water drain plug on water pump side of block
   - Timing chain tensioner cover

- Water pump cover
- Primary timing chain tensioner
- Water pump mounting bolts

4. Turn the crankshaft pulley counterclockwise until the timing chain slack on the water pump pulley is at maximum.
5. Place M8 bolts in the upper and lower M8 threaded holes of the water pump.
6. Tighten each bolt by turning alternately ½ turn until they reach the timing chain rear case. Be sure to turn each bolt ½ turn at a time to prevent damage.
7. Lift up the water pump and remove it.
8. When removing the water pump, do not allow the water pump gear to hit the timing chain.
9. Remove and discard the O-rings from the water pump.
10. Clean all traces of liquid gasket from the water pump and covers.

To install:

11. Install the water pump using new O-rings to the engine block. Lubricate the inner O-ring with clean engine oil and the

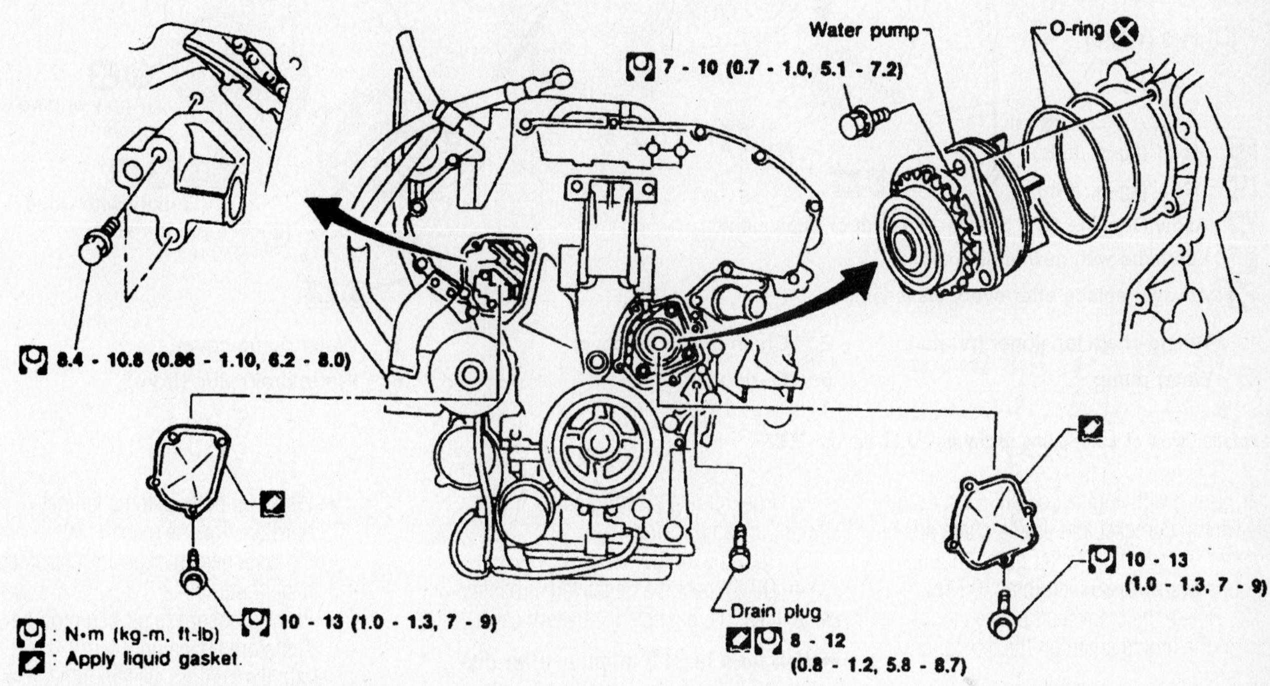

Water pump and timing cover assembly—3.0L engine

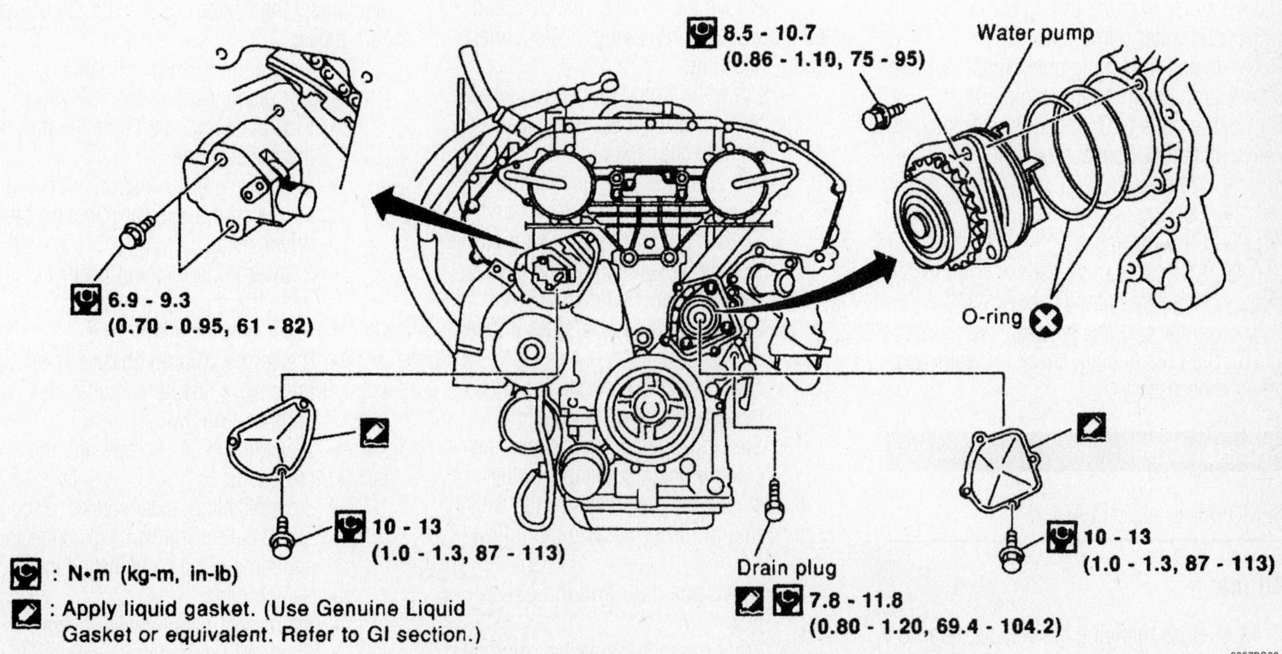

Water pump and timing cover assembly—3.5L engine–Altima and Maxima

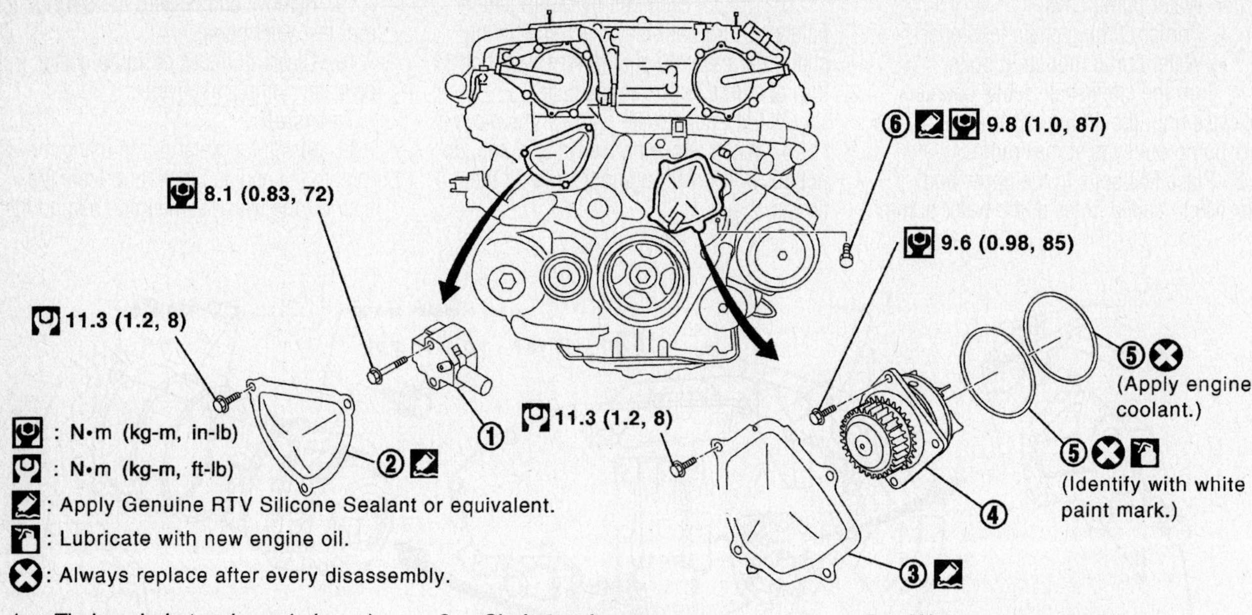

Exploded view of water pump mounting—3.5L engine–350Z

outer O-ring with engine coolant. Ensure the water pump sprocket and timing chain are engaged. Torque the 3 water pump mounting bolts evenly to 85 inch lbs. (10 Nm).

12. Rotate the crankshaft pulley clockwise so the timing chain on the tensioner side is loose.
13. Install the primary timing chain tensioner.
14. Apply a continuous 0.091–0.130 in. (2.3–3.3mm) bead of liquid sealant to the mating surfaces of the timing chain tensioner and water pump covers.
15. Install the timing chain tensioner and water pump covers to the engine block. Torque the bolts to 97 inch lbs. (11 Nm).
16. Install or connect the following:
    - Water drain plug
    - Cooling fan
    - Radiator hoses
    - Accessory drive belts
    - Negative battery cable
17. Fill the cooling system.
18. Start the engine, check for leaks and repair if necessary.

Heater Core

REMOVAL & INSTALLATION

Altima

1. Before servicing the vehicle, refer to the precautions in the beginning of this section.
2. Position the steering wheel in the straight-ahead position.
3. Turn the ignition switch OFF.
4. Disconnect the negative (–) battery cable; then, the positive (+) battery cable.

➡Wait for a least 3 minutes after disconnecting the battery cables for the charge in the air bag circuit to dissipate before working on the air bag module(s).

5. Remove the driver's side SRS and steering wheel by removing or disconnecting the following:
    - Lower lid from the steering wheel and disconnect the driver's air bag module connector
    - Left and right side lids from the steering wheel
    - Special bolts from both side of the steering wheel using a tamper resistant Torx® wrench (T50)
    - Air bag module and store it face up
    - Horn's electrical connector and remove the steering wheel nut
    - Steering wheel from the steering column using a suitable puller
6. Remove the passenger's side SRS by removing or disconnecting the following:
    - Glove box door and the glove box
    - Front passenger's air bag module connector
    - 2 special bolts using a tamper resistant Torx® wrench (T50)
    - 4 passenger's air bag-to-instrument panel nuts
    - Front passenger's air bag module and store it face up.
7. Drain the cooling system into a clean container for reuse.
8. Discharge and recover the air conditioning system refrigerant.
9. Working in the engine compartment, disconnect the heater hoses from the heater core tubes.
10. Remove the instrument panel by removing or disconnecting the following:
    - Kick plate and dash side finisher on the driver's side
    - 2 lower panel-to-instrument panel screws and the lower panel on the driver's side
    - 2 lower reinforcement panel-to-instrument panel screws and the lower reinforcement panel
    - 6 steering column cover screws, the covers, the spiral cable and combination switch
    - 2 cluster lid "A" screws and the cluster lid "A"
    - 3 combination meter screws, disconnect the electrical harness connector and remove the combination meter
    - Switch panel
    - Instrument panel lower covers
    - Snap out the transmission shifter finisher (boot)

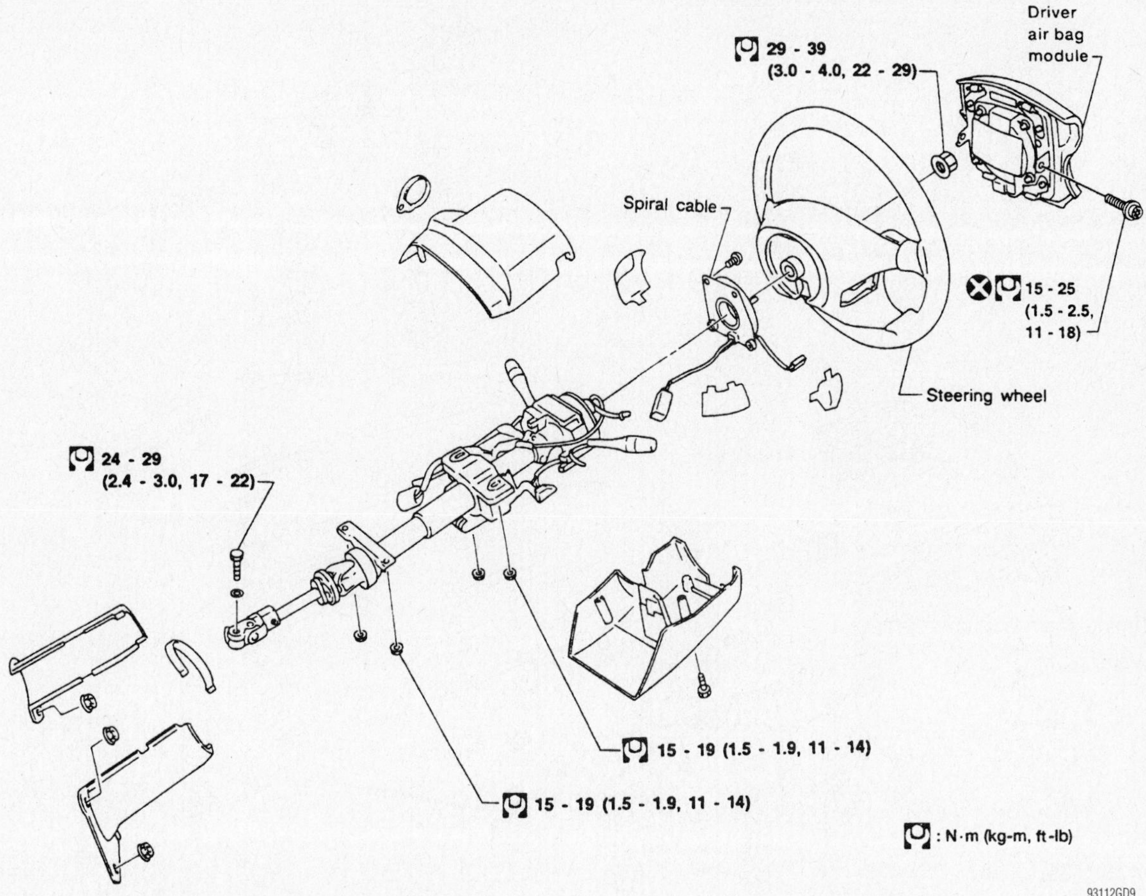

Exploded view of the steering wheel and air bag module—Altima

- 4 cluster lid "C" screws and the cluster lid "C"
- 4 audio and deck pocket-to-instrument panel screws and the audio and deck pocket
- 5 center console screws and the center console
- 2 center instrument panel screws and the center panel
- front defroster grilles
- Front pillar garnish
- Instrument panel 3 nuts/4 screws and the instrument panel
- 8 instrument stay assembly nuts and the stay
- Steering member assembly 5 nuts/1 bolt and the steering member

11. Remove the air conditioning housing assembly by removing or disconnecting the following:
    - Refrigerant lines from the air conditioning housing assembly
    - Thermo control amp
    - Air conditioning housing assembly
12. Remove or disconnect the following:
    - Heater unit
    - Heater core from the heater unit

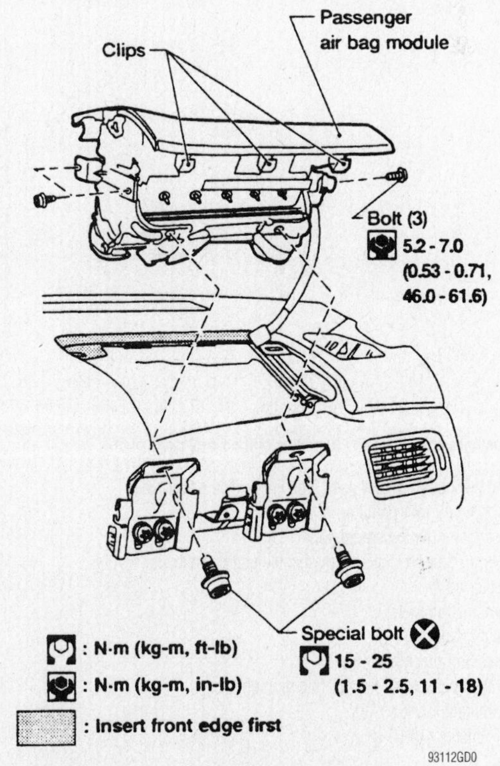

Exploded view of the passenger's side air bag module—Altima

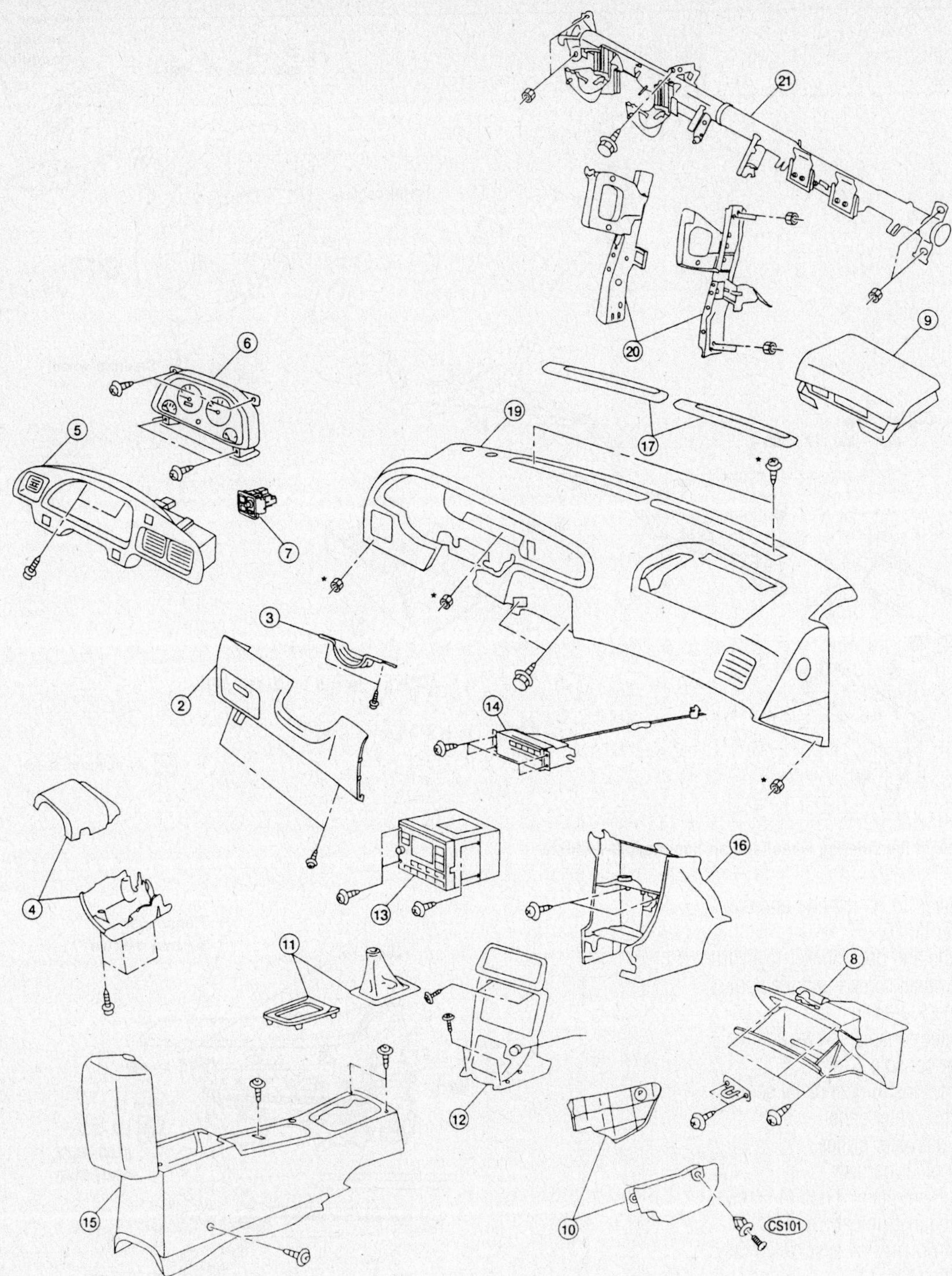

*: Instrument panel assembly mounting bolts, screws and nuts.

1. Remove kick plate and dash side finisher on driver side
2. Instrument lower panel on driver side
3. Dash lower reinforcement panel
4. Steering column covers, spiral cable and combination switch
5. Cluster lid A
6. Combination meter
7. Switch panel
8. Glove box assembly
9. Remove passenger side air bag moldule
10. Instrument lower covers
11. A/T finisher or M/T boot
12. Cluster lid C
13. Audio and deck pocket
14. A/C & heater control
15. Center console assembly
16. Instrument center panel
17. Front defroster grilles
18. Front pillar garnish
19. Instrument panel assembly
20. Instrument stay assemblies, if necesary
21. Steering member assembly, if necessary

93112GE1

Exploded view of the instrument panel assembly—Altima

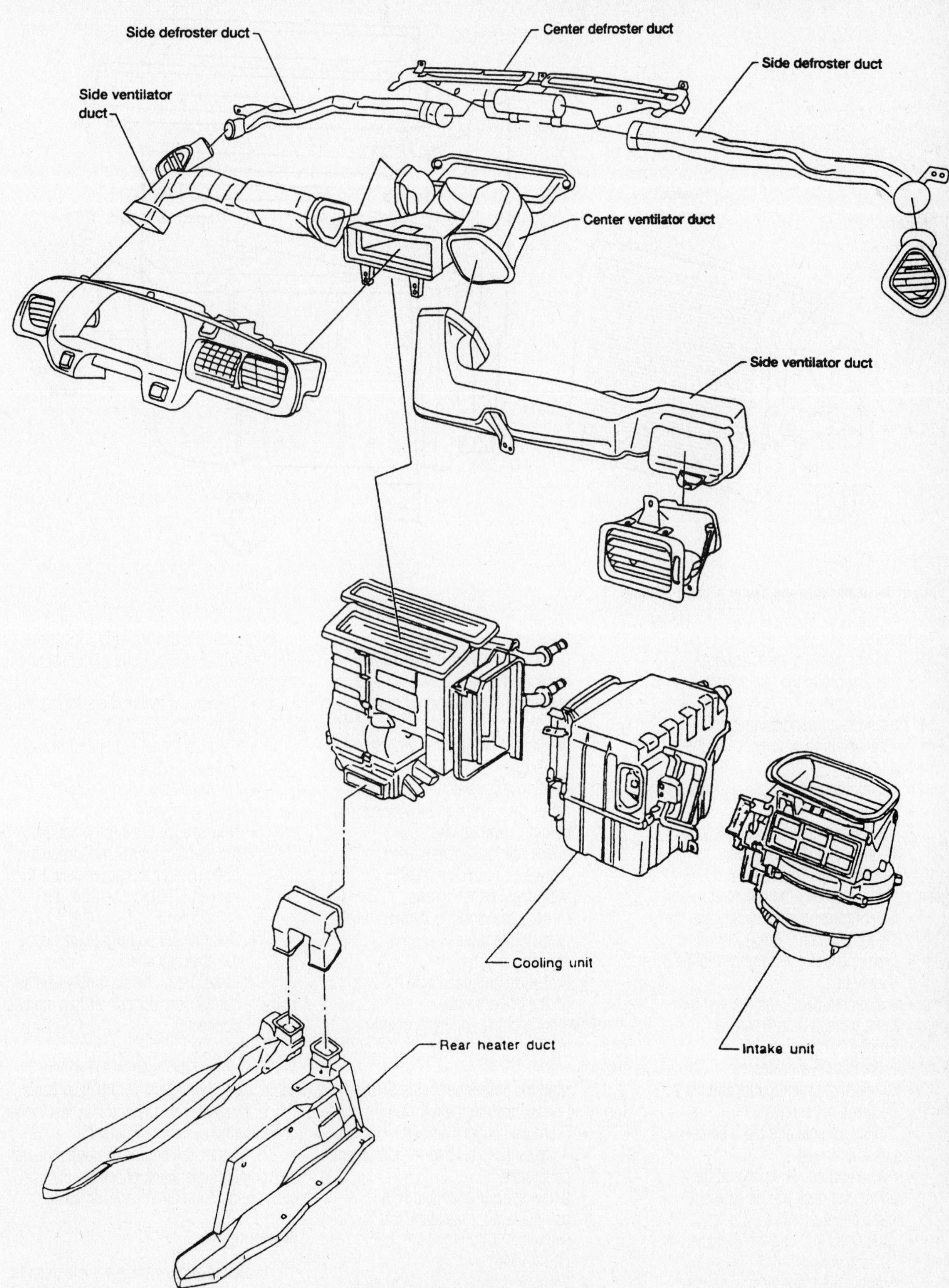

Exploded view of the heater housing assembly and related components—Altima

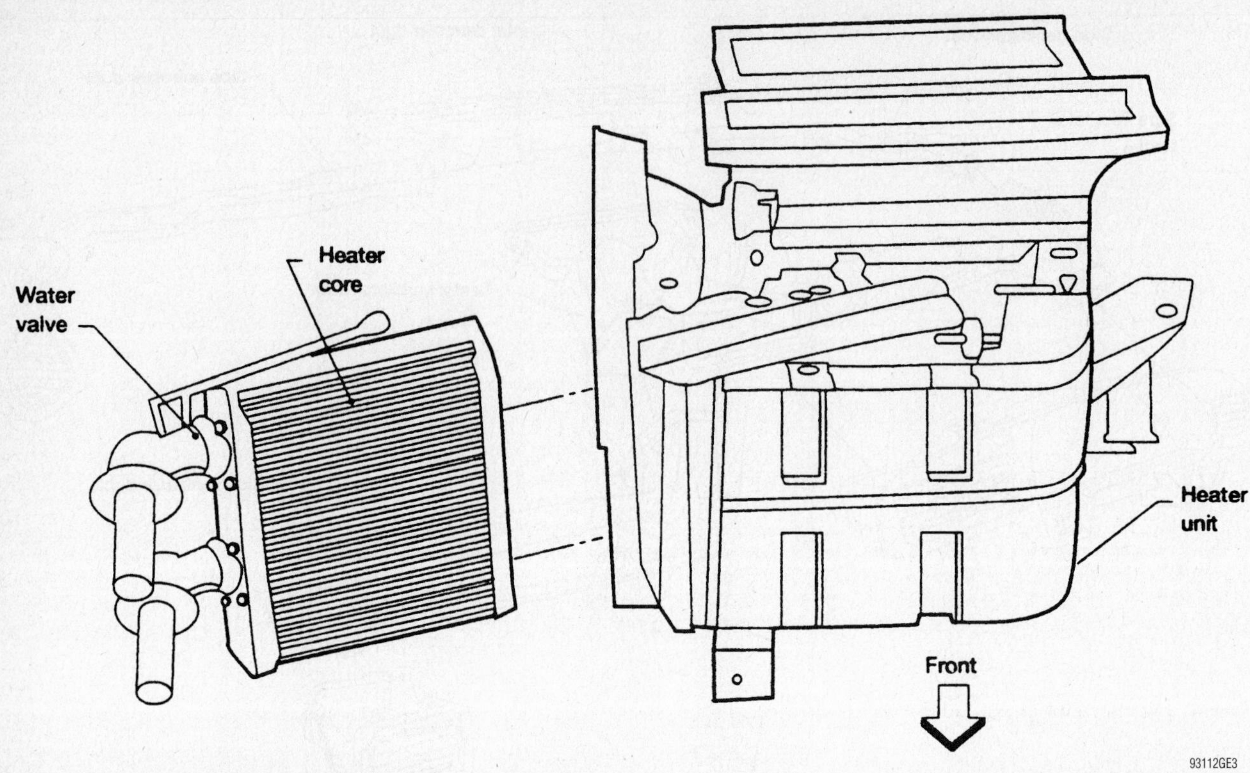

View of the heater core and heater housing—Altima

To install:

13. Install or connect the following:
    - Heater core to the heater unit
    - Heater unit

14. Install the air conditioning housing assembly by installing or connecting the following:
    - Air conditioning housing assembly
    - Thermo control amp
    - Refrigerant lines to the air conditioning housing assembly

15. Install the instrument panel by installing or connecting the following:
    - Steering member assembly and the steering member 5 nuts/1 bolt
    - Instrument stay assembly and the 8 stay nuts
    - Instrument panel and the instrument panel 3 nuts/4 screws
    - Front pillar garnish
    - Front defroster grilles
    - Center instrument panel and the 2 center panel screws
    - Center console and the 5 center console screws
    - Audio and deck pocket and the 4 audio and deck pocket-to-instrument panel screws
    - Cluster lid "C" and the 4 cluster lid "C" screws
    - Snap in the transmission shifter finisher (boot)
    - Instrument panel lower covers
    - Switch panel
    - Combination meter, connect the electrical harness connector and install the 3 combination meter screws
    - Cluster lid "A" and the 2 cluster lid "A" screw
    - Combination switch, the spiral cable, the covers and the 6 steering column cover screws
    - Lower reinforcement panel and the 2 lower reinforcement panel-to-instrument panel screws
    - Lower panel and the 2 lower panel-to-instrument panel screws, on the driver's side
    - Kick plate and dash side finisher, on the driver's side

16. Working in the engine compartment, connect the heater hoses to the heater core tubes.

17. Install the passenger's side SRS by installing or connecting the following:
    - Front passenger's air bag module
    - 4 passenger's air bag-to-instrument panel nuts
    - 2 new special bolts and torque using a tamper resistant Torx® wrench (T50) to 11–18 ft. lbs. (15–25 Nm)
    - Front passenger's air bag module connector
    - Glove box door and the glove box

18. Install the driver's side SRS and steering wheel by installing or connecting the following:
    - Steering wheel to the steering column
    - Steering wheel nut and torque the nut to 22–29 ft. lbs.
    - Horn's electrical connector
    - Air bag module
    - New special bolts to both sides of the steering wheel and torque the bolts using a tamper resistant Torx® wrench (T50) to 11–18 ft. lbs. (15–25 Nm).
    - Both the left and right side lids to the steering wheel
    - Lower lid to the steering wheel and connect the driver's air bag module connector

19. Refill the cooling system.

20. Connect the positive (+) battery cable; then, the negative (–) battery cable.

21. Evacuate, charge and leak test the air conditioning system refrigerant.

22. Operate the engine to normal operating temperatures; then, check the climate control operation and check for leaks.

Maxima

1. Before servicing the vehicle, refer to the precautions in the beginning of this section.

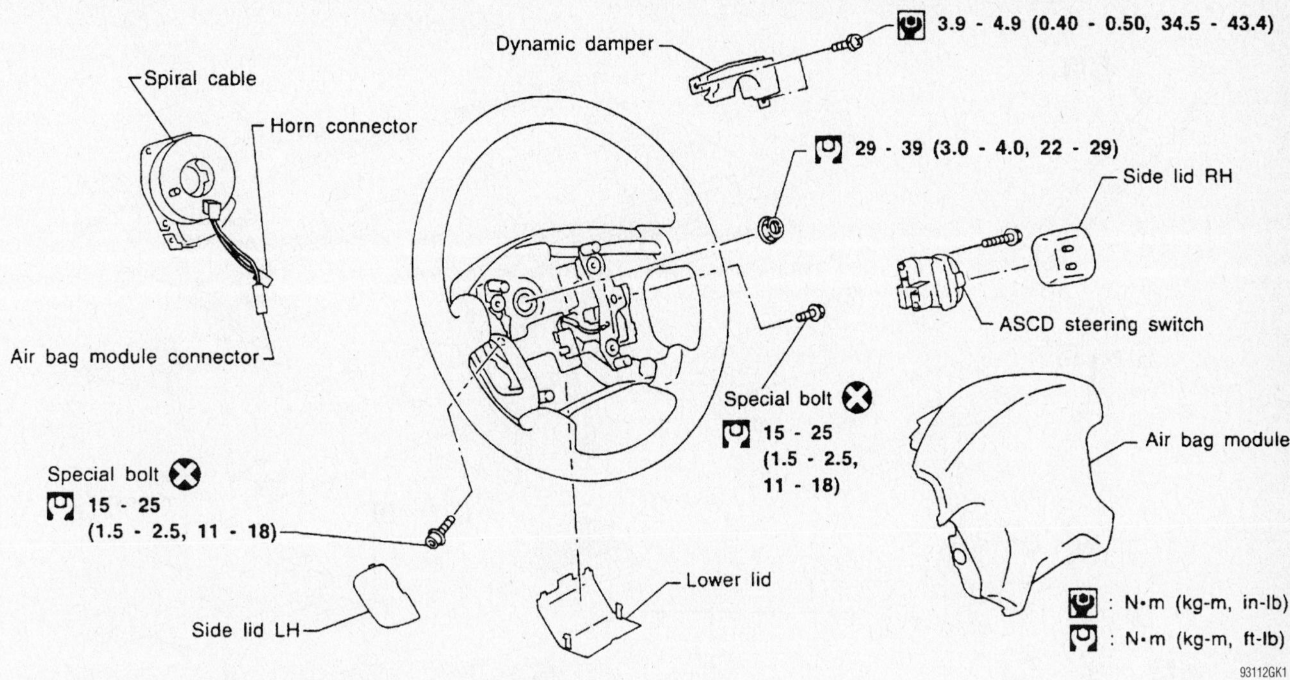

Exploded view of the air bag module and steering wheel—Maxima

2. Disconnect the negative battery terminal.

CAUTION

After disconnecting the negative battery cable, wait for at least 3 minutes for the SRS modules to deplete its energy.

3. Drain the cooling system into a clean container for reuse.
4. Remove the air bag module and steering wheel by removing or disconnecting the following:
   - Place the front wheels in the straight-ahead position
   - Lower lid and disconnect the air bag electrical connector at the bottom of the steering wheel
   - Side lids from both sides of the steering wheel
   - Torx® bolts using a Torx® wrench T50 from both side of the steering wheel; then, discard the bolts
   - Air bag module from the steering wheel
   - Steering wheel nut
   - Steering wheel from the steering column using a suitable puller
5. Disarm the passenger's side air bag by removing or disconnecting the following:
   - Glove box lid
   - Passenger's air bag electrical connector
6. Remove the instrument panel by removing or disconnecting the following:
   - Upper and lower glove box screws and remove the glove box
   - Lower instrument panel screws and the panel at the driver's side
   - Knee protector screws and the knee protector
   - Steering column cover screws and the cover
   - Combination switch-to-steering column screws, disconnect the electrical connector and the combination switch
   - Cluster lid "A" screws and the lid
   - Combination meter screws, disconnect the electrical connectors and the combination meter
   - Center ventilator with the switch panel using a suitable prytool
   - Cover plate (automatic transmission) or the shifter cover plate (manual transmission)
   - Ashtray
   - Upper and lower audio/air conditioning control unit assembly screws and the assembly
   - Console box screws and the console box (under the shifter cover plate); be sure to remove the rear screws
   - Front pillar garnish
   - Left and right lower cover and the center lower cover at the instrument panel dash
   - Defroster grille
   - Instrument panel-to-chassis nuts/bolts and the instrument panel
7. Remove or disconnect the following:
   - Rear heater ducts
   - Side ventilator ducts
   - Center defroster duct and the center ventilator duct
   - Heater housing-to-chassis fasteners and remove the heater housing
   - Heater core from the heater housing

To install:

8. Install or connect the following:
   - Heater core to the heater housing
   - Heater housing and the heater housing-to-chassis fasteners
   - Center ventilator duct and the center defroster duct
   - Side ventilator ducts
   - Rear heater ducts
9. Install the instrument panel by installing or connecting the following:
   - Instrument panel and the instrument panel-to-chassis nuts/bolts
   - Defroster grille
   - Left and right lower cover and the center lower cover
   - Front pillar garnish
   - Console box and the console box screws under the shifter cover plate; be sure to install the rear screws
   - Audio/air conditioning control unit assembly and the upper and lower assembly screws
   - Ashtray
   - Cover plate (automatic transmission) or the shifter cover plate (manual transmission)

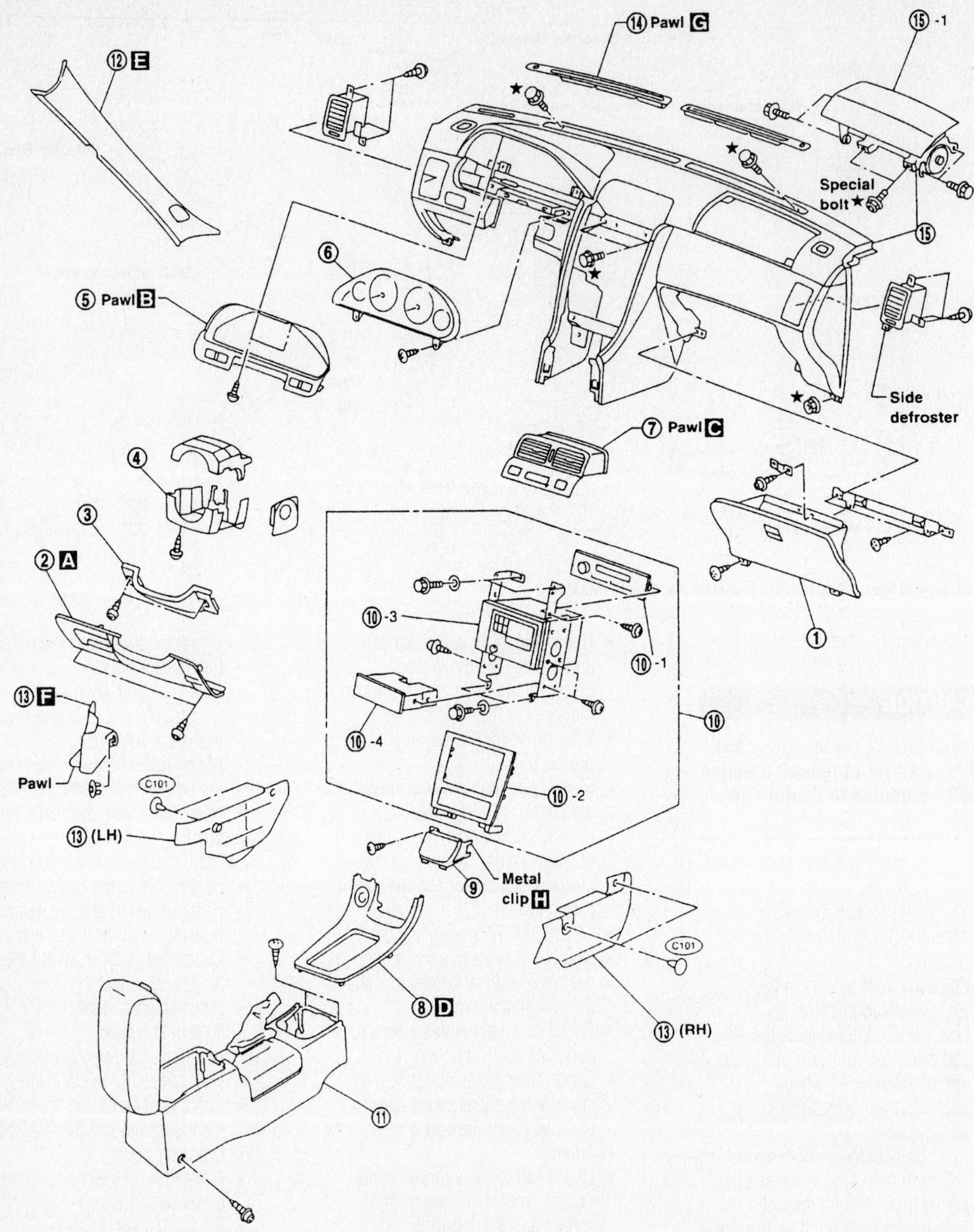

1. Glove box assembly
2. Instrument lower panel on driver side
3. Knee protector assembly
4. Steering column cover & combination switch
5. Cluster lid A
6. Combination meter
7. Center ventilator with switch panel
8. A/T shifter cover plate or M/T shifter cover plate
9. Ashtray
10. Audio & A/C control unit assembly
11. Console box
12. Front pillar garnish
13. Instrument dash: lower cover and center lower cover on LH, RH
14. Defroster grille
15. Instrument panel assembly
15. -1 Passenger air bag module

93112GK2

Exploded view of the instrument panel, console and related components—Maxima

- Center ventilator with the switch panel
- Combination meter, connect the electrical connectors and the combination meter screws
- Cluster lid "A" and the lid screws
- Combination switch, connect the electrical connector and the combination switch-to-steering column screws
- Steering column cover and the cover screws
- Knee protector and the knee protector screws
- Lower instrument panel and the panel screws on the driver's side
- Glove box and the upper and lower glove box screws

10. Arm the passenger's side air bag by installing or connecting the following:
    - Passenger's air bag electrical connector
    - Glove box lid
11. Install the air bag module and steering wheel by installing or connecting the following:
    - Steering wheel to the steering column
    - Steering wheel nut and torque it to 22–29 ft. lbs. (29–39 Nm)
    - Air bag module to the steering wheel
    - Torque the new Torx® bolts (using a Torx® wrench T50), at both side of the steering wheel to 11–18 ft. lbs. (15–25 Nm).
    - Side lids to both sides of the steering wheel
    - Air bag electrical connector and install the lower lid at the bottom of the steering wheel
12. Refill the cooling system.

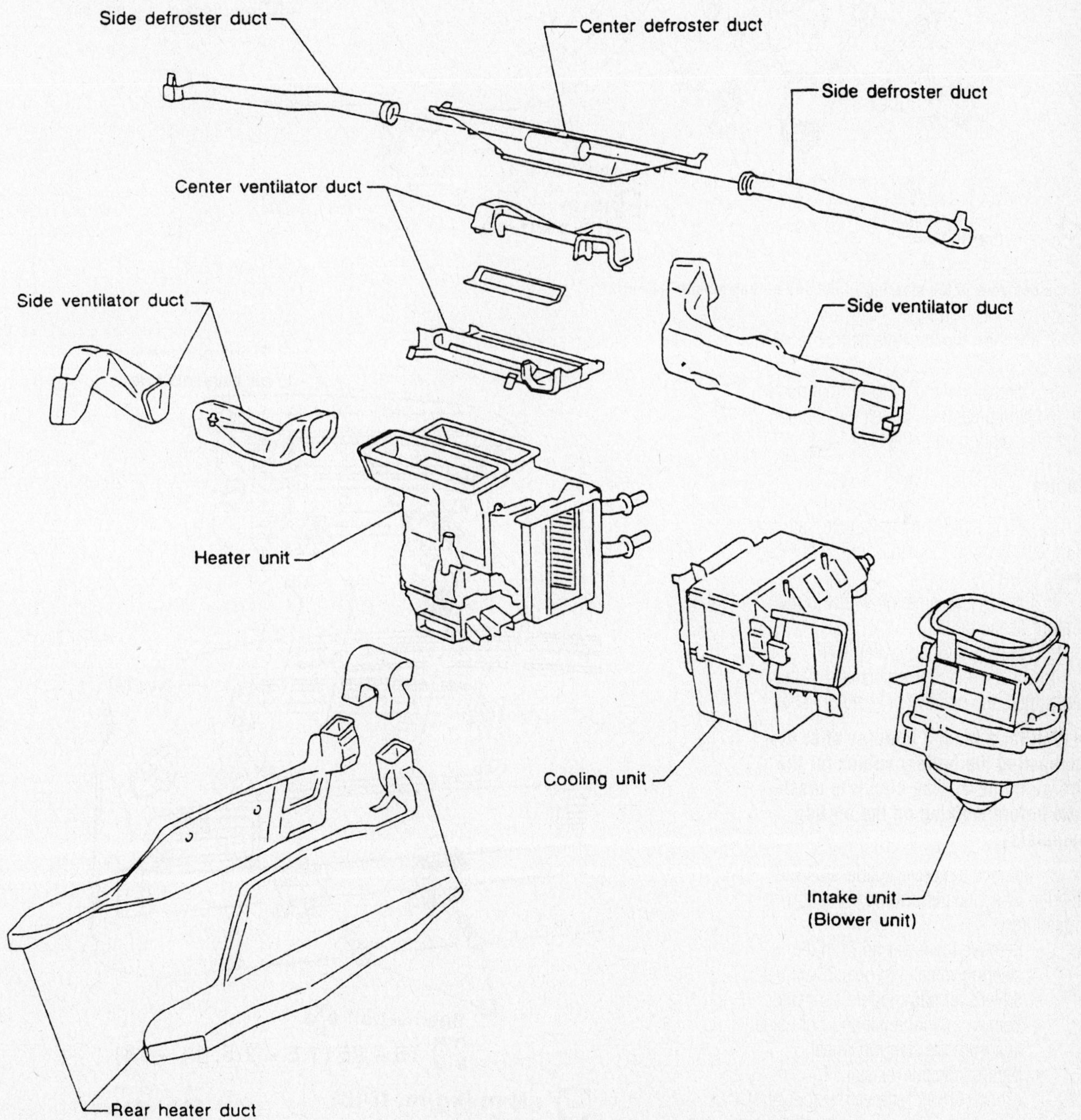

Exploded view of the heater housing, evaporator housing, ventilator system and related components—Maxima

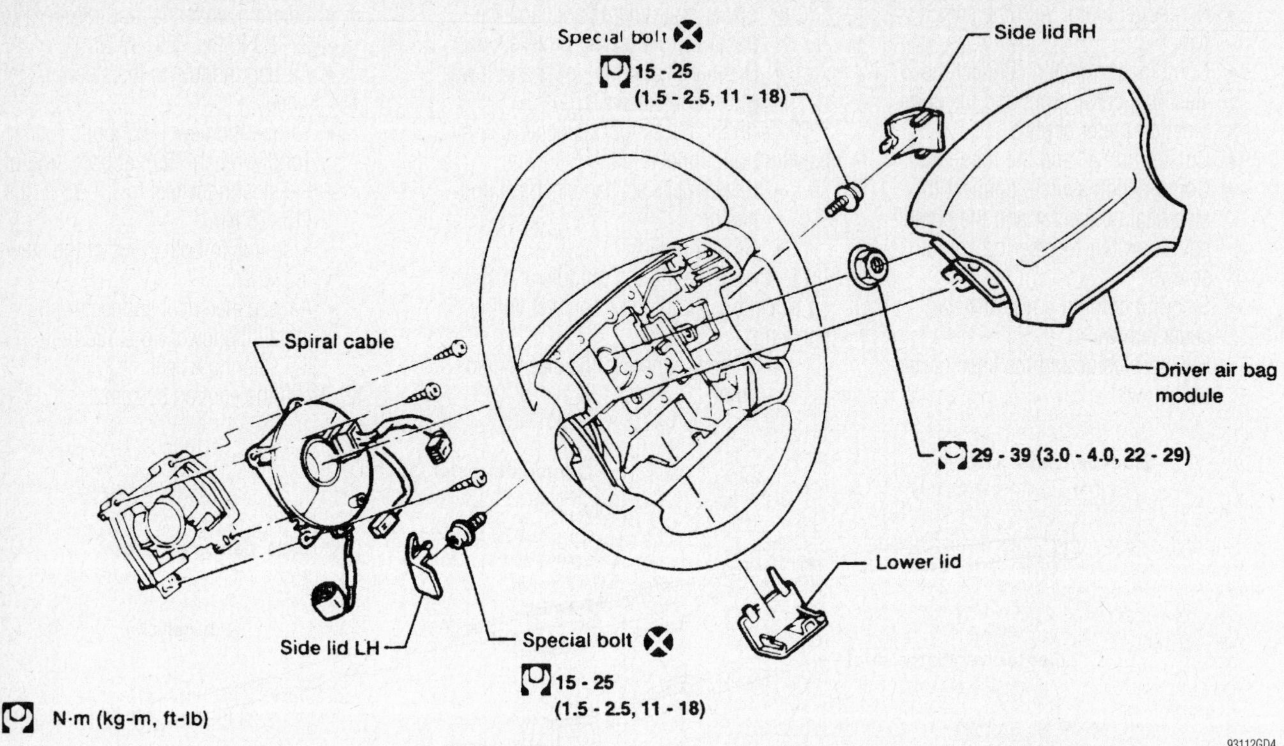

Exploded view of the steering wheel and air bag module—Sentra

13. Connect the negative battery terminal.
14. Operate the engine to normal operating temperatures; then, check the climate control operation and check for leaks.

Sentra

1. Before servicing the vehicle, refer to the precautions in the beginning of this section.
2. Position the steering wheel in the straight-ahead position.
3. Turn the ignition switch OFF.
4. Disconnect the negative (–) battery cable; then, the positive (+) battery cable.

➡Wait for a least 3 minutes after disconnecting the battery cables for the charge in the air bag circuit to dissipate before working on the air bag module(s).

5. Remove the driver's side SRS and steering wheel by performing the following procedure:
   - Remove the lower lid from the steering wheel and disconnect the driver's air bag module connector.
   - Remove both the left and right side lids from the steering wheel.
   - Using a tamper resistant Torx® wrench (T50), remove the special bolts from both side of the steering wheel.

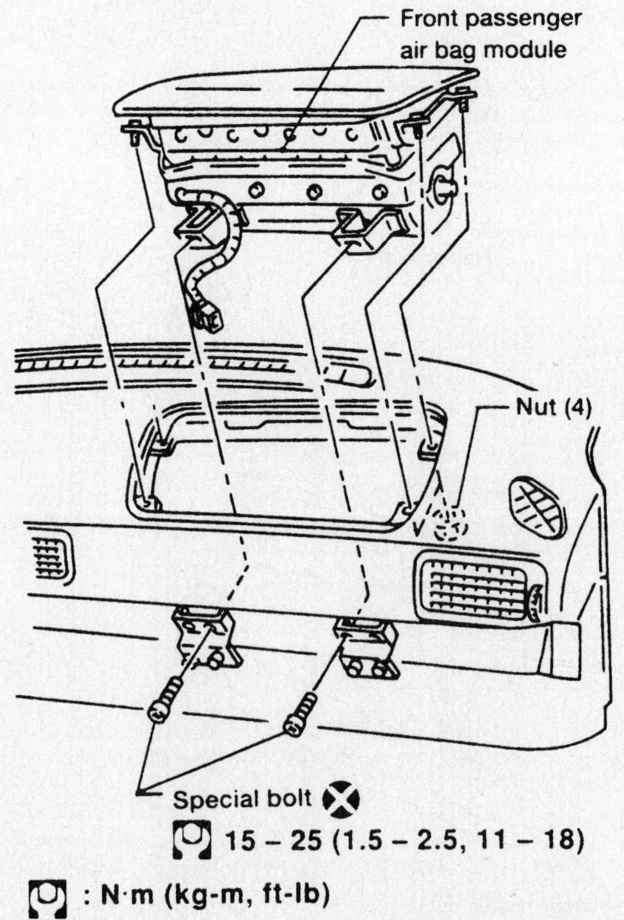

Exploded view of the passenger's side air bag module—Sentra

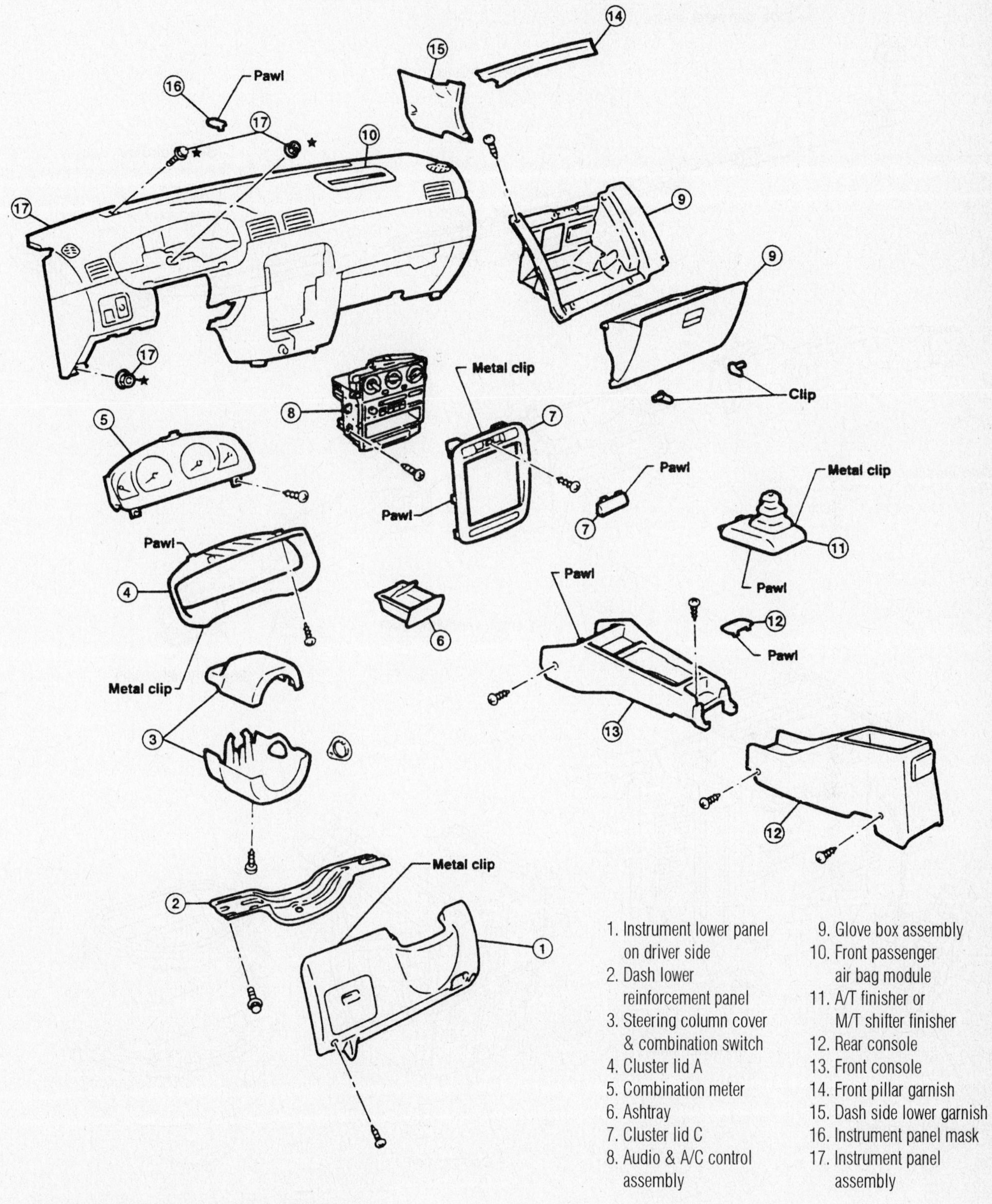

Exploded view of the instrument panel assembly—Sentra

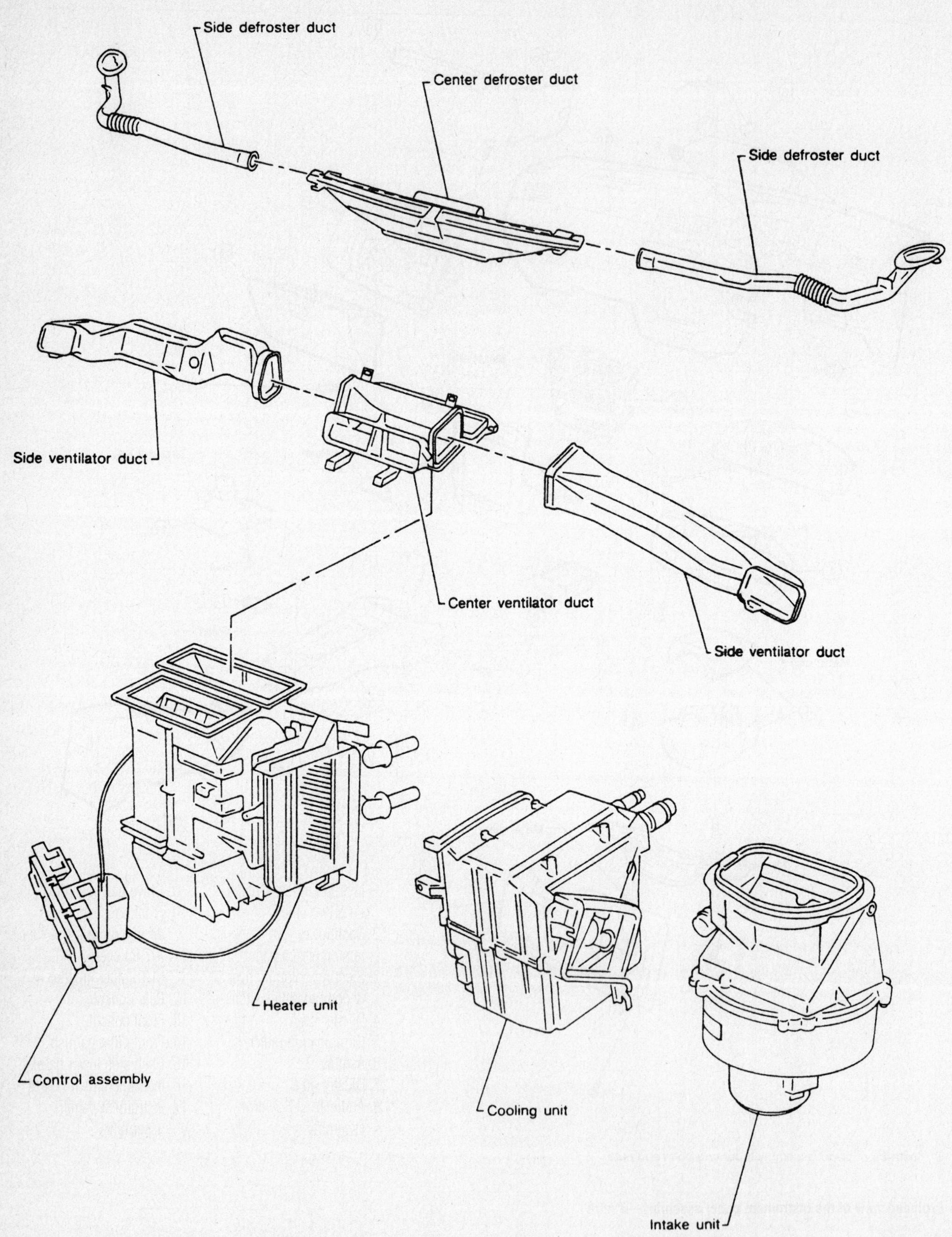

Exploded view of the heater housing assembly and related components—Sentra

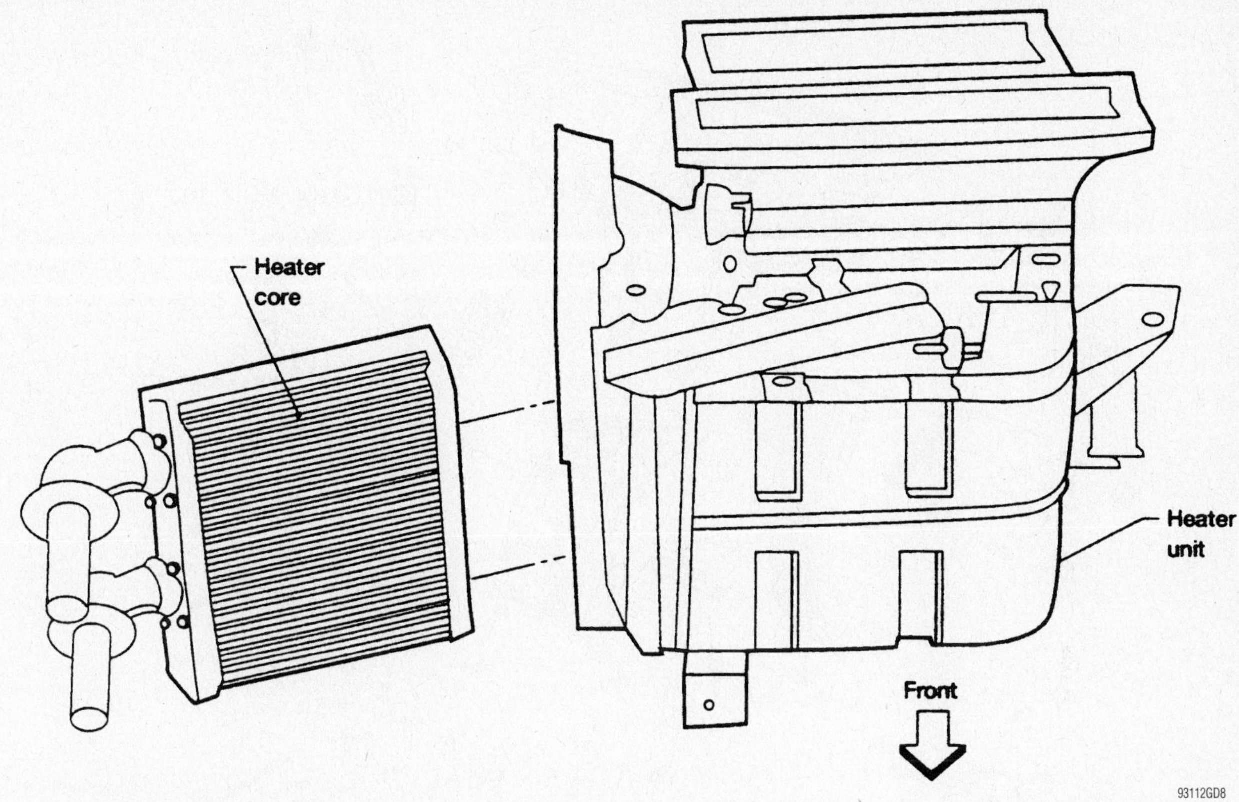

View of the heater core and heater housing—Sentra

- Carefully, remove the air bag module and store it face up.
- Disconnect the horn's electrical connector and remove the steering wheel nut.
- Using a steering wheel puller, press the steering wheel from the steering column.

6. Remove the passenger's side SRS by performing the following procedure:
   - Remove the glove box door and the glove box.
   - Disconnect the front passenger's air bag module connector.
   - Using a tamper resistant Torx® wrench (T50), remove the 2 special bolts.
   - Remove the 4 passenger's air bag-to-instrument panel nuts.
   - Carefully, remove the front passenger's air bag module and store it face up.
7. Drain the cooling system into a clean container for reuse.
8. Discharge and recover the air conditioning system refrigerant.
9. Working in the engine compartment, disconnect the heater hoses from the heater core tubes.
10. Remove the instrument panel by performing the following procedures:
    - On the driver's side, remove the 2 lower panel-to-instrument panel screws and the lower panel.
    - Remove the 2 lower reinforcement panel-to-instrument panel screws and the lower reinforcement panel.
    - Remove the 6 steering column cover screws, the cover and combination switch.
    - Remove the 2 cluster lid "A" screws and the cluster lid "A".
    - Remove the 3 combination meter screws, disconnect the electrical harness connector and remove the combination meter.
    - Remove the ashtray.
    - Remove the cluster lid "C" mask, screw and the cluster lid "C".
    - Remove the 8 audio and air conditioning control assembly-to-instrument panel screws, the electrical connectors and the audio and air conditioning control assembly.
    - Remove the transmission shifter finisher.
    - Remove the rear console mask, the 4 screws and the rear console.
    - Remove the 4 front console screws and the front console.
    - Remove the front pillar garnish.
    - Remove the lower dash side garnish.
    - Remove the instrument panel mask.
    - Remove the instrument panel-to-chassis nuts/bolts and the instrument panel.
11. Remove the air conditioning housing assembly by performing the following procedure:
    - Disconnect the refrigerant lines from the air conditioning housing assembly.
    - Disconnect the thermo control amp.
    - Remove the air conditioning housing assembly.
12. Remove the heater unit.
13. Remove the heater core from the heater unit.

To install:

14. Install or connect the following:
    - Heater core to the heater unit
    - Heater unit
15. Install the air conditioning housing assembly by performing the following procedure:
    - Install the air conditioning housing assembly.
    - Connect the thermo control amp.
    - Connect the refrigerant lines to the air conditioning housing assembly.
16. Install the instrument panel by installing or connecting the following:
    - Instrument panel and the instrument panel-to-chassis nuts/bolts
    - Instrument panel mask

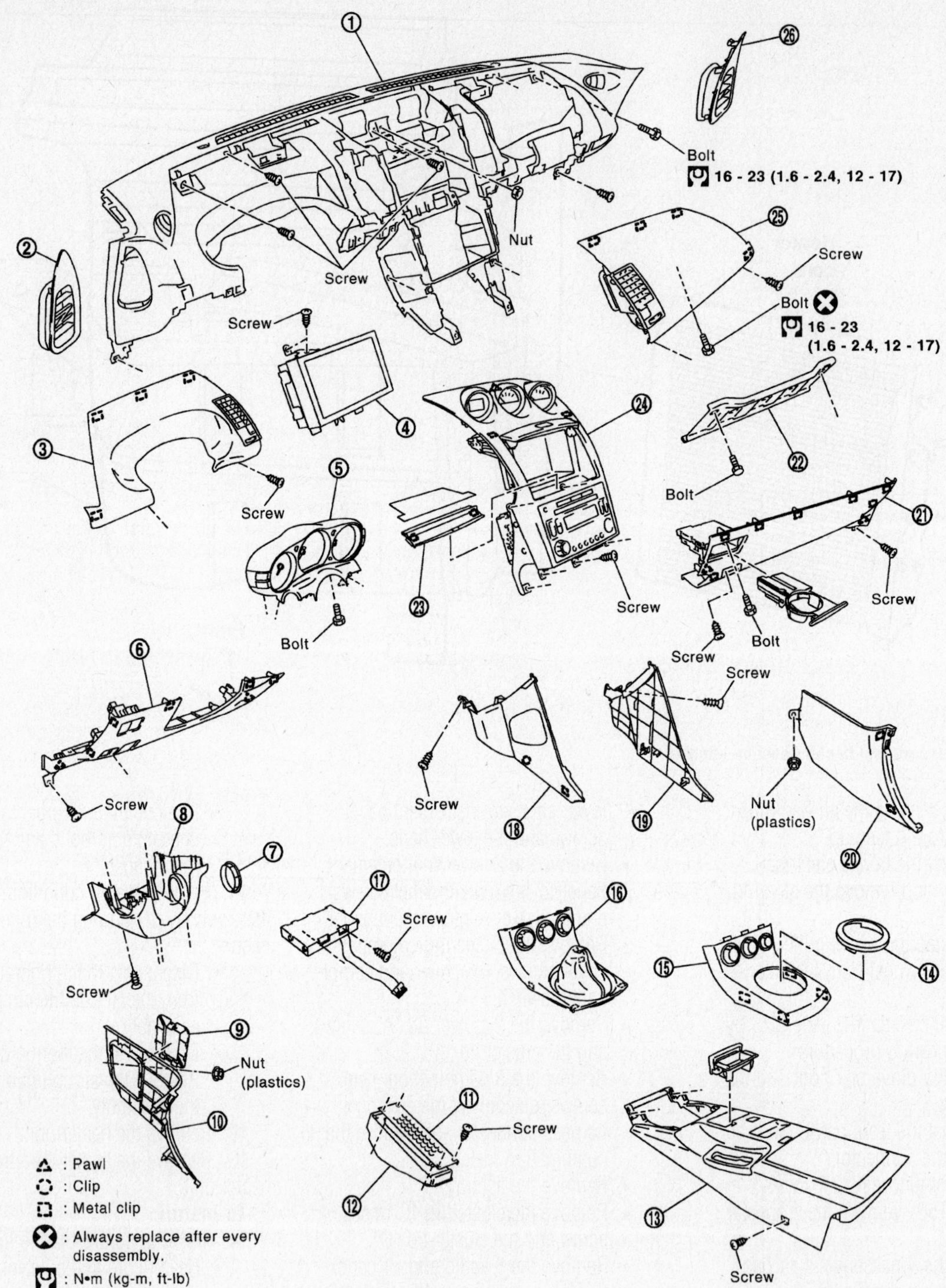

1.	Instrument panel	2.	Instrument side finisher (LH)	3.	Instrument driver panel upper
4.	Display	5.	Combination meter	6.	Instrument driver panel lower
7.	Steering lock escutcheon	8.	Steering column lower cover	9.	Fuse cover
10.	Dash side finisher (LH)	11.	Foot rest plate	12.	Foot rest
13.	Center console	14.	Console finisher (A/T ring)	15.	Console finisher (A/T)
16.	Console boot (M/T)	17.	Unified meter and A/C amp	18.	Instrument side panel (LH)
19.	Instrument side panel (RH)	20.	Dash side finisher (RH)	21.	Instrument passenger panel lower
22.	Knee protector	23.	NAVI switch / Switch mask	24.	Cluster lid C
25.	Instrument passenger panel upper	26.	Instrument side finisher (RH)		

67170-NISS-G010

Exploded view of the instrument panel assembly—350Z

- Lower dash side garnish
- Front pillar garnish
- Front console and the 4 front console screws
- Rear console, the 4 screws and the rear console mask
- Transmission shifter finisher
- Air conditioning control assembly, the electrical connectors and the audio and the 8 audio and air conditioning control assembly-to-instrument panel screws.
- Cluster lid "C", the screw and the cluster lid "C" mask
- Ashtray
- Combination meter, connect the electrical harness connector and install the 3 combination meter screws
- Cluster lid "A" and the 2 cluster lid "A" screws
- Combination switch, the cover and the 6 steering column cover screws
- Lower reinforcement panel and the 2 lower reinforcement panel-to-instrument panel screws
- Lower panel and the 2 lower panel-to-instrument panel screws, on the driver's side

17. Working in the engine compartment, connect the heater hoses to the heater core tubes.

18. Install the passenger's side SRS by installing or connecting the following:

- Front passenger's air bag module
- 4 passenger's air bag-to-instrument panel nuts
- 2 new special bolts and torque to 11–18 ft. lbs. (15–25 Nm), using a tamper resistant Torx® wrench (T50)
- Front passenger's air bag module connector
- Glove box door and the glove box

19. Install the driver's side SRS and steering wheel by installing or connecting the following:

- Steering wheel to the steering column
- Steering wheel nut and torque the nut to 22–29 ft. lbs. (30–39 Nm)
- Horn's electrical connector
- Air bag module
- New special bolts to both side of the steering wheel and torque the bolts to 11–18 ft. lbs. (15–25 Nm), using a tamper resistant Torx® wrench (T50)
- Left and right side lids to the steering wheel
- Lower lid to the steering wheel
- Driver's air bag module connector.

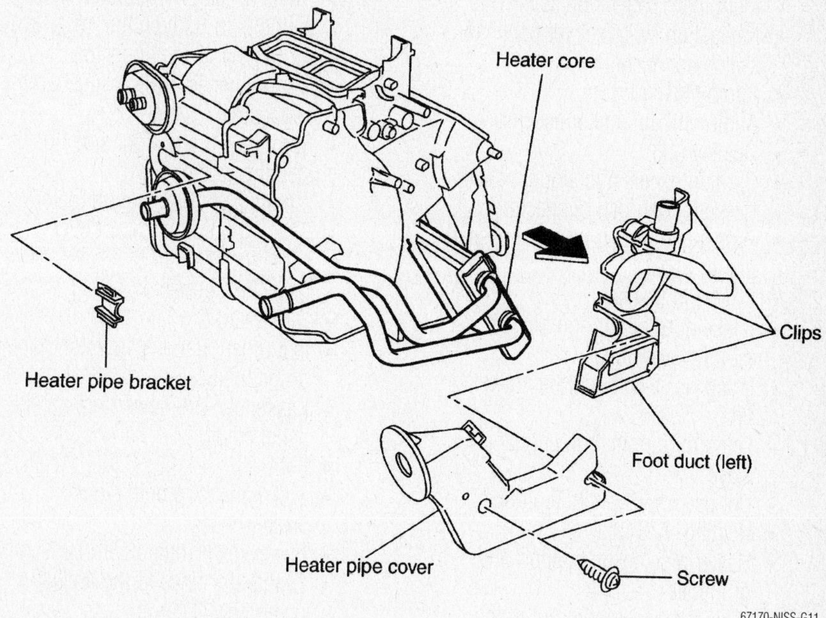

View of the heater core and heater housing—350Z

20. Refill the cooling system.

21. Connect the positive (+) battery cable; then, the negative (–) battery cable.

22. Evacuate, charge and leak test the air conditioning system refrigerant.

23. Operate the engine to normal operating temperatures; then, check the climate control operation and check for leaks.

350Z

1. Before servicing the vehicle, refer to the precautions in the beginning of this section.

2. Discharge the A/C system using approved recycling equipment.

3. Drain the cooling system.

4. Remove or disconnect the following:

- Hood ledge cover
- Both wiper arms
- Cowl rubber seal
- Cowl top cover and washer hose
- Evaporator lines from the firewall and cap openings
- Electronic throttle control assembly
- Heater hoses
- Kick panels on both sides
- Foot rests
- Passenger side lower instrument panel cover
- Instrument panel side finish panels on both sides
- Cluster Lid C
- Data link connector
- Hood lock cable
- Steering column lower cover
- 4 bolts and combination meter

5. Position the steering wheel in the straight-ahead position.

6. Turn the ignition switch OFF.

7. Disconnect the negative (–) battery cable; then, the positive (+) battery cable.

➡Wait for a least 3 minutes after disconnecting the battery cables for the charge in the air bag circuit to dissipate before working on the air bag module(s).

8. Remove the driver's side SRS and steering wheel by performing the following procedure:

- Remove both the left and right side lids from the steering wheel.
- Using a tamper resistant Torx® wrench (T50), remove the special bolts from both sides of the steering wheel.
- Steering wheel switch sub-harness connector
- Air bag harness connector
- Carefully remove the air bag module and store it face up.

9. Remove or disconnect the following:

- Steering wheel
- Steering column upper cover
- Spiral cable connector
- Combination switch
- Automatic transmission console finisher panel, if equipped
- Manual transmission shift knob, if equipped
- Center console
- Cup holder
- Passenger side lower instrument panel cover

- Instrument panel side cover
- Navigation switch cover panel and switch connector
- Audio cluster lid
- Audio unit and meter assembly
- Display unit
- Garnish panels and side finishers
- Passenger air bag connector
- Passenger air bag bolt and passenger air bag
- ECM and bracket
- Intake door motor and blower motor connectors
- 2 screws, 1 bolt and the blower unit
- Left and right instrument panel stays
- Defroster and ventilation ducts
- Heating–A/C unit
- Heater pipe cover, support and grommet
- Slide the heater core out of the heating–A/C unit

To install:

10. Install or connect the following:
    - Heater core to the heater–A/C unit
    - Heater pipe cover, support and grommet
    - Heater–A/C unit and tighten the bolts to 61 inch lbs. (7 Nm)
    - Defroster and ventilation ducts
    - Left and right instrument panel stays
    - 2 screws, 1 bolt and the blower unit
    - Intake door motor and blower motor connectors
    - ECM and bracket
    - Passenger air bag and tighten the bolt to 15–21 ft. lbs. (20–29 Nm)
    - Passenger air bag connector
    - Garnish panels and side finishers
    - Display unit
    - Audio unit and meter assembly
    - Audio cluster lid
    - Navigation switch cover panel and switch connector
    - Instrument panel side cover
    - Passenger side lower instrument panel cover
    - Cup holder
    - Center console
    - Manual transmission shift knob, if equipped
    - Automatic transmission console finisher panel, if equipped
    - Combination switch
    - Spiral cable connector
    - Steering column upper cover
    - Steering wheel and tighten the bolt to 22–28 ft. lbs. (30–39 Nm)
    - Driver air bag module and tighten the bolts to 83 inch lbs. (9.4 Nm)
    - Air bag harness connector
    - Steering wheel switch sub-harness connector
    - Both the left and right side lids to steering wheel
    - 4 bolts and combination meter
    - Steering column lower cover
    - Hood lock cable
    - Data link connector
    - Cluster lid C
    - Instrument panel side finish panels on both sides
    - Passenger side lower instrument panel cover
    - Foot rests
    - Kick panels on both sides
    - Heater hoses
    - Electronic throttle control assembly
    - Evaporator lines to the firewall using new O-rings
    - Cowl top cover and washer hose
    - Cowl rubber seal
    - Both wiper arms
    - Hood ledge cover
11. Refill the cooling system.
12. Connect the positive (+) battery cable; then, the negative (–) battery cable.
13. Evacuate, charge and leak test the air conditioning system refrigerant.
14. Operate the engine to normal operating temperatures; then, check the climate control operation and check for leaks.

Cylinder Head

REMOVAL & INSTALLATION

1.8L Engines

1. Before servicing the vehicle, refer to the precautions in the beginning of this section.
2. Drain the cooling system.
3. Properly relieve the fuel system pressure.
4. Remove or disconnect the following:
    - Negative battery cable
    - Engine drive belts
    - Power steering pulley
    - Oil pump and bracket
    - Air duct to the intake manifold collector
    - Right front wheel
    - Engine side and under covers
    - Front exhaust tube
    - Cylinder head front mounting bracket
    - Rocker cover by loosening the bolts in numerical order
    - Distributor, plug wires and spark plugs
    - Spark plugs
    - Intake manifold support and set the No. 1 cylinder at the Top Dead Center (TDC) position
    - Idler pulley, camshaft sprockets and timing chains
    - Camshafts
5. Loosen the cylinder head bolts in 2–3 steps in the reverse order of the tightening sequence to prevent warpage or cracking of the cylinder head assembly.
    - Cylinder head (carefully), from the block, pulling the head up evenly from both ends. If the head seems stuck, do not pry it off. Tap lightly around the lower perimeter of the head with a rubber mallet to help break the seal. The cylinder head and the intake and exhaust manifolds are removed together.
    - Cylinder head gasket(s)

To install:

6. Thoroughly clean both the cylinder block and head mating surfaces. Avoid scratching either surface.
7. Coat the threads and the seating surface of the head bolts with clean engine oil. Install the cylinder head assembly (always replace the head gasket). Install head bolts (with washers) in their proper locations.
8. For 1.8L engines, tighten the bolts in sequence, as follows:
    a. Step 1: Bolts 1–10 to 22 ft. lbs. (29 Nm).
    b. Step 2: Bolts 1–10 to 43 ft. lbs. (59 Nm).
    c. Step 3: Loosen bolts 1–10 completely.
    d. Step 4: Bolts 1–10 to 22 ft. lbs. (29 Nm).
    e. Step 5: Bolts 1–10 plus 50–55 degrees.
    f. Step 5: Bolts 11–14 to 74 inch lbs. (8 Nm).
9. Install or connect the following:
    - Camshafts
    - Idler pulley, camshaft sprockets and timing chains
    - Intake manifold support
    - Distributor
    - Spark plugs and wires
    - Distributor cap
    - Rocker arm cover and torque the bolts to 34 inch lbs. (4 Nm)
    - Cylinder head front mounting bracket
    - Front exhaust tube
    - Engine side and under covers
    - Right front wheel

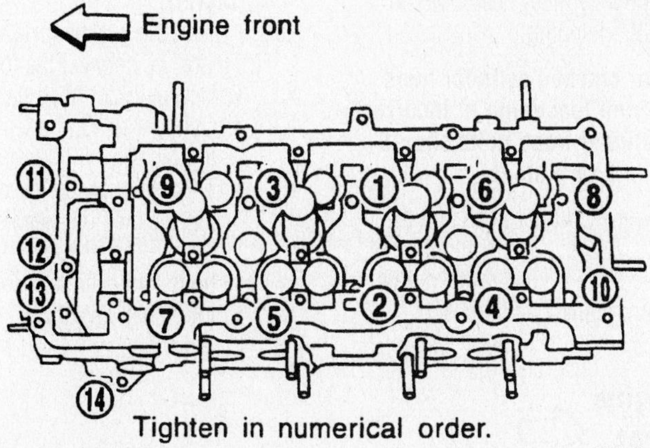

Cylinder head torque sequence—1.8L engine

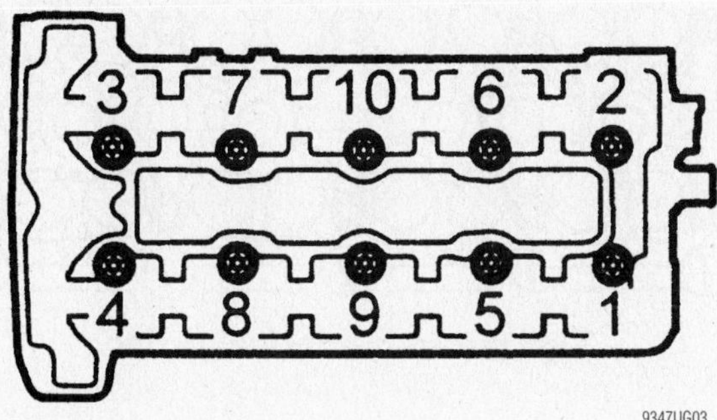

Tighten the rocker cover bolts in sequence—1.8L engines

- Air duct to the intake manifold collector
- Oil pump and bracket
- Power steering pulley
- Drive belts
- Negative battery cable

10. Fill the cooling system.
11. Start the vehicle, check for leaks and repair if necessary.

2.0L Engine

1. Before servicing the vehicle, refer to the precautions in the beginning of this section.
2. Release the fuel system pressure.
3. Drain the cooling system.
4. Remove or disconnect the following:

- Negative battery cable
- Engine under covers
- Right front wheel and engine side cover
- Radiator assembly
- Air duct and intake manifold
- Drive belts and water pump pulley
- Alternator
- Power steering pump
- Cylinder head cover and oil separator
- Oil filter and power steering pump brackets
- Front exhaust pipe from the exhaust manifold
- Distributor assembly
- Timing chain, tensioner, chain guide and camshaft sprockets
- Camshafts
- Water hose from the cylinder block and water hose from the heater
- Starter motor
- Water pipe bolt
- Knock Sensor (KS) harness connector and the Exhaust Gas Recirculation (EGR) tube
- Cylinder outside bolts. Remove the cylinder head bolts in 2 or 3 steps.
- Cylinder head completely with manifolds attached

To install:

5. Check all components for wear. Replace as necessary. Clean all mating surfaces and replace the cylinder head gasket.

➡If the length of any cylinder head bolt exceeds 6.22 in. (158.2mm), replace the bolt.

6. Install cylinder head. Torque the cylinder head bolts in the following sequence:

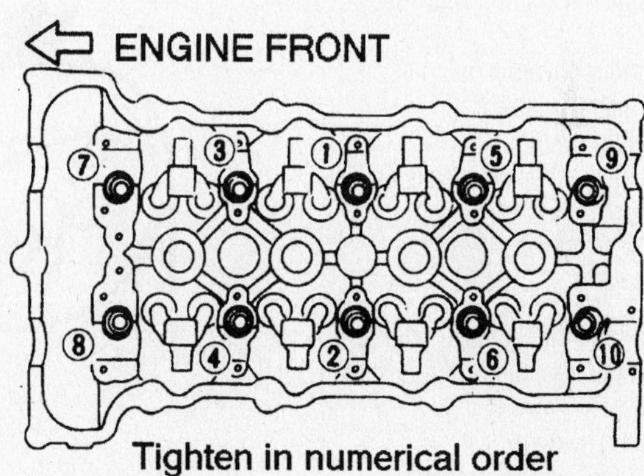

Cylinder head bolt torque sequence—2.0L engine

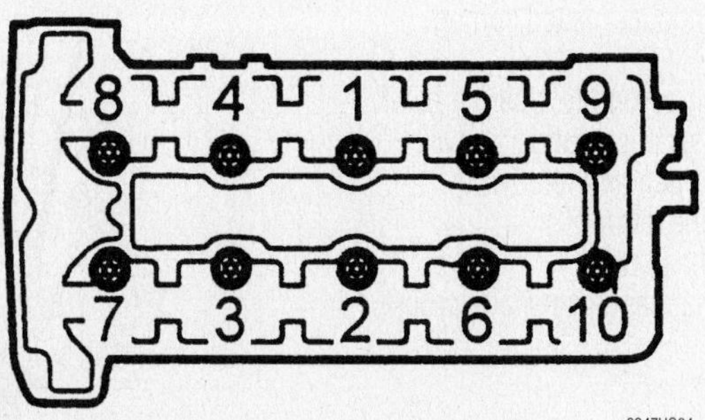

Tighten the rocker cover bolts in sequence—2.0L engines

a. Step 1: 29 ft. lbs. (39 Nm).
b. Step 2: 58 ft. lbs. (78 Nm).
c. Step 3: Loosen all bolts in sequence completely.
d. Step 4: 25–33 ft. lbs. (34–44 Nm).
e. Step 5: Plus 90–95 degrees clockwise in sequence.
f. Step 6: Plus additional 90–95 degrees.

➡Do not turn any bolt 180–200 degrees clockwise all at once.

7. Install or connect the following:
   - KS connector and EGR tube
   - Starter motor and the wiring
   - Water hoses to the engine block and heater
   - Camshafts
   - Camshaft sprockets timing chain guide, tensioner and timing chain
   - Distributor assembly
   - Front exhaust pipe to the exhaust manifold
   - Oil filter and power steering pump brackets
   - Cylinder head cover and oil separator
   - Power steering pump
   - Alternator
   - Drive belts
   - Air duct and intake manifold
   - Radiator
   - Engine side cover and right front wheel
   - Engine under covers
   - Negative battery cable
8. Fill the cooling system.
9. Start the vehicle, check for leaks and repair if necessary.

2.4L Engine

1. Before servicing the vehicle, refer to the precautions in the beginning of this section.
2. Drain cooling system.
3. Relieve the fuel system pressure.
4. Remove or disconnect the following:
   - Negative battery cable
   - Intake manifold collector, exhaust manifold and all related components
   - Distributor assembly. Using a block of wood, set a jack under the aluminum oil pan and remove the front engine mount.
   - Cylinder head cover
   - Timing chain and camshaft sprockets
   - Camshafts

➡The valvetrain components must be reassembled in their original positions.

5. Loosen the cylinder head bolts in reverse order of tightening.

➡A warped or cracked cylinder head could result from loosening in incorrect order. The cylinder head bolts should be loosened in 2 or 3 steps.

6. Remove the cylinder head and the intake manifold. Remove the cylinder head gasket. The lower timing chain will not be disengaged from crankshaft sprocket.

To install:

7. Clean the gasket surfaces.
8. Install or connect the following:
   - New cylinder head gasket
   - Cylinder head and temporarily tighten the cylinder head bolts. This is necessary to avoid damaging the cylinder head gasket. Be sure to install washers between the bolts and cylinder head.
   - Idler shaft assembly

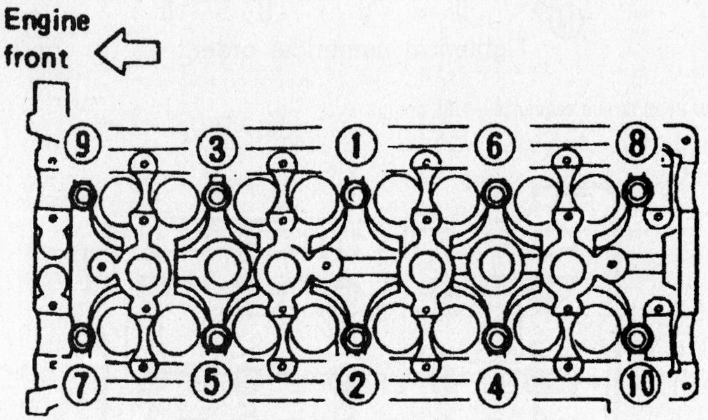

Cylinder head bolt torque sequence—2.4L engine

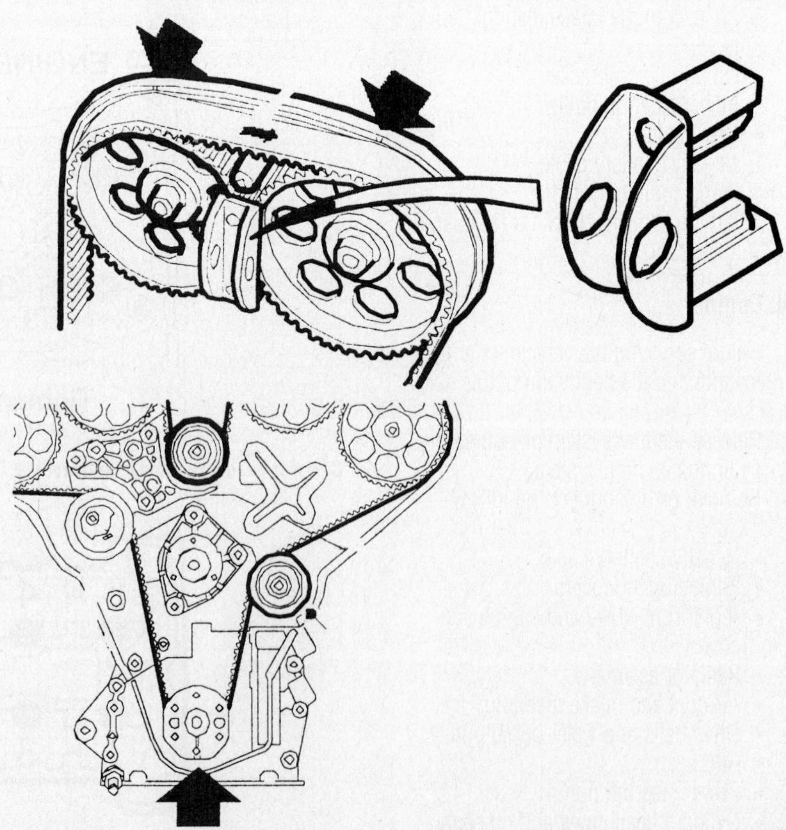

Tighten the rocker arm bolts in sequence—2.4L engines

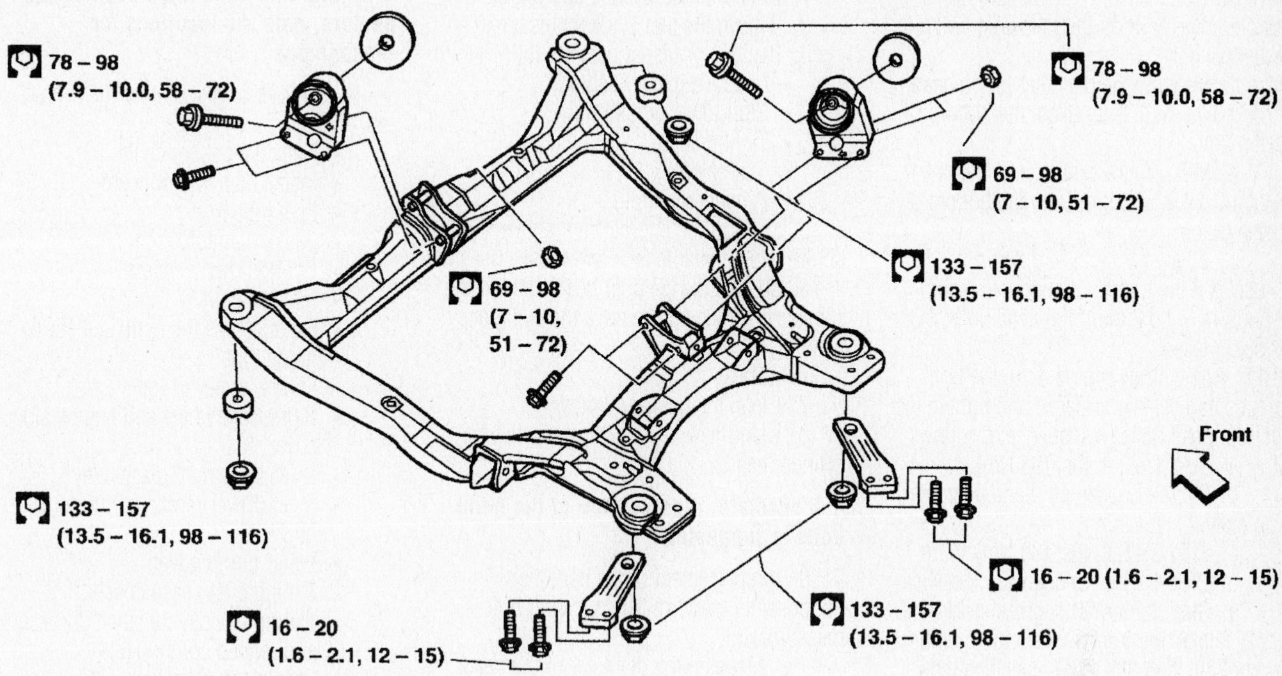

Removing the front suspension member—Altima

- Upper timing chain and cover

9. Tighten the cylinder head bolts in sequence as follows:
   a. Step 1: 22 ft. lbs. (29 Nm).
   b. Step 2: 59 ft. lbs. (79 Nm).
   c. Step 3: Loosen all the bolts completely.
   d. Step 4: 18–25 ft. lbs. (25–34 Nm).
   e. Step 5: Plus 86–91 degrees clockwise.
10. Install or connect the following:
   - Camshafts
   - Timing chains, chain tensioner and camshaft sprockets
   - Cylinder head cover
   - Distributor assembly
   - Intake manifold collector
   - Negative battery cable
11. Fill the cooling system.
12. Start the vehicle, check for leaks and repair if necessary.

2.5L Engine

ALTIMA

1. Before servicing the vehicle, refer to the precautions in the beginning of this section.
2. Drain cooling system and engine oil.
3. Relieve the fuel system pressure.
4. Remove or disconnect the following:
   - Negative battery cable
   - Upper radiator hose and heater hose
   - Right side fuse/relay box and move aside
   - Engine undercover
   - Resonator and air cleaner case
   - Engine top cover
   - All electrical and vacuum lines at the head
   - Fuse hose quick connector at fuel tube side
   - Engine harness and power steering hose bracket
   - Timing chain
   - Drive belt tensioner
   - Camshafts
   - Spark plugs
   - Exhaust manifold
5. Support the engine from above and below with a suitable jack and hoist.
6. Remove or disconnect the following:
   - Power steering pump and reservoir and wire aside
   - Auxiliary drive belts
   - A/C compressor and wire aside
   - Upper sway bar links
   - Front and rear engine mount through bolts
   - Lower ball joints
   - 2 steering gear housing mounting bolts
   - Front suspension member bolts and front suspension member
7. Loosen the cylinder head bolts in the sequence shown and remove the cylinder head.

To install:

8. Clean the gasket surfaces.
9. Using new head gaskets, install the cylinder heads.

➡If possible, replacement of the head bolts is suggested.

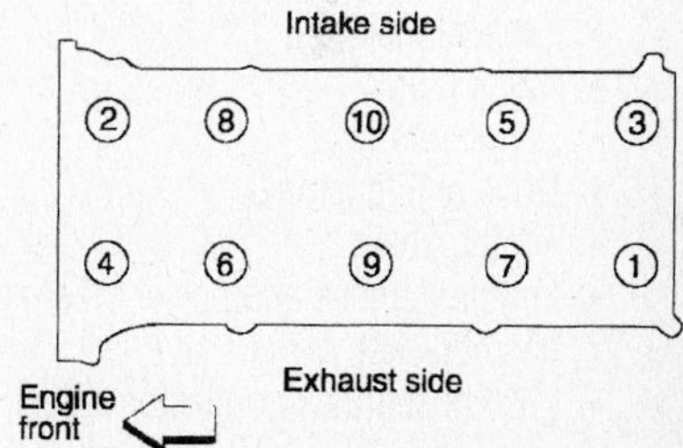

Cylinder head bolt removal sequence—2.5L engine–Altima and Sentra

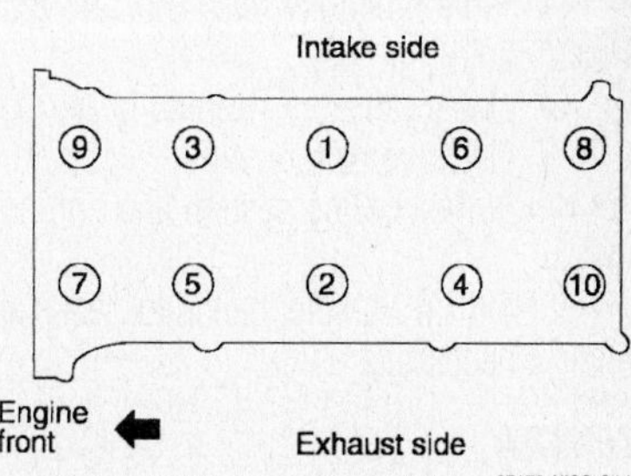

Cylinder head bolt tightening sequence—2.5L engine–Altima and Sentra

10. If replacement of the head bolts is not possible, perform the following bolt measurement:
   a. Measure the diameter of the head bolt 0.43 in. (11mm) from the bottom of the bolt.
   b. Measure the diameter of the head bolt 2.17 in. (55mm) from the bottom of the bolt.
   c. Whenever the size difference between the 2 measurements exceeds 0.0091 in. (0.23mm) the head bolts must be replaced.
11. Apply clean engine oil to the bolt threads and seating surfaces and tighten the cylinder head bolts in sequence as follows:
   a. Step 1: 72 ft. lbs. (98 Nm).
   b. Step 2: Loosen all the bolts completely.
   c. Step 2: 29 ft. lbs. (33 Nm).
   d. Step 4: Plus 90 degrees clockwise.
   e. Step 5: Plus 90 degrees clockwise.
12. Install the front suspension member and tighten the bolts to the specifications shown in the illustration.
13. Install or connect the following:
   - 2 steering gear housing mounting bolts
   - Lower ball joints
   - Front and rear engine mount through bolts
   - Upper sway bar links
   - A/C compressor
   - Auxiliary drive belts
   - Power steering pump and reservoir
   - Exhaust manifold
   - Spark plugs
   - Camshafts
   - Drive belt tensioner
   - Timing chain
   - Engine harness and power steering hose bracket
   - Fuse hose quick connector at fuel tube side
   - All electrical and vacuum lines at the head
   - Engine top cover
   - Resonator and air cleaner case
   - Engine undercover
   - Right side fuse/relay box and move aside
   - Upper radiator hose and heater hose
   - Negative battery cable
14. Fill the cooling system and engine oil
15. Start the vehicle, check for leaks and repair if necessary.

SENTRA

1. Before servicing the vehicle, refer to the precautions in the beginning of this section.
2. Drain cooling system and engine oil.
3. Relieve the fuel system pressure.
4. Remove or disconnect the following:
   - Negative battery cable
   - Strut tower cross brace
   - Timing chain
   - Camshafts
   - Exhaust manifold
5. Support the engine from above and below with a suitable jack and hoist.
6. Loosen the cylinder head bolts in the sequence shown and remove the cylinder head.

To install:

7. Clean the gasket surfaces.
8. Using new head gaskets, install the cylinder heads.

➡If possible, replacement of the head bolts is suggested.

9. If replacement of the head bolts is not possible, perform the following bolt measurement:
   a. Measure the diameter of the head bolt 0.47 in. (12mm) from the bottom of the bolt.
   b. Measure the diameter of the head bolt 2.17 in. (55mm) from the bottom of the bolt.
   c. Whenever the size difference between the 2 measurements exceeds 0.0091 in. (0.23mm) the head bolts must be replaced.
10. Apply clean engine oil to the bolt threads and seating surfaces and tighten the cylinder head bolts in sequence as follows:
   a. Step 1: 72 ft. lbs. (98 Nm).
   b. Step 2: Loosen all the bolts completely.
   c. Step 2: 29 ft. lbs. (33 Nm).
   d. Step 4: Plus 75 degrees clockwise.
   e. Step 5: Plus 75 degrees clockwise.
11. Install or connect the following:
   - Exhaust manifold
   - Camshafts
   - Timing chain
   - Strut tower cross brace
   - Negative battery cable
12. Fill the cooling system and engine oil
13. Start the vehicle, check for leaks and repair if necessary.

3.0L Engine

1. Before servicing the vehicle, refer to the precautions in the beginning of this section.
2. Relieve the fuel system pressure.
3. Drain the engine oil.
4. Drain the cooling system.

➡Before detaching any hoses or connectors, note the locations for reassembly.

5. Remove or disconnect the following:
   - Negative battery cable
   - Intake manifold collector
   - Fuel tube
   - Intake manifold
   - Cylinder head covers
   - Ignition coils
   - Exhaust Gas Recirculation (EGR) guide tube
   - Engine under cover
   - Right front wheel and engine side cover
   - Drive belts and idler pulley
   - Steel (lower) and aluminum (upper) oil pans
   - Water pump cover
   - Timing chain case cover
   - Timing chains, camshaft sprockets and related components
   - Crankshaft sprocket
6. Loosen the bolts that secure the rear timing chain case. The bolts must be loosened in the reverse order of installation sequence.
   - Rear timing case cover using seal cutter tool

➡Remove the O-rings from the front of the engine block.

   - Camshafts
   - Cylinder head bolts in the reverse order of the tightening sequence. The bolts should be loosened in 2–3 steps.

➡A warped or cracked cylinder head could result from removing the bolts in incorrect order.

   - Cylinder heads from the vehicle
   - Discard the head gaskets
7. Remove all traces of liquid gasket from the timing chain case and from the water pump covers.
8. Remove all traces of liquid gasket from the engine block.
9. Inspect the timing chain for excessive wear or damage and replace as necessary.

To install:

10. Turn the crankshaft until the No. 1 piston is set 240 degrees before Top Dead Center (TDC) on compression stroke.
11. Using new head gaskets, install the cylinder heads.

➡If possible, replacement of the head bolts is suggested.

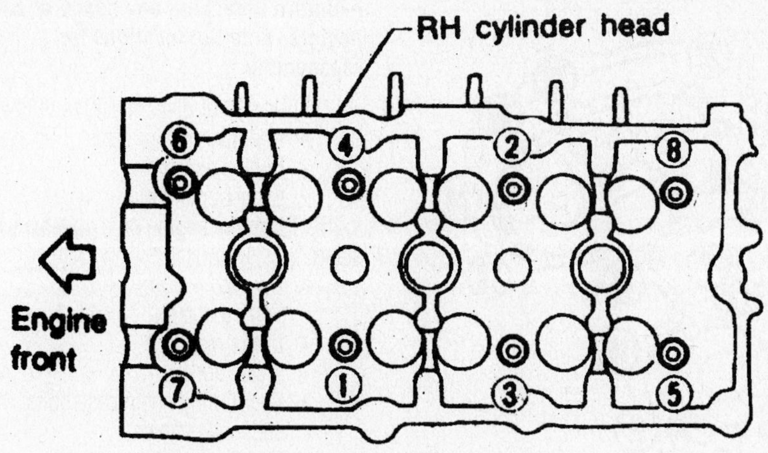

Right cylinder head bolt torque sequence—3.0L engine

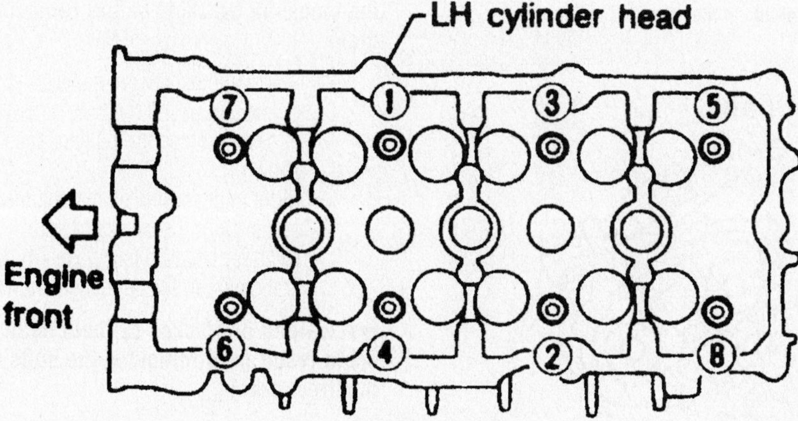

Left cylinder head bolt torque sequence—3.0L engine

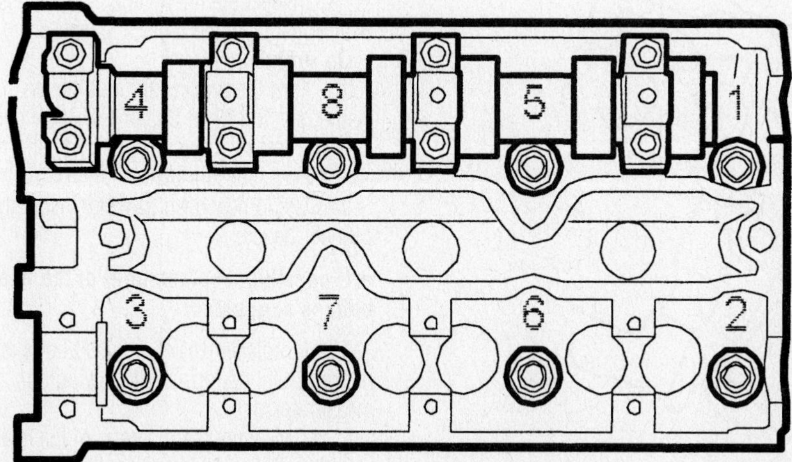

Tighten the rocker arm bolts in sequence—3.0L and 3.5L engines

12. If replacement of the head bolts is not possible, perform the following bolt measurement:
    a. Measure the diameter of the head bolt 0.43 in. (11mm) from the bottom of the bolt.
    b. Measure the diameter of the head bolt 1.89 in. (48mm) from the bottom of the bolt.
    c. Whenever the size difference between the 2 measurements exceeds 0.0043 in. (0.11mm) the head bolts must be replaced.
13. Install the cylinder head bolts and torque in sequence as follows:
    a. Step 1: 72 ft. lbs. (98 Nm).
    b. Step 2: Completely loosen all bolts.
    c. Step 3: 25–33 ft. lbs. (34–44 Nm).
    d. Step 4: plus 90–95 degrees clockwise.
    e. Step 5: plus 90–95 degrees clockwise.
14. Install or connect the following:
    - Camshafts and related components
    - New O-rings to the front of the engine block
15. Apply sealant to the hatched portion of the of the rear timing chain case.
16. Align the rear timing chain case with the dowel pins and install onto the cylinder heads and engine block.
17. Torque the rear timing chain case mounting bolts in sequence to 105–121 inch lbs. (11.8–13.7 Nm).
18. Install or connect the following:
    - Water pump cover
    - Upper and lower oil pans
    - Idler pulley and drive belts
    - Cylinder head covers
    - Intake manifold
    - Fuel tube
    - Intake manifold collector
    - Negative battery cable
19. Fill the cooling system.
20. Fill the engine with clean oil.
21. Start the vehicle, check for leaks and repair if necessary.

3.5L Engine

ALTIMA AND MAXIMA

➡You must remove the engine from the vehicle in order to remove the cylinder head, for this procedure.

1. Before servicing the vehicle, refer to the precautions in the beginning of this section.
2. Relieve the fuel system pressure.
3. Drain the engine oil.
4. Drain the cooling system.

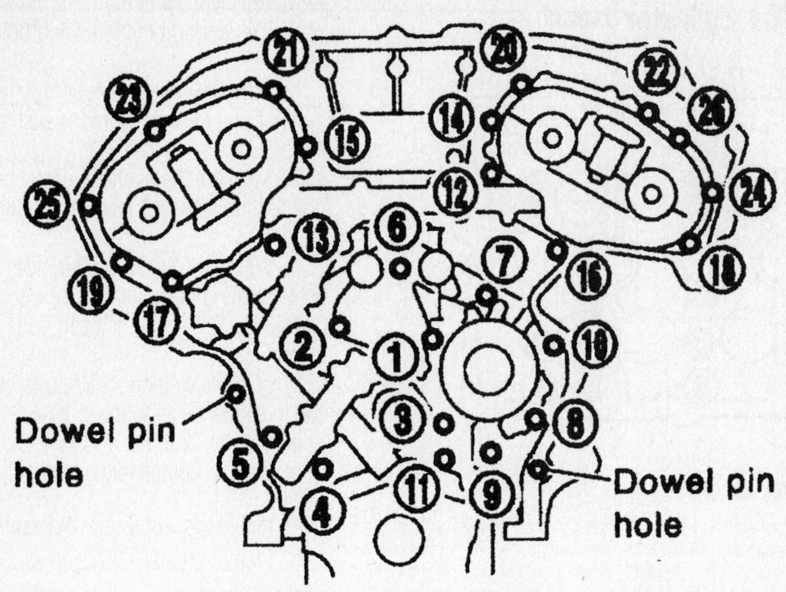

Loosen the rear timing chain case bolts in sequence—Altima & Maxima 3.5L engine

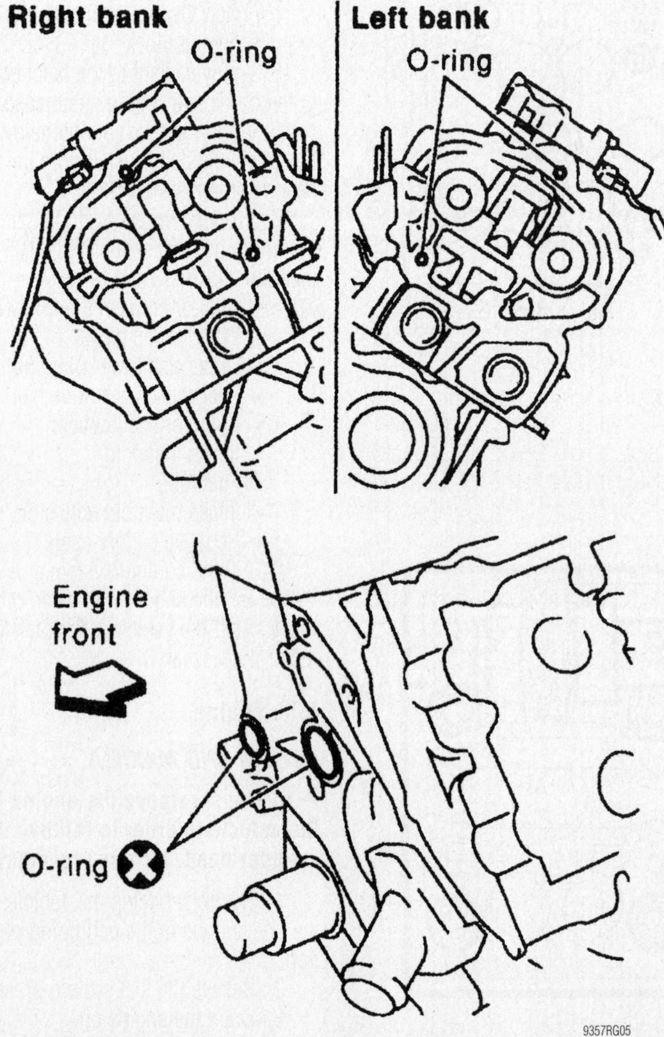

Remove the O-rings from the cylinder head and block— Altima & Maxima 3.5L engine

➡Before detaching any hoses or connectors, note the locations for reassembly.

5. Remove or disconnect the following:
   - Negative battery cable
   - Engine assembly
   - Exhaust manifold

6. Place the engine on a suitable workstand.
   - Oil pan
   - Timing chain
   - Intake manifold
   - Water outlet
   - Rear timing chain case bolts, in the sequence shown
   - Rear timing chain case
   - O-rings from the cylinder head and block
   - Intake valve timing control solenoid valves

➡For installation purposes, matchmark the camshaft brackets before removing them.

- Intake and exhaust camshafts and brackets. Loosen the bracket bolts in several steps, in the sequence shown.
- Right and left side cam chain tensioner from the cylinder head
- Cylinder head bolts. Loosen in several steps, in the sequence shown.

➡A warped or cracked cylinder head could result from removing the bolts in incorrect order.

- Cylinder heads from the vehicle
- Discard the head gaskets

7. Remove all traces of liquid gasket from the timing chain case and from the water pump covers.

8. Remove all traces of liquid gasket from the engine block.

9. Inspect the timing chain for excessive wear or damage and replace as necessary.

To install:

10. Turn the crankshaft until the No. 1 piston is a Top Dead Center (TDC) on compression stroke. The crankshaft key should face toward the right bank.

11. Using new head gaskets, install the cylinder heads.

➡If possible, replacement of the head bolts is suggested.

12. If replacement of the head bolts is not possible, perform the following bolt measurement:
    a. Measure the diameter of the head bolt 0.43 in. (11mm) from the bottom of the bolt.

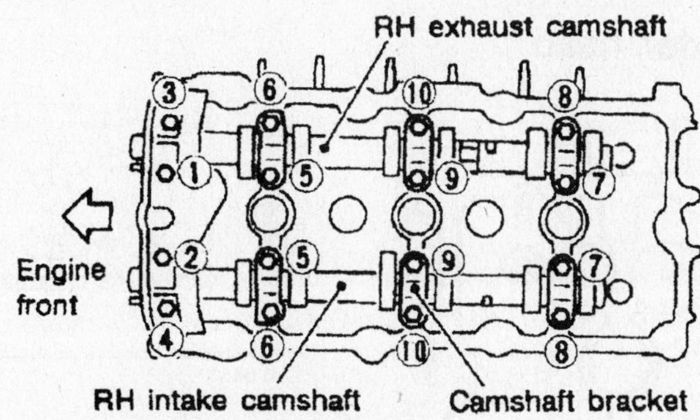

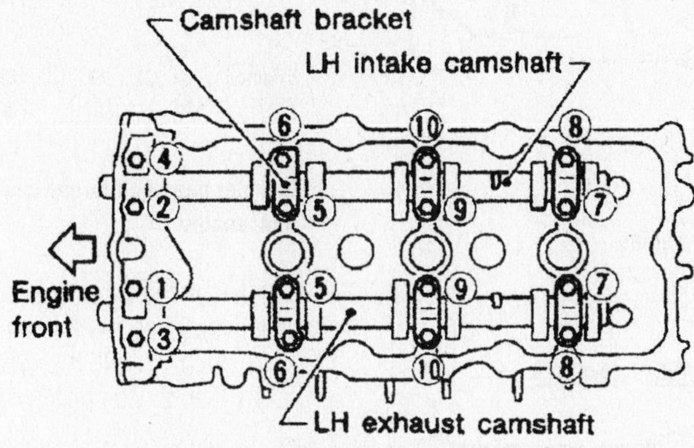

Right and left camshaft bracket bolt loosening sequence— Altima & Maxima 3.5L engine

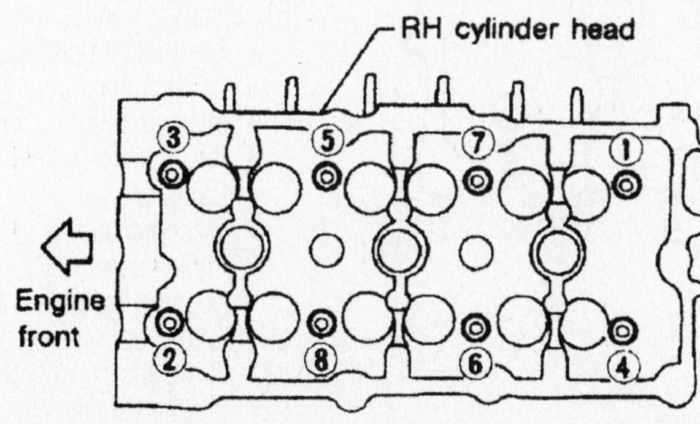

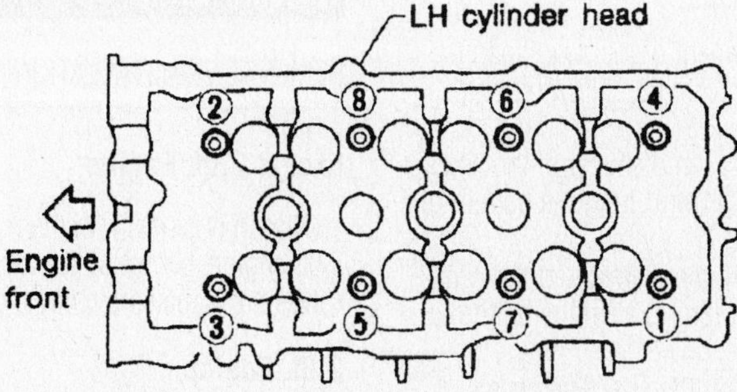

Cylinder head bolt loosening sequence— Altima & Maxima 3.5L engine

b. Measure the diameter of the head bolt 1.89 in. (48mm) from the bottom of the bolt.

c. Whenever the size difference between the 2 measurements exceeds 0.0043 in. (0.11mm) the head bolts must be replaced.

13. Install the cylinder head bolts and torque in sequence as follows:

a. Step 1: 72 ft. lbs. (98 Nm).

b. Step 2: Completely loosen all bolts.

c. Step 3: 26–32 ft. lbs. (34–44 Nm).

d. Step 4: plus 90–95 degrees clockwise.

e. Step 5: plus 90–95 degrees clockwise.

14. Install or connect the following:

- Camshafts and related components
- Intake valve timing control solenoid valves
- New O-rings to the front of the engine block and cylinder head

15. Apply sealant to the hatched portion of the of the rear timing chain case.

16. Align the rear timing chain case with the dowel pins and install onto the cylinder heads and engine block.

17. Torque the rear timing chain case mounting bolts in sequence to 105–121 inch lbs. (11.8–13.7 Nm).

18. Install or connect the following:

- Water outlet
- Intake manifold
- Timing chain
- Oil pan
- Exhaust manifold
- Engine assembly into the vehicle
- Negative battery cable

19. Fill the cooling system.

20. Fill the engine with clean oil.

21. Start the vehicle, check for leaks and repair if necessary.

350Z

➡You must remove the engine from the vehicle in order to remove the cylinder head, for this procedure.

1. Before servicing the vehicle, refer to the precautions in the beginning of this section.
2. Relieve the fuel system pressure.
3. Drain the engine oil.
4. Drain the cooling system.

➡Before detaching any hoses or connectors, note the locations for reassembly.

5. Remove or disconnect the following:

- Negative battery cable
- Engine assembly

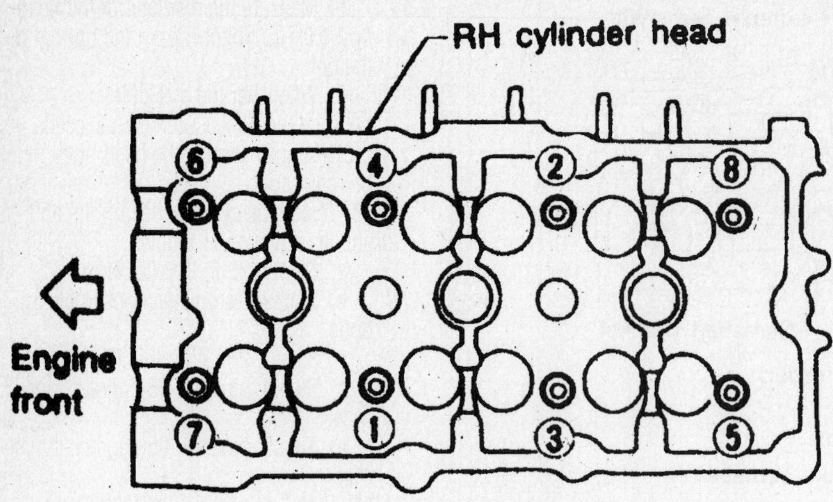

Right cylinder head bolt torque sequence— Altima & Maxima 3.5L engine

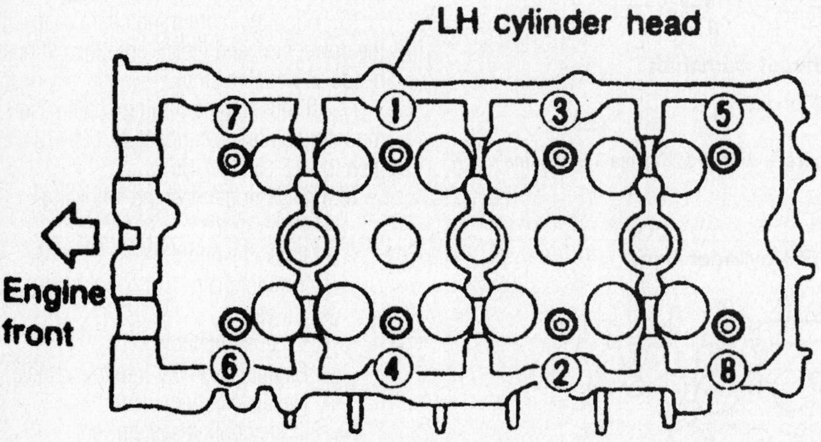

Left cylinder head bolt torque sequence— Altima & Maxima 3.5L engine

- Timing chain
- Camshafts
- Fuel injector assembly
- Intake manifold
- Exhaust manifold
- Thermostat housing
- Cylinder head bolts in the reverse of the tightening sequence
- Cylinder head gaskets

To install:

6. Turn the crankshaft until the No. 1 piston is a Top Dead Center (TDC) on compression stroke. The crankshaft key should face toward the right bank.
7. Using new head gaskets, install the cylinder heads.

➡If possible, replacement of the head bolts is suggested.

8. If replacement of the head bolts is not possible, perform the following bolt measurement:
   a. Measure the diameter of the head bolt 0.43 in. (11mm) from the bottom of the bolt.
   b. Measure the diameter of the head bolt 1.89 in. (48mm) from the bottom of the bolt.
   c. Whenever the size difference between the 2 measurements exceeds 0.0043 in. (0.11mm) the head bolts must be replaced.

Right bank

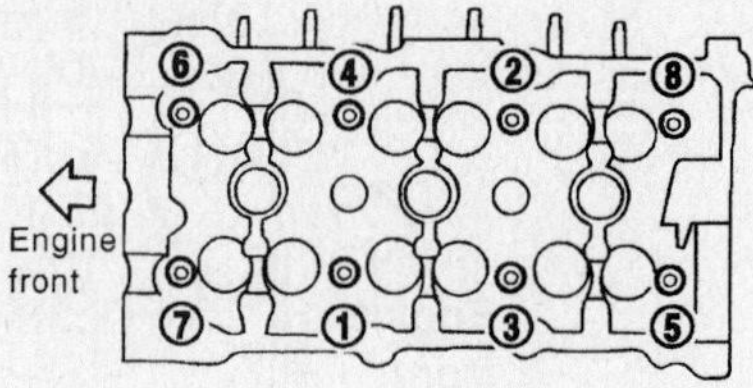

Left bank

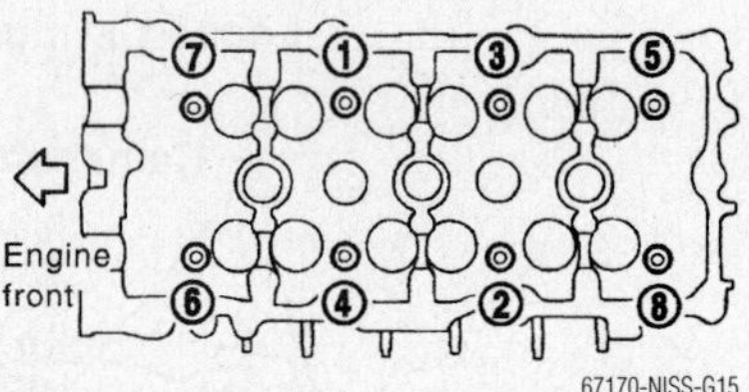

Cylinder head bolt torque sequence—350Z 3.5L engine

9. Install the cylinder head bolts and torque in sequence as follows:
   a. Step 1: 72 ft. lbs. (98 Nm).
   b. Step 2: Completely loosen all bolts.
   c. Step 3: 26–32 ft. lbs. (34–44 Nm).
   d. Step 4: plus 90–95 degrees clockwise.
   e. Step 5: plus 90–95 degrees clockwise.
10. Install or connect the following:
    - Thermostat housing
    - Exhaust manifold
    - Intake manifold
    - Fuel injector assembly
    - Camshafts
    - Timing chain
    - Engine assembly
    - Negative battery cable
11. Fill the cooling system.
12. Fill the engine with clean oil.
13. Start the vehicle, check for leaks and repair if necessary.

Rocker Arms

REMOVAL & INSTALLATION

Except 2.0L Engine

Nissan engines, with the exception of the 2.0L engine, do not utilize rocker arms. The valves are actuated directly by the camshafts.

2.0L Engine

1. Before servicing the vehicle, refer to the precautions in the beginning of this section.

2. Release the fuel pressure following the proper procedure.

3. Remove or disconnect the following:

- Negative battery cable
- Rocker arm cover, gasket and the oil separator
- Intake manifold supports, oil filter bracket and the power steering pump

4. Set the No. 1 cylinder at Top Dead Center (TDC) on the compression stroke.

- Timing chain tensioner from the side of the head

5. Matchmark the position of the rotor and housing and remove the distributor.

- Timing chain guide
- Camshaft sprockets while holding the camshaft stationary with a large wrench. Secure the timing chain with wire so the timing is not lost. The front cover will have to be removed if the chain timing is lost.

➡When removing the camshafts, loosen the journal caps in the opposite sequence of tightening. Camshaft damage may result if this step is not followed.

- Camshafts, brackets, oil tubes and the baffle plate. Label all components for proper installation.

➡It is essential that all parts be kept in the same order and orientation for reinstallation. Be sure to mark and separate parts to keep them from getting mixed. This will aid assembly.

- Rocker arms, shims, rocker arm guides and the hydraulic lash adjusters. Label all components for proper installation.

➡The valve lifters must be stored in the vertical position or submersed in clean oil to prevent air from entering the lifters.

6. Inspect the surfaces of the rockers and replace if there are any signs of damage.

To install:

7. Lubricate the rocker arms, shims, rocker arm guides and the hydraulic lash adjusters.

8. Install or connect the following:

- Rocker arms, shims, rocker arm guides and the hydraulic lash adjusters in their original locations
- Camshafts, brackets, oil tubes and the baffle plate in the proper location

9. Tighten the bolts in sequence as follows:

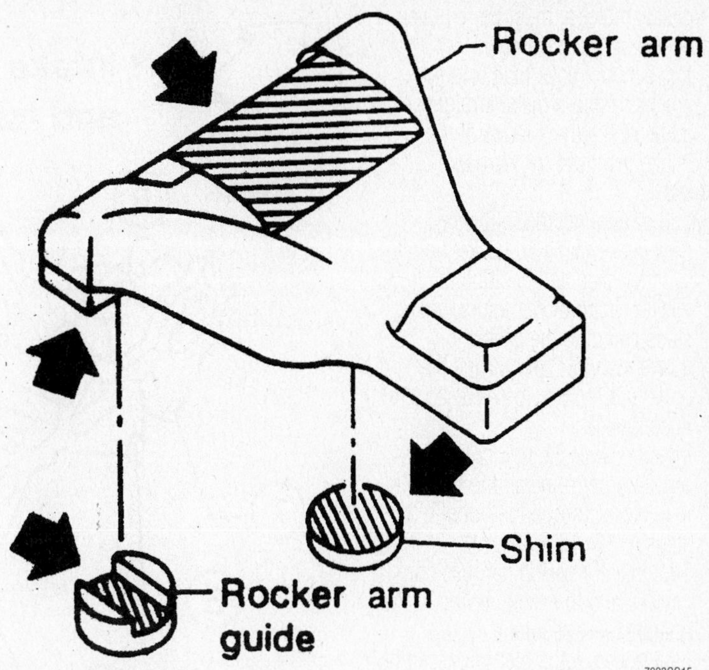

Rocker arm, guide and shim—2.0L engine

a. Right camshaft bolts Nos. 9 and No. 10: 17 inch lbs. (2 Nm).

b. Right camshaft bolts No. 1 through 8: 17 inch lbs. (2 Nm).

c. Left camshaft bolts Nos. 11 and 12: 17 inch lbs. (2 Nm).

d. Left camshaft bolts Nos. 1 through 10: 17 inch lbs. (2 Nm).

e. All camshaft bolts in numerical sequence: 52 inch lbs. (6 Nm).

f. All camshaft bolts in numerical sequence: 87–104 inch lbs. (10–11 Nm).

g. Rear 2 bolts of the left-hand camshaft: 13–19 ft. lbs. (18–25 Nm).

10. Install or connect the following:

- Camshaft sprockets while holding the camshaft stationary with a large wrench
- Remaining components in the reverse order from which they were removed
- Negative battery cable

11. Check and adjust the ignition and valve timing. If there is air in the lifters, bleed the air by running the engine at 1000 rpm for 10 minutes.

Intake Manifold

REMOVAL & INSTALLATION

1.8L Engine

1. Before servicing the vehicle, refer to the precautions in the beginning of this section.

2. Relieve the fuel system pressure.

3. Drain the cooling system.

4. Remove or disconnect the following:

- Negative battery cable
- Air cleaner assembly
- Throttle linkage, electrical connections and vacuum lines from the throttle body
- Intake manifold collector support brackets

5. The throttle body can be removed from the manifold at this point or can be removed as an assembly with the intake manifold.

- Bolts holding the upper portion of the intake to the lower portion. Remove the bolts in reverse order of the tightening sequence.
- Upper portion of the intake
- Fuel injector wiring harness connectors and the vacuum line from the fuel pressure regulator
- Fuel hoses from the fuel rail assembly
- Bolts that secure the fuel rail to the intake
- Injectors with the fuel rail assembly
- Intake manifold retaining bolts in the proper sequence and separate the manifold from the cylinder head. Remove the bolts in reverse order of the tightening sequence.
- Intake manifold gasket and clean all the gasket contact surfaces thoroughly with a gasket scraper and suitable solvent. All traces of old

gasket material must be removed to ensure proper sealing.
- Inspect the intake manifold for cracks. Using a metal straightedge, check the surface of the intake manifold for warpage.

To install:

6. Install or connect the following:
   - New intake manifold gasket onto the cylinder head and position the lower intake manifold over the mounting studs and onto the gasket.
   - Intake manifold and torque the bolts to 13–15 ft. lbs. (18–21 Nm) in sequence
   - Injectors with the fuel rail assembly. Be sure to install the fuel rail insulators. Torque the bolts in 2 steps to 13–15 ft. lbs. (18–21 Nm).
   - Fuel injector wiring harness connectors and the vacuum line to the fuel pressure regulator
   - Fuel hoses to the fuel rail assembly using new hose clamps
   - Intake manifold collector using a new gasket. Torque the bolts to 13–15 ft. lbs. (18–21 Nm) in sequence.
   - Throttle body or throttle chamber, if removed. Torque the bolts in a crisscross pattern to 13–16 ft. lbs. (18–22 Nm).

➡Be sure to properly position the throttle body gasket with the cut out facing down.

   - Intake manifold collector support brackets
   - Throttle linkage, electrical connections and vacuum lines
   - Air cleaner
   - Negative battery cable
7. Fill the cooling system to the proper level.
8. Start the engine, check for leaks and repair if necessary.

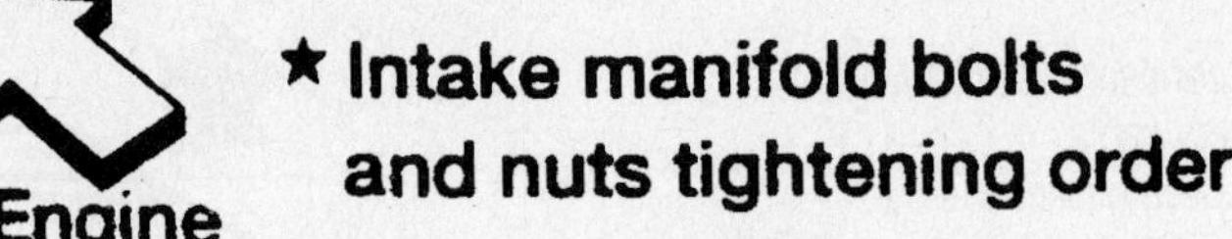
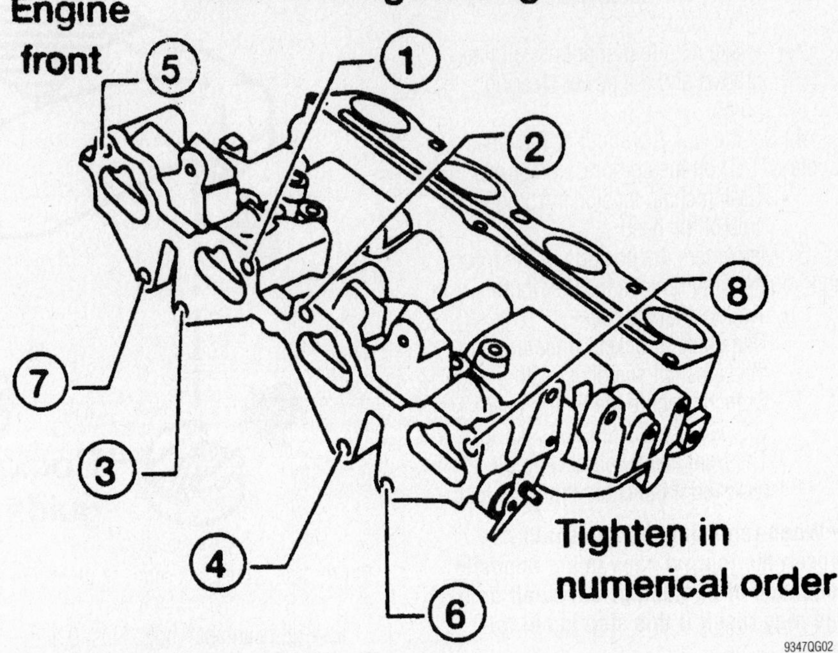

Lower intake manifold torque sequence—1.8L engine

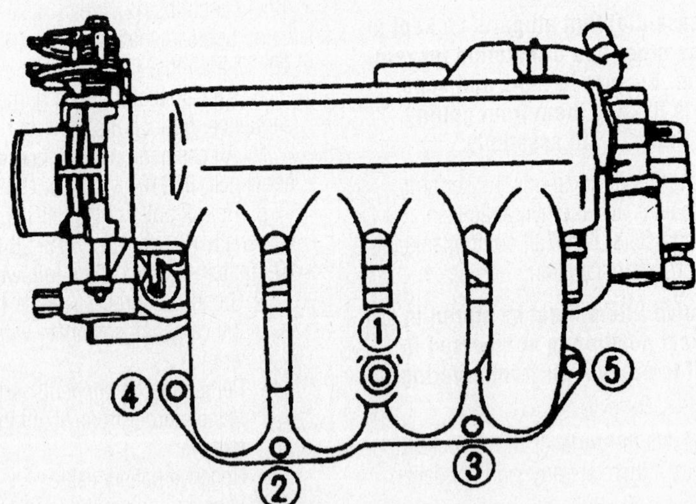

Upper intake manifold torque sequence—1.8L engine

2.0L Engine

1. Before servicing the vehicle, refer to the precautions in the beginning of this section.
2. Relieve the fuel system pressure.
3. Drain the cooling system.
4. Remove or disconnect the following:
   - Negative battery cable
   - Air cleaner assembly
   - Manifold support brackets
   - Throttle linkage, electrical connections and vacuum lines from the throttle body. Be sure to note the locations of all connections.
   - Exhaust Gas Recirculation (EGR) tube from the manifold
   - Fuel rail assembly
   - Drive belts and water pump pulley
   - Alternator and power steering pump
   - Oil filter bracket and power steering bracket
   - Intake manifold collector retaining bolts in the reverse of installation sequence and separate the collector from the manifold
   - Intake manifold assembly retaining bolts in the reverse order of the tightening sequence and separate the manifold from the cylinder head

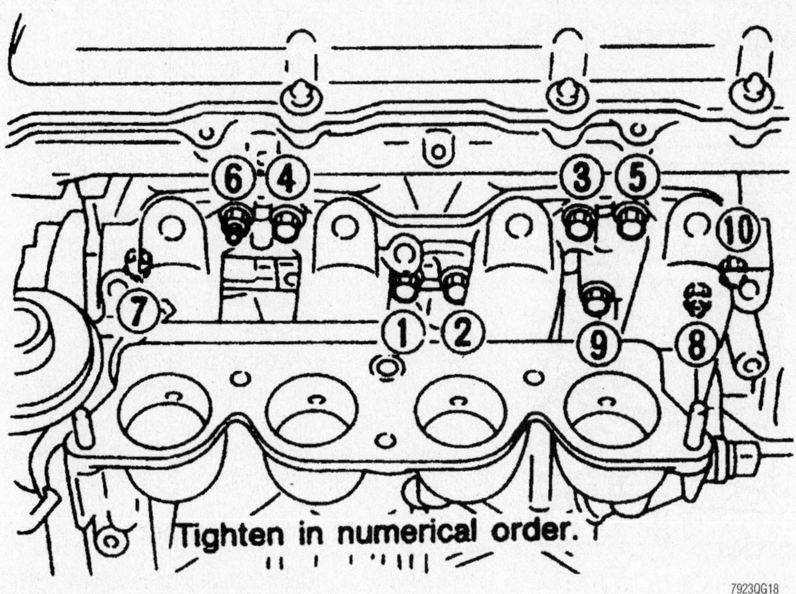

Lower intake manifold torque sequence—2.0L engine

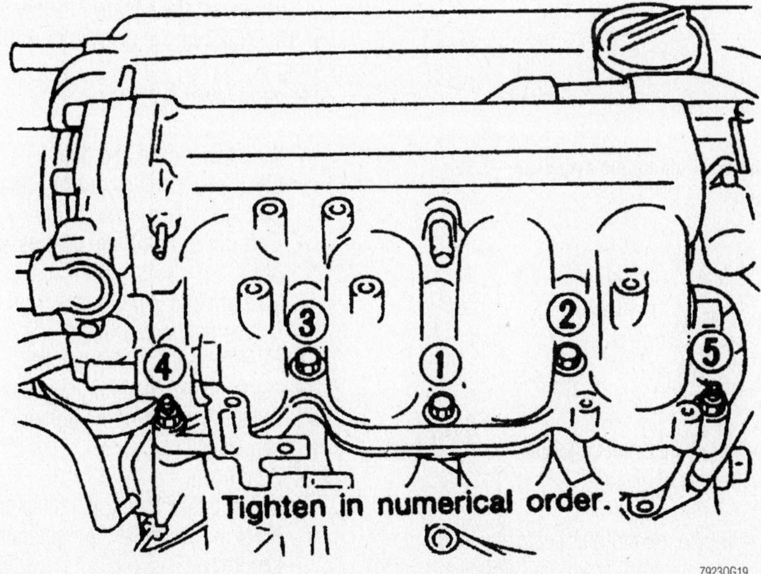

Upper intake manifold torque sequence—2.0L and 1.8L engines

- All gasket material and clean all the gasket contact surfaces thoroughly with a gasket scraper and a suitable solvent. All traces of old gasket material must be removed to ensure proper sealing.
- Inspect the intake manifold for cracks. Using a metal straightedge, check the surface of the intake manifold for warpage.

To install:

5. Install or connect the following:
   - Intake manifold with new gaskets and torque the bolts, in sequence, to 13–15 ft. lbs. (18–21 Nm)
   - Intake manifold collector with new gaskets and torque the bolts, in sequence, to 13–15 ft. lbs. (18–21 Nm)
   - Oil filter bracket
   - Power steering bracket
   - Alternator and power steering pump
   - Water pump pulley and drive belts
   - Fuel rail assembly with new insulators and torque the bolts to 35 inch lbs. (4 Nm)
   - EGR tube
   - Throttle linkage, vacuum lines and electrical connections to the throttle body
   - Manifold support brackets with new gaskets and torque the bolts to 20 ft. lbs. (26 Nm)
   - Air cleaner assembly
   - Negative battery cable
6. Fill the cooling system to the proper level.
7. Start the vehicle, check for leaks and repair if necessary.

2.4L Engine

1. Before servicing the vehicle, refer to the precautions in the beginning of this section.
2. Relieve the fuel system pressure.
3. Drain the cooling system.
4. Remove or disconnect the following:
   - Negative battery cable
   - Air duct between the air flow meter and the throttle body
   - Throttle cable and the cruise control cable, if equipped
   - Fuel supply and return lines from the fuel injector assembly. Plug the lines to prevent leakage.
   - Electrical connectors and the vacuum hoses to the throttle body and intake manifold/collector assembly
   - Spark plug wires from the spark plugs
   - Throttle body assembly from the intake manifold
   - Exhaust Gas Recirculation (EGR) valve tube from the exhaust manifold
   - Intake manifold mounting brackets
   - Intake manifold collector-to-intake manifold bolts/nuts in the reverse sequence of the tightening procedure and separate the intake manifold from the intake manifold collector
   - Bolts that secure the intake manifold to the cylinder head
   - Manifold. Be sure to loosen the bolts in the reverse sequence of the tightening procedure.
5. Using a putty knife, clean the gasket mounting surfaces. Check the intake manifold/collector for cracks and warpage.

To install:

6. Install or connect the following:
   - Intake manifold with new gaskets and torque the bolts, in sequence, to 12–14 ft. lbs. (16–19 Nm)
   - Intake manifold collector using new gaskets and torque the bolts in sequence to 12–14 ft. lbs. (16–19 Nm)
   - Intake manifold mounting brackets
   - EGR valve tube to the exhaust manifold

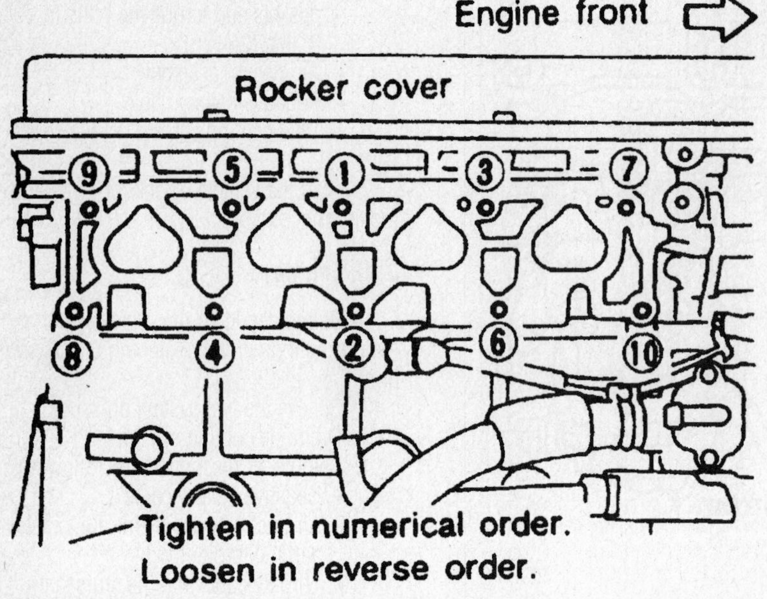

Intake manifold torque sequence—2.4L engine

- Throttle body using a new gasket and torque the bolts in a crisscross pattern to 13–16 ft. lbs. (18–22 Nm). Be sure to tighten the bolts in 2 progressive steps.
- Spark plug wires
- Vacuum hoses and electrical connectors to the throttle body
- EGR valve to the exhaust manifold
- Fuel return and supply lines
- Throttle and cruise control cables, if equipped
- Air duct
- Negative battery cable

7. Fill the cooling system to the proper level.
8. Start the vehicle, check for leaks and repair if necessary.

2.5L Engine

1. Before servicing the vehicle, refer to the precautions in the beginning of this section.
2. Relieve the fuel system pressure.
3. Drain the cooling system.
4. Remove or disconnect the following:
   - Negative battery cable
   - Mass Air Flow (MAF) sensor connector
   - Air cleaner case and duct
   - PCV hose
   - EVAP canister purge solenoid
   - Electronic throttle control actuator
   - Brake booster vacuum hose
   - Fuel line quick connector
   - Intake manifold collector harness and vacuum hose

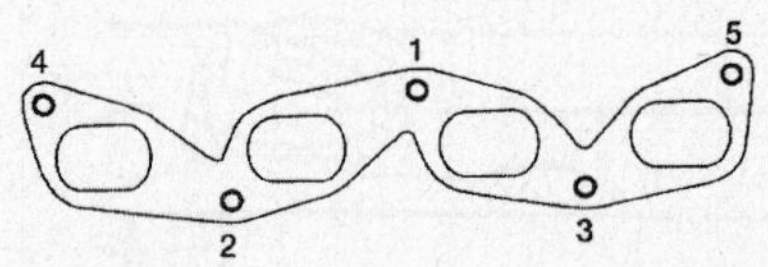

Intake manifold torque sequence—2.5L engine

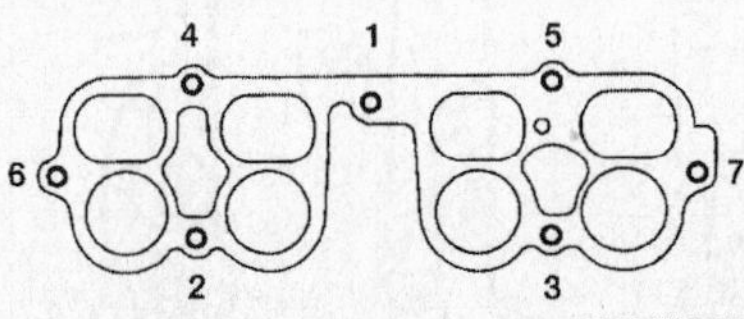

Intake manifold collector torque sequence—2.5L engine

   - Intake manifold collector-to-intake manifold bolts/nuts in the reverse sequence of the tightening procedure and separate the intake manifold from the intake manifold collector
   - Intake manifold in the reverse sequence of the tightening procedure.
5. Using a putty knife, clean the gasket mounting surfaces. Check the intake manifold/collector for cracks and warpage.

To install:

6. Install or connect the following:
   - Intake manifold with new gaskets and torque the bolts, in sequence, to 13–15 ft. lbs. (18–22 Nm)
   - Intake manifold collector using new gaskets and torque the bolts in sequence to 13–15 ft. lbs. (18–22 Nm)
   - Intake manifold collector harness and vacuum hose
   - Fuel line quick connector
   - Brake booster vacuum hose
   - Electronic throttle control actuator
   - EVAP canister purge solenoid
   - PCV hose
   - Air cleaner case and duct
   - Mass Air Flow (MAF) sensor connector
   - Negative battery cable
7. Fill the cooling system to the proper level.
8. Start the vehicle, check for leaks and repair if necessary.

3.0L and 3.5L Engines

1. Before servicing the vehicle, refer to the precautions in the beginning of this section.
2. Drain the cooling system.
3. Release the fuel system pressure.
4. Disconnect the negative battery cable.
5. On 350Z remove or disconnect the following:
   - Strut tower bar
   - Engine cover
   - Air cleaner case and duct
   - Electronic throttle control actuator bolt in the sequence shown
   - Fuel injector and fuel tube assembly
   - Vacuum and water hoses from intake manifold collector
   - EVAP solenoid valve bracket
   - Upper intake manifold collector bolts in reverse of the tightening sequence
   - PCV hose
   - Lower intake manifold collector bolts in reverse of the tightening sequence
6. On Altima and Maxima remove or disconnect the following:
   - Throttle body coolant hoses
   - Electrical connectors from the Throttle Position (TP) sensor
   - Hoses from the throttle body, the Exhaust Gas Recirculation (EGR) valve, intake manifold collector, Idle Air Control (IAC) valve and the fuel pressure regulator
   - Canister purge hose and blow-by hose
   - EGR guide tube
   - Accelerator cable from the throttle body
   - Intake manifold collector support brackets

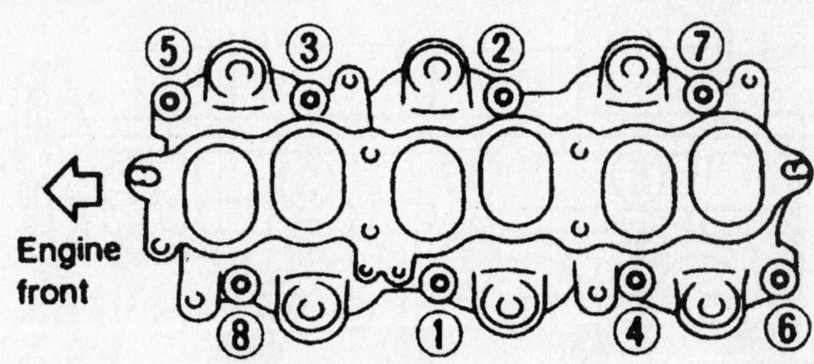

Intake manifold torque sequence—3.0L and 3.5L engines

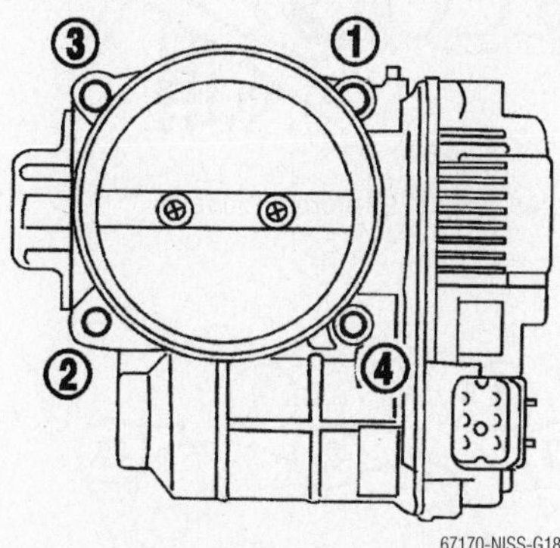

Electronic throttle actuator removal and installation sequence—350Z 3.5L engine

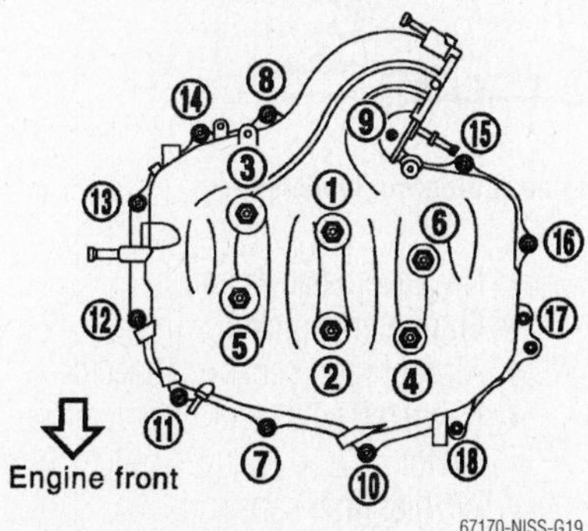

Upper intake manifold collector tightening sequence—350Z 3.5L engine

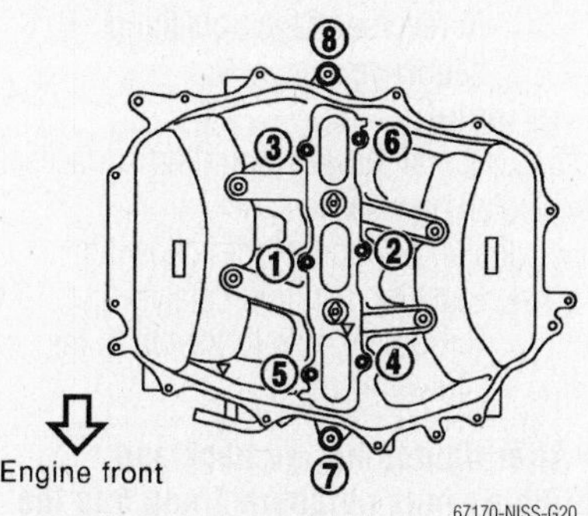

Lower intake manifold collector tightening sequence—350Z 3.5L engine

- Right side electrical connectors from the ignition coils
- Electrical connector from the crank angle sensor and the power transistor, if necessary
- Intake manifold collector-to-intake manifold bolts/nuts and the intake manifold collector

7. Remove the fuel injector assembly by performing the following procedures:
   a. Detach the electrical connectors from the fuel injectors.
   b. Disconnect the fuel lines from the fuel injector assembly.
   c. Remove the fuel rail-to-cylinder head bolts.
   d. Remove the fuel rail assembly from the engine.
8. On all models, remove or disconnect the following:
   - Intake manifold bolts/nuts in the reverse of the installation sequence
   - Intake manifold from the engine and discard the gaskets
9. Clean all gasket mounting surfaces.

To install:

10. Using new gaskets, install the intake manifold to the engine.
11. Torque the bolts in sequence as follows:
    a. Step 1: 44–89 inch lbs. (5–10 Nm).
    b. Step 2: 20–23 ft. lbs. (26–31 Nm).
12. On 350Z install or connect the following:
    - Lower intake manifold collector bolts in reverse of the tightening sequence and tighten to 10 ft. lbs. (14 Nm).
    - PCV hose
    - Upper intake manifold collector bolts in reverse of the tightening sequence and tighten to 10 ft. lbs. (14 Nm).
    - EVAP solenoid valve bracket
    - Vacuum and water hoses from intake manifold collector
    - Fuel injector and fuel tube assembly
    - Electronic throttle control actuator bolts in the sequence shown and tighten to 64–85 inch lbs. (7–10 Nm)
    - Air cleaner case and duct
    - Engine cover
    - Strut tower bar and tighten to 24 ft. lbs. (32 Nm).
13. On Altima and Maxima install or connect the following:
14. Install the fuel injector assembly by performing the following procedures:
    a. Install the fuel rail assembly to the engine.
    b. Install the fuel rail-to-cylinder head bolts and tighten the bolts to 15–20 ft. lbs. (21–26 Nm) in 2 progressive steps.
    c. Connect the fuel lines to the fuel injector assembly.
    d. Connect the electrical connectors to the fuel injectors.
15. Install the intake manifold collector. Torque the bolts to 8–11 ft. lbs. (11–15 Nm).
16. Install or connect the following:
    - Crank angle sensor and transmitter electrical connectors
    - Right side ignition coil electrical connectors
    - Intake manifold collector support brackets
    - Accelerator cable to the throttle body
    - EGR guide tube
    - Canister purge and blow by hoses
    - Throttle body, EGR valve and intake manifold collector hoses
    - IAC valve and fuel pressure regulator hoses
    - TP sensor electrical connector
    - Throttle body coolant hose
17. On all models, connect the negative battery cable.
18. Fill the cooling system.
19. Start the vehicle, check for leaks and repair if necessary.

Exhaust Manifold

REMOVAL & INSTALLATION

1.8L and 2.0L Engines

1. Before servicing the vehicle, refer to the precautions in the beginning of this section.
2. Remove or disconnect the following:
   - Negative battery cable
   - Engine undercovers

- Air cleaner or collector assembly
- Heat shields from the manifold and front exhaust pipe
- Front exhaust pipe from the exhaust manifold
- Temperature sensors, Oxygen (O_2) sensors and air induction pipes from the manifold
- Manifold support brackets
- Exhaust manifold attaching nuts and the manifold from the block. Discard the exhaust manifold gaskets.

3. Clean the gasket surfaces and check the manifold for cracks and warpage.

To install:

4. Install the exhaust manifold with a new gasket.
5. On 1.8L engines, tighten the mounting nuts with washers in sequence to 19–21 ft. lbs. (26–29 Nm).
6. On 2.0L engines, torque the fasteners from the center outward in several stages to 27–35 ft. lbs. (37–48 Nm).
7. Install or connect the following:
   - Temperature sensors, O_2sensors and air induction pipes
   - Manifold support brackets
   - Exhaust pipe to the manifold using a new gasket. Torque the nuts to 21–25 ft. lbs. (28–33 Nm) for 1.8L engines, or 32–37 ft. lbs. (43–50 Nm) for 2.0L engine models.
   - Heat shields
   - Air cleaner or collector assembly
   - Engine undercovers
   - Negative battery cable
8. Start the engine and check for exhaust leaks.

2.4L Engine

1. Before servicing the vehicle, refer to the precautions in the beginning of this section.
2. Remove or disconnect the following:
   - Negative battery cable
   - Exhaust pipe from the exhaust manifold
   - Oxygen (O_2) sensor electrical connector
   - Exhaust manifold cover
   - Exhaust Gas Recirculation (EGR) tube from the exhaust manifold
   - Exhaust manifold-to-engine bolts and nuts and discard the gaskets
   - Retaining bolts and nuts in reverse of the tightening sequence
   - Exhaust manifold from the vehicle

To install:

3. Clean all gasket mounting surfaces and install new gaskets.

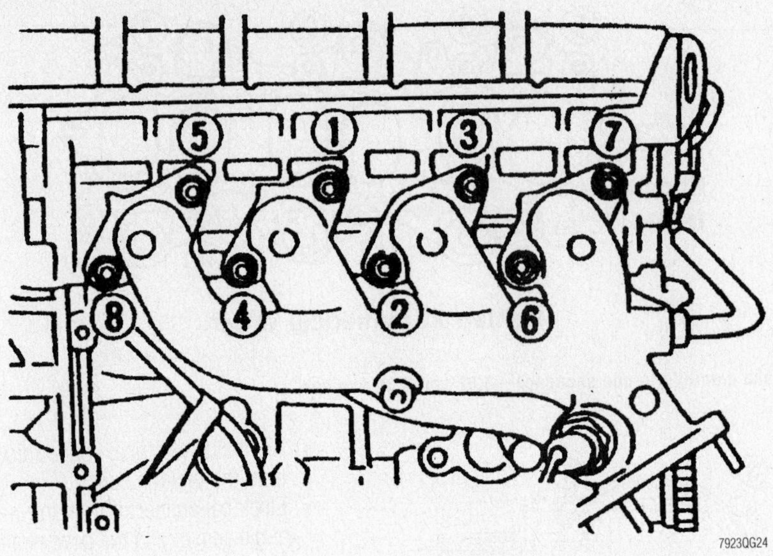

Exhaust manifold bolt tightening sequence—2.4L engine (except California models)

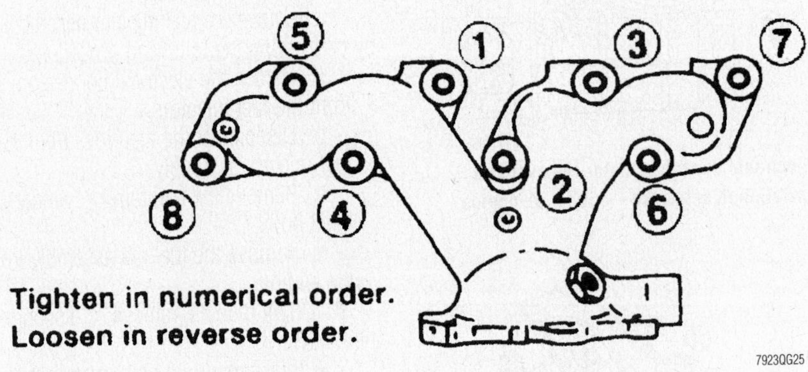

Exhaust manifold bolt tightening sequence—2.4L engine (California models)

4. Install or connect the following:
   - Exhaust manifold to the engine and torque the nuts to 27–35 ft. lbs. (37–48 Nm)
   - EGR tube to the exhaust manifold and torque the nuts to 29–36 ft. lbs. (39–49 Nm)
   - Exhaust manifold cover and torque the bolts to 45–57 inch lbs. (5–7 Nm)
   - O_2 sensor electrical connector
   - Exhaust pipe to the exhaust manifold and torque the bolts to 33–44 ft. lbs. (45–60 Nm)
   - Negative battery cable
5. Start the engine and check for exhaust leaks.

2.5L Engine

1. Before servicing the vehicle, refer to the precautions in the beginning of this section.
2. Remove or disconnect the following:
   - Negative battery cable
   - Engine undercover
   - Air/fuel ratio sensor connector
   - Oxygen (O_2) sensor electrical connector
   - Air/fuel ratio sensor
   - Oxygen (O_2) sensor
   - Upper and lower exhaust manifold covers
   - Exhaust manifold/catalyst assembly in reverse of the tightening sequence

To install:

3. Clean all gasket mounting surfaces and install new gaskets.
4. Install or connect the following:
   - Exhaust manifold/catalyst and torque the nuts to 29–32 ft. lbs. (40–44 Nm)

➡After tightening, go back and retighten nuts numbers 1 and 3 to the specification.

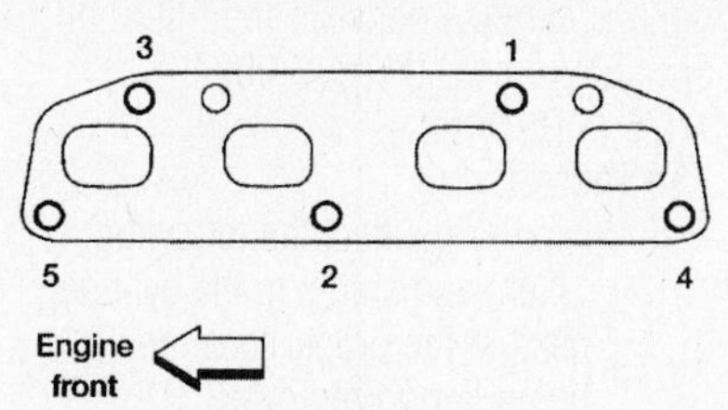

Exhaust manifold bolt tightening sequence—2.5L engine

- Air/fuel ratio sensor after coating the threads with anti-sieze
- Oxygen (O_2) sensor after coating the threads with anti-sieze
- Oxygen (O_2) sensor electrical connector
- Air/fuel ratio sensor connector
- Engine undercover
- Negative battery cable

5. Start the engine and check for exhaust leaks.

3.0L and Altima and Maxima 3.5L Engines

1. Before servicing the vehicle, refer to the precautions in the beginning of this section.
2. Remove or disconnect the following:
   - Negative battery cable
   - Exhaust manifolds from the exhaust pipes
   - Protective covers from the manifolds
   - Heated Oxygen Sensor (HO_2S) from the manifold, if equipped
   - Exhaust manifold-to-engine mounting nuts
   - Manifolds from the engine and discard the gaskets

To install:

3. Clean all gasket mounting surfaces. Install new gaskets.
4. Install or connect the following:
   - Exhaust manifold and torque the nuts in steps to 22–24 ft. lbs. (30–32 Nm)
   - Protective shields and torque the bolts in steps to 46–57 inch lbs. (5–7 Nm)
   - Exhaust manifolds to the exhaust pipes and torque the nuts to 32–37 ft. lbs. (43–50 Nm)
   - HO_2S sensor to the manifold and torque the fastener to 30–44 ft. lbs. (40–60 Nm), if equipped
   - Negative battery cable
5. Start the engine and check for exhaust leaks.

350Z 3.5L Engine

1. Before servicing the vehicle, refer to the precautions in the beginning of this section.
2. Drain the engine coolant
3. Remove or disconnect the following:
   - Negative battery cable
   - Strut tower bar
   - Engine cover
   - Air cleaner case and duct
   - Heated oxygen sensor No. 2 connectors and sensors
   - Exhaust mounting bracket between transmission and catalytic converters
   - Catalytic converters
   - Heated oxygen sensor No. 1 connectors and sensors
   - Water and heater pipe on both sides Heat shield
   - Exhaust manifold bolts in reverse of the tightening sequence
   - Manifold gaskets

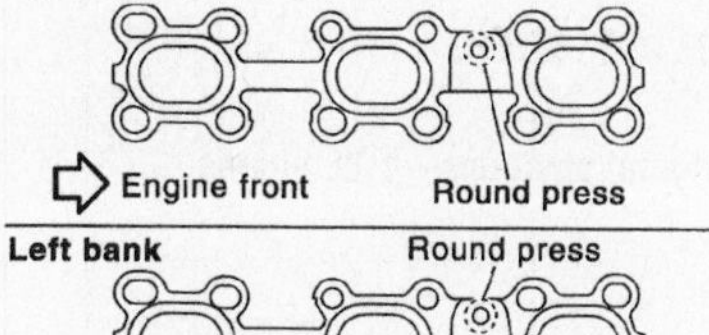

Exhaust manifold gasket identification—350Z 3.5L engine

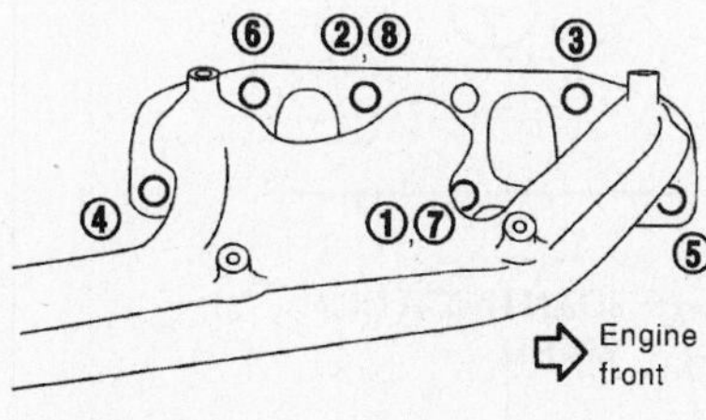

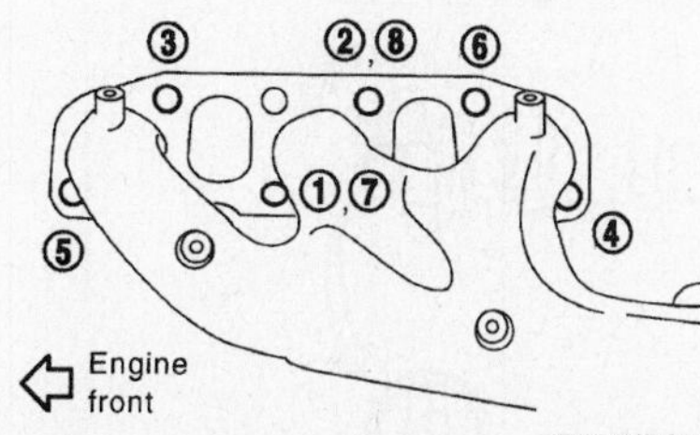

Exhaust manifold bolt tightening sequence—350Z 3.5L engine

To install:

4. Clean all gasket mounting surfaces. Install new gaskets noting the correct placement.
5. Installation is the reverse of the removal procedure noting the following:
   - Exhaust manifold and torque the nuts in sequence to 22 ft. lbs. (30 Nm)
   - Heat shields and torque the bolts in steps to 51 inch lbs. (6 Nm)
   - Exhaust manifolds to the exhaust pipes and torque the nuts to 46 ft. lbs. (63 Nm)
   - Oxygen sensors and torque the fastener to 33 ft. lbs. (45 Nm)

Front Crankshaft Seal

REMOVAL & INSTALLATION

➡The front crankshaft seal procedure is applicable to timing belt-equipped engines only. For the front seal on engines equipped with timing chains, refer to the timing chain, sprockets, front cover and seal procedure later in this section.

Camshaft and Valve Lifters

REMOVAL & INSTALLATION

1.8L Engines

1. Before servicing the vehicle, refer to the precautions in the beginning of this section.
2. Drain the cooling system.
3. Relieve the fuel system pressure.
4. Remove or disconnect the following:
   - Negative battery cable
   - All engine drive belts
   - Exhaust pipe from the exhaust manifold
   - Power steering pulley and pump with the mounting bracket
   - Cylinder head cover
   - Distributor assembly
   - Timing chain tensioners and camshaft sprocket

➡Before the camshafts are removed from the cylinder head, note the positioning of the pins at the end of the camshafts for reassembly purposes.

   - Camshaft bearing caps in sequence
   - Camshafts from the cylinder head
   - Idler sprocket bolt. These parts should be reassembled in their original position.

- Shims from the tops of the lifters. Be sure to note the position of each shim.
- Valve lifters from the bores in the cylinder head. Note the positioning of the lifters for reassembly.

5. Measure the diameter of the lifters. The diameter should be 1.1795–1.1801 in. (29.960–29.975mm).
6. Measure the diameter of the lifter bores. The diameter should be 1.1811–1.1819 in. (30.000–30.021mm).

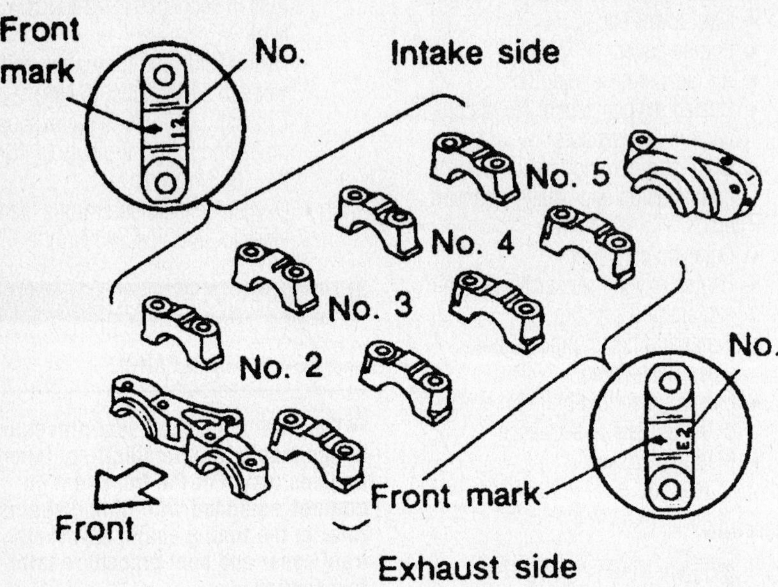

Be sure to install the camshaft bearing caps in their original positions—1.8L engine

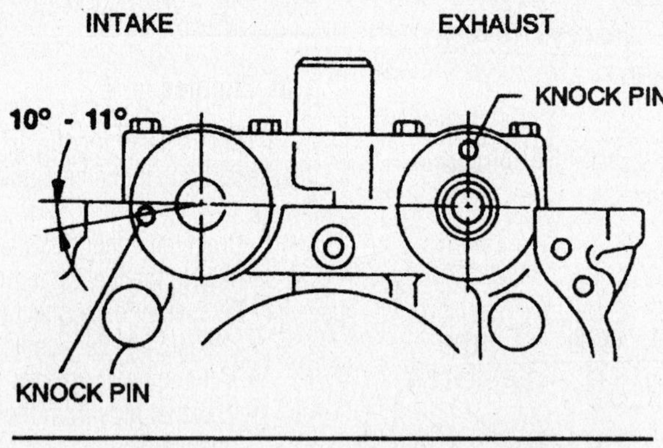

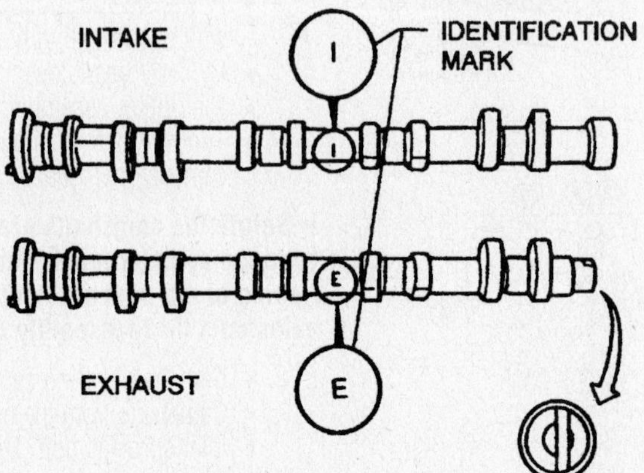

Positioning and identification of the camshafts—1.8L engine

7. Clearance between the lifter and bore should be 0.0010–0.0024 in. (0.025–0.061mm).

To install:

8. Install or connect the following:
   - Lifters and shims to the cylinder head in the proper locations as noted during removal

➡The exhaust and intake camshafts are marked with identification stamps. (E for exhaust and I for intake).

   - Camshafts to the cylinder head and position the intake camshaft knock pin at the 9 o'clock position and the exhaust camshaft at the 12 o'clock position
9. Install the camshaft bearing caps and tighten the mounting bolts as follows:
   a. Bolts 11 through 15, then bolts 1 through 10: 18 inch lbs. (2 Nm).
   b. Bolts 1 through 15: 53 inch lbs. (6 Nm).
   c. Bolts 1 through 14: 87–105 inch lbs. (10–12 Nm).
   d. Bolt 15: 56–73 inch lbs. (7–8 Nm).

➡If any part of the valvetrain has been has been replaced, the valve adjustment must be checked. DO NOT adjust the valves or rotate the camshafts at this point. Internal engine damage will result.

10. Install or connect the following:
    - Camshaft sprockets with timing chains
    - Distributor assembly
11. Check and adjust the valve clearance.
    - Cylinder head cover
    - Power steering pulley pump
    - Exhaust pipe from the exhaust manifold
    - All engine drive belts
    - Negative battery cable
12. Fill the cooling system.
13. Start the vehicle, check for leaks and repair if necessary.

2.0L Engine

1. Before servicing the vehicle, refer to the precautions in the beginning of this section.
2. Properly relieve the fuel system pressure.
3. Remove or disconnect the following:
   - Negative battery cable
   - Rocker cover and oil separator
4. Rotate the crankshaft until the No.1 piston is at Top Dead Center (TDC) on the compression stroke. Then, rotate the crank-

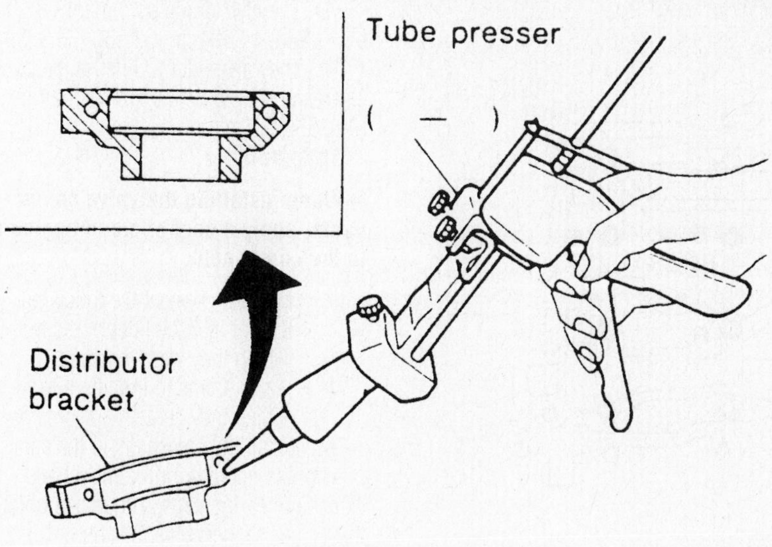

If removed, apply liquid gasket to the distributor bracket as shown—1.8L engine

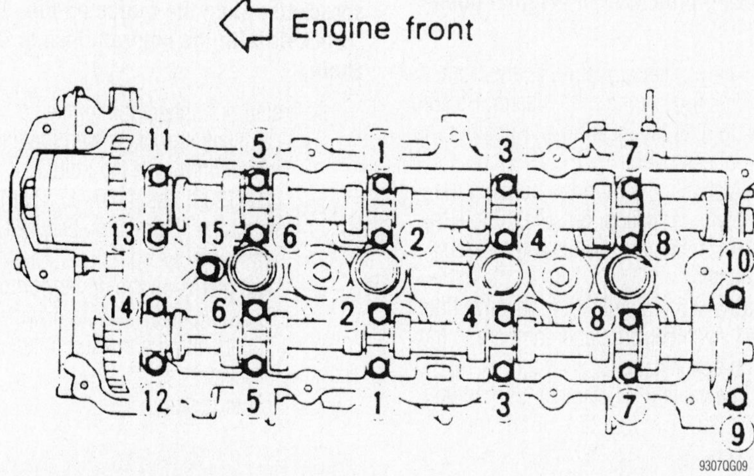

Camshaft bolt torque sequence—1.8L engine

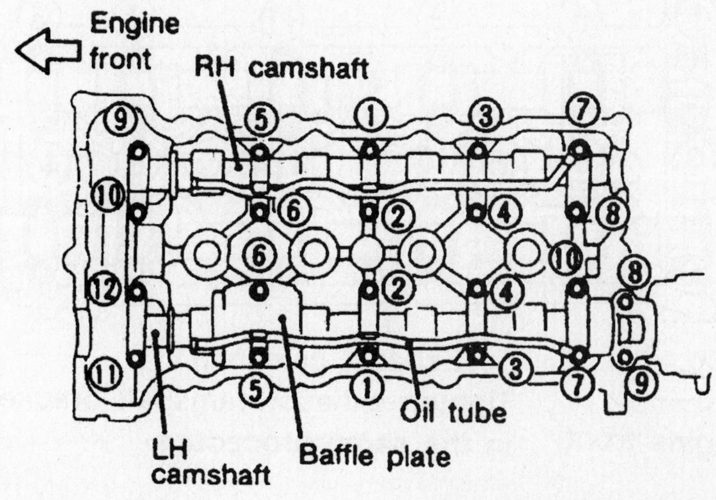

To prevent damage to the camshafts, torque the bearing caps in the numbered . . .

shaft until the mating marks on the camshaft sprockets line up with the mating marks on the timing chain.

- Timing chain tensioner
- Distributor
- Timing chain guide
- Camshaft sprockets. Use a wrench to hold the camshaft while loosening the sprocket bolt.
- Camshaft bracket bolts in the opposite order of the tightening sequence
- Camshaft

To install:

5. Clean the left-hand camshaft end bracket and coat the mating surface with liquid gasket. Install the camshafts, camshaft brackets, oil tubes and baffle plate. Ensure the left camshaft key is at 12 o'clock and the right camshaft key is at 10 o'clock.
6. The procedure for tightening camshaft bolts must be followed exactly to prevent camshaft damage. Torque the bolts as follows:
   a. Right bolts 9 and 10 (in that order): 18 inch lbs. (2 Nm).
   b. Right bolts 1–8 (in that order): 18 inch lbs. (2 Nm).
   c. Left bolts 11 and 12 (in that order): 18 inch lbs. (2 Nm).
   d. Left bolts 1–10 (in that order): 18 inch lbs. (2 Nm).
   e. All bolts in sequence: 52 inch lbs. (6 Nm).
   f. All bolts in sequence: 7–9 ft. lbs. (9–12 Nm) for type A, B and C bolts and 13–19 ft. lbs. (18–25 Nm) for type D bolts.
7. Line up the mating marks on the timing chain and camshaft sprockets and install the sprockets. Torque the sprocket bolts to 101–116 ft. lbs. (137–157 Nm).
8. Install or connect the following:
   - Timing chain guide
   - Distributor
   - Timing chain tensioner
   - Rocker cover and oil separator
   - Negative battery cable
9. Start the vehicle, check for leaks and repair if necessary.

2.4L Engine

1. Before servicing the vehicle, refer to the precautions in the beginning of this section.
2. Relieve the fuel system pressure.
3. Drain coolant from the engine and radiator.
4. Remove or disconnect the following:
   - Negative battery cable
   - Air intake ducts and the air cleaner assembly

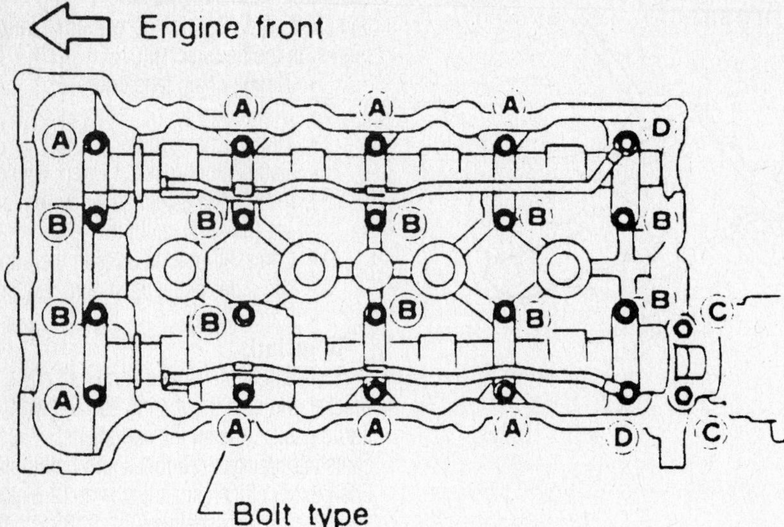

. . . and lettered sequence shown. Refer to the text for the proper torque—2.0L engines

- Vacuum hoses, fuel hoses, wires, harness and connectors that are necessary for removal of the rocker cover
- Alternator and mounting bracket
- Upper radiator hose and cooling fan

5. Set the No. 1 piston at Top Dead Center (TDC) on its compression stroke.

- Spark plug wires from the spark plugs
- Distributor assembly. Matchmark and note the positioning of the distributor rotor and housing to the engine block before removing the distributor.
- Rocker cover by loosening the bolts in the reverse order of installation

6. Wire the chain to the sprocket so the chain does not fall off during sprocket removal. Hold the flats of the camshaft with a wrench just behind the first camshaft bearing cap. Loosen the bolts and remove the sprockets.

- Camshaft sprockets

➡The stoppers on camshaft covers prevent the upper timing chain from disengaging from the idle sprocket. Also, after removal of the camshaft sprockets, note the positioning of the pins at the end of the camshafts for reinstallation purposes.

- Camshaft bearing caps in reverse order of installation
- Camshafts. The camshaft brackets must be loosened in the correct sequence to prevent damage to the camshaft.

➡All the valvetrain components must be reassembled in their original positions.

- Valve lifter adjusting shims from the tops of the of the lifters. Be sure to note the location and positioning of each shim.
- Valve lifters from the bores in the cylinder heads. Be sure to note the location and positioning of each lifter.

7. Check the diameter of the valve lifter and the valve lifter bore and compare to the following specifications.

8. The valve lifter diameter should be 1.3370–1.3376 in. (33.960–33.975mm).

9. The lifter guide bore diameter should be 1.3386–1.3394 in. (33.960–33.975mm).

10. The valve lifter to lifter guide bore clearance should be 0.0010–0.0024 in. (0.025–0.061mm).

To install:

➡When installing the valve components, apply a coat of clean engine oil to the component.

11. Install or connect the following:

- Lifters into the lifter bores from which they were removed
- Valve shims to the lifters from which they came

12. Install the camshafts in the same position as noted during removal and camshaft bearing caps; torque cap bolts in the proper sequence as follows:

a. Step 1: 17 inch lbs. (2 Nm).

b. Step 2: 81–104 inch lbs. (9–12 Nm).

➡When installing the timing chain and sprockets, align the marks on the sprockets with the colored links of the chain.

13. Install or connect the following:

- Camshaft sprockets with the timing chain and torque the bolts to 123–130 ft. lbs. (167–177 Nm). Install the chain guide between both camshaft sprockets. The alignment marks on the upper portion of the timing chain should now be aligned with the marks on the sprockets.
- Rocker cover

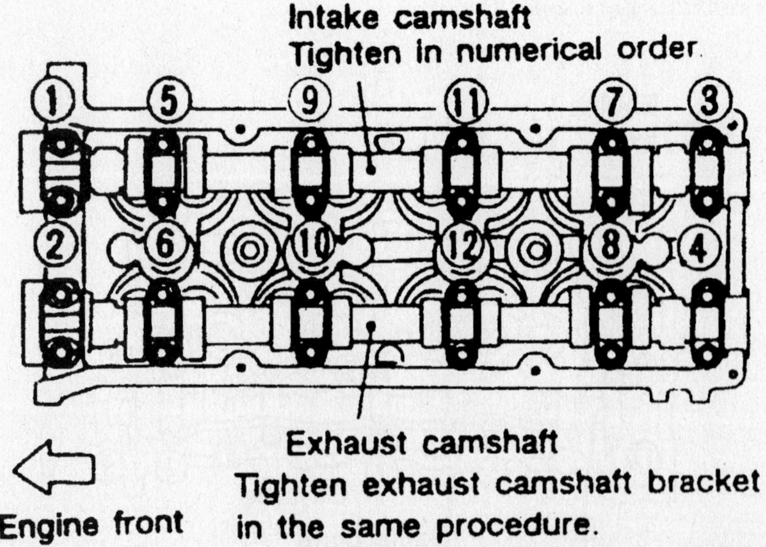

Tighten the camshaft bearing caps in sequence to prevent damage to the camshaft and cylinder head—2.4L engines

- Distributor
- Spark plugs and wires
- Upper radiator hose and cooling fan
- Alternator
- Vacuum hoses, fuel lines and electrical connectors
- Air cleaner assembly
- Negative battery cable

14. Fill the cooling system.
15. Start the vehicle, check for leaks and repair if necessary.

2.5L Engine

1. Before servicing the vehicle, refer to the precautions in the beginning of this section.
2. Relieve the fuel system pressure.
3. Drain coolant from the engine and radiator.
4. Remove or disconnect the following:
   - Negative battery cable
   - Engine undercover
   - Right front wheel
   - Valve cover
   - Drive belt
   - Coolant reservoir
   - Variable timing control solenoid connector
   - Camshaft position sensor
   - Intake valve timing control cover in reverse of the tightening sequence
   - Set the No. 1 piston at Top Dead Center (TDC) on its compression stroke
5. Check that the mating marks on the camshaft sprockets are lined up with the Yellow links on the timing chain. If not, rotate the crankshaft one revolution until the links line up.
6. Remove the timing chain guide out between the camshaft sprockets through the front cover.
7. Push in the plunger on the timing chain tensioner, then insert a pin to hold the tensioner retracted.
8. Remove the timing chain tensioner.

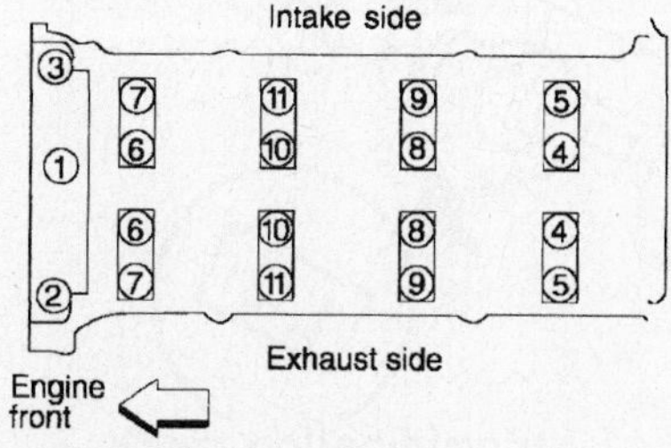

Camshaft bearing cap removal sequence—2.5L engines

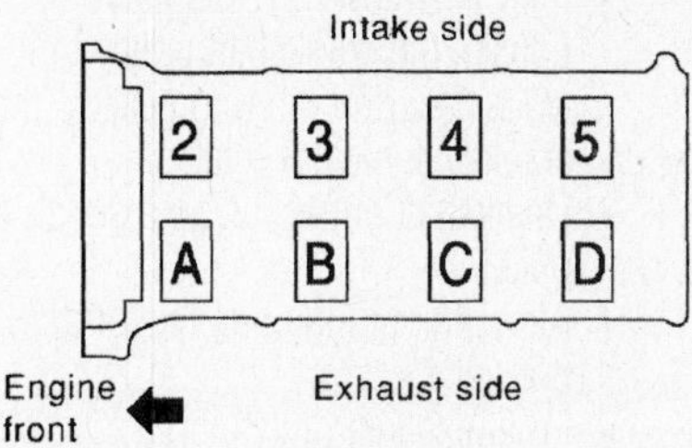

Identifiying camshaft bearing cap installation marks—2.5L engines

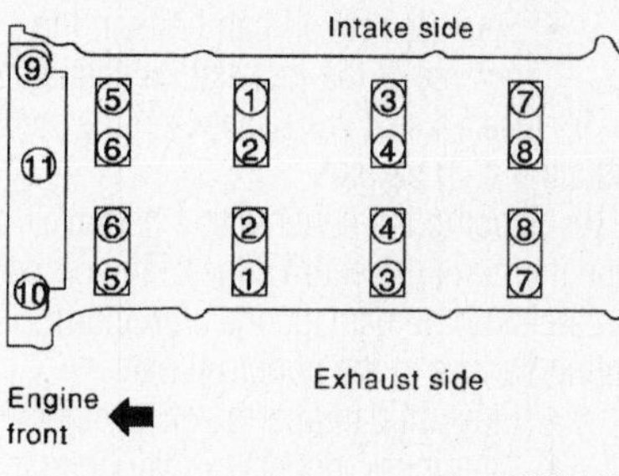

Camshaft bearing cap tightening sequence—2.5L engines

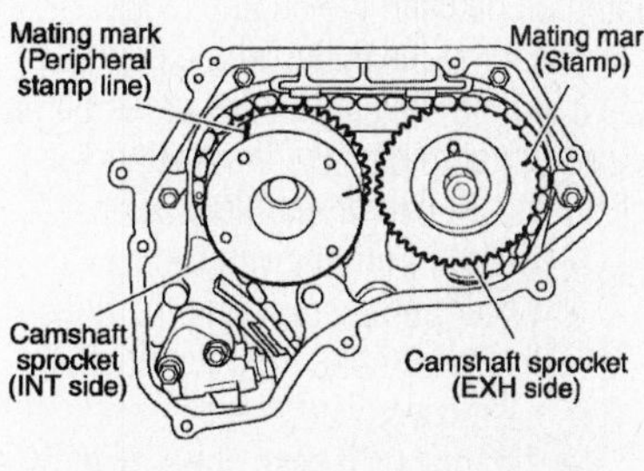

Camshaft sprocket alignment marks—2.5L engines

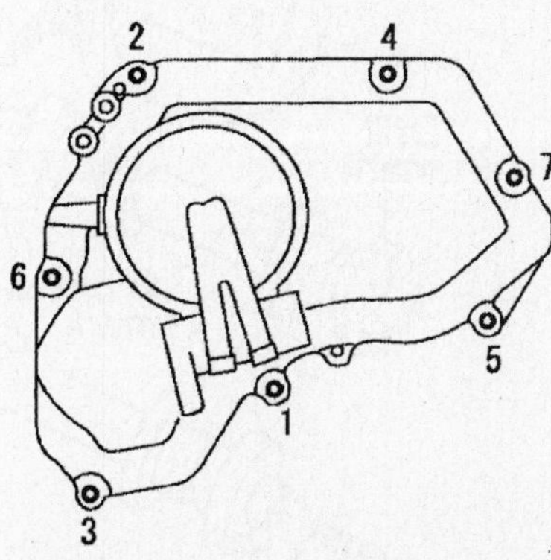

Intake valve timing cover tightening sequence—2.5L engines

9. While holding the flat of the camshaft with an open end wrench, loosen and remove the camshaft sprockets.
10. Loosen the camshaft brackets in the order shown.
11. Remove the camshafts.
12. Remove the valve lifters.

To install:

➡When installing the valve components, apply a coat of clean engine oil to the component.

13. Install or connect the following:
    - Lifters into the lifter bores from which they were removed
    - Valve shims to the lifters from which they came
14. Install the camshafts so the dowel pin on the intake camshaft is placed at the three o'clock position and the exhaust camshaft pin is placed at the twelve o'clock position.
15. Install the camshaft caps with the identifying marks placed in the correct position as shown.
16. Place a bead of sealant of the No. 1 bearing cap bottom edge.
17. Place a bead of sealant on the back side of the front cover where the No. 1 bearing cap lines up.
18. Apply sealant to the bolt hole on the front cover.
19. Install the No. 1 bearing cap so the sealant joints line up.
20. Tighten the camshaft bearing caps bolts in the proper sequence as follows:
    a. Step 1 bolts 9–11: 17 inch lbs. (2 Nm).
    b. Step 2 bolts 1–8: 17 inch lbs. (2 Nm).
    c. Step 3 bolts 1–11: 52 inch lbs. (6 Nm).
    d. Step 4 bolts 1–11: 80–104 inch lbs. (9–12 Nm).
21. Install the camshaft sprockets so the mating marks are aligned as shown.
22. Install or connect the following:
    - Timing chain tensioner and remove the pin
    - Intake valve timing control cover and tighten the bolts in the sequence shown.
    - Check and adjust the valve clearance as necessary.
    - Timing chain guide between the camshaft sprockets through the front cover
    - Camshaft position sensor
    - Variable timing control solenoid connector
    - Coolant reservoir
    - Drive belt

- Valve cover
- Right front wheel
- Negative battery cable
- Engine undercover

23. Fill the cooling system.
24. Start the vehicle, check for leaks and repair if necessary.

3.0L and Altima and Maxima 3.5L Engines

1. Before servicing the vehicle, refer to the precautions in the beginning of this section.
2. Relieve the fuel system pressure.
3. Drain the engine oil.
4. Drain the cooling system.
5. Remove or disconnect the following:
   - Negative battery cable
   - Left side rocker cover ornament

➡Before detaching any hoses or connectors, note the locations for reassembly.

   - Air duct to intake manifold hose, collector hose, blow-by hose and vacuum hoses
   - Fuel hoses and detach the harness connections
   - Canister purge hoses
   - Water hoses from the cylinder head and intake manifold
   - All 6 ignition coils from the spark plugs
   - Spark plugs
   - Bolts that secure the Exhaust Gas Recirculation (EGR) tube
   - EGR tube
   - Intake manifold collector supports and the collector
   - Bolts that secure the fuel tube and the fuel tube
   - Bolts that secure the intake manifold to the engine block and the manifold. Loosen the bolts in the reverse sequence of the tightening procedure.
   - Left-hand and right-hand rocker covers from the cylinder head
   - Engine undercovers
   - Right front wheel and engine side covers
   - Drive belts and idler pulley
   - Power steering oil pump belt and the power steering oil pump assembly
   - Camshaft Position (CMP) sensor (PHASE) and Crankshaft Position (CKP) sensors (REF)/(POS)
6. Set the No. 1 piston to Top Dead Center (TDC) of compression stroke by rotating the crankshaft.
   - Ring gear cover access plate. Loosen the crankshaft pulley bolt while securing the ring gear so the crankshaft cannot rotate
   - Crankshaft pulley, using a suitable puller
   - Air conditioning compressor and bracket
   - Front exhaust pipe and its support
7. Hang the engine at the right and left side engine slingers with a suitable hoist.
8. Support the transaxle with jack.
   - Right side engine mounting, mounting bracket and nuts
   - Center crossmember assembly
   - Steel (lower) oil pan bolts in the reverse of the installation sequence
9. Insert a seal cutter between the steel and aluminum oil pan
10. Tapping the cutter with a hammer, slide it around the entire edge of the oil pan. Be careful not to damage the aluminum mating surface of the upper oil pan.
   - Steel oil pan and the oil strainer
   - Aluminum (upper) oil pan bolts in the reverse of the installation sequence
   - Transaxle bolts that secure the oil pan
11. Insert a seal cutter between the aluminum oil pan and the engine block.
12. Tapping the cutter with a hammer, slide it around the entire edge of the oil pan. Be careful not to damage the mating surfaces of the oil pan or engine block.
   - Oil pan from the vehicle
   - Water pump cover and the bolts that secure the front timing chain case cover
   - Timing chain case cover, using the seal cutter
   - Internal timing chain guide and the upper chain guide
   - Timing chain tensioner and slack side chain guide
   - Left and right intake camshaft sprockets first. Be sure to hold the flats of the camshafts while removing the sprocket bolts.
   - Lower timing chain assembly. Be sure to note the aligning marks of the chain before removal.
13. Insert a suitable stopper pin for the left and right camshaft tensioners.
   - Left and right exhaust camshaft sprocket bolts. Be sure to hold the flats of the camshafts while removing the sprocket bolts.
   - Upper timing chain assembly. Be sure to note the aligning marks of the chain before removal.
   - Lower timing chain guide
   - Crankshaft sprocket
   - Bolts that secure the rear timing chain case. The bolts must be loosened in sequence.
   - Rear timing case cover, using the seal cutter

➡Remove the O-rings from the front of the engine block.

   - Camshaft bearing caps in several steps. The bearing caps MUST be loosened in sequence.

➡Keep all bearing caps and camshafts in proper order for installation.

   - Left-hand and right-hand camshaft tensioners from the cylinder head
   - Camshafts from the cylinder heads

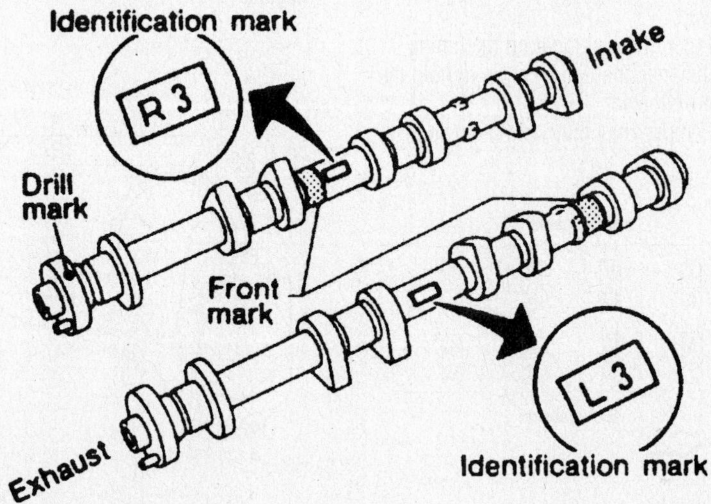

Camshaft identification marks—3.0L and 3.5L engines

➡The valve lifters have a replaceable shim on the top of the lifter. Note the proper locations of each shim to lifter and remove the shims from the lifters.

- Valve adjusting shim from the lifter, using a magnet
- Lifter assembly from the bore. Be sure to note the locations from where each lifter came.

14. Check the diameter of the valve lifter and the valve lifter guide bore.

15. The diameter of the lifter should be 1.3764–1.3770 in. (34.960–34.975mm) and the diameter of the bore should be 1.3780–1.3788 in. (35.000–35.021mm).

16. Remove all traces of liquid gasket from the timing chain case and from the water pump covers.

17. Remove all traces of liquid gasket from the engine block.

18. Inspect the camshafts for excessive wear or damage and replace as necessary.

To install:

➡Before installing the camshaft brackets, apply RTV sealant to the mating surface of the No. 1 journal head.

19. Lubricate the valve lifters with clean engine oil and install the lifters into the bore from which they were removed.

20. Lubricate the valve lifter shims with clean engine oil and install the shims into the lifter from which they were removed.

21. Turn the crankshaft clockwise until the No. 1 piston is set 240 degrees before TDC on compression stroke.

22. Install or connect the following:

- Camshaft tensioners on both sides of the cylinder heads and torque the bolts to 75–96 inch lbs. (8.4–10.8 Nm)

➡The camshafts can be identified by the paint marks on the camshaft. The left cylinder head camshafts have a YELLOW paint mark and the right cylinder head camshafts have a WHITE paint mark.

- Exhaust and intake camshafts and install the bearing caps. Before installing the No. 1 bearing cap, apply liquid gasket to the corners of the cap.

➡When installing the camshafts, position the camshaft keys at the 12 o'clock position in respect to the cylinder head angle.

23. Torque the camshaft bearing caps as follows:

    a. Bolts No. 7–10: 17 inch lbs. (2 Nm).

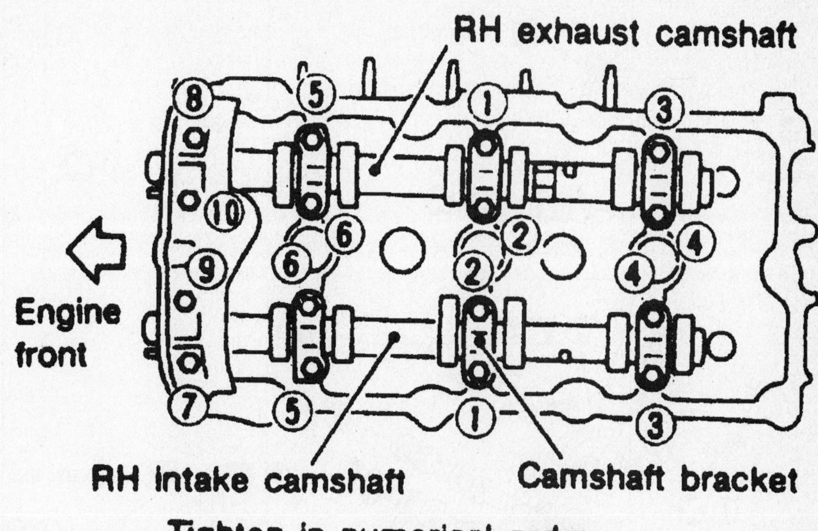

Right cylinder head camshaft bearing cap tightening sequence—3.0L and 3.5L engines

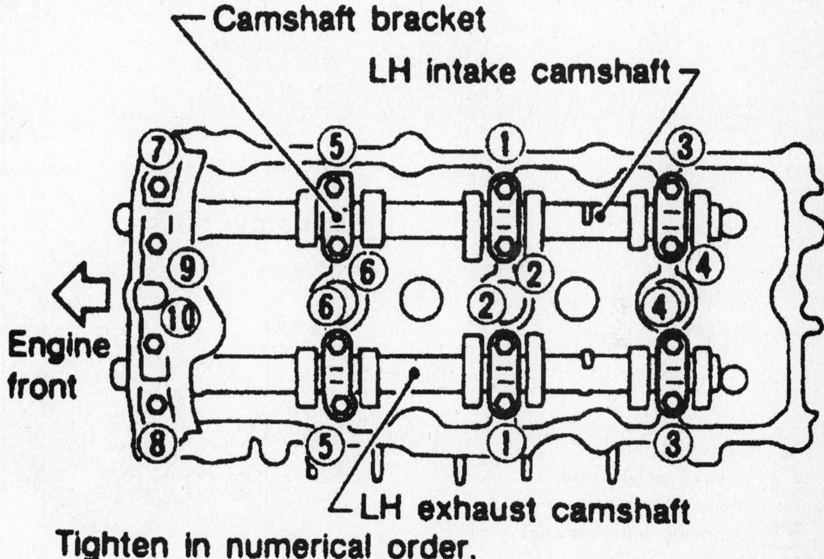

Left cylinder head camshaft bearing cap tightening sequence—3.0L and 3.5L engines

    b. Bolts No. 1–6: 17 inch lbs. (2 Nm).
    c. Bolts No. 1–10: 52 inch lbs. (6 Nm).
    d. Bolts No. 1–10: 81–104 inch lbs. (9–11 Nm).

24. Install new O-rings to the front of the engine block.

25. Apply sealant to the hatched portion of the of the rear timing chain case.

26. Align the rear timing chain case with the dowel pins and install onto the cylinder heads and engine block.

27. Torque the rear timing chain case mounting bolts in sequence to 105–121 inch lbs. (11.8–13.7 Nm).

28. Install the crankshaft sprocket with the mating mark facing out.

29. Rotate the crankshaft clockwise and position the crankshaft to TDC of compression stroke and align the dowels of the camshaft sprockets to the 12 o'clock position in respect to the cylinder head.

30. Install the lower chain guide on the dowel pin with the front mark on the guide facing upward.

31. On a workbench, align the marks on the intake and exhaust camshaft sprockets with the marks of the chain.

32. Put the exhaust camshaft sprockets onto the dowel pin and torque the bolts to

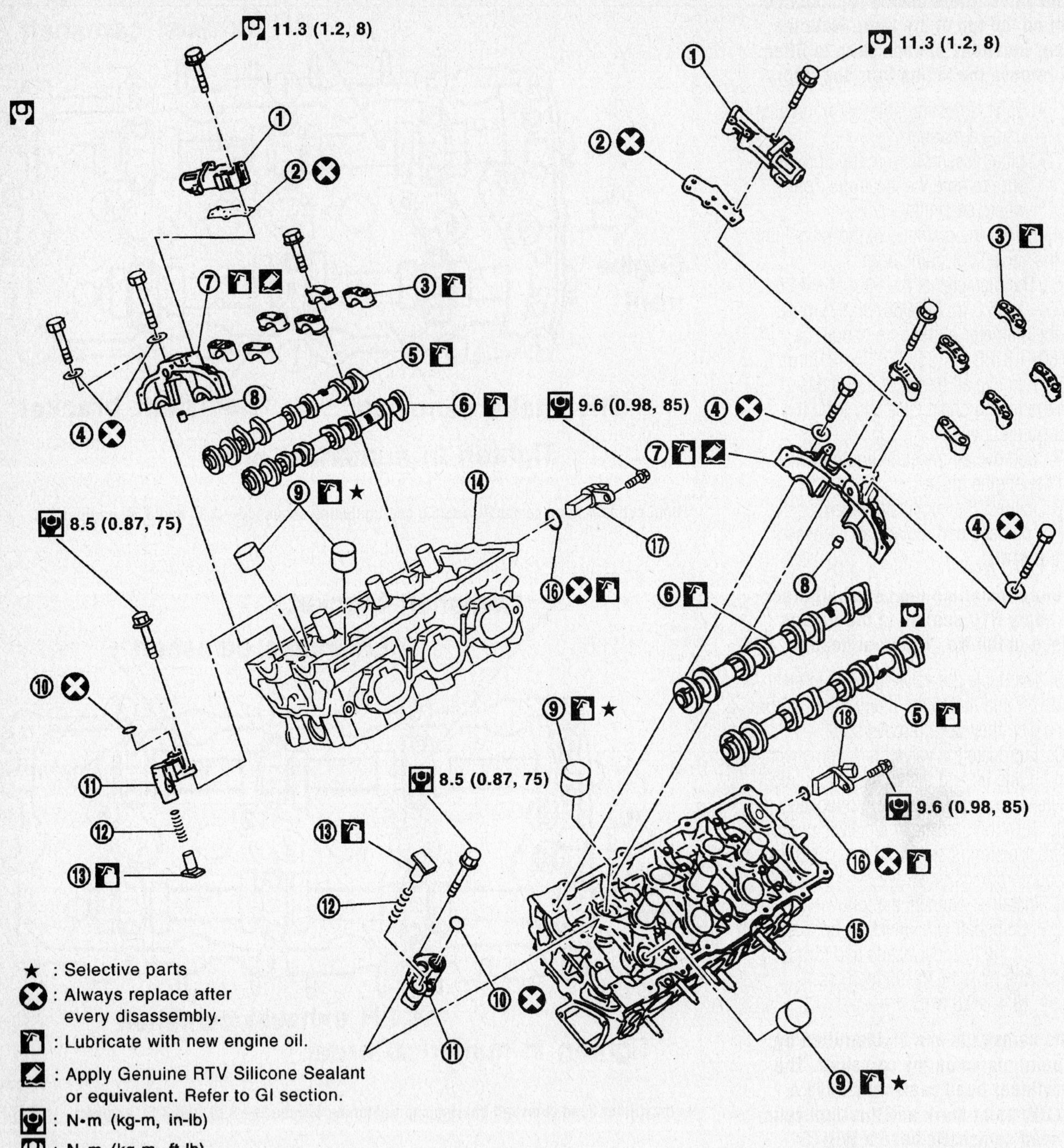

1.	Intake valve timing control solenoid valve	2.	Gasket	3.	Camshaft bracket (No. 2 to No. 4)
4.	Seal washer	5.	Camshaft (EXH)	6.	Camshaft (INT)
7.	Camshaft bracket (No. 1)	8.	Dowel pin	9.	Valve lifter
10.	O-ring	11.	Timing chain tensioner (Secondary)	12.	Spring
13.	Plunger	14.	Cylinder head (right bank)	15.	Cylinder head (left bank)
16.	O-ring	17.	Camshaft position sensor (PHASE) (right bank)	18.	Camshaft position sensor (PHASE) (left bank)

67170-NISS-G29

Exploded view of camshaft assemblies—350Z 3.5L engine

88–95 ft. lbs. (119–128 Nm). Be sure to secure the camshafts while tightening the bolts.

33. Install or connect the following:
- Timing chains, sprockets and related components
- Transaxle bolts that secure the oil pan
- Oil pan strainer and torque the bolts to 12–14 ft. lbs. (16–19 Nm)

34. Apply a 0.177–0.217 in. (4.5–5.5mm) continuous bead of liquid gasket to the lower oil pan mating surface and install the oil pan. Torque the bolts in sequence to 57–66 inch lbs. (6.4–7.5 Nm).
- Center crossmember assembly
- Right side engine mounting bracket and mount assembly

35. Remove the engine slinger assembly.
- Front exhaust pipe and its support
- Air conditioning compressor and bracket
- Crankshaft pulley to the crankshaft and install the mounting bolt. Torque the bolt to 14–22 ft. lbs. (20–29 Nm). Torque the crankshaft bolt an additional 60–66 degrees clockwise. This is about the angle from one hexagon bolt head corner to another
- CMP sensor, PHASE and CKP sensors
- Power steering pump
- Idler pulley and all belts
- Engine side and under covers
- Right front wheel
- Intake manifold
- Rocker covers
- Fuel tube
- Intake manifold support and collector
- EGR tube
- Spark plugs and ignition coils
- Coolant hoses
- Canister purge hoses
- Fuel feed and return lines
- Vacuum hoses
- Negative battery cable

36. Fill the cooling system.
37. Fill the engine with clean oil.
38. Start the vehicle, check for leaks and repair if necessary.

350Z 3.5L Engine

1. Before servicing the vehicle, refer to the precautions in the beginning of this section.
2. Relieve the fuel system pressure.
3. Drain the engine oil.
4. Drain the cooling system.

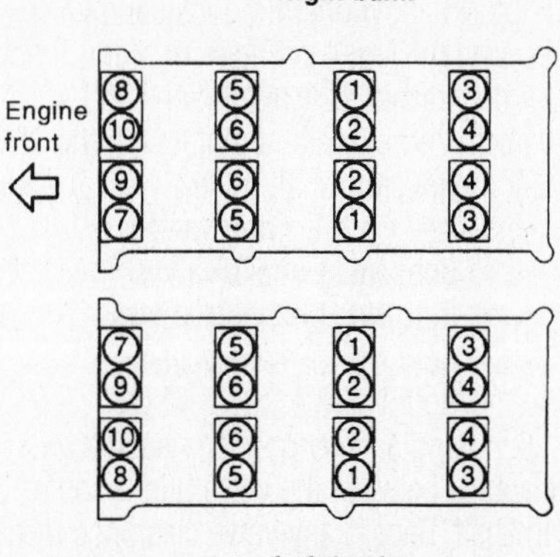

Camshaft bearing cap tightening sequence—350Z 3.5L engine

5. Disconnect the negative battery cable.
6. Remove the timing chain case, camshaft sprockets, timing chain and rear timing chain case.
7. Remove or disconnect the following:
- Camshaft position sensors (PHASE) from the back of the cylinder heads
- Intake valve timing control solenoid valves and discard the gaskets
- Camshaft bearing caps in the reverse of the tightening sequence
- Camshafts
- Valve lifters
- Secondary timing chain tensioners

To install:

➡Before installing the camshaft brackets, apply RTV sealant to the mating surface of the No. 1 journal head.

8. Lubricate the valve lifters with clean engine oil and install the lifters into the bore from which they were removed.
9. Lubricate the valve lifter shims with clean engine oil and install the shims into the lifter from which they were removed.
10. Ensure the crankshaft is set to TDC for the No. 1 cylinder.
11. Install the camshaft tensioners using new O-rings on both sides of the cylinder heads. The sliding part faces downward on the right head and upward on the left head. Torque the bolts to 75 inch lbs. (8.4 Nm).

➡The camshafts can be identified by the paint marks on the camshaft. The intake camshafts have a PINK paint mark and the exhaust camshafts have a ORANGE paint mark.

- Install the camshafts so the large and small pin holes are located on

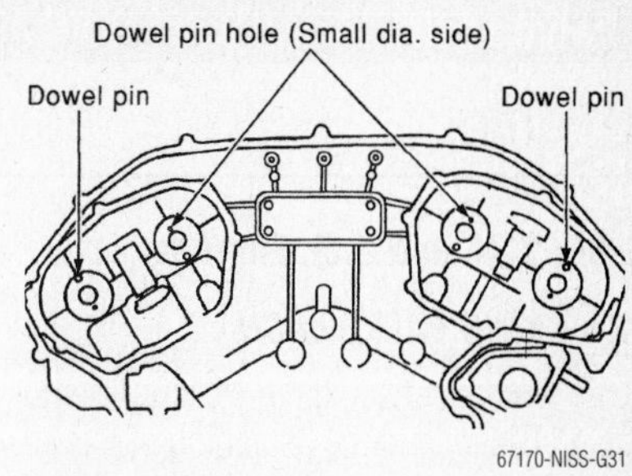

Camshaft dowel pin installation location—350Z 3.5L engine

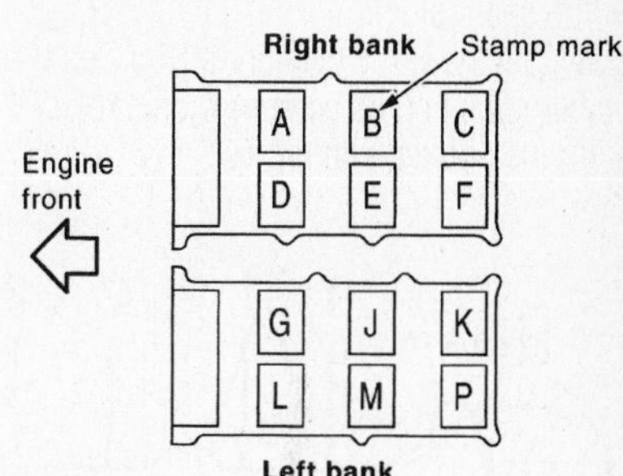

Camshaft bearing cap stamp positioning—350Z 3.5L engine

the front face of the camshafts at 180° intervals.

12. Install the bearing caps aligning the stamp marks on the caps as shown.
13. Torque the camshaft bearing caps as follows:
    a. Bolts No. 7–10: 17 inch lbs. (2 Nm).
    b. Bolts No. 1–6: 17 inch lbs. (2 Nm).
    c. Bolts No. 1–10: 52 inch lbs. (6 Nm).
    d. Bolts No. 1–6: 81–104 inch lbs. (9–11 Nm).
    e. Bolts No. 7–10: 74–91 inch lbs. (8.3–10.3 Nm).
14. Check and adjust the valve clearance.
15. Install or connect the following:
- Intake valve timing control solenoid valves using new gaskets and tighten the bolts to 8 ft. lbs. (11.3 Nm)
- Camshaft position sensors (PHASE) and tighten the bolts to 85 inch lbs. (9.6 Nm)
- Timing chain case, camshaft sprockets, timing chain and rear timing chain case
- Negative battery cable

16. Fill the cooling system.
17. Fill the engine with clean oil.
18. Start the vehicle, check for leaks and repair if necessary.

Valve Lash

ADJUSTMENT

1.8L, 2.4L and 2.5L Engines

CHECKING VALVE LASH

1. Before servicing the vehicle, refer to the precautions in the beginning of this section.
2. Run the engine until it reaches normal operating temperature and shut if off.
3. Remove the cylinder head cover and all the spark plugs.
4. Set the No. 1 cylinder at Top Dead Center (TDC) on its compression stroke. Align the pointer with the TDC mark on the crankshaft pulley. Check that the valve lifters on the No. 1 cylinder are loose and valve lifters on the No. 4 cylinder are tight. If not, turn the crankshaft 1 revolution (360 degrees) and align the pointer with the TDC mark on the crankshaft pulley.
5. Check the following valves:
   - Both No. 1 intake valves
   - Both No. 1 exhaust valves
   - Both No. 2 intake valves
   - Both No. 3 exhaust valves
6. Using a feeler gauge, measure the clearance between the valve lifter and the camshaft. Record any valve clearance measurements which are out of specification.
7. Turn the crankshaft 1 revolution (360 degrees) and align the mark on the crankshaft pulley with the pointer. Check the following valves:
   - Both No. 2 exhaust valves
   - Both No. 3 intake valves
   - Both No. 4 intake valves
   - Both No. 4 exhaust valves
8. Using a feeler gauge, measure the clearance between the valve lifter and the camshaft. Record any valve clearance measurements which are out of specification.
9. If all the valve clearances are within specification, install the cylinder head cover and the spark plugs.

ADJUSTING VALVE LASH

1. Before servicing the vehicle, refer to the precautions in the beginning of this section.
2. If an adjustment is necessary, adjust the valve clearance while engine is cold by removing the adjusting shim. The adjusting shim can be removed by using the following procedures:
   a. Turn the crankshaft so the camshaft lobe of the valve to be adjusted is pointed straight up.
   b. Turn the lifter so the notch is pointed towards the center of the cylinder head; this will facilitate the shim removal process.
   c. Using a depressor tool, push down on the lifter and insert a keeper tool on the edge of the lifter to keep the lifter in the depressed position.
   d. Remove the depressor tool and remove the shim with a magnet.
3. Determine the replacement adjusting shim size by using the following procedures and formula:
   a. Using a micrometer determine thickness of the removed shim.
   b. Calculate the thickness of a new adjusting shim so valve clearance is within the specified values.
   c. R = thickness of the removed shim.
   d. N = thickness of the new shim.
   e. M = measured valve clearance.
   f. 1.8L engine: Intake shim determination formula: N = R + (M – 0.0146 in. or 0.37mm)
   g. 1.8L engine: Exhaust shim determination formula: N = R + (M – 0.0157 in. or 0.40mm)
   h. 2.4L engine: Intake shim determination formula: N = R + (M – 0.0138 in. or 0.35mm)
   i. 2.4L engine: Exhaust shim determination formula: N = R + (M – 0.0146 in. or 0.37mm)

Shims are available in different sizes from 0.0772–0.1055 in. (1.96–2.68mm) in increments of 0.0008 in. (0.02mm). The

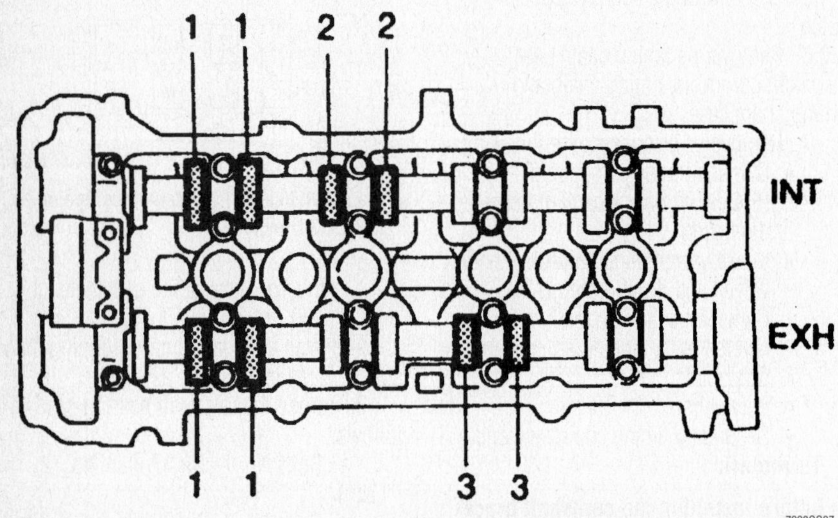

Measure the clearance of the valves indicated when the No. 1 piston is at TDC on compression—1.8L and 2.4L engines

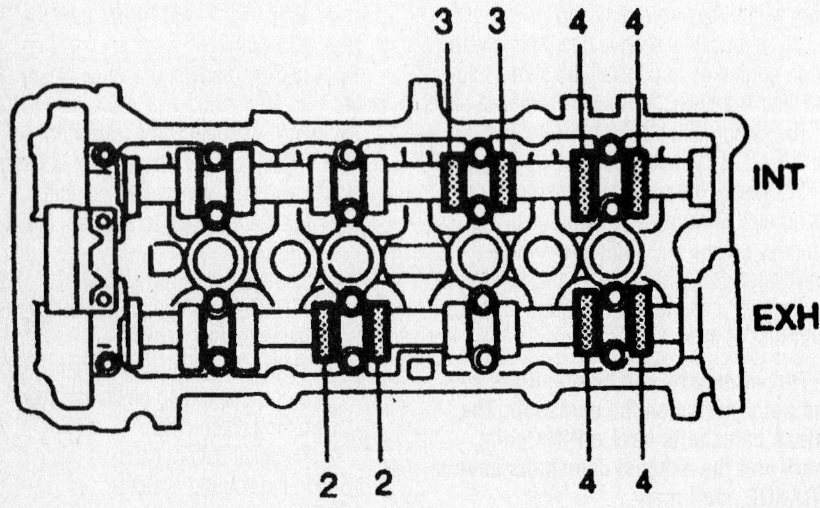

Measure the clearance of the valves indicated when the No. 4 piston is at TDC on compression—1.8L and 2.4L engines

thickness is stamped on the shim; this side is always installed facing down. Select new shims with thickness as close as possible to calculated valve and install it in the lifter.

4. Install the new shim onto the lifter.
5. Depress the lifter and remove the keeper tool. Remove the depressor tool and recheck the valve clearance. Repeat this procedure for any other valves requiring adjustment.
6. Install the cylinder head cover and spark plugs when all valve adjustments are finished.

2.0L Engines

The engine is equipped with hydraulic lash adjusters. The valve lash is not adjustable.

3.0L and 3.5L Engines

➡Check and adjust the valve clearances while the engine is cold and not running.

CHECKING VALVE LASH

1. Before servicing the vehicle, refer to the precautions in the beginning of this section.
2. Remove or disconnect the following:
   - Intake manifold collector
   - Left and right rocker covers
   - Spark plugs
3. Set the No. 1 cylinder at Top Dead Center (TDC) on its compression stroke. Align the pointer with the TDC mark on the crankshaft pulley. Check that the valve lifters on the No. 1 cylinder are loose and valve lifters on the No. 4 cylinder are tight. If not, turn the crankshaft 1 revolution (360 degrees) and align the pointer with the TDC mark on the crankshaft pulley.
4. Check the following valves:
   - Both No. 1 intake valves
   - Both No. 2 exhaust valves
   - Both No. 3 exhaust valves
   - Both No. 6 intake valves
5. Using a feeler gauge, measure the clearance between the valve lifter and the camshaft. Record any valve clearance measurements that are out of specification. Intake valve clearance (cold) is 0.010–0.013 in. (0.26–0.34mm) and exhaust valve clearance (cold) is 0.011–0.015 in. (0.29–0.37mm).
6. Turn the crankshaft 240 degrees and set the No. 3 cylinder to TDC of its compression stroke.
7. Check the following valves:
   - Both No. 2 intake valves
   - Both No. 3 intake valves

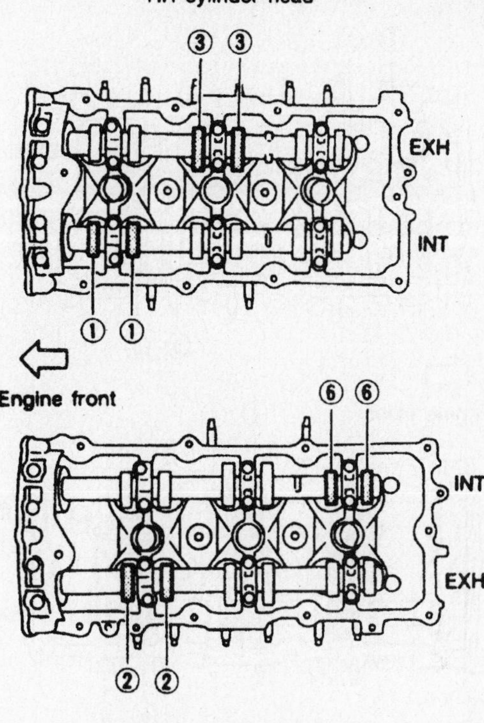

Measure the valves indicated while the No. 1 piston is at TDC on the compression stroke—3.0L and 3.5L engines

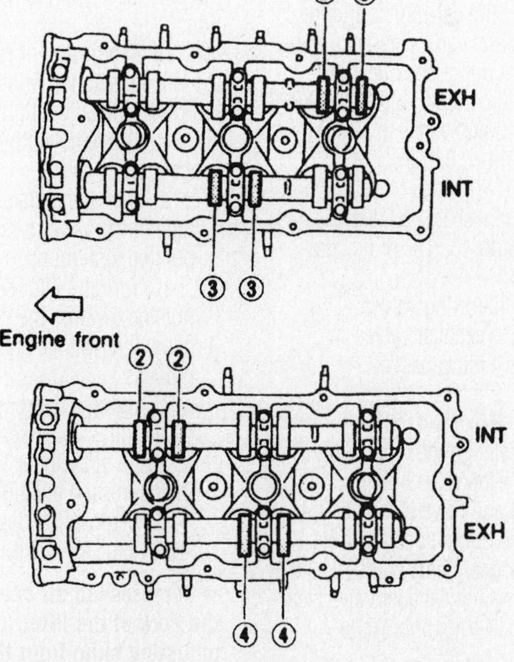

Measure the valves indicated while the No. 3 piston is at TDC on the compression stroke—3.0L and 3.5L engines

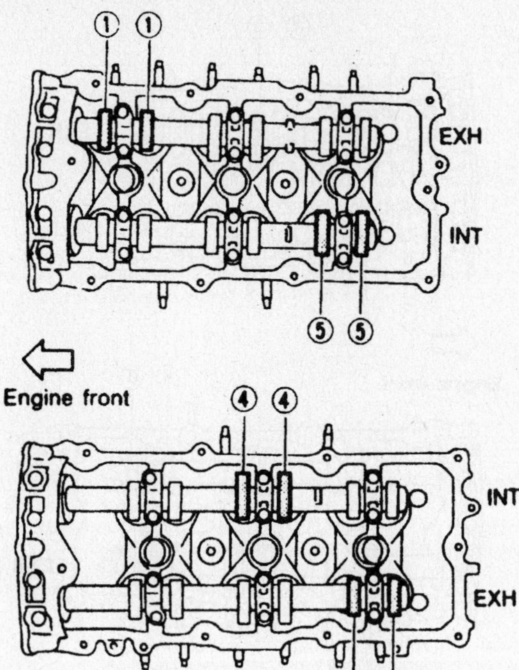

Measure the valves indicated while the No. 5 piston is at TDC on compression—3.0L and 3.5L engines

- Both No. 4 exhaust valves
- Both No. 5 exhaust valves

8. Using a feeler gauge, measure the clearance between the valve lifter and the camshaft. Record any valve clearance measurements that are out of specification. Intake valve clearance (cold) is 0.010–0.013 in. (0.26–0.34mm) and exhaust valve clearance (cold) is 0.011–0.015 in. (0.29–0.37mm).
9. Turn the crankshaft 240 degrees and set the No. 5 cylinder to TDC of its compression stroke.
10. Check the following valves:
    - Both No. 1 exhaust valves
    - Both No. 4 intake valves
    - Both No. 5 intake valves
    - Both No. 6 exhaust valves
11. Using a feeler gauge, measure the clearance between the valve lifter and the camshaft. Record any valve clearance measurements that are out of specification. Intake valve clearance (cold) is 0.010–0.013 in. (0.26–0.34mm) and exhaust valve clearance (cold) is 0.011–0.015 in. (0.29–0.37mm).
12. If all the valve clearances are within specification, install the cylinder head cover, spark plugs and the intake manifold collector.

ADJUSTING VALVE LASH

1. Before servicing the vehicle, refer to the precautions in the beginning of this section.
2. If an adjustment is necessary, adjust the valve clearance while engine is cold by removing the adjusting shim. The adjusting shim can be removed by using the following procedures:
    a. Turn the crankshaft so the camshaft lobe of the valve to be adjusted is pointed straight up.
    b. Turn the lifter so the notch is pointed towards the center of the cylinder head; this will facilitate the shim removal process.
    c. Using a depressor tool, push down on the lifter and insert a keeper tool on the edge of the lifter to keep the lifter in the depressed position.
    d. Remove the depressor tool and remove the shim with a magnet.

➡Compressed air can be blown into the hole of the lifter to separate the adjusting shim from the lifter.

3. Determine the replacement adjusting shim size by using the following procedures and formula:
    a. Using a micrometer determine thickness of the removed shim.
    b. Calculate the thickness of a new adjusting shim so valve clearance is within the specified values.
    c. R = thickness of the removed shim.
    d. N = thickness of the new shim.
    e. M = measured valve clearance.
    - Intake shim determination formula: N = R + (M – 0.0118 in. or 0.30mm)
    - Exhaust shim determination formula: N = R + (M – 0.0130 in. or 0.33mm)
4. Shims are available in 64 sizes from 0.0913–0.1161 in. (2.32–2.95mm) in steps of 0.004 in. (0.01mm). The thickness is stamped on the shim; this side is always installed facing down. Select new shims with thickness as close as possible to calculated valve and install it in the lifter.
5. Install the new shim onto the lifter.
6. Depress the lifter and remove the keeper tool. Remove the depressor tool and recheck the valve clearance. Repeat this procedure for any other valves requiring adjustment.
7. When all valve adjustments are finished, install the cylinder head cover, spark plugs and the intake manifold collector.

Starter Motor

REMOVAL & INSTALLATION

1.8L and 2.0L Engines

1. Before servicing the vehicle, refer to the precautions in the beginning of this section.
2. Remove or disconnect the following:
    - Negative battery cable
    - Wiring at the starter
    - Bolts attaching the starter to the engine
    - Starter

To install:

3. Install or connect the following:
    - Starter and torque the bolts to 70 inch lbs. (8 Nm)
    - Starter electrical connections
    - Negative battery cable

2.4L Engines

1. Remove or disconnect the following:
    - Negative battery cable
    - Air inlet tube
    - Harness bracket
    - Wiring at the starter
    - Starter

To install:

2. Install or connect the following:
    - Starter and torque the bolts to 60 inch lbs. (7 Nm)
    - Starter electrical connectors
    - Harness bracket
    - Air inlet tube
    - Negative battery cable

2.5L Engines

1. Remove or disconnect the following:
    - Negative battery cable
    - Air cleaner cover and duct
    - Harness bracket
    - Wiring at the starter
    - Starter

To install:

2. Install or connect the following:
    - Starter and torque the bolts to 60 inch lbs. (7 Nm)
    - Starter electrical connectors
    - Harness bracket
    - Air inlet tube
    - Negative battery cable

3.0L and 3.5L Engines

1. Remove or disconnect the following:
    - Negative battery cable
    - Air duct
    - Harness protector from the harness
    - Starter wiring at the starter
    - Starter-to-engine bolts
    - Starter from the vehicle

To install:

2. Install or connect the following:
    - Starter and torque the long bolt to 57–72 ft. lbs. (77–98 Nm) on Altima and Maxima and 41 ft. lbs. (55 Nm) on 350 Z. torque the short bolt to 22–30 ft. lbs. (30–41 Nm)
    - Starter wiring
    - Harness protector
    - Air duct
    - Negative battery cable

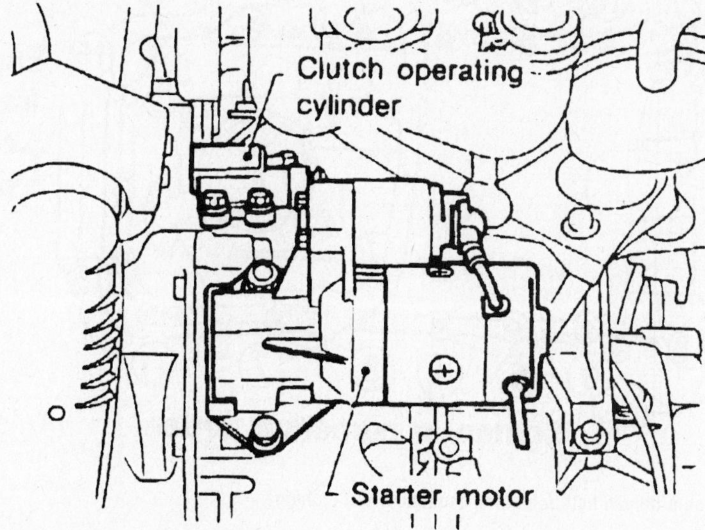

Starter location and mounting detail—3.0L Maxima with a manual transaxle

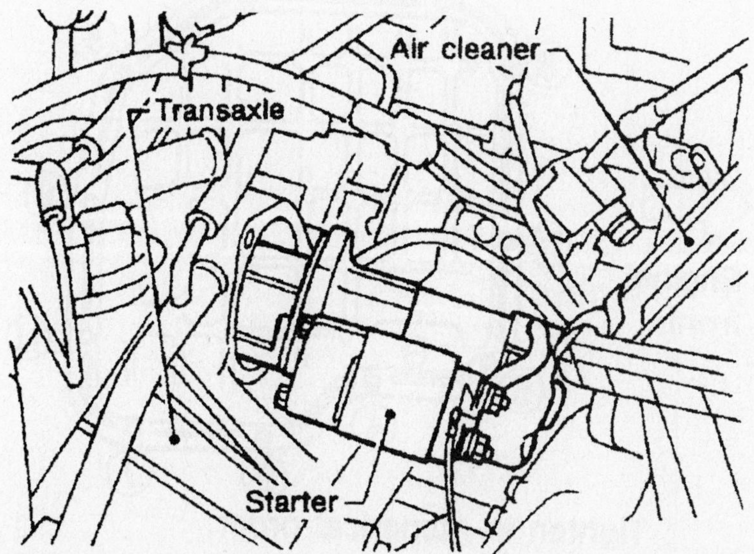

Starter location and mounting detail—3.5L Maxima

Oil Pan

REMOVAL & INSTALLATION

1.8L Engine

1. Before servicing the vehicle, refer to the precautions in the beginning of this section.
2. Drain the engine oil.
3. Remove or disconnect the following:
    - Negative battery cable
    - Engine undercovers
    - Front exhaust tube and properly support the transaxle assembly
    - Center crossmember
    - Support brackets from the sides of the oil pan
    - Rear cover plate, models equipped with a automatic transaxle
    - Oil pan mounting bolts
4. Using an oil pan seal cutter, separate the oil pan from the engine.

**** WARNING**

Do not drive the seal cutter into the oil pump or rear oil seal retainer portion, for the aluminum mating surfaces will be damaged. Do not use a prybar to remove the oil pan; the flange will be deformed.

5. Clean all the sealing surfaces.

To install:

6. Apply sealant to the rear oil seal retainer.
7. Apply a 0.128–0.177 in. (3.5–4.5mm) continuous bead of liquid gasket to the oil pan mating surface.
8. Install or connect the following:
    - Oil pan and torque bolts, in sequence, to 56–73 inch lbs. (6.3–8.3 Nm)
    - Rear cover plate, models equipped with a automatic transaxle
    - Oil pan support brackets
    - Center crossmember

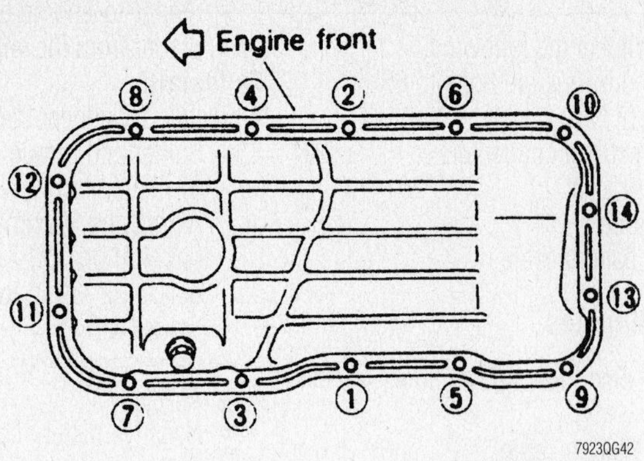

Tighten the oil pan bolts in the correct sequence to prevent oil leakage—1.8L engines

- Front exhaust tube
- Engine undercovers
- Oil pan plug, using a new gasket and tighten the plug to 21–28 ft. lbs. (7–8 Nm)
- Negative battery cable

9. After 30 minutes of gasket curing time, refill the oil pan with the specified quantity of clean oil.
10. Start the vehicle, check for leaks and repair if necessary.

2.0L Engine

1. Before servicing the vehicle, refer to the precautions in the beginning of this section.
2. Drain the engine oil.
3. Remove or disconnect the following:
   - Negative battery cable
   - Engine undercover
   - Lower steel oil pan bolts in the reverse of installation sequence
   - Steel oil pan. Insert a cutting tool between steel oil pan and aluminum oil pan. Tap the tool around the perimeter of the pan to cut the gasket material.
   - Oil baffle bolts and oil baffle
   - Front exhaust tube and set a suitable jack under the transaxle and raise the engine
   - Center crossmember from the vehicle
   - Transaxle shift control cable, if equipped with an automatic transaxle
   - Compressor gussets and the rear cover plate
   - Aluminum oil pan bolts. Loosen aluminum oil pan bolts in reverse order of the tightening sequence.
   - 2 transaxle mounting bolts and refit the them into vacant holes at the bottom of the oil pan. Use a cutting tool to cut the gasket material.
   - 2 transaxle mounting bolts that were relocated and the pan from the vehicle

To install:

4. Clean the oil pan rail of all liquid gasket and apply a new bead of 5⁄32 in. (4.5mm) thickness to the aluminum oil pan rail.
5. Install the aluminum oil pan. Torque the bolts in the opposite order of removal as follows:
   a. Bolts No. 1–16: 12–14 ft. lbs. (16–19 Nm).
   b. Bolts No. 17–18: 60–72 inch lbs. (6–8 Nm).
6. Install or connect the following:

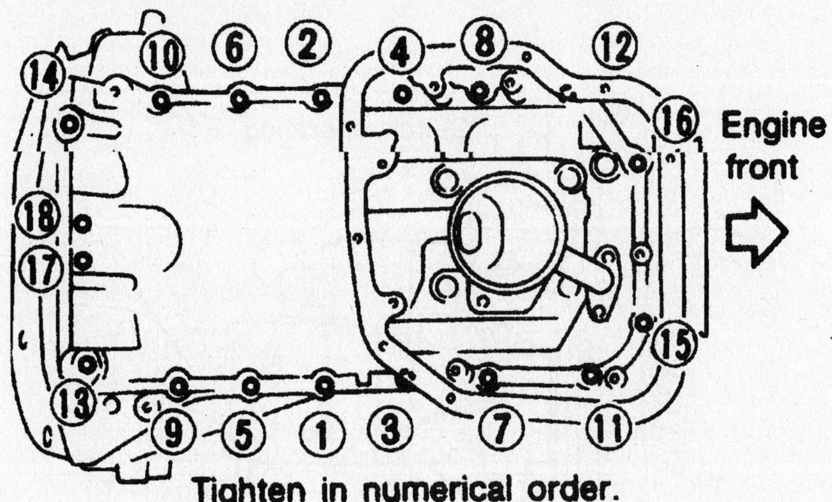

Aluminum oil pan bolt tightening sequence—2.0L engine

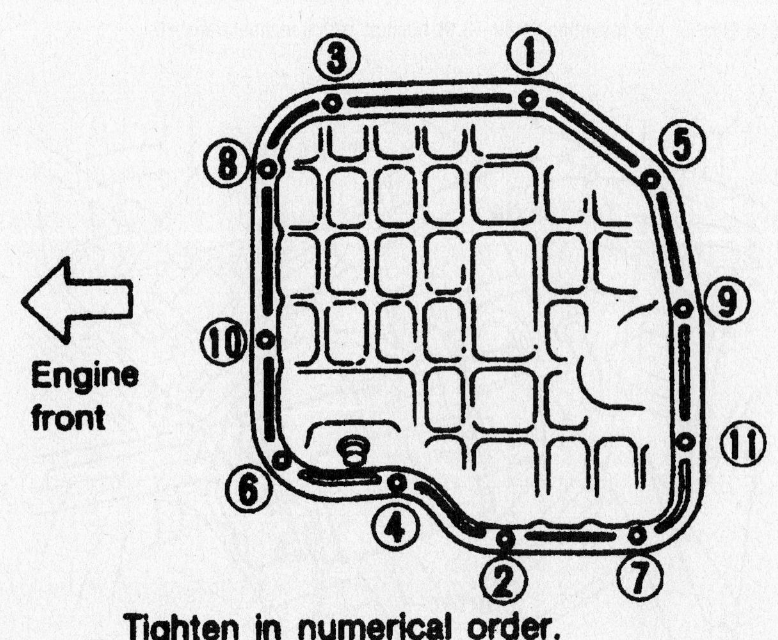

Steel oil pan bolt tightening sequence—2.0L engine

- 2 transaxle mounting bolts, rear cover plate and the compressor gussets
- Automatic transmission shift control cable, if equipped
- Center crossmember member, front exhaust tube and the baffle plate and torque the bolts to 56–66 inch lbs. (6.4–7.5 Nm)

7. Clean the steel oil pan rail of all liquid gasket and apply a new bead of 5⁄32 in. (4.5mm) thickness to the steel oil pan rail.
   - Steel oil pan and torque the bolts in the proper sequence to 56–66 inch lbs. (6.4–7.5 Nm)
   - Negative battery cable

8. After 30 minutes, refill the engine with clean oil

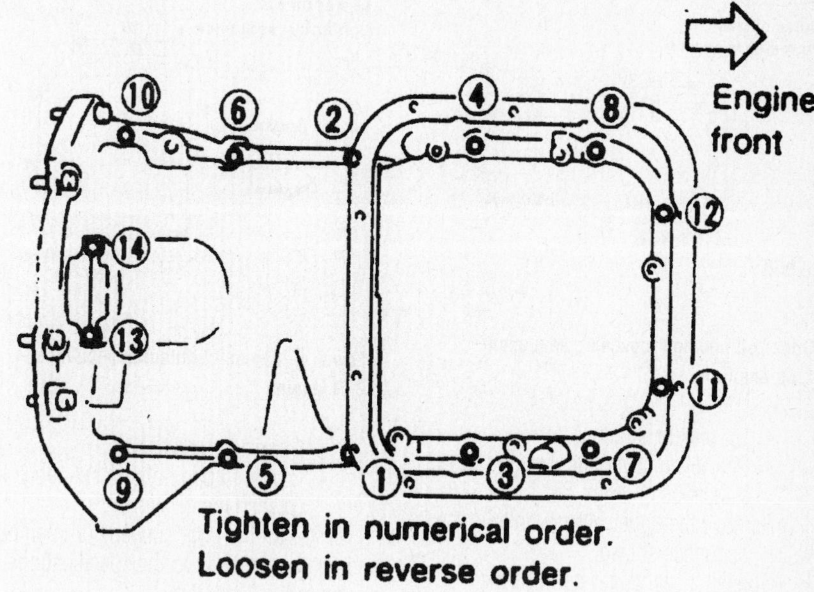

Oil pan bolt loosening and tightening sequence—2.4L engine

2.4L Engine

1. Before servicing the vehicle, refer to the precautions in the beginning of this section.
2. Drain the engine oil.
3. Remove or disconnect the following:
   - Negative battery cable
   - Engine undercover
   - Bolts securing the steel oil pan to the aluminum oil pan in reverse order of the tightening sequence
4. Install a seal cutter between the steel oil pan and the aluminum oil pan
5. Tapping the cutter with a hammer, slide it around the entire edge of the oil pan. Take care not to damage the aluminum oil pan.
6. Remove or disconnect the following:
   - Steel oil pan
   - Baffle plate and oil strainer
   - Front exhaust tube
   - Front suspension member
   - A/C compressor gussets
   - Rear cover plate
   - Aluminum oil pan retaining bolts in reverse order of the tightening sequence
7. Insert a seal cutter between the oil pan and the cylinder block.
8. Tapping the cutter with a hammer, slide it around the entire edge of the oil pan. Take care not to damage the aluminum oil pan.
9. Lower the oil pan from the cylinder block and remove it from the engine.

To install:

10. Carefully scrape the old gasket material away from the pan and cylinder block mounting surfaces, then apply a continuous 3.5–4.5mm wide bead of liquid gasket around the oil pan. Install the pan within 5 minutes or else this step will have to be repeated.
11. Install or connect the following:
    - Aluminum oil pan and torque the bolts, in sequence, to 13 ft. lbs. (17.5 Nm)
    - Baffle plate
    - Steel oil pan. Torque the bolts in sequence to 61 inch lbs. (7 Nm).
    - Rear cover plate
    - Front suspension member
    - Front exhaust tube
    - A/C compressor gussets
    - Front suspension member
    - Engine undercovers
    - Negative battery cable
12. Wait 30 minutes before refilling the crankcase with clean oil.
13. Fill the engine with clean oil.
14. Start the vehicle, checks for leaks and repair if necessary.

2.5L Engine

ALTIMA

1. Before servicing the vehicle, refer to the precautions in the beginning of this section.
2. Drain the engine oil.
3. Remove or disconnect the following:
   - Negative battery cable
   - Engine undercover
   - Front exhaust pipe
   - Power steering hose bracket from the collector
4. Support the engine from above and below with a suitable jack and hoist.
5. Remove or disconnect the following:
   - Power steering pump and reservoir and wire aside
   - Auxiliary drive belts
   - A/C compressor and wire aside
   - Upper sway bar links
   - Front and rear engine mount through bolts
   - Lower ball joints
   - 2 steering gear housing mounting bolts
   - Front suspension member bolts and front suspension member

   Remove the lower oil pan bolts in the sequence shown
6. Install a seal cutter between the steel oil pan and the aluminum oil pan
7. Tapping the cutter with a hammer, slide it around the entire edge of the oil pan. Take care not to damage the aluminum oil pan.
8. Remove or disconnect the following:
   - Steel oil pan

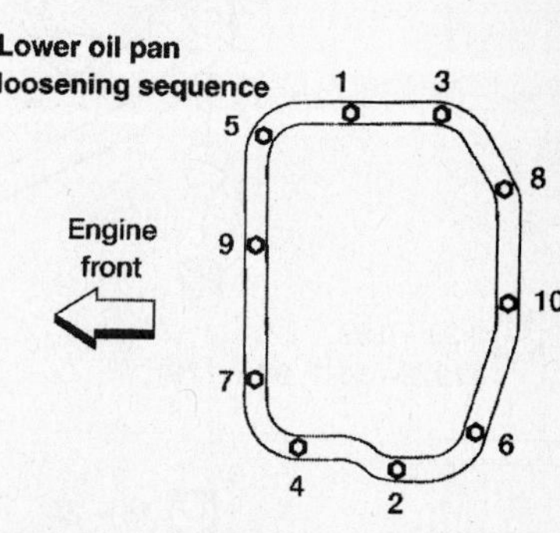

Lower oil pan bolt loosening sequence—2.5L engine

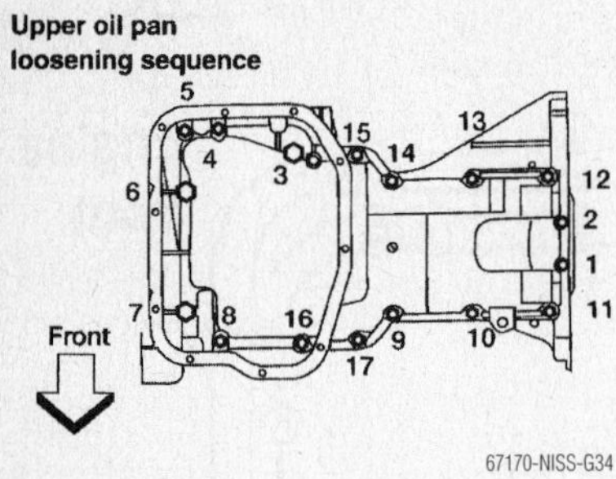

Upper oil pan bolt loosening sequence—2.5L engine

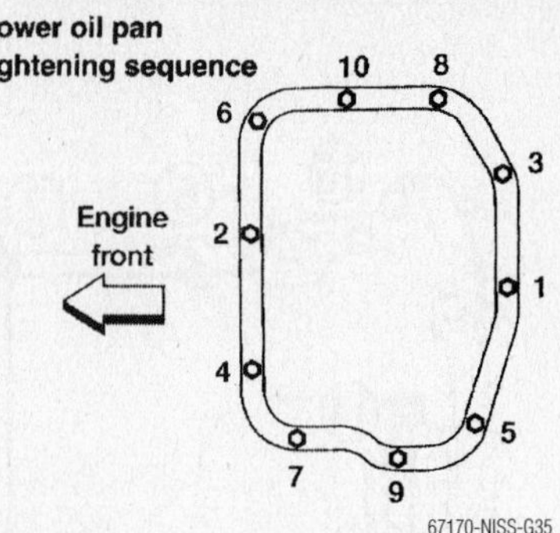

Lower oil pan bolt tightening sequence—2.5L engine

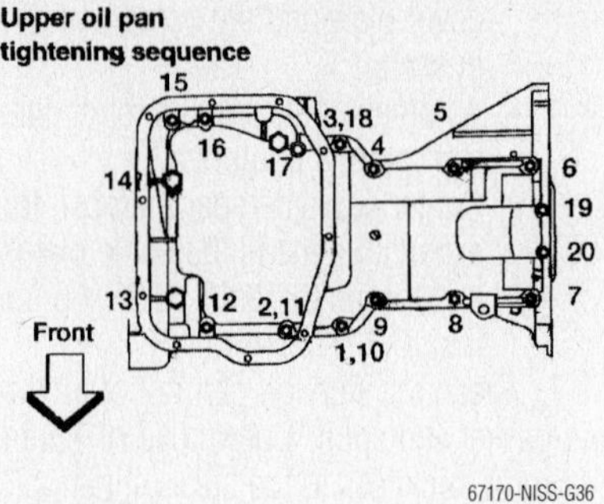

Upper oil pan bolt tightening sequence—2.5L engine

- Oil pickup screen
- Rear plate cover and 4 engine-to-transmission bolts

Remove the upper oil pan bolts in the sequence shown

9. Install a seal cutter between the block and the upper oil pan
10. Remove the upper oil pan.

To install:

11. Carefully scrape the old gasket material away from the pan and cylinder block mounting surfaces, then apply a continuous 3.5–4.5mm wide bead of liquid gasket around the oil pan. Install the pan within 5 minutes or else this step will have to be repeated.
12. Install or connect the following:
    - Upper oil pan and torque the bolts in sequence to 16 ft. lbs. (22 Nm)
    - Rear plate cover and 4 engine-to-transmission bolts
    - Oil pickup screen
    - Lower oil pan bolts in sequence to 61 inch lbs. (7 Nm).
13. Install the front suspension member and tighten the bolts to the specifications shown in the illustration.
14. Install or connect the following:
    - 2 steering gear housing mounting bolts
    - Lower ball joints
    - Front and rear engine mount through bolts
    - Upper sway bar links
    - A/C compressor
    - Auxiliary drive belts
    - Power steering pump and reservoir
    - Power steering hose bracket to the collector
    - Front exhaust pipe
    - Engine undercover
    - Negative battery cable
15. Wait 30 minutes before refilling the crankcase with clean oil.
16. Fill the engine with clean oil.
17. Start the vehicle, checks for leaks and repair if necessary.

SENTRA

1. Before servicing the vehicle, refer to the precautions in the beginning of this section.
2. Drain the engine oil.
3. Remove or disconnect the following:
    - Negative battery cable
    - Engine undercover
    - Front exhaust pipe
4. Support the engine from above and below with a suitable jack and hoist.
5. Remove or disconnect the following:

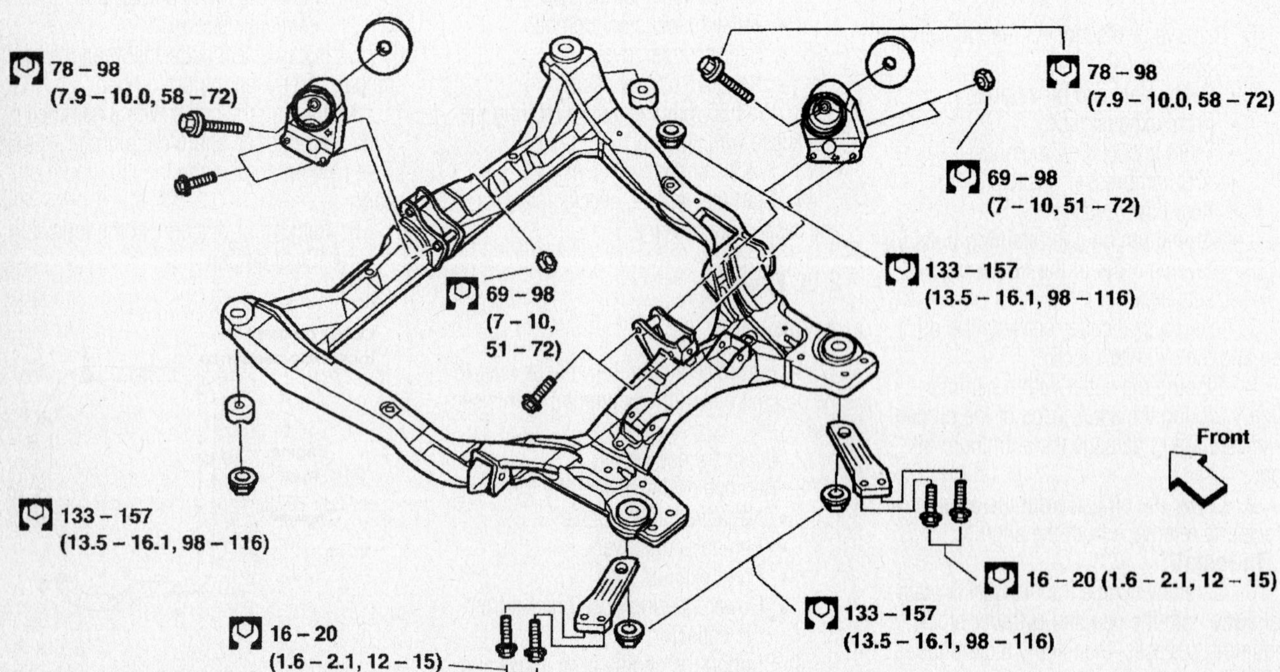

Exploded view of the front suspension member—Altima

- Front and rear engine mount through bolts
- Center crossmember
- A/C compressor and wire aside
- Remove the lower oil pan bolts in the sequence shown

6. Install a seal cutter between the steel oil pan and the aluminum oil pan
7. Tapping the cutter with a hammer, slide it around the entire edge of the oil pan. Take care not to damage the aluminum oil pan.
8. Remove or disconnect the following:
   - Steel oil pan
   - Oil pickup screen
   - Rear plate cover and 4 engine-to-transmission bolts

   Remove the upper oil pan bolts in the sequence shown
9. Install a seal cutter between the block and the upper oil pan
10. Remove the upper oil pan.

To install:

11. Carefully scrape the old gasket material away from the pan and cylinder block mounting surfaces, then apply a continuous 3.5–4.5mm wide bead of liquid gasket around the oil pan. Install the pan within 5 minutes or else this step will have to be repeated.
12. Install or connect the following:
    - Upper oil pan and torque the bolts in sequence to 16 ft. lbs. (22 Nm)
    - Rear plate cover and 4 engine-to-transmission bolts
    - Oil pickup screen
    - Lower oil pan bolts in sequence to 61 inch lbs. (7 Nm).
    - A/C compressor
    - Center crossmember
    - Front and rear engine mount through bolts
    - Front exhaust pipe
    - Engine undercover
    - Negative battery cable
13. Wait 30 minutes before refilling the crankcase with clean oil.
14. Fill the engine with clean oil.
15. Start the vehicle, checks for leaks and repair if necessary.

3.0L and 3.5L Engines

1. Before servicing the vehicle, refer to the precautions in the beginning of this section.
2. Drain the engine oil
3. Remove or disconnect the following:
   - Negative battery cable
   - Engine undercovers
   - Steel (lower) oil pan bolts in the reverse of the installation sequence
4. Insert a seal cutter between the steel and aluminum oil pan.
5. Tapping the cutter with a hammer, slide it around the entire edge of the oil pan. Be careful not to damage the aluminum mating surface of the upper oil pan.
   - Steel oil pan and the oil strainer
   - Front exhaust pipe and its support
6. Hang the engine at the right and left side engine slingers with a suitable hoist.
7. Position a suitable jack under the transaxle.
   - Crankshaft Position (CKP) sensors (REFERENCE and POSITION) from the oil pan
   - Front and rear engine mounting nuts and bolts
   - Center crossmember assembly
   - Engine drive belts
   - A/C compressor and mounting bracket
   - Rear cover plate and the lower transaxle bolts
   - Aluminum (upper) oil pan bolts in

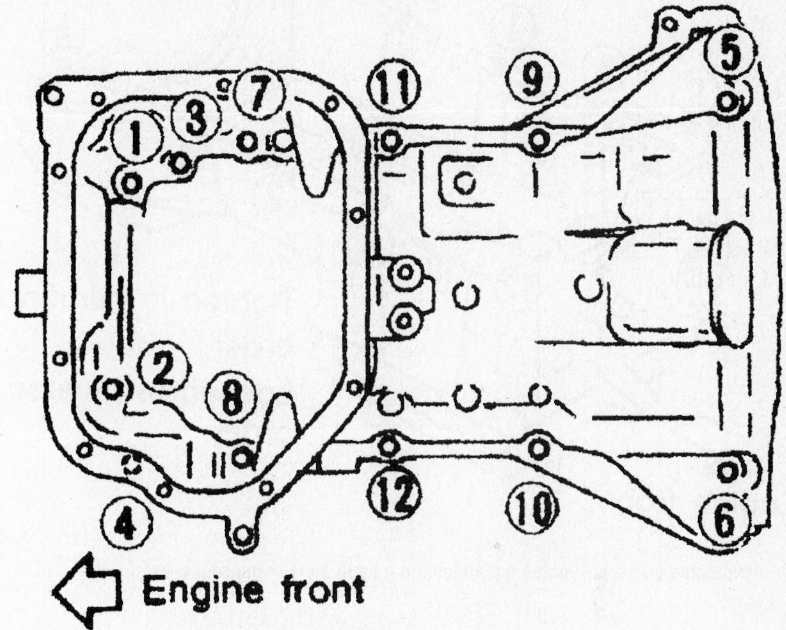

Aluminum (upper) oil pan bolt loosening sequence—3.0L and 3.5L engine

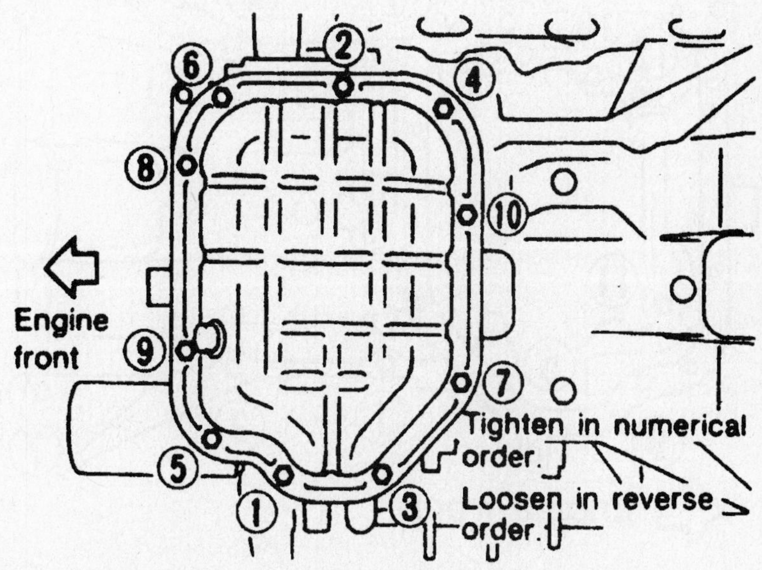

Bolt tightening sequence for the steel oil pan—3.0L engines

the reverse of the installation sequence

8. Insert a seal cutter between the aluminum oil pan and the engine block.

9. Tapping the cutter with a hammer, slide it around the entire edge of the oil pan. Be careful not to damage the mating surfaces of the oil pan or engine block.

10. Remove or disconnect the following:
- Oil pan assembly
- Bolts that secure the baffle plate and the baffle plate
- O-rings from the cylinder block and oil pump body

To install:

11. Install or connect the following:
- Baffle plate to the oil pan and torque the bolts to 22–27 inch lbs. (2.5–3.1 Nm) and apply sealant to the front and rear seal of the oil pan
- New O-rings to the cylinder block and the oil pump body

12. Apply a 4.5–5.5mm wide continuous bead of liquid gasket to the upper oil pan mating surface and install the oil pan. Torque the bolts in sequence to 12–14 ft. lbs. (16–19 Nm).

13. Install or connect the following:
- Oil pan strainer and torque the bolts to 12–14 ft. lbs. (16–19 Nm)
- Rear cover plate and lower transaxle bolts
- A/C compressor and bracket
- Drive belts
- Center crossmember
- Front and rear engine mount hardware
- CKP sensors
- Front exhaust tube and support
- Oil strainer
- Steel oil pan and torque the bolts, in sequence, to 66 inch lbs. (7.5 Nm)
- Engine under covers
- Negative battery cable

14. After waiting approximately 30 minutes, fill the engine with clean oil.

15. Start the vehicle, check for leaks and repair if necessary.

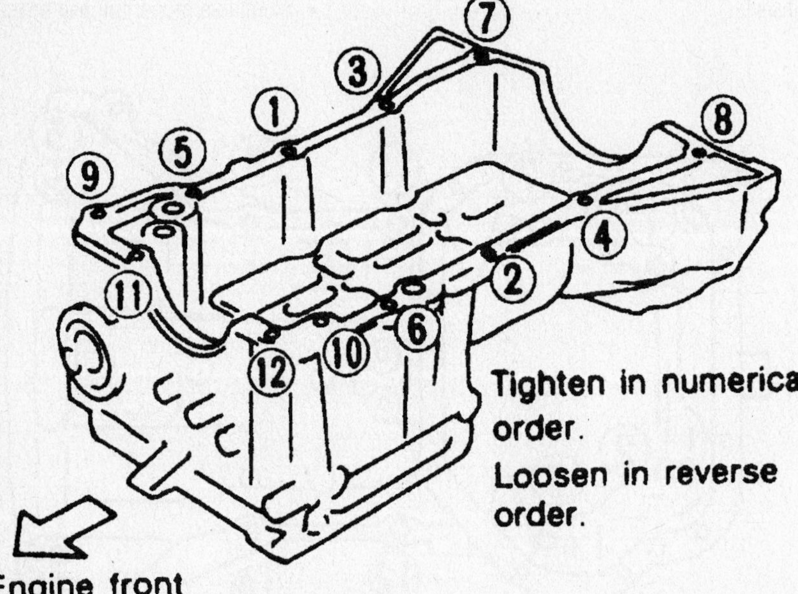

To prevent pan warpage, tighten the aluminum oil pan bolts in the sequence shown—3.0L engines

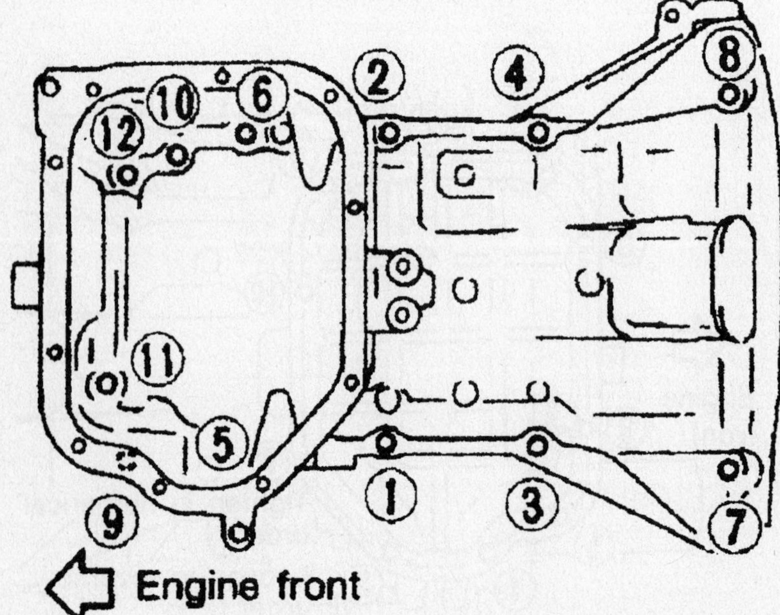

To prevent pan warpage, tighten the aluminum oil pan bolts in the sequence shown—3.5L engines

Oil Pump

REMOVAL & INSTALLATION

1.8L and 2.0L Engines

1. Before servicing the vehicle, refer to the precautions in the beginning of this section.
2. Drain the engine oil
3. Remove or disconnect the following:
   - Negative battery cable
   - Drive belts
   - Cylinder head
   - Oil pan and strainer
   - Engine front cover
   - Oil pump from the front cover

To install:

4. Install or connect the following:
   - Oil pump cover and torque the long bolt to 70 inch lbs. (8 Nm) and the short bolt to 44 inch lbs. (5 Nm)
   - Front cover and torque the bolts to 43 ft. lbs. (58 Nm)
   - Oil strainer and oil pan
   - Cylinder head
   - Drive belts
   - Negative battery cable
5. Fill the engine with clean oil.

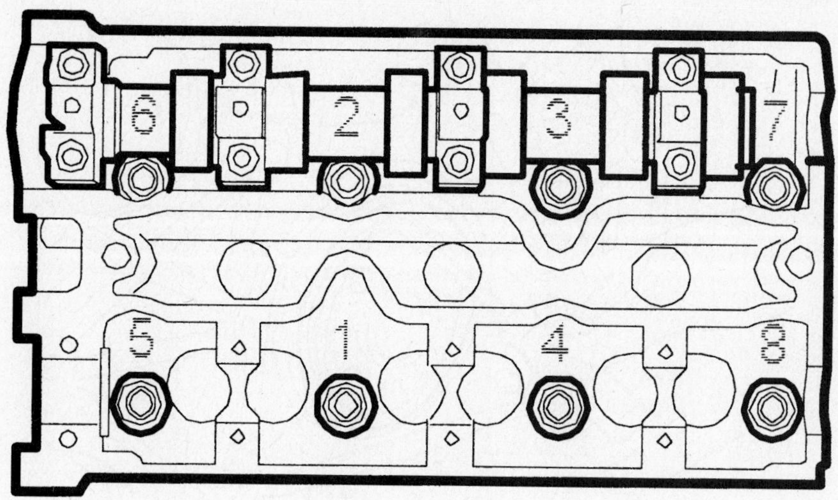

Exploded view of the oil pump assembly–2.0L engine

6. Start the vehicle, check for leaks and repair if necessary.

2.4L Engine

1. Before servicing the vehicle, refer to the precautions in the beginning of this section.
2. Drain the engine oil
3. Remove or disconnect the following:
   - Negative battery cable
   - Engine front cover
   - Oil pump cover and pump

To install:

4. Install or connect the following:
   - Oil pump
   - Oil pump cover and torque the long bolt to 15 ft. lbs. (20 Nm) and shorter bolt to 69 inch lbs. (8 Nm)
   - Front cover
   - Negative battery cable
5. Fill the engine with clean oil.
6. Start the vehicle, check for leaks and repair if necessary.

2.5L Engine

1. Before servicing the vehicle, refer to the precautions in the beginning of this section.
2. Drain the engine oil
3. Remove or disconnect the following:
   - Negative battery cable
   - Engine front cover
   - Oil pump cover and pump

To install:

4. Install or connect the following:
   - Oil pump
   - Oil pump cover and torque the bolts to 61 inch lbs. (7 Nm) and shorter bolt to 69 inch lbs. (8 Nm)
   - Front cover
   - Negative battery cable
5. Fill the engine with clean oil.
6. Start the vehicle, check for leaks and repair if necessary

3.0L and 3.5L Engines

1. Before servicing the vehicle, refer to the precautions in the beginning of this section.
2. Drain the engine oil
3. Remove or disconnect the following:
   - Negative battery cable
   - Drive belts
   - Camshaft Position (CMP) sensor (PHASE) and the Crankshaft Position (CKP) sensor (REF)/(POS)
   - Engine lower covers
   - Crankshaft pulley
   - Front exhaust tube and support
   - Right side mounting insulator and bracket
   - Center member
   - A/C compressor and move it aside
   - Oil pans
   - Water pump cover
   - Front cover
   - Timing chain
   - Oil pump assembly
4. Clean all mating surfaces.

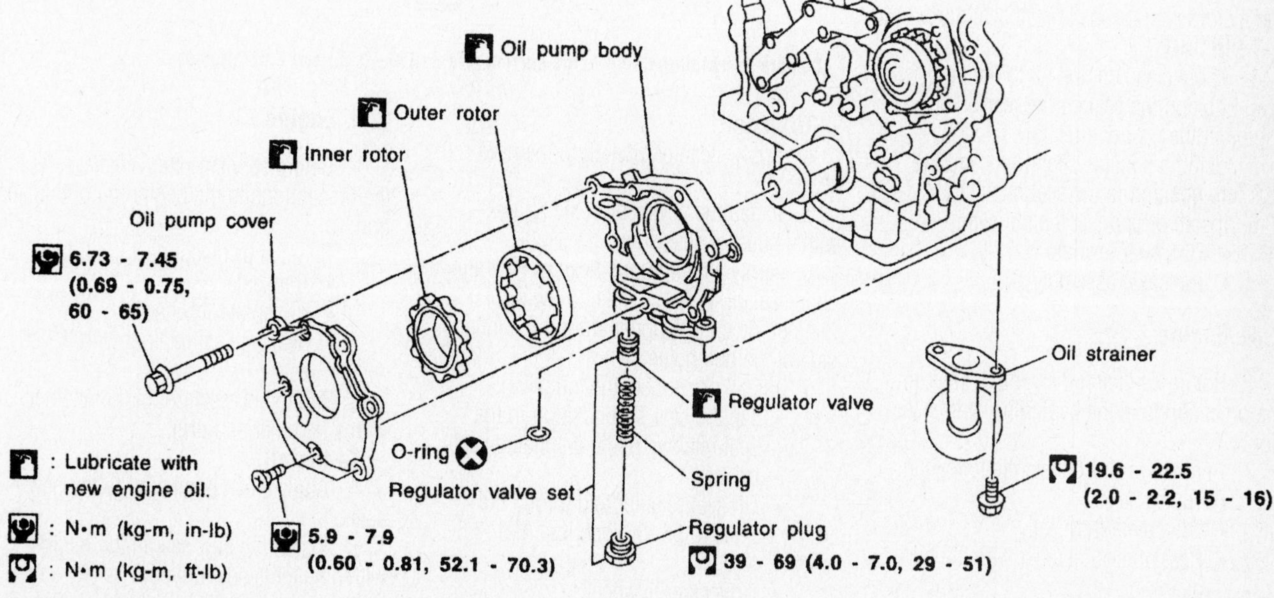

Exploded view of the oil pump assembly–3.0L and 3.5L engines

To install:

5. Install or connect the following:
   - Oil pump
   - Timing chain
   - Front cover and torque the long bolt to 95 inch lbs. (11 Nm) and the short bolt to 71 inch lbs. (8 Nm)
   - Water pump cover
   - Oil pans
   - A/C compressor
   - Center member
   - Right side mounting insulator and bracket
   - Front exhaust tube and support
   - Crankshaft pulley
   - CMP and CKP sensors
   - Engine lower covers
   - Drive belts
   - Negative battery cable
6. Fill the engine with clean oil
7. Start the vehicle, check for leaks and repair if necessary.

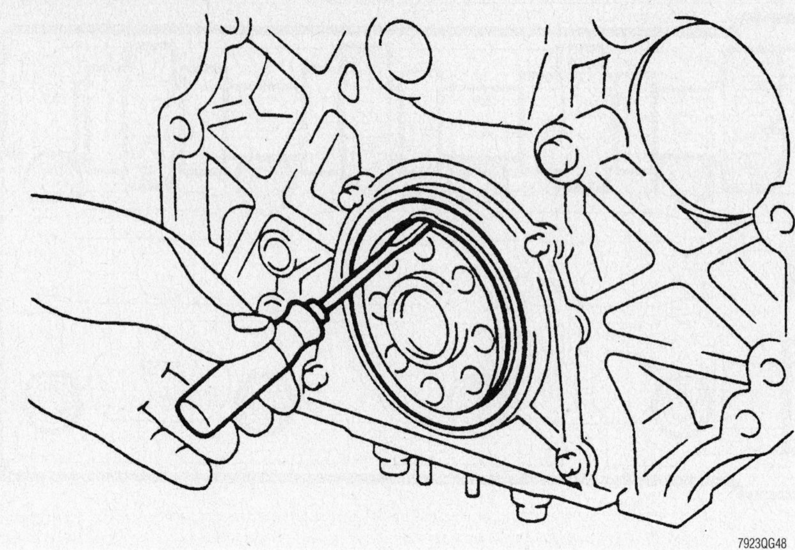

Carefully pry the rear main seal out of the retainer on the rear of the engine—1.8L and 2.0L engines

Rear Main Seal

REMOVAL & INSTALLATION

1.8L and 2.0L Engines

1. Before servicing the vehicle, refer to the precautions in the beginning of this section.
2. Remove or disconnect the following:
   - Transaxle
   - Driveplate/ flywheel
   - Oil seal retainer
3. Carefully pry the seal from the retainer. Be sure not to scratch the sealing surface of the crankshaft or oil seal bore.

To install:

4. Apply clean engine oil to the new seal. Position the seal on the rear of the engine in the proper direction.
5. Using a suitable seal driver, tap the seal into position in the seal retainer.
6. Install or connect the following:
   - Flywheel/flexplate
   - Transaxle assembly

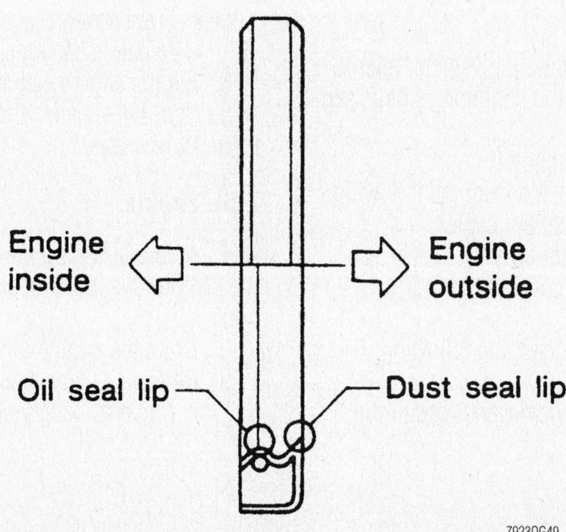

Be sure to install the seal in the correct orientation—1.8L and 2.0L engines

2.4L Engine

1. Before servicing the vehicle, refer to the precautions in the beginning of this section.
2. Remove or disconnect the following:
   - Transaxle
   - Driveplate/flywheel
   - Rear oil seal retainer with the oil seal
3. Tap the oil seal out of the retainer with a hammer and drift.

To install:

4. Apply clean engine oil to the new seal.
5. Install the new seal in the retainer with a suitable seal driver.
6. Apply a continuous bead of RTV silicone sealant, 2–3mm wide, to the seal retainer. Be sure to apply around the inner side of the bolt holes.
7. Install or connect the following:
   - Tap the seal into position in the seal retainer, using a suitable seal driver
   - Oil seal retainer and torque the bolts to 56–66 inch lbs. (6.5–7.5 Nm)
   - Driveplate/flywheel
   - Transaxle

2.5L Engine

1. Before servicing the vehicle, refer to the precautions in the beginning of this section.
2. Remove or disconnect the following:
   - Transaxle
   - Driveplate/flywheel
   - Rear oil seal retainer with the oil seal
3. Tap the oil seal out of the retainer with a hammer and drift.

To install:

4. Apply clean engine oil to the new seal.
5. Install the new seal in the retainer with a suitable seal driver.
6. Apply a continuous bead of RTV silicone sealant, 2–3mm wide, to the seal

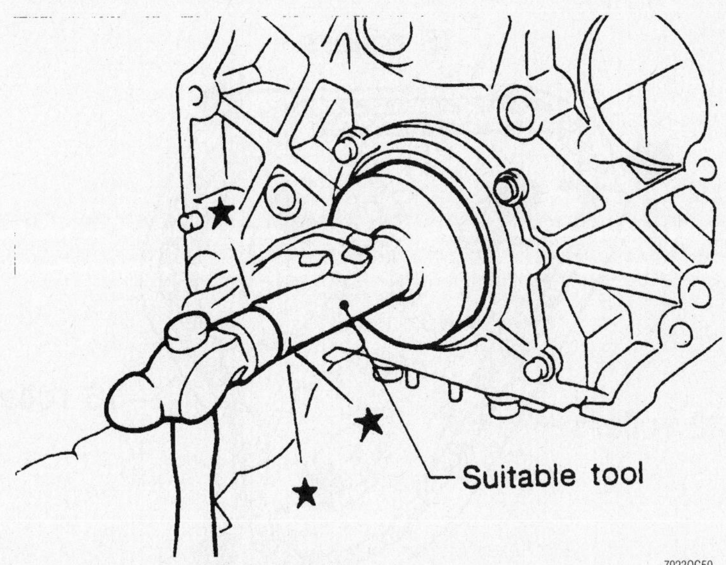

Install the rear main seal using a suitable driver—1.8L and 2.0L engines

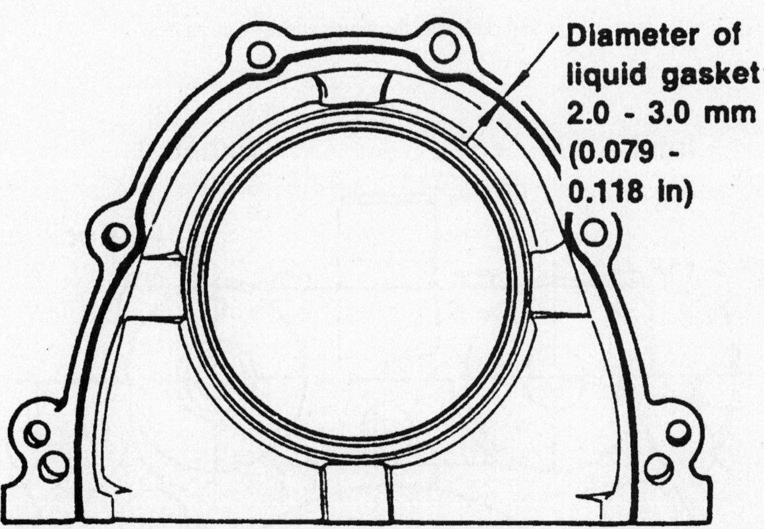

Apply sealant to the seal retainer as shown—2.4L engine

retainer. Be sure to apply around the inner side of the bolt holes.

7. Install or connect the following:
   - Tap the seal into position in the seal retainer, using a suitable seal driver
   - Oil seal retainer and torque the bolts to 56–66 inch lbs. (6.5–7.5 Nm)
   - Driveplate/flywheel
   - Transaxle

3.0L and 3.5L Engines

1. Before servicing the vehicle, refer to the precautions in the beginning of this section.
2. Drain the engine oil.
3. Remove or disconnect the following:
   - Transaxle
   - Driveplate/flywheel
   - Oil pan
   - Oil seal retainer
4. Tap the oil seal out of the retainer with a hammer and drift.
5. Clean all mating surfaces of any residual liquid gasket.

To install:

6. Install or connect the following:
   - New seal into the retainer
   - Oil seal retainer
   - Oil pan
   - Driveplate/flywheel
   - Transaxle
7. Fill the engine with clean oil.
8. Start the vehicle, check for leaks and repair if necessary.

Timing Chain, Sprockets, Front Cover and Seal

REMOVAL & INSTALLATION

1.8L Engine

1. Before servicing the vehicle, refer to the precautions in the beginning of this section.
2. Relieve the fuel system pressure.
3. Drain the cooling system.
4. Remove or disconnect the following:
   - Negative battery cable
   - Upper radiator hose
   - Engine drive belts
   - Power steering pulley and the pump with bracket
   - Air duct from the intake manifold collector
   - Right front wheel and engine side covers
   - Engine undercovers
   - Front exhaust pipe
   - Cylinder head front mounting bracket
   - Cylinder head cover from the engine
   - Rocker cover
   - Distributor cap
   - Spark plugs
   - Intake manifold support and set the No. 1 piston at the Top Dead Center (TDC) compression stroke
   - Distributor
   - Cylinder head front cover
   - Water pump pulley
   - Thermostat housing
   - Lower timing chain tensioner
   - Upper timing chain tensioner and slack side timing chain guide
   - Idler sprocket bolt
   - Camshaft sprocket bolts and the sprockets from the camshafts. Be sure to mark the sprockets for proper reinstallation.
   - Camshaft mounting caps by loosening the bolts in 2 or 3 steps
   - Camshafts from the engine
   - Idler sprocket bolt
   - Cylinder head with the manifolds
   - Idler sprocket shaft from the rear side
   - Upper timing chain and support the engine assembly
   - Center crossmember
   - Oil pan and strainer assembly
   - Crankshaft pulley

- Engine front mount and bracket
- Bolts that secure the front timing cover and the cover from the engine. Once the timing chain cover is removed, drive out the old oil seal.
- Idler sprocket and the lower timing chain
- Oil pump drive spacer and the crankshaft sprocket
- Timing chain guide

To install:

5. Drive a new oil seal into the front cover. Lubricate the oil seal lip with clean engine oil.
6. Confirm that No. 1 piston is set at Top Dead Center (TDC) on compression stroke.
7. Install or connect the following:
   - Crankshaft sprocket with the marks of the sprocket facing the front of the engine
   - Oil pump drive spacer and the chain guide
   - Lower timing chain. Set the chain by aligning its mating mark with the one on the crankshaft sprocket. Be sure the sprocket's mating mark faces the front of the engine.

➡**The number of links between the alignment marks are the same for the left and the right side.**

   - Crankshaft sprocket and the lower timing chain. Set the timing chain by aligning its mating mark with the one on the crankshaft sprocket. Be sure sprocket's mating mark faces engine front.
   - Front cover assembly, using liquid gasket
   - Engine front mounting bracket and the engine mount
   - Oil strainer, oil pan assembly and the crankshaft pulley
   - Center crossmember
8. Set the idler sprocket by aligning the mating mark on the larger sprocket with the silver mating mark on the lower timing chain.
   - Upper timing chain and set it by aligning the mating mark on the smaller sprocket with the silver mating marks on the upper timing chain. Be sure sprocket marks face engine front.
   - Idler sprocket shaft to the rear side
   - Cylinder head assembly
   - Idler sprocket bolt. Be sure to lubricate the bolt with clean engine oil.
   - Exhaust and intake camshafts. The camshafts and marked **I** for intake and **E** for exhaust.

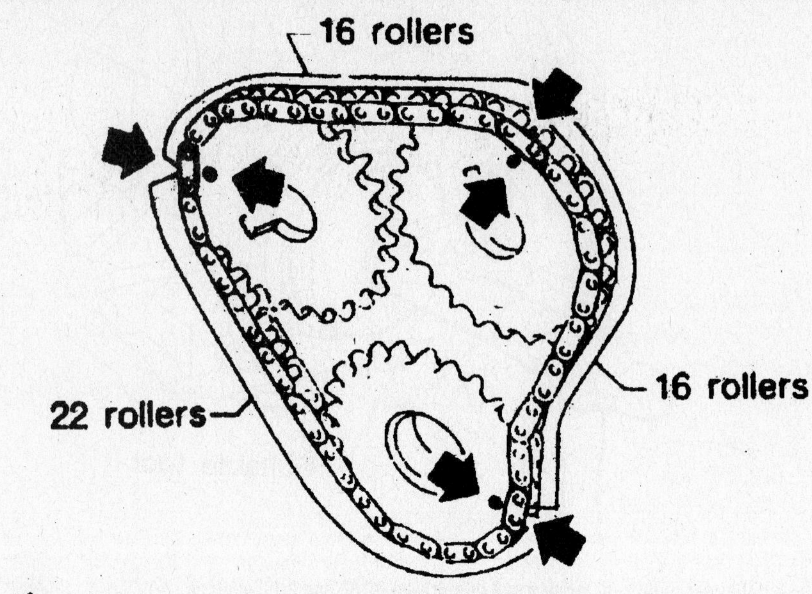

Be sure to align the camshaft sprockets with the timing chain—1.8L engine

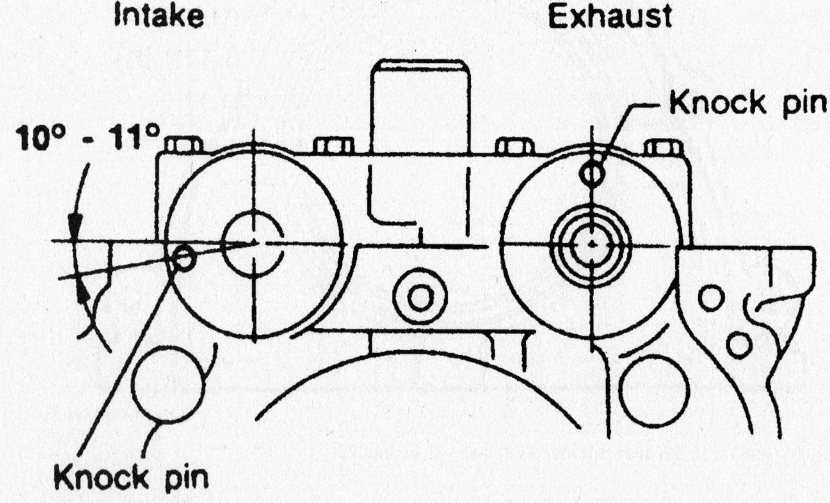

Positioning of camshaft knock pins during assembly—1.8L engine

9. Position the intake camshaft knock pin at the 9 o'clock position and the exhaust camshaft knock pin at the 12 o'clock position.
10. Install or connect the following:
    - Camshaft bearing caps and distributor bracket. Apply liquid sealant to the distributor bracket.
11. Torque the mounting bolts in sequence as follows:
    a. Bolts 11–15, then bolts 1–10: 18 inch lbs. (2.0 Nm).
    b. Bolts 1–15: 52 inch lbs. (6 Nm).
    c. Bolts 1–14: 104 inch lbs. (12 Nm).
    d. Bolt 15: 73 inch lbs. (8 Nm).
12. Install or connect the following:
    - Camshaft sprockets with timing chain. Set the camshaft sprockets by aligning the mating marks of the timing chain with the marks on the camshaft sprockets.
    - Camshaft sprocket bolts. Torque the bolts to 86 ft. lbs. (117 Nm). Be sure to lubricate the bolts with clean engine oil.
    - Upper timing chain tensioner. Before installation of the tensioner, install a suitable pin to hold the tensioner in the relaxed position.

After installing the chain tensioner, remove the pin.
- Lower timing chain tensioner. Be sure the notch of the gasket is positioned down.
- Thermostat housing
- Water pump pulley
- Cylinder head front cover
- Distributor
- Intake manifold support
- Spark plugs and leads
- Distributor cap
- Rocker cover
- Cylinder head cover
- Front exhaust pipe
- Engine under covers
- Engine side covers and the right front wheel
- Air duct to the intake manifold collector
- Power steering pulley and oil pump
- Drive belts
- Upper radiator hose
- Negative battery cable

13. Fill the cooling system.
14. Start the vehicle, check for leaks and repair if necessary.

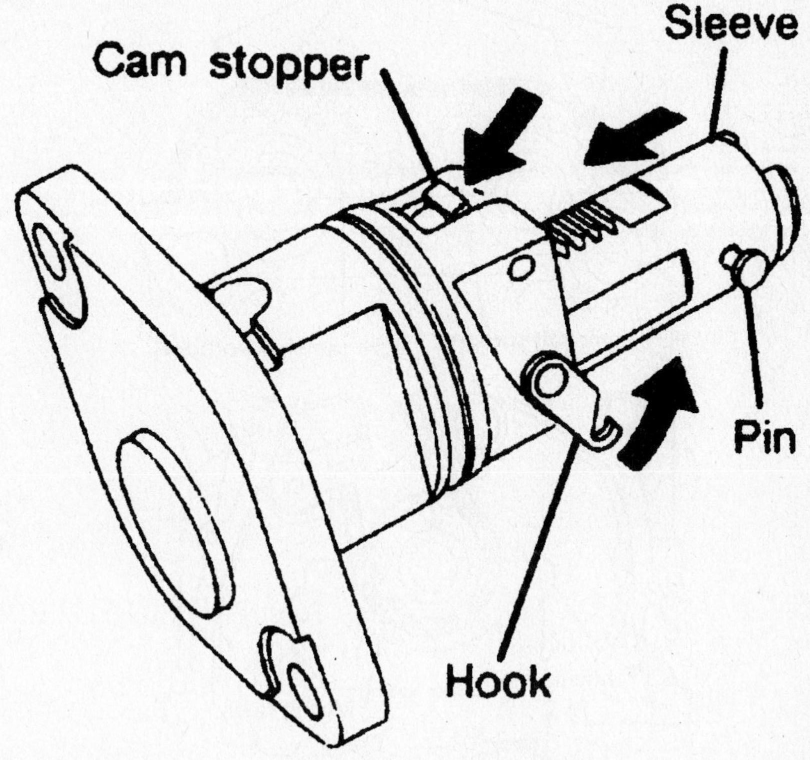

Timing chain tensioner—2.0L engines

2.0L Engine

1. Before servicing the vehicle, refer to the precautions in the beginning of this section.
2. Relieve the fuel system pressure.
3. Drain the cooling system.
4. Remove or disconnect the following:
   - Negative battery cable
   - Radiator
   - Right front wheel and engine side cover
   - Spark plugs and rotate the engine and position the No. 1 cylinder to Top Dead Center (TDC)
   - Air duct to the intake manifold
   - Drive belts and the water pump pulley
   - Alternator and the power steering pump from the engine
   - Vacuum hoses, fuel hoses and the wiring harness connectors
   - Cylinder head cover
   - Intake manifold supports
   - Oil filter bracket and the power steering pump bracket
   - Timing chain tensioner
   - Distributor
   - Timing chain guide and holding the flats of the camshaft sprockets, remove the bolts that secure the sprockets
   - Timing chain sprockets from the camshafts
   - Oil tubes, baffle plate and camshaft brackets
   - Camshafts from the cylinder head
   - Starter motor
   - Coolant hoses from the engine block
   - Knock sensor (KS) harness connector
   - Exhaust Gas Recirculation (EGR) tube
   - Cylinder head
   - Oil pan, oil strainer and the baffle plate
   - Crankshaft pulley using a suitable puller
   - Engine front mount
   - Front cover and oil pump drive spacer
   - Timing chain guides and timing chain. Check the timing chain for excessive wear at the roller links.

To install:

5. Clean all gasket mating surfaces.
6. Install or connect the following:
   - Crankshaft sprocket. Position the crankshaft so that No. 1 piston is set at TDC (keyway at 12 o'clock, mating mark at 4 o'clock) fit timing chain to crankshaft sprocket so the mating mark is in line with mating mark on crankshaft sprocket. The mating marks on timing chain for the camshaft sprockets should be silver. The mating mark on the timing chain for the crankshaft sprocket should be gold.
   - Timing chain to the crankshaft sprocket and install the timing chain guides. Tighten the timing chain guides to 10–14 ft. lbs. (13–19 Nm). Drape the timing chain over the left chain guide.
   - Oil pump drive spacer to the crankshaft
7. Apply a continuous bead of liquid sealant to the front timing cover and install the cover. Tighten the front cover mounting bolts to 57–66 inch lbs. (6.4–7.5 Nm).
   - Right front engine mount
   - Crankshaft pulley and torque the bolt to 105–112 ft. lbs. (142–152 Nm). Be sure the No. 1 piston is at TDC.
   - Oil strainer, baffle plate and the oil pan assembly
   - Cylinder head assembly. Be sure to apply a bead of sealant to the joint of the block and front timing cover.
   - EGR tube
   - KS harness connector

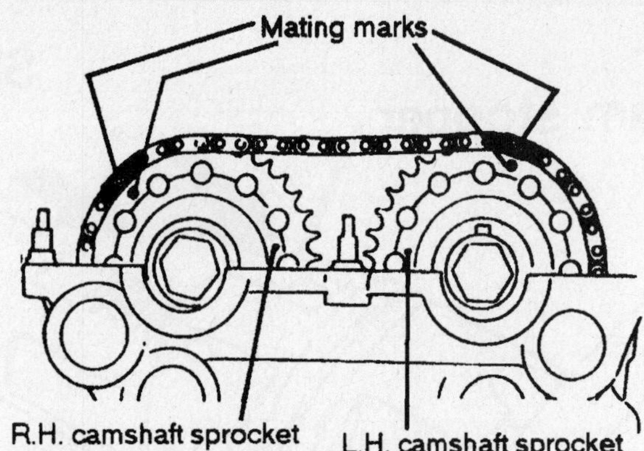

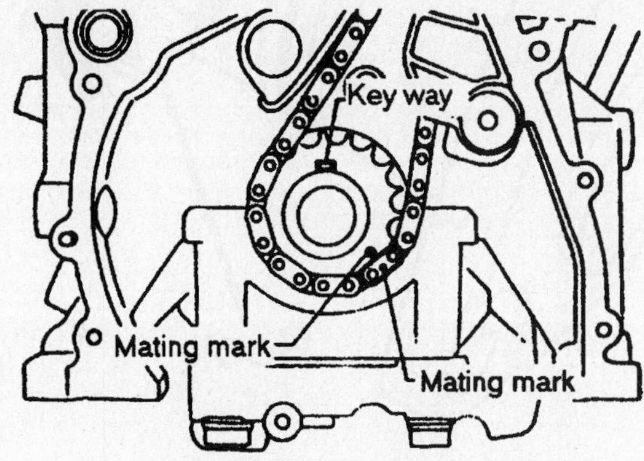

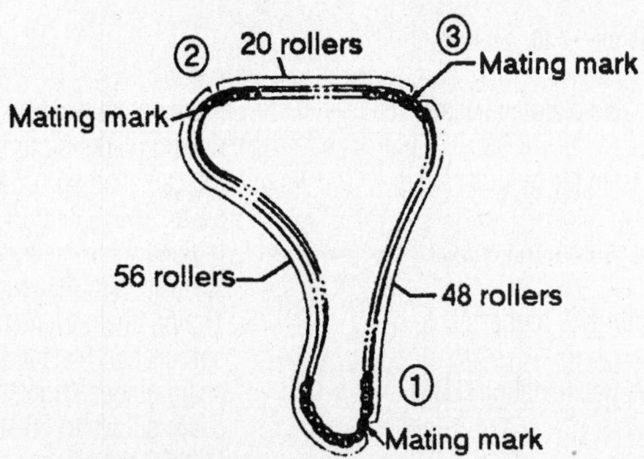

7923QG55

Timing chain sprocket alignment marks—2.0L engines

- Coolant hoses to the engine block, using new hose clamps
- Starter motor
- Camshafts, camshaft bearing caps, oil tubes and the baffle plate

➡**When installing the camshafts, be sure to position the left-hand and right-hand camshaft keys at 12 o'clock. Also be sure the camshaft brackets are facing in the correct direction.**

- Camshaft sprockets by lining up the mating marks on the timing chain with the mating marks on the camshaft sprockets and torque the bolts to 101–116 ft. lbs. (137–157 Nm)
- Timing chain guide and distributor
- Chain tensioner. Press the cam stopper down and the press-in sleeve until the hook can be engaged on the pin. When tensioner is bolted in position the hook will release automatically.

➡**Ensure the arrow on the outside of the tensioner faces the front of the engine.**

- Oil filter bracket and power steering pump bracket
- Intake manifold supports
- Cylinder head cover
- Vacuum hoses and fuel lines
- Alternator and power steering pump
- Water pump pulley and drive belts
- Air duct
- Spark plugs after making certain that the No. 1 piston is at the TDC position
- Engine side cover and right front wheel
- Radiator
- Negative battery cable

8. Fill the cooling system.
9. Start the engine, check for leaks and repair if necessary.

2.4L Engine

1. Before servicing the vehicle, refer to the precautions in the beginning of this section.
2. Drain the cooling system.
3. Drain the engine oil.
4. Remove or disconnect the following:
   - Negative battery cable
   - Spark plug wires and set the No. 1 piston at the Top Dead Center (TDC) of the compression stroke
   - Engine undercover
   - Vacuum hoses, fuel hoses, wires, harness and connectors
   - Drive belts
   - Power steering reservoir
   - Alternator and bracket, the upper radiator hose, the air duct and the front exhaust tube
   - Intake manifold collector supports, intake manifold collector and the exhaust manifold
   - Distributor
5. Using a block of wood, set a transmission jack under the aluminum oil pan and remove the front engine mounting.
   - Rocker cover. Remove the rocker cover bolts in the proper sequence.
   - Camshaft sprockets

➡**The stoppers on camshaft covers prevent the upper timing chain from disengaging from the idle sprocket.**

- Cam bearing caps in sequence
- Camshafts. The camshaft brackets must be loosened in reverse order of tightening to prevent damage to the camshaft.

➡**These parts must be reassembled in their original positions.**

- Cylinder head bolts in the reverse order of installation

➡**A warped or cracked cylinder head could result from loosening in incorrect order. The cylinder head bolts should be loosened in 2 or 3 steps.**

- Cam sprocket cover
- Upper chain tensioner and upper chain guides
- Upper timing chain
- Idler sprocket bolt
- Cylinder head and the intake manifold
- Cylinder head gasket. The lower timing chain will not be disengaged from crankshaft sprocket.

➡**The cast portion of the front cover is located on the lower side of the crankshaft sprocket, so the lower timing chain need not be disengaged from idler sprocket.**

- Steel oil pan bolts in the reverse sequence of the tightening procedure

6. Install a seal cutter between the steel oil pan and the aluminum oil pan.
7. Tapping the cutter with a hammer, slide it around the entire edge of the oil pan. Take care not to damage the aluminum oil pan.
8. Remove or disconnect the following:
   - Steel oil pan
   - Baffle plate, oil strainer and the front tube
9. Support the transaxle with a jack and the engine with a engine hoist.
   - Front suspension member
   - A/C compressor gussets
   - Rear cover plate
   - Aluminum oil pan retaining bolts in sequence
10. Insert a seal cutter between the oil pan and the cylinder block.
11. Tapping the cutter with a hammer, slide it around the entire edge of the oil pan. Take care not to damage the aluminum oil pan.
   - Oil pan from the engine
   - Crankshaft pulley
   - Front timing chain cover
   - Oil pump drive spacer
   - Lower timing chain tensioner, tensioner arm and lower timing chain guide
   - Lower timing chain and idler sprocket

To install:

12. Install or connect the following:
   - Crankshaft sprocket and oil pump drive spacer
   - Idler sprocket and lower timing chain
13. Set the lower timing chain on the sprockets, aligning the mating marks. The mating marks on the timing chain assembly will be silver.
   - Chain tension arm and chain guide
   - Lower timing chain tensioner
14. Apply a continuous bead of liquid gasket to the front cover and install the front cover. Install a new oil seal.
   - Crankshaft pulley and torque tighten bolt to 105–112 ft. lbs. (142–152 Nm)
15. Carefully scrape the old gasket material away from the pan and cylinder block mounting surfaces, then apply a continuous 3.5–4.5mm wide bead of liquid gasket around the oil pan and the cylinder block.
   - Aluminum oil pan and torque the bolts, in sequence, to 13 ft. lbs. (17.5 Nm)
   - Baffle plate, oil strainer and the front tube
   - Steel oil pan and torque the bolts, in sequence, to 61 inch lbs. (7 Nm)
   - Rear cover plate
   - A/C compressor gussets
   - Front suspension member
   - Front engine mounting and remove the engine hoist and transaxle support
   - New cylinder head gasket
   - Cylinder head and temporarily tighten the cylinder head bolts when installing the front cover. This is necessary to avoid damaging the cylinder head gasket. Be sure to install washers between the bolts and cylinder head.
   - Upper timing chain, chain tensioner and chain guide
16. Set the upper timing chain on idler sprockets, aligning the mating marks.
   - Cam sprocket cover. Apply a continuous bead of liquid gasket to front cover. Be careful not to damage the cylinder head gasket. Be careful that the upper timing chain does not slip or jump when installing cam sprocket cover.
17. Tighten cylinder head bolts.
   - Camshafts and camshaft bearing caps
   - Camshaft sprockets, then torque the sprocket bolts to 123–130 ft. lbs. (167–176 Nm). Install the chain guide between both camshaft sprockets. The alignment marks on the upper portion of the timing chain should now be aligned.
   - Rocker cover

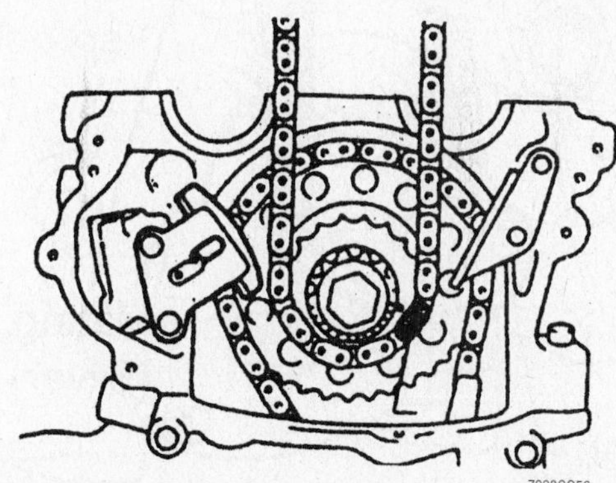

Be sure to align the mark on the idler sprocket with the mark on the upper chain—2.4L engines

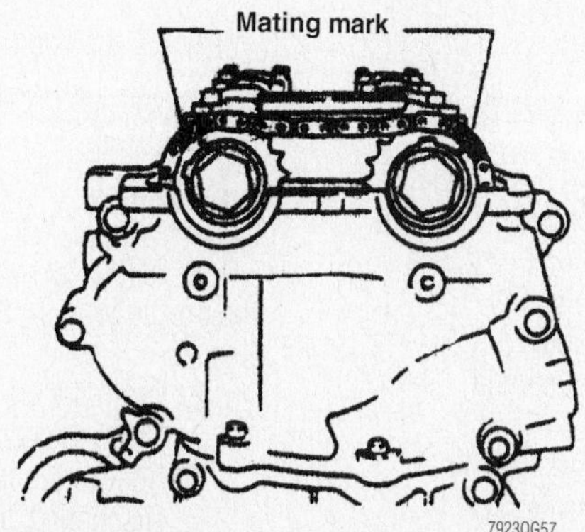

Align the marks on the camshaft sprockets with the upper portion of the upper timing chain and mating marks—2.4L engine

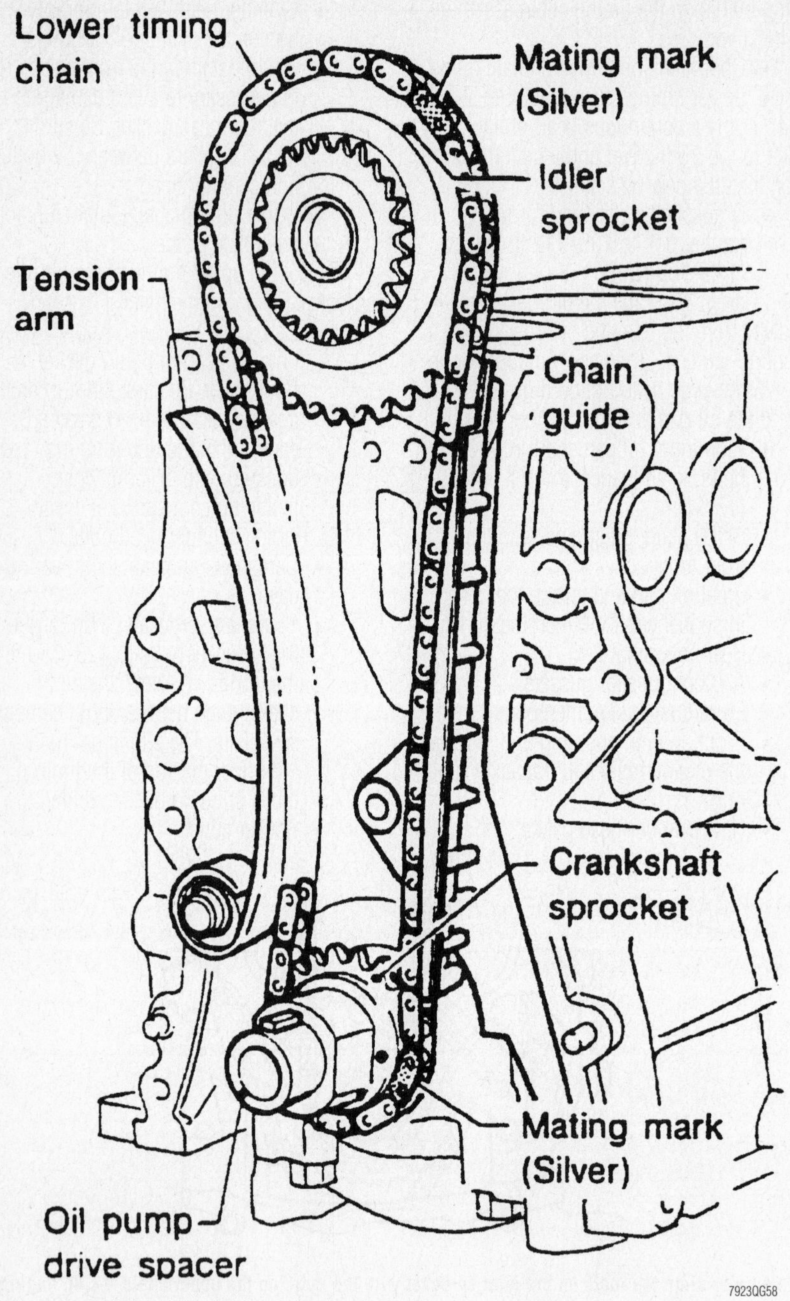

Lower timing chain alignment marks—2.4L engine

- Distributor
- Intake manifold collector supports and collector
- Exhaust manifold
- Alternator and bracket
- Upper radiator hose
- Air duct assembly
- Front exhaust tube
- Power steering reservoir
- Drive belts
- Vacuum hoses and fuel lines
- Engine under cover
- Spark plug wires
- Negative battery cable

18. Fill the cooling system.
19. Fill the engine with clean oil.
20. Start the vehicle, check for leaks and repair if necessary.

2.5L Engine

1. Before servicing the vehicle, refer to the precautions in the beginning of this section.
2. Drain the engine oil.
3. Drain the cooling system.
4. Relieve the fuel system pressure.
5. Disconnect the negative battery cable.
6. Support the engine and transaxle with suitable tools.
7. On Altima, remove or disconnect the following:
   - Engine under cover
   - Upper and lower oil pans
   - Alternator
   - Engine cover
   - Variable timing control solenoid conector
   - Engine ground
   - Coolant reservoir tank
   - Right side fuse and relay box and move aside
   - Right engine mount and bracket
8. On Sentra, remove or disconnect the following:
   - Air cleaner and duct
   - Spark plugs
   - Valve cover
   - Coolant reservoir tank
   - Auxiliary drive belt tensioner
   - Alternator
   - Strut tower cross brace
   - A/C compressor and wire aside
   - Power steering pump and reservoir and wire aside
   - Upper and lower oil pans
9. On all models, remove or disconnect the following:
   - Intake valve timing control cover in reverse of the tightening sequence
   - Timing chain guide between the camshaft sprockets and through the front cover
   - Rotate the crankshaft clockwise and set the No. 1 piston to TDC of the compression stroke
   - Crankshaft pulley using a suitable puller
   - Front timing chain cover bolts in reverse order of the tightening sequence
   - Front timing chain case
   - Insert a suitable stopper pin in the timing chain tensioner
   - Chain tensioner
10. While holding the camshaft with an open end wrench on the camshaft flat, remove the camshaft sprocket bolts and remove the camshaft sprockets.
11. Remove the timing chain slack guide, tension guide, timing chain and oil pump drive spacer.
12. Lift up on the timing chain tensioner lever for the balancer and release the ratchet.
13. Push the tensioner sleeve in and hold.
14. Insert a stopper pin to secure the tensioner sleeve
15. Remove the balancer tensioner.
16. Remove the timing chain for the balancer unit.

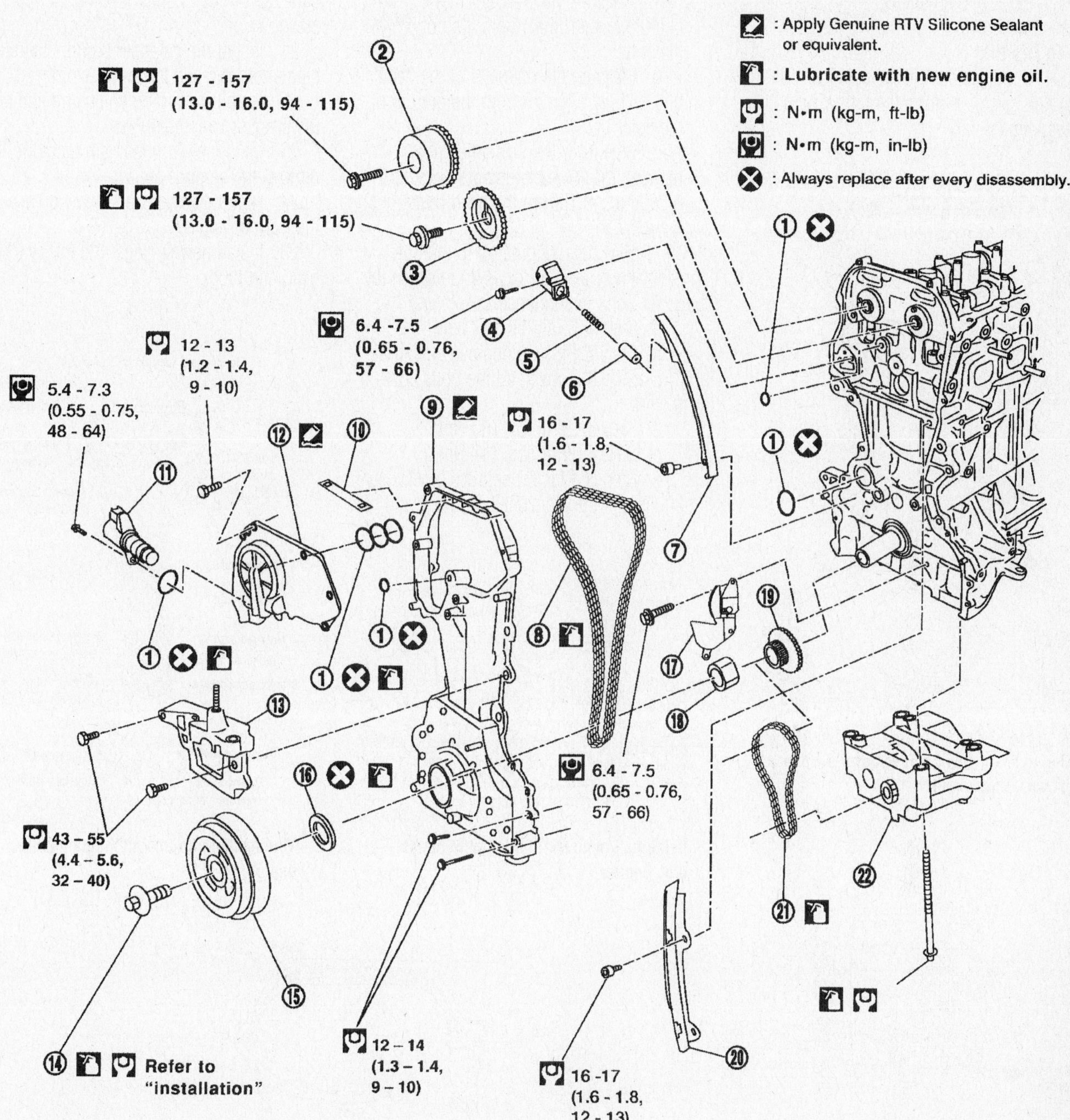

1. O rings
2. Camshaft sprocket (INT)
3. Camshaft sprocket (EXH)
4. Chain tensioner
5. Spring
6. Chain tensioner plunger
7. Timing chain slack guide
8. Timing chain
9. Front cover
10. Chain guide
11. IVT control solenoid valve
12. IVT control cover
13. Engine mounting bracket
14. Crankshaft pulley bolt
15. Crankshaft pulley
16. Front oil seal
17. Balancer unit timing chain tensioner
18. Oil pump drive spacer
19. Crankshaft sprocket
20. Timing chain tension guide
21. Balancer unit timing chain
22. Balancer unit

67170-NISS-G36A

Exploded view of the timing chain assembly—2.5L engine

17. Remove the balancer unit using the sequence shown.

To install:

18. If replacement of the balancer bolts is not possible, perform the following bolt measurement:

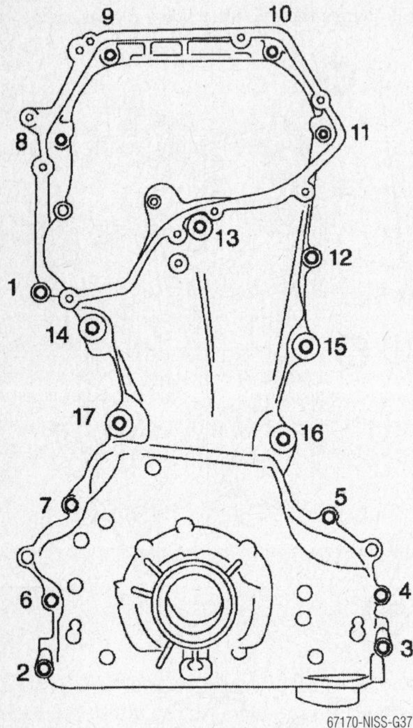

Timing chain cover bolt removal sequence shown—2.5L engine

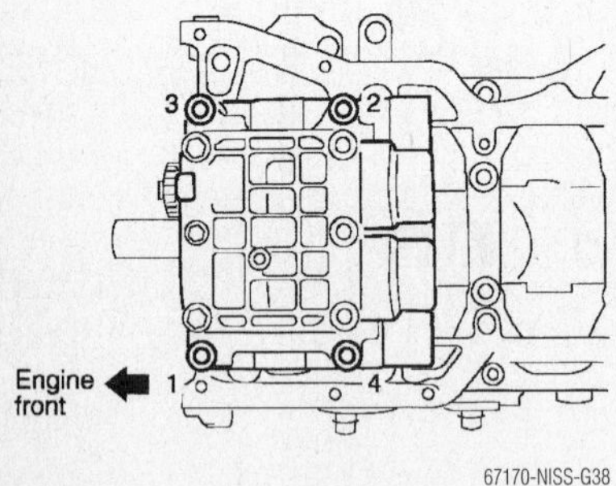

Balancer unit mounting bolts removal sequence shown—2.5L engine

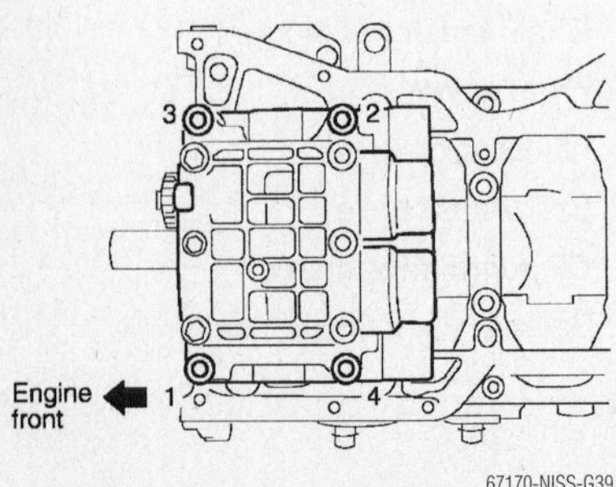

Balancer unit mounting bolts tightening sequence shown—2.5L engine

a. Measure the diameter of the head bolt 0.39 in. (10mm) from the bottom of the bolt.

b. Measure the diameter of the head bolt 2.75 in. (70mm) from the bottom of the bolt.

c. Whenever the size difference between the 2 measurements exceeds 0.0059 in. (0.15mm) the head bolts must be replaced.

19. Apply clean engine oil to the bolt threads and seating surfaces and tighten the balancer bolts in sequence as follows:

a. Step 1: 35 ft. lbs. (47 Nm).

b. Step 2: Plus 90 degrees clockwise

c. Step 3: Loosen all the bolts completely.

d. Step 2: 29 ft. lbs. (33 Nm).

e. Step 4: 35 ft. lbs. (47 Nm).

f. Step 5: Plus 90 degrees clockwise

20. Install the crankshaft sprocket and balancer timing chain and align the marks as shown.

21. Install the balancer timing chain tensioner and remove the stopper pin.

22. Install the timing chain and and align the timing marks as shown.

23. Install the timing chain tensioner and remove the stopper pin.

24. Using adrift, install a new oil seal in the front cover.

25. Install new O-rings into the cylinder head and block.

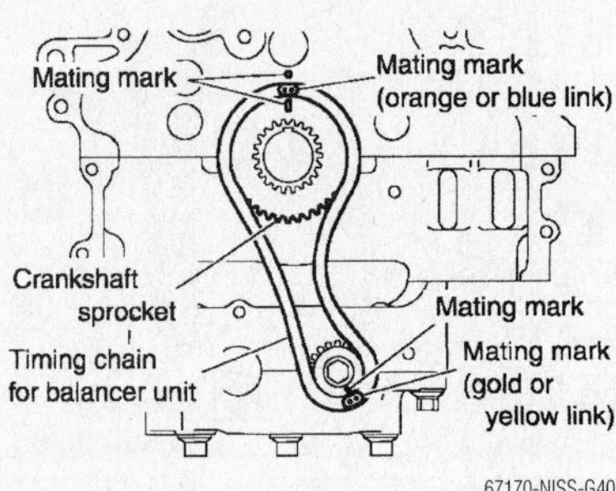

Balancer timing chain alignment marks—2.5L engine

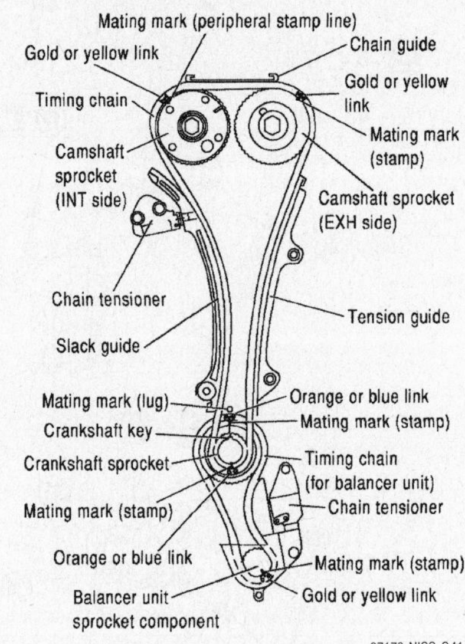

Timing chain alignment marks—2.5L engine

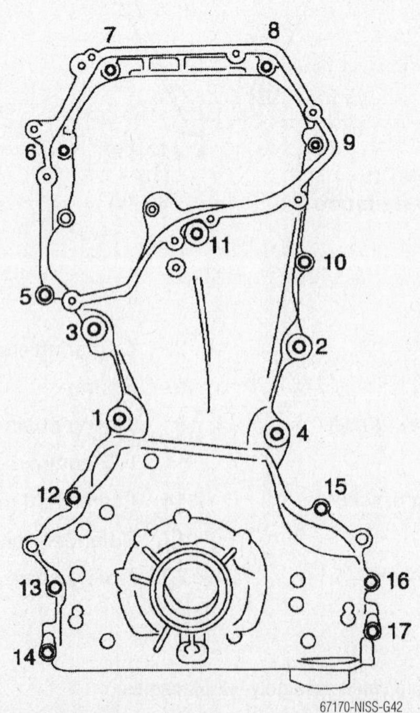

Timing chain cover bolt removal sequence shown—2.5L engine

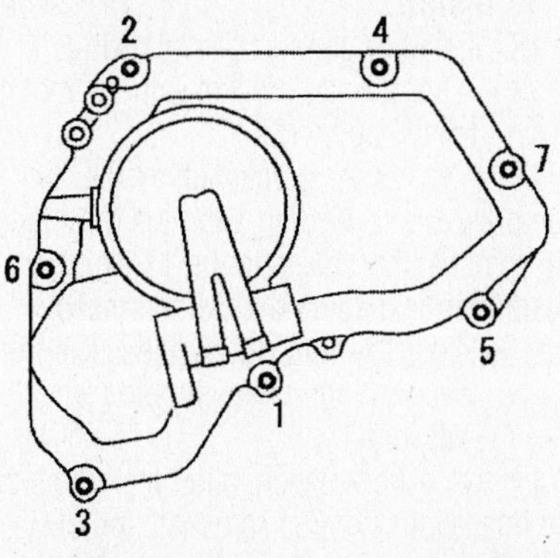

Intake valve timing control cover tightening sequence shown—2.5L engine

26. Apply a bead of sealant around the inside of the front cover sealing surface.
27. Install the front cover and tighten the bolts in the sequence shown.
28. Install the chain guide between camshaft sprockets.
29. Install the intake valve timing control cover in the sequence shown and tighten the bolts to 10 ft. lbs. (13 Nm).
30. Install the crankshaft pulley and tighten the bolt to 31 ft. lbs. (42 Nm), plus an additional 60°.
31. The remainder of the installation is the reverse of the removal procedure.

3.0L and Altima and Maxima 3.5L Engines

1. Before servicing the vehicle, refer to the precautions in the beginning of this section.
2. Drain the engine oil.
3. Drain the cooling system.
4. Relieve the fuel system pressure.
5. Remove or disconnect the following:
    - Negative battery cable
    - Left side rocker cover ornament

➡**Before detaching any hoses or connectors, note the locations for reassembly.**

- Air duct to intake manifold hose, collector hose, blow-by hose and vacuum hoses
- Fuel hoses and detach the harness connections
- Canister purge hoses
- Water hoses from the cylinder head and intake manifold
- All 6 ignition coils from the spark plugs
- Spark plugs
- Bolts that secure the Exhaust Gas Recirculation (EGR) tube and remove the tube

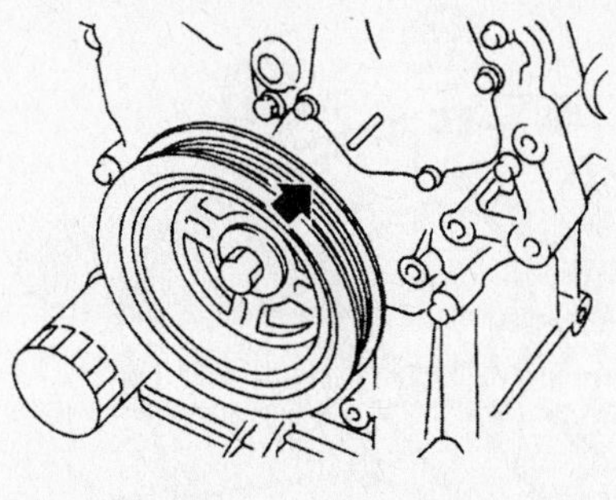

Set the No. 1 piston to Top Dead Center (TDC)—3.0L and 3.5L engines

- Intake manifold collector supports and the collector
- Fuel tube assembly
- Intake manifold. Loosen the bolts in the reverse sequence of the tightening procedure.
- Left-hand and right-hand intake valve timing control solenoid valves, 3.5L engine only
- Left-hand and right-hand rocker covers from the cylinder head
- Engine undercovers
- Right front wheel and the engine side covers
- Drive belts and the idler pulley

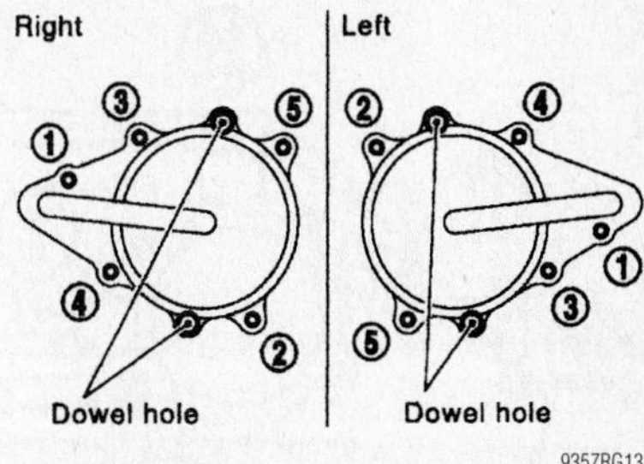

Loosen the intake valve timing control valve cover bolts in the reverse of the order shown—3.5L engine

- Power steering oil pump belt and the power steering oil pump assembly
- Camshaft Position (CMP) sensor (PHASE) and Crankshaft Position (CKP) sensors (REF)/(POS)

6. Set the No. 1 piston to Top Dead Center (TDC) of compression stroke by rotating the crankshaft.
7. Loosen the crankshaft pulley bolt while securing the ring gear so the crankshaft cannot rotate.
    - Ring gear cover access plate

➡**Use care not to damage the ring gear teeth.**

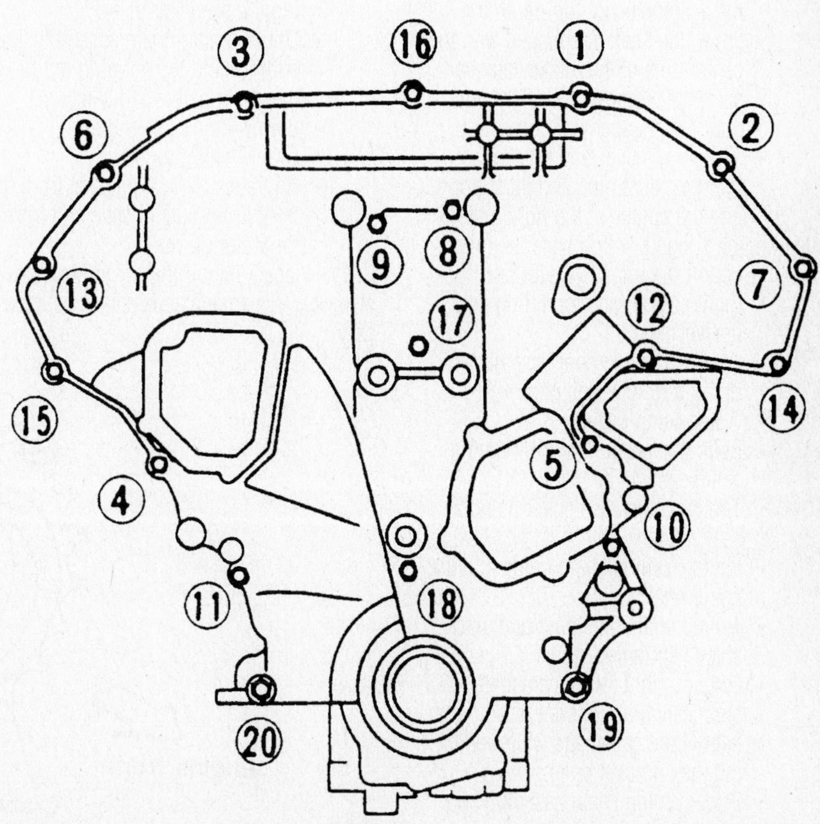

Remove the front timing chain case mounting bolts in the sequence shown—3.0L engines

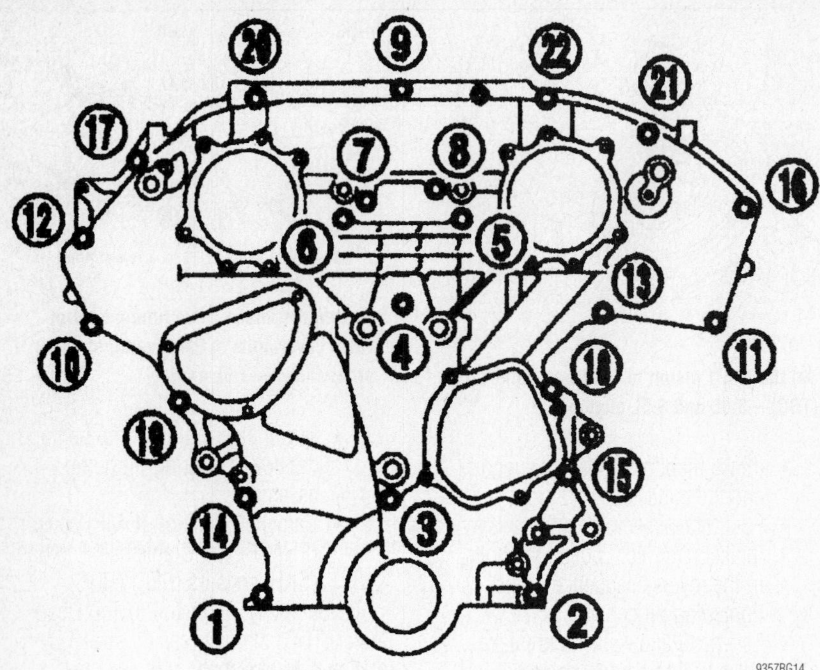

Remove the front timing chain case mounting bolts in the reverse of the sequence shown—3.5L engines

- Crankshaft pulley using a suitable puller
- Intake valve timing control valve cover, 3.5L engine only. Loosen the bolts in the reverse order shown in the accompanying figure. In the cover, the shaft is engaged with the center hole of the intake camshaft sprocket. Remove it straight out until the engagement comes off.
- A/C compressor and bracket
- Front exhaust pipe and its support

8. Hang the engine at the right and left side engine slingers with a suitable hoist.
9. Support the transaxle with jack.
   - Right side engine mounting and bracket
   - Center crossmember assembly
   - Upper and lower oil pans
   - Water pump cover
   - Bolts that secure the front timing chain case, in sequence
   - Timing chain case cover using a seal cutter
   - Internal timing chain guide and the upper chain guide
   - Timing chain tensioner and slack side chain guide
   - Left and right intake camshaft sprockets first. Be sure to hold the flats of the camshafts while removing the sprocket bolts.
   - Lower timing chain assembly. Be sure to note the aligning marks of the chain before removal.
10. Insert a suitable stopper pin for the left and right camshaft tensioners.
    - Left and right exhaust camshaft sprocket bolts. Be sure to hold the flats of the camshafts while removing the sprocket bolts.
    - Upper timing chain assembly. Be sure to note the aligning marks of the chain before removal.
    - Lower timing chain guide
    - Crankshaft sprocket
    - All traces of liquid gasket from the front timing chain case and from the water pump
11. Inspect the timing chain for excessive wear or damage and replace as necessary.

To install:

12. Install or connect the following:
    - Crankshaft sprocket with the mating mark facing out
13. Position the crankshaft to TDC of compression stroke and align the dowels of the camshaft sprockets to the 12 o'clock position in respect to the cylinder head.
    - Lower timing chain guide. The front mark on the guide should face upwards.
14. On a work bench, align the marks on the intake and exhaust camshaft sprockets with the marks of the chain.
    - Exhaust camshaft sprockets onto the dowel pin and torque the mounting bolts to 88–95 ft. lbs. (119–128 Nm). Be sure to secure the camshafts while tightening the bolts.
    - Timing chains and sprockets to the intake camshafts. Be sure to align the timing chain and sprocket mating marks.
    - Left and right camshaft tensioner stopper pins
15. Align the mating mark on the crankshaft with the matchmark (gold link) on the lower timing chain.
16. Install the lower timing chain to the water pump sprocket.
17. Working counterclockwise, install the lower timing chain camshaft sprockets. Be sure to align the sprocket marks with the blue links of the timing chain during installation.
18. Install or connect the following:
    - Intake sprocket and torque the bolts to 88–95 ft. lbs. (119–128 Nm). Be sure to secure the camshafts while tightening the bolts.
    - Internal timing chain guide, upper timing chain guide, lower timing

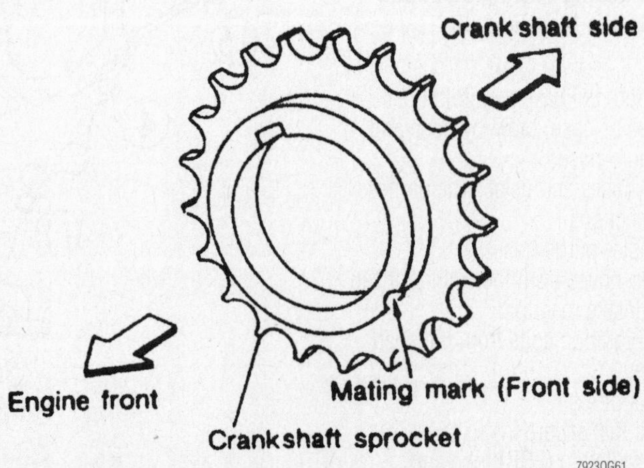

Crankshaft sprocket with mating marks—3.0L engine

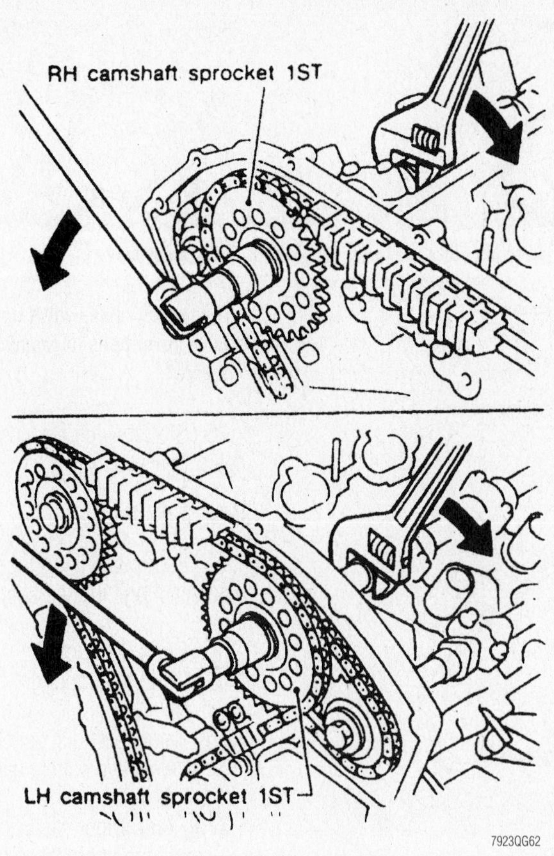

Hold the camshaft with a wrench while removing the sprocket bolts—3.0L and 3.5L engines

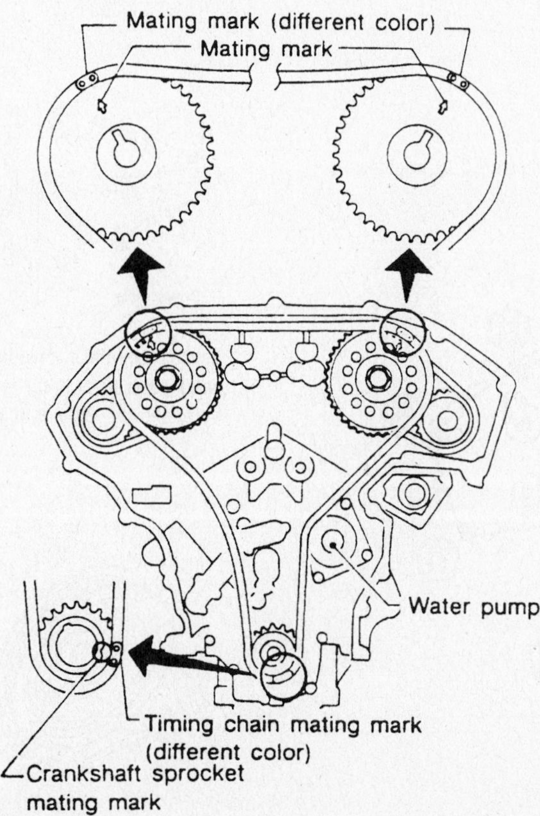

Timing chain alignment marks—3.0L and 3.5L engines

chain tensioner and slack side timing chain guide

19. Torque the tensioner mounting bolt to 75–96 inch lbs. (8.4–10.8 Nm) and the guide bolts to 108–168 inch lbs. (13–19 Nm).

20. Apply a 0.102–0.142 in. (2.6–3.6mm) continuous bead of liquid gasket to all necessary areas as shown on the front timing cover.

- Timing cover evenly and gently. Be sure to align the dowel pin holes.

21. Torque the mounting bolts in sequence as follows:

a. Bolts No. 1 and 2: 19–23 ft. lbs. (26–31 Nm).

b. Bolts No. 3–20: 105–121 inch lbs. (11.8–13.7 Nm).

➡Leave the bolts unattended for 30 minutes or more after tightening. This will allow the liquid gasket to cure sufficiently.

22. Apply a 0.091–0.130 in. (2.3–3.3mm) continuous bead of liquid gasket to the water pump cover and install the cover. Torque the bolts to 84–108 inch lbs. (10–13 Nm).

23. Install or connect the following:

- Oil pans
- Center crossmember
- Right side engine mount and bracket
- Front exhaust pipe and remove the transaxle support
- A/C compressor and bracket
- Crankshaft pulley
- Ring gear access cover plate
- CMP sensor and CKP sensors
- Power steering pump
- Idler pulley and drive belts
- Engine side cover and right front wheel
- Engine under covers
- Intake valve timing control solenoid valves with new covers, 3.5L engine only. Tighten the cover bolts in the proper sequence.
- Rocker covers
- Intake manifold
- Fuel tube assembly
- Intake manifold collector and support
- EGR tube
- Spark plugs and ignition coils
- Coolant hoses
- Fuel hoses
- Air duct assembly and hoses
- Left side rocker cover ornament
- Negative battery cable

24. Fill the cooling system.

25. Fill the engine with clean oil.

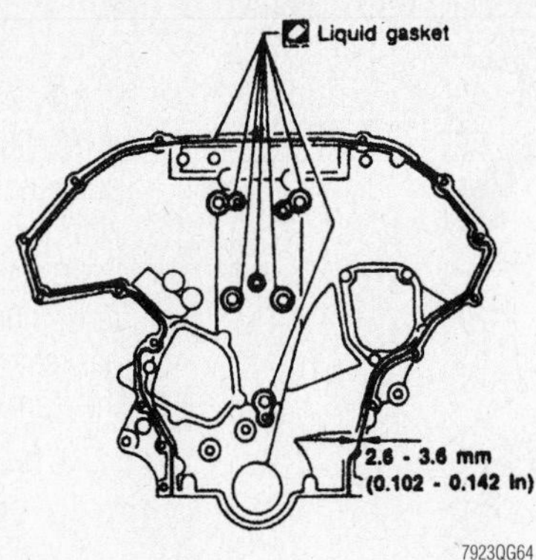

Application of liquid gasket to the front timing case—3.0L and 3.5L engines

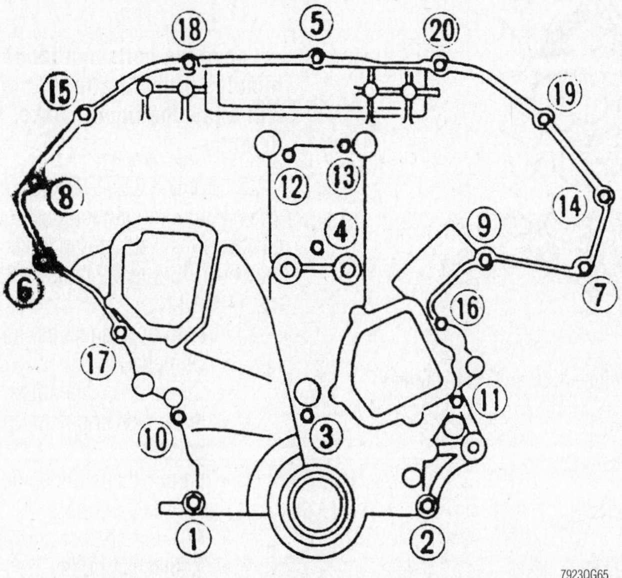

Tighten the front timing chain case bolts according to the sequence shown—3.0L engines

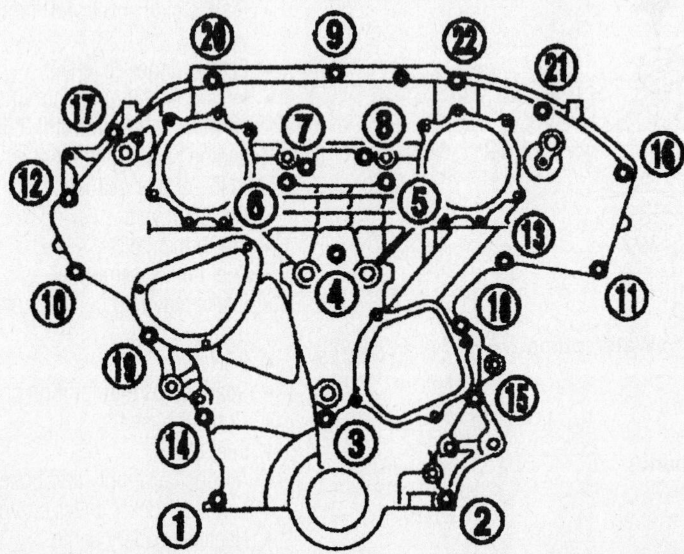

Tighten the front timing chain case bolts according to the sequence shown—3.5L engines

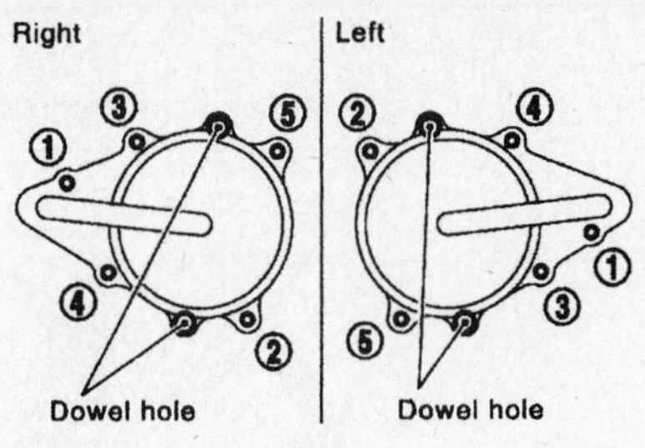

Tighten the intake valve timing control valve cover bolts in sequence—3.5L engine

26. Start the vehicle, check for leaks and repair if necessary.

350Z 3.5L Engine

1. Before servicing the vehicle, refer to the precautions in the beginning of this section.
2. Drain the engine oil.
3. Drain the cooling system.
4. Relieve the fuel system pressure.
5. Remove or disconnect the following:
   - Negative battery cable
   - Engine cover
   - Air cleaner case assembly
   - Engine harnesses from timing chain case
   - Upper and lower intake manifold collectors

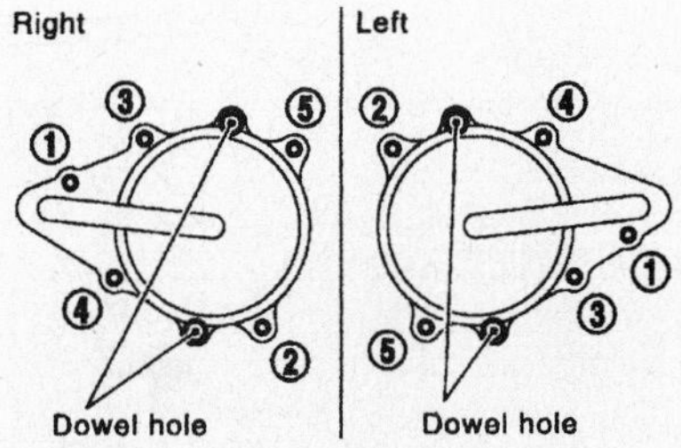

Loosen the intake valve timing control valve cover bolts in the reverse of the order shown—350Z 3.5L engine

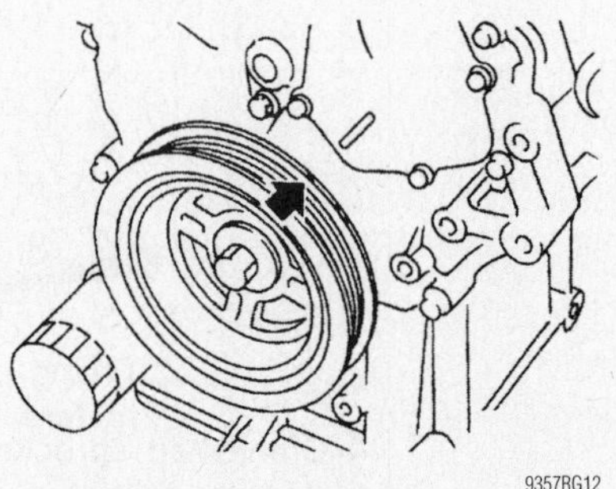

Set the No. 1 piston to Top Dead Center (TDC)—350Z 3.5L engine

- Cooling fan
- Drive belts
- A/C compressor and wire aside
- Power steering pump and wire aside
- Power steering pump bracket
- Alternator
- Water bypass hose, clamp and idler pulley bracket from timing chain case
- Intake valve timing control valve covers. Loosen the bolts in the reverse order shown in the accompanying figure. In the cover, the shaft is engaged with the center hole of the intake camshaft sprocket. Remove it straight out until the engagement comes off.
- O-ring from timing chain case on both sides
- Both valve covers
- Rotate the crankshaft clockwise and set the No. 1 piston to TDC of the compression stroke

6. Make sure the intake and exhaust camshaft lobes are facing toward the inside of the cylinder head
7. Remove the starter and lock the flywheel through the starter mounting hole.
8. Loosen the crankshaft pulley bolt.
   - Crankshaft pulley using a suitable puller
   - Upper and lower oil pans
   - Front timing chain cover bolts in reverse order of the tightening sequence
   - Front timing chain case
   - O-rings from rear timing chain cover
   - Water pump and chain tensioner cover from rear cover
   - Pry out front oil seal
   - Insert a suitable stopper pin for the

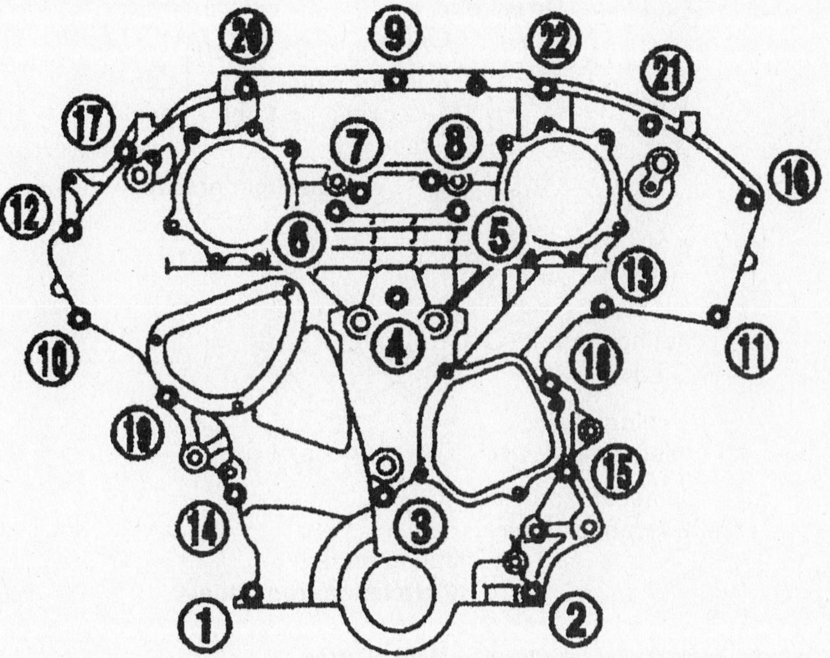

Remove the front timing chain case mounting bolts in the reverse of the sequence shown—350Z 3.5L engine

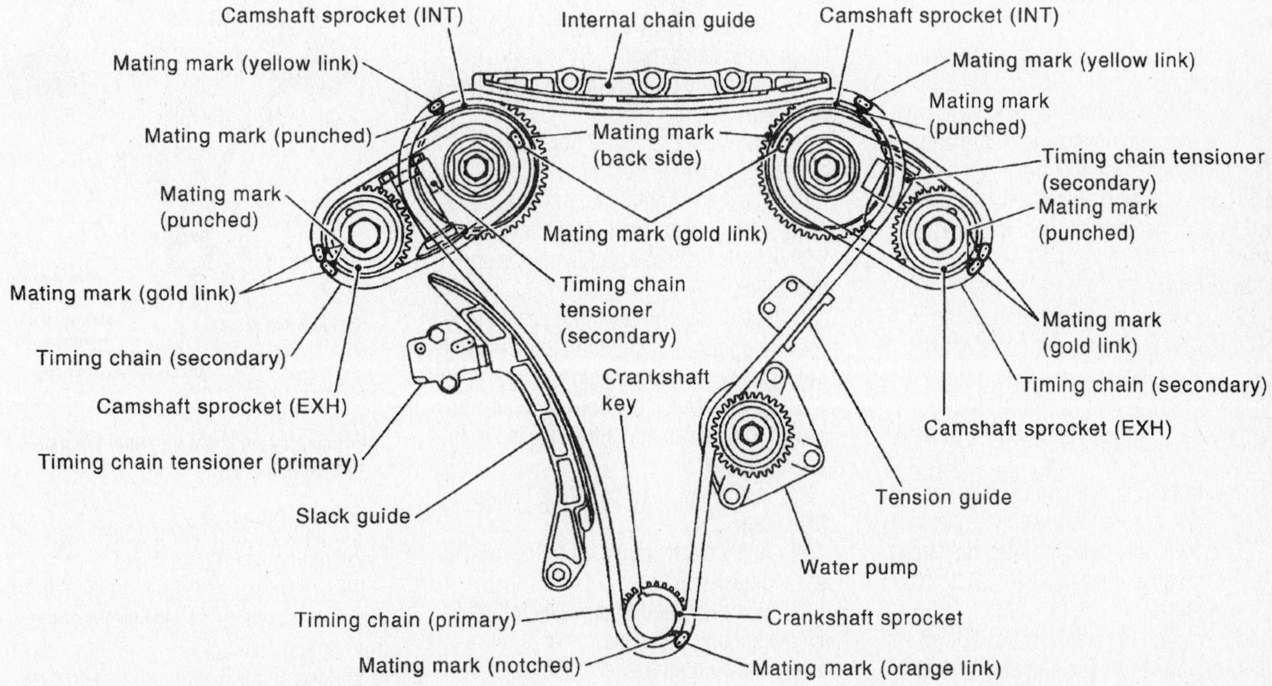

Timing chain alignment marks—350Z 3.5L engine

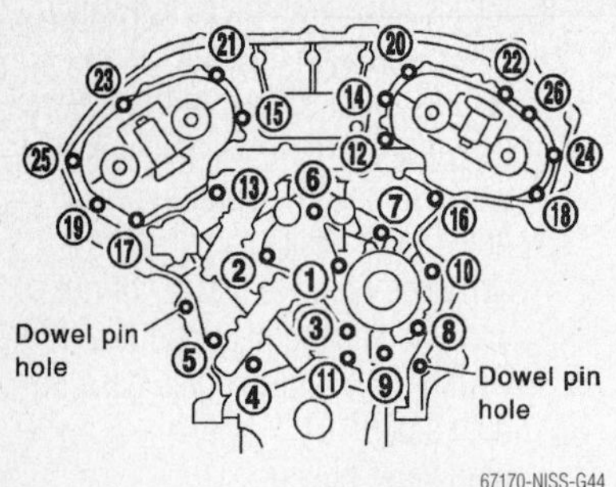

Rear timing chain case tightening sequence—350Z 3.5L engine

left and right primary camshaft tensioners

- Primary chain tensioners
- Timing chain guide, slack guide and tension guide
- Primary timing chain and crankshaft sprocket
- Insert a suitable stopper pin for the left and right secondary camshaft tensioners
- Secondary chain tensioners
- Left and right intake camshaft sprocket bolts first, then exhaust sprocket bolts. Be sure to hold the flats of the camshafts while removing the sprocket bolts.
- Lower timing chain assembly with camshaft sprockets. Be sure to note the aligning marks of the chain before removal.

9. Remove the rear timing chain case bolts in the reverse order of the tightening sequence.

10. Remove the O-rings from the cylinder heads and block.

- All traces of liquid gasket from the front and timing chain case and from the water pump

11. Inspect the timing chain for excessive wear or damage and replace as necessary.

To install:

12. Install or connect the following:

- New O-rings to the cylinder heads and block
- Apply a bead of sealant to the back side of the rear timing chain case
- Install the rear timing chain case and tighten the bolts in sequence to 10 ft. lbs. (14 Nm). After tightening, retighten them again to 10 ft. lbs. (14 Nm).

13. Position the crankshaft to TDC of compression stroke and align the dowels of the camshaft sprockets to the 12 o'clock position in respect to the cylinder head.

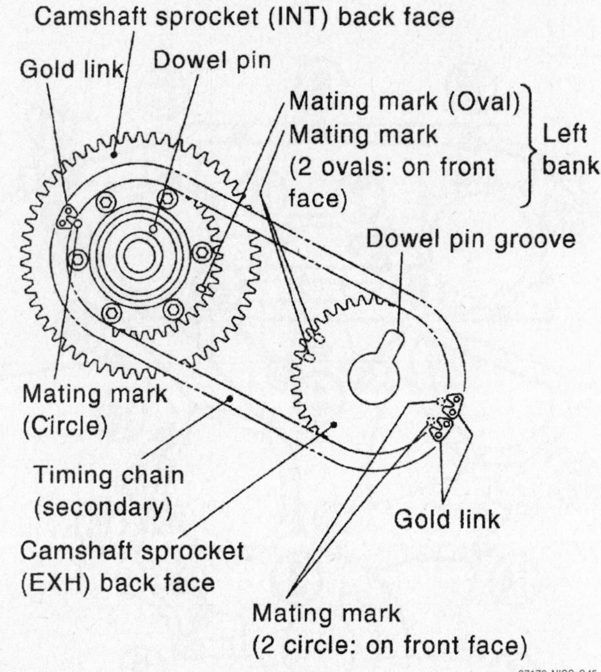

Secondary timing chain alignment marks—350Z 3.5L engine

14. Install the secondary timing chain and camshaft sprockets aligning the timing marks and timing chain links as shown.

15. Tighten the camshaft sprocket bolts to 76 ft. lbs. (103 Nm).

16. Remove the pins from the secondary timing chain tensioners.

17. Install the primary timing chain tension guide.

18. Install the crankshaft sprocket with the mating marks on the front side.

19. Install the primary timing chain and so the mating marks are aligned as shown.

20. Install the internal chain guide and slack guide.

21. Install the primary timing chain tensioner and tighten the bolt to 72 inch lbs. (8 Nm). Remove the stopper pin

22. Double check that the mating marks on the timing chain and sprockets are in the correct locations.

23. Install new O-rings on the timing chain case.

24. Apply clean engine oil to the front oil seal and dust seal lips.

25. Use a drift and press fit the oil seal into the timing chain case.

26. Apply liquid gasket to the water pump and chain tensioner cover openings, then install the covers.

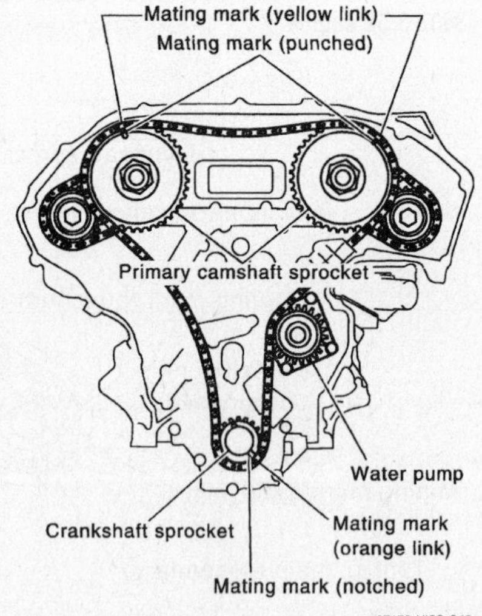

Primary timing chain alignment marks—350Z 3.5L engine

27. Apply liquid gasket to the back side of the timing chain case cover.

28. Install the dowel pin on the rear timing chain case into the dowel pin of the front timing chain case.

29. Install the front timing chain case and tighten the bolts in sequence to 19–23 ft. lbs. (26–31 Nm for M8 bolts and 9–10 ft.

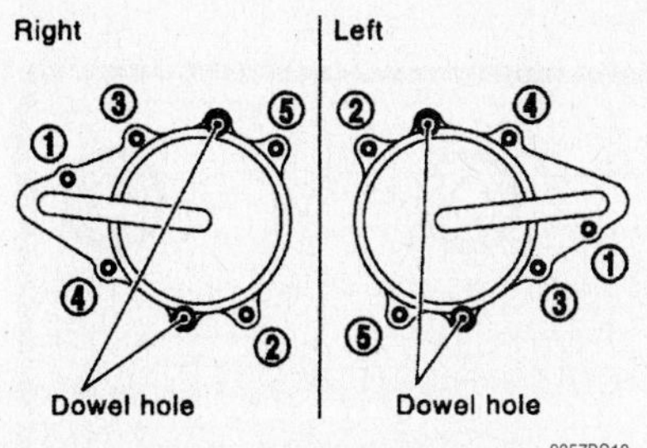

Tighten the intake valve timing control valve cover bolts in sequence—3.5L engine

lbs. (12–14 Nm for M6 bolts. After tightening, retighten them again to the same specification.

30. Install seal rings in the timing control cover shaft grooves and apply liquid gasket to the covers.

31. Install new O-rings in the timing chain case oil holes.

32. Install the timing control covers and tighten the bolt in sequence to 72 inch lbs. (8 Nm).

33. Install or connect the following:
- Upper and lower oil pans
- Valve covers
- Crankshaft pulley and tighten the bolt to 29–36 ft. lbs. (40–49 Nm), plus an additional 60°.

34. The remainder of the installation is the reverse of the removal procedure.

35. Fill the cooling system.

36. Fill the engine with clean oil.

37. Start the vehicle, check for leaks and repair if necessary.

Piston and Ring

POSITIONING

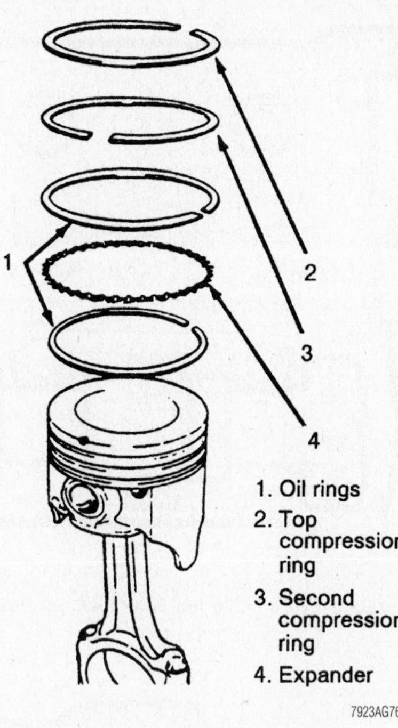

Exploded view of common piston ring mounting

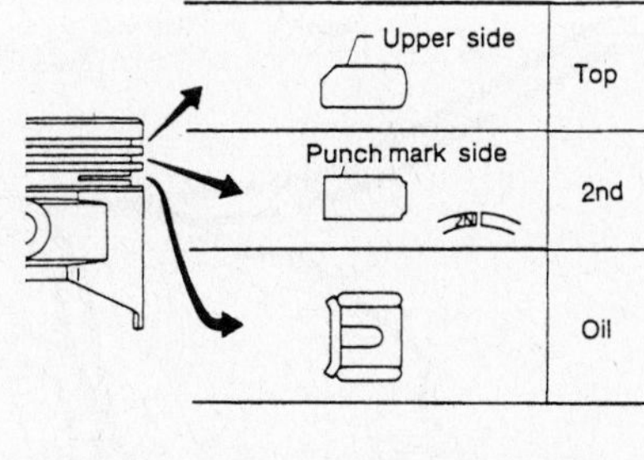

Piston ring positioning—2.0L and 2.4L engines

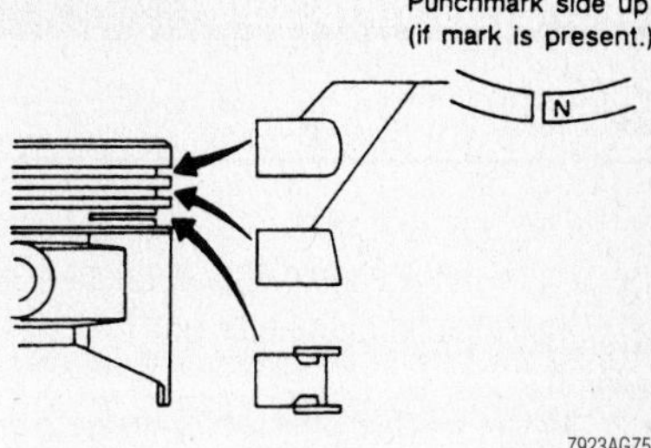

Piston ring positioning—3.0L and 3.5L engines

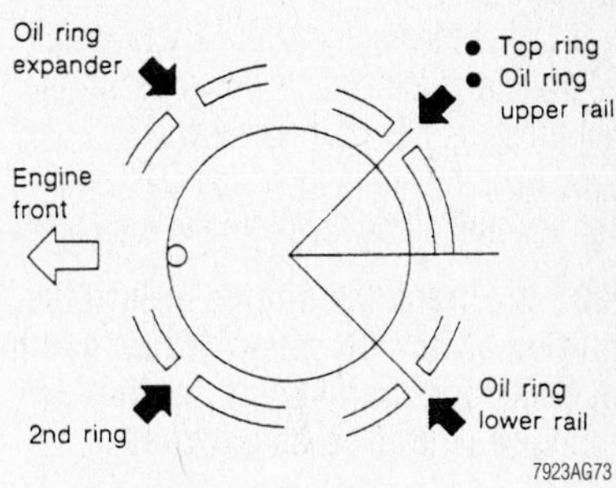

Piston ring end-gap spacing

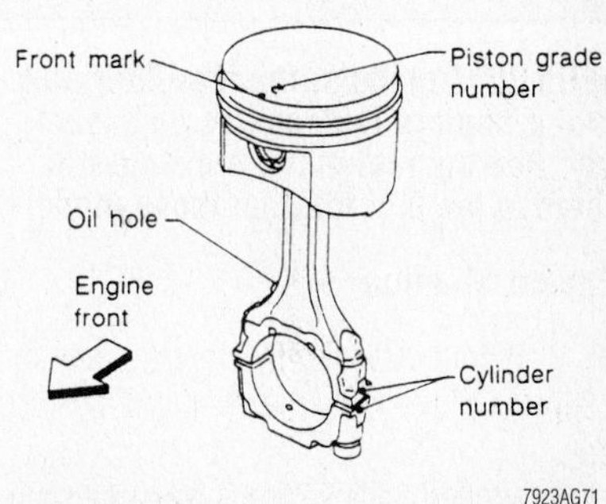

Piston and connecting rod assembly positioning

FUEL SYSTEM

Fuel System Service Precautions

Safety is the most important factor when performing not only fuel system maintenance but any type of maintenance. Failure to conduct maintenance and repairs in a safe manner may result in serious personal injury or death. Maintenance and testing of the vehicle's fuel system components can be accomplished safely and effectively by adhering to the following rules and guidelines.

- To avoid the possibility of fire and personal injury, always disconnect the negative battery cable unless the repair or test procedure requires that battery voltage be applied.
- Always relieve the fuel system pressure prior to disconnecting any fuel system component (injector, fuel rail, pressure regulator, etc.), fitting or fuel line connection. Exercise extreme caution whenever relieving fuel system pressure, to avoid exposing skin, face and eyes to fuel spray. Please be advised that fuel under pressure may penetrate the skin or any part of the body that it contacts.
- Always place a shop towel or cloth around the fitting or connection prior to loosening to absorb any excess fuel due to spillage. Ensure that all fuel spillage (should it occur) is quickly removed from engine surfaces. Ensure that all fuel soaked cloths or towels are deposited into a suitable waste container.
- Always keep a dry chemical (Class B) fire extinguisher near the work area.
- Do not allow fuel spray or fuel vapors to come into contact with a spark or open flame.
- Always use a back-up wrench when loosening and tightening fuel line connection fittings. This will prevent unnecessary stress and torsion to fuel line piping. Always follow the proper torque specifications.
- Always replace worn fuel fitting O-rings with new. Do not substitute fuel hose where fuel pipe is installed.

Fuel System Pressure

RELIEVING

The fuel pump fuse is located in the dash fuse box or in the engine compartment fuse box. Check the lid of the fuse box for exact location.

1. Before servicing the vehicle, refer to the precautions in the beginning of this section.
2. Remove the fuel pump fuse.
3. Start the engine.
4. Start the engine and run until the engine stalls.
5. After the engine stalls, try to restart the engine; if the engine will not start, the fuel pressure has been released.
6. Turn the ignition switch **OFF**. Reinstall the fuel pump fuse into the fuse block.

➡Do not crank the engine or turn the ignition switch ON after the fuel pump fuse has been reinstalled, or the fuel pressure will be re-established.

Fuel Filter

REMOVAL & INSTALLATION

➡On 350Z models, the fuel filter and fuel pump are replaced as an assembly. See the fuel pump procedure to remove the fuel filter on these models.

Except Maxima

1. Before servicing the vehicle, refer to the precautions in the beginning of this section.
2. Properly relieve fuel system pressure.
3. Remove or disconnect the following:
   - Negative battery cable
   - Fuel hose clamps
   - Hoses from the fuel filter
   - Bolt securing the filter to the bracket or the filter from the bracket clips
   - Filter

To install:

4. Install or connect the following:
   - New filter and secure the filter in the bracket
5. If necessary, replace the fuel line hoses and hose clamps
   - Fuel hoses and tighten the clamps
   - Negative battery cable
   - Fuel pump fuse
6. Start the engine and check for leaks.

Maxima

1. Before servicing the vehicle, refer to the precautions in the beginning of this section.

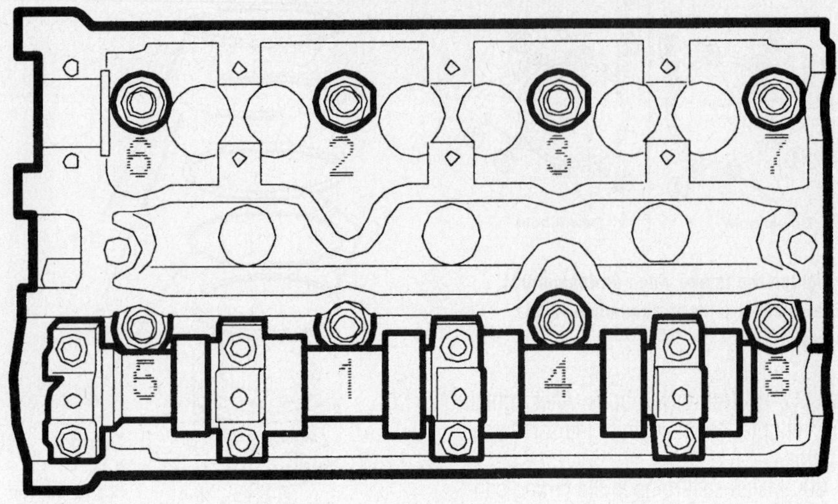

Exploded view of the fuel filter location—except Maxima

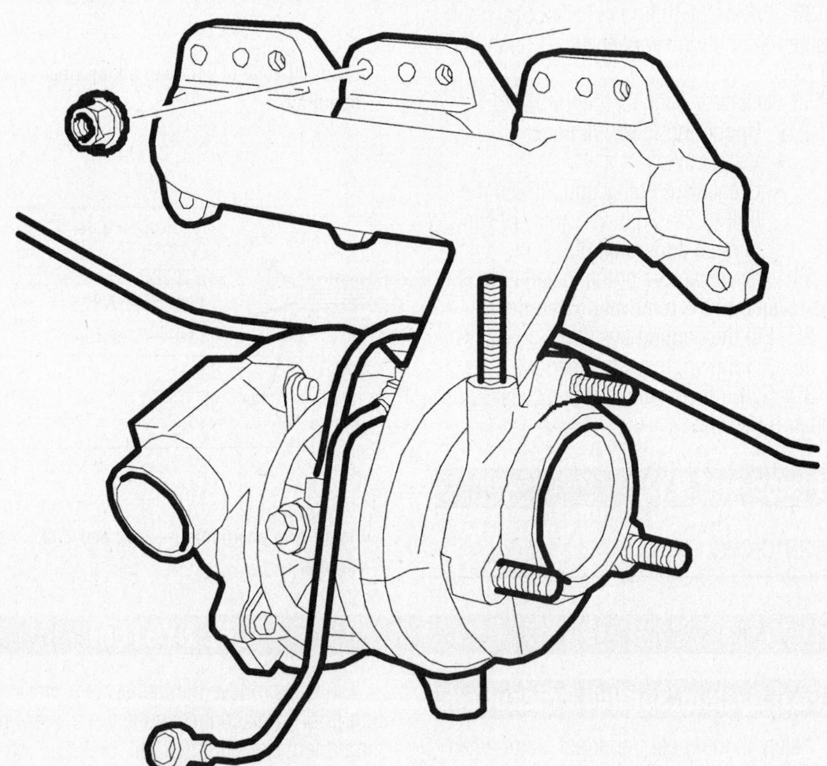

Remove the fuel filter from the fuel chamber–Maxima

2. Properly relieve fuel system pressure.
3. Remove or disconnect the following:
   - Negative battery cable
   - Rear seat bottom
   - Inspection hole cover
   - Electrical and quick connectors
   - Six screws
   - Fuel level sensor unit and fuel pump assembly
   - Flange and snap fit portion of the fuel pump
   - Fuel tank temperature sensor harness
   - Fuel level sensor flange
   - Fuel pump connector
   - Quick connectors from the fuel level sensor
   - Fuel level sensor from the chamber
   - Fuel filter from the chamber

To install:

4. Install or connect the following:
   - Fuel filter to the chamber
   - Fuel level sensor to the chamber
   - Quick connectors to the fuel level sensor
   - Fuel pump connector
   - Fuel level sensor flange
   - Fuel tank temperature sensor harness
   - Fuel pump assembly to the fuel tank
   - Screws and electrical connectors
   - Quick connectors
   - Negative battery cable
5. Start the vehicle, check for leaks and repair if necessary.
   - Inspection hole cover
   - Rear seat bottom

Fuel Pump

REMOVAL & INSTALLATION

Sentra

The fuel pump is located in the fuel tank on all vehicles. In-tank fuel pumps are accessible by lifting up the rear seat to gain access to the inspection cover.

1. Before servicing the vehicle, refer to the precautions in the beginning of this section.
2. Relieve the fuel system pressure.
3. Remove or disconnect the following:
   - Negative battery cable
   - Rear seat from the vehicle
   - Inspection cover that is located under the rear seat
   - Inlet and outlet fuel lines from the fuel pump assembly
   - Fuel pump and gauge wiring connections
   - 6 mounting bolts that secure the fuel pump assembly to the top of the fuel tank
4. Raise up the fuel pump assembly and detach the fuel tubes and connector.
   - Fuel gauge assembly
   - Fuel pump with the fuel chamber
5. Pull up the front of the fuel pump chamber and slide the chamber forward.
   - Fuel pump from the chamber
   - O-ring seal or gasket

To install:

6. Install or connect the following:
   - Fuel pump to the fuel pump chamber and slide chamber rearward
   - Fuel pump with the fuel pump chamber
   - Fuel gauge assembly using a new O-ring
   - Fuel tubes and connector. Use new hoses and clamps.
   - 6 mounting bolts to the top of the fuel gauge unit and torque the bolts to 17–22 inch lbs. (2–3 Nm)
   - Fuel pump and gauge wiring connections
   - New inlet and outlet fuel lines to the fuel pump assembly
   - Negative battery cable
7. Start the vehicle, check for leaks and repair if necessary.
   - Inspection cover and the rear seat

Altima

1. Before servicing the vehicle, refer to the precautions in the beginning of this section.
2. Relieve the pressure from the fuel system.
3. Remove or disconnect the following:
   - Negative battery cable
   - Rear seat and the access cover
   - Fuel pump electrical connector
   - Fuel lines from the fuel pump assembly
   - Locking ring
   - Fuel gauge assembly
   - Fuel tube and connector from the fuel gauge

➡When the fuel sending unit needs to be removed, pull the tab upwards. The tab is located on the sending unit, opposite the end of the float. After the tab is pulled, the sending unit will lift straight out of the tank bracket.

   - Fuel pump by pinching the 2 locking tabs together. Lift the fuel pump assembly straight upward and out of fuel tank.
   - O-ring and discard
4. Place a clean rag in the hole to keep out dirt.

To install:

5. Remove the rag
6. Install or connect the following:
   - New O-ring and fuel pump
   - Electrical connection and fuel tube to the fuel gauge sending unit
   - Fuel sending unit into the tank

➡Verify that the mark on the fuel tank and the components are aligned when installing the pump and fuel gauge sending unit.

   - Locking ring and torque the ring to 22–26 ft. lbs. (30–35 Nm)
   - Fuel lines and fuel pump electrical connector. Always install new clamps on the fuel lines.
   - Negative battery cable
7. Start the engine, check for leaks and repair if necessary.
   - Fuel pump access cover
   - Rear seat

Maxima

1. Before servicing the vehicle, refer to the precautions in the beginning of this section.
2. Relieve the fuel system pressure
3. Remove or disconnect the following:
   - Negative battery cable
   - Rear seat or open the access panel in the trunk
   - Fuel gauge electrical connector and pump electrical connector
   - Fuel outlet and the return hoses
   - Fuel tank, if necessary
4. On some models you need to remove the fuel pump assembly-to-fuel tank bolts and lift the fuel pump assembly from the fuel tank.
5. On other models you need to remove the locking ring and raise the fuel pump from the tank. Disconnect the feed tube while raising the pump.
6. Discard the O-ring. Plug the fuel tank opening with a clean rag to prevent dirt from entering the system.

➡When removing or installing the fuel pump assembly, be careful not to damage or deform it and always install a new O-ring.

To install:

7. Remove the rag
8. Install or connect the following:
   - Fuel pump assembly into the fuel tank using a new O-ring
   - Fuel pump assembly-to-fuel tank bolts and torque the bolts to 17–22 inch lbs. (2.0–2.5 Nm)
   - Locking ring assembly and tighten
   - Fuel tank assembly, if removed
   - Fuel lines and the electrical connectors. Always use new clamps when reconnecting fuel line hoses.

➡When installing the upper plate, be sure to align the mark with the center marks on the fuel tank.

   - Negative battery cable
9. Start the engine, check for fuel leaks and repair if necessary.
10. Install the fuel pump access cover.

➡On some models, the Check Engine Light will stay ON after installation is completed. The memory code in the control unit must be erased. This code is stored for an open fuel pump circuit, this is caused when the fuel pressure is

released. To erase the code, disconnect the battery cable for 10 seconds, then reconnect after installation of fuel pump.

350Z

1. Before servicing the vehicle, refer to the precautions in the beginning of this section.

➡If the fuel tank is more than three quarter full, some fuel will have to be drained before removing the fuel pump/fuel filter assembly.

2. Relieve the fuel system pressure
3. Remove or disconnect the following:
   - Negative battery cable
   - Rear floor box
   - Fuel pump inspection cover
   - Harness connector and fuel feed tube
   - Fuel feed tube quick connector
   - Retainer
   - Raise the fuel pump assembly and remove the fuel hose connector
   - Remove the fuel pump assembly
   - Reverse the removal procedure to install

Fuel Injector

REMOVAL & INSTALLATION

Except Maxima And 350Z

1. Before servicing the vehicle, refer to the precautions in the beginning of this section.
2. Relieve the fuel system pressure
3. Remove or disconnect the following:
   - Negative battery cable
   - Intake manifold collector
   - Vacuum hose from the pressure regulator
   - Fuel hoses from the rail
   - Injector electrical connectors
   - Fuel rail bolts
   - Injector rail assembly with injectors from the intake manifold
   - Injector from the rail by pushing on the injector tail piece
   - Discard injector O-rings

To install:

4. Clean the injector tail piece and lubricate new O-rings with a smear of clean engine oil.
5. Install or connect the following:
   - New O-rings
   - Injector to the fuel rail
   - Fuel rail and the injectors as an assembly to the intake manifold
6. Install the fuel rail bolts and tighten in 2 steps as follows:
   a. Step 1: Bolts to 84–96 inch lbs. (9–10 Nm).
   b. Step 2: Bolts to 15–20 ft. lbs. (21–26 Nm).
7. Install or connect the following:
   - Injector electrical connectors
   - Vacuum hose to the pressure regulator
   - Fuel hoses to the rail
   - All remaining components in the reverse order of removal

Maxima

1. Before servicing the vehicle, refer to the precautions in the beginning of this section.
2. Relieve the fuel system pressure
3. Remove or disconnect the following:
   - Negative battery cable
   - Intake manifold collector
   - Vacuum hose from the pressure regulator
   - Fuel hoses from the rail
   - Injector electrical connectors
   - Fuel rail bolts
4. To remove the fuel injector from the fuel rail, expand and remove the clips securing the injectors and press the fuel injector out from the fuel rail. Discard the O-rings.

To install:

5. Apply a thin coat of engine oil to the new O-rings, install them on the injectors, then press the injector into the fuel rail.
6. Install or connect the following:
   - New injector retaining clips
   - New injector gaskets onto the manifold
   - Fuel rail assembly to the engine
   - Fuel rail-to-cylinder head bolts and torque the bolts to 84–96 inch lbs. (9.3–10.8 Nm). Then tighten them again to 16–19 ft. lbs. (21–26 Nm).
   - Fuel lines to the rail assembly
   - Vacuum hose to the fuel pressure regulator
   - Electrical connectors to the fuel injectors
   - Intake manifold collector
   - Negative battery cable.
7. Start the engine and check for leaks.

350Z

1. Before servicing the vehicle, refer to the precautions in the beginning of this section.
2. Relieve the fuel system pressure
3. Remove or disconnect the following:
   - Negative battery cable
   - Engine cover
   - Fuel feed hose and damper
   - Under vehicle fuel line quick connectors
   - Upper and lower intake manifold collectors
   - Fuel injector harness connectors
   - Fuel rail mounting bolts in reverse of the tightening sequence
   - Intake manifold spacers
4. To remove the fuel injector from the fuel rail, expand and remove the clips securing the injectors and press the fuel injector out from the fuel rail. Discard the O-rings.

To install:

➡Upper and lower O-rings are different. The Blue rings go on the fuel tube side and the Brown rings go on the nozzle side.

5. Apply a thin coat of engine oil to the new O-rings, install them on the injectors, then press the injector into the fuel rail.
6. Install or connect the following:
   - New injector retaining clips
   - Intake manifold spacers
   - Fuel rail mounting bolts in sequence and tighten to 8 ft. lbs. (11 Nm), then to 16–19 ft. lbs. (21–27 Nm).
   - Fuel injector harness connectors
   - Upper and lower intake manifold collectors
   - Under vehicle fuel line quick connectors
   - Fuel feed hose and damper
   - Engine cover
   - Negative battery cable
7. Start the engine and check for leaks.

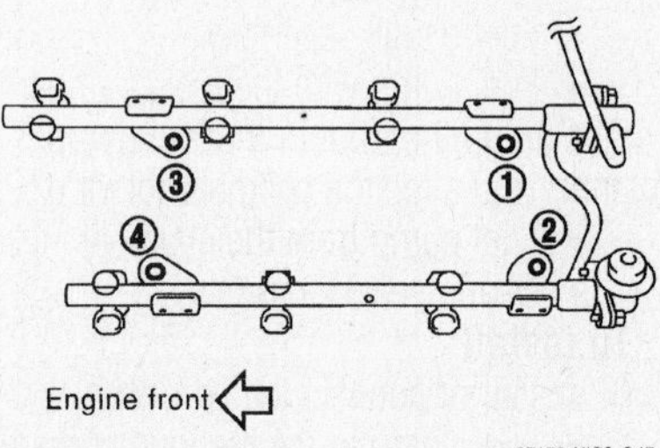

Fuel rail tightening sequence–350Z

DRIVE TRAIN

Transaxle Assembly

REMOVAL & INSTALLATION

Manual

SENTRA

1. Before servicing the vehicle, refer to the precautions in the beginning of this section.
2. Drain the fluid from the transaxle.
3. Remove or disconnect the following:
   - Both battery cables
   - Battery and bracket from the vehicle
   - Crankshaft Position (CKP) sensor from the transaxle
   - Air cleaner assembly
   - All electrical connectors from the transaxles
   - Control cable from the transaxle
   - Speed sensor, OD position switch and Back-up lamp switch
   - Neutral position switch
   - Starter
   - Shift control rod
   - Halfshafts and properly support the transmission
   - Left side and rear engine to transmission mounts
4. Slide the transmission away from the engine and lower the transmission assembly.

To install:

5. Install the transaxle mounting bolts in the proper location as noted during removal.
   a. On 1.8L engines: torque the 2 bottom bolts to 12–15 ft. lbs. (16–21 Nm) and all other bolts to 22–30 ft. lbs. (30–40 Nm).
   b. On 2.0L engines: torque the 2 bottom bolts to 23–31 ft. lbs. (31–42 Nm) and all other bolts to 51–59 ft. lbs. (70–79 Nm).
6. Install or connect the following:
   - Left side and rear engine to transmission mounts
   - Halfshafts
   - Shift control rod
   - Starter
   - Neutral position switch
   - Speed sensor, OD position switch and back-up lamp switch
   - Clutch control cable
   - CKP sensor
   - Air cleaner assembly
   - Battery and both cables
7. Fill the transmission with clean oil to the proper level.
8. Start the vehicle, check for leaks and repair if necessary.

ALTIMA

1. Before servicing the vehicle, refer to the precautions in the beginning of this section.
2. Drain the transmission fluid.
3. Remove or disconnect the following:
   - Battery cables
   - Battery and tray
   - Air cleaner box with the Mass Air Flow (MAF) sensor
   - Air duct
   - Clutch operating cylinder

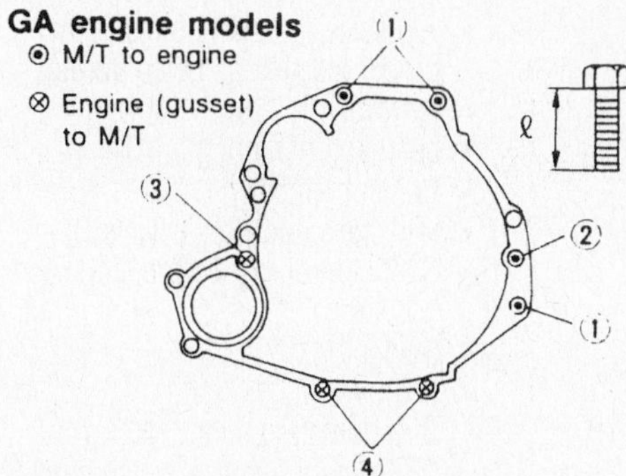

Bolt No.	Tightening torque N·m (kg-m, ft-lb)	"ℓ" mm (in)
①	30 - 40 (3.1 - 4.1, 22 - 30)	70 (2.76)
②	30 - 40 (3.1 - 4.1, 22 - 30)	85 (3.35)
③	30 - 40 (3.1 - 4.1, 22 - 30)	30 (1.18)
④	16 - 21 (1.6 - 2.1, 12 - 15)	25 (0.98)
Front gusset to engine	30 - 40 (3.1 - 4.1, 22 - 30)	20 (0.79)
Rear gusset to engine	16 - 21 (1.6 - 2.1, 12 - 15)	16 (0.63)

9307QG13

Bolt locations and torque specifications—1.8L engine with manual transaxle

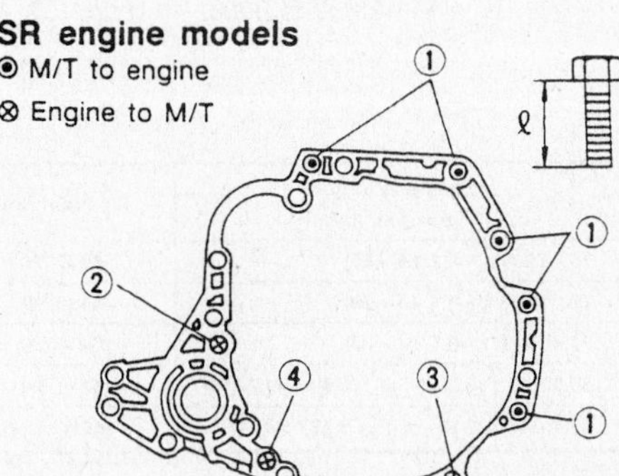

Bolt No.	Tightening torque N·m (kg-m, ft-lb)	"ℓ" mm (in)
①	70 - 79 (7.1 - 8.1, 51 - 59)	55 (2.17)
②	70 - 79 (7.1 - 8.1, 51 - 59)	65 (2.56)
③	31 - 42 (3.2 - 4.3, 23 - 31)	35 (1.38)
④	31 - 42 (3.2 - 4.3, 23 - 31)	45 (1.77)

9307QG14

Bolt locations and torque specifications—2.0L engine with manual transaxle

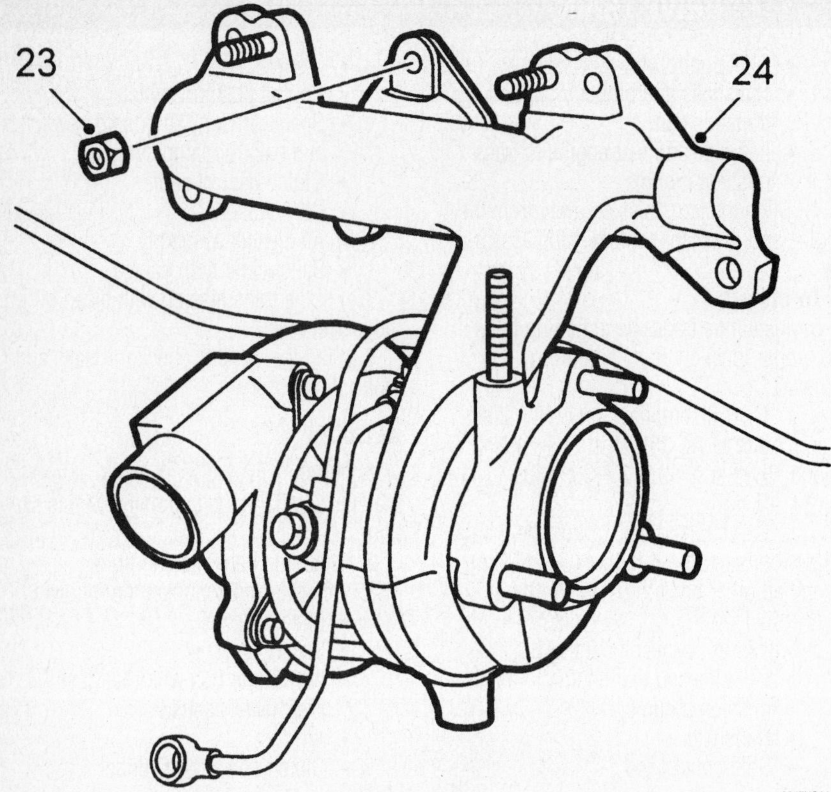

Bolt locations and torque specifications for the Altima with a manual transaxle

- Speedometer pinion electrical connectors
- Park/Neutral Position (PNP) switch electrical connectors
- Starter
- Crankshaft Position (CKP) sensor
- Shift control rod
- Front wheels
- Halfshafts and properly support the engine
- Rear and left side engine mounts
- Transaxle assembly

To install:

4. Install the transaxle assembly into the vehicle.
5. Torque the 4 lower mounting bolts to 22–30 ft. lbs. (30–40 Nm) and all remaining bolts to 29–36 ft. lbs. (39–49 Nm).
6. Install or connect the following:
   - Rear and left side engine mounts
   - Halfshafts and remove the engine support
   - Both front wheels
   - Shift control rod
   - CKP sensor
   - Starter and torque the bolts to 30 ft. lbs. (41 Nm)
   - PNP switch electrical connectors
   - Speedometer pinion connectors
   - Clutch operating cylinder
   - Air duct and air cleaner assembly
   - Battery tray and battery
   - Both battery cables
7. Fill the transmission with clean fluid.
8. Start the vehicle, check for leaks and repair if necessary.

MAXIMA

1. Before servicing the vehicle, refer to the precautions in the beginning of this section.
2. Drain the fluid from the transaxle.
3. Remove or disconnect the following:
   - Battery cables
   - Battery and the battery tray
   - Air cleaner and Mass Air Flow (MAF) sensor
   - Air breather hose
   - Control cable and cable mounting bracket
   - Clutch operating cylinder and hose clamps
   - Speedometer pinion, Park/Neutral Position (PNP) and ground harness switch connectors
   - Starter
   - Crankshaft Position (CKP) sensor (POS) from the transaxle
   - Shift control rod and support rod bracket
   - Front wheels
   - Halfshafts and properly support the engine with a jack under the oil pan

WARNING

Do not place the jack under the oil pan drain plug.

   - Center member
   - Left-hand mounting bracket from the transaxle and body
   - Transaxle

To install:

➡The transaxle mounting bolts are different lengths and require special torque specifications. Use care when installing and tightening these bolts.

4. Install or connect the following:
   - Transaxle assembly into the vehicle
   - Transaxle mounting bolts in the proper location as noted during removal
   - Left hand mounting bracket
   - Center member
   - Halfshafts and front wheels
   - Shift control rod and support rod bracket
   - Starter
   - CKP sensor
   - Speedometer pinion, PNP switch and ground harness switch connectors
   - Clutch operating cylinder and hose clamp
   - Control cable and cable mounting bracket
   - Air breather hose
   - Air cleaner and MAF sensor

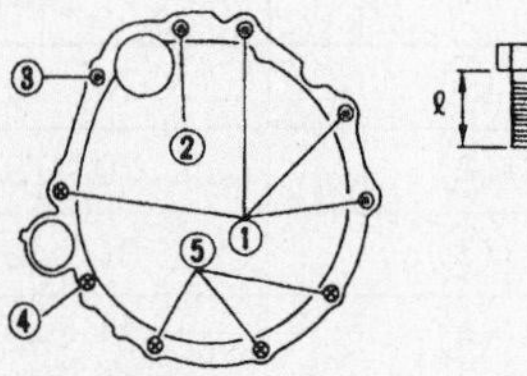

Bolt No.	Tightening torque N·m (kg-m, ft-lb)	"ℓ" mm (in)
①	70 - 79 (7.1 - 8.1, 51 - 59)	52 (2.05)
②	70 - 79 (7.1 - 8.1, 51 - 59)	65 (2.56)
③	70 - 79 (7.1 - 8.1, 51 - 59)	124 (4.88)
④	35.1 - 47.1 (3.58 - 4.80, 25.89 - 34.74)	40 (1.57)
⑤	35.1 - 47.1 (3.58 - 4.80, 25.89 - 34.74)	40 (1.57)

③ with starter
④ with support rod bracket

7923QG73

Manual transaxle bolt torque specifications and locations—Maxima with 3.0L engine

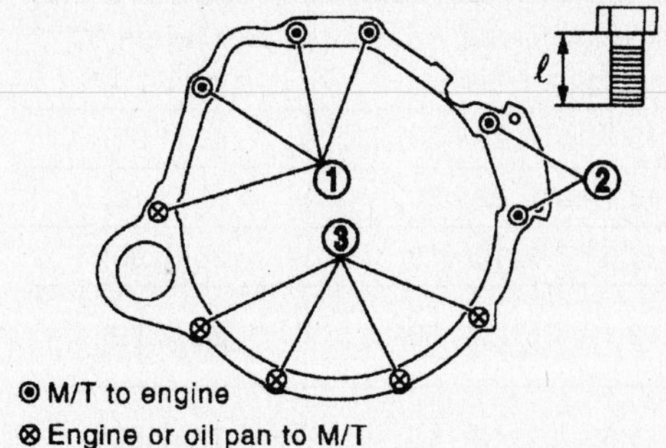

Bolt No.	Tightening torque N·m (kg-m, ft-lb)	"ℓ" mm (in)
1	69.6 - 79.4 (7.1 - 8.0, 52 - 58)	52 (2.05)
2	69.6 - 79.4 (7.1 - 8.0, 52 - 58)	113 (4.45)
3	36 - 47 (3.7 - 4.7, 27 - 34)	40 (1.57)

9357RG15

Manual transaxle bolt torque specifications and locations—Maxima with 3.5L engine

Bolt No.	1	2	3	4	5
Quantity	1	5	2	2	2
"ℓ" mm (in)	55 (2.17)	65 (2.56)	50 (1.97)	35 (1.38)	65 (2.56)
Tightening torque N·m (kg-m, ft-lb)	75 (7.7, 55)		55.4 (5.7, 41)	46.6 (4.8, 34)	55.4 (5.7,41)

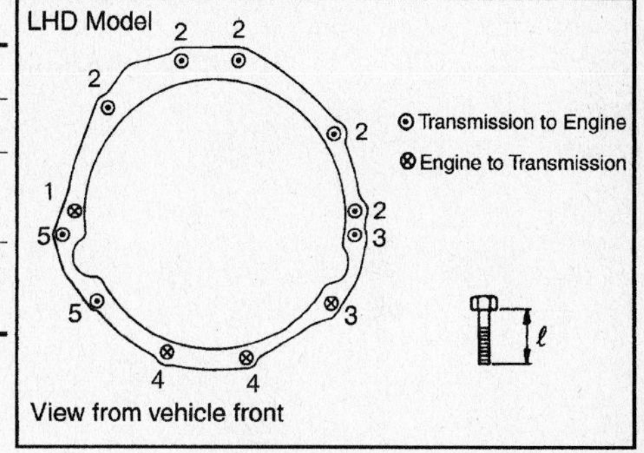

67170-NISS-G48

Manual transaxle bolt torque specifications and locations—350Z

- Battery and tray
- Battery cables

5. Fill the transaxle with clean fluid.
6. Start the vehicle, check for leaks and repair if necessary.

350Z

1. Before servicing the vehicle, refer to the precautions in the beginning of this section.
2. Drain the fluid from the transaxle.
3. Remove or disconnect the following:
   - Negative battery cable
   - Engine undercover
   - Strut tower bar
   - Front under vehicle cross bar
   - Catalytic converter and front exhaust pipe
   - Drive shaft
   - Shift lever assembly from the control rod
   - Shifter console and boot
   - Shift lever from shift housing
   - Clutch slave cylinder
   - Crankshaft position sensor
   - Back-up light and neutral safety switch
   - Wire harnesses from transmission
   - Starter
   - Transmission cover plate
4. Place a transmission jack under the transmission.
5. Remove the rear engine crossmember.
6. Remove the transmission mounting bolts and the transmission.

To install:

➡The transaxle mounting bolts are different lengths and require special torque specifications. Use care when installing and tightening these bolts.

7. Installation is the reverse of the removal procedure noting the following:
   a. Tighten the rear engine crossmember bolts to 36 ft. lbs. (49 Nm).
8. Fill the transaxle with clean fluid.
9. Start the vehicle, check for leaks and repair if necessary.

Automatic

SENTRA

1. Before servicing the vehicle, refer to the precautions in the beginning of this section.
2. Drain the fluid from the transaxle.
3. Remove or disconnect the following:
   - Both battery cables
   - Battery and bracket from the vehicle
   - Crankshaft Position (CKP) sensor from the transaxle
   - Air cleaner assembly
   - Torque converter clutch solenoid valve electrical connector
   - Inhibitor switch and Vehicle Speed Sensor (VSS) electrical connectors
   - Throttle wire from the engine side
   - Control cable
   - Oil cooler hoses
   - Halfshafts
   - Intake manifold support bracket
   - Starter
   - Upper engine to transmission bolts and properly support the transmission
   - Center member
   - Front and rear gussets and engine rear plate
   - Rear transaxle to engine bracket
   - Rear transaxle mount
   - Transaxle assembly

To install:

When connecting the torque converter to the transaxle, be sure to measure the distance between the mounting lug of the converter and the front edge of the transaxle.

4. The measured distance between the converter and the front of the transaxle should be:
   a. 1.8L engines: 0.831 in. (21.1mm) or more.
   b. 2.0L engines: 0.626 in. (15.9mm) or more.

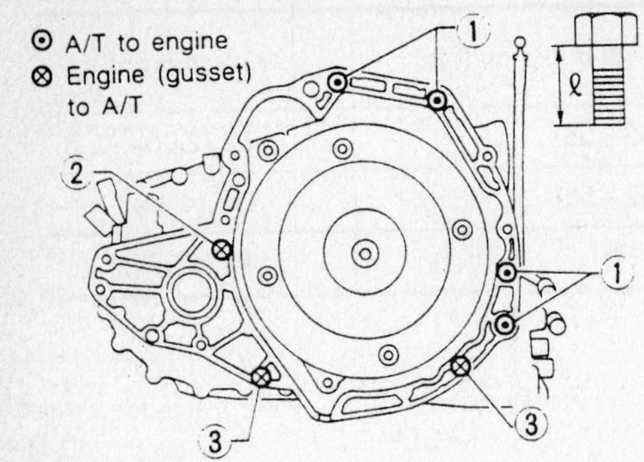

Bolt No.	Tightening torque N·m (kg-m, ft-lb)	Bolt length "ℓ" mm (in)
①	30 - 40 (3.1 - 4.1, 22 - 30)	50 (1.97)
②	30 - 40 (3.1 - 4.1, 22 - 30)	30 (1.18)
③	16 - 21 (1.6 - 2.1, 12 - 15)	25 (0.98)
Front gusset to engine	30 - 40 (3.1 - 4.1, 22 - 30)	20 (0.79)
Rear gusset to engine	16 - 21 (1.6 - 2.1, 12 - 15)	16 (0.63)

9307QG11

Bolt locations and torque specifications—1.8L engine with automatic transaxle

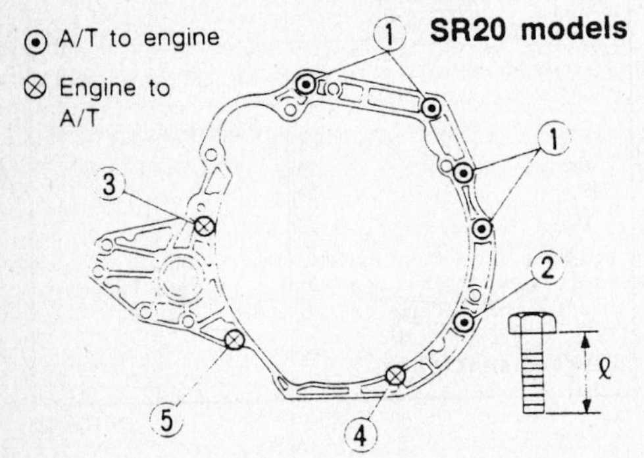

Bolt No.	Tightening torque N·m (kg-m, ft-lb)	Bolt length "ℓ" mm (in)
①	70 - 79 (7.1 - 8.1, 51 - 59)	55 (2.17)
②	70 - 79 (7.1 - 8.1, 51 - 59)	50 (1.97)
③	70 - 79 (7.1 - 8.1, 51 - 59)	65 (2.56)
④	16 - 21 (1.6 - 2.1, 12 - 15)	35 (1.38)
⑤	16 - 21 (1.6 - 2.1, 12 - 15)	45 (1.77)

9307QG12

Bolt locations and torque specifications—2.0L engine with automatic transaxle

5. Raise the transaxle and install to engine drive plate.

6. Install the transaxle mounting bolts in the proper location as noted during removal.

7. On 1.8L engines torque the 2 bottom bolts to 12–15 ft. lbs. (16–21 Nm) and all other bolts to 22–30 ft. lbs. (30–40 Nm).

8. On 2.0L engines torque the 2 bottom bolts to 12–15 ft. lbs. (16–21 Nm) and all other bolts to 51–59 ft. lbs. (70–79 Nm).

9. Install or connect the following:

- Torque converter to the drive plate and torque the bolts to 33–43 ft. lbs. (44–59 Nm)
- Rear transmission mount
- Rear transmission bracket
- Front and rear gussets and the rear engine plate
- Center member
- Starter and torque the bolts to 31 ft. lbs. (42 Nm)
- Intake manifold support bracket
- Both half shafts
- Control cable
- Throttle wire
- CKP sensor
- Torque converter clutch solenoid valve, VSS and inhibitor switch electrical connectors
- Air duct
- Battery and both cables

10. Fill the transmission to the proper level.

11. Start the vehicle, check for leaks and repair if necessary.

ALTIMA

1. Before servicing the vehicle, refer to the precautions in the beginning of this section.
2. Drain the transmission fluid.
3. Remove or disconnect the following:

- Battery cables
- Battery and tray
- Air cleaner and resonator
- Park/Neutral Position (PNP) switch
- Revolution sensor and Vehicle Speed Sensor (VSS) electrical connectors
- Crankshaft Position (CKP) sensor
- Left hand mounting bracket from the transaxle and body
- Control cable
- Both front wheels
- Halfshafts
- Oil cooler pipes
- Starter and properly support the engine
- Center member
- Rear cover plate
- Torque converter
- Transaxle assembly

➡When removing the torque converter, turn the crankshaft for access to the bolts. Place alignment marks on the converter and drive plate, so the converter can be installed in its original position.

➡The transaxle mounting bolts are different lengths. Tagging the bolts upon removal will facilitate proper tightening during installation.

To install:

➡When installing the torque converter to the transaxle, measure the depth of the converter to ensure proper installation.

4. Using a straight edge across the mounting flange, measure the depth of the converter. The measurement is to the bolt mounting flange of the converter.

5. The depth measurement of the converter should be 0.75 in. (19mm) or more.

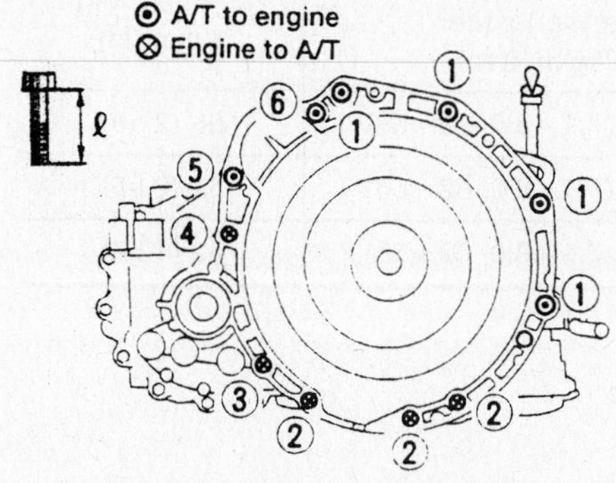

Bolt No.	Tightening torque N·m (kg-m, ft-lb)	ℓ mm (in)
①	39 - 49 (4.0 - 5.0, 29 - 36)	45 (1.77)
②	30 - 36 (3.1 - 3.7, 22 - 27)	30 (1.18)
③	30 - 36 (3.1 - 3.7, 22 - 27)	40 (1.57)
④	74 - 83 (7.5 - 8.5, 54 - 61)	45 (1.77)
⑤	30 - 36 (3.1 - 3.7, 22 - 27)	80 (3.15)
⑥	30 - 36 (3.1 - 3.7, 22 - 27)	65 (2.56)

7923QG72

Be sure to install the bolts in the correct location and tighten them to specification—Altima with automatic transaxle

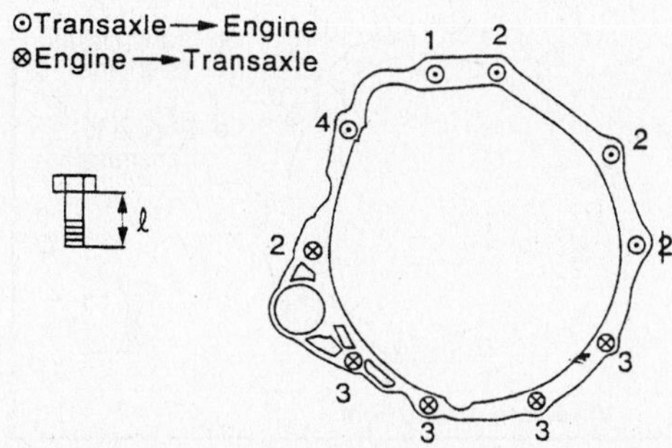

Bolt No.	Tightening torque N·m (kg-m, ft-lb)	ℓ mm (in)
1	70 - 79 (7.1 - 8.1, 52 - 58)	65 (2.56)
2	70 - 79 (7.1 - 8.1, 52 - 58)	52 (2.05)
3	70 - 79 (7.1 - 8.1, 52 - 58)	40 (1.57)
4	78 - 98 (7.9 - 10.0, 58 - 72)	124 (4.88)

9307QG20

Automatic transaxle bolt torque specifications and locations—Maxima with 3.0L engine

➡The transaxle mounting bolts are different lengths and require special torque specifications. Use care when installing and tightening these bolts.

6. Install the transaxle assembly into the vehicle.
7. Refer to the diagram for the automatic transaxle mounting bolt torque specifications.
8. Torque the bolts holding the converter to the flexplate to 33–43 ft. lbs. (44–59 Nm).
9. Install or connect the following:
   - Rear cover plate
   - Center member
   - Starter and remove the engine support
   - Oil cooler pipes
   - Halfshafts and both front wheels
   - Control cable
   - Left hand mounting bracket
   - CKP sensor
   - Revolution and VSS sensor electrical connectors
   - PNP switch
   - Air cleaner and resonator
   - Battery, tray and both cables
10. Fill the transaxle with the proper type and amount of fluid.
11. Start the vehicle, check for leaks and repair if necessary.

MAXIMA

1. Before servicing the vehicle, refer to the precautions in the beginning of this section.
2. Drain the transaxle fluid.
3. Remove or disconnect the following:
   - Battery cables
   - Battery and tray
   - Air cleaner and resonator
   - Park/Neutral Position (PNP) switch
   - Revolution sensor and Vehicle Speed Sensor (VSS) electrical connectors
   - Crankshaft Position (CKP) sensor
   - Left hand mounting bracket from the transaxle and body
   - Control cable
   - Both front wheels
   - Halfshafts
   - Oil cooler pipes
   - Starter and properly support the engine
   - Center member
   - Rear cover plate
   - Torque converter
   - Transaxle assembly

➡When removing the torque converter, turn the crankshaft for access to the bolts. Place alignment marks on the converter and drive plate, so the converter can be installed in its original position.

➡The transaxle mounting bolts are different lengths. Tagging the bolts upon removal will facilitate proper tightening during installation.

To install:

➡When installing the torque converter to the transaxle, measure the depth of the converter to ensure proper installation.

4. Using a straight edge across the mounting flange, measure the depth of the converter. The measurement is to the bolt mounting flange of the converter.
5. The depth measurement of the converter should be 0.75 in. (19mm) or more.

➡The transaxle mounting bolts are different lengths and require special torque specifications. Use care when installing and tightening these bolts.

6. Install the transaxle assembly into the vehicle.

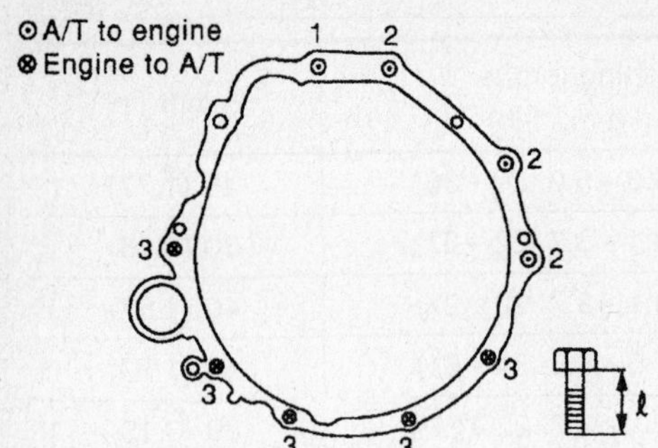

Bolt No.	Tightening torque N·m (kg-m, ft-lb)	ℓ mm (in)
1	69.6 - 79.4 (7.1 - 8.0, 52 - 58)	65 (2.56)
2	69.6 - 79.4 (7.1 - 8.0, 52 - 58)	52 (2.05)
3	69.6 - 79.4 (7.1 - 8.0, 52 - 58)	40 (1.57)

9357RG16

Automatic transaxle bolt torque specifications and locations—Maxima with 3.5L engine

Bolt No.	1	2	3	4
Number of bolts	1	5	2	2
Bolt length "ℓ"mm (in)	55 (2.17)	65 (2.56)	56 (2.20)	35 (1.38)
Tightening torque N·m (kg-m, ft-lb)	70 - 80 (7.2 - 8.1, 52 - 59)		49.0 - 61.8 (5.0 - 6.3, 37 - 45)	41.2 - 52.0 (4.2 - 5.3, 31 - 38)

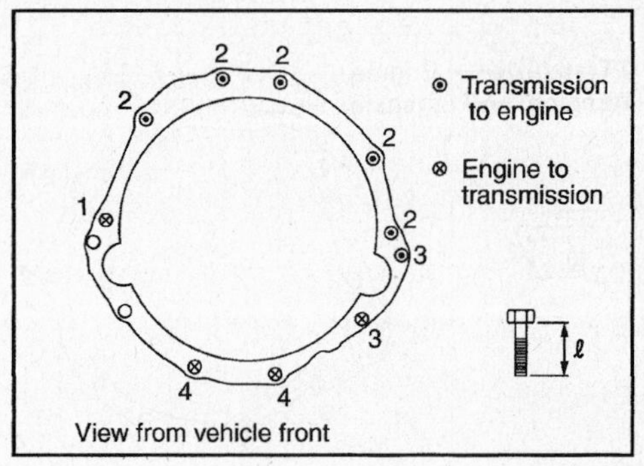

67170-NISS-G49

Automatic transaxle bolt torque specifications and locations—350Z

7. Refer to the diagram for the automatic transaxle mounting bolt torque specifications.

8. Torque the bolts holding the converter to the flexplate to 33–43 ft. lbs. (44–59 Nm).

9. Install or connect the following:

- Rear cover plate
- Center member
- Starter and remove the engine support
- Oil cooler pipes
- Halfshafts and both front wheels
- Control cable
- Left hand mounting bracket
- CKP sensor
- Revolution and VSS sensor electrical connectors
- PNP switch
- Air cleaner and resonator
- Battery, tray and both cables

10. Fill the transaxle with the proper type and amount of fluid.

11. Start the vehicle, check for leaks and repair if necessary.

350Z

1. Before servicing the vehicle, refer to the precautions in the beginning of this section.

2. Drain the fluid from the transaxle.

3. Remove or disconnect the following:

- Negative battery cable
- Engine undercover
- Strut tower bar
- Front under vehicle cross bar
- Catalytic converter and front exhaust pipe
- Drive shaft
- Control rod
- Transmission harness connector
- Crankshaft position sensor
- Cooler lines and cap openings
- Air breather hose
- Starter
- Torque converter dust cover
- Torque converter bolts

4. Place a transmission jack under the transmission.

5. Remove the rear engine crossmember.

6. Remove the transmission mounting bolts and the transmission.

To install:

➡The transaxle mounting bolts are different lengths and require special torque specifications. Use care when installing and tightening these bolts.

7. Installation is the reverse of the removal procedure noting the following:

a. Tighten the rear engine crossmember bolts to 36 ft. lbs. (49 Nm).

b. Tighten the torque converter bolts to 33—42 ft. lbs. (44–58 Nm).

8. Fill the transaxle with clean fluid.

9. Start the vehicle, check for leaks and repair if necessary.

Clutch

REMOVAL & INSTALLATION

1. Before servicing the vehicle, refer to the precautions in the beginning of this section.

2. Remove or disconnect the following:

- Transmission/transaxle assembly

3. Insert a clutch disc centering tool into the clutch disc hub for support.

- Pressure plate bolts evenly in reverse order of the tightening sequence, a little at a time to prevent distortion
- Clutch assembly
- Throw-out bearing from the clutch lever

To install:

4. Apply a light coating of chassis lube to the clutch disc spleens, input shaft and pilot bearing. Use a disc centering tool to aid installation.

5. Install or connect the following:
   - Disc and pressure plate
6. On all except Maxima and 350Z, torque the pressure plate bolts in a crisscross pattern and in several steps to 16–22 ft. lbs. (20–26 Nm).
7. On Maxima, torque the pressure plate bolts in a crisscross pattern in the following 2 steps:
   a. Step 1: 7–14 ft. lbs. (10–20 Nm).
   b. Step 2: 25–33 ft. lbs. (34–44 Nm).
8. On 350Z, torque the pressure plate bolts in a crisscross pattern in the following 2 steps:
   a. Step 1: 11 ft. lbs. (115 Nm).
   b. Step 2: 29 ft. lbs. (340 Nm).
9. Install or connect the following:
   - New throw-out bearing in the clutch release lever. Remove the clutch disc centering tool.
   - Transaxle into the vehicle. If the mating surfaces will not come together, do not force the units together. Remove the transaxle and recheck that the disc is centered.

➡DO NOT draw the transaxle to the engine with the bolts. This may damage the clutch and/or transaxle. Also, be careful not to move the throw-out bearing when installing the transaxle.

10. After the transaxle is installed, connect the clutch cable and check operation before complete reassembly.
11. Adjust the clutch pedal as necessary.

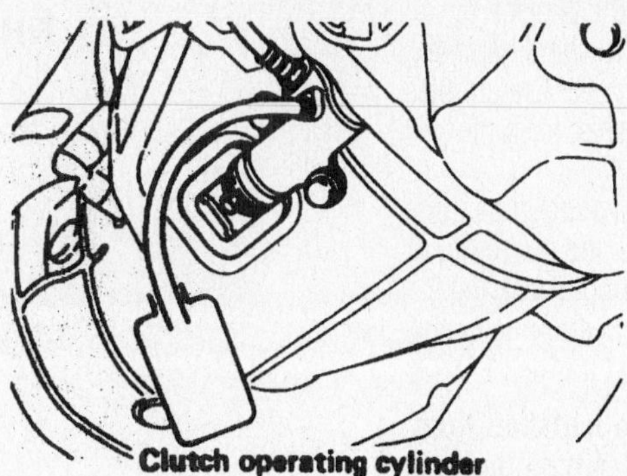

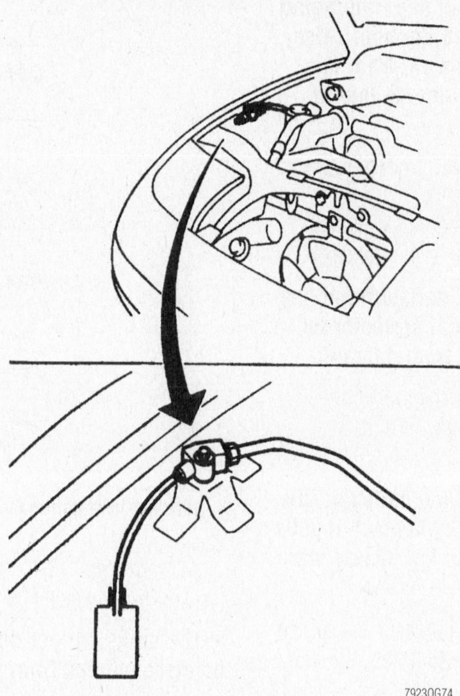

Clutch system bleeding points—Altima and Maxima

Hydraulic Clutch System

BLEEDING

Bleeding is required to remove air trapped in the hydraulic system. The bleed screw is located on the clutch slave (operating) cylinder.

Some models are also equipped with a clutch damper mechanism. The clutch damper mechanism is bled in exactly the same manner as the operating cylinder. It should be bled along with the operating cylinder.

1. Before servicing the vehicle, refer to the precautions in the beginning of this section.
2. Remove the bleed screw dust cap.
3. Attach a transparent vinyl tube to the bleed screw, immersing the free end in a clean container of clean brake fluid.
4. Fill the master cylinder with the proper fluid.
5. Open the bleed screw about ¾ turn.
6. Depress the clutch pedal quickly. Hold it down. Have an assistant tighten the bleed screw. Allow the pedal to return slowly.
7. Repeat the above procedure until no more air bubbles are seen in the fluid container.
8. Remove the bleed tube.
9. Replace the dust cap and refill the master cylinder.
10. Bleed the clutch damper, if equipped.

Halfshaft

REMOVAL & INSTALLATION

Sentra

➡The halfshafts will require a special tool for the spline alignment of the halfshaft end into the transaxle case. Do not perform this procedure without access to this tool. The Kent Moore tool Number is J-34296 and J-34297

1. Before servicing the vehicle, refer to the precautions in the beginning of this section.
2. Raise the front of the vehicle and support it on jackstands, then remove the wheel and the tire assembly.
3. Remove or disconnect the following:
   - Wheel
   - Hub nut using a bar to hold the wheel from turning
   - Clip and separate the brake hose from the strut
   - Caliper assembly and support it with a wire. Do not allow the caliper to hang from the brake hose.
   - Bolts that secure the strut to the steering knuckle

➡Cover the halfshaft boots with shop towels to protect them during removal of the shaft.

- Halfshaft from the knuckle by lightly tapping it with a hammer. If it is hard to remove, use a puller.

4. Remove the halfshaft from the transaxle as follows:
   a. Models without support bearing: Pry the halfshaft from the transaxle.
   b. Models with support bearing: Remove the support bearing bolts and pull the halfshaft from transaxle.

➡When removing the halfshaft from the transaxle, do not pull on the halfshaft. The halfshaft will separate at the sliding joint (damaging the boot). Use a small prybar to remove it from the transaxle. Be sure to replace the oil seal in the transaxle.

5. Remove the halfshaft from the vehicle.

To install:

6. Use a new circlip on the halfshaft and install a new oil seal to the transaxle.

➡When installing the halfshaft into the transaxle, use a oil seal protector tool to protect the oil seal from damage.

7. Install or connect the following:
   - Halfshaft assembly into the transaxle

➡After installation of the halfshaft, try to pull the flange out by hand. If it pulls out, the circular clip is not locked into the transaxle.

   - Support bearing bracket and torque the bolts to 19–26 ft. lbs. (25–35 Nm)

8. Lubricate the splines of the halfshaft and insert the shaft through the steering knuckle.
9. Align the steering knuckle with the lower strut mount. Torque the bolts to 68–82 ft. lbs. (92–111 Nm).
   - Disc brake caliper and the brake hose to the strut with the clip
   - Washer and hub nut to the halfshaft and torque the nut to 145–202 ft. lbs. (197–274 Nm)
   - Adjusting cap and a new cotter pin in drive axle
   - Wheel and tire assembly and lower the vehicle

10. Road test the vehicle for proper operation.

Altima

1. Before servicing the vehicle, refer to the precautions in the beginning of this section.
2. Raise and safely support the vehicle with the front wheels hanging freely.

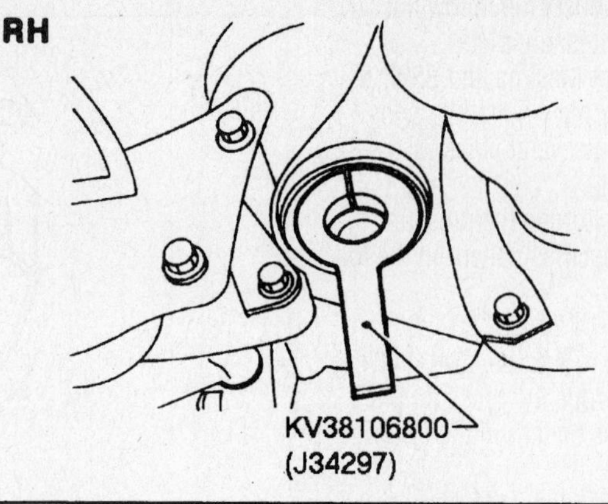

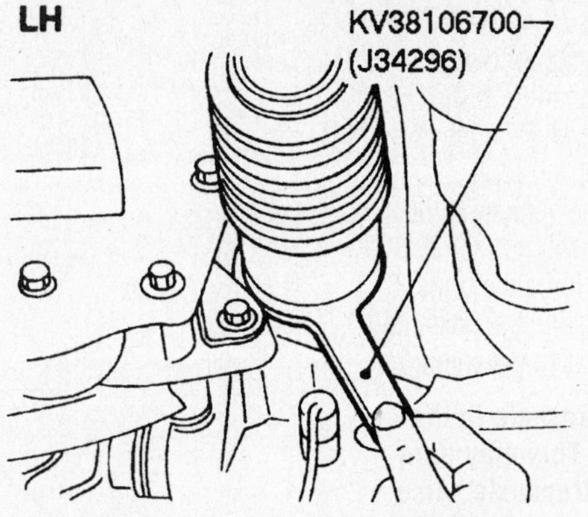

Halfshaft installation tools—Sentra

3. Remove or disconnect the following:
   - Front wheels from the vehicle

➡The brake caliper does not need to be disconnected from the knuckle.

   - Cotter pin from the castellated nut on the wheel hub
   - Wheel bearing locknut

➡Cover the CV-joint boots with a shop towel or waste cloth so not to damage them when removing the halfshaft.

   - Cotter pin and castle nut from the lower ball joint

4. Strike the knuckle with a hammer and pull down on the transverse link to separate the lower ball joint from the knuckle.
   - Tie rod end from the steering knuckle
   - Halfshaft from the steering knuckle by tapping it with a block of wood and a mallet

5. Using a prybar, reach through the engine crossmember and carefully pry the right inner CV-joint from the transaxle.
6. If equipped with manual transaxle, carefully pry the left inner CV-joint from the transaxle.
7. If equipped with automatic transaxle, insert a long tool into the opening for the right halfshaft and strike the tool to with a hammer.
8. Remove the left halfshaft from the transaxle.

To install:

➡Whenever the halfshafts are removed, the axle seals should be replaced.

9. When installing the shafts into the transaxle, use a new oil seal and install an alignment tool along the inner circumference of the oil seal.
10. Install or connect the following:
    - Halfshaft into the transaxle, align the serration's and remove the alignment tool

11. Push the halfshaft, then press-fit the circular clip on the shaft into the clip groove on the side gear.

➡After insertion, attempt to pull the flange out of the side joint to be sure the circular clip is properly seated in the side gear and will not come out.

   - Halfshaft into the steering knuckle
   - Lower ball joint and tie rod end and torque the lower ball joint-to-control arm nuts to 52–64 ft. lbs. (71–86 Nm) and the tie rod end-to-

steering knuckle nut to 22–29 ft. lbs. (29–39 Nm)
- New cotter pins to the castle nuts
- Axle nut and torque the locknut to 174–231 ft. lbs. (235–314 Nm)
- New cotter pin on the wheel hub and install the wheel
- Front wheels to the vehicle

12. Road test the vehicle for proper operation.
13. Check the transaxle fluid level and top off as necessary.

Maxima

1. Before servicing the vehicle, refer to the precautions in the beginning of this section.
2. Raise and support the front of the vehicle safely and remove the wheels.
3. Remove or disconnect the following:
   - Anti-Lock Brake (ABS) wheel sensor and move it out of the way
   - Brake hose from the strut
   - Wheel bearing locknut
   - Bolts attaching the steering knuckle to the strut. Matchmark the bolts before removal.

➡Cover axle boots with waste cloth so as not to damage them when removing halfshaft.

- Halfshaft from the knuckle by slightly tapping it
- Bolts attaching the support bearing to the support bearing bracket
- Halfshaft from the transaxle with a flat-bladed tool, if equipped with a manual transaxle

4. If equipped with a automatic transaxle perform the following:
   a. Remove the right halfshaft from the vehicle.
   b. Insert a flat-bladed tool into the transaxle where the right halfshaft was, place the end of the tool on the halfshaft, then, drive the left shaft from the pinion side gear.
5. Remove or disconnect the following:
   - Support bearing bolts and the halfshaft from the vehicle
   - Discard the circlip on the end of the halfshaft
   - Seal from the transaxle

To install:

6. Install or connect the following:
   - New seal into the transaxle and install a halfshaft alignment tool into the transaxle seal
   - New circlip to the halfshaft, then insert the halfshaft into the transaxle

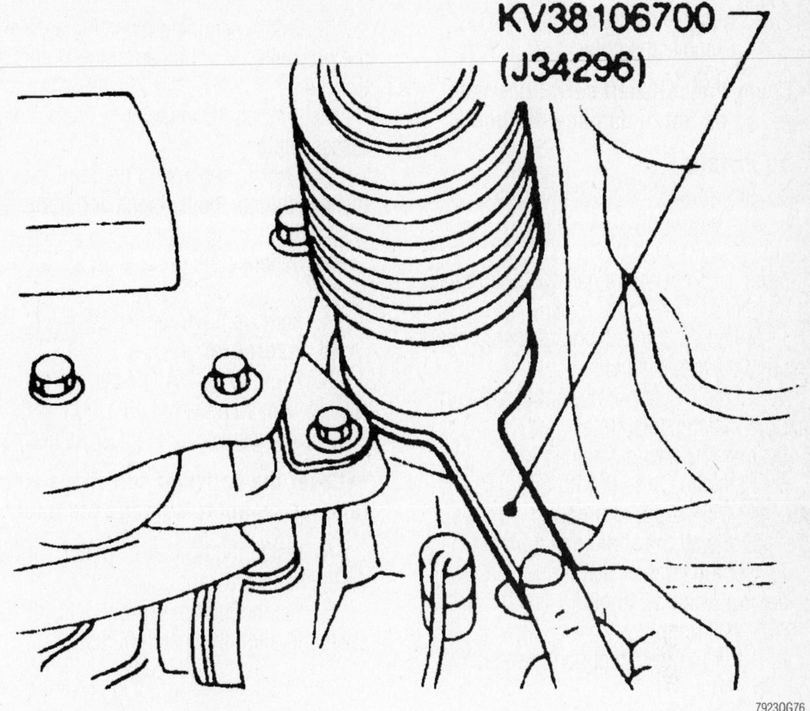

Left halfshaft alignment tool—Maxima

7. With the serration's aligned remove the alignment tool.
8. Push the halfshaft fully into the transaxle to seat the circlip. Try to pull the halfshaft from the transaxle by hand to verify that the circlip is properly seated.
   - Support bearing and torque the bolts to 10–14 ft. lbs. (13–19 Nm)
   - Halfshaft into the steering knuckle and install the hub locknut, do not tighten the hub nut
   - Steering knuckle to the strut
   - Strut mounting bolts to the matchmarks and torque the bolts to 103–117 ft. lbs. (140–159 Nm)
   - Brake hose to the strut
   - ABS wheel sensor and torque the bolt to 13–17 ft. lbs. (18–24 Nm)
   - Front wheels and torque hub locknut to 174–231 ft. lbs. (235–314 Nm)
9. Check and/or adjust the wheel alignment as necessary.

350Z

1. Before servicing the vehicle, refer to the precautions in the beginning of this section.
2. Raise and support the rear of the vehicle safely and remove the wheels.
3. Remove or disconnect the following:
   - Cotter pin and axle nut
   - Stabilizer bar connecting rod
4. Remove the nuts and bolts between the side flange and the drive shaft.
5. Use a suitable puller and remove the drive shaft from the axle.

To install:

6. Install or connect the following:
   - New seal into the transaxle and install halfshaft
   - Tighten the side flange bolts to 47–58 ft. lbs. (63–79 Nm)
   - Stabilizer bar connecting rod
   - Axle nut and new cotter pin. Tighten the axle nut to 152–202 ft. lbs. (206–274 Nm)
   - Rear wheels

CV-Joints

OVERHAUL

Sentra

TRANSAXLE SIDE

1. Before servicing the vehicle, refer to the precautions in the beginning of this section.
2. Disassemble the joint as follows:
   a. Remove the boot bands.
   b. Matchmark the slide joint housing and inner race before separating the assembly.
   c. Matchmark the spider assembly and drive shaft.

d. Remove the snapring.
e. Remove the spider assembly.
f. Remove the boot.

➡Cover the halfshaft serrations with tape, so as not to damage the boot.

To install:

3. Assemble the joint as follows:
a. Thoroughly clean all parts in solvent and dry with compressed air. Check parts for evidence of damage and replace as necessary.
b. Install the boot and new small boot band on the halfshaft.
4. Install the spider assembly. Confirm that the matchmarks are aligned.
5. Install a new outer snapring.
6. Pack the CV joint with 5.5–5.8 ounces (155–165g) of grease.
a. Install the slide joint housing.
7. Set the boot so that it does not swell or deform when its length is 4–4.07 in. (101.5–103.5mm).
8. Lock the new boot bands securely.

WHEEL SIDE

1. Before servicing the vehicle, refer to the precautions in the beginning of this section.
The joint on the wheel side cannot be disassembled.
2. Prior to separating the joint assembly, matchmark the halfshaft and joint assembly.
3. Separate the joint using a slide hammer.
4. Remove the boot bands.

To assemble:

5. Thoroughly clean all parts in solvent and dry with compressed air. Check parts for evidence of damage and replace as necessary.

➡Cover the halfshaft serrations with tape, so as not to damage the boot.

6. Install the boot and small boot band on the halfshaft.
7. Set the joint assembly onto the halfshaft and align the matchmarks.
8. Attach the joint assembly to the halfshaft by lightly tapping the serrated end with a plastic hammer.

➡Using a metal hammer may damage the threads on the end of the joint.

9. Pack the CV joint with 4.6–4.41 ounces (115–125g) of grease.
10. Ensure that the boot is properly installed on the halfshaft groove.
11. Set the boot so that it does not swell or deform when its length is 3.78–3.86 in. (96–98mm).
12. Lock the new boot bands securely.

Altima

TRANSAXLE SIDE

1. Remove the boot bands.
2. Matchmark the slide joint housing and inner race, prior to separating the joint assembly.
3. Pry off the snapring and remove the ball cage, inner race and balls as a unit.

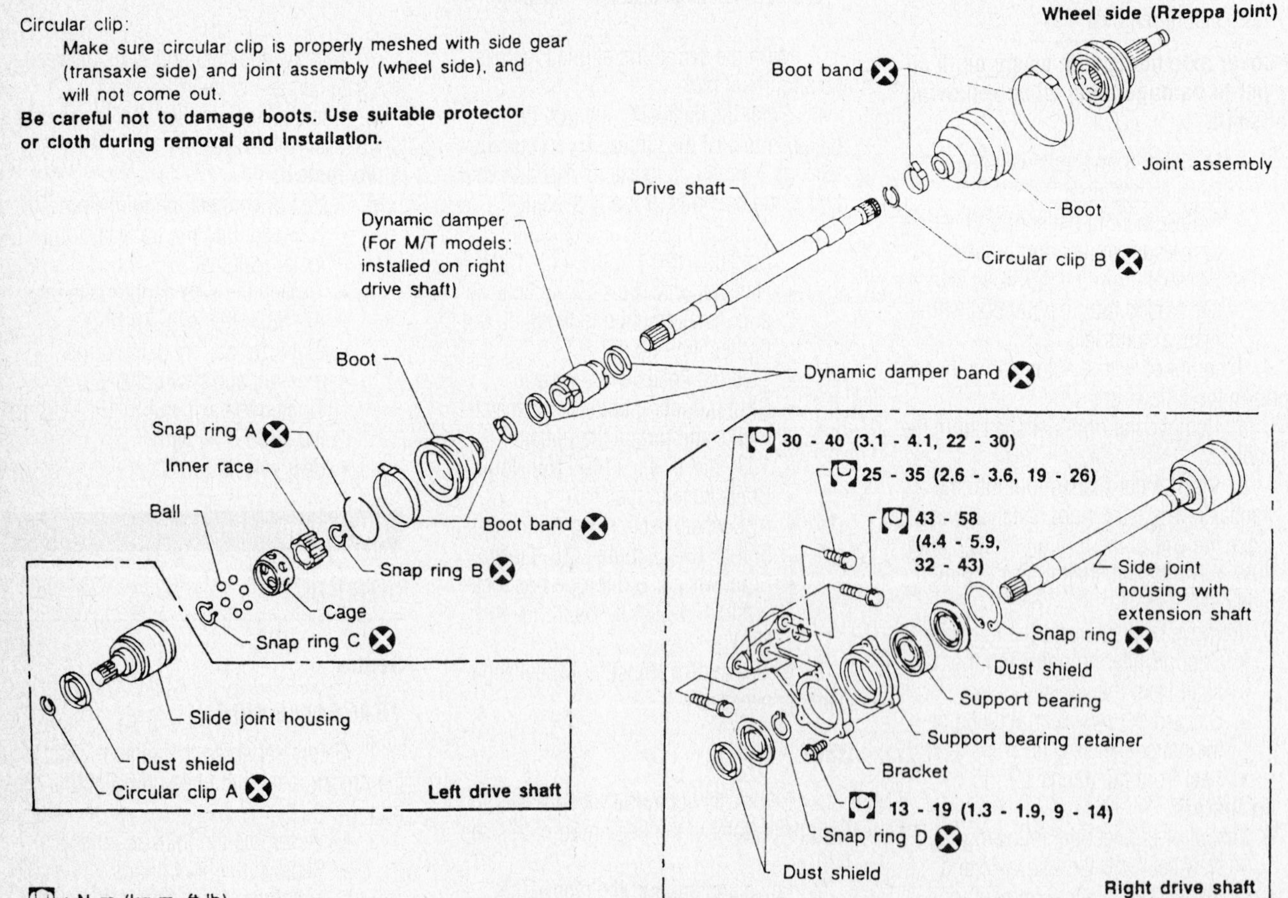

Exploded view of the halfshafts and related components

4. Remove the snapring and withdraw the boot.

➡Cover the halfshaft serrations with tape, so as not to damage the boot.

To install:

5. Thoroughly clean all parts in solvent and dry with compressed air. Check parts for evidence of damage and replace as necessary.

6. Install or connect the following:

- Boot and new boot band on the halfshaft
- New inner snapring
- Ball cage, inner race and balls as a unit. Ensure that the matchmarks are aligned.
- New outer snapring

7. Pack the halfshaft with 5.0–6.0 ounces (165–175 g) of grease.

8. Ensure that the boot is properly installed on the halfshaft groove.

9. Set the boot so that it does not swell or deform when its length is 3.82–3.90 in. (97–99mm).

10. Lock the new boot bands securely.

WHEEL SIDE

The joint on the wheel side cannot be disassembled.

1. Prior to separating the joint assembly, matchmark the halfshaft and joint assembly.

2. Separate the joint using a slide hammer.

3. Remove the boot bands.

To assemble:

4. Thoroughly clean all parts in solvent and dry with compressed air. Check parts for evidence of damage and replace as necessary.

➡Cover the halfshaft serrations with tape, so as not to damage the boot.

5. Install the boot and small boot band on the halfshaft.

6. Set the joint assembly onto the halfshaft and align the matchmarks.

7. Attach the joint assembly to the halfshaft by lightly tapping the serrated end with a plastic hammer.

➡Using a metal hammer may damage the threads on the end of the joint.

8. Pack the halfshaft with 3.5–4.0 ounces (100–115 g) of grease.

9. Ensure that the boot is properly installed on the halfshaft groove.

10. Set the boot so that it does not swell or deform when its length is 3.327–3.406 in. (84.5–86.5mm).

11. Lock the new boot bands securely.

Maxima

TRANSAXLE SIDE

1. Remove the boot bands.

2. Matchmark the slide joint housing and inner race, prior to separating the joint assembly.

3.Pry off the snapring and remove the ball cage, inner race and balls as a unit.

4. Remove the snapring and withdraw the boot.

To install:

➡Cover the halfshaft serrations with tape, so as not to damage the boot.

5. Thoroughly clean all parts in solvent and dry with compressed air. Check parts for evidence of damage and replace as necessary.

6. Install or connect the following:

- Boot and new boot band on the halfshaft
- New inner snapring
- Ball cage, inner race and balls as a unit. Confirm that the matchmarks are aligned.
- New outer snapring

7. Pack the CV joint with 5.0–6.0 ounces (165–175 g) of grease.

8. Ensure that the boot is properly installed on the halfshaft groove.

9. Set the boot so that it does not swell or deform when its length is 3.86 in. (98mm).

10. Lock the new boot bands securely.

WHEEL SIDE

The joint on the wheel side cannot be disassembled.

The inner CV joint uses a large C-clip to retain the ball and cage assembly in the outer housing

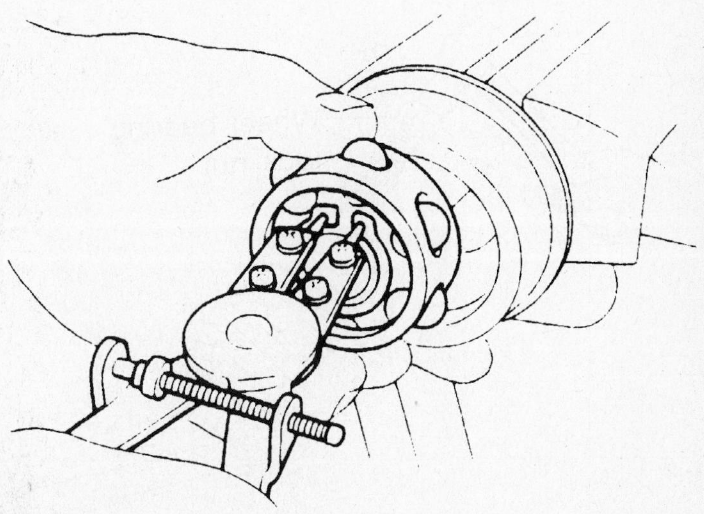

After the outer housing is removed, the ball and cage assembly can slide from the shaft by removing the C-clip

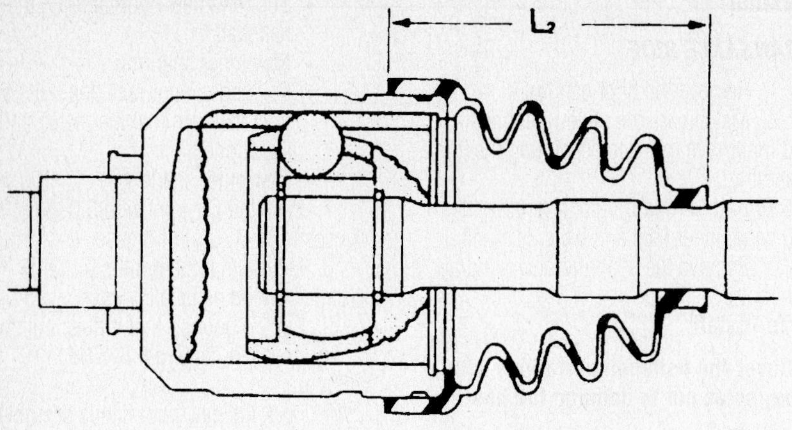

Make sure to properly position the boot before tightening the boot clamps

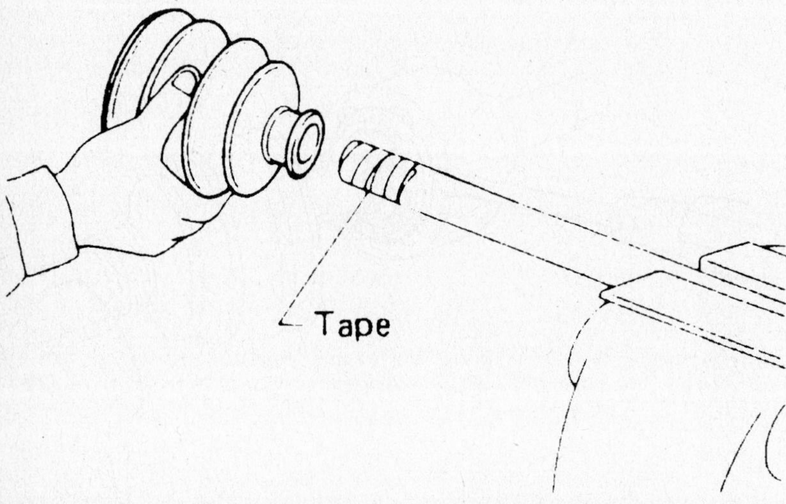

Use vinyl tape and wrap the end of the shaft to protect the boot during installation

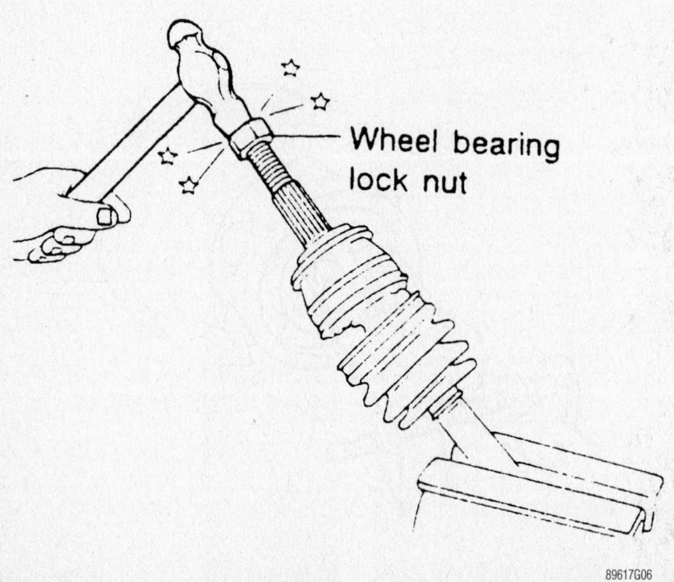

Use an old nut to protect the threads when tapping the outer CV joint onto the shaft

1. Prior to separating the joint assembly, matchmark the halfshaft and joint assembly.
2. Separate the joint using a slide hammer.
3. Remove the boot bands and the boot.

To install:

4. Thoroughly clean all parts in solvent and dry with compressed air. Check parts for evidence of damage and replace as necessary.

➡Cover the halfshaft serrations with tape, so as not to damage the boot.

5. Install the boot and small boot band on the halfshaft.
6. Set the joint assembly onto the halfshaft and align the matchmarks.
7. Attach the joint assembly to the halfshaft by lightly tapping the serrated end with a plastic hammer.

➡Using a metal hammer may damage the threads on the end of the joint.

8. Pack the CV joint with 4.76–5.11 ounces (135–145 g) of grease.
9. Ensure that the boot is properly installed on the halfshaft groove.
10. Set the boot so that it does not swell or deform when its length is 3.82 in. (97mm).
11. Lock the new boot bands securely.

350Z

FINAL DRIVE SIDE

1. Remove the boot bands.
2. Remove the stopper ring and pull out of the housing.
3. Remove the snap ring and remove the ball cage and inner race assembly using a puller.
4. Remove the boot from the shaft.

To assemble:

5. Install or connect the following:
   - Boot and new boot band on the halfshaft
   - Ball cage and inner race as a unit and install a new snap ring.
6. Pack the CV joint with 4.4–4.7 ounces (124–134 g) of grease.
7. Install a new stopper ring in the housing.
8. Ensure that the boot is properly installed on the halfshaft groove.
9. Set the boot so that it does not swell or deform when its length is 3.70 in. (94mm).
10. Lock the new boot bands securely.

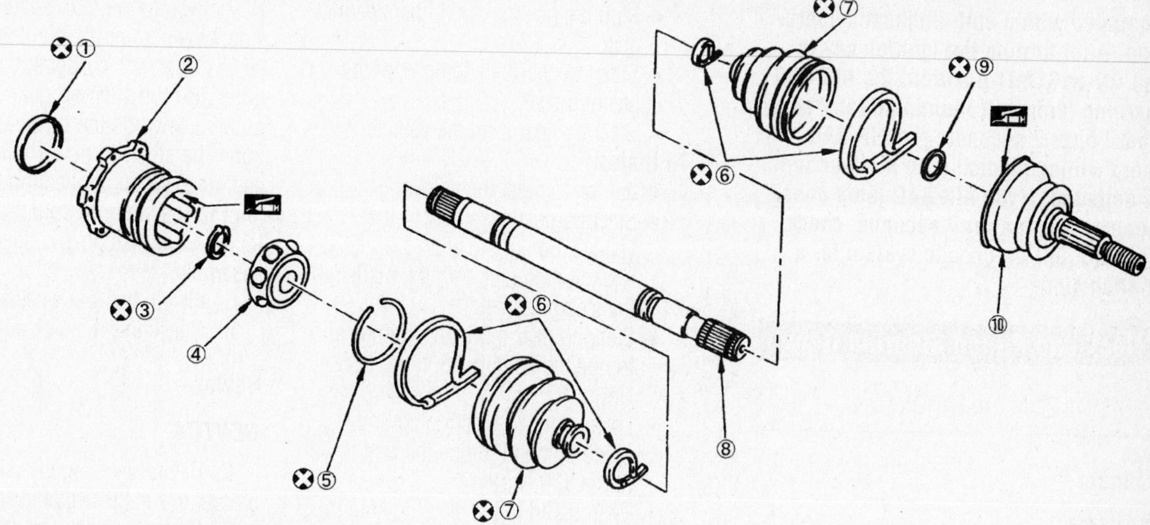

Exploded view of the rear halfshaft–350Z

WHEEL SIDE

The joint on the wheel side cannot be disassembled.

1. Prior to separating the joint assembly, matchmark the halfshaft and joint assembly.
2. Separate the joint using a slide hammer.
3. Remove the boot bands.

To assemble:

4. Thoroughly clean all parts in solvent and dry with compressed air. Check parts for evidence of damage and replace as necessary.

➡Cover the halfshaft serrations with tape, so as not to damage the boot.

5. Install the boot and small boot band on the halfshaft.
6. Set the joint assembly onto the halfshaft and align the matchmarks.
7. Attach the joint assembly to the halfshaft by lightly tapping the serrated end with a plastic hammer.

➡Using a metal hammer may damage the threads on the end of the joint.

8. Pack the halfshaft with 3.0–3.4 ounces (86–96 g) of grease.
9. Ensure that the boot is properly installed on the halfshaft groove.
10. Set the boot so that it does not swell or deform when its length is 3.82 in. (0.97mm).
11. Lock the new boot bands securely.

STEERING AND SUSPENSION

Air Bag

PRECAUTIONS

Several precautions must be observed when handling the inflator module to avoid accidental deployment and possible personal injury.

1. Never carry the inflator module by the wires or connector on the underside of the module.
2. When carrying a live inflator module, hold securely with both hands and ensure that the bag and trim cover are pointed away.
3. Place the inflator module on a bench or other surface with the bag and trim cover facing up.
4. With the inflator module on the bench, never place anything on or close to the module that may be thrown in the event of an accidental deployment.

DISARMING

➡All Supplemental Restraint System (SRS) electrical wiring harnesses and connectors are covered with YELLOW outer insulation. Do not use electrical test equipment on any circuit related to the SRS (air bag) sensors. When installing SRS components, always install with the arrow marks facing the front of the vehicle.

To disarm the SRS system turn the ignition switch to **OFF** position. Then, disconnect the both battery cables starting with the negative cable first and wait at least 10 minutes after the cables are disconnected. Be sure to insulate the battery terminal ends.

REARMING

To arm the Supplemental Restraint System (SRS) system turn the ignition switch to **OFF** position. Connect the both battery cables starting with the positive cable first.

➡The SRS or air bag system is equipped with a self-diagnostic operation. After turning the ignition key to the ON or START position, the AIR BAG warning lamp will illuminate for 7 seconds. After 7 seconds, the AIR BAG lamp will extinguish if no malfunction is detected. If the AIR BAG lamp does not extinguish after 7 seconds, check the SRS self-diagnostic system for a malfunction.

Rack and Pinion Steering Gear

REMOVAL & INSTALLATION

Manual

SENTRA

1. Before servicing the vehicle, refer to the precautions in the beginning of this section.
2. Remove or disconnect the following:
   - Front wheels
   - Both tie rod ends from the steering knuckles
3. Matchmark the steering column shaft to the lower joint and remove the pinch bolt from the joint.
   - Steering gear mounting bolts
   - Mounting clamps from the steering gear
   - Steering gear by sliding it off the steering shaft
   - Steering gear from the vehicle

To install:

4. Install or connect the following:
   - Steering gear assembly to the vehicle. Be sure to align the matchmarks of the rack with the marks on the steering shaft.
   - Steering gear mounting clamps and torque the bolts to 58 ft. lbs. (78 Nm)
   - Lower joint-to-steering column pinch bolt and torque the bolt to 22 ft. lbs. (29 Nm)
   - Tie rod end to the steering knuckle and torque the castle nut to 29 ft. lbs. (39 Nm) and install a new cotter pin

➡If installing a new rack and pinion assembly, transfer the lower steering joint to the new rack and pinion prior to installation. When installing the lower steering joint to the steering gear, be sure that the wheels are aligned with the vehicle (straight-ahead position).

5. To center the steering gear, turn it all the way to the lock position on one side. Now, count the number of turns it takes to get to the opposite side lock position. Turn the steering gear ½ the number of turns towards the original starting position. The steering rack should now be centered. When connecting the steering joint to the steering column shaft, be sure to align the matchmarks made during disassembly.
6. Install the front wheels
7. Check the vehicle's alignment.

Power

SENTRA

1. Before servicing the vehicle, refer to the precautions in the beginning of this section.
2. Remove or disconnect the following:
   - Low pressure hose clamp
   - Low pressure hose at the steering gear. Be sure to use a pan to catch the fluid.
   - Flare nut and the high pressure tube at the steering gear, then drain the fluid from the gear
   - Tie rod ends from the steering knuckle

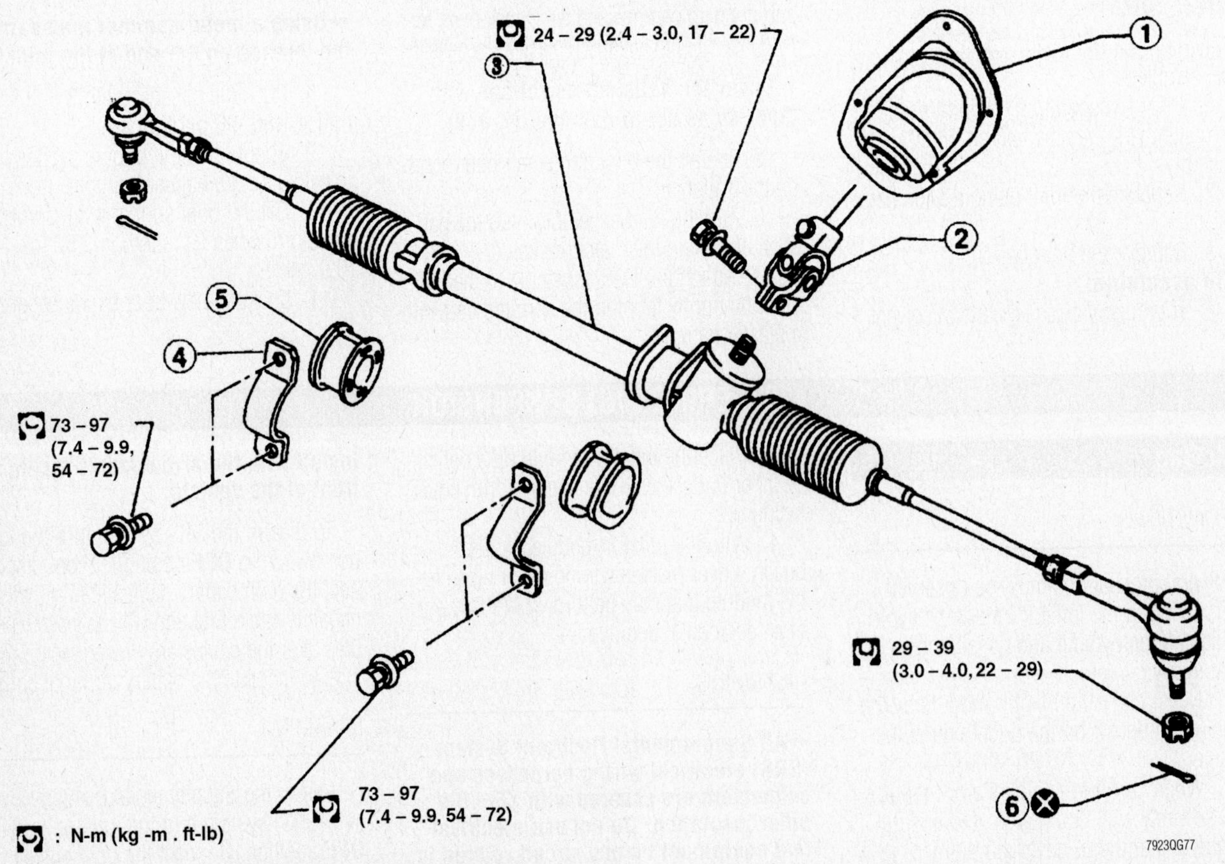

Exploded view of the manual rack and pinion steering gear mounting—Sentra

3. Place a floor jack under the transaxle and support it.
   - Front exhaust pipe and the rear engine mount
4. Position the front wheels so they are pointing straight ahead.
5. Matchmark the steering column lower joint to the steering gear.

➡The steering gear splines have a flat spot or keyway. Be sure to note this during removal.

6. Remove or disconnect the following:
   - Bolt that secures steering column lower joint
   - Bolts, steering gear unit and the linkage

To install:

7. Install or connect the following:
   - Power steering gear assembly to the vehicle. Align the steering column to the steering gear.

➡Be sure to align the flat spot or keyway during installation.

   - Steering gear mounts and torque the bolts in sequence to 54–72 ft. lbs. (73–97 Nm)
   - Pinch bolt for the steering column-to-gear connection and torque the bolt to 17–22 ft. lbs. (24–29 Nm)

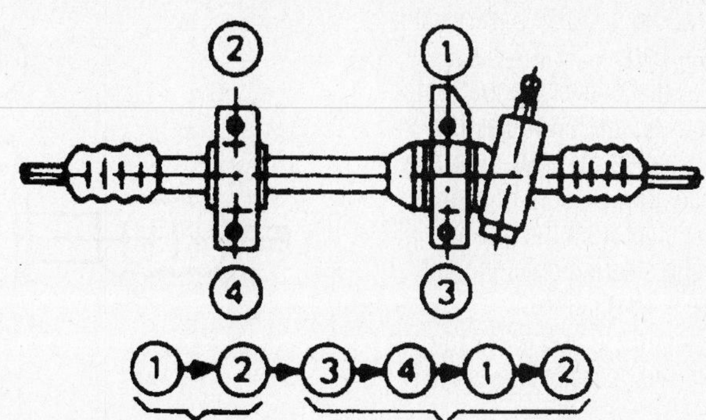

Tighten the power steering gear mounting bolts according to the sequence shown—Sentra

   - Tie rod ends to the steering knuckle and torque the nut to 22–29 ft. lbs. (29–39 Nm). Tighten the tie rod mounting nut further so the groves in the nut align with first cotter pin hole. Install a new cotter pin.
   - Power steering low pressure hose and torque the fitting to 20–29 ft. lbs. (27–39 Nm)
   - Power steering high pressure and torque the fitting to 11–18 ft. lbs. (15–25 Nm)
   - Rear engine mount and remove the floor jack
   - Front exhaust pipe assembly using new gaskets
8. Fill the power steering system and start the engine.
9. Check the wheel alignment.

ALTIMA

1. Before servicing the vehicle, refer to the precautions in the beginning of this section.

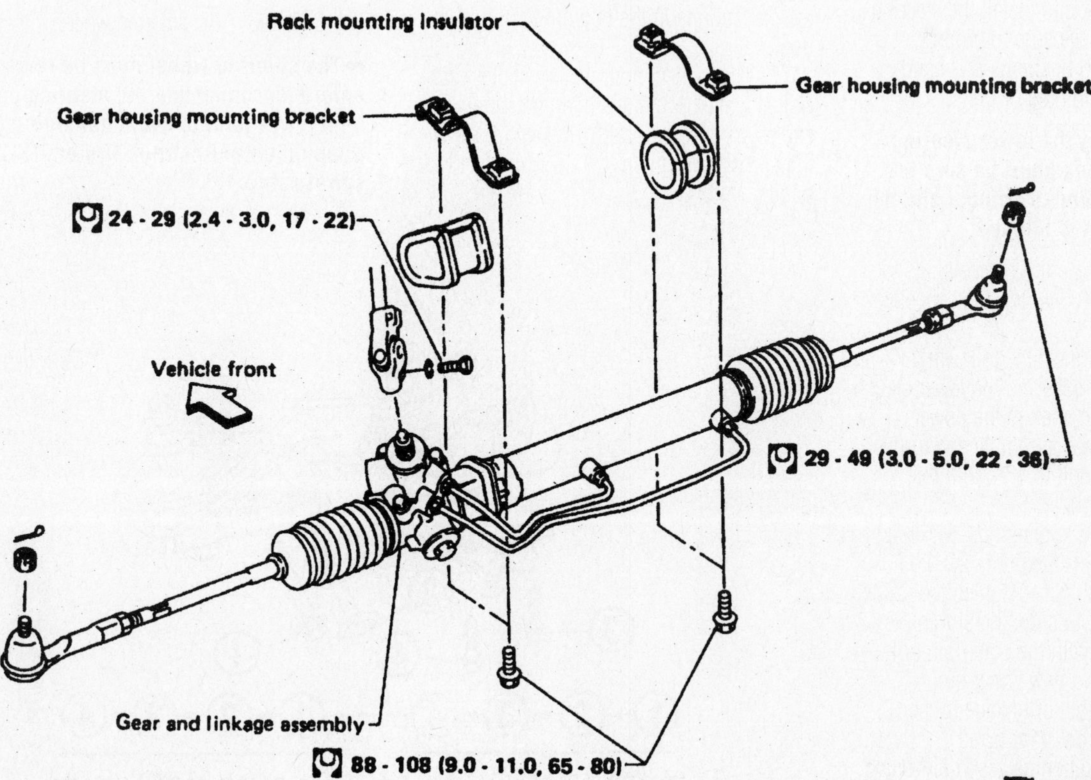

Exploded view of the power steering gear mounting—Altima

2. Disconnect the negative battery cable and disarm the air bag.
3. Remove or disconnect the following:
- Bolt securing the lower steering column shaft to the power steering gear assembly. Be sure to matchmark the shaft from the steering gear to the steering column joint for correct installation.
- Hoses from the power steering gear and plug the hoses to prevent leakage
- Cotter pins and castle nuts from the tie rod ends
- Tie rod ends from the steering knuckle, using a ball joint separating tool
- Front exhaust pipe mounting nuts and bolts
- Front exhaust pipe from the vehicle
- Control cable or linkage from the transmission and position it out of the way, if necessary
- Power steering gear mounting bolts or nuts
- Steering gear from the vehicle. Use care when separating the steering column joint.

4. Inspect the steering gear mount bushings and replace as necessary.

To install:

5. Align the steering column-to-steering gear matchmark and install the steering gear to the vehicle. Be sure to properly install the mounting bushings and hand-tighten the mounting nuts or bolts.

➡When installing the lower steering joint to the steering gear, be sure that the wheels are aligned straight and the steering joint slot is aligned.

6. Torque the steering gear mounts to 54–72 ft. lbs. (73–97 Nm) in the sequence illustrated.
7. Install or connect the following:
- Pinch bolt securing the lower steering column shaft to the power steering gear assembly and torque the pinch bolt to 17–22 ft. lbs. (24–29 Nm)
- Tie rod end to steering knuckle and torque the castle nut to 22–29 ft. lbs. (29–39 Nm). Tighten the castle nut further to align the slot in the castle nut with the cotter pin hole and install a new cotter pin.
- Control cable or linkage to the transmission, if removed
- Front exhaust pipe assembly, using new gaskets
- Power steering hoses to the steering gear

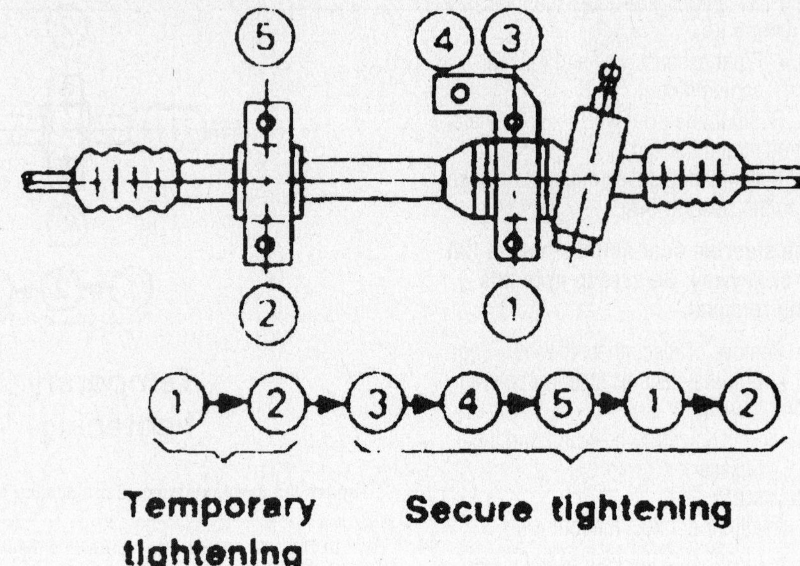

Tighten the mounting bolts using the illustrated procedure—Altima

8. Start the engine and fill the power steering reservoir.
9. Perform a front end alignment.
10. Connect the negative battery cable.
11. If equipped, enable the air bag system.

MAXIMA

1. Before servicing the vehicle, refer to the precautions in the beginning of this section.
2. Disconnect both battery cables and wait at least 10 minutes after the battery cables are disconnected. This will disarm the air bag system so the steering wheel can be removed.
3. Point the front tires straight ahead and lock the steering in this position.

⁂ WARNING

Do not turn the steering wheel or column with the lower joint removed from the steering column or the spiral cable may be damaged.

4. Remove the steering wheel.

➡The steering wheel must be removed before disconnecting the steering column lower joint to avoid damaging the Supplemental Restraint System (SRS) spiral cable.

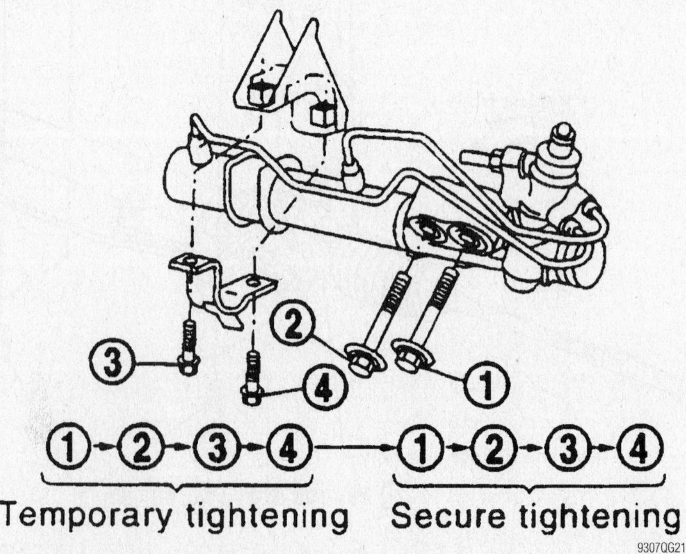

Tighten the mounting bolts using the illustrated procedure—Maxima

5. Raise and support the vehicle safely and remove the front wheels.

6. Remove or disconnect the following:
- Tie rod ends from the steering knuckles
- Front exhaust tube and properly support the engine
- Bolts attaching the engine mounts to the engine mounting center member
- Engine mounting center member and rear engine mount
- Front stabilizer bar from the vehicle
- Nuts attaching the hole cover to the bulkhead

7. Move the hole cover aside and disconnect the lower joint from the rack and pinion. Matchmark the pinion shaft and the pinion housing to record the steering neutral position.
- Power steering fluid pipes from the rack and pinion
- Bolts attaching the mounting brackets
- Rack and pinion from the vehicle

To install:

8. Position the rack and pinion in the vehicle and install the mounting brackets. Torque the mounting nuts and bolts in the proper sequence to 54–72 ft. lbs. (73–97 Nm).

9. Install or connect the following:
- New O-rings to the power steering fluid pipes and connect them to the rack and pinion. Torque the low pressure line 20–29 ft. lbs. (27–39 Nm) and the high pressure line to 11–18 ft. lbs. (15–25 Nm).

10. Align the lower steering joint to the pinion shaft and install the joint onto the pinion shaft. Torque the bolt to 17–22 ft. lbs. (24–29 Nm).
- Hole cover and torque the nuts to 36–43 inch lbs. (4–5 Nm)
- Front stabilizer
- Engine mounting center member and torque the bolts to 57–72 ft. lbs. (77–98 Nm)
- Engine mounts to the center member. Torque the bolts to 57–72 ft. lbs. (77–98 Nm). Remove the support from the engine.
- Remaining components in the reverse order of removal

11. Torque the tie rod end nuts to 22–29 ft. lbs. (29–39 Nm), then install a new cotter pin.

12. Fill the power steering reservoir with fluid and bleed the air from the power steering system.

13. Check the vehicle front end alignment and adjust as necessary.

350Z

1. Before servicing the vehicle, refer to the precautions in the beginning of this section.

⁂ WARNING

Do not turn the steering wheel or column with the lower joint removed from the steering column or the spiral cable may be damaged.

2. Raise and support the vehicle safely and remove the front wheels.

3. Remove or disconnect the following:
- Engine undercover
- Front crossbar
- Cotter pin at steering outer socket, then loosen mounting nut
- Separate outer socket from steering knuckle
- Power steering fluid pipes from the rack and pinion

4. Loosen upper yoke bolt and remove lower yoke bolt, then slide lower joint onto lower shaft.

5. Separate the steering gear from the lower shaft.

6. Remove the fixing bolt and remove the steering gear, rack mounting bracket and insulator from the vehicle.

To install:

7. Installation is the reverse of the removal procedure noting the following:

a. Tighten the rack mounting bracket bolt to 46–56 ft. lbs. (62–76 Nm).

b. Tighten the steering gear bolt to 89–103 ft. lbs. (120–140 Nm).

c. Tighten the outer socket nut to 22–28 ft. lbs. (30–40 Nm).

d. Orientate the steering column yoke to the steering gear to the specifications shown.

e. Tighten the steering column lower joint bolts to gear bolt to 18–21 ft. lbs. (24–34 Nm).

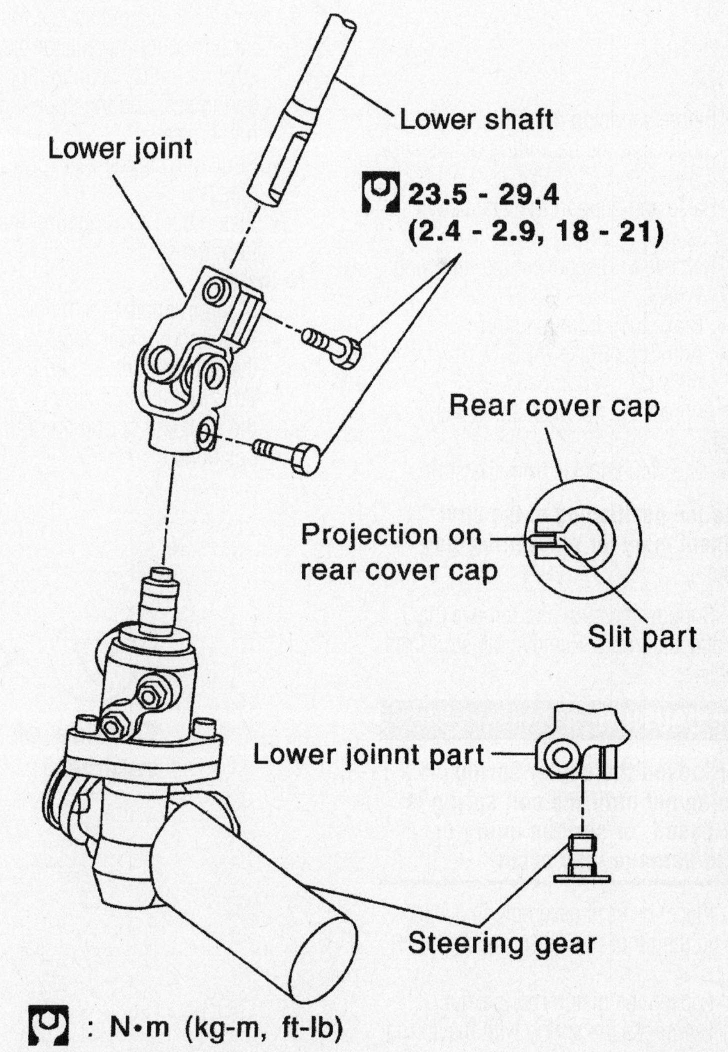

Separating steering shaft from steering gear—350Z

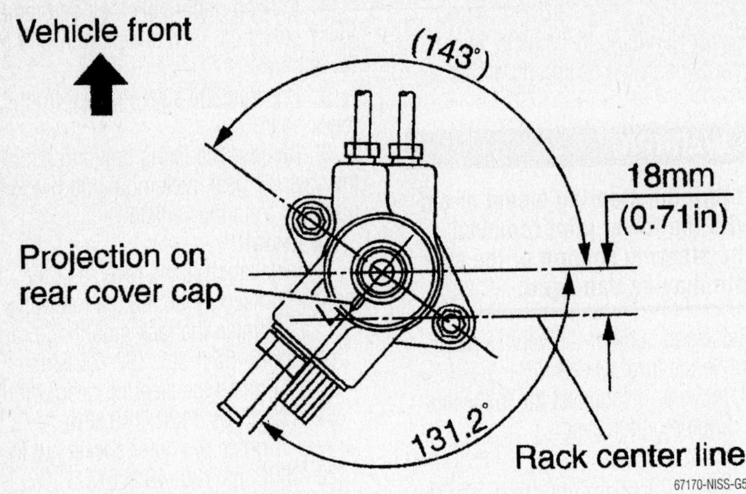

Proper orientation of the steering lower joint-to-gear—350Z

Strut and Coil Spring

REMOVAL & INSTALLATION

Front

SENTRA

1. Before servicing the vehicle, refer to the precautions in the beginning of this section.
2. Raise and support the vehicle on jackstands.
3. Remove or disconnect the following:
   - Wheel
   - Brake tube from the strut
   - Anti-Lock Brake (ABS) wiring from the strut, if equipped
4. Support the transverse link with a jackstand.
   - Steering knuckle from the strut

➡Note the positioning of the strut alignment mark for reassembly purposes.

5. Support the strut and remove the 3 upper attaching nuts. Remove the strut from the vehicle.

⁂ CAUTION

Never loosen the center spring retaining nut until the coil spring is compressed, or serious injury or vehicle damage may occur.

6. Place the strut assembly in a vise with a holding tool or in a spring compressor.
7. Loosen the piston rod locknut.
8. Compress the spring with the spring compressor, then remove the piston rod locknut.

➡Before removing the strut from the coil spring, note the positioning of the strut in relationship to the coil spring for reassembly.

9. Remove or disconnect the following:
   - Strut mounting insulator bracket, strut mounting bearing, upper spring seat and the upper spring rubber seat
   - Strut, leaving the coil spring compressed
   - Piston boot and rebound bumper from the strut

To install:

10. Install or connect the following:
    - Rebound bumper and the boot to the strut piston
    - Strut into the coil spring, be sure the strut and spring are properly positioned
    - Upper spring rubber seat, upper spring seat, strut mounting bearing and the strut mounting insulator bracket. Be sure that the cutout on the upper spring seat is facing the outside of the vehicle.
    - Piston rod locknut. Remove the tool and torque the piston rod locknut to 43–54 ft. lbs. (59–74 Nm).

➡When installing the strut, be sure to position the alignment mark toward the outside of the vehicle.

   - Strut to the vehicle
   - 3 upper attaching nuts and torque the nuts to 18–22 ft. lbs. (25–29 Nm)
   - Steering knuckle to the strut and torque the bolts to 68–82 ft. lbs. (92–111 Nm)
   - Brake tube to the strut and the ABS wiring to the strut, if it was removed
11. Bleed the brake system and install the wheel.
12. Perform a front end alignment.

ALTIMA

1. Before servicing the vehicle, refer to the precautions in the beginning of this section.
2. Raise and support the vehicle on jackstands.
3. Remove or disconnect the following:
   - Wheel
   - Brake tube from the strut
   - Anti-Lock Brake (ABS) wiring from the strut, if equipped with ABS
4. Support the transverse link with a jackstand.

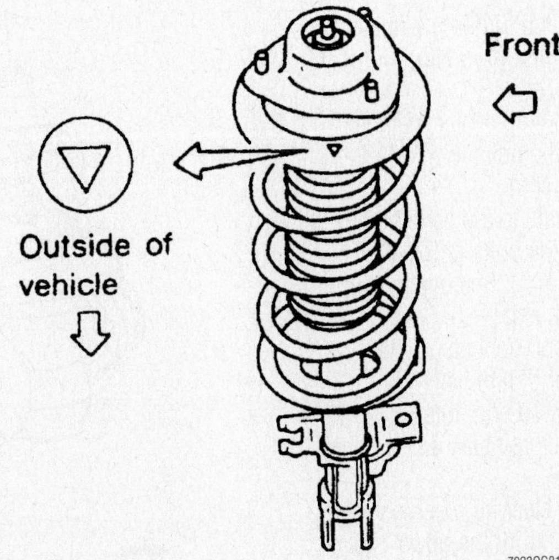

During assembly, be sure to point the alignment mark toward the outside of the vehicle—Sentra

- Steering knuckle from the strut and properly support the strut
- 3 upper attaching nuts
- Strut from the vehicle

⁂ WARNING

Never loosen the center spring retaining nut until the coil spring is compressed, or serious injury or vehicle damage may occur.

To install:

➡When installing the strut, be sure to position the alignment mark toward the outside of the vehicle.

5. Install or connect the following:
   - Strut to the vehicle
   - 3 upper attaching nuts and torque the nuts to 29–40 ft. lbs. (39–54 Nm)
   - Steering knuckle to the strut and torque the nuts to 123–137 ft. lbs. (167–186 Nm)
   - Brake tube to the strut and the ABS wiring to the strut, if removed
6. Bleed the brake system and install the wheel.
7. Lower the vehicle and perform a front end alignment.

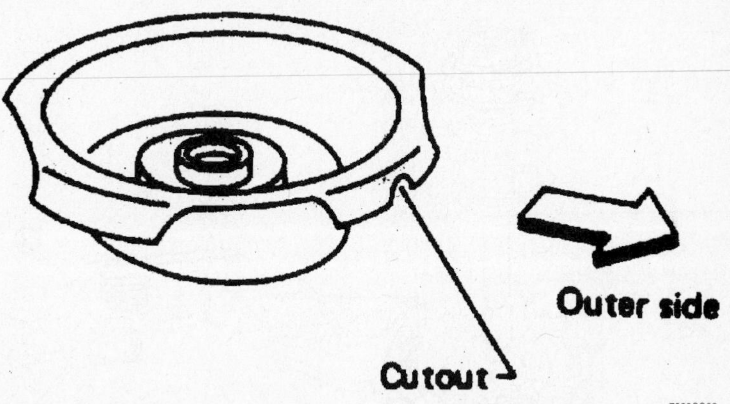

Position the alignment mark toward the outside of the vehicle—Altima

MAXIMA

1. Before servicing the vehicle, refer to the precautions in the beginning of this section.
2. Raise and safely support the vehicle.
3. Remove or disconnect the following:
   - Wheel
4. Matchmark the position of the strut-to-steering knuckle location.
   - Brake hose from the strut
   - Anti-Lock Brake (ABS) wheel sensor and move it out of the way
   - Bolts attaching the steering knuckle to the strut. Matchmark the bolts before removal.
5. Open the hood and remove the strut attaching nuts while holding the strut.

⁂ CAUTION

Do not remove the center locknut from the strut assembly until the strut is safely compressed.

- Strut from the vehicle

6. Place the strut assembly in a vise with a holding tool or in a spring compressor.
7. Loosen the piston rod locknut.

⁂ CAUTION

Do not remove the piston rod locknut, the spring is under tension and can cause serious personal injury.

8. Compress the spring with the spring compressor, then remove the piston rod locknut.

➡Before removing the strut from the coil spring, note the positioning of the strut in relationship to the coil spring for reassembly.

9. Remove or disconnect the following:
   - Strut mounting insulator bracket, strut mounting bearing, upper spring seat and the upper spring rubber seat
   - Strut, leaving the coil spring compressed
   - Piston boot and rebound bumper from the strut

To install:

10. Install or connect the following:
    - Rebound bumper and the boot to the strut piston
    - Strut into the coil spring, be sure the strut and spring are properly positioned
    - Upper spring rubber seat, upper spring seat, strut mounting bearing and the strut mounting insulator bracket. Be sure that the cutout on the upper spring seat is facing the outside of the vehicle.
    - Piston rod locknut. Remove the tool and torque the piston rod locknut to 44–65 ft. lbs. (59–88 Nm).
    - Strut into the strut tower
    - New attaching nuts and torque to 32–38 ft. lbs. (43–51 Nm)
    - Bolts attaching the steering knuckle to the strut and align the matchmarks and torque to 103–117 ft. lbs. (140–159 Nm)
    - ABS wheel sensor and torque to 13–17 ft. lbs. (18–24 Nm)
    - Brake hose to the strut
    - Front wheels
11. Lower the vehicle.
12. Check and/or adjust the wheel alignment as necessary.

350Z

1. Before servicing the vehicle, refer to the precautions in the beginning of this section.
2. Raise and safely support the vehicle.
3. Remove or disconnect the following:
   - Wheel
   - Engine undercover
   - Anti-Lock Brake (ABS) wheel sensor and move it out of the way
   - Brake hose from the strut
   - Strut-to-transverse link nut and bolt
4. Open the hood and remove the strut attaching nuts while holding the strut.

⁂ CAUTION

Do not remove the center locknut from the strut assembly until the strut is safely compressed.

- Strut from the vehicle

5. Place the strut assembly in a vise with a holding tool or in a spring compressor.
6. Loosen the piston rod locknut.

⁂ CAUTION

Do not remove the piston rod locknut, the spring is under tension and can cause serious personal injury.

7. Compress the spring with the spring compressor, then remove the piston rod locknut

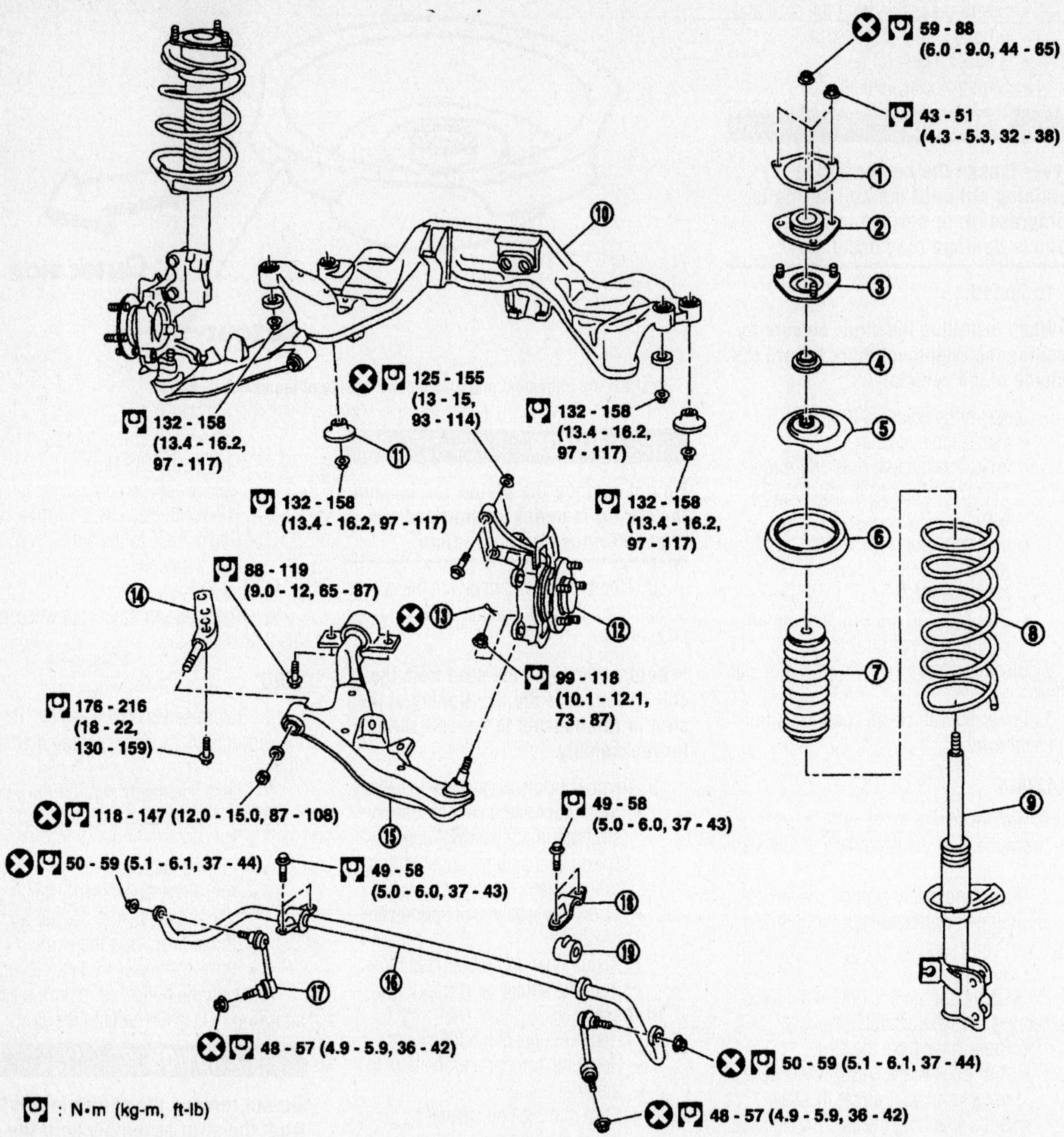

1. Strut spacer
2. Strut mount insulator
3. Strut mount bracket
4. Strut mount bearing
5. Spring upper seat
6. Spring rubber seat
7. Bound bumper rubber
8. Coil spring
9. Shock absorber
10. Suspension member
11. Rebound stopper
12. Wheel hub and steering knuckle
13. Cotter pin
14. Bush link pin
15. Transverse link
16. Stabilizer
17. Connecting rod
18. Stabilizer clamp
19. Bushing

9357RG17

Exploded view of the front suspension—2002 Maxima shown

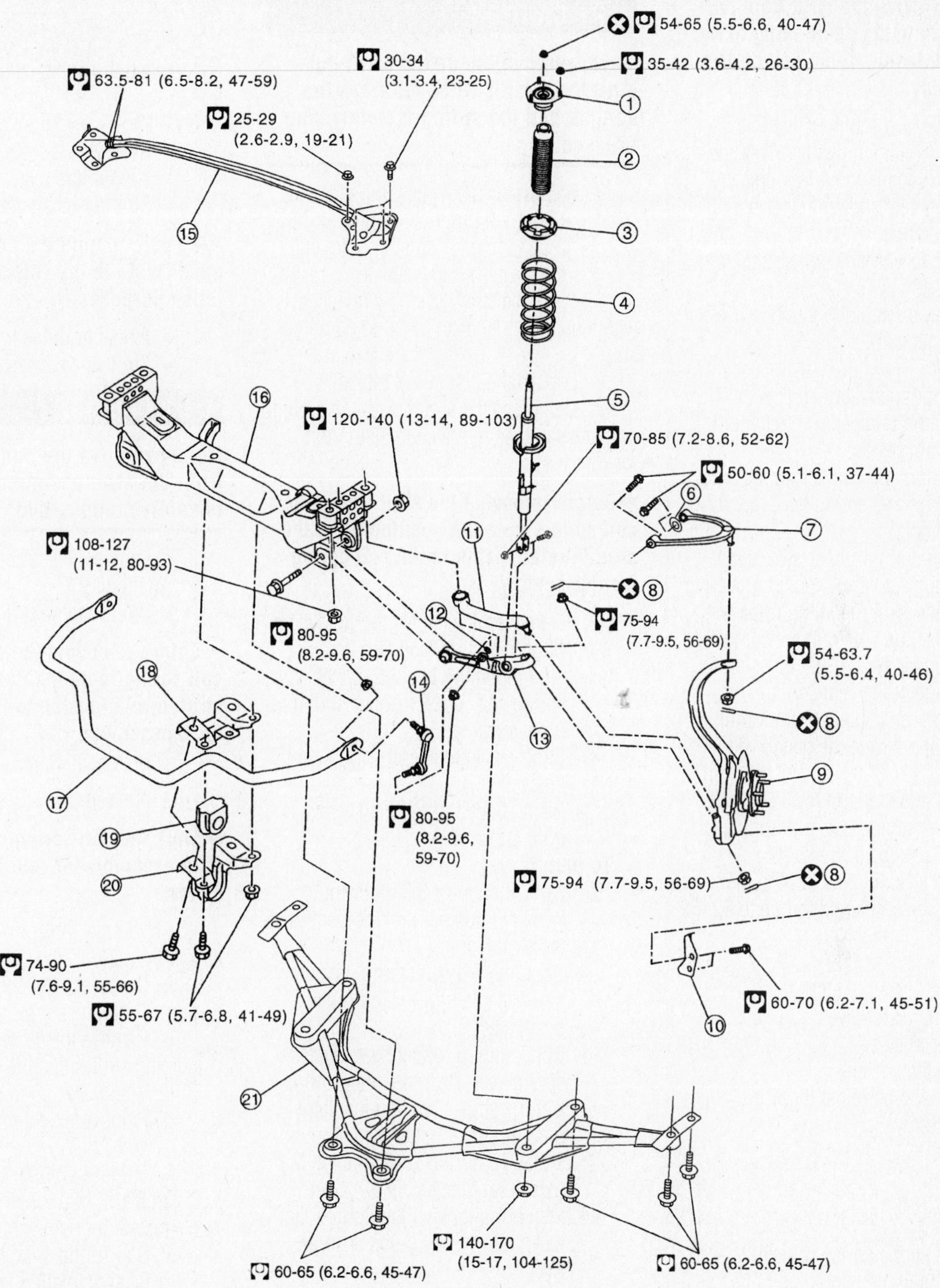
54-65 (5.5-6.6, 40-47)
35-42 (3.6-4.2, 26-30)
30-34
(3.1-3.4, 23-25)
63.5-81 (6.5-8.2, 47-59)
25-29
(2.6-2.9, 19-21)
120-140 (13-14, 89-103)
70-85 (7.2-8.6, 52-62)
50-60 (5.1-6.1, 37-44)
108-127
(11-12, 80-93)
80-95
(8.2-9.6, 59-70)
75-94
(7.7-9.5, 56-69)
54-63.7
(5.5-6.4, 40-46)
80-95
(8.2-9.6,
59-70)
75-94 (7.7-9.5, 56-69)
74-90
(7.6-9.1, 55-66)
55-67 (5.7-6.8, 41-49)
60-70 (6.2-7.1, 45-51)
140-170
(15-17, 104-125)
60-65 (6.2-6.6, 45-47)
60-65 (6.2-6.6, 45-47)
: N•m (kg-m, ft-lb)
: Always replace after every disassembly
1. Mounting insulator
2. Bound bumper
3. Spring upper seat
4. Coil spring
5. Shock absorber
6. Stopper rubber
7. Upper link
8. Cotter pin
9. Front axle
10. Steering stopper bracket
11. Compression rod
12. Washer
13. Transverse link
14. Stabilizer connecting rod
15. Tower bar
16. Front suspension member
17. Stabilizer bar
18. Stabilizer clamp bracket
19. Stabilizer bushing
20. Stabilizer clamp
21. Front cross bar
67170-NISS-G53

Exploded view of the front suspension—350Z

➡Before removing the strut from the coil spring, note the positioning of the strut in relationship to the coil spring for reassembly.

8. Remove or disconnect the following:
   - Strut mounting insulator bracket, strut mounting bearing, upper spring seat and the upper spring rubber seat
   - Strut, leaving the coil spring compressed
   - Piston boot and rebound bumper from the strut

To install:

9. Install or connect the following:
   - Rebound bumper and the boot to the strut piston
   - Strut into the coil spring, be sure the strut and spring are properly positioned
   - Upper spring rubber seat, upper spring seat, strut mounting bearing and the strut mounting insulator bracket. Be sure that the cutout on the upper spring seat is facing the outside of the vehicle.
   - Piston rod locknut. Remove the tool and torque the piston rod locknut to 40–47 ft. lbs. (54–65 Nm).
   - Strut into the strut tower
   - New attaching nuts and torque to 26–30 ft. lbs. (35–42 Nm)
   - Bolts attaching the strut to the transverse link and torque to 52–62 ft. lbs. (70–80 Nm)
   - ABS wheel sensor
   - Brake hose to the strut
   - Engine undercover
   - Front wheels
10. Lower the vehicle.
11. Check and/or adjust the wheel alignment as necessary.

Rear

SENTRA

1. Before servicing the vehicle, refer to the precautions in the beginning of this section.
2. Remove or disconnect the following:
   - Rear wheel
   - Trim panel from the trunk to gain access to the upper mounting nuts of the strut
   - Protective cap from the upper portion of the strut
3. Position a floor jack under the rear axle for support.

➡Note and mark the positioning of the upper strut plate to the vehicle body.

⁂ CAUTION

Never remove the center strut nut until the strut is removed from the vehicle and the spring is safely compressed.

4. Remove or disconnect the following:
   - Lower strut mounting through-bolt
   - 2 upper strut mounting nuts and the strut from the vehicle
5. Place the strut assembly in a vise with a holding tool or in a spring compressor.
6. Loosen the piston rod locknut.
7. Compress the spring with the spring compressor, then remove the piston rod locknut.

➡Before removing the strut from the coil spring, note the positioning of the strut in relationship to the coil spring for reassembly.

8. Remove or disconnect the following:
   - Strut mounting insulator bracket, strut mounting bearing, upper spring seat and the upper spring rubber seat
   - Strut, leaving the coil spring compressed
   - Piston boot and rebound bumper from the strut

To install:

9. Install or connect the following:
   - Rebound bumper and the boot to the strut piston
   - Strut into the coil spring, be sure the strut and spring are properly positioned
   - Upper spring rubber seat, upper spring seat, strut mounting bearing and the strut mounting insulator bracket. Be sure that the cutout on the upper spring seat is facing the outside of the vehicle.
   - Piston rod locknut and torque the locknut to 13–17 ft. lbs. (18–24 Nm)
10. Remove the spring compressor from the coil spring.
   - Strut and torque the 2 upper mounting nuts to 12–14 ft. lbs. (16–19 Nm)
   - Upper mount protective cap
   - Through-bolt to the lower mount of the strut and torque the bolt to 72–87 ft. lbs. (98–118 Nm)
   - Trunk trim panel
   - Rear wheel
11. Lower the vehicle and perform an alignment.

ALTIMA

1. Before servicing the vehicle, refer to the precautions in the beginning of this section.
2. Remove or disconnect the following:
   - Rear wheels from the vehicle and support the rear axle with a jack
   - Strut lower mounting through-bolts

➡Be sure to note the position the strut upper plate to the vehicle for reinstallation purposes.

   - 2 nuts from the top of the strut
   - Strut as an assembly

⁂ CAUTION

Do not remove the center locknut from the strut assembly until the strut is safely compressed.

3. Compress the strut coil spring with a spring compressor.
   - Strut assembly center locknut

➡Before removing the strut from the coil spring, note the positioning of the strut in relationship to the coil spring for reassembly.

   - Strut leaving the coil spring compressed

➡Mark the coil spring position to the strut assembly for reinstallation purposes.

4. To remove the spring from the strut assembly, perform the following steps:
   a. Compress the coil spring with the proper compressor tool.
   b. Remove the center retaining nut holding strut mounting insulator.
   c. Slowly decompress the coil spring.
   d. Remove the strut mounting insulator.
   e. Remove coil spring.

To install:

5. Install or connect the following:
   - Coil spring onto the strut assembly. Be sure to align the matchmarks made during the removal procedure.
   - Strut mounting insulator and compress the coil spring assembly

➡It will be necessary to use a new locknut for the center retaining nut of the coil spring.

   - Center retaining nut and torque to 43–58 ft. lbs. (59–78 Nm). Be sure the spring is seated properly on the strut and in the mounting insulator.
6. Slowly remove the spring compressor tool.

- Strut assembly and torque the upper nuts to 31–40 ft. lbs. (42–54 Nm)
- Lower strut through-bolt and torque to 123–137 ft. lbs. (167–186 Nm)

➡**Be sure to hold the through-bolt and tighten the nuts.**

7. Install the wheels, lower the vehicle and perform a front end alignment.

MAXIMA

1.Before servicing the vehicle, refer to the precautions in the beginning of this section.

2. Remove the rear wheels.
3. Support the rear torsion beam assembly with a jack.
4. Open the trunk and remove the 2 nuts attaching the strut to the vehicle.

CAUTION

Do not remove the center locknut from the strut assembly until the strut is safely compressed.

5. Remove the bolt attaching the strut to the rear torsion beam assembly and remove the strut.
6. Place the strut assembly in a vise

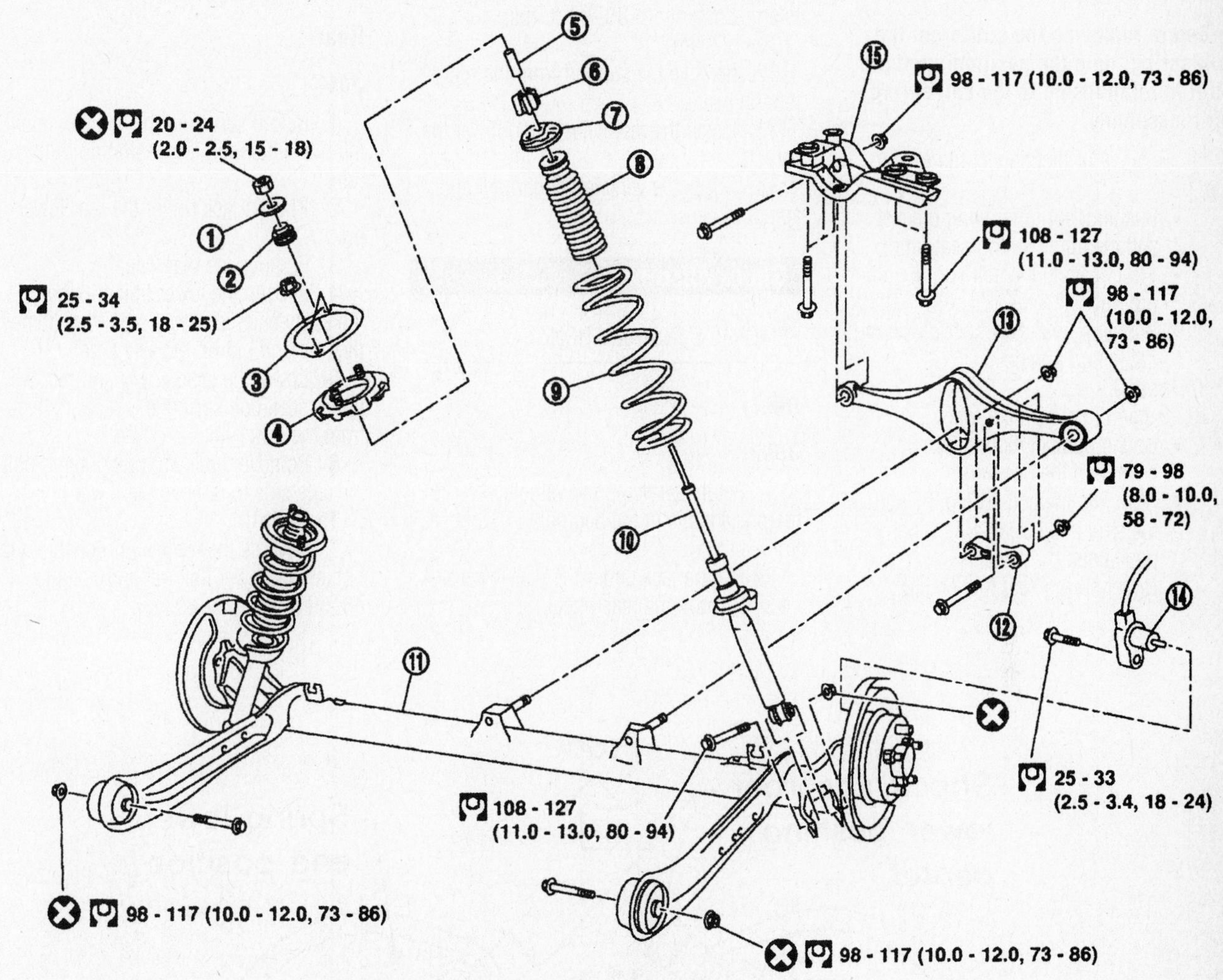

1. Washer
2. Bushing
3. Shock absorber mounting seal
4. Shock absorber mounting bracket
5. Distance tube
6. Bushing
7. Bound bumper cover
8. Bound bumper
9. Coil spring
10. Shock absorber
11. Torsion beam
12. Control rod
13. Lateral link
14. ABS sensor
15. Suspension member

9357RG18

Exploded view of the rear suspension—2002 Maxima shown

with a holding tool or in a spring compressor.

7. Loosen the piston rod locknut.

**** CAUTION**

Do not remove the piston rod locknut, the spring is under tension and can cause serious personal injury.

8. Compress the spring with the spring compressor, then remove the piston rod locknut.

➡Before removing the strut from the coil spring, note the positioning of the strut in relationship to the coil spring for reassembly.

9. Remove or disconnect the following:
   - Bushing, strut mounting bracket and the upper spring seat rubber
   - Strut, leaving the coil spring compressed
   - Bushing, bound bumper cover and the bound bumper

To install:

10. Install or connect the following:
   - Bound bumper, bound bumper cover and the bushing
   - Strut into the coil spring, be sure the strut and spring are properly positioned
   - Upper spring seat rubber, strut mounting bracket and the bushing. Be sure that the mounting bracket is properly positioned.
   - Piston rod locknut. Remove the tool and torque the piston rod locknut to 15–18 ft. lbs. (20–24 Nm).
   - Strut and torque the new nuts to 18–25 ft. lbs. (25–34 Nm)

11. Position the strut on the rear torsion beam and install the bolt. Torque the bolt attaching the strut to the torsion beam assembly to 80–94 ft. lbs. (108–127 Nm).

12. Remove the support from the rear torsion beam.

13. Install the rear wheels and lower the vehicle.

14. Check the vehicle's alignment and adjust as necessary.

Shock Absorber

REMOVAL & INSTALLATION

Rear

350Z

1. Before servicing the vehicle, refer to the precautions in the beginning of this section.
2. Place a jack under the rear lower link.
3. Remove the rear wheel.
4. Remove the upper and lower mounting bolts and remove the shock absorber.

To install:

5. Reverse the removal procedure and tighten the upper shock nuts to 21 ft. lbs. (28 Nm) and the lower bolt to 74–88 ft. lbs. (100–120 Nm).

Coil Spring & Rear Lower Link

REMOVAL & INSTALLATION

Rear

350Z

1. Before servicing the vehicle, refer to the precautions in the beginning of this section.
2. Place a jack under the rear lower link.
3. Remove the rear wheel.
4. Loosen the lower link nut and bolt on the suspension member side, then remove the bolt and nut on the axle side.
5. Lower the jack slowly and remove the upper seat, coil spring and rubber sheet from the lower link.
6. Remove the lower link nut and bolt on the axle side to remove the lower link.

To install:

7. Reverse the removal procedure and tighten the lower link nut and bolts to 48–59 ft. lbs. (65–80 Nm).

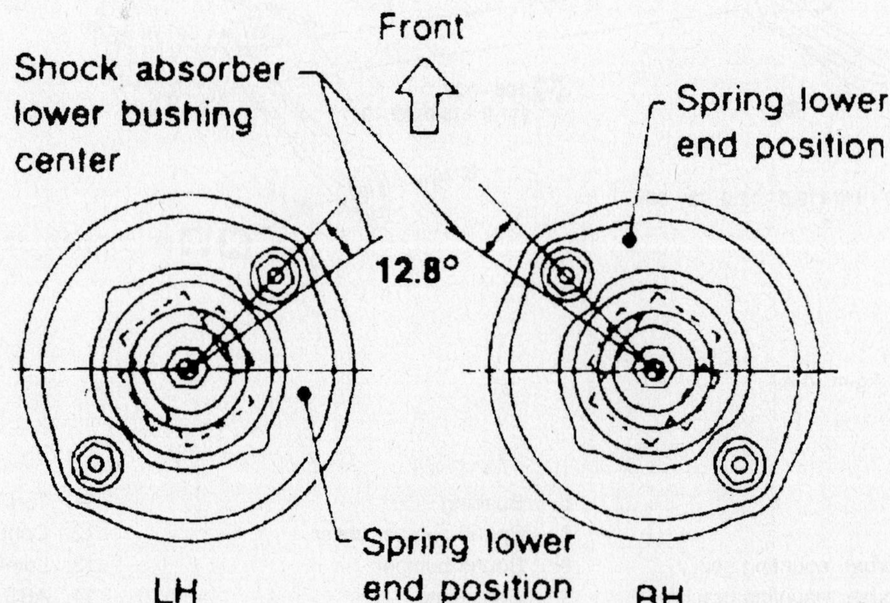

Positioning of the strut mounting brackets—Maxima

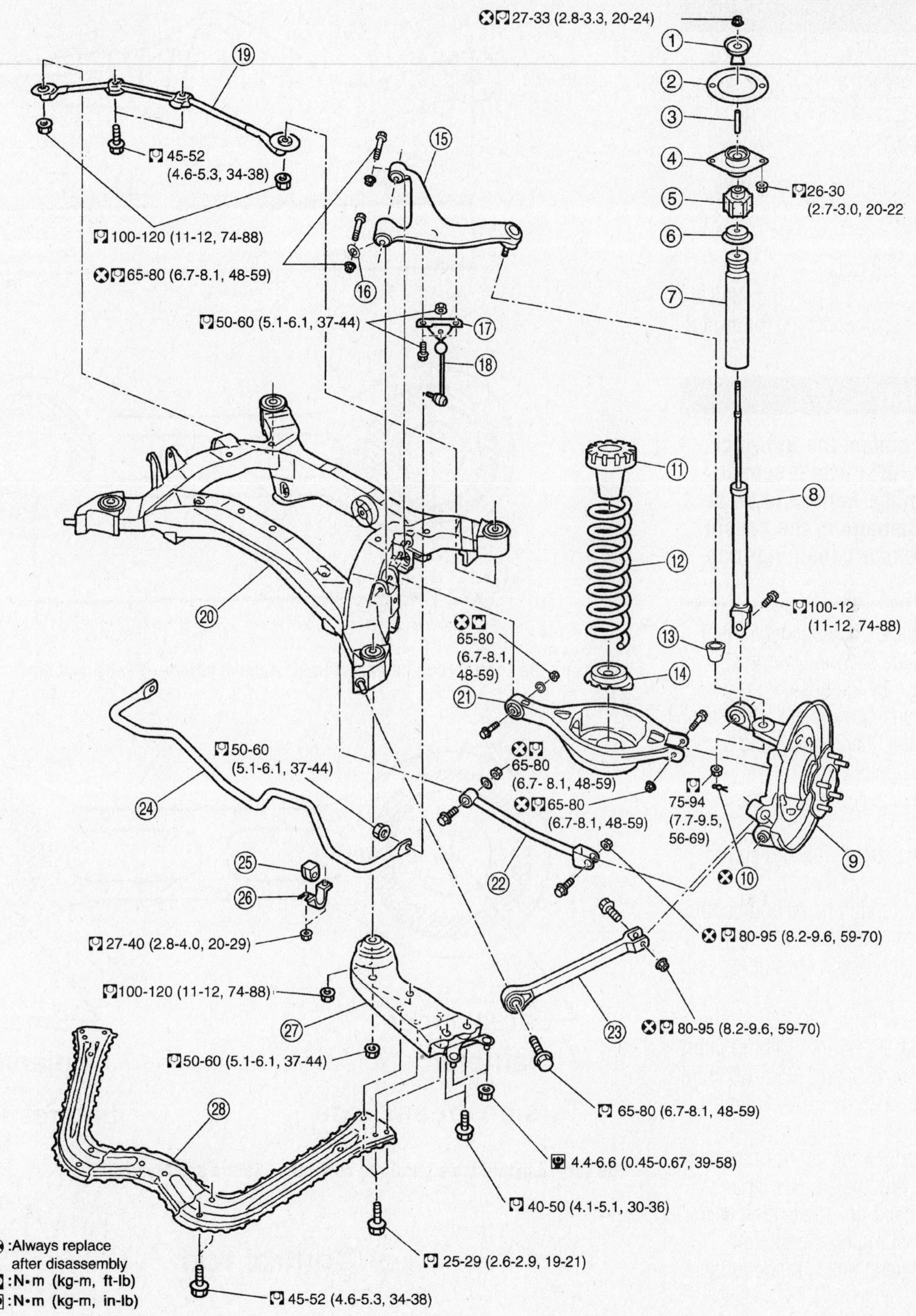

1. Bushing	2. Mounting seal	3. Distance tube
4. Mounting seal bracket	5. Bushing	6. Bound bumper cover
7. Bound bumper	8. Shock absorber	9. Axle
10. Cotter pin	11. Upper seat	12. Coil spring
13. Ball seat	14. Rubber seat	15. Suspension arm
16. Stopper rubber	17. Stabilizer connecting rod mounting bracket	18. Stabilizer connecting rod
19. Rear pin stay	20. Rear suspension member	21. Rear lower link
22. Front lower link	23. Radius rod	24. Stabilizer bar
25. Stabilizer bushing	26. Stabilizer clamp	27. Member stay
28. Tunnel stay		

67170-NISS-G54

Exploded view of the rear suspension—350Z

Torsion Bars

REMOVAL & INSTALLATION

Sentra and Maxima

1. Before servicing the vehicle, refer to the precautions in the beginning of this section.
2. Loosen the lug nuts.
3. Raise and safely support the vehicle
4. Remove or disconnect the following:
   - Wheels

WARNING

Be sure to disconnect the Anti-lock Brake System (ABS) wheel sensor from the assembly. Failure to do so may result in damage to the sensor wire and the sensor becoming inoperative.

   - Brake calipers and suspend them with a piece of wire. Do not let them hang by the hose.
5. Using a transmission jack, raise the torsion beam a little, then remove the suspension mounting bolts.
6. Lower the jack and remove the suspension assembly.
7. Remove the lateral link and control rod.
8. Inspect the torsion beam and control rod for cracks, wear and deformation. The length of the lateral link and control rod is as follows:
   - A—8.15–8.19 in. (207–208mm)
   - B—15.51–15.55 in. (394–395mm)
   - C—23.66–23.74 in. (601–603mm)
   - D—4.17–4.25 in. (106–108mm)

To install:

9. When installing the control rod, connect the bushing with the smaller inner diameter to the lateral link. Install the lateral link and the control rod on the torsion beam. Place the lateral link with the arrow topside.
10. Place the lateral link and control rod horizontally against the beam and tighten the bolts. Refer to the illustration.
11. Secure the torsion beam to the vehicle. Make sure the lateral link is horizontal, then tighten the link to the chassis.
12. Attach the struts to the torsion beam and tighten the fasteners.
13. Tighten the torsion beam-to-chassis bolts.
14. Install the calipers, ABS sensor and wheels. Lower the vehicle to the ground.

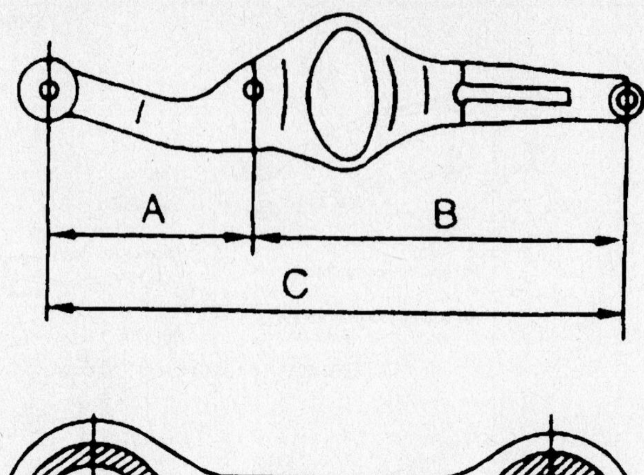

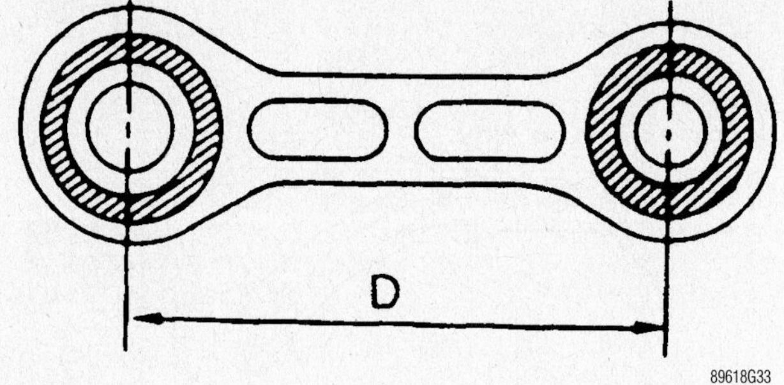

Measure the control rod and lateral links at these points—Sentra and Maxima

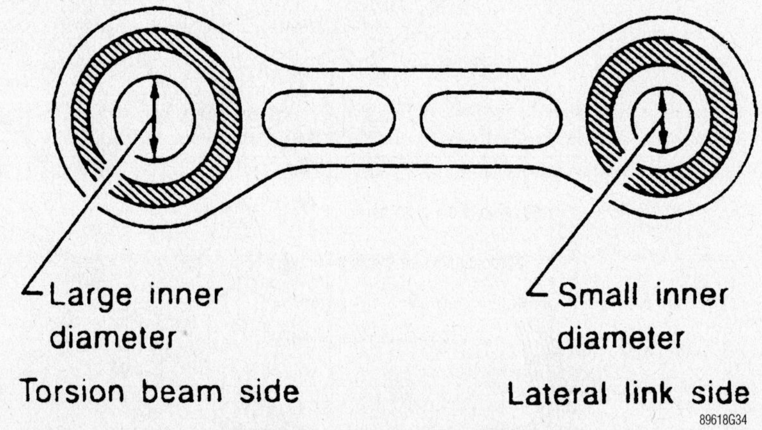

Be sure to install the control rod correctly—Sentra and Maxima

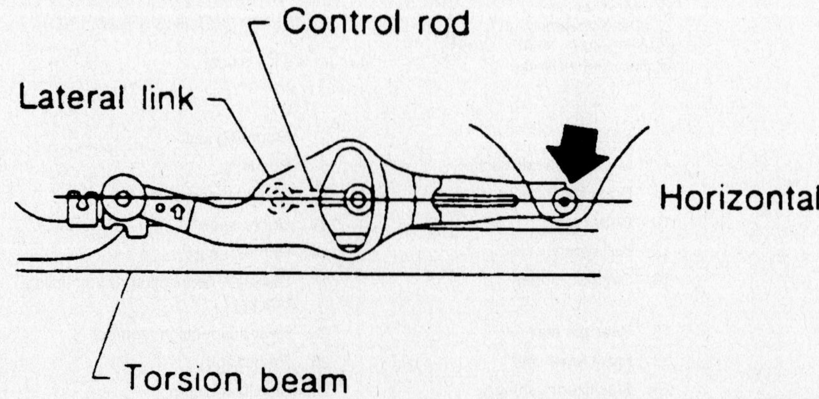

The lateral link must be in the horizontal position when tightening the bolts—Sentra and Maxima

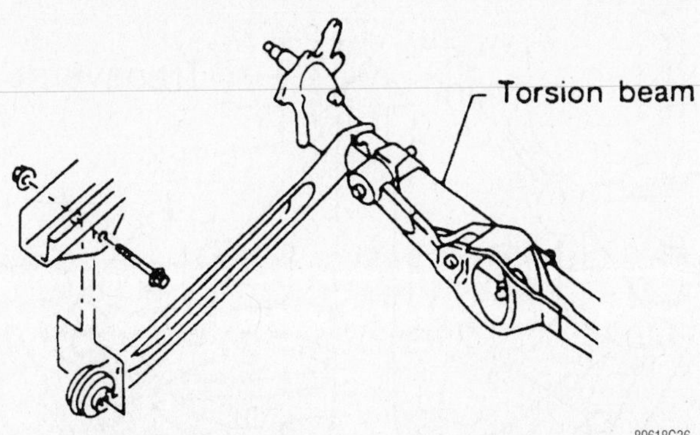

Tighten the torsion beam-to-chassis bolts with the suspension unloaded—Sentra and Maxima

Lower Ball Joint

REMOVAL & INSTALLATION

The ball joint is an integral part of the lower control arm. If the ball joint is defective the control arm must be replaced.

Lower Control Arm (Transverse Link)

REMOVAL & INSTALLATION

Sentra

1. Before servicing the vehicle, refer to the precautions in the beginning of this section.
2. Remove or disconnect the following:
   - Front wheels
   - Disc brake caliper from the steering knuckle

⁂ WARNING

DO NOT allow the disc brake caliper to hang from the brake hose. Support the disc caliper with safety wire.

   - Cotter pin and loosen the wheel bearing locknut
   - Cotter pin and the castle nut from the tie rod ball joint. Separate the tie rod with a suitable puller.
   - 2 bolts that secure the lower portion of the strut to the steering knuckle
3. Using a plastic or rubber mallet, tap on the loosened wheel bearing locknut to loosen the halfshaft in the knuckle. Remove the locknut and remove the halfshaft from the steering knuckle. Be sure to cover the CV-joints with a shop rag.

➡Support the halfshaft assembly with wire. Do not allow the halfshaft to hang by the inner joint.

   - Nut that secures the stabilizer link to the lower control arm
   - Link from the control arm. Note the positioning of the washers and spacers for reassembly.
   - Cotter pin and castle nut from the lower ball joint
   - Lower ball joint from the knuckle
   - Knuckle from the vehicle
   - Mounting nuts/bolts that secure the lower control arm to the frame
   - Control arm from the vehicle

To install:

➡Final tightening of all suspension components should take place with the weight of the vehicle on the wheels.

4. Install the lower control arm assembly and torque mounting bolts/nuts as follows:
   a. Through bolt and nut: 76–90 ft. lbs. (103–123 Nm).
   b. 2 saddle bracket mounting bolts: 58–72 ft. lbs. (78–98 Nm).
5. Install or connect the following:
   - Steering knuckle to the lower ball joint and torque the castle nut to 43–54 ft. lbs. (59–74 Nm). Install a new cotter pin.
   - Stabilizer link to the lower control arm and torque the nut to 12–16 ft. lbs. (16–22 Nm)
   - Halfshaft through the wheel bearing
   - Wheel bearing locknut. Do not torque the locknut at this time.
   - Steering knuckle to the strut and torque the bolts to 68–82 ft. lbs. (92–111 Nm)
   - Tie rod end and torque the castle nut to 22–29 ft. lbs. (29–39 Nm). Install a new cotter pin.
   - Disc brake caliper to the steering knuckle
6. Tighten the halfshaft mounting nut (hub nut) and torque the nut to 145–202 ft. lbs. (196–274 Nm). It may be necessary to have an assistant hold the brake pedal while tightening the locknut. Install the adjusting cap and a new cotter pin.
7. Install the front wheels, lower the vehicle and perform a front end alignment.

Altima

➡The lower ball joint is integral with the lower control arm (transverse link). They are removed and replaced as an assembly.

1. Before servicing the vehicle, refer to the precautions in the beginning of this section.
2. Remove or disconnect the following:
   - Front wheels
   - Stabilizer bar. The bar is removed by unfastening the nut that hold the bar to the transverse link gusset plate.

➡Take note of position of marks on clamp face and stabilizer bar for reassembling.

   - Lower ball joint to knuckle cotter pin and nut
   - Ball joint stud from knuckle using the proper tool
   - Transverse link mounting bolts and nuts
   - Link

To install:

3. Install or connect the following:
   - Transverse link with mounting bolts and torque nuts and bolts to 87–108 ft. lbs. (118–147 Nm)

➡The final tightening of suspension components must be done with wheels on the ground and vehicle at curb weight.

   - Lower ball joint to the knuckle and torque the nut to 52–64 ft. lbs. (71–86 Nm) and install a new cotter pin
   - Stabilizer bar link to the transverse link and torque the nuts to 30–35 ft. lbs. (41–47 Nm)
4. Install wheels and safely lower vehicle to ground.
5. Check the front end alignment.

Maxima

1. Before servicing the vehicle, refer to the precautions in the beginning of this section.
2. Remove or disconnect the following:
   - Front wheels
   - Anti-Lock Brake (ABS) wheel sensor and move it out of the way
   - Wheel bearing locknut
   - Tie rod from the steering knuckle

- Bolts attaching the strut to the steering knuckle. Matchmark the bolts before removal.
- Halfshaft from the steering knuckle by lightly tapping the end of the shaft
- Steering knuckle and the lower ball joint
- Stabilizer bar from the lower control arm
- Bolts attaching the link bushing pin to the chassis
- Nut attaching the link to the control arm and the link, if necessary
- Bolts attaching the compression rod bushing clamp
- Lower control arm/traverse link

To install:

3. Install or connect the following:
   - Lower control arm and the compression rod bushing clamp into the vehicle
   - Link bushing pin, if removed from the control arm
4. Tighten all bolts and nuts until they are snug enough to support the weight of the vehicle but not fully tight, the bolts should be torqued to specification with the vehicle on the floor.

➡Always use a new nut when installing the ball joint to the control arm.

   - Steering knuckle to the strut and to the halfshaft
   - Strut mounting bolts and torque the bolts to 103–117 ft. lbs. (140–159 Nm)
   - Tie rod ball joint and torque the nut to 22–29 ft. lbs. (29–39 Nm)
   - Wheel bearing locknut
   - ABS wheel sensor and torque the bolt to 13–17 ft. lbs. (18–24 Nm)
   - Front wheels, lower the vehicle and torque hub locknut to 174–231 ft. lbs. (235–314 Nm)
5. Torque the bolts attaching the compression rod bushing clamp and the link bushing pin, in the proper sequence to 87–108 ft. lbs. (118–147 Nm).
6. If the link bushing pin was removed from the control arm torque the attaching nut to 87–108 ft. lbs. (118–147 Nm).
7. Torque the sway bar attaching nut to 30–35 ft. lbs. (41–47 Nm).
8. Check the vehicle alignment.

CONTROL ARM BUSHING REPLACEMENT

The bushing is an integral part of the lower control arm. If the bushing is defective the control arm must be replaced.

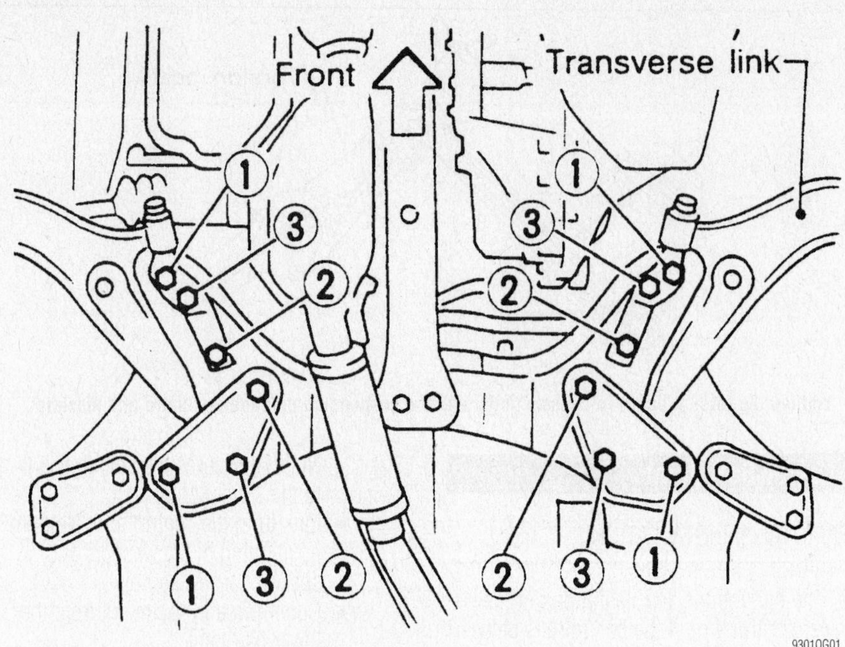

Bolt tightening sequence for the lower control arms—2000 Maxima

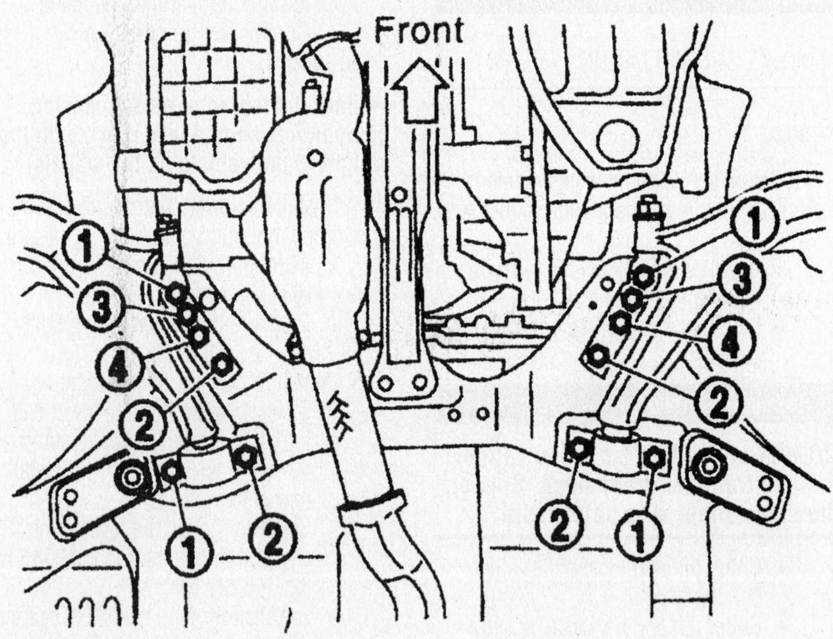

Bolt tightening sequence for the lower control arms—2001–03 Maxima

Wheel Bearings

ADJUSTMENT

Front

➡Whenever the hub or bearing assemblies are removed, the wheel bearing must be replaced. Never reuse the old bearing assembly.

The wheel bearings are sealed and are not adjustable. If defective, replacement is the only option.

Rear

If the wheel hub bearing assembly is removed, it must be replaced.

➡The wheel hub bearing assembly is not repairable; it must be replaced when defective.

1. Before servicing the vehicle, refer to the precautions in the beginning of this section.
2. Torque the wheel bearing locknut to 138–188 ft. lbs. (187–255 Nm).
3. Verify that the wheel bearings operate smoothly.
4. Install a new cotter pin into the spindle to hold the wheel bearing locknut.
5. Install a dial indicator to the rear wheel hub bearing assembly and check the axial end-play; it should be less than 0.0020 in. (0.05mm).
6. Install the grease cap.
7. If the axial end-play exceeds specifications, the wheel bearing must be replaced.

REMOVAL & INSTALLATION

Front

SENTRA

➡Whenever the hub or bearing assembly is removed, the wheel bearing assembly must be replaced. Never reuse the old bearing assembly.

1. Before servicing the vehicle, refer to the precautions in the beginning of this section.
2. Remove or disconnect the following:
   - Front wheel
   - Wheel bearing/axle shaft locknut while depressing the brake pedal
   - Brake caliper and support it with a piece of wire. It is not necessary to disconnect the brake line from the caliper.
   - Anti-Lock Brake System (ABS) sensor from the steering knuckle

➡Do not depress the brake pedal or twist the brake line.

   - Tie rod end
   - Halfshaft from the knuckle by slightly tapping with a soft hammer. Position the axle shaft nut on the threads of the shaft to protect them when lightly tapping.
   - Lower ball joint nut and separate
   - 2 strut-to-knuckle retaining bolts and separate
   - Steering knuckle from the vehicle
3. Place the assembly in a vise. Drive the hub with the inner race from the knuckle with a suitable tool. Remove the inner and outer grease seals.
   - Bearing inner race and outer grease seal from the hub
   - Snapring and press the bearing outer race to remove the bearing from the steering knuckle

To install:

4. Press a new wheel bearing into the knuckle assembly not exceeding 3.3 tons (2994 kg) pressure.
5. Install or connect the following:
   - Snapring and pack the grease seal lips with chassis grease
   - Inner and outer grease seals
6. Press the wheel hub into the knuckle not exceeding 3.3 tons (2994 kg) pressure.
7. Check bearing operation and by applying 3.9–5.5 tons (3538–4990 kg) pressure to the hub assembly. Spin the hub several times in both directions.
8. Be sure the bearings rotate freely. If the bearings do not rotate freely, replace the bearings.
9. Install or connect the following:
   - Knuckle and wheel hub assembly
   - Lower ball joint and torque the nut to 43–54 ft. lbs. (59–74 Nm). Install a new cotter pin.
   - Strut and torque the bolts to 68–82 ft. lbs. (92–118 Nm)
   - Tie rod end and tighten the nut to 22–29 ft. lbs. (29–39 Nm). Install a new cotter pin.
   - Disc brake caliper
   - Torque the wheel bearing locknut to 145–203 ft. lbs. (196–275 Nm). Install a new cotter pin.
   - Front wheels and lower the vehicle
10. Check the vehicle's alignment.
11. Road test the vehicle and verify proper operation.

MAXIMA AND ALTIMA

➡Whenever the hub or bearing assembly is removed, the wheel bearing assembly must be replaced. Never reuse the old bearing assembly.

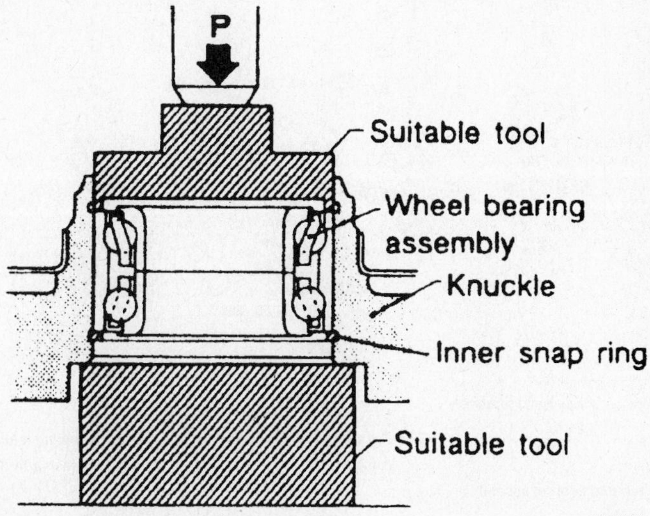

Typical method of installing the wheel bearing

1. Before servicing the vehicle, refer to the precautions in the beginning of this section.
2. Remove or disconnect the following:
   - Knuckle assembly from the vehicle
   - Hub with the inner race from the steering knuckle, using a shop press and a suitable tool
   - Bearing inner race from the hub, using a shop press and a suitable tool
   - Outer grease seal
   - Inner grease seal from the steering knuckle, using a prybar
   - Inner and outer snaprings from the steering knuckle, using snapring pliers
   - Sealed bearing assembly from the steering knuckle, using a shop press and a suitable tool
3. Inspect the hub, steering knuckle and snaprings for cracks and/or wear; if necessary, replace the damaged part(s).

To install:

4. Install or connect the following:
   - Inner snapring in the steering knuckle groove
   - New wheel bearing assembly into the steering knuckle, using a shop press and a suitable tool, until it seats, using a maximum pressure of 3 tons (2722 kg)
   - Outer snapring
5. Pack the new grease seal lips with multi-purpose grease.
   - New outer grease seal into the steering knuckle, using a shop press and a suitable tool
   - Hub into the steering knuckle, using a shop press and a suitable tool, until it seats, using a maxi-

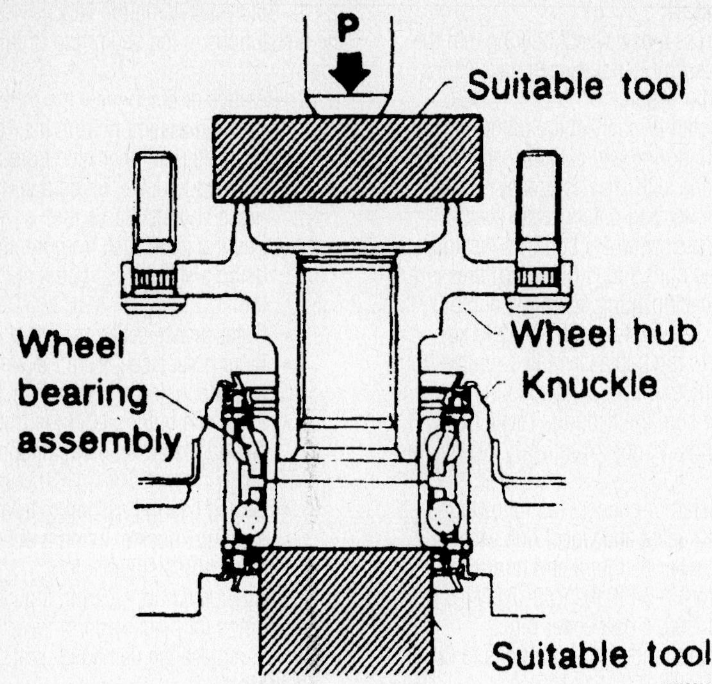

Use a press to install the hub into the knuckle assembly

mum pressure of 5.5 tons (4990 kg); be careful not to damage the grease seal

6. To check the bearing operation, perform the following procedures:
   a. Increase the press pressure to 3.5–5.0 tons (3175–4536 kg).
   b. Spin the steering knuckle, several turns, in both directions.
   c. Be sure the wheel bearings operate smoothly.
7. If the wheel bearings do not operate smoothly, replace the wheel bearing assembly.
8. Install the knuckle assembly.
9. Install the halfshaft into the hub. Torque the locknut to 174–231 ft. lbs. (235–314 Nm).
10. Install the wheel assembly and lower the vehicle.
11. Road test the vehicle and verify proper operation.

350Z

➡If the wheel bearing is damaged, the steering knuckle and bearing must be replaced as an assembly.

54 - 63.7 (5.5 - 6.4, 40 - 46)
78.5 - 98.1 (8.0 - 10, 58 - 72)
29.5 - 39.2 (3.0 - 3.9, 22 - 28)
75 - 94 (7.7 - 9.5, 56 - 59)
75 - 94 (7.7 - 9.5, 56 - 59)

: N•m (kg-m, ft-lb)
: Always replace after disassembly

1. Hub cap
2. Ball seat
3. Steering knuckle
4. Cotter pin
5. Washer
6. Splash guard
7. Wheel hub and bearing assembly

67170-NISS-G55

Exploded view of the front steering knuckle and wheel bearing assembly—350Z

1. Before servicing the vehicle, refer to the precautions in the beginning of this section.
2. Remove or disconnect the following:
   - Front wheel
   - Engine undercover
   - Brake caliper and wire aside
   - Brake rotor
   - ABS sensor
   - Brake hose bracket
   - Loosen steering outer socket nut
   - Separate outer socket from steering knuckle
   - Upper link from knuckle
   - Transverse link from knuckle
   - Compression rod from knuckle
   - Loosen steering knuckle nut
   - Knuckle and hub assembly from the vehicle
   - Separate the wheel hub from the knuckle

To install:

3. Install the wheel hub to the knuckle.
4. Install or connect the following:
   - Knuckle and hub assembly
   - Tighten the steering knuckle/hub nut to 58–72 ft. lbs. (79–98 Nm)
   - Compression rod to knuckle and tighten the nut to 56–69 ft. lbs. (75–94 Nm)
   - Transverse link to knuckle and tighten the nut to 56–69 ft. lbs. (75–94 Nm)
   - Upper link to knuckle and tighten the nut to 40–46 ft. lbs. (54–64 Nm)
   - Install outer socket to steering knuckle
   - Steering outer socket nut
   - Brake hose bracket
   - ABS sensor
   - Brake rotor
   - Brake caliper
   - Engine undercover
   - Front wheel

Rear

If the wheel hub bearing assembly is removed, it must be replaced.

➡If the vehicle is equipped with Anti-Lock Brake (ABS), the sensor must be removed to protect the sensor and its wiring.

1. Before servicing the vehicle, refer to the precautions in the beginning of this section.
2. Raise and safely support the vehicle. Remove the rear wheel(s).
3. If equipped with disc brakes, remove or disconnect the following:

- Brake caliper and hang it by a piece of wire
- Brake caliper support
- Disc brake pads
- Brake rotor

4. If equipped with drum brakes, remove or disconnect the following:
   - Brake drum
   - Brake shoe assembly, if necessary
   - Grease cap
5. Remove the cotter pin, wheel bearing locknut, washer and the wheel hub bearing assembly. A slide hammer may be needed to remove the hub bearing assembly.

➡The wheel hub bearing assembly is not repairable; it must be replaced when defective.

To install:

➡If the vehicle is equipped with ABS, the sensor ring must be removed and installed on the new hub.

6. Apply oil to the threaded portion of the spindle and both sides of the plain washer.
7. Install the wheel hub bearing assembly, the washer and the wheel bearing locknut. Torque the wheel bearing locknut to 138–188 ft. lbs. (187–255 Nm).
8. Verify that the wheel bearings operate smoothly.
9. Install or connect the following:
   - New cotter pin into the spindle to hold the wheel bearing locknut
10. Install a dial micrometer to the rear wheel hub bearing assembly and check the axial end-play. It should be less than 0.0020 in. (0.05mm).
    - Grease cap
    - ABS sensor and its wiring, if removed
    - Brake assembly and the wheels

BRAKES

Brake Caliper

REMOVAL & INSTALLATION

Front

ALTIMA, MAXIMA AND SENTRA

1. Before servicing the vehicle, refer to the precautions in the beginning of this section.
2. Remove or disconnect the following:
   - Front wheels
   - Brake fluid hose
   - Pin bolts
   - Caliper assembly from the vehicle

To install:

3. Use a large C-clamp to press the caliper piston back into the caliper.
4. Install or connect the following:
   - New pads, new shims and pad retainers
   - Brake caliper and torque the pin bolts to 23 ft. lbs. (31 Nm)
   - Brake line to the caliper, using new copper washers, and torque the connecting bolt to 12–14 ft. lbs. (17–20 Nm).
   - Wheels
5. Bleed the brake system and top off the master cylinder as necessary.

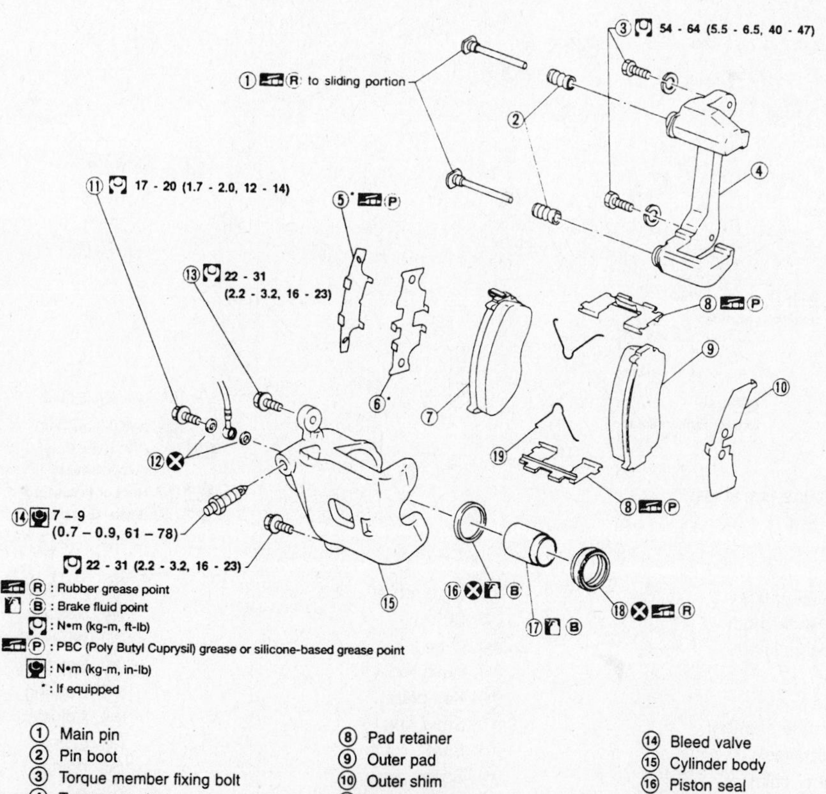

Front brake caliper—Sentra

350Z WITH CLZ25VD CALIPER

1. Before servicing the vehicle, refer to the precautions in the beginning of this section.
2. Remove or disconnect the following:
   - Front wheels
   - Brake fluid hose
   - Sliding pin bolts
   - Caliper assembly from the vehicle

To install:

3. Use a large C-clamp to press the caliper piston back into the caliper.
4. Install or connect the following:
   - New pads, new shims and pad retainers
   - Brake caliper and torque the sliding pin bolts to17–22 ft. lbs. (22–31 Nm).
   - Brake line to the caliper, using new copper washers, and torque the connecting bolt to 12–14 ft. lbs. (17–20 Nm).
   - Wheels
5. Bleed the brake system and top off the master cylinder as necessary.

350Z WITH OPB27VACALIPER

1. Before servicing the vehicle, refer to the precautions in the beginning of this section.
2. Remove or disconnect the following:
   - Front wheels
   - Caliper bolts
   - Caliper assembly from the vehicle

To install:

3. Use a large C-clamp to press the caliper piston back into the caliper.
4. Install or connect the following:
   - New pads, new shims and pad retainers
   - Brake caliper and torque the mounting bolts to 112 ft. lbs. (152 Nm).
   - Wheels
5. Bleed the brake system and top off the master cylinder as necessary.

Rear

EXCEPT 350Z

1. Before servicing the vehicle, refer to the precautions in the beginning of this section.
2. Remove or disconnect the following:
   - Rear wheels
   - Parking brake cable and the lock spring
   - Brake fluid hose from the caliper

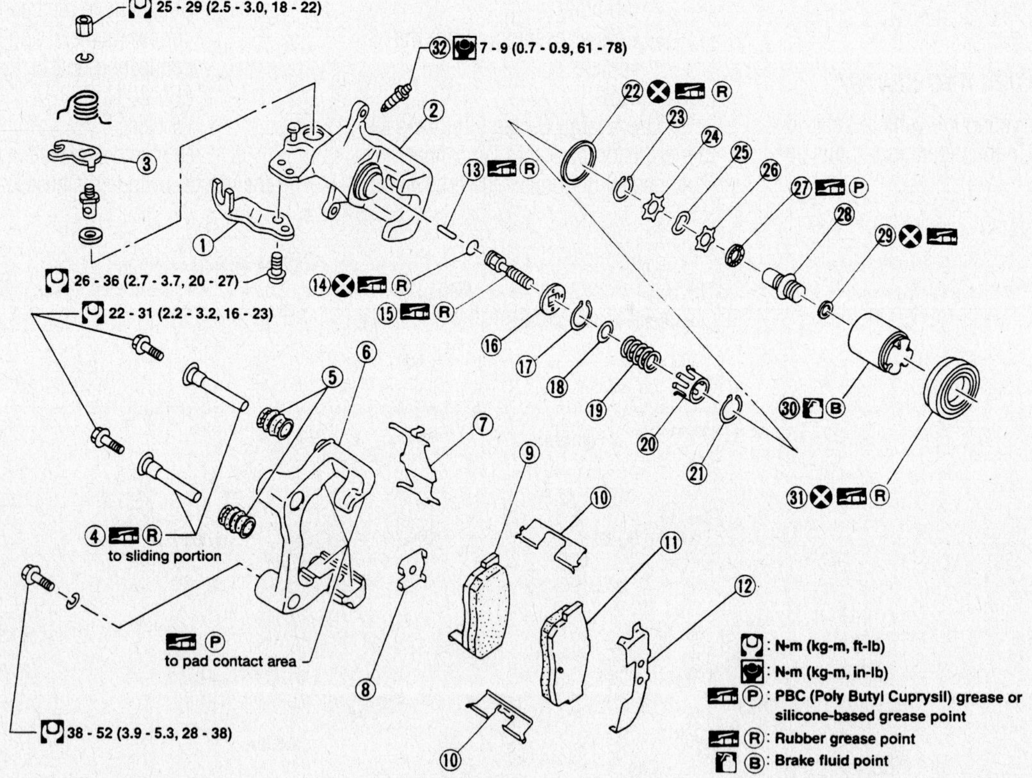

Rear disc brakes—Sentra shown

- Caliper pin bolts and remove the caliper

To install:

3. Turn the piston clockwise back into the caliper body. Remove some brake fluid from the master cylinder, if necessary. Take care not to damage the piston boot.
4. Coat the pad contact area on the mounting support with a silicone based grease.
5. Install or connect the following:
   - New pads, shims and the pad springs
   - Caliper body into position and torque the caliper pin bolts to 16–23 ft. lbs. (22–31 Nm)
   - Brake fluid hose, using new copper washers, and tighten the flare nut to 12–14 ft. lbs. (17–20 Nm)
   - Lock spring and the parking brake cable
6. Bleed the brake system and top off the master cylinder as necessary.
7. Replace the wheels.

350Z WITH AD14VE CALIPER

1. Before servicing the vehicle, refer to the precautions in the beginning of this section.
2. Remove or disconnect the following:
   - Rear wheels
   - Brake fluid hose from the caliper
   - Caliper pin bolts and remove the caliper

To install:

3. Install or connect the following:
   - New pads, shims and the pad springs
   - Caliper body into position and torque the caliper pin bolts to 16–23 ft. lbs. (22–31 Nm)
   - Brake fluid hose, using new copper washers, and tighten the flare nut to 12–14 ft. lbs. (17–20 Nm)
4. Bleed the brake system and top off the master cylinder as necessary.
5. Replace the wheels.

350Z WITH OPB13VB CALIPER

1. Before servicing the vehicle, refer to the precautions in the beginning of this section.
2. Remove or disconnect the following:
   - Rear wheels
   - Caliper pin bolts and remove the caliper

To install:

3. Install or connect the following:
   - New pads, shims and the pad springs
   - Caliper body into position and torque the caliper mounting bolts to 53–71 ft. lbs. (72–97 Nm)

Disc Brake Pads

REMOVAL & INSTALLATION

Sentra

FRONT

1. Before servicing the vehicle, refer to the precautions in the beginning of this section.
2. Remove or disconnect the following:
   - Wheels
   - Bottom guide pin from the caliper and swing the caliper cylinder body up
   - Brake pad retainers and the pads

To install:

3. Compress the piston of the disc brake caliper.
4. Install or connect the following:
   - Brake pads, shims, and retainers
   - Caliper assembly. Torque the guide pin to 23–30 ft. lbs. (31–41 Nm).
   - Wheels
5. Check the master cylinder and add fluid if necessary.

REAR

1. Before servicing the vehicle, refer to the precautions in the beginning of this section.
2. Remove or disconnect the following:
   - Wheels
   - Parking brake cable bracket bolt
   - Pin bolts and lift off the caliper body
   - Pad springs and the pads and shims

To install:

3. Turn the piston clockwise back into the caliper body. Take care not to damage the piston boot.
4. Coat the pad contact area on the mounting support with a silicone based grease.
5. Install or connect the following:
   - Pads, shims and retainer springs
   - Caliper body into position in the mounting support and tighten the pin bolts to 28–38 ft. lbs. (38–52 Nm)
   - Wheels and bleed the system if necessary.

Altima and Maxima

FRONT

1. Before servicing the vehicle, refer to the precautions in the beginning of this section.
2. Remove or disconnect the following:
   - Wheels
   - Bottom guide pin from the caliper and swing the caliper cylinder body up
   - Brake pad retainers and the pads

To install:

3. Compress the piston of the disc brake caliper.
4. Install or connect the following:
   - Brake pads, retainers, and caliper assembly. Torque the guide pin to 16–23 ft. lbs. (22–31 Nm).
   - Wheels
5. Check the master cylinder and add fluid if necessary.

REAR

1. Before servicing the vehicle, refer to the precautions in the beginning of this section.
2. Remove or disconnect the following:
   - Rear wheels
   - Parking brake cable bracket bolt
   - Pin bolts and lift off the caliper body
   - Pad springs by pulling them out

To install:

3. Turn the piston clockwise back into the caliper body. Take care not to damage the piston boot.
4. Coat the pad contact area on the mounting support with a silicone based grease.
5. Install or connect the following:
   - Pads, shims, and the pad springs
   - Caliper body into position in the mounting support and tighten the pin bolts to 16–23 ft. lbs. (22–31 Nm)
   - Wheels
6. Check the master cylinder and add fluid if necessary.

350Z

1. Before servicing the vehicle, refer to the precautions in the beginning of this section.
2. Remove or disconnect the following:
   - Front wheels
   - Sliding pin bolts
   - Rotate the caliper up and remove the brake pads

To install:

3. Install or connect the following:
   - New pads, new shims and pad retainers
   - Brake caliper
   - Wheels
4. Bleed the brake system and top off the master cylinder as necessary.

Brake Drums

REMOVAL & INSTALLATION

1. Before servicing the vehicle, refer to the precautions in the beginning of this section.
2. Remove the wheels.
3. Remove the brake drum from the brake shoes. Two 8mm x 1.25 bolts can be used to press the drum from the hub.

To install:

4. Install the drum assembly to the vehicle.
5. Install the wheels.
6. Adjust the rear brakes.

Brake Shoes

REMOVAL & INSTALLATION

1. Before servicing the vehicle, refer to the precautions in the beginning of this section.
2. Remove or disconnect the following:
   - Brake drum
   - Return springs, adjuster assembly, hold-down springs, and brake shoes
   - Parking brake cable from the toggle lever

To install:

3. Install or connect the following:
   - Parking brake cable
   - Shoes with the hold-down springs
   - Return springs, by hooking them into the new shoes
   - Adjuster assembly
   - Drums and wheels. Adjust the brakes and bleed the hydraulic system, if necessary.
4. Check the parking brake adjustment.

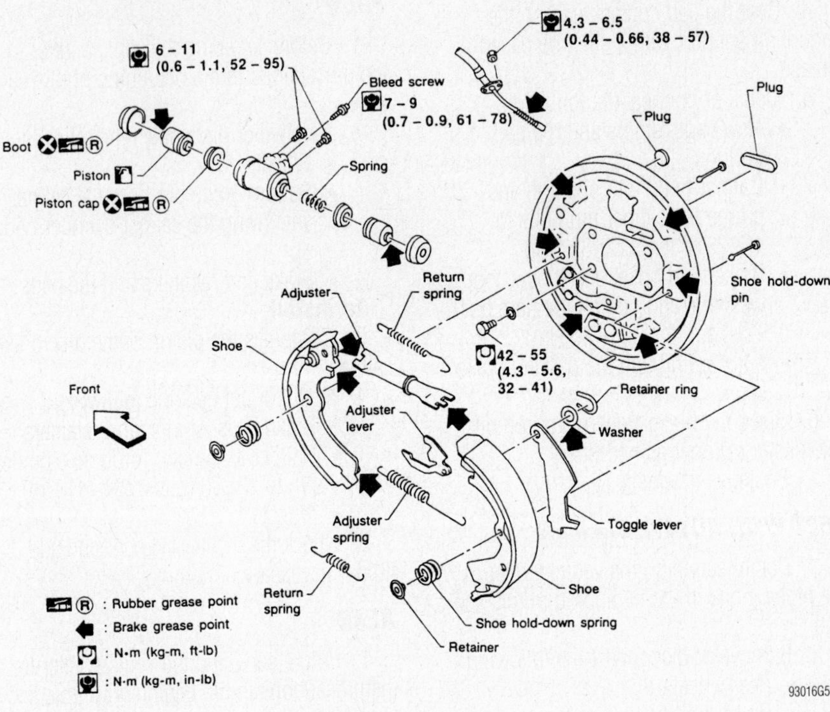

Exploded view of the rear drum brakes—Sentra

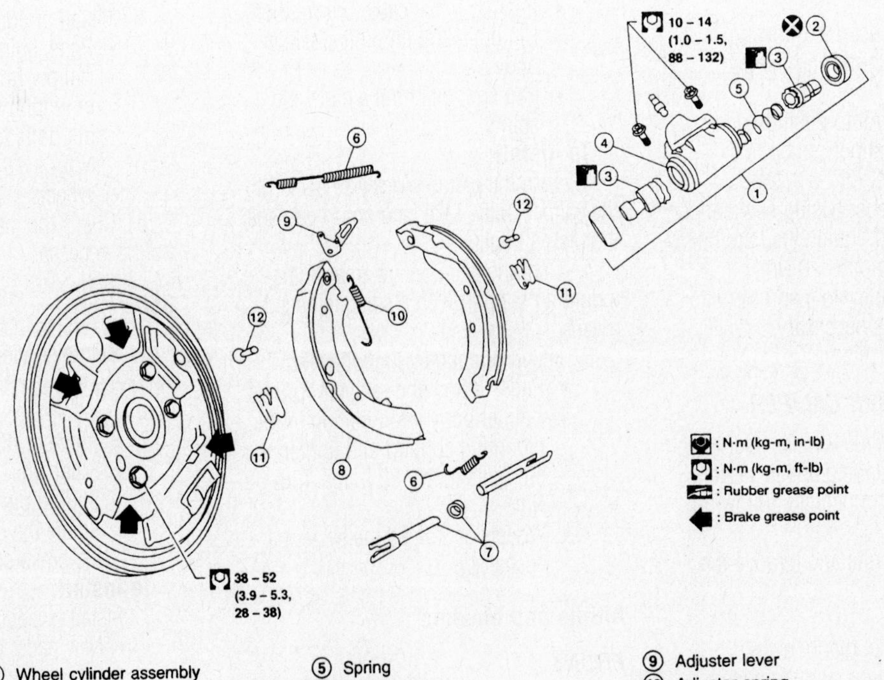

Rear drum brakes—Altima

NISSAN

6

Armada

SPECIFICATIONS AND MAINTENANCE CHARTS

ENGINE AND VEHICLE IDENTIFICATION

Engine						
Code ①	Liters (cc)	Cu. In.	Cyl.	Fuel Sys.	Engine	Eng. Mfg.
VK56DE	5.6 (5552)	338.8	8	MFI	DOHC	Nissan

Model Year	
Code ②	Year
4	2004
5	2005

MFI: Multi-port Fuel Injection

DOHC: Double Overhead Camshafts

① Located on the timing belt cover

② 10th digit of the Vehicle Identification Number (VIN)

67170-ARMA-C01

GENERAL ENGINE SPECIFICATIONS

Year	Model	Engine Displacement Liters	Engine ID	Net Horsepower @ rpm	Net Torque @ rpm (ft. lbs.)	Bore x Stroke (in.)	Com-pression Ratio	Oil Pressure @ rpm
2004	Armada	5.6	VK56DE	305@4900	385@3600	3.86X3.62	9.8:1	43@2000

67170-ARMA-C02

ENGINE TUNE-UP SPECIFICATIONS

Year	Engine Displacement Liters	Engine ID	Spark Plug Gap (in.)	Ignition Timing	Fuel Pump (psi) ①	Idle Speed ②	Valve Clearance (in.)	
							In.	Ex.
2004	5.6	VK56DE	0.043	15B	51	650	0.010-0.013	0.011-0.015

NOTE: The Vehicle Emission Control Information label often reflects specification changes made during production. The label figures must be used if they differ from those in this chart.

B: Before top dead center

① System pressure at idle

② Automatic transmission in Neutral

67170-ARMA-C03

CAPACITIES

Year	Model	Engine Displacement Liters	Engine ID	Engine Oil with Filter (qts.)	Transmission (pts.)	Transfer Case (pts.)	Drive Axle Front (pts.)	Drive Axle Rear (pts.)	Fuel Tank (gal.)	Cooling System (qts.)
2004	Armada	5.6	VK56DE	6.2	22.5	6.25	3.375	3.75	28.0	14.4

NOTE: All capacities are approximate. Add fluid gradually and check to be sure a proper fluid level is obtained.

67170-ARMA-C04

VALVE SPECIFICATIONS

Year	Engine Displacement Liters	Engine ID	Seat Angle (deg.)	Face Angle (deg.)	Spring Test Pressure (lbs. @ in.)	Spring Installed Height (in.)	Stem-to-Guide Clearance (in.) Intake	Stem-to-Guide Clearance (in.) Exhaust	Stem Diameter (in.) Intake	Stem Diameter (in.) Exhaust
2004	5.6	VK56DE	45.15-45.45	45	37.0@1.457	1.9913	0.0008-0.0021	0.0012-0.0025	0.2348-0.2354	0.2344-0.2350

67170-ARMA-C05

CRANKSHAFT AND CONNECTING ROD SPECIFICATIONS

All measurements are given in inches.

Year	Engine Displ. Liters	Engine ID	Crankshaft Main Brg. Journal Dia.	Crankshaft Main Brg. Oil Clearance	Crankshaft Shaft End-play	Crankshaft Thrust on No.	Connecting Rod Journal Diameter	Connecting Rod Oil Clearance	Connecting Rod Side Clearance
2004	5.6	VK56DE	①	②	0.0118	3	③	0.0008-0.0015	0.0079-0.0157

NA - Not Available

① There are 17 different grades, ranging from 0 (2.483) to 78 (2.510)

② No. 1 and 5: 0.00004-0.0004

No. 2, 3 and 4: 0.0003-0.0007

③ Grade 0: 2.0441-2.0441

Grade 1: 2.0441-2.0442

Grade 2: 2.0442-2.0442

Grade 3: 2.0442-2.0443

Grade 4: 2.0443-2.0443

Grade 5: 2.0443-2.0443

Grade 6: 2.0443-2.0444

Grade 7: 2.0444-2.0444

Grade 8: 2.0444-2.0444

Grade 9: 2.0444-2.0445

Grade A: 2.0445-2.0445

Grade B: 2.0445-2.0446

Grade C: 2.0446-2.0446

67170-ARMA-C06

PISTON AND RING SPECIFICATIONS

All measurements are given in inches.

	Engine			Ring Gap			Ring Side Clearance		
	Displacement	Engine	Piston	Top	Bottom	Oil	Top	Bottom	Oil
Year	Liters	ID	Clearance	Comp.	Comp.	Control	Comp.	Comp.	Control
2004	5.6	VK56DE	0.0004-0.0012	0.0091-0.0110	0.0189-0.0217	0.0079-0.0197	0.0014-0.0033	0.0012-0.0028	0.0006-0.0073

67170-ARMA-C07

TORQUE SPECIFICATIONS

All readings in ft. lbs.

	Engine		Cylinder	Main	Rod	Crankshaft					Oil Pan
	Displacement	Engine	Head	Bearing	Bearing	Damper	Flywheel	Manifold		Spark	Drain
Year	Liters	ID	Bolts	Bolts	Bolts	Bolts	Bolts	Intake	Exhaust	Plugs	Plug
2004	5.6	VK56DE	①	②	③	④	65	6	21	18	25

NA: Information not available

① Step 1: 72 ft. lbs
Step 2: Loosen all bolts completely
Step 3: 33 ft. lbs.
Step 4: +60 degrees
Step 5: +60 degrees

② Step 1: Main Bolts to 29 ft. lbs.
Step 2: Sub-bolts to 22 ft. lbs.
Step 3: Main Bolts +40 degrees
Step 4: Sub-Bolts +30 degrees
Step 5: Side Bolts to 36 ft. lbs.

③ Step 1: 11 ft. lbs.
Step 2: +90 degrees

④ Step 1: 65 ft. lbs.
Step 2: +90 degre

67170-ARMA-C08

WHEEL ALIGNMENT

		Caster		Camber		
			Preferred		Preferred	
		Range	Setting	Range	Setting	Toe-in
Year	Model	(+/-Deg.)	(Deg.)	(+/-Deg.)	(Deg.)	(in.)
2004	Armada ①	0.75	②	0.75	③	0.08+/-0.11

① Assumes P285/70R17 tire

② 4x2: +3.47
4x4: +3.05

③ 4x2: -0.10
4x4: +0.18

67170-ARMA-C09

TIRE, WHEEL AND BALL JOINT SPECIFICATIONS

Year	Model	OEM Tires		Tire Pressures (psi)		Wheel Size	Ball Joint Inspection	Lugnut Torque (ft. lbs.)
		Standard	Optional	Front	Rear			
2004	Armada	P285/70R17	P265/70R18	35	35	std.: 17, opt: 18	①	98

OEM: Original Equipment Manufacturer

PSI: Pounds Per Square Inch

① Axial play

Upper: 0

67170-ARMA-C10

BRAKE SPECIFICATIONS

All measurements in inches unless noted

Year	Model	Brake Disc			Minimum Lining Thickness	Brake Caliper	
		Original Thickness	Minimum Thickness	Maximum Runout		Bracket Bolts (ft. lbs.)	Mounting Bolts (ft. lbs.)
2004	Armada	1.024	0.965	0.0016	0.039	155	20
		0.551	0.472	0.002	0.039	NA	32

NA: Not applicable

67170-ARMA-C11

SCHEDULED MAINTENANCE INTERVALS

Nissan Armada

TO BE SERVICED	TYPE OF SERVICE	7.5	15	22.5	30	37.5	45	52.5	60
Engine oil & filter	R	✓	✓	✓	✓	✓	✓	✓	✓
Brake lines & cables	S/I		✓		✓		✓		✓
Brake pads and rotors	I		✓		✓		✓		✓
Driveshaft boots & propeller shaft (4x4)	L/I		✓		✓		✓		✓
Transmission, transfer & differential gear oil	I		✓		✓		✓		✓
Air cleaner filter	R				✓				✓
Engine coolant ①	R								✓
Spark plugs (Platinum)	R	Replace every 105,000 miles							
Drive belt(s) ②	S/I								✓
Cabin air filter	R		✓		✓		✓		✓
Exhaust system	I				✓				✓
Fuel lines	S/I				✓				✓
Steering gear, linkage, axle & suspension parts	I				✓				✓
Vapor lines	S/I				✓				✓

R: Replace S/I: Service or Inspect L: Lubricate

① Coolant: After 60,000 miles, inspect every 30,000 miles.

② Drive Belts: After 60,000 miles, inspect every 15,000 miles. Replace belts if damaged.

③ Fuel Filter: Maintenance free item.

FREQUENT OPERATION MAINTENANCE (SEVERE SERVICE)

If a vehicle is operated under any of the following conditions it is considered severe service:

- Extremely dusty areas.

- Rough, muddy, or salt spread roads.

- 50% or more of the vehicle constant operation is in 32°C (90°F) or higher temperatures, or temperatures below 0°C (32°F).

- Prolonged idling (vehicle operation in stop and go traffic).

- Frequent short running periods (engine does not warm to normal operating temperatures).

- Police, taxi, delivery usage or trailer towing usage.

Oil & oil filter: replace every 3750 miles.

Brake pads, discs, drums & linings: service or inspect every 7500 miles.

Driveshaft boots & propeller shaft: service or inspect every 7500 miles.

Exhaust system: service or inspect every 7500 miles.

Steering gear (box) & linkage, (steering damper-4x4), axle & suspension parts: service or inspect every 7500 miles.

Steering linkage ball joints & front suspension ball joints: service or inspect every 7500 miles.

67170-ARMA-C12

PRECAUTIONS

Before servicing any vehicle, please be sure to read all of the following precautions, which deal with personal safety, prevention of component damage, and important points to take into consideration when servicing a motor vehicle:

- Never open, service or drain the radiator or cooling system when the engine is hot; serious burns can occur from the steam and hot coolant.
- Observe all applicable safety precautions when working around fuel. Whenever servicing the fuel system, always work in a well-ventilated area. Do not allow fuel spray or vapors to come in contact with a spark, open flame, or excessive heat (a hot drop light, for example). Keep a dry chemical fire extinguisher near the work area. Always keep fuel in a container specifically designed for fuel storage; also, always properly seal fuel containers to avoid the possibility of fire or explosion. Refer to the additional fuel system precautions later in this section.
- Fuel injection systems often remain pressurized, even after the engine has been turned **OFF**. The fuel system pressure must be relieved before disconnecting any fuel lines. Failure to do so may result in fire and/or personal injury.
- Brake fluid often contains polyglycol ethers and polyglycols. Avoid contact with the eyes and wash your hands thoroughly after handling brake fluid. If you do get brake fluid in your eyes, flush your eyes with clean, running water for 15 minutes. If eye irritation persists, or if you have taken brake fluid internally, IMMEDIATELY seek medical assistance.
- The EPA warns that prolonged contact with used engine oil may cause a number of skin disorders, including cancer. You should make every effort to minimize your exposure to used engine oil. Protective gloves should be worn when changing oil. Wash your hands and any other exposed skin areas as soon as possible after exposure to used engine oil. Soap and water, or waterless hand cleaner should be used.
- All new vehicles are now equipped with an air bag system, often referred to as a Supplemental Restraint System (SRS) or Supplemental Inflatable Restraint (SIR) system. The system must be disabled before performing service on or around system components, steering column, instrument panel components, wiring and sensors. Failure to follow safety and disabling procedures could result in accidental air bag deployment, possible personal injury, and unnecessary system repairs.
- Always wear safety goggles when working with, or around, the air bag system. When carrying a non-deployed air bag, be sure the bag and trim cover are pointed away from your body. When placing a non-deployed air bag on a work surface, always face the bag and trim cover upward, away from the surface. This will reduce the motion of the module if it is accidentally deployed. Refer to the additional air bag system precautions later in this section.
- Clean, high quality brake fluid from a sealed container is essential to the safe and proper operation of the brake system. You should always buy the correct type of brake fluid for your vehicle. If the brake fluid becomes contaminated, completely flush the system with new fluid. Never reuse any brake fluid. Any brake fluid that is removed from the system should be discarded. Also, do not allow any brake fluid to come in contact with a painted surface; it will damage the paint.
- Never operate the engine without the proper amount and type of engine oil; doing so will result in severe engine damage.
- Timing belt maintenance is extremely important. Many models utilize an interference-type, non-freewheeling engine. If the timing belt breaks, the valves in the cylinder head may strike the pistons, causing potentially serious (also time-consuming and expensive) engine damage. Refer to the maintenance interval charts in the front of this manual for the recommended replacement interval for the timing belt, and to the timing belt section for belt replacement and inspection.
- Disconnecting the negative battery cable on some vehicles may interfere with the functions of the on-board computer system(s) and may require the computer to undergo a relearning process once the negative battery cable is reconnected.
- When servicing drum brakes, only disassemble and assemble one side at a time, leaving the remaining side intact for reference.

ENGINE REPAIR

Distributor

REMOVAL& INSTALLATION

These models use a Distributorless Ignition System (DIS) controlled by the Powertrain Control Module (PCM).

Alternator

REMOVAL & INSTALLATION

1. Before servicing the vehicle, refer to the precautions in the beginning of this section.
2. Remove or disconnect the following:
   - Negative battery cable
   - Fan shroud

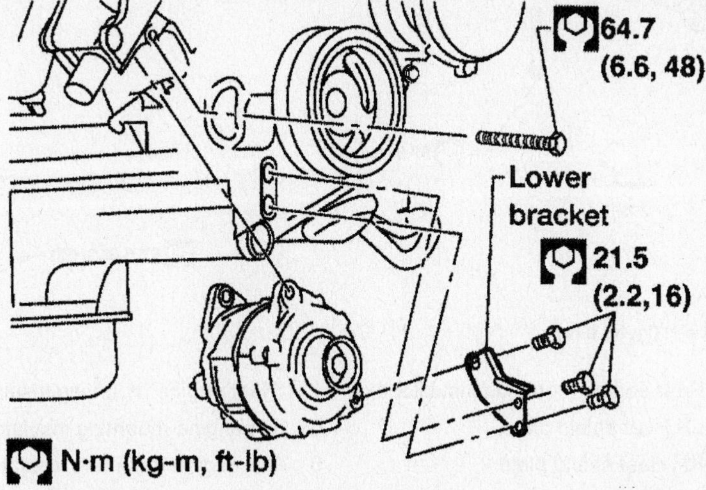

Alternator mounting

- Drive belt
- Lower alternator bracket
- Alternator upper bolt
- Alternator harness connectors
- Alternator

To install:

3. Install or connect the following:
   - Alternator
   - Alternator harness connectors
   - Upper bolt, tighten to 48 ft. lbs. (65 Nm)
   - Lower bracket, tighten to 16 ft. lbs (22 Nm)
   - Drive belt
   - Fan shroud
   - Negative battery cable

Ignition Timing

ADJUSTMENT

Ignition timing is controlled by the ECM and manual adjustment is not possible.

Engine Assembly

REMOVAL & INSTALLATION

1. Before servicing the vehicle, refer to the precautions in the beginning of this section.
2. Drain the cooling system.
3. Partially drain the automatic transmission fluid.
4. Relieve the fuel system pressure.
5. Remove or disconnect the following:
   - Hood
   - Cowl extension
   - Engine cover
   - Air intake assembly
   - Vacuum hose between vehicle and engine
   - Radiator hoses
   - Radiator
   - Drive belts
   - Engine fan
   - Wiring harness
   - ECM
   - Power steering reservoir tank and oil pump
   - A/C compressor
   - Brake booster vacuum line
   - EVAP line
   - Fuel hose
   - Heater hoses
   - Exhaust manifolds
   - Front final drive assembly
   - Automatic transmission dipstick tube assembly
   - Automatic transmission
6. Install engine slings onto the left and right cylinder heads and tighten to 33 ft. lbs. (45 Nm).
7. Attach an engine hoist to slings and lift engine out of the vehicle

To install:

8. Lower engine into the vehicle
9. Install or connect the following:
   - Automatic transmission
   - Automatic transmission dipstick tube assembly

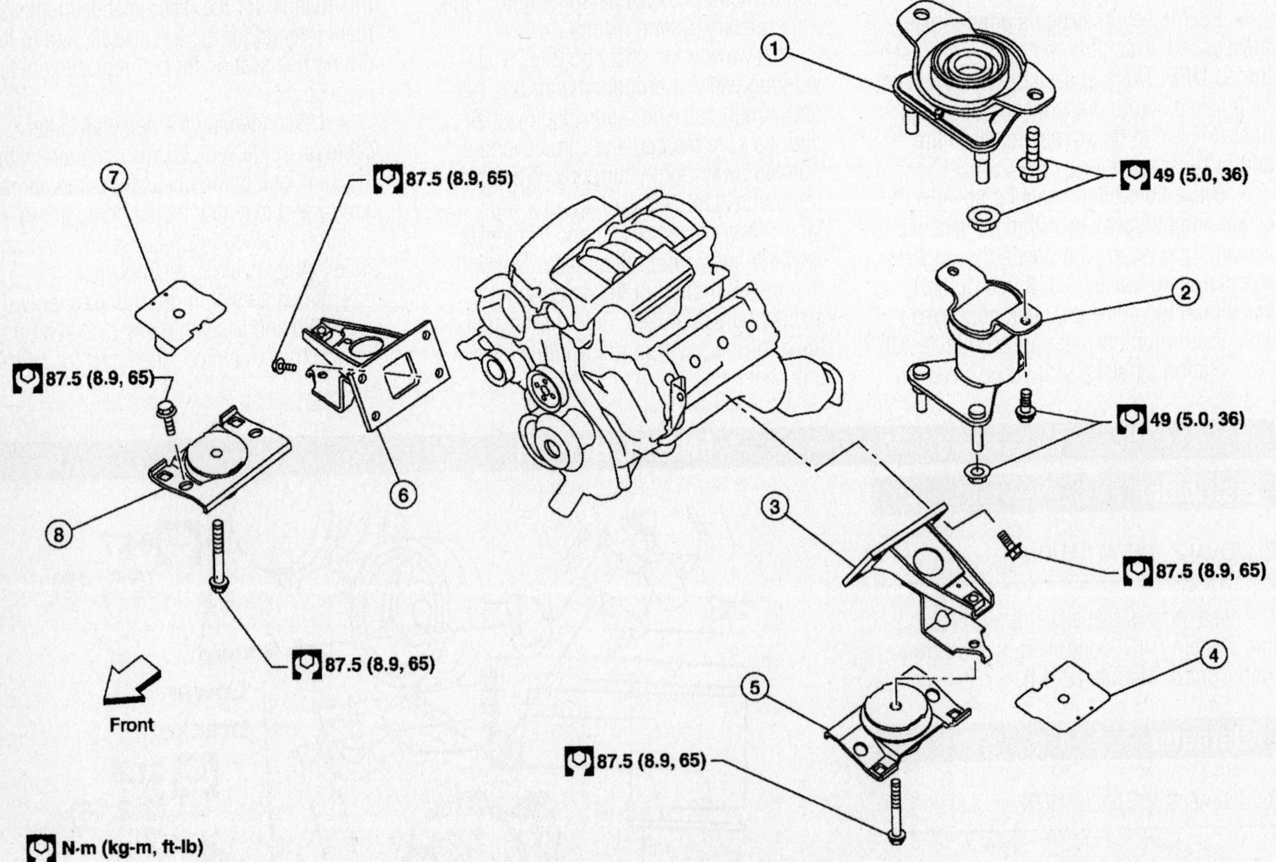

Engine mounts

- Front final drive assembly
- Exhaust manifolds
- Heater hoses
- Fuel hose
- EVAP line
- Brake booster vacuum line
- A/C compressor
- Power steering reservoir tank and oil pump
- ECM
- Wiring harness
- Engine fan
- Drive belts
- Radiator and radiator hoses
- Vacuum hose between vehicle and engine
- Air intake assembly
- Engine cover
- Cowl extension
- Hood

10. Refill the automatic transmission fluid.
11. Refill the cooling system.
12. Start the engine and check for leaks.

Water Pump

REMOVAL & INSTALLATION

1. Before servicing the vehicle, refer to the precautions in the beginning of this section.
2. Drain the cooling system.
3. Remove or disconnect the following:
   - Engine splash guard
   - Air intake assembly
   - Accessory drive belt

➡**Leave tensioner pulley in its fixed position.**

   - Water pump pulley
   - Water pump

To install:

4. Install or connect the following:
   - Water pump with a new gasket. Tighten bolts to 18 ft. lbs. (25 Nm).
   - Water pump pulley and tighten bolts to 87 in. lbs. (10 Nm).
   - Accessory drive belt
   - Air intake assembly
   - Engine splash guard
5. Refill the cooling system.
6. Start the engine and check for leaks.

Heater Core

REMOVAL & INSTALLATION

Front Heater Assembly

1. Discharge the A/C system.
2. Drain the cooling system.
3. Remove or disconnect the following:
   - Cowl top extension
   - Exhaust system
   - Front heater hoses
   - High/Low pressure pipes
   - Instrument and console panels
   - Steering column
   - Instrument panel wiring harness
   - Steering member from body
   - Heater assembly

To install:

4. Install or connect the following:
   - Heater assembly
   - Steering member
   - Instrument panel wiring harness
   - Steering column
   - Console panel
   - Instrument panel
   - High/Low pressure pipes
   - Heater hoses
   - Exhaust system
   - Cowl top extension
5. Refill the cooling system.
6. Recharge the A/C system.

Rear Heater Assembly

1. Discharge the A/C system.
2. Drain the cooling system.
3. Remove or disconnect the following:
   - Rear heater hoses
   - Rear A/C pipes
   - Rear right-hand interior trim panel
   - Rear blower motor electrical connection
   - Rear blower motor resistor electrical connection
   - Rear air mix door motor electrical connection
   - Ducts from the heater assembly
   - Heater assembly

To install:

4. Install or connect the following:
   - Heater assembly and ducts
   - Rear air mix door motor electrical connection

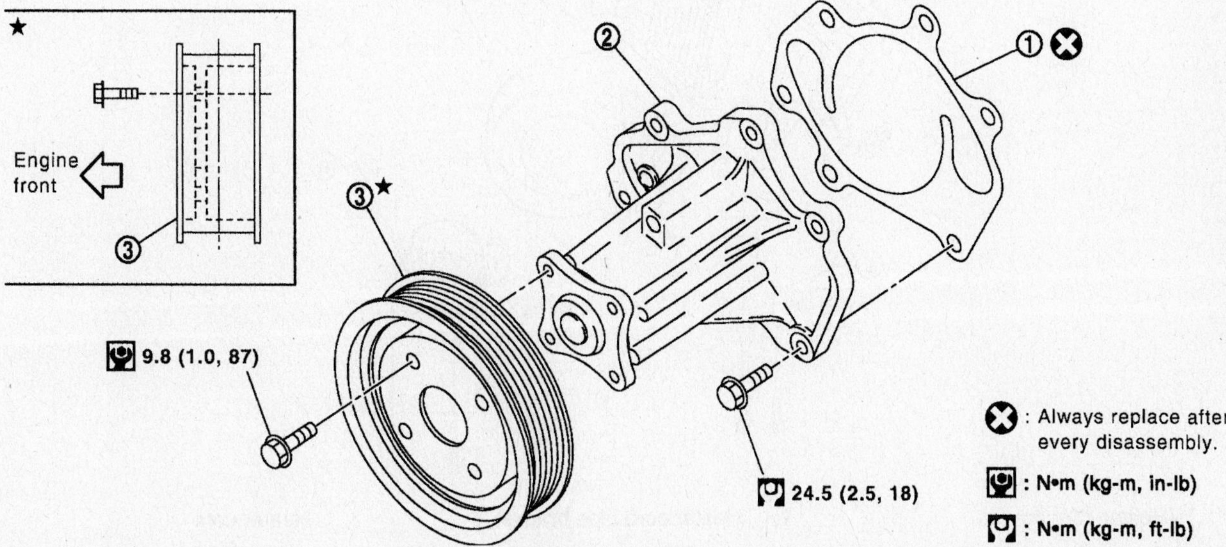

Water pump mounting

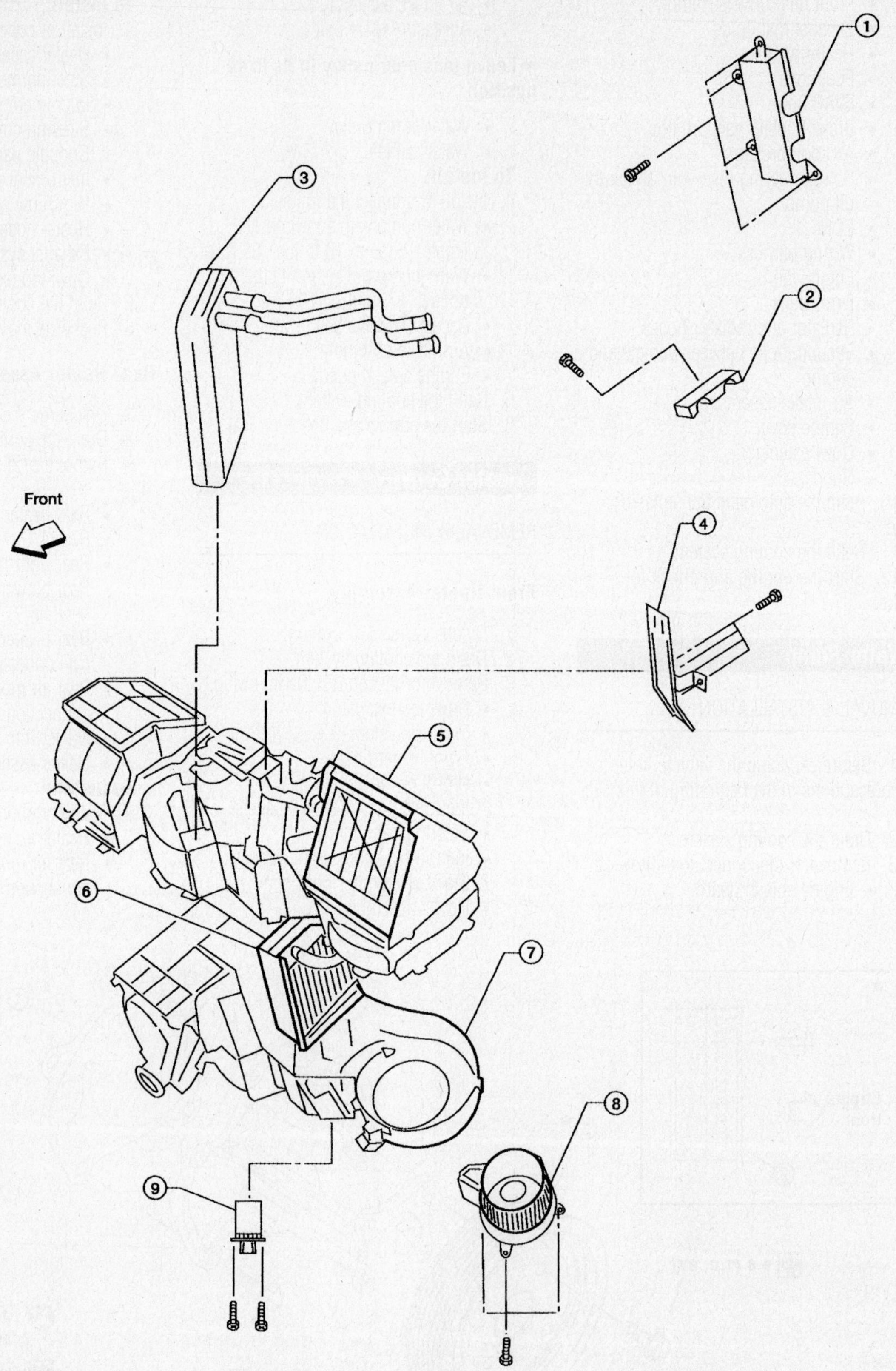

1. Heater core cover
2. Heater core pipe bracket
3. Heater core
4. Upper bracket
5. Upper heater and cooling unit case
6. A/C evaporator
7. Lower heater and cooling unit case
8. Blower motor
9. Variable blower control

67170-ARMA-G26

Front heater/AC assembly

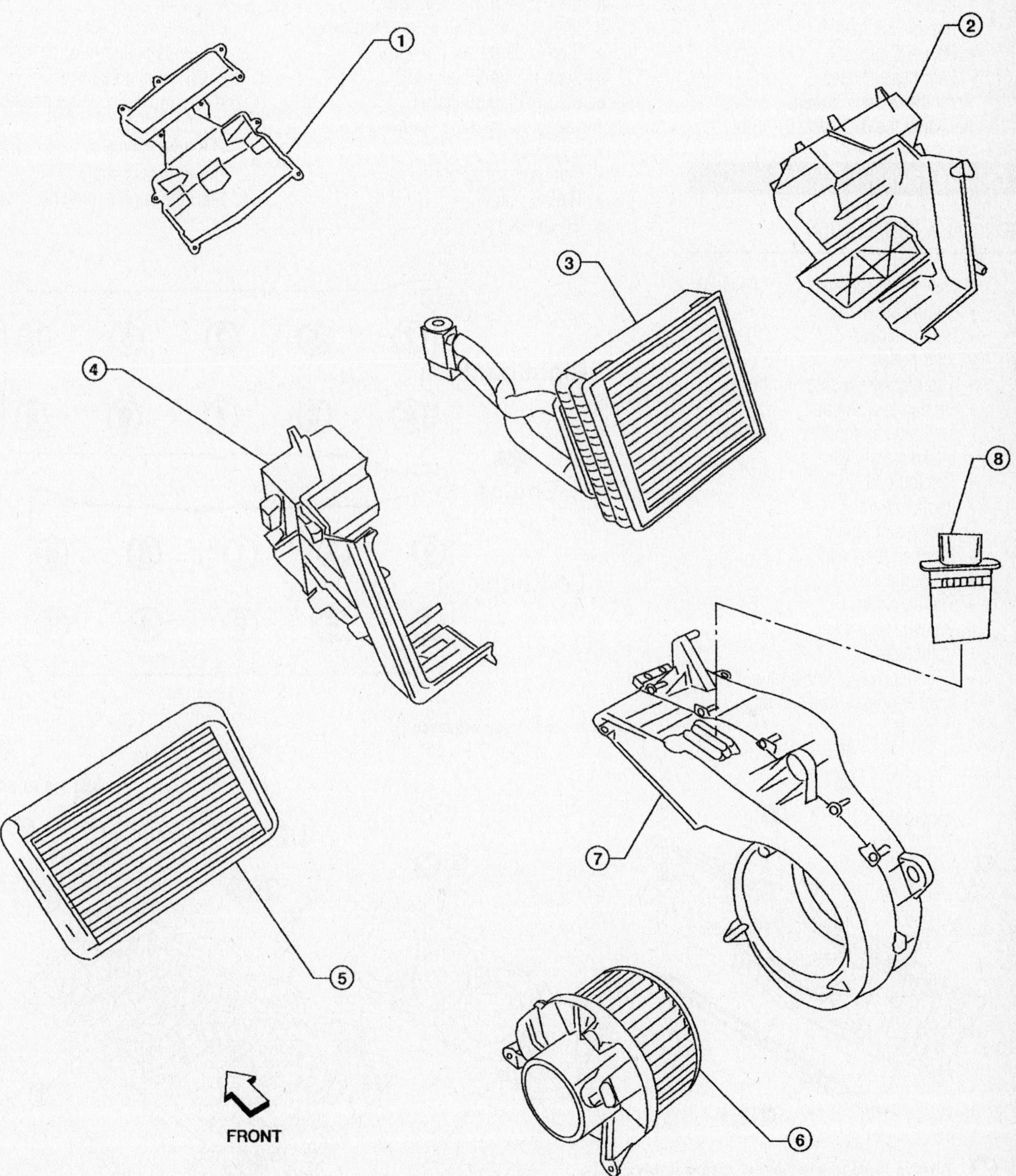

1. Front cover	2. Evaporator and heater core case	3. Evaporator
4. Side cover	5. Heater core	6. Blower motor
7. Blower motor case	8. Blower motor resistor	

67170-ARMA-G27

Rear heater/AC assembly

- Rear blower motor resistor electrical connection
- Rear blower motor electrical connection
- Interior trim panel
- Rear A/C pipes
- Rear heater hoses

5. Refill the cooling system.
6. Recharge the A/C system.

Cylinder Head

REMOVAL & INSTALLATION

1. Remove or disconnect the following:
   - Engine assembly
   - Belt tensioner
   - Idler pulley
   - Thermostat housing and hose
   - Oil pan and strainer
   - Fuel tube and injector assembly
   - Intake manifold
   - Ignition coil
   - Rocker cover
   - Crankshaft pulley
   - Front engine cover
   - Oil pump
   - Timing chain
   - Camshaft sprockets
   - Camshafts
   - Cylinder head, removing bolts in reverse order shown in figure
2. Install the cylinder head with a new gasket. Tighten the bolts in sequence as follows:
   a. Step 1: 72 ft. lbs. (98 Nm)
   b. Step 2: Loosen all bolts completely
   c. Step 3: 33 ft. lbs. (44 Nm)
   d. Step 4: Plus 60 degrees
   e. Step 5: Plus 60 degrees
3. Install or connect the following:
   - Camshaft
   - Camshaft sprockets
   - Timing chain
   - Oil pump
   - Front engine cover
   - Crankshaft pulley
   - Rocker cover
   - Ignition coil
   - Intake manifold
   - Fuel tube and injector assembly
   - Oil pain and strainer
   - Thermostat housing and hose
   - Idler pulley
   - Belt tensioner
   - Engine assembly
4. Start the engine and check for leaks

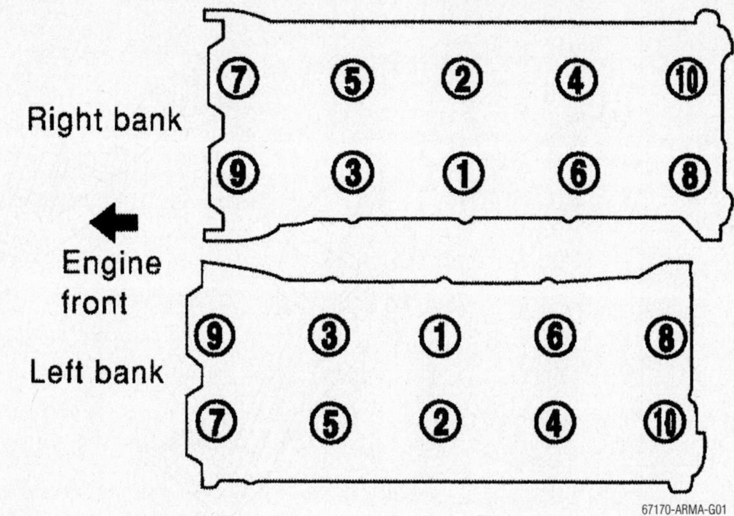

Cylinder head torque sequence

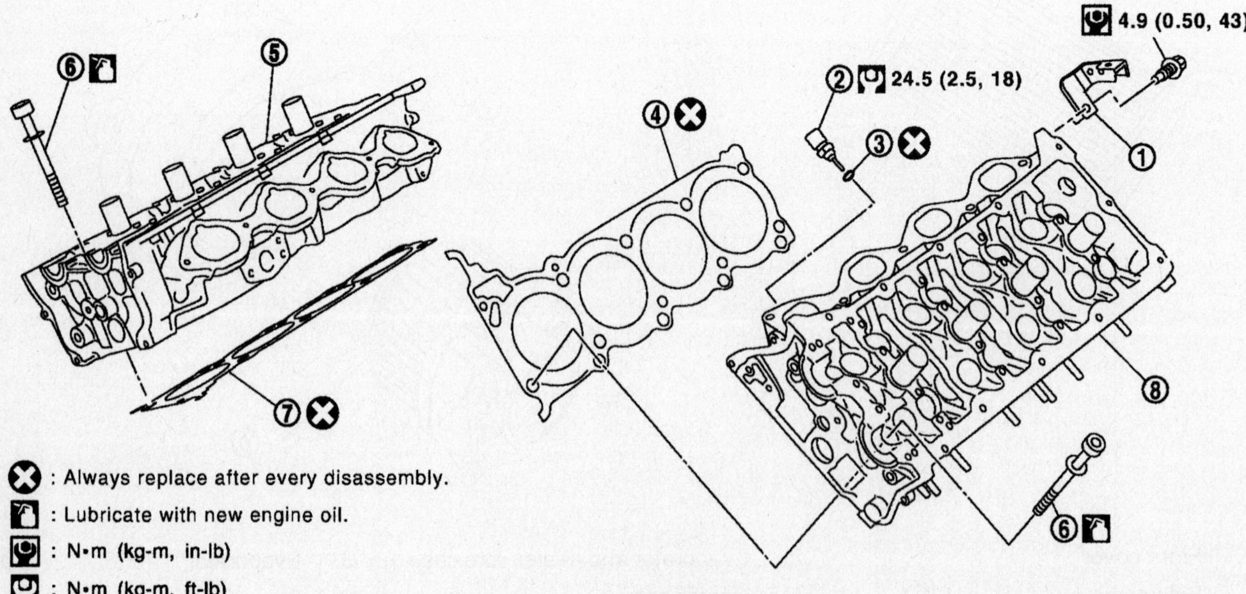

Cylinder heads and gaskets

Intake Manifold

REMOVAL & INSTALLATION

1. Drain the cooling system.
2. Relieve the fuel system pressure.
3. Remove or disconnect the following:
   - Engine cover
   - Air intake assembly
   - Fuel tube quick connector using special tool J-45488
   - Wiring harnesses and brackets from manifold
   - Vacuum hoses
   - PCV hose and tube
   - Electric throttle control actuator, loosening bolts diagonally
   - Fuel injectors
   - Fuel tube assembly
   - Intake manifold, removing bolts in reverse order shown in figure

To install:

4. Install the intake manifold with new gaskets. Tighten the bolts in order as shown.
5. Install or connect the following:
   - Fuel tube assembly
   - Fuel injectors
   - Electronic throttle control actuator, tightening the bolts in several steps
   - PCV hose
   - Vacuum hoses
   - Wiring harnesses
6. Connect the fuel tube as follows:
   a. Apply a thin layer of engine oil on the tube from tip end to spool end.
   b. Insert tube into quick connector past the white identification mark
   c. Insert tube into quick connector

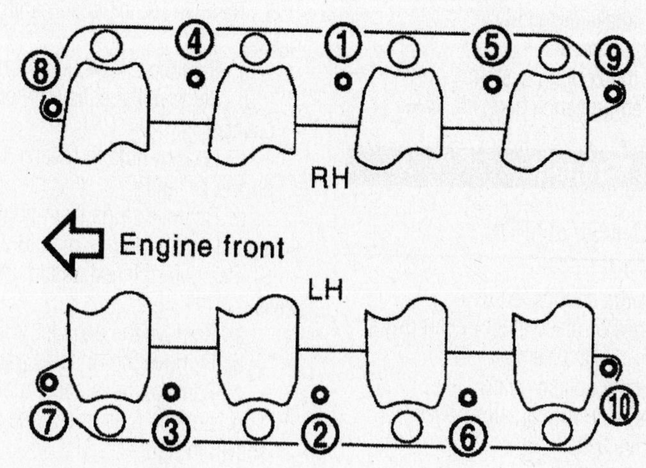

Intake manifold torque sequence

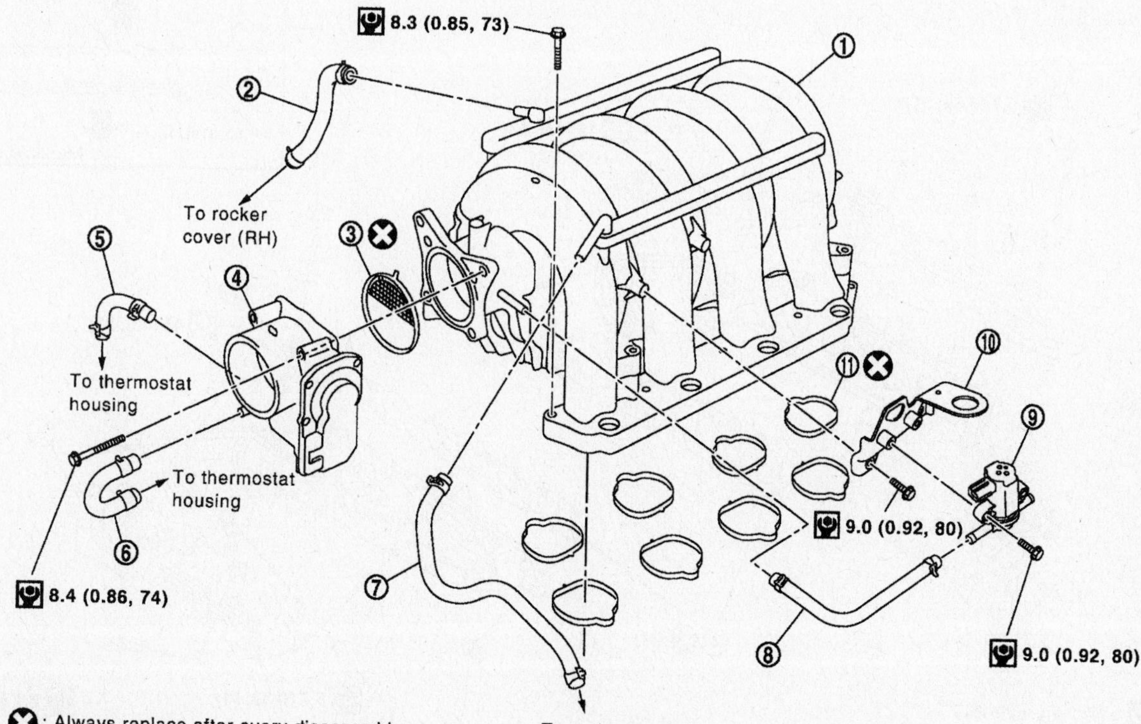

Intake manifold and related parts

until top spool is completely inside the connector and 2nd level spool is exposed right below the connector.

d. Pull slightly on the quick connector to ensure it is fully engaged.

e. Install quick connector cap on quick connector joint.

7. Install or connect the following:
   - Air intake assembly
   - Engine cover
8. Refill the cooling system.
9. Start engine and check for leaks.

Exhaust Manifold

REMOVAL & INSTALLATION

1. Drain the cooling system.
2. Remove or disconnect the following:
   - Air intake assembly
   - Engine splash guard
   - Radiator and radiator hoses
   - Drive belts
3. Remove air fuel ratio sensors as follows:
   a. Remove engine cover.
   b. Remove wiring harness from each sensor
   c. Remove sensors, using special tool J-38356
4. Remove front cross bar
5. Remove left exhaust manifold as follows:
   a. Remove the exhaust front tube.
   b. Remove the exhaust manifold cover.
   c. Loosen nuts in reverse order shown in figure.
   d. Remove studs from position 2, 4, 6, and 8 and remove manifold.
6. Remove right exhaust manifold as follows:
   a. Remove the exhaust front tube.
   b. Remove the oil level gauge guide.
   c. Remove the exhaust manifold cover.
   d. Loosen nuts in reverse order shown in figure.
   e. Remove studs from position 2, 4, 6, and 8 and remove manifold.

To install:

7. Install or connect the following:
   - Exhaust manifold gasket with triangle mark facing up and coated (gray) face toward exhaust manifold.
   - Exhaust manifold, tightening the nuts as shown in figure
   - Exhaust manifold cover
   - Oil level gauge guide (right side only)
   - Exhaust front tube
   - Front cross bar
   - Air fuel ratio sensors, with anti-seize lubricant
   - Engine cover
   - Drive belts
   - Radiator and radiator hoses
   - Engine splash guard
   - Air intake assembly
8. Refill the cooling system
9. Start engine and check for leaks.

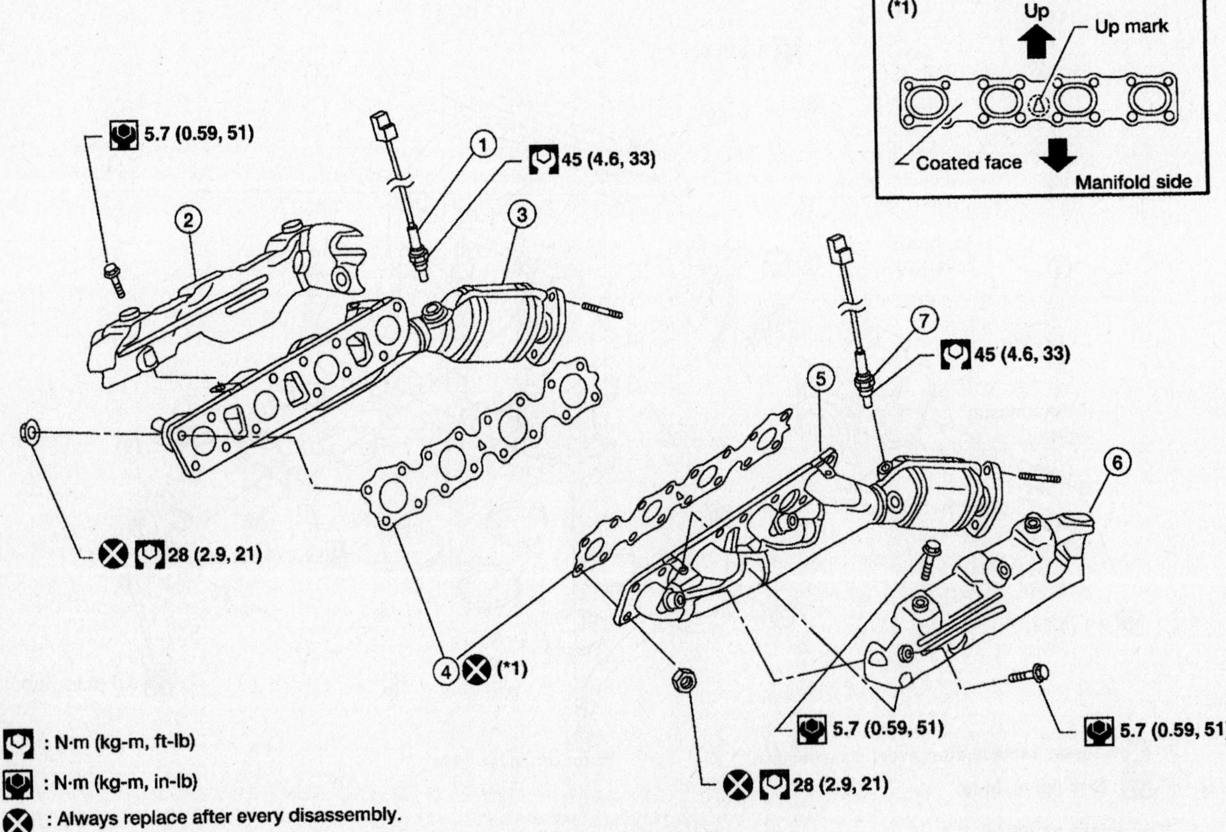

Exhaust manifolds and related parts

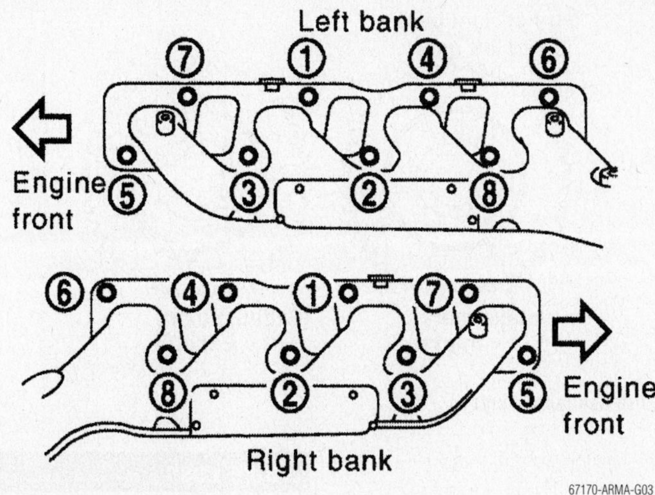

Exhaust manifold torque sequence

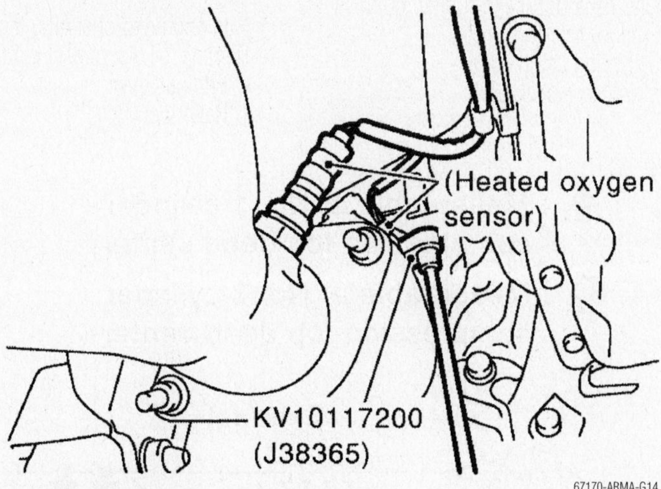

Removing Air-Fuel ratio sensors

Camshaft and Valve Lifters

REMOVAL & INSTALLATION

1. Remove rocker cover.
2. Obtain compression Top Dead Center (TDC) of No. 1 cylinder.
3. Remove timing chain case cover.
4. Matchmark the timing chain, aligning with the camshaft sprocket marks.
5. Remove chain tensioner from left bank as follows:
   a. Squeeze end clips and push plunger into tensioner body.
   b. Secure plunger using stopper pin.
   c. Remove chain tensioner.
6. Remove chain tensioner from right bank as follows:
   a. Remove chain tensioner cover using special tool J-37228.
   b. Squeeze end clips and push plunger into tensioner body.

67170-ARMA-G04

Camshaft torque sequence

   c. Secure plunger using stopper pin.
   d. Remove chain tensioner.
7. With camshaft locked with a wrench, loosen bolts to remove camshaft sprocket.
8. Remove front cover bolts.
9. Remove camshaft brackets, removing bolts in reverse order shown in figure.
10. Remove camshaft.
11. Remove valve lifters.

To install:

12. Install valve lifters.
13. Install camshaft, refer to table for correct placement.

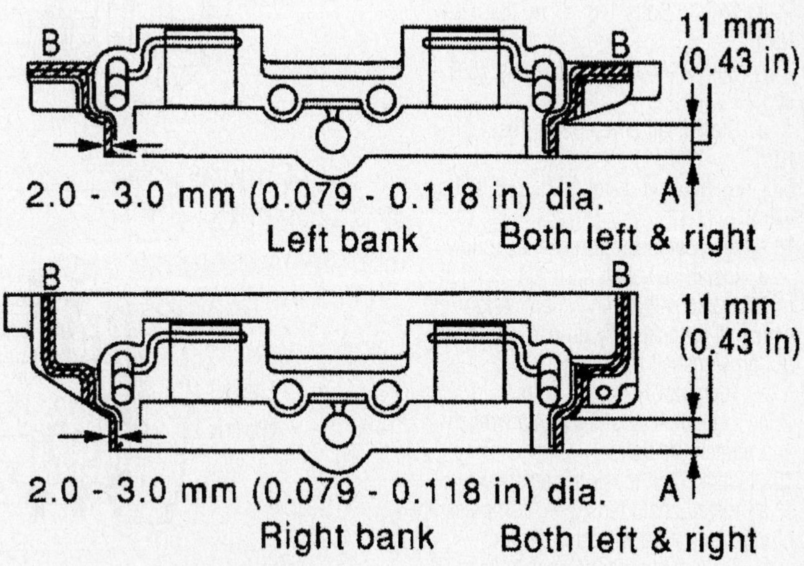

Gasket application for camshaft bracket

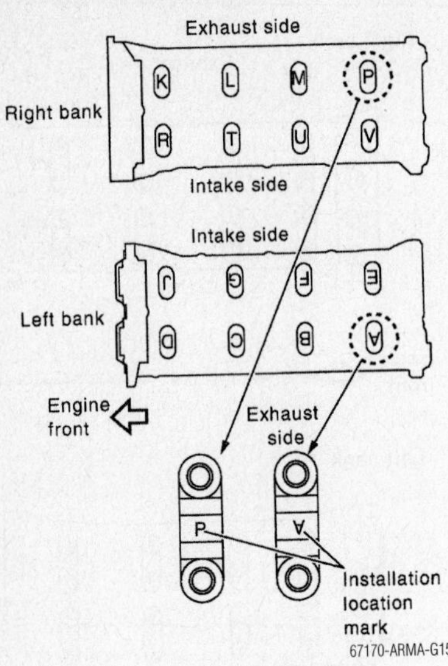

Camshaft bracket installation markings

14. Install camshaft brackets as follows:
   a. Refer to location mark on upper surface of bracket.
   b. Installation mark should be correctly read when viewed from intake side.
15. Install camshaft bracket #1 as follows:
   a. Apply liquid gasket to bracket and backside of front cover as shown in figure.
   b. Carefully position and mount camshaft bracket #1.
   c. Temporarily tighten front cover bolts
16. Tighten fixing bolts for camshaft brackets as follows:
   a. Step 1: Bolts 9-12: 17 in. lbs. (1.9 Nm)
   b. Step 2: Bolts 1-8: 17 in. lbs. (1.9 Nm)
   c. Step 3: All bolts: 52 in. lbs. (5.9 Nm)
   d. Step 4: All bolts: 92 in. lbs. (10 Nm)
17. Tighten front cover bolts to 8 ft. lbs. (11 Nm)
18. Install camshaft sprocket as follows:
   a. Install camshaft sprocket aligning matchmarks with timing chain. Align camshaft sprocket key groove with dowel pin on camshaft front edge.
   b. Temporarily tighten bolts.
   c. Lock the camshaft with a wrench and tighten the bolts.
19. Install chain tensioner as shown:
   a. Install chain tensioner, compress plunger and hold with stopper pin.
   b. Tighten chain tensioner bolts to 61 in. lbs. (7 Nm)

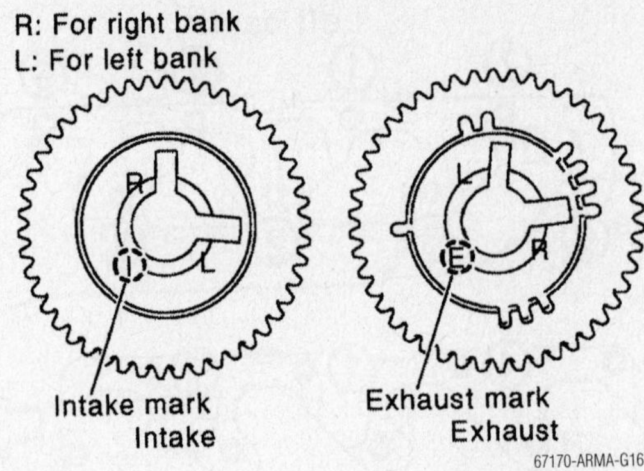

Camshaft installation markings

   c. Remove stopper pin, release plunger and apply tension to timing chain.
   d. Install chain tensioner front cover (Right-hand bank only) and tighten bolts to 80 in. lbs. (9 Nm).
20. Install timing chain cover.
21. Install rocker cover.

Valve Lash

INSPECTION

1. Run engine to operating temperature.
2. Remove or disconnect the following:
   - Engine cover
   - Battery cover

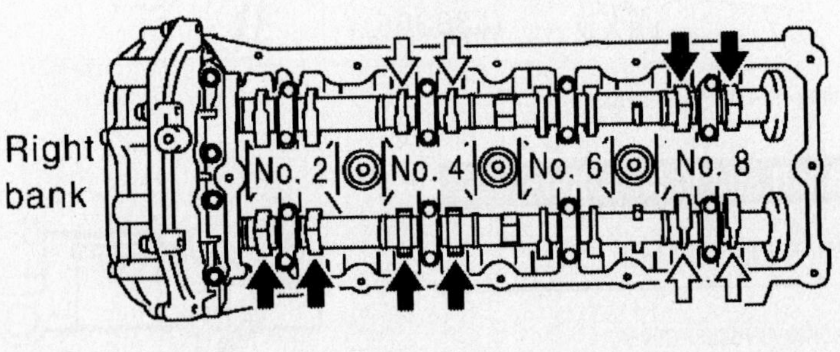

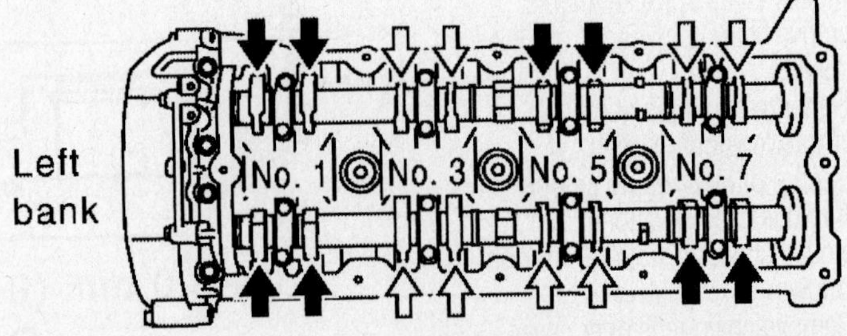

Locations to measure clearance with No. 1 cylinder at TDC

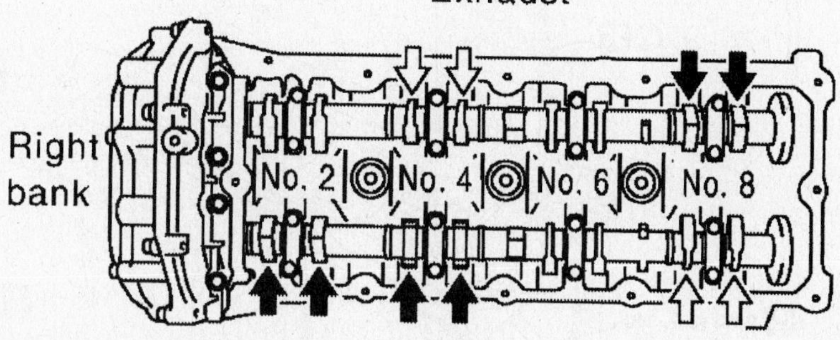

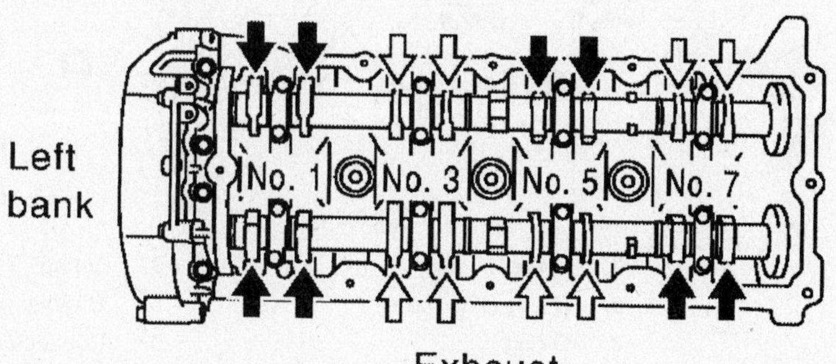

Locations to measure clearance with No. 3 cylinder at TDC

- Air intake assembly
- Left and right rocker covers

3. Turn the crankshaft pulley clockwise to Top Dead Center (TDC) identification notch with timing indicator.
4. Ensure that both the intake and exhaust cam noses of the No. 1 cylinder face outside.
5. Measure the valve clearances at locations marked 'x' shown in figure.
6. Turn the crankshaft pulley clockwise 270 degrees from the position of No. 1 cylinder compression to obtain No. 3 cylinder compression TDC.
7. Measure the valve clearances at locations marked 'x' shown in next figure.
8. Turn crankshaft pulley clockwise 90 degrees and measure the intake and exhaust valve clearance of No. 6 cylinder and exhaust valve clearance of No. 2 cylinder.

ADJUSTMENT

1. Remove camshaft and valve lifter(s) out of specification.
2. Install replacement valve lifter(s).
3. Install the camshaft.
4. Manually turn the crankshaft pulley several turns.
5. Recheck valve clearances with engine at operating temperature.

Oil Pan

REMOVAL & INSTALLATION

1. Before servicing the vehicle, refer to the precautions in the beginning of this section.
2. Remove engine assembly.
3. Remove lower oil pan, loosening bolts in reverse order shown in figure.
4. Remove oil strainer from upper oil pan.
5. Gently pry and remove upper oil pan from engine block.

To install:

6. Apply liquid gasket to upper oil pan mating surfaces.
7. Install new O-rings to oil pump and front cover side.
8. Tighten upper oil pan bolts in following numerical order:
   a. No. 15, 16
   b. No. 1, 3, 5, 7, 11, 13
   c. No. 2, 4, 6, 8, 10, 14
   d. No. 9, 12
9. Install or connect the following:
   - Rear plate cover
   - Oil strainer to upper oil pan
   - Lower oil pan, tightening bolts in order shown in figure

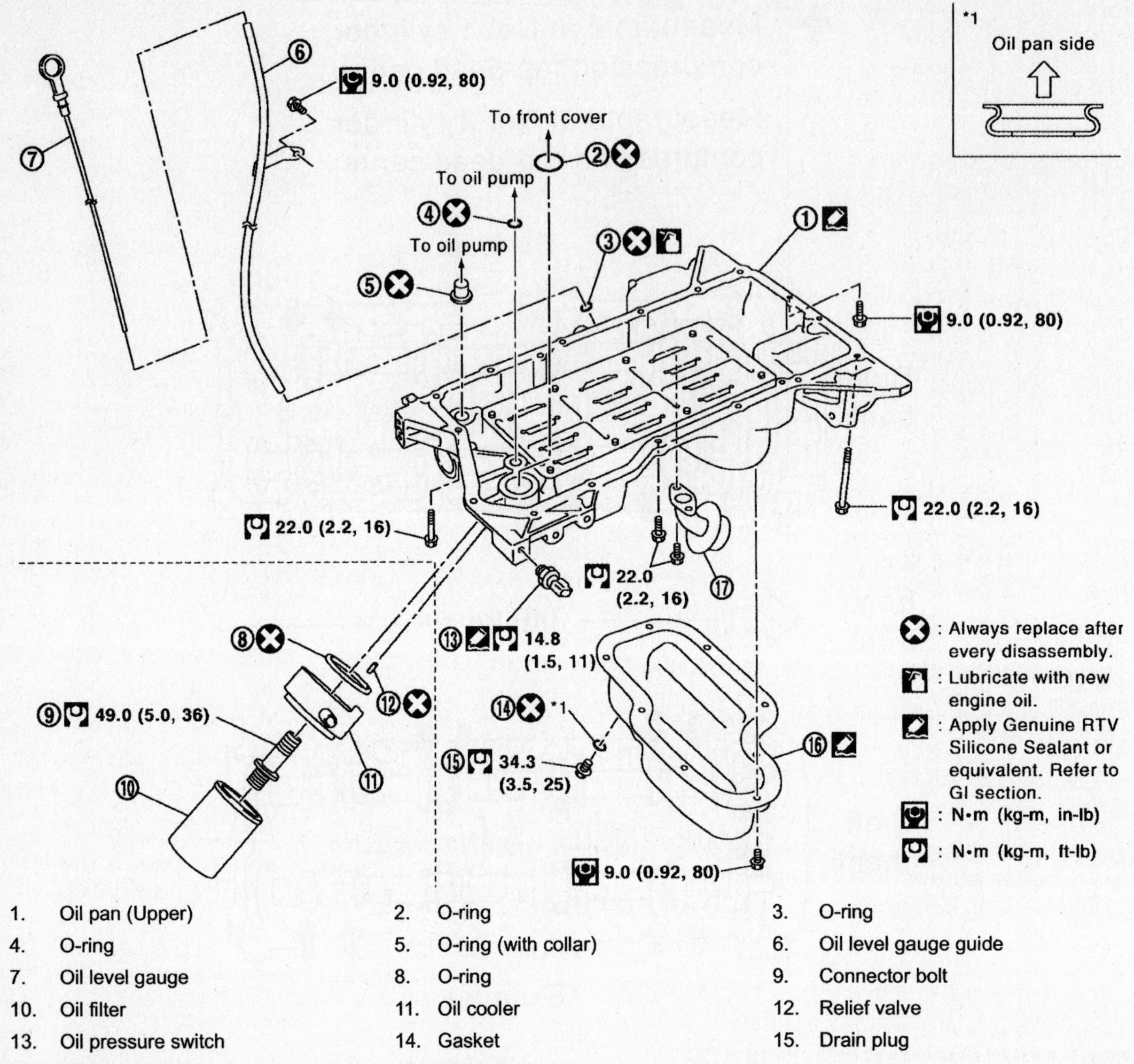

1. Oil pan (Upper)
2. O-ring
3. O-ring
4. O-ring
5. O-ring (with collar)
6. Oil level gauge guide
7. Oil level gauge
8. O-ring
9. Connector bolt
10. Oil filter
11. Oil cooler
12. Relief valve
13. Oil pressure switch
14. Gasket
15. Drain plug
16. Oil pan (Lower)
17. Oil strainer

67170-ARMA-G31

Oil pan and related parts

Engine front

67170-ARMA-G08

Lower oil pan torque sequence

Engine front

67170-ARMA-G09

Upper oil pan torque sequence

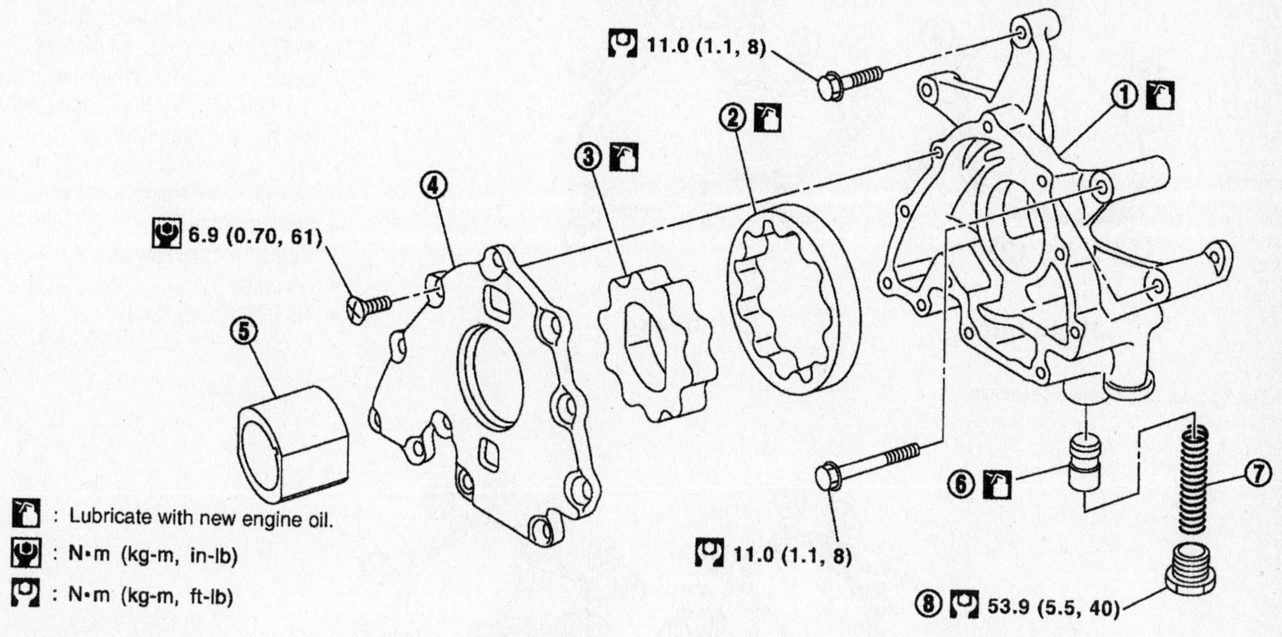

Oil pump exploded view

Oil Pump

REMOVAL & INSTALLATION

1. Remove or disconnect the following:
   - Timing chain cover
   - Oil pump drive spacer
   - Oil pump

To install:

2. Install or connect the following:
   - Oil pump
   - Oil pump drive spacer
   - Timing chain cover

Rear Main Seal

REMOVAL & INSTALLATION

1. Remove or disconnect the following:
   - Transmission assembly
   - Drive plate
   - Engine rear plate
   - Rear main seal using suitable tool

To install:

2. Install or connect the following:
   - Rear main seal using suitable tool
   - Engine rear plate
   - Drive plate
   - Transmission assembly

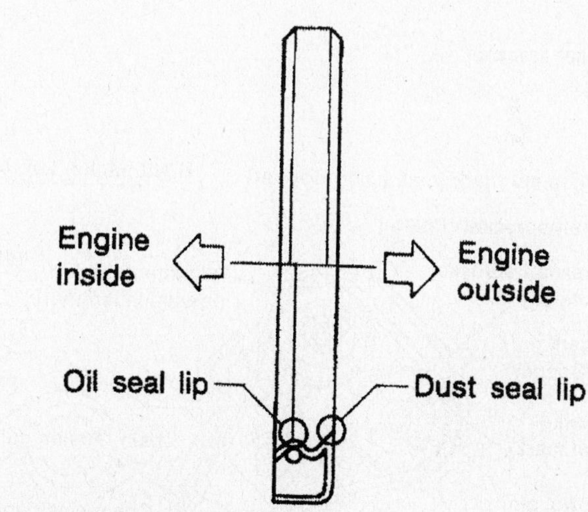

Proper seal installation direction

Timing Chain, Sprockets, Front Cover and Seal

REMOVAL & INSTALLATION

1. Remove or disconnect the following:
   - Engine assembly
   - Drive belt auto tensioner
   - Idler pulley
   - Thermostat housing and water hose
   - Power steering pump bracket
   - Oil pan (upper and lower)
   - Oil strainer
   - Ignition coil
   - Rocker cover
   - Timing chain case cover, loosening bolts in reverse order shown in figure
2. Obtain compression TDC of No. 1 cylinder as follows:
   a. Turn crankshaft pulley to align the

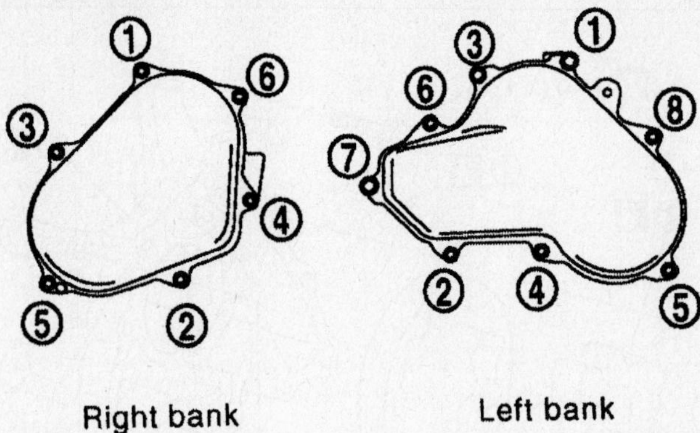

Timing case cover torque sequence

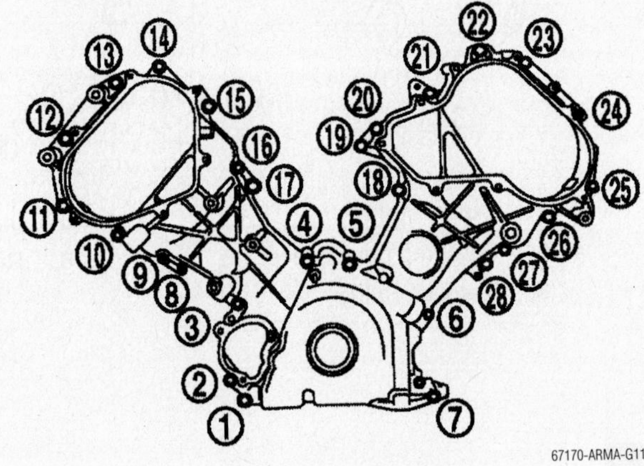

Front cover torque sequence

TDC identification notch with timing indicator on front cover.

b. Ensure intake and exhaust cam lobes of No. 1 cylinder point outside.

3. Remove or disconnect the following:
   - Crankshaft pulley from crankshaft using a suitable puller
   - Front cover, loosening bolts in reverse order shown in figure
   - Front oil seal
   - Oil pump drive spacer
   - Oil pump
   - Timing chain tensioner
   - Chain tension guide and slack guide
   - Timing chain

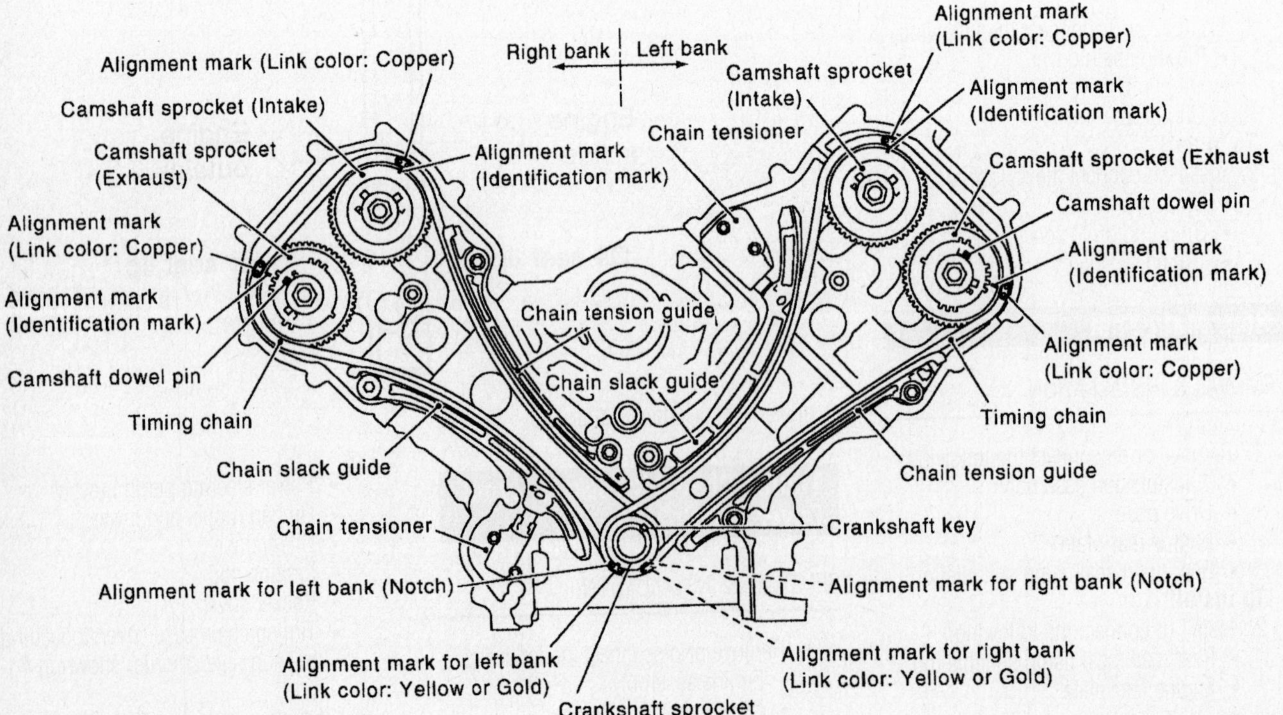

Timing mark alignment

- Camshaft sprocket

To install:

4. Ensure that the crankshaft key and dowel pin of each camshaft are facing the same direction.
5. Install or connect the following:
   - Camshaft sprockets
   - Timing chain
   - Chain tension guide and slack guide
   - Oil pump
   - Oil pump drive spacer
   - Front oil seal, using suitable tool
   - Front cover, using new O-rings and tighten bolts in order shown in figure
   - Chain case cover, and tighten bolts in order shown in figure
   - Crankshaft pulley and tighten bolt to 69 ft. lbs. (93 Nm) plus 90 degrees
   - Ignition coil
   - Oil strainer
   - Lower and upper oil pan
   - Power steering pump bracket
   - Thermostat housing and water hose
   - Idler pulley
   - Drive belt auto tensioner
   - Engine assembly

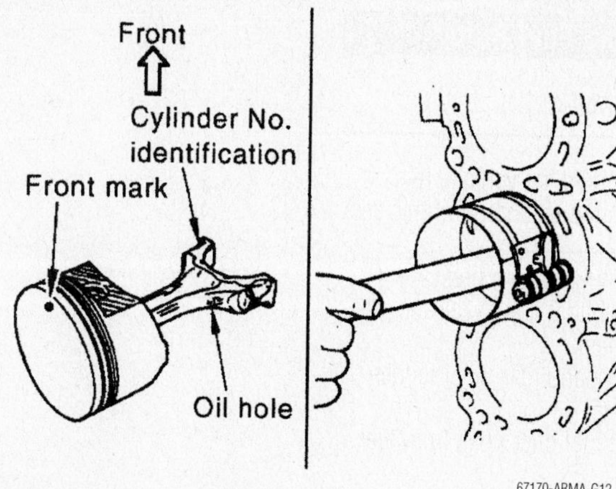

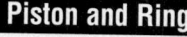

Piston and rod positioning and identification

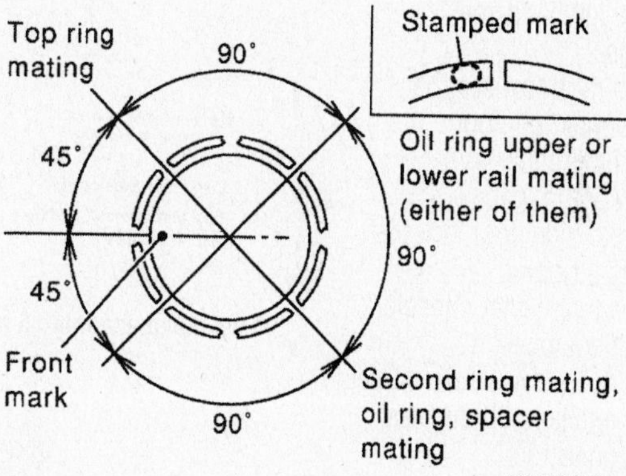

Piston ring installation

Piston and Ring

POSITIONING

FUEL SYSTEM

Fuel System Service Precautions

Safety is the most important factor when performing not only fuel system maintenance but any type of maintenance. Failure to conduct maintenance and repairs in a safe manner may result in serious personal injury or death. Maintenance and testing of the vehicle fuel system components can be accomplished safely and effectively by adhering to the following rules and guidelines.

- To avoid the possibility of fire and personal injury, always disconnect the negative battery cable unless the repair or test procedure requires that battery voltage be applied.
- Always relieve the fuel system pressure prior to disconnecting any fuel system component (injector, fuel rail, pressure regulator, etc.), fitting or fuel line connection. Exercise extreme caution whenever relieving fuel system pressure to avoid exposing skin, face and eyes to fuel spray. Please be advised that fuel under pressure may penetrate the skin or any part of the body that it contacts.
- Always place a shop towel or cloth around the fitting or connection prior to loosening to absorb any excess fuel due to spillage. Ensure that all fuel spillage (should it occur) is quickly removed from engine surfaces. Ensure that all fuel soaked cloths or towels are deposited into a suitable waste container.
- Always keep a dry chemical (Class B) fire extinguisher near the work area.
- Do not allow fuel spray or fuel vapors to come into contact with a spark or open flame.
- Always use a back-up wrench when loosening and tightening fuel line connection fittings. This will prevent unnecessary stress and torsion to fuel line piping. Always follow the proper tighten specifications.
- Always replace worn fuel fitting O-rings with new. Do not substitute fuel hose or equivalent, where fuel pipe is installed.

Relieving Fuel System Pressure

With CONSULT-II

1. Turn ignition switch **ON**.
2. Perform "FUEL PRESSURE RELEASE" in "WORK SUPPORT" mode with CONSULT-II.
3. Start engine.
4. After engine stalls, turn over the engine two or three times to release all fuel pressure.
5. Turn ignition switch **OFF**.

Without CONSULT-II

1. Remove fuel pump fuse located in IPDM E/R.
2. Start engine.
3. After engine stalls, turn over engine two or three times to release all fuel pressure.
4. Turn ignition switch **OFF**.
5. Reinstall fuel pump fuse after servicing fuel system.

Fuel Filter and Fuel Pump

REMOVAL & INSTALLATION

1. Before servicing the vehicle, refer to the precautions in the beginning of this section.
2. Relieve the fuel system pressure.
3. Remove fuel filler cap to release pressure from inside tank.
4. Remove left hand rear inner fender liner.
5. Disconnect fuel filler hose from fuel filler pipe.
6. Drain fuel tank through the fuel filler hose using a suitable hose.
7. Remove or disconnect the following:
   - Second row left hand seat
   - Third row seat
   - Second and third row seat belt buckles mounted on floor
   - Left hand center pillar trim
   - Left hand rear trim panel
   - Left hand rear side door kick plate and weather stripping
   - Second row rear center console and base, if equipped
   - Inspection hole cover under carpet by turning retainers 90 degrees
   - Electrical connectors
   - EVAP hose
   - Fuel supply hose
   - Lock ring using special tool J-46214
   - Fuel level sensor
   - Fuel filter
   - Fuel pump assembly
8. Install or connect the following:
   - Fuel pump assembly
   - Fuel filter
   - Fuel level sensor
   - Lock ring using special tool J-46214
   - Fuel supply hose
   - EVAP hose
   - Electrical connectors
   - Inspection hole cover
   - Second row rear center console and base, if equipped
   - Left hand rear side door kick plate and weather stripping
   - Left hand rear trim panel
   - Left hand center pillar trim
   - Second and third row seat belt buckles
   - Third row seat
   - Second row left hand seat
   - Fuel filler hose to fuel filler pipe
   - Left hand rear inner fender liner
9. Start the engine and check for leaks.

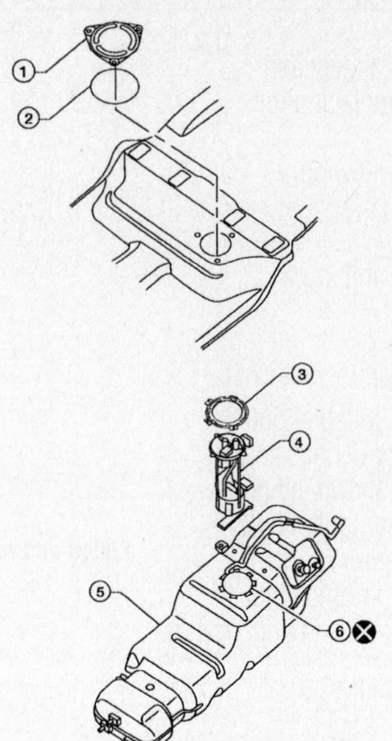

Fuel pump and related parts

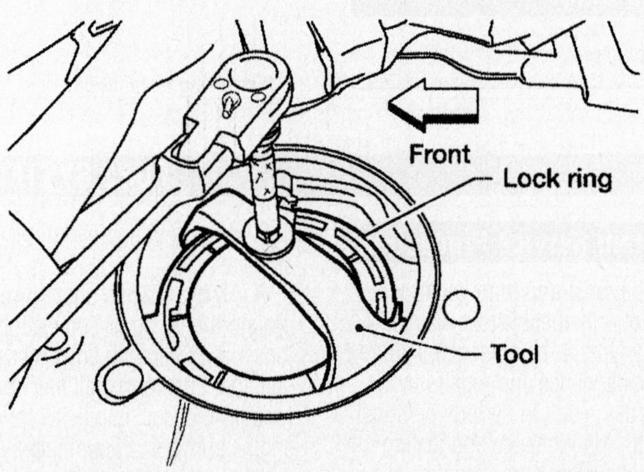

Removing fuel assembly lock ring

Fuel Injectors

REMOVAL & INSTALLATION

1. Remove engine cover
2. Relieve fuel system pressure.
3. Remove or disconnect the following:
   - Negative battery cable
   - Fuel injector harness connectors
   - Fuel hose assembly from right and left fuel rails
   - Fuel injectors with fuel rail as an assembly
   - Fuel injector from fuel rail

To install:

4. Install or connect the following:
   - New clip onto the fuel injector
   - Fuel injector to fuel rail
   - Fuel injectors and fuel rail as an assembly to the intake manifold
   - Fuel hose assembly
   - Fuel injector harness connectors
   - Negative battery cable
   - Engine cover
5. Start engine and check for leaks.

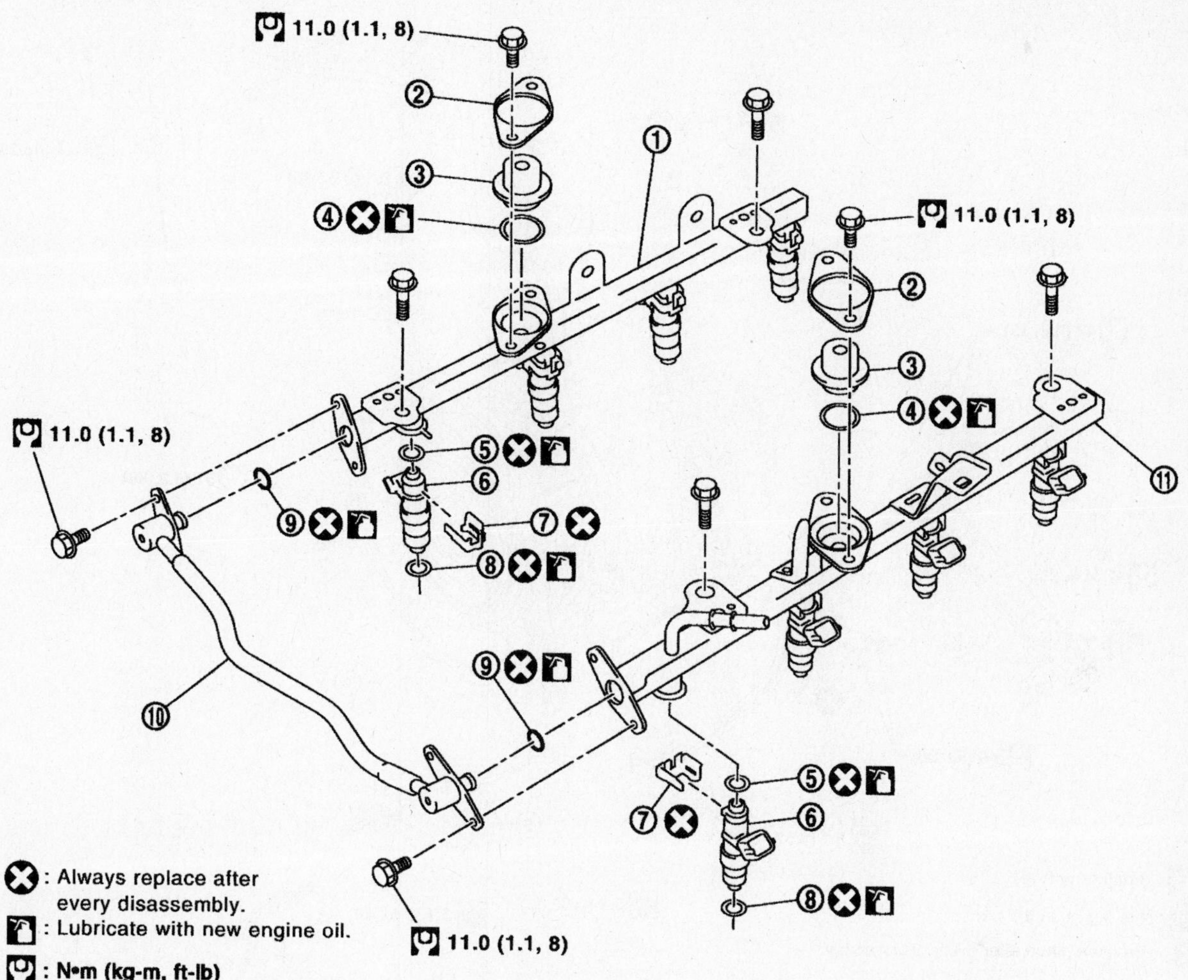

Fuel injectors and related parts

DRIVE TRAIN

Transmission Assembly

REMOVAL & INSTALLATION

2-Wheel Drive

1. Remove or disconnect the following:
   - Negative battery cable
   - Engine cover
   - Transmission fluid indicator gauge
   - Engine splash guard
   - Exhaust front pipe
   - Center muffler
   - Rear drive shaft
   - Transmission control cable
   - Crankshaft position sensor
   - Fluid cooler tube
   - Dust cover from converter housing
2. Turning crankshaft clockwise, remove the four tightening bolts for drive plate and torque converter
3. Support the transmission with a suitable jack.
4. Remove or disconnect the following:
   - Transmission cross member
   - Air breather hose
   - Transmission assembly connector
   - Fluid indicator tube from transmission assembly
   - Transmission assembly to engine bolts
   - Transmission assembly from vehicle

To install:

5. Install or connect the following:
   - Transmission assembly into vehicle
   - Transmission assembly to engine bolts tightening to 83 ft. lbs. (113 Nm)
   - Fluid indicator tube to transmission assembly
   - Transmission assembly connector
   - Air breather hose
   - Transmission cross member
6. Turning crankshaft clockwise, install the torque converter to drive plate.

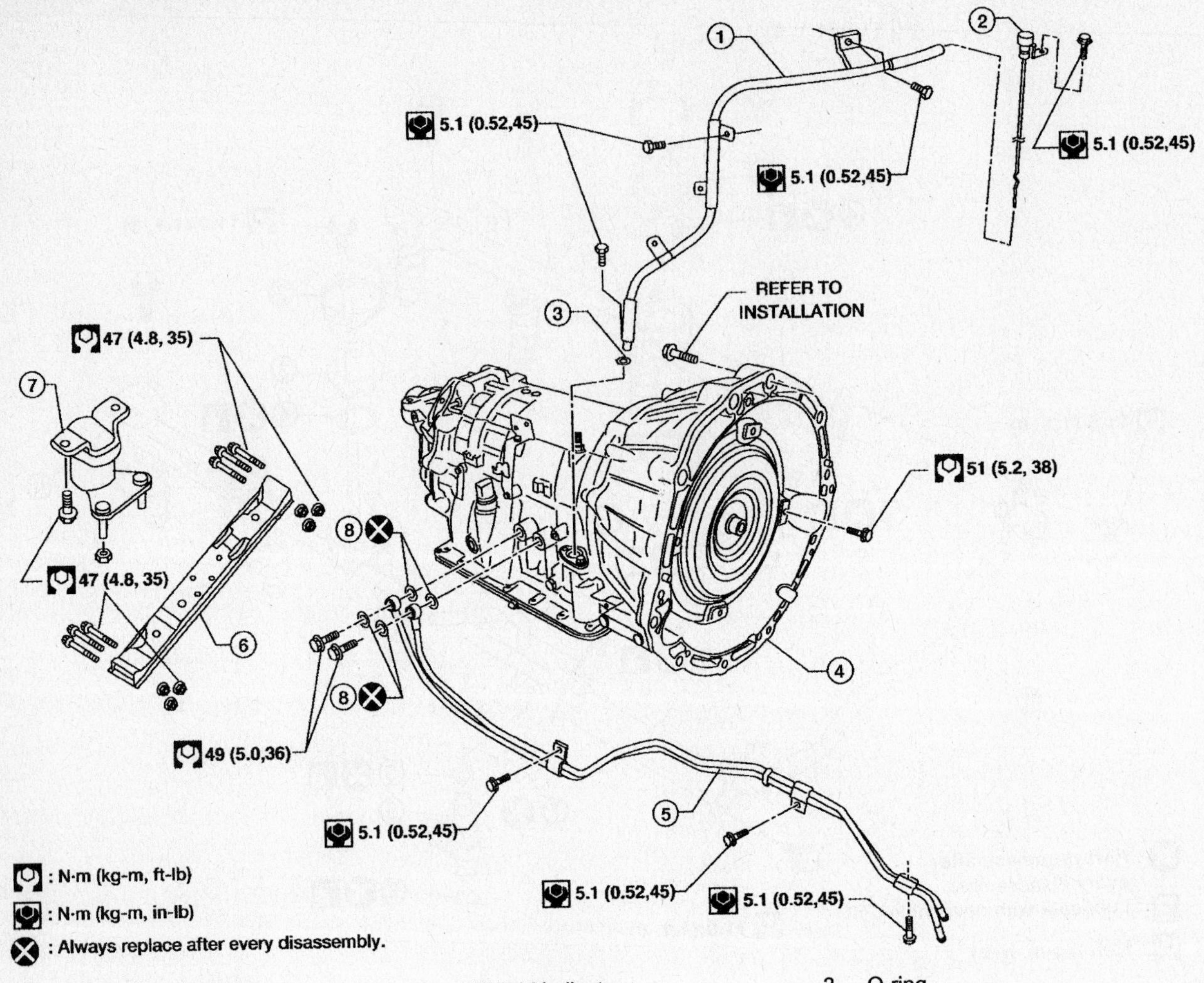

Transmission and related parts—2-wheel drive

➡**After torque converter is installed, rotate the crankshaft to ensure transmission rotates freely.**

7. Install or connect the following:
   - Dust cover for converter housing
   - Fluid cooler tube
   - Crankshaft position sensor
   - Transmission control cable
   - Rear drive shaft
   - Center muffler
   - Exhaust front pipe
   - Engine splash guard
   - Transmission fluid indicator gauge
   - Engine cover
   - Negative battery cable
8. Start engine and check for leaks.

4-Wheel Drive

1. Remove or disconnect the following:
   - Negative battery cable
   - Engine cover
   - Transmission fluid indicator gauge
   - Engine splash guard
   - Exhaust front pipe
   - Center muffler
   - Drive shaft
   - Transmission control cable
   - Crankshaft position sensor
   - Fluid cooler tube
   - Dust housing for torque converter
2. Turning the crankshaft clockwise, remove the four tightening bolts for drive plate and torque converter.
3. Support the transmission assembly with a suitable jack.
4. Remove transmission cross member.
5. Tilt the transmission slightly to keep clearance between the body and the transmission assembly, then disconnect the air breather hose.
6. Remove or disconnect the following:
   - Transmission assembly connector and transfer case connector
   - Fluid indicator pipe
   - Transmission assembly to engine bolts
   - Transmission assembly, with transfer case attached, from vehicle
   - Transmission assembly from transfer case

To install:

7. Install or connect the following:
   - Transfer case to transmission assembly
   - Transmission assembly into vehicle
   - Transmission assembly to engine bolts tightening to 83 ft. lbs. (113 Nm)
8. With the transmission slightly tilted

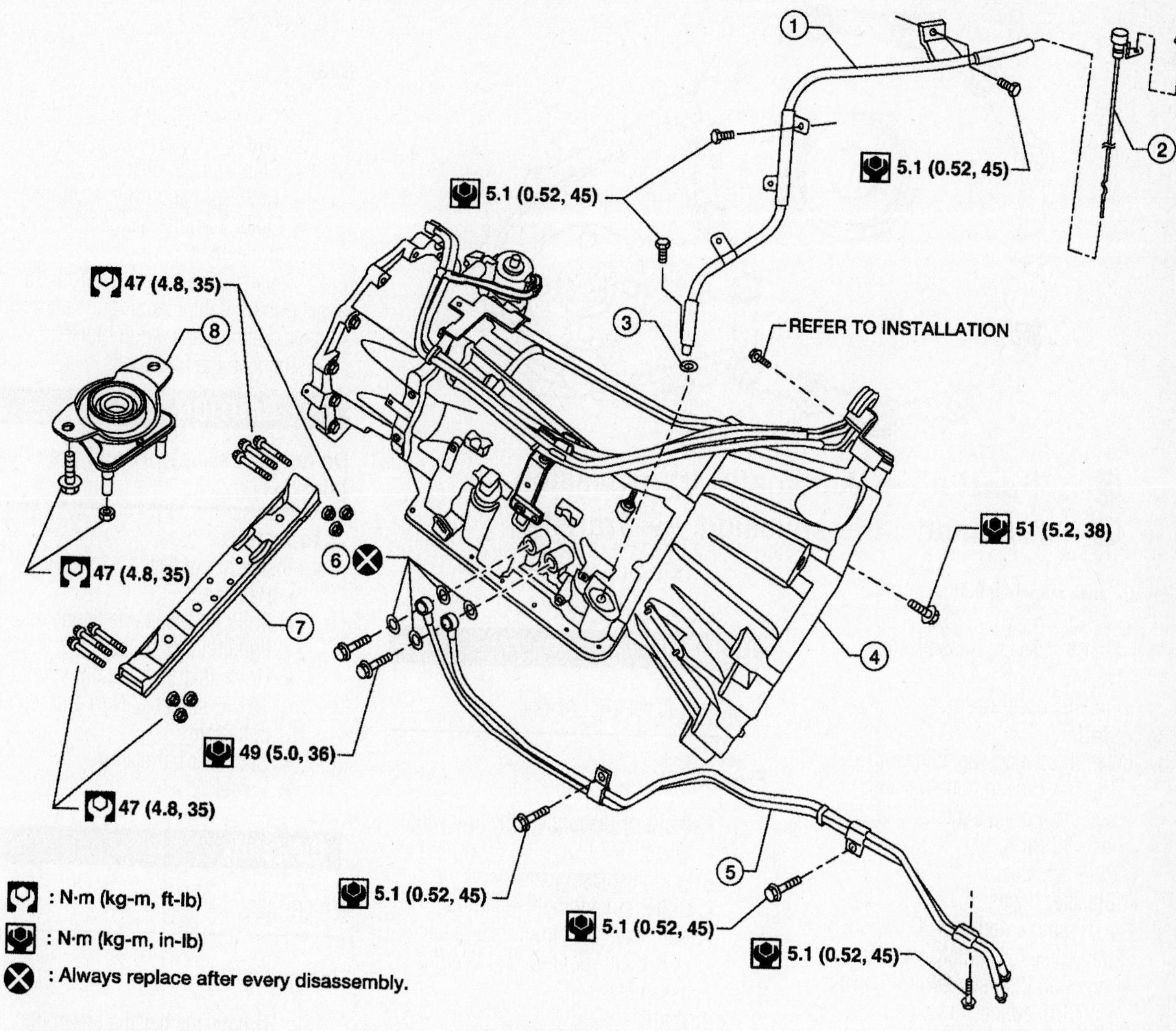

Transmission and related parts—with 4-wheel drive

to allow clearance between body and transmission, connect the air breather hose.

9. Install the transmission cross member.

10. Turning crankshaft clockwise, install the torque converter to drive plate.

➡After torque converter is installed, rotate the crankshaft to ensure transmission rotates freely.

11. Install or connect the following:
- Dust housing for torque converter
- Fluid cooler tube
- Crankshaft position sensor
- Transmission control cable
- Drive shaft
- Center muffler
- Front exhaust pipe
- Engine splash guard
- Transmission fluid indicator gauge
- Engine cover
- Negative battery cable

12. Start engine and check for leaks.

Transfer Case Assembly

REMOVAL & INSTALLATION

1. Remove or disconnect the following:
- Transmission splash guard
- Center exhaust pipe and muffler
- Front and rear drive shafts

➡Plug rear oil seal after removing rear drive shaft.

- Transmission assembly mounting bolts

2. Support the transmission assembly with a suitable jack and remove the crossmember.

3. Remove or disconnect the following:
- ATP switch, neutral 4LO switch, wait detection switch, transfer motor and transfer control device electrical connectors
- Breather hoses
- Shift actuator from the extension housing

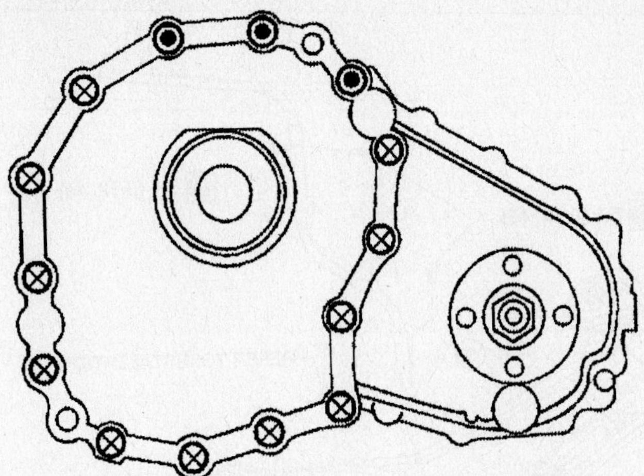

Transfer case mounting bolt locations

- Transfer case to transmission assembly bolts
- Transfer case assembly

To install:

4. Install or connect the following:
   - Transfer case to transmission assembly bolts tightening to 26 ft. lbs. (36 Nm)
   - Shift actuator
   - Breather hoses
   - ATP switch, neutral 4LO switch, wait detection switch, transfer motor and transfer control device electrical connectors
   - Support crossmember
   - Transmission mounting bolts
   - Drive shafts
   - Muffler and center exhaust pipe
   - Transmission splash guard

Halfshaft

REMOVAL & INSTALLATION

Front

1. Remove or disconnect the following:
   - Wheel
   - Engine splash guard
   - Cotter pin and half shaft nut
   - Half shaft from front differential
   - Half shaft from hub and bearing assembly

To install:

2. Install or connect the following:
   - Half shaft into hub
   - Half shaft into front differential
   - Half shaft nut and tighten to 101 ft. lbs. and replace cotter pin
   - Engine splash guard
   - Wheel

Rear

1. Remove or disconnect the following:
   - Wheel
   - Stabilizer bar clamp
   - Cotter pin and drive shaft nut
   - Bolts from the inside flange of the drive shaft
2. Separate the drive shaft from the wheel hub by lightly tapping the end with suitable hammer and wood block.
3. Remove the half shaft.

⁂ CAUTION

Do not excessively extend the slide joint.

To install:

4. Install or connect the following:
   - Half shaft
   - Bolts for the inside flange and tighten to 87 ft. lbs. (118 Nm)
   - Drive shaft nut and tighten nut to 101 ft. lbs. (137 Nm) and replace cotter pin
   - Stabilizer bar clamp
   - Wheel

CV-Joints

OVERHAUL

Inner

1. Remove the halfshaft from the vehicle.
2. Mount halfshaft in a vise.
3. Remove the dust boot bands.
4. Remove the stopper ring with a flat-bladed screwdriver or suitable tool.

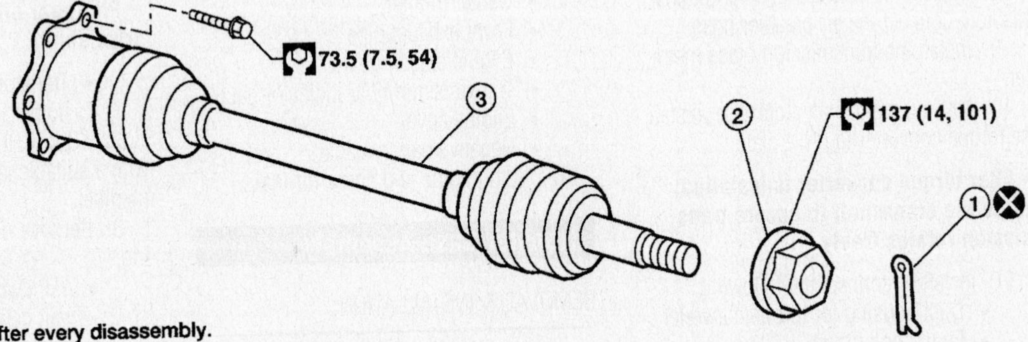

Front halfshaft

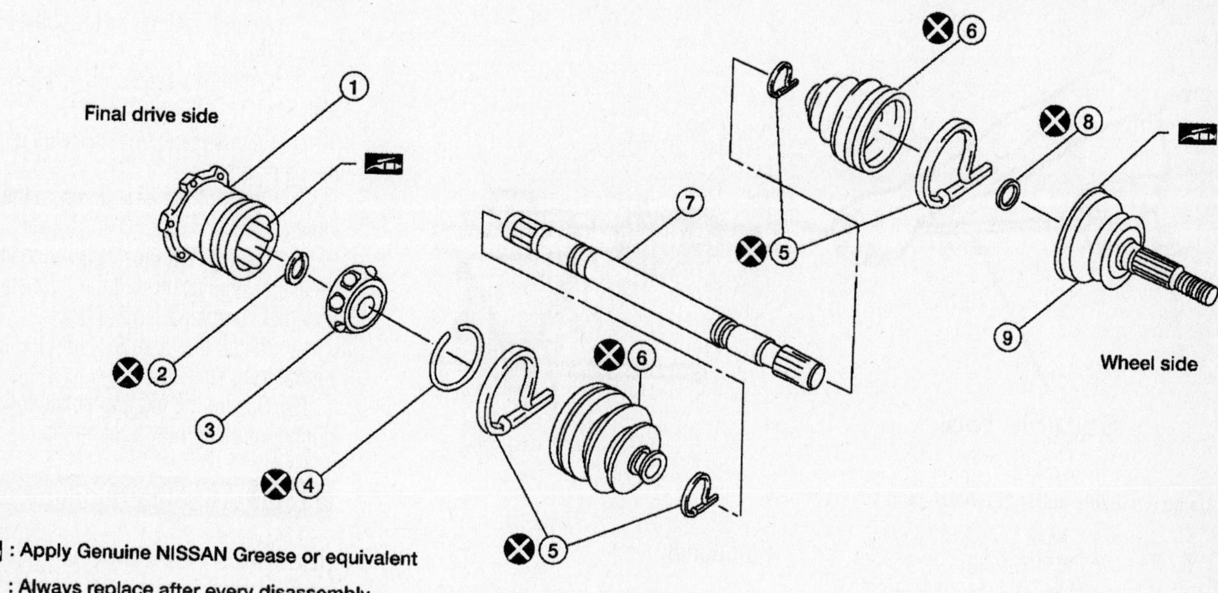

1. Housing
2. Snap ring
3. Ball cage, steel ball, iiner race assembly
4. Stopper ring
5. Boot band
6. Boot
7. Shaft
8. Circlip
9. Joint sub-assembly

Front halfshaft—exploded view

67170-ARMA-G43

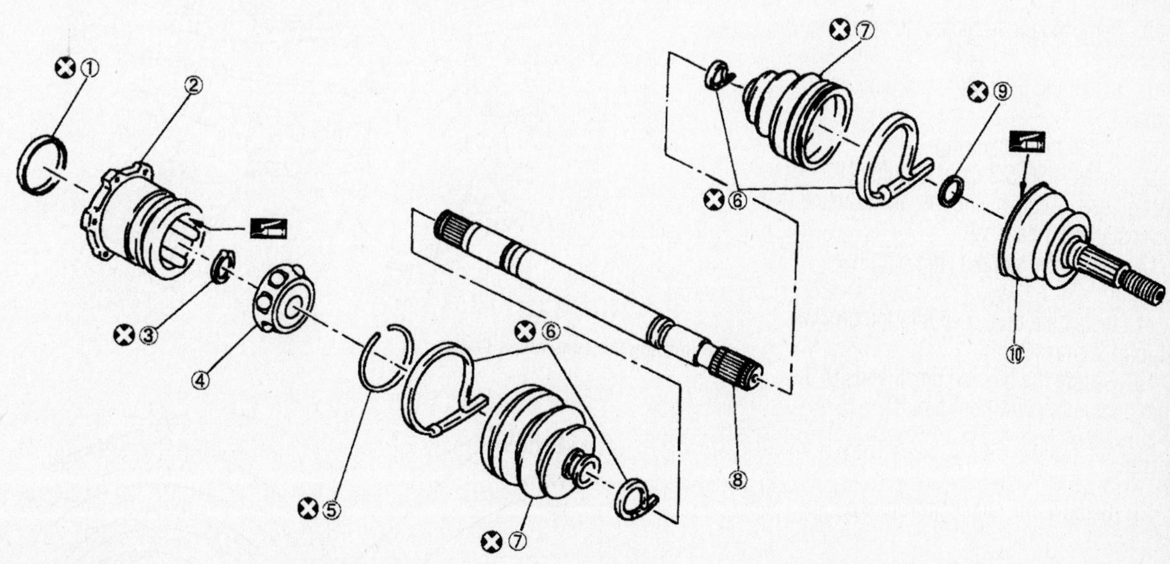

1. Plug
2. Housing
3. Snap ring
4. Ball cage, steel ball, liner race assembly
5. Stopper ring
6. Boot band
7. Boot
8. Shaft
9. Circlip
10. Joint sub-assembly

Rear halfshaft—exploded view

67170-ARMA-G44

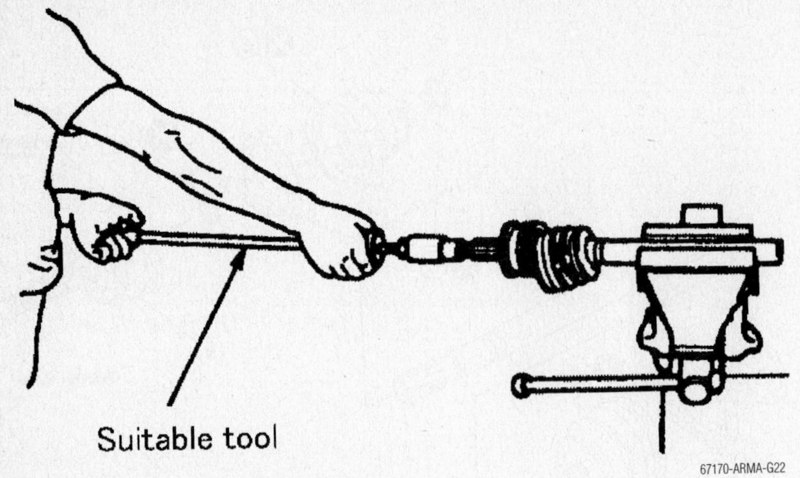

Using a suitable puller to remove joint sub-assembly.

5. Remove the snap ring.
6. Disassemble the cage, ball and inner race assembly and dust boot for cleaning and inspection.

To install:

➡Discard old dust boot, dust boot bands and snap ring and use new ones for assembly.

7. Wrap the serrated part of the halfshaft with tape.
8. Install new dust boot and band onto halfshaft.
9. Remove tape from serrated part of halfshaft.
10. Install the cage, ball and inner race assembly.
11. Install new snap ring.
12. Insert 4.50-5.3 oz of genuine NISSAN grease or equivalent onto the housing and install onto halfshaft.
13. Install the stopper ring onto the housing.
14. Install the dust boot into the grooves on joint sub-assembly.
15. Secure the big and small ends of the dust boot using new boot bands.

Outer

1. Remove the halfshaft from the vehicle.
2. Mount halfshaft in a vise.
3. Remove the dust boot bands and dust boot from joint sub-assembly.
4. Insert a suitable puller into the threaded part of the halfshaft. Pull the joint sub-assembly off of the halfshaft as shown in figure.
5. Remove dust boot and circlip from halfshaft for cleaning and inspection.

To install:

➡Discard old dust boot, dust boot bands and circlip and use new ones for assembly.

6. Insert genuine NISSAN grease or equivalent into the joint sub-assembly until grease oozes from the ball groove and serration hole.
7. Wrap the serrated part of the halfshaft with tape.
8. Install new dust boot and band onto halfshaft.
9. Remove tape from serrated part of the halfshaft.
10. Press-fit the new circlip to the halfshaft.
11. Insert 5.1-5.8 oz of genuine NISSAN grease or equivalent into the joint sub-assembly and large end of boot.
12. Install the dust boot into the grooves on the joint sub-assembly.
13. Secure the big and small ends of the dust boot using new boot bands.

Front Differential Pinion Seal

REMOVAL & INSTALLATION

1. Remove or disconnect the following:
   - Front drive shaft
   - Halfshafts
2. Measure and record the pinion bearing preload using special tool J-25765-A.

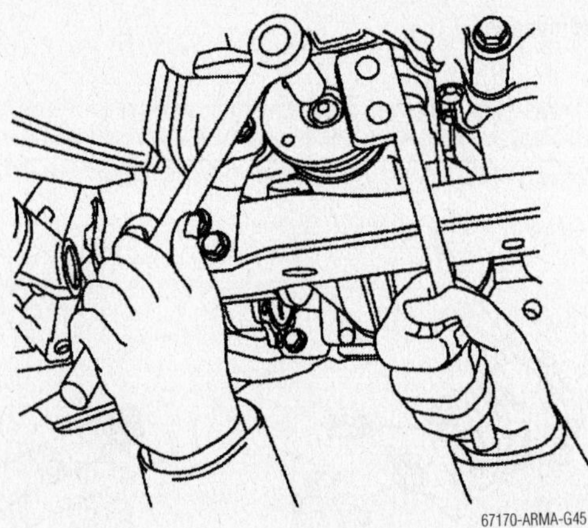

Removing the companion flange

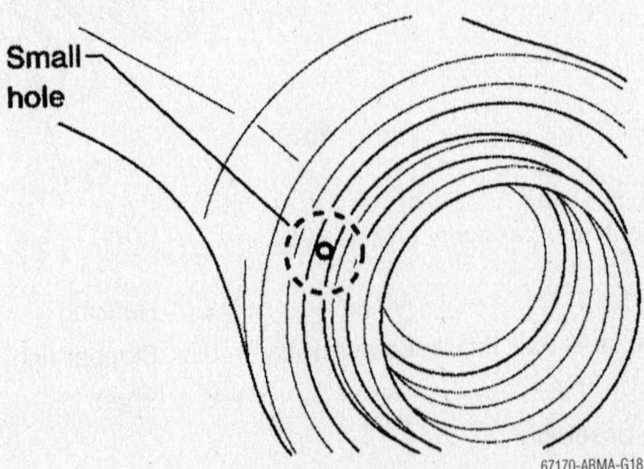

Small hole in casing

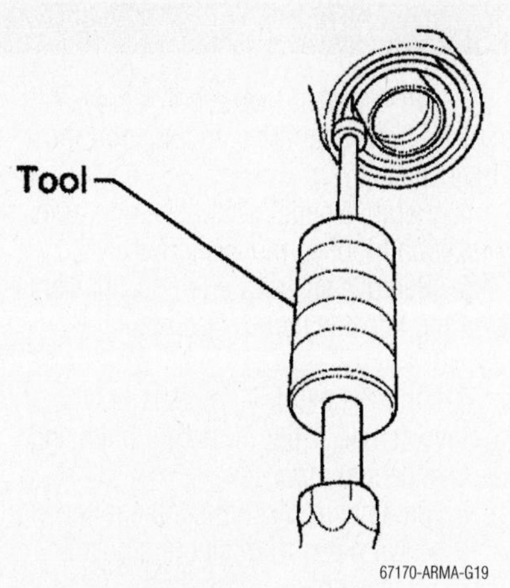

Removing pinion seal

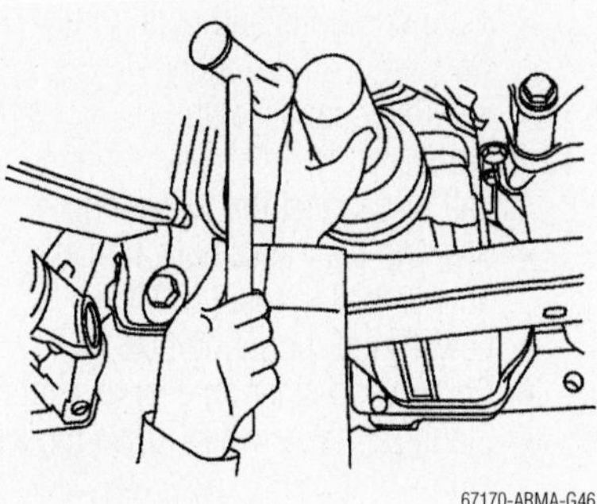

Seal installation

3. Loosen the pinion nut while holding the companion flange using special tool J-44195.
4. Remove the companion flange using a suitable tool.
5. Using a punch or drill, place a small hole in the case.
6. Remove the seal using special tool SP8P.

To install:

7. Press front seal into carrier using a suitable tool.
8. Install companion flange and new pinion nut. Tighten pinion nut until there is no end play and until recorded pinion bearing preload is met plus an additional 5 inch lbs. (0.5 Nm).
9. Install or connect the following:
   - Halfshafts
   - Front drive shaft

Rear Differential Pinion Seal

REMOVAL & INSTALLATION

1. Remove the rear drive shaft.
2. Measure and record the total preload.
3. Matchmark the drive pinion to position 'B' on the companion flange.
4. Remove the drive pinion nut using suitable tool.
5. Remove the companion flange using suitable tool.
6. Remove the rear pinion seal using special tool J-34286.

To install:

7. Press the rear pinion seal into the carrier using suitable tool.
8. Align the matchmark on the companion flange to the drive pinion and install the companion flange.
9. Lubricate the drive pinion threads and seating surfaces of the drive pinion nut with grease.

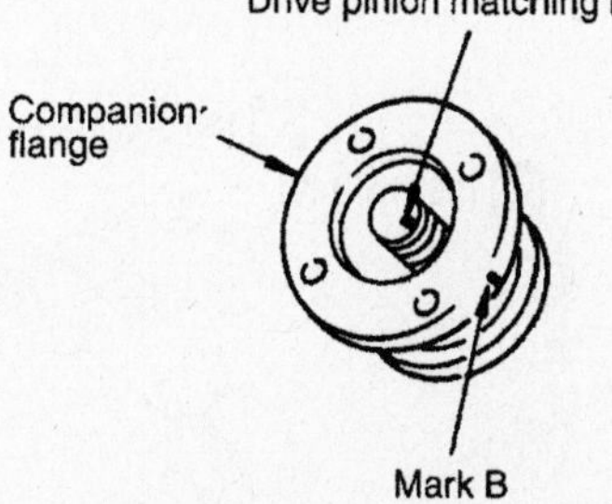

Companion flange marking

10. Using a new drive pinion nut, tighten to 124-274 ft. lbs. (167-372 Nm).

➡Final torque is determined when adjusting total preload using special tool J-25765-A.

11. Install rear drive shaft.

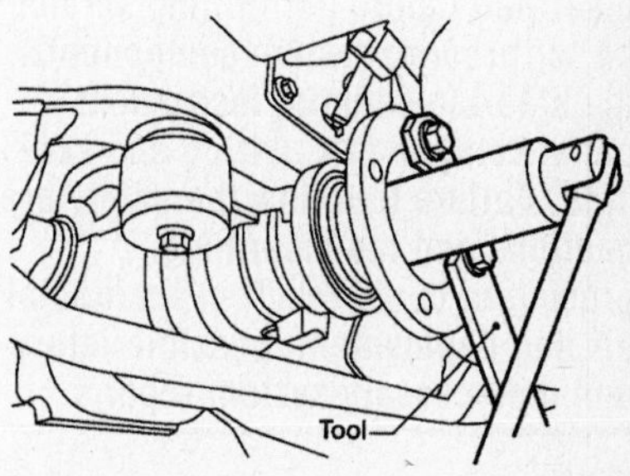

Loosening the flange nut

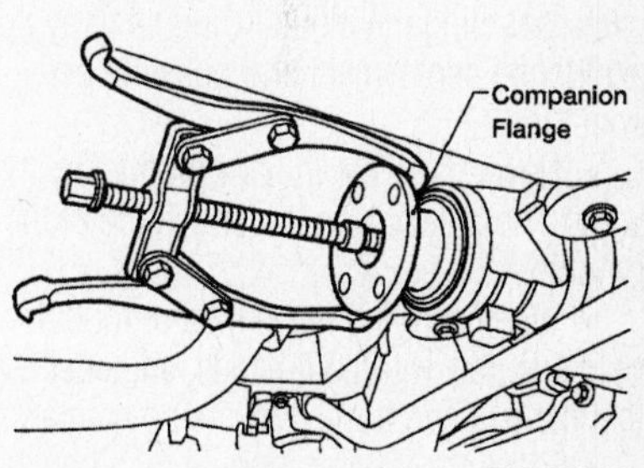

Removing the companion flange

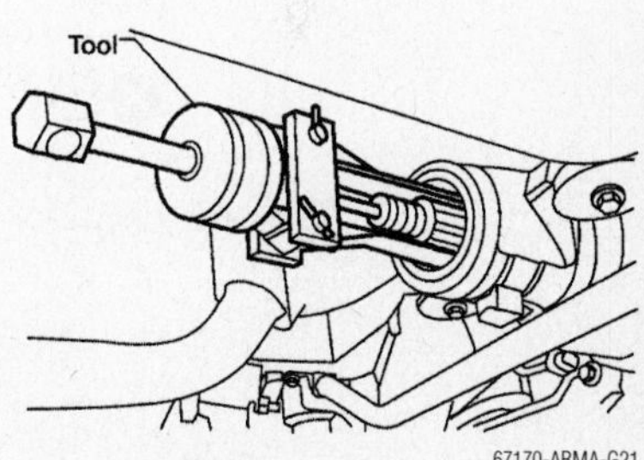

Removing the pinion seal

STEERING AND SUSPENSION

Air Bag

** CAUTION

Some vehicles are equipped with an air bag system. The system must be disarmed before performing service on, or around, system components, the steering column, instrument panel components, wiring and sensors. Failure to follow the safety precautions and the disarming procedure could result in accidental air bag deployment, possible injury and unnecessary system repairs.

PRECAUTIONS

Several precautions must be observed when handling the inflator module to avoid accidental deployment and possible personal injury.

- Never carry the inflator module by the wires or connector on the underside of the module.
- When carrying a live inflator module, hold securely with both hands, and ensure that the bag and trim cover are pointed away.
- Place the inflator module on a bench or other surface with the bag and trim cover facing up.
- With the inflator module on the bench, never place anything on or close to the module which may be thrown in the event of an accidental deployment.

DISARMING THE SYSTEM

1. Before servicing the vehicle, refer to the precautions in the beginning of this section.
2. Disconnect both battery cables.
3. Wait at least 3 minutes before working on the vehicle. The air bag system is designed to retain enough power to deploy the air bag for a short time after the battery has been disconnected.
4. After repairs are complete, connect the negative battery cable. Turn the ignition switch to the **ON** position and check the air bag warning light blinks for proper operation.

Power Steering Gear

REMOVAL & INSTALLATION

1. Ensure the wheels are in the straight-ahead position.
2. Remove or disconnect the following:
   - Wheels
   - Engine splash guard
3. On 4-wheel drive models only, remove front final drive and support the drive shafts.
4. Remove cotter pin at steering outer socket and loosen mounting nut.
5. Remove steering outer socket from steering knuckle using special tool J-25730-A.
6. On 2-wheel drive models only, remove stabilizer bar mounting bolts and secure the stabilizer bar.
7. Remove or disconnect the following:
   - Oil pipes from steering gear assembly
   - Lower joint mounting bolt from lower shaft
   - Mounting bolts and nuts from steering gear assembly
   - Steering gear assembly

To install:

8. Install or connect the following:
   - Steering gear assembly, tighten nuts to 133 ft. lbs. (180 Nm)
   - Lower joint mounting bolt
   - Oil pipes to steering gear assembly
   - Stabilizer bar, 2 wheel-drive models only
   - Steering outer socket to steering knuckle, tighten nut to 63 ft. lbs. (86 Nm)
   - Front final drive, 4-wheel drive models only

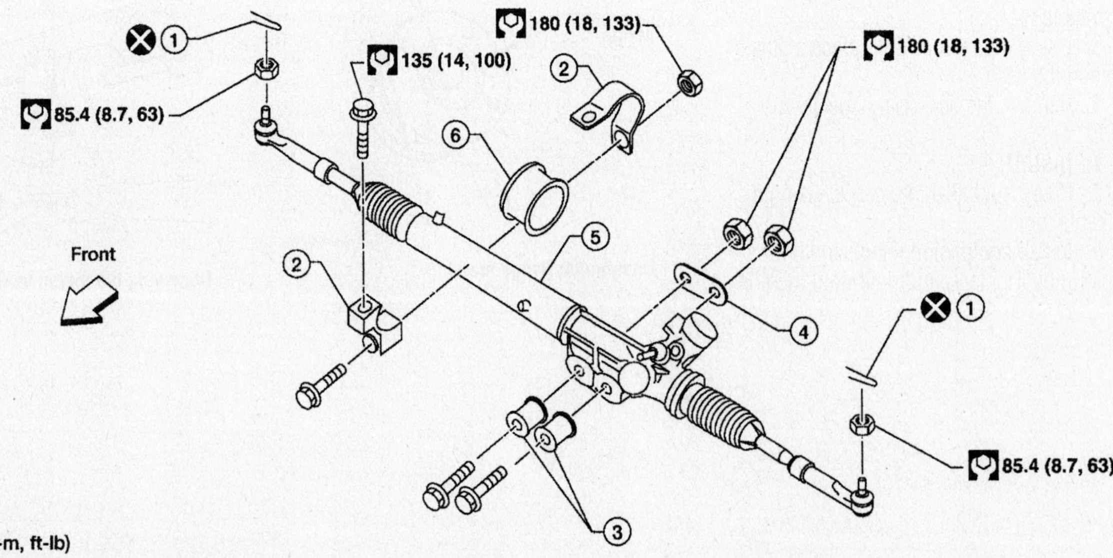

Steering gear and related parts

- Engine splash guard
- Wheels

9. Check the wheel alignment and adjust as necessary.

Shock Absorber

REMOVAL & INSTALLATION

Front

1. Remove or disconnect the following:
   - Wheel
   - Lower shock absorber bolt
   - Upper shock absorber bolts
   - Coil spring and shock absorber assembly
2. Secure the shock absorber in a vice and loosen (without removing) the piston rod lock nut.
3. Install a spring compressor and tighten until the shock absorber mounting insulator can be turned by hand.
4. Remove piston rod lock nut and remove shock absorber.

To install:

5. Install upper mounting insulator in line with the lower shock absorber mount and step in shock absorber lower seat as shown in figure.
6. Tighten the new piston rod lock nut to 40 ft. lbs. (54 Nm).
7. Install or connect the following:
   - Coil spring and shock absorber assembly
   - Upper shock absorber bolts and tighten to 22 ft. lbs (30 Nm)
   - Lower shock absorber bolt and tighten to 99 ft. lbs. (134 Nm)
   - Wheel
8. Check wheel alignment and adjust as necessary.

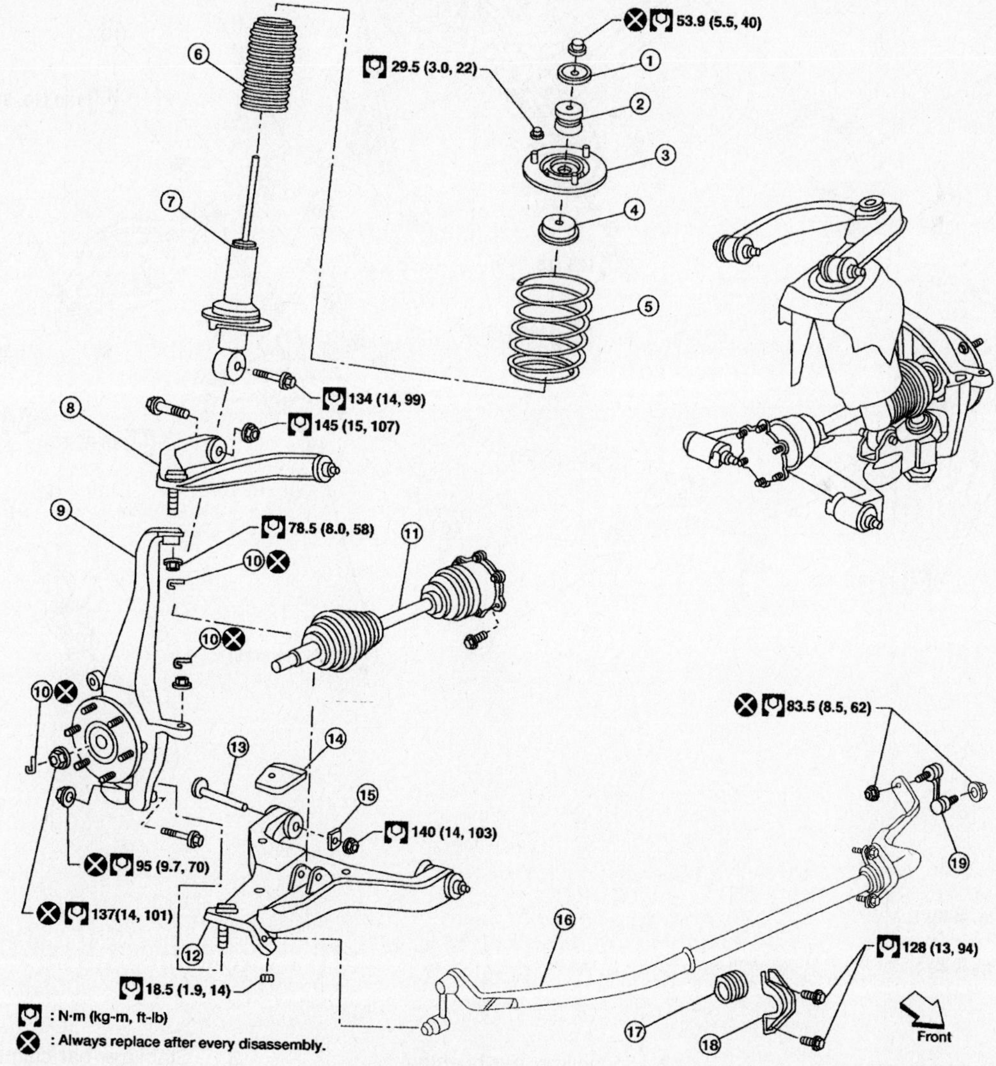

1. Washer
2. Shock absorber bushing
3. Shock absorber mounting insulator
4. Upper seat
5. Coil spring
6. Dust cover
7. Shock absorber
8. Upper link
9. Steering knuckle
10. Cotter pin
11. Drive shaft
12. Lower link
13. Cam bolt
14. Jounce bumper
15. Cam washer
16. Stabilizer bar
17. Stabilizer bar bushing
18. Stabilizer bar mounting bracket
19. Connecting rod

67170-ARMA-G13

Front suspension

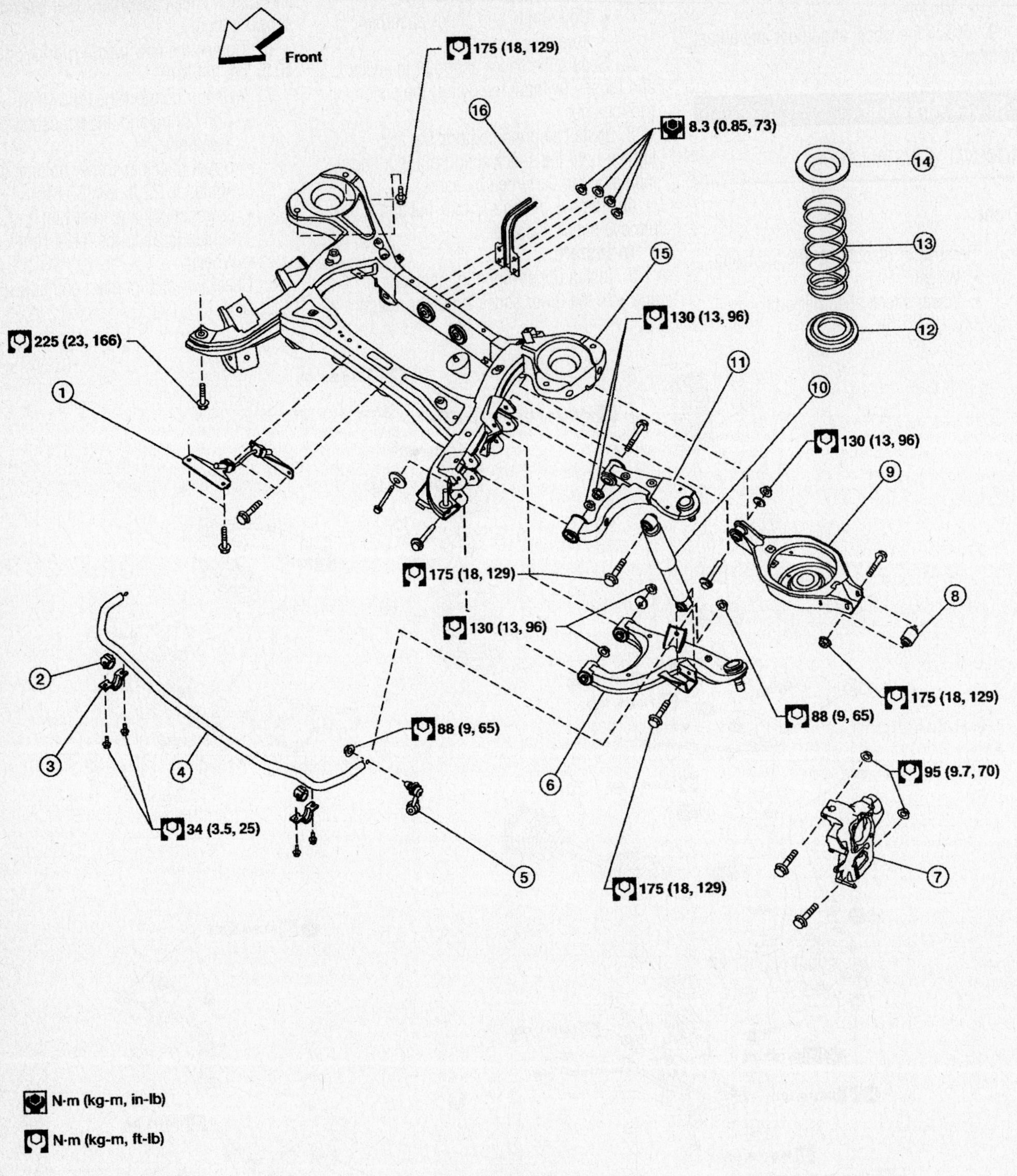

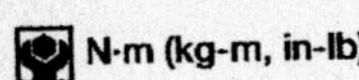

N·m (kg-m, in-lb)

N·m (kg-m, ft-lb)

1. Seat belt latch anchor
2. Stabilizer bar bushing
3. Stabilizer bar clamp
4. Stabilizer bar
5. Connecting rod
6. Front lower link
7. Wheel hub and spindle assembly
8. Bushing
9. Rear lower link
10. Shock absorber
11. Suspension arm
12. Lower rubber seat
13. Coil spring
14. Upper rubber seat
15. Rear suspension member
16. Spare tire bracket

67170-ARMA-G50

Standard rear suspension

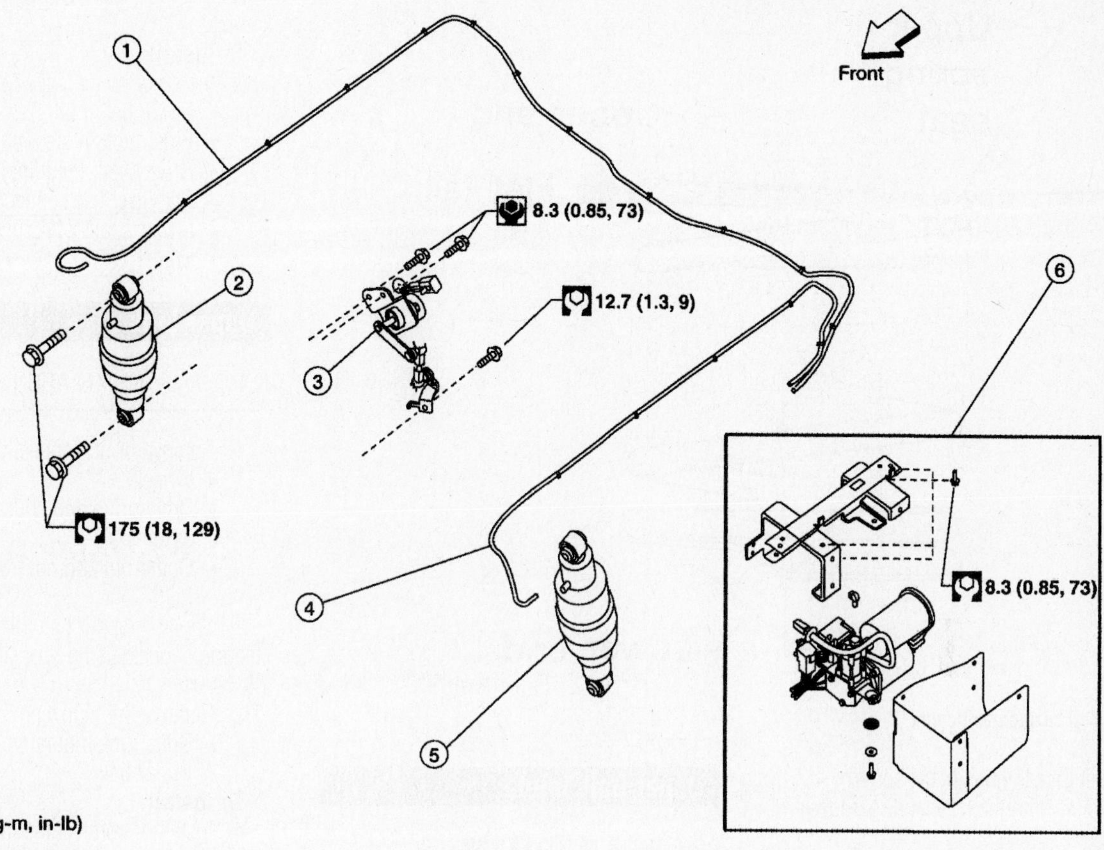

Rear load leveling air suspension

Rear

1. Remove the rear wheel.
2. Release the air pressure from the rear load leveling air suspension system using the CONSULT-II "EXHAUST SOLENOID" active test.
3. Remove or disconnect the following:
   - Rear fender protector
   - Rear load leveling air suspension hose from the shock absorber
   - Shock absorber upper and lower end bolts
   - Shock absorber

To install:

4. Install or connect the following:
   - Shock absorber and tighten end bolts to 129 ft. lbs. (175 Nm)
   - Rear load leveling air suspension hose
   - Rear fender protector
   - Rear wheel

Coil Spring

REMOVAL & INSTALLATION

Front

1. Remove or disconnect the following:
   - Wheel
   - Lower shock absorber bolt
   - Upper shock absorber bolts
   - Coil spring and shock absorber assembly
2. Secure the shock absorber in a vice and loosen (without removing) the piston rod lock nut.
3. Install a spring compressor and tighten until the shock absorber mounting insulator can be turned by hand.
4. Remove piston rod lock nut and remove shock absorber from the coil spring.

To install:

5. Install upper mounting insulator in line with the lower shock absorber mount and step in shock absorber lower seat as shown in figure.
6. Tighten the new piston rod lock nut to 40 ft. lbs. (54 Nm).
7. Install or connect the following:
   - Coil spring and shock absorber assembly
   - Upper shock absorber bolts and tighten to 22 ft. lbs (30 Nm)
   - Lower shock absorber bolt and tighten to 99 ft. lbs. (134 Nm)
   - Wheel
8. Check wheel alignment and adjust as necessary.

Rear

1. Remove the rear wheel.
2. Release the air pressure from the rear load leveling air suspension system using the CONSULT-II "EXHAUST SOLENOID" active test.

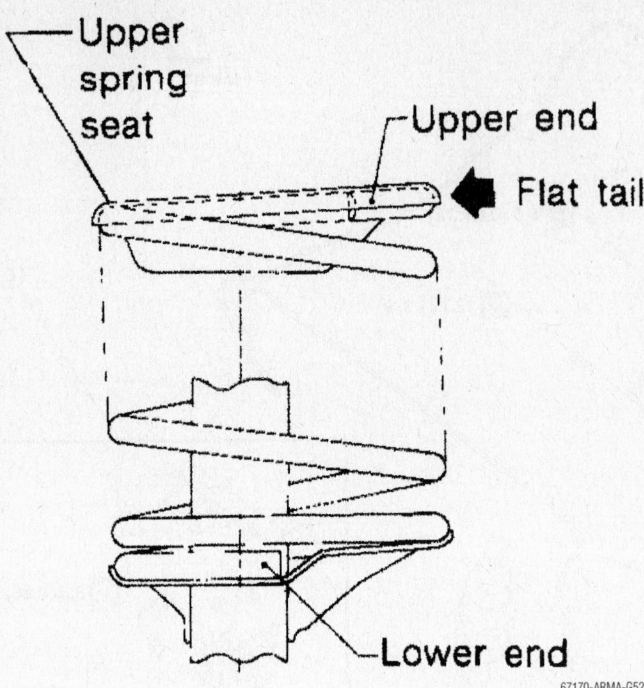

Front coil spring positioning

3. Remove the height sensor arm bracket bolt from the left-hand rear lower link.
4. Place a suitable jack under the rear lower link and relieve the coil spring tension.
5. Loosen the rear lower link adjusting bolt and nut connected to the rear suspension member.
6. Remove the rear lower link bolt and nut from the knuckle.
7. Slowly lower the jack to relieve the coil spring tension.
8. Remove the coil spring.

To install:

9. Install or connect the following:
   - Coil spring

➡When installing the rubber seats for the coil spring, ensure the embossed arrows point outward towards the wheel.

   - Rear lower link bolt to knuckle and tighten nut to 70 ft. lbs. (95 Nm)
   - Rear lower link adjusting bolt to rear suspension member and tighten nut to 101 ft. lbs. (137 Nm)
   - Height sensor arm bracket bolt to left-head rear lower link and tighten to 9 ft. lbs. (12 Nm)
   - Rear wheel

Upper Ball Joint

REMOVAL & INSTALLATION

1. Remove or disconnect the following:
   - Wheel
   - Coil spring and shock absorber assembly
   - Cotter pin and nut from upper ball joint
2. Separate upper ball joint from steering knuckle using special tool J-24319-01

To install:

3. Install or connect the following:
   - Upper ball joint
   - New cotter pin and tighten nut to 58 ft. lbs. (79 Nm)
   - Coil spring and shock absorber assembly
   - Wheel

Lower Ball Joint

REMOVAL & INSTALLATION

1. Remove or disconnect the following:
   - Wheel
   - Lower shock absorber bolt
   - Stabilizer bar connecting rod
   - Drive shaft, if equipped
   - Pinch bolt from steering knuckle
2. Separate lower ball joint from steering knuckle

To install:

3. Install or connect the following:
   - Lower ball joint
   - Pinch bolt to steering knuckle
   - Drive shaft, if equipped
   - Stabilizer bar connecting rod
   - Lower shock absorber bolt
   - Wheel

Upper Link

REMOVAL & INSTALLATION

1. Remove or disconnect the following:
   - Wheel
   - Coil spring and shock absorber assembly
   - Cotter pin and nut from upper ball joint
2. Separate upper ball joint stud from steering knuckle using special tool J-24319-01.
3. Remove the following:
   - Upper link mounting bolts
   - Upper link

To install:

4. Install or connect the following:
   - Upper link and tighten bolts to 107 ft. lbs. (145 Nm)
   - Upper ball joint with new cotter pin and tighten nut to 58 ft. lbs. (79 Nm)
   - Coil spring and shock absorber assembly
   - Wheel

Lower Control Arm

REMOVAL & INSTALLATION

1. Remove or disconnect the following:
   - Wheel
   - Lower shock absorber bolt
   - Stabilizer bar connecting rod
   - Drive shaft, if equipped
   - Pinch bolt from steering knuckle
2. Separate lower ball joint from steering knuckle.
3. Remove the following:
   - Lower link adjusting bolts
   - Lower link

To install:

4. Install or connect the following:
   - Lower link and tighten adjusting bolts to 98 ft. lbs. (133 Nm)
   - Lower ball joint

- Pinch bolt
- Drive shaft, if equipped
- Stabilizer bar connected rod
- Lower shock absorber bolt
- Wheel

Wheel Bearings

REMOVAL & INSTALLATION

Front

1. Remove or disconnect the following:
   - Wheel
   - Engine splash guard
   - Brake caliper without disconnecting the hydraulic lines, and reposition aside with wire
2. Matchmark the brake rotor to the wheel hub and remove the brake rotor.
3. Remove or disconnect the following:
   - Cotter pin and lock nut from drive shaft
   - Drive shaft from wheel hub and bearing assembly
   - ABS sensor
   - Wheel hub and bearing assembly bolts
   - Wheel hub and bearing assembly

To install:

4. Install or connect the following:
   - Wheel hub and bearing assembly, using new bolts and tighten to 155 ft. lbs. (210 Nm)
   - ABS sensor
   - Drive shaft to wheel hub and bearing assembly
   - Cotter pin and lock nut and tighten to 101 ft. lbs. (137 Nm)
   - Brake rotor
   - Brake caliper
   - Engine splash guard
   - Wheel

Rear

1. Remove or disconnect the following:
   - Wheel
   - Brake caliper without disconnecting the hydraulic lines, and reposition aside with wire
   - Brake rotor
   - Cotter pin and nut from drive shaft
   - Drive shaft
   - Wheel hub and bearing assembly bolts

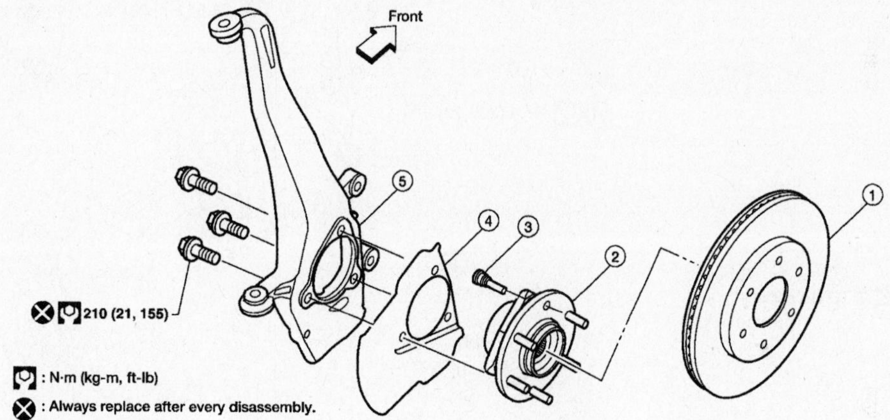

Front hub/bearing assembly

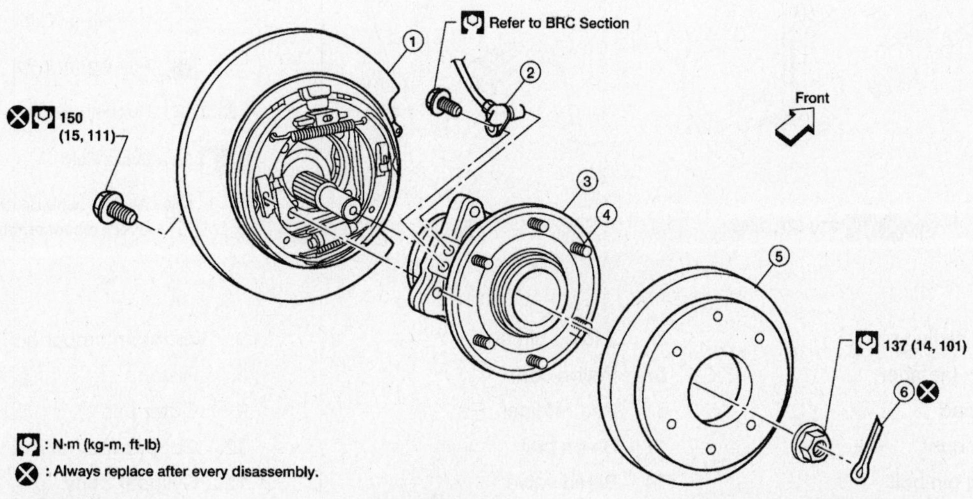

Rear hub/bearing assembly

2. Pulling out the wheel hub and bearing assembly slightly, remove the ABS sensor.

3. Remove the wheel hub and bearing assembly.

To install:

4. Install or connect the following:
   - ABS sensor
   - Wheel hub and bearing assembly, using new bolts and tighten to 111 ft. lbs. (150 Nm)
   - Drive shaft
   - Lock nut and tighten to 101 ft. lbs. (137 Nm) and new cotter pin
   - Brake rotor
   - Brake caliper
   - Wheel

BRAKES

Brake Caliper

REMOVAL & INSTALLATION

Front

1. Drain brake fluid as necessary.
2. Remove or disconnect the following:
   - Wheel
   - Union bolt
   - Caliper-to-torque member slide pins, or remove the caliper and torque member as an assembly.
   - Brake caliper

To install:

3. Install or connect the following:
   - Brake caliper, tighten torque member bolts to 155 ft. lbs. (210 Nm); the caliper slide pins to 20 ft. lbs. (27 Nm)
   - Union bolt and tighten to 13 ft. lbs. (18 Nm)
4. Fill the master cylinder and bleed the brake system.
5. Install the wheels.

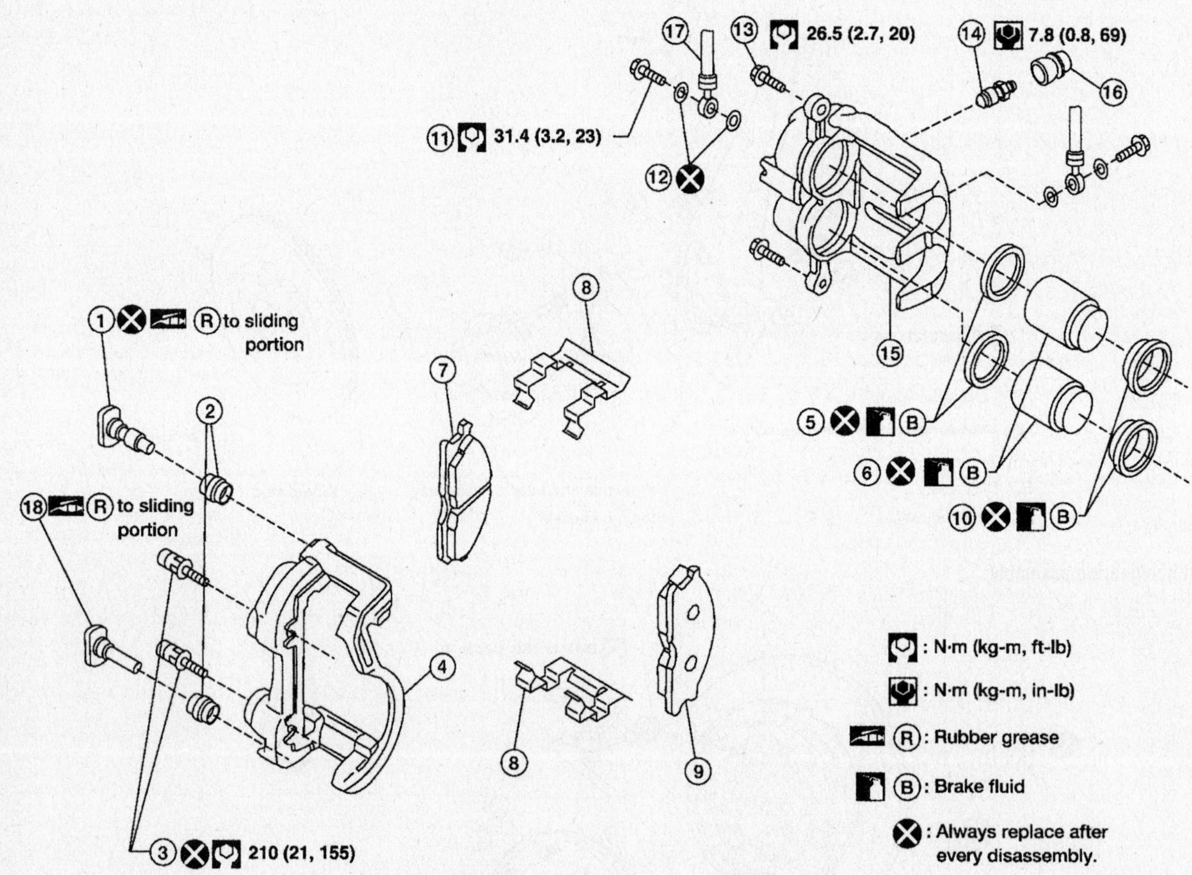

1.	Upper sliding pin	2.	Sliding pin boot	3.	Torque member bolt
4.	Torque member	5.	Piston seal	6.	Piston
7.	Inner pad	8.	Pad retainer	9.	Outer pad
10.	Piston boot	11.	Union bolt	12.	Copper washer
13.	Sliding pin bolt	14.	Bleed valve	15.	Cylinder body
16.	Cap	17.	Brake hose	18.	Lower sliding pin

67170-ARMA-G55

Front brake components

Rear

1. Drain brake fluid as necessary.
2. Remove or disconnect the following:
   - Wheel
   - Union bolt
   - Mounting bolts
   - Brake caliper assembly

To install:

3. Install or connect the following:
   - Brake caliper assembly and tighten mounting bolts to 23ft. lbs. (44 Nm)
   - Union bolt and tighten to 13 ft. lbs. (18 Nm)
4. Fill the master cylinder and bleed the brake system.
5. Install the wheels.

Disc Brake Pads

REMOVAL& INSTALLATION

Front

1. Remove the wheel.
2. Remove lower sliding pin bolt.

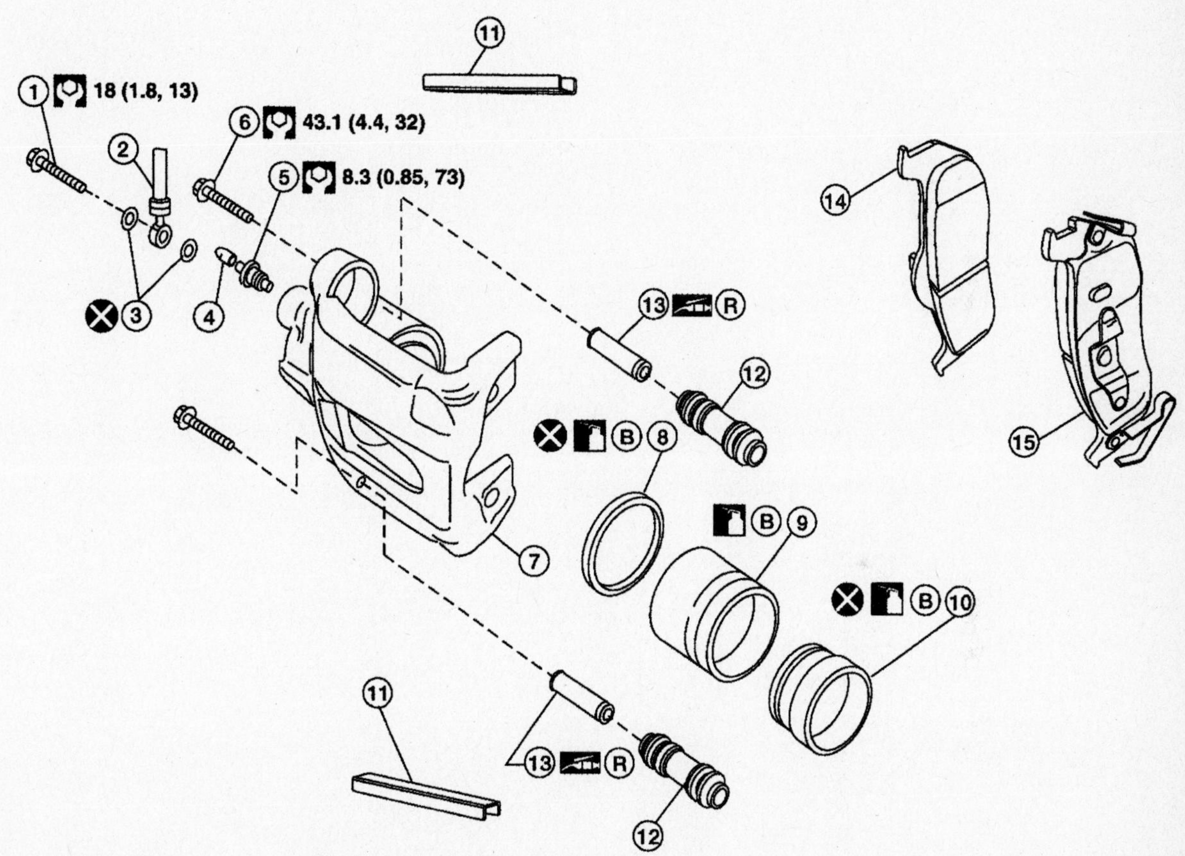

1. Union bolt	2. Brake hose	3. Copper washer
4. Cap	5. Bleed valve	6. Mounting bolt
7. Cylinder body	8. Piston seal	9. Piston
10. Piston boot	11. Knuckle slide	12. Sliding sleeve boot
13. Sliding sleeve	14. Inner pad	15. Outer pad

67170-ARMA-G56

Rear brake components

3. Suspend brake caliper with a remove and remove brake pad and shim from torque member.

To install:

4. Push pistons in so that the pad is firmly installed, using a suitable tool.
5. Mount the brake caliper to torque member.
6. Attach pad retainer to torque member.
7. Lubricate lower sliding pin bolt with a thin layer of silicone grease and install. Torque to 20 ft. lbs. (27 Nm).
8. Install the wheel.

Rear

1. Remove the wheel.
2. Remove mounting bolt from the top mount.
3. Swing brake caliper open and remove the brake pads.

To install:

4. Push pistons in so that the pad is firmly installed, using a suitable tool.
5. Install pads to the brake caliper.
6. Install top mounting bolt and tighten to 32 ft. lbs. (44 Nm).
7. Install the wheel.

NISSAN

Frontier • Xterra

7

SPECIFICATION AND MAINTENANCE CHARTS

ENGINE AND VEHICLE IDENTIFICATION

Engine						
Code ①	Liters (cc)	Cu. In.	Cyl.	Fuel Sys.	Engine Type	Eng. Mfg.
KA24DE	2.4 (2389)	146	4	MFI	DOHC	Nissan
VG33E	3.3 (3277)	199	6	MFI	SOHC	Nissan
VG33ER	3.3 (3277)	199	6	MFI	SOHC	Nissan

Model Year	
Code ②	Year
1	2001
2	2002
3	2003
4	2004
5	2005

MFI: Multi-port Fuel Injection

SOHC: Single Overhead Camshaft

DOHC: Double Overhead Camshafts

① Located on the timing belt cover

② 10th digit of the Vehicle Identification Number (VIN)

67170-FRON-C01

GENERAL ENGINE SPECIFICATIONS

Year	Model	Engine Displacement Liters	Engine ID	Net Horsepower @ rpm	Net Torque @ rpm (ft. lbs.)	Bore x Stroke (in.)	Com-pression Ratio	Oil Pressure @ rpm
2001	Frontier	2.4	KA24DE	143@5200	154@4000	3.50x3.78	9.2:1	60-70@3000
	Frontier	3.3	VG33E	170@4800	200@2800	3.60X3.27	8.9:1	60-65@2000
	Xterra	2.4	KA24DE	143@5200	154@4000	3.50X3.78	9.2:1	60-70@3000
	Xterra	3.3	VG33E	170@4800	200@2800	3.60X3.27	8.9:1	60-65@2000
2002	Frontier	2.4	KA24DE	143@5200	154@4000	3.50x3.78	9.2:1	60-70@3000
	Frontier	3.3	VG33E	170@4800	200@2800	3.60X3.27	8.9:1	60-65@2000
	Frontier	3.3	VG33ER	210@4800	231@2800	3.60X3.27	8.3:1	60-65@2000
	Xterra	2.4	KA24DE	143@5200	154@4000	3.50X3.78	9.2:1	60-70@3000
	Xterra	3.3	VG33E	170@4800	200@2800	3.60X3.27	8.9:1	60-65@2000
	Xterra	3.3	VG33ER	210@4800	231@2800	3.60X3.27	8.3:1	60-65@2000
2003	Frontier	2.4	KA24DE	143@5200	154@4000	3.50x3.78	9.2:1	60-70@3000
	Frontier	3.3	VG33E	170@4800	200@2800	3.60X3.27	8.9:1	60-65@2000
	Frontier	3.3	VG33ER	210@4800	246@2800	3.60X3.27	8.3:1	60-65@2000
	Xterra	2.4	KA24DE	143@5200	154@4000	3.50X3.78	9.2:1	60-70@3000
	Xterra	3.3	VG33E	170@4800	200@2800	3.60X3.27	8.9:1	60-65@2000
	Xterra	3.3	VG33ER	210@4800	231@2800	3.60X3.27	8.3:1	60-65@2000
2004	Frontier	2.4	KA24DE	143@5200	154@4000	3.50x3.78	9.2:1	60-70@3000
	Frontier	3.3	VG33E	170@4800	200@2800	3.60X3.27	8.9:1	60-65@2000
	Frontier	3.3	VG33ER	210@4800	246@2800	3.60X3.27	8.3:1	60-65@2000
	Xterra	2.4	KA24DE	143@5200	154@4000	3.50X3.78	9.2:1	60-70@3000
	Xterra	3.3	VG33E	170@4800	200@2800	3.60X3.27	8.9:1	60-65@2000
	Xterra	3.3	VG33ER	210@4800	231@2800	3.60X3.27	8.3:1	60-65@2000

MFI: Multi-port Fuel Injection

67170-FRON-C02

ENGINE TUNE-UP SPECIFICATIONS

Year	Engine Displ. Liters	Engine ID	Spark Plug Gap (in.)	Ignition Timing (deg.)		Fuel Pump (psi) ①	Idle Speed (rpm)		Valve Clearance (in.)	
				MT	AT		MT	AT ②	In.	Ex.
2001	2.4	KA24DE	0.039-0.043	18-22B	18-22B	34	750-850	750-850	0.012-0.015	0.013-0.016
	3.3	VG33E	0.039-0.043	13-17B	13-17B	34	700-800	700-800	HYD	HYD
2002	2.4	KA24DE	0.039-0.043	18-22B	18-22B	34	750-850	750-850	0.012-0.015	0.013-0.016
	3.3	VG33E	0.039-0.043	13-17B	13-17B	34	700-800	700-800	HYD	HYD
	3.3	VG33ER	0.043	10B	10B	35	700-800	700-800	HYD	HYD
2003	2.4	KA24DE	0.039-0.043	18-22B	18-22B	34	750-850	750-850	0.012-0.015	0.013-0.016
	3.3	VG33E	0.039-0.043	13-17B	13-17B	34	700-800	700-800	HYD	HYD
	3.3	VG33ER	0.043	10B	10B	35	700-800	700-800	HYD	HYD
2004	2.4	KA24DE	0.043	18-22B	18-22B	34	750-850	750-850	0.012-0.015	0.013-0.016
	3.3	VG33E	0.043	13-17B	13-17B	34	700-800	700-800	HYD	HYD
	3.3	VG33ER	0.043	10B	10B	35	700-800	700-800	HYD	HYD

NOTE: The Vehicle Emission Control Information label often reflects specification changes made during production. The label figures must be used if they differ from those in this chart.

B: Before top dead center

HYD: Hydraulic

① System pressure at idle with vacuum hose connected
Should increase to 43 psi when disconnected

② Automatic transmission in Neutral

67170-FRON-C03

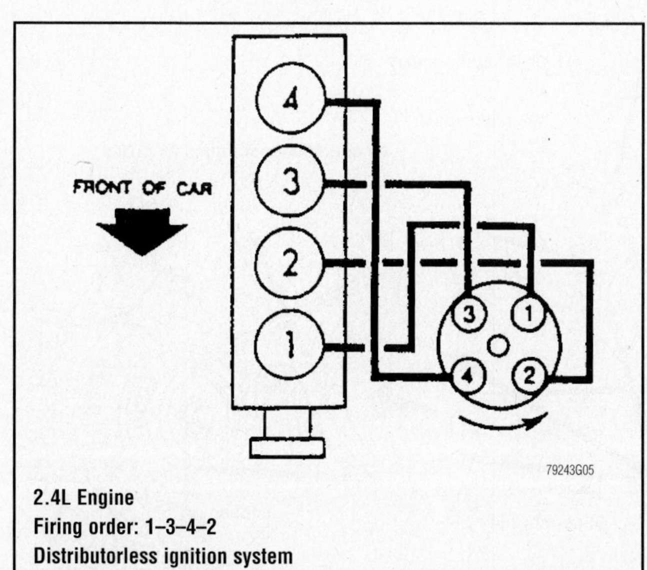

2.4L Engine
Firing order: 1–3–4–2
Distributorless ignition system

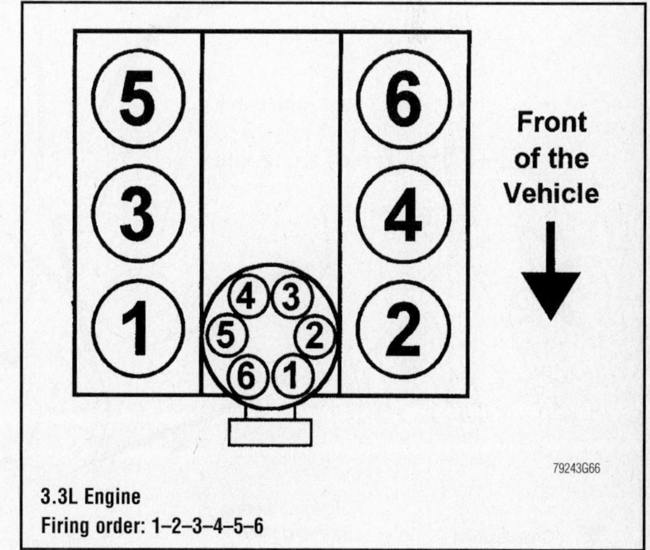

3.3L Engine
Firing order: 1–2–3–4–5–6

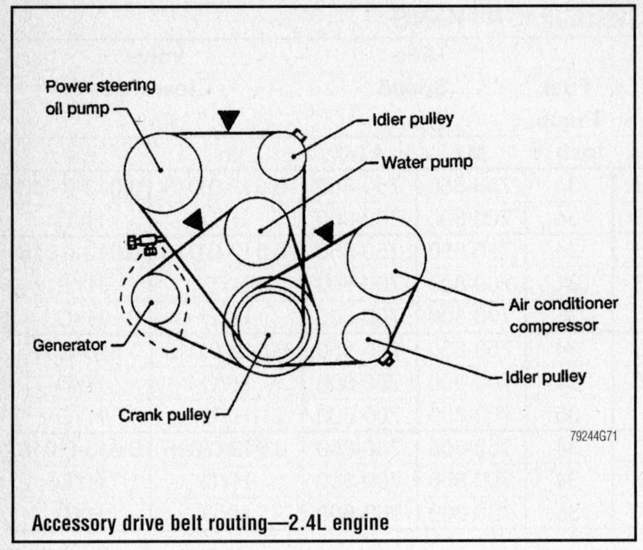

Power steering oil pump
Idler pulley
Water pump
Air conditioner compressor
Generator
Idler pulley
Crank pulley
79244G71

Accessory drive belt routing—2.4L engine

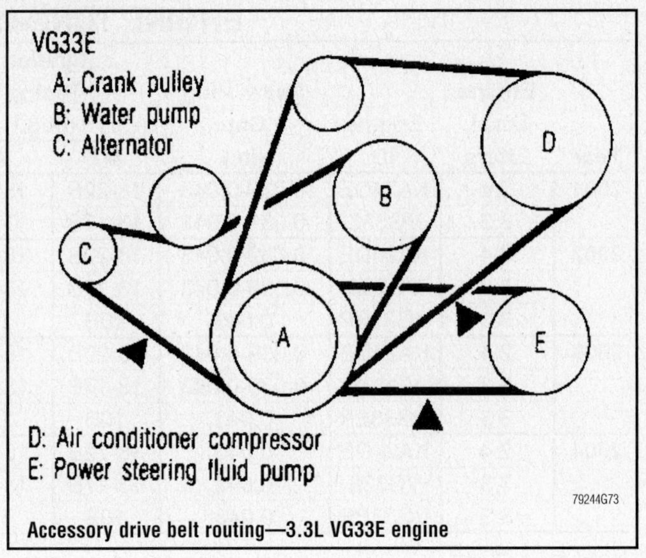

VG33E
A: Crank pulley
B: Water pump
C: Alternator
D: Air conditioner compressor
E: Power steering fluid pump
79244G73

Accessory drive belt routing—3.3L VG33E engine

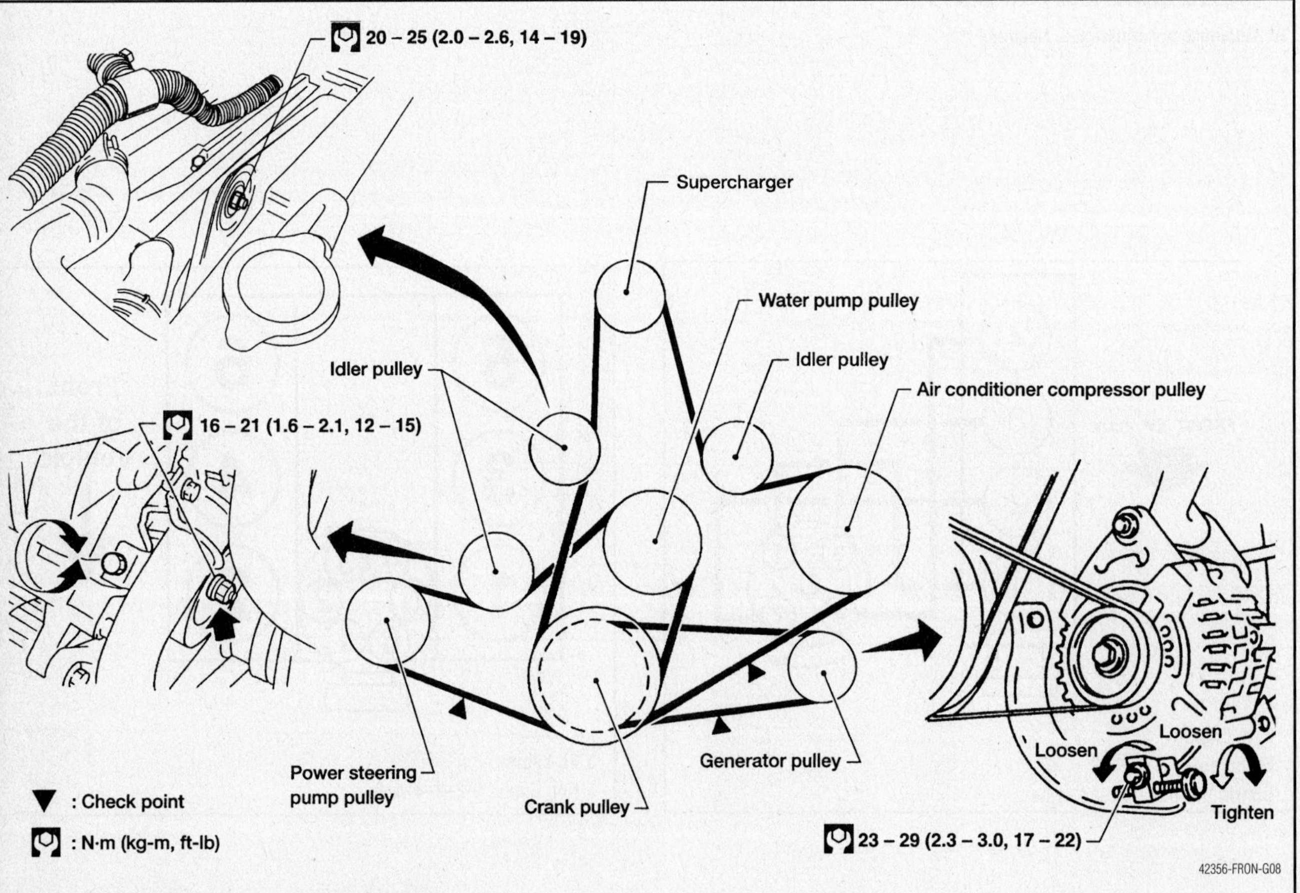

20 – 25 (2.0 – 2.6, 14 – 19)
Supercharger
Water pump pulley
Idler pulley
Idler pulley
Air conditioner compressor pulley
16 – 21 (1.6 – 2.1, 12 – 15)
Loosen
Loosen
Tighten
Generator pulley
Power steering pump pulley
Crank pulley
▼ : Check point
: N·m (kg-m, ft-lb)
23 – 29 (2.3 – 3.0, 17 – 22)
42356-FRON-G08

Accessory drive belt routing—3.3L VG33ER engine

CAPACITIES

Year	Model	Engine Displacement Liters	Engine ID	Engine Oil with Filter (qts.)	Transmission (pts.) 5-Spd	Transmission (pts.) Auto.	Transfer Case (pts.)	Drive Axle Front (pts.)	Drive Axle Rear (pts.)	Fuel Tank (gal.)	Cooling System (qts.)
2001	Frontier	2.4	KA24DE	①	4.25	—	—	—	②	15.9	7.75
	Frontier	3.3	VG33E	①	③	16.8	4.8	2.8	②	19.4	④
	Xterra	2.4	KA24DE	3.75	4.25	—	—	—	3.1	19.4	7.75
	Xterra	3.3	VG33E	3.8	⑤	⑥	4.8	4.4	4.9	19.4	11.25
2002	Frontier	2.4	KA24DE	3.5	4.25	—	—	—	2.4	15.9	7.75
	Frontier	3.3	VG33E	3.5	⑤	⑥	4.8	5.8	3.12	19.4	11.6
	Xterra	2.4	KA24DE	3.75	4.25	—	—	—	3.1	19.4	7.75
	Xterra	3.3	VG33E	3.5	⑤	⑥	4.8	3.1	5.9	19.4	11.6
	Xterra	3.3	VG33ER	3.5	⑤	⑥	4.8	3.1	5.9	19.4	11.6
2003	Frontier	2.4	KA24DE	3.5	4.25	—	—	—	2.4	15.9	7.75
	Frontier	3.3	VG33E	3.5	⑤	⑥	4.8	5.8	3.12	19.4	11.6
	Frontier	3.3	VG33ER	3.5	⑤	⑥	4.8	5.8	3.12	19.4	11.6
	Xterra	2.4	KA24DE	3.75	4.25	—	—	—	3.1	19.4	7.75
	Xterra	3.3	VG33E	3.5	⑤	⑥	4.8	3.1	5.9	19.4	11.6
	Xterra	3.3	VG33ER	3.5	⑤	⑥	4.8	3.1	5.9	19.4	11.6
2004	Frontier	2.4	KA24DE	3.75	4.25	16.75	—	—	2.4	15.9	7.75
	Frontier	3.3	VG33E	3.5	⑤	⑥	4.8	3.75	5.8	19.4	11.6
	Frontier	3.3	VG33ER	3.5	⑤	⑥	4.8	3.75	5.8	19.4	11.6
	Xterra	2.4	KA24DE	3.75	4.25	16.75	—	—	5.8	19.4	7.75
	Xterra	3.3	VG33E	3.5	⑤	⑥	4.8	3.75	5.9	19.4	11.6
	Xterra	3.3	VG33ER	3.5	⑤	⑥	4.8	3.75	5.9	19.4	11.6

NOTE: All capacities are approximate. Add fluid gradually and check to be sure a proper fluid level is obtained.

① 2WD: 3.75 qts.
4WD: 4.125 qts.

② H190A: 3.1 pts.
C200: 2.75 pts.
H233B: 5.9 pts.

③ 2WD: 4.25 pts.
4WD: 10.4 pts.

④ 2WD: 8.6 qts.; 4WD: 9.5 qts.

⑤ MT: 5.8; AT: 10.75

⑥ MT: 17.5; AT: 18 pts.

67170-FRON-C04

VALVE SPECIFICATIONS

Year	Engine Displacement Liters	Engine ID	Seat Angle (deg.)	Face Angle (deg.)	Spring Test Pressure (lbs. @ in.)	Spring Installed Height (in.)	Stem-to-Guide Clearance (in.) Intake	Stem-to-Guide Clearance (in.) Exhaust	Stem Diameter (in.) Intake	Stem Diameter (in.) Exhaust
2001	2.4	KA24DE	45	45.5	93.9@1.15	NA	0.0008-0.0021	0.0016-0.0029	0.2742-0.2748	0.2734-0.2740
	3.3	VG33E	45	45.25-46.75	①	NA	0.0008-0.0021	0.0016-0.0029	0.2742-0.2748	0.3135-0.3138
2002	2.4	KA24DE	45	45.5	93.9@1.15	NA	0.0008-0.0021	0.0016-0.0029	0.2742-0.2748	0.2734-0.2740
	3.3	VG33E	45	45.25-46.75	①	NA	0.0008-0.0021	0.0016-0.0029	0.2742-0.2748	0.3135-0.3138
	3.3	VG33ER	45	45.25-46.75	①	NA	0.0008-0.0021	0.0012-0.0019	0.2742-0.2748	0.3135-0.3138
2003	2.4	KA24DE	45	45.5	93.9@1.15	NA	0.0008-0.0021	0.0016-0.0029	0.2742-0.2748	0.2734-0.2740
	3.3	VG33E	45	45.25-46.75	①	NA	0.0008-0.0021	0.0016-0.0029	0.2742-0.2748	0.3135-0.3138
	3.3	VG33ER	45	45.25-46.75	①	NA	0.0008-0.0021	0.0012-0.0019	0.2742-0.2748	0.3135-0.3138
2004	2.4	KA24DE	45	45.5	93.9@1.15	NA	0.0008-0.0021	0.0016-0.0029	0.2742-0.2748	0.2734-0.2740
	3.3	VG33E	45	45.25-46.75	①	NA	0.0008-0.0021	0.0016-0.0029	0.2742-0.2748	0.3135-0.3138
	3.3	VG33ER	45	45.25-46.75	①	NA	0.0008-0.0021	0.0012-0.0019	0.2742-0.2748	0.3135-0.3138

NA: Not Available

① Inner: 57.3 @ 0.984
Outer: 117.7 @ 1.181

67170-FRON-C05

CRANKSHAFT AND CONNECTING ROD SPECIFICATIONS

All measurements are given in inches.

Year	Engine Displacement Liters	Engine ID	Crankshaft Main Brg. Journal Dia.	Crankshaft Main Brg. Oil Clearance	Crankshaft Shaft End-play	Crankshaft Thrust on No.	Connecting Rod Journal Diameter	Connecting Rod Oil Clearance	Connecting Rod Side Clearance
2001	2.4	KA24DE	2.3609-2.3612	0.0008-0.0019	0.0020-0.0071	3	1.9672-1.9675	0.0004-0.0014	0.0080-0.0160
	3.3	VG33E	2.4790-2.4793	0.0011-0.0022	0.0020-0.0067	4	1.9967-1.9675	0.0006-0.0021	0.0079-0.0138
2002	2.4	KA24DE	2.3609-2.3612	0.0008-0.0019	0.0020-0.0071	3	1.9672-1.9675	0.0004-0.0014	0.0080-0.0160
	3.3	VG33E	2.4790-2.4793	0.0011-0.0022	0.0020-0.0067	4	1.9967-1.9675	0.0006-0.0021	0.0079-0.0138
	3.3	VG33ER	①	②	0.0020-0.0067	4	1.9667-1.9675	0.0009-0.0025	0.0079-0.0138
2003	2.4	KA24DE	2.3609-2.3612	0.0008-0.0019	0.0020-0.0071	3	1.9672-1.9675	0.0004-0.0014	0.0080-0.0160
	3.3	VG33E	2.4790-2.4793	0.0011-0.0022	0.0020-0.0067	4	1.9967-1.9675	0.0006-0.0021	0.0079-0.0138
	3.3	VG33ER	①	②	0.0020-0.0067	4	1.9667-1.9675	0.0009-0.0025	0.0079-0.0138
2004	2.4	KA24DE	2.3609-2.3612	0.0008-0.0019	0.0020-0.0071	3	1.9672-1.9675	0.0004-0.0014	0.0080-0.0160
	3.3	VG33E	2.4790-2.4793	0.0011-0.0022	0.0020-0.0067	4	1.9967-1.9675	0.0006-0.0021	0.0079-0.0138
	3.3	VG33ER	①	②	0.0020-0.0067	4	1.9667-1.9675	0.0009-0.0025	0.0079-0.0138

① Except No. 1
Grade 0: 2.4790-2.4793
Grade 1: 2.4787-2,4790
Grade 2: 2.4784-2.4787
No. 1
Grade 3: 2.4683-2.4793
Grade 4: 2.4789-2.4791
Grade 5: 2.4786-2.4789
Grade 6: 2.4784-2.4786

② No. 1: 0.0012-0.0019
Nos. 2, 3, 4: 0.0015-0.0026

67170-FRON-C06

PISTON AND RING SPECIFICATIONS

All measurements are given in inches.

Year	Engine Displacement Liters	Engine ID	Piston Clearance	Ring Gap Top Comp.	Ring Gap Bottom Comp.	Ring Gap Oil Control	Ring Side Clearance Top Comp.	Ring Side Clearance Bottom Comp.	Ring Side Clearance Oil Control
2001	2.4	KA24E	0.0008-0.0016	0.011-0.021	0.018-0.027	0.008-0.027	0.0016-0.0031	0.0012-0.0028	0.0026-0.0053
	3.3	VG33E	①	0.0083-0.0157	0.0197-0.0272	0.0079-0.0272	0.0009-0.0030	0.0012-0.0028	0.0006-0.0073
2002	2.4	KA24E	0.0008-0.0016	0.011-0.021	0.018-0.027	0.008-0.027	0.0016-0.0031	0.0012-0.0028	0.0026-0.0053
	3.3	VG33E	①	0.0083-0.0157	0.0197-0.0272	0.0079-0.0272	0.0009-0.0030	0.0012-0.0028	0.0006-0.0073
	3.3	VG33ER	②	0.0083-0.0122	0.0197-0.0236	0.0079-0.0236	0.0016-0.0031	0.0012-0.0028	0.0006-0.0073
2003	2.4	KA24E	0.0008-0.0016	0.011-0.021	0.018-0.027	0.008-0.027	0.0016-0.0031	0.0012-0.0028	0.0026-0.0053
	3.3	VG33E	①	0.0083-0.0157	0.0197-0.0272	0.0079-0.0272	0.0009-0.0030	0.0012-0.0028	0.0006-0.0073
	3.3	VG33ER	②	0.0083-0.0122	0.0197-0.0236	0.0079-0.0236	0.0016-0.0031	0.0012-0.0028	0.0006-0.0073
2004	2.4	KA24E	0.0008-0.0016	0.011-0.021	0.018-0.027	0.008-0.027	0.0016-0.0031	0.0012-0.0028	0.0026-0.0053
	3.3	VG33E	①	0.0083-0.0157	0.0197-0.0272	0.0079-0.0272	0.0009-0.0030	0.0012-0.0028	0.0006-0.0073
	3.3	VG33ER	②	0.0083-0.0122	0.0197-0.0236	0.0079-0.0236	0.0016-0.0031	0.0012-0.0028	0.0006-0.0073

① Except cylinders 3 and 4: 0.0010 - 0.0018 in.
Cylinders 3 and 4: 0.0006 - 0.0010 in.

② Cylinders 3, 4: 0.0006-0.0010 in.
Cylinders 1, 2, 5, 6: 0.0010-0.0018 in.

67170-FRON-C07

TORQUE SPECIFICATIONS

All readings in ft. lbs.

Year	Engine Displacement Liters	Engine ID	Cylinder Head Bolts	Main Bearing Bolts	Rod Bearing Bolts	Crankshaft Damper Bolts	Flywheel Bolts	Manifold Intake	Manifold Exhaust	Spark Plugs	Oil Pan Drain Plug
2001	2.4	KA24DE	①	34-41	②	105-112	105-112	12-14	27-35	14-22	22-29
	3.3	VG33E	③	67-74	②	141-156	61-69	③	21-25	14-22	22-29
2002	2.4	KA24DE	①	34-41	②	105-112	105-112	12-14	27-35	14-22	22-29
	3.3	VG33E	③	67-74	②	141-156	61-69	③	21-25	14-22	22-29
	3.3	VG33ER	③	67-74	④	141-156	61-69	③	21-25	14-22	22-29
2003	2.4	KA24DE	①	34-41	②	105-112	105-112	12-14	27-35	14-22	22-29
	3.3	VG33E	③	67-74	②	141-156	61-69	③	21-25	14-22	22-29
	3.3	VG33ER	③	67-74	④	141-156	61-69	③	21-25	14-22	22-29
2004	2.4	KA24DE	①	34-41	②	105-112	105-112	12-14	27-35	14-22	22-29
	3.3	VG33E	③	67-74	②	141-156	61-69	③	21-25	14-22	22-29
	3.3	VG33ER	③	67-74	④	141-156	61-69	③	21-25	14-22	22-29

① Step 1: 22 ft. lbs.
Step 2: 59 ft. lbs.
Step 3: Loosen completely then retorque to 22 ft. lbs.
Step 4: 18-25 ft. lbs.
Step 5: Plus 86-91 degrees

② 10-12 ft. lbs. plus 60-65 degrees or 28-33 ft. lbs.

③ The cylinder heads and the lower intake manifold are installed together
Step 1: Tighten the cylinder head bolts to 22 ft. lbs.
Step 2: Tighten the cylinder head bolts to 43 ft. lbs.
Step 3: Loosen the cylinder head bolts completely
Step 4: Tighten the cylinder head bolts to 84 inch lbs.
Step 5: Tighten the intake manifold fasteners to 35 inch lbs.
Step 6: Tighten the intake manifold fasteners to 13 ft. lbs.
Step 7: Tighten the intake manifold fasteners to 12-14 ft. lbs.
Step 8: Loosen all intake manifold fasteners completely
Step 9: Tighten the cylinder head bolts to 22 ft. lbs.
Step 10: Tighten the cylinder head bolts 60-65 degrees
Step 11: Tighten the cylinder head sub-bolts to 80-105 inch lbs.
Step 12: Tighten the intake manifold fasteners to 35 inch lbs.
Step 13: Tighten the intake manifold fasteners to 78 inch lbs.
Step 14: Tighten the intake manifold fasteners to 70-84 inch lbs.

④ 10-12 ft. lbs. +60-65 degrees

67170-FRON-C08

WHEEL ALIGNMENT

			Caster		Camber		
Year	Model		Range (+/-Deg.)	Preferred Setting (Deg.)	Range (+/-Deg.)	Preferred Setting (Deg.)	Toe-in (in.)
2001	Frontier	2.4L	0.50	+0.60	0.50	+0.42	0.12+/-0.04
		3.3L	0.50	+2.17	0.50	+0.60	0.16+/-0.04
	Xterra	ALL	0.75	+0.60	0.50	+0.42	0.12+/-0.04
2002	Frontier	2.4L	0.50	+0.60	0.50	+0.42	0.12+/-0.04
		3.3L	0.50	+2.17	0.50	+0.60	0.16+/-0.04
	Xterra	2WD	0.50	+2.57	0.50	+0.33	0.16+/-0.04
	Xterra	4WD	0.50	+2.10	0.50	+0.60	0.16+/-0.04
2003	Frontier	2.4L	0.50	+0.60	0.50	+0.42	0.12+/-0.04
		3.3L	0.50	+2.17	0.50	+0.60	0.16+/-0.04
	Xterra	2WD	0.50	+2.57	0.50	+0.33	0.16+/-0.04
	Xterra	4WD	0.50	+2.10	0.50	+0.60	0.16+/-0.04
2004	Frontier	2.4L	0.50	+0.60	0.50	+0.42	0.12+/-0.04
		3.3L	0.50	+2.17	0.50	+0.60	0.16+/-0.04
	Xterra	2WD	0.50	+2.57	0.50	+0.33	0.16+/-0.04
	Xterra	4WD	0.50	+2.10	0.50	+0.60	0.16+/-0.04

67170-FRON-C09

TIRE, WHEEL AND BALL JOINT SPECIFICATIONS

Year	Model	OEM Tires		Tire Pressures (psi)		Wheel Size	Ball Joint Inspection	Lugnut Torque (ft. lbs.)
		Standard	Optional	Front	Rear			
2001	Frontier 2wd 4-Cyl.	P215/65R15	None	26	29	6JJ	U: 0.020 in. L: ①	87-108
	Frontier 2wd 6-Cyl.	P235/70R15	P265/70R15	26	26	6.5JJ/7JJ	U: 0.020 in. L: ①	87-108
	Frontier 4wd XE	P215/75R15	P235/70R15	30	30	7-JJ	U: 0.020 in. L: ①	87-108
	Frontier 4wd SE	P255/65R16	None	26	26	7JJ	U: 0.020 in. L: ①	87-108
	Xterra	P235/70R15	P265/70R15	26	26	7JJ	U: 0.020 in. L: ①	87-108
2002	Frontier 2wd 4-Cyl.	P225/70R15	None	②	②	NA	U: 0.020 in. L: ①	87-108
	Frontier 4wd 6-Cyl.	P255/65R16	None	②	②	NA	U: 0.020 in. L: ①	87-108
	Frontier 2wd SC 6-Cyl.	P265/55R17	None	②	②	NA	U: 0.020 in. L: ①	87-108
	Frontier 4wd SC 6-Cyl	P265/65R17	None	②	②	NA	U: 0.020 in. L: ①	87-108
	Frontier 4wd XE 6-cyl.	P265/70R15	None	②	②	NA	U: 0.020 in. L: ①	87-108
	Frontier 6-Cyl. XE Desert Runner	P265/70R15	None	②	②	NA	U: 0.020 in. L: ①	87-108
	Frontier 6-Cyl. SE Desert Runner	P255/65R16	None	②	②	NA	U: 0.020 in. L: ①	87-108
	Frontier SE V6	P265/70R16	None	②	②	NA	U: 0.020 in. L: ①	87-108
	Frontier 2wd SC V6 Crew Cab	P265/55R17	None	②	②	NA	U: 0.020 in. L: ①	87-108
	Frontier 4wd SC V6 Crew Cab	P265/65R17	None	②	②	NA	U: 0.020 in. L: ①	87-108
	Xterra SE, SE S/C, and XE S/C	P265/70R16	None	②	②	7JJ	③	87-108
	Xterra XE, XE V6	P265/70R15	None	②	②	7JJ	③	87-108
2003	Frontier 2wd 4-Cyl.	P225/70R15	None	②	②	NA	U: 0.020 in. L: ①	87-108
	Frontier 4wd 6-Cyl.	P255/65R16	None	②	②	NA	U: 0.020 in. L: ①	87-108
	Frontier 2wd SC 6-Cyl.	P265/55R17	None	②	②	NA	U: 0.020 in. L: ①	87-108
	Frontier 4wd SC 6-Cyl	P265/65R17	None	②	②	NA	U: 0.020 in. L: ①	87-108
	Frontier 4wd XE 6-cyl.	P265/70R15	None	②	②	NA	U: 0.020 in. L: ①	87-108
	Frontier 6-Cyl. XE Desert Runner	P265/70R15	None	②	②	NA	U: 0.020 in. L: ①	87-108

67170-FRON-C10

TIRE, WHEEL AND BALL JOINT SPECIFICATIONS

Year	Model	OEM Tires		Tire Pressures (psi)		Wheel	Ball Joint	Lugnut Torque
		Standard	Optional	Front	Rear	Size	Inspection	(ft. lbs.)
2003 cont.	Frontier 6-Cyl. SE Desert Runner	P255/65R16	None	②	②	NA	U: 0.020 in. L: ①	87-108
	Frontier SE V6	P265/70R16	None	②	②	NA	U: 0.020 in. L: ①	87-108
	Frontier 2wd SC V6 Crew Cab	P265/55R17	None	②	②	NA	U: 0.020 in. L: ①	87-108
	Frontier 4wd SC V6 Crew Cab	P265/65R17	None	②	②	NA	U: 0.020 in. L: ①	87-108
	Xterra SE, SE S/C, and XE S/C	P265/70R16	None	②	②	7JJ	③	87-108
	Xterra XE, XE V6	P265/70R15	None	②	②	7JJ	③	87-108
2004	Frontier 2wd 4-Cyl.	P225/70R15	None	②	②	NA	U: 0.020 in. L: ①	87-108
	Frontier 4wd 6-Cyl.	P255/65R16	None	②	②	NA	U: 0.020 in. L: ①	87-108
	Frontier 2wd SC 6-Cyl.	P265/55R17	None	②	②	NA	U: 0.020 in. L: ①	87-108
	Frontier 4wd SC 6-Cyl	P265/65R17	None	②	②	NA	U: 0.020 in. L: ①	87-108
	Frontier 4wd XE 6-cyl.	P265/70R15	None	②	②	NA	U: 0.020 in. L: ①	87-108
	Frontier 6-Cyl. XE Desert Runner	P265/70R15	None	②	②	NA	U: 0.020 in. L: ①	87-108
	Frontier 6-Cyl. SE Desert Runner	P255/65R16	None	②	②	NA	U: 0.020 in. L: ①	87-108
	Frontier SE V6	P265/70R16	None	②	②	NA	U: 0.020 in. L: ①	87-108
	Frontier 2wd SC V6 Crew Cab	P265/55R17	None	②	②	NA	U: 0.020 in. L: ①	87-108
	Frontier 4wd SC V6 Crew Cab	P265/65R17	None	②	②	NA	U: 0.020 in. L: ①	87-108
	Xterra SE, SE S/C, and XE S/C	P265/70R16	None	②	②	7JJ	③	87-108
	Xterra XE, XE V6	P265/70R15	None	②	②	7JJ	③	87-108

OEM: Original Equipment Manufacturer

PSI: Pounds Per Square Inch

STD: Standard

OPT: Optional

L: Lower

U: Upper

① Replace if any measurable movement is found.

② See placard on vehicle

③ Axial play

Upper: 0

Lower: 0.008 in.

67170-FRON-C11

BRAKE SPECIFICATIONS

All measurements in inches unless noted

		Brake Disc			Brake Drum Diameter			Minimum Lining Thickness		Brake Caliper	
Year	Model	Original Thickness	Minimum Thickness	Maximum Runout	Original Inside Diameter	Max. Wear Limit	Maximum Machine Diameter	Front	Rear	Bracket Bolts (ft. lbs.)	Mounting Bolts (ft. lbs.)
2001	Frontier	①	②	0.003	③	NA	④	0.079	0.059	⑦	24-31
	Xterra	①	②	0.003	③	NA	④	0.079	0.059	53-72	24-31
2002	Frontier	①	②	0.003	③	NA	④	0.079	0.059	⑦	17-22
	Xterra	1.100	1.024	0.003	11.61	NA	11.67	0.079	0.059	101-130	17-22
2003	Frontier	NA	⑤	0.003	③	NA	⑥	0.079	0.059	⑦	17-22
	Xterra	1.100	1.024	0.003	11.61	NA	11.67	0.079	0.059	101-130	17-22
2004	Frontier	NA	⑤	0.003	③	NA	⑥	0.079	0.059	⑦	17-22
	Xterra	1.100	1.024	0.003	11.61	NA	11.67	0.079	0.059	101-130	17-22

NA: Information not available

① 2WD: 0.870
4WD: 1.020

② 2WD: 0.787
4WD: 0.945

③ 2WD: 10.20
4WD: 11.60

④ 2WD: 10.30
4WD: 11.67

⑤ 4-cyl.: 0.945
6-cyl.: 1.024

⑥ 4-cyl.: 10.30
6-cyl.: 11.67

⑦ 4-cyl.: 53-72 ft. lbs.
6-cyl.: 101-130 ft. lbs.

67170-FRON-C12

SCHEDULED MAINTENANCE INTERVALS

Nissan Frontier and Xterra

TO BE SERVICED	TYPE OF SERVICE	VEHICLE MILEAGE INTERVAL (x1000)												
		7.5	15	22.5	30	37.5	45	52.5	60	67.5	75	82.5	90	97.5
Engine oil & filter	R	✓	✓	✓	✓	✓	✓	✓	✓	✓	✓	✓	✓	✓
Brake lines & cables	S/I		✓		✓		✓		✓		✓		✓	
Brake pads, discs, drums & linings	S/I		✓		✓		✓		✓		✓		✓	
Driveshaft boots & propeller shaft	S/I				✓				✓				✓	
Front wheel bearings (4x2)	S/I				✓				✓				✓	
Front wheel bearings (4x4)	S/I				✓				✓				✓	
Automatic & manual transmission, transfer & differential gear oil ①	S/I		✓		✓		✓		✓		✓		✓	
Air cleaner filter	R				✓				✓				✓	
Engine coolant	R				✓				✓				✓	
PCV filter (KA24E)	R				✓				✓				✓	
Spark plugs	R				✓				✓				✓	
Drive belt(s)	S/I				✓				✓				✓	
Exhaust system	S/I				✓				✓				✓	
Fuel lines	S/I				✓				✓				✓	
Steering gear (box) & linkage, axle & suspension parts	S/I				✓				✓				✓	
Vapor lines	S/I				✓				✓				✓	
Timing belt ②	R													

R: Replace   S/I: Service or Inspect

① Differential (w/limited-slip differential) oil: replace oil every 30,000 miles.

② Timing belt: replace at 105,000 miles.

FREQUENT OPERATION MAINTENANCE (SEVERE SERVICE)

If a vehicle is operated under any of the following conditions it is considered severe service:

- Extremely dusty areas.

- 50% or more of the vehicle operation is in 32°C (90°F) or higher temperatures, or constant operation in temperatures below 0°C (32°F).

- Prolonged idling (vehicle operation in stop and go traffic).

- Frequent short running periods (engine does not warm to normal operating temperatures).

- Police, taxi, delivery usage or trailer towing usage.

Oil & oil filter: replace every 3750 miles.

Brake pads, discs, drums & linings: service or inspect every 7500 miles.

Driveshaft boots & propeller shaft: service or inspect every 7500 miles.

Exhaust system: service or inspect every 7500 miles.

Steering gear (box) & linkage, (steering damper-4x4), axle & suspension parts: service or inspect every 7500 miles.

Steering linkage ball joints & front suspension ball joints: service or inspect every 7500 miles.

67170-FRON-C13

PRECAUTIONS

Before servicing any vehicle, please be sure to read all of the following precautions, which deal with personal safety, prevention of component damage, and important points to take into consideration when servicing a motor vehicle:

- Never open, service or drain the radiator or cooling system when the engine is hot; serious burns can occur from the steam and hot coolant.
- Observe all applicable safety precautions when working around fuel. Whenever servicing the fuel system, always work in a well-ventilated area. Do not allow fuel spray or vapors to come in contact with a spark, open flame, or excessive heat (a hot drop light, for example). Keep a dry chemical fire extinguisher near the work area. Always keep fuel in a container specifically designed for fuel storage; also, always properly seal fuel containers to avoid the possibility of fire or explosion. Refer to the additional fuel system precautions later in this section.
- Fuel injection systems often remain pressurized, even after the engine has been turned **OFF**. The fuel system pressure must be relieved before disconnecting any fuel lines. Failure to do so may result in fire and/or personal injury.
- Brake fluid often contains polyglycol ethers and polyglycols. Avoid contact with the eyes and wash your hands thoroughly after handling brake fluid. If you do get brake fluid in your eyes, flush your eyes with clean, running water for 15 minutes. If eye irritation persists, or if you have taken brake fluid internally, IMMEDIATELY seek medical assistance.
- The EPA warns that prolonged contact with used engine oil may cause a number of skin disorders, including cancer! You should make every effort to minimize your exposure to used engine oil. Protective gloves should be worn when changing oil. Wash your hands and any other exposed skin areas as soon as possible after exposure to used engine oil. Soap and water, or waterless hand cleaner should be used.
- All new vehicles are now equipped with an air bag system. The system must be disabled before performing service on or around system components, steering column, instrument panel components, wiring and sensors. Failure to follow safety and disabling procedures could result in accidental air bag deployment, possible personal injury and unnecessary system repairs.
- Always wear safety goggles when working with, or around, the air bag system. When carrying a non-deployed air bag, be sure the bag and trim cover are pointed away from your body. When placing a non-deployed air bag on a work surface, always face the bag and trim cover upward, away from the surface. This will reduce the motion of the module if it is accidentally deployed. Refer to the additional air bag system precautions later in this section.
- Clean, high quality brake fluid from a sealed container is essential to the safe and proper operation of the brake system. You should always buy the correct type of brake fluid for your vehicle. If the brake fluid becomes contaminated, completely flush the system with new fluid. Never reuse any brake fluid. Any brake fluid that is removed from the system should be discarded. Also, do not allow any brake fluid to come in contact with a painted surface; it will damage the paint.
- Never operate the engine without the proper amount and type of engine oil; doing so WILL result in severe engine damage.
- Timing belt maintenance is extremely important! Many models utilize an interference-type, non-freewheeling engine. If the timing belt breaks, the valves in the cylinder head may strike the pistons, causing potentially serious (also time-consuming and expensive) engine damage. Refer to the maintenance interval charts in the front of this manual for the recommended replacement interval for the timing belt, and to the timing belt section for belt replacement and inspection.
- Disconnecting the negative battery cable on some vehicles may interfere with the functions of the on-board computer system(s) and may require the computer to undergo a relearning process once the negative battery cable is reconnected.
- When servicing drum brakes, only disassemble and assemble one side at a time, leaving the remaining side intact for reference.

ENGINE REPAIR

➡Disconnecting the negative battery cable on some vehicles may interfere with the functions of the on board computer system. The computer may undergo a relearning process once the negative battery cable is reconnected.

Distributor

REMOVAL

1. Before servicing the vehicle, refer to the precautions in the beginning of this section.
2. Remove or disconnect the following:
   - Negative battery cable
   - Distributor cap
   - Distributor wiring harness connector
3. Matchmark the rotor to the distributor housing and the distributor housing to the cylinder head.
4. Remove the distributor.

INSTALLATION

Timing Not Disturbed

1. Install or connect the following:
   - Distributor by aligning the matchmarks made during removal
   - Distributor wiring harness connector
   - Distributor cap
   - Negative battery cable
2. Check the ignition timing and adjust, as necessary.

Timing Disturbed

2.4L ENGINE

1. Set the engine to Top Dead Center (TDC) of the compression stroke for the No. 1 cylinder.
2. Install the distributor so that the distributor shaft engages the oil pump driveshaft.
3. Check that the distributor rotor is aligned, as shown.
4. Install or connect the following:
   - Distributor cap
   - Distributor harness connector
5. Check the ignition timing and adjust, as necessary.

3.3L ENGINE

1. Set the engine to Top Dead Center (TDC) of the compression stroke for the No. 1 cylinder.

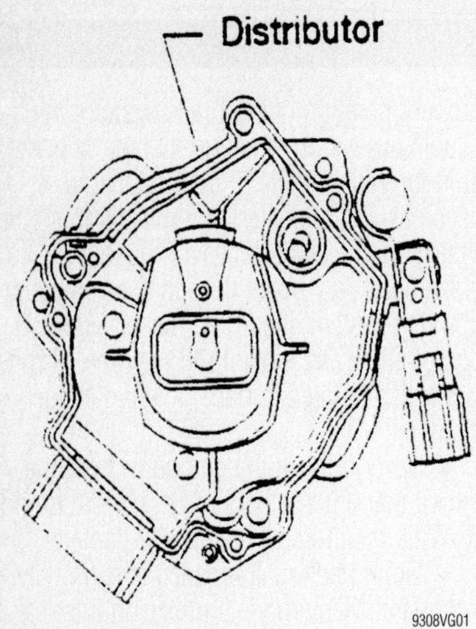

Distributor rotor alignment with the engine at Top Dead Center (TDC)—2.4L engine

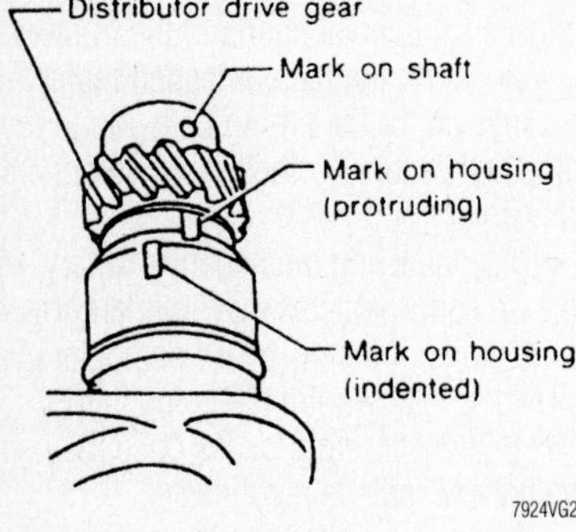

Distributor shaft alignment—3.3L engine

2. Align the index mark on the distributor shaft with the protrusion on the distributor housing.
3. Install the distributor and check that the distributor rotor is aligned.

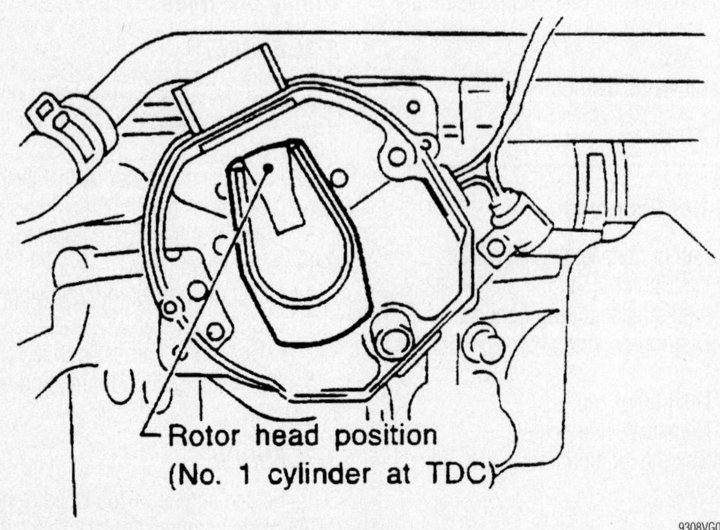

Distributor rotor alignment—3.3L engine

4. Install or connect the following:
   - Distributor cap
   - Distributor harness connector
5. Check the ignition timing and adjust, as necessary.

Alternator

REMOVAL

2.4L Engine

1. Before servicing the vehicle, refer to the precautions in the beginning of this section.
2. Remove or disconnect the following:
   - Negative battery cable
   - Engine under cover
   - Right splash shield
   - Alternator harness connectors
   - Alternator belt
   - Alternator

3.3L Engine

1. Before servicing the vehicle, refer to the precautions in the beginning of this section.
2. Remove or disconnect the following:
   - Negative battery cable
   - Alternator harness connectors
   - Engine under cover
   - Alternator belt
   - Alternator

INSTALLATION

2.4L Engine

1. Install or connect the following:
   - Alternator
   - Alternator belt. Tighten the adjustment bolt to 12–14 ft. lbs. (16–19 Nm) and the pivot bolt to 32–38 ft. lbs. (44–52 Nm).
   - Alternator harness connectors
   - Right splash shield
   - Engine under cover
   - Negative battery cable

3.3L Engine

1. Install or connect the following:
   - Alternator
   - Alternator belt. Tighten the adjustment bolt to 12–14 ft. lbs. (16–19 Nm) and the pivot bolts to 16–22 ft. lbs. (22–30 Nm).
   - Engine under cover
   - Alternator harness connectors
   - Negative battery cable

Ignition Timing

ADJUSTMENT

➡Ignition timing is set with the engine at operating temperature, transmission in Neutral and all electrical accessories OFF.

1. Before servicing the vehicle, refer to the precautions in the beginning of this section.
2. Attach a timing light to the No. 1 spark plug wire.
3. Start the engine and allow it to reach normal operating temperature.
4. Check that the idle speed is less than 1000 rpm.
5. Run the engine at 2000 rpm for 2 minutes.
6. Rev the engine to 3000 rpm 2–3 times and allow it to idle for 1 minute.
7. Check for the presence of Diagnostic Trouble Codes (DTC) and service as necessary.
8. Run the engine at 2000 rpm for 2 minutes.
9. Stop the engine and disconnect the Throttle Position (TP) sensor.
10. Start the engine and rev it to 3000 rpm 2–3 times and allow it to idle.
11. Set the base timing to 8–12 degrees Before Top Dead Center (BTDC).
12. Tighten the distributor lockbolt to 83–113 inch lbs. (9–13 Nm).
13. Set the base idle speed to 700–800 rpm.
14. Stop the engine and connect the TP sensor.

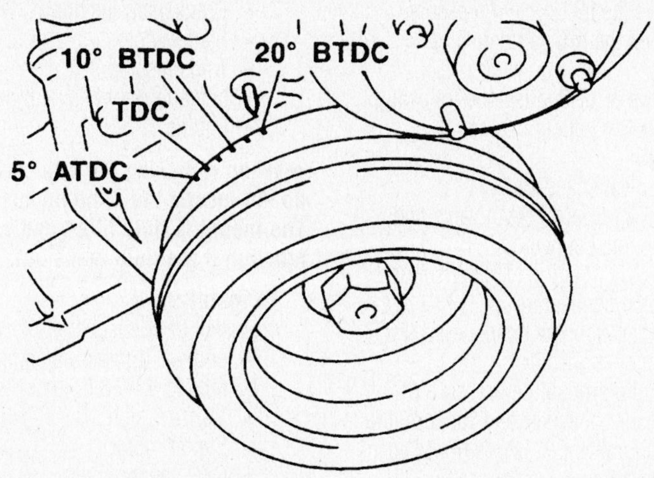

Timing indicator—3.3L engine shown

Engine Assembly

REMOVAL & INSTALLATION

Frontier

2.4L ENGINE

1. Before servicing the vehicle, refer to the precautions in the beginning of this section.
2. Drain the cooling system.
3. Relieve the fuel system pressure.
4. Remove or disconnect the following:
   - Negative battery cable
   - Hood
   - Air cleaner assembly
   - Idle Air Control (IAC) valve and solenoid connectors
   - Throttle Position (TP) sensor and switch connectors
   - Engine Coolant Temperature (ECT) sensor connector
   - Manifold Absolute Pressure (MAP) sensor connector and vacuum line
   - Evaporative Emissions (EVAP) canister purge valve connector and vacuum line
   - Mass Air Flow (MAF) sensor connector
   - Brake booster vacuum line
   - Fuel lines
   - Exhaust Gas Recirculation (EGR) temperature sensor connector
   - Throttle cable
   - Accessory drive belts
   - Radiator and hoses
   - Heater hoses
   - Exhaust manifold heat shield
   - Heated Oxygen (HO_2S) sensor connectors
   - Exhaust front pipe
   - A/C compressor, if equipped
   - Power steering pump, if equipped
   - Crankshaft Position (CKP) sensor
   - Starter motor
   - Transmission
   - Left and right engine mounts
   - Engine

To install:

5. Install or connect the following:
   - Engine. Tighten the engine mount nuts to 30–38 ft. lbs. (41–52 Nm).

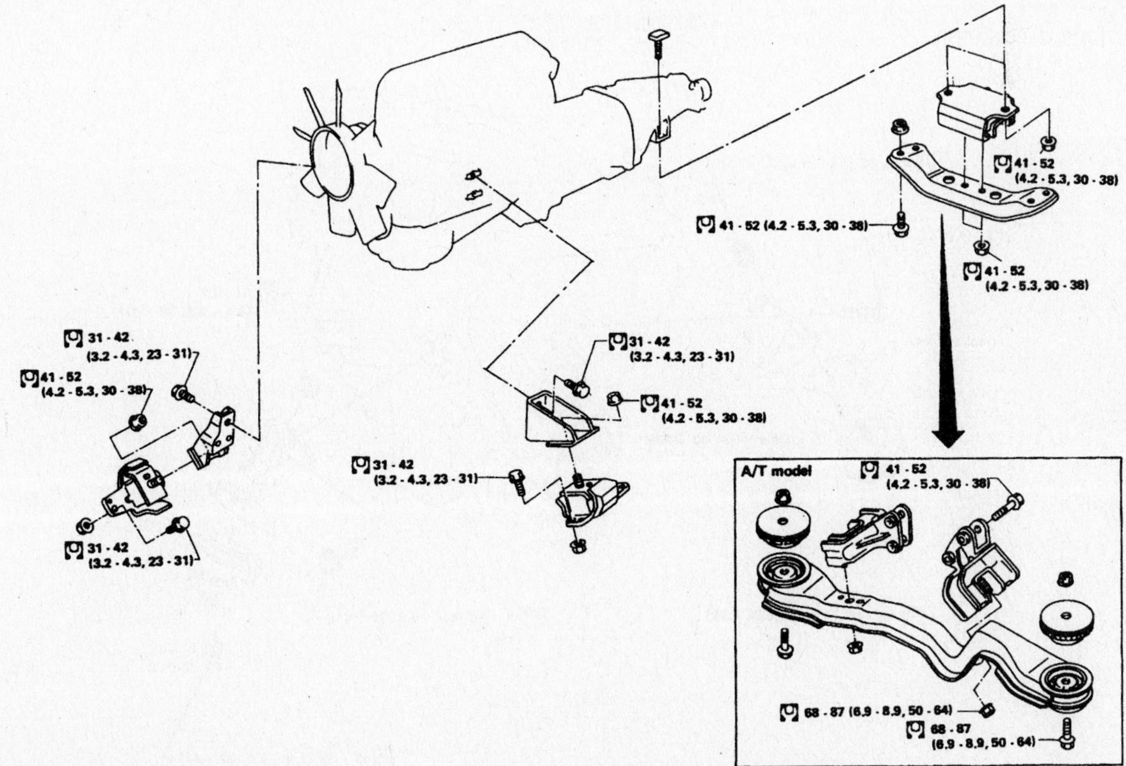

Engine and transmission mounts—2.4L engine

- Transmission
- Starter motor
- CKP sensor
- Power steering pump, if equipped
- A/C compressor, if equipped
- Exhaust front pipe
- HO_2S sensor connectors
- Exhaust manifold heat shield
- Heater hoses
- Radiator and hoses
- Accessory drive belts
- Throttle cable
- EGR temperature sensor connector
- Fuel lines
- Brake booster vacuum line
- MAF sensor connector
- EVAP canister purge valve connector and vacuum line
- MAP sensor connector and vacuum line
- ECT sensor connector
- TP sensor and switch connectors
- IAC valve and solenoid connectors
- Air cleaner assembly
- Hood
- Negative battery cable

6. Fill the cooling system.
7. Start the engine and check for leaks.

3.3L Engine

1. Before servicing the vehicle, refer to the precautions in the beginning of this section.
2. Drain the cooling system.
3. Relieve the fuel system pressure.
4. Recover the A/C refrigerant, if equipped.
5. Remove or disconnect the following:
   - Negative battery cable
   - Hood
   - Air cleaner assembly
   - Idle Air Control (IAC) valve and solenoid connectors
   - Throttle Position (TP) sensor and switch connectors
   - Engine Coolant Temperature (ECT) sensor connector
   - Manifold Absolute Pressure (MAP) sensor connector and vacuum line
   - Evaporative Emissions (EVAP) canister purge valve connector and vacuum line
   - Mass Air Flow (MAF) sensor connector
   - Brake booster vacuum line
   - Fuel lines
   - Exhaust Gas Recirculation (EGR) temperature sensor connector
   - Throttle cable
   - Accessory drive belts
   - Cooling fan and shroud
   - Radiator and hoses
   - Engine under cover
   - A/C compressor manifold
   - Power steering pump
   - Heated Oxygen (HO_2S) sensor connectors
   - Exhaust front pipes
   - Crankshaft Position (CKP) sensor
   - Starter motor
   - Transmission
   - Left and right engine mounts
   - Engine

➡When removing the engine mounts, do not loosen the 4 mount cover nuts. The mount is fluid filled and will not function if the fluid leaks out.

To install:

6. Install or connect the following:
   - Engine. Tighten the engine mount nuts to 43–58 ft. lbs. (59–78 Nm).
   - Transmission
   - Starter motor
   - CKP sensor
   - Exhaust front pipes
   - HO_2S sensor connectors
   - Power steering pump
   - A/C compressor manifold
   - Engine under cover
   - Radiator and hoses
   - Cooling fan and shroud
   - Accessory drive belts
   - Throttle cable
   - EGR temperature sensor connector
   - Fuel lines
   - Brake booster vacuum line
   - MAF sensor connector
   - EVAP canister purge valve connector and vacuum line
   - MAP sensor connector and vacuum line

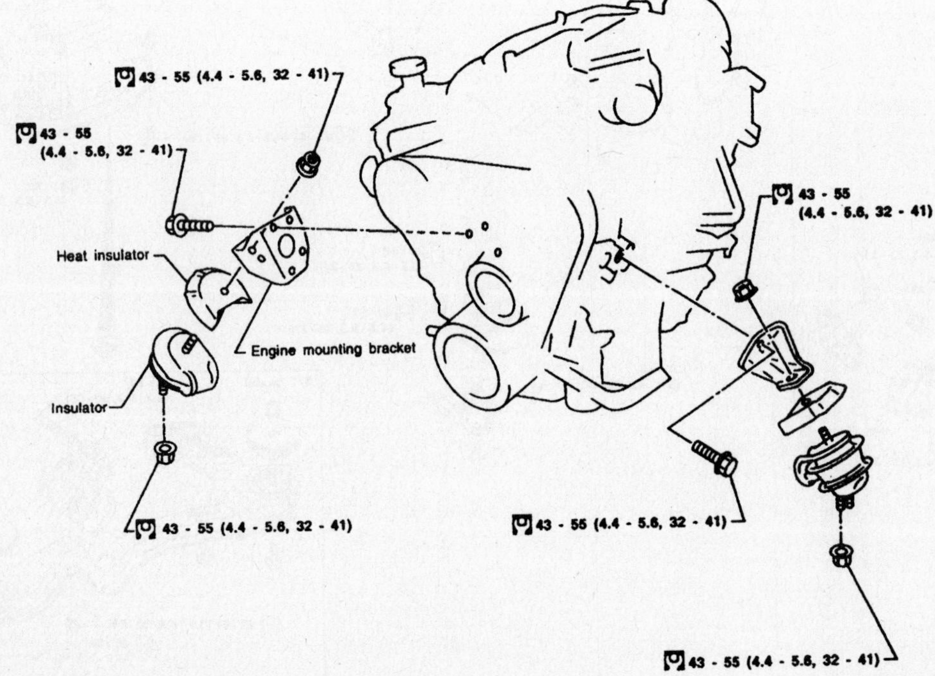

Engine mounts and related components—3.3L engine

- ECT sensor connector
- TP sensor and switch connectors
- IAC valve and solenoid connectors
- Air cleaner assembly
- Hood
- Negative battery cable

7. Fill the cooling system.
8. Recharge the A/C system, if equipped.
9. Start the engine and check for leaks.

Xterra

➡Do not loosen front engine mounting insulator cover securing bolts. When cover is removed, damper oil flows out and mounting insulator will not function.

1. Remove engine undercover and hood.
2. Drain coolant from cylinder block and radiator.
3. Remove vacuum hoses, fuel tubes, wires, harnesses and connectors.
4. Before disconnecting fuel hose, release fuel pressure from fuel line.
5. Remove radiator with shroud and cooling fan.
6. Remove drive belts.
7. Discharge refrigerant.
8. Remove A/C compressor manifold.
9. Remove power steering oil pump from engine.
10. Remove front exhaust tubes.
11. Remove transmission from vehicle.
12. Install engine slingers. Tighten the slinger bolts to 15–20 ft. lbs. (20–26 Nm).
13. Hoist engine with engine slingers and remove engine mounting nuts from both sides.
14. Lift and remove engine from vehicle.
15. Installation is the reverse of removal. See the accompanying illustration for installation torques.

Heater Core

REMOVAL & INSTALLATION

2001–02 Frontier and Xterra

1. Disconnect both the negative (1st) and positive (2nd) battery cables.
2. Remove the steering wheel by performing the following procedure:
   a. Turn the ignition switch to the OFF position.

⁂ CAUTION

Wait 3 minutes after disconnecting the battery cables and turning the ignition switch to the OFF position before servicing the air bag system.

   b. Remove the lower lid and disconnect the driver's air bag module connector.

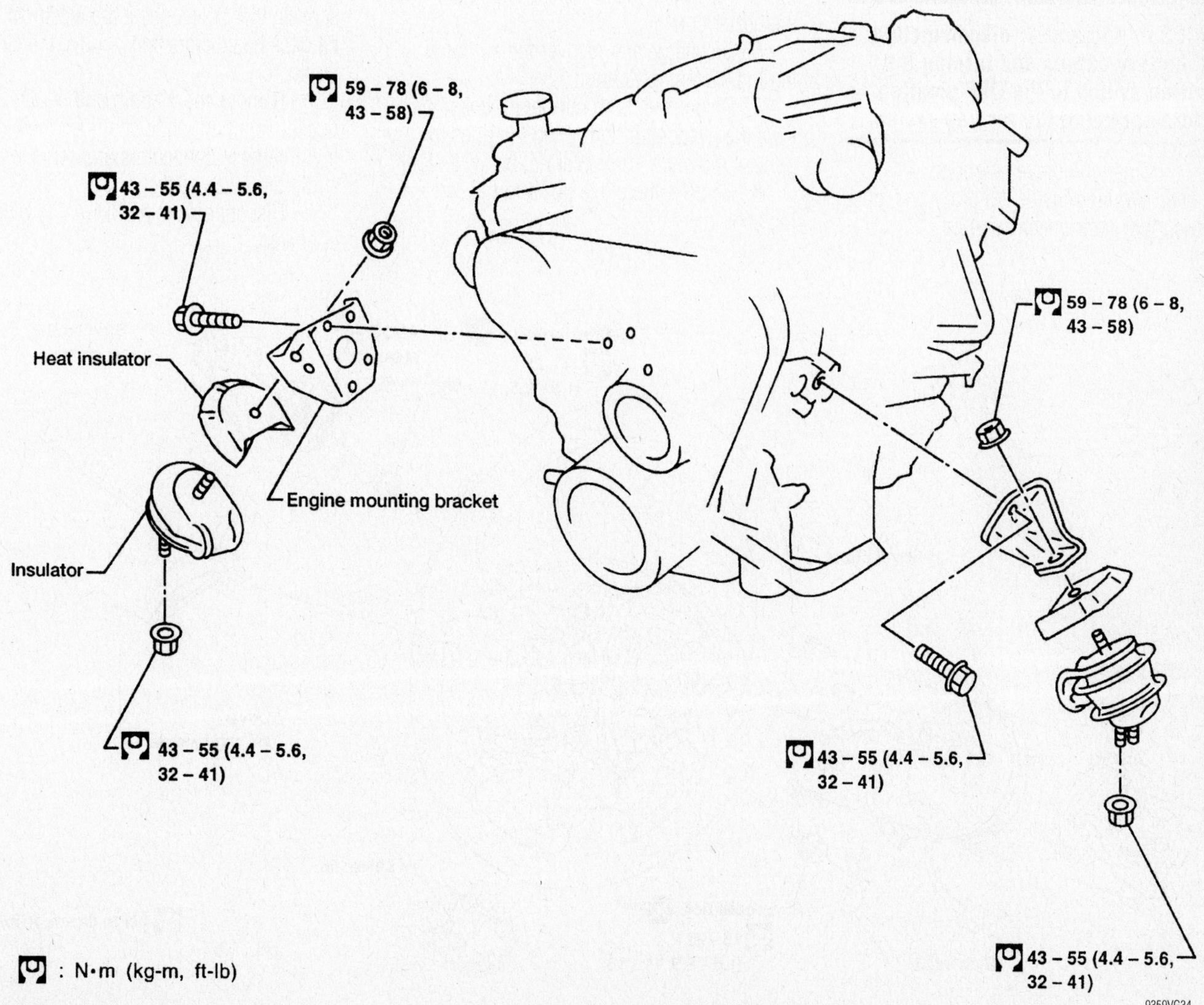

Engine mounting—Xterra 3.3L engine

c. Remove both side lids.

d. Using the Tamper Resistant Torx® Wrench size T50, remove the special bolts from both sides of the steering wheel and discard them.

e. Remove the SRS module from the steering wheel.

⁂ CAUTION

Always store the SRS module face up.

f. Position the steering wheel in the straight-ahead position.

g. Disconnect the horn connector and remove the steering wheel nut.

h. Using a steering wheel puller, press the steering wheel from the steering column.

3. Remove the passenger's side air bag by disconnecting or removing the following items:

a. Turn the ignition switch to the OFF position.

⁂ CAUTION

Wait 3 minutes after disconnecting the battery cables and turning the ignition switch to the OFF position before servicing the air bag system.

b. Open the glove box door.

c. Working inside the glove box, open the lower instrument panel lid.

d. Remove the passenger's air bag module connector clip from the lid.

e. Disconnect the passenger's SRS module connector.

f. Remove the glove box and the lower passenger's side instrument panel.

g. Using the Tamper Resistant Torx® Wrench size T50, remove the SRS module-to-instrument panel special bolts and discard them.

h. Remove the 4 SRS module-to-instrument panel mounting nuts.

i. Release the SRS module-to-instrument panel clips and remove the SRS module.

⁂ CAUTION

Always store the SRS module face up.

4. Drain the cooling system into a clean container for reuse.

5. Working inside the engine compartment, disconnect the 2 heater hoses from the heater core.

6. Discharge and recover the air conditioning system refrigerant.

7. Disconnect both refrigerant lines from the evaporator core. Plug the lines to prevent moisture from entering the system.

8. Remove the glove box and the mating trim.

9. Disconnect the thermal amp connector.

10. Remove the air conditioning housing assembly from the vehicle.

11. Remove the instrument panel assembly by performing the following procedure:

a. Remove the 4 steering column cover screws; then, separate and remove the steering column covers.

b. Remove the 2 driver's side lower instrument panel screws and the lower instrument panel.

c. Remove the 4 cluster cover screws and the cluster cover.

d. Remove the 6 combination meter screws; then, disconnect the combination meter electrical connector and remove the meter.

e. Remove the 2 glove box screws and the glove box.

f. Remove the 2 instrument stay cover screws; then, disconnect the electrical harness connectors and remove the stay cover.

g. Remove the 2 cluster lid "C" screws; then, disconnect the electrical harness connectors and remove the cluster lid "C".

h. Remove the 4 audio and deck pocket screws; then, disconnect the electrical harness connectors and remove the audio and deck pocket.

i. Disconnect the ASCD main switch connector.

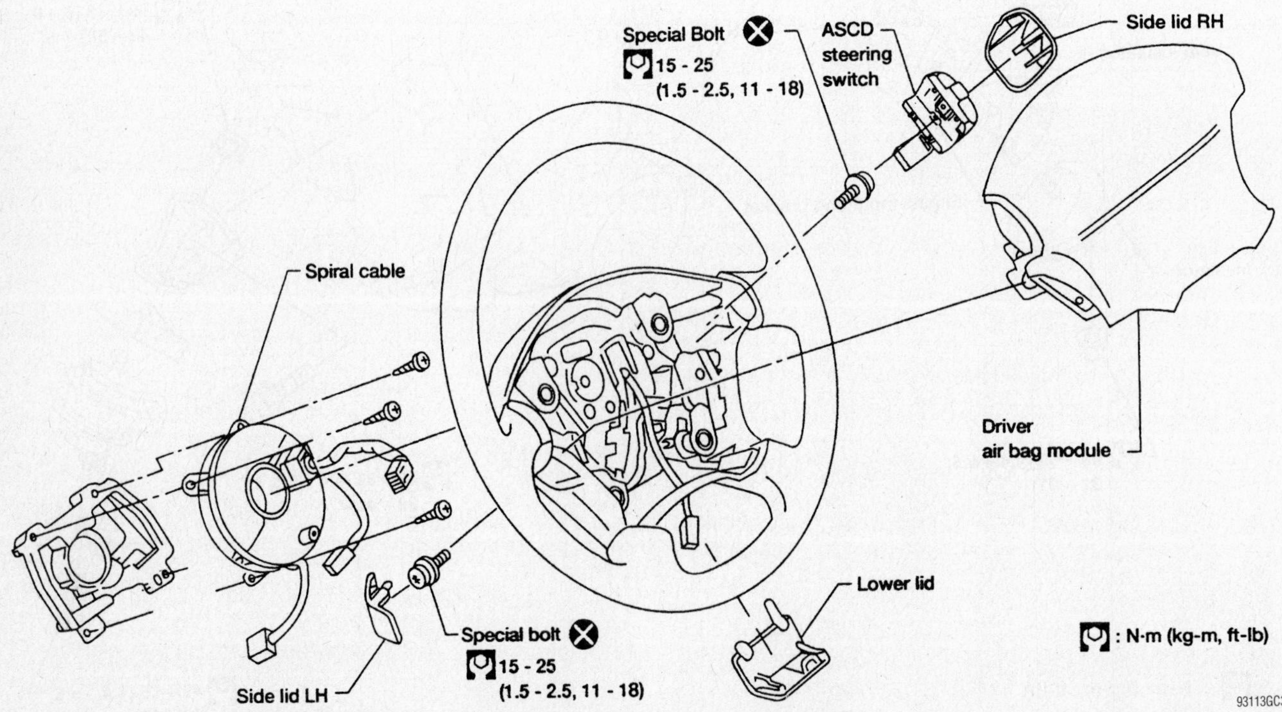

Exploded view of the steering wheel and SRS module and related components—2001–02 Frontier

j. Remove the 2 meter cover screws; then, disconnect the electrical harness connectors and remove the meter cover.

k. Remove the 4 air conditioning-heater control screws; then, disconnect the control cables and remove the air conditioning-heater control.

l. Remove the front pillar garnish.

m. Remove the 3 instrument panel assembly nuts and 2 bolts; then, remove the instrument panel.

12. Remove the heater housing assembly.

13. Remove the heater core from the heater housing assembly.

To install:

14. Install the heater core to the heater housing assembly.

15. Install the heater housing assembly.

16. Install the instrument panel assembly by performing the following procedure:

a. Install the instrument panel and the 3 instrument panel assembly nuts and 2 bolts.

b. Install the front pillar garnish.

c. Install the air conditioning-heater control, connect the control cables and install the 4 air conditioning/heater control screws.

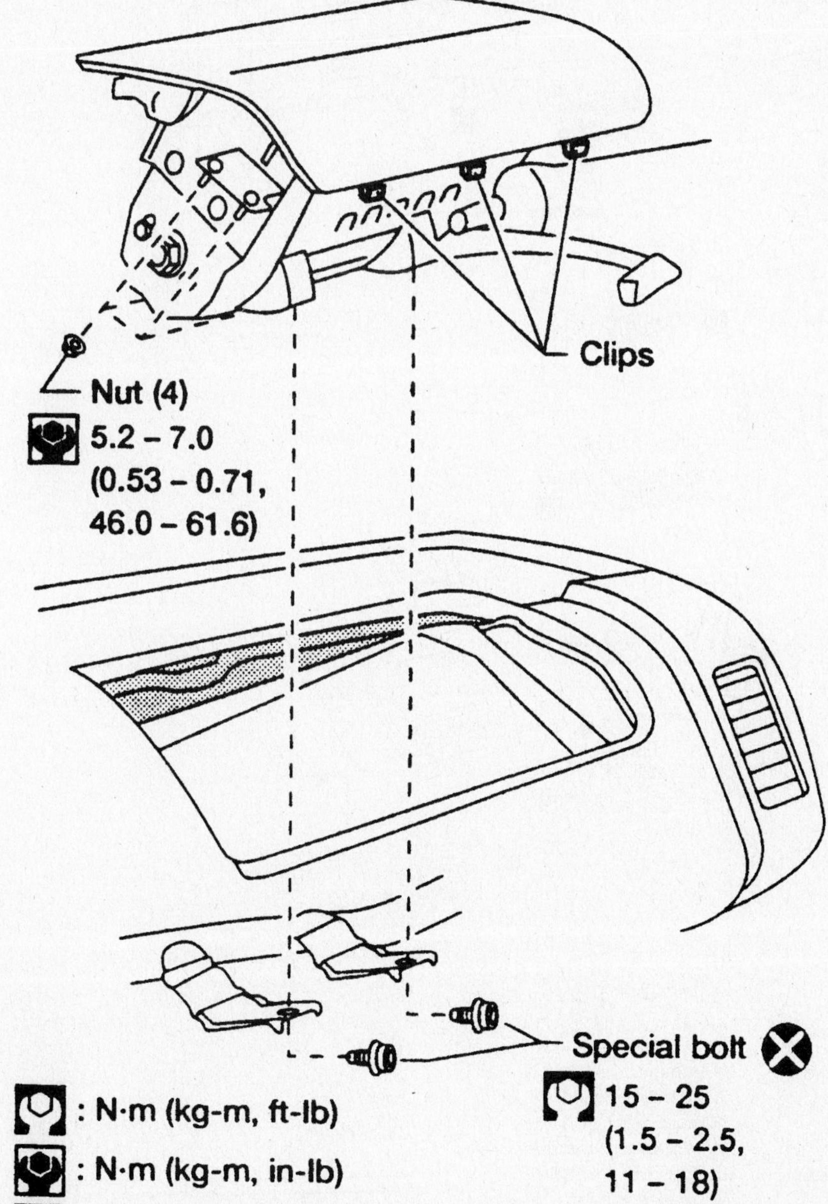

View of the passenger's side SRS module and related components—2001–02 Frontier

d. Install the meter cover, connect the electrical harness connectors and install the 2 meter cover screws.

e. Connect the ASCD main switch connector.

f. Install the audio and deck pocket, connect the electrical harness connectors and install the 4 audio and deck pocket screws.

g. Install the cluster lid "C", connect the electrical harness connectors and install the 2 cluster lid "C" screws.

h. Install the stay cover, connect the electrical harness connectors and the 2 instrument stay cover screws.

i. Install the glove box and the 2 glove box screws.

j. Install the meter, connect the combination meter electrical connector and install the 6 combination meter screws.

k. Install the cluster cover and the 4 cluster cover screws.

l. Install the lower instrument panel and the 2 driver's side lower instrument panel screws.

m. Install the steering column covers and the 4 steering column cover screws.

17. Install the air conditioning housing assembly to the vehicle.

18. Connect the thermal amp connector.

19. Install the glove box and the mating trim.

20. Connect both refrigerant lines to the evaporator core.

21. Inside the engine compartment, connect the heater hoses to the heater core.

22. Refill the cooling system.

23. Install the passenger's side air bag by performing the following procedure:

a. Install the SRS module and secure the SRS module-to-instrument panel clips.

b. Install the 4 SRS module-to-instrument panel mounting nuts.

c. Using the Tamper Resistant Torx® Wrench size T50, install the new special SRS module-to-instrument panel bolts.

d. Install the lower passenger's side instrument panel and the glove box.

e. Connect the passenger's SRS module connector.

f. Install the passenger's air bag module connector clip to the lid.

g. Inside the glove box, close the lower instrument panel lid.

h. Close the glove box door.

24. Install the steering wheel by performing the following procedure:

a. Align the spiral cable pin guide and install the steering wheel by pulling the spiral cable connectors through it.

b. Connect the horn connector and

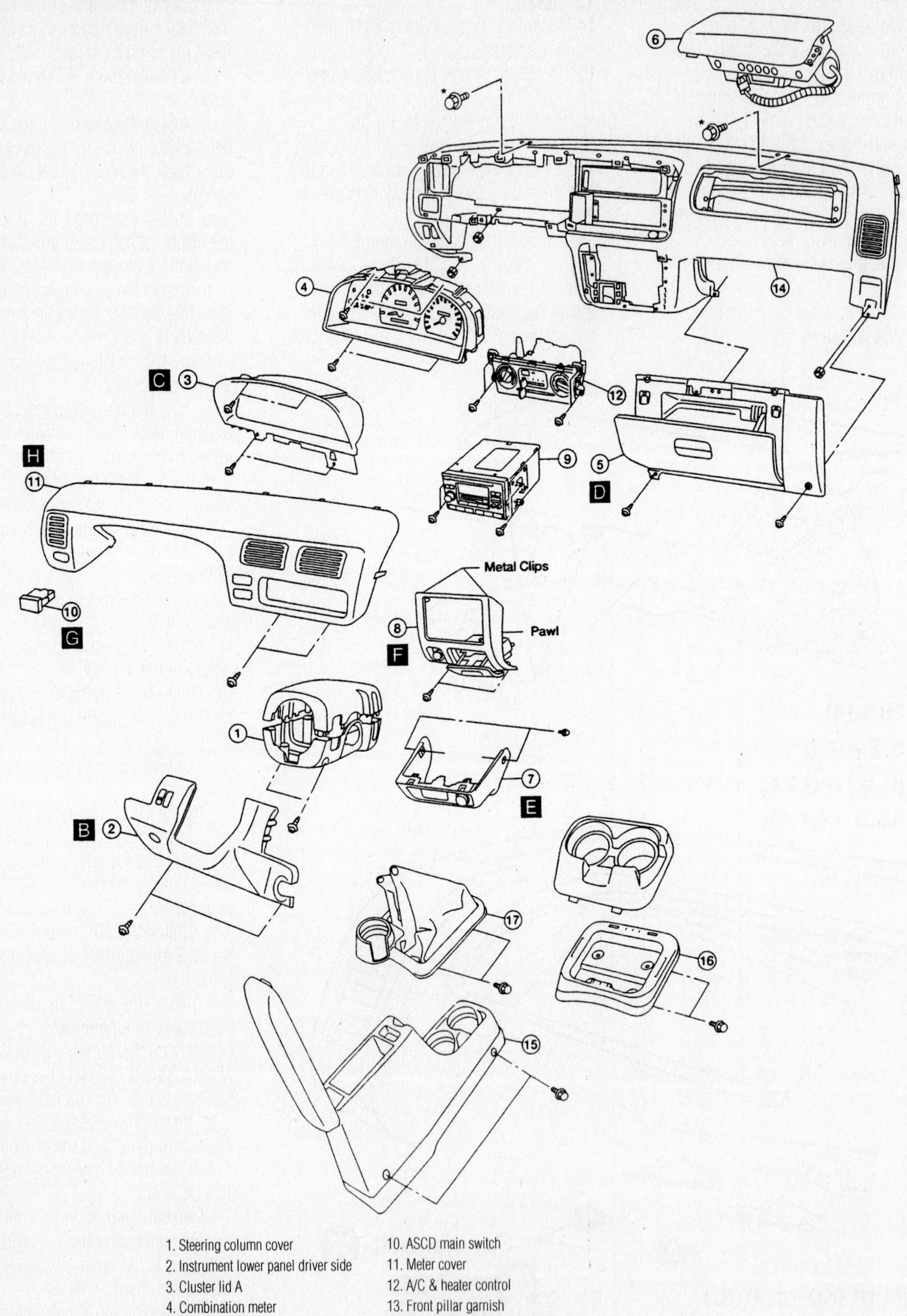

1. Steering column cover
2. Instrument lower panel driver side
3. Cluster lid A
4. Combination meter
5. Glove box assembly
6. Passenger side air bag module
7. Instrument stay cover
8. Cluster lid C
9. Audio and deck pocket
10. ASCD main switch
11. Meter cover
12. A/C & heater control
13. Front pillar garnish
14. Instrument panel assembly
15. Center console assembly
16. Cup holder assembly
17. M/T boot assembly

93113GC5

Exploded view of the instrument panel and related components—2001–02 Frontier

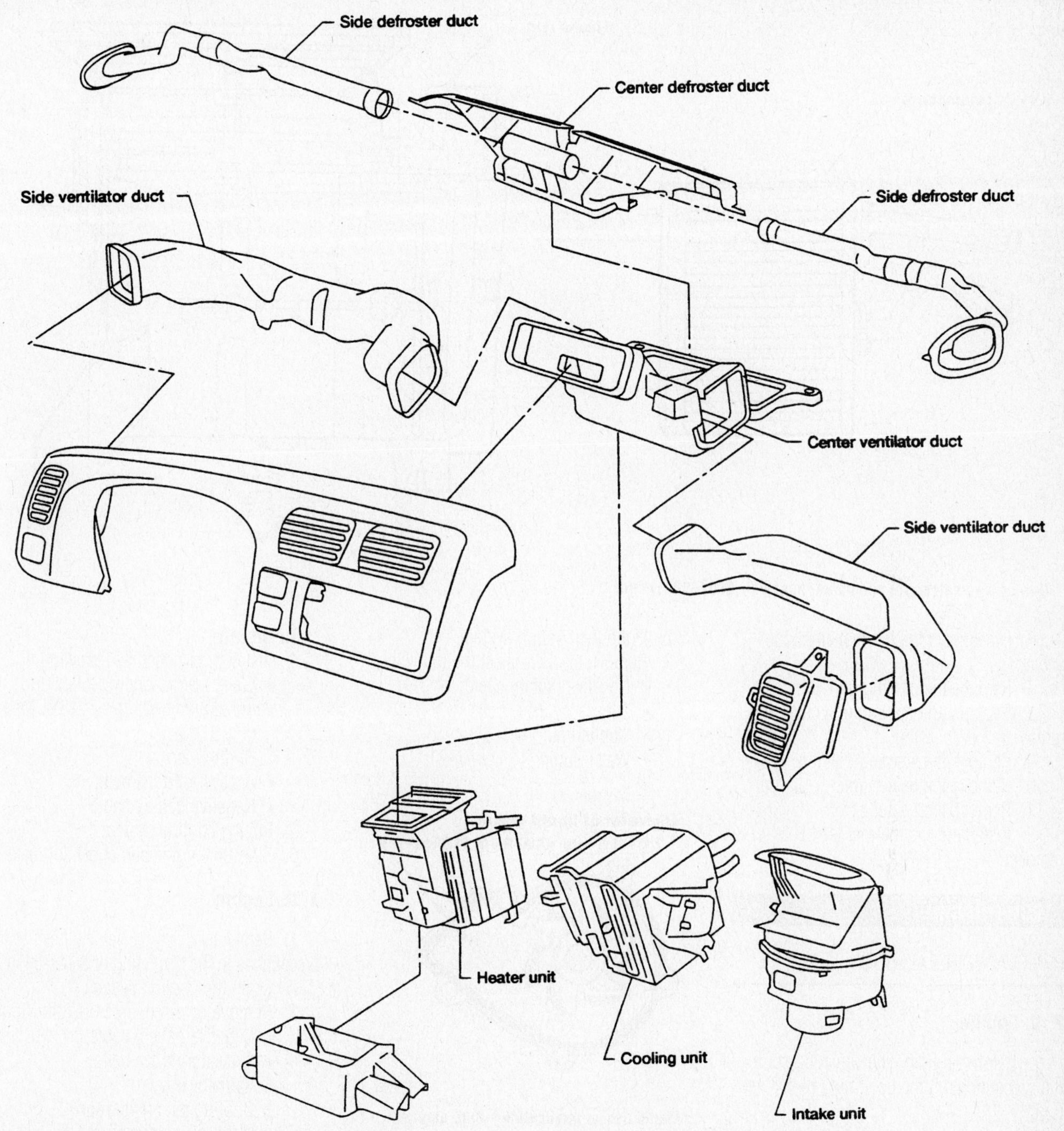

Exploded view of the heater housing, air conditioning housing and related components—2001–02 Frontier

connect the spiral cable by aligning the pawls in the steering wheel.

c. Install the steering wheel nut and torque it to 22–29 ft. lbs. (29–39 Nm).

d. Install the SRS module to the steering wheel.

e. Using the Tamper Resistant Torx® Wrench size T50, install the new special bolts to both sides of the steering wheel.

f. Install the lower lid and disconnect the driver's air bag module connector.

g. Install both side lids.

h. Rotate the steering wheel fully right and left to make sure that the spiral cable is set in the neutral position.

25. Connect both the positive (1st) and negative (2nd) battery cables.

26. Evacuate and charge the air conditioning system refrigerant.

27. Run the engine to normal operating temperatures; then, check the climate control operation and check for leaks.

2003–04 Frontier and Xterra

1. Drain cooling system.
2. Disconnect the two heater hoses from the engine compartment side.
3. Discharge the A/C system.
4. Disconnect the two evaporator core refrigerant lines from the engine compartment side. Cap the refrigerant lines to prevent moisture from entering the system.
5. Remove the glove box and mating trim.

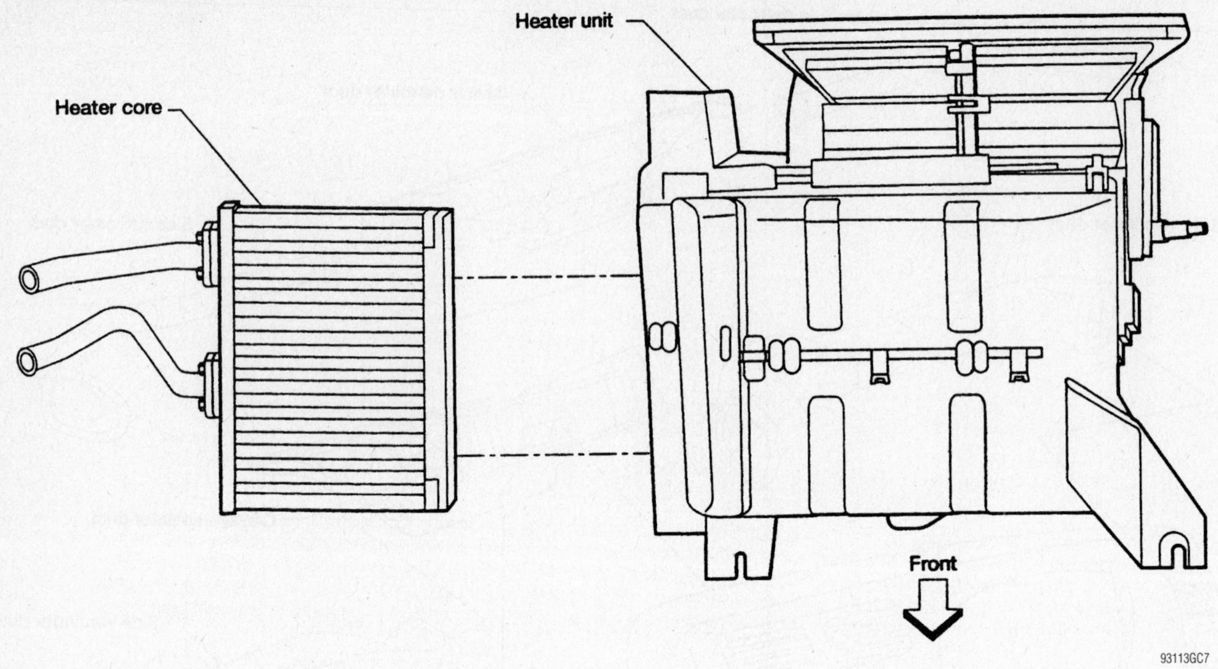

View of the heater core and heater housing—2001–02 Frontier

6. Disconnect the thermal amp. connector.
7. Remove the cooling unit.
8. Separate the cooling unit case, and remove the evaporator.
9. Remove the steering column.
10. Remove the heater unit.
11. Remove the heater core.
12. Installation is the reverse of removal.

Water Pump

REMOVAL & INSTALLATION

2.4L Engine

1. Before servicing the vehicle, refer to the precautions in the beginning of this section.
2. Drain the cooling system.
3. Remove or disconnect the following:
   - Negative battery cable
   - Accessory drive belts
   - Cooling fan
   - Water pump

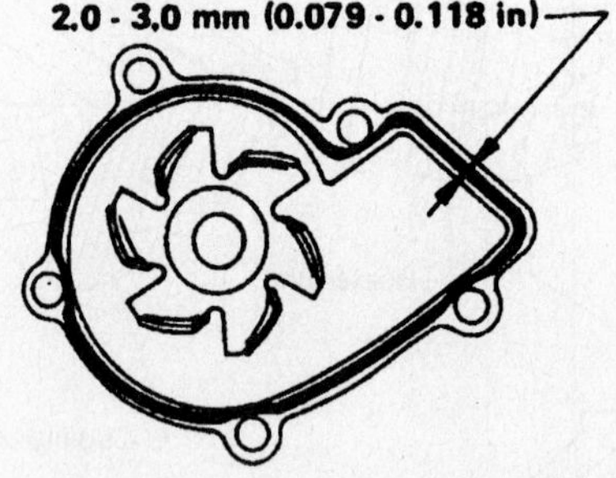

Liquid gasket application—2.4L engine

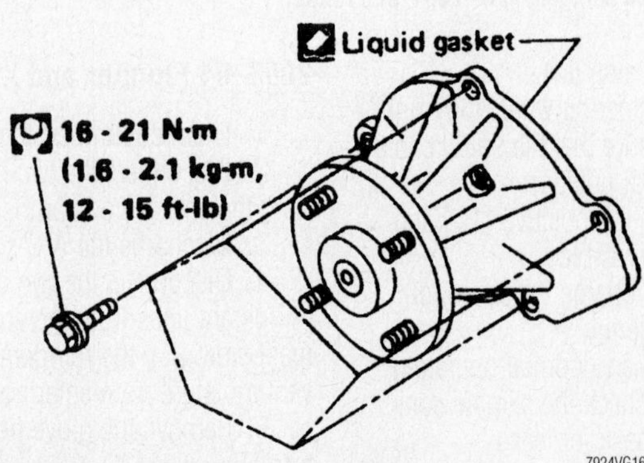

Water pump assembly—2.4L engine

To install:

4. Install or connect the following:
   - Water pump. Apply sealant and tighten the bolts to 12–15 ft. lbs. (16–21 Nm).
   - Cooling fan
   - Accessory drive belts
   - Negative battery cable
5. Fill the cooling system.
6. Start the engine and check for leaks.

3.3L Engine

1. Before servicing the vehicle, refer to the precautions in the beginning of this section.
2. Drain the cooling system.
3. Remove or disconnect the following:
   - Negative battery cable
   - Accessory drive belts
   - Radiator hoses
   - Cooling fan and shroud
   - Water pump pulley
   - Front cover
   - Timing belt. Refer to the Timing Belt unit repair section.
   - Water pump

To install:

4. Install or connect the following:
   - Water pump. Tighten the bolts to 12–15 ft. lbs. (16–21 Nm).
   - Timing belt
   - Front cover
   - Water pump pulley
   - Cooling fan and shroud
   - Radiator hoses
   - Accessory drive belts
   - Negative battery cable

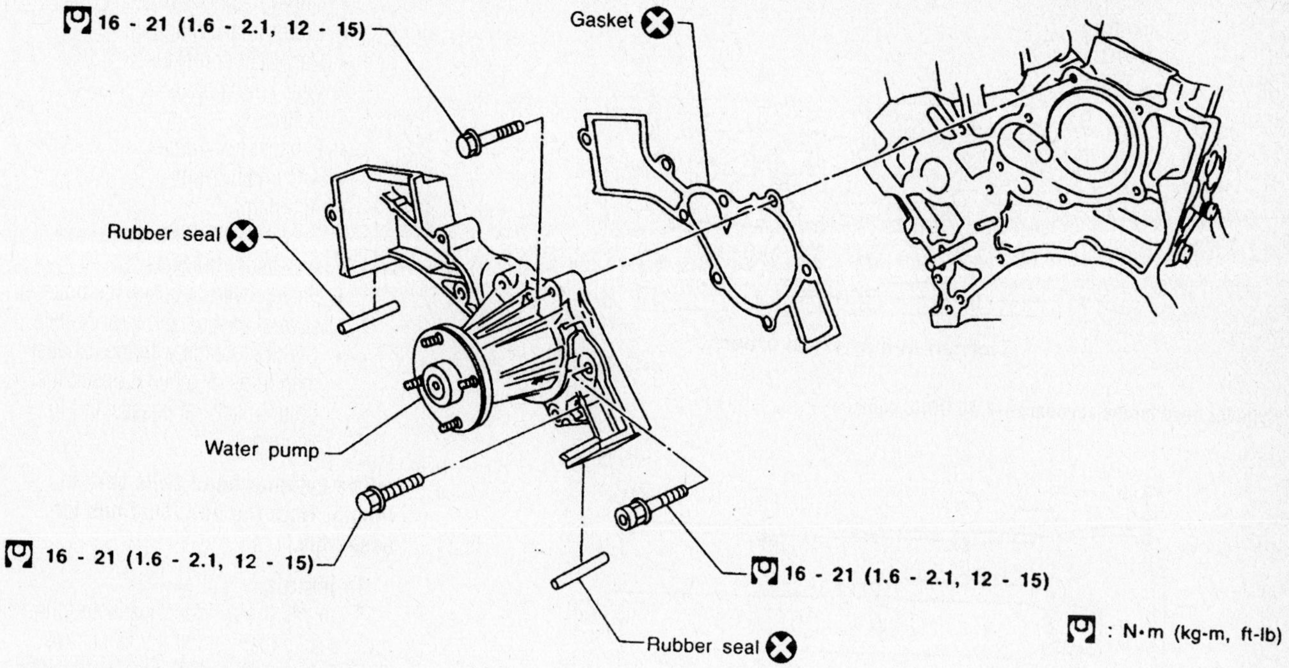

16 - 21 (1.6 - 2.1, 12 - 15)
Gasket
Rubber seal
Water pump
16 - 21 (1.6 - 2.1, 12 - 15)
16 - 21 (1.6 - 2.1, 12 - 15)
Rubber seal
: N•m (kg-m, ft-lb)
7924VG20

Exploded view of the water pump assembly—3.3L engine

5. Fill the cooling system.
6. Start the engine and check for leaks.

Cylinder Head

REMOVAL & INSTALLATION

2.4L Engine

1. Before servicing the vehicle, refer to the precautions in the beginning of this section.
2. Drain the cooling system.
3. Relieve the fuel system pressure.
4. Remove or disconnect the following:
   - Negative battery cable
   - Air cleaner assembly
   - Spark plug wires
   - Radiator hoses
   - Accessory drive belts
   - Fuel lines
   - Intake manifold
   - Exhaust manifold
   - Valve cover. Remove the bolts in the sequence shown.
   - Camshaft sprocket cover
   - Camshafts sprockets and upper timing chain
5. Wedge the lower timing chain in place to prevent the chain tensioner from expanding.
6. Remove or disconnect the following:
   - Timing chain idler sprocket
   - Camshafts
   - Cylinder head. Loosen the bolts in several passes and in sequence as shown.

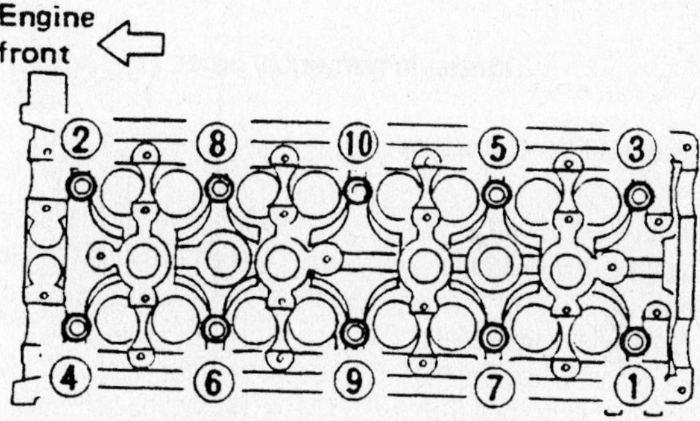

Cylinder head loosening sequence—2.4L engine

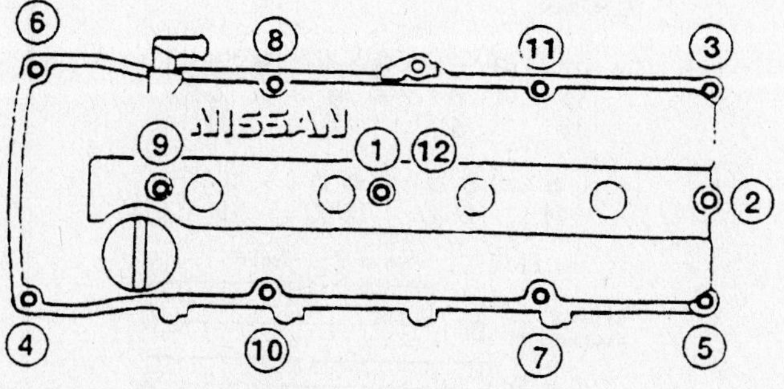

Valve cover loosening sequence—2.4L engine

To install:

7. Install the cylinder head with a new gasket. Tighten the bolts in sequence as follows:
   a. Step 1: 22 ft. lbs. (30 Nm)
   b. Step 2: 59 ft. lbs. (79 Nm)
   c. Step 3: Loosen all bolts completely
   d. Step 4: 18–25 ft. lbs. (25–34 Nm)

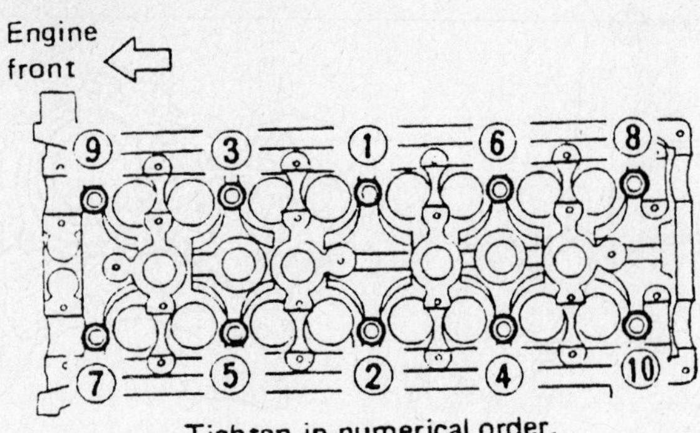

Cylinder head torque sequence—2.4L DOHC engine

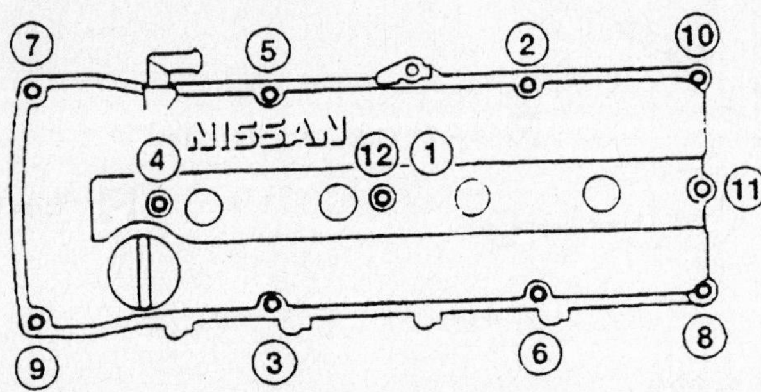

Valve cover torque sequence—2.4L engine

e. Step 5: Plus 86–91 degrees

8. Install or connect the following:
   - Camshafts
   - Timing chain idler sprocket and lower timing chain. Remove the wedge and tighten the bolt to 48–61 ft. lbs. (66–83 Nm).
   - Camshaft sprockets and upper timing chain. Tighten the bolts to 123–130 ft. lbs. (167–177 Nm).
   - Camshaft sprocket cover
   - Valve cover. Tighten the bolts in sequence to 69–95 ft. lbs. (8–11 Nm).
   - Exhaust manifold
   - Intake manifold
   - Fuel lines
   - Accessory drive belts
   - Radiator hoses
   - Spark plug wires
   - Air cleaner assembly
   - Negative battery cable
9. Fill the cooling system.
10. Start the engine and check for leaks.

3.3L Engine

1. Before servicing the vehicle, refer to the precautions in the beginning of this section.
2. Drain the cooling system.
3. Relieve the fuel system pressure.
4. Remove or disconnect the following:
   - Negative battery cable
   - Accessory drive belts
   - Front cover
   - Timing belt. Refer to the Timing Belt unit repair section.
   - Upper intake manifold
   - Lower intake manifold
   - Camshaft sprockets
   - Rear timing cover
   - Distributor
   - Exhaust front pipes
   - A/C compressor
   - Alternator
   - Power steering pump
   - Accessory brackets
   - Valve covers. Loosen the bolts in several passes and in sequence.
   - Cylinder heads with the exhaust manifolds attached. Loosen the bolts in several passes and in sequence.

➡The cylinder head bolts vary in length. Note the bolt locations for assembly.

To install:

5. Install the cylinder heads and the lower intake manifold at the same time. Tighten the bolts in sequence as follows:
   a. Step 1: Tighten the cylinder head bolts to 22 ft. lbs. (29 Nm)
   b. Step 2: Tighten the cylinder head bolts to 43 ft. lbs. (59 Nm)
   c. Step 3: Loosen all cylinder head bolts completely
   d. Step 4: Tighten the cylinder head bolts to 84 inch lbs. (10 Nm)
   e. Step 5: Tighten the intake manifold fasteners to 35 inch lbs. (4 Nm)
   f. Step 6: Tighten the intake manifold fasteners to 13 ft. lbs. (18 Nm)
   g. Step 7: Tighten the intake manifold fasteners to 12–14 ft. lbs. (16–20 Nm)

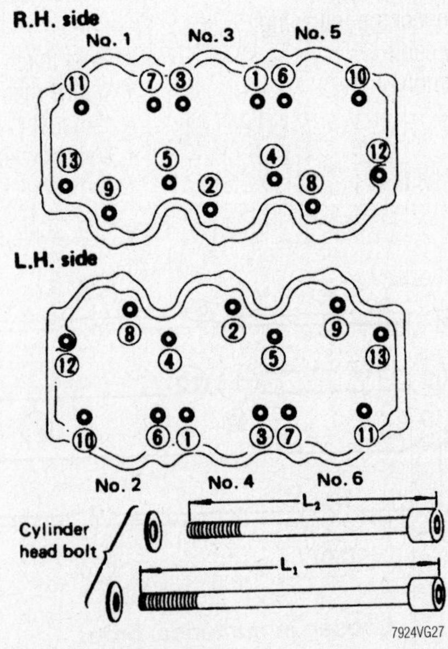

Cylinder head torque sequence—3.3L engine

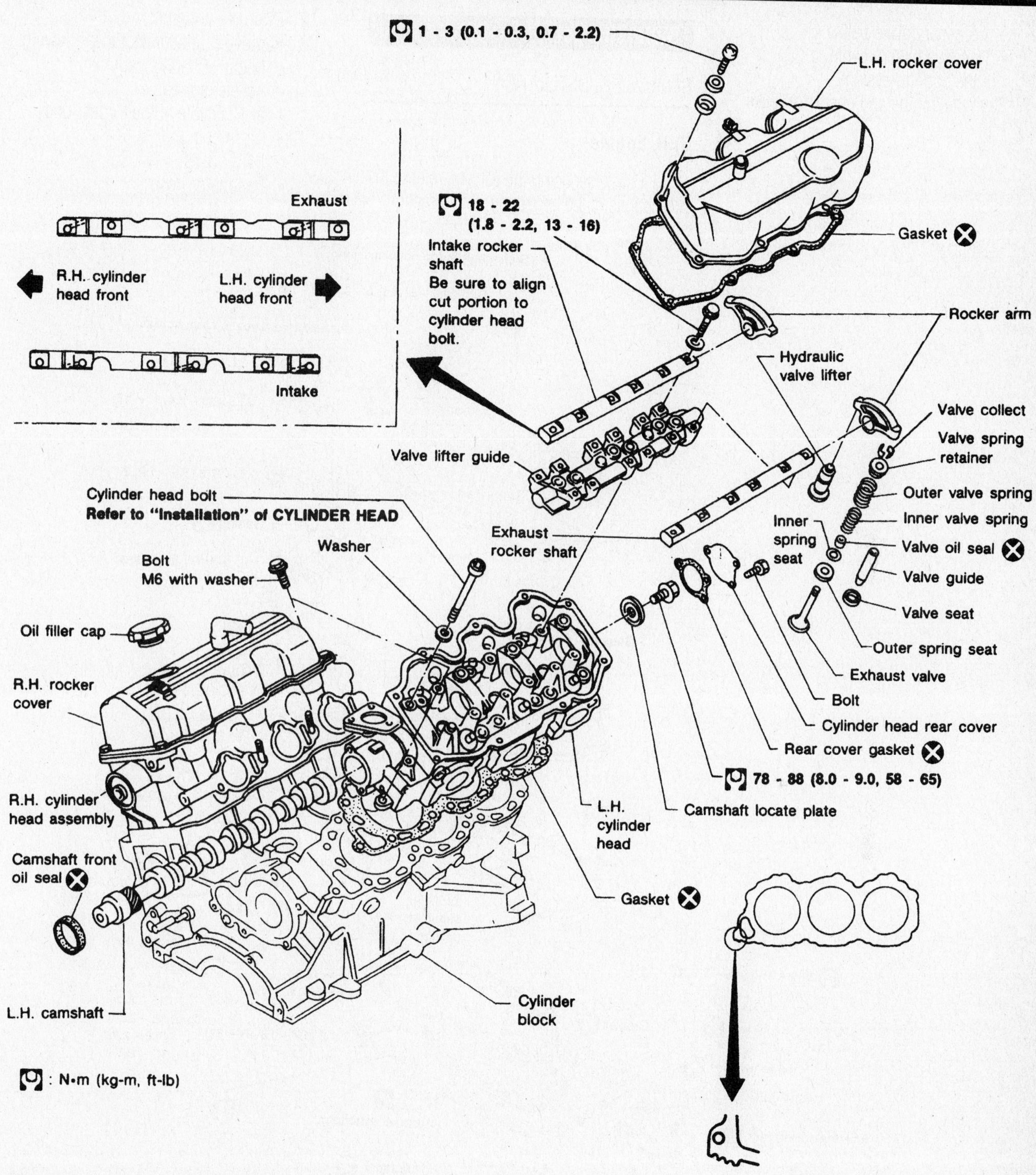

Exploded view of the cylinder head assembly—3.3L engine

h. Step 8: Loosen all intake fasteners completely

i. Step 9: Tighten the cylinder head bolts to 22 ft. lbs. (29 Nm)

j. Step 10: Tighten the cylinder head bolts 60–65 degrees **OR** tighten to 40–47 ft. lbs. (54–64 Nm)

k. Step 11: Tighten the cylinder head sub-bolts to 80–105 inch lbs. (9–12 Nm)

l. Step 12: Tighten the intake manifold fasteners to 35 inch lbs. (4 Nm)

m. Step 13: Tighten the intake manifold fasteners to 78 inch lbs. (9 Nm)

n. Step 14: Tighten the intake manifold fasteners to 70–84 inch lbs. (6–7 Nm)

6. Install or connect the following:
   - Valve covers
   - Accessory brackets
   - Power steering pump
   - Alternator
   - A/C compressor
   - Exhaust front pipes
   - Distributor
   - Rear timing cover
   - Camshaft sprockets
   - Upper intake manifold
   - Timing belt
   - Front cover

- Accessory drive belts
- Negative battery cable

7. Fill the cooling system.
8. Start the engine and check for leaks.

Supercharger

REMOVAL & INSTLLATION

3.3L Engine

1. Before servicing the vehicle, refer to the precautions in the beginning of this section.
2. Drain the coolant.
3. Remove or disconnect the following:
   - Negative battery cable
   - Accelerator cable
   - ASCD cable at the throttle body
   - Air inlet duct
   - PCV hoses
   - Resonator hose

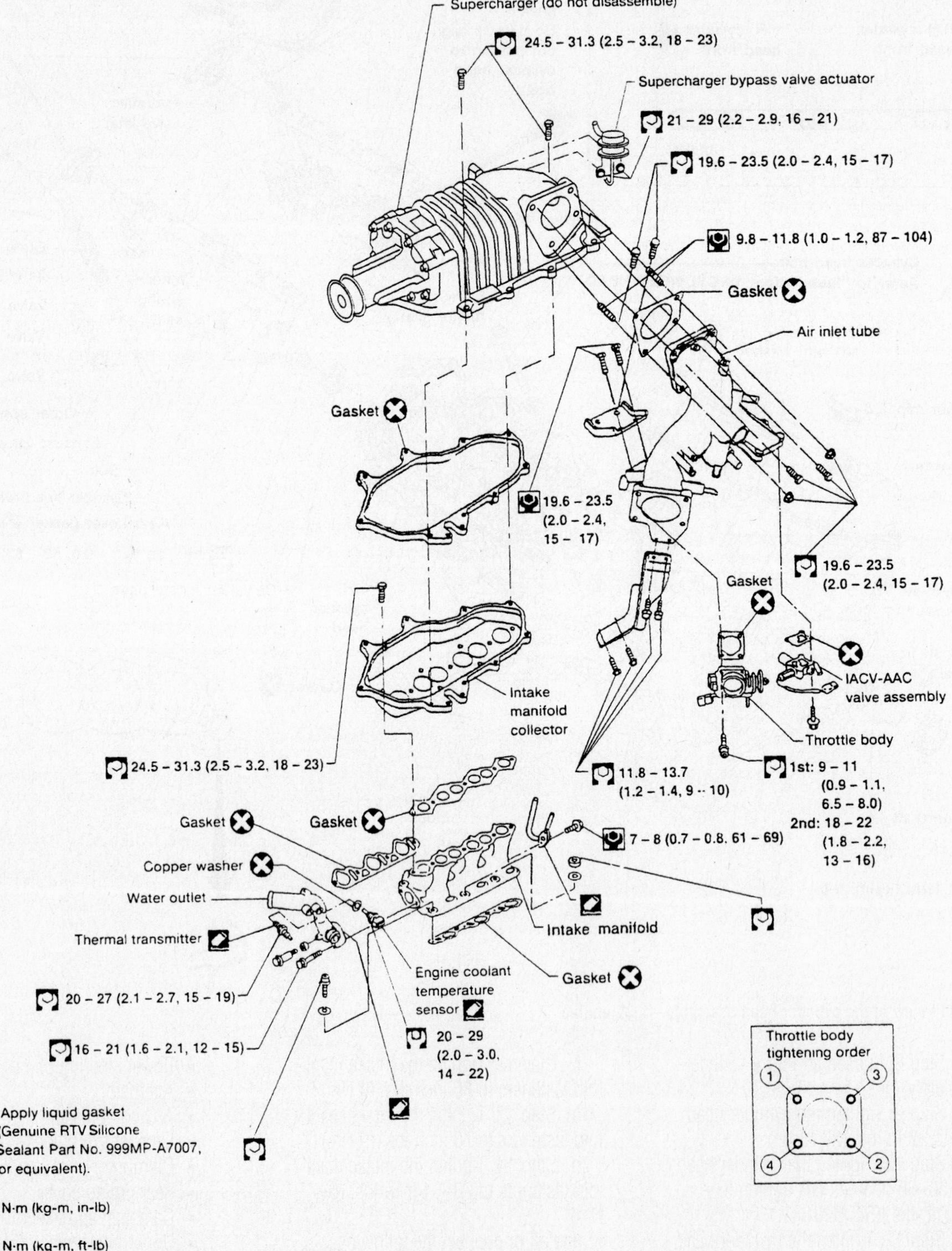

Supercharger and related parts—3.3L engine

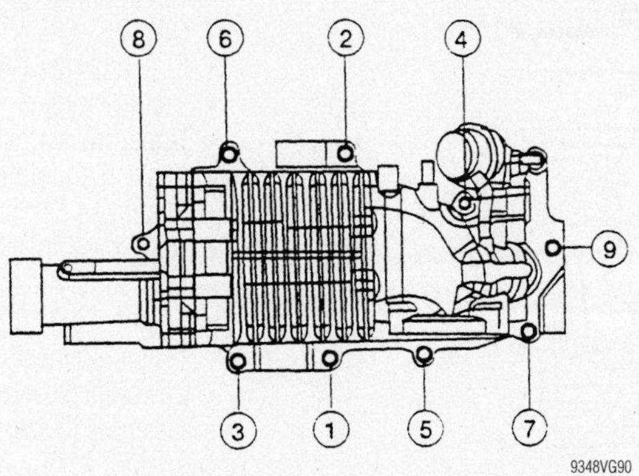

Supercharger torque sequence. Loosen in reverse order—3.3L engine

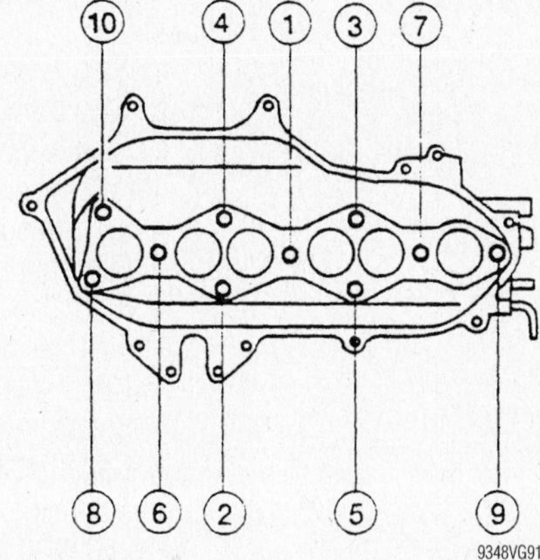

Intake manifold collector torque sequence. Loosen in reverse order—3.3L supercharged engine

- Supercharger pulley cover
- Supercharger drive belt
- Air inlet tube supports
- Air inlet tube
- EVAP vacuum hose
- Brake booster hose
- All remaining hoses and wires in the way of removal
- Intake manifold collector
- Heater hoses
- Supercharger

4. Installation is the reverse of removal. Observe the following torques:
   - Supercharger mounting bolts: 18–23 ft. lbs. (24–31 Nm).
   - Air inlet tube-to-supercharger: 15–17 ft. lbs. (20–24 Nm).

Rocker Arms/Shafts

REMOVAL & INSTALLATION

2.4L Engine

This engine is not equipped with rocker arms. The camshafts act directly on the valve lifters.

3.3L Engine

1. Before servicing the vehicle, refer to the precautions in the beginning of this section.
2. Remove or disconnect the following:
   - Negative battery cable
   - Supercharger or upper intake manifold
   - Valve covers
   - Rocker arm and shaft assemblies
   - Rocker arms from the shafts

➡Keep all valvetrain components in order for assembly.

To install:

3. Lubricate all contact points with clean engine oil and assemble the rocker arms to the shafts in their original positions.
4. Install or connect the following:
   - Rocker arm and shaft assemblies. Tighten the bolts to 13–16 ft. lbs. (18–22 Nm).
   - Valve covers
   - Upper intake manifold
   - Negative battery cable
5. Start the engine and check for leaks.

Intake Manifold

REMOVAL & INSTALLATION

2.4L Engine

1. Before servicing the vehicle, refer to the precautions in the beginning of this section.
2. Drain the cooling system.
3. Relieve the fuel system pressure.
4. Remove or disconnect the following:
   - Negative battery cable
   - Air cleaner assembly
   - Coolant hoses
   - Fuel lines
   - Accelerator cable
   - Cruise control cable, if equipped
   - Positive Crankcase Ventilation (PCV) valve and hose
   - Exhaust Gas Recirculation (EGR) tube
   - EGR temperature sensor connector
   - Idle Air Control (IAC) valve and solenoid connectors
   - Throttle Position (TP) sensor and switch connectors
   - Engine Coolant Temperature (ECT) sensor connector
   - Manifold Absolute Pressure (MAP) sensor connector and vacuum line
   - Evaporative Emissions (EVAP) canister purge valve vacuum line
   - Brake booster vacuum line
   - Fuel injector connectors
   - Intake manifold bracket
   - Intake manifold. Loosen the fasteners in reverse of the torque sequence.

To install:

5. Install or connect the following:
   - Intake manifold. Tighten the bolts to 12–14 ft. lbs. (16–19 Nm).
   - Intake manifold bracket. Tighten the bolts to 24–28 ft. lbs. (32–38 Nm).
   - Fuel injector connectors
   - Brake booster vacuum line
   - EVAP canister purge valve vacuum line
   - MAP sensor connector and vacuum line
   - ECT sensor connector
   - TP sensor and switch connectors
   - IAC valve and solenoid connectors
   - EGR temperature sensor connector
   - EGR tube
   - PCV valve and hose
   - Cruise control cable, if equipped
   - Accelerator cable
   - Fuel lines
   - Coolant hoses

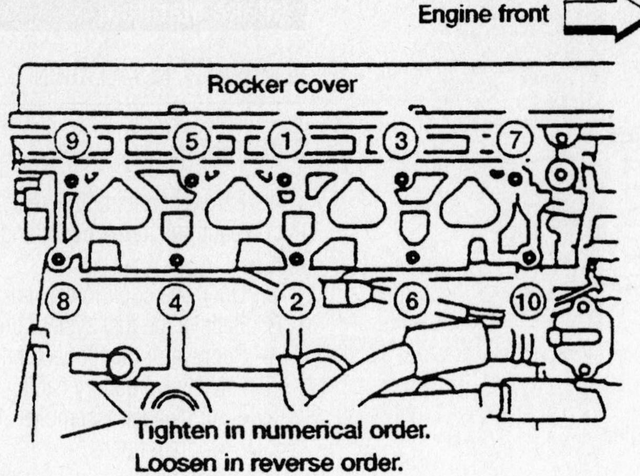

Intake manifold torque sequence—2.4L engines

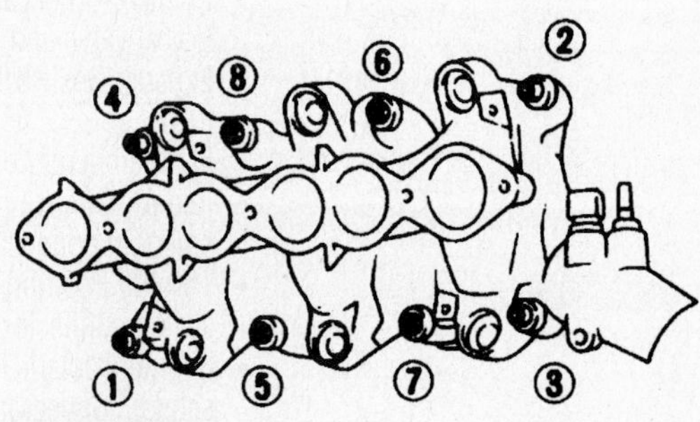

Intake manifold loosening sequence—3.3L engine

- Air cleaner assembly
- Negative battery cable

6. Fill the cooling system.
7. Start the engine and check for leaks.

3.3L Engine

1. Before servicing the vehicle, refer to the precautions in the beginning of this section.
2. Drain the cooling system.
3. Relieve the fuel system pressure.
4. Remove or disconnect the following:
   - Negative battery cable
   - Air intake duct
   - Accelerator cable
   - Cruise control cable
   - Idle Air Control (IAC) valve connector
   - Throttle Position (TP) sensor and switch connectors
   - Ignition coil and power transistor connectors
   - Exhaust Gas Recirculation (EGR) Solenoid valve connector
   - EGR temperature sensor connector
   - Radiator hoses
   - Heater hoses
   - Positive Crankcase Ventilation (PCV) valve and hose
   - Evaporative Emissions (EVAP) canister vacuum and purge hoses
   - Brake booster vacuum hose
   - Fuel pressure regulator vacuum hose
   - EGR tube
   - Spark plug wires
   - Distributor
   - Left bank injector connectors
   - Thermal transmitter
   - Upper intake manifold ground cable (VG33)
   - Supercharger (VG33ER)
   - Breather pipe
   - Upper intake manifold (VG33)
   - Intake manifold collector (VG33ER)
   - Fuel lines
   - Right bank injector connectors
   - Fuel supply manifold
   - Engine Coolant Temperature (ECT) sensor connector
   - Lower intake manifold. Loosen the fasteners in the sequence shown.

To install:

5. Install the lower intake manifold with a new gasket.
6. Tighten the fasteners in sequence as follows:
   a. Step 1: 35 inch lbs. (4 Nm)
   b. Step 2: 78 inch lbs. (9 Nm)
   c. Step 3: 70–84 inch lbs. (8–10 Nm)
7. Install or connect the following:
   - ECT sensor connector
   - Fuel supply manifold
   - Right bank injector connectors
   - Fuel lines
   - Upper intake manifold
   - Breather pipe
   - Upper intake manifold ground cable
   - Thermal transmitter
   - Left bank injector connectors
   - Distributor
   - Spark plug wires
   - EGR tube
   - Fuel pressure regulator vacuum hose
   - Brake booster vacuum hose
   - EVAP canister vacuum and purge hoses
   - PCV valve and hose
   - Heater hoses
   - Radiator hoses
   - EGR temperature sensor connector
   - EGR Solenoid valve connector
   - Ignition coil and power transistor connectors

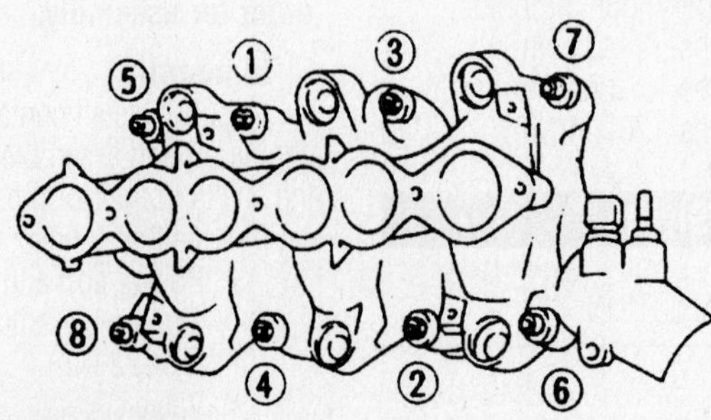

Intake manifold torque sequence—3.3L engine

- TP sensor and switch connectors
- IAC valve connector
- Cruise control cable
- Accelerator cable
- Air intake duct
- Negative battery cable
8. Fill the cooling system.
9. Start the engine and check for leaks.

Exhaust Manifold

REMOVAL & INSTALLATION

2.4L Engine

1. Before servicing the vehicle, refer to the precautions in the beginning of this section.
2. Remove or disconnect the following:
   - Negative battery cable
   - Heated Oxygen (HO_2S) sensor connector
   - Exhaust manifold heat shield
   - Exhaust Gas Recirculation (EGR) tube
   - Exhaust front pipe
   - Exhaust manifold. Loosen the nuts in reverse of the torque sequence.

To install:

3. Install or connect the following:
   - Exhaust manifold. Tighten the nuts in sequence to 28–35 ft. lbs. (37–48 Nm).
   - Exhaust front pipe. Tighten the fasteners to 32–37 ft. lbs. (43–50 Nm).
   - EGR tube. Tighten the flange fittings to 29–36 ft. lbs. (39–49 Nm).
   - Exhaust manifold heat shield. Tighten the bolts to 45–57 inch lbs. (5–7 Nm).
   - HO_2S sensor connector
   - Negative battery cable
4. Start the engine and check for leaks.

3.3L Engine

1. Before servicing the vehicle, refer to the precautions in the beginning of this section.

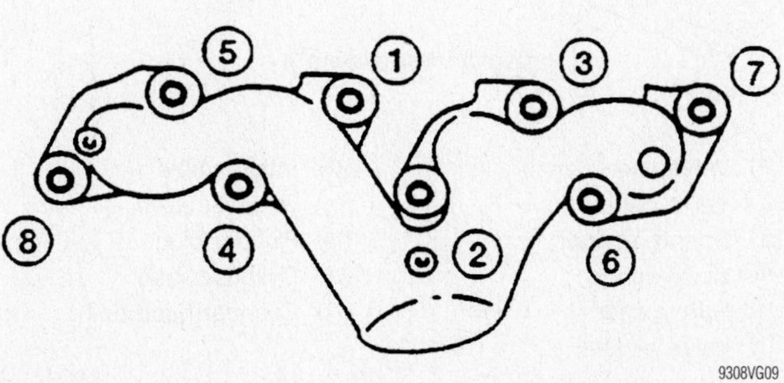

Exhaust manifold torque sequence—2.4L engine

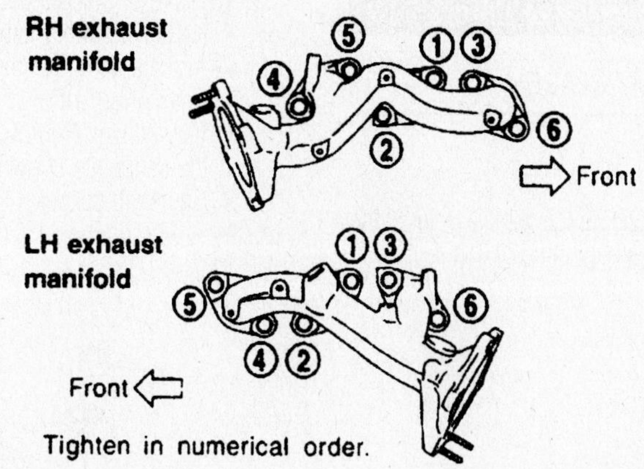

Exhaust manifold torque sequence—3.3L engine

2. Remove or disconnect the following:
   - Negative battery cable
   - Exhaust manifold heat shields
   - Exhaust Gas Recirculation (EGR) tube
   - Heated Oxygen (HO_2S) sensor connectors
   - Exhaust front pipes
   - Exhaust manifolds with catalytic converters attached. Loosen the nuts in the reverse of the torque sequence.

To install:

3. Install or connect the following:
   - Exhaust manifolds with catalytic converters attached. Tighten the nuts in sequence to 21–25 ft. lbs. (28–33 Nm).
   - Exhaust front pipes. Tighten the bolts to 21–25 ft. lbs. (28–33 Nm).
   - Heated Oxygen (HO_2S) sensor connectors
   - EGR tube. Tighten the flange fittings to 29–36 ft. lbs. (39–49 Nm).
   - Exhaust manifold heat shields. Tighten the bolts to 84–96 inch lbs. (9–11 Nm)
   - Negative battery cable
4. Start the engine and check for leaks.

Front Crankshaft Seal

REMOVAL & INSTALLATION

2.4L Engine

Refer to the Timing Chain, Sprockets, Front Cover and Seal procedure in this section.

3.3L Engine

1. Before servicing the vehicle, refer to the precautions in the beginning of this section.
2. Drain the cooling system.
3. Remove or disconnect the following:
   - Negative battery cable
   - Accessory drive belts
   - Radiator hoses
   - Crankshaft pulley
   - Front cover
   - Timing belt. Refer to the Timing Belt unit repair section.
   - Crankshaft timing sprocket
   - Front crankshaft seal

To install:

4. Install or connect the following:
   - Front crankshaft seal flush with the oil pump housing
   - Crankshaft timing sprocket
   - Timing belt
   - Front cover. Tighten the bolts to 26–43 inch lbs. (3–5 Nm).
   - Crankshaft pulley. Tighten the bolt to 141–156 ft. lbs. (191–211 Nm).
   - Radiator hoses
   - Accessory drive belts
   - Negative battery cable
5. Fill the cooling system.
6. Start the engine and check for leaks.

Camshaft and Valve Lifters

REMOVAL & INSTALLATION

2.4L Engine

1. Before servicing the vehicle, refer to the precautions in the beginning of this section.
2. Remove or disconnect the following:
   - Negative battery cable
   - Air cleaner assembly
   - Spark plug wires
   - Valve cover. Remove the bolts in the sequence shown.
   - Camshaft sprocket cover
   - Camshaft sprockets and upper timing chain

➡Keep all valvetrain components in order for assembly.

   - Camshaft bearing caps. Loosen the bolts in several passes in reverse of the torque sequence.

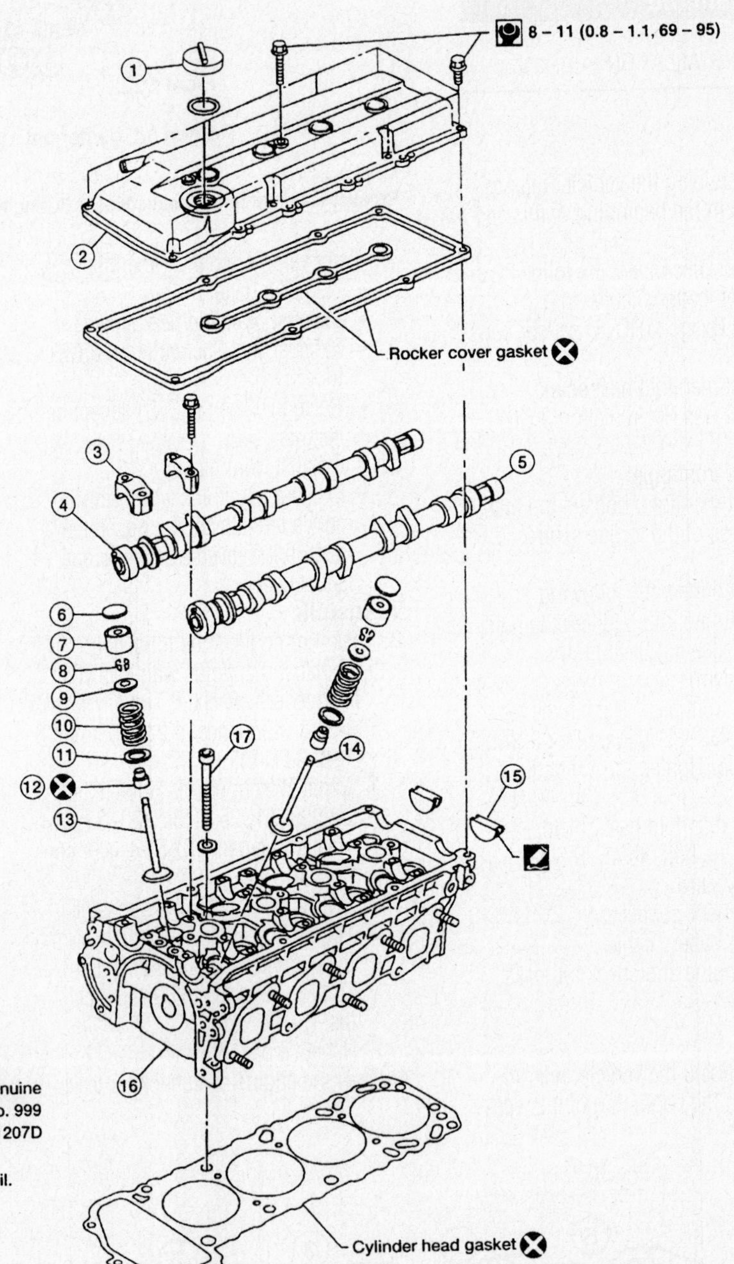

: Apply liquid gasket. Use Genuine RTV silicone sealant, Part No. 999 MP-A7007, Three Bond TB 1207D or equivalent.
: Lubricate with new engine oil.
: N·m (kg-m, in-lb)
: N·m (kg-m, ft-lb)

1. Oil filler cap
2. Rocker cover
3. Camshaft bracket
4. Intake camshaft
5. Exhaust camshaft
6. Shim
7. Valve lifter
8. Valve cotter
9. Spring retainer
10. Valve spring
11. Spring seat
12. Valve oil seal
13. Intake valve
14. Exhaust valve
15. Rubber plug
16. Cylinder head
17. Cylinder head bolt

7924VG53

Exploded view of the camshafts and related components—2.4L engine

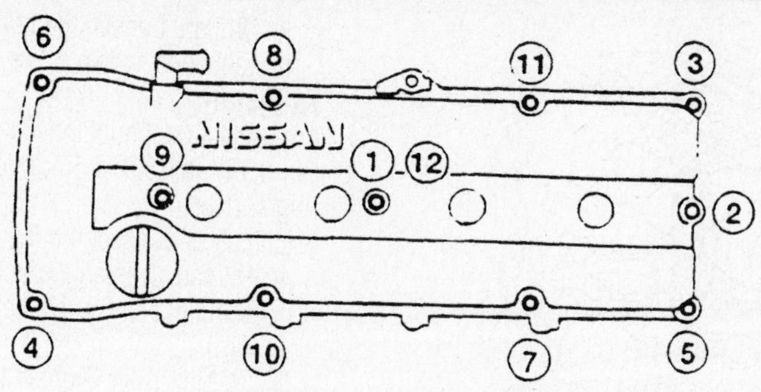

Valve cover loosening sequence—2.4L engine

- Camshafts
- Valve lifters and shims

To install:

3. Install or connect the following:
   - Valve lifters and shims in their original positions
   - Camshafts
4. Install the bearing caps. Tighten the bolts in sequence as follows:
   a. Step 1: 17 inch lbs. (2 Nm)
   b. Step 2: 80–104 inch lbs. (9–12 Nm)
5. Install or connect the following:
   - Camshaft sprockets and upper timing chain. Tighten the sprocket bolts to 123–130 ft. lbs. (167–177 Nm).
   - Camshaft sprocket cover
   - Valve cover. Tighten the bolts in sequence to 69–95 inch lbs. (8–11 Nm).
   - Spark plug wires
   - Air cleaner assembly
   - Negative battery cable

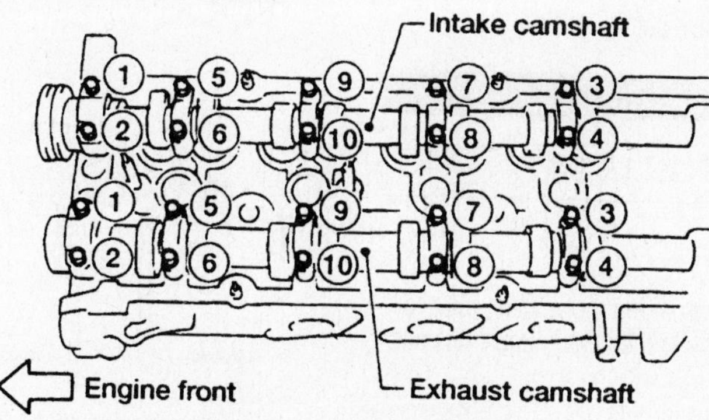

Bearing cap bolt torque sequence—2.4L engine

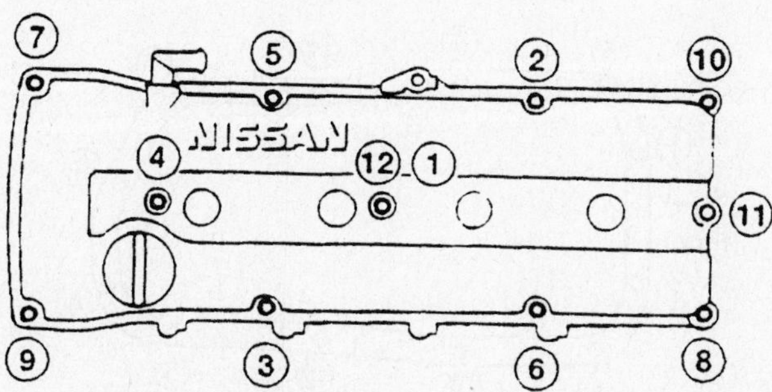

Valve cover torque sequence—2.4L engine

3.3L Engines

1. Before servicing the vehicle, refer to the precautions in the beginning of this section.
2. Drain the cooling system.
3. Remove or disconnect the following:
   - Negative battery cable
   - Upper intake manifold
   - Valve covers

➡Keep all valvetrain components in order for assembly.

   - Rocker arm and shaft assemblies
   - Valve lifter guide and valve lifters. Attach a wire to the top of the lifters so that they will not drop from the lifter guide.
   - Radiator
   - Accessory drive belts
   - Front cover
   - Timing belt. Refer to the Timing Belt unit repair section.
   - Camshaft sprockets
   - Camshaft seals
   - Rear timing cover
   - Distributor
   - Cylinder head rear covers
   - Camshaft locating plates
   - Camshafts

To install:

4. Install or connect the following:
   - Camshafts
   - Camshaft locating plates. Tighten the bolts to 58–65 ft. lbs. (78–88 Nm).
   - Cylinder head rear covers
   - Distributor
   - Rear timing cover
   - Camshaft seals
   - Camshaft sprockets. Tighten the bolts to 58–65 ft. lbs. (78–88 Nm).
   - Timing belt
   - Front cover
   - Accessory drive belts
   - Radiator
   - Valve lifter guide and valve lifters
   - Rocker arm and shaft assemblies. Tighten the bolts to 13–16 ft. lbs. (18–22 Nm).
   - Valve covers
   - Upper intake manifold
   - Negative battery cable
5. Fill the cooling system.
6. Start the engine and check for leaks.

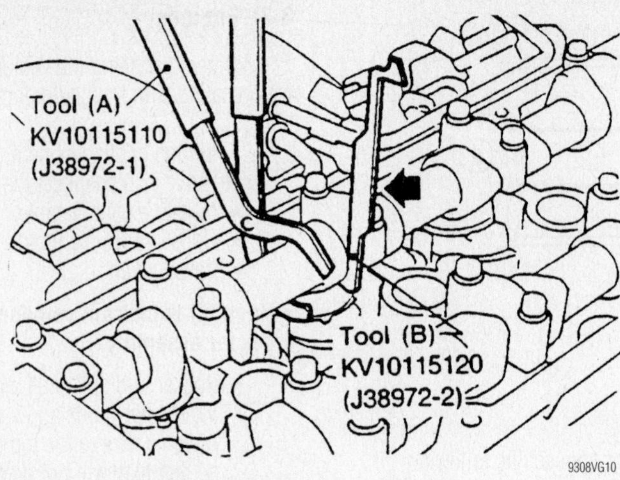

Valve adjustment tools (A) and (B)—2.4L engine

Valve Lash

ADJUSTMENT

3.3L Engines

These engines are equipped with hydraulic valve lifters that do not require periodic adjustment.

2.4L Engine

➡Measure valve clearance with the engine warm.

1. Before servicing the vehicle, refer to the precautions in the beginning of this section.
2. Remove the valve cover.
3. Set the engine to the top of the compression stroke with the valves closed for the cylinder to be measured.
4. Check the valve clearance. The valve clearance specifications are as follows:
   - Intake: 0.012–0.015 in. (0.31–0.39mm)
   - Exhaust: 0.013–0.016 in. (0.33–0.41mm)
5. If adjustment is necessary, compress the valve spring with Tool **A** and insert Tool **B** to hold the valve in the open position as shown.
6. Replace the shims as necessary to achieve the correct valve clearance.
7. Repeat for each valve to be adjusted.

Starter Motor

REMOVAL & INSTALLATION

1. Before servicing the vehicle, refer to the precautions in the beginning of this section.
2. Remove or disconnect the following:
   - Negative battery cable
   - Engine under cover
   - Starter harness connectors
   - Starter motor

To install:

3. Install or connect the following:
   - Starter motor. Tighten the bolts to 22–27 ft. lbs. (30–36 Nm).
   - Starter harness connectors
   - Engine under cover
   - Negative battery cable

Oil Pan

REMOVAL & INSTALLATION

2.4L Engine

1. Before servicing the vehicle, refer to the precautions in the beginning of this section.
2. Drain the engine oil.
3. Remove or disconnect the following:
   - Negative battery cable
   - Engine under cover
   - Stabilizer bar
   - Oil pan. Loosen the bolts in the sequence shown.

To install:

4. Apply a continuous bead of sealant 0.138–0.177 in. (3.5–4.5mm) to the oil pan mating surface.
5. Install or connect the following:
   - Oil pan. Tighten the bolts in sequence to 60–72 inch lbs. (7–8 Nm).
   - Stabilizer bar. Tighten the bracket bolts to 38–45 ft. lbs. (51–61 Nm) and the link nuts to 12–16 ft. lbs. (16–22 Nm).
   - Engine under cover
   - Negative battery cable

➡Wait 30 minutes after installation of the oil pan to allow the sealant to cure before adding oil.

6. Fill the crankcase to the correct level.
7. Start the engine and check for leaks.

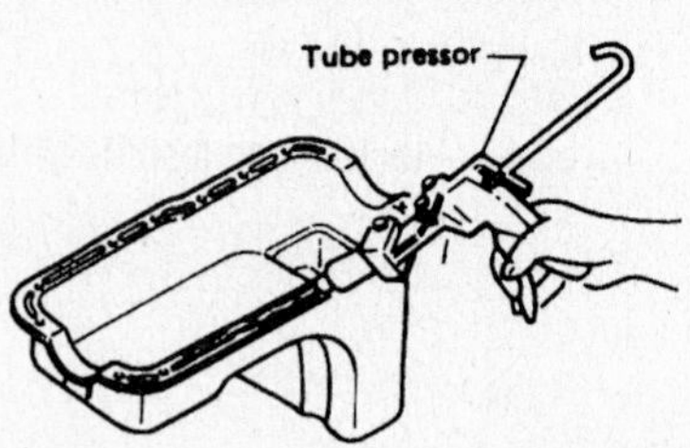

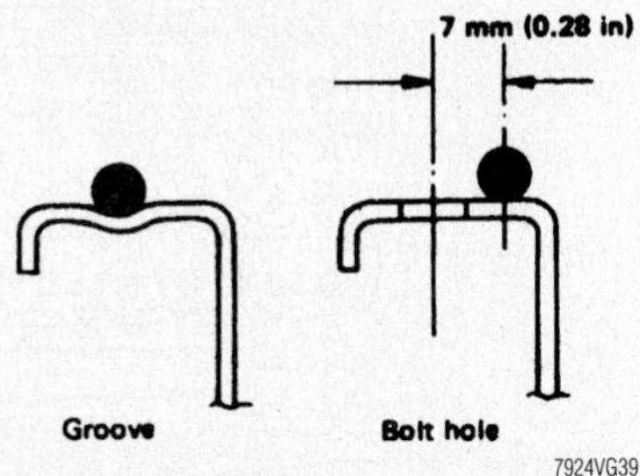

Oil pan sealant application—2.4L engine shown

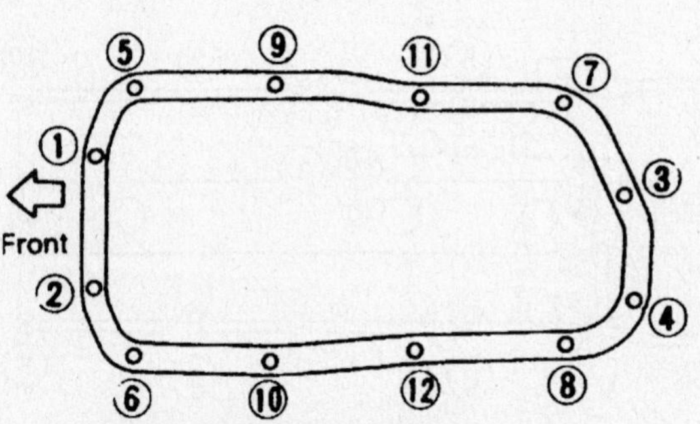

Oil pan bolt removal sequence—2.4L engine

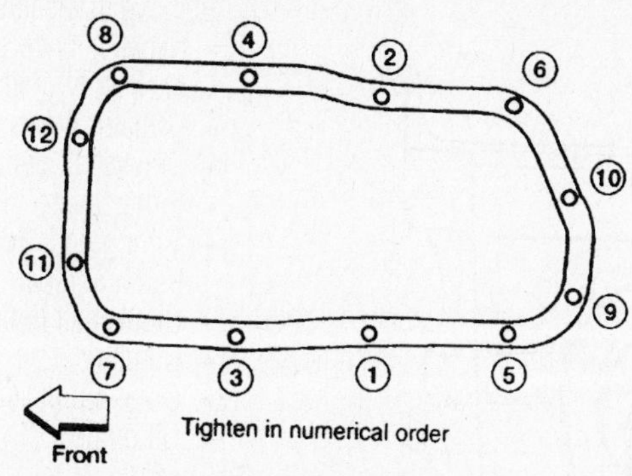

Oil pan bolt installation sequence—2.4L engine

3.3L Engine

2WD MODELS

1. Before servicing the vehicle, refer to the precautions in the beginning of this section.
2. Drain the engine oil.
3. Remove or disconnect the following:
   - Negative battery cable
   - Engine under cover
   - Stabilizer bar
   - Front crossmember
   - Starter motor
   - Transmission mount
   - Left and right motor mounts
   - Power steering gear
4. Raise and support the engine for clearance.
5. Remove or disconnect the following:
   - Oil pan bolts in the sequence
   - Oil pan

To install:

6. Apply a continuous bead of sealant 0.138–0.177 in. (3.5–4.5mm) to the oil pan mating surface.
7. Install or connect the following:
   - Oil pan. Tighten the bolts in reverse of the removal sequence to 62 inch lbs. (7 Nm).
   - Power steering gear
   - Left and right motor mounts
   - Transmission mount
   - Starter motor
   - Front crossmember
   - Stabilizer bar
   - Engine under cover
   - Negative battery cable

➡Wait 30 minutes after installation of the oil pan to allow the sealant to cure before adding oil.

8. Fill the crankcase to the correct level.
9. Start the engine and check for leaks.

4WD MODELS

1. Before servicing the vehicle, refer to the precautions in the beginning of this section.
2. Drain the engine oil.
3. Remove or disconnect the following:
   - Negative battery cable
   - Engine under cover
   - Stabilizer bar brackets
   - Front driveshaft
   - Axle halfshafts
   - Front suspension crossmember
   - Front differential and mounting bracket
   - Starter motor
   - Transmission mount
   - Left and right motor mounts
   - Power steering gear
   - Relay rod
4. Raise and support the engine for clearance.
5. Remove or disconnect the following:
   - Oil pan bolts in the sequence
   - Oil pan

To install:

6. Apply a continuous bead of sealant 0.138–0.177 in. (3.5–4.5mm) to the oil pan mating surface.
7. Install or connect the following:
   - Oil pan. Tighten the bolts in reverse of the removal sequence to 62 inch lbs. (7 Nm).
   - Relay rod
   - Power steering gear
   - Left and right motor mounts
   - Transmission mount
   - Starter motor
   - Front differential and mounting bracket
   - Front suspension crossmember
   - Axle halfshafts
   - Front driveshaft
   - Stabilizer bar brackets
   - Engine under cover
   - Negative battery cable

➡Wait 30 minutes after installation of the oil pan to allow the sealant to cure before adding oil.

8. Fill the crankcase to the correct level.
9. Start the engine and check for leaks.

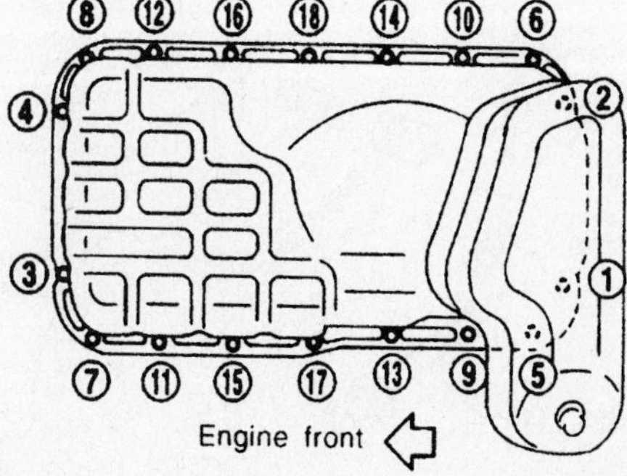

Oil pan bolt removal sequence—3.3L engine

Oil Pump

REMOVAL & INSTALLATION

2.4L Engine

1. Before servicing the vehicle, refer to the precautions in the beginning of this section.
2. Set the engine to Top Dead Center (TDC) of the compression stroke for the No. 1 cylinder.
3. Remove or disconnect the following:
   - Distributor cap
   - Distributor
   - Engine under cover
   - Stabilizer bar
   - Oil pump and drive spindle

To install:

4. Fill the pump housing with engine oil, then align the punch mark on the spindle with the hole in the oil pump as shown.
5. Install or connect the following:
   - Oil pump and drive spindle. Tighten the mounting bolts to 96–132 inch lbs. (11–15 Nm).
   - Stabilizer bar

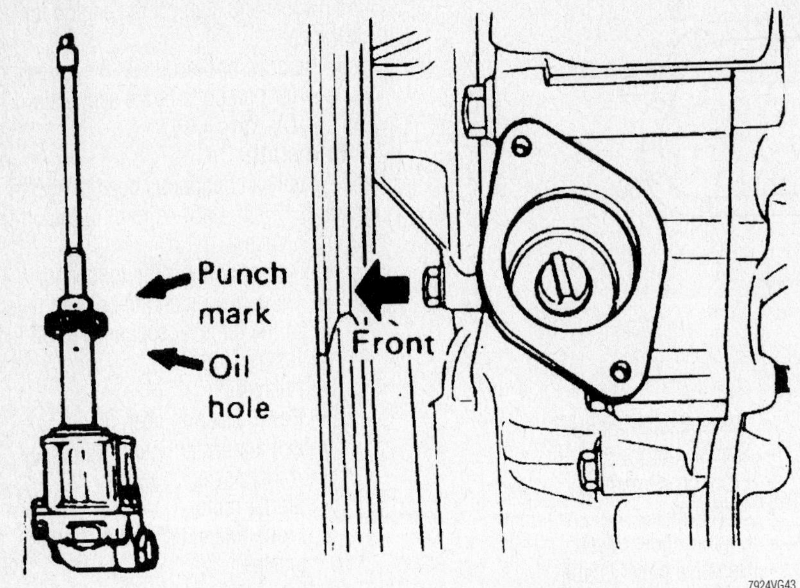

Align the punch mark with the oil hole before oil pump installation—2.4L engine

- Engine under cover
- Distributor
- Distributor cap

6. Start the engine and check for leaks.
7. Check the ignition timing and adjust, as necessary.

3.3L Engine

1. Before servicing the vehicle, refer to the precautions in the beginning of this section.
2. Drain the engine oil.
3. Drain the cooling system.
4. Remove or disconnect the following:
   - Negative battery cable
   - Accessory drive belts
   - Radiator hoses
   - Crankshaft pulley
   - Front cover
   - Timing belt. Refer to the Timing Belt unit repair section.
   - Crankshaft timing sprocket
   - Oil pan
   - Oil pump pickup tube
   - Oil pump

To install:

5. Install or connect the following:
   - Oil pump. Tighten the large bolts to 16–22 ft. lbs. (22–29 Nm) and the small bolts to 55–74 inch lbs. (6–8 Nm).
   - Oil pump pickup tube. Tighten the flange bolts to 12 ft. lbs. (16 Nm) and the bracket bolt to 55–74 inch lbs. (6–8 Nm).
   - Oil pan
   - Crankshaft timing sprocket
   - Timing belt
   - Front cover
   - Crankshaft pulley
   - Radiator hoses
   - Accessory drive belts
   - Negative battery cable
6. Fill the cooling system.

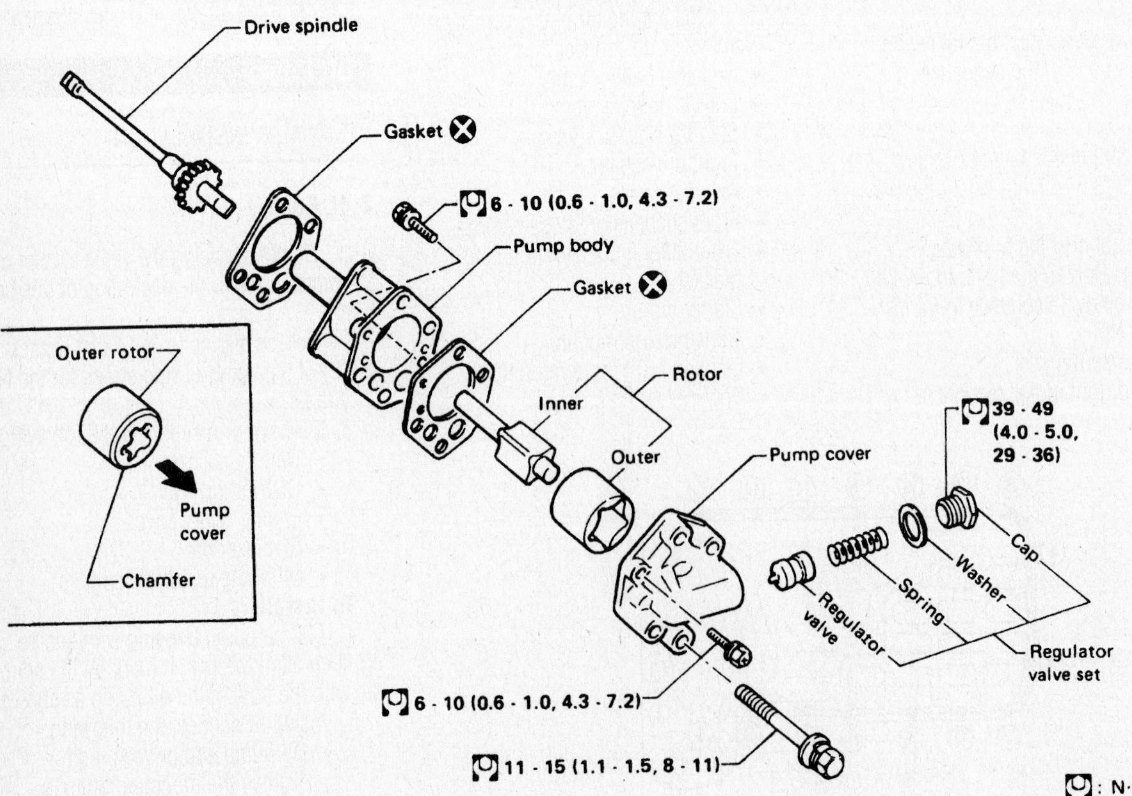

Exploded view of oil pump assembly—2.4L engine

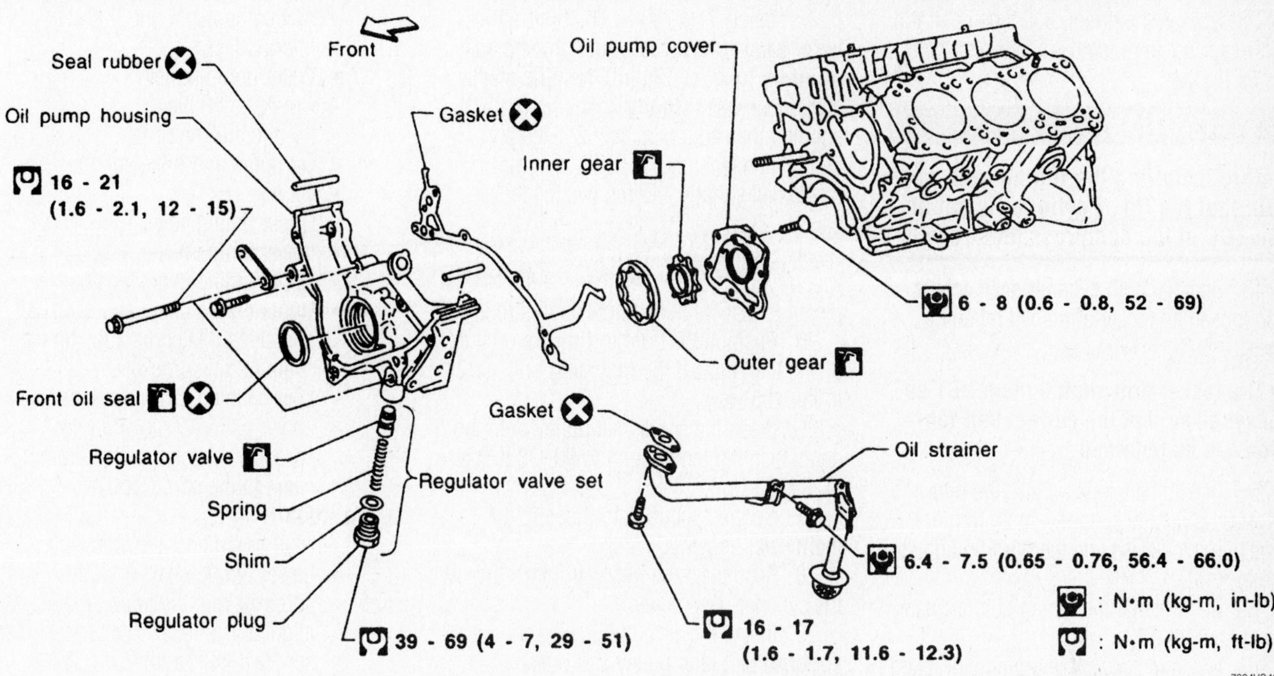

Oil pump assembly exploded view—3.3L engine

7. Fill the crankcase to the correct level.
8. Start the engine and check for leaks.

Rear Main Seal

REMOVAL & INSTALLATION

1. Before servicing the vehicle, refer to the precautions in the beginning of this section.
2. Remove or disconnect the following:
   - Transmission
   - Flywheel
   - Rear main seal

To install:

3. Install the seal so that it is flush with the retainer housing.
4. Install or connect the following:
   - Flywheel. Tighten the bolts to 61–69 ft. lbs. (83–93 Nm).
   - Transmission

Timing Belt

REMOVAL & INSTALLATION

3.3L Engines

1. Remove the engine undercover.
2. Remove the radiator shroud, the fan and the pulleys.
3. Drain the coolant from the radiator and remove the water pump hose.

**** CAUTION**

When draining the coolant, keep in mind that cats and dogs are attracted by the ethylene glycol antifreeze, and are quite likely to drink any that is left in an uncovered container or in puddles on the ground. This will prove fatal in sufficient quantity. Always drain the coolant into a sealable container. Coolant should be reused unless it is contaminated or several years old.

4. Remove the radiator.
5. Remove the power steering, air conditioning compressor and alternator drive belts.
6. Remove the spark plugs.
7. Remove the distributor protector (dust shield).
8. Remove the air conditioning compressor drive belt idler pulley and bracket.
9. Remove the fresh air intake tube at the cylinder head cover.
10. Disconnect the radiator hose at the thermostat housing.
11. Remove the crankshaft pulley bolt, then pull off the pulley with a suitable puller.
12. Remove the bolts, then remove the front upper and lower timing belt covers.
13. Set the No. 1 piston at Top Dead Center (TDC) of its compression stroke. Align the punchmark on the left camshaft sprocket with the punchmark on the timing belt upper rear cover. Align the punchmark on the crankshaft sprocket with the notch on the oil pump housing. Temporarily install the crank pulley bolt so the crankshaft can be rotated if necessary.

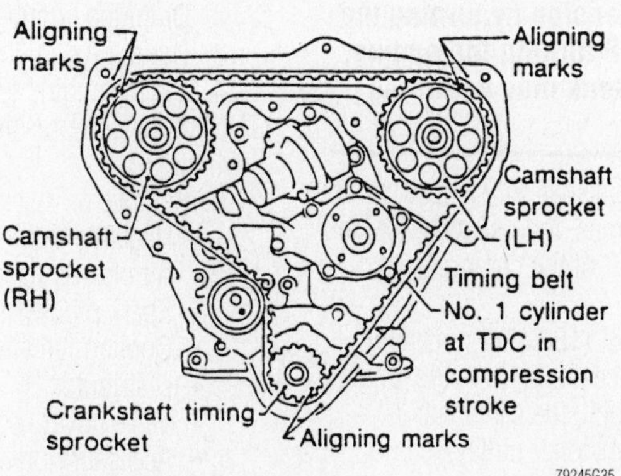

Timing belt alignment mark locations—3.3L engines

14. Loosen the timing belt tensioner and return spring, then remove the timing belt.

To install:

✻✻ CAUTION

Before installing the timing belt, confirm that the No. 1 cylinder is set at the TDC of the compression stroke.

15. Remove both cylinder head covers and loosen all rocker arm shaft retaining bolts.

➡The rocker arm shaft bolts MUST be loosened so that the correct belt tension can be obtained.

16. Install the tensioner and the return spring. Using a hexagon wrench, turn the tensioner clockwise and temporarily tighten the locknut.
17. Be sure that the timing belt is clean and free from oil or water.
18. When installing the timing belt, align the white lines on the belt with the punch-marks on the camshaft and crankshaft sprockets. Have the arrow on the timing belt pointing toward the front belt covers.

➡A good way (although rather tedious!) to check for proper timing belt installation is to count the number of belt teeth between the timing marks. There are 133 teeth on the belt; there should be 40 teeth between the timing marks on the left and right side camshaft sprockets, and 43 teeth between the timing marks on the left side camshaft sprocket and the crankshaft sprocket.

19. While keeping the tensioner steady, loosen the locknut with a hex wrench.
20. Turn the tensioner approximately 70–80 degrees clockwise with the wrench, then tighten the locknut.

✻✻ WARNING

If any binding is felt when adjusting the timing belt tension by turning the crankshaft, STOP turning the engine, because the pistons may be hitting the valves.

21. Turn the crankshaft in a clockwise direction several times, then **slowly** set the No. 1 piston to TDC of the compression stroke.
22. Apply 22 lbs. (10 kg) of pressure (push it in!) to the center span of the timing belt between the right side camshaft sprocket and the tensioner pulley, then loosen the tensioner locknut.
23. Using a 0.0138 in. (0.35mm) thick feeler gauge (the actual width of the blade **must** be ½ in. or 13mm!), turn the crankshaft clockwise (**slowly!**). The timing belt should move approximately 2½ teeth. Tighten the tensioner locknut, turn the crankshaft slightly and remove the feeler gauge.
24. Slowly rotate the crankshaft clockwise several more times, then set the No. 1 piston to TDC of the compression stroke.
25. Position the 2 timing covers on the block, then tighten the mounting bolts to 24 ft. lbs. (35 Nm).
26. Press the crankshaft pulley onto the shaft, then tighten the bolt to 90–98 ft. lbs. (123–132 Nm).
27. Connect the radiator hose to the thermostat housing.
28. Reconnect the fresh air intake tube at the cylinder head cover.
29. Install the air conditioning compressor drive belt idler pulley and bracket.
30. Install the distributor protector (dust shield).
31. Install the spark plugs.
32. Install the power steering, air conditioning compressor and alternator drive belts.
33. Install the radiator.
34. Reconnect the water pump hose and fill the engine with coolant. Install the fan shroud and pulleys.
35. Install the engine undercover.
36. Start the engine and check for any leaks.

Timing Chain, Sprockets, Front Cover and Seal

REMOVAL & INSTALLATION

2.4L Engine

1. Before servicing the vehicle, refer to the precautions in the beginning of this section.
2. Drain the cooling system.
3. Drain the engine oil.
4. Set the engine to Top Dead Center (TDC) of the compression stroke for the No. 1 cylinder.
5. Remove or disconnect the following:
   - Negative battery cable
   - Air cleaner assembly
   - Spark plug wires
   - Cooling fan and shroud
   - Distributor
   - Valve cover
   - Accessory drive belts
   - Power steering pump and brackets
   - A/C compressor and bracket
   - Idler pulleys
   - Water pump pulley
   - Crankshaft pulley
   - Front crankshaft seal
   - Oil pump and drive spindle
   - Oil pan
   - Upper timing cover
   - Lower timing cover
   - Upper timing chain tensioner
   - Upper timing chain and camshaft sprockets. Matchmark the timing chain to the sprockets.
   - Lower timing chain tensioner
   - Lower timing chain and idler sprocket. Matchmark the timing chain to the sprockets.

To install:

6. Install or connect the following:
   - Lower timing chain and idler sprocket with the timing marks aligned as shown. Tighten the idler sprocket bolt to 48–61 ft. lbs. (66–83 Nm).
   - Lower timing chain tensioner. Tighten the bolts to 56–66 inch lbs. (6.5–7.5 Nm).
   - Upper timing chain and camshaft sprockets with the timing marks aligned as shown. Tighten the camshaft sprocket bolts to 123–130 ft. lbs. (167–177 Nm).
   - Upper timing chain tensioner. Tighten the bolts to 56–66 inch lbs. (6.5–7.5 Nm).
   - Lower timing cover. Tighten the large bolts to 12–14 ft. lbs. (16–19

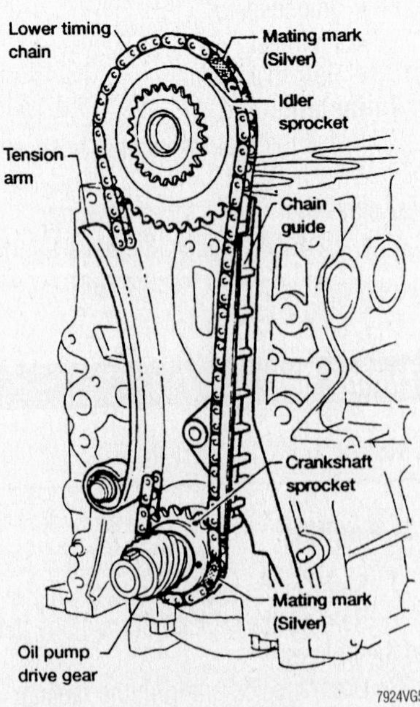

Lower timing chain alignment—2.4L engine

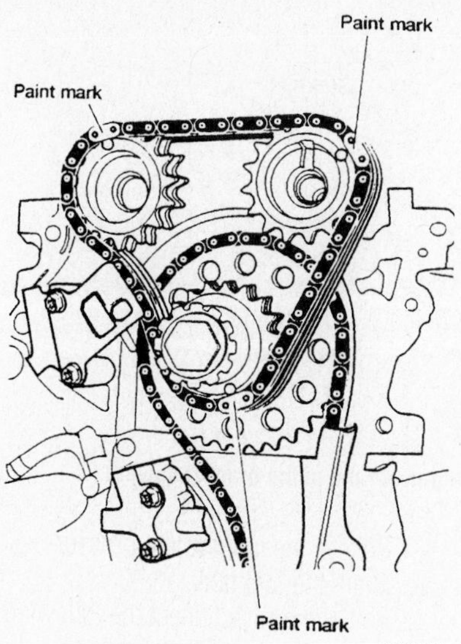

Upper timing chain alignment—2.4L engine

Nm) and the small bolts to 56–66 inch lbs. (6.5–7.5 Nm).

- Upper timing cover. Tighten the large bolts to 12–14 ft. lbs. (16–19 Nm) and the small bolts to 56–66 inch lbs. (6.5–7.5 Nm).
- Oil pan
- Oil pump and drive spindle
- Front crankshaft seal
- Crankshaft pulley. Tighten the bolt to 105–112 ft. lbs. (142–152 Nm).
- Water pump pulley
- Idler pulleys
- A/C compressor and bracket
- Power steering pump and brackets
- Accessory drive belts
- Valve cover
- Distributor
- Cooling fan and shroud
- Spark plug wires
- Air cleaner assembly
- Negative battery cable

7. Fill the cooling system.
8. Fill the crankcase to the correct level.
9. Start the engine and check for leaks.
10. Check the ignition timing and adjust, as necessary.

Piston and Ring

POSITIONING

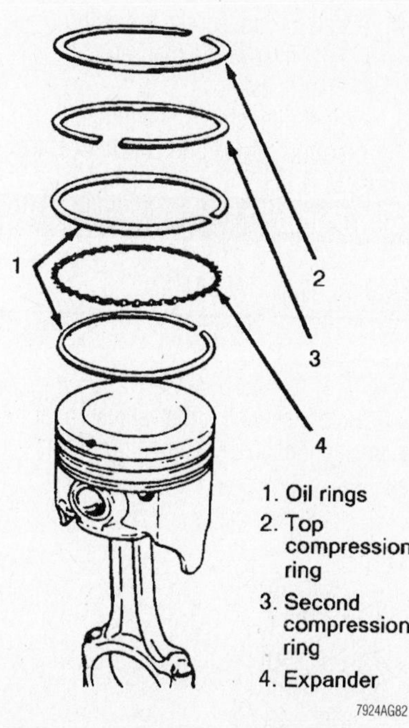

Piston ring positioning—2.4L engine

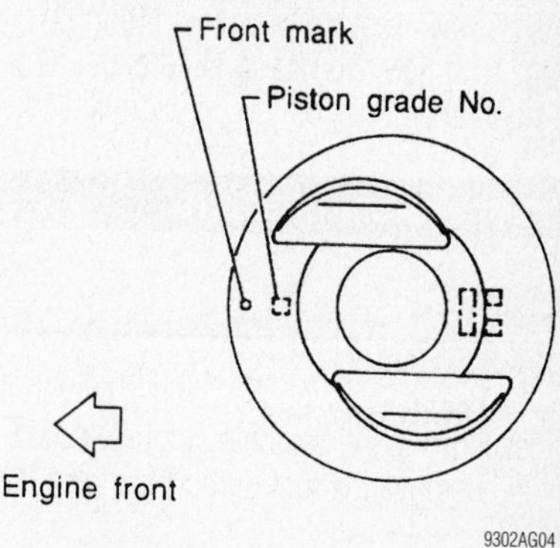

Piston ring positioning—3.3L engine

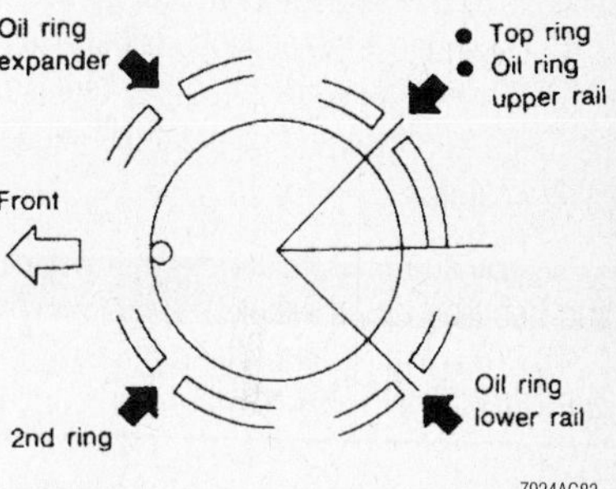

Piston ring end-gap spacing

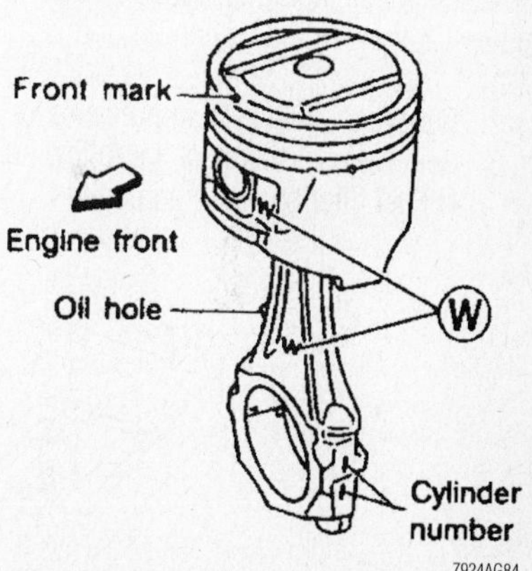

Piston and connecting rod positioning

FUEL SYSTEM

Fuel System Service Precautions

Safety is the most important factor when performing not only fuel system maintenance but any type of maintenance. Failure to conduct maintenance and repairs in a safe manner may result in serious personal injury or death. Maintenance and testing of the vehicle's fuel system components can be accomplished safely and effectively by adhering to the following rules and guidelines.

- To avoid the possibility of fire and personal injury, always disconnect the negative battery cable unless the repair or test procedure requires that battery voltage be applied.
- Always relieve the fuel system pressure prior to disconnecting any fuel system component (injector, fuel rail, pressure regulator, etc.), fitting or fuel line connection. Exercise extreme caution whenever relieving fuel system pressure, to avoid exposing skin, face and eyes to fuel spray. Please be advised that fuel under pressure may penetrate the skin or any part of the body that it contacts.
- Always place a shop towel or cloth around the fitting or connection prior to loosening to absorb any excess fuel due to spillage. Ensure that all fuel spillage (should it occur) is quickly removed from engine surfaces. Ensure that all fuel soaked cloths or towels are deposited into a suitable waste container.
- Always keep a dry chemical (Class B) fire extinguisher near the work area.
- Do not allow fuel spray or fuel vapors to come into contact with a spark or open flame.
- Always use a back-up wrench when loosening and tightening fuel line connection fittings. This will prevent unnecessary stress and torsion to fuel line piping. Always follow the proper torque specifications.

- Always replace worn fuel fitting O-rings with new. Do not substitute fuel hose or equivalent, where fuel pipe is installed.

Fuel System Pressure

RELIEVING

1. Before servicing the vehicle, refer to the precautions in the beginning of this section.
2. Remove the fuel pump fuse from the panel.
3. Start the engine and allow it to run until it stalls. Crank the engine for a few seconds to relieve additional fuel pressure.
4. Disconnect the negative battery cable.
5. When repairs are complete, replace the fuel pump fuse and connect the negative battery cable.

Fuel Filter

REMOVAL & INSTALLATION

➡**The fuel filter is located under the vehicle near the fuel tank.**

1. Before servicing the vehicle, refer to the precautions in the beginning of this section.
2. Relieve the fuel system pressure.
3. Remove or disconnect the following:
   - Fuel filter shield, if equipped
   - Fuel lines
   - Fuel filter from the bracket

To install:

4. Install or connect the following:
   - Fuel filter to the bracket
   - Fuel lines
   - Fuel filter shield, if equipped
5. Start the engine and check for leaks.

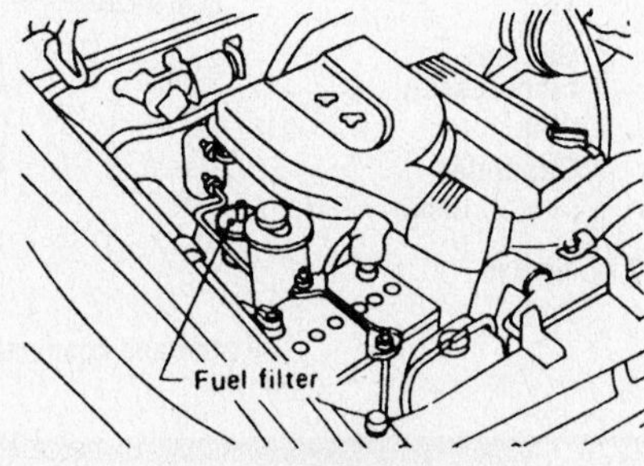

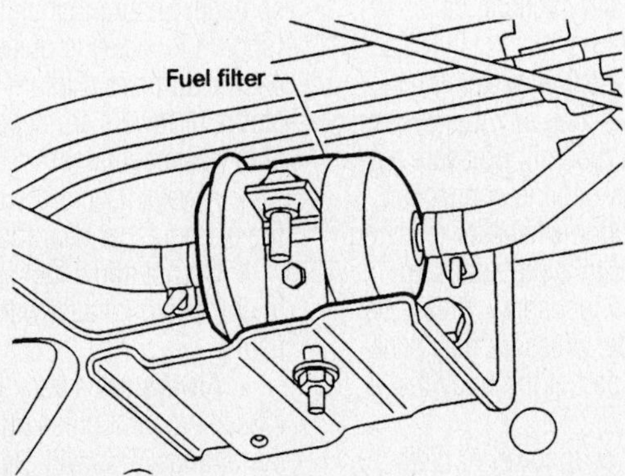

Typical fuel filter locations

Fuel Pump

REMOVAL & INSTALLATION

Frontier

1. Before servicing the vehicle, refer to the precautions in the beginning of this section.

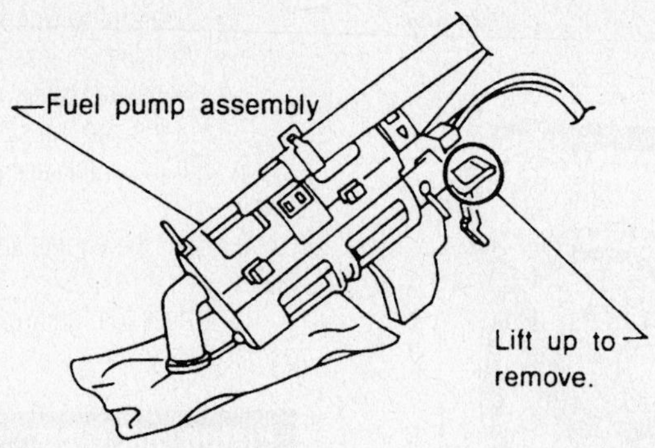

Remove the fuel pump with bracket while lifting the pawl of the pump bracket upward

2. Relieve the fuel system pressure.
3. Drain the fuel tank.
4. Remove or disconnect the following:
   - Negative battery cable
   - Fuel pump module harness connectors
   - Filler hose shield
   - Fuel pressure and return lines
   - Filler hose
   - Vent hose
   - Evaporative Emissions (EVAP) hose
   - Fuel tank skid plate
   - Fuel tank
   - Fuel level sender
   - Fuel pump

To install:

5. Install or connect the following:
   - Fuel pump
   - Fuel level sender. Tighten the screws to 17–23 inch lbs. (2.0–2.5 Nm).
   - Fuel tank. Tighten the bolts to 27–36 ft. lbs. (37–49 Nm).
   - Fuel tank skid plate. Tighten the bolts to 27–36 ft. lbs. (37–49 Nm).
   - EVAP hose
   - Vent hose
   - Filler hose
   - Fuel pressure and return lines
   - Filler hose shield
   - Fuel pump module harness connectors
   - Negative battery cable
6. Fill the fuel tank.
7. Start the engine and check for leaks.

Xterra

1. Before servicing the vehicle, refer to the precautions in the beginning of this section.
2. Relieve the fuel system pressure.
3. Remove the rear seat cushion and the access panel.
4. Remove or disconnect the following:
   - Negative battery cable
   - Fuel pump electrical connectors
5. Matchmark the installed position of

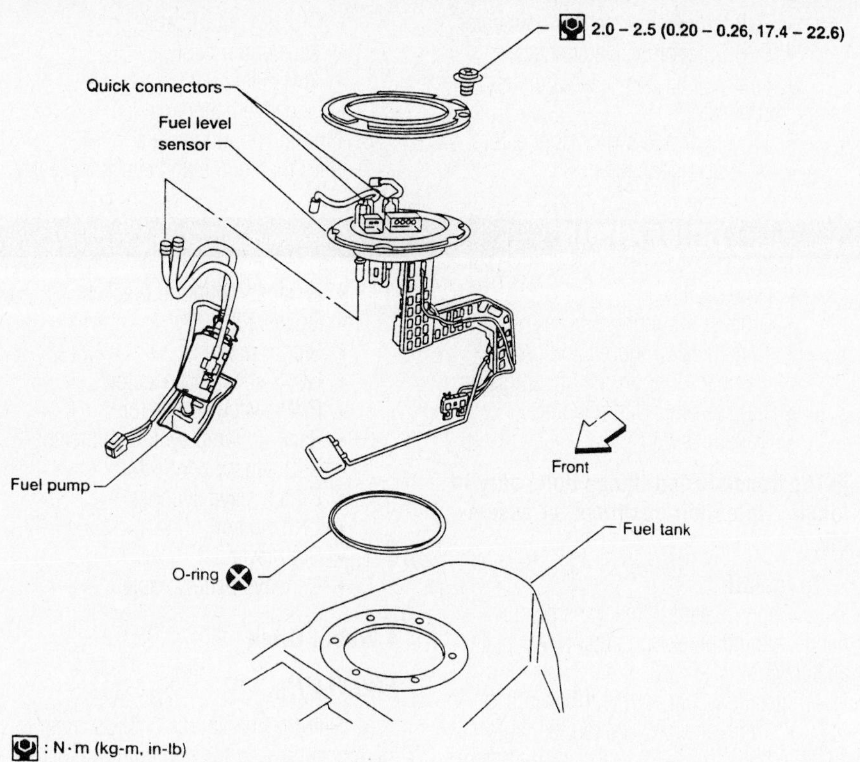

Exploded view of the fuel pump—Xterra shown

the fuel line quick connect fittings, then disconnect the fittings by holding the sides of the connector, push in the tabs and pull out the tube inserted in the retainer.

➡The tube can be removed when the tabs are completely pushed in. Do NOT use any tools to remove the quick connector.

- Six screws
- Fuel level sensor retainer and fuel level sensor
- Fuel pump with the bracket, while lifting the pawl of the fuel bracket upward
- Fuel level sensor

To install:

6. Installation is the reverse of the removal procedure.
7. Start the engine and check for leaks.

Fuel Injectors

REMOVAL & INSTALLATION

2.4L Engine

1. Before servicing the vehicle, refer to the precautions in the beginning of this section.
2. Relieve the fuel system pressure.
3. Remove or disconnect the following:
   - Negative battery cable
   - Air cleaner assembly
   - Fuel lines
   - Fuel pressure regulator vacuum line
   - Fuel injector connectors
   - Fuel supply manifold with the injectors attached
   - Fuel injector caps
   - Fuel injectors

To install:

➡Use new insulators and O-ring seals for assembly.

4. Install or connect the following:
   - Fuel injectors
   - Fuel injector caps. Tighten the screws to 26–34 inch lbs. (3–4 Nm).
   - Fuel supply manifold with the injectors attached. Tighten the bolts to 96–132 inch lbs. (11–15 Nm).
   - Fuel injector connectors
   - Fuel pressure regulator vacuum line
   - Fuel lines
   - Air cleaner assembly
   - Negative battery cable
5. Start the engine and check for leaks.

3.3L Engine

1. Before servicing the vehicle, refer to the precautions in the beginning of this section.
2. Drain the cooling system.
3. Relieve the fuel system pressure.
4. Remove or disconnect the following:
   - Negative battery cable
   - Air intake duct
   - Accelerator cable
   - Cruise control cable
   - Idle Air Control (IAC) valve connector
   - Throttle Position (TP) sensor and switch connectors
   - Ignition coil and power transistor connectors
   - Exhaust Gas Recirculation (EGR) Solenoid valve connector
   - EGR temperature sensor connector
   - Radiator hoses
   - Heater hoses
   - Positive Crankcase Ventilation (PCV) valve and hose
   - Evaporative Emissions (EVAP) canister vacuum and purge hoses
   - Brake booster vacuum hose
   - Fuel pressure regulator vacuum hose
   - EGR tube
   - Left bank injector connectors
   - Thermal transmitter
   - Upper intake manifold ground cable
   - Breather pipe
   - Supercharger or upper intake manifold
   - Fuel lines
   - Right bank injector connectors
   - Fuel supply manifold with the injectors attached
   - Fuel injector caps
   - Fuel injectors

To install:

➡Use new insulators and O-ring seals for assembly.

5. Install or connect the following:
   - Fuel injectors
   - Fuel injector caps. Tighten the screws to 26–34 inch lbs. (3–4 Nm).
   - Fuel supply manifold with the injectors attached. Tighten the bolts to 96–132 inch lbs. (11–15 Nm).
   - Right bank injector connectors
   - Fuel lines
   - Upper intake manifold
   - Breather pipe
   - Upper intake manifold ground cable
   - Thermal transmitter
   - Left bank injector connectors
   - EGR tube
   - Fuel pressure regulator vacuum hose

- Brake booster vacuum hose
- EVAP canister vacuum and purge hoses
- PCV valve and hose
- Heater hoses
- Radiator hoses
- EGR temperature sensor connector
- EGR Solenoid valve connector
- Ignition coil and power transistor connectors
- TP sensor and switch connectors
- IAC valve connector
- Cruise control cable
- Accelerator cable
- Air intake duct
- Negative battery cable

6. Fill the cooling system.
7. Start the engine and check for leaks.

DRIVE TRAIN

Manual Transmission

REMOVAL & INSTALLATION

2 Wheel Drive

1. Before servicing the vehicle, refer to the precautions in the beginning of this section.
2. Remove or disconnect the following:
   - Negative battery cable
   - Shift lever
   - Crankshaft Position (CKP) sensor
   - Clutch slave cylinder
   - Vehicle Speed (VSS) sensor connector
   - Back-up lamp switch connector
   - Park/Neutral Position (PNP) switch connector
   - Rear Heated Oxygen (HO_2S) sensor connector
   - Starter motor
   - Driveshaft
   - Exhaust mounting bracket
   - Transmission mount and crossmember. Support the transmission.
   - Transmission flange bolts
   - Transmission

➡The transmission flange bolts vary in length. Note their positions for assembly.

To install:

3. Apply sealant to the transmission flange, engine block and engine rear plate as shown.
4. Install or connect the following:
   - Transmission. Tighten the large bolts to 29–36 ft. lbs. (39–49 Nm) and the small bolts to 12–16 ft. lbs. (16–22 Nm).
   - Transmission mount and crossmember. Tighten the mount and crossmember fasteners to 30–38 ft. lbs. (41–52 Nm).
   - Exhaust mounting bracket
   - Driveshaft
   - Starter motor
   - HO_2S sensor connector
   - PNP switch connector
   - Back-up lamp switch connector
   - VSS sensor connector
   - Clutch slave cylinder
   - CKP sensor
   - Shift lever
   - Negative battery cable

4 Wheel Drive

FRONTIER

1. Before servicing the vehicle, refer to the precautions in the beginning of this section.
2. Remove or disconnect the following:
   - Negative battery cable
   - Shift lever
   - Transfer case select lever
   - Crankshaft Position (CKP) sensor

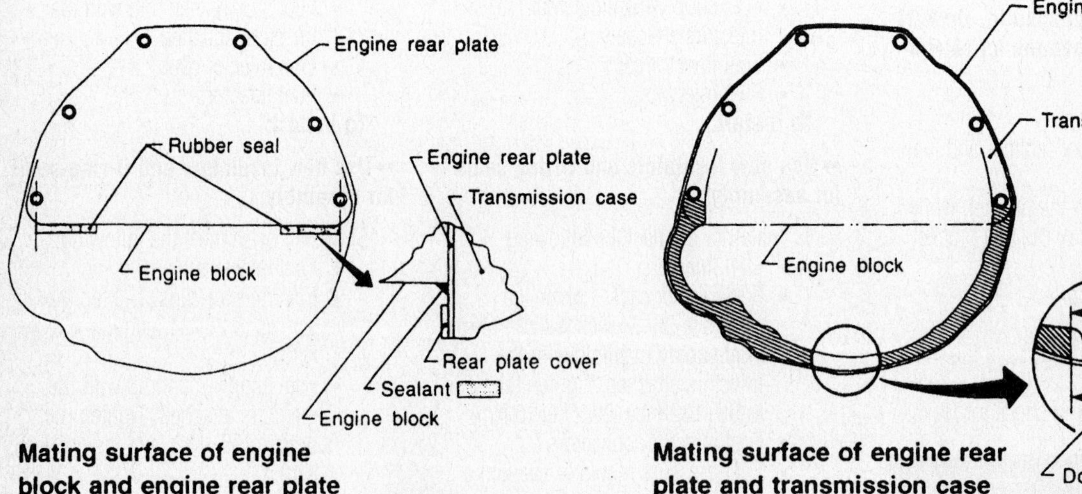

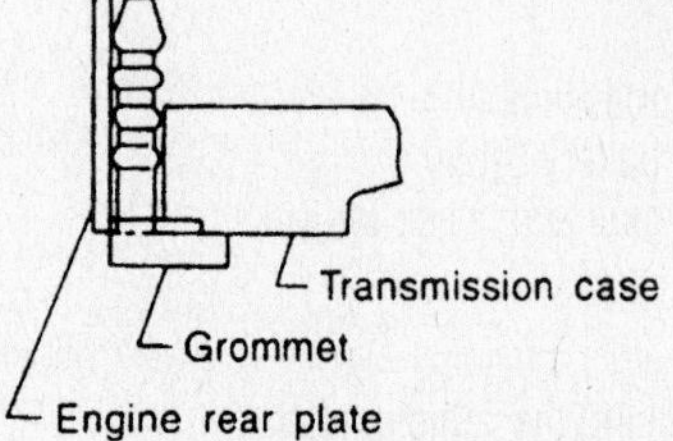

Apply sealant to the indicated areas between the engine block, transmission and engine rear plate—4 Wheel Drive shown

- Clutch slave cylinder
- Vehicle Speed (VSS) sensor connector
- Back-up lamp switch connector
- Park/Neutral Position (PNP) switch connector
- Rear Heated Oxygen (HO_2S) sensor connector
- Starter motor
- Front and rear driveshafts
- Exhaust front pipes
- Exhaust center pipe
- Torsion bars and mounts
- Rear torsion bar cross mount
- Transmission mount and crossmember. Support the transmission.
- Transmission flange bolts
- Transmission

➡The transmission flange bolts vary in length. Note their positions for assembly.

To install:

3. Apply sealant to the transmission flange, engine block, and engine rear plate as shown.
4. Install or connect the following:
   - Transmission. Tighten the large bolts to 29–36 ft. lbs. (39–49 Nm) and the small bolts to 22–29 ft. lbs. (29–39 Nm).
   - Transmission mount and crossmember. Tighten the mount and crossmember fasteners to 30–38 ft. lbs. (41–52 Nm).
   - Rear torsion bar cross mount
   - Torsion bars and mounts
   - Exhaust center pipe
   - Exhaust front pipes
   - Front and rear driveshafts
   - Starter motor
   - HO_2S sensor connector
   - PNP switch connector
   - Back-up lamp switch connector
   - VSS sensor connector
   - Clutch slave cylinder
   - CKP sensor
   - Transfer case select lever
   - Shift lever
   - Negative battery cable

XTERRA

1. Before servicing the vehicle, refer to the precautions in the beginning of this section.
2. Remove or disconnect the following:
   - Negative battery cable
   - Shift lever
   - Transfer case select lever
   - Crankshaft Position (CKP) sensor
   - Clutch slave cylinder
   - Vehicle Speed (VSS) sensor connector
   - Back-up lamp switch connector
   - Park/Neutral Position (PNP) switch connector
   - Rear Heated Oxygen (HO_2S) sensor connector
   - Starter motor
   - Front and rear driveshafts
   - Exhaust front pipes
   - Exhaust center pipe
   - Torsion bars and mounts
   - Rear torsion bar cross mount
   - Transmission mount and crossmember. Support the transmission.
   - Transmission flange bolts
   - Transmission

➡The transmission flange bolts vary in length. Note their positions for assembly.

To install:

3. Apply sealant to the transmission flange, engine block, and engine rear plate as shown.
4. Install or connect the following:
   - Transmission. Tighten the large bolts to 29–36 ft. lbs. (39–49 Nm) and the small bolts to 22–29 ft. lbs. (29–39 Nm).
   - Transmission mount and crossmember. Tighten the mount and crossmember fasteners to 30–38 ft. lbs. (41–52 Nm).
   - Rear torsion bar cross mount
   - Torsion bars and mounts
   - Exhaust center pipe
   - Exhaust front pipes
   - Front and rear driveshafts
   - Starter motor
   - HO_2S sensor connector
   - PNP switch connector
   - Back-up lamp switch connector
   - VSS sensor connector
   - Clutch slave cylinder
   - CKP sensor
   - Transfer case select lever
   - Shift lever
   - Negative battery cable

Automatic Transmission

REMOVAL & INSTALLATION

2 Wheel Drive

1. Before servicing the vehicle, refer to the precautions in the beginning of this section.
2. Remove or disconnect the following:
   - Negative battery cable
   - Crankshaft Position (CKP) sensor
   - Exhaust front pipes
   - Exhaust rear pipes
   - Transmission dipstick tube
   - Transmission oil cooler lines
   - Driveshaft
   - Shift cable
   - Transmission control harness connectors
   - Vehicle Speed (VSS) sensor connector
   - Starter motor
   - Torque converter
   - Transmission mount and crossmember. Support the transmission.
   - Transmission flange bolts
   - Transmission

➡The transmission flange bolts vary in length. Note their positions for assembly.

To install:

3. Install or connect the following:
   - Transmission. Tighten the large bolts to 29–36 ft. lbs. (39–49 Nm) and the small bolts to 22–29 ft. lbs. (29–39 Nm).
   - Transmission mount and crossmember. Tighten the mount and crossmember fasteners to 30–38 ft. lbs. (41–52 Nm).
   - Torque converter. Tighten the bolts to 33–43 ft. lbs. (44–59 Nm).
   - Starter motor
   - VSS sensor connector
   - Transmission control harness connectors
   - Shift cable
   - Driveshaft
   - Transmission oil cooler lines
   - Transmission dipstick tube
   - Exhaust rear pipes
   - Exhaust front pipes
   - CKP sensor
   - Negative battery cable

4 Wheel Drive

1. Before servicing the vehicle, refer to the precautions in the beginning of this section.
2. Remove or disconnect the following:
   - Negative battery cable
   - Crankshaft Position (CKP) sensor
   - Exhaust front pipes
   - Exhaust rear pipes
   - Transmission dipstick tube
   - Transmission oil cooler lines
   - Front and rear driveshafts
   - Transfer case linkage
   - Shift cable
   - Transmission control harness connectors
   - Vehicle Speed (VSS) sensor connector

- Starter motor
- Torque converter
- Transmission mount and crossmember. Support the transmission.
- Transmission flange bolts
- Transmission

➡The transmission flange bolts vary in length. Note their positions for assembly.

To install:

3. Install or connect the following:
   - Transmission. Tighten the large bolts to 29–36 ft. lbs. (39–49 Nm) and the small bolts to 22–29 ft. lbs. (29–39 Nm).
   - Transmission mount and crossmember. Tighten the mount and crossmember fasteners to 30–38 ft. lbs. (41–52 Nm).
   - Torque converter. Tighten the bolts to 33–43 ft. lbs. (44–59 Nm).
   - Starter motor
   - VSS sensor connector
   - Transmission control harness connectors
   - Shift cable
   - Transfer case linkage
   - Front and rear driveshafts
   - Transmission oil cooler lines
   - Transmission dipstick tube
   - Exhaust rear pipes
   - Exhaust front pipes
   - CKP sensor
   - Negative battery cable

Clutch

REMOVAL & INSTALLATION

1. Before servicing the vehicle, refer to the precautions in the beginning of this section.
2. Remove or disconnect the following:
   - Negative battery cable
   - Transmission
   - Pressure plate. Loosen the bolts evenly in ½ turn steps.
   - Clutch disc

To install:

3. Install or connect the following:
   - Clutch disc and pressure plate. Tighten the pressure plate bolts evenly in ½ turns to 16–22 ft. lbs. (22–29 Nm).
   - Transmission
   - Negative battery cable

Hydraulic Clutch System

BLEEDING

1. Before servicing the vehicle, refer to the precautions in the beginning of this section.
2. Have an assistant pump the clutch pedal slowly several times and hold it depressed.
3. Open the slave cylinder bleeder screw and allow air to escape.
4. Close the bleeder screw before releasing the clutch pedal.
5. Repeat until all air is purged from the clutch hydraulic system.
6. Refill the reservoir to the full mark.

Transfer Case Assembly

REMOVAL & INSTALLATION

1. Before servicing the vehicle, refer to the precautions in the beginning of this section.
2. Remove or disconnect the following:
   - Negative battery cable
   - Front and rear driveshafts
   - Torsion bars and mounts
   - Rear torsion bar crossmember
   - Exhaust front pipes
   - Exhaust rear pipes
   - Vehicle Speed (VSS) sensor connector
   - Transfer case shift linkage
   - Transfer case neutral switch connector
   - 4 wheel drive switch connector
   - Vent hose
   - Transfer case flange bolts
   - Transfer case

To install:

3. Install or connect the following:
   - Transfer case. Tighten the flange bolts to 23–30 ft. lbs. (31–41 Nm).
   - Vent hose
   - 4 wheel drive switch connector

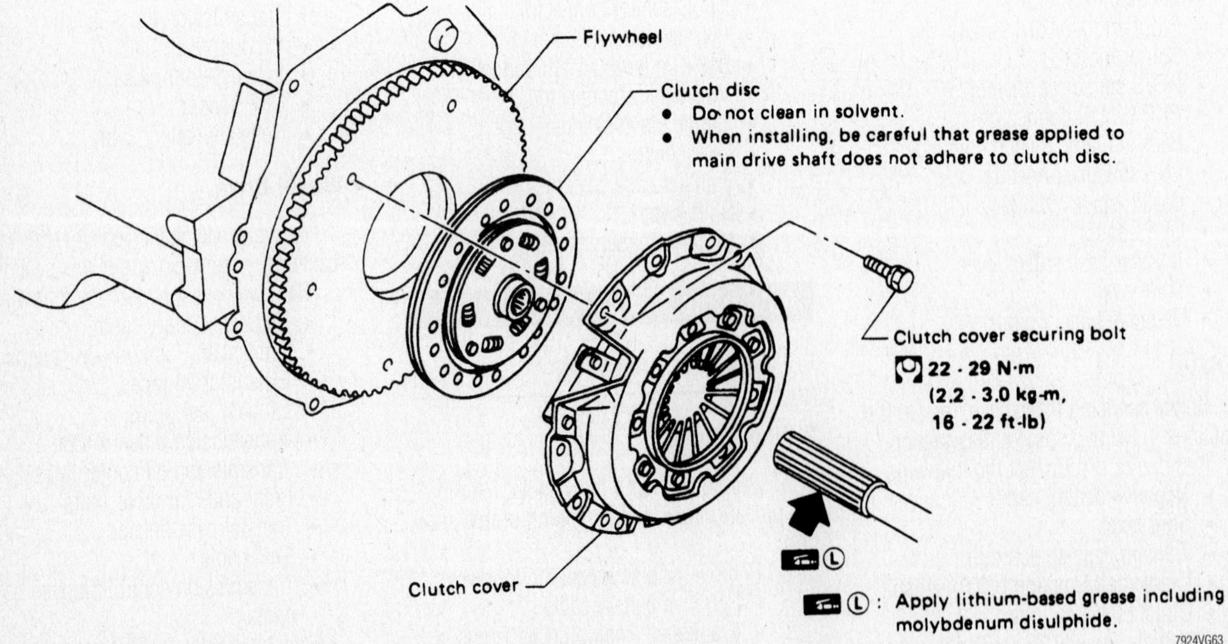

Exploded view of the pressure plate and clutch disc and related components—all models

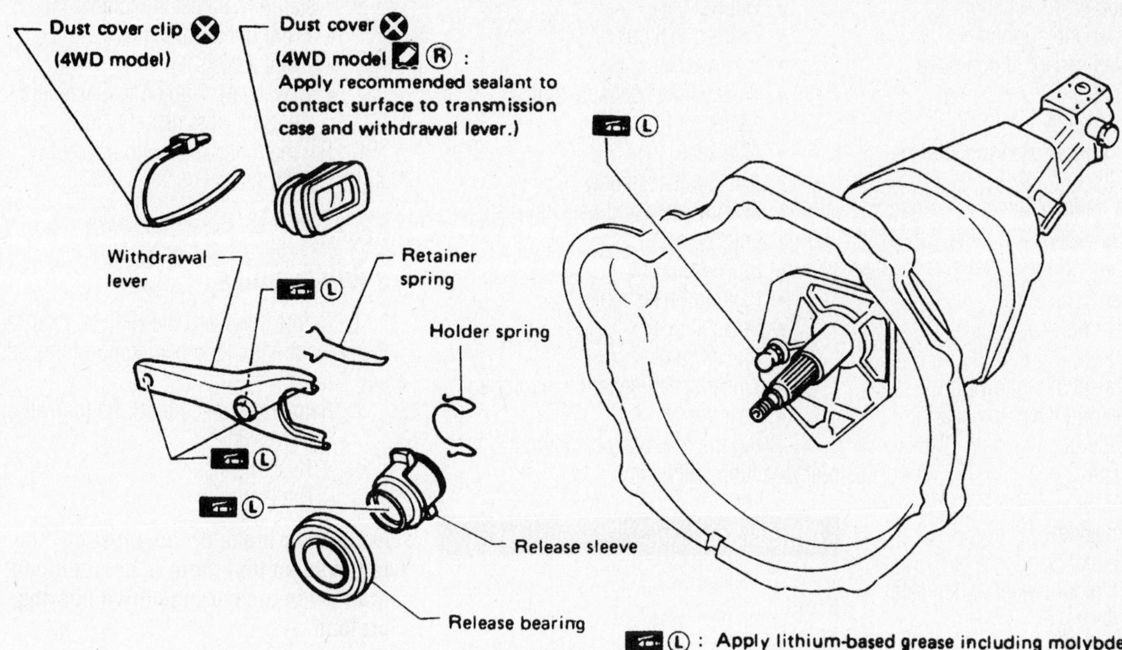

Clutch release mechanism exploded view—all models

- Transfer case neutral switch connector
- Transfer case shift linkage
- VSS sensor connector
- Exhaust rear pipes
- Exhaust front pipes
- Rear torsion bar crossmember
- Torsion bars and mounts
- Front and rear driveshafts
- Negative battery cable

Halfshaft

REMOVAL & INSTALLATION

1. Before servicing the vehicle, refer to the precautions in the beginning of this section.
2. Remove or disconnect the following:
   - Front wheel
   - Wheel speed sensor, if equipped
   - Locking hub or drive flange
   - Snapring
   - Spindle washer
   - Thrust washer
   - Inner CV-joint bolts
   - Axle halfshaft. Separate the stub shaft from the spindle by tapping with a plastic hammer.

To install:

3. Install or connect the following:
   - Axle halfshaft. Guide the stub shaft into the spindle and tighten the inner CV-joint bolts to 25–33 ft. lbs. (34–44 Nm).
   - Thrust washer
   - Spindle washer
   - Snapring
   - Locking hub or drive flange
   - Wheel speed sensor, if equipped
   - Front wheel

CV-Joints

OVERHAUL

Outer CV-Joint

1. Before servicing the vehicle, refer to the precautions in the beginning of this section.
2. Remove the axle halfshaft from the vehicle.
3. Remove the CV-joint boot clamps and push the boot away from the joint.
4. Remove the CV-joint from the axle shaft by tapping it with a brass hammer.

To install:

➡Use new circlips and boot clamps for assembly.

5. Install the CV-joint to the axle shaft by tapping it with a brass hammer.
6. Pack the joint with grease.
7. Install the boot clamps.
8. Install the axle halfshaft to the vehicle.

Inner Tri-Pot Joint

1. Before servicing the vehicle, refer to the precautions in the beginning of this section.
2. Remove the axle halfshaft from the vehicle.
3. Remove the plug seal by tapping around the joint housing flange with a brass hammer.
4. Remove or disconnect the following:
   - CV-joint boot clamps
   - Snapring
   - Spider assembly
   - CV-joint housing
   - CV-joint boot

To install:

➡Use new snaprings and plug seals for assembly.

5. Install or connect the following:
   - CV-joint boot
   - CV-joint housing
   - Spider assembly
   - Snapring. Pack the joint with grease.
   - CV-joint boot clamps
   - Plug seal
6. Install the axle halfshaft to the vehicle.

Spindle Bearings

REMOVAL, PACKING AND INSTALLATION

1. Before servicing the vehicle, refer to the precautions in the beginning of this section.
2. Remove or disconnect the following:
   - Front wheel

- Locking hub or drive flange
- Brake caliper and support
- Wheel speed sensor, if equipped
- Axle halfshaft
- Outer tie rod ends
- Upper ball joint or steering knuckle bracket bolts
- Lower ball joint
- Steering knuckle
- Inner seal
- Thrust washer
- Spindle bearing

To install:

3. Install or connect the following:
   - Spindle bearing. Coat the bearing with multi-purpose grease.
   - Thrust washer
   - Inner seal
   - Steering knuckle
   - Lower ball joint
   - Upper ball joint or steering knuckle bracket bolts
   - Outer tie rod ends
   - Axle halfshaft
   - Wheel speed sensor, if equipped
   - Brake caliper and support
   - Locking hub or drive flange
   - Front wheel

Axle Shaft, Bearing and Seal

REMOVAL & INSTALLATION

1. Before servicing the vehicle, refer to the precautions in the beginning of this section.
2. Remove or disconnect the following:
   - Rear wheel
   - Wheel speed sensor, if equipped
   - Brake drum
   - Brake shoes
   - Parking brake cable
   - Brake fluid line
   - Bearing cage and backing plate bolts
   - Axle shaft assembly
   - Axle seal
   - Wheel speed sensor rotor, if equipped
   - Lockwasher
   - Bearing locknut
   - Flat washer
   - Wheel bearing
   - Wheel bearing cage grease seal

To install:

➡**Use new lockwashers, seals and bearings for assembly.**

3. Install or connect the following:
   - Wheel bearing cage grease seal
   - Wheel bearing
   - Flat washer
   - Bearing locknut
   - Lockwasher
   - Wheel speed sensor rotor, if equipped
   - Axle seal
   - Axle shaft assembly
   - Bearing cage and backing plate bolts
   - Brake fluid line
   - Parking brake cable
   - Brake shoes
   - Brake drum
   - Wheel speed sensor, if equipped
   - Rear wheel
4. Bleed the rear brakes and check the rear axle lubricant level.

Pinion Seal

Front

1. Before servicing the vehicle, refer to the precautions in the beginning of this section.
2. Remove or disconnect the following:
   - Driveshaft
   - Front wheels
   - Front brake calipers

➡**The front brake calipers must be removed so that there is no additional drag when measuring pinion bearing preload.**

3. Use an inch lb. torque wrench and measure the amount of torque required to maintain pinion rotation through several revolutions.
4. Remove or disconnect the following:
   - Pinion flange
   - Oil seal

To install:

5. Install or connect the following:
   - Pinion seal
   - Pinion flange
6. Rotate the pinion flange occasionally while tightening the flange nut to make sure the pinion bearings seat correctly.
7. Take frequent bearing preload torque readings. Tighten the flange nut to achieve the preload torque readings originally recorded. Do not exceed 137–180 ft. lbs. (186–245 Nm) torque when tightening the pinion flange nut.

✲✲ CAUTION

If the bearing preload can not be achieved at the specified torque, remove the pinion bearing and install a new adjustment spacer.

8. Install or connect the following:
   - Front brake calipers
   - Front wheels
   - Driveshaft. Tighten the fasteners to 29–33 ft. lbs. (39–44 Nm).
9. Fill the differential with gear lubricant and check for leaks.

Rear

2 WHEEL DRIVE

1. Before servicing the vehicle, refer to the precautions in the beginning of this section.
2. Remove or disconnect the following:
   - Driveshaft
   - Rear wheels
   - Brake drums

➡**The rear brake drums must be removed so that there is no additional drag when measuring pinion bearing preload.**

3. Use an inch lb. torque wrench and measure the amount of torque required to maintain pinion rotation through several revolutions.
4. Remove or disconnect the following:
   - Pinion flange
   - Wheel speed sensor and rotor, if equipped
   - Oil seal
   - Pinion bearing
   - Collapsible spacer

To install:

➡**Use a new collapsible spacer and wheel speed sensor rotor for assembly.**

5. Install or connect the following:
   - Collapsible spacer
   - Pinion bearing
   - Pinion seal
   - Pinion flange
6. Rotate the pinion flange occasionally while tightening the flange nut to make sure the pinion bearings seat correctly.
7. Take frequent bearing preload torque readings. Tighten the flange nut to achieve the preload torque readings originally recorded. Do not exceed 137–180 ft. lbs. (186–245 Nm) torque when tightening the pinion flange nut.

✲✲ CAUTION

Never loosen the pinion nut to reduce bearing preload. If it is necessary to reduce bearing preload, install a new collapsible spacer.

8. Install or connect the following:
   - Brake drums
   - Rear wheels

- Driveshaft. Tighten the fasteners to 58–65 ft. lbs. (78–88 Nm).

9. Fill the differential with gear lubricant and check for leaks.

4 WHEEL DRIVE

1. Before servicing the vehicle, refer to the precautions in the beginning of this section.
2. Remove or disconnect the following:
   - Driveshaft
   - Rear wheels
   - Brake drums

➡The rear brake drums must be removed so that there is no additional drag when measuring pinion bearing preload.

3. Use an inch lb. torque wrench and measure the amount of torque required to maintain pinion rotation through several revolutions.
4. Remove or disconnect the following:
   - Pinion flange
   - Oil seal

To install:

5. Install or connect the following:
   - Pinion seal
   - Pinion flange
6. Rotate the pinion flange occasionally while tightening the flange nut to make sure the pinion bearings seat correctly.
7. Take frequent bearing preload torque readings. Tighten the flange nut to achieve the preload torque readings originally recorded. Do not exceed 137–180 ft. lbs. (186–245 Nm) torque when tightening the pinion flange nut.

**** CAUTION**

If the bearing preload can not be achieved at the specified torque, remove the pinion bearing and install a new adjustment spacer.

8. Install or connect the following:
   - Brake drums
   - Rear wheels
   - Driveshaft. Tighten the fasteners to 58–65 ft. lbs. (78–88 Nm).
9. Fill the differential with gear lubricant and check for leaks.

STEERING AND SUSPENSION

Air Bag

**** CAUTION**

Some vehicles are equipped with an air bag system. The system must be disarmed before performing service on, or around, system components, the steering column, instrument panel components, wiring and sensors. Failure to follow the safety precautions and the disarming procedure could result in accidental air bag deployment, possible injury and unnecessary system repairs.

PRECAUTIONS

Several precautions must be observed when handling the inflator module to avoid accidental deployment and possible personal injury.

- Never carry the inflator module by the wires or connector on the underside of the module.
- When carrying a live inflator module, hold securely with both hands, and ensure that the bag and trim cover are pointed away.
- Place the inflator module on a bench or other surface with the bag and trim cover facing up.
- With the inflator module on the bench, never place anything on or close to the module which may be thrown in the event of an accidental deployment.

DISARMING

To disarm the **SRS** system turn the ignition switch to the **OFF** position. Then, disconnect both battery cables starting with the negative cable first and wait at least 3 minutes after the cables are disconnected.

To rearm the **SRS** system, turn the ignition switch to the **OFF** position. Connect both battery cables starting with the positive cable first.

Recirculating Ball Power Steering Gear

REMOVAL & INSTALLATION

1. Before servicing the vehicle, refer to the precautions in the beginning of this section.
2. Remove or disconnect the following:
   - Pitman arm
   - Steering column intermediate shaft
   - Power steering hoses
   - Steering gear

To install:

3. Install or connect the following:
   - Steering gear. Tighten the bolts to 62–71 ft. lbs. (84–96 Nm).
   - Power steering hoses. Tighten the banjo fittings to 29–38 ft. lbs. (39–51 Nm).
   - Steering column intermediate shaft. Tighten the pinch bolt to 17–22 ft. lbs. (24–29 Nm).
   - Pitman arm. Tighten the nut to 102-130 ft. lbs. (138-176 Nm) (4-cyl.) or 174–195 ft. lbs. (235–265 Nm) (6-cyl.).
4. Check the wheel alignment and adjust, as necessary.

Shock Absorber

REMOVAL & INSTALLATION

Front

FRONTIER

1. Before servicing the vehicle, refer to the precautions in the beginning of this section.
2. Support the lower control arm.
3. Remove or disconnect the following:
   - Front wheel
   - Lower shock absorber mounting bolt
   - Upper shock absorber mounting nut
   - Shock absorber

To install:

4. Install or connect the following:
   - Shock absorber
   - Upper shock absorber mounting nut. Tighten the nut to 12–16 ft. lbs. (16–22 Nm).
   - Lower shock absorber mounting bolt. Tighten the bolt to 87–106 ft. lbs. (118–147 Nm).
   - Front wheel

XTERRA

1. Before servicing the vehicle, refer to the precautions in the beginning of this section.
2. Support the lower control arm.
3. Remove or disconnect the following:
   - Front wheel
   - Lower shock absorber mounting bolt
   - Upper shock absorber mounting nut
   - Shock absorber

To install:

4. Install or connect the following:
   - Shock absorber
   - Upper shock absorber mounting nut. Tighten the nut to 12–16 ft. lbs. (16–22 Nm).
   - Lower shock absorber mounting bolt. Tighten the bolt to 87–108 ft. lbs. (118–147 Nm).
   - Front wheel

Rear

FRONTIER

1. Before servicing the vehicle, refer to the precautions in the beginning of this section.
2. Support the rear axle.
3. Remove or disconnect the following:
   - Lower shock absorber bolt
   - Upper shock absorber bolt
   - Shock absorber

To install:

➡Use new fasteners for assembly.

4. Install the shock absorber and tighten the bolts to 49–65 ft. lbs. (67–88 Nm).
5. Before servicing the vehicle, refer to the precautions in the beginning of this section.
6. Remove or disconnect the following:
   - Upper and lower shock absorber nuts
   - Shock absorber

To install:

➡Use new nuts for assembly.

7. Install the shock absorber and tighten the nuts to 30–37 ft. lbs. (40–50 Nm).

XTERRA

1. Before servicing the vehicle, refer to the precautions in the beginning of this section.
2. Remove or disconnect the following:
   - Upper and lower shock absorber nuts
   - Shock absorber

To install:

➡Use new nuts for assembly.

3. Install the shock absorber and tighten the nuts to 30–37 ft. lbs. (40–50 Nm).

Leaf Springs

REMOVAL & INSTALLATION

1. Before servicing the vehicle, refer to the precautions in the beginning of this section.
2. Support the vehicle at the frame.
3. Support the axle with a floor jack.
4. Remove or disconnect the following:
   - Rear wheels
   - Shock absorbers
   - Axle U-bolts and spring pad
   - Spring shackle
   - Front mount bolt
   - Leaf spring

To install:

➡Use new fasteners for assembly.

5. Install or connect the following:
   - Leaf spring. Tighten the front mount bolt to 86–108 ft. lbs. (117–147 Nm).
   - Spring shackle. Tighten the nuts to 58–72 ft. lbs. (78–98 Nm).
   - Axle U-bolts and spring pad. Tighten the nuts to 72–80 ft. lbs. (98–108 Nm).
   - Shock absorbers
   - Rear wheels

Torsion Bar

1. Before servicing the vehicle, refer to the precautions in the beginning of this section.
2. Matchmark the torsion bar to the control arm mount and the anchor arm.
3. Measure the adjustment bolt protrusion as shown and note the length (L) for assembly.
4. Loosen the adjustment bolt so that all tension is released.
5. Remove the torsion bar mount from the control arm and remove the torsion bar.

To install:

6. Align the matchmarks and install the torsion bar. Tighten the large mount nut to 66–87 ft. lbs. (89–118 Nm) and the small nut to 33–44 ft. lbs. (45–60 Nm).
7. Tighten the adjustment bolt to achieve the measurement (L) noted earlier. Tighten the locknut to 22–30 ft. lbs. (30–40 Nm).
8. If a new torsion bar is being installed, set length (L) to 2.68 inches.

Upper Ball Joint

REMOVAL & INSTALLATION

The upper ball joint is serviced with the upper control arm as an assembly.

Lower Ball Joint

REMOVAL & INSTALLATION

The lower ball joint is serviced with the lower control arm as an assembly.

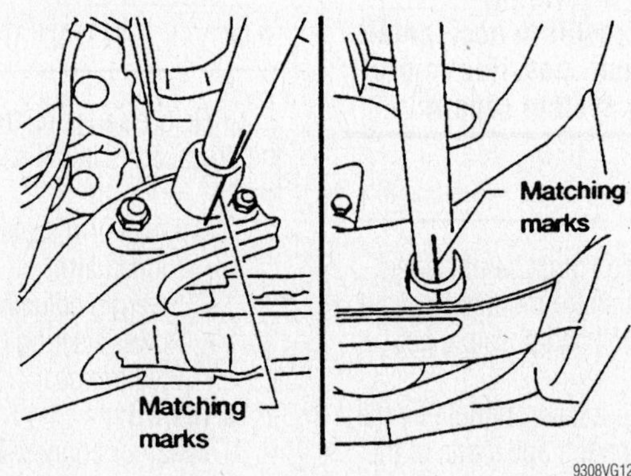

Torsion bar matchmarks

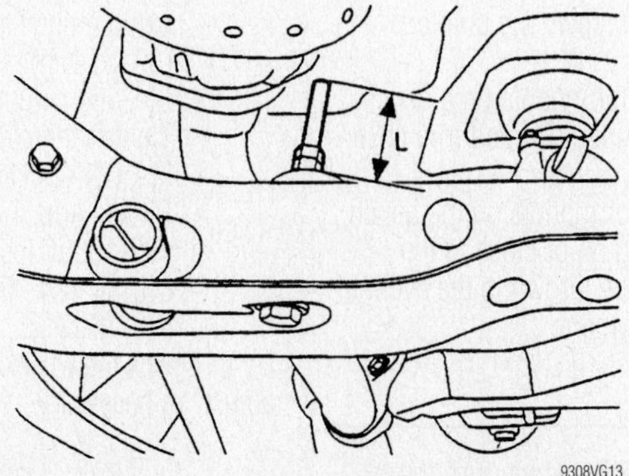

Adjustment bolt measurement (L)

Upper Control Arm

REMOVAL & INSTALLATION

2001–02 Frontier and Xterra

1. Before servicing the vehicle, refer to the precautions in the beginning of this section.
2. Support the lower control arm.
3. Remove or disconnect the following:
   - Front wheel
   - Shock absorber
   - Upper ball joint
   - Control arm mounting bolts
   - Upper control arm

To install:

4. Install or connect the following:
   - Upper control arm. Tighten the mounting bolts to 72–87 ft. lbs. (98–118 Nm).
   - Upper ball joint. Tighten the nut to 58–108 ft. lbs. (78–147 Nm).
   - Shock absorber
   - Front wheel
5. Check the wheel alignment and adjust, as necessary.

2003–04 Frontier and Xterra

1. Remove shock absorber.
2. Separate upper ball joint stud from knuckle spindle.

✻✻ CAUTION

Support lower link with jack.

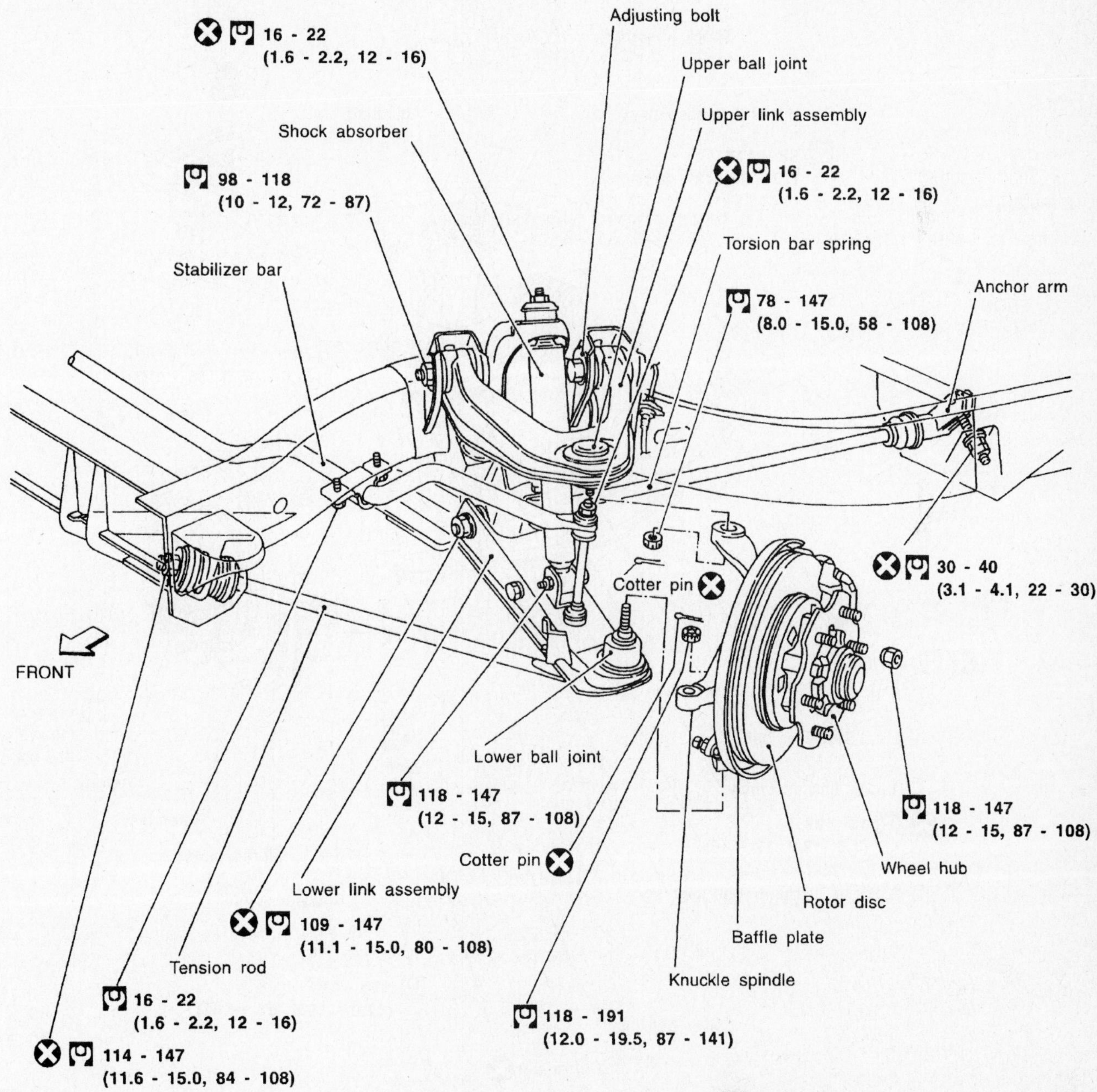

Front suspension components—2wd 4-cyl. 2003–04 Frontier

3. Put matching marks on adjusting bolts and remove adjusting bolts.

To install:

4. While aligning the adjusting bolts with the matching marks, install the upper link. If a new upper link or any other suspension part is installed, align the matching mark with the slit as indicated in the accompanying figure, then install the upper link.

5. Install shock absorber.

6. Tighten adjusting bolts under unladen condition (fuel, radiator coolant, and engine oil full; with spare tire, jack, hand tools, and mats in designated positions) with tires on ground. See the accompanying illustrations for the proper torques.

7. After installing, check wheel alignment. Adjust if necessary.

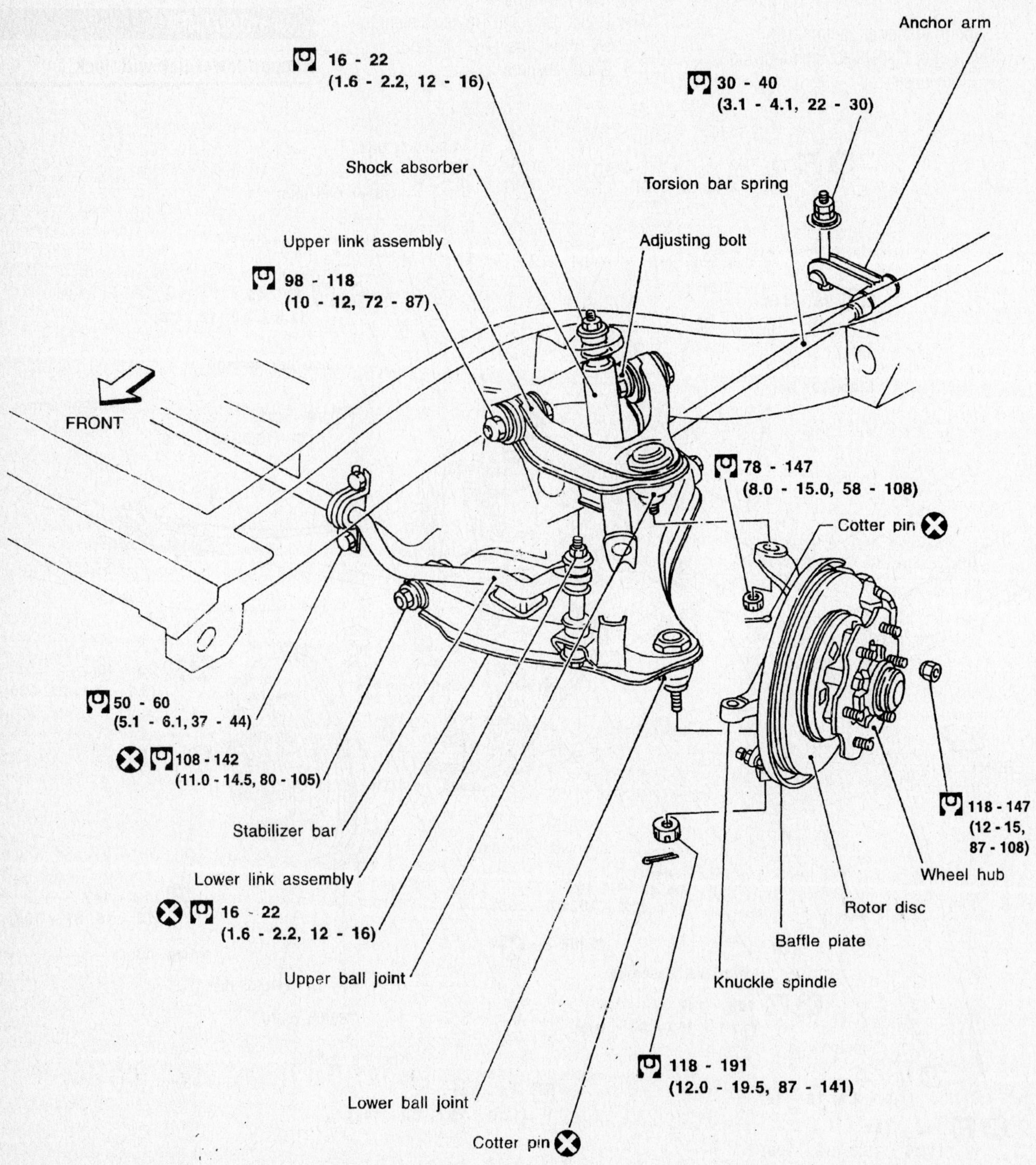

Front suspension components—2wd 6-cyl. 2003–04 Frontier and Xterra

CONTROL ARM BUSHING REPLACEMENT

1. Before servicing the vehicle, refer to the precautions in the beginning of this section.
2. Remove the control arm from the vehicle.
3. Remove the control arm bushing with a press.

To install:

4. Lubricate the control arm bushings with liquid soap.
5. Install the bushings with a press.
6. Install the control arm to the vehicle.
7. Check the wheel alignment and adjust, as necessary.

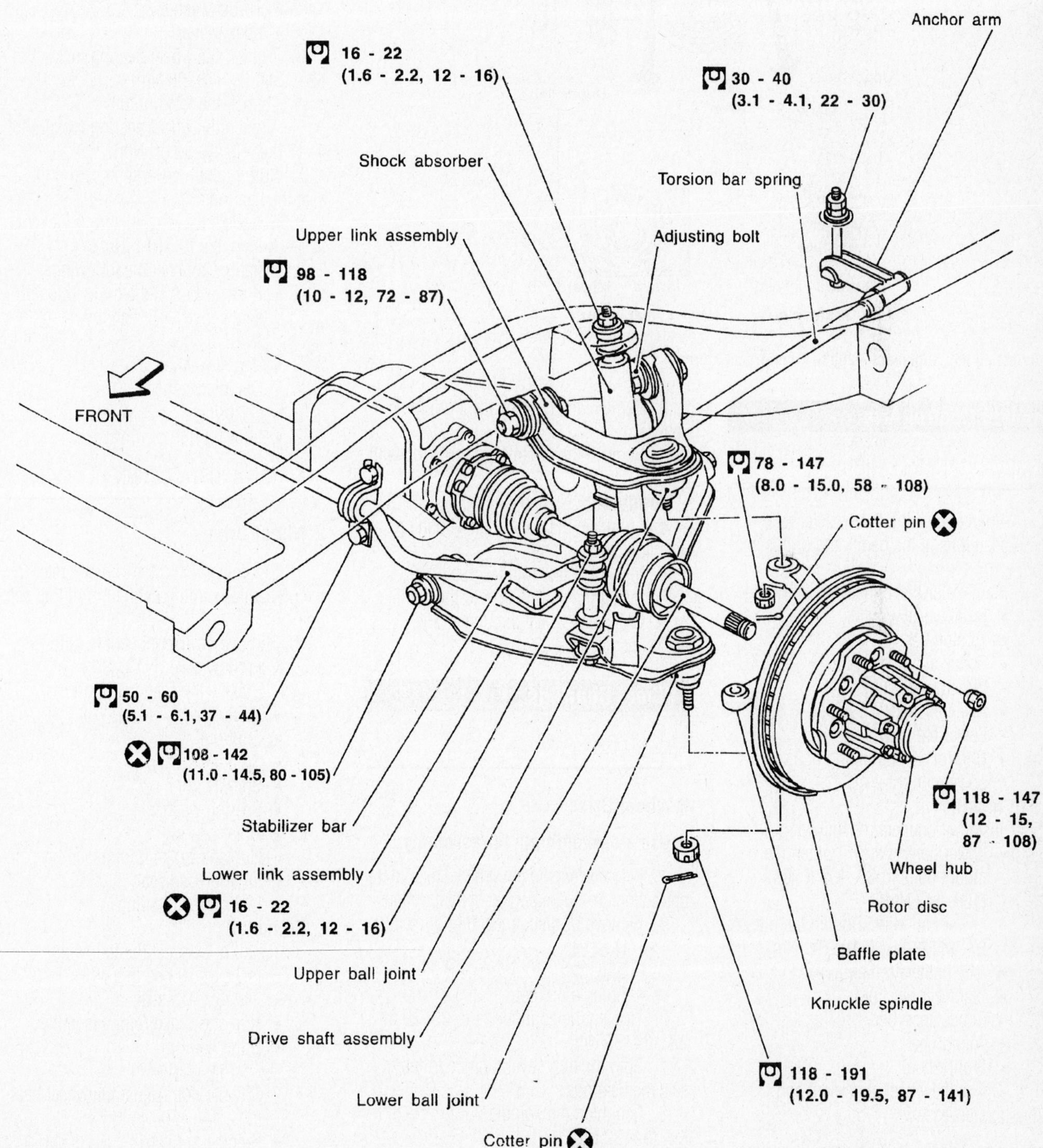

Front suspension components—4wd 2003–04 Frontier and Xterra

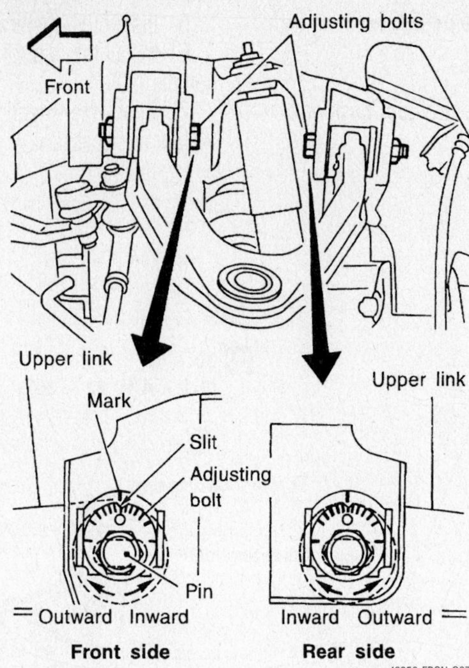

Adjusting bolt alignment—2003– Frontier and Xterra

Lower Control Arm

REMOVAL & INSTALLATION

1. Before servicing the vehicle, refer to the precautions in the beginning of this section.
2. Remove or disconnect the following:
   - Front wheel
   - Torsion bar
   - Shock absorber
   - Stabilizer bar link
   - Axle halfshaft, if equipped
   - Lower ball joint
   - Control arm mounting bolts
   - Lower control arm

To install:

3. Install or connect the following:
   - Lower control arm. Tighten the mount bolts to 80–105 ft. lbs. (108–142 Nm).
   - Lower ball joint. Tighten the nut to 87–141 ft. lbs. (118–191 Nm).
   - Axle halfshaft, if equipped
   - Stabilizer bar link
   - Shock absorber
   - Torsion bar
   - Front wheel
4. Check the wheel alignment and adjust, as necessary.

CONTROL ARM BUSHING REPLACEMENT

1. Before servicing the vehicle, refer to the precautions in the beginning of this section.
2. Remove the control arm from the vehicle.
3. Remove the control arm bushing with a press.

To install:

4. Lubricate the control arm bushings with liquid soap.
5. Install the bushings with a press.
6. Install the control arm to the vehicle.
7. Check the wheel alignment and adjust, as necessary.

Wheel Bearings

ADJUSTMENT

2 Wheel Drive

➡**Use a new split pin for assembly.**

1. Before servicing the vehicle, refer to the precautions in the beginning of this section.
2. Remove or disconnect the following:
   - Dust cap
   - Split pin
   - Spindle nut cap
3. Tighten the spindle nut to 25–29 ft. lbs. (34–39 Nm).
4. Spin the hub several times to fully seat the bearings.
5. Retighten the spindle nut to 25–29 ft. lbs. (34–39 Nm).
6. Loosen the spindle nut 45–60 degrees and install the spindle nut cap and split pin.
7. Install the dust cap.

4 Wheel Drive

1. Before servicing the vehicle, refer to the precautions in the beginning of this section.
2. Remove or disconnect the following:
   - Locking hub or driveplate
   - Snapring
   - Spindle washer
   - Thrust washer
   - Lockwasher
3. Tighten the wheel bearing locknut to 58–72 ft. lbs. (78–98 Nm).
4. Loosen the locknut fully.
5. Tighten the wheel bearing locknut to 4–13 inch lbs. (0.5–1.5 Nm).
6. Spin the hub several times to fully seat the bearings.
7. Retighten the wheel bearing locknut to 4–13 inch lbs. (0.5–1.5 Nm).
8. Install or connect the following:
   - Lockwasher. Tighten the retaining screw to 10–16 inch lbs. (1–2 Nm).
   - Thrust washer
   - Spindle washer
   - Snapring
   - Locking hub or driveplate

REMOVAL & INSTALLATION

2 Wheel Drive

1. Before servicing the vehicle, refer to the precautions in the beginning of this section.
2. Remove or disconnect the following:
   - Front wheel
   - Brake caliper and support
   - Dust cap
   - Split pin
   - Spindle nut cap
   - Spindle nut
   - Bearing washer
   - Outer bearing
   - Hub and brake rotor assembly
   - Inner grease seal
   - Inner wheel bearing

To install:

3. Install or connect the following:
   - Inner wheel bearing
   - Inner grease seal
   - Hub and brake rotor assembly
   - Outer bearing
   - Bearing washer
   - Spindle nut. Adjust the wheel bearings.
   - Spindle nut cap
   - Split pin
   - Dust cap
   - Brake caliper and support
   - Front wheel

4 Wheel Drive

1. Before servicing the vehicle, refer to the precautions in the beginning of this section.
2. Remove or disconnect the following:
   - Front wheel
   - Brake caliper and support
   - Locking hub or driveplate
   - Snapring
   - Spindle washer
   - Thrust washer
   - Lockwasher
   - Wheel bearing locknut
   - Outer bearing
   - Hub and brake rotor assembly
   - Inner grease seal
   - Inner wheel bearing

To install:

3. Install or connect the following:
   - Inner wheel bearing
   - Inner wheel bearing
   - Inner grease seal
   - Hub and brake rotor assembly
   - Outer bearing
   - Wheel bearing locknut. Adjust the wheel bearings.
   - Lockwasher
   - Thrust washer
   - Spindle washer
   - Snapring
   - Locking hub or driveplate
   - Brake caliper and support
   - Front wheel

BRAKES

Brake Caliper

REMOVAL & INSTALLATION

1. Raise the vehicle and support safely.
2. Remove the appropriate tire and wheel assembly.
3. Remove the bolt attaching the brake hose to the caliper. Plug the brake hose to prevent brake fluid loss.
4. Remove the caliper support mounting bolts and lift the caliper assembly from the knuckle.

To install

5. Position the caliper assembly onto the knuckle and install the bolts. Make sure the rotor fits between the brake pads. Torque the bolts to 53–72 ft. lbs. (72–92 Nm) for 2001–03 4-cyl. Frontier; 101–103 ft. lbs. (137–177 Nm) for all others.
6. Using new copper washers, connect the brake hose to the caliper. Torque the brake hose attaching bolt to 12–14 ft. lbs. (17–20 Nm).
7. Bleed the brake system.
8. Apply the brake pedal and inspect the system. Ensure proper operation and no leakage.
9. Install tire and wheel assembly. Lower the vehicle and road-test.

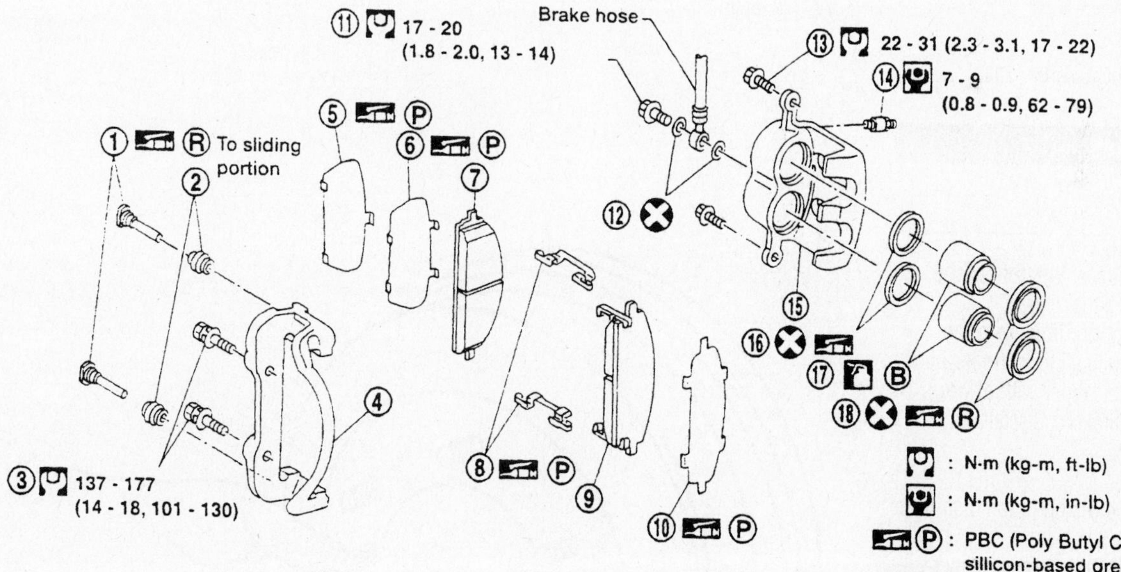

1. Main pin
2. Pin boot
3. Torque member fixing bolt
4. Torque member
5. Shim cover (if so equipped)
6. Inner shim
7. Inner pad
8. Pad retainer
9. Outer pad
10. Outer shim
11. Connecting bolt
12. Copper washer
13. Main pin bolt
14. Bleed valve
15. Cylinder body
16. Piston seal
17. Piston
18. Piston boot

Front disc brake components—2001–02

9348VG95

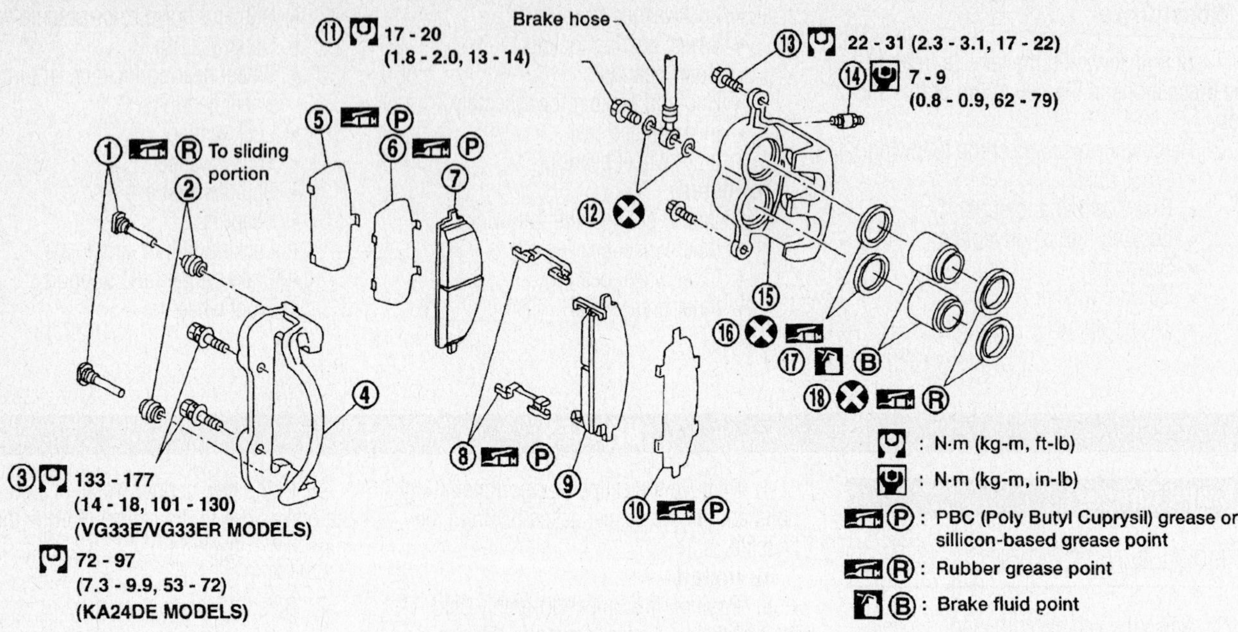

1. Main pin	2. Pin boot	3. Torque member fixing bolt
4. Torque member	5. Shim cover (if equipped)	6. Inner shim
7. Inner pad	8. Pad retainer	9. Outer pad
10. Outer shim	11. Connecting bolt	12. Copper washer
13. Main pin bolt	14. Bleed valve	15. Cylinder body
16. Piston seal	17. Piston	18. Piston boot

42356-FRON-G01

Front disc brake components—2003

Disc Brake Pads

REMOVAL & INSTALLATION

1. Raise and support the front of the vehicle, then remove the wheels.
2. Remove the bottom pin from the caliper and swing the caliper cylinder body upward; support the caliper with a wire.
3. Remove the brake pad retainers, shims and the pads.

To install:

4. Compress the piston of the disc brake caliper.
5. Install the brake pads and caliper assembly. Torque the guide pin to 17–22 ft. lbs. (22–31 Nm).

Brake Drums

REMOVAL & INSTALLATION

1. Remove the hub cap and loosen the lug nuts.
2. Raise the rear of the vehicle and support it on jackstands.
3. Remove the lug nuts, tire and wheel.

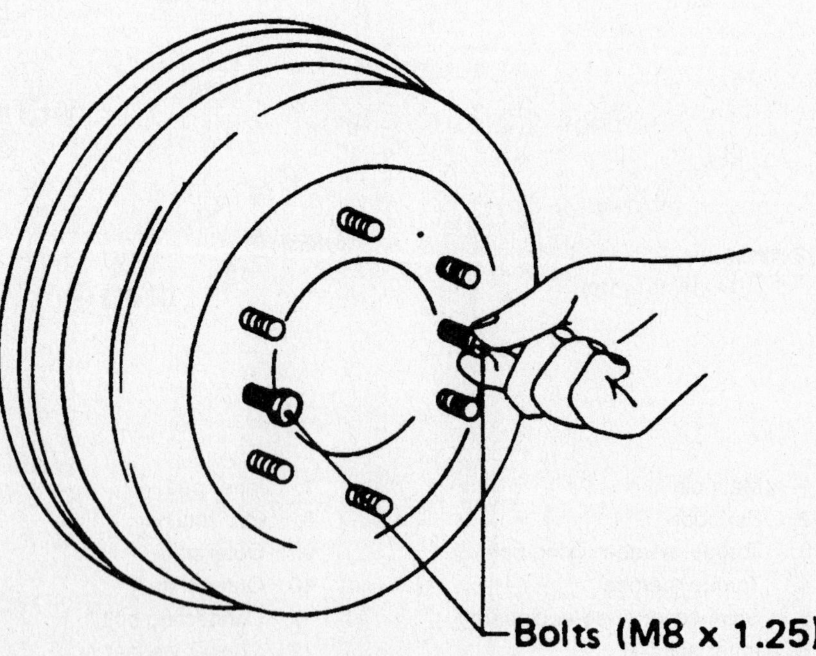

93026G66

Install and tighten 2 bolts to remove a stubborn brake drum

4. Release the parking brake.
5. Pull the brake drum from the hub. If difficult to remove try the following:
   a. Strike the face of the drum with a plastic or rubber mallet. This will break free any rust that may develop between the drum and the hub.
   b. Install 2, M8x1.25mm bolts into the holes in the drum and gradually tighten them to pull the drum off the hub.

To install:

6. Install the brake drum to the hub.
7. Install the wheel.
8. Remove the jackstands and lower the vehicle.
9. Road-test the vehicle to ensure that the brakes are working properly.

Brake Shoes

REMOVAL & INSTALLATION

4-Cyl. Frontier and 2001–02 6-Cyl. Frontier and Xterra

1. Release the parking brake.
2. Safely raise and support the vehicle.
3. Remove the rear wheel and drum.
4. Remove the hold-down pin retainers.
5. Remove the leading shoe and then the trailing shoe.
6. Remove the adjuster.
7. Disconnect the parking brake cable from the toggle lever on the rear shoe.

To install:

8. Transfer the toggle lever to the new rear shoe.
9. Apply a small amount of brake grease to the tips of the shoes and the 6 pads on the backing plate that contact the brake shoe.
10. Shorten the adjuster by turning it.
11. Connect the parking brake cable to the toggle lever on the rear shoe.
12. Install the lower return spring to both shoes and install the shoes on the backing plate with the hold down pins and retainers.
13. Install the adjuster and the remaining springs. Pay attention to the direction of the adjuster assembly.
14. Inspect the complete assembly and install the brake drum.
15. Adjust the shoe to drum clearance.
16. Install the wheel assembly and lower the vehicle to the floor.

2003 6-Cyl. Frontier and Xterra

1. Remove the drum.
2. After removing shoe hold pin by rotating retainer, remove leading shoe then remove trailing shoe. Remove spring by rotating shoes in direction arrow.

⁂ WARNING

Be careful not to damage wheel cylinder piston boots. Be careful not to damage parking brake cable when separating it.

3. Remove the adjuster.
4. Disconnect the parking brake cable from toggle lever.
5. Remove retainer clip with a suitable tool. Then separate toggle lever and brake shoe (trailing side).
6. Installation is the reverse of removal.

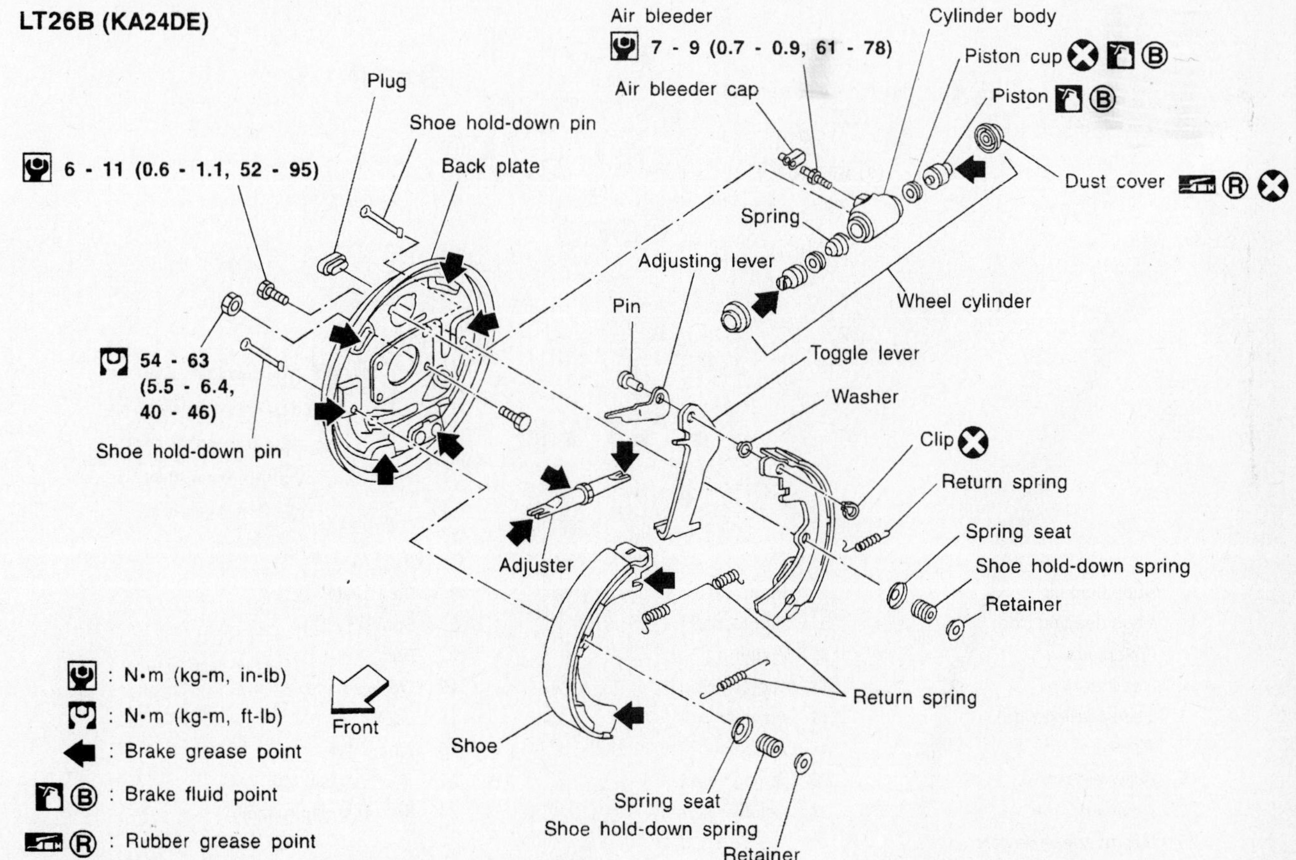

Rear drum brake assembly and related components—4-cyl. Frontier

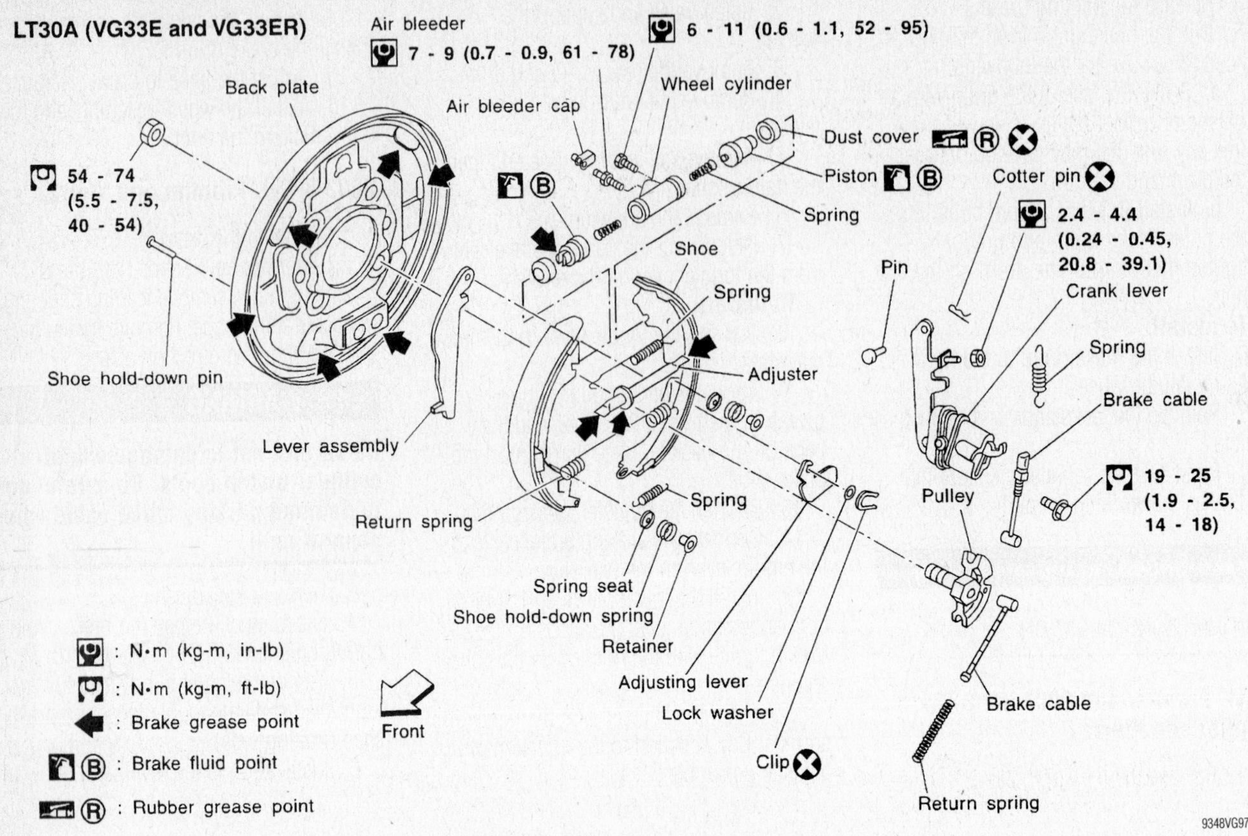

Rear drum brake assembly and related components—2001–02 6-cyl. Frontier and Xterra

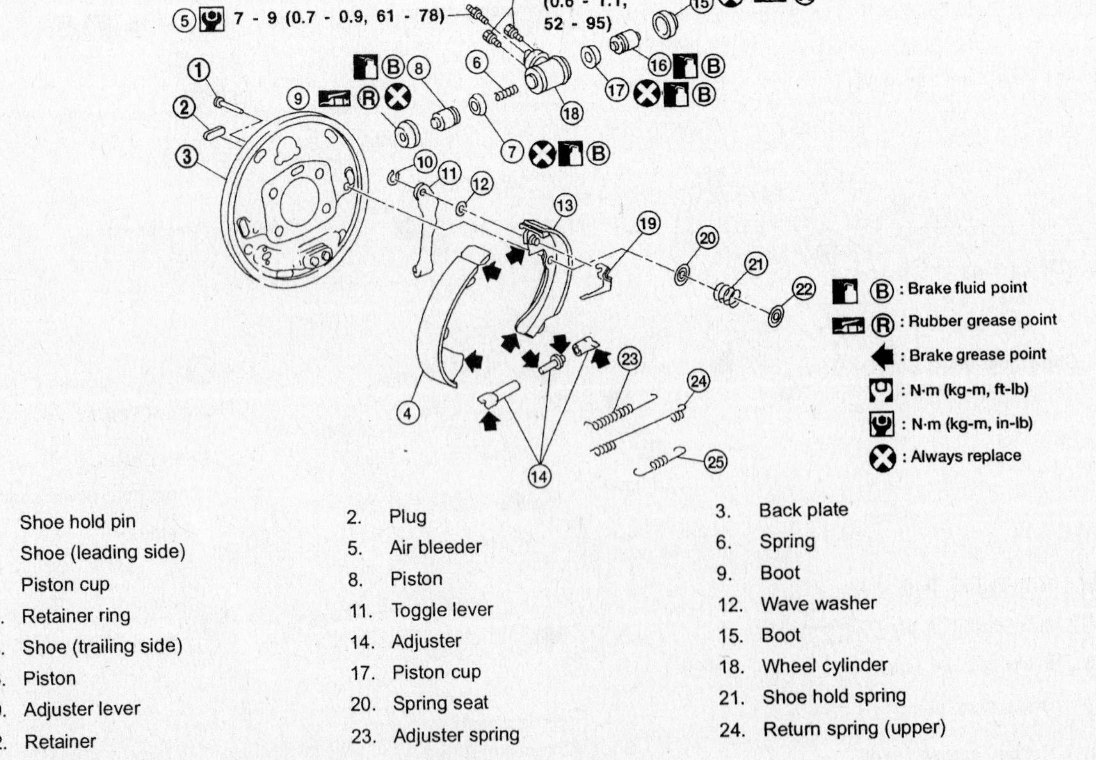

Rear drum brake assembly and related components—2003 6-cyl. Frontier and Xterra

NISSAN

Murano

8

SPECIFICATION AND MAINTENANCE CHARTS

ENGINE AND VEHICLE IDENTIFICATION

Engine						
Code ①	Liters (cc)	Cu. In.	Cyl.	Fuel Sys.	Engine	Eng. Mfg.
VQ35DE	3.5 (3498)	213	6	MFI	DOHC	Nissan

MFI: Multi-port Fuel Injection

DOHC: Double Overhead Camshafts

Model Year	
Code	Year
3	2003
4	2004
5	2005

67170-MURA-C01

GENERAL ENGINE SPECIFICATIONS

Year	Model	Engine Displacement Liters	Engine ID	Net Horsepower @ rpm	Net Torque @ rpm (ft. lbs.)	Bore x Stroke (in.)	Com-pression Ratio	Oil Pressure @ rpm
2003	Murano	3.5	VQ35DE	240@6000	265@3200	3.76X3.20	10.0:1	43@2000
2004	Murano	3.5	VQ35DE	240@6000	265@3200	3.76X3.20	10.0:1	43@2000

67170-MURA-C02

ENGINE TUNE-UP SPECIFICATIONS

Year	Engine Displacement Liters	Engine ID	Spark Plug Gap (in.)	Ignition Timing	Fuel Pump (psi)	Idle Speed RPM	Valve Clearance (in.) In.	Ex.
2003	3.5	VQ35DE	0.043	15B	51 ①	600-700	HYD	HYD
2004	3.5	VQ35DE	0.043	15B	51 ①	600-700	HYD	HYD

NOTE: The Vehicle Emission Control Information label often reflects specification changes made during production. The label figures must be used if they differ from those in this chart.

B: Before top dead center

HYD: Hydraulic

① At idle

67170-MURA-C03

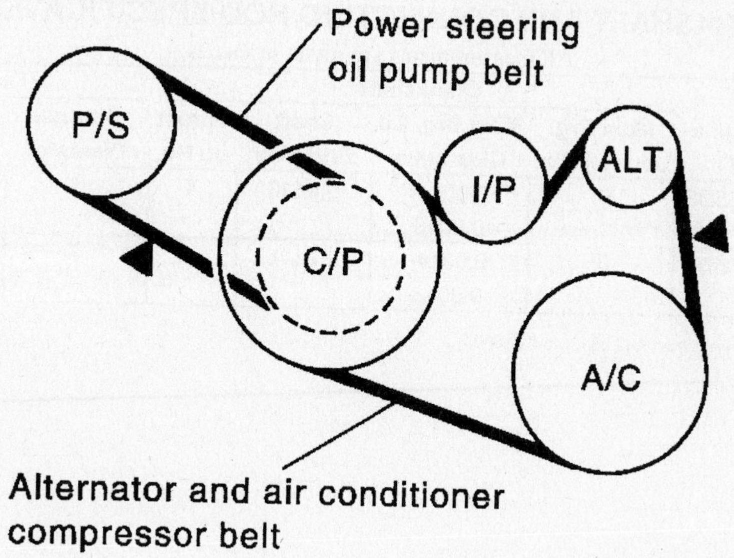

Accessory drive belt routing —3.5L engine

42356-MURA-G75

CAPACITIES

Year	Model	Engine Displacement Liters	Engine ID	Engine Oil with Filter (qts.)	Transmission (pts.)	Transfer Case (pts.)	Drive Axle Front (pts.)	Drive Axle Rear (pts.)	Fuel Tank (gal.)	Cooling System (qts.)
2003	Murano	3.5	VQ35DE	4.25	21.5	0.32	—	1.5	21.6	9.75
2004	Murano	3.5	VQ35DE	4.25	21.5	0.63	—	1.1	21.6	9.75

NOTE: All capacities are approximate. Add fluid gradually and check to be sure a proper fluid level is obtained.

67170-MURA-C04

VALVE SPECIFICATIONS

Year	Engine Displacement Liters	Engine ID	Seat Angle (deg.)	Face Angle (deg.)	Spring Test Pressure (lbs. @ in.)	Spring Installed Height (in.)	Stem-to-Guide Clearance (in.) Intake	Stem-to-Guide Clearance (in.) Exhaust	Stem Diameter (in.) Intake	Stem Diameter (in.) Exhaust
2003	3.5	VQ35DE	45.15-45.45	45	45.4@1.457	1.457	0.0008-0.0021	0.0016-0.0029	0.2348-0.2354	0.2341-0.2346
2004	3.5	VQ35DE	45.15-45.45	45	45.4@1.457	1.457	0.0008-0.0021	0.0016-0.0029	0.2348-0.2354	0.2341-0.2346

67170-MURA-C05

CRANKSHAFT AND CONNECTING ROD SPECIFICATIONS

All measurements are given in inches.

Year	Engine Displacement Liters	Engine ID	Crankshaft Main Brg. Journal Dia.	Crankshaft Main Brg. Oil Clearance	Crankshaft Shaft End-play	Crankshaft Thrust on No.	Connecting Rod Journal Diameter	Connecting Rod Oil Clearance	Connecting Rod Side Clearance
2003	3.5	VQ35DE	①	0.0014-0.0018	0.0118	4	②	0.0013-0.0023	0.0079-0.0138
2004	3.5	VQ35DE	①	0.0014-0.0018	0.0118	4	②	0.0013-0.0023	0.0079-0.0138

① There are 24 different grades, ranging from A (2.3612) to 7 (2.3603)

② Grade 0: 2.0460-2.0462
Grade 1: 2.0457-2.0460
Grade 2: 2.0445-2.0457

67170-MURA-C06

PISTON AND RING SPECIFICATIONS

All measurements are given in inches.

Year	Engine Displacement Liters	Engine ID	Piston Clearance	Ring Gap Top Comp.	Ring Gap Bottom Comp.	Ring Gap Oil Control	Ring Side Clearance Top Comp.	Ring Side Clearance Bottom Comp.	Ring Side Clearance Oil Control
2003	3.5	VQ35DE	0.0004-0.0012	0.0091-0.0130	0.0130-0.0189	0.0079-0.0236	0.0016-0.0031	0.0012-0.0028	0.0006-0.0020
2004	3.5	VQ35DE	0.0004-0.0012	0.0091-0.0130	0.0130-0.0189	0.0079-0.0236	0.0016-0.0031	0.0012-0.0028	0.0006-0.0020

67170-MURA-C07

TORQUE SPECIFICATIONS

All readings in ft. lbs.

Year	Engine Displacement Liters	Engine ID	Cylinder Head Bolts	Main Bearing Bolts	Rod Bearing Bolts	Crankshaft Damper Bolts	Driveplate Bolts	Manifold Intake	Manifold Exhaust	Spark Plugs	Oil Pan Drain Plug
2003	3.5	VQ35DE	①	②	③	④	61-69	⑤	21-24	14-22	25
2004	3.5	VQ35DE	①	②	③	④	61-69	⑤	21-24	14-22	25

① Step 1: 72 ft. lbs.
Step 2: Loosen all bolts completely
Step 3: 25-33 ft. lbs.
Step 4: +90 degrees
Step 5: +90 degrees

② Step 1: 24-28 ft. lbs.
Step 2: +90 degrees

③ Step 1: 15 ft. lbs.
Step 2: +90 degrees

④ 29-36 ft. lbs. +60-66 degrees

⑤ Step 1: 44-86 inch lbs.
Step 2: 20-23 ft. lbs.

67170-MURA-C08

WHEEL ALIGNMENT

Year	Model		Caster Range (+/-Deg.)	Caster Preferred Setting (Deg.)	Camber Range (+/-Deg.)	Camber Preferred Setting (Deg.)	Toe-in (in.)	Kingpin Inclination (Deg.)
2003	Murano	F	0.75	+2.58	0.25	-0.33	0.02+/-0.04	14.33
		R	—	—	0.50	-0.80	0.14+/-0.70	—
2004	Murano	F	0.75	+2.58	0.25	-0.33	0.02+/-0.04	14.33
		R	—	—	0.50	-0.80	0.14+/-0.70	—

67170-MURA-C09

TIRE, WHEEL AND BALL JOINT SPECIFICATIONS

Year	Model	OEM Tires Standard	OEM Tires Optional	Tire Pressures (psi) Front	Tire Pressures (psi) Rear	Wheel Size	Ball Joint Inspection	Lugnut Torque (ft. lbs.)
2003	Murano	P235/65SR18	none	33	33	7.5	①	90
2004	Murano	P235/65SR18	none	33	33	7.5	①	90

OEM: Original Equipment Manufacturer

PSI: Pounds Per Square Inch

NA: Not available

① Rotating torque: 5-30 inch lbs.

67170-MURA-C10

BRAKE SPECIFICATIONS

All measurements in inches unless noted

Year	Model		Brake Disc Original Thickness	Brake Disc Minimum Thickness	Brake Disc Maximum Runout	Minimum Lining Thickness	Brake Caliper Bracket Bolts (ft. lbs.)	Brake Caliper Mounting Bolts (ft. lbs.)
2003	Murano	F	1.102	1.024	0.0016	0.079	101-129	17-22
		R	0.630	0.551	0.002	0.079	53-71	28-35
2004	Murano	F	1.102	1.024	0.0016	0.079	101-129	17-22
		R	0.630	0.551	0.002	0.079	53-71	28-35

F: Front

R: Rear

67170-MURA-C11

SCHEDULED MAINTENANCE INTERVALS

Nissan—Murano

TO BE SERVICED	TYPE OF SERVICE	7.5	15	22.5	30	37.5	45	52.5	60
Engine oil & filter	R	✓	✓	✓	✓	✓	✓	✓	✓
Brake lines & cables	S/I		✓		✓		✓		✓
Brake pads, discs	I		✓		✓		✓		✓
Driveshaft boots & propeller shaft	L/I		✓		✓		✓		✓
CVT, transfer case and differential fluid	I		✓		✓		✓		✓
Air cleaner filter	R				✓				✓
Drive belt(s) ①	S/I								
Engine coolant ②	R								✓
Spark plugs	R	Platinum plugs, every 100,000 miles							
Cabin air filter	R		✓		✓		✓		✓
Exhaust system	I			✓					✓
Fuel lines	S/I				✓				✓
Steering gear, linkage, axle & suspension parts	I		✓		✓		✓		✓
Vapor lines	S/I				✓				✓

R: Replace S/I: Service or Inspect L: Lubricate

① First a 60,000, then every 15,000 miles

② After 60,000, replace every 30,000

FREQUENT OPERATION MAINTENANCE (SEVERE SERVICE)

If a vehicle is operated under any of the following conditions it is considered severe service:

- Extremely dusty areas.

- 50% or more of the vehicle operation is in 32°C (90°F) or higher temperatures, or constant temperatures below 0°C (32°F).

- Prolonged idling (vehicle operation in stop and go traffic).

- Frequent short running periods (engine does not warm to normal operating temperatures).

- Police, taxi, delivery usage or trailer towing usage.

Oil & oil filter: replace every 3750 miles.

Brake pads, discs, drums & linings: service or inspect every 7500 miles.

Driveshaft boots & propeller shaft: service or inspect every 7500 miles.

Exhaust system: service or inspect every 7500 miles.

Steering gear (box) & linkage, (steering damper-4x4), axle & suspension parts: service or inspect every 7500 miles.

Steering linkage ball joints & front suspension ball joints: service or inspect every 7500 miles.

67170-MURA-C12

PRECAUTIONS

Before servicing any vehicle, please be sure to read all of the following precautions, which deal with personal safety, prevention of component damage, and important points to take into consideration when servicing a motor vehicle:

- Never open, service or drain the radiator or cooling system when the engine is hot; serious burns can occur from the steam and hot coolant.
- Observe all applicable safety precautions when working around fuel. Whenever servicing the fuel system, always work in a well-ventilated area. Do not allow fuel spray or vapors to come in contact with a spark, open flame, or excessive heat (a hot drop light, for example). Keep a dry chemical fire extinguisher near the work area. Always keep fuel in a container specifically designed for fuel storage; also, always properly seal fuel containers to avoid the possibility of fire or explosion. Refer to the additional fuel system precautions later in this section.
- Fuel injection systems often remain pressurized, even after the engine has been turned **OFF**. The fuel system pressure must be relieved before disconnecting any fuel lines. Failure to do so may result in fire and/or personal injury.
- Brake fluid often contains polyglycol ethers and polyglycols. Avoid contact with the eyes and wash your hands thoroughly after handling brake fluid. If you do get brake fluid in your eyes, flush your eyes with clean, running water for 15 minutes. If eye irritation persists, or if you have taken brake fluid internally, IMMEDIATELY seek medical assistance.
- The EPA warns that prolonged contact with used engine oil may cause a number of skin disorders, including cancer! You should make every effort to minimize your exposure to used engine oil. Protective gloves should be worn when changing oil. Wash your hands and any other exposed skin areas as soon as possible after exposure to used engine oil. Soap and water, or waterless hand cleaner should be used.
- All new vehicles are now equipped with an air bag system. The system must be disabled before performing service on or around system components, steering column, instrument panel components, wiring and sensors. Failure to follow safety and disabling procedures could result in accidental air bag deployment, possible personal injury and unnecessary system repairs.
- Always wear safety goggles when working with, or around, the air bag system. When carrying a non-deployed air bag, be sure the bag and trim cover are pointed away from your body. When placing a non-deployed air bag on a work surface, always face the bag and trim cover upward, away from the surface. This will reduce the motion of the module if it is accidentally deployed. Refer to the additional air bag system precautions later in this section.
- Clean, high quality brake fluid from a sealed container is essential to the safe and proper operation of the brake system. You should always buy the correct type of brake fluid for your vehicle. If the brake fluid becomes contaminated, completely flush the system with new fluid. Never reuse any brake fluid. Any brake fluid that is removed from the system should be discarded. Also, do not allow any brake fluid to come in contact with a painted surface; it will damage the paint.
- Never operate the engine without the proper amount and type of engine oil; doing so WILL result in severe engine damage.
- Timing belt maintenance is extremely important! Many models utilize an interference-type, non-freewheeling engine. If the timing belt breaks, the valves in the cylinder head may strike the pistons, causing potentially serious (also time-consuming and expensive) engine damage. Refer to the maintenance interval charts in the front of this manual for the recommended replacement interval for the timing belt, and to the timing belt section for belt replacement and inspection.
- Disconnecting the negative battery cable on some vehicles may interfere with the functions of the on-board computer system(s) and may require the computer to undergo a relearning process once the negative battery cable is reconnected.
- When servicing drum brakes, only disassemble and assemble one side at a time, leaving the remaining side intact for reference.

ENGINE REPAIR

➡Disconnecting the negative battery cable on some vehicles may interfere with the functions of the on board computer system. The computer may undergo a relearning process once the negative battery cable is reconnected.

Alternator

REMOVAL & INSTALLATOIN

1. Before servicing the vehicle, refer to the precautions in the beginning of this section.
2. Remove or disconnect the following:
   - Negative battery cable
   - Alternator harness connectors
   - Engine right side under cover
   - Radiator
   - Remove alternator and air conditioner compressor belt.

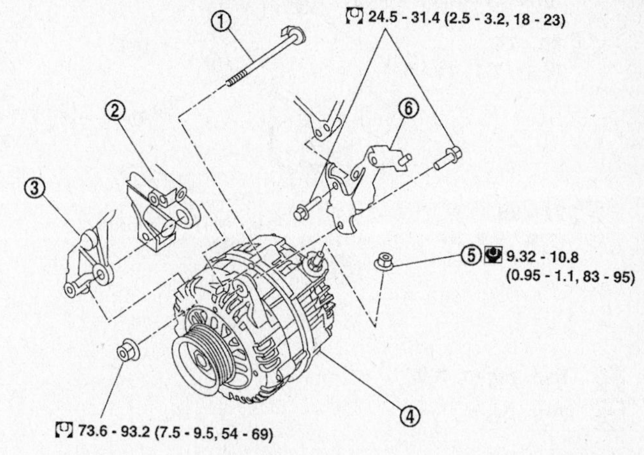

Alternator exploded view

- Idler pulley
- Alternator

To install:

3. Install or connect the following:
   - Alternator
   - Idler pulley
   - Alternator belt. Tighten the through-bolts to 18–23 ft. lbs. (25–31 Nm).
   - Engine under cover
   - Radiator
   - Alternator harness connectors
   - Negative battery cable

Ignition Timing

ADJUSTMENT

Timing is not adjustable.

Engine Assembly

REMOVAL & INSTALLATION

1. Release fuel pressure.
2. Remove the engine cover, and the splash guards.
3. Drain engine coolant.
4. Remove or disconnect the following:
   - Battery and tray
   - Air inlet duct
   - Air duct and air cleaner case (upper) assembly with mass air flow sensor
   - Power brake booster vacuum hose
   - Drive belts
   - Radiator assembly, coolant reservoir, and system hoses
   - RH windshield wiper arm and right font cowl top cover
   - Engine room harness from the ECM side
   - Heater hoses
   - Wheel and tires
   - A/C compressor with piping connected, and temporarily secure it aside
   - Fuel hose quick connector at vehicle piping side
   - Transaxle shift control cable.
   - Starter motor
   - Front exhaust tube
   - Reservoir tank for the power steering from engine compartment bracket and position it aside.
   - Power steering gear from steering lower joint
   - Steering outer socket from steering knuckle
   - Stabilizer connecting rod
   - Propeller shaft (AWD models)

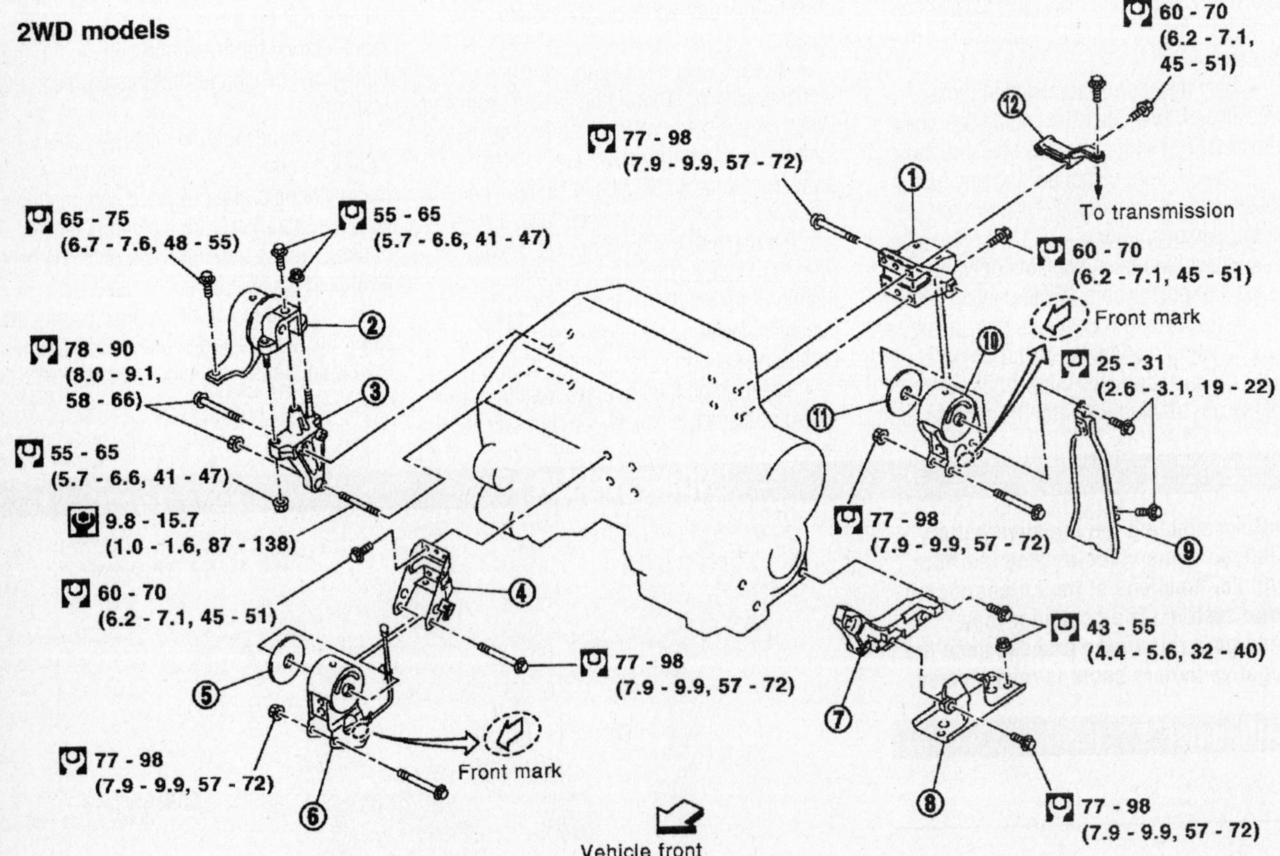

Engine mounting—2wd

AWD models

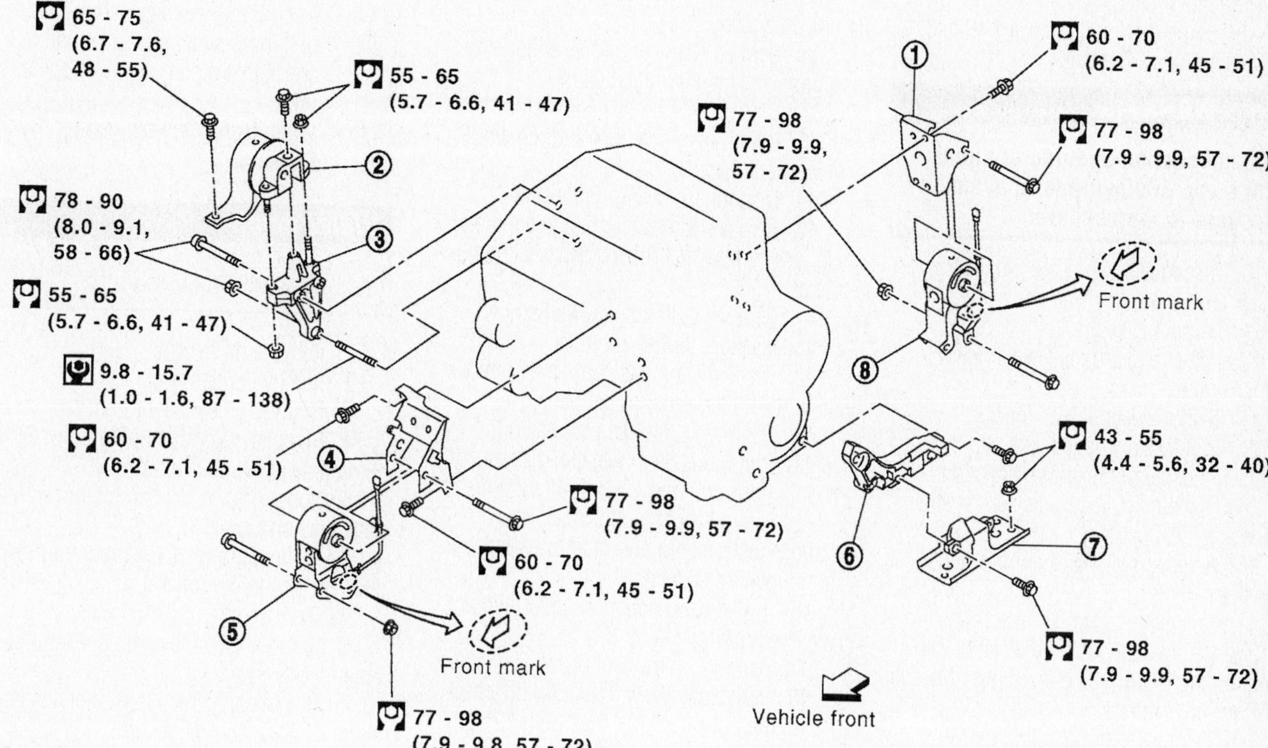

Engine mounting—AWD

- Left and right front halfshafts
- Power steering piping from power steering oil cooler

5. Position a manual lift table caddy under the engine and transaxle assembly.
6. Remove the right engine mounting insulator
7. Remove mounting bolt between transverse link and front suspension member
8. Carefully lower the engine, transaxle, transfer case (AWD models) and front suspension member assembly with the manual lift table caddy, avoiding interference with the vehicle body.

✲✲ WARNING

Before and during this procedure, always check if any harnesses are left connected. Avoid any damage to, or any oil or grease smearing or spills onto the engine mounting insulators.

9. Remove the crankshaft position sensor (POS).
10. Disconnect front suspension mounting nuts and bolts to remove engine, transaxle, transfer case (AWD models) and front suspension member assembly as a unit.
11. Separate the engine, transaxle and transfer case (AWD models) assembly and front suspension member.
12. Installation is the reverse order of removal. See the accompanying illustrations for the proper torque values.

Heater Core

REMOVAL & INSTALLATION

1. Use a refrigerant collecting equipment (for HFC-134a) to discharge refrigerant.
2. Drain coolant from cooling system.
3. Remove both right/left wiper arms.
4. Remove cowl top seal rubber.
5. Remove clips from cowl top cover (right) and remove cowl top cover (right).
6. Remove clips from cowl top cover (left) and remove cowl top cover (left)
7. Remove washer nozzles and hose from cowl top cover.
8. Remove cowl top cover.
9. Disconnect evaporator-side one-touch joints.

a. Install a disconnector tool (High-pressure side: 92530-89908, Low pressure side: 92530-89916) on A/C piping.

b. Slide a disconnector toward vehicle front until it clicks.

c. Slide A/C piping toward vehicle front and disconnect it.

⁂ WARNING

Seal connection opening of piping with a cap or vinyl tape to avoid exposure to atmosphere.

10. Disconnect two heater hoses from heater core.
11. Remove fuse lid.
12. Remove instrument driver lower panel screws.
13. Remove data link connector.
14. Pull to disengage metal clip by removing panel in horizontal direction.
15. Disconnect in-vehicle sensor and each electrical parts.
16. Remove bolts, and remove hood lock opener.
17. Remove tilt lever knob screws.
18. Remove the knob by picking it up and pulling it out. Using a remover, ply and remove tilt lever mask.
19. Remove steering column cover screws.
20. Disengage the tab, and remove steering column cover. After removing combination meter screws, remove harness connector.
21. Using a remover, pry and remove side ventilator assembly (left/right).
22. Disconnect aiming switch harness connector and VDC switch harness connector only left side.
23. Insert a remover into lower space of instrument side finisher (left/right) and remove by lifting.
24. Remove screws and glove box striker, disconnect connectors, and remove instrument passenger lower panel assembly.
25. Detach the damper from glove box right side.
26. Remove glove box pins, and remove glove box.
27. Insert a remover into lower space of center ventilator and remove by lifting.
28. Insert a remover into upper space of center ventilator and the upper clip is removed.
29. Remove screws. Disconnect harness connector, and remove tweeter with right and left part.
30. Insert a remover into front space of instrument stay cover (left/right) and detach.
31. Disconnect the left side harness connector only.
32. Remove cluster lid screws.
33. Disconnect A/C and AV harness connectors, and remove cluster lid C.
34. Remove display unit screws.
35. Disconnect harness connector and remove display.
36. Using a remover and disengage the ignition key finisher metal clips
37. Disconnect harness connector.
38. Using a remover, pry and remove instrument mask.
39. Disconnect harness connector.
40. Remove screw. Disengage the metal clip and remove instrument driver upper panel.
41. Remove bolt and screws Remove front passenger air bag module.
42. Disconnect metal clips, then remove instrument passenger upper panel.
43. Disconnect harness connector.
44. Pull to inside of vehicle, disconnect metal clips and remove front pillar garnish.
45. Remove bolts and screws, and remove instrument panel from passenger door opening portion.
46. Tweeter and sensor harness clip are removed from the duct.
47. Remove instrument panel assembly.
48. Remove ECM with bracket attached.
49. Remove nuts (2), then bolts (2) and screw (1), then remove blower unit.

➡Move blower unit to the right and remove locating pin (1) and joint. Then remove blower unit downward.

50. Disconnect intake door motor connector and blower fan motor connector.
51. Remove harness clips (2) from blower unit.
52. Remove blower unit.
53. Remove clips from vehicle harness from steering member.
54. Remove instrument stays (driver side and passenger side).
55. Remove rear ventilator duct1 and front floor duct.
56. Remove mounting screws from heater & cooling unit.
57. Remove the steering member, and then remove heater & cooling unit.
58. Remove foot duct (right).
59. Remove heater core cover.
60. Remove heater pipe support and heater pipe grommet.
61. Slide heater core to passenger side.
62. Installation is the reverse of removal. Note the following points:
    - Replace O-rings for A/C piping with new ones, coated with compressor oil.
    - Connection point for female-side piping is thin. So, when inserting male-side piping, take care not to deform female-side piping. Slowly insert in axial direction.
    - Insert one-touch joint connection point securely until it clicks.
    - After piping has been connected, pull male-side piping by hand to check that piping does not come off.
    - When recharging refrigerant, check for leaks.

Water Pump

REMOVAL & INSTALLATION

1. Remove drive belts.
2. Remove undercover.
3. Drain engine coolant from radiator.
4. Remove water drain plug on water pump side of cylinder block.
5. Support lower oil pan bottom with a transmission jack.
6. Remove right engine mounting insulator and mounting bracket.
7. Remove idler pulley bracket.
8. Remove chain tensioner cover and water pump cover.
9. Remove the chain tensioner assembly in the following procedure.

    a. Pull the lever down and release the plunger stopper tab.

    b. Insert the stopper pin into the tensioner body hole to hold the lever and keep the stopper tab released.

    c. Insert the plunger into the tensioner body by pressing the timing chain slack guide.

    d. Keep the slack guide pressed and hold the plunger in by pushing the stopper pin deeper through the lever and into the tensioner body hole.

    e. Turn crankshaft pulley approximately 20 degrees clockwise so that the timing chain on the chain tensioner side is loose.
10. Remove chain tensioner.
11. Remove the 3 water pump bolts. Secure a gap between water pump gear and timing chain, by turning crankshaft pulley approximately 20 degrees counterclockwise.
12. Screw M8 bolts, pitch: 1.25mm (0.049in) length: Approx. 50 mm (1.97in) into water pump's upper and lower mounting-bolt holes until they reach timing chain case. Then, alternately tighten each bolt for a half turn, and pull out water pump.
13. Pull straight out while preventing vane from contacting socket in installation area. Remove water pump without causing sprocket to contact timing chain.

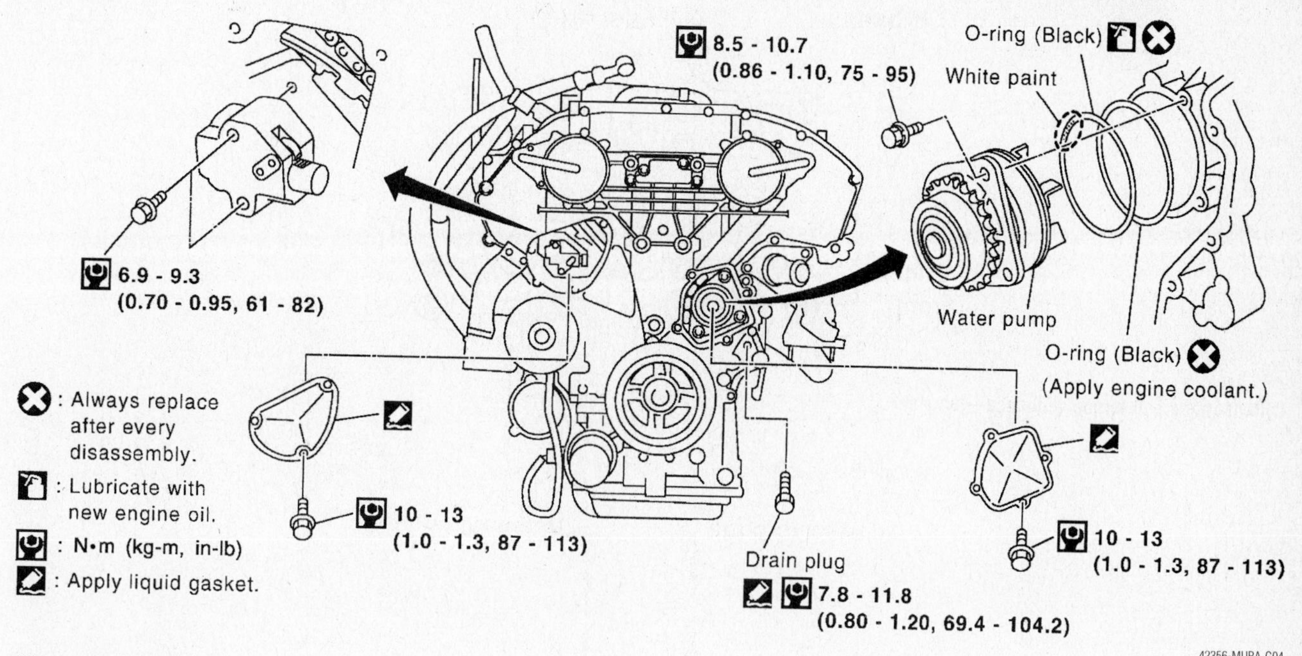

Water pump mounting

14. Remove M8 bolts and O-rings from water pump.

To install:

15. Install new O-rings to water pump.
16. Apply engine oil and engine coolant to the O-rings as shown. Locate the O-ring with white paint mark to engine front side.
17. Install the water pump.

➡**Do not allow cylinder block to nip the O-rings when installing the water pump.**

18. Check that timing chain and water pump sprocket are engaged.
19. Insert water pump by tightening mounting bolts alternately and evenly.
20. Remove dust and foreign material completely from backside of chain tensioner and from installation area of rear timing chain case.
21. Turn the crankshaft pulley clockwise so that the timing chain on the timing chain tensioner side is loose.

➡**When installing the timing chain tensioner, engine oil should be applied to the oil hole and tensioner.**

22. Install the timing chain tensioner.
23. Remove the stopper pin.
24. Install chain tensioner cover and water pump cover.
    a. Before installing, remove all traces of liquid gasket from mating surface of water pump cover and chain tensioner cover using a scraper. Also remove traces of liquid gasket from the mating surface of the front cover.
    b. Apply a continuous bead of liquid gasket, to mating surface of chain tensioner cover and water pump cover. Use RTV silicon sealant or equivalent
25. Install water drain plug on water pump side of cylinder block.
26. Installation is in the reverse order of removal for remaining parts.
27. After starting engine, let idle for three minutes, then rev engine up to 3,000 rpm under no load to purge air from the high-pressure chamber of the chain tensioner. The engine may produce a rattling noise. This indicates that air still remains in the chamber and is not a matter of concern.

Cylinder Head

REMOVAL & INSTALLATION

1. Before servicing the vehicle, refer to the precautions in the beginning of this section.
2. Remove or disconnect the following:
   - Negative battery cable
   - Fuel tube and fuel injector assembly
   - Intake manifold
   - Exhaust manifold

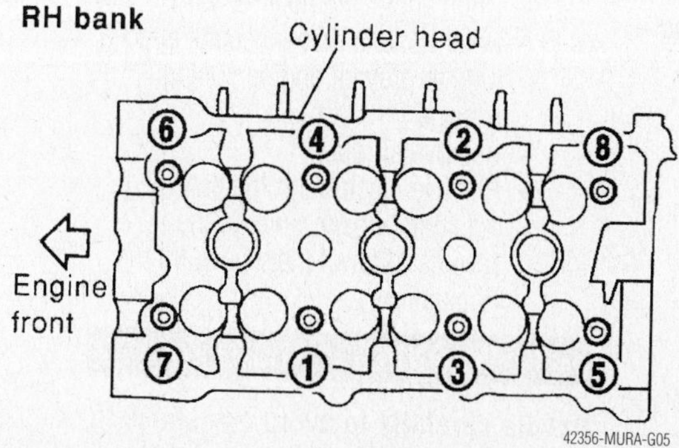

Cylinder head bolt torque sequence—right side

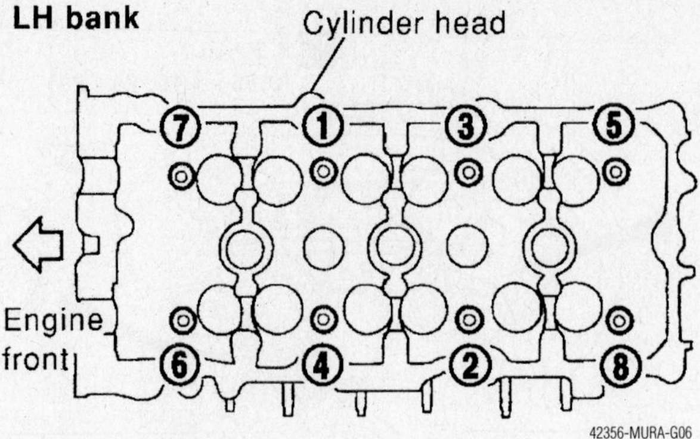

Cylinder head bolt torque sequence—left side

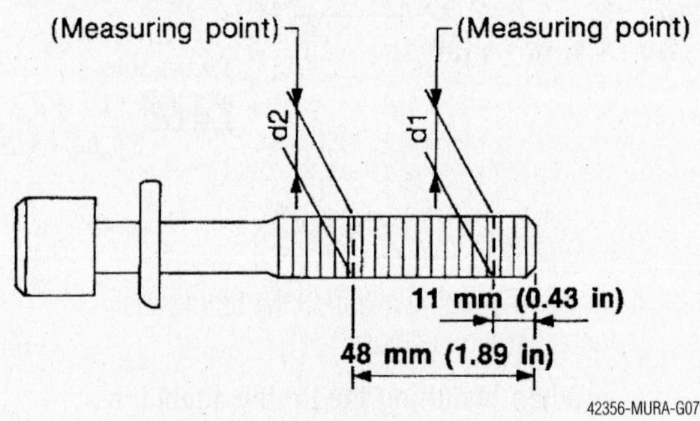

Cylinder head bolt measurement

- Water inlet and thermostat assembly
- Water outlet and water piping
- Camshaft
- Cylinder head bolts in reverse of the tightening sequence

3. Inspect the bolts. Cylinder head bolts are tightened by plastic zone tightening method. Whenever the size difference between d1 and d2 exceeds the limit, replace them with new one. If reduction of outer diameter appears in a position other than d2, use it as d2 point.

To install:

4. Install cylinder head gasket.
5. Turn the crankshaft until No. 1 piston is set at TDC on the compression stroke. The crankshaft key should line up with the right bank cylinder center line as shown.
6. Install cylinder head. Tighten the head bolts in the order shown in illustration.
    a. Step 1: Tighten all bolts to 98.1 Nm (72 ft. lbs.).
    b. Step 2: Completely loosen to 0 in the reverse order.
    c. Step 3: Tighten all bolts to 34.3–44.1 Nm (26–32 ft. lbs.).
    d. Step 4: Turn all bolts 90 degrees clockwise.
    e. Step 5: Turn all bolts an additional 90 degrees clockwise.
7. After installing cylinder head, measure distance between front end faces of cylinder block and cylinder head (left and right banks). If measurement is outside the specified range, reinstall cylinder head.
8. The remainder of installation is the reverse of removal.

Intake Manifold

REMOVAL & INSTALLATION

1. Before servicing the vehicle, refer to the precautions in the beginning of this section.
2. Remove or disconnect the following:
    - Negative battery cable
    - Remove engine cover.
    - Drain engine coolant, or when water hose is disconnected, attach plug to prevent engine coolant leakage.
    - Remove air duct.
    - Remove electric throttle control actuator. Loosen bolts in the reverse order of that shown in the illustration.

WARNING

Handle carefully to avoid any shock to the electric throttle control actuator.

- Disconnect vacuum hose and water hose from intake manifold collector (upper and lower).
- Disconnect EVAP canister purge volume control solenoid valve mounting bolt from intake manifold collector (lower).
- Remove VIAS control solenoid valve and vacuum tank.
- Remove the right windshield wiper arm and right front cowl top cover.
- Disconnect the power steering hose bracket.
- Remove intake manifold collector support bracket.
- Remove PCV hose
- Loosen bolts in reverse order of illustration, and remove intake manifold collector (upper and lower) assembly.
- Loosen bolts in reverse order of illustration to remove intake manifold collector (upper)
- Remove power valve in reverse order of illustration.
- Remove fuel tube and fuel injector assembly.
- Loosen bolts and nuts in reverse order of illustration to remove intake manifold assembly

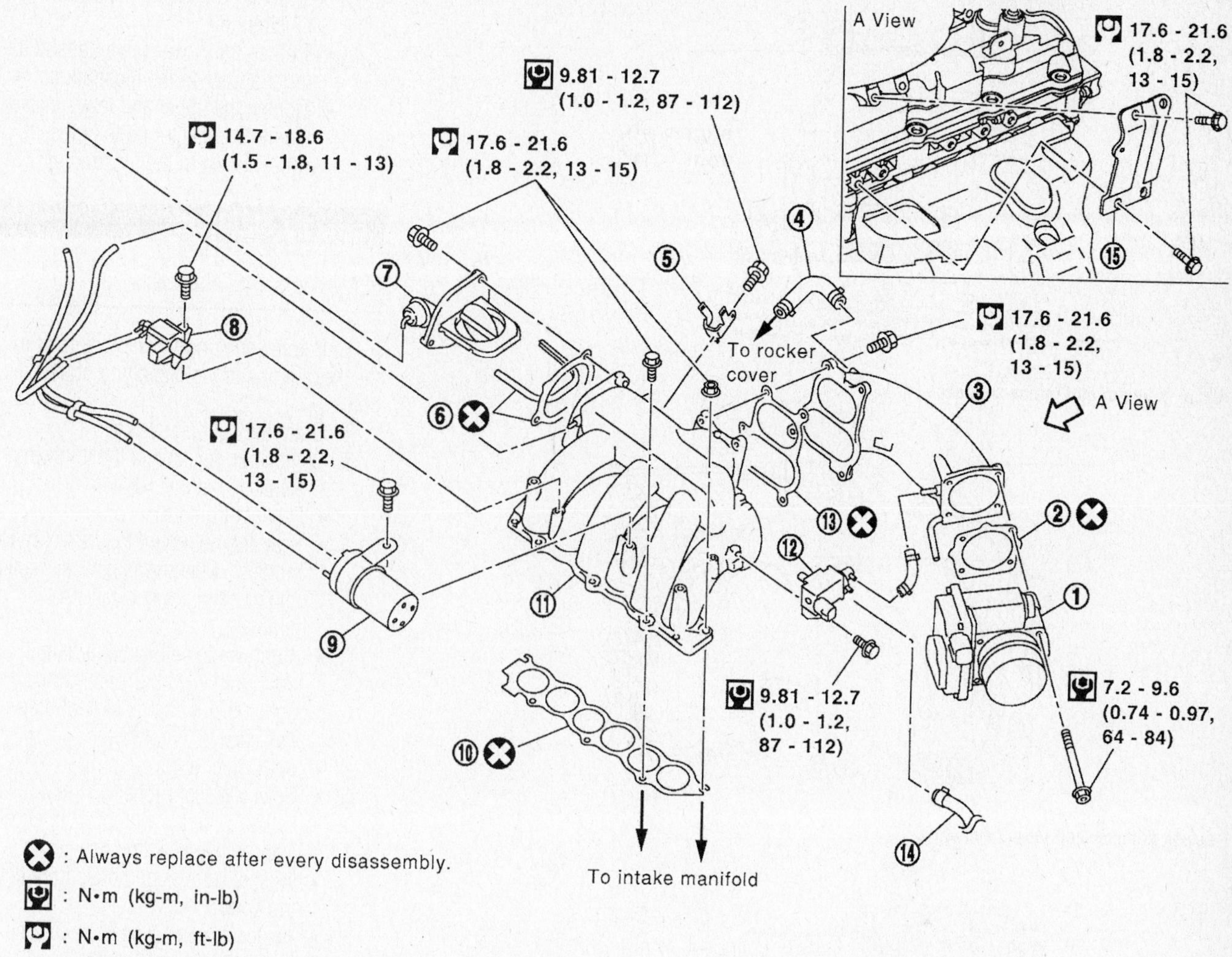

⊗ : Always replace after every disassembly.

: N•m (kg-m, in-lb)

: N•m (kg-m, ft-lb)

To intake manifold

To rocker cover

A View

1. Electric throttle control actuator
2. Gasket
3. Intake manifold collector (upper)
4. PCV hose
5. Harness bracket
6. Gasket
7. Power valve
8. VIAS control solenoid valve
9. Vacuum tank
10. Gasket
11. Intake manifold collector (lower)
12. EVAP canister purge volume control solenoid valve
13. Gasket
14. EVAP hose
15. Support bracket

42356-MURA-G08

Upper and lower intake manifold collectors

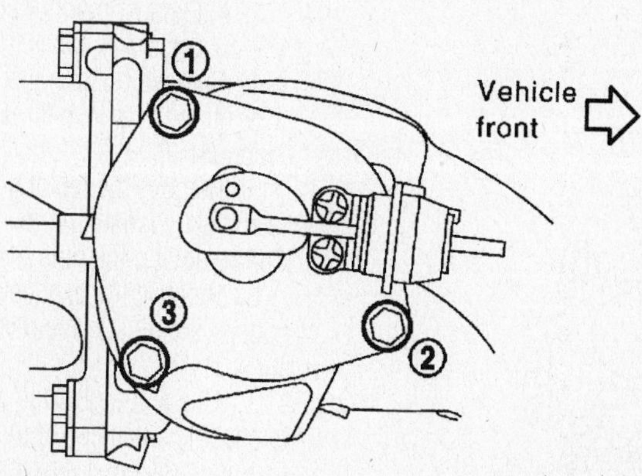

Power valve installation

3. Using straightedge and feeler gauge, inspect the surface distortion of each surface on intake manifold. If it exceeds the limit, replace the intake manifold.

4. Using straightedge and feeler gauge, inspect the surface distortion of intake manifold collector (lower). If it exceeds the limit, replace the intake manifold collector.

5. Installation is the reverse of removal paying attention to the following.

- If the intake manifold stud bolts were removed, install them and tighten to the torque specified below.
- Tighten all mounting bolts and nuts to specified torque in two or more

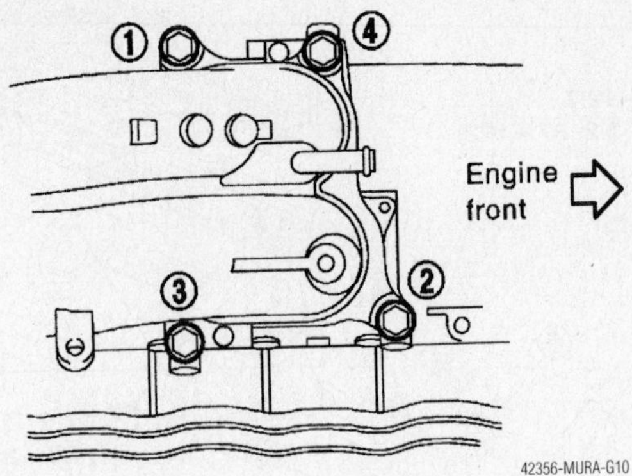

Upper collector bolt torque sequence

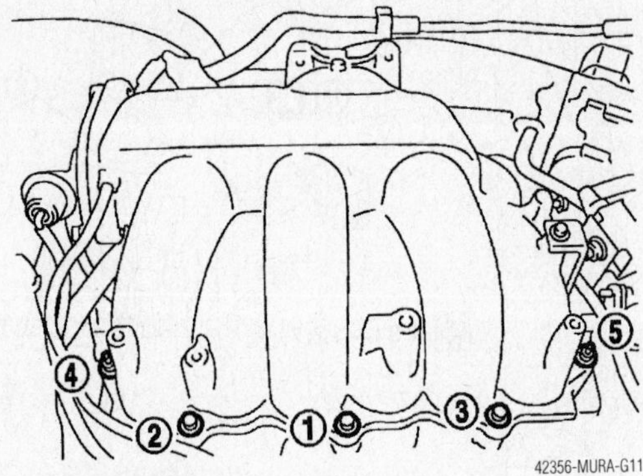

Lower collector bolt torque sequence

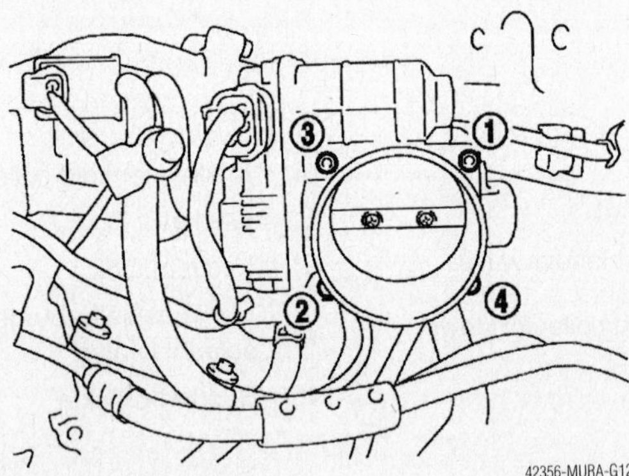

Throttle control actuator bolt torque sequence

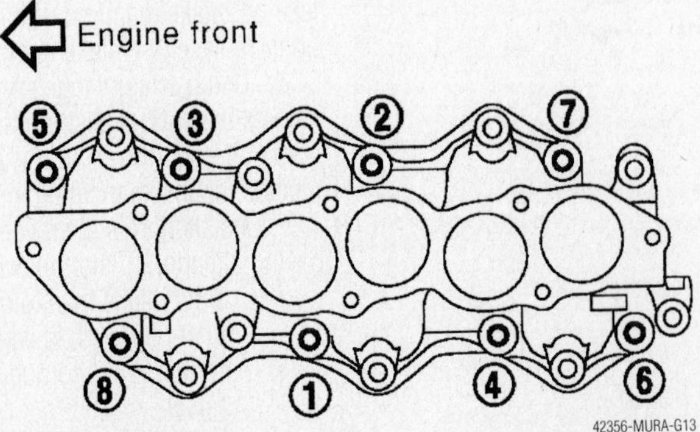

Intake manifold bolt torque sequence

steps in numerical order shown in illustration.
- Tighten the power valve bolts in the order shown in the illustration.
- Tighten the upper and lower intake manifold collector bolts in the order shown in the illustration.

Exhaust Manifold

REMOVAL & INSTALLATION

1. Before servicing the vehicle, refer to the precautions in the beginning of this section.
2. Drain the cooling system.
3. Remove or disconnect the following:
   - Negative battery cable
   - Exhaust front tube
   - Rear engine mount insulator (2WD models), when right exhaust manifold and three way catalyst is removed
   - Right windshield wiper arm and right front cowl top cover, when right exhaust manifold and three way catalyst is removed
   - Both wiper arms
   - Cowl top seal rubber
   - Left and right cowl top cover clips and remove cowl top covers
   - Washer nozzles and hose from cowl top cover
   - Heated oxygen sensor 1 and 2 on both left and right bank.

WARNING

Be careful not to damage heated oxygen sensor. Discard any heated oxygen sensor which has been dropped from a height of more than 0.5 m (19.7 inches) onto a hard surface such as a concrete floor; replace with a new sensor.

   - Exhaust manifold covers and the three way catalyst heat shields
4. Remove bolts in the reverse order of illustration to remove three way catalyst supports.
5. Remove the left and right three way catalysts by loosening the bolts first and then removing the nuts.
6. Remove the exhaust manifolds. Loosen the nuts in the reverse order as shown.
7. Use a straightedge and feeler gauge to check the flatness of the exhaust manifold mating surfaces. If it exceeds the limit (0.012 inch), replace the exhaust manifold.
8. Installation is in the reverse of

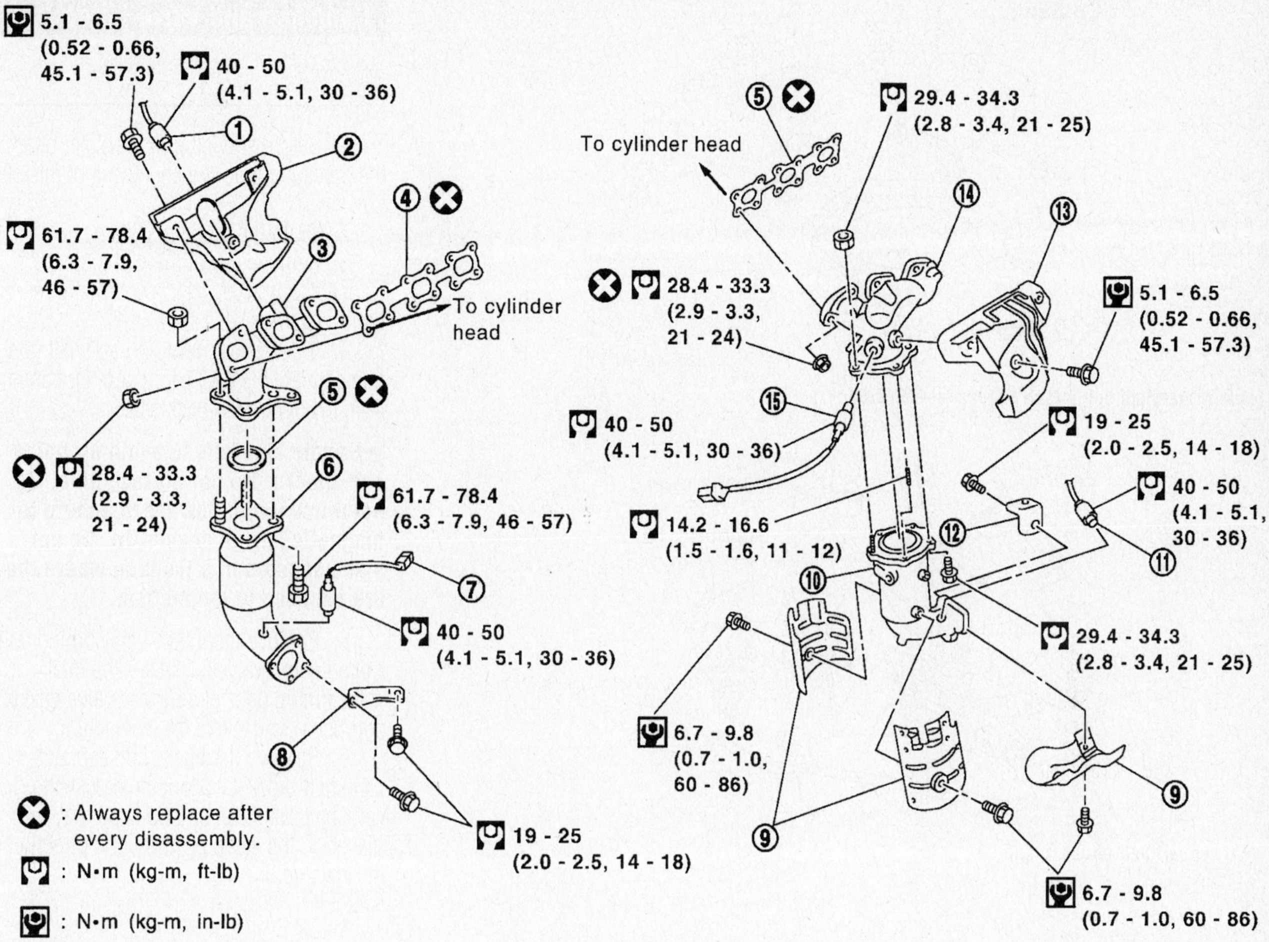

5.1 - 6.5
(0.52 - 0.66,
45.1 - 57.3)
40 - 50
(4.1 - 5.1, 30 - 36)
61.7 - 78.4
(6.3 - 7.9,
46 - 57)
To cylinder head
28.4 - 33.3
(2.9 - 3.3,
21 - 24)
61.7 - 78.4
(6.3 - 7.9, 46 - 57)
40 - 50
(4.1 - 5.1, 30 - 36)
19 - 25
(2.0 - 2.5, 14 - 18)
: Always replace after every disassembly.
: N•m (kg-m, ft-lb)
: N•m (kg-m, in-lb)
To cylinder head
29.4 - 34.3
(2.8 - 3.4, 21 - 25)
28.4 - 33.3
(2.9 - 3.3,
21 - 24)
5.1 - 6.5
(0.52 - 0.66,
45.1 - 57.3)
40 - 50
(4.1 - 5.1, 30 - 36)
19 - 25
(2.0 - 2.5, 14 - 18)
40 - 50
(4.1 - 5.1,
30 - 36)
14.2 - 16.6
(1.5 - 1.6, 11 - 12)
29.4 - 34.3
(2.8 - 3.4, 21 - 25)
6.7 - 9.8
(0.7 - 1.0,
60 - 86)
6.7 - 9.8
(0.7 - 1.0, 60 - 86)

1. Heated oxygen sensor 1 (bank 1)	2. Exhaust manifold cover	3. Exhaust manifold (RH bank)
4. Gasket	5. Gasket	6. Three way catalyst (manifold) (RH bank)
7. Heated oxygen sensor 2 (bank 1)	8. Support (RH)	9. Three way catalyst heat shield
10. Three way catalyst (manifold) (LH bank)	11. Heated oxygen sensor 2 (bank 2)	12. Support (LH)
13. Exhaust manifold cover	14. Exhaust manifold (LH bank)	15. Heated oxygen sensor 1 (bank 2)

42356-MURA-G14

Exhaust manifold removal and installation

removal. Check the accompanying illustrations for the proper torque values. Note the following:

- When installing the heated oxygen sensor, tighten to the middle of specified torque range, because the length of the torque wrench may increase the actual tightness. Do not tighten to the maximum specified torque range.
- Install the exhaust manifold nuts in the order shown.
- When installing the Three Way Catalyst Supports, install in the order shown.

RH bank
Engine front

42356-MURA-G15

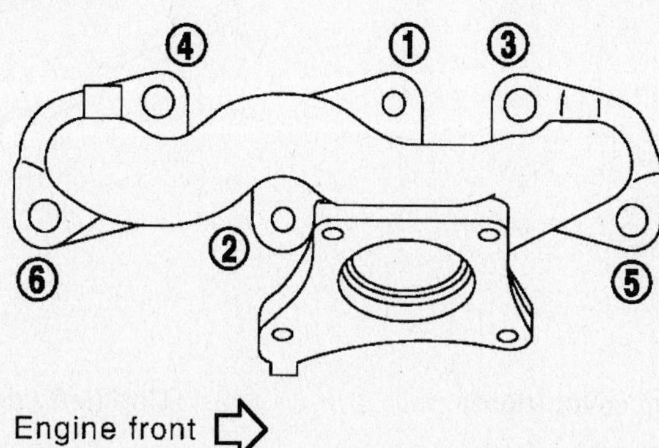

Exhaust manifold bolt torque sequence—right side

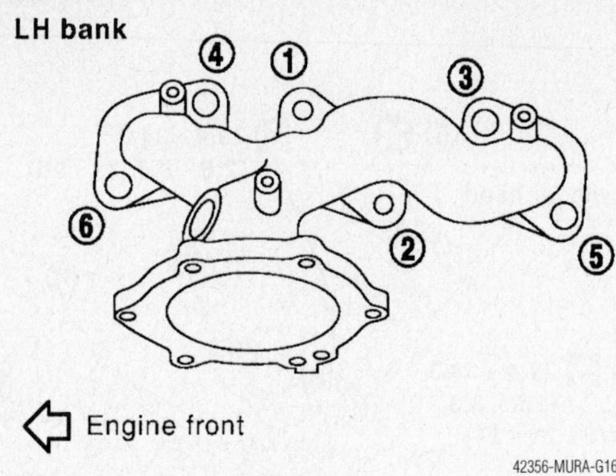

Exhaust manifold bolt torque sequence—left side

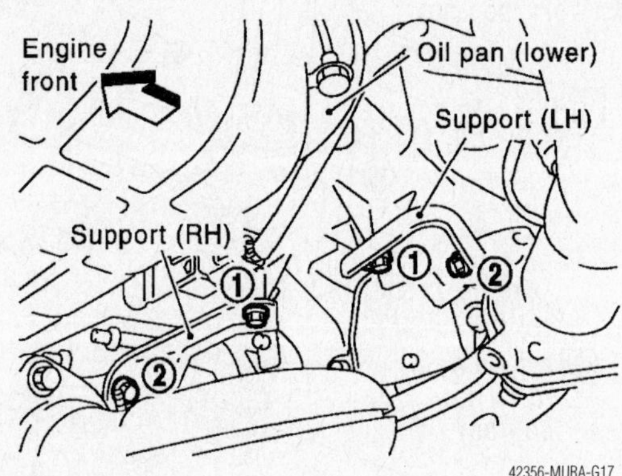

TWC support bolt torque sequence

Camshaft and Valve Lifters

REMOVAL

1. Before servicing the vehicle, refer to the precautions in the beginning of this section.
2. Drain the cooling system.
3. Remove front timing chain case, camshaft sprocket, timing chain and rear timing chain case.
4. If necessary, remove camshaft position sensor (PHASE) (right and left banks) from cylinder head back side.

➡Handle carefully to avoid dropping and shocks. Do not disassemble. Do not allow metal powder to adhere to magnetic part at sensor tip. Do not place sensors in a location where they are exposed to magnetism.

5. Remove the intake valve timing control solenoid valves. Discard the intake valve timing control solenoid valve gaskets and use new gaskets for installation.
6. Remove the intake and exhaust camshaft brackets. Mark the camshafts, camshaft brackets, and bolts so they are placed in the same position and direction for installation.

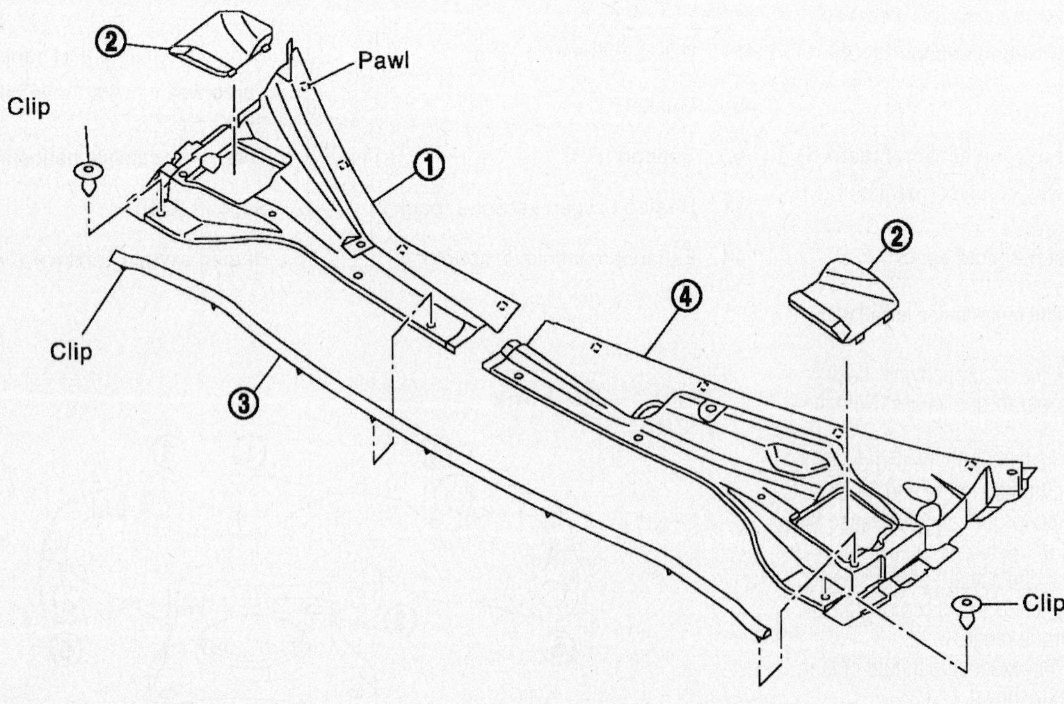

1. Cowl top cover (right) 2. Cap (left / right) 3. Cowl top seal rubber
4. Cowl top cover (left)

42356-MURA-G18

Cowl top removal

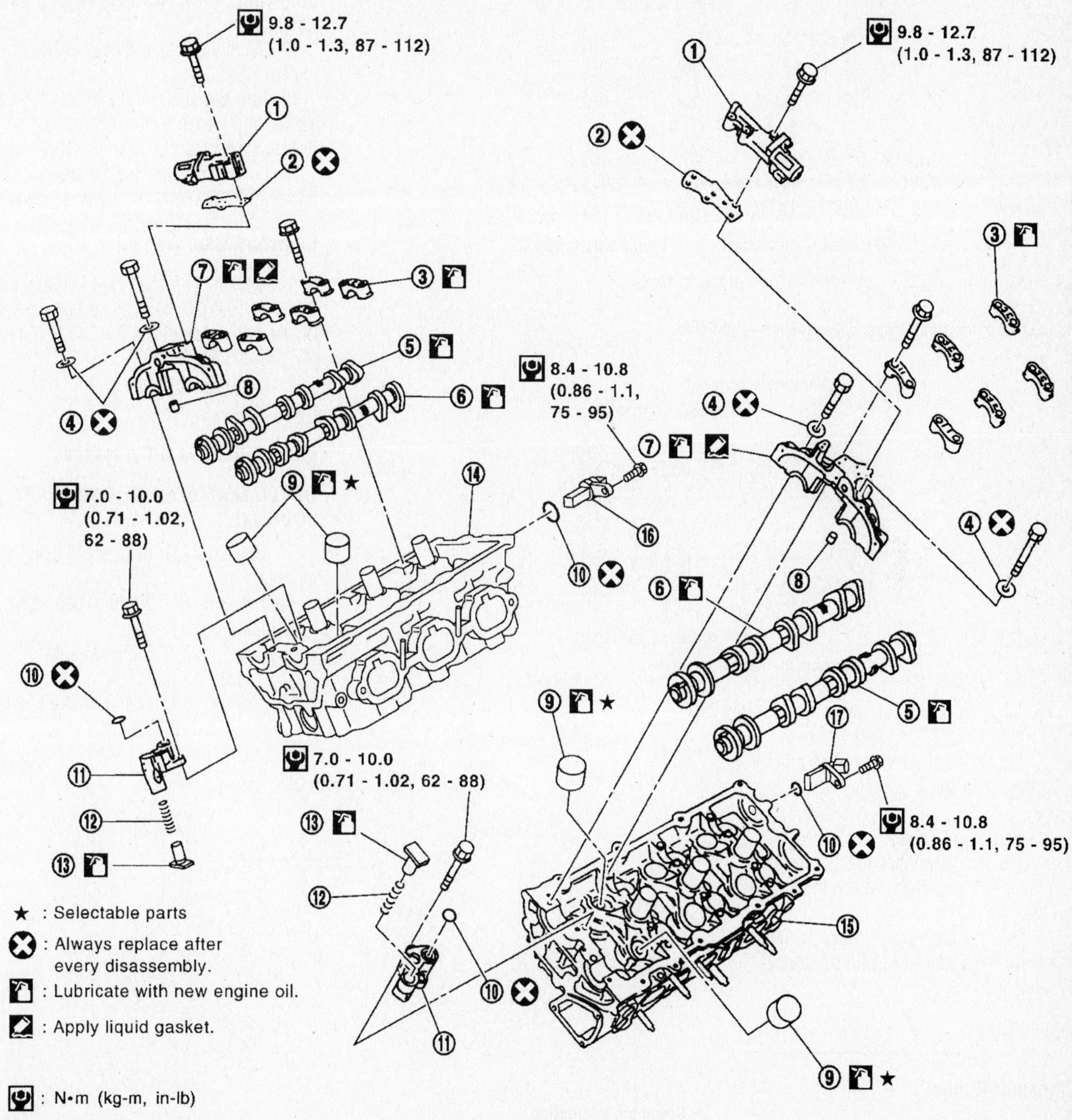

1.	Intake valve timing control solenoid valve	2.	Gasket	3.	Camshaft bracket (No.2 to No.4)
4.	Seal washer	5.	Camshaft (EXH)	6.	Camshaft (INT)
7.	Camshaft bracket (No.1)	8.	Dowel pin	9.	Valve lifter
10.	O-ring	11.	Chain tensioner	12.	Spring
13.	Plunger	14.	Cylinder head (RH bank)	15.	Cylinder head (LH bank)
16.	Camshaft position sensor (PHASE) (RH bank)	17.	Camshaft position sensor (PHASE) (LH bank)		

42356-MURA-G19

Camshaft removal and installation

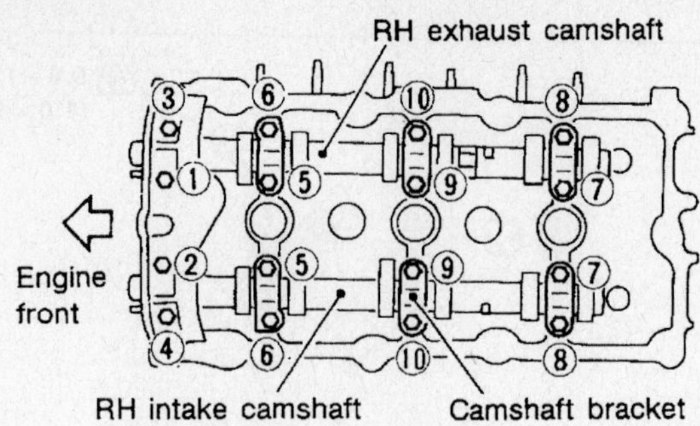

Camshaft bracket bolt loosening sequence—right side

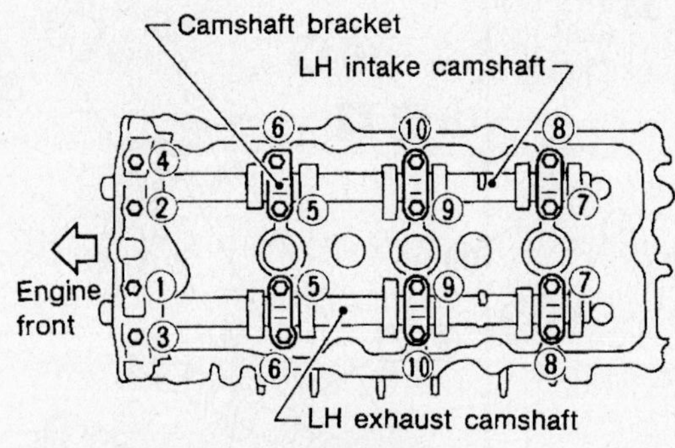

Camshaft bracket bolt loosening sequence—left side

7. Equally loosen the camshaft bracket bolts in several steps in the numerical order shown.
8. Remove camshaft.
9. Remove valve lifter. Identify installation positions, and store them without mixing them up.
10. Remove secondary timing chain tensioner from cylinder head with its stopper pin attached.

INSPECTION

Camshaft Runout

1. Put V block on precise flat bed, and support No. 2 and No. 4 journal of camshaft.
2. Set dial gauge vertically to No. 3 journal.
3. Turn camshaft to one direction with hands, and measure camshaft runout on dial gauge. (Total indicator reading)
4. If it exceeds the limit, replace camshaft.

Camshaft Cam Height

1. Measure camshaft cam height. Limit: 0.05 mm (0.0020 in). Standard cam height (intake and exhaust): 44.865–45.055 mm (1.7663–1.7738 in)
2. If wear is beyond the limit, replace camshaft.

Camshaft Journal Clearance

OUTER DIAMETER OF CAMSHAFT JOURNAL

1. Measure outer diameter of camshaft journal.
   Cam wear limit: 0.2 mm (0.008 inch)
   Standard outer diameter:
   - No. 1: 25.935–25.955 mm (1.0211–1.0218 in)
   - No. 2, 3, 4: 23.445–23.465 mm (0.9230–0.9238 in)

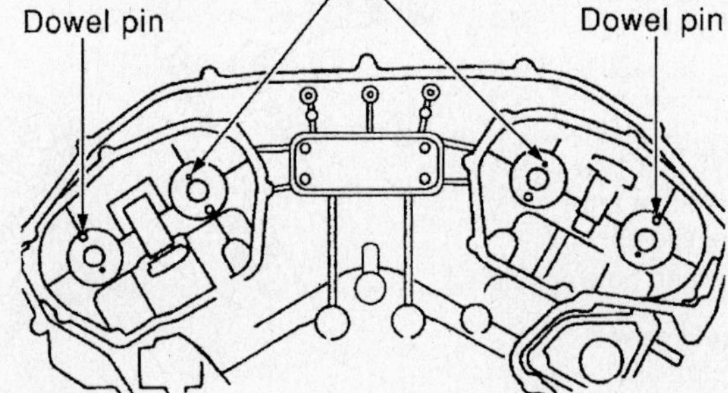

Dowel pin orientation

Bank	INT/EXH	Dowel pin	Paint marks		ID mark
			M1	M2	
RH	INT	No	Pink	No	RE
	EXH	Yes	No	Orange	RE
LH	INT	No	Pink	No	LH
	EXH	Yes	No	Orange	LH

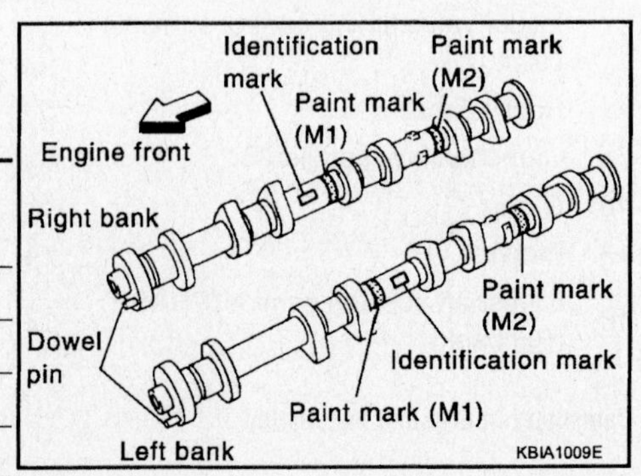

Camshaft identification

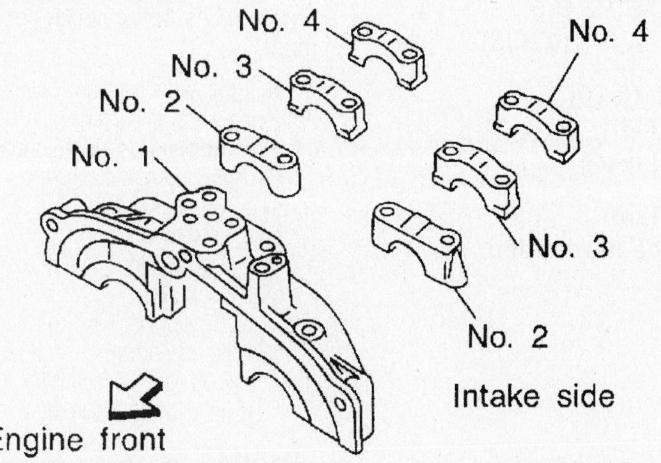

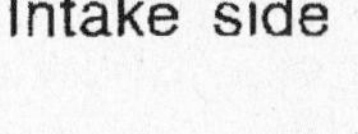

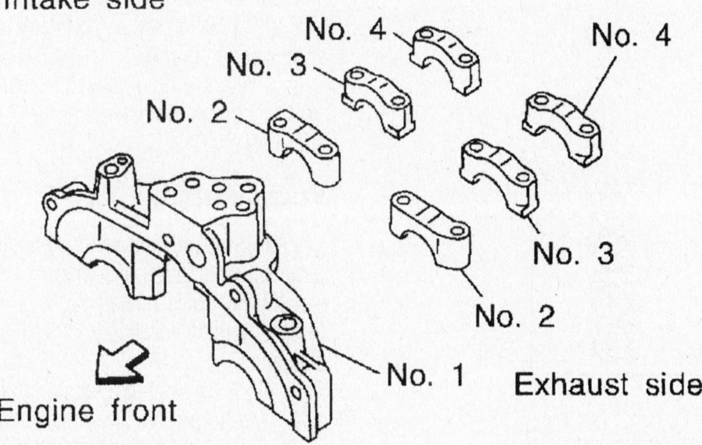

Camshaft bracket identification

INNER DIAMETER OF CAMSHAFT BRACKET

1. Tighten camshaft bracket bolt with specified torque.
2. Using inside micrometer, measure inner diameter "A" of camshaft bracket.

CALCULATION OF CAMSHAFT JOURNAL CLEARANCE

Journal clearance = inner diameter of camshaft bracket—outer diameter of camshaft journal. When outside the limit, replace either or both camshaft and cylinder head.

➡Inner diameter of camshaft bracket is manufactured together with cylinder head. Replace the whole cylinder head assembly.

Camshaft End Play

1. Install dial gauge in thrust direction on front end of camshaft. Measure end play of dial gauge when camshaft is moved forward and backward.
2. When out of the limit, replace with new camshaft and measure again.
3. When out of the limit again, replace with new cylinder head.

INSTALLATION

1. Install secondary chain tensioners on both sides of cylinder head.
   a. Install chain tensioner with its stopper pin attached.
   b. Install tensioner with sliding part facing downward on right side cylinder head, and with sliding part facing upward on left side cylinder head.
   c. Install O-ring as shown.
2. Install valve lifters in their original position.

   Valve lifter outer diameter (Intake and exhaust)
   - 33.977–33.987mm (1.3377–1.3381 inch)

   Standard (Intake and exhaust)
   - 34.000–34.016 mm (1.3386–1.3392 inch)

   Standard (Intake and exhaust)
   - 0.013–0.039 mm (0.0005–0.0015 in)
3. Install camshafts.
   a. Install camshaft with dowel pin attached to its front end face on the exhaust side.
   b. Follow your identification marks made during removal, or follow the iden-

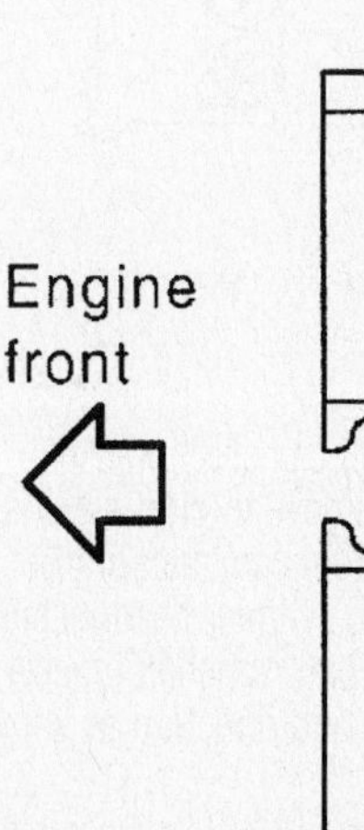

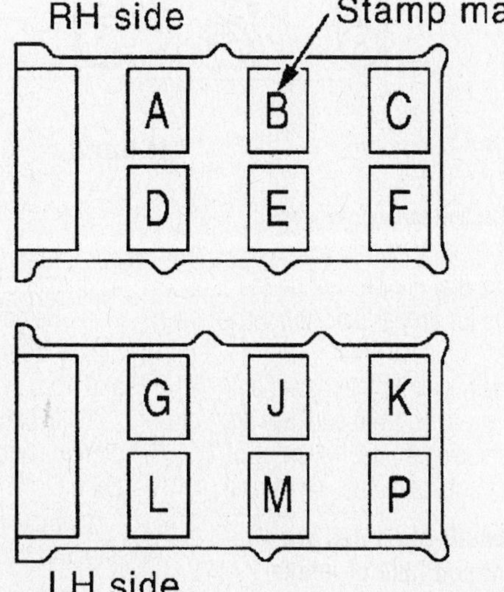

Align the stamp marks as shown

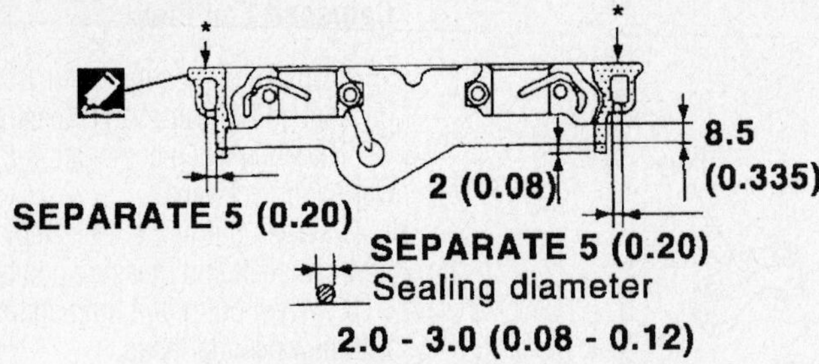

RTV sealer application

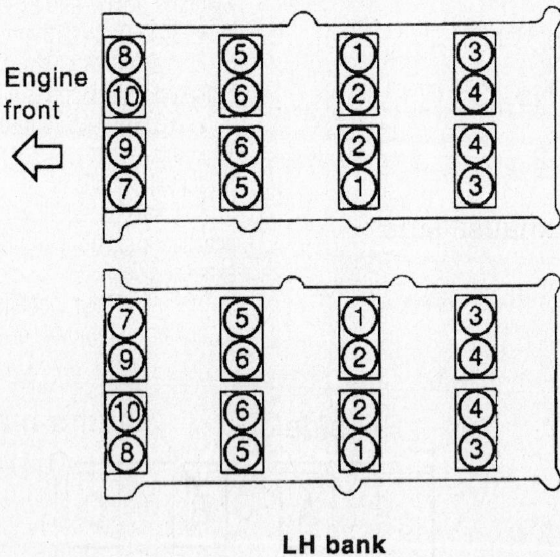

Camshaft bracket bolt torque sequence

tification marks that are present on the new camshafts for proper placement and direction.

c. Install camshaft so that dowel pin hole and dowel pin on front end face are positioned as shown in illustration. (No. 1 cylinder TDC on its compression stroke)

➡Large- and small-pin holes are located on front end face of intake camshaft, at intervals of 180 degrees. Face small dia. side pin hole upward (in cylinder head upper face direction).

4. Install camshaft brackets.

a. Remove foreign material completely from camshaft bracket backside and from cylinder head installation face.

b. Install camshaft bracket in original position and direction as shown in illustration.

c. Install No.2 to 4 camshaft brackets aligning the stamp marks as shown.

➡There are no identification marks indicating left and right for No. 1 camshaft bracket.

d. Apply sealant to mating surface of No.1 camshaft bracket as shown on right and left banks.

➡Use RTV silicone sealant or equivalent.

5. Tighten the camshaft brackets in the following steps, in numerical order as shown.

a. Tighten No. 7 to 10, then tighten No.1 to 6 in order as shown.

b. Tighten No.1 to 10 in numerical order as shown.

c. Tighten No. 1 to 6 in the numerical order as shown.

d. Tighten No. 7 to 10 in the numerical order as shown.

- 1.96 Nm (17 inch lbs.)
- 5.88 Nm (52 inch lbs.)
- 9.02–11.8 Nm (80–104 inch lbs.)
- 8.3–10.3 Nm (74–91 inch lbs.)

6. Measure difference in levels between front end faces of No. 1 camshaft bracket and cylinder head. If measurement is outside the specified range, re-install camshaft and camshaft bracket.

7. Inspect and adjust valve clearance.

8. Install in the reverse order of removal after this step.

Valve Clearance

Perform inspection as follows after removal, installation or replacement of camshaft or valve-related parts, or if there is unusual engine conditions regarding valve clearance.

1. Remove right and left rocker covers.
2. Measure valve clearance as below:

a. Set No.1 cylinder at TDC of its compression stroke. Align crankshaft pulley timing mark (grooved line without color) with timing indicator. Check that No. 1 cylinder intake and exhaust cam nose is facing in direction shown in illustration. If not, rotate crankshaft pulley 360 degrees clockwise (when viewed from front).

b. Using a feeler gauge, measure valve clearance. Standard: -0.14 to 0.14 mm (-0.0055 to 0.0055 in)

➡If inspection was carried out with cold engine, check that values with fully warmed up engine are still within specifications.

c. Rotate crankshaft by 240 degrees clockwise (when viewed from front) to align No. 3 cylinder at TDC of its compression stroke.

➡Crankshaft pulley mounting bolt flange has a stamped line every 60 degrees. They can be used as a guide to rotation angle.

d. Turn crankshaft pulley clockwise by 240 degrees from the position of No. 5 cylinder at compression TDC.

3. For measurements that are outside the specified range, perform adjustment below.

ADJUSTMENT

Perform adjustment depending on selected head thickness of valve lifter. The specified valve lifter thickness is the dimension at normal temperatures. Ignore dimensional differences caused by temperature. Use the specifications for hot engine condition to adjust.

1. Remove camshaft.
2. Remove the valve lifters at the locations that are outside the standard.

Measuring position (RH bank)		No.1 CYL.	No.3 CYL.	No.5 CYL.
No.1 cylinder at TDC	EXH		×	
	INT	×		
Measuring position (LH bank)		No.2 CYL.	No.4 CYL.	No.6 CYL.
No.1 cylinder at TDC	INT			×
	EXH	×		

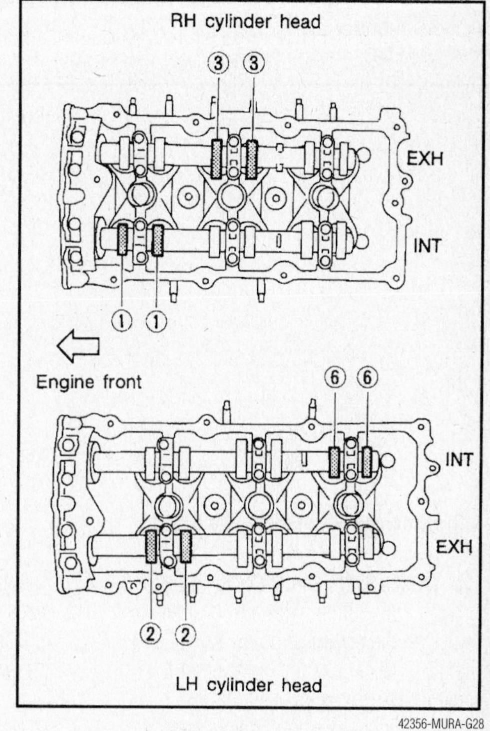

Valve clearance inspection—No.1 TDC

Measuring position (RH bank)		No.1 CYL.	No.3 CYL.	No.5 CYL.
No.3 cylinder at TDC	EXH			×
	INT		×	
Measuring position (LH bank)		No.2 CYL.	No.4 CYL.	No.6 CYL.
No.3 cylinder at TDC	INT	×		
	EXH		×	

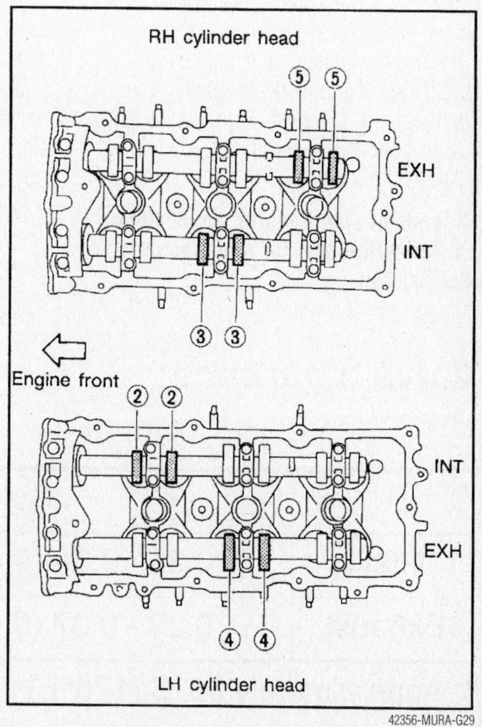

Valve clearance inspection—No.3 TDC

Measuring position (RH bank)		No.1 CYL.	No.3 CYL.	No.5 CYL.
No.5 cylinder at TDC	EXH	×		
	INT			×
Measuring position (LH bank)		No.2 CYL.	No.4 CYL.	No.6 CYL.
No.5 cylinder at TDC	INT		×	
	EXH			×

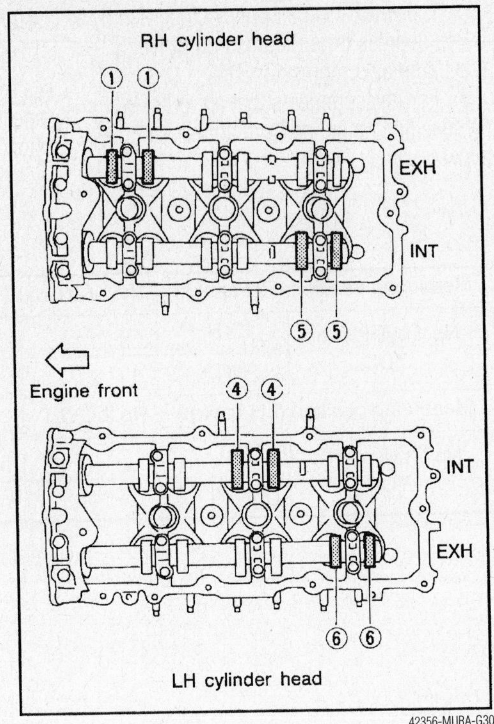

Valve clearance inspection—No.5 TDC

3. Measure the center thickness of the removed valve lifters with a micrometer.
4. Use the equation below to calculate valve lifter thickness for replacement.

Valve lifter thickness calculation:

- Thickness of replacement valve lifter = t1 + (C1—C2)
- t1 = Thickness of removed valve lifter
- C1 = Measured valve clearance
- C2=Standard valve clearance:

Thickness of a new valve lifter can be identified by stamp marks on the reverse side (inside the cylinder). Stamp mark 788U or 788R indicates 7.88 mm (0.3102 in) in thickness.

➡2 types of stamp marks are used for parallel setting and for manufacturer identification.

Available thickness of valve lifter: 27 sizes with range 7.88 to 8.40 mm (0.3102 to 0.3307 in) in steps of 0.02 mm (0.0008 in) (when manufactured at factory).

5. Install the selected valve lifter.
6. Install camshaft.
7. Manually turn crankshaft pulley a few turns.
8. Check that valve clearances for cold engine are within specifications by referring to the specified values.
9. After completing the repair, check valve clearances again with the specifications for warmed engine. Make sure the values are within specifications.

Valve clearance:

- Intake: 0.30 mm (0.012 in)
- Exhaust: 0.33 mm (0.013 in)

Starter Motor

REMOVAL & INSTALLATION

1. Before servicing the vehicle, refer to the precautions in the beginning of this section.
2. Remove or disconnect the following:
   - Battery
   - Air intake duct
   - Battery bracket
   - Fluid charging pipe
   - S connector
   - B terminal nut
   - Starter motor mounting bolts
   - Starter motor to the direction of upper side the vehicle

Valve clearance:

Unit: mm (in)

	Cold	Hot * (reference data)
Intake	0.26 - 0.34 (0.010 - 0.013)	0.304 - 0.416 (0.012 - 0.016)
Exhaust	0.29 - 0.37 (0.011 - 0.015)	0.308 - 0.432 (0.012 - 0.016)

*: Approximately 80°C (176°F)

42356-MURA-G31

Valve clearance specifications

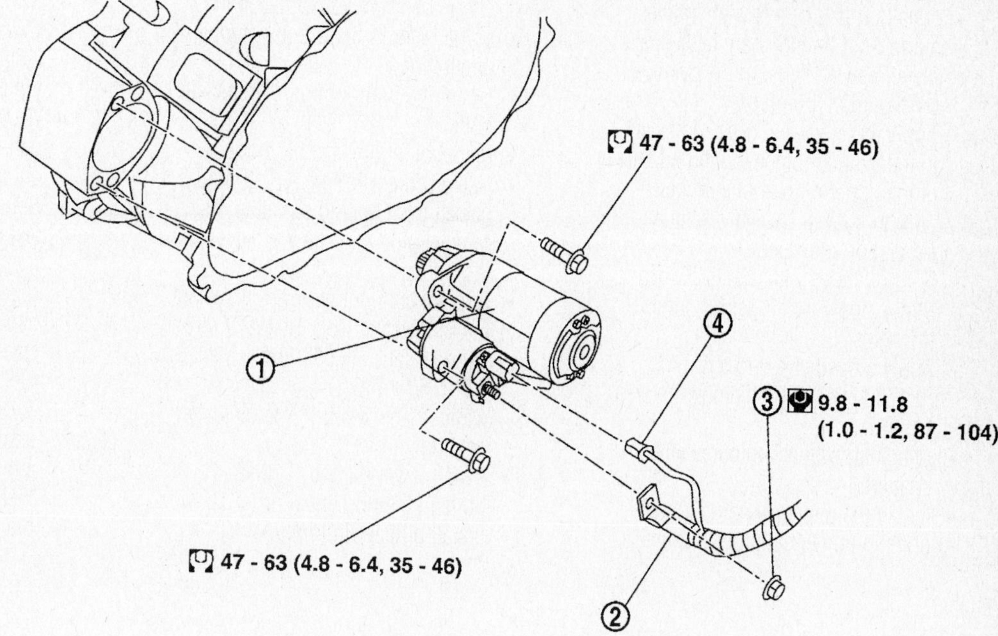

Starter removal

3. Install in the reverse order of removal. Observe the following toques:

- B terminal nut: 9.8–11.8 Nm (87–104 inch lbs.)
- Starter motor mounting bolt: 47–63 Nm (35–46 ft. lbs.)
- Battery bracket mounting bolt: 14–20 Nm (10–15 ft. lbs.)

Oil Pan

REMOVAL & INSTALLATION

2WD Models

1. Before servicing the vehicle, refer to the precautions in the beginning of this section.

➡When removing the upper oil pan from the engine, first remove the crankshaft position sensor (POS). Be careful not to damage sensor edges or signal plate teeth.

2. Remove engine cover.
3. Remove right splash guard.
4. Remove the front right road wheel and tire.
5. Drain engine oil.
6. Drain engine coolant.
7. Remove oil filter.
8. Remove oil cooler and water pipes.
9. Remove all drive belts.
10. Remove A/C compressor with piping connected, and temporarily secure it aside.
11. Remove exhaust front tube.
12. Remove the heated oxygen sensor 2 (bank 2) and remove the three way catalyst (manifold) (bank 2) from the exhaust manifold.
13. Loosen lower oil pan bolts in reverse order of illustration.
14. Insert a seal cutter (special service tool) between the lower oil pan and the upper oil pan.

➡Be careful not to damage the mating surface. Do not insert a screwdriver; this will damage the mating surfaces.

15. Slide seal cutter (special service tool) by tapping on the side of the tool with a hammer.
16. Remove lower oil pan.
17. Remove oil strainer.
18. Remove the oil pressure switch.
19. Remove crankshaft position sensor (POS).

➡Handle carefully to avoid dropping and shocks. Do not disassemble. Do not allow metal powder to adhere to magnetic part at sensor tip. Do not place sensors in a location where they are exposed to magnetism.

20. Remove the four engine-to-transaxle bolts.
21. Remove upper oil pan. Loosen bolts in reverse order shown.
22. Insert an appropriate size tool into the notch of the upper oil pan shown (1). Pry off the upper oil pan by moving the tool up and down shown (2).
23. Remove O-rings from the bottom of the cylinder block and oil pump body.
24. Remove oil pan gasket.

To install:

Installation is the reverse of removal. Note the following:

- Use a scraper to remove old liquid gasket from mating surfaces.
- Also remove the old liquid gasket from mating surface of the cylinder block.
- Remove the old liquid gasket from the bolt holes and threads.

➡Do not scratch or damage the mating surfaces when cleaning off the old liquid gasket.

- Apply Genuine RTV Silicone Sealant or equivalent, to the front timing chain case gasket and the rear oil seal retainer gasket shown.
- To install, align protrusion of oil pan gasket with notches of front timing chain case and rear oil seal retainer.

- Install oil pan gasket with smaller arc to front timing chain case side.
- Install new O-rings on the cylinder block and oil pump side.
- Apply a continuous bead of sealant to the cylinder block mating surface of the upper oil pan to a limited portion shown. Use RTV silicone sealant or equivalent.
- For bolt holes with marks (5 locations), apply liquid gasket outside the holes.
- Apply a bead of 4.5 to 5.5 mm (0.177 to 0.217 in) diameter to area "A".
- Attaching within 5 minutes after coating.
- Install the upper oil pan. Tighten bolts in numerical order shown.

There are two types of mounting bolts. Refer to the accompanying illustration.

- Install the four engine-to-transaxle bolts.
- Install oil strainer to oil pump.
- Use a scraper to remove old liquid gasket from mating surfaces. Also remove old liquid gasket from mating surface of upper oil pan.
- Apply a continuous bead of sealant to the lower oil pan. Use RTV silicone sealant. Be sure the sealant is 4.5–5.5 mm (0.177–0.217 inch) wide. Attach within 5 minutes after coating.
- Install lower oil pan. Tighten the bolts in the numerical order shown.
- Install oil pan drain plug.
- Refer to illustration for installation of washer.
- Install in the reverse order of removal after this step.
- Wait at least 30 minutes after oil pan is installed, before adding engine oil.
- Before starting engine, check the levels of engine coolant, engine oil and working fluid. If less than required quantity, fill to the specified level.
- Use procedure below to check for fuel leakage.
- Turn ignition switch ON (with engine stopped). With fuel pressure applied to fuel piping, check for fuel leakage at connection points.
- Start engine. With engine speed

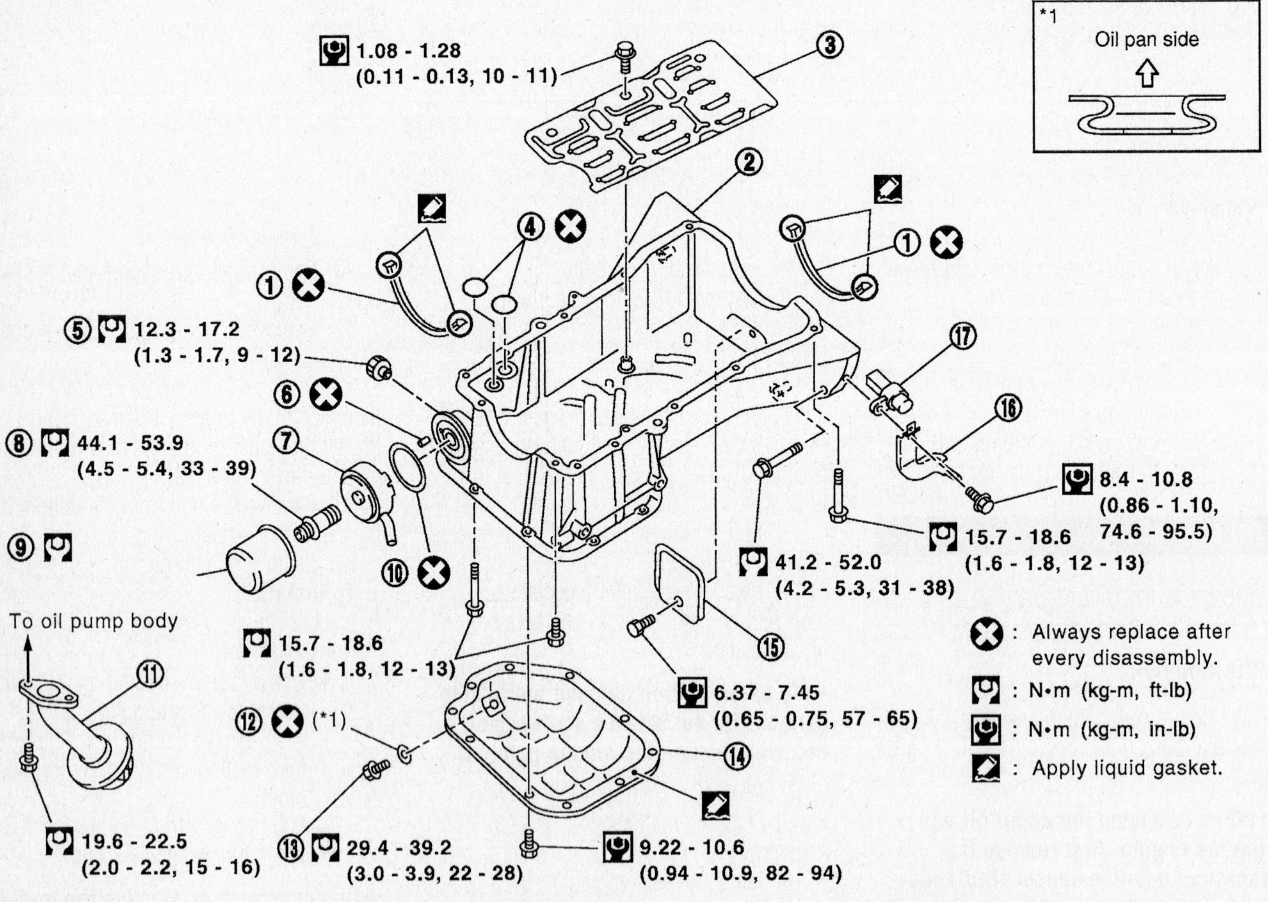

1. Gasket
2. Upper oil pan
3. Baffle plate
4. O-ring
5. Oil pressure switch
6. Relief valve
7. Oil cooler
8. Oil cooler connector
9. Oil filter
10. Gasket
11. Oil strainer
12. Gasket
13. Drain plug
14. Lower oil pan
15. Rear cover plate
16. Heated oxygen sensor (bank 2) harness clamp (2WD models)
17. Crankshaft position sensor (POS)

42356-MURA-G33

Oil pan exploded view

42356-MURA-G34

RTV sealer application at the timing case

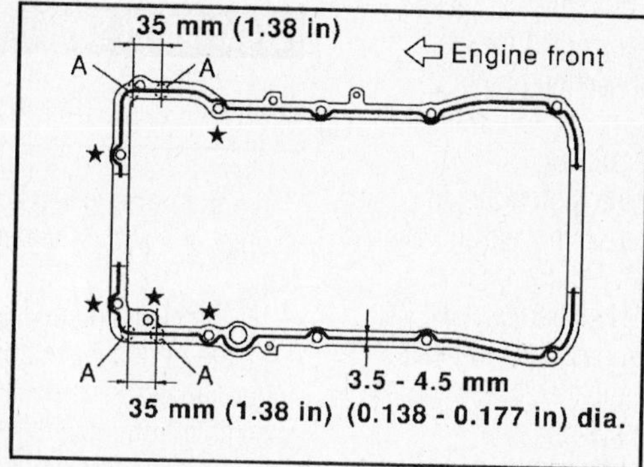

RTV sealer application on the pan

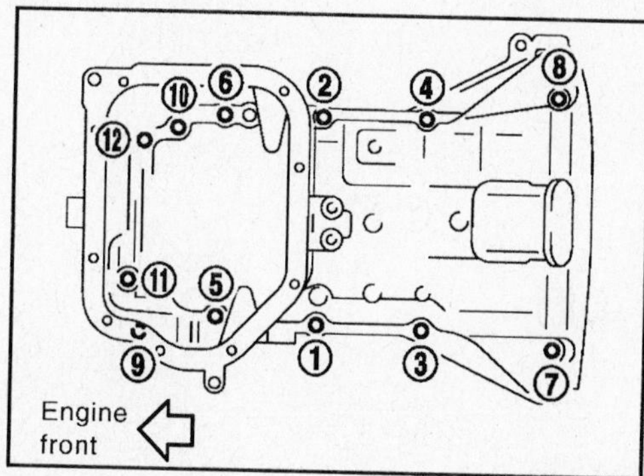

Oil pan bolt torque sequence

increased, check again for fuel leakage at connection points.

- Run engine to check for unusual noise and vibration.
- Warm up engine thoroughly to make sure there is no leakage of engine coolant, engine oil and working fluid, fuel and exhaust gas.
- Bleed air from passages in pipes and tubes of applicable lines, such as in cooling system.
- After cooling down engine, again check amounts of engine coolant, engine oil and working fluid. Refill to specified level, if necessary.

AWD Models

1. Before servicing the vehicle, refer to the precautions in the beginning of this section.

➡When removing the upper oil pan from the engine, first remove the crankshaft position sensor (POS). Be careful not to damage sensor edges or signal plate teeth.

2. Remove engine assembly from vehicle, and separate front suspension member, transaxle and transfer case assembly from engine.
3. Install an engine crane.
4. Remove the engine and mount it onto the engine stand.
5. Drain engine oil.
6. Remove oil filter.
7. Remove oil cooler and water pipes.
8. Remove the heated oxygen sensor 2 (bank 2) and remove the three way catalyst (manifold) (bank 2) from the exhaust manifold.
9. Loosen lower oil pan bolts in reverse order of illustration.
10. Insert a seal cutter (special service tool) between the lower oil pan and the upper oil pan.

➡Be careful not to damage the mating surface. Do not insert a screwdriver, this will damage the mating surfaces.

11. Slide seal cutter (special service tool) by tapping on the side of the tool with a hammer. Remove lower oil pan.
12. Remove oil strainer.
13. Remove the oil pressure switch.
14. Remove upper oil pan. Loosen bolts in reverse order shown.
15. Insert an appropriate size tool into the notch of the upper oil pan shown (1). Pry off the upper oil pan by moving the tool up and down shown (2).
16. Remove O-rings from the bottom of the cylinder block and oil pump body.
17. Remove oil pan gasket.

To install:

Installation is the reverse of removal. Note the following:

- Use a scraper to remove old liquid gasket from mating surfaces.
- Also remove the old liquid gasket from mating surface of the cylinder block.
- Remove the old liquid gasket from the bolt holes and threads.

➡Do not scratch or damage the mating surfaces when cleaning off the old liquid gasket.

- Apply Genuine RTV Silicone Sealant or equivalent, to the front timing chain case gasket and the rear oil seal retainer gasket shown.
- To install, align protrusion of oil pan gasket with notches of front timing chain case and rear oil seal retainer.
- Install oil pan gasket with smaller arc to front timing chain case side.
- Install new O-rings on the cylinder block and oil pump side.
- Apply a continuous bead of sealant to the cylinder block mating surface of the upper oil pan to a limited

portion shown. Use RTV silicone sealant or equivalent.

- For bolt holes with marks (5 locations), apply liquid gasket outside the holes.
- Apply a bead of 4.5 to 5.5 mm (0.177 to 0.217 in) diameter to area "A".
- Attaching within 5 minutes after coating.
- Install the upper oil pan. Tighten bolts in numerical order shown. There are two types of mounting bolts. Refer to the accompanying illustration.
- Install oil strainer to oil pump.
- Use a scraper to remove old liquid gasket from mating surfaces. Also remove old liquid gasket from mating surface of upper oil pan.
- Apply a continuous bead of sealant to the lower oil pan. Use RTV silicone sealant. Be sure the sealant is 4.5–5.5 mm (0.177–0.217 inch) wide. Attach within 5 minutes after coating.
- Install lower oil pan. Tighten the bolts in the numerical order shown.
- Install oil pan drain plug.
- Refer to illustration for installation of washer.
- Install in the reverse order of removal.
- Wait at least 30 minutes after oil pan is installed, before adding oil.
- Before starting engine, check the levels of engine coolant, engine oil and working fluid. If less than required quantity, fill to the specified level.
- Use procedure below to check for fuel leakage.
- Turn ignition switch ON (with engine stopped). With fuel pressure applied to fuel piping, check for fuel leakage at connection points.
- Start engine. With engine speed increased, check again for fuel leakage at connection points.
- Run engine to check for unusual noise and vibration.
- Warm up engine thoroughly to make sure there is no leakage of engine coolant, engine oil and working fluid, fuel and exhaust gas.
- Bleed air from passages in pipes and tubes of applicable lines, such as in cooling system.
- After cooling down engine, again check amounts of engine coolant, engine oil and working fluid. Refill to specified level, if necessary.

Oil Pump

REMOVAL & INSTALLATION

1. Before servicing the vehicle, refer to the precautions in the beginning of this section.
2. Remove oil pan and oil strainer.
3. Remove front timing chain case and timing chain (primary).
4. Remove oil pump assembly.
5. Installation is the reverse of removal. When installing, align crankshaft flat faces with inner rotor flat faces.

Rear Main Seal

REMOVAL & INSTALLATION

1. Before servicing the vehicle, refer to the precautions in the beginning of this section.
2. Remove engine from vehicle, and separate front suspension member, transaxle and transfer case (AWD models) assembly from engine.
3. Remove drive plate.
4. Remove upper oil pan.
5. Use a seal cutter (special service tool) to cut away liquid gasket and remove rear oil seal retainer.

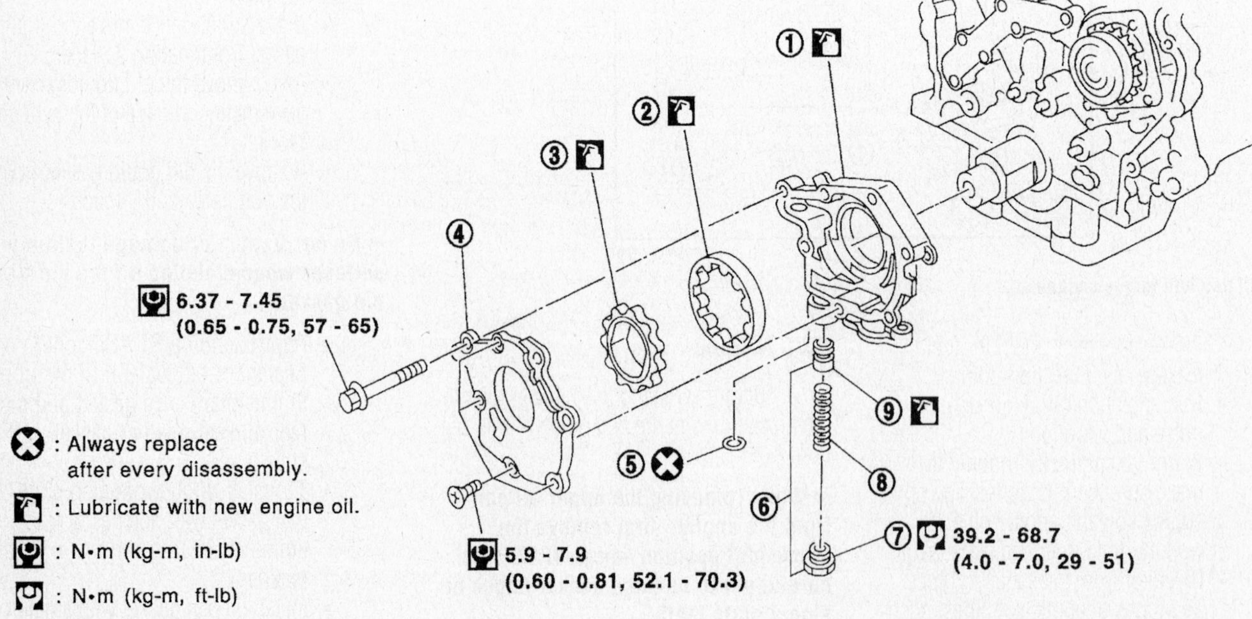

Oil pump exploded view

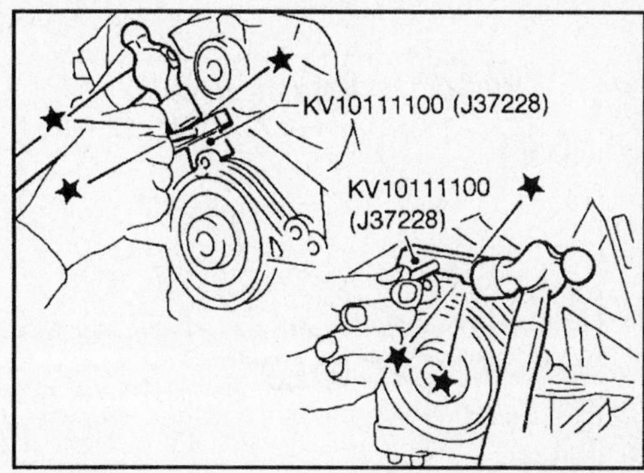

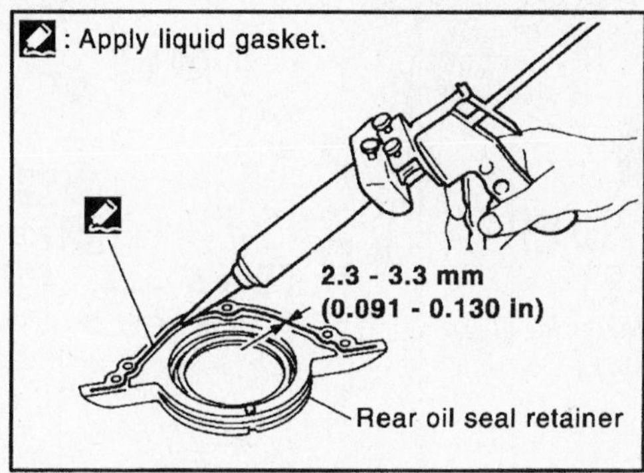

Applying sealer to the rear main seal

To install:

6. Remove old liquid gasket from mating surface of cylinder block and oil pan using scraper.
7. Apply liquid gasket to rear oil seal retainer as shown in the illustration. Use RTV silicone sealant. Assembly should be done within 5 minutes after coating.
8. Install rear oil seal retainer to cylinder block.
9. Perform the remaining steps in the reverse order of removal.

Front Case and Seal, Timing Chain, Sprockets, Rear Case

REMOVAL & INSTALLATION

1. Before servicing the vehicle, refer to the precautions in the beginning of this section.
2. Remove engine assembly from vehicle, and separate front suspension member, transaxle and transfer case (AWD models) assembly from engine.
3. Drain engine oil.
4. Remove engine harnesses.
5. Remove water hoses.
6. Remove EVAP canister purge volume control solenoid valve..
7. Remove drive belts and idler pulley bracket.
8. Remove power steering oil pump assembly.
9. Remove alternator.
10. Remove right and left rocker covers.
11. Remove crankshaft position sensor (POS).

➡Handle carefully to avoid dropping and shocks. Do not disassemble. Do not allow metal powder to adhere to magnetic part at sensor tip. Do not place sensors in a location where they are exposed to magnetism.

12. Obtain compression TDC of No.1 cylinder as follows. Rotate crankshaft pulley clockwise to align timing mark (grooved line without color) with timing indicator.
13. Remove lower and upper oil pans.
14. Remove crankshaft pulley as follows:
    a. Lock crankshaft with a hammer handle or similar tool to loosen bolts.
    b. Remove crankshaft pulley with a suitable puller.
15. Remove the right and left intake valve timing control covers. Loosen bolts in reverse order as shown. Use seal cutter to cut liquid gasket for removal.

➡Shaft is internally jointed with intake camshaft sprocket center hole. When removing, keep it horizontal until it is completely disconnected.

16. Remove right engine mounting bracket.
17. Remove front timing chain case.
    a. Loosen mounting bolts in reverse order as shown.
    b. Insert the appropriate size tool into the notch at the top of the front timing chain case as shown (1).
    c. Pry off the case by moving the tool as shown (2). Use seal cutter to cut liquid gasket for removal.
18. Remove water pump cover and chain tensioner cover from front timing chain case. Use seal cutter to cut liquid gasket for removal.
19. Remove the front oil seal from the front timing chain case using a suitable tool. Use screwdriver for removal. Exercise care not to damage front timing chain case.
20. Remove internal chain guide, timing chain tensioner, tension guide and slack guide.
21. Remove timing chain tensioner as follows:
    a. Pull lever down and release plunger stopper tab. Plunger stopper tab can be pushed up to release (coaxial structure with lever).
    b. Insert stopper pin into tensioner body hole to hold lever, and keep the tab released.

➡An Allen wrench [2.5 mm (0.098 in)] is used for a stopper pin as an example.

    c. Insert plunger into tensioner body by pressing the slack guide.
    d. Keep the slack guide pressed and hold it by pushing the stopper pin through the lever hole and body hole.
    e. Remove the mounting bolts and remove the timing chain tensioner.
22. Remove timing chain (primary) and crankshaft sprocket.

➡After removing timing chain, do not turn the crankshaft and camshaft separately, or the valves will strike the piston heads.

23. Attach a suitable stopper pin to the right and left camshaft chain tensioners (for secondary timing chains).
24. Remove intake and exhaust camshaft sprocket bolts. Apply paint to timing chain and camshaft sprockets for alignment during installation. Secure the hexagonal portion of the camshaft using a wrench to loosen the mounting bolts.

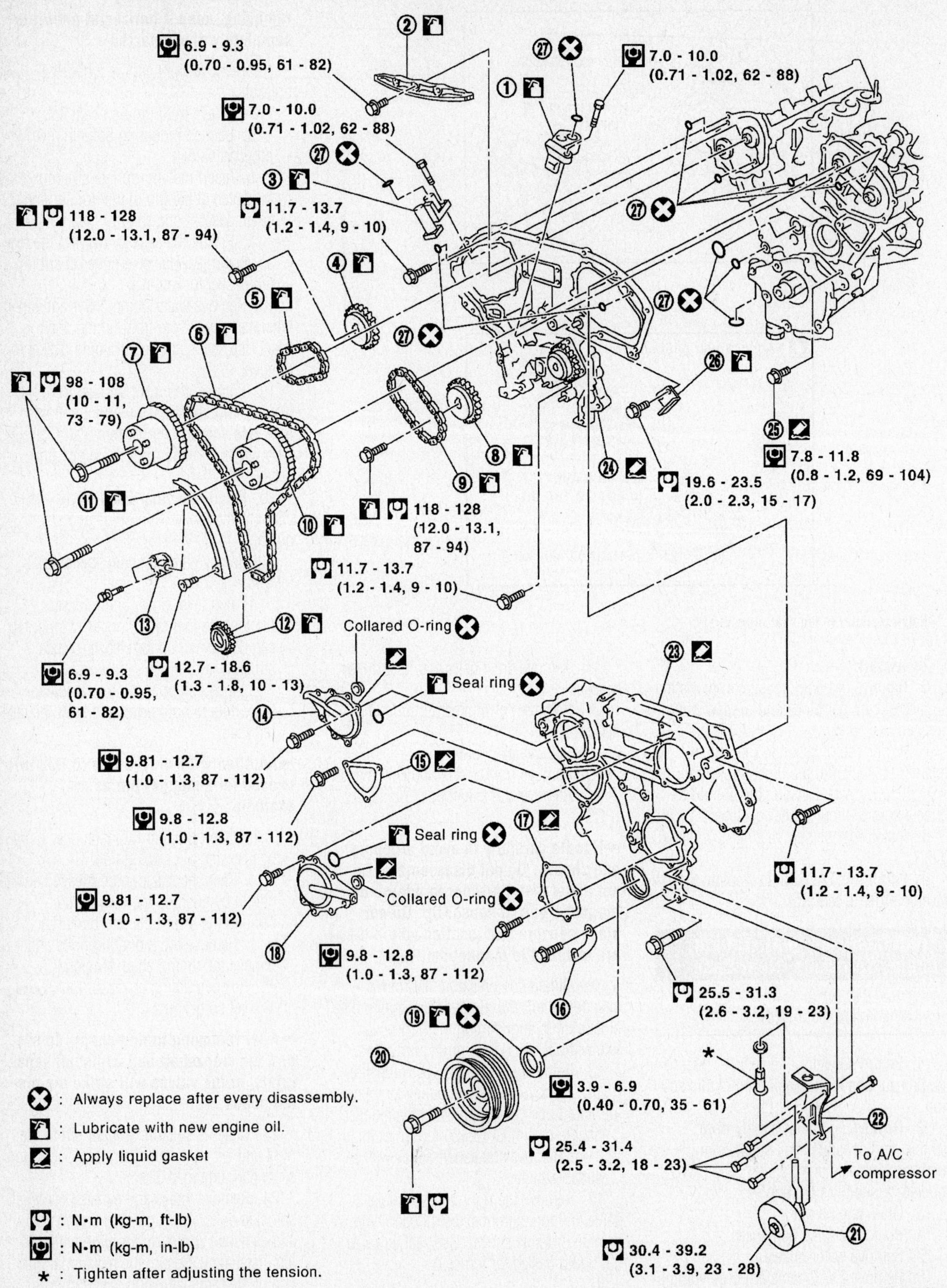

6.9 - 9.3 (0.70 - 0.95, 61 - 82)
7.0 - 10.0 (0.71 - 1.02, 62 - 88)
7.0 - 10.0 (0.71 - 1.02, 62 - 88)
118 - 128 (12.0 - 13.1, 87 - 94)
11.7 - 13.7 (1.2 - 1.4, 9 - 10)
98 - 108 (10 - 11, 73 - 79)
19.6 - 23.5 (2.0 - 2.3, 15 - 17)
7.8 - 11.8 (0.8 - 1.2, 69 - 104)
118 - 128 (12.0 - 13.1, 87 - 94)
11.7 - 13.7 (1.2 - 1.4, 9 - 10)
Collared O-ring
Seal ring
6.9 - 9.3 (0.70 - 0.95, 61 - 82)
12.7 - 18.6 (1.3 - 1.8, 10 - 13)
9.81 - 12.7 (1.0 - 1.3, 87 - 112)
9.8 - 12.8 (1.0 - 1.3, 87 - 112)
Seal ring
11.7 - 13.7 (1.2 - 1.4, 9 - 10)
9.81 - 12.7 (1.0 - 1.3, 87 - 112)
Collared O-ring
9.8 - 12.8 (1.0 - 1.3, 87 - 112)
25.5 - 31.3 (2.6 - 3.2, 19 - 23)
3.9 - 6.9 (0.40 - 0.70, 35 - 61)
25.4 - 31.4 (2.5 - 3.2, 18 - 23)
To A/C compressor
30.4 - 39.2 (3.1 - 3.9, 23 - 28)
: Always replace after every disassembly.
: Lubricate with new engine oil.
: Apply liquid gasket
: N•m (kg-m, ft-lb)
: N•m (kg-m, in-lb)
* : Tighten after adjusting the tension.
42356-MURA-G39

Timing chain and related components

1. Timing chain tensioner (secondary)
2. Internal chain guide
3. Timing chain tensioner (secondary)
4. Camshaft sprocket (EXH)
5. Timing chain (secondary)
6. Timing chain (primary)
7. Camshaft sprocket (INT)
8. Camshaft sprocket (EXH)
9. Timing chain (secondary)
10. Camshaft sprocket (INT)
11. Slack guide
12. Crankshaft sprocket
13. Timing chain tensioner (primary)
14. Intake valve timing control cover
15. Chain tensioner cover
16. Water hose clamp
17. Water pump cover
18. Intake valve timing control cover
19. Front oil seal
20. Crankshaft pulley
21. Idler pulley
22. Idler pulley bracket
23. Front timing chain case
24. Rear timing chain case
25. Water drain plug
26. Tension guide
27. O-ring

25. Remove secondary timing chain together with camshaft sprockets. Turn camshaft slightly to secure slackness of timing chain on chain tensioner side. Insert 0.5 mm (0.020 in) thick metal or resin plate between timing chain and chain tensioner plunger (guide). Remove secondary timing chain together with camshaft sprockets from guide groove.

➡Be careful of plunger coming off when removing timing chain. This is because plunger of chain tensioner moves during operation, leading to coming off of fixed stopper pin.

➡Camshaft sprocket (INT) is two-for-one structure of primary and secondary sprockets. When handling camshaft sprocket (INT), handle carefully to avoid any shock to camshaft sprocket. Do not disassemble. (Never loosen bolts "A" and "B" as shown).

26. Remove chain tension guide.

27. Remove rear timing chain case as follows:

a. Loosen and remove mounting bolts in reverse order as shown.

b. Cut the sealant using a seal cutter and remove rear timing chain case.

➡Do not remove plate metal cover of oil passage. After removing chain case, do not apply any load which affects flatness.

28. Remove right and left camshaft chain tensioners from cylinder head as follows if necessary.

a. Remove No.1 camshaft brackets.

b. Remove secondary chain tensioners with stopper pin attached.

29. Use a scraper to remove all traces of liquid gasket from front and rear timing chain cases, and opposite mating surfaces. Remove old liquid gasket from the bolt hole and thread.

30. Use a scraper to remove all traces of liquid gasket from water pump cover, chain tensioner cover and intake valve timing control covers.

31. Check for cracks and any excessive wear at the roller links of the timing chain. Replace the timing chain as necessary.

To install:

32. Install right and left camshaft chain tensioners to cylinder head as follows if removed.

a. Install secondary chain tensioners with stopper pin attached and new O-ring.

b. Install No.1 camshaft brackets.

33. Install O-rings onto cylinder block.

34. Install O-rings to cylinder head.

35. Apply liquid gasket to rear timing chain case back side as shown. Use RTV silicone sealant or equivalent.

36. Align the rear timing chain case and water pump assembly with the dowel pins (right and left) on the cylinder block and install the case. Make sure the O-rings stay in place during installation to cylinder block and cylinder head.

a. Tighten the mounting bolts in the numerical order as shown. There are two bolt lengths used.

- Bolt length: Bolt position 20 mm (0.79 inch)
- 1, 2, 3, 6, 7, 8, 9, 10 16 mm (0.63 in): Except the above 11.7–13.7 Nm (9–10 ft. lbs.)

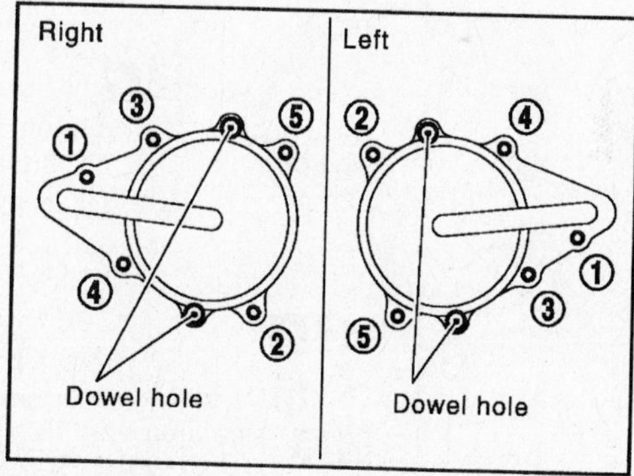

Timing control cover bolt torque sequence

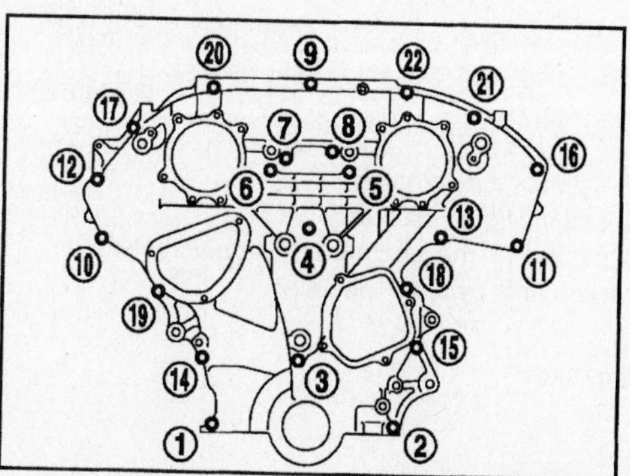

Front timing chain case bolt torque sequence

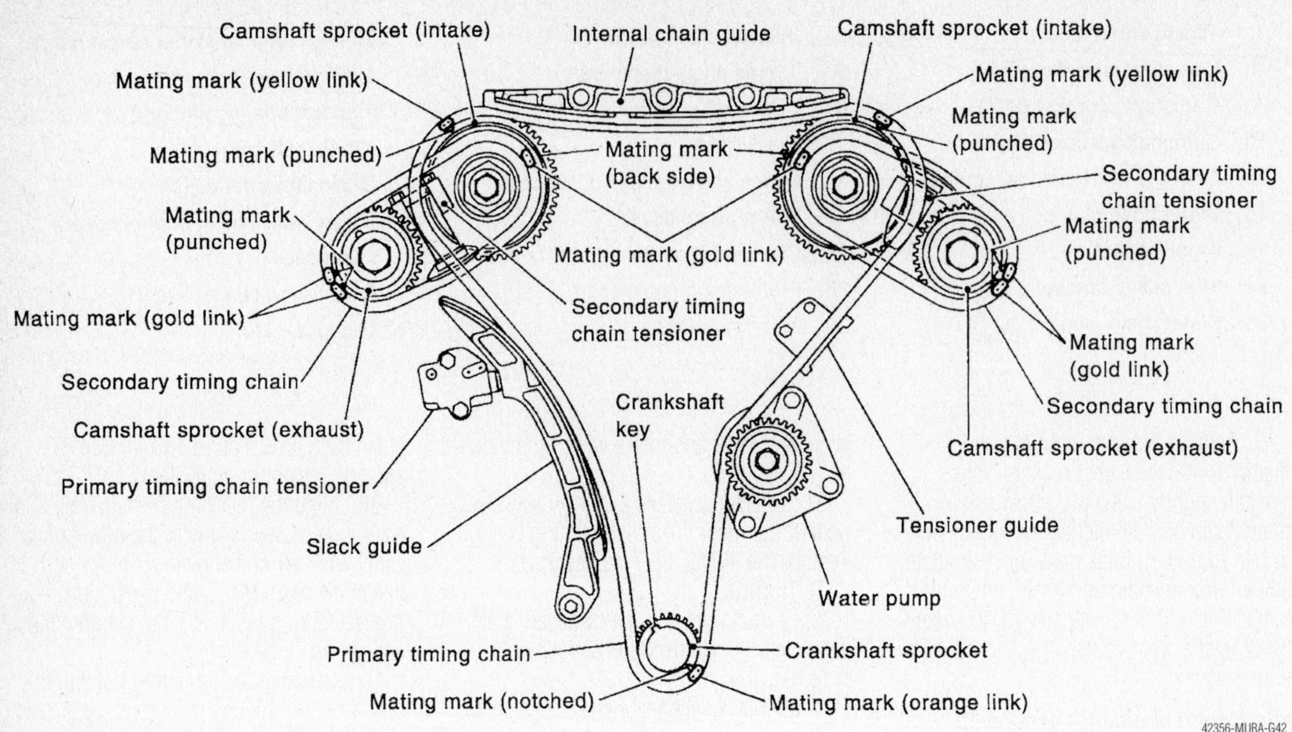

Timing chain mating marks

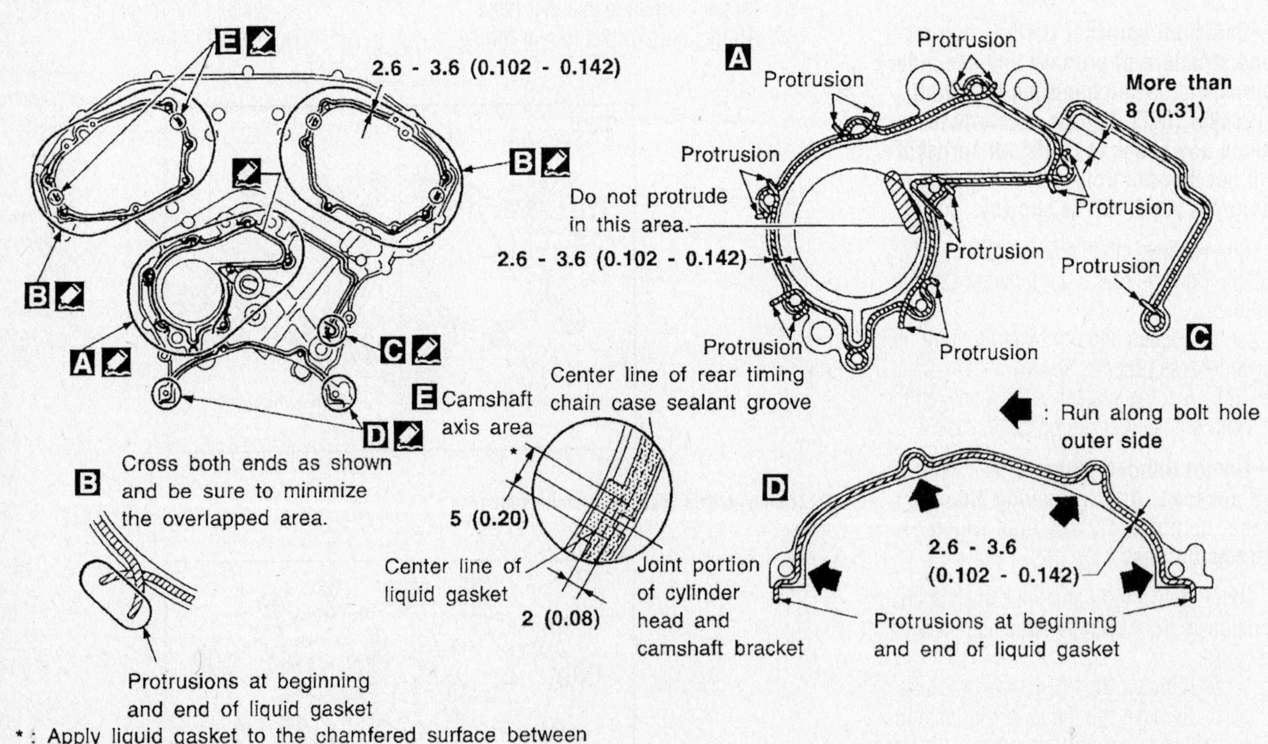

Rear timing case sealer application

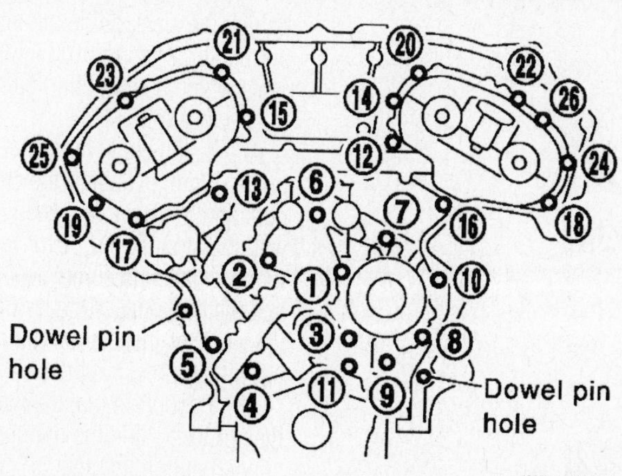

Rear timing case bolt torque sequence

Standard Rear timing chain case to cylinder block: -0.24 to 0.14 mm (-0.0094 to 0.0055 inch)

b. After all bolts are temporarily tightened, retighten them to the specification in the numerical order as shown. If the RTV protrudes, wipe it off immediately.

37. After installing rear timing chain case, check surface height difference between following parts on oil pan mounting surface. If not within standard, repeat above installation procedure.

38. Install chain tension guide.

39. Position the crankshaft so No. 1 piston is set at TDC on the compression stroke. Make sure that the dowel pin hole, dowel pin and crankshaft key are located as shown.

➡Hole on small dia. side must be used for intake side dowel pin hole. Do not misidentify.

40. Install the timing chains (secondary) and camshaft sprockets.

a. Install secondary timing chains and camshaft sprockets. Align the mating marks on the secondary timing chain (gold link) with the ones on the intake and exhaust sprockets (stamped), and install them.

➡Mating marks for the intake sprocket are on the back side of the secondary camshaft sprocket. There are two types of mating marks, circle and oval types. They should be used for the right and left banks, respectively.

41. Align the dowel pin and pin hole on the camshaft with the groove and dowel pin on the sprocket, and install them.

a. On the intake side, align the pin hole on the small diameter side of the camshaft front end with the dowel pin on the back side of the camshaft sprocket, and install them.

b. On the exhaust side, align the dowel pin on the camshaft front end with the pin groove on the camshaft sprocket, and install them.

c. Mounting bolts for the camshaft sprockets must be tightened in the next step. Tightening them by hand is enough to prevent the dislocation of the dowel pins.

d. It may be difficult to visually check the dislocation of mating marks during and after installation. To make the matching easier, make a mating mark on the top of sprocket teeth and its extended line in advance with paint.

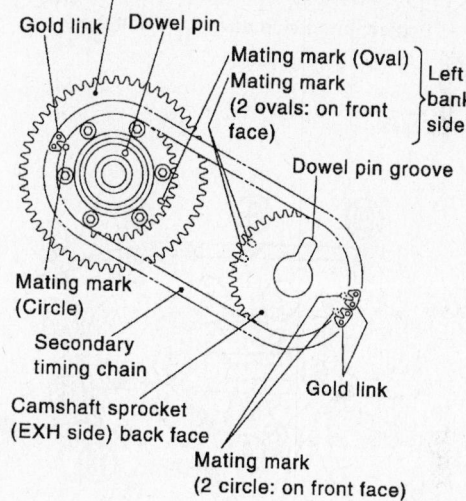

Camshaft sprocket (INT side) back face
Gold link
Dowel pin
Mating mark (Oval)
Mating mark (2 ovals: on front face)
Left bank side
Dowel pin groove
Mating mark (Circle)
Secondary timing chain
Gold link
Camshaft sprocket (EXH side) back face
Mating mark (2 circle: on front face)
42356-MURA-G45

Secondary timing chain alignment

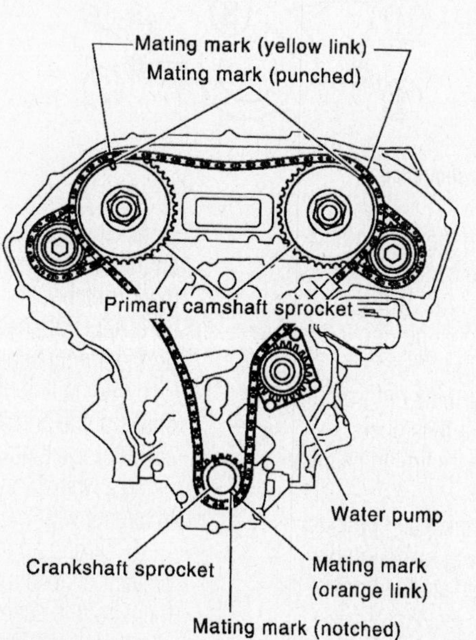

Primary timing chain alignment

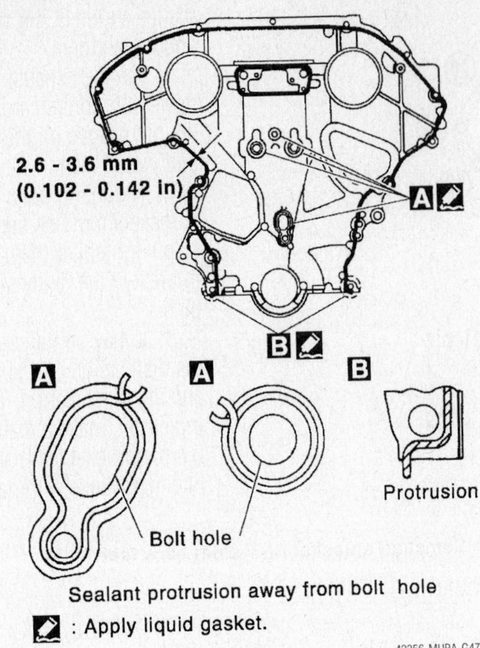

Sealer application on front case

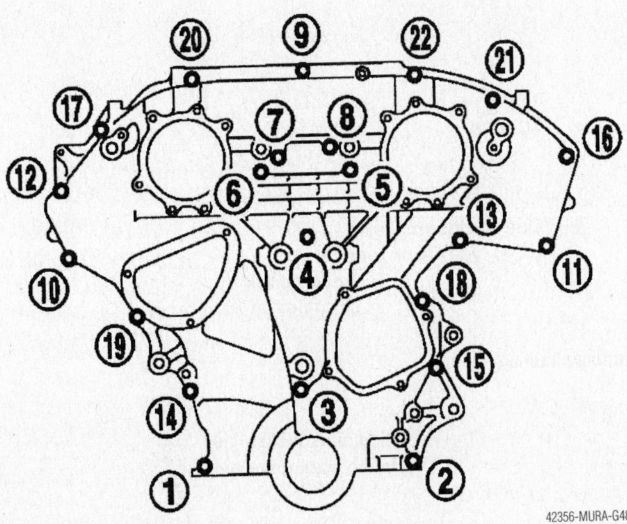

Front case bolt torque sequence

42. After confirming the mating marks are aligned, tighten the camshaft sprocket mounting bolts. Secure the camshaft using a wrench at the hexagonal portion to tighten the mounting bolts.

43. Pull the stopper pins out from the secondary timing chain tensioners.

44. Install the primary timing chain as follows:

a. Install the crankshaft sprocket. Make sure the mating marks on the crankshaft sprocket face the front of the engine.

b. Install the primary timing chain. Install primary timing chain so the mating mark (punched) on camshaft sprocket is aligned with the yellow link on the timing chain, while the mating mark (notched) on the crankshaft sprocket is aligned with the orange one on the timing chain, as shown.

c. When it is difficult to align mating marks of the primary timing chain with each sprocket, gradually turn the camshaft using a wrench on the hexagonal portion to align it with the mating marks.

d. During alignment, be careful to prevent dislocation of mating mark alignments of the secondary timing chains.

45. Install the internal chain guide and tension guide.

46. Install slack guide. Do not over tighten the slack guide mounting bolts. It is normal for a gap to exist under the bolt seats when the mounting bolts are tightened to specification.

47. Install chain tensioner for slack guide. When installing the chain tensioner, push in the sleeve and keep it pressed in with the stopper pin. Remove any dirt and foreign materials completely from the back and the mounting surfaces of the chain tensioner. After installation, pull out the stopper pin by pressing the slack guide.

48. Reconfirm that the mating marks on the sprockets and the timing chain have not slipped out of alignment.

49. Install new O-rings on the rear timing chain case.

50. Install the front oil seal on the front timing chain case. Apply new engine oil to the oil seal edges. Install it so that each seal lip is oriented as shown in illustration. Using a suitable drift, press-fit oil seal until it becomes flush with timing chain case end face. Make sure the garter spring is in position and seal lip is not inverted.

51. Install the water pump cover and the chain tensioner cover to front cover. Apply RTV silicone sealant.

52. Install front timing chain case as follows:

a. Apply liquid gasket to front timing chain case back side as shown.

b. Install dowel pin on the rear timing chain case into dowel pin hole on front timing chain case.

c. Tighten bolts to the specified torque in order shown in the illustration.

- 8 mm (0.31 in) dia. bolts 1, 2: 25.5–31.3 Nm (19–23 ft. lbs.)
- 6 mm (0.24 in) dia. bolts Except the above: 11.7–13.7 Nm (9–10 ft. lbs.)

d. After tightening, retighten them to specified torque in numerical order shown in illustration.

e. After tightening, retighten them to specified torque in numerical order shown in illustration.

53. After installing the front timing chain case, check the surface height difference between the following parts on the oil pan mounting surface. If not within specification, repeat the installation procedure.

54. Install right and left intake valve timing control covers as follows:

a. Install seal rings in shaft grooves.

b. Apply liquid gasket to the intake valve timing control covers. Use RTV Silicone Sealant.

c. Install collared O-ring in front cover oil hole (left and right sides).

d. Being careful not to move the seal

ring from the installation groove, align the dowel pins on the chain case with the holes to install the intake valve timing control covers.

e. Tighten bolts in the numerical order as shown.

55. Install right and left rocker covers.
56. Install crankshaft pulley as follows:

a. Fix crankshaft using a hammer shaft or an equivalent tool.

b. Install crankshaft pulley, taking care not to damage front oil seal. When press-fitting crankshaft pulley with a plastic hammer, tap on its center portion (not circumference).

c. Tighten bolt to 39.2 to 49.0 Nm (29 to 36 ft. lbs.).

d. Put a paint mark on crankshaft pulley aligning with angle mark on crankshaft pulley bolt. Then, further retighten bolt by 60 to 65 degrees.

57. Rotate crankshaft pulley in normal direction (clockwise when viewed from front) to confirm it turns smoothly.
58. For the following operations, perform steps in the reverse order of removal.

➡If hydraulic pressure inside chain tensioner drops after removal/installation, slack in the guide may generate a pounding noise during and just after engine start. However, this is not unusual. Noise will stop after hydraulic pressure rises.

Piston and Ring

POSITIONING

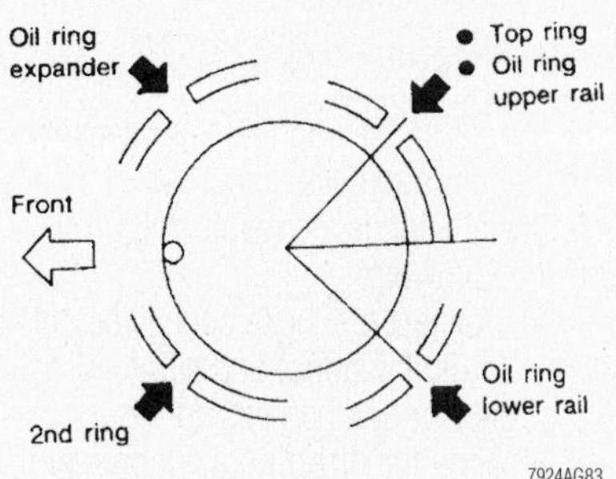

Piston ring end-gap spacing

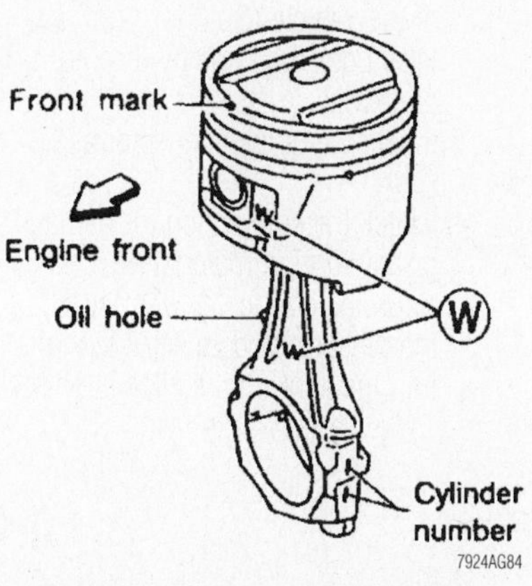

Piston and connecting rod positioning

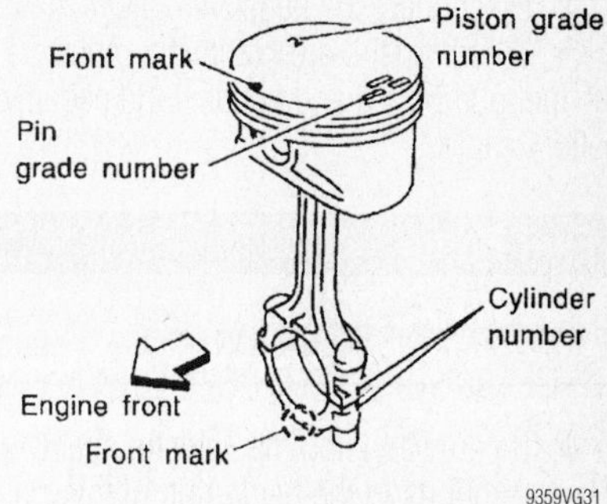

Piston and connecting rod positioning

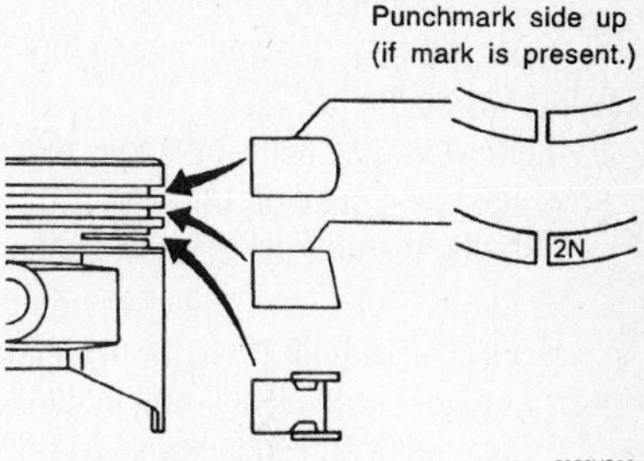

Piston ring positioning

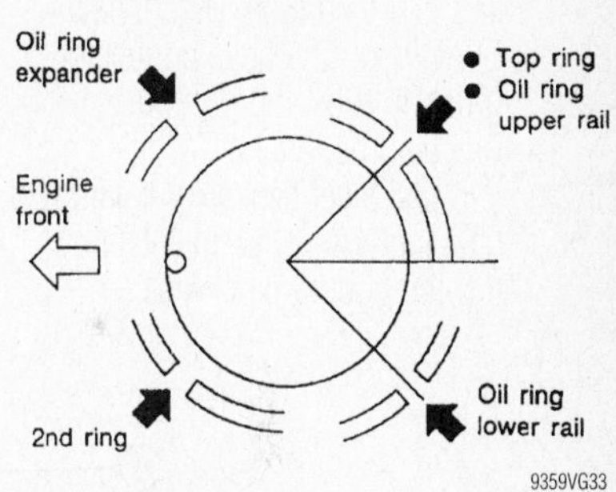

Piston ring positioning

FUEL SYSTEM

Fuel System Service Precautions

Safety is the most important factor when performing not only fuel system maintenance but any type of maintenance. Failure to conduct maintenance and repairs in a safe manner may result in serious personal injury or death. Maintenance and testing of the vehicle's fuel system components can be accomplished safely and effectively by adhering to the following rules and guidelines.

• To avoid the possibility of fire and personal injury, always disconnect the negative battery cable unless the repair or test procedure requires that battery voltage be applied.

• Always relieve the fuel system pressure prior to disconnecting any fuel system component (injector, fuel rail, pressure regulator, etc.), fitting or fuel line connection. Exercise extreme caution whenever relieving fuel system pressure, to avoid exposing skin, face and eyes to fuel spray. Please be advised that fuel under pressure may penetrate the skin or any part of the body that it contacts.

• Always place a shop towel or cloth around the fitting or connection prior to loosening to absorb any excess fuel due to spillage. Ensure that all fuel spillage (should it occur) is quickly removed from engine surfaces. Ensure that all fuel soaked cloths or towels are deposited into a suitable waste container.

• Always keep a dry chemical (Class B) fire extinguisher near the work area.

• Do not allow fuel spray or fuel vapors to come into contact with a spark or open flame.

• Always use a back-up wrench when loosening and tightening fuel line connection fittings. This will prevent unnecessary stress and torsion to fuel line piping. Always follow the proper torque specifications.

• Always replace worn fuel fitting O-rings with new. Do not substitute fuel hose or equivalent, where fuel pipe is installed.

Fuel System Pressure

RELIEVING

1. Before servicing the vehicle, refer to the precautions in the beginning of this section.
2. Remove the fuel pump fuse from the panel.
3. Start the engine and allow it to run until it stalls. Crank the engine for a few seconds to relieve additional fuel pressure.

4. Disconnect the negative battery cable.
5. When repairs are complete, replace the fuel pump fuse and connect the negative battery cable.

Fuel Pump

REMOVAL & INSTALLATION

1. Before servicing the vehicle, refer to the precautions in the beginning of this section.
2. Relieve the fuel system pressure.
3. Open the fuel filler lid.
4. Open the filler cap and release the pressure inside the fuel tank.
5. Remove or disconnect the following:
   - Rear seat cushion trim and pad bolts, then lift up rear seat cushion.
   - Inspection hole cover for main and sub fuel level sensor unit by turning clips counterclockwise by 90°.
   - Harness connector and quick connectors for EVAP/Vent line hose and fuel feed tube. Disconnect EVAP/Vent line hose connector (push in tubs and pull out).
   - Remove the retainer for main and sub fuel level sensor unit with fuel tank lock ring wrench (SST) by turning counterclockwise.
   - Remove main fuel level sensor unit, fuel filter and fuel pump assembly, and sub fuel level sensor unit. Raise the main fuel level sensor unit, fuel filter and fuel pump assembly, and disconnect the fuel hose connector (push in tabs and pull out) and sub fuel level sensor unit harness connector. Raise and release the sub fuel level sensor unit to remove.
6. Installation is the reverse of removal. Note the following:
   - Connect fuel hose connector (push in until it stops) and sub fuel level sensor unit harness connector.
   - Align the direction mark on main and sub fuel level sensor unit with that on fuel tank as shown in the illustration.
   - Install the inspection hole cover with front mark (arrow) facing front of the vehicle (Both for right and left). Lock the clips by turning clockwise.
7. Connect the quick connector as follows.
   - Check the connection for damage or any foreign materials.
   - Align the connector with the tube, then insert the connector straight into the tube until a click is heard.
   - After connecting, make sure that the connection is secure by following method.
   - Pull the tube and the connector to make sure they are securely connected. Visually confirm that the two retainer tabs are connected to the connector.
8. Turn ignition switch "ON" (with engine stopped), then check connections for leaks by applying fuel pressure to fuel piping.
9. Start the engine and let it idle and make sure there are no fuel leaks at the fuel system connections.

Fuel Injectors

REMOVAL & INSTALLATION

1. Before servicing the vehicle, refer to the precautions in the beginning of this section.
2. Remove the engine cover.
3. Release the fuel pressure.
4. Remove the right windshield wiper arm and the right cowl top cover.
5. Remove radiator cover grille, air duct (inlet), air cleaner case, air duct assembly and mass air flow sensor.
6. Disconnect electric throttle control actuator and engine coolant hoses.

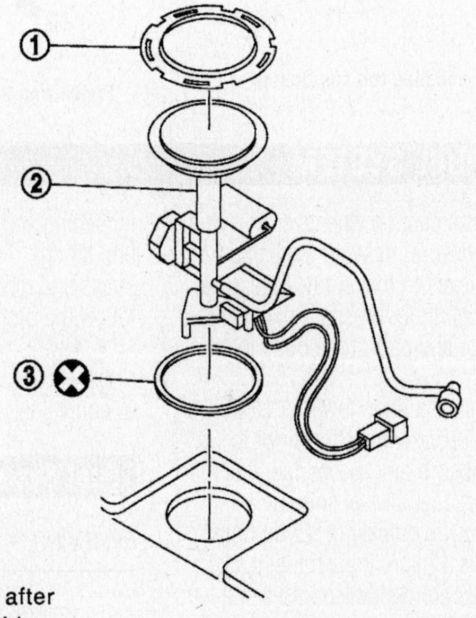
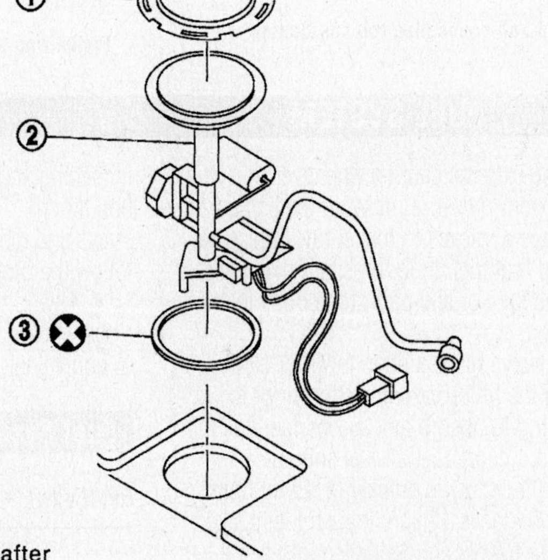

Left side

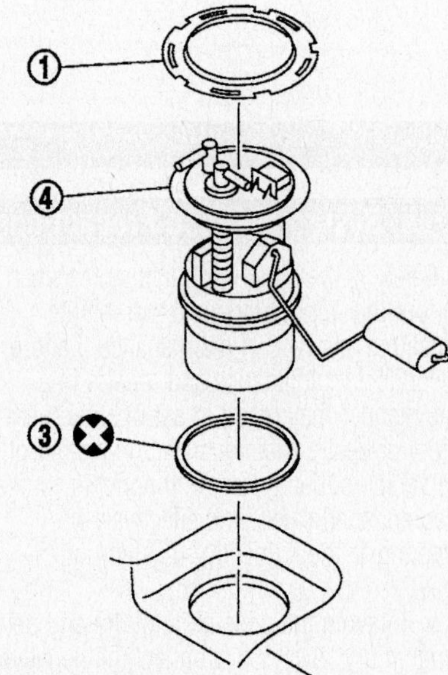

Ⓧ: Always replace after every disassembly.

1. Retainer
2. Sub fuel level sensor unit
3. O-ring
4. Main fuel level sensor unit, fuel filter and fuel pump assembly

42356-MURA-G49

Fuel level sensor unit, filter and pump

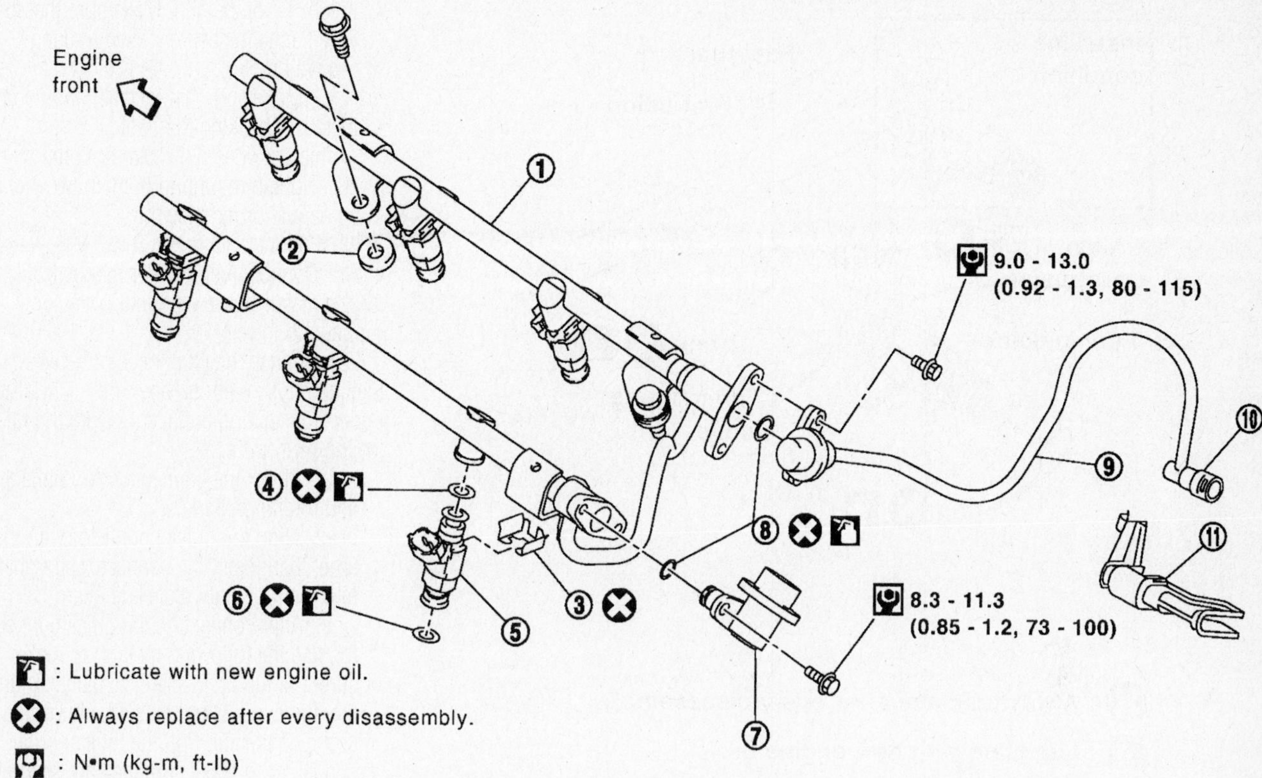

Fuel injector and fuel rail

7. Disconnect vacuum hose, fuel injectors electrical connectors, and PCV hose.

8. Remove the vacuum tank from intake manifold collector (lower).

9. Disconnect the power steering hose bracket.

10. Remove the intake manifold collector (upper and lower).

The intake manifold collector (upper) should be moved aside with water hoses connected.

11. Remove fuel feed hose (with damper) from fuel tube.

12. Disconnect fuel feed hose (with damper) quick connector at vehicle piping side. When separating fuel feed hose and centralized under-floor piping connection, disconnect quick connector with the following procedure.

a. Remove quick connector cap from quick connector.

b. Disconnect quick connector from centralized under-floor piping.

13. Remove harness connector from fuel injector.

14. Loosen mounting bolts in numerical order in the illustration, and remove fuel tube and fuel injector assembly.

15. Remove fuel injector from fuel tube with following procedure.

a. Open and remove clip.

b. Remove fuel injector from the fuel tube by pulling straight.

16. Remove fuel damper from fuel tube.

To install:

17. Install fuel damper. Insert fuel damper straight into fuel tube. Tighten mounting bolts evenly in turn. After tightening mounting bolts, make sure that there is

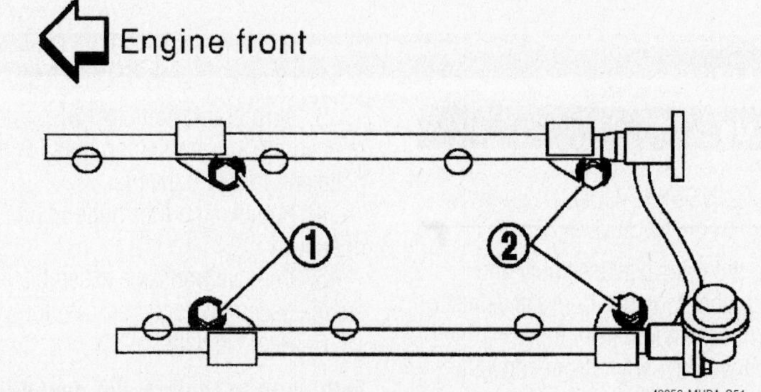

Fuel rail bolt loosening sequence

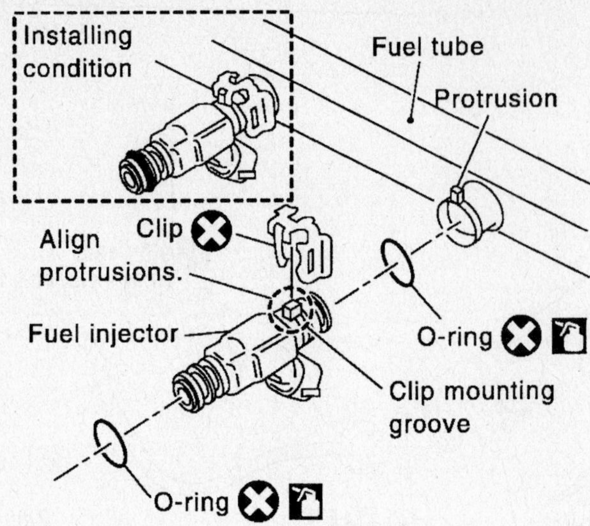

Fuel injector installation

no gap between flange and fuel tube. When handling O-rings, be careful of the following:

- Handle O-ring with bare hands. Never wear gloves.
- Lubricate O-ring with new engine oil.
- Do not clean O-ring with solvent.
- Make sure that O-ring and its mating part are free of foreign material.
- When installing O-ring, be careful not to scratch it with tool or fingernails. Also be careful not to twist or stretch O-ring. If O-ring was stretched while it was being attached, do not insert it quickly into fuel tube.

18. Install O-rings on the fuel injector. Upper and lower O-ring are different.
19. Install fuel injector to fuel tube with the following procedure.
    a. Insert clip into clip mounting groove on fuel injector. Insert clip so that lug "A" of fuel injector matches notch "A" of the clip.

➡Do not reuse clip. Replace it with a new one. Be careful to keep clip from interfering with O-ring. If interference occurs, replace O-ring.

    b. Insert fuel injector into fuel tube with clip attached. Insert it while matching it to the axial center. Insert fuel injector so that lug "B" of fuel tube matches notch "B" of the clip. Make sure that fuel tube flange is securely fixed in flange groove on clip.
    c. Make sure that installation is complete by checking that fuel injector does not rotate or come off.
20. Tighten mounting bolts in two steps in numerical order shown in illustration.
21. Connect fuel injector harness.
22. Install intake manifold collector (upper and lower).
23. Connect quick connector between fuel feed hose (with damper) and centralized under-floor piping connection with the following procedure:
    a. Check the connection for damage and foreign materials.
    b. Align the quick connector with the tube, then insert the connector straight into the tube until a click is heard.
    c. After connecting the quick connector, use the following method to make sure it is full connected. Visually confirm that the two retainer tabs are connected to the connector. Pull the tube and the connector to make sure they are securely connected.
    d. Install quick connector cap to quick connector connection. Install quick connector cap with arrow on surface facing in direction of quick connector.

➡If cap cannot be installed smoothly, quick connector may have not been installed correctly. Check connection again.

24. The remainder of installation is the reverse of removal.
25. Turn ignition switch ON (with engine stopped). With fuel pressure applied to fuel piping, check for fuel leakage at connection points.
26. Start engine. With engine speed increased, check again for fuel leakage at connection points.

DRIVE TRAIN

Transaxle

REMOVAL & INSTALLATION

Remove the transaxle assembly and engine assembly together from the vehicle.

1. Remove exhaust front tube.
2. Remove dust cover from converter housing part.
3. Turn crankshaft clockwise and remove the four tightening nuts for drive plate and torque converter.
4. Remove the four bolts in the illustration.
5. Remove transaxle assembly and engine assembly together from the vehicle.
6. Remove halfshaft.

➡Be sure to replace the new differential side oil seal every removal of halfshaft.

7. Remove transfer case gusset. (AWD models)
8. Remove transfer case assembly.

➡Be sure to replace the new differential side oil seal (converter housing side only) whenever the transfer case is removed.

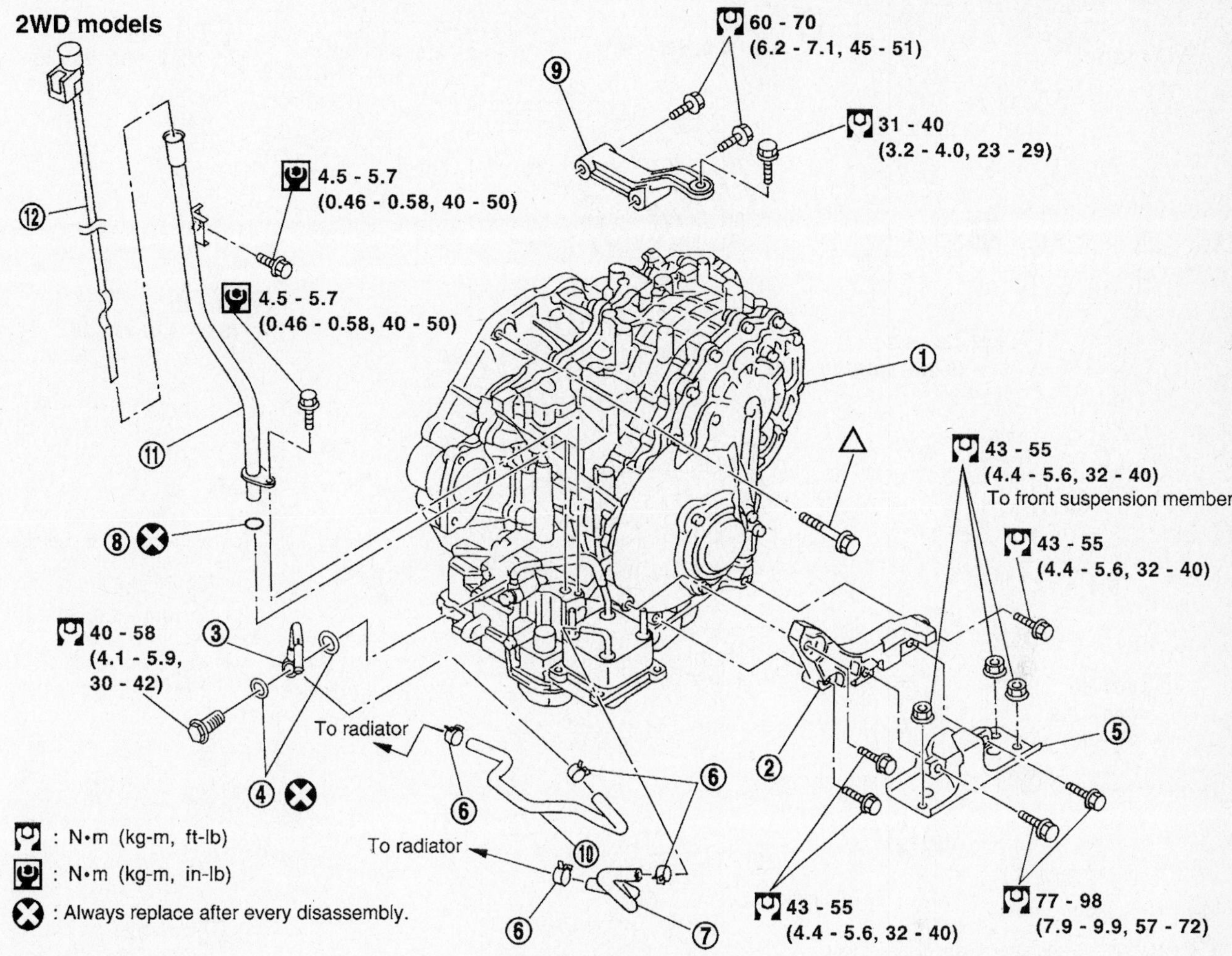

1. Transaxle assembly
2. LH engine mounting bracket
3. Fluid cooler tube
4. Copper washer
5. LH engine mounting insulator
6. Hose clamp
7. CVT fluid cooler hose
8. O-ring
9. Rear gusset
10. CVT fluid cooler hose
11. CVT fluid charging pipe
12. CVT fluid level gauge

42356-MURA-G53

Transaxle and related parts—2wd

9. Remove filler pipe.
10. Disconnect harness connector and wire harness.
11. Remove POS sensor, from engine assembly.
12. Remove starter motor.
13. Remove CVT fluid cooler valve assembly. (With CVT fluid cooler tube assembly and heater hose).
14. Install slinger to transaxle assembly.
15. Remove rear gusset.
16. Remove left engine mounting bracket and left engine mounting insulator.
17. Remove front suspension member from transaxle assembly and engine assembly.
18. Remove transaxle assembly bolts.
19. Remove transaxle assembly from engine assembly with a hoist. Secure torque converter to prevent it from dropping.

➡After installing a torque converter to a transaxle, be sure to check dimension A to ensure it is within the reference value limit.

20. Installation is the reverse of the removal. Note the following:
   - Screw and set the locator into the stud bolts for the torque converter locate.
   - Rotate the torque converter to allow the locator to go down.
   - Rotate the drive plate so that the hole of the drive plate locator faces down.
   - Installing transaxle assembly from engine assembly with a hoist.
   - When installing fluid cooler tube to transaxle assembly, transaxle assembly the part with the tube aligned with the rib.
   - When installing CVT fluid cooler valve assembly to the engine, torque the bolts to: 28–32 Nm (21–23 ft. lbs.)
   - Align the positions of tightening nuts for drive plate with those of the torque converter, and temporarily tighten the nuts. Then, tighten the nuts to the specified torque.
   - Install POS sensor.
   - After completing installation, check for fluid leakage, fluid level, and the positions of CVT.
   - When replacing the CVT assembly, erase EEP ROM in TCM.

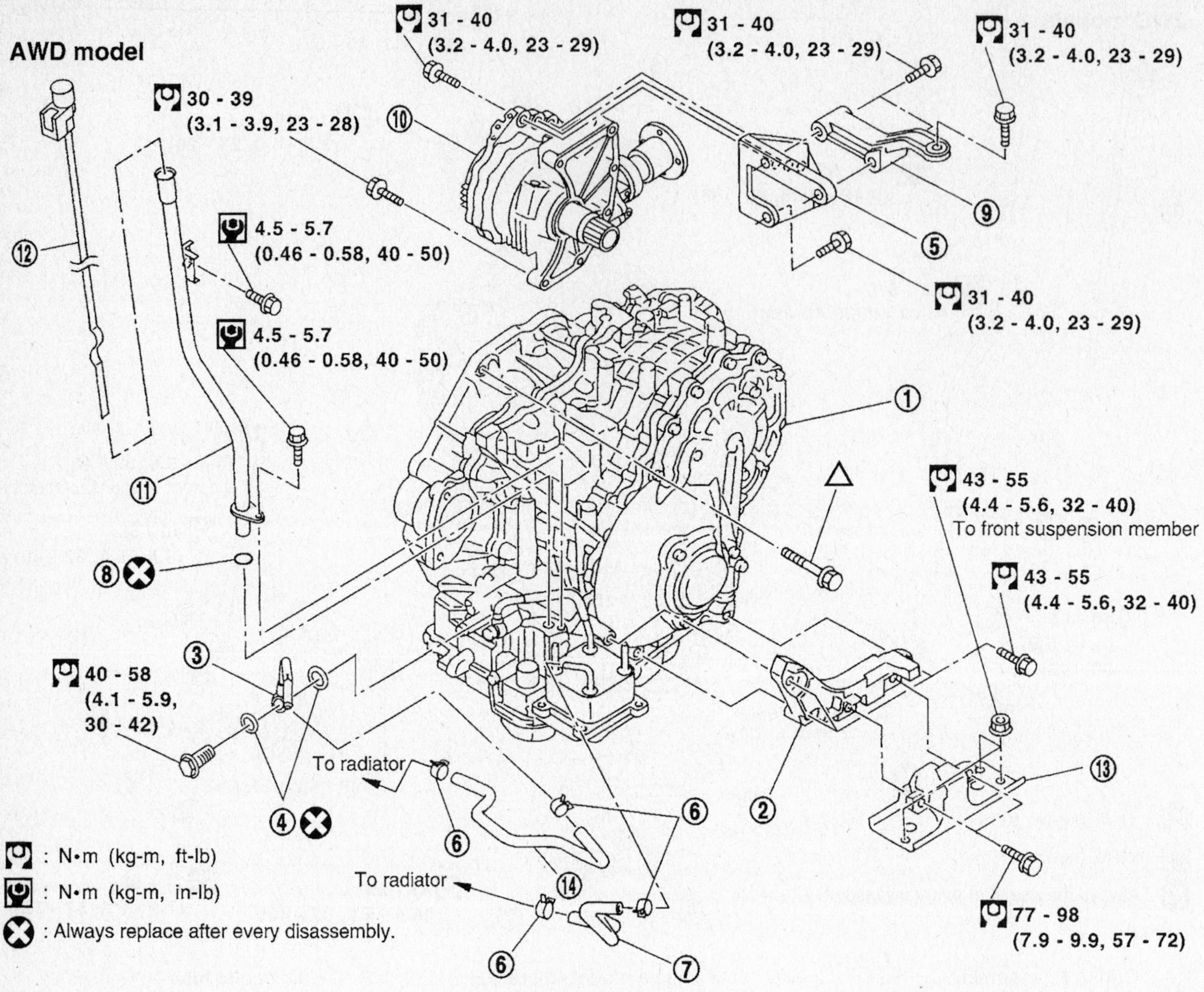

AWD model

: N•m (kg-m, ft-lb)

: N•m (kg-m, in-lb)

: Always replace after every disassembly.

1. Transaxle assembly	2. LH engine mounting bracket	3. Fluid cooler tube
4. Copper washer	5. Transfer gusset	6. Hose clamp
7. CVT fluid cooler hose	8. O-ring	9. Rear gusset
10. Transfer assembly	11. CVT fluid charging pipe	12. CVT fluid level gauge
13. LH engine mounting insulator	14. CVT fluid cooler hose	

42356-MURA-G54

Transaxle and related parts—AWD

Bolt No.	1	2	3	4
Number of bolts	1	2	2	4
Bolt length "ℓ"mm (in)	52 (2.05)	36 (1.42)	105 (4.13)	35 (1.38)
Tightening torque N-m (kg-m, ft-lb)	70 - 79 (7.1 - 8.0, 51 - 58)			42-52 (4.3 - 5.3, 31 - 38)

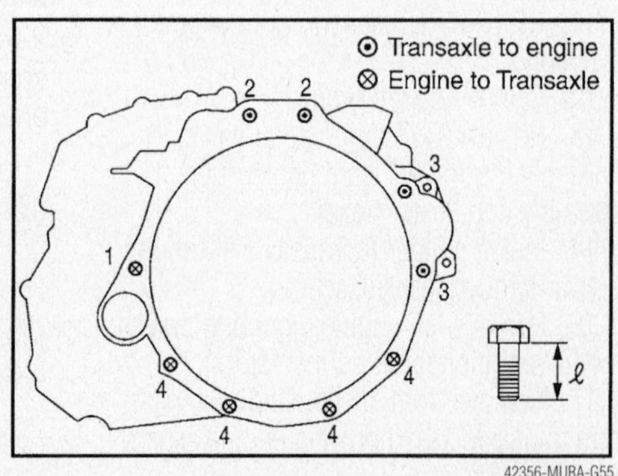

42356-MURA-G55

Transaxle bolt location and torque sequence

A
Straightedge
Distance"A"
Scale
42356-MURA-G56

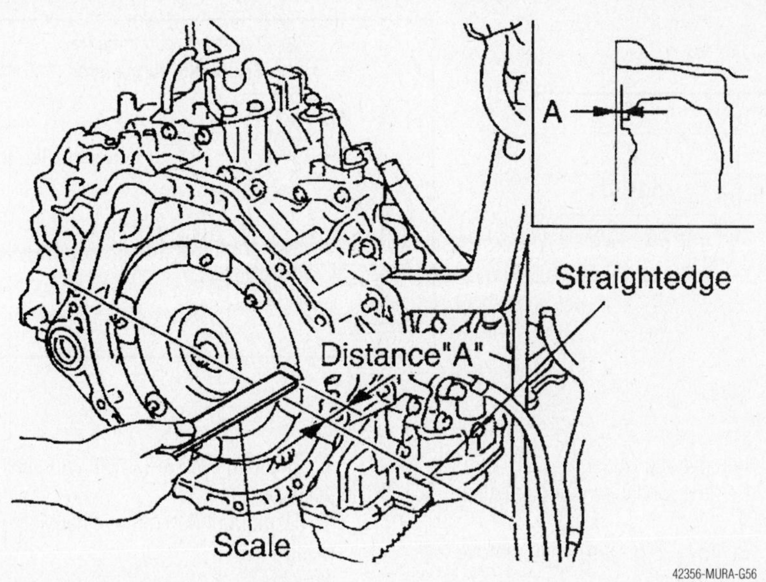

Torque converter installation inspection

Transfer Case Assembly

REMOVAL & INSTALLATION

1. Before servicing the vehicle, refer to the precautions in the beginning of this section.
2. Remove the engine assembly.
3. Remove gusset mounting bolts, and then remove gusset from engine and transaxle.
4. Remove transfer case mounting bolts and separate transfer case from transaxle.

➡After removing transfer case from transaxle, be sure to replace differential side oil seal of the transaxle side with new one.

5. Installation is the reverse of removal.

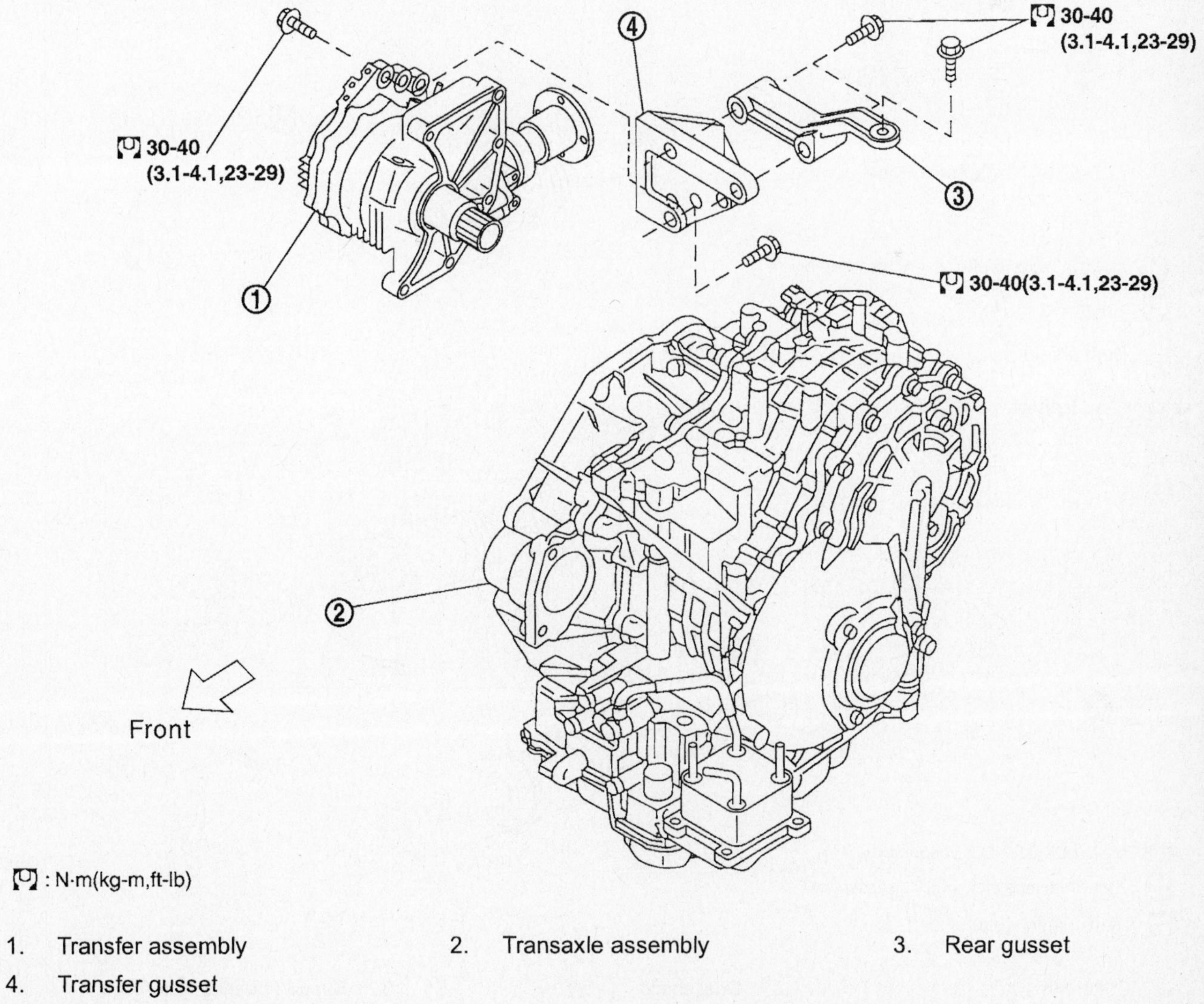

30-40
(3.1-4.1,23-29)
30-40
(3.1-4.1,23-29)
30-40(3.1-4.1,23-29)
④
③
①
②
Front
: N·m(kg-m,ft-lb)
1. Transfer assembly
2. Transaxle assembly
3. Rear gusset
4. Transfer gusset
42356-MURA-G57

Transfer case and related parts

Bolt No.	1	2
Quantity	4	2
Nominal length mm (in)	65 (2.56)	40 (1.57)
Tightening torque [N·m (kg·m, ft.-lb.)]	29.4 - 39.2 (3.0 - 3.9, 22 - 28)	

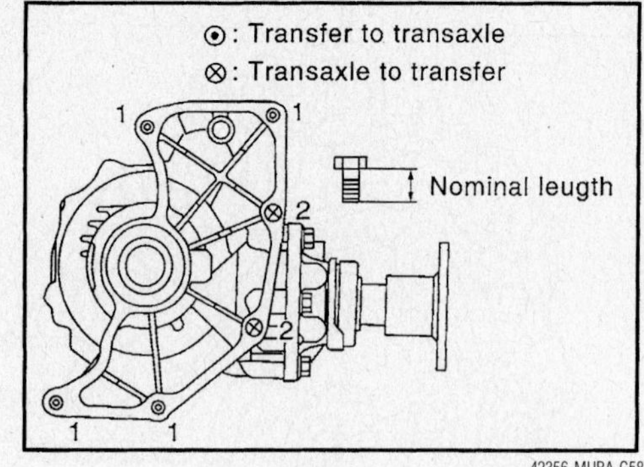

Transfer case bolt location and torque sequence

Front Halfshaft

REMOVAL & INSTALLATION

Left Side

1. Before servicing the vehicle, refer to the precautions in the beginning of this section.
2. Remove the wheel.
3. Remove wheel sensor from steering knuckle.
4. Remove cotter pin. Then remove lock nut from halfshaft.
5. Remove brake hose lock plate. Then remove brake hose from strut assembly.
6. Remove strut assembly and steering knuckle bolt and nut.
7. Using a puller, remove halfshaft from steering knuckle.
8. Remove halfshaft from transaxle.
9. Installation is the reverse of removal.

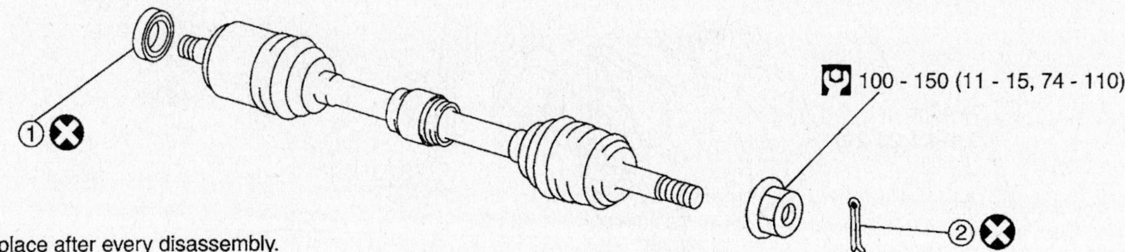

Left side front halfshaft

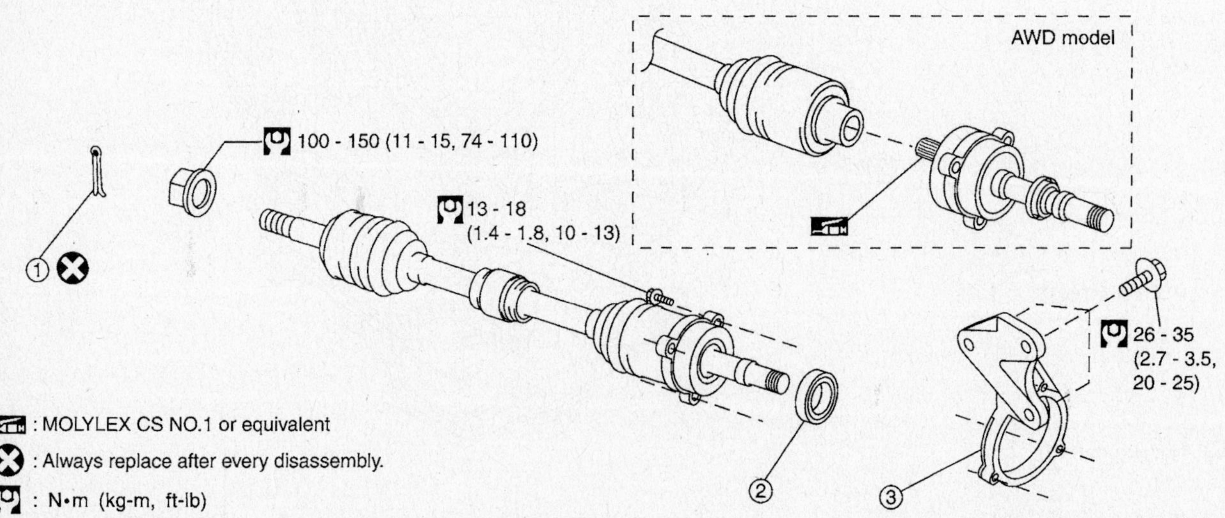

Right side front halfshaft

Right Side

2WD MODELS

1. Before servicing the vehicle, refer to the precautions in the beginning of this section.
2. Remove the wheel.
3. Remove wheel sensor from steering knuckle.
4. Remove cotter pin. Then remove lock nut from halfshaft.
5. Remove brake hose lock plate. Then remove brake hose from strut assembly.
6. Remove strut assembly and steering knuckle bolt and nut.
7. Using a puller, remove halfshaft from axle.
8. Remove support bearing bolts, and pull halfshaft from transaxle. Pry off halfshaft from transaxle.
9. Installation is the reverse of removal.

AWD MODELS

1. Before servicing the vehicle, refer to the precautions in the beginning of this section.
2. Remove the wheel.
3. Remove wheel sensor from steering knuckle.
4. Remove brake hose lock plate. Then remove brake hose from strut assembly.
5. Remove cotter pin. Then remove lock nut from halfshaft.
6. Remove strut assembly and steering knuckle bolt and nut.
7. Using a puller, remove halfshaft from knuckle.
8. Remove halfshaft from transaxle.
9. Installation is the reverse of removal.

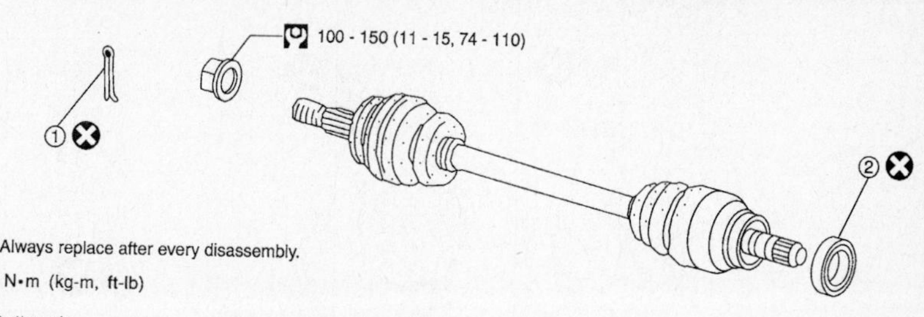

Rear halfshaft

Rear Halfshaft

REMOVAL & INSTALLATION

1. Remove the wheel.
2. Remove wheel sensor from axle.
3. Remove cotter pin. Then remove lock nut from halfshaft.
4. Remove parking cable and parking brake shoe from back plate.
5. Remove wheel hub and bearing assembly bolts, then remove wheel hub and bearing assembly from axle.
6. Use a wheel wrench or other tool to remove halfshaft from final drive.
7. Installation is the reverse of removal.

Front Halfshaft CV-Joints

OVERHAUL

Outer CV-Joint

1. Before servicing the vehicle, refer to the precautions in the beginning of this section.

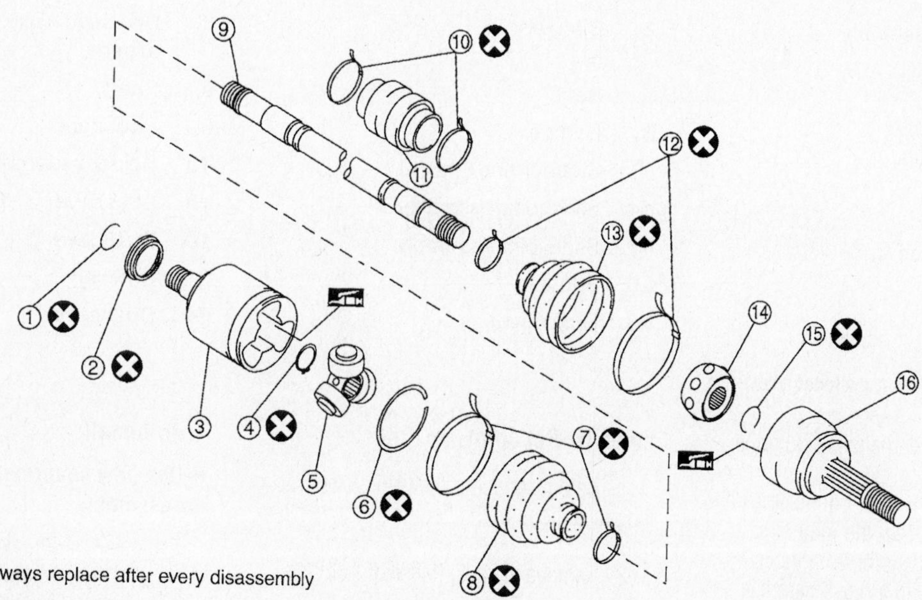

Left side front halfshaft exploded view

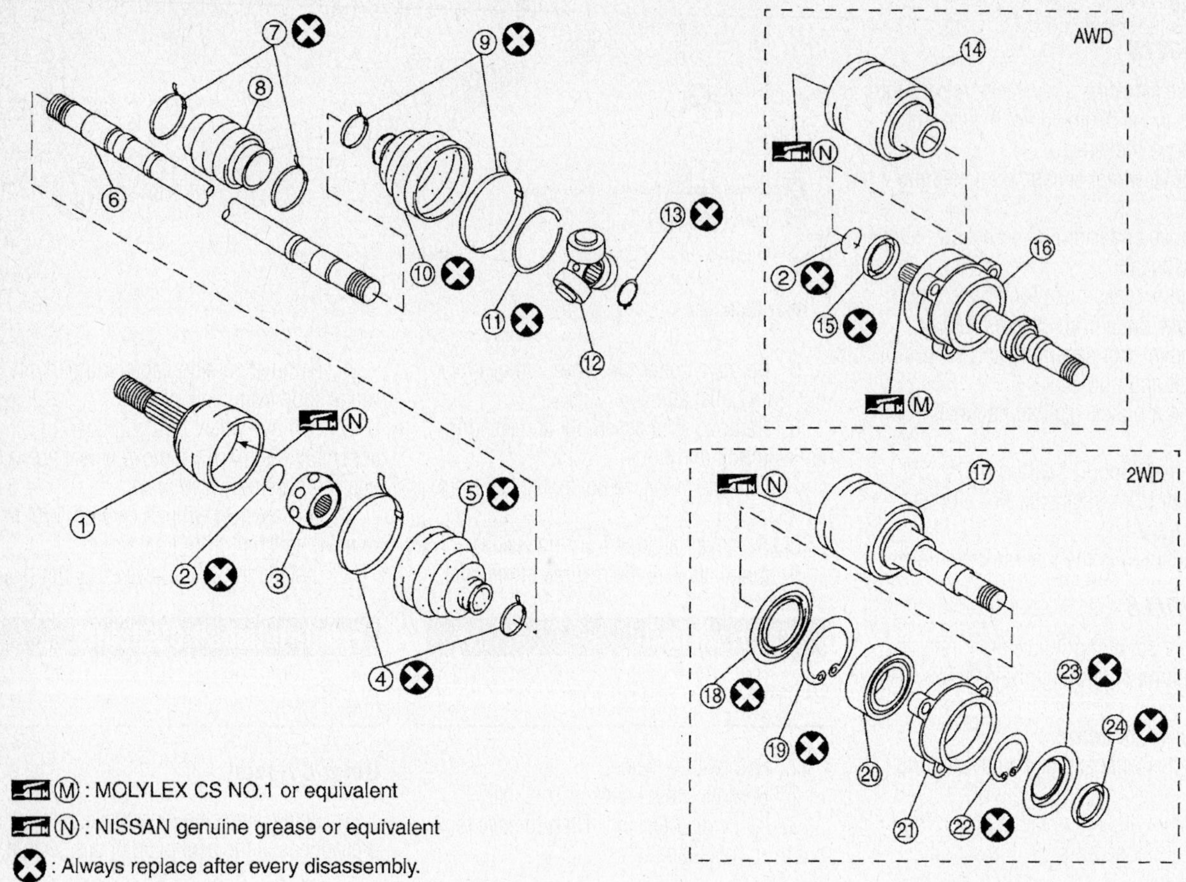

1. Joint sub-assembly
2. Circular clip
3. Ball cage / Steel ball / Inner race assembly
4. Boot band
5. Boot
6. Shaft
7. Damper band
8. Damper
9. Boot band
10. Boot
11. Stopper ring
12. Spider assembly
13. Circular clip
14. Slide joint assembly
15. Dust cover
16. Support bearing
17. Slide joint assembly
18. Dust cover
19. Snap ring
20. Bearing
21. Bracket
22. Snap ring
23. Dust cover
24. Dust cover

42356-MURA-G62

Right side front halfshaft exploded view

2. Remove the axle halfshaft from the vehicle.
3. Remove the CV-joint boot clamps and push the boot away from the joint.
4. Remove the CV-joint from the axle shaft by tapping it with a brass hammer.

To install:

➡Use new circlips and boot clamps for assembly.

5. Install the CV-joint to the axle shaft by tapping it with a brass hammer.
6. Pack the joint with grease.
7. Install the boot clamps.
8. Install the axle halfshaft to the vehicle.

Inner Tri-Pot Joint

1. Before servicing the vehicle, refer to the precautions in the beginning of this section.
2. Remove the axle halfshaft from the vehicle.
3. Remove the plug seal by tapping around the joint housing flange with a brass hammer.
4. Remove or disconnect the following:
   - CV-joint boot clamps
   - Snapring
   - Spider assembly
   - CV-joint housing
   - CV-joint boot

To install:

➡Use new snaprings and plug seals for assembly.

5. Install or connect the following:
   - CV-joint boot
   - CV-joint housing
   - Spider assembly
   - Snapring. Pack the joint with grease.
   - CV-joint boot clamps
   - Plug seal
6. Install the axle halfshaft to the vehicle.

Rear Halfshaft CV-Joints

DISASSEMBLY

Final Drive Side

1. Press shaft in a vise.
2. Remove boot band.

➡When retaining shaft in a vice, always use copper or aluminum plates between vise and shaft.

3. Put matching marks on slide joint assembly and shaft before separating slide joint assembly.
4. Put matching marks on spider assembly and shaft.
5. Remove snap ring, then remove spider assembly from shaft.
6. Remove boot from shaft.
7. Remove old grease on slide joint assembly with paper towels.
8. Remove circular clip and dust shield from slide joint assembly.

Wheel Side

1. Place shaft in a vise.

➡When retaining shaft in a vise, always use copper or aluminum plates between vise and shaft.

2. Remove boot bands. Then remove boot from joint sub-assembly.
3. Screw a halfshaft puller (suitable tool) 30 mm (1.18 in) or more into threaded part of joint sub-assembly. Pull joint sub-assembly out of shaft.

➡If joint sub-assembly cannot be removed after five or more unsuccessful attempts, replace the entire halfshaft assembly. Align sliding hammer and halfshaft and remove them by pulling directly.

4. Remove boot from shaft.
5. Remove circular clip from shaft.
6. While rotating ball cage, remove old grease on joint sub-assembly with paper towels.
7. Replace halfshaft if there is any runout, cracking, or other damage.

➡If there are any irregular conditions of joint sub-assembly components, replace the entire joint sub-assembly.

ASSEMBLY

Final Drive Side

1. Install new boot and new small boot band on shaft.

➡Cover shaft serration with tape to prevent damage to boot during installation.

2. Remove protective tape wound around serrated part of shaft.
3. Install spider assembly securely, making sure the matching marks which were made during disassembly are properly aligned.
4. Install new snap ring.
5. Insert the amount of new grease listed below into housing from large end of boot. Grease amount: 85–95 g (3.0–3.35 oz)
6. Install slide joint assembly.
7. Install boot securely into grooves (indicated by * marks) shown in the illustration.

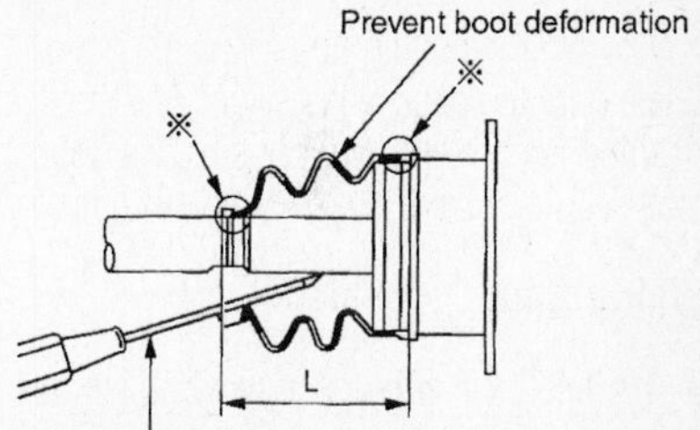

Checking boot length

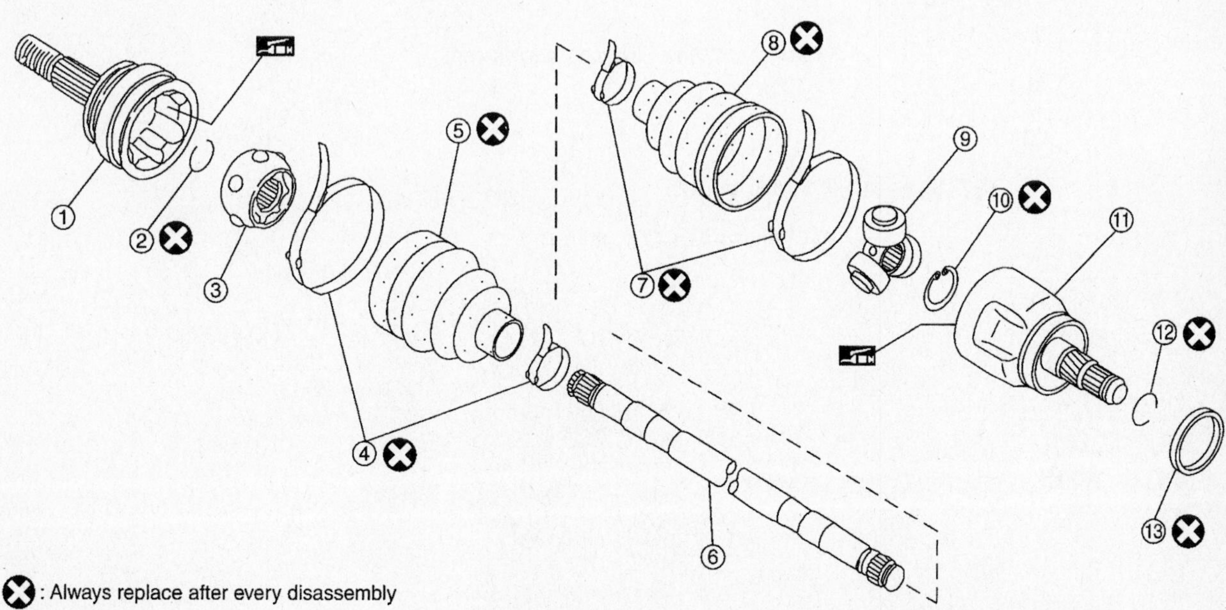

1. Joint sub-assembly	2. Circular clip	3. Ball cage/Steel ball/Inner race assembly
4. Boot band (Wheel side)	5. Boot (Wheel side)	6. Shaft
7. Boot band (Final drive side)	8. Boot (Final drive side)	9. Spider assembly
10. Snap ring	11. Slide joint assembly	12. Circular clip
13. Dust shield		

42356-MURA-G64

Rear halfshaft exploded view

➡If there is grease on boot mounting surfaces (indicated by * marks) of shaft and housing, boot may come off. Remove all grease from surfaces.

8. Make sure boot installation length "L" is the length indicated below. Insert a flat-bladed screwdriver or similar tool into smaller side of boot. Bleed air from boot to prevent boot deformation. Boot installation length "L ": 79.6 mm (3.13 in)

➡Boot may break if boot installation length is less than standard value. Take care not to touch the tip of screwdriver to inside of boot.

9. Secure big and small ends of boot with new boot bands as shown in the illustration.

➡Discard old boot bands; replace with new ones.

10. After installing housing and shaft, rotate boot to check whether or not the actual position is correct. If boot position is not correct, secure boot with new boot band again.

Wheel Side

1. Insert the amount of new grease into joint sub-assembly serration hole until grease begins to ooze from ball groove and serration hole. After inserting grease, use a shop cloth to wipe off old grease that has oozed out.
2. Wind serrated part of shaft with tape. Install new boot band and boot to shaft. Be careful not to damage boot.

➡Discard old boot band and boot; replace with new ones.

3. Remove protective tape wound around serrated part of shaft.
4. Install new circular clip to shaft. At this time, circular clip must fit securely into shaft groove. Attach nut to joint sub-assembly. Use a wooden hammer to press-fit.

➡Discard old circular clip; replace with new one.

5. Insert the amount of new grease listed below into housing from large end of boot. Grease amount: 75–85 g (2.65–3.0 oz)
6. Install boot securely into grooves (indicated by * marks) shown in the illustration.

➡If there is grease on boot mounting surfaces (indicated by * marks) of shaft and housing, boot may come off. Remove all grease from surfaces.

7. Make sure boot installation length "L" is the length indicated below. Insert a flat-bladed screwdriver or similar tool into smaller side of boot. Bleed air from boot to prevent boot deformation. Boot installation length "L": 67.7mm (2.67 in)

➡Boot may break if boot installation length is less than standard value. Be careful that screwdriver tip does not contact inside surface of boot.

8. Secure big and small ends of boot with new boot bands as shown in the illustration.

➡Discard old boot bands; replace with new ones.

9. Check installation status of boot. Rotate joint to make sure boot is securely in place. If not, reinstall using a new boot band.

Rear Final Drive

REMOVAL & INSTALLATION

1. Before servicing the vehicle, refer to the precautions in the beginning of this section.

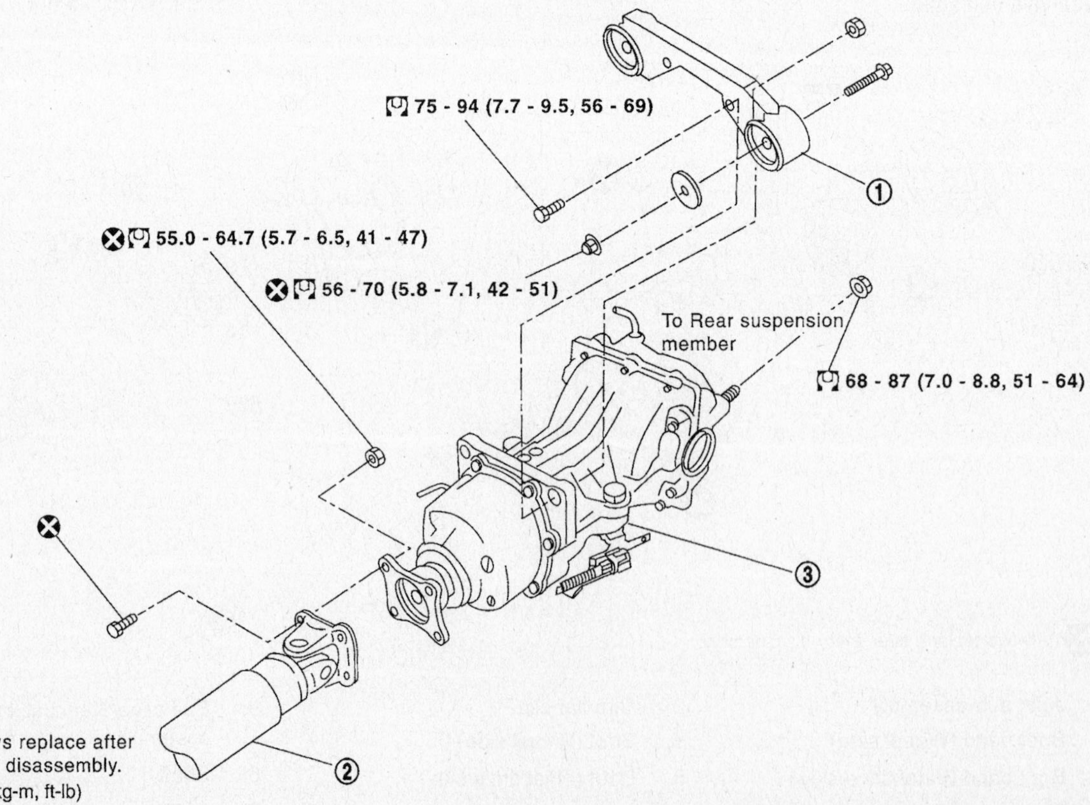

Rear final drive

2. Remove rear propeller shaft.
3. Remove rear stabilizer mounting bracket.
4. Remove wheel sensor.
5. Remove rear halfshaft.
6. Remove electric controlled coupling connector.
7. Remove electric controlled coupling breather hose and rear final drive breather hose.
8. Remove canister.
9. Set Transmission Jack to rear final drive assembly, and then remove nuts from rear suspension member.

➡Do not place a transmission jack on the rear cover (aluminum case).

10. Remove bolt and nut from final drive mount bracket, and then remove rear final drive assembly from vehicle.
11. Installation is the reverse of removal. Supporting rear final drive assembly securely with transmission jack, install it to final drive mount bracket and rear suspension member with bolt and nut.

Rear Final Drive Seals

REMOVAL & INSTALLATION

Pinion Seal

1. Remove the propeller shaft. Refer
2. Put a mark on the end of the drive pinion corresponding to the position mark on the final drive companion flange.
3. Using the drive pinion flange wrench, Remove companion flange nut.
4. Using the puller, remove the companion flange.
5. Using the side bearing outer race puller, remove front oil seal.

To install:

6. Apply multi-purpose grease to sealing lips of oil seal. Press front oil seal into carrier with tool.
7. Align the matching mark of drive pinion with the matching mark of companion flange, then install the companion flange.
8. Apply oil or grease on the screw part of drive pinion and the seating surface of companion flange nut.
9. Install companion flange nut with tool.

➡Never reuse companion flange nut.

10. Install propeller shaft.

Side Seal

1. Remove rear wheel sensor.
2. Remove rear axle assembly.
3. Remove rear halfshaft.
4. Using flat tip screwdriver, remove side oil seal.

To install:

5. Apply multi-purpose grease to sealing lips of oil seal.
6. Using the drift, press-fit oil seal so that its surface comes face to face with the end surface of the case.
7. Install rear halfshaft.
8. Install rear axle assembly.

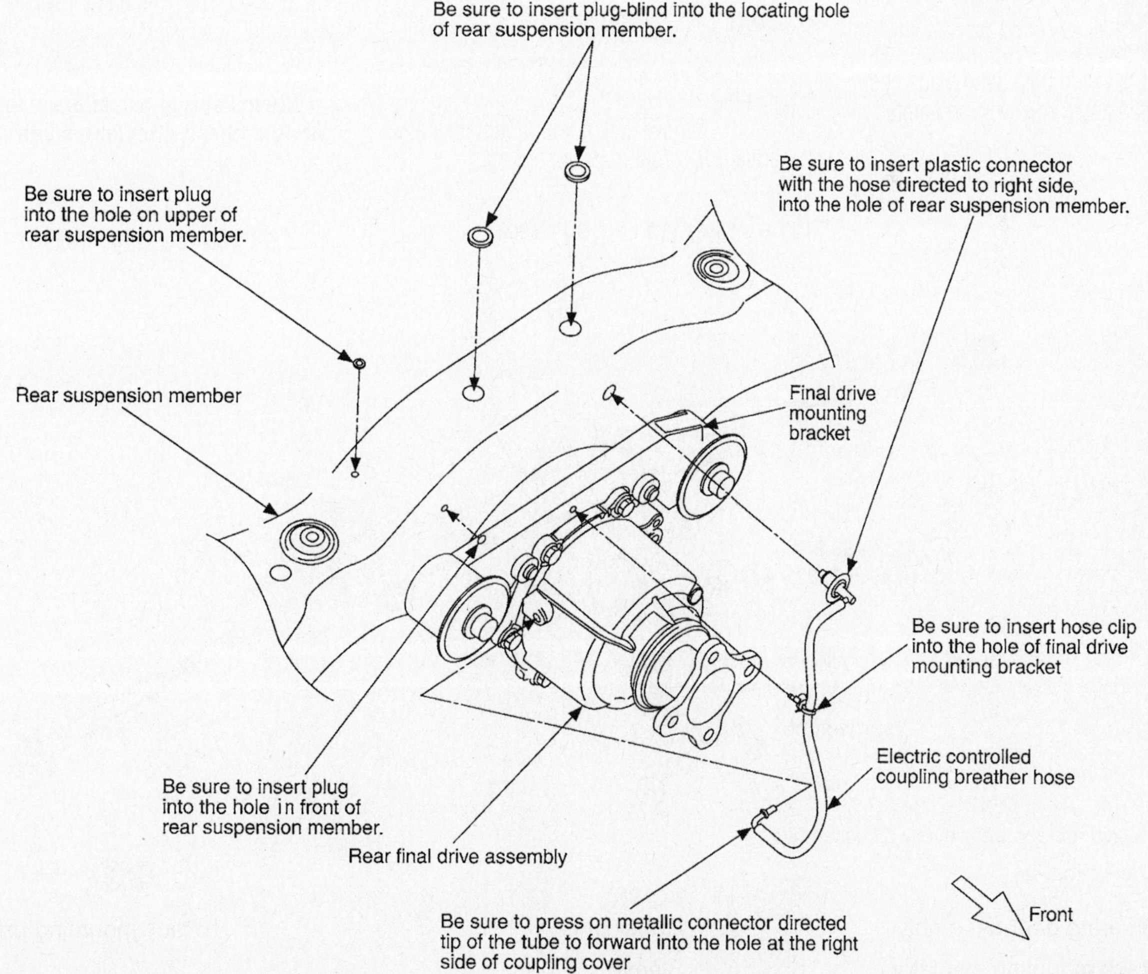
Be sure to insert plug-blind into the locating hole of rear suspension member.
Be sure to insert plug into the hole on upper of rear suspension member.
Be sure to insert plastic connector with the hose directed to right side, into the hole of rear suspension member.
Rear suspension member
Final drive mounting bracket
Be sure to insert hose clip into the hole of final drive mounting bracket
Electric controlled coupling breather hose
Be sure to insert plug into the hole in front of rear suspension member.
Rear final drive assembly
Be sure to press on metallic connector directed tip of the tube to forward into the hole at the right side of coupling cover
Front
42356-MURA-G67

Electronically controlled coupling breather hose

STEERING AND SUSPENSION

Air Bag

✻✻ CAUTION

Some vehicles are equipped with an air bag system. The system must be disarmed before performing service on, or around, system components, the steering column, instrument panel components, wiring and sensors. Failure to follow the safety precautions and the disarming procedure could result in accidental air bag deployment, possible injury and unnecessary system repairs.

PRECAUTIONS

Several precautions must be observed when handling the inflator module to avoid accidental deployment and possible personal injury.

- Never carry the inflator module by the wires or connector on the underside of the module.
- When carrying a live inflator module, hold securely with both hands, and ensure that the bag and trim cover are pointed away.
- Place the inflator module on a bench or other surface with the bag and trim cover facing up.
- With the inflator module on the bench, never place anything on or close to the module which may be thrown in the event of an accidental deployment.

DISARMING

To disarm the **SRS** system turn the ignition switch to the **OFF** position. Then, disconnect both battery cables starting with the negative cable first and wait at least 3 minutes after the cables are disconnected.

To rearm the **SRS** system, turn the ignition switch to the **OFF** position. Connect both battery cables starting with the positive cable first.

Rack and Pinion Steering Gear

REMOVAL & INSTALLATION

2WD

1. Set wheels in the straight ahead position.
2. Remove lock nut and bolt, then separate lower joint from upper joint.
3. Remove wheels.
4. Confirm that the slit on lower joint fits with the projection on rear cover cap, also the matchmark on steering gear assembly fits with the projection on rear cover cap.
5. Remove cotter pin at steering knuckle, then loosen mounting nut.
6. Use a ball joint remover to remove steering outer socket from steering knuckle. Be careful not to damage ball joint boot.

➡To prevent damage to threads and to prevent ball joint remover from coming off, temporarily tighten mounting nut.

7. Remove oil pipes (high pressure side and low pressure side) from steering gear assembly, then drain fluid from pipes.
8. Remove mounting bolt (lower side) from lower joint.
9. Remove mounting bolts and nut from steering gear assembly, and then remove steering gear assembly, rack mounting bracket, rack mounting insulator and sleeve from vehicle.
10. Installation is the reverse of removal.

➡When steering wheel is set in the straight ahead direction, confirm slit of

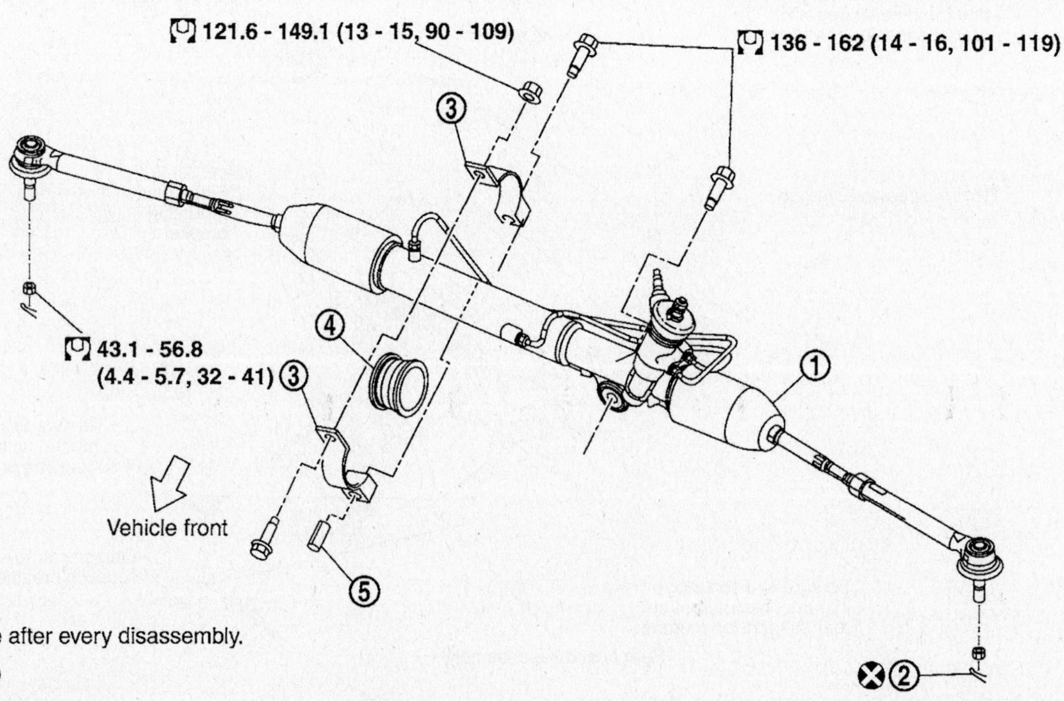

1. Steering gear assembly
2. Cotter pin
3. Rack mounting bracket
4. Rack mounting insulator
5. Sleeve

Steering gear and linkage

lower joint fits with the projection on rear cover cap, also the matchmarks on steering gear assembly fit with the projection on rear cover cap.

11. After installation, bleed air from piping.
12. Check if steering wheel turns smoothly when it is turned several times fully to the end of the left and right.

AWD

1. Set wheels in the straight-ahead position.
2. Remove lock nut and bolt, then separate lower joint from upper joint.
3. Remove tires from vehicle.
4. Remove undercover from vehicle.
5. Confirm slit of lower joint fits with the projection on rear cover cap, furthermore marking position on steering gear assembly nearly fits with the projection on rear cover cap.
6. Remove oil pipes (high pressure side and low pressure side) from steering gear assembly, then drain fluid from pipes.
7. Remove cotter pin at steering knuckle, then loosen mounting nut.
8. Use a ball joint remover to remove steering outer socket from steering knuckle. Be careful not to damage ball joint boot.

➡To prevent damage to threads and to prevent ball joint remover from coming off, and temporarily tighten mounting nut.

9. Remove mounting bolt (lower side) from lower joint.
10. Remove front exhaust tube.
11. Remove rear propeller shaft.
12. Remove mounting nuts on lower position from stabilizer connecting rod.
13. Remove mounting bolts from stabilizer clamp and hang stabilizer on vehicle.
14. Remove steering hydraulic piping bracket from front suspension member.
15. Disconnect electrical rear engine mounting actuator harness connector.
16. Set jack under engine and front suspension member.
17. Remove mounting bolts from rear engine mounting insulator.
18. Loosen mounting nuts of front suspension member (front side).
19. Remove mounting bolts from member stay (body side), then loosen mounting nuts of member stay (front suspension member side).
20. Move jack down slowly (front suspension member side) to remove rear engine mounting insulator from engine and front suspension member.
21. Remove mounting bolts and nut from steering gear assembly, and then remove steering gear assembly, rack mounting bracket, rack mounting insulator and sleeve from vehicle.
22. Installation is the reverse of removal.
23. When steering wheel is set in the straight ahead direction, confirm slit of lower joint fits with the projection on rear cover cap, also the matchmarks on steering gear assembly nearly fits with the projection on rear cover cap.
24. After installation, bleed air from piping.
25. Check if steering wheel turns smoothly when it is turned several times fully to the end of the left and right.

Strut

REMOVAL & INSTALLATION

1. Before servicing the vehicle, refer to the precautions in the beginning of this section.
2. Remove wheel.
3. Remove cowl top grille.
4. Remove brake caliper. Hang it in a place where it will not interfere with work.
5. Remove lock plate from brake hose from strut assembly.
6. Remove harness from wheel sensor from strut assembly. Do not pull on wheel sensor harness.
7. Remove mounting nut between strut assembly and connecting rod.
8. Remove mounting bolt and nut between strut assembly and steering knuckle.
9. Remove mounting nuts on mounting insulator bracket, then remove strut assembly from vehicle.
10. Installation is the reverse of removal. Perform final tightening of strut assembly lower side (rubber bushing) under unladen conditions with tires on level ground. Check wheel alignment.

Shock Absorber

REMOVAL & INSTALLATION

1. Before servicing the vehicle, refer to the precautions in the beginning of this section.
2. Remove wheel.
3. Remove bolt in lower side of shock absorber assembly.
4. Remove mounting seal bracket nuts from shock absorber upper side and remove shock absorber assembly from vehicle.
5. Installation is the reverse of removal. Perform final tightening of shock absorber lower side (rubber bushing) under unladen conditions with tires on level ground.

Front Coil Spring

REMOVAL & INSTALLATION

1. Before servicing the vehicle, refer to the precautions in the beginning of this section.
2. Remove the strut assembly.
3. Compress the coil spring and remove the piston rod nut.
4. Remove or disconnect the following:
   - Upper strut mount
   - Strut mount bracket
   - Upper strut bearing
   - Spring upper seat
   - Coil spring

To install:

➡Use new fasteners for assembly.

5. Install or connect the following:
   - Coil spring
   - Spring upper seat
   - Upper strut bearing
   - Strut mount bracket
   - Upper strut mount. Tighten the piston rod nut to 43–58 ft. lbs. (59–78 Nm).
6. Remove the spring compressor and install the strut assembly to the vehicle.
7. Check the wheel alignment and adjust, as necessary.

Rear Coil Spring and Lower Member

REMOVAL & INSTALLATION

1. Before servicing the vehicle, refer to the precautions in the beginning of this section.
2. Remove wheel.
3. Set jack under rear lower link.
4. Loosen bolt and nut between rear lower link and suspension member, and then remove bolt and nut between rear axle and rear lower link.
5. Slowly lower jack, then remove upper seat, coil spring and rubber seat from rear lower link.
6. Remove bolt and nut between rear suspension member and rear lower link.
7. Installation is the reverse of removal.

➡Insert bracket tabs (3) and the inside protrusion on upper seat into each

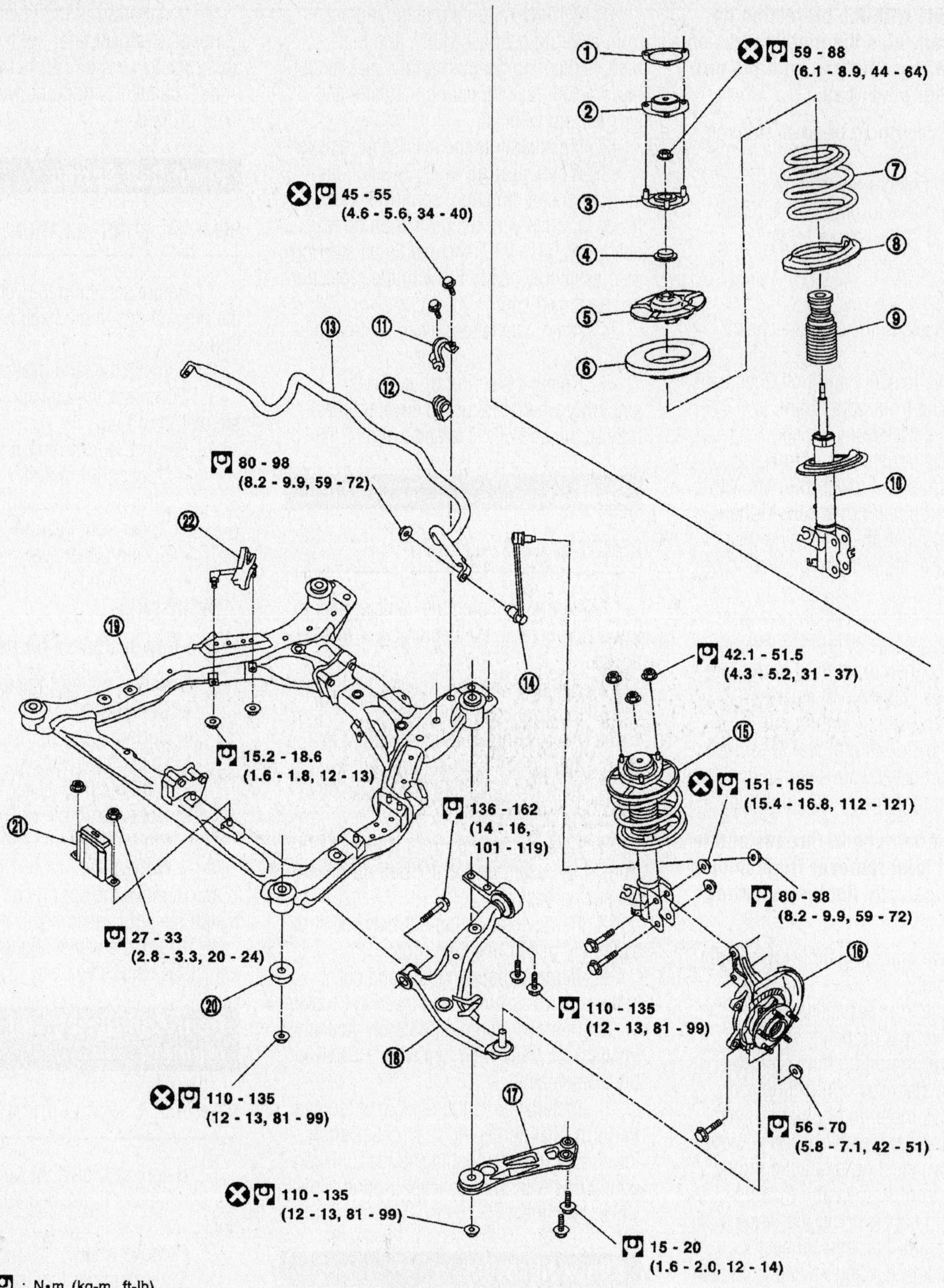

1. Upper mounting plate	2. Mounting insulator	3. Mounting insulator bracket
4. Mounting bearing	5. Spring upper seat	6. Spring upper rubber seat
7. Coil spring	8. Spring lower rubber seat	9. Bound bumper
10. Strut	11. Stabilizer clamp	12. Stabilizer bushing
13. Stabilizer	14. Connecting rod	15. Strut assembly
16. Front axle	17. Member stay	18. Transverse link
19. Front suspension member	20. Rebound stopper	21. Damper assembly
22. Air guide		

42356-MURA-G69

Front suspension

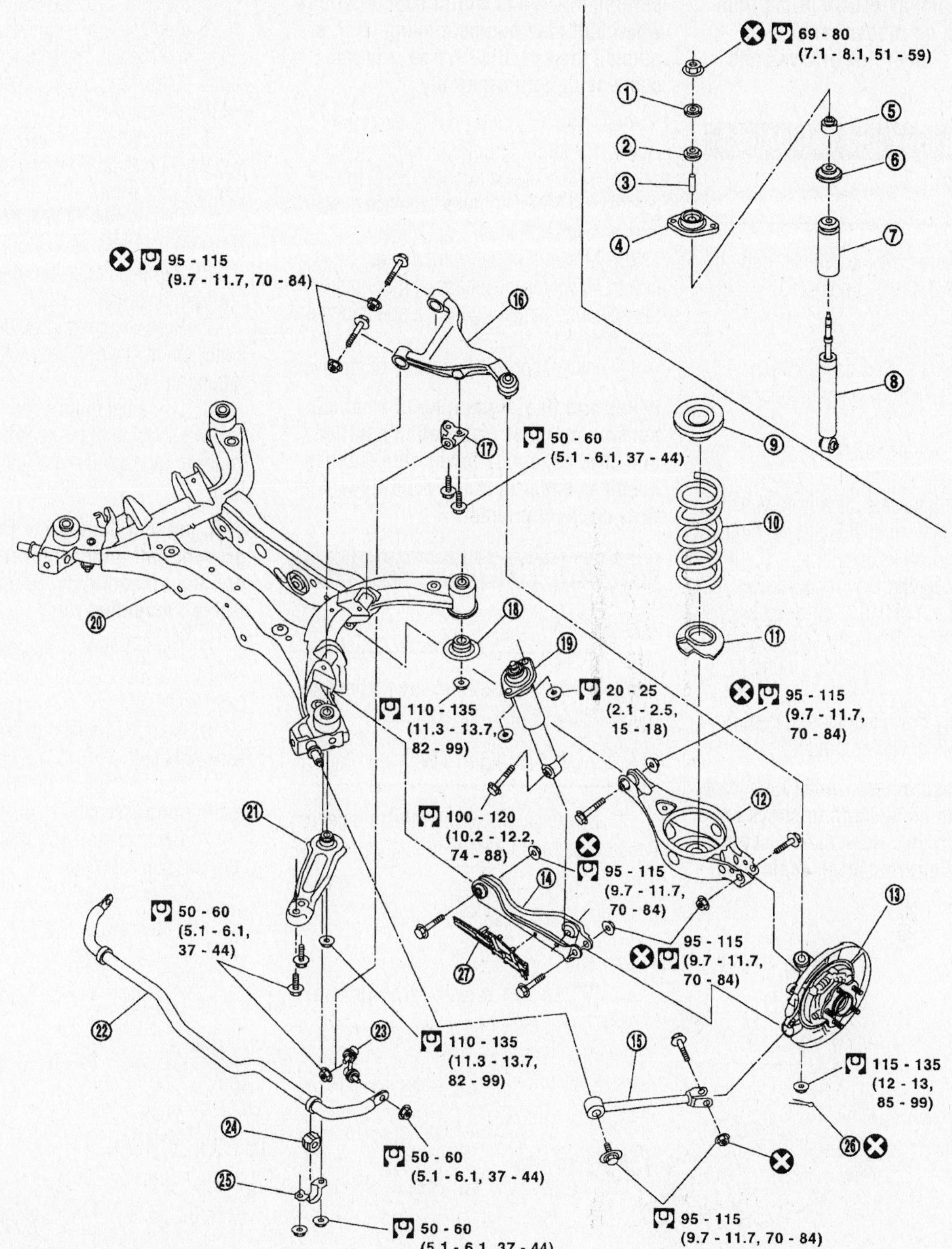

1. Outer washer	2. Bushing A	3. Distance tube
4. Mounting seal bracket	5. Bushing B	6. Bound bumper cover
7. Bound bumper	8. Shock absorber	9. Upper seat
10. Coil spring	11. Rubber seat	12. Rear lower link
13. Axle	14. Front lower link	15. Radius rod
16. Suspension arm	17. Stabilizer connecting rod mount bracket	18. Rebound stopper
19. Shock absorber assembly	20. Rear suspension member	21. Member stay
22. Stabilizer bar	23. Stabilizer connecting rod	24. Stabilizer bushing
25. Stabilizer clamp	26. Cotter pin	27. Front lower link protector

42356-MURA-G70

Rear suspension

other beforehand as shown in the illustration. Match up rubber seat indentions and rear lower link grooves and attach.

Transverse Link and Ball Joint

REMOVAL & INSTALLATION

1. Before servicing the vehicle, refer to the precautions in the beginning of this section.
2. Remove wheel.
3. Remove mounting bolt between transverse link and front suspension member.
4. Remove transverse link from steering knuckle.
5. Remove transverse link from vehicle.
6. Check transverse link and bushing for deformation, cracks, or damage. If any non-standard condition is found, replace it.
7. Check boot of ball joint for cracks or other damage, and also for grease leakage. If any non-standard condition is found, replace it.
8. Manually move ball stud to confirm it moves smoothly with no binding.

➡Before measurement, move ball joint at least ten times by hand to check for smooth movement. Hook spring scale at ball stud. Confirm spring scale measurement value is within specifications when ball stud begins moving. If it is outside the specified range, replace suspension arm assembly.

9. Attach mounting nut to ball stud. Check that sliding torque is within specifications with a preload gauge (SST). If it is outside the specified range, replace suspension arm assembly.
10. Move tip of ball joint in axial direction to check for looseness. If it is outside the specified range, replace suspension arm assembly.
11. Installation is the reverse of removal.

➡Perform final tightening of front suspension member installation position and strut assembly lower side (rubber bushing) under unladen conditions with tires on level ground.

Front Wheel Bearings

ADJUSTMENT

The front wheel bearings are part of a unitized hub and are not adjustable.

REMOVAL & INSTALLATION

1. Before servicing the vehicle, refer to the precautions in the beginning of this section.
2. Remove tire from vehicle.
3. Remove brake caliper. Hang it in a place where it will not interfere with work. Avoid depressing brake pedal while brake caliper is removed.
4. Put alignment marks on disc rotor and wheel hub and bearing assembly, then remove disc rotor.
5. Remove wheel sensor from steering knuckle.
6. Remove cotter pin, then remove lock nut from halfshaft.
7. Remove steering outer socket and cotter pin at steering knuckle, then loosen mounting nut.
8. Use a ball joint remover (SST) to remove steering outer socket from steering knuckle. Be careful not to damage ball joint boot.

➡To prevent damage to threads and to prevent ball joint remover (SST) from coming off suddenly, temporarily tighten mounting nut.

9. Using a puller (suitable tool), remove wheel hub and bearing assembly from halfshaft.
10. Remove wheel hub and bearing assembly bolt.
11. Remove splash guard and wheel hub and bearing assembly from steering knuckle.
12. Remove strut assembly and steering knuckle bolts and nuts.

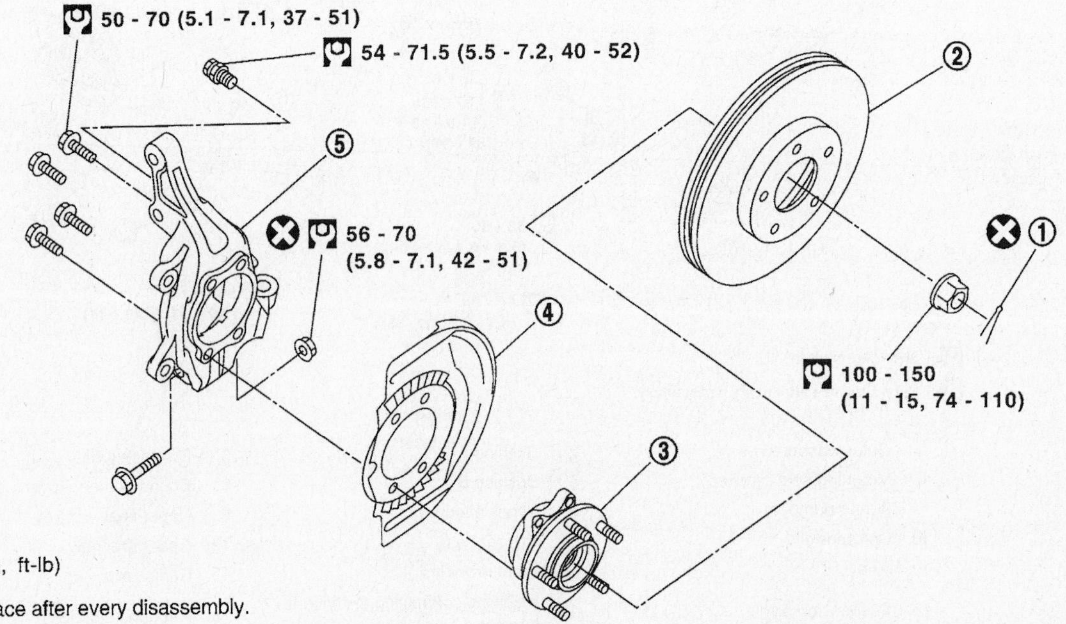

Front wheel hub/bearing assembly

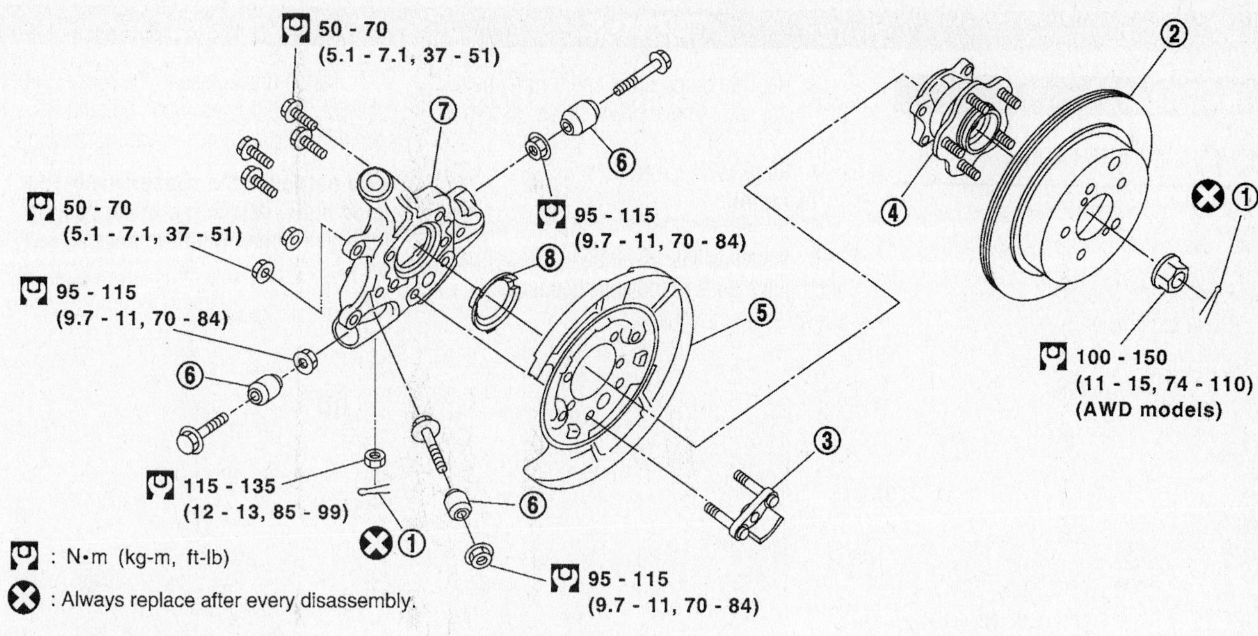

Rear hub/bearing assembly

13. Remove transverse link and steering knuckle bolt and nut.
14. Remove steering knuckle from vehicle.
15. Check for deformity, cracks and damage on each parts, replace if necessary.
16. Check for boot breakage, axial looseness, and torque of transverse link ball joint.
17. Installation is the reverse of removal.

Rear Wheel Bearings

ADJUSTMENT

The front wheel bearings are part of a unitized hub and are not adjustable.

REMOVAL & INSTALLATION

1. Before servicing the vehicle, refer to the precautions in the beginning of this section.
2. Remove the wheel.
3. Remove brake caliper. Hang it in a place where it will not interfere with work.
4. Put alignment marks on disc rotor and wheel hub and bearing assembly, then remove disc rotor.
5. Remove wheel sensor from axle.
6. Remove parking cable and parking brake shoe from back plate.
7. Remove cotter pin. Then remove lock nut from halfshaft. (AWD models)
8. Using a puller (suitable tool), remove wheel hub and bearing assembly from halfshaft. (AWD models)
9. Remove wheel hub and bearing assembly from axle.
10. Loosen bolts and nuts of front lower link, radius rod and rear lower link in side of suspension member.
11. Remove shock absorber bolt (lower), front lower link bolt and nut (axle-side) while supporting rear lower link with jack.
12. Remove bolt and nut in axle side of rear lower link. Then remove coil spring.
13. Remove bolt and nut in axle side of radius rod.

Remove suspension arm and cotter pin at axle, then loosen mounting nut.

14. Use a ball joint remover (suitable tool) to remove suspension arm from axle. Be careful not to damage ball joint boot.

➡To prevent damage to threads and to prevent ball joint remover (suitable tool) from coming off suddenly, and temporarily tighten mounting nut.

15. Remove axle from vehicle.
16. Remove nuts from anchor block, then remove anchor block and back plate from axle.
17. Remove axle cap (2WD) or dust shield (AWD) from axle.
18. Check for deformity, cracks and damage on each parts, replace if necessary.
19. Check for boot breakage, axial looseness, and torque of suspension arm ball joint.
20. Installation is the reverse of removal.
21. Perform final tightening of installation position of suspension links (rubber bushing) under unladen conditions with tires on level ground.

BRAKES

Brake Caliper

REMOVAL AND INSTALLATION

Front

1. Remove the wheels.
2. Drain brake fluid.
3. Remove union bolts and torque member bolts, and remove brake caliper assembly.
4. Remove disc rotor.

To install:

5. Install disc rotor.
6. Install caliper assembly to the vehicle, and tighten bolts to the specified torque shown in the illustration.
7. Install brake hose to the brake caliper assembly, and tighten union bolts to the specified torque shown in the illustration.

➡Do not reuse the copper washer for union bolts. Attach the brake hose to the brake hose mounting boss.

8. Refill new brake fluid and bleed air.
9. Install the wheels.

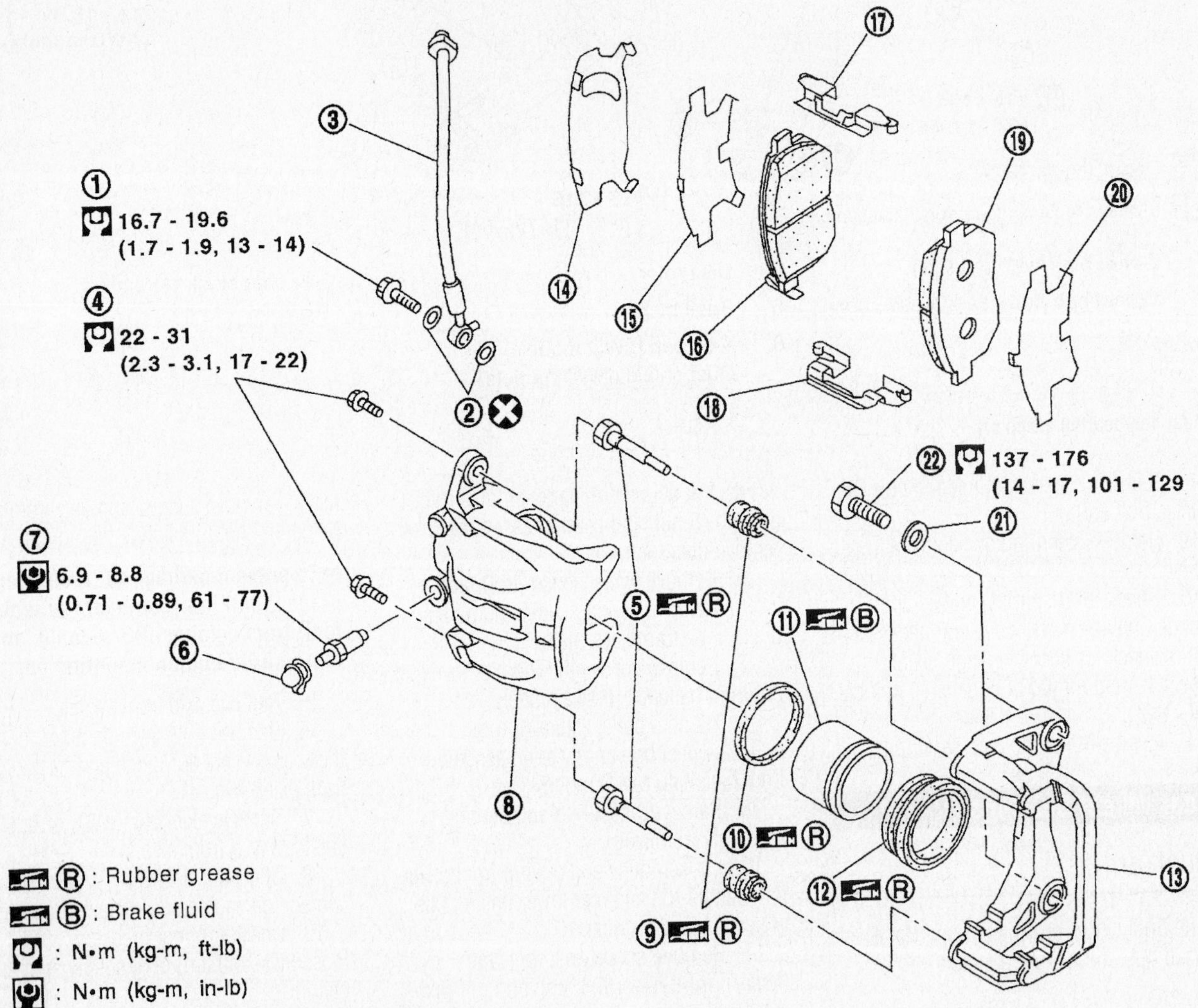

1. Union bolt
2. Copper washer
3. Brake hose
4. Sliding pin bolt
5. Sliding pin
6. Cap
7. Bleed valve
8. Cylinder body
9. Sliding pin boot
10. Piston seal
11. Piston
12. Piston boot
13. Torque member
14. Inner shim cover
15. Inner shim
16. Inner pad
17. Pad retainer (Upper)
18. Pad retainer (Lower)
19. Outer pad
20. Outer shim
21. Washer
22. Torque member fixing bolt

42356-MURA-G73

Front disc brake assembly

Rear

1. Remove wheels from vehicle.
2. Drain brake fluid.
3. Remove union bolts and torque member bolts, and remove brake caliper assembly.

To install:

4. Install disc rotor.
5. Install caliper assembly to the vehicle, and tighten bolts to the specified torque shown in the illustration.
6. Install brake hose to caliper assembly and tighten union bolts to the specified torque shown in the illustration.

➡Do not reuse the copper washer for union bolts. Attach brake hose to the brake hose mounting boss.

7. Refill new brake fluid and bleed air. Refer
8. Install the tires to the vehicle.

Disc Brake Pads

REMOVAL AND INSTALLATION

Front

1. Remove the wheels.
2. Remove sliding pin bolt (top).
3. Suspend cylinder body with a wire, and remove pads, pad retainers, shims from torque member.

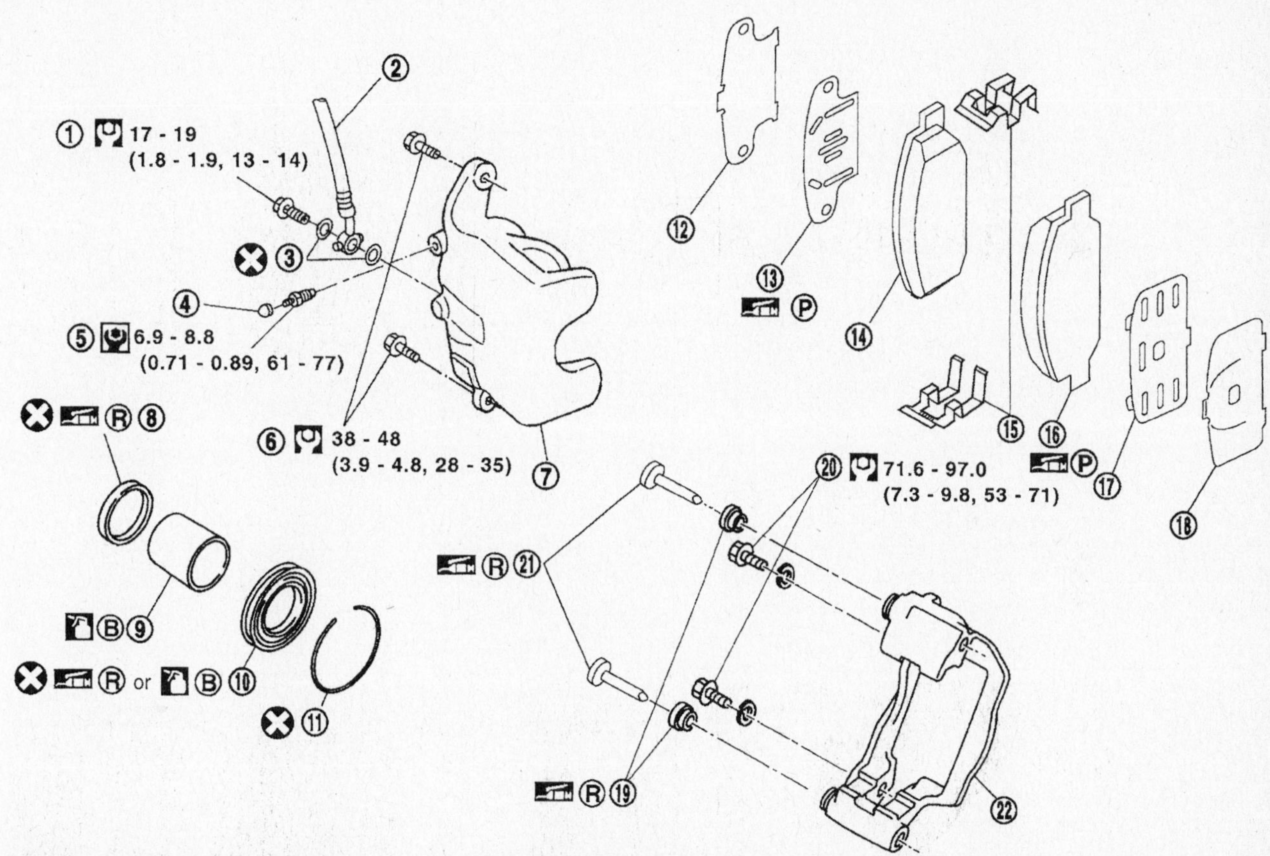

1.	Union bolt	2.	Brake hose	3.	Copper washer
4.	Cap	5.	Bleed valve	6.	Sliding pin bolt
7.	Cylinder body	8.	Piston seal	9.	Piston
10.	Piston boot	11.	Retaining ring	12.	Inner shim cover
13.	Inner shim	14.	Inner pad	15.	Pad retainer
16.	Outer pad	17.	Outer shim	18.	Outer shim cover
19.	Slide pin boot	20.	Torque member fixing bolt	21.	Sliding pin
22.	Torque member				

42356-MURA-G74

Rear disc brake assembly

To install:

4. Apply PBC (Poly Butyl Cuprysil) grease or silicon-based grease to the rear of the pad and to both sides of the shim, and attach the inner shim and shim cover to the inner pad, and the outer shim and outer shim cover to the outer pad.
5. Attach the pad retainer and pad to the torque member.
6. Push the piston in so that the pad is attached and attach the cylinder body to the torque member.
7. Install the sliding pin bolt (top) and tighten to the specified torque.
8. Check brake for drag.
9. Install the tires to the vehicle.

Rear

1. Remove tires from vehicle.
2. Remove sliding pin bolt (top).
3. Suspend cylinder body with a wire, and remove pads, pad retainers, shims from torque member.

To install:

4. Apply PBC (Poly Butyl Cuprysil) grease or silicon based grease to the rear of the pad and to both sides of the shim, and attach the inner shim and shim cover to the inner pad, and the outer shim and outer shim cover to the outer pad.
5. Attach the pad retainer and pad to the torque member.
6. Push the piston in so that the pad is attached and attach the cylinder body to the torque member.
7. Install the sliding pin bolt (one on top) and tighten to the specified torque.
8. Check brake for drag.
9. Install the tires to the vehicle.

NISSAN

Pathfinder

9

SPECIFICATIONS AND MAINTENANCE CHARTS

ENGINE AND VEHICLE IDENTIFICATION

Engine						
Code ①	Liters (cc)	Cu. In.	Cyl.	Fuel Sys.	Engine	Eng. Mfg.
VG33E	3.3 (3277)	199.8	6	MFI	SOHC	Nissan
VQ35DE	3.5 (3498)	213	6	MFI	DOHC	Nissan

Model Year	
Code ②	Year
1	2001
2	2002
3	2003
4	2004
5	2005

MFI: Multi-port Fuel Injection

SOHC: Single Overhead Camshaft

DOHC: Double Overhead Camshafts

① Located on the timing belt cover

② 10th digit of the Vehicle Identification Number (VIN)

67170-PATH-C01

GENERAL ENGINE SPECIFICATIONS

Year	Model	Engine Displacement Liters	Engine ID	Net Horsepower @ rpm	Net Torque @ rpm (ft. lbs.)	Bore x Stroke (in.)	Com-pression Ratio	Oil Pressure @ rpm
2001	Pathfinder	3.3	VG33E	170@4800	200@2800	3.60X3.27	8.9:1	60-65@2000
2002	Pathfinder	3.5	VQ35DE	240@6000	265@3200	3.76X3.20	10.0:1	43@2000
2003	Pathfinder	3.5	VQ35DE	240@6000	265@3200	3.76X3.20	10.0:1	43@2000
2004	Pathfinder	3.5	VQ35DE	240@6000	265@3200	3.76X3.20	10.0:1	43@2000

67170-PATH-C02

ENGINE TUNE-UP SPECIFICATIONS

Year	Engine Displacement Liters	Engine ID	Spark Plug Gap (in.)	Ignition Timing (deg.)		Fuel Pump (psi) ①	Idle Speed (rpm)		Valve Clearance (in.)	
				MT	AT		MT	AT ②	In.	Ex.
2001	3.3	VG33E	0.039-0.043	13-17B	13-17B	34	700-800	700-800	HYD	HYD
2002	3.5	VQ35DE	0.044	15B	15B	35	700-800	700-800	HYD	HYD
2003	3.5	VQ35DE	0.044	15B	15B	35	700-800	700-800	HYD	HYD
2004	3.5	VQ35DE	0.044	15B	15B	35	700-800	700-800	HYD	HYD

NOTE: The Vehicle Emission Control Information label often reflects specification changes made during production. The label figures must be used if they differ from those in this chart.

B: Before top dead center

HYD: Hydraulic

① System pressure at idle with vacuum hose connected
Should increase to 43 psi when disconnected

② Automatic transmission in Neutral

67170-PATH-C03

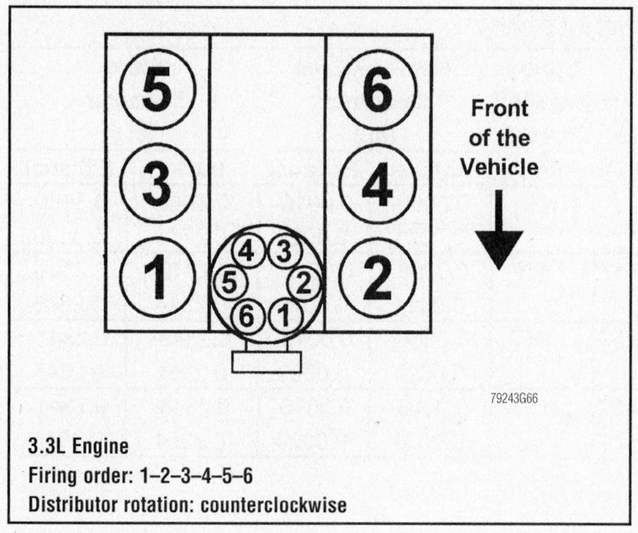

3.3L Engine
Firing order: 1–2–3–4–5–6
Distributor rotation: counterclockwise

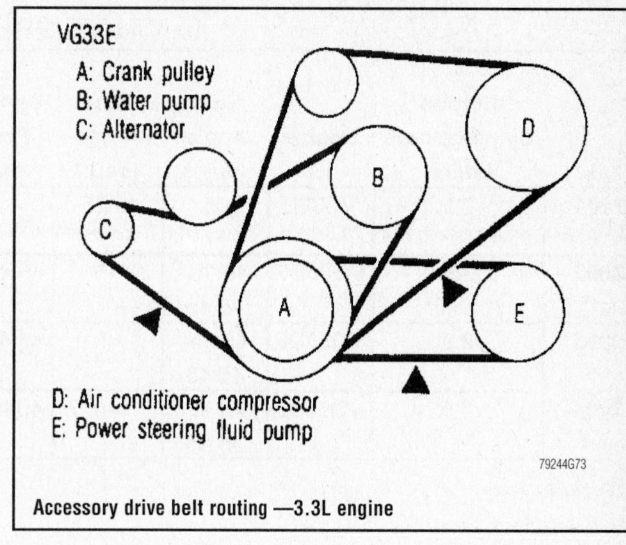

Accessory drive belt routing —3.3L engine

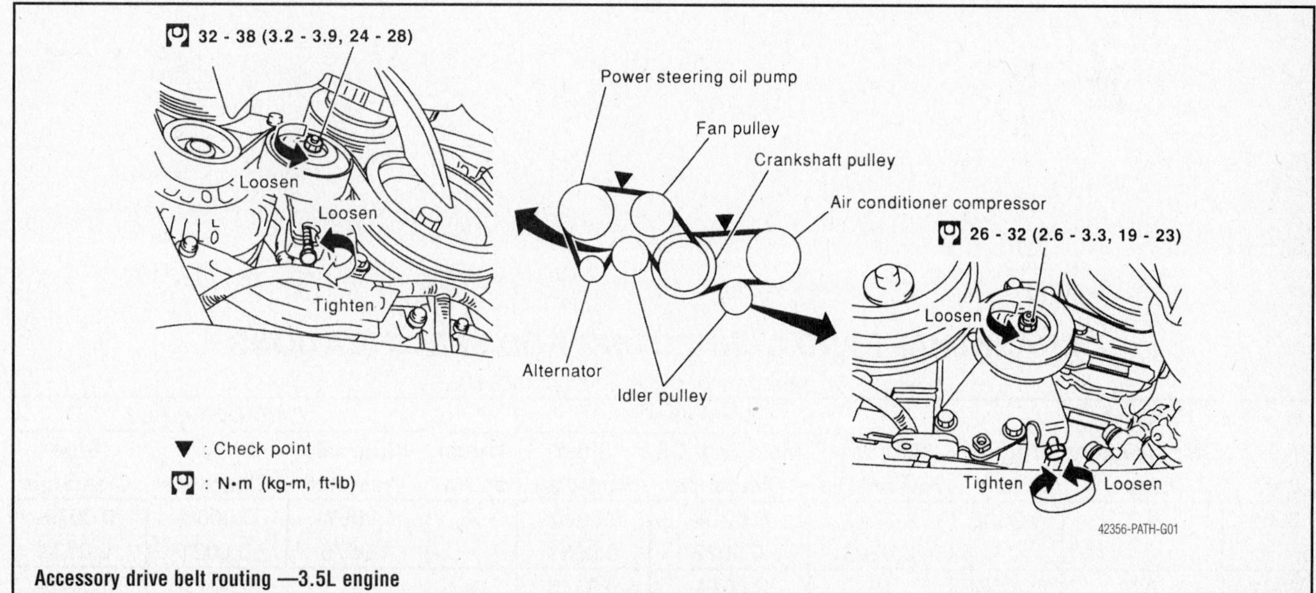

Accessory drive belt routing —3.5L engine

CAPACITIES

Year	Model	Engine Displacement Liters	Engine ID	Engine Oil with Filter (qts.)	Transmission (pts.) 5-Spd	Transmission (pts.) Auto.	Transfer Case (pts.)	Drive Axle Front (pts.)	Drive Axle Rear (pts.)	Fuel Tank (gal.)	Cooling System (qts.)
2001	Pathfinder	3.3	VG33E	3.8	①	②	4.8	4.4	4.9	21.1	11.25
2002	Pathfinder	3.5	VQ35DE	5.25	10.75	18.0	③	3.9	5.9	21.1	9.75
2003	Pathfinder	3.5	VQ35DE	5.25	10.75	18.0	③	3.9	5.9	21.1	9.75
2004	Pathfinder	3.5	VQ35DE	5.25	10.75	18.0	④	3.9	5.9	21.1	9.75

NOTE: All capacities are approximate. Add fluid gradually and check to be sure a proper fluid level is obtained.

① 2WD: 5.125 pts.
4WD: 10.75 pts.

② 2WD: 17.5 pts.
4WD: 18 pts.

③ Part time: 2.375; full time: 2.625 pts.

④ Part time: 2.375; full time: 3.125

67170-PATH-C04

VALVE SPECIFICATIONS

Year	Engine Displacement Liters	Engine ID	Seat Angle (deg.)	Face Angle (deg.)	Spring Test Pressure (lbs. @ in.)	Spring Installed Height (in.)	Stem-to-Guide Clearance (in.) Intake	Stem-to-Guide Clearance (in.) Exhaust	Stem Diameter (in.) Intake	Stem Diameter (in.) Exhaust
2001	3.3	VG33E	45	45.25-46.75	①	NA	0.0008-0.0021	0.0016-0.0029	0.2742-0.2748	0.3135-0.3138
2002	3.5	VQ35DE	45.15-45.45	45	45.4@1.457	1.457	0.0008-0.0021	0.0016-0.0029	0.2348-0.2354	0.2341-0.2346
2003	3.5	VQ35DE	45.15-45.45	45	45.4@1.457	1.457	0.0008-0.0021	0.0016-0.0029	0.2348-0.2354	0.2341-0.2346
2004	3.5	VQ35DE	45.15-45.45	45	45.4@1.457	1.457	0.0008-0.0021	0.0016-0.0029	0.2348-0.2354	0.2341-0.2346

NA: Not Available

① Inner: 57.3 @ 0.984

Outer: 117.7 @ 1.181

67170-PATH-C05

CRANKSHAFT AND CONNECTING ROD SPECIFICATIONS

All measurements are given in inches.

Year	Engine Displacement Liters	Engine ID	Crankshaft Main Brg. Journal Dia.	Crankshaft Main Brg. Oil Clearance	Crankshaft Shaft End-play	Crankshaft Thrust on No.	Connecting Rod Journal Diameter	Connecting Rod Oil Clearance	Connecting Rod Side Clearance
2001	3.3	VG33E	2.4790-2.4793	0.0011-0.0022	0.0020-0.0067	4	1.9967-1.9675	0.0006-0.0021	0.0079-0.0138
2002	3.5	VQ35DE	①	0.0014-0.0018	0.0118	4	②	0.0013-0.0023	0.0079-0.0138
2003	3.5	VQ35DE	①	0.0014-0.0018	0.0118	4	②	0.0013-0.0023	0.0079-0.0138
2004	3.5	VQ35DE	①	0.0014-0.0018	0.0118	4	②	0.0013-0.0023	0.0079-0.0138

NA - Not Available

① There are 24 different grades, ranging from A (2.3612) to 7 (2.3603)

② Grade 0: 2.0460-2.0462

Grade 1: 2.0457-2.0460

Grade 2: 2.0445-2.0457

67170-PATH-C06

PISTON AND RING SPECIFICATIONS

All measurements are given in inches.

Year	Engine Displacement Liters	Engine ID	Piston Clearance	Ring Gap Top Comp.	Ring Gap Bottom Comp.	Ring Gap Oil Control	Ring Side Clearance Top Comp.	Ring Side Clearance Bottom Comp.	Ring Side Clearance Oil Control
2001	3.3	VG33E	①	0.0083-0.0157	0.0197-0.0272	0.0079-0.0272	0.0009-0.0030	0.0012-0.0028	0.0006-0.0073
2002	3.5	VQ35DE	0.0004-0.0012	0.0091-0.0130	0.0130-0.0189	0.0079-0.0236	0.0016-0.0031	0.0012-0.0028	0.0006-0.0020
2003	3.5	VQ35DE	0.0004-0.0012	0.0091-0.0130	0.0130-0.0189	0.0079-0.0236	0.0016-0.0031	0.0012-0.0028	0.0006-0.0020
2004	3.5	VQ35DE	0.0004-0.0012	0.0091-0.0130	0.0130-0.0189	0.0079-0.0236	0.0016-0.0031	0.0012-0.0028	0.0006-0.0020

① Cylinders 1, 2, 6: 0.0010 - 0.0018 in.
Cylinders 3 and 4: 0.0006 - 0.0010 in.
Cylinder 5: 0.0012-0.0016 in.

67170-PATH-C07

TORQUE SPECIFICATIONS

All readings in ft. lbs.

Year	Engine Displacement Liters	Engine ID	Cylinder Head Bolts	Main Bearing Bolts	Rod Bearing Bolts	Crankshaft Damper Bolts	Flywheel Bolts	Manifold Intake	Manifold Exhaust	Spark Plugs	Oil Pan Drain Plug
2001	3.3	VG33E	①	67-74	②	141-156	61-69	①	21-25	14-22	21-28
2002	3.5	VQ35DE	③	④	⑤	⑥	61-69	⑦	21-24	14-22	21-28
2003	3.5	VQ35DE	③	④	⑤	⑥	61-69	⑦	21-24	14-22	21-28
2004	3.5	VQ35DE	③	④	⑤	⑥	61-69	⑦	21-24	14-22	21-28

① The cylinder heads and the lower intake manifold are installed together
Step 1: Tighten the cylinder head bolts to 22 ft. lbs.
Step 2: Tighten the cylinder head bolts to 43 ft. lbs.
Step 3: Loosen the cylinder head bolts completely
Step 4: Tighten the cylinder head bolts to 84 inch lbs.
Step 5: Tighten the intake manifold fasteners to 35 inch lbs.
Step 6: Tighten the intake manifold fasteners to 13 ft. lbs.
Step 7: Tighten the intake manifold fasteners to 12-14 ft. lbs.
Step 8: Loosen all intake manifold fasteners completely
Step 9: Tighten the cylinder head bolts to 22 ft. lbs.
Step 10: Tighten the cylinder head bolts 60-65 degrees
Step 11: Tighten the cylinder head sub-bolts to 80-105 inch lbs.
Step 12: Tighten the intake manifold fasteners to 35 inch lbs.
Step 13: Tighten the intake manifold fasteners to 78 inch lbs.
Step 14: Tighten the intake manifold fasteners to 70-84 inch lbs.

② 10-12 ft. lbs. plus 60-65 degrees or 28-33 ft. lbs.

③ Step 1: 72 ft. lbs.
Step 2: Loosen all bolts completely
Step 3: 25-33 ft. lbs.
Step 4: +90 degrees
Step 5: +90 degrees

④ Step 1: 24-28 ft. lbs.
Step 2: +90 degrees

⑤ Step 1: 15 ft. lbs.
Step 2: +90 degrees

⑥ 29-36 ft. lbs. +60-66 degrees

⑦ Step 1: 44-86 inch lbs.
Step 2: 20-23 ft. lbs.

67170-PATH-C08

WHEEL ALIGNMENT

		Caster		Camber			
Year	Model	Range (+/-Deg.)	Preferred Setting (Deg.)	Range (+/-Deg.)	Preferred Setting (Deg.)	Toe-in (in.)	Kingpin Inclination (Deg.)
2001	Pathfinder	0.75	+3.00	0.75	+0.17	0.08+/-0.04	14.33
2002	Pathfinder	0.75	+3.00	0.75	+0.17	0.08+/-0.04	14.33
2003	Pathfinder	0.75	+3.00	0.75	+0.17	0.08+/-0.04	14.33
2004	Pathfinder	0.75	+3.00	0.75	+0.17	0.08+/-0.04	14.33

67170-PATH-C09

TIRE, WHEEL AND BALL JOINT SPECIFICATIONS

		OEM Tires		Tire Pressures (psi)				
Year	Model	Standard	Optional	Front	Rear	Wheel Size	Ball Joint Inspection	Lugnut Torque (ft. lbs.)
2001	Pathfinder	P235/70R15	P265/70R15	30	30	Std: 6.5-JJ Opt: 7-JJ	U: 0.020 in. L: ①	87-108
2002	Pathfinder LE	P245/65SR17	none	②	②	8J	③	87-108
	Pathfinder SE	P255/65SR16	None	②	②	7JJ	③	87-108
2003	Pathfinder LE	P245/65SR17	none	②	②	8J	③	87-108
	Pathfinder SE	P255/65SR16	None	②	②	7JJ	③	87-108
2004	Pathfinder LE	P245/65SR17	none	②	②	8J	③	87-108
	Pathfinder SE	P255/65SR16	None	②	②	7JJ	③	87-108

OEM: Original Equipment Manufacturer

PSI: Pounds Per Square Inch

STD: Standard

OPT: Optional

L: Lower

U: Upper

① Replace if any measurable movement is found.

② See placard on vehicle

③ Axial play
Upper: 0
Lower: 0.008 in.

67170-PATH-C10

BRAKE SPECIFICATIONS

All measurements in inches unless noted

		Brake Disc			Brake Drum Diameter			Minimum Lining Thickness		Brake Caliper	
Year	Model	Original Thickness	Minimum Thickness	Maximum Runout	Original Inside Diameter	Max. Wear Limit	Maximum Machine Diameter	Front	Rear	Bracket Bolts (ft. lbs.)	Mounting Bolts (ft. lbs.)
2001	Pathfinder	1.100	1.024	0.004	11.60	NA	11.67	0.079	0.059	53-72	16-23
2002	Pathfinder	1.100	1.024	0.003	11.61	NA	11.67	0.079	0.059	①	①
2003	Pathfinder	1.100	1.024	0.003	11.61	NA	11.67	0.079	0.059	①	①
2004	Pathfinder	1.100	1.024	0.003	11.61	NA	11.67	0.079	0.059	②	②

NA: Not Available

① Torque member mounting bolt: 127-134
Main pin bolt: 24-31

② Torque member mounting bolt: 107-136
Main pin bolt: 24-31

67170-PATH-C11

SCHEDULED MAINTENANCE INTERVALS

Normal Driving Conditions

Nissan—Pathfinder

		VEHICLE MILEAGE INTERVAL (x1000)															
TO BE SERVICED	TYPE OF SERVICE	3.8	7.5	11	15	19	22.5	26	30	34	37.5	41	45	49	52.5	56	60
Engine oil & filter	R	✓	✓	✓	✓	✓	✓	✓	✓	✓	✓	✓	✓	✓	✓	✓	✓
Brake lines & cables	S/I				✓				✓				✓				✓
Brake pads, discs, drums & linings	I		✓		✓		✓		✓		✓		✓		✓		✓
Driveshaft boots & propeller shaft	L/I		✓		✓		✓		✓		✓		✓		✓		✓
Front wheel bearings (4x2)	I								✓								✓
Automatic & manual transmission, transfer & differential gear oil ①	I				✓				✓				✓				✓
LSD gear oil	R								✓								✓
Front wheel bearing grease (4x4)	R								✓								✓
Timing belt ②	R																
Air cleaner filter	R								✓								✓
Engine coolant ③	R																✓
Spark plugs	R	platinum tipped plugs every 105,000 miles															
Drive belt(s)	S/I								✓								✓
Cabin air filter	I/R		I		R		I		R		I		R		I		R
Exhaust system	I		✓		✓		✓		✓		✓		✓		✓		✓
Fuel lines	S/I								✓								✓
Steering gear (box) & linkage, axle & suspension parts	I		✓		✓		✓		✓		✓		✓		✓		✓
Vapor lines	S/I								✓								✓

R: Replace S/I: Service or Inspect L: Lubricate

① Differential (w/limited-slip differential) oil: replace oil every 30,000 miles.

② Timing belt: replace at 105,000 miles.

③ After 60,000, replace every 30,000

67170-PATH-C12

PRECAUTIONS

Before servicing any vehicle, please be sure to read all of the following precautions, which deal with personal safety, prevention of component damage, and important points to take into consideration when servicing a motor vehicle:

- Never open, service or drain the radiator or cooling system when the engine is hot; serious burns can occur from the steam and hot coolant.
- Observe all applicable safety precautions when working around fuel. Whenever servicing the fuel system, always work in a well-ventilated area. Do not allow fuel spray or vapors to come in contact with a spark, open flame, or excessive heat (a hot drop light, for example). Keep a dry chemical fire extinguisher near the work area. Always keep fuel in a container specifically designed for fuel storage; also, always properly seal fuel containers to avoid the possibility of fire or explosion. Refer to the additional fuel system precautions later in this section.
- Fuel injection systems often remain pressurized, even after the engine has been turned **OFF**. The fuel system pressure must be relieved before disconnecting any fuel lines. Failure to do so may result in fire and/or personal injury.
- Brake fluid often contains polyglycol ethers and polyglycols. Avoid contact with the eyes and wash your hands thoroughly after handling brake fluid. If you do get brake fluid in your eyes, flush your eyes with clean, running water for 15 minutes. If eye irritation persists, or if you have taken brake fluid internally, IMMEDIATELY seek medical assistance.
- The EPA warns that prolonged contact with used engine oil may cause a number of skin disorders, including cancer! You should make every effort to minimize your exposure to used engine oil. Protective gloves should be worn when changing oil. Wash your hands and any other exposed skin areas as soon as possible after exposure to used engine oil. Soap and water, or waterless hand cleaner should be used.
- All new vehicles are now equipped with an air bag system. The system must be disabled before performing service on or around system components, steering column, instrument panel components, wiring and sensors. Failure to follow safety and disabling procedures could result in accidental air bag deployment, possible personal injury and unnecessary system repairs.
- Always wear safety goggles when working with, or around, the air bag system. When carrying a non-deployed air bag, be sure the bag and trim cover are pointed away from your body. When placing a non-deployed air bag on a work surface, always face the bag and trim cover upward, away from the surface. This will reduce the motion of the module if it is accidentally deployed. Refer to the additional air bag system precautions later in this section.
- Clean, high quality brake fluid from a sealed container is essential to the safe and proper operation of the brake system. You should always buy the correct type of brake fluid for your vehicle. If the brake fluid becomes contaminated, completely flush the system with new fluid. Never reuse any brake fluid. Any brake fluid that is removed from the system should be discarded. Also, do not allow any brake fluid to come in contact with a painted surface; it will damage the paint.
- Never operate the engine without the proper amount and type of engine oil; doing so WILL result in severe engine damage.
- Timing belt maintenance is extremely important! Many models utilize an interference-type, non-freewheeling engine. If the timing belt breaks, the valves in the cylinder head may strike the pistons, causing potentially serious (also time-consuming and expensive) engine damage. Refer to the maintenance interval charts in the front of this manual for the recommended replacement interval for the timing belt, and to the timing belt section for belt replacement and inspection.
- Disconnecting the negative battery cable on some vehicles may interfere with the functions of the on-board computer system(s) and may require the computer to undergo a relearning process once the negative battery cable is reconnected.
- When servicing drum brakes, only disassemble and assemble one side at a time, leaving the remaining side intact for reference.

ENGINE REPAIR

➡Disconnecting the negative battery cable on some vehicles may interfere with the functions of the on board computer system. The computer may undergo a relearning process once the negative battery cable is reconnected.

Distributor

REMOVAL

1. Before servicing the vehicle, refer to the precautions in the beginning of this section.
2. Remove or disconnect the following:
   - Negative battery cable
   - Distributor cap
   - Distributor wiring harness connector
3. Matchmark the rotor to the distributor housing and the distributor housing to the cylinder head.
4. Remove the distributor.

INSTALLATION

Timing Not Disturbed

1. Install or connect the following:
   - Distributor by aligning the matchmarks made during removal
   - Distributor wiring harness connector
   - Distributor cap
   - Negative battery cable
2. Check the ignition timing and adjust, as necessary.

Timing Disturbed

3.3L ENGINE

1. Set the engine to Top Dead Center (TDC) of the compression stroke for the No. 1 cylinder.
2. Align the index mark on the distributor shaft with the protrusion on the distributor housing.
3. Install the distributor and check that the distributor rotor is aligned.
4. Install or connect the following:

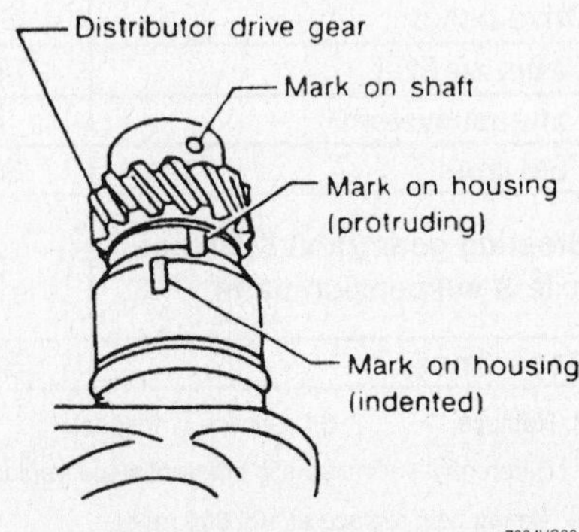

Distributor shaft alignment—3.3L engine

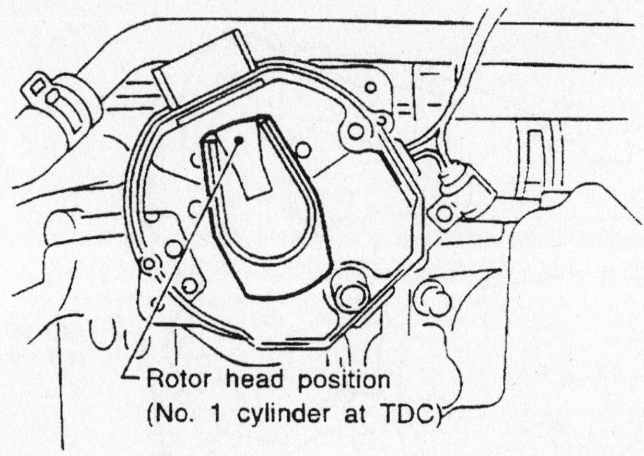

Distributor rotor alignment—3.3L engine

- Distributor cap
- Distributor harness connector

5. Check the ignition timing and adjust, as necessary.

Alternator

REMOVAL & INSTALLATOIN

3.3L and 3.5L Engines

1. Before servicing the vehicle, refer to the precautions in the beginning of this section.
2. Remove or disconnect the following:
   - Negative battery cable
   - Alternator harness connectors
   - Engine under cover
   - Alternator belt
   - Alternator

To install:

3. Install or connect the following:
   - Alternator
   - Alternator belt. Tighten the adjustment bolts as shown in the accompanying illustration.
   - Engine under cover
   - Alternator harness connectors
   - Negative battery cable

Engine Assembly

REMOVAL & INSTALLATION

3.3L Engine

1. Before servicing the vehicle, refer to the precautions in the beginning of this section.
2. Drain the cooling system.

Alternator mounting

3. Relieve the fuel system pressure.
4. Recover the A/C refrigerant, if equipped.
5. Remove or disconnect the following:
   - Negative battery cable
   - Hood
   - Air cleaner assembly
   - Idle Air Control (IAC) valve and solenoid connectors
   - Throttle Position (TP) sensor and switch connectors
   - Engine Coolant Temperature (ECT) sensor connector
   - Manifold Absolute Pressure (MAP) sensor connector and vacuum line
   - Evaporative Emissions (EVAP) canister purge valve connector and vacuum line
   - Mass Air Flow (MAF) sensor connector
   - Brake booster vacuum line
   - Fuel lines
   - Exhaust Gas Recirculation (EGR) temperature sensor connector
   - Throttle cable
   - Accessory drive belts
   - Cooling fan and shroud
   - Radiator and hoses
   - Engine under cover
   - A/C compressor manifold
   - Power steering pump
   - Heated Oxygen (HO_2S) sensor connectors
   - Exhaust front pipes
   - Crankshaft Position (CKP) sensor
   - Starter motor
   - Transmission
   - Left and right engine mounts
   - Engine

➡When removing the engine mounts, do not loosen the 4 mount cover nuts. The mount is fluid filled and will not function if the fluid leaks out.

To install:

6. Install or connect the following:
   - Engine. Tighten the engine mount nuts to 43–58 ft. lbs. (59–78 Nm).
   - Transmission
   - Starter motor
   - CKP sensor
   - Exhaust front pipes
   - HO_2S sensor connectors
   - Power steering pump
   - A/C compressor manifold
   - Engine under cover
   - Radiator and hoses
   - Cooling fan and shroud
   - Accessory drive belts
   - Throttle cable

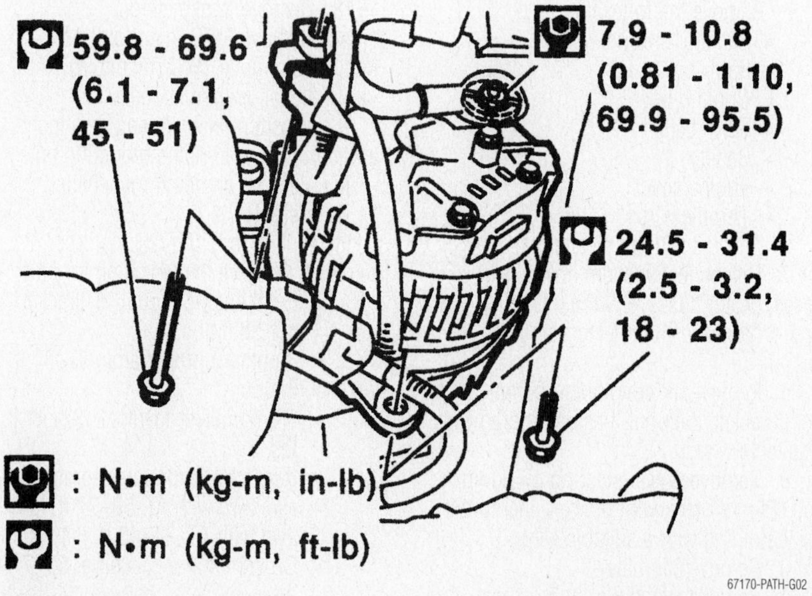

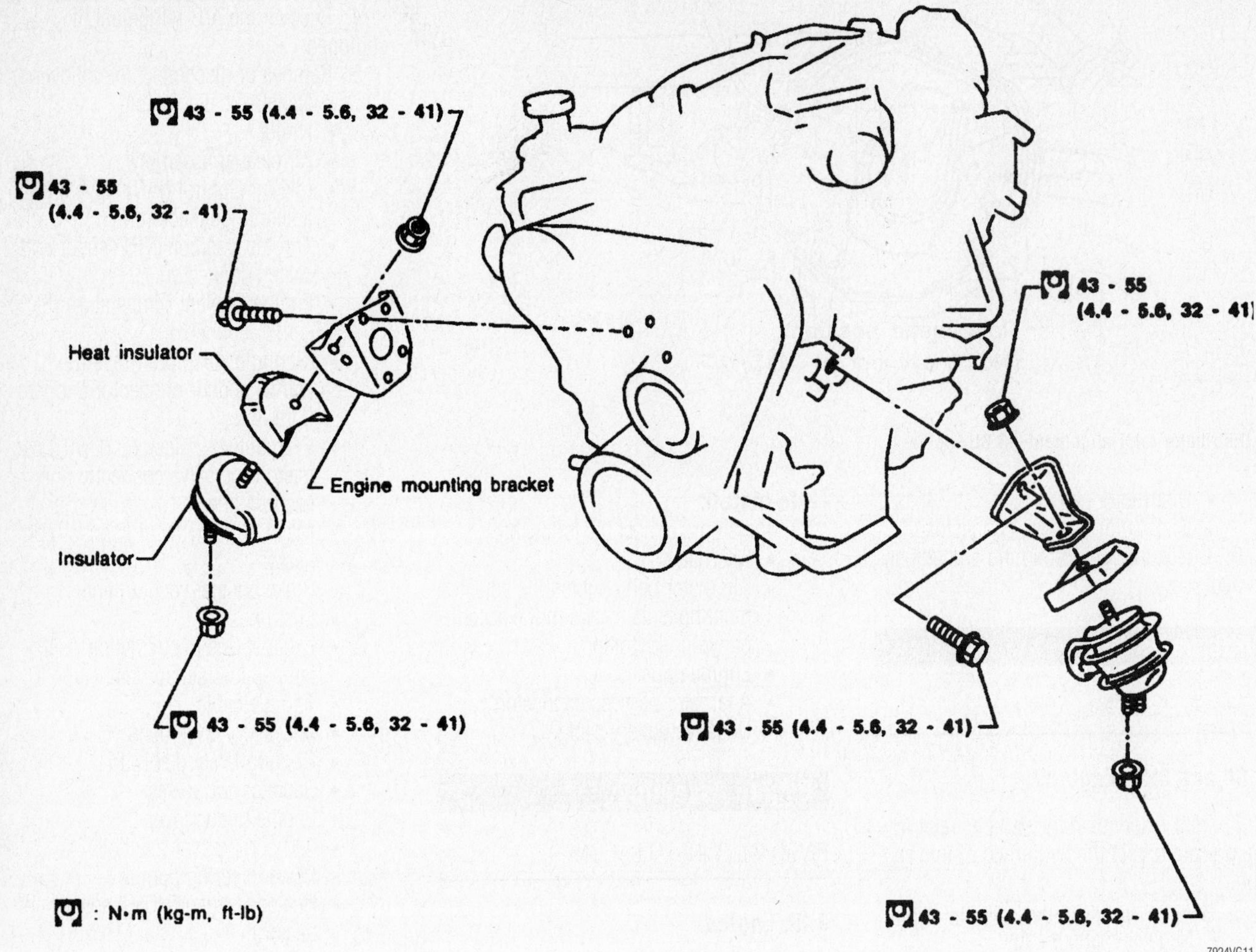

Engine mounts and related components—3.3L engine

- EGR temperature sensor connector
- Fuel lines
- Brake booster vacuum line
- MAF sensor connector
- EVAP canister purge valve connector and vacuum line
- MAP sensor connector and vacuum line
- ECT sensor connector
- TP sensor and switch connectors
- IAC valve and solenoid connectors
- Air cleaner assembly
- Hood
- Negative battery cable

7. Fill the cooling system.

8. Recharge the A/C system, if equipped.

9. Start the engine and check for leaks.

3.5L Engine

1. Release fuel pressure.

2. Remove engine hood and front RH and LH wheels.

3. Remove engine undercover and suspension member stay.

4. Drain coolant from radiator.

5. Remove the following parts.
   - Radiator shroud
   - Radiator
   - Cooling fan
   - Drive belts
   - Battery
   - Engine cover
   - Throttle wires

6. Air duct with air cleaner case.

7. Disconnect vacuum hoses, fuel hoses, heater hoses, EVAP canister hoses, harnesses, connectors and so on.

8. Remove air conditioner compressor from bracket, then put it aside holding with a suitable wire.

9. Remove power steering oil pump and reservoir tank with bracket, then put it aside holding with a suitable wire.

10. Remove alternator.

11. Remove exhaust front tube heat insulators, then remove rear heated oxygen sensors.

12. Remove exhaust front and rear tubes.

13. Remove transmission.

14. Remove TWC (manifold) heat insulators, then remove TWC (manifold).

15. Install engine slingers.

16. Hoist engine with engine slingers and remove front engine mounting nuts.

17. Remove engine from vehicle.

To install:

Installation is in the reverse order of removal. Observe the following torques:

- Front engine mount-to-bracket: 43–58 ft. lbs.
- Front mount-to-frame: 32–41 ft. lbs.
- Front bracket-to-block: 32–41 ft. lbs.
- Rear engine mount-to-bracket: all exc. 2wd with AT: 58–77 ft. lbs.; 2wd with AT: 32–40 ft. lbs.
- Crossmember-to-frame: 58–77 ft. lbs.

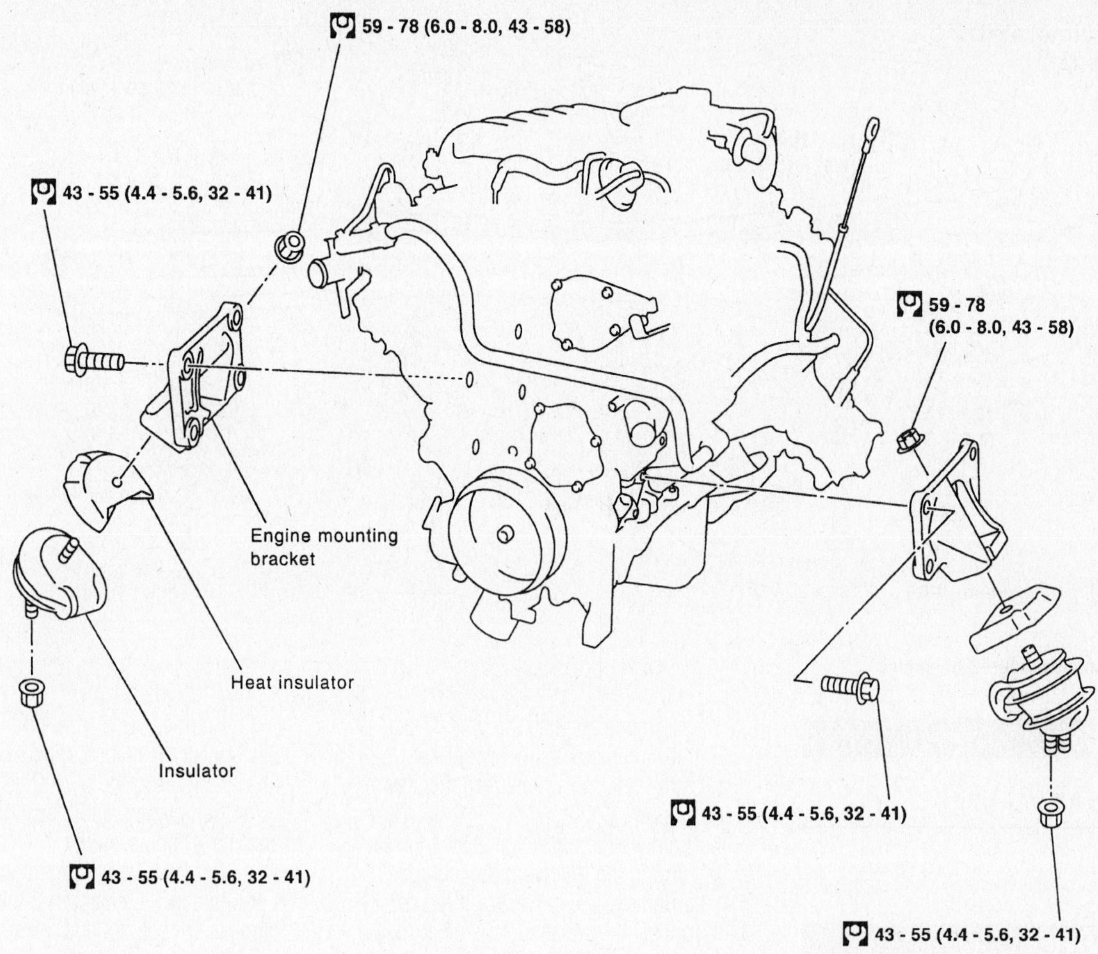

Front engine mounting—3.5L engine

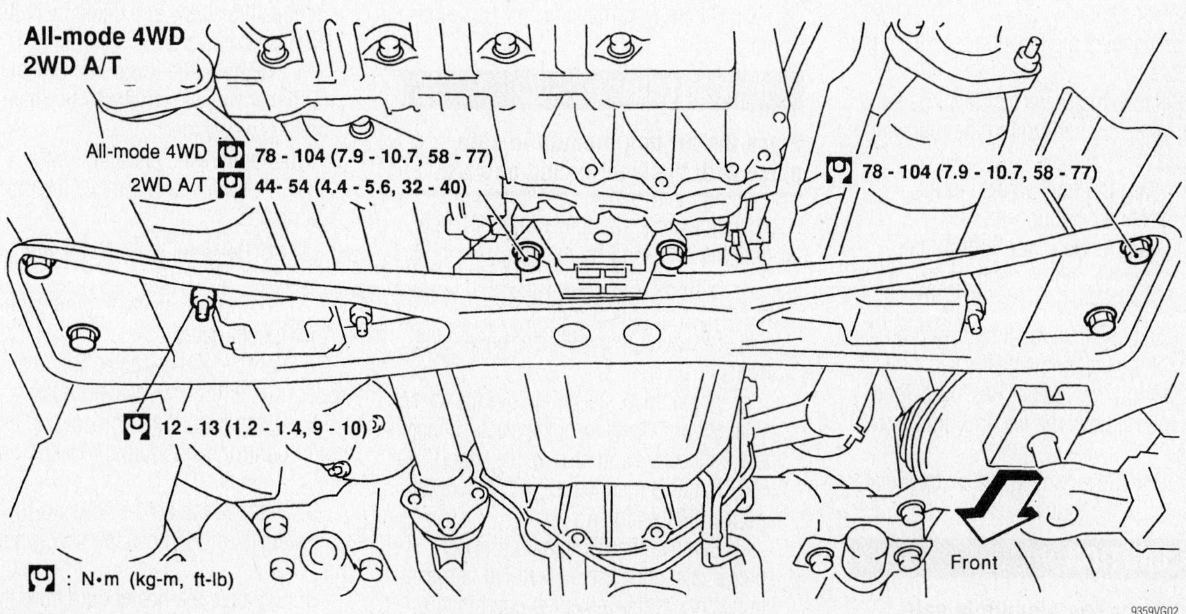

Rear engine mounting—3.5L engine

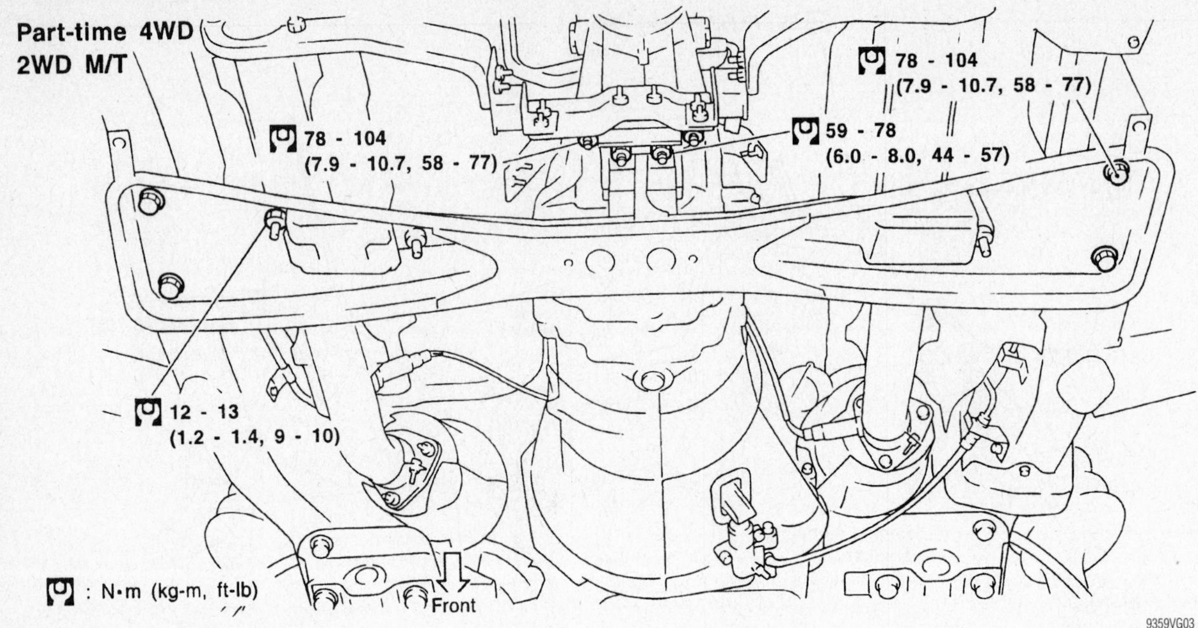

Rear engine mounting—3.5L engine

Heater Core

REMOVAL & INSTALLATION

1. Disconnect the negative battery cable.

✻✻ CAUTION

After disconnecting the negative battery cable, wait for at least 3 minutes before working on the steering column or instrument panel.

2. Drain the cooling system into a clean container for reuse.
3. Disconnect the heater hoses from the heater core.
4. Remove the driver's side air bag and steering wheel by performing the following procedure:
   a. Place the front wheels in the straight-ahead position.
   b. Remove the lower lid from the steering wheel and disconnect the air bag module connector.
   c. Remove the side lids from both sides of the steering wheel.
   d. Using the Tamper Resistant Torx® tool T50, remove the left and right Torx® bolts.
   e. Carefully, remove the air bag module.

✻✻ CAUTION

Place the air bag module in safe place with the front facing upward.

   f. Remove the steering wheel nut.
   g. Using a steering wheel puller, press the steering wheel from the steering column.
5. Remove the passenger's side air bag by performing the following procedure:
   a. Remove the glove box clips and disconnect the passenger's side air bag module connector.
   b. Remove the lower panel screws; then, disconnect the harness connector and remove the air bag module bracket.
   c. Using the Tamper Resistant Torx® tool T50, remove the passenger's side air bag module bolts.
   d. Carefully, remove the air bag module.

✻✻ CAUTION

Place the air bag module in safe place with the front facing upward.

6. Remove the instrument panel by performing the following procedure:
   a. Remove the steering column cover and the combination switch.
   b. Remove the instrument panel side lower finisher.
   c. At the driver's side, remove the lower panel screws, disconnect the electrical harness connectors and remove the panel.
   d. Remove the cluster lid "A" screws and the cluster lid "A".
   e. Remove the combination meter screws, disconnect the electrical harness connectors and remove the combination meter.
   f. Remove the cluster lid "C" screws, disconnect the electrical harness connectors and remove the cluster lid "C".
   g. Remove the audio assembly screws and the audio assembly.
   h. Remove the air conditioning control unit screws, disconnect the electrical harness connectors and the air conditioning control unit.
   i. Remove the ashtray.
   j. Remove the shifter (automatic transmission) or shift lever boot (manual transmission); then, remove the screw and disconnect the harness connector.
   k. Remove the console box; then, remove the screw and disconnect the harness connector.
   l. Remove the lower instrument center panel screws and the lower instrument center panel.
   m. Remove the defroster grille.
   n. At both sides, remove the pillar garnishes.
   o. Remove the instrument panel and pads nuts and bolts.
   p. Using an assistant, remove the instrument panel.
7. Remove the defroster nozzle and the heater nozzle from the heater housing.
8. Disconnect the electrical connector and/or control cable from the heater housing.
9. Remove the heater housing-to-chassis fasteners and remove the heater housing.
10. Separate the heater core from the heater housing and remove the heater core.

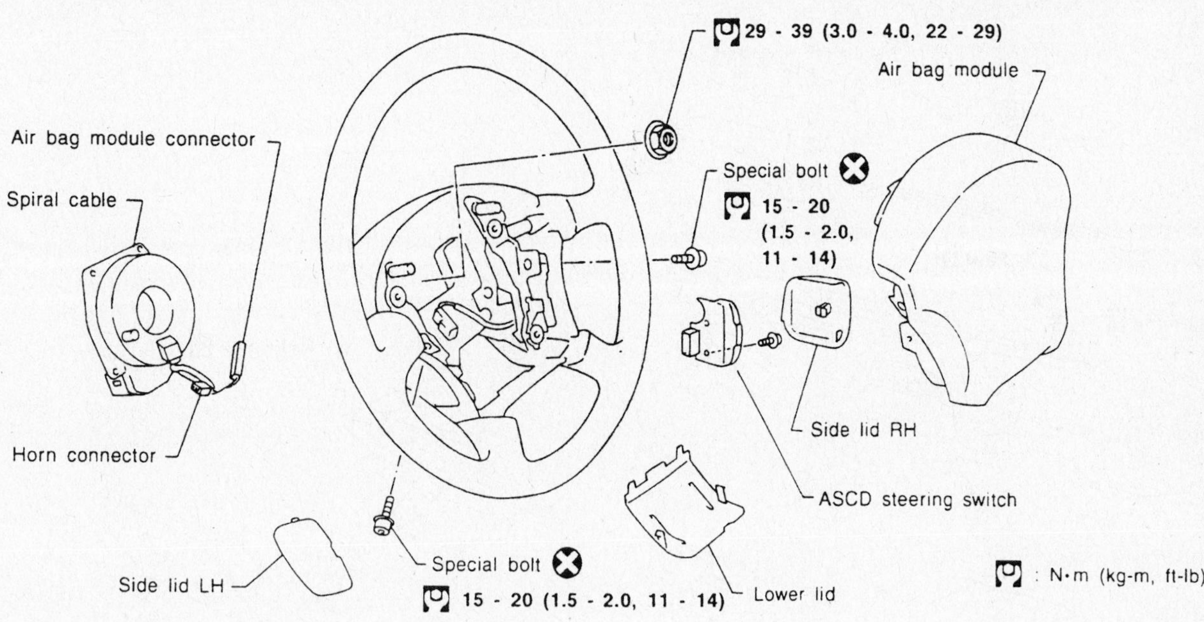

Exploded view of the driver's side air bag module and steering wheel

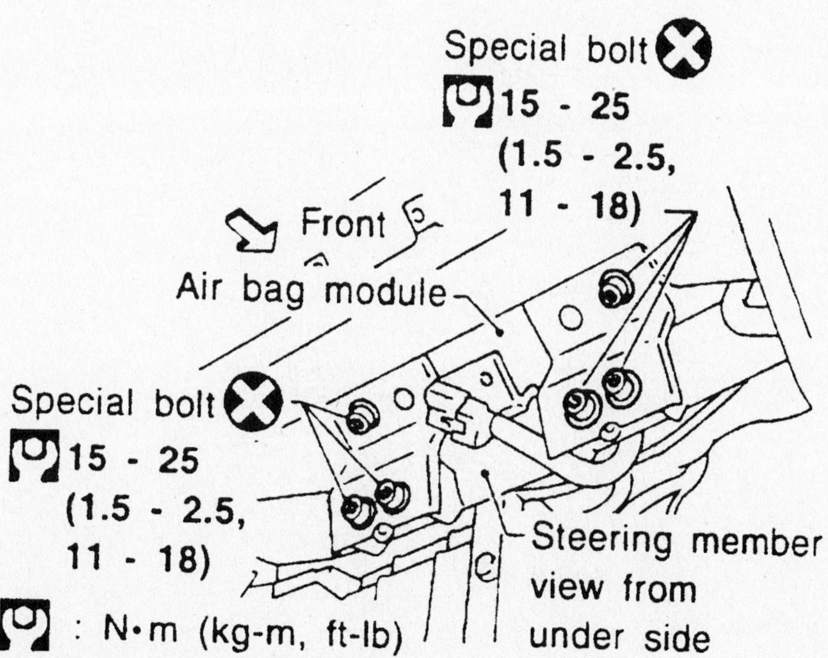

Exploded view of the passenger's side air bag module

To install:

11. Install the heater core and assemble the heater housing.
12. Install the heater housing and the heater housing-to-chassis fasteners.
13. Connect the electrical connector and/or control cable to the heater housing.
14. Install the defroster nozzle and the heater nozzle to the heater housing.
15. Install the passenger's side air bag by performing the following procedure:
   a. Carefully, install the air bag module.
   b. Using the Tamper Resistant Torx® tool T50, install the passenger's side air bag module bolts. Torque the bolts to 11–18 ft. lbs. (15 –25 Nm).
   c. Connect the harness connector and install the air bag module bracket; then, install the lower panel screws.
   d. Connect the passenger's side air bag module connector and install the glove box clips.
16. Install the instrument panel by performing the following procedure:
   a. Using an assistant, position the instrument panel.
   b. Install the instrument pads, nuts and bolts.
   c. At both sides, install the pillar garnishes.
   d. Install the defroster grille.
   e. Install the lower instrument center panel and the lower instrument center panel screws.
   f. Install the console box; then, install the screw and connect the harness connector.
   g. Connect the harness connector and install the screw; then, install the shifter (automatic transmission) or shift lever boot (manual transmission).
   h. Install the ashtray.
   i. Install the air conditioning control unit, connect the electrical harness connectors and the air conditioning control unit screws.
   j. Install the audio assembly and the audio assembly screws.
   k. Install the cluster lid "C", connect the electrical harness connectors and install the cluster lid "C" screws.
   l. Install the combination meter, connect the electrical harness connectors and install the combination meter screws.

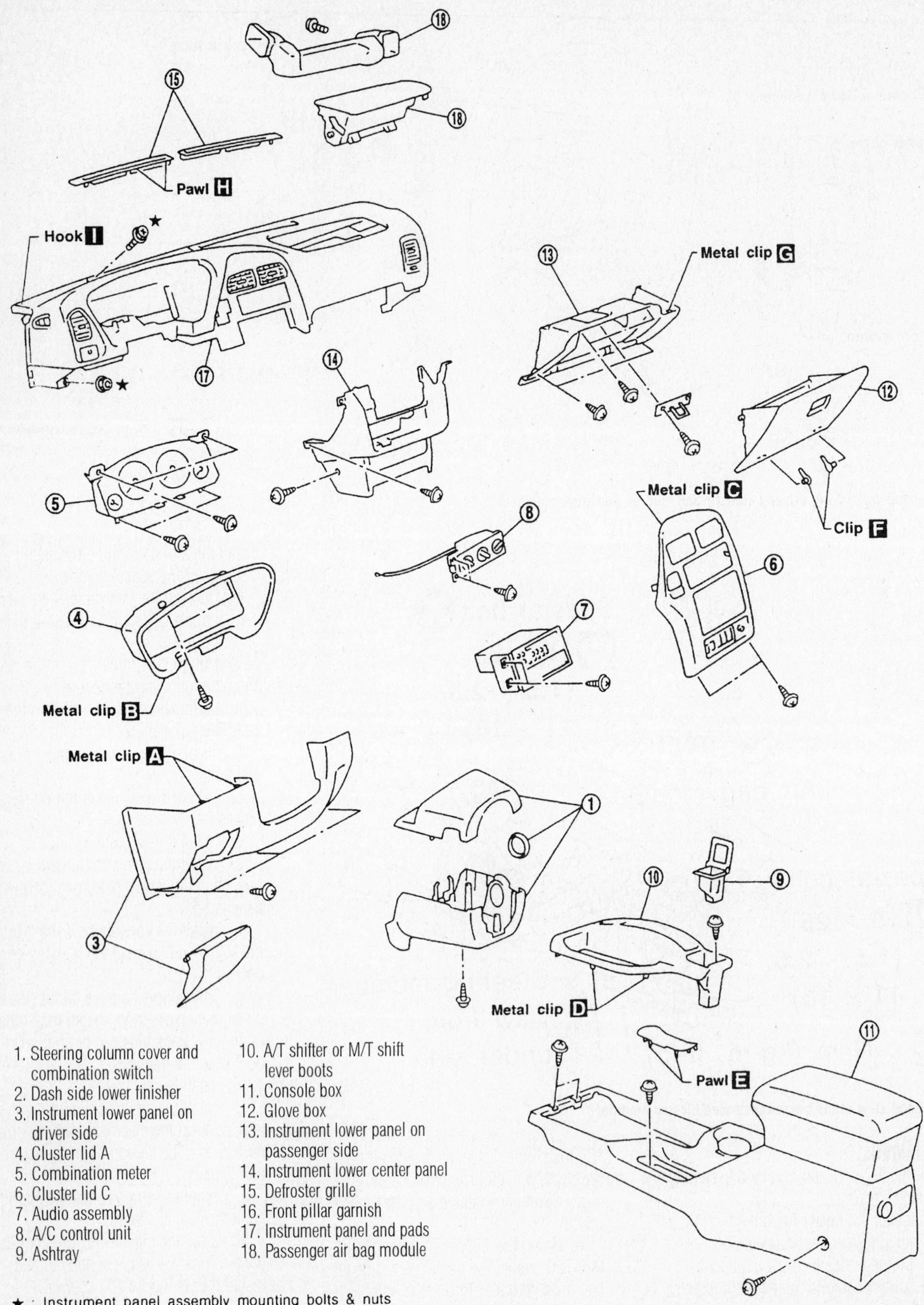
18
15
18
Pawl H
Hook I
17
13
Metal clip G
14
12
5
Metal clip C
Clip F
8
6
4
7
Metal clip B
Metal clip A
1
10
9
3
Metal clip D
11
Pawl E
1. Steering column cover and combination switch
2. Dash side lower finisher
3. Instrument lower panel on driver side
4. Cluster lid A
5. Combination meter
6. Cluster lid C
7. Audio assembly
8. A/C control unit
9. Ashtray
10. A/T shifter or M/T shift lever boots
11. Console box
12. Glove box
13. Instrument lower panel on passenger side
14. Instrument lower center panel
15. Defroster grille
16. Front pillar garnish
17. Instrument panel and pads
18. Passenger air bag module
★ : Instrument panel assembly mounting bolts & nuts
93113GH0

Exploded view of the instrument panel and related accessories

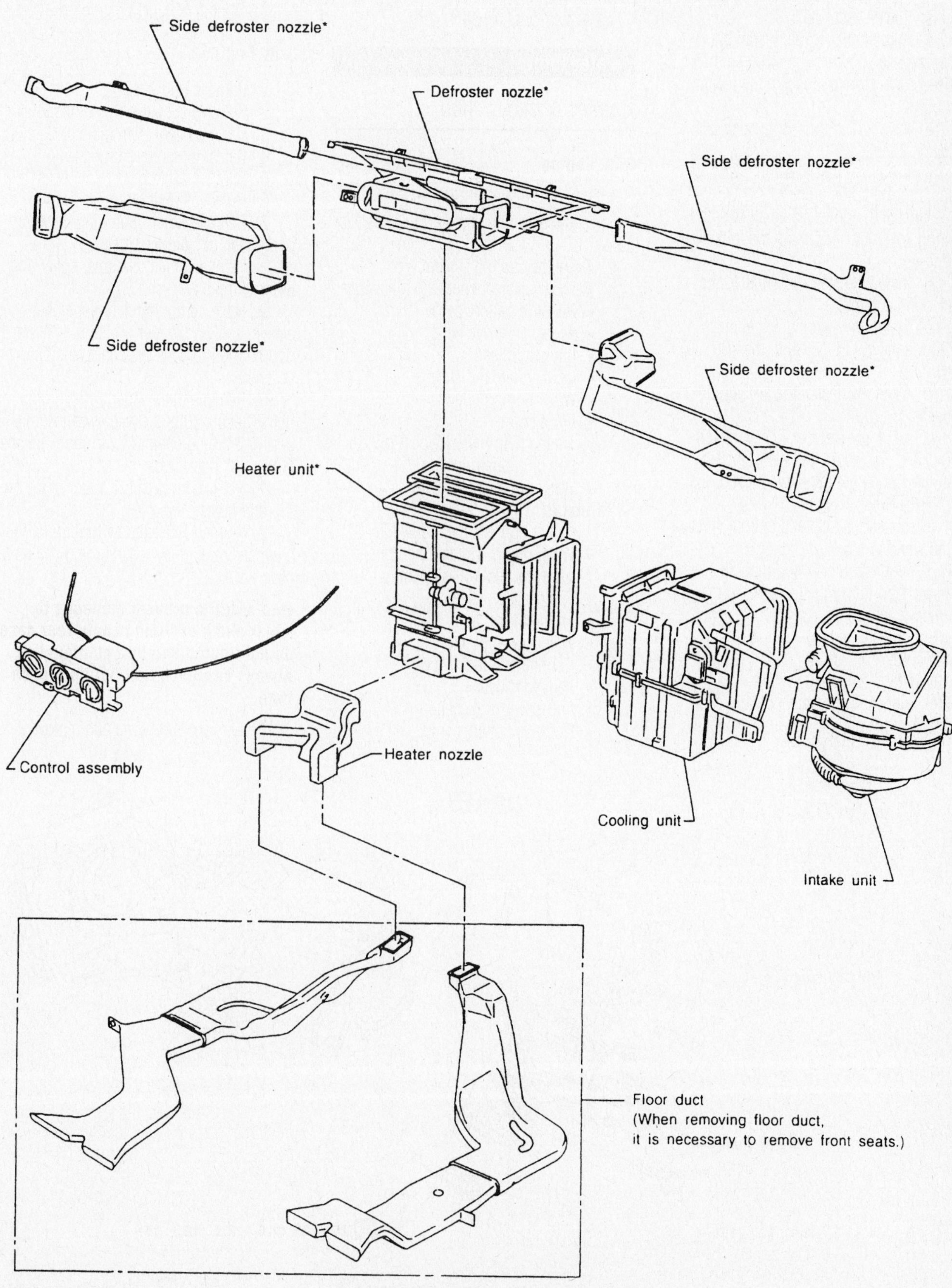

Exploded view of the heater housing, the evaporator housing, the ventilation dusts and related accessories

m. Install the cluster lid "A" and the cluster lid "A" screws.

n. At the driver's side, install the lower panel, connect the electrical harness connectors and install the panel screws.

o. Install the instrument panel side lower finisher.

p. Install the combination switch and the steering column cover.

17. Install the driver's side air bag and steering wheel by performing the following procedure:

a. Install the steering wheel to the steering column.

b. Install the steering wheel nut. Torque the nut to 22–29 ft. lbs. (29–39 Nm).

c. Carefully, install the air bag module.

d. Using the Tamper Resistant Torx® tool T50, install the left and right Torx® bolts. Torque the bolts to 11–14 ft. lbs. (15–20 Nm).

e. Install the side lids to both sides of the steering wheel.

f. Connect the air bag module connector and install the lower lid to the steering wheel.

18. Connect the heater hoses to the heater core.

19. Refill the cooling system.

20. Connect the negative battery cable.

21. Run the engine to normal operating temperatures; then, check the climate control operation and check for leaks.

Water Pump

REMOVAL & INSTALLATION

3.3L Engine

1. Before servicing the vehicle, refer to the precautions in the beginning of this section.
2. Drain the cooling system.
3. Remove or disconnect the following:
   - Negative battery cable
   - Accessory drive belts
   - Radiator hoses
   - Cooling fan and shroud
   - Water pump pulley
   - Front cover
   - Timing belt. Refer to the Timing Belt unit repair section.
   - Water pump

To install:

4. Install or connect the following:
   - Water pump. Tighten the bolts to 12–15 ft. lbs. (16–21 Nm).
   - Timing belt
   - Front cover
   - Water pump pulley
   - Cooling fan and shroud
   - Radiator hoses
   - Accessory drive belts
   - Negative battery cable
5. Fill the cooling system.
6. Start the engine and check for leaks.

3.5L Engine

1. Remove undercover.
2. Remove suspension member stay.
3. Drain coolant from radiator.
4. Remove radiator shrouds.
5. Remove drive belts.
6. Remove cooling fan.
7. Remove water drain plug on water pump side of cylinder block.
8. Remove chain tensioner cover and water pump cover.
9. Pushing timing chain tensioner sleeve, apply a stopper pin so it does not return. Then remove the chain tensioner assembly.
10. Remove the 3 water pump fixing bolts. Secure a gap between water pump gear and timing chain, by turning crankshaft pulley 20° backwards.
11. Put M8 bolts to two water pump fixing bolt holes.
12. Tighten M8 bolts by turning half turn alternately until they reach timing chain rear case.

➡In order to prevent damages to water pump or timing chain rear case, do not tighten one bolt continuously. Always turn each bolt half turn each time.

13. Lift up water pump and remove it.

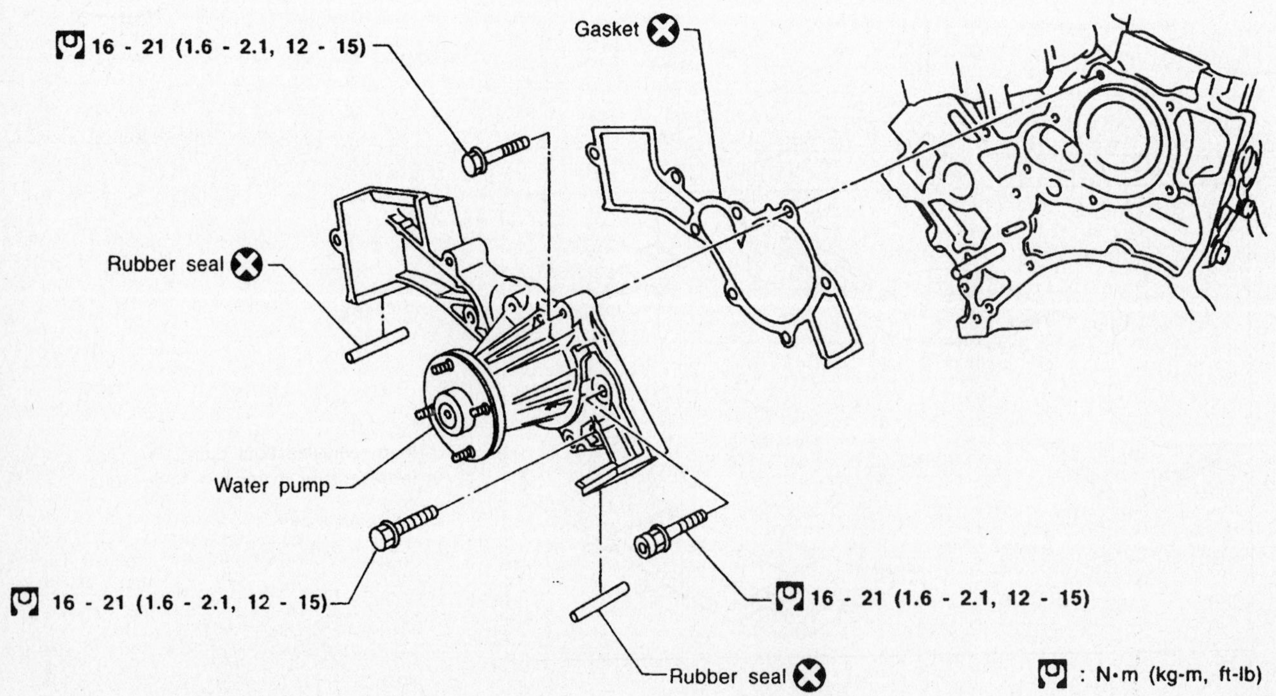

Exploded view of the water pump assembly—3.3L engine

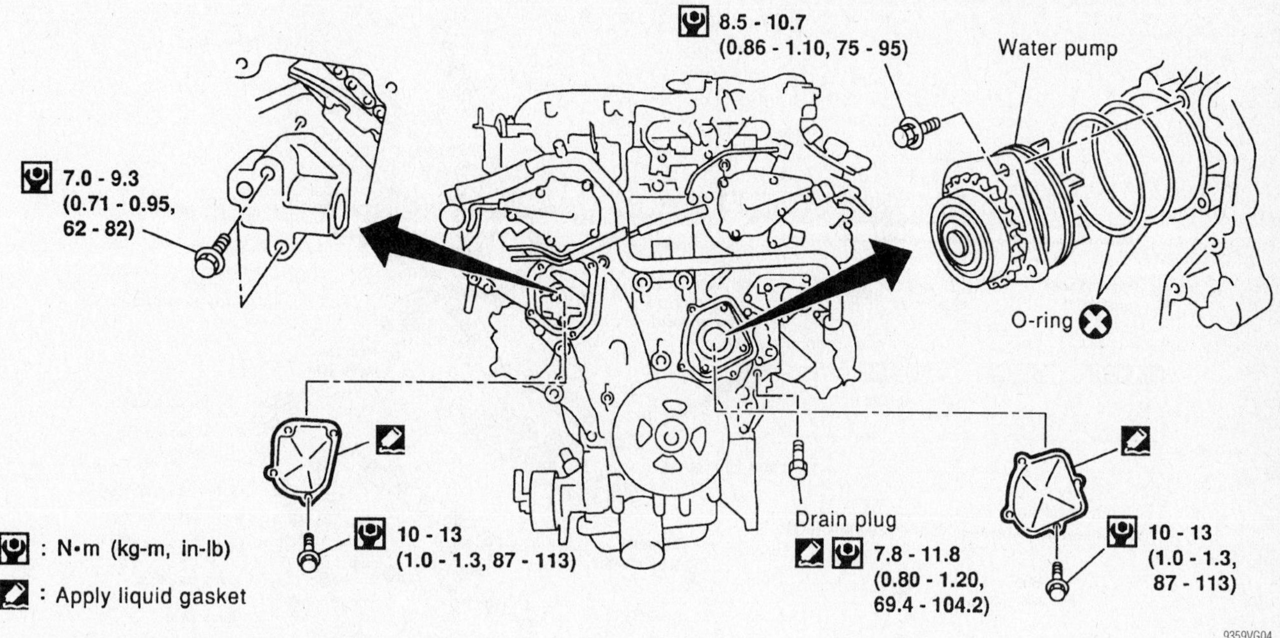

Exploded view of the water pump assembly—3.5L engine

➡When lifting up water pump, do not allow water pump gear to hit timing chain.

To install:

14. Apply engine oil and coolant to O-rings as shown in the figure.
15. Install water pump.

➡Do not allow cylinder block to nip O-rings when installing water pump.

16. Before installing, remove all traces of liquid gasket from mating surface of water pump cover and chain tensioner cover using a scraper. Also remove traces of liquid gasket from mating surface of front cover.
17. Apply a continuous bead of liquid gasket to mating surface of chain tensioner cover and water pump cover. Use Genuine RTV silicone sealant or equivalent.
18. Return the crankshaft pulley to its original position by turning it 20° forward.
19. Install timing chain tensioner, then remove the stopper pin.

➡When installing the timing chain tensioner, engine oil should be applied to the oil hole and tensioner.

➡After starting engine, let idle for three minutes, then rev engine up to 3,000 rpm under no load to purge air from the high-pressure chamber of the chain tensioners. The engine may produce a rattling noise. This indicates that air still remains in the chamber and is not a matter of concern.

20. Reinstall any parts removed in reverse order of removal.

Cylinder Head

REMOVAL & INSTALLATION

3.3L Engine

1. Before servicing the vehicle, refer to the precautions in the beginning of this section.
2. Drain the cooling system.
3. Relieve the fuel system pressure.
4. Remove or disconnect the following:
   - Negative battery cable
   - Accessory drive belts
   - Front cover
   - Timing belt. Refer to the Timing Belt unit repair section.
   - Upper intake manifold
   - Lower intake manifold
   - Camshaft sprockets
   - Rear timing cover
   - Distributor
   - Exhaust front pipes
   - A/C compressor
   - Alternator
   - Power steering pump
   - Accessory brackets
   - Valve covers. Loosen the bolts in several passes and in sequence.
   - Cylinder heads with the exhaust manifolds attached. Loosen the bolts in several passes and in sequence.

➡The cylinder head bolts vary in length. Note the bolt locations for assembly.

To install:

5. Install the cylinder heads and the lower intake manifold at the same time. Tighten the bolts in sequence as follows:
   a. Step 1: Tighten the cylinder head bolts to 22 ft. lbs. (29 Nm)
   b. Step 2: Tighten the cylinder head bolts to 43 ft. lbs. (59 Nm)
   c. Step 3: Loosen all cylinder head bolts completely
   d. Step 4: Tighten the cylinder head bolts to 84 inch lbs. (10 Nm)
   e. Step 5: Tighten the intake manifold fasteners to 35 inch lbs. (4 Nm)
   f. Step 6: Tighten the intake manifold fasteners to 13 ft. lbs. (18 Nm)
   g. Step 7: Tighten the intake manifold fasteners to 12–14 ft. lbs. (16–20 Nm)
   h. Step 8: Loosen all intake fasteners completely
   i. Step 9: Tighten the cylinder head bolts to 22 ft. lbs. (29 Nm)
   j. Step 10: Tighten the cylinder head bolts 60–65 degrees **OR** tighten to 40–47 ft. lbs. (54–64 Nm)
   k. Step 11: Tighten the cylinder head sub-bolts to 80–105 inch lbs. (9–12 Nm)
   l. Step 12: Tighten the intake manifold fasteners to 35 inch lbs. (4 Nm)
   m. Step 13: Tighten the intake manifold fasteners to 78 inch lbs. (9 Nm)

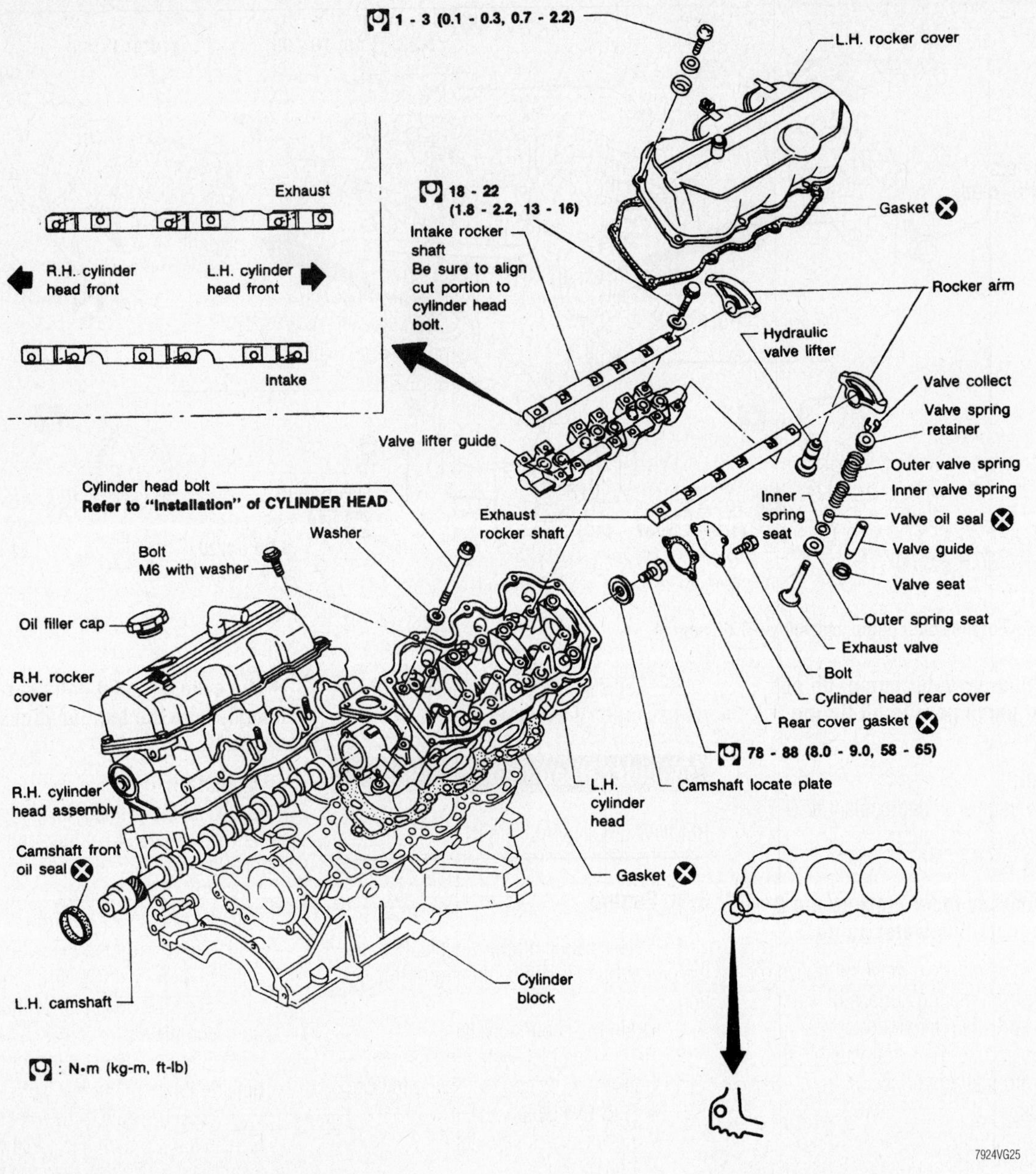

1 - 3 (0.1 - 0.3, 0.7 - 2.2)
L.H. rocker cover
Exhaust
R.H. cylinder head front
L.H. cylinder head front
Intake
18 - 22 (1.8 - 2.2, 13 - 16)
Intake rocker shaft
Be sure to align cut portion to cylinder head bolt.
Gasket
Rocker arm
Hydraulic valve lifter
Valve collect
Valve spring retainer
Outer valve spring
Inner valve spring
Valve oil seal
Valve guide
Valve seat
Outer spring seat
Exhaust valve
Bolt
Cylinder head rear cover
Rear cover gasket
78 - 88 (8.0 - 9.0, 58 - 65)
Camshaft locate plate
Valve lifter guide
Cylinder head bolt
Refer to "Installation" of CYLINDER HEAD
Washer
Exhaust rocker shaft
Inner spring seat
Bolt
M6 with washer
Oil filler cap
R.H. rocker cover
R.H. cylinder head assembly
Camshaft front oil seal
L.H. camshaft
L.H. cylinder head
Gasket
Cylinder block
: N•m (kg-m, ft-lb)
7924VG25

Exploded view of the cylinder head assembly—3.3L engine

Right bank
Engine front
Left bank
9308VG08

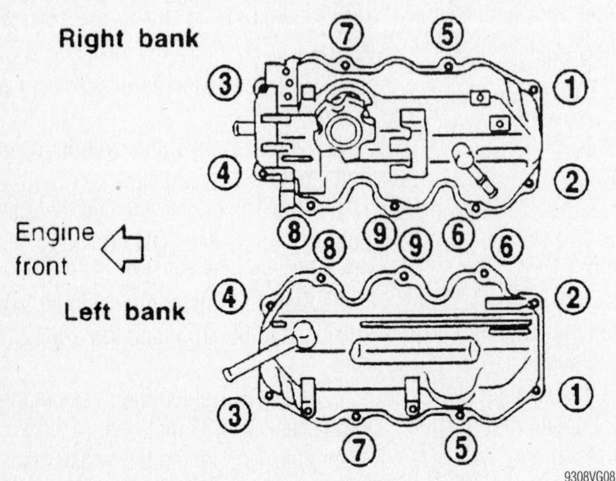

Valve cover bolt loosening sequence—3.3L engine

For L.H. cylinder head
For R.H. cylinder head
No. 1
No. 3
No. 5
ENGINE FRONT
7924VG26

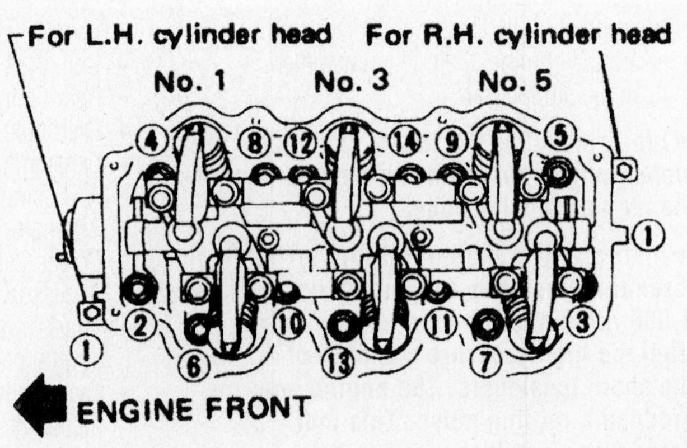

Cylinder head loosening sequence—3.3L engine

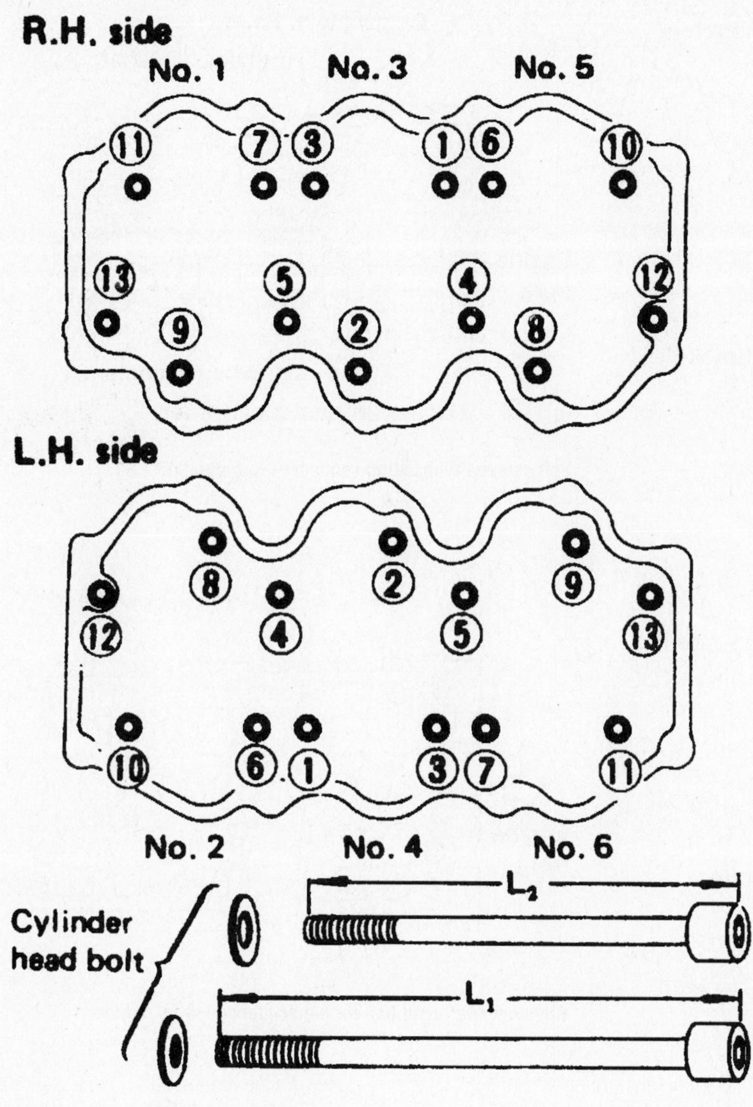

Cylinder head torque sequence—3.3L engine

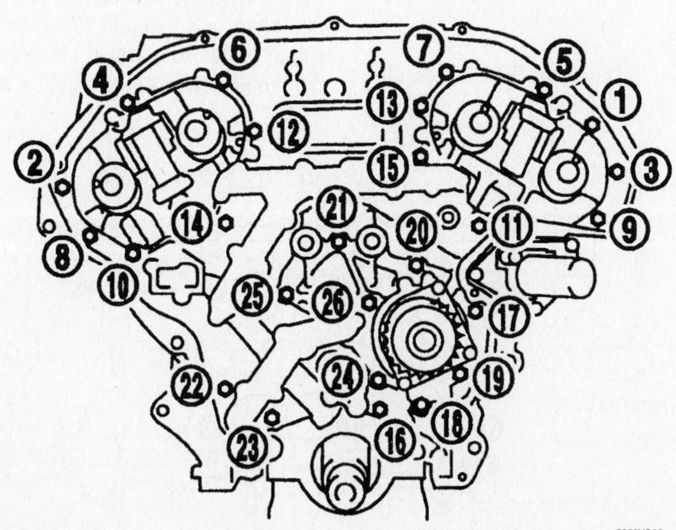

Rear timing case loosening sequence—3.5L engine

n. Step 14: Tighten the intake manifold fasteners to 70–84 inch lbs. (6–7 Nm)

6. Install or connect the following:
   - Valve covers
   - Accessory brackets
   - Power steering pump
   - Alternator
   - A/C compressor
   - Exhaust front pipes
   - Distributor
   - Rear timing cover
   - Camshaft sprockets
   - Upper intake manifold
   - Timing belt
   - Front cover
   - Accessory drive belts
   - Negative battery cable
7. Fill the cooling system.
8. Start the engine and check for leaks.

3.5L Engine

1. Remove engine from vehicle.
2. Remove exhaust manifolds in reverse order of installation.
3. Place engine on a work stand.
4. Remove aluminum oil pan
5. Remove timing chain.
6. Remove intake manifold in reverse order of installation.
7. Remove water outlet.
8. Remove rear timing chain case bolts. Loosen in numerical order as shown in the figure.
9. Remove rear timing chain case.
10. Remove O-rings to cylinder head.
11. Remove O-rings to cylinder block.
12. Remove intake valve timing control solenoid valves.
13. Remove intake and exhaust camshafts and camshaft brackets. Equally loosen camshaft bracket bolts in several steps in the numerical order shown in the figure. For reinstallation, be sure to put marks on camshaft bracket before removal.
14. Remove RH and LH camshaft chain tensioners from cylinder head.
15. Remove cylinder head bolts. Cylinder head bolts should be loosened in two or three steps.
16. Remove cylinder head.

To install:

17. Before installing rear timing chain case, remove old liquid gasket from mating surface using a scraper. Also remove old liquid gasket from mating surface of cylinder block. Remove old liquid gasket from the bolt hole and thread.
18. Before installing cam bracket, remove old liquid gasket from mating surface using a scraper.
19. Before installing the cylinder head

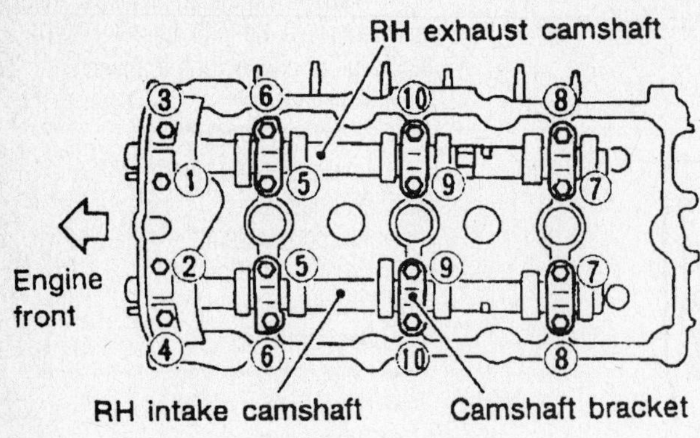

Right camshaft loosening sequence—3.5L engine

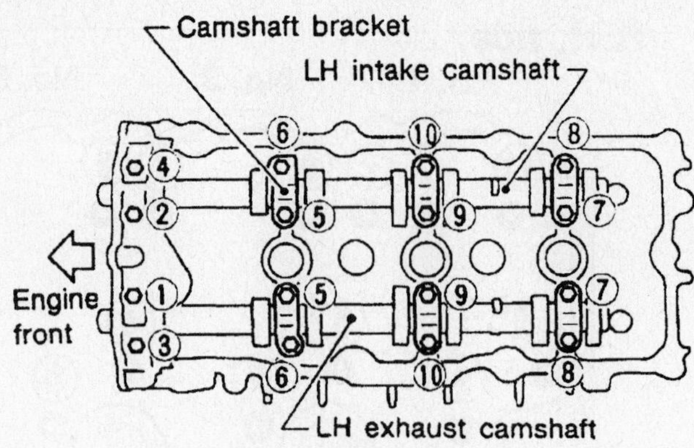

Left camshaft loosening sequence—3.5L engine

Cylinder head bolt

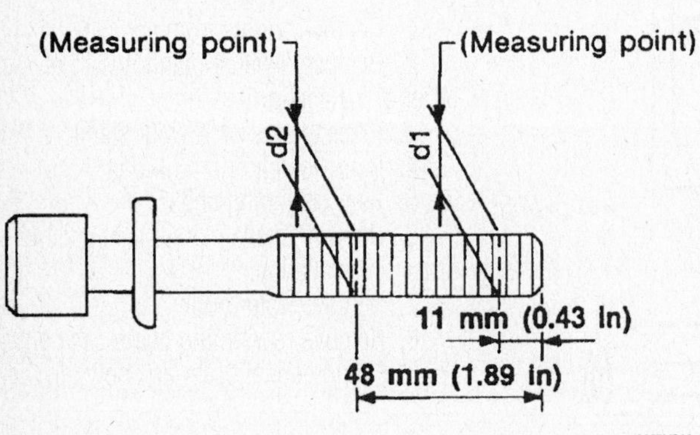

Head bolt checking—3.5L engine

RH cylinder head

Engine front

9359VG09

Right cylinder head bolt torque sequence—3.5L engine

gasket, be sure that No. 1 cylinder is at TDC. At this time, the crankshaft key should face toward the right bank.

20. Install cylinder heads with new gaskets.

➡Do not rotate crankshaft and camshaft separately, or valves will strike piston heads.

⁂ CAUTION

Cylinder head bolts are tightened by plastic zone tightening method. Whenever the size difference between d1 and d2 exceeds the limit, replace them with new ones. Limit (d1–d2): 0.0043 in. Lubricate threads and seat surfaces of the bolts with new engine oil.

21. Install cylinder head outside bolts Tighten in numerical order shown in the figure. Tightening procedure:

a. Tighten all bolts to 98 Nm (10 kg-m, 72 ft-lb).

b. Completely loosen all bolts.

c. Tighten all bolts to 34 to 44 Nm (3.5 to 4.5 kg-m, 25 to 33 ft-lb).

d. Turn all bolts 90 to 95 degrees clockwise.

e. Turn all bolts 90 to 95 degrees clockwise.

22. Install camshaft chain tensioners on both sides of cylinder head.

23. Install exhaust and intake camshafts and camshaft brackets.

➡Intake camshaft has a drill mark on camshaft sprocket mounting flange. Install it on the intake side. Position camshaft. RH exhaust camshaft dowel pin at about 10 o'clock; LH exhaust camshaft dowel pin at about 2 o'clock

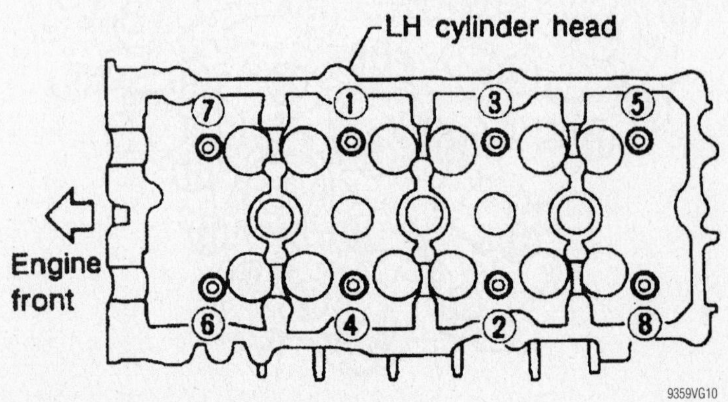

Left cylinder head bolt torque sequence—3.5L engine

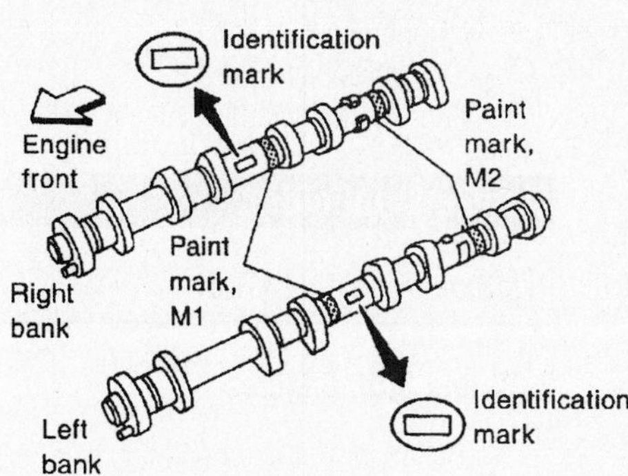

- **Identification marks are present on camshafts.**

Bank	INT/EXH	ID mark	Drill mark	Paint mark	
				M1	M2
RH	INT	R3	Yes	Yes	No
	EXH	R3	No	No	Yes
LH	INT	L3	Yes	Yes	No
	EXH	L3	No	No	Yes

9359VG11

Camshaft identification—3.5L engine

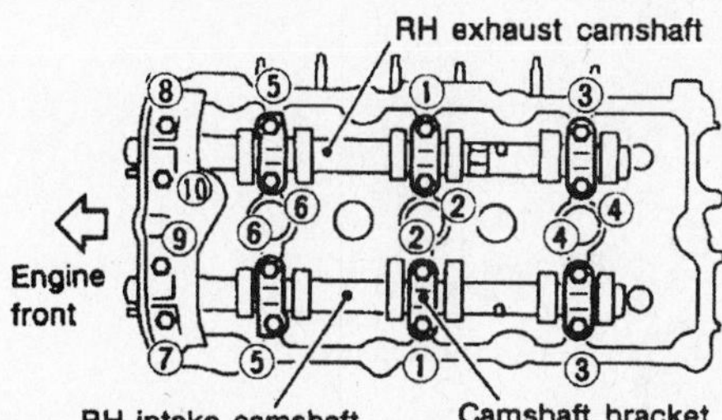

- Tighten the camshaft brackets in the following steps.

Step	Tightening torque	Tightening order
1	1.96 N·m (0.2 kg-m, 17 in-lb)	Tighten in the order of 7 to 10, then tighten 1 to 6.
2	5.88 N·m (0.6 kg-m, 52 in-lb)	Tighten in the numerical order.
3	9.02 - 11.8 N·m (0.92 - 1.20 kg-m, 79.9 - 104.2 in-lb)	Tighten in the order of 1 to 6.
	8.3 - 10.3 N·m (0.9 - 1.0 kg-m, 74 - 91 in-lb)	Tighten in the order of 7 to 10.

9359VG12

Right camshaft bolt torque sequence—3.5L engine

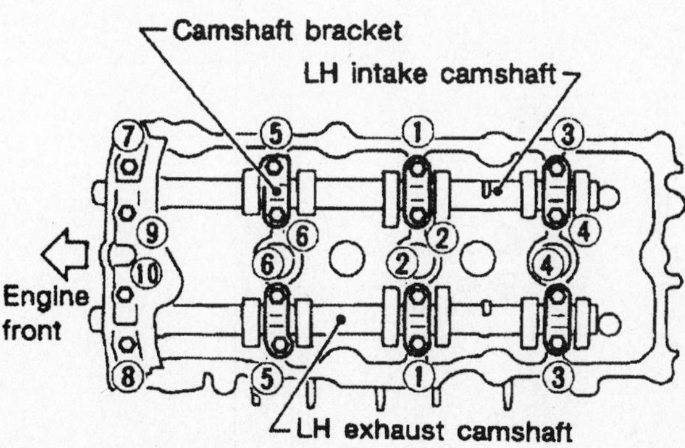

9359VG13

Left camshaft bolt torque sequence—3.5L engine

12 - 13 N•m (1.2 - 1.4 kg-m, 9 - 10 ft-lb)

9359VG14

Rear timing case bolt torque sequence—3.5L engine

24. Before installing camshaft brackets, apply sealant to mating surface of No. 1 journal head. Use Genuine RTV silicone sealant or equivalent. Install camshaft brackets in their original positions. Align stamp mark as shown in the figure. If any part of valve assembly or camshaft is replaced, check valve clearance according to reference data. After completing assembly check valve clearance. Valve clearance (Cold):

- Intake 0.26—0.34 mm (0.010—0.013 in)
- Exhaust 0.29—0.37 mm (0.011—0.015 in)

➡Lubricate threads and seat surfaces of camshaft bracket bolts with new engine oil before installing them.

25. Install intake valve timing control solenoid valves.
26. Install O-rings to cylinder block.
27. Install O-rings to cylinder head.
28. Apply sealant to the hatched portion of rear timing chain case. Apply continuous bead of liquid gasket to mating surface of rear timing chain case. Before installation, wipe off the protruding sealant.
29. Align rear timing chain case with dowel pins, then install on cylinder head and block.
30. Tighten rear chain case bolts.

a. Tighten bolts in numerical order shown in the figure.

b. Repeat above step a.

31. Reinstall all removed parts in reverse order of removal.

Rocker Arms/Shafts

REMOVAL & INSTALLATION

3.3L Engine

1. Before servicing the vehicle, refer to the precautions in the beginning of this section.
2. Remove or disconnect the following:
   - Negative battery cable
   - Upper intake manifold
   - Valve covers
   - Rocker arm and shaft assemblies
   - Rocker arms from the shafts

➡**Keep all valvetrain components in order for assembly.**

To install:

3. Lubricate all contact points with clean engine oil and assemble the rocker arms to the shafts in their original positions.
4. Install or connect the following:
   - Rocker arm and shaft assemblies. Tighten the bolts to 13–16 ft. lbs. (18–22 Nm).
   - Valve covers
   - Upper intake manifold
   - Negative battery cable
5. Start the engine and check for leaks.

Intake Manifold

REMOVAL & INSTALLATION

3.3L and 3.5L Engines

1. Before servicing the vehicle, refer to the precautions in the beginning of this section.
2. Drain the cooling system.

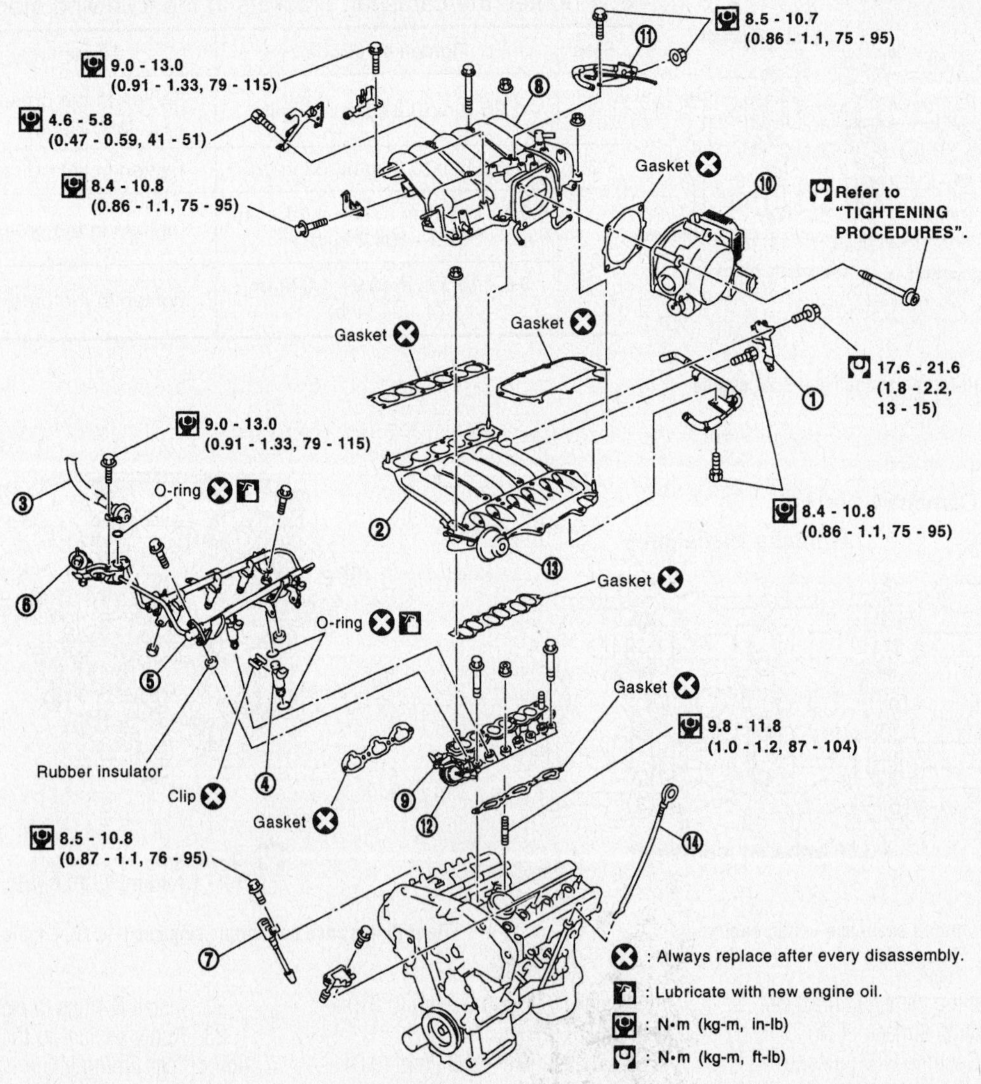

1. Intake manifold collector support
2. Lower intake manifold collector
3. Fuel damper and fuel feed hose assembly
4. Fuel injector
5. [illegible]
6. Fuel pressure regulator
7. Ignition coil with power transistor
8. Upper intake manifold collector
9. Intake manifold
10. Electric throttle control actuator
11. EVAP canister purge volume control solenoid valve
12. Swirl control valve actuator
13. Power valve actuator (A/T)
14. Oil level gauge

67170-PATH-G03

Intake manifold exploded view—3.5L engine

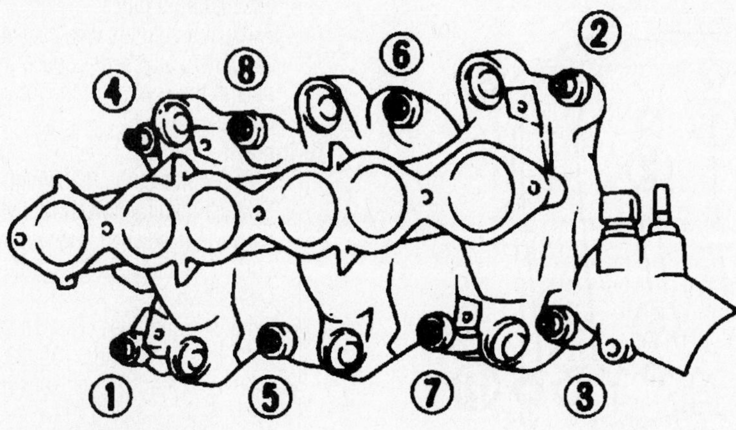

Intake manifold loosening sequence—3.3L engine

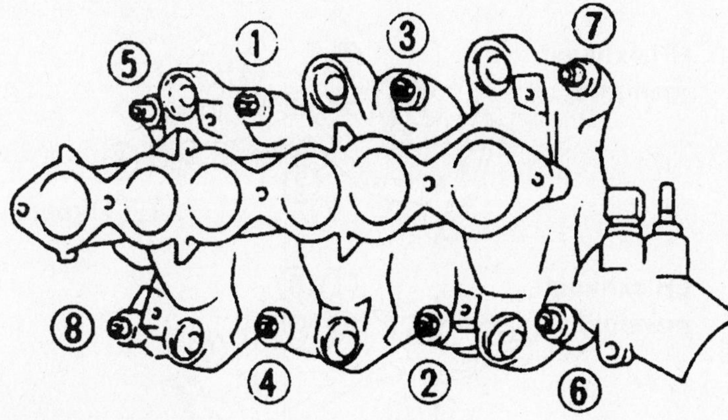

Intake manifold torque sequence—3.3L engine

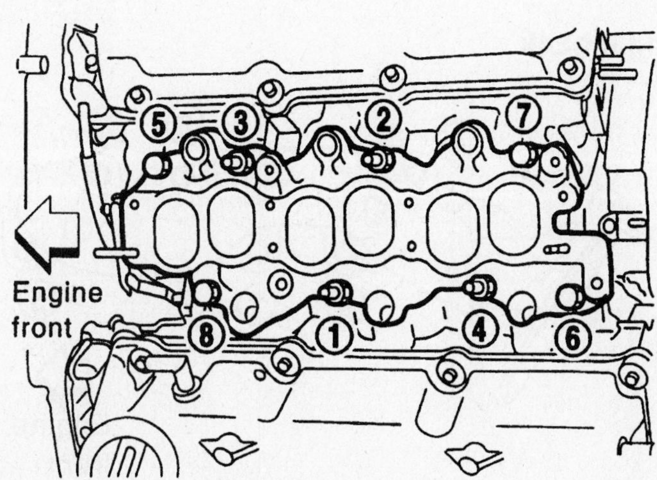

Lower intake manifold torque sequence—3.5L engine

3. Relieve the fuel system pressure.
4. Remove or disconnect the following:
   - Negative battery cable
   - Air intake duct
   - Accelerator cable
   - Cruise control cable
   - Idle Air Control (IAC) valve connector
   - Throttle Position (TP) sensor and switch connectors
   - Ignition coil and power transistor connectors
   - Exhaust Gas Recirculation (EGR) Solenoid valve connector
   - EGR temperature sensor connector
   - Radiator hoses
   - Heater hoses
   - Positive Crankcase Ventilation (PCV) valve and hose
   - Evaporative Emissions (EVAP) canister vacuum and purge hoses
   - Brake booster vacuum hose
   - Fuel pressure regulator vacuum hose
   - EGR tube
   - Spark plug wires
   - Distributor
   - Left bank injector connectors
   - Thermal transmitter
   - Upper intake manifold ground cable
   - Breather pipe
   - Upper intake manifold
   - Fuel lines
   - Right bank injector connectors
   - Fuel supply manifold
   - Engine Coolant Temperature (ECT) sensor connector
   - Lower intake manifold. Loosen the fasteners in the sequence shown.

To install:

5. Install the lower intake manifold with a new gasket.
6. For 3.3L engines, tighten the fasteners in sequence as follows:
   a. Step 1: 35 inch lbs. (4 Nm)
   b. Step 2: 78 inch lbs. (9 Nm)
   c. Step 3: 70–84 inch lbs. (8–10 Nm)
7. For 3.5L engines, tighten the fasteners in sequence as follows:
   a. Step 1: 86 inch lbs. (4 Nm)
   b. Step 2: 23 ft. lbs. (9 Nm)
8. Install or connect the following:
   - ECT sensor connector
   - Fuel supply manifold
   - Right bank injector connectors
   - Fuel lines
   - Upper intake manifold
   - Breather pipe
   - Upper intake manifold ground cable
   - Thermal transmitter
   - Left bank injector connectors
   - Distributor

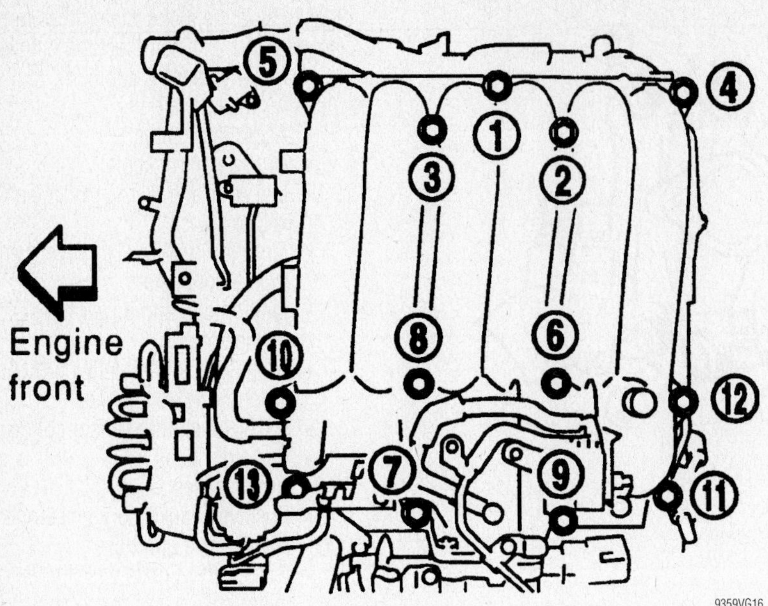

Upper intake manifold torque sequence—3.5L engine

- Spark plug wires
- EGR tube
- Fuel pressure regulator vacuum hose
- Brake booster vacuum hose
- EVAP canister vacuum and purge hoses
- PCV valve and hose
- Heater hoses
- Radiator hoses
- EGR temperature sensor connector
- EGR Solenoid valve connector
- Ignition coil and power transistor connectors
- TP sensor and switch connectors
- IAC valve connector
- Cruise control cable
- Accelerator cable
- Air intake duct
- Negative battery cable

9. Fill the cooling system.
10. Start the engine and check for leaks.

Exhaust Manifold

REMOVAL & INSTALLATION

3.3L and 3.5L Engines

1. Before servicing the vehicle, refer to the precautions in the beginning of this section.
2. Remove or disconnect the following:
    - Negative battery cable
    - Exhaust manifold heat shields
    - Exhaust Gas Recirculation (EGR) tube
    - Heated Oxygen (HO_2S) sensor connectors
    - Exhaust front pipes
    - Exhaust manifolds with catalytic converters attached. Loosen the nuts in the reverse of the torque sequence.

To install:

3. Install or connect the following:
    - Exhaust manifolds with catalytic converters attached. Tighten the nuts in sequence to 21–25 ft. lbs. (28–33 Nm).
    - Exhaust front pipes. Tighten the bolts to 21–25 ft. lbs. (28–33 Nm).
    - Heated Oxygen (HO_2S) sensor connectors
    - EGR tube. Tighten the flange fittings to 29–36 ft. lbs. (39–49 Nm).
    - Exhaust manifold heat shields. Tighten the bolts to 84–96 inch lbs. (9–11 Nm)
    - Negative battery cable
4. Start the engine and check for leaks.

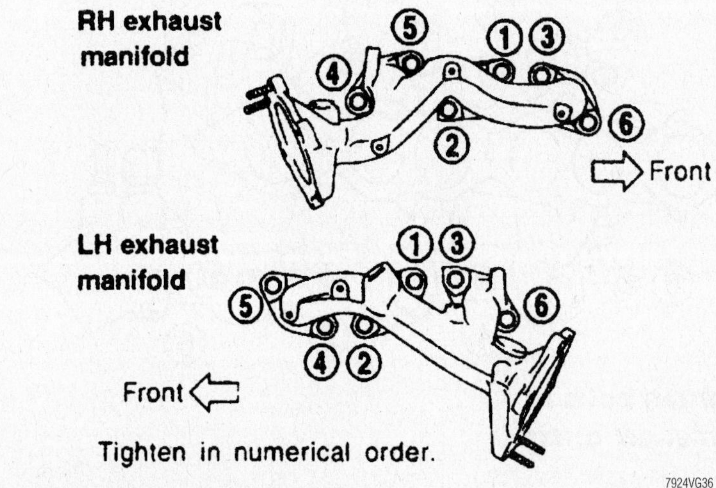

Exhaust manifold torque sequence—3.3L engine

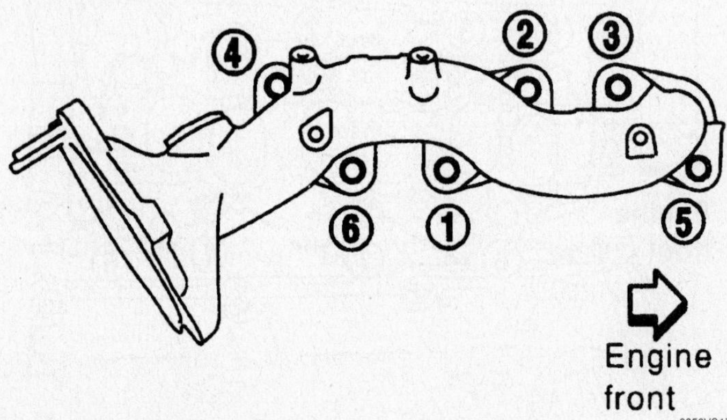

Right exhaust manifold torque sequence—3.5L engine

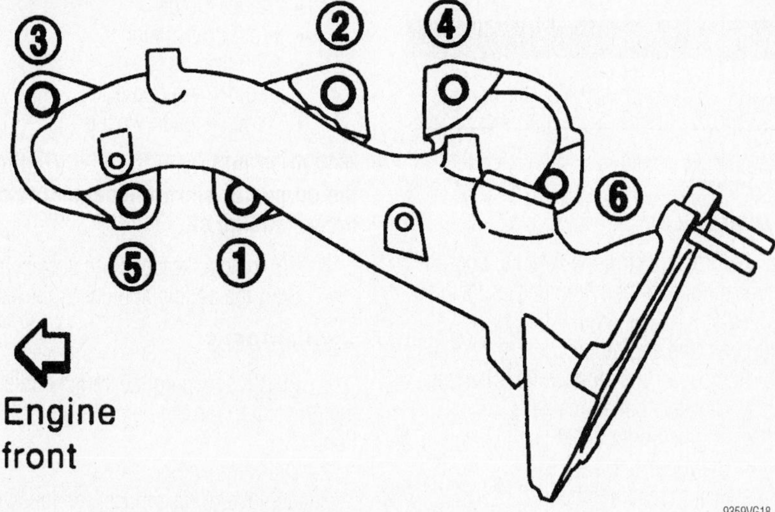

Exhaust manifold torque sequence—3.5L engine

Front Crankshaft Seal

REMOVAL & INSTALLATION

For the 3.5L engine, see the Timing Chain procedure.

3.3L Engine

1. Before servicing the vehicle, refer to the precautions in the beginning of this section.
2. Drain the cooling system.
3. Remove or disconnect the following:
   - Negative battery cable
   - Accessory drive belts
   - Radiator hoses
   - Crankshaft pulley
   - Front cover
   - Timing belt. Refer to the Timing Belt unit repair section.
   - Crankshaft timing sprocket
   - Front crankshaft seal

To install:

4. Install or connect the following:
   - Front crankshaft seal flush with the oil pump housing
   - Crankshaft timing sprocket
   - Timing belt
   - Front cover. Tighten the bolts to 26–43 inch lbs. (3–5 Nm).
   - Crankshaft pulley. Tighten the bolt to 141–156 ft. lbs. (191–211 Nm).
   - Radiator hoses
   - Accessory drive belts
   - Negative battery cable
5. Fill the cooling system.
6. Start the engine and check for leaks.

Camshaft and Valve Lifters

REMOVAL & INSTALLATION

3.3L Engines

1. Before servicing the vehicle, refer to the precautions in the beginning of this section.
2. Drain the cooling system.
3. Remove or disconnect the following:
   - Negative battery cable
   - Upper intake manifold
   - Valve covers

➡Keep all valvetrain components in order for assembly.

   - Rocker arm and shaft assemblies
   - Valve lifter guide and valve lifters. Attach a wire to the top of the lifters so that they will not drop from the lifter guide.
   - Radiator
   - Accessory drive belts
   - Front cover
   - Timing belt. Refer to the Timing Belt unit repair section.
   - Camshaft sprockets
   - Camshaft seals
   - Rear timing cover
   - Distributor
   - Cylinder head rear covers
   - Camshaft locating plates
   - Camshafts

To install:

4. Install or connect the following:
   - Camshafts
   - Camshaft locating plates. Tighten the bolts to 58–65 ft. lbs. (78–88 Nm).
   - Cylinder head rear covers
   - Distributor
   - Rear timing cover
   - Camshaft seals
   - Camshaft sprockets. Tighten the bolts to 58–65 ft. lbs. (78–88 Nm).
   - Timing belt
   - Front cover
   - Accessory drive belts
   - Radiator
   - Valve lifter guide and valve lifters
   - Rocker arm and shaft assemblies. Tighten the bolts to 13–16 ft. lbs. (18–22 Nm).
   - Valve covers
   - Upper intake manifold
   - Negative battery cable
5. Fill the cooling system.
6. Start the engine and check for leaks.

3.5L Engines

See the Cylinder Head Removal and Installation procedure.

Valve Lash

ADJUSTMENT

3.3L Engines

These engines are equipped with hydraulic valve lifters that do not require periodic adjustment.

3.5L Engines

➡Adjust valve clearance while engine is cold.

1. Turn crankshaft, to position cam lobe on camshaft of valve that must be adjusted upward.
2. Thoroughly wipe off engine oil around adjusting shim using a rag.
3. Using an extra-fine screwdriver, turn the round hole of the adjusting shim in the direction of the arrow.
4. Place Tool (A) around camshaft as shown in figure.

Before placing Tool (A), rotate notch toward center of cylinder head (See figure.), to simplify shim removal later.

✲✲ CAUTION

Be careful not to damage cam surface with Tool (A).

5. Rotate Tool (A) (See figure.) so that valve lifter is pushed down.
6. Place Tool (B) between camshaft and the edge of the valve lifter to retain valve lifter.

⁂ CAUTION

Tool (B) must be placed as close to camshaft bracket as possible. Be careful not to damage cam surface with Tool (B).

7. Remove Tool (A).
8. Blow air into the hole to separate adjusting shim from valve lifter.
9. Remove adjusting shim using a small screwdriver and a magnetic finger.
10. Determine replacement adjusting shim size following formula. Using a micrometer determine thickness of removed shim. Calculate thickness of new adjusting shim so valve clearance comes within specified values.
    - R = Thickness of removed shim
    - N = Thickness of new shim
    - M = Measured valve clearance
    - Intake: N = R + [M—0.30 mm (0.0118 in)]
    - Exhaust: N = R + [M—0.33 mm (0.0130 in)]

Shims are available in 64 sizes from 2.32 mm (0.0913 in) to 2.95 mm (0.1161 in), in steps of 0.01 mm (0.0004 in). Select new shim with thickness as close as possible to calculated value.

11. Install new shim using a suitable tool. Install with the surface on which the thickness is stamped facing down.
12. Place Tool (A) as mentioned in steps 2 and 3.
13. Remove Tool (B).
14. Remove Tool (A).
15. Recheck valve clearance.

Valve clearance (Cold)
- Intake: 0.010—0.013
- Exhaust: 0.011—0.015

Starter Motor

REMOVAL & INSTALLATION

1. Before servicing the vehicle, refer to the precautions in the beginning of this section.
2. Remove or disconnect the following:
    - Negative battery cable
    - Engine under cover
    - Starter harness connectors
    - Starter motor

To install:

3. Install or connect the following:
    - Starter motor. Tighten the bolts to 22–27 ft. lbs. (30–36 Nm) on the 3.3L; 37–45 ft. lbs. (61-69NM) on the 3.5L.
    - Starter harness connectors
    - Engine under cover
    - Negative battery cable

Oil Pan

REMOVAL & INSTALLATION

3.3L Engine

2WD MODELS

1. Before servicing the vehicle, refer to the precautions in the beginning of this section.
2. Drain the engine oil.
3. Remove or disconnect the following:
    - Negative battery cable
    - Engine under cover
    - Stabilizer bar
    - Front crossmember
    - Starter motor
    - Transmission mount
    - Left and right motor mounts
    - Power steering gear
4. Raise and support the engine for clearance.
5. Remove or disconnect the following:
    - Oil pan bolts in the sequence
    - Oil pan

To install:

6. Apply a continuous bead of sealant 0.138–0.177 in. (3.5–4.5mm) to the oil pan mating surface.
7. Install or connect the following:
    - Oil pan. Tighten the bolts in reverse of the removal sequence to 62 inch lbs. (7 Nm).
    - Power steering gear
    - Left and right motor mounts
    - Transmission mount
    - Starter motor
    - Front crossmember
    - Stabilizer bar
    - Engine under cover
    - Negative battery cable

➡Wait 30 minutes after installation of the oil pan to allow the sealant to cure before adding oil.

8. Fill the crankcase to the correct level.
9. Start the engine and check for leaks.

4WD MODELS

1. Before servicing the vehicle, refer to the precautions in the beginning of this section.
2. Drain the engine oil.
3. Remove or disconnect the following:
    - Negative battery cable
    - Engine under cover
    - Stabilizer bar brackets
    - Front driveshaft
    - Axle halfshafts
    - Front suspension crossmember
    - Front differential and mounting bracket
    - Starter motor
    - Transmission mount
    - Left and right motor mounts
    - Power steering gear
    - Relay rod
4. Raise and support the engine for clearance.
5. Remove or disconnect the following:
    - Oil pan bolts in the sequence
    - Oil pan

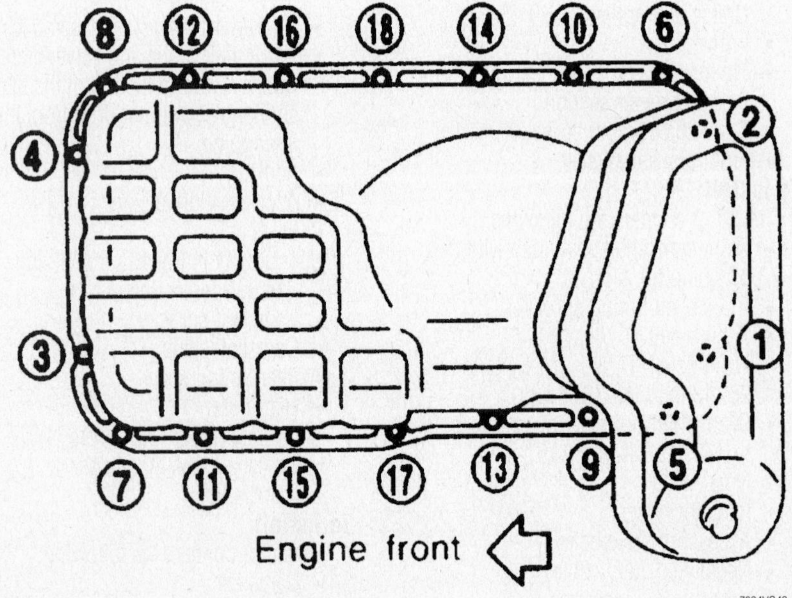

Oil pan bolt removal sequence—3.3L engine

To install:

6. Apply a continuous bead of sealant 0.138–0.177 in. (3.5–4.5mm) to the oil pan mating surface.
7. Install or connect the following:
   - Oil pan. Tighten the bolts in reverse of the removal sequence to 62 inch lbs. (7 Nm).
   - Relay rod
   - Power steering gear
   - Left and right motor mounts
   - Transmission mount
   - Starter motor
   - Front differential and mounting bracket
   - Front suspension crossmember
   - Axle halfshafts
   - Front driveshaft
   - Stabilizer bar brackets
   - Engine under cover
   - Negative battery cable

➡**Wait 30 minutes after installation of the oil pan to allow the sealant to cure before adding oil.**

8. Fill the crankcase to the correct level.
9. Start the engine and check for leaks.

3.5L Engines

1. Remove front RH and LH wheels.
2. Remove battery.
3. Remove oil level gauge.
4. Remove engine undercover.
5. Remove suspension member stay.
6. Drain engine coolant from radiator drain plug.
7. Disconnect A/T oil cooler hoses. (A/T)
8. Drain engine oil.
9. Remove the crankshaft position sensors (REF and POS).
10. Remove drive belts and idler pulley with bracket.
11. Remove power steering oil pump, then put it aside holding with a suitable wire.
12. Remove alternator.
13. Install engine slingers.
14. Remove front propeller shaft. (4WD)
15. Remove exhaust front tube heat insulators, then remove rear heat oxygen sensors.
16. Remove exhaust front tube from both sides.
17. Remove front final drive. (4WD)
18. Remove starter motor.
19. Disconnect oil pressure switch harness connector.
20. Loosen and disconnect the bolts fixing the steering column assembly lower joint and the power steering gear.
21. Set a suitable transmission jack under the front suspension member and hoist engine with engine slingers.
22. Remove front engine mounting nuts from both sides.
23. Remove front suspension member bolts.
24. Lower the transmission jack carefully to secure clearance between the oil pan and suspension member.
25. Remove A/T oil cooler tube. (A/T)
26. Remove water hose and tube. (A/T)
27. Remove the four engine-to-transmission bolts.
28. Remove aluminum oil pan bolts in numerical order.
29. Remove aluminum oil pan.
   a. Insert tool between aluminum oil pan and cylinder block.

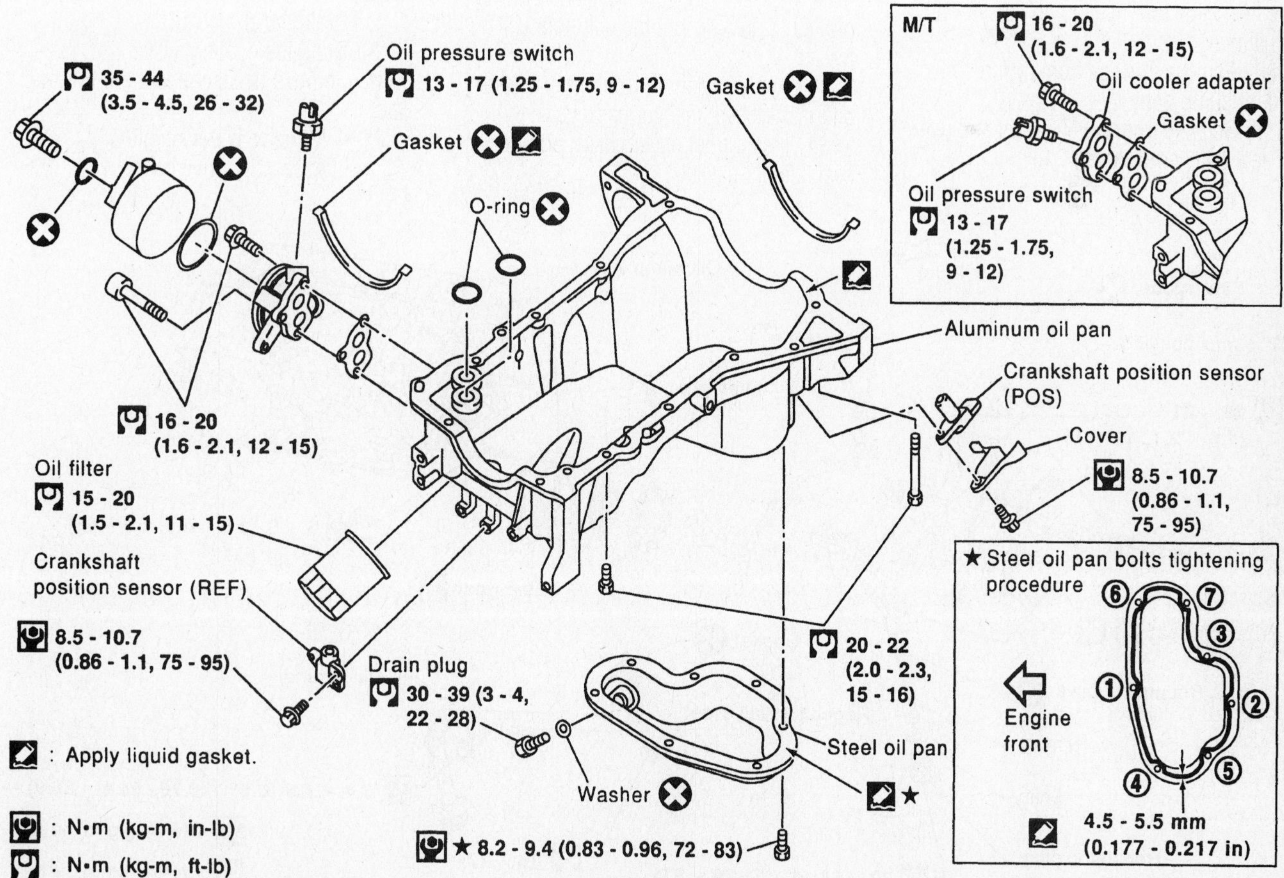

Oil pan exploded view—3.5L engine

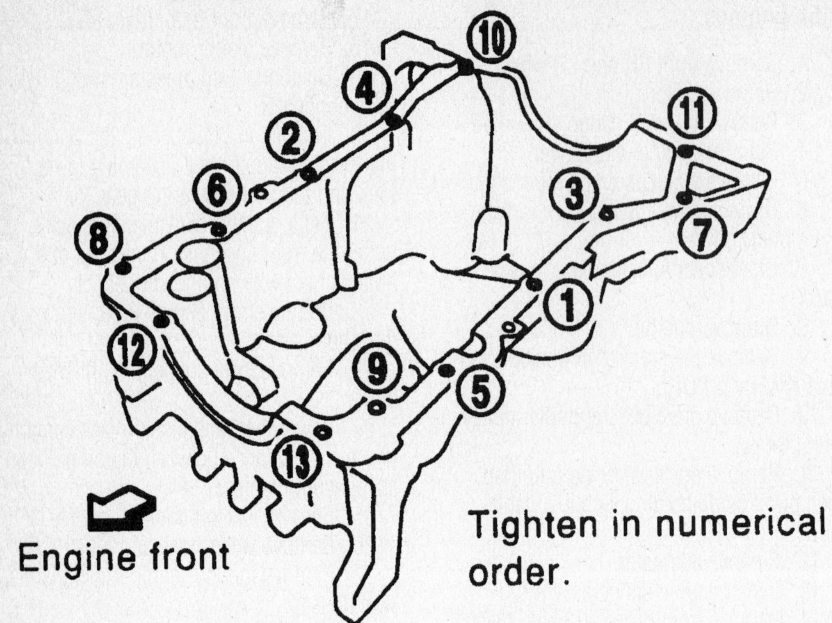

Oil pan bolt torque sequence—3.5L engine

➡Be careful not to damage aluminum mating surface. I Do not insert screwdriver, or oil pan flange will be deformed.

b. Slide tool by tapping its side with a hammer.

30. Remove O-rings from cylinder block and oil pump body.

31. Remove front cover gasket and rear oil seal retainer gasket.

To install:

32. Before installing oil pan, remove old liquid gasket from mating surface using a scraper. Also remove old liquid gasket from mating surface of cylinder block. Remove old liquid gasket from the bolt hole and thread.

33. Apply sealant to front cover gasket and rear oil seal retainer gasket.

34. Install front cover gasket and rear oil seal retainer gasket.

35. Apply a continuous bead of liquid gasket to mating surface of aluminum oil pan. Use RTV silicone sealant or equivalent.

36. Apply liquid gasket to inner sealing surface as shown in figure. Be sure liquid gasket is 4.0 to 5.0 mm (0.157 to 0.197 in) or 4.5 to 5.5 mm (0.177 to 0.217 in) wide. Attaching should be done within 5 minutes after coating.

37. Install O-rings, cylinder block and oil pump body.

38. Install aluminum oil pan. Tighten bolts in numerical order. Wait at least 30 minutes before refilling engine oil.

39. Install the four engine-to-transmission bolts.

40. Reinstall in the reverse order of removal.

Oil Pump

REMOVAL & INSTALLATION

3.3L Engine

1. Before servicing the vehicle, refer to the precautions in the beginning of this section.
2. Drain the engine oil.
3. Drain the cooling system.
4. Remove or disconnect the following:
   - Negative battery cable
   - Accessory drive belts

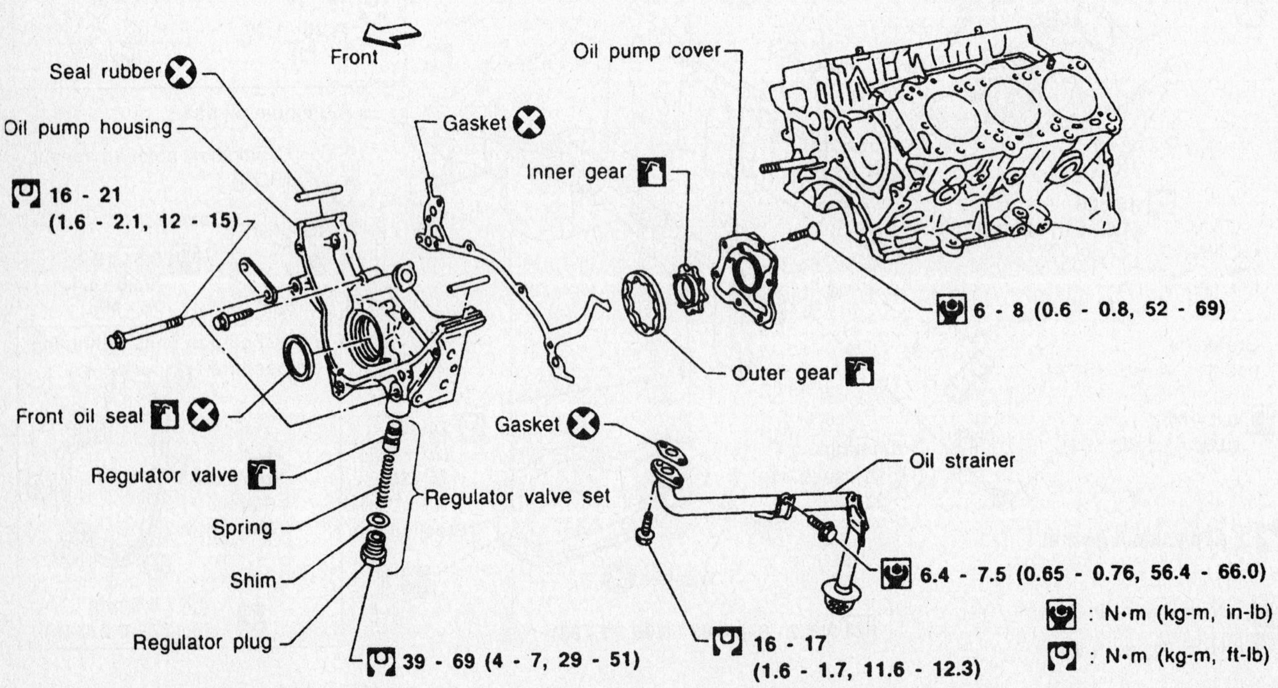

Oil pump assembly exploded view—3.3L engine

- Radiator hoses
- Crankshaft pulley
- Front cover
- Timing belt. Refer to the Timing Belt unit repair section.
- Crankshaft timing sprocket
- Oil pan
- Oil pump pickup tube
- Oil pump

To install:

5. Install or connect the following:
   - Oil pump. Tighten the large bolts to 16–22 ft. lbs. (22–29 Nm) and the small bolts to 55–74 inch lbs. (6–8 Nm).
   - Oil pump pickup tube. Tighten the flange bolts to 12 ft. lbs. (16 Nm) and the bracket bolt to 55–74 inch lbs. (6–8 Nm).
   - Oil pan
   - Crankshaft timing sprocket
   - Timing belt
   - Front cover
   - Crankshaft pulley
   - Radiator hoses
   - Accessory drive belts
   - Negative battery cable
6. Fill the cooling system.
7. Fill the crankcase to the correct level.
8. Start the engine and check for leaks.

3.5L Engine

1. Remove timing chain.
2. Remove oil pump assembly.
3. Reinstall any parts removed in reverse order of removal.

Rear Main Seal

REMOVAL & INSTALLATION

3.3L Engines

1. Before servicing the vehicle, refer to the precautions in the beginning of this section.
2. Remove or disconnect the following:
   - Transmission
   - Flywheel
   - Rear main seal

To install:

3. Install the seal so that it is flush with the retainer housing.
4. Install or connect the following:
   - Flywheel. Tighten the bolts to 61–69 ft. lbs. (83–93 Nm).
   - Transmission

3.5L Engine

1. Remove transmission.
2. Remove flywheel or drive plate.
3. Remove oil pan.
4. Remove rear oil seal retainer.
5. Remove old liquid gasket using scraper. Remove old liquid gasket from the bolt hole and thread.
6. Apply liquid gasket to rear oil seal retainer.

Timing Belt

REMOVAL & INSTALLATION

3.3L Engines

1. Remove the engine undercover.
2. Remove the radiator shroud, the fan and the pulleys.
3. Drain the coolant from the radiator and remove the water pump hose.

✱✱ CAUTION

When draining the coolant, keep in mind that cats and dogs are attracted by the ethylene glycol antifreeze, and are quite likely to drink any that is left in an uncovered container or in puddles on the ground. This will prove fatal in sufficient quantity. Always drain the coolant into a sealable container. Coolant should be reused unless it is contaminated or several years old.

4. Remove the radiator.
5. Remove the power steering, air conditioning compressor and alternator drive belts.
6. Remove the spark plugs.
7. Remove the distributor protector (dust shield).
8. Remove the air conditioning compressor drive belt idler pulley and bracket.

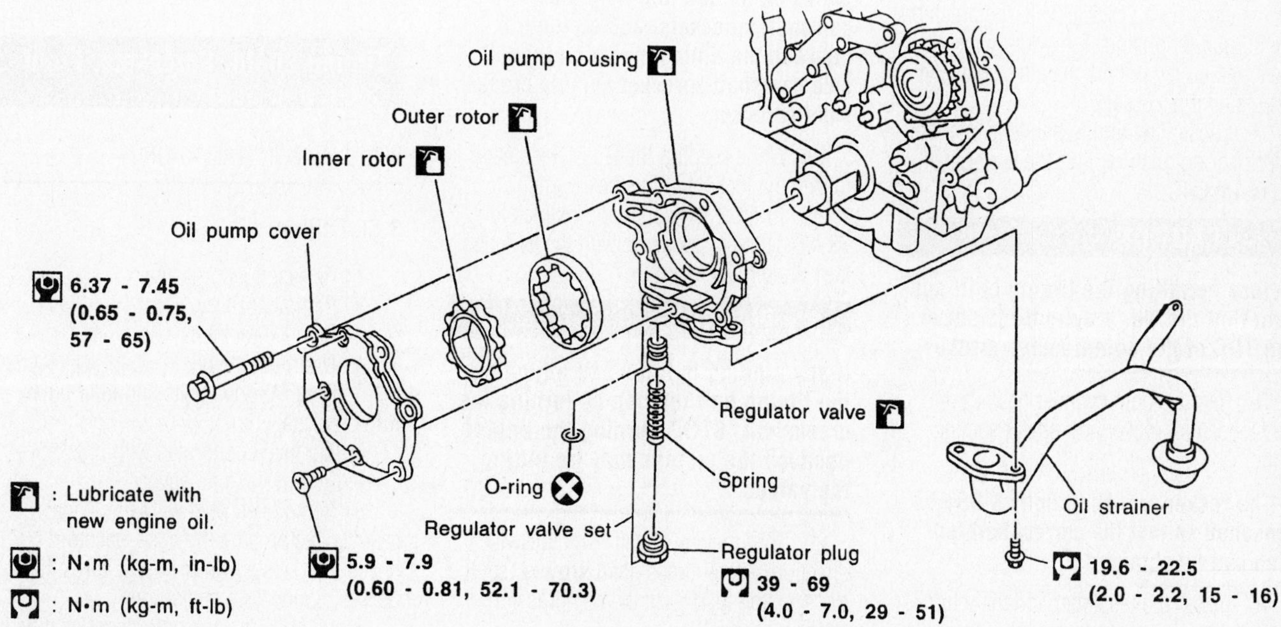

Oil pump assembly exploded view—3.5L engine

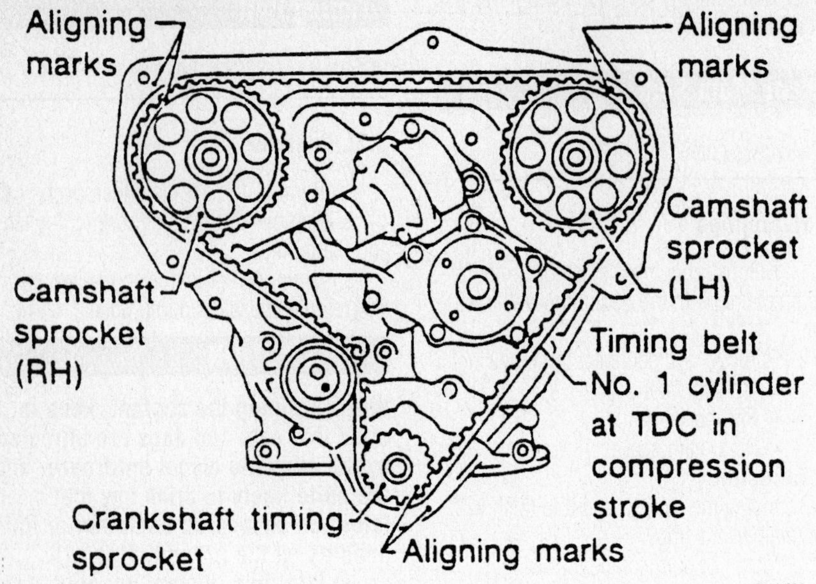

Timing belt alignment mark locations —3.0L and 3.3L engines

9. Remove the fresh air intake tube at the cylinder head cover.
10. Disconnect the radiator hose at the thermostat housing.
11. Remove the crankshaft pulley bolt, then pull off the pulley with a suitable puller.
12. Remove the bolts, then remove the front upper and lower timing belt covers.
13. Set the No. 1 piston at Top Dead Center (TDC) of its compression stroke. Align the punchmark on the left camshaft sprocket with the punchmark on the timing belt upper rear cover. Align the punchmark on the crankshaft sprocket with the notch on the oil pump housing. Temporarily install the crank pulley bolt so the crankshaft can be rotated if necessary.
14. Loosen the timing belt tensioner and return spring, then remove the timing belt.

To install:

⁂ CAUTION

Before installing the timing belt, confirm that the No. 1 cylinder is set at the TDC of the compression stroke.

15. Remove both cylinder head covers and loosen all rocker arm shaft retaining bolts.

➡The rocker arm shaft bolts MUST be loosened so that the correct belt tension can be obtained.

16. Install the tensioner and the return spring. Using a hexagon wrench, turn the tensioner clockwise and temporarily tighten the locknut.
17. Be sure that the timing belt is clean and free from oil or water.
18. When installing the timing belt, align the white lines on the belt with the punchmarks on the camshaft and crankshaft sprockets. Have the arrow on the timing belt pointing toward the front belt covers.

➡A good way (although rather tedious!) to check for proper timing belt installation is to count the number of belt teeth between the timing marks. There are 133 teeth on the belt; there should be 40 teeth between the timing marks on the left and right side camshaft sprockets, and 43 teeth between the timing marks on the left side camshaft sprocket and the crankshaft sprocket.

19. While keeping the tensioner steady, loosen the locknut with a hex wrench.
20. Turn the tensioner approximately 70–80 degrees clockwise with the wrench, then tighten the locknut.

⁂ WARNING

If any binding is felt when adjusting the timing belt tension by turning the crankshaft, STOP turning the engine, because the pistons may be hitting the valves.

21. Turn the crankshaft in a clockwise direction several times, then **slowly** set the No. 1 piston to TDC of the compression stroke.
22. Apply 22 lbs. (10 kg) of pressure (push it in!) to the center span of the timing belt between the right side camshaft sprocket and the tensioner pulley, then loosen the tensioner locknut.
23. Using a 0.0138 in. (0.35mm) thick feeler gauge (the actual width of the blade **must** be ½ in. or 13mm!), turn the crankshaft clockwise (**slowly!**). The timing belt should move approximately 2½ teeth. Tighten the tensioner locknut, turn the crankshaft slightly and remove the feeler gauge.
24. Slowly rotate the crankshaft clockwise several more times, then set the No. 1 piston to TDC of the compression stroke.
25. Position the 2 timing covers on the block, then tighten the mounting bolts to 24 ft. lbs. (35 Nm).
26. Press the crankshaft pulley onto the shaft, then tighten the bolt to 90–98 ft. lbs. (123–132 Nm).
27. Connect the radiator hose to the thermostat housing.
28. Reconnect the fresh air intake tube at the cylinder head cover.
29. Install the air conditioning compressor drive belt idler pulley and bracket.
30. Install the distributor protector (dust shield).
31. Install the spark plugs.
32. Install the power steering, air conditioning compressor and alternator drive belts.
33. Install the radiator.
34. Reconnect the water pump hose and fill the engine with coolant. Install the fan shroud and pulleys.
35. Install the engine undercover.
36. Start the engine and check for any leaks.

Timing Chain, Sprockets, Front Cover and Seal

REMOVAL & INSTALLATION

3.5L Engine

1. Release fuel pressure.
2. Remove battery.
3. Remove radiator.
4. Drain engine oil.
5. Remove drive belts and idler pulley with brackets.
6. Remove cooling fan with bracket.
7. Remove engine cover.
8. Remove air duct with air cleaner case, collector, blow-by hose, vacuum hoses, fuel hoses, water hoses, wires, harnesses, connectors and so on.
9. Remove the air compressor, and tie it down using rope or the like to keep it from interfering.

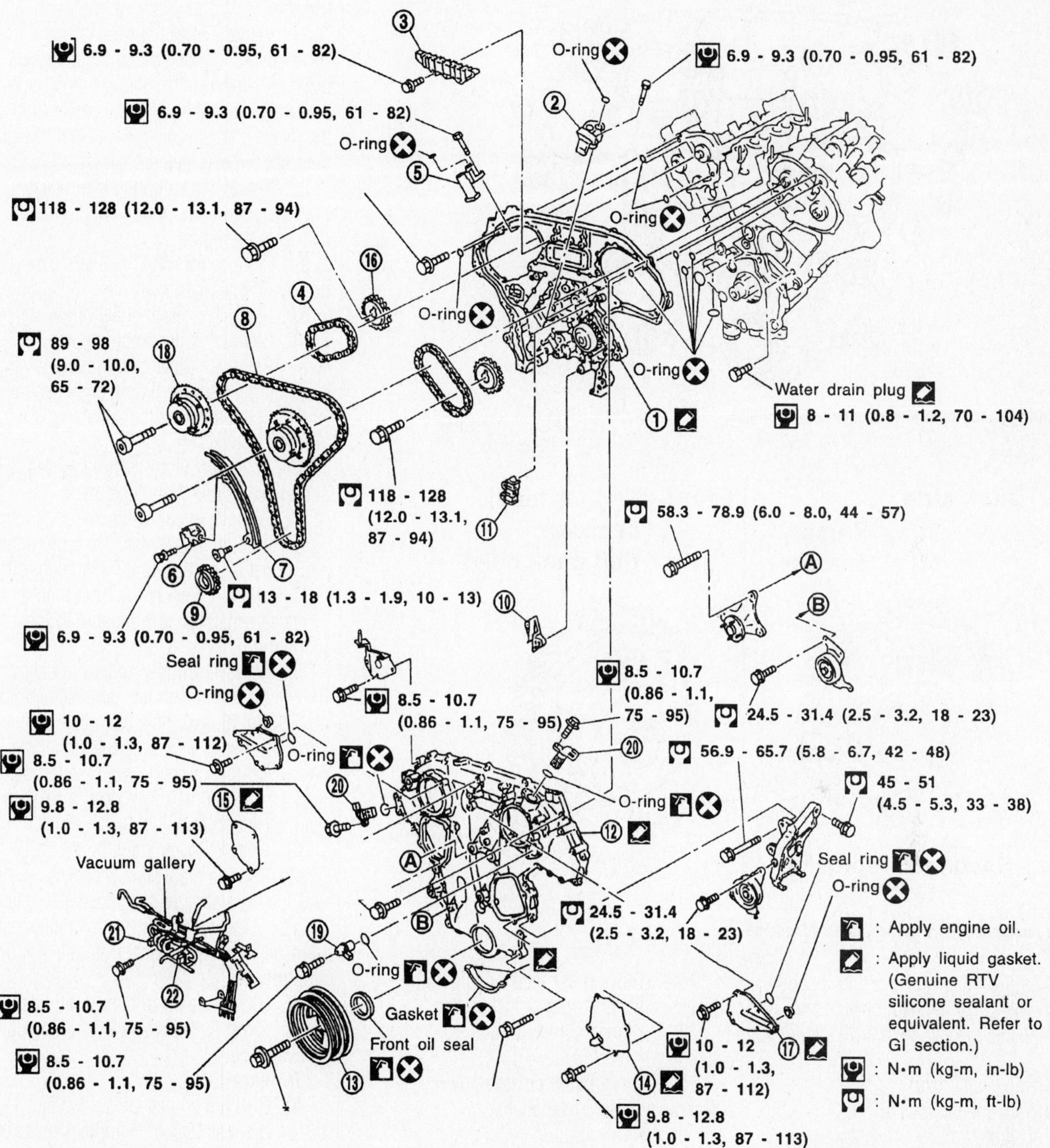

1. Rear timing chain case
2. Left camshaft chain tensioner
3. Internal guide
4. Timing chain (Secondary)
5. Right camshaft chain tensioner
6. Timing chain tensioner
7. Slack guide
8. Timing chain (Primary)
9. Crankshaft sprocket
10. Lower tension guide
11. Upper tension guide
12. Front timing chain case
13. Crankshaft pulley
14. Water pump cover
15. Chain tensioner cover
16. Exhaust camshaft sprocket
17. Intake valve timing control valve cover
18. Intake camshaft sprocket
19. Camshaft position sensor (PHASE)
20. Intake valve timing control position sensor
21. Power valve actuator (A/T)
22. Swirl control valve control solenoid valve

9359VG30

Timing chain components—3.5L engine

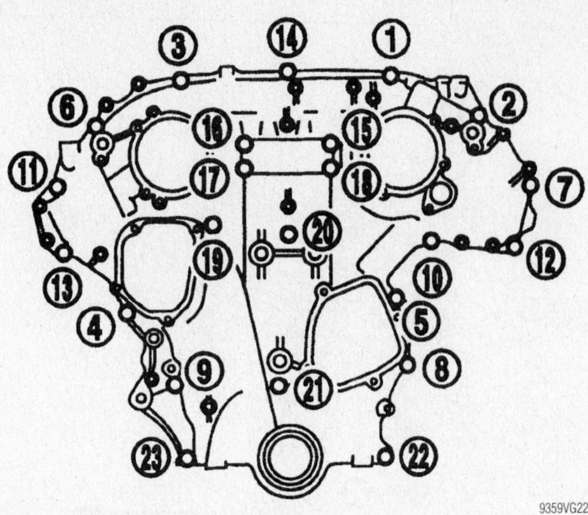

Rear timing case removal sequence—3.5L engine

Primary and secondary sprockets—3.5L engine

10. Remove the power steering oil pump and reservoir tank. Tie them down using rope or the like to keep them from interfering.

11. Remove alternator.

12. Remove the following.

- Vacuum gallery
- Water bypass pipe
- Brackets

13. Remove camshaft position sensor (PHASE), intake valve timing control position sensors and crankshaft position sensor.

➡Avoid impact such as dropping. Do not disassemble the components. Do not place them on areas where iron powder may adhere. Keep away from the objects susceptible to magnetism.

14. Remove upper intake manifold collector in reverse order of installation.

15. Remove intake manifold collector support bolts.

16. Remove lower intake manifold collector in reverse order of installation.

17. Disconnect injector harness connectors.

18. Remove fuel tube assembly in reverse order of installation.

19. Remove ignition coils.

20. Remove RH and LH rocker covers from cylinder head.

21. Set No. 1 piston at TDC on the compression stroke by rotating crankshaft. Align pointer with TDC mark on crankshaft pulley. Check that intake and exhaust cam nose on No. 1 cylinder are installed as shown left. If not, turn the crankshaft one revolution (360°) and align as above.

22. Remove starter motor, and set ring gear stopper using the mounting bolt hole. Be careful not to damage the signal plate teeth.

23. Loosen the crankshaft pulley bolt.

24. Remove crankshaft pulley with a suitable puller.

25. Remove aluminum oil pan.

26. Temporarily install the suspension member bolts and engine mounting nuts.

27. Remove intake valve timing control valve covers. Loosen bolts in numerical order as shown in the figure. In the cover, the shaft is engaged with the center hole of the intake cam sprocket. Remove it straight out until the engagement comes off.

28. Remove front timing chain case bolts. Loosen bolts in numerical order as shown in the figure.

29. Remove front timing chain case. Do not scratch sealing surfaces.

30. Remove internal chain guide.

31. Remove upper tension guide.

32. Remove timing chain tensioner and slack guide. Remove timing chain tensioner. (Push piston and insert a suitable pin into pinhole.)

33. Attach a suitable stopper pin to RH and LH camshaft chain tensioners.

34. Remove intake and exhaust camshaft sprocket bolts. I Apply paint to timing chain and camshaft sprockets for alignment during installation. Secure the hexagonal head of the camshaft using a spanner to loosen mounting bolts.

35. Remove primary and secondary timing chains along with the camshaft sprockets. Do not disassemble the intake camshaft sprocket. Avoid damaging the signal mark protrusion area at the front of the left bank intake camshaft sprocket. Keep it away from magnetized objects.

36. Remove lower chain guide.

37. Remove crankshaft sprocket.

38. Use a scraper to remove all traces of liquid gasket from front timing chain case. Remove old liquid gasket from the bolt hole and thread.

39. Use a scraper to remove all traces of liquid gasket from intake valve timing control valve cover.

To install:

40. Position crankshaft so that No. 1 piston is set at TDC on compression stroke.

41. Install crankshaft sprocket on crankshaft. Make sure that mating marks on crankshaft sprocket face

front of engine.

42. Install lower chain guide on dowel pin, with front mark on the guide facing upside.

43. Press and shrink the secondary chain tensioner sleeve, and fix it using stopper pins. Lubricate threads and seat surfaces of camshaft sprocket bolts with new engine oil.

44. Install secondary timing chain and sprocket to one of the banks (Right bank shown in the figure) as described below.

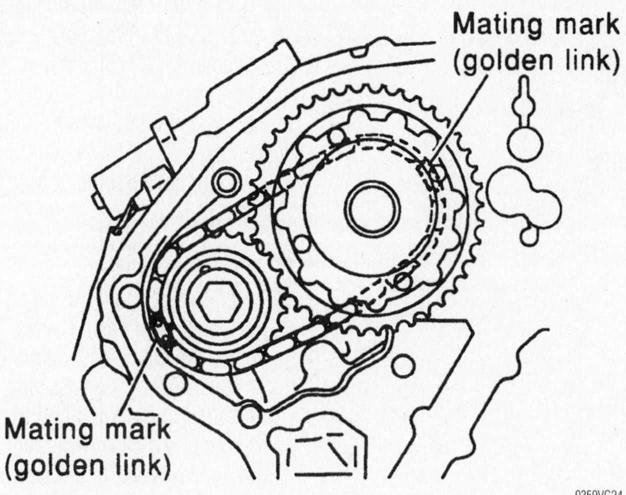

Secondary timing chain installed—3.5L engine

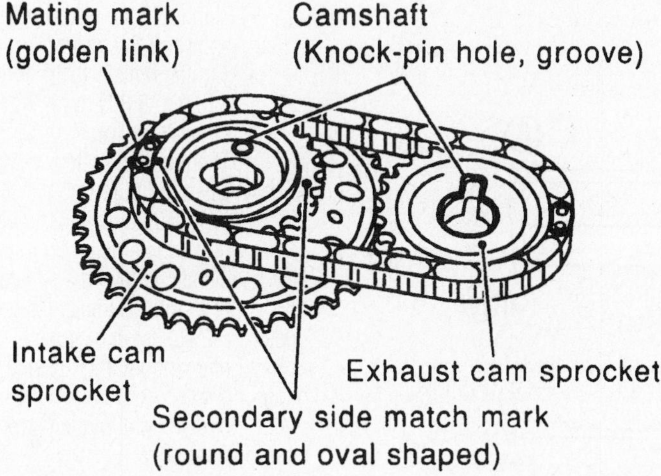

Intake sprocket mating marks—3.5L engine

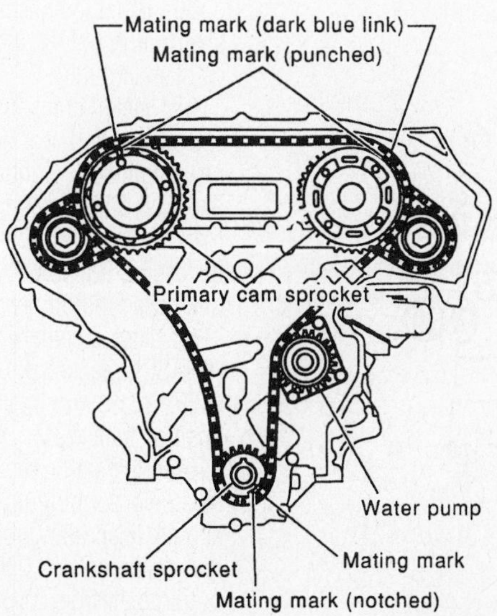

Primary timing chain installation—3.5L engine

a. Align mating marks (golden links) on secondary timing chain with those (punched marks) on the intake and exhaust sprockets.

b. Align camshaft knock pins with the sprocket groove and hole. Because camshaft sprocket mounting bolts are tightened in step 7, perform manual tightening to the extent necessary to keep camshaft knock pin from dislocating. Matching marks of the intake sprocket are on the back side of the secondary sprockets. There are two types of the marks; round and oval types, which should be used for right and left banks respectively.

- Right bank: Round
- Left bank: Oval

It may be difficult to visually check the dislocation of mating marks during and after installation. To make the matching easier, make a mating mark on the sprocket teeth in advance using paint.

45. Install secondary timing chain and sprocket to the other bank. Install primary timing chain at the same time. Installation of the secondary timing chain follows the procedure described in step 5.

46. Install primary timing chain so that mating mark (punched) on camshaft sprocket is aligned with that (dark blue link) on the timing chain, and mating mark (notched) on crankshaft sprocket is aligned with that on the timing chain, respectively.

47. When it is difficult to align mating marks of the primary timing chain with each sprocket, gradually turn the camshaft hexagonal head using a spanner so it is aligned with the mating mark.

48. During alignment, be careful to prevent dislocation of mating marks on the secondary timing chain.

49. After confirming the mating marks are aligned, tighten the camshaft sprocket mounting bolts. Secure the camshaft hexagonal head using a spanner to tighten mounting bolts.

50. Pull out the stopper pin from the secondary timing chain tensioner.

51. Install internal guide.

52. Install upper tension guide and slack guide.

53. Install timing chain tensioner, then remove the stopper pin. When installing the timing chain tensioner, engine oil should be applied to the oil hole and tensioner.

54. Install O-rings on rear timing chain case.

55. Apply liquid gasket to front timing chain case. Before installation, wipe off the protruding sealant.

56. Install rear case pin into dowel pin hole on front timing chain case.

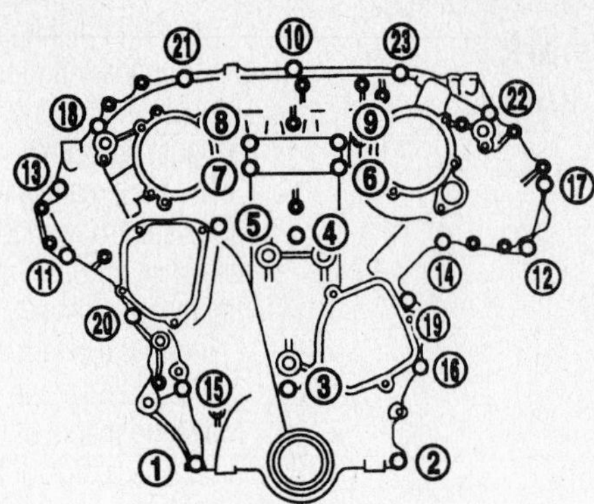

Rear timing case installation—3.5L engine

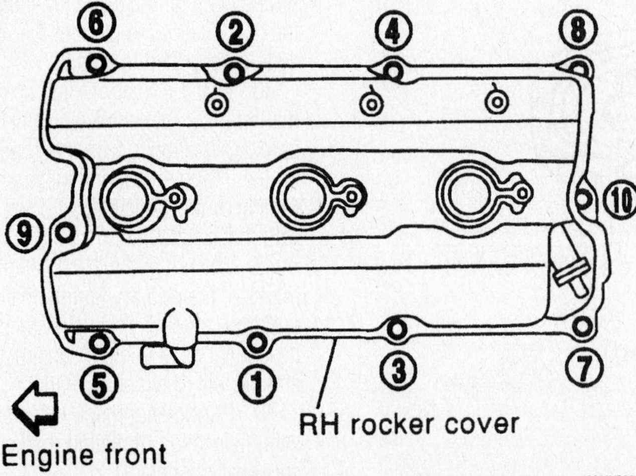

Right rocker cover installation—3.5L engine

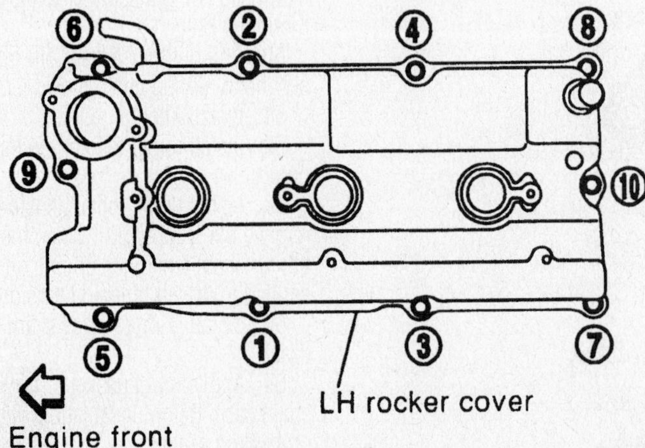

Left rocker cover installation—3.5L engine

57. Tighten bolts to the specified torque in order shown in the figure. Leave the bolts unattended for 30 minutes or more after tightening.

58. Install intake valve timing control valve cover.

a. Install O-rings at front timing chain case.

b. Install seal ring at intake valve timing control valve covers.

c. Apply liquid gasket to intake valve timing control valve covers. Use RTV silicone sealant or equivalent. I Being careful not to move the seal ring from the installation groove, align the dowel pins on the chain case with the holes to install the intake valve timing control valve cover. Tighten in numerical order as shown in the figure.

59. Install RH and LH rocker covers. Rocker cover tightening procedure:

- Tighten in numerical order as shown in the figure.
- Tighten bolts 1 to 10 in that order to 6.9 to 8.8 N·m (0.7 to 0.9 kg-m, 61 to 78 in-lb).
- Then tighten bolts 1 to 10 as indicated in figure to 6.9 to 8.8 N·m (0.7 to 0.9 kg-m, 61 to 78 in-lb).

60. Hang engine using the right and left side engine slingers with a suitable hoist.

61. Set a suitable transmission jack under the suspension member.

62. Remove right and left side engine mounting nuts.

63. Remove right and left side suspension member bolts.

64. Install aluminum oil pan.

65. Set ring gear stopper using the mounting bolt hole. Be careful not to damage the signal plate teeth.

66. Install crankshaft pulley to crankshaft. Align pointer with TDC mark on crankshaft pulley.

67. Install crankshaft pulley bolt. Lubricate thread and seat surface of the bolt with new engine oil. Tighten to 39 to 49 N·m (4.0 to 5.0 kg-m, 29 to 36 ft-lb). Put a paint mark on the crankshaft pulley. Again tighten by turning 60° to 66°, about the angle from one hexagon bolt head corner to another.

68. Install camshaft position sensor (PHASE), crankshaft position sensors (REF)/(POS) and intake valve timing control position sensors.

69. Reinstall removed parts in the reverse order of removal. After starting engine, keep idling for three minutes. Then rev engine up to 3,000 rpm under no load to purge air from the high-pressure chamber of the chain tensioners. The engine may produce a rattling noise. This indicates that

air still remains in the chamber and is not a matter of concern.

Piston and Ring

POSITIONING

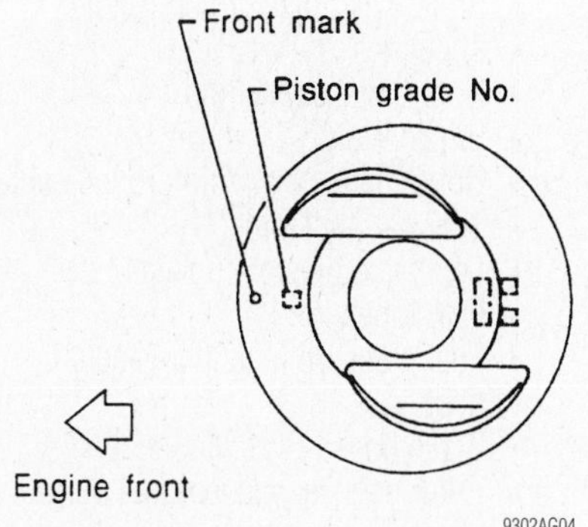

Piston ring positioning—3.3L engine

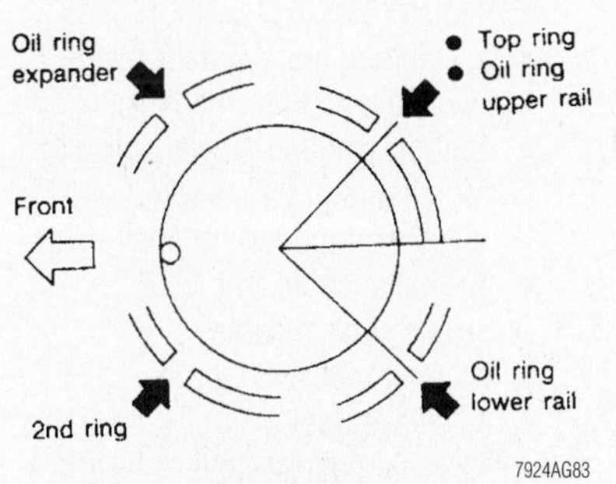

Piston ring end-gap spacing

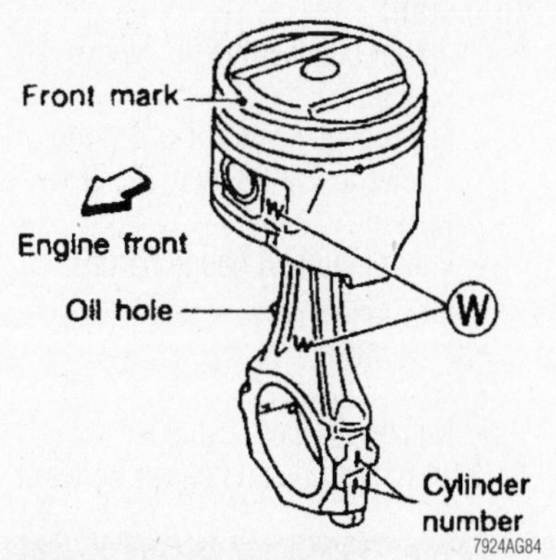

Piston and connecting rod positioning

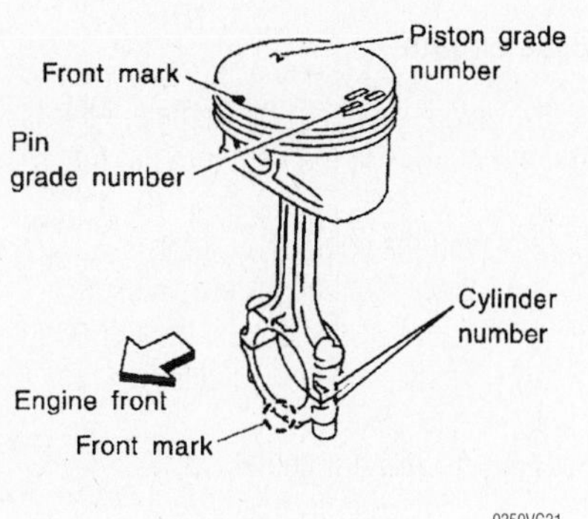

Piston and connecting rod positioning–3.5L

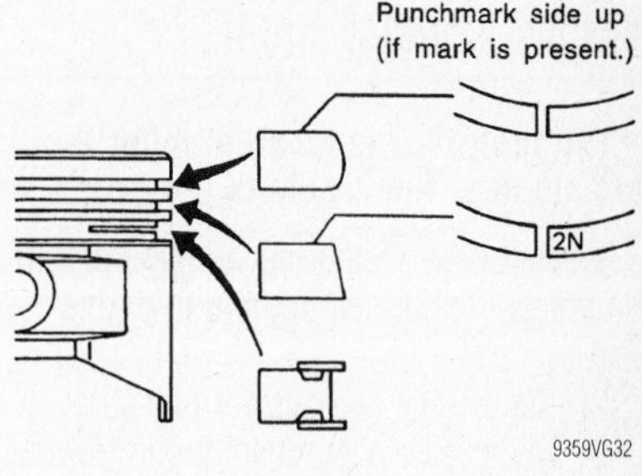

Piston ring positioning–3.5L

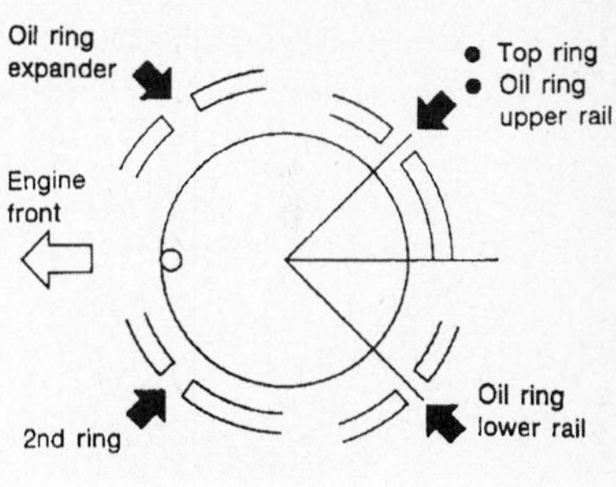

Piston ring positioning–3.5L

FUEL SYSTEM

Fuel System Service Precautions

Safety is the most important factor when performing not only fuel system maintenance but any type of maintenance. Failure to conduct maintenance and repairs in a safe manner may result in serious personal injury or death. Maintenance and testing of the vehicle's fuel system components can be accomplished safely and effectively by adhering to the following rules and guidelines.

- To avoid the possibility of fire and personal injury, always disconnect the negative battery cable unless the repair or test procedure requires that battery voltage be applied.
- Always relieve the fuel system pressure prior to disconnecting any fuel system component (injector, fuel rail, pressure regulator, etc.), fitting or fuel line connection. Exercise extreme caution whenever relieving fuel system pressure, to avoid exposing skin, face and eyes to fuel spray. Please be advised that fuel under pressure may penetrate the skin or any part of the body that it contacts.
- Always place a shop towel or cloth around the fitting or connection prior to loosening to absorb any excess fuel due to spillage. Ensure that all fuel spillage (should it occur) is quickly removed from engine surfaces. Ensure that all fuel soaked cloths or towels are deposited into a suitable waste container.
- Always keep a dry chemical (Class B) fire extinguisher near the work area.
- Do not allow fuel spray or fuel vapors to come into contact with a spark or open flame.
- Always use a back-up wrench when loosening and tightening fuel line connection fittings. This will prevent unnecessary stress and torsion to fuel line piping. Always follow the proper torque specifications.
- Always replace worn fuel fitting O-rings with new. Do not substitute fuel hose or equivalent, where fuel pipe is installed.

Fuel System Pressure

RELIEVING

1. Before servicing the vehicle, refer to the precautions in the beginning of this section.
2. Remove the fuel pump fuse from the panel.
3. Start the engine and allow it to run until it stalls. Crank the engine for a few seconds to relieve additional fuel pressure.
4. Disconnect the negative battery cable.
5. When repairs are complete, replace the fuel pump fuse and connect the negative battery cable.

Fuel Filter

REMOVAL & INSTALLATION

➡The fuel filter is located under the vehicle near the fuel tank.

1. Before servicing the vehicle, refer to the precautions in the beginning of this section.
2. Relieve the fuel system pressure.
3. Remove or disconnect the following:
   - Fuel filter shield, if equipped
   - Fuel lines
   - Fuel filter from the bracket

To install:

4. Install or connect the following:
   - Fuel filter to the bracket
   - Fuel lines
   - Fuel filter shield, if equipped
5. Start the engine and check for leaks.

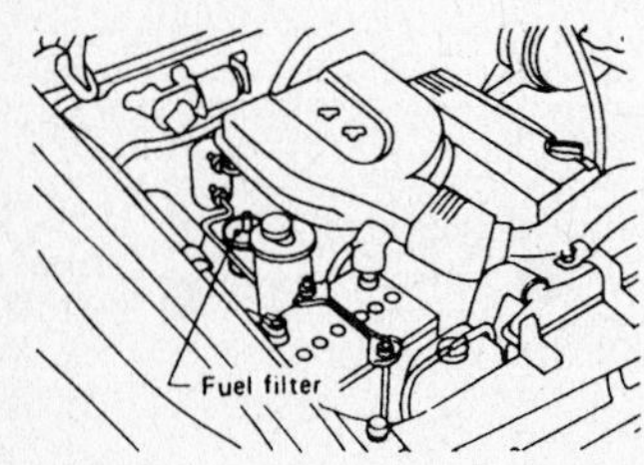

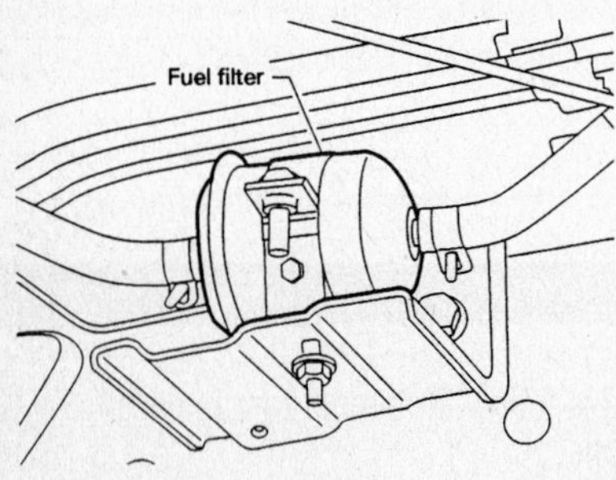

Typical fuel filter locations

Fuel Pump

REMOVAL & INSTALLATION

1. Before servicing the vehicle, refer to the precautions in the beginning of this section.
2. Relieve the fuel system pressure.
3. Remove or disconnect the following:
   - Negative battery cable
   - Access panel behind the rear seat
   - Fuel lines
   - Fuel pump and gauge harness connectors
   - Fuel gauge sender
   - Fuel pump

To install:

4. Install or connect the following:
   - Fuel pump
   - Fuel gauge sender. Tighten the screws to 17–23 inch lbs. (2.0–2.5 Nm).
   - Fuel pump and gauge harness connectors
   - Fuel lines
   - Access panel
   - Negative battery cable
5. Start the engine and check for leaks.

Fuel Injectors

REMOVAL & INSTALLATION

3.3L Engine

1. Before servicing the vehicle, refer to the precautions in the beginning of this section.
2. Drain the cooling system.
3. Relieve the fuel system pressure.
4. Remove or disconnect the following:
   - Negative battery cable
   - Air intake duct
   - Accelerator cable
   - Cruise control cable
   - Idle Air Control (IAC) valve connector
   - Throttle Position (TP) sensor and switch connectors
   - Ignition coil and power transistor connectors
   - Exhaust Gas Recirculation (EGR) Solenoid valve connector
   - EGR temperature sensor connector
   - Radiator hoses
   - Heater hoses
   - Positive Crankcase Ventilation (PCV) valve and hose
   - Evaporative Emissions (EVAP) canister vacuum and purge hoses
   - Brake booster vacuum hose
   - Fuel pressure regulator vacuum hose
   - EGR tube
   - Left bank injector connectors
   - Thermal transmitter
   - Upper intake manifold ground cable
   - Breather pipe
   - Upper intake manifold
   - Fuel lines
   - Right bank injector connectors
   - Fuel supply manifold with the injectors attached
   - Fuel injector caps
   - Fuel injectors

To install:

➡Use new insulators and O-ring seals for assembly.

5. Install or connect the following:
   - Fuel injectors
   - Fuel injector caps. Tighten the screws to 26–34 inch lbs. (3–4 Nm).
   - Fuel supply manifold with the injectors attached. Tighten the bolts to 96–132 inch lbs. (11–15 Nm).
   - Right bank injector connectors
   - Fuel lines
   - Upper intake manifold
   - Breather pipe
   - Upper intake manifold ground cable
   - Thermal transmitter
   - Left bank injector connectors
   - EGR tube
   - Fuel pressure regulator vacuum hose
   - Brake booster vacuum hose
   - EVAP canister vacuum and purge hoses
   - PCV valve and hose
   - Heater hoses
   - Radiator hoses
   - EGR temperature sensor connector
   - EGR Solenoid valve connector
   - Ignition coil and power transistor connectors
   - TP sensor and switch connectors
   - IAC valve connector
   - Cruise control cable
   - Accelerator cable
   - Air intake duct
   - Negative battery cable
6. Fill the cooling system.
7. Start the engine and check for leaks.

3.5L

1. Release fuel pressure to zero.
2. Remove intake manifold collector.
3. Remove fuel tube assemblies in numerical sequence as shown in the figure at left.
4. Expand and remove clips securing fuel injectors.
5. Extract fuel injectors straight from fuel tubes.

➡Be careful not to damage injector nozzles during removal. Do not bump or drop fuel injectors.

6. Carefully install O-rings, including the one used with the pressure regulator. Lubricate O-rings with a smear of engine oil.

➡Be careful not to damage O-rings with service tools, finger nails or clips. Do not expand or twist O-rings. Discard old clips; replace with new ones.

7. Position clips in grooves on fuel injectors. Make sure that protrusions of fuel injectors are aligned with cutouts of clips after installation.

8. Align protrusions of fuel tubes with those of fuel injectors. Insert fuel injectors straight into fuel tubes.

9. After properly inserting fuel injectors, check to make sure that fuel tube protrusions are engaged with those of fuel injectors, and that flanges of fuel tubes are engaged with clips.

10. Tighten fuel tube assembly mounting nuts in numerical sequence (indicated in the figure at left) and in two stages. Tighten to:

Step 1: 84–96 inch lbs.

Step 2: 16–19 ft. lbs.

11. Install all parts removed in reverse order of removal.

DRIVE TRAIN

Manual Transmission

REMOVAL & INSTALLATION

2 Wheel Drive

1. Before servicing the vehicle, refer to the precautions in the beginning of this section.
2. Remove or disconnect the following:
   - Negative battery cable
   - Shift lever
   - Crankshaft Position (CKP) sensor
   - Clutch slave cylinder
   - Vehicle Speed (VSS) sensor connector
   - Back-up lamp switch connector
   - Park/Neutral Position (PNP) switch connector
   - Rear Heated Oxygen (HO_2S) sensor connector
   - Starter motor
   - Driveshaft
   - Exhaust mounting bracket
   - Transmission mount and crossmember. Support the transmission.
   - Transmission flange bolts
   - Transmission

➡The transmission flange bolts vary in length. Note their positions for assembly.

To install:

3. Apply sealant to the transmission flange, engine block and engine rear plate as shown.
4. Install or connect the following:
   - Transmission. Tighten the large bolts to 29–36 ft. lbs. (39–49 Nm) and the small bolts to 12–16 ft. lbs. (16–22 Nm).
   - Transmission mount and crossmember. Tighten the mount and crossmember fasteners to 30–38 ft. lbs. (41–52 Nm).
   - Exhaust mounting bracket
   - Driveshaft
   - Starter motor
   - HO_2S sensor connector
   - PNP switch connector
   - Back-up lamp switch connector
   - VSS sensor connector
   - Clutch slave cylinder
   - CKP sensor
   - Shift lever
   - Negative battery cable

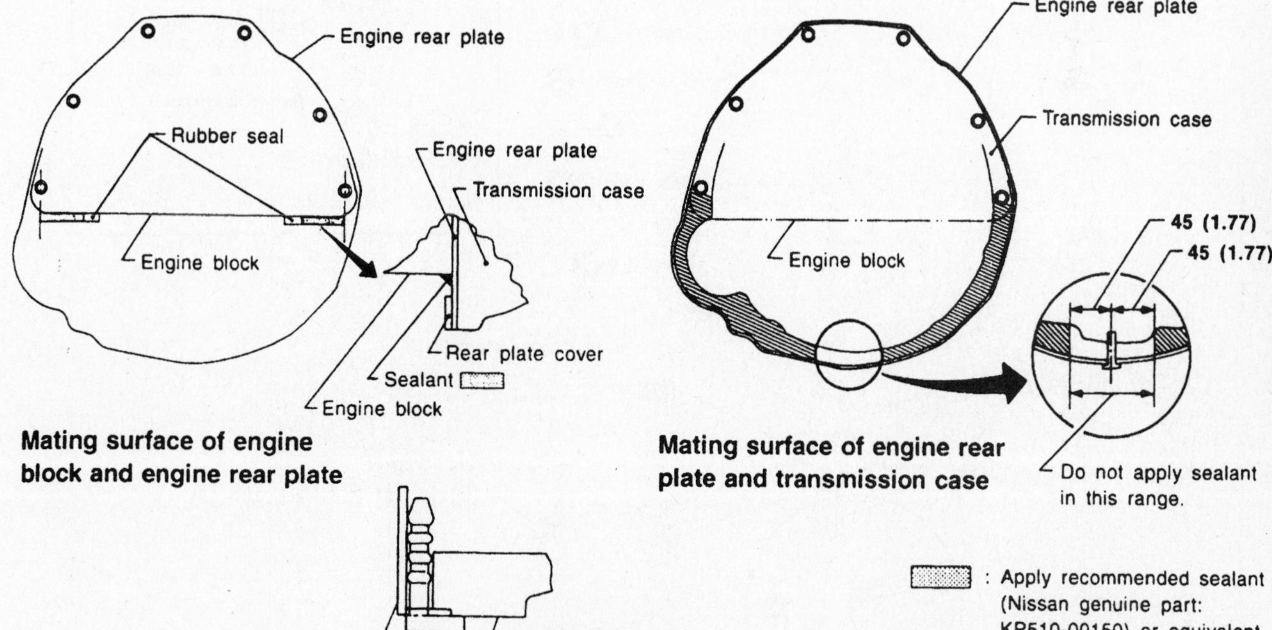

Apply sealant to the indicated areas between the engine block, transmission and engine rear plate—4 Wheel Drive shown

4 Wheel Drive

PATHFINDER

1. Before servicing the vehicle, refer to the precautions in the beginning of this section.
2. Remove or disconnect the following:

- Negative battery cable
- Shift lever
- Transfer case select lever
- Crankshaft Position (CKP) sensor
- Clutch slave cylinder
- Vehicle Speed (VSS) sensor connector
- Back-up lamp switch connector
- Park/Neutral Position (PNP) switch connector
- Rear Heated Oxygen (HO_2S) sensor connector
- Starter motor
- Front and rear driveshafts

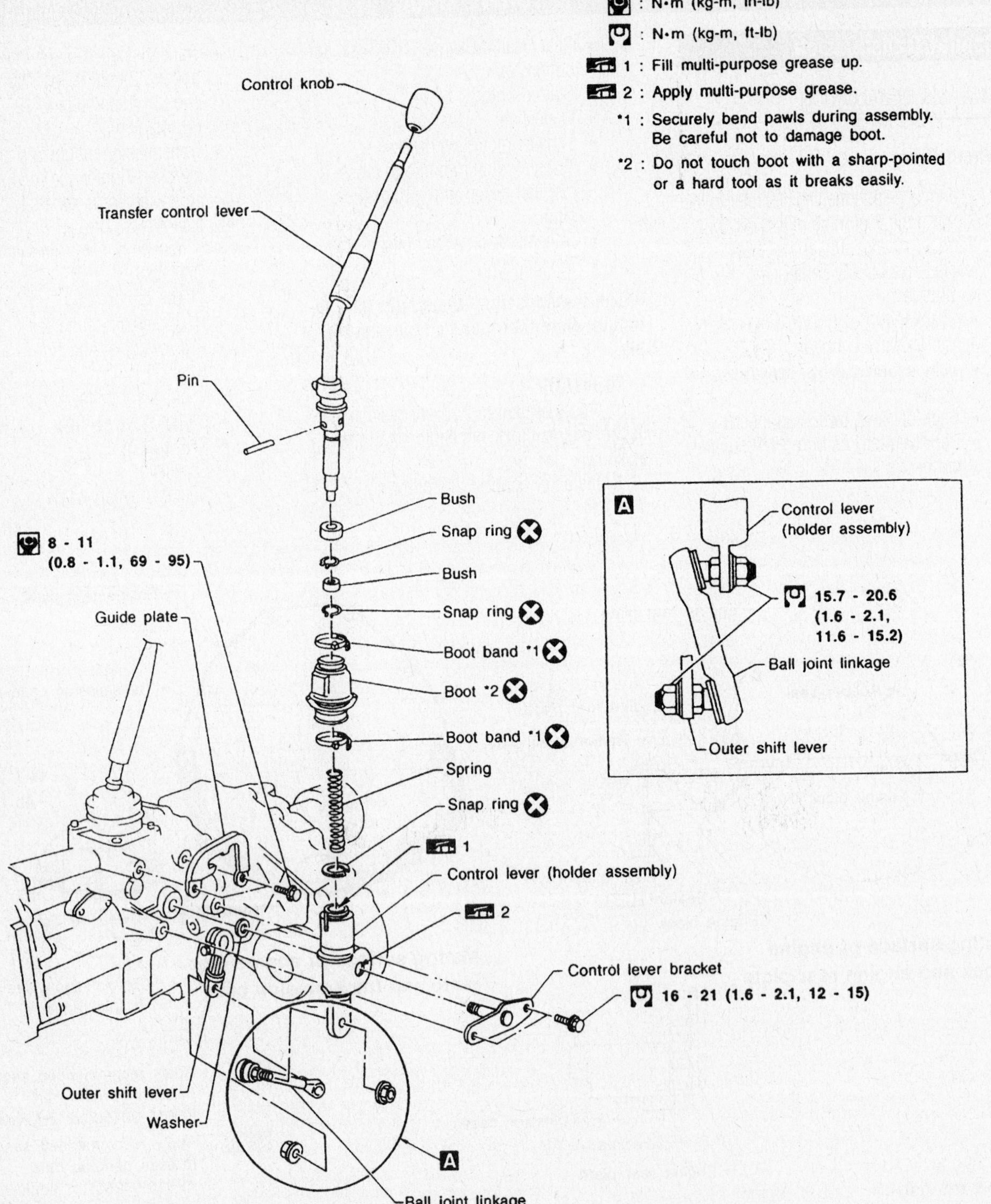

Exploded view of the transfer case shifter lever and related components—Pathfinder 4WD

- Exhaust front pipes
- Exhaust center pipe
- Transmission mount and crossmember. Support the transmission.
- Transmission flange bolts
- Transmission

➡The transmission flange bolts vary in length. Note their positions for assembly.

To install:

3. Apply sealant to the transmission flange, engine block, and engine rear plate as shown.
4. Install or connect the following:
   - Transmission. Tighten the large bolts to 29–36 ft. lbs. (39–49 Nm) and the small bolts to 22–29 ft. lbs. (29–39 Nm).
   - Transmission mount and crossmember. Tighten the mount and crossmember fasteners to 30–38 ft. lbs. (41–52 Nm).
   - Exhaust center pipe
   - Exhaust front pipes
   - Front and rear driveshafts
   - Starter motor
   - HO_2S sensor connector
   - PNP switch connector
   - Back-up lamp switch connector
   - VSS sensor connector
   - Clutch slave cylinder
   - CKP sensor
   - Transfer case select lever
   - Shift lever
   - Negative battery cable

Automatic Transmission

REMOVAL & INSTALLATION

2 Wheel Drive

1. Before servicing the vehicle, refer to the precautions in the beginning of this section.
2. Remove or disconnect the following:
   - Negative battery cable
   - Crankshaft Position (CKP) sensor
   - Exhaust front pipes
   - Exhaust rear pipes
   - Transmission dipstick tube
   - Transmission oil cooler lines
   - Driveshaft
   - Shift cable
   - Transmission control harness connectors
   - Vehicle Speed (VSS) sensor connector
   - Starter motor
   - Torque converter
   - Transmission mount and crossmember. Support the transmission.
   - Transmission flange bolts
   - Transmission

➡The transmission flange bolts vary in length. Note their positions for assembly.

To install:

3. Install or connect the following:
   - Transmission. Tighten the large bolts to 29–36 ft. lbs. (39–49 Nm) and the small bolts to 22–29 ft. lbs. (29–39 Nm).
   - Transmission mount and crossmember. Tighten the mount and crossmember fasteners to 30–38 ft. lbs. (41–52 Nm).
   - Torque converter. Tighten the bolts to 33–43 ft. lbs. (44–59 Nm).
   - Starter motor
   - VSS sensor connector
   - Transmission control harness connectors
   - Shift cable
   - Driveshaft
   - Transmission oil cooler lines
   - Transmission dipstick tube
   - Exhaust rear pipes
   - Exhaust front pipes
   - CKP sensor
   - Negative battery cable

4 Wheel Drive

1. Before servicing the vehicle, refer to the precautions in the beginning of this section.
2. Remove or disconnect the following:
   - Negative battery cable
   - Crankshaft Position (CKP) sensor
   - Exhaust front pipes
   - Exhaust rear pipes
   - Transmission dipstick tube
   - Transmission oil cooler lines
   - Front and rear driveshafts
   - Transfer case linkage
   - Shift cable
   - Transmission control harness connectors
   - Vehicle Speed (VSS) sensor connector
   - Starter motor
   - Torque converter
   - Transmission mount and crossmember. Support the transmission.
   - Transmission flange bolts
   - Transmission

➡The transmission flange bolts vary in length. Note their positions for assembly.

To install:

3. Install or connect the following:
   - Transmission. Tighten the large bolts to 29–36 ft. lbs. (39–49 Nm) and the small bolts to 22–29 ft. lbs. (29–39 Nm).
   - Transmission mount and crossmember. Tighten the mount and crossmember fasteners to 30–38 ft. lbs. (41–52 Nm).
   - Torque converter. Tighten the bolts to 33–43 ft. lbs. (44–59 Nm).
   - Starter motor
   - VSS sensor connector
   - Transmission control harness connectors
   - Shift cable
   - Transfer case linkage
   - Front and rear driveshafts
   - Transmission oil cooler lines
   - Transmission dipstick tube
   - Exhaust rear pipes
   - Exhaust front pipes
   - CKP sensor
   - Negative battery cable

Clutch

REMOVAL & INSTALLATION

1. Before servicing the vehicle, refer to the precautions in the beginning of this section.
2. Remove or disconnect the following:
   - Negative battery cable
   - Transmission
   - Pressure plate. Loosen the bolts evenly in ½ turn steps.
   - Clutch disc

To install:

3. Install or connect the following:
   - Clutch disc and pressure plate. Tighten the pressure plate bolts evenly in ½ turns to 16–22 ft. lbs. (22–29 Nm).
   - Transmission
   - Negative battery cable

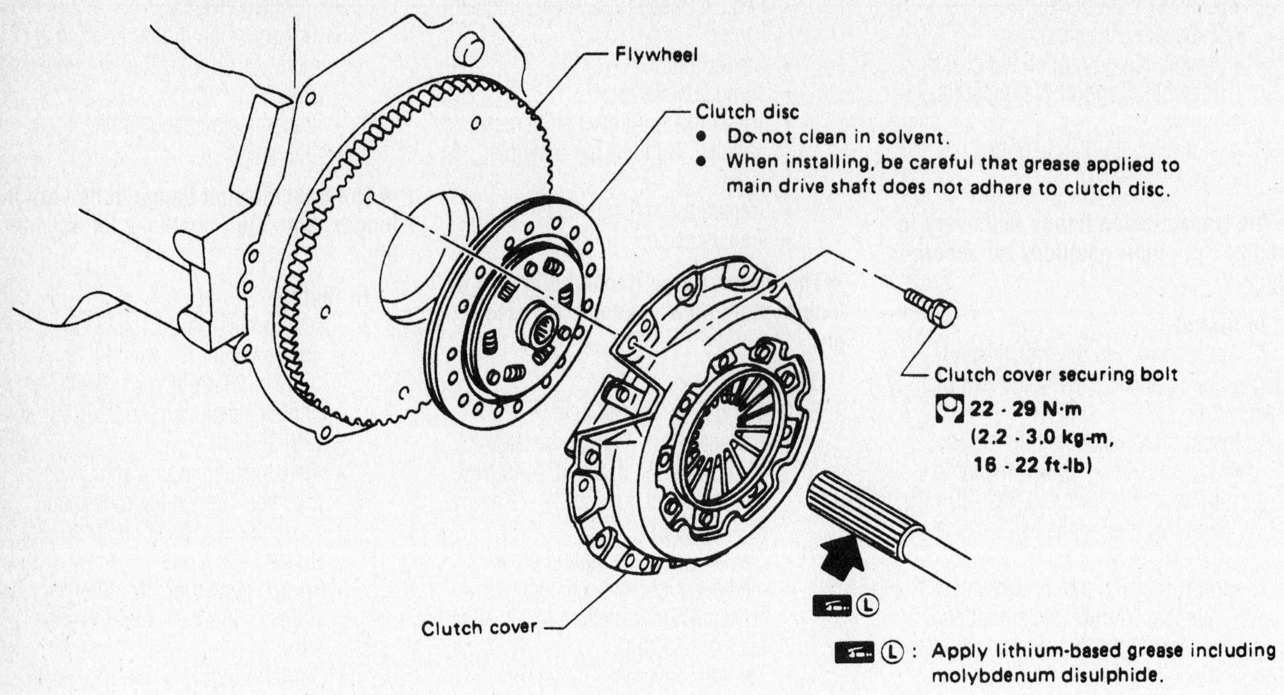

Exploded view of the pressure plate and clutch disc and related components—all models

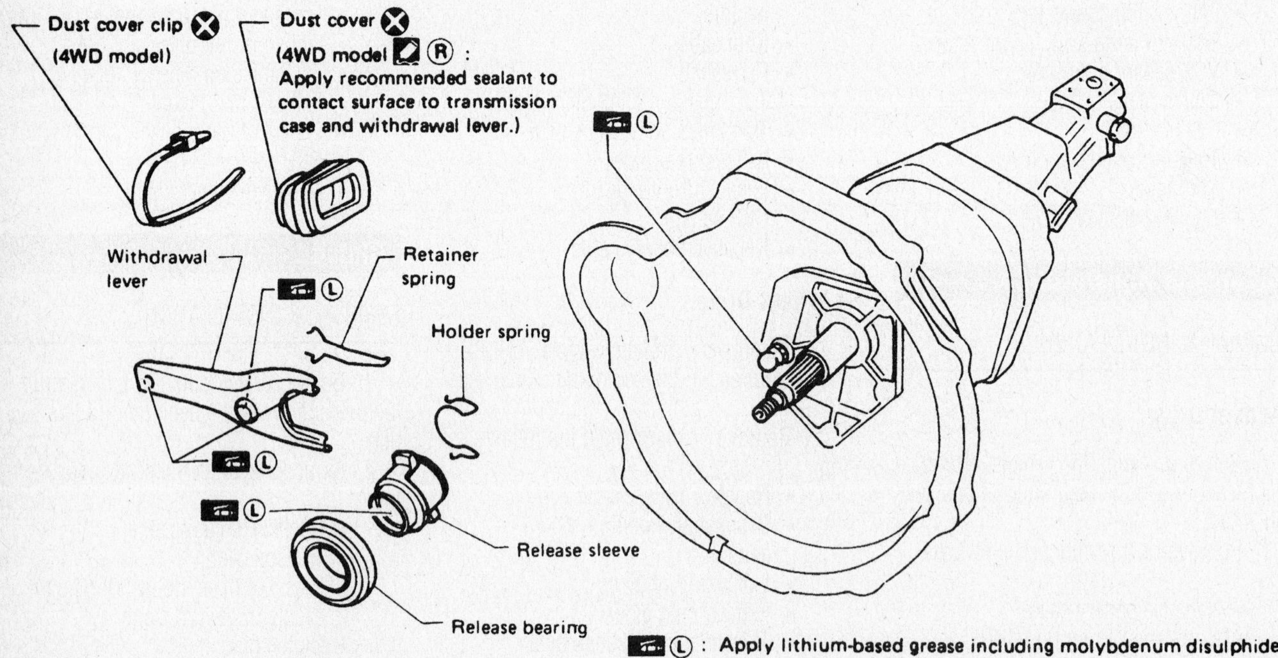

Clutch release mechanism exploded view—all models

Hydraulic Clutch System

BLEEDING

1. Before servicing the vehicle, refer to the precautions in the beginning of this section.
2. Fill the clutch master cylinder reservoir with fresh clean brake fluid.
3. Connect a clear plastic hose to the air bleeder.
4. Have an assistant pump the clutch pedal slowly several times and hold it depressed.
5. Open the slave cylinder bleeder screw and allow air to escape.
6. Close the bleeder screw before releasing the clutch pedal.
7. Repeat until all air is purged from the clutch hydraulic system.
8. Refill the reservoir to the full mark.

Transfer Case Assembly

REMOVAL & INSTALLATION

1. Before servicing the vehicle, refer to the precautions in the beginning of this section.
2. Remove or disconnect the following:
   - Negative battery cable
   - Front and rear driveshafts
   - Torsion bars and mounts
   - Rear torsion bar crossmember
   - Exhaust front pipes
   - Exhaust rear pipes
   - Vehicle Speed (VSS) sensor connector
   - Transfer case shift linkage
   - Transfer case neutral switch connector
   - 4 wheel drive switch connector
   - Vent hose
   - Transfer case flange bolts
   - Transfer case

To install:

3. Install or connect the following:
   - Transfer case. Tighten the flange bolts to 23–30 ft. lbs. (31–41 Nm).
   - Vent hose
   - 4 wheel drive switch connector
   - Transfer case neutral switch connector
   - Transfer case shift linkage
   - VSS sensor connector
   - Exhaust rear pipes
   - Exhaust front pipes
   - Rear torsion bar crossmember
   - Torsion bars and mounts
   - Front and rear driveshafts
   - Negative battery cable

Halfshaft

REMOVAL & INSTALLATION

1. Before servicing the vehicle, refer to the precautions in the beginning of this section.
2. Remove or disconnect the following:
   - Front wheel
   - Wheel speed sensor, if equipped
   - Locking hub or drive flange
   - Snapring
   - Spindle washer
   - Thrust washer
   - Inner CV-joint bolts
   - Axle halfshaft. Separate the stub shaft from the spindle by tapping with a plastic hammer.

To install:

3. Install or connect the following:
   - Axle halfshaft. Guide the stub shaft into the spindle and tighten the inner CV-joint bolts to 25–33 ft. lbs. (34–44 Nm).
   - Thrust washer
   - Spindle washer
   - Snapring
   - Locking hub or drive flange
   - Wheel speed sensor, if equipped
   - Front wheel

CV-Joints

OVERHAUL

Outer CV-Joint

1. Before servicing the vehicle, refer to the precautions in the beginning of this section.
2. Remove the axle halfshaft from the vehicle.
3. Remove the CV-joint boot clamps and push the boot away from the joint.
4. Remove the CV-joint from the axle shaft by tapping it with a brass hammer.

To install:

➡Use new circlips and boot clamps for assembly.

5. Install the CV-joint to the axle shaft by tapping it with a brass hammer.
6. Pack the joint with grease.
7. Install the boot clamps.
8. Install the axle halfshaft to the vehicle.

Inner Tri-Pot Joint

1. Before servicing the vehicle, refer to the precautions in the beginning of this section.
2. Remove the axle halfshaft from the vehicle.
3. Remove the plug seal by tapping around the joint housing flange with a brass hammer.
4. Remove or disconnect the following:
   - CV-joint boot clamps
   - Snapring
   - Spider assembly
   - CV-joint housing
   - CV-joint boot

To install:

➡Use new snaprings and plug seals for assembly.

5. Install or connect the following:
   - CV-joint boot
   - CV-joint housing
   - Spider assembly
   - Snapring. Pack the joint with grease.
   - CV-joint boot clamps
   - Plug seal
6. Install the axle halfshaft to the vehicle.

Axle Shaft, Bearing and Seal

REMOVAL & INSTALLATION

1. Before servicing the vehicle, refer to the precautions in the beginning of this section.
2. Remove or disconnect the following:
   - Rear wheel
   - Wheel speed sensor, if equipped
   - Brake drum
   - Brake shoes
   - Parking brake cable
   - Brake fluid line
   - Bearing cage and backing plate bolts
   - Axle shaft assembly
   - Axle seal
   - Wheel speed sensor rotor, if equipped
   - Lockwasher
   - Bearing locknut
   - Flat washer
   - Wheel bearing
   - Wheel bearing cage grease seal

To install:

➡Use new lockwashers, seals and bearings for assembly.

3. Install or connect the following:
   - Wheel bearing cage grease seal
   - Wheel bearing
   - Flat washer
   - Bearing locknut
   - Lockwasher
   - Wheel speed sensor rotor, if equipped
   - Axle seal
   - Axle shaft assembly
   - Bearing cage and backing plate bolts
   - Brake fluid line
   - Parking brake cable
   - Brake shoes
   - Brake drum
   - Wheel speed sensor, if equipped
   - Rear wheel
4. Bleed the rear brakes and check the rear axle lubricant level.

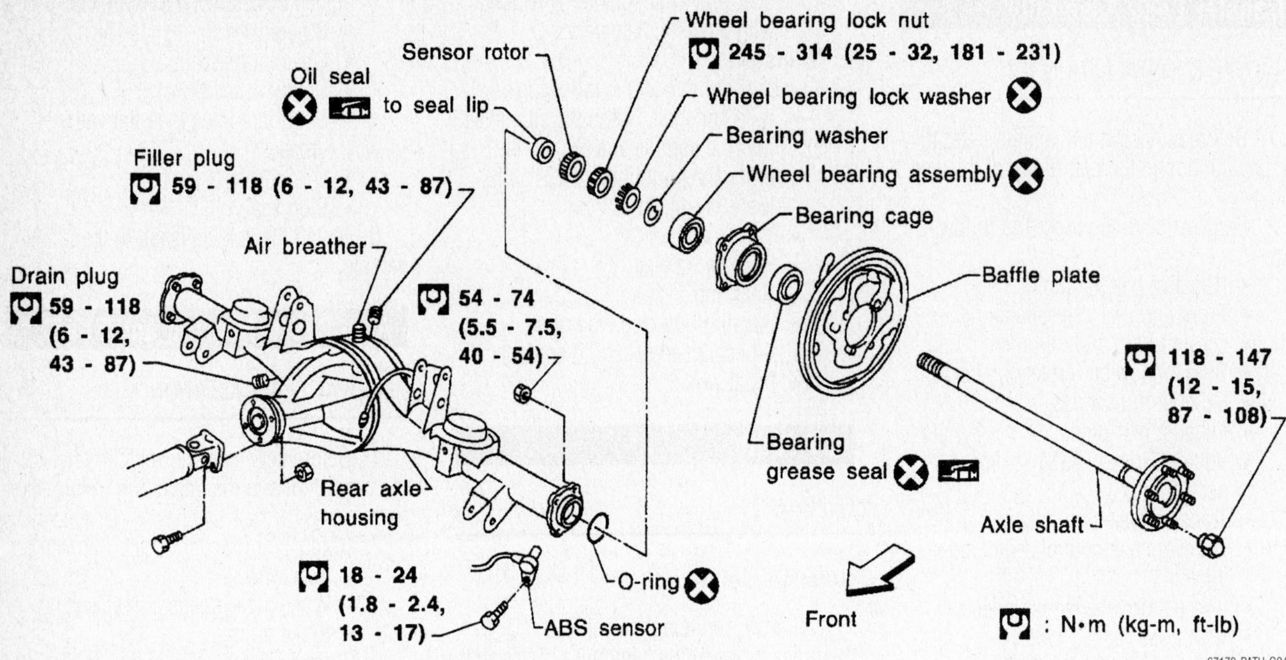

Rear axle components

Pinion Seal

Front

1. Before servicing the vehicle, refer to the precautions in the beginning of this section.
2. Remove or disconnect the following:
   - Driveshaft
   - Front wheels
   - Front brake calipers

➡The front brake calipers must be removed so that there is no additional drag when measuring pinion bearing preload.

3. Use an inch lb. torque wrench and measure the amount of torque required to maintain pinion rotation through several revolutions.
4. Remove or disconnect the following:
   - Pinion flange
   - Oil seal

To install:

5. Install or connect the following:
   - Pinion seal
   - Pinion flange
6. Rotate the pinion flange occasionally while tightening the flange nut to make sure the pinion bearings seat correctly.
7. Take frequent bearing preload torque readings. Tighten the flange nut to achieve the preload torque readings originally recorded. Do not exceed 137–217 ft. lbs. (186–294 Nm) torque when tightening the pinion flange nut.

✲✲ CAUTION

If the bearing preload can not be achieved at the specified torque, remove the pinion bearing and install a new adjustment spacer.

8. Install or connect the following:
   - Front brake calipers
   - Front wheels
   - Driveshaft. Tighten the fasteners to 29–33 ft. lbs. (39–44 Nm).
9. Fill the differential with gear lubricant and check for leaks.

Rear

2 WHEEL DRIVE

1. Before servicing the vehicle, refer to the precautions in the beginning of this section.
2. Remove or disconnect the following:
   - Driveshaft
   - Rear wheels
   - Brake drums

➡The rear brake drums must be removed so that there is no additional drag when measuring pinion bearing preload.

3. Use an inch lb. torque wrench and measure the amount of torque required to maintain pinion rotation through several revolutions.
4. Remove or disconnect the following:
   - Pinion flange
   - Wheel speed sensor and rotor, if equipped
   - Oil seal
   - Pinion bearing
   - Collapsible spacer

To install:

➡Use a new collapsible spacer and wheel speed sensor rotor for assembly.

5. Install or connect the following:
   - Collapsible spacer
   - Pinion bearing
   - Pinion seal
   - Pinion flange
6. Rotate the pinion flange occasionally while tightening the flange nut to make sure the pinion bearings seat correctly.
7. Take frequent bearing preload torque readings. Tighten the flange nut to achieve the preload torque readings originally recorded. Do not exceed 137–217 ft. lbs. (186–294 Nm) torque when tightening the pinion flange nut.

✲✲ CAUTION

Never loosen the pinion nut to reduce bearing preload. If it is necessary to reduce bearing preload, install a new collapsible spacer.

8. Install or connect the following:
   - Brake drums
   - Rear wheels
   - Driveshaft. Tighten the fasteners to 58–65 ft. lbs. (78–88 Nm).

9. Fill the differential with gear lubricant and check for leaks.

4 WHEEL DRIVE

1. Before servicing the vehicle, refer to the precautions in the beginning of this section.
2. Remove or disconnect the following:
   - Driveshaft
   - Rear wheels
   - Brake drums

➡The rear brake drums must be removed so that there is no additional drag when measuring pinion bearing preload.

3. Use an inch lb. torque wrench and measure the amount of torque required to maintain pinion rotation through several revolutions.
4. Remove or disconnect the following:
   - Pinion flange
   - Oil seal

To install:

5. Install or connect the following:
   - Pinion seal
   - Pinion flange
6. Rotate the pinion flange occasionally while tightening the flange nut to make sure the pinion bearings seat correctly.
7. Take frequent bearing preload torque readings. Tighten the flange nut to achieve the preload torque readings originally recorded. Do not exceed 137–217 ft. lbs. (186–294 Nm) torque when tightening the pinion flange nut.

⁂ CAUTION

If the bearing preload can not be achieved at the specified torque, remove the pinion bearing and install a new adjustment spacer.

8. Install or connect the following:
   - Brake drums
   - Rear wheels
   - Driveshaft. Tighten the fasteners to 58–65 ft. lbs. (78–88 Nm).
9. Fill the differential with gear lubricant and check for leaks.

STEERING AND SUSPENSION

Air Bag

⁂ CAUTION

Some vehicles are equipped with an air bag system. The system must be disarmed before performing service on, or around, system components, the steering column, instrument panel components, wiring and sensors. Failure to follow the safety precautions and the disarming procedure could result in accidental air bag deployment, possible injury and unnecessary system repairs.

PRECAUTIONS

Several precautions must be observed when handling the inflator module to avoid accidental deployment and possible personal injury.

- Never carry the inflator module by the wires or connector on the underside of the module.
- When carrying a live inflator module, hold securely with both hands, and ensure that the bag and trim cover are pointed away.
- Place the inflator module on a bench or other surface with the bag and trim cover facing up.
- With the inflator module on the bench, never place anything on or close to the module which may be thrown in the event of an accidental deployment.

DISARMING

To disarm the **SRS** system turn the ignition switch to the **OFF** position. Then, disconnect both battery cables starting with the negative cable first and wait at least 3 minutes after the cables are disconnected.

To rearm the **SRS** system, turn the ignition switch to the **OFF** position. Connect both battery cables starting with the positive cable first.

Recirculating Ball Power Steering Gear

REMOVAL & INSTALLATION

1. Before servicing the vehicle, refer to the precautions in the beginning of this section.
2. Remove or disconnect the following:
   - Pitman arm
   - Steering column intermediate shaft
   - Power steering hoses
   - Steering gear

To install:

3. Install or connect the following:
   - Steering gear. Tighten the bolts to 62–71 ft. lbs. (84–96 Nm).
   - Power steering hoses. Tighten the banjo fittings to 29–38 ft. lbs. (39–51 Nm).
   - Steering column intermediate shaft. Tighten the pinch bolt to 17–22 ft. lbs. (24–29 Nm).
   - Pitman arm. Tighten the nut to 174–195 ft. lbs. (235–265 Nm).
4. Check the wheel alignment and adjust, as necessary.

Rack and Pinion Steering Gear

REMOVAL & INSTALLATION

1. Before servicing the vehicle, refer to the precautions in the beginning of this section.
2. Remove or disconnect the following:
   - Front wheels
   - Outer tie rod ends
   - Steering shaft coupler
   - Power steering hoses
   - Steering gear

To install:

3. Install or connect the following:
   - Steering gear. Tighten the bolts to 101 ft. lbs. (137 Nm) for 2001–02; 116–137 ft. lbs. (157–186 Nm) for 2003–04 models.
   - Power steering hoses. Tighten the fittings to 25 ft. lbs. (35 Nm).
   - Steering shaft coupler. Tighten the bolt to 22 ft. lbs. (29 Nm).
   - Outer tie rod ends. Tighten the nuts to 65 ft. lbs. (88 Nm).
   - Front wheels

Front Strut

REMOVAL & INSTALLATION

1. Before servicing the vehicle, refer to the precautions in the beginning of this section.

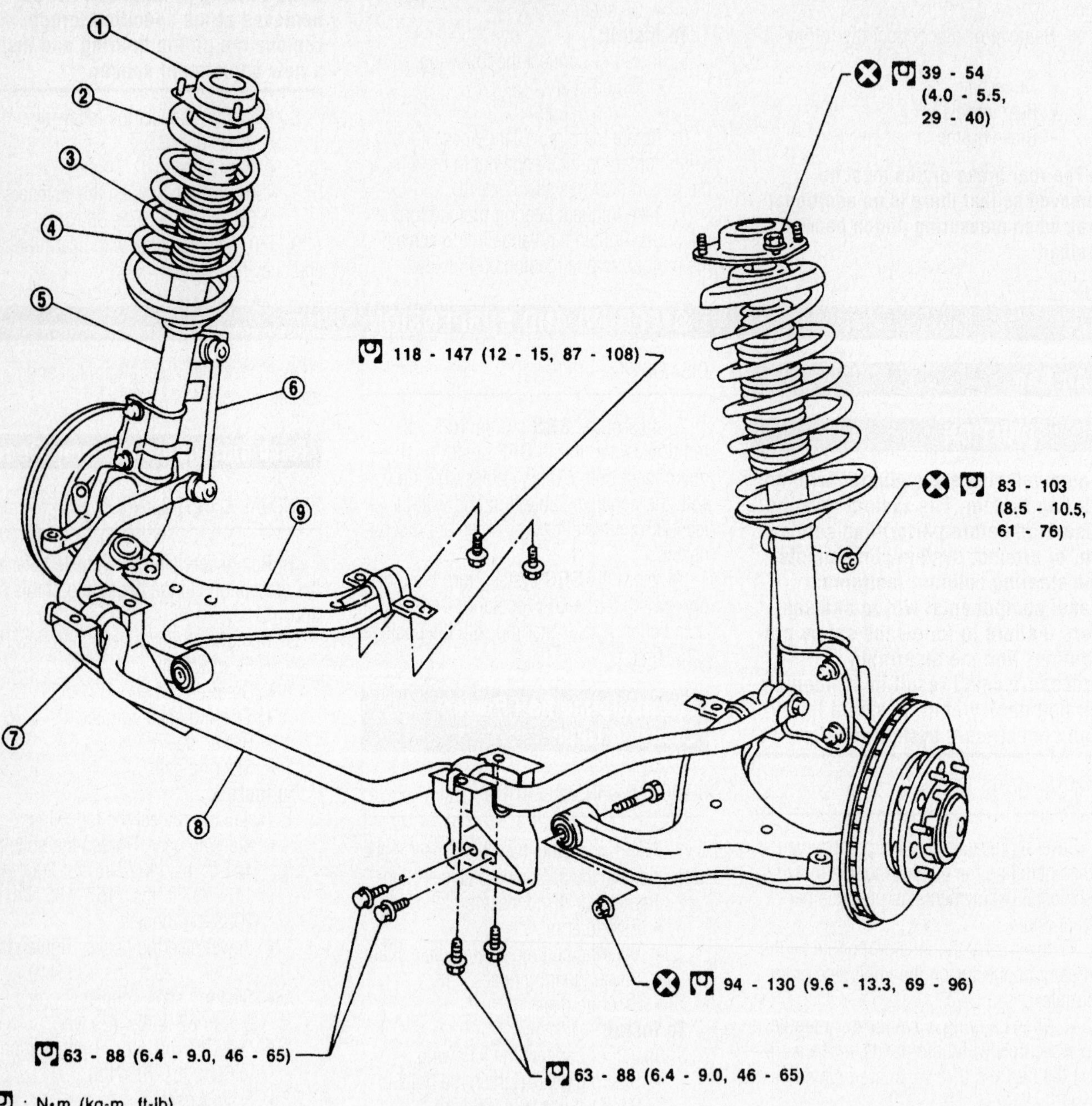
When installing rubber parts, final tightening must be carried out under unladen condition* with tires on ground.
Fuel, radiator coolant and engine oil full.
Spare tire, jack, hand tools and mats in designated positions.
39 - 54 (4.0 - 5.5, 29 - 40)
118 - 147 (12 - 15, 87 - 108)
83 - 103 (8.5 - 10.5, 61 - 76)
94 - 130 (9.6 - 13.3, 69 - 96)
63 - 88 (6.4 - 9.0, 46 - 65)
63 - 88 (6.4 - 9.0, 46 - 65)
: N•m (kg-m, ft-lb)
① Strut mounting insulator
② Spring upper seat
③ Bound bumper
④ Coil spring
⑤ Strut assembly
⑥ Stabilizer connecting rod
⑦ Bracket
⑧ Stabilizer bar
⑨ Transverse link
7924VG66

Exploded view of the front suspension—2WD shown

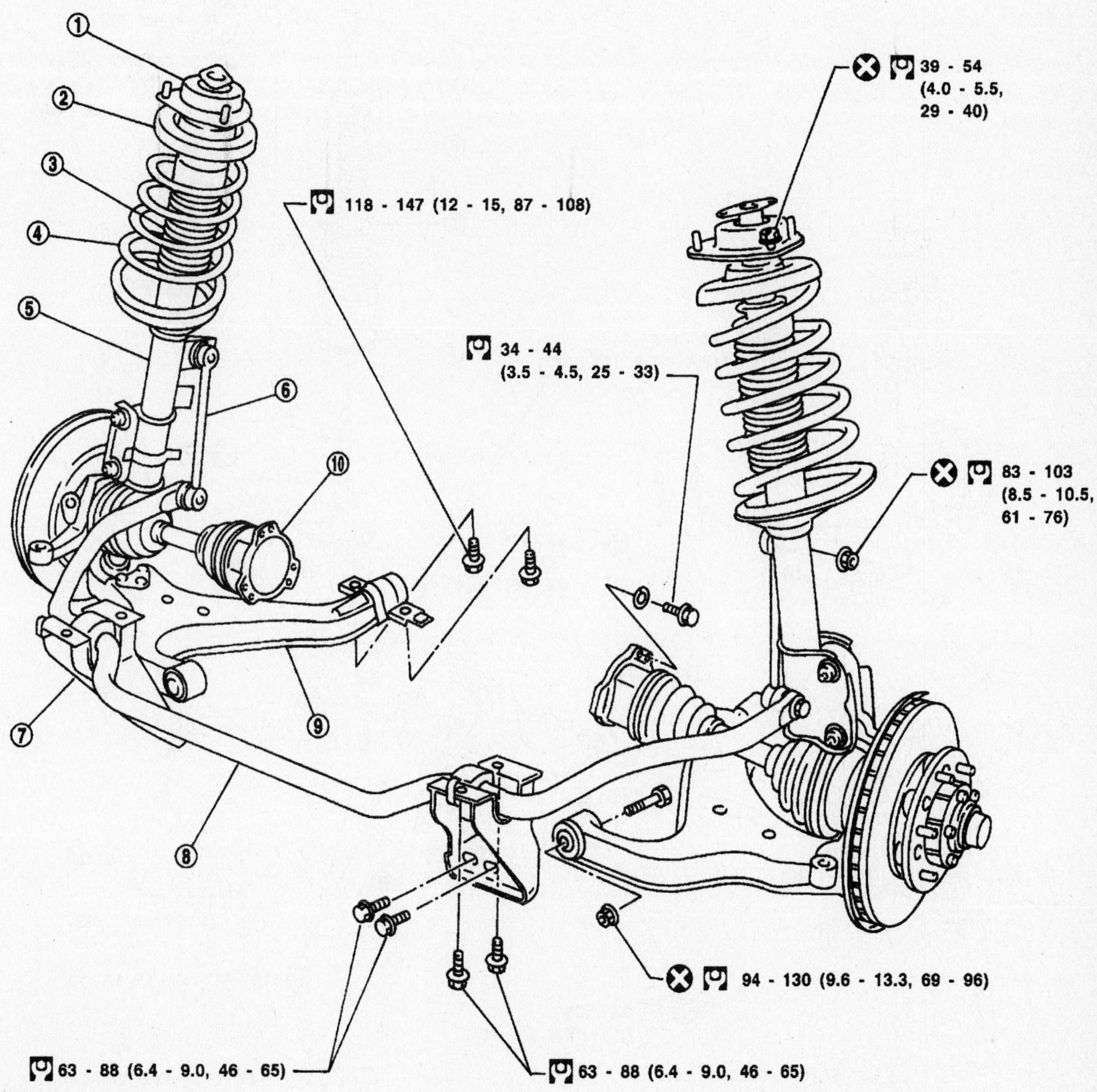

Exploded view of the front suspension—4WD shown

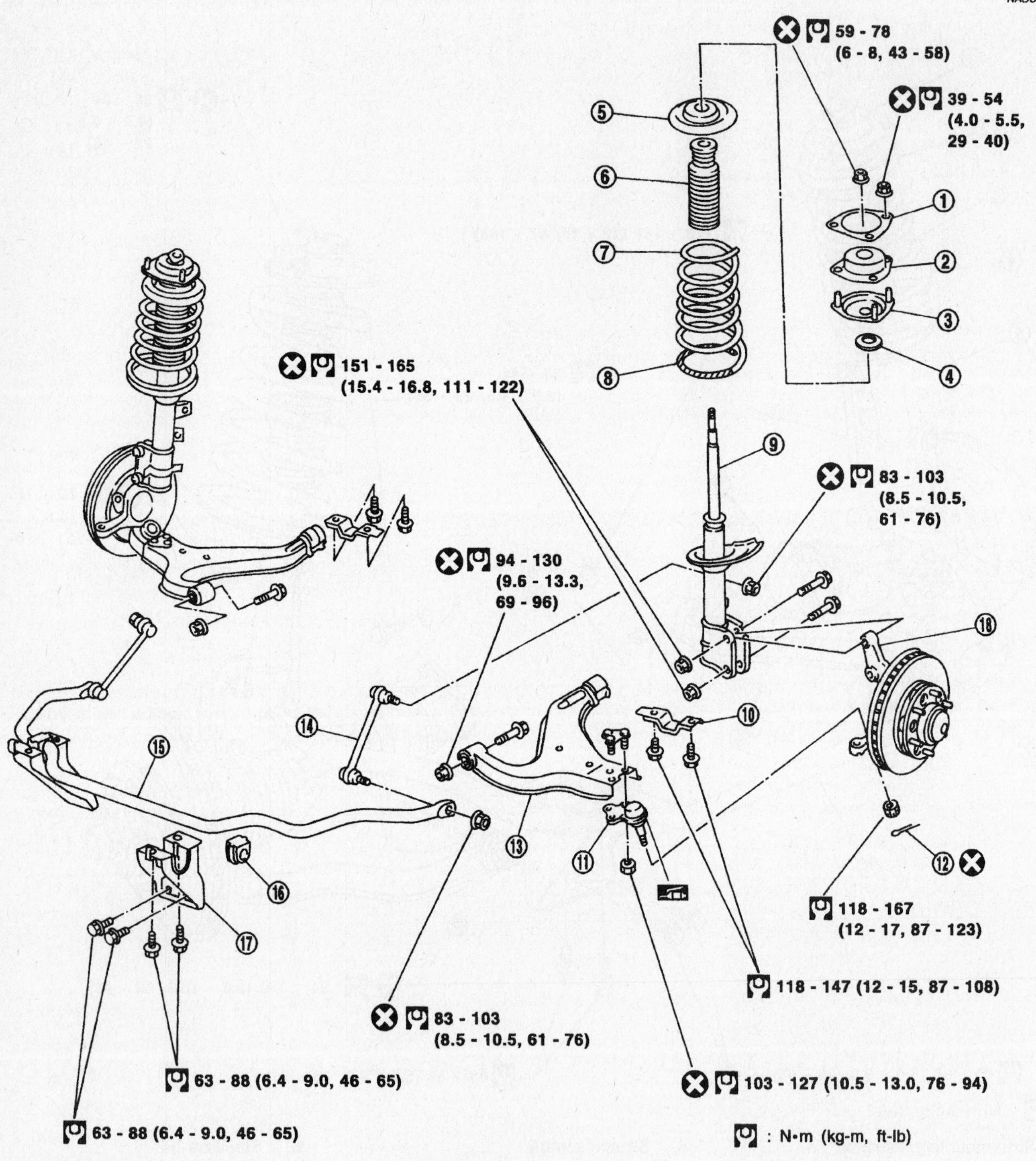

1. Spacer
2. Strut mounting insulator
3. Bracket
4. Strut mounting bearing
5. Spring upper seat
6. Bound bumper
7. Coil spring
8. (Polyurethane tube)
9. Strut assembly
10. Bracket
11. Lower ball joint assembly
12. Cotter pin
13. Transverse link
14. Stabilizer connecting rod
15. Stabilizer bar
16. Bushing
17. Bracket
18. Knuckle spindle

67170-PATH-G06

Front strut and related parts—2002–04 2WD shown

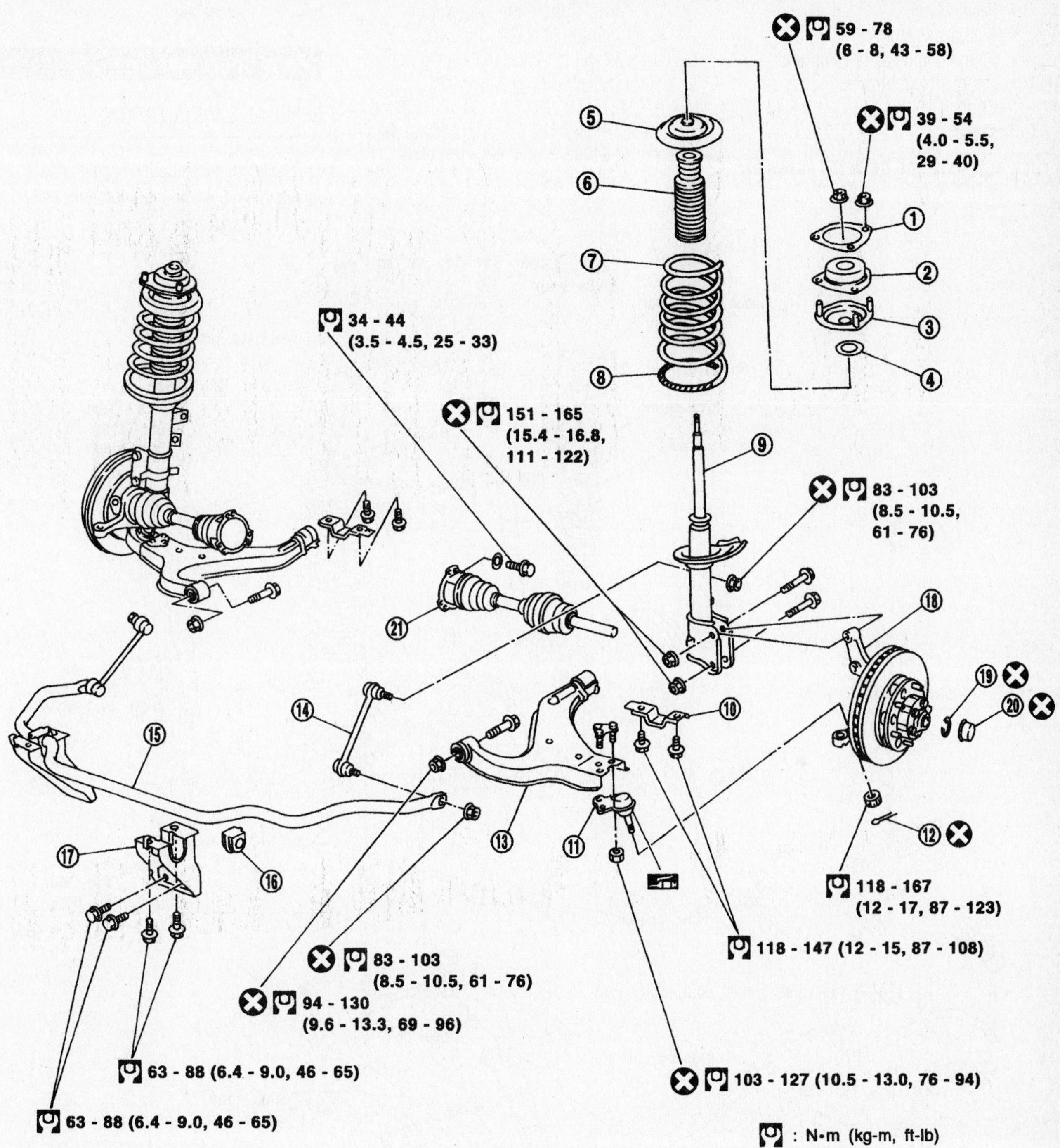

1. Spacer
2. Strut mounting insulator
3. Bracket
4. Strut mounting bearing
5. Spring upper seat
6. Bound bumper
7. Coil spring
8. (Polyurethane tube)
9. Strut assembly
10. Bracket
11. Lower ball joint assembly
12. Cotter pin
13. Transverse link
14. Stabilizer connecting rod
15. Stabilizer bar
16. Bushing
17. Bracket
18. Knuckle spindle
19. Snap ring
20. Hub cap
21. Drive shaft

67170-PATH-G07

Front strut and related parts—2002–04 4WD shown

2. Remove or disconnect the following:
   - Front wheel
   - Stabilizer bar link
   - Steering knuckle bracket bolts
   - Upper strut mount nuts
   - Strut

To install:

➡Use new nuts and bolts for assembly.

3. Install or connect the following:
   - Strut. Tighten the upper strut mount nuts to 29–40 ft. lbs. (39–54 Nm) and the knuckle bracket bolts to 111–122 ft. lbs. (151–165 Nm).
   - Stabilizer bar link. Tighten the nut to 61–76 ft. lbs. (83–103 Nm).
   - Front wheel

4. Check the wheel alignment and adjust, as necessary.

Rear Shock Absorber

REMOVAL & INSTALLATION

1. Before servicing the vehicle, refer to the precautions in the beginning of this section.

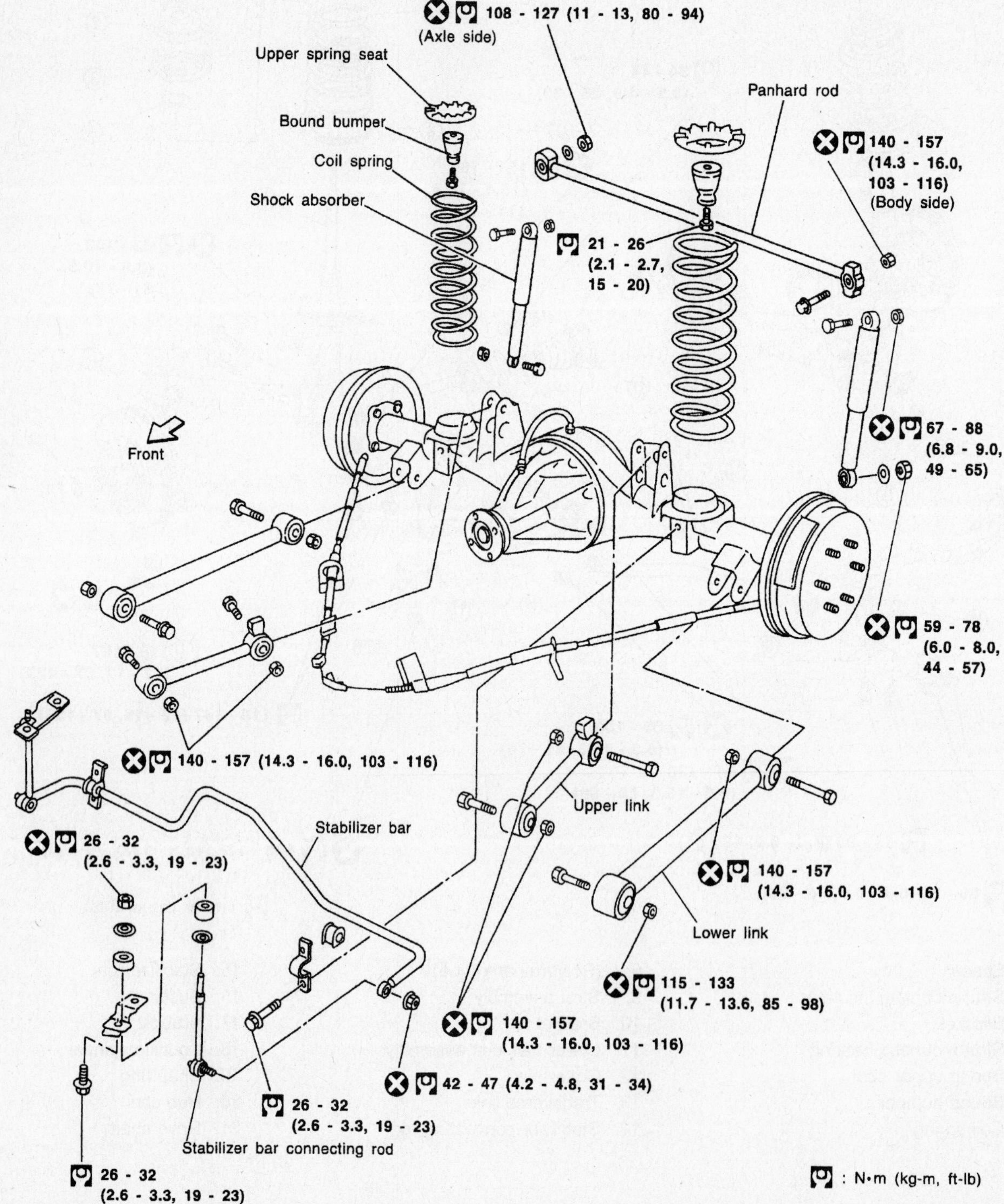

Rear suspension components

2. Support the rear axle.
3. Remove or disconnect the following:
   - Lower shock absorber bolt
   - Upper shock absorber bolt
   - Shock absorber

To install:

➡Use new fasteners for assembly.

4. Install the shock absorber and tighten the upper bolt to 49–65 ft. lbs. (67–88 Nm); the lower bolt to 44–57 ft. lbs. (59–78 Nm).

Coil Spring

REMOVAL & INSTALLATION

Front

1. Before servicing the vehicle, refer to the precautions in the beginning of this section.
2. Remove the strut assembly.
3. Compress the coil spring and remove the piston rod nut.
4. Remove or disconnect the following:
   - Upper strut mount
   - Strut mount bracket
   - Upper strut bearing
   - Spring upper seat
   - Coil spring

To install:

➡Use new fasteners for assembly.

5. Install or connect the following:
   - Coil spring
   - Spring upper seat
   - Upper strut bearing
   - Strut mount bracket
   - Upper strut mount. Tighten the piston rod nut to 43–58 ft. lbs. (59–78 Nm).
6. Remove the spring compressor and install the strut assembly to the vehicle.
7. Check the wheel alignment and adjust, as necessary.

Rear

1. Before servicing the vehicle, refer to the precautions in the beginning of this section.
2. Support the vehicle at the frame.
3. Support the axle with a floor jack.
4. Remove or disconnect the following:
   - Rear wheels
   - Shock absorbers
   - Stabilizer bar links
   - Lateral control rod
   - Coil springs

To install:

➡Use new fasteners for assembly.

5. Install or connect the following:
   - Coil springs
   - Lateral control rod.
   - Stabilizer bar links.
   - Shock absorbers
   - Rear wheels
6. See the accompanying illustration for relevant torques.

Lower Ball Joint

REMOVAL & INSTALLATION

1. Before servicing the vehicle, refer to the precautions in the beginning of this section.
2. Support the lower control arm.
3. Remove or disconnect the following:
   - Front wheel
   - Lower ball joint

To install:

4. Install or connect the following:
   - Lower ball joint. Tighten the control arm-to-ball joint bolts to 76–94 ft. lbs. (103–127 Nm) and the stud nut to 87–123 ft. lbs. (118–167 Nm).
   - Front wheel

Lower Control Arm

REMOVAL & INSTALLATION

1. Before servicing the vehicle, refer to the precautions in the beginning of this section.
2. Remove or disconnect the following:
   - Front wheel
   - Torsion bar
   - Shock absorber
   - Stabilizer bar link
   - Axle halfshaft, if equipped
   - Lower ball joint
   - Control arm mounting bolts
   - Lower control arm

To install:

3. Install or connect the following:
   - Lower control arm. Tighten the mount bolts to 80–105 ft. lbs. (108–142 Nm) for 2001; 69–96 ft. lbs. (94–130 Nm) for 2002–04 models.
   - Lower ball joint. Tighten the nut to 87–141 ft. lbs. (118–191 Nm) for 2001; 87–123 ft. lbs. (118–167 Nm) for 2002–04 models.
   - Axle halfshaft, if equipped
   - Stabilizer bar link
   - Shock absorber
   - Torsion bar
   - Front wheel
4. Check the wheel alignment and adjust, as necessary.

CONTROL ARM BUSHING REPLACEMENT

1. Before servicing the vehicle, refer to the precautions in the beginning of this section.
2. Remove the control arm from the vehicle.
3. Remove the control arm bushing with a press.

To install:

4. Lubricate the control arm bushings with liquid soap.
5. Install the bushings with a press.
6. Install the control arm to the vehicle.
7. Check the wheel alignment and adjust, as necessary.

Wheel Bearings

ADJUSTMENT

2 Wheel Drive

➡Use a new split pin for assembly.

1. Before servicing the vehicle, refer to the precautions in the beginning of this section.
2. Remove or disconnect the following:
   - Dust cap
   - Split pin
   - Spindle nut cap
3. Tighten the spindle nut to 25–29 ft. lbs. (34–39 Nm).
4. Spin the hub several times to fully seat the bearings.
5. Retighten the spindle nut to 25–29 ft. lbs. (34–39 Nm).
6. Loosen the spindle nut 45–60 degrees and install the spindle nut cap and split pin.
7. Install the dust cap.

4 Wheel Drive

1. Before servicing the vehicle, refer to the precautions in the beginning of this section.
2. Remove or disconnect the following:
   - Locking hub or driveplate
   - Snapring
   - Spindle washer

- Thrust washer
- Lockwasher

3. Tighten the wheel bearing locknut to 58–72 ft. lbs. (78–98 Nm).
4. Loosen the locknut fully.
5. Tighten the wheel bearing locknut to 4–13 inch lbs. (0.5–1.5 Nm).
6. Spin the hub several times to fully seat the bearings.
7. Retighten the wheel bearing locknut to 4–13 inch lbs. (0.5–1.5 Nm).
8. Install or connect the following:
   - Lockwasher. Tighten the retaining screw to 10–16 inch lbs. (1–2 Nm).
   - Thrust washer
   - Spindle washer
   - Snapring
   - Locking hub or driveplate

REMOVAL & INSTALLATION

2 Wheel Drive

1. Before servicing the vehicle, refer to the precautions in the beginning of this section.
2. Remove or disconnect the following:
   - Front wheel
   - Brake caliper and support
   - Dust cap
   - Split pin
   - Spindle nut cap
   - Spindle nut
   - Bearing washer
   - Outer bearing
   - Hub and brake rotor assembly
   - Inner grease seal
   - Inner wheel bearing

To install:

3. Install or connect the following:
   - Inner wheel bearing
   - Inner grease seal
   - Hub and brake rotor assembly
   - Outer bearing
   - Bearing washer
   - Spindle nut. Adjust the wheel bearings.
   - Spindle nut cap
   - Split pin
   - Dust cap
   - Brake caliper and support
   - Front wheel

4 Wheel Drive

1. Before servicing the vehicle, refer to the precautions in the beginning of this section.
2. Remove or disconnect the following:
   - Front wheel

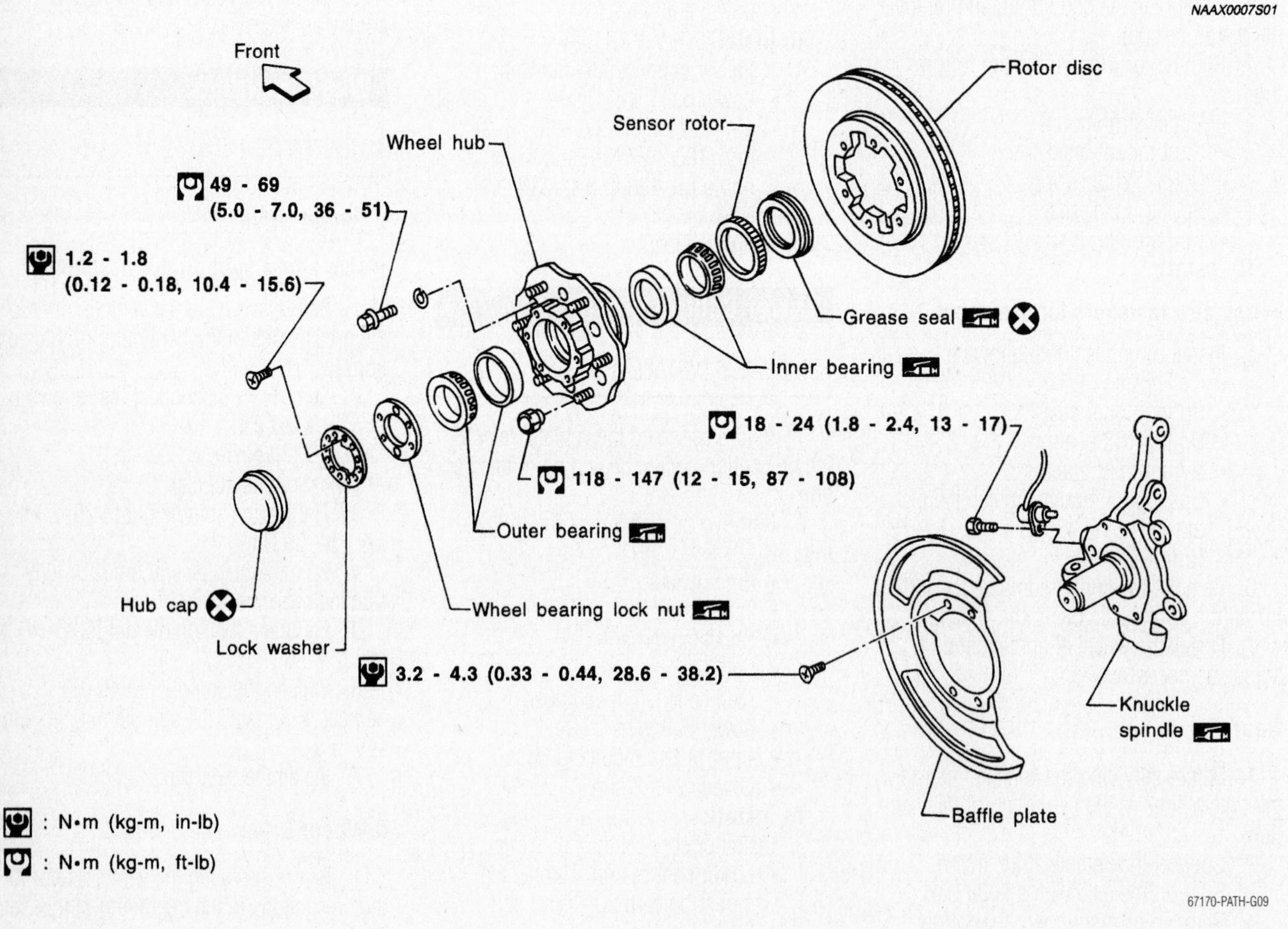

2WD front hub and related parts

- Brake caliper and support
- Locking hub or driveplate
- Snapring
- Spindle washer
- Thrust washer
- Lockwasher
- Wheel bearing locknut
- Outer bearing
- Hub and brake rotor assembly
- Inner grease seal
- Inner wheel bearing

To install:

3. Install or connect the following:
   - Inner wheel bearing
   - Inner wheel bearing
   - Inner grease seal
   - Hub and brake rotor assembly
   - Outer bearing
   - Wheel bearing locknut. Adjust the wheel bearings.
   - Lockwasher
   - Thrust washer
   - Spindle washer
   - Snapring
   - Locking hub or driveplate
   - Brake caliper and support
   - Front wheel

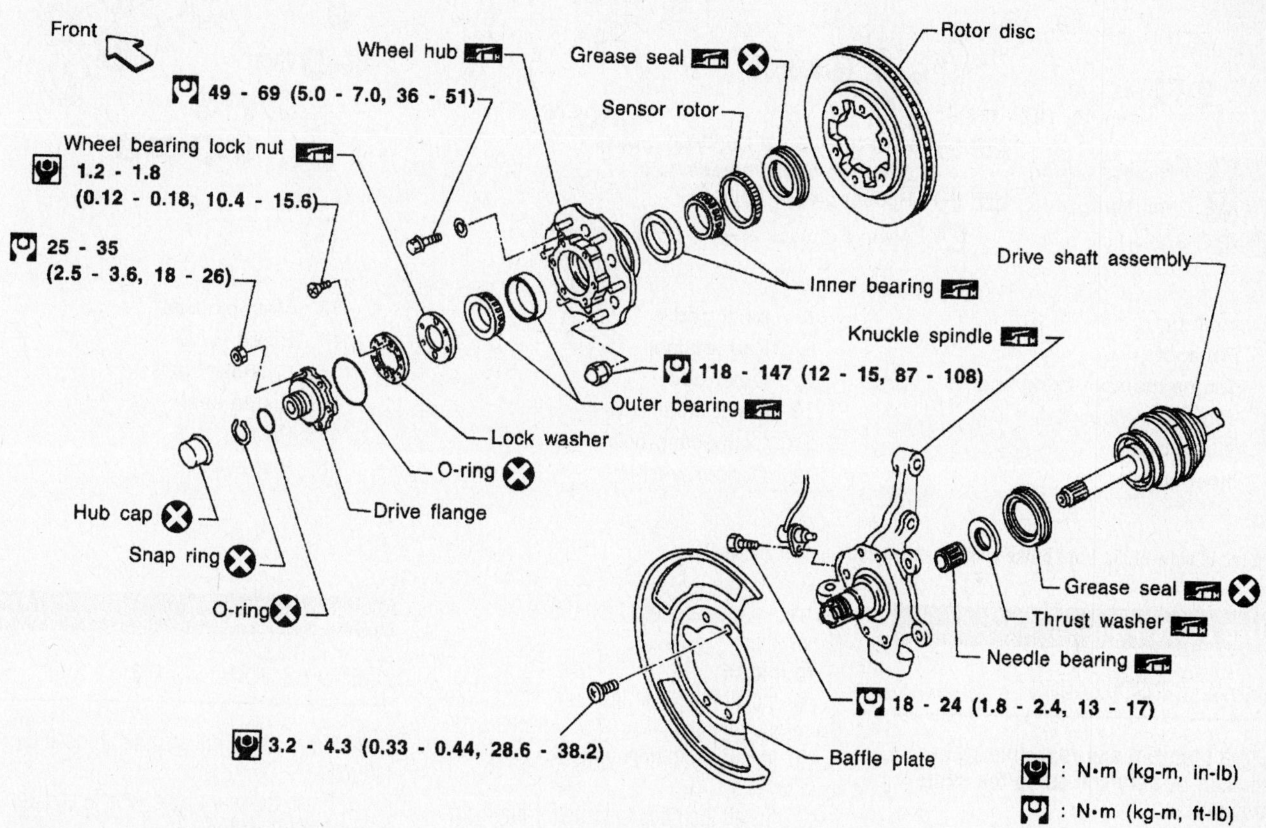

4WD front hub and related parts

BRAKES

Brake Caliper

REMOVAL & INSTALLATION

1. Raise the vehicle and support safely.
2. Remove the appropriate tire and wheel assembly.
3. Remove the bolt attaching the brake hose to the caliper. Plug the brake hose to prevent brake fluid loss.
4. Remove the caliper support mounting bolts and lift the caliper/support assembly from the knuckle.

To install:

5. Position the caliper assembly onto the knuckle and install the bolts. Make sure the rotor fits between the brake pads. Torque the bolts to:
   - 2001: 53–72 ft. lbs. (72–97 Nm).
   - 2002–03: 124–137 ft. lbs. (169–186 Nm)
   - 2004: 107–136 ft. lbs. (145–185 Nm)
6. Using new copper washers, connect the brake hose to the caliper. Torque the brake hose attaching bolt to 12–14 ft. lbs. (17–20 Nm).
7. Bleed the brake system.
8. Apply the brake pedal and inspect the system. Ensure proper operation and no leakage.
9. Install tire and wheel assembly. Lower the vehicle and road-test.

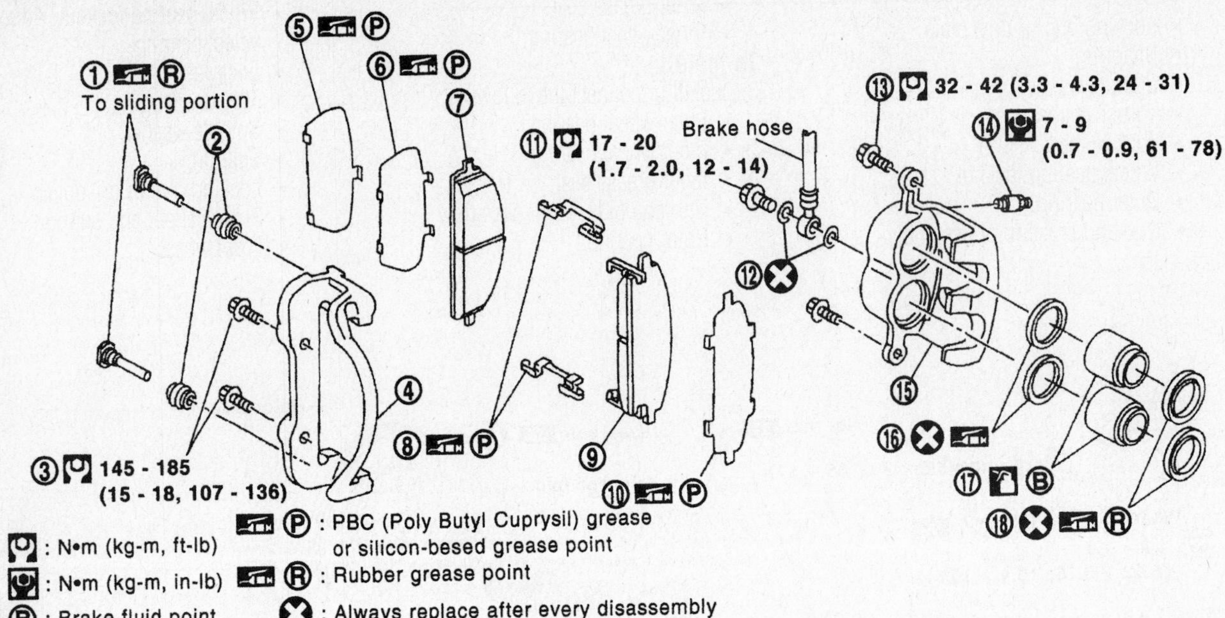

1. Main pin
2. Pin boot
3. Torque member fixing bolt
4. Torque member
5. Shim cover
6. Inner shim
7. Inner pad
8. Pad retainer
9. Outer pad
10. Outer shim
11. Connecting bolt
12. Copper washer
13. Main pin bolt
14. Bleed valve
15. Cylinder body
16. Piston seal
17. Piston
18. Piston boot

67170-PATH-G01

Exploded view of the front brake components—2004 model shown

Disc Brake Pads

REMOVAL & INSTALLATION

➡Both the front and rear disc brake pads can be serviced using the same procedure.

1. Using a syringe, siphon brake fluid from the reservoir, leaving reservoir approximately ½ full.
2. Raise and properly support the vehicle.
3. Remove the wheel assemblies.
4. Remove the lower pin bolt from the brake caliper.
5. Swivel the caliper up and away from the torque member. Tie the caliper to a suspension member so that it is out of the way.
6. Lift the 2 brake pads out of the torque member.
7. Remove the inner and outer shims. Remove the 2 pad retainers if they are not attached to the pads.
8. Check the pad thickness and replace the pads if they are less than 0.079 in. (2mm) thick.

To install:

9. Install the inner and outer shims into the torque member.
10. Install a pad retainer to the bottom of each pad.
11. Install the pads into the torque member.
12. Use a C-clamp or hammer handle and press the caliper piston(s) back into the housing.
13. Untie the caliper and swivel it back into position over the torque plate so that the dust boot is not pinched. Install the pin bolt and torque it to 16–23 ft. lbs. (22–31 Nm) for 2001 models; 24–31 ft. lbs. (32–42 Nm) for 2002–04 models.
14. Check the condition of the pin boot. Gently pull on it to expel any trapped air.
15. Install the wheel and lower the vehicle.
16. Pump the brakes until the pedal is firm and check the level of brake fluid. Road-test the vehicle.

Brake Drums

REMOVAL & INSTALLATION

1. Remove the hub cap and loosen the lug nuts.
2. Raise the rear of the vehicle and support it on jackstands.
3. Remove the lug nuts, tire and wheel.
4. Release the parking brake.
5. Pull the brake drum from the hub. If difficult to remove try the following:
   a. Strike the face of the drum with a plastic or rubber mallet. This will break free any rust that may develop between the drum and the hub.
   b. Install 2, M8x1.25mm bolts into the holes in the drum and gradually tighten them to pull the drum off the hub.

To install:

6. Install the brake drum to the hub.
7. Install the wheel.
8. Remove the jackstands and lower the vehicle.

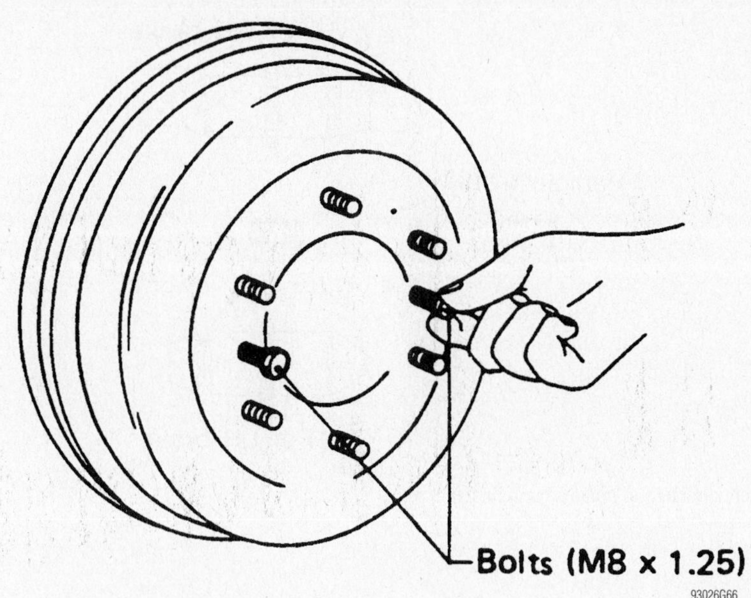

Install and tighten 2 bolts to remove a stubborn brake drum

9. Road-test the vehicle to ensure that the brakes are working properly.

Brake Shoes

REMOVAL & INSTALLATION

Pathfinder

1. Release the parking brake.
2. Safely raise and support the vehicle.
3. Remove the rear wheel and drum.
4. Remove the hold-down pin retainers.
5. Remove the leading shoe and then the trailing shoe.
6. Remove the adjuster.
7. Disconnect the parking brake cable from the toggle lever on the rear shoe.

To install:

8. Transfer the toggle lever to the new rear shoe.
9. Apply a small amount of brake

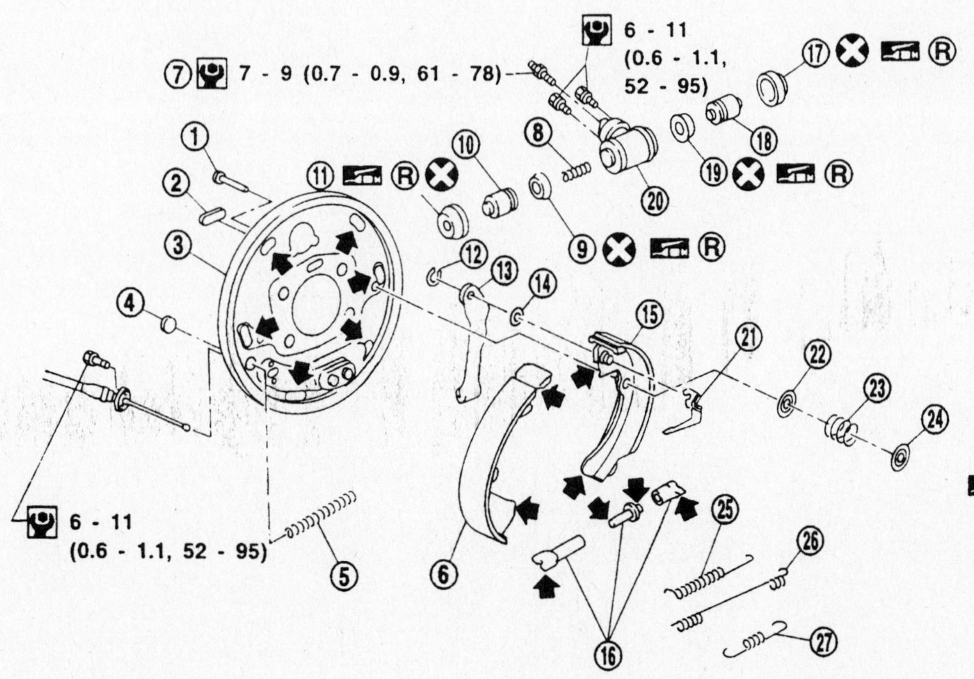

1. Shoe hold pin
2. Plug
3. Back plate
4. Check plug
5. Spring
6. Shoe (leading side)
7. Air bleeder
8. Spring
9. Piston cup
10. Piston
11. Boot
12. Retainer ring
13. Toggle lever
14. Wave washer
15. Shoe (trailing side)
16. Adjuster
17. Boot
18. Piston
19. Piston cup
20. Wheel cylinder
21. Adjuster lever
22. Spring seat
23. Shoe hold spring
24. Retainer
25. Adjuster spring
26. Return spring (upper)
27. Return spring (lower)

93026G69

Drum brake assembly exploded view

grease to the tips of the shoes and the 6 pads on the backing plate that contact the brake shoe.

10. Shorten the adjuster by turning it.
11. Connect the parking brake cable to the toggle lever on the rear shoe.
12. Install the lower return spring to both shoes and install the shoes on the backing plate with the hold down pins and retainers.
13. Install the adjuster and the remaining springs. Pay attention to the direction of the adjuster assembly.
14. Inspect the complete assembly and install the brake drum.
15. Adjust the shoe to drum clearance.
16. Install the wheel assembly and lower the vehicle to the floor.

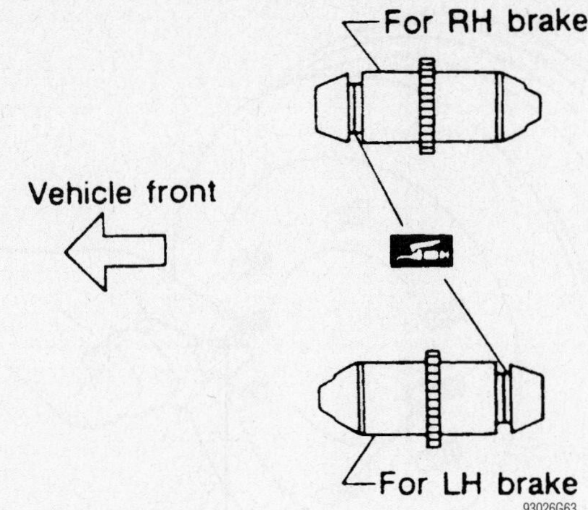

Correct direction of brake shoe adjuster

NISSAN

QUEST

10

SPECIFICATION AND MAINTENANCE CHARTS

VEHICLE AND ENGINE IDENTIFICATION CHART

Engine Code						
Code	Liters (cc)	Cu. In.	Cyl.	Fuel Sys.	Engine Type	Eng. Mfg.
T	3.3 (3275)	200	6	SEFI	SOHC	Nissan
VQ35DE	3.5 (3498)	213	6	MFI	DOHC	Nissan

MFI: Multi-port Fuel Injection

SEFI: Sequential Multi-port Fuel Injection

Model Year	
Code	Year
1	2001
2	2002
3	2003
4	2004
5	2005

67170-NIQU-C01

GENERAL ENGINE SPECIFICATIONS

Year	Engine Displacement Liters (VIN)	Net Horsepower @ rpm	Net Torque @ rpm (ft. lbs.)	Bore x Stroke (in.)	Com-pression Ratio	Oil Pressure @ rpm
2000	3.3 (T)	195@4500	190@3800	3.60x3.27	8.9:1	60-65@2000
2001	3.3 (T)	195@4500	190@3800	3.60x3.27	8.9:1	60-65@2000
2002	3.3 (T)	195@4500	190@3800	3.60x3.27	8.9:1	60-65@2000
2004-05	3.5 (VQ35DE)	240@5800	242@4400	3.76X3.20	10.3:1	43@2000

MFI: Multiport fuel injection

SEFI: Sequential Multi-port Fuel Injection

67170-NIQU-C02

ENGINE TUNE-UP SPECIFICATIONS

Year	Engine Displacement Liters (VIN)	Spark Plug Gap (in.)	Ignition Timing (deg.)		Fuel Pump (psi)	Idle Speed (rpm)		Valve Clearance	
			MT	AT		MT	AT ②	In.	Ex.
2000	3.3 (T)	0.043	—	13-17B	34 ①	—	650-750	HYD	HYD
2001	3.3 (T)	0.043	—	13-17B	34 ①	—	650-750	HYD	HYD
2002	3.3 (T)	0.043	—	13-17B	34 ①	—	650-750	HYD	HYD
2004-05	3.5 (VQ35DE)	0.043	—	10-20B	51 ③	—	650-750	HYD	HYD

NOTE: The Vehicle Emission Control Information label must be used if they differ from those in this chart.

B: Before top dead center

HYD: Hydraulic

① System pressure at idle with vacuum hose connected should increase to 43 psi when disconnected

② Transmission in Neutral

③ System pressure at idle

67170-NIQU-C03

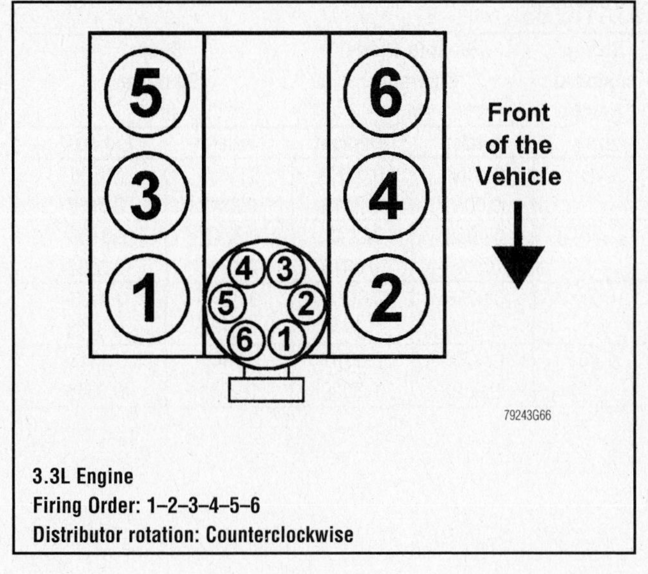

3.3L Engine
Firing Order: 1–2–3–4–5–6
Distributor rotation: Counterclockwise

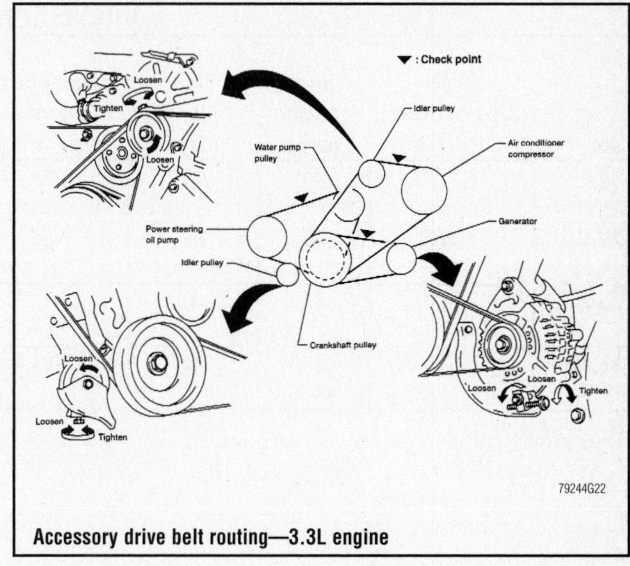

Accessory drive belt routing—3.3L engine

CAPACITIES

Year	Model	Engine Displacement Liters (VIN)	Engine Oil with Filter (qts.)	Transmission (pts.) 4-Spd	Transmission (pts.) 5-Spd	Transmission (pts.) Auto.	Drive Axle Front (pts.)	Drive Axle Rear (pts.)	Fuel Tank (gal.)	Cooling System (qts.) ②
2000	Quest	3.3 (T)	4.0	—	—	20.0	①	—	20	11.25
2001	Quest	3.3 (T)	4.0	—	—	20.0	①	—	20	11.25
2002	Quest	3.3 (T)	4.0	—	—	20.0	①	—	20	11.25
2004-05	Quest	3.5 (VQ35DE)	3.78	—	—	③	—	—	20	10.50

NOTE: All capacities are approximate. Add fluid gradually and check to be sure a proper fluid level is obtained.

① Included in transaxle capacity

② Includes reservoir tank.

③ 4 speed: 9 qts.
5 speed: 7 7/8 qts.

67170-NIQU-C04

VALVE SPECIFICATIONS

Year	Engine Displacement Liters (VIN)	Seat Angle (deg.)	Face Angle (deg.)	Spring Test Pressure (lbs. @ in.)	Spring Installed Height (in.)	Stem-to-Guide Clearance (in.)		Stem Diameter (in.)	
						Intake	Exhaust	Intake	Exhaust
2000	3.3 (T)	45	45	①	②	0.0008-0.0021	0.0012-0.0019	0.2742-0.2748	0.3136-0.3138
2001	3.3 (T)	45	45	①	②	0.0008-0.0021	0.0012-0.0019	0.2742-0.2748	0.3136-0.3138
2002	3.3 (T)	45	45	①	②	0.0008-0.0021	0.0012-0.0019	0.2742-0.2748	0.3136-0.3138
2004-05	3.5 (VQ35DE)	45.15-45.45	45	45.4@1.457	1.457	0.0008-0.0021	0.0016-0.0029	0.2348-0.2354	0.2341-0.2346

① Outer spring: 118@1.81
Inner spring: 57.3@0.984

② Spring height measured unloaded
Minimum length. outer spring: 2.016
Minimum length. inner spring: 1.736

67170-NIQU-C05

CRANKSHAFT AND CONNECTING ROD SPECIFICATIONS

All measurements are given in inches.

Year	Engine Displacement Liters (VIN)	Crankshaft				Connecting Rod		
		Main Brg. Journal Dia.	Main Brg. Oil Clearance	Shaft End-play	Thrust on No.	Journal Diameter	Oil Clearance	Side Clearance
2000	3.3 (T)	2.4791-2.4793	0.0011-0.0022	0.0020-0.0067	3	1.9667-1.9675	0.0006-0.0021	0.0079-0.00138
2001	3.3 (T)	2.4791-2.4793	0.0011-0.0022	0.0020-0.0067	3	1.9667-1.9675	0.0006-0.0021	0.0079-0.00138
2002	3.3 (T)	2.4791-2.4793	0.0011-0.0022	0.0020-0.0067	3	1.9667-1.9675	0.0006-0.0021	0.0079-0.00138
2004-05	3.5 (VQ35DE)	①	0.0014-0.0018	0.0118	4	②	0.0013-0.0023	0.0079-0.0138

① There are 24 different grades, ranging from A (2.3612) to 7 (2.3603)

② Grade 0: 0.0591-0.0592
Grade 1: 0.0592-0.0593
Grade 2: 0.0593-0.0594

67170-NIQU-C06

PISTON AND RING SPECIFICATIONS

All measurements are given in inches.

Year	Engine Displacement Liters (VIN)	Piston Clearance	Ring Gap Top Compression	Ring Gap Bottom Compression	Ring Gap Oil Control	Ring Side Clearance Top Compression	Ring Side Clearance Bottom Compression	Ring Side Clearance Oil Control
2000	3.3 (T)	①	0.0083-0.0122	0.0197-0.0236	0.0079-0.0236	0.0016-0.0031	0.0012-0.0028	0.0006-0.0073
2001	3.3 (T)	①	0.0083-0.0122	0.0197-0.0236	0.0079-0.0236	0.0016-0.0031	0.0012-0.0028	0.0006-0.0073
2002	3.3 (T)	①	0.0083-0.0122	0.0197-0.0236	0.0079-0.0236	0.0016-0.0031	0.0012-0.0028	0.0006-0.0073
2004-05	3.5 (VQ35DE)	0.0004-0.0012	0.0091-0.0130	0.0130-0.0189	0.0079-0.0197	0.0018-0.0031	0.0012-0.0028	0.0026-0.0053

① Journals 1, 2 and 6: 0.0010 - 0.0018 in.
Journals 3 and 4: 0.0006 - 0.0010 in.
Journal 5: 0.0012 - 0.0016 in.

67170-NIQU-C07

TORQUE SPECIFICATIONS

All readings in ft. lbs.

Year	Engine Displacement Liters (VIN)	Cylinder Head Bolts	Main Bearing Bolts	Rod Bearing Bolts	Crankshaft Damper Bolts	Flywheel Bolts	Manifold Intake	Manifold Exhaust	Spark Plugs	Oil Pan Drain Plug
2000	3.3 (T)	①	②	②	141-156	61-69	①	13-16	14-22	24
2001	3.3 (T)	①	②	②	141-156	61-69	①	13-16	14-22	24
2002	3.3 (T)	①	②	②	141-156	61-69	①	13-16	14-22	24
2004-05	3.5 (VQ35DE)	③	④	⑤	⑥	61-69	14	23	18	25

① Intake manifold and cylinder heads are installed at the same time.
Step 1: cylinder head bolts to 22 ft. lbs.
Step 2: cylinder head bolts to 43 ft. lbs.
Step 3: Loosen all bolts completely
Step 4: cylinder head bolts to 7 ft. lbs.
Step 5: Intake manifold bolts to 2.9 (4Nm) ft. lbs.
Step 6: Intake manifold bolts to 13 ft. lbs.
Step 7: Intake manifold bolts to 14 ft. lbs.
Step 8: Loosen all maifold bolts completely
Step 9: Cylinder head bolts to 26 inch lbs.
Step 10: cylinder head bolts to 47 ft. lbs. Or, an additional 65 degrees
Step 11: cylinder head sub-bolts to 104 inch lbs.
Step 12: Intake manifold bolts to 36 inch lbs.
Step 13: Intake manifold bolts to 78 inch lbs.
Step 14: Intake manifold bolts to 84 inch lbs.

② Step 1: 34-37 ft. lbs.
Step 2: 67-74 ft. lbs.

③ Step 1: 72 ft. lbs.
Step 2: Loosen all bolts completely
Step 3: 29 ft. lbs.
Step 4: +90 degrees
Step 5: +90 degrees

④ Step 1: 26 ft. lbs.
Step 2: +90-95 degrees

⑤ Step 1: 15 ft. lbs.
Step 2: +90-95 degrees

⑥ 32 ft. lbs. +60-65 degrees

67170-NIQU-C08

WHEEL ALIGNMENT

			Caster		Camber		
Year	Model		Range (Deg.)	Preferred Setting (Deg.)	Range (Deg.)	Preferred Setting (Deg.)	Toe-in (in.)
2000	Quest	F	0.75	2.75	0.75	-0.25	0.04 +/- 0.04
		R	—	—	0.75	-1.00	0.04 +/- 0.16
2001	Quest	F	0.75	2.75	0.75	-0.25	0.04 +/- 0.04
		R	—	—	0.75	-1.00	0.04 +/- 0.16
2002	Quest	F	0.75	2.75	0.75	-0.25	0.04 +/- 0.04
		R	—	—	0.75	-1.00	0.04 +/- 0.16
2004-05	Quest ① ②	F	③	③	④	④	0.02 +/- 0.04
		R	—	—	⑤	⑤	0.02 +/- 0.16

① Vehicle unladen

② Specifications are decimal degrees

③ Minimum: 2.17
Nominal: 2.92
Maximum: 3.67
Left and right difference 0.75 or less

④ Minimum: - 0.92
Nominal: - 0.17
Maximum: 0.58
Left and right difference 0.75 or less

⑤ Minimum: - 0.24
Nominal: - 0.26
Maximum: - 0.76

67170-NIQU-C09

TIRE, WHEEL AND BALL JOINT SPECIFICATIONS

		OEM Tires		Tire Pressures (psi)		Wheel	Ball Joint	Lug
Year	Model	Standard	Optional	Front	Rear	Size	Inspection	Nut
2000	Quest	P215/70R15	P225/60R16	30	30	Std: 5.5-JJ Opt: 6.5-JJ	①	80
2001	Quest	P215/70R15	P225/60R16	30	30	Std: 5.5-JJ Opt: 6.5-JJ	①	80
2002	Quest	P215/70R15	P225/60R16	30	30	Std: 5.5-JJ Opt: 6.5-JJ	①	80
2004-05	Quest	P225/65R16	P225/60R17	35	35	NA	①	83

NA: Not Available

OEM: Original Equipment Manufacturer

PSI: Pounds Per Square Inch

① Replace if any measurable movement is found.

67170-NIQU-C10

BRAKE SPECIFICATIONS

All measurements in inches unless noted

			Brake Disc			Brake Drum Diameter				Brake Caliper	
Year	Model		Original Thickness	Minimum Thickness	Maximum Runout	Original Inside Diameter	Max, Wear Limit	Maximum Machine Diameter	Minimum Lining Thickness	Bracket-to-Hub Bolt (ft. lbs.)	Mounting Pin or Bolt (ft. lbs.)
2000	Quest	F	1.002	0.940	0.0028	—	—	—	0.079	—	18-25
		R	—	—	—	9.84	9.90	9.86	0.079	—	—
2001	Quest	F	1.002	0.940	0.0028	—	—	—	0.079	—	18-25
		R	—	—	—	9.84	9.90	9.86	0.079	—	—
2002	Quest	F	1.002	0.940	0.0028	—	—	—	0.079	—	18-25
		R	—	—	—	9.84	9.90	9.86	0.079	—	—
2004-05	Quest	F	1.100	1.02	0.0006	—	—	—	0.079	115	20
		R	0.630	0.551	0.0006	—	—	—	0.079	62	32

NOTE: Due to changes made during production, refer to the manufacturer's specifications if they differ from those in this chart

F: Front

R: Rear

67170-NIQU-C11

SCHEDULED MAINTENANCE INTERVALS
2001—02 NISSAN QUEST

TO BE SERVICED	TYPE OF SERVICE	VEHICLE MILEAGE INTERVAL (x1000)												
		5	10	15	20	25	30	35	40	45	50	55	60	65
Engine oil & filter	R	✓	✓	✓	✓	✓	✓	✓	✓	✓	✓	✓	✓	✓
Rotate tires	S/I	✓		✓		✓		✓		✓		✓		✓
Engine coolant strength hoses & clamps	S/I			✓			✓			✓			✓	
Air cleaner filter	R						✓						✓	
Automatic transmission fluid & filter	R						✓						✓	
Engine coolant ①	R						✓						✓	
PCV valve	R												✓	
Spark plugs ②	R													
Drive belts ③	S/I			✓			✓			✓			✓	
Timing belt ④	S/I													
Exhaust system & heat shields	S/I			✓			✓			✓			✓	
Drive shaft boots	S/I			✓			✓			✓		✓		
Front & rear brake components	S/I	✓	✓	✓	✓	✓	✓	✓	✓	✓	✓	✓	✓	✓

R: Replace S/I: Service or Inspect

① Engine coolant: change every 30,000 miles or 36 months.

② Replace every 105,000 miles

③ Inspect every 15, 000 miles or 12 months and replace every 60,000 miles or 48 months.

④ Replace every 105,000 miles

FREQUENT OPERATION MAINTENANCE (SEVERE SERVICE)

If a vehicle is operated under any of the following conditions it is considered severe service:

- Extremely dusty areas.

- 50% or more of the vehicle operation is in 32°C (90°F) or higher temperatures, or constant operation in temperatures below 0°C (32°F).

- Prolonged idling (vehicle operation in stop and go traffic.

- Frequent short running periods (engine does not warm to normal operating temperatures).

- Police, taxi, delivery usage or trailer towing usage.

Engine oil & filter: replace every 3000 miles.

Rotate tires initially at 6000 miles and every 9000 miles thereafter.

Air cleaner filter: change every 15,000 miles.

Engine coolant strength, hoses & clamps: check every 15,000 miles.

Exhaust system: check every 15,000 miles.

Automatic transmission fluid & filter: change every 21,000 miles.

67170-NIQU-C12

SCHEDULED MAINTENANCE INTERVALS
2004—05 NISSAN QUEST

TO BE SERVICED	TYPE OF SERVICE	VEHICLE MILEAGE INTERVAL (x1000)															
		3.75	7.5	11.3	15	18.8	22.5	26.3	30	33.8	37.5	41.3	45	48.8	52.5	56.3	60
Engine oil & filter	R	✓	✓	✓	✓	✓	✓	✓	✓	✓	✓	✓	✓	✓	✓	✓	✓
Brake lines & cables	S/I				✓				✓				✓				✓
Brake pads and rotors	S/I		✓		✓		✓		✓		✓		✓		✓		✓
Driveshaft boots & propeller shaft	L/I		✓		✓		✓		✓		✓		✓		✓		✓
Transaxle fluid ①	I																✓
Air cleaner filter ②	R																✓
Drive belt(s) ③	S/I																✓
Engine coolant ④	R																✓
Spark plugs	R	platinum tipped plugs every 105,000 miles															
Cabin air filter	R				✓				✓				✓				✓
Exhaust system	I		✓		✓		✓		✓		✓		✓		✓		✓
Fuel lines	S/I								✓								✓
Steering gear, linkage, axle & suspension parts	I		✓		✓		✓		✓		✓		✓		✓		✓
Valve clearance ⑤																	
Vapor lines	S/I																✓

R: Replace S/I: Service or Inspect L: Lubricate

① Replace first a 60,000 or 48 months, inspect every 15,000 miles

② If operating in dusty conditions more frequent maintenance may be required

③ Replace first a 60,000 or 48 months, inspect every 15,000 miles, replace if damaged during any inspection

④ Replace first after 60,000 or 48 months, replace every 30,000 or 24 months

⑤ If valve noise increases inspect clearance

FREQUENT OPERATION MAINTENANCE (SEVERE SERVICE)

If a vehicle is operated under any of the following conditions it is considered severe service:

- Extremely dusty areas.

- 50% or more of the vehicle operation is in 32°C (90°F) or higher temperatures, or constant operation in temperatures below 0°C (32°F).

- Prolonged idling (vehicle operation in stop and go traffic).

- Frequent short running periods (engine does not warm to normal operating temperatures).

- Police, taxi, delivery usage or trailer towing usage.

Oil & oil filter: replace every 3750 miles.

Brake pads, discs, drums & linings: service or inspect every 7500 miles.

Driveshaft boots & propeller shaft: service or inspect every 7500 miles.

Exhaust system: service or inspect every 7500 miles.

Steering gear (box) & linkage, (steering damper-4x4), axle & suspension parts: service or inspect every 7500 miles.

Steering linkage ball joints & front suspension ball joints: service or inspect every 7500 miles.

67170-NIQU-C13

PRECAUTIONS

Before servicing any vehicle, please be sure to read all of the following precautions, which deal with personal safety, prevention of component damage, and important points to take into consideration when servicing a motor vehicle:

- Never open, service or drain the radiator or cooling system when the engine is hot; serious burns can occur from the steam and hot coolant.
- Observe all applicable safety precautions when working around fuel. Whenever servicing the fuel system, always work in a well-ventilated area. Do not allow fuel spray or vapors to come in contact with a spark, open flame or excessive heat (a hot drop light, for example). Keep a dry chemical fire extinguisher near the work area. Always keep fuel in a container specifically designed for fuel storage; also, always properly seal fuel containers to avoid the possibility of fire or explosion. Refer to the additional fuel system precautions later in this section.
- Fuel injection systems often remain pressurized, even after the engine has been turned **OFF**. The fuel system pressure must be relieved before disconnecting any fuel lines. Failure to do so may result in fire and/or personal injury.
- Brake fluid often contains polyglycol ethers and polyglycols. Avoid contact with the eyes and wash your hands thoroughly after handling brake fluid. If you do get brake fluid in your eyes, flush your eyes with clean, running water for 15 minutes. If eye irritation persists, or if you have taken brake fluid internally, IMMEDIATELY seek medical assistance.
- The EPA warns that prolonged contact with used engine oil may cause a number of skin disorders, including cancer! You should make every effort to minimize your exposure to used engine oil. Protective gloves should be worn when changing oil. Wash your hands and any other exposed skin areas as soon as possible after exposure to used engine oil. Soap and water, or waterless hand cleaner should be used.
- All new vehicles are now equipped with an air bag system. The system must be disabled before performing service on or around system components, steering column, instrument panel components, wiring and sensors. Failure to follow safety and disabling procedures could result in accidental air bag deployment, possible personal injury and unnecessary system repairs.
- Always wear safety goggles when working with, or around, the air bag system. When carrying a non-deployed air bag, be sure the bag and trim cover are pointed away from your body. When placing a non-deployed air bag on a work surface, always face the bag and trim cover upward, away from the surface. This will reduce the motion of the module if it is accidentally deployed. Refer to the additional air bag system precautions later in this section.
- Clean, high quality brake fluid from a sealed container is essential to the safe and proper operation of the brake system. You should always buy the correct type of brake fluid for your vehicle. If the brake fluid becomes contaminated, completely flush the system with new fluid. Never reuse any brake fluid. Any brake fluid that is removed from the system should be discarded. Also, do not allow any brake fluid to come in contact with a painted surface; it will damage the paint.
- Never operate the engine without the proper amount and type of engine oil; doing so WILL result in severe engine damage.
- Timing belt maintenance is extremely important! Many models utilize an interference-type, non-freewheeling engine. If the timing belt breaks, the valves in the cylinder head may strike the pistons, causing potentially serious (also time-consuming and expensive) engine damage.
- Disconnecting the negative battery cable on some vehicles may interfere with the functions of the on-board computer system(s) and may require the computer to undergo a relearning process once the negative battery cable is reconnected.
- When servicing drum brakes, only disassemble and assemble one side at a time, leaving the remaining side intact for reference.
- Only an MVAC-trained, EPA-certified automotive technician should service the air conditioning system or its components.

ENGINE REPAIR

Distributor

REMOVAL

This applies to the 3.3L engine only.

1. Before servicing the vehicle, refer to the precautions in the beginning of this section.
2. Remove or disconnect the following:
   - Negative battery cable
   - Distributor cap
   - Distributor wiring harness connector
3. Matchmark the rotor to the distributor housing and the distributor housing to the cylinder head.
4. Remove the distributor hold-down bolt and the distributor.

INSTALLATION

Timing Not Disturbed

1. Install or connect the following:
   - Distributor and align the matchmarks made during removal. Tighten the hold-down bolt to 10–12 ft. lbs. (14–17 Nm).
   - Distributor wiring harness connector
   - Distributor cap
   - Negative battery cable
2. Check the ignition timing and adjust, as necessary.

Timing Disturbed

1. Set the engine to Top Dead Center (TDC) of the compression stroke for the No. 1 cylinder.
2. Align the index mark on the distributor shaft with the protrusion on the distributor housing.
3. Install the distributor and check that the distributor rotor is aligned.
4. Install or connect the following:

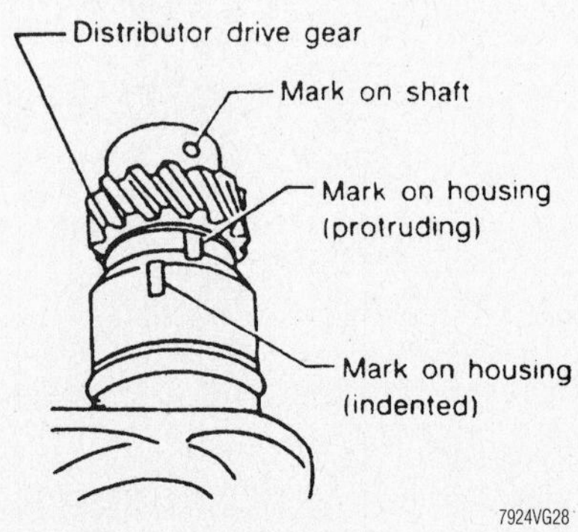

Distributor shaft alignment

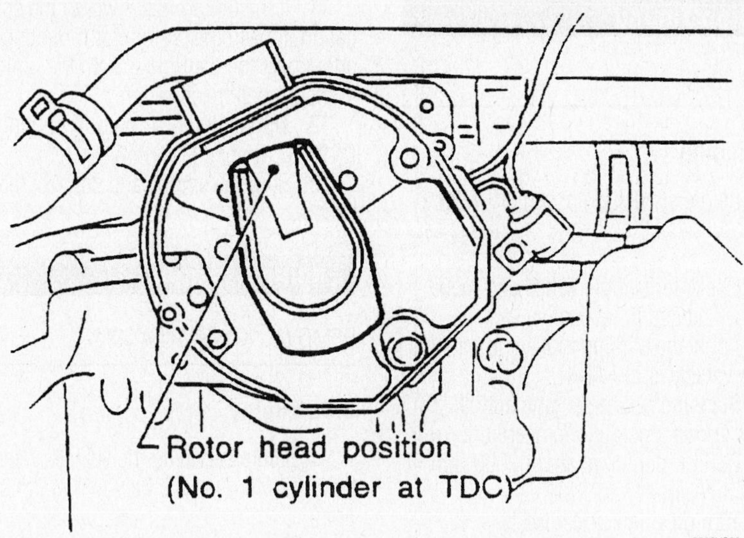

Distributor rotor alignment

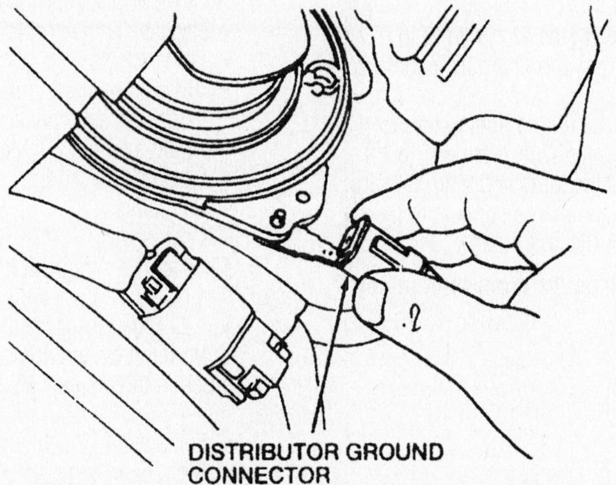

Disengage the distributor ground connector when removing the distributor

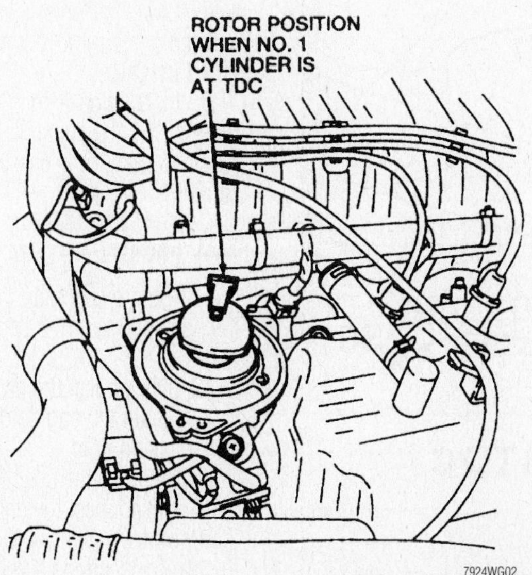

Note the position of the rotor when the No. 1 piston is at TDC on the compression stroke

- Distributor. Tighten the hold-down bolt to 10–12 ft. lbs. (14–17 Nm).
- Distributor cap
- Distributor harness connector

5. Check the ignition timing and adjust, as necessary.

Alternator

REMOVAL

3.3L Engine

1. Before servicing the vehicle, refer to the precautions in the beginning of this section.
2. Remove or disconnect the following:
   - Negative battery cable
   - Idler adjusting bolt, loosen
   - A/C belt
   - Engine undercover
   - Alternator electrical connectors and bracket
   - Alternator mounting bolts
   - Alternator belt
   - Alternator

3.5L Engine

1. Before servicing the vehicle, refer to the precautions in the beginning of this section.
2. Disconnect the negative battery terminal.
3. Remove radiator.
4. Remove the drive belt.
5. Remove idler pulley.
6. Remove the alternator adjustable top mount.
7. Remove the alternator lower bolt and nut.
8. Disconnect the alternator harness connectors.
9. Remove the alternator upper bolt.
10. Remove the alternator.

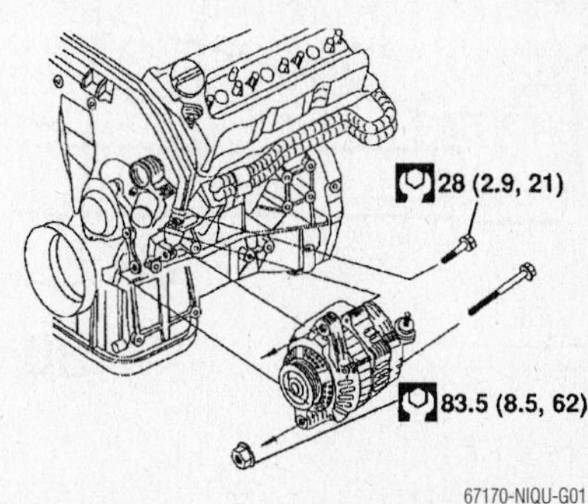

Alternator exploded view–3.5L engine

INSTALLATION

3.3L Engine

1. Before servicing the vehicle, refer to the precautions in the beginning of this section.
2. Install the components in the reverse order of removal. Tighten the fasteners to the following specifications:
    a. Alternator mounting bolts to 16–22 ft. lbs. (22–29 Nm).
    b. Alternator bracket bolt to 12–15 ft. lbs. (16–20 Nm).

3.5L Engine

1. Before servicing the vehicle, refer to the precautions in the beginning of this section.
2. Install the alternator.
3. Install the alternator upper bolt. Tighten to 21 ft. lbs. (28 Nm).
4. Connect the alternator harness connectors.
5. Install the alternator lower bolt and nut. Tighten to 62 ft. lbs. (83 Nm).
6. Install the alternator adjustable top mount.
7. Install idler pulley.
8. Install the drive belt.
9. Install radiator.
10. Connect the negative battery terminal.

Ignition Timing

ADJUSTMENT

3.3L Engine

1. Before servicing the vehicle, refer to the precautions in the beginning of this section.
2. Check for trouble codes and make necessary repairs if needed.
3. Apply the parking brake and be sure that the vehicle is in PARK.
4. Start and run the engine until it reaches normal operating temperature.
5. Run the engine at about 2000 rpm for 2 minutes under no-load.
6. Turn off all electrical loads.
7. Disconnect the Throttle Position (TP) sensor electrical connector.
8. Be sure the engine speed is 700–800 rpm.
9. Rev the engine 2 or 3 times to 2,000–3,000 rpm and return the engine to idle speed.
10. Connect a timing light to the No. 1 cylinder spark plug wire at the distributor end and check the ignition timing. Be sure that the timing pointer is pointing to the 15° BTDC mark on the crankshaft pulley.

➡**Each notch on the crankshaft pulley represents 5°.**

11. If the timing is not within the specification, loosen the distributor mounting bolt and adjust the distributor until the timing is at the proper specification.
12. Tighten the distributor mounting bolt to 10–12 ft. lbs., (14–17 Nm).
13. Stop the engine and connect the TP sensor.

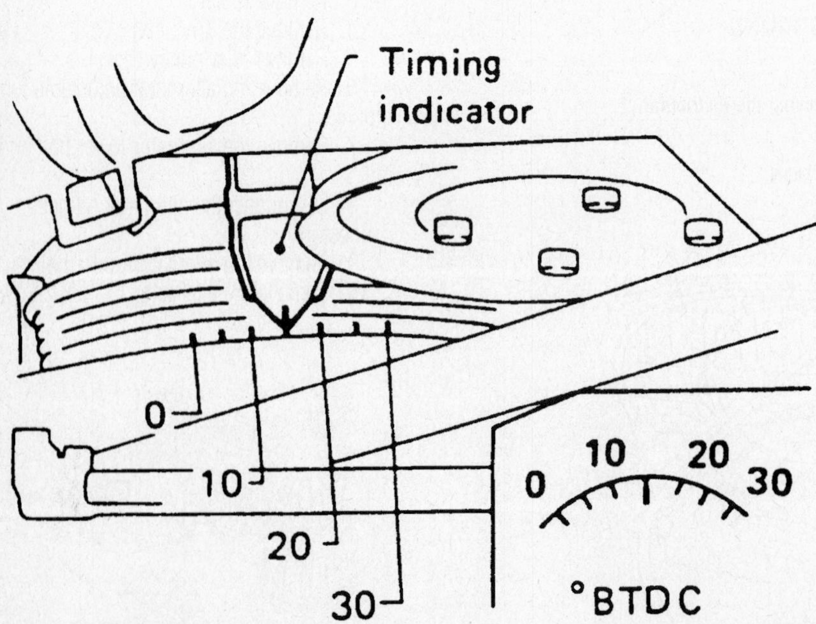

Adjust the timing so the pointer on the engine indicates 15° before top dead center (3 notches from TDC) on the crankshaft pulley.

Engine Assembly

REMOVAL & INSTALLATION

3.3L Engine

1. Before servicing the vehicle, refer to the precautions in the beginning of this section.
2. Properly relieve the fuel system pressure.
3. Drain the coolant and crankcase.
4. Remove or disconnect the following:
    - Negative battery cable
    - Front wheels
    - All vacuum hoses, fuel lines, wires, harnesses and connectors that would interfere with engine removal
    - Exhaust tube
    - Ball joints
    - Drive shafts
5. Recover the refrigerant from the A/C system
    - A/C compressor manifold
    - Power steering pump
6. Support the engine using a suitable lift.
    - Left hand engine mount bolts
    - Right hand engine mount
    - Rear A/C refrigerant line bracket, if equipped
    - Crossmember
7. Lower the engine and transaxle assembly and remove it from the vehicle.

To install:

8. Installation is the reverse of removal. Refer to the accompanying engine mounting illustration for all necessary torque specifications.

3.5L Engine

1. Before servicing the vehicle, refer to the precautions in the beginning of this section.
2. Disconnect the battery cables.
3. Drain the engine oil, coolant and transmission fluid.
4. Remove the cowl top extension.
5. Disconnect the engine harness from the Powertrain Control Module (PCM) and the two connections at the right hand strut tower.

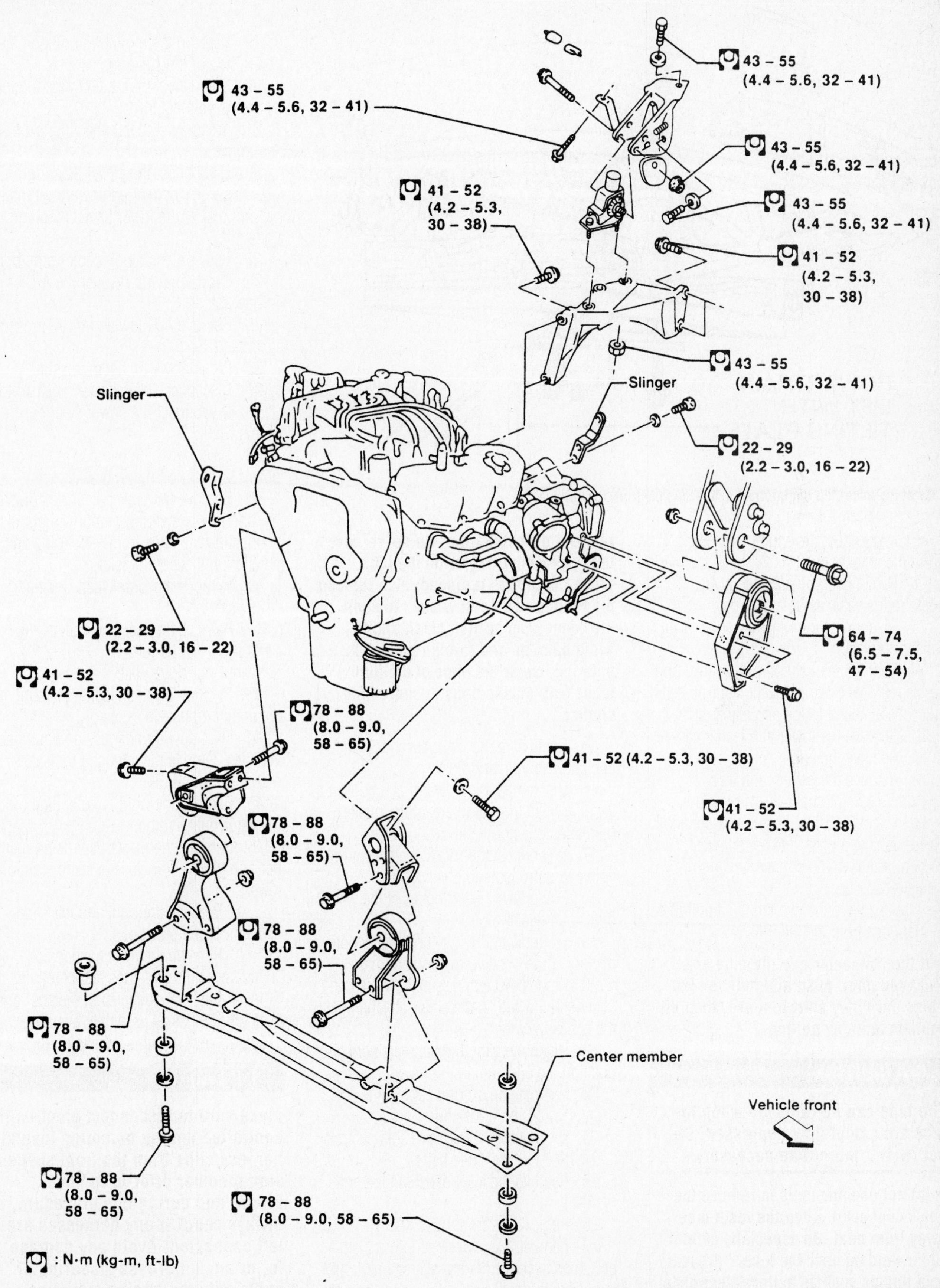

Engine mounting components and specifications—3.3L engines

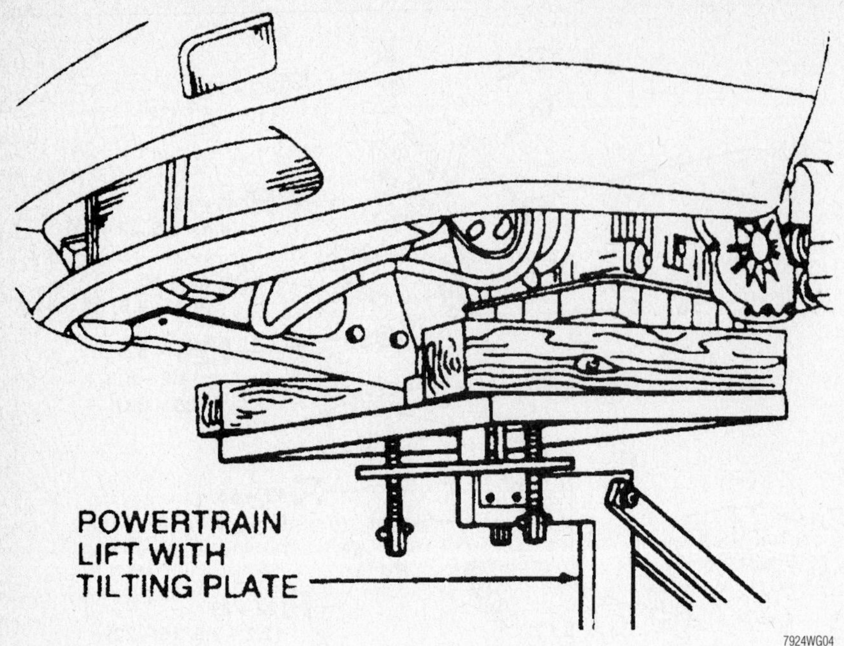

Carefully lower the engine/transaxle assembly from the vehicle.

6. Disconnect the engine harness ground connections.
7. Disconnect the coolant valve and position the valve aside.
8. Disconnect the Mass Air Flow (MAF) sensor connector.
9. Remove the fresh air intake tube and air cleaner–to–electric throttle control actuator tube attached to the air cleaner lid.
10. Remove the lower air cleaner case.
11. Remove the engine cover.
12. Remove the battery and tray.
13. Relieve the fuel system pressure.
14. Disconnect fuel hose quick connection at vehicle piping side as follows:
    a. Remove the connector cap from the fuel hose.
    b. Squeeze the two tabs and pull the fuel hose from the fuel line.

➡If the connector and the tube are stuck together, push and pull several times until they start to move, then disconnect them by pulling.

✻✻ CAUTION

The tube can be removed when the tabs are completely depressed. Do not twist it more than necessary.

➡Do not use any tools to remove the quick connector. Keep the resin tube away from heat. Be especially careful when welding near the tube.? Prevent acid liquids such as battery electrolyte, etc. from getting on the resin tube. Do not bend or twist the tube during removal or installation. Do not remove the remaining retainer on the tube. When the tube is replaced, also replace the retainer with a new one. To keep the connecting portion clean and to avoid damage and foreign materials entering, cover the ends of the fuel tubes with plastic bags or something similar.

15. Remove the radiator assembly, coolant reservoir, and hoses.
16. Disconnect the power brake booster vacuum hose from the back of the intake manifold collector.
17. Disconnect the EVAP canister purge volume control solenoid valve hose.
18. Disconnect heater hoses at the water outlet and heater pipe.
19. Disconnect the two fusible link connectors at the positive battery terminal.
20. Disconnect two engine harness connectors below the MAF sensor attached to the shock tower.
21. Disconnect the harness retainers and position the engine harness aside.
22. Remove the ground cable and ground wire from transaxle.
23. Disconnect the transaxle shift controls.
24. Remove the drive belts.
25. Remove the front exhaust tube and hanger.
26. Remove the front drive shafts.
27. Remove the lower ball joint pinch bolt, then separate the transverse link from the steering knuckle.
28. Remove the power steering line bracket from the front suspension member.
29. Remove the mounting bolts on the lower side of the steering gear.
30. Disconnect the front engine mount electrical connector.
31. Disconnect the connecting rod from the front.
32. Disconnect the power steering line brackets from the rear engine mount insulator and rear of the lower intake manifold collector.
33. Remove engine oil cooler pipe bolts.
34. Discharge and recover the A/C refrigerant.
35. Remove A/C low and high–pressure flexible hoses.
36. Remove the A/C compressor.
37. Disconnect the transaxle breather hose.
38. Disconnect the power steering pressure switch.
39. Disconnect the harness retainer from power steering oil pump bracket.
40. Remove the power steering adjusting bar and power steering pump, without disconnecting the lines from the engine and position and secure it aside.
41. Remove the Crankshaft Position (CKP) sensor.
42. Remove the rear cover plate and bolts securing the torque converter–to–drive plate.
43. Position a transmission jack under the engine/transaxle assembly.
44. On 4 speed transmission equipped vehicles, remove the left hand transaxle mount through bolt.
45. Remove the right hand engine mount insulator nuts and bolt.
46. Remove the front suspension member and engine/transaxle assembly as follows:
    a. Remove the right and left hand member pin stay bolts.
    b. Remove the front suspension member nuts and cups and carefully lower the front suspension member and engine/transaxle assembly avoiding interference with the vehicle body.

✻✻ CAUTION

Make sure to disconnect electrically controlled engine mounting insulator harness clips from the front suspension member prior to removal. Before and during this procedure, always check if any harnesses are left connected. Avoid any damage to, or any oil/grease smearing or spills onto the engine mounting insulators.

47. Remove the starter motor.

48. Disconnect the electrical connectors, harness retainers and remove the harnesses.

49. Disconnect the transmission cooler hoses and remove the cooler.

50. Remove the front and rear engine mount through bolts.

51. On 5 speed transmission equipped models, remove the left hand transaxle mount bolts.

52. Raise the engine/transaxle and remove the front suspension member.

53. Remove the transmission cooler valve from the engine with the hoses attached.

54. Separate the engine and transaxle and mount the engine on a suitable engine stand.

To install:

55. Installation is in the reverse order of removal. Tighten the bolts to the specifications shown in the engine mounting and front suspension member exploded view illustrations and note the following steps:

a. With converter installed, rotate crankshaft several turns to check that transaxle rotates freely without binding. Tighten the transmission–to–engine bolts for the 4 or 5 speed transmission vehicles as shown in the accompanying illustrations.

b. Tighten the converter–to–drive plate bolts to 40 ft. lbs. (54 Nm).

c. Tighten the rear cover plate bolt to 61 inch lbs. (7 Nm).

d. Tighten the CKP sensor to 85 inch lbs. (10 Nm).

e. Fill the engine with oil, and coolant and recharge the A/C system.

f. Check for any leaks and for proper vehicle operation.

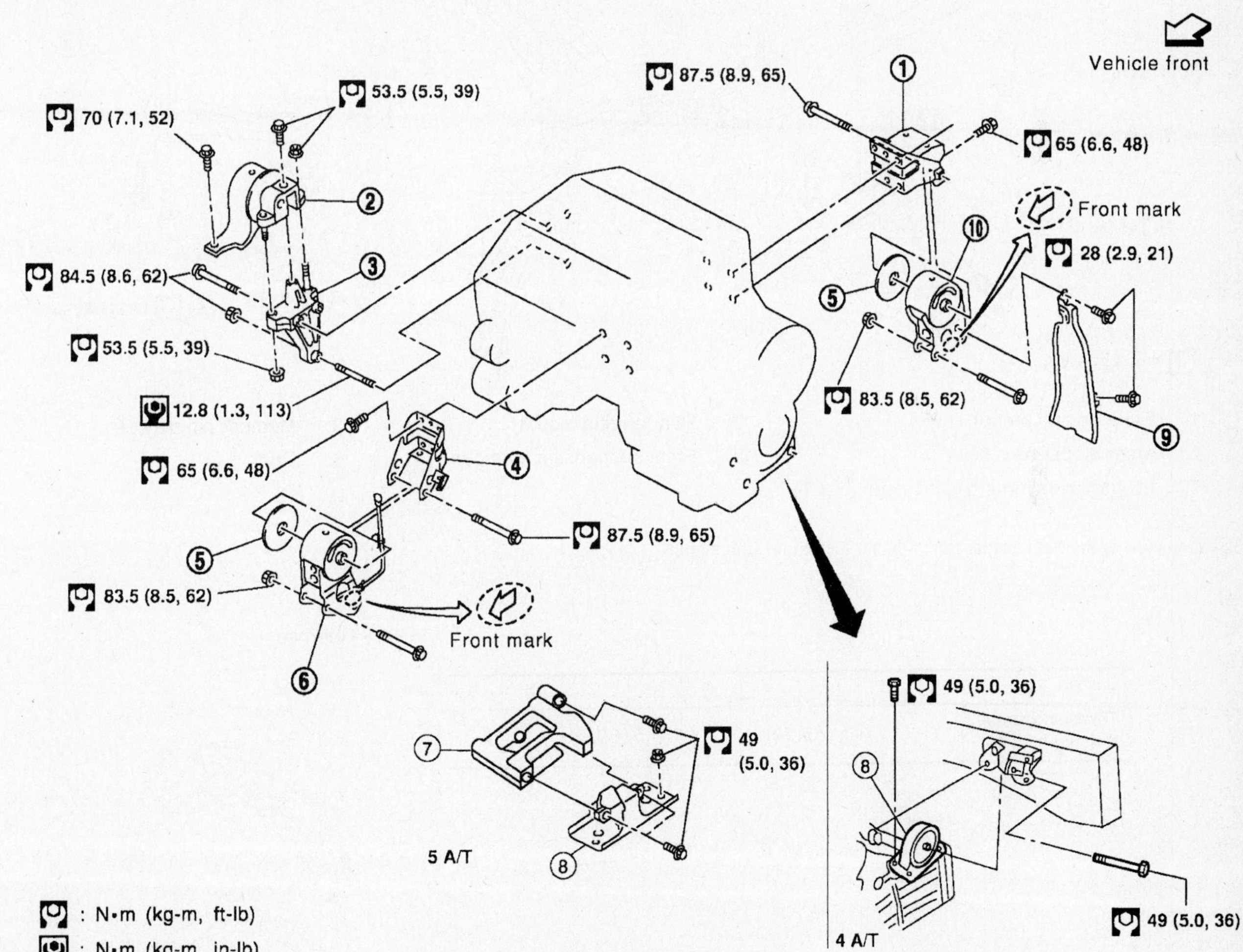

1. Rear engine mounting bracket
2. RH engine mounting insulator
3. RH engine mount
4. Front engine mount
5. Stopper
6. Front engine mounting insulator
7. LH transaxle mount
8. LH transaxle mounting insulator
9. Air guide
10. Rear engine mounting insulator

67170-NIQU-G73

Exploded view of the engine mounting–3.5L engine

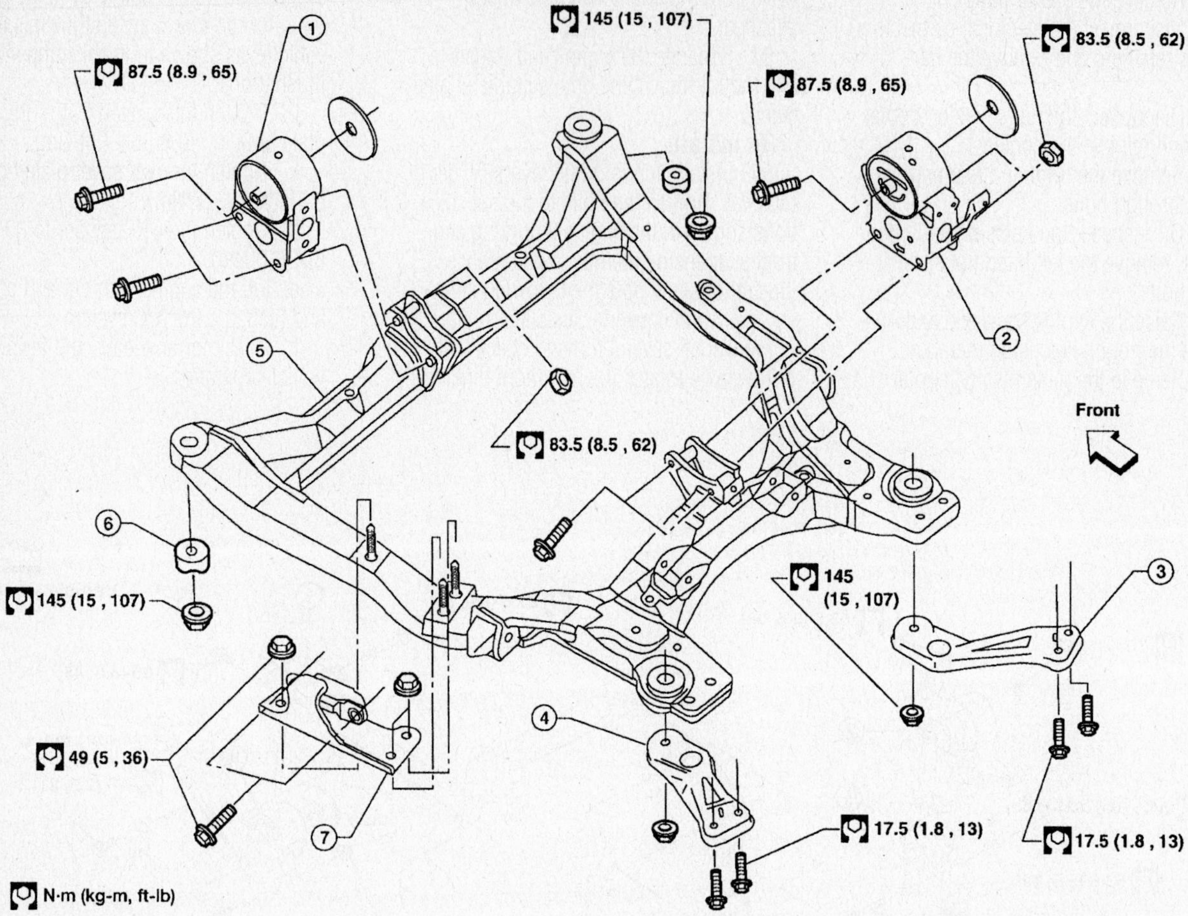

145 (15 , 107)
87.5 (8.9 , 65)
83.5 (8.5 , 62)
87.5 (8.9 , 65)
83.5 (8.5 , 62)
Front
145 (15 , 107)
145 (15 , 107)
49 (5 , 36)
17.5 (1.8 , 13)
17.5 (1.8 , 13)
N·m (kg-m, ft-lb)

1. Front engine mount
2. Rear engine mount
3. Member pin stay, RH
4. Member pin stay, LH
5. Front suspension member
6. Cup
7. LH transaxle mounting insulator (5 A/T)

67170-NIQU-G02

Exploded view of the front suspension member mounting–3.5L engine

Bolt No.	1	2	3	4	5	6	7	8	9
Tightening torque N·m (kg-m, ft-lb)	74.5 (7.6, 55)					41.5 (4.2, 31)			

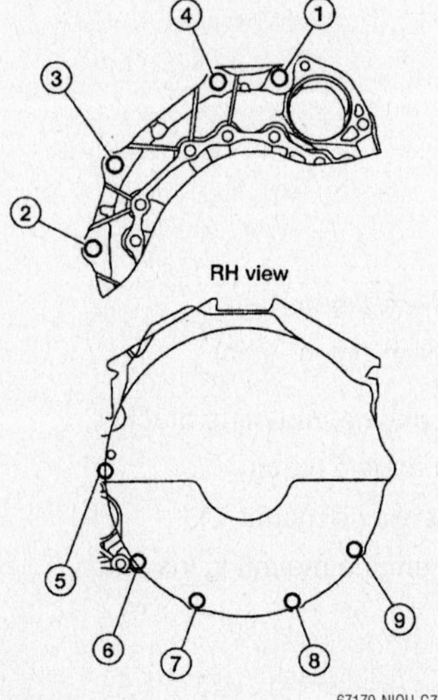

4 A/T bolt tightening sequence
LH view
RH view

67170-NIQU-G74

Exploded view of the 4 speed transmission–to–engine torque sequence and specifications–3.5L engine

Bolt No.	1	2	3	4	5	6	7	8	9
Tightening torque N·m (kg-m, ft-lb)	74.5 (7.6, 55)					41.5 (4.2, 31)			

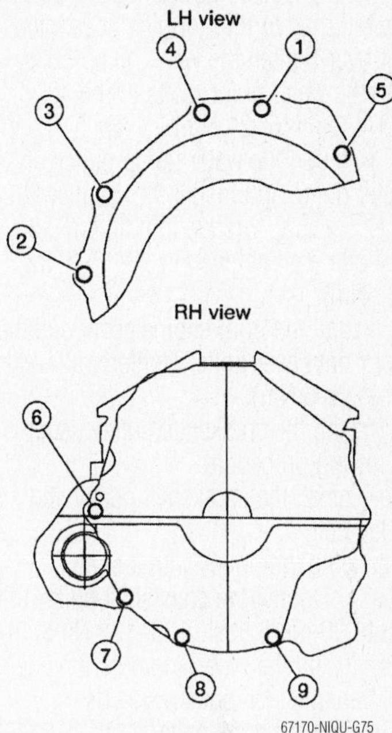

Exploded view of the 5 speed transmission–to–engine torque sequence and specifications–3.5L engine

Water Pump

REMOVAL & INSTALLATION

3.3L Engine

1. Before servicing the vehicle, refer to the precautions in the beginning of this section.
2. Drain the coolant.
3. Remove or disconnect the following:
   - Negative battery cable
   - Radiator hoses and fan shroud
   - Drive belts
   - Water pump pulley using strap wrench 303–D055–(D85L–6000–A) to hold the pulley while removing the bolts
4. Remove the crankshaft pulley using the following procedure:
   a. Raise and safely support the vehicle.
   b. Remove the 5 right side inner engine and transmission splash shield bolts and 2 screws and remove the inner engine and transmission shield.
   c. Remove the 4 right side outer engine and transmission splash shield bolts and 2 screws and remove the right side outer engine and transmission splash shields.
   d. Use a strap wrench to hold the crankshaft pulley while removing the crankshaft pulley bolt.
   e. Use a crankshaft damper remover to draw the crankshaft pulley off the front of the crankshaft.
5. Remove the 5 lower engine front cover bolts and take of the front cover.
6. Remove the 6 water pump bolts. Make note of the locations of the bolts since one should be a stud/bolt and must be returned to its original location. Remove the water pump.

To install:

7. Clean all parts well. The bolt threads should be cleaned of any old sealer or corrosion. Be sure the mating surfaces between the water pump and the engine block are cleaned of any old sealant. Apply a continuous bead of gasket maker type sealer approximately ⅛ inch (3mm) wide onto the water pump and position the water pump on the engine block.

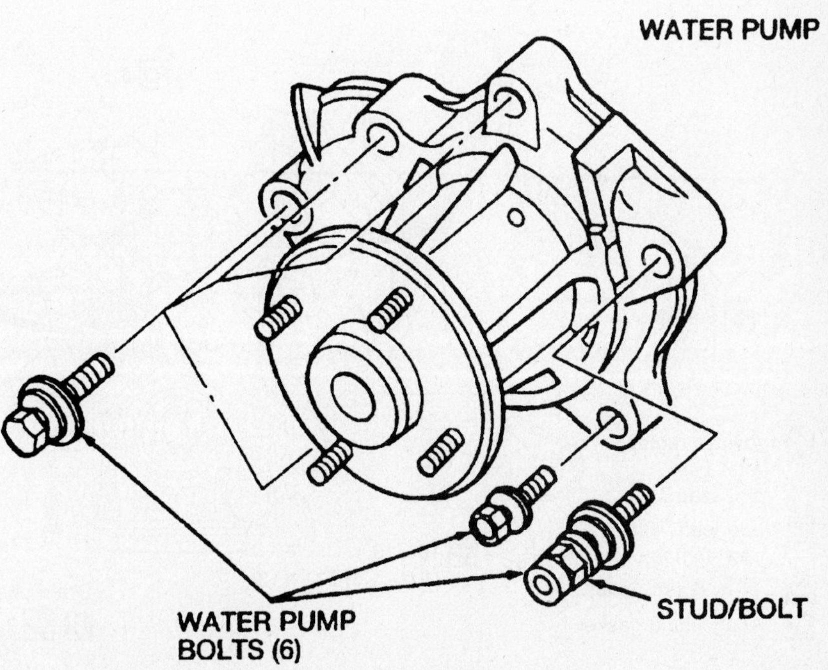

Water pump mounting. Note the location of the stud/bolt—3.3L engine

8. Install the 6 water pump bolts. Refer to any notes made at removal so the bolts can be returned to their original locations. Do not over-tighten the water pump bolts. Tighten the water pump bolts evenly to 12–15 ft. lbs. (16–21 Nm).

9. Position the water pump pulley on the water pump and install the 4 pulley bolts. Use a strap wrench to hold the pulley as the bolts are tightened to 12–15 ft. lbs. (16–21 Nm).

10. Install the front engine cover and the 5 lower front cover bolts. Tighten to 27–44 inch lbs. (3–5 Nm).

11. Install the crankshaft pulley using the following procedure:

a. Install the crankshaft pulley and pulley bolt.

b. Hold the pulley with a strap wrench. Tighten the crankshaft pulley bolt to 90–98 ft. lbs. (123–132 Nm).

c. Install the inner and outer engine and transmission splash shields.

12. Install the drive belts.

13. Connect the negative battery cable.

14. Refill the cooling system.

15. Start the engine and check for leaks.

3.5L Engine

1. Before servicing the vehicle, refer to the precautions in the beginning of this section.

2. Remove drive belts.

3. Remove undercover.

4. Drain engine coolant from radiator.

5. Remove water drain plug on water pump side of cylinder block.

6. Support lower oil pan bottom with a transmission jack.

7. Remove right engine mounting insulator and mounting bracket.

8. Remove idler pulley bracket.

9. Remove chain tensioner cover and water pump cover.

10. Remove the chain tensioner assembly in the following procedure.

a. Pull the lever down and release the plunger stopper tab.

b. Insert the stopper pin into the tensioner body hole to hold the lever and keep the stopper tab released.

c. Insert the plunger into the tensioner body by pressing the timing chain slack guide.

d. Keep the slack guide pressed and hold the plunger in by pushing the stopper pin deeper through the lever and into the tensioner body hole.

e. Turn crankshaft pulley approximately 20 degrees clockwise so that the timing chain on the chain tensioner side is loose.

11. Remove chain tensioner.

12. Remove the 3 water pump bolts. Secure a gap between water pump gear and timing chain, by turning crankshaft pulley approximately 20 degrees counterclockwise.

13. Screw M8 bolts, pitch: 1.25mm (0.049in) length: Approx. 50 mm (1.97in) into water pump's upper and lower mounting-bolt holes until they reach timing chain case. Then, alternately tighten each bolt for a half turn, and pull out water pump.

14. Pull straight out while preventing the vane from contacting socket in the installation area. Remove the water pump without the sprocket contacting the timing chain.

15. Remove the M8 bolts and O-rings from water pump.

To install:

16. Install new O-rings to water pump.

17. Apply engine oil and engine coolant to the O-rings as shown. Locate the O-ring with white paint mark to engine front side.

18. Install the water pump.

➡Do not allow cylinder block to nip the O-rings when installing the water pump.

19. Check that timing chain and water pump sprocket are engaged.

20. Insert the water pump by tightening the mounting bolts alternately and evenly.

21. Remove the dust and foreign material completely from the backside of the chain tensioner and from the installation area of the rear timing chain case.

22. Turn the crankshaft pulley clockwise so that the timing chain on the timing chain tensioner side is loose.

➡When installing the timing chain tensioner, engine oil should be applied to the oil hole and tensioner.

23. Install the timing chain tensioner.

24. Remove the stopper pin.

25. Install chain tensioner cover and water pump cover.

a. Before installing, remove all traces

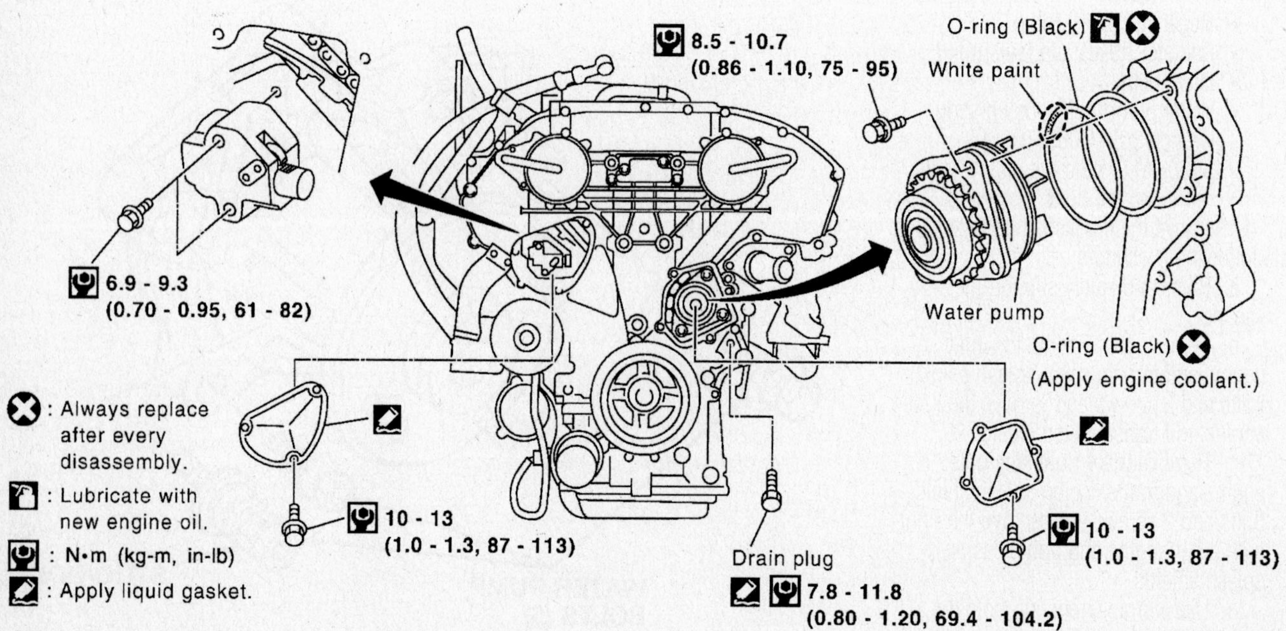

Water pump mounting—3.5L engine

of liquid gasket from mating surface of water pump cover and chain tensioner cover using a scraper. Also remove traces of liquid gasket from the mating surface of the front cover.

b. Apply a continuous bead of liquid gasket, to the mating surface of the chain tensioner cover and the water pump cover. Use RTV silicon sealant or equivalent

26. Install the water drain plug on the water pump side of the cylinder block.
27. Installation is in the reverse order of removal for any remaining parts.
28. After starting the engine, let it idle for three minutes, then rev the engine up to 3,000 rpm under no load to purge the air from the high-pressure chamber of the chain tensioner. The engine may produce a rattling noise. This indicates that air still remains in the chamber and is not a matter of concern.

Heater Core

REMOVAL & INSTALLATION

Front System

2001–02 MODELS

1. Before servicing the vehicle, refer to the precautions in the beginning of this section.
2. Disconnect the negative battery cable.
3. Drain the cooling system into a clean container for reuse.
4. Remove or disconnect the following:
   - Heater hoses at the bulkhead and plug
   - Storage bin, then both side covers by the bin and the footlamp, if equipped
   - Control console bezel (1 screw in the center), then the ashtray assembly
   - Climate control console screws, pull the console rearward and detach the electrical connectors

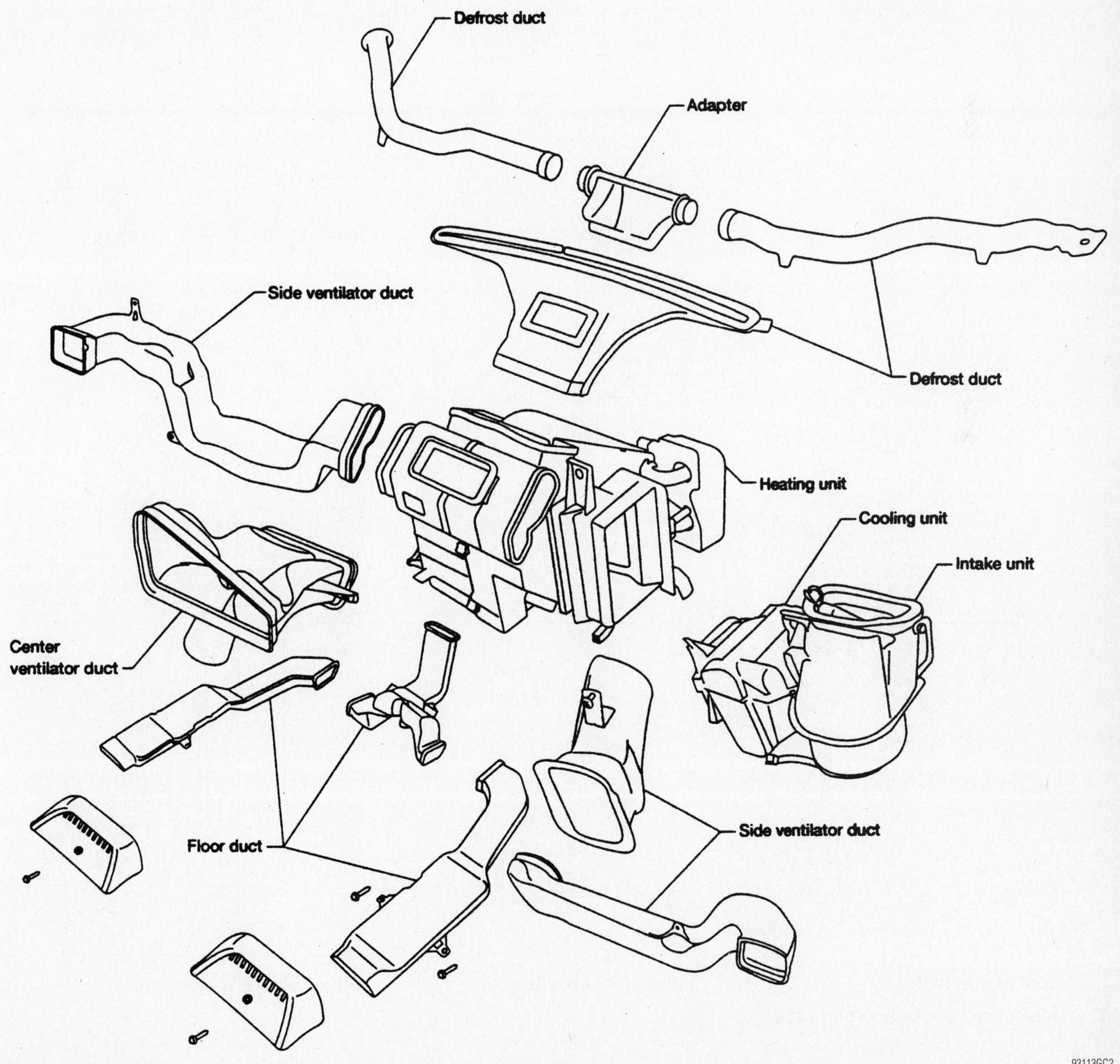

Exploded view the front heater/air conditioning assembly

HEATER UNIT
SEAL
PLATE
GROMMET
HEATER CORE
93113G58

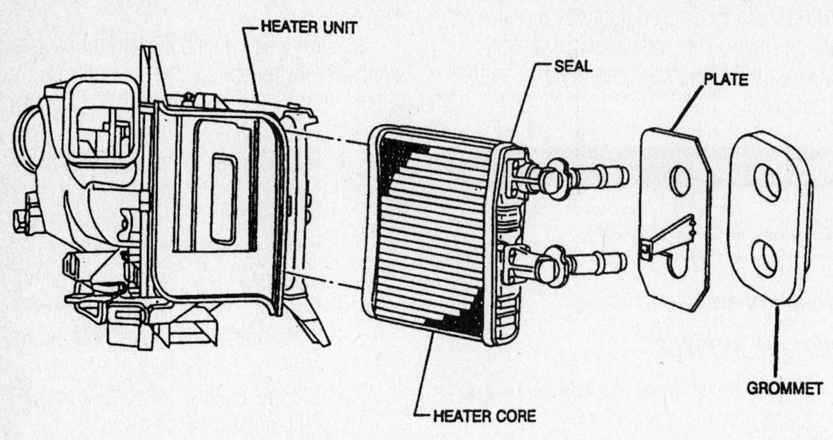

View the front heater core and heater housing assembly

- 4 radio assembly screws and take the radio out of the vehicle
- Floor duct and the right and left knee reinforcement plates
- ABS control module.

5. The speed control module, keyless entry module (if equipped) and the passive restraint (air bag) module are all located behind the center console and can be removed after detaching the respective connectors and removing the retaining nuts or screws.

WARNING

The control modules are very sensitive to static electricity and can be damaged if exposed to static or stray electrical impulses.

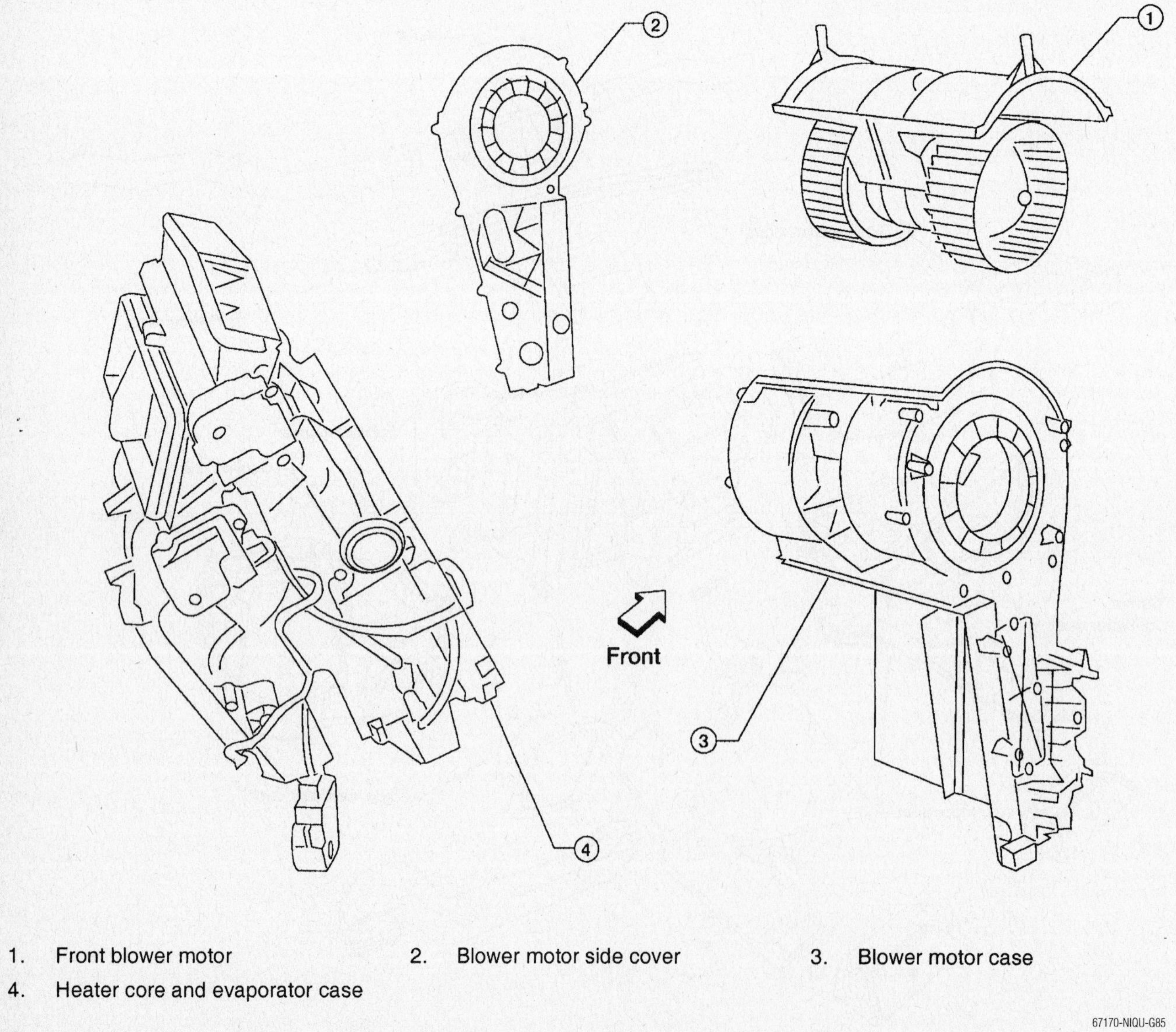

1
2
3
4
Front
1. Front blower motor
2. Blower motor side cover
3. Blower motor case
4. Heater core and evaporator case
67170-NIQU-G85

Exploded view of the front heater core assembly and related components—2004–05 models

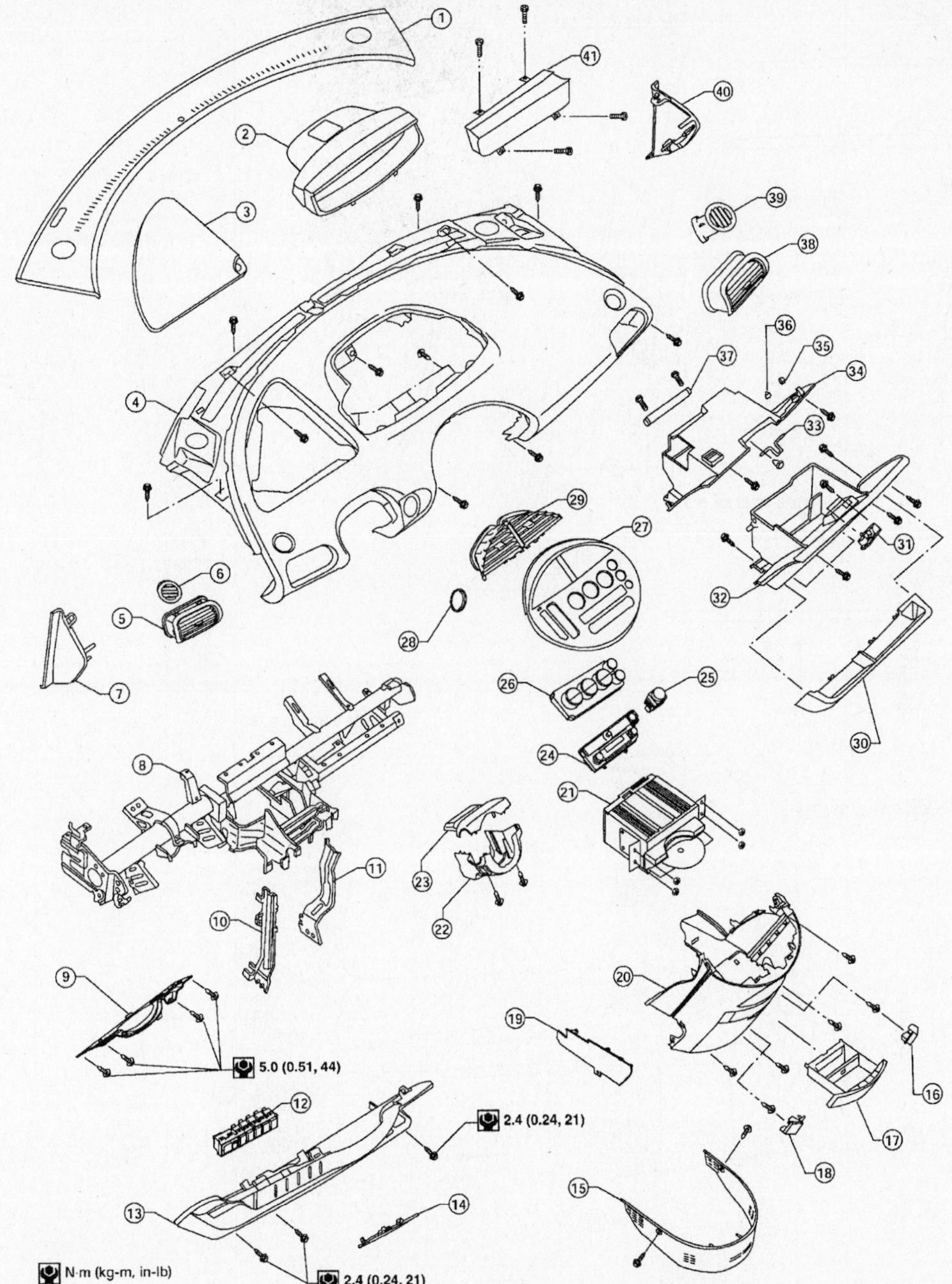

1. Defrost grille
2. Combination meter cover
3. Instrument panel storage bin
4. Instrument panel
5. Side ventilator assembly LH
6. Side defroster assembly LH
7. Instrument panel side cover LH
8. Steering member assembly
9. Knee protector
10. Instrument stay LH
11. Instrument stay RH
12. Switch assembly
13. Instrument lower panel LH
14. Fuse block cover
15. Center console lower cover
16. Console mask RH
17. Tray assembly
18. Console mask LH
19. Center console side finisher LH
20. Center console
21. Audio unit
22. Steering column lower cover
23. Steering column upper cover
24. AV switch
25. Hazard switch
26. Front air control
27. Cluster lid C
28. Steering lock escutcheon
29. Center ventilator assembly
30. Storage tray
31. Glove box latch assembly
32. Instrument lower panel RH
33. Glove box striker
34. Glove box housing
35. Glove box lamp receptacle
36. Glove box lamp
37. Glove box damper
38. Side ventilator assembly RH
39. Side defroster assembly RH
40. Instrument panel side cover RH
41. Combination meter

67170-NIQU-G86

Exploded view of the instrument panel components—2004–05 models

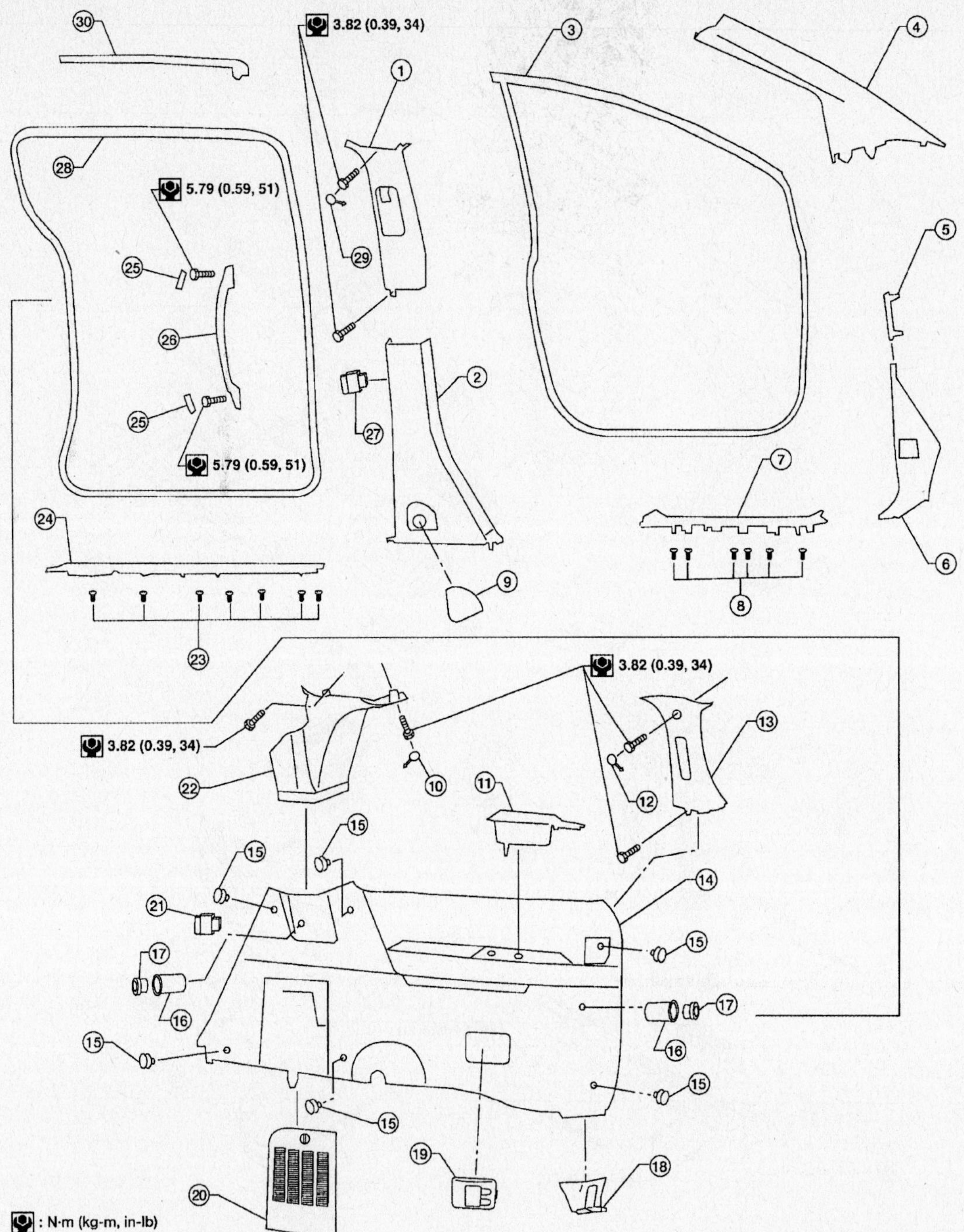

1.	Center pillar upper finisher LH	2.	Center pillar lower finisher LH	3.	Front door welt
4.	Front pillar finisher LH	5.	Upper dash side finisher LH	6.	Lower dash side finisher LH
7.	Front kicking plate	8.	Trim clips	9.	Center pillar lower escutcheon LH
10.	Rear finisher cover	11.	Luggage side cup holder LH	12.	Center finisher cover
13.	Rear center pillar finisher LH	14.	Rear lower finisher assembly LH	15.	Cargo net hooks
16.	Power point	17.	Power point cap	18.	Rear kick escutcheon LH
19.	Luggage side lower escutcheon LH	20.	Luggage side panel LH	21.	Power sliding door switch assembly
22.	Rear pillar upper finisher LH	23.	Trim clips	24.	Rear kicking plate
25.	Assist grip end caps	26.	Center pillar assist grip	27.	Back door open/close switch
28.	Sliding door welt	29.	Center pillar upper cover	30.	Sliding door drip weatherstrip LH

67170-NIQU-G87

Exploded view of the left hand center console side finisher components—2004–05 models

6. Remove or disconnect the following:
   - Center air duct
   - 2 ground wire bolts, the U-bracket and the 2 console brackets
   - Glove box and lamp
   - Accelerator pedal and pedal stop
   - Floor air duct
   - Temperature blend sir door actuator and mode door actuator by unfastening the attaching bracket bolts and detaching the electrical connections
   - Center distribution duct
   - 4 evaporator/blower assembly screws, the 4 heater assembly screws and the heater assembly
   - Hater pipe plate from the assembly
   - Hater core retainer, disengage the shut-off valve control rod
   - Hater core from the assembly

To install:

7. Install or connect the following:
   - Hater core to the case, the retainer and pipe plate
   - Hater assembly in the vehicle and attach the 4 retaining screws
   - Center distribution duct, the blend air and mode door actuators
   - Floor air duct
   - Accelerator stop and pedal
   - Glove box and lamp, then the center console and U-brackets
   - Center air duct, the passive restraint, the keyless entry, the speed control and the ABS modules, as removed
   - Remaining center console components
   - Heater hoses to the heater core
8. Refill the cooling system.
9. Connect the negative battery cable.
10. Run the engine to normal operating temperatures; then, check the climate control operation and check for leaks.

2004–05 MODELS

1. Before servicing the vehicle, refer to the precautions in the beginning of this section.
2. Discharge the refrigerant from the A/C system.

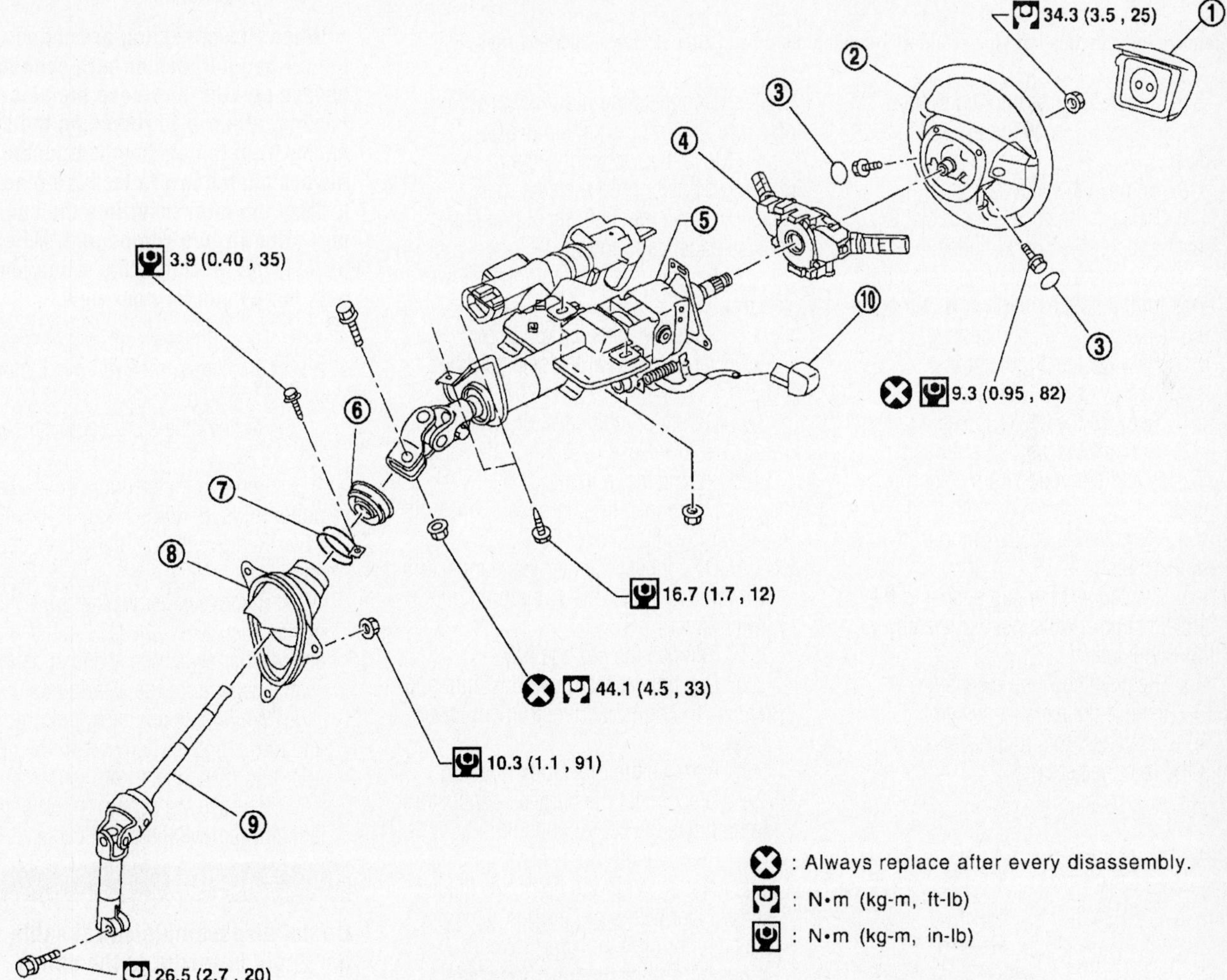

Exploded view of the steering column components—2004–05 models

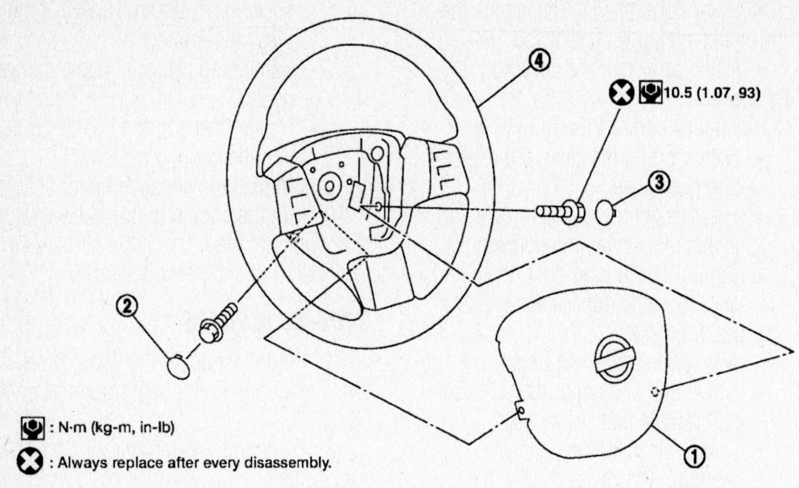

Remove the side lids from the retainers retaining the drivers side air bag—2004–05 models

3. Drain the engine cooling system.

4. Remove the cowl top extension as follows:

a. Remove the wiper arms.

b. Release the clips under cowl top cover and clips on hood–ledge.

c. Remove the seal by releasing the ends from tabs the on the cowl top cover and releasing the plastic clips.

d. Remove the weatherstrip.

e. Remove the cowl top cover.

f. Remove the cowl top seal.

g. Remove the windshield washer nozzles and the hoses from cowl top cover.

h. Remove the heater pump from cowl top extension.

i. Disconnect the clip attaching the coolant control valve hose to the cowl top extension.

j. Remove cowl top extension.

5. Remove the exhaust system.

6. Disconnect the front heater hoses from the front heater core.

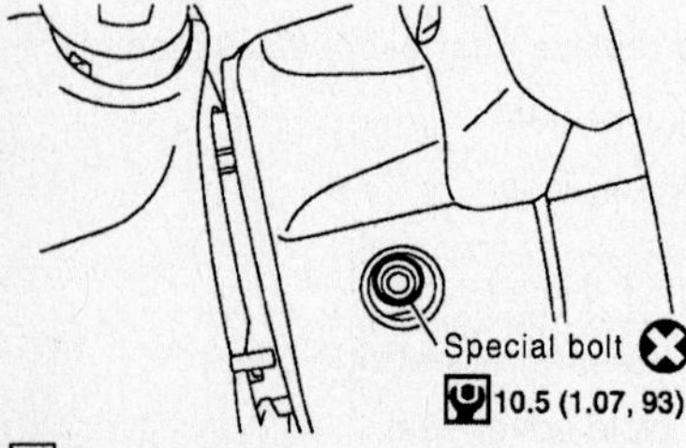

Remove the left and right side air bag bolts—2004–05 models

7. Disconnect the high/low pressure pipe from the front expansion valve.

8. Move the two front seats to the rear–most position on the seat track.

9. Remove the instrument panel and console panel as follows:

a. Remove the center console lower cover.

b. Slide the selector knob cover downwards to reveal the selector knob latch.

c. Gently pry the selector knob latch outward to release it, then lift the selector knob up to remove it.

d. Disconnect the harness connectors and remove the cluster lid C.

e. Disconnect the harness connectors and remove the left hand instrument lower panel.

f. Remove the storage tray.

g. Disconnect the glove box light and remove the right hand instrument lower panel.

h. Remove the glove box housing.

i. Disconnect the harness connectors and remove the center console.

j. Remove left hand center console side finisher. Wrap the tip of flat-bladed screwdriver with a cloth when removing the clips from the finishers. When removing or installing the body side welts, do not allow butyl seal to come in contact with the pillar finisher.

k. Disconnect the harness connectors and remove the combination meter cover and the combination meter.

l. Remove the instrument panel storage bin.

m. Remove the knee protector.

n. Remove the upper and lower steering column covers.

CAUTION

Before servicing the air bag system, turn ignition switch OFF, disconnect both battery cables and wait at least 3 minutes. When servicing the SRS, do not work from directly in front of air bag module.

o. Set the front wheels in the straight-ahead position.

p. Remove the side lids from the retainers retaining the driver's side air bag.

q. Remove the left and right side bolts.

r. Lift the driver's side air bag from the steering wheel.

➡When Disconnecting or connecting the air bag harness or horn connectors, always pull up to release the black locking tab prior to removing the connector from the air bag component. Always push down to lock the black locking tab after installing the connector to the air bag component. When locked, the black locking tab is level with the connector housing.

s. Disconnect the air bag harness and horn connectors, then remove the driver air bag module.

t. Remove the steering wheel center nut.

u. Remove the steering wheel using puller tools J–1859–A and J–42578.

v. Remove the column cover upper and lower.

w. Remove wiper washer switch connector, then pinch the tabs at the wiper and washer switch base and slide the switch away from the steering column.

x. While pressing tabs, pull the light and turn signal switch toward the driver door and disconnect from the base.

y. Remove the screws, release the tab, and remove the spiral cable.

CAUTION

Do not disassemble spiral cable. Do not apply lubricant to the spiral cable.

z. Remove the spiral cable connectors.

CAUTION

With the steering linkage disconnected, the spiral cable may snap by

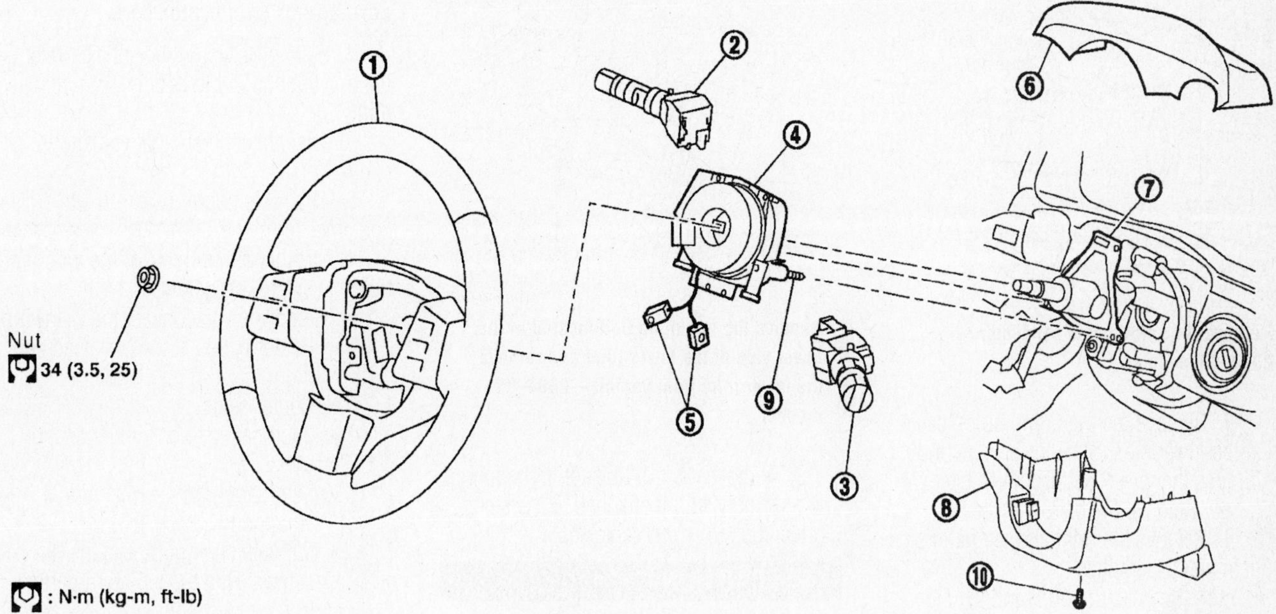

Exploded view of the steering wheel, spiral cable and air bag components—2004-05 models

turning the steering wheel beyond the limited number of turns. The spiral cable can be turned counterclockwise about 2.5 turns from the right end position.

aa. Inspect the steering wheel near the puller holes for damage. If damaged, replace the steering wheel.

bb. Remove the tilt lever knob from the tilt lever by inserting a suitable tool into the slot of the tilt knob, then depress the tab and withdraw the lever knob.

cc. Remove the instrument panel driver's side lower panel.

dd. Remove the steering column cover and ignition key finisher.

ee. Remove the mounting screws of the knee protector, then remove the knee protector.

ff. Remove the lock nut and bolt, then separate the lower joint from the upper joint.

gg. Remove the nuts from steering member, remove the steering column assembly from the steering member.

hh. Remove the hole cover seal and clamp.

ii. Remove the nuts, then remove the hole cover from the dash panel.

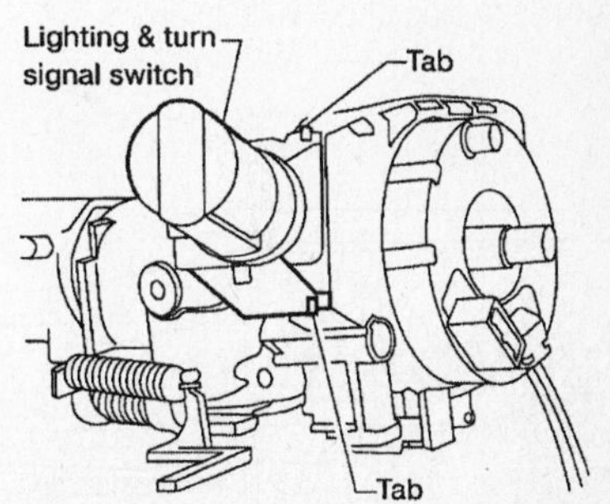

While pressing tabs, pull the light and turn signal switch toward the driver door and disconnect from the base—2004-05 models

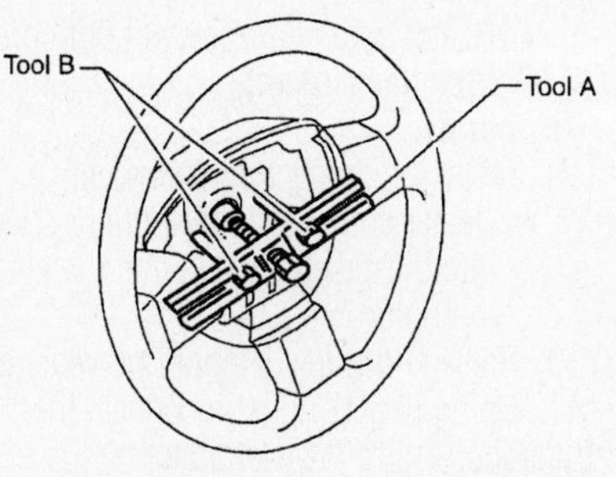

Remove the steering wheel using puller tools J-1859-A and J-42578—2004-05 models

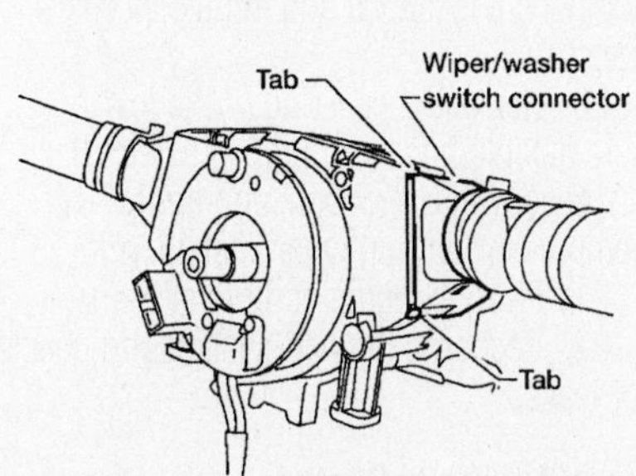

Remove wiper washer switch connector, pinch the tabs at the wiper and washer switch base and slide the switch away from the column—2004-05 models

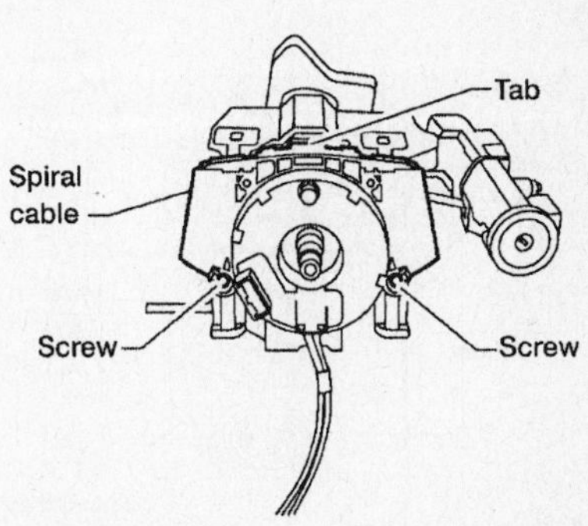

Remove the screws, release the tab, and remove the spiral cable—2004-05 models

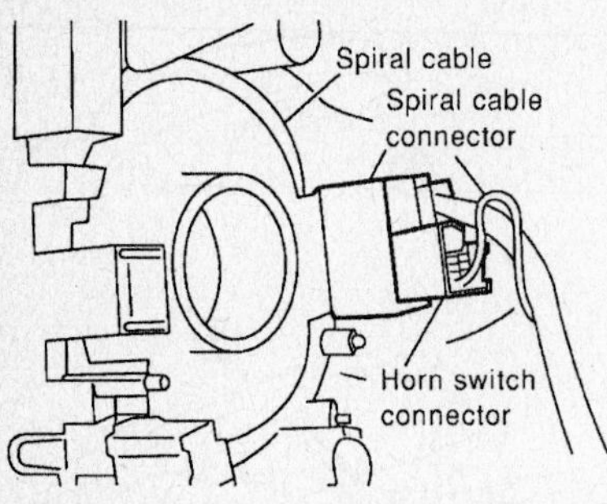

Remove the spiral cable connectors—2004–05 models

jj. Remove the mounting bolt located at the lower side of the lower joint and remove the lower joint from vehicle.

kk. Remove the defrost grille.

ll. Remove the right and left hand side defroster assemblies.

mm. Remove the right and left side ventilator assemblies.

nn. Remove the right and left front pillar finishers.

oo. Remove the instrument panel.

10. Disconnect the instrument panel wire harness at the right and left hand in–line connector brackets, and the fuse block electrical connectors.

11. Disconnect the steering member from each side of the vehicle body.

Remove the knee protector—2004–05 models

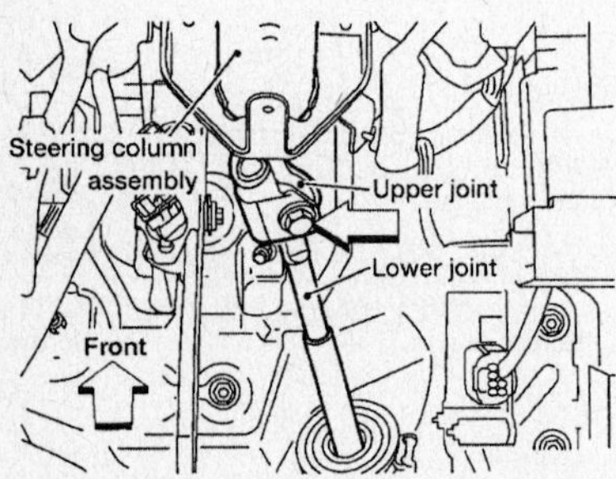

Remove the lock nut and bolt, then separate the lower joint from the upper joint—2004–05 models

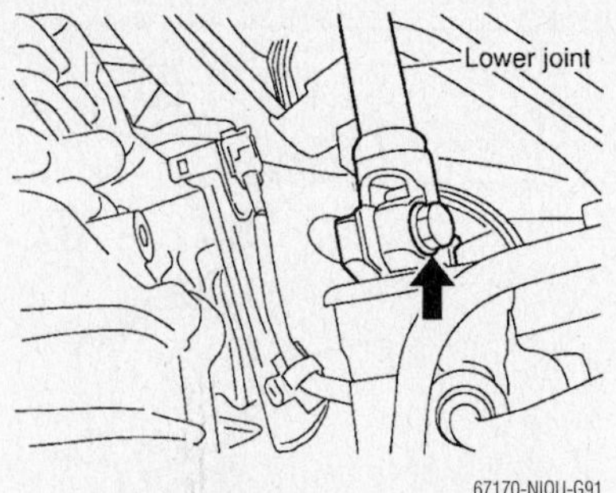

Remove the mounting bolt located at the lower side of the lower joint and remove the lower joint from vehicle—2004–05 models

12. Remove the front heater and cooling unit assembly with it attached to the steering member, from the vehicle.

⁂ CAUTION

Use care not to damage the seats and interior trim panels when removing the front heater and cooling unit assembly with it attached to the steering member.

13. Remove the front heater and cooling unit assembly from the steering member.

14. Remove the blower motor side cover.

15. Remove the front blower motor.

16. Remove heater core and evaporator case bottom cover.

17. Remove the blower motor case.

18. Remove the front heater core.

➡If the in-cabin microfilters are contaminated from coolant leaking from the heater core, replace the microfilters before installing the new front heater core.

To install:

19. Installation is the reverse of removal, refer to the component exploded views for component locations and torque specifications.

20. Replace the O-ring of the low-pressure flexible hose and high-pressure flexible hose with a new one, and apply compressor oil to it when installing it.

21. Fill the engine cooling system.

22. Recharge the A/C system and check for leaks.

Rear Auxiliary System

2001–02 MODELS

➡The rear heater/air conditioning assembly must be removed as a complete unit in order to remove the heater core and/or evaporator core.

1. Before servicing the vehicle, refer to the precautions in the beginning of this section.
2. Disconnect the negative battery cable.
3. Drain the cooling system into a clean container for reuse.
4. Discharge and recover the air conditioning system refrigerant.
5. Remove or disconnect the following:
    - Heater hoses at the bulkhead and plug
    - Center seats
    - 2 left half seat belt lower anchor bolts
    - Left rear cargo net retainers, if equipped
    - Lift gate scuff plate and the 3 screws from the left rear quarter trim panel. Gently pry the rear seat remote control (if equipped) from the trim panel. Disconnect the remote control wiring connector and remove the rear radio control panel. Pull the top of the trim panel away from the body.
    - Rear climate control panel wiring, if equipped
    - Left front lap belt guide from the left quarter trim panel and pass the belt through the trim panel
    - Trim panel from the vehicle
    - Upper duct from the assembly (6 screws)
    - Blower motor and resistor wiring
    - Temperature blend and vent door actuator connectors
6. Raise and safely support the vehicle. Use the spring lock coupling tool to disconnect and plug the refrigerant line connections from beneath the vehicle.
7. Lower the vehicle.
8. Remove or disconnect the following:
    - 4 heater/air conditioning assembly bolts and the assembly from the vehicle
    - Heater core and/or evaporator core from the assembly

To install:

9. Install or connect the following:
    - Heater core and/or evaporator core into the assembly and the 4 retaining bolts
10. Raise and safely support the vehicle.
11. Using new O-rings, reconnect the refrigerant lines to the evaporator.
12. Lower the vehicle.
13. Install or connect the following:
    - All wiring connectors

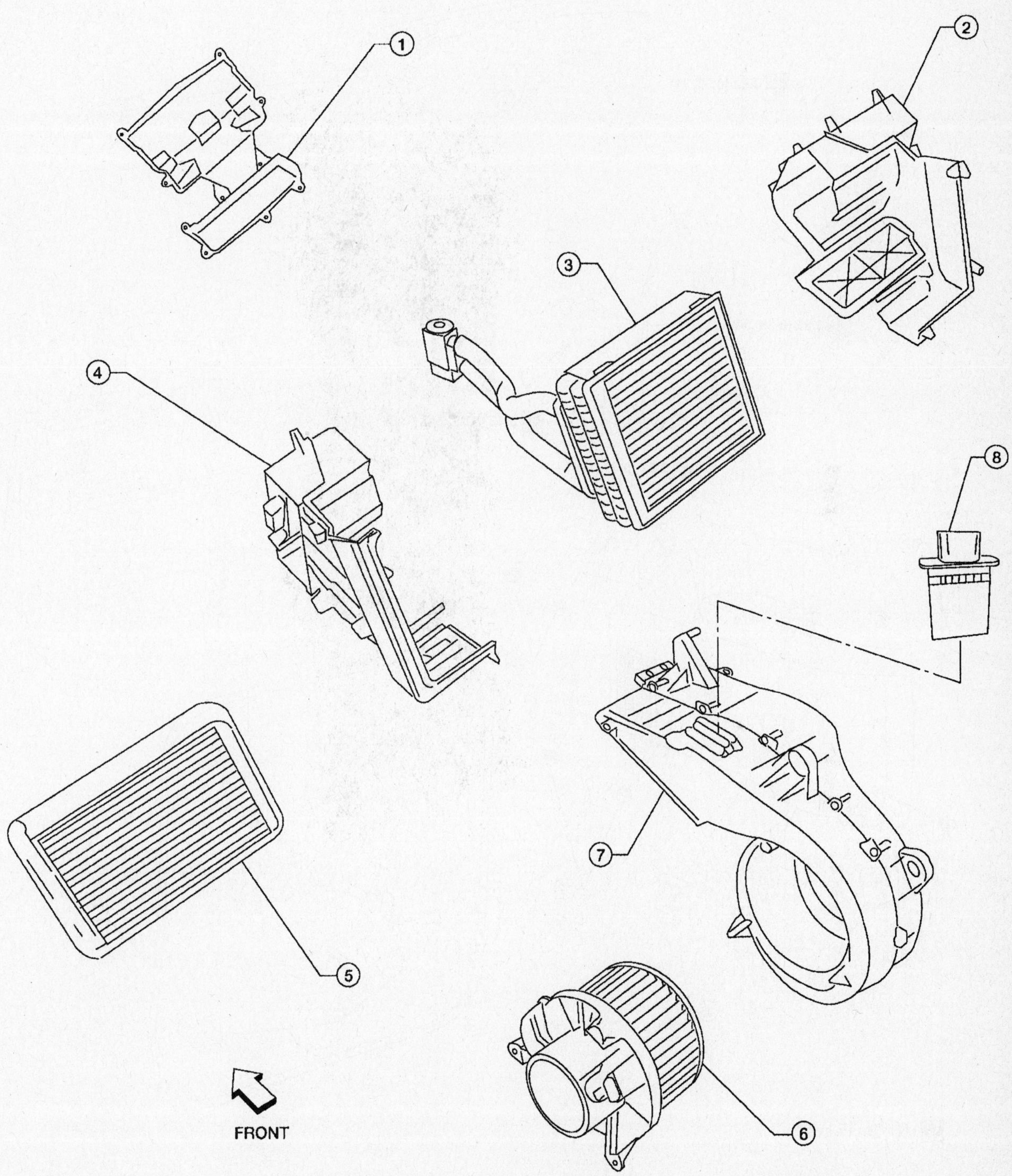

1.	Front cover	2.	Evaporator and heater core case	3.	Evaporator
4.	Side cover	5.	Heater core	6.	Rear blower motor
7.	Blower motor case	8.	Rear blower motor resistor		

67170-NIQU-GAA

Exploded view of the rear heater core assembly—2004–05 models

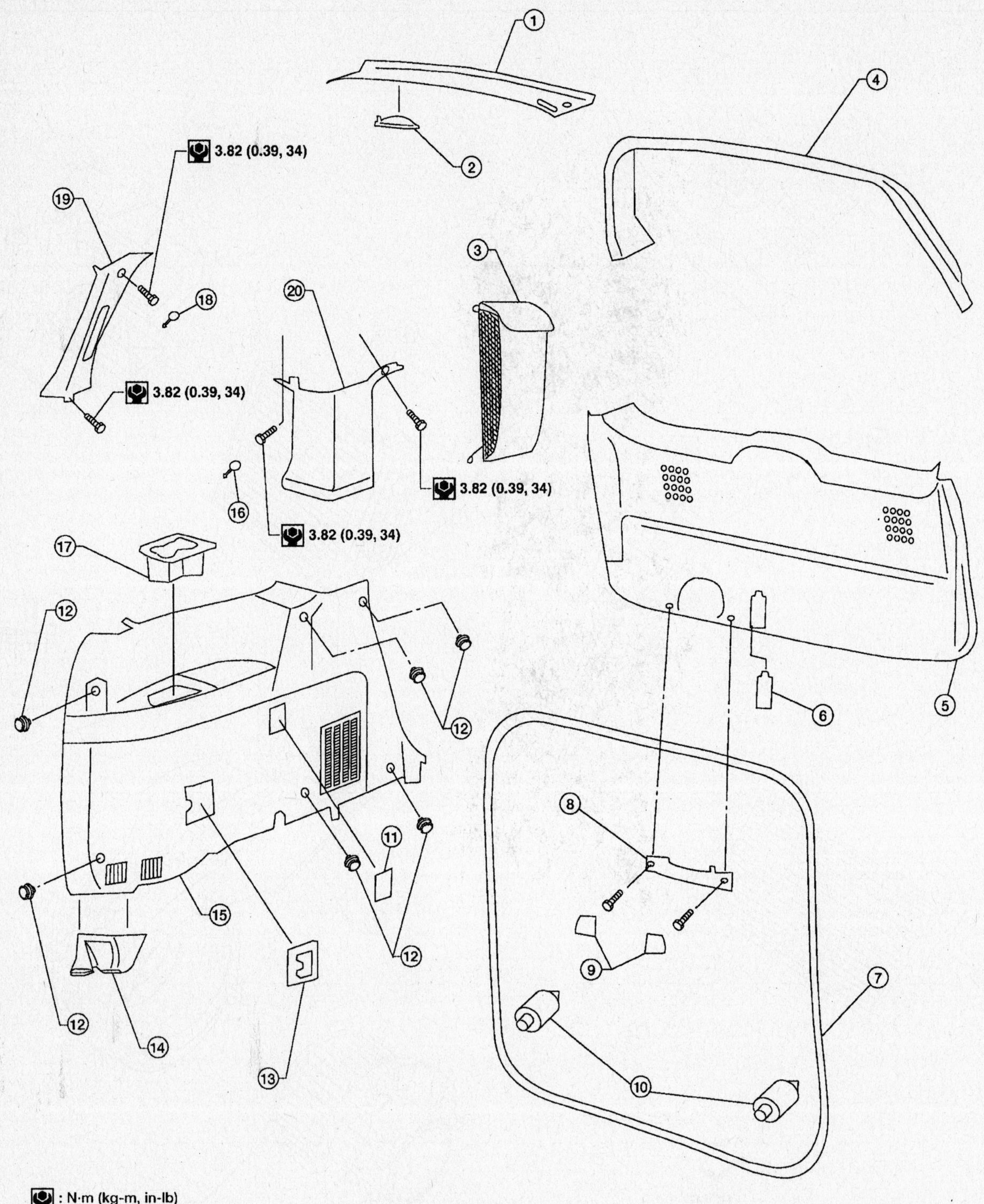

1. Roof finisher
2. Luggage room lamp
3. Cargo net
4. Back door upper finisher
5. Back door lower finisher
6. Back door mask
7. Back door welt
8. Back door pull handle
9. Pull handle covers RH/LH
10. Back door bumpers
11. Gas leak check lid
12. Cargo net hooks
13. Luggage side lower escutcheon RH
14. Rear kick escutcheon RH
15. Rear lower finisher assembly RH
16. Rear pillar cover RH
17. Cup holder luggage side RH
18. Center pillar cover RH
19. Rear center pillar finisher RH
20. Rear pillar upper finisher RH

67170-NIQU-GAB

Exploded view of the rear trim components—2004–05 models

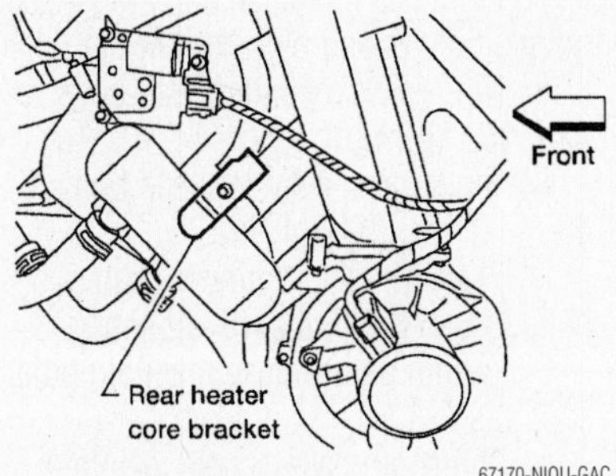

Remove the rear heater core bracket—2004–05 models

- Upper air duct with the 6 screws
- Trim panel and pass the lap seat belt through the panel slot
- Rear climate control panel
- Rear radio and rear remote control
- Remaining trim panel and components

14. Refill the cooling system.
15. Connect the negative battery cable.
16. Evacuate and charge the air conditioning system.
17. Run the engine to normal operating temperatures; then, check the climate control operation and check for leaks.

2004–05 MODELS

1. Before servicing the vehicle, refer to the precautions in the beginning of this section.
2. Partially drain the engine cooling system.
3. Remove the rear right hand interior trim panel.
4. Disconnect the rear heater core hoses from the rear heater core.
5. Remove the rear heater core bracket.
6. Remove the rear heater core.

Cylinder Head

REMOVAL & INSTALLATION

3.3L Engine

The factory specifies that the cylinder head bolts ARE NOT to be reused. Obtain the proper replacement parts before beginning this procedure. Check carefully that all bolts are removed before attempting to remove a cylinder head. A tab, part of the head, contains 1 lightly tightened head bolt that is external to the valve cover. Do not overlook this "hidden" bolt or the head will be damaged.

1. Before servicing the vehicle, refer to the precautions in the beginning of this section.

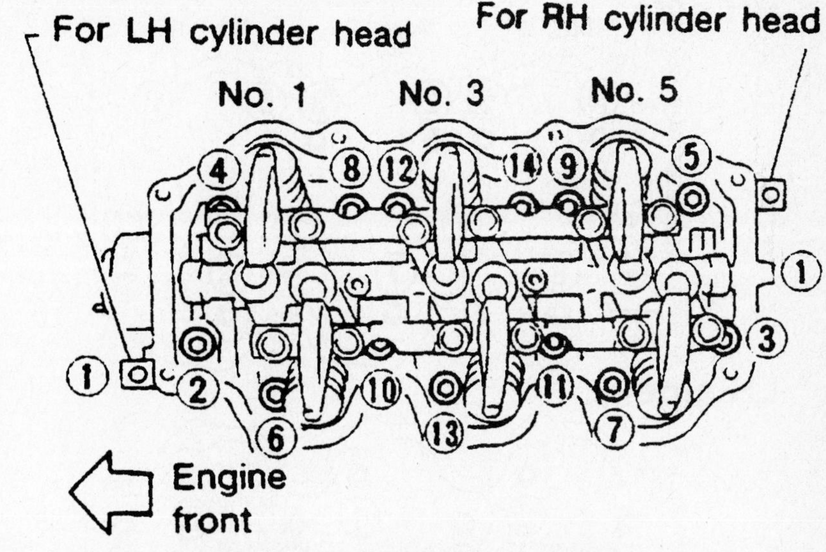

Remove the cylinder head bolts in the sequence shown—3.3L engines

2. Properly relieve the fuel system pressure.
3. Drain the coolant.
4. Remove or disconnect the following:
   - Negative battery cable
   - Air intake tube
   - Timing belt
   - Upper intake manifold (plenum)
   - Fuel feed and return hoses from the fuel rail
   - Fuel injector's electrical connections
   - Fuel rail and injectors as an assembly
   - Intake manifold (lower)
   - Camshaft sprockets
   - Rear timing belt cover
   - Distributor
   - Harness clamp from the right hand rocker cover
   - Exhaust tube from the left hand manifold
   - Left hand exhaust manifold from the right hand exhaust manifold
   - Left hand manifold-to-bracket bolt
   - A/C compressor, alternator and their brackets
   - Rocker covers

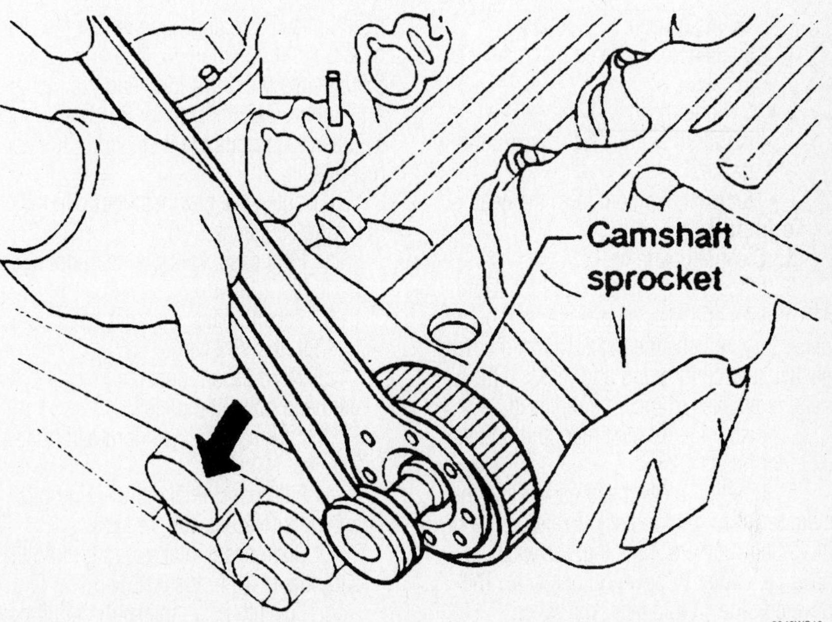

Hold the camshaft sprocket while removing the sprocket retaining bolt—3.3L engines

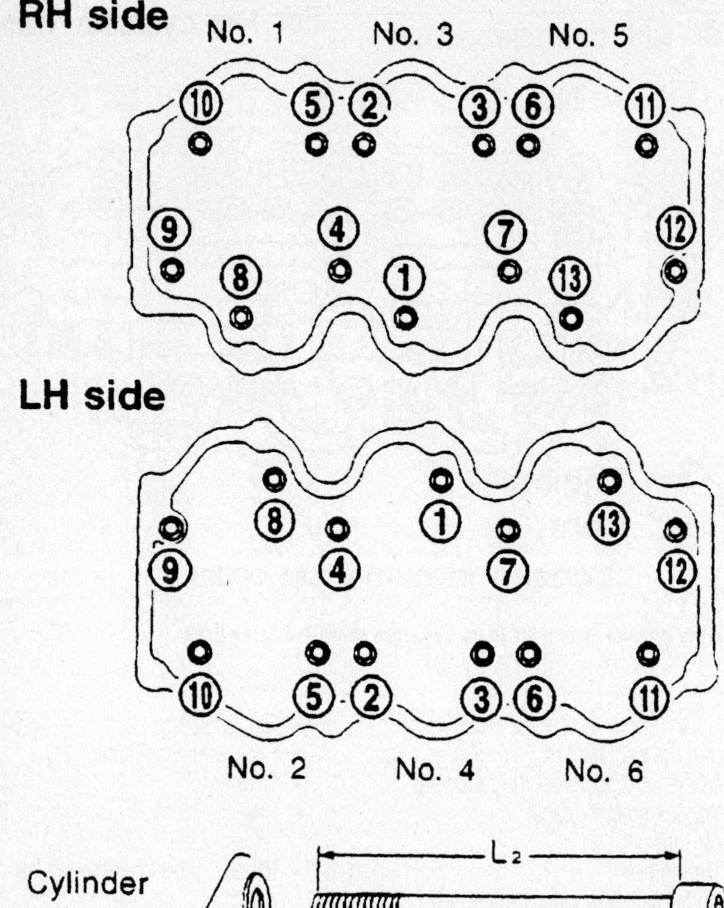

Tighten the cylinder head bolts in sequence as shown—3.3L engines

- Cylinder head bolts in the sequence illustrated using several passes
- Cylinder head with the exhaust manifold and gasket. Discard the gasket.
- Exhaust manifold from the head

To install:

5. Clean all parts well.
6. Inspect the cylinder head for damage, cracks and leakage of water and oil. If necessary, replace the head. Check the head gasket surface for burrs and nicks. If the head is cracked, it must be replaced.
7. Install the exhaust manifold on the cylinder head.
8. Position a new head gasket and the cylinder head on the block. Examine the head bolt washers. Note that the washers have a chamfer or bevel on one side. The beveled side should face "up" when installed. Examine the new replacement head bolts. There are different lengths. The head bolts in positions 4, 7, 9 and 12 are 5.00 inches (127mm) long and the rest are 4.17 inches (106mm) long. Be sure the new cylinder head bolts are installed in the correct positions.
9. Tighten the new head bolts in the following sequence:
   a. First pass: cylinder head bolts to 22 ft. lbs. (29 Nm).
   b. Second pass: cylinder head bolts to 43 ft. lbs. (59 Nm).
   c. Third pass: Loosen all of the cylinder head bolts completely.
   d. Fourth pass: cylinder head bolts to 7 ft. lbs. (10 Nm).
   e. Fifth pass: intake manifold bolts and nuts to 2.9 ft. lbs. (4 Nm).
   f. Sixth pass: intake manifold bolts and nuts to 13 ft. lbs. (18 Nm).
   g. Seventh pass: intake manifold bolts and nuts to 12–14 ft. lbs. (16–20 Nm).
   h. Eight pass: Loosen all of the intake manifold bolts and nuts completely.
   i. Ninth pass: cylinder head bolts to 22 ft. lbs. (29 Nm).
   j. Tenth pass: cylinder head bolts to 40–47 ft. lbs. (54–64 Nm).
   k. Eleventh pass: cylinder head sub-bolts to 6.7–8.7 ft. lbs. (9–12 Nm).
   l. Twelfth pass: intake manifold bolts and nuts to 2.9 ft. lbs. (4 Nm).
   m. Thirteenth pass: intake manifold bolts and nuts to 6.5 ft. lbs. (9 Nm).
   n. Fourteenth pass: intake manifold bolts and nuts to 6–7 ft. lbs. (8–10 Nm).
   - Rocker covers
   - A/C compressor, alternator brackets
   - A/C compressor and alternator
   - Left hand manifold-to-bracket bolt
   - Left hand exhaust manifold to the right hand exhaust manifold
   - Exhaust tube to the left hand manifold
   - Harness clamp to the right hand rocker cover
   - Distributor
   - Rear timing belt cover
   - Camshaft sprockets
   - Intake manifold (lower)
   - Fuel rail and injectors as an assembly
   - Fuel injector's electrical connections
   - Fuel feed and return hoses to the fuel rail
   - Upper intake manifold (plenum)
   - Timing belt
   - Air intake tube
   - Negative battery cable
10. Fill the cooling system. An oil and filter change is recommended.
11. Start the vehicle and check for leaks. Check the ignition timing and adjust as required.

3.5L Engine

1. Before servicing the vehicle, refer to the precautions in the beginning of this section.
2. Disconnect the negative battery cable.
3. Remove the fuel tube and fuel injector assembly.
4. Remove the intake manifold.
5. Remove the exhaust manifold.
6. Remove the water inlet and thermostat assembly.
7. Remove the water outlet and water piping.
8. Remove the camshaft.

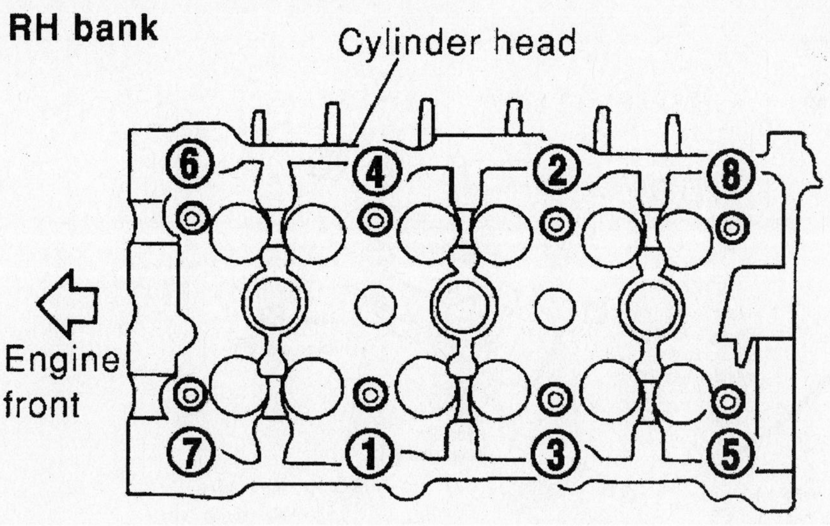

Cylinder head bolt torque sequence—3.5L engine–right side

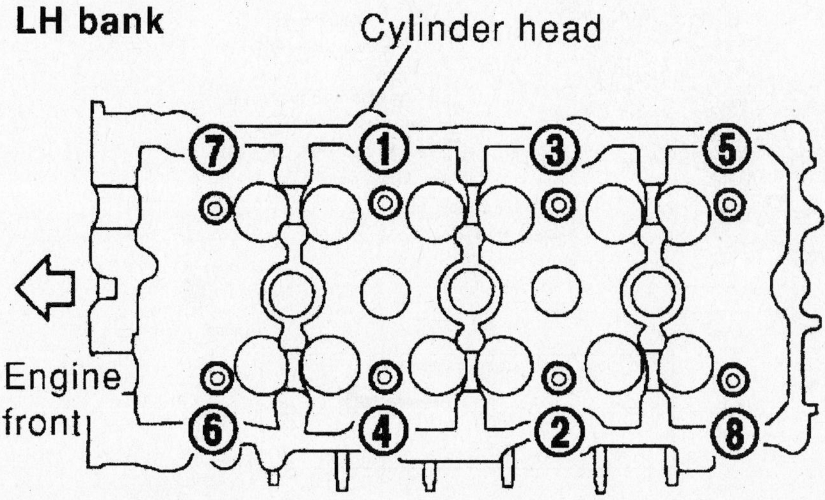

Cylinder head bolt torque sequence—3.5L engine–left side

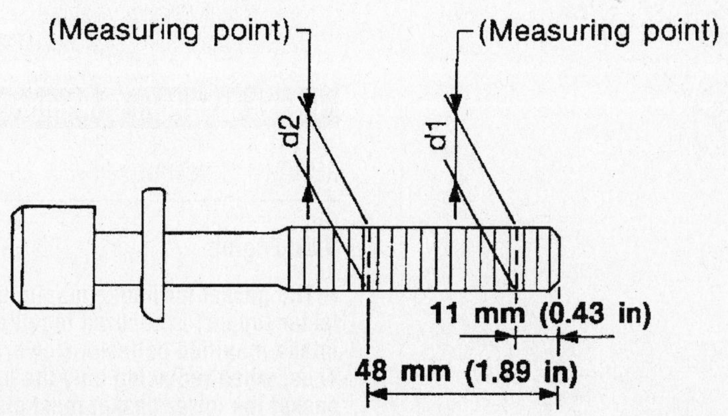

Cylinder head bolt measurement—3.5L engine

9. Remove the cylinder head bolts in reverse of the tightening sequence.
10. Inspect the bolts. Cylinder head bolts are tightened by plastic zone tightening method. Whenever the size difference between d1 and d2 exceeds the limit, replace them with new one. If reduction of outer diameter appears in a position other than d2, use it as d2 point.

To install:

11. Install cylinder head gasket.
12. Turn the crankshaft until No. 1 piston is set at TDC on the compression stroke. The crankshaft key should line up with the right bank cylinder center line as shown.
13. Install cylinder head. Tighten the head bolts in the order shown in illustration.
    a. Step 1: Tighten all bolts to 72 ft. lbs. (98 Nm).
    b. Step 2: Completely loosen to 0 in the reverse order.
    c. Step 3: Tighten all bolts to 26–32 ft. lbs. (34.3–44.1 Nm).
    d. Step 4: Turn all bolts 90 degrees clockwise.
    e. Step 5: Turn all bolts an additional 90 degrees clockwise.
14. After installing cylinder head, measure distance between front end faces of cylinder block and cylinder head (left and right banks). If measurement is outside the specified range, reinstall cylinder head.
15. The remainder of installation is the reverse of removal.

Rocker Arms/Shafts

REMOVAL & INSTALLATION

3.3L Engine

1. Before servicing the vehicle, refer to the precautions in the beginning of this section.
2. Remove or disconnect the following:
    - Negative battery cable
    - Upper intake manifold
    - Valve covers
    - Rocker arm and shaft assemblies
    - Rocker arms from the shafts

➡Keep all valvetrain components in order for assembly.

To install:

3. Lubricate all contact points with clean engine oil and assemble the rocker arms to the shafts in their original positions.
4. Install or connect the following:
    - Rocker arm and shaft assemblies.

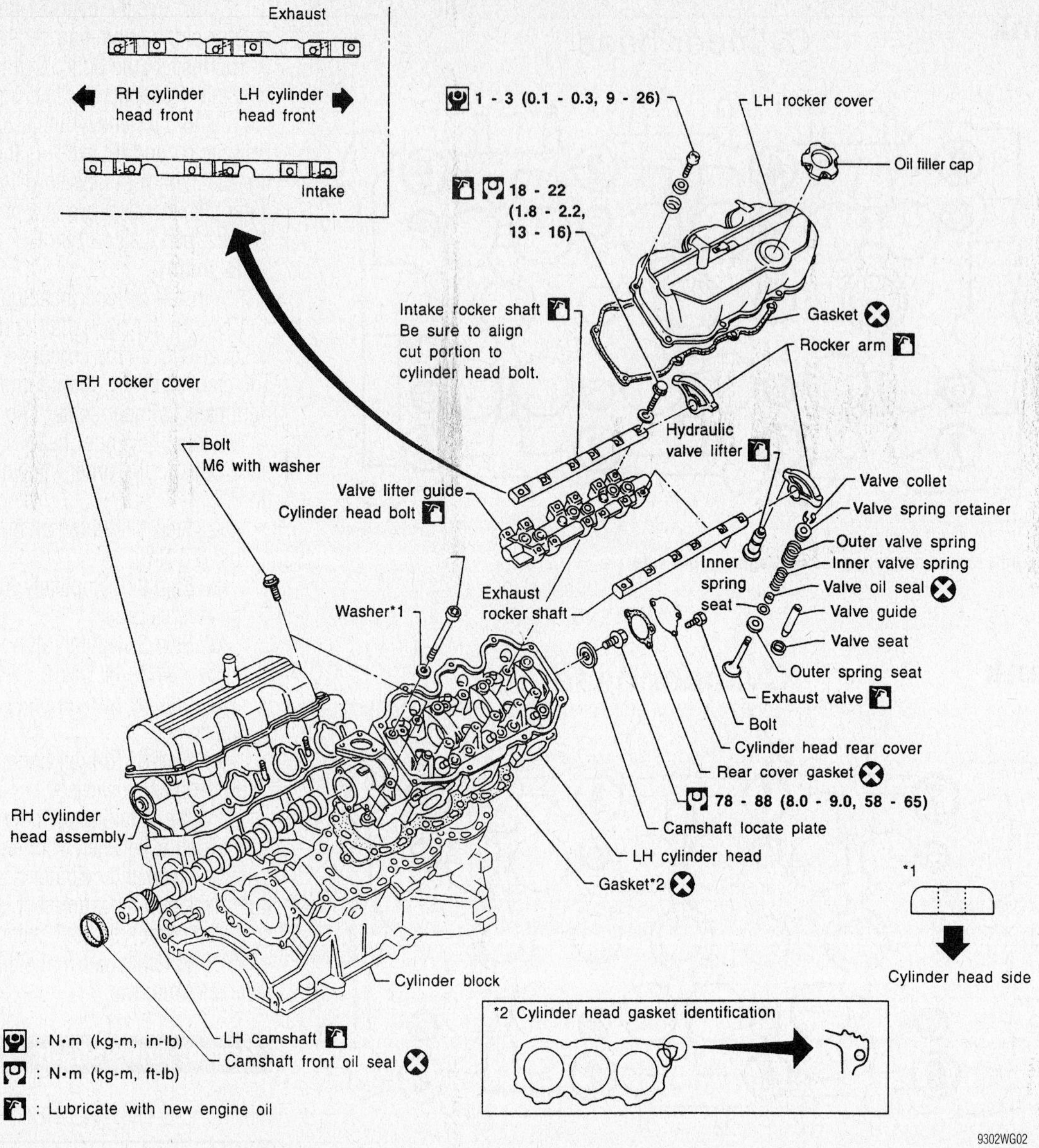

Exhaust

RH cylinder head front

LH cylinder head front

Intake

1 - 3 (0.1 - 0.3, 9 - 26)

LH rocker cover

Oil filler cap

18 - 22 (1.8 - 2.2, 13 - 16)

Intake rocker shaft Be sure to align cut portion to cylinder head bolt.

Gasket

Rocker arm

RH rocker cover

Hydraulic valve lifter

Bolt M6 with washer

Valve lifter guide

Cylinder head bolt

Valve collet

Valve spring retainer

Outer valve spring

Inner valve spring

Inner spring seat

Valve oil seal

Valve guide

Valve seat

Exhaust rocker shaft

Washer*1

Outer spring seat

Exhaust valve

Bolt

Cylinder head rear cover

Rear cover gasket

78 - 88 (8.0 - 9.0, 58 - 65)

Camshaft locate plate

RH cylinder head assembly

LH cylinder head

Gasket*2

*1

Cylinder head side

Cylinder block

*2 Cylinder head gasket identification

: N•m (kg-m, in-lb)

: N•m (kg-m, ft-lb)

: Lubricate with new engine oil

LH camshaft

Camshaft front oil seal

9302WG02

Rocker arm and shaft components—3.3L engine

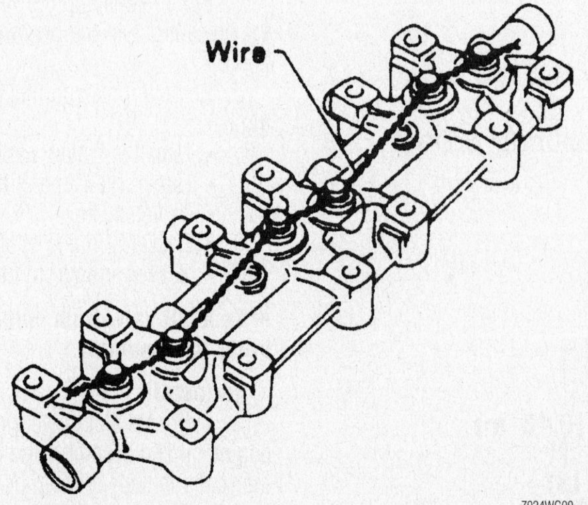

Wire the lifters on top of the guide so they won't fall out when the guide is removed from the head—3.3L engines

Tighten the bolts to 13–16 ft. lbs. (18–22 Nm).

- Valve covers
- Upper intake manifold
- Negative battery cable

5. Start the engine and check for leaks.

Intake Manifold Collector

REMOVAL & INSTALLATION

3.5L Engine

➡The gasket for intake manifold collector (upper) is secured together with intake manifold collector (lower) bolt. Thus, when replacing only the upper gasket the lower gasket must also be replaced.

1. Before servicing the vehicle, refer to the precautions in the beginning of this section.
2. Remove the cowl top and cowl top extension.
3. Remove the engine cover using power tool.
4. Remove upper air cleaner case, Mass Air Flow (MAF) sensor, and air cleaner–to–electric throttle control actuator tube as an assembly.
5. Partially drain the coolant when the engine is cool.
6. Disconnect the following:
   a. Power brake booster vacuum hose.
   b. Coolant hoses from the intake manifold collector
   c. Vacuum lines from the upper intake manifold collector and power valve.
   d. Fuel injector electrical connectors.
   e. Positive Crankcase Ventilation (PCV) hose.
   f. Electric throttle control actuator electrical connectors.
   g. EVAP canister purge hose.
   h. Exhaust Gas Recirculation (EGR) temperature sensor electrical connector.

➡Cover any engine openings to avoid the entry of any foreign material.

7. Remove the EGR tube–to–lower intake manifold collector nuts.
8. Disconnect the power steering hose bracket from the back of the intake manifold collector.
9. Remove the EVAP canister purge volume solenoid valve bracket bolt and position the valve aside.
10. Remove the VIAS control solenoid valve bracket bolt and position the valve aside.
11. Remove the vacuum tank.
12. Remove the intake manifold collector support bracket from the back of the intake manifold collector.
13. Loosen the intake manifold collector bolts in the order illustrated and remove the intake manifold collector and gasket.

To install:

14. Install a new gasket and the collector. Tighten the bolts in sequence to 14 ft. lbs. (19 Nm).
15. Install the intake manifold collector support bracket to the back of the intake manifold collector.
16. Install the vacuum tank.
17. Install the VIAS control solenoid valve and the bracket bolt.
18. Install the EVAP canister purge volume solenoid valve and bracket bolt.
19. Connect the power steering hose bracket to the back of the intake manifold collector.
20. Install the EGR tube–to–lower intake manifold collector nuts.
21. Connect the following:
   a. EGR temperature sensor electrical connector
   b. EVAP canister purge hose.
   c. Electric throttle control actuator electrical connectors.
   d. PCV hose.
   e. Fuel injector electrical connectors.
   f. Vacuum lines to the upper intake manifold collector and power valve.
   g. Coolant hoses to the intake manifold collector.
   h. Power brake booster vacuum hose.
22. Install upper air cleaner case, MAF sensor, and air cleaner–to–electric throttle control actuator tube as an assembly.
23. Install the engine cover.
24. After installation, it is necessary to re-calibrate the electric throttle control actuator as follows:
   a. Perform the "Throttle Valve Closed Position Learning" when the harness connector of the electric throttle control actuator is disconnected.

Intake Manifold

REMOVAL & INSTALLATION

3.3L Engine

1. Before servicing the vehicle, refer to the precautions in the beginning of this section.
2. Drain the cooling system.
3. Relieve the fuel system pressure.
4. Remove or disconnect the following:
   - Negative battery cable
   - Air intake duct
   - Idle Air Control (IAC) valve connectors
   - Throttle Position (TP) sensor and switch connectors
   - Exhaust Gas Recirculation (EGR) solenoid valve connector
   - Evaporative Emissions (EVAP) canister vacuum and purge hoses
   - Water, heater and Positive Crankcase Ventilation (PCV) valve hoses
   - Vacuum hoses from the EVAP canister, brake cylinder, pressure regulator and EGR tube
   - Spark plug wires
   - Distributor cap
   - 3 left bank injector connectors
   - Thermal transmitter
   - Ground harness
   - Breather pipe
   - Upper manifold
   - Fuel feed and return lines from the fuel rail
   - Right injector harness connectors
   - Fuel rail and injectors
   - Coolant temperature switch harness connector
   - Water hose from the thermostat
   - Lower manifold bolts in the sequence illustrated.
   - Manifold gasket and discard

To install:

5. Install the lower intake manifold with a new gasket.
   a. Step 1: 35 inch lbs. (4 Nm)
   b. Step 2: 78 inch lbs. (9 Nm)
   c. Step 3: 70–84 inch lbs. (8–10 Nm)
6. Install or connect the following:
   - ECT sensor connector
   - Fuel supply manifold
   - Right bank injector connectors
   - Fuel lines
   - Upper intake manifold
   - Breather pipe
   - Upper intake manifold ground cable
   - Thermal transmitter

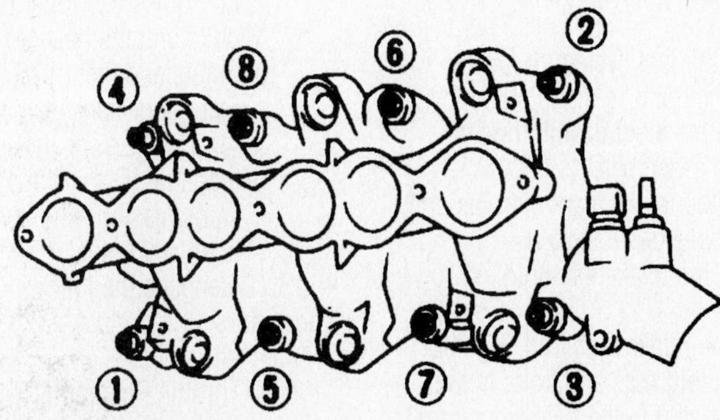

Intake manifold loosening sequence—3.3L engine

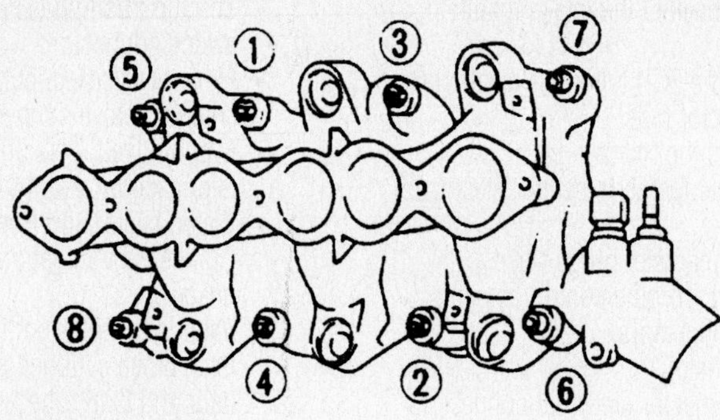

Tighten bolts in numerical order.
7924VG33

Intake manifold tightening sequence—3.3L engine

- Left bank injector connectors
- Distributor
- Spark plug wires
- EGR tube
- Fuel pressure regulator vacuum hose
- Brake booster vacuum hose
- EVAP canister vacuum and purge hoses
- PCV valve and hose
- Heater hoses
- Radiator hoses
- EGR temperature sensor connector
- EGR solenoid valve connector
- Ignition coil and power transistor connectors
- TP sensor and switch connectors
- IAC valve connector
- Cruise control cable
- Accelerator cable
- Air intake duct
- Negative battery cable

7. Fill the cooling system.
8. Start the engine and check for leaks.

3.5L Engine

1. Before servicing the vehicle, refer to the precautions in the beginning of this section.
2. Relieve the fuel system pressure.
3. Remove the intake manifold collector.
4. Remove the fuel rail with the fuel injectors as an assembly.
5. Loosen the intake manifold nuts and bolts in the sequence illustrated.
6. Remove the intake manifold.

To install:

7. Install the intake manifold. Tighten the intake manifold nuts and bolts in the sequence illustrated in the following steps:?
   a. Stud bolts: 96 inch lbs. (10 Nm).
   b. Manifold bolts first pass: 65 inch lbs. (7 Nm).
   c. Manifold bolts second pass: 21 ft. lbs. (29 Nm).
8. Install the fuel rail with the fuel injectors as an assembly.
9. Install the intake manifold collector.

Exhaust Manifold

REMOVAL & INSTALLATION

3.3L Engine

REAR (RIGHT-HAND) EXHAUST MANIFOLD

1. Before servicing the vehicle, refer to the precautions in the beginning of this section.
2. Remove or disconnect the following:
   - Negative battery cable
   - Radiator overflow hose from the radiator
   - Radiator coolant-recovery reservoir off of the bracket
   - Reservoir
   - Air cleaner intake tube and the engine air intake resonator
   - 6 rear (right-hand) exhaust manifold crossover tube heat-shield bolts and the heat shields
   - 2 nuts and the 1 bolt securing the rear (right-hand) exhaust manifold tube to the front (left-hand) exhaust manifold. Discard the gasket.
   - Transmission fluid level indicator tube heat shield
3. Disengage the following electrical connectors:
   - Idle switch
   - Throttle Position (TP) sensor
   - Exhaust Gas Recirculation (EGR) control solenoid
4. Raise and safely support the vehicle.
5. Remove or disconnect the following:
   - EGR valve-to-back-pressure transducer valve tube nut and position it out of the way
   - 2 EGR valve-to-exhaust manifold tube nuts and tube
   - 6 rear exhaust manifold nuts in the reverse order of the tightening sequence
6. Safely lower the vehicle, remove the exhaust manifold and discard the exhaust manifold gasket.

To install:

7. Raise and safely support the vehicle.
8. Be sure that both the exhaust manifold and the cylinder head mating surfaces are clean of any old gasket material.
9. Install or connect the following:
   - Rear (right-hand) exhaust manifold gasket onto the exhaust manifold mounting studs
10. Lower the vehicle safely.
    - Rear (right-hand) exhaust manifold onto the studs
    - 6 rear (right-hand) exhaust manifold nuts. Tighten the nuts in sequence to 13–16 ft. lbs. (18–22 Nm).
    - EGR valve-to-exhaust manifold tube and tube nuts
    - EGR valve-to-back-pressure transducer valve tube nut
11. Lower the vehicle carefully.
12. Reconnect the following electrical connectors:
    - EGR solenoid
    - TP sensor
    - Idle switch
    - Transmission fluid level indicator tube heat shield
    - New gasket between the front (left-hand) exhaust manifold and the rear exhaust manifold crossover tube
    - 2 nuts and the 1 bolt securing the rear (right-hand) exhaust manifold crossover tube to the front (left-hand) exhaust manifold. Tighten the rear exhaust manifold crossover tube-to-front (left-hand) exhaust manifold nuts and bolt.
    - Rear (right-hand) exhaust manifold crossover tube heat shield with the 6 mounting bolts
    - Rear (right-hand) exhaust manifold crossover tube bolts
    - Air cleaner intake tube and the engine air intake resonator
    - Radiator coolant recovery reservoir
    - Radiator overflow hose to the radiator
    - Negative battery cable
13. Start the engine and check for leaks and proper operation.

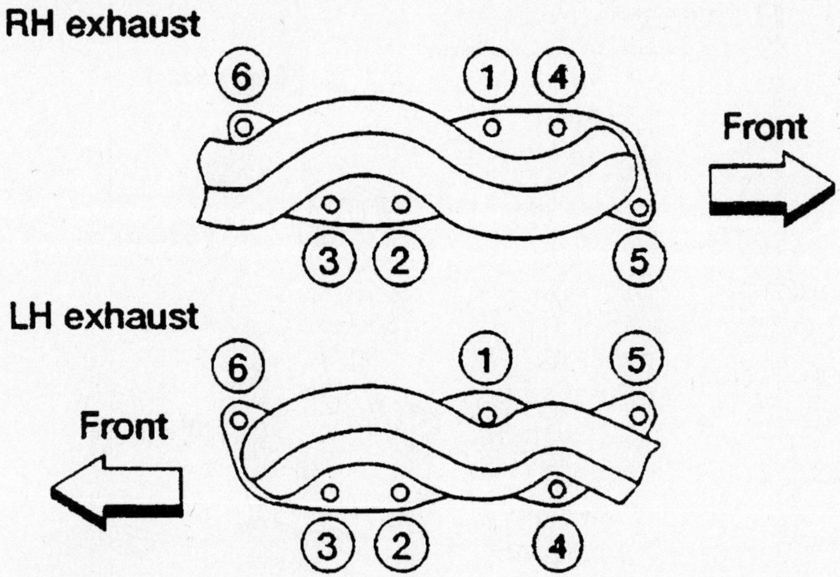

To avoid warping the exhaust manifolds, use this sequence when loosening the bolts—3.3L engine

FRONT (LEFT-HAND) EXHAUST MANIFOLD

1. Before servicing the vehicle, refer to the precautions in the beginning of this section.
2. Remove or disconnect the following:
   - Negative battery cable and wait at least 90 seconds before performing any work. This allows time for the SRS or air bag system to deplete its back up energy supply.
   - 2 nuts and the 1 bolt securing the front (left-hand) exhaust manifold to the rear (right-hand) exhaust manifold crossover tube. Discard the gasket.
3. Remove the transmission fluid level indicator tube heat shield.
   - 6 front (left-hand) exhaust manifold nuts in 2 steps in the reverse order of the tightening sequence. Do not remove the 3 lower front (left-hand) exhaust manifold nuts.
   - Front (left-hand) exhaust manifold-to-mounting bracket bolt
4. Raise and safely support the vehicle.
   - Heated Oxygen Sensor (HO_2S) electrical connector
   - 3 front (left-hand) exhaust manifold-to-inlet pipe nuts
   - Exhaust system flex tube bracket bolt
   - Left-hand inner engine and transmission splash shield bolts and screws
   - Left-hand inner engine and transmission splash shield
   - 3 lower exhaust manifold nuts
   - Front (left-hand) exhaust manifold and discard the exhaust manifold gasket

To install:

5. Be sure that both the exhaust manifold and the cylinder head mating surfaces are clean of any old gasket material.
6. Install or connect the following:
   - New front exhaust manifold gasket in place
   - Front (left-hand) exhaust manifold
   - 3 lower exhaust manifold mounting nuts. Do not tighten the nuts at this time.
   - Left-hand inner engine and transmission splash shield with their mounting bolts and screws
   - Exhaust system flex tube bracket bolt
   - 3 exhaust manifold-to-exhaust inlet pipe nuts
   - HO_2S electrical connector
7. Lower the vehicle.
   - Front (left-hand) exhaust manifold-to-mounting bracket bolt
   - 3 upper exhaust manifold mounting bolts and tighten all 6 exhaust manifold mounting bolts in sequence to 13–16 ft. lbs. (18–22 Nm)
   - Transmission fluid level indicator tube heat shield
   - 2 nuts and the 1 bolt securing the front (left-hand) exhaust manifold to the rear (right-hand) exhaust manifold crossover tube
   - Negative battery cable
8. Start the engine, check for leaks and road test for proper operation.

3.5L Engine

1. Before servicing the vehicle, refer to the precautions in the beginning of this section.

➡When removing the front and rear engine mounting through bolts and nuts, lift the engine up slightly for safety.

2. Remove the front wheels.
3. Remove the engine undercover.
4. Remove the wheel splash shields.
5. If removing the left hand exhaust manifold, remove the radiator and cooling fan assembly.
6. If removing the right hand exhaust manifold, remove the front suspension member as follows:
   a. Remove the left hand transaxle mount insulator nuts, if equipped with a 5 speed transmission.
   b. Attach an engine lifting bracket to the transaxle at the location illustrated.
   c. Install an engine support tool. Make sure the tool is securely resting on the hood ledge.
   d. Remove the three transaxle mount insulator nuts.
   e. Remove the lower ball joint bolt and separate the transverse link from the steering knuckle.
   f. Remove the front exhaust tube.
   g. Remove the power steering line bracket.
   h. Remove the mounting bolts from the lower side of the steering gear.
   i. Disconnect the front engine mount electrical connector.
   j. Disconnect the connecting rod from the front strut.
   k. Place a transmission jack under the front suspension member and remove the mounting nuts from the front suspension member.
   l. Remove the front suspension member bolts from the pin stay on the vehicle body side.
   m. Remove the through bolts from the front and rear engine mounts.
   n. Lower the transmission jack to remove the front member. It may be necessary to remove the exhaust hanger bracket, the front and rear engine mounts

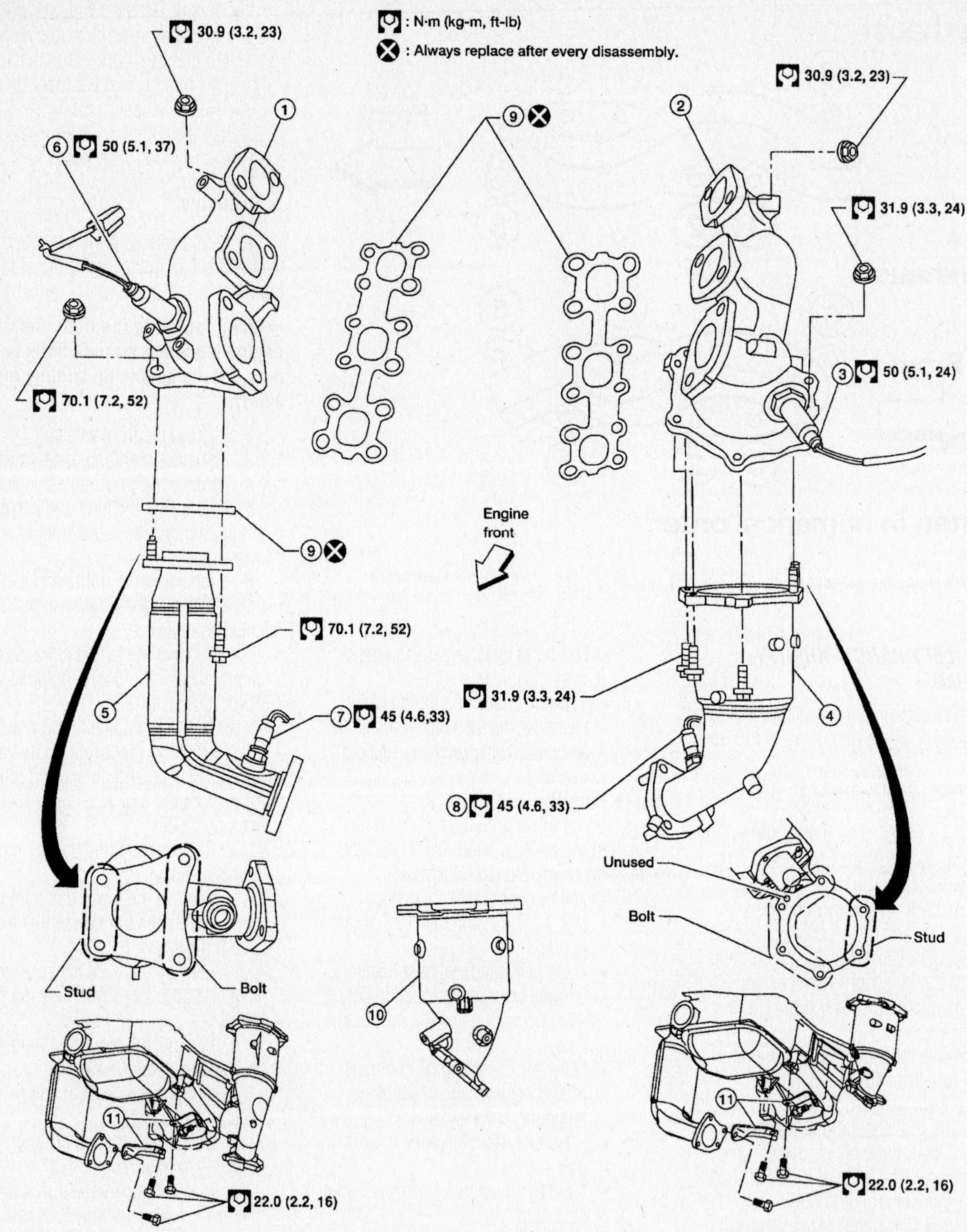

1. Exhaust manifold (RH bank)
2. Exhaust manifold (LH bank)
3. Air fuel ratio (A/F) sensor 1 (bank
4. Three way catalyst (manifold) (bank 2) (4 A/T only)
5. Three way catalyst (manifold) (bank 1)
6. Air fuel ratio (A/F) sensor 1 (bank
7. Heated oxygen sensor 2 (bank 1)
8. Heated oxygen sensor 2 (bank 2) (4 A/T only)
9. Gasket
10. Three way catalyst (manifold) (bank 2) (5 A/T only)
11. Three way catalyst supports

67170-NIQU-G04

Exploded view of the exhaust manifold and catalyst assembly–3.5L engine

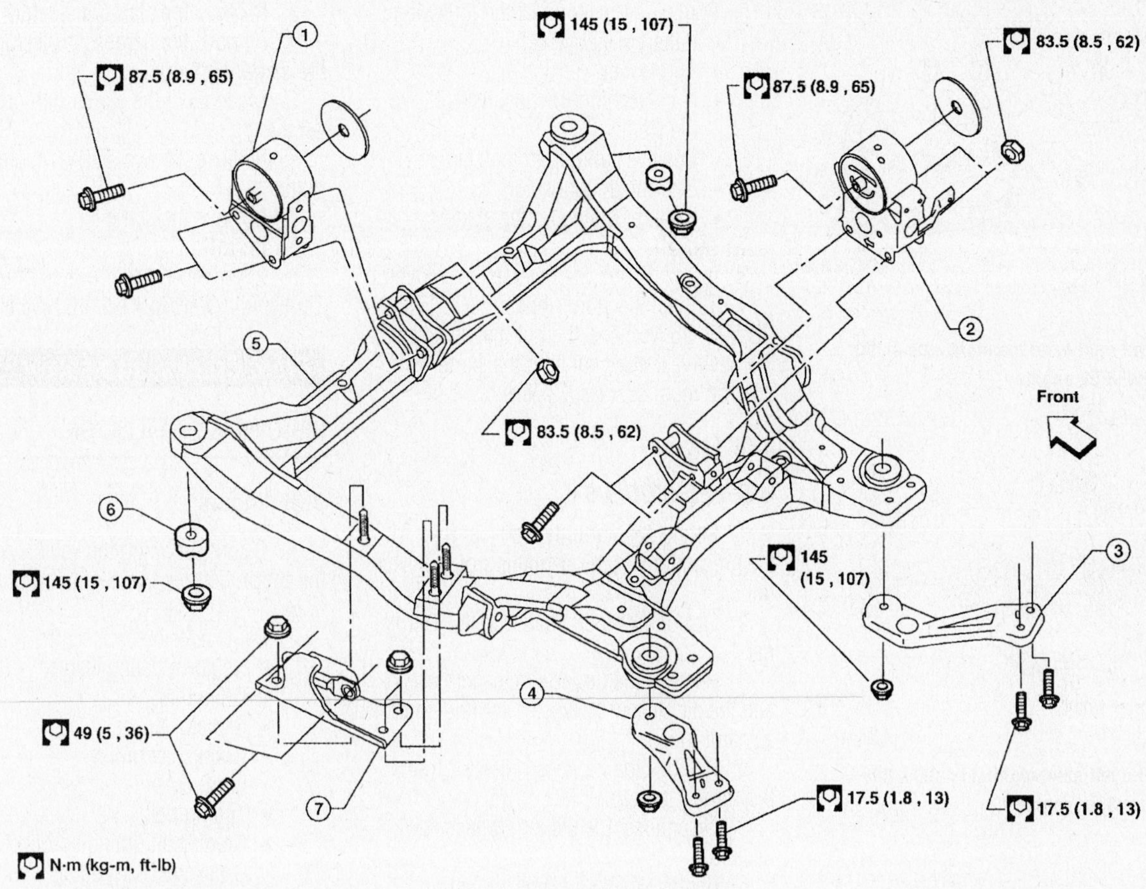

Exploded view of the front suspension member mounting–3.5L engine

and the transverse link to enable removal.

7. Remove the right and left hand three way catalyst support bolts in the sequence illustrated.

8. Remove the Heated Oxygen (HO_2S) sensors, Air Fuel Ratio (AFR) sensors.

9. Remove exhaust manifold and three way catalyst heat shields.

10. Remove the three way catalyst by loosening the bolts first and then removing the nuts and through bolts.

11. Remove the exhaust manifolds by loosening the nuts in the sequence illustrated.

To install:

12. Install the manifolds and tighten the nuts in sequence to 23 ft. lbs. (30 Nm).

13. Install the three way catalyst by tightening the nuts first and then the bolts and tighten to 16 ft. lbs. (22 Nm) in the sequence illustrated.

14. Install the exhaust manifold and three way catalyst heat shields.

15. Install the Heated Oxygen (HO_2S) sensors and Air Fuel Ratio (AFR) sensors. Tighten to the specifications shown in the exploded view of the exhaust manifold assembly illustration.

16. Install the front support member in

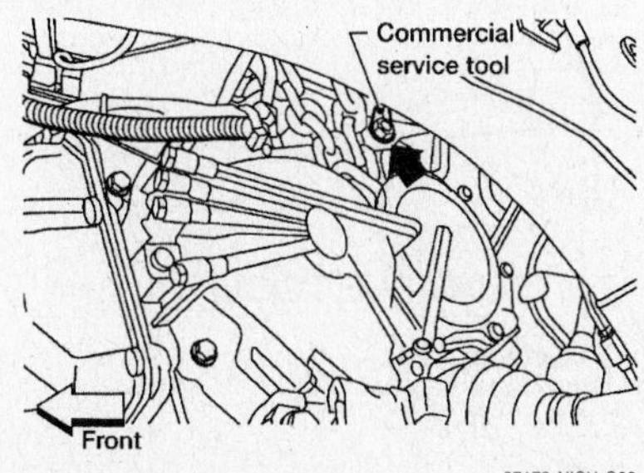

Attach a engine lifting bracket to the tranaxle–3.5L engine

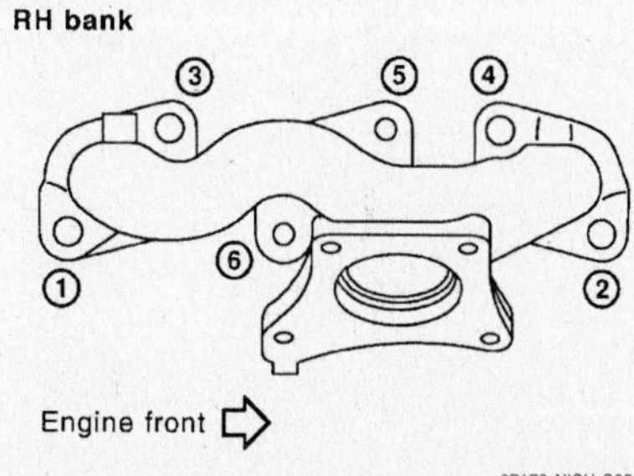

Remove the right hand manifold nuts in this sequence–3.5L engine

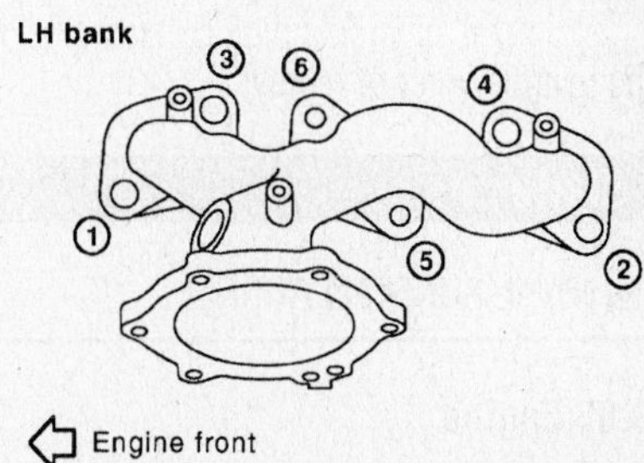

Remove the left hand manifold nuts in this sequence–3.5L engine

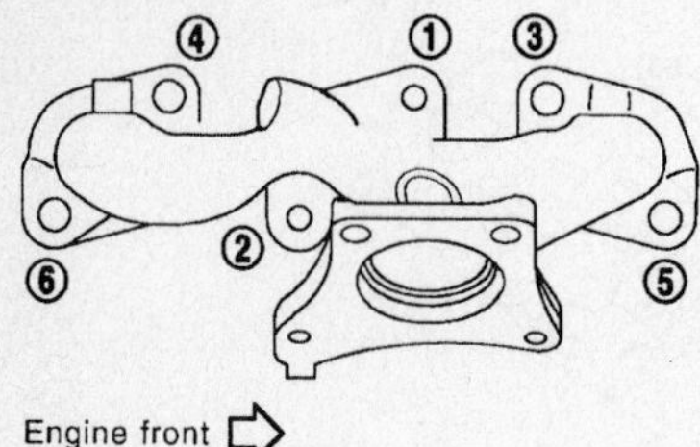

Install the right hand manifold nuts in this sequence–3.5L engine

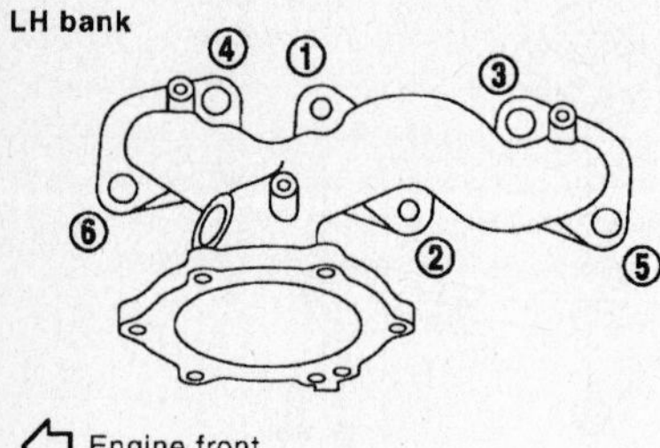

Install the left hand manifold nuts in this sequence–3.5L engine

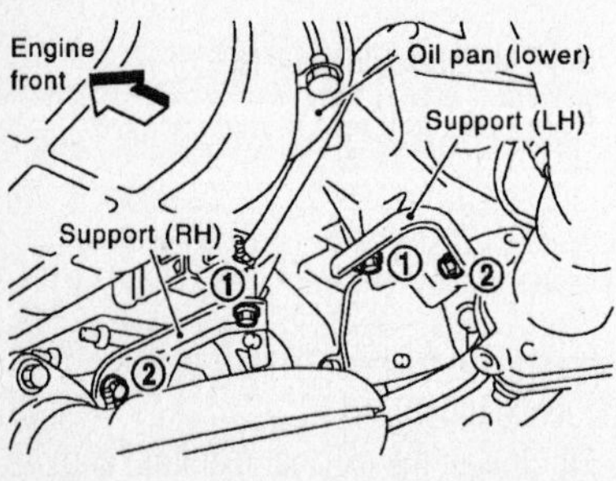

Install the three way catalyst by tightening the nuts first and then the bolts in the sequence shown–3.5L engine

the reverse order of removal and tighten the retainers to the specifications shown in the front suspension member mounting exploded view illustration.

17. Install any remaining components in the reverse order of removal.

Starter

REMOVAL & INSTALLATION

3.3L Engine

1. Before servicing the vehicle, refer to the precautions in the beginning of this section.
2. Remove or disconnect the following:
   - Battery negative cable
   - Air cleaner
   - Nut attaching the positive cable to the starter
   - Positive cable from the starter
   - S-terminal connector
   - 2 starter bolts and the starter

To install:

3. Installation is the reverse of removal.
4. Tighten the starter bolts to 17–19 ft. lbs. (23–26 Nm) and the nut that attaches the positive battery cable to the starter to 87–104 inch lbs. (10–12 Nm).

3.5L Engine

4 SPEED MODELS

1. Before servicing the vehicle, refer to the precautions in the beginning of this section.
2. Disconnect the negative battery terminal.
3. Remove the upper air cleaner case and the air cleaner to electric throttle control actuator tube.
4. Remove the harness protector from the starter harness.
5. Disconnect the starter harness connectors.
6. Remove the two starter mounting bolts.
7. Remove the starter.

To install:

8. Installation is the reverse of removal. Tighten the upper bolt to 41 ft. lbs. (55 Nm) and the lower bolt to 55 ft. lbs. (74 Nm).

5 SPEED MODELS

1. Before servicing the vehicle, refer to the precautions in the beginning of this section.
2. Disconnect the negative battery terminal.
3. Remove the starter insulator.
4. Remove the harness protector from the starter harness.
5. Disconnect the starter harness connectors.
6. Remove the two starter mounting bolts.
7. Remove the starter.

To install:

8. Installation is the reverse of removal. Tighten the bolts to 41 ft. lbs. (55 Nm).

Front Crankshaft Seal

REMOVAL & INSTALLATION

3.3L Engine

1. Before servicing the vehicle, refer to the precautions in the beginning of this section.
2. Remove or disconnect the following:
   - Negative battery cable
   - Drive belts
   - Radiator hoses
   - Crankshaft pulley
   - Front cover
   - Timing belt
   - Crankshaft timing sprocket
   - Crankshaft seal using a suitable prytool

To install:

3. Install or connect the following:
   - Crankshaft seal using a driver and a hammer until its flush with the housing
   - Crankshaft timing sprocket
   - Timing belt
   - Front cover and tighten the bolts to 26–43 inch lbs. (3–5 Nm)
   - Crankshaft pulley and tighten the bolt to 141–156 ft. lbs. (191–211 Nm)
   - Radiator hoses

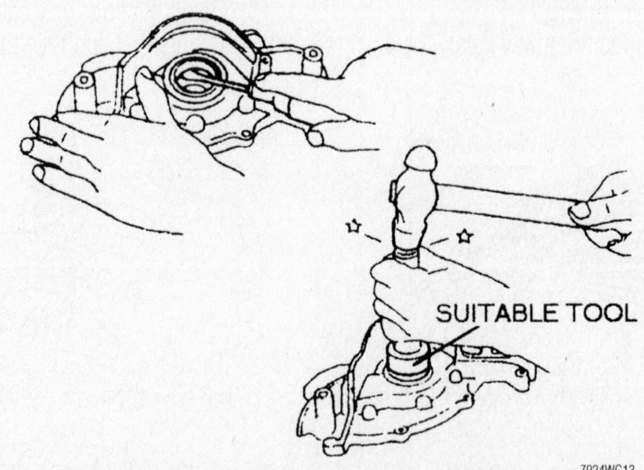

Removal and installation of the front crankshaft seal—3.3L engines

- Drive belts
- Negative battery cable

4. Fill the cooling system, start the vehicle and check for leaks.

3.3L Engine

Refer to the front timing cover case removal and installation for seal removal.

Camshaft And Valve Lifters

REMOVAL & INSTALLATION

3.3L Engine

1. Before servicing the vehicle, refer to the precautions in the beginning of this section.
2. Drain the cooling system.
3. Remove or disconnect the following:
   - Negative battery cable
   - Upper intake manifold
   - Valve covers

➡Keep all valvetrain components in order for assembly.

- Rocker arm and shaft assemblies
- Valve lifter guide and valve lifters.

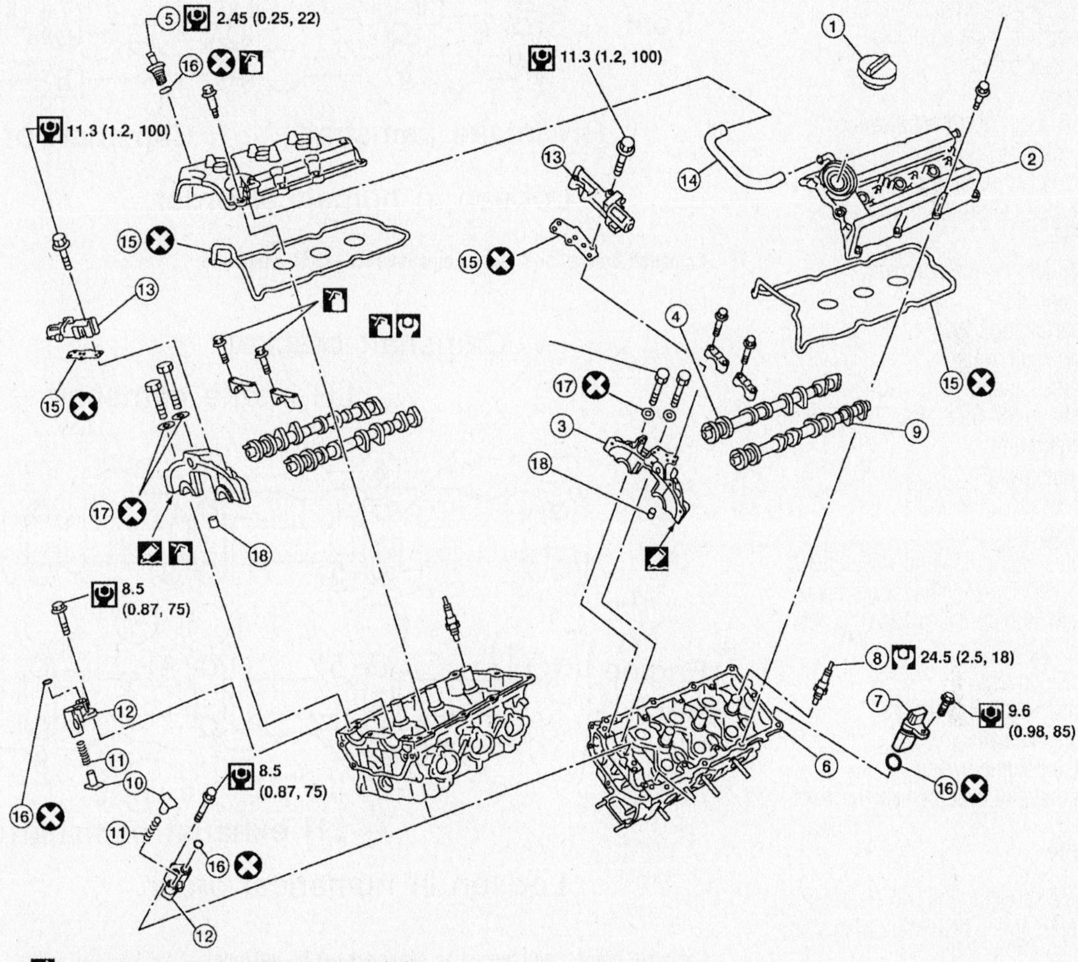

Exploded view of the camshaft and related components–3.5L engine

Attach a wire to the top of the lifters so that they will not drop from the lifter guide.

- Radiator
- Accessory drive belts
- Front cover
- Timing belt
- Camshaft sprockets
- Camshaft seals
- Rear timing cover
- Distributor
- Cylinder head rear covers
- Camshaft locating plates
- Camshafts

To install:

4. Install or connect the following:
   - Camshafts
   - Camshaft locating plates. Tighten the bolts to 58–65 ft. lbs. (78–88 Nm).
   - Cylinder head rear covers
   - Distributor
   - Rear timing cover
   - Camshaft seals
   - Camshaft sprockets. Tighten the bolts to 58–65 ft. lbs. (78–88 Nm).
   - Timing belt
   - Front cover
   - Accessory drive belts
   - Radiator
   - Valve lifter guide and valve lifters
   - Rocker arm and shaft assemblies. Tighten the bolts to 13–16 ft. lbs. (18–22 Nm).
   - Valve covers
   - Upper intake manifold
   - Negative battery cable
5. Fill the cooling system.
6. Start the engine and check for leaks.

3.5L Engine

1. Before servicing the vehicle, refer to the precautions in the beginning of this section.

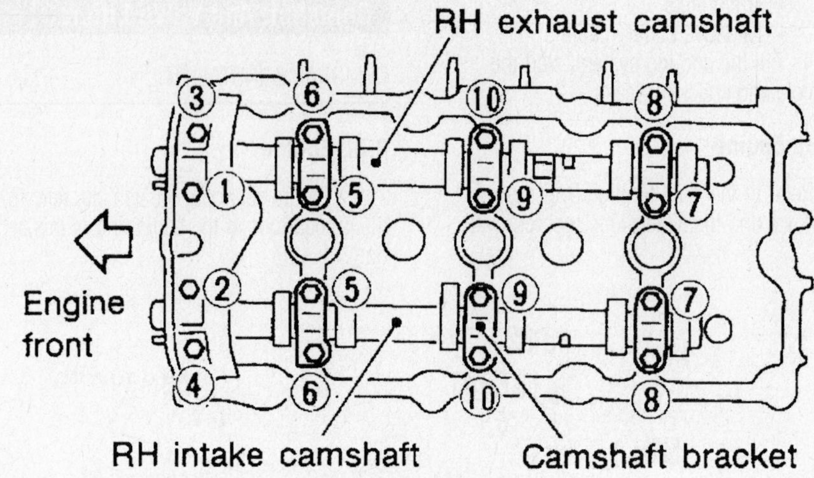

Camshaft bracket bolt loosening sequence—3.5L right side

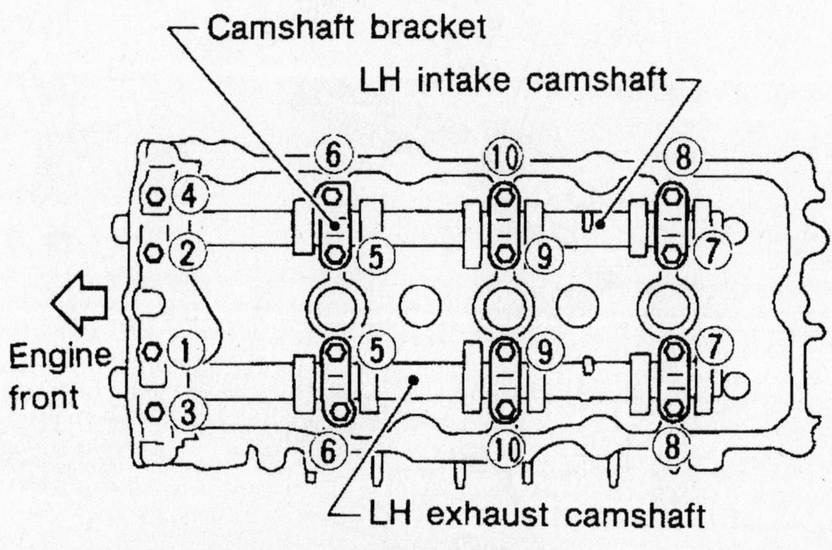

Camshaft bracket bolt loosening sequence—3.5L left side

Bank	INT/EXH	ID mark	Drill mark	Paint marks	
				M1	M2
RH	INT	RE	Yes	Yes	No
	EXH	RE	No	No	Yes
LH	INT	LH	Yes	Yes	No
	EXH	LH	No	No	Yes

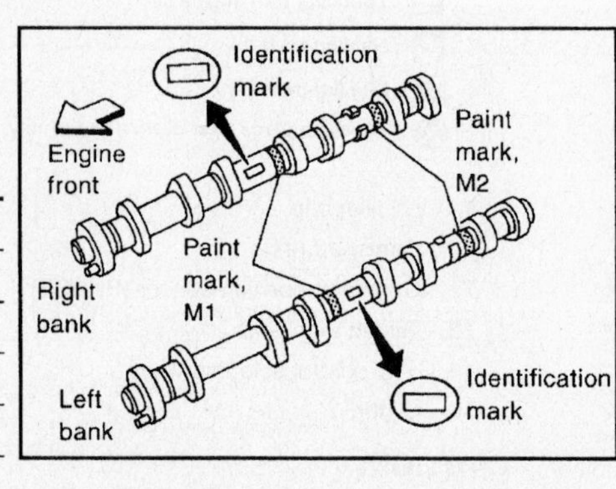

Camshaft identification–3.5L engine

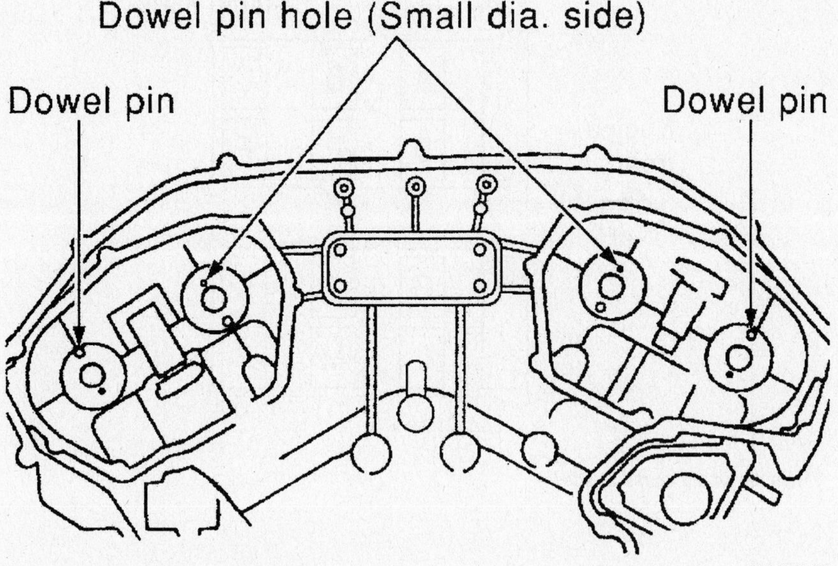

Dowel pin orientation–3.5L engine

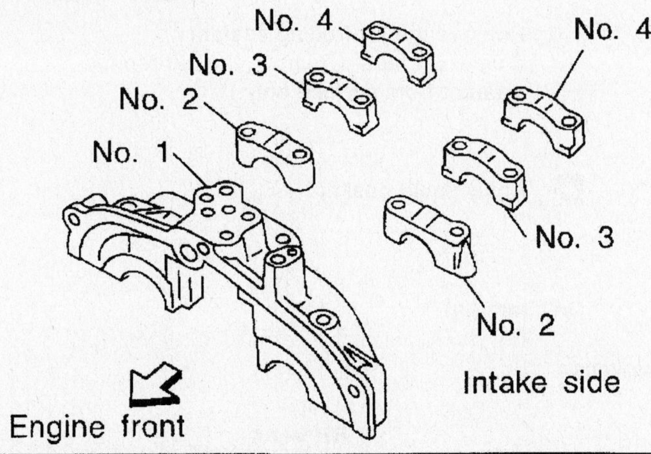

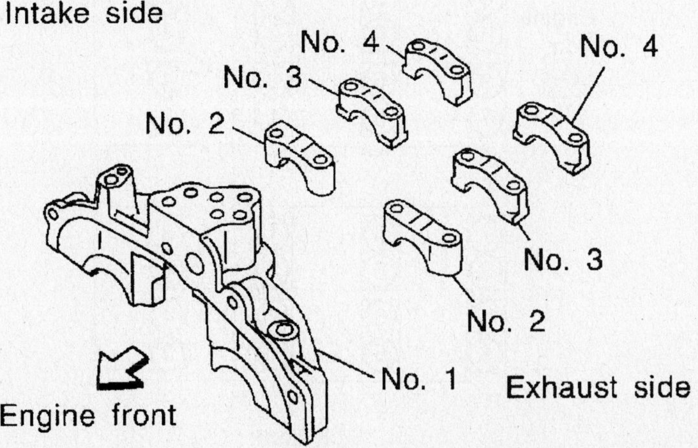

Camshaft bracket identification–3.5L engine

2. Drain the cooling system.
3. Remove the front timing chain case, camshaft sprocket, timing chain and the rear timing chain case.
4. If necessary, remove both the Camshaft Position (CMP) sensors from cylinder head back side.

➡Handle carefully to avoid dropping and shocks. Do not disassemble. Do not allow metal powder to adhere to magnetic part at sensor tip. Do not place sensors in a location where they are exposed to magnetism.

5. Remove the intake valve timing control solenoid valves. Discard the intake valve timing control solenoid valve gaskets and use new gaskets for installation.
6. Remove the intake and exhaust camshaft brackets. Mark the camshafts, camshaft brackets, and bolts so they are placed in the same position and direction for installation.
7. Equally loosen the camshaft bracket bolts in several steps in the numerical order shown.
8. Remove the camshaft.
9. Remove the valve lifter. Identify installation positions, and store them without mixing them up.
10. Remove the secondary timing chain tensioner from cylinder head with its stopper pin attached.
11. Inspect the Camshaft Runout as follows:
    a. Put V block on precise flat bed, and support No. 2 and No. 4 journal of camshaft.
    b. Set dial gauge vertically to No. 3 journal.
    c. Turn camshaft to one direction with hands, and measure camshaft runout on dial gauge. (Total indicator reading)
    d. If it exceeds the limit, replace camshaft.
12. Inspect the camshaft cam height as follows:
    a. Measure the camshaft cam height. Limit: 0.0020 inch (0.05 mm). Standard cam height (intake and exhaust): 1.7663–1.7738 inch (44.865–45.055 mm)
    b. If wear is beyond the limit, replace camshaft.
13. Inspect the camshaft journal clearance of the outer diameter of the camshaft journal as follows:
    a. Measure outer diameter of camshaft journal. Cam wear limit: 0.008 inch (0.2 mm)

Standard outer diameter:

- No. 1: 1.0211–1.0218 inch (25.935–25.955 mm)
- No. 2, 3, 4: 0.9230–0.9238 inch (23.445–23.465 mm)

14. Inspect the camshaft journal clearance of the inner diameter of the camshaft bracket as follows:

a. Tighten camshaft bracket bolt with specified torque.

b. Using inside micrometer, measure inner diameter "A" of camshaft bracket.

15. Calculate the camshaft journal clearance as follows:

Journal clearance = inner diameter of camshaft bracket—outer diameter of camshaft journal. When outside the limit, replace either or both camshaft and cylinder head.

➡Inner diameter of camshaft bracket is manufactured together with cylinder head. Replace the whole cylinder head assembly.

16. Inspect the camshaft end play as follows:

a. Install dial gauge in thrust direction on front end of camshaft. Measure end play of dial gauge when camshaft is moved forward and backward.

b. When out of the limit, replace with new camshaft and measure again.

c. When out of the limit again, replace with new cylinder head.

To install:

17. Install secondary chain tensioners on both sides of cylinder head.

a. Install chain tensioner with its stopper pin attached.

b. Install tensioner with sliding part facing downward on right side cylinder head, and with sliding part facing upward on left side cylinder head.

c. Install O-ring as shown.

18. Install valve lifters in their original position.

Valve lifter outer diameter (Intake and exhaust)

- 1.3377–1.3381 inch (33.977–33.987 mm)

Standard (Intake and exhaust)

- 1.3386–1.3392 inch (34.000–34.016 mm)

Standard (Intake and exhaust)

- 0.0005–0.0015 inch (0.013–0.039 mm)

19. Install camshafts.

a. Install camshaft with dowel pin attached to its front end face on the exhaust side.

b. Follow your identification marks

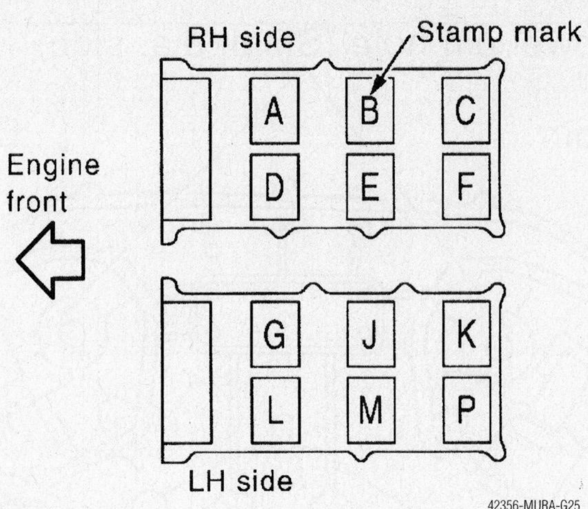

Align the stamp marks as shown

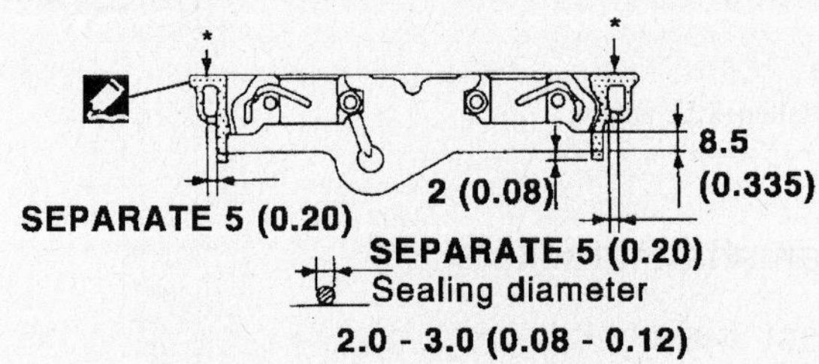

RTV sealer application

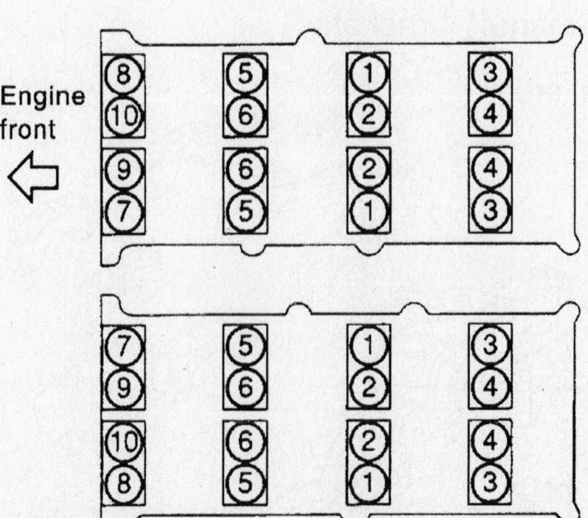

Camshaft bracket bolt torque sequence

made during removal, or follow the identification marks that are present on the new camshafts for proper placement and direction.

c. Install camshaft so that dowel pin hole and dowel pin on front end face are positioned as shown in illustration. (No. 1 cylinder TDC on its compression stroke)

➡Large- and small-pin holes are located on front end face of intake camshaft, at intervals of 180 degrees. Face small dia. side pin hole upward (in cylinder head upper face direction).

20. Install camshaft brackets.

a. Remove foreign material completely from camshaft bracket backside and from cylinder head installation face.

b. Install camshaft bracket in original position and direction as shown in illustration.

c. Install No.2 to 4 camshaft brackets aligning the stamp marks as shown.

➡There are no identification marks indicating left and right for No. 1 camshaft bracket.

d. Apply sealant to mating surface of No.1 camshaft bracket as shown on right and left banks.

➡Use RTV silicone sealant or equivalent.

21. Tighten the camshaft brackets in the following steps, in numerical order as shown.

a. Tighten No. 7 to 10, then tighten No.1 to 6 in order as shown.

b. Tighten No.1 to 10 in numerical order as shown.

c. Tighten No. 1 to 6 in the numerical order as shown.

d. Tighten No. 7 to 10 in the numerical order as shown.

- 17 inch lbs. (1.96 Nm)
- 52 inch lbs. (5.88 Nm)
- 80–104 inch lbs. (9.02–11.8 Nm)
- 74–91 inch lbs. (8.3–10.3 Nm)

22. Measure difference in levels between front end faces of No. 1 camshaft bracket and cylinder head. If measurement is outside the specified range, re-install camshaft and camshaft bracket.

23. Inspect and adjust valve clearance.

24. Install in the reverse order of removal after this step.

25. Inspect the valve clearance as follows:

Perform inspection as follows after removal, installation or replacement of camshaft or valve-related parts, or if there is unusual engine conditions regarding valve clearance.

a. Remove right and left rocker covers.

b. Measure valve clearance as below:

- Set No.1 cylinder at TDC of its compression stroke. Align crankshaft pulley timing mark (grooved line without color) with timing indicator. Check that No. 1 cylinder intake and exhaust cam nose is facing in direction shown in illustration. If not, rotate crankshaft pulley 360 degrees clockwise (when viewed from front).
- Using a feeler gauge, measure valve clearance. Standard: -0.0055–0.0055 inch (-0.14 to 0.14 mm)

➡If inspection was carried out with cold engine, check that values with fully warmed up engine are still within specifications.

- Rotate crankshaft by 240 degrees clockwise (when viewed from front) to align No. 3 cylinder at TDC of its compression stroke.

➡Crankshaft pulley mounting bolt flange has a stamped line every 60? degrees. They can be used as a guide to rotation angle.

- Turn crankshaft pulley clockwise by 240 degrees from the position of No. 5 cylinder at compression TDC.

c. For measurements that are outside the specified range, perform adjustment below.

26. Adjust the clearance as follows:

Perform adjustment depending on selected head thickness of valve lifter. The specified valve lifter thickness is the dimension at normal temperatures. Ignore dimensional differences caused by temperature. Use the specifications for hot engine condition to adjust.

a. Remove camshaft.

b. Remove the valve lifters at the locations that are outside the standard.

c. Measure the center thickness of the removed valve lifters with a micrometer.

d. Use the equation below to calculate valve lifter thickness for replacement.

Valve lifter thickness calculation:

- Thickness of replacement valve lifter = t1 + (C1–C2)
- t1 = Thickness of removed valve lifter
- C1 = Measured valve clearance
- C2 = Standard valve clearance:

Thickness of a new valve lifter can be identified by stamp marks on the reverse side (inside the cylinder). Stamp mark 788U or 788R indicates 0.3102 inch (7.88 mm) in thickness.

➡2 types of stamp marks are used for parallel setting and for manufacturer identification.

Available thickness of valve lifter: 27 sizes with range 0.3102–0.3307 inch (7.88–8.40 mm) in steps of 0.0008 inch (0.02 mm) (when manufactured at factory).

Crank Position	Valve No. 1	Valve No. 2	Valve No. 3	Valve No. 6
No. 1 TDC	Intake	Exhaust	Exhaust	Intake

67170-NIQU-G70

Valve clearance inspection—No.1 TDC on 3.5L engine

Crank Position	Valve No. 2	Valve No. 3	Valve No. 4	Valve No. 5
No. 3 TDC	Intake	Intake	Exhaust	Exhaust

67170-NIQU-G71

Valve clearance inspection—No.3 TDC on 3.5L engine

Crank Position	Valve No. 1	Valve No. 4	Valve No. 5	Valve No. 6
No. 5 TDC	Exhaust	Intake	Intake	Exhaust

67170-NIQU-G72

Valve clearance inspection—No.5 TDC on 3.5L engine

Valve clearance:

Unit: mm (in)

	Cold	Hot * (reference data)
Intake	0.26 - 0.34 (0.010 - 0.013)	0.304 - 0.416 (0.012 - 0.016)
Exhaust	0.29 - 0.37 (0.011 - 0.015)	0.308 - 0.432 (0.012 - 0.016)

*: Approximately 80°C (176°F)

42356-MURA-G31

Valve clearance specifications–3.5L engines

a. Install the selected valve lifter.
b. Install camshaft.
c. Manually turn crankshaft pulley a few turns.
d. Check that valve clearances for cold engine are within specifications by referring to the specified values.
e. After completing the repair, check valve clearances again with the specifications for warmed engine. Make sure the values are within specifications.
Valve clearance:

- Intake: 0.012 inch (0.30 mm)
- Exhaust: 0.013 inch (0.33 mm)

Valve Lash

ADJUSTMENT

The engines covered in this section use hydraulic valve lifters that automatically adjust the valve lash. No periodic adjustment is needed.

Oil Pan

REMOVAL & INSTALLATION

3.3L Engine

1. Before servicing the vehicle, refer to the precautions in the beginning of this section.
2. Drain the engine oil.
3. Remove or disconnect the following:
   - Negative battery cable
   - Front engine mount (support) insulator through-bolt
   - Rear engine mount (support) through-bolt
   - 2 rear refrigerant/heater pipe hold down bracket bolts
   - 4 crossmember (also called a transverse member) bolts, and remove the crossmember.
   - Exhaust inlet pipe
   - 4 rear transaxle-to-engine brace bolts and the 5 front transaxle-to-engine brace bolts
   - Front transaxle-to-engine brace
   - Low oil level sensor electrical connector
   - 18 oil pan bolts in the reverse order of the tightening sequence, working from the outside, towards the center bolts.
   - Oil pan and discard the seals

To install:

4. Clean all parts well. Be sure that all old sealing material is removed from the oil pan and engine mating surfaces.
5. Position new oil pan seals. Apply Loctite® Ultra Gray 599 Silicone Sealer, or equivalent, to the ends of the oil pan seals.
6. Apply a bead of Loctite® Ultra Gray 599 Silicone Sealer or equivalent to the oil pan gasket rail inboard of the bolt holes.
7. Install or connect the following:
   - Oil pan on the engine block. Tighten the 18 oil pan bolts in sequence, working from the inside, towards the outer bolts. Do not over-tighten. Tighten to 62–70 inch lbs. (7–8 Nm).
   - Low oil level sensor electrical connector.
   - Front and rear transaxle braces. Tighten all bolts to 22–30 ft. lbs. (30–40 Nm).
   - Exhaust inlet pipe
   - Crossmember and tighten the bolts to 58–65 ft. lbs. (78–88 Nm).
   - Both engine support through-bolts and tighten to 58–65 ft. lbs. (78–88 Nm).
8. Remove the support jack from under the crankshaft pulley.
9. Lower the vehicle.
10. Fill the engine with the specified engine oil to the required level.
11. Connect the negative battery cable. Start the engine and check for leaks.

3.5L Engine

➡When removing the front and rear engine mounting through bolts and nuts, lift the engine up slightly for safety.

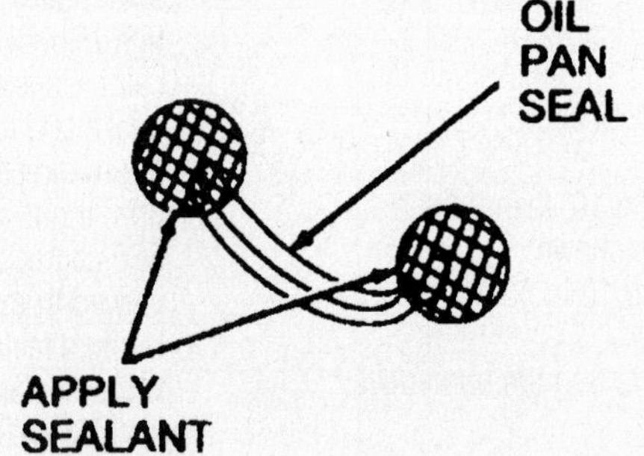

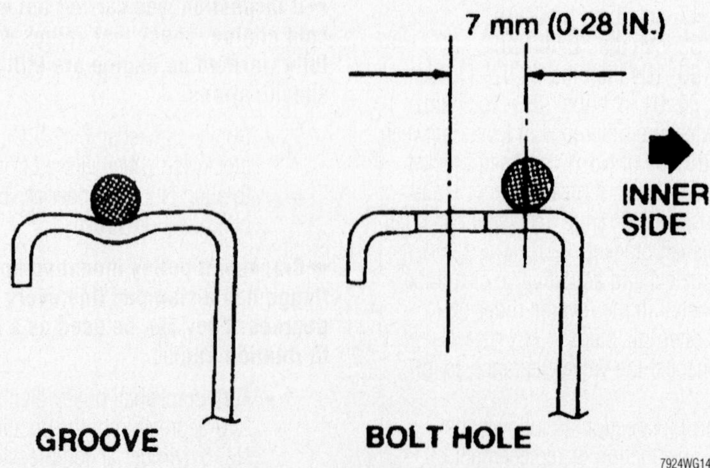

Apply RTV silicone sealer to the seal ends and to the oil pan gasket rail

TIGHTENING SEQUENCE

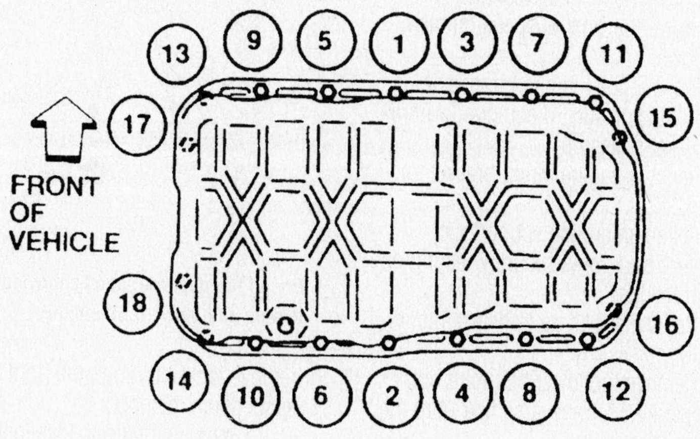

Tighten the 18 oil pan bolts in sequence, working from the inside, towards the outer bolts

CAUTION

When removing the upper oil pan from the engine, first remove the Crankshaft Position (CKP) sensor.

1. Before servicing the vehicle, refer to the precautions in the beginning of this section.
2. Remove the front right hand wheel.
3. Drain engine oil and coolant.
4. Remove the oil dipstick.
5. Remove the engine undercover.
6. Remove the right hand inner fender splash shield.
7. Remove the A/C drive belt.
8. Remove the front exhaust tube.
9. Remove the coolant pipe bolts.
10. Discharge and recover the A/C refrigerant.

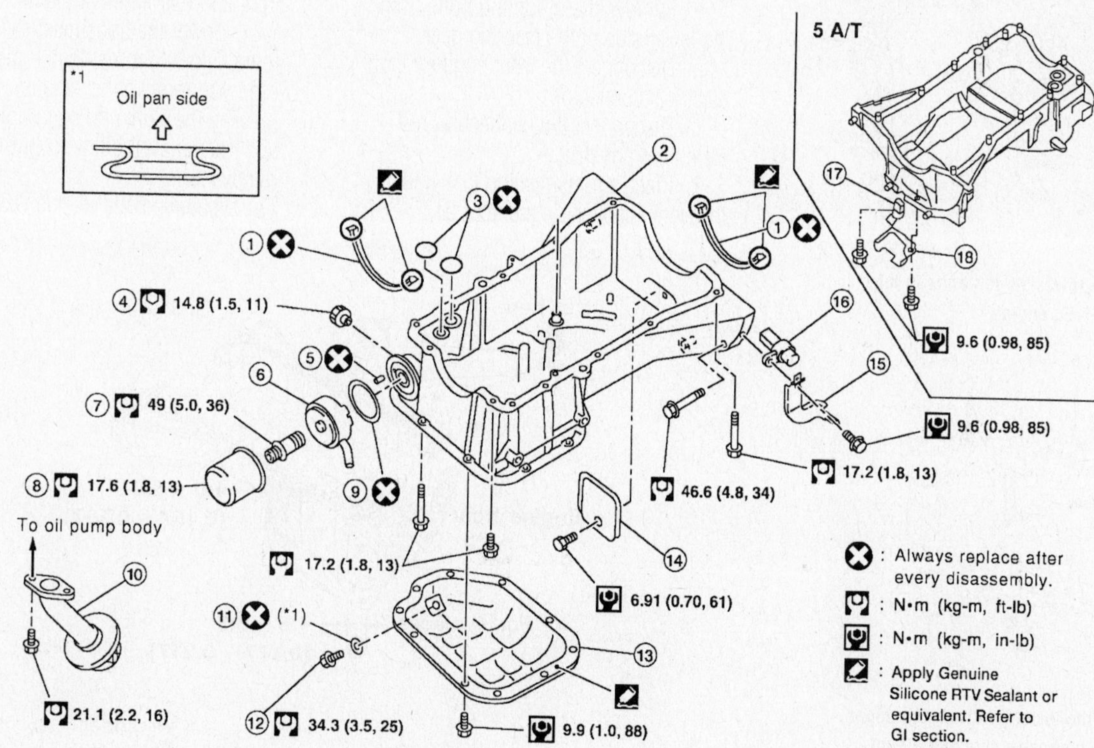

1. Gasket
2. Upper oil pan
3. O-ring
4. Oil pressure switch
5. Relief valve
6. Oil cooler
7. Oil cooler connection
8. Oil filter
9. Gasket
10. Oil strainer
11. Gasket
12. Drain plug
13. Lower oil pan
14. Rear plate cover
15. Heated oxygen sensor (bank 2) harness clamp (4 A/T only)
16. Crankshaft position sensor (POS) (4 A/T)
17. Crankshaft position sensor (POS) (5 A/T)
18. Crankshaft position sensor (POS) (5 A/T) shield

67170-NIQU-G10

Exploded view of the oil pan assembly–3.5L engine

11. Remove the A/C compressor.

12. Disconnect the coolant lines from the engine oil cooler and plug the lines them to prevent coolant loss.

13. Remove the oil filter and engine oil cooler from the upper oil pan.

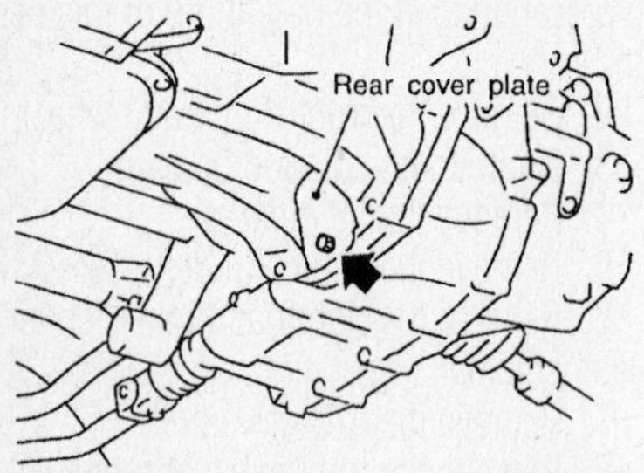

Location of the rear plate–3.5L engine

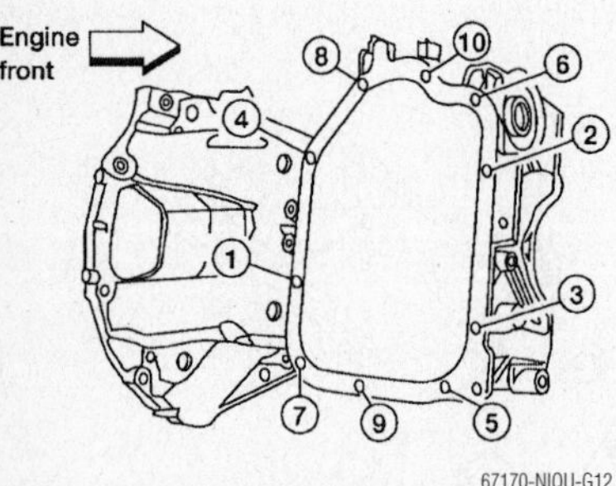

Loosen the lower oil pan bolts in this sequence–3.5L engine

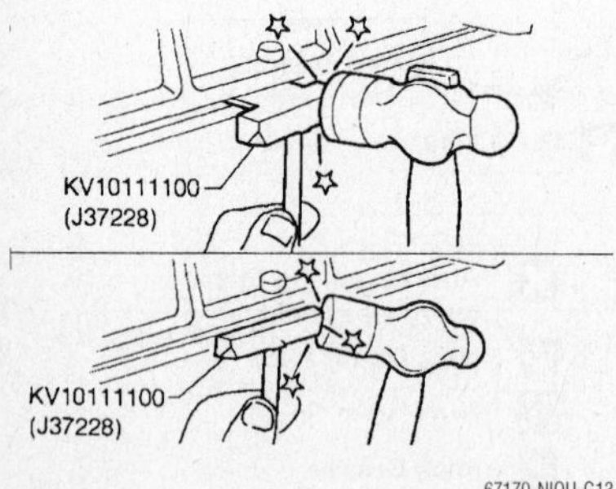

Separate the lower pan from the upper pan using a suitable tool and a hammer–3.5L engine

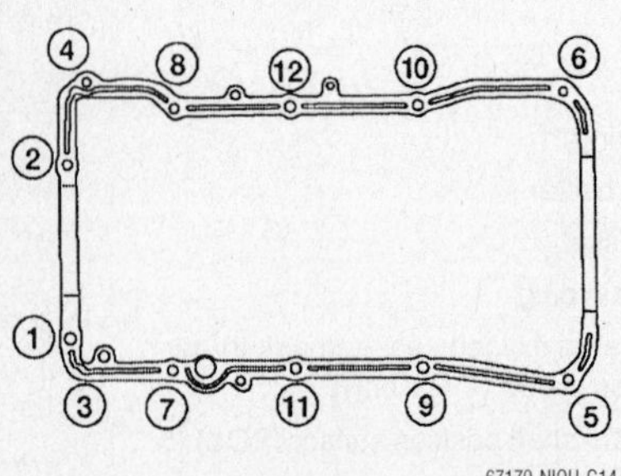

Loosen the upper oil pan bolts in this sequence–3.5L engine

14. Remove the oil pressure switch, and the CKP sensor from the upper oil pan.

15. Remove the front driveshafts.

16. Remove the front suspension member as follows:

a. Remove the left hand transaxle mount insulator nuts, if equipped with a 5 speed transmission.

b. Attach a engine lifting bracket to the tranaxle at the location illustrated.

c. Install an engine support tool. Make sure the tool is securely resting on the hood ledge.

d. Remove the three transaxle mount insulator nuts.

e. Remove the lower ball joint bolt and separate the transverse link from the steering knuckle.

f. Remove the front exhaust tube.

g. Remove the power steering line bracket.

h. Remove the mounting bolts from the lower side of the steering gear.

i. Disconnect the front engine mount electrical connector.

j. Disconnect the connecting rod from the front strut.

k. Place a transmission jack under the front suspension member and remove the mounting nuts from the front suspension member.

l. Remove the front suspension member bolts from the pin stay on the vehicle body side.

m. Remove the through bolts from the front and rear engine mounts.

n. Lower the transmission jack to remove the front member. It may be necessary to remove the exhaust hanger bracket, the front and rear engine mounts and the transverse link to enable removal.

17. Disconnect the Heated Oxygen

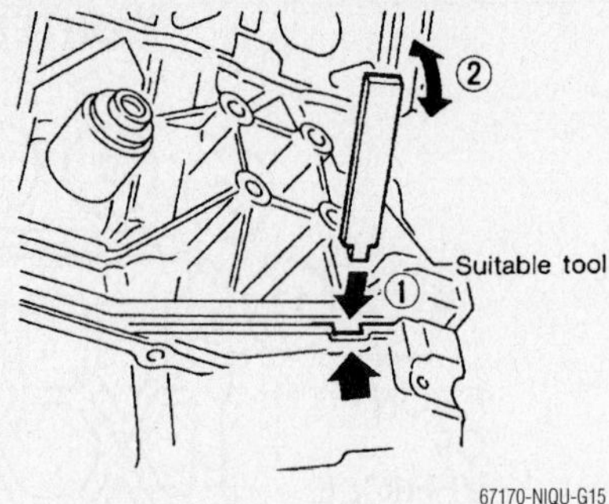

Pry off the upper oil pan by moving the tool up and down–3.5L engine

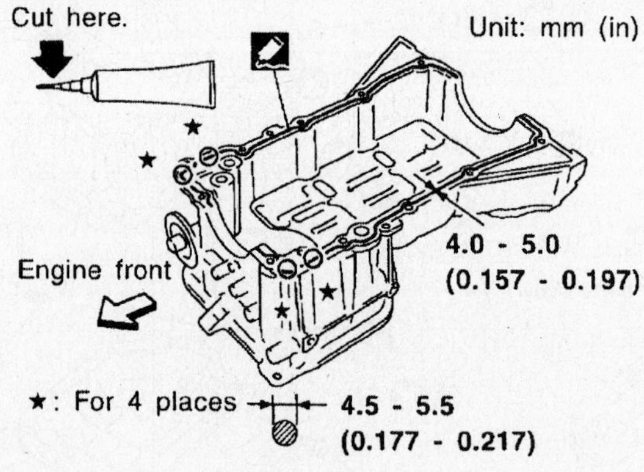

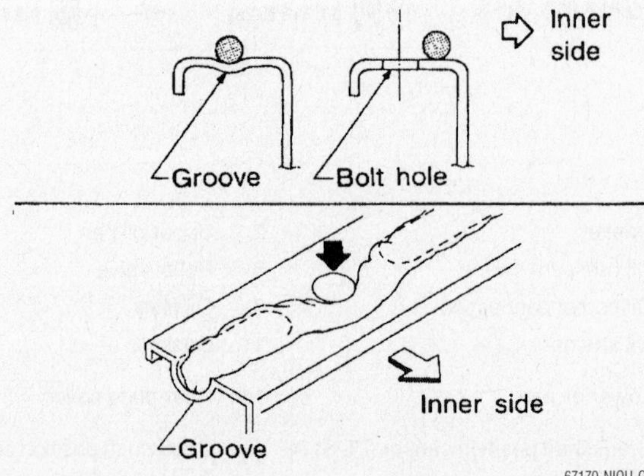

Apply a 0.15–0.197 inch (4–5mm) bead of sealant to the cylinder block mating surface of the upper oil pan–3.5L engine

(HO_2S) sensors and Air Flow Ratio (AFR) sensors and remove the two three way catalysts from the exhaust manifolds.

18. Remove the rear plate cover from the upper oil pan.
19. Loosen the lower oil pan bolts in the sequence illustrated.
20. Remove the lower oil pan by inserting a suitable prytool between the lower and the upper pans.

➡Do not insert a screwdriver, this will damage the mating surfaces. Slide the Tool by tapping its side with a hammer to remove the lower oil pan from the upper oil pan.

21. Remove the four upper oil pan–to–transaxle bolts.

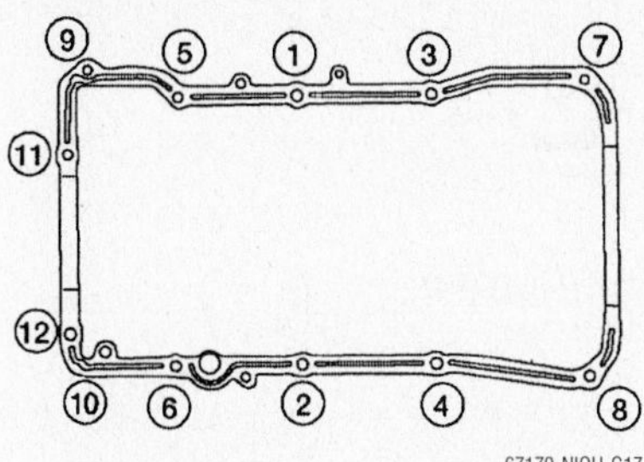

Tighten the upper oil pan bolts in this sequence–3.5L engine

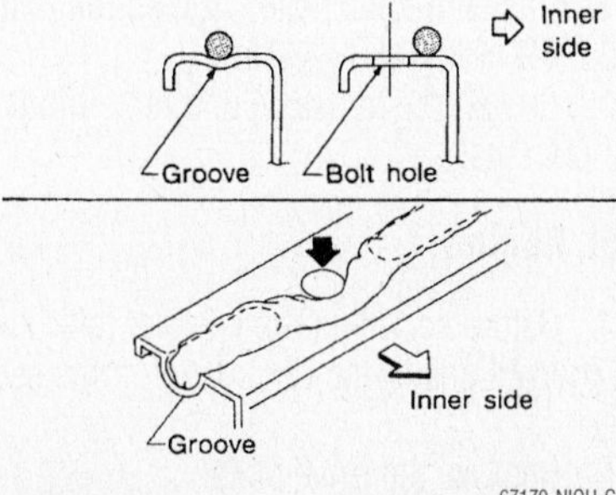

Apply a 0.177–0.217 inch (4.5–5.5mm) bead of sealant to the lower oil pan–3.5L engine

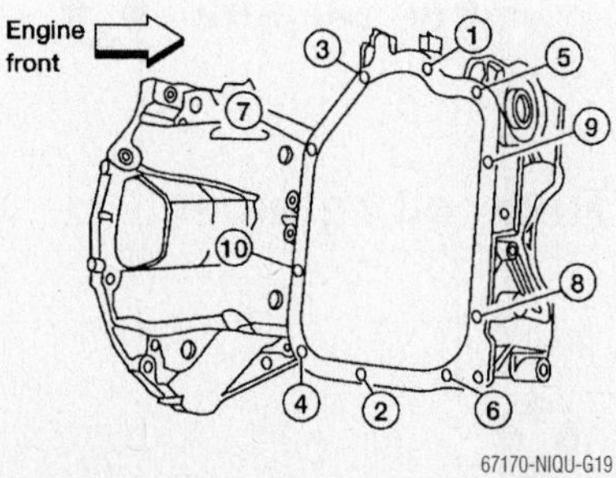

Tighten the lower oil pan bolts in this sequence–3.5L engine

22. Remove O_2S sensor wire retainer bracket.
23. Remove the upper oil pan by loosening the bolts in the sequence illustrated.
24. Insert an appropriate size tool into the notch of the upper oil pan as shown in the accompanying illustration.
25. Pry off the upper oil pan by moving the tool up and down as illustrated.
26. If re-installing the original oil pan, remove the old sealant from the mating surfaces using a scraper.

To install:

⁂ CAUTION

Wait at least 30 minutes before refilling the engine with oil.

27. Apply a 0.15–0.197 inch (4–5mm) bead of sealant to the cylinder block mating surface of the upper oil pan as illustrated. Attach the pan within 5 minutes of applying the sealant.
28. Install new O-rings on the cylinder block and oil pump body.
29. Install the upper oil pan. Tighten upper oil pan bolts in the order illustrated to 13 ft. lbs. (17 Nm).
30. Install the four upper oil pan to transaxle bolts.
31. Apply a 0.177–0.217 inch (4.5–5.5mm) bead of sealant to the lower oil pan. Attach the pan within 5 minutes of applying the sealant.
32. Install the lower oil pan. Tighten the lower oil pan bolts in sequence to 88 inch lbs. (10 Nm).
33. Install rear plate cover.
34. Install the remaining components is in the reverse order of removal.
35. Wait at least 30 minutes before refilling the engine with oil.
36. Start the engine and check for leaks.
37. Inspect the engine oil level.

Oil Pump

REMOVAL & INSTALLATION

3.3L Engine

1. Before servicing the vehicle, refer to the precautions in the beginning of this section.
2. Drain the engine oil.
3. Drain the cooling system.
4. Remove or disconnect the following:
   - Negative battery cable
   - Oil pan
5. After removing the oil pan, reinstall the crossmember and the mount bolts.
   - Timing belt
   - Crankshaft sprocket and timing belt plate
   - Oil pump

To install:

6. Install or connect the following:
   - Oil pump-to-body bolts to 52–69 inch lbs. (6–8 Nm)
   - Timing belt plate and tighten the bolts to 52–69 inch lbs. (6–8 Nm)
   - Crankshaft sprocket
   - Timing belt
   - Oil pan
   - Negative battery cable

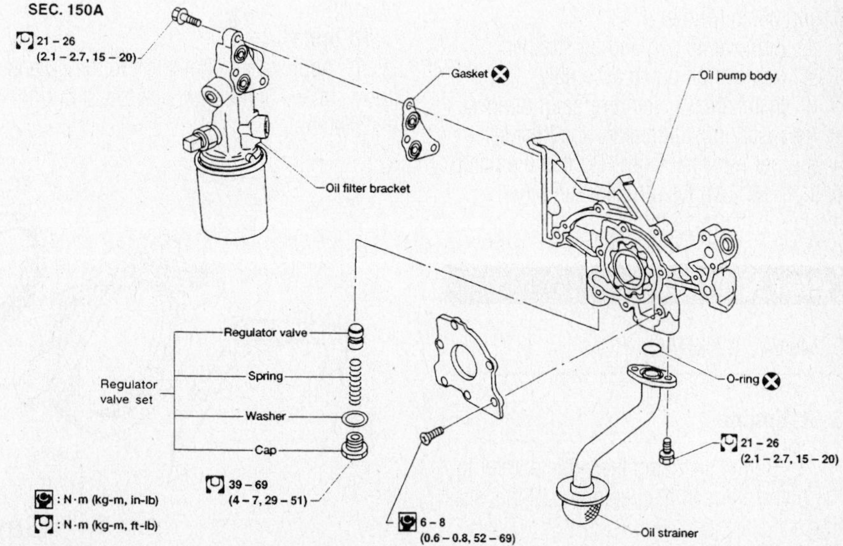

Exploded view of the oil pump assembly—3.3L engine

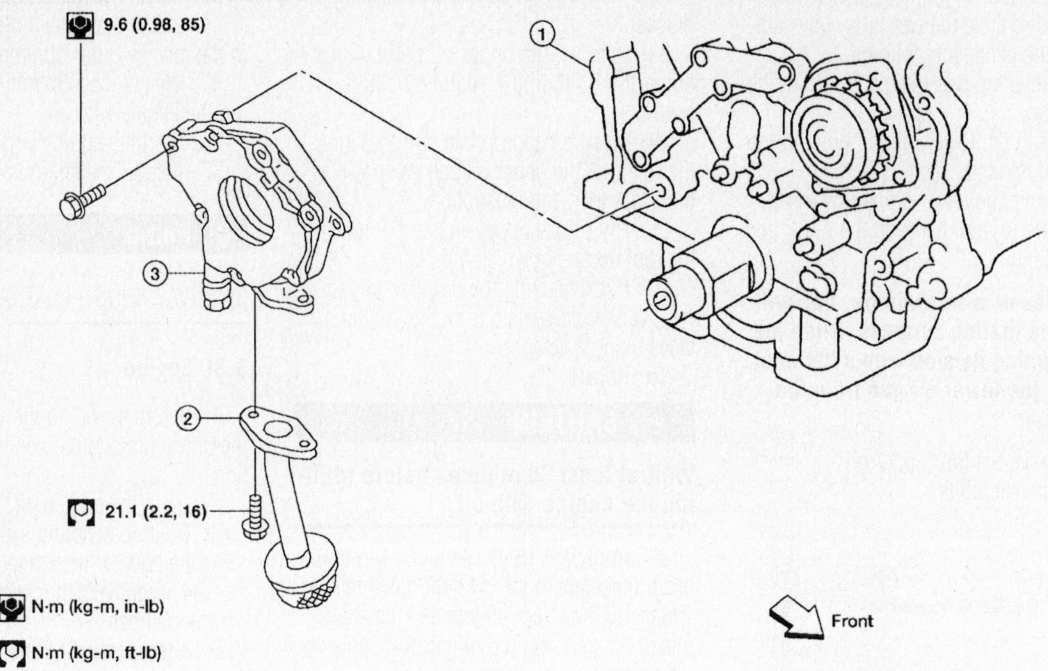

Oil pump exploded view—3.5L engine

7. Fill the cooling system.
8. Fill the crankcase to the correct level.
9. Start the engine and check for leaks.

3.5L Engine

1. Before servicing the vehicle, refer to the precautions in the beginning of this section.
2. If equipped with a 5 speed transmission, remove the engine/transmission and front suspension member.
3. Remove front timing chain case and timing chain (primary).
4. Remove oil pan and oil strainer.
5. Remove oil pump assembly.
6. Installation is the reverse of removal. When installing, align crankshaft flat faces with inner rotor flat faces. Tighten the pump bolts to 85 inch lbs. (10 Nm) and the strainer bolts to 16 ft. lbs. (21 Nm).

Rear Main Seal

REMOVAL & INSTALLATION

3.3L Engine

1. Before servicing the vehicle, refer to the precautions in the beginning of this section.
2. Disconnect the negative battery cable.
3. Remove the transaxle from the vehicle.
4. Remove the flexplate from the crankshaft.
5. Remove the rear oil seal retainer.

WARNING

Do not scratch the seal bore of the oil seal retainer when removing the oil seal.

6. Remove the oil seal from the seal retainer.

To install:

7. Apply clean engine oil to the lip and outer surface of the new seal to aid during installation.
8. Install the seal in the retainer using a suitable seal driver.
9. Using a new gasket install the retainer on the engine. Tighten the bolts to 52–61 inch lbs. (6–7 Nm).
10. Install the flexplate. Tighten the bolts to 61–69 ft. lbs. (83–93 Nm).
11. Install the transaxle and remaining components.

3.5L Engine

1. Before servicing the vehicle, refer to the precautions in the beginning of this section.
2. Remove the drive plate.
3. Remove the engine from the vehicle.
4. Remove the upper oil pan.

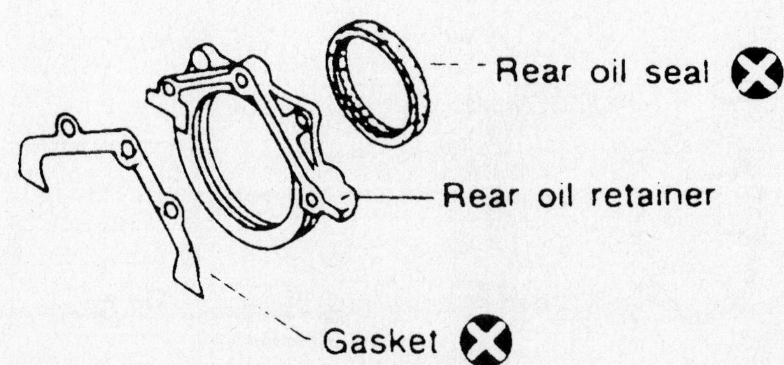

Exploded view of the oil seal, retainer and gasket—3.3L engine

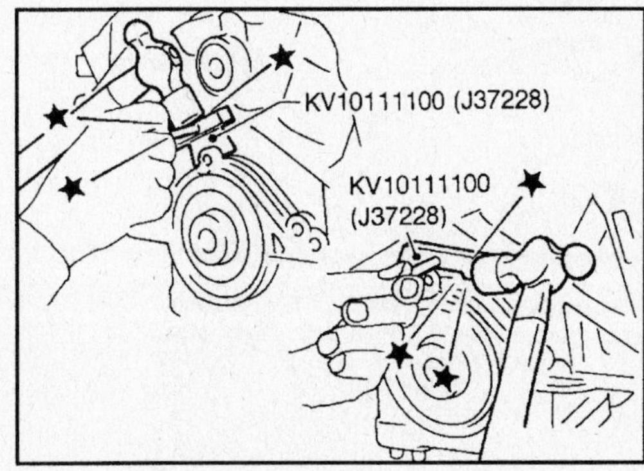

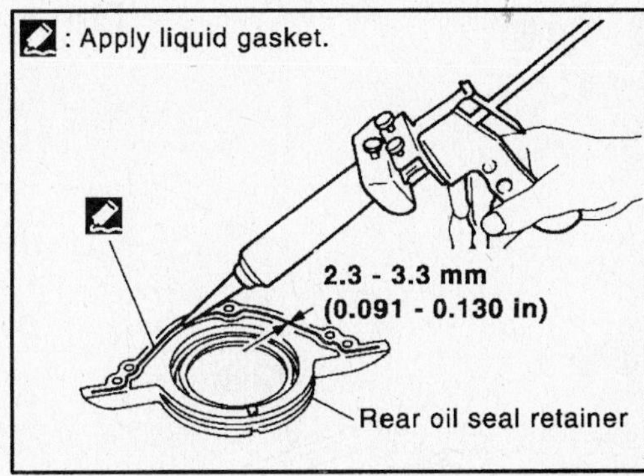

Applying sealer to the rear main seal

5. Use a seal cutter (special service tool) to cut away the liquid gasket and remove the rear oil seal retainer.

To install:

6. Remove the old liquid gasket from the mating surface of the cylinder block and the oil pan ausing scraper.
7. Apply liquid gasket to the rear oil seal retainer as shown in the illustration. Use RTV silicone sealant. Assembly should be done within 5 minutes after coating.
8. Install the rear oil seal retainer to the cylinder block.
9. Perform the remaining steps in the reverse order of removal.

Front Timing Cover

REMOVAL & INSTALLATION

3.5L Engine

1. Before servicing the vehicle, refer to the precautions in the beginning of this section.

➡This section describes procedures for removal/installation procedure of the front timing chain case and timing chain related parts without removing the upper oil pan from the vehicle. If the upper oil needs to be removed or installed, or when rear timing chain case is removed or installed, remove both oil pans first. Then remove front timing chain case and timing chain related parts.

2. Disconnect the negative battery cable.
3. Drain the engine cooling system and engine oil.
4. Remove the engine cover.
5. Remove the upper air cleaner case, Mass Air Flow (MAF) sensor and air cleaner–to–electric throttle control actuator tube.
6. Remove the engine coolant reservoir.
7. Remove the cowl top and cowl top extension.
8. Remove the IPDM E/R and position aside. Remove the bracket.
9. Remove the front right hand wheel.
10. Remove the engine undercover.
11. Remove the right hand inner splash shield.
12. Remove the drive belts and idler pulley.
13. Recover the A/C refrigerant and remove the A/C compressor.
14. Removev engine oil cooler pipe bolts.
15. Remove the power steering oil pump and reservoir tank with lines attached and position them aside.
16. Remove the lower oil pan.
17. Remove the alternator.
18. Disconnect the engine harness and position aside.
19. Remove the A/C low-pressure flexible hose.
20. Support the engine using a suitable support device and remove the right hand engine mount insulator, mount and bracket.
21. Remove the chain tensioner cover and water pump cover.
22. Remove the left and right IVT control covers. Loosen the IVT control cover bolts in the order illustrated.

➡The shaft in the cover is inserted into the center hole of the intake camshaft sprocket. Remove the cover by pulling straight out until the cover disengages from the camshaft sprocket.

23. Remove the starter.
24. Remove the intake manifold collector.
25. Remove the ignition coils. Make sure to note the locations for installation purposes.
26. Remove the spark plugs.
27. Remove the engine oil dipstick.
28. If removing the secondary timing chains, remove the rocker covers as follows:
    a. Remove the engine cover.
    b. Remove side engine covers.
    c. If removing right hand rocker cover, disconnect the Mass Air Flow (MAF) sensor electrical connector and remove the air cleaner–to–electric throttle control actuator tube and air cleaner lid.
    d. If removing right hand rocker cover, remove the following:
    e. Front cowl panel.
    f. Windshield wiper arms and motor assembly.
    g. Intake manifold collector.
    h. If removing left hand rocker cover, disconnect the Air Fuel Ratio (AFR) sensor.
    i. Remove the ignition coils.
    j. Position engine harness aside.
    k. Disconnect Positive Crankcase Ventilation (PCV) hose.
    l. Remove the dipstick.
    m. Remove the rocker cover bolts in the sequence illustrated.
29. Remove the IVT control solenoid valves and discard the gaskets.

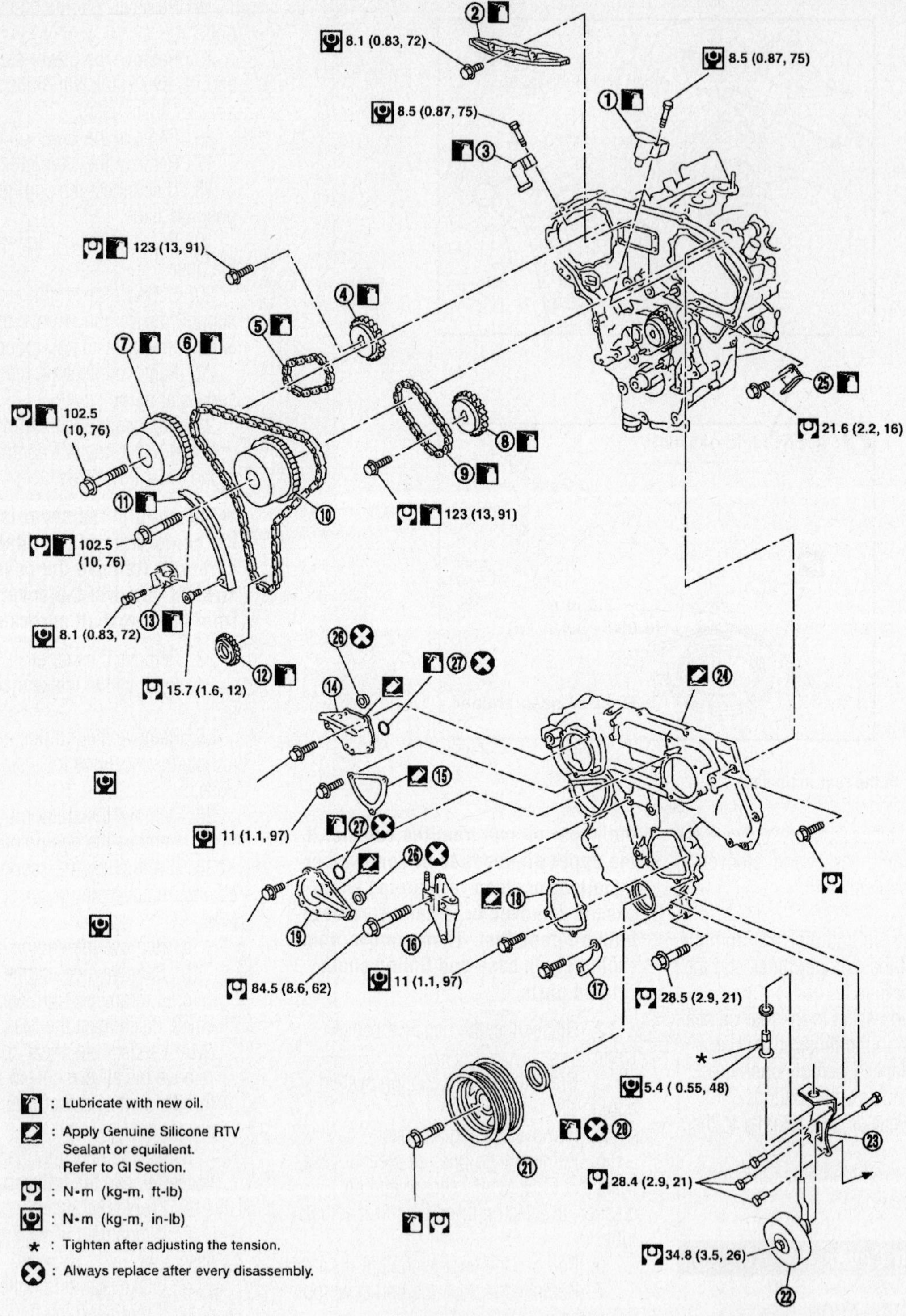

1. Timing chain tensioner
2. Internal chain guide
3. Timing chain tensioner
4. Camshaft sprocket (EXH)
5. Timing chain (secondary)
6. Timing chain (primary)
7. Camshaft sprocket (INT)
8. Camshaft sprocket (EXH)
9. Timing chain (secondary)
10. Camshaft sprocket (INT)
11. Slack guide
12. Crankshaft sprocket
13. Timing chain tensioner
14. IVT control valve cover - right
15. Chain tensioner cover
16. RH engine mounting bracket
17. Water hose clamp
18. Water pump cover
19. IVT control valve cover - left
20. Front oil seal
21. Crankshaft pulley
22. Idler pulley
23. Idler pulley bracket
24. Front timing chain case
25. Timing tension guide
26. Collared O-ring
27. Seal ring

67170-NIQU-G48

Exploded view of the timing chain, cover and related components–3.5L engine

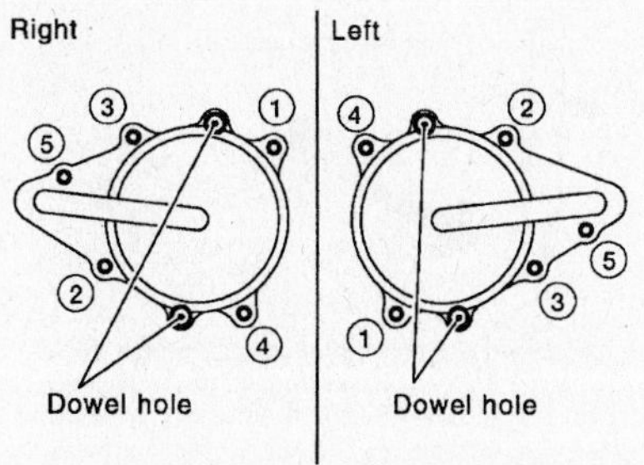

Loosen the IVT control cover bolts in this sequence–3.5L engine

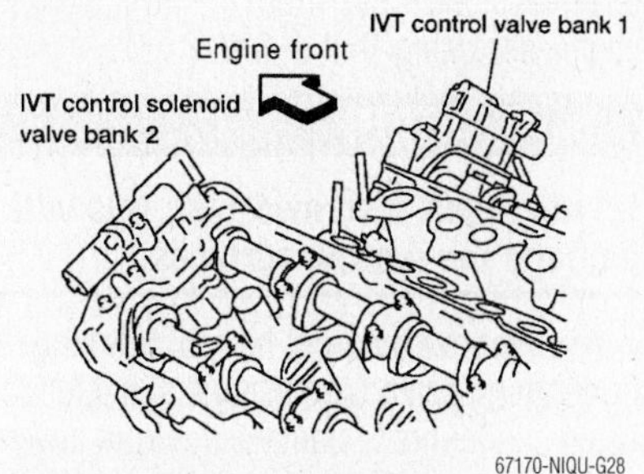

Remove the IVT control solenoid valves–3.5L engine

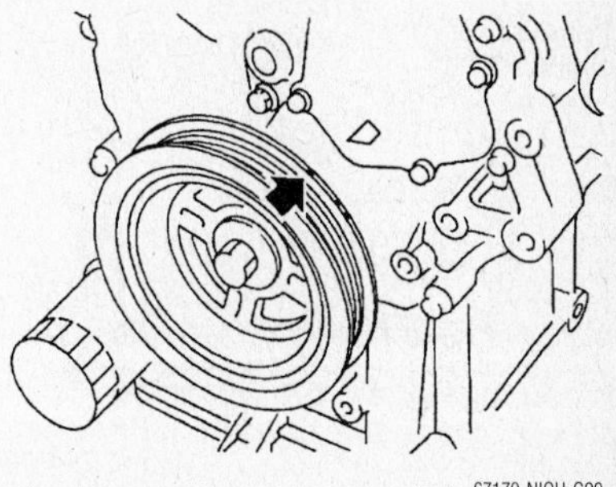

Position the engine at Top Dead Center (TDC) of No. 1 cylinder–3.5L engine

30. Position the engine at Top Dead Center (TDC) of No. 1 cylinder as follows:

a. Rotate crankshaft pulley clockwise to align timing mark (grooved line without color) with timing indicator.

b. Check that intake and exhaust camshaft lobes on No. 1 cylinder on the right bank of engine are located as illustrated. If not, turn the crankshaft one revolution (360 degrees) and align as illustrated.

31. Lock the ring gear using tool J–44716 attached to the starter bolt hole.

➡Do not damage the ring gear teeth, or the signal plate teeth behind the ring gear, when setting the tool J–44716.

32. Remove the crankshaft pulley as follows:

a. Loosen crankshaft pulley bolt using tool KV10109300 and locate bolt seating surface at 0.39 inch (10 mm) from its original position. surface at 0.39 inch (10mm) from its original position.

33. Position a pulley puller at recess

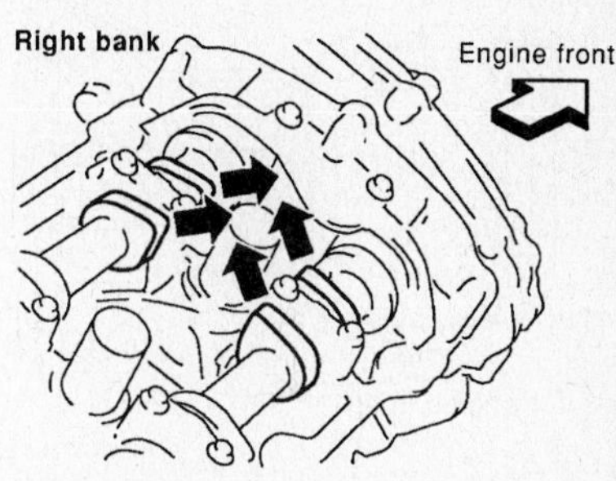

Check that intake and exhaust camshaft lobes on No. 1 cylinder on the right bank of engine are located as shown–3.5L engine

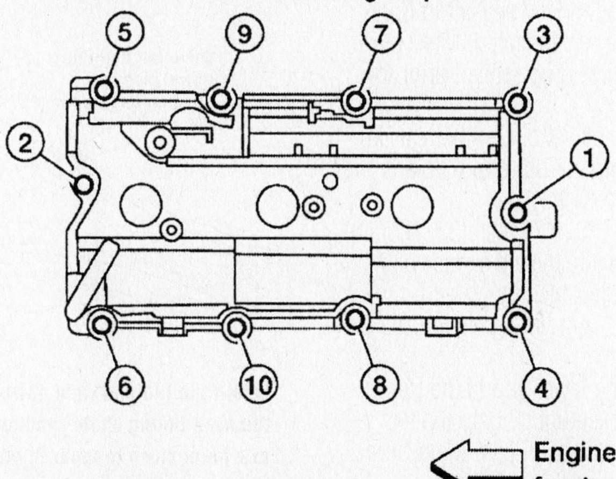

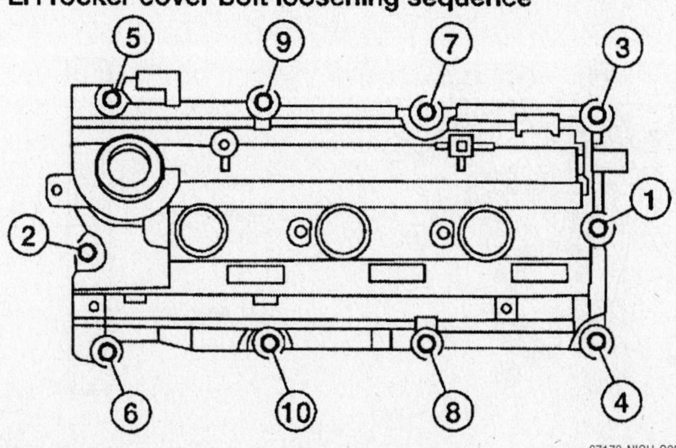

Loosen the rocker cover bolts in this sequence–3.5L engine

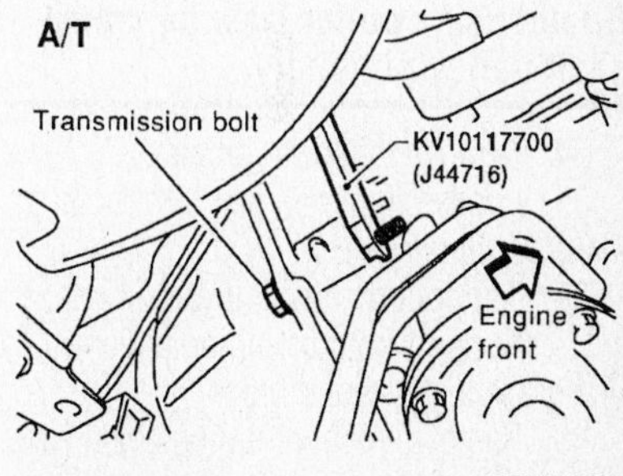

Lock the ring gear using tool J–44716 attached to the starter bolt hole–3.5L engine

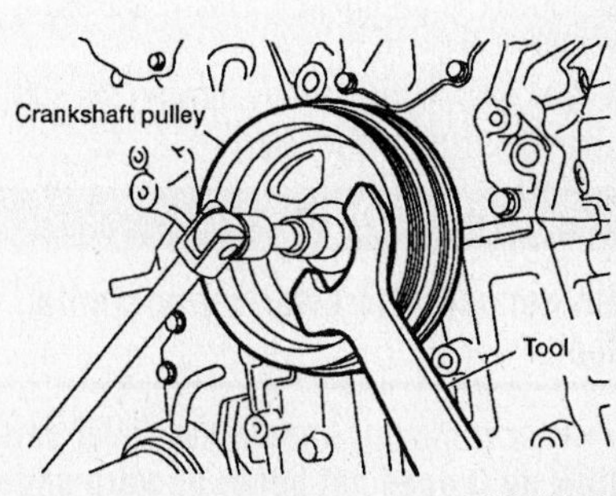

Loosen crankshaft pulley bolt using tool KV10109300 and locate bolt seating

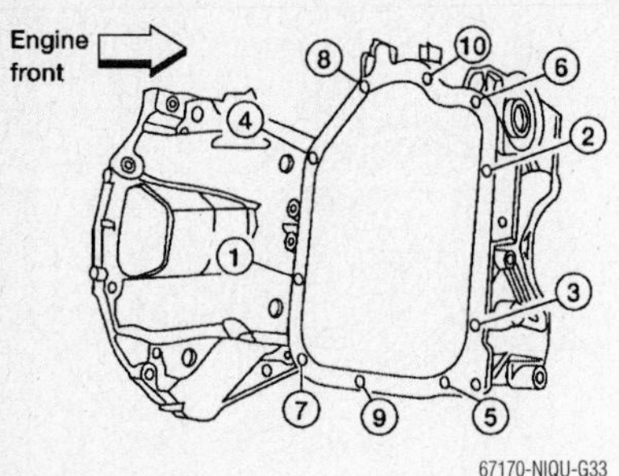

Loosen lower oil pan front bolts in this order–3.5L engine

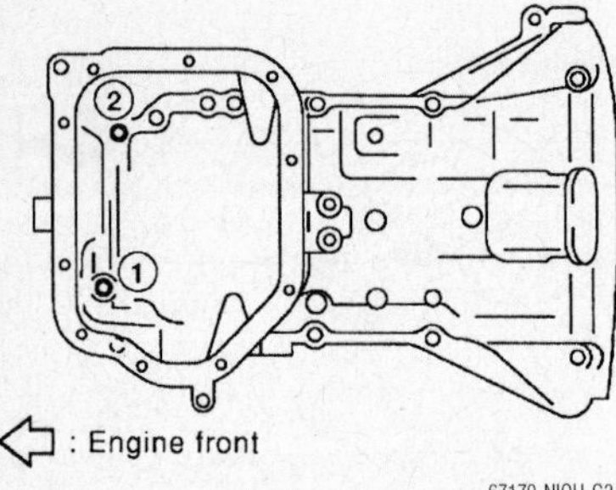

Loosen upper oil pan front bolts in this order–3.5L engine

hole of crankshaft pulley to remove crankshaft pulley.

⁂ CAUTION

Do not use a puller claw on crankshaft pulley periphery.

34. Remove the lower oil pan.
35. Loosen upper oil pan front bolts in the order illustrated.
36. Temporarily install lower oil pan.
37. Support front of engine under oil pan using a jack and wood block.
38. Remove the front timing chain case as follows:
    a. Loosen the front timing chain case bolts in the order illustrated.
    b. Insert the appropriate size tool such as J–37228 into the notch at the top of the front timing chain case as illustrated.
39. Pry off the case by moving the suitable tool back and forward.

⁂ CAUTION

Do not use a screwdriver or similar tool.

➡After removal, handle the cover carefully so it does not bend, or warp under a load.

40. Remove the water pump cover and chain tensioner cover from the front timing chain case using Tool J–37228.

⁂ CAUTION

Do not insert a screwdriver, this will damage the mating surfaces.

41. Remove the front oil seal from the front timing chain case using a suitable tool being careful not to damage the front cover.
42. If necessary, remove timing chain and related parts.
43. Use a scraper to remove all of the old sealant from the front timing chain mating surfaces. Be careful not to damage the mating surfaces while cleaning.

To install:

44. Install the timing chain and related parts, if removed.
45. Install the left and right dowel pins into the front timing chain case up to a point close to taper in order to shorten protrusion length.
46. Install the front oil seal on the front timing chain case. Apply new engine oil to the oil seal edges. Install it so that each seal lip is oriented as illustrated.
47. Make sure the garter spring is in position and seal lip is not inverted.
48. Apply a 0.102–0.142 inch (2.6–3.6mm) bead of RTV sealant to front timing chain case as illustrated. Make sure to wipe off the protruding sealant prior to installation.
49. Install the dowel pin on the rear timing chain case into the dowel pin hole in the front timing chain case.
50. Apply a 0.138–0.77 inch (3.5–4.5mm) bead of RTV sealant to the top surface of the upper oil pan as illustrated.
51. Install the front timing chain case as follows:
    a. Install the lower end of the front timing chain case tightly onto the top surface of the upper oil pan. Make sure that the oil pan gasket is in place.

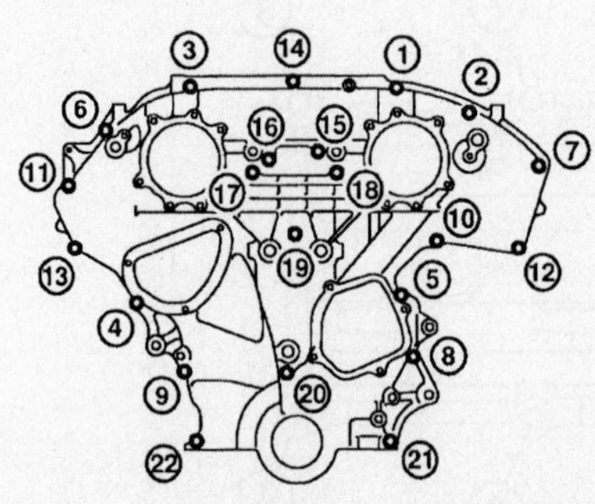

Loosen the front timing chain case bolts in this order–3.5L engine

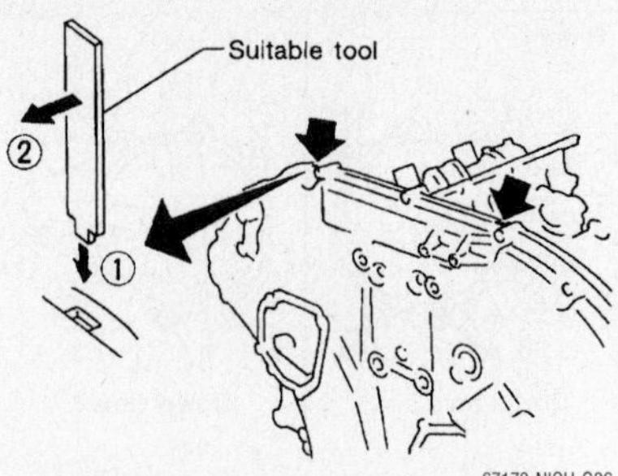

Insert the appropriate size tool such as J–37228 into the notch at the top of the front timing chain case–3.5L engine

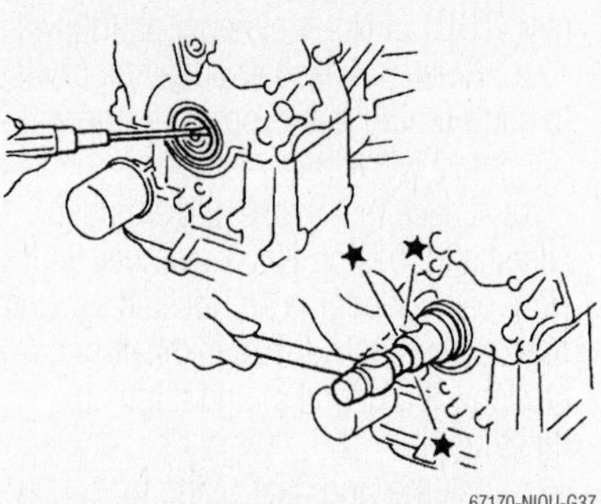

Remove the front oil seal from the front timing chain case using a suitable tool–3.5L engine

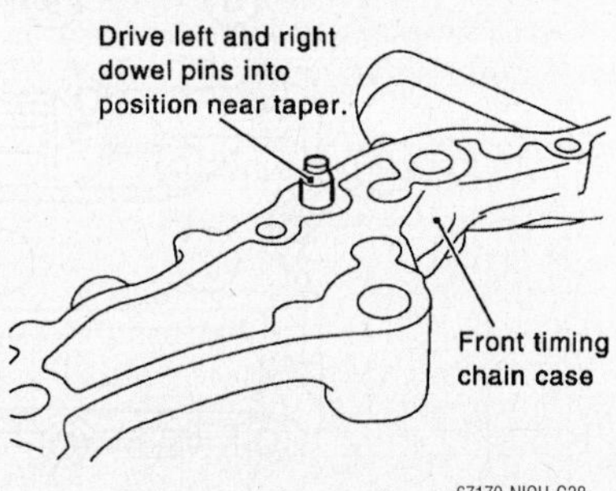

Install the left and right dowel pins into the front timing chain case up to a point close to taper in order to shorten protrusion length–3.5L engine

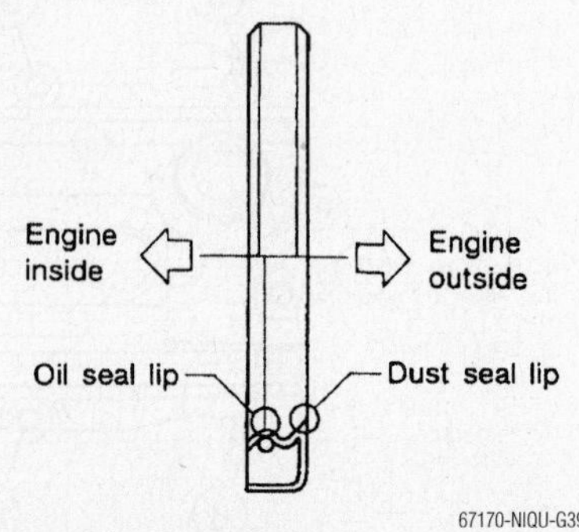

Install the front oil seal on the front timing chain case–3.5L engine

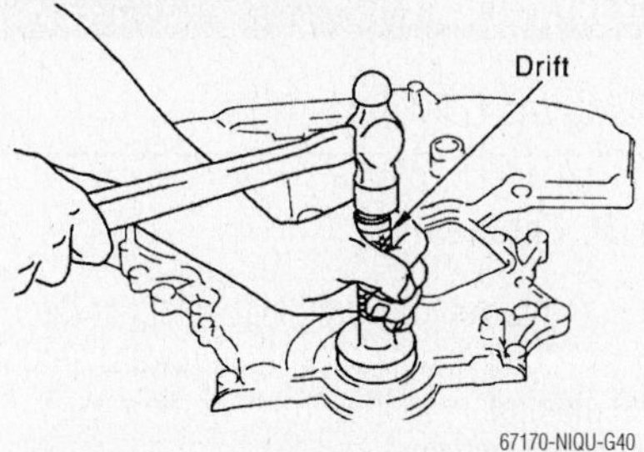

Make sure the garter spring is in position and seal lip is not inverted when installing the front oil seal–3.5L engine

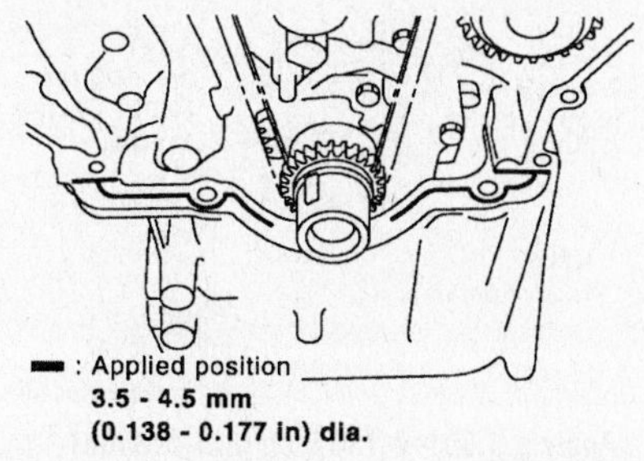

Apply a 0.138–0.77 inch (3.5–4.5mm) bead of RTV sealant to top surface of the upper oil pan–3.5L engine

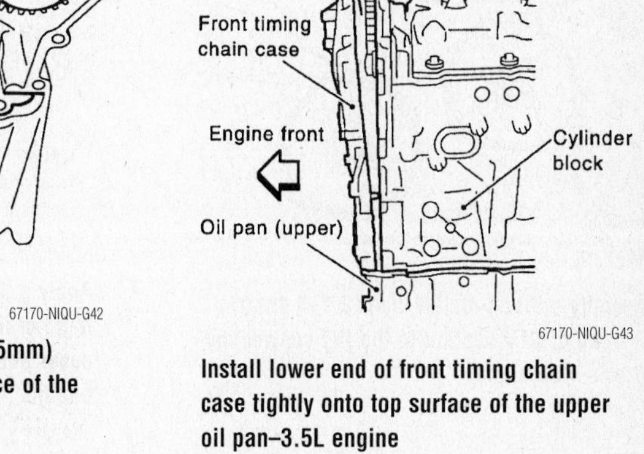

Install lower end of front timing chain case tightly onto top surface of the upper oil pan–3.5L engine

b. While pressing the front timing chain case from its front and top as illustrated, hammer the dowel pin until the outer end becomes flush with the surface.

52. Loosely install the front timing chain case bolts, refer to the illustration for bolt location.

53. Tighten the front timing chain case bolts in the order illustrated. Bolt positions in the illustration and torques are as follows:

a. Bolt location 1 and 2 with a diameter of 0.31 inch (8 mm) are torqued to 21 ft. lbs. (28 Nm).

b. Bolt locations for bolts 3 through 22 with a diameter of 0.24 inch (6 mm) are torqued to 9 ft. lbs. (13 Nm).

54. Install the upper oil pan front bolts in the order illustrated and tighten to 13 ft. lbs. (17 Nm).

55. Install the IVT control valve covers as follows:

a. Install new collared O-rings in front cover oil hole on both sides.

b. Install new seal rings on the IVT control covers.

c. Apply a 0.083–0.122 inch (2.1–3.1mm) bead of RTV sealant to the IVT control covers as illustrated.

56. Being careful not to move the seal ring from the installation groove, align the dowel pins on the chain case with the holes to install the IVT control covers.

57. Tighten the intake valve timing control cover bolts in the order illustrated to 100 inch lbs. (11 Nm).

58. Apply a 0.091–0.130 inch (2.3–3.3mm) bead of liquid gasket to the water pump cover and the chain tensioner cover and install covers. Tighten the bolts on the pump and cover to 97 inch lbs. (11 Nm).

59. Install the crankshaft pulley, lubricate the pulley bolt with clean engine oil and tighten the bolt in two steps:

a. Step 1: 32 ft. lbs. (44 Nm).

b. Step 2: 60–65 degrees clockwise.

60. Rotate crankshaft pulley in a clockwise direction as viewed from front of the engine to confirm it turns smoothly.

61. Install the rocker covers in the reverse order of removal, make sure to perform the following:

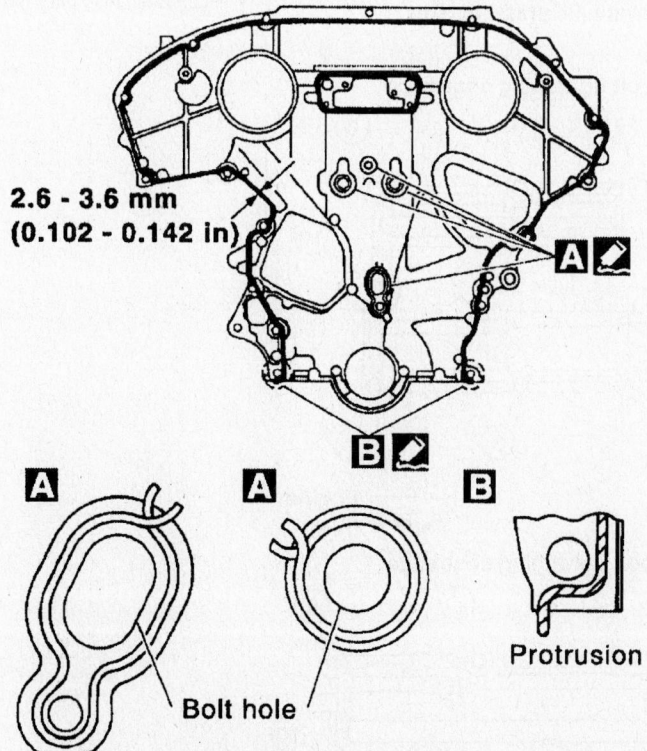

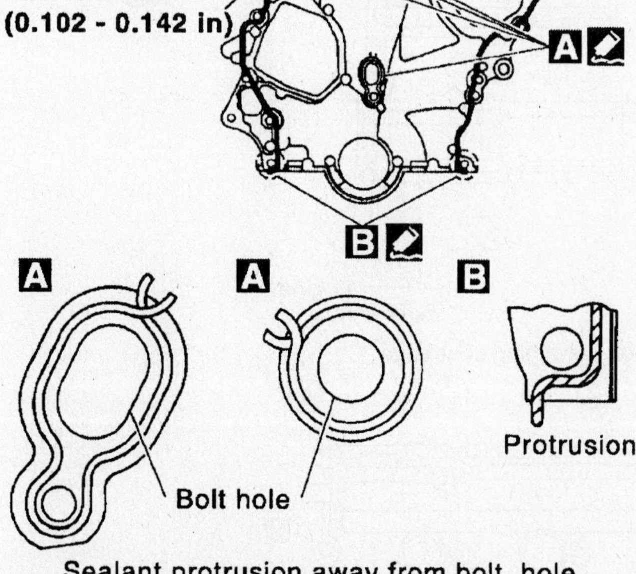

Apply a 0.102–0.142 inch (2.6–3.6mm) bead of RTV sealant to front timing chain case–3.5L engine

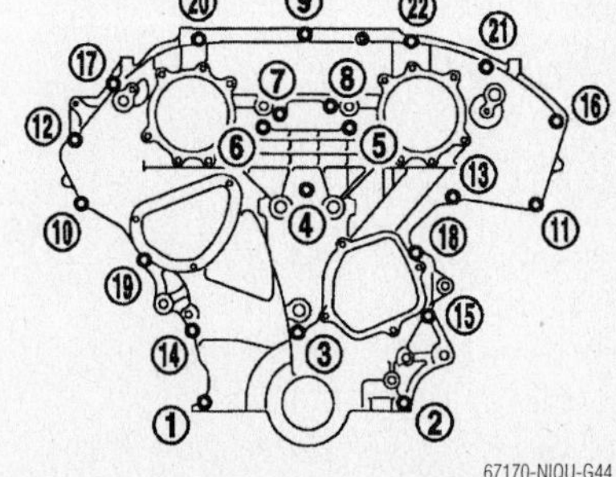

Tighten the front timing chain case bolts in the order shown–3.5L engine

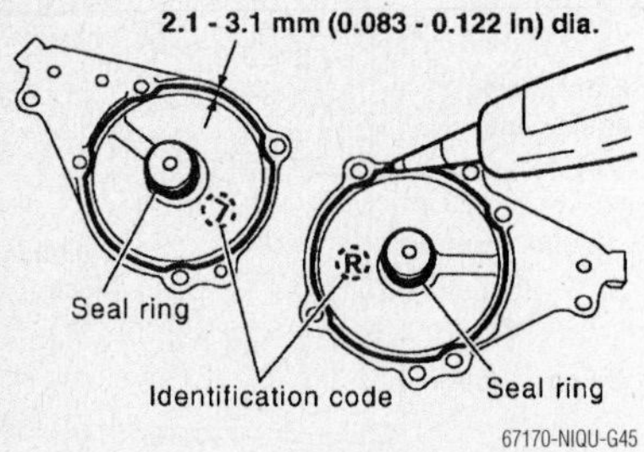

Apply a 0.083–0.122 inch (2.1–3.1mm) bead of RTV sealant to the IVT control covers–3.5L engine

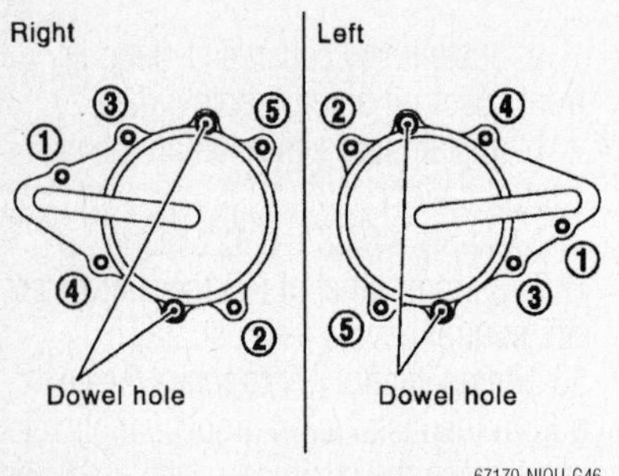

Tighten the intake valve timing control cover bolts in this order–3.5L engine

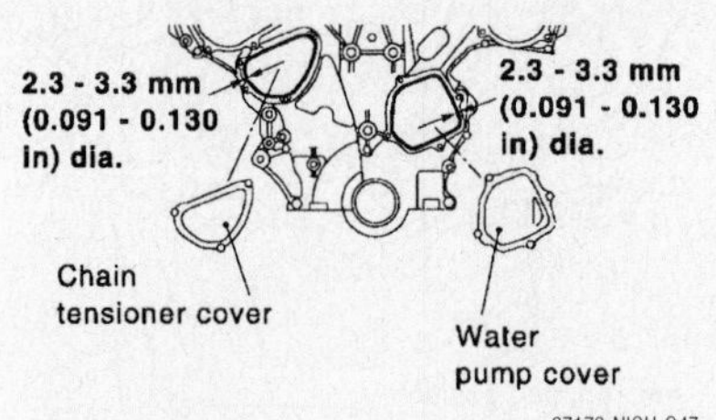

Apply a 0.091–0.130 inch (2.3–3.3mm) bead of liquid gasket to the water pump cover and the chain tensioner cover–3.5L engine

62. Apply sealant to the areas on the front corners.
63. Tighten the rocker cover bolts in two steps in the sequence illustrated:
    a. Step 1: 17 inch lbs. (1.96 Nm).
    b. Step 2: 74 inch lbs. (8 Nm).
64. Install the remaining components is in reverse order of removal.

➡If hydraulic pressure inside chain tensioner drops after removal/installation, slack in the guide may generate a pounding noise during and just after engine start. This is normal. Noise will stop after hydraulic pressure rises.

Timing Belt

REMOVAL & INSTALLATION

3.3L Engines

On this vehicle, right side refers to the "rear" components (near the firewall) and left side refers to the "front" components (near the radiator).

1. If the timing belt is to be removed, it is good practice to turn the crankshaft until the engine is at Top Dead Center (TDC) of the No. 1 cylinder, compression stroke (firing position), before beginning work. This should align all timing marks and serve as a reference for all work that follows. After verifying that the engine is at TDC for the No. 1 cylinder, do not crank the engine or allow the crankshaft or camshaft sprockets to be turned otherwise engine timing will be lost.
2. Before servicing the vehicle, refer to the precautions in the beginning of this section.
3. Drain the cooling system.
4. Remove or disconnect the following:
    - Negative battery cable
    - Alternator drive belt, water pump

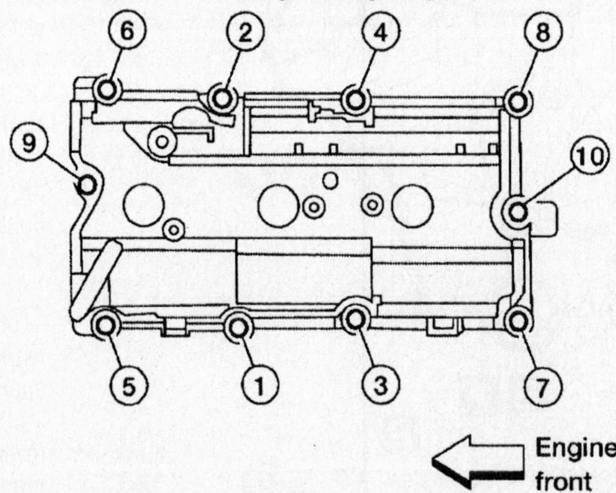

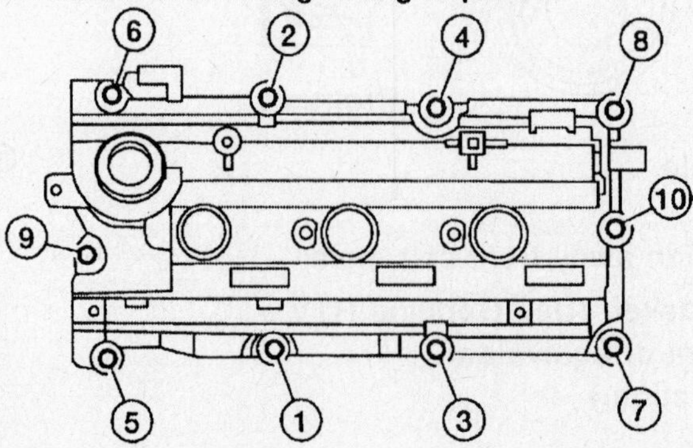

Tighten the rocker cover bolts in this sequence–3.5L engine

and power steering pump belt and the air conditioning compressor belt, if equipped
- 3 air conditioning compressor drive belt idler pulley bolts and the idler pulley, if equipped with air conditioning
- Upper radiator hose bracket bolt
- Upper hose with the bracket from the vehicle
- Water bypass hose from between the thermostat housing and the lower water hose connection
- Main wiring harness from the upper engine front cover
- 8 upper engine front cover bolts and the upper cover
- Right side front wheel and tire assembly
- 4 right side engine and transmission splash shield bolts and 2 screws, and right side outer engine and transaxle splash shield

5. Use a strap wrench to hold the water pump pulley. Remove the 4 pulley bolts, and the water pump pulley.

6. Use a strap wrench to hold the crankshaft pulley. Remove the center pulley bolt, and the crankshaft pulley using a harmonic balancer (damper) puller to draw the pulley from the front of the crankshaft.
- 5 lower engine front cover bolts, then remove the lower engine front cover

7. Be sure that the timing marks between the crankshaft sprocket and the oil pump housing align.

8. If the timing belt is to be reused, mark an arrow on the belt indicating the direction of rotation. The directional arrow is necessary to ensure that the timing belt, if it to be reused, can reinstalled in the same direction.

9. Loosen the timing belt tensioner nut and slip the timing belt off of the sprockets.

10. If necessary, the camshaft sprockets can be removed. A special spanner tool is designed to hold the sprocket to keep it from turning while the center bolt is being loosened. Use care if using substitutes.

➡The sprockets are not interchangeable.

11. If necessary, the crankshaft sprocket can be removed. The outer timing belt guide (looks like a large washer) and the crankshaft sprocket simply pull off the front of the crankshaft.

➡Be careful, there are 2 crankshaft keys. Use care not to loose them.

To install:

12. Clean all parts well. If removed, inspect the crankshaft sprocket for warping or abnormal wear. Check the sprocket teeth for wear, deformation, chipping or other damage. Replace as necessary. Clean the sprocket mounting surface to ease installation. Install the key. Slip the sprocket onto the crankshaft. Tap it in place with a suitably-sized socket.

13. If removed, inspect the camshaft sprockets for damage and wear. Replace as required. The sprockets should be marked **L3** to designate the front, or left side camshaft and **R3** to designate the rear, or right side camshaft. Use care to install the sprockets properly. A special spanner tool is designed to hold the sprocket to keep it from turning while the center bolt is being tightened. Use care if using a substitute. Tighten the camshaft sprocket center bolts to 61 ft. lbs. (83 Nm). Verify that the timing marks on the camshaft sprockets and the timing marks on the rear cover (called the seal plate) are aligned.

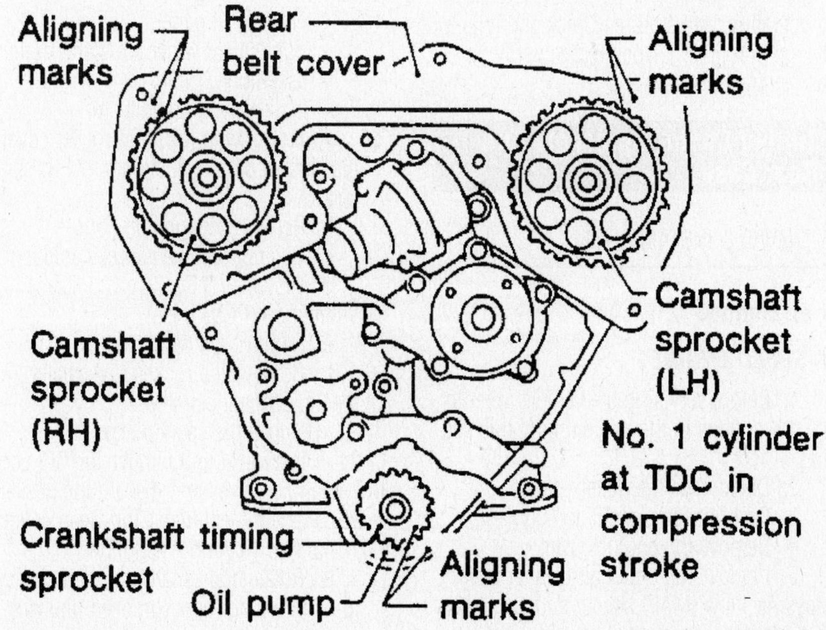

Use a shop rag to clean the alignment marks for the timing belt— 3.3L engines

14. Use an Allen wrench to turn the timing belt tensioner clockwise until the belt tensioner spring is fully extended. Temporarily tighten the tensioner nut to 32–43 ft. lbs. (43–58 Nm).

15. If a new timing belt is to be installed, look for a printed arrow on the belt. Be sure the arrow is pointing away from the engine. If the original timing belt is to be reused, be sure that the directional arrow that was marked at disassembly is facing the correct direction.

16. A new Original Equipment Manufacture (OEM) timing belt should have 3 white timing marks on it that indicate the correct timing positions of the camshafts and the crankshaft. These marks are to help ensure that the engine is properly timed. When the engine is properly timed, each white timing mark on the timing belt will be aligned with the corresponding camshaft and crankshaft timing mark on the sprocket. Because the white timing marks are not evenly spaced, the technician needs to use care in installing the belt. There should be 40 timing belt teeth between the timing marks on the front and rear camshaft sprockets and 43 teeth between the timing mark on the front camshaft sprocket and the timing mark on the crankshaft sprocket.

17. Verify that the camshaft timing marks are aligned with the timing marks on the rear cover (seal plate) and that the crankshaft sprocket timing mark is aligned with the timing mark on the oil pump housing.

18. Install the timing belt starting at the crankshaft sprocket and moving around the camshaft sprockets following a counterclockwise path. Do not allow any slack in the timing belt between the sprockets. After all of the timing marks are aligned with the timing belt installed, slip the timing belt onto the belt tensioner.

19. While holding the timing belt tensioner with an Allen wrench, loosen the tensioner nut. Allow the tensioner to put pressure on the timing belt. Use an Allen wrench to turn the timing belt tensioner 70–80 degrees clockwise and tighten the timing belt tensioner nut to 32–43 ft. lbs. (43–58 Nm).

✲✲ WARNING

If any binding is felt when adjusting the timing belt tension by turning the crankshaft, STOP turning the engine, because the pistons may be hitting the valves.

20. Rotate the crankshaft clockwise twice and align the No. 1 piston to TDC on the compression stroke (firing position).
21. Apply 22 lbs. (10kg) of force on the timing belt between the rear camshaft sprocket and the timing belt tensioner. An assistant may be needed. While holding the timing belt tensioner steady with an Allen wrench, loosen the timing belt tensioner nut. Remove the Allen wrench and adjust the timing belt tensioner using the following procedure:
    a. Install a 0.0138 in. (0.35mm) thick and 0.500 in. (12.7mm) wide feeler gauge where the timing belt just starts to go around the tensioner (approximately the 4 o'clock position, looking at the tensioner).
    b. Turn the crankshaft sprocket clockwise, which should force the feeler gauge between the timing belt and the tensioner, up to a position on the tensioner of about 1 o'clock.
    c. Tighten the timing belt tensioner nut to 61 ft. lbs. (83 Nm).
    d. Turn the crankshaft clockwise to rotate the feeler gauge out from between the timing belt tensioner and the timing belt.
22. Rotate the crankshaft clockwise twice, and once again align the No. 1 piston to TDC on the compression stroke (firing position).
23. Apply 22 lbs. (10kg) of force on the timing belt between the front and rear camshaft sprockets. Measure the amount of belt deflection. Belt deflection should be between 0.51–0.59 in. (13–15mm). If belt deflection is out of specification, repeat Steps 29 through 33. If the timing belt deflection cannot be adjusted into specification, the timing belt will have to be replaced.
24. Install or connect the following:
    - Lower engine front cover and the 5 lower cover bolts. Do not over tighten. Tighten to 27–44 inch lbs. (3–5 Nm).
    - Outer timing belt guide next to the crankshaft sprocket with the dished side facing away from the cylinder block. Install the crankshaft pulley. Use a strap wrench to keep the crankshaft pulley from turning and tighten the center bolt to 148 ft. lbs. (201 Nm).
    - Water pump pulley on the pump. Install the 4 bolts. Use a strap wrench to keep the water pump pulley from turning and tighten the 4 water pump pulley bolts to 89 inch lbs. (10 Nm).
    - Right side outer engine and transaxle splash shield, and secure with the 4 bolts and 2 screws
    - Right side front wheel. Tighten the lug nuts to 72–87 ft. lbs. (98–118 Nm).
    - Upper engine timing belt front cover, and tighten the 8 bolts to 27–44 inch lbs. (3–5 Nm)
    - Main wiring harness on the upper engine front cover
    - Water bypass hose between the thermostat housing and water connection
    - Upper radiator hose between the radiator and the water hose connection. Secure the hoses with clamps. Install the upper radiator hose bracket. Tighten the bracket bolt to 34–58 ft. lbs. (46–65 Nm).
    - Air conditioning compressor drive belt idler pulley and install the 3 bolts. Tighten to 15 ft. lbs. (21 Nm), if equipped
    - Alternator drive belt, the water pump and power steering pump drive belt and the air conditioning compressor drive belt, if equipped.
    - Negative battery cable
25. Fill the cooling system.
26. Start the engine and allow it to warm to operating temperature. Check and adjust the ignition timing. Road test to verify correct engine operation.

Timing Chain, Sprockets And Rear Case

REMOVAL & INSTALLATION

3.5L Engines

4–SPEED ENGINE

1. Before servicing the vehicle, refer to the precautions in the beginning of this section.
2. Disconnect the battery cables.
3. Drain the engine oil and coolant.
4. Remove the engine cover.
5. Remove the upper air cleaner case, Mass Air Flow (MAF) sensor and air cleaner to electric throttle control actuator tube.
6. Remove the battery and tray.
7. Disconnect the heater pump and position aside.
8. Remove the cowl top and cowl top extension.
9. Disconnect the engine room harness from the Powertrain Control Module (PCM) and the two connections at the right hand strut tower.
10. Disconnect the engine harness ground connections.
11. Remove the radiator assembly, coolant reservoir, and all system hoses.
12. Remove the idler pulley and bracket.
13. Remove the upper and lower oil pans.
14. Remove the alternator.
15. Disconnect the engine harness and position it aside.
16. Support the engine using a suitable support device and remove the right hand engine mount insulator, mount and bracket.
17. Remove the timing chain tensioner cover.
18. Remove both the IVT control covers, remove the bolts in the sequence illustrated.

➡The shaft in the cover is inserted into the center hole of the intake camshaft sprocket. Remove the cover by pulling straight out until the cover disengages from the camshaft sprocket.

19. Remove the starter.
20. Remove the intake manifold collector.
21. Remove the rocker covers as follows:
    a. Remove the engine cover.
    b. Remove side engine covers.
22. If removing right hand rocker cover, remove the following:
    - Front cowl panel
    - Windshield wiper arms and motor assembly
    - Intake manifold collector.

    a. If removing left hand rocker cover, disconnect the Air Fuel Ratio (AFR) sensor.
    b. Remove the ignition coils.
    c. Position engine harness aside.
    d. Disconnect Positive Crankcase Ventilation (PCV) hose.
    e. Remove the dipstick.
    f. Remove the rocker cover bolts in the sequence illustrated.
23. Remove the spark plugs.
24. Disconnect and remove the IVT control solenoid valves and discard the gaskets
25. Place the engine at Top Dead center (TDC) of No. 1 cylinder as follows:
    a. Rotate crankshaft pulley clockwise to align timing mark (grooved line without color) with the timing indicator.
    b. Check that intake and exhaust

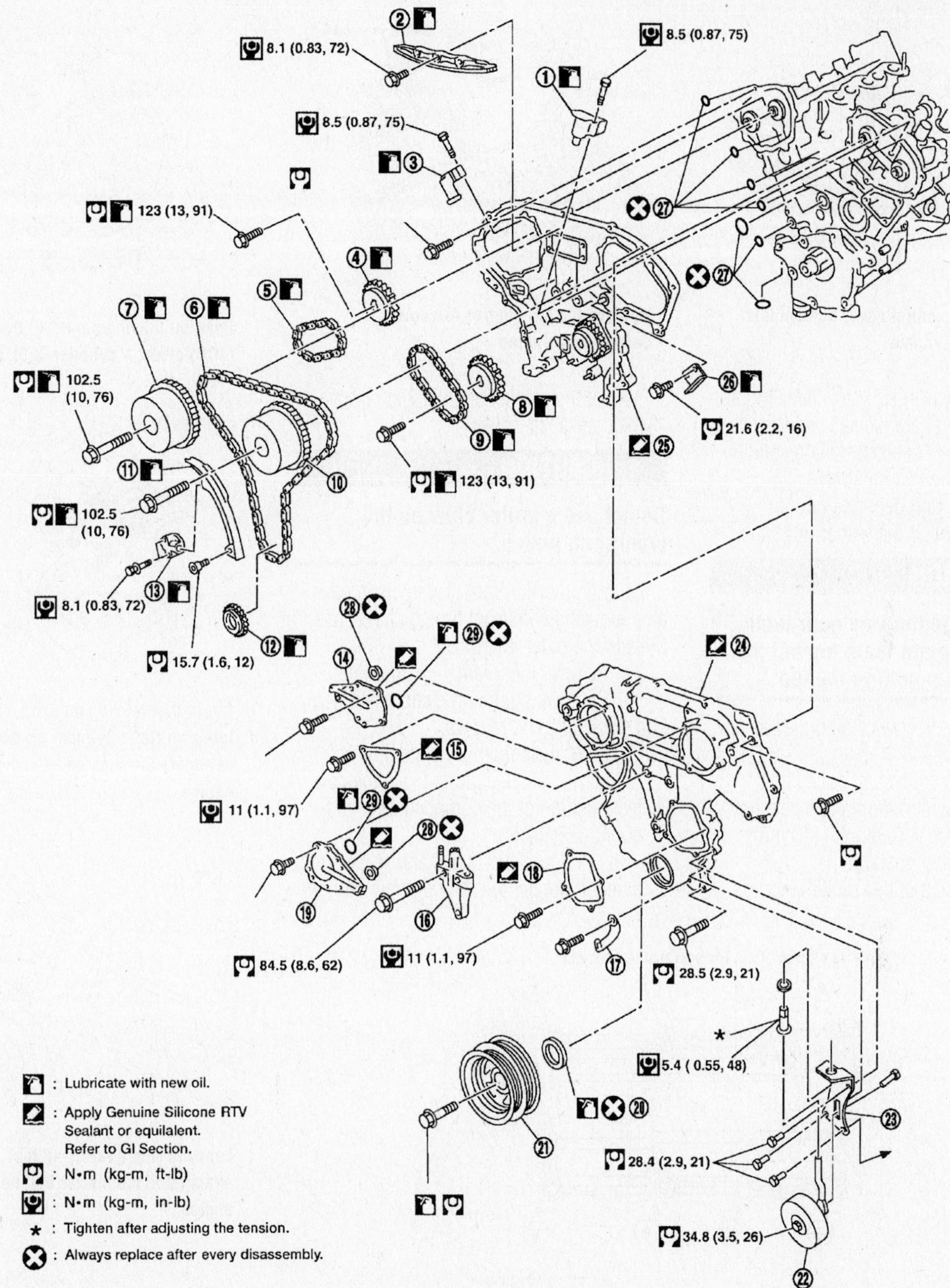

1. Timing chain tensioner
2. Internal chain guide
3. Timing chain tensioner
4. Camshaft sprocket (EXH)
5. Timing chain (secondary)
6. Timing chain (primary)
7. Camshaft sprocket (INT)
8. Camshaft sprocket (EXH)
9. Timing chain (secondary)
10. Camshaft sprocket (INT)
11. Slack guide
12. Crankshaft sprocket
13. Timing chain tensioner
14. IVT control valve cover - right
15. Chain tensioner cover
16. RH engine mounting bracket
17. Water hose clamp
18. Water pump cover
19. IVT control valve cover - left
20. Front oil seal
21. Crankshaft pulley
22. Idler pulley
23. Idler pulley bracket
24. Front timing chain case
25. Rear timing chain case
26. Timing chain tension guide
27. O-ring
28. Collared O-ring
29. Seal ring

67170-NIQU-G49

Exploded view of the timing chain and related components–3.5L engine

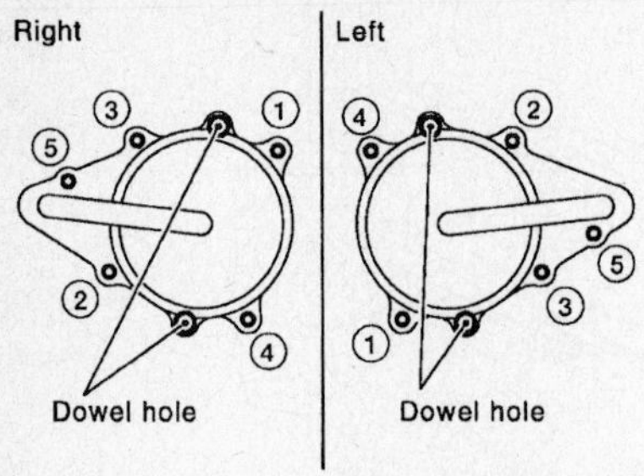

Loosen the IVT control cover bolts in this sequence–3.5L engine

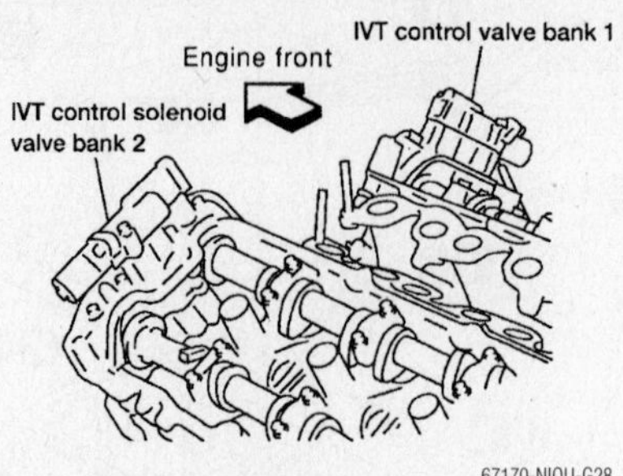

Remove the IVT control solenoid valves–3.5L engine

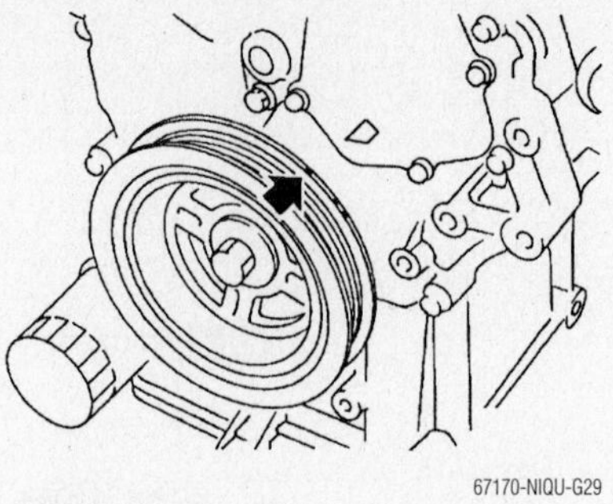

Position the engine at Top Dead Center (TDC) of No. 1 cylinder–3.5L engine

camshaft lobes on No. 1 cylinder (right bank of engine) are located as illustrated. If not, turn the crankshaft one revolution (360°) and align as illustrated.

26. Lock the ring gear using tool J–44716 attached to the starter bolt hole.

CAUTION

Do not damage the ring gear teeth, or the signal plate teeth behind the ring gear, when setting the tool.

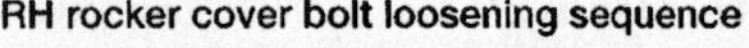

27. Remove the crankshaft pulley as follows:
    a. Loosen crankshaft pulley bolt using tool KV10109300 and locate bolt seating surface at 0.39 inch (10 mm) from its original position.
    b. Position a pulley puller at recess hole of the crankshaft pulley to remove the crankshaft pulley.

CAUTION

Do not use a puller claw on the crankshaft pulley.

28. Remove the front timing chain case after loosening the front timing chain case bolts in the order illustrated.
29. Remove the internal chain guide.
30. Remove the timing chain tensioner and slack guide as follows:
    a. Put matchmarks on the timing chain and sprockets to indicate the correct position of the components to aid during installation.
    b. Pull the lever down and release the plunger stopper tab. The plunger stopper

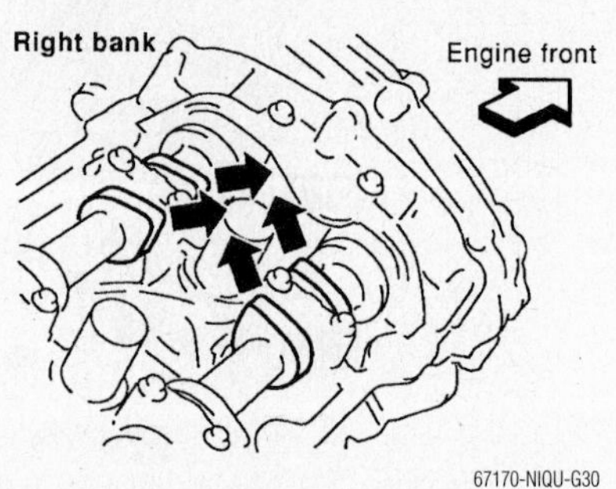

Check that intake and exhaust camshaft lobes on No. 1 cylinder on the right bank of engine are located as shown–3.5L engine

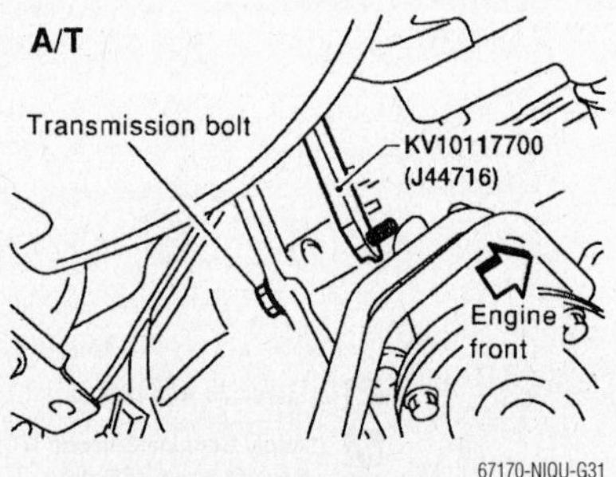

Lock the ring gear using tool J–44716 attached to the starter bolt hole–3.5L engine

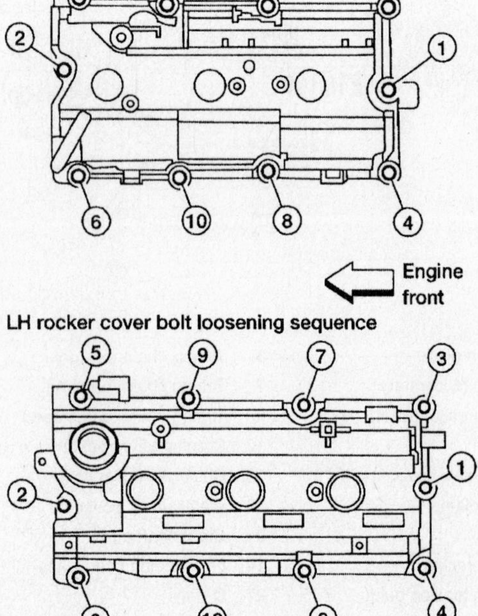

Loosen the rocker cover bolts in this sequence–3.5L engine

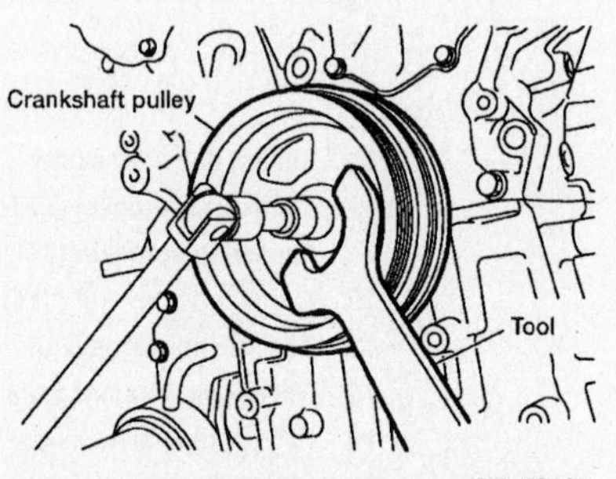

Loosen crankshaft pulley bolt using tool KV10109300 and locate bolt seating surface at 0.39 inch (10mm) from its original position–3.5L engine

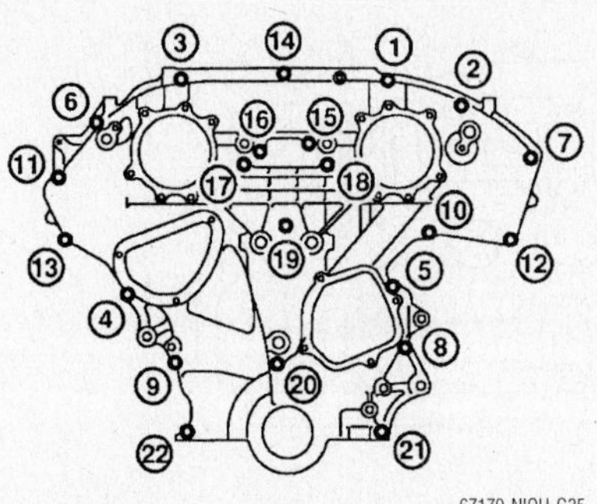

Loosen the front timing chain case bolts in this order–3.5L engine

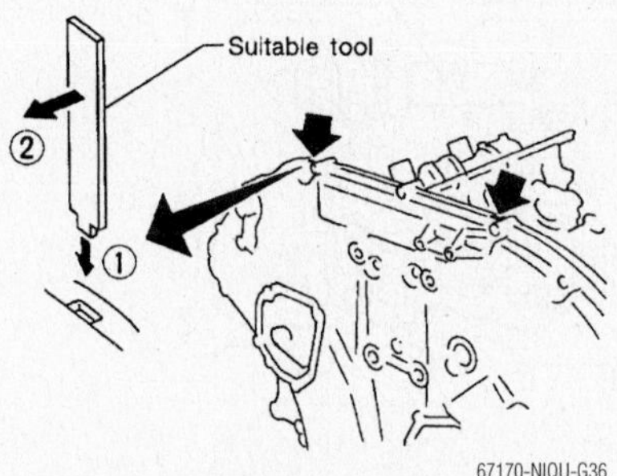

Insert the appropriate size tool such as J–37228 into the notch at the top of the front timing chain case–3.5L engine

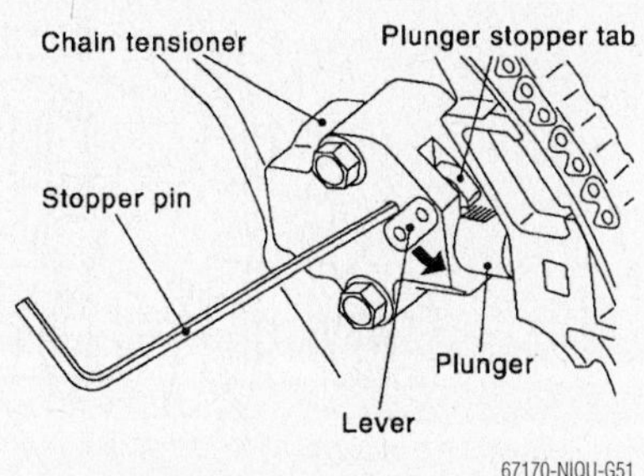

Insert a 0.098 inch (2.5mm) Allen wrench into the tensioner body hole to hold the chain tensioner lever–3.5L engine

tab can be pushed up to release (coaxial structure with lever).

c. Insert a 0.098 inch (2.5mm) Allen wrench into the tensioner body hole to hold the lever, and keep the tab released.

d. Insert the plunger into tensioner body by pressing the slack side chain guide.

e. Keep the slack side chain guide pressed and hold it by pushing a stopper pin such as a 0.098 inch (2.5mm) Allen wrench through the lever hole and body hole.

f. Remove the timing chain tensioner bolts and the timing chain tensioner.

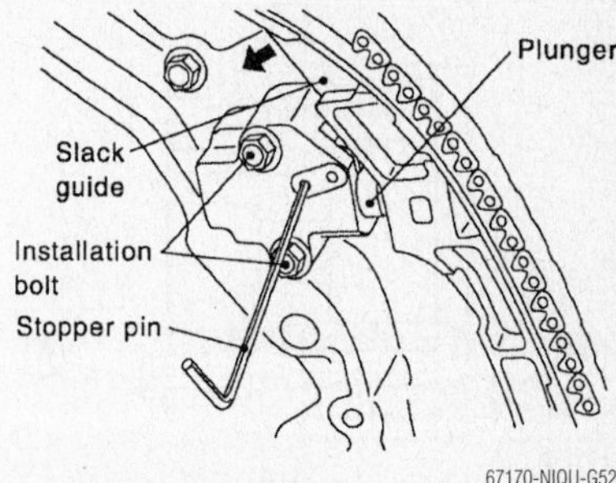

Keep the slack side chain guide pressed and hold it by pushing a stopper pin–3.5L engine

g. Remove the slack guide bolt and the guide.

31. Remove primary timing chain and crankshaft sprocket.

CAUTION

After removing timing chain, do not turn the crankshaft and camshaft separately, or the valves will strike the pistons.

32. Attach a suitable stopper pin such as a 0.098 inch (2.5mm) Allen wrench to the right and left camshaft chain tensioners, for secondary timing chains.

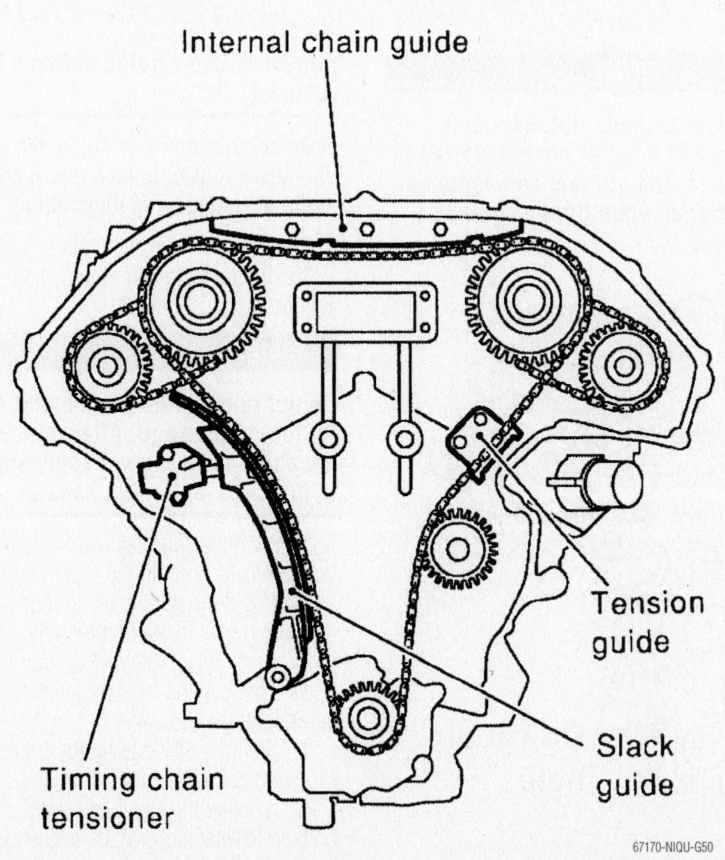

Location of the internal chain guide–3.5L engine

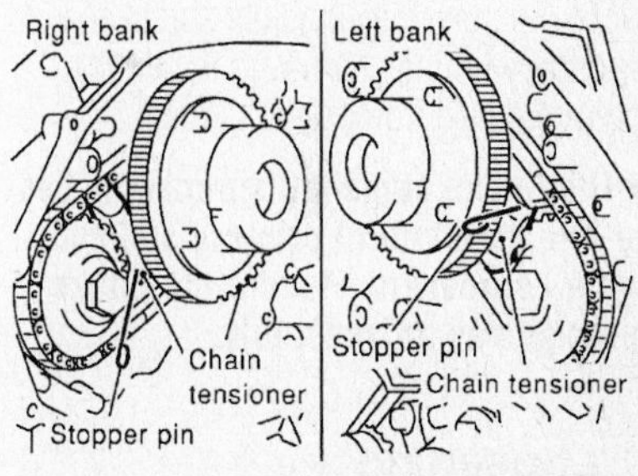

Attach a suitable stopper pin such as a 0.098 inch (2.5mm) Allen wrench to the right and left camshaft chain tensioners–3.5L engine

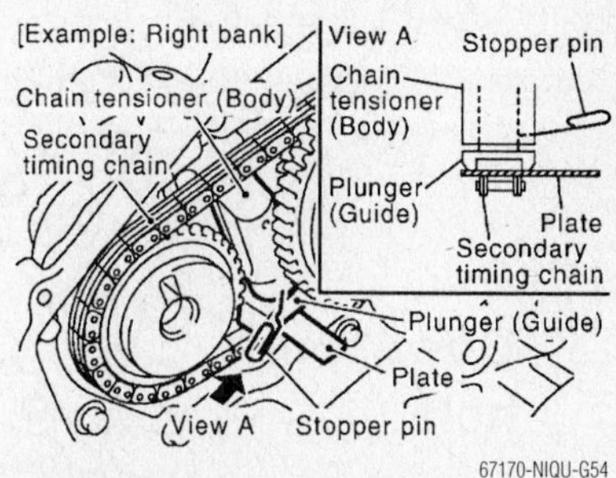

Insert a 0.020 inch (0.5 mm) metal or resin plate into the guide between timing chain and chain tensioner plunger–3.5L engine

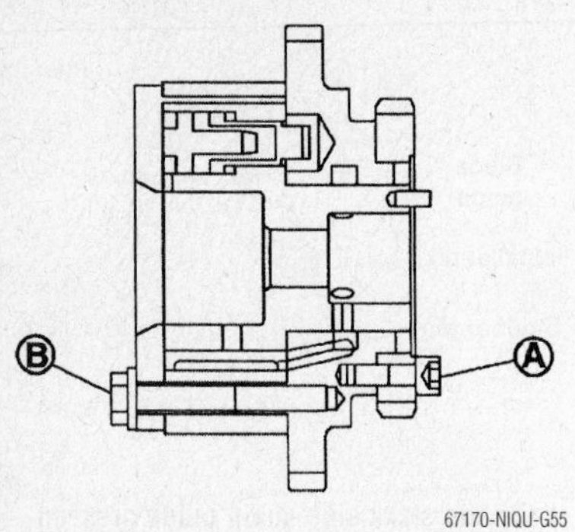

Do not disassemble the intake sprockets (never loosen bolts A and B)–3.5L engine

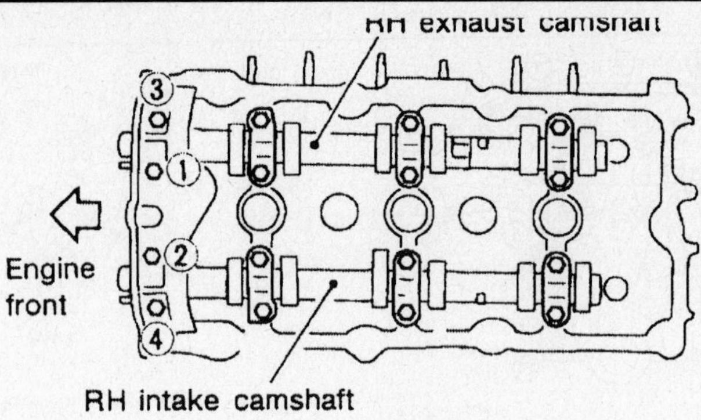

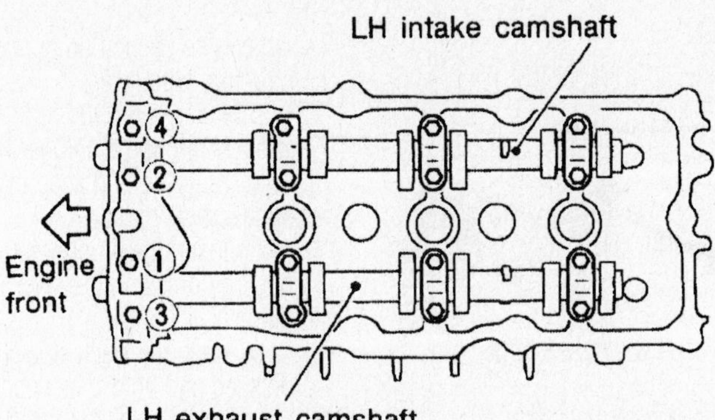

Loosen the No. 1 camshaft bracket bolts in several steps in the sequence illustrated–3.5L engine

33. Remove the intake and exhaust camshaft sprocket bolts. Put matchmarks on the timing chain and camshaft sprockets to aid alignment during installation.

34. Use a wrench on the hexagonal portion of the camshaft and loosen the bolts as illustrated.

35. Remove the secondary timing chains with camshaft sprockets as follows:

a. Rotate camshaft slightly, and slacken timing chain of timing chain tensioner side.

b. Insert a 0.020 inch (0.5 mm) metal or resin plate into the guide between timing chain and chain tensioner plunger.

36. Remove the camshaft sprocket and secondary timing chain with the timing chain removed from guide groove.

➡The intake camshaft sprocket is two-for-one structure of primary and secondary sprockets. Handle the intake sprockets as an assembly.

⁂ CAUTION

The chain tensioner plunger can move while stopper pin is inserted in tensioner. The plunger can come out of tensioner when timing chain is removed. Use caution during removal.

➡Do not disassemble the intake sprockets (never loosen bolts A and B shown in the accompanying illustration).

37. Remove the timing chain tension guide.

38. Remove the rear timing chain case as follows:

⁂ CAUTION

Do not remove the plate metal cover for the oil passage. After removing the chain case, do not apply any load to the case that might bend it.

a. Loosen and remove the rear timing chain case bolts in the order illustrated.

b. Cut the sealant using tool J–37228 and remove the rear timing chain case.

39. Disconnect the inlet coolant hose.

40. Remove the inlet coolant housing, gasket and thermostat.

41. Remove O–rings on the cylinder head and cylinder block.

42. Loosen the No. 1 camshaft bracket bolts in several steps in the sequence illustrated and remove No. 1 camshaft brackets.

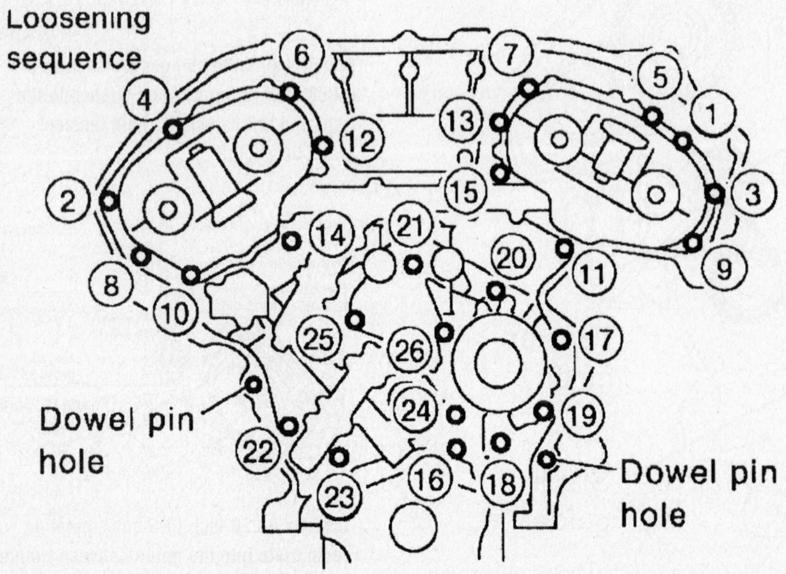

Remove the rear timing chain case bolts in the order illustrated–3.5L engine

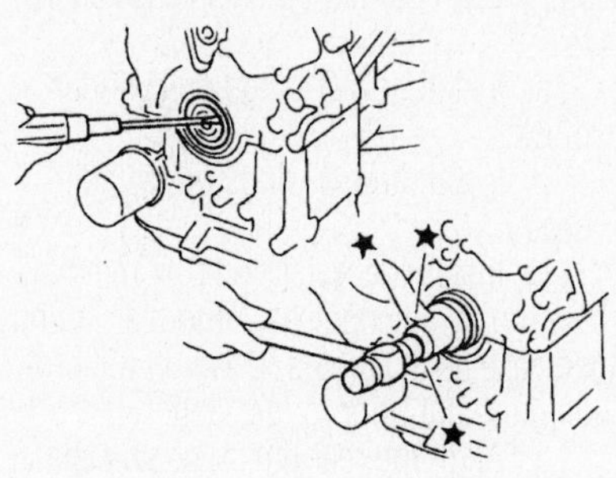

67170-NIQU-G37

Remove the front oil seal from the front timing chain case using a suitable tool–3.5L engine

43. Remove the camshaft chain tensioners, for secondary timing chains.
44. If necessary, remove the water pump.
45. Use a scraper to remove all of the old sealant from the front and rear timing chain case mating surfaces and the bolt holes, bolts, camshaft No. 1 bracket and water pump cover. Being careful not to damage the mating surfaces.
46. Remove the front oil seal from the front timing chain case using a suitable tool.
47. Check for cracks and any excessive wear at the roller links of the timing chain. Replace the timing chain as necessary.

To install:

48. Install the camshaft chain tensioners, for secondary timing chain Tighten the bolts to 75 inch lbs. (9 Nm).
49. Before installing the No. 1 camshaft bracket, apply sealant to mating surface and wipe off any excess sealant before installation.
50. Tighten the No. 1 camshaft bracket in three steps, in the order illustrated as follows:
51. Step 1: 17 inch lbs. (2 Nm).

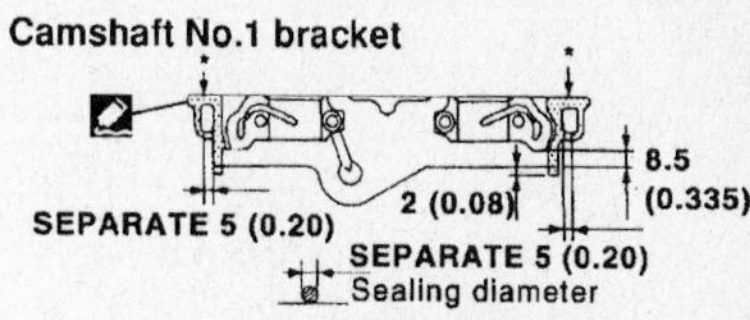

67170-NIQU-G59

Before installing the No. 1 camshaft bracket, apply sealant to mating surface and wipe off any excess sealant–3.5L engine

a. Step 2: 52 inch lbs. (6 Nm).
b. Step 3: 92 inch lbs. (10 Nm).

52. Install the thermostat, gasket and coolant inlet housing. Tighten the bolts to 87 inch lbs. (10 Nm).

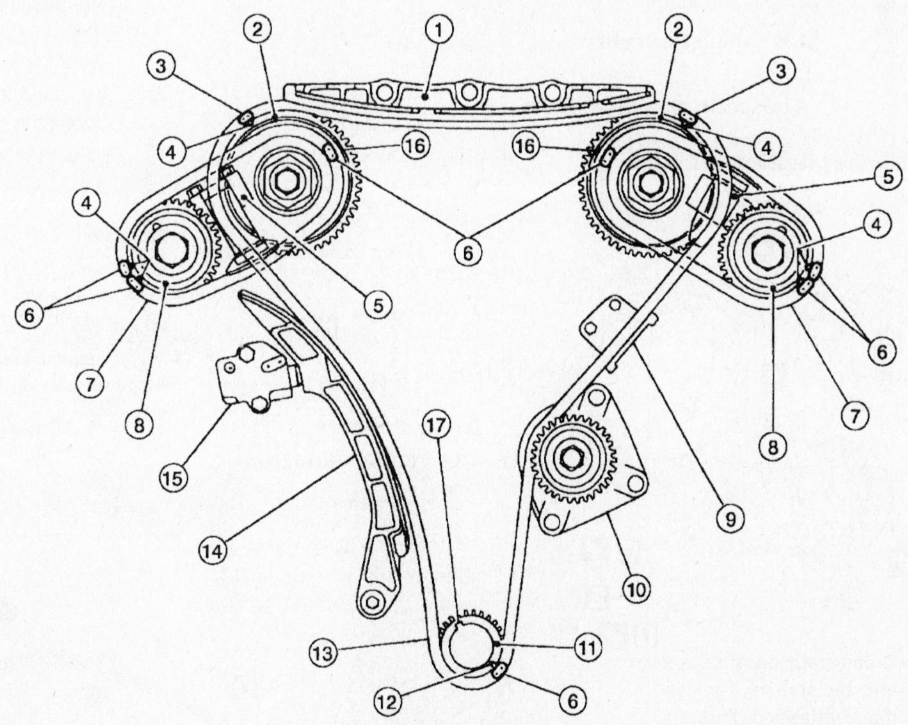

1. Internal chain guide
2. Camshaft sprocket (intake)
3. Mating mark (copper link)
4. Mating mark (punched)
5. Secondary timing chain tensioner
6. Mating mark (gold link)
7. Secondary timing chain
8. Camshaft sprocket (exhaust)
9. Tensioner guide
10. Water pump
11. Crankshaft sprocket
12. Mating mark (notched)
13. Primary timing chain
14. Slack guide
15. Primary timing chain tensioner
16. Mating mark (back side)
17. Crankshaft key

67170-NIQU-G58

Make sure all timing marks are aligned as shown when the chain is installed–3.5L engine

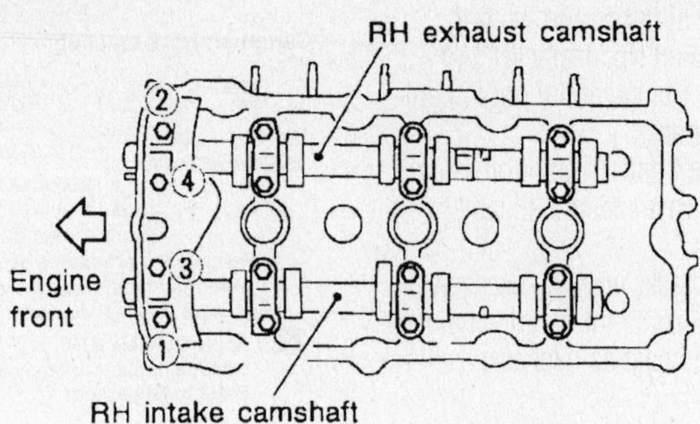

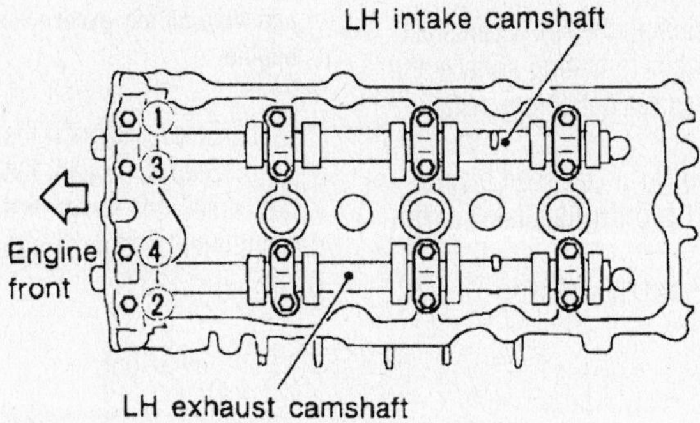

Camshaft bolt tightening sequence–3.5L engine

53. Install rear timing chain case as follows:
 a. Install new O–rings on cylinder block.
 b. Install new O–rings on cylinder head.
 c. Install the water pump, if removed.
54. Apply silicone sealant to rear timing chain case as illustrated, clean off the protruding sealant prior to installation.
 a. Align the rear timing chain case with the dowel pins on the cylinder block and install the case. Make sure the O–rings stay in place during installation.
 b. Tighten the rear timing chain case bolts in the order illustrated.
 c. Install the 0.79 inch (20 mm) bolts in locations 1, 2, 3, 6, 7, 8, 9 and 10. Tighten to 9 ft. lbs. (13 Nm). Install the 0.63 inch (16 mm) bolts in locations 4, 5, 11 through 26 and tighten to 9 ft. lbs. (13 Nm).
 d. After all bolts are initially tightened, retighten them to the specification in the order illustrated.
 e. After installing rear timing chain case, check surface height difference between the rear timing chain case to cylinder block. The measurement should

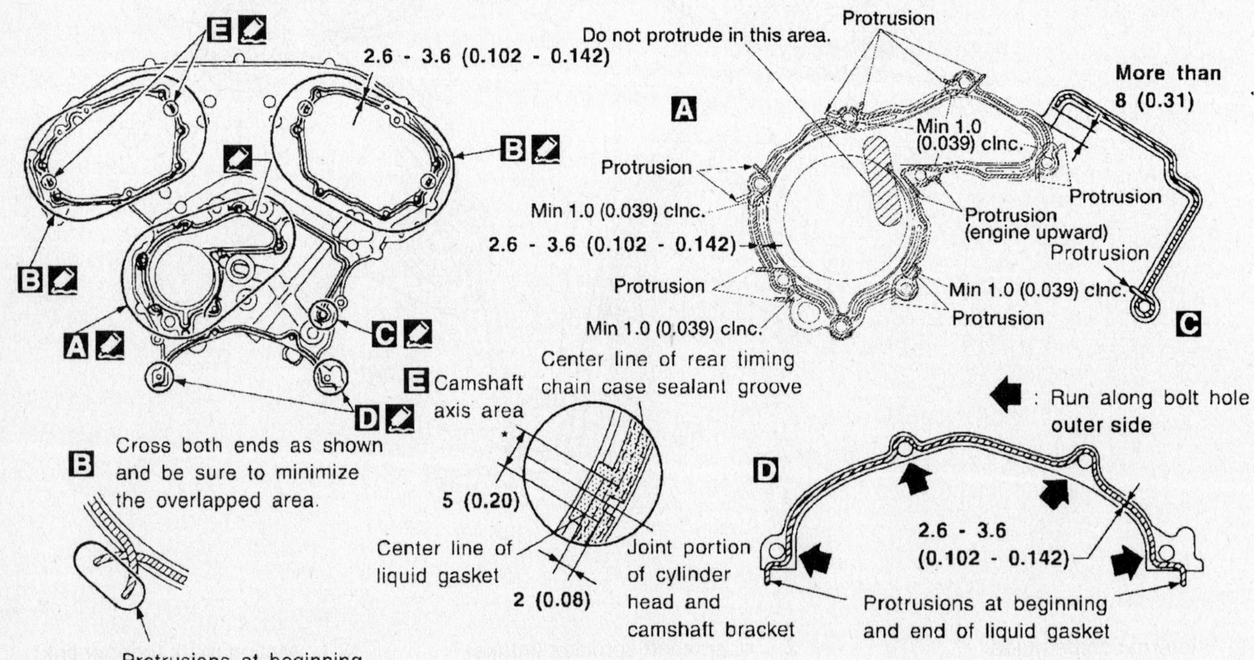

Apply silicone sealant to rear timing chain case as illustrated–3.5L engine

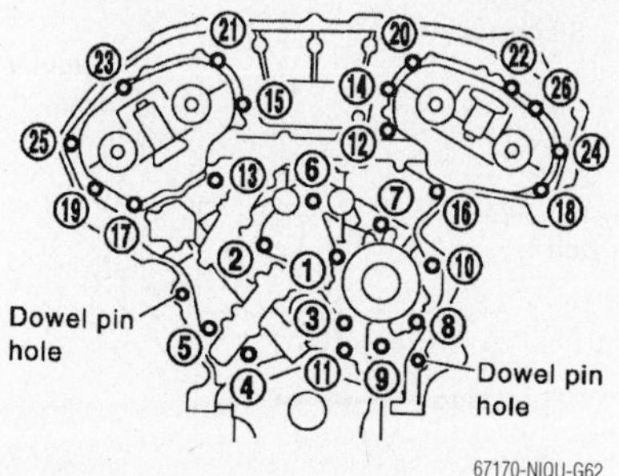

Rear timing chain case torque sequence–3.5L engine

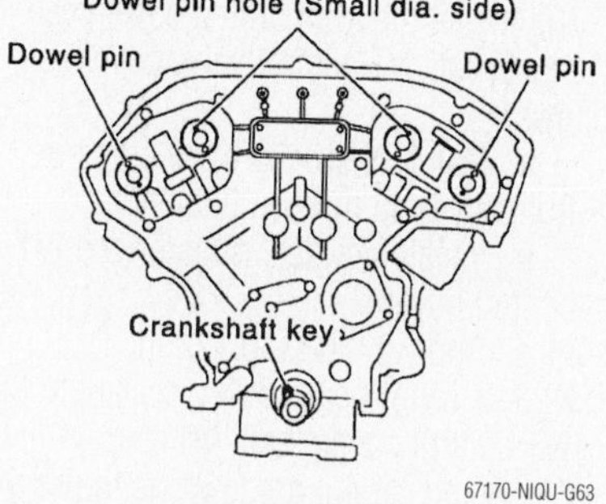

Make sure that the dowel pin hole, dowel pin and crankshaft key are located as illustrated–3.5L engine

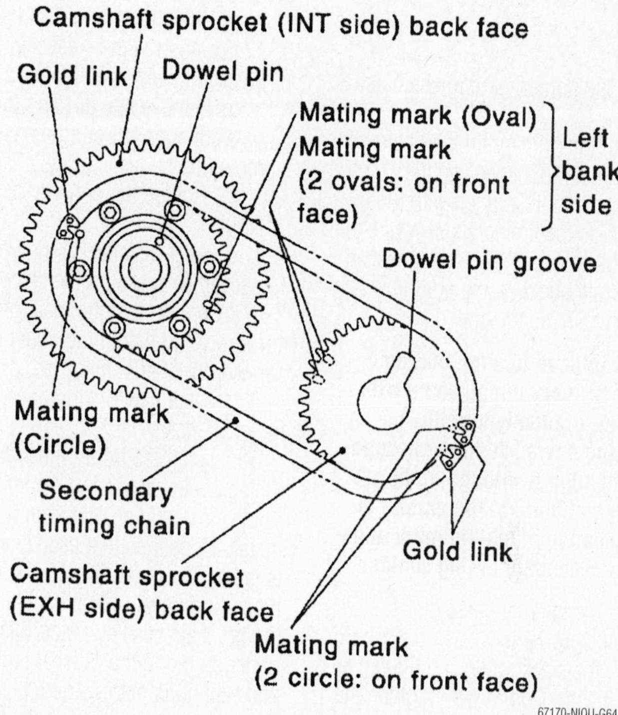

Match marks for the intake sprocket are on the back side of the secondary sprocket. There are two types of match marks, round and oval types–3.5L engine

be 0.0094–0.0055 inch (0.24–0.14 mm). If not within specification, repeat the cover installation procedure.

55. Install the timing chain tension guide. Tighten the bolts to 9 ft. lbs. (13 Nm).

56. Position the crankshaft so the No. 1 piston is set at TDC on the compression stroke.

57. Make sure that the dowel pin hole, dowel pin and crankshaft key are located as illustrated. The camshaft dowel pin hole (intake side): at cylinder head upper face side in each bank. The camshaft dowel pin (exhaust side): at cylinder head upper face side in each bank. The crankshaft key: at cylinder head side of right hand bank.

**** CAUTION**

The hole on small diameter side must be used for the intake camshaft sprocket dowel pin. Do not misidentify (ignore the big diameter side).

58. Install the secondary timing chains and camshaft sprockets as follows:

**** CAUTION**

Match marks between the timing chain and sprockets can slip easily. Check all match mark positions repeatedly during the installation process. Push the sleeve of the secondary chain tensioner and keep it pressed in with a stopper pin.

a. Align the match marks on the secondary timing chain (gold link) with the ones on the intake and exhaust sprockets (stamped), and install them.

➡Match marks for the intake sprocket are on the back side of the secondary sprocket. There are two types of match marks, round and oval types. They should be used for the right hand and left hand banks, respectively. The right hand banks use round type and the left hand bank use oval type.

b. Align the dowel pin and pin hole on the camshaft with the groove and dowel pin on the sprocket, and install them.

c. On the intake side, align the pin hole on the small diameter side of the camshaft front end with the dowel pin on the back side of the camshaft sprocket, and install them.

d. On the exhaust side, align the dowel pin on the camshaft front end with the pin groove on the camshaft sprocket, and install them.

e. Tighten the camshaft sprocket bolts by hand to prevent the dislocation of the dowel pins.

59. Tighten the timing chain tension guide bolts to 16 ft. lbs. (21 Nm).

➡It may be difficult to visually check the dislocation of mating marks during and after installation. To make the matching easier, make a mating mark on the sprocket teeth in advance with paint.

60. After confirming the mating marks are aligned, tighten the camshaft sprocket bolts to 76 ft. lbs. (102 Nm).

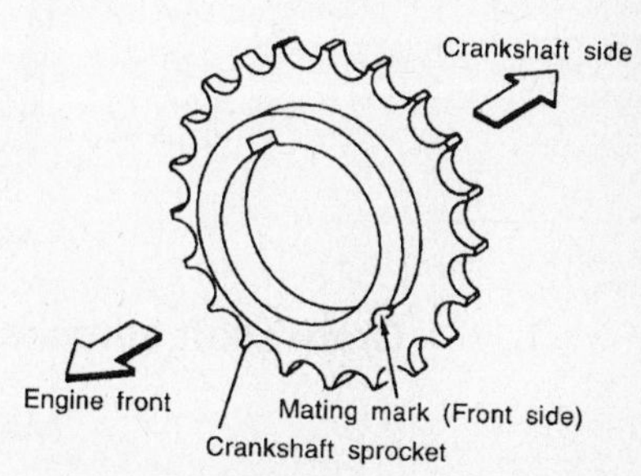

Make sure the mating marks on the crankshaft sprocket face the front of the engine–3.5L engine

61. Remove the stopper pins out from the timing chain tensioners, on secondary timing chains.

62. Install the crankshaft sprocket on the crankshaft. Make sure the mating marks on the crankshaft sprocket face the front of the engine.

63. Install the primary timing chain as follows:

a. Install the primary timing chain so the mating mark (punched) on the camshaft sprocket is aligned with the copper link on the timing chain, while the mating mark (notched) on the crankshaft sprocket is aligned with the gold link on the timing chain, as illustrated.

➡When it is difficult to align mating marks of the primary timing chain with each sprocket, gradually turn the camshaft using a wrench on the hexagonal portion to align it with the mating marks. During alignment, be careful to prevent dislocation of mating mark alignments of the secondary timing chains.

64. Install the internal chain guide. Tighten the bolts to 72 inch lbs. (8 Nm).

65. Install the slack guide and tighten to 12 ft. lbs. (15 Nm). Do not over–tighten the slack guide installation bolt. It is normal for a gap to exist under the bolt seats when the bolt is tightened to specification.

66. Install the timing chain tensioner for the slack guide as follows:

a. When installing the chain tensioner, push in the sleeve and keep it pressed in with the stopper pin.

b. Remove any dirt and foreign materials completely from the back and the mounting surfaces of the chain tensioner.

c. Tighten the bolts to 72 inch lbs. (8 Nm).

d. After installation, pull out the stopper pin by pressing the slack guide.

67. Confirm that the match marks on the sprockets and the timing chain have not slipped out of alignment.

68. Install new O–rings on the rear timing chain case.

69. Install the front oil seal on the front timing chain case using a suitable tool. Apply clean engine oil to the oil seal edges. Install the seal so that each seal lip is oriented as illustrated. Using a suitable drift, press-fit oil seal until it becomes flush with timing chain case end face. Make sure the garter spring in the oil seal is in position and seal lip is not inverted.

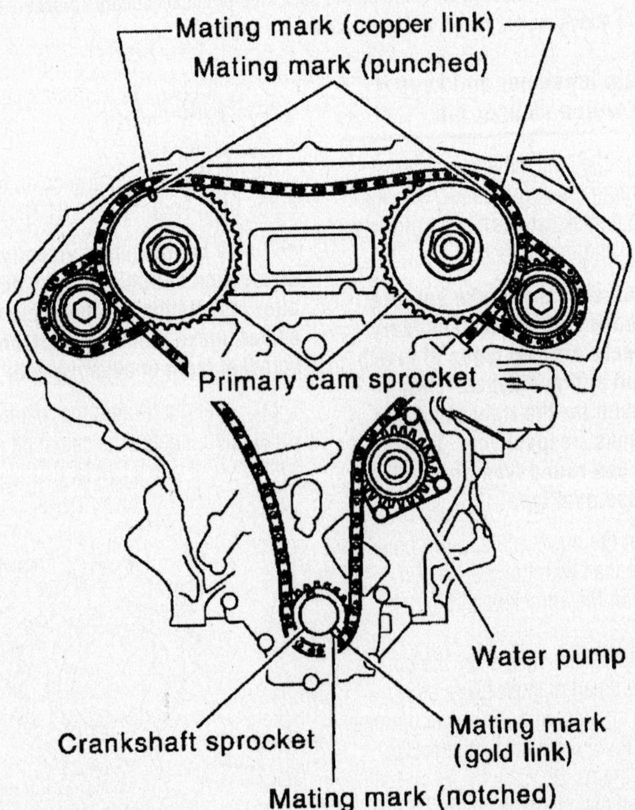

Install the primary timing chain so the match mark (punched) on the cam sprocket is aligned with the copper link on the timing chain, while the mating mark (notched) on the crank sprocket is aligned with the gold link on the timing chain–3.5L engine

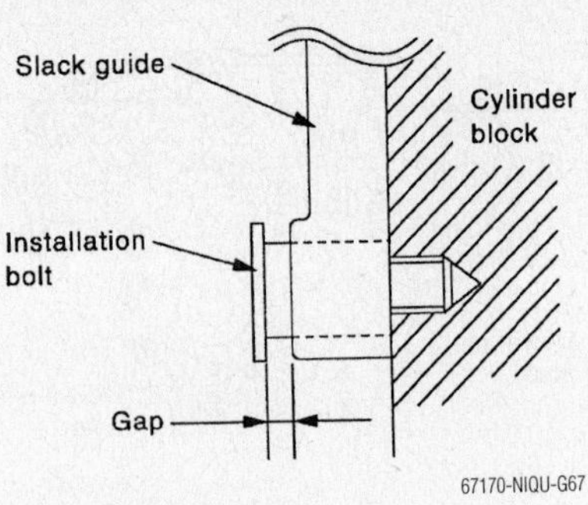

It is normal for a gap to exist under the bolt seats when the internal chain guide bolt is tightened –3.5L engine

70. Install the front timing cover case.

71. Install IVT control valve covers as follows:

a. Install new collared O–rings in front cover oil hole on both sides.

b. Install new seal rings on the IVT control covers.

c. Apply a 0.083–0.122 inch (2.1–3.1mm) bead of RTV sealant to the IVT control covers as illustrated.

72. Being careful not to move the seal ring from the installation groove, align the dowel pins on the chain case with the holes to install the IVT control covers.

73. Tighten the intake valve timing control cover bolts in the order illustrated to 100 inch lbs. (11 Nm).

74. Apply a 0.091–0.130 inch (2.3–3.3mm) bead of liquid gasket to the water pump cover and the chain tensioner cover and install covers. Tighten the bolts on the pump and cover to 97 inch lbs. (11 Nm).

75. Install crankshaft pulley lubricate the pulley bolt with clean engine oil and tighten the bolt in two steps:

a. Step 1: 32 ft. lbs. (44 Nm).

b. Step 2: 60–65 degrees clockwise.

76. Rotate crankshaft pulley in a clockwise direction as viewed from front of the engine to confirm it turns smoothly.

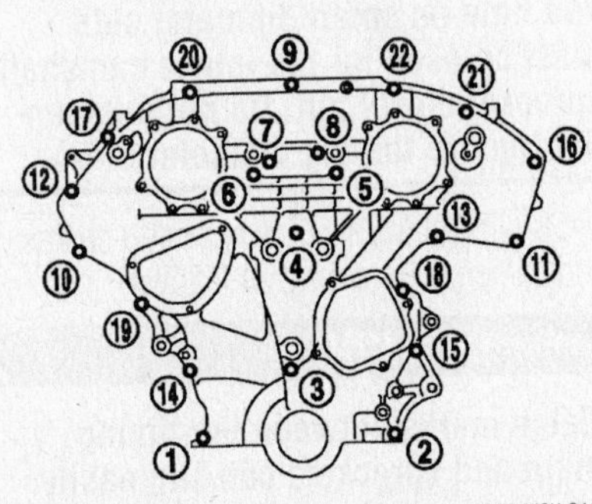

Tighten the front timing chain case bolts in the order shown–3.5L engine

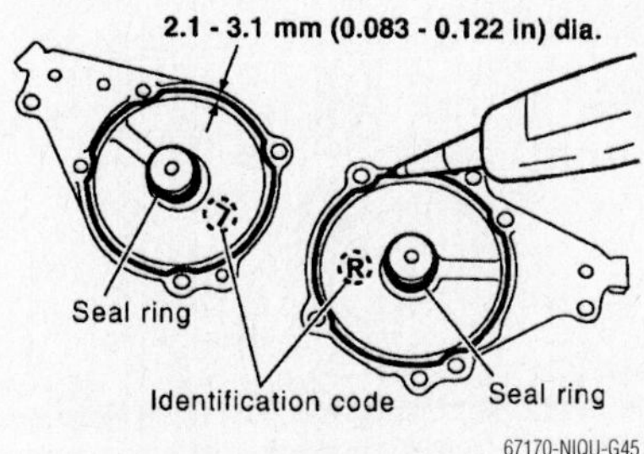

Apply a 0.083–0.122 inch (2.1–3.1mm) bead of RTV sealant to the IVT control covers–3.5L engine

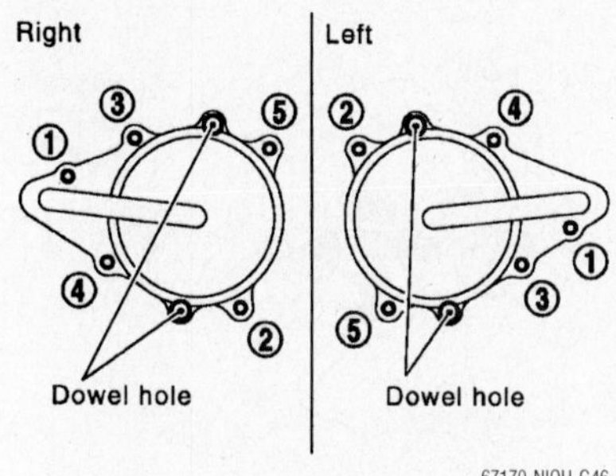

Tighten the intake valve timing control cover bolts in this order–3.5L engine

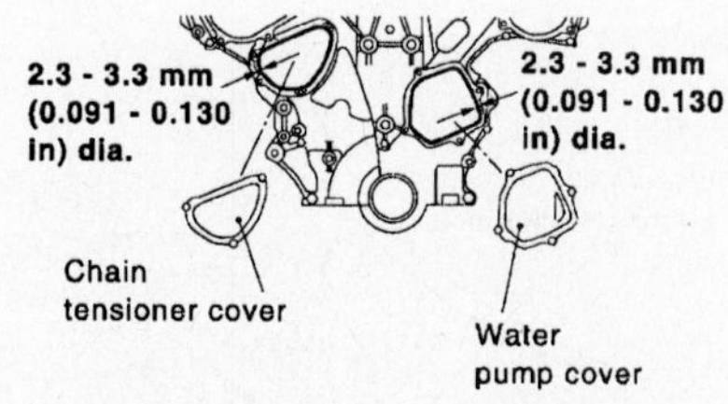

Apply a 0.091–0.130 inch (2.3–3.3mm) bead of liquid gasket to the water pump cover and the chain tensioner cover–3.5L engine

77. Install the right hand engine mount.
78. Install the rocker covers in the reverse order of removal, make sure to perform the following:
79. Apply sealant to the areas on the front corners.
80. Tighten the rocker cover bolts in two steps in the sequence illustrated:
    a. Step 1: 17 inch lbs. (1.96 Nm).
    b. Step 2: 74 inch lbs. (8 Nm).
81. Install the remaining components is in reverse order of removal.

➡If hydraulic pressure inside chain tensioner drops after removal/installation, slack in the guide may generate a pounding noise during and just after engine start. This is normal. Noise will stop after hydraulic pressure rises.

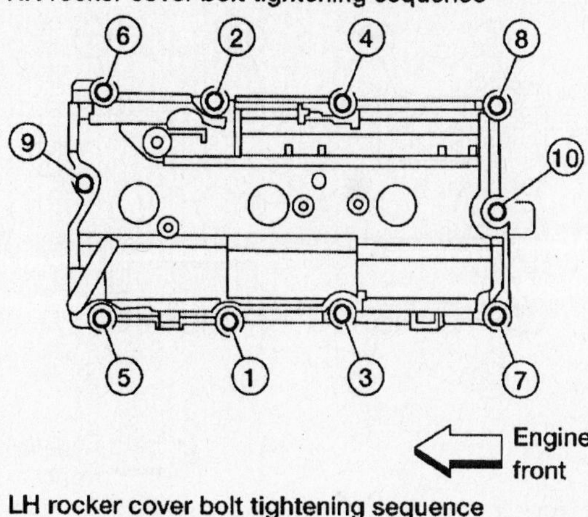

Tighten the rocker cover bolts in this sequence–3.5L engine

5-SPEED MODELS

1. Before servicing the vehicle, refer to the precautions in the beginning of this section.
2. Remove the engine from the vehicle.
3. Remove the idler pulley and bracket.
4. Remove the upper and lower oil pans.
5. Remove the alternator.
6. Remove the timing chain tensioner cover.
7. Remove both the IVT control covers, remove the bolts in the sequence illustrated.

➡The shaft in the cover is inserted into the center hole of the intake camshaft sprocket. Remove the cover by pulling straight out until the cover disengages from the camshaft sprocket.

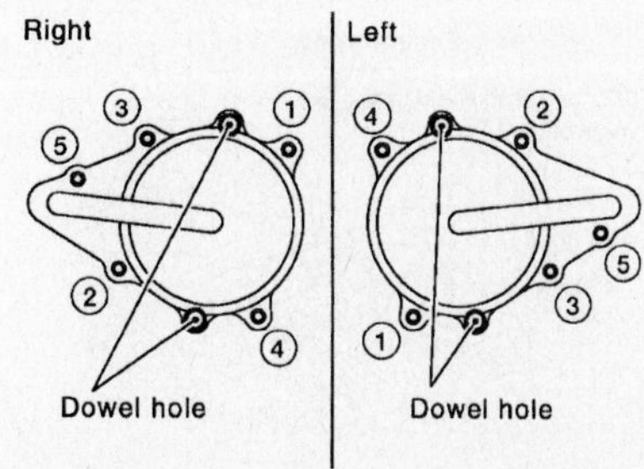

Loosen the IVT control cover bolts in this sequence–3.5L engine

8. Remove the starter.
9. Remove the intake manifold collector.
10. Remove the rocker covers as follows:
    a. Remove the engine cover.
    b. Remove side engine covers.
11. If removing right hand rocker cover, remove the following:
    - Front cowl panel
    - Windshield wiper arms and motor assembly
    - Intake manifold collector.
    a. If removing left hand rocker cover, disconnect the Air Fuel Ratio (AFR) sensor.
    b. Remove the ignition coils.
    c. Position engine harness aside.
    d. Disconnect Positive Crankcase Ventilation (PCV) hose.
    e. Remove the dipstick.
    f. Remove the rocker cover bolts in the sequence illustrated.
12. Remove the spark plugs.
13. Disconnect and remove the IVT control solenoid valves and discard the gaskets
14. Place the engine at Top Dead center (TDC) of No. 1 cylinder as follows:
    a. Rotate crankshaft pulley clockwise

RH rocker cover bolt loosening sequence

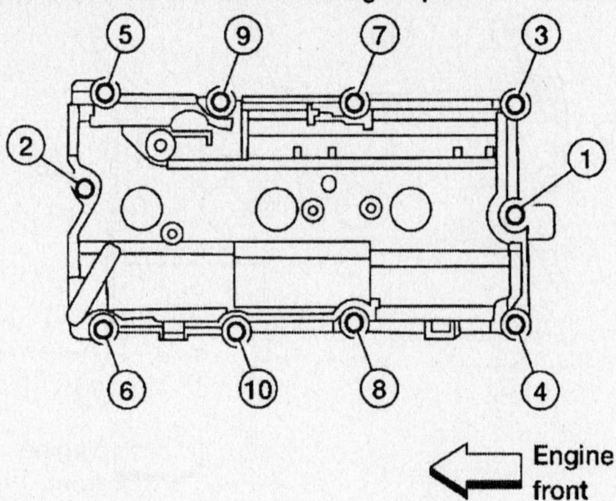

LH rocker cover bolt loosening sequence

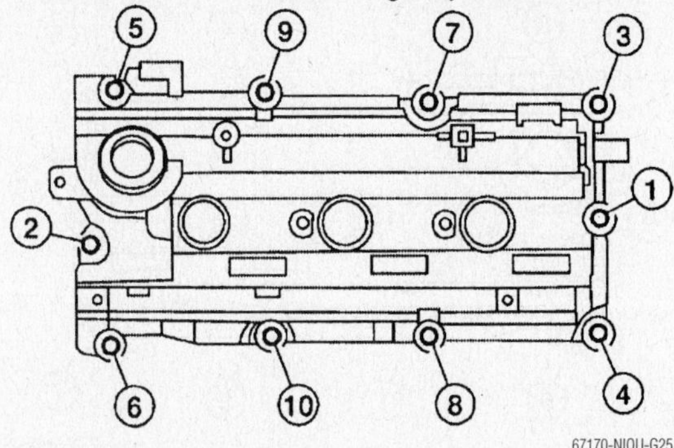

Loosen the rocker cover bolts in this sequence–3.5L engine

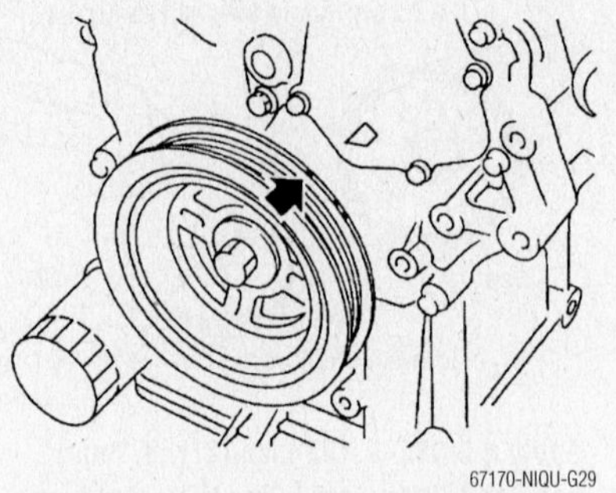

Position the engine at Top Dead Center (TDC) of No. 1 cylinder–3.5L engine

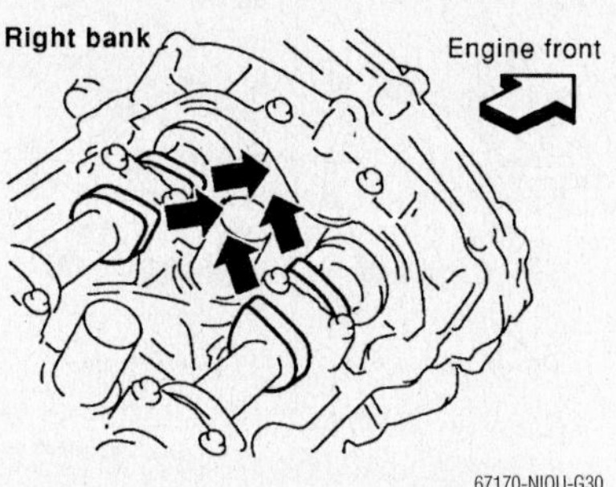

Check that intake and exhaust camshaft lobes on No. 1 cylinder on the right bank of engine are located as shown–3.5L

to align timing mark (grooved line without color) with the timing indicator.

b. Check that intake and exhaust camshaft lobes on No. 1 cylinder (right bank of engine) are located as illustrated. If not, turn the crankshaft one revolution (360 degrees) and align as illustrated.

15. Lock the ring gear using tool J–44716 attached to the starter bolt hole.

⁂ CAUTION

Do not damage the ring gear teeth, or the signal plate teeth behind the ring gear, when setting the tool.

16. Remove the crankshaft pulley as folows:

a. Loosen crankshaft pulley bolt using tool KV10109300 and locate bolt seating surface at 0.39 inch (10 mm) from its original position.

b. Position a pulley puller at recess hole of crankshaft pulley to remove crankshaft pulley.

⁂ CAUTION

Do not use a puller claw on crankshaft pulley.

17. Remove the front timing chain case after loosening the front timing chain case bolts in the order illustrated.

18. Remove the internal chain guide.

19. Remove the timing chain tensioner and slack guide as follows:

a. Put matchmarks on the timing chain and sprockets to indicate the correct position of the components to aid during installation.

b. Pull the lever down and release the plunger stopper tab. The plunger stopper tab can be pushed up to release (coaxial structure with lever).

c. Insert a 0.098 inch (2.5mm) Allen

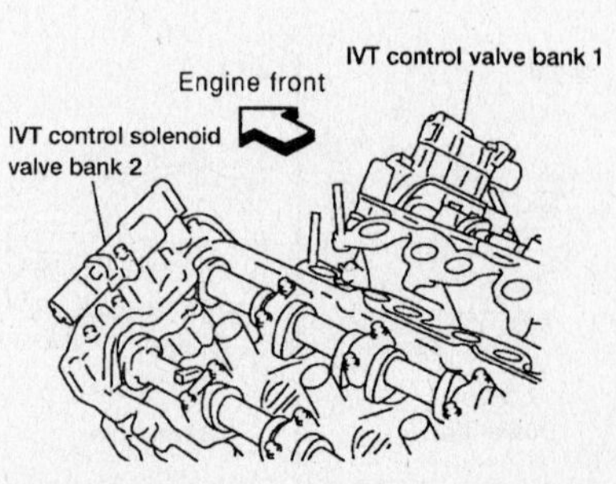

Remove the IVT control solenoid valves–3.5L engine

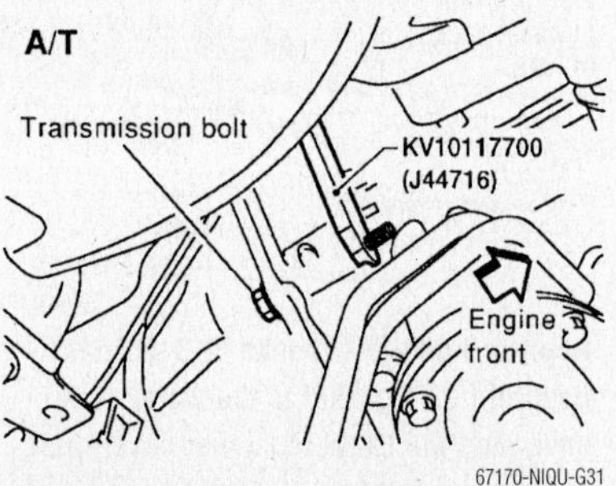

Lock the ring gear using tool J–44716 attached to the starter bolt hole–3.5L engine

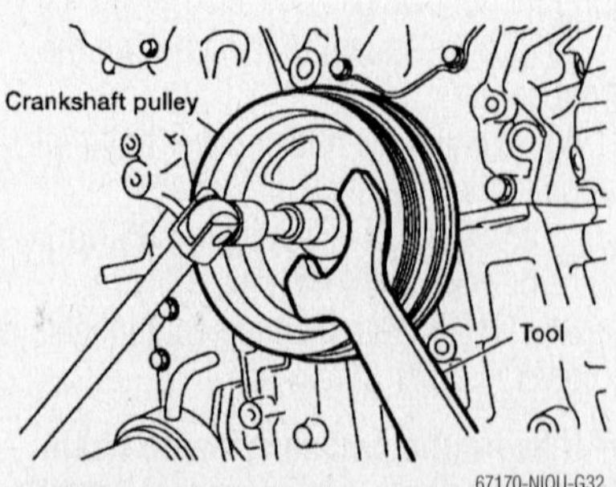

Loosen crankshaft pulley bolt using tool KV10109300 and locate bolt seating surface at 0.39 inch (10mm) from its original position–3.5L engine

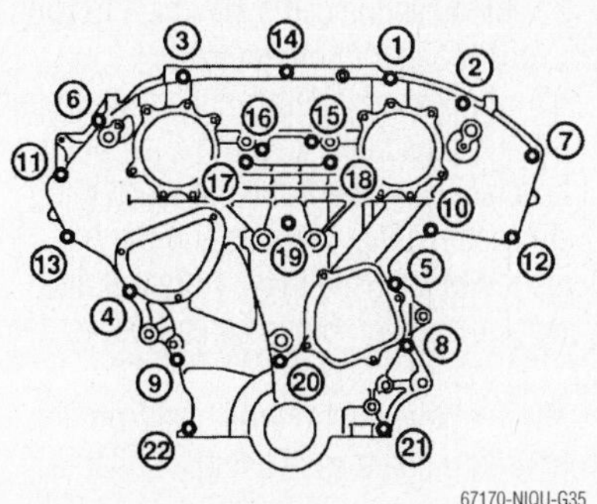

Loosen the front timing chain case bolts in this order–3.5L engine

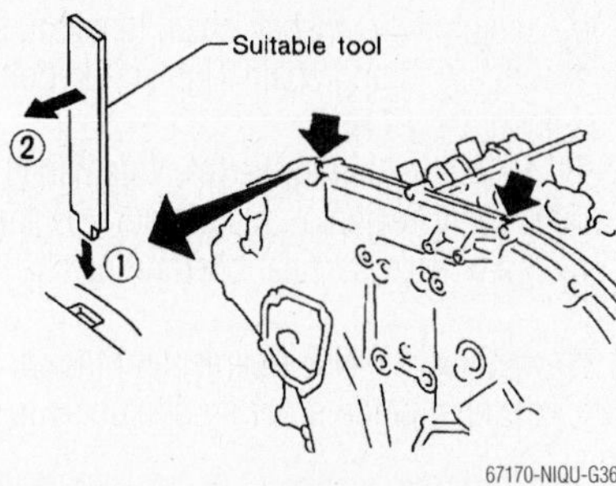

Insert the appropriate size tool such as J-37228 into the notch at the top of the front timing chain case–3.5L engine

wrench into the tensioner body hole to hold the lever, and keep the tab released.

d. Insert the plunger into tensioner body by pressing the slack side chain guide.

e. Keep the slack side chain guide pressed and hold it by pushing a stopper pin such as a 0.098 inch (2.5mm) Allen wrench through the lever hole and body hole.

f. Remove the timing chain tensioner bolts and the timing chain tensioner.

g. Remove the slack guide bolt and the guide.

20. Remove primary timing chain and crankshaft sprocket.

CAUTION

After removing timing chain, do not turn the crankshaft and camshaft separately, or the valves will strike the pistons.

21. Attach a suitable stopper pin such as a 0.098 inch (2.5mm) Allen wrench to the right and left camshaft chain tensioners, for secondary timing chains.

22. Remove the intake and exhaust

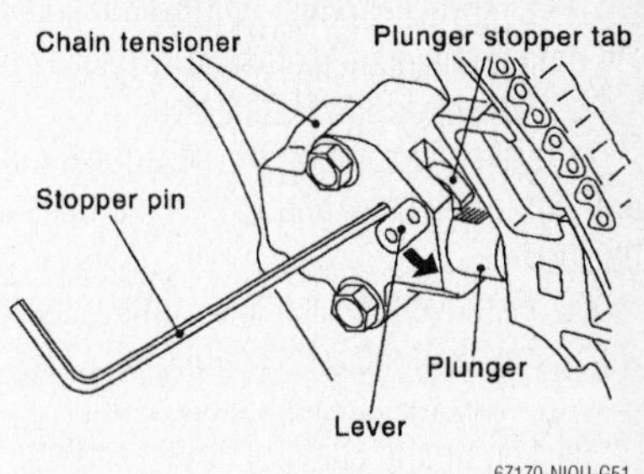

Insert a 0.098 inch (2.5mm) Allen wrench into the tensioner body hole to hold the chain tensioner lever–3.5L engine

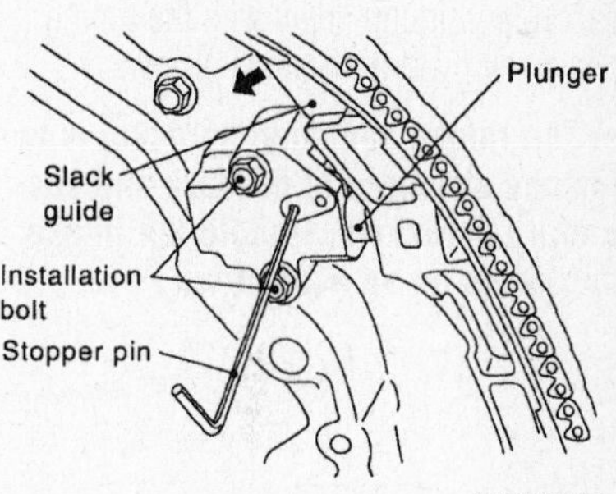

Keep the slack side chain guide pressed and hold it by pushing a stopper pin–3.5L engine

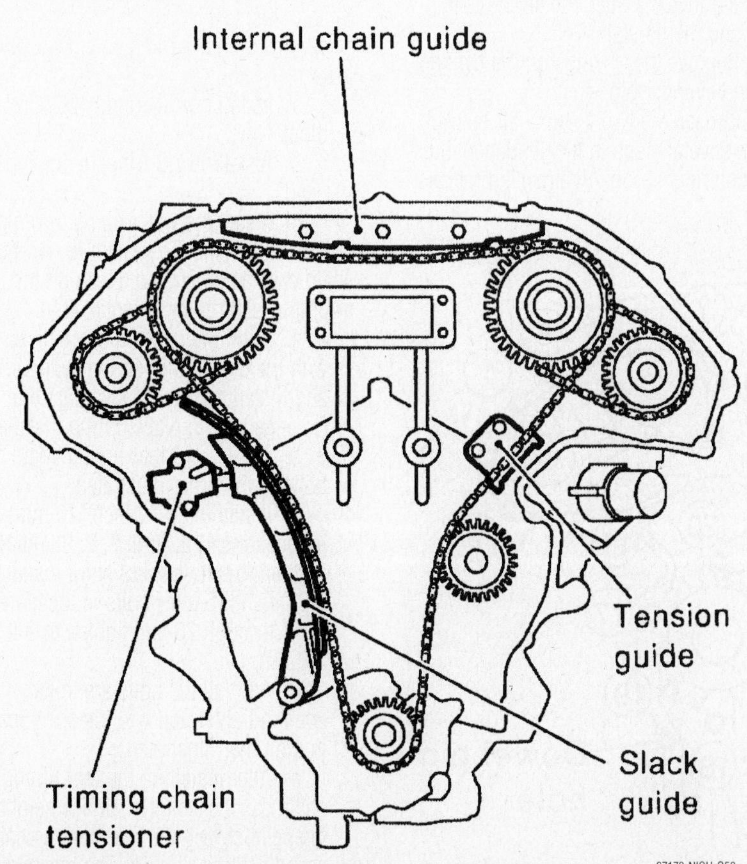

Location of the internal chain guide–3.5L engine

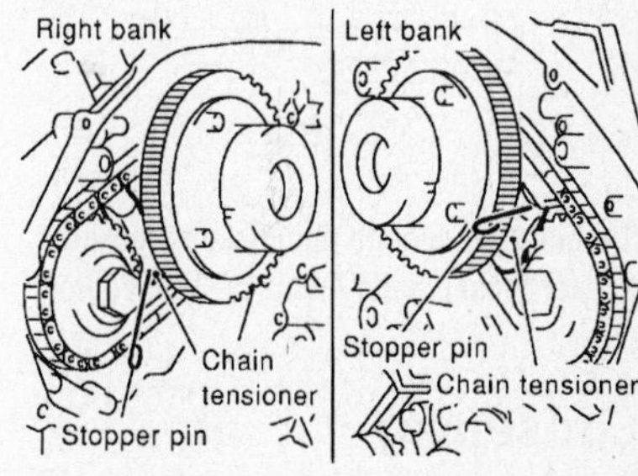

Attach a suitable stopper pin such as a 0.098 inch (2.5mm) Allen wrench to the right and left camshaft chain tensioners–3.5L engine

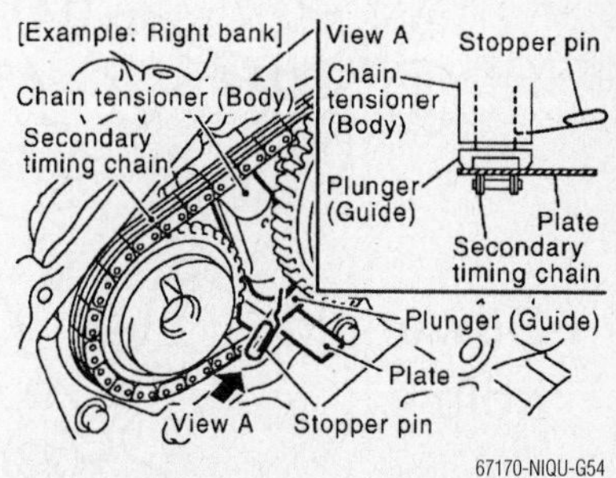

Insert a 0.020 inch (0.5 mm) metal or resin plate into the guide between timing chain and chain tensioner plunger–3.5L engine

camshaft sprocket bolts. Put matchmarks on the timing chain and camshaft sprockets to aid alignment during installation.

23. Use a wrench on the hexagonal portion of the camshaft and loosen the bolts as illustrated.

24. Remove the secondary timing chains with camshaft sprockets as follows:

a. Rotate camshaft slightly, and slacken timing chain of timing chain tensioner side.

b. Insert a 0.020 inch (0.5 mm) metal or resin plate into the guide between timing chain and chain tensioner plunger.

25. Remove the camshaft sprocket and secondary timing chain with the timing chain removed from guide groove.

➡**The intake camshaft sprocket is two-for-one structure of primary and secondary sprockets. Handle the intake sprockets as an assembly.**

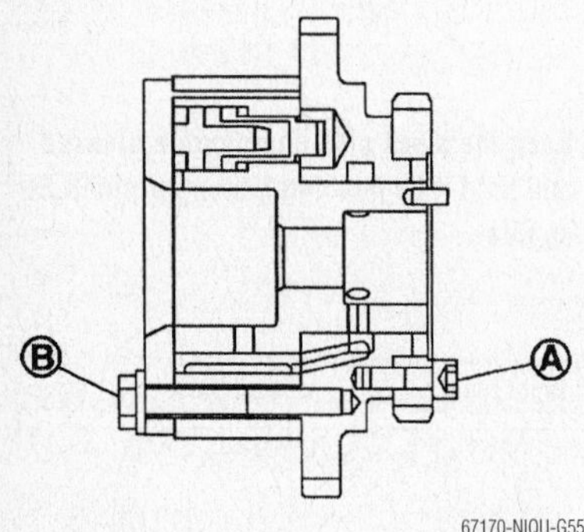

Do not disassemble the intake sprockets (never loosen bolts A and B)–3.5L engine

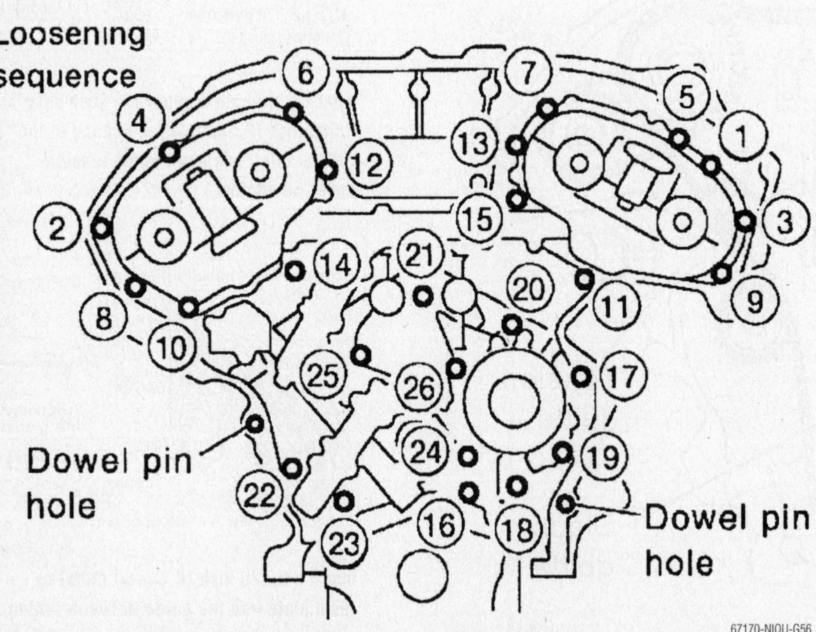

Remove the rear timing chain case bolts in the order illustrated–3.5L engine

⁂ CAUTION

The chain tensioner plunger can move while stopper pin is inserted in tensioner. The plunger can come out of tensioner when timing chain is removed. Use caution during removal.

➡**Do not disassemble the intake sprockets (never loosen bolts A and B shown in the accompanying illustration).**

26. Remove the timing chain tension guide.

27. Remove the rear timing chain case as follows:

⁂ CAUTION

Do not remove the plate metal cover for the oil passage. After removing the chain case, do not apply any load to the case that might bend it.

a. Loosen and remove the rear timing chain case bolts in the order illustrated.

b. Cut the sealant using tool J–37228 and remove the rear timing chain case.

28. Disconnect the inlet coolant hose.

29. Remove the inlet coolant housing, gasket and thermostat.

30. Remove the O–rings on the cylinder head and cylinder block.

31. Loosen the No. 1 camshaft bracket bolts in several steps in the sequence illustrated and remove No. 1 camshaft brackets.

32. Remove the camshaft chain tensioners, for secondary timing chains.

33. If necessary, remove the water pump.

34. Use a scraper to remove all of the old sealant from the front and rear timing chain case mating surfaces and the bolt holes, bolts, camshaft No. 1 bracket and water pump cover. Being careful not to damage the mating surfaces.

35. Remove the front oil seal from the front timing chain case using a suitable tool.

36. Check for cracks and any excessive wear at the roller links of the timing chain. Replace the timing chain as necessary.

To install:

37. Install the camshaft chain tensioners, for secondary timing chain Tighten the bolts to 75 inch lbs. (9 Nm).

38. Before installing the No. 1 camshaft bracket, apply sealant to mating surface and wipe off any excess sealant before installation.

39. Tighten the No. 1 camshaft bracket in three steps, in the order illustrated as follows:

40. Step 1: 17 inch lbs. (2 Nm).

a. Step 2: 52 inch lbs. (6 Nm).

b. Step 3: 92 inch lbs. (10 Nm).

41. Install the thermostat, gasket and coolant inlet housing. Tighten the bolts to 87 inch lbs. (10 Nm).

42. Install rear timing chain case as follows:

a. Install new O–rings on cylinder block.

b. Install new O–rings on cylinder head.

c. Install the water pump, if removed.

43. Apply silicone sealant to rear timing chain case as illustrated, clean off the protruding sealant prior to installation.

a. Align the rear timing chain case with the dowel pins on the cylinder block and install the case. Make sure the O–rings stay in place during installation.

b. Tighten the rear timing chain case bolts in the order illustrated.

c. Install the 0.79 inch (20 mm) bolts in locations 1, 2, 3, 6, 7, 8, 9 and 10. Tighten to 9 ft. lbs. (13 Nm). Install the 0.63 inch (16 mm) bolts in locations 4, 5, 11 through 26 and tighten to 9 ft. lbs. (13 Nm).

d. After all the bolts are initially tightened, retighten them to the specification in the order illustrated.

e. After installing the rear timing chain case, check the surface height difference between the rear timing chain case to cylinder block. The measurement should be 0.0094–0.0055 inch

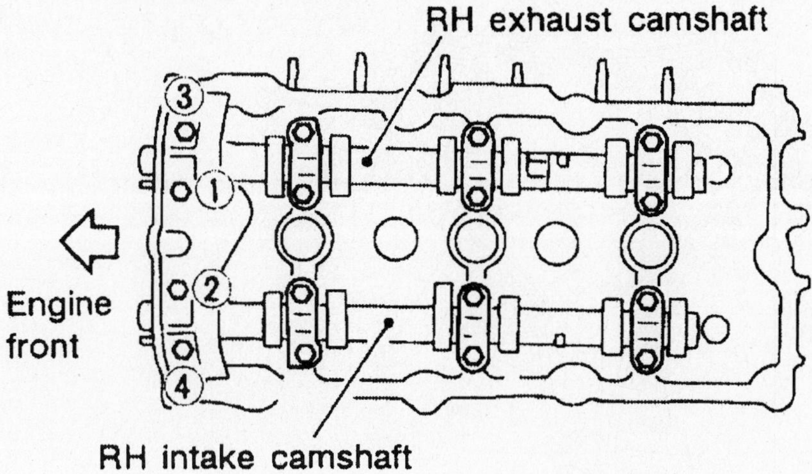

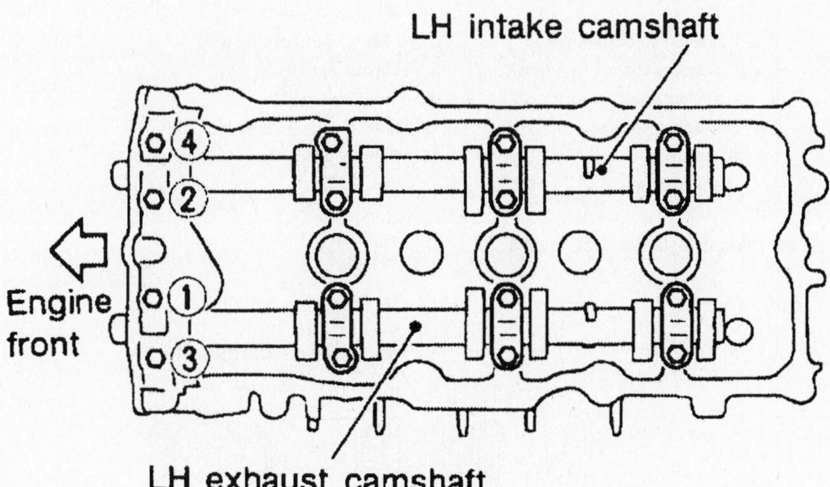

Loosen the No. 1 camshaft bracket bolts in several steps in the sequence illustrated–3.5L engine

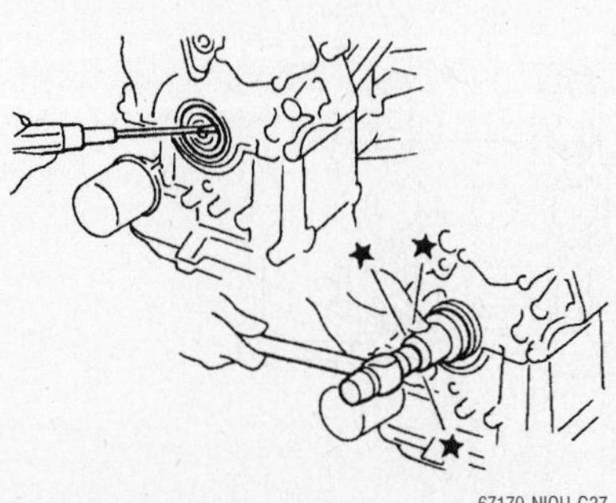

Remove the front oil seal from the front timing chain case using a suitable tool–3.5L engine

(0.24–0.14 mm). If not within specification, repeat the cover installation procedure.

44. Install the timing chain tension guide. Tighten the bolts to 9 ft. lbs. (13 Nm).

45. Position the crankshaft so the No. 1 piston is set at TDC on the compression stroke.

46. Make sure that the dowel pin hole, dowel pin and crankshaft key are located as illustrated. The camshaft dowel pin hole (intake side): at cylinder head upper face side in each bank. The camshaft dowel pin (exhaust side): at cylinder head upper face side in each bank. The crankshaft key: at cylinder head side of right hand bank.

** CAUTION**

The hole on small diameter side must be used for the intake camshaft sprocket dowel pin. Do not misidentify (ignore the big diameter side).

47. Install the secondary timing chains and camshaft sprockets as follows:

** CAUTION**

Match marks between the timing chain and sprockets can slip easily. Check all match mark positions repeatedly during the installation process. Push the sleeve of the secondary chain tensioner and keep it pressed in with a stopper pin.

a. Align the match marks on the secondary timing chain (gold link) with the ones on the intake and exhaust sprockets (stamped), and install them.

➡Match marks for the intake sprocket are on the back side of the secondary sprocket. There are two types of match marks, round and oval types. They should be used for the right hand and left hand banks, respectively. The right hand banks use round type and the left hand bank use oval type.

b. Align the dowel pin and pin hole on the camshaft with the groove and dowel pin on the sprocket, and install them.

c. On the intake side, align the pin hole on the small diameter side of the camshaft front end with the dowel pin on the back side of the camshaft sprocket, and install them.

d. On the exhaust side, align the dowel pin on the camshaft front end with the pin groove on the camshaft sprocket, and install them.

e. Tighten the camshaft sprocket bolts by hand to prevent the dislocation of the dowel pins.

48. Tighten the timing chain tension guide bolts to 16 ft. lbs. (21 Nm).

➡It may be difficult to visually check the dislocation of mating marks during and after installation. To make the matching easier, make a mating mark on the sprocket teeth in advance with paint.

49. After confirming the mating marks are aligned, tighten the camshaft sprocket bolts to 76 ft. lbs. (102 Nm).

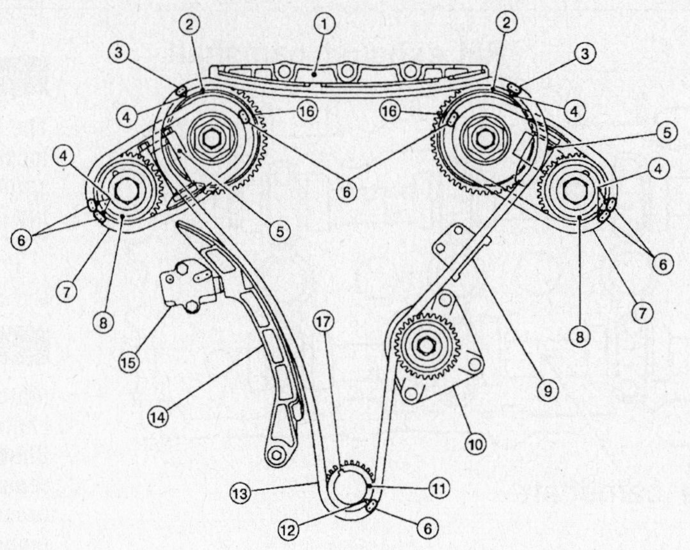

1. Internal chain guide
2. Camshaft sprocket (intake)
3. Mating mark (copper link)
4. Mating mark (punched)
5. Secondary timing chain tensioner
6. Mating mark (gold link)
7. Secondary timing chain
8. Camshaft sprocket (exhaust)
9. Tensioner guide
10. Water pump
11. Crankshaft sprocket
12. Mating mark (notched)
13. Primary timing chain
14. Slack guide
15. Primary timing chain tensioner
16. Mating mark (back side)
17. Crankshaft key

67170-NIQU-G58

Make sure all timing marks are aligned as shown when the chain is installed–3.5L engine

50. Remove the stopper pins out from the timing chain tensioners, on secondary timing chains.

51. Install the crankshaft sprocket on the crankshaft. Make sure the mating marks on the crankshaft sprocket face the front of the engine.

52. Install the primary timing chain as follows:

a. Install the primary timing chain so the mating mark (punched) on the camshaft sprocket is aligned with the copper link on the timing chain, while the

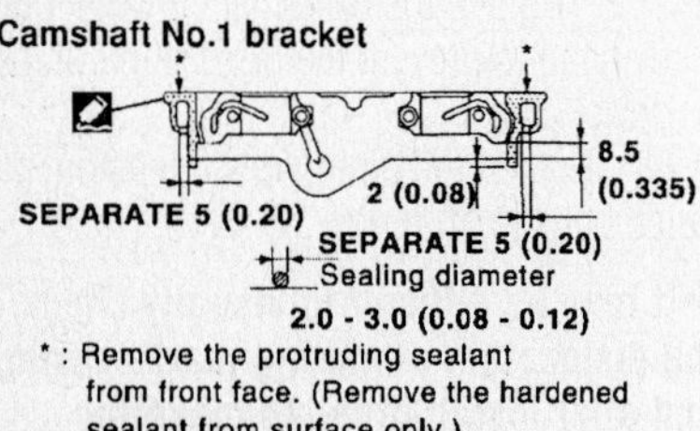

* : Remove the protruding sealant from front face. (Remove the hardened sealant from surface only.)

: Apply Genuine RTV Silicone Sealant or equivalent. Refer to GI section.

67170-NIQU-G59

Before installing the No. 1 camshaft bracket, apply sealant to mating surface and wipe off any excess sealant–3.5L engine

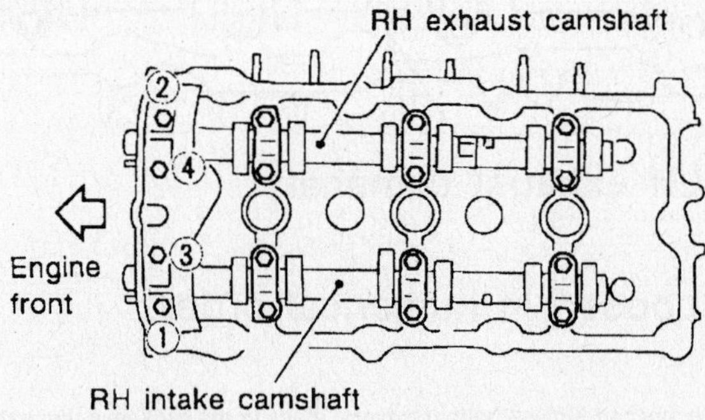

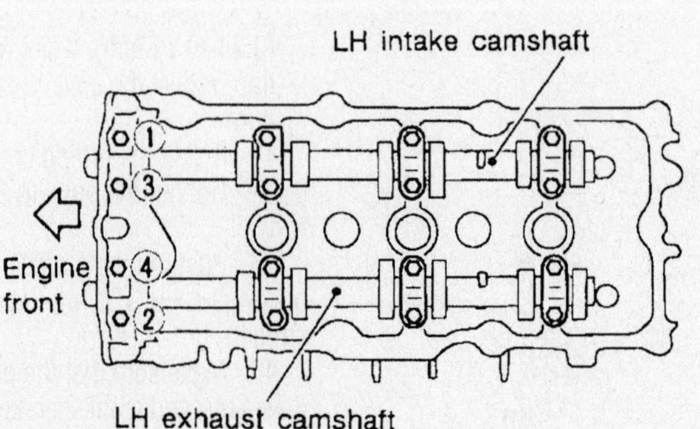

Loosen in numerical order.

67170-NIQU-G60

Camshaft bolt tightening sequence–3.5L engine

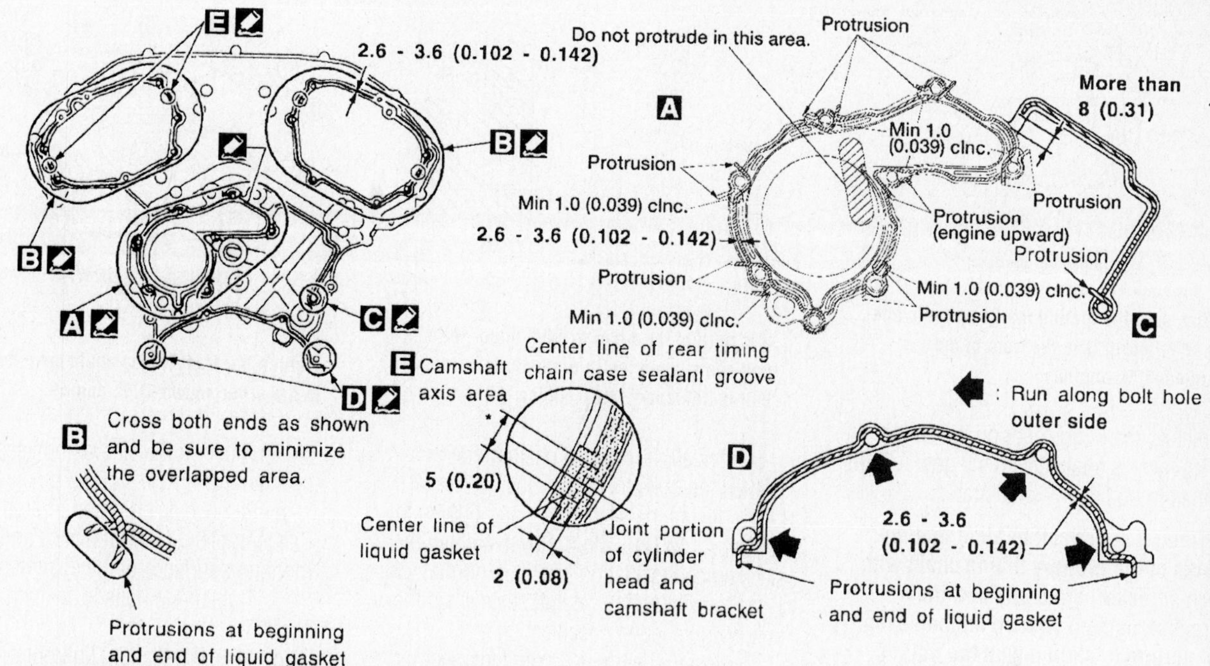

Apply silicone sealant to rear timing chain case as illustrated–3.5L engine

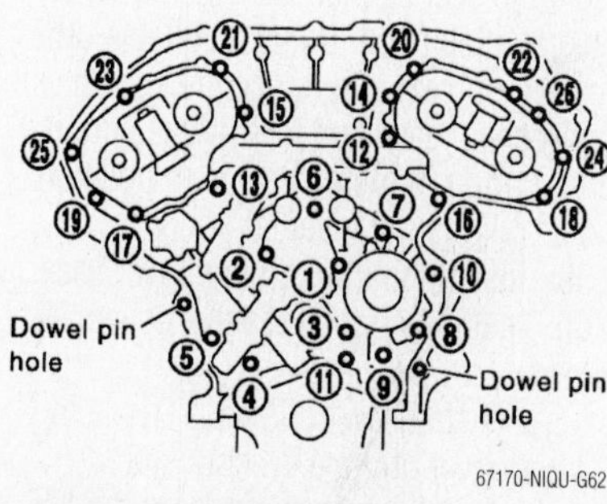

Rear timing chain case torque sequence–3.5L engine

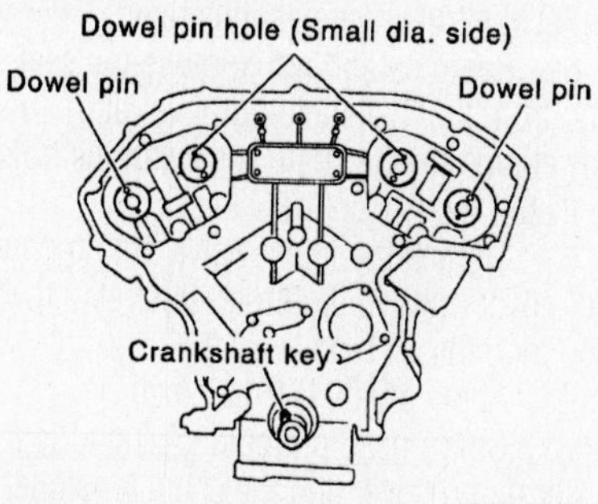

Make sure that the dowel pin hole, dowel pin and crankshaft key are located as illustrated–3.5L engine

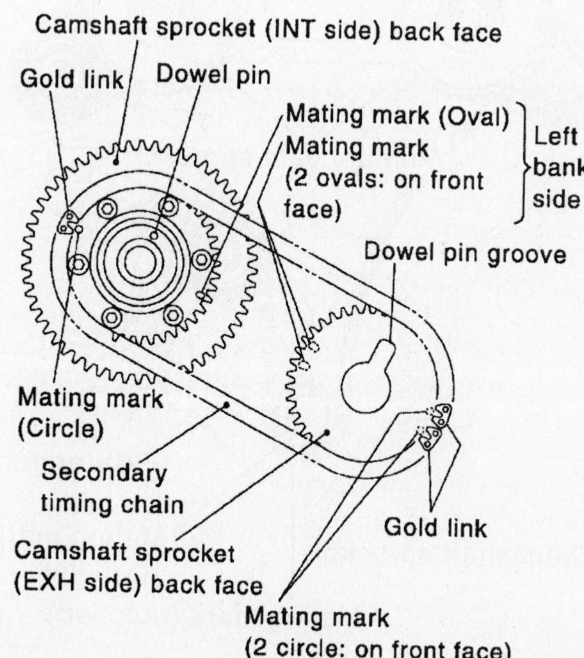

Match marks for the intake sprocket are on the back side of the secondary sprocket. There are two types of match marks, round and oval types–3.5L engine

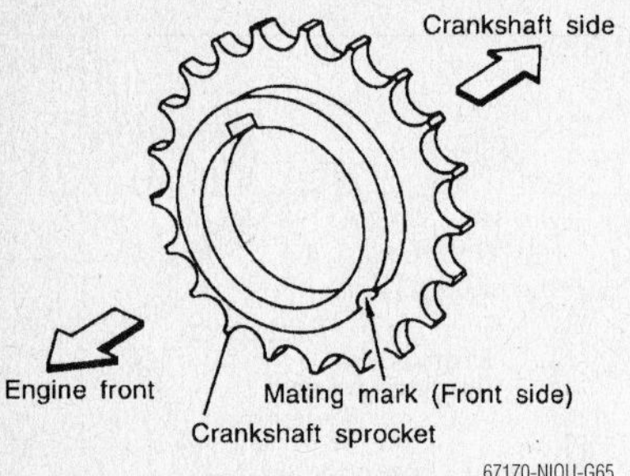

Make sure the mating marks on the crankshaft sprocket face the front of the engine–3.5L engine

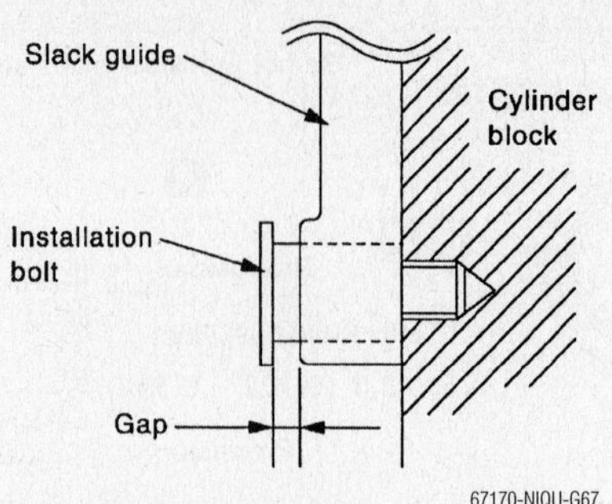

It is normal for a gap to exist under the bolt seats when the internal chain guide bolt is tightened –3.5L engine

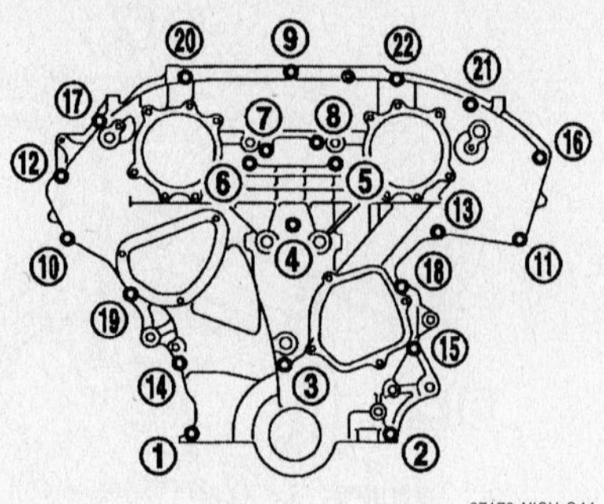

Tighten the front timing chain case bolts in the order shown–3.5L engine

mating mark (notched) on the crankshaft sprocket is aligned with the gold link on the timing chain, as illustrated.

➡When it is difficult to align mating marks of the primary timing chain with each sprocket, gradually turn the camshaft using a wrench on the hexagonal portion to align it with the mating marks. During alignment, be careful to prevent dislocation of mating mark alignments of the secondary timing chains.

53. Install the internal chain guide. Tighten the bolts to 72 inch lbs. (8 Nm).

54. Install the slack guide and tighten to 12 ft. lbs. (15 Nm). Do not over–tighten the slack guide installation bolt. It is normal for a gap to exist under the bolt seats when the bolt is tightened to specification.

55. Install the timing chain tensioner for the slack guide as follows:

a. When installing the chain tensioner, push in the sleeve and keep it pressed in with the stopper pin.

b. Remove any dirt and foreign materials completely from the back and the mounting surfaces of the chain tensioner.

c. Tighten the bolts to 72 inch lbs. (8 Nm).

d. After installation, pull out the stopper pin by pressing the slack guide.

56. Confirm that the match marks on the sprockets and the timing chain have not slipped out of alignment.

57. Install new O–rings on the rear timing chain case.

58. Install the front oil seal on the front timing chain case using a suitable tool. Apply clean engine oil to the oil seal edges. Install the seal so that each seal lip is oriented as illustrated. Using a suitable drift, press-fit oil seal until it becomes flush with timing chain case end face. Make sure the garter spring in the oil seal is in position and seal lip is not inverted.

59. Install the front timing cover case.

60. Install IVT control valve covers as follows:

a. Install new collared O–rings in front cover oil hole on both sides.

b. Install new seal rings on the IVT control covers.

c. Apply a 0.083–0.122 inch (2.1–3.1mm) bead of RTV sealant to the IVT control covers as illustrated.

61. Being careful not to move the seal ring from the installation groove, align the dowel pins on the chain case with the holes to install the IVT control covers.

62. Tighten the intake valve timing control cover bolts in the order illustrated to 100 inch lbs. (11 Nm).

63. Apply a 0.091–0.130 inch (2.3–3.3mm) bead of liquid gasket to the water pump cover and the chain tensioner cover and install covers. Tighten the bolts on the pump and cover to 97 inch lbs. (11 Nm).

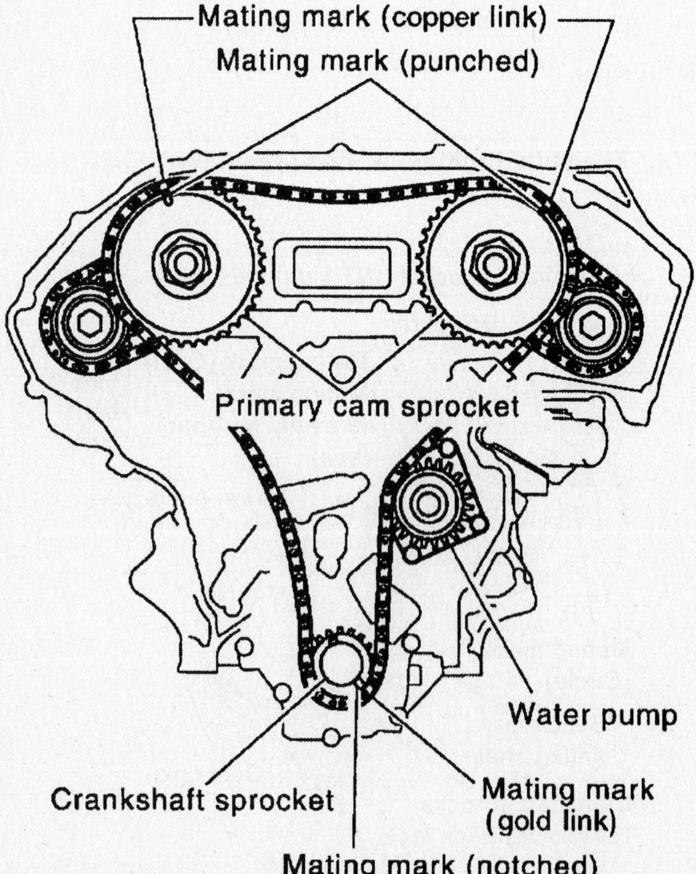

Install the primary timing chain so the match mark (punched) on the cam sprocket is aligned with the copper link on the timing chain, while the mating mark (notched) on the crank sprocket is aligned with the gold link on the timing chain–3.5L engine

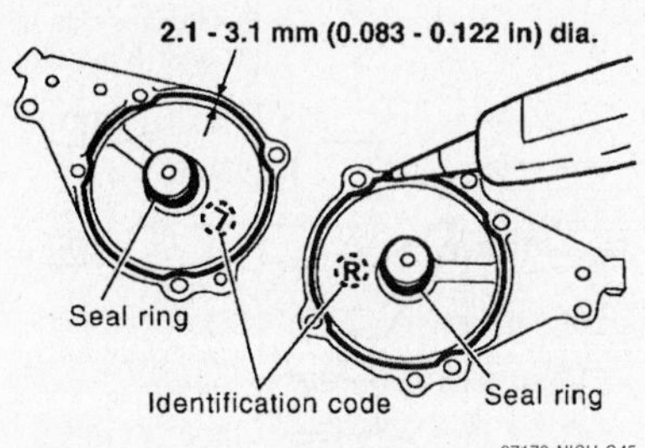

Apply a 0.083–0.122 inch (2.1–3.1mm) bead of RTV sealant to the IVT control covers–3.5L engine

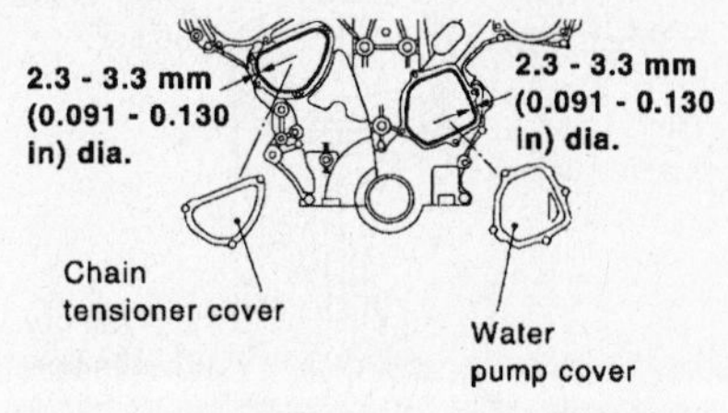

Apply a 0.091–0.130 inch (2.3–3.3mm) bead of liquid gasket to the water pump cover and the chain tensioner cover–3.5L engine

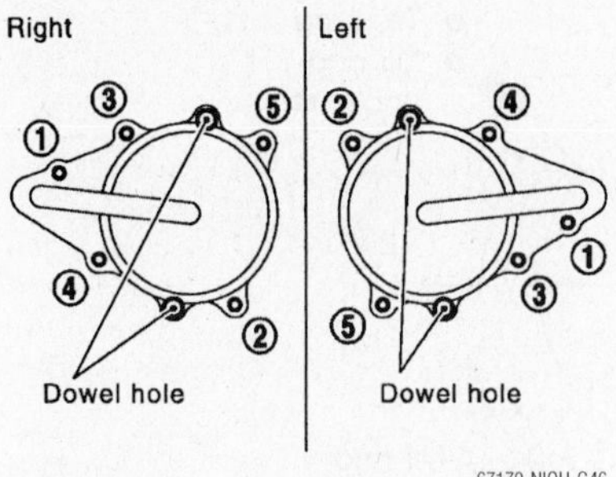

Tighten the intake valve timing control cover bolts in this order–3.5L engine

64. Install crankshaft pulley lubricate the pulley bolt with clean engine oil and tighten the bolt in two steps:
   a. Step 1: 32 ft. lbs. (44 Nm).
   b. Step 2: 60–65 degrees clockwise.
65. Rotate crankshaft pulley in a clockwise direction as viewed from front of the engine to confirm it turns smoothly.
66. Install the rocker covers in the reverse order of removal, make sure to perform the following:
67. Apply sealant to the areas on the front corners.
68. Tighten the rocker cover bolts in two steps in the sequence illustrated:
   a. Step 1: 17 inch lbs. (1.96 Nm).
   b. Step 2: 74 inch lbs. (8 Nm).
69. Install the remaining components is in reverse order of removal.

➡**If hydraulic pressure inside chain tensioner drops after removal/installation, slack in the guide may generate a pounding noise during and just after engine start. This is normal. Noise will stop after hydraulic pressure rises.**

Piston and Ring Positioning

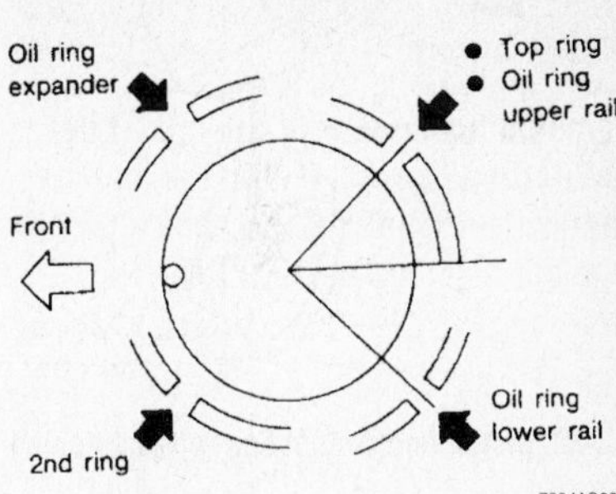

3.3L engines piston ring end-gap spacing

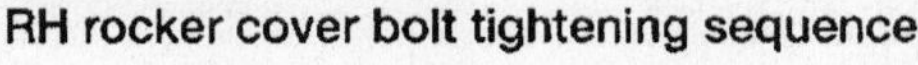

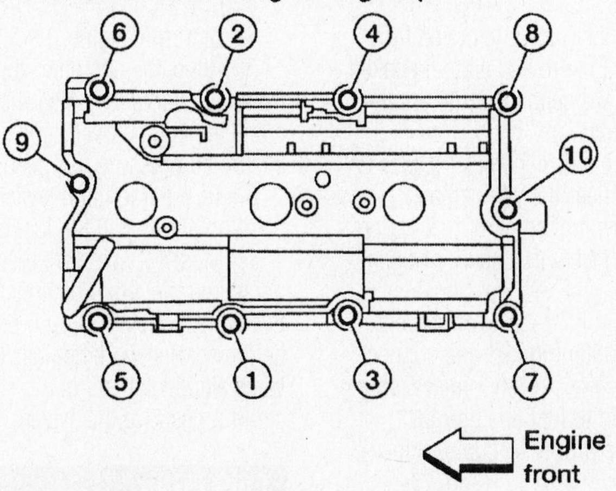

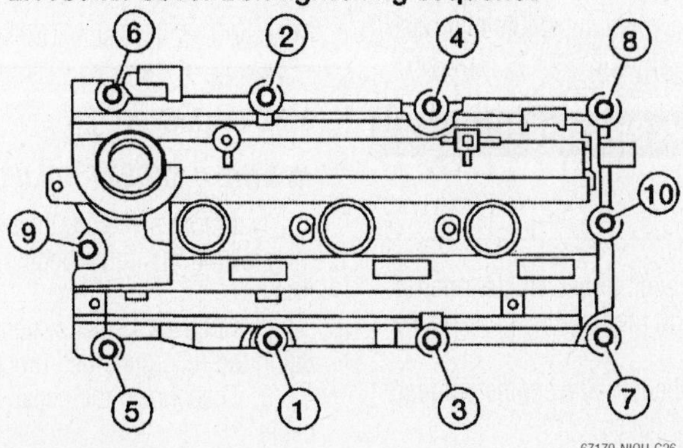

Tighten the rocker cover bolts in this sequence–3.5L engine

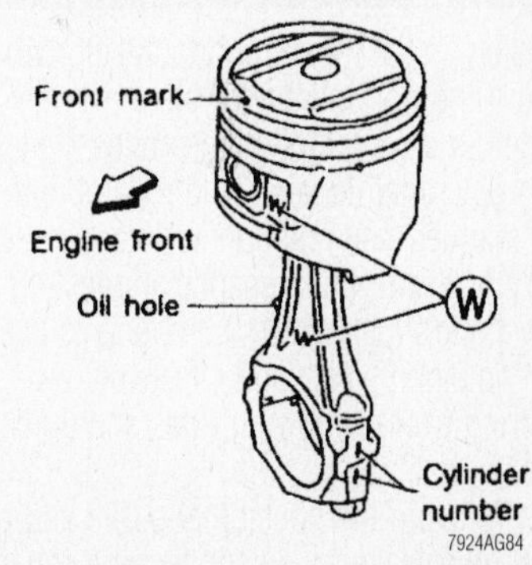

3.3L engines piston and connecting rod assembly positioning

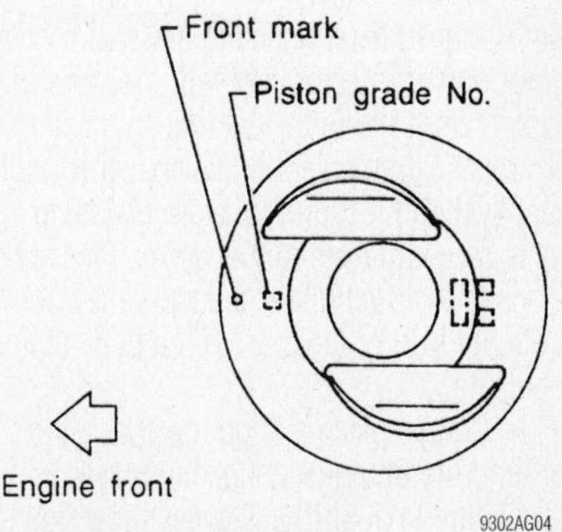

3.3L engine piston positioning

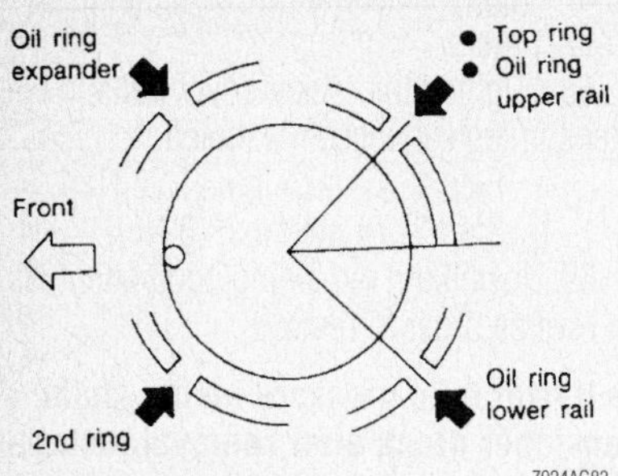

Piston ring end-gap spacing

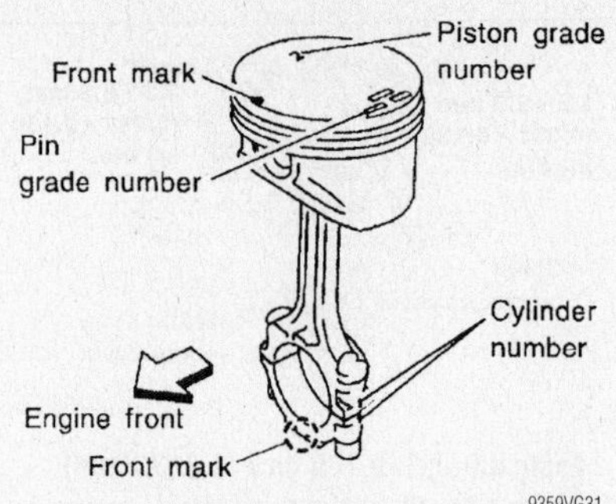

3.5L piston and connecting rod positioning

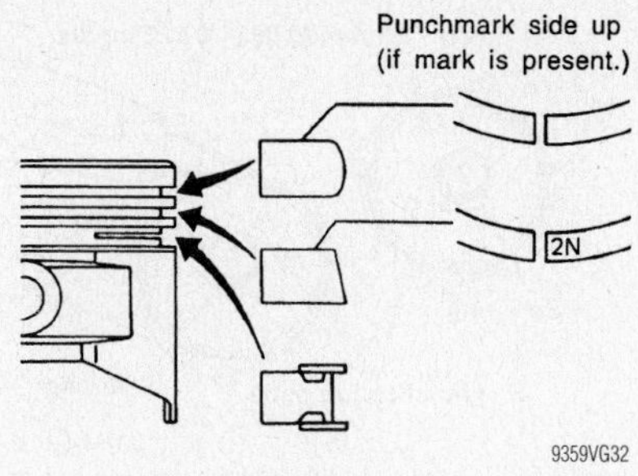

3.5L piston ring positioning

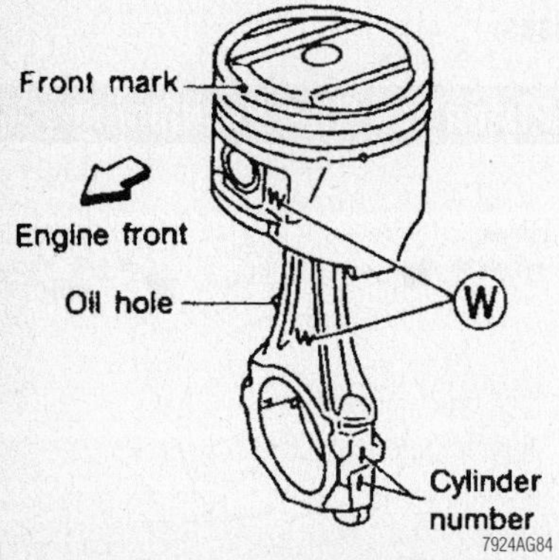

3.5L piston and connecting rod positioning

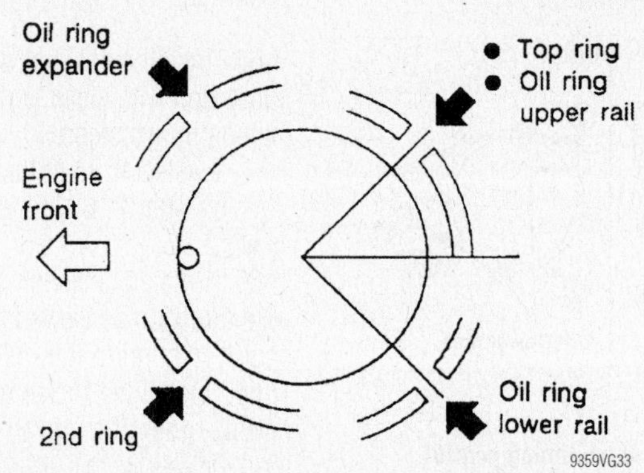

3.5L piston ring positioning

FUEL SYSTEM

Fuel System Service Precautions

Safety is the most important factor when performing not only fuel system maintenance but any type of maintenance. Failure to conduct maintenance and repairs in a safe manner may result in serious personal injury or death. Maintenance and testing of the vehicle's fuel system components can be accomplished safely and effectively by adhering to the following rules and guidelines.

- To avoid the possibility of fire and personal injury, always disconnect the negative battery cable unless the repair or test procedure requires that battery voltage be applied.
- Always relieve the fuel system pressure prior to disconnecting any fuel system component (injector, fuel rail, pressure regulator, etc.), fitting or fuel line connection. Exercise extreme caution whenever relieving fuel system pressure, to avoid exposing skin, face and eyes to fuel spray. Please be advised that fuel under pressure may penetrate the skin or any part of the body that it contacts.
- Always place a shop towel or cloth around the fitting or connection prior to loosening to absorb any excess fuel due to spillage. Ensure that all fuel spillage (should it occur) is quickly removed from engine surfaces. Ensure that all fuel soaked cloths or towels are deposited into a suitable waste container.
- Always keep a dry chemical (Class B) fire extinguisher near the work area.
- Do not allow fuel spray or fuel vapors to come into contact with a spark or open flame.
- Always use a back-up wrench when loosening and tightening fuel line connection fittings. This will prevent unnecessary stress and torsion to fuel line piping. Always follow the proper torque specifications.
- Always replace worn fuel fitting O-rings with new. Do not substitute fuel hose or equivalent, where fuel pipe is installed.

Fuel System Pressure

RELIEVING

1. Before servicing the vehicle, refer to the precautions in the beginning of this section.
2. Remove the left side engine compartment relay panel cover.
3. Locate and remove the fuel pump relay from the relay panel.
4. Start the engine.
5. Allow the engine to run until it stalls from fuel starvation. After the engine stalls, crank the engine over 2 more times to ensure all pressure has been released.
6. Turn the ignition switch to the **OFF** position and install the fuel pump relay.
7. Most service work that follows fuel pressure relief also requires that the negative battery cable (ground) be disconnected before service work begins. This also prevents accidental fuel pump energizing that could pressurize the system.

Fuel Filter

REMOVAL & INSTALLATION

3.3L Engine

IN-LINE—EXCEPT CALIFORNIA

1. Before servicing the vehicle, refer to the precautions in the beginning of this section.
2. Relieve the fuel system pressure using the recommended procedure.
3. Disconnect the negative battery cable.
4. Raise and safely support the vehicle.
5. Remove the fuel hose clamps.

6. Disconnect and plug the hoses to prevent leakage.

7. Remove the fuel filter from the bracket.

To install:

8. Install the fuel filter into the bracket with the arrow facing up, in the direction of the fuel travel to the engine.

9. Reconnect the fuel hoses.

10. Install and tighten the hose clamps. Verify that the clamps are properly tightened. System operating pressure is approximately 36 psi (248 kPa) and fuel will leak is connections are not properly made.

11. Lower the vehicle.

12. Reconnect the negative battery cable.

13. Check for leaks.

IN-LINE—CALIFORNIA

1. Before servicing the vehicle, refer to the precautions in the beginning of this section.

2. Relieve the fuel system pressure using the recommended procedure.

3. Disconnect the negative battery cable.

4. Raise and safely support the vehicle.

5. Remove the filter splash shield bolts.

6. Disconnect the lines from each end of the filter and plug the hoses to prevent leakage.

7. Loosen the filter bracket nuts.

8. Remove the fuel filter from the bracket.

To install:

9. Install the fuel filter into the bracket with the arrow facing forward. Tighten the bracket bolts to 44 inch lbs. (5 Nm).

10. Reconnect the fuel hoses.

11. Lower the vehicle.

12. Reconnect the negative battery cable.

13. Check for leaks.

Fuel Pump

REMOVAL & INSTALLATION

3.3L Engine

1. Before servicing the vehicle, refer to the precautions in the beginning of this section.

2. Properly relieve the fuel system pressure.

3. Disconnect the negative battery cable.

4. Raise and safely support the vehicle.

5. Remove the fuel tank as follows:

   a. Drain the fuel from the tank.

   b. Remove the filler protector.

   c. Disconnect the filler tube.

   d. Detach any electrical connectors related to the fuel pump and fuel level sending unit.

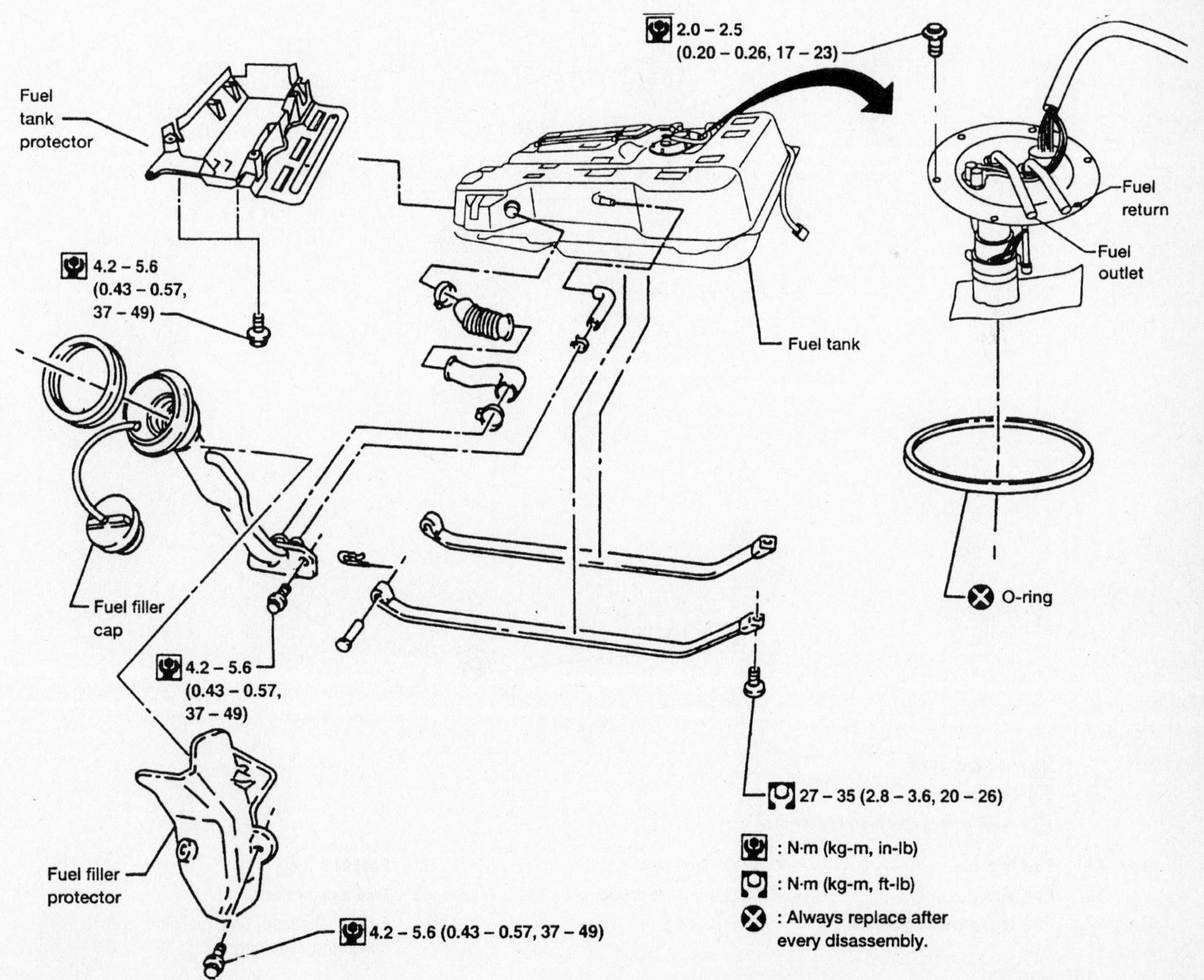

Fuel tank and related components

e. Detach the fuel line quick connectors.

f. Safely support the fuel tank.

g. Remove the tank mounting straps, then lower the tank out of the vehicle.

6. Remove the 6 fuel pump bolts.

7. Lift the fuel pump out of the fuel tank. Use care. The fuel level sensor and fuel pump and bracket must be tipped to remove it from the fuel tank. Do not lift the fuel sensor and pump assembly straight out of the fuel tank or damage to the level sensor may occur.

8. Remove the 2 bolts attaching the level sensor to the fuel pump.

9. Remove the fuel pump level sensor and the gasket.

10. Discard the gasket.

11. Remove the fuel pump from the bracket.

To install:

12. Position the fuel level sensor on the fuel pump and bracket and install the 2 bolts.

13. Install a new level sensor gasket. Carefully install the level sensor and pump assembly.

14. Install the 6 fuel pump bolts. Do not over-tighten the bolts. Tighten the bolts to just 17–23 inch lbs. (2–3 Nm).

15. Install the fuel tank in the reverse order of removal, be sure to tighten the tank mounting straps to 20–26 ft. lbs. (27–35 Nm).

16. Lower the vehicle. Refill the fuel tank as required.

17. Connect the negative battery cable. Verify that the fuel pump relay has been properly installed. Start the engine and check for proper operation.

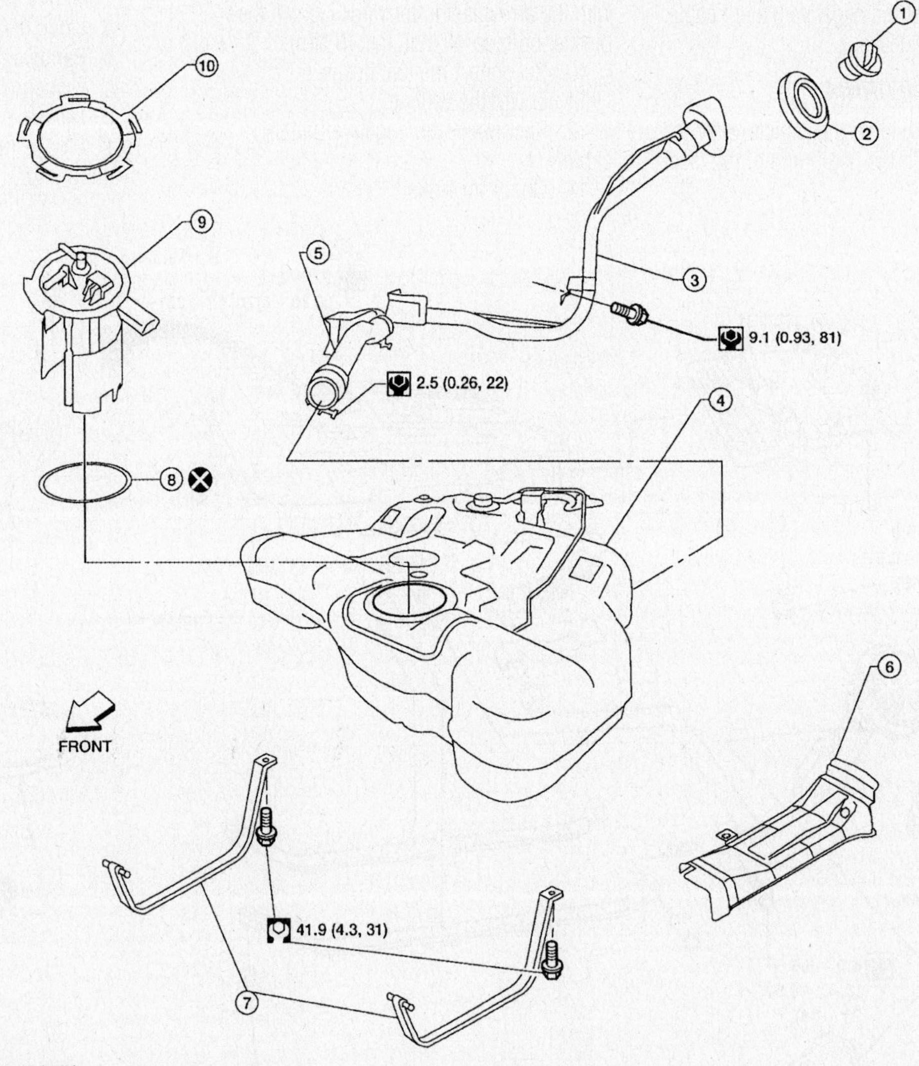

1. Fuel filler cap
2. Grommet
3. Fuel filler tube
4. Fuel tank
5. Fuel filler hose
6. Fuel tank protector
7. Fuel tank mounting straps
8. O-ring
9. Fuel level sensor unit, fuel filter, and fuel pump assembly
10. Lock ring

67170-NIQU-G77

Exploded view of the fuel tank and pump assembly–3.5L engine

3.5L Engine

1. Before servicing the vehicle, refer to the precautions in the beginning of this section.
2. Disconnect the negative battery cable.
3. Drain the fuel tank.
4. Open the fuel door and unscrew the fuel filler cap to release the pressure inside the fuel tank.
5. Release the fuel system pressure
6. Remove the center exhaust tube, with mufflers attached.
7. Disconnect the parking brake cables from the equalizer, then disconnect the three parking brake cable mounting brackets on each cable and position the cables out of the way.
8. Remove the fuel tank protector.
9. Disconnect the fuel filler hose, recirculation hose and EVAP canister hose at the fuel tank.
10. Disconnect the fuel tank mounting straps while supporting the fuel tank.
11. Lower the fuel tank to access the top of the fuel level sensor unit, fuel filter, and fuel pump assembly.
12. Disconnect the fuel level sensor unit, fuel filter, and fuel pump assembly electrical connector, and the fuel feed hose from the fuel level sensor unit, fuel filter, and fuel pump assembly.
13. Remove the fuel tank.
14. Remove the lock ring using a socket drive handle and tool J–16214.
15. Remove the fuel level sensor, fuel filter, and fuel pump assembly from the fuel tank.

CAUTION

Do not bend the float arm during removal.

16. Make sure the fuel level sensor, fuel filter, and fuel pump is free from defects and foreign materials.

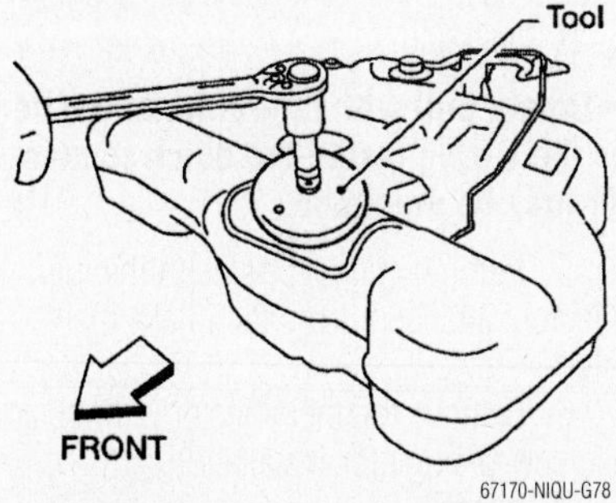

Remove the lock ring using a socket drive handle and tool J–16214–3.5L engine

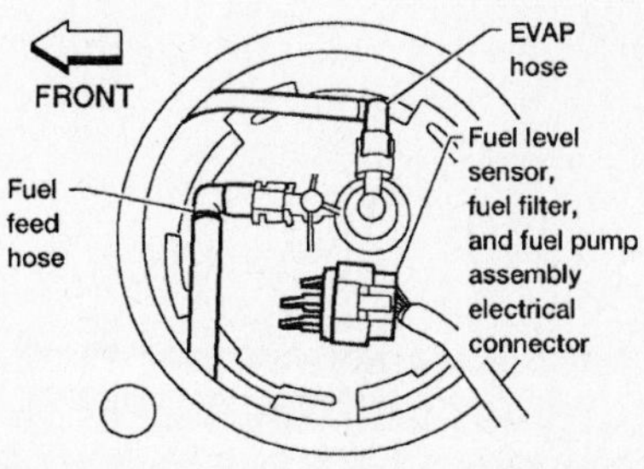

Install the fuel level sensor, fuel filter, and fuel pump assembly with the fuel feed hose facing the front of the vehicle–3.5L engine

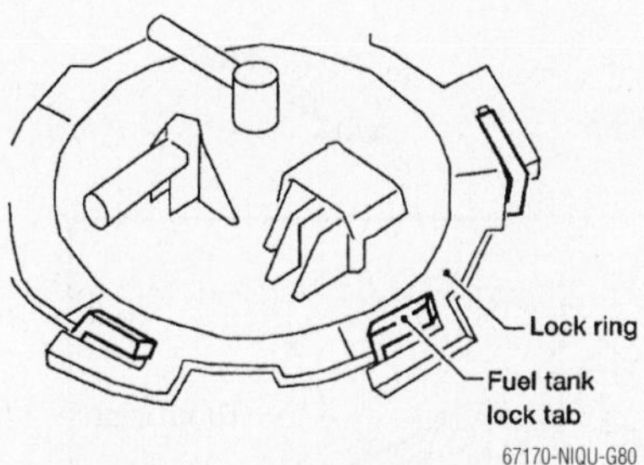

Turn the lock ring until the lock ring is fully rotated into the fuel tank lock tabs–3.5L engine

To install:

17. Install the fuel level sensor, fuel filter, and fuel pump assembly with the fuel feed hose facing the front of the vehicle.
18. Turn the lock ring until the lock ring is fully rotated into the fuel tank lock tabs as illustrated.
19. Install the fuel tank in the reverse order of removal.
20. Before tightening the fuel tank mounting straps, temporarily install the filler hose, recirculation hose, and signal hose. Tighten the straps to 31 ft. lbs. (41 Nm).
21. Connect the quick connector as follows:
    a. Check the connection for damage or any foreign materials.
    b. Align the connector with the tube, then insert the connector straight into the tube until a click is heard.
    c. After the tube is connected, make sure the connection is secure by pulling on the tube and the connector to make sure they are securely connected.
22. Turn the ignition switch to ON with the engine OFF to check the connections for fuel leaks with the electric fuel pump applying fuel pressure to the fuel lines.
23. Start the engine and let it idle to check that there are no fuel leaks at the fuel system tube and hose connections.

Fuel Injector

REMOVAL & INSTALLATION

3.3L Engine

1. Before servicing the vehicle, refer to the precautions in the beginning of this section.
2. Disconnect the negative battery cable.
3. If removing a rear injector, remove the upper intake manifold.
4. Disengage the injector electrical connection.

➡When removing the fuel injectors, use a screwdriver head socket to remove the injector cap screws.

5. Remove the injector cap screws and the cap.
6. Pull the injector from the fuel rail.
7. Remove and discard the injector O-rings.

To install:

➡Use new insulators and O-ring seals for assembly.

8. Install or connect the following:
    - Fuel injectors with the rail and tighten the fasteners to 8–11 ft. lbs. (11–15 Nm)
    - Fuel injector caps. Tighten the screws to 26–33 inch lbs. (3–4 Nm).
    - Injector electrical connections.
    - Intake manifold, if removed.
    - Negative battery cable.
9. Start the vehicle and check for leaks.

3.5L Engine

1. Before servicing the vehicle, refer to the precautions in the beginning of this section.
2. Remove the intake manifold collector.
3. Disconnect the fuel quick connector using tool J–45488 as follows:
    a. Remove the quick connector cap.
    b. With the sleeve side of tool facing quick connector, install the tool on to fuel tube.
4. Insert Tool into quick connector until sleeve contacts and goes no further. Hold the tool in that position.

CAUTION

Inserting the tool hard will not disconnect quick connector. Hold tool where it contacts and will go no further.

    a. Pull the quick connector straight

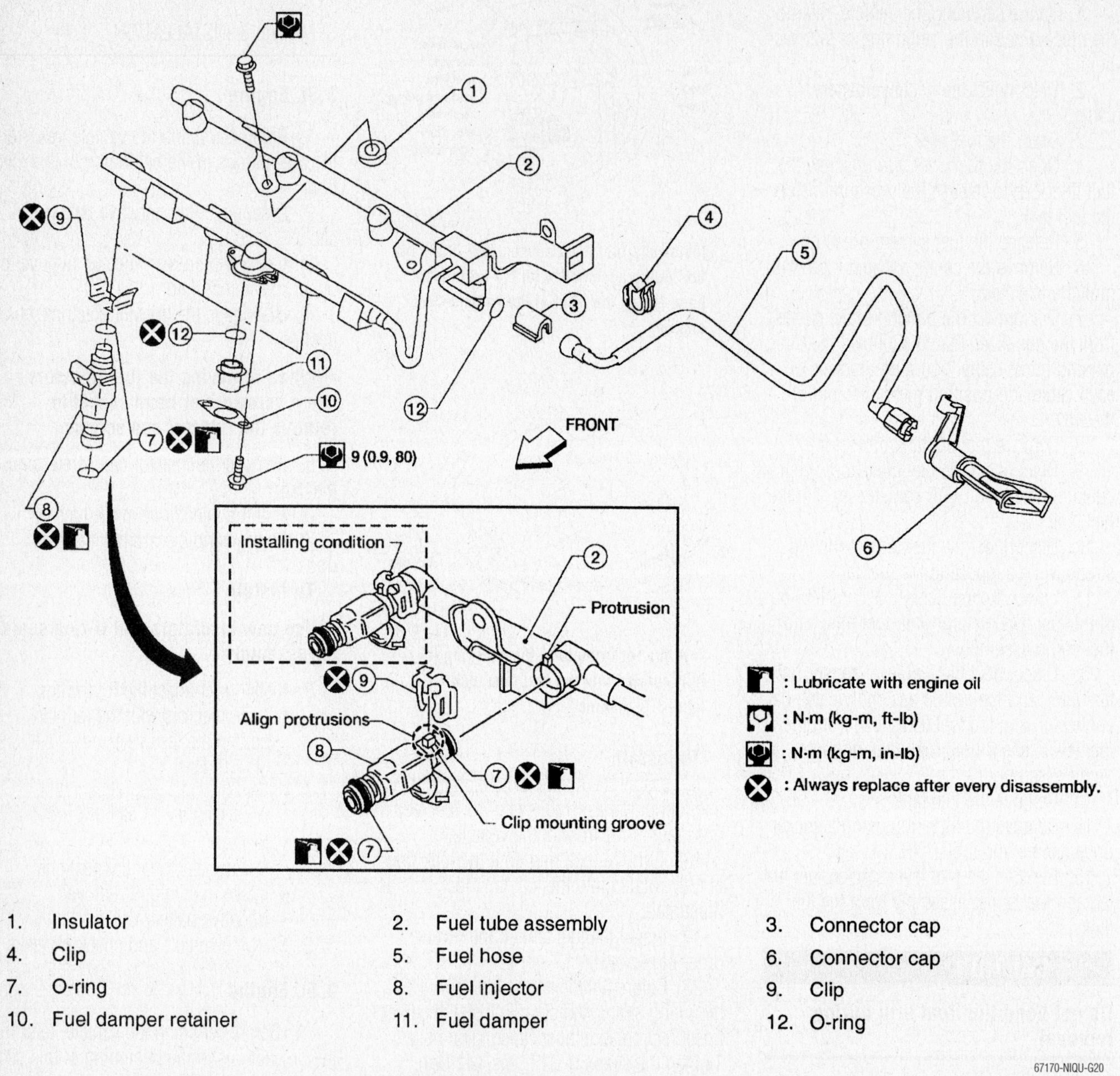

FRONT
9 (0.9, 80)
Installing condition
Protrusion
Align protrusions
Clip mounting groove
: Lubricate with engine oil
: N·m (kg-m, ft-lb)
: N·m (kg-m, in-lb)
: Always replace after every disassembly.
1. Insulator
2. Fuel tube assembly
3. Connector cap
4. Clip
5. Fuel hose
6. Connector cap
7. O-ring
8. Fuel injector
9. Clip
10. Fuel damper retainer
11. Fuel damper
12. O-ring
67170-NIQU-G20

Exploded view of the fuel rail and injector assembly–3.5L engine

Quick connector
Quick connector cap
Fuel tube
67170-NIQU-G21

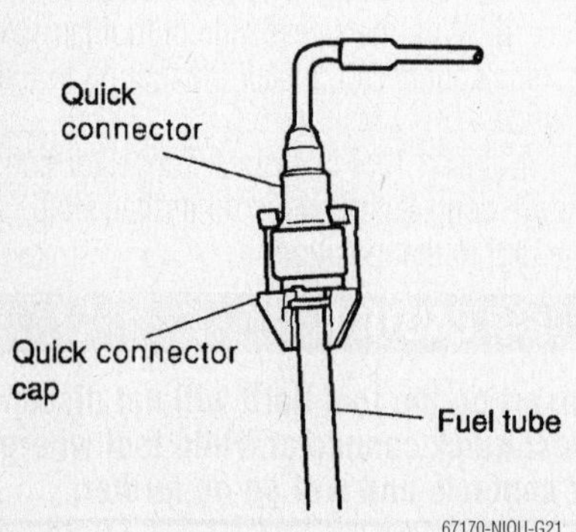

Quick connector components–3.5L engine

J-45488
Pull quick connector
Sleeve
Quick connector
A
Insert and retain
J-45488
Fuel tube
67170-NIQU-G22

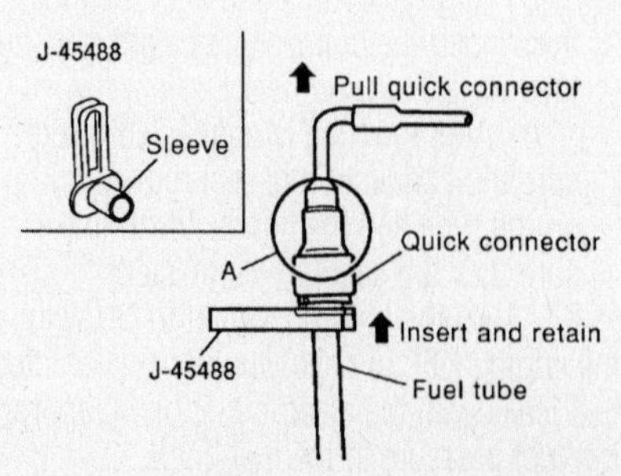

Use tool J-45488 to disconnect the quick connector–3.5L engine

out from the fuel tube. Pull the quick connector holding it at the (A) position, as illustrated.

➡**Do not pull with lateral force applied as the O-ring inside the quick connector may be damaged.**

5. Remove the fuel rail with the fuel injectors attached, from the intake manifold.

6. Remove the fuel injector O-rings and use new O-rings for installation.

To install:

7. Install the fuel rails with fuel injectors attached.

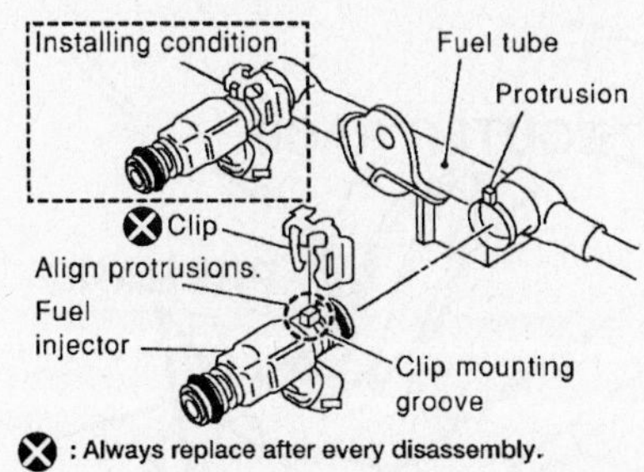

Position clips in grooves on the fuel injectors. Make sure that protrusions of fuel injectors are aligned with cutouts of clips after installation–3.5L engine

8. Install new O-rings. Lubricate the O-rings by lightly coating with new engine oil.

➡**Be careful not to damage the O-rings and surfaces for O-ring sealing surfaces. Do not expand or twist O-rings.**

9. Install new clips, position the clips in grooves on the fuel injectors. Make sure that protrusions of the fuel injectors are

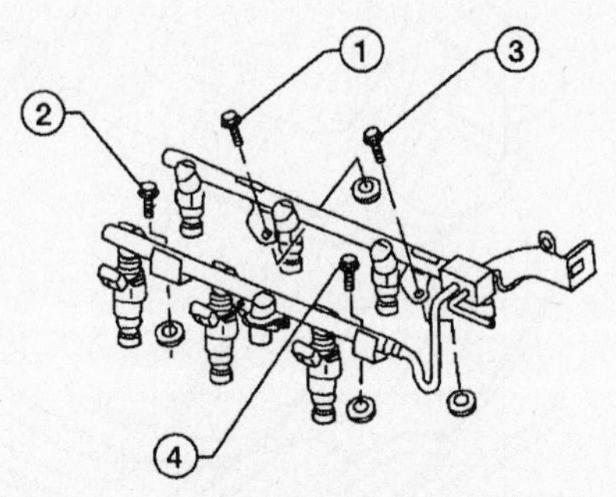

Tighten the fuel tube assembly bolts in this sequence–3.5L engine

aligned with the cutouts of clips after installation.

10. After properly inserting the fuel injectors onto the fuel tube assembly, check that the fuel tube protrusions are engaged with those of fuel injectors, and the flanges of the fuel tube assembly are fully engaged with the clips.

11. Tighten the fuel tube assembly bolts in the sequence illustrated, in two steps:

a. Step 1: 89 inch lbs. (10 Nm).

b. Step 2: 16 ft. lbs. (22 Nm).

12. Install the quick connector as follows:

a. Make sure no dirt or foreign objects are around the tube and quick connector to avoid damage.

b. Align the center to insert the quick connector straight onto the fuel tube.

c. Insert the fuel tube until a click is heard.

13. Install the remaining components is in the reverse of removal.

14. Make sure there is no fuel leakage at connections as follows:

a. Apply fuel pressure to fuel lines by turning ignition switch **ON** with the engine **OFF** and check for fuel leaks at connections.

b. Start the engine and rev it up and check for fuel leaks at connections. Use mirrors for checking on connections out of the direct line of sight.

DRIVE TRAIN

Automatic Transaxle Assembly

REMOVAL & INSTALLATION

2000–02 Models

1. Before servicing the vehicle, refer to the precautions in the beginning of this section.
2. Remove or disconnect the following:
   - Negative battery cable
   - Battery and tray
   - Resonator
   - Terminal cord assembly harness connector
   - Vacuum lines
   - Starter motor
   - Transaxle fluid from the unit
   - Halfshafts
   - Transaxle cooler hose and control cable
   - Front exhaust manifold
   - Crankshaft Position (CKP) sensor
   - Engine gusset and torque converter undercover
   - Bolts from the drive plate from the torque converter. Rotate the crankshaft to access all the bolts.
3. Support the transaxle with a suitable jack.
   - Front mount
   - Rear mount
   - Bolts attaching the transaxle to the engine
4. Carefully separate the transaxle assembly from the engine assembly. Lower the assembly from the vehicle.

To install:

5. Be sure that the transaxle is secured firmly to the transaxle jack.
6. Carefully raise the transaxle into the vehicle and align the transaxle to the engine assembly, making sure that the alignment dowels are positioned properly.
7. Install or connect the remaining components in the reverse order of removal. Refer to the accompanying transaxle torque

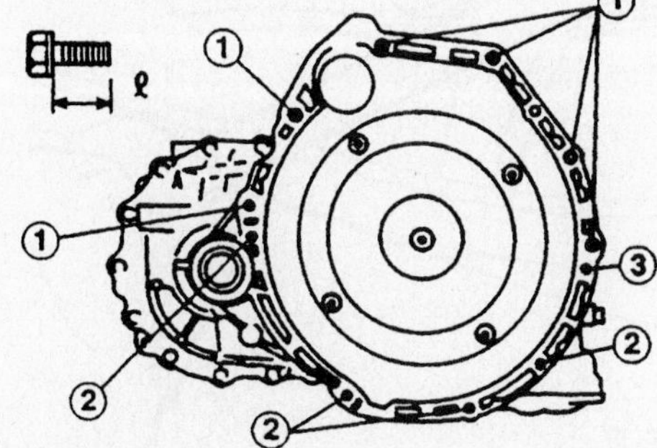

Bolt No.	Tightening torque N·m (kg-m, ft-lb)	ℓ mm (in)
1	39 - 49 (4.0 - 5.0, 29 - 36)	60 (2.36)
2	30 - 40 (3.1 - 4.1, 22 - 30)	25 (0.98)
3*	30 - 40 (3.1 - 4.1, 22 - 30)	25 (0.98)

*: TORX bolt

9348WG17

Transaxle torque specification and locations—3.3L engine

specification illustration for bolt locations and their specifications.

8. Connect the negative battery cable.
9. Fill the transaxle with the correct amount and type of fluid.
10. Start the engine.
11. Check for leaks and proper operation.

2004–05 Models

The transmission is removed along with the engine as an assembly from the vehicle. Please refer to the engine removal and installation procedure in this section for transmission removal and installation

Halfshaft

REMOVAL & INSTALLATION

2001–02 Models

1. Before servicing the vehicle, refer to the precautions in the beginning of this section.
2. Raise and safely support the vehicle.
3. Remove or disconnect the following:
   - Wheel
   - Fender splash shield
   - Cotter pin, nut retainer, and the hub retainer washers from the front hub assembly
   - Lower ball joint from the knuckle
   - Sway bar from the lower control arm at the sway bar link nut
   - Halfshaft and CV-joint from the wheel hub
4. Position a drain pan under the transaxle since some fluid may run out when the inner joint is disengaged from the transaxle.
5. A prybar is used to separate the inner CV-joint from the transaxle. Use great care that the prybar does not damage the transaxle case, differential oil seal, outer race or boot. If removing the left side halfshaft, position prybars on both sides of the outer race, between the outer race and the transaxle case. Gently pry outward to unseat the circlip.
6. When removing the right side halfshaft, it is not be necessary to remove the halfshaft bearing retainer bracket from the cylinder block. Remove the 3 bearing retainer bolts and pull the right side halfshaft CV-joint with the bearing retainer from the differential side gear.
7. Support the halfshafts and remove them from the vehicle. Use care not to damage the boots. Place the halfshafts on a flat, protected work area.

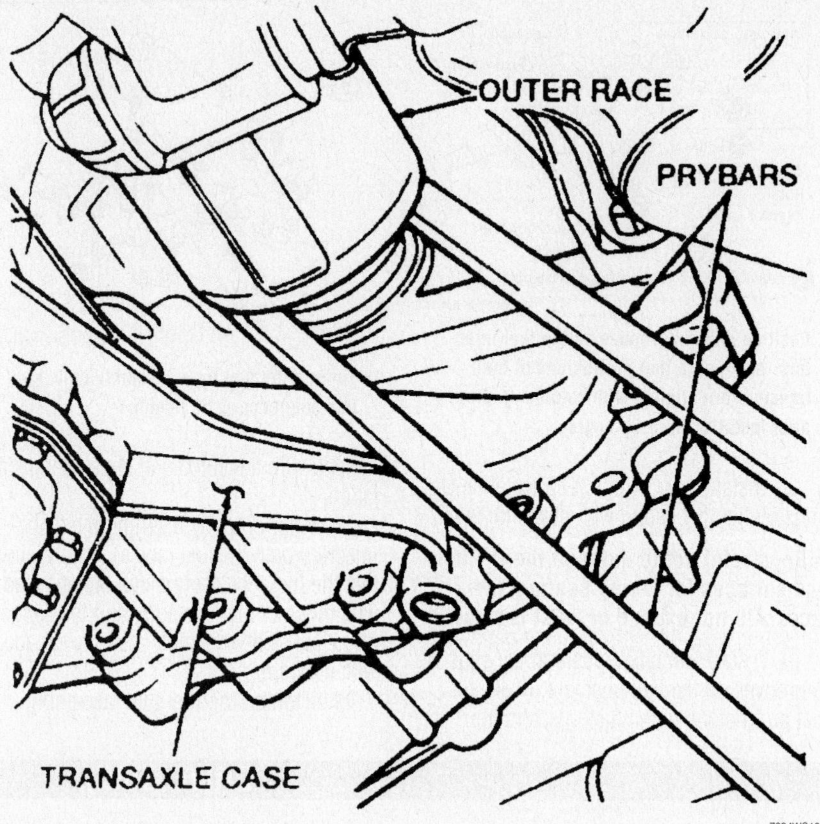

Removing the left side halfshaft by gently prying with 2 prybars to unseat the circlip

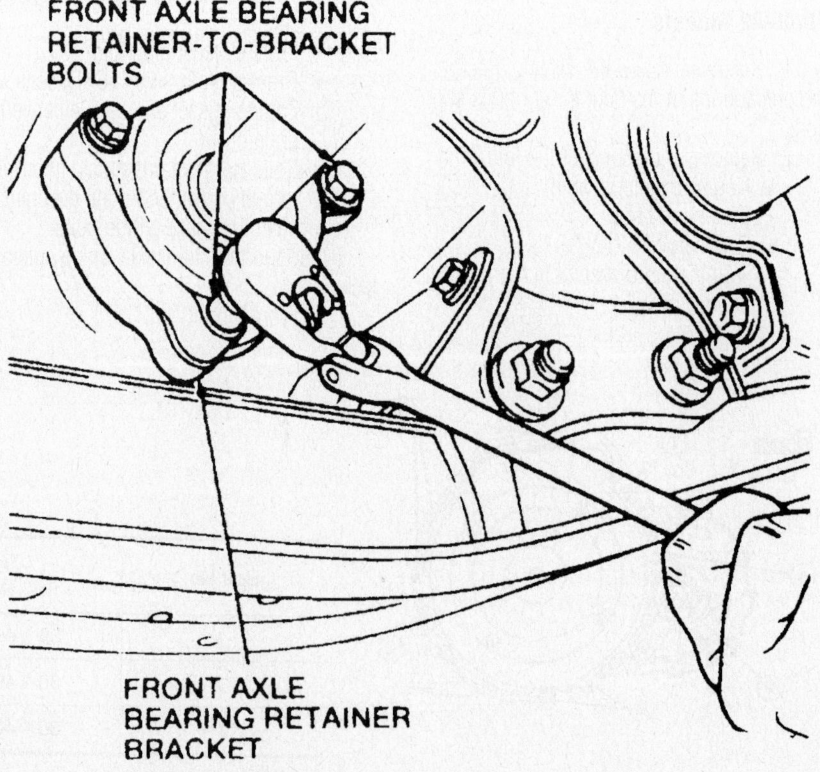

Right side halfshaft bearing retainer bracket

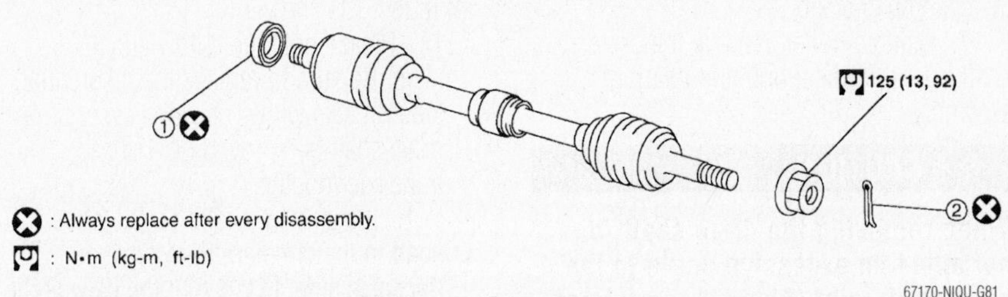

Exploded view of the left front halfshaft—2004–05 models

To install:

⁂ CAUTION

Do not reuse the circlip used on the left side halfshaft.

8. To prevent over-expanding the circlip, install the circlip carefully, starting one end in the shaft groove, then working the circlip over the CV-joint splined end. Always use a new circlip. No circlip is used on the right side halfshaft.

9. Inspect the CV-joint boots. If service is required, replace the CV-joint boots.

10. Inspect the differential oil seals. If damaged, the factory recommends using a hook-type puller and slide hammer arrangement to remove the seals. A seal driver is used to install the replacement differential oil seals.

11. If installing the left side halfshaft and CV-joint assembly, position the CV-joint so the splines are aligned with the differential side gear splines, then push the halfshaft joint into the differential case. As the circlip locks into the differential side gear groove, a click will be felt.

12. If installing the right side halfshaft and CV-joint assembly, simply push the CV-joint into the differential side gear. Position the bearing retainer onto the bearing retainer bracket that should still be on the cylinder block. Install the 3 bolts and tighten to 8–14 ft. lbs. (13–19 Nm).

13. Install or connect the following:
- Halfshaft
- Lower ball joint and tighten the lower ball joint stud nut to 52–63 ft. lbs. (71–86 Nm). Secure the nut with a new cotter pin.
- Sway bar link to the lower control arm and tighten the link nut to 12–16 ft. lbs. (16–22 Nm).
- Wheel outer bearing retainer, washer and axle nut. Tighten the hub nut to 174–231 ft. lbs. (235–314 Nm). Install the nut retainer and secure with a new cotter pin.
- Splash shield
- Wheel. Tighten the lug nuts to 72–87 ft. lbs. (98–118 Nm).

14. Lower the vehicle.

15. Check the transaxle fluid level.

16. Road test the vehicle to verify correct operation and no noise or vibration.

2004–05 Models

LEFT SIDE

1. Before servicing the vehicle, refer to the precautions in the beginning of this section.

2. Remove the wheel.

3. Remove the ABS sensor from the steering knuckle.

4. Remove the cotter pin, then remove the lock nut from the drive shaft.

5. Remove the brake hose lock plate, then remove the brake hose from the strut.

6. Remove the lower ball joint pinch bolt, then separate the lower ball joint from the steering knuckle.

7. Using a puller, remove the drive shaft from the wheel hub and bearing assembly.

⁂ CAUTION

When removing the drive shaft, do not apply an excessive angle to the drive shaft joint. Also be careful not to excessively extend the slide joint.

8. Remove the drive shaft from the transaxle using a suitable tool and drive shaft puller KV40107500.

9. Move the joint up and down, left and right, and in an axial direction. Check for any rough movement or significant looseness.

10. Check the boot for cracks or other damage, and for grease leakage. If damaged, disassemble the drive shaft to verify damage, and repair or replace as necessary.

To install:

11. Installation is in the reverse order of removal. Tighten the axle nut to 92 ft. lbs. (125 Nm).

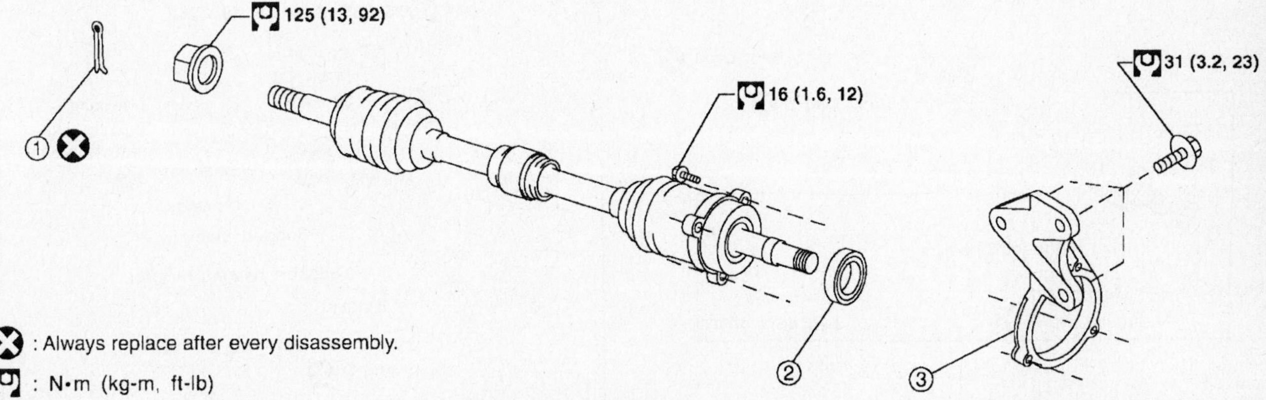

Exploded view of the right front halfshaft—2004–05 models

12. Please note the following:

a. Be sure to replace the differential side oil seal with a new one every time drive shaft is removed on 4 speed transaxle models.

b. Install new circlip on the drive shaft in the circular clip groove on the transaxle side. Make sure the new circlip on the drive shaft is securely fastened. After its insertion, try to pull the flange out of the slide joint by hand. If it pulls out, the circlip is not properly meshed with the transaxle side gear.

RIGHT SIDE

1. Before servicing the vehicle, refer to the precautions in the beginning of this section.
2. Remove the wheel.
3. Remove the ABS sensor from the steering knuckle.
4. Remove the cotter pin, then remove the lock nut from the drive shaft.
5. Remove the brake hose lock plate, then remove the brake hose from the strut.
6. Remove the lower ball joint pinch bolt, then separate the lower ball joint from the steering knuckle.
7. Using a puller, remove the drive shaft from the wheel hub and bearing assembly.

⁂ CAUTION

When removing the drive shaft, do not apply an excessive angle to the drive shaft joint. Also be careful not to excessively extend the slide joint.

8. Remove the support bearing bolts.
9. Remove the drive shaft from the transaxle using a suitable prytool.
10. Move the joint up and down, left and right, and in an axial direction. Check for any rough movement or significant looseness.
11. Check the boot for cracks or other damage, and for grease leakage. If damaged, disassemble drive shaft to verify damage, and repair or replace as necessary.

To install:

12. Installation is in the reverse order of removal. Tighten the axle nut to 92 ft. lbs. (125 Nm) and the support bearing bolts to 23 ft. lbs. (31 Nm).
13. Please note the following:

a. Be sure to replace the differential side oil seal with a new one every time drive shaft is removed on 4 speed transaxle models.

b. Install new circlip on the drive shaft in the circular clip groove on transaxle side. Make sure the new circlip on the drive shaft is securely fastened. After its insertion, try to pull the flange out of the slide joint by hand. If it pulls out, the circlip is not properly meshed with the transaxle side gear.

CV-Joint

OVERHAUL

2000–02 Models

INNER

1. Before servicing the vehicle, refer to the precautions in the beginning of this section.

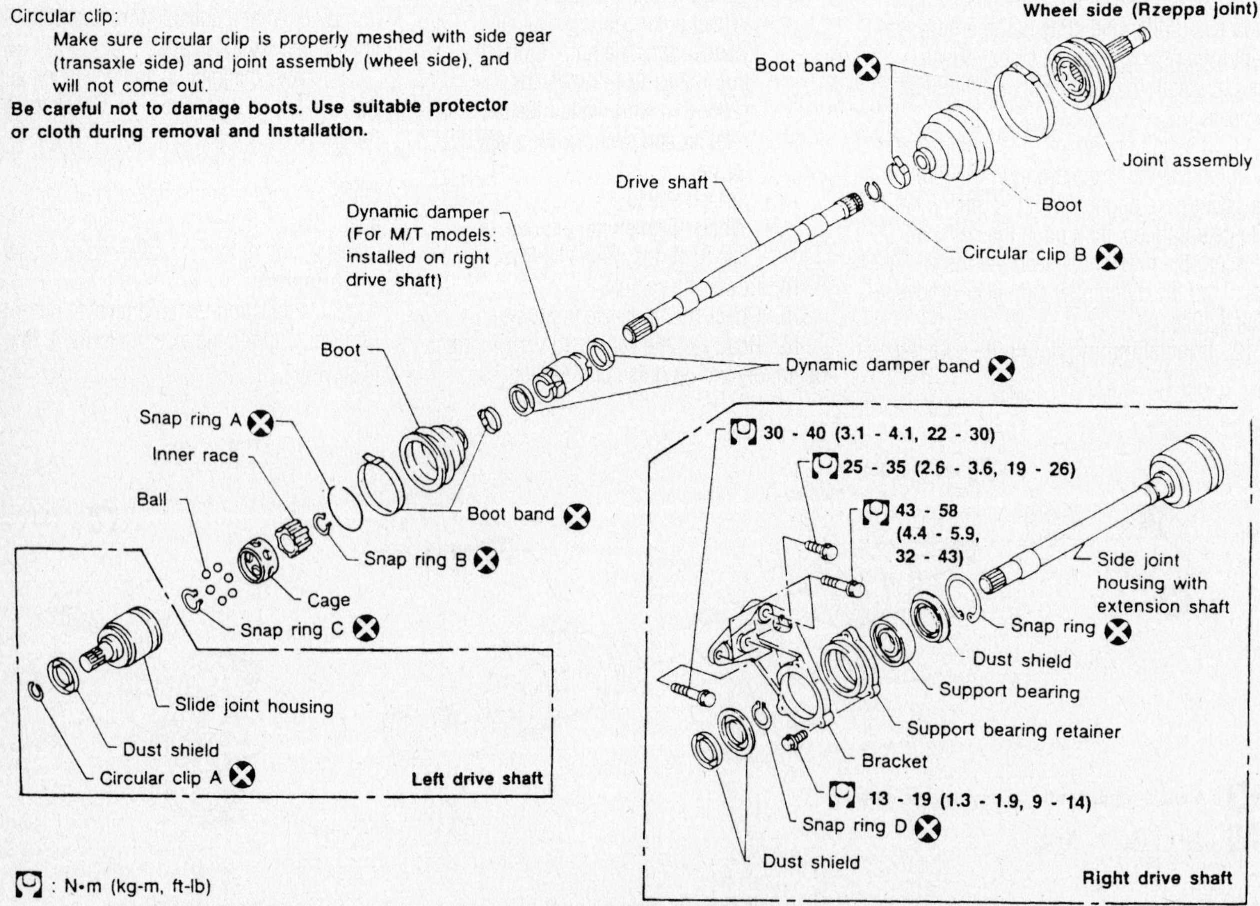

Exploded view of the halfshafts and related components

89617G07

The inner CV-joint uses a large C-clip to retain the ball and cage assembly in the outer housing

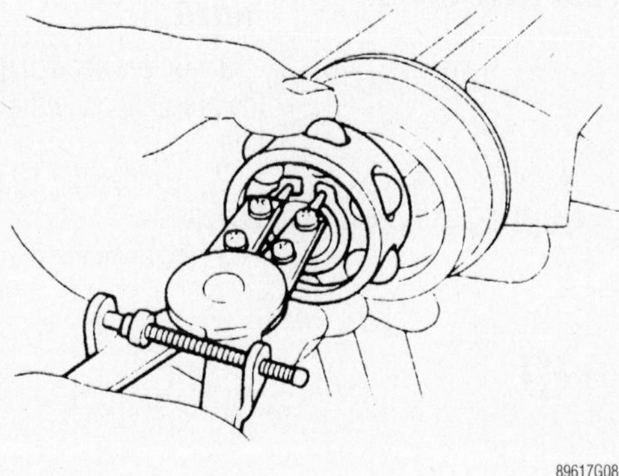
89617G08

After the outer housing is removed, the ball and cage assembly can slide from the shaft by removing the C-clip

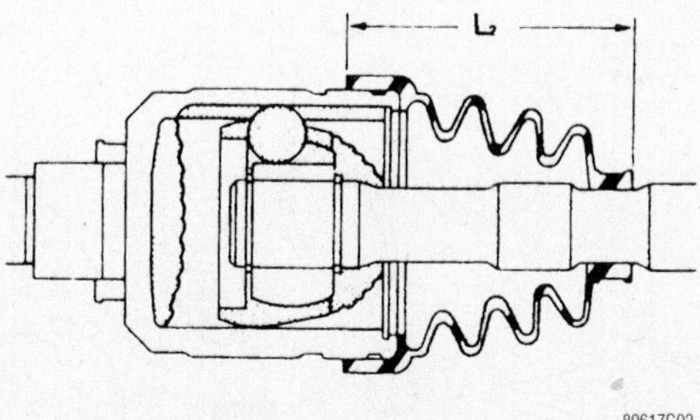

89617G02

Make sure to properly position the boot before tightening the boot clamps

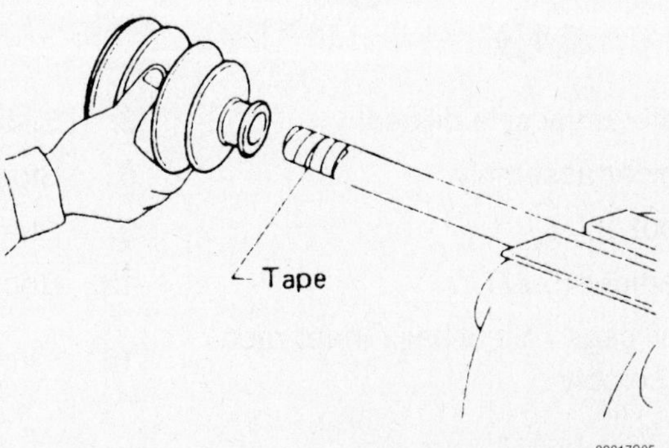

89617G05

Use vinyl tape and wrap the end of the shaft to protect the boot during installation

2. Remove the boot bands.
3. Matchmark the slide joint housing and inner race, prior to separating the joint assembly.
4. Pry off the snapring and remove the ball cage, inner race and balls as a unit.
5. Remove the snapring and withdraw the boot.

To install:

➡Cover the halfshaft serrations with tape, so as not to damage the boot.

6. Thoroughly clean all parts in solvent and dry with compressed air. Check parts for evidence of damage and replace as necessary.
7. Install the boot and new boot band on the halfshaft.
8. Install a new inner snapring.
9. Install the ball cage, inner race and balls as a unit. Confirm that the matchmarks are aligned.
10. Install a new outer snapring.
11. Pack the CV-joint with 5.0–6.0 ounces (165–175 g) of grease.
12. Ensure that the boot is properly installed on the halfshaft groove.
13. Set the boot so that it does not swell or deform when its length is 3.86 in. (98mm).
14. Lock the new boot bands securely.

OUTER

1. Before servicing the vehicle, refer to the precautions in the beginning of this section.

The joint on the wheel side cannot be disassembled.

2. Prior to separating the joint assembly, matchmark the halfshaft and joint assembly.
3. Separate the joint using a slide hammer.
4. Remove the boot bands and the boot.

To install:

5. Thoroughly clean all parts in solvent and dry with compressed air. Check parts for evidence of damage and replace as necessary.

➡Cover the halfshaft serrations with tape, so as not to damage the boot.

6. Install the boot and small boot band on the halfshaft.
7. Set the joint assembly onto the halfshaft and align the matchmarks.
8. Attach the joint assembly to the halfshaft by lightly tapping the serrated end with a plastic hammer.

➡Using a metal hammer may damage the threads on the end of the joint.

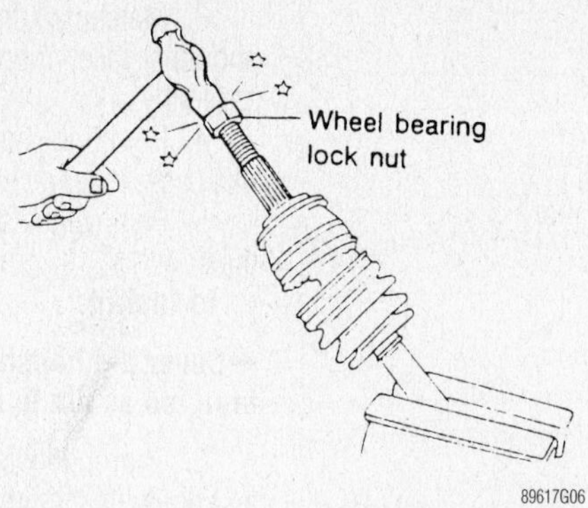

Use an old nut to protect the threads when tapping the outer CV-joint onto the shaft

9. Pack the CV-joint with 4.76–5.11 ounces (135–145 g) of grease.
10. Ensure that the boot is properly installed on the halfshaft groove.
11. Set the boot so that it does not swell or deform when its length is 3.82 in. (97mm).
12. Lock the new boot bands securely.

2004–05 Models

OUTER

1. Before servicing the vehicle, refer to the precautions in the beginning of this section.
2. Remove the axle halfshaft from the vehicle.
3. Remove the CV-joint boot clamps and push the boot away from the joint.
4. Remove the CV-joint from the axle shaft by tapping it with a brass hammer.

To install:

➡Use new circlips and boot clamps for assembly.

5. Install the CV-joint to the axle shaft by tapping it with a brass hammer.
6. Pack the joint with grease.
7. Install the boot clamps.
8. Install the axle halfshaft to the vehicle.

INNER

1. Before servicing the vehicle, refer to the precautions in the beginning of this section.
2. Remove the axle halfshaft from the vehicle.
3. Remove the plug seal by tapping

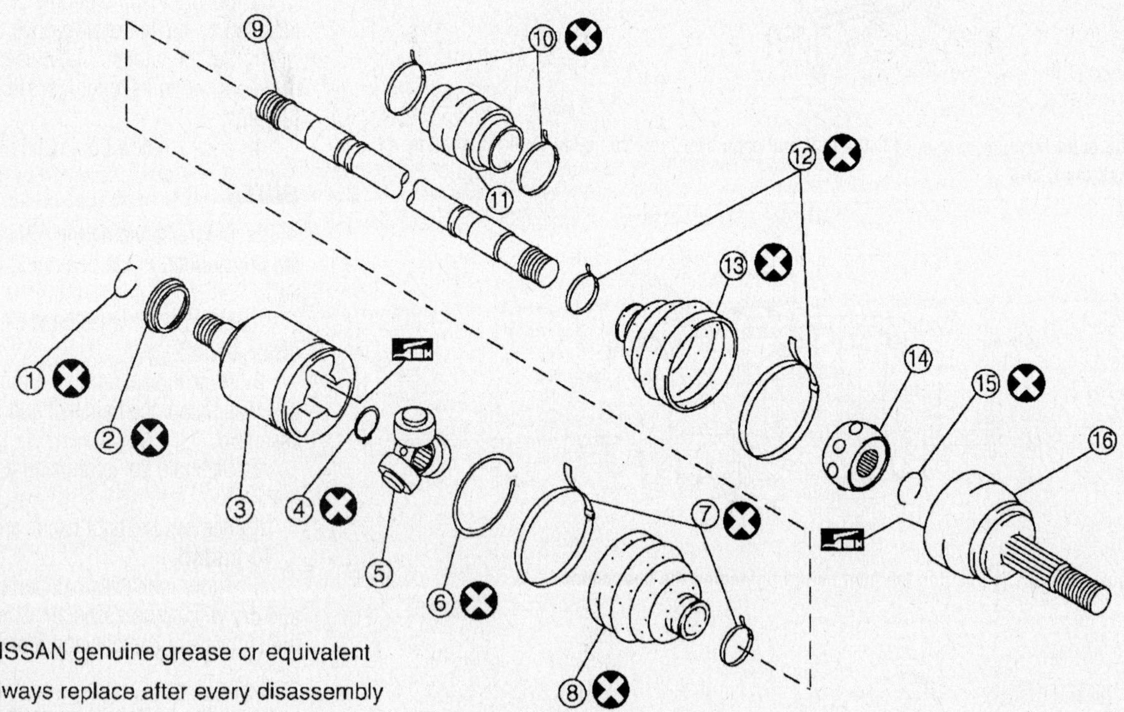

1. Circlip
2. Differential side oil seal
3. Slide joint housing
4. Snap ring
5. Spider assembly
6. Stopper ring
7. Boot band
8. Boot
9. Shaft
10. Damper band (5 A/T)
11. Damper (5 A/T)
12. Boot band
13. Boot
14. Ball cage / Steel ball / Inner race assembly
15. Circlip
16. Joint sub-assembly

67170-NIQU-G83

Left side front halfshaft components exploded view—2004–05 models

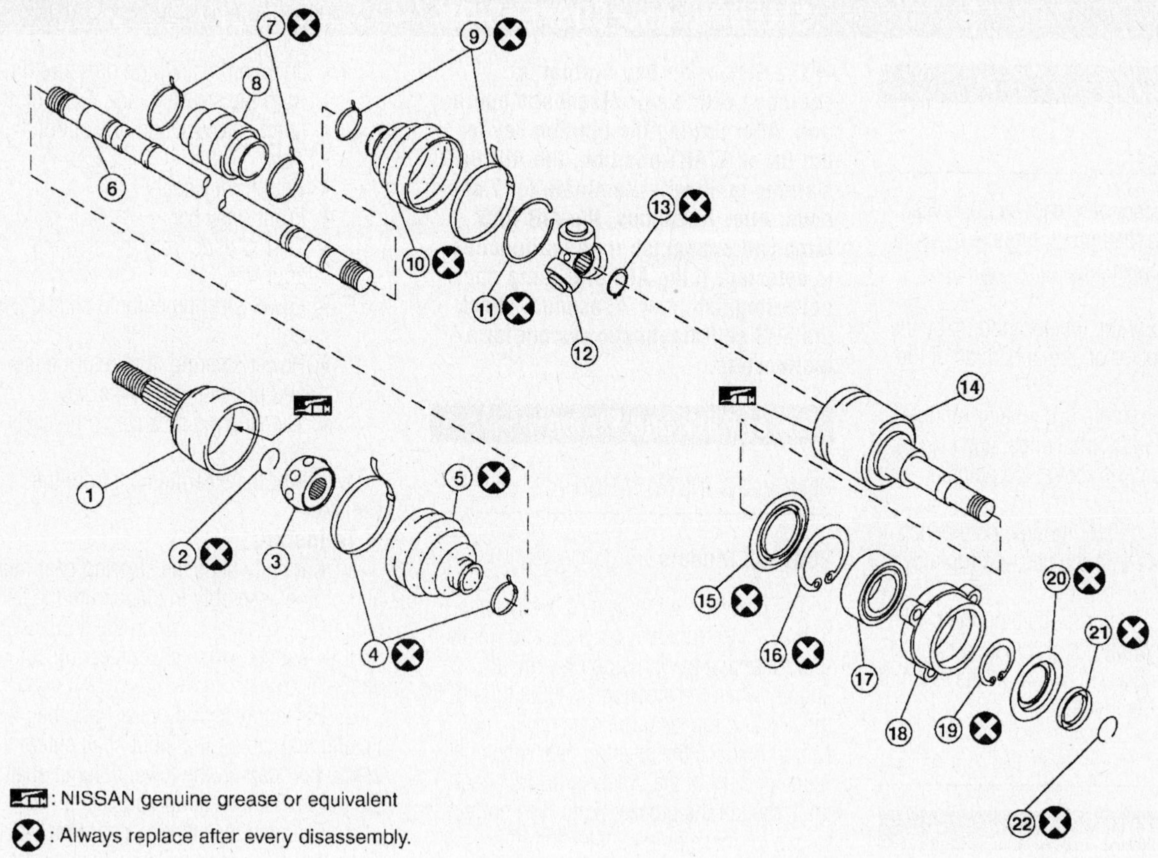

1. Joint sub-assembly	2. Circlip	3. Ball cage / Steel ball / Inner race assembly
4. Boot bands	5. Boot	6. Shaft
7. Damper bands (5 A/T)	8. Damper (5 A/T)	9. Boot band
10. Boot	11. Stopper ring	12 Spider assembly
13. Snap ring	14. Slide joint housing	15. Dust cover
16. Snap ring	17. Bearing	18. Bracket
19. Snap ring	20. Dust cover	21. Differential side oil seal
22. Circlip		

67170-NIQU-G84

Right side front halfshaft components exploded view—2004–05 models

around the joint housing flange with a brass hammer.

4. Remove or disconnect the following:
   - CV-joint boot clamps
   - Snapring
   - Spider assembly
   - CV-joint housing
   - CV-joint boot

To install:

➡Use new snaprings and plug seals for assembly.

5. Install or connect the following:
   - CV-joint boot
   - CV-joint housing
   - Spider assembly
   - Snapring. Pack the joint with grease.
   - CV-joint boot clamps
   - Plug seal
6. Install the axle halfshaft to the vehicle.

STEERING AND SUSPENSION

Air Bag

PRECAUTIONS

Several precautions must be observed when handling the inflator module to avoid accidental deployment and possible personal injury.

- Never carry the inflator module by the wires or connector on the underside of the module.
- When carrying a live inflator module, hold securely with both hands, and ensure that the bag and trim cover are pointed away.
- Place the inflator module on a bench or other surface with the bag and trim cover facing up.
- With the inflator module on the bench, never place anything on or close to the module which may be thrown in the event of an accidental deployment.

DISARMING

CAUTION

To avoid rendering the Supplemental Restraint System (SRS) inoperative, which could lead to personal injury or death in the event of a severe frontal collision, extreme caution must be taken when servicing the electrical related systems.

➡All SRS electrical wiring harnesses and connectors are covered with YELLOW outer insulation. Do not use electrical test equipment on any circuit related to the SRS (air bag) sensors. When installing SRS components, always install with the arrow marks facing the front of the vehicle.

Disarming

To disarm the Supplemental Restraint System (SRS) system turn the ignition switch to the **OFF** position. Then, disconnect the both battery cables starting with the negative cable first and wait at least 10 minutes after the cables are disconnected. Be sure to insulate the battery terminal ends.

Arming

To arm the Supplemental Restraint System (SRS) system turn the ignition switch to **OFF** position. Connect the both battery cables starting with the positive cable first.

➡The SRS or air bag system is equipped with a self-diagnostic operation. After turning the ignition key to the ON or START position, the AIR BAG warning lamp will illuminate for 7 seconds. After 7 seconds, the AIR BAG lamp will extinguish if no malfunction is detected. If the AIR BAG lamp does not extinguish after 7 seconds, check the SRS self-diagnostic system for a malfunction.

Power Rack and Pinion

REMOVAL & INSTALLATION

2001–02 Models

The power steering gear is held in position by 2 steering gear brackets and insulators. Note that the housing may move slightly when the steering wheel is turned. If the housing moves more than 0.080 inch (2mm), replace the steering gear insulators. If one or both of the brackets move, check the torque of the bracket bolts. The correct torque for these bolts is 54–72 ft. lbs. (73–97 Nm).

1. Before servicing the vehicle, refer to the precautions in the beginning of this section.
2. Place a drain pan under the steering rack.
3. Remove or disconnect the following:
   - Brake master cylinder remote reservoir bracket screws. Position the reservoir out of the way and secure with wire.
   - Junction block/high pressure line from the steering rack. Position the junction block and line out of the way.
   - Both front wheels
   - Front sway bar
   - Tie rod ends from the steering knuckles
   - Lower steering column shaft clamp bolt
   - Power steering fluid return hose and position out of the way.
   - The steering rack clamp bracket bolts
4. Lower the steering rack from the vehicle.

To install:

5. Carefully slide the steering gear rack and pinion assembly in place from the left side of the vehicle. Position the input shaft so it is just below the lower steering column shaft clamp.
6. Raise the steering gear until the plastic aligning tab on the input shaft enters the clamp bolt gap on the lower column shaft. Do not install the clamp bolt yet.
7. Examine the steering gear brackets. They should be marked UP with arrows pointing to one end of the bracket. Be sure the brackets are installed correctly. Tighten the steering gear bracket bolts to 54–72 ft. lbs. (73–97Nm) in sequence, working counterclockwise from the number 1 bolt (upper right side).
8. Install or connect the following:
   - Fluid return line to the steering gear
   - Steering column shaft clamp bolt.

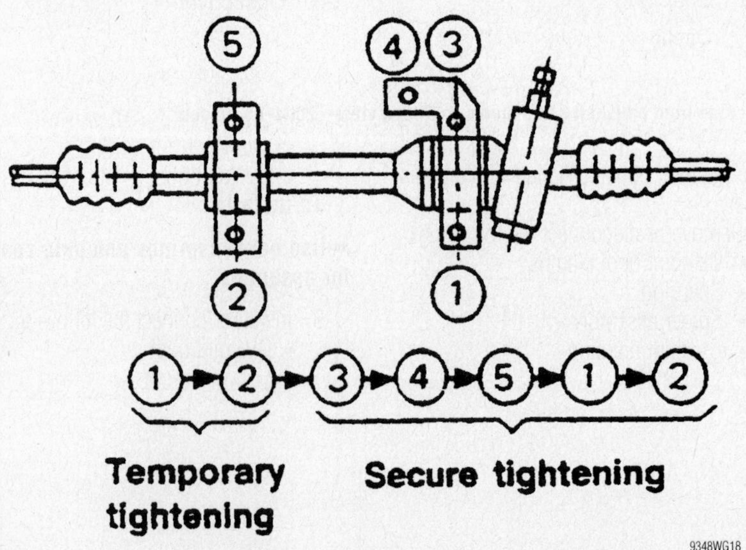

Tighten the power steering rack mounting bolts in the sequence shown

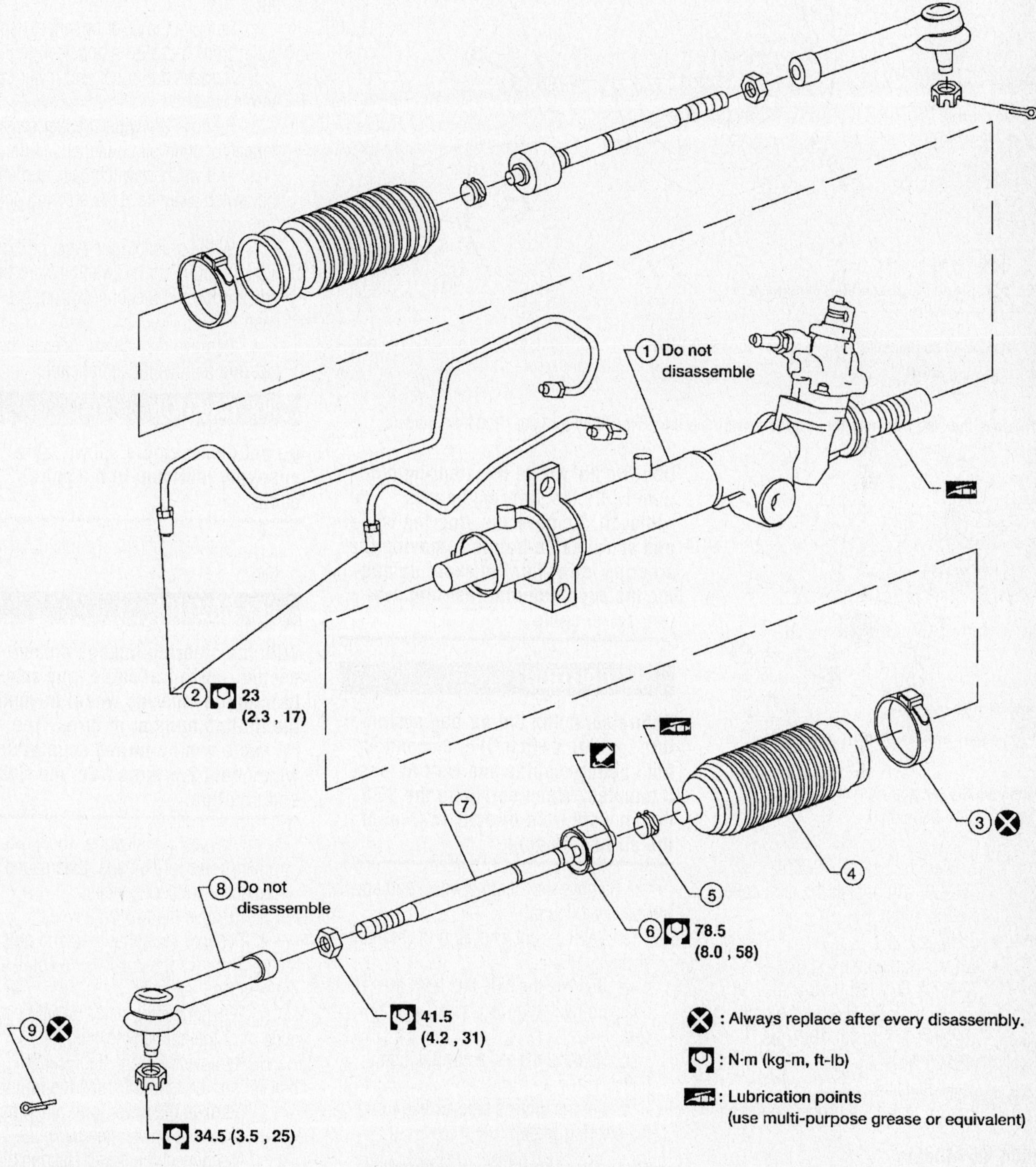

Exploded view of the steering gear mounting with torque specifications—2004–05 models

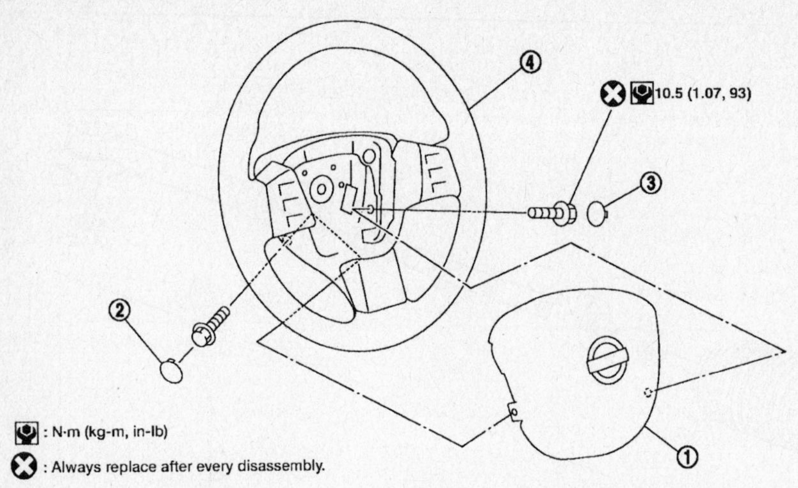

Remove the side lids from the retainers retaining the drivers side air bag—2004–05 models

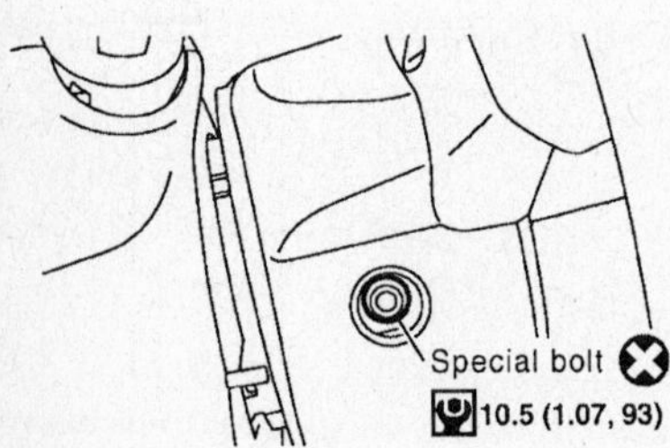

Remove the left and right side air bag bolts—2004–05 models

Tighten the bolt to 17–22 ft lbs. (24–29 Nm). Install the dust cover.

- Tie rod ends
- Stabilizer bar
- Wheel. Tighten the lug nuts to 72–87 ft. lbs. (98–118 Nm).
- Junction block. Tighten the high-pressure line to 11–18 ft. lbs. (15–25 Nm).
- Brake master cylinder reservoir

9. Check for leaks and proper operation.

2004–05 Models

1. Before servicing the vehicle, refer to the precautions in the beginning of this section.

CAUTION

The rotation of the driver air bag spiral cable is limited. If the steering gear must be removed, set the front wheels in the straight-ahead direction. Do not rotate the steering column while the steering gear is removed. Remove the steering wheel and spiral cable before removing the steering lower joint to avoid damaging the supplemental restraint system spiral cable.

CAUTION

Before servicing the air bag system, turn ignition switch OFF, disconnect both battery cables and wait at least 3 minutes. When servicing the SRS, do not work from directly in front of the air bag module.

2. Remove the steering wheel and spiral gear as follows:
   a. Set the front wheels in the straight-ahead position.
   b. Remove the side lids from the retainers retaining the driver's side air bag.
   c. Remove the left and right side bolts.
   d. Lift the driver's side air bag from the steering wheel.

➡When Disconnecting or connecting the air bag harness or horn connectors, always pull up to release the black locking tab prior to removing the connector from the air bag component. Always push down to lock the black locking tab after installing the connector to the air bag component. When locked, the black locking tab is level with the connector housing.

   e. Disconnect the air bag harness and horn connectors, then remove the driver air bag module.
   f. Remove the steering wheel center nut.
   g. Remove the steering wheel using puller tools J–1859–A and J–42578.
   h. Remove the upper and lower column covers.
   i. Remove the wiper washer switch connector, then pinch the tabs at the wiper and washer switch base and slide the switch away from the steering column.
   j. While pressing the tabs, pull the light and turn signal switch toward the driver door and disconnect from the base.
   k. Remove the screws, release the tab, and remove the spiral cable.

CAUTION

Do not disassemble spiral cable. Do not apply lubricant to the spiral cable.

   l. Remove the spiral cable connectors.

CAUTION

With the steering linkage disconnected, the spiral cable may snap by turning the steering wheel beyond the limited number of turns. The spiral cable can be turned counterclockwise about 2.5 turns from the right end position.

   m. Inspect the steering wheel near the puller holes for damage. If damaged, replace the steering wheel.
3. Remove the two front wheels.
4. Remove the cotter pins and nuts, then disconnect the outer tie-rod ends from the knuckle.
5. Disconnect the outer stabilizer bar ends from the connecting rods.
6. Remove the front stabilizer bar bracket rear bolts and loosen the front bolt.
7. Remove the lower joint pinch bolt.
8. Drain the power steering fluid.
9. Disconnect the power steering high and low pressure lines from the steering gear.
10. Position the stabilizer bar up and out of the way.
11. Remove the two gear housing mounting bolts. Do not remove the gear housing mounting bracket from the gear housing.
12. Remove the power steering gear and linkage assembly.

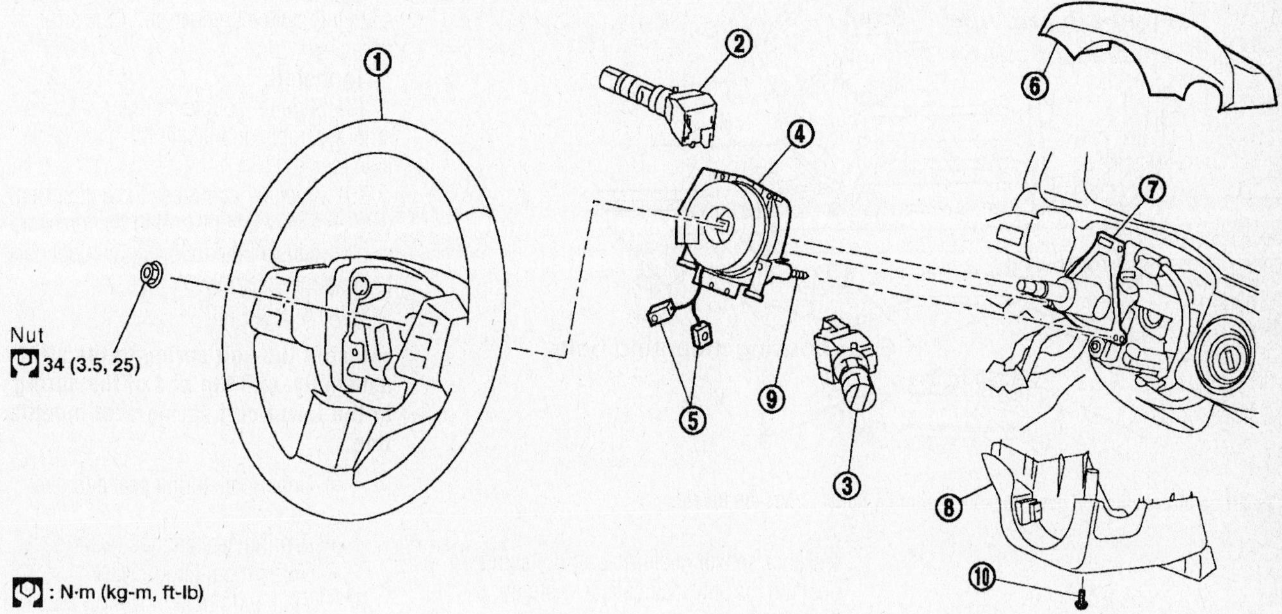

Exploded view of the steering wheel, spiral cable and air bag components—2004–05 models

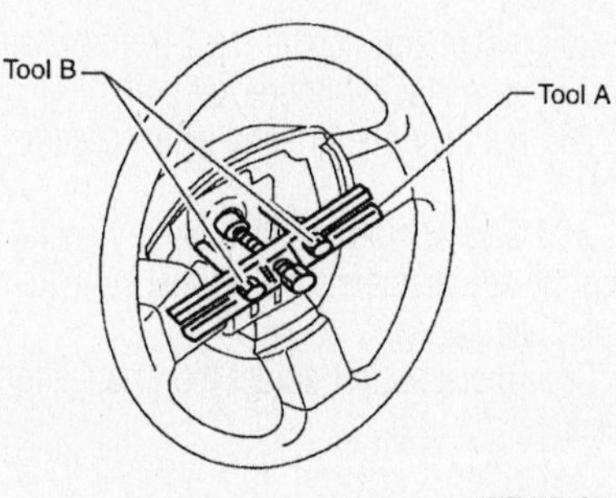

Remove the steering wheel using puller tools J-1859-A and J-42578—2004-05 models

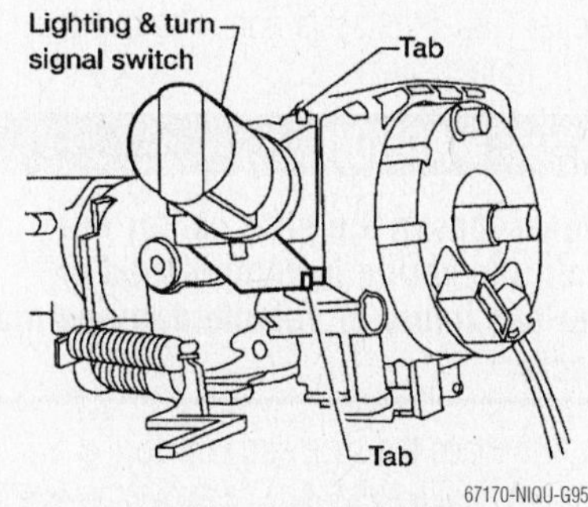

While pressing tabs, pull the light and turn signal switch toward the driver door and disconnect from the base—2004–05 models

To install:

13. Installation is in the reverse order of removal, please note the following:

a. Use the specified tightening torque when installing the high pressure and low-pressure hose connections.

✲✲ CAUTION

Excessive tightening will damage threads of connection or O-ring.

b. The O-ring in low-pressure hose connector is larger than that in high-pressure connector. Take care to install the proper O-ring.

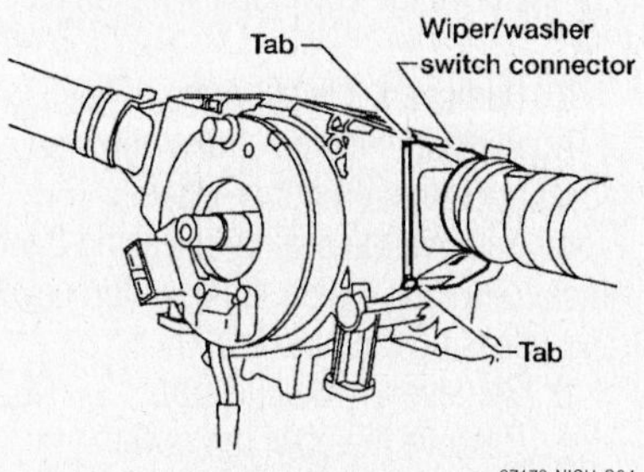

Remove wiper washer switch connector, pinch the tabs at the wiper and washer switch base and slide the switch away from the column—2004–05 models

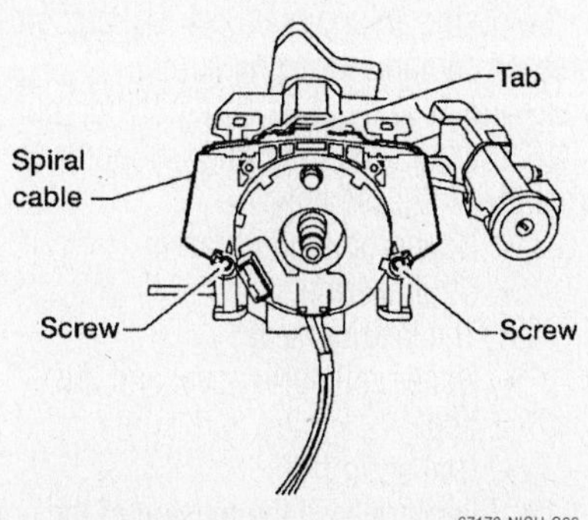

Remove the screws, release the tab, and remove the spiral cable—2004–05 models

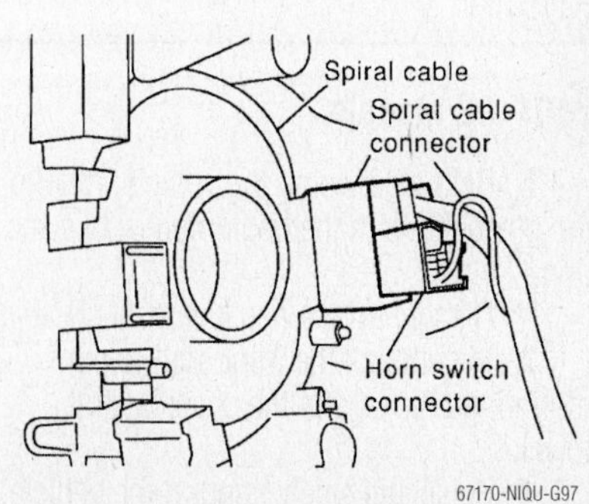

Remove the spiral cable connectors—2004–05 models

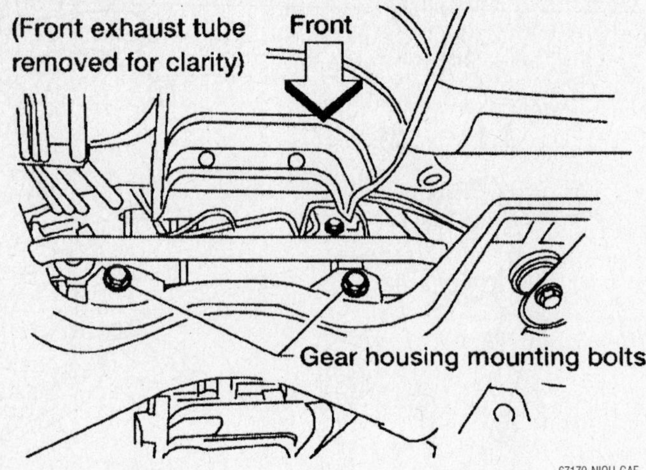

Location of the steering gear housing mounting bolts—2004–05 models

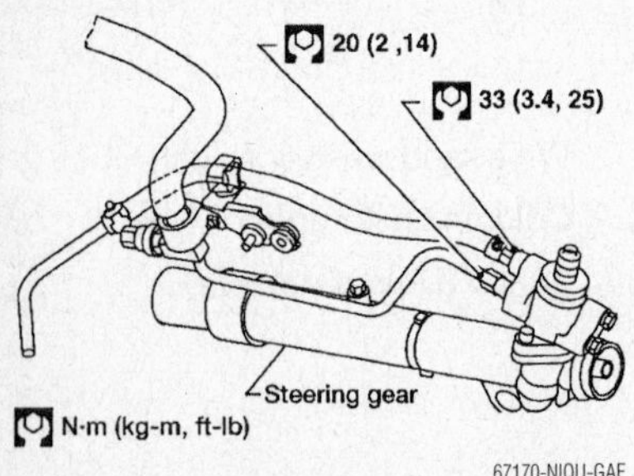

Use the specified tightening torque when installing the high pressure and low-pressure hose connections—2004–05 models

c. Initially, tighten the nut on the tie-rod outer socket and knuckle arm to the specification shown in the exploded view illustration of 25 ft. lbs. (34 Nm). Then tighten further to align nut groove with the first pin hole so that the cotter pin can be installed. The tightening torque must not exceed 36 ft. lbs. (49 Nm).

14. Refill the power steering system and bleed after installation.

MacPherson Strut

REMOVAL & INSTALLATION

2001–02 Models

1. Before servicing the vehicle, refer to the precautions in the beginning of this section.
2. Disconnect the negative battery cable.
3. Matchmark the front strut upper mounting bracket and the chassis strut tower.
4. Raise and safely support the vehicle.
5. Remove the front wheel.
6. If equipped, remove the 2 front brake anti-lock sensor cable bracket bolts and position the anti-lock sensor cable out of the way.
7. Detach the brake tube from the strut.
8. Support the control arm.
9. Matchmark the knuckle to the strut so it can installed in the same position. This is important for the camber angle of the front wheel.
10. Remove the strut-to-steering knuckle bolts.
11. Support the strut and remove the 3 upper strut-to-chassis nuts. Remove the strut from the vehicle.

⁂ WARNING

Never loosen the strut center nut until the spring is compressed or serious injury or vehicle damage may occur.

12. Place the strut and coil spring assembly in a suitable vise and remove the strut nut cover.
13. Slightly loosen, but **do not** remove the front strut nut.

If desired, use the following steps to remove the coil spring from the strut.

14. Using an approved coil spring compressor, compress the coil spring.
15. Remove the strut assembly top nut.
16. Remove the following components from the strut assembly:
    - Upper mounting bracket
    - Strut bearing
    - The bearing seat.
    - Upper coil spring seat and dust boot
    - Coil spring
17. Slowly release the tension of the coil spring compressor and remove the coil spring from the compressor tool.
18. Remove the coil spring insulator and slide the jounce bumper off of the strut assembly.

To install:

19. Slide the jounce bumper onto the strut assembly and install the coil spring insulator.
20. Carefully compress the coil spring with an approved coil spring compressor.
21. Reinstall the following components to the strut assembly:
    - Coil spring

➡Install the coil spring to the strut assembly with the end of the spring in the lower coil spring seat indentation.

    - Upper coil spring seat and dust boot
    - Bearing seat and the bearing
    - Upper mounting bracket
22. Install and tighten the strut assembly nut and tighten the nut to 43–58 ft. lbs. (59–78 Nm).
23. Install the strut assembly onto the vehicle and tighten the following:
    - Strut-to-body nuts: 29–40 ft. lbs. (39–54 Nm)
    - Strut-to-knuckle bolts: 101–116 ft. lbs. (137–157 Nm)
24. Reattach the brake tube to the strut assembly.
25. Install and tighten the 2 front brake anti-lock sensor cable bracket bolts.
26. Reinstall the tire and wheel assembly.
27. Connect the negative battery cable and the adjustable strut electrical connectors, if equipped.
28. Check and/or adjust the wheel alignment.

2004–05 Models

1. Before servicing the vehicle, refer to the precautions in the beginning of this section.
2. Remove the wheels.
3. Remove the cowl top extension as follows:
    a. Remove the wiper arms.
    b. Release the clips under cowl top cover and clips on hood–ledge.
    c. Remove the seal by releasing the ends from tabs the on the cowl top cover and releasing the plastic clips.
    d. Remove the weatherstrip.
    e. Remove the cowl top cover.
    f. Remove the cowl top seal.
    g. Remove the windshield washer nozzles and the hoses from cowl top cover.

When installing rubber parts, final tightening must be carried out under unladen condition* with tires on ground.
*: Fuel, radiator coolant and engine oil full. Spare tire, jack, hand tools and mats in designated positions.

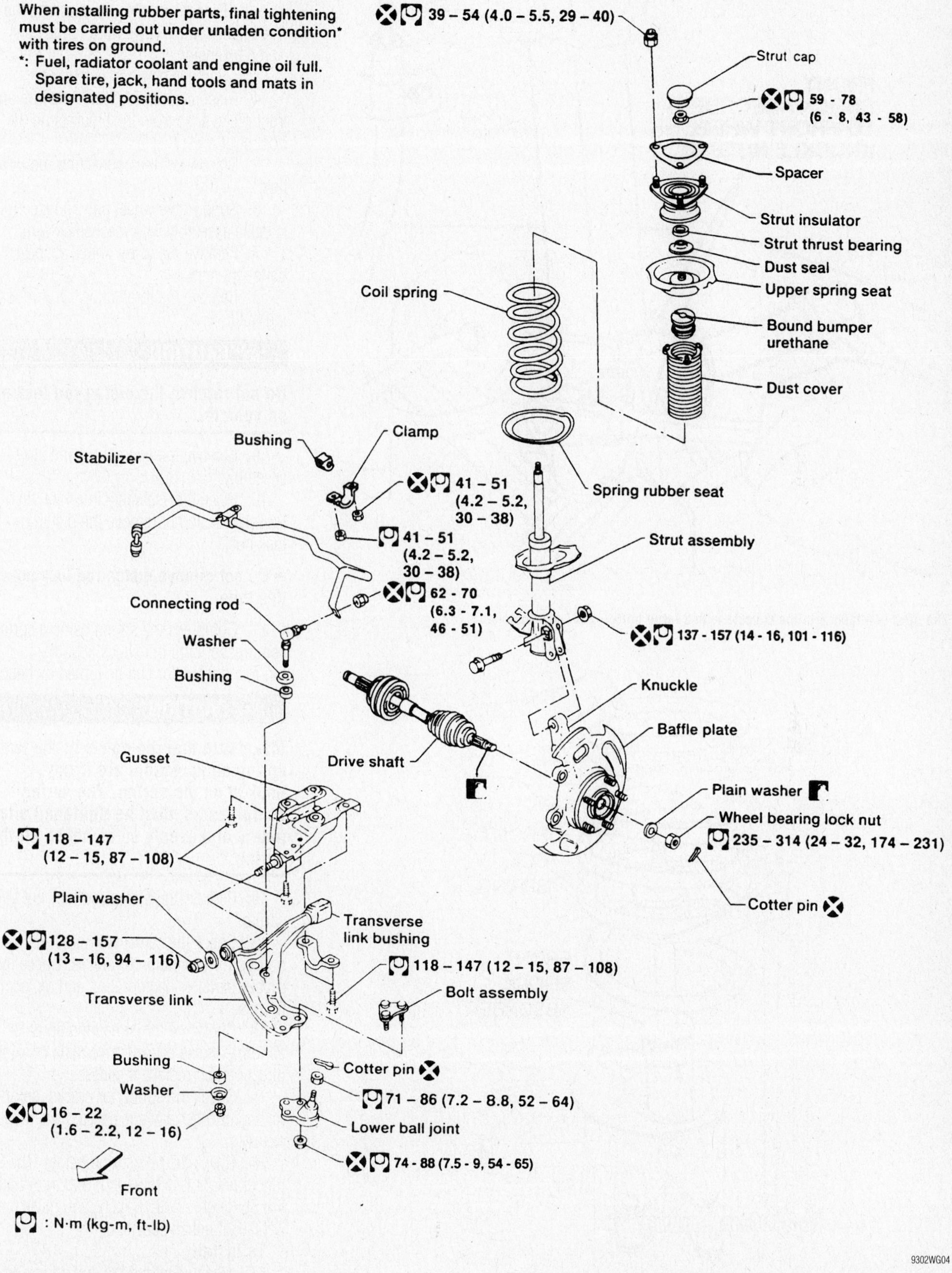

39 – 54 (4.0 – 5.5, 29 – 40)
Strut cap
59 - 78 (6 - 8, 43 - 58)
Spacer
Strut insulator
Strut thrust bearing
Dust seal
Upper spring seat
Coil spring
Bound bumper urethane
Dust cover
Clamp
Bushing
Stabilizer
41 – 51 (4.2 – 5.2, 30 – 38)
Spring rubber seat
41 – 51 (4.2 – 5.2, 30 – 38)
Strut assembly
62 - 70 (6.3 - 7.1, 46 - 51)
Connecting rod
137 - 157 (14 - 16, 101 - 116)
Washer
Bushing
Knuckle
Baffle plate
Gusset
Drive shaft
Plain washer
Wheel bearing lock nut
118 – 147 (12 – 15, 87 – 108)
235 – 314 (24 – 32, 174 – 231)
Plain washer
Cotter pin
Transverse link bushing
128 – 157 (13 – 16, 94 – 116)
118 – 147 (12 – 15, 87 – 108)
Bolt assembly
Transverse link
Bushing
Cotter pin
Washer
71 – 86 (7.2 – 8.8, 52 – 64)
16 – 22 (1.6 – 2.2, 12 – 16)
Lower ball joint
74 - 88 (7.5 - 9, 54 - 65)
Front

: N·m (kg-m, ft-lb)

9302WG04

Coil spring and strut assembly—2001–02 models

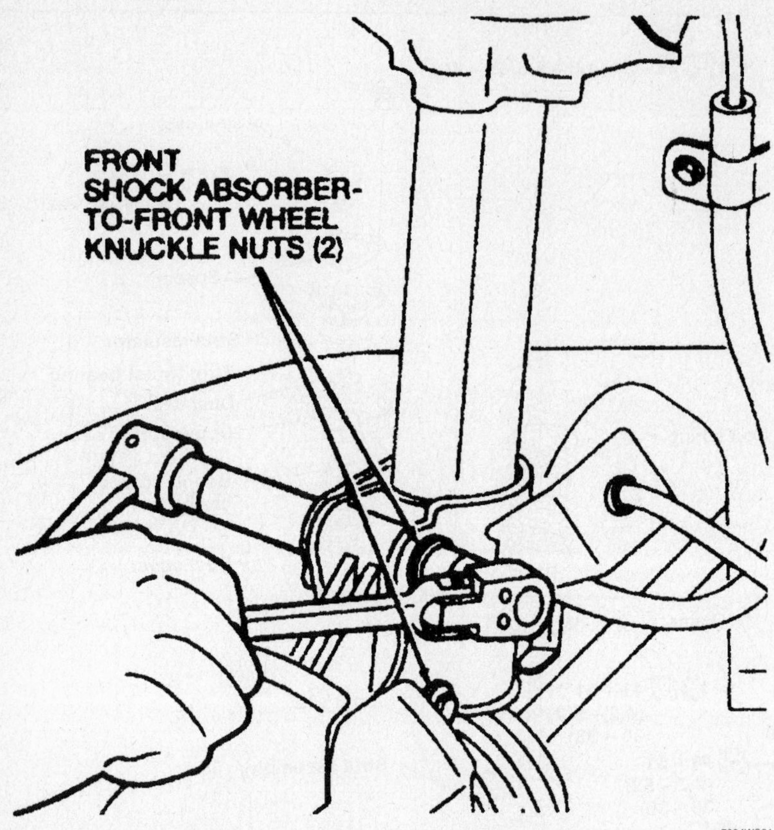

The strut is attached to the knuckle with 2 large bolts—2001–02 models

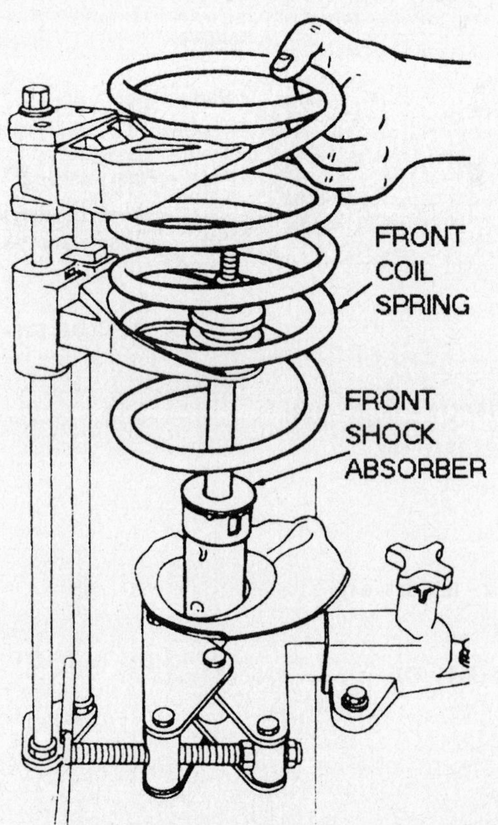

Compress the coil spring in an approved spring compressor—2001–02 models

h. Remove the heater pump from cowl top extension.

i. Disconnect the clip attaching the coolant control valve hose to the cowl top extension.

j. Remove cowl top extension.

4. Disconnect the ABS sensor wire and front brake hose from the brackets on the front strut.

5. Disconnect the connecting rod upper link.

6. Support the wheel hub and steering knuckle assembly with mechanics wire.

7. Remove the lower shock absorber bolts and nuts.

8. Remove the three upper strut mounting nuts.

**** CAUTION**

Do not remove the piston rod lock nut on vehicle.

9. Remove the coil spring and strut assembly.

10. Place the assembly in a vise, then loosen, but do not remove the piston rod lock nut.

➡Do not remove piston rod lock nut at this time.

11. Compress the spring using a spring compressor until the shock absorber mounting insulator can be turned by hand.

**** CAUTION**

Make sure that the pawls of the two spring compressors are firmly hooked on the spring. The spring compressors must be tightened alternately and evenly so as not to tilt the spring.

12. Remove the piston rod lock nut and spring.

13. Check the cemented rubber-to-metal portion for separation or cracks. Check the rubber parts for deterioration and replace if necessary.

14. Check the thrust bearing parts for abnormal noise or excessive rattle in axial direction and replace if necessary.

15. Check the spring for cracks, deformation or other damage and replace if necessary.

16. Check the free spring height. The SE model height is 13.39 inch (340 mm) and the SL height is 13.78 inch (350 mm). Replace if not as specified.

To install:

17. When installing the coil spring on strut, it must be positioned as illustrated.

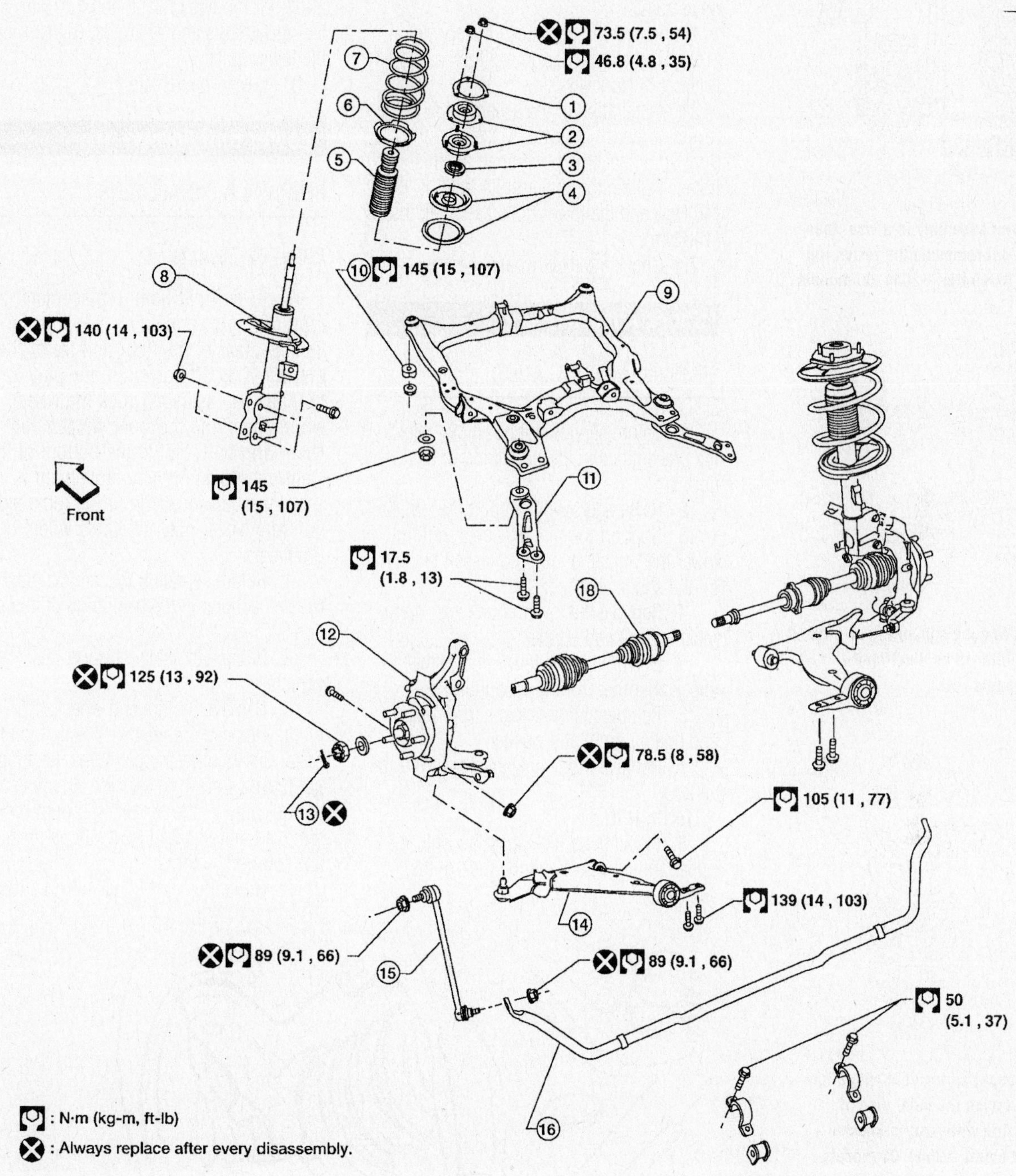

1.	Gasket	2.	Shock absorber mounting insulator	3.	Upper rubber seat
4.	Shock absorber bushing	5.	Dust cover	6.	Coil spring
7.	Lower rubber seat	8.	Shock absorber	9.	Front suspension member
10.	Cup	11.	Member pin stay	12.	Wheel hub and steering knuckle assembly
13.	Cotter pin	14.	Transverse link	15.	Connecting rod
16.	Stabilizer bar				

67170-NIQU-GAG

Exploded view of the front suspension assembly—2004-05 models

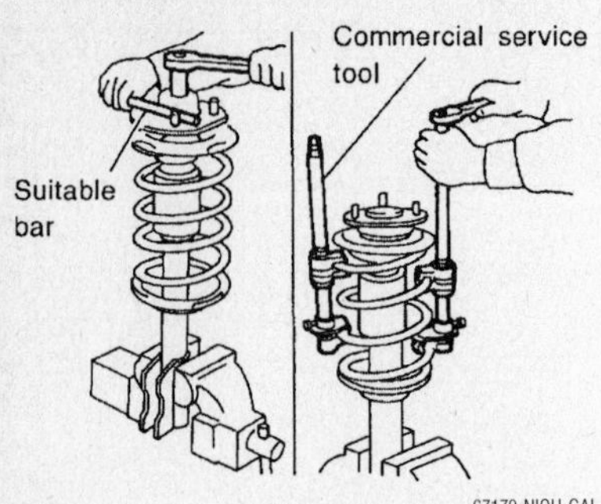

Place the strut assembly in a vise, then loosen, but not removing the piston rod lock nut as illustrated—2004–05 models

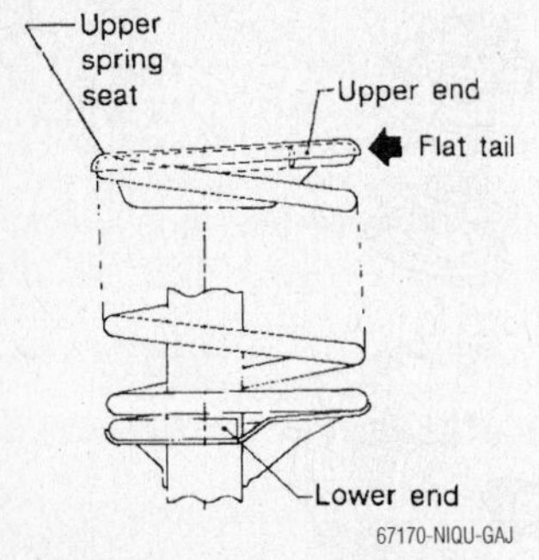

When installing the coil spring on strut, it must be positioned as illustrated—2004–05 models

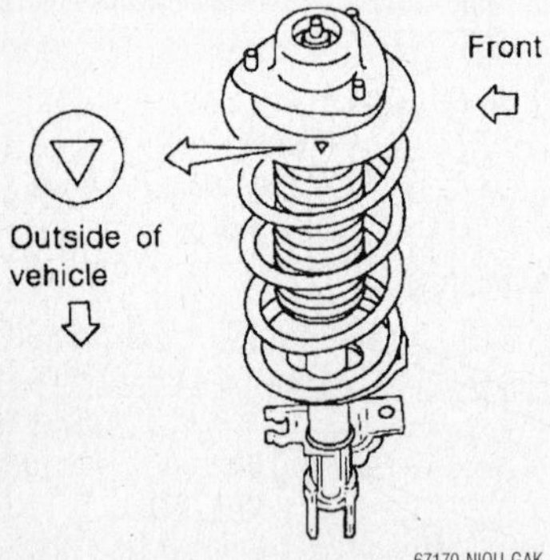

Install the upper spring seat with alignment mark facing the outer side of vehicle, in line with strut-to-knuckle attachment points—2004–05 models

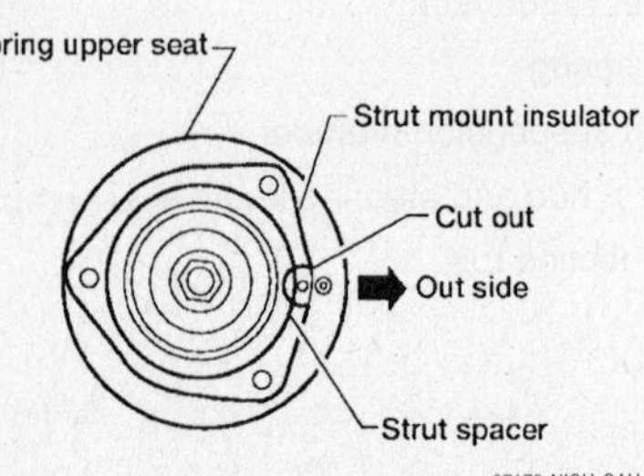

Make sure the front strut spacer is positioned as illustrated—2004–05 models

18. Install the upper spring seat with alignment mark facing the outer side of vehicle, in line with strut-to-knuckle attachment points.
19. Install and compress the spring.
20. Tighten the piston rod lock nut to 54. ft. lbs. (73 Nm).
21. Installation of the assembly is in the reverse order of removal. Make sure the front strut spacer is positioned as illustrated. Tighten the upper bolts to 35 ft. lbs. (46 Nm) and the lower bolts to 103 ft. lbs. (140 Nm).
22. Check the front wheel alignment

Shock Absorber

REMOVAL & INSTALLATION

1. Before servicing the vehicle, refer to the precautions in the beginning of this section.
2. Raise and safely support the vehicle.
3. Support the rear axle and slightly lower the vehicle enough to lessen tension on the shock absorber.
4. Remove the lower shock absorber retaining nut and washer.
5. Disconnect the lower end of the shock absorber from the mounting stud.
6. Remove the shock absorber upper end retaining nut and washer.
7. Remove the shock absorber from the vehicle.

To install:

8. Install the shock absorber onto the upper and lower mounting studs of the vehicle.
9. Install the washers and retaining nuts. Tighten the upper and lower retaining nuts to 22–30 ft. lbs. (30–41 Nm) on 2001–02 models or 22 ft. lbs. (30 Nm) for the upper bolts and 47 ft. lbs. (64 Nm) on the lower bolts.
10. Lower the vehicle.

Lower Ball Joints

REMOVAL & INSTALLATION

2001–02 Models

To check if ball joint replacement is required, raise and safely support the vehicle clear of the floor and try to rock the wheel up and down. If any play is felt, have an assistant rock the wheel while observing the front suspension lower arm ball joint at the bottom of the steering knuckle. If any movement is seen, the ball joint should be replaced. If not, any wheel play indicates wheel bearing wear.

1. Before servicing the vehicle, refer to the precautions in the beginning of this section.
2. Raise and safely support the vehicle.
3. Remove the tire and wheel.
4. Remove and discard the ball joint cotter pin. Loosen the ball joint attaching nut from the steering knuckle. Because of tight clearance, the nut likely cannot be removed until the ball joint stud is loosened and lowered slightly.
5. Strike the front knuckle with a ham-

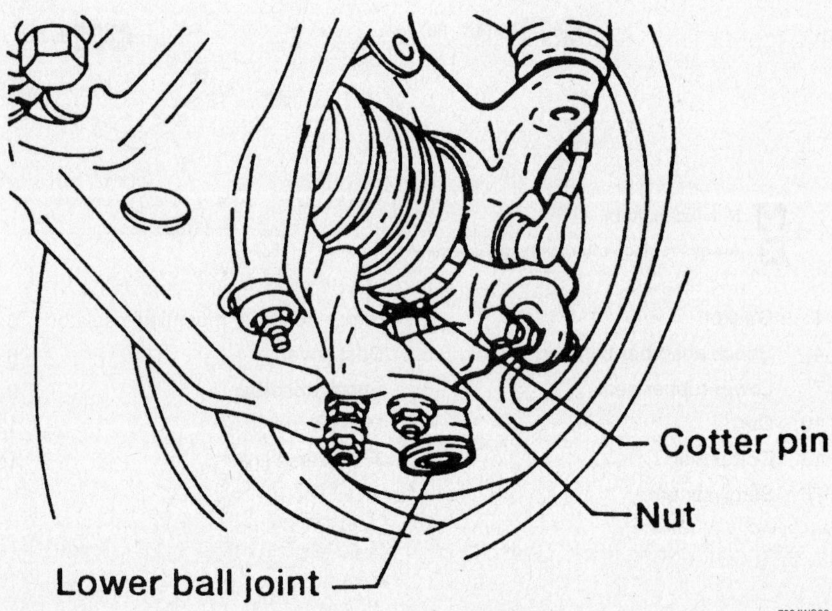

Loosen the nut on the lower ball joint stud—2001–02 models

mer while pulling down on the lower control arm. There should now be enough clearance to allow removal of the ball joint stud nut. Separate the ball joint from the steering knuckle.

6. Remove the 3 bolts attaching the ball joint to the control arm.
7. Remove the ball joint from the control arm.

To install:

8. Install the ball joint to the control arm and install the attaching bolts.
9. Tighten the bolts to 54–65 ft. lbs. (74–88 Nm).
10. Install the ball joint into the steering knuckle, just enough to get the nut started on the stud. Then, push the ball joint stud fully in place. Tighten the nut to 52–63 ft. lbs. (71–86 Nm). Secure the nut with a new cotter pin.
11. Install the tire and wheel.
12. Lower the vehicle.
13. A front end alignment check is recommended.

2004–05 Models

The ball joint on 2004–05 models is an integral part of the lower control arm (transverse link) assembly. If the ball joint is found to be defective, replace the transverse link.

Lower Control Arm

REMOVAL & INSTALLATION

2001–02 Models

1. Before servicing the vehicle, refer to the precautions in the beginning of this section.
2. Remove the wheel.
3. Disconnect the ball joint.
4. Disconnect the stabilizer bar from the control arm.
5. Remove the 2 rear arm bolts and the mounting bracket.
6. Remove the lower arm nut.
7. Pull the rear of the arm down and gently pry the arm forward and off the gusset.
8. Installation is the reverse of removal. Observe the following torques:
    - Stabilizer bar-to-lower arm: 12–16 ft. lbs. (16–22 Nm)
    - Lower arm rear bolts: 87–108 ft. lbs. (118–147 Nm)
    - Lower arm nuts: 94–115 ft. lbs. (128–156 Nm)
    - Ball stud nut: 56–80 ft. lbs. (76–109 Nm)

2004–05 Models

1. Before servicing the vehicle, refer to the precautions in the beginning of this section.
2. Remove the wheel.
3. Remove the lower ball joint pinch bolt.
4. Separate the transverse link from the steering knuckle assembly.
5. Remove the two transverse link pivot bolts.
6. Remove the transverse link from the front suspension member.
7. Check the transverse link for damage, cracks or any deformities and replace as necessary.
8. Check the bushing for damage, cracks or any deformities. Replace the transverse link if necessary.
9. Check the ball joint for excessive play. Replace the transverse link assembly if the lower ball joint stud is worn, the ball joint is hard to swing, the ball joint play in axial directions or end play is excessive.

➡Before checking the axial forces and end play, turn the lower ball joint at least 10 revolutions so that the ball joint is properly broken in.

10. The swinging force (measuring from the cotter pin hole of the ball stud) is 1.8–12.3 ft. lbs. (7.8–54.9 N). Refer to the illustration for the location of the swinging force measurement point identified by the letter **A**.
11. The turning force is 4.3–30.4 inch lbs. (0.49–3.43 Nm). Refer to the illustration for the location of the turning force measurement point identified by the letter **B**.
12. Check the transverse link vertical end play, it should be zero. Refer to the illustration for the location of the vertical end play measurement point identified by the letter **C**.Check the dust cover for damage. Replace it and the cover clamp if necessary.

To install:

13. Hand tighten the transverse link mounting bolts.
14. Once all components are installed final tighten the transverse link bolts with the vehicle at curb weight and the tires on the ground. Tighten the ball joint pinch bolt to 58 ft. lbs. (78 Nm), the two outer link bolts to 103 ft. lbs. (139 Nm) and the single inner bolt to 77 ft. lbs. (105 Nm). Refer the exploded view of the front suspension assembly illustration for bolt locations.

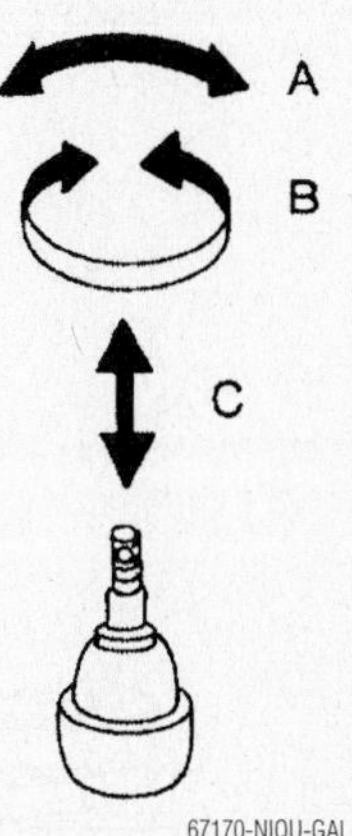

Ball joint swinging force (A), turning force (B) and transverse link vertical end play (C) measuring locations—2004-05 models

15. Installation of the remaining components is in the reverse order of removal.

BUSHING REPLACEMENT

2001–02 Models

The bushings are press-fit types. Support the arm in a press, using the proper adapters. Ford tool numbers are: T93P-5493-A, T75L-1165-B and -DA.

Wheel Bearings

ADJUSTMENT

The wheel bearings are not adjustable. If the bearings become loose or make noise, they must be replaced using the following procedure.

REMOVAL & INSTALLATION

Front

2001–02 MODELS

1. Before servicing the vehicle, refer to the precautions in the beginning of this section.
2. Raise and safely support the vehicle.
3. Remove the wheel and tire.
4. Remove the brake caliper assembly. DO NOT disconnect the brake hose. Hang the caliper on a piece of wire from a near by support such as the strut.
5. Remove the brake rotor.
6. Remove and discard the cotter pin from the end of the outboard CV-joint stub shaft. Remove the hub nut retainer, washer and the hub nut. There should be another washer under the hub nut that acts as a front wheel bearing outer bearing retainer.

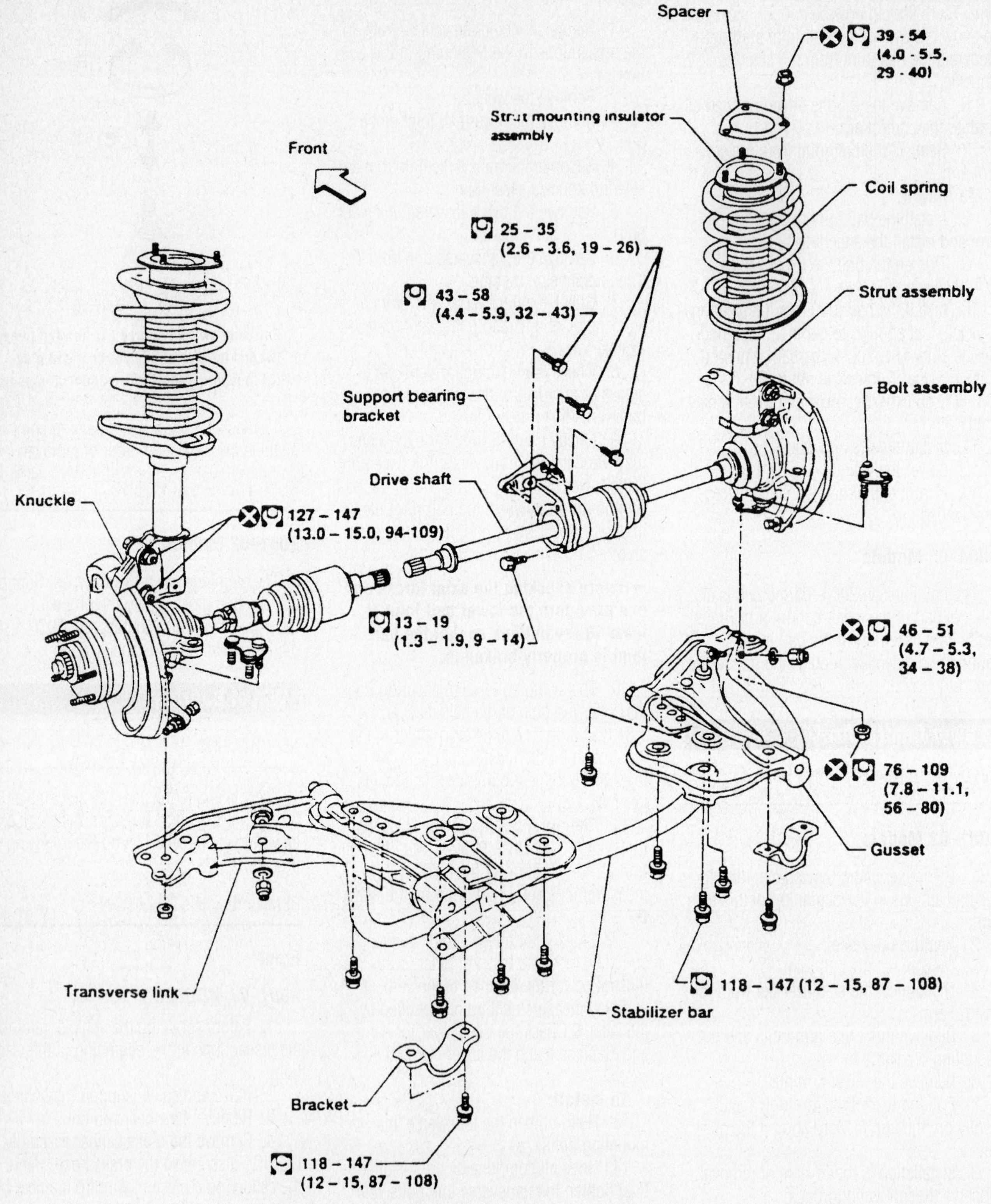

Exploded view of the front suspension and drive axles—2001–02 models

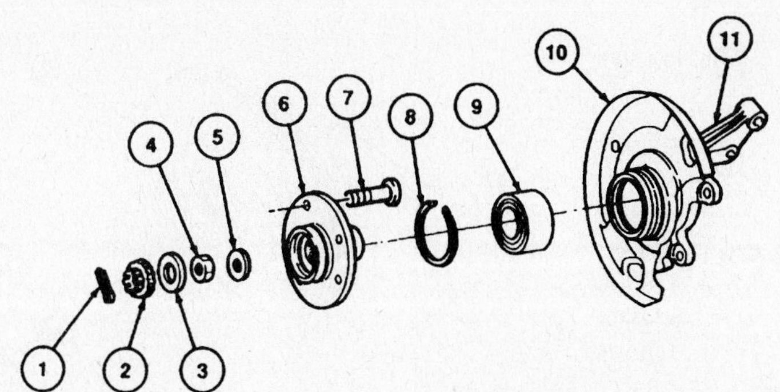

1. Cotter pin
2. Nut retainer
3. Insulator
4. Front axle wheel hub retainer
5. Front wheel outer bearing retainer washer
6. Wheel hub
7. Wheel hub bolt
8. Snap ring
9. Front wheel bearing
10. Front disc brake rotor shield
11. Front wheel knuckle

7924WG23

Exploded view of the knuckle, hub and bearing—2001–02 models

7. Disengage the lower ball joint stud from the steering knuckle using the following procedure.
   a. Remove and discard the cotter pin from the front lower ball joint.
   b. Loosen the lower ball joint nut until it contacts the front halfshaft joint.
   c. Strike the front knuckle with a hammer while pulling down on the lower control arm until the ball joint stud separates from the knuckle.
   d. Remove the ball joint nut.
   e. Disengage the lower ball joint stud from the steering knuckle.
8. Disengage the outer tie rod end stud from the steering knuckle using the following procedure.
   a. Remove and discard the cotter pin from the outer tie rod end stud.
   b. Remove the outer tie rod end retaining nut.
   c. Use a tie rod end puller to carefully

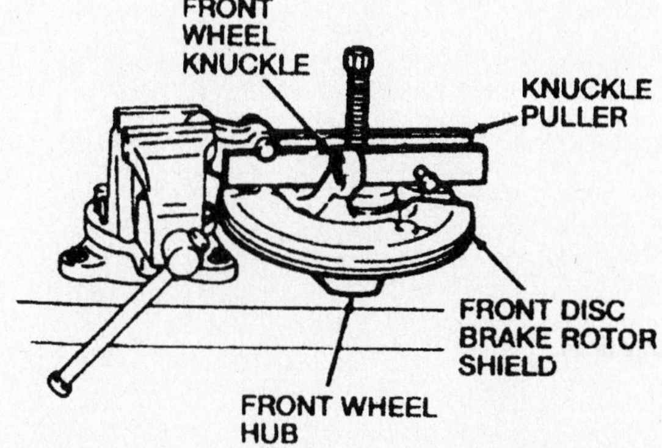

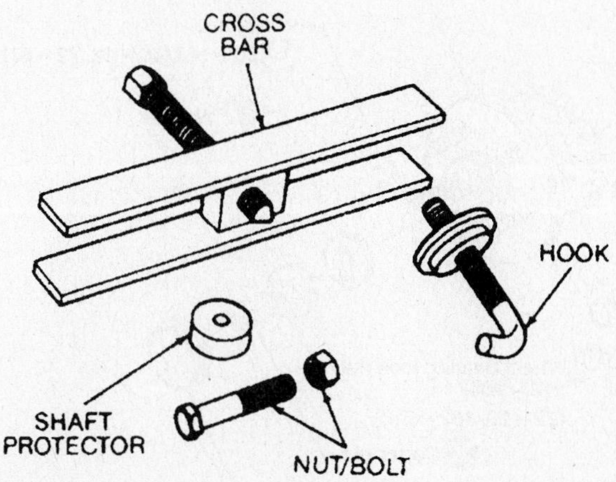

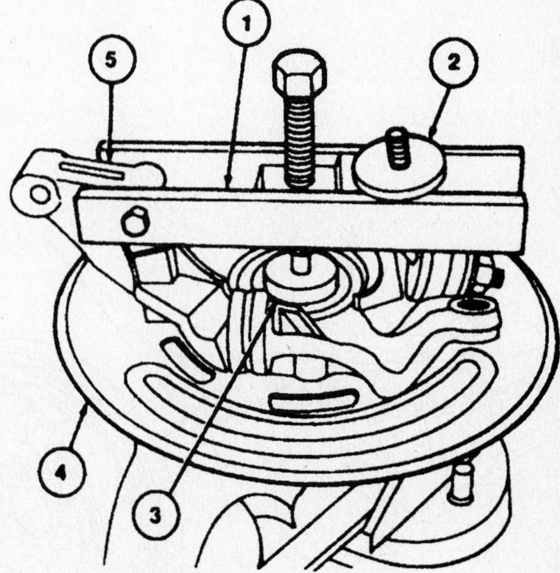

1. Knuckle puller
2. Knuckle puller adapter
3. Step plate adapter
4. Front disc brake rotor shield
5. Front wheel knuckle

7924WG27

Example of a puller set up to bear against the front wheel bearing inner race—2001–02 models

60 (6.1, 44)
63 (6.4, 46)
63 (6.4, 46)
125 (13, 92)
: N•m (kg-m, ft-lb)
: Always replace after every disassembly.

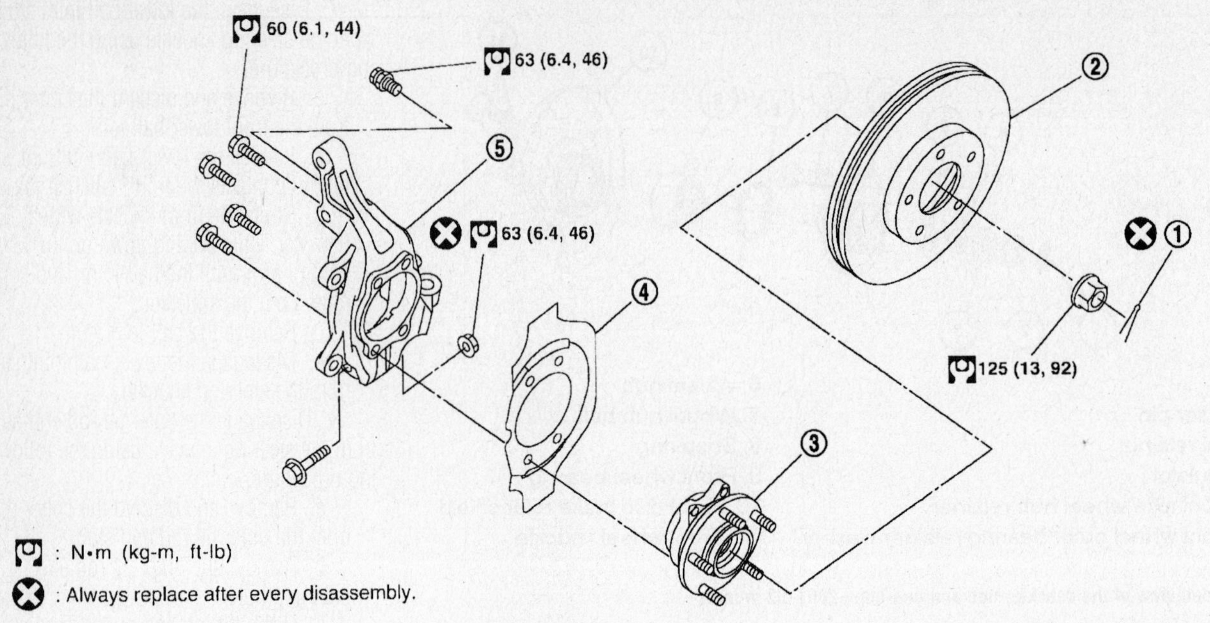

1. Cotter pin
2. Disc rotor
3. Wheel hub and bearing assembly
4. Splash guard
5. Steering knuckle

67170-NIQU-GAM

Exploded view of the front wheel hub and knuckle components—2004–05 models

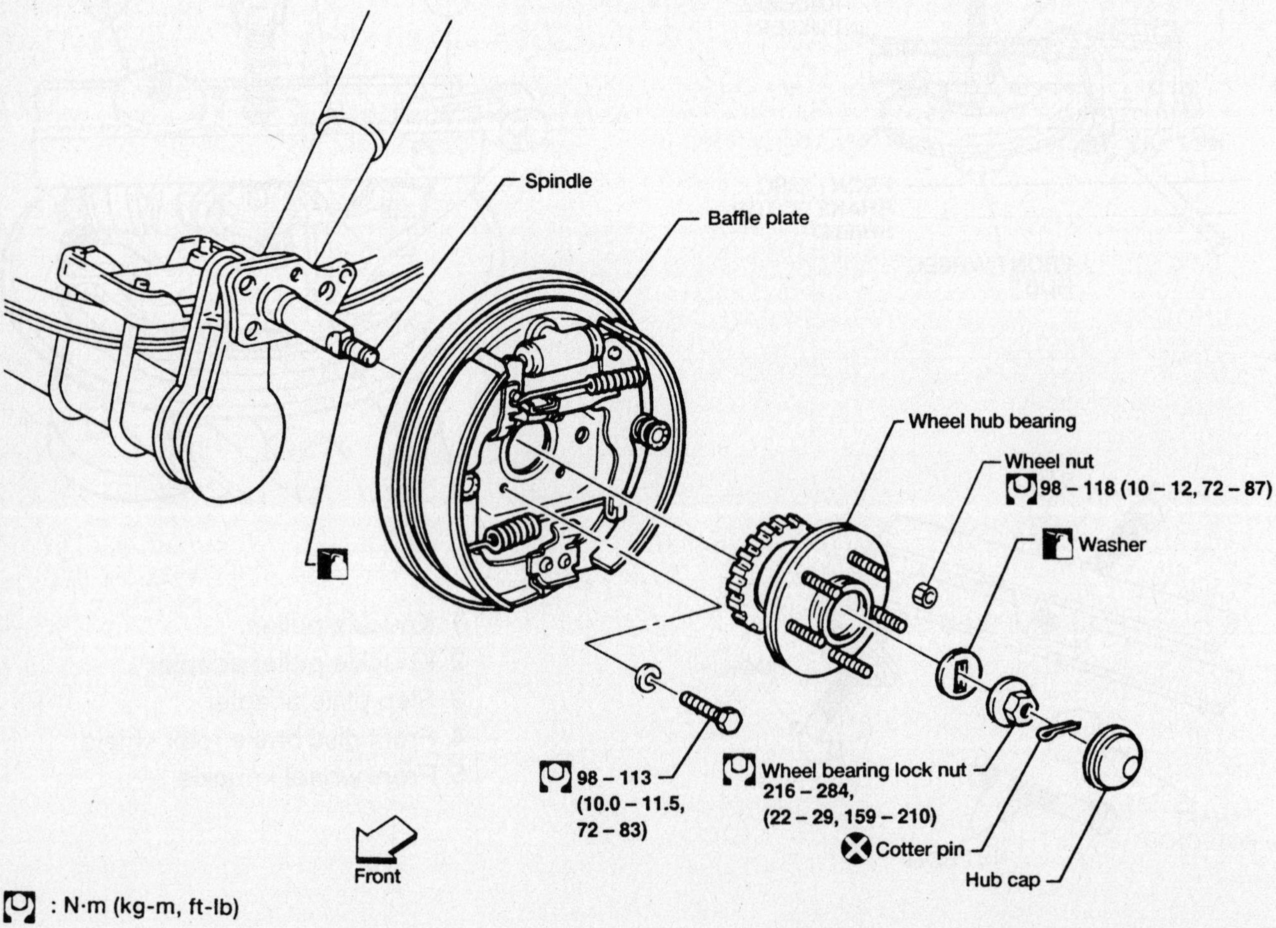

Spindle
Baffle plate
Wheel hub bearing
Wheel nut
98 – 118 (10 – 12, 72 – 87)
Washer
98 – 113
(10.0 – 11.5,
72 – 83)
Wheel bearing lock nut
216 – 284,
(22 – 29, 159 – 210)
Cotter pin
Hub cap
Front
: N·m (kg-m, ft-lb)

9302WG05

Rear hub assembly—2001–02 models

press the tie rod end from the steering knuckle.

9. Remove the front ABS sensor bolt.

10. Remove the 2 front strut-to-front knuckle nuts and remove the 2 bolts. Disengage the strut from the steering knuckle.

11. Use a 2-jaw puller to separate the front halfshaft outboard CV-joint stub shaft from the knuckle/bearing assembly.

12. Remove the front wheel hub, knuckle and wheel bearing assembly from the vehicle.

13. If the knuckle is being replaced with a service part, change over the steering stop bolt and jam nut from the old knuckle to the replacement part.

14. To remove the front wheel bearing, jig up a puller to bear against the front wheel bearing inner race and pull the race from the hub/knuckle assembly.

15. Use a shop press to press out damaged wheel studs and also to press out the outer bearing race.

16. Use a shop press to press out the inner bearing race.

To install:

17. If the front wheel bearings were removed, assemble the ABS sensing ring, if removed and the disc brake dust shield under the steering knuckle. Use a shop press to push in new front wheel bearing inner and outer races. Support the knuckle and press the front wheel bearing into the knuckle and install the snapring retainer. Support the bearing assemblies and press the hub onto the knuckle and wheel bearing assembly.

18. Install the hub, knuckle and bearings as an assembly. Position the assembly on the halfshaft outer CV-joint stub axle end. Guide the knuckle into the front strut and install the 2 knuckle-to-strut bolts and nuts. Tighten the nuts to 83–91 ft. lbs. (113–123 Nm).

19. Install the ABS sensor bolt. Do not over-tighten. Tighten to just 16–21 inch lbs. (1.8–2.4 Nm).

20. Install the outer tie rod end to the steering knuckle. Tighten the nut to 22–29 ft. lbs. (29–39 Nm). If the cotter pin holes do not align, tighten the nut slightly until they do. Never loosen the nut to align the holes. Secure the nut with a new cotter pin.

21. Start the lower ball joint stud to the steering knuckle and partially install the nut, then push the ball joint stud fully in place. Tighten the ball joint stud nut to 52–63 ft. lbs. (71–86 Nm). Secure the nut with a new cotter pin.

22. Install the front wheel outer bearing retaining washer and the hub retainer nut. Tighten to 174–231 ft. lbs. (235–314 Nm). Install the nut retainer, insulator and a new cotter pin.

23. Install the front brake rotor and install the disc brake caliper.

24. If removed, install the steering stop bolt.

25. Install the tire and wheel assembly. Tighten the lug nuts to 72–87 ft. lbs. (98 to 118 Nm).

26. Lower the vehicle. Pump the brake pedal slowly to seat the front brake pads. Do not move the vehicle until a firm pedal is obtained.

27. A front end alignment is recommended.

2004–05 Models

1. Before servicing the vehicle, refer to the precautions in the beginning of this section.

2. Remove the wheel.

3. Remove the brake caliper, leaving the line attached and suspend the caliper using a piece of wire.

4. Place alignment marks on the brake rotor, wheel hub and bearing assembly, then remove the rotor.

5. Remove the ABS sensor from the steering knuckle.

6. Remove the cotter pin and lock nut from the drive shaft.

7. Remove the steering outer socket cotter pin at the steering knuckle, then loosen the mounting nut.

8. Disconnect the steering outer socket from the steering knuckle using tool J25730–A. Be careful not to damage ball joint boot.

✱✱ CAUTION

To prevent damage to the threads and to prevent the tool from coming off suddenly, temporarily tighten mounting nut.

9. Remove the transverse link and steering knuckle pinch bolt and nut.

10. Remove the wheel hub and bearing assembly from drive shaft using a suitable puller.

➡When removing the wheel hub and bearing assembly, do not apply an excessive angle to drive shaft joint. Be careful not to excessively extend the slide joint and support the driveshaft when removing.

11. Remove the wheel hub and bearing assembly bolts.

12. Remove the splash guard and the wheel hub and bearing assembly from steering knuckle.

To install:

13. Installation is in the reverse order of removal. Tighten the bolts to the specifications shown in the exploded view of the front wheel hub and knuckle illustration. Also please note the following:

a. Replace the differential side oil seal with a new one every time driveshaft is removed on 4 speed transmission models

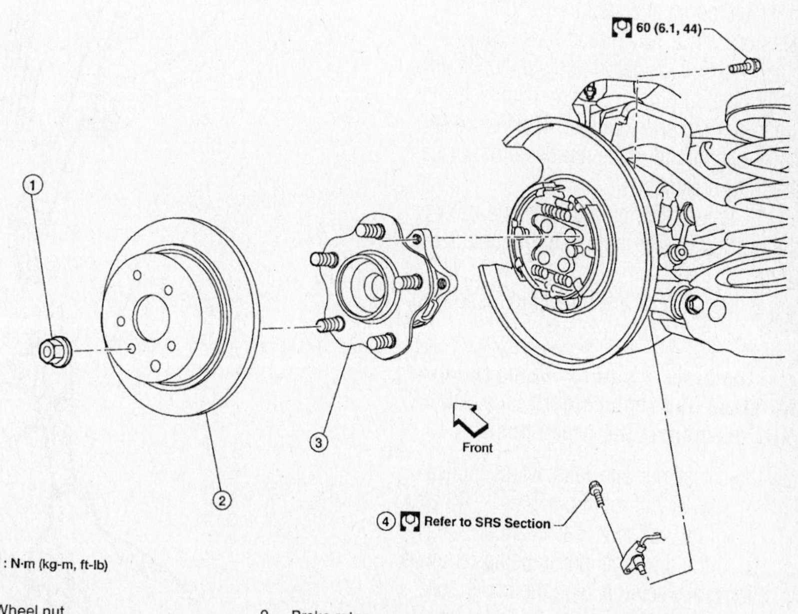

Exploded view of the rear wheel hub assembly—2004–05 models

b. When installing the wheel hub and bearing assembly to steering knuckle, align cutout in toner ring cover with ABS sensor mounting hole in steering knuckle.

c. When installing the rotor on the wheel hub and bearing assembly, align the marks made prior to removal.

d. A front end alignment is recommended.

Rear

2001–02 MODELS

1. Before servicing the vehicle, refer to the precautions in the beginning of this section.
2. Raise and safely support the vehicle.
3. Remove the rear wheel(s).
4. Remove the brake drum.
5. Remove the grease cap for the hub.
6. Remove and discard the cotter pin.
7. Remove the wheel bearing nut and washer.
8. Remove the rear wheel hub and bearing assembly.

To install:

9. Install the rear wheel hub and bearing assembly onto the vehicle.
10. Install the rear wheel bearing washer and nut and tighten the bearing nut to 159–210 ft. lbs. (216–284 Nm). Install a new cotter pin.
11. Install the wheel hub grease cap. Install the brake drum.
12. Install the rear wheel(s) and lug nuts. Tighten the lug nuts, in a star sequence, to 72–87 ft. lbs. (98–118 Nm).
13. Lower the vehicle.

2004–05 Models

1. Before servicing the vehicle, refer to the precautions in the beginning of this section.
2. Remove the rear wheel.
3. Remove the brake caliper assembly without disconnecting the hydraulic line and suspend the caliper aside using wire.
4. Release the parking brake and remove the rotor.
5. Remove the rear ABS sensor, then position it aside using wire.
6. Remove the wheel hub assembly from the knuckle.
7. Check for cracks, and damage on the wheel hub assembly and replace as necessary.

To install:

8. Installation is in the reverse order of removal. Tighten the wheel hub bolts to 44 ft. lbs. (60 Nm).
9. Check that the wheel bearing operates smoothly.
10. Check that the wheel hub bearing axial end play is within 0.004 inch (0.1mm) or less using a dial indicator.

BRAKES

Brake Caliper

REMOVAL & INSTALLATION

2001–02 Models

The front disc brake caliper slides on 2 stainless steel locating pins. The front disc brakes use a conventional pin slider-type front disc brake caliper with a 10.875 inch (27.6cm) front disc rotor. The front disc brake caliper is attached to the front suspension with 2 Torx® head brake caliper bolts. Rubber insulators isolate the stainless steel locating pins from direct contact with the front disc brake caliper. The front disc brake calipers must be removed to replace the front brake pads.

1. Before servicing the vehicle, refer to the precautions in the beginning of this section.
2. Remove or disconnect the following:
   - Wheel and tire

➡If the brake caliper is being removed for brake pad replacement only, DO NOT disconnect the brake hose.

   - 2 caliper pin bolts. Most applications will require a Torx® T-40 bit to remove the 2 brake caliper bolts.
3. If the brake caliper is being removed just for brake service, with the brake hose still attached to the caliper, use a length of wire to support the caliper from the front shock absorber. Do not let the caliper hang by the brake hose. If the caliper is being completely removed from the vehicle for overhaul, use care not to drip brake fluid on the paint.

➡If both calipers are being completely removed from the vehicle at the same time, mark them Left and Right so the calipers can be reinstalled to their original locations. The reason for this is that the bleeder screws must be positioned on the top of the front disc brake caliper when installed on the vehicle.

To install:

4. Clean all parts well. Use a C-clamp and a used brake pad to push the caliper

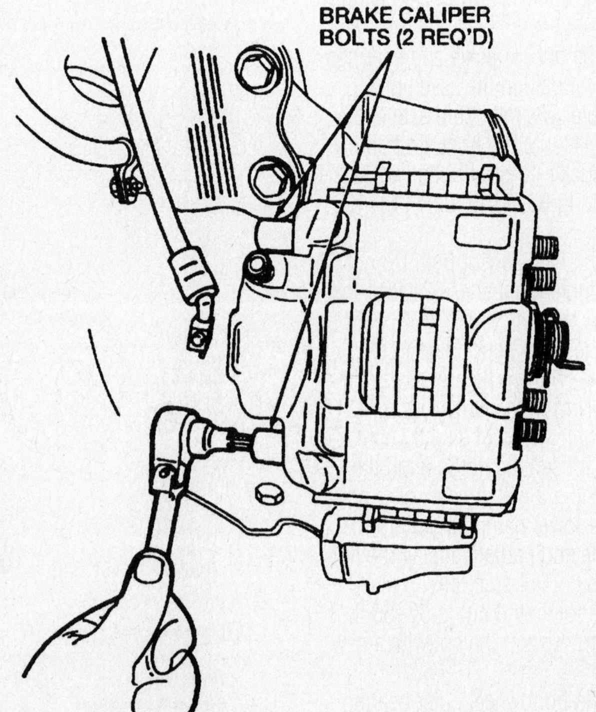

Caliper pin bolt removal—2001–02 models

piston fully in the piston bore. Inspect the caliper pins and clean any dirt and debris.

5. Install the caliper onto the rotor. Make sure the inboard and outboard brake pads are properly positioned.

6. Lubricate the stainless steel locating pins with a Silicone Dielectric Compound such as Ford DZAZ-19A331-A or equivalent silicone grease. Install the 2 caliper pin bolts and torque to 18–25 ft. lbs. (24–34 Nm).

7. If disconnected, install the brake hose using a new replacement copper washer, install the banjo bolt and torque to 12–14 ft. lbs. (17–20 Nm).

8. If the brake hose had been disconnected, bleed the brake system.

9. Install the wheel and tire.

10. Torque the lug nuts to 72–87 ft. lbs. (98–118 Nm).

11. Check the master cylinder reservoir and add fresh DOT 3 brake fluid as required.

12. Lower the vehicle. Pump the brake pedal slowly until a firm brake pedal is obtained, indicating that the brake pads are properly seated, before attempting to move the vehicle. Road-test and check for proper brake operation.

2004–05 Models

1. Before servicing the vehicle, refer to the precautions in the beginning of this section.

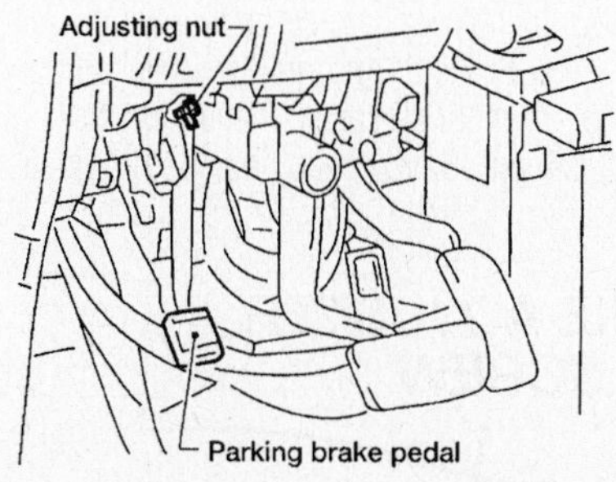

Location of the brake pedal adjusting nut—2004–05 models

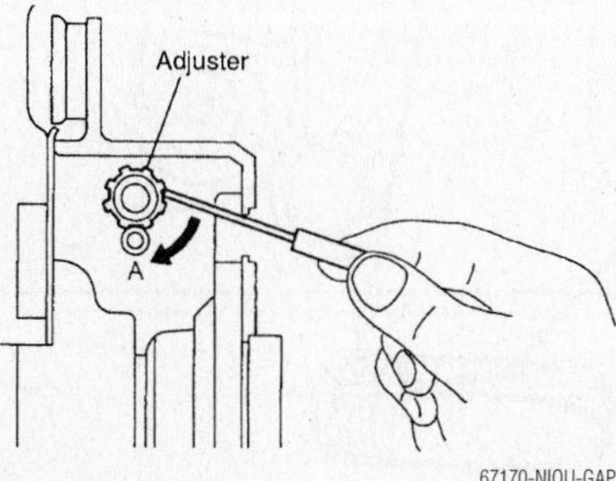

Location of the parking brake adjuster—2004–05 models

2. Remove the wheel.

3. Drain the brake fluid.

4. Remove brake hose bolt and caliper bolts, then remove caliper assembly.

To install:

5. Install caliper assembly, and tighten bolts to 115 ft. lbs. (156 Nm) on the front caliper or 62 ft. lbs. (84 Nm) on the rear caliper.

6. Install brake hose to the caliper assembly using a new copper washer, and tighten the bolt to 13 ft. lbs. (18 Nm).

7. Refill the system with new brake fluid and bleed the brakes.

8. If the rear caliper was removed, adjust the parking brake as follows:

a. Insert a deep socket wrench to rotate adjusting nut and loosen the cable sufficiently. Then, return the pedal to the free height.

b. Using the lug nuts, secure the rotor to the hub and prevent it from tilting.

c. Remove the adjusting hole plug installed on the rotor. Using a suitable tool, rotate the adjuster in direction **A** as illustrated, until the rotor is locked. After locking, turn the adjuster in the opposite direction by 5 or 6 notches.

d. Rotate the rotor to make sure there is no drag and install the adjusting hole plug.

e. Pump the brake pedal 10 or more times with a force of 66 lbs. (294 N).

f. Rotate the adjusting nut with a deep socket to adjust the pedal stroke. The stroke should be 5 to 6 notches

➡Do not reuse the adjusting nut after removing it.

g. When parking brake pedal is operated at specified force, make sure the force is 44 lbs. (196 N).

h. With the pedal completely returned, make sure there is no drag on the rear brake.

9. Install the wheel.

Disc Brake Pads

REMOVAL & INSTALLATION

Front

2001–02 MODELS

1. Before servicing the vehicle, refer to the precautions in the beginning of this section.

2. Remove or disconnect the following:
- Wheels
- Bottom guide pin from the caliper and swing the caliper cylinder body upward; support the caliper with a wire
- Brake pad retainers and the pads

To install:

3. Compress the piston of the disc brake caliper.

4. Install or connect the following:
- Brake pads and caliper assembly. Torque the guide pin to 23–30 ft. lbs. (31–41 Nm).
- Wheels

5. Apply the brakes a few times to seat the pads. Check the master cylinder and add fluid if necessary. Bleed the brakes, if necessary.

2004–05 Models

1. Before servicing the vehicle, refer to the precautions in the beginning of this section.

2. Remove the wheels.

3. Remove the bottom guide pin from the caliper and swing the caliper cylinder body upward; support the caliper with a wire.

4. Remove the brake pad retainers and the pads.

To install:

5. Compress the piston of the disc brake caliper.

6. Install the retainers and brake pads.

7. Install the caliper assembly. Torque the guide pin to 20 ft. lbs. (26 Nm).

8. Install the wheels.

9. Apply the brakes a few times to seat the pads. Check the master cylinder and add fluid if necessary. Bleed the brakes, if necessary.

Rear

2001–02 MODELS

1. Before servicing the vehicle, refer to the precautions in the beginning of this section.

➡Do not press the piston into the bore as performed on the front disc brakes. Due to the parking brake mechanism, the caliper piston must be turned into the bore using a special tool.

2. Remove or disconnect the following:
- Rear wheels

3. Release the parking brake.
- Parking brake cable bracket bolt
- Pin bolts and lift off the caliper body
- Pad springs
- Pads and shims

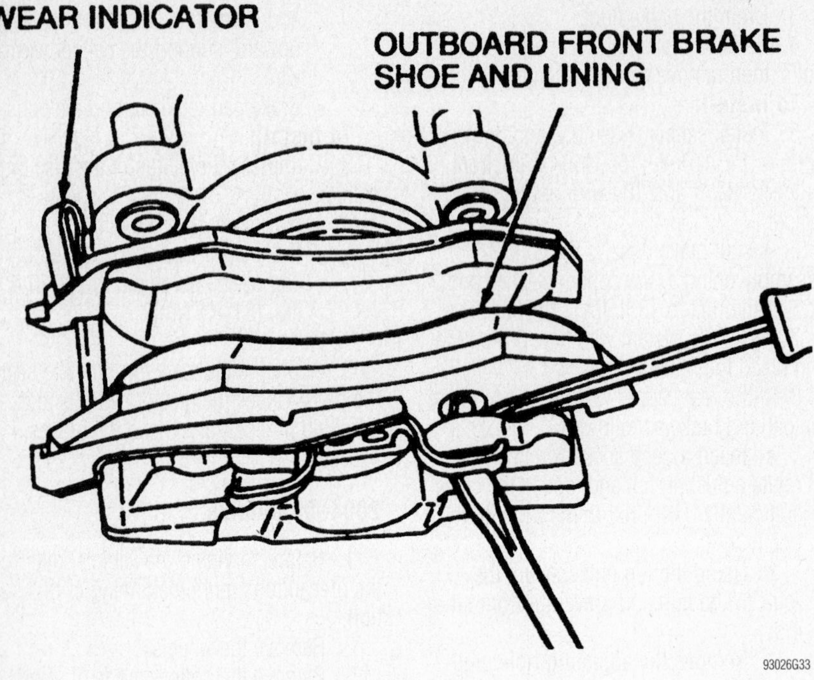

Replacing the disc brake pads—2001–02 models

To install:

4. Clean the piston end of the caliper body and the area around the pin holes. Be careful not to get oil on the rotor.
5. Using the proper tool, carefully turn the piston clockwise back into the caliper body. Take care not to damage the piston boot.
6. Coat the pad contact area on the mounting support with a silicone based grease.
7. Install or connect the following:
   - Pads, shims, and the pad springs. Always use new shims.
   - Caliper body in the mounting support and tighten the pin bolts to 28–38 ft. lbs. (38–52 Nm)
   - Wheels
8. Bleed the system if necessary.

2004–05 Models

1. Before servicing the vehicle, refer to the precautions in the beginning of this section.
2. Remove the wheels.
3. Remove the bottom guide pin from the caliper and swing the caliper cylinder body upward; support the caliper with a wire.
4. Remove the brake pad retainers, shims and the pads.

To install:

5. Compress the piston of the disc brake caliper.
6. Coat the pad contact area on the mounting support with a silicone based grease.
7. Install the retainers and brake pads.
8. Install the caliper assembly. Torque the guide pin to 32 ft. lbs. (43 Nm).
9. Install the wheels.
10. Apply the brakes a few times to seat the pads. Check the master cylinder and add fluid if necessary. Bleed the brakes, if necessary.

Brake Drums

REMOVAL & INSTALLATION

2001–02 Models

1. Before servicing the vehicle, refer to the precautions in the beginning of this section.

The rear drum brakes used on these vehicles are conventional expanding shoe-type with the brake shoe lining applied to the inside of the rotating drum. An incremental brake adjuster screw is designed to actuate whenever sufficient wear occurs.

2. Remove or disconnect the following:
   - Wheel and tire
   - Brake drum by pulling it from the wheel studs
3. If necessary for brake drum removal, pry off the access hole plug from the access hole. Insert a screwdriver and a brake adjustment tool. Press the screwdriver against the adjusting lever to disengage it from the adjuster. Loosen the adjuster using the brake adjusting tool.

To install:

4. Clean all parts well. It is good practice to inspect the wheel cylinder for leaks anytime the brake drum is removed. If a new replacement brake drum is being installed, inspect it for a protective coating on the machined inside braking surface. Remove any coating with suitable solvent.
5. Install or connect the following:
   - Brake drum onto the wheel studs
6. In most all cases, manual brake adjustment IS NOT recommended. Adjust-

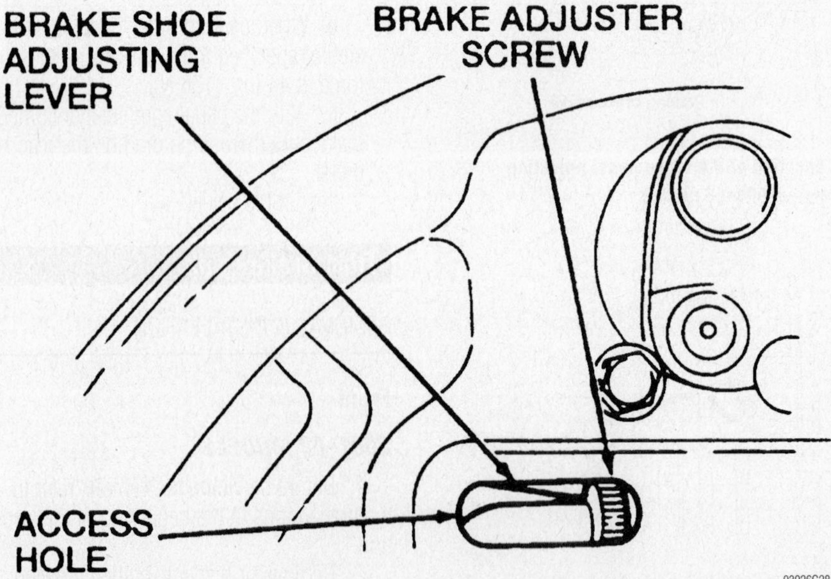

Brake shoe adjustment may need to be loosened to remove the brake drum—2001–02 models

ment is performed by driving the vehicle and applying the brakes.

- Tire and wheel and torque the fasteners to 72–87 ft. lbs. (98–118 Nm)

7. Adjust the rear brake shoes by sharply applying the brakes several times while driving the vehicle alternately forwards and backwards. Check the brake operation by making several stops while driving forward.

Brake Shoes

REMOVAL & INSTALLATION

2001–02 Models

1. Before servicing the vehicle, refer to the precautions in the beginning of this section.

The rear drum brakes use an internal rear wheel cylinder with expanding shoes and lining that are applied against a rotating

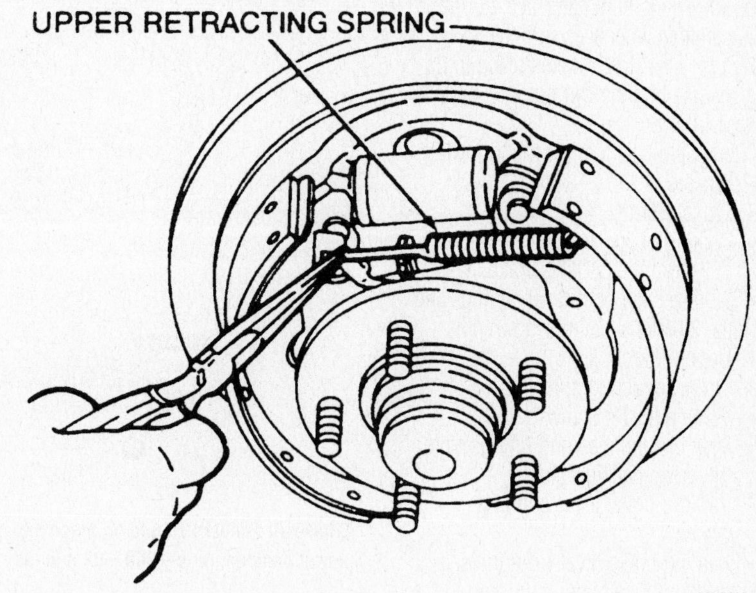

Remove the upper retracting spring—2001–02 models

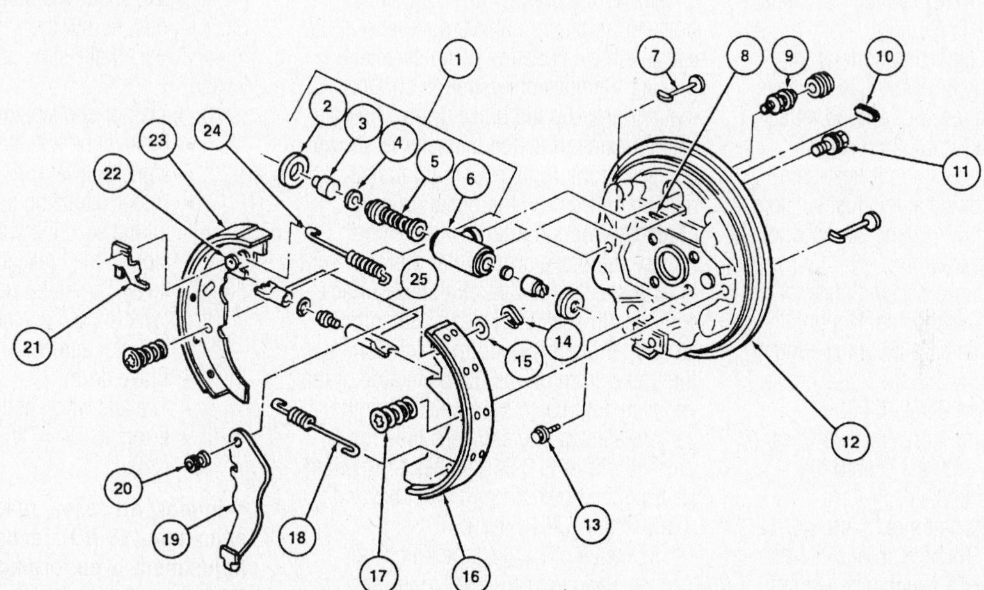

1 Rear Wheel Cylinder
2 Dust Boot (2 Req'd)
3 Wheel Cylinder Piston (2 Req'd)
4 Cup (2 Req'd)
5 Wheel Cylinder Piston Cup Spring
6 Wheel Cylinder Housing
7 Brake Shoe Hold-Down Pin (2 Req'd)
8 Access Hole
9 Rear Brake Bleeder Screw
10 Access Hole Plug
11 Rear Wheel Cylinder Bolt (2 Req'd)
12 Rear Brake Backing Plate
13 Rear Brake Backing Plate Bolts (4 Req'd)
14 Parking Brake Lever Clip
15 Spring Washer
16 Secondary Brake Shoe and Lining
17 Brake Shoe Hold-Down Spring
18 Lower Retracting Spring
19 Parking Brake Lever
20 Parking Brake Lever Pin
21 Brake Shoe Adjusting Lever
22 Adjuster Lever Pin
23 Primary Brake Shoe and Lining
24 Upper Retracting Spring
29 Brake Adjuster Screw

93026G29

Rear drum brake assembly and related components—2001–02 models

brake drum. An incremental brake adjuster screw is actuated whenever sufficient wear occurs. Brake adjustment takes place in forward or reverse braking but not with parking brake application.

2. Remove or disconnect the following:
- Wheels
- Brake drum
- Parking brake rear cable and conduit from the parking brake lever
- 2 brake shoe hold-down springs and the 2 brake shoe hold-down pins
- Upper retracting spring
- Lower retracting spring
- Brake adjuster screw
- Rear brake shoes and linings from the brake backing plate
- Parking brake lever clip and washer
- Parking brake lever from the secondary brake shoe and lining

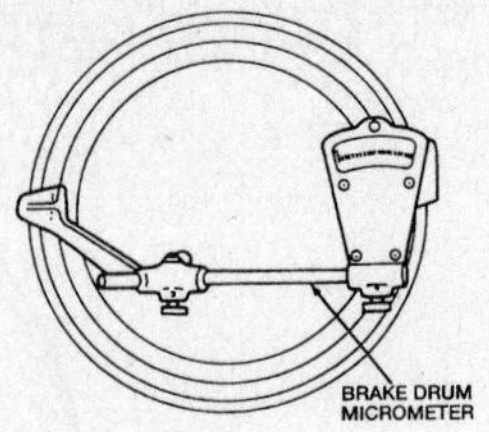

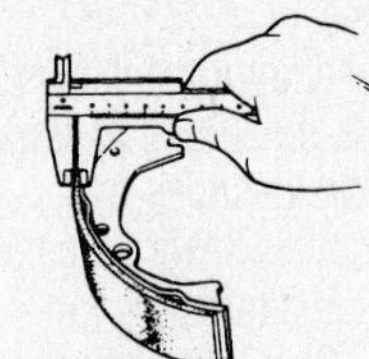

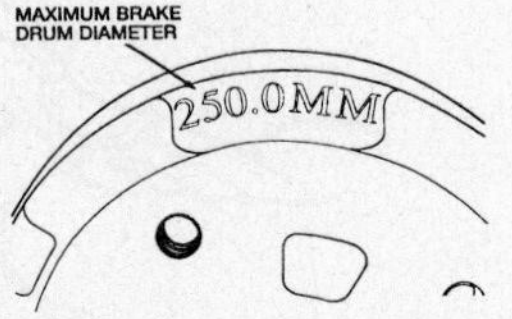

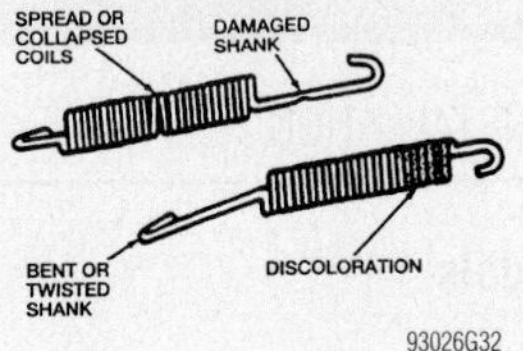

These size checks should be made and the retracting springs' condition checked. Replace questionable parts—2001–02 models

To install:

3. Clean all parts well.

4. Inspect the wheel cylinder for signs of leaking. Service as required.

5. Inspect the retracting springs for heat damage, bends or damage to the coils or shank or loss of tension. A good retracting spring will make a full thud when dropped on a concrete floor. A heat-damaged retracting spring that has lost tension will make a distinctive ringing sound when dropped on a concrete floor.

6. Check the brake backing plate for signs of scoring. The shoe contact points must be smooth and have a light coating of lithium grease. Verify that the brake lining thickness is between 0.059–0.232 in. (1.5–5.9mm). Failure to replace worn rear brake shoes will result in a scored drum.

7. Inspect the brake drum for scratches, scoring, bell mouth and out-of-round conditions. Remove minor scores on a brake drum with sandpaper. Do not refinish brake drums to remove scoring marks. A brake drum surface that is highly polished can cause the brakes to lock up. Remove polished surfaces with sandpaper or refinish the brake drum. Refinish a brake drum that is out-of-round enough to cause vehicle vibration or noise when braking. Remove only enough surface metal to true-up the brake drum. Brake drum maximum inside diameter is shown on each drum. If the maximum inside diameter shown on the brake drum is exceeded through wear or refinishing, replace the brake drum. After a brake drum is refinished, wipe the refinished surface with a cloth soaked in clean denatured alcohol. If one brake drum is refinished, the brake drum on the opposite side of the vehicle should also be refinished to the same diameter. The standard inner brake drum diameter is 9.840 inches (250.0mm). Replace the brake drum if worn beyond 9.900 inches (251.5mm).

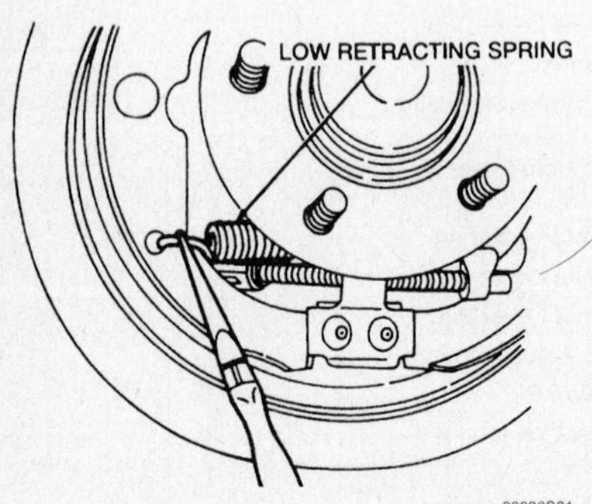

Remove the lower retracting spring—2001–02 models

8. Install or connect the following:
- Parking brake lever to the secondary brake shoe and lining with a new parking brake lever clip
- Secondary (rear) shoe on the backing plate and the brake shoe hold-down spring and pin
- Primary (front) shoe on the backing plate and the brake shoe hold-down spring and pin
- Parking brake rear cable and conduit to the parking brake lever
- Lower retracting spring to the rear brake shoes

9. Apply a light coat of high-quality grease to the threaded areas of the adjuster nut and adjuster socket. Turn the adjuster nut all the way down on the brake adjuster screw, then loosen the adjuster ½ turn. Install the adjuster screw in the slots on the rear brake shoes. The wider slot on the socket must fit in the slot on the primary (front) brake shoe. The slot on the adjuster nut end must fit into the slots in the secondary (rear) brake shoe and parking brake lever.

10. Install or connect the following:
- Brake shoe adjusting lever on the adjuster lever pin
- Upper retracting spring in the slot on the secondary shoe and in the slot on the brake shoe adjusting lever. The brake shoe adjusting lever should contact the brake adjuster screw.
- Brake drum
- Tire and wheel and torque the fasteners to 72–87 ft. lbs. (98–118 Nm).

➡In most all cases, manual brake adjustment IS NOT recommended. Adjustment is performed by driving the vehicle and applying the brakes.

11. The rear brakes do not require adjustment when being serviced to obtain a firm brake pedal feel. To achieve a firm brake pedal after servicing the rear brakes, sharply apply the brake pedal several times while driving the vehicle alternately forwards and backwards. Check the brake operation by making several stops while driving forward. The self-adjusting mechanism will sufficiently adjust the rear brake shoes without any manual tightening at the brake shoe adjuster. If the rear brake shoes are manually adjusted, the additional action of the brake shoe adjuster can cause the brakes to become over-tightened and result in binding or overheated rear brakes.

NISSAN

Titan

11

SPECIFICATIONS AND MAINTENANCE CHARTS

ENGINE AND VEHICLE IDENTIFICATION

Engine						
Code ①	Liters (cc)	Cu. In.	Cyl.	Fuel Sys.	Engine	Eng. Mfg.
VK56DE	5.6 (5552)	338.8	8	MFI	DOHC	Nissan

Model Year	
Code ②	Year
4	2004
5	2005

MFI: Multi-port Fuel Injection

DOHC: Double Overhead Camshafts

② 10th digit of the Vehicle Identification Number (VIN)

67170-TITA-C01

GENERAL ENGINE SPECIFICATIONS

Year	Model	Engine Displacement Liters	Engine ID	Net Horsepower @ rpm	Net Torque @ rpm (ft. lbs.)	Bore x Stroke (in.)	Com-pression Ratio	Oil Pressure @ rpm
2004	Titan	5.6	VK56DE	305@4900	379@3600	3.86X3.62	9.8:1	43@2000

67170-TITA-C02

ENGINE TUNE-UP SPECIFICATIONS

Year	Engine Displacement Liters	Engine ID	Spark Plug Gap (in.)	Ignition Timing	Fuel Pump (psi) ①	Idle Speed ②	Valve Clearance (in.) In.	Valve Clearance (in.) Ex.
2004	5.6	VK56DE	0.043	15B	51	650	0.010-0.013	0.011-0.015

NOTE: The Vehicle Emission Control Information label often reflects specification changes made during production. The label figures must be used if they differ from those in this chart.

B: Before top dead center

① System pressure at idle

② Automatic transmission in Neutral

67170-TITA-C03

CAPACITIES

Year	Model	Engine Displacement Liters	Engine ID	Engine Oil with Filter (qts.)	Transmission (pts.)	Transfer Case (pts.)	Drive Axle Front (pts.)	Drive Axle Rear (pts.)	Fuel Tank (gal.)	Cooling System (qts.)
2004	Titan	5.6	VK56DE	6.2	22.5	6.25	3.375	4.25	28.0	13

NOTE: All capacities are approximate. Add fluid gradually and check to be sure a proper fluid level is obtained.

67170-TITA-C04

VALVE SPECIFICATIONS

Year	Engine Displacement Liters	Engine ID	Seat Angle (deg.)	Face Angle (deg.)	Spring Test Pressure (lbs. @ in.)	Spring Installed Height (in.)	Stem-to-Guide Clearance (in.) Intake	Stem-to-Guide Clearance (in.) Exhaust	Stem Diameter (in.) Intake	Stem Diameter (in.) Exhaust
2004	5.6	VK56DE	45.15-45.45	45	37.0@1.457	1.9913	0.0008-0.0021	0.0012-0.0025	0.2348-0.2354	0.2344-0.2350

67170-TITA-C06

CRANKSHAFT AND CONNECTING ROD SPECIFICATIONS

All measurements are given in inches.

Year	Engine Displ. Liters	Engine ID	Crankshaft Main Brg. Journal Dia.	Crankshaft Main Brg. Oil Clearance	Crankshaft Shaft End-play	Crankshaft Thrust on No.	Connecting Rod Journal Diameter	Connecting Rod Oil Clearance	Connecting Rod Side Clearance
2004	5.6	VK56DE	①	②	0.0118	3	③	0.0008-0.0015	0.0079-0.0157

① There are 24 different grades, ranging from G (2.5183) to 9 (2.5173)

② No. 1 and 5: 0.00004-0.0004

No. 2, 3 and 4: 0.0003-0.0007

③ Grade 0: 2.2441-2.2441

Grade 1: 2.2441-2.2442

Grade 2: 2.2442-2.2442

Grade 3: 2.2442-2.2443

Grade 4: 2.2443-2.2443

Grade 5: 2.2443-2.2443

Grade 6: 2.2443-2.2444

Grade 7: 2.2444-2.2444

Grade 8: 2.2444-2.2444

Grade 9: 2.2444-2.2445

Grade A: 2.2445-2.2445

Grade B: 2.2445-2.22446

Grade C: 2.2446-2.2446

67170-TITA-C05

PISTON AND RING SPECIFICATIONS

All measurements are given in inches.

Year	Engine Displacement Liters	Engine ID	Piston Clearance	Ring Gap Top Comp.	Ring Gap Bottom Comp.	Ring Gap Oil Control	Ring Side Clearance Top Comp.	Ring Side Clearance Bottom Comp.	Ring Side Clearance Oil Control
2004	5.6	VK56DE	0.0004-0.0012	0.0091-0.0130	0.0091-0.0130	0.0079-0.0236	0.0014-0.0033	0.0012-0.0028	0.0006-0.0073

67170-TITA-C07

TORQUE SPECIFICATIONS

All readings in ft. lbs.

Year	Engine Displacement Liters	Engine ID	Cylinder Head Bolts	Main Bearing Bolts	Rod Bearing Bolts	Crankshaft Damper Bolts	Flywheel Bolts	Manifold Intake	Manifold Exhaust	Spark Plugs	Oil Pan Drain Plug
2004	5.6	VK56DE	①	②	③	④	65	73	21	18	25

NA: Information not available

① Step 1: 72 ft. lbs
Step 2: Loosen all bolts completely
Step 3: 33 ft. lbs.
Step 4: +60 degrees
Step 5: +60 degrees

② Step 1: Main Bolts to 29 ft. lbs.
Step 2: Sub-bolts to 22 ft. lbs.
Step 3: Main Bolts +40 degrees
Step 4: Sub-Bolts +30 degrees
Step 5: Side Bolts to 36 ft. lbs.

③ Step 1: 11 ft. lbs.
Step 2: +90 degrees

④ Step 1: 65 ft. lbs.
Step 2: +90 degrees

67170-TITA-C08

WHEEL ALIGNMENT

Year	Model	Caster Range (+/-Deg.)	Caster Preferred Setting (Deg.)	Camber Range (+/-Deg.)	Camber Preferred Setting (Deg.)	Toe-in (in.)
2004	Titan ①	0.75	②	0.75	③	0.11+/-0.05

① Assumes P245/70R17 tire

② 4x2: +3.27
4x4: +2.37

③ 4x2: -0.12
4x4: +0.43

67170-TITA-C09

TIRE, WHEEL AND BALL JOINT SPECIFICATIONS

		OEM Tires		Tire Pressures (psi)		Wheel	Ball Joint	Lugnut Torque
Year	Model	Standard	Optional	Front	Rear	Size	Inspection	(ft. lbs.)
2004	Titan	P245/70R17	①	35	35	17	②	98

OEM: Original Equipment Manufacturer

PSI: Pounds Per Square Inch

① P285/70R17

P265/70R18

② Axial play

Upper: 0

67170-TITA-C10

BRAKE SPECIFICATIONS

All measurements in inches unless noted

			Brake Disc			Minimum	Brake Caliper	
			Original	Minimum	Maximum	Lining	Bracket Bolts	Mounting Bolts
Year	Model		Thickness	Thickness	Runout	Thickness	(ft. lbs.)	(ft. lbs.)
2004	Titan	F	1.024	0.965	0.0016	0.039	155	32
		R	0.551	0.472	0.002	0.039	NA	24

NA: Not applicable

67170-TITA-C11

SCHEDULED MAINTENANCE INTERVALS

Nissan - Titan

TO BE SERVICED	TYPE OF SERVICE	7.5	15	22.5	30	37.5	45	52.5	60
Engine oil & filter	R	✓	✓	✓	✓	✓	✓	✓	✓
Brake lines & cables	S/I		✓		✓		✓		✓
Brake pads and rotors	I		✓		✓		✓		✓
Driveshaft boots & propeller shaft (4x4)	L/I		✓		✓		✓		✓
Transmission, transfer & differential gear oil	I		✓		✓		✓		✓
Air cleaner filter	R				✓				✓
Engine coolant ①	R								✓
Spark plugs (Platinum)	R	Replace every 105,000 miles							
Drive belt(s) ②	S/I								✓
Cabin air filter	R		✓		✓		✓		✓
Exhaust system	I				✓				✓
Fuel lines	S/I				✓				✓
Fuel Filter ③									
Steering gear (box) & linkage, axle & suspension parts	I				✓				✓
Vapor lines	S/I				✓				✓

R: Replace S/I: Service or Inspect L: Lubricate

① Coolant: After 60,000 miles, inspect every 30,000 miles.

② Drive Belts: After 60,000 miles, inspect every 15,000 miles. Replace belts if damaged.

③ Fuel Filter: Maintenance free item.

FREQUENT OPERATION MAINTENANCE (SEVERE SERVICE)

If a vehicle is operated under any of the following conditions it is considered severe service:

- Extremely dusty areas.

- Rough, muddy, or salt spread roads.

- 50% or more of the vehicle constant operation is in 32°C (90°F) or higher temperatures, or temperatures below 0°C (32°F).

- Prolonged idling (vehicle operation in stop and go traffic).

- Frequent short running periods (engine does not warm to normal operating temperatures).

- Police, taxi, delivery usage or trailer towing usage.

Oil & oil filter: replace every 3750 miles.

Brake pads, discs, drums & linings: service or inspect every 7500 miles.

Driveshaft boots & propeller shaft: service or inspect every 7500 miles.

Exhaust system: service or inspect every 7500 miles.

Steering gear (box) & linkage, (steering damper-4x4), axle & suspension parts: service or inspect every 7500 miles.

Steering linkage ball joints & front suspension ball joints: service or inspect every 7500 miles.

67170-TITA-C12

PRECAUTIONS

Before servicing any vehicle, please be sure to read all of the following precautions, which deal with personal safety, prevention of component damage, and important points to take into consideration when servicing a motor vehicle:

- Never open, service or drain the radiator or cooling system when the engine is hot; serious burns can occur from the steam and hot coolant.
- Observe all applicable safety precautions when working around fuel. Whenever servicing the fuel system, always work in a well-ventilated area. Do not allow fuel spray or vapors to come in contact with a spark, open flame, or excessive heat (a hot drop light, for example). Keep a dry chemical fire extinguisher near the work area. Always keep fuel in a container specifically designed for fuel storage; also, always properly seal fuel containers to avoid the possibility of fire or explosion. Refer to the additional fuel system precautions later in this section.
- Fuel injection systems often remain pressurized, even after the engine has been turned **OFF**. The fuel system pressure must be relieved before disconnecting any fuel lines. Failure to do so may result in fire and/or personal injury.
- Brake fluid often contains polyglycol ethers and polyglycols. Avoid contact with the eyes and wash your hands thoroughly after handling brake fluid. If you do get brake fluid in your eyes, flush your eyes with clean, running water for 15 minutes. If eye irritation persists, or if you have taken brake fluid internally, IMMEDIATELY seek medical assistance.
- The EPA warns that prolonged contact with used engine oil may cause a number of skin disorders, including cancer. You should make every effort to minimize your exposure to used engine oil. Protective gloves should be worn when changing oil. Wash your hands and any other exposed skin areas as soon as possible after exposure to used engine oil. Soap and water, or waterless hand cleaner should be used.
- All new vehicles are now equipped with an air bag system, often referred to as a Supplemental Restraint System (SRS) or Supplemental Inflatable Restraint (SIR) system. The system must be disabled before performing service on or around system components, steering column, instrument panel components, wiring and sensors. Failure to follow safety and disabling procedures could result in accidental air bag deployment, possible personal injury, and unnecessary system repairs.
- Always wear safety goggles when working with, or around, the air bag system. When carrying a non-deployed air bag, be sure the bag and trim cover are pointed away from your body. When placing a non-deployed air bag on a work surface, always face the bag and trim cover upward, away from the surface. This will reduce the motion of the module if it is accidentally deployed. Refer to the additional air bag system precautions later in this section.
- Clean, high quality brake fluid from a sealed container is essential to the safe and proper operation of the brake system. You should always buy the correct type of brake fluid for your vehicle. If the brake fluid becomes contaminated, completely flush the system with new fluid. Never reuse any brake fluid. Any brake fluid that is removed from the system should be discarded. Also, do not allow any brake fluid to come in contact with a painted surface; it will damage the paint.
- Never operate the engine without the proper amount and type of engine oil; doing so will result in severe engine damage.
- Timing belt maintenance is extremely important. Many models utilize an interference-type, non-freewheeling engine. If the timing belt breaks, the valves in the cylinder head may strike the pistons, causing potentially serious (also time-consuming and expensive) engine damage. Refer to the maintenance interval charts in the front of this manual for the recommended replacement interval for the timing belt, and to the timing belt section for belt replacement and inspection.
- Disconnecting the negative battery cable on some vehicles may interfere with the functions of the on-board computer system(s) and may require the computer to undergo a relearning process once the negative battery cable is reconnected.
- When servicing drum brakes, only disassemble and assemble one side at a time, leaving the remaining side intact for reference.

ENGINE REPAIR

Distributor

REMOVAL & INSTALLATION

These models use a Distributorless Ignition System (DIS) controlled by the Powertrain Control Module (PCM).

Alternator

REMOVAL & INSTALLATION

1. Before servicing the vehicle, refer to the precautions in the beginning of this section.
2. Remove or disconnect the following:
    - Negative battery cable
    - Fan shroud
    - Drive belt

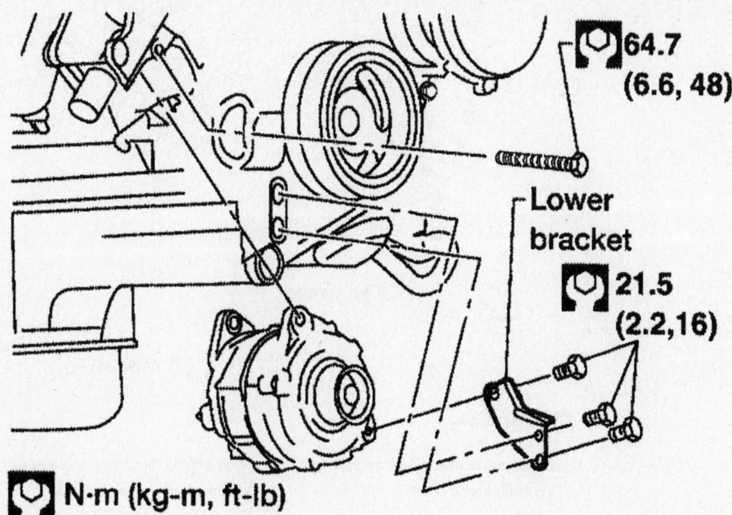

Alternator mounting

- Lower alternator bracket
- Alternator upper bolt
- Alternator harness connectors
- Alternator

To install:

3. Install or connect the following:
   - Alternator
   - Alternator harness connectors
   - Upper bolt, tighten to 48 ft. lbs. (65 Nm)
   - Lower bracket, tighten to 16 ft. lbs (22 Nm)
   - Drive belt
   - Fan shroud
   - Negative battery cable

Ignition Timing

ADJUSTMENT

Ignition timing is controlled by the ECM and manual adjustment is not possible.

Engine Assembly

REMOVAL & INSTALLATION

1. Before servicing the vehicle, refer to the precautions in the beginning of this section.
2. Drain the cooling system.
3. Partially drain the automatic transmission fluid.
4. Relieve the fuel system pressure.
5. Remove or disconnect the following:
   - Hood
   - Cowl extension
   - Engine cover
   - Air intake assembly
   - Vacuum hose between vehicle and engine
   - Radiator hoses
   - Radiator
   - Drive belts
   - Engine fan
   - Wiring harness
   - ECM
   - Power steering reservoir tank and oil pump
   - A/C compressor
   - Brake booster vacuum line
   - EVAP line
   - Fuel hose
   - Heater hoses
   - Transmission dipstick assembly
   - Front final drive assembly (4WD models only)
   - Exhaust manifolds
6. Install engine slings onto the left and right cylinder heads and tighten to 33 ft. lbs. (45 Nm).
7. Attach an engine hoist to slings and lift engine out of the vehicle

To install:

8. Lower engine into the vehicle and remove the engine slings.
9. Install or connect the following:
   - Exhaust manifolds
   - Front final drive assembly (4WD models only)
   - Transmission dipstick assembly
   - Heater hoses
   - Fuel hose
   - EVAP line
   - Brake booster vacuum line
   - A/C compressor
   - Power steering oil pump and reservoir tank
   - ECM
   - Wiring harnesses
   - Engine fan
   - Drive belts
   - Radiator
   - Radiator hoses
   - Vacuum hose between engine and vehicle
   - Air intake assembly
   - Engine cover
   - Cowl extension
   - Hood
10. Refill the automatic transmission fluid.
11. Refill the cooling system.
12. Start the engine and check for leaks.

Water Pump

REMOVAL & INSTALLATION

1. Before servicing the vehicle, refer to the precautions in the beginning of this section.

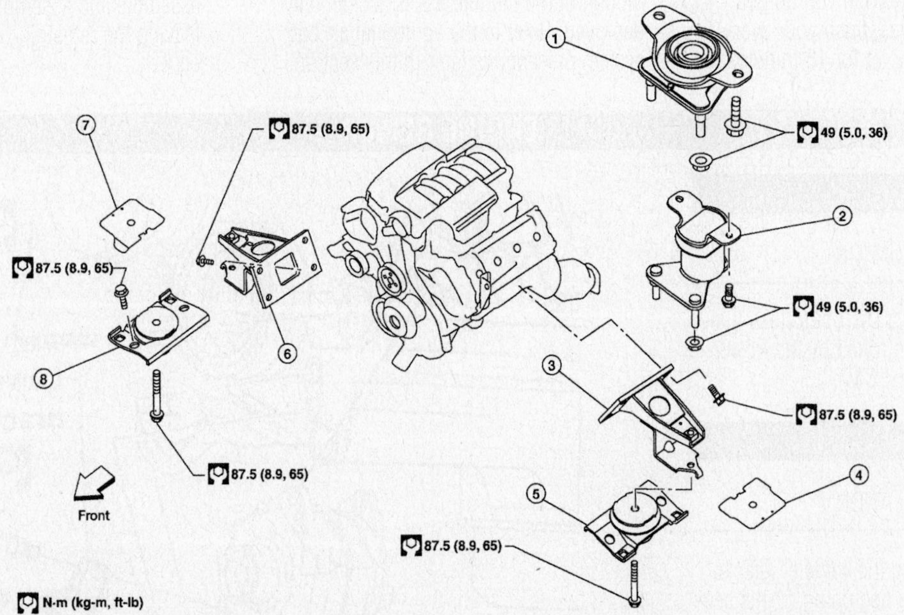

Engine mounts

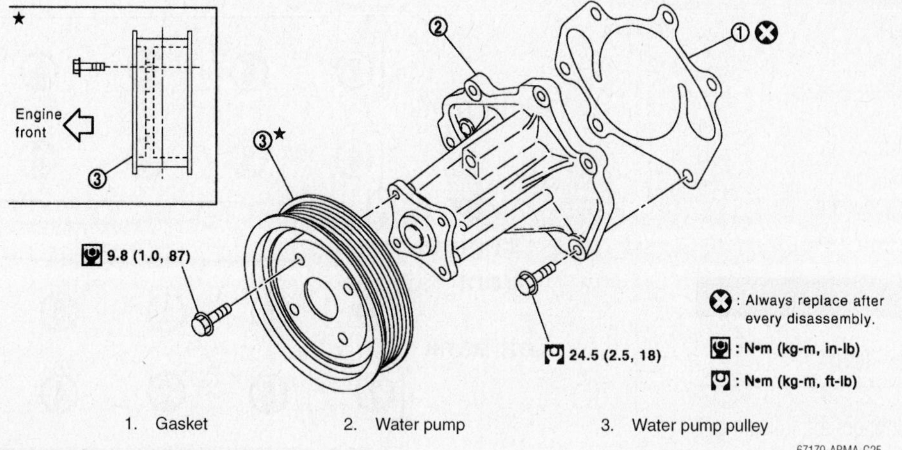

Water pump mounting

2. Drain the cooling system.
3. Remove or disconnect the following:
   - Engine splash guard
   - Air intake assembly
   - Accessory drive belt

➡Leave tensioner pulley in its fixed position.

   - Water pump pulley
   - Water pump

To install:

4. Install or connect the following:
   - Water pump with a new gasket. Tighten bolts to 18 ft. lbs. (25 Nm).
   - Water pump pulley and tighten bolts to 87 in. lbs. (10 Nm).
   - Accessory drive belt
   - Air intake assembly
   - Engine splash guard
5. Refill the cooling system.
6. Start the engine and check for leaks.

Heater Core

REMOVAL & INSTALLATION

1. Discharge the A/C system.
2. Drain the cooling system.
3. Remove or disconnect the following:
   - Cowl top extension
   - Exhaust system
   - Front heater hoses
   - High/Low pressure pipes
   - Instrument and console panels
   - Steering column
   - Instrument panel wiring harness
   - Steering member from body
   - Heater assembly

To install:

4. Install or connect the following:
   - Heater assembly
   - Steering member
   - Instrument panel wiring harness

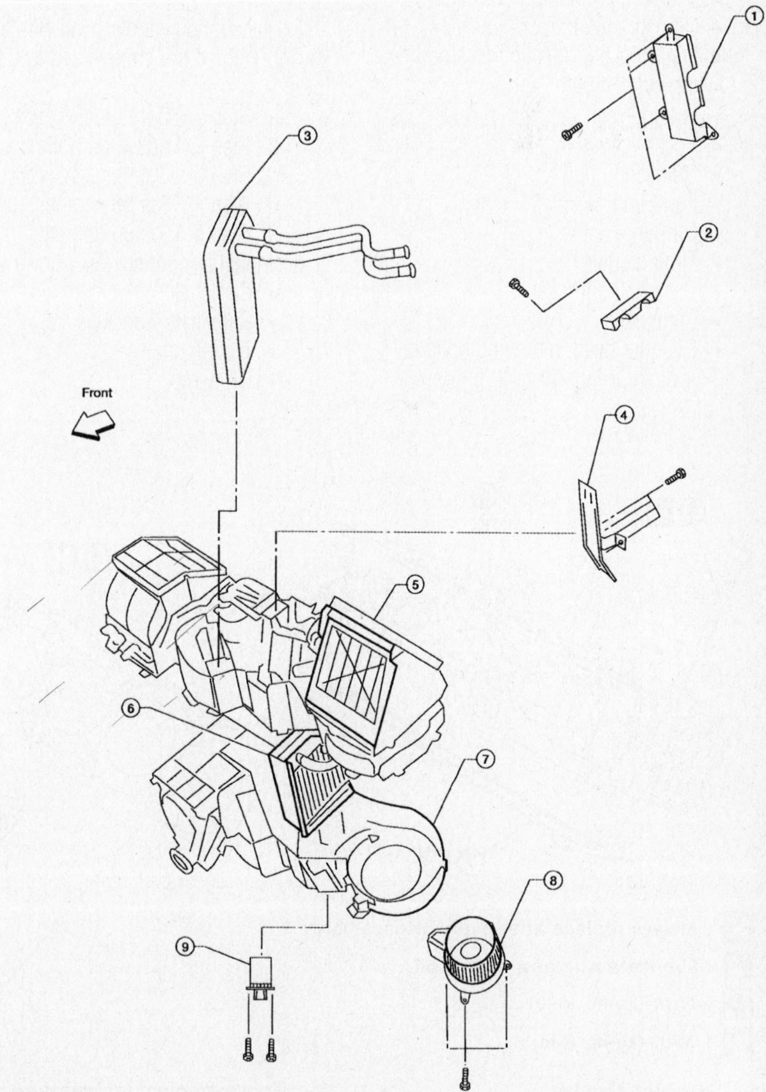

Front heater/AC assembly

- Steering column
- Console panel
- Instrument panel
- High/Low pressure pipes
- Heater hoses
- Exhaust system
- Cowl top extension

5. Refill the cooling system.
6. Recharge the A/C system.

Cylinder Head

REMOVAL & INSTALLATION

1. Remove or disconnect the following:
   - Engine assembly
   - Belt tensioner
   - Idler pulley
   - Thermostat housing and hose
   - Oil pan and strainer
   - Fuel tube and injector assembly
   - Intake manifold
   - Ignition coil
   - Rocker cover
   - Crankshaft pulley
   - Front engine cover
   - Oil pump
   - Timing chain
   - Camshaft sprockets
   - Camshafts
   - Cylinder head, removing bolts in reverse order shown in figure

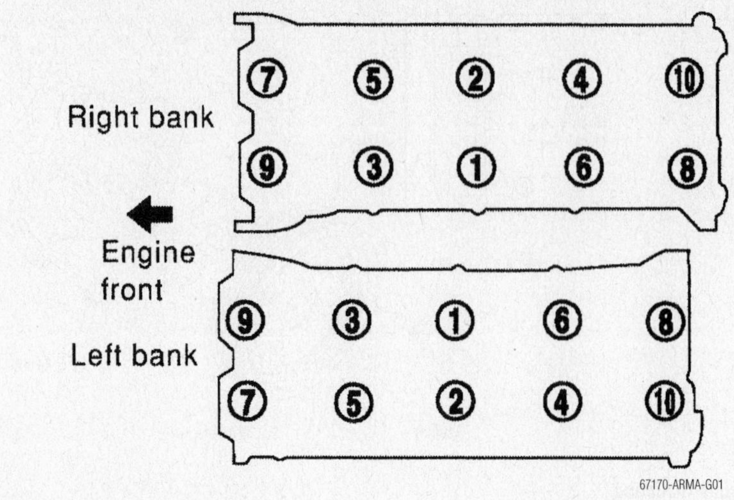

Cylinder head torque sequence

2. Install the cylinder head with a new gasket. Tighten the bolts in sequence as follows:
   a. Step 1: 72 ft. lbs. (98 Nm)
   b. Step 2: Loosen all bolts completely
   c. Step 3: 33 ft. lbs. (44 Nm)
   d. Step 4: Plus 60 degrees
   e. Step 5: Plus 60 degrees
3. Install or connect the following:
   - Camshaft
   - Camshaft sprockets
   - Timing chain
   - Oil pump
   - Front engine cover
   - Crankshaft pulley
   - Rocker cover
   - Ignition coil
   - Intake manifold
   - Fuel tube and injector assembly
   - Oil pain and strainer
   - Thermostat housing and hose
   - Idler pulley
   - Belt tensioner
   - Engine assembly
4. Start the engine and check for leaks

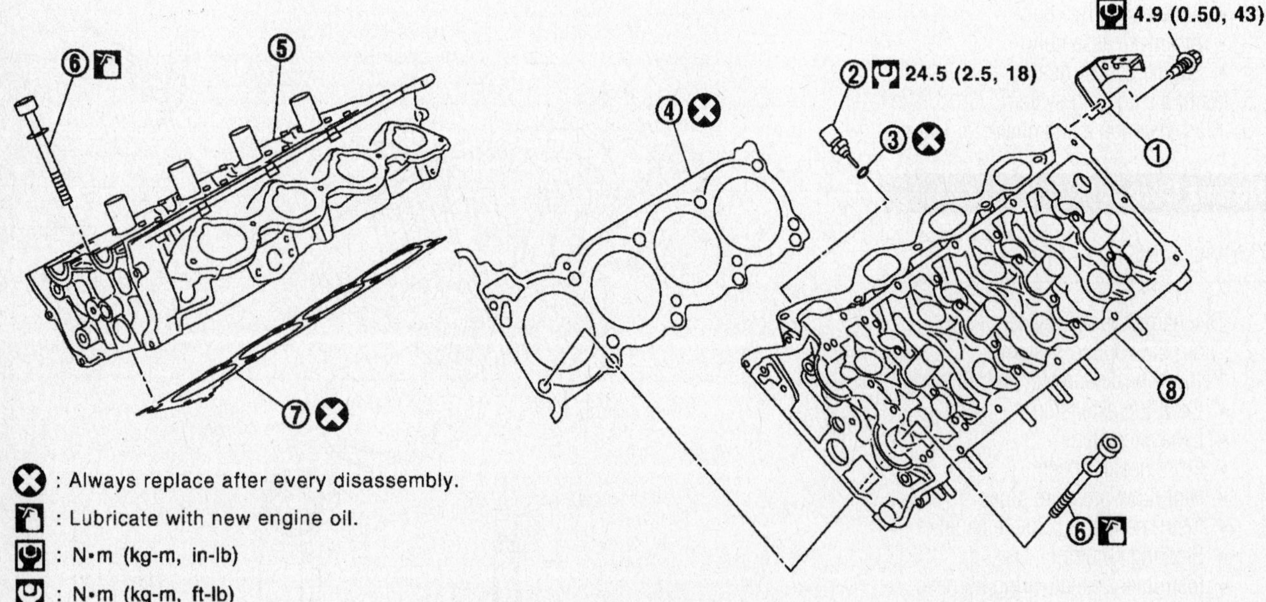

Cylinder heads and gaskets

Intake Manifold

REMOVAL & INSTALLATION

1. Drain the cooling system.
2. Relieve the fuel system pressure.
3. Remove or disconnect the following:
   - Engine cover
   - Air intake assembly
   - Fuel tube quick connector using special tool J-45488
   - Wiring harnesses and brackets from manifold
   - Vacuum hoses
   - PCV hose and tube
   - Electric throttle control actuator, loosening bolts diagonally
   - Fuel injectors
   - Fuel tube assembly
   - Intake manifold, removing bolts in reverse order shown in figure

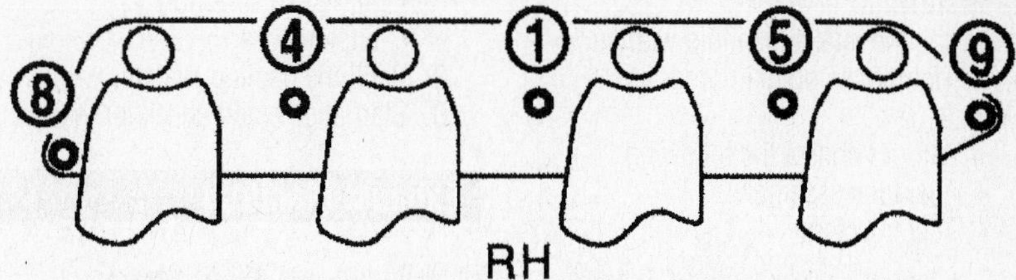

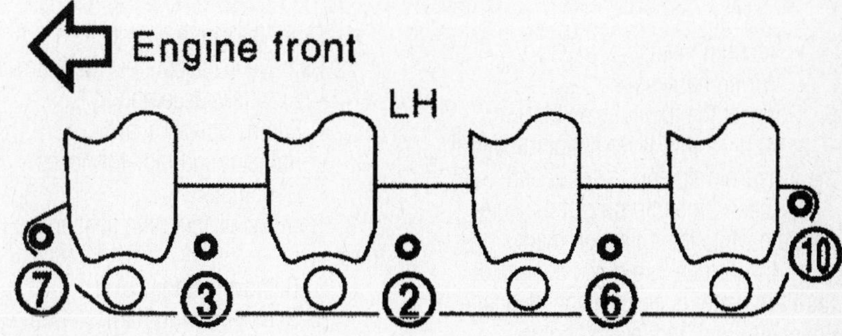

Intake manifold torque sequence

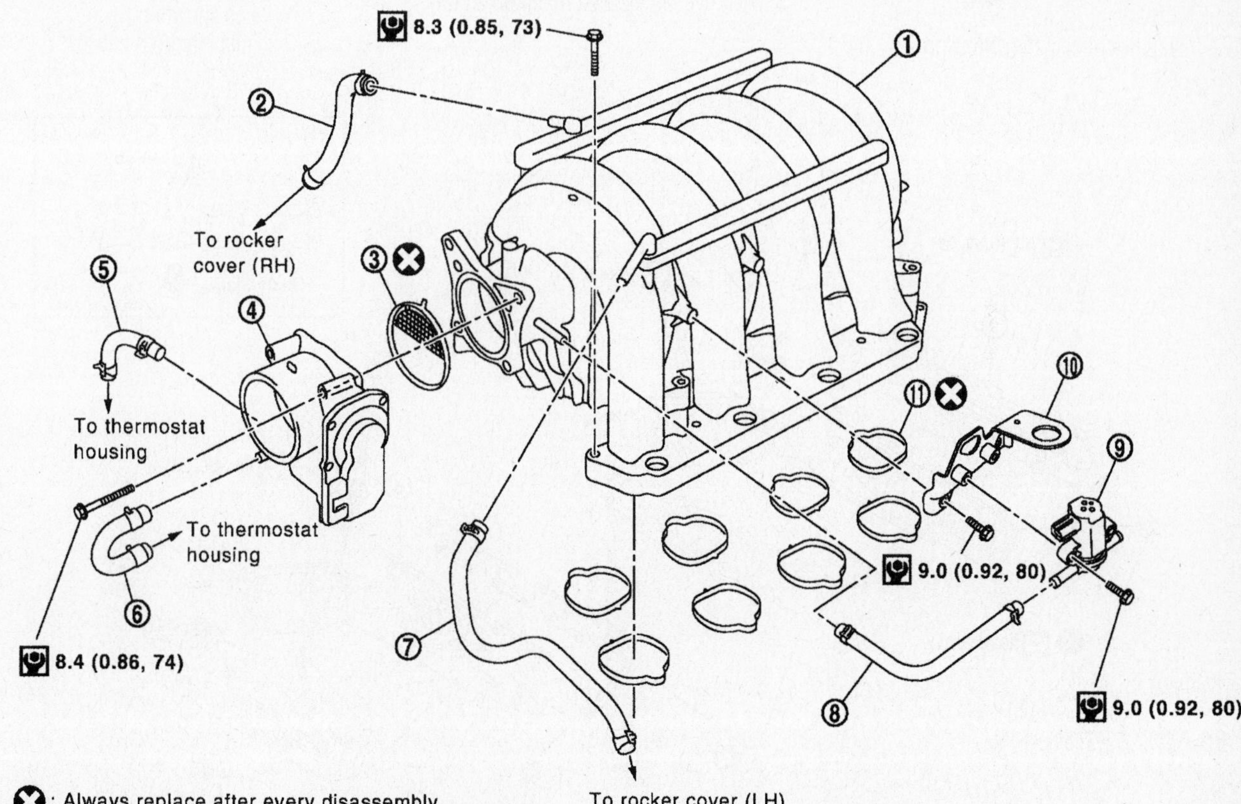

1.	Intake manifold	2.	PCV hose	3.	Gasket
4.	Electric throttle control actuator	5.	Water hose	6.	Water hose
7.	PCV hose	8.	EVAP hose	9.	EVAP canister purge control solenoid valve
10.	Bracket	11.	Gasket		

67170-ARMA-G29

Intake manifold and related parts

To install:

4. Install the intake manifold with new gaskets. Tighten the bolts in order as shown.
5. Install or connect the following:
   - Fuel tube assembly
   - Fuel injectors
   - Electronic throttle control actuator, tightening the bolts in several steps
   - PCV hose
   - Vacuum hoses
   - Wiring harnesses
6. Connect the fuel tube as follows:
   a. Apply a thin layer of engine oil on the tube from tip end to spool end.
   b. Insert tube into quick connector past the white identification mark
   c. Insert tube into quick connector until top spool is completely inside the connector and 2nd level spool is exposed right below the connector.
   d. Pull slightly on the quick connector to ensure it is fully engaged.
   e. Install quick connector cap on quick connector joint.
7. Install or connect the following:
   - Air intake assembly
   - Engine cover
8. Refill the cooling system.
9. Start engine and check for leaks.

Exhaust Manifold

REMOVAL & INSTALLATION

1. Drain the cooling system.
2. Remove or disconnect the following:
   - Air intake assembly
   - Engine splash guard
   - Radiator and radiator hoses
   - Drive belts
3. Remove air fuel ratio sensors as follows:
   a. Remove engine cover.
   b. Remove wiring harness from each sensor
   c. Remove sensors, using special tool J-38356
4. Remove front cross bar
5. Remove left exhaust manifold as follows:
   a. Remove the exhaust front tube.
   b. Remove the exhaust manifold cover.
   c. Loosen nuts in reverse order shown in figure.
   d. Remove studs from position 2, 4, 6, and 8 and remove manifold.
6. Remove right exhaust manifold as follows:
   a. Remove the exhaust front tube.
   b. Remove the oil level gauge guide.
   c. Remove the exhaust manifold cover.
   d. Loosen nuts in reverse order shown in figure.
   e. Remove studs from position 2, 4, 6, and 8 and remove manifold.

To install:

7. Install or connect the following:
   - Exhaust manifold gasket with triangle mark facing up and coated (gray) face toward exhaust manifold.
   - Exhaust manifold, tightening the nuts as shown in figure
   - Exhaust manifold cover

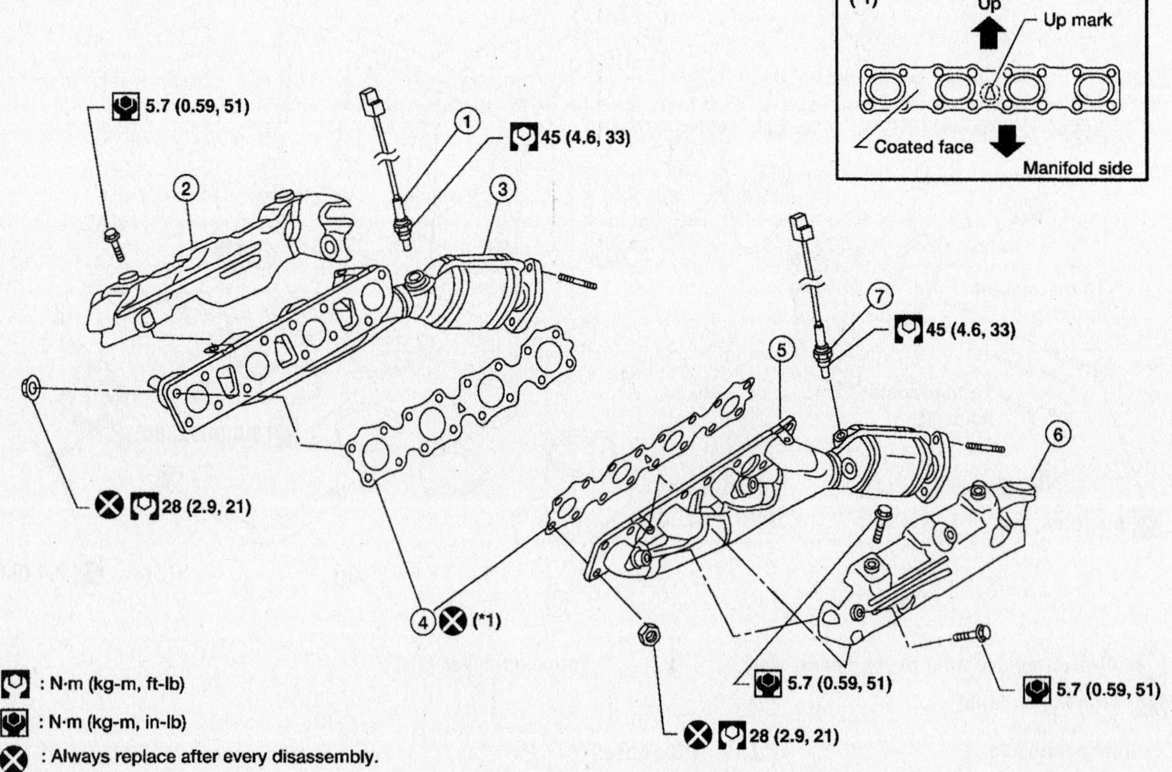

Exhaust manifolds and related parts

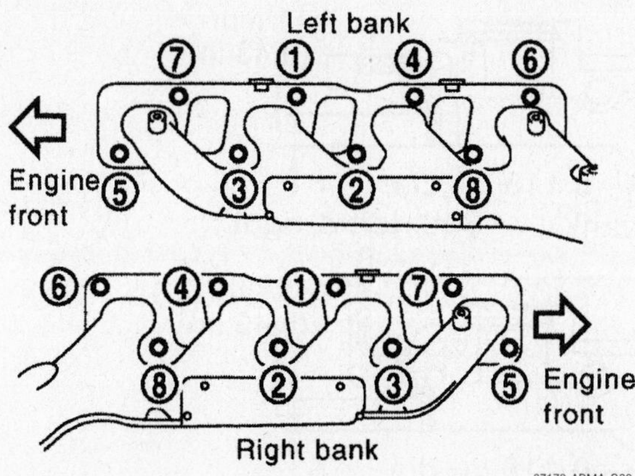

Exhaust manifold torque sequence

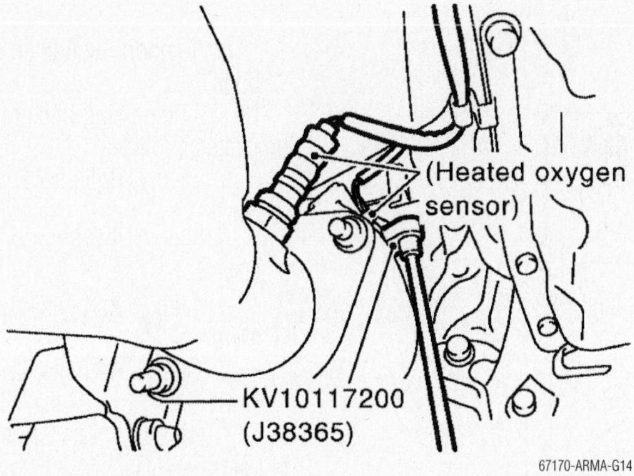

Removing Air-Fuel ratio sensors

- Oil level gauge guide (right side only)
- Exhaust front tube
- Front cross bar
- Air fuel ratio sensors, with anti-seize lubricant
- Engine cover
- Drive belts
- Radiator and radiator hoses
- Engine splash guard
- Air intake assembly

8. Refill the cooling system
9. Start engine and check for leaks.

Camshaft and Valve Lifters

REMOVAL & INSTALLATION

1. Remove rocker cover.
2. Obtain compression Top Dead Center (TDC) of No. 1 cylinder.
3. Remove timing chain case cover.
4. Matchmark the timing chain, aligning with the camshaft sprocket marks.
5. Remove chain tensioner from left bank as follows:
    a. Squeeze end clips and push plunger into tensioner body.
    b. Secure plunger using stopper pin.
    c. Remove chain tensioner.
6. Remove chain tensioner from right bank as follows:
    a. Remove chain tensioner cover using special tool J-37228.
    b. Squeeze end clips and push plunger into tensioner body.
    c. Secure plunger using stopper pin.
    d. Remove chain tensioner.
7. With camshaft locked with a wrench, loosen bolts to remove camshaft sprocket.
8. Remove front cover bolts.
9. Remove camshaft brackets, removing bolts in reverse order shown in figure.
10. Remove camshaft.
11. Remove valve lifters.

To install:

12. Install valve lifters.
13. Install camshaft, refer to table for correct placement.
14. Install camshaft brackets as follows:
    a. Refer to location mark on upper surface of bracket.
    b. Installation mark should be correctly read when viewed from intake side.
15. Install camshaft bracket #1 as follows:
    a. Apply liquid gasket to bracket and backside of front cover as shown in figure.
    b. Carefully position and mount camshaft bracket #1.

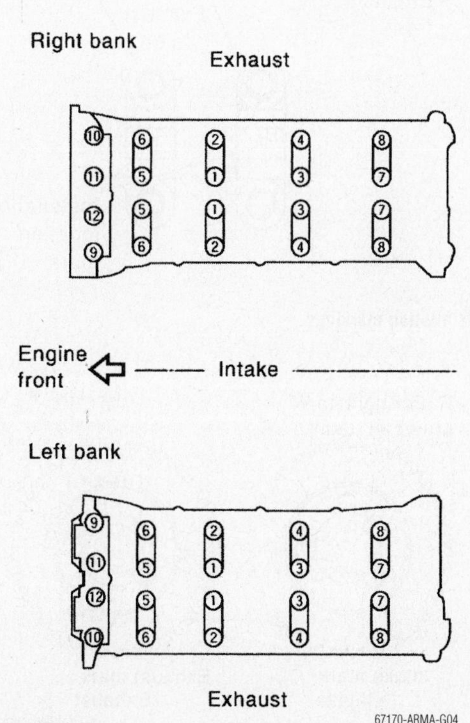

Camshaft torque sequence

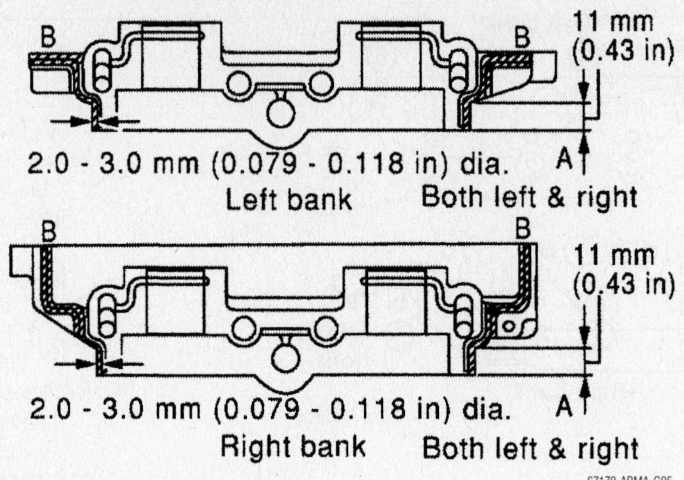

Gasket application for camshaft bracket

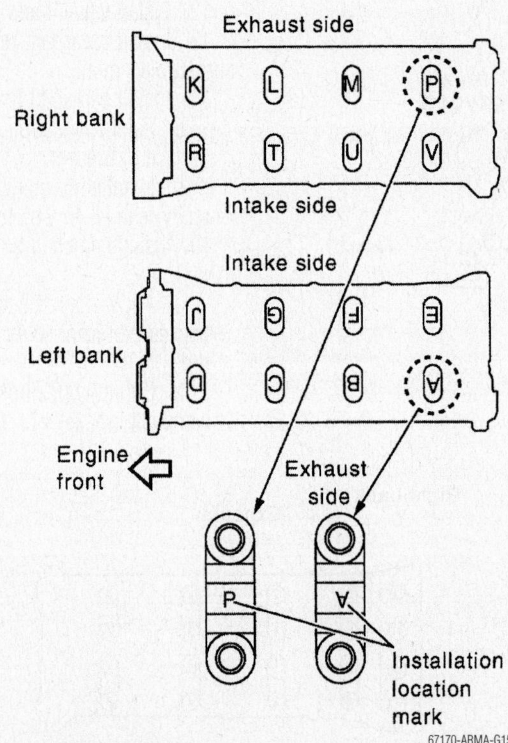

Camshaft bracket installation markings

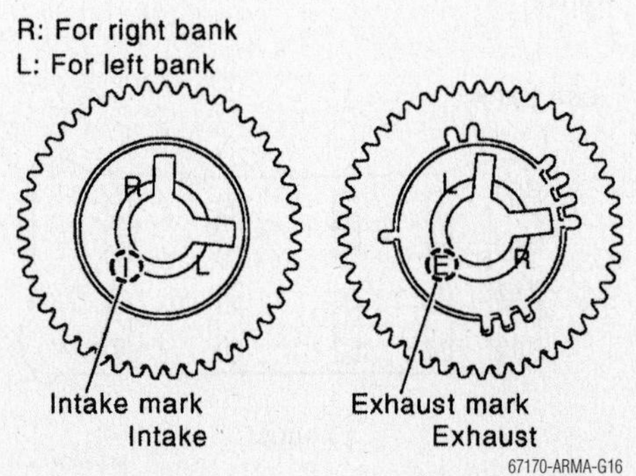

Camshaft installation markings

c. Temporarily tighten front cover bolts

16. Tighten fixing bolts for camshaft brackets as follows:

a. Step 1: Bolts 9-12: 17 in. lbs. (1.9 Nm)

b. Step 2: Bolts 1-8: 17 in. lbs. (1.9 Nm)

c. Step 3: All bolts: 52 in. lbs. (5.9 Nm)

d. Step 4: All bolts: 92 in. lbs. (10 Nm)

17. Tighten front cover bolts to 8 ft. lbs. (11 Nm)

18. Install camshaft sprocket as follows:

a. Install camshaft sprocket aligning matchmarks with timing chain. Align camshaft sprocket key groove with dowel pin on camshaft front edge.

b. Temporarily tighten bolts.

c. Lock the camshaft with a wrench and tighten the bolts.

19. Install chain tensioner as shown:

a. Install chain tensioner, compress plunger and hold with stopper pin.

b. Tighten chain tensioner bolts to 61 in. lbs. (7 Nm)

c. Remove stopper pin, release plunger and apply tension to timing chain.

d. Install chain tensioner front cover (Right-hand bank only) and tighten bolts to 80 in. lbs. (9 Nm).

20. Install timing chain cover.

21. Install rocker cover.

Valve Lash

INSPECTION

1. Run engine to operating temperature.
2. Remove or disconnect the following:

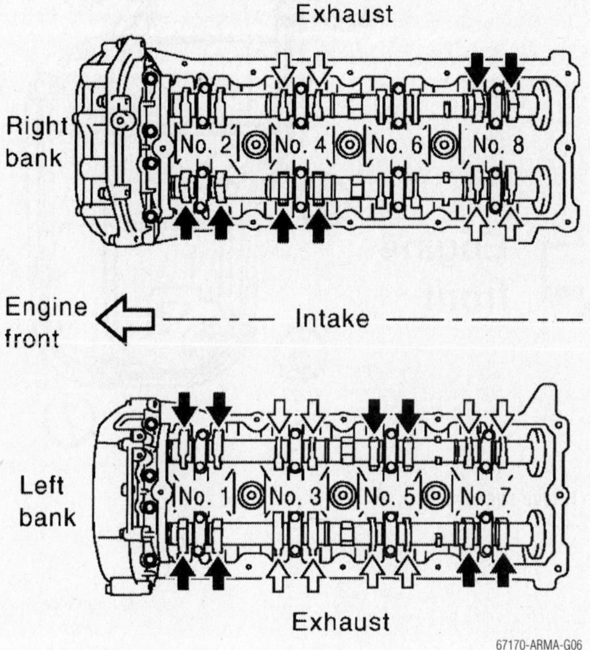

Locations to measure clearance with No. 1 cylinder at TDC

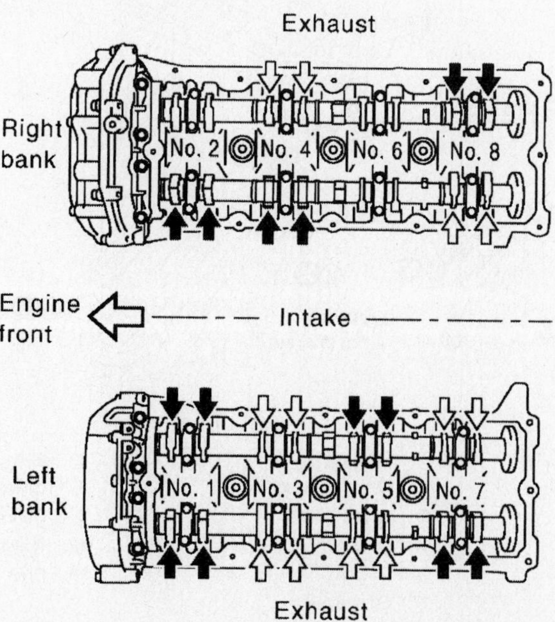

Locations to measure clearance with No. 3 cylinder at TDC

- Engine cover
- Battery cover
- Air intake assembly
- Left and right rocker covers

3. Turn the crankshaft pulley clockwise to Top Dead Center (TDC) identification notch with timing indicator.
4. Ensure that both the intake and exhaust cam noses of the No. 1 cylinder face outside.
5. Measure the valve clearances at locations marked 'x' shown in figure.
6. Turn the crankshaft pulley clockwise 270 degrees from the position of No. 1 cylinder compression to obtain No. 3 cylinder compression TDC.
7. Measure the valve clearances at locations marked 'x' shown in next figure.
8. Turn crankshaft pulley clockwise 90 degrees and measure the intake and exhaust valve clearance of No. 6 cylinder and exhaust valve clearance of No. 2 cylinder.

ADJUSTMENT

1. Remove camshaft and valve lifter(s) out of specification.
2. Install replacement valve lifter(s).
3. Install the camshaft.
4. Manually turn the crankshaft pulley several turns.

5. Recheck valve clearances with engine at operating temperature.

Oil Pan

REMOVAL & INSTALLATION

1. Before servicing the vehicle, refer to the precautions in the beginning of this section.
2. Remove engine assembly.
3. Remove lower oil pan, loosening bolts in reverse order shown in figure.
4. Remove oil strainer from upper oil pan.
5. Gently pry and remove upper oil pan from engine block.

To install:

6. Apply liquid gasket to upper oil pan mating surfaces.
7. Install new O-rings to oil pump and front cover side.
8. Tighten upper oil pan bolts in following numerical order:

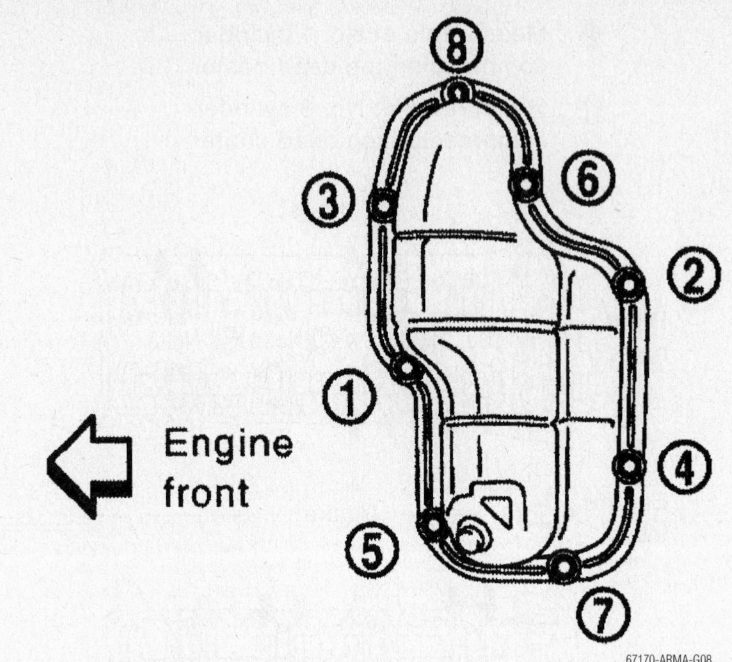

Lower oil pan torque sequence

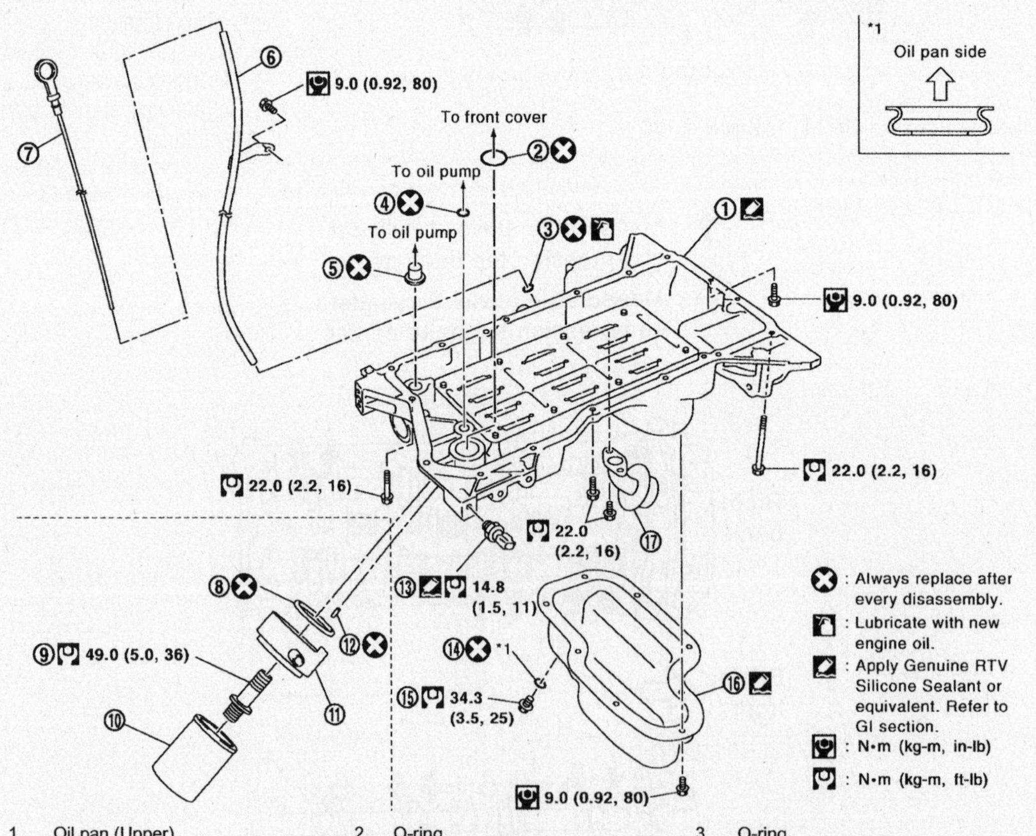

Oil pan and related parts

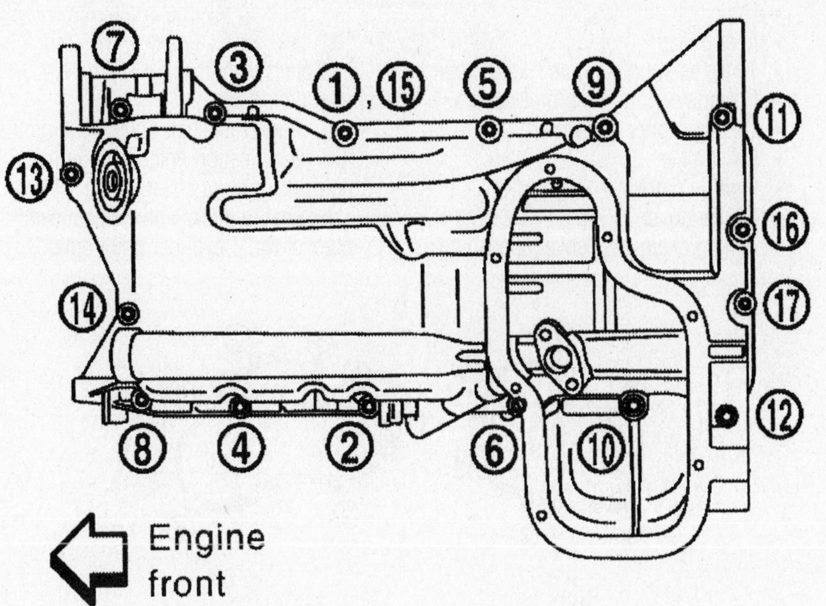

Upper oil pan torque sequence

a. No. 15, 16
b. No. 1, 3, 5, 7, 11, 13
c. No. 2, 4, 6, 8, 10, 14
d. No. 9, 12

9. Install or connect the following:
   - Rear plate cover
   - Oil strainer to upper oil pan
   - Lower oil pan, tightening bolts in order shown in figure

Oil Pump

REMOVAL & INSTALLATION

1. Remove or disconnect the following:
   - Timing chain cover
   - Oil pump drive spacer
   - Oil pump

To install:

2. Install or connect the following:
   - Oil pump
   - Oil pump drive spacer
   - Timing chain cover

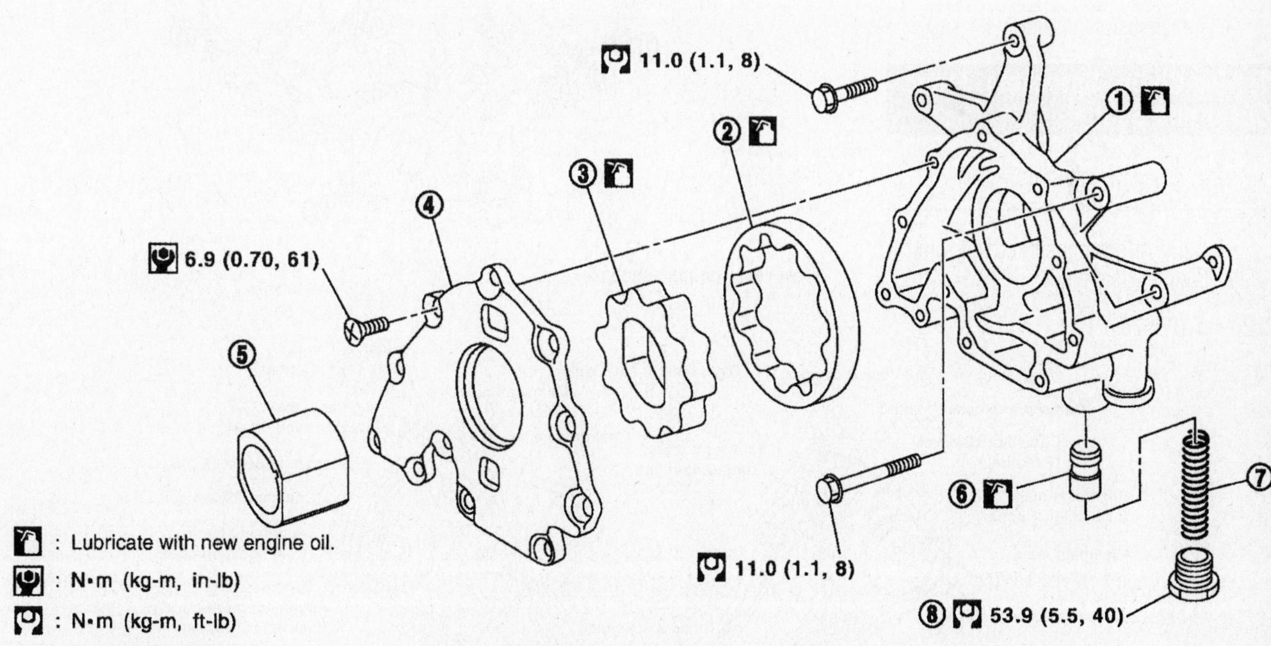

1. Oil pump body
2. Outer rotor
3. Inner rotor
4. Oil pump cover
5. Oil pump drive spacer
6. Regulator valve
7. Regulator spring
8. Regulator plug

67170-ARMA-G32

Oil pump exploded view

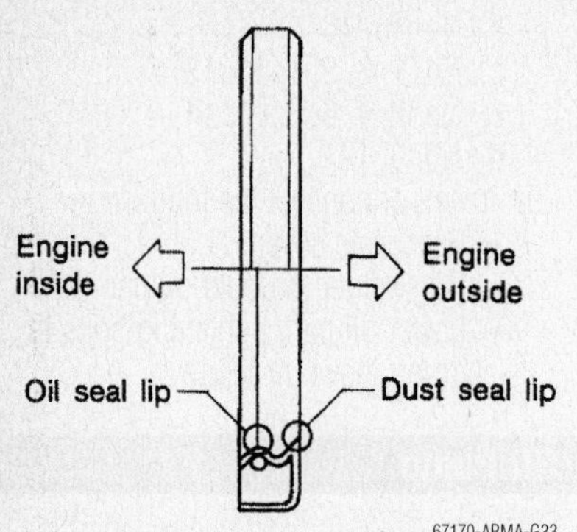

Proper seal installation direction

Rear Main Seal

REMOVAL & INSTALLATION

1. Remove or disconnect the following:
   - Transmission assembly
   - Pressure plate
   - Engine rear plate
   - Rear main seal using suitable tool

To install:

2. Install or connect the following:
   - Rear main seal using suitable tool
   - Engine rear plate
   - Pressure plate
   - Transmission assembly

Timing Chain, Sprockets, Front Cover and Seal

REMOVAL & INSTALLATION

1. Remove or disconnect the following:
   - Engine assembly
   - Drive belt auto tensioner
   - Idler pulley
   - Thermostat housing and water hose
   - Power steering pump bracket
   - Oil pan (upper and lower)
   - Oil strainer
   - Ignition coil
   - Rocker cover
   - Timing chain case cover, loosening bolts in reverse order shown in figure
2. Obtain compression TDC of No. 1 cylinder as follows:
   a. Turn crankshaft pulley to align the TDC identification notch with timing indicator on front cover.
   b. Ensure intake and exhaust cam lobes of No. 1 cylinder point outside.

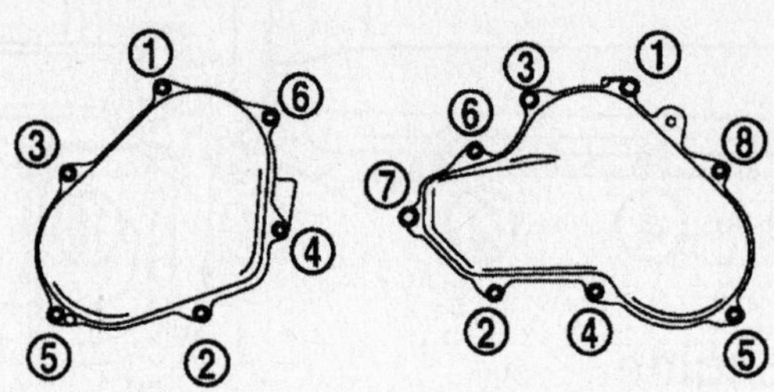

Timing case cover torque sequence

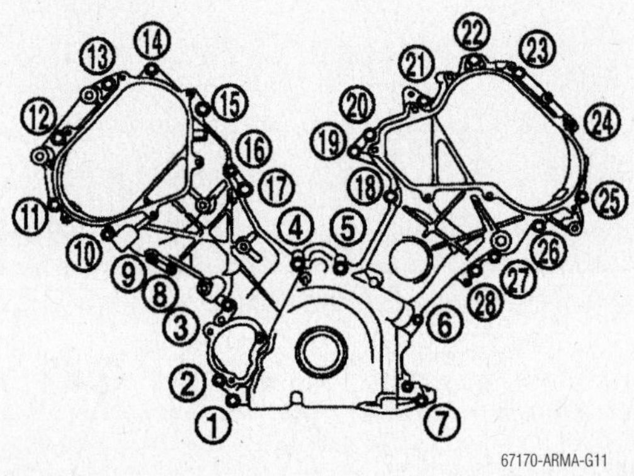

Front cover torque sequence

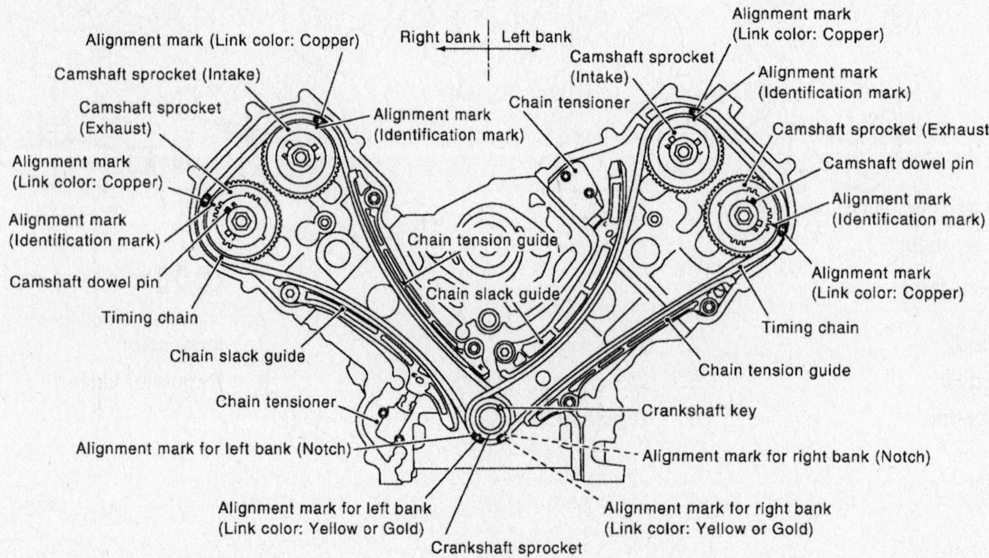

Timing mark alignment

3. Remove or disconnect the following:
    - Crankshaft pulley from crankshaft using a suitable puller
    - Front cover, loosening bolts in reverse order shown in figure
    - Front oil seal
    - Oil pump drive spacer
    - Oil pump
    - Timing chain tensioner
    - Chain tension guide and slack guide
    - Timing chain
    - Camshaft sprocket

To install:

4. Ensure that the crankshaft key and dowel pin of each camshaft are facing the same direction.
5. Install or connect the following:
    - Camshaft sprockets
    - Timing chain
    - Chain tension guide and slack guide
    - Oil pump
    - Oil pump drive spacer
    - Front oil seal, using suitable tool
    - Front cover, using new O-rings and tighten bolts in order shown in figure
    - Chain case cover, and tighten bolts in order shown in figure
    - Crankshaft pulley and tighten bolt to 69 ft. lbs. (93 Nm) plus 90 degrees
    - Ignition coil
    - Oil strainer
    - Lower and upper oil pan
    - Power steering pump bracket
    - Thermostat housing and water hose
    - Idler pulley
    - Drive belt auto tensioner
    - Engine assembly

Piston and Ring

POSITIONING

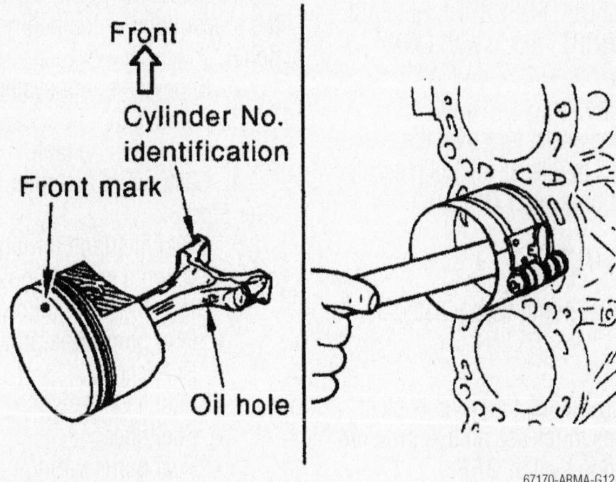

Piston and rod positioning and identification

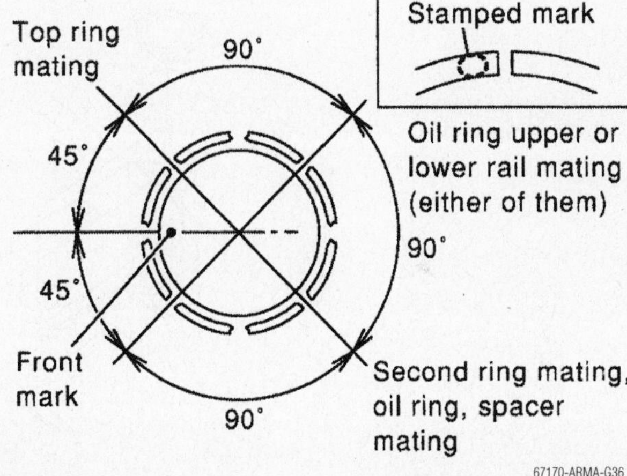

Piston ring installation

FUEL SYSTEM

Fuel System Service Precautions

Safety is the most important factor when performing not only fuel system maintenance but any type of maintenance. Failure to conduct maintenance and repairs in a safe manner may result in serious personal injury or death. Maintenance and testing of the vehicle fuel system components can be accomplished safely and effectively by adhering to the following rules and guidelines.

- To avoid the possibility of fire and personal injury, always disconnect the negative battery cable unless the repair or test procedure requires that battery voltage be applied.
- Always relieve the fuel system pressure prior to disconnecting any fuel system component (injector, fuel rail, pressure regulator, etc.), fitting or fuel line connection. Exercise extreme caution whenever relieving fuel system pressure to avoid exposing skin, face and eyes to fuel spray. Please be advised that fuel under pressure may penetrate the skin or any part of the body that it contacts.
- Always place a shop towel or cloth around the fitting or connection prior to loosening to absorb any excess fuel due to spillage. Ensure that all fuel spillage (should it occur) is quickly removed from engine surfaces. Ensure that all fuel soaked cloths or towels are deposited into a suitable waste container.
- Always keep a dry chemical (Class B) fire extinguisher near the work area.
- Do not allow fuel spray or fuel vapors to come into contact with a spark or open flame.
- Always use a back-up wrench when loosening and tightening fuel line connection fittings. This will prevent unnecessary stress and torsion to fuel line piping. Always follow the proper tighten specifications.
- Always replace worn fuel fitting O-rings with new. Do not substitute fuel hose or equivalent, where fuel pipe is installed.

Relieving Fuel System Pressure

With CONSULT-II

1. Turn ignition switch **ON**.
2. Perform "FUEL PRESSURE RELEASE" in "WORK SUPPORT" mode with CONSULT-II.
3. Start engine.
4. After engine stalls, turn over the engine two or three times to release all fuel pressure.
5. Turn ignition switch **OFF**.

Without CONSULT-II

1. Remove fuel pump fuse located in IPDM E/R.
2. Start engine.
3. After engine stalls, turn over engine two or three times to release all fuel pressure.
4. Turn ignition switch **OFF**.
5. Reinstall fuel pump fuse after servicing fuel system.

Fuel Filter and Fuel Pump

REMOVAL & INSTALLATION

1. Before servicing the vehicle, refer to the precautions in the beginning of this section.
2. Relieve the fuel system pressure.
3. Remove fuel filler cap to release pressure from inside tank.
4. Disconnect fuel filler hose from fuel filler pipe.
5. Drain fuel tank through the fuel filler opening using a suitable hose.
6. Disconnect the following:
   - Fuel pump line protector
   - EVAP hose
   - Fuel level sensor
   - Fuel filter
   - Fuel pump wiring harness
   - Fuel supply hose
7. Using a suitable jack to support the fuel tank, remove the strap bolts and remove the fuel tank from the vehicle.
8. Remove the lock ring using special tool J-46536.
9. Remove the following:
   - Fuel level sensor
   - Fuel filter
   - Fuel pump assembly

To install:

10. Install or connect the following:
    - Fuel pump assembly, using new O-ring
    - Fuel filter, using new filter
    - Fuel level sensor, using new sensor
    - Fuel pump assembly lock ring
    - Fuel tank
    - Fuel supply hose
    - Fuel pump wiring harness
    - EVAP hose
    - Fuel pump line protector
    - Fuel filler pipe
11. Start engine and check for leaks.

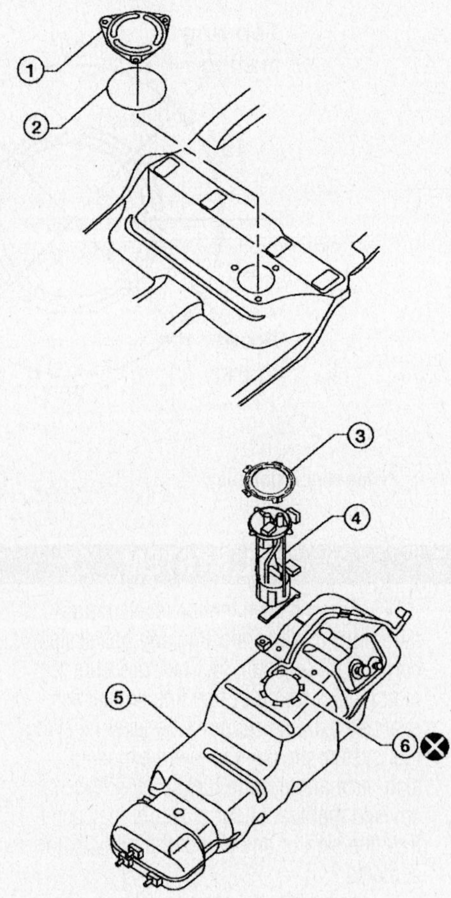

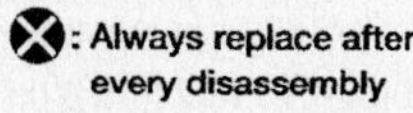

1. Inspection hole cover
2. Inspection hole cover O-ring
3. Lock ring
4. Fuel level sensor, fuel filter, and fuel pump assembly
5. Fuel tank
6. Fuel level sensor, fuel filter, and fuel pump assembly O-ring

67170-ARMA-G37

Fuel pump and related parts

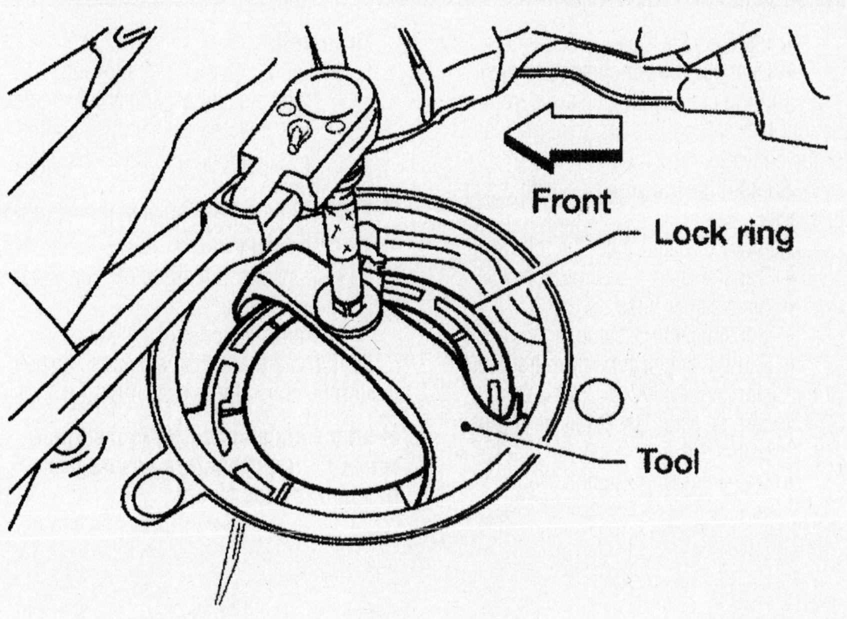

Removing fuel assembly lock ring

Fuel Injectors

REMOVAL & INSTALLATION

1. Remove engine cover
2. Relieve fuel system pressure.
3. Remove or disconnect the following:
   - Negative battery cable
   - Fuel injector harness connectors
   - Fuel hose assembly from right and left fuel rails
   - Fuel injectors with fuel rail as an assembly
   - Fuel injector from fuel rail

To install:

4. Install or connect the following:
   - New clip onto the fuel injector
   - Fuel injector to fuel rail
   - Fuel injectors and fuel rail as an assembly to the intake manifold
   - Fuel hose assembly
   - Fuel injector harness connectors
   - Negative battery cable
   - Engine cover
5. Start engine and check for leaks.

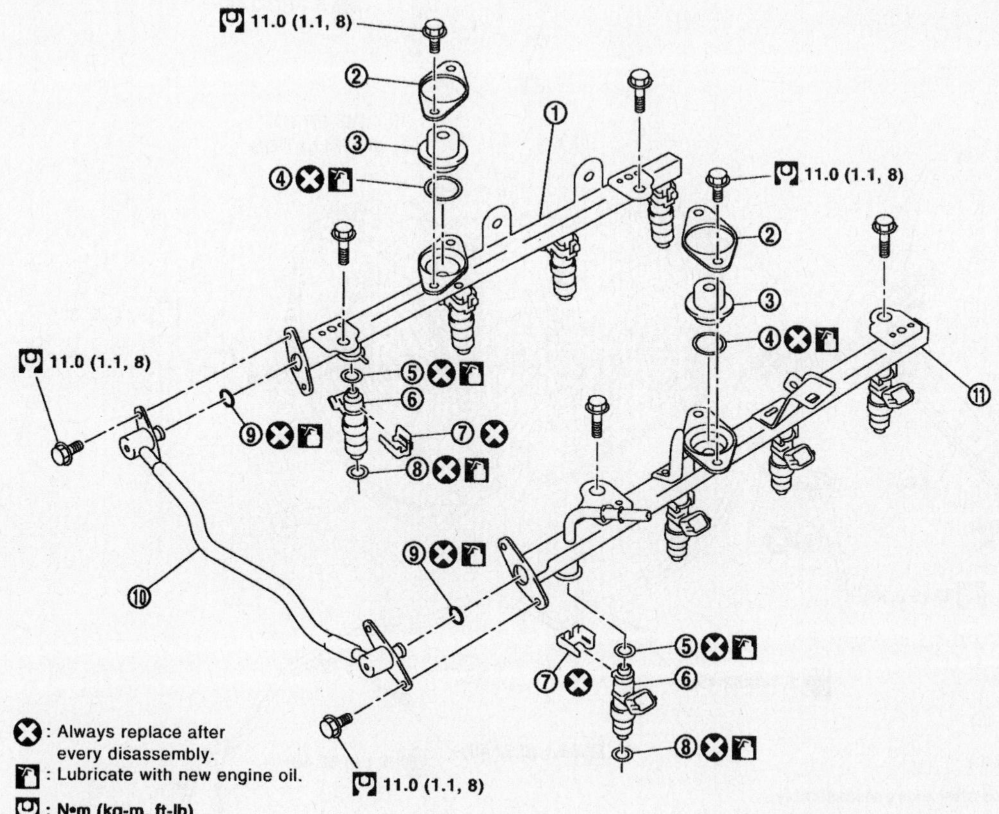

1.	Fuel tube (right bank)	2.	Cap	3.	Fuel damper
4.	O-ring	5.	O-ring (Blue)	6.	Fuel injector
7.	Clip	8.	O-ring (Brown)	9.	O-ring
10.	Fuel hose assembly	11.	Fuel tube (left bank)		

67170-ARMA-G38

Fuel injectors and related parts

DRIVE TRAIN

Transmission Assembly

REMOVAL & INSTALLATION

2-Wheel Drive

1. Remove or disconnect the following:
   - Negative battery cable
   - Engine cover
   - Transmission fluid indicator gauge
   - Engine splash guard
   - Exhaust front pipe
   - Center muffler
   - Rear drive shaft
   - Transmission control cable
   - Crankshaft position sensor
   - Fluid cooler tube
   - Dust cover from converter housing
2. Turning crankshaft clockwise, remove the four tightening bolts for drive plate and torque converter
3. Support the transmission with a suitable jack.
4. Remove or disconnect the following:
   - Transmission cross member
   - Air breather hose
   - Transmission assembly connector
   - Fluid indicator tube from transmission assembly
   - Transmission assembly to engine bolts
   - Transmission assembly from vehicle

To install:

5. Install or connect the following:
   - Transmission assembly into vehicle
   - Transmission assembly to engine bolts tightening to 83 ft. lbs. (113 Nm)
   - Fluid indicator tube to transmission assembly
   - Transmission assembly connector
   - Air breather hose
   - Transmission cross member
6. Turning crankshaft clockwise, install the torque converter to drive plate.

➡After torque converter is installed, rotate the crankshaft to ensure transmission rotates freely.

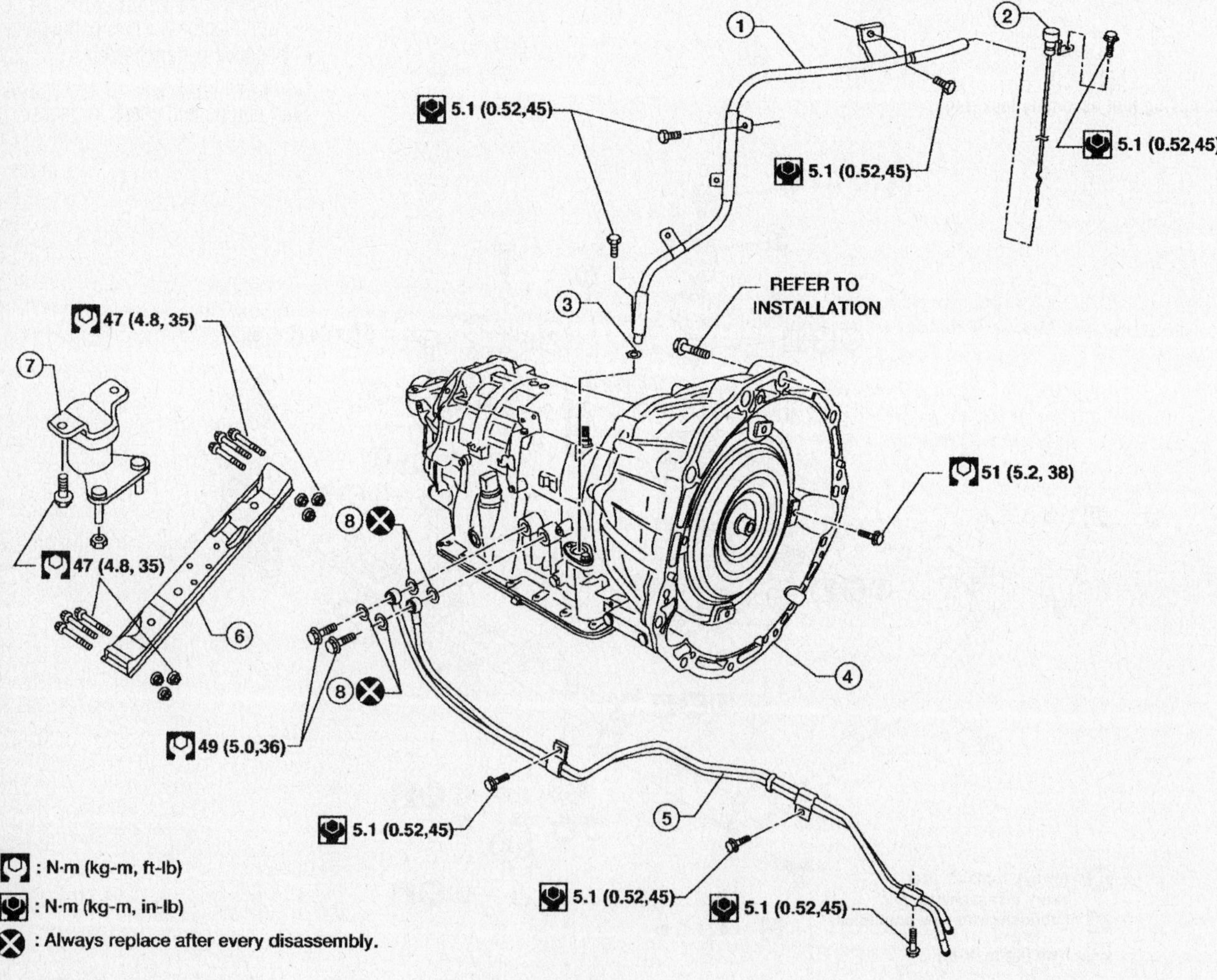

Transmission and related parts—2-wheel drive

7. Install or connect the following:
   - Dust cover for converter housing
   - Fluid cooler tube
   - Crankshaft position sensor
   - Transmission control cable
   - Rear drive shaft
   - Center muffler
   - Exhaust front pipe
   - Engine splash guard
   - Transmission fluid indicator gauge
   - Engine cover
   - Negative battery cable
8. Start engine and check for leaks.

4-Wheel Drive

1. Remove or disconnect the following:
   - Negative battery cable
   - Engine cover
   - Transmission fluid indicator gauge
   - Engine splash guard
   - Exhaust front pipe
   - Center muffler
   - Drive shaft
   - Transmission control cable
   - Crankshaft position sensor
   - Fluid cooler tube
   - Dust housing for torque converter

2. Turning the crankshaft clockwise, remove the four tightening bolts for drive plate and torque converter.

3. Support the transmission assembly with a suitable jack.

4. Remove transmission cross member.

5. Tilt the transmission slightly to keep clearance between the body and the transmission assembly, then disconnect the air breather hose.

6. Remove or disconnect the following:
   - Transmission assembly connector and transfer case connector
   - Fluid indicator pipe
   - Transmission assembly to engine bolts
   - Transmission assembly, with transfer case attached, from vehicle
   - Transmission assembly from transfer case

To install:

7. Install or connect the following:
   - Transfer case to transmission assembly
   - Transmission assembly into vehicle
   - Transmission assembly to engine bolts tightening to 83 ft. lbs. (113 Nm)
8. With the transmission slightly tilted

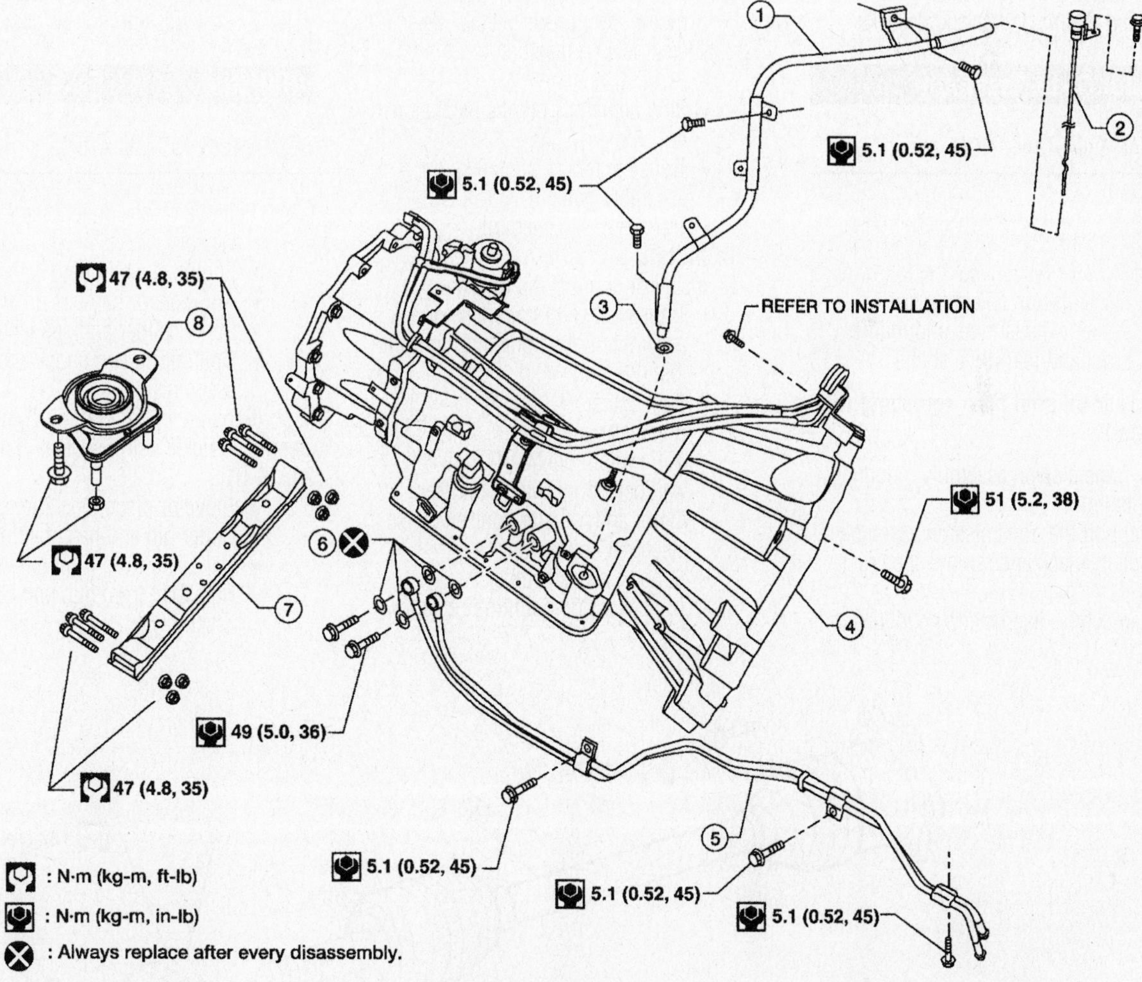

Transmission and related parts—with 4-wheel drive

to allow clearance between body and transmission, connect the air breather hose.

9. Install the transmission cross member.

10. Turning crankshaft clockwise, install the torque converter to drive plate.

➡After torque converter is installed, rotate the crankshaft to ensure transmission rotates freely.

11. Install or connect the following:
    - Dust housing for torque converter
    - Fluid cooler tube
    - Crankshaft position sensor
    - Transmission control cable
    - Drive shaft
    - Center muffler
    - Front exhaust pipe
    - Engine splash guard
    - Transmission fluid indicator gauge
    - Engine cover
    - Negative battery cable

12. Start engine and check for leaks.

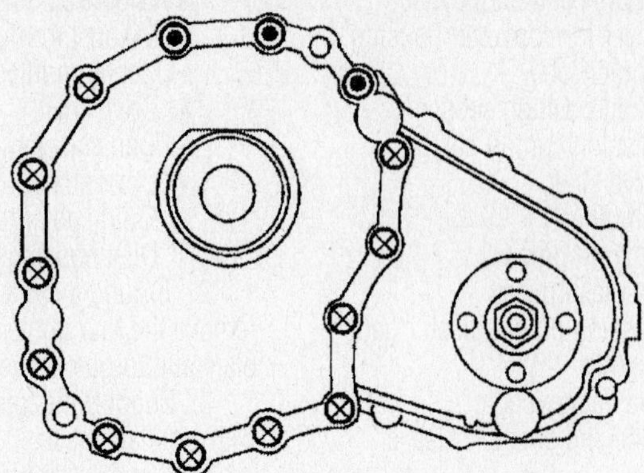

Transfer case mounting bolt locations

Transfer Case Assembly

REMOVAL & INSTALLATION

1. Ensure the transfer case is set to 2WD.
2. Remove or disconnect the following:
    - Transmission splash guard
    - Center exhaust pipe and muffler
    - Front and rear drive shafts

➡Plug rear oil seal after removing rear drive shaft.

    - Transmission assembly mounting bolts

3. Support the transmission assembly with a suitable jack and remove the crossmember.
4. Remove or disconnect the following:
    - ATP switch, neutral 4LO switch, wait detection switch, transfer motor and transfer control device electrical connectors
    - Breather hoses
    - Shift actuator from the extension housing
    - Transfer case to transmission assembly bolts
    - Transfer case assembly

To install:

5. Install or connect the following:
    - Transfer case to transmission assembly bolts tightening to 26 ft. lbs. (36 Nm)
    - Shift actuator
    - Breather hoses
    - ATP switch, neutral 4LO switch, wait detection switch, transfer motor and transfer control device electrical connectors
    - Support crossmember
    - Transmission mounting bolts
    - Drive shafts
    - Muffler and center exhaust pipe
    - Transmission splash guard

Halfshaft

REMOVAL & INSTALLATION

1. Remove or disconnect the following:
    - Wheel
    - Engine splash guard
    - ABS sensor harness on knuckle
    - Brake caliper and suspend it aside
    - Coil spring and shock absorber assembly
2. Separate upper ball joint stud from steering knuckle using special tool J-24319-01.
3. Remove or disconnect the following:
    - Cotter pin and half shaft nut
    - Half shaft from front differential
    - Half shaft from hub and bearing assembly

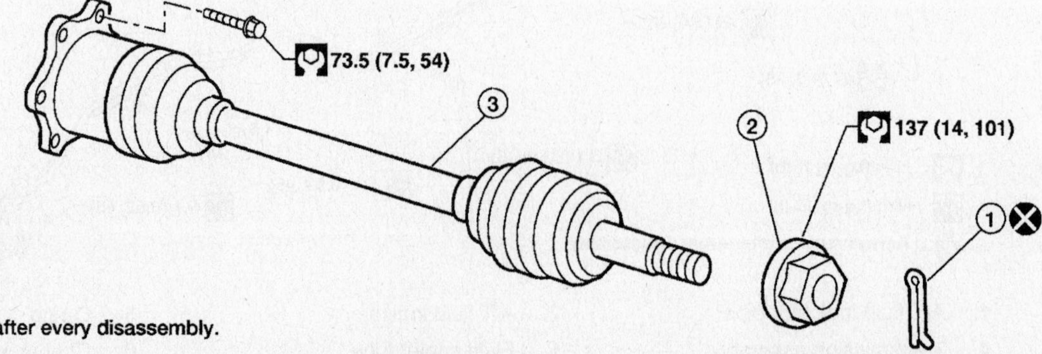

Front halfshaft

To install:

4. Install or connect the following:
   - Half shaft into hub
   - Halt shaft into front differential
   - Half shaft nut and tighten to 101 ft. lbs. and replace cotter pin
   - Upper ball joint to steering knuckle
   - Coil spring and shock absorber assembly
   - Brake caliper
   - ABS sensor
   - Engine splash guard
   - Wheel

CV-Joints

OVERHAUL

Inner

1. Remove the halfshaft from the vehicle.
2. Mount halfshaft in a vise.
3. Remove the dust boot bands.
4. Remove the stopper ring with a flat-bladed screwdriver or suitable tool.
5. Remove the snap ring.
6. Disassemble the cage, ball and inner race assembly and dust boot for cleaning and inspection.

To install:

➡Discard old dust boot, dust boot bands and snap ring and use new ones for assembly.

7. Wrap the serrated part of the half-shaft with tape.
8. Install new dust boot and band onto halfshaft.
9. Remove tape from serrated part of halfshaft.
10. Install the cage, ball and inner race assembly.
11. Install new snap ring.
12. Insert 4.50-5.3 oz of genuine NISSAN grease or equivalent onto the housing and install onto halfshaft.
13. Install the stopper ring onto the housing.
14. Install the dust boot into the grooves on joint sub-assembly.
15. Secure the big and small ends of the dust boot using new boot bands.

Outer

1. Remove the halfshaft from the vehicle.
2. Mount halfshaft in a vise.
3. Remove the dust boot bands and dust boot from joint sub-assembly.
4. Insert a suitable puller into the threaded part of the halfshaft. Pull the joint sub-assembly off of the halfshaft as shown in figure.

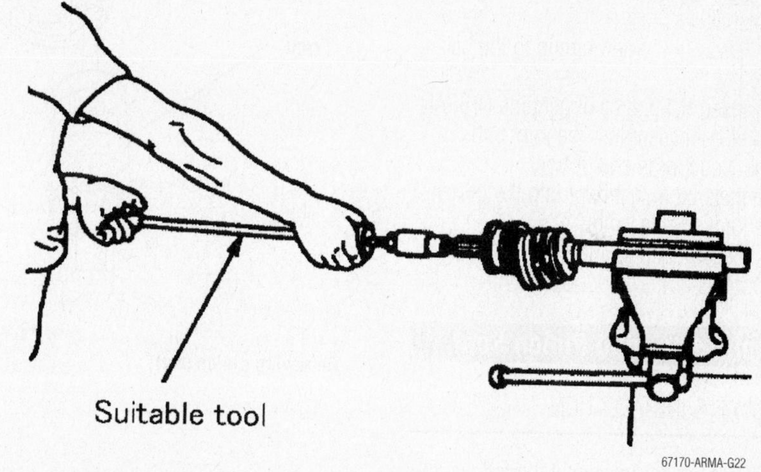

Using a suitable puller to remove joint sub-assembly.

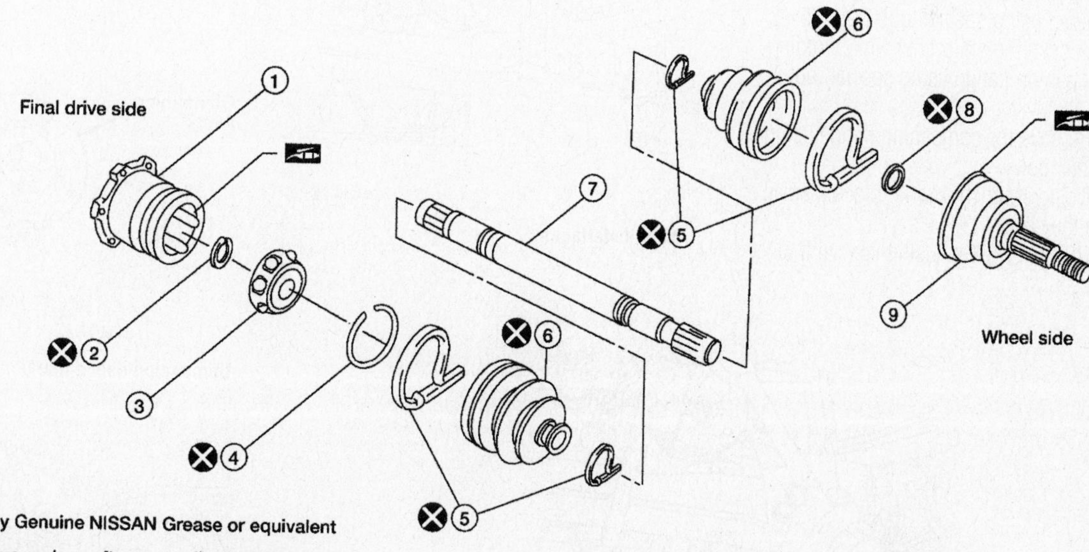

1. Housing
2. Snap ring
3. Ball cage, steel ball, iner race assembly
4. Stopper ring
5. Boot band
6. Boot
7. Shaft
8. Circlip
9. Joint sub-assembly

67170-ARMA-G43

Front halfshaft—exploded view

5. Remove dust boot and circlip from halfshaft for cleaning and inspection.

To install:

➡Discard old dust boot, dust boot bands and circlip and use new ones for assembly.

6. Insert genuine NISSAN grease or equivalent into the joint sub-assembly until grease oozes from the ball groove and serration hole.
7. Wrap the serrated part of the halfshaft with tape.
8. Install new dust boot and band onto halfshaft.
9. Remove tape from serrated part of the halfshaft.
10. Press-fit the new circlip to the halfshaft.
11. Insert 5.1-5.8 oz of genuine NISSAN grease or equivalent into the joint sub-assembly and large end of boot.
12. Install the dust boot into the grooves on the joint sub-assembly.
13. Secure the big and small ends of the dust boot using new boot bands.

Front Differential Pinion Seal

REMOVAL & INSTALLATION

1. Remove or disconnect the following:
   - Front drive shaft
   - Halfshafts
2. Measure and record the pinion bearing preload using special tool J-25765-A.
3. Loosen the pinion nut while holding the companion flange using special tool J-44195.
4. Remove the companion flange using a suitable tool.
5. Using a punch or drill, place a small hole in the case.
6. Remove the seal using special tool SP8P.

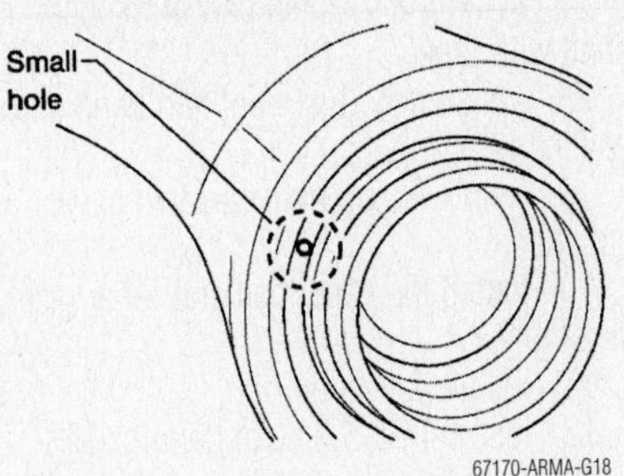

Small hole in casing

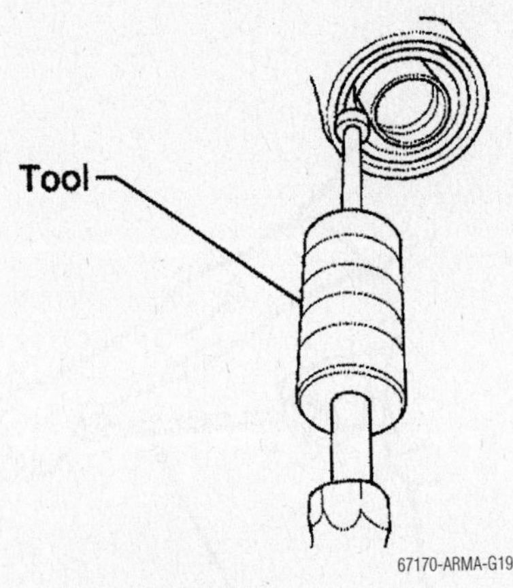

Removing pinion seal

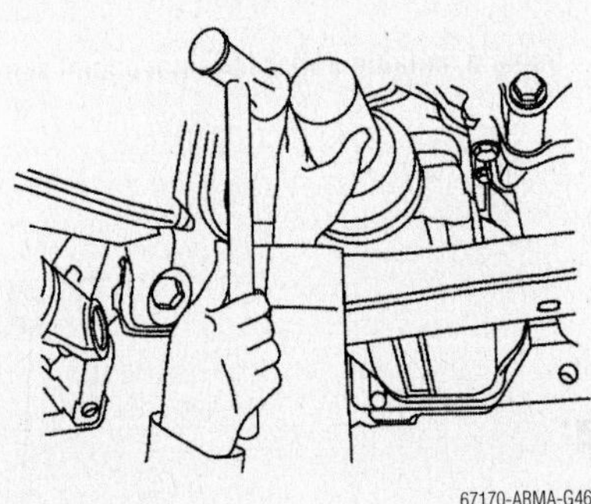

Seal installation

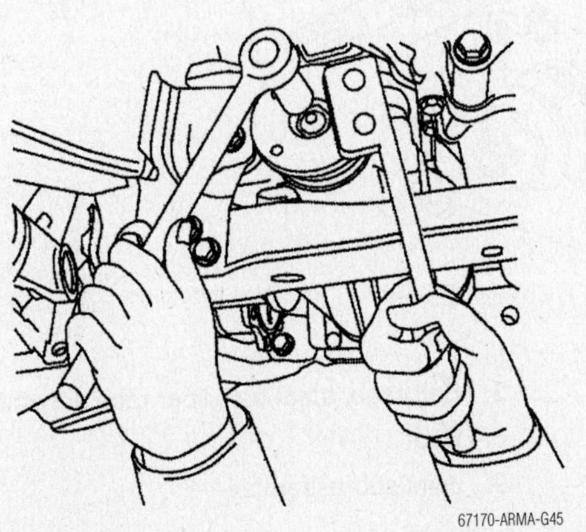

Removing the companion flange

To install:

7. Press front seal into carrier using a suitable tool.
8. Install companion flange and new pinion nut. Tighten pinion nut until there is no end play and until recorded pinion bearing preload is met plus an additional 5 inch lbs. (0.5 Nm).
9. Install or connect the following:
   - Halfshafts
   - Front drive shaft

Rear Differential Pinion Seal

REMOVAL & INSTALLATION

1. Remove the rear drive shaft.
2. Remove brake calipers and rotors.
3. Measure and record the total preload.
4. Matchmark the drive pinion to position 'B' on the companion flange.
5. Remove the drive pinion nut using suitable tool.
6. Remove the companion flange using suitable tool.
7. Remove the rear pinion seal using special tool J-34286.

To install:

8. Press the rear pinion seal into the carrier using suitable tool.
9. Align the matchmark on the companion flange to the drive pinion and install the companion flange.

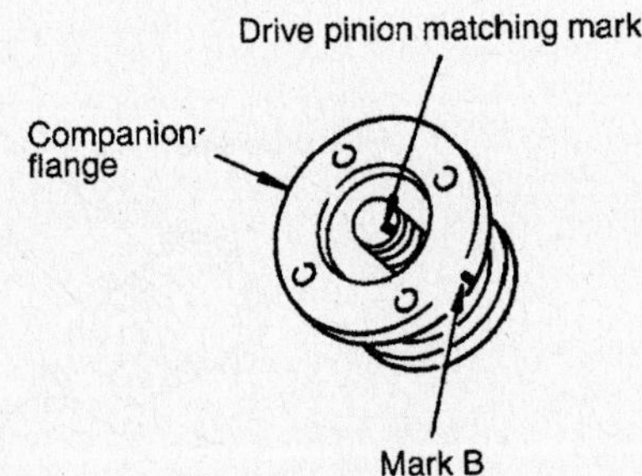

Companion flange marking

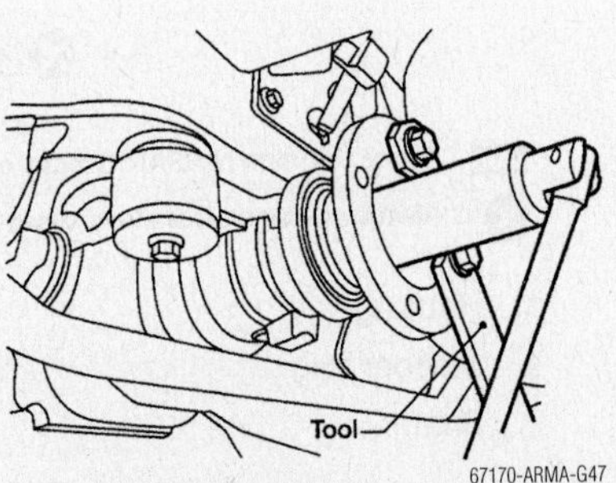

Loosening the flange nut

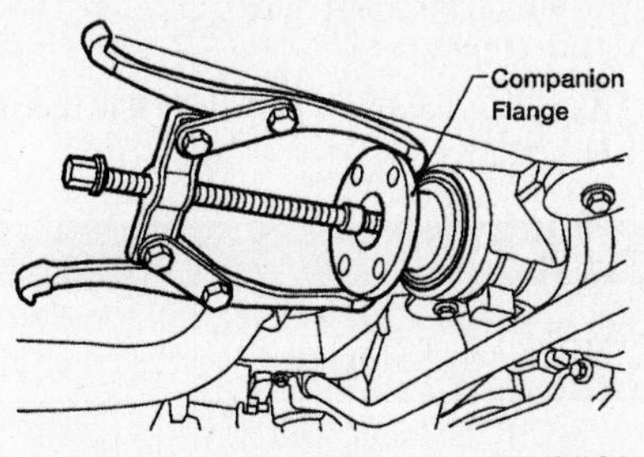

Removing the companion flange

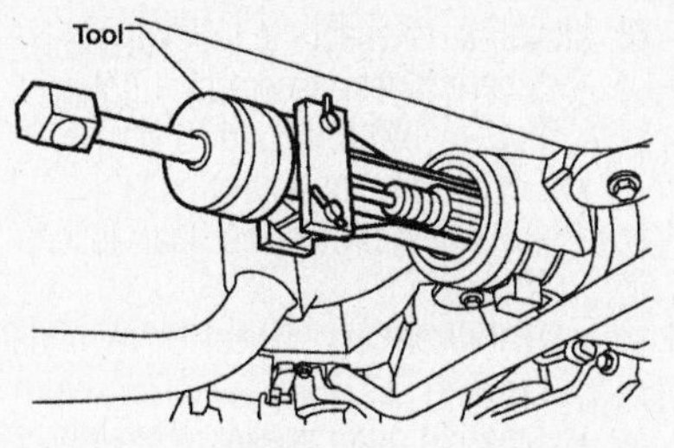

Removing the pinion seal

10. Lubricate the drive pinion threads and seating surfaces of the drive pinion nut with grease.

11. Using a new drive pinion nut, tighten to 124-274 ft. lbs. (167-372 Nm).

➡Final torque is determined when adjusting total preload using special tool J-25765-A.

12. Install brake calipers and rotors.

13. Install rear drive shaft.

STEERING AND SUSPENSION

Air Bag

⁂ CAUTION

Some vehicles are equipped with an air bag system. The system must be disarmed before performing service on, or around, system components, the steering column, instrument panel components, wiring and sensors. Failure to follow the safety precautions and the disarming procedure could result in accidental air bag deployment, possible injury and unnecessary system repairs.

PRECAUTIONS

Several precautions must be observed when handling the inflator module to avoid accidental deployment and possible personal injury.

- Never carry the inflator module by the wires or connector on the underside of the module.
- When carrying a live inflator module, hold securely with both hands, and ensure that the bag and trim cover are pointed away.
- Place the inflator module on a bench or other surface with the bag and trim cover facing up.
- With the inflator module on the bench, never place anything on or close to the module which may be thrown in the event of an accidental deployment.

DISARMING THE SYSTEM

1. Before servicing the vehicle, refer to the precautions in the beginning of this section.
2. Disconnect both battery cables.
3. Wait at least 3 minutes before working on the vehicle. The air bag system is designed to retain enough power to deploy the air bag for a short time after the battery has been disconnected.
4. After repairs are complete, connect the negative battery cable. Turn the ignition switch to the **ON** position and check the air bag warning light blinks for proper operation.

Power Steering Gear

REMOVAL & INSTALLATION

1. Ensure the wheels are in the straight-ahead position.
2. Remove or disconnect the following:
   - Wheels
   - Engine splash guard
3. On 4-wheel drive models only, remove front final drive and support the drive shafts.
4. Remove cotter pin at steering outer socket and loosen mounting nut.

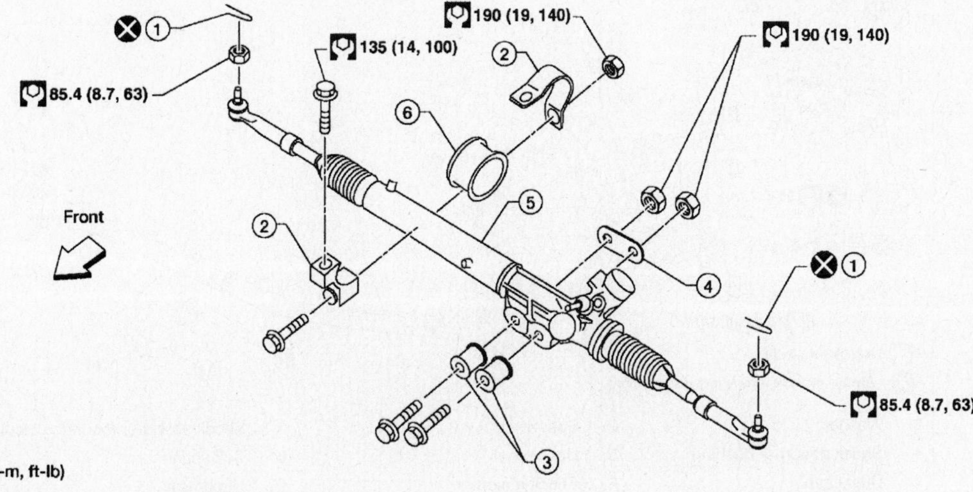

1.	Cotter pin	2.	Mounting bracket	3.	Bushing
4.	Washer	5.	Steering gear assembly	6.	Mounting insulator

Steering gear and related parts

5. Remove steering outer socket from steering knuckle using special tool J-25730-A.

6. On 2-wheel drive models only, remove stabilizer bar mounting bolts and secure the stabilizer bar.

7. Remove or disconnect the following:

- Oil pipes from steering gear assembly
- Lower joint mounting bolt from lower shaft
- Mounting bolts and nuts from steering gear assembly
- Steering gear assembly

To install:

8. Install or connect the following:

- Steering gear assembly, tighten nuts to 140 ft. lbs. (190 Nm)
- Lower joint mounting bolt
- Oil pipes to steering gear assembly
- Stabilizer bar, 2 wheel-drive models only
- Steering outer socket to steering knuckle, tighten nut to 63 ft. lbs. (86 Nm)
- Front final drive, 4-wheel drive models only
- Engine splash guard
- Wheels

9. Check the wheel alignment and adjust as necessary.

Shock Absorber

REMOVAL & INSTALLATION

Front

1. Remove or disconnect the following:

- Wheel

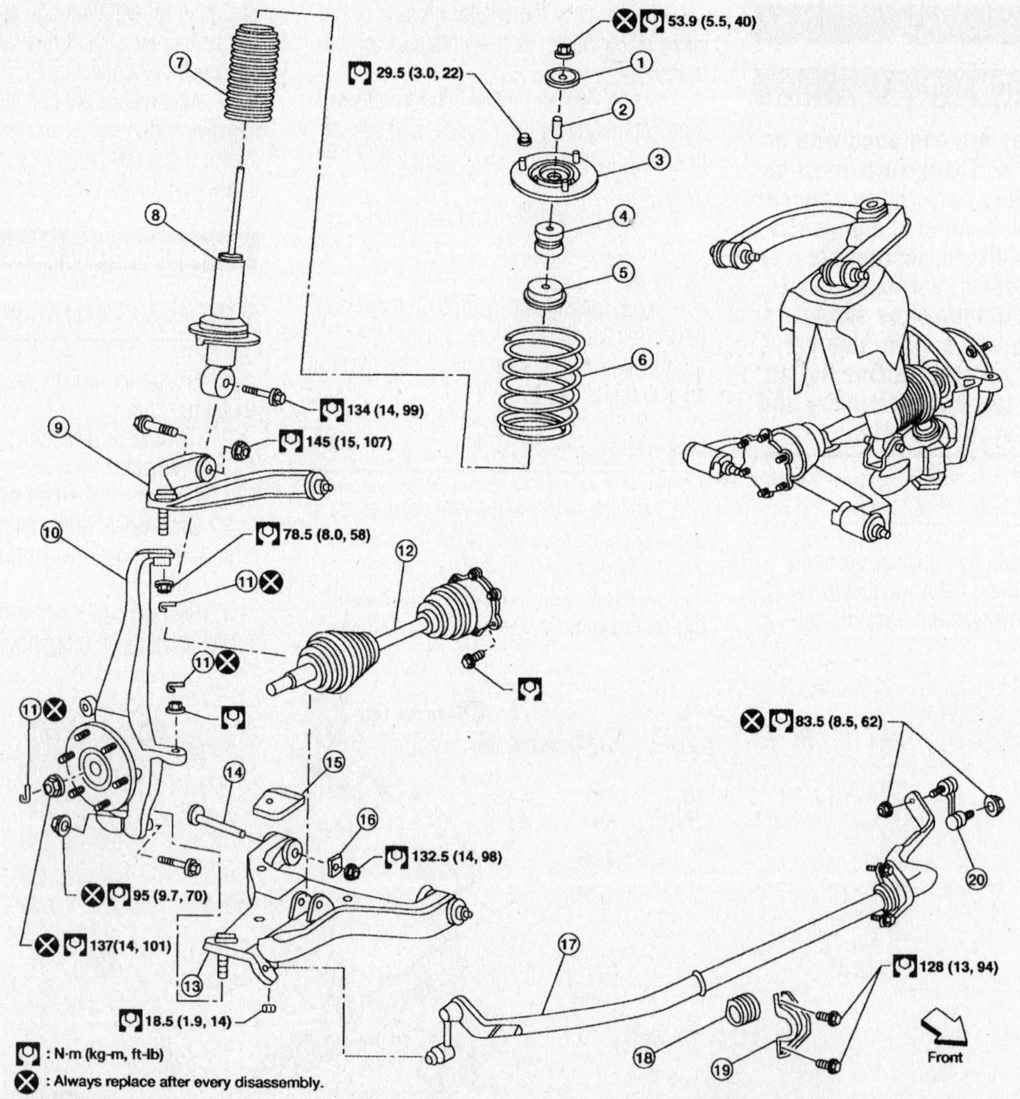

1. Washer
2. Spacer
3. Shock absorber mounting insulator
4. Shock absorber bushing
5. Upper seat
6. Coil spring
7. Dust cover
8. Shock absorber
9. Upper link
10. Steering knuckle
11. Cotter pin
12. Drive shaft
13. Lower link
14. Cam bolt
15. Jounce bumper
16. Cam washer
17. Stabilizer bar
18. Stabilizer bar bushing
19. Stabilizer bar mounting bracket
20. Connecting rod

67170-TITA-G98

Front suspension

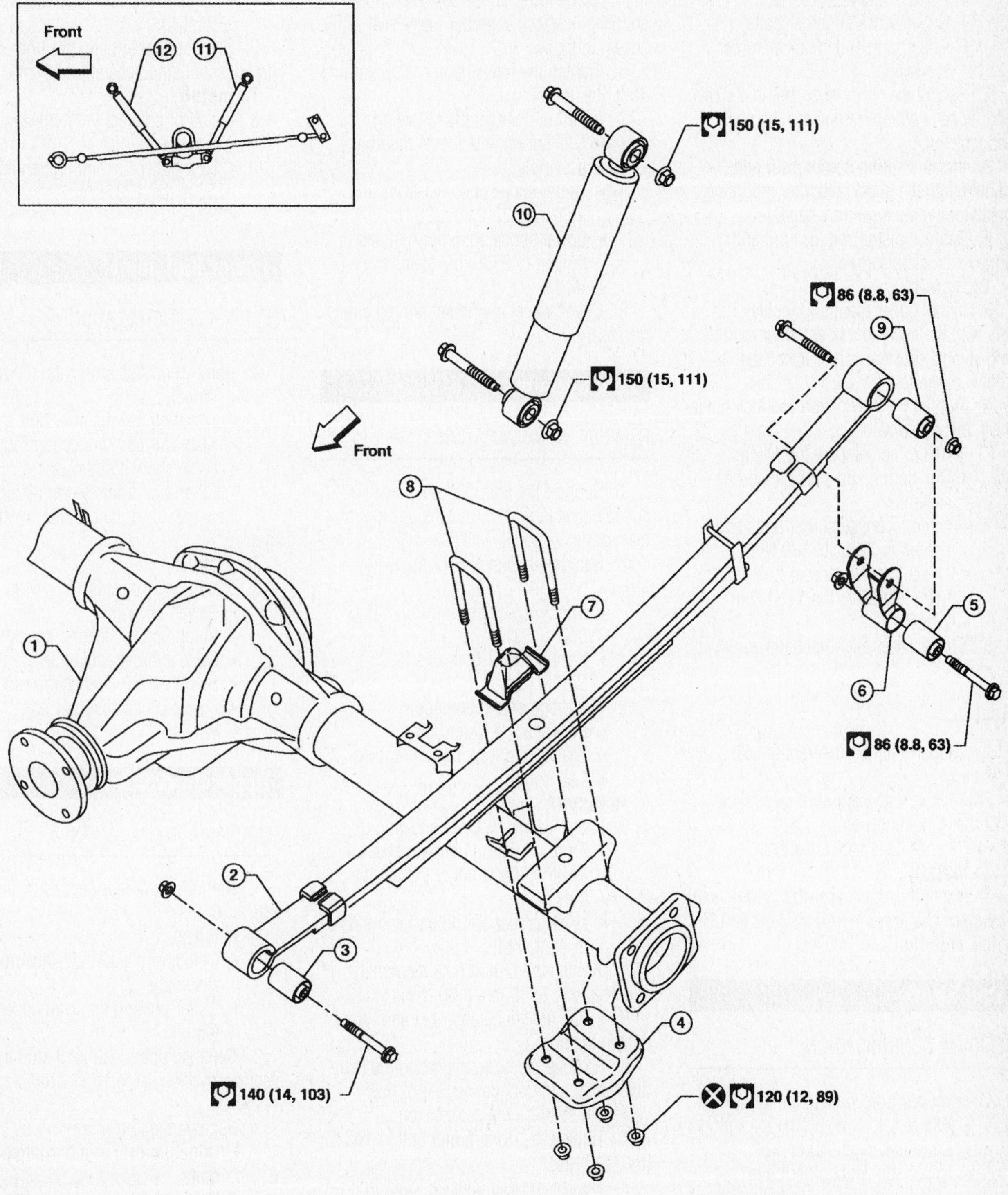

1.	Rear final drive	2.	Rear leaf spring	3.	Rear spring bushing (front)
4.	Rear spring pad	5.	Rear spring shackle bushing	6.	Rear spring shackle
7.	Bumper	8.	Rear spring clip U-bolts	9.	Rear spring bushing (rear)
10.	Shock absorber	11.	Shock absorber (left side)	12.	Shock absorber (right side)

67170-TITA-G97

Rear suspension

- Lower shock absorber bolt
- Upper shock absorber bolts
- Coil spring and shock absorber assembly

2. Secure the shock absorber in a vice and loosen (without removing) the piston rod lock nut.
3. Install a spring compressor and tighten until the shock absorber mounting insulator can be turned by hand.
4. Remove piston rod lock nut and remove shock absorber.

To install:

5. Install upper mounting insulator in line with the lower shock absorber mount and step in shock absorber lower seat as shown in figure.
6. Tighten the new piston rod lock nut to 40 ft. lbs. (54 Nm).
7. Install or connect the following:
   - Coil spring and shock absorber assembly
   - Upper shock absorber bolts and tighten to 22 ft. lbs (30 Nm)
   - Lower shock absorber bolt and tighten to 99 ft. lbs. (134 Nm)
   - Wheel
8. Check wheel alignment and adjust as necessary.

Rear

1. Support the rear differential with a suitable jack.
2. Remove the upper and lower shock absorber mounting bolts.
3. Remove the shock absorber.

To install:

4. Install the shock absorber and tighten the upper and lower mounting bolts to 111 ft. lbs. (150 Nm).

Coil Spring

REMOVAL & INSTALLATION

1. Remove or disconnect the following:
   - Wheel
   - Lower shock absorber bolt
   - Upper shock absorber bolts
   - Coil spring and shock absorber assembly
2. Secure the shock absorber in a vice and loosen (without removing) the piston rod lock nut.
3. Install a spring compressor and tighten until the shock absorber mounting insulator can be turned by hand.
4. Remove piston rod lock nut and remove shock absorber from the coil spring.

To install:

5. Install upper mounting insulator in line with the lower shock absorber mount and step in shock absorber lower seat as shown in figure.
6. Tighten the new piston rod lock nut to 40 ft. lbs. (54 Nm).
7. Install or connect the following:
   - Coil spring and shock absorber assembly
   - Upper shock absorber bolts and tighten to 22 ft. lbs (30 Nm)
   - Lower shock absorber bolt and tighten to 99 ft. lbs. (134 Nm)
   - Wheel
8. Check wheel alignment and adjust as necessary.

Leaf Springs

REMOVAL & INSTALLATION

1. Support the rear differential with a suitable jack to relieve the tension from the leaf spring.
2. Remove or disconnect the following:
   - Shock absorber lower mounting bolt
   - Spring clip U-bolt nuts
   - Spring pad
   - Storage box, if equipped
   - Rear shackle lower bolt
   - Leaf spring front mounting bolt
   - Leaf spring

To install:

3. Install or connect the following:
   - Front mounting bolt and shackle lower bolt and finger tighten the nuts
   - U-bolts, rear spring pad and nuts or the U-bolts
4. Tighten the U-bolt nuts diagonally and evenly to 89 ft. lbs. (120 Nm).
5. Install the shock absorber and finger tighten the nuts.
6. Remove the jack supporting the rear differential and bounce the rear of the vehicle to stabilize the suspension.
7. Tighten the front mount bolt to 103 ft. lbs. (140 Nm).
8. Tighten the rear shackle lower bolt to 63 ft. lbs. (86 Nm).
9. Tighten the shock absorber lower mounting bolt to 111 ft. lbs. (150 Nm).

Upper Ball Joint

REMOVAL & INSTALLATION

1. Remove or disconnect the following:
   - Wheel
   - Cotter pin and nut from upper ball joint
2. Separate upper ball joint from steering knuckle using special tool J-24319-01

To install:

3. Install or connect the following:
   - Upper ball joint
   - New cotter pin and tighten nut to 58 ft. lbs. (79 Nm)
   - Wheel

Lower Ball Joint

REMOVAL & INSTALLATION

1. Remove or disconnect the following:
   - Wheel
   - Lower shock absorber bolt
   - Stabilizer bar connecting rod
   - Drive shaft, if equipped
   - Pinch bolt from steering knuckle
2. Separate lower ball joint from steering knuckle

To install:

3. Install or connect the following:
   - Lower ball joint
   - Pinch bolt to steering knuckle
   - Drive shaft, if equipped
   - Stabilizer bar connecting rod
   - Lower shock absorber bolt
   - Wheel

Upper Control Arm

REMOVAL & INSTALLATION

1. Remove or disconnect the following:
   - Wheel
   - Coil spring and shock absorber assembly
   - Cotter pin and nut from upper ball joint
2. Separate upper ball joint stud from steering knuckle using special tool J-24319-01.
3. Remove the following:
   - Upper control arm mounting bolts
   - Upper arm

To install:

4. Install or connect the following:
   - Upper control arm and tighten bolts to 107 ft. lbs. (145 Nm)
   - Upper ball joint with new cotter pin and tighten nut to 58 ft. lbs. (79 Nm)
   - Coil spring and shock absorber assembly
   - Wheel

Lower Control Arm

REMOVAL AND & INSTALLATION

1. Remove or disconnect the following:
   - Wheel
   - Lower shock absorber bolt
   - Stabilizer bar connecting rod
   - Drive shaft, if equipped
   - Pinch bolt from steering knuckle
2. Separate lower ball joint from steering knuckle.
3. Remove the following:
   - Lower control arm adjusting bolts
   - Lower control arm

To install:

4. Install or connect the following:
   - Lower control arm and tighten adjusting bolts to 98 ft. lbs. (133 Nm)
   - Lower ball joint
   - Pinch bolt
   - Drive shaft, if equipped
   - Stabilizer bar connected rod
   - Lower shock absorber bolt
   - Wheel

Wheel Bearings

REMOVAL & INSTALLATION

Front

1. Remove or disconnect the following:
   - Wheel
   - Engine splash guard
   - Brake caliper without disconnecting the hydraulic lines, and reposition aside with wire
2. Matchmark the brake rotor to the wheel hub and remove the brake rotor.
3. Remove or disconnect the following:
   - Cotter pin and lock nut from drive shaft
   - Drive shaft from wheel hub and bearing assembly
   - ABS sensor
   - Wheel hub and bearing assembly bolts
   - Wheel hub and bearing assembly

To install:

4. Install or connect the following:
   - Wheel hub and bearing assembly, using new bolts and tighten to 155 ft. lbs. (210 Nm)
   - ABS sensor
   - Drive shaft to wheel hub and bearing assembly
   - Cotter pin and lock nut and tighten to 101 ft. lbs. (137 Nm)
   - Brake rotor
   - Brake caliper
   - Engine splash guard
   - Wheel

Rear

1. Remove or disconnect the following:
   - ABS sensor
   - Rear brake rotor
   - Parking brake assembly
   - Four axle shaft bearing cage nuts and lock washers
   - Axle shaft assembly using special tool J-25604-01 and J-25840-A
   - Snap ring from axle shaft
2. Remove the bearing ring retainer by drilling ¾ of thickness of the ring and using a hammer and chisel to break the ring free.
3. Remove the axle shaft bearing from the axle shaft using special tool 205-D002.

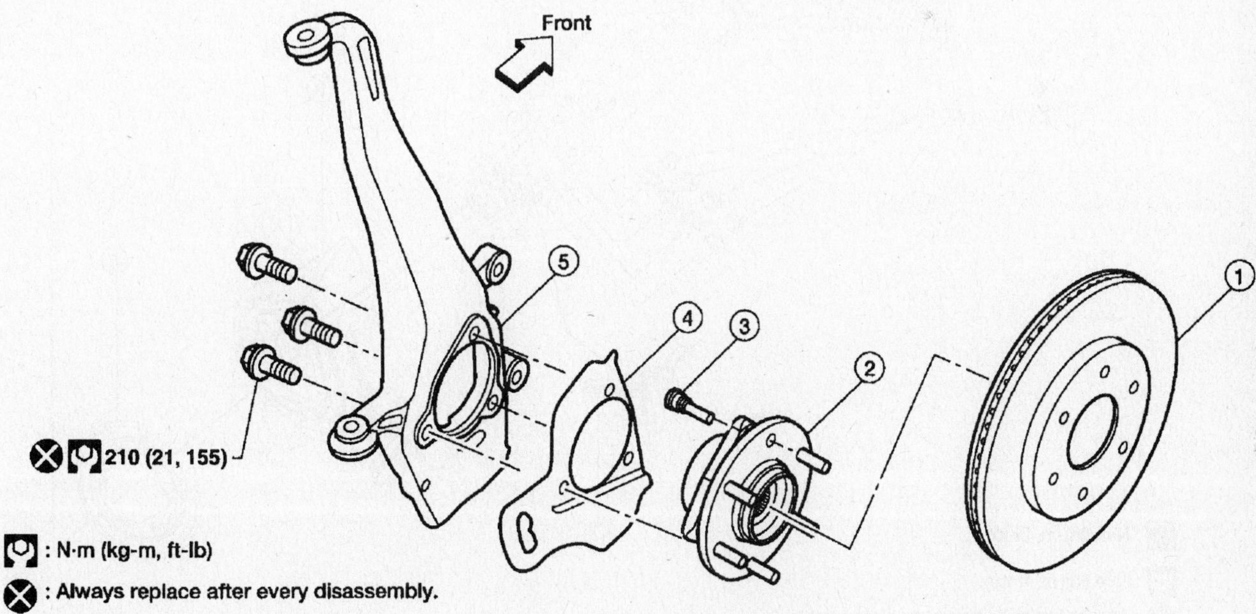

1. Disc rotor
2. Wheel hub and bearing assembly
3. Wheel stud
4. Splash guard
5. Steering knuckle

67170-TITA-G96

Front hub

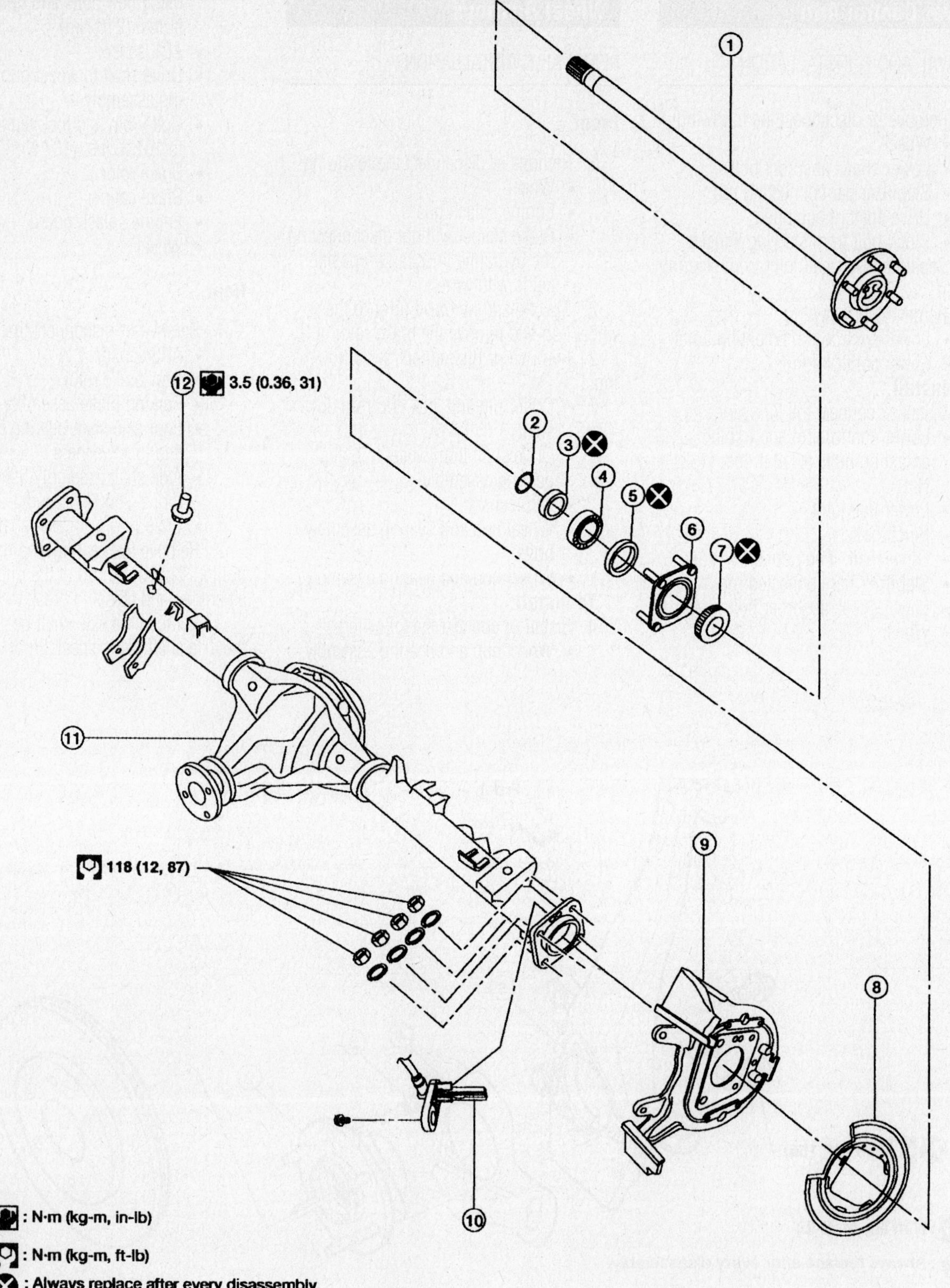

1. Axle shaft
2. Snap ring
3. Bearing ring retainer
4. Axle shaft bearing and cup
5. Axle oil seal
6. Axle shaft bearing cage
7. ABS sensor rotor
8. Back plate
9. Torque member
10. ABS sensor
11. Rear final drive
12. Breather

67170-TITA-G95

Rear axle shaft and related parts

4. Remove and discard the axle oil seal.
5. Remove the wheel bearing assembly.

To install:

6. Install or connect the following:
   - Wheel bearing assembly
   - New axle oil seal
   - Axle shaft bearing on axle shaft
   - Bearing ring retainer onto the axle shaft
   - Snap ring
   - Axle shaft assembly into the axle shaft housing
   - Axle shaft bearing cage lock washers and nuts and tighten to 87 ft. lbs. (118 Nm)
   - Parking brake assembly
   - Rear brake rotor
   - ABS sensor

BRAKES

Brake Caliper

REMOVAL & INSTALLATION

Front

1. Drain brake fluid as necessary.
2. Remove or disconnect the following:
   - Wheel
   - Union bolt
   - Caliper-to-torque member slide pins, or remove the caliper and torque member as an assembly.
   - Brake caliper

To install:

3. Install or connect the following:
   - Brake caliper, tighten torque member bolts to 155 ft. lbs. (210 Nm); the caliper slide pins to 32 ft. lbs. (44 Nm)
   - Union bolt and tighten to 13 ft. lbs. (18 Nm)
4. Fill the master cylinder and bleed the brake system.
5. Install the wheels.

Rear

1. Remove brake hose and mounting bolts.
2. Remove brake caliper assembly.

To install:

3. Install caliper assembly and tighten mounting bolts to 24 ft. lbs. (32 Nm).
4. Install brake hose and tighten to 13 ft. lbs. (18 Nm).

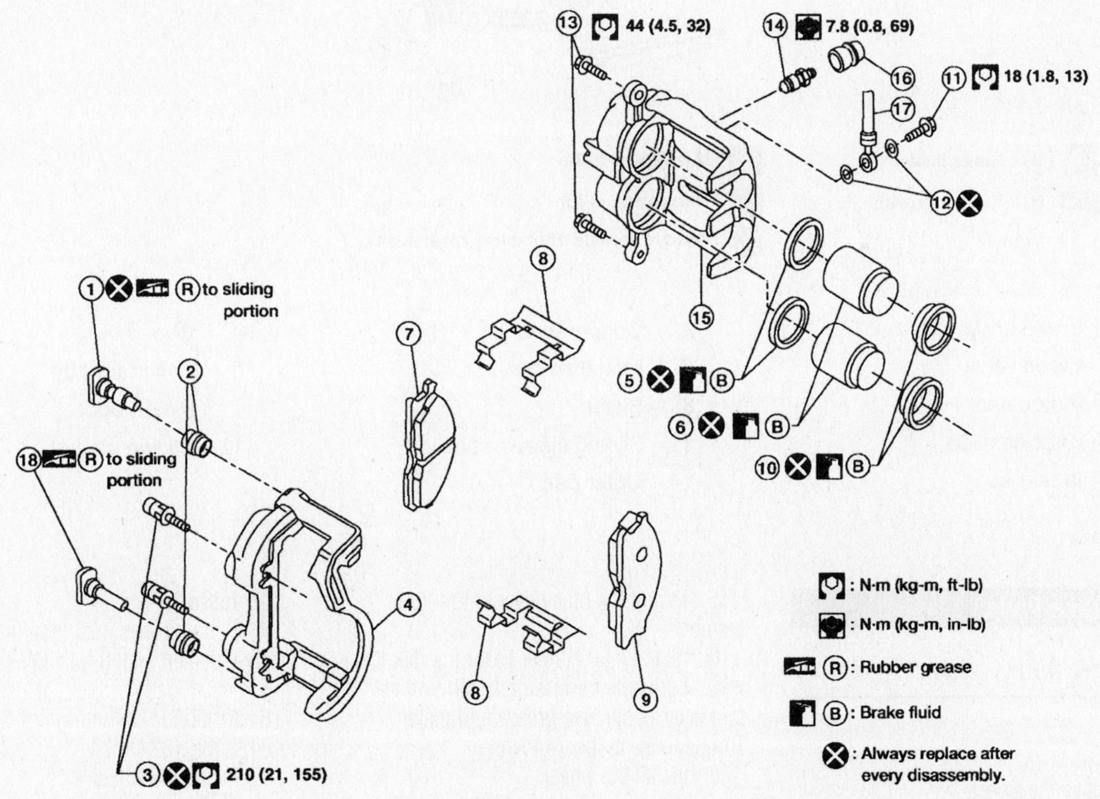

1.	Upper sliding pin	2.	Sliding pin boot	3.	Torque member bolt
4.	Torque member	5.	Piston seal	6.	Piston
7.	Inner pad	8.	Pad retainer	9.	Outer pad
10.	Piston boot	11.	Union bolt	12.	Copper washer
13.	Sliding pin bolt	14.	Bleed valve	15.	Cylinder body
16.	Cap	17.	Brake hose	18.	Lower sliding pin

67170-TITA-G94

Front disc brake

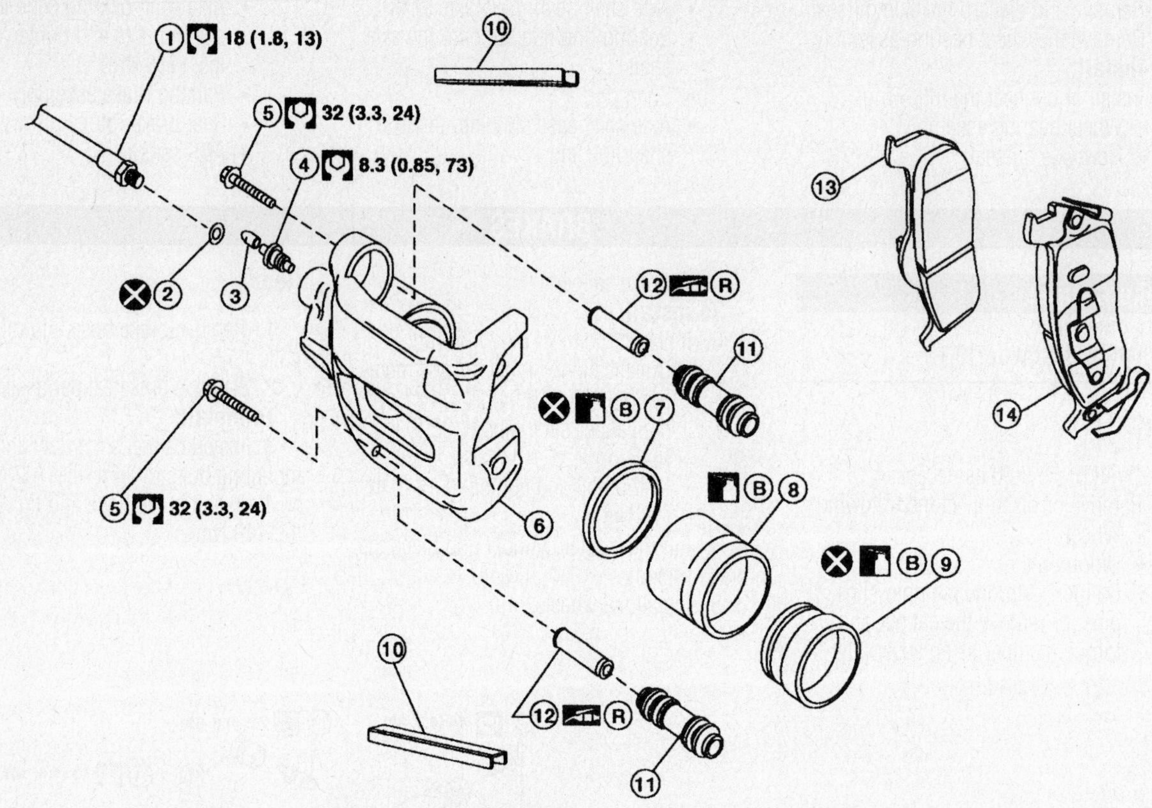

1. Brake hose	2. Copper washer	3. Cap
4. Bleed valve	5. Mounting bolt	6. Cylinder body
7. Piston seal	8. Piston	9. Piston boot
10. Knuckle slide	11. Sliding sleeve boot	12. Sliding sleeve
13. Inner pad	14. Outer pad	

67170-TITA-G93

Rear brakes

Disc Brake Pads

REMOVAL & INSTALLATION

Front

1. Remove the wheel.
2. Remove lower sliding pin bolt.
3. Suspend brake caliper with a remove and remove brake pad and shim from torque member.

To install:

4. Push pistons in so that the pad is firmly installed, using a suitable tool.
5. Mount the brake caliper to torque member.
6. Attach pad retainer to torque member.
7. Lubricate lower sliding pin bolt with a thin layer of silicone grease and install. Torque to 32 ft. lbs. (44 Nm).
8. Install the wheel.

Rear

1. Remove the wheel.
2. Remove mounting bolt from the top mount.
3. Swing brake caliper open and remove the brake pads.

To install:

4. Push pistons in so that the pad is firmly installed, using a suitable tool.
5. Install pads to the brake caliper.
6. Install top mounting bolt and tighten to 24 ft. lbs. (32 Nm).
7. Install the wheel.

SUBARU

12

Baja • Impreza • Impreza Outback • Impreza Outback Sport •
Legacy • Legacy Outback • Legacy SUS • WRX

SPECIFICATION AND MAINTENANCE CHARTS

ENGINE AND VEHICLE IDENTIFICATION CHART

Engine Code							Model Year	
Code ①	Liters (cc)	Cu. In.	Cyl.	Fuel Sys.	Type	Eng. Mfg.	Code ②	Year
4	2.2 (2212)	135	4	MFI	SOHC	Subaru	1	2001
6	2.5 (2457)	150	4	MFI	DOHC	Subaru	2	2002
6	2.5 (2457)	150	4	MFI	SOHC	Subaru	3	2003
8	3.0 (3000)	183	6	MFI	DOHC	Subaru	4	2004
2	2.0 (1994)	121	4	MFI	DOHC	Subaru	5	2005

MFI: Multiport Fuel Injection

SOHC: Single Overhead Camshaft

DOHC: Double Overhead Camshaft

① 6th digit of the VIN

② 10th digit of the VIN

67170-SBCR-C01

GENERAL ENGINE SPECIFICATIONS

Year	Model	Engine Displacement Liters (VIN)	Net Horsepower @ rpm	Net Torque @ rpm (ft. lbs.)	Bore x Stroke (in.)	Com-pression Ratio	Oil Pressure @ rpm
2001	Impreza ①	2.2 (4)	142@5600	149@3600	3.82x2.95	10.0:1	14@600
	Impreza RS	2.5 (6)	165@5600	166@4000	3.92x3.11	10.0:1	14@600
	Outback	2.5 (6)	165@5600	166@4000	3.92x3.11	10.0:1	14@600
	Outback	3.0 (8)	212@6000	210@4400	3.51x3.15	10.7:1	14@600
	Legacy	2.5 (6)	165@5600	166@4000	3.92x3.11	10.0:1	14@600
2002	Impreza RS	2.5 (6)	165@5600	166@4000	3.92x3.11	10.0:1	14@600
	WRX	2.0 (2)	227@6000	217@4000	3.62x2.95	8.0:1	14@800
	Outback	2.5 (6)	165@5600	166@4000	3.92x3.11	10.0:1	14@600
	Outback	3.0 (8)	212@6000	210@4400	3.51x3.15	10.7:1	14@600
	Legacy	2.5 (6)	165@5600	166@4000	3.92x3.11	10.0:1	14@600
2003-04	Impreza RS	2.5 (6)	165@5600	166@4000	3.92x3.11	10.0:1	14@600
	WRX	2.0 (2)	227@6000	217@4000	3.62x2.95	8.0:1	14@800
	Outback	2.5 (6)	165@5600	166@4000	3.92x3.11	10.0:1	14@600
	Outback	3.0 (8)	212@6000	210@4400	3.51x3.15	10.7:1	14@600
	Baja	2.5 (6)	165@5600	166@4000	3.92x3.11	10.0:1	14@600
	Legacy	2.5 (6)	165@5600	166@4000	3.92x3.11	10.0:1	14@600
2004-05	Impreza RS	2.5 (6)	165@5600	166@4000	3.92x3.11	10.0:1	14@600
	WRX	2.0 (2)	227@6000	217@4000	3.62x2.95	8.0:1	14@800
	Outback	2.5 (6)	165@5600	166@4000	3.92x3.11	10.0:1	14@600
	Outback	3.0 (8)	212@6000	210@4400	3.51x3.15	10.7:1	14@600
	Baja	2.5 (6)	165@5600	166@4000	3.92x3.11	10.0:1	14@600
	Legacy	2.5 (6)	165@5600	166@4000	3.92x3.11	10.0:1	14@600

MFI: Multi-port Fuel Injection

Note: All capacities are approximate. Add fluid gradually and check to be sure a proper fluid level is obtained.

① Includes Outback

67170-SBCR-C02

ENGINE TUNE-UP SPECIFICATIONS

Year	Liters (VIN)	Engine Displacement Liters (VIN)	Spark Plug Gap (in.)	Ignition Timing (deg.) ① MT	Ignition Timing (deg.) ① AT	Fuel Pump (psi)	Idle Speed (rpm) MT	Idle Speed (rpm) AT	Valve Clearance ② In.	Valve Clearance ② Ex.
2001	Impreza/ Outback	2.2 (4)	0.039-0.043	6-22	12-28	34-38	600-800	600-800	0.0071-0.0087	0.0090-0.0106
	Impreza RS	2.5 (6)	0.039-0.043	7-23	7-23	34-38	600-800	600-800	0.0071-0.0087	0.0090-0.0106
	Outback/SUS ③	3.0 (8)	0.039-0.043	7-23	7-23	34-38	600-800	600-800	0.0063-0.0095	0.0078-0.0118
	Legacy	2.5 (6)	0.039-0.043	7-23	7-23	34-38	600-800	600-800	0.0071-0.0087	0.0090-0.0106
2002	Impreza RS	2.5 (6)	0.039-0.043	7-23	7-23	34-38	600-800	600-800	0.0071-0.0087	0.0090-0.0106
	Outback/SUS ③	2.5 (6)	0.039-0.043	7-23	7-23	34-38	600-800	600-800	0.0071-0.0087	0.0090-0.0106
	Outback/SUS ③	3.0 (8)	0.039-0.043	2-18 ④	2-18 ④	34-38	600	600	0.0063-0.0095	0.0078-0.0118
	WRX	2.0 (2)	0.028-0.031	2-22	2-22	33-38	650-850	650-850	0.0071-0.0087	0.0090-0.0106
	Legacy	2.5 (6)	0.039-0.043	7-23	7-23	34-38	600-800	600-800	0.0071-0.0087	0.0090-0.0106
2003	Impreza RS	2.5 (6) 2.0 (2)	0.039-0.043	7-23	7-23	34-38	600-800	600-800	0.0071-0.0087	0.0090-0.0106
	Outback/SUS ③	2.5 (6)	0.039-0.043	7-23	7-23	34-38	600-800	600-800	0.0071-0.0087	0.0090-0.0106
	Outback/SUS ③	3.0 (8)	0.039-0.043	7-23	7-23	34-38	600-800	600-800	0.0063-0.0095	0.0078-0.0118
	WRX	2.0 (2)	0.028-0.031	2-22	2-22	33-38	650-850	650-850	0.0071-0.0087	0.0090-0.0106
	Baja	2.5 (6)	0.039-0.043	7-23	7-23	34-38	600-800	600-800	0.0071-0.0087	0.0090-0.0106
	Legacy	2.5 (6)	0.039-0.043	7-23	7-23	34-38	600-800	600-800	0.0071-0.0087	0.0090-0.0106
2004-05	Impreza RS	2.5 (6)	0.039-0.043	7-23	7-23	34-38	600-800	600-800	0.0071-0.0087	0.0090-0.0106
	Outback/SUS ③	2.5 (6)	0.039-0.043	7-23	7-23	34-38	600-800	600-800	0.0071-0.0087	0.0090-0.0106
	Outback/SUS ③	3.0 (8)	0.039-0.043	7-23	7-23	34-38	600-800	600-800	0.0063-0.0095	0.0078-0.0118
	WRX	2.0 (2)	0.028-0.031	2-22	2-22	33-38	650-850	650-850	0.0071-0.0087	0.0090-0.0106
	Baja	2.5 (6)	0.039-0.043	7-23	7-23	34-38	600-800	600-800	0.0071-0.0087	0.0090-0.0106
	Legacy	2.5 (6)	0.039-0.043	7-23	7-23	34-38	600-800	600-800	0.0071-0.0087	0.0090-0.0106

Note: The Vehicle Emission Control Information label often reflects specification changes made during production. The lable figures mudst be used if they differ from those in this chart.

① Before Top Dead Center

② Measured with engine cold

③ Sport Utility Sedan

67170-SBCR-C03

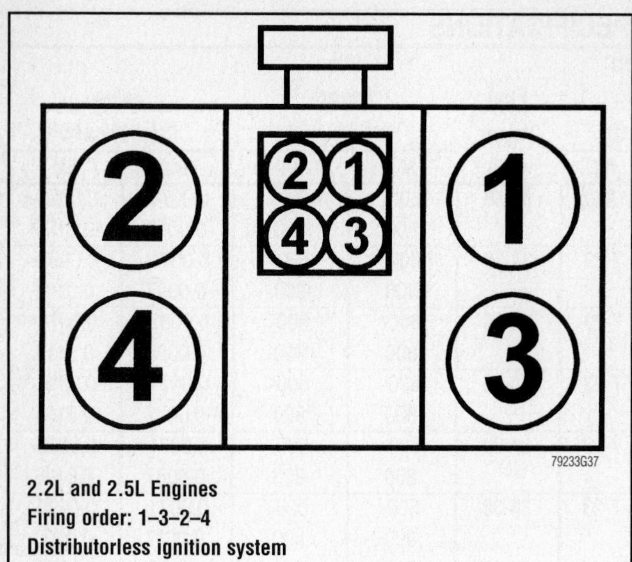

2.2L and 2.5L Engines
Firing order: 1–3–2–4
Distributorless ignition system

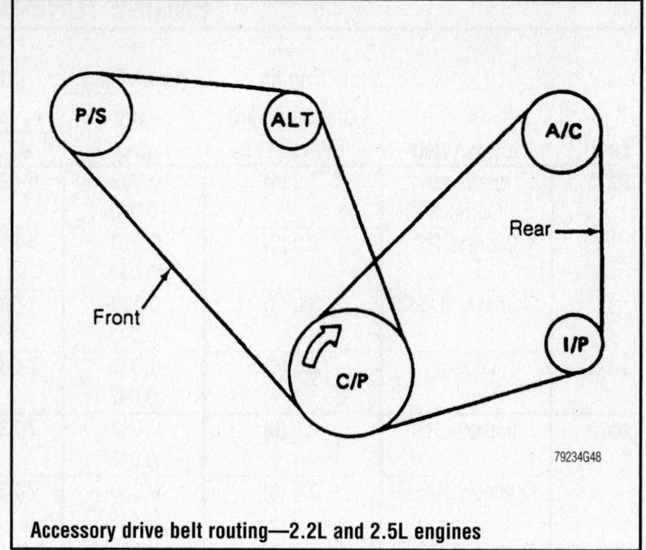

Accessory drive belt routing—2.2L and 2.5L engines

CAPACITIES

Year	Liters (VIN)	Engine Displacement Liters (VIN)	Engine Oil with Filter (qts.)	Transmission (pts.) 5-Spd	Transmission (pts.) Auto.	Transfer Case (pts.)	Drive Axle Front ① (pts.)	Drive Axle Rear (pts.)	Fuel Tank (gal.)	Cooling System (qts.)
2001	Impreza ②	2.2 (4)	4.4	7.4	16.8	–	2.6	1.6	13.2	6.6
	Impreza RS	2.5 (6)	4.2	7.4	20.0	–	2.6	1.6	13.2	6.6
	Outback	3.0 (8)	5.9	7.4	20.2	–	2.6	1.6	16.9	8.4
	Legacy	2.5 (6)	4.2	7.4	20.0	–	2.6	1.6	16.9	7.1
2002	Impreza RS	2.5 (6)	4.2	7.4	20.0	–	2.6	1.6	13.2	6.6
	WRX	2.0 (2)	4.8	7.4	19.6	–	2.6	1.6	15.9	8.0
	Outback	3.0 (8)	5.9	7.4	20.2	–	2.6	1.6	16.9	8.4
	Legacy	2.5 (6)	4.2	7.4	20.0	–	2.6	1.6	16.9	7.1
2003	Impreza RS	2.5 (6)	4.2	7.4	20.0	–	2.6	1.6	13.2	6.6
	WRX	2.0 (2)	4.8	7.4	19.6	–	2.6	1.6	15.9	8.0
	Baja	2.5 (6)	4.2	7.4	20.0	–	2.6	1.6	16.9	7.1
	Outback	3.0 (8)	5.9	7.4	20.2	–	2.6	1.6	16.9	8.4
	Legacy	2.5 (6)	4.2	7.4	20.0	–	2.6	1.6	16.9	7.1
2003	Impreza RS	2.5 (6)	4.2	7.4	20.0	–	2.6	1.6	13.2	6.6
	WRX	2.0 (2)	4.8	7.4	19.6	–	2.6	1.6	15.9	8.0
	Baja	2.5 (6)	4.2	7.4	20.0	–	2.6	1.6	16.9	7.1
	Outback	3.0 (8)	5.9	7.4	20.2	–	2.6	1.6	16.9	8.4
	Legacy	2.5 (6)	4.2	7.4	20.0	–	2.6	1.6	16.9	7.1
2004-05	Impreza RS	2.5 (6)	4.2	7.4	20.0	–	2.6	1.6	13.2	6.6
	WRX	2.0 (2)	4.8	7.4	19.6	–	2.6	1.6	15.9	8.0
	Baja	2.5 (6)	4.2	7.4	20.0	–	2.6	1.6	16.9	7.1
	Outback	3.0 (8)	5.9	7.4	20.2	–	2.6	1.6	16.9	8.4
	Legacy	2.5 (6)	4.2	7.4	20.0	–	2.6	1.6	16.9	7.1

Note: All capacities are approximate. Add fluid gradually and check to be sure a proper fluid level is obtained.

① A/T differential only

② Includes Outback

67170-SBCR-C04

VALVE SPECIFICATIONS

Year	Engine Displacement Liters (VIN)	Seat Angle (deg.)	Face Angle (deg.)	Spring Test Pressure (lbs. @ in.)	Spring Installed Height (in.)	Stem-to-Guide Clearance (in.)		Stem Diameter (in.)	
						Intake	Exhaust	Intake	Exhaust
2001	2.2 (4)	45	45	91 - 103@ 1.110	①	0.0014-0.0059	0.0016-0.0059	0.2344-0.2350	0.2341-0.2346
	2.5 (6)	45	45	102 - 118@ 1.315	②	0.0014-0.0024	0.0016-0.0026	0.2343-0.2348	0.2341-0.2346
	3.0 (8)	45	45	102 - 118@ 1.315	③	0.0012-0.0022	0.0016-0.0026	002148-0.2154	0.2148-0.2150
2002	2.0 (2)	45	45	91 - 103@ 1.110	④	0.0014-0.0059	0.0016-0.0059	0.2343-0.2348	0.2341-0.2346
	2.5 (6)	45	45	102 - 118@ 1.315	②	0.0014-0.0024	0.0016-0.0026	0.2343-0.2348	0.2341-0.2346
	3.0 (8)	45	45	102 - 118@ 1.315	③	0.0012-0.0022	0.0016-0.0026	002148-0.2154	0.2148-0.2150
2003	2.0 (2)	45	45	91 - 103@ 1.110	④	0.0014-0.0059	0.0016-0.0059	0.2343-0.2348	0.2341-0.2346
	2.5 (6)	45	45	102 - 118@ 1.315	②	0.0014-0.0024	0.0016-0.0026	0.2343-0.2348	0.2341-0.2346
	3.0 (8)	45	45	102 - 118@ 1.315	③	0.0012-0.0022	0.0016-0.0026	002148-0.2154	0.2148-0.2150
2004-05	2.0 (2)	45	45	91 - 103@ 1.110	④	0.0014-0.0059	0.0016-0.0059	0.2343-0.2348	0.2341-0.2346
	2.5 (6)	45	45	102 - 118@ 1.315	②	0.0014-0.0024	0.0016-0.0026	0.2343-0.2348	0.2341-0.2346
	3.0 (8)	45	45	102 - 118@ 1.315	③	0.0012-0.0022	0.0016-0.0026	002148-0.2154	0.2148-0.2150

① Free length: 1.7342 in.

② Free length: 1.8913 in.

③ Free length: 1.8421 in.

④ Free length: 1.7587 in.

67170-SBCR-C05

CRANKSHAFT AND CONNECTING ROD SPECIFICATIONS

All measurements are given in inches.

Year	Engine Displacement Liters (VIN)	Crankshaft Main Brg. Journal Dia.	Crankshaft Main Brg. Oil Clearance	Crankshaft Shaft End-play	Crankshaft Thrust on No.	Connecting Rod Journal Diameter	Connecting Rod Oil Clearance	Connecting Rod Side Clearance
2001	2.2 (4)	2.3619-2.3625	①	0.0012-0.0098	3	2.0466-2.0472	0.0006-0.0020	0.0028-0.0160
	2.5 (6)	2.3619-2.3625	②	0.0012-0.0098	3	1.8891-1.8898	0.0005-0.0015	0.0028-0.0130
	3.0 (8)	2.3619-2.3625	③	0.0012-0.0098	3	1.8891-1.8898	0.0009-0.0020	0.0028-0.0130
2002	2.0 (2)	2.3619-2.3625	0.0004 0.0012	0.0012-0.0048	3	1.8891-1.8898	0.0008-0.0018	0.0028-0.0130
	2.5 (6)	2.5194-2.5200	②	0.0012-0.0098	3	1.8891-1.8898	0.0005-0.0015	0.0028-0.0130
	3.0 (8)	2.3619-2.3625	③	0.0012-0.0098	3	1.8891-1.8898	0.0009-0.0020	0.0028-0.0130
2003	2.0 (2)	2.3619-2.3625	0.0004 0.0012	0.0012-0.0048	3	1.8891-1.8898	0.0008-0.0018	0.0028-0.0130
	2.5 (6)	2.3619-2.3625	②	0.0012-0.0098	3	1.8891-1.8898	0.0005-0.0015	0.0028-0.0130
	3.0 (8)	2.3619-2.3625	③	0.0012-0.0098	3	1.8891-1.8898	0.0009-0.0020	0.0028-0.0130
2004-05	2.0 (2)	2.3619-2.3625	0.0004 0.0012	0.0012-0.0048	3	1.8891-1.8898	0.0008-0.0018	0.0028-0.0130
	2.5 (6)	2.3619-2.3625	②	0.0012-0.0098	3	1.8891-1.8898	0.0005-0.0015	0.0028-0.0130
	3.0 (8)	2.3619-2.3625	③	0.0012-0.0098	3	1.8891-1.8898	0.0009-0.0020	0.0028-0.0130

① Journals 1 and 5: 0.0001 - 0.0016 in.
Journals 2 and 4: 0.0004 - 0.0014 in.
Journal 3: 0.0004 - 0.0014 in.

② Journals 1 and 5: 0.0001 - 0.0016 in.
Journals 2 and 4: 0.0004 - 0.0018 in.
Journal 3: 0.0004 - 0.0016 in.

③ 0.0006 - 0.0012 in.

67170-SBCR-C06

PISTON AND RING SPECIFICATIONS

All measurements are given in inches.

Year	Engine Displacement Liters (VIN)	Piston Clearance	Ring Gap			Ring Side Clearance		
			Top Compression	Bottom Compression	Oil Control	Top Compression	Bottom Compression	Oil Control
2001	2.2 (4)	0.0004-0.0020	0.0079-0.0390	0.0079-0.0390	0.0079-0.0590	0.0016-0.0590	0.0012-0.0590	NA
	2.5 (6)	0.0004-0.0012	0.0079-0.0138	0.0146-0.0250	0.0079-0.0197	0.0016-0.0031	0.0012-0.0028	NA
	3.0 (8)	0.0004-0.0012	0.0079-0.0138	0.0138-0.0197	0.0079-0.0236	0.0016-0.0031	0.0012-0.0028	NA
2002	2.0 (2)	0.0004-0.0012	0.0079-0.0138	0.0138-0.0197	0.0079-0.0276	0.0016-0.0031	0.0012-0.0028	NA
	2.5 (6)	0.0004-0.0012	0.0079-0.0138	0.0146-0.0250	0.0079-0.0197	0.0016-0.0031	0.0012-0.0028	NA
	3.0 (8)	0.0004-0.0012	0.0079-0.0138	0.0138-0.0197	0.0079-0.0236	0.0016-0.0031	0.0012-0.0028	NA
2003	2.0 (2)	0.0004-0.0012	0.0079-0.0138	0.0138-0.0197	0.0079-0.0276	0.0016-0.0031	0.0012-0.0028	NA
	2.5 (6)	0.0004-0.0012	0.0079-0.0138	0.0146-0.0250	0.0079-0.0197	0.0016-0.0031	0.0012-0.0028	NA
	3.0 (8)	0.0004-0.0012	0.0079-0.0138	0.0138-0.0197	0.0079-0.0236	0.0016-0.0031	0.0012-0.0028	NA
2004-05	2.0 (2)	0.0004-0.0012	0.0079-0.0138	0.0138-0.0197	0.0079-0.0276	0.0016-0.0031	0.0012-0.0028	NA
	2.5 (6)	0.0004-0.0012	0.0079-0.0138	0.0146-0.0250	0.0079-0.0197	0.0016-0.0031	0.0012-0.0028	NA
	3.0 (8)	0.0004-0.0012	0.0079-0.0138	0.0138-0.0197	0.0079-0.0236	0.0016-0.0031	0.0012-0.0028	NA

NA: Not Available

67170-SBCR-C07

TORQUE SPECIFICATIONS

All readings in ft. lbs.

Year	Engine Displacement Liters (VIN)	Cylinder Head Bolts	Main ① Bearing Bolts	Rod Bearing Bolts	Crankshaft Damper Bolts	Flywheel Bolts	Manifold Intake	Manifold Exhaust	Spark Plugs	Oil Pan Drain Plug
2001	2.2 (4)	②	③	31 - 44	87 - 101	51 - 55	17 - 20	19 - 26	13 - 17	33
	2.5 (6)	②	③	31 - 44	123 - 137	51 - 55	17 - 20	19 - 26	13 - 17	33
	3.0 (8)	④	14	39	131	60	18	22	15	33
2002	2.0 (2)	②	③	31 - 44	94	51 - 55	18	26	13 - 17	33
	2.5 (6)	②	③	31 - 44	123 - 137	51 - 55	17 - 20	19 - 26	13 - 17	33
	3.0 (8)	④	14	39	131	60	18	22	15	33
2003	2.0 (2)	②	③	31 - 44	94	51 - 55	18	26	13 - 17	33
	2.5 (6)	②	③	31 - 44	123 - 137	51 - 55	17 - 20	19 - 26	13 - 17	33
	3.0 (8)	④	14	39	131	60	18	22	15	33
2004-05	2.0 (2)	②	③	31 - 44	94	51 - 55	18	26	13 - 17	33
	2.5 (6)	②	③	31 - 44	123 - 137	51 - 55	17 - 20	19 - 26	13 - 17	33
	3.0 (8)	④	14	39	131	60	18	22	15	33

① Engine block connecting bolts

② Step 1: Tighten all bolts to 22 ft. lbs.
Step 2: Tighten all bolts to 51 ft. lbs.
Step 3: Loosen all botls 180 degrees.
Step 4: Repeat Step 3.
Step 5: Tighten bolts 1 and 2 to 25 ft. lbs.
Step 6: Tighten bolts 3, 4, 5 and 6 to 11 ft. lbs.
Step 7: Tighten all bolts 80 to 90 degrees.
Step 8: Repeat Step 7. Do not exceed 180 degrees total tightening.

③ Split engine case connecting bolts:
Short bolts: 17-20 ft. lbs.
Long bolts: 33-37 ft. lbs.
Smaller short bolts (if used) 5 ft. lbs.

④ Step 1: Tighten all bolts to 14 ft. lbs.
Step 2: Tighten all bolts to 37 ft. lbs.
Step 3: Loosen all bolts 180 degrees the an additional 180 degrees in two steps in the reverse order of tightening sequence.
Step 4: Tighten all bolts to 18 ft. lbs.
Step 5: Tighten all bolts to 18 ft. lbs.
Step 6: Tighten bolts 90 degrees
Step 7: Tighten bolts 1, 2, 3 and 4 90 degrees.
Step 8: Tighten bolts 5, 6, 7 and 8 45 degrees.

67170-SBCR-C08

WHEEL ALIGNMENT

Year	Model		Caster Range (+/-Deg.)	Caster Preferred Setting (Deg.)	Camber Range (+/-Deg.)	Camber Preferred Setting (Deg.)	Toe-in (in.)
2001	Impreza 2.2L	F	1.00	+3.00	0.50	0	0+/-0.12
		R	—	—	0.75	-0.92	0+/-0.12
	Impreza 2.5L	F	1.00	+3.05	0.50	-0.42	0+/-0.12
		R	—	—	0.75	-1.17	0+/-0.12
	Legacy Sedan	F	0.75	+3.05	0.50	-0.05	0+/-0.12
		R	—		0.75	-0.50	0+/-0.12
	Legacy Wagon	F	0.75	+2.05	0.50	-0.05	0+/-0.12
		R	—		0.75	-0.20	0+/-0.12
	Outback	F	0.75	+2.40	0.50	+0.33	0+/-0.12
		R	—		0.75	-0.17	0+/-0.12
2002	Impreza 2.5L	F	1.00	+3.05	0.50	-0.42	0+/-0.12
		R	—	—	0.75	-1.17	0+/-0.12
	Legacy Sedan	F	0.75	+3.05	0.50	-0.05	0+/-0.12
		R	—		0.75	-0.50	0+/-0.12
	Legacy Wagon	F	0.75	+2.05	0.50	-0.05	0+/-0.12
		R	—		0.75	-0.20	0+/-0.12
	Outback	F	0.75	+2.40	0.50	+0.33	0+/-0.12
		R	—		0.75	-0.17	0+/-0.12
	WRX	F	0.75	+2.40	0.50	+0.33	0+/-0.12
		R	—		0.75	-0.17	0+/-0.12
2003	Impreza 2.5L	F	1.00	+3.05	0.50	-0.42	0+/-0.12
		R	—	—	0.75	-1.17	0+/-0.12
	Legacy Sedan	F	0.75	+3.05	0.50	-0.05	0+/-0.12
		R	—		0.75	-0.50	0+/-0.12
	Legacy Wagon	F	0.75	+2.05	0.50	-0.05	0+/-0.12
		R	—		0.75	-0.20	0+/-0.12
	Baja	F	1.00	+3.05	0.50	-0.42	0+/-0.12
		R	—	—	0.75	-1.17	0+/-0.12
	Outback	F	0.75	+2.40	0.50	+0.33	0+/-0.12
		R	—		0.75	-0.17	0+/-0.12
	WRX	F	0.75	+2.40	0.50	+0.33	0+/-0.12
		R	—		0.75	-0.17	0+/-0.12
2004-05	Impreza 2.5L	F	1.00	+3.05	0.50	-0.42	0+/-0.12
		R	—	—	0.75	-1.17	0+/-0.12
	Legacy Sedan	F	0.75	+3.05	0.50	-0.05	0+/-0.12
		R	—		0.75	-0.50	0+/-0.12
	Legacy Wagon	F	0.75	+2.05	0.50	-0.05	0+/-0.12
		R	—		0.75	-0.20	0+/-0.12
	Baja	F	1.00	+3.05	0.50	-0.42	0+/-0.12
		R	—	—	0.75	-1.17	0+/-0.12
	Outback	F	0.75	+2.40	0.50	+0.33	0+/-0.12
		R	—		0.75	-0.17	0+/-0.12
	WRX	F	0.75	+2.40	0.50	+0.33	0+/-0.12
		R	—		0.75	-0.17	0+/-0.12

67170-SBCR-C09

TIRE, WHEEL AND BALL JOINT SPECIFICATIONS

Year	Model	OEM Tires		Tire Pressures (psi)		Wheel	Ball Joint	Lug
		Standard	Optional	Front	Rear	Size	Inspection	Nut
2001	Impreza	P195/60R15	None	32	29	6-JJ	0.012 in.	58 - 72
	Impreza RS	P205/55R16	None	32	29	7-JJ	0.012 in.	58 - 72
	Impreza Outback	P205/60R15	None	32	29	7-JJ	0.012 in.	65
	Legacy	P185/70R14	P195/60R15 P205/55R16 P205/70R15	32	30	6.5-JJ	0.012 in.	58 - 72
2002	Impreza	P195/60R15	None	32	29	6-JJ	0.012 in.	58 - 72
	Impreza RS	P205/55R16	None	32	29	7-JJ	0.012 in.	58 - 72
	Impreza Outback	P225/60R16	T145/80R16	32	30	16x16 1/2JJ	0.012 in.	65
	Impreza WRX	P195/60R15	P205/55R16 P215/45R17	32	29	6-JJ 7-JJ	0.012 in.	58 - 72
	Legacy	P185/70R14	P195/60R15 P205/55R16 P205/70R15	32	30	6.5-JJ	0.012 in.	58 - 72
2003	Impreza	P195/60R15	None	32	29	6-JJ	0.012 in.	58 - 72
	Impreza RS	P205/55R16	None	32	29	7-JJ	0.012 in.	58 - 72
	Impreza Outback	P225/60R16	T145/80R16	32	30	16x16 1/2JJ	0.012 in.	65
	Impreza WRX	P195/60R15	P205/55R16 P215/45R17	32	29	6-JJ 7-JJ	0.012 in.	58 - 72
	Baja	P185/70R14	P195/60R15 P205/55R16 P205/70R15	32	30	6.5-JJ	0.012 in.	58 - 72
	Legacy	P185/70R14	P195/60R15 P205/55R16 P205/70R15	32	30	6.5-JJ	0.012 in.	58 - 72
2004-05	Impreza	P195/60R15	None	32	29	6-JJ	0.012 in.	58 - 72
	Impreza RS	P205/55R16	None	32	29	7-JJ	0.012 in.	58 - 72
	Impreza Outback	P225/60R16	T145/80R16	32	30	16x16 1/2JJ	0.012 in.	65
	Impreza WRX	P195/60R15	P205/55R16 P215/45R17	32	29	6-JJ 7-JJ	0.012 in.	58 - 72
	Baja	P185/70R14	P195/60R15 P205/55R16 P205/70R15	32	30	6.5-JJ	0.012 in.	58 - 72
	Legacy	P185/70R14	P195/60R15 P205/55R16 P205/70R15	32	30	6.5-JJ	0.012 in.	58 - 72

OEM: Original Equipment Manufacturer

PSI: Pounds Per Square Inch

67170-SBCR-C10

BRAKE SPECIFICATIONS

All measurements in inches unless noted

			Brake Disc			Brake Drum Diameter			Minimum Lining Thickness		Brake Caliper	
Year	Model		Original Thickness	Minimum Thickness	Maximum Runout	Original Inside Diameter	Max. Wear Limit	Maximum Machine Diameter	Front	Rear	Bracket Bolts (ft. lbs.)	Mounting Bolts (ft. lbs.)
2001	Impreza	F	0.940	0.870	0.003	—	—	—	0.059	—	59	23-30
		R	0.390	0.335	0.004	9.000 ①	9.079 ②	NA	—	0.059	—	23-30
	Outback	F	0.940	0.870	0.003	—	—	—	0.059	—	59	23-30
		R	0.390	0.335	0.004	9.000 ①	9.079 ②	NA	—	0.059	59	23-30
	Legacy	F	0.940	0.870	0.003	—	—	—	0.059	—	59	23-30
		R	0.390	0.335	0.004	9.000 ①	9.079 ②	NA	—	0.059	—	23-30
2002	Impreza	F	0.940	0.870	0.003	—	—	—	0.059	—	59	23-30
		R	0.390	0.335	0.004	9.000 ①	9.079 ②	NA	—	0.059	—	23-30
	WRX	F	0.940	0.870	0.003	—	—	—	0.295	—	59	19.5
		R	0.390	0.335	0.004	9.000 ①	9.080 ②	9.000 ①	—	0.059	59	27.5
	Outback	F	0.945	0.866	0.003	—	—	—	0.295	—	59	29
		R	0.390	0.335	0.003	9.000 ①	9.079 ②	NA	—	0.256	59	29
	Legacy	F	0.940	0.870	0.003	—	—	—	0.059	—	59	23-30
		R	0.390	0.335	0.004	9.000 ①	9.079 ②	NA	—	0.059	—	23-30
2003	Impreza	F	0.940	0.870	0.003	—	—	—	0.059	—	59	23-30
		R	0.390	0.335	0.004	9.000 ①	9.079 ②	NA	—	0.059	—	23-30
	WRX	F	0.940	0.870	0.003	—	—	—	0.295	—	59	19.5
		R	0.390	0.335	0.004	9.000 ①	9.080 ②	9.000 ①	—	0.059	59	27.5
	Outback	F	0.945	0.866	0.003	—	—	—	0.295	—	59	29
		R	0.390	0.335	0.003	9.000 ①	9.079 ②	NA	—	0.256	59	29
	Baja	F	0.940	0.870	0.003	—	—	—	0.059	—	59	23-30
		R	0.390	0.335	0.004	9.000 ①	9.079 ②	NA	—	0.059	—	23-30
	Legacy	F	0.940	0.870	0.003	—	—	—	0.059	—	59	23-30
		R	0.390	0.335	0.004	9.000 ①	9.079 ②	NA	—	0.059	—	23-30
2004-05	Impreza	F	0.940	0.870	0.003	—	—	—	0.059	—	59	23-30
		R	0.390	0.335	0.004	9.000 ①	9.079 ②	NA	—	0.059	—	23-30
	WRX	F	0.940	0.870	0.003	—	—	—	0.295	—	59	19.5
		R	0.390	0.335	0.004	9.000 ①	9.080 ②	9.000 ①	—	0.059	59	27.5
	Outback	F	0.945	0.866	0.003	—	—	—	0.295	—	59	29
		R	0.390	0.335	0.003	9.000 ①	9.079 ②	NA	—	0.256	59	29
	Baja	F	0.940	0.870	0.003	—	—	—	0.059	—	59	23-30
		R	0.390	0.335	0.004	9.000 ①	9.079 ②	NA	—	0.059	—	23-30
	Legacy	F	0.940	0.870	0.003	—	—	—	0.059	—	59	23-30
		R	0.390	0.335	0.004	9.000 ①	9.079 ②	NA	—	0.059	—	23-30

NA: Not Available

① Parking brake drum on vehicles with rear disc brakes: 6.69 in.

② Specification is for the parking brake drum.

67170-SBCR-C11

SCHEDULED MAINTENANCE INTERVALS

SUBARU—IMPREZA, OUTBACK, WRX, LEGACY & BAJA

TO BE SERVICED	TYPE OF SERVICE	VEHICLE MILEAGE INTERVAL (x1000)												
		7.5	15	22.5	30	37.5	45	52.5	60	67.5	75	82.5	90	97.5
Engine oil & filter	R	✓	✓	✓	✓	✓	✓	✓	✓	✓	✓	✓	✓	✓
Brake lines	S/I		✓		✓		✓		✓		✓		✓	
Clutch & hill holder system	S/I		✓		✓		✓		✓		✓		✓	
Disc brake pads & discs, front & rear axle boots & axle shaft joint portions	S/I		✓		✓		✓		✓		✓		✓	
Parking brake	S/I		✓		✓		✓		✓		✓		✓	
Steering & suspension	S/I		✓		✓		✓		✓		✓		✓	
Air filter element	R				✓				✓				✓	
Engine coolant	R				✓				✓				✓	
Fuel filter	R				✓				✓				✓	
Spark plugs	R								✓					
Automatic transmission fluid & filter	S/I				✓				✓				✓	
Brake fluid	S/I				✓				✓				✓	
Brake linings & drums	S/I				✓				✓				✓	
Camshaft drive belt ①	S/I				✓				✓				✓	
Coolant level, hoses & clamps	S/I				✓				✓				✓	
Drive belts	S/I				✓				✓				✓	
Fuel system, hoses & connections	S/I				✓				✓				✓	
Transmission and/or differential gear fluid	S/I				✓								✓	
Front & rear wheel bearing repack	S/I								✓					

R: Replace S/I: Service or Inspect

① Non-California vehicles: replace every 60,000 miles.

FREQUENT OPERATION MAINTENANCE (SEVERE SERVICE)

If a vehicle is operated under any of the following conditions it is considered severe service:

- Extremely dusty areas.

- 50% or more of the vehicle operation is in 32°C (90°F) or higher temperatures, or constant operation in temperatures below 0°C (32°F).

- Prolonged idling (vehicle operation in stop and go traffic).

- Frequent short running periods (engine does not warm to normal operating temperatures).

- Police, taxi, delivery usage or trailer towing usage.

Oil & oil filter change: change every 3750 miles.

Clutch & hill holder system: service or inspect every 7500 miles.

Disc brake pads & discs, front & rear axle boots & axle shaft joint portions: service or inspect every 7500 miles.

Steering & suspension: service or inspect every 7500 miles.

Air filter element: service or inspect every 15,000 miles.

Automatic transmission fluid: service or inspect every 15,000 miles.

Brake linings & drums: service or inspect every 15,000 miles.

Coolant level, hoses & clamps: service or inspect every 15,000 miles.

Drive belts: service or inspect every 15,000 miles.

Transmission/differential gear oil (except SVX): service or inspect every 15,000 miles.

Front & rear wheel bearing repack: service or inspect every 30,000 miles.

67170-SBCR-C12

PRECAUTIONS

Before servicing any vehicle, please be sure to read all of the following precautions, which deal with personal safety, prevention of component damage, and important points to take into consideration when servicing a motor vehicle:

- Never open, service or drain the radiator or cooling system when the engine is hot; serious burns can occur from the steam and hot coolant.
- Observe all applicable safety precautions when working around fuel. Whenever servicing the fuel system, always work in a well-ventilated area. Do not allow fuel spray or vapors to come in contact with a spark, open flame or excessive heat (a hot drop light, for example). Keep a dry chemical fire extinguisher near the work area. Always keep fuel in a container specifically designed for fuel storage; also, always properly seal fuel containers to avoid the possibility of fire or explosion. Refer to the additional fuel system precautions later in this section.
- Fuel injection systems often remain pressurized, even after the engine has been turned **OFF**. The fuel system pressure must be relieved before disconnecting any fuel lines. Failure to do so may result in fire and/or personal injury.
- Brake fluid often contains polyglycol ethers and polyglycols. Avoid contact with the eyes and wash your hands thoroughly after handling brake fluid. If you do get brake fluid in your eyes, flush your eyes with clean, running water for 15 minutes. If eye irritation persists, or if you have taken brake fluid internally, IMMEDIATELY seek medical assistance.
- The EPA warns that prolonged contact with used engine oil may cause a number of skin disorders, including cancer! You should make every effort to minimize your exposure to used engine oil. Protective gloves should be worn when changing oil. Wash your hands and any other exposed skin areas as soon as possible after exposure to used engine oil. Soap and water, or waterless hand cleaner should be used.
- All new vehicles are now equipped with an air bag system. The system must be disabled before performing service on or around system components, steering column, instrument panel components, wiring and sensors. Failure to follow safety and disabling procedures could result in accidental air bag deployment, possible personal injury and unnecessary system repairs.
- Always wear safety goggles when working with, or around, the air bag system. When carrying a non-deployed air bag, be sure the bag and trim cover are pointed away from your body. When placing a non-deployed air bag on a work surface, always face the bag and trim cover upward, away from the surface. This will reduce the motion of the module if it is accidentally deployed. Refer to the additional air bag system precautions later in this section.
- Clean, high quality brake fluid from a sealed container is essential to the safe and proper operation of the brake system. You should always buy the correct type of brake fluid for your vehicle. If the brake fluid becomes contaminated, completely flush the system with new fluid. Never reuse any brake fluid. Any brake fluid that is removed from the system should be discarded. Also, do not allow any brake fluid to come in contact with a painted surface; it will damage the paint.
- Never operate the engine without the proper amount and type of engine oil; doing so WILL result in severe engine damage.
- Timing belt maintenance is extremely important! Many models utilize an interference-type, non-freewheeling engine. If the timing belt breaks, the valves in the cylinder head may strike the pistons, causing potentially serious (also time-consuming and expensive) engine damage.
- Disconnecting the negative battery cable on some vehicles may interfere with the functions of the on-board computer system(s) and may require the computer to undergo a relearning process once the negative battery cable is reconnected.
- When servicing drum brakes, only disassemble and assemble one side at a time, leaving the remaining side intact for reference.
- Only an MVAC-trained, EPA-certified automotive technician should service the air conditioning system or its components.

ENGINE REPAIR

➡Disconnecting the negative battery cable on some vehicles may interfere with the functions of the on board computer systems and may require the computer to undergo a relearning process, once the negative battery cable is reconnected.

Alternator

REMOVAL & INSTALLATION

1. Remove or disconnect the following:
   - Negative battery cable
   - Connector and terminal from the alternator
   - V-belt cover, if equipped
   - Front side V-belt
   - Alternator to bracket bolts
   - Alternator from the vehicle

To install:

2. Install or connect the following:
   - Alternator into the vehicle
   - Alternator to bracket bolts
   - Front side V-belt
   - V-belt cover, if equipped
   - Connector and terminal to the alternator
   - Negative battery cable
3. Check and adjust the belt tension.

Ignition Timing

ADJUSTMENT

All Subaru models are equipped with Distributorless Ignition System (DIS). The ignition timing is controlled by the Powertrain Control Module (PCM) and is not adjustable.

Engine Assembly

⁂ CAUTION

Some models covered by this manual may be equipped with an air bag. Whenever working near any of the Supplemental Restraint System (SRS) components, such as the impact sensors, the air bag module, steering column and instrument panel, properly disable the SRS.

REMOVAL & INSTALLATION

2.0L Engines

1. Before servicing the vehicle, refer to the precautions in the beginning of this section.

2. Relieve the fuel system pressure.

3. Drain the engine oil and coolant into suitable containers.

4. Raise the rear seat and turn the floor mat up.

5. Disconnect the fuel pump relay connector, start the engine and let it stall. Once the engine stalls, crank it for a further 5 seconds to ensure the fuel system is properly relieved.

6. Remove the fuel filler cap.

7. Remove or disconnect the following:

- Air cleaner cover and element
- Battery cables
- Radiator
- Coolant filler tank

8. If equipped with air conditioning, discharge the system using an approved recovery/recycling machine. Disconnect and cap the lines from the compressor.

- Intercooler

9. Disconnect the following electrical connections:

- Engine harness connector
- Engine ground terminal
- Alternator connector, terminal and A/C compressor connections

10. Remove or disconnect the following:

- Accelerator cable
- Clutch release spring
- Brake booster hose
- Heater inlet and outlet hoses

11. Remove the power steering pump from the bracket by performing the following steps:

a. Loosen the lock and slider bolts.

b. Remove the V-belt.

c. Disconnect the power steering switch connection.

d. Remove the pipe with bracket from the intake manifold.

e. Remove the power steering pump from the engine.

f. Remove the power steering tank from the bracket by pulling it upwards.

g. Place the power steering pump on the wheel apron on the right.

12. Remove or disconnect the following:

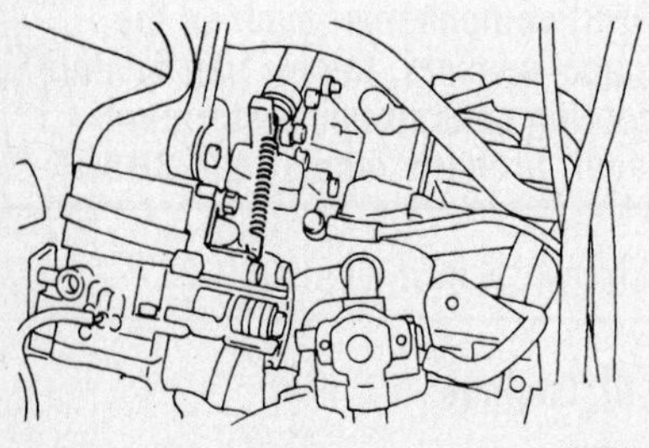

View of the clutch release spring location—2.0L engines

- Center exhaust pipe
- Nuts that attach the lower side of the engine to the transmission
- Nuts that attach the front cushion rubber onto the crossmember

13. Disconnect the clutch release fork from the release bearing as follows:

a. Remove the clutch cylinder from the transmission.

b. Using a 10mm wrench, remove the plug.

c. Screw a 6mm diameter bolt into the release fork and remove it.

d. Raise the release fork and unfasten the release tabs to free the release fork.

14. Disconnect the torque converter clutch from the drive plate on models equipped with an automatic transaxle as follows:

a. Remove the service hole plug.

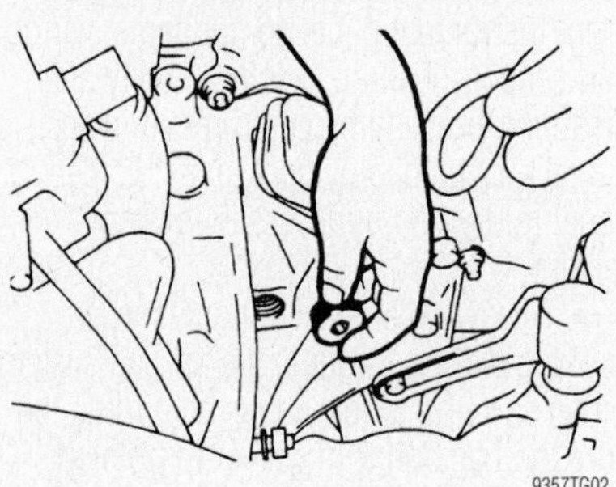

Using a 10mm wrench, remove the plug—2.0L engines

b. Remove the torque converter clutch-to-drive plate bolts.

c. Remove the remaining bolts while rotating the engine using a crankshaft pulley wrench.

15. Remove or disconnect the following:

- Pitching stopper
- Fuel delivery, return and evaporation hoses
- Fuel filter and bracket

16. Attach a lifting device to the engine.

17. Using a floor jack, support the transmission.

18. Remove the starter.

19. Separate the engine from the transmission.

20. Remove the upper right transmission-to-engine bolts.

21. Remove the engine as follows:

a. Raise the engine slightly.

b. Using the floor jack, raise the transmission.

c. Move the engine horizontally until the mainshaft is withdrawn from the clutch cover.

d. Remove the engine.

To install:

22. Installation is the reverse of removal, please note the following torques:

- Clutch release fork plug: 32 ft. lbs. (44 Nm)
- Front cushion rubbers: 25 ft. lbs. (34 Nm)

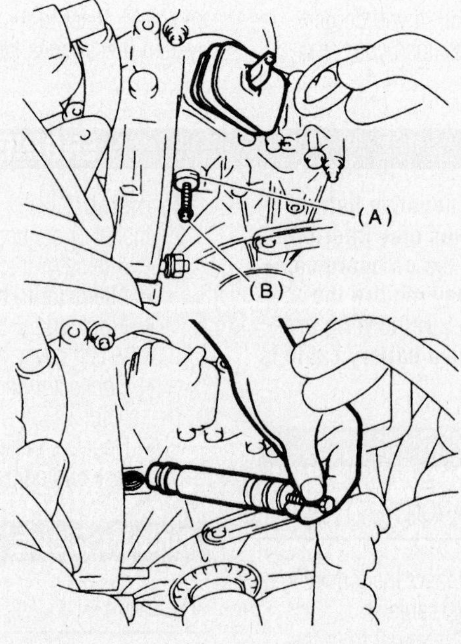

Screw a 6mm diameter bolt into the release fork and remove it—2.0L engines

- Bolts attaching the upper right side of the transmission to the engine: 37 ft. lbs. (50 Nm)
- Pitching stopper-to-fender bolt: 37 ft. lbs. (50 Nm)
- Pitching stopper-to-engine bolt: 43 ft. lbs. (58 Nm)
- Torque converter clutch-to-drive plate bolts, while rotating the engine: 18 ft. lbs. (25 Nm)
- Power steering pump bolts: 15 ft. lbs. (20 Nm)
- Bolts attaching the lower side of the transmission to the engine: 37 ft. lbs. (50 Nm)
- Front cushion rubber-to-cross-member bolts: 61 ft. lbs. (83 Nm)

23. Fill the engine with the recommended oil.
24. Fill and bleed the cooling system.
25. Charge the air conditioning system using an approved recovery/recycling machine.
26. Adjust the clutch cable.
27. If equipped, check the automatic transaxle fluid level and add Dexron®II if necessary.
28. Start the engine and allow it to reach normal operating temperature. Check for leaks.

2.2L and 2.5L Engines

1. Before servicing the vehicle, refer to the precautions in the beginning of this section.
2. Relieve the fuel system pressure.
3. Drain the engine oil and coolant into suitable containers.
4. Remove or disconnect the following:
   - Battery cables
   - Battery from the vehicle
   - Radiator hoses
   - Fan motor harness
   - Radiator
5. If equipped with air conditioning, discharge the system using an approved recovery/recycling machine. Disconnect and cap the lines from the compressor.
   - Air intake duct
   - Air cleaner element and upper cover
   - Evaporator canister and bracket
   - Oxygen Sensor (O_2S) connector
   - Engine ground terminal
   - Crankshaft Position (CKP) sensor connector
   - Camshaft Position (CMP) sensor connector
   - Knock Sensor (KS) connector
   - Alternator connector and terminal
   - Air conditioning compressor connectors, if equipped
   - Accelerator cable
   - Cruise control cable, if equipped
   - Clutch release spring, clutch cable and hill holder cable, if equipped with a manual transaxle
   - Brake booster hose(s)
   - Heater inlet and outlet hoses
   - Alternator drive belt
   - Spark plug wires from left side of engine
   - Power steering pump line bracket
   - Power steering pump, leave the lines connected and position aside
   - Exhaust Y-pipe
   - Lower starter nuts
   - Lower engine-to-transaxle nuts
   - Front engine mount-to-crossmember nuts
   - Starter
6. If equipped with an automatic transaxle, perform the following:
   a. Remove the torque converter service hole plug.
   b. Rotate the engine. Remove the torque converter-to-drive plate bolts as they become accessible.
7. Remove or disconnect the following:
   - Pitching stopper
   - Fuel delivery, return and evaporation hoses
8. Support the engine with a suitable lifting device attached to the engine lifting eyes.
9. Slightly raise the engine.
10. Raise the transaxle with a floor jack.
11. Slowly remove the engine from the vehicle.

To install:

12. Apply a small amount of grease to the splines of the mainshaft.
13. Position the engine in the engine compartment and align it with the transaxle.
14. Install or connect the following:
    - Engine upper bolts and tighten to 34–40 ft. lbs. (44–54 Nm)
15. Remove the lifting device and floor jack.
16. Install the pitching stopper and tighten the bolts to the following specifications:
    a. Body side: 49 ft. lbs. (67 Nm).
    b. Bracket side: 40 ft. lbs. (54 Nm).
17. If equipped with an automatic transaxle, perform the following:
    a. Install the torque converter-to-drive plate bolts while rotating the engine, and tighten to 20 ft. lbs. (26 Nm).
    b. Install the service hole cover.
18. Install or connect the following:
    - Evaporator canister and bracket
    - Power steering pump. Torque retainers to 22–36 ft. lbs. (29–47 Nm).
    - Drive belt, adjust tension
    - Starter. Tighten bolts to 34–40 ft. lbs. (44–52 Nm).
    - Lower engine-to-transaxle nuts. Tighten to 34–40 ft. lbs. (44–52 Nm).
    - Lower engine mounting nuts. Tighten to 61 ft. lbs. (83 Nm) in the inner most elliptical hole in the front crossmember so the clearance is 0.16–0.24 in. (4–6mm).
    - Exhaust Y-pipe with new gaskets and nuts
    - Brake booster hose
    - Heater inlet and outlet hoses
    - Accelerator and the cruise control cables, if equipped
19. If equipped with a manual transaxle, install the following:
    - Clutch release spring
    - Clutch cable
    - Hill holder cable
20. Install or connect the following:
    - Engine harness connectors
    - O_2 sensor connector
    - Engine ground terminal
    - CKP sensor connector
    - CMP sensor connector
    - Knock sensor connector
    - Alternator connector and terminal
    - Air conditioning compressor connectors, if equipped
    - Air cleaner element and cover
    - Air conditioning lines, if equipped, with new O-rings. Tighten the bolts to 23 ft. lbs. (31 Nm).
    - Radiator
    - Engine cover
    - Battery
21. Fill the engine with the recommended oil.
22. Fill and bleed the cooling system.
23. Charge the air conditioning system using an approved recovery/recycling machine.
24. Adjust the clutch cable.
25. If equipped, check the automatic transaxle fluid level and add Dexron®II if necessary.
26. Start the engine and allow it to reach normal operating temperature. Check for leaks.

3.0L Engines

1. Before servicing the vehicle, refer to the precautions in the beginning of this section.
2. Relieve the fuel system pressure.
3. Drain the engine oil and coolant into suitable containers.

4. Remove or disconnect the following:
   - Battery cables
   - Battery
   - Air intake duct
   - Engine undercover
   - Radiator
   - Drive belt

5. If equipped with air conditioning, discharge the system using an approved recovery/recycling machine. Disconnect and cap the lines from the compressor.
   - Engine ground terminal
   - Engine harness connectors
   - Alternator connector and terminal
   - Air conditioning compressor connectors, if equipped
   - Accelerator and the cruise control cables, if equipped
   - Brake booster hose
   - Heater inlet and outlet hoses
   - Power steering pump line bracket
   - Power steering pump, leave the lines connected and position aside
   - Exhaust Y-pipe
   - Lower engine-to-transaxle nuts
   - Front engine mount-to-crossmember nuts

6. If equipped with an automatic transaxle, perform the following:
   a. Remove the torque converter service hole plug.
   b. Rotate the engine. Remove the torque converter-to-drive plate bolts as they become accessible.

7. Remove or disconnect the following:
   - Pitching stopper
   - Fuel delivery, return and evaporation hoses

8. Support the engine with a suitable lifting device attached to the engine lifting eyes.

9. Support the transaxle with a floor jack.

10. Remove or disconnect the following:
    - Starter
    - Upper engine-to-transmission bolts

11. Slightly raise the engine.

12. Raise the transaxle with a floor jack.

13. Slowly remove the engine from the vehicle.

To install:

14. Apply a small amount of grease to the splines of the mainshaft.

15. Position the engine in the engine compartment and align it with the transaxle.

16. Install or connect the following:
    - Engine upper bolts and tighten to 36 ft. lbs. (50 Nm)

17. Remove the lifting device and floor jack.

18. Install the pitching stopper and tighten the bolts to the following specifications:
    a. Body side: 42 ft. lbs. (57 Nm).
    b. Bracket side: 42 ft. lbs. (49 Nm).

19. Install the starter. Tighten bolts to 37 ft. lbs. (50 Nm).

20. If equipped with an automatic transaxle, perform the following:
    a. Install the torque converter-to-drive plate bolts while rotating the engine, and tighten to 18 ft. lbs. (25 Nm).
    b. Install the service hole cover.

21. Install or connect the following:
    - Power steering pump. Torque retainers to 14 ft. lbs. (20 Nm).
    - Lower engine-to-transaxle nuts. Tighten to 36 ft. lbs. (50 Nm).
    - Lower engine mounting nuts. Tighten to 54 ft. lbs. (74 Nm).
    - Exhaust Y-pipe with new gaskets and nuts
    - Fuel delivery, return and evaporation hoses
    - Heater inlet and outlet hoses
    - Brake booster hose
    - Engine ground terminal
    - Engine harness connectors
    - Alternator connector and terminal
    - Air conditioning compressor connectors, if equipped
    - Accelerator and the cruise control cables, if equipped
    - Air conditioning lines, if equipped, with new O-rings. Tighten the bolts to 10 ft. lbs. (15 Nm).
    - Radiator
    - Drive belt, adjust tension
    - Air cleaner element and cover
    - Engine under cover
    - Battery
    - Battery cables

22. Fill the engine with the recommended oil.

23. Fill and bleed the cooling system.

24. Charge the air conditioning system using an approved recovery/recycling machine.

25. Check the automatic transaxle fluid level and add Dexron®II if necessary.

26. Start the engine and allow it to reach normal operating temperature. Check for leaks.

Water Pump

REMOVAL & INSTALLATION

Except 3.0L Engines

1. Before servicing the vehicle, refer to the precautions in the beginning of this section.

2. Remove or disconnect the following:
   - Negative battery cable
   - Engine undercover, if equipped

3. Drain the coolant into a suitable container.

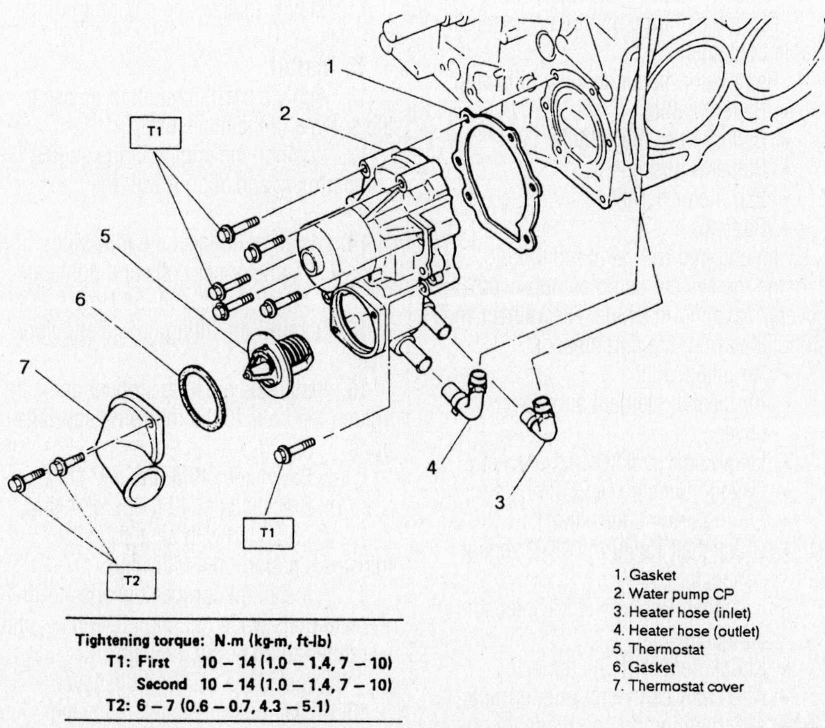

Exploded view of the water pump assembly—except 3.0L engines

4. Remove or disconnect the following:
   - Radiator fan connector(s)
   - Radiator outlet and heater hoses
   - Heater bypass hose or overflow hose, if equipped
   - Reservoir tank, on Legacy models
   - Radiator fan motor assembly(ies)
   - Accessory drive belts
   - Timing belt
   - Belt tension adjuster
   - Belt idler No. 2
   - Camshaft Position (CMP) sensor
   - Left side camshaft pulley(s)
   - Left side rear timing belt cover
   - Tensioner bracket
   - Radiator and heater hoses from water pump
   - Water pump retainer bolts
   - Water pump
5. Inspect the radiator hoses for deterioration and replace as necessary.

To install:

6. Clean the gasket mating surfaces thoroughly. Always use new gaskets during installation.
7. Install or connect the following:
   - Water pump, tighten the bolts in sequence to 10 ft. lbs. (13 Nm). After tightening the bolts once, retighten to the same specification again.
   - Radiator heater hoses to water pump
   - Tensioner bracket and tighten to 18 ft. lbs. (25 Nm)
   - Left side rear timing belt cover
   - Left side camshaft pulley(s). Tighten to 58 ft. lbs. (78 Nm) on non-turbo models and 72 ft. lbs. (98 Nm) on turbo models.
   - CMP sensor

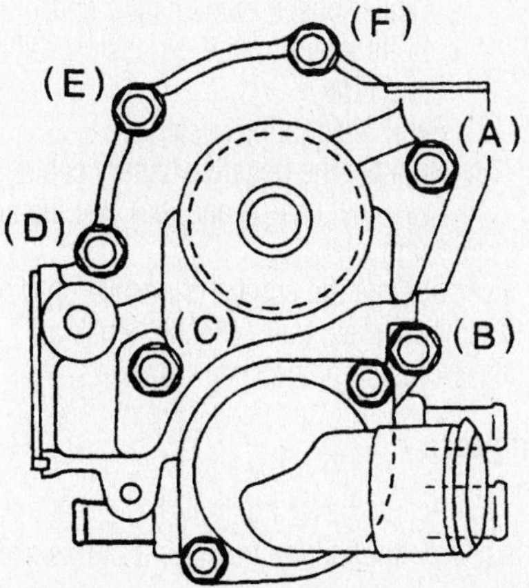

Tighten the water pump bolts is two steps using the following sequence—2.0L engine Turbo model shown, others similar

   - Belt idler No. 2 and tighten to 29 ft. lbs. (39 Nm)
   - Belt tension adjuster
   - Timing belt
   - Accessory drive belts
   - Radiator fan assembly(ies)
   - Reservoir tank, if removed
   - Heater bypass hose or overflow hose, if equipped
   - Air intake duct
   - Radiator outlet and heater hoses
   - Radiator fan connector(s)
   - Engine undercover, if removed
8. Fill the system with coolant and connect the negative battery cable.
9. Start the engine and allow it to reach operating temperature.
10. Check for leaks.

3.0L Engines

1. Before servicing the vehicle, refer to the precautions in the beginning of this section.
2. Remove or disconnect the following:
   - Negative battery cable
   - Engine undercover, if equipped
3. Drain the coolant into a suitable container.
4. Remove or disconnect the following:
   - Radiator
   - Accessory drive belts
   - Timing chain
   - Water pump retainer bolts
   - Water pump
5. Inspect the radiator hoses for deterioration and replace as necessary.

To install:

6. Clean the gasket mating surfaces thoroughly. Always use new gaskets during installation.
7. Apply coolant to the new O-ring before installation
8. Install or connect the following:
   - Water pump with a new O-ring, tighten the bolts to 5 ft. lbs. (7 Nm).
   - Timing chain
   - Front chain cover
   - Accessory drive belts

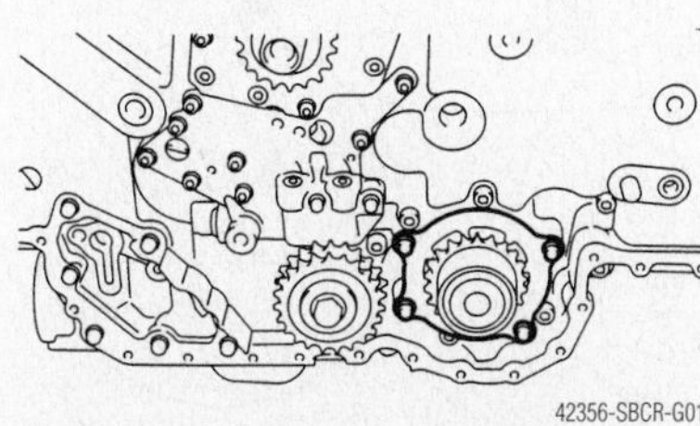

View of the water pump assembly—3.0L engines

   - Radiator
   - Engine undercover, if removed
9. Fill the system with coolant and connect the negative battery cable.
10. Start the engine and allow it to reach operating temperature.
11. Check for leaks.

Heater Core

REMOVAL & INSTALLATION

WRX

1. Before servicing the vehicle, refer to the precautions in the beginning of this section.
2. Disconnect the negative battery cable.

CAUTION

If equipped with an air bag system, wait 10 minutes after disconnecting the negative battery cable before performing any further work while the system fully de-energizes in order to avoid accidental deployment. All air bag system wiring is yellow. Do not use electrically powered test equipment on these circuits.

3. Drain the cooling system into a clean container for reuse.
4. Remove or disconnect the following:
   - Bolts retaining the expansion valve and pipe
   - Heater hoses from the heater core. Plug the heater core and heater hoses.
   - Lower cover from the instrument panel
   - Glove box
5. Remove the center console panel as follows:
   a. Remove lower panel and control wires.
   b. Pull the control panel out and disconnect the connections.
6. Remove the passenger side air bag module as follows:
   a. Disconnect the airbag connector from the support beam bracket and then unplug the connector.
   b. Unfasten the 3 airbag module bolts and remove the module.
7. Remove or disconnect the following:
   - 4 screws and two nuts then remove the lower console panel
   - Hooks that retain the defroster panel
   - Nuts and remove the electrical con-

nections as shown in the illustration
- Instrument panel bolts and remove the panel
- Support beam
- Blower motor
- Heater core retainers
- Heater core

To install:

8. Install or connect the following:
   - Heater core
   - Heater core retainers
   - Blower motor
   - Support beam
   - Instrument panel and tighten the bolts
   - Electrical connections and tighten the nuts
   - Hooks to the defroster panel
   - Lower console panel and tighten the 4 screws and two nuts
9. Install the passenger side air bag module as follows:
   a. Install the module and tighten the 3 airbag module bolts.

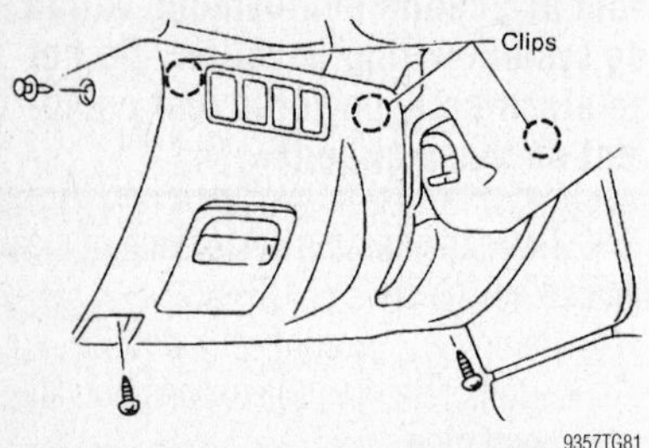

Location of the lower cover retainers—WRX models

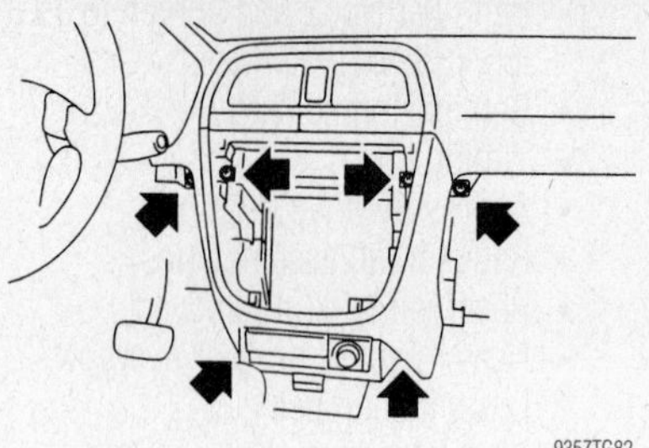

Location of the lower console panel retainers—WRX models

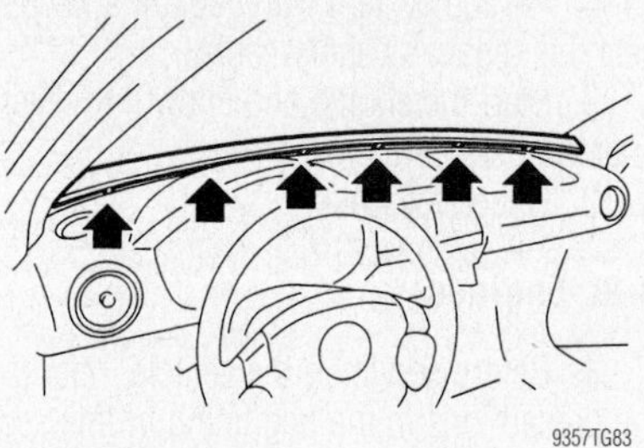

Loosen the hooks that retain the defroster panel—WRX models

Disconnect the two electrical connections after removing the console panel—WRX models

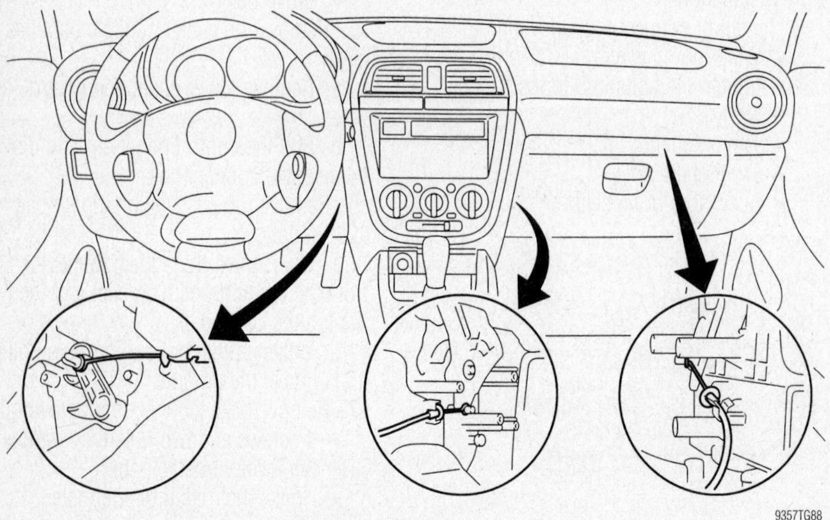

Location of the control unit control wires—WRX models

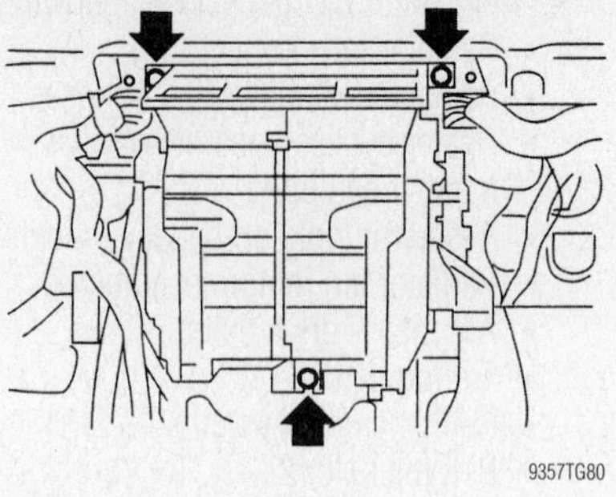

9357TG80

Location of the heater core retainers—WRX models

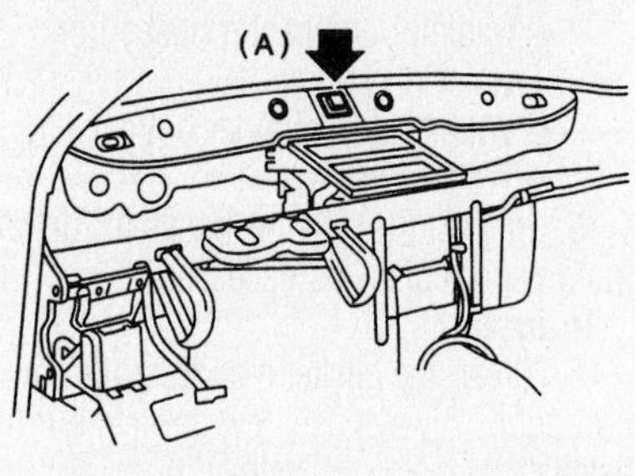

When installing the instrument panel, push the hook into grommet (A) on the body panel—WRX models

   b. Connect the airbag connector and then attach it to the support beam bracket.
10. Install the center console panel as follows:
   a. Attach the control panel connections and insert the panel into position.
   b. Install the lower panel and control wires.
11. Install or connect the following:
   - Glove box
   - Lower cover to the instrument panel
   - Heater hoses to the heater core
   - Bolts retaining the expansion valve and pipe
12. Refill the cooling system.
13. Connect the negative battery cable.
14. Evacuate, charge and leak test the air conditioning system.
15. Operate the engine to normal operating temperatures; then, check the climate control operation and check for leaks.

Impreza

1. Before servicing the vehicle, refer to the precautions in the beginning of this section.
2. Disconnect the negative battery cable.

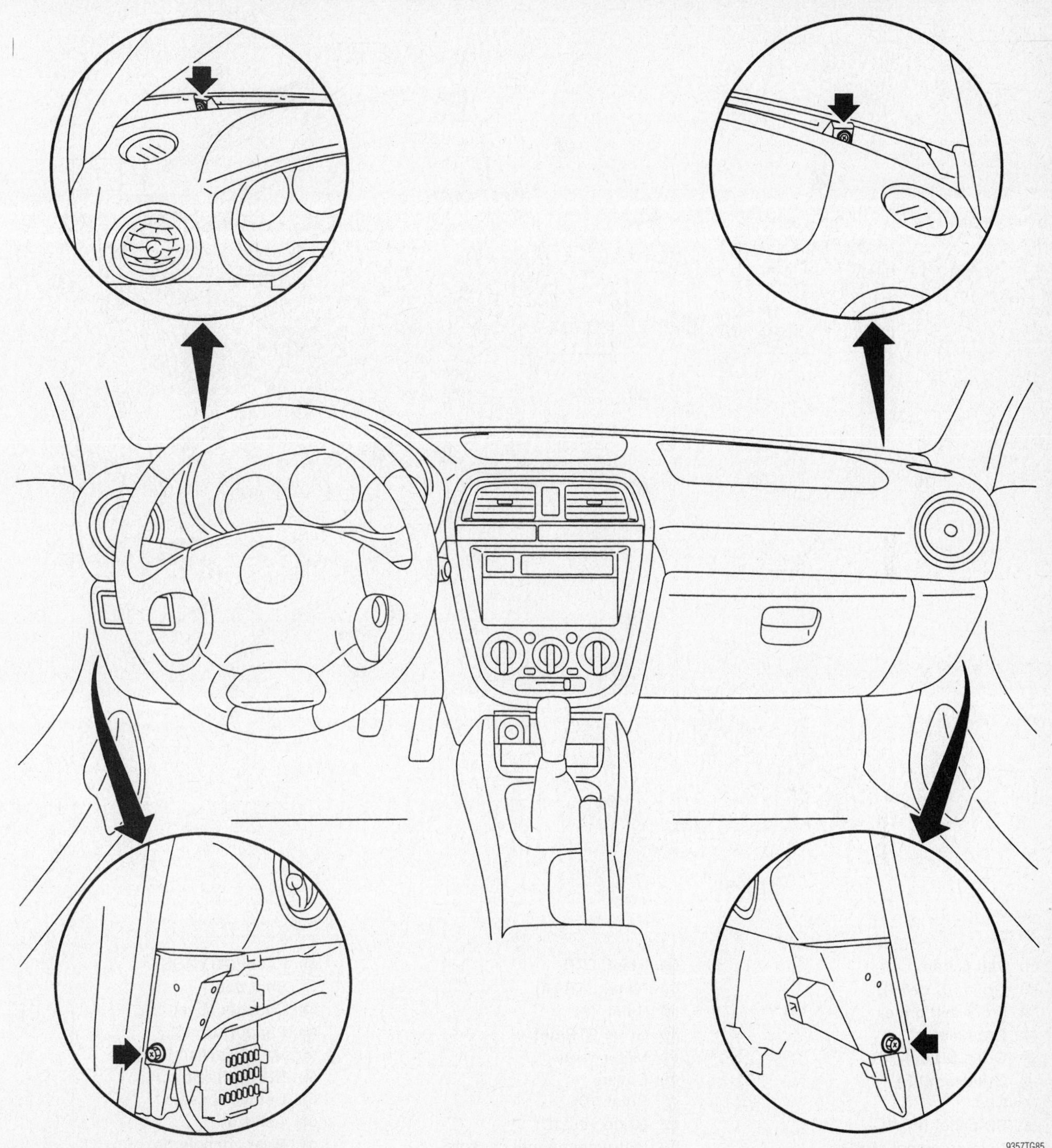

Location of the instrument panel bolts—WRX models

**** CAUTION**

If equipped with an air bag system, wait 10 minutes after disconnecting the negative battery cable before performing any further work while the system fully de-energizes in order to avoid accidental deployment. All air bag system wiring is yellow. Do not use electrically powered test equipment on these circuits.

3. Drain the cooling system into a clean container for reuse.
4. Remove or disconnect the following:
   - Heater hoses from the heater core. Plug the heater core and heater hoses.
   - Radio box or console
5. Remove the instrument panel as follows:
   a. Remove the rear console box.
   b. Pull the cup holder.
   c. Turn over the shift lever boot (manual transaxle models) or remove select lever cover (automatic transaxle models).
   d. Remove the console cover.
   e. Remove the audio assembly and disconnect the antenna cable and connectors.
   f. Remove the lower cover and then disconnect the seat belt timer connector.
   g. Remove the glove box.
   h. Remove the instrument panel console.

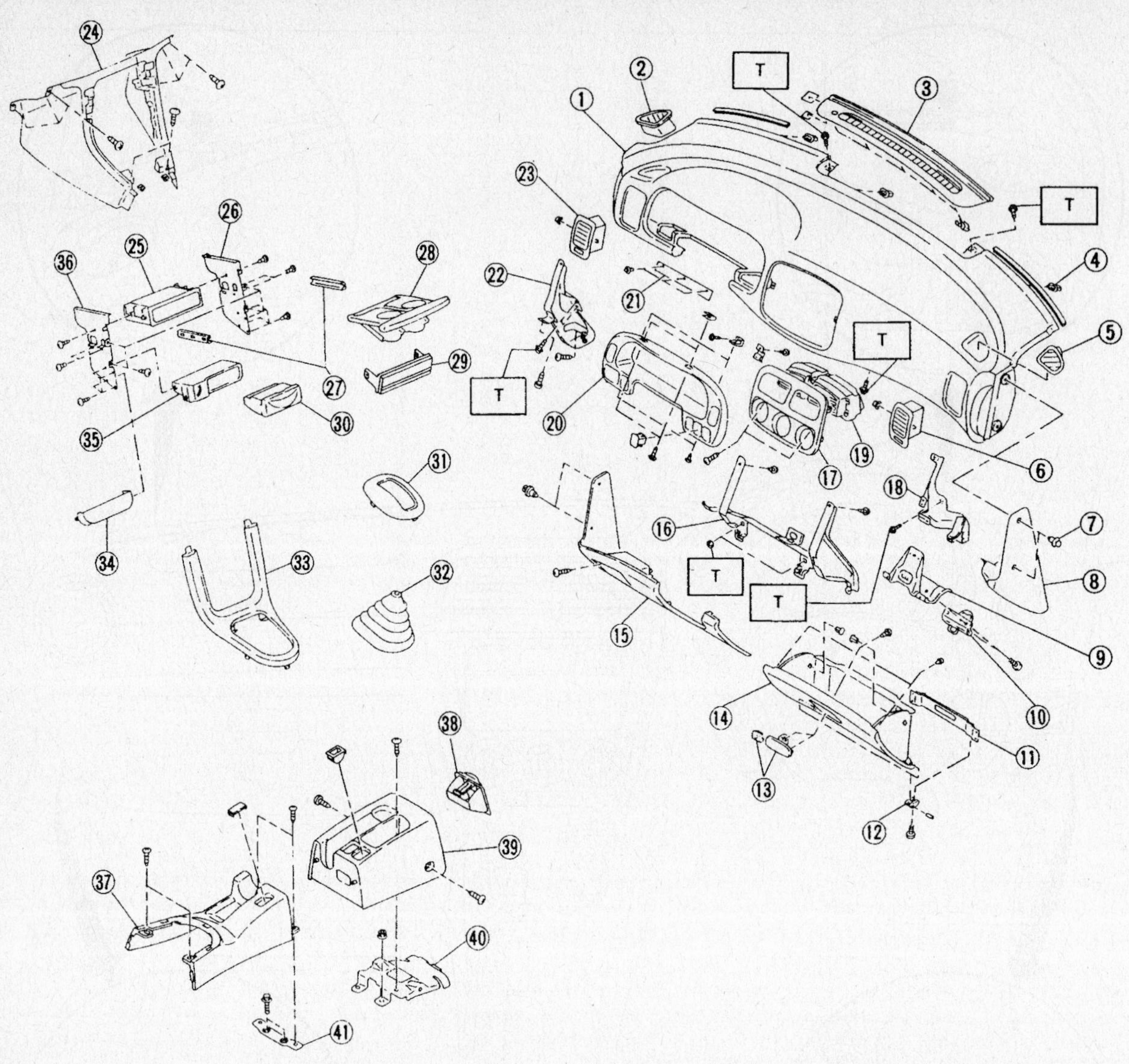

1. Pad & frame
2. Grille SD def. (D)
3. Front def. grille
4. Grommet
5. Grille SD def. (P)
6. Grille vent (P)
7. Clip
8. SD panel (P)
9. Reinforcement striker
10. Striker
11. Frame pocket
12. Hinge
13. Lock ASSY
14. Pocket ASSY
15. Lower cover ASSY
16. Reinf. CTR
17. Panel CTR (A)
18. Reinf. (P)
19. Grille CTR def.
20. Meter visor
21. Cover
22. Reinf. (D)
23. Grille vent (D)
24. Instrument panel console
25. Pocket CTR
26. BRKT (Radio) RH
27. Rail (Cup holder)
28. Cup holder
29. Panel (Radio)
30. Ash tray
31. Panel (AT) ASSY
32. Shift boot
33. Console cover
34. Panel (Airbag)
35. Housing (Ash tray)
36. BRKT (Radio) LH
37. Center console
38. Ash tray
39. Rear console box
40. Rear console BRKT
41. Center console BRKT

Tightening torque: N·m (kg-cm, in-lb)
T: 6.9 ± 1.0 (70 ± 10, 60.8 ± 8.7)

87970G59

Exploded view of the instrument panel and center console—Impreza

i. Remove the 2 bolts and lower the steering column.

j. Remove the column cover.

k. Remove the hood opening lever.

l. Set the temperature control switch to MAX COLD, and mode selector switch to the defroster position.

m. Disconnect both the temperature control cable and the mode selector cable from the link.

➡Do not move the switch and link when installing.

n. Tag or match mark the wiring connectors, then disconnect by holding the connectors and not the wiring.

o. Remove the 6 instrument panel retaining bolts and nuts.

p. Remove the front defroster grille and 2 bolts.

q. Carefully remove the instrument panel from the body and then disconnect the speedometer cable from the back of the combination meter.

6. Remove or disconnect the following:
   - Heater control cables and the fan motor wiring harness
   - Duct between the heater unit and the blower heater unit. Lift up and out on the heater unit and remove it.

7. With the heater assembly out of the vehicle, remove the heater core tube retaining clamps and lift the core from the heater case.

8. Remove the heater core.

To install:

9. Install or connect the following:
   - Heater core into the heater case. Secure it in place with the retaining clamps and screws.
   - Heater assembly to its mounting position under the dash. Torque the mounting bolts to 48–84 inch lbs. (5–9 Nm).
   - Heater control cables and fan motor wiring harness connectors
   - Instrument panel
   - Radio and console assemblies
   - Heater hoses in the engine compartment

10. Refill the cooling system.

11. Connect the negative battery cable.

12. Evacuate, charge and leak test the air conditioning system.

13. Operate the engine to normal operating temperatures; then, check the climate control operation and check for leaks.

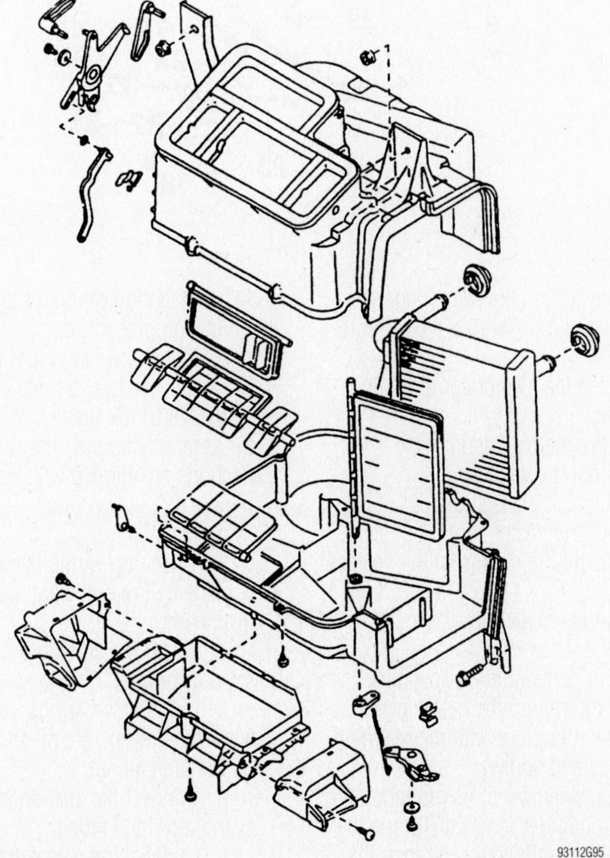

Exploded view of the heater assembly—Impreza

Legacy and Baja

1. Before servicing the vehicle, refer to the precautions in the beginning of this section.

2. Disconnect the negative battery cable.

CAUTION

If equipped with an air bag system, wait 10 minutes after disconnecting the negative battery cable before performing any further work while the system fully de-energizes in order to avoid accidental deployment. All air bag system wiring is yellow. Do not use electrically powered test equipment on these circuits.

3. Drain the cooling system into a clean container for reuse.

4. Remove or disconnect the following:
   - Heater hoses from the heater core. Plug the heater core and heater hoses.
   - Radio box or console

5. Remove the instrument panel as follows:

a. Remove the center console retaining screws and remove the center console assembly.

b. Remove the instrument panel retaining bolt covers by prying them from the panel.

c. Remove the lower part of the front A pillar trim. Remove the instrument panel under covers from the driver's and passenger's sides.

d. Remove the hood release cable from the hood release lever.

e. Disconnect the wiring harness connectors under the instrument panel.

f. Remove the instrument cluster assembly. Remove the glove box assembly

g. Disconnect the ventilation control cables and electrical connectors at the heater unit. Disconnect the vacuum line at the blower housing.

h. Disconnect the radio antenna feeder wire. Disconnect the main harness connector at the fuse box.

i. Remove the lower steering column covers. Remove the steering column retaining bolts and allow the column to hang down.

j. Remove the instrument panel retaining bolts.

➡When removing the instrument panel, check that all wiring and cables are disconnected before pulling it completely away from the firewall.

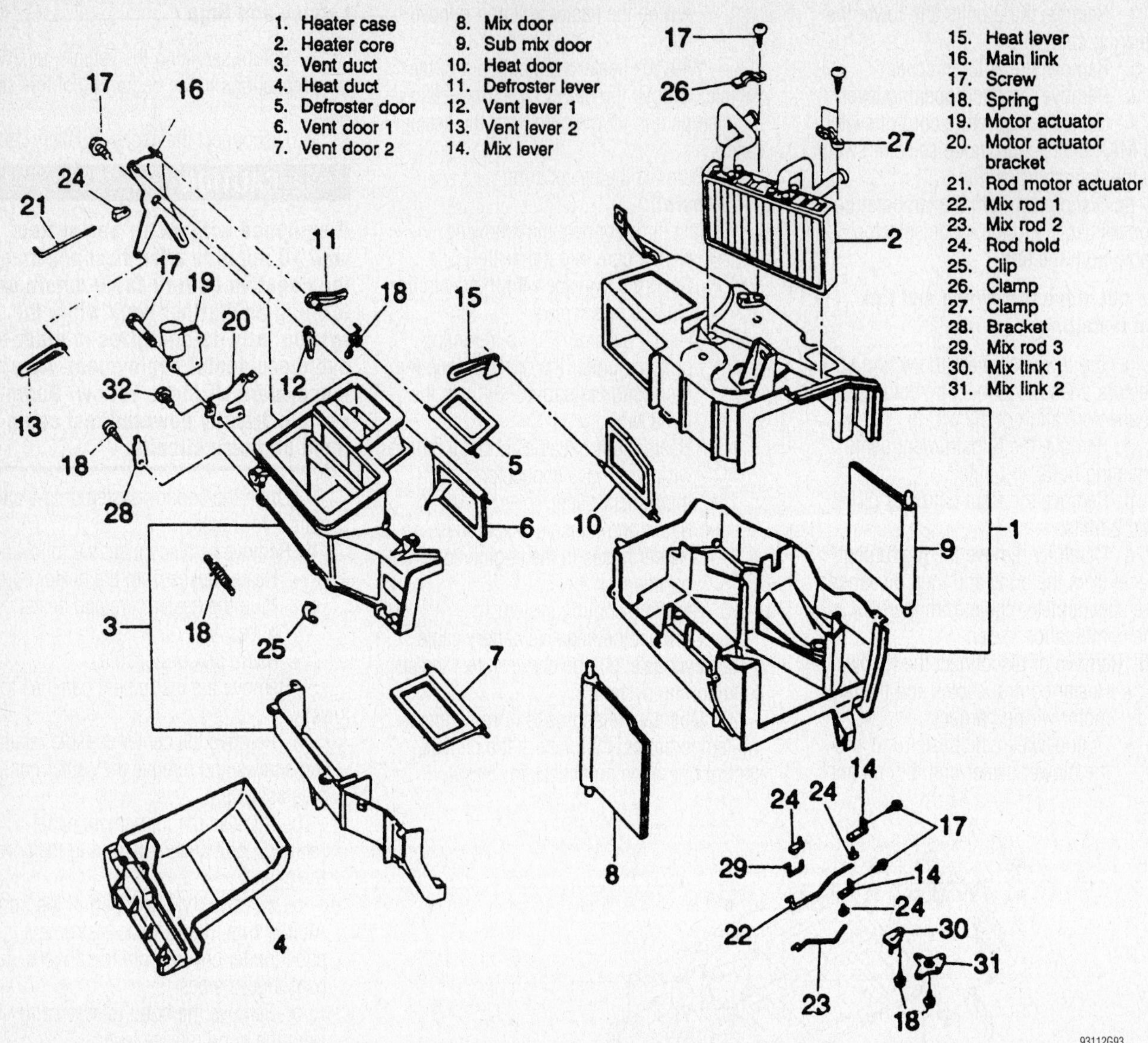

Exploded view of the heater assembly—Legacy

k. With the help of an assistant, lift and remove the instrument panel from the vehicle.

6. Remove or disconnect the following:
- Heater control cables and the fan motor wiring harness
- Duct between the heater unit and the blower heater unit. Lift up and out on the heater unit and remove it.

7. Remove the evaporator case assembly by performing the following procedure:

a. Discharge and recover the air conditioning system refrigerant.

b. Disconnect the low and high pressure line from the evaporator outlet and cap the fittings.

c. Remove the inlet and outlet pipe grommets.

d. Remove the glove box and support bracket.

e. Disconnect the air conditioning wiring harness from the evaporator. Disconnect the drain hose from the evaporator.

f. Remove the evaporator mounting nut and bolt.

g. Remove the evaporator case assembly from the vehicle.

8. With the heater assembly out of the vehicle, remove the heater core tube retaining clamps and lift the core from the heater case.

9. Remove the heater core.

To install:

10. Install or connect the following:
- Heater core into the heater case. Secure it in place with the retaining clamps and screws.
- Heater assembly to its mounting position under the dash Torque the mounting bolts to 48–84 inch lbs. (5–9 Nm).

11. Install the evaporator by performing the following procedure:

a. Install the evaporator case assembly and the nuts and bolts.

b. Adjust the position of the evaporator assembly so the inlet and outlet connections are aligned with the heater and blower unit connections.

c. Install the drain hose. Connect the air conditioning wiring harness.

d. Install the inlet and outlet pipe grommets.

e. Install the glove box and the lower support bracket.

f. Using new O-rings, lubricate them with clean refrigerant oil and install them on the pipe fittings.

g. Connect the suction hose to the evaporator inlet fitting.

h. Connect the discharge hose to the evaporator outlet fitting.

i. Connect the heater control cables and fan motor wiring harness connectors.

12. Install or connect the following:
- Instrument panel
- Radio and console assemblies
- Heater hoses in the engine compartment

13. Refill the cooling system.
14. Connect the negative battery cable.
15. Evacuate, charge and leak test the air conditioning system.
16. Operate the engine to normal operating temperatures; then, check the climate control operation and check for leaks.

Outback

1. Before servicing the vehicle, refer to the precautions in the beginning of this section.
2. Disconnect the negative battery cable.

⁂ CAUTION

If equipped with an air bag system, wait 10 minutes after disconnecting the negative battery cable before performing any further work while the system fully de-energizes in order to avoid accidental deployment. All air bag system wiring is yellow. Do not use electrically powered test equipment on these circuits.

3. Drain the cooling system into a clean container for reuse.
4. Remove or disconnect the following:
- Negative battery cable
- LLC
- Heater hoses from the heater core. Plug the heater core and heater hoses.
- Bolts attaching the expansion valve and pipe in the engine compartment

5. Remove the instrument panel as follows:

a. Remove the lower cover.

b. Remove the lower column cover and disconnect the harness.

c. Set the tires in the straight ahead position.

d. Using a T30 Torx® bit, remove the Torx bolts from either side of the steering wheel.

e. Slide the airbag module forward and disconnect the electrical connection. Remove the module and store in a safe area.

➡Matchmark the steering wheel to shaft location, prior to removal.

f. Steering wheel nut and use a puller to draw the wheel off of the shaft.

g. Remove the universal joint bolts and remove the joint.

h. If not already removed, remove the trim from under the instrument panel.

i. Remove the lower steering column cover screw.

j. Disconnect all electrical connectors from the column.

k. Remove the 2 bolts that secure the column to the instrument panel.

l. Remove the steering column.

m. Remove the glove box stopper and the glove box.

n. Remove the side panels from both sides.

o. Disconnect passenger air bag module from the support beam.

p. Remove the 3 bolts securing the passenger air bag module and remove the module.

q. Remove the shift knob.

r. Remove the tray the console box cover and the console box.

s. Remove the front trim pillar from both sides.

t. Remove the front pillar lower trim from the passenger side.

u. Set the temperature control switch to **FULL HOT** and disconnect the control cable from the bottom of the heater unit. Do not move the switch and link.

v. Remove the instrument panel bolts, harness connectors and the panel.

6. Remove or disconnect the following:
- Keyless and CRU units
- Sunroof connector
- Servo motor connector
- Heater blower transistor connector
- Blower motor and in-vehicle temperature sensor connectors
- Intake unit bolts and nuts
- Drain hose from the intake unit
- Intake unit
- Heater unit case screws
- Heater core from the case

To install:

7. Install or connect the following:
- Heater core into the heater case. Secure it in place with the retaining screws.
- Drain hose to the intake unit
- Intake unit to its mounting position. Torque the mounting bolts to 48–84 inch lbs. (5–9 Nm).
- Blower motor and in-vehicle temperature sensor connectors
- Heater blower transistor connector
- Servo motor connector
- Sunroof connector
- Keyless and CRU units

8. Install the instrument panel as follows:

a. Install the instrument panel, tighten the bolts and connect the harness connectors

b. Connect the control cable to the bottom of the heater unit. Do not move the switch and link during installation.

c. Install the front pillar lower trim to the passenger side.

d. Install the front trim pillar on both sides.

e. Install the console box, box cover and tray.

f. Install the shift knob.

g. Passenger side airbag module and tighten the 3 bolts securing the module.

h. Connect passenger air bag module to the support beam.

i. Install the side panels on both sides.

j. Install the glove box and stopper.

k. Install the steering column.

l. Install the 2 bolts that secure the column to the instrument panel. Tighten the bolts to 18 ft. lbs. (25 Nm).

m. Connect all electrical connectors to the column.

n. Install the lower steering column cover with the tilt lever held in the lowered position.

o. Install the universal joint. Align the bolt hole on the long yoke side of the joint with the cutout at the serrated section of the shaft end and insert the joint. Align the bolt hole on the short yoke side of the joint at the serrated section of the gearbox assembly and lower the joint completely. Temporarily tighten the bolt on the short yoke side and raise the joint to ensure the bolt is passing properly through the cutout. Tighten the bolt on the long yoke side, then the short yoke side to 17 ft. lbs. (24 Nm).

⁂ CAUTION

Make sure the joint bolt is tightened through the notch in the shaft serration. Do not over tighten the joint bolts as this can lead to heavy steering wheel operation. Make sure the clearance between the gearbox is over 0.59 inch (15mm).

p. Making sure the wheels are still in the straight ahead position, turn the roll connector (A) clockwise until it stops. Then turn the roll connector pin (A) counterclockwise approximately 2.65 turns until the marks are aligned. Refer to the accompanying illustration for more detail.

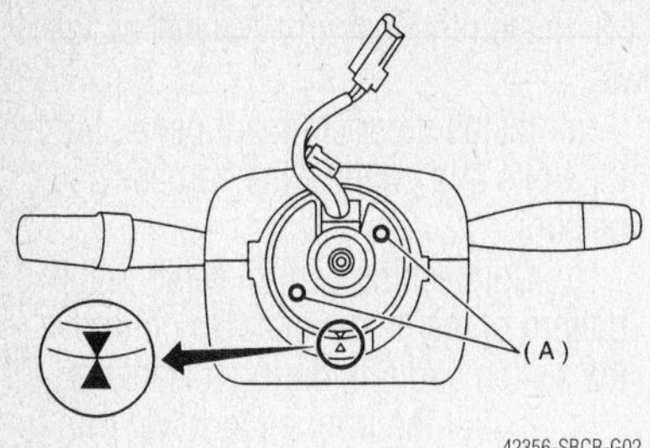

Make sure the roll pin is properly aligned—Outback

q. Steering wheel, align the matchmarks made during removal and install the nut. Tighten the nut to 32 ft. lbs. (45 Nm). make sure the column cover-to-steering wheel clearance is 0.08–0.16 inch (2–4mm).

r. Drivers side airbag, engage the connector and tighten the fasteners.

⁂ CAUTION

Insert the roll connector guide pin into the guide hole on the lower end of the surface of the steering wheel to prevent any damage. Draw out the airbag, horn and cruise control connectors from the guide hole of the steering wheel lower end.

s. Connect any remaining connections and install the lower column cover.

t. Install the lower cover.

9. Install or connect the following:
- Bolts attaching the expansion valve and pipe in the engine compartment
- Heater hoses to the heater core after removing the plugs.
- LLC
- Negative battery cable

10. Refill the cooling system.

11. Connect the negative battery cable.

12. Evacuate, charge and leak test the air conditioning system.

13. Operate the engine to normal operating temperatures; then, check the climate control operation and check for leaks.

Cylinder Head

➡On some models, engine compartment room is limited, so it may be necessary to remove the engine to service the cylinder heads.

REMOVAL & INSTALLATION

2.0L Engines

1. Before servicing the vehicle, refer to the precautions in the beginning of this section.
2. Remove or disconnect the following:
- Negative battery cable
- Drive belt
- Crankshaft pulley
- Belt cover
- Timing belt assembly
- Camshaft sprocket
- Intake manifold
- Bolt that attaches the A/C compressor bracket to the head
- Camshaft
- Cylinder head bolts in the proper sequence. Leave bolts A and D installed loosely to prevent the cylinder head from falling.
- Cylinder head from the block, use a plastic-faced hammer, if needed, to separate the head from the cylinder block
- Bolts A and D
- Cylinder head and gasket

3. Clean all gasket material from both mating surfaces.

To install:

4. Inspect the cylinder head for warpage. Warpage should not exceed 0.0020 in. (0.05mm).
5. Install a new head gasket and the cylinder head.
6. Secure the head in place with the mounting bolts. Coat each bolts with clean engine oil, and hand-tighten. Tighten the cylinder head bolts, in sequence, to the following specifications:

a. Step 1: 22 ft. lbs. (29 Nm).

b. Step 2: 51 ft. lbs. (69 Nm).

c. Step 3: loosen all bolts by 180 degrees, then loosen an additional 180 degrees.

d. Step 4: bolts 1 and 2 to 25 ft. lbs. (34 Nm).

e. Step 5: bolts 3, 4, 5 and 6 to 11 ft. lbs. (15 Nm).

f. Step 6: all bolts plus 80–90 degrees.

g. Step 7: all bolts plus 80–90 degrees.

➡Do not exceed 180 degrees total tightening.

7. Install or connect the following:
- Camshaft

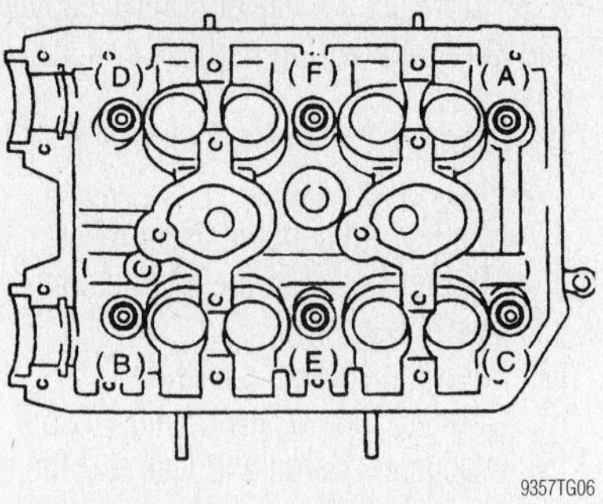

Cylinder head bolt loosening sequence (except bolts A and D which are left in place at this time)—2.0L engines

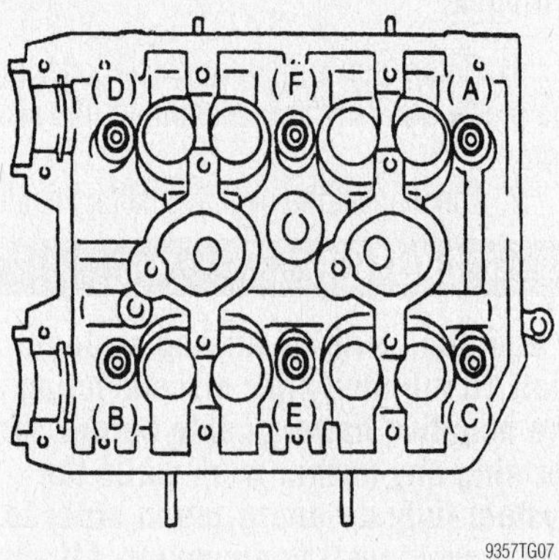

Tap on the block with a rubber mallet prior to removing cylinder head bolts A and D—2.0L engines

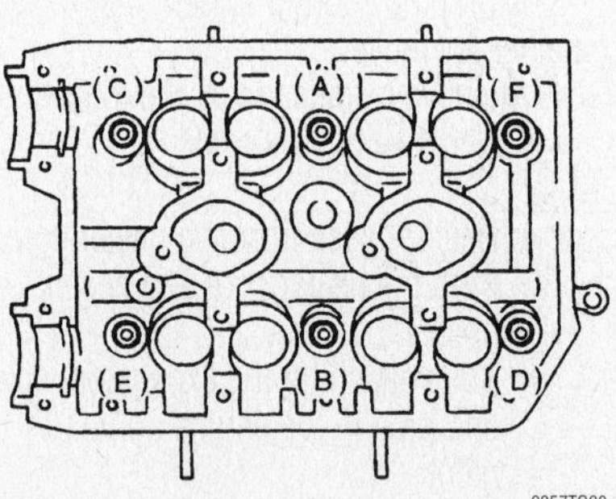

Cylinder head bolt tightening sequence—2.0L engines

- Bolt that attaches the A/C compressor bracket to the head
- Camshaft sprocket
- Intake manifold
- Timing belt assembly
- Belt cover
- Crankshaft pulley
- Drive belt
- Negative battery cable

8. Start the engine and allow it to reach operating temperature. Check for leaks.

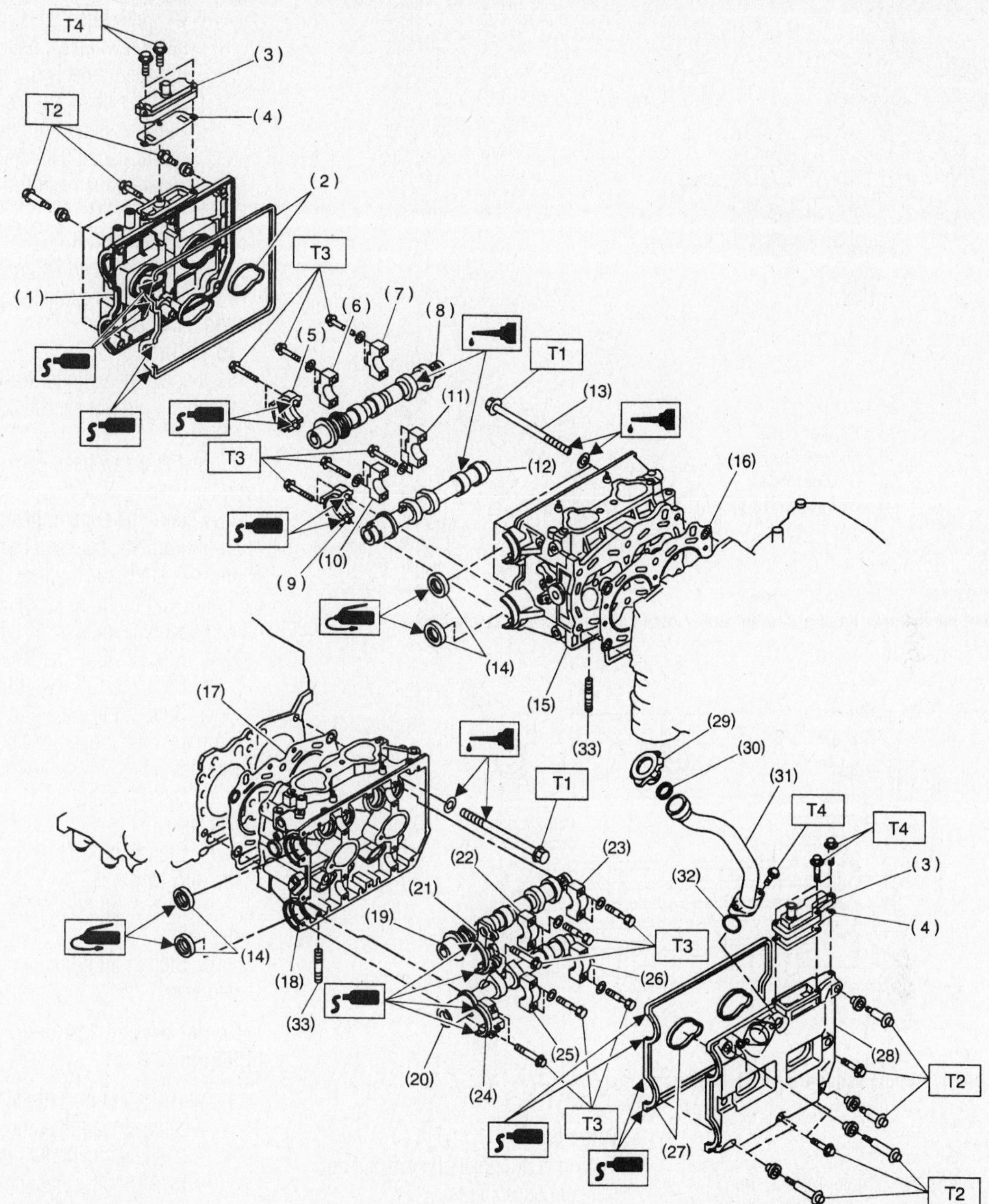

(1) Rocker cover (RH)
(2) Rocker cover gasket (RH)
(3) Oil separator cover
(4) Gasket
(5) Intake camshaft cap (Front RH)
(6) Intake camshaft cap (Center RH)
(7) Intake camshaft cap (Rear RH)
(8) Intake camshaft (RH)
(9) Exhaust camshaft cap (Front RH)
(10) Exhaust camshaft cap (Center RH)
(11) Exhaust camshaft cap (Rear RH)
(12) Exhaust camshaft (RH)
(13) Cylinder head bolt
(14) Oil seal
(15) Cylinder head (RH)
(16) Cylinder head gasket (RH)
(17) Cylinder head gasket (LH)
(18) Cylinder head (LH)
(19) Intake camshaft (LH)
(20) Exhaust camshaft (LH)
(21) Intake camshaft cap (Front LH)
(22) Intake camshaft cap (Center LH)
(23) Intake camshaft cap (Rear LH)
(24) Exhaust camshaft (Front LH)
(25) Exhaust camshaft cap (Center LH)
(26) Exhaust camshaft cap (Rear LH)
(27) Rocker cover gasket (LH)
(28) Rocker cover (LH)
(29) Oil filler cap
(30) Gasket
(31) Oil filler duct
(32) O-ring
(33) Stud bolt

Tightening torque: N·m (kgf-m, ft-lb)

T1: ***<Ref. to ME(DOHC TURBO)-64, INSTALLATION, Cylinder Head Assembly.>***

T2: ***5 (0.5, 3.6)***

T3: ***10 (1.0, 7)***

T4: ***6.4 (0.65, 4.7)***

9357TG05

Exploded view of the cylinder head assembly and related components—2.0L engines

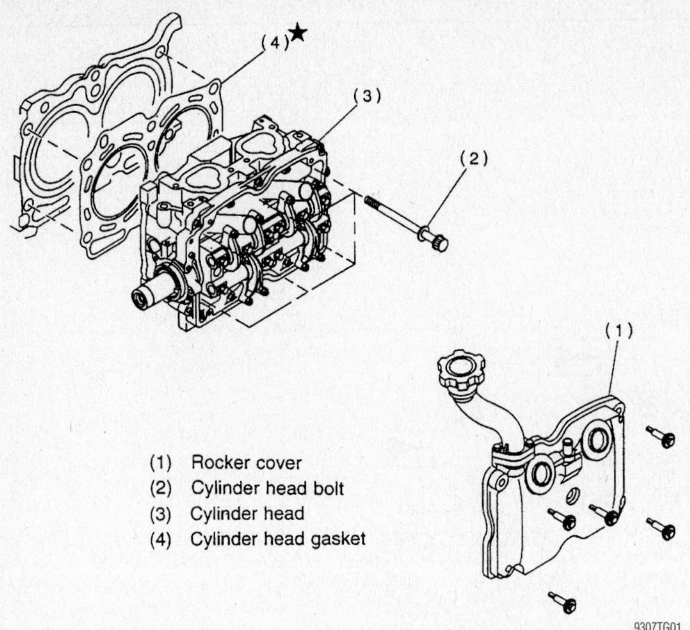

Exploded view of the cylinder head assembly and related components—2.2L engines

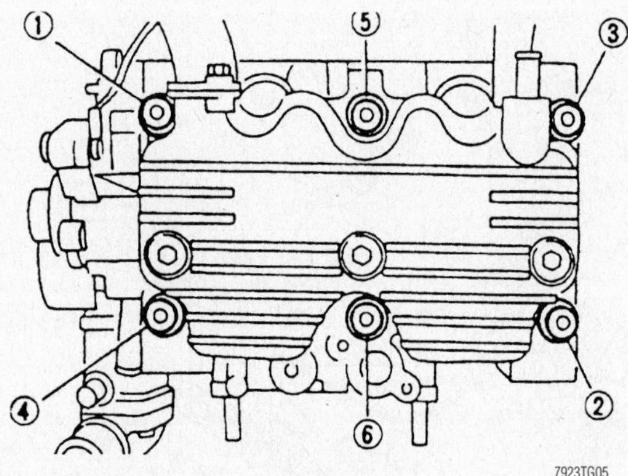

Cylinder head bolt loosening sequence—2.2L engines

2.2L Engines

1. Before servicing the vehicle, refer to the precautions in the beginning of this section.
2. Remove or disconnect the following:
   - Negative battery cable
   - Drive belt(s)
   - Power steering pump
   - Alternator and bracket
   - Valve rocker cover
   - Positive Crankcase Ventilation (PCV) hose and spark plug wires
   - Connector bracket attaching bolt
   - Crankshaft Position (CKP) and Camshaft Position (CMP) ensors
   - Oil pressure switch
   - Knock sensor
   - Blow-by hose
3. Relieve the fuel system pressure and disconnect the fuel pipes.
   - Intake manifold and gasket
   - Water pipe
   - Timing belt, camshaft sprocket and related components
   - Oil level gauge guide attaching bolt on the left cylinder head
   - Cylinder head bolts in the proper sequence. Leave bolts 1 and 3 installed loosely to prevent the cylinder head from falling.
   - Cylinder head from the block, use a plastic-faced hammer, if needed, to separate the head from the cylinder block
   - Bolts 1 and 3
   - Cylinder head and gasket
4. Clean all gasket material from both mating surfaces.

To install:

5. Inspect the cylinder head for warpage. Warpage should not exceed 0.0020 in. (0.05mm).
6. Install a new head gasket and the cylinder head.
7. Secure the head in place with the mounting bolts. Coat each bolts with clean engine oil, and hand-tighten. Tighten the cylinder head bolts, in sequence, to the following specifications:
   a. Step 1: 22 ft. lbs. (29 Nm).
   b. Step 2: 51 ft. lbs. (69 Nm).
   c. Step 3: loosen all bolts by 180 degrees, then loosen an additional 180 degrees.
   d. Step 4: bolts 1 and 2 to 25 ft. lbs. (34 Nm).
   e. Step 5: bolts 3, 4, 5 and 6 to 11 ft. lbs. (15 Nm).
   f. Step 6: all bolts plus 80–90 degrees.
   g. Step 7: all bolts plus 80–90 degrees.

➡Do not exceed 180 degrees total tightening.

8. Install or connect the following:
   - Oil level gauge guide attaching bolt on the left cylinder head
   - Timing belt, camshaft sprocket and related components
   - Water pipe

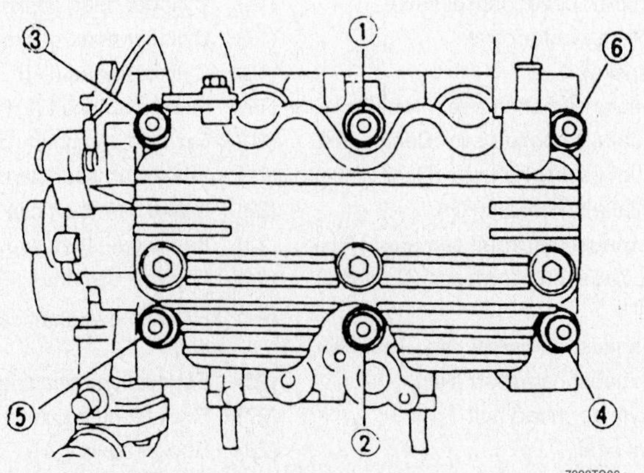

Cylinder head bolt tightening sequence—2.2L engines

- Intake manifold and tighten the bolts to 21–25 ft. lbs. (28–34 Nm)
- Fuel delivery pipes
- Blow-by hose
- Knock sensor
- Oil pressure switch connector
- CKP and CMP sensors
- Connector bracket attaching bolt
- Spark plug wires
- PCV hose
- Valve rocker cover and tighten the bolts to 44 inch lbs. (5 Nm)
- Alternator
- Power steering pump
- Accessory drive belt
- Negative battery cable

9. Start the engine and allow it to reach operating temperature. Check for leaks.

2.5L Engine

1. Before servicing the vehicle, refer to the precautions in the beginning of this section.

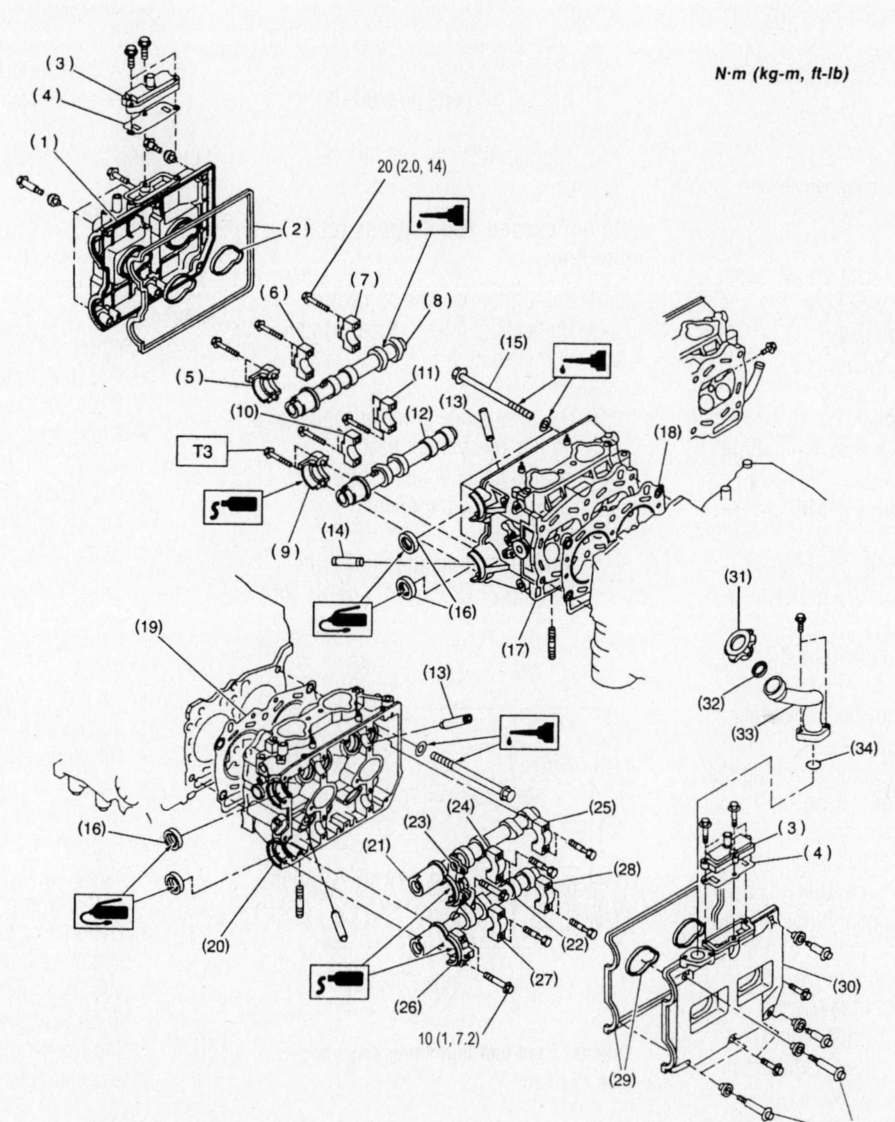

(1) Rocker cover (RH)
(2) Rocker cover gasket (RH)
(3) Oil separator cover
(4) Gasket
(5) Intake camshaft cap (Front RH)
(6) Intake camshaft cap (Center RH)
(7) Intake camshaft cap (Rear RH)
(8) Intake camshaft (RH)
(9) Exhaust camshaft cap (Front RH)
(10) Exhaust camshaft cap (Center RH)
(11) Exhaust camshaft cap (Rear RH)
(12) Exhaust camshaft (RH)
(13) Intake valve guide
(14) Exhaust valve guide
(15) Cylinder head bolt
(16) Oil seal
(17) Cylinder head (RH)
(18) Cylinder head gasket (RH)
(19) Cylinder head gasket (LH)
(20) Cylinder head (LH)
(21) Intake camshaft (LH)
(22) Exhaust camshaft (LH)
(23) Intake camshaft cap (Front LH)
(24) Intake camshaft cap (Center LH)
(25) Intake camshaft cap (Rear LH)
(26) Exhaust camshaft (Front LH)
(27) Exhaust camshaft cap (Center LH)
(28) Exhaust camshaft cap (Rear LH)
(29) Rocker cover gasket (LH)
(30) Rocker cover (LH)
(31) Oil filler cap
(32) Gasket
(33) Oil filler duct
(34) O-ring

7923TG07

Exploded view of the cylinder head and related components—2.5L engine

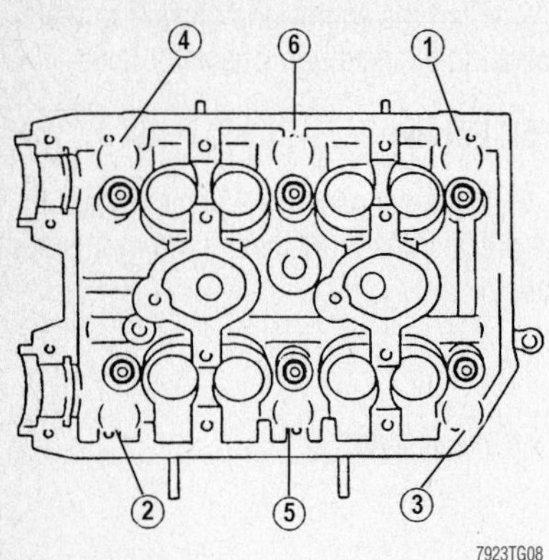

Cylinder head bolt loosening sequence—2.5L engine

2. Remove or disconnect the following:
   - Negative battery cable
   - Accessory drive belts
   - Power steering pump
   - Alternator and bracket
   - Valve rocker cover
   - Connector bracket attaching bolt
   - Crankshaft Position (CKP) and Camshaft Position (CMP) sensors
   - Coolant filler tank
3. Relieve the fuel system pressure and disconnect the fuel pipes.
   - Intake manifold and gasket
   - Water pipe
   - Timing belt, camshaft sprocket and related components
   - Oil level gauge guide attaching bolt on the left cylinder head
   - Valve covers
   - Camshafts, refer to the camshaft procedure in this section
   - Cylinder head bolts, in the proper sequence. Leave bolts 1 and 3 installed loosely to prevent the cylinder head from falling.
   - Cylinder head from the block using a plastic-faced hammer, if needed
   - Bolts 1 and 3
   - Cylinder head and gasket
4. Clean all gasket material from both mating surfaces.

To install:

5. Inspect the cylinder head for warpage. Warpage should not exceed 0.0020 in. (0.05mm).
6. Install a new head gasket and the cylinder head.
7. Secure the head in place with the mounting bolts. Coat each bolts with clean engine oil, and hand-tighten. Tighten the cylinder head bolts to the following specifications:
   a. Step 1: all bolts to 22 ft. lbs. (29 Nm).
   b. Step 2: all bolts to 51 ft. lbs. (69 Nm).
   c. Step 3: loosen all bolts by 180 degrees, then loosen an additional 180 degrees.
   d. Step 4: bolts 1 and 2 to 25 ft. lbs. (24 Nm).
   e. Step 5: bolts 3, 4, 5 and 6 to 11 ft. lbs. (15 Nm).
   f. Step 6: all bolts plus 80–90 degrees.
   g. Step 7: All bolts plus 80–90 degrees.

➡Do not exceed 180 degrees total tightening.

8. Install or connect the following:
   - Camshafts, refer to the procedure in this section
   - Valve covers
   - Oil level gauge guide attaching bolt on the left cylinder head
   - Timing belt, camshaft sprockets and related components
   - Water pipe
   - Intake manifold and tighten the bolts to 21–25 ft. lbs. (28–34 Nm)

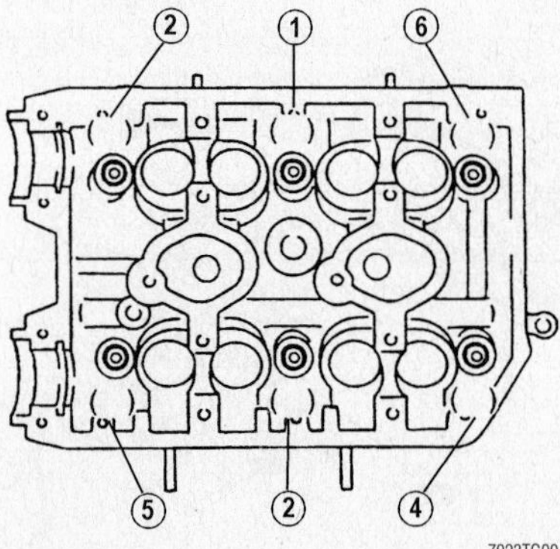

Cylinder head bolt tightening sequence—2.5L engines

   - Fuel delivery pipes
   - Blow-by hose
   - Knock sensor
   - CKP and CMP sensors
   - Connector bracket attaching bolt
   - Spark plug wires
   - Valve rocker cover and tighten the bolts to 48 inch lbs. (9 Nm)
   - Alternator
   - Power steering pump
   - Accessory drive belt
   - Negative battery cable
9. Start the engine and allow it to reach operating temperature. Check for leaks.

3.0L Engine

1. Before servicing the vehicle, refer to the precautions in the beginning of this section.
2. Remove or disconnect the following:
   - Negative battery cable
   - Crankshaft pulley cover
   - Crankshaft pulley bolt using tool 499977100 to hold the crankshaft
   - Crankshaft pulley
   - Timing chain cover and the chain. Refer to the procedure in this section.
   - Camshaft and crankshaft sprockets. Refer to the procedure in this section.
   - Oil pump. Refer to the procedure in this section.
   - Oil pump relief valve
   - Water pump. Refer to the procedure in this section.
   - Rear chain cover
   - Camshafts. Refer to the procedure in this section.
   - Cylinder head bolts in the sequence illustrated. Leave bolts 2 and 4 connected by a few threads to prevent the head from falling. Tap the head with a plastic mallet to separate it from the block.

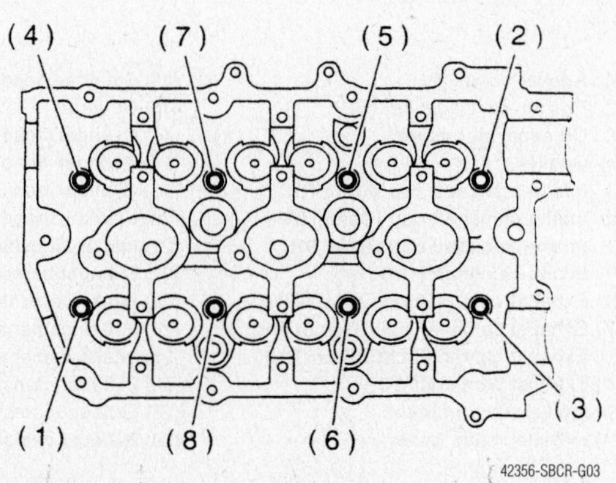

Cylinder head bolt loosening sequence—3.0L engines

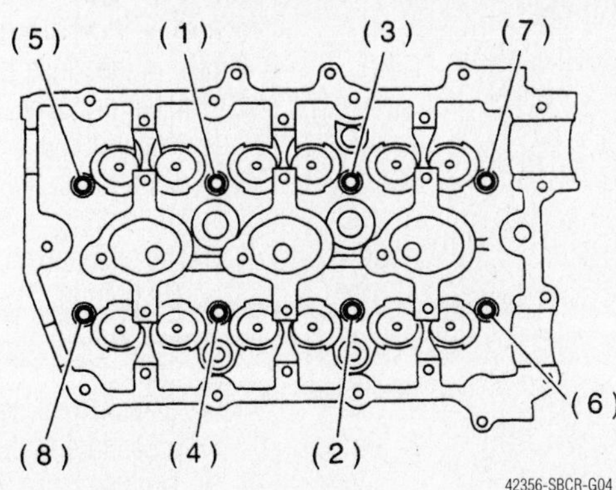

Cylinder head bolt tightening sequence—3.0L engines

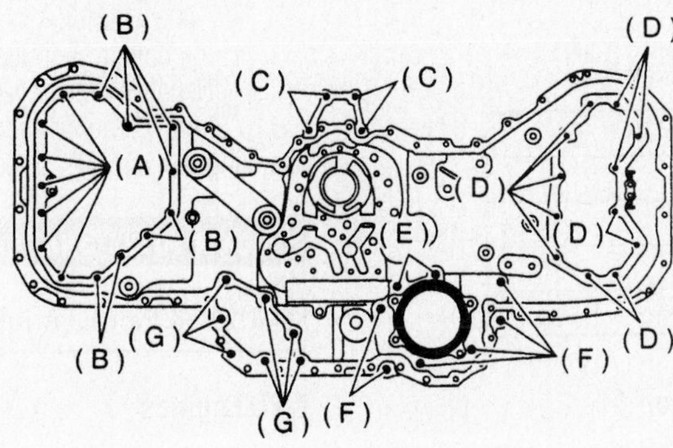

Rear timing chain bolt sizes and locations—3.0L engine

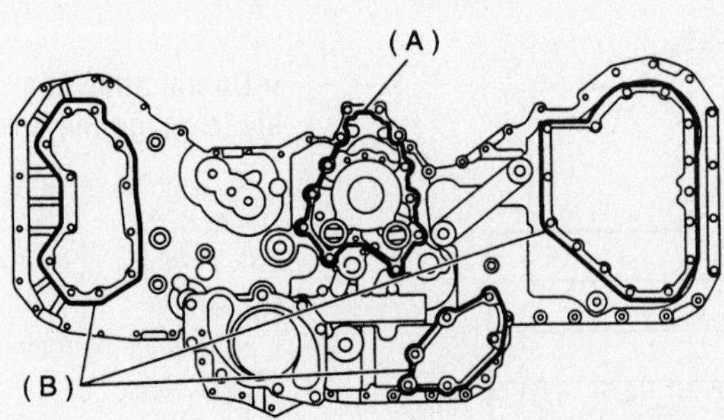

Apply liquid gasket maker to the mating surfaces of the cover as shown—3.0L engine

- Bolts 2 and 4, the cylinder head and gasket

3. Clean all gasket material from both mating surfaces.

To install:

4. Inspect the cylinder head for warpage. Warpage should not exceed 0.0020 in. (0.05mm).
5. Make sure not to scratch or damage the mating surfaces of the cylinder head, block and oil pump.
6. Install a new head gasket and the cylinder head.
7. Secure the head in place with the mounting bolts. Coat each bolts with clean engine oil, and hand-tighten. Tighten the cylinder head bolts to the following specifications:
   a. Step 1: all bolts to 14 ft. lbs. (20 Nm).
   b. Step 2: all bolts to 37 ft. lbs. (50 Nm).
   c. Step 3: loosen all bolts by 180 degrees, then loosen an additional 180 degrees.
   d. Step 4: all bolts to 18 ft. lbs. (25 Nm).
   e. Step 5: all bolts to 18 ft. lbs. (25 Nm).
   f. Step 6: all bolts plus 90 degrees.
   g. Step 7: bolts 1, 2, 3 and 4 plus 80–90 degrees.
   h. Step 8: bolts 5, 6, 7 and 8 plus 45 degrees.

➡Do not exceed 180 degrees total tightening.

8. Install or connect the following:
   - Camshafts. Refer to the procedure in this section.
9. Install the rear chain cover as follows:

➡There are several size bolts used, refer to the illustration for size and locations.

   a. Rear chain cover gasket and clean the mating surfaces.
   b. Apply liquid gasket maker to the mating surfaces of the cover. Refer to the illustration for gasket maker application and diameter.
   c. Install new O-rings. Refer to the illustration for O-ring location and size
   d. Install the rear chain cover and temporarily tighten the bolts, refer to the illustration for size and locations.
   e. Replace mounting bolts **G** with new bolts, refer to the illustration for location.
   f. Tighten the cover bolts in the

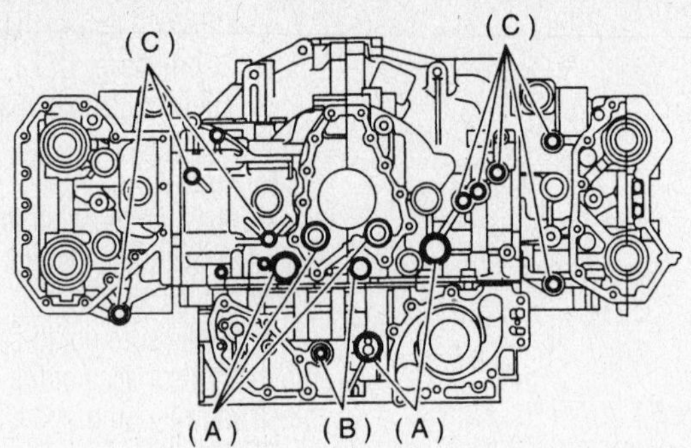

(A) O-ring (Large)
(B) O-ring (Medium)
(C) O-ring (Small)

42356-SBCR-G07

Rear cover O-ring sizes and locations—3.0L engine

(1) to (11)	9 N·m (0.9 kgf-m, 6.5 ft-lb)
(12) to (19)	20 N·m (2.0 kgf-m, 14 ft-lb)
(20) to (31)	9 N·m (0.9 kgf-m, 6.5 ft-lb)
(32) to (39)	12 N·m (1.2 kgf-m, 8.7 ft-lb)
(40) to (46)	9 N·m (0.9 kgf-m, 6.5 ft-lb)

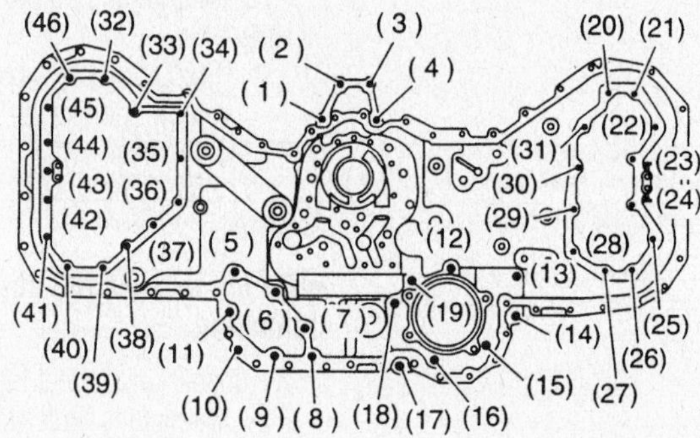

42356-SBCR-G08

Rear cover bolt tightening sequence and torque specifications—3.0L engine

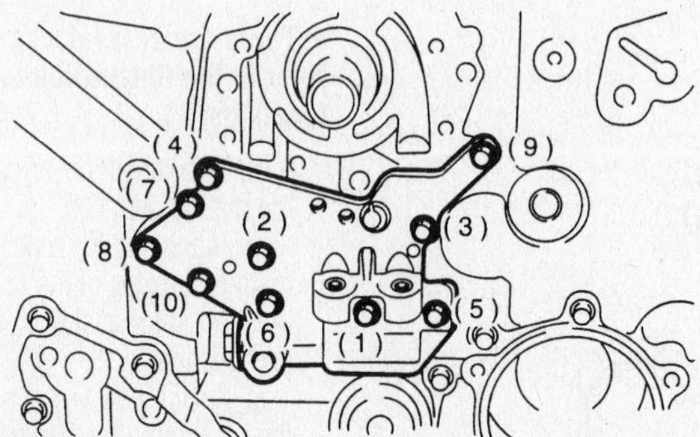

Bolt installation position	Bolt dimension
(1) and (5)	6 x 26
(2), (3), (4) and (9)	6 x 35
(6), (7), (8) and (10)	6 x 16

42356-SBCR-G09

Oil pump relief valve bolt tightening sequence and bolt specifications—3.0L engine

sequence illustrated to the specifications shown in the illustration.

10. Install or connect the following:
    - Water pump
    - Oil pump relief valve and tighten the bolts to 4.7 ft. lbs. (6.4 Nm) in the sequence illustrated.
    - Oil pump
    - Crankshaft sprocket. Refer to the procedure in this section.
    - Camshaft sprocket. Refer to the procedure in this section.
    - Timing chain. Refer to the procedure in this section.
    - Crankshaft pulley, apply clean oil to the bolt threads and tighten the bolt to 131 ft. lbs. (178 Nm)
    - Camshaft pulley cover and tighten the bolts to 5 ft. lbs. (7 Nm)
    - Negative battery cable
11. Start the engine and allow it to reach operating temperature. Check for leaks.

Rocker Arms/Shafts

REMOVAL & INSTALLATION

2.2L Engines

1. Before servicing the vehicle, refer to the precautions in the beginning of this section.
2. Remove or disconnect the following:
    - Positive Crankcase Ventilation (PCV) hose
    - Rocker cover
    - Valve rocker assembly by removing bolts 2 through 4 in numerical sequence
3. Loosen bolt 1, but leave it engaged to retain the valve rocker assembly.
4. Remove or disconnect the following:
    - Bolts 5 through 8, taking care not to gouge the dowel pin
    - Valve rocker assembly

To install:

5. Install the valve rocker assembly on the cylinder head.
6. Temporarily tighten bolts 1 through 4 equally.

➡Do not allow the valve rocker assembly to gouge the dowel pins.

7. Tighten bolts 5–8 to 108 inch lbs. (12 Nm).
8. Tighten bolts 1–4 to 108 inch lbs. (12 Nm).
9. Install the rocker cover and connect the PCV hose.

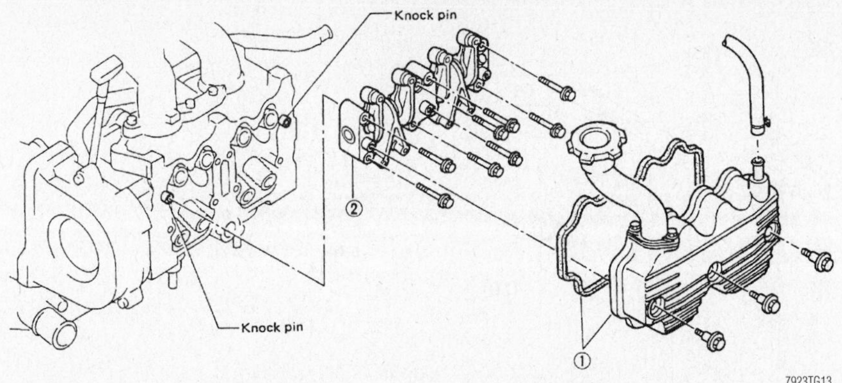

Exploded view of the cylinder head and rocker assembly—2.2L engines

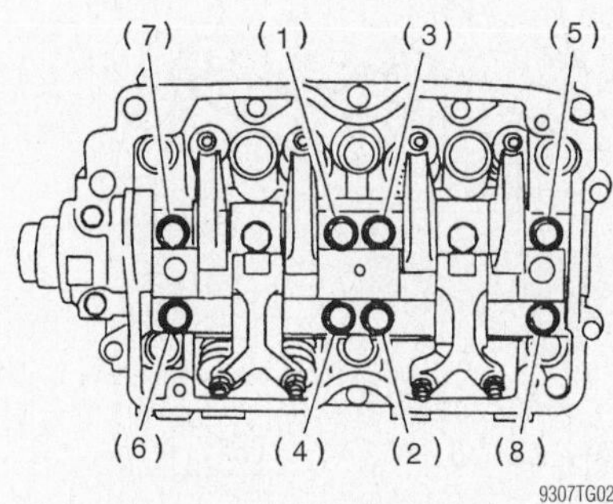

Rocker shaft bolt loosening/tightening sequence—2000–01 2.2L engines

Except 2.2L Engines

These engines are not equipped with either rocker shafts or rocker arms. Instead, the camshafts act directly on the individual valves.

Intake Manifold

REMOVAL & INSTALLATION

2.0L Engines

1. Before servicing the vehicle, refer to the precautions in the beginning of this section.
2. Release the fuel system pressure.
3. Remove or disconnect the following:
   - Negative battery cable
   - Engine cover, if necessary
   - Air cleaner upper cover and boot
   - Air cleaner filter
   - Intercooler
   - Accelerator cable
   - Coolant filler tank
4. Remove the power steering pump from the bracket by performing the following steps:
   a. Loosen the lock and slider bolts.
   b. Remove the V-belt.
   c. Disconnect the power steering switch connection.
   d. Remove the bolts that attach the power steering pump pipe brackets to the intake manifold. Do not disconnect the hose.
   e. Bolts that attach the power steering pump bracket.
   f. Remove the power steering tank from the bracket by pulling it upwards.
   g. Place the power steering pump on the wheel apron on the right.
5. Remove or disconnect the following:
   - Emission hose from the Positive Crankcase Ventilation (PCV) valve
   - Engine coolant temperature hoses from the throttle body
   - Brake booster hose
   - Pressure hose from the intake duct
   - Engine harness connectors from the bulkhead connections
6. Disconnect the following electrical connections:
   - Engine Coolant Temperature (ECT) sensor
   - Oil pressure switch
   - Crankshaft Position (CKP) sensor
   - Knock Sensor (KS)
   - Camshaft Position (CMP) sensor
   - Ignition coil
7. Remove or disconnect the following:
   - Engine harness fixed clip from the bracket
   - Fuel delivery, return and evaporative hoses
   - Intake manifold bolts
   - Intake manifold and gasket

To install:

8. Install or connect the following:
   - Intake manifold and gasket
   - Intake manifold bolts and tighten to 18 ft. lbs. (25 Nm)
   - Fuel delivery, return and evaporative hoses
   - Engine harness fixed clip from the bracket

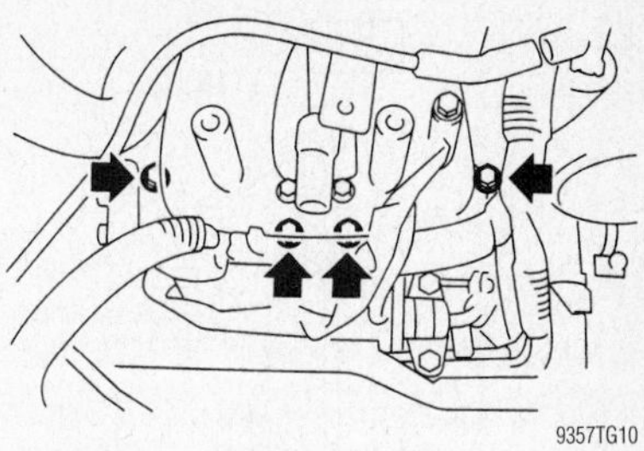

Location of the intake manifold bolts—2.0L engine

9. Connect the following electrical connections:
   - Oil pressure switch
   - CKP sensor
   - ECT sensor
   - KS sensor
   - CMP sensor
   - Ignition coil
   - Engine harness connectors to the bulkhead connections
10. Install or connect the following:
   - Brake booster hose
   - Engine coolant temperature hoses to the throttle body
   - Emission hose to the PCV valve
   - Pressure hose to the intake duct
11. Install the power steering pump on the bracket by performing the following steps:
   a. Install the power steering tank on the bracket.
   b. Attach the power steering switch connection.
   c. Install the bolts that attach the power steering pump bracket and tighten to 16 ft. lbs. (22 Nm).
   d. Attach the power steering pump pipe brackets to the intake manifold.
   e. Install the V-belt.
12. Install or connect the following:
   - Coolant filler tank
   - Accelerator cable
   - Intercooler
   - Air cleaner filter
   - Air cleaner upper cover and boot
   - Engine undercover
   - Negative battery cable
13. Fill the cooling system.
14. Start the engine and allow it to reach operating temperature. Check for leaks and test drive the vehicle.

2.2L Engines

1. Before servicing the vehicle, refer to the precautions in the beginning of this section.
2. Release the fuel system pressure.
3. Remove or disconnect the following:

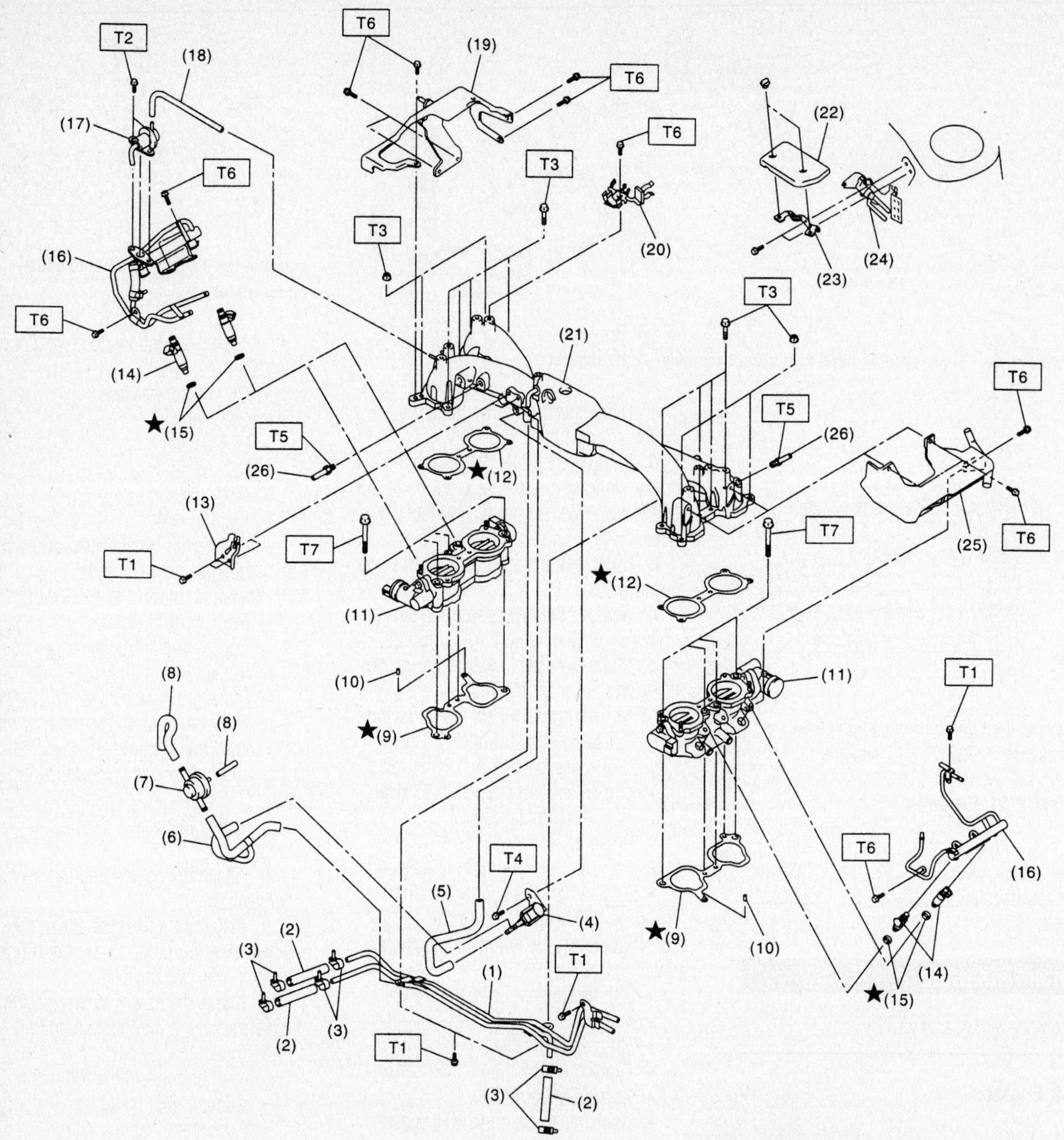

(1) Fuel pipe ASSY
(2) Fuel hose
(3) Clip
(4) Purge control solenoid valve
(5) Vacuum hose
(6) Vacuum control hose
(7) Purge valve
(8) Purge hose
(9) Intake manifold gasket
(10) Guide pin
(11) Tumble generator valve ASSY
(12) Tumble generator valve gasket
(13) Accelerator cable bracket
(14) Fuel injector
(15) Insulator
(16) Fuel injector pipe
(17) Pressure regulator
(18) Pressure regulator hose
(19) Fuel pipe protector RH
(20) Blow-by hose stay
(21) Intake manifold
(22) Solenoid valve cover
(23) Solenoid valve cover stay
(24) Wastegate control solenoid valve ASSY
(25) Fuel pipe protector LH
(26) Nipple

Tightening torque: N·m (kgf-m, ft-lb)

T1: 4.9 (0.5, 3.6)
T2: 6.4 (0.65, 4.7)
T3: 8.25 (0.84, 6.1)
T4: 16 (1.6, 11.8)
T5: 17 (1.73, 12.5)
T6: 19 (1.94, 13.7)
T7: 25 (2.5, 18.1)

9357TG09

Exploded view of the intake manifold assembly—2.0L engine

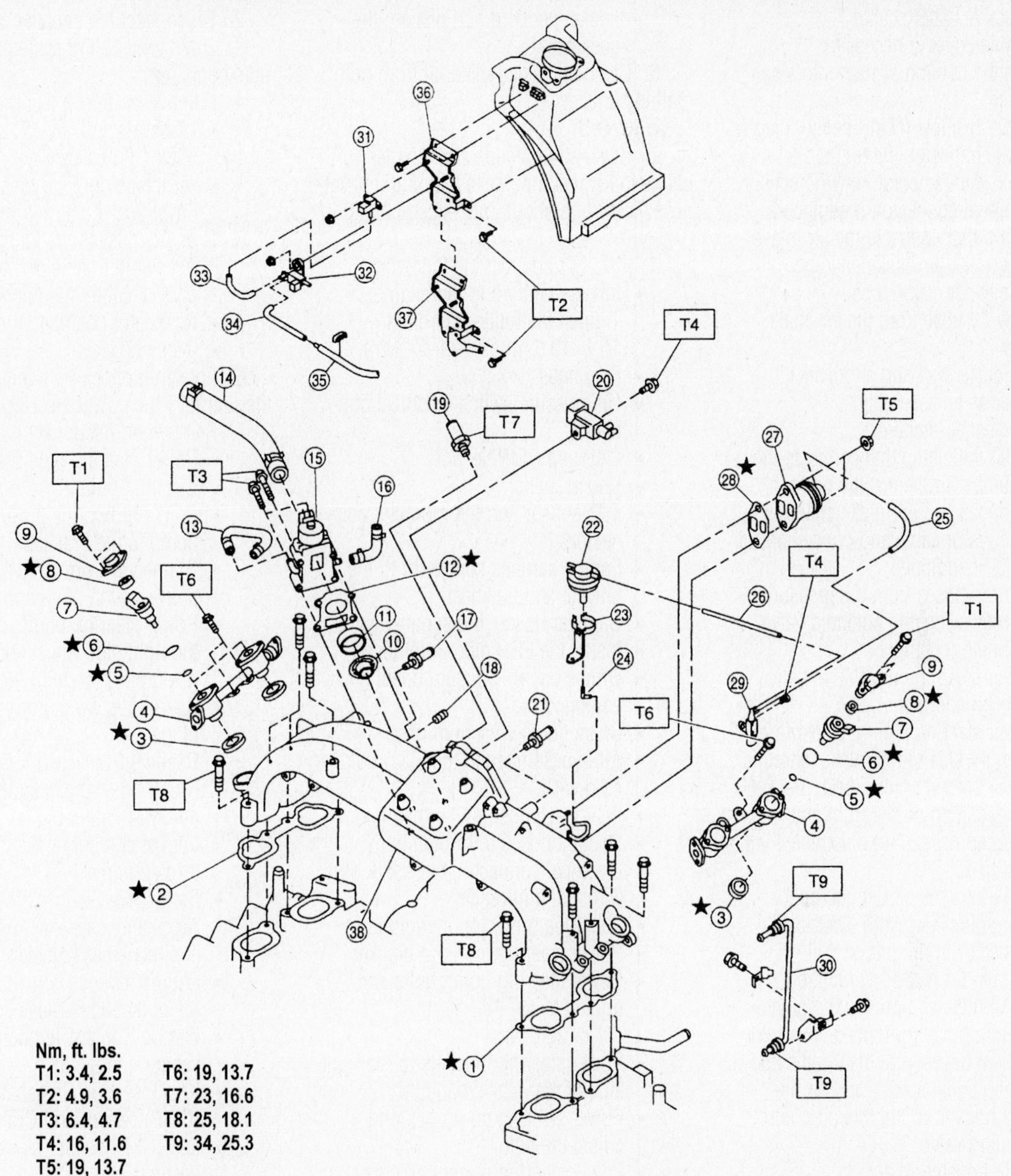

1. Intake manifold gasket LH
2. Intake manifold gasket RH
3. Fuel injector pipe insulator
4. Fuel injector pipe
5. O-ring A
6. O-ring B
7. Fuel injector
8. Insulator
9. Fuel injector cap
10. Plate
11. Sealing
12. Gasket
13. Engine coolant hose B
14. Air by-pass hose
15. Idle air control solenoid valve
16. Engine coolant hose A
17. Nipple (AT vehicles)
18. Plug
19. PCV valve
20. Purge control solenoid valve
21. Nipple
22. BPT (AT vehicles)
23. BPT holder bracket (AT vehicles)
24. Back pressure hose (AT vehicles)
25. EGR vacuum hose A (AT vehicles)
26. EGR vacuum hose B (AT vehicles)
27. EGR valve (AT vehicles)
28. Gasket (AT vehicles)
29. EGR solenoid valve (AT vehicles)
30. EGR pipe (AT vehicles)
31. Pressure sensor
32. Pressure sources switching solenoid valve
33. Vacuum hose A
34. Vacuum hose B
35. Vacuum hose C
36. Bracket (Except Canada spec. vehicles)
37. Bracket (For Canada spec. vehicles)
38. Intake manifold

7923TG16

Exploded view of the intake manifold assembly—2.2L engine

- Negative battery cable
- Engine cover, if necessary

4. Drain the cooling system into a suitable container.

- Mass Air Flow (MAF) sensor electrical connector, if equipped
- Intake Air Temperature (IAT) sensor electrical connector, if equipped
- Clamp that connects the air intake duct to the air intake chamber
- Air cleaner cover clips
- Blow-by hose from the air intake duct
- Air intake duct and air cleaner cover as an assembly
- Air cleaner element

5. Loosen the clamp that connects the air intake chamber to the throttle body

- Air hoses and air intake chamber
- Accelerator cable and cruise control cable, if equipped
- Vacuum hoses from the pressure sources switching solenoid valve
- Resonator chamber
- Power steering pump drive belt cover(s) and belt
- Power steering pipe brackets from the right side of the intake manifold
- Power steering pump from bracket, then position on the right side wheel apron; do not disconnect the fluid lines
- Fuel pipe protector, if equipped
- Spark plug wires and electrical connector from ignition coil
- Positive Crankcase Ventilation (PCV) hose and pressure regulator vacuum hose from intake manifold
- Coolant hoses from the throttle body
- Engine coolant hose and air bypass hose from the idle air control solenoid valve
- Brake booster hose
- Cruise control vacuum hose, if equipped
- Exhaust Gas Recirculation (EGR) pipe, on 2.2L engines with automatic transaxle
- Canister hoses from the pipes
- Engine harness connectors from bulkhead harness connectors, then remove from bracket
- Engine Coolant Temperature (ECT) sensor connector and thermometer connector, if equipped
- Knock sensor connector
- Crankshaft Position (CKP) and Camshaft Position (CMP) sensor connectors
- Oil pressure switch connector
- Fuel supply lines
- Intake manifold bolts
- Intake manifold and discard the gaskets

6. Clean all gasket material from both mating surfaces.

To install:

7. Use a straightedge and a feeler gauge to inspect the intake manifold for flatness. Distortion should not exceed 0.020 in. (0.5mm).
8. Install or connect the following:

- Intake manifold to the engine. Tighten the retaining bolts to 16.7–19.5 ft. lbs. (23–27 Nm).
- Fuel lines
- Oil pressure switch electrical connector
- CKP and CMP sensors
- Knock sensor
- ECT sensor and thermometer connectors
- Engine harness connector to the bracket and bulkhead
- Canister hoses, if disconnected
- Install the EGR pipe, if removed
- Cruise control vacuum hose, if equipped
- Brake booster vacuum hose
- Air bypass hose to the FICD solenoid valve
- PCV valve hose to the manifold
- Coolant hoses to throttle body
- Electrical connector and spark plug wires to ignition coil
- Fuel pipe protector, if equipped
- Power steering pump to bracket
- Power steering pump bolts and tighten to 13–16.6 ft. lbs. (17.6–22.6 Nm)
- Power steering brackets to the right side of the intake manifold
- Power steering pump belt and adjust the belt
- Power steering pump belt cover
- Vacuum hoses to pressure sources switching solenoid valve
- Accelerator cable and cruise control cable, if equipped
- Air intake chamber and hoses
- Air cleaner element, and the upper cover and duct as a unit
- MAF sensor electrical connector
- Engine cover, if equipped
- Negative battery cable

9. Fill the cooling system.
10. Start the engine and allow it to reach operating temperature. Check for leaks and test drive the vehicle.

2.5L Engine

1. Before servicing the vehicle, refer to the precautions in the beginning of this section.
2. Disconnect the negative battery cable.
3. Drain the cooling system into a suitable container.
4. Remove or disconnect the following:

- Mass Air Flow (MAF) sensor connector, if necessary
- Air intake duct, air cleaner upper cover and the air cleaner element

5. Properly release the fuel pressure.

- Accelerator cable and the cruise control cable, if equipped
- Resonator chamber, if equipped
- V-belt cover(s)

6. Loosen the lock bolt and slider bolt, then remove the power steering belt
7. Remove or disconnect the following:

- Power steering pipe bracket-to-manifold bolts
- Bolts holding the power steering pump to the bracket
- Connector from the power steering pump switch, if equipped
- Power steering pump, and place on the right side wheel apron. Do NOT disconnect the fluid lines.
- Spark plug wires from the spark plugs
- Positive Crankcase Ventilation (PCV) hose and vacuum hose from the intake manifold
- Engine coolant hoses from the throttle body
- Brake booster hose
- Air cleaner case stay (right side) and engine harness bracket, if necessary
- Engine harness connectors form the bulkhead harness connectors
- Engine Coolant Temperature (ECT) sensor connector
- Knock sensor, Camshaft Position (CMP) sensor and Crankshaft Position (CKP) sensor electrical connectors
- Oil pressure switch
- Fuel hoses from the fuel pipes
- Intake manifold mounting bolts
- Intake manifold and discard the gaskets

➡The intake manifold sits on pins that protrude from the cylinder heads. Be sure the pins remain in the cylinder heads.

To install:

8. Install or connect the following:

- New gaskets
- Intake manifold to the engine. Tighten the mounting bolts to 16.7–19.5 ft. lbs. (23–27 Nm).
- Fuel hoses to the fuel pipes, be sure to secure the hoses with new clamps

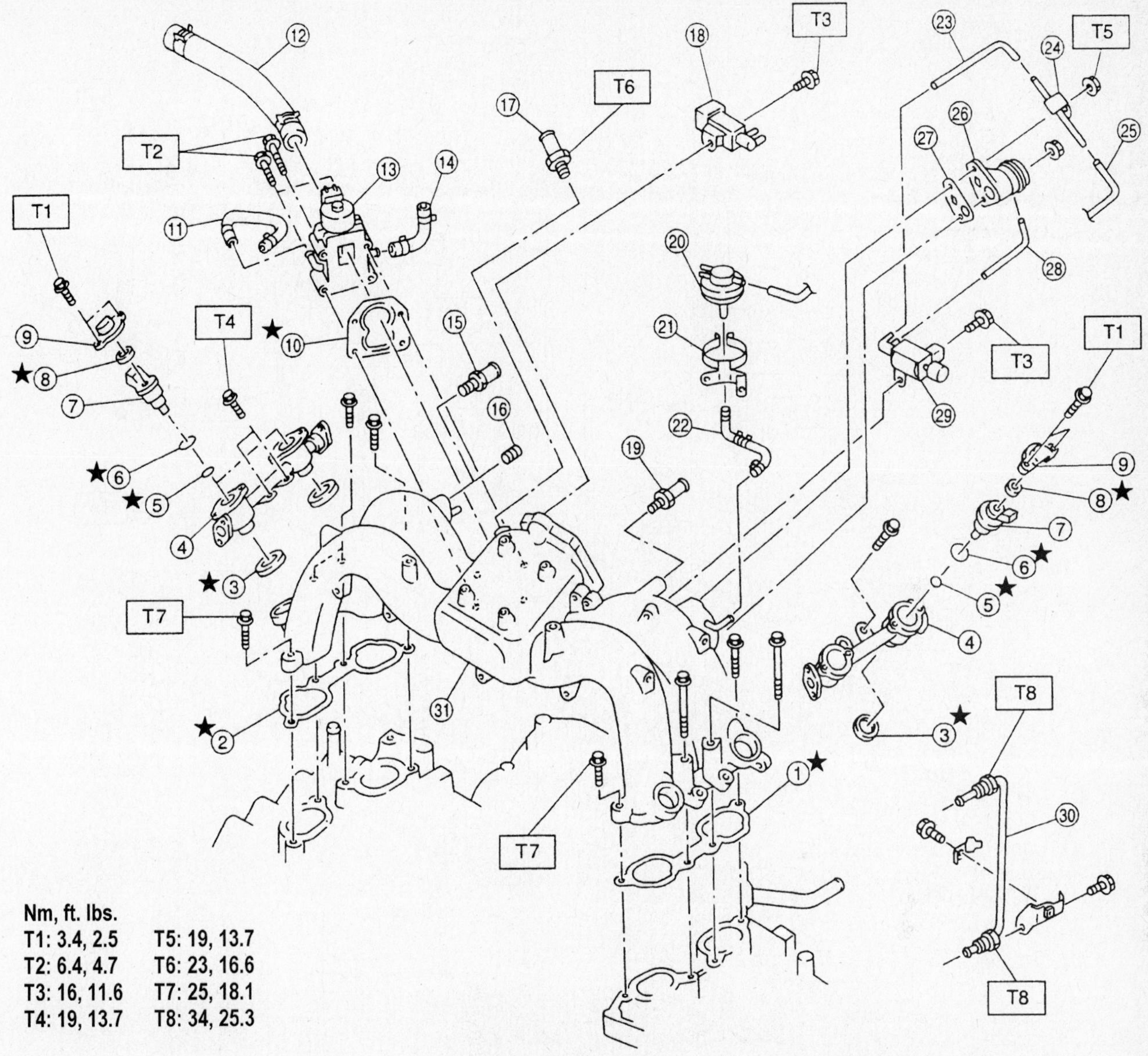

1. Intake manifold gasket RH
2. Intake manifold gasket LH
3. Fuel injector pipe insulator
4. Fuel injector pipe
5. O-ring A
6. O-ring B
7. Fuel injector
8. Insulator
9. Fuel injector cap
10. Gasket
11. Engine coolant hose B
12. Air by-pass hose
13. Idle air control solenoid valve
14. Engine coolant hose A
15. Nipple (AT model)
16. Plug
17. PCV valve
18. Purge control solenoid valve
19. Nipple
20. BPT
21. BPT holder bracket
22. Back pressure hose
23. EGR vacuum hose A
24. EGR vacuum pipe
25. EGR vacuum hose C
26. EGR valve
27. Gasket
28. EGR vacuum hose B
29. EGR solenoid valve
30. EGR pipe
31. Intake manifold

7923TG17

Exploded view of the intake manifold and related components—2.5L engine

- Knock sensor, CMP sensor, CKP sensor, oil pressure switch and ECT sensor wiring
- Engine harness bracket and engine harness connectors to bulkhead connectors
- EGR pipe, if removed
- Brake booster hose
- Engine coolant hose and air bypass hose to idle air control solenoid valve, if removed
- Engine coolant hoses to the throttle body
- PCV hoses to intake manifold
- Spark plug wires to ignition coil or plugs, as applicable
- Power steering pump to bracket. Tighten the bolts to 13–16.6 ft. lbs. (17.6–22.6 Nm).
- Power steering pipe brackets on right side of the intake manifold

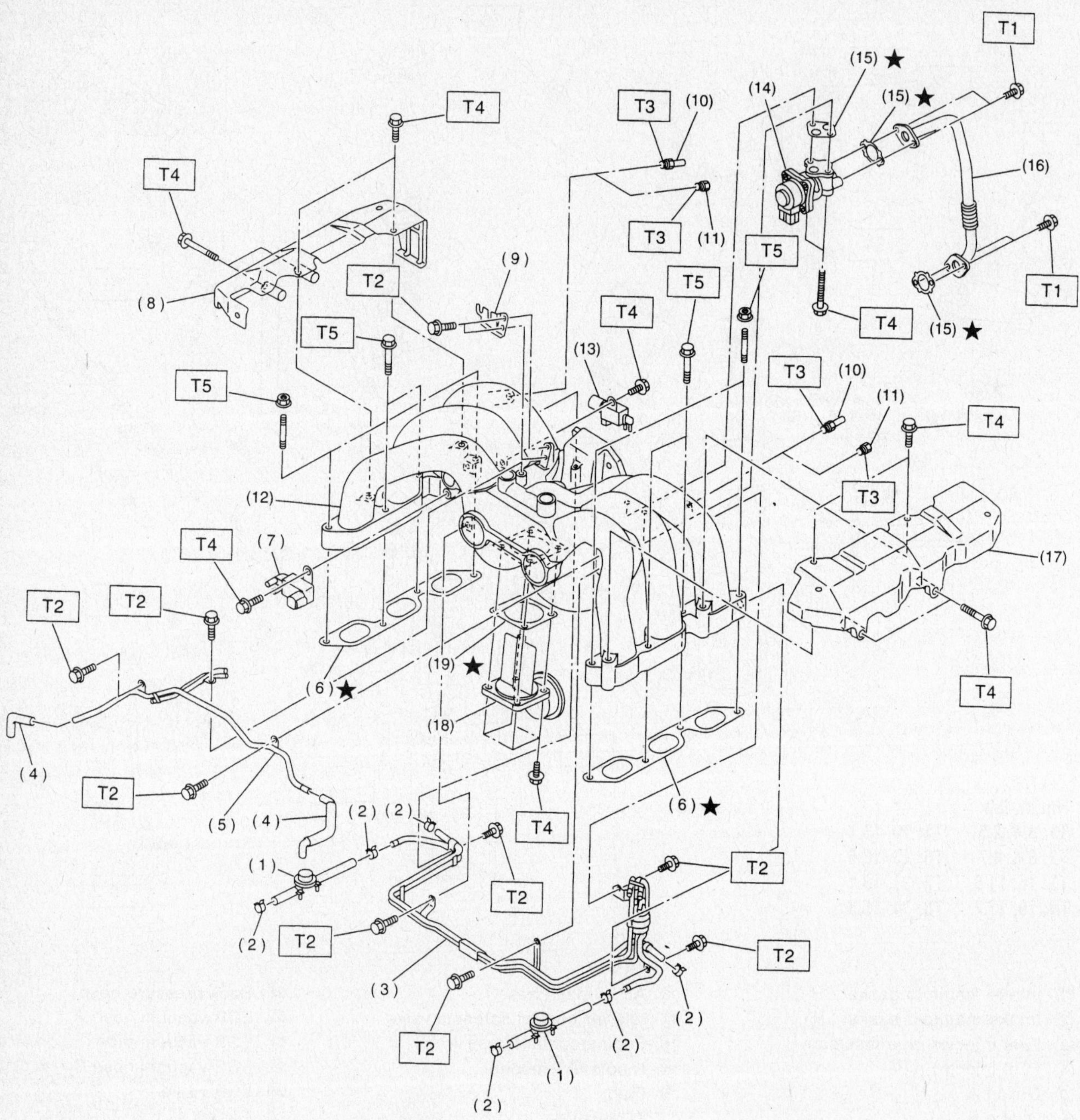

(1) Fuel damper valve
(2) Clamp
(3) Fuel pipe ASSY
(4) Air assist hose
(5) Air assist and purge pipe ASSY
(6) Gasket
(7) Purge control solenoid valve
(8) Fuel pipe protector RH
(9) Accelerator cable bracket
(10) Nipple
(11) Plug
(12) Intake manifold
(13) Induction valve control solenoid
(14) EGR valve
(15) Gasket
(16) EGR pipe
(17) Fuel pipe protector LH
(18) Induction valve
(19) Gasket

Tightening torque: N·m (kgf-m, ft-lb)

T1: 6.4 (0.65, 4.7)
T2: 5.0 (0.51, 3.7)
T3: 17 (1.7, 12)
T4: 19 (1.9, 14)
T5: 25 (2.5, 18)

42356-SBCR-G10

Exploded view of the intake manifold and related components—3.0L engine

- Bolt, which installs the power steering pump stiffener on the engine block, to 14.4–17.3 ft. lbs. (20–24 Nm)
- Power steering pump belt and adjust the belt as necessary
- V-belt cover(s)
- Resonator chamber, if equipped
- Accelerator cable and the cruise control cable, if equipped
- Air cleaner assembly
- MAF sensor connector, if disconnected

9. Connect the negative battery cable and refill the cooling system. Start the engine, and bleed the cooling system. Check for leaks.

3.0L Engine

1. Before servicing the vehicle, refer to the precautions in the beginning of this section.
2. Disconnect the negative battery cable.
3. Properly release the fuel pressure.
4. Remove or disconnect the following:
   - Air intake duct, air cleaner upper cover and the air cleaner element
   - Resonator chamber
   - Accelerator cable and the cruise control cable, if equipped
   - Power steering pipe bracket-to-manifold bolts
   - V-belt
   - Bolts holding the power steering pump to the bracket
   - Connector from the power steering pump switch
   - Power steering pump, and place on the right side wheel apron. Do NOT disconnect the fluid lines.
   - Washer reservoir bolts, hose from the front washer motor and connector from the rear washer motor.
   - Rear washer hose from the motor and plug the hose
   - Washer reservoir aside
   - Positive Crankcase Ventilation (PCV) hose from the cylinder head cover
   - Engine coolant hoses from the throttle body
   - Brake booster hose
   - Exhaust Gas Recirculation (EGR) pipe from the valve being careful not to drop the gaskets
   - Engine harness connectors from the bulkhead connectors
   - Engine ground terminal from the manifold
   - Fuel hoses from the fuel pipes
   - Ground cable from the fuel pipe protector on the left hand side
   - Fuel pipe protector on the left hand side
   - Air assist hose on the left hand side
   - Bolt securing the fuel injector pipe to the cylinder head on the left hand side
   - Ground cable from the fuel pipe protector on the right hand side
   - Fuel pipe protector on the right hand side
   - Air assist hose on the right hand side
   - Bolt securing the fuel injector pipe to the cylinder head on the right hand side
   - Intake manifold mounting bolts
   - Intake manifold and discard the gaskets

To install:

5. Install or connect the following:
   - New gaskets
   - Intake manifold to the engine. Tighten the mounting bolts to 18 ft. lbs. (25 Nm).
   - Bolt securing the fuel injector pipe to the cylinder head on the right hand side. Tighten to 14 ft. lbs. (19 Nm).
   - Air assist hose on the right hand side
   - Fuel pipe protector on the right hand side. Tighten to 14 ft. lbs. (19 Nm).
   - Ground cable from the fuel pipe protector on the right hand side
   - Bolt securing the fuel injector pipe to the cylinder head on the left hand side. Tighten to 14 ft. lbs. (19 Nm).
   - Air assist hose on the left hand side
   - Fuel pipe protector on the left hand side. Tighten to 14 ft. lbs. (19 Nm).
   - Ground cable to the fuel pipe protector on the left hand side
   - EGR pipe to the valve using new gaskets and tighten to 5 ft. lbs. (7 Nm)
   - Fuel hoses to the fuel pipes
   - Engine ground terminal to the manifold
   - Engine harness connectors to the bulkhead connectors
   - Brake booster hose
   - Engine coolant hoses to the throttle body
   - PCV hose to the cylinder head cover
   - Power steering pump
   - Connector to the power steering pump switch
   - Bolts holding the power steering pump to the bracket and tighten to 15 ft. lbs. (20 Nm)
   - V-belt
   - Rear washer hose to the motor
   - Washer reservoir hose to the front washer motor and connector to the rear washer motor.
   - Washer reservoir
   - Accelerator cable and the cruise control cable, if equipped
   - Resonator chamber
   - Air intake duct, air cleaner upper cover and the air cleaner element

6. Connect the negative battery cable and refill the cooling system. Start the engine, and bleed the cooling system. Check for leaks.

Exhaust Manifold

Due to the unique design of the Subaru engine, an exhaust manifold is not used. The exhaust enters directly into the front Y-pipe.

REMOVAL & INSTALLATION

Except 2.0L Engines

1. Before servicing the vehicle, refer to the precautions in the beginning of this section.
2. Remove or disconnect the following:
   - Negative battery cable
   - Air cleaner case, if necessary
   - Front Oxygen Sensor (O_2S)
   - Front undercover
   - Rear O_2S electrical connector
   - Y-pipe-to-rear pipe mounting nuts and separate the Y-pipe from the rear pipe
   - Bolts that secure the front Y-pipe to the cylinder head
   - Y-pipe from the hanger bracket
   - Front exhaust pipe from the catalytic converter and discard the gaskets

To install:

3. Clean all gasket surfaces completely.
4. Install or connect the following:
   - New gaskets
   - Front catalytic converter to front exhaust pipe. Tighten the bolts to 18.8–26 ft. lbs. (25–35 Nm) on 2.2L and 2.5L engines and 22 ft. lbs. (30 Nm) on 3.0L engines.
   - Y-pipe. Temporarily tighten the bolt that holds the center exhaust pipe to the hanger bracket.
   - Y-pipe, to the cylinder head. Tighten the retainers to 18.8–26 ft.

lbs. (25–35 Nm)) on 2.2L and 2.5L engines and 22 ft. lbs. (30 Nm) on 3.0L engines.
- Y-pipe to the rear pipe. Tighten the retainers to 9.4–16.6 ft. lbs. (13–23 Nm)) on 2.2L and 2.5L engines and 13 ft. lbs. (18 Nm) on 3.0L engines.

5. Tighten the center exhaust pipe to hanger bracket bolt to 22.4–29.6 ft. lbs. (30–40 Nm)) on 2.2L and 2.5L engines and 26 ft. lbs. (35 Nm) on 3.0L engines..
- Rear O_2S electrical connector
- Front O_2S electrical connector
- Front undercover, if equipped
- Air cleaner case, if removed
- Negative battery cable

6. Start the engine and check for exhaust leaks.

2.0L Engines

1. Before servicing the vehicle, refer to the precautions in the beginning of this section.
2. Remove or disconnect the following:
- Negative battery cable
- Front Oxygen Sensor (O_2S)
- Front undercover, if equipped
- Lower exhaust manifold cover on the right hand side
- Upper and lower exhaust manifold covers on the left hand side
- Nuts and bolts that attach the front exhaust pipe to the turbocharger joint pipe
- Nuts that attach the front exhaust pipe to the to the cylinder head while holding the front pipe
- Front exhaust pipe assembly
- Covers from the front exhaust pipe and manifold
- Front exhaust pipe from the manifolds and discard the gaskets

To install:

3. Clean all gasket surfaces completely.
4. Install or connect the following:
- New gaskets
- Front exhaust pipe to the manifolds

9357TG11

Location of the front exhaust pipe retainers—2.0L engines

and tighten the retainers to 26 ft. lbs. (35 Nm)
- Covers to the front exhaust pipe and tighten to 18 ft. lbs. (25 Nm)
- Upper exhaust manifold cover on the right hand side and tighten the retainers to 13 ft. lbs. (19 Nm)
- Front exhaust pipe assembly and tighten the retainers to 26 ft. lbs. (35 Nm)
- Right hand side manifold to the turbocharger joint pipe and tighten the retainers to 13 ft. lbs. (19 Nm)
- Upper and lower manifold covers on the left hand side to 13 ft. lbs. (19 Nm)
- Front O_2S
- Front undercover, if equipped
- Negative battery cable

Turbocharger

REMOVAL & INSTALLATION

2.0L Engines

1. Before servicing the vehicle, refer to the precautions in the beginning of this section.
2. Remove or disconnect the following:
- Negative battery cable
- Center exhaust pipe
- Turbocharger joint pipe from the turbocharger
- Engine coolant hose from the filler tank
- Clamp that attaches the turbocharger to the inlet duct
- Bolt that attaches the bracket of the oil pipe to the turbocharger
- Oil pipe from the turbocharger
- Turbocharger bracket
- Oil outlet hose from the pipe
- Turbocharger.

To install:

3. Installation is the reverse of removal.
4. Tighten the front pipe to the turbocharger retainers to 22 ft. lbs. (30 Nm).

2.5L Engines

1. Before servicing the vehicle, refer to the precautions in the beginning of this section.
2. Disconnect the negative battery cable.
3. Remove the center exhaust pipe.
4. Separate the turbocharger joint pipe from the turbocharger.
5. Disconnect the engine coolant hose from the coolant filler tank.
6. Loosen the clamp that attaches the turbocharger to the intake duct.
7. Remove the oil pipe–to–turbocharger bolt and remove the oil pipe from the turbocharger.
8. Remove the turbocharger bracket and disconnect the oil outlet hose from the pipe.
9. Remove the turbocharger.

To install:

10. Installation is the reverse of removal, please note the following torque specs:
    a. Turbocharger–to–intake duct: 2 ft. lbs. (3 Nm).
    b. Oil pipe–to–turbocharger: 11 ft. lbs. (16 Nm).
    c. Joint pipe–to–turbocharger using a new gasket: 25 ft. lbs. (35 Nm).
    d. Turbocharger bracket: 24 ft. lbs. (33 Nm).

Front Crankshaft Seal

REMOVAL & INSTALLATION

2.0L Engines

1. Before servicing the vehicle, refer to the precautions in the beginning of this section.
2. Remove or disconnect the following:
- Negative battery cable
- Drive belt

3. Secure the crankshaft pulley with tool No. 499977300.
4. Remove or disconnect the following:
- Crankshaft pulley bolt and pulley
- Left, right and center timing belt cover(s) mounting bolts
- Belt covers
- Timing belt
- Crankshaft seal from the oil pump housing

To install:

5. Using a suitable seal driver, install a new crankshaft seal.
6. Install or connect the following:
- Timing belt crankshaft sprocket and timing belt
- Belt covers and tighten the bolts to 36–48 inch lbs. (4–5 Nm)
- Crankshaft pulley and tighten the bolt to 33 ft. lbs. (44 Nm)
- Drive belt
- Negative battery cable

2.2L Engines

1. Before servicing the vehicle, refer to the precautions in the beginning of this section.
2. Remove or disconnect the following:
- Negative battery cable
- Accessory drive belts
- Power steering pump and alternator

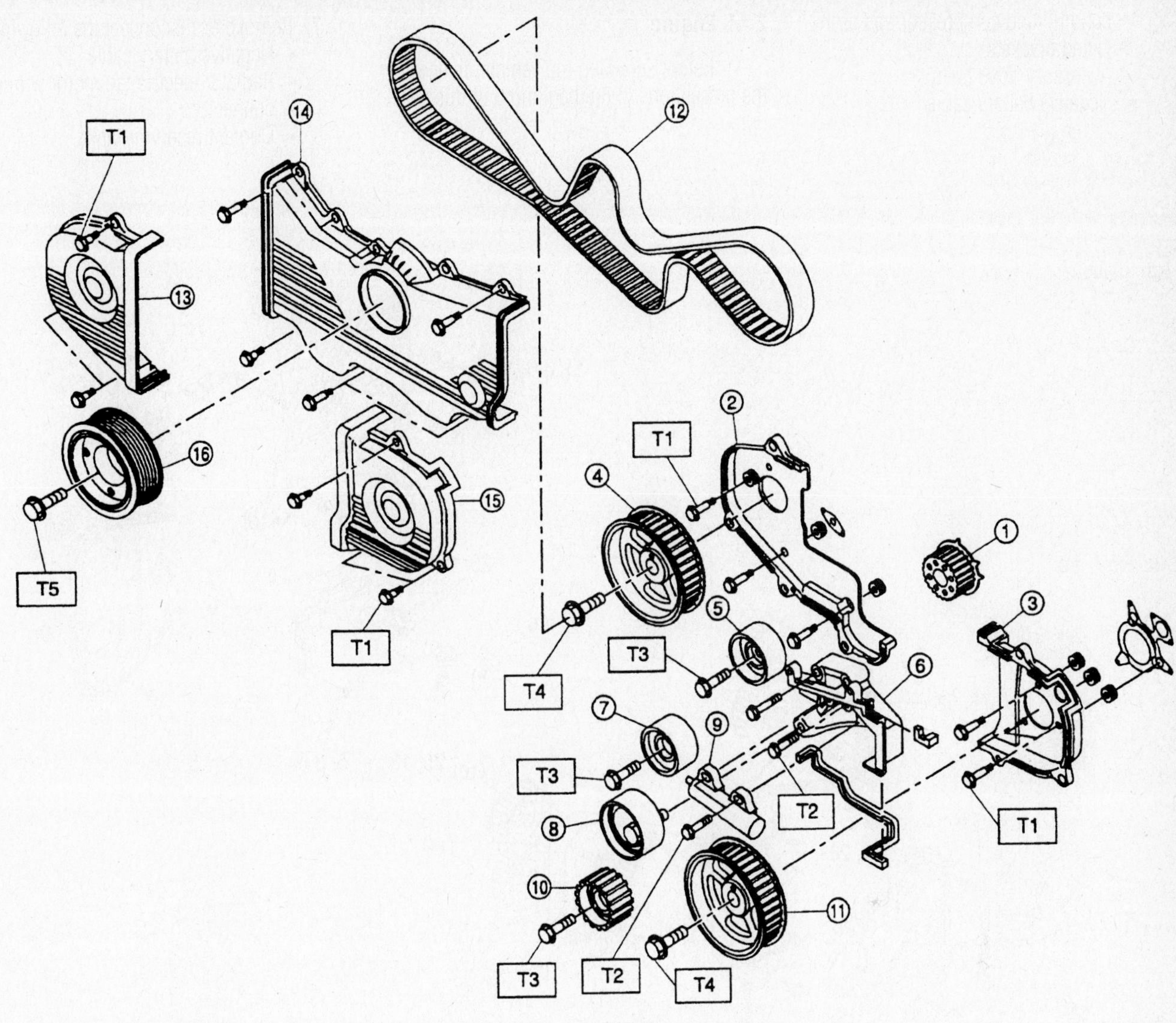

1. Crankshaft sprocket
2. Belt cover No. 2 (RH)
3. Belt cover No. 2 (LH)
4. Camshaft sprocket (RH)
5. Belt idler
6. Tensioner bracket
7. Belt idler
8. Belt tensioner
9. Tensioner adjuster
10. Belt idler No. 2
11. Camshaft sprocket (LH)
12. Timing belt
13. Belt cover (RH)
14. Front belt cover
15. Belt cover (LH)
16. Crankshaft pulley

Tightening torque: N·m (kg-m, ft-lb)
T1: 5 ± 1 (0.5 ± 0.1, 3.6 ± 0.7)
T2: 25 ± 2 (2.5 ± 0.2, 18.1 ± 1.4)
T3: 39 ± 4 (4.0 ± 0.4, 28.9 ± 2.9)
T4: 78 ± 5 (8.0 ± 0.5, 57.9 ± 3.6)
T5: 108^{+10}_{-5} ($11^{+1.0}_{-0.5}$, $79.6^{+7.2}_{-3.6}$)

7923TG20

Exploded view of the timing belt covers and components—2.2L engine

- Air conditioner compressor brackets

3. Secure the crankshaft pulley with tool No. 499977000.
4. Remove or disconnect the following:
   - Crankshaft pulley bolt and pulley
   - Timing belt cover mounting bolts
   - Belt covers
   - Timing belt
   - Timing belt crankshaft sprocket
   - Crankshaft seal from the oil pump housing

To install:

5. Using a suitable seal driver, install a new crankshaft seal.
6. Install or connect the following:
   - Timing belt crankshaft sprocket and timing belt
   - Belt covers and tighten the bolts to 36–48 inch lbs. (4–5 Nm)
   - Crankshaft pulley and tighten the bolt to 69–76 ft. lbs. (93–103 Nm)
   - Power steering pump, alternator, air

conditioning compressor and associated brackets
- Accessory drive belts
- Negative battery cable

2.5L Engine

1. Before servicing the vehicle, refer to the precautions in the beginning of this section.
2. Remove or disconnect the following:
   - Negative battery cable
   - Radiator electric fan motor wiring connectors
   - Coolant reservoir tank

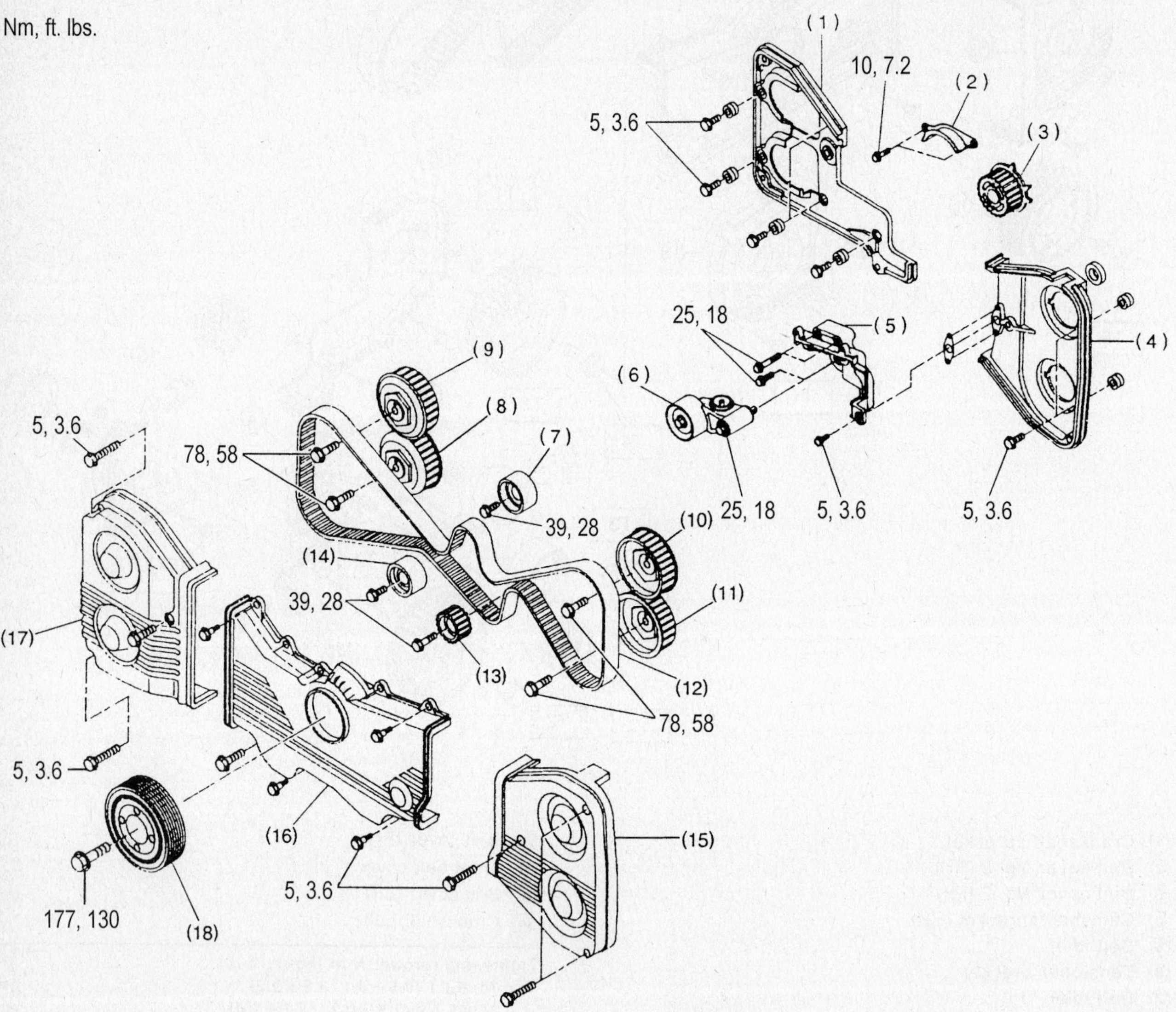

(1) Right-hand belt cover No. 2
(2) Timing belt guide (MT vehicles only)
(3) Crankshaft sprocket
(4) Left-hand belt cover No. 2
(5) Tensioner bracket
(6) Automatic belt tension adjuster ASSY
(7) Belt idler
(8) Right-hand exhaust camshaft sprocket
(9) Right-hand intake camshaft sprocket
(10) Left-hand intake camshaft sprocket
(11) Left-hand exhaust camshaft sprocket
(12) Timing belt
(13) Belt idler No. 2
(14) Belt idler
(15) Left-hand belt cover
(16) Front belt cover
(17) Right-hand belt cover
(18) Crankshaft pulley

7923TG21

Exploded view of the timing belt covers and components—2.5L engine

- 4 bolts that secure the radiator shroud, then remove the fan assembly

3. Position the No. 1 piston to Top Dead Center (TDC) of its compression stroke.
   - Accessory drive belt cover
   - Air conditioning compressor drive belt and tensioner
4. Secure the crankshaft pulley with tool No. ST499977000.
   - Crankshaft pulley bolt and pulley
   - Left timing belt cover mounting bolts and the left cover
   - Right timing belt cover mounting bolts and the right cover
   - Center timing belt cover mounting bolts and the center cover
   - Timing belt
   - Timing belt crankshaft sprocket
   - Crankshaft seal from the oil pump housing

To install:

5. Install or connect the following:
   - New crankshaft seal, using a suitable seal driver
   - Timing belt crankshaft sprocket and the timing belt
   - Center, right, then the left timing belt covers. Tighten the bolts to 44 inch lbs. (5 Nm).
   - Crankshaft pulley and tighten the bolt to 94 ft. lbs. (127 Nm)
   - Air conditioning compressor drive belt tensioner and the drive belts
   - Fan shroud and fan motor assembly
   - Accessory drive belt cover
   - Negative battery cable

3.0L Engine

1. Before servicing the vehicle, refer to the precautions in the beginning of this section.
2. Remove or disconnect the following:
   - Negative battery cable
   - Drive belt
3. Secure the crankshaft pulley with tool No. 499977100.
4. Remove or disconnect the following:
   - Crankshaft pulley bolt and pulley
   - Front chain cover. Refer to the timing chain procedure in this section.
   - Timing chain
   - Crankshaft seal

To install:

5. Using a suitable seal driver, install a new crankshaft seal.
6. Install or connect the following:
   - Timing chain
   - Front chain cover. Refer to the timing chain procedure in this section.
   - Crankshaft pulley and tighten the bolt to 131 ft. lbs. (178 Nm)
   - Drive belt
   - Negative battery cable

Camshaft

On some models, it may be necessary to remove the engine from the vehicle to perform this service.

REMOVAL & INSTALLATION

2.0L Engines

1. Before servicing the vehicle, refer to the precautions in the beginning of this section.
2. Remove or disconnect the following:
   - Negative battery cable
   - Drive belt
3. Secure the crankshaft pulley with tool No. 499977300.
4. Remove or disconnect the following:
   - Crankshaft pulley bolt and pulley
   - Belt covers

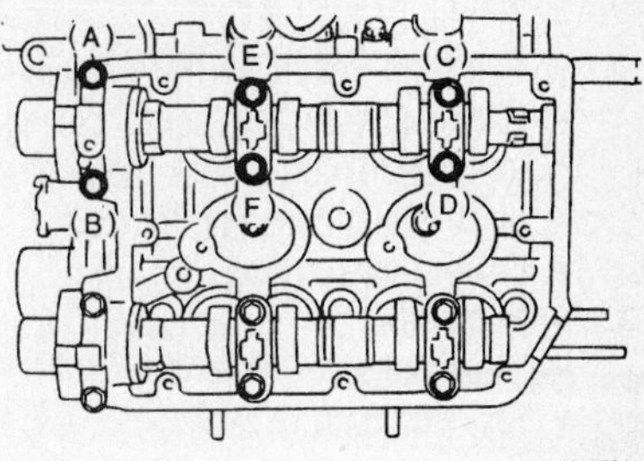

Loosen the intake camshaft caps as shown—2.0L engines

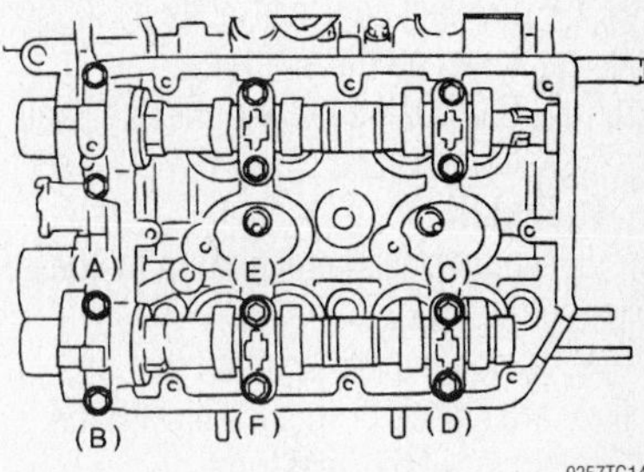

Loosen the exhaust camshaft caps as shown—2.0L engines

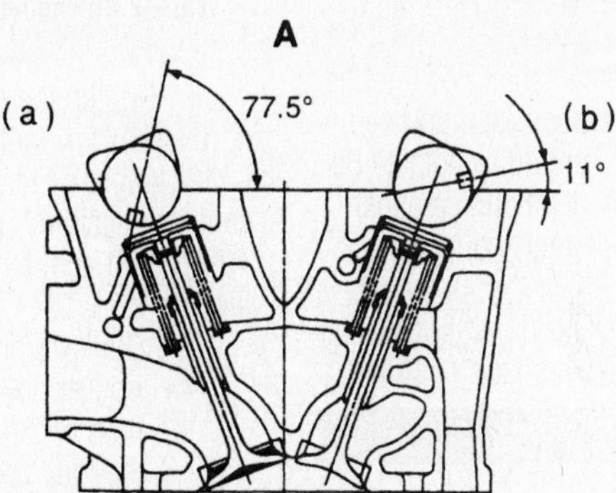

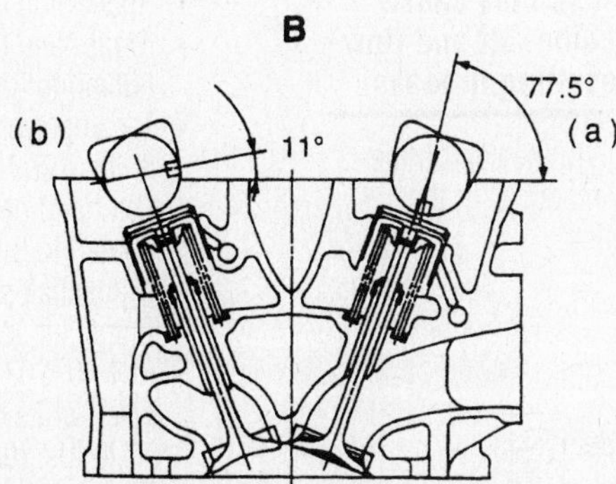

A Left side cylinder head
B Right side cylinder head
(a) Intake camshaft
(b) Exhaust camshaft

9357TG12

The right hand camshaft need not be rotated when set at the position illustrated, the left hand camshaft should be rotated 80 degrees clockwise. The left hand camshaft should be rotated 45 degrees counterclockwise—2.0L engines

- Timing belt
- Camshaft Position (CMP) sensor
- Camshaft sprockets using locking tool 499207400 to lock them in place while loosening the bolt
- Crankshaft sprocket
- Right hand belt cover No. 2
- Dipstick tube
- Spark plug cord
- Rocker cover and gasket
- Intake camshaft bolts as illustrated a little at a time
- Camshaft caps and the camshaft
- Exhaust camshaft bolts as illustrated a little at a time
- Camshaft caps and the camshaft
- Camshafts

To install:

5. Apply clean engine oil to the bearings on the head.
6. Install the camshaft so that the valves are closed or in contact with the "base circle" of the cam lobe.
7. If the camshafts are positioned as shown in the accompanying illustration, the camshafts need to be rotated at a minimum to align the with the timing belt during installation.
8. The right hand camshaft need not be rotated when set at the position illustrated, the left hand camshaft should be rotated 80 degrees clockwise. The left hand camshaft should be rotated 45 degrees counterclockwise.
9. Apply fluid packing (three bond 1215) sparingly to the cap mating surface.

⁂ WARNING

Do not apply fluid packing excessively. Failure to do so may cause excess packing to come out and flow towards oil seal, resulting in leaks.

10. Apply engine oil to the cap on the camshaft as shown by mark "A" in the accompanying illustration.

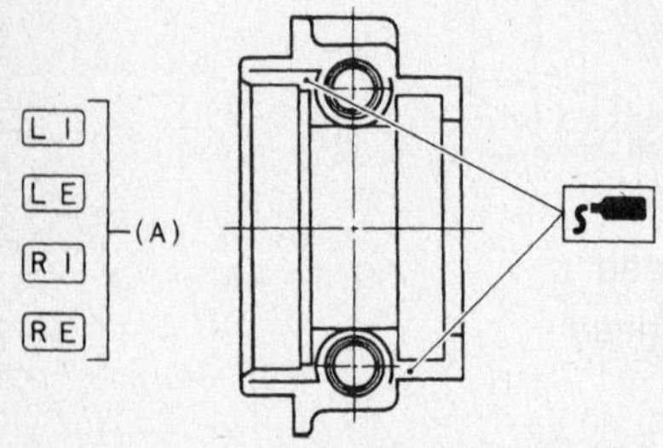

Apply engine oil to the cap on the camshaft as shown by mark "A"—2.0L engines

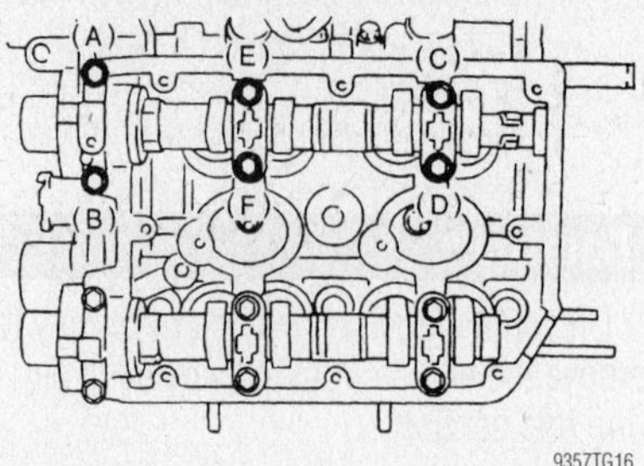

Tighten the cap in two stages in the sequence shown—2.0L engines

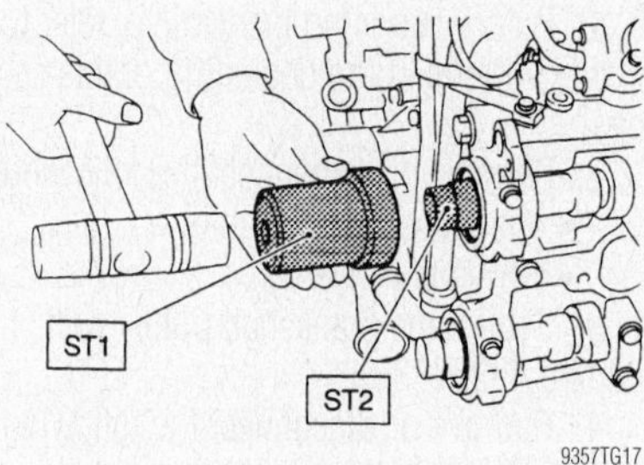

Using guide tool 499597200 and installer 499587600, install the new camshaft seal—2.0L engines

11. Tighten the cap in two stages to 14.5 ft. lbs. (20 Nm) in the sequence illustrated.
12. Apply oil to the lip of the new camshaft seal and using guide tool 499597200 and installer 499587600, install the new seal.
13. Install or connect the following:
    - Rocker cover, making sure to apply fluid packing (three bond 1215) to the four front open edges of the gasket
    - Spark plug cord
    - Right hand belt cover No. 2 and tighten to 3 ft. lbs. (4 Nm)
    - Tensioner bracket and tighten to 18 ft. lbs. (25 Nm)
    - Left hand belt cover No. 2 and tighten to 3 ft. lbs. (4 Nm)
    - Crankshaft sprocket
    - Camshaft sprockets using locking tool 499207400 to lock them in place while tightening the bolt to 72 ft. lbs. (98 Nm)
    - Timing belt
    - Belt cover(s)
    - Crankshaft pulley and tighten the bolt to 33 ft. lbs. (44 Nm)
    - Drive belt
    - Negative battery cable

2.2L Engines

1. Before servicing the vehicle, refer to the precautions in the beginning of this section.
2. Remove or disconnect the following:
    - Negative battery cable
    - Timing belt covers, timing belt and camshaft sprockets
    - Valve rocker covers
    - Rocker arm assemblies. Refer to the rocker arms/shafts procedure in this section.
    - Camshaft cap bolts in the proper sequence
    - Camshaft cap
3. To remove the left camshaft, remove or disconnect the following:
    - Camshaft Position (CMP) sensor
    - Oil dipstick tube attaching bolt and the dipstick tube
    - CMP sensor support
4. To remove the right camshaft, remove the camshaft support on the right side.
5. Remove or disconnect the following:
    - Camshaft O-ring
    - Camshaft and rear seal
    - Oil seal from the camshaft support

To install:

➡Lubricate the camshaft journals with clean engine oil prior to installation.

6. Install or connect the following:
    - Rear oil seal, then the camshaft into the cylinder head
7. Apply a bead of sealant in the camshaft cap. Position the camshaft cap, then tighten the bolts 7–10, in sequence, temporarily
8. Install the rocker arm assemblies. Refer to the rocker arms/shafts procedure in this section.
9. Tighten the remaining camshaft cap bolts, in sequence, as follows:
    a. Step 1: bolts 1–8 to 17–20 ft. lbs. (23–27 Nm).
    b. Step 2: bolts 9–14 to 12–15 ft. lbs. (16–20 Nm).
    c. Step 3: bolts 15–22 to 6–9 ft. lbs. (8–12 Nm), using SST 499497000.
    d. Step 4: bolts 23–24 to 6–9 ft. lbs. (8–12 Nm).
10. To install the left camshaft, install or connect the following:
    - O-ring into the camshaft support and install the support. Tighten the front retainer bolts to 84 inch lbs. (9 Nm), and the rear bolts to 12 ft. lbs. (16 Nm).
    - Oil seal into the camshaft support
    - Dipstick tube. Tighten the retaining bolt to 10 ft. lbs. (13 Nm).
    - CMP sensor
11. To install the right camshaft, install or connect the following:
    - O-ring into the camshaft support

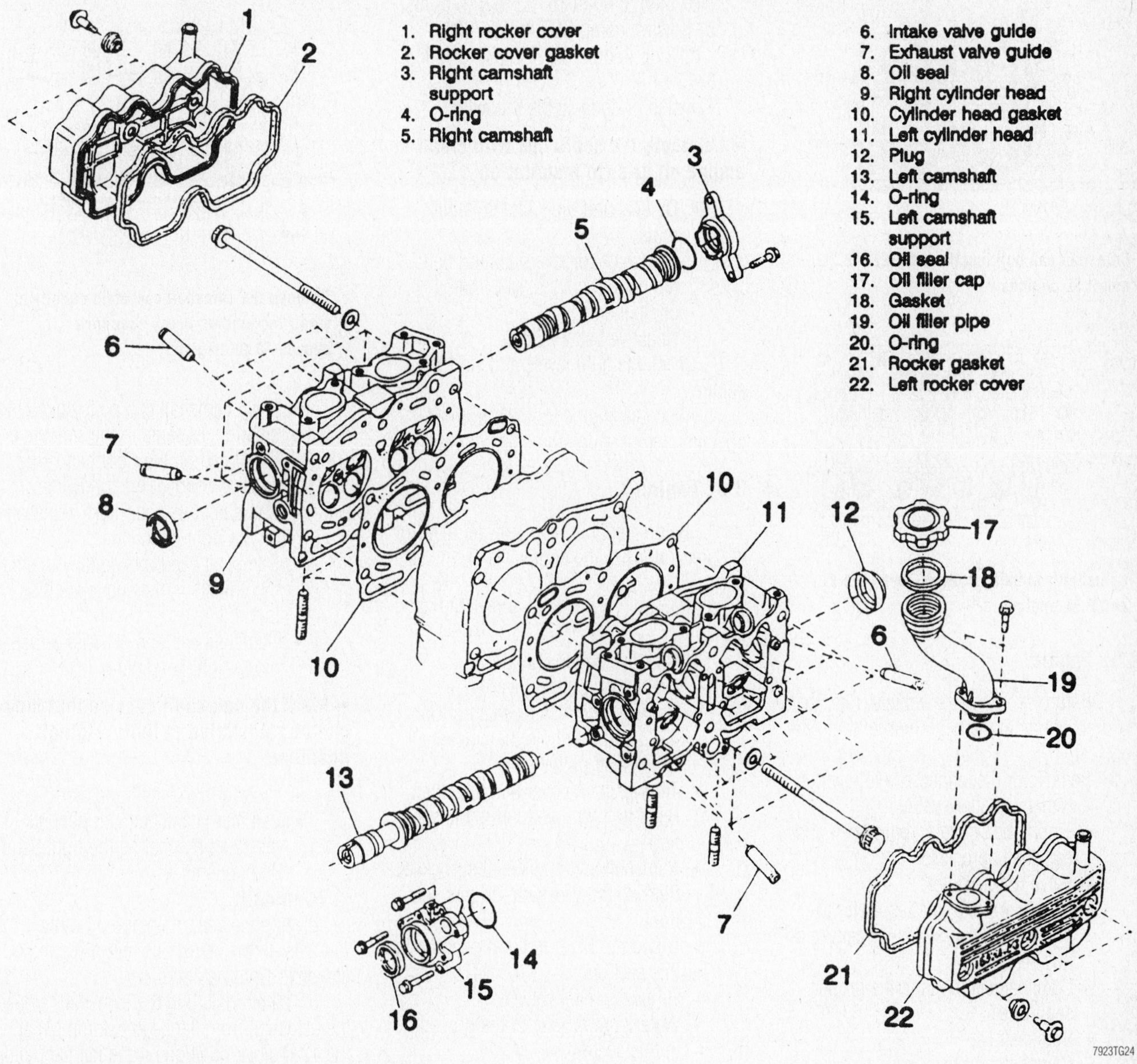

1. Right rocker cover
2. Rocker cover gasket
3. Right camshaft support
4. O-ring
5. Right camshaft
6. Intake valve guide
7. Exhaust valve guide
8. Oil seal
9. Right cylinder head
10. Cylinder head gasket
11. Left cylinder head
12. Plug
13. Left camshaft
14. O-ring
15. Left camshaft support
16. Oil seal
17. Oil filler cap
18. Gasket
19. Oil filler pipe
20. O-ring
21. Rocker gasket
22. Left rocker cover

7923TG24

Exploded view of the camshaft assembly—2.2L engine

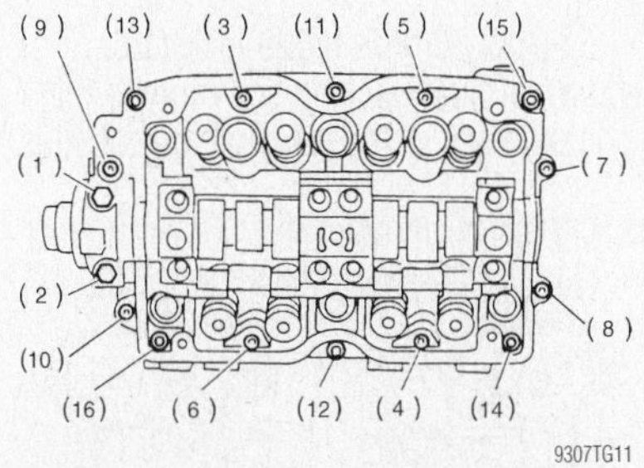

Camshaft cap bolt loosening sequence—2000–01 2.2L and 2.5L engines

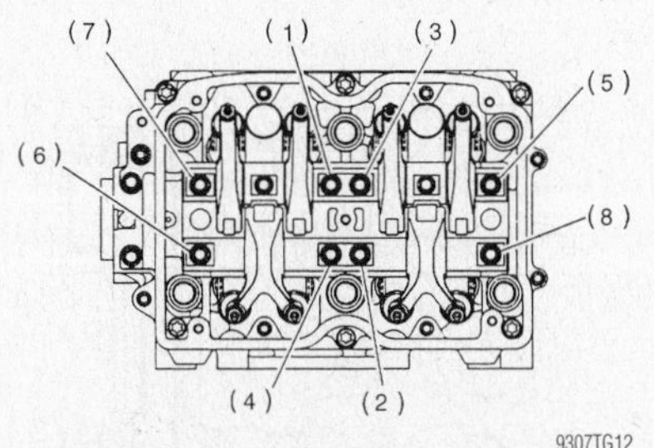

Camshaft cap bolt locations 1–8—2.2L and 2.5L engines

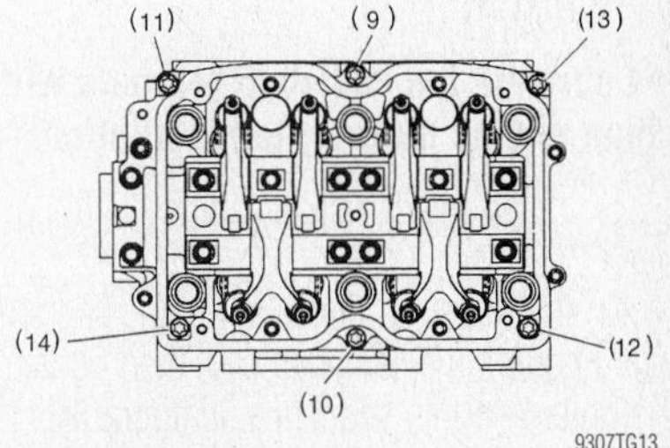

Camshaft cap bolt locations 9–14—2.2L and 2.5L engines

and install the support. Tighten the retainer bolts to 12 ft. lbs. (16 Nm).

- New oil seal in the rear of the cylinder head

12. Install or connect the following:
    - Camshaft sprockets, timing belt, timing belt covers, and related components
    - Negative battery cable
13. Check the fluid levels and start the engine.
14. Allow the engine to reach normal operating temperature and check for leaks.

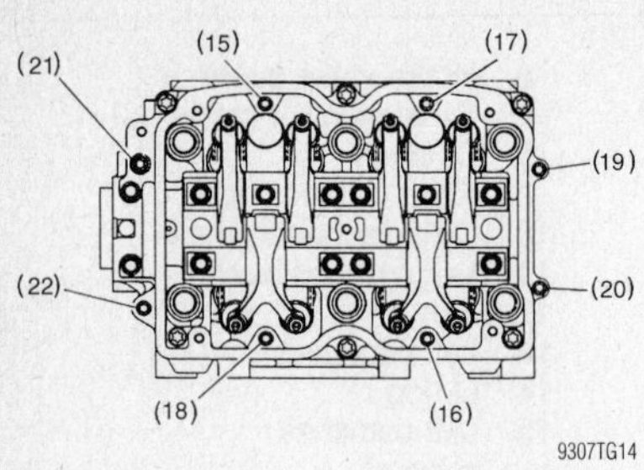

Camshaft cap bolt locations 15–22—2.2L and 2.5L engines

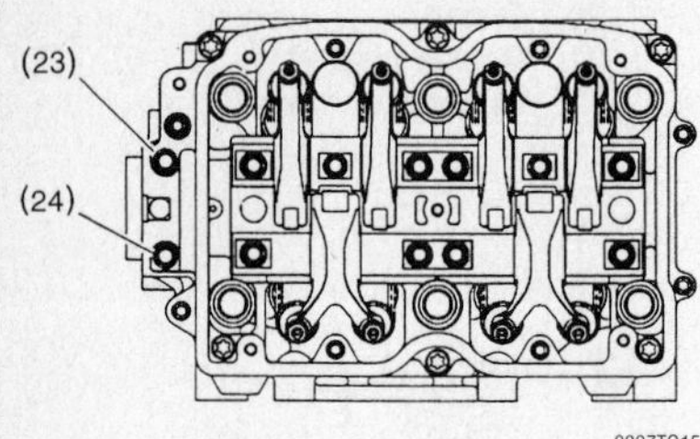

Camshaft cap bolt locations 23–24—2.2L and 2.5L engines

2.5L Engine

1. Before servicing the vehicle, refer to the precautions in the beginning of this section.
2. Remove or disconnect the following:
    - Negative battery cable
    - Timing belt covers, timing belt and camshaft sprockets
    - Valve rocker covers
    - Rocker arm assemblies. Refer to the rocker arms/shafts procedure in this section.
    - Camshaft cap bolts in the proper sequence
    - Camshaft cap
    - Camshaft and rear seal
    - Oil seal from the rear side of the camshaft

To install:

➡Lubricate the camshaft journals with clean engine oil prior to installation.

3. Install the camshaft into the cylinder head
4. Apply a bead of sealant in the camshaft cap. Position the camshaft cap, then tighten the bolts 7–10, in sequence, temporarily
5. Install the rocker arm assemblies. Refer to the rocker arms/shafts procedure in this section.
6. Tighten the remaining camshaft cap bolts, in sequence, as follows:
    a. Step 1: bolts 1–8 to 17–20 ft. lbs. (23–27 Nm).
    b. Step 2: bolts 9–14 to 12–15 ft. lbs. (16–20 Nm).
    c. Step 3: bolts 15–22 to 6–9 ft. lbs. (8–12 Nm), using SST 499497000.
    d. Step 4: bolts 23–24 to 6–9 ft. lbs. (8–12 Nm).
7. Install or connect the following:

➡Lubricate the seals lips with clean engine oil prior to installation

- Oil seal and plug with suitable tools
- Camshaft sprockets, timing belt, timing belt covers, and related components
- Negative battery cable

8. Check the fluid levels and start the engine.
9. Allow the engine to reach normal operating temperature and check for leaks.

3.0L Engine

1. Before servicing the vehicle, refer to the precautions in the beginning of this section.
2. Remove or disconnect the following:
    - Negative battery cable
    - Crankshaft pulley cover
    - Crankshaft pulley bolt using tool 499977100 to hold the crankshaft
    - Crankshaft pulley
    - Timing chain cover and the chain. Refer to the procedure in this section.
    - Camshaft and crankshaft sprockets. Refer to the procedure in this section.
    - Oil pump. Refer to the procedure in this section.
    - Oil pump relief valve
    - Water pump. Refer to the procedure in this section.
    - Rear chain cover
    - Right hand valve cover

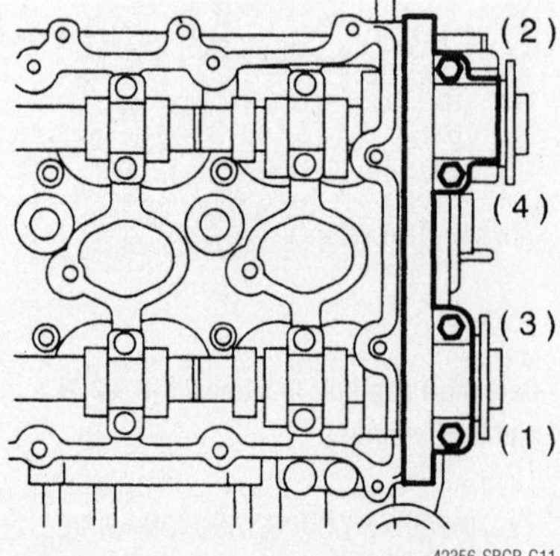

Remove the front camshaft cap bolts equally in small increments in the sequence shown—3.0L engine

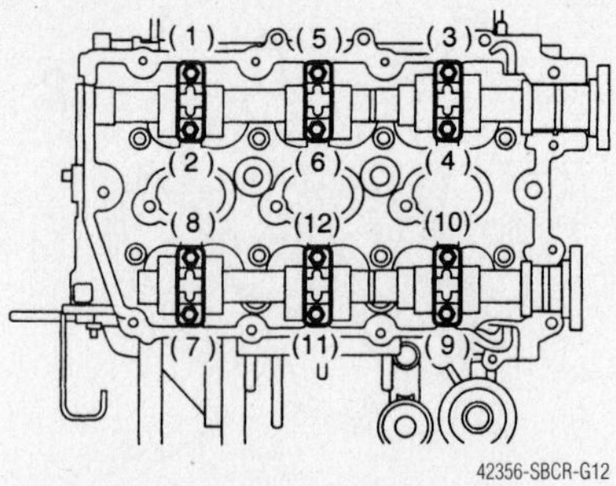

Remove the camshaft cap bolts equally in small increments in the sequence shown—3.0L engine

- Front camshaft cap bolts equally in small increments in the sequence illustrated on the right hand side

3. Install or connect the following:
    - Camshaft cap and intake camshaft on the right hand side
    - Camshaft cap bolts equally in small increments in the sequence illustrated
    - Camshaft cap and exhaust camshaft of the right hand side

➡Mark the camshaft caps so that they can be reinstalled in their original positions.

- Plug from the left hand side
- Left hand camshaft components in the same manner as the right hand components

To install:

4. Apply a coat of engine oil to the journals on the camshafts and place the camshafts into position.
5. When installing the camshaft, adjust the camshaft front flange knock pin (A) at the 12 O'clock position on the left hand side and the 10 O'clock position on the right hand side. Refer to the illustration for more detail.
6. Apply 0.059–0.099 inch (1.5–2.5mm) of fluid packing sparingly to the front and back of the camshaft cap as illus-

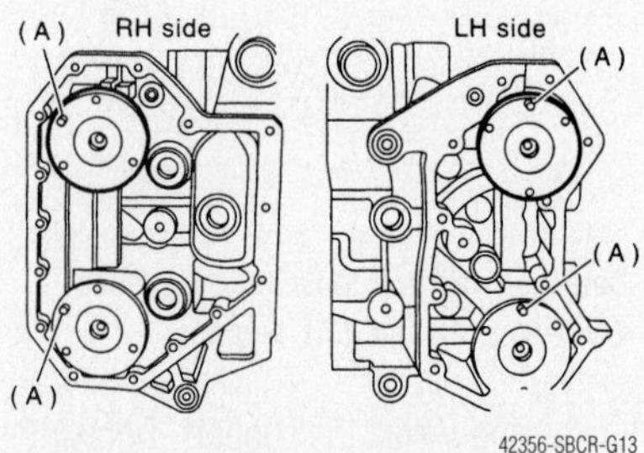

Install the camshafts so the flange knock pin (A) is positioned as shown—3.0L engine

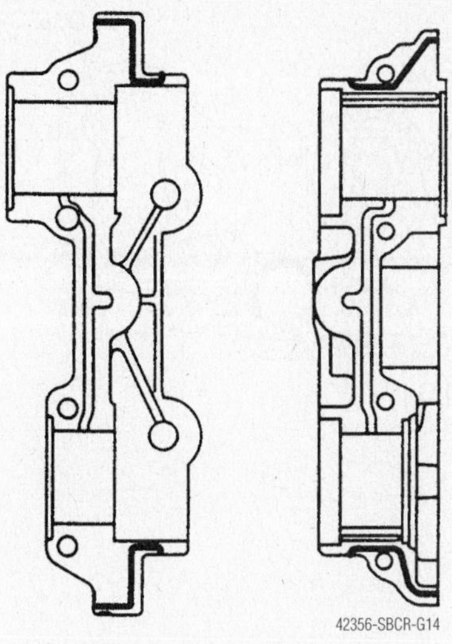

Apply 0.059–0.099 inch (1.5–2.5mm) of fluid gasket sparingly to the front and back of the camshaft cap where indicated—3.0L engine

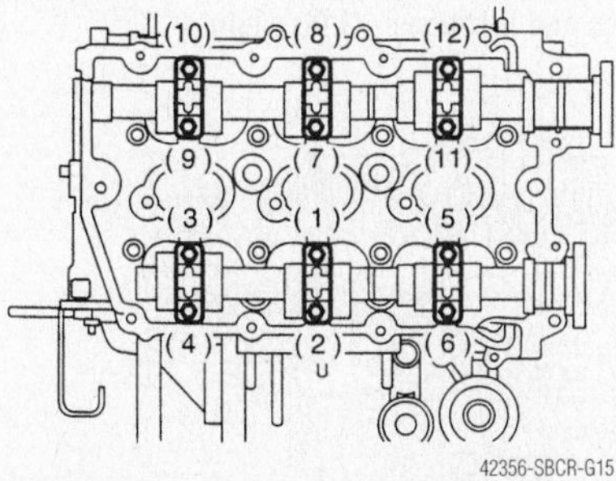

Tighten the camshaft cap bolts in the sequence shown to the proper specification—3.0L engine

trated. Subaru recommends Three Bond 1280B for this purpose.

✲✲ CAUTION

Do not apply too much gasket maker. This may cause excess fluid gasket maker to come out and flow towards the camshaft journal resulting in engine damage.

7. Apply oil to the cap bearing surface.
8. Install or connect the following:
   - Camshaft cap in the their original location
   - Camshaft cap bolts and tighten to 11.6 ft. lbs. (16 Nm) in the sequence illustrated
   - Front camshaft cap bolts and tighten to 7 ft. lbs. (10 Nm) in the sequence illustrated
9. Apply fluid gasket maker of the of the cylinder heads and valve covers as illustrated. Subaru recommends Three Bond 1280B for this purpose.

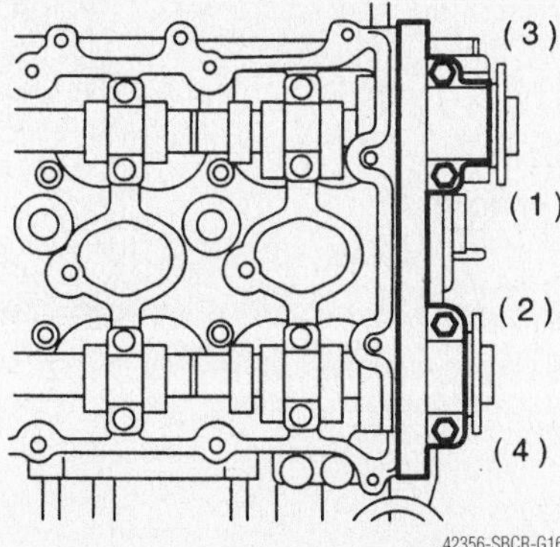

Tighten the front camshaft cap bolts in the sequence shown—3.0L engine

✲✲ CAUTION

Do not apply too much gasket maker. This may cause excess fluid gasket maker to come out and flow towards the camshaft journal resulting in engine damage.

10. Install or connect the following:
    - Valve cover bolts and tighten to 5 ft. lbs. (7 Nm) in the sequence illustrated.
11. Install the rear chain cover as follows:

➡There are several size bolts used, refer to the illustration for size and locations.

a. Rear chain cover gasket and clean the mating surfaces.

b. Apply liquid gasket maker to the mating surfaces of the cover. Refer to the illustration for gasket maker application and diameter.

c. Install new O-rings. Refer to the illustration for O-ring location and size

d. Install the rear chain cover and temporarily tighten the bolts, refer to the illustration for size and locations.

e. Replace mounting bolts **G** with new bolts, refer to the illustration for location.

f. Tighten the cover bolts in the sequence illustrated to the specifications shown in the illustration.

12. Install or connect the following:
    - Water pump
    - Oil pump relief valve and tighten the bolts to 4.7 ft. lbs. (6.4 Nm) in the sequence illustrated.
    - Oil pump
    - Crankshaft sprocket. Refer to the procedure in this section.
    - Camshaft sprocket. Refer to the procedure in this section.
    - Timing chain. Refer to the procedure in this section.
    - Crankshaft pulley, apply clean oil to the bolt threads and tighten the bolt to 131 ft. lbs. (178 Nm)
    - Camshaft pulley cover and tighten the bolts to 5 ft. lbs. (7 Nm)

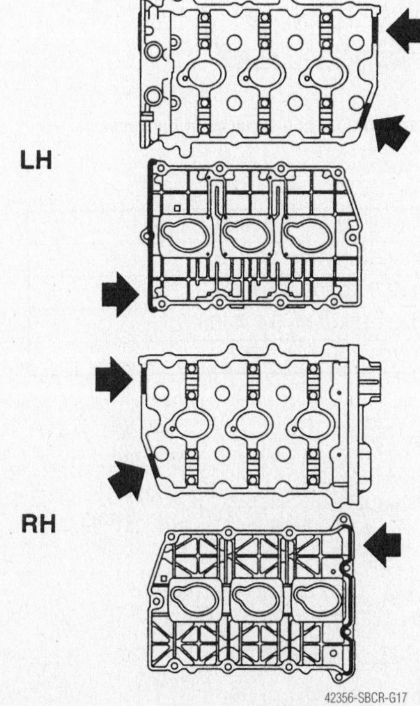

Apply fluid gasket maker of the of the cylinder heads and valve covers as shown—3.0L engine

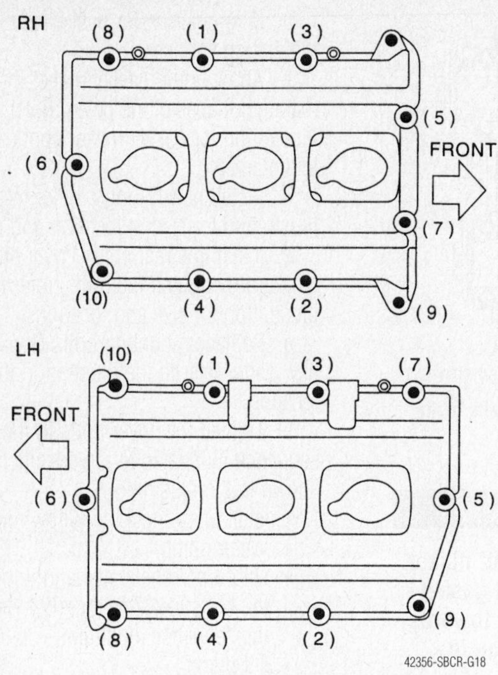

Tighten the valve cover bolts in this sequence—3.0L engine

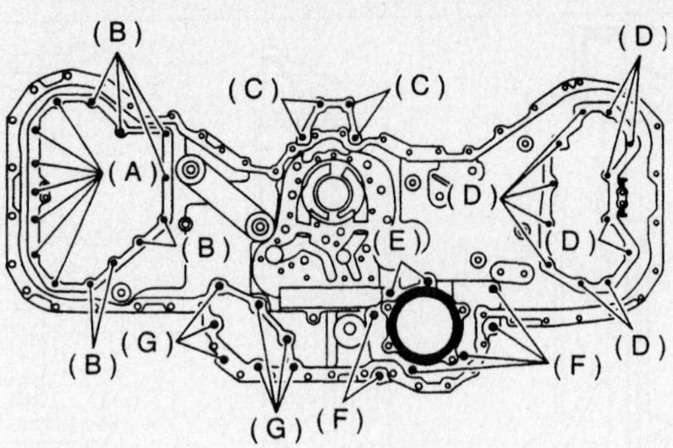

(A) 6 × 14
(B) 6 × 18 (Silver)
(C) 6 × 30
(D) 6 × 18
(E) 6 × 40
(F) 6 × 30
(G) 6 × 22

42356-SBCR-G05

Rear timing chain bolt sizes and locations—3.0L engine

Fluid gasket application diameter:
(A) 1.0±0.5 mm (0.039±0.020 in)
(B) 3.0±1.0 mm (0.118±0.039 in)

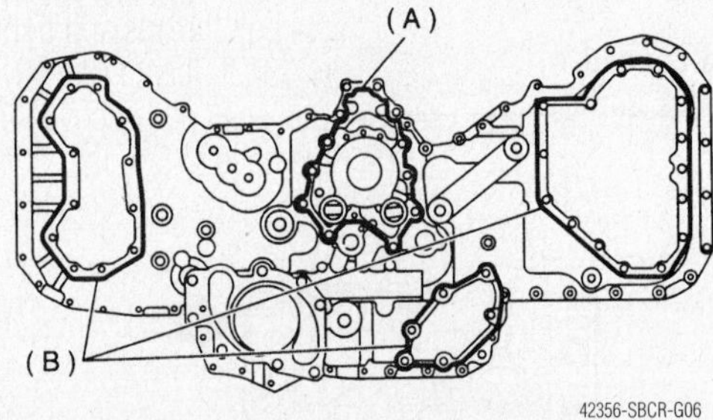

Apply liquid gasket maker to the mating surfaces of the cover as shown—3.0L engine

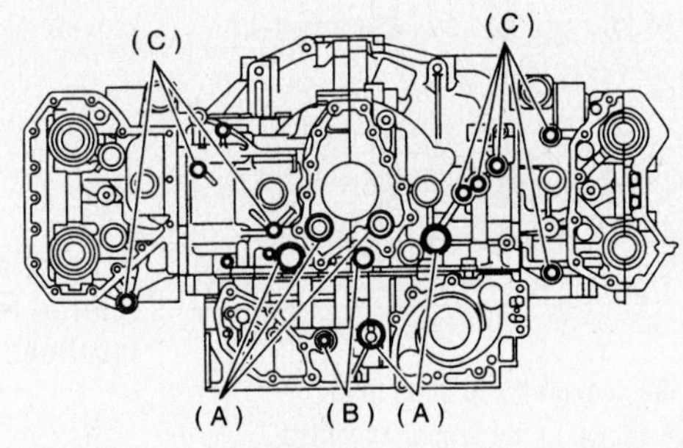

(A) O-ring (Large)
(B) O-ring (Medium)
(C) O-ring (Small)

42356-SBCR-G07

Rear cover O-ring sizes and locations—3.0L engine

(1) to (11)	9 N·m (0.9 kgf-m, 6.5 ft-lb)
(12) to (19)	20 N·m (2.0 kgf-m, 14 ft-lb)
(20) to (31)	9 N·m (0.9 kgf-m, 6.5 ft-lb)
(32) to (39)	12 N·m (1.2 kgf-m, 8.7 ft-lb)
(40) to (46)	9 N·m (0.9 kgf-m, 6.5 ft-lb)

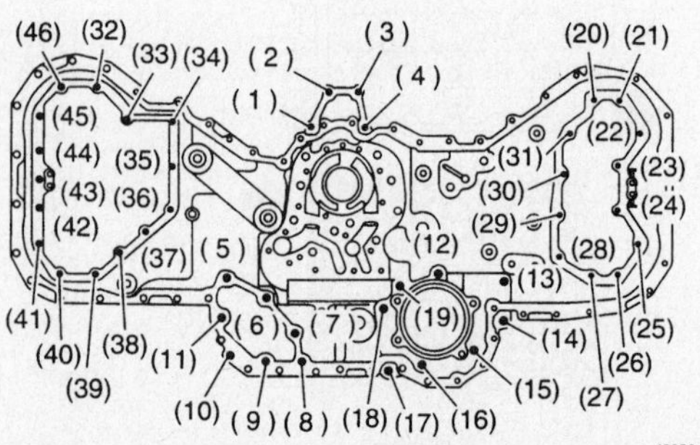

Rear cover bolt tightening sequence and torque specifications—3.0L engine

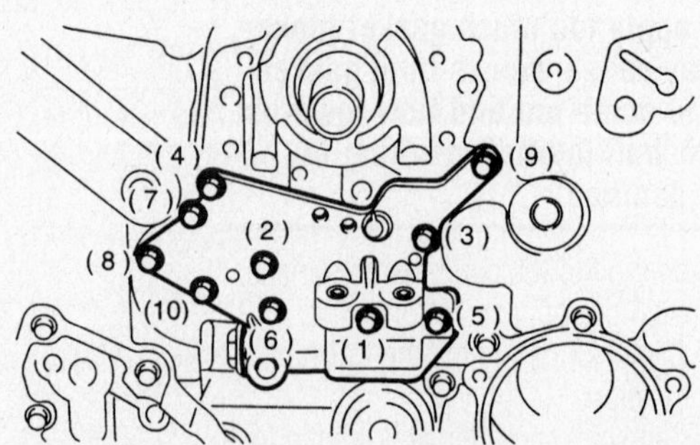

Bolt installation position	Bolt dimension
(1) and (5)	6 x 26
(2), (3), (4) and (9)	6 x 35
(6), (7), (8) and (10)	6 x 16

42356-SBCR-G09

Oil pump relief valve bolt tightening sequence and bolt specifications—3.0L engine

- Negative battery cable

13. Start the engine and allow it to reach operating temperature. Check for leaks.

Valve Lash

INSPECTION & ADJUSTMENT

2.2L Engine

1. Before servicing the vehicle, refer to the precautions in the beginning of this section.
2. With the engine cold, rotate the engine so that the No. 1 piston is at Top Dead Center (TDC) of its compression stroke.
3. Check the clearance of both the intake and exhaust valves of the No. 1 cylinder by inserting a feeler gauge between each valve stem and rocker arm.
4. If the clearance is not within specifications, loosen the locknut with the proper size wrench and turn the adjusting stud either in or out until the valve clearance is correct.

➡Proper valve clearance is obtained when the feeler gauge slides between the valve stem and the rocker arm with a minimum amount of resistance.

5. Tighten the locknut and recheck the valve stem-to-rocker clearance.
6. The rest of the valves are adjusted in the same way. Bring each piston to TDC of its compression stroke, then check and adjust the valves for that cylinder. The proper valve adjustment sequence is 1–3–2–4.
7. Rotate the crankshaft at least 2 revolutions, then recheck the valve clearance.
8. Tighten the rocker arm locknuts to 10–13 ft. lbs. (14–18 Nm).
9. Install the valve covers using new gaskets. Tighten the retaining nuts to 24–36 inch lbs. (3–4 Nm).

2.5L Engine

➡The valve adjustment should be performed while the engine is cold. A Shim Replace Kit 498187100 will be needed to perform the valve adjustment.

1. Before servicing the vehicle, refer to the precautions in the beginning of this section.
2. Adjustment should be performed when engine is cold.
3. Remove or disconnect the following:
    - Negative battery cable
    - Engine coolant reservoir tank
    - Timing belt cover on the left hand side
4. When inspecting the No. 1 and 3 cylinders remove the following:
    - Air intake duct as a unit
    - Resonator chamber
    - Spark plug wires from the No. 1 and 3 cylinders
    - Blow-by house from valve cover
    - Engine undercover
    - Timing belt cover on the right hand side
    - Valve cover on the right hand side
5. When inspecting the No. 2 and 4 cylinders remove the following:
    - Battery and battery tray
    - Window washer motor connectors front and rear
    - Rear gate glass washer hose from the washer motor
    - Washer tank mounting bolts and secure out of the way
    - Spark plug wires from the No. 2 and 4 cylinders
    - Blow by house from valve cover
    - Timing belt cover on the right hand side
    - Valve cover on the left hand side
6. Set No. 1 cylinder to Top Dead Center (TDC).

➡When arrow mark on the left hand side comes exactly to the top, No. 1 cylinder piston is brought to TDC of the compression stroke.

7. Check the valve clearance:
    - Intake valve: 0.0071–0.0087 in. (0.18–0.22mm)
    - Exhaust valve: 0.0090–0.0106 in. (0.23–0.27mm)
8. If any valve needs adjustment, perform the following:
    a. Loosen the valve rocker nut and screw.
    b. Place a thickness gage in at as horizontal a direction as possible with respect to the valve stem and face.
    c. Adjust the screw until proper clearance is obtained.
    d. Tighten the rocker nut after adjusted.
9. Install or connect the following:
    - Valve covers left and right
    - Timing belt covers
    - Blow-by houses to valve covers
    - Spark plug wires
    - Washer tank
    - Rear gate glass washer hose to the washer motor
    - Washer motor connectors
    - Battery and battery tray
    - Engine undercover

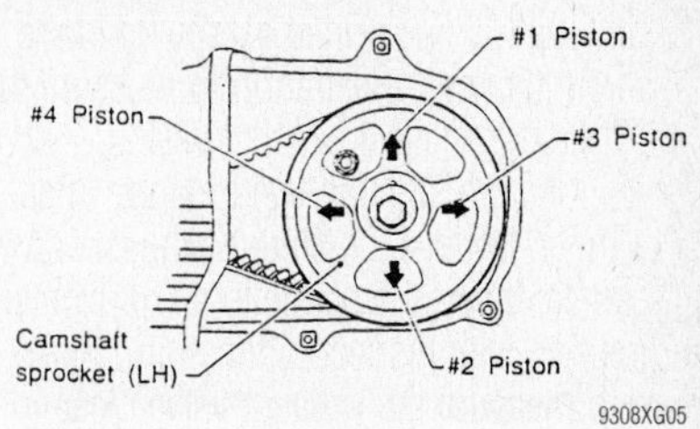

Position the camshaft for adjustment to valves

- Resonator chamber
- Air intake duct unit
- Engine coolant reservoir tank

2.0L Engine

➡Inspection and adjustment of the valve clearance should be performed with the engine cold.

1. Before servicing the vehicle, refer to the precautions in the beginning of this section.
2. Remove or disconnect the following:
    - Negative battery cable
    - Air intake duct
    - Bolt that attaches the right hand timing cover
    - Engine undercover
    - Remaining bolts attaching the right hand timing belt cover and the cover
3. When inspecting the # 1 and # 3 cylinders:
    a. Pull out the engine harness connector with the bracket from the air cleaner upper cover.
    b. Remove the air cleaner case.
    c. Disconnect the spark plug wires from the # 1 and # 3 cylinders.
    d. Disconnect the Positive Crankcase Ventilation (PCV) hose from the right hand rocker cover.
    e. Remove the right hand rocker cover.
4. When inspecting the # 2 and # 4 cylinders:
    a. Remove the battery and tray.
    b. Remove the bolt that attaches the engine harness onto the body.
    c. Disconnect the washer motor connectors.
    d. Remove the washer tank bolts and lift the tank upwards.
    e. Disconnect the spark plug wires from the # 2 and # 4 cylinders.
    f. Disconnect the PCV hose from the left hand rocker cover.
    g. Remove the left hand rocker cover.
    h. Turn the crankshaft pulley clockwise until the arrow mark on the

camshaft is positioned as shown in the accompanying illustration to measure the # 1 intake and # 3 exhaust valves.

5. Using a suitable feeler gauge, measure the # 1 and # 3 cylinder exhaust valve clearance. Insert the gauge in as horizontal a direction with respect to the shim. Make sure to measure the exhaust valve clearances while lifting up the vehicle.

6. The intake valve clearance should be 0.0071–0.0087 inch (0.18–0.22mm). The exhaust valve clearance should be 0.0090–0.0106 inch (0.23–0.27mm).

7. If not within specification, adjust the valve as outlined in the adjustment steps.

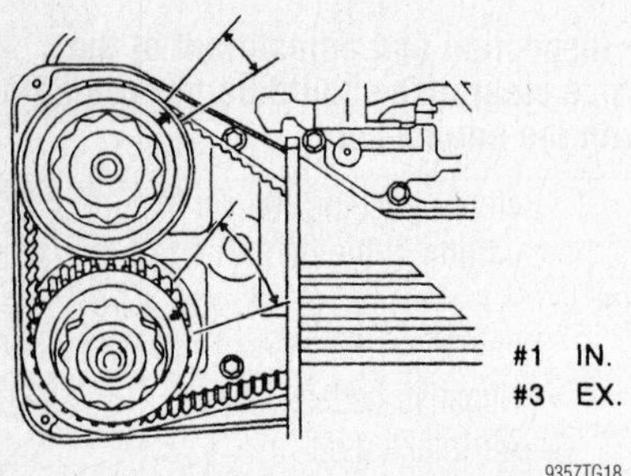

Turn the crankshaft pulley clockwise until the arrow mark on the camshaft is positioned as shown to measure the # 1 intake and # 3 exhaust valves—2.0L engines

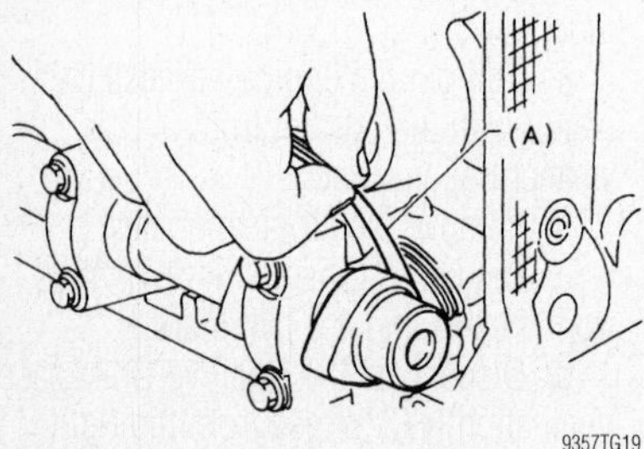

Use a feeler gauge to inspect the valve clearance—2.0L engines

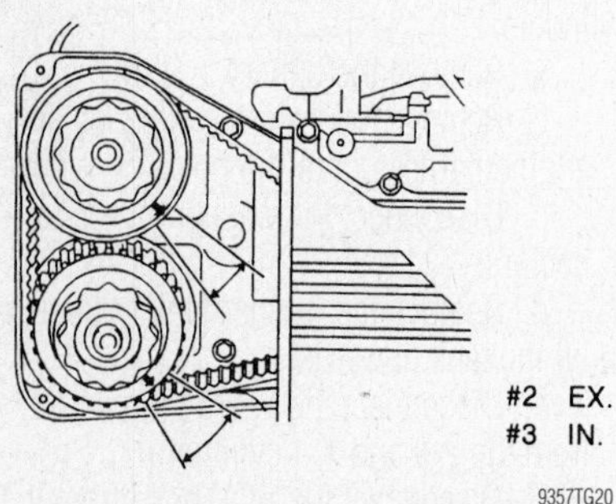

Turn the crankshaft pulley clockwise until the arrow mark on the camshaft is positioned as shown to measure the # 2 exhaust and # 3 intake valves—2.0L engines

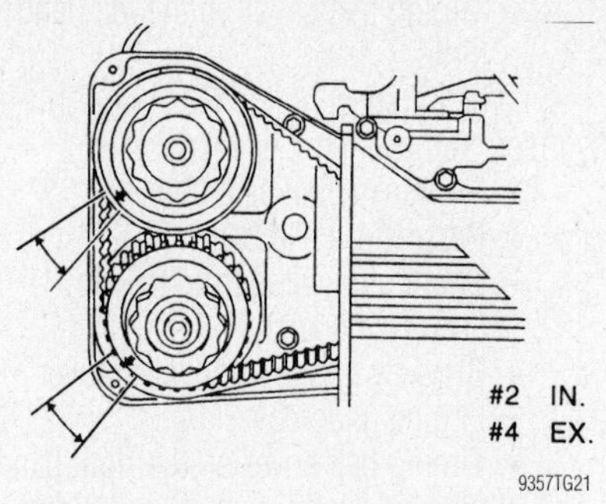

Turn the crankshaft pulley clockwise until the arrow mark on the camshaft is positioned as shown to measure the # 2 intake and # 4 exhaust valves—2.0L engines

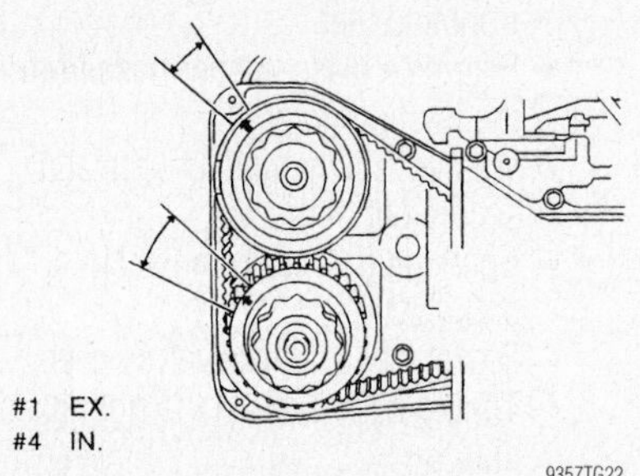

Turn the crankshaft pulley clockwise until the arrow mark on the camshaft is positioned as shown to measure the # 1 exhaust and # 4 intake valves—2.0L engines

8. Turn the crankshaft pulley clockwise to measure the valve clearance for the # 2 exhaust and # 3 intake valves as shown in the accompanying illustration.

9. Turn the crankshaft pulley clockwise to measure the valve clearance for the # 2 intake and # 4 exhaust valves as shown in the accompanying illustration.

10. Turn the crankshaft pulley clockwise to measure the valve clearance for the # 1 exhaust and # 4 intake valves as shown in the accompanying illustration.

11. Adjust the valve clearance as follows:

a. Measure and record all valve clearances using the procedures outlined in the inspection steps in this section.

b. Prepare shim replacer tool 498187200.

c. Rotate the notch of the valve lifter outwards 45 degrees.

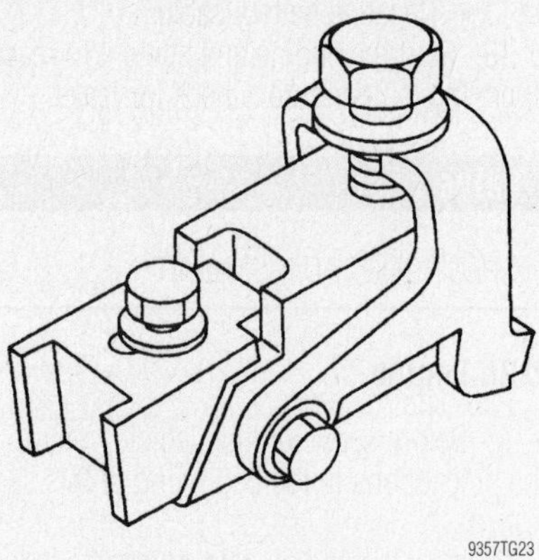

Shim replacer tool 498187200 is required to adjust the valves—2.0L engines

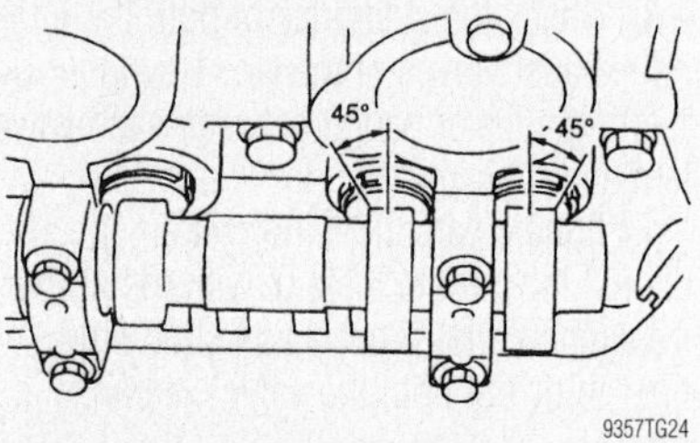

Rotate the notch of the valve lifter outwards 45 degrees—2.0L engines

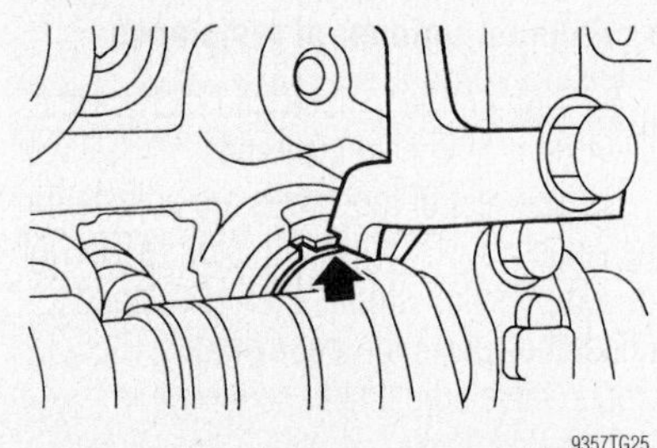

Adjust the shim replacer tool notch to the lifter and set it—2.0L engines

Unit: mm
Intake valve:S =(V + T) - 0.20 Exhaust valve:S =(V + T) - 0.25
S: Shim thickness to be used V: Measured valve clearance T: Shim thickness required

9357TG28

Use this table to help you select a suitable shim—2.0L engines

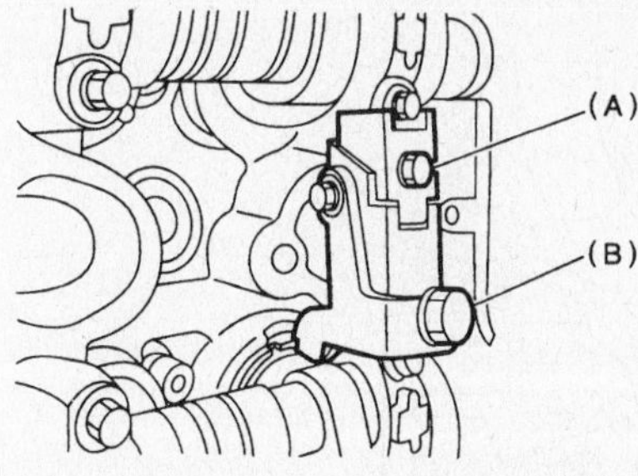

Location of bolts "A" and "B" on the shim replacer tool—2.0L engines

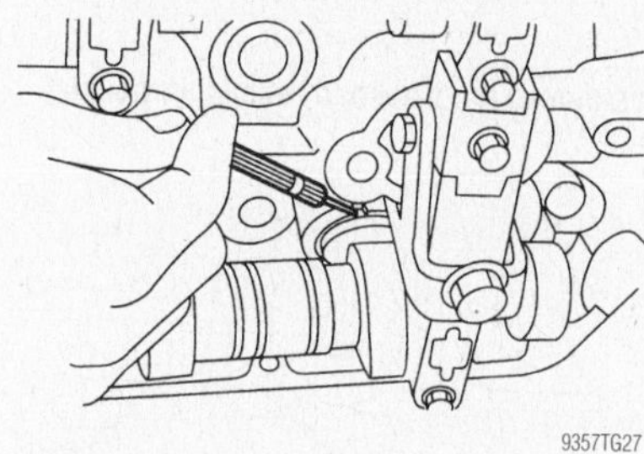

Remove the shim from the lifter—2.0L engines

d. Adjust the shim replacer tool notch to the lifter and set it.

➡**Make sure when setting the tool that the edge does not touch the shim.**

e. Tighten bolt "A" and attach it to the cylinder head. Refer to the accompanying illustration for bolt locations.

f. Tighten bolt "B" and insert the lifter. Refer to the accompanying illustration for bolt locations.

g. Use tweezers and remove the shim from the lifter. A magnet can also be used to remove the shim.

h. Measure the shim thickness using a micrometer.

i. Using the table supplied, select a suitable shim using measured valve clearance and shim thickness.

j. Install the replacement shim to the lifter.

k. After all shims have been adjusted, inspect the valve clearances again.

l. After completion, install all removed components.

3.0L Engine

➡**The valve adjustment should be performed while the engine is cold.**

1. Before servicing the vehicle, refer to the precautions in the beginning of this section.
2. Adjustment should be performed when engine is cold.
3. Remove or disconnect the following:

Part No.	Thickness mm (in)
13218 AK010	2.00 (0.0787)
13218 AK020	2.02 (0.0795)
13218 AK030	2.04 (0.0803)
13218 AK040	2.06 (0.0811)
13218 AK050	2.08 (0.0819)
13218 AK060	2.10 (0.0827)
13218 AK070	2.12 (0.0835)
13218 AK080	2.14 (0.0843)
13218 AK090	2.16 (0.0850)
13218 AK100	2.18 (0.0858)
13218 AK110	2.20 (0.0866)
13218 AE710	2.22 (0.0874)
13218 AE730	2.24 (0.0882)
13218 AE750	2.26 (0.0890)
13218 AE770	2.28 (0.0898)
13218 AE790	2.30 (0.0906)
13218 AE810	2.32 (0.0913)
13218 AE830	2.34 (0.0921)
13218 AE850	2.36 (0.0929)
13218 AE870	2.38 (0.0937)
13218 AE890	2.40 (0.0945)
13218 AE910	2.42 (0.0953)
13218 AE920	2.43 (0.0957)
13218 AE930	2.44 (0.0961)
13218 AE940	2.45 (0.0965)
13218 AE950	2.46 (0.0969)
13218 AE960	2.47 (0.0972)
13218 AE970	2.48 (0.0976)
13218 AE980	2.49 (0.0980)
13218 AE990	2.50 (0.0984)
13218 AF000	2.51 (0.0988)
13218 AF010	2.52 (0.0992)
13218 AF020	2.53 (0.0996)
13218 AF030	2.54 (0.1000)
13218 AF040	2.55 (0.1004)
13218 AF050	2.56 (0.1008)
13218 AF060	2.57 (0.1012)
13218 AF070	2.58 (0.1016)
13218 AF090	2.60 (0.1024)
13218 AF110	2.62 (0.1031)
13218 AF130	2.64 (0.1039)
13218 AF150	2.66 (0.1047)
13218 AF170	2.68 (0.1055)
13218 AF190	2.70 (0.1063)

9357TG30

Valve adjusting shim chart—2.0L engines

- Negative battery cable
- Engine undercover

4. To inspect the right hand side perform remove the following:
   - Drive belt
   - Power steering hose from the bracket
   - Power steering pump bracket bolts and place the pump assembly on the right side wheel apron with the lines still attached
   - Fuel pipe protector from the right hand side
   - Fuel injector connections
   - Front Oxygen Sensor (O_2S) connector
   - Oil pressure switch connector
   - Ignition coils
   - Valve cover on the right hand side
5. To inspect the left hand side perform remove the following:
   - Battery
   - Window washer motor connectors front and rear
   - Rear gate glass washer hose from the washer motor
   - Washer tank mounting bolts and secure out of the way
   - Positive Crankcase Ventilation (PCV) and blow–by hose from the left hand rocker cover
   - Fuel pipe protector from the left hand side
   - Fuel injector connections
   - Front O_2S connector
   - Ignition coils
   - Valve cover on the left hand side
6. Using crankshaft socket tool ST 18252AA000, turn the crankshaft clockwise. Adjust the camshaft position so the camshaft lobe is perpendicular to the shim as illustrated.
7. Check the valve clearance:
   - Intake valve: 0.0063–0.0095 in. (0.16–0.24mm).
   - Exhaust valve: 0.0078–0.0118 in. (0.20–0.25mm).
8. If any valve needs adjustment, perform the following:
   a. Record each valve measurement after it has been measured.
   b. Using shim replacer tool ST 18329AA000, remove the shim from the lifter.
   c. Rotate the notch of the valve lifter outwards by 45 degrees.
   d. Adjust the shim replacer notch to the valve lifter and set it as illustrated. Make sure when setting the replacer, the edge does not touch the shim.
   e. Tighten bolt **A** on the tool and attach it to the cylinder head, then tighten

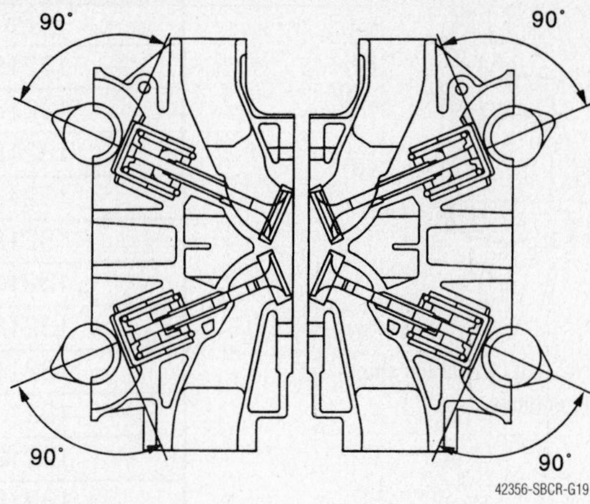

Turn the crankshaft clockwise. Adjust the camshaft position so the camshaft lobe is perpendicular to the shim —3.0L engine

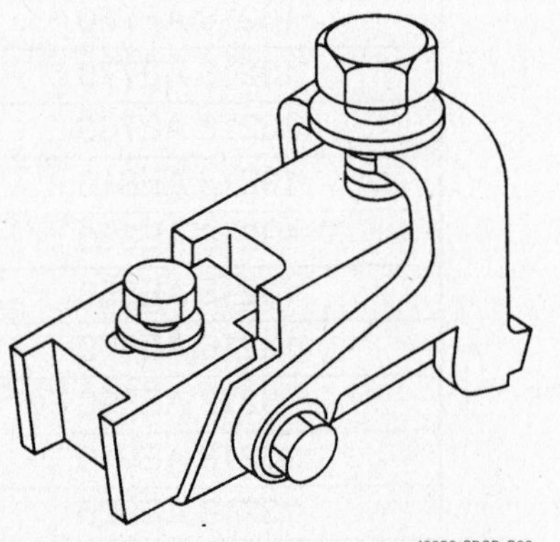

Use shim replacer tool ST 18329AA000 to remove the shim from the lifter —3.0L engine

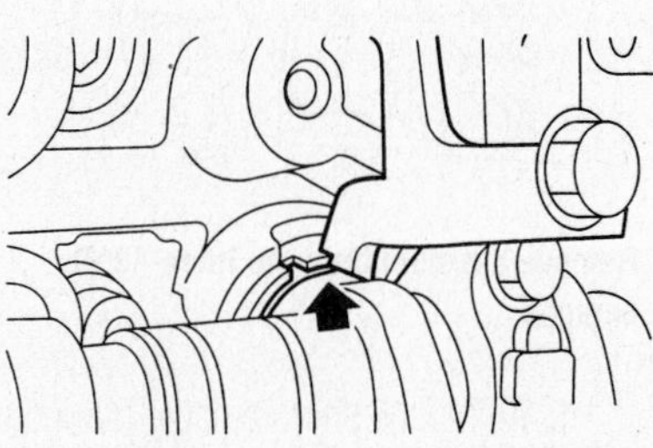

Adjust the shim replacer notch to the valve lifter and set it as shown—3.0L engine

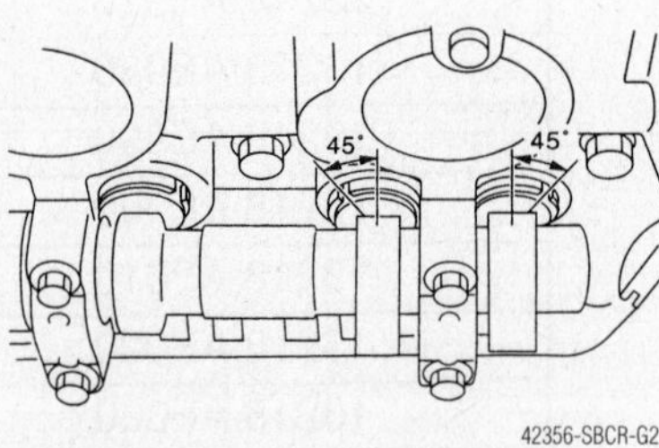

Rotate the notch of the valve lifter outwards by 45 degrees—3.0L engine

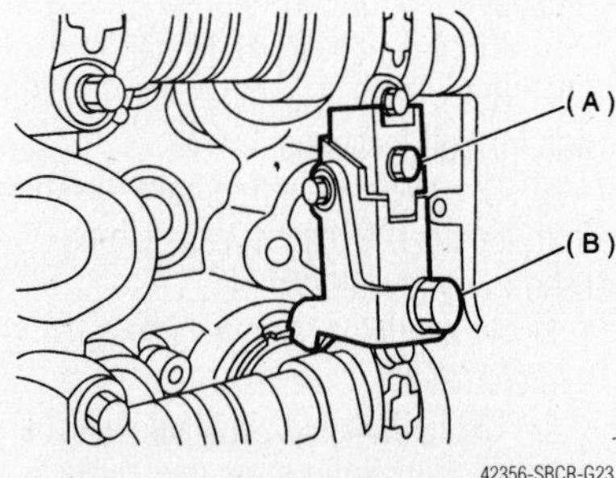

Tighten bolt A on the tool and attach it to the cylinder head, then tighten bolt B and insert the valve lifter—3.0L engine

Unit: mm
Intake valve:S =(V + T) - 0.20 Exhaust valve:S =(V + T) - 0.25
S: Shim thickness to be used V: Measured valve clearance T: Shim thickness required

9357TG28

Use this table to help you select a suitable shim—3.0L engines

Part No.	Thickness mm (in)
13218 AK010	2.00 (0.0787)
13218 AK020	2.02 (0.0795)
13218 AK030	2.04 (0.0803)
13218 AK040	2.06 (0.0811)
13218 AK050	2.08 (0.0819)
13218 AK060	2.10 (0.0827)
13218 AK070	2.12 (0.0835)
13218 AK080	2.14 (0.0843)
13218 AK090	2.16 (0.0850)
13218 AK100	2.18 (0.0858)
13218 AK110	2.20 (0.0866)
13218 AE710	2.22 (0.0874)
13218 AE720	2.23 (0.0878)
13218 AE730	2.24 (0.0882)
13218 AE740	2.25 (0.0886)
13218 AE750	2.26 (0.0890)
13218 AE760	2.27 (0.0894)
13218 AE770	2.28 (0.0898)
13218 AE780	2.29 (0.0902)
13218 AE790	2.30 (0.0906)
13218 AE800	2.31 (0.0909)
13218 AE810	2.32 (0.0913)
13218 AE820	2.33 (0.0917)
13218 AE830	2.34 (0.0921)
13218 AE840	2.35 (0.0925)
13218 AE850	2.36 (0.0929)
13218 AE860	2.37 (0.0933)
13218 AE870	2.38 (0.0937)
13218 AE880	2.39 (0.0941)
13218 AE890	2.40 (0.0945)
13218 AE900	2.41 (0.0949)
13218 AE910	2.42 (0.0953)

42356-SBCR-G24

Valve adjusting shim chart (1 of 2)—3.0L engines

Part No.	Thickness mm (in)
13218 AE920	2.43 (0.0957)
13218 AE930	2.44 (0.0961)
13218 AE940	2.45 (0.0965)
13218 AE950	2.46 (0.0969)
13218 AE960	2.47 (0.0972)
13218 AE970	2.48 (0.0976)
13218 AE980	2.49 (0.0980)
13218 AE990	2.50 (0.0984)
13218 AF000	2.51 (0.0988)
13218 AF010	2.52 (0.0992)
13218 AF020	2.53 (0.0996)
13218 AF030	2.54 (0.1000)
13218 AF040	2.55 (0.1004)
13218 AF050	2.56 (0.1008)
13218 AF060	2.57 (0.1012)
13218 AF070	2.58 (0.1016)
13218 AF090	2.60 (0.1024)
13218 AF100	2.61 (0.1028)
13218 AF110	2.62 (0.1031)
13218 AF120	2.63 (0.1035)
13218 AF130	2.64 (0.1039)
13218 AF140	2.65 (0.1043)
13218 AF150	2.66 (0.1047)
13218 AF160	2.67 (0.1051)
13218 AF170	2.68 (0.1055)
13218 AF180	2.69 (0.1059)
13218 AF190	2.70 (0.1063)
13218 AF200	2.71 (0.1067)
13218 AF210	2.72 (0.1071)
13218 AF220	2.73 (0.1075)
13218 AF230	2.74 (0.1079)
13218 AF240	2.75 (0.1083)
13218 AF250	2.76 (0.1087)
13218 AF260	2.77 (0.1091)
13218 AF270	2.78 (0.1094)
13218 AF280	2.79 (0.1098)
13218 AF290	2.80 (0.1102)
13218 AF300	2.81 (0.1106)

42356-SBCR-G25

Valve adjusting shim chart (2 of 2)—3.0L engines

bolt **B** and insert the valve lifter. Refer to the accompanying illustration.

f. Use tweezers to remove the shim. A magnet may be used as well to remove the shim without dropping it.

g. Use a micrometer to measure the shim.

9. Measure the shim thickness using a micrometer.

10. Using the table supplied, select a suitable shim using measured valve clearance and shim thickness.

11. Install the replacement shim to the lifter.

12. After all shims have been adjusted, inspect the valve clearances again.

13. After completion, install all removed components.

Starter Motor

REMOVAL & INSTALLATION

1. Before servicing the vehicle, refer to the precautions in the beginning of this section.
2. Remove or disconnect the following:
   - Negative battery cable
   - Intake Air Temperature (IAT) connector, on Legacy models equipped with a manual transaxle
   - Air cleaner case and duct
   - Air cleaner case stay, on Legacy models
   - Connector and terminal from starter
   - Retaining bolts and/or nuts
   - Starter from transmission

To install:

3. Install or connect the following:
   - Starter to the transmission
   - Starter retaining bolts and/or nuts and tighten to 34–40 ft. lbs. (46–54 Nm)
   - Connector and terminal to starter
   - Air cleaner case stay, on Legacy models
   - Air cleaner case and duct
   - IAT connector, on Legacy models equipped with a manual transaxle
   - Negative battery cable

Oil Pan

REMOVAL & INSTALLATION

1. Before servicing the vehicle, refer to the precautions in the beginning of this section.

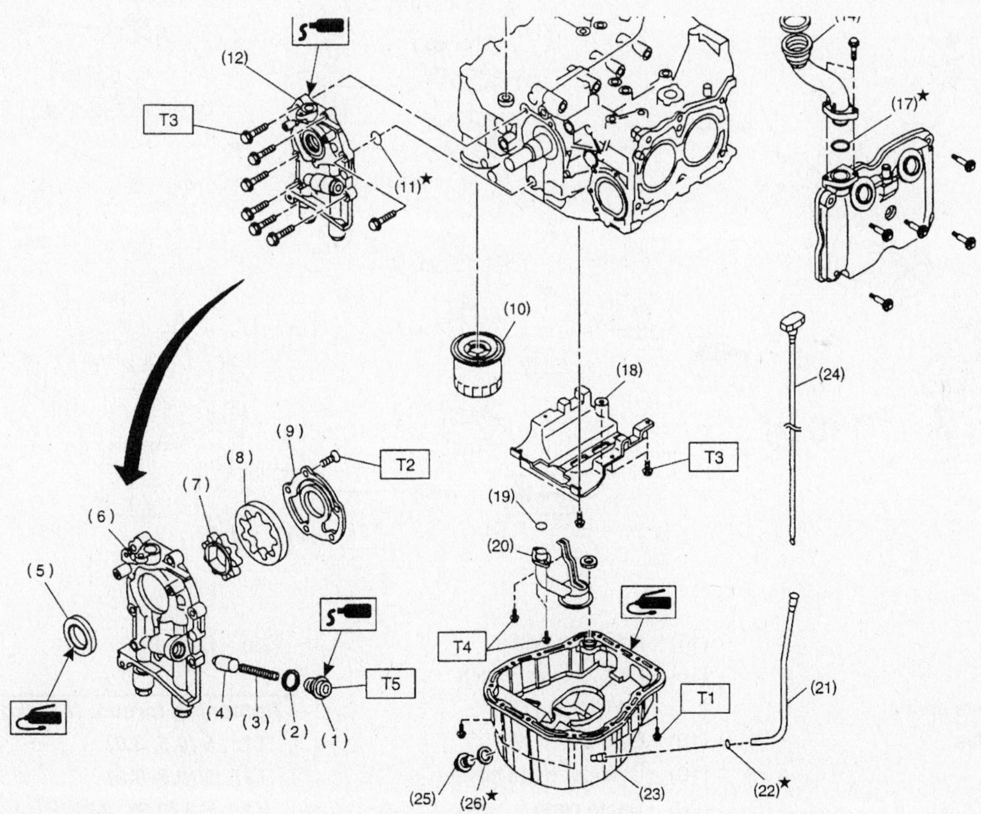

1) Plug
2) Washer
3) Relief valve spring
4) Relief valve
5) Oil seal
6) Oil pump case
7) Inner rotor
8) Outer rotor
9) Oil pump cover
10) Oil filter
11) O-ring
12) Oil pump ASSY
(13) Oil pressure switch
(14) Oil filler duct
(15) O-ring
(16) Oil filler cap
(17) O-ring
(18) Baffle plate
(19) O-ring
(20) Oil strainer
(21) Oil level gauge guide
(22) O-ring
(23) Oil pan
(24) Oil level gauge
(25) Drain plug
(26) Metal gasket

Tightening torque: N·m (kg-m, ft-lb)
T1: 5 (0.5, 3.6)
T2: 5^{+1}_{-0} $(0.5^{+0.1}_{-0}, 3.6^{+0.7}_{-0})$
T3: 6.4 (0.65, 4.7)
T4: 10 (1.0, 7.2)
T5: 44.1±3.4 (4.5±0.35, 32.5±2.5)

9307TG03

Exploded view of the oil pan and lubrication components—2.2L and 2.5L engines

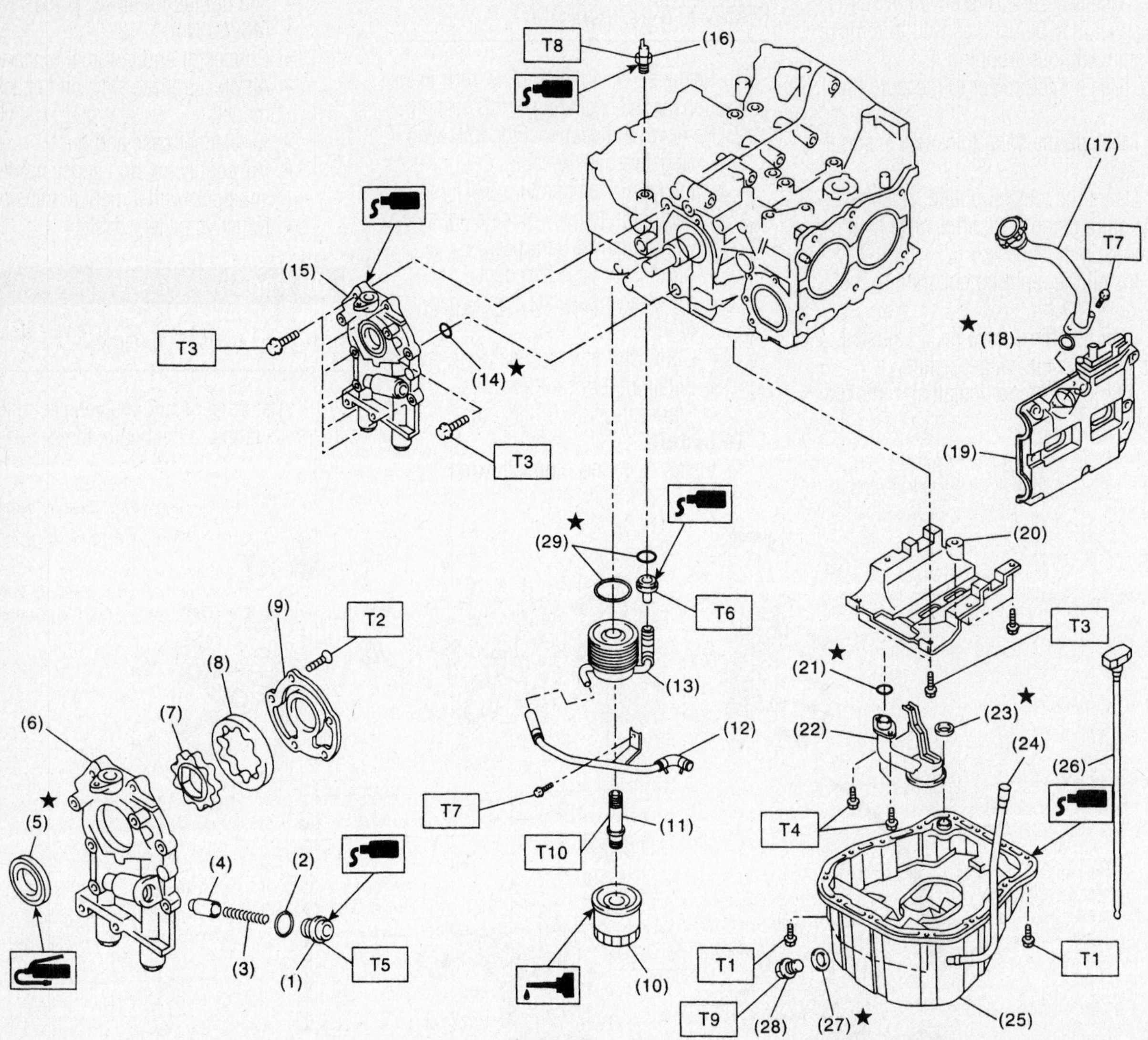

(1) Plug
(2) Washer
(3) Relief valve spring
(4) Relief valve
(5) Oil seal
(6) Oil pump case
(7) Inner rotor
(8) Outer rotor
(9) Oil pump cover
(10) Oil filter
(11) Connector
(12) Water by-pass pipe
(13) Oil cooler
(14) O-ring
(15) Oil pump ASSY
(16) Oil pressure switch
(17) Oil filler duct
(18) O-ring
(19) Cylinder head cover
(20) Baffle plate
(21) O-ring
(22) Oil strainer
(23) Gasket
(24) Oil level gauge guide
(25) Oil pan
(26) Oil level gauge
(27) Metal gasket
(28) Drain plug
(29) O-ring

Tightening torque: N·m (kgf-m, ft-lb)

T1: 5 (0.5, 3.6)
T2: 5 (0.5, 3.6)
T3: 6.4 (0.65, 4.7)
T4: 10 (1.0, 7.0)
T5: 44.1 (4.5, 32.5)
T6: 69 (7.0, 4.7)
T7: 6.4 (0.65, 50.6)
T8: 25 (2.5, 18.1)
T9: 44 (4.5, 33)
T10: 54 (5.5, 40)

9357TG31

Exploded view of the oil pan and lubrication components—2.0L turbocharged engines

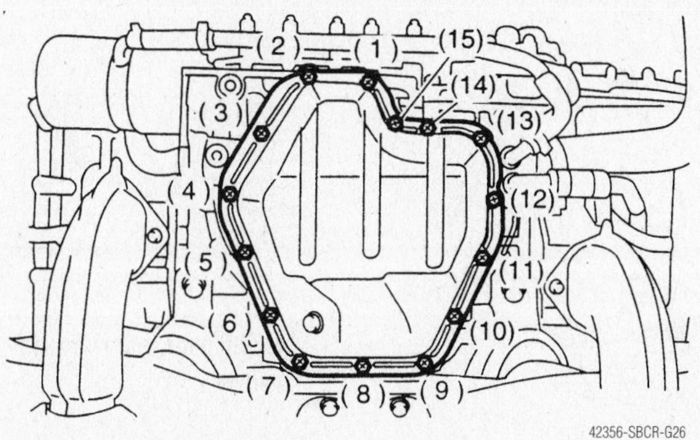

Oil pan bolt torque sequence—3.0L engines

2. Remove or disconnect the following:
    - Negative battery cable
    - Air intake duct
    - Mass Air Flow (MAF) sensor on turbo models
    - Air intake boot and air cleaner upper cover on turbo models
    - Intercooler on turbo models
    - Front Oxygen Sensor (O_2S) electrical connector
    - Pitching stopper
    - Upper radiator brackets
3. Support the engine with a suitable lifting device.
    - Front wheel and tire assemblies
4. Lift up the engine slightly.
    - Engine undercover
5. Drain the oil from the engine into a suitable container.
6. Install the drain plug with a new gasket and tighten it to 33–36 ft. lbs. (43–47 Nm).
7. Remove or disconnect the following:
    - Rear O_2S electrical connector
    - Exhaust Y-pipe
    - Nuts that secure the front engine mounts to the front crossmember
    - Oil pan mounting bolts
8. Insert an oil pan cutter blade between the upper and lower pans, on 3.0L engines
9. While supporting the oil pan, use a rubber mallet and tap the oil pan to free it from the engine.
10. Clean all gasket material from both mating surfaces.

To install:

11. Apply a continuous bead of sealer to a new oil pan gasket.
12. Install the oil pan assembly. Tighten the bolts to 36–48 inch lbs. (4–5 Nm) on all models except 3.0L engines. On 3.0L engines tighten the bolts to 5 ft. lbs. (7 Nm).
13. Lower the engine onto the front crossmember.
14. Install or connect the following:
    - Front engine mount nuts and tighten to 61 ft. lbs. (83 Nm)
    - Y-pipe with new gaskets. Tighten the pipe-to-engine nuts to 23 ft. lbs. (30 Nm)
    - Rear O_2S electrical connector
    - Engine undercover
    - Front wheel and tire assemblies
15. Remove the engine lifting device.
    - Front O_2S sensor electrical connector
    - Pitching stopper. Tighten the front bolt to 40 ft. lbs. (54 Nm) and the rear bolt to 49 ft. lbs. (67 Nm).
    - Upper radiator brackets
    - MAF sensor on turbo models
    - Air intake boot and air cleaner upper cover on turbo models
    - Intercooler on turbo models
    - Air intake duct
    - Negative battery cable
16. Fill the engine to the proper level with the recommended oil and run the engine. Check for leaks.

Oil Pump

REMOVAL & INSTALLATION

2.2L and 2.5L Engines

1. Before servicing the vehicle, refer to the precautions in the beginning of this section.
2. Remove or disconnect the following:
    - Negative battery cable

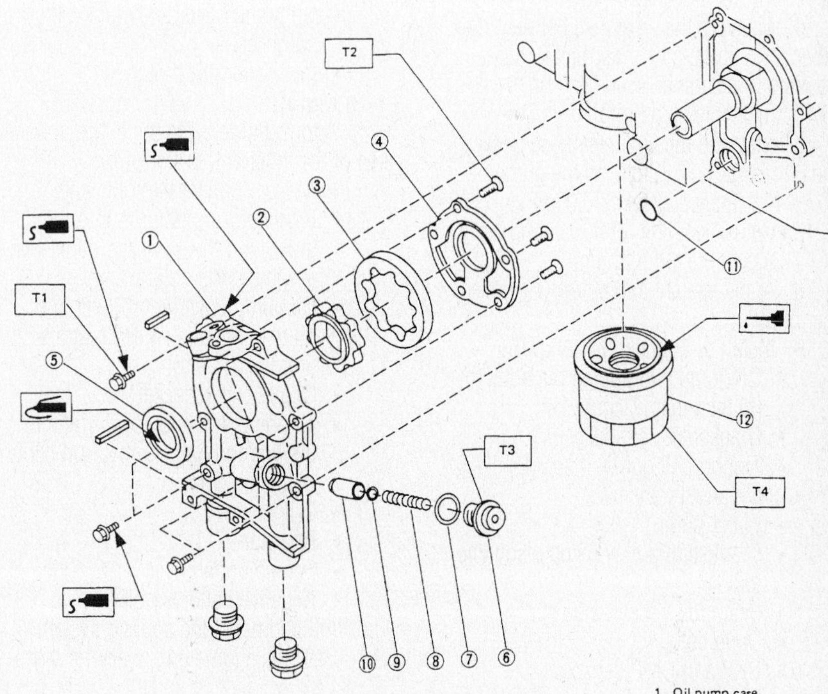

Tightening torque: N·m (kg-m, ft-lb)
T1: 6 – 7 (0.6 – 0.7, 4.3 – 5.1)
T2: 4 – 7 (0.4 – 0.7, 2.9 – 5.1)
T3: 40 – 48 (4.1 – 4.9, 30 – 35)
T4: 12 – 16 (1.2 – 1.6, 9 – 12)

1 Oil pump case
2 Inner rotor
3 Outer rotor
4 Oil pump cover
5 Front oil seal
6 Plug
7 Washer
8 Relief spring
9 Washer
10 Relief valve
11 O-ring
12 Oil filter

7923TG37

Oil pump and components

- Engine undercover

3. Drain the coolant into a suitable separate container.
   - Radiator main fan and sub fan assemblies, on Impreza models
   - Radiator, on Legacy models
   - Crankshaft Position (CKP) and Camshaft Position (CMP) sensors
   - Drive belts
   - Rear side V-belt tensioner
   - Crankshaft pulley using a suitable tool
   - Water pump
   - Timing belt guide, if equipped with a manual transaxle
   - Crankshaft sprocket
   - Oil pump mounting bolts
   - Oil pump by carefully prying it from the engine block

WARNING

Use extreme care not to damage the engine block or the oil pump during removal of the pump.

To install:

4. Measure the tip clearance of the rotors. If clearance is greater than 0.0071 in. (0.18mm), replace the rotors.
5. Measure the clearance between the outer rotor and the cylinder block rotor housing. If clearance exceeds 0.0079 in. (0.20mm), replace the rotor.
6. Measure the side clearance between the oil pump inner rotor and the pump cover. If clearance exceeds 0.0059 in. (0.15mm), replace the rotor or pump body.
7. Assemble the oil pump.
8. Apply sealant and a new O-ring to the oil pump.
9. Install or connect the following:
   - Oil pump and tighten the bolts to 60 inch lbs. (7 Nm)
   - Crankshaft sprocket
   - Timing belt guide, if equipped with a manual transaxle
   - Water pump
   - Crankshaft pulley using a suitable tool
   - Rear side V-belt tensioner
   - Drive belts
   - CKP and CMP sensors
   - Radiator, on Legacy models
   - Radiator main fan and sub fan assemblies, on Impreza models
   - Engine undercover
   - Negative battery cable
10. Fill and bleed the cooling system.
11. Start the engine and check for leaks.

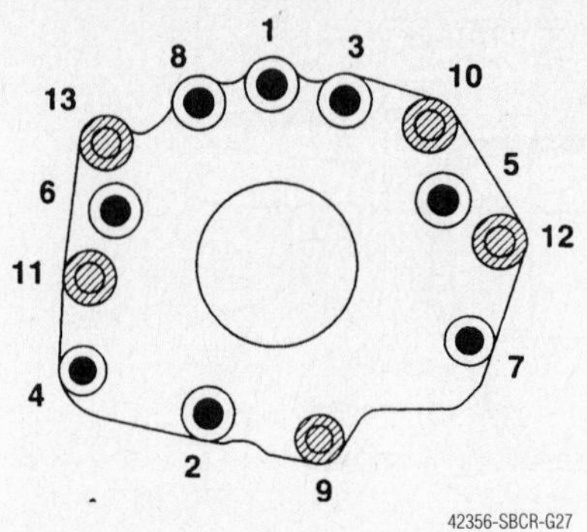

Oil pump cover torque sequence—3.0L engines

3.0L Engine

1. Before servicing the vehicle, refer to the precautions in the beginning of this section.
2. Remove or disconnect the following:
   - Negative battery cable
   - Engine under cover
3. Drain the coolant.
   - Radiator
   - Drive belt
   - Front timing chain cover and chain. Refer to the timing chain removal procedure in this section.
   - Oil pump cover and crankshaft sprocket
   - Inner and outer rotor

To install:

4. Apply engine oil to the entire surface area of the inner and outer rotor
5. Install or connect the following:
   - Inner rotor by fitting it into the groove on the crankshaft and then assemble the outer rotor
   - Oil pump cover and tighten the bolts in the sequence illustrated to 5 ft. lbs. (7 Nm)
   - Crankshaft sprocket
   - Timing chain and cover. Refer to the timing chain removal procedure in this section.
   - Drive belt
   - Radiator
   - Engine undercover
   - Negative battery cable
6. Fill and bleed the cooling system.
7. Start the engine and check for leaks.

Rear Main Seal

REMOVAL & INSTALLATION

1. Before servicing the vehicle, refer to the precautions in the beginning of this section.
2. Remove or disconnect the following:
   - Engine from the vehicle

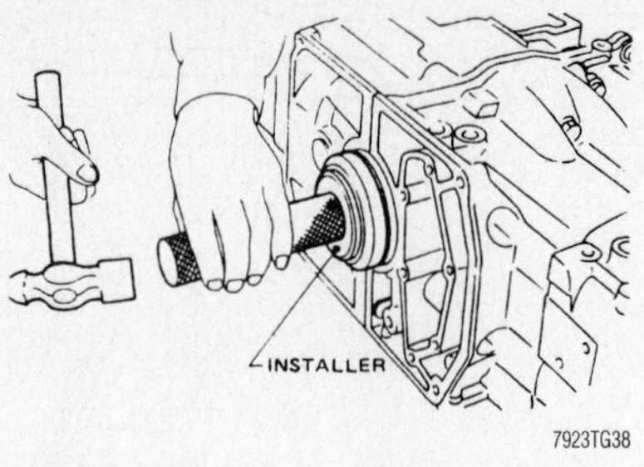

Installing the rear main seal—except 3.0L engines

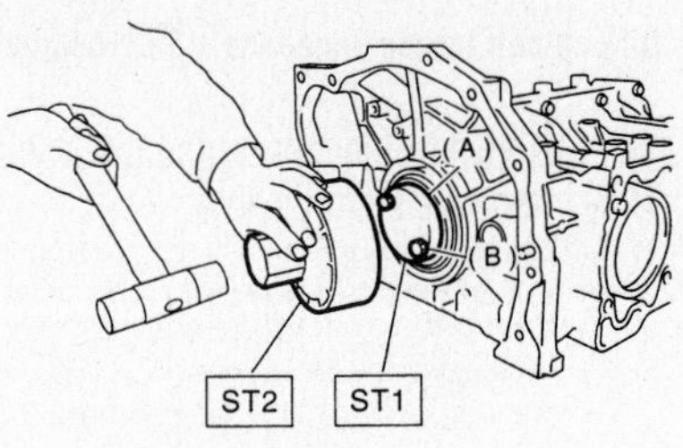

Installing the rear main seal using oil seal guide ST1 499597100 and ST2 499598200—3.0L engines

- Clutch assembly/flywheel using the Clutch Disc Guide tool 499747000, if equipped with a manual transmission
- Torque converter flexplate from the crankshaft, if equipped with an automatic transmission
- Oil seal from the cylinder block using a small prybar

To install:

3. Install or connect the following:
   - New oil seal by pressing it into the cylinder block using the appropriate driver and hammer
   - Flywheel housing using new gaskets and sealant where necessary.
   - Flywheel and tighten the bolts to 50–54 ft. lbs. (69–75 Nm) on all models except 3.0L engines. On 3.0L engines tighten the bolts to 60 ft. lbs. (81 Nm).
   - Engine

Timing Belt

REMOVAL & INSTALLATION

2.0L Engine

1. Before servicing the vehicle, refer to the precautions in the beginning of this section.

2. Disconnect the negative battery cable.
3. Remove the V-belt.
4. Remove the crankshaft pulley.
5. Remove the belt cover as follows:
   a. Crankshaft pulley.
   b. Left hand belt cover.
   c. Right hand belt cover.
   d. Front hand belt cover.
6. Remove the timing belt guides on models equipped with a manual transmission.
7. If the alignment marks that indicate rotation are faded, put new marks on the belt before removal as follows:
   a. Turn the crankshaft using crankshaft sprocket tool 499987500 and a breaker bar to align the crankshaft sprocket, left hand intake camshaft sprocket, left hand exhaust camshaft, right hand intake camshaft sprocket and

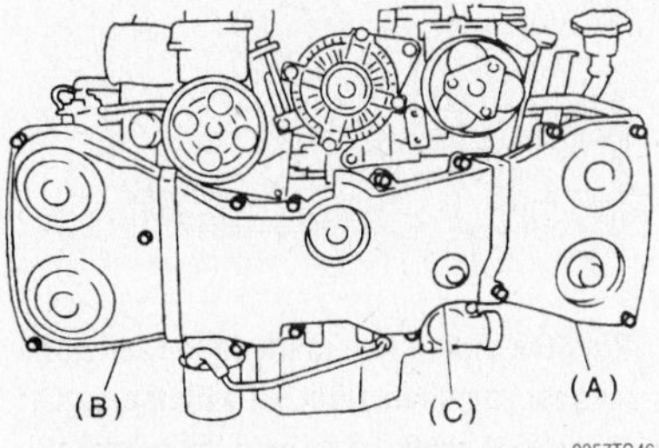

Remove the left hand (A), right hand (B) and front (C) belt covers–2.0L engines

Location of the upper belt guide–2.0L engines

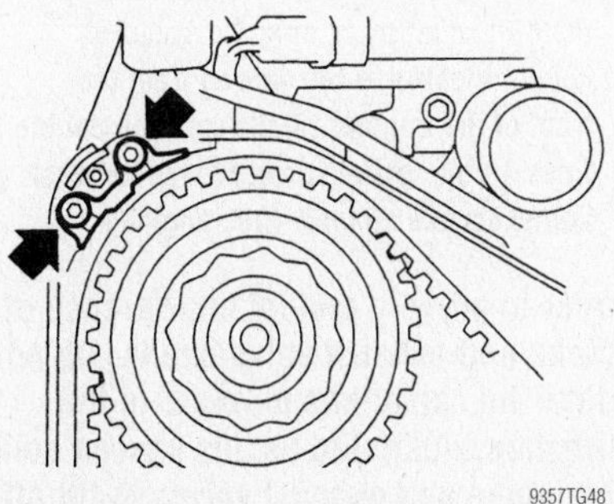

Location of the upper left belt guide–2.0L engines

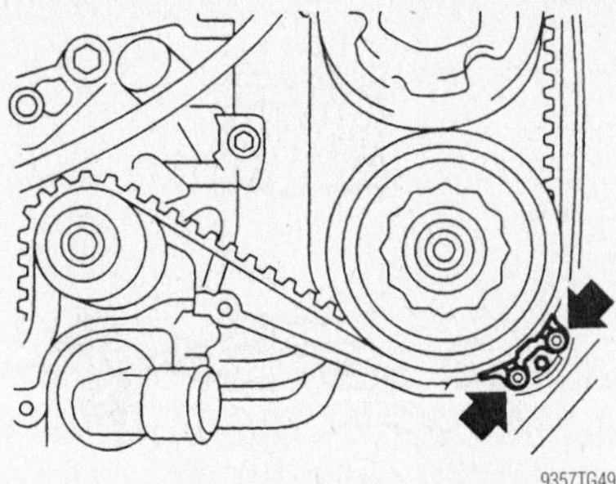

Location of the lower right belt guide–2.0L engines

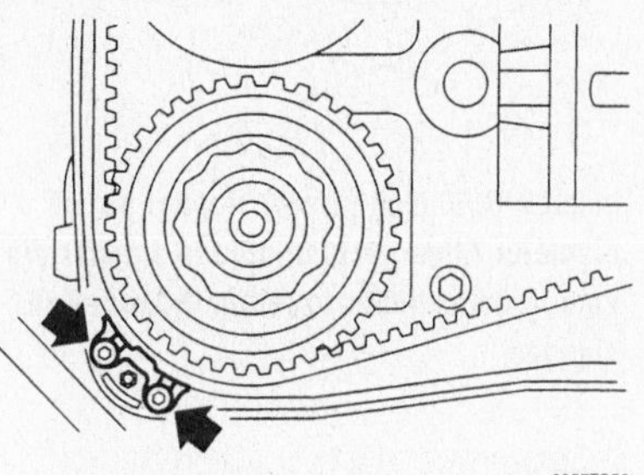

Location of the lower left belt guide–2.0L engines

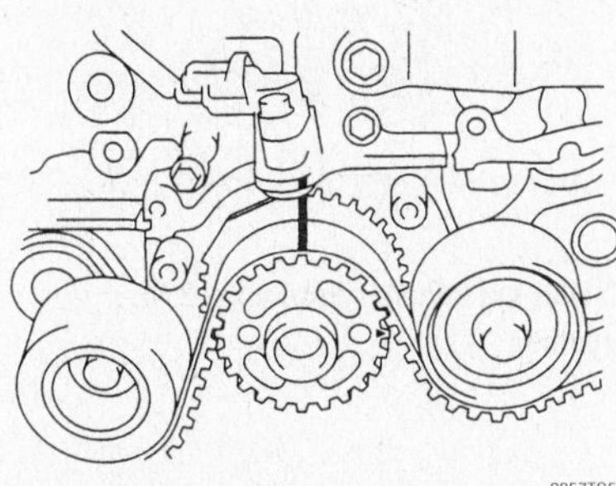

Mark the upper belt-to-sprocket alignment–2.0L engines

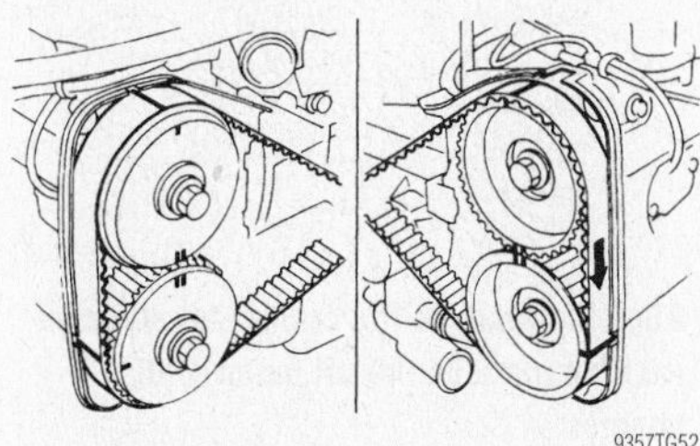

Mark the lower belt-to-sprocket alignment–2.0L engines

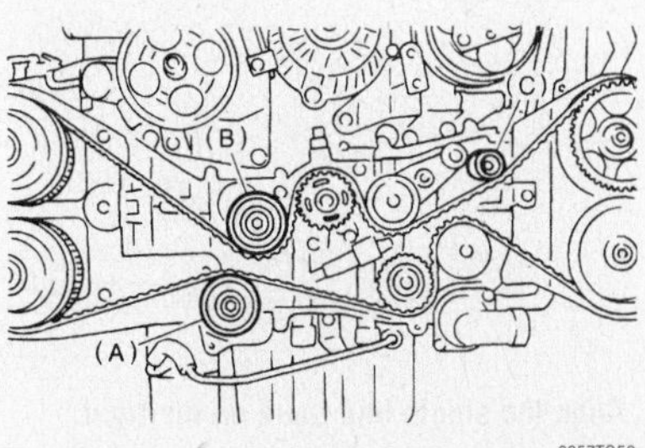

Remove the belt idler (A)–2.0L engines

right hand exhaust camshaft sprocket with the on the cover and cylinder block.
   b. Using white paint such as white out, place alignment marks on the belts in relation to the sprockets.
8. Remove the belt idler (A), illustrated in the accompanying illustration.
9. Remove the timing belt.
10. If necessary, remove belt idlers (B) and (C).
11. Remove the belt idler 2.
12. Remove the automatic belt tension adjuster assembly.

To install:

13. To prepare the automatic belt tensioner for assembly, perform the following steps:
   a. Always use a vertical type pressing tool to move the adjuster rod down.
   b. Do not use a lateral type vise.
   c. Always push the adjuster rod vertically.
   d. Make sure to slowly move the adjuster rod down applying a pressure of 66 lbs. (294 N).
   e. Press in the push adjuster rod gradually taking more than 3 minutes.
   f. Never allow the press pressure to exceed 2,205 lbs. (9,807 N).
   g. Press the adjuster rod as far as the end surface of the cylinder. Do not press the rod into the cylinder as doing so may damage the cylinder.
   h. Never release the press pressure until the stopper pin has been fully inserted.
14. Attach the automatic belt tension adjuster assembly to the vertical pressing tool.
15. Move the adjuster rod down slowly using a pressure of 66 lbs. (294 N) until the rod is aligned with the stopper pin hole in the cylinder.
16. Insert a 0.08 inch (2mm) stopper pin or diameter Allen wrench into the stopper pin hole in the cylinder to retain the rod.
17. Install the adjuster assembly and tighten the retainers to 29 ft. lbs. (39 Nm).
18. Install belt idle 2 and tighten the retainers to 29 ft. lbs. (39 Nm).
19. Install the belt idler and tighten to 29 ft. lbs. (39 Nm).
20. Align the mark on the crankshaft sprocket with the mark on the oil pump.
21. Align the single line mark on the right hand exhaust camshaft sprocket with the notch on the belt cover.
22. Align the single line mark on the right hand intake camshaft with the notch on the belt cover. Make sure the double lines on the intake camshaft and exhaust sprockets are aligned as shown in the accompanying illustration.

23. Align the single line mark on the left hand exhaust camshaft sprocket with the notch on the belt cover by turning the sprocket counterclockwise (as viewed from the front of the engine).

24. Align the single line mark on the left hand intake camshaft sprocket with the notch on the belt cover by turning the sprocket counterclockwise (as viewed from the front of the engine). Make sure the double lines on the intake camshaft and exhaust sprockets are aligned as shown in the accompanying illustration.

➡Make sure the camshaft and crankshaft sprockets are positioned correctly. The intake and exhaust camshafts on this engine can be rotated independently with the timing belt removed. By looking at the accompanying illustration it will show you that if the intake and exhaust valve are lift together the heads will hit each other and bend.

➡When the timing belts are not installed, 4 camshafts are held at "zero lift" position, where all cams on the camshafts do not push the intake and exhaust valves down (under this condition all valves remain unlifted). When the camshafts are rotated to install the timing belts, # 2 intake and # 4 exhaust cam of the left hand camshafts are held to push their corresponding valves down. Under this condition these valves are held lifted. The right side camshafts are held in so that their cams do not push the valves down. The left hand camshafts must be rotated from the "zero lift" position to the position where the timing belt is to be installed at as small an angle as possible, in

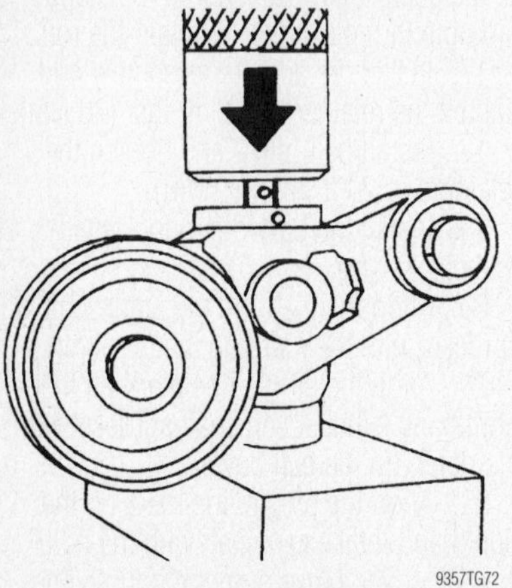

Attach the automatic belt tension adjuster assembly to the vertical pressing tool–2.0L engines

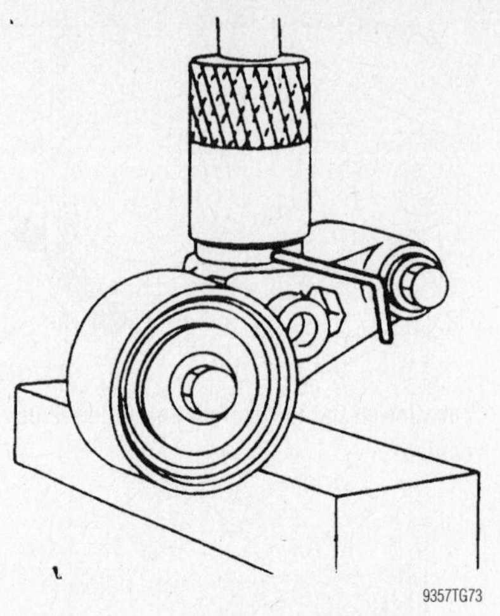

Insert a 0.08 inch (2mm) stopper pin or diameter Allen wrench into the stopper pin hole in the cylinder to retain the rod–2.0L engines

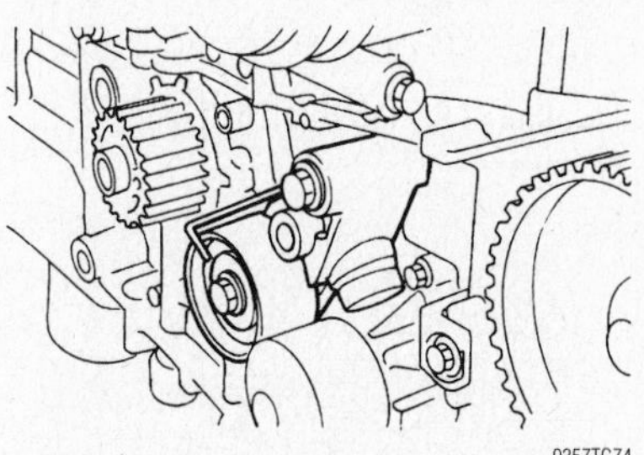

Install the adjuster assembly–2.0L engines

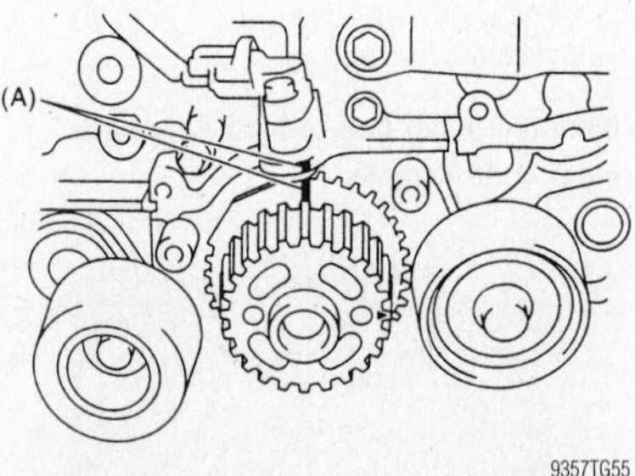

Align the mark on the crankshaft sprocket with the mark on the oil pump–2.0L engines

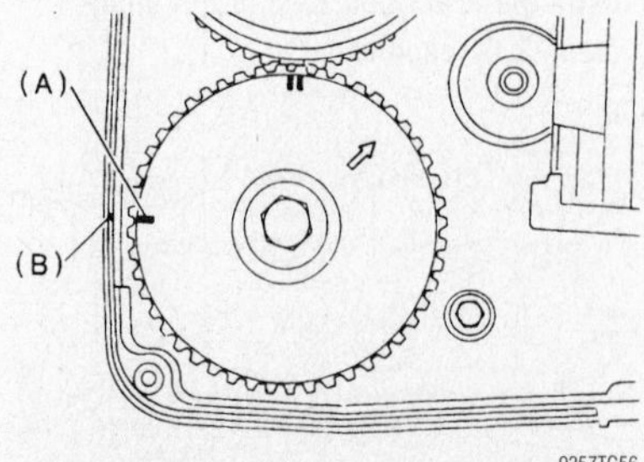

Align the single line mark on the right hand exhaust camshaft sprocket with the notch on the belt cover–2.0L engines

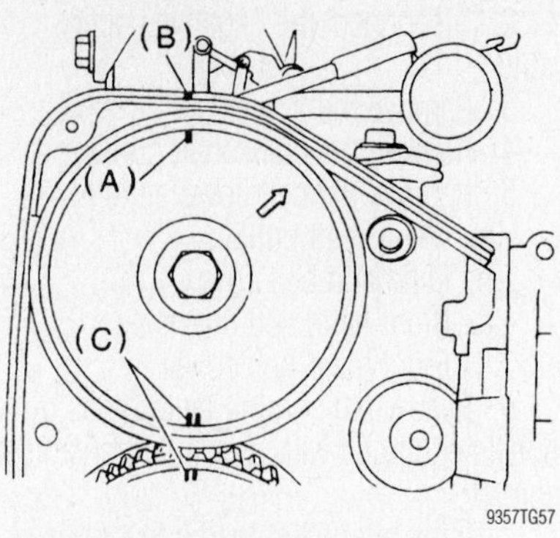

Align the single line mark on the right hand intake camshaft with the notch on the belt cover. Make sure the double lines on the intake camshaft and exhaust sprockets are aligned–2.0L engines

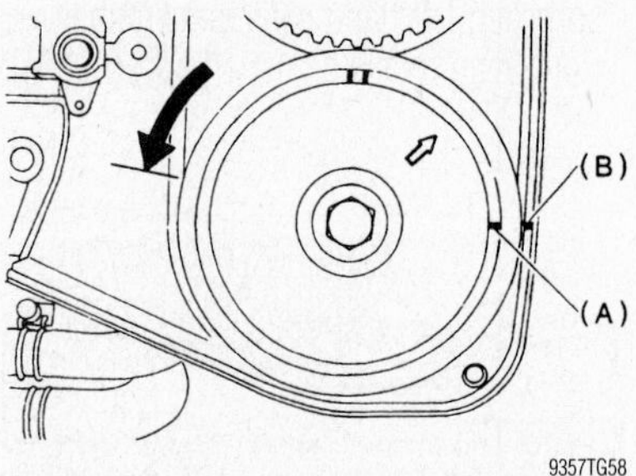

Align the single line mark on the left hand exhaust camshaft sprocket with the notch on the belt cover by turning the sprocket counterclockwise (as viewed from the front of the engine)–2.0L engines

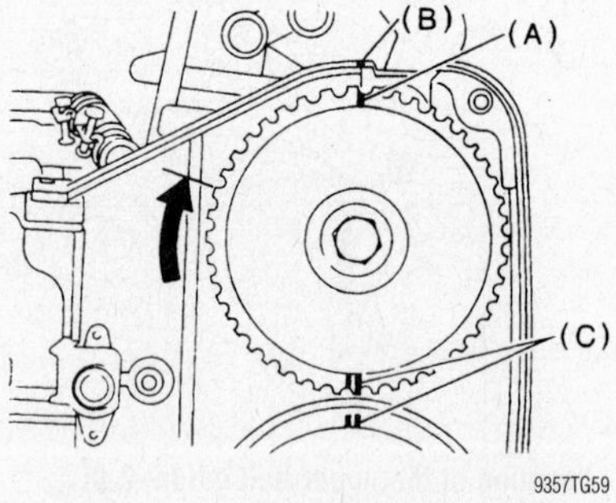

Align the single line mark on the left hand intake camshaft sprocket with the notch on the belt cover by turning the sprocket counterclockwise (as viewed from the front of the engine). Make sure the double lines on the intake camshaft and exhaust sprockets are aligned–2.0L engines

order to prevent mutual interference of intake and exhaust valve heads. Do not allow the camshafts to rotate in the direction illustrated as this causes both the intake and exhaust valves to lift off at the same time with will cause valve damage.

25. When installing the belt, make sure to align the marks made during removal or if using a new belt, align the in alphabetical order as shown in the accompanying illustration.

⁂ WARNING

Disengagement of more than 3 timing belt teeth may result in contact between the valve and piston. Always make sure the belts rotation is correct.

26. Install the belt idlers and tighten to 29 ft. lbs. (39 Nm).

⁂ WARNING

Make sure the marks on the belt and sprockets are properly aligned.

27. Once the marks on the belt and sprockets are aligned, remove the stopper pin from the tensioner adjuster.
28. Install the timing belt guide on models with manual transmission. measure the clearance between the belt and guide. the clearance should be 0.019–0.059 inch (0.5–1.5mm) and tighten the retainers to 7 ft. lbs. (10 Nm).
29. Install the belt covers and tighten to 3.5 ft. lbs. (5 Nm).
30. Install the crankshaft pulley and tighten the bolt to 94 ft. lbs. (127 Nm).
31. Install the V-belt.

2.2L and 2.5L Engine

1. Before servicing the vehicle, refer to the precautions in the beginning of this section.

When servicing the timing belt, note the following:

a. The intake and exhaust camshafts can be rotated independently when the timing belt is removed. If the intake and exhaust valves are lifted off of their seats simultaneously, their heads will contact each other, possibly causing damage.
b. When the timing belt is removed, the camshafts are positioned so that none of the valves are lifted off of their seats, resulting in a "zero-lift" position.
c. The left-hand cylinder head camshafts must be rotated from the "zero-lift" position as little as possible when orienting it for timing belt installation, otherwise possible valve head interference may occur.
d. Never allow the camshafts to rotate in the direction shown in the accompanying illustration, which would cause both the intake and exhaust valves to lift simultaneously, causing interference.

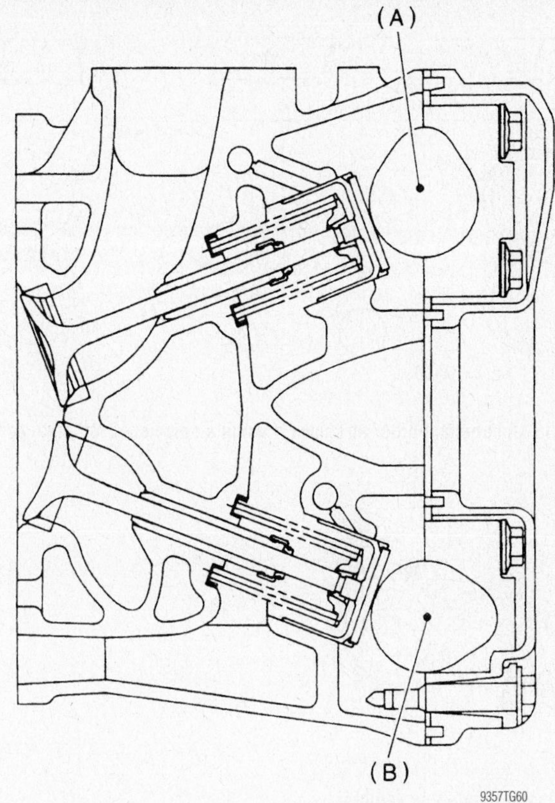

If the intake and exhaust valve are lift together the heads will hit each other and bend –2.0L engines

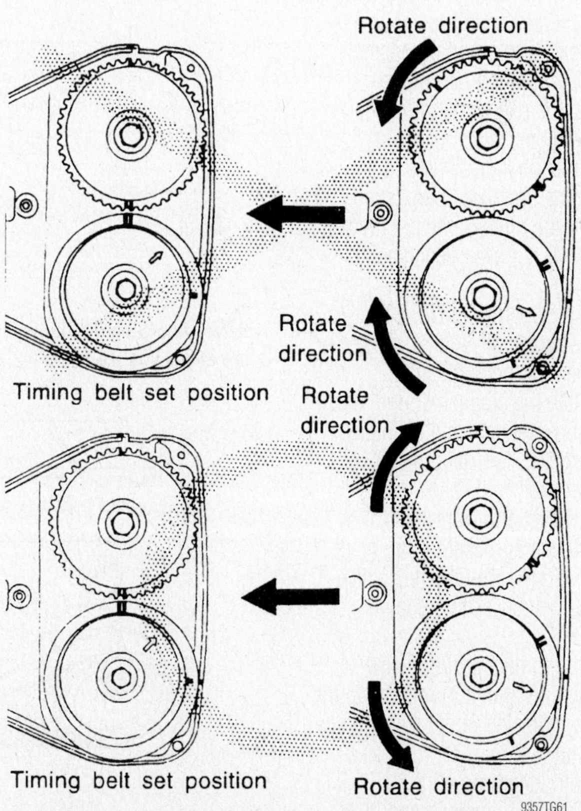

Do not allow the camshafts to rotate in the direction shown as this causes both the intake and exhaust valves to lift off at the same time with will cause valve damage–2.0L engines

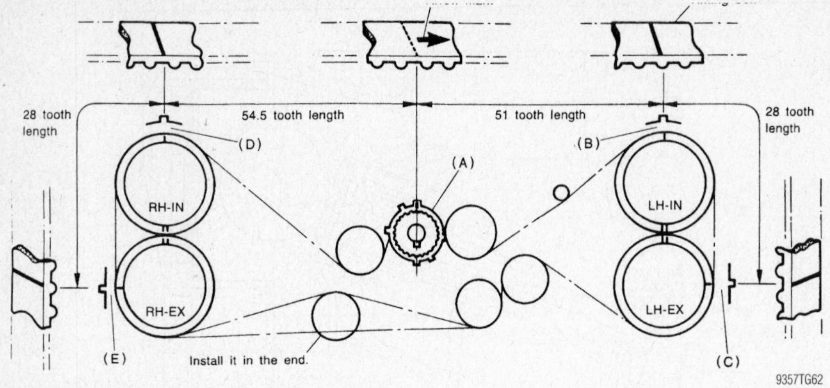

Align the marks in alphabetical order as shown if using a new belt–2.0L engines

Remove the timing belt guide on vehicles equipped with manual transmission—2.5L engine shown, 2.2L similar

2. Remove all necessary components to gain access to the timing belt.

3. If equipped with manual transmissions, loosen the 2 timing belt guide mounting bolts, then separate the guide from the engine block.

4. If the directional arrow and alignment marks on the timing belt are faded, and the belt is to be reused, remark the belt with white paint or a grease pencil as follows:

a. Using a Subaru tool No. ST-499987500 Crankshaft Socket, or equivalent, installed on the crankshaft sprocket, rotate the crankshaft until the crankshaft sprocket, left-hand exhaust camshaft sprocket, left-hand intake camshaft sprocket, right-hand intake camshaft sprocket and right-hand exhaust camshaft sprocket timing mark notches are aligned with the respective marks on the belt cover and engine block.

b. Make alignment and/or arrow marks on the timing belt in relation to the sprockets as indicated in the accompanying illustration.

- Z1: 44 tooth length on all 2.2L engines and 2000 2.5L engines and 46.8 tooth length on 2001–04 2.5L engines
- Z2: 40.5 tooth length on all 2.2L engines and 2000 2.5L engines or 43.7 tooth length on 2001–04 2.5L engines

5. Loosen the center bolt from the timing belt idler pulley, then remove the idler pulley from the engine block.

⁂ WARNING

After removing the timing belt, DO NOT rotate the camshafts. Damage to the valves may occur.

6. Carefully remove the timing belt from all of the sprockets.

7. Remove the automatic belt tension adjuster assembly as follows:

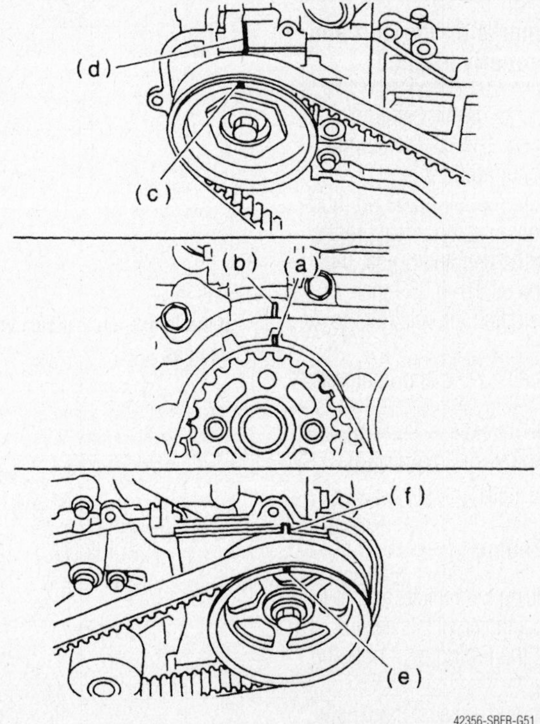

Before removing the timing belt, turn the crankshaft sprocket until all of the alignment marks are aligned as indicated—2.5L engine shown, 2.2L similar

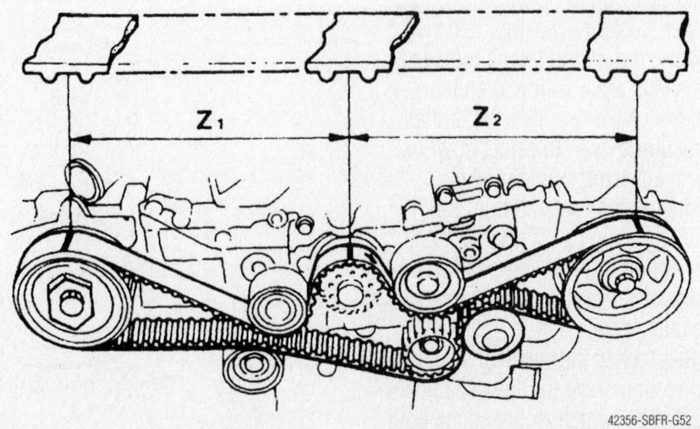

If the original marks on the timing belt are worn or faded, make new alignment marks in the positions indicated—2.5L engine shown, 2.2L similar

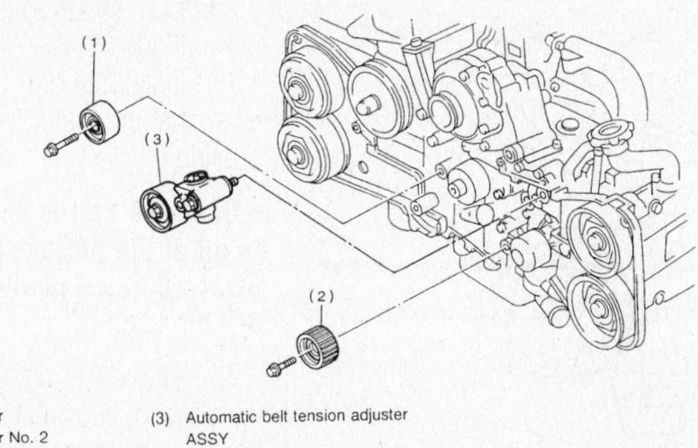

It is necessary to remove the automatic adjuster assembly and reset the pushrod for timing belt installation—2.5L engine shown, 2.2L similar

a. Remove the 2 timing belt idler pulleys, as indicated in the accompanying illustration.

b. Loosen the automatic tension adjuster assembly mounting bolts, then separate the adjuster assembly from the engine block.

To install:

WARNING

Do not allow oil, grease, or coolant to come in contact with the timing belt. If this occurs, quickly and thoroughly remove all traces of the compound. Also, never bend the timing belt sharply; the minimum bending radius is 2.36 in. (60mm).

8. Inspect the camshaft and crankshaft sprocket teeth for abnormal or excessive wear or scratches. Ensure there is no free-play between the sprocket and the key. Inspect the crankshaft sprocket sensor notch for damage or contamination with debris or dirt.

➡When preparing the automatic tension adjuster assembly for installation, adhere to the following points:

- Always use a vertical press, rather than a horizontal press or vise, to depress the adjuster assembly rod

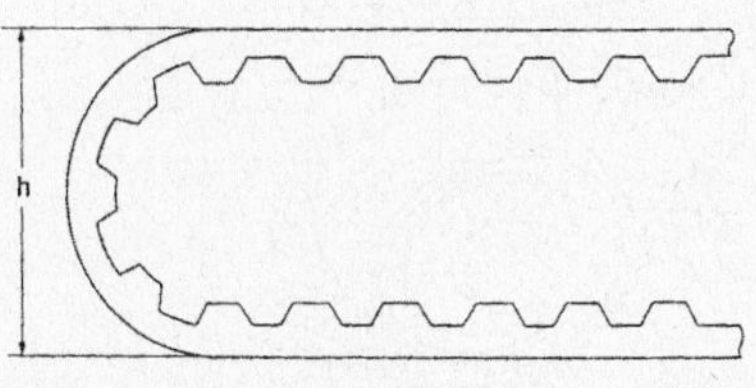

Never bend the timing belt into a radius tighter than 2.36 in./60mm (h), otherwise it will be damaged beyond—2.5L engine shown, 2.2L similar

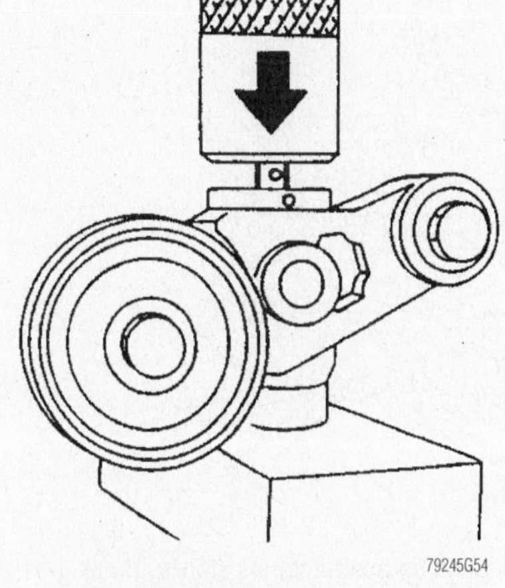

Use a vertical press to push the adjuster rod into its housing until it is flush with the assembly's outer surface . . .

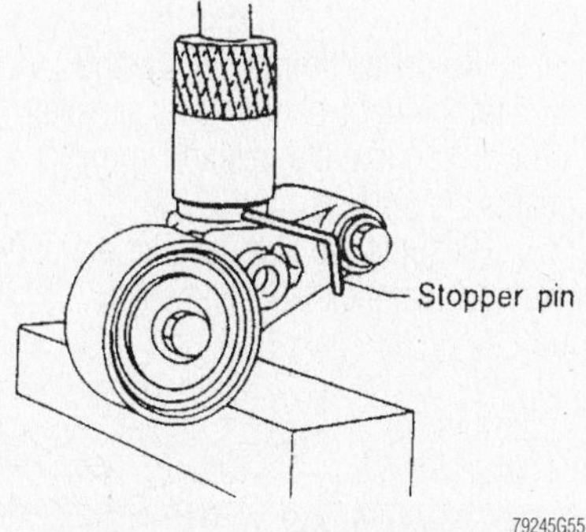

. . . then insert a 0.08 in. (2mm) diameter pin or Allen wrench into the housing and rod holes to hold it in position—2.5L engine shown, 2.2L similar

- Depress the adjuster rod in a vertical position ONLY
- Depress the adjuster rod slowly (taking more than 3 minutes) with a force of 66 lbs. (30 kg)
- Do not allow the press force to exceed 2205 lbs. (1000 kg)
- Press the adjuster rod in as far as the end surface of the cylinder — do not press the rod into the cylinder, which may cause damage to the assembly
- Do not release the press force from the rod until the stopper pin is completely inserted in the cylinder

9. Prepare the automatic timing belt tension adjuster assembly for installation as follows:

a. Position the adjuster assembly in a vertical press.

b. Slowly depress the adjuster rod with a force of 66 lbs. (30 kg) until the hole in the rod is aligned with the hole in the adjuster cylinder housing.

c. Insert a 0.08 in. (2mm) diameter stopper pin or Allen wrench through the hole in the cylinder housing and rod, then slowly release the press force from the adjuster rod.

10. Install the adjuster assembly onto the engine block.

11. Install timing belt idler pulley No. 2 on the engine block. Tighten the bolts to 28 ft. lbs. (39 Nm).

12. Install the timing belt idler pulley No. 1 on the engine block. Tighten the bolts to 28 ft. lbs. (39 Nm).

13. If the camshaft and crankshaft timing marks are no longer aligned, perform the following:

a. Position the crankshaft sprocket so that its mark is aligned with the mark on the oil pump cover on the engine block.

b. Align the single line mark on the right-hand exhaust camshaft sprocket with the notch on the belt cover.

c. Rotate the right-hand intake camshaft so that the single line mark is aligned with the notch on the belt cover.

➡At this point, the double line marks on both right-hand camshaft sprockets should be aligned.

d. Turn the left-hand exhaust (lower) camshaft counterclockwise (as viewed from the front of the engine) until the single line mark is aligned with the notch on the belt cover.

e. Position the single line mark on the left-hand intake camshaft sprocket so that it is aligned with the notch on the belt cover. When rotating the camshaft, do so only in a clockwise direction (as viewed from the front of the engine).

➡At this point, the double line marks on both left-hand camshaft sprockets should be aligned.

f. Ensure the timing marks are

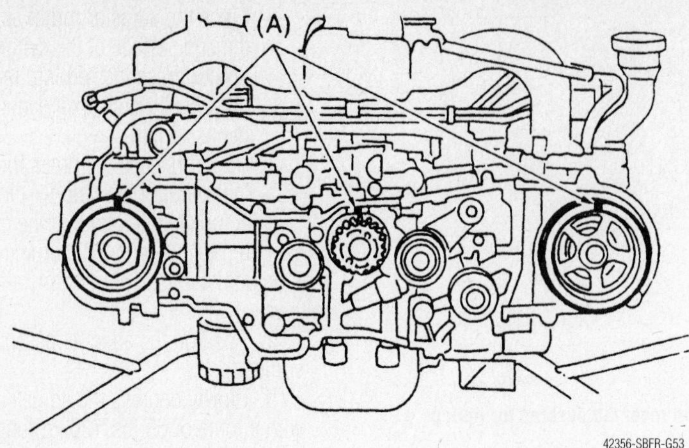

Make sure the sprockets are aligned as shown—2.5L engine shown, 2.2L similar

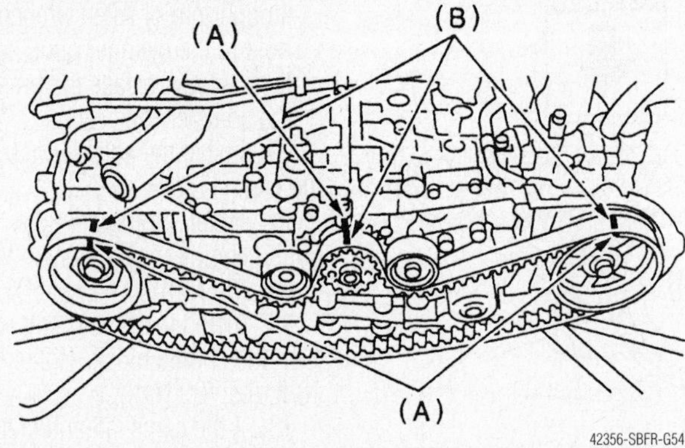

After installing the timing belt, the alignment marks should be positioned as shown. If not, remove the belt and align the sprockets and reinstall the belt—2.5L engine shown, 2.2L similar

aligned as shown in the accompanying illustration. If they are not, repeat Sub steps 12a through 12e until they are properly aligned.

14. Install the timing belt around the camshaft, crankshaft and idler pulleys so that the positioning marks on the timing belt are aligned with the marks on the sprockets as follows:

a. Position the timing belt on the crankshaft sprocket so that the marks are aligned.

b. Route the belt down and under the left-hand, upper idler pulley, then up and around the left-hand intake camshaft sprocket, ensuring the camshaft sprocket mark is aligned with the mark on the belt.

c. Route the belt down and around the left-hand exhaust camshaft sprocket, making sure the marks are properly aligned, then up and over the first lower idler pulley and down and around the second lower idler pulley.

d. While holding the timing belt on the inner, left-hand, lower idler pulley, route the other side of the timing belt (from the crankshaft sprocket) down and under the right-hand upper idler pulley.

e. Route the timing belt up and around the right-hand intake camshaft sprocket so that the belt and sprocket marks are aligned.

f. Position the belt down and around the right-hand exhaust camshaft sprocket, ensuring the positioning marks are aligned.

15. Install the right-hand lower idler pulley so that the timing belt is routed over the top side of it.

➡Once the belt is completely installed on all of the pulleys and sprockets, ensure that the positioning marks are still all aligned.

16. After ensuring all of the marks are still aligned, use a pair of pliers to withdraw the stopper pin or Allen wrench from the adjuster assembly housing.

17. On models with manual transmissions, perform the following:

a. Install the timing belt guide by temporarily tightening the mounting bolts.

b. Position the timing belt guide so that there is 0.019–0.059 in. (0.5–1.5mm) clearance between the timing belt and the belt guide.

c. Tighten the guide mounting bolts securely, then double check the guide clearance.

18. Install the timing belt covers and all remaining engine components.

On models equipped with manual transmissions, ensure the timing belt-to-guide clearance (arrows) is correct before tightening the mounting bolts—2.5L engine shown, 2.2L similar

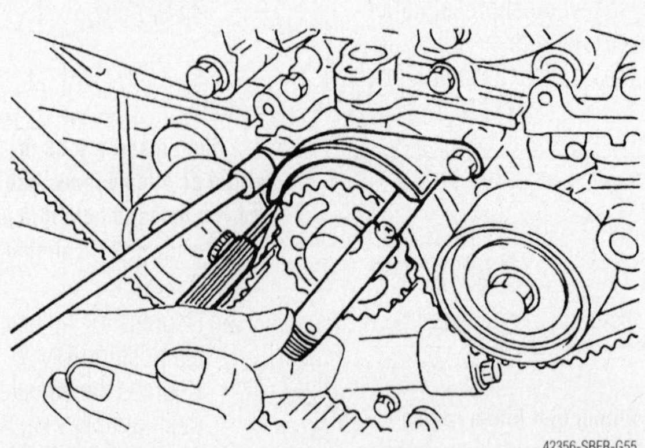

On models equipped with manual transmissions, use a feeler gauges to adjust the clearance between the timing belt and the belt guide—2.5L engine shown, 2.2L similar

Timing Chain

REMOVAL & INSTALLATION

3.0L Engines

1. Before servicing the vehicle, refer to the precautions in the beginning of this section.
2. Remove or disconnect the following:
   - Negative battery cable
   - Crankshaft pulley cover
   - Crankshaft pulley bolt using tool 499977100 to hold the crankshaft
   - Crankshaft pulley

➡There are 4 different types of front cover bolts. Note their sizes and keep them separate to avoid a problem during installation.

   - Front timing cover bolts. Note the location and the size of the bolts as you remove them, this will help during installation
   - Chain tensioner on the right hand side. The plunger **A** does not come out. Refer to illustration for plunger location.
   - Chain guide on the right hand side located between the cams
   - Chain guide from the right hand side
   - Chain tensioner lever from the right hand side
   - Timing chain from the right hand side
   - Chain tensioner on the left hand side. The plunger **A** does not come out. Refer to illustration for plunger location.
   - Chain tensioner lever from the left hand side
   - Chain guide on the left hand side located between the cams
   - Chain guide from the left hand side
   - Center chain guide
   - Upper idler sprocket
   - Timing chain from the left hand side
   - Lower idler sprocket

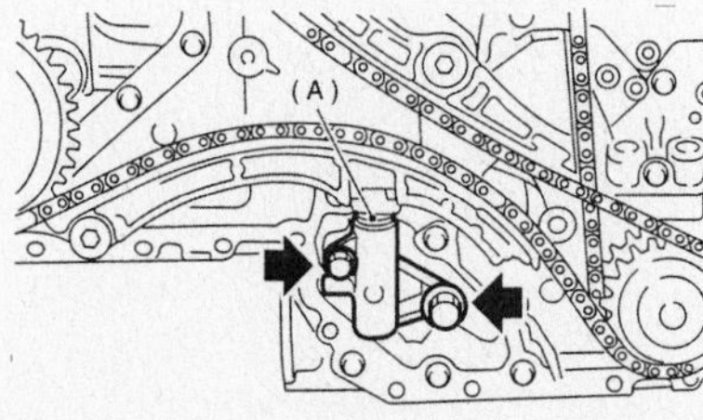

The chain tensioner plunger A does not come out—right hand side—3.0L engines

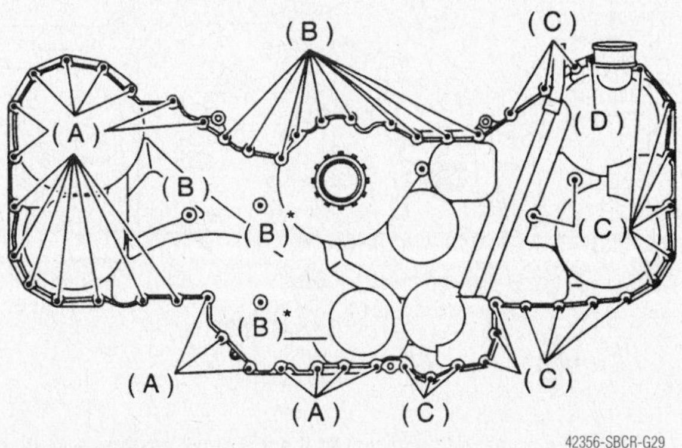

Front timing cover bolt sizes and locations—3.0L engines

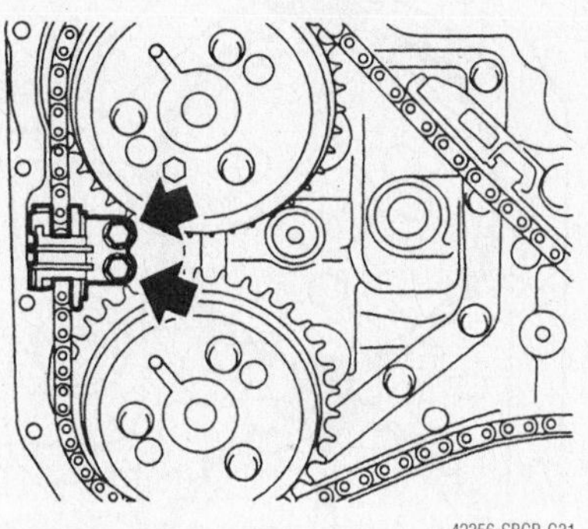

Location of the chain guide between the cams—right hand side—3.0L engines

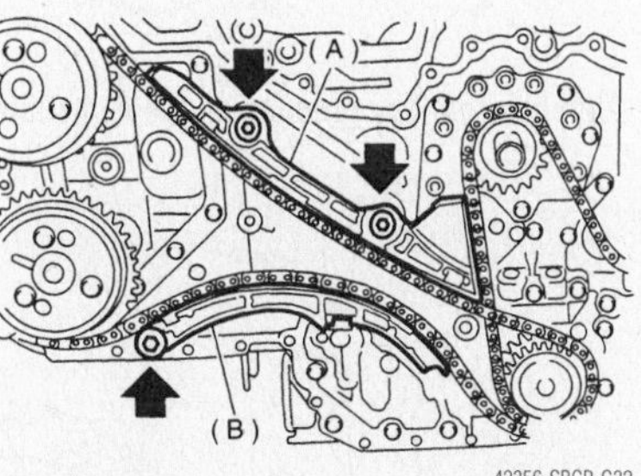

Location of the chain guide—right hand side—3.0L engines

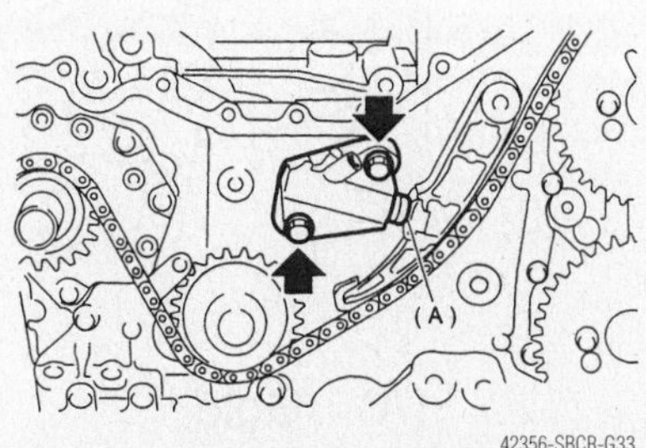

The chain tensioner plunger A does not come out—left hand side—3.0L engines

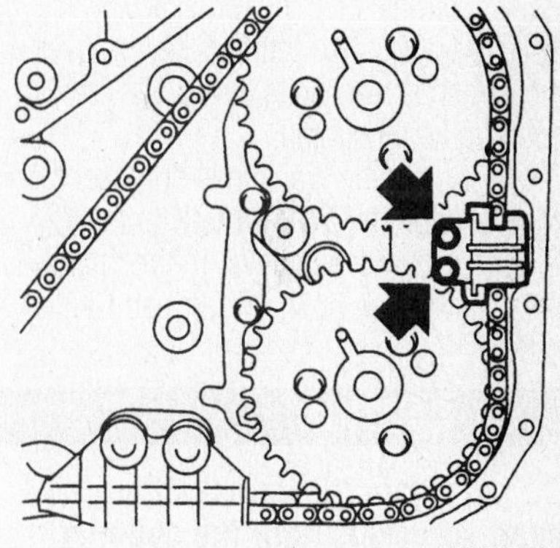

Location of the chain guide between the cams—left hand side—3.0L engines

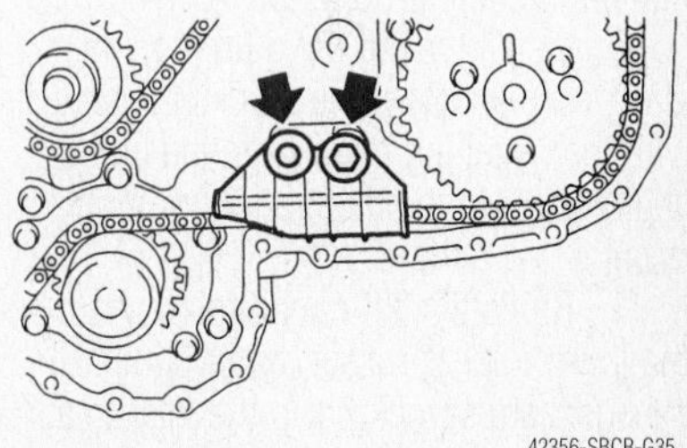

Location of the chain guide—left hand side—3.0L engines

To install:

3. Make sure all components are clean. Apply oil to the chain guide, tensioner lever and idler sprockets.
4. Place the screw, spring, pin and tension rod into the tensioner body.
5. While pressing the tensioner onto a rubber mat, twist it to the left and right to shorten the rod. Place a thin pin into the holes between the rod and body to hold it in place. Always perform this task on a rubber mat.
6. Using the crankshaft socket tool, align the **TOP MARK** on the crankshaft

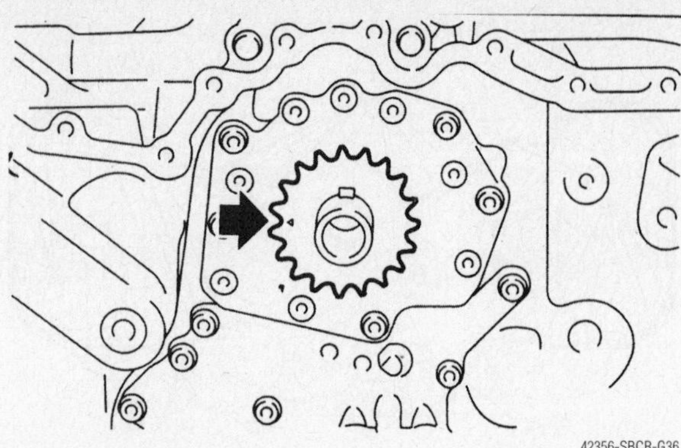

Align the TOP MARK on the crankshaft sprocket to the 9 O'clock position —3.0L engines

sprocket to the 9 O'clock position as shown in the accompanying illustration.

7. Using camshaft sprocket wrench ST 18231AA000, align the four key grooves on the camshaft sprockets to the 12 O'clock position as illustrated.

8. Rotate the crankshaft sprocket clockwise to align the **TOP MARK** to the 12 O'clock position as shown in the illustration. Piston 1 is now at Top Dead Center (TDC).

CAUTION

Do not rotate the camshaft or crankshaft sprockets until the chain is completely routed or damage will occur.

9. Install the lower idler sprocket and tighten the bolt to 50 ft. lbs. (69 Nm).

10. Install the left hand timing chain, align the mark **B** on the crankshaft sprocket with the matching mark **A** on the timing chain. Refer to the illustration for more detail

11. Route the left hand timing chain onto the lower idler sprocket, water pump, exhaust cam sprocket and the intake cam sprocket in that order.

12. Make sure the mark **A** on the chain and the mark **B** camshaft sprocket are aligned the same way as the one on the crankshaft sprocket or damage will occur.

13. Install or connect the following:

- Upper chain idler and tighten the bolt to 50 ft. lbs. (69 Nm)
- Chain guide on the left hand side between the cams and tighten the bolt to 4.6 ft. lbs. (6 Nm) using a NEW bolt
- Chain guide on the left hand side and tighten the bolts to 11 ft. lbs. (16 Nm)
- Tensioner lever on the left hand side and tighten the bolt to 11 ft. lbs. (16 Nm)
- Chain tensioner on the left hand side and tighten the bolts to 11 ft. lbs. (16 Nm)
- Right hand timing chain. On the lower idler sprocket align the match marks on the timing chain on the left and right hand sides as illustrated. Route the chain onto the intake cam sprocket and the exhaust cam sprocket.

14. Make sure the mark **A** on the chain and the mark **B** camshaft sprocket are aligned the same way as the one on the crankshaft sprocket or damage will occur.

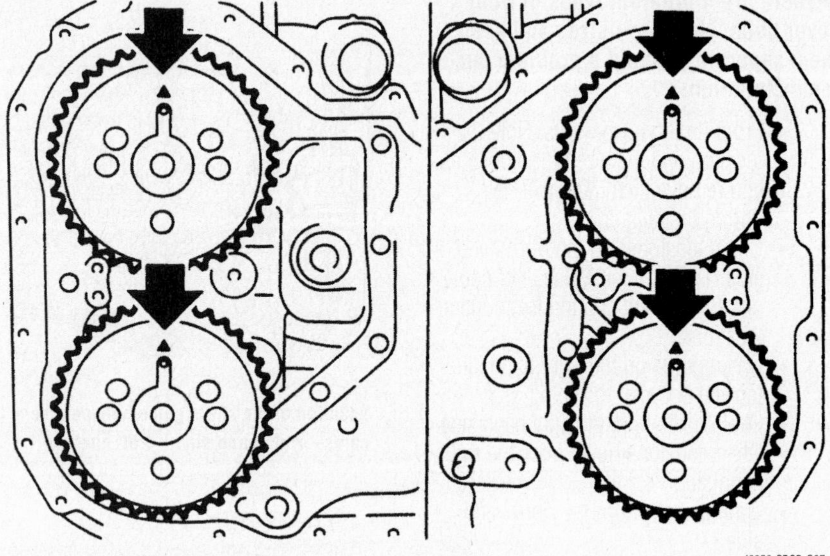

Align the four key grooves on the camshaft sprockets to the 12 O'clock position—3.0L engines

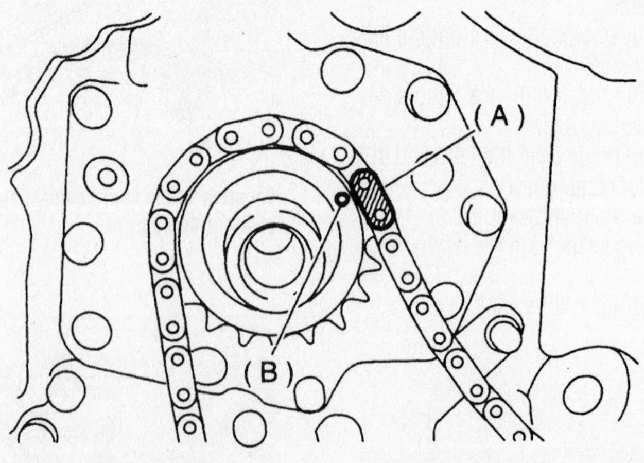

Align the mark B on the crankshaft sprocket with the matching mark A on the left hand timing chain—3.0L engines

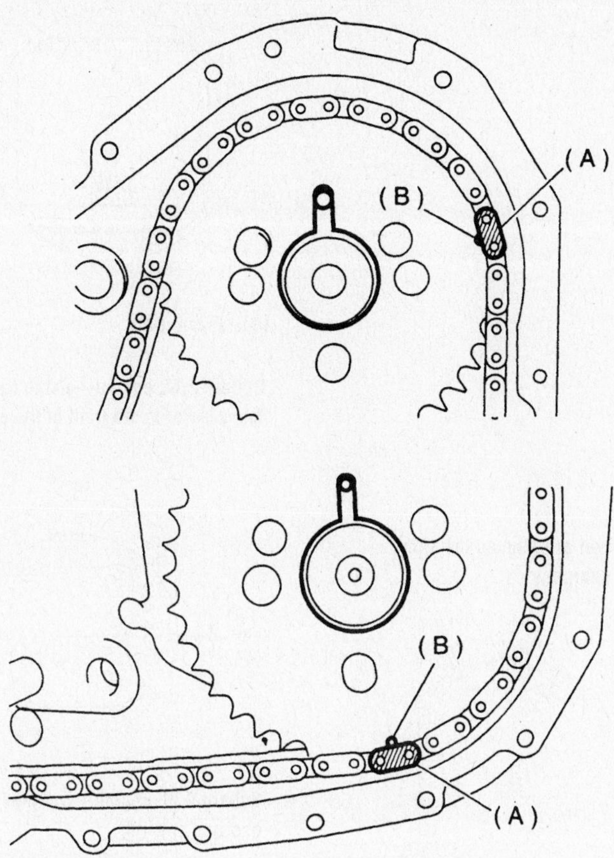

Make sure the mark A on the chain and the mark B camshaft sprocket are aligned the same way as the one on the crankshaft sprocket–left hand side—3.0L engines

15. Install or connect the following:
    - Right hand chain guide
    - Right hand chain tensioner lever and tighten the bolts to 11 ft. lbs. (16 Nm)
    - Right hand chain guide and tighten the NEW bolt to 4.6 ft. lbs. (6 Nm)
    - Right hand chain tensioner and tighten the bolts to 11 ft. lbs. (16 Nm)
16. Adjust the clearance between the chain guide on the right hand side and the center chain guide so that there is range between 0.331–0.339 inch (8.4–8.6mm).
    - Center chain guide and tighten the NEW bolt to 6 ft. lbs. (8 Nm)
17. Check the match marks on each sprocket and corresponding timing chain are correct, remove the stopper from the tensioner.
18. Clean the mating surfaces on the front timing cover. Apply a 0.078–0.126 inch (2–5mm) bead of gasket maker to the mating surface of the front cover. Subaru recommends Three Bond 1280B for this procedure.
19. Install the front chain cover and temporarily tighten the bolts. use the illustration showing the bolt sizes and location for proper installation.
20. Tighten the front timing cover bolts in the sequence illustrated to 5 ft. lbs. (7 Nm).
21. Install or connect the following:
    - Crankshaft pulley, apply clean oil to the bolt threads and tighten the bolt to 131 ft. lbs. (178 Nm)
    - Camshaft pulley cover and tighten the bolts to 5 ft. lbs. (7 Nm)

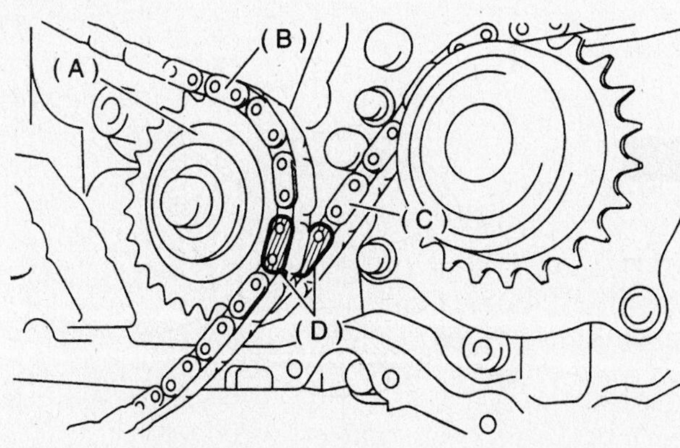

On the lower idler sprocket align the match marks on the timing chain on the left and right hand sides–right hand side chain installation—3.0L engines

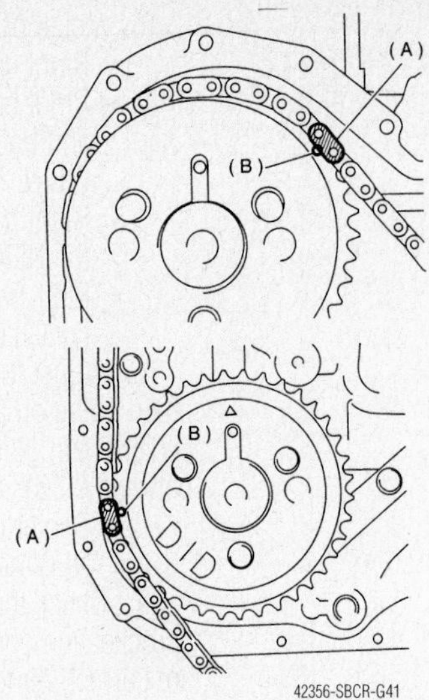

Make sure the mark A on the chain and the mark B camshaft sprocket are aligned the same way as the one on the crankshaft sprocket–right hand side—3.0L engines

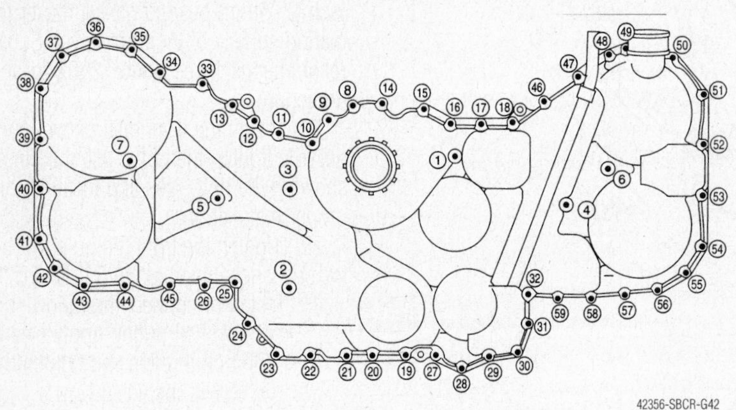

Tighten the front timing cover bolts in the sequence illustrated to the correct specification—3.0L engines

- Negative battery cable

22. Start the engine and allow it to reach operating temperature. Check for leaks.

Piston and Ring

POSITIONING

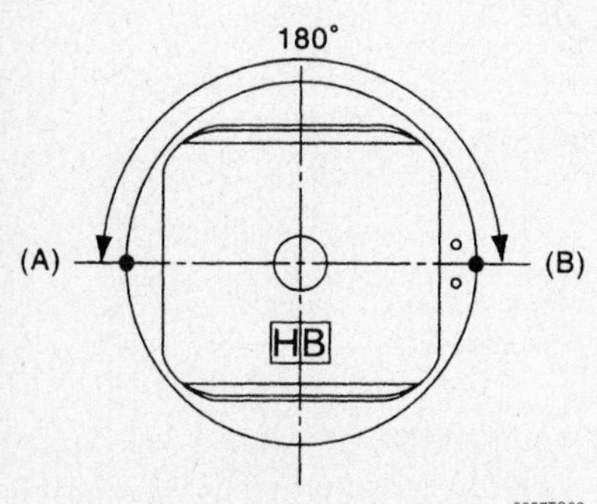

Subaru 2.0L engine—top ring end-gap spacing

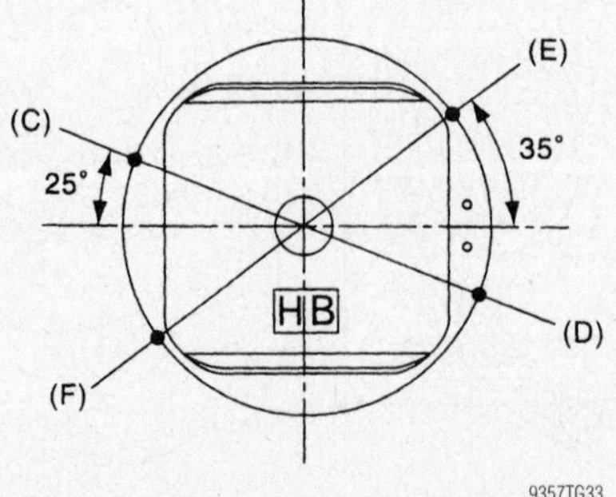

Subaru 2.0L engine—upper rail end-gap spacing

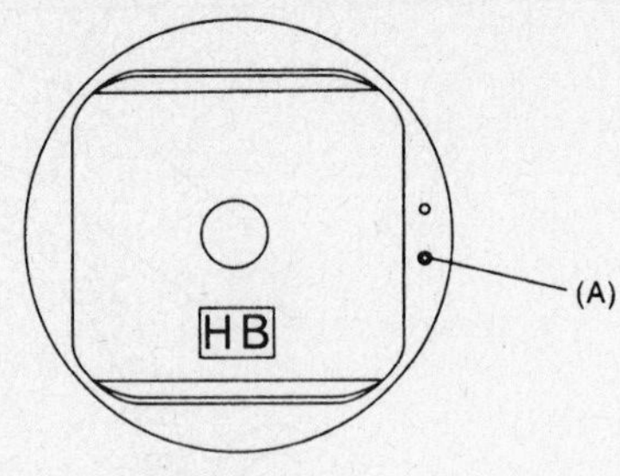

(A) Front mark

9357TG34

Subaru 2.0L engine—piston front mark faces towards the front of the engine

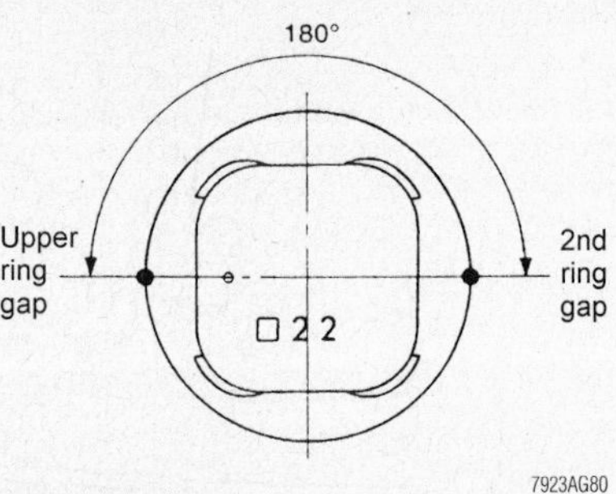

Subaru 2.2L engine—compression ring end-gap spacing

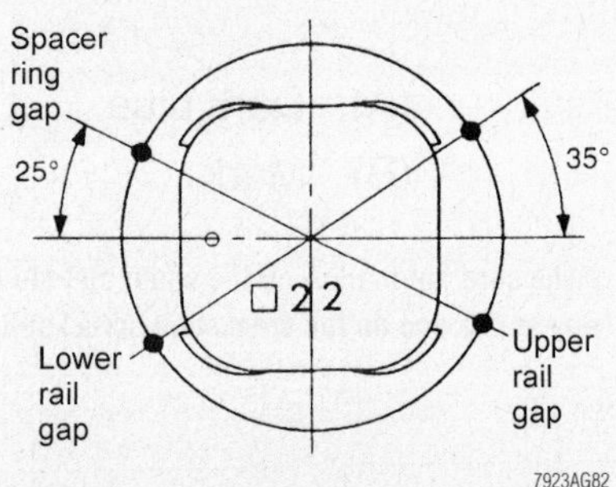

Subaru 2.2L engine—upper, spacer and lower oil ring end-gap spacing

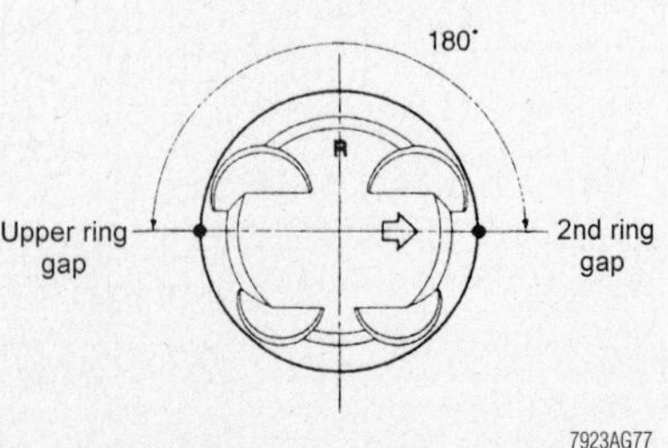

Subaru 2.5L engine—compression ring end-gap spacing

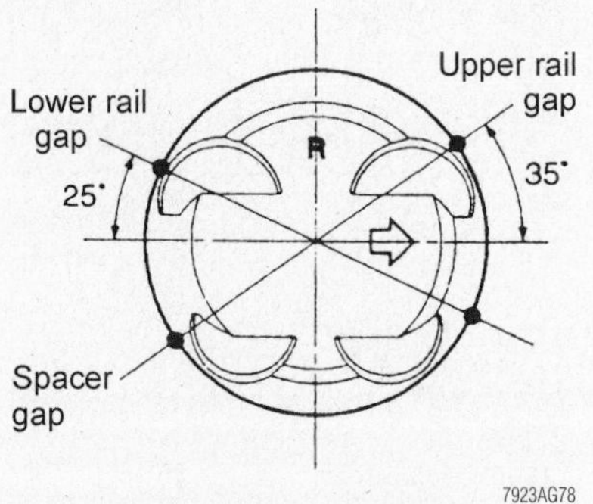

Subaru 2.5L engine—upper, spacer and lower oil ring end-gap spacing

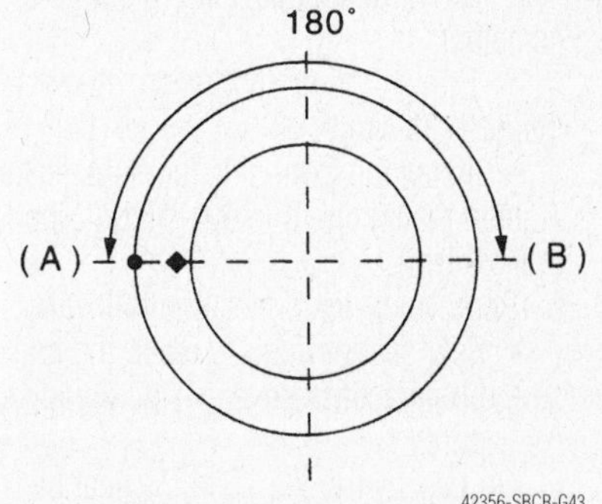

Top ring end-gap (A), second ring gap (B)—3.0L engines

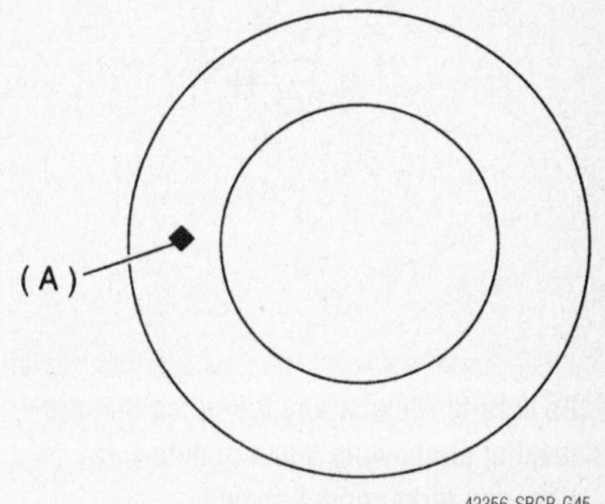

Piston front mark (A) faces towards the front of the engine—3.0L engine

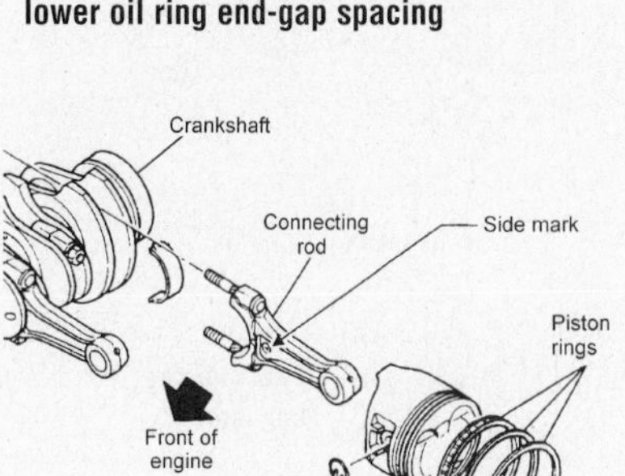

Subaru engines—piston and connecting rod assembly positioning

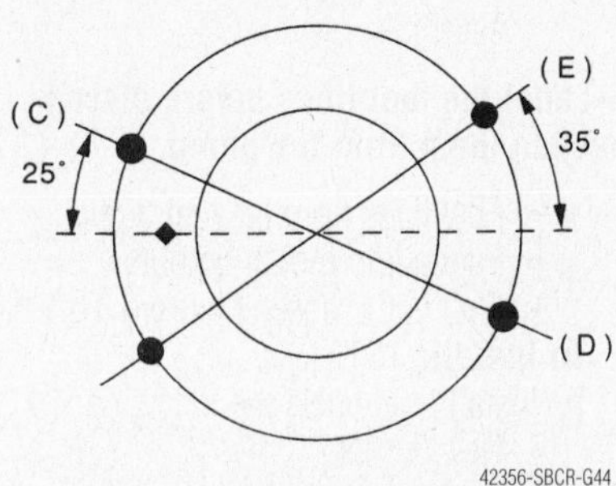

Upper rail end-gap (C) , expander gap (D) and lower rail gap (E)—3.0L engines

FUEL SYSTEM

Fuel System Service Precautions

Safety is the most important factor when performing not only fuel system maintenance, but any type of maintenance. Failure to conduct maintenance and repairs in a safe manner may result in serious personal injury or death. Maintenance and testing of the vehicle's fuel system components can be accomplished safely and effectively by adhering to the following rules and guidelines.

- To avoid the possibility of fire and personal injury, always disconnect the negative battery cable unless the repair or test procedure requires that battery voltage be applied.
- Always relieve the fuel system pressure prior to disconnecting any fuel system component (injector, fuel rail, pressure regulator, etc.), fitting or fuel line connection. Exercise extreme caution whenever relieving fuel system pressure, to avoid exposing skin, face and eyes to fuel spray. Please be advised that fuel under pressure may penetrate the skin or any part of the body that it contacts.
- Always place a shop towel or rag around the fitting or connection prior to loosening to absorb any excess fuel due to spillage. Ensure that all fuel spillage (should it occur) is quickly removed from engine surfaces. Ensure that all fuel soaked cloths or towels are deposited into a suitable waste container.
- Always keep a dry chemical (Class B) fire extinguisher near the work area.
- Do not allow fuel spray or fuel vapors to come into contact with a spark or open flame.
- Always use a back-up wrench when loosening and tightening fuel line connection fittings. This will prevent unnecessary stress and torsion to fuel line piping.
- Always replace worn fuel fitting O-rings with new. Do not substitute fuel hose or equivalent, where fuel pipe is installed.

Fuel System Pressure

RELIEVING

➡This procedure must be performed prior to servicing any component of the fuel injection system.

1. Before servicing the vehicle, refer to the precautions in the beginning of this section.
2. Disconnect the fuel pump connector from the fuel pump relay.
3. Start the engine and let it stall.
4. Crank the engine for 5 seconds or more to ensure the fuel pressure is properly relieved. If the engine starts during this time, allow it to run until it stalls.
5. After performing the required service, connect the fuel pump harness.

Fuel Filter

REMOVAL & INSTALLATION

1. Before servicing the vehicle, refer to the precautions in the beginning of this section.
2. Locate the fuel filter in the engine compartment.
3. Properly relieve the fuel system pressure.
4. Remove or disconnect the following:
   - Negative battery cable
   - Hose clamp screws and slide the hoses off the filter
   - Filter from the bracket

To install:

5. Inspect the hoses for wear or cracks, and replace if needed.
6. Install or connect the following:
   - New filter into the bracket and tighten the hose clamp screws

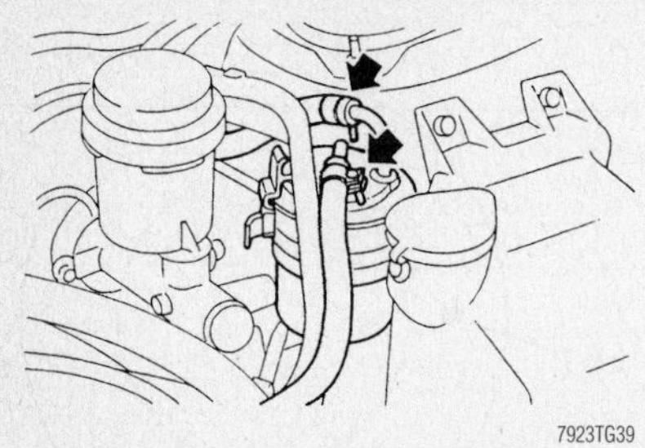

Be sure to replace any fuel lines that are leaking or showing signs of deterioration–non turbo model shown

- Negative battery cable

7. Start the engine and check for leaks.

Fuel Pump

REMOVAL & INSTALLATION

1. Before servicing the vehicle, refer to the precautions in the beginning of this section.
2. Relieve the fuel system pressure.
3. Disconnect the negative battery cable.
4. On the 2000–03 Legacy, Outback and turbocharged models, perform the following:
   a. Raise and safely support the vehicle.
   b. Remove the front side fuel tank cover.

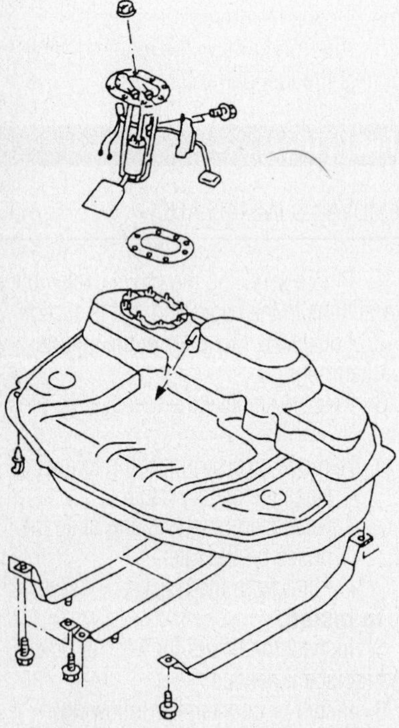

Exploded view of the fuel pump assembly

   c. Drain the fuel tank into a suitable container.
   d. Tighten the drain plug to 14–24 ft. lbs. (19–33 Nm).
   e. Install the front side fuel tank cover Tighten the retainers to 9.4–16.6 ft. lbs. (13–23 Nm).
5. Remove or disconnect the following:
   - Rear seat bottom, to reach the fuel pump access cover, if not already done
6. On Legacy models, fold the seat back, then roll the floor mat back.
   - Fuel pump cover mounting bolts and the cover
   - Electrical harness from the pump assembly

➡Label the fuel lines before disconnecting them from the pump.

- Fuel lines from the fuel pump
- Fuel pump mounting nuts
- Fuel pump assembly from the tank

To install:

7. Install or connect the following:
   - New gasket
   - Fuel pump assembly into the fuel tank and secure with the mounting nuts. Tighten the nuts to 24–48 inch lbs. (3–6 Nm) on non turbo models and 3 ft. lbs. (4 Nm) on turbo models.
   - Electrical harness to the fuel pump assembly
   - Fuel lines to the pump assembly, then tighten the clamps and fittings
   - Fuel pump service cover and cover mounting bolts
   - Rear seat bottom
   - Negative battery cable
8. Start the engine and check for leaks.

Fuel Injector

REMOVAL & INSTALLATION

Except 2.0L Engines

1. Before servicing the vehicle, refer to the precautions in the beginning of this section.
2. Relieve the fuel system pressure.
3. To remove the right side injectors, remove or disconnect the following:
   - Air cleaner ducts and resonator chamber, on California vehicles
   - Mass Air Flow (MAF) sensor connector and air intake duct and air cleaner upper cover as a unit, on non-California vehicles
   - Air cleaner element

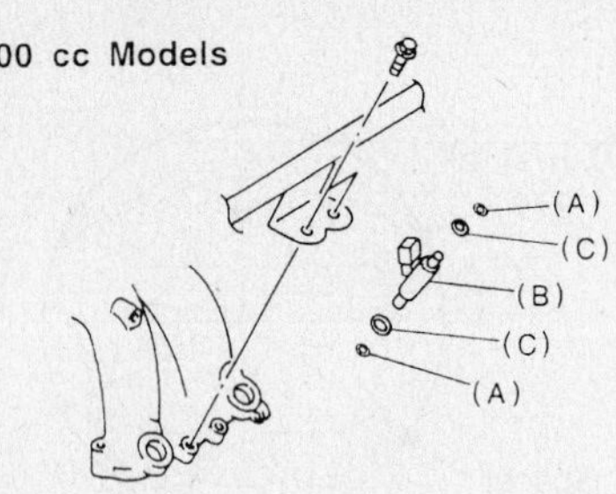

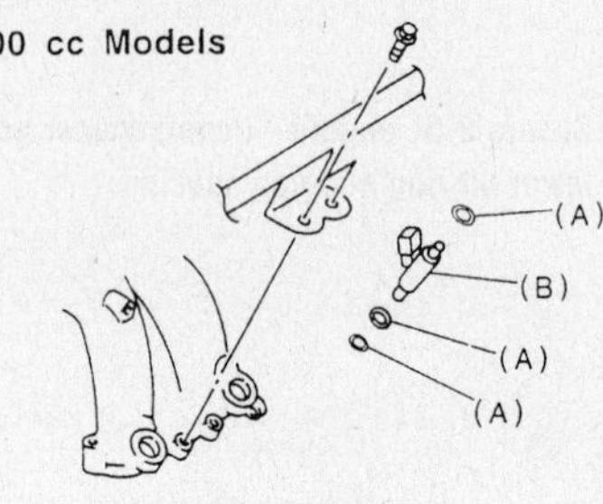

Exploded view of the fuel injector

   - Spark plug wires from the right side spark plugs
   - V-belt covers and power steering pump belt
   - Power steering pump brackets-to-intake manifold bolts
   - Power steering pump-to-bracket bolts, then position the pump on the right side wheel apron
4. To remove the injectors on the left side, remove or disconnect the following:
   - Windshield washer motor electrical connector
   - Electrical connector from the rear window washer, on station wagon only
   - Rear window washer hose from the washer motor and plug or cap the line
   - Two bolts that secure the washer tank to the body
   - Washer tank and secure it out of the way
   - Spark plug wires from the left side spark plugs
   - Fuel pipe protector
5. Remove or disconnect the following (for either side):
   - Band that secures the engine harness to the fuel injector pipe, if equipped
   - Intake manifold protector, if equipped
   - Fuel injector electrical connector(s)
   - Bolts that hold the fuel injector pipe

(fuel rail) to the intake manifold, if applicable

➡**Automatic transaxle equipped Legacy's may have a retaining clip that must be removed before the injector can be removed.**

6. Pull up on the injector pipe (fuel rail), then remove the fuel injector(s) from the intake manifold. Remove and discard the injector O-rings.

To install:

- Install or connect the following:
- New injector O-rings
- Fuel injector(s) into the intake manifold
- Retaining clips, if applicable
- Injector pipe (fuel rail) and secure with the retaining bolts. Tighten the bolts to 14 ft. lbs. (19 Nm).
- Fuel injector electrical connector(s)
- Intake manifold protector, if equipped
- Band that secures the engine harness to the fuel injector pipe, if equipped

7. To install the injectors on the left side, install or connect the following:
   - Fuel pipe protector
   - Spark plug wires to the left side spark plugs
   - Washer tank and secure with the two mounting bolts
   - Rear window washer hose to the washer motor
   - Electrical connector to the rear window washer, on station wagon only
   - Windshield washer motor electrical connector
8. To install the right side injectors, install or connect the following:
   - Power steering pump into position
   - Pump-to-bracket bolts
   - Power steering pump brackets-to-intake manifold bolts
   - Power steering pump belt and V-belt covers
   - Spark plug wires to the right side spark plugs
   - Air cleaner element
   - Air intake duct and upper cover and the MAF sensor connector
   - Air cleaner ducts and resonator chamber, on California vehicles

2.0L Engines

1. Before servicing the vehicle, refer to the precautions in the beginning of this section.
2. Relieve the fuel system pressure.
3. Remove or disconnect the following:
   - Intake manifold
   - Fuel pipe protector
   - Electrical connector from the fuel injector
   - Bolts that attach the injector pipe to the intake manifold
   - Fuel injector while lifting up the fuel injector pipe

To install:

- Install or connect the following:
- New injector O-rings
- Remaining components in the reverse order of removal. Tighten the injector pipe bolts to 13 ft. lbs. (19 Nm).

DRIVE TRAIN

Transaxle Assembly

REMOVAL & INSTALLATION

Manual

1. Before servicing the vehicle, refer to the precautions in the beginning of this section.
2. Remove or disconnect the following:
   - Negative battery cable
   - Air intake duct and cleaner case
   - Air cleaner stay, if equipped
   - Intercooler on turbo models
   - Front Oxygen Sensor (O_2S) connector
   - Neutral position switch connector
   - Back-up light switch connector
   - Vehicle Speed Sensor (VSS) connector, if equipped
   - Transmission ground cable, if necessary
   - Clutch cable, if equipped
   - Clutch release spring, if equipped
   - Starter
   - Operating cylinder from the transmission
   - Pitching stopper
3. On turbo models, perform the following to disconnect the clutch release fork from the release bearing:
   a. Remove the clutch cylinder from the transmission.
   b. Using a 10mm wrench, remove the plug.
   c. Screw a 6mm diameter bolt into the release fork and remove it.
   d. Raise the release fork and unfasten the release tabs to free the release fork.
4. Remove or disconnect the following:
   - Drive belt cover, if necessary
   - Slave (operating) cylinder, on 2.5L engines
5. Install engine support assembly 927670000, on 3.0L engines, install support assembly 41099AA00.
   - Bolt securing the right upper side of the transaxle to the engine
   - Engine undercover, if equipped
   - Rear O_2S connector
   - Front Y-pipe
   - Rear exhaust pipe and muffler, on Legacy models
   - Heat shield cover
   - Hanger bracket from the right side of the transaxle
   - Driveshaft
   - Spring, and disconnect the shifter stay and rod from the transaxle
   - Bolts securing the sway bar clamps to the crossmember
   - Ball joints from the steering knuckle
   - Halfshafts from the transaxle
   - Nuts securing the lower side of the transaxle to the engine
6. Support the transaxle with a jack.
   - Rear transaxle crossmember
   - Transaxle from the vehicle. Move the jack rearward until the mainshaft is withdrawn from the clutch cover.

To install:

7. Install or connect the following:
   - Transaxle assembly and secure it to the engine block
   - Crossmember
8. Tighten the crossmember retainers to the following specifications:
   a. Step 1: T1 to 40–62 ft. lbs. (54–84 Nm), on all engines except the 3.0L. On the 3.0L engine, tighten the T1 bolts to 55 ft. lbs. (75 Nm)
   b. Step 2: T2 to 87–115 ft. lbs. (117–157 Nm), on all engines except the 3.0L. On the 3.0L engine, tighten the T2 bolts to 103 ft. lbs. (140 Nm)
9. Remove the transmission jack.
10. Install or connect the following:
    - Nuts securing the lower portion of the engine to the transaxle and tighten to 40 ft. lbs. (54 Nm) on all except turbo and 3.0L models. On turbo and 3.0L models, tighten to 37 ft. lbs. (50 Nm).

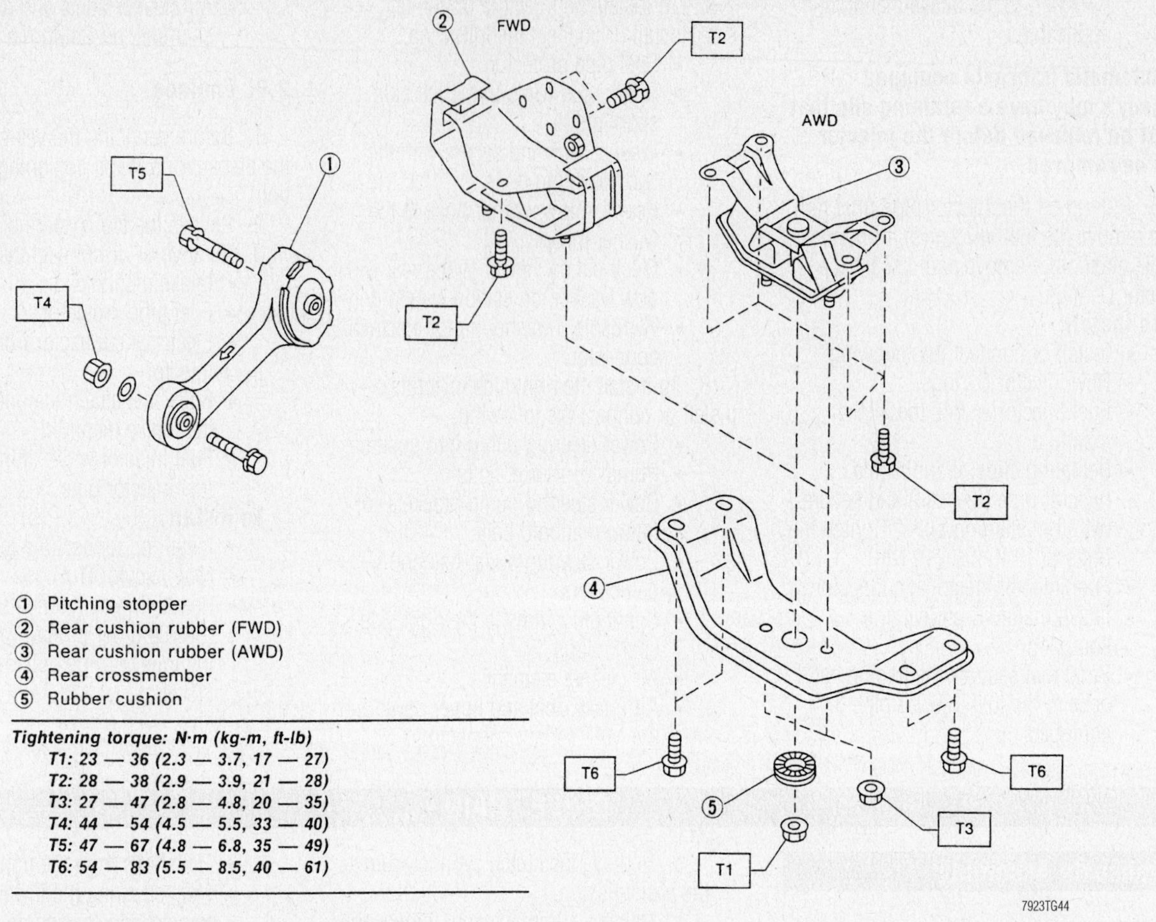

Exploded view of the transaxle mounting—Impreza models

- Bolt securing the right upper side of the transaxle to the engine and tighten it to 40 ft. lbs. (54 Nm) on all except turbo and 3.0L models. On turbo and 3.0L models, tighten to 37 ft. lbs. (50 Nm).

11. Remove the engine support.
    - Drive belt cover
    - Slave (operating) cylinder, on 2.5L engines

12. Install the pitching stopper and tighten the bolts to the following specifications:

    a. Step 1: T1 to 33–40 ft. lbs. (44–54 Nm) on all except turbo and 3.0L models. On turbo and 3.0L models, tighten to 37 ft. lbs. (50 Nm).

    b. Step 2: T2 to 35–49 ft. lbs. (47–67 Nm). on all except turbo and 3.0L models. On turbo and 3.0L models, tighten to 43 ft. lbs. (58 Nm).

13. Install or connect the following:
    - Halfshafts into the transaxle with new roll pins
    - Ball joint to the steering knuckle and tighten the bolt to 29–43 ft. lbs. (39–59 Nm) on all except turbo and 3.0L models. On turbo and 3.0L models, tighten to 36 ft. lbs. (49 Nm).
    - Sway bar to the crossmember and tighten the clamp bolts to 15–21 ft. lbs. (21–29 Nm) on all except turbo and 3.0L models. On turbo models, tighten to 33 ft. lbs. (45 Nm). On 3.0L models, tighten the clamp bolts to 22 ft. lbs. (30 Nm).
    - Shift control rod and stay to the transaxle and install the spring
    - Driveshaft
    - Heat shield cover, if removed
    - Rear exhaust pipe and muffler, if removed
    - Y-pipe with new gaskets and nuts
    - Hanger bracket on the right side of the transaxle, if removed
    - Rear O_2S connector
    - Engine undercover, if removed
    - Transaxle connectors bracket
    - Drive belt cover
    - Pitching stopper
    - Starter
    - Front O_2S connector
    - VSS connector, if equipped
    - Neutral position switch connector
    - Back-up light switch connector
    - Clutch cable (if equipped)
    - Clutch release spring
    - Air cleaner case stay and case
    - Air intake duct and attach the airflow sensor connector
    - Negative battery cable

Automatic

EXCEPT WRX MODELS

1. Before servicing the vehicle, refer to the precautions in the beginning of this section.
2. Drain the transaxle fluid.
3. Remove or disconnect the following:
    - Negative battery cable
    - Air intake duct with air cleaner case
    - Air cleaner case stay
    - Front Oxygen Sensor (O_2S) connector
    - Speedometer cable or electronic wiring connector from the speed sensor
    - Transaxle harness connector
    - Inhibitor switch connector, if equipped
    - Revolution sensor connector, if equipped

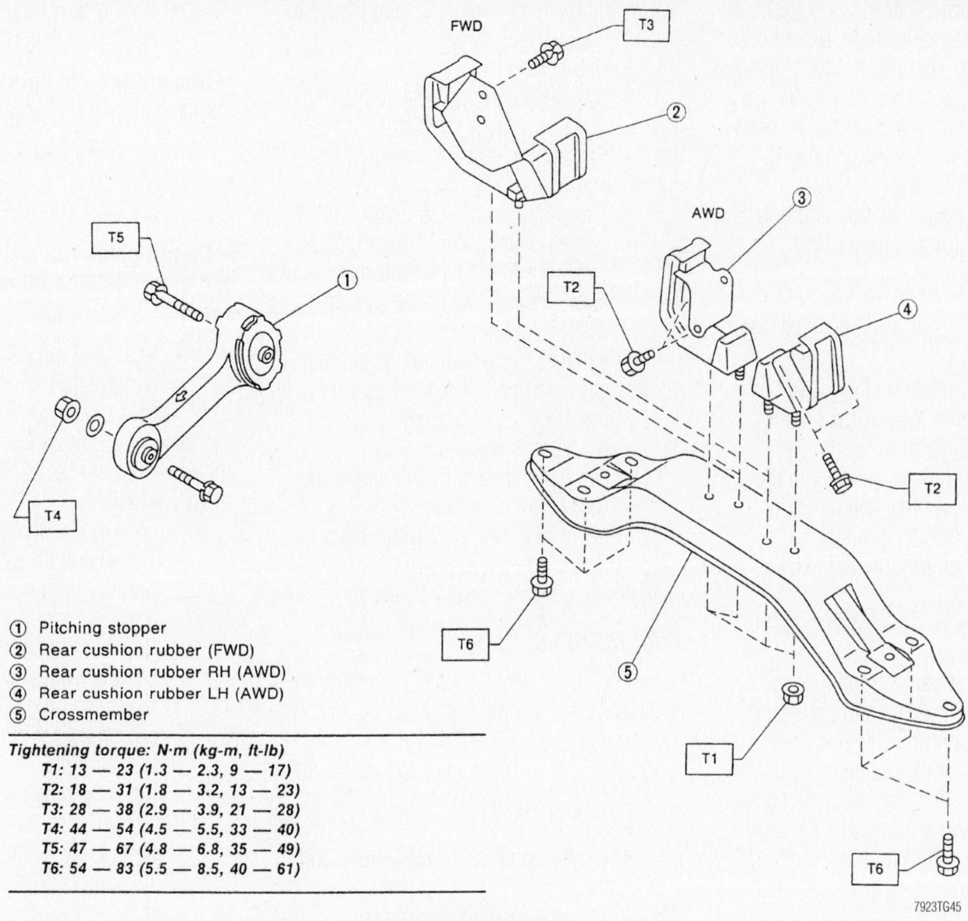

Exploded view of the engine and transaxle mounts—Impreza and Legacy models

- Transaxle ground terminal
- Clip band that secures the air breather hose to the pitching stopper, if equipped
- Starter and air intake boot
- Pitching stopper
- Timing hole inspection plug
- 4 bolts that hold the torque converter to the driveplate
- Automatic Transaxle Fluid (ATF) level gauge
- Engine-to-transaxle mounting nut and bolt on the right side
- Buffer rod

4. Support the engine assembly with special engine support tool.
   - Exhaust system
   - Exhaust brackets or hangers that attach to the transaxle, as necessary
5. Drain the transmission fluid.
   - ATF cooler hoses from the pipes of the transmission side
   - ATF level gauge guide

➡Matchmark the installed position of the driveshaft before removal.

- Driveshaft. Plug the opening at the rear of extension housing to prevent oil from flowing out.
- Gearshift cable from the transaxle select lever
- Stabilizer from the transverse link
- Parking brake cable bracket from the transverse link
- Transverse link bolts and lower the link
- Spring pins
- Halfshafts from the transaxle

➡Discard the old spring pin and always install a new pin.

- Oil cooler hoses

6. Place a transaxle jack under the transaxle.
   - Engine to transaxle mounting nuts

➡Do not place the jack under the oil pan otherwise the oil pan may be damaged.

- Rear cushion rubber mounting nuts and the rear crossmember

7. Move the torque converter and transaxle as a unit away from the engine and lower it from the vehicle.

To install:

8. Install or connect the following:
   - Transaxle to the engine and temporarily tighten the engine-to-transaxle mounting nuts
   - Rear crossmember to the rear cushion rubber mounts. Align the rear cushion guide with the rear crossmember guide hole and tighten nuts.

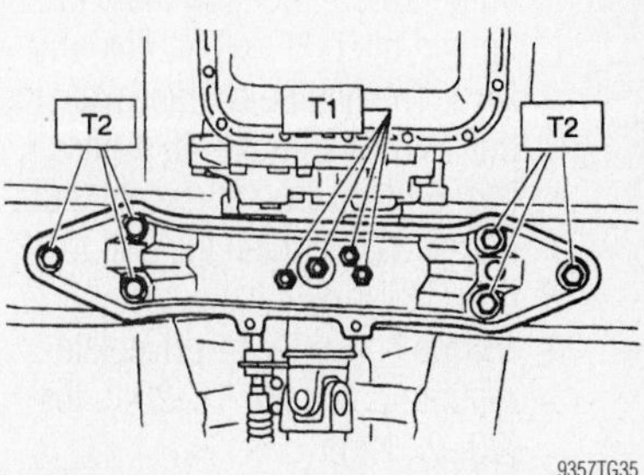

Location of the rear crossmember bolts—3.0L engine

- Rear crossmember to the chassis. Tighten the rear crossmember bolts to 39–49 ft. lbs. (53–66 Nm) on all models except the 3.0L engine. On the 3.0L engine tighten the T1 bolts to 26 ft. lbs. (35 Nm) and the T2 bolts to 55 ft. lbs. (75 Nm).
- Engine to transaxle retaining nuts to 34–40 ft. lbs. (46–54 Nm)

9. Remove the transaxle jack from the vehicle.

10. Remove the engine support tool and install the buffer rod.

11. Install or connect the following:
    - Axle shafts to the transaxle using new spring pins
    - Transverse link temporarily to the front crossmember. Do not complete final torque at this point.
    - Stabilizer temporarily to the transverse link
    - Parking brake cable bracket to the transverse link
    - Transverse link-to-front crossmember mounting bolts and transverse link-to-stabilizer mounting bolts, with the tires placed on the ground
    - Transverse link to front crossmember (self-locking nuts) to 40–62 ft. lbs. (54–84 Nm) and the transverse link to stabilizer to 18–32 ft. lbs. (24–44 Nm).
    - Gearshift cable to the select lever. Be sure the lever operates smoothly all across the operating range.
    - Driveshaft. Tighten the driveshaft-to-rear differential retaining bolts to 17–24 ft. lbs. (23–33 Nm) and center bearing location retaining bolts to 25–33 ft. lbs. (34–45 Nm).
    - Oil cooler hoses
    - Engine to transaxle bolts to 34–40 ft. lbs. (46–54 Nm)
    - Starter
    - Pitching stopper. Be sure to tighten the bolt for the body side first. Tightening torque for the body side bolt is 35–49 ft. lbs. (47–67 Nm) on all except 3.0L; on the 3.0L tighten the bolt to 37 ft. lbs. (50 Nm). The engine or transaxle side bolt is torque to 33–40 ft. lbs. (44–54 Nm)) on all except 3.0L; on the 3.0L tighten the bolt to 43 ft. lbs. (58 Nm).
    - Torque converter-to-driveplate mounting bolts to 17–20 ft. lbs. (23–27 Nm)
    - ATF level gauge guide
    - ATF cooler hoses to the pipes of the transmission side
    - Timing hole inspection plug, air intake boot and air breather hose to the pitching stopper
    - O_2S connector
    - Transaxle harness connector
    - Inhibitor switch connector
    - Revolution sensor connector, if equipped
    - Transaxle ground terminal
    - Speedometer cable. Tighten the cable nut by hand, then turn it approximately 30 degrees more with a tool.
    - Exhaust system and exhaust brackets or hangers that attach to the transaxle, as necessary
    - Air cleaner case stay
    - Air intake duct with air cleaner case
    - Battery ground cable

12. Refill and check transaxle oil level.

13. Road test the vehicle for proper operation across all operating ranges.

WRX MODELS

1. Before servicing the vehicle, refer to the precautions in the beginning of this section.

2. Drain the transaxle fluid.

3. Remove or disconnect the following:
    - Negative battery cable
    - Intercooler
    - Center and rear exhaust pipes and the muffler
    - Transaxle harness connector
    - Transaxle ground terminal
    - Starter
    - Pitching stopper
    - Torque converter service hole plug
    - Bolts that hold the torque converter to the driveplate
    - Automatic Transaxle Fluid (ATF) level gauge

4. Support the engine assembly with special engine support tool.
    - Engine-to-transaxle mounting and bolt(s) on the right side
    - Undercover
    - Heat shield cover
    - Buffer rod

5. Drain the transmission fluid.
    - ATF cooler hoses from the pipes of the transmission side
    - ATF level gauge guide

➡Matchmark the installed position of the driveshaft before removal.

- Driveshaft. Plug the opening at the rear of extension housing to prevent oil from flowing out.
- Gearshift cable from the transaxle select lever
- Stabilizer from the transverse link
- Transverse link bolts and lower the link
- Spring pins
- Halfshafts from the transaxle

➡Discard the old spring pin and always install a new pin.

6. Place a transaxle jack under the transaxle.
    - Engine to transaxle mounting nuts

➡Do not place the jack under the oil pan otherwise the oil pan may be damaged.

- Rear cushion rubber mounting nuts and the rear crossmember

7. Move the torque converter and transaxle as a unit away from the engine and lower it from the vehicle.

To install:

8. Install or connect the following:
    - Transaxle to the engine and temporarily tighten the engine-to-transaxle mounting nuts
    - Rear crossmember to the rear cushion rubber mounts. Align the rear cushion guide with the rear crossmember guide hole and tighten nuts.
    - Rear crossmember to the chassis. Tighten the rear crossmember T1 bolts to 26 ft. lbs. (35 Nm) and the T2 bolts to 51 ft. lbs. (70 Nm). Refer to the accompanying illustration for bolt location.
    - Engine to transaxle retaining nuts to 36 ft. lbs. (50 Nm)
    - Starter
    - Torque converter clutch plate bolts to 18 ft. lbs. (25 Nm0
    - Torque converter plug
    - Pitching stopper. Be sure to tighten the bolt for the body side first. Tightening torque for the body side bolt is 43 ft. lbs. (58 Nm). The engine or transaxle side bolt is torque to 37 ft. lbs. (50 Nm).
    - Axle shafts to the transaxle using new spring pins
    - Transverse link temporarily to the

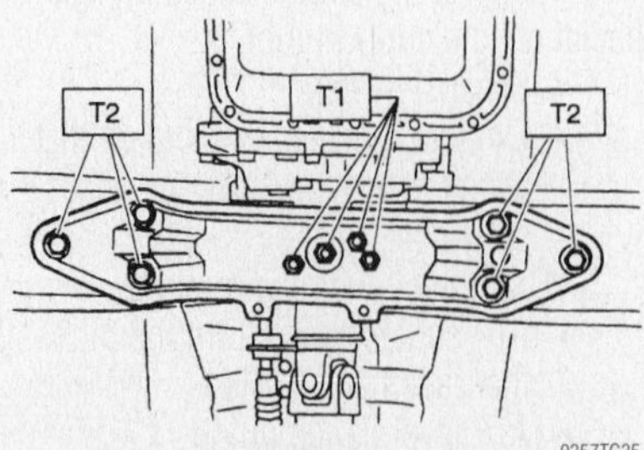

Location of the rear crossmember bolts—WRX models

front crossmember. Do not complete final torque at this point.

- Stabilizer temporarily to the transverse link
- Transverse link-to-front crossmember mounting bolts and transverse link-to-stabilizer mounting bolts, with the tires placed on the ground
- Transverse link to front crossmember (self-locking nuts) to 22 ft. lbs. (30 Nm) and the transverse link to stabilizer to 37 ft. lbs. (50 Nm).
- Gearshift cable to the select lever. Be sure the lever operates smoothly all across the operating range.
- ATF level gauge guide
- Oil cooler hoses
- Driveshaft
- Heat shield cover
- Center, rear exhaust pipes and the muffler
- Undercover
- ATF fluid level gauge
- Transmission harness connectors
- Transmission ground terminal
- Air cleaner case stay
- Intercooler
- Battery ground cable

9. Refill and check transaxle oil level.
10. Road test the vehicle for proper operation across all operating ranges.

Clutch

ADJUSTMENT

Some models are equipped with a mechanical clutch system that is adjustable. Other models are equipped with a hydraulic system that is not adjustable.

Cable

The clutch cable can be adjusted at the cable bracket where the cable is attached to the side of the transaxle housing.

1. Before servicing the vehicle, refer to the precautions in the beginning of this section.
2. Remove the circlip and clamp.
3. Slide the cable end in the direction desired, then replace the circlip and clamp into the nearest gutters on the cable end.

➡The cable should not be stretched out straight nor should it have right angle kinks in it. Any straightening should be gradual.

4. Check the clutch for proper operation.

Pedal Height

Adjust the pedal with the return stop bolt, so that its pad is on the same level as the brake pedal pad.

Check to be sure that the stroke of the pedal is 5.12–5.31 in. (130–135mm). Check the clutch release fork stroke. It should be 0.67 in. (17mm).

Free-Play

1. Before servicing the vehicle, refer to the precautions in the beginning of this section.
2. Remove the clutch release lever return spring from the lever, and loosen the locknut on the fork adjusting nut.

➡Be careful not to twist the cable during adjustment

3. Turn the adjusting nut (spherical nut)

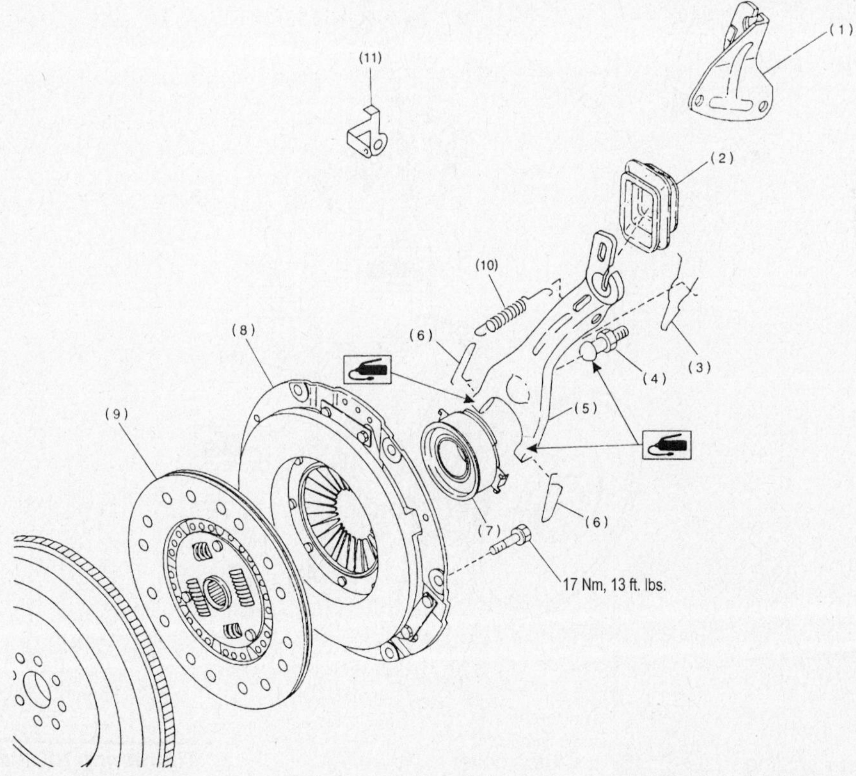

Exploded view of the clutch system components—mechanical clutch, non-turbo models

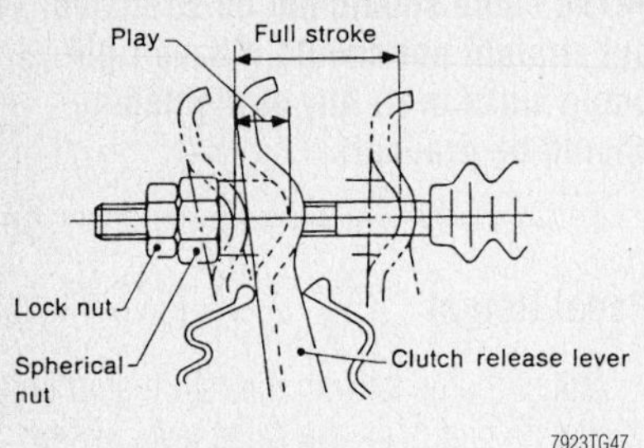

Be sure to tighten the locknut after making the necessary adjustments—mechanical clutch, non-turbo models

until a release fork free-play of 0.14–0.18 in. (3.5–4.5mm) is obtained.

4. Tighten the locknut.
5. Install the return spring on the lever. Hook the long hook side of the return spring with the lever.
6. Check the pedal free-play. It should be 0.12–0.16 in. (3.0–4.0mm).
7. Adjust the pedal free-play, as necessary, with the pedal adjusting bolt.

REMOVAL & INSTALLATION

✲✲ CAUTION

The clutch driven disc may contain asbestos that has been determined to be a cancer causing agent. Never clean clutch surfaces with compressed air. Avoid inhaling any dust from any clutch surface. When cleaning clutch surfaces, use a commercially available brake cleaning fluid.

1. Before servicing the vehicle, refer to the precautions in the beginning of this section.
2. Remove or disconnect the following:
   - Negative battery cable
   - Transaxle
3. Gradually unscrew the bolts which hold the pressure plate assembly on the flywheel. Loosen the bolts only 1 turn at a time, working around the pressure plate.
4. When all of the bolts have been removed, remove the clutch plate and disc.

✲✲ WARNING

Do not get oil or grease on the clutch facing.

5. Remove the 2 retaining springs and remove the throwout bearing and the release fork.

➡Do not disassemble either the clutch cover or disc. Inspect the parts for wear or damage and replace any parts as necessary. Replace the clutch disc if there is any oil or grease on the facing. Do not wash or attempt to lubricate the throwout bearing. If it requires replacement, the bearing may be removed and a new one installed in the holder by means of a press.

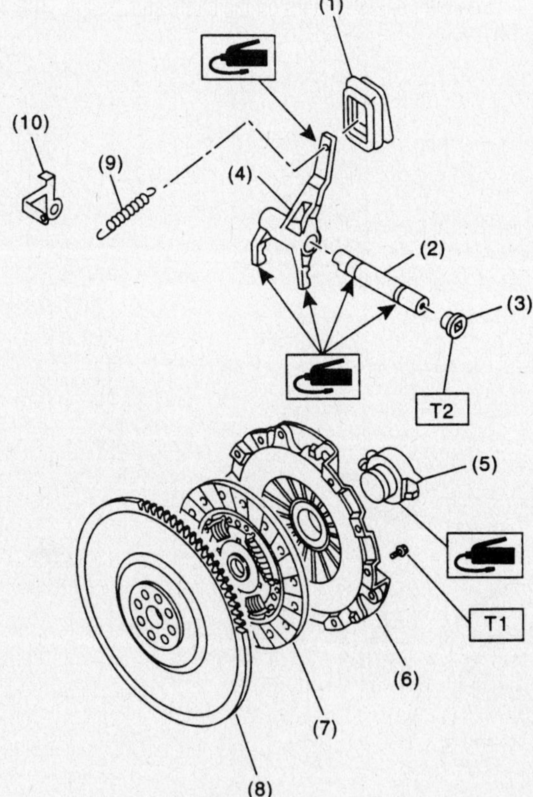

Clutch release lever sealing	(6)	Clutch cover
Release lever shaft	(7)	Clutch disc
Plug	(8)	Flywheel
Release lever	(9)	Spring
Release bearing	(10)	Bracket

Tightening torque: N·m (kgf-m, ft-lb)
T1: 15.7 (1.6, 11.6)
T2: 44 (4.5, 32.5)

9357TG36

Exploded view of the clutch system components—mechanical clutch, turbo models

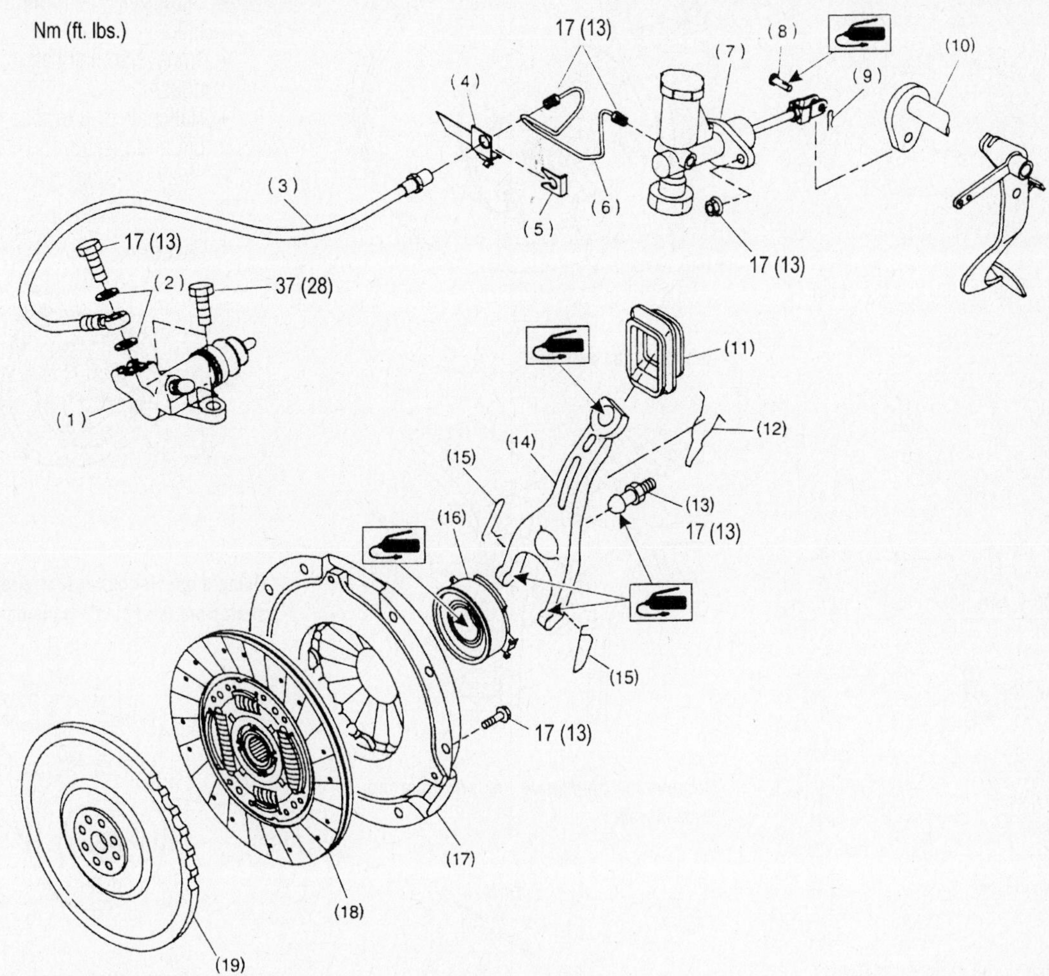

Exploded view of the clutch system components—hydraulic clutch

To install:

6. Fit the release fork boot on the front of the transaxle housing.
7. Install or connect the following:
    - Release fork
    - Throwout bearing assembly and secure it with the 2 springs. Coat the inside diameter of the bearing holder and the fork-to-holder contact points with grease.
8. Insert a pilot shaft through the clutch cover and disc, then insert the end of the pilot into the needle bearing.
9. If equipped, position the **O** marks on the clutch cover and flywheel 120 degrees apart.
10. Tighten the pressure plate bolts gradually, 1 turn at a time, until the proper torque is reached. Tighten to 11 ft. lbs. (15 Nm).

WARNING

When installing the clutch pressure plate assembly, be sure that the O marks on the flywheel and the clutch pressure plate assembly are at least 120 degrees apart. These marks indicate the direction of residual unbalance. Also, be sure that the clutch disc is installed properly, noting the FRONT and REAR markings.

11. After installation of the transaxle in the car, perform the adjustments outlined above.

Hydraulic Clutch System

BLEEDING

➡To properly bleed the system, it must be bled at the slave cylinder and at the damper.

1. Before servicing the vehicle, refer to the precautions in the beginning of this section.
2. Connect a vinyl tube to the air bleeder on the clutch operating (slave) cylinder. Put the other end in a jar with clean clutch fluid.

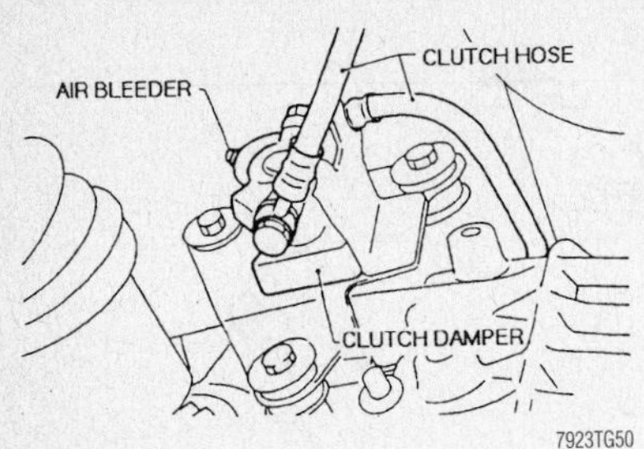

Bleeding the hydraulic clutch at the clutch damper

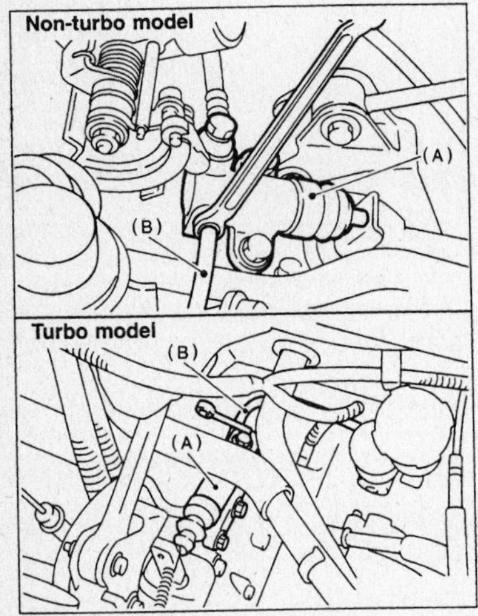

(A) Operating cylinder
(B) Vinyl tube

9357TG37

Bleeding the hydraulic clutch at the slave cylinder

3. With the help of an assistant depressing the clutch pedal, slowly open the bleeder valve. Close the bleeder valve and release the pedal. Repeat this process until no air bubbles appear in the jar.

4. Move the tube to the bleeder on the slave cylinder and repeat the process. Check the operation of the clutch after the bleed procedure is complete.

Transfer Case Assembly

REMOVAL & INSTALLATION

The transfer case must be removed as an assembly with the transaxle.

Halfshaft

REMOVAL & INSTALLATION

Front

1. Before servicing the vehicle, refer to the precautions in the beginning of this section.

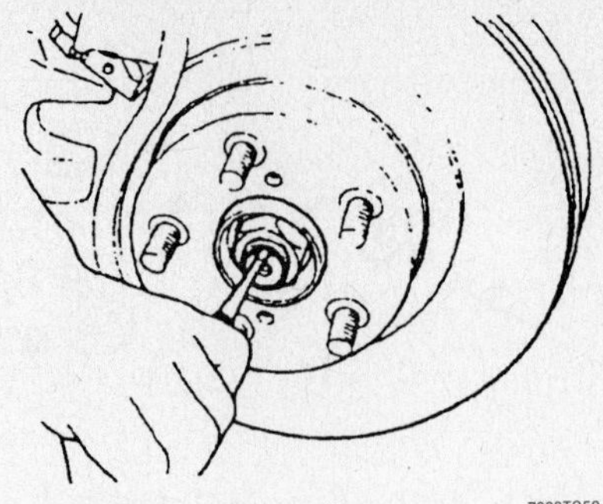

Unstaking the axle nut

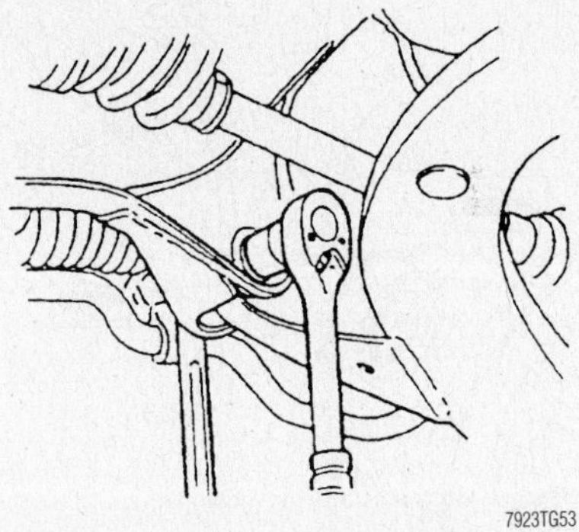

Remove the transverse link arm from the crossmember

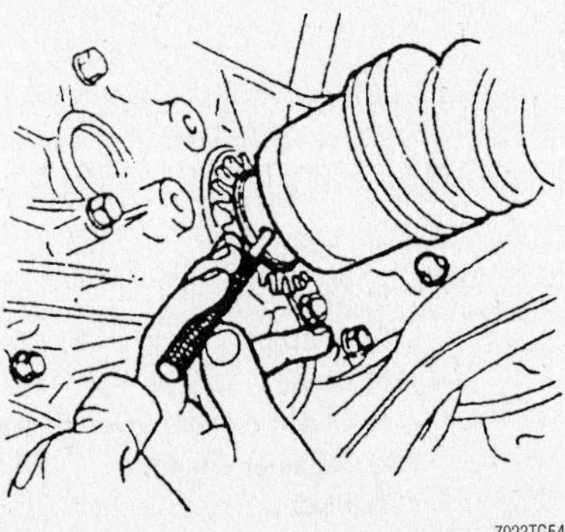

Drive out the halfshaft-to-transaxle roll pin

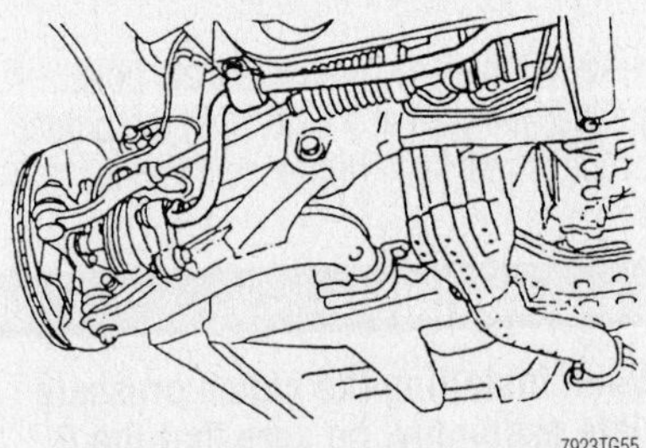

Remove the sway bar bracket

2. Remove or disconnect the following:
   - Negative battery cable
   - Wheel
   - Axle nut, unstake the nut before attempting removal
   - Stabilizer link from the transverse link
   - Transverse link ball joint from the housing
   - Halfshaft-to-transaxle roll/spring pin and discard it
   - Sway bar bracket

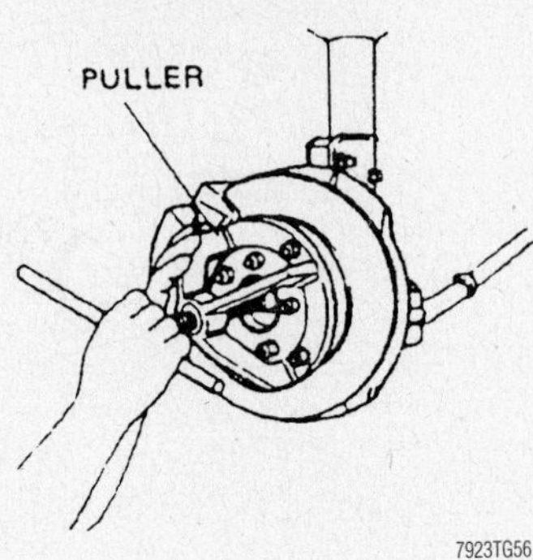

Using a special puller tool, press the axle shaft from the spindle housing

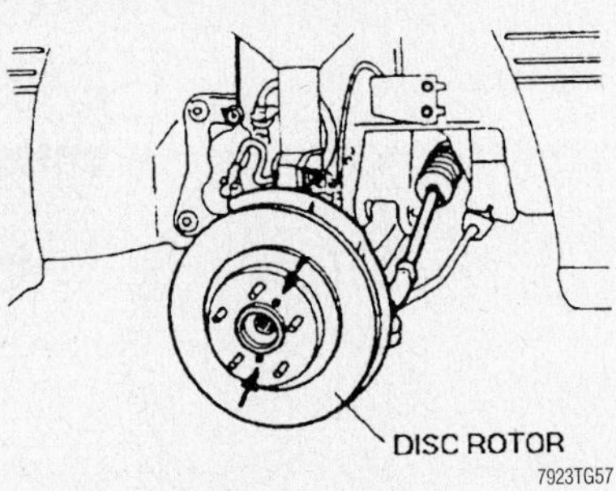

Use two 8mm bolts (arrows) to loosen the rotor from the spindle housing

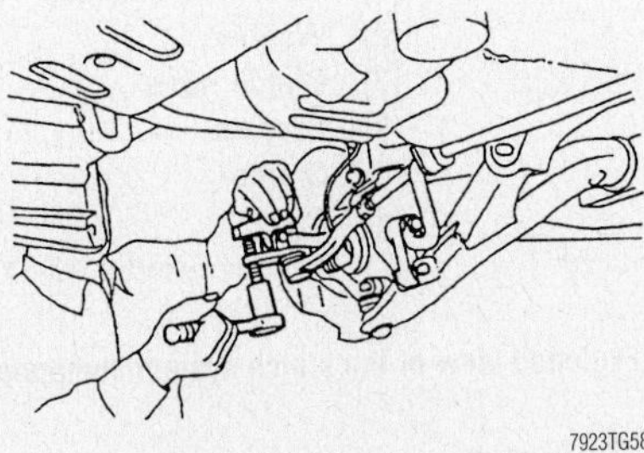

Using a special tool to separate the tie rod end from the steering knuckle

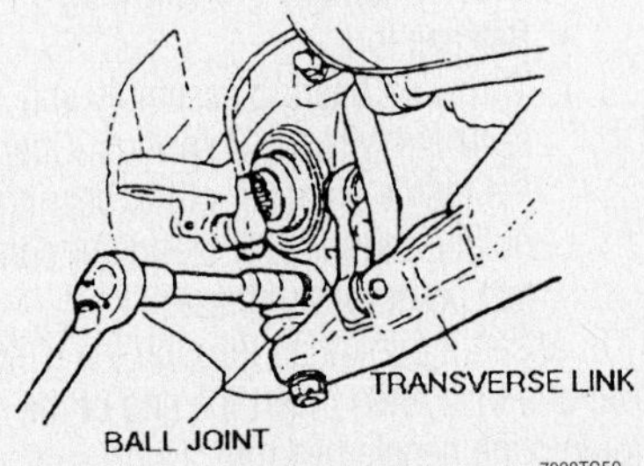

Remove the transverse link arm from the spindle housing

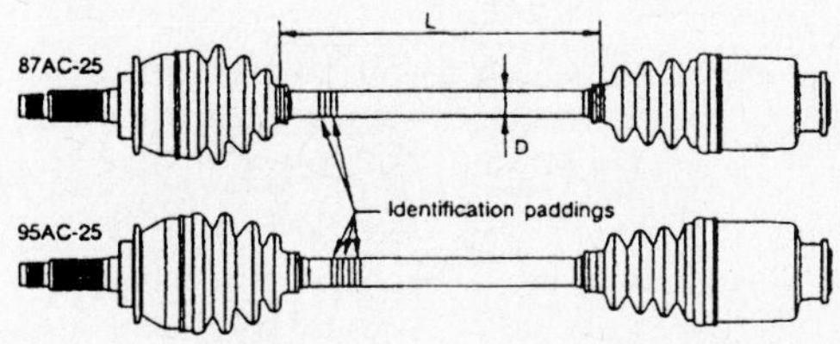

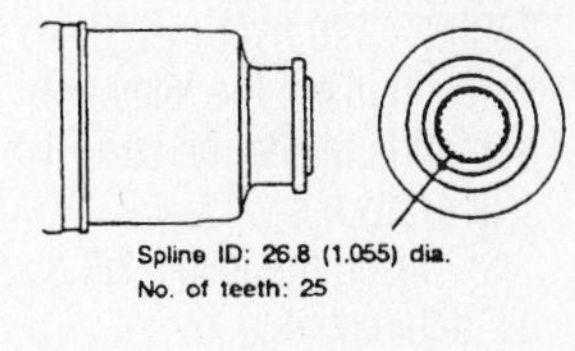

Be sure to identify the correct halfshaft—Except WRX models

- Halfshaft from the transaxle
- Halfshaft from the hub using puller 92707000

To install:

3. Install or connect the following:
   - Halfshaft into the hub
4. Using installer 922431000 and adapter 927390000, pull the halfshaft through the hub.
5. Install or connect the following:
   - Temporarily tighten a new axle nut
   - Align the halfshaft roll/spring pin hole
   - Halfshaft onto the transaxle
   - New pin.
   - Transverse link to the knuckle and tighten a new self-locking nut to 36 ft. lbs. (50 Nm)
   - Sway bar bracket
   - New axle nut to 152 ft. lbs. (206 Nm) on all models except WRX and models equipped with the 3.0L engines. On WRX models, tighten the nut to 137 ft. lbs. (186 Nm). On models equipped with the 3.0L engines, tighten the nut to 159 ft. lbs. (216 Nm). Stake the nut
   - Wheel
   - Negative battery cable

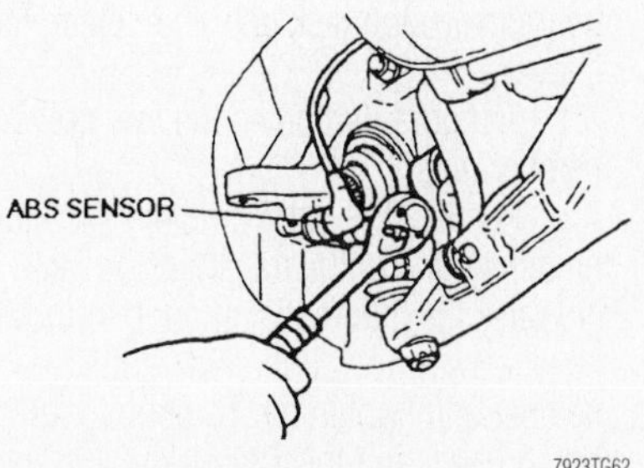
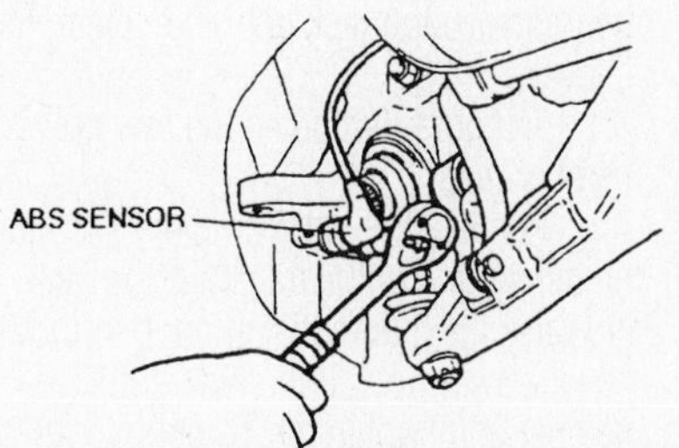

Removing the ABS sensor

Rear

EXCEPT IMPREZA

1. Before servicing the vehicle, refer to the precautions in the beginning of this section.
2. Remove or disconnect the following:
   - Negative battery cable
   - Wheel
   - Axle nut, unstake the nut before attempting removal
3. Remove the rear differential from models equipped with a T–type as follows:

a. Place the gear shifter into the **Neutral** position.

b. Release the parking brake.

c. Remove the rear exhaust pipe and muffler.

d. Remove the heat shield cover.

e. Remove the driveshaft (propeller).

f. Remove the rear differential protector, if equipped.

g. Place a transmission jack under the differential and loose the nuts that attach the differential to the rear crossmember.

h. Remove the driveshaft joint from the differential using a suitable shaft removal tool such as ST 28099PA100.

i. Remove the protector nut.

j. Remove the differential front member and support the differential assembly with the transmission jack making sure to securely attach the differential to the jack with a chain or strap.

k. Remove the nuts attaching the differential to the crossmember.

l. Remove the differential stud bolt from the rear crossmember bushing. You may have to carefully adjust the angle and position of the jack to facilitate stud removal.

m. Once the stud bolt has been removed, lower the jack, making sure the rear drive shaft does not strike the lateral link bolt.

n. Pull the driveshaft out of the differential and remove the differential.

4. Remove the rear differential from models equipped with a VA–type as follows:

a. Place the gear shifter into the **Neutral** position.

b. Release the parking brake.

c. Remove the rear exhaust pipe and muffler.

d. Remove the heat shield cover.

e. Remove the driveshaft (propeller).

f. Place a transmission jack under the differential and loosen the nuts that attach the differential to the rear crossmember.

g. Remove the driveshaft joint from the differential using a suitable shaft removal tool such as ST 28099PA100.

h. Remove the protector nut.

i. Remove the nuts with attach the differential to the front member. Support the differential assembly with the transmission jack.

j. Remove the differential front member and securely attach the differential to the jack with a chain or strap.

k. Remove the nuts attaching the differential to the crossmember.

l. Remove the differential stud bolt from the rear crossmember bushing. You may have to carefully adjust the angle and position of the jack to facilitate stud removal.

m. Once the stud bolt has been removed, lower the jack, making sure the

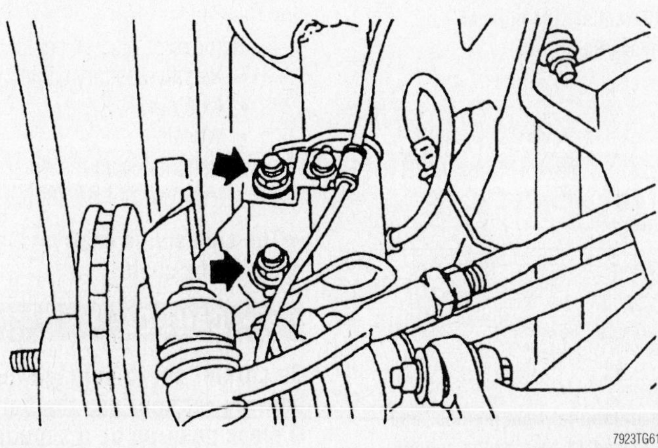

Before loosening the strut-to-housing bolts (arrows), matchmark the camber adjustment bolt and strut

rear drive shaft does not strike the lateral link bolt.

n. Pull the driveshaft out of the differential and remove the differential.

5. Remove the axle nut and using a suitable puller, remove the halfshaft from the being careful not to damage the tone wheel using puller ST 926470000 and plate ST 927140000.

To install:

6. Insert the axle into the hub splines being careful not to damage the tone wheel using adapter installer ST1 922431000 and adapter ST2 927390000. Tighten the axle nut until snug.

7. Install the rear differential from models equipped with a VA–type as follows:

a. Place the differential on a transmission jack and fasten securely with a chain or band.

b. Place a seal protector tool such as ST28099PA090 on the differential.

c. Insert the splined shaft of the halfshaft until the spline portion is inside the oil seal.

d. Remove the seal protector.

e. Insert the driveshaft into the differential until it is fully seated.

f. Adjust the transmission jack as necessary so the stud bolt is inserted correctly into the crossmember bushing.

g. Once the stud bolt is inserted, raise the transmission jack until the differential is level.

h. Temporarily tighten the rear crossmember nuts.

i. Remove the band or chain securing the differential to the transmission jack and raise the differential enough to move the jack away. Install the differential member and two inner bolts to 81 ft. lbs. (110 Nm) and the outer bolt to 48 ft. lbs. (65 Nm).

j. Tighten the crossmember self locking nut to 51 ft. lbs. (70 Nm) and the protector nut to 47 ft. lbs. (64 Nm).

k. Remove the jack and install the driveshaft (propeller).

l. Install the heat shield cover.

m. Install the rear exhaust pipe and muffler.

n. Apply the parking brake.

o. Place the gear shifter into the **Park** position.

8. Remove the filler plug and fill the differential to the level. Tighten the filler plug to 25 ft. lbs. (34 Nm).

9. Install the rear differential from models equipped with a T–type as follows:

a. Place the differential on a transmission jack and fasten securely with a chain or band.

b. Place seal protector tool ST 28099PA090 over the differential side oil seal and insert the shaft until the spline portion is inside the seal. Remove the seal protector.

c. Insert the axle shaft completely into the differential

d. Adjust the transmission jack as necessary so the stud bolt is inserted correctly into the crossmember bushing.

e. Once the stud bolt is inserted, raise the transmission jack until the differential is level.

f. Temporarily tighten the rear crossmember nuts.

g. Remove the band or chain securing the differential to the transmission jack and raise the differential enough to move the jack away. Install the differential member and tighten the T1 bolts to 48 ft. lbs. (65 Nm) and the T2 bolt to 81 ft. lbs. (110 Nm).

h. Tighten the rear crossmember self locking nut to 51 ft. lbs. (70 Nm) and the protector nut to 47 ft. lbs. (64 Nm).

i. Remove the jack and install the driveshaft (propeller).

j. Install the heat shield cover.

k. Install the rear exhaust pipe and muffler.

l. Apply the parking brake.

m. Place the gear shifter into the **Park** position.

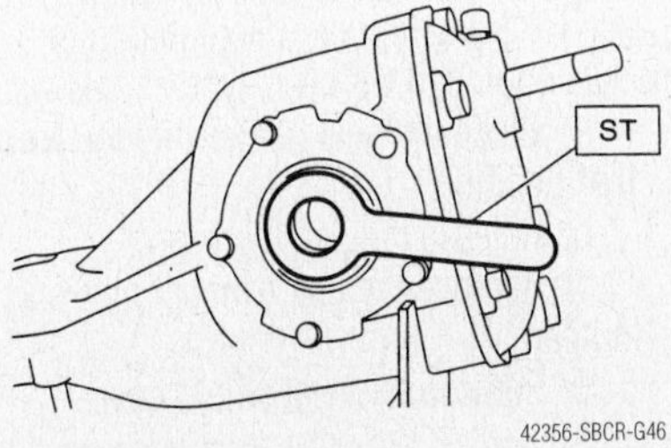

Place seal protector tool ST 28099PA090 over the differential side oil seal—Legacy/Outback/Baja

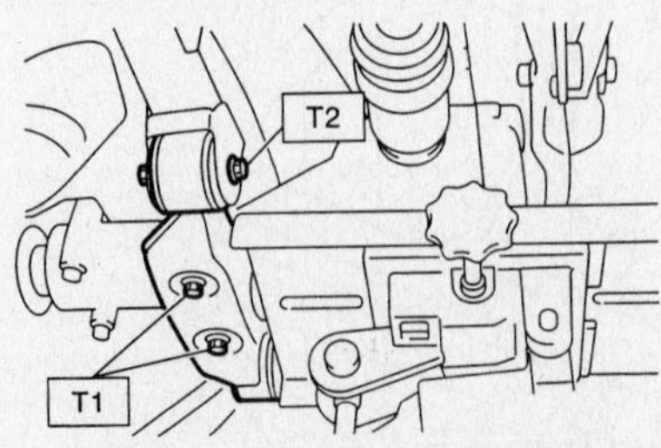

Tighten the differential front member and tighten the T1 and T2 bolts to the specifications outlined in the procedure—Legacy/Outback/Baja

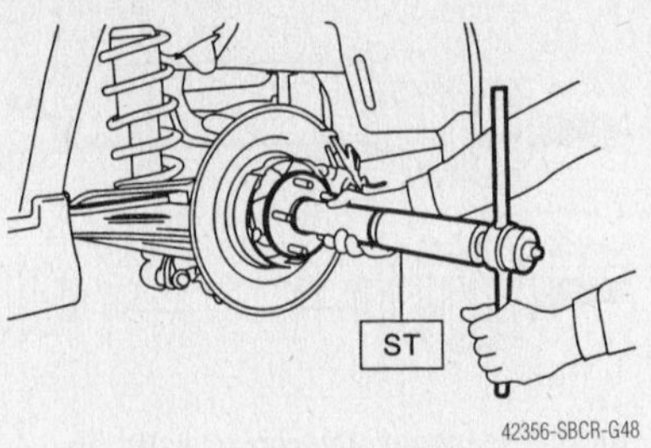

Install the axle into the hub splines using adapter installer ST1 922431000 and adapter ST2 927390000—Legacy/Outback/Baja

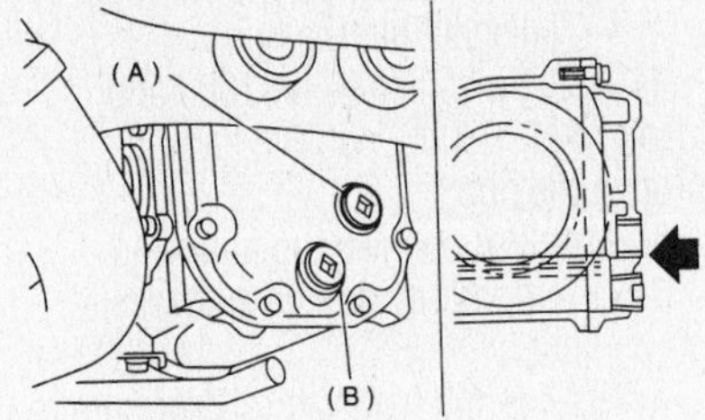

(A) Filler plug
(B) Drain plug

42356-SBCR-G49

Location of the rear differential drain and fill plugs—Legacy/Outback/Baja

10. Remove the filler plug and fill the differential to the level. Tighten the filler plug to 36 ft. lbs. (49 Nm).

11. Install or connect the following:

- Axle nut and tighten to 174 ft. lbs. (235 Nm). Stake the nut.
- Wheels
- Negative battery cable

12. Check all fluids and road test the vehicle.

IMPREZA

1. Before servicing the vehicle, refer to the precautions in the beginning of this section.

2. Remove or disconnect the following:

- Negative battery cable
- Axle nut
- Wheels
- Halfshaft from the differential using remover tool ST 28099PA100

➡The side spline shaft circlip comes out with the shaft.

CAUTION

Be careful not to damage the side bearing retainer use the bolt at the 5 O'clock position as a supporting point when using the removal tool.

- Axle shaft from the hub using puller ST 9266470000 and puller plate 927140000

To install:

3. Install the axle shaft into the hub. Use installer ST 922431000 and adapter ST 927390000 to pull the drive shaft into place and hand tighten the axle nut.
4. Place a seal protector tool such as ST28099PA090 on the differential.
5. Insert the splined shaft of the half-shaft until the spline portion is inside the oil seal.
6. Remove the seal protector.
7. Insert the driveshaft into the differential until it is fully seated.
8. Insert the splined shaft of the half-shaft until the spline portion is inside the oil seal.
9. Torque the new axle nut to 137 ft. lbs. (186 Nm) on 2000–01 models or 174 ft. lbs. (235 Nm) on 2002–03 models and stake the nut.

- Wheel
- Negative battery cable

CV-Joints

OVERHAUL

Front

1. Before servicing the vehicle, refer to the precautions in the beginning of this section.
2. Place alignment marks on the shaft and outer race.
3. Remove the inner boot band and boot.
4. Remove the circlip from the inner joint outer race using a suitable prytool.
5. Outer race from the shaft assembly and wipe off the grease.
6. Place alignment marks on the free ring and trunnion as shown in the illustration.
7. Remove the free ring from the trunnion.
8. Place alignment marks on the trunnion and shaft as shown in the illustration.
9. Remove the snapring and trunnion.
10. Place the shaft in a vise between wooden blocks.
11. Using a suitable prytool, raise the outer boot band claws.
12. Cut and remove the boot.
13. Only the boot can be replaced, the joint is not serviceable and must be replaced if damaged.

To install:

14. Place the half shaft in a vise.

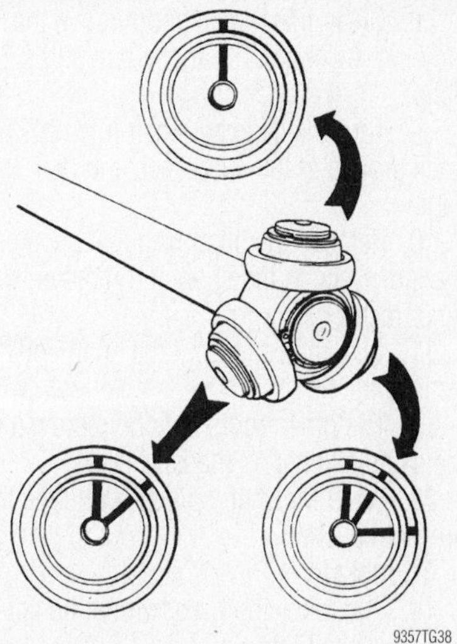

Place alignment marks on the free ring and trunnion as shown

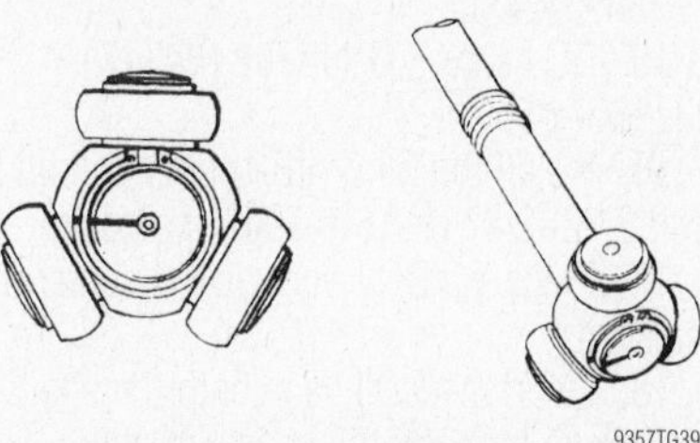

Place alignment marks on the trunnion and shaft as shown

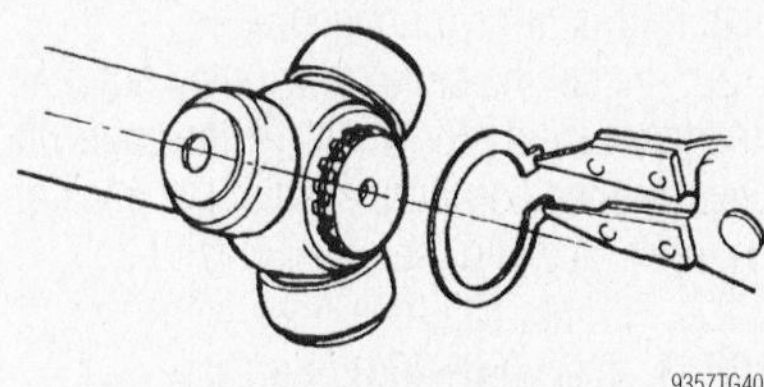

Remove the snapring and trunnion

15. Place the outer boot and small band on the shaft.
16. Apply 2.12–2.47 oz. (60–70g) of supplied grease to the joint.
17. Apply 0.71–1.06 oz. (20–30g) of supplied grease to the whole inner surface of the boot, and apply some grease to the shaft.
18. Install the boot to the joint groove, and attach the large boot band as shown.
19. Install the boot to the shaft groove, and attach the small boot band as shown.
20. Using boot band plier tool 28099A000 to tighten the large band to 116 ft. lbs. (157 Nm) and the small band to 98 ft. lbs. (133 Nm).
21. Place the inner boot on the center of the shaft.
22. Align the alignment marks from earlier and install the trunnion and snapring. Make sure the snapring is fully engaged.
23. Apply 3.53–3.88 oz. (100–110g) of supplied grease to the joint outer race.
24. Apply a coat of supplied grease to the free ring and trunnion.

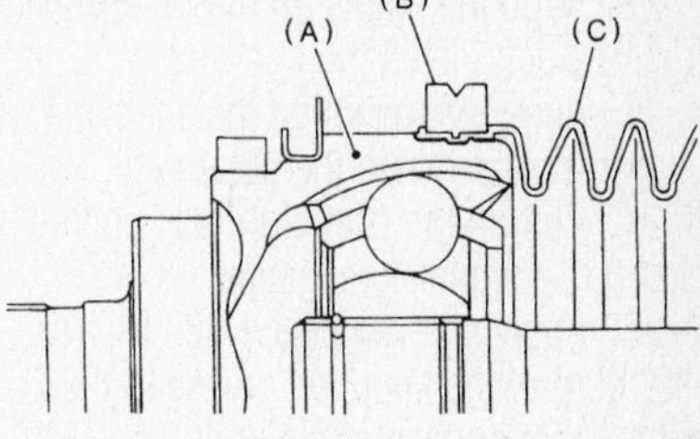

Position the boot to the joint groove, and attach the large boot band as shown

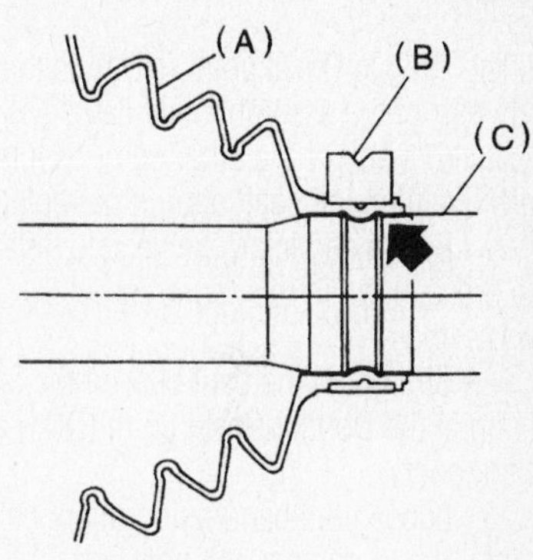

Position the boot to the shaft groove, and attach the small boot band as shown

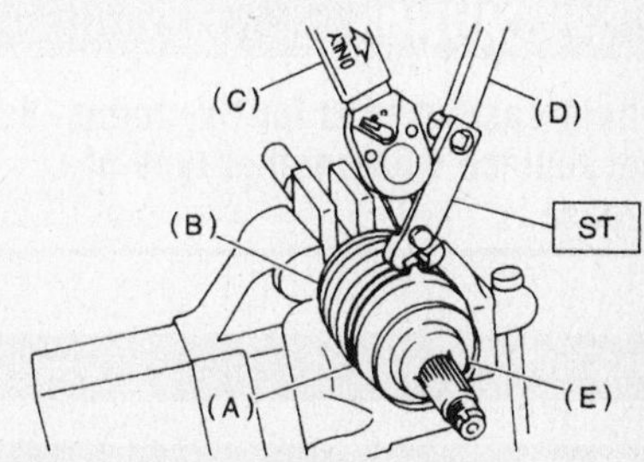

Using boot band plier tool 28099A000 tighten the boot bands

25. Align the marks on the free ring and trunnion and install the free ring.
26. Align the marks on the shaft and outer race and install the outer race.
27. Pull on the shaft lightly to ensure the circlip is completely engaged.
28. Apply an even coat 1.06–1.41 oz. (30–40g) of the supplied grease to the entire inner surface of the boot.
29. Install the boot and band.
30. Once the band is properly tightened, cut off any excess to leave only 0.39 inch (10mm) and bend it over.
31. Install the shaft.

Rear

The Double Offset Joint (DOJ), is the only part of the assembly that can be replaced, if any of the other components are defective then the shaft should be replaced.

1. Before servicing the vehicle, refer to the precautions in the beginning of this section.
2. Straighten the bent claw of the large clamp at the Double Offset Joint (DOJ) end of the boot.
3. Loosen the band using pliers being careful not to damage the bolt.
4. Remove the small boot band from the DOJ using the same technique.
5. Remove the boot from the large end of the DOJ outer race.
6. Remove the round circlip using a suitable prytool from the neck of the joint outer race.
7. Remove the joint outer race.

⁂ CAUTION

The grease used is for CV–Joints, do not replace with another type of grease

8. Clean the grease and remove the balls. Be careful not to loose any of the 6 balls.
9. Turn the cage by a half pitch to the track groove of the inner race and shift the cage.
10. Remove the snap ring that secures the inner race to the shaft and remove the inner race.
11. Take the cage from the shaft and remove the boot.
12. The other boots may be removed in the same manner as the DOJ boot.
13. Wrap the shaft splines with tape to prevent damage.

To install:

14. The following grease must be used during assembly:
    a. BJ side (Non-Turbo): Molylex No. 2 #723223010.
    b. EBJ side (Turbo): NTG2218 # 28093AA000.
15. DOJ side: VU–3A702 (Yellow) # 23223GA050.
16. Install the BJ or EBJ boots and fill it with 2.12–2.47 oz. (60–70g) of grease.
17. Place the DOJ boot on the center of the shaft.
18. Insert the DOJ cage onto the shaft. Make sure to insert the cage with the cut–out portion facing shaft end.
19. Install the inner race on the shaft and fasten with the snap ring. Make sure the snap ring is firmly engaged.
20. Install the cage (previously fitted) to the inner race on the shaft. Fit the cage with the protruded part aligned with the track on the inner race and then turn a half pitch.
21. Fill the DOJ inner race with 2.82–3.17 oz. (80–90g) of grease.
22. Apply a coat of grease to the to the cage pocket and 6 balls.
23. Insert the 6 balls.
24. Align the outer race track and ball positions and place where the shaft, inner race, cage and balls were located prior to removal and then install the outer race.
25. Install the circlip into the groove on the outer race.

➡Make sure the balls, cage and inner race are fully seated. make sure not to place the matched position of the circlip in the ball groove of the outer race. Pull the shaft lightly to make sure the circlip is fully engaged.

26. Apply an even coat 0.71–1.06 oz. (20–30g) of grease to the entire inner surface of the boot and to the shaft.
27. Make sure the boot is free from any dirt or foreign materials prior to installation.
28. Place the outer race of the boot at the center of its travel.
29. Put a band through the boot clip and wind twice in alignment with the groove on the boot.
30. Pinch the end of the band using tool ST 9250910000 and tighten securely until it cannot be moved by hand. Make sure there is appropriate air inside the boot.
31. Tap on the clip with a punch to lock it making sure not to damage the damaged while tapping. Cut any excess off the band leaving 0.39 inch (10mm) and bend the remaining portion over the clip. Make sure the end of the band is close to the clip.
32. Install the remaining boot clamps in the same manner.

STEERING AND SUSPENSION

Air Bag

⁂ CAUTION

Some vehicles are equipped with an air bag system. The system must be disabled before performing service on or around system components, steering column, instrument panel components, wiring and sensors. Failure to follow safety and disabling procedures could result in accidental air bag deployment, possible personal injury and unnecessary system repairs.

PRECAUTIONS

Several precautions must be observed when handling the inflator module to avoid accidental deployment and possible personal injury.

- Never carry the inflator module by the wires or connector on the underside of the module.
- When carrying a live inflator module, hold it securely with both hands, and ensure that the bag and trim cover are pointed away.
- Place the inflator module on a bench or other surface with the bag and trim cover facing up.
- With the inflator module on the bench, never place anything on or close to the module that may be thrown in the event of an accidental deployment.

DISARMING

1. Before servicing the vehicle, refer to the precautions in the beginning of this section.
2. Disconnect the negative battery cable.
3. Disconnect the positive battery cable.
4. Wait more than 20 seconds before starting work.

Power Rack and Pinion Steering Gear

REMOVAL & INSTALLATION

Legacy, Baja and Outback Models

1. Before servicing the vehicle, refer to the precautions in the beginning of this section.
2. Remove or disconnect the following:
   - Negative battery cable
   - Air intake duct
   - Front axle nut, loosen only at this time
   - Front tire and wheel assemblies
   - Electrical connector from the Oxygen Sensor (O_2S)
   - Front exhaust pipe assembly
   - Tie rod end cotter pin and loosen the castle nut
   - Tie rod ends from the steering knuckle arm using a ball joint puller
   - Jack up plate and the front stabilizer bar
3. From the power steering rack, remove the center pressure pipe, connect a vinyl hose to the pipe and joint, then turn the steering wheel to discharge the fluid into a container.

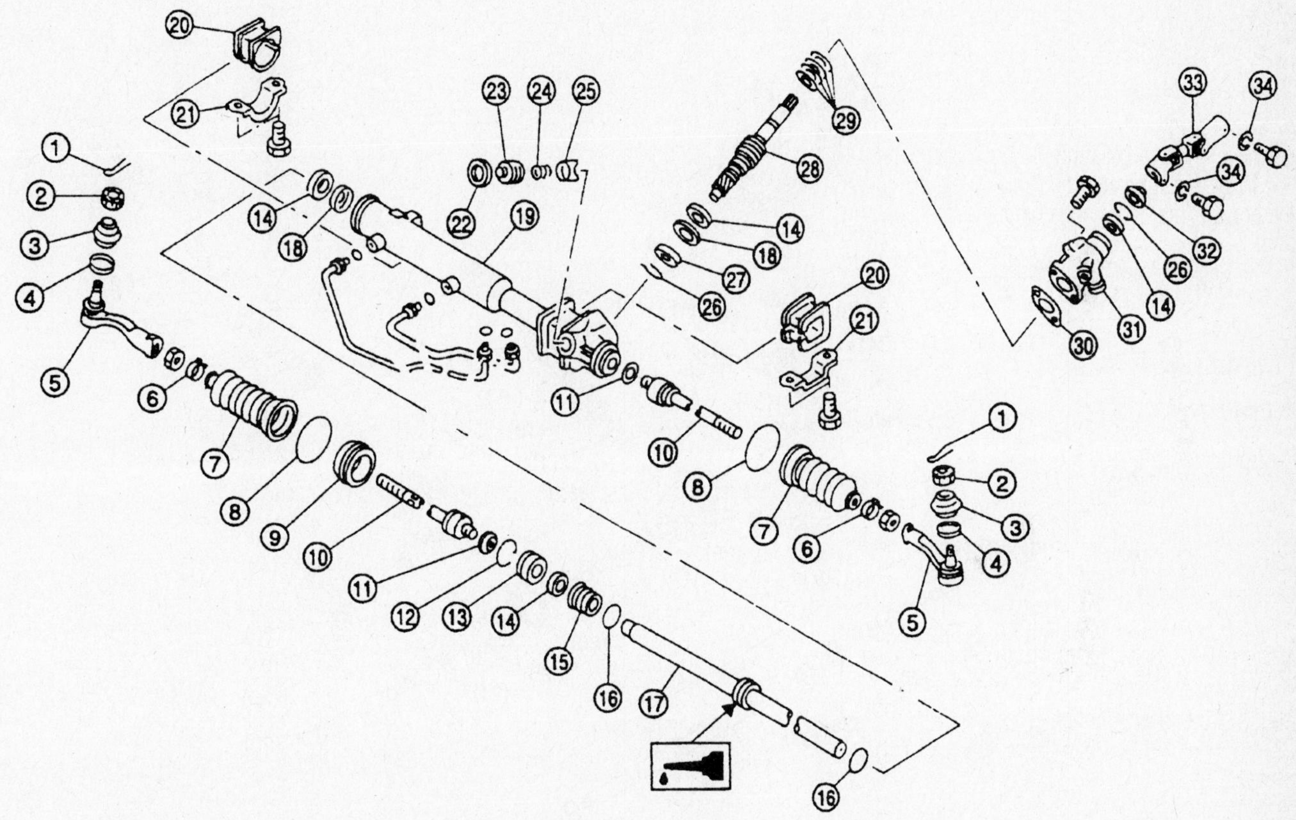

1. Cotter pin
2. Castle nut
3. Dust cover
4. Clip
5. Tie-rod end
6. Clip
7. Boot
8. Clip
9. Spacer
10. Tie-rod
11. Lock washer
12. Circlip
13. Rack stopper
14. Oil seal
15. Rack bushing
16. O-ring
17. Rack
18. Back-up washer
19. Rack housing
20. Adapter
21. Clamp
22. Lock nut
23. Adjusting screw
24. Spring
25. Sleeve
26. C-ring
27. Ball bearing
28. Valve
29. Seal ring
30. Packing
31. Valve housing
32. Dust seal
33. Universal joint
34. Spring washer

7923TG63

Exploded view of the steering rack assembly—Legacy models

➡When discharging the power steering fluid (line A and B), turn the steering wheel fully, left and right. Be sure to disconnect the other pipe and drain the fluid in the same manner.

4. From the control valve of the gearbox assembly, remove the power steering **C** and **D** pressure pipes. Remove pipe **D** first and pipe **C** second.

5. If not disconnected when draining the fluid from the control valve of the gearbox assembly, remove the power steering **A** and **B** pressure pipes. Remove pipe **A** first and pipe **B** second.

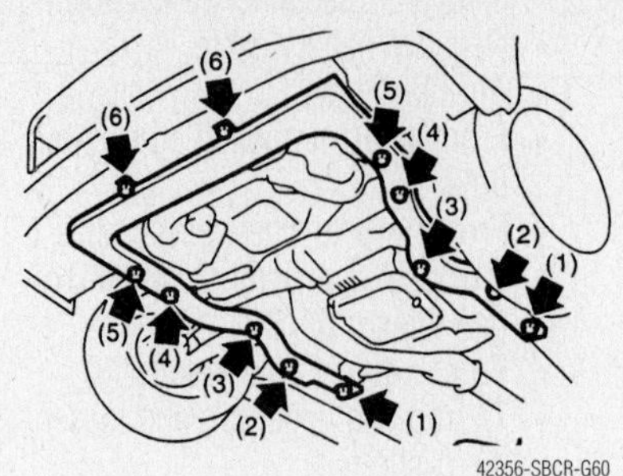

Remove the sub-frame bolts as illustrated—Impreza models

6. Remove or disconnect the following:
   - Universal joint assembly. Matchmark the assembly before removal.
   - Power steering gearbox-to-crossmember assembly bolts
   - Gearbox assembly from the vehicle

To install:

7. Install or connect the following:
   - Power steering rack and tighten the rack to crossmember bolts to 35–52 ft. lbs. (47–70 Nm)
   - Universal joint assembly making sure to align the matchmarks

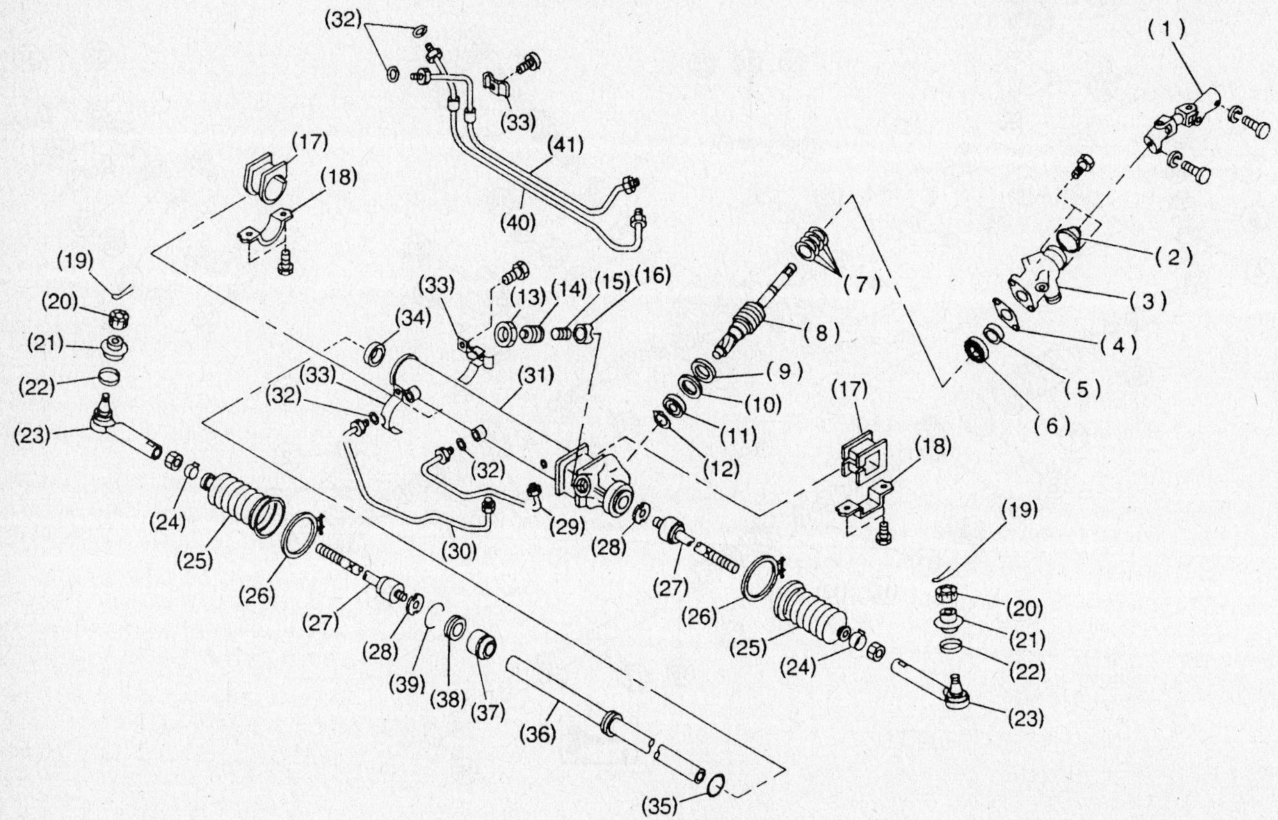

Exploded view of the steering rack assembly—Impreza models

- Power steering pressure pipes and tighten to 7–12 ft. lbs. (10–16 Nm)
- Universal joint assembly-to-power steering gearbox bolts and tighten to 16–19 ft. lbs. (22–24 Nm) and the universal joint assembly-to-steering shaft bolts 16–19 ft. lbs. (22–24 Nm)
- Tie rod end to steering knuckle nut and tighten to 18–22 ft. lbs. (25–29 Nm). After tightening this nut, turn it no more than 60 degrees further to align the cotter pin hole.
- New cotter pin
- Front stabilizer into the vehicle
- Exhaust Y-pipe and O_2S connector
- Tires and tighten the wheel lug nuts to specification

8. Partially lower the vehicle, then refill and bleed the power steering system.
9. Check for fluid leaks and the fluid level, then install the jack up plate.
10. Check and adjust the toe-in and the steering angle.

Impreza

1. Before servicing the vehicle, refer to the precautions in the beginning of this section.
2. Remove or disconnect the following:
   - Negative battery cable
   - Front wheels
   - Under cover
   - Bolt, clip on the sub-frame
   - Sub-frame bolts. Leave bolt (1) connected by a few threads and remove the bolts in the sequence illustrated. Once the other bolts are removed, remove bolt (1) and the sub-frame. Refer to the illustration for bolt location.
   - Front Y-pipe
   - Tie rod end cotter pin and nut
   - Tie rod ends from the steering knuckle using a puller
   - Jack-up plate and front sway bar
   - Fluid lines from the rack and pinion
3. Matchmark the universal joint to the serration in the steering rack for installation reference.
   - Lower and upper universal joint bolts and lift the joint upward, disconnecting it from the rack and pinion shaft
   - Clamp bolts securing the rack and pinion to the crossmember
   - Rack and pinion

To install:

4. Install the rack and pinion. Tighten the clamp bolts to 43 ft. lbs. (59 Nm).
   - Fluid pipes and tighten to 9 ft. lbs. (13 Nm)
5. Align the steering rack to the universal joint. Push the long yoke of the joint all the way into the serrated position of the steering shaft, setting the bolt hole in the cut-out. Pull the short yoke all the way out of the serrated portion of the rack and pinion, setting the bolt hole in the cut-out. Insert the bolt through the short yoke. Pull the yoke and ensure the bolt is properly engaged in the cut-out. Fasten the short yoke side with the spring washer and bolt, then fasten the yoke side. Tighten the bolts to 17 ft. lbs. (24 Nm).
6. Install or connect the following:
   - Tie rod ends to the steering knuckle
   - Sway bar and jack-up plate
   - Y-pipe with new gaskets and nuts
   - Sub-frame. Tighten the T1 bolts to 25 ft. lbs. (34 Nm), the T2 bolts to 41 ft. lbs. (55 Nm) and the T3 bolts to 52 ft. lbs. (71 Nm).
   - Wheels
7. Fill and bleed the steering system.

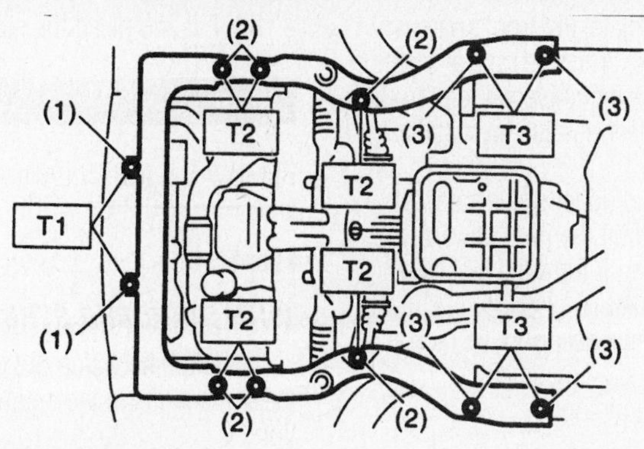

Tighten the sub-frame bolts to the sequences outlined in the procedure—Impreza models

WRX Models

1. Before servicing the vehicle, refer to the precautions in the beginning of this section.
2. Remove or disconnect the following:
   - Negative battery cable
   - Front wheels
   - Engine undercover
3. Remove the sub-frame as follows:
   a. Remove the bolt cover.
   b. Remove the clip.
   c. Loosen the sub-frame bolt (1) and leave it screwed in a few threads. Remove the remaining bolts in the following order: 2, 3, 4, 5 and 6. See illustration for bolt location.
4. Remove or disconnect the following:
   - Front exhaust pipe
   - Tie rod end cotter pin and nut
   - Tie rod ends from the steering knuckle using a puller

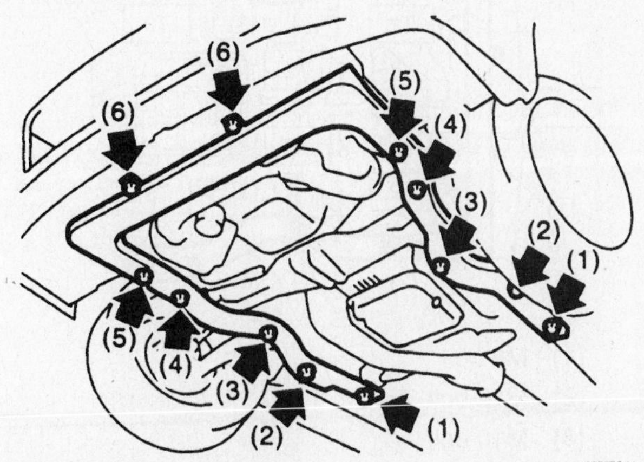

Location of the sub-frame bolts–WRX models

- Jack-up plate and front sway bar
- Fluid lines from the rack and pinion

5. Matchmark the universal joint to the serration in the steering rack for installation reference.

- Lower and upper universal joint bolts and lift the joint upward, disconnecting it from the rack and pinion shaft
- Clamp bolts securing the rack and pinion to the crossmember
- Rack and pinion

To install:

6. Install the rack and pinion. Tighten the clamp bolts to 43 ft. lbs. (59 Nm).

7. Align the steering rack to the universal joint. Push the long yoke of the joint all the way into the serrated position of the steering shaft, setting the bolt hole in the cut-out. Pull the short yoke all the way out of the serrated portion of the rack and pinion, setting the bolt hole in the cut-out. Insert the bolt through the short yoke. Pull the yoke and ensure the bolt is properly engaged in the cut-out. Fasten the short yoke side with the spring washer and bolt, then fasten the yoke side. Tighten the bolts to 19 ft. lbs. (27 Nm).

8. Install or connect the following:

- Tie rod ends to the steering knuckle
- Sway bar and jack-up plate

9. Install the sub-frame and bolts. Tighten the bolts as follows, referring to the illustration for location:

a. T1: 25 ft. lbs. (34 Nm).
b. T2: 41 ft. lbs. (55 Nm).
c. T3: 52 ft. lbs. (71 Nm).

- Bolt and clip
- Exhaust pipe with new gaskets and nuts
- Engine undercover
- Wheels

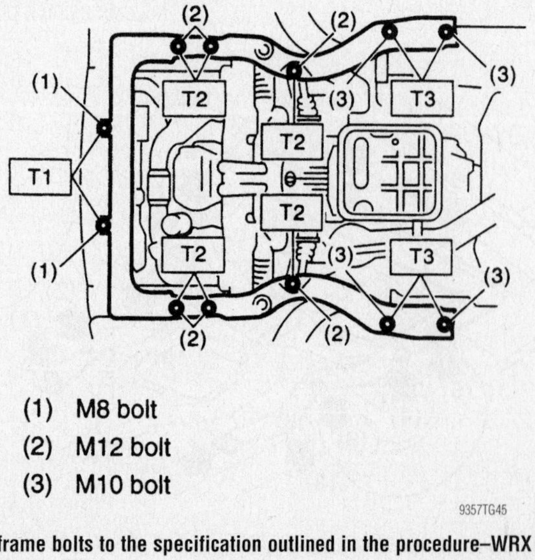

10. Fill and bleed the steering system.

Strut

REMOVAL & INSTALLATION

Front

WITH STANDARD STRUTS

1. Before servicing the vehicle, refer to the precautions in the beginning of this section.

2. Remove or disconnect the following:

- Negative battery cable
- Front wheel assembly
- Caliper, if necessary, leaving the line connected, and suspend it out of the way with a piece of wire or string
- Clip or bolt attaching the brake line to the strut housing
- Bolt securing the Anti-lock Brake System (ABS) sensor harness, if equipped

➡Scribe a matchmark on the camber adjusting bolt which secures the strut to the housing.

- 2 bolts and nuts securing the strut to the steering knuckle. Notice that the shaft of the top bolt is not round. This bolt is used for camber adjustment, and must always be installed in the top hole.
- 3 nuts securing the strut to the body in the engine compartment
- Strut and coil spring assembly from the vehicle

To install:

3. Install or connect the following:

- Strut assembly into the vehicle
- Upper strut retainer nuts and tighten the nuts to 15 ft. lbs. (20 Nm)
- Lower strut nuts and bolts. Be sure the alignment adjustment bolt is installed in the top mounting hole. Tighten the nuts to 112 ft. lbs. (152 Nm) for Impreza and Outback models, or to 130 ft. lbs. (177 Nm) for Legacy and WRX models.
- ABS sensor harness, if equipped and tighten the bolt to 24 ft. lbs. (32 Nm)
- Caliper, if removed
- Brake line to the strut and install the clip or bolt
- Front wheel
- Negative battery cable

4. Check the front end alignment and adjust as necessary.

WITH PNEUMATIC STRUTS

1. Before servicing the vehicle, refer to the precautions in the beginning of this section.

2. Remove or disconnect the following:

- Negative battery cable
- Air line and height sensor harness (from inside the engine compartment) from the strut assembly
- Front wheel assembly
- Anti-lock Brake System (ABS) sensor, if equipped
- Caliper, leaving the line connected, and suspend it out of the way with a piece of wire or string
- Clip attaching the brake line to the strut housing

3. Matchmark the camber adjustment bolt to the strut housing as reference for installation.

- Bolt securing the ABS sensor harness, if equipped
- 2 bolts and nuts securing the strut to the steering knuckle. Notice that the shaft of the top bolt is not round. This bolt is used for camber adjustment, and must always be installed in the top hole.
- 3 nuts securing the strut to the body in the engine compartment
- Strut and coil spring assembly

To install:

4. Install or connect the following:

- Strut assembly
- Upper strut retainer nuts and tighten to 15 ft. lbs. (20 Nm)
- ABS sensor harness (if equipped), and tighten the bolt to 15 ft. lbs. (20 Nm)
- Lower strut nuts and bolts. Be sure the alignment adjustment bolt is installed in the top mounting hole.

(1) M8 bolt
(2) M12 bolt
(3) M10 bolt

9357TG45

Tighten the sub-frame bolts to the specification outlined in the procedure–WRX models

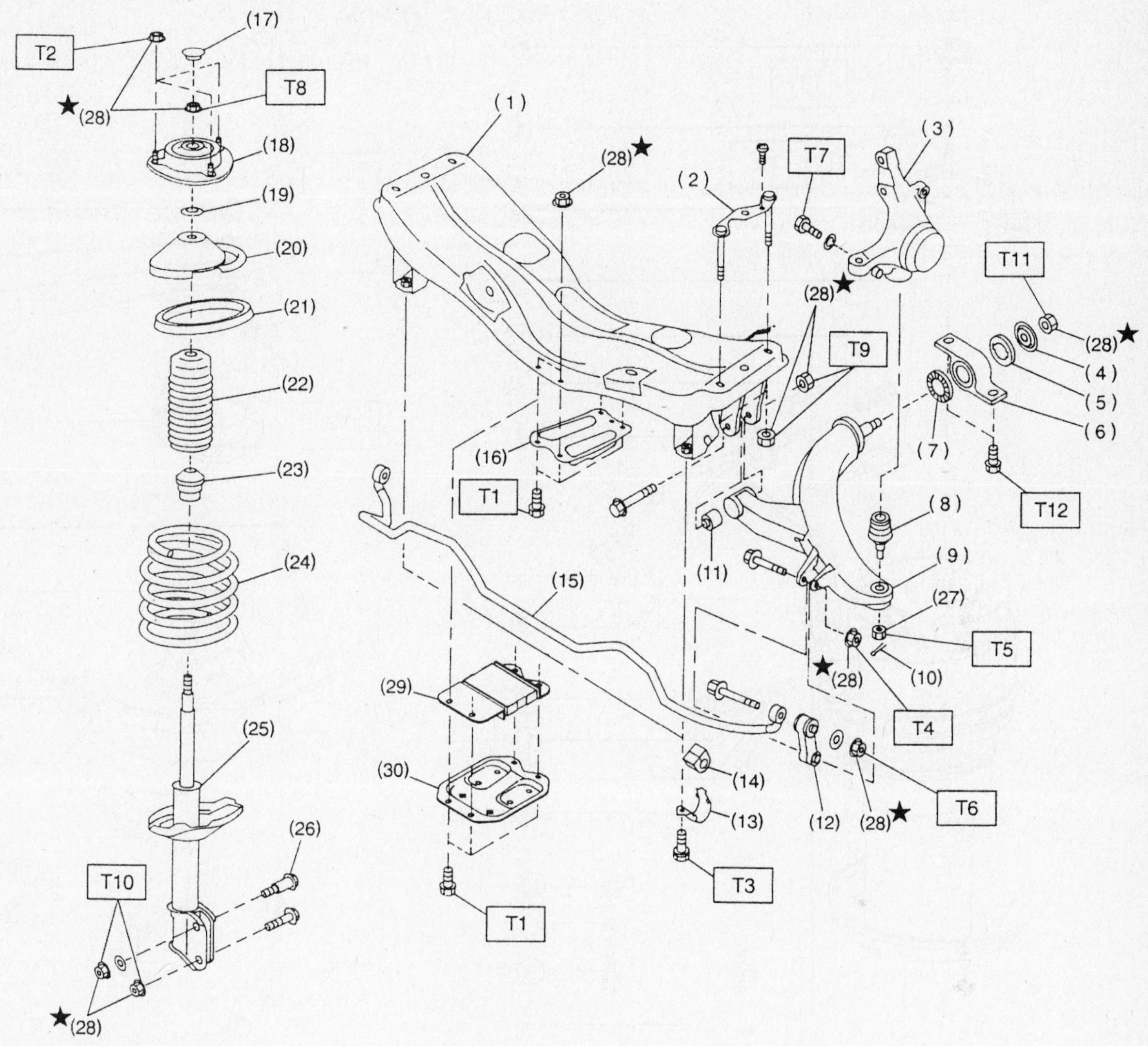

(1) Crossmember
(2) Bolt ASSY
(3) Housing
(4) Washer
(5) Stop rubber (Rear)
(6) Rear bushing
(7) Stop rubber (Front)
(8) Ball joint
(9) Transverse link
(10) Cotter pin
(11) Front bushing
(12) Stabilizer link
(13) Clamp
(14) Bushing
(15) Stabilizer
(16) Jack-up plate (Except 2500 cc MT model)
(17) Dust seal
(18) Strut mount
(19) Spacer
(20) Upper spring seat
(21) Rubber seat
(22) Dust cover
(23) Helper
(24) Coil spring
(25) Damper strut
(26) Adjusting bolt
(27) Castle nut
(28) Self-locking nut
(29) Dynamic damper (2500 cc MT model)
(30) Jack-up plate (2500 cc MT model)

Tightening torque: N·m (kg-m, ft-lb)

T1: 18±5 (1.8±0.5, 13.0±3.6)
T2: 20±6 (2.0±0.6, 14.5±4.3)
T3: 25±4 (2.5±0.4, 18.1±2.9)
T4: 29±5 (3.0±0.5, 21.7±3.6)
T5: 39 (4, 29)
T6: 44±6 (4.5±0.6, 32.5±4.3)
T7: 49±10 (5.0±1.0, 36±7)
T8: 54±5 (5.5±0.5, 39.8±3.6)
T9: 98±15 (10.0±1.5, 72±11)
T10: 152±20 (15.5±2.0, 112±14)
T11: 186±10 (19.0±1.0, 137±7)
T12: 245±49 (25.0±5.0, 181±36)

9307TG06

Exploded view of the front suspension assembly—Legacy and Impreza models

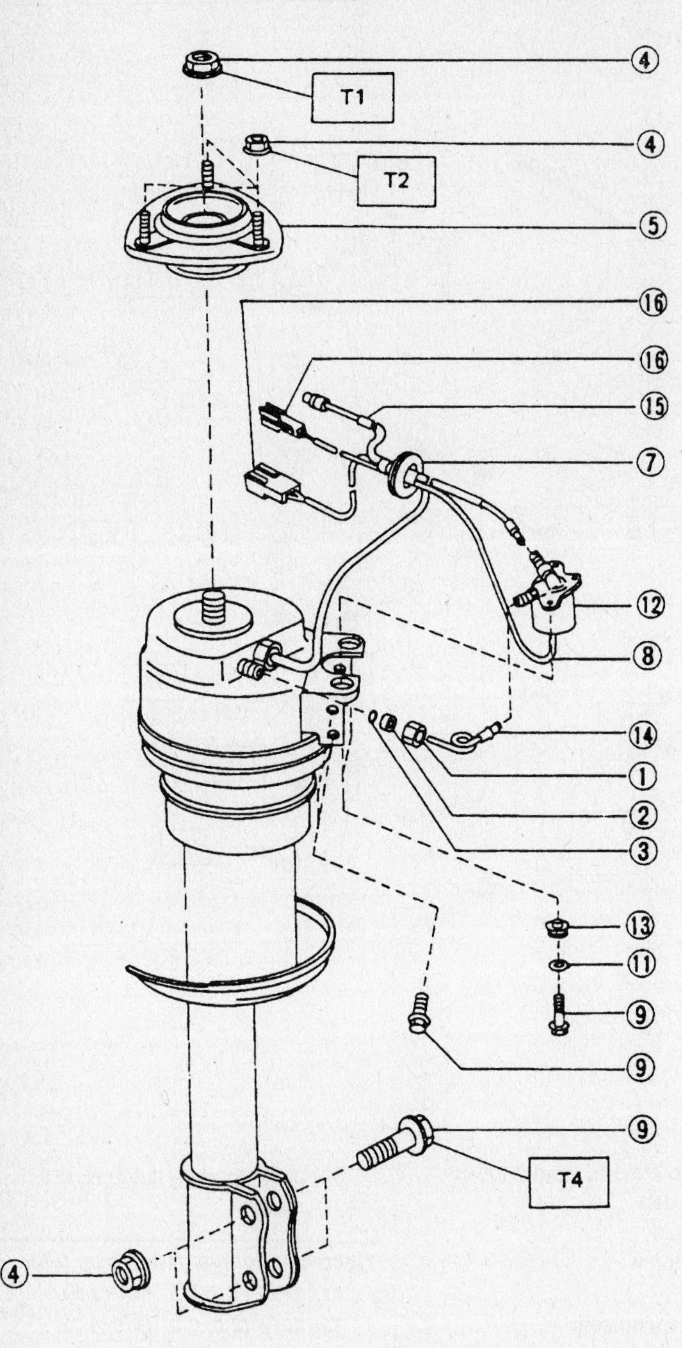

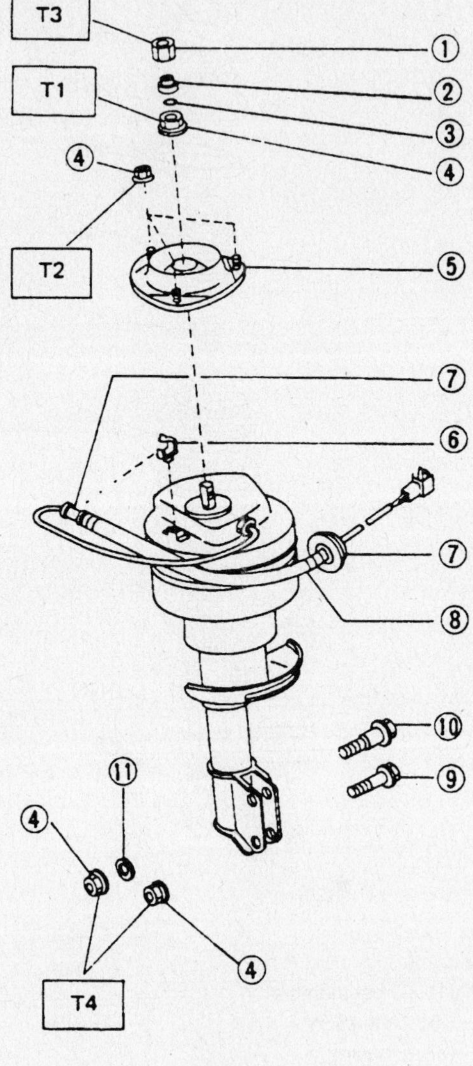

1 Cap
2 Air bushing
3 O-ring
4 Self lock nut
5 Strut mount
6 Clip
7 Grommet
8 Corrugate tube
9 Flange bolt
10 Adjusting bolt
11 Washer
12 Solenoid valve
13 Insulator
14 Air pipe for solenoid valve
15 Air pipe
16 Connector

Tightening torque: N·m (kg-m, ft-lb)

T1: 49 – 69 (5 – 7, 36 – 51)

T2: 14 – 25 (1.4 – 2.6, 10 – 19)

T3: 7 – 17 (0.7 – 1.7, 5.1 –12.3)

T4: 186 – 235 (19 – 24, 137 – 174)

7923TG69

Exploded view of the front and rear pneumatic suspension assembly—Legacy and Impreza models

Tighten the nuts to 112 ft. lbs. (152 Nm).

- Caliper
- Brake line to the strut and install the clip
- Height sensor harness and air line
- Front wheel
- Negative battery cable

5. Start the vehicle and allow enough time for the struts to pressurize before driving.

6. Check the front end alignment and adjust as necessary.

Rear

WITH STANDARD STRUTS

1. Before servicing the vehicle, refer to the precautions in the beginning of this section.

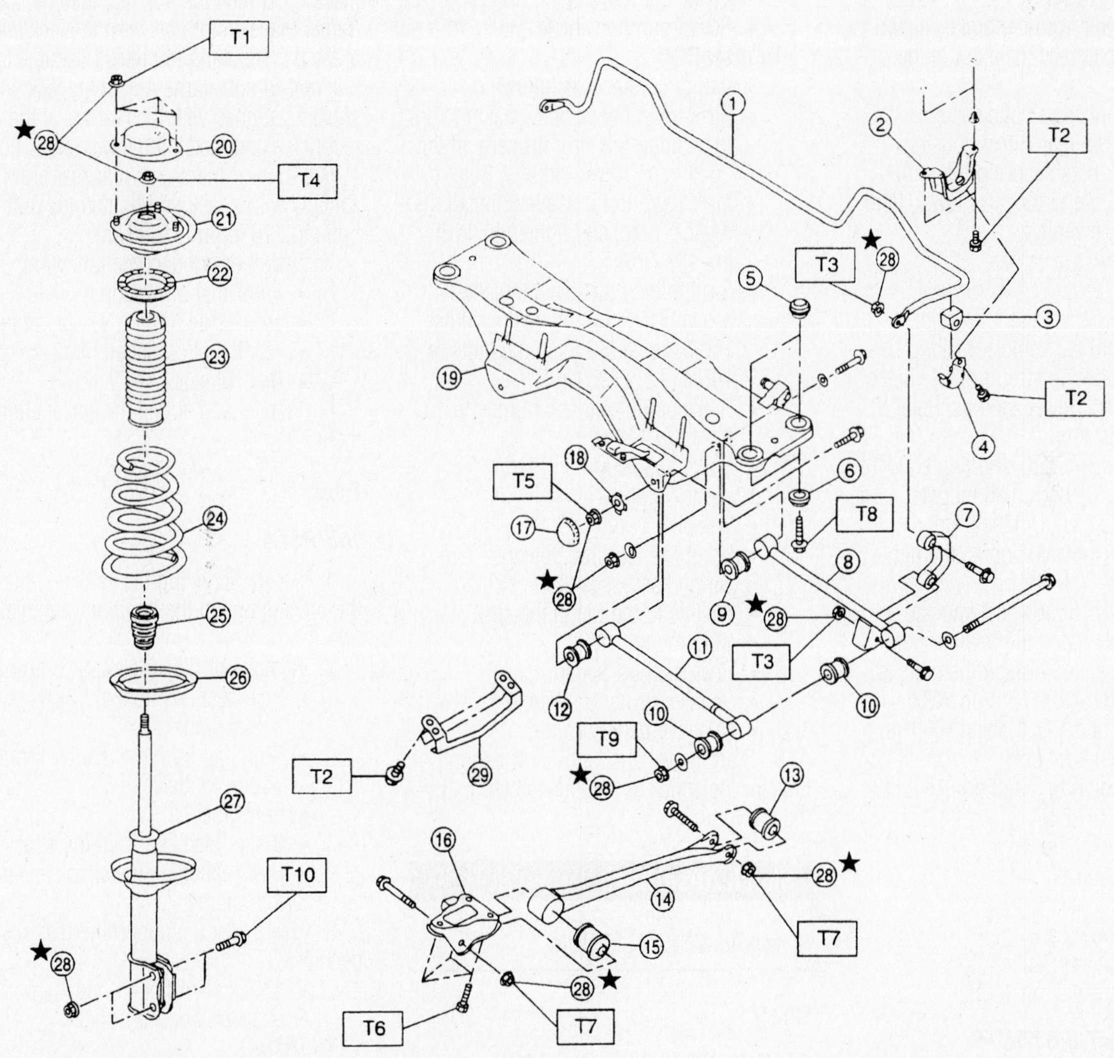

① Stabilizer
② Stabilizer bracket
③ Stabilizer bushing
④ Clamp
⑤ Floating bushing
⑥ Stopper
⑦ Stabilizer link
⑧ Rear lateral link
⑨ Bushing (C)
⑩ Bushing (A)
⑪ Front lateral link
⑫ Bushing (B)
⑬ Trailing link rear bushing
⑭ Trailing link
⑮ Trailing link front bushing
⑯ Trailing link bracket
⑰ Cap (Protection)
⑱ Washer
⑲ Crossmember
⑳ Strut mount cap
㉑ Strut mount
㉒ Rubber seat upper
㉓ Dust cover
㉔ Coil spring
㉕ Helper
㉖ Rubber seat lower
㉗ Damper strut
㉘ Self-locking nut
㉙ Crossmember reinforcement lower (Sedan model only)

Tightening torque: N·m (kg-m, ft-lb)

T1: 20 ± 6 (2.0 ± 0.6, 14.5 ± 4.3)
T2: 25 ± 7 (2.5 ± 0.7, 18.1 ± 5.1)
T3: 44 ± 6 (4.5 ± 0.6, 32.5 ± 4.3)
T4: 59 ± 10 (6.0 ± 1.0, 43 ± 7)
T5: 98 ± 15 (10.0 ± 1.5, 72 ± 11)
T6: 98 ± 20 (10.0 ± 2.0, 72 ± 14)
T7: 113 ± 15 (11.5 ± 1.5, 83 ± 11)
T8: 127 ± 20 (13.0 ± 2.0, 94 ± 14)
T9: 137 ± 20 (14.0 ± 2.0, 101 ± 14)
T10: 196^{+39}_{-10} $(20.0^{+4.0}_{-1.0}, 145^{+29}_{-7})$

7923TG71

Exploded view of the rear standard strut assembly—Legacy and Impreza models

2. Remove or disconnect the following:
   - Rear seat assembly, on Sedan only
   - Rear speaker grille and service hole cap, on Wagon only
   - Strut mount cap
   - Wheel and tire assembly
   - Brake hose clip
   - Union bolt from the brake caliper. Move the brake hose out of the way.
   - Lower nuts and bolts securing the strut to the rear wheel housing
   - Retainer nuts securing the strut bearing cap to the strut tower, from inside the vehicle
   - Strut from the vehicle

To install:

3. Install or connect the following:
   - Strut onto the vehicle, making sure to position the strut properly in the upper strut tower mounts. Refer to the illustration.
   - Strut retainer nuts and tighten to 14 ft. lbs. (20 Nm) on all models except Legacy/Outback. On Legacy/Outback, tighten the nut to 22 ft. lbs. (30 Nm).
   - Strut to the rear wheel knuckle assembly using the retainer nuts and bolts, and tighten the bolts to 162 ft. lbs. (220 Nm) on WRX models and 145 ft. lbs. (196 Nm) on all other models
   - Brake union bolt and tighten to 13 ft. lbs. (18 Nm)
   - Brake hose clip
4. Bleed the brakes.
   - Wheel
   - Strut mount cap
   - Rear seat, on Sedan
   - Speaker grille, on Wagon

WITH PNEUMATIC STRUTS

1. Before servicing the vehicle, refer to the precautions in the beginning of this section.
2. Remove or disconnect the following:
   - Negative battery cable
   - Rear seat assembly, on the Sedan
   - Rear speaker grille and service hole cap, on the Wagon
   - Strut mount cap
   - Air line from the top of the strut assembly
   - Height sensor and solenoid valve wiring harnesses from the strut assembly
   - Wheel and tire assembly
   - Brake hose clip
   - Union bolt from the brake caliper. Move the brake hose out of the way.
   - Lower nuts and bolts securing the strut to the rear wheel housing
   - Retainer nuts securing the strut bearing cap to the strut tower (from inside the vehicle)
   - Strut from the vehicle

To install:

3. Install or connect the following:
   - Strut on to the vehicle, making sure to position the strut properly in the upper strut tower mounts. Refer to the illustration if needed. Install the retainer nuts, and tighten to 11 ft. lbs. (15 Nm).
   - Strut to the rear wheel knuckle assembly, using the retainer nuts and bolts, and tighten the bolts to 145 ft. lbs. (196 Nm)
   - Brake union bolt and tighten to 13 ft. lbs. (18 Nm)
   - Brake hose clip
4. Bleed the brakes.
   - Wheel
   - Height sensor and solenoid valve wiring harnesses to the strut
   - Air line to the top of the strut
   - Strut mount cap
   - Rear seat, on Sedan
   - Speaker grille, on Wagon
   - Negative battery cable
5. Start the vehicle, and allow enough time for the shock to pressurize before driving the vehicle.

Shock Absorber

REMOVAL & INSTALLATION

Rear

LEGACY, BAJA AND OUTBACK

1. Before servicing the vehicle, refer to the precautions in the beginning of this section.
2. Remove or disconnect the following:
   - Rear wheels
   - Floor mat on wagon models
   - Trunk mat on sedan models
   - Roll up the trunk side trim on sedan models
   - Bolt attaching the shock absorber to the rear arm
3. Use a jack to support the rear suspension.
   - Shock absorber-to-body upper retainers and the shock

To install:

4. Install or connect the following:
   - Shock absorber and the shock-to-body using NEW upper retainers and tighten to 22 ft. lbs. (30 Nm).
5. Place the vehicle jack (the one supplied with the vehicle) upside down and place it between the link rear and the sub-frame. Place a cloth between the jack and areas it is touching to prevent damage to the link rear or sub-frame Adjust the jack so the shock is aligned with the rear are at the correct holes and install the lower shock bolts.
6. Support the shock/rear rear arm horizontally with a jack and tighten the nuts and bolts to 118 ft. lbs. (160 Nm).
7. Install or connect the following:
   - Floor mat on wagon models
   - Trunk side trim on sedan models
   - Trunk mat on sedan models
   - Rear wheels
8. Inspect and adjust the wheel alignment.

Rear

IMPREZA

1. Before servicing the vehicle, refer to the precautions in the beginning of this section.
2. Remove or disconnect the following:
   - Rear seat cushion and backrest on sedan models
   - Strut cap from the quarter trim on wagon models
   - Rear wheels
   - Brake hose clip from the strut
   - Bolts attaching the shock absorber to housing
3. Use a jack to support the rear suspension.
   - Shock absorber-to-body upper retainers and the shock

To install:

4. Install or connect the following:
   - Shock absorber and the shock-to-body using NEW upper retainers and tighten to 14 ft. lbs. (20 Nm).
   - Bolts attaching the shock absorber to housing and tighten to 162 ft. lbs. (220 Nm)
   - Brake hose clip to the strut
   - Rear wheels
   - Floor mat on wagon models
   - Rear seat cushion and backrest on sedan models
   - Strut cap to the quarter trim on wagon models
5. Inspect and adjust the wheel alignment.

Coil Spring

REMOVAL & INSTALLATION

Front

1. Before servicing the vehicle, refer to the precautions in the beginning of this section.
2. Remove the strut assembly from the vehicle.
3. Place the strut assembly in a vise with a holding tool and install a spring compressor.
4. Compress the spring slightly.
5. Loosen but do not remove the bearing cap locknut.
6. Compress the spring with the spring compressor, then remove the locknut.
7. Remove or disconnect the following:
   - Strut bearing cap, mounting insulator bracket and upper spring seat
   - Coil assembly, leaving the spring compressed
   - Strut boot and rebound bumper from the strut. Inspect and replace if worn.
   - Strut retainer nut using a suitable wrench
   - Strut insert from the assembly

To install:

8. Install or connect the following:
   - Strut into the chamber and install the retainer nut. Tighten the nut snugly.
   - Rebound bumper and the boot to the strut piston rod
   - Coil spring on the strut assembly. Be sure the spring is properly positioned on the lower bracket.
   - Upper spring seat, mounting insulator and bearing cap. Be sure the upper spring seat is facing the proper direction.
   - Locknut and tighten to 36–43 ft. lbs. (49–59 Nm) on all models except Legacy/Outback or to 41 ft. lbs. (55Nm) on Legacy/Outback models.
9. Loosen and remove the spring compressor from the coil spring.
10. Install the strut to the vehicle.

Rear

1. Before servicing the vehicle, refer to the precautions in the beginning of this section.
2. Remove the strut assembly from the vehicle and secure in a soft jawed vise.
3. Compress the coil spring with a spring compressor until the upper spring seat can be turned by hand.
4. Remove the self-locking nut on the top of the strut assembly, then remove the upper spring seat.
5. Remove the coil spring and compressor. If the spring is being replaced, slowly release the spring from the compressor and compress the new coil spring.

To install:

6. Place the proper end of the coil spring on the lower spring seat on the strut.
7. Install the insulator, upper spring seat and strut mount on the strut piston. Install a new self-locking nut. Tighten the nut to 36–43 ft. lbs. (49–59 Nm).
8. Slowly release the spring compressor.
9. Install the strut on to the vehicle.

Lower Ball Joint

REMOVAL & INSTALLATION

1. Before servicing the vehicle, refer to the precautions in the beginning of this section.
2. Remove or disconnect the following:
   - Negative battery cable
   - Front wheel and tire assembly
   - Ball joint castle nut cotter pin and discard the cotter pin
   - Castle nut
   - Ball joint from the lower control arm using a suitable puller or prytool
   - Bolt securing the ball joint to the steering knuckle
   - Ball joint using a suitable wedge to expand the steering knuckle connection point

To install:

3. Install or connect the following:
   - Ball joint to the steering knuckle
   - Retaining bolt and tighten to 37 ft. lbs. (50 Nm)
   - Ball joint to the lower control arm and tighten the castle nut on all models except WRX to 29 ft. lbs. (39 Nm). On WRX sedan models, tighten to 22 ft. lbs. (30 Nm) and 33 ft. lbs. (45 Nm) on all other WRX models. Then, tighten the castle nut an additional 60 degrees until the slot in the castle nut is aligned with the cotter pin hole in the ball joint.
   - New cotter pin
   - Wheel
   - Negative battery cable

Front Lower Control Arm

REMOVAL & INSTALLATION

Except WRX Models

1. Before servicing the vehicle, refer to the precautions in the beginning of this section.
2. Remove or disconnect the following:
   - Tire and wheel assembly
3. On Impreza models, remove the sub-frame as follows:
   a. Remove the bolt cover.
   b. Remove the clip.
   c. Loosen the sub-frame bolt (1) and leave it screwed in a few threads.

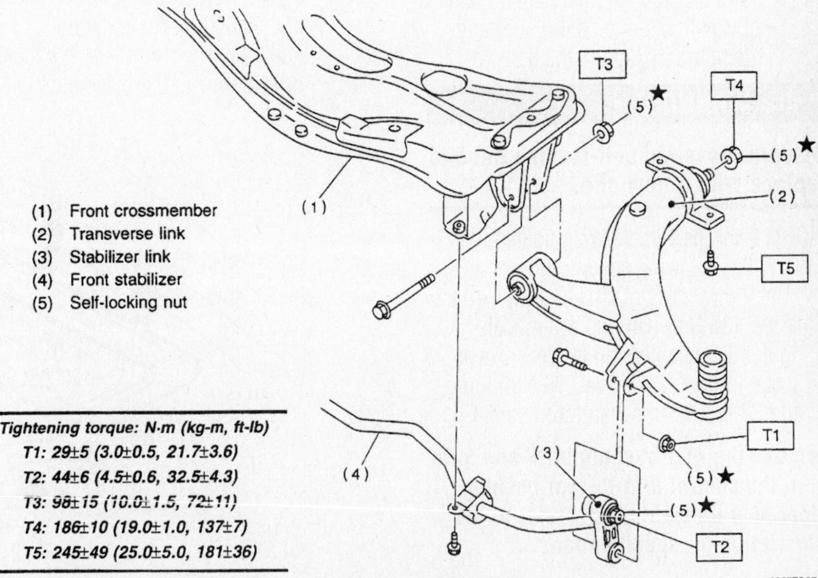

Exploded view of the lower control arm (transverse link)—Impreza and Legacy

Remove the remaining bolts in the following order: 2, 3, 4, 5 and 6. See illustration for bolt location.

- Sway link from the lower control arm
- Bolt securing the ball joint to the steering knuckle
- Nuts (NOT the bolts) securing the lower control arm to the crossmember
- 2 bolts holding the bushing bracket of the control arm to the body
- Ball joint from the steering knuckle
- Bolts securing the lower control arm to the crossmember, then the lower control

To install:

4. Install or connect the following:
   - Lower control arm; temporarily tighten the 2 bolts used to secure the rear bushing of the lower control arm to the body

➡These bolts should be tightened so they can still move back and forth in the oblong shaped hole in the bracket that holds the bushing.

5. On Impreza models, install the sub-frame as follows:
6. Install the sub-frame and bolts. Tighten the bolts as follows, referring to the illustration for location:
   a. T1: 25 ft. lbs. (34 Nm).
   b. T2: 41 ft. lbs. (55 Nm).
   c. T3: 52 ft. lbs. (71 Nm).
7. Install or connect the following:
   - Bolts used to secure the lower control arm to the crossmember and temporarily tighten the nuts
   - Ball joint into the steering knuckle and secure with the retaining bolt
   - Sway link to the control arm and temporarily tighten the bolts

⁂ WARNING

Discard loosened self-locking nut and replace with a new one.

8. Lower the vehicle, then tighten the bolts to the following specifications:
   a. Lower control arm-to-sway bar: 22 ft. lbs. (30 Nm). On Impreza models tighten the retainers on models except sedan turbo to 22 ft. lbs. (30 Nm) turbo and 33 ft. lbs. on sedan turbo models.

➡Move the rear bushing back and forth until the control arm-to-rear bushing clearance if established. Refer to the illustration for specifications.

   b. Lower control arm-to-crossmember: 74 ft. lbs. (100 Nm).

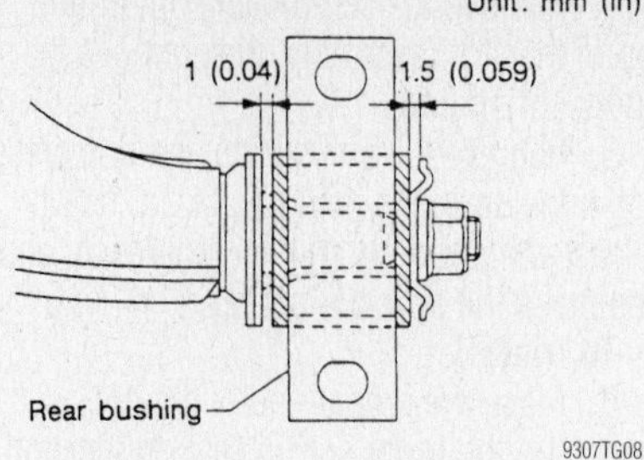

Proper control arm-to-rear bushing clearance specifications—all models

   c. Lower control arm-to-rear link bushing-to-body: 184 ft. lbs. (250 Nm).
9. Check the wheel alignment and adjust if necessary.

WRX Models

1. Before servicing the vehicle, refer to the precautions in the beginning of this section.
2. Remove or disconnect the following:
   - Tire and wheel assembly
3. Remove the sub-frame as follows:
   a. Remove the bolt cover.
   b. Remove the clip.
   c. Loosen the sub-frame bolt (1) and leave it screwed in a few threads. Remove the remaining bolts in the following order: 2, 3, 4, 5 and 6. See illustration for bolt location.
   - Sway link from the lower control arm
   - Bolt securing the ball joint to the steering knuckle
   - Nuts (NOT the bolts) securing the lower control arm to the crossmember
   - 2 bolts holding the bushing bracket of the control arm to the body
   - Ball joint from the steering knuckle
   - Bolts securing the lower control arm to the crossmember, then the lower control

To install:

4. Install or connect the following:
   - Lower control arm; temporarily tighten the 2 bolts used to secure the rear bushing of the lower control arm to the body

➡These bolts should be tightened so they can still move back and forth in the oblong shaped hole in the bracket that holds the bushing.

   - Bolts used to secure the lower control arm to the crossmember and temporarily tighten the nuts
   - Ball joint into the steering knuckle and secure with the retaining bolt
   - Sway link to the control arm and temporarily tighten the bolts

⁂ WARNING

Discard loosened self-locking nut and replace with a new one.

5. Lower the vehicle, then tighten the bolts to the following specifications:
   a. Lower control arm-to-sway bar: 33 ft. lbs. (45 Nm) on sedan models or 22 ft. lbs. (30 Nm) on all except sedan models.
   b. Lower control arm and crossmember to 74 ft. lbs. (100 Nm).
   c. Lower control arm rear bushing and body to 184 ft. lbs. (250 Nm).

➡Move the rear bushing back and forth until the control arm-to-rear bushing clearance if established. Refer to the illustration for specifications.

6. Install the sub-frame and bolts. Tighten the bolts as follows referring to the illustration for location:
   a. T1: 25 ft. lbs. (34 Nm).

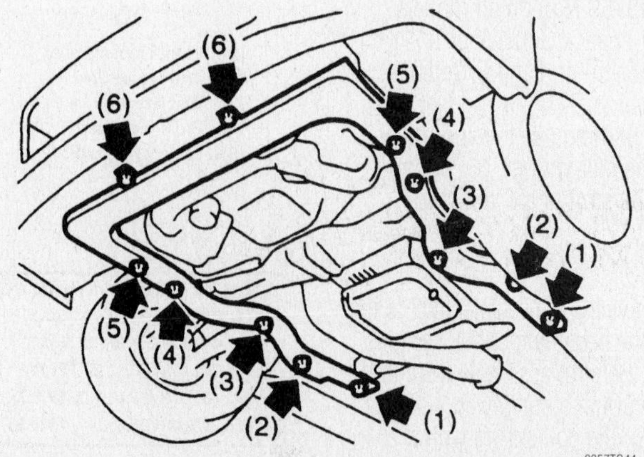

Location of the sub-frame bolts–WRX models

b. T2: 41 ft. lbs. (55 Nm).
c. T3: 52 ft. lbs. (71 Nm).

7. Check the wheel alignment and adjust if necessary.

CONTROL ARM BUSHING REPLACEMENT

Front

1. Remove the control arm from the vehicle.
2. Mount the control arm in a soft jawed vise.
3. Use either a press or a control arm bushing fixture (C-clamp like tool) along with a slotted washer and a piece of pipe (slightly larger than the bushing) and press out the old bushing.
4. Clean the inside bushing contact surfaces of rust and old rubber.

To install:

5. Apply a light coating of grease to both the replacement busing and bushing contact surfaces on the control arm.
6. Align the bushing according to the illustration.
7. Install the bushing using the press tool. A bushing install clamp can also be used to compress the bushing into the control arm.
8. Install the control arm on the vehicle.

Rear

1. Remove the control arm from the vehicle.
2. Scribe a matchmark on the control arm and rear bushing.
3. Loosen the nut and remove the rear bushing. Discard the nut.

To install:

4. Install the rear bushing to the control arm, making sure to align the marks made during removal.
5. Install a new nut and tighten to 140 ft. lbs. (190 Nm) on WRX models or 137 ft. lbs. (168 Nm) on all models except WRX.

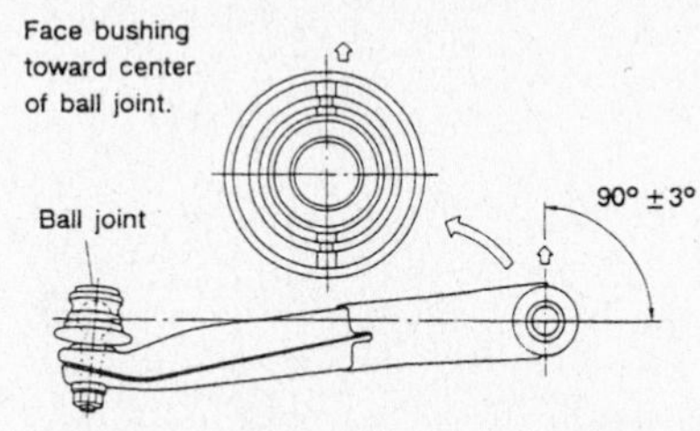

The front control arm bushing must be installed in the proper direction

Rear Lower Control Arm

REMOVAL & INSTALLATION

Impreza

TRAILING LINK

1. Before servicing the vehicle, refer to the precautions in the beginning of this section.
2. Remove or disconnect the following:
   - Tire and wheel assembly
   - Rear parking bracket clamp and Anti-lock Brake System (ABS) sensor harness, if equipped
   - Bolts that secure the trailing link to the bracket
   - Bolt that secures the trailing link to the rear housing
   - Trailing link from the vehicle

To install:

3. Install or connect the following:
   - Trailing link and the through-bolts and nuts. DO NOT tighten the nuts and bolts at this time.
   - ABS sensor bracket, if equipped, and parking brake cable to the trailing link.
   - Tire and wheel assembly
4. Tighten the trailing link-to-bracket bolt to 72 ft. lbs. (98 Nm) and the nut to 83 ft. lbs. (113 Nm).
5. Tighten the trailing link-to-rear housing to 83 ft. lbs. (113 Nm).
6. Check the wheel alignment and adjust if necessary.

LATERAL LINK

1. Before servicing the vehicle, refer to the precautions in the beginning of this section.
2. Remove or disconnect the following:
   - Tire and wheel assembly
   - Stabilizers on turbo models
   - Anti-lock Brake System (ABS) sensor harness from the trailing link, if equipped.
   - Bolts that secure the lateral link to the rear housing.

➡Discard the old self-locking nuts and replace with new ones during installation.

   - Bolts which secure the trailing link to the rear housing
   - Halfshaft from the rear differential using a suitable tool.

➡On all except 2.2L engines, do not remove the circlip attached to the inside of the differential. On 2.2L engines, the side spline circlip comes out together with the shaft. Be careful not to damage the side bearing retainer.

3. Scribe an alignment mark on the rear lateral link adjusting bolt and crossmember.
4. Remove or disconnect the following:
   - Outer lateral link bolt securing the lateral link to the housing
   - Bolts securing the front and rear lateral links to the crossmember
   - Lateral links from the vehicle

To install:

5. Install or connect the following:
   - Bolts securing the front and rear lateral links to the crossmember and hand-tighten
   - Outer lateral link bolt securing the lateral link to the housing and hand-tighten
   - Halfshaft to the rear differential

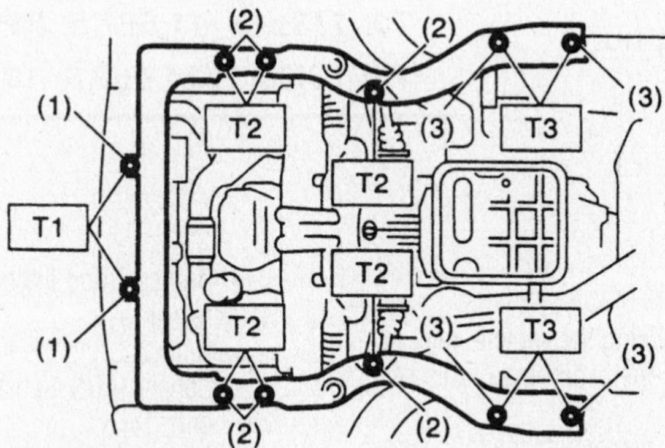

Tighten the sub-frame bolts to the specification outlined in the procedure–WRX models

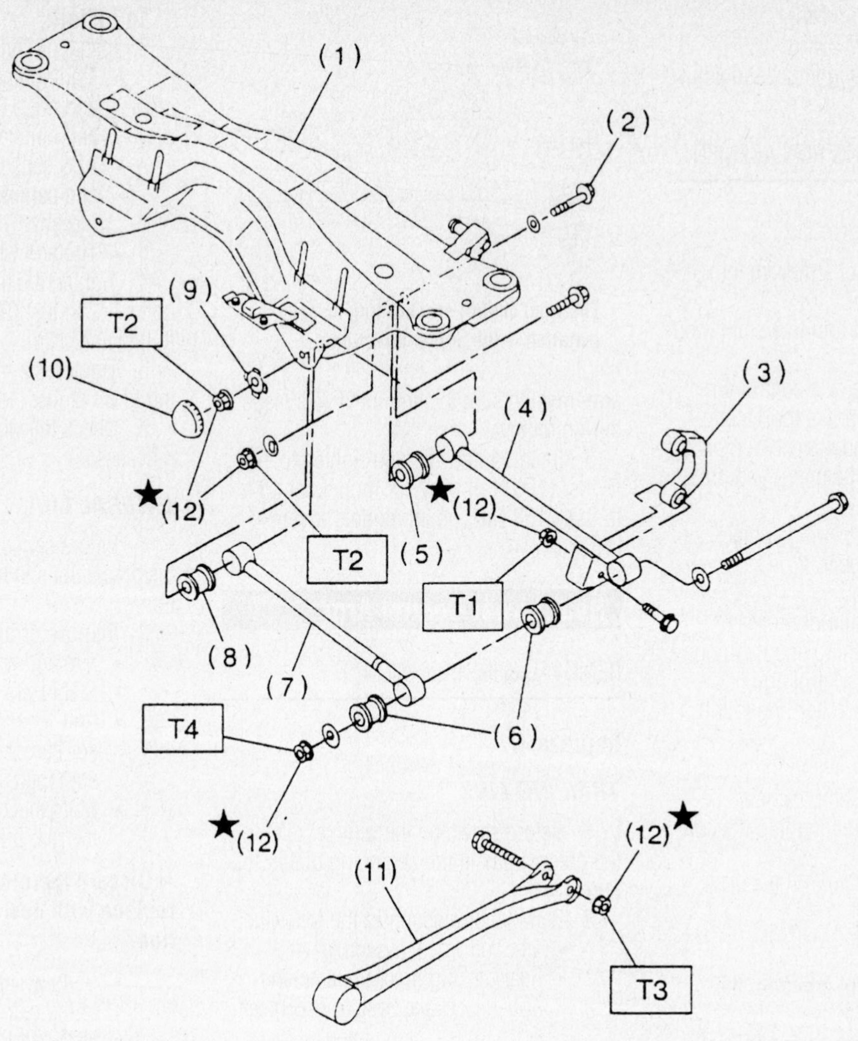

(1) Crossmember
(2) Adjusting bolt
(3) Stabilizer link
(4) Rear lateral link
(5) Bushing (C)
(6) Bushing (A)
(7) Front lateral link
(8) Bushing (B)
(9) Washer
(10) Cap
(11) Trailing link
(12) Self-locking nut

Tightening torque: N·m (kg-m, ft-lb)
T1: 44±6 (4.5±0.6, 32.5±4.3)
T2: 98±15 (10.0±1.5, 72±11)
T3: 113±15 (11.5±1.5, 83±11)
T4: 137±20 (14.0±2.0, 101±14)

9307TG10

Lateral link mounting and tightening specifications—Impreza

- Bolts that secure the lateral link to the rear housing
- Bolts which secure the trailing link to the rear housing
- ABS sensor harness to the trailing link, if equipped
- Tire and wheel assembly

6. Tighten the lateral link bolts as shown in the illustration.
7. Check the wheel alignment and adjust if necessary.

Legacy/Baja

1. Before servicing the vehicle, refer to the precautions in the beginning of this section.
2. Remove or disconnect the following:
   - Tire and wheel assembly
   - Wheel bearing assembly, refer to the procedure in this section
   - Bolt holding the parking brake cable to the control arm
   - Bolt securing the brake hose to the rear arm
   - Bolt securing the Anti-lock Brake Sensor (ABS) sensor to the rear arm
   - Brake line from the wheel cylinder with a flare nut wrench, if equipped with drum brakes. Plug the line to avoid contaminating the system. Suspend the brake backing plate from the sub-frame.

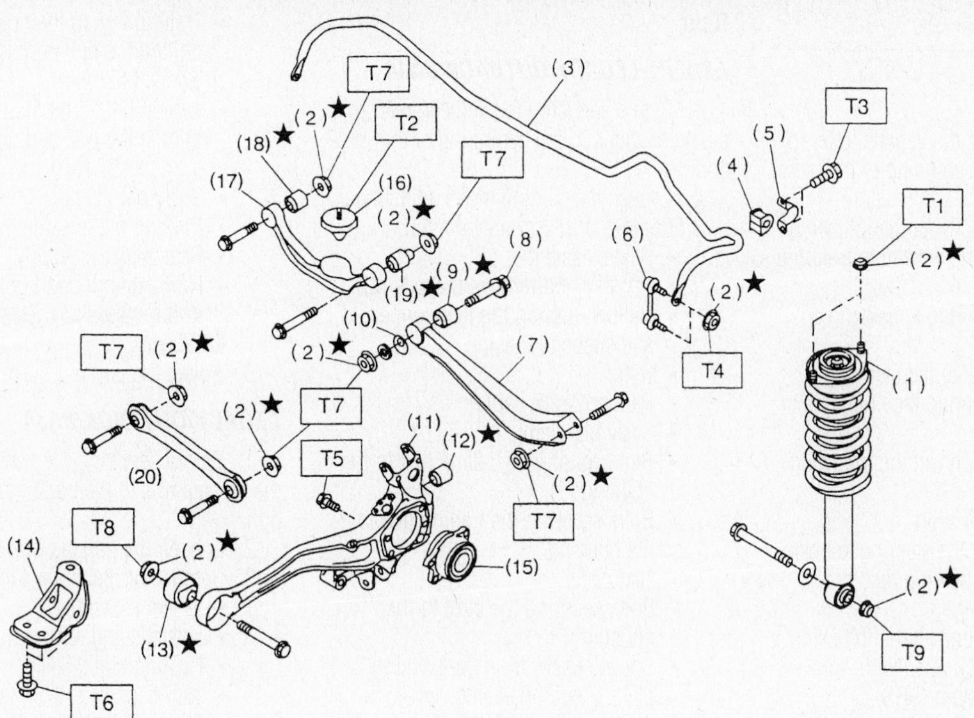

1) Shock absorber
2) Self-locking nut
3) Stabilizer
4) Stabilizer bushing
5) Clamp
6) Stabilizer link
7) Link rear
8) Adjusting bolt
9) Link rear bushing
10) Adjusting washer
11) Rear arm
(12) Rear arm rear bushing
(13) Rear arm front bushing
(14) Rear arm bracket
(15) Hub bearing unit
(16) Helper
(17) Link upper
(18) Link upper bushing (Inside)
(19) Link upper bushing (Outside)
(20) Link front

Tightening torque: N·m (kg-m, ft-lb)
T1: 30±7 (3.1±0.7, 22.4±5.1)
T2: 32±10 (3.3±1.0, 23.9±7.2)
T3: 39±7 (4.0±0.7, 28.9±5.1)
T4: 44±6 (4.5±0.6, 32.5±4.3)
T5: 66±10 (6.7±1.0, 48.5±7.2)
T6: 108±15 (11±1.5, 80±11)
T7: 123±15 (12.5±1.5, 90±11)
T8: 147±20 (15±2, 108±14)
T9: 157±20 (16±2, 116±14)

9307TG16

Rear control arm mounting and tightening specifications—Legacy

- Nut securing the stabilizer link to the rear arm
- Bolt holding the shock absorber to the rear arm

3. Use a suitable transaxle jack to support the rear arm horizontally.
4. Remove or disconnect the following:
   - Bolt securing the rear arm to the body
   - Nut securing the front link to the rear arm, loosen
   - Nut securing the rear link to the rear arm, loosen
   - Bolts holding the rear arm to the links
   - Rear arm

To install:

5. Use a transaxle jack to support the rear arm.
6. Install or connect the following:
   - Rear arm and temporarily tighten the bolts securing the rear arm to the link
   - Wheel bearing unit
   - Bolt securing the ABS sensor to the rear arm
   - Brake hose-to-rear arm bolt
   - Parking brake cable clamp-to-rear arm bolt

➡Place a rag or cloth between the jack and its mating area to avoid scratching the rear link and sub-frame.

7. Place the tire changing jack (supplied with the car) upside down and position between the rear link and sub-frame. Adjust the jack position so the rear shock absorber is aligned with the rear arm at their corresponding holes. Install the lower shock absorber bolts.
8. Using the transmission jack, support the rear arm horizontally, then tighten the nuts and bolts holding the rear arm, front and rear links, upper link and shock absorber. Refer to the specifications in the illustration.
9. Install the tire and wheel assembly.
10. Check and adjust the alignment, if necessary.

BUSHING REPLACEMENT

1. Use a suitable press to remove and install the bushings.

Wheel Bearings

ADJUSTMENT

The wheel bearings are not adjustable.

REMOVAL & INSTALLATION

Front

1. Before servicing the vehicle, refer to the precautions in the beginning of this section.
2. Remove or disconnect the following:
   - Steering knuckle assembly from the vehicle
3. Position the steering knuckle in a soft-jawed vise.
4. Press the hub from the steering knuckle. If the inner bearing race remains in the hub, press it out.
5. Remove or disconnect the following:
   - Rotor shield
   - Inner and outer seals
   - Snapring from the steering knuckle
6. Press the inner bearing race to remove the outer bearing.
7. Remove or disconnect the following:
   - Tone ring, if equipped with Anti-lock Brake System (ABS)
   - Wheel lugs from the hub using a suitable press

➡To prevent deforming the hub, do not hammer the lugs out.

To install:

8. Install or connect the following:
   - Wheel lugs into the hub using a suitable press
9. If equipped with ABS, clean all foreign material from the hub and tone ring.
10. Install or connect the following:
    - Tone ring
11. Clean the inside of the steering knuckle.
12. Remove the plastic lock from the inner race and press a new greased bearing into the hub by pressing the outer race.
13. Install or connect the following:
    - Snapring into its groove
    - New outer oil seal using a press, until it contacts the bottom of the housing
    - New inner oil seal using a press, until it contacts the circlip
14. Apply grease to the oil seal lips.
15. Install or connect the following:
    - Rotor shield and tighten the bolts to 10 ft. lbs. (14 Nm)
    - Hub to the steering knuckle
16. Press a new bearing into the hub by driving the inner race.
17. Install the steering knuckle on the vehicle.

Rear

EXCEPT LEGACY/OUTBACK/BAJA

1. Before servicing the vehicle, refer to the precautions in the beginning of this section.
2. Loosen the parking brake adjustment.
3. Remove or disconnect the following:
   - Wheel assembly
   - Unstake and remove the axle nut
   - Caliper, leaving the line connected, and suspend it aside
   - Rotor
   - Parking brake cable
   - Sway bar clamp
   - Bolt securing the lateral link to the housing
   - Bolts securing the trailing link to the housing
   - Halfshaft
   - Bolts securing the strut to the housing
   - Speed sensor from the backing plate, if equipped with Anti-lock Brake System (ABS)
   - Housing assembly
   - Hub from the rear housing using hub stand 92708000 and puller 927420000
   - Backing plate from the housing
   - Outer, inner and sub oil seals
   - Snapring
   - Bearing by pressing the inner race

To install:

4. Clean the housing thoroughly.

➡Do not remove the plastic lock from the inner race when installing the bearing.

5. Install or connect the following:
   - New bearing into the housing by pressing the outer race and pack the bearing with grease
   - Snapring and ensure that it fits properly
   - New outer seal until it contacts the snapring using a press
   - New inner seal until it contacts the bottom using a press
   - New sub oil seal and apply grease to the oil seal lip
   - Backing plate and tighten the bolts to 43 ft. lbs. (58 Nm)
   - Hub into the housing using installer 927450000 to press it into position
   - Housing to the strut and tighten the bolts to 119 ft. lbs. (162 Nm)
   - Speed sensor, if equipped with ABS
   - Halfshaft
   - Trailing link to the housing and tighten the bolt and new nut to 94 ft. lbs. (127 Nm)
   - Lateral link to the housing and tighten the bolt and new nut to 116 ft. lbs. (157 Nm)
   - Sway bar clamp
   - Parking brake cable
   - Rear brake assembly
   - New axle nut and tighten it to 152 ft. lbs. (206 Nm). Stake the nut.
   - Wheel
6. Adjust the parking brake cable.

LEGACY/OUTBACK/BAJA

1. Before servicing the vehicle, refer to the precautions in the beginning of this section.
2. Loosen the parking brake adjustment.
3. Remove or disconnect the following:
   - Wheel assembly
   - Unstake and remove the axle nut
   - Parking brake lever
   - ABS sensor
   - Caliper, leaving the line connected, and suspend it aside
   - Rotor
   - Four bolts from the rear arm and the hub and bearing assembly
4. Disassemble the hub/bearing assembly as follows:
   a. Place the hub/bearing in a press.
   b. Using dummy collar tool ST 398507703 to press the bearing from the hub.
   c. Place the hub assembly on hub stand ST 927080000 and using a com-

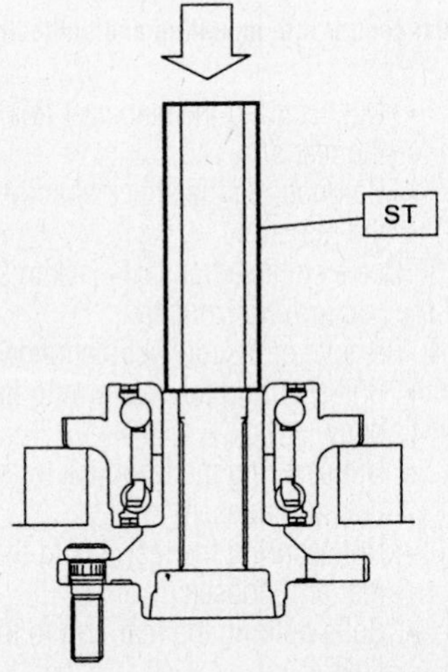

Use dummy collar tool ST 398507703 to press the bearing from the hub—Legacy/Outback/Baja

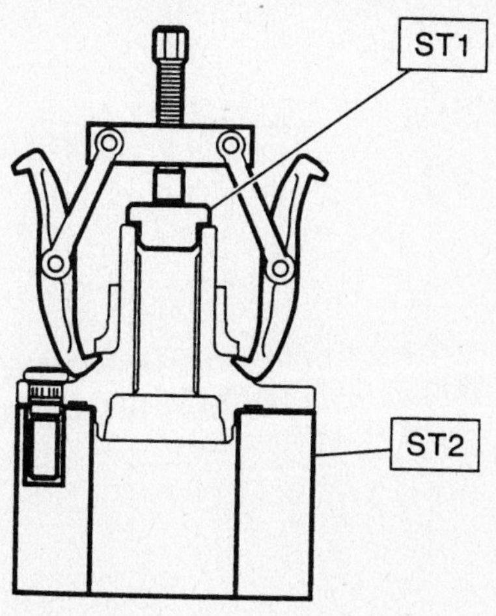

Place the hub assembly on stand ST 927080000 and use a common puller and tool ST 399520105, remove the bearing inner race—Legacy/Outback/Baja

mon puller and tool ST 399520105, remove the inner race of the bearing. Discard the bearing as it should not be reused.

d. Using the hub stand tool, press out the hub bolt.

To install:

5. Install the bearing assembly to the hub as follows:

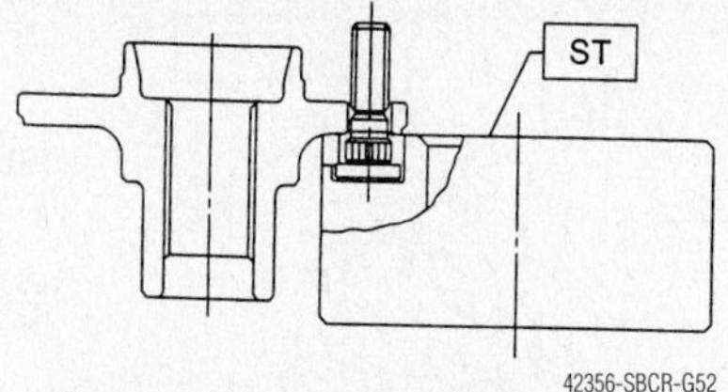

Using the hub stand tool, press out the hub bolt—Legacy/Outback/Baja

a. Press a new hub bolt into place using the hub stand. Make sure the bolt closely contacts the hub. use a 0.47 inch (12mm) hole in the hub stand to prevent the bolt from tilting while installing.

b. Using the hub stand, hub installer ST 927450000 and spacer ST 28499AE000, press the NEW bearing assembly into the hub. Make sure to always press on the inner race while installing.

6. Install or connect the following:
- Hub assembly with the mounting holes on the backing plate and install the hub assembly and backing plate. Temporarily tighten the axle nuts. be careful not to damage the tone ring.
- Four bolts and tighten to 48 ft. lbs. (66 Nm).

7. Remove the axle nut, use axle shaft installer tool ST 922431000 and adapter

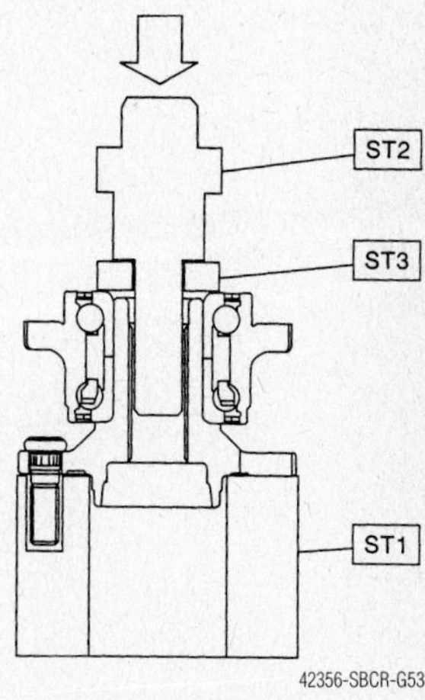

Installing the bearing into the hub—Legacy/Outback/Baja

927390000 to pull the axle shaft into position and hand tighten the axle nut
- Rotor
- Caliper
- ABS sensor
- Parking brake lever
- Tighten the axle nut to 174 ft. lbs. (235 Nm) and stake the nut.
- Wheel

8. Adjust the parking brake cable.

BRAKES

Brake Caliper

REMOVAL & INSTALLATION

WRX

FRONT

1. Before servicing the vehicle, refer to the precautions in the beginning of this section.
2. Remove or disconnect the following:
- Front wheels
- Brake hose from the caliper body
- Caliper retainer bolts and the caliper
- Caliper bracket, if necessary

To install:

3. Compress the piston assembly into the cylinder bore.
4. Install or connect the following:
- Caliper bracket to the spindle assembly, and secure in place with the retainer bolts. Tighten the retainer bolts to 59 ft. lbs. (80 Nm).
- Caliper and tighten the retainers to 19 ft. lbs. (26 Nm)
- Brake hose using new sealing washers, and tighten the fitting to 13 ft. lbs. (18 Nm)

5. Bleed the brake system.
6. Install the wheels and check the fluid level in the master cylinder.

REAR

1. Before servicing the vehicle, refer to the precautions in the beginning of this section.
2. Remove or disconnect the following:
- Rear wheels
- Brake hose from the caliper body
- Caliper bracket retainer bolts
- Caliper and bracket assembly off the rotor

To install:

3. Compress the piston assembly into the cylinder bore.
4. Install or connect the following:
- Caliper bracket to the spindle assembly, and secure in place with the retainer bolts. Tighten the retainer bolts to 27 ft. lbs. (37 Nm).
- Brake hose using new sealing washers, and tighten the fitting to 13 ft. lbs. (18 Nm)

5. Bleed the brake system.
6. Install the wheels and check the fluid level in the master cylinder.

Except WRX

FRONT

1. Before servicing the vehicle, refer to the precautions in the beginning of this section.
2. Remove or disconnect the following:
- Front wheels
- Brake hose from the caliper body
- Caliper retainer bolts and the caliper
- Caliper bracket, if necessary

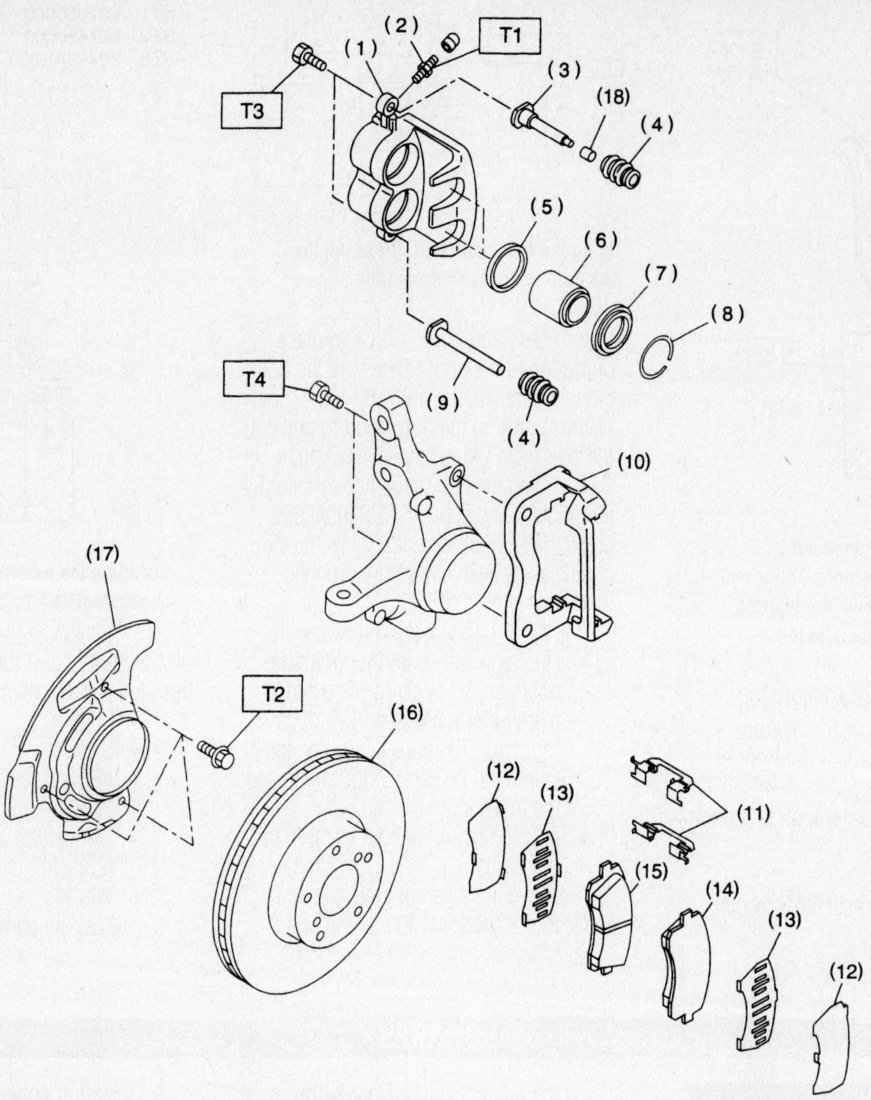

(1) Caliper body
(2) Air bleeder screw
(3) Guide pin (Green)
(4) Pin boot
(5) Piston seal
(6) Piston
(7) Piston boot
(8) Boot ring
(9) Lock pin (Yellow)
(10) Support
(11) Pad clip
(12) Outer shim
(13) Inner shim
(14) Pad (Outside)
(15) Pad (Inside)
(16) Disc rotor
(17) Disc cover
(18) Bush

Tightening torque: N·m (kgf-m, ft-lb)
T1: 8 (0.8, 5.8)
T2: 18 (1.8, 13.0)
T3: 37 (3.8, 27.5)
T4: 80 (8.2, 59)

9357TG65

Exploded view of the front disc brakes—Subaru WRX

To install:

3. Compress the piston assembly into the cylinder bore.
4. Install or connect the following:
   - Caliper bracket to the spindle assembly, and secure in place with the retainer bolts. Tighten the retainer bolts to 58 ft. lbs. (78 Nm).
   - Caliper and tighten the retainers to 29 ft. lbs. (40 Nm) on 2000–02 models and Outback models 19 ft. lbs. (26 Nm)
   - Brake hose using new sealing washers, and tighten the fitting to 13 ft. lbs. (18 Nm)
5. Bleed the brake system.
6. Install the wheels and check the fluid level in the master cylinder.
7. Pump the brake pedal several times to seat the brakes before attempting to move the vehicle and road test the vehicle.

REAR

1. Before servicing the vehicle, refer to the precautions in the beginning of this section.
2. Install or connect the following:
   - Rear wheels

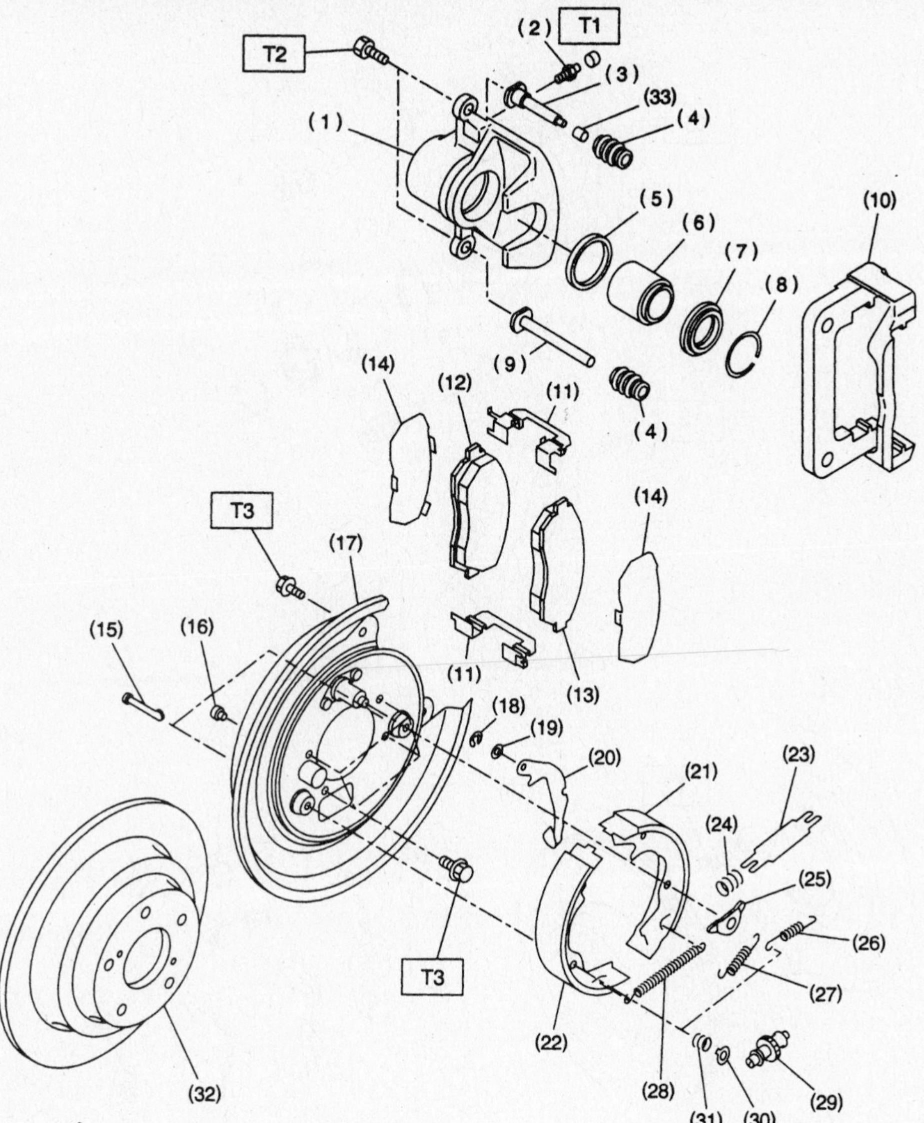

(1)	Caliper body	(14)	Shim	(27)	Primary shoe return spring
(2)	Air bleeder screw	(15)	Shoe hold-down pin	(28)	Adjusting spring
(3)	Guide pin (Green)	(16)	Cover	(29)	Adjuster
(4)	Pin boot	(17)	Back plate	(30)	Shoe hold-down cup
(5)	Piston seal	(18)	Retainer	(31)	Shoe hold-down spring
(6)	Piston	(19)	Spring washer	(32)	Disc rotor
(7)	Piston boot	(20)	Parking brake lever	(33)	Bush
(8)	Boot ring	(21)	Parking brake shoe (Secondary)		
(9)	Lock pin (Yellow)	(22)	Parking brake shoe (Primary)		
(10)	Support	(23)	Strut		
(11)	Pad clip	(24)	Strut shoe spring		
(12)	Inner pad	(25)	Shoe guide plate		
(13)	Outer pad	(26)	Secondary shoe return spring		

Tightening torque: N·m (kgf-m, ft-lb)

T1: *8 (0.8, 5.8)*
T2: *37 (3.8, 27.5)*
T3: *53 (5.4, 39.1)*

9357TG66

Exploded view of the rear disc brakes—Subaru WRX

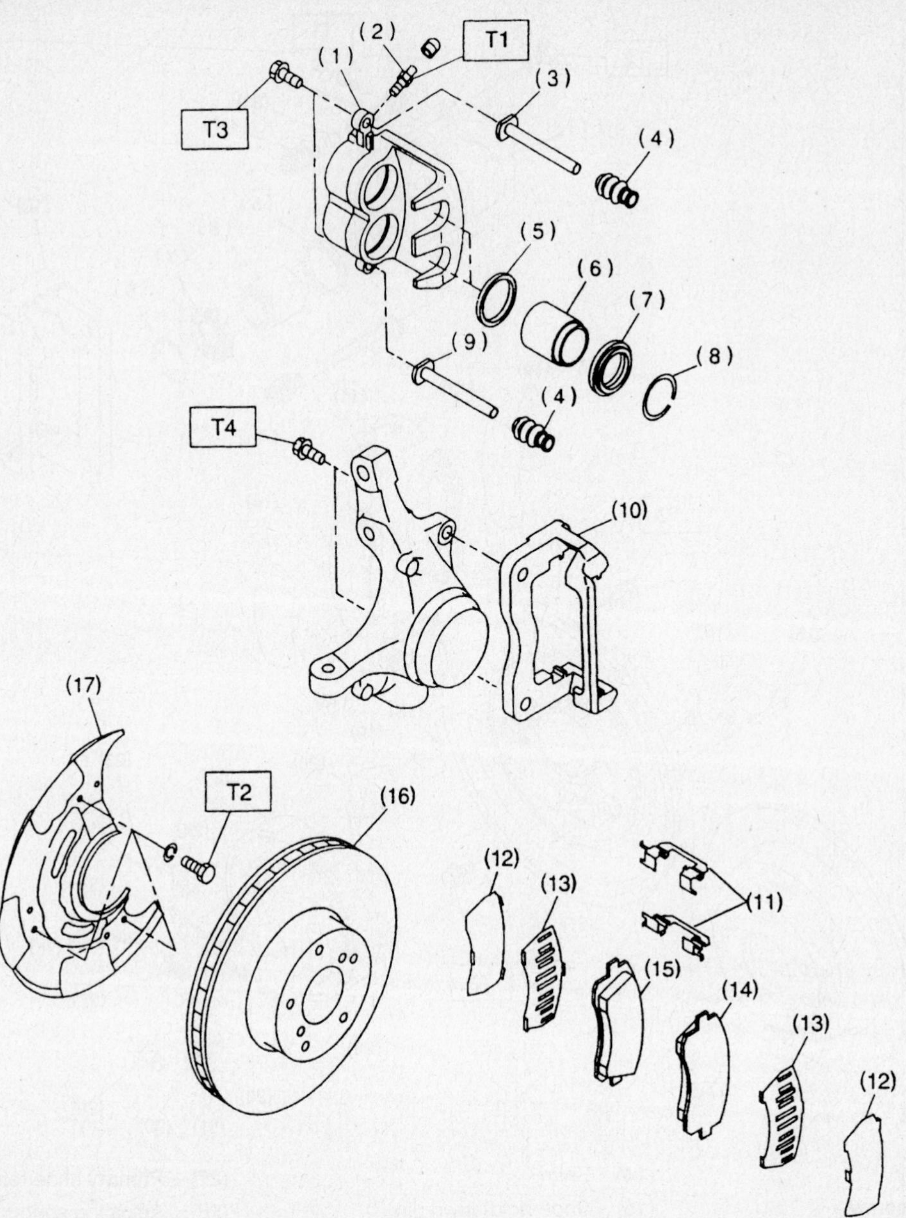

(1) Caliper body
(2) Air bleeder screw
(3) Guide pin (Green)
(4) Pin boot
(5) Piston seal
(6) Piston
(7) Piston boot
(8) Boot ring
(9) Lock pin (Yellow)
(10) Support
(11) Pad clip
(12) Outer shim
(13) Inner shim
(14) Pad (Outside)
(15) Pad (Inside)
(16) Disc rotor
(17) Disc cover

Tightening torque: N·m (kg-m, ft-lb)
T1: 8±1 (0.8±0.1, 5.8±0.7)
T2: 18±5 (1.8±0.5, 13.0±3.6)
T3: 37±5 (3.8±0.5, 27.5±3.6)
T4: 78±10 (8.0±1.0, 58±7)

93016G60

Exploded view of the front disc brakes—Subaru models, except WRX

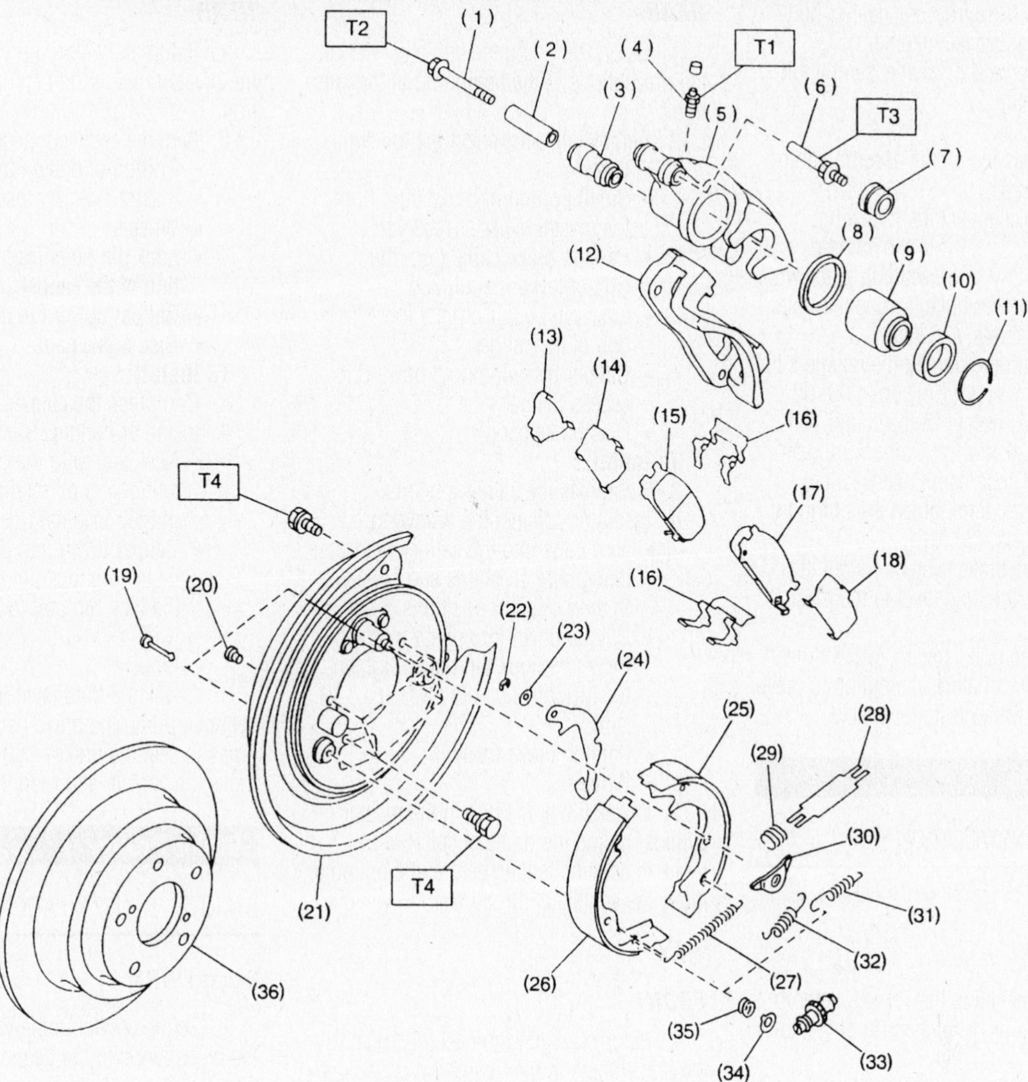

(1) Lock pin
(2) Lock pin sleeve
(3) Lock pin boot
(4) Air bleeder screw
(5) Caliper body
(6) Guide pin
(7) Guide pin boot
(8) Piston seal
(9) Piston
(10) Piston boot
(11) Boot ring
(12) Support
(13) Shim
(14) Inner shim
(15) Inner pad
(16) Pad clip
(17) Outer pad
(18) Outer shim
(19) Shoe hold-down pin
(20) Cover
(21) Back plate
(22) Retainer
(23) Spring washer
(24) Parking brake lever
(25) Parking brake shoe (Secondary)
(26) Parking brake shoe (Primary)
(27) Adjusting spring
(28) Strut
(29) Strut shoe spring
(30) Shoe guide plate
(31) Secondary shoe return spring
(32) Primary shoe return spring
(33) Adjuster
(34) Shoe hold-down cup
(35) Shoe hold-down spring
(36) Disc rotor

Tightening torque: N·m (kg-m, ft-lb)
T1: 8 ± 1 (0.8 ± 0.1, 5.8 ± 0.7)
T2: 20 ± 4 (2.0 ± 0.4, 14.5 ± 2.9)
T3: 26 ± 5 (2.7 ± 0.5, 19.5 ± 3.6)
T4: 52 ± 6 (5.3 ± 0.6, 38.3 ± 4.3)

93016G61

Exploded view of the rear disc brakes—Subaru models, except WRX

- Brake hose from the caliper body
- Caliper bracket retainer bolts
- Caliper and bracket assembly off the rotor

To install:

3. Compress the piston assembly into the cylinder bore.
4. Install or connect the following:
   - Caliper bracket to the spindle assembly, and secure in place with the retainer bolts, torque the bolts to 58 ft. lbs. (78 Nm)
   - Caliper and tighten the retainers to 29 ft. lbs. (40 Nm) on 2000–02 models except Outback and 19 ft. lbs. (26 Nm) on Outback models
   - Brake hose using new sealing washers, and tighten the fitting to 13 ft. lbs. (18 Nm)
5. Bleed the brake system. Install the wheels. Check the fluid level in the master cylinder.
6. Pump the brake pedal several times to seat the brakes before attempting to move the vehicle and road test the vehicle.

Brake Pads

REMOVAL & INSTALLATION

Except WRX

FRONT

1. Before servicing the vehicle, refer to the precautions in the beginning of this section.
2. Remove or disconnect the following:
   - Wheels
   - Lock pin bolts from the lower portion of the caliper
   - Caliper by swinging it upward to access the pads
   - Disc brake pads

To install:

3. Compress the caliper pistons.
4. Install or connect the following:
   - New pads into the caliper brackets, being sure all shims and clips are in their original positions
   - Caliper and tighten the retainers to 29 ft. lbs. (40 Nm) on 2000–02 models except Outback and 19 ft. lbs. (26 Nm) on Outback models
   - Wheels
5. Check the fluid level in the master cylinder, pump the brake pedal several times to seat the brakes before attempting to move the vehicle and road test the vehicle.

REAR

1. Before servicing the vehicle, refer to the precautions in the beginning of this section.
2. Remove or disconnect the following:
   - Wheels
   - Small portion of brake fluid from the master cylinder reservoir
   - Parking brake cable from the caliper lever, if equipped
   - Lock pin bolts from the lower portion of the caliper
   - Caliper by swinging it upward to access the pads
   - Disc brake pads

To install:

3. Compress the caliper pistons.
4. Install or connect the following:
   - New pads into the caliper brackets, being sure all shims and clips are in their original positions
   - Caliper down into position and install the lock pin bolts. Tighten the lock pin bolts to 29 ft. lbs. (40 Nm).
   - Parking brake cable
   - Wheels
5. Check the fluid level in the master cylinder, pump the brake pedal several times to seat the brakes before attempting to move the vehicle and road test the vehicle.

WRX

FRONT

1. Before servicing the vehicle, refer to the precautions in the beginning of this section.
2. Remove or disconnect the following:
   - Wheels
   - Lock pin bolts from the lower portion of the caliper
   - Caliper upward to access the pads
   - Disc brake pads

To install:

3. Compress the caliper piston.
4. Install or connect the following:
   - New pads into the caliper brackets, being sure all shims and clips are in their original positions
   - Calipers down into position and install the lock pin bolts. Tighten the lock pin bolts to 19 ft. lbs. (26 Nm).
   - Wheels
5. Check the fluid level in the master cylinder, pump the brake pedal several times to seat the brakes before attempting to move the vehicle and road test the vehicle.

REAR

1. Before servicing the vehicle, refer to the precautions in the beginning of this section.
2. Remove or disconnect the following:
   - Portion of brake fluid from the master cylinder reservoir
   - Wheels
   - Lock pin bolts from the lower portion of the caliper
   - Caliper upward to access the pads
   - Disc brake pads

To install:

3. Compress the caliper piston.
4. Install or connect the following:
   - New pads into the caliper bracket, being sure all shims and clips are in their original positions
   - Caliper down into position and install the lock pin bolts. Tighten the lock pin bolt to 27 ft. lbs. (37 Nm).
   - Wheels
5. Check the fluid level in the master cylinder, pump the brake pedal several times to seat the brakes before attempting to move the vehicle and road test the vehicle.

Brake Drums

REMOVAL & INSTALLATION

Except WRX

1. Before servicing the vehicle, refer to the precautions in the beginning of this section.
2. Remove or disconnect the following:
   - Rear wheels
   - Center cap by prying it off
   - Cotter pin, castle nut, and center retainer washer
   - Drum

To install:

3. Install or connect the following:
   - Drum
   - Center retainer washer and castle nut. Tighten the castle nut to 108 ft. lbs. (147 Nm). Install a new cotter pin.
   - Center cap
4. Adjust the brake shoes.
5. Install the rear wheels.

WRX

1. Before servicing the vehicle, refer to the precautions in the beginning of this section.
2. Remove the rear wheels.

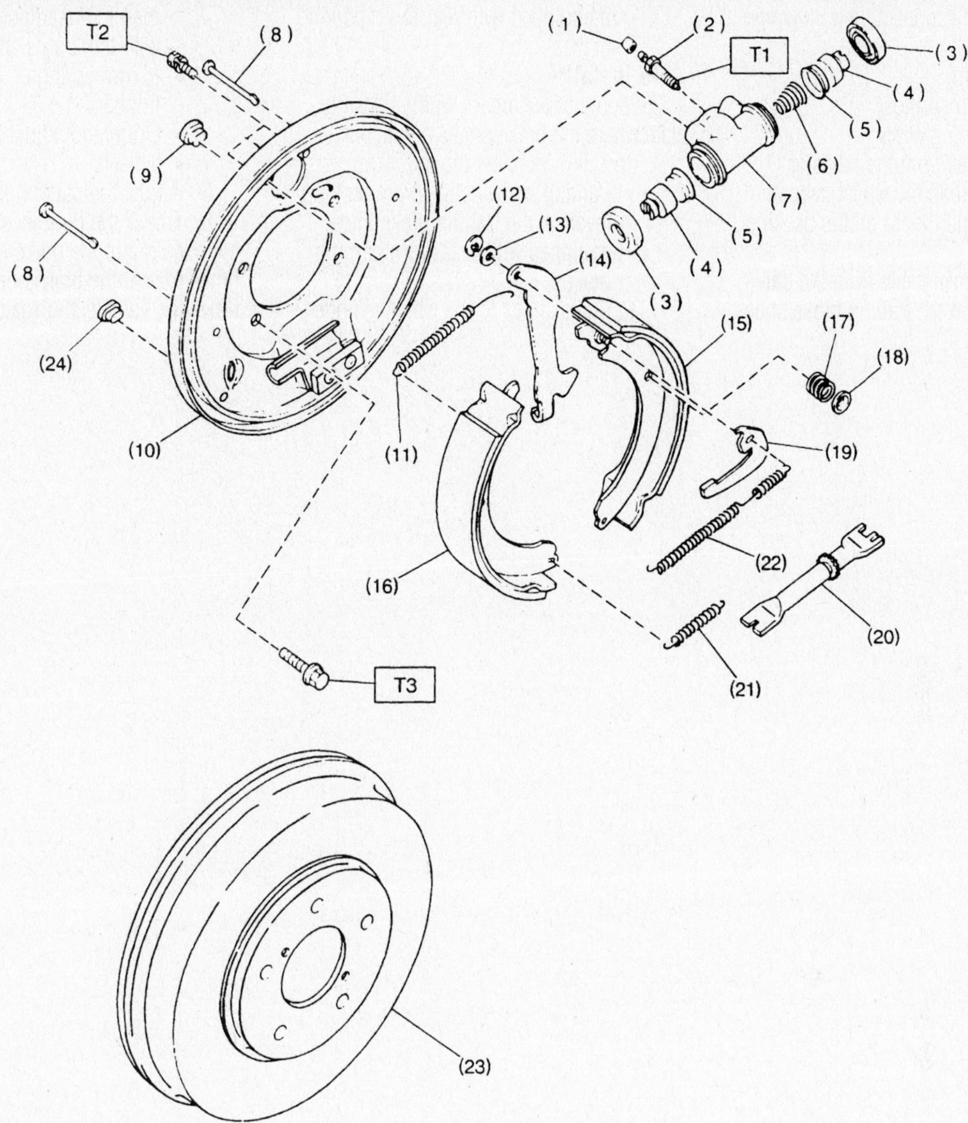

(1) Air bleeder cap
(2) Air bleeder screw
(3) Boot
(4) Piston
(5) Cup
(6) Spring
(7) Wheel cylinder body
(8) Pin
(9) Plug
(10) Back plate
(11) Upper shoe return spring
(12) Retainer
(13) Washer
(14) Parking brake lever
(15) Brake shoe (Trailing)
(16) Brake shoe (Leading)
(17) Shoe hold-down spring
(18) Cup
(19) Adjusting lever
(20) Adjuster
(21) Lower shoe return spring
(22) Adjusting spring
(23) Drum
(24) Plug

Tightening torque: N·m (kg-m, ft-lb)
T1: 8±1 (0.8±0.1, 5.8±0.7)
T2: 10±2 (1.0±0.2, 7.2±1.4)
T3: 52±6 (5.3±0.6, 38.3±4.3)

93016G62

Exploded view of a typical rear drum brake assembly

3. Release the parking brake.
4. If necessary, remove the adjusting hole cover from the backing plate and using a suitable tool, back off the shoe adjuster.
5. If the drum is difficult to remove, insert an 8mm bolt into the hole on the drum to push it off.
6. Remove the drum.

To install:

7. Install the drum.
8. Adjust the brake shoes.
9. Install the rear wheels.

Brake Shoes

REMOVAL & INSTALLATION

1. Before servicing the vehicle, refer to the precautions in the beginning of this section.

2. Remove or disconnect the following:
   - Wheels
   - Brake drum
   - Both return springs
   - Both retaining clips
   - Brake shoes from the adjuster side first, then the wheel cylinder side, and pull them off the backing plate
   - Parking brake cable from the parking lever on the trailing brake shoe, if equipped with rear drum parking brakes

To install:

3. Apply brake grease to the backing plate where the brake shoes contact it.
4. Install or connect the following:
   - Parking brake cable to the parking lever on the trailing brake shoe, if equipped with rear drum parking brakes
   - Brake shoes to the wheel cylinder, then to the adjuster. Secure in place with the 2 pins and retaining clips.
   - Return springs. The upper spring is thinner.
   - Drum and adjust the brake shoes
   - Wheels
5. Adjust the parking brake.
6. Check the fluid level in the master cylinder, pump the brake pedal several times to seat the brakes before attempting to move the vehicle and road test the vehicle.

SUBARU

Forester

13

SPECIFICATION AND MAINTENANCE CHARTS

ENGINE AND VEHICLE IDENTIFICATION CHART

Engine Code						
Code ①	Liters (cc)	Cu. In.	Cyl.	Fuel Sys.	Type	Eng. Mfg.
6 ③	2.5 (2457)	150	4	MFI	DOHC	Subaru
6	2.5 (2457)	150	4	MFI	SOHC	Subaru

Model Year	
Code ②	Year
1	2001
2	2002
3	2003
4	2004
5	2005

MFI: Multiport Fuel Injection

DOHC: Double Overhead Camshafts

SOHC: Singlee Overhead Camshaft

① 6th digit of the VIN.

② 10th digit of the VIN.

③ Turbo

67170-SBFR-C01

GENERAL ENGINE SPECIFICATIONS

Year	Model	Engine Displacement Liters (VIN)	Net Horsepower @ rpm	Net Torque @ rpm (ft. lbs.)	Bore x Stroke (in.)	Com-pression Ratio	Oil Pressure psi @ rpm
2001	Forester	2.5 (6)	165@5600	162@4000	3.92x3.11	9.7:1	14@600
2002	Forester	2.5 (6)	165@5600	162@4000	3.92x3.11	9.7:1	14@600
2003-04	Forester	2.5 (6) ①	210@5600	235@4600	3.92x3.11	8.2:1	14@600
		2.5 (6)	165@5600	162@4000	3.92x3.11	9.7:1	14@600

MFI: Multi-port Fuel Injection

① DOHC Turbo

67170-SBFR-C02

ENGINE TUNE-UP SPECIFICATIONS

Year	Engine Displacement Liters (VIN)	Spark Plugs Gap (in.)	Ignition Timing (deg.) ① MT	Ignition Timing (deg.) ① AT	Fuel Pump (psi)	Idle Speed (rpm) ② MT	Idle Speed (rpm) ② AT	Valve Clearance ③ In.	Valve Clearance ③ Ex.
2001	2.5 (6) ④	0.039-0.043	7-23 BTDC	7-23 BTDC	34-38	600-800	600-800	0.0071-0.0087	0.0090-0.0106
2002	2.5 (6) ④	0.039-0.043	7-23 BTDC	7-23 BTDC	34-38	600-800	600-800	0.0071-0.0087	0.0090-0.0106
2003-04	2.5 (6) ④	0.039-0.043	7-23 BTDC	7-23 BTDC	34-38	600-800	600-800	0.0071-0.0087	0.0090-0.0106
	2.5 (6) ⑤	0.028-0.031	7-24 BTDC	7-24 BTDC	34-38	600-800	600-800	0.0071-0.0087	0.0090-0.0106

BTDC: Before Top Dead Center

① At idle speed.

② With engine under no load.

③ With engine cold.

④ SOHC Non-Turbo

⑤ DOHC Turbo

67170-SBFR-C03

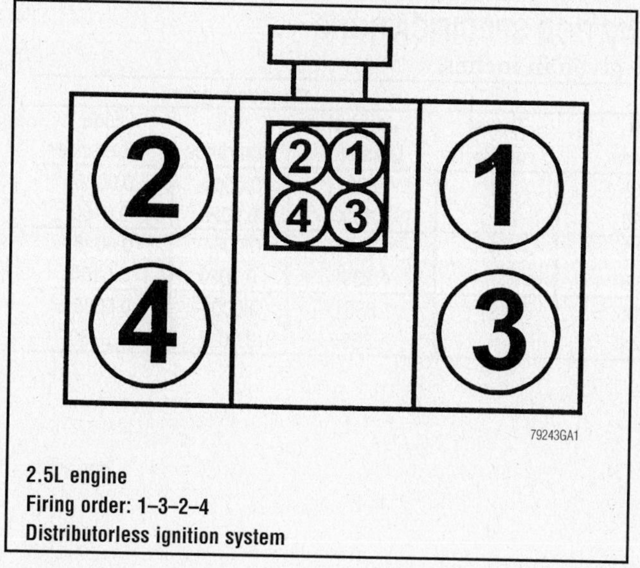

2.5L engine
Firing order: 1–3–2–4
Distributorless ignition system

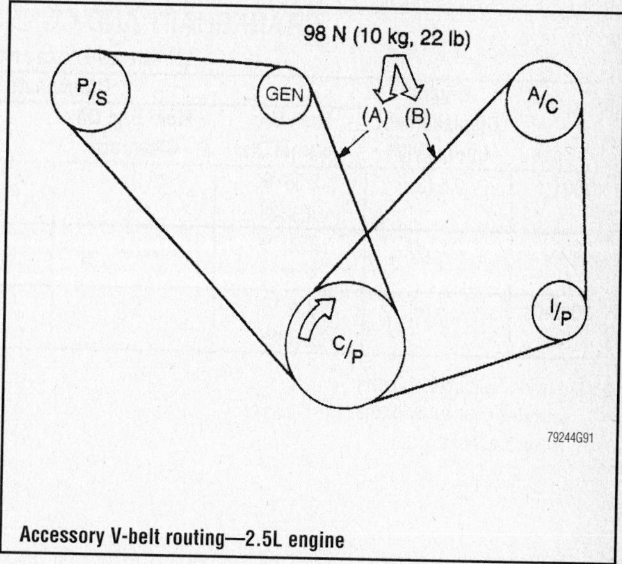

Accessory V-belt routing—2.5L engine

CAPACITIES

Year	Model	Engine Displacement Liters (VIN)	Engine Oil with Filter (qts.)	Transmission (pts.) 4-Spd	5-Spd	Auto.	Transfer Case (pts.)	Drive Axle Front (pts.)	Rear (pts.)	Fuel Tank (gal.)	Cooling System (qts.)
2001	Forester	2.5 (6)	4.2	—	7.4	20	—	2.6 ①	1.6	15.9	6.3
2002	Forester	2.5 (6)	4.2	—	7.4	20	—	2.6 ①	1.6	15.9	6.3
2003-04	Forester	2.5 (6)	4.2	—	7.4	20	—	2.6 ①	1.6	15.9	②

① A/T differential only.
② SOHC: 7.2 quarts
DOHC: 7.8 quarts

67170-SBFR-C04

VALVE SPECIFICATIONS

Year	Engine Displacement Liters (VIN)	Seat Angle (deg.)	Face Angle (deg.)	Spring Test Pressure (lbs. @ in.)	Spring Installed Height (in.)	Stem-to-Guide Clearance (in.) Intake	Exhaust	Stem Diameter (in.) Intake	Exhaust
2001	2.5 (6)	①	①	33-38@ 1.654 ②	③	0.0014- ③ 0.0024	0.0016- ④ 0.0026	0.2343- 0.2348	0.2343- 0.2348
2002	2.5 (6)	①	①	33-38@ 1.654 ②	③	0.0014- ③ 0.0024	0.0016- ④ 0.0026	0.2343- 0.2348	0.2343- 0.2348
2003-04	2.5 (6)	①	①	33-38@ 1.654 ②	③	0.0014- ③ 0.0024	0.0012- 0.0022	⑤	⑥

① Refacing angle: 90 degrees
② 102-118 lbs. @ 1.315 in.
③ Free length: DOHC: 1.863 in. SOHC: 2.1387 in.
④ Wear limit: 0.0059 in.
⑤ SOHC: 0.2343-0.2348
DOHC: 0.2374-0.2350 in.
⑥ SOHC: 0.2341-0.2346
DOHC: 0.2341-0.2346

67170-SBFR-C05

CRANKSHAFT AND CONNECTING ROD SPECIFICATIONS

All measurements are given in inches.

	Engine	Crankshaft				Connecting Rod		
Year	Displacement Liters (VIN)	Main Brg. Journal Dia.	Main Brg. Oil Clearance	Shaft End-play	Thrust on No.	Journal Diameter	Oil Clearance	Side Clearance
2001	2.5 (6)	2.3619-2.3625	①	0.0012-0.0098	3	1.8891-1.8898	0.0004-0.0020	0.0028-0.0160
2002	2.5 (6)	2.3619-2.3625	①	0.0012-0.0098	3	1.8891-1.8898	0.0004-0.0020	0.0028-0.0160
2003-04	2.5 (6)	2.3619-2.3625	①	0.0012-0.0045	3	1.8891-1.8898	0.0006-0.0017	0.0028-0.0130

① Journals 1 and 5: 0.0001-0.0016
Journals 2 and 4: 0.0004-0.0018
Journal 3: 0.0004-0.0016

67170-SBFR-C06

PISTON AND RING SPECIFICATIONS

All measurements are given in inches.

	Engine		Ring Gap			Ring Side Clearance		
Year	Displacement Liters (VIN)	Piston Clearance	Top Compression	Bottom Compression	Oil Control	Top Compression	Bottom Compression	Oil Control
2001	2.5 (6)	0.0004-0.0020	0.0079-0.0390	0.0146-0.0390	0.0079-0.0590	0.0016-0.0059	0.0012-0.0059	NA
2002	2.5 (6)	0.0004-0.0020	0.0079-0.0390	0.0146-0.0390	0.0079-0.0590	0.0016-0.0059	0.0012-0.0059	NA
2003-04	2.5 (6)	①	②	③	④	0.0016-0.0031	0.0012-0.0028	NA

NA: Not Available

① SOHC: 0.0004-0.0012 in.
DOHC: -0.0004-0.0004 in.

② SOHC:0.0079-0.0138 in.
DOHC: 0.0079-0.0098 in.

③ SOHC: 0.0146-0.0205 in.
DOHC: 0.015-0.020 in.

④ SOHC: 0.0079-0.0197 in.
DOHC: 0.015-0.020 in.

67170-SBFR-C07

TORQUE SPECIFICATIONS

All readings in ft. lbs.

Year	Engine Displacement Liters (VIN)	Cylinder Head Bolts	Main Bearing Bolts	Rod Bearing Bolts	Crankshaft Damper Bolts	Flywheel Bolts	Manifold Intake	Manifold Exhaust	Spark Plugs	Oil Pan Drain Plug
2001	2.5 (6)	①	②	31-34	123-137	51-55	14-17	③	13-17	32
2002	2.5 (6)	①	②	31-34	123-137	51-55	14-17	③	13-17	32
2003-04	2.5 (6)	④ ⑤	⑥	33-38	131	53	18	③	15	32

① Step 1: Tighten all bolts, in sequence, to 22 ft. lbs.
Step 2: Tighten all bolts, in sequence, to 51 ft. lbs.
Step 3: Loosen all bolts 180 degrees (one-half turn)
Step 4: Loosen all bolts another 180 degrees (one-half turn)
Step 5: Tighten all bolts in sequence, 40-45 degrees
Step 6: Tighten bolts A and B, in sequence, 40-45 degrees

② Split engine case bolts:
10mm bolts: 33-37 ft. lbs.
8mm bolts: A thru G to 17-20 ft. lbs. and H to 5 ft. lbs.

③ No separate exhaust manifold is used, the front pipe bolts directly to the cylinder heads
Tighten the bolts to 22 ft. lbs. on SOHC engines or 26 ft. lbs. on DOHC engines

④ SHOC: Step 1: Tighten all bolts, in sequence, to 22 ft. lbs.
Step 2: Tighten all bolts, in sequence, to 51 ft. lbs.
Step 3: Loosen all bolts 180 degrees (one-half turn)
Step 4: Loosen all bolts another 180 degrees (one-half turn)
Step 5: Tighten all bolts in sequence, to 40-45 degrees
Step 6: Tighten bolts A and B, in sequence 40-45 degrees

⑤ DHOC: Step 1: Tighten all bolts, in sequence, to 22 ft. lbs.
Step 2: Tighten all bolts, in sequence, to 51 ft. lbs.
Step 3: Loosen all bolts 180 degrees (one-half turn)
Step 4: Loosen all bolts another 180 degrees (one-half turn)
Step 5: Tighten all bolts in sequence, to 36 ft. lbs. (49Nm)
Step 6: Tighten all bolts in sequence, to 80-90 degrees
Step 7: Tighten all bolts in sequence, to 40-45 degrees
Step 8: Tighten bolts A and B, in sequence 40-45 degrees

⑥ Split engine case bolts SOHC:
10mm bolts: 11 ft. lbs.
8mm bolts: Left hand A thru D to 90 dregrees
8mm bolts: Right hand E thru J to 90 dregrees
6 and 8mm bolts: A thru G to 18 ft. lbs. and H 5 ft. lbs.
Split engine case bolts DOHC:
10mm bolts: 7 ft. lbs.
8mm bolts: Left hand A thru D. Bolts A and C 15 ft. lbs Bolts B and D 11 ft. lbs.
8mm bolts: Right hand E thru J. Bolts Ethru G 15 ft. lbs Bolts H and J 13 ft. lbs.
8mm bolts: Left hand A thru D another 90 degrees
8mm bolts: Right hand E thru J another 90 degrees
6 and 8mm bolts: A thru G to 18 ft. lbs. and H 5 ft. lbs.

67170-SBFR-C08

WHEEL ALIGNMENT

Year	Model		Caster Range (+/-Deg.)	Caster Preferred Setting (Deg.)	Camber Range (+/-Deg.)	Camber Preferred Setting (Deg.)	Toe-in (in.)
2001	Forester	F	0.75	+2.58	0.50	-0.25	0+/-0.12
		R	—	—	0.75	-0.58	0.08+/-0.04
2002	Forester	F	0.75	+2.58	0.50	-0.25	0+/-0.12
		R	—	—	0.75	-0.58	0.08+/-0.04
2003-04	Forester	F	0.75	+2.58	0.50	-0.25	0+/-0.12
		R	—	—	0.75	-0.58	0.08+/-0.04

67170-SBFR-C09

TIRE, WHEEL AND BALL JOINT SPECIFICATIONS

Year	Model	OEM Tires Standard	OEM Tires Optional	Tire Pressures (psi) Front	Tire Pressures (psi) Rear	Wheel Size	Ball Joint Inspection	Lug Nut
2001	Forester	P205/70R15 95S	P215/60R16 94H	29	26 ①	②	0.012 in. ③	58-72
2002	Forester	P205/70R15 95S	P215/60R16 94H	29	26 ①	②	0.012 in. ③	58-72
2003-04	Forester	P215/60R16 94H	—	29	28 ①	16X6.5 JJ	0.012 in. ③	68

OEM: Original Equipment Manufacturer

PSI: Pounds Per Square Inch

STD: Standard

OPT: Optional

① With full load: 36 psi.

② With standard tires: 6-JJ
With optional tires: 6.5-JJ

③ Apply 154 lbs. vertical force

67170-SBFR-C10

BRAKE SPECIFICATIONS

All measurements in inches unless noted

Year	Model		Brake Disc Original Thickness	Brake Disc Minimum Thickness	Brake Disc Maximum Runout	Brake Drum Diameter Original Inside Diameter	Brake Drum Diameter Max. Wear Limit	Brake Drum Diameter Maximum Machine Diameter	Minimum Lining Thickness Front	Minimum Lining Thickness Rear	Brake Caliper Bracket Bolts (ft. lbs.)	Brake Caliper Mounting Bolts (ft. lbs.)
2001	Forester	F	0.945	0.660	0.003	—	—	—	0.295	0.256	59	28
		R	0.390	0.335	0.0028	9.00 ①	9.08 ②	NA	—	0.059	38	28
2002	Forester	F	0.945	0.660	0.003	—	—	—	0.295	0.256	59	28
		R	0.390	0.335	0.0028	9.00 ①	9.08 ②	NA	—	0.059	38	28
2003-04	Forester	F	0.940	0.870	0.003	—	—	—	0.59	0.059	59	28
		R	0.390	0.335	0.0028	9.00 ①	9.08 ②	NA	—	0.059	38	28

NA: Not Available

① Parking brake drum on vehicles with rear disc brakes: 6.69 in.

② Parking brake drum on vehicles with rear disc brakes: 6.73 in.

67170-SBFR-C11

SCHEDULED MAINTENANCE INTERVALS

SUBARU—FORESTER

TO BE SERVICED	TYPE OF SERVICE	VEHICLE MILEAGE INTERVAL (x1000)																
		3	7.5	15	22.5	30	37.5	45	52.5	60	67.5	75	82.5	90	97.5	105	112.5	120
Accessory drive belts	R									✓								✓
Accessory drive belts	S/I					✓								✓				
Air cleaner filter	R					✓				✓				✓				✓
Automatic transmission fluid	S/I					✓				✓				✓				✓
Axle shaft joints	S/I			✓		✓		✓		✓		✓		✓		✓		✓
Brake fluid	R					✓				✓				✓				✓
Brake system lines	S/I			✓		✓		✓		✓		✓		✓		✓		✓
Clutch operation	S/I			✓		✓		✓		✓		✓		✓		✓		✓
Disc brake pads & rotors	S/I			✓		✓		✓		✓		✓		✓		✓		✓
Drums brake linings & drums	S/I					✓				✓				✓				✓
Engine coolant	R					✓				✓				✓				✓
Engine cooling system, hoses & connections	S/I					✓				✓				✓				✓
Engine oil & filter	R	✓	✓	✓	✓	✓	✓	✓	✓	✓	✓	✓	✓	✓	✓	✓	✓	✓
Front & rear axle boots	S/I			✓		✓		✓		✓		✓		✓		✓		✓
Front & rear wheel bearings	S/I & L									✓								✓
Fuel filter	R					✓				✓				✓				✓
Parking & service brake systems' operation	S/I			✓		✓		✓		✓		✓		✓		✓		✓
Spark plugs	R									✓								✓
Steering & suspension	S/I			✓		✓		✓		✓		✓		✓		✓		✓
Timing belt	R															✓		
Timing belt	S/I					✓				✓				✓				
Transmission & differential fluid levels	S/I					✓				✓				✓				✓
Valve clearance	S/I															✓		

R: Replace S/I: Inspect and service, if needed L: Lubricate

FREQUENT OPERATION MAINTENANCE (SEVERE SERVICE)

If a vehicle is operated under any of the following conditions it is considered severe service:

- Towing a trailer or using a camper or car-top carrier.

- Repeated short trips of less than 5 miles in temperatures below freezing, or trips of less than 10 miles in any temperature.

- Extensive idling or low-speed driving for long distances as in heavy commercial use, such as delivery, taxi or police cars.

- Operating on rough, muddy or salt-covered roads, or extensive mountain driving.

- Operating on unpaved or dusty roads.

- Driving in extremely hot (over 90°) conditions.

Engine oil and filter: replace every 3000 miles or 3 months, whichever occurs first.

Fuel filter: replace every 7500 miles or 7.5 months, whichever occurs first.

Fuel system, hoses & connections: inspect every 7500 miles or 7.5 months, whichever occurs first.

Transmission & differential fluid: replace every 15,000 miles.

Automatic transmission fluid: replace every 15,000 miles.

Brake fluid: replace every 15,000 miles.

Disc brake pads & rotors: inspect every 7500 miles or 7.5 months, whichever occurs first.

Front & rear axle boots: inspect every 7500 miles or 7.5 months, whichever occurs first.

Axle shaft boots: inspect every 7500 miles or 7.5 months, whichever occurs first.

Drum brake linings & drums: inspect every 7500 miles or 7.5 months, whichever occurs first.

Brake lines: inspect every 7500 miles or 7.5 months, whichever occurs first.

Parking & service brake system operation: inspect every 7500 miles or 7.5 months, whichever occurs first.

Clutch operation: inspect every 7500 miles or 7.5 months, whichever occurs first.

PRECAUTIONS

Before servicing any vehicle, please be sure to read all of the following precautions, which deal with personal safety, prevention of component damage, and important points to take into consideration when servicing a motor vehicle:

- Never open, service or drain the radiator or cooling system when the engine is hot; serious burns can occur from the steam and hot coolant.
- Observe all applicable safety precautions when working around fuel. Whenever servicing the fuel system, always work in a well-ventilated area. Do not allow fuel spray or vapors to come in contact with a spark, open flame or excessive heat (a hot drop light, for example). Keep a dry chemical fire extinguisher near the work area. Always keep fuel in a container specifically designed for fuel storage; also, always properly seal fuel containers to avoid the possibility of fire or explosion. Refer to the additional fuel system precautions later in this section.
- Fuel injection systems often remain pressurized, even after the engine has been turned **OFF**. The fuel system pressure must be relieved before disconnecting any fuel lines. Failure to do so may result in fire and/or personal injury.
- Brake fluid often contains polyglycol ethers and polyglycols. Avoid contact with the eyes and wash your hands thoroughly after handling brake fluid. If you do get brake fluid in your eyes, flush your eyes with clean, running water for 15 minutes. If eye irritation persists, or if you have taken brake fluid internally, IMMEDIATELY seek medical assistance.
- The EPA warns that prolonged contact with used engine oil may cause a number of skin disorders, including cancer! You should make every effort to minimize your exposure to used engine oil. Protective gloves should be worn when changing oil. Wash your hands and any other exposed skin areas as soon as possible after exposure to used engine oil. Soap and water, or waterless hand cleaner should be used.
- All new vehicles are now equipped with an air bag system. The system must be disabled before performing service on or around system components, steering column, instrument panel components, wiring and sensors. Failure to follow safety and disabling procedures could result in accidental air bag deployment, possible personal injury and unnecessary system repairs.
- Always wear safety goggles when working with, or around, the air bag system. When carrying a non-deployed air bag, be sure the bag and trim cover are pointed away from your body. When placing a non-deployed air bag on a work surface, always face the bag and trim cover upward, away from the surface. This will reduce the motion of the module if it is accidentally deployed. Refer to the additional air bag system precautions later in this section.
- Clean, high quality brake fluid from a sealed container is essential to the safe and proper operation of the brake system. You should always buy the correct type of brake fluid for your vehicle. If the brake fluid becomes contaminated, completely flush the system with new fluid. Never reuse any brake fluid. Any brake fluid that is removed from the system should be discarded. Also, do not allow any brake fluid to come in contact with a painted surface; it will damage the paint.
- Never operate the engine without the proper amount and type of engine oil; doing so WILL result in severe engine damage.
- Timing belt maintenance is extremely important! Many models utilize an interference-type, non-freewheeling engine. If the timing belt breaks, the valves in the cylinder head may strike the pistons, causing potentially serious (also time-consuming and expensive) engine damage.
- Disconnecting the negative battery cable on some vehicles may interfere with the functions of the on-board computer system(s) and may require the computer to undergo a relearning process once the negative battery cable is reconnected.
- When servicing drum brakes, only disassemble and assemble one side at a time, leaving the remaining side intact for reference.
- Only an MVAC-trained, EPA-certified automotive technician should service the air conditioning system or its components.

ENGINE REPAIR

➡Disconnecting the negative battery cable on some vehicles may interfere with the functions of the on board computer systems and may require the computer to undergo a relearning process, once the negative battery cable is reconnected.

Distributor

The Forester is equipped with a distributorless ignition system.

Alternator

REMOVAL & INSTALLATION

1. Before servicing the vehicle, refer to the precautions in the beginning of this section.
2. Remove or disconnect the following:

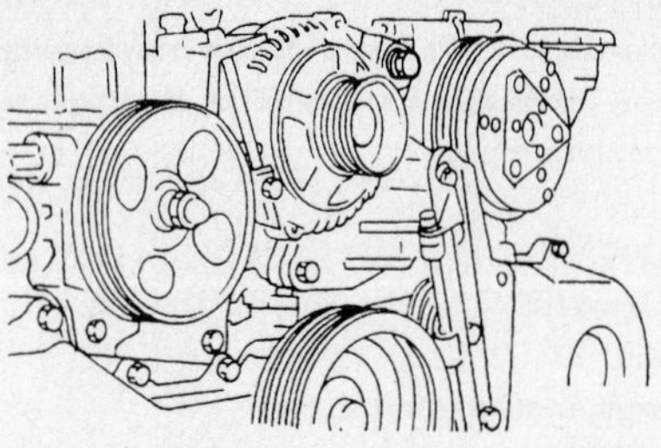

View of alternator mounting

- Negative battery cable
- Wires
- Belt cover
- Drive belt
- Mounting bolts
- Alternator

To install:

Install or connect the following:

- Alternator
- Mounting bolts
- Drive belt. Adjust the tension to 0.276–0.354 in. (7–9mm) for new or 0.354–0.433 in. (9–11mm) for used. Torque the slider bolt to 4–6 ft. lbs. (6–10 Nm) and the lockbolt to 16.5 ft. lbs. (19.5 Nm).
- Belt cover
- Wires
- Negative battery cable

Ignition Timing

ADJUSTMENT

The ignition timing is controlled by the engine control computer and is not adjustable. To check the ignition timing proceed as follows:

1. Before servicing the vehicle, refer to the precautions in the beginning of this section.

2. Warm up the engine, then turn the ignition **OFF**.

3. Connect a timing light to the No. 1 spark plug wire according to the manufactures directions.

4. Start the engine. With the vehicle at idle check the timing.

5. The timing should be 7–23 degrees Before Top Dead Center (BTDC) at 700 RPM on vehicles with an automatic transmission and 2–18 degrees at 650 RPM on models with a manual transmission.

6. If the timing is not correct, there could be a problem in the ignition control system.

Engine Assembly

REMOVAL & INSTALLATION

SOHC Engines

2001 MODELS

✱✱ CAUTION

The fuel injection system remains under pressure after the engine has been turned OFF. Properly relieve fuel pressure before disconnecting any fuel lines. Failure to do so may result in fire or personal injury.

1. Before servicing the vehicle, refer to the precautions in the beginning of this section.
2. Relieve the fuel system pressure.
3. Drain the engine oil and coolant.
4. Discharge and recover the air conditioning system.
5. Remove or disconnect the following:
   - Negative battery cables and battery
   - Engine undercover
   - Radiator hoses and fan motor harness
   - Radiator
   - Air conditioning compressor and cap the lines
   - Air intake duct
   - Air cleaner element and upper cover
   - Evaporative Emissions (EVAP) canister and bracket
   - Front Oxygen (O_2S) sensor
6. If equipped with California emissions specifications, disconnect the rear O_2S sensor.
7. Remove or disconnect the following:
   - Engine ground terminal
   - Crankshaft Position (CKP) sensor connector
   - Camshaft Position (CMP) sensor connector
   - Knock Sensor (KS) connector
   - Alternator connector and terminal
   - Air conditioning compressor connectors, if equipped
   - Accelerator cable
   - Cruise control cable, if equipped
   - Brake booster hose
   - Heater inlet and outlet hoses
   - Alternator drive belt
   - Wires from the spark plugs on the left side of the engine
   - Power steering pump line bracket
   - Power steering pump, leaving the lines connected and position it aside
   - Exhaust Y-pipe
   - Lower starter nuts
   - Lower engine-to-transmission nuts
   - Front engine mount-to-crossmember nuts
   - Starter
8. If equipped with an automatic transmission, perform the following:
   a. Remove the torque converter service hole plug.
   b. Matchmark the torque converter-to-driveplate.
   c. Rotate the engine to remove the torque converter-to-driveplate bolts as they become accessible.
9. Remove or disconnect the following:
   - Flywheel cover, if equipped with a manual transmission
   - Pitching stopper
   - Fuel delivery, return and evaporation hoses
10. Support the engine with a suitable lifting device attached to the engine lifting eyes.
11. Slightly raise the engine.
12. Raise the transmission with a floor jack.
13. If equipped with a manual transmission, pull the engine forward then up and out of the vehicle to clear the transmission mainshaft.
14. If equipped with an automatic transmission, pull the engine forward then up and out of the vehicle.

To install:

15. If equipped with a manual transmission, apply a small amount of grease to the splines of the mainshaft.
16. Position the engine in the engine compartment and align it with the transmission.
17. Install the engine. Torque the upper bolts to 37 ft. lbs. (50 Nm).
18. Remove the lifting device and floor jack.
19. Install or connect the following:
   - Pitching stopper. Torque the bolts to 43 ft. lbs. (58 Nm) on the body side and 37 ft. lbs. (50 Nm) on the bracket side.
   - Flywheel cover, if equipped with a manual transmission
20. If equipped with an automatic transmission, perform the following:
   a. Align the matchmarks, install the torque converter-to-driveplate bolts while rotating the engine and tighten to 18 ft. lbs. (25 Nm).
   b. Install the service hole cover.
21. Install or connect the following:
   - EVAP canister and bracket
   - Power steering pump. Tighten the retainer bolts to 14 ft. lbs. (20 Nm).
   - Accessory drive belt
   - Starter. Torque the bolts to 34–40 ft. lbs. (44–52 Nm).
   - Lower engine-to-transmission nuts. Tighten them to 36 ft. lbs. (50 Nm).

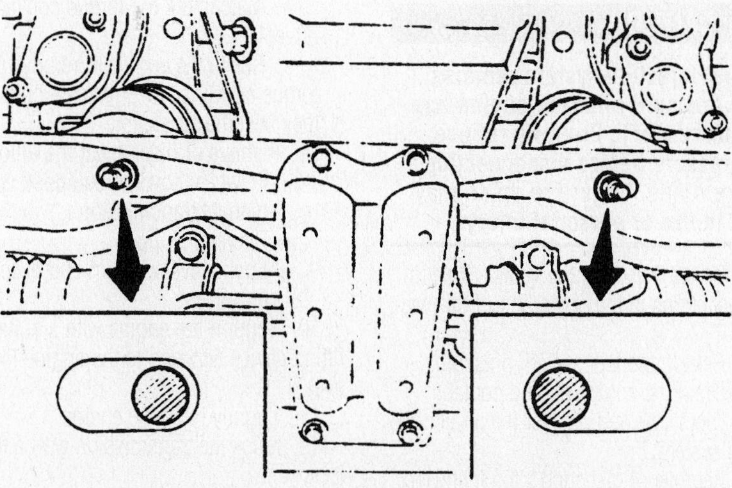

Be sure to tighten the front cushion rubber mounting bolts in the innermost elliptical hole in the front crossmember—2001 Models

- Lower engine mounting nuts. Tighten them to 61 ft. lbs. (83 Nm) in the inner most elliptical hole in the front crossmember so the clearance is 0.16–0.24 in. (4–6mm).
- Exhaust Y-pipe with new gaskets and nuts
- Brake booster hose
- Heater inlet and outlet hoses
- Accelerator cable
- Cruise control cable, if equipped
- Engine harness connectors
- Engine ground terminal
- CKP sensor connector
- CMP sensor connector
- Knock sensor connector
- Alternator connector and terminal
- Air conditioning compressor connectors, if equipped
- Front O_2S sensor, and if removed, the rear O_2S sensor.
- Air cleaner element and cover
- Air conditioning lines with new O-rings, if equipped. Torque the bolts to 23 ft. lbs. (31 Nm).
- Radiator
- Engine undercover
- Negative battery cable

22. Fill the crankcase to the proper level with clean engine oil.
23. Fill and bleed the cooling system.
24. Charge the air conditioning system using an approved recovery/recycling machine.
25. If equipped, check the automatic transmission fluid level and add Dexron®II if necessary.
26. Start the engine and allow it to reach normal operating temperature. Check for leaks.

2002–04 MODELS

⁂ CAUTION

The fuel injection system remains under pressure after the engine has been turned OFF. Properly relieve fuel pressure before disconnecting any fuel lines. Failure to do so may result in fire or personal injury.

1. Before servicing the vehicle, refer to the precautions in the beginning of this section.
2. Relieve the fuel system pressure.
3. Drain the engine oil and coolant.
4. Discharge and recover the air conditioning system.
5. Remove or disconnect the following:
   - Negative battery cables and battery
   - Air cleaner assembly
   - Engine undercover
   - Radiator hoses and fan motor harness
   - Radiator
   - Air conditioning compressor and cap the lines
   - Air cleaner case stay
   - Front Oxygen (O_2S) sensor
6. If equipped with California emissions specifications, disconnect the rear O_2S sensor.
7. Remove or disconnect the following:
   - Engine ground terminal
   - Crankshaft Position (CKP) sensor connector
   - Camshaft Position (CMP) sensor connector
   - Knock Sensor (KS) connector
   - Alternator connector and terminal
   - Air conditioning compressor connectors, if equipped
   - Accelerator cable
   - Cruise control cable, if equipped
   - Pressure Switch
   - Brake booster hose
   - Heater inlet and outlet hoses
   - Resonator chamber
   - Front side V–belt
   - Pipe with bracket from the intake manifold
   - Power steering pump, leaving the lines connected and position it aside
   - Exhaust Y-pipe
   - Lower starter nuts
   - Lower engine-to-transmission nuts
   - Front engine mount-to-crossmember nuts
8. If equipped with an automatic transmission, perform the following:
   a. Remove the torque converter service hole plug.
   b. Matchmark the torque converter-to-driveplate.
   c. Rotate the engine to remove the torque converter-to-driveplate bolts as they become accessible.
9. Remove or disconnect the following:
   - Flywheel cover, if equipped with a manual transmission
   - Pitching stopper
   - Fuel delivery, return and evaporation hoses
10. Support the engine with a suitable lifting device attached to the engine lifting eyes.
11. Slightly raise the engine.
12. Raise the transmission with a floor jack.
    - Starter
    - Upper engine-to-transmission bolts
13. If equipped with a manual transmission, pull the engine forward then up and out of the vehicle to clear the transmission mainshaft.
14. If equipped with an automatic transmission, pull the engine forward then up and out of the vehicle.

To install:

15. If equipped with a manual transmission, apply a small amount of grease to the splines of the mainshaft.
16. Position the engine in the engine compartment and align it with the transmission.
17. Install the engine. Torque the upper bolts to 37 ft. lbs. (50 Nm).
18. Remove the lifting device and floor jack.
19. Install or connect the following:
    - Pitching stopper. Torque the bolts to 43 ft. lbs. (58 Nm) on the body side and 37 ft. lbs. (50 Nm) on the bracket side.
    - Starter
    - Flywheel cover, if equipped with a manual transmission
20. If equipped with an automatic transmission, perform the following:
    a. Align the matchmarks, install the torque converter-to-driveplate bolts while rotating the engine and tighten to 18 ft. lbs. (25 Nm).
    b. Install the service hole cover.
21. Install or connect the following:
    - Power steering pump. Tighten the retainer bolts to 14 ft. lbs. (20 Nm).
    - Power steering switch connector
    - Accessory drive belt
    - Resontaor chamber and tighten the bolts to 24 ft. lbs. (33 Nm)
    - Lower engine-to-transmission nuts. Tighten them to 36 ft. lbs. (50 Nm).
    - Lower engine mounting nuts. Tighten them to 63 ft. lbs. (85 Nm) in the inner most elliptical hole in the front crossmember so the clearance is 0.16–0.24 in. (4–6mm).
    - Exhaust Y-pipe with new gaskets and nuts
    - Brake booster hose
    - Heater inlet and outlet hoses
    - Accelerator cable
    - Cruise control cable, if equipped
    - Engine harness connectors
    - Engine ground terminal
    - CKP sensor connector
    - CMP sensor connector
    - Knock sensor connector
    - Alternator connector and terminal
    - Air conditioning compressor connectors, if equipped
    - Front O_2S sensor, and if removed, the rear O_2S sensor.

42356-SBFR-G01

Attach a suitable lift device to the engine

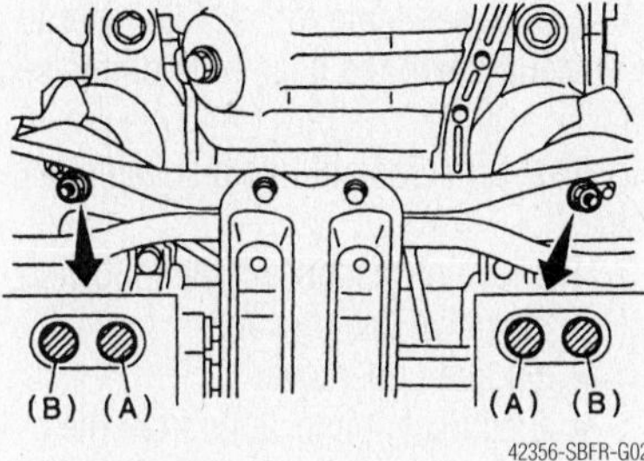

42356-SBFR-G02

Be sure to tighten the front cushion rubber mounting bolts in the innermost elliptical hole in the front crossmember—2002–04 Models

- Air cleaner element and cover
- Air conditioning lines with new O-rings, if equipped. Torque the bolts to 18 ft. lbs. (25 Nm).
- Radiator
- Engine undercover
- Negative battery cable

22. Fill the crankcase to the proper level with clean engine oil.
23. Fill and bleed the cooling system.
24. Charge the air conditioning system using an approved recovery/recycling machine.
25. If equipped, check the automatic transmission fluid level and add Dexron®II if necessary.
26. Start the engine and allow it to reach normal operating temperature. Check for leaks.

DOHC Engines

2002–04 MODELS

⁂ CAUTION

The fuel injection system remains under pressure after the engine has been turned OFF. Properly relieve fuel pressure before disconnecting any fuel lines. Failure to do so may result in fire or personal injury.

1. Before servicing the vehicle, refer to the precautions in the beginning of this section.
2. Relieve the fuel system pressure.
3. Drain the engine oil and coolant.
4. Discharge and recover the air conditioning system.
5. Disconnect the negative battery cable.
6. Remove the radiator.
7. Remove the coolant filler tank.
8. Disconnect the air conditioning hoses from the compressor and cap the lines.
9. Remove the intercooler as follows:
   a. Disconnect the Positive Crankcase Ventilation (PCV) hose from the pipe.
   b. Remove the PCV pipe from the intercooler.
   c. Disconnect the air by–pass valve hose.
   d. Remove the intercooler–to–bracket bolts.
   e. Loosen the clamps that connect the intercooler to the turbocharger and throttle body.
   f. Disconnect the intercooler duct from the turbocharger and remove the intercooler.
10. Disconnect the following:
    - Engine harness connector
    - Engine ground terminal
    - Left and right engine ground cables

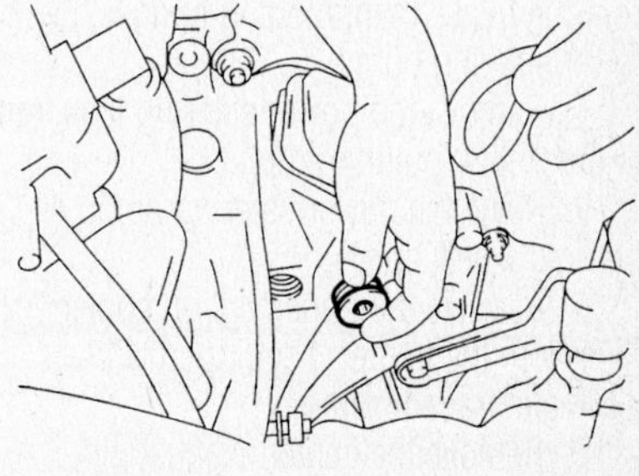
67170-SBFR-G01

Remove the plug using a 10mm wrench–DOHC engines

- Alternator connector and terminal
- Air conditioning compressor connectors, if equipped
- Accelerator cable, on models equipped with a manual transmission
- Clutch release spring, if equipped with a manual transmission
- Brake booster hose
- Heater inlet and outlet hoses

11. Loosen the power steering pump lock bolt and slider bolt. Remove the power steering belt.
12. Disconnect the power steering switch connector.

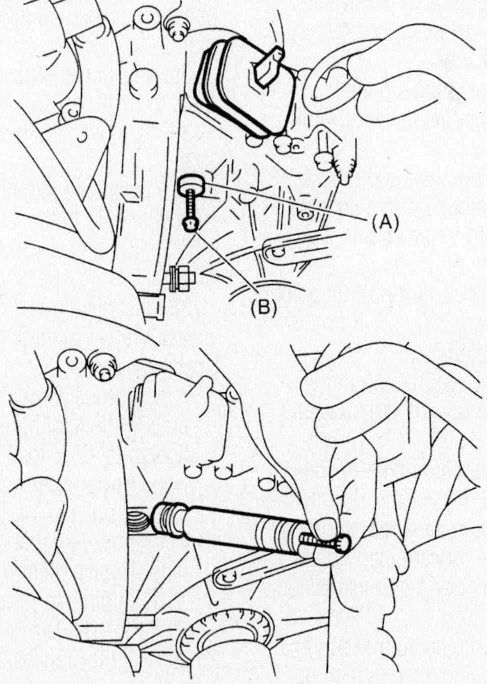

(A) Shaft
(B) Bolt

67170-SBFR-G02

Screw the 6mm bolt into the release fork and remove it–DOHC engines

13. Remove the pipe with the bracket from the intake manifold and remove the power steering pump.
14. Remove the power steering tank from its bracket by pulling it up.
15. Place the power steering pump on the right wheel apron.
16. Remove the transmission cooler lines from the frame, if equipped with an automatic transmission.
17. Remove the center exhaust pipe.
18. Remove the lower transmission–to–engine nuts.
19. Remove the nuts which attach the front mount rubber to the crossmember.
20. On models equipped with a manual transmission, perform the following:
 a. Remove the clutch operating cylinder from the transmission.
 b. Remove the plug using a 10mm wrench
 c. Screw the 6mm bolt into the release fork and remove it.
 d. Raise the release fork and unfasten the release bearing tabs to free the fork.
21. If equipped with an automatic transmission, perform the following:
 a. Remove the torque converter service hole plug.
 b. Matchmark the torque converter-to-driveplate.
 c. Rotate the engine to remove the torque converter-to-driveplate bolts as they become accessible.
22. Remove the pitching stopper.
23. Disconnect the fuel delivery, return and evaporation hoses
24. Remove the fuel filter and bracket.
25. Support the engine with a suitable lifting device attached to the engine lifting eyes.
26. Support the transmission with a floor jack.
27. Remove the starter.
28. Install tool ST 498277200 to the converter housing to prevent the converter from falling.
29. Remove the upper engine-to-transmission bolts.
30. If equipped with a manual transmission, pull the engine forward then up and out of the vehicle to clear the transmission mainshaft.
31. If equipped with an automatic transmission, pull the engine forward then up and out of the vehicle.

To install:

32. If equipped with a manual transmission, perform the following:
 a. Remove the release bearing from the clutch cover using a flat bladed tool.

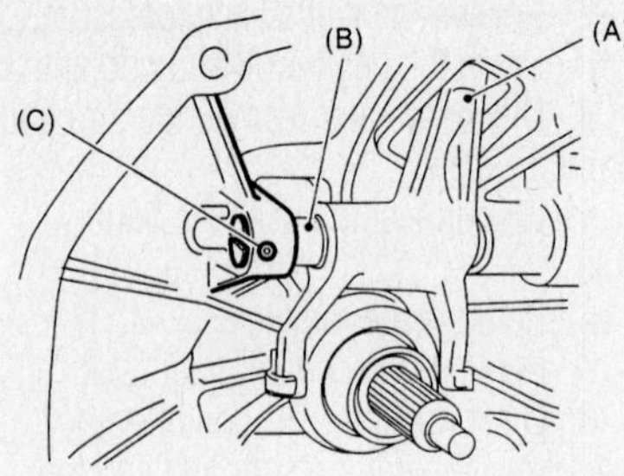

(A) Release fork
(B) Release shaft
(C) Spring pin

67170-SBFR-G03

Release fork and related components–DOHC engines

 b. Install the release bearing onto the transmission.
 c. Install the release fork into the release bearing tab.
 d. Apply a small amount of grease to the splines of the mainshaft.
 e. Insert the release fork shaft into the release fork. Make sure the cutout portion of the release fork shaft contacts the spring pin.
 f. Install the plug and tighten to 32 ft. lbs. (44 Nm).
33. Install the front mount rubber to the engine and tighten to 26 ft. lbs. (35 Nm).
34. Position the engine in the engine compartment and align it with the transmission.
35. Install the engine. Torque the upper bolts to 37 ft. lbs. (50 Nm).
36. Remove the lifting device and floor jack.
37. Install the pitching stopper. Torque the bolts to 43 ft. lbs. (58 Nm) on the body side and 37 ft. lbs. (50 Nm) on the bracket side.
38. Installation is the reverse of removal, please keep in mind the following torque specifications:
 a. Torque converter-to-driveplate bolts while rotating the engine and tighten to 18 ft. lbs. (25 Nm).
 b. Power steering pump. Tighten the retainer bolts to 14 ft. lbs. (20 Nm).
 c. Lower engine-to-transmission nuts. Tighten them to 37 ft. lbs. (50 Nm).
 d. Engine–to–crossmember nuts to 63 ft. lbs. (85 Nm).
39. Fill the crankcase to the proper level with clean engine oil.
40. Fill and bleed the cooling system.
41. Charge the air conditioning system using an approved recovery/recycling machine.
42. If equipped, check the automatic transmission fluid level and add Dexron®II if necessary.
43. Start the engine and allow it to reach normal operating temperature. Check for leaks.

Water Pump

REMOVAL & INSTALLATION

SOHC Engine

1. Before servicing the vehicle, refer to the precautions in the beginning of this section.
2. Drain the coolant into a suitable container.
3. Remove or disconnect the following:
 - Negative battery cable
 - Engine undercover
 - Electrical connections from the radiator fan motor and sub motor(s)
 - Bolt that retains the water by-pass pipe of the oil cooler onto the oil pump on vehicles equipped with an automatic transmission
 - Radiator outlet hose
 - Radiator fan motor assembly
 - Accessory drive belts
 - Timing belt and tensioner
 - Belt idler number 2
 - Camshaft Position (CMP) sensor
 - Left side camshaft pulleys and left side rear timing belt cover
 - Tensioner bracket
 - Radiator hose and heater hose from the water pump
 - Water pump retainer bolts
 - Water pump

To install:

4. Clean the gasket mating surfaces thoroughly. Always use new gaskets during installation.
5. Install or connect the following:
 - Water pump. Torque the bolts, in sequence, to 9 ft. lbs. 12 Nm). After tightening the bolts once, retighten to the same specification again.
 - Radiator hose and heater hose to the water pump
 - Left side rear timing belt cover, left side camshaft pulleys and tensioner bracket
 - CMP sensor
 - Timing belt and tensioner
 - Accessory drive belts
 - Water pipe bypass pipe retaining bolt
 - Radiator fan motor assembly
 - Radiator outlet hose

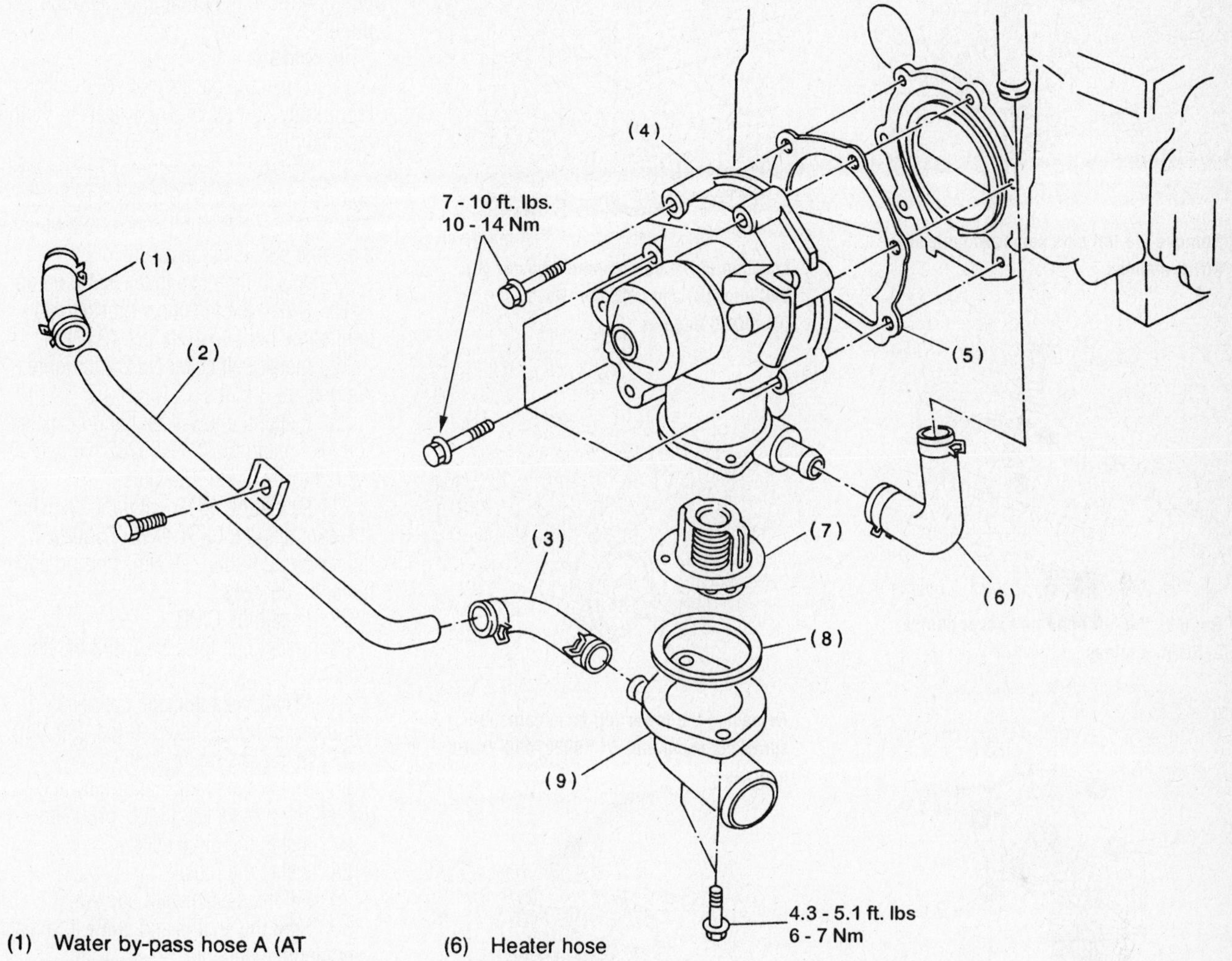

7 - 10 ft. lbs.
10 - 14 Nm

4.3 - 5.1 ft. lbs
6 - 7 Nm

(1) Water by-pass hose A (AT vehicles)
(2) Water by-pass pipe (AT vehicles)
(3) Water by-pass hose B (AT vehicles)
(4) Water pump ASSY
(5) Gasket
(6) Heater hose
(7) Thermostat
(8) Gasket
(9) Thermostat case

7924XG01

Exploded view of the water pump mounting and related components–SOHC engines

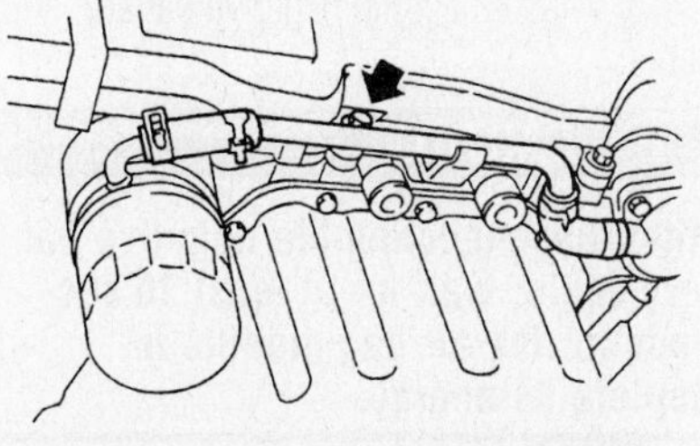

42356-SBFR-G03

Remove the bolt retaining the water by-pass pipe of the oil cooler onto the oil pump on vehicles equipped with A/T–SOHC engines

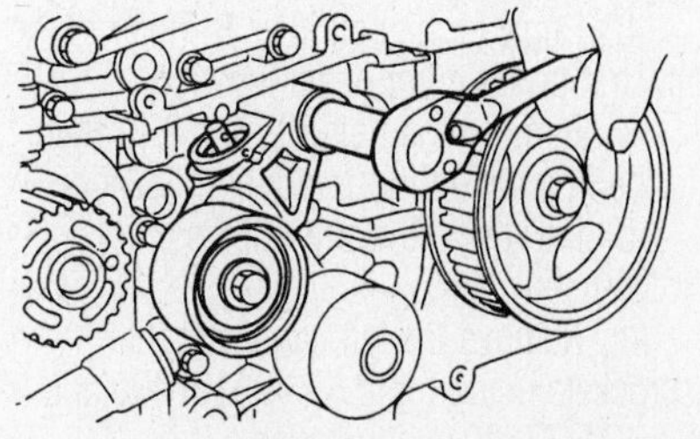

42356-SBFR-G04

Remove the automatic belt tensioner–SOHC engines

42356-SBFR-G05

Remove belt idler number 2–SOHC engines

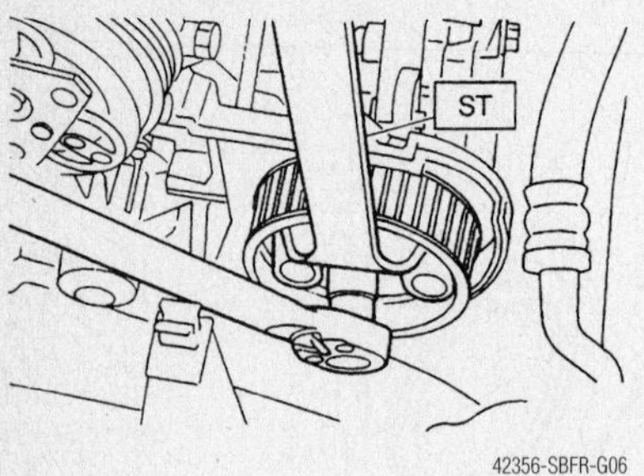

Remove the left side camshaft pulleys–SOHC engines

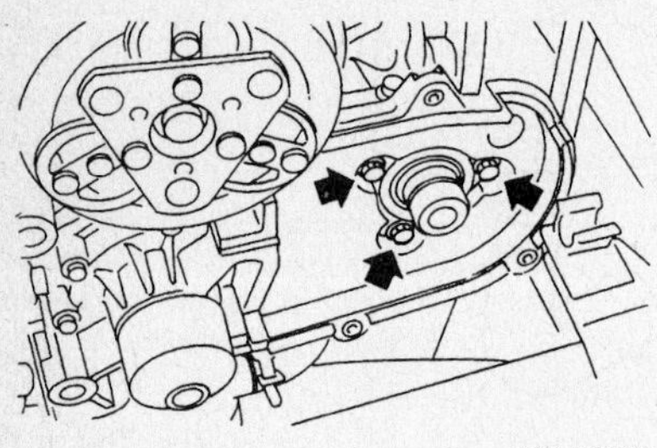

Remove the left hand belt cover number 2–SOHC engines

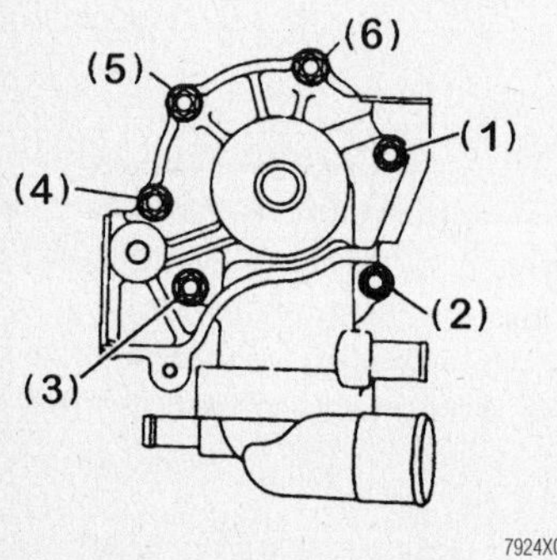

Water pump bolt tightening sequence–SOHC and DOHC engines

- Engine undercover
- Negative battery cable

6. Fill the system with coolant.
7. Start the engine and allow it to reach operating temperature.
8. Check for leaks.

DOHC Engines

1. Before servicing the vehicle, refer to the precautions in the beginning of this section.
2. Drain the coolant into a suitable container.
3. Disconnect the negative battery cable.
4. Remove the radiator.
5. Remove the drive belts.
6. Remove the crankshaft pulley and the timing belt.

Location of the belt tension adjuster (A), belt idler (B) and belt idler No. 2 (C)–DOHC engines

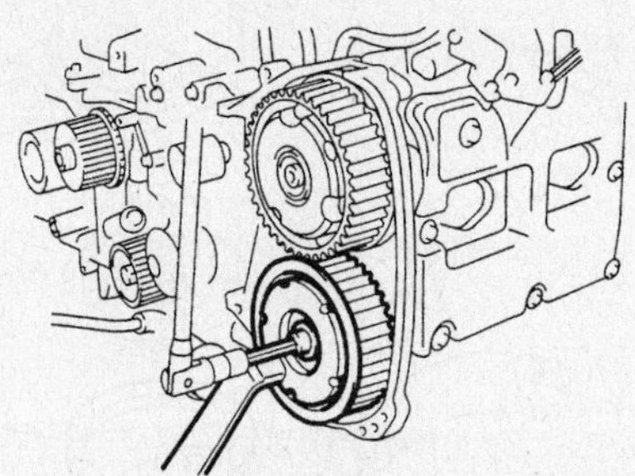

Removing the lower left hand camshaft sprockets using tool ST 499207400–DOHC engines

Removing the upper left hand camshaft sprockets using tool ST 499977500–DOHC engines

7. Remove the automatic belt tension adjuster **A**. Refer to the illustration for component location.
8. Remove the belt idler **B**. Refer to the illustration for component location.
9. Remove the belt idle No. 2 **C**. Refer to the illustration for component location.
10. Remove the Camshaft Position (CMP) sensor.
11. Remove the left hand camshaft sprockets using tools ST 499207400 and ST 499977500 as illustrated.
12. Remove belt cover No.2 from the left hand side.
13. Remove the tensioner bracket.
14. Disconnect the hose from the water pump, remove the pump bolts and the pump.

To install:

15. Clean the gasket mating surfaces thoroughly. Always use new gaskets during installation.
16. Install the water pump. Torque the bolts, in sequence, to 9 ft. lbs. (12 Nm). After tightening the bolts once, retighten to the same specification again.
17. Attach the hose to the water pump.
18. Install the tensioner bracket and tighten the bolts to 18 ft. lbs. (25 Nm).
19. Install belt cover No.2 and tighten the bolts to 3 ft. lbs. (5 Nm).
20. Install the lower left hand camshaft sprocket using tool ST 499207400 and tighten to 72 ft. lbs. (98 Nm)
21. Install the upper left hand camshaft sprocket using tool ST 499977500 and tighten to 22 ft. lbs. (29 Nm) plus an additional 45 degrees.
22. Install the CMP.
23. Install belt idler No.2 and the belt idler.
24. Install the automatic belt tension adjuster and tighten to 29 ft. lbs. (39 Nm).
25. Install the timing belt.
26. Install the crankshaft pulley and tighten the bolt to 132 ft. lbs. (180 Nm).
27. Install the drive belts.
28. Install the radiator.
29. Fill the system with coolant.
30. Start the engine and allow it to reach operating temperature.
31. Check for leaks.

Heater Core

REMOVAL & INSTALLATION

2001–02 Models

1. Before servicing the vehicle, refer to the precautions in the beginning of this section.
2. Disconnect the negative battery cable.

**** CAUTION**

After disconnecting the negative battery cable, wait for at least 20 seconds for the air bag module to deplete its energy.

3. Drain the engine coolant into a clean container for reuse.
4. Disconnect the heater hoses from the heater core.

5. Remove the instrument panel by performing the following procedure:

a. If equipped with a manual transmission, remove the shift knob.

b. Remove both the front and rear console covers.

c. Remove the console box-to-chassis screws and the console box.

d. Remove the 3 lower left side cover assembly screws, disengage the 3 upper clips, and remove the cover assembly.

e. Using a screwdriver, disconnect the data link connector from the lower cover.

f. Remove the knee panel.

g. At the glove box, remove the right side cover screw, the clip and the side cover.

h. Remove the glove box screws and remove the glove box.

i. Remove the center panel bezel.

j. Remove the audio assembly screws, disconnect the electrical connectors and remove the audio assembly.

k. Remove the 2 steering column-to-instrument panel bolts and lower the steering column.

l. Move the temperature control switch to FULL HOT, the mode selector switch to DEF and the recirculation switch to FRESH positions.

m. Disconnect the temperature control cable and the mode control cable from the

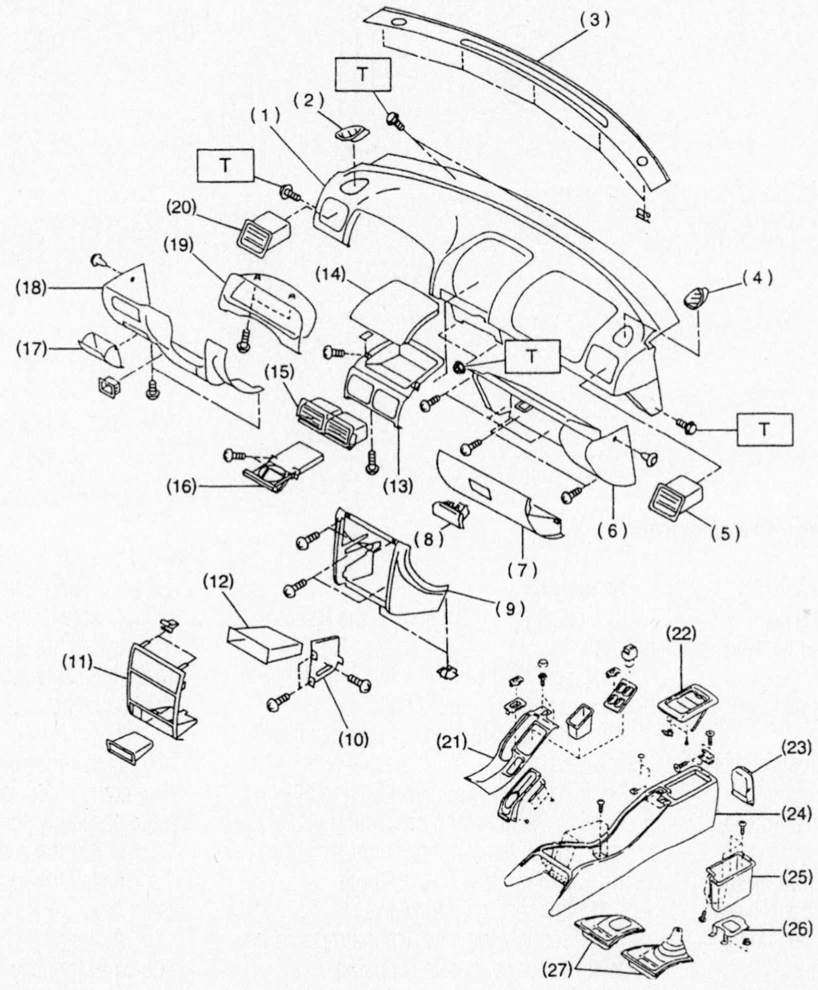

1 Pad & frame
2 Grille side (D)
3 Front def. grille
4 Grille side (P)
5 Grille vent (P)
6 Glove box panel
7 Glove box lid
8 Knob
9 Instrument panel center console
10 BRKT (Radio)
11 Center console cover
12 Pocket
13 Panel center
14 Center pocket lid
15 Grille center
16 Cup holder
17 Side pocket
18 Lower cover ASSY
19 Meter visor
20 Grille vent (D)
21 Console cover
22 Console lid
23 Rear cup holder
24 Console box
25 Console pocket
26 Rear console BRKT
27 Front cover

Tightening torque: N·m (kg-m, ft-lb)

T: 7±1 (0.7±0.1, 5.1±0.7)

93113G18

Exploded view of the instrument panel assembly

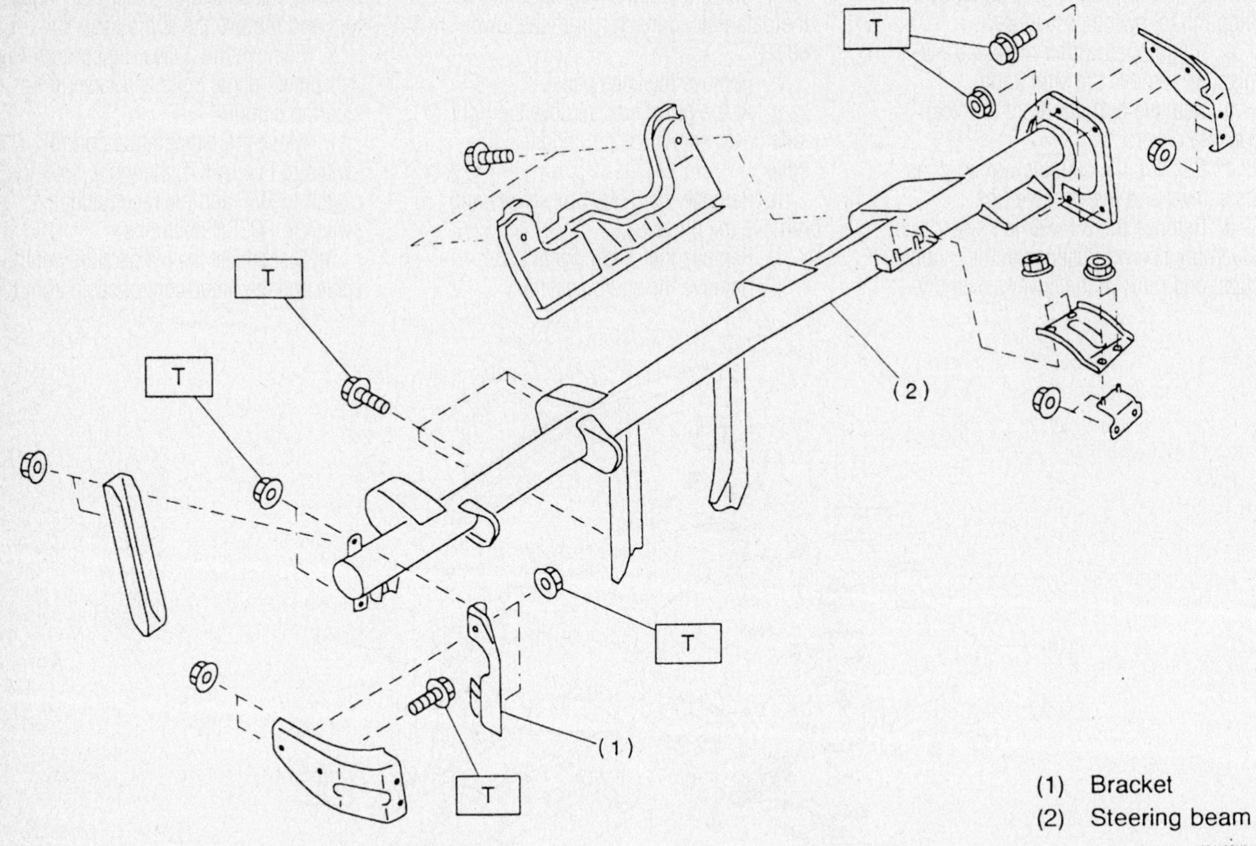

Exploded view of the steering support beam assembly

heater housing.; then, the recirculation control cable from the intake housing.

n. Disconnect the electrical harness connectors.

o. Remove the instrument panel-to-chassis bolts.

p. Remove the 2 front defroster grille bolts.

q. Carefully, remove the instrument panel.

6. Remove the steering support beam bracket nuts and the steering support beam.

7. Remove the evaporator housing by performing the following procedure:

a. Discharge and recover the air conditioning system refrigerant.

b. Remove the refrigerant line-to-cowl connector bolt, separate the lines, discard the O-rings and plug the openings to prevent contamination.

c. Disconnect the electrical harness connector from the evaporator housing.

d. Disconnect the drain hose.

e. Remove the evaporator housing nut/bolts and the evaporator housing.

8. Remove the heater housing-to-chassis bolts and the heater housing.

9. Remove the heater core from the heater housing.

To install:

10. Install the heater core to the heater housing.

11. Install the heater housing and the heater housing-to-chassis bolts.

12. Install the steering support beam and the steering support beam bracket nuts.

13. Install the evaporator housing by performing the following procedure:

a. Install the evaporator housing and the evaporator housing nut/bolts.

b. Connect the drain hose.

c. Connect the electrical harness connector to the evaporator housing.

d. Using new O-rings, assemble the refrigerant lines and install the refrigerant line-to-cowl connector bolt.

14. Install the instrument panel by performing the following procedure:

a. Carefully, install the instrument panel.

b. Install the 2 front defroster grille bolts.

c. Install the instrument panel-to-chassis bolts.

d. Connect the electrical harness connectors.

e. Connect the temperature control cable and the mode control cable to the heater housing. Then, the recirculation control cable to the intake housing.

f. Install the steering column and lower the 2 steering column-to-instrument panel bolts and torque to 14–21 ft. lbs. (20–30 Nm).

g. Install the audio assembly, connect the electrical connectors and install the audio assembly screws.

h. Install the center panel bezel.

i. Install the glove box and the glove box screws.

j. At the glove box, install the right side cover, the clip and the side cover screw.

k. Install the knee panel.

l. Connect the data link connector to the lower cover.

m. Install the lower left side cover assembly, engage the 3 upper clips and install the cover assembly screws.

n. Install the console box and the console box-to-chassis screws.

o. Install both the front and rear console covers.

p. If equipped with a manual transmission, install the shift knob.

15. Connect the heater hoses to the heater core.

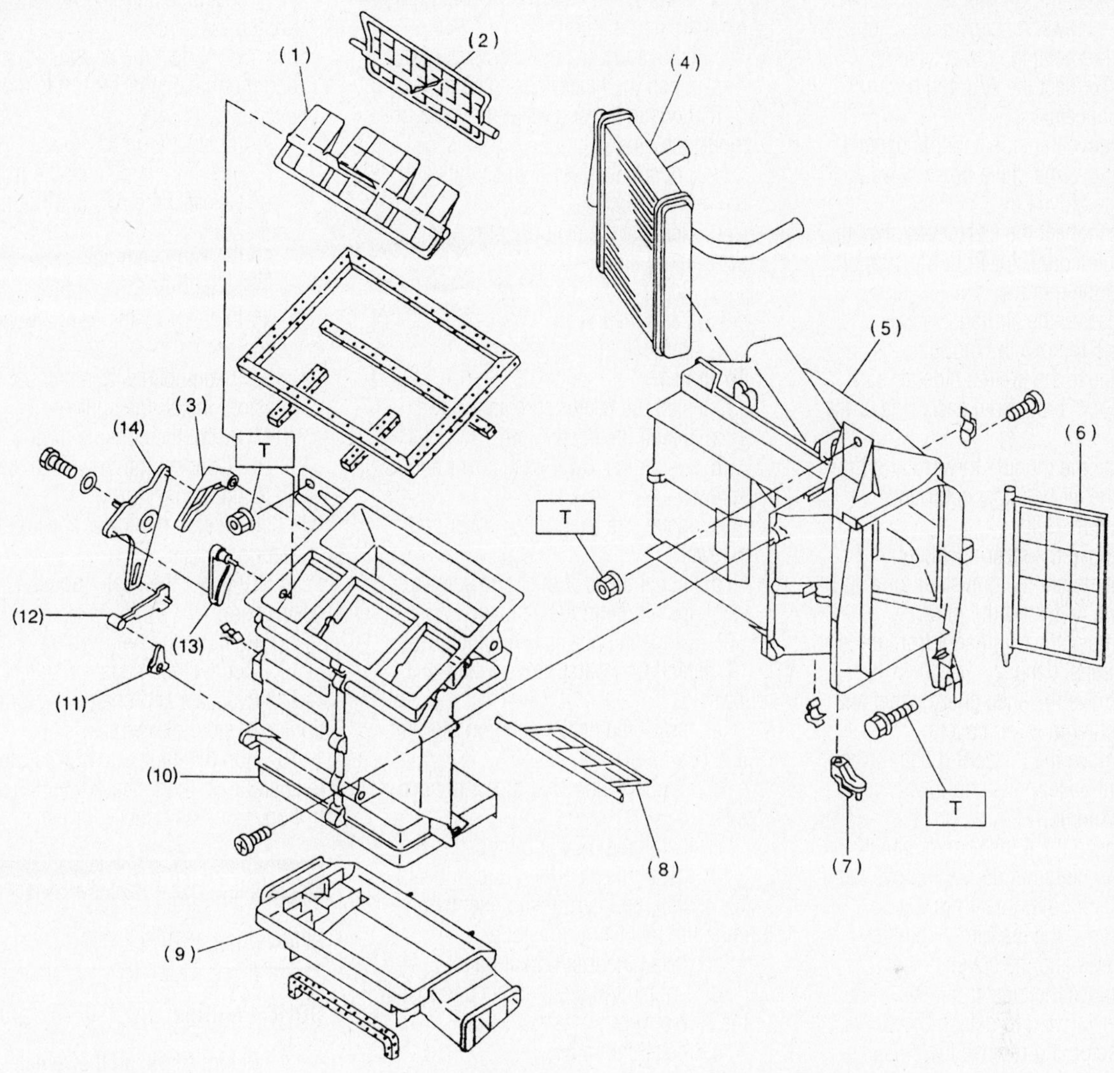

1 Vent door
2 DEF door
3 DEF lever
4 Heater core
5 Heater case FRONT
6 Mix door
7 Mix lever
8 Foot door
9 Foot duct
10 Heater case REAR
11 Foot lever lower
12 Foot lever upper
13 Vent lever
14 Side link

Tightening torque: N·m (kg-m, ft-lb)
T: 7.35±1.96
(0.750±0.200, 5.421±1.446)

93113GI7

Exploded view of the heater core, heater housing and related components

16. Refill the cooling system.

17. Connect the negative battery cable.

18. Evacuate and charge the air conditioning system refrigerant.

19. Run the engine to normal operating temperatures; then, check the climate control operation and check for leaks.

2003–04 Models

1. Before servicing the vehicle, refer to the precautions in the beginning of this section.

2. Disconnect the negative battery cable.

CAUTION

After disconnecting the negative battery cable, wait for at least 20 seconds for the air bag module to deplete its energy.

3. Drain the engine coolant into a clean container for reuse.

4. Disconnect the heater hoses from the heater core.

5. Remove the instrument panel by performing the following procedure:

a. Remove the lower cover from below the steering wheel.

b. Disconnect the in–vehicle sensor hose and connector.

c. Remove the console front cover.

d. If equipped with a manual transmission, remove the shift knob.

e. Remove both the front and rear console covers.

f. Remove the console box-to-chassis screws and the console box.

g. Remove the 3 lower left side cover assembly screws, disengage the 3 upper clips, and remove the cover assembly.
h. Disconnect the A/C and hazard switch connectors.
i. Remove the side console panel.
j. Remove the glove box screws and remove the glove box.
k. Disconnect the passenger side air bag module connector from the support beam and the module.
l. Unfasten the air bag module screws and remove the module.
m. Remove the drivers side air bag module Torx® bolts from each side of the module.
n. Slide the module forward and disconnect the air bag connector and remove the module.
o. Remove the steering wheel.
p. Matchmark the universal joint and remove the bolts and the joint.
q. Remove the trim panel from under the instrument panel.
r. Remove the knee guard panel and steering column lower covers.
s. Remove the 2 steering column-to-instrument panel bolts and lower the steering column.
t. Loosen the 4 instrument panel installation bolts but do not remove the lower bolts for alignment purposes.
u. Remove the instrument cluster cover screws and the cover.
v. Remove the cluster screws, slide the cluster forward to detach the electrical connector and remove the cluster.
w. Loosen the instrument panel screws.
x. Using a flat bladed tool, pry the instrument panel center compartment up to disengage the clips and remove the compartment.
y. Remove the center panel screws and clips at the radio, remove the panel.
z. Disconnect the center panel connector.
aa. Loosen the radio screws, slide the radio forward and detach the antenna and electrical connectors and remove the radio.
bb. Remove the side covers from the instrument panel and loosen the two bolts.
cc. Remove the instrument panel.
6. Remove the heater and cooling unit as follows:
a. Remove the steering support beam bracket nuts and the steering support beam.
b. Remove the blower motor.
c. Disconnect the servo motor connectors.
d. Remove the heater and cooling unit retainers and the unit.
7. Remove the heater core as follows:
a. Open the heater core pipe cover.
b. Loosen the screws and remove the mode actuator.
c. Loosen the foot duct screws and remove the duct.
d. Remove the evaporator cover screws and the cover.
e. Remove the lower case cover screws and the cover
f. Remove the heater core.

To install:

8. Install the heater core as follows:
a. Install the heater core.
b. Install the lower case cover and screws
c. Install the evaporator cover and screws.
d. Install the foot duct and screws.
e. Install the mode actuator.
f. Close the heater core pipe cover.
9. Install the heater and cooling unit as follows:
a. Install the heater and cooling unit and retainers.
b. Connect the servo motor connectors.
c. Install the blower motor.
d. Install the steering support beam.
10. Install the instrument panel by performing the following procedure:
a. Install the instrument panel.
b. Tighten the 2 side bolts and install the side covers.
c. Install the radio.
d. Connect the center panel connector.
e. Install the center panel.
f. Install the instrument panel center compartment.
g. Tighten the instrument panel screws.
h. Install the cluster, screws and the cover.
i. Tighten the 4 instrument panel installation bolts.
j. Raise the steering column and tighten the bolts to 18 ft. lbs. 925 Nm).
k. Install the knee guard panel and steering column lower covers.
l. Install the trim panel under the instrument panel.
m. Install the universal joint and tighten the bolts to 17 ft. lbs. (24 Nm).
n. Install the steering wheel.
o. Connect the driver's side air bag connector and install the module.
p. Install the drivers side air bag module Torx® bolts on each side of the module.
q. Install the air bag module and tighten the screws.
r. Connnect the passenger side air bag module connector to the module and the support beam.
s. Install the glove box.
t. Install the side console panel.
u. Connect the A/C and hazard switch connectors.
v. Install the left side cover assembly.
w. Install the console box.
x. Install both the front and rear console covers.
y. If equipped with a manual transmission, install the shift knob.
z. Install the console front cover.
aa. Connect the in–vehicle sensor hose and connector.
bb. Install the lower cover below the steering wheel.
11. Connect the heater hoses to the heater core.
12. Refill the cooling system.
13. Connect the negative battery cable.
14. Evacuate and charge the air conditioning system refrigerant.
15. Run the engine to normal operating temperatures; then, check the climate control operation and check for leaks.

Cylinder Head

REMOVAL & INSTALLATION

SOHC Engines

1. Before servicing the vehicle, refer to the precautions in the beginning of this section.
2. Properly relieve the fuel system pressure.
3. Remove or disconnect the following:
   - Negative battery cable
   - Oxygen (O_2S) sensor. If equipped with California emissions, disconnect the rear O_2S sensor.
   - Engine undercover
   - Exhaust Y-pipe and lower it just enough to clear the studs in the heads. Do not allow the Y-pipe to hang without support.
   - Accessory drive belts
   - Engine accessories and brackets from the side of the engine the cylinder head is being removed
   - Connector bracket attaching bolt, if necessary
4. On the left cylinder head, remove the CMP sensor.
5. Remove or disconnect the following:
   - Fuel pipes

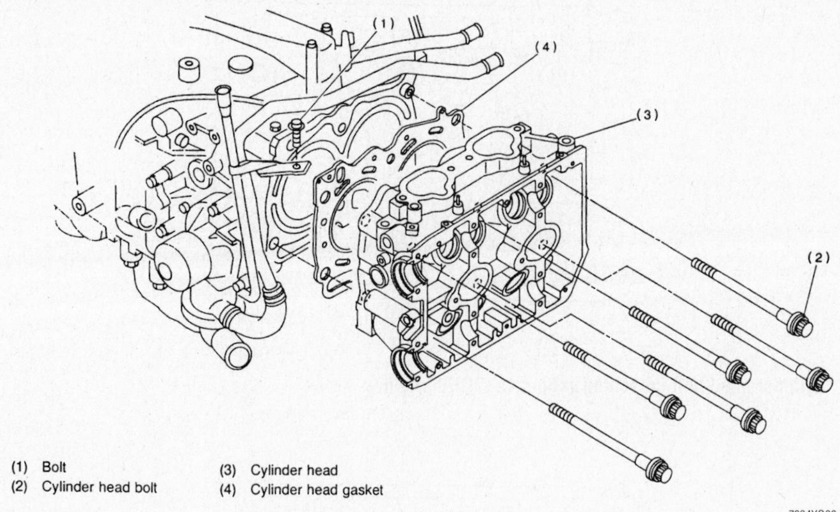

Exploded view of the cylinder head mounting–SOHC engine

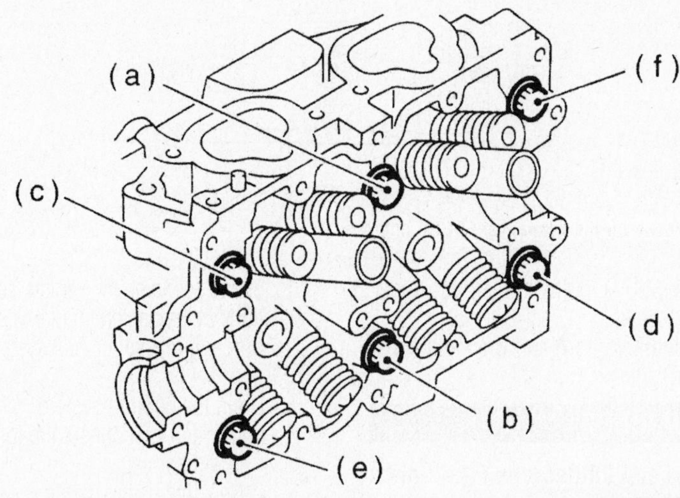

Cylinder head bolt loosening sequence–SOHC engine

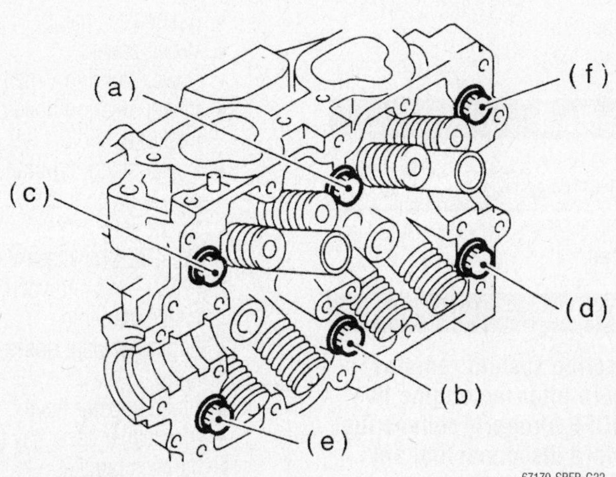

Cylinder head torque sequence—SOHC engine

- Intake manifold and gasket
- Timing belt, camshaft sprockets, and related components
- Valve covers, camshafts and related components
- Oil dipstick tube attaching bolt on the left cylinder head
- Bolt attaching the A/C compressor to the head
- Cylinder head bolts in the proper sequence. Leave bolts C and F installed loosely to prevent the cylinder head from falling.

6. Separate the cylinder head from the block. Use a plastic-faced hammer, if needed.
7. Remove bolts C and F. Remove the cylinder head and gasket.
8. Clean all gasket material from both mating surfaces.

To install:

9. Inspect the cylinder head for warpage. Warpage should not exceed 0.0020 in. (0.05mm).
10. Install the cylinder head(s) on the block using new gaskets. Secure in place with the mounting bolts. Coat each bolts with clean engine oil, and hand-tighten.
11. Tighten the cylinder head bolts as follows:
    a. Step 1: Torque all the bolts in sequence to 22 ft. lbs. (29 Nm).
    b. Step 2: Torque all the bolts in sequence to 51 ft. lbs. (69 Nm).
    c. Step 3: Loosen all bolts in sequence by 180 degrees, then loosen an additional 180 degrees.
    d. Step 4: Tighten all bolts in sequence: 40–45 degrees.

✱✱ CAUTION

Do not tighten the bolts more that 45 degrees in step 4.

    e. Step 5: Tighten bolts A and B in sequence: 40–45 degrees.

✱✱ WARNING

Do not exceed 90 degrees total tightening.

12. Install or connect the following:
    - Oil dipstick tube attaching bolt on the left cylinder head
    - Valve covers, camshafts and related components
    - Camshaft sprocket, timing belt, and related components
    - Intake manifold
    - Fuel delivery pipes
13. On the left cylinder head, install the CMP sensor.

14. Install or connect the following:
- Connector bracket attaching bolt
- Spark plug wires
- Engine accessories and brackets
- Accessory drive belts
- Exhaust Y-pipe. Torque the fasteners to 19–26 ft. lbs. (25–35 Nm).
- Engine undercover
- Front O_2S and rear O_2S, if removed.
- Negative battery cable

15. Start the engine and allow it to reach operating temperature.
16. Check for leaks.

DOHC Engines

1. Before servicing the vehicle, refer to the precautions in the beginning of this section.
2. Properly relieve the fuel system pressure.
3. Remove or disconnect the following:
4. Disconnect the negative battery cable.
5. Remove the drive belts.
6. Remove the crankshaft pulley.
7. Remove the timing belt cover, timing belt assembly and camshaft sprockets.
8. Remove the intake manifold.
9. Remove the bolt attaching the A/C compressor bracket to the head.
10. Remove the camshaft.
11. Remove the bolts in the sequence illustrated. Leave bolts **A** and **D** installed loosely to prevent the cylinder head from falling.
12. Separate the cylinder head from the block. Use a plastic-faced hammer, if needed.
13. Remove bolts A and D. Remove the cylinder head and gasket.
14. Clean all gasket material from both mating surfaces.

To install:

15. Inspect the cylinder head for warpage. Warpage should not exceed 0.0020 in. (0.05mm).
16. Install the cylinder head(s) on the block using new gaskets. Secure in place with the mounting bolts. Coat each bolts with clean engine oil, and hand-tighten.
17. Tighten the cylinder head bolts as follows:
    a. Step 1: Torque all the bolts in sequence to 22 ft. lbs. (29 Nm).
    b. Step 2: Torque all the bolts in sequence to 51 ft. lbs. (69 Nm).
    c. Step 3: Loosen all bolts in sequence by 180 degrees, then loosen an additional 180 degrees.
    d. Step 4: Tighten all bolts in sequence to 36 ft. lbs. (49 Nm).
    e. Step 5: Tighten all bolts in sequence an additional 80–90 degrees

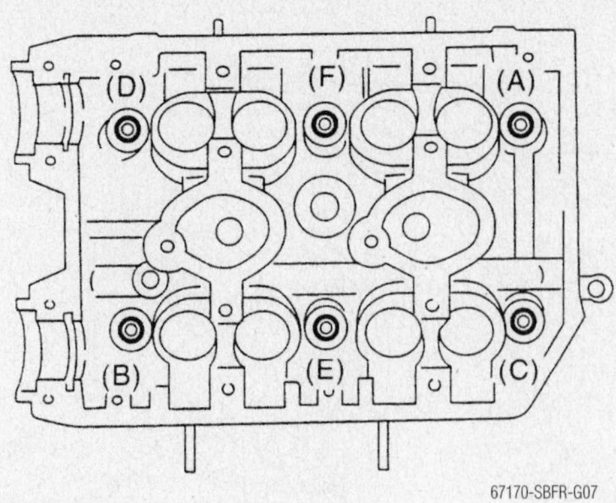

Cylinder head bolt loosening sequence–DOHC engines

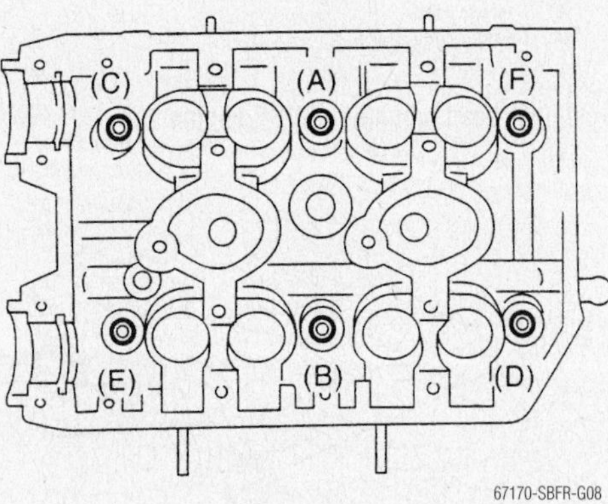

Cylinder head torque sequence—DOHC engines

    f. Step 6: All bolts: turn an additional 40–45 degrees.
    g. Step 7: Bolts **A** and **B** an additional 40–45 degrees.

⁂ WARNING

Do not exceed 90 degrees total tightening in the previous two steps.

18. Install the remaining components in the reverse order of removal.
19. Start the engine and allow it to reach operating temperature.
20. Check for leaks.

Intake Manifold

REMOVAL & INSTALLATION

SOHC Engines

⁂ CAUTION

The fuel injection system remains under pressure after the engine has been turnedOFF. Properly relieve fuel pressure before disconnecting any fuel lines. Failure to do so may result in fire or personal injury.

1. Before servicing the vehicle, refer to the precautions in the beginning of this section.
2. Properly relieve the fuel system pressure.
3. Drain the cooling system.
4. Remove or disconnect the following:
- Fuel filler cap
- Negative battery cable
- Air intake duct, air cleaner upper cover and the air cleaner element
- Accelerator cable and the cruise control cable, if equipped
- Resonator chamber
- V-belt covers
- Power steering V-belt
- Power steering hose brackets from intake manifold
- Power steering pump and set aside, do not disconnect hoses
- Spark plug wires
- Positive Crankcase Ventilation (PCV) hose and vacuum hose from intake manifold
- Engine coolant hoses from throttle body
- Brake booster hose
- Air cleaner case stay and engine harness bracket
- Engine Coolant Temperature (ECT) sensor connector

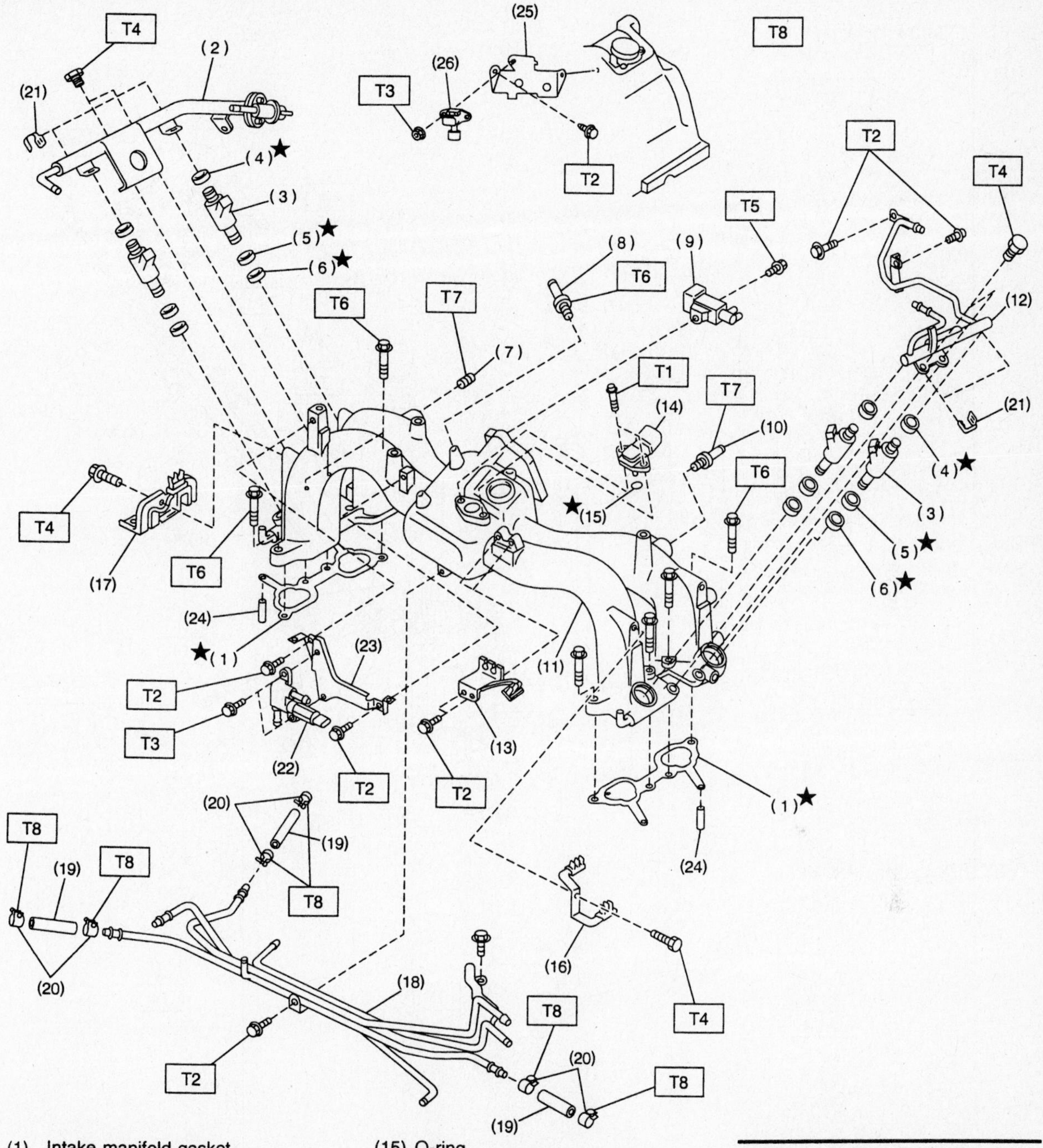

(1) Intake manifold gasket
(2) Fuel injector pipe RH
(3) Fuel injector
(4) O-ring
(5) O-ring
(6) O-ring
(7) Plug
(8) PCV valve
(9) Purge control solenoid valve
(10) Nipple
(11) Intake manifold
(12) Fuel injector pipe LH
(13) Accelerator cable bracket
(14) Intake air temperature and pressure sensor
(15) O-ring
(16) Plug cord holder LH
(17) Plug cord holder RH
(18) Fuel pipe ASSY
(19) Fuel hose
(20) Clip
(21) Clip
(22) Air assist injector solenoid valve
(23) Air assist injector solenoid valve bracket
(24) Guide pin
(25) Atmospheric pressure sensor bracket
(26) Atmospheric pressure sensor

Tightening torque: N·m (kgf-m, ft-lb)
T1: 3.4 (0.35, 2.5)
T2: 5.0 (0.51, 3.7)
T3: 6.4 (0.65, 4.7)
T4: 19 (1.9, 13.7)
T5: 16 (1.6, 12)
T6: 25 (2.6, 18.8)
T7: 17 (1.7, 12)
T8: 1.5 (0.15, 1.1)

Exploded view of the intake manifold—2001–03 SOHC models shown

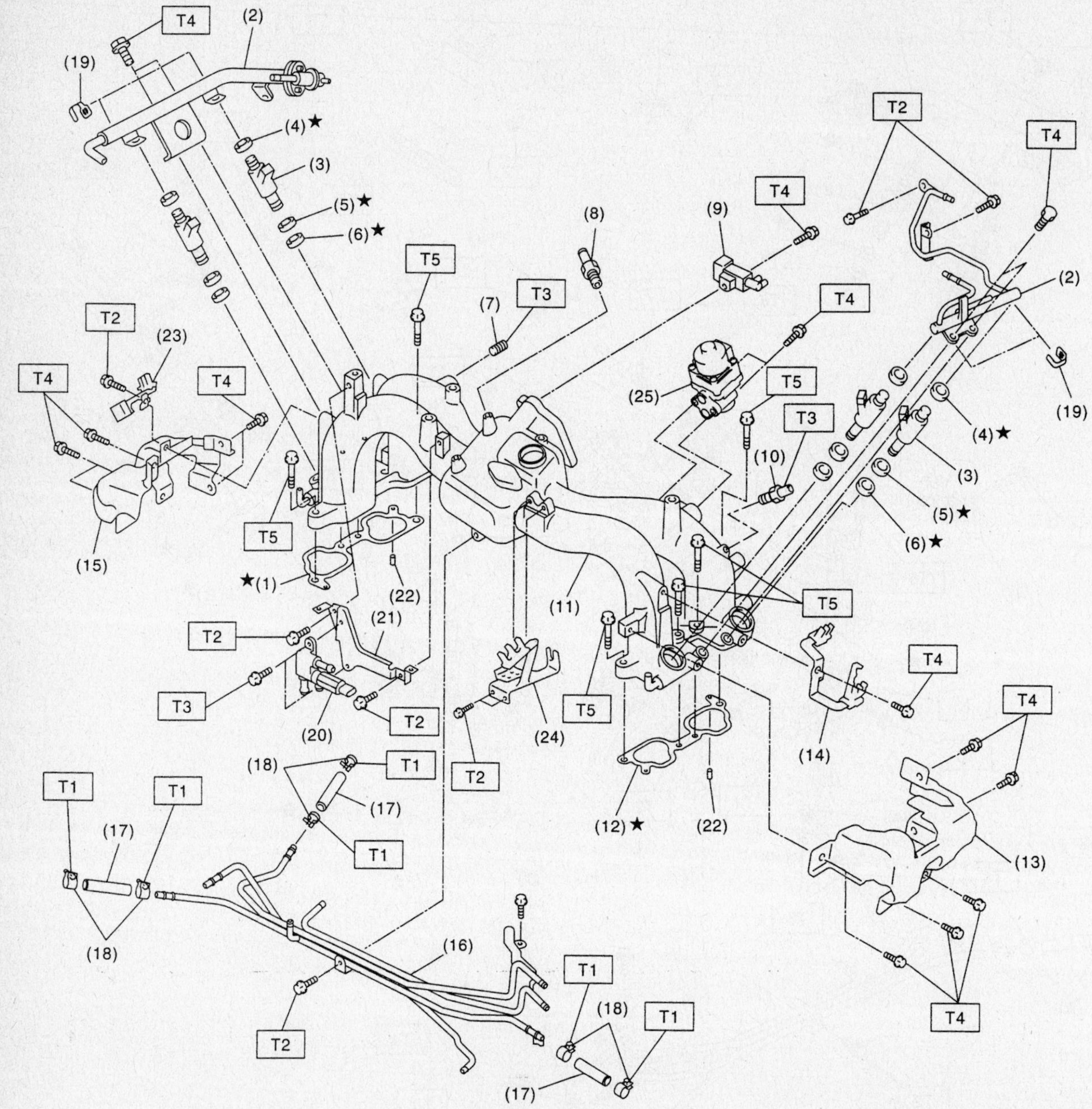

(1) Intake manifold gasket RH
(2) Fuel injector pipe
(3) Fuel injector
(4) O-ring
(5) O-ring
(6) O-ring
(7) Plug
(8) Nipple
(9) Purge control solenoid valve
(10) Nipple
(11) Intake manifold
(12) Intake manifold gasket LH
(13) Fuel pipe protector LH
(14) Plug cord holder LH
(15) Fuel pipe protector RH
(16) Fuel pipe ASSY
(17) Fuel hose
(18) Clamp
(19) Clip
(20) Air assist injector solenoid valve
(21) Air assist injector solenoid valve bracket
(22) Guide pin
(23) Plug cord holder RH
(24) Accelerator cable bracket
(25) EGR valve

Tightening torque: N·m (kgf-m, ft-lb)

T1: 1.5 (0.15, 1.1)
T2: 5.0 (0.51, 3.7)
T3: 17 (1.7, 12.5)
T4: 19 (1.9, 13.7)
T5: 25 (2.5, 18.1)

67170-SBFR-G33

Exploded view of the intake manifold—2004 SOHC models shown

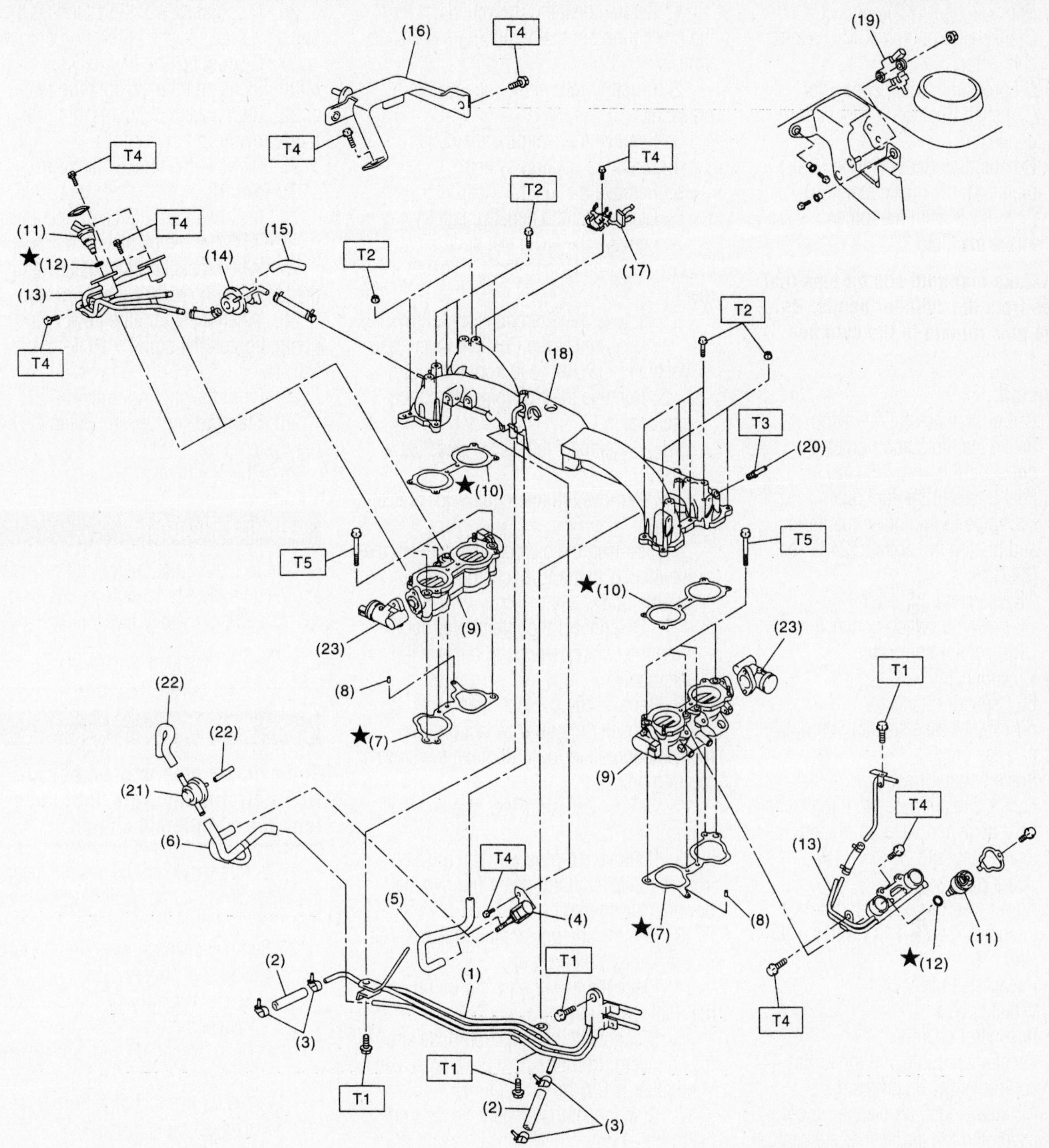

(1)	Fuel pipe ASSY	(13)	Fuel injector pipe
(2)	Fuel hose	(14)	Pressure regulator
(3)	Clamp	(15)	Pressure regulator hose
(4)	Purge control solenoid valve	(16)	Fuel pipe protector RH
(5)	Vacuum hose	(17)	Blow-by hose stay
(6)	Vacuum control hose	(18)	Intake manifold
(7)	Intake manifold gasket	(19)	Wastegate control solenoid valve ASSY
(8)	Guide pin		
(9)	Tumble generator valve ASSY	(20)	Nipple
(10)	Tumble generator valve gasket	(21)	Purge valve
(11)	Fuel injector	(22)	Purge hose
(12)	O-ring	(23)	Tumble generator valve actuator

Tightening torque: N·m (kgf-m, ft-lb)

T1: 5 (0.5, 3.7)
T2: 8.25 (0.84, 6.1)
T3: 17 (1.73, 12.5)
T4: 19 (1.94, 13.7)
T5: 25 (2.5, 18.1)

67170-SBFR-G09

Exploded view of the intake manifold—DOHC models

- Knock Sensor (KS) connector
- Crankshaft Position (CMP) sensor connector
- Oil pressure switch connector
- Camshaft Position (CKP) sensor connector
- Exhaust Gas Recirculation (EGR) pipe from the intake manifold
- Fuel hoses from fuel pipes
- Intake manifold

➡The intake manifold sits on pins that protrude from the cylinder heads. Be sure the pins remain in the cylinder heads.

To install:

5. Install or connect the following:
   - Intake manifold and tighten the bolts to 18 ft. lbs. (25 Nm)
   - Fuel hoses to the fuel pipes
   - EGR pipe to the intake manifold and tighten the bolts to 24 ft. lbs. (34 Nm)
   - CKP sensor connector
   - Oil pressure switch connector
   - CMP sensor connector
   - KS connector
   - ECT sensor connector
   - Air cleaner case stay and engine harness bracket
   - Brake booster hose
   - Engine coolant hoses to throttle body
   - PCV hose and vacuum hose to the intake manifold
   - Spark plug wires
   - Power steering pump and tighten the bracket bolts to 14 ft. lbs. (20 Nm)
   - Power steering V-belt
   - V-belt covers
   - Resonator chamber
   - Accelerator cable and the cruise control cable, if equipped
   - Air intake duct, air cleaner upper cover and the air cleaner element
   - Negative battery cable
   - Fuel filler cap
6. Refill the cooling system.
7. Start the engine, and bleed the cooling system.
8. Check for leaks.

DOHC Engines

⁂ CAUTION

The fuel injection system remains under pressure after the engine has been turnedOFF. Properly relieve fuel pressure before disconnecting any fuel lines. Failure to do so may result in fire or personal injury.

1. Before servicing the vehicle, refer to the precautions in the beginning of this section.
2. Properly relieve the fuel system pressure.
3. Remove the engine undercover.
4. Drain the cooling system.
5. Remove the fuel filler cap.
6. Disconnect the negative battery cable.
7. Remove the air intake duct, air cleaner upper cover and the air cleaner element.
8. Remove the intercooler as follows:
   a. Disconnect the Positive Crankcase Ventilation (PCV) hose from the pipe.
   b. Remove the PCV pipe from the intercooler.
   c. Disconnect the air by–pass valve hose.
   d. Remove the intercooler–to–bracket bolts.
   e. Loosen the clamps that connect the intercooler to the turbocharger and throttle body.
   f. Disconnect the intercooler duct from the turbocharger and remove the intercooler.
9. Remove the coolant filler tank.
10. Loosen the power steering pump lock bolt and slider bolt. Remove the power steering belt.
11. Disconnect the power steering switch connector.
12. Remove the pipe with the bracket from the intake manifold and remove the power steering pump.
13. Remove the power steering tank from its bracket by pulling it up.
14. Place the power steering pump on the right wheel apron.
15. Disconnect the emission hose and the connector from the Positive Crankcase Ventilation (PCV) hose assembly.
16. Disconnect the coolant hoses from the throttle body.
17. Disconnect the brake booster hose.
18. Disconnect the pressure hose from the intake duct.
19. Disconnect the following electrical connectors:
   - Engine harness connectors from the bulkhead connector
   - Engine Coolant Temperature (ECT) sensor connector
   - Crankshaft Position (CMP) sensor connector
   - Oil pressure switch connector
   - Knock Sensor (KS) connector
   - Camshaft Position (CKP) sensor connector
   - Ignition coil connectors on both sides
20. Disconnect the clip attaching the engine harness to the bracket on both sides.
21. Disconnect the fuel delivery and return hoses and the evaporation hose.
22. Remove the bolts attaching the intake manifold to the heads.
23. Remove the intake manifold.

To install:

24. Installation of the components is the reverse order of removal.
25. When installing the intake manifold, tighten the bolts to 18 ft. lbs. (25 Nm)
26. When installing the power steering pump, tighten the bolts to 16 ft. lbs. (22 Nm)
27. Refill the cooling system.
28. Start the engine, and bleed the cooling system.
29. Check for leaks.

Exhaust Manifold

Due to the unique design of the Subaru engine an exhaust manifold is not used. The exhaust enters directly into the front Y-pipe.

REMOVAL & INSTALLATION

⁂ CAUTION

The exhaust pipe may be hot; DO NOT perform any work until the system has completely cooled.

1. Before servicing the vehicle, refer to the precautions in the beginning of this section.
2. Remove or disconnect the following:
   - Negative battery cable
   - Front Oxygen (O_2) sensor electrical connectors
   - Rear Oxygen (O_2) sensor electrical connectors
   - Center exhaust pipe from front exhaust pipe
   - Nuts that secure the exhaust pipe to the cylinder head
   - Front pipe-to-front catalytic converter mounting nuts
3. Discard the gaskets.

To install:

4. Clean all gasket surfaces completely.
5. Install or connect the following:
   - Catalytic converter to front exhaust pipe using new gasket. Torque the bolts to 22 ft. lbs. (30 Nm).
   - Exhaust pipe to the cylinder head using new gaskets. Torque the mounting nuts to 22 ft. lbs. (30 Nm).
   - Exhaust pipe to the center pipe using new gaskets. Torque the

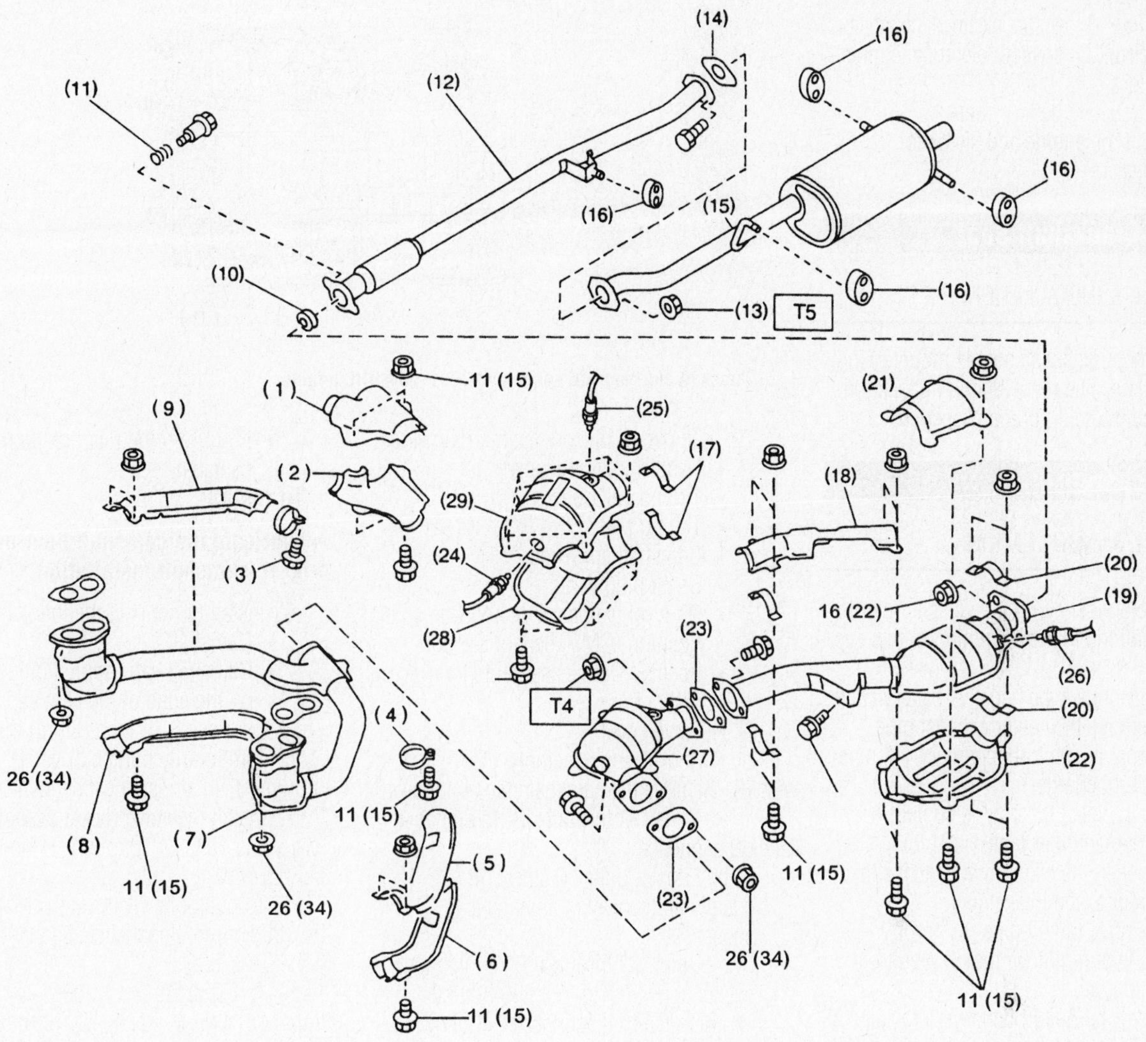

ft. lbs. (Nm)

(1) Upper front exhaust pipe cover CTR
(2) Lower front exhaust pipe cover CTR
(3) Band RH
(4) Band LH
(5) Upper front exhaust pipe cover LH
(6) Lower front exhaust pipe cover LH
(7) Front exhaust pipe
(8) Lower front exhaust pipe cover RH
(9) Upper front exhaust pipe cover RH
(10) Gasket
(11) Spring
(12) Rear exhaust pipe
(13) Self-locking nut
(14) Gasket
(15) Muffler
(16) Cushion rubber
(17) Clamp
(18) Upper center exhaust pipe cover
(19) Center exhaust pipe
(20) Clamp B
(21) Upper rear catalytic converter cover
(22) Lower rear catalytic converter cover
(23) Gasket
(24) Front oxygen sensor
(25) Rear oxygen sensor (California spec. vehicles)
(26) Rear oxygen sensor (Except California spec. vehicles)
(27) Front catalytic converter
(28) Lower front catalytic converter cover
(29) Upper front catalytic converter cover

7924XG07

Exploded view of the exhaust system and related components

mounting nuts to 26 ft. lbs. (35 Nm).
- Rear O_2 sensors electrical connectors
- Front O_2 sensors electrical connectors
- Negative battery cable

6. Start the engine and check for exhaust leaks.

Front Crankshaft Seal

REMOVAL & INSTALLATION

The front crankshaft seal is mounted in the oil pump. The removal and installation is covered in the oil pump procedure.

Turbocharger

REMOVAL & INSTALLATION

1. Before servicing the vehicle, refer to the precautions in the beginning of this section.
2. Disconnect the negative battery cable.
3. Remove the center exhaust pipe.
4. Separate the turbocharger joint pipe from the turbocharger.
5. Disconnect the engine coolant hose from the coolant filler tank.
6. Loosen the clamp that attaches the turbocharger to the intake duct.
7. Remove the oil pipe–to–turbocharger bolt and remove the oil pipe from the turbocharger.
8. Remove the turbocharger bracket and disconnect the oil outlet hose from the pipe.
9. Remove the turbocharger.

To install:

10. Installation is the reverse of removal, please note the following torque specs:
    a. Turbocharger–to–intake duct: 2 ft. lbs. (2 Nm).
    b. Oil pipe–to–turbocharger: 11 ft. lbs. (16 Nm).
    c. Joint pipe–to–turbocharger using a new gasket: 22 ft. lbs. (30 Nm).
    d. Turbocharger bracket: 24 ft. lbs. (33 Nm).

Camshaft and Valve Lifters

REMOVAL & INSTALLATION

SOHC Engine

2001–03 MODELS

1. Before servicing the vehicle, refer to the precautions in the beginning of this section.

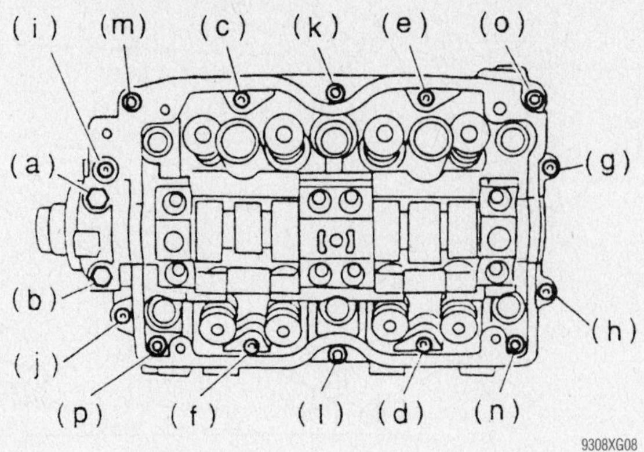

Camshaft cap removal sequence–2001–03 SOHC engines

2. Remove or disconnect the following:
   - Negative battery cable
   - Timing belt covers
   - Timing belt
   - Camshaft sprockets
   - Spark plug wires
   - Oil level gauge guide and Camshaft Position (CMP) sensor support
   - Positive Crankcase Ventilation (PCV) hose
   - Valve cover
   - Rocker arm assembly
3. Remove the camshaft cap as follows:
   a. Bolts "A" through "B" in alphabetical sequence
   b. Loosen bolts "C" through "J" equally all the way in alphabetical sequence
   c. Bolts "K" through "P" in alphabetical sequence using tool 499497000
4. Remove or disconnect the following:
   - Camshaft
   - Oil seal, if necessary
   - Plug from the rear side of the camshaft

To install:

➡Lubricate the camshaft bearings prior to camshaft installation.

5. Install or connect the following:
   - Camshaft
   - Camshaft cap. Apply liquid gasket on the edge of the cam cap mating surface 0.12 inch (3mm) thick
6. Temporarily tighten bolts "G" through "J" in alphabetical sequence
7. Install the valve rocker assembly. Torque the bolts "A" through "H" to 18 ft. lbs. (25 Nm).
8. Torque bolts "I" through "N" to 13 ft. lbs. (25 Nm) in alphabetical sequence using tool 499497000
9. Torque bolts "O" through "X" to 7.2 ft. lbs. (10 Nm) in alphabetical sequence
10. Install or connect the following:
    - Oil seal to the camshaft using tool

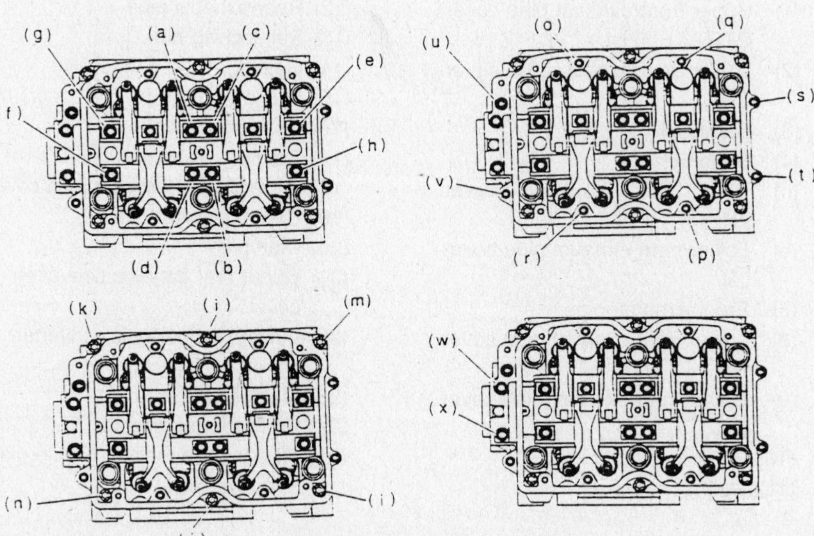

Camshaft cap tightening sequence–2001–03 SOHC engines

499597000 oil seal guide and 49958700 oil seal installer

- Plug to the rear side of the camshaft using tool 499587700 oil seal installer
- Valve cover
- Oil level gauge guide and CMP sensor support
- PCV house
- Spark plug wires
- Camshaft sprockets. Torque the bolts to 58 ft. lbs. (78 Nm).
- Timing belt
- Timing belt covers. Torque the bolts to 3.6 ft. lbs. (5 Nm).
- Negative battery cable

2004 MODELS

1. Before servicing the vehicle, refer to the precautions in the beginning of this section.
2. Disconnect the negative battery cable.
3. Remove the drive belts.
4. Remove the crankshaft pulley.
5. Remove the timing belt covers.
6. Remove the timing belt.
7. Remove the camshaft sprocket.
8. Remove the crankshaft sprocket
9. Remove timing belt cover No.2 from the left and right hand sides.
10. Remove the tensioner bracket.
11. Remove the Camshaft Position (CMP) sensor support from the left hand side.
12. Remove the oil level gauge guide from the left hand side.
13. Remove the spark plug wires.
14. Remove the valve cover and gasket.
15. Remove the valve rocker assembly.

➡Before removing the camshaft cap bolts mark and note their location so they may be reinstalled in their original positions.

16. Loosen the camshaft cap bolts using several passes in the sequences illustrated in the following order:
    a. Remove bolts A and B.
    b. Loosen bolts C through J in sequence.
    c. Using a suitable size torque bit, remove bolts K through P in sequence.
17. Remove the camshaft caps, camshaft and oil seal.
18. Remove the plug from the rear right hand side of the camshaft.

⁂ CAUTION

Do not remove the oil seal unless needed. Be careful not to scratch or damage the journal mating surfaces.

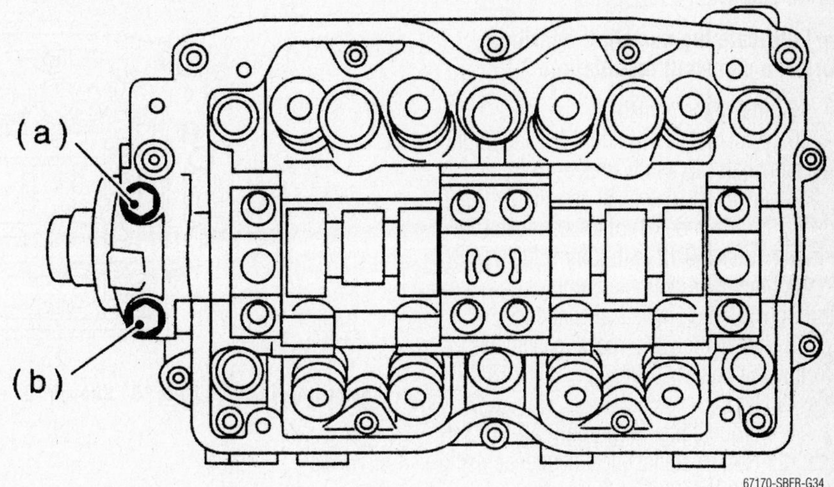

Remove camshaft cap bolts A and B using several passes–2004 SOHC engines

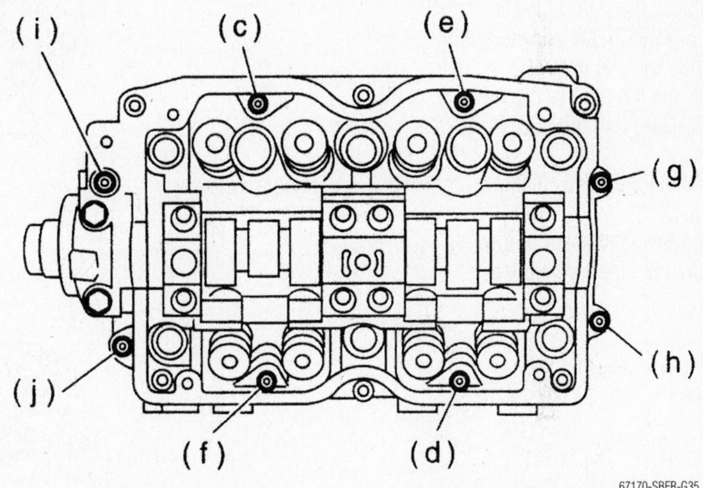

Remove camshaft cap bolts C through J using several passes–2004 SOHC engines

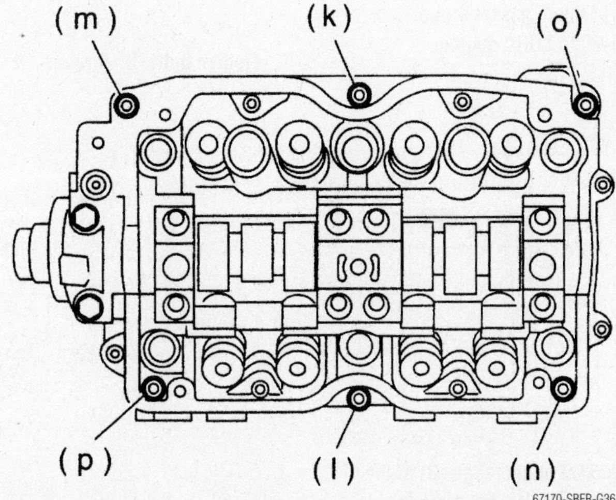

Remove camshaft cap bolts K through P using several passes–2004 SOHC engines

To install:

➡Lubricate the camshaft bearings prior to camshaft installation.

19. Install the camshaft.
20. Apply a bead of liquid gasket around the camshaft cap as illustrated.
21. Install the camshaft cap. Apply liquid gasket on the edge (B) of the cam cap (C) mating surface 0.12 inch (3mm) thick (A).
22. Temporarily tighten bolts "G" through "J" in alphabetical sequence.
23. Install the valve rocker assembly. Torque the bolts "A" through "H" to 18 ft. lbs. (25 Nm).
24. Torque bolts "I" through "N" to 13 ft. lbs. (25 Nm) in alphabetical sequence using Torx tool 499497000.
25. Torque bolts "O" through "V" to 7.2 ft. lbs. (10 Nm) in alphabetical sequence.
26. Torque bolts "W" through "X" to 7.2 ft. lbs. (10 Nm) in alphabetical sequence.
27. Install the oil seal to the camshaft using tool 499597000 oil seal guide and 499587500 oil seal installer.
28. Install the plug to the rear side of the camshaft using tool 499587700 oil seal installer.
29. Adjust the valve clearance.
30. Install the valve cover.
31. Connect the spark plug wires.
32. Connect the PVC house.

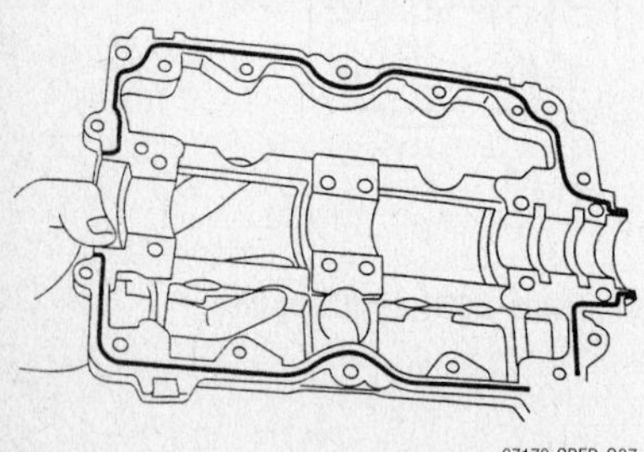

Apply a bead of liquid gasket around the camshaft cap–2004 SOHC engines

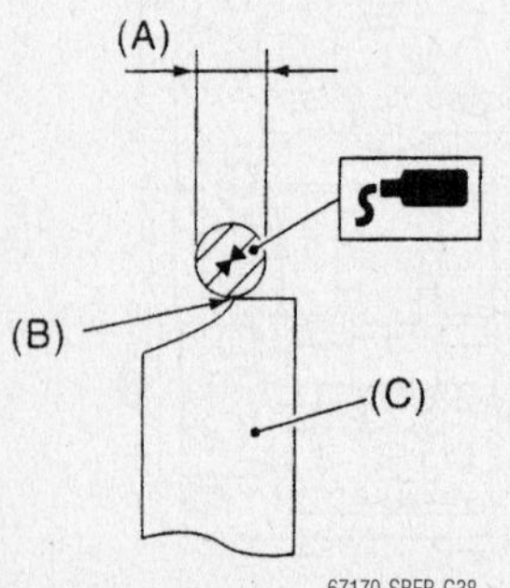

Apply liquid gasket on the edge (B) of the cam cap (C) mating surface 0.12 inch (3mm) thick (A)–2004 SOHC engines

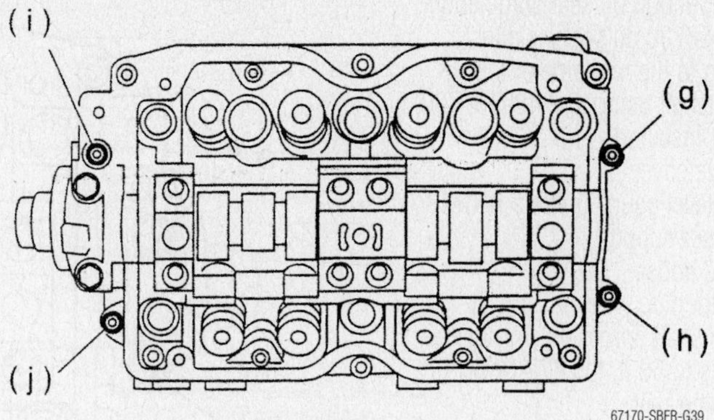

Temporarily tighten bolts "G" through "J" in sequence–2004 SOHC engines

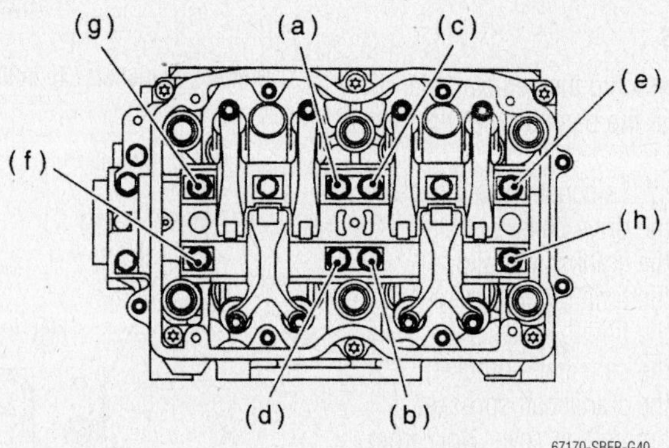

Tighten bolts "A" through "H" in sequence–2004 SOHC engines

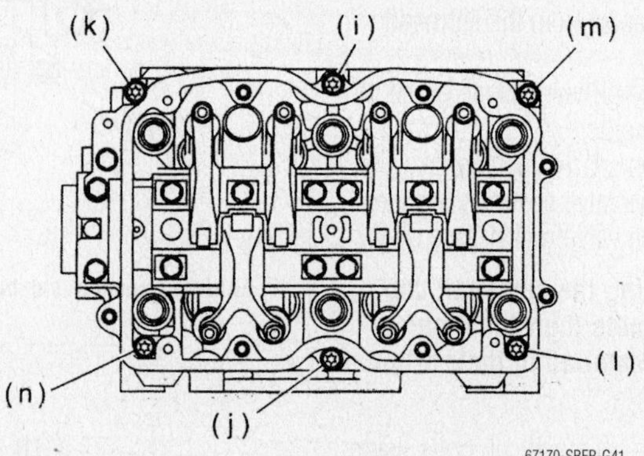

Tighten bolts "I" through "N" in sequence–2004 SOHC engines

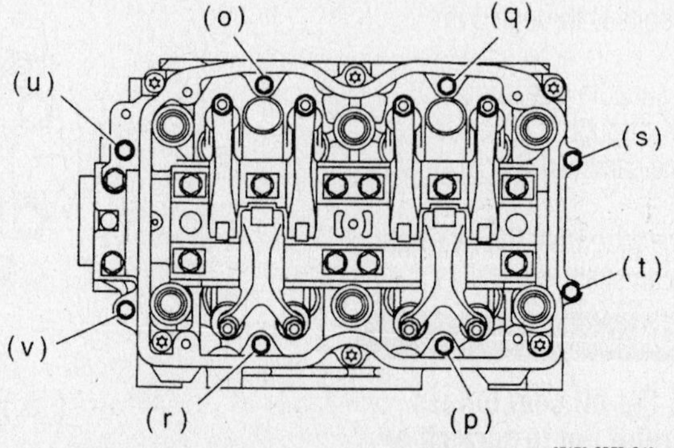

Tighten bolts "O" through "V" in sequence–2004 SOHC engines

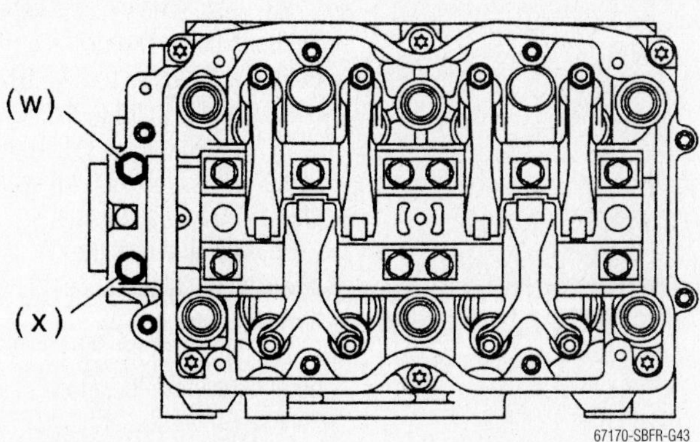

Tighten bolts "W" through "X" in sequence–2004 SOHC engines

33. Install the oil level gauge guide and CMP sensor support
34. Install the tensioner bracket and tighten to 18 ft. lbs. (25 Nm).
35. Install the right and left hand timing belt cover No.2 and tighten the bolts to 3 ft. lbs. (5 Nm).
36. Install the crankshaft and camshaft sprockets.
37. Install the timing belt and cover.
38. Install the crankshaft pulley and tighten the bolt to 130 ft. lbs. (177 Nm).
39. Install the drive belts.
40. Connect the negative battery cable.

DOHC Engines

1. Before servicing the vehicle, refer to the precautions in the beginning of this section.
2. Disconnect the negative battery cable.
3. Remove the drive belts.
4. Remove the crankshaft pulley.
5. Remove the timing belt covers.
6. Remove the timing belt.
7. Remove the camshaft sprockets.
8. Remove the crankshaft sprocket
9. Disconnect the oil flow control solenoid valve connector.
10. Remove the tensioner bracket.
11. Remove timing belt cover No.2 from the left hand side and the timing belt cover from the right hand side.
12. Remove the spark plug wires.
13. Remove the oil level gauge guide from the left hand side.
14. Remove the valve cover and gasket.
15. Remove the oil pipe.

➡Before removing the camshaft cap bolts mark and note their location so they may be reinstalled in their original positions.

16. Loosen the oil flow control solenoid valve assembly and intake camshaft cap bolts using several passes in the sequence illustrated.
17. Remove the oil flow control solenoid valve assembly the camshaft caps and the camshaft.
18. Loosen the exhaust camshaft cap bolts using several passes in the sequence illustrated. Remove the caps and camshaft.

To install:

19. Lubricate the camshaft bearings prior to camshaft installation. Install the camshaft so that each valve is close to or in contact with the base circle of the cam lobe.

➡When the camshafts are positioned as shown in the accompanying illustration, the camshafts need to be rotated at a minimum to align with the timing belt during installation. The right hand camshaft will not need to be rotated when set at the position show in the accompanying illustration. The left hand intake camshaft should be rotated 80 degrees clockwise and the left hand exhaust camshaft should be rotated 45 degrees counterclockwise.

20. Install the intake camshaft caps and oil flow control solenoid valve assembly as follows:
    a. Apply liquid gasket sparingly to the cap mating surfaces as illustrated. Do not use an excessive amount of gasket maker as this can cause excess packing to come out and block the oil seal which will cause an oil leak.
21. Apply clean engine oil to the cap bearing surface and install the cap on the camshaft indicated by the letter **A** in the accompanying illustration.
22. Tighten the camshaft cap and oil control valve assembly bolts in two to three passes. Apply clean engine oil to the cap bearing surface and install the cap on the camshaft indicated by the letter **A** n the sequence illustrated. Tighten bolts C, D E and F to 7 ft. lbs. (10 Nm) and bolts A and B to 14 ft. lbs. (20 Nm).

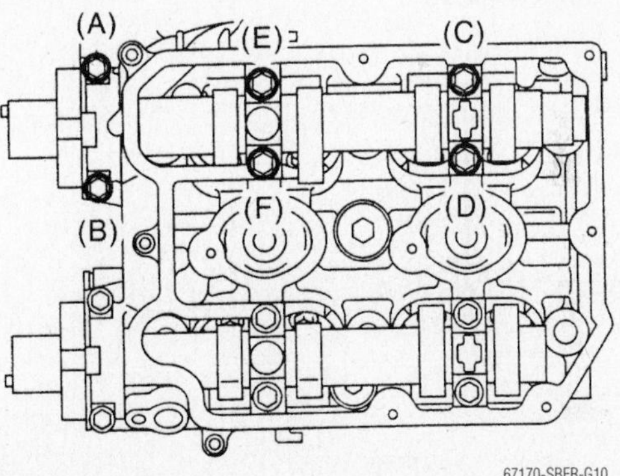

Loosen the oil flow control solenoid valve assembly and intake camshaft cap bolts using several passes–DOHC engines

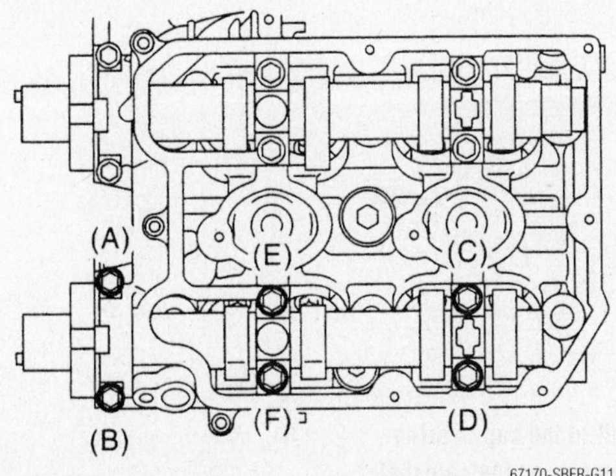

Loosen exhaust camshaft cap bolts using several passes–DOHC engines

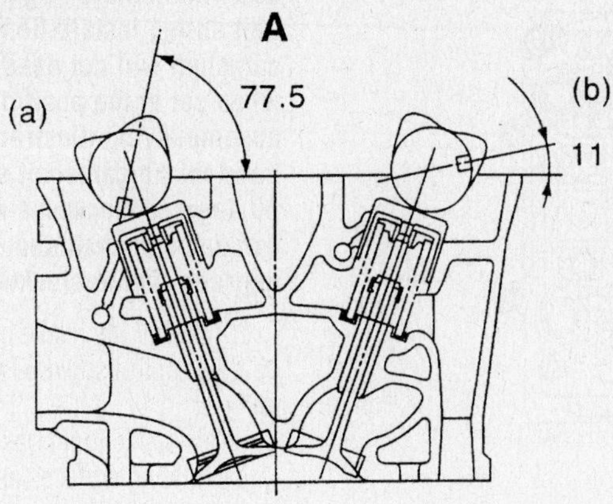

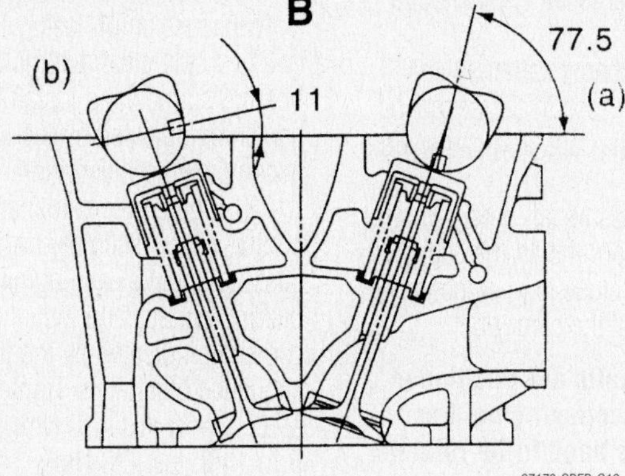

Camshaft alignment–DOHC engines

23. Install the exhaust caps and tighten the bolts using two or three passes. Tighten bolts C, D E and F to 7 ft. lbs. (10 Nm) and bolts A and B to 14 ft. lbs. (20 Nm).

24. Install the oil seal to the camshaft using tool 499587600 oil seal guide and 499597200 oil seal installer

25. Install a new gasket on the valve cover, also install the peripheral and ignition coil gaskets.

26. Apply liquid gasket maker to the points shown in the accompanying illustration.

27. Install the valve cover and make sure the gasket are positioned correctly. Tighten the valve cover bolts in the sequence illustrated to 5 ft. lbs. (7 Nm).

28. Install the oil pipe and tighten to 21 ft. lbs. (29 Nm).

29. Connect the oil flow control solenoid valve connector.

30. Connect the spark plug wires.

31. Install the right and left hand timing belt cover No.2 and tighten the bolts to 3 ft. lbs. (5 Nm).

32. Install the tensioner bracket and tighten to 18 ft. lbs. (25 Nm).

33. Install the crankshaft and camshaft sprockets.

34. Install the timing belt and cover.

35. Install the crankshaft pulley and tighten the bolt to 132 ft. lbs. (180 Nm).

67170-SBFR-G13

Apply liquid gasket sparingly to the cap mating surfaces–DOHC engines

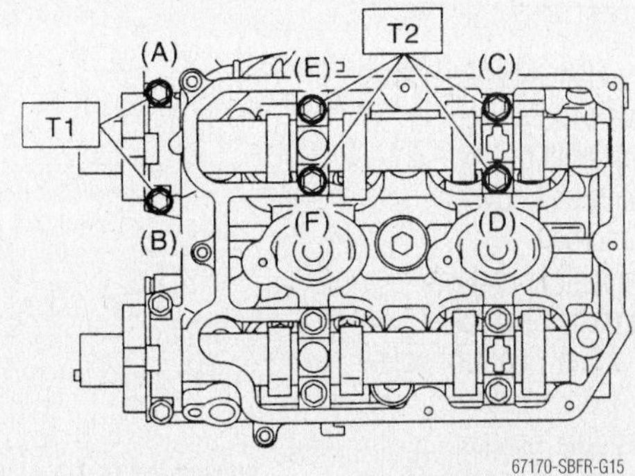

Intake camshaft and oil control valve bearing cap torque sequence–DOHC engines

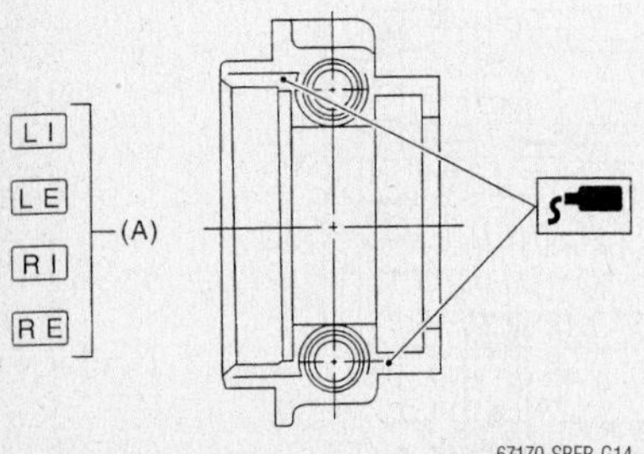

Apply clean engine oil to the cap bearing surface and install the cap on the camshaft indicated by the letter A–DOHC engines

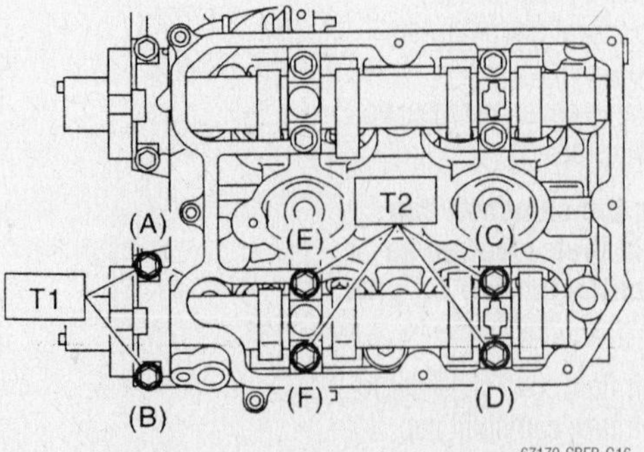

Exhaust camshaft bearing cap torque sequence–DOHC engines

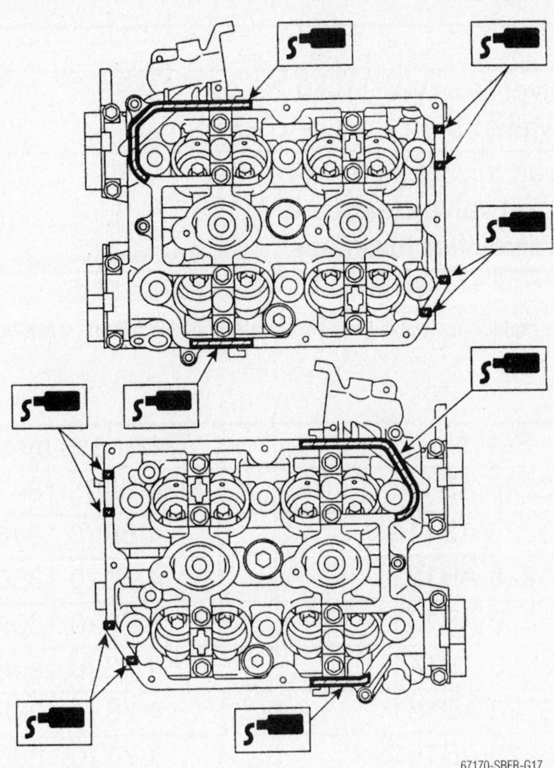

Apply liquid gasket maker to the areas shown before installing the valve cover–DOHC engines

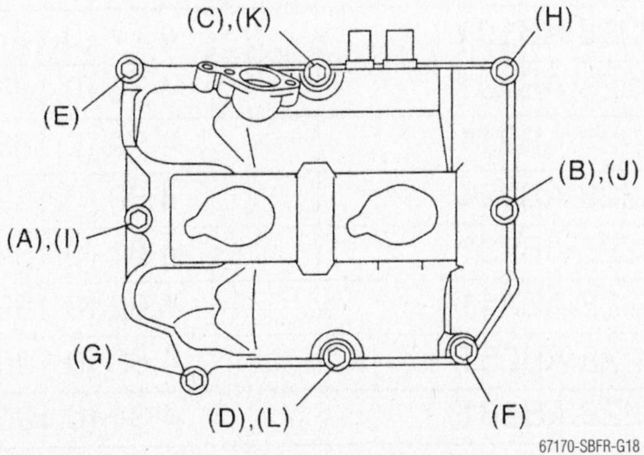

Tighten the valve cover bolts in this sequence–DOHC engines

36. Install the drive belts.
37. Connect the negative battery cable.

Valve Lash

ADJUSTMENT

SOHC Engines

➡The valve adjustment should be performed while the engine is cold. A Shim Replace Kit 498187100 will be needed to perform the valve adjustment.

1. Before servicing the vehicle, refer to the precautions in the beginning of this section.
2. Adjustment should be performed when engine is cold.
3. Remove or disconnect the following:
   - Negative battery cable
   - Engine coolant reservoir tank
   - Timing belt cover on the left hand side
4. When inspecting Nos. 1 and 3 cylinders remove the following:
   - Air intake duct as a unit
   - Resonator chamber
   - Spark plug wires from Nos. 1 and 3 cylinders
   - Blow-by house from valve cover
   - Engine undercover
   - Timing belt cover on the right hand side
   - Valve cover on the right hand side
5. When inspecting Nos. 2 and 4 cylinders remove the following:
   - Battery and battery tray
   - Window washer motor connectors front and rear
   - Rear gate glass washer hose from the washer motor
   - Washer tank mounting bolts and secure out of the way
   - Spark plug wires from Nos. 2 and 4 cylinders
   - Blow by hose from valve cover
   - Timing belt cover on the right hand side
   - Valve cover on the left hand side
6. Set No. 1 cylinder to Top Dead Center (TDC).

➡When arrow mark on the left hand side comes exactly to the top, No. 1 cylinder piston is brought to TDC of the compression stroke.

7. Check the valve clearance:
   - Intake valve: 0.0071–0.0087 in. (0.18–0.22mm).

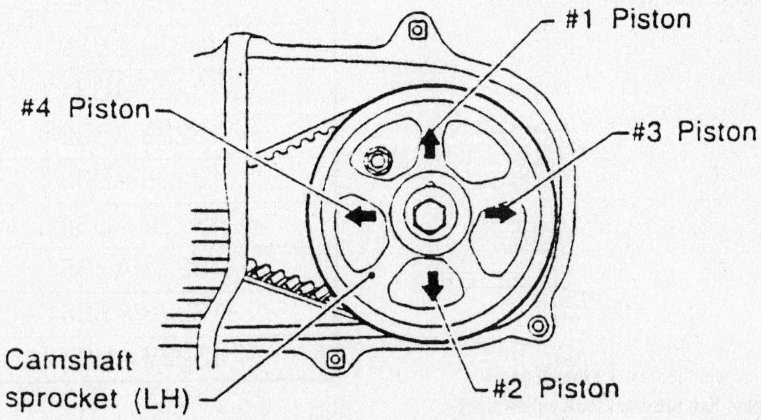

Position the camshaft for adjustment to valves

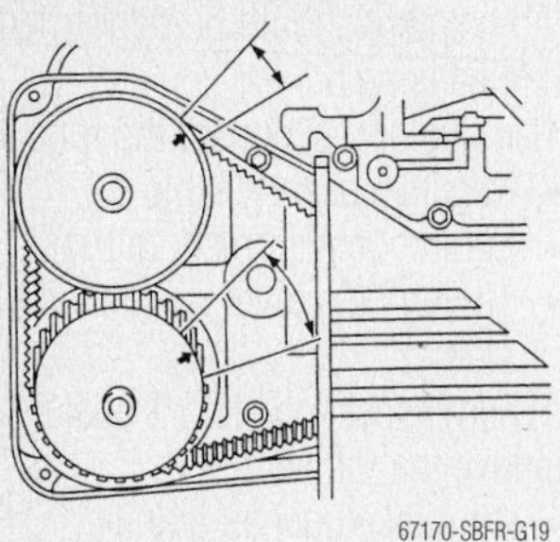

67170-SBFR-G19

To inspect the Nos. 1 intake and 3 exhaust, turn the crankshaft pulley clockwise until the arrow mark on the camshaft sprocket is set as shown–DOHC engine

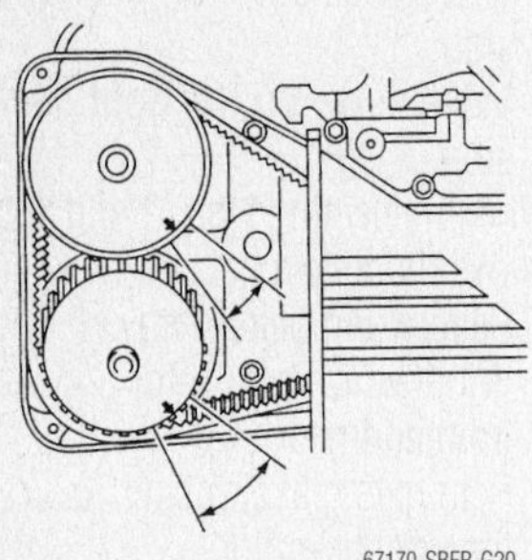

67170-SBFR-G20

To inspect the Nos. 2 exhaust and 3 intake, turn the crankshaft pulley clockwise until the arrow mark on the camshaft sprocket is set as shown–DOHC engine

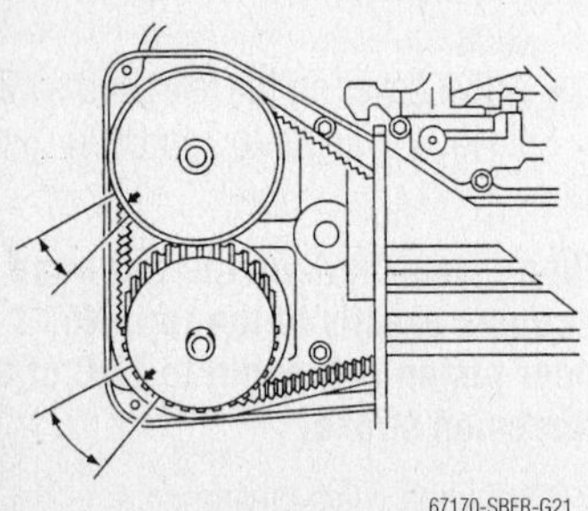

67170-SBFR-G21

To inspect the Nos. 2 intake and 4 exhaust, turn the crankshaft pulley clockwise until the arrow mark on the camshaft sprocket is set as shown–DOHC engine

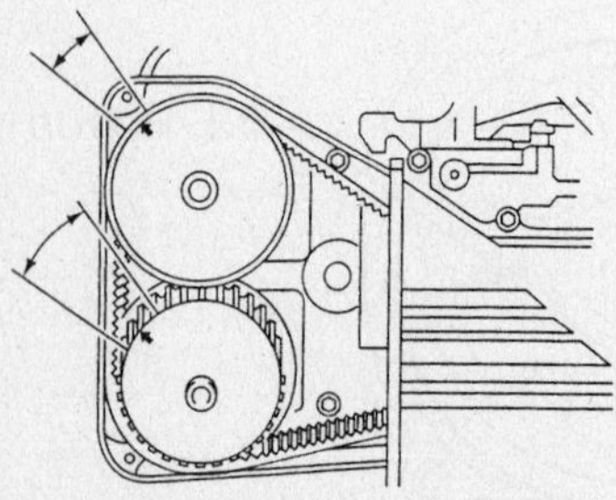

67170-SBFR-G22

To inspect the Nos. 1 exhaust and 4 intake, turn the crankshaft pulley clockwise until the arrow mark on the camshaft sprocket is set as shown–DOHC engine

Unit: mm
Intake valve: S = (V + T) – 0.20 Exhaust valve: S = (V + T) – 0.35
S: Required thickness of valve lifter V: Measured valve clearance T: Used valve lifter thickness

67170-SBFR-G23

Select a suitable replacement valve lifter after referring to this formula table–DOHC engines

Part No.	Thickness mm (in)
13228 AB101	4.68 (0.1843)
13228 AB111	4.69 (0.1846)
13228 AB121	4.70 (0.1850)
13228 AB131	4.71 (0.1854)
13228 AB141	4.72 (0.1858)
13228 AB151	4.73 (0.1862)
13228 AB161	4.74 (0.1866)
13228 AB171	4.75 (0.1870)
13228 AB181	4.76 (0.1874)
13228 AB191	4.77 (0.1878)
13228 AB201	4.78 (0.1882)
13228 AB211	4.79 (0.1886)
13228 AB221	4.80 (0.1890)
13228 AB231	4.81 (0.1894)
13228 AB241	4.82 (0.1898)
13228 AB251	4.83 (0.1902)
13228 AB261	4.84 (0.1906)
13228 AB271	4.85 (0.1909)
13228 AB281	4.86 (0.1913)
13228 AB291	4.87 (0.1917)
13228 AB301	4.88 (0.1921)
13228 AB311	4.89 (0.1925)
13228 AB321	4.90 (0.1929)
13228 AB331	4.91 (0.1933)
13228 AB341	4.92 (0.1937)
13228 AB351	4.93 (0.1941)
13228 AB361	4.94 (0.1945)
13228 AB371	4.95 (0.1949)

67170-SBFR-G24

Valve lifter thickness table–DOHC engines (1 of 3)

Part No.	Thickness mm (in)
13228 AB381	4.96 (0.1953)
13228 AB391	4.97 (0.1957)
13228 AB401	4.98 (0.1961)
13228 AB411	4.99 (0.1965)
13228 AB421	5.00 (0.1969)
13228 AB431	5.01 (0.1972)
13228 AB441	5.02 (0.1976)
13228 AB451	5.03 (0.1980)
13228 AB461	5.04 (0.1984)
13228 AB471	5.05 (0.1988)
13228 AB481	5.06 (0.1992)
13228 AB491	5.07 (0.1996)
13228 AB501	5.08 (0.2000)
13228 AB511	5.09 (0.2004)
13228 AB521	5.10 (0.2008)
13228 AB531	5.11 (0.2012)
13228 AB541	5.12 (0.2016)
13228 AB551	5.13 (0.2020)
13228 AB561	5.14 (0.2024)
13228 AB571	5.15 (0.2028)
13228 AB581	5.16 (0.2031)
13228 AB591	5.17 (0.2035)
13228 AB601	5.18 (0.2039)
13228 AB611	5.19 (0.2043)
13228 AB621	5.20 (0.2047)
13228 AB631	5.21 (0.2051)
13228 AB641	5.22 (0.2055)
13228 AB651	5.23 (0.2059)
13228 AB661	5.24 (0.2063)
13228 AB671	5.25 (0.2067)
13228 AB681	5.26 (0.2071)
13228 AB691	5.27 (0.2075)
13228 AB701	4.38 (0.1724)
13228 AB711	4.40 (0.1732)
13228 AB721	4.42 (0.1740)
13228 AB731	4.44 (0.1748)
13228 AB741	4.46 (0.1756)
13228 AB751	4.48 (0.1764)
13228 AB761	4.50 (0.1771)
13228 AB771	4.52 (0.1780)
13228 AB781	4.54 (0.1787)
13228 AB791	4.56 (0.1795)
13228 AB801	4.58 (0.1803)
13228 AB811	4.60 (0.1811)
13228 AB821	4.62 (0.1819)
13228 AB831	4.64 (0.1827)
13228 AB841	4.66 (0.1835)
13228 AB851	5.29 (0.2083)
13228 AB861	5.31 (0.2091)
13228 AB871	5.33 (0.2098)
13228 AB881	5.35 (0.2106)

67170-SBFR-G25

Valve lifter thickness table–DOHC engines (2 of 3)

- Exhaust valve: 0.0090–0.0106 in. (0.23–0.27mm).

8. If any valve needs adjustment, perform the following:
   a. Loosen the valve rocker nut and screw.
   b. Place a thickness gage in at as horizontal a direction as possible with respect to the vale stem and face.
   c. Adjust the screw until proper clearance is obtained.
   d. Tighten the rocker nut after adjusted to 7 ft. lbs. (10 Nm).
9. Install or connect the following:
   - Valve covers left and right
   - Timing belt covers
   - Blow-by hoses to valve covers
   - Spark plug wires
   - Washer tank
   - Rear gate glass washer hose to the washer motor
   - Washer motor connectors
   - Battery and battery tray
   - Engine undercover
   - Resonator chamber
   - Air intake duct unit
   - Engine coolant reservoir tank

DOHC Engines

➡The valve adjustment should be performed while the engine is cold.

1. Before servicing the vehicle, refer to the precautions in the beginning of this section.
2. Disconnect the negative battery cable.
3. Remove the air intake duct.
4. Remove the right hand timing belt cover.
5. When inspecting Nos. 1 and 3 cylinders remove the following:
   a. Pull out the engine harness connector with the bracket from the air cleaner upper cover.
   b. Remove the air cleaner case and the ignition coil.
   c. Disconnect the Positive Crankcase ventilation (PCV) hose from the valve cover.
   d. Remove the right hand rocker cover.
6. When inspecting Nos. 2 and 4 cylinders remove the following:
   a. Remove the battery and battery tray.
   b. Remove the bolt that attaches the engine harness bracket to the body.
   c. Disconnect the window washer motor connectors.
   d. Remove the washer tank mounting bolts and secure out of the way.
   e. Remove the ignition coil.

Part No.	Thickness mm (in)
13228 AB891	5.37 (0.2114)
13228 AB901	5.39 (0.2122)
13228 AB911	5.41 (0.2123)
13228 AB921	5.43 (0.2138)
13228 AB931	5.45 (0.2146)
13228 AB941	5.47 (0.2154)
13228 AB951	5.49 (0.2161)
13228 AB961	5.51 (0.2169)
13228 AB971	5.53 (0.2177)
13228 AB981	5.55 (0.2185)
13228 AB991	5.57 (0.2193)
13228 AC001	5.59 (0.2201)
13228 AC011	5.61 (0.2209)
13228 AC021	5.63 (0.2217)
13228 AC031	5.65 (0.2224)

67170-SBFR-G26

Valve lifter thickness table–DOHC engines (3 of 3)

7. Disconnect the Positive Crankcase ventilation (PCV) hose from the valve cover.
   a. Remove the left hand rocker cover.
8. To inspect the Nos. 1 intake and 3 exhaust, turn the crankshaft pulley clockwise until the arrow mark on the camshaft sprocket is set as shown in the accompanying illustration.
9. To inspect the Nos. 2 exhaust and 3 intake, turn the crankshaft pulley clockwise until the arrow mark on the camshaft sprocket is set as shown in the accompanying illustration.
10. To inspect the Nos. 2 intake and 4 exhaust, turn the crankshaft pulley clockwise until the arrow mark on the camshaft sprocket is set as shown in the accompanying illustration.
11. To inspect the Nos. 1 exhaust and 4 intake, turn the crankshaft pulley clockwise until the arrow mark on the camshaft sprocket is set as shown in the accompanying illustration.
12. Check the valve clearance:
    - Intake valve: 0.0063–0.0095 in. (0.14–0.24mm).
    - Exhaust valve: 0.0118–0.0158 in. (0.30–0.40mm).
13. If any valve needs adjustment, perform the following:
    a. Remove the camshaft.
    b. Remove the valve lifter and measure the thickness of the lifter using a micrometer.
    c. Select a suitable replacement valve lifter after referring to the formula table and thickness tables illustrated.
    d. Re-inspect valve clearances again and adjust if necessary.
14. Once valve adjustment has been performed and rechecked, install the removed components in the reverse order of removal. Refer to camshaft removal and installation procedure in this section for camshaft and valve cover installation procedures.

Starter

REMOVAL & INSTALLATION

1. Before servicing the vehicle, refer to the precautions in the beginning of this section.
2. Remove or disconnect the following:
   - Negative battery
   - Air intake duct and assembly
   - Wires
   - Starter

To install:

3. Install or connect the following:
   - Starter. Torque the mounting bolts to 37 ft. lbs. (50 Nm).
   - Wires
   - Air intake duct and assembly
   - Negative battery

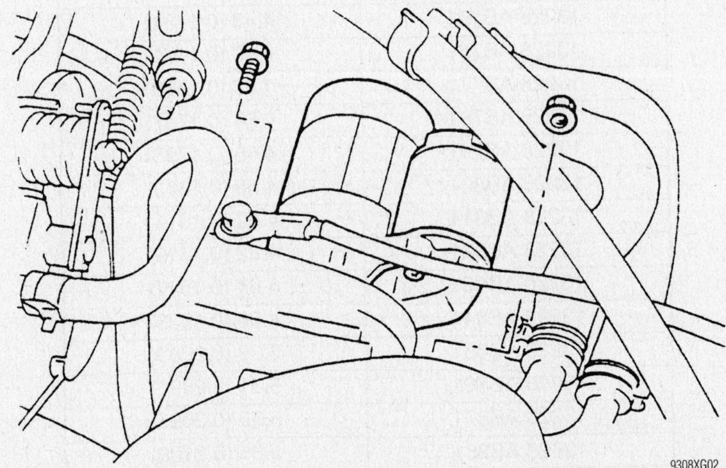

View of the starter mounting

Oil Pan

REMOVAL & INSTALLATION

1. Before servicing the vehicle, refer to the precautions in the beginning of this section.
2. Drain the oil from the engine.
3. Remove or disconnect the following:
   - Front wheels
   - Negative battery cable
   - Air intake duct
   - Mass Air Flow (MAF) connector, on turbo models
4. Remove the intercooler as follows:
   a. Disconnect the Positive Crankcase Ventilation (PCV) hose from the pipe.
   b. Remove the PCV pipe from the intercooler.
   c. Disconnect the air by–pass valve hose.
   d. Remove the intercooler–to–bracket bolts.
   e. Loosen the clamps that connect the intercooler to the turbocharger and throttle body.
   f. Disconnect the intercooler duct from the turbocharger and remove the intercooler.
5. Remove or disconnect the following:

- Oxygen (O_2S) sensor connectors, if necessary
- Pitching stopper
- Upper radiator brackets

6. Support the engine with a suitable lifting device and lift the engine slightly.

- Engine undercover
- Rear O_2S sensor connector, if equipped
- Exhaust front Y-pipe and center pipe, on non turbo models
- Nuts which secure the front engine mounts to the front crossmember

7. Remove the oil pan.

8. Clean all gasket material from both mating surfaces.

To install:

9. Install the drain plug with a new gasket. Torque the bolt to 32 ft. lbs. (44 Nm).

10. Apply a continuous bead of sealer to a new oil pan gasket.

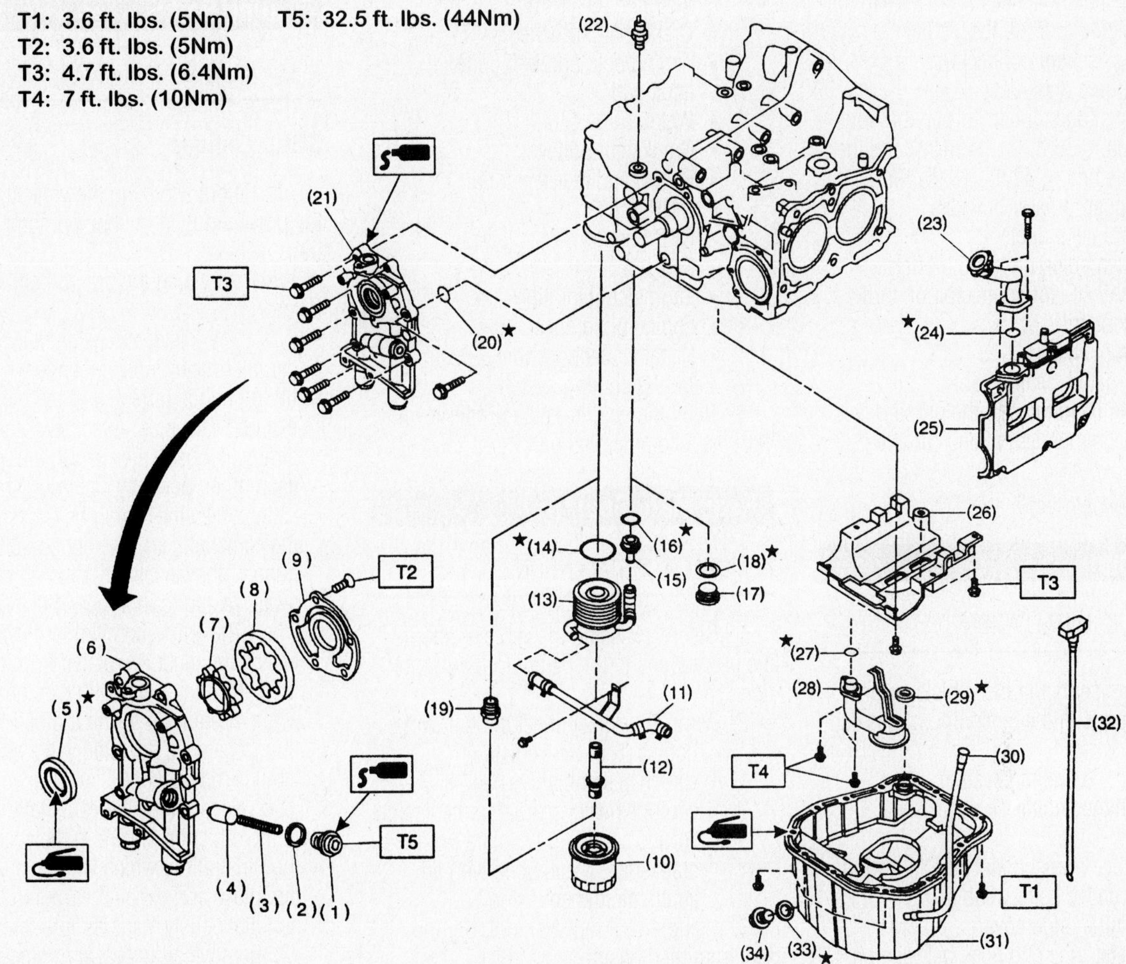

Oil pan and lubrication components

11. Install the oil pan assembly. Torque the bolts to 36–48 inch lbs. (4–5 Nm).
12. Lower the engine onto the front crossmember.
13. Install or connect the following:
- Front engine mount nuts. Torque the nuts to 51 ft. lbs. (69 Nm).
- Exhaust front Y-pipe with new gaskets, on non turbo models. Torque the nuts that secure the pipe to the engine to 23 ft. lbs. (30 Nm).
- O_2S sensor connectors
- Engine undercover
- Pitching stopper. Torque the front bolt to 36 ft. lbs. (49 Nm) and the rear bolt to 42 ft. lbs. (57 Nm).
- Upper radiator brackets
- Air intake duct
- Intercooler, on turbo models
- MAF sensor connector, on turbo models
- Front wheels
- Negative battery cable

14. Refill the engine to the proper level with the recommended oil and run the engine.
15. Check for leaks.

Oil Pump

REMOVAL & INSTALLATION

1. Before servicing the vehicle, refer to the precautions in the beginning of this section.
2. Drain the cooling system.
3. Drain the engine oil into a separate container.
4. Remove or disconnect the following:
- Negative battery cable
- Engine undercover
- Water pipe and hose between oil cooler and water pump
- Radiator
- Crankcase Position (CKP) Sensor
- Camshaft Position (CMP) Sensor
- Belt(s) and rear tensioner
- Crankshaft pulley
- Water pump
- Timing belt, and belt guide, if equipped
- Crankshaft sprocket
- Oil pump mounting bolts and carefully pry the pump from the engine block

⁂ WARNING

Use extreme care not to damage the engine block or the oil pump during removal of the pump.

To install:

5. Apply a continuous bead sealant to the mating surfaces of the oil pump.
6. Install or connect the following:
- New front seal to the oil pump coat the inside of the seal with engine oil
- New O-ring to the oil pump
- Oil pump. Torque the bolts to 56 inch lbs. (6.4 Nm).
- Crankshaft sprocket
- Timing belt, and belt guide, if equipped
- Water pump
- Crankshaft pulley
- Belt(s) and rear tensioner
- CMP Sensor
- CKP Sensor
- Radiator
- Engine coolant pipe
- Engine undercover
- Negative battery cable

7. Refill the cooling system.
8. Refill the engine to the proper level with the recommended oil.

Rear Main Seal

REMOVAL & INSTALLATION

1. Before servicing the vehicle, refer to the precautions in the beginning of this section.
2. Remove or disconnect the following:
- Engine from the vehicle
- Clutch assembly/flywheel, if equipped with manual transmission
- Torque converter flexplate from the crankshaft, if equipped with an automatic transmission

3. Using a seal removal tool, pry the oil seal from the housing.

To install:

4. Utilizing the appropriate seal installer, install or connect the following:
- New oil seal and press it into the housing using oil seal guide ST 499597100 and installer ST 499587200.
- Clutch assembly/flywheel, if equipped
- Flywheel/flexplate and tighten the bolts to 53 ft. lbs. (72 Nm), if equipped
- Engine into the vehicle

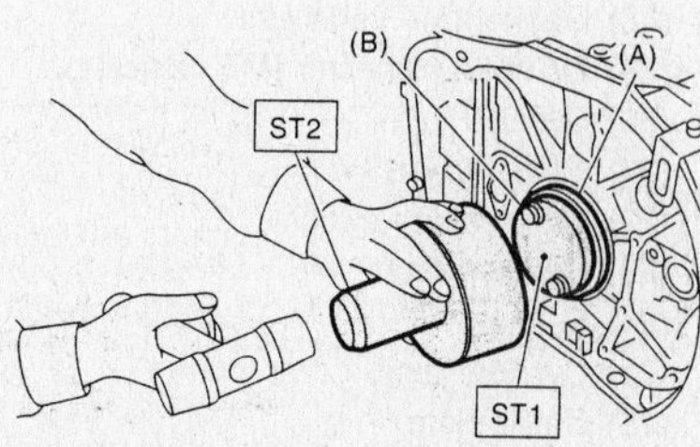

(A) Rear oil seal
(B) Flywheel attaching bolt

67170-SBFR-G27

Installing the rear main seal

Timing Belt

REMOVAL & INSTALLATION

SOHC Engines

1. Before servicing the vehicle, refer to the precautions in the beginning of this section.

When servicing the timing belt, note the following:

a. The intake and exhaust camshafts can be rotated independently when the timing belt is removed. If the intake and exhaust valves are lifted off of their seats simultaneously, their heads will contact each other, possibly causing damage.

b. When the timing belt is removed, the camshafts are positioned so that none of the valves are lifted off of their seats, resulting in a "zero-lift" position.

c. The left-hand cylinder head camshafts must be rotated from the "zero-lift" position as little as possible when orienting it for timing belt installation, otherwise possible valve head interference may occur.

d. Never allow the camshafts to rotate in the direction shown in the accompanying illustration, which would cause both the intake and exhaust valves to lift simultaneously, causing interference.

2. Remove all necessary components to gain access to the timing belt.
3. If equipped with manual transmissions, loosen the 2 timing belt guide mounting bolts, then separate the guide from the engine block.

42356-SBFR-G50

Remove the timing belt guide on vehicles equipped with manual transmission–SOHC engine

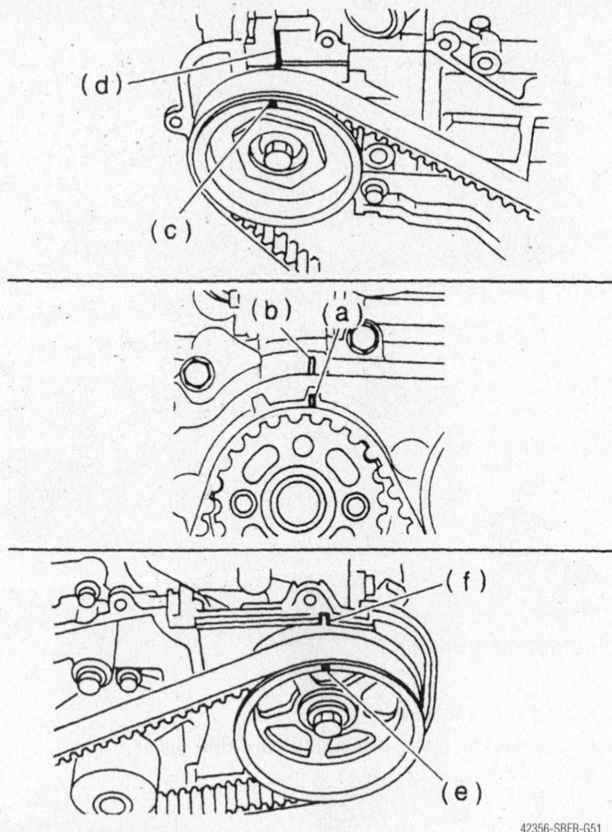

Before removing the timing belt, turn the crankshaft sprocket until all of the alignment marks are aligned as indicated–SOHC engine

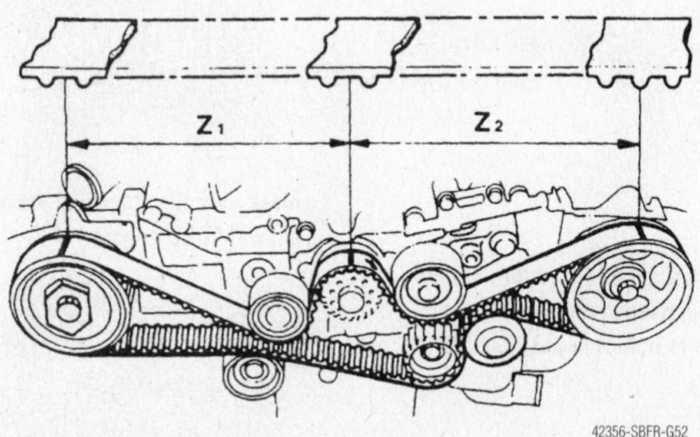

If the original marks on the timing belt are worn or faded, make new alignment marks in the positions indicated—SOHC engine

4. If the directional arrow and alignment marks on the timing belt are faded, and the belt is to be reused, remark the belt with white paint or a grease pencil as follows:

a. Using a Subaru tool No. ST-499987500 Crankshaft Socket, or equivalent, installed on the crankshaft sprocket, rotate the crankshaft until the crankshaft sprocket, left-hand exhaust camshaft sprocket, left-hand intake camshaft sprocket, right-hand intake camshaft sprocket and right-hand exhaust camshaft sprocket timing mark notches are aligned with the respective marks on the belt cover and engine block.

b. Make alignment and/or arrow marks on the timing belt in relation to the sprockets as indicated in the accompanying illustration.

- Z1: 46.8 tooth length
- Z2: 43.7 tooth length

5. Loosen the center bolt from the timing belt idler pulley, then remove the idler pulley from the engine block.

✲✲ WARNING

After removing the timing belt, DO NOT rotate the camshafts. Damage to the valves may occur.

6. Carefully remove the timing belt from all of the sprockets.

7.Remove the automatic belt tension adjuster assembly as follows:

a. Remove the 2 timing belt idler pulleys, as indicated in the accompanying illustration.

b. Loosen the automatic tension adjuster assembly mounting bolts, then separate the adjuster assembly from the engine block.

To install:

✲✲ WARNING

Do not allow oil, grease, or coolant to come in contact with the timing belt. If this occurs, quickly and thoroughly remove all traces of the compound. Also, never bend the timing belt sharply; the minimum bending radius is 2.36 in. (60mm).

8. Inspect the camshaft and crankshaft sprocket teeth for abnormal or excessive wear or scratches. Ensure there is no free-play between the sprocket and the key. Inspect the crankshaft sprocket sensor notch for damage or contamination with debris or dirt.

➡When preparing the automatic tension adjuster assembly for installation, adhere to the following points:

- Always use a vertical press, rather than a horizontal press or vise, to depress the adjuster assembly rod
- Depress the adjuster rod in a vertical position ONLY
- Depress the adjuster rod slowly (taking more than 3 minutes) with a force of 66 lbs. (30 kg)
- Do not allow the press force to exceed 2205 lbs. (1000 kg)
- Press the adjuster rod in as far as the end surface of the cylinder — do not press the rod into the cylinder, which may cause damage to the assembly
- Do not release the press force from the rod until the stopper pin is completely inserted in the cylinder

9. Prepare the automatic timing belt tension adjuster assembly for installation as follows:

a. Position the adjuster assembly in a vertical press.

b. Slowly depress the adjuster rod

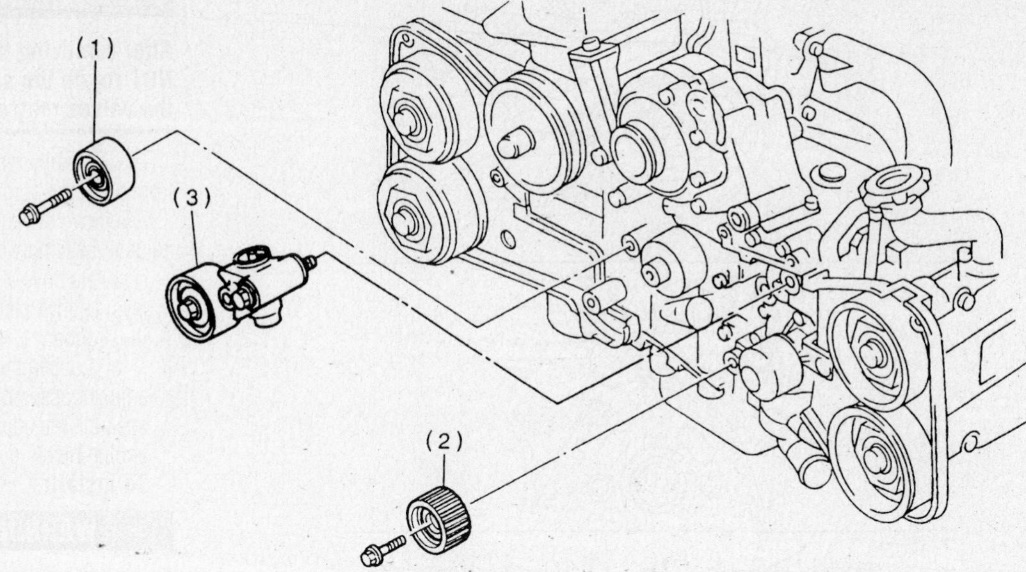

It is necessary to remove the automatic adjuster assembly and reset the pushrod for timing belt installation–SOHC engine

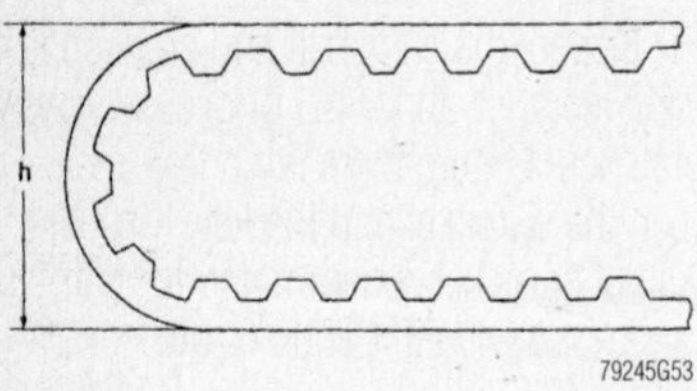

Never bend the timing belt into a radius tighter than 2.36 in./60mm (h), otherwise it will be damaged beyond–SOHC engine

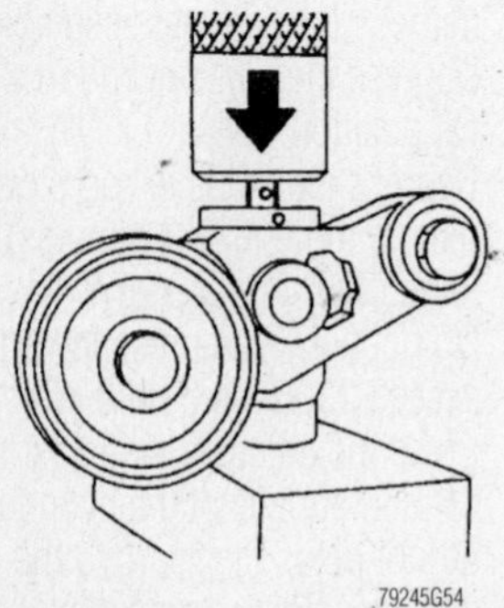

Use a vertical press to push the adjuster rod into its housing until it is flush with the assembly's outer surface . . .

with a force of 66 lbs. (30 kg) until the hole in the rod is aligned with the hole in the adjuster cylinder housing.

c. Insert a 0.08 in. (2mm) diameter stopper pin or Allen wrench through the hole in the cylinder housing and rod,

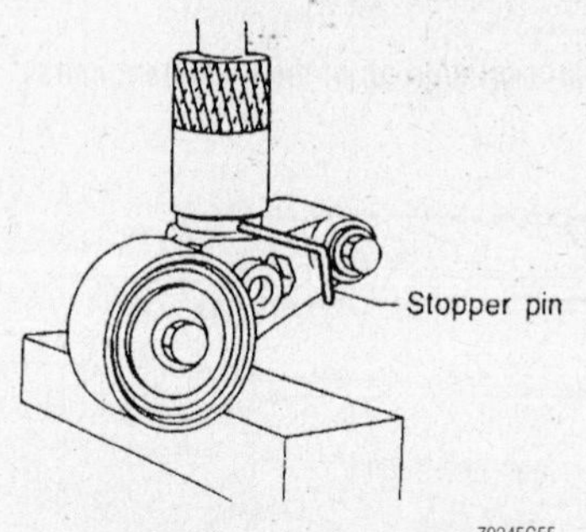

. . . then insert a 0.08 in. (2mm) diameter pin or Allen wrench into the housing and rod holes to hold it in position–SOHC engine

then slowly release the press force from the adjuster rod.

10. Install the adjuster assembly onto the engine block.

11. Install timing belt idler pulley No. 2 on the engine block. Tighten the bolts to 28 ft. lbs. (39 Nm).

12. Install the timing belt idler pulley No. 1 on the engine block. Tighten the bolts to 28 ft. lbs. (39 Nm).

13. If the camshaft and crankshaft timing marks are no longer aligned, perform the following:

a. Position the crankshaft sprocket so that its mark is aligned with the mark on the oil pump cover on the engine block.

b. Align the single line mark on the

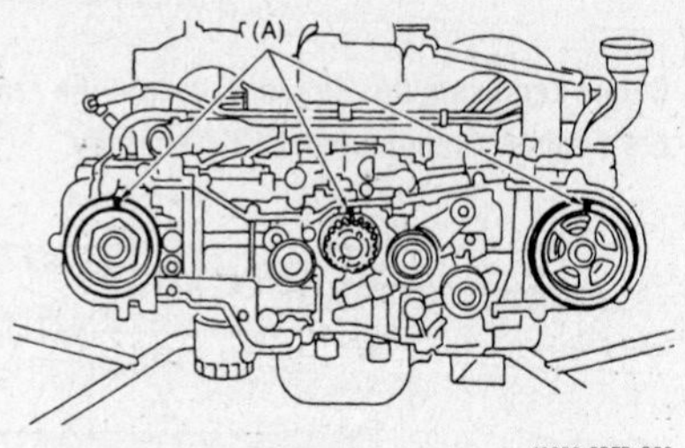

Make sure the sprockets are aligned as shown–SOHC engine

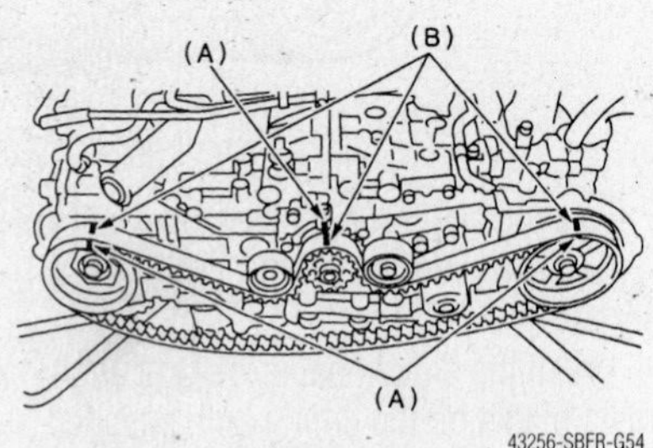

After installing the timing belt, the alignment marks should be positioned as shown. If not, remove the belt and align the sprockets and reinstall the belt–SOHC engine

right-hand exhaust camshaft sprocket with the notch on the belt cover.

c. Rotate the right-hand intake camshaft so that the single line mark is aligned with the notch on the belt cover.

➡At this point, the double line marks on both right-hand camshaft sprockets should be aligned.

d. Turn the left-hand exhaust (lower) camshaft counterclockwise (as viewed from the front of the engine) until the single line mark is aligned with the notch on the belt cover.

e. Position the single line mark on the left-hand intake camshaft sprocket so that it is aligned with the notch on the belt cover. When rotating the camshaft, do so only in a clockwise direction (as viewed from the front of the engine).

➡At this point, the double line marks on both left-hand camshaft sprockets should be aligned.

f. Ensure the timing marks are aligned as shown in the accompanying illustration. If they are not, repeat Substeps 12a through 12e until they are properly aligned.

14. Install the timing belt around the camshaft, crankshaft and idler pulleys so that the positioning marks on the timing belt are aligned with the marks on the sprockets as follows:

a. Position the timing belt on the crankshaft sprocket so that the marks are aligned.

b. Route the belt down and under the left-hand, upper idler pulley, then up and around the left-hand intake camshaft sprocket, ensuring the camshaft sprocket mark is aligned with the mark on the belt.

c. Route the belt down and around the left-hand exhaust camshaft sprocket, making sure the marks are properly aligned, then up and over the first lower idler pulley and down and around the second lower idler pulley.

d. While holding the timing belt on the inner, left-hand, lower idler pulley, route the other side of the timing belt (from the crankshaft sprocket) down and under the right-hand upper idler pulley.

e. Route the timing belt up and around the right-hand intake camshaft sprocket so that the belt and sprocket marks are aligned.

f. Position the belt down and around the right-hand exhaust camshaft sprocket, ensuring the positioning marks are aligned.

15. Install the right-hand lower idler pulley so that the timing belt is routed over the top side of it.

➡Once the belt is completely installed on all of the pulleys and sprockets, ensure that the positioning marks are still all aligned.

On models equipped with manual transmissions, ensure the timing belt-to-guide clearance (arrows) is correct before tightening the mounting bolts–SOHC engine

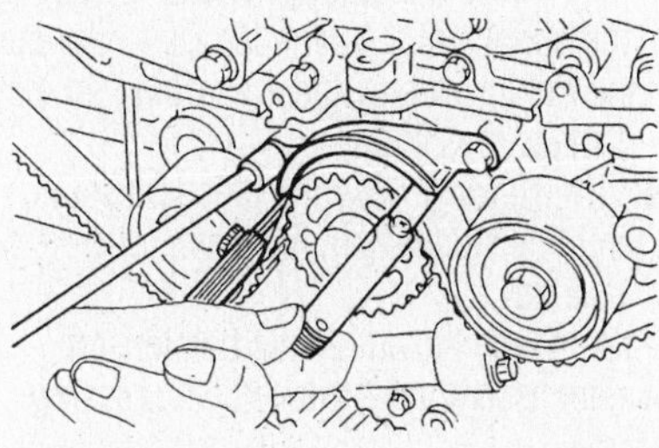

On models equipped with manual transmissions, use a feeler gauges to adjust the clearance between the timing belt and the belt guide–SOHC engine

16. After ensuring all of the marks are still aligned, use a pair of pliers to withdraw the stopper pin or Allen wrench from the adjuster assembly housing.

17. On models with manual transmissions, perform the following:

a. Install the timing belt guide by temporarily tightening the mounting bolts.

b. Position the timing belt guide so that there is 0.019–0.059 in. (0.5–1.5mm) clearance between the timing belt and the belt guide.

c. Tighten the guide mounting bolts securely, then double check the guide clearance.

18. Install the timing belt covers and all remaining engine components.

DOHC Engines

1. Before servicing the vehicle, refer to the precautions in the beginning of this section.

2. Disconnect the negative battery cable.

3. Remove the belt cover as follows:

a. Remove the V-belt.

b. Crankshaft pulley.

c. Left hand belt cover.

d. Right hand belt cover.

e. Front belt cover.

4. Remove the timing belt guides.

5. If the alignment marks that indicate rotation are faded, put new marks on the belt before removal as follows:

a. Turn the crankshaft using crankshaft sprocket tool 499987500 and a breaker bar to align the crankshaft sprocket, left hand intake camshaft sprocket, left hand exhaust camshaft, right hand intake camshaft sprocket and right hand exhaust camshaft sprocket with the on the cover and cylinder block.

b. Using white paint such as white out, place alignment marks on the belts in relation to the sprockets.

6. Remove the belt idler (A), shown in the accompanying illustration.

Location of the upper belt guide–DOHC engines

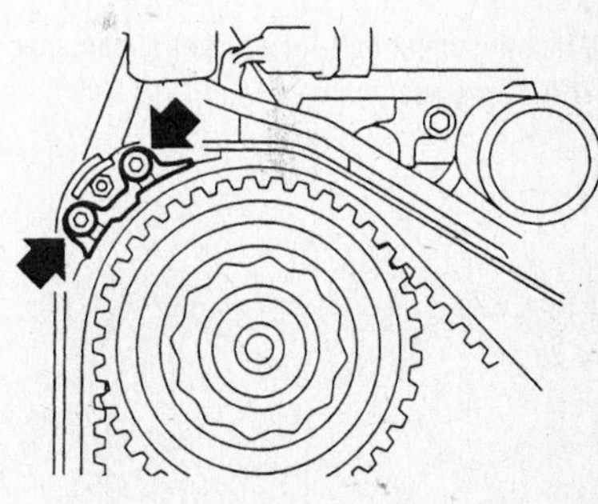

Location of the upper left belt guide–DOHC engines

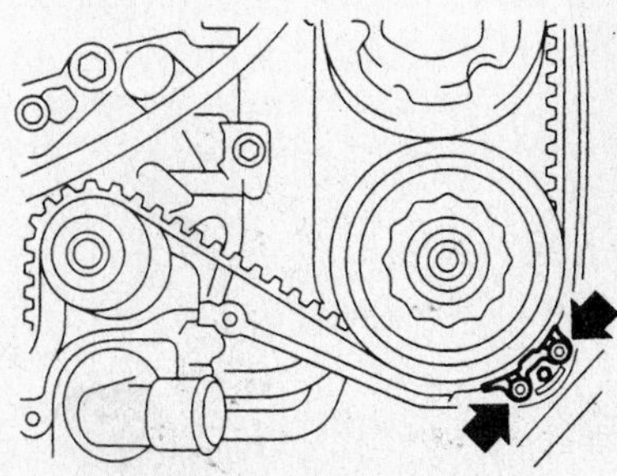

Location of the lower right belt guide–DOHC engines

7. Remove the timing belt.

8. If necessary, remove belt idlers (B) and (C).

9. Remove the belt idler 2.

10. Remove the automatic belt tension adjuster assembly.

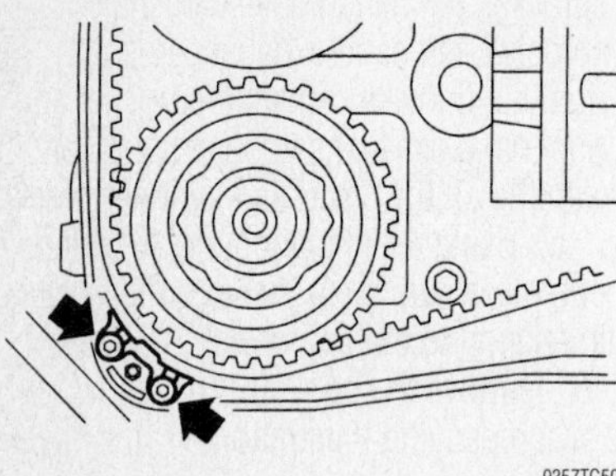

Location of the lower left belt guide–DOHC engines

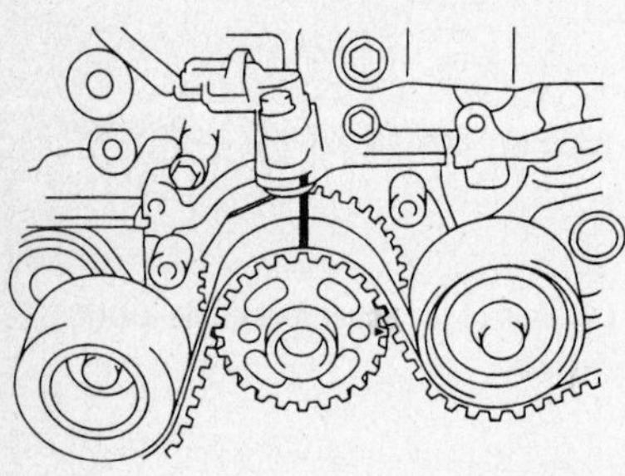

Mark the upper belt-to-sprocket alignment–DOHC engines

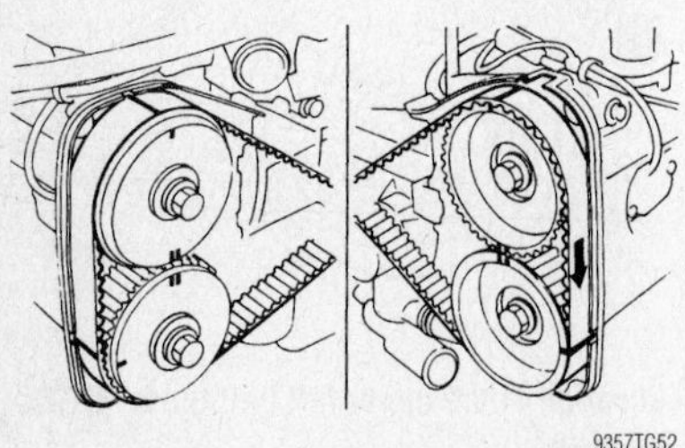

Mark the lower belt-to-sprocket alignment–DOHC engines

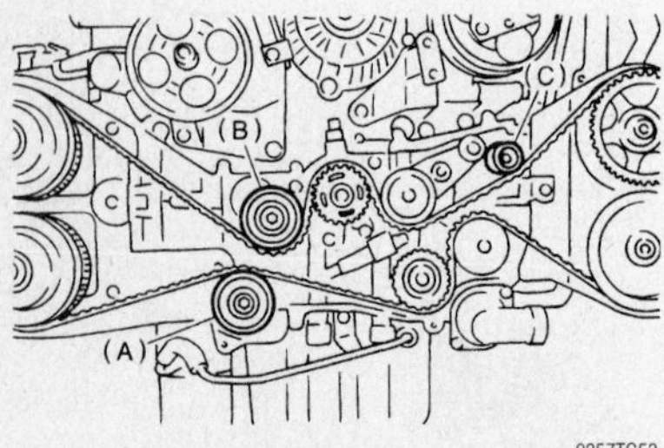

Remove the belt idler (A)–DOHC engines

To install:

11. To prepare the automatic belt tensioner for assembly, perform the following steps:

a. Always use a vertical type pressing tool to move the adjuster rod down.

b. Do not use a lateral type vise.

c. Always push the adjuster rod vertically.

d. Make sure to slowly move the adjuster rod down applying a pressure of 66 lbs. (294 N).

e. Press in the push adjuster rod gradually taking more than 3 minutes.

f. Never allow the press pressure to exceed 2,205 lbs. (9,807 N).

g. Press the adjuster rod as far as the end surface of the cylinder. Do not press the rod into the cylinder as doing so may damage the cylinder.

h. Never release the press pressure until the stopper pin has been fully inserted.

12. Attach the automatic belt tension adjuster assembly to the vertical pressing tool.

13. Move the adjuster rod down slowly using a pressure of 66 lbs. (294 N) until the rod is aligned with the stopper pin hole in the cylinder.

14. Insert a 0.08 inch (2mm) stopper pin or diameter Allen wrench into the stopper pin hole in the cylinder to retain the rod.

15. Install the adjuster assembly and tighten the retainers to 29 ft. lbs. (39 Nm).

16. Install belt idle 2 and tighten the retainers to 29 ft. lbs. (39 Nm).

17. Install the belt idler and tighten to 29 ft. lbs. (39 Nm).

18. Align the mark on the crankshaft sprocket with the mark on the oil pump.

19. Align the single line mark on the right hand exhaust camshaft sprocket with the notch on the belt cover.

20. Align the single line mark on the right hand intake camshaft with the notch on the belt cover. Make sure the double lines on the intake camshaft and exhaust sprockets are aligned as shown in the accompanying illustration.

21. Align the single line mark on the left hand exhaust camshaft sprocket with the notch on the belt cover by turning the sprocket counterclockwise (as viewed from the front of the engine).

22. Align the single line mark on the left hand intake camshaft sprocket with the notch on the belt cover by turning the sprocket counterclockwise (as viewed from the front of the engine). Make sure the double lines on the intake camshaft and exhaust sprockets are aligned as shown in the accompanying illustration.

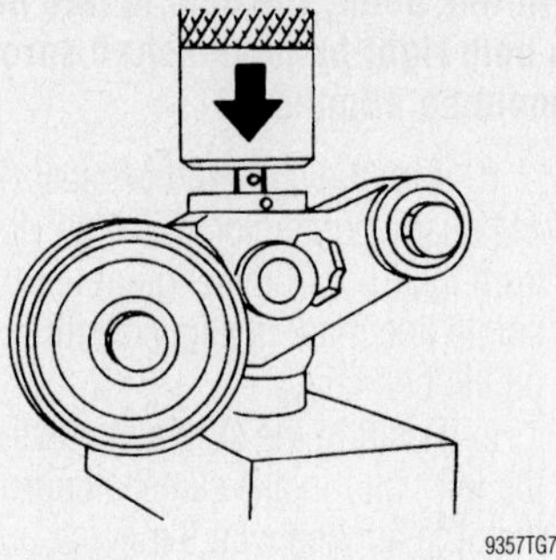

Attach the automatic belt tension adjuster assembly to the vertical pressing tool–DOHC engines

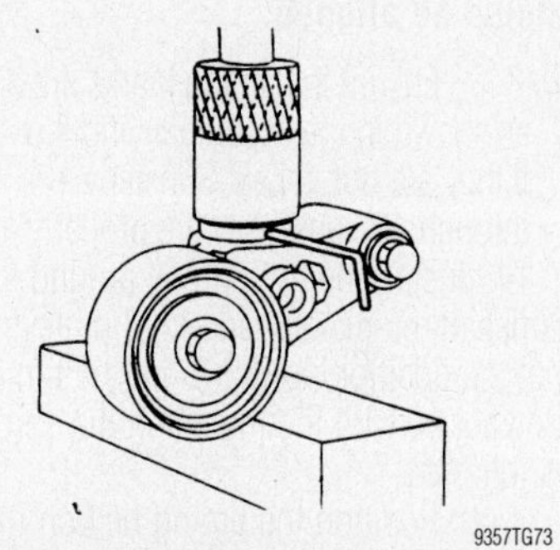

Insert a 0.08 inch (2mm) stopper pin or diameter Allen wrench into the stopper pin hole in the cylinder to retain the rod–DOHC engines

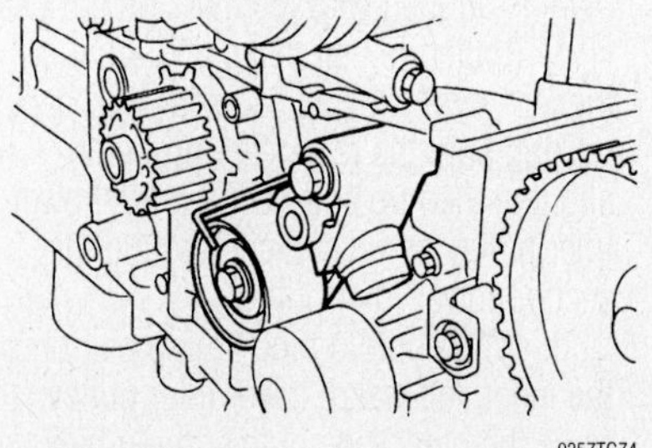

Install the adjuster assembly–DOHC engines

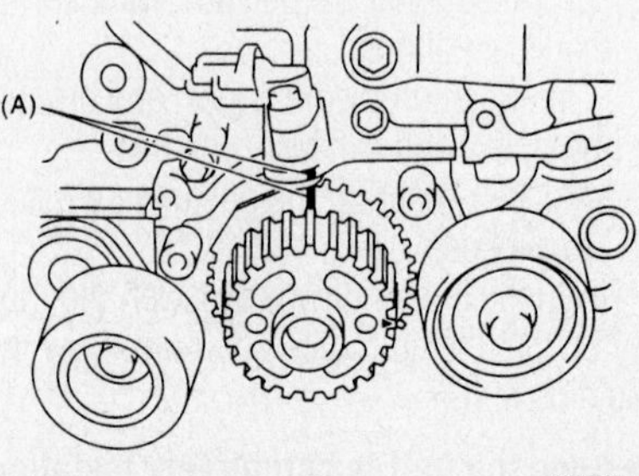

Align the mark on the crankshaft sprocket with the mark on the oil pump–DOHC engines

➡Make sure the camshaft and crankshaft sprockets are positioned correctly. The intake and exhaust camshafts on this engine can be rotated independently with the timing belt removed. By looking at the accompanying illustration it will show you that if the intake and exhaust valve are lifted together the heads will hit each other and bend.

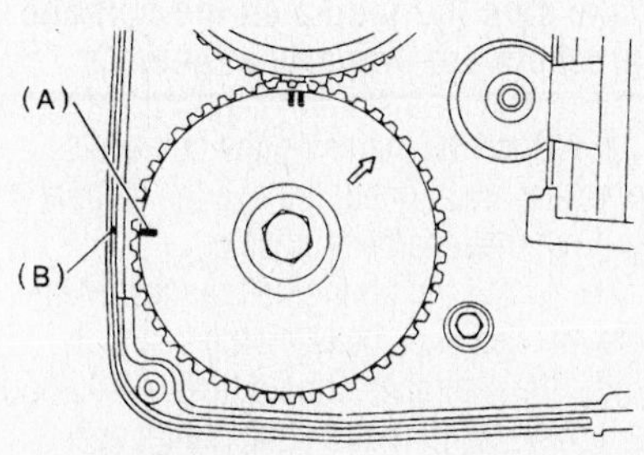

Align the single line mark on the right hand exhaust camshaft sprocket with the notch on the belt cover–DOHC engines

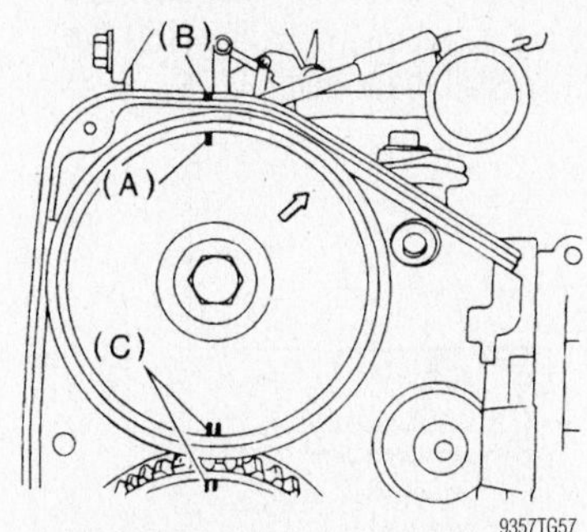

Align the single line mark on the right hand intake camshaft with the notch on the belt cover. Make sure the double lines on the intake camshaft and exhaust sprockets are aligned–DOHC engines

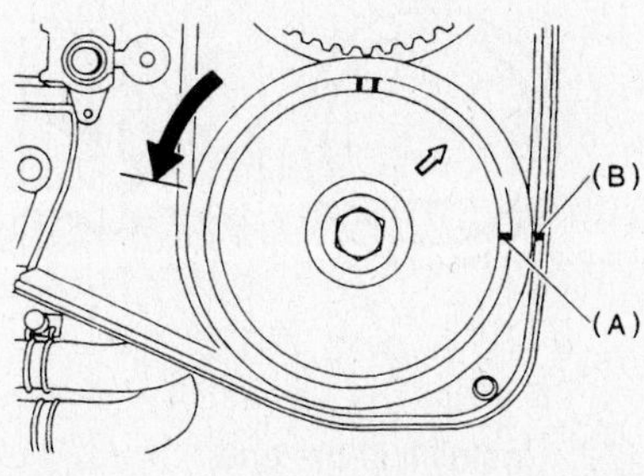

Align the single line mark on the left hand exhaust camshaft sprocket with the notch on the belt cover by turning the sprocket counterclockwise (as viewed from the front of the engine)–DOHC engines

➡When the timing belts are not installed, 4 camshafts are held at "zero lift" position, where all cams on the camshafts do not push the intake and exhaust valves down (under this condition all valves remain unlifted). When the camshafts are rotated to install the timing belts, # 2 intake and # 4 exhaust cam of the left hand camshafts are held to push their corresponding valves down. Under this condition these valves are held lifted. The right side camshafts are held in so that their cams do not push the valves down. The left hand camshafts must be rotated from the "zero lift" position to the position where the timing belt is to be installed at as small an angle as possible, in order to prevent mutual interference of intake and exhaust valve heads. Do not allow the camshafts to rotate in the direction illustrated as this causes both the intake and exhaust valves to lift off at the same time with will cause valve damage.

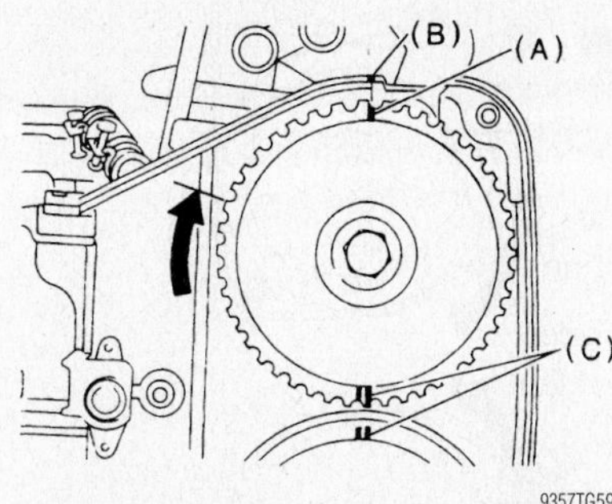

Align the single line mark on the left hand intake camshaft sprocket with the notch on the belt cover by turning the sprocket counterclockwise (as viewed from the front of the engine). Make sure the double lines on the intake camshaft and exhaust sprockets are aligned–DOHC engines

23. When installing the belt, make sure to align the marks made during removal or if using a new belt, align the in alphabetical order as shown in the accompanying illustration.

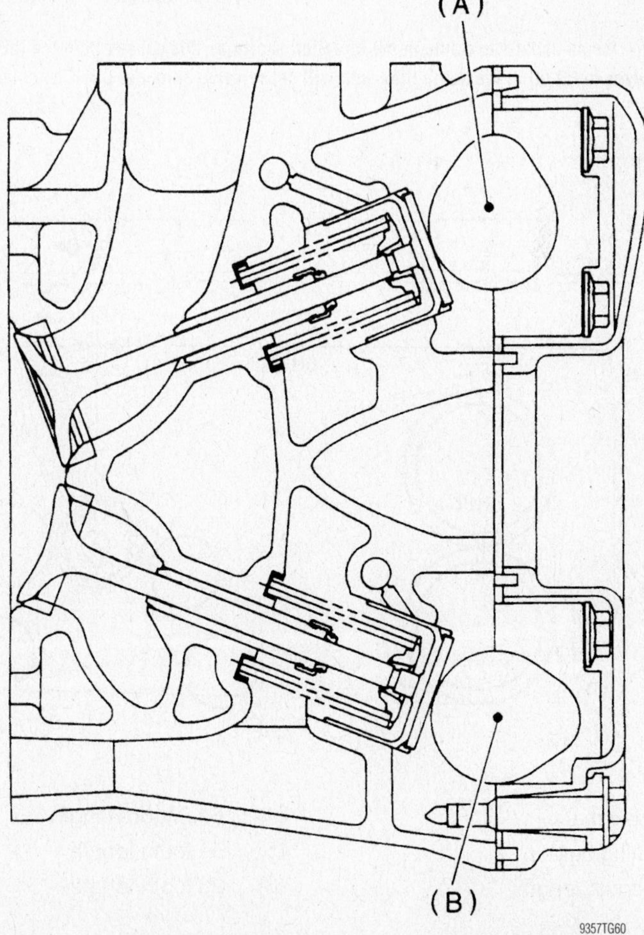

If the intake and exhaust valve are lifted together the heads will hit each other and bend–DOHC engines

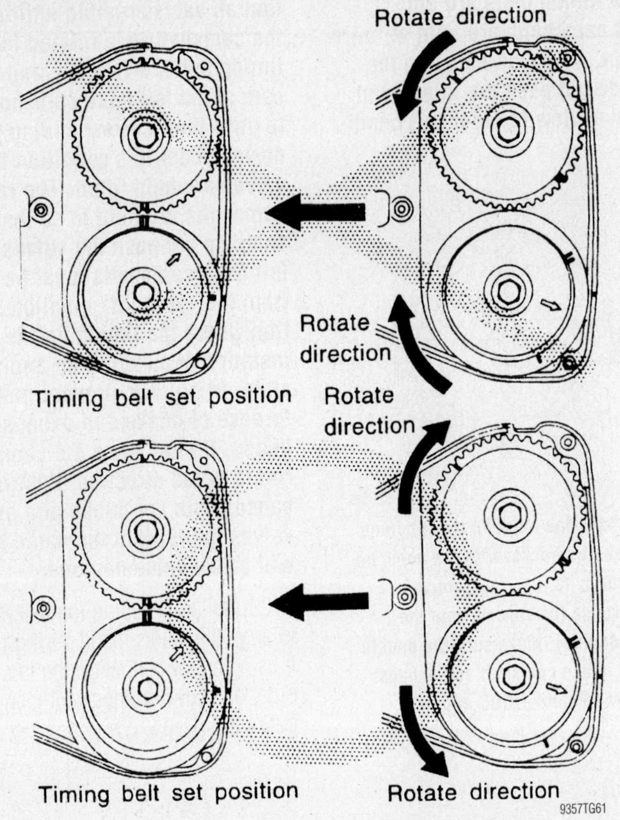

Do not allow the camshafts to rotate in the direction shown as this causes both the intake and exhaust valves to lift off at the same time with will cause valve damage–DOHC engines

WARNING

Disengagement of more than 3 timing belt teeth may result in contact between the valve and piston. Always make sure the belts rotation is correct.

24. Install the belt idlers and tighten to 29 ft. lbs. (39 Nm).

WARNING

Make sure the marks on the belt and sprockets are properly aligned.

25. Once the marks on the belt and sprockets are aligned, remove the stopper pin from the tensioner adjuster.
26. Install the timing belt guides. Measure the clearance between the belt and guide. the clearance should be 0.019–0.059 inch (0.5–1.5mm) and tighten the upper belt guide retainers to 7 ft. lbs. (10 Nm) and the remaining belt guides, tighten the retainers to 5 ft. lbs. (6.6 Nm).
27. Install the belt covers and tighten to 3.5 ft. lbs. (5 Nm).
28. Install the crankshaft pulley and tighten the bolt to 132 ft. lbs. (180 Nm).
29. Install the V-belt.

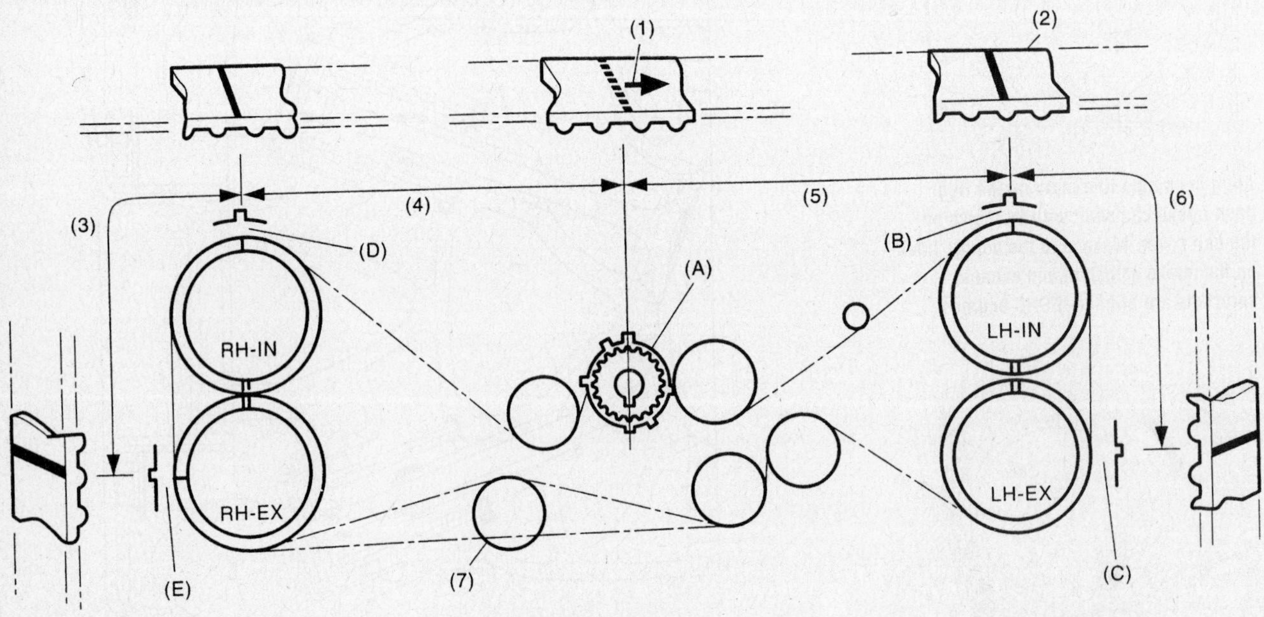

Align the marks in alphabetical order as shown if using a new belt–DOHC engines

Piston and Ring

POSITIONING

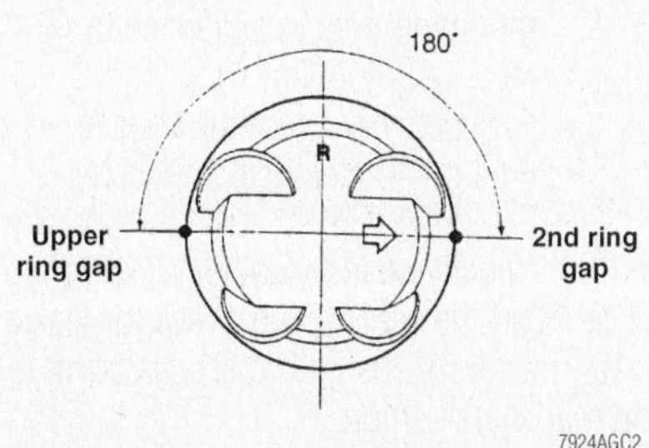

Compression ring end-gap spacing—2.5L engine

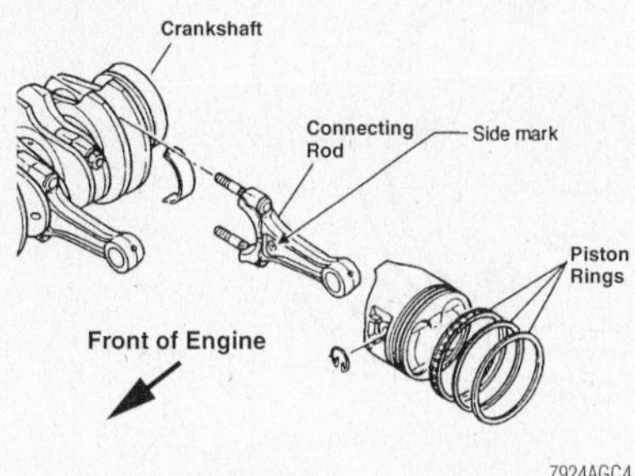

Piston and connecting rod assembly positioning—2.5L engine

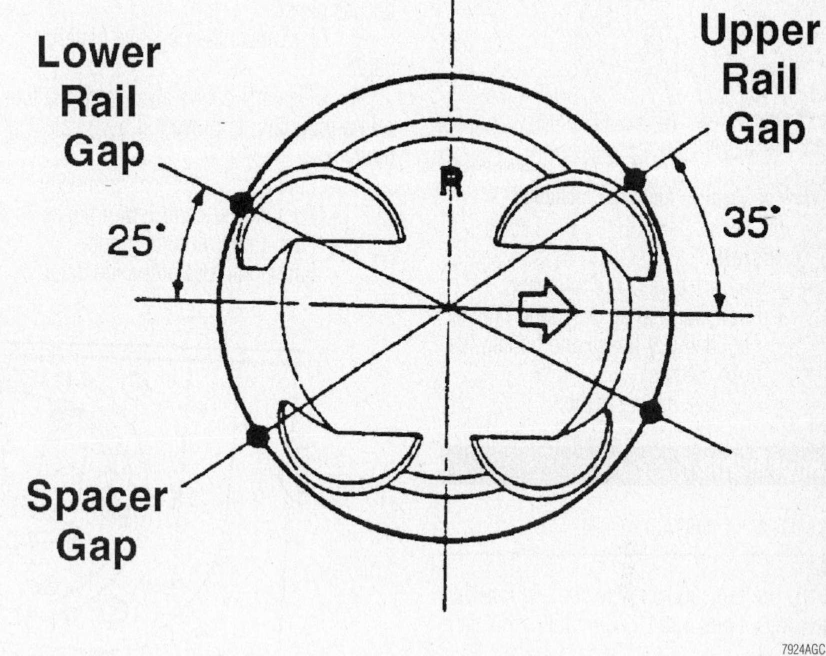

Upper, spacer and lower oil ring end-gap spacing—2.5L engine

FUEL SYSTEM

Fuel System Service Precautions

Safety is the most important factor when performing not only fuel system maintenance but any type of maintenance. Failure to conduct maintenance and repairs in a safe manner may result in serious personal injury or death. Maintenance and testing of the vehicle's fuel system components can be accomplished safely and effectively by adhering to the following rules and guidelines.

- To avoid the possibility of fire and personal injury, always disconnect the negative battery cable unless the repair or test procedure requires that battery voltage be applied.
- Always relieve the fuel system pressure prior to disconnecting any fuel system component (injector, fuel rail, pressure regulator, etc.), fitting or fuel line connection. Exercise extreme caution whenever relieving fuel system pressure, to avoid exposing skin, face and eyes to fuel spray. Please be advised that fuel under pressure may penetrate the skin or any part of the body that it contacts.
- Always place a shop towel or cloth around the fitting or connection prior to loosening to absorb any excess fuel due to spillage. Ensure that all fuel spillage (should it occur) is quickly removed from engine surfaces. Ensure that all fuel soaked cloths or towels are deposited into a suitable waste container.
- Always keep a dry chemical (Class B) fire extinguisher near the work area.
- Do not allow fuel spray or fuel vapors to come into contact with a spark or open flame.
- Always use a backup wrench when loosening and tightening fuel line connection fittings. This will prevent unnecessary stress and torsion to fuel line piping. Always follow the proper torque specifications.
- Always replace worn fuel fitting O-rings with new. Do not substitute fuel hose or equivalent, where fuel pipe is installed.

Fuel System Pressure

RELIEVING

➡This procedure must be performed prior to servicing any component of the fuel injection system.

1. Before servicing the vehicle, refer to the precautions in the beginning of this section.
2. Disconnect the connector from the fuel pump relay located behind the passenger side front sill cover.
3. Crank the engine for 5 seconds or more to relieve the fuel pressure. If the engine starts during this time, allow it to run until it stalls.
4. Connect the fuel pump relay connector after repairs are completed.

Fuel Filter

REMOVAL & INSTALLATION

1. Before servicing the vehicle, refer to the precautions in the beginning of this section.
2. Properly relieve the fuel system pressure.
3. Remove or disconnect the following:
   - Negative battery cable
   - Fuel delivery hoses from the fuel filter
   - Fuel filter from its holder

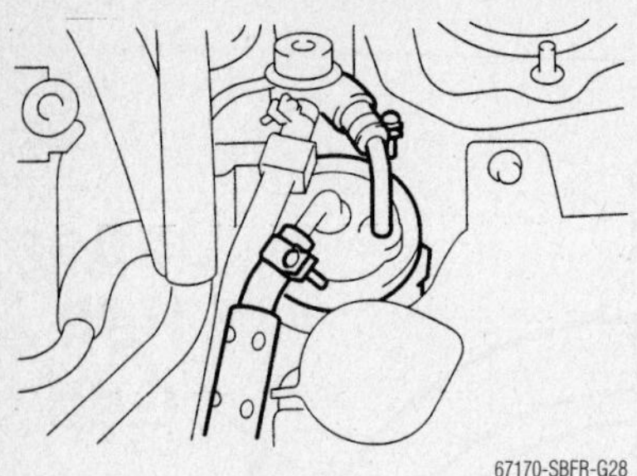

View of a typical fuel filter mounting

To install:

4. Install or connect the following:
   - Fuel filter into its mounting bracket
   - Fuel delivery hoses and tighten the hose clamps
   - Negative battery cable

Fuel Pump

REMOVAL & INSTALLATION

1. Before servicing the vehicle, refer to the precautions in the beginning of this section.
2. Properly relieve the fuel system pressure.
3. Drain the fuel tank by removing the drain plugs from the tank and draining into an approved container. Once the fuel has been drained, replace the plugs and tighten to 19 ft. lbs. (26 Nm).
4. Remove the rear seat cushion and access panel.
5. Disconnect the negative battery cable.
6. Clean any debris away from the fuel pump mounting to prevent it from entering the tank.
   - Fuel pump electrical connector
   - Fuel delivery and return hoses
   - Fuel pump mounting nuts
   - Fuel pump out of the fuel tank

To install:

7. Replace the sealing gaskets for the fuel pump.
8. Install or connect the following:
   - Fuel pump into the tank. Torque the mounting nuts in sequence to 39 inch lbs. (4.4 Nm).
   - Fuel delivery and return hoses
   - Fuel pump electrical connector
   - Fuel filler cap
   - Negative battery cable
9. Start the vehicle and check for leaks.
10. Install the fuel pump access cover and rear seat cushion.

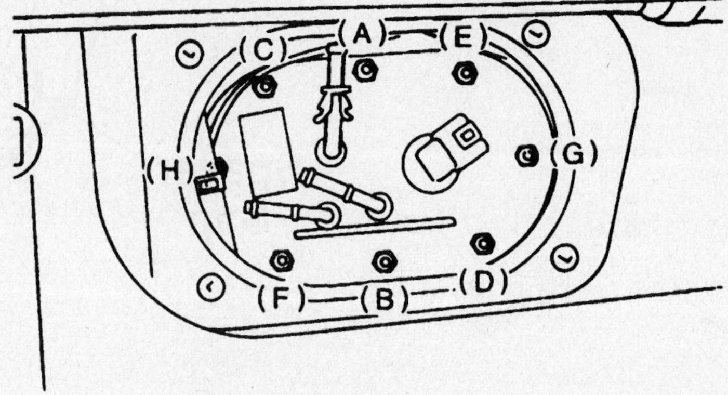

Fuel pump mounting nut tightening sequence

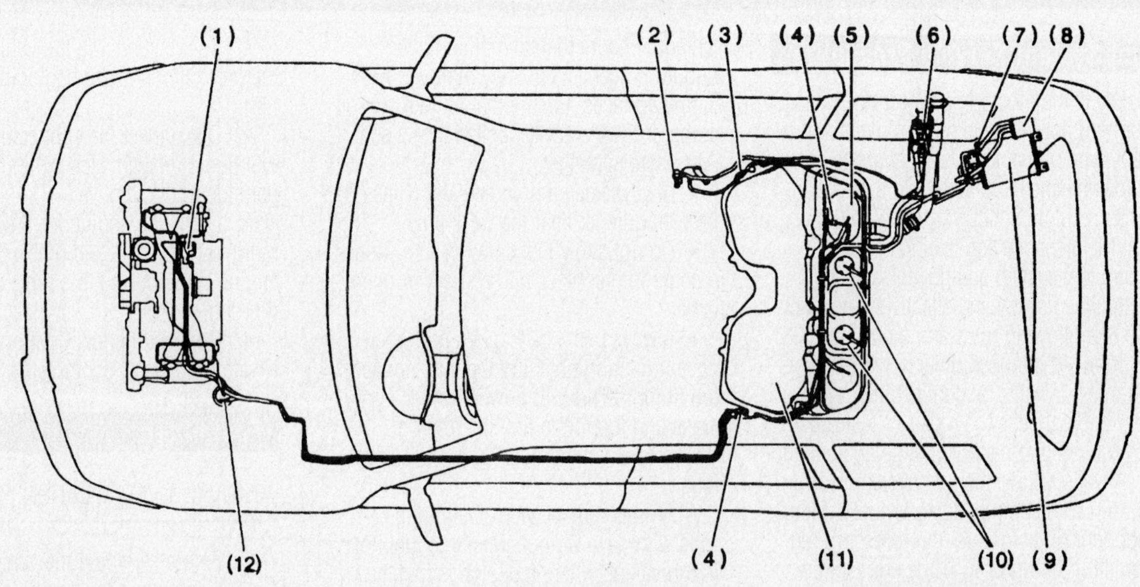

Typical fuel system component locations

Fuel Injector

REMOVAL & INSTALLATION

Right hand Side

SOHC ENGINES

1. Before servicing the vehicle, refer to the precautions in the beginning of this section.
2. Properly relieve the fuel system pressure.
3. Remove or disconnect the following:
   - Negative battery cable
   - Air duct and cleaner assembly
   - Resonator chamber
   - Spark plug wires No. 1 and 3
   - V-belt covers
   - Power steering pump belt
   - Power steering pipe bracket from manifold
   - Power steering pump and set aside
   - Fuel pipe protector, if equipped
   - Wires from fuel injector
   - Injector pipe from intake manifold
   - Fuel injector

To install:

4. Install or connect the following:
   - Fuel injector
   - Injector pipe to intake manifold
   - Wires to fuel injector
   - Fuel pipe protector, if equipped
   - Power steering pump
   - Power steering pipe bracket to manifold
   - Power steering pump belt
   - V-belt covers
   - Spark plug wires No. 1 and 3
   - Resonator camber
   - Air duct and cleaner assembly
   - Negative battery cable

View of fuel injector removal–SOHC engines

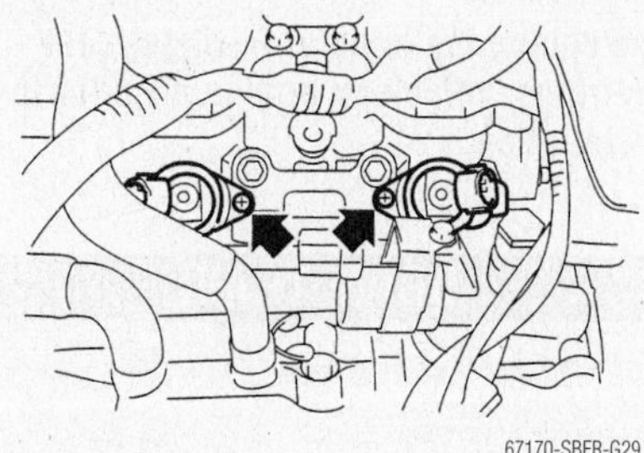

The fuel injectors are retained by a screw–DOHC engines

DOHC ENGINES

1. Before servicing the vehicle, refer to the precautions in the beginning of this section.
2. Properly relieve the fuel system pressure.
3. Disconnect the negative battery cable.
4. Remove the right hand fuel pipe protector.
5. Disconnect the connector from the injector.
6. Remove the retaining screw and the injector.

To install:

➡Replace the injector O–rings, coat the rings with clean engine oil prior to installation.

7. Install the injector and the retaining screw
8. Connect the injector connector.
9. Install the right hand fuel pipe protector.
10. Connect the negative battery cable.

Left hand Side

SOHC ENGINES

1. Before servicing the vehicle, refer to the precautions in the beginning of this section.
2. Properly relieve the fuel system pressure.
3. Remove or disconnect the following:
   - Negative battery cable
   - Air duct and cleaner assembly
   - Resonator chamber
   - Two bolts that attach the washer reservoir to the body
   - Connector from the front window washer motor
   - Rear window washer hose from the motor and plug the line
   - Washer reservoir and secure it to one side
   - Spark plug wires No. 2 and 4
   - Left hand fuel pipe connector
   - Wires from fuel injector
   - Injector pipe from intake manifold
   - Fuel injector

To install:

4. Install or connect the following:
   - Fuel injector
   - Injector pipe to intake manifold
   - Wires to fuel injector
   - Spark plug wires No. 2 and 4
   - Washer reservoir
   - Rear window washer hose to the motor
   - Connector to the front window washer motor
   - Two bolts that attach the washer reservoir to the body
   - Resonator camber
   - Air duct and cleaner assembly
   - Negative battery cable

DOHC ENGINES

1. Before servicing the vehicle, refer to the precautions in the beginning of this section.
2. Properly relieve the fuel system pressure.
3. Disconnect the negative battery cable.

4. Remove the intake manifold.
5. Disconnect the connector from the injector.
6. Remove the retaining screw and the injector.

To install:

➡Replace the injector O–rings, coat the rings with clean engine oil prior to installation.

7. Install the injector and the retaining screw
8. Connect the injector connector.
9. Install the intake manifold.
10. Connect the negative battery cable.

DRIVE TRAIN

Transmission Assembly

REMOVAL & INSTALLATION

1. Before servicing the vehicle, refer to the precautions in the beginning of this section.
2. On automatic transmission vehicles, drain the automatic transmission fluid and the front differential.
3. Remove or disconnect the following:
   - Negative battery cable
   - Air intake and chamber, then the chamber stays
   - Intercooler on turbo models
   - Air cleaner case stay
   - Front and rear Oxygen (O_2S) sensor connectors, if equipped
   - Transmission harness connector, if equipped with an automatic transmission
   - Transmission ground terminal
   - Neutral position switch connector, if equipped with a manual transmission
   - Backup light switch connector, if equipped with a manual transmission
   - 2 Vehicle Speed (VSS) sensor connectors
   - Starter
   - Clutch slave cylinder and hold it aside with wire, on non–turbo engines with a manual transmission
   - Pitching stopper
4. On automatic transmission vehicles, remove the timing hole inspection plug. Matchmark the torque converter-to-driveplate and remove the 4 bolts which hold torque converter to driveplate.
5. Remove or disconnect the following:
   - Automatic transmission fluid dipstick and tube
6. Install Engine Support Assembly ST 41099AA020. (Also available as part no. 927670000).
   - Clutch operating cylinder from the transmission by removing the plug using a 10mm wrench, on turbo models with a manual transmission
   - Screw the 6mm bolt into the release fork and remove it

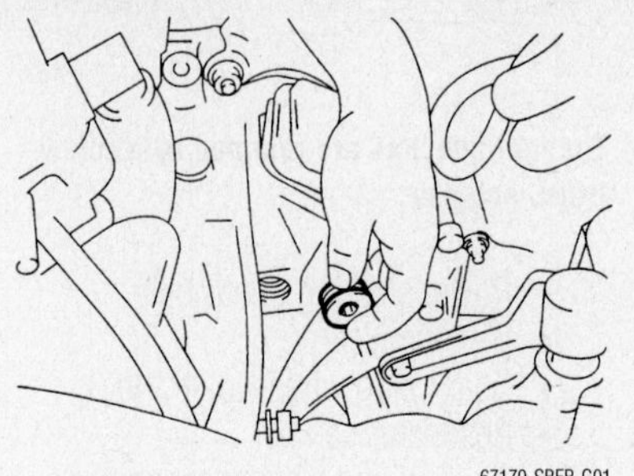

Remove the plug using a 10mm wrench–DOHC engines

   - Raise the release fork and unfasten the release bearing tabs to free the fork.
7. Remove or disconnect the following:
   - Bolt securing the right upper side of the transmission to the engine
   - Engine undercover
   - Front Y-pipe
8. Disconnect connector from rear O_2S sensor
9. Remove or disconnect the following:
   - Center exhaust pipe from rear pipe and hanger bolt
   - Rear exhaust pipe and if equipped the heat shield cover
   - Hanger bracket from the right side of the transmission
   - Transmission cooler lines
   - Rear driveshaft, matchmark for reassembly
   - Center bearing bracket

➡Plug the opening at the rear of the extension housing to prevent oil from flowing out.

   - Shifter stay and rod from the transmission, on manual transmission
   - Gear shift cable from the transmission select lever, on automatic transmission

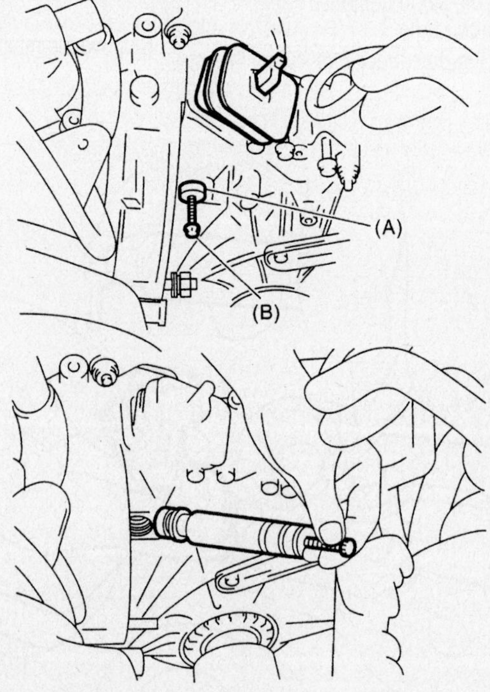

Screw the 6mm bolt into the release fork and remove it–DOHC engines

- Sway bar from the transverse link
- Parking brake cable bracket from the transverse link and bolt holding the transverse link to the cross-member on each side, lower the transverse link
- Lower ball joint from knuckle
- Spring pin and separate the half-shaft from the transmission on each side

➡Use a small punch to remove the spring pin. Discard old spring pin and always install a new pin.

10. Disconnect the halfshaft from transmission on each side. Be sure to remove the axle shaft from the transmission by pushing the rear of the tire outward.

11. Remove the engine-to-transmission mounting nuts

12. Support the transmission with a jack.

➡Do not place jack under the transmission oil pan, otherwise the oil pan may be damaged.

13. Remove or disconnect the following:
    - Rear transmission crossmember
    - Transmission

To install:

14. If equipped with a manual transmission, perform the following:

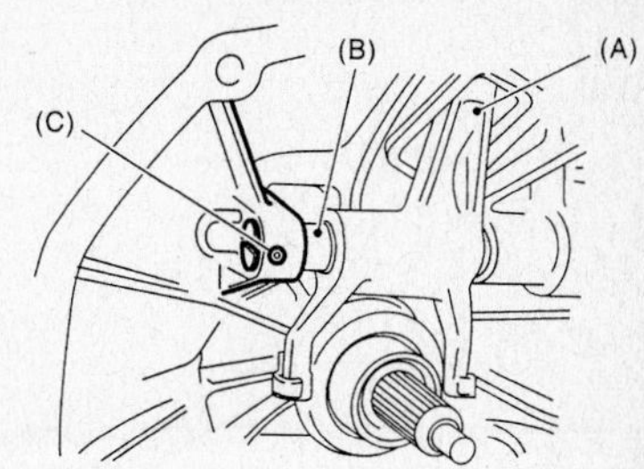

(A) Release fork
(B) Release shaft
(C) Spring pin

67170-SBFR-G03

Release fork and related components–DOHC engines

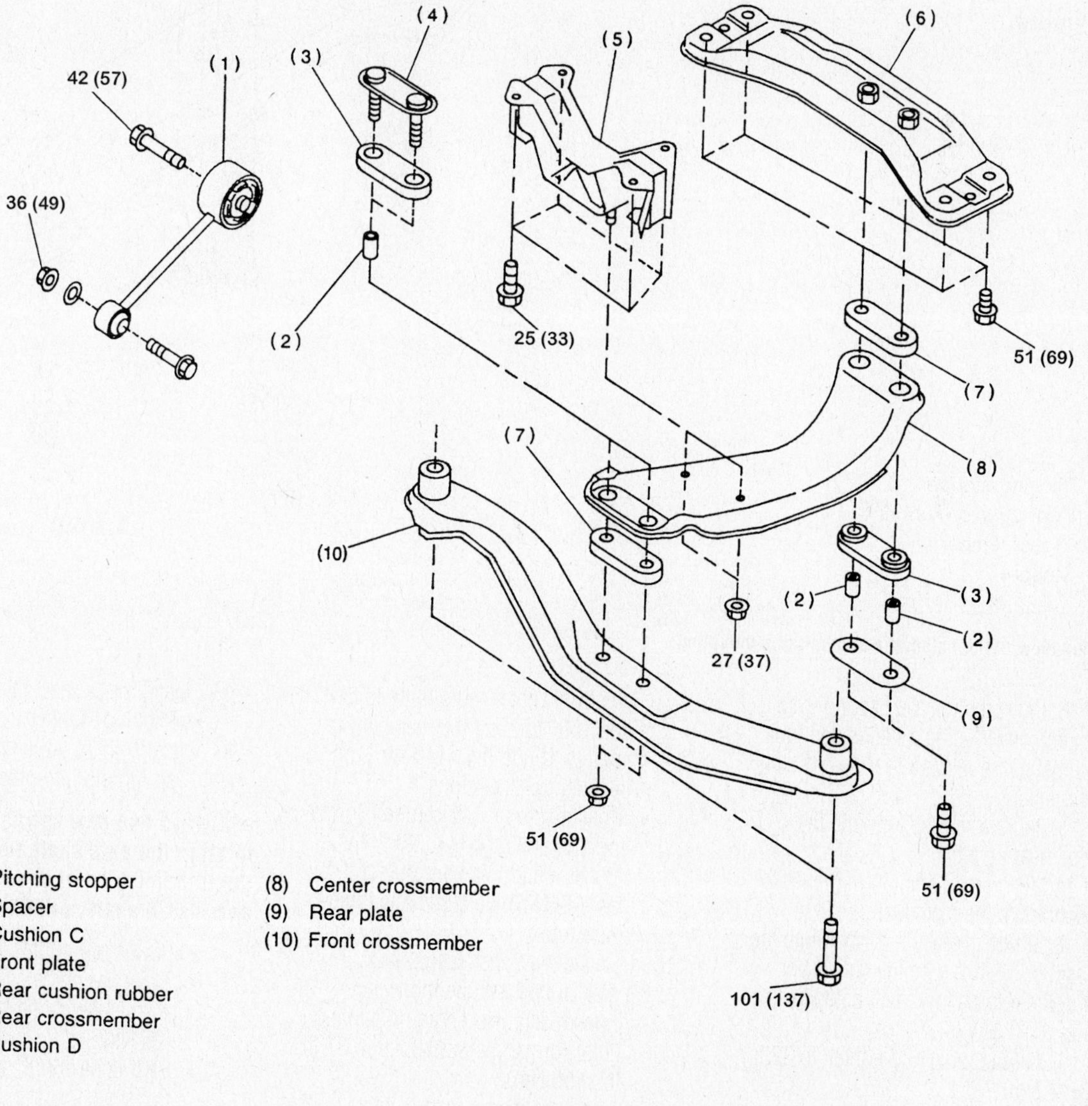

(1) Pitching stopper
(2) Spacer
(3) Cushion C
(4) Front plate
(5) Rear cushion rubber
(6) Rear crossmember
(7) Cushion D
(8) Center crossmember
(9) Rear plate
(10) Front crossmember

42 (57)
36 (49)
25 (33)
51 (69)
27 (37)
51 (69)
51 (69)
101 (137)

ft. lbs. (Nm)

7924XG14

Exploded view of the manual transmission mounting

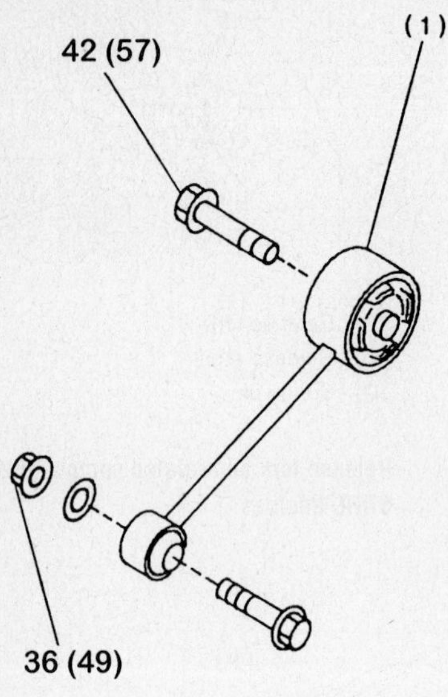

(1) Pitching stopper
(2) Rear cushion rubber
(3) Crossmember
(4) Stopper

Exploded view of the automatic transmission mounting

a. Remove the release bearing from the clutch cover using a flat bladed tool.

b. Install the release bearing onto the transmission.

c. Install the release fork into the release bearing tab.

d. Apply a small amount of grease to the splines of the mainshaft.

e. Insert the release fork shaft into the release fork. Make sure the cutout portion of the release fork shaft contacts the spring pin.

f. Install the plug and tighten to 32 ft. lbs. (44 Nm).

15. Install or connect the following:
    - Special tool 498277200 to torque converter clutch case
    - Transmission to the engine
    - Transmission crossmember. Torque the front (outer) nuts/bolts to 51 ft. lbs. (69 Nm) and the rear (inner) nuts to 103 ft. lbs. (140 Nm), on manual transmissions
    - Transmission crossmember. Torque the inner nuts/bolts to 26 ft. lbs. (35 Nm) and the outer nuts to 51 ft. lbs. (69 Nm), on automatic transmissions
16. Remove the transmission jack.
17. Install or connect the following:
    - Transmission-to-engine mounting nuts. Torque the nuts/bolts to 37 ft. lbs. (50 Nm).
    - Torque converter-to-driveplate bolts on automatic transmission vehicles. Torque the bolts to 18 ft. lbs. (25 Nm).
18. Remove special tool 927670000.
19. Install or connect the following:
    - Pitching stopper
    - Halfshaft to transmission and spring pin into place

➡Always use new spring pin. Be sure to align the axle shaft and shaft from the transmission at chamfered holes and install shaft splines correctly.

- Lower ball joint to knuckle
- Sway bar to the crossmember. Torque the clamp bolts to 18 ft. lbs. (25 Nm).
- Shift control rod, shifter stay to the transmission and the spring, on manual transmission vehicles
- Gear shift cable to the select lever, on automatic transmission vehicles
- Fluid cooler lines

- Driveshaft. Tighten the bolts to 23 ft. lbs. (31 Nm).
- Center bearing bracket. Torque the bolt to 38 ft. lbs. (52 Nm).
- Heat shield cover, if removed
- Center exhaust pipe from rear pipe and hanger bolt
- Rear exhaust pipe and heat shield cover
- Hanger bracket from the right side of the transmission
- Y-pipe with new gaskets and nuts
- Rear O_2S sensor connector
- Clutch slave cylinder on manual transmission vehicles. Torque the mounting bolts to 27 ft. lbs. (37 Nm).
- Automatic transmission fluid dipstick tube
- Transmission connector bracket
- Starter. Torque the mounting bolts to 37 ft. lbs. (50 Nm).
- Front and rear O_2S sensor connectors
- Transmission harness connector, on automatic transmission vehicles
- Transmission ground terminal
- Neutral position switch connector, on manual transmission vehicles
- Backup light switch connector, on manual transmission vehicles
- 2 VSS connectors
- Air intake and chamber, and the camber stays
- Intercooler, on turbo models
- Negative battery cable

20. On automatic transmission vehicles, fill the automatic transmission fluid with Dexron®II or III or equivalent.

21. On manual transmission vehicles, check and fill the transmission with 75W-90 gear oil.

22. Road test the vehicle.

Clutch

ADJUSTMENT

This vehicle is equipped with a hydraulic clutch that is self-adjusting, therefore no adjustment is possible or necessary.

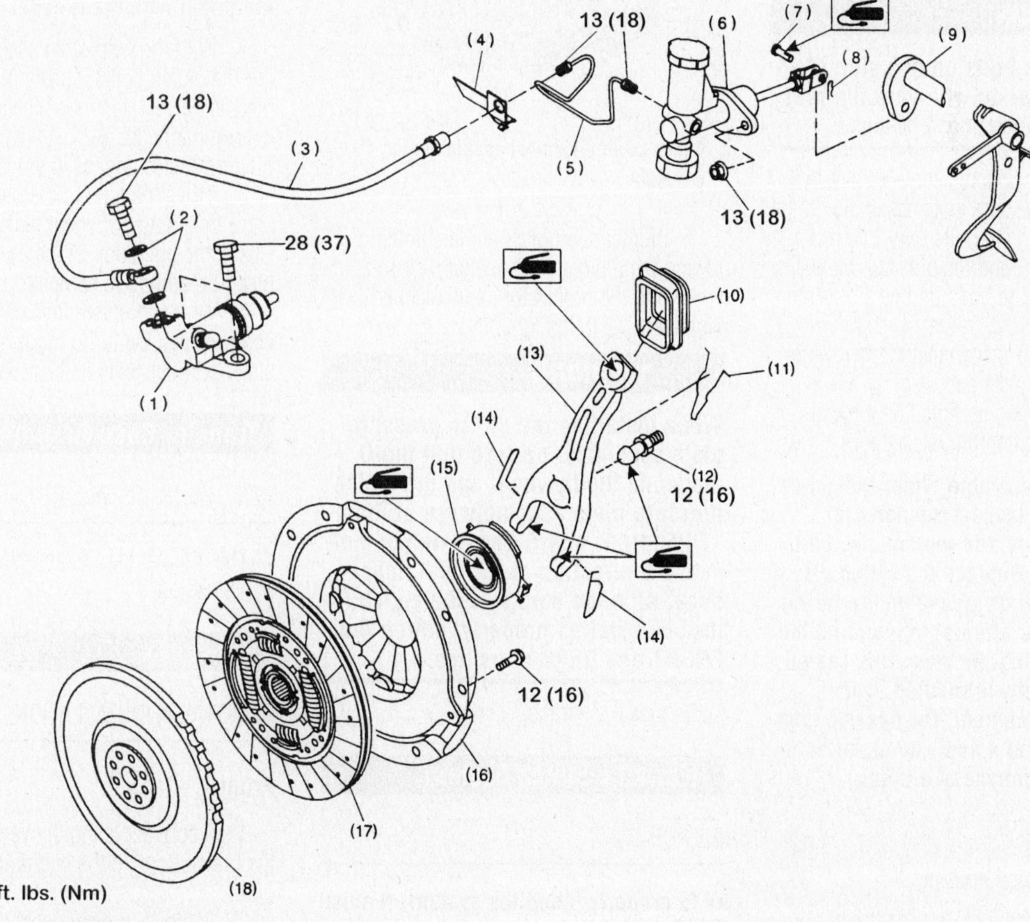

Exploded view of clutch system

REMOVAL & INSTALLATION

**** CAUTION**

The clutch driven disc may contain asbestos, which has been determined to be a cancer-causing agent. Never clean clutch surfaces with compressed air. Avoid inhaling any dust from any clutch surface. When cleaning clutch surfaces, use a commercially available brake cleaning fluid.

1. Before servicing the vehicle, refer to the precautions in the beginning of this section.
2. Remove or disconnect the following:
   - Negative battery cable
   - Transmission

**** WARNING**

Removing the bolts on one side of the pressure plate will warp the pressure plate, rendering it useless.

3. Gradually unscrew the six 6mm bolts that hold the pressure plate assembly on the flywheel. Loosen the bolts only 1 turn at a time, working around the pressure plate. Do not unscrew all the bolts on one side at one time.
4. Remove or disconnect the following:
   - Clutch plate and disc
   - 2 retaining springs, the throwout bearing and the release fork

➡Do not disassemble either the clutch cover or disc. Inspect the parts for wear or damage and replace any parts as necessary. Replace the clutch disc if there is any oil or grease on the facing. Do not wash or attempt to lubricate the throwout bearing, because it is sealed and permanently lubricated. If it requires replacement, the bearing may be removed and a new one installed in the holder by means of a press.

To install:

5. Fit the release fork boot on the front of the transmission housing.
6. Install or connect the following:
   - Release fork
   - Throwout bearing assembly and secure it with the 2 springs

➡Coat the inside diameter of the throwout bearing and the release lever contact points with grease.

- Clutch alignment tool through the clutch cover and disc, then insert the end of the tool into the needle bearing

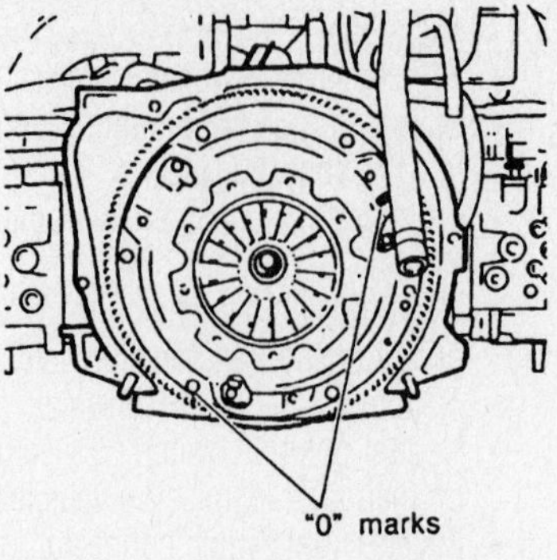

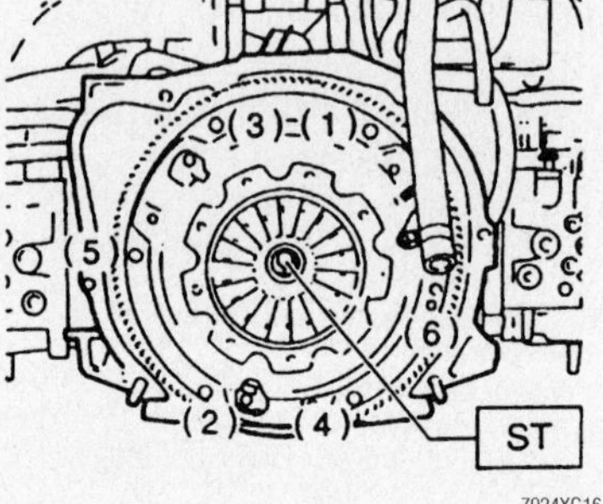

Clutch cover alignment and tightening sequence

7. Tighten the pressure plate bolts following the illustrated sequence, 1 turn at a time, until the proper torque is reached. Tighten to 12 ft. lbs. (16 Nm).

**** WARNING**

When installing the clutch pressure plate assembly, be sure that the O marks on the flywheel and the clutch pressure plate assembly are at least 120 degrees apart. These marks indicate the direction of residual unbalance. Also, be sure that the clutch disc is installed properly, noting the FRONT and REAR markings.

8. Install the transmission.

Hydraulic Clutch System

BLEEDING

➡To properly bleed the system, it must be bled at the slave cylinder and at the damper. Each of these has an air bleeder on it.

1. Before servicing the vehicle, refer to the precautions in the beginning of this section.
2. Remove any components necessary to access the slave cylinder.
3. On turbo models, remove the cylinder without disconnecting the clutch hose.

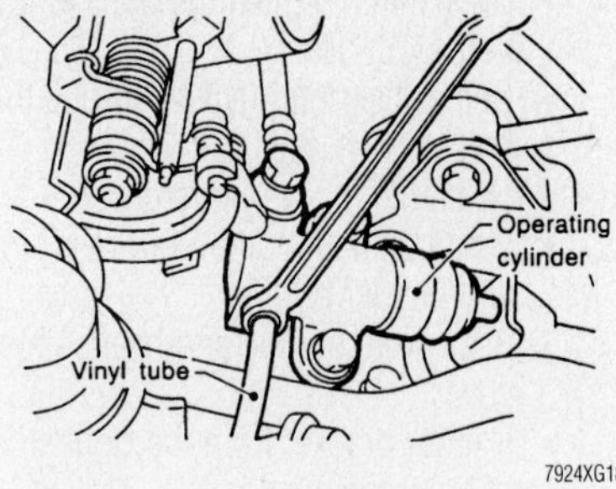

Bleeding the hydraulic clutch at the slave cylinder

4. Connect a vinyl tube to the air bleeder on the damper and put the other end in a jar with clean clutch fluid.

➡Do not let the fluid level fall too low in the master cylinder. Do not release the pedal with the bleeder open.

5. With the help of an assistant depressing the clutch pedal, slowly open the bleeder valve. Close the bleeder valve and release the pedal. Repeat this process until no air bubbles appear in the jar.
6. Move the tube to the bleeder on the slave cylinder and repeat the process. Check the operation of the clutch after the bleed procedure is complete.
7. When complete, reinstall all removed components. On turbo models tighten the slave cylinder bolts to 27 ft. lbs. (37 Nm).

Transfer Case Assembly

REMOVAL & INSTALLATION

The transfer case is an integral part of the transmission.

Halfshafts

REMOVAL & INSTALLATION

Front

1. Before servicing the vehicle, refer to the precautions in the beginning of this section.
2. Remove or disconnect the following:
   - Negative battery cable
   - Wheel
   - Axle nut
   - Stabilizer link from transverse link
   - Brake caliper and suspend from the housing using wire
   - Brake rotor
   - Tie rod from the knuckle
   - Anti-Lock Brake (ABS) sensor and harness, if equipped

- Transverse link ball joint from the knuckle
- Halfshaft-to-transmission roll pin and discard it
- Halfshaft from the transmission

3. Using Axle Shaft Puller 926470000 and Plate 927140000, remove the halfshaft from the hub.

To install:

4. Install the halfshaft into the hub.
5. Using halfshaft Installer 922431000 and Adapter 927390000, pull the halfshaft through the hub.
6. Install and temporarily tighten a new axle nut.
7. Install or connect the following:
   - Halfshaft onto the transmission, by aligning the halfshaft roll pin hole
   - New roll pin
   - Transverse link to housing. Torque the nut to 37 ft. lbs. (50 Nm).
   - Tie rod and tighten the castellated nut to 20 ft. lbs. (27 Nm) and then up to an additional 60 degrees to align the cotter pin hole with the slot on the nut, then install a new cotter pin
   - Stabilizer link
   - Brake rotor
   - Brake caliper
   - ABS sensor and harness, if equipped

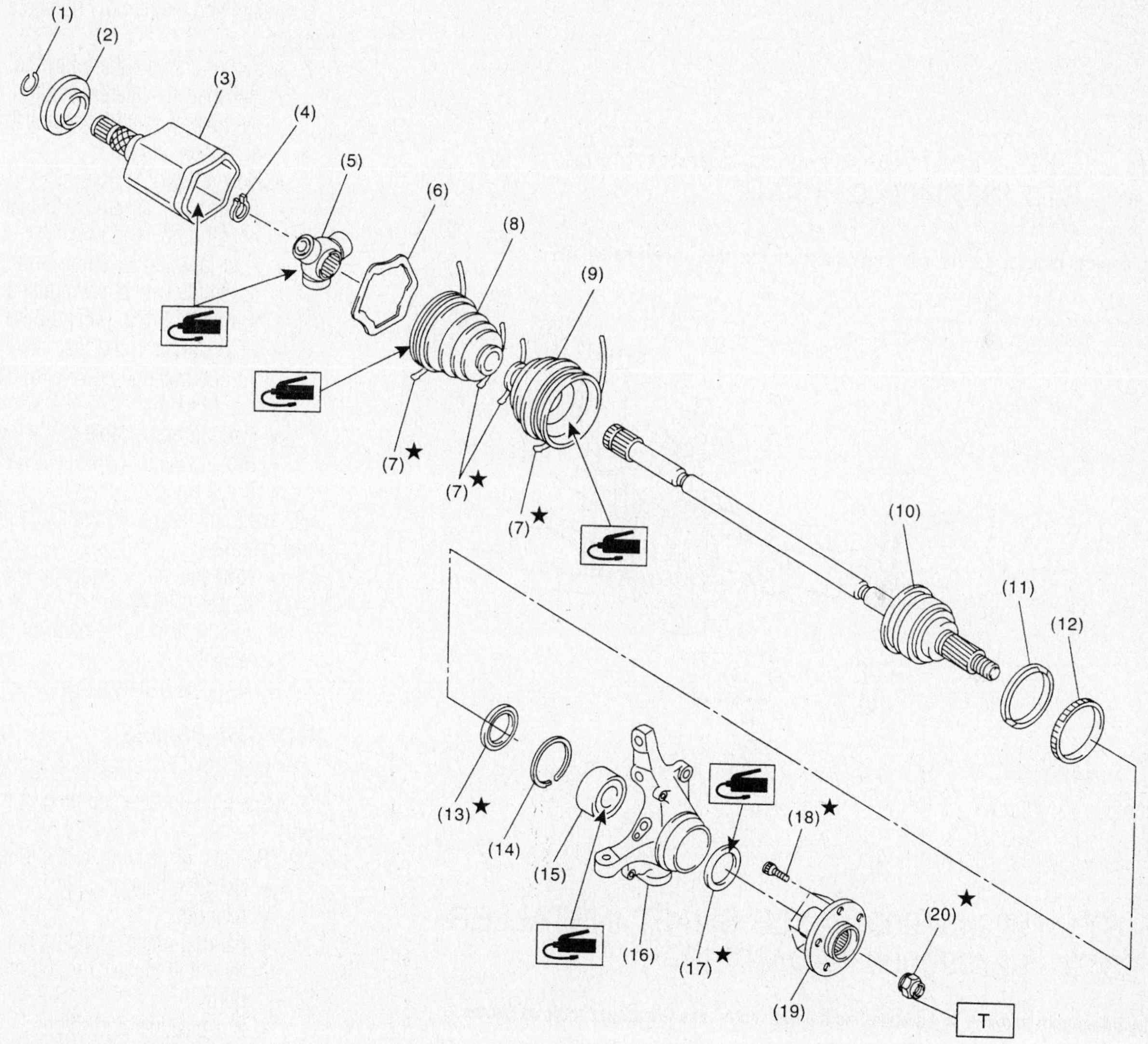

(1)	Spring pin	(9)	Boot (AC)	(17)	Oil seal (OUT)
(2)	Baffle plate (SFJ)	(10)	AC ASSY	(18)	Hub bolt
(3)	Outer race (SFJ)	(11)	Tone wheel	(19)	Hub
(4)	Snap ring	(12)	Baffle plate	(20)	Axle nut
(5)	Trunnion	(13)	Oil seal (IN)		
(6)	Retainer	(14)	Snap ring		
(7)	Boot band	(15)	Bearing		
(8)	Boot (AARi)	(16)	Housing		

Tightening torque: N·m (kgf-m, ft-lb)
T: 190 (19.4, 140)

67170-SBFR-G30

Exploded view of the front halfshaft

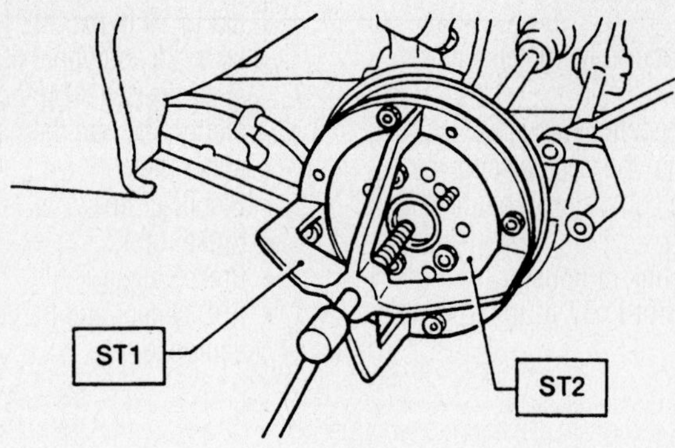

ST1 926470000 AXLE SHAFT PULLER
ST2 927140000 PLATE

7924XG29

Be sure not to damage the threads when removing the front or rear halfshafts

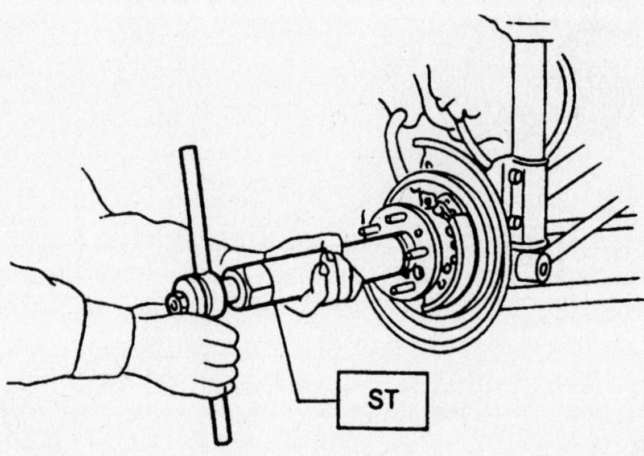

ST1 922431000 AXLE SHAFT INSTALLER
ST2 927390000 ADAPTER

7924XG30

To avoid using a hammer when installing the halfshafts, use the proper tools as shown

- New axle nut. Torque the nut to 140 ft. lbs. (190 Nm) and stake the nut.
- Wheel
- Negative battery cable

Rear

WITH DRUM BRAKES

1. Before servicing the vehicle, refer to the precautions in the beginning of this section.
2. Remove or disconnect the following:
   - Negative battery cable
   - Axle nut
   - Parking brake adjusting nut after returning the lever to the off position
   - Brake drum
   - Brake hose from the wheel cylinder and plug the line and wheel cylinder
   - Parking brake cable from the lever
   - Lateral link assembly from rear housing
   - ABS sensor from the backing plate
   - Halfshaft from the differential
3. Using Axle Shaft Puller 926470000 and Plate 927140000, remove the halfshaft from the hub.

To install:

4. Install the halfshaft into the rear housing.
5. Using halfshaft Installer 922431000 and Adapter 927390000, pull the halfshaft into place.
6. Install and temporarily tighten a new axle nut.
7. Install or connect the following:
   - Halfshaft-to-differential align roll pin holes and slide the halfshaft onto the splines
   - New roll pin
   - Trailing link assembly-to-rear housing bolt and nut.
   - Rear housing assembly and strut assembly with a new nut and torque to 145 ft. lbs. (196 Nm).
   - Rear stabilizer and rear lateral link. Torque bolt and new nut to 32 ft. lbs. (44 Nm).
   - Parking brake cable to the lever
   - Brake hose to the wheel cylinder
   - Brake drum
8. Bleed the brake system and adjust the parking brake.
   - New axle nut. Torque the nut to 137 ft. lbs. (186 Nm).
   - ABS sensor to the backing plate
   - Wheel
   - Negative battery cable

WITH DISC BRAKES

1. Before servicing the vehicle, refer to the precautions in the beginning of this section.
2. Remove or disconnect the following:
   - Negative battery cable
   - Axle nut
   - Parking brake adjusting nut after returning the lever to the off position
   - Brake caliper and rotor
   - Parking brake cable from the lever
   - Lateral link assembly to rear housing
   - ABS sensor from the backing plate
   - Halfshaft from the differential
3. Using Axle Shaft Puller 926470000 and Plate 927140000, remove the halfshaft from the hub.

To install:

4. Install the halfshaft into the rear housing.
5. Using halfshaft Installer 922431000

and Adapter 927390000, pull the halfshaft into place.

6. Install and temporarily tighten a new axle nut.

7. Install or connect the following:

- Halfshaft-to-differential align roll pin holes and slide the halfshaft onto the splines
- New roll pin
- Rear housing assembly and strut assembly with a new nut and torque to 145 ft. lbs. (196 Nm).
- Rear stabilizer and rear lateral link. Torque bolt and new nut to 32 ft. lbs. (44 Nm).
- Parking brake cable to the lever
- Brake rotor and caliper and adjust the parking brake
- New axle nut. Torque the nut to 137 ft. lbs. (186 Nm).
- ABS sensor to the backing plate
- Wheel
- Negative battery cable

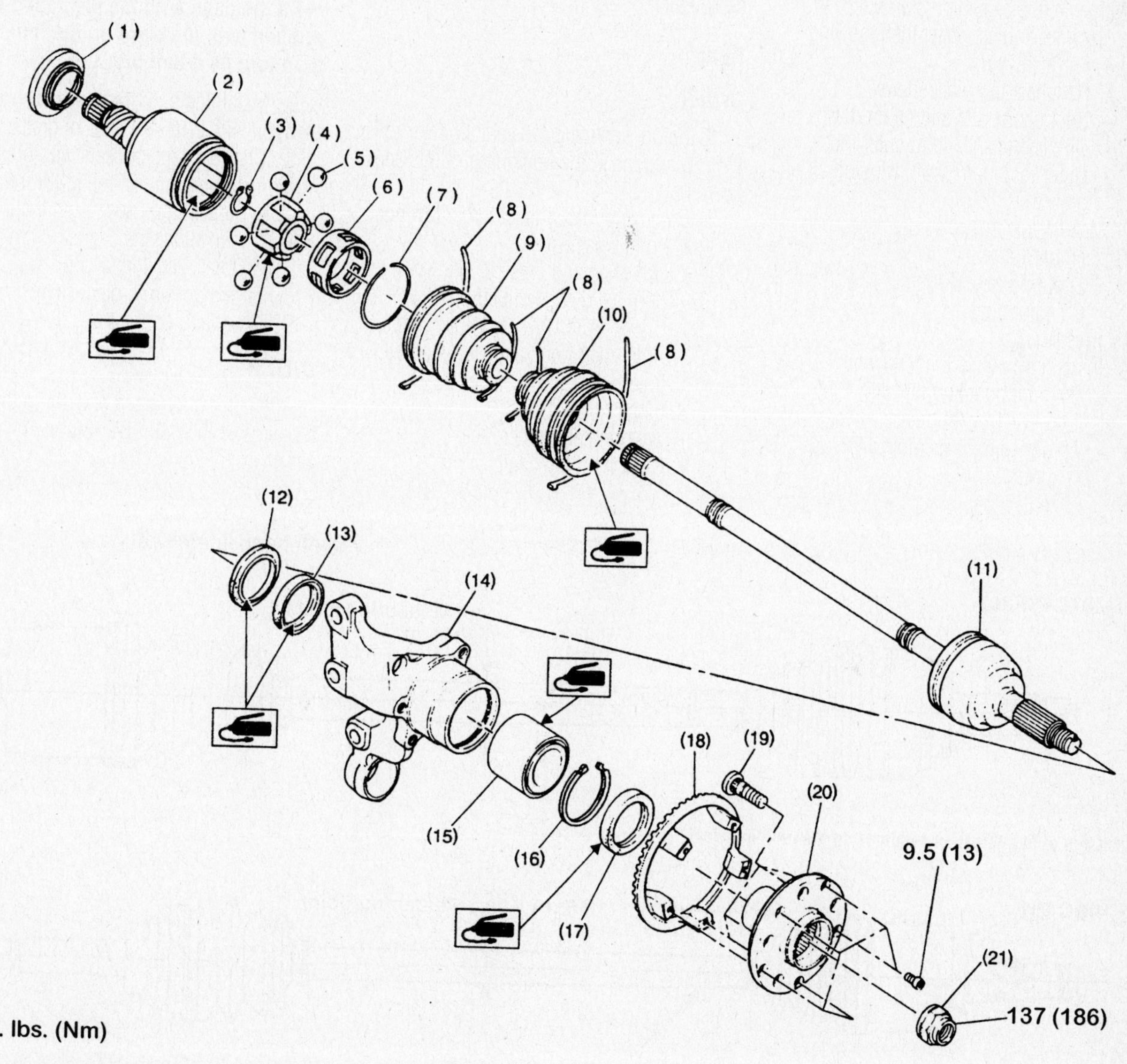

Exploded view of the rear halfshaft

CV-Joints

REMOVAL & INSTALLATION

Front

INNER

1. Before servicing the vehicle, refer to the precautions in the beginning of this section.
2. Remove or disconnect the following:
   - Front wheel(s)
   - Halfshaft and place in vise
   - Boot bands and slide boot down
   - Circlip from CV-joint outer race
   - Outer race from shaft, wipe off all grease
3. Matchmark Tri-pot spider assembly for reassembly.
   - Snapring and Tri-pot
   - CV-joint boot

To install:

4. Install or connect the following:
   - CV-joint boot and fill with 1.06–1.41 oz. of grease
   - Tri-pot spider assembly and snapring
   - Outer race, fill with 3.53–3.88 oz. of grease
   - Circlip to CV-joint outer race
5. Slide boot onto outer race
6. New band on the boot

OUTER

The outer boot is the only thing that is replaceable the outer CV-joint is not disassembled.

Rear

INNER

1. Before servicing the vehicle, refer to the precautions in the beginning of this section.
2. Remove or disconnect the following:
   - Front wheel(s)
   - Half shaft and place in vise
   - Boot bands and slide boot down
   - Circlip from CV- joint outer race
   - Outer race from shaft, wipe off all grease
   - 6 balls
   - Snapring and inner race
   - Cage from the shaft
   - CV boot

To install:

3. Install or connect the following:
   - CV-joint boot and fill with 2.12–2.47 oz. of grease
   - Cage to the shaft with the cut out portion facing the shaft end
   - Inner race and snapring
4. Install the cage to the inner race.

➡Fit the cage with the protruded part aligned with the track on the inner race then turn by a half pitch.

5. Install the outer race and snapring; then, fill with 2.82–3.17 oz. of grease
6. Coat the cage pockets with grease.
7. Install or connect the following:
   - 6 balls into cage, align the inner race and cage
   - Outer race and circlip
8. Slide boot on to outer race
9. New band on the boot

OUTER

The outer boot is the only thing that is replaceable the outer CV-joint is not disassembled.

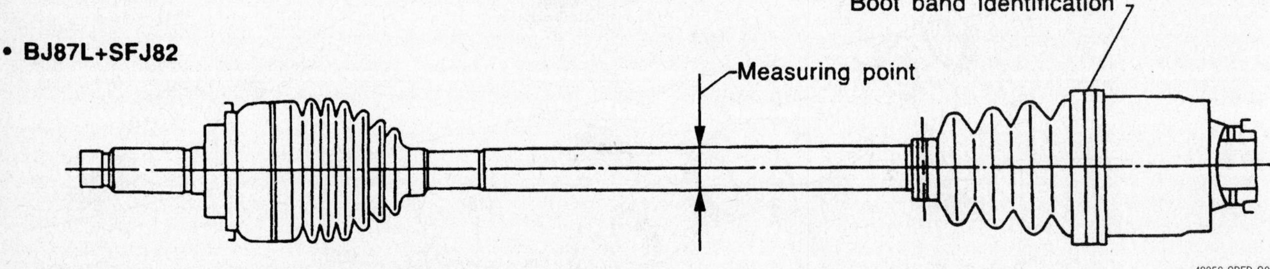

Location of identification paddings on the front halfshaft

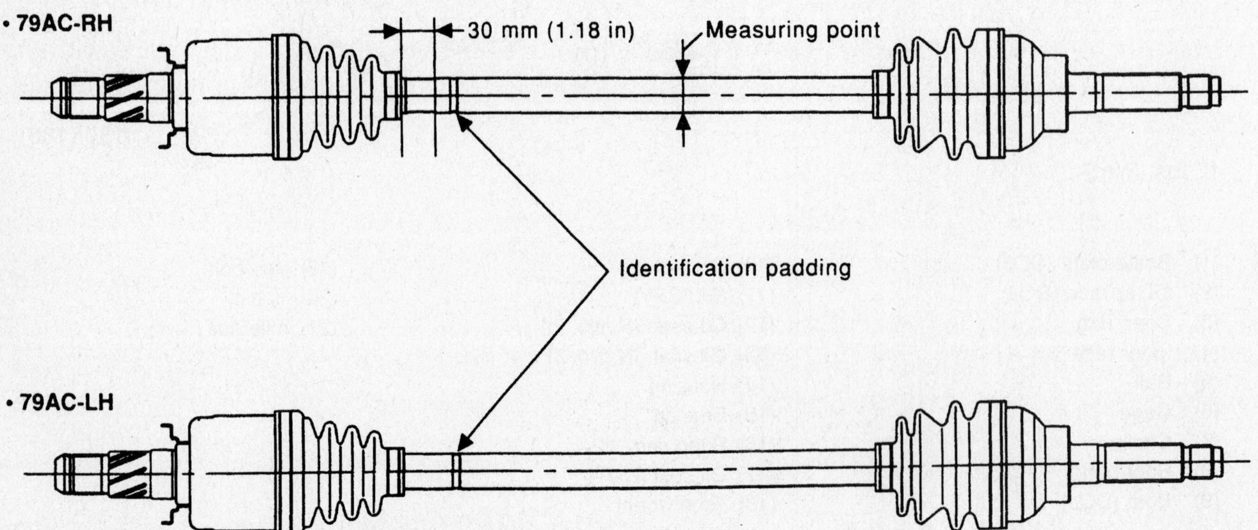

Location of identification paddings on the rear halfshaft

STEERING AND SUSPENSION

Air Bag

✲✲ CAUTION

All vehicles are equipped with an air bag system. The system must be disabled before performing service on or around system components, steering column, instrument panel components, wiring and sensors. Failure to follow safety and disabling procedures could result in accidental air bag deployment, possible personal injury and unnecessary system repairs.

PRECAUTIONS

Several precautions must be observed when handling the inflator module to avoid accidental deployment and possible personal injury.

- Never carry the inflator module by the wires or connector on the underside of the module.
- When carrying a live inflator module, hold securely with both hands, and ensure that the bag and trim cover are pointed away.
- Place the inflator module on a bench or other surface with the bag and trim cover facing up.
- With the inflator module on the bench, never place anything on or close to the module, which may be thrown in the event of an accidental deployment.

DISARMING

1. Before servicing the vehicle, refer to the precautions in the beginning of this section.
2. Disconnect the negative battery cable.
3. Disconnect the positive battery cable.
4. Wait more than 20 seconds to allow the air bag system to deplete its backup power before starting work.
5. To rearm the air bag system, reconnect the positive, then the negative battery cables.

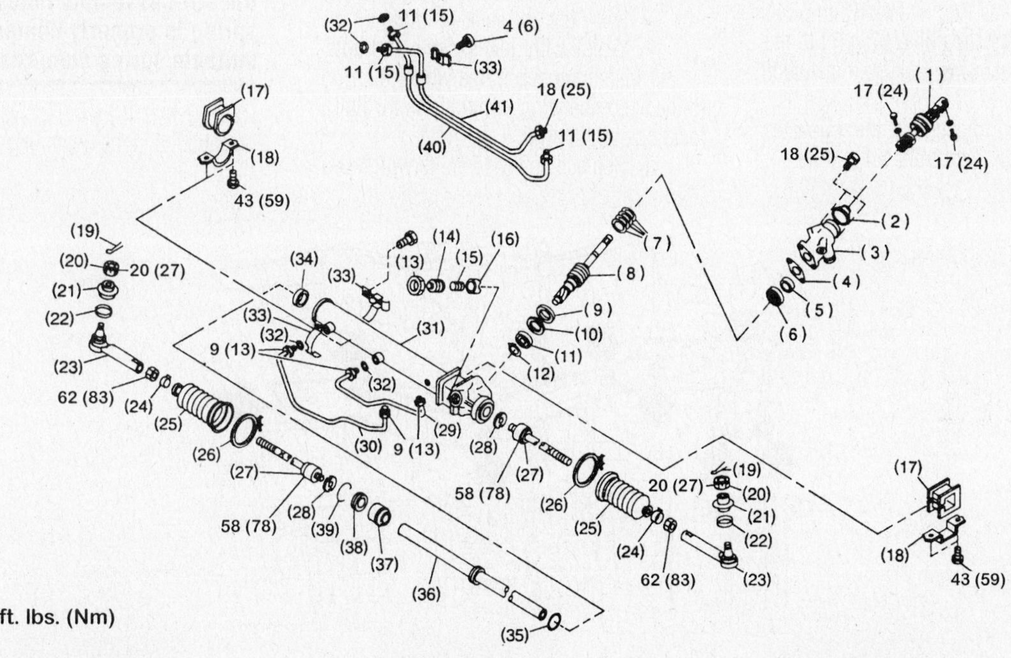

(1) Universal joint
(2) Dust cover
(3) Valve housing
(4) Gasket
(5) Oil seal
(6) Special bearing
(7) Seal ring
(8) Pinion and valve ASSY
(9) Oil seal
(10) Back-up washer
(11) Ball bearing
(12) Snap ring
(13) Lock nut
(14) Adjusting screw
(15) Spring
(16) Sleeve
(17) Adapter
(18) Clamp
(19) Cotter pin
(20) Castle nut
(21) Dust cover
(22) Clip
(23) Tie-rod end
(24) Clip
(25) Boot
(26) Band
(27) Tie-rod
(28) Lock washer
(29) Pipe B
(30) Pipe A
(31) Housing ASSY
(32) O-ring
(33) Clamp
(34) Oil seal
(35) Piston ring
(36) Rack
(37) Rack bushing
(38) Rack stopper
(39) Circlip
(40) Pipe E
(41) Pipe F

7924XG22

Exploded view of the rack and pinion steering gear

Rack and Pinion Steering Gear

REMOVAL & INSTALLATION

1. Before servicing the vehicle, refer to the precautions in the beginning of this section.
2. Remove or disconnect the following:
   - Negative battery cable
   - Front wheels
   - Engine undercover
3. Remove the sub frame while referring to the accompanying illustrations for bolt locations as follows:
   a. Loosen bolt (1) but leave it connected by a few threads.
   b. Remove bolts 2, 3, 4, 5 and 6 (in that order) and remove the subframe.
   - Y-pipe
   - Tie rod end cotter pin and nut
   - Jack-up plate and front sway bar
   - Fluid lines from the rack and pinion
4. Matchmark the universal joint to the serration in the steering rack for installation reference.
5. Remove or disconnect the following:
   - Universal joint bolts and lift the joint upward disconnecting it from the rack and pinion shaft.
   - Rack and pinion

To install:

6. Install the rack and pinion. Torque the clamp bolts to 43 ft. lbs. (59 Nm).
7. Align the steering rack to the universal joint. Push the long yoke of the joint all the way into the serrated position of the steering shaft, setting the bolt hole in the cut-out. Pull the short yoke all the way out of the serrated portion of the rack and pinion, setting the bolt hole in the cut-out. Insert the bolt through the short yoke. Pull the yoke and ensure the bolt is properly engaged in the cut-out. Fasten the short yoke side with the spring washer and bolt, then fasten the yoke side. Tighten the bolts to 17 ft. lbs. (24 Nm).
8. Install or connect the following:
   - Tie rod ends to the steering knuckle
   - Sway bar and jack-up plate
   - Y-pipe with new gaskets and nuts
9. Install the sub frame while referring to the accompanying illustrations for bolt locations as follows:
   a. Replace any M12 bolts with new ones. tighten the bolts marked in the illustration as (T1) to 41 ft. lbs. (55 Nm) and the bolts marked (T2) to 52 ft. lbs. (71 Nm).
   b. Inspect all the bolts and make sure they are torqued to the proper specification.
10. Install or connect the following:
   - Engine undercover
   - Wheels
11. Fill and bleed the steering system.

Strut

REMOVAL & INSTALLATION

Front

CAUTION

Do not remove the large nut on top of the strut assembly unless the coil spring is properly compressed with a suitable spring compressor.

1. Before servicing the vehicle, refer to the precautions in the beginning of this section.

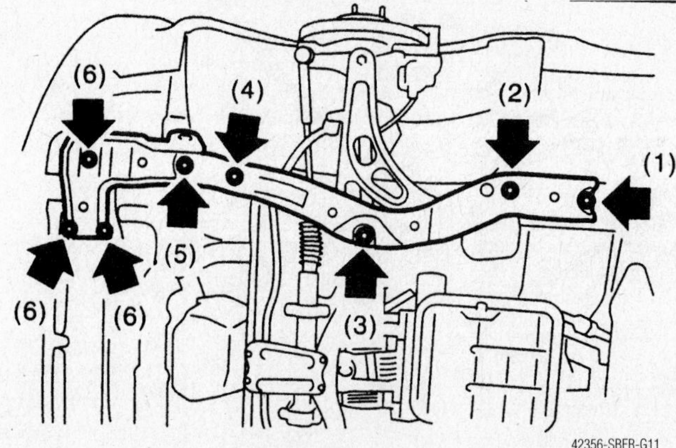

Remove the sub frame bolts in the sequence illustrated following the procedure specified in the text

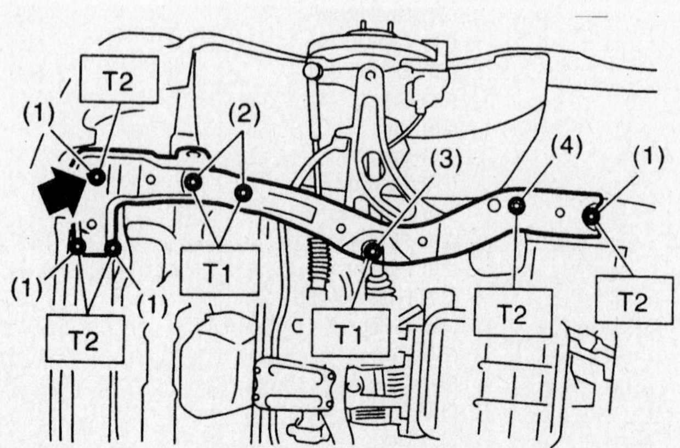

Install the sub frame bolts and tighten to the specifications listed in the procedure

2. Remove or disconnect the following:
- Negative battery cable
- Front wheel assembly
- Brake line to the strut housing bolt
- Caliper, leaving the line connected and suspend it out of the way

3. Matchmark the camber adjustment bolt to the strut housing as reference for installation.

4. Remove or disconnect the following:
- Anti-locking Brakes System (ABS) sensor and harness, if equipped
- Strut from the steering knuckle. Note that the shaft of the top bolt is not round.
- Strut from the body in the engine compartment
- Strut and coil spring assembly

To install:

5. Install the strut and coil assembly. Torque the upper strut retainer nuts to 15 ft. lbs. (20 Nm).

6. Align matchmark on camber adjustment bolt and strut housing.

7. Install or connect the following:
- Lower strut nuts and bolts. Tighten the nuts, while securing the bolts to 112 ft. lbs. (152 Nm) on 2001 models and 129 ft. lbs. (175 Nm) on 2002–04 models.
- ABS sensor and harness. Torque the bolt to 24 ft. lbs. (32 Nm), if equipped.
- Brake line to the strut bolt. Torque the bolt to 24 ft. lbs. (32 Nm).
- Caliper
- Front wheel
- Negative battery cable

8. Check and adjust the front end alignment.

Rear

⁂ CAUTION

Do not remove the large nut on top of the strut assembly unless the coil spring is properly retained with a spring compressor.

1. Before servicing the vehicle, refer to the precautions in the beginning of this section.

2. Remove or disconnect the following:
- Strut mount cap located at the rear interior quarter trim
- Wheel
- Brake hose clip
- Union bolt from the brake caliper, if equipped with disc brakes and move the brake hose out of the way
- Brake hose and pipe from strut and drum, if equipped with drum brakes

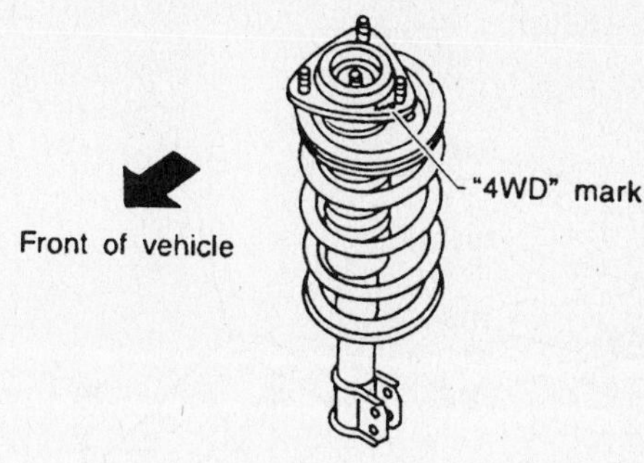

Position the upper strut bearing as shown—rear strut assembly

3. Support rear with jack.

4. Remove or disconnect the following:

5. Remove or disconnect the following:
- Retainer nuts securing the strut bearing cap to the strut tower, from inside the vehicle
- Lower nuts and bolts securing the strut to the rear wheel housing
- Strut

To install:

6. Install or connect the following:
- Strut on to the vehicle, making sure to position the strut with the "4WD" mark on the strut mount facing the outside of the vehicle as shown in the illustration. Torque the retaining nuts to 15 ft. lbs. (20 Nm).
- Strut and mount cap. Torque the strut mount cap bolts to 14.5 ft. lbs. (20 Nm).
- Strut to the rear wheel knuckle assembly. Torque the retainer nuts/bolts to 145 ft. lbs. (196 Nm) on 2001 models and 148 ft. lbs. (200 Nm) on 2002–04 models.
- Union bolt, if equipped with disc brakes. Torque the bolt and to 13 ft. lbs. (18 Nm).
- Brake hose to brake pipe, if equipped with drum brakes. Torque to 10 ft. lbs. (15 Nm).
- Brake hose clip
- Wheel
- Strut mount cap

7. Bleed the brakes.

Coil Spring

REMOVAL & INSTALLATION

Front and Rear

⁂ CAUTION

Do not remove the large nut on top of the strut assembly unless the coil spring is properly compressed with a spring compressor.

1. Before servicing the vehicle, refer to the precautions in the beginning of this section.

2. Remove the strut assembly.

3. Place the strut assembly in a vise with a holding tool and install a spring compressor.

4. Compress the spring slightly.

5. Loosen but do not remove the bearing cap locknut.

6. Unload the spring seat using the spring compressor, then remove the locknut.

7. Remove or disconnect the following:
- Strut bearing cap, mounting insulator bracket and upper spring seat
- Coil spring and compressor. If the spring is being replaced, slowly release the spring from the compressor and compress the new coil spring.
- Strut boot and rebound bumper from the strut, inspect and replace if worn
- Strut retainer nut and the strut insert from the assembly

To install:

8. Install or connect the following:
- Strut into the chamber
- Retainer nut. Tighten the nut until snug.
- Rebound bumper and the boot to the strut piston rod
- Coil spring on the strut assembly
- Upper spring seat, mounting insulator and bearing cap
- Locknut. Tighten it to 41 ft. lbs. (55 Nm).
- Spring compressor from the coil spring
- Strut

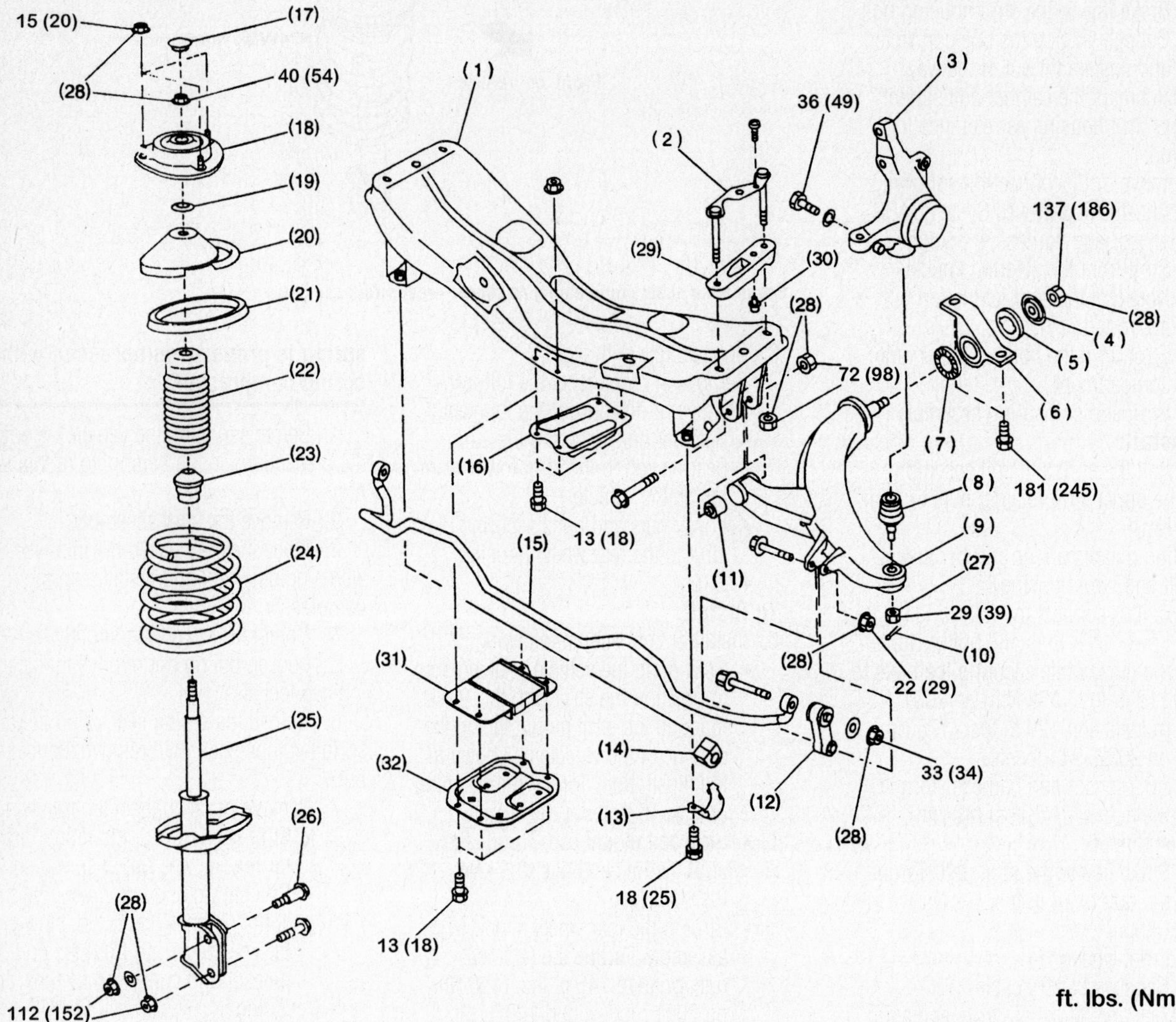

(1) Front crossmember
(2) Bolt ASSY
(3) Housing
(4) Washer
(5) Stopper rubber (Rear)
(6) Rear bushing
(7) Stopper rubber (Front)
(8) Ball joint
(9) Transverse link
(10) Cotter pin
(11) Front bushing
(12) Stabilizer link
(13) Clamp
(14) Bushing
(15) Stabilizer
(16) Jack-up plate (Except MT model)
(17) Dust seal
(18) Strut mount
(19) Spacer
(20) Upper spring seat
(21) Rubber seat
(22) Dust cover
(23) Helper
(24) Coil spring
(25) Damper strut
(26) Adjusting bolt
(27) Castle nut
(28) Self-locking nut
(29) Adapter front crossmember
(30) Clip
(31) Dynamic damper (MT model)
(32) Jack-up plate (MT model)

7924XG23

Exploded view of the front suspension

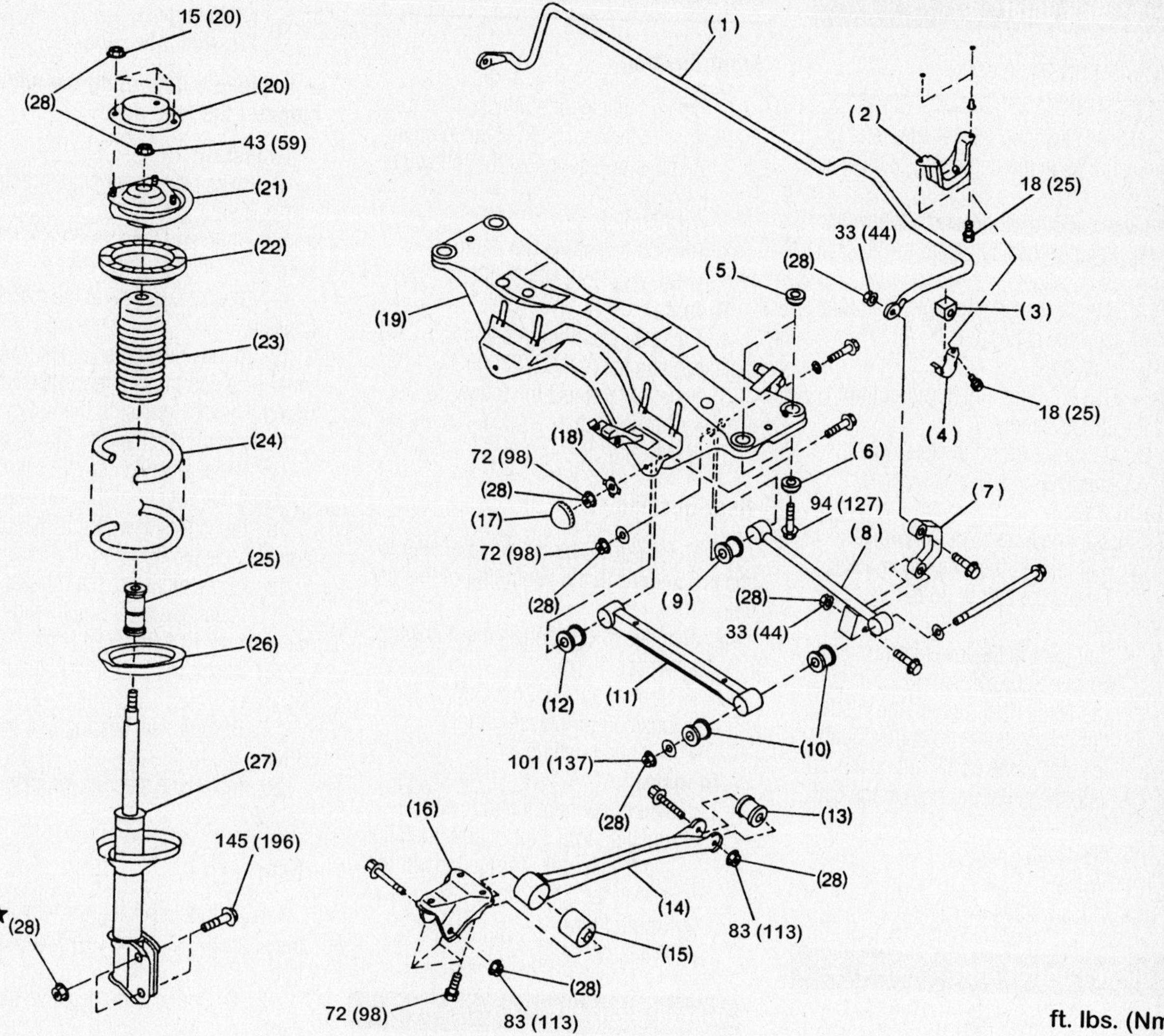

(1) Stabilizer
(2) Stabilizer bracket
(3) Stabilizer bushing
(4) Clamp
(5) Floating bushing
(6) Stopper
(7) Stabilizer link
(8) Rear lateral link
(9) Bushing (C)
(10) Bushing (A)
(11) Front lateral link
(12) Bushing (B)
(13) Trailing link rear bushing
(14) Trailing link
(15) Trailing link front bushing
(16) Trailing link bracket
(17) Cap (Protection)
(18) Washer
(19) Rear crossmember
(20) Strut mount cap
(21) Strut mount
(22) Rubber seat upper
(23) Dust cover
(24) Coil spring
(25) Helper
(26) Rubber seat lower
(27) Damper strut
(28) Self-locking nut

7924XG24

Exploded view of the rear suspension

Lower Ball Joint

REMOVAL & INSTALLATION

1. Before servicing the vehicle, refer to the precautions in the beginning of this section.
2. Remove or disconnect the following:
   - Negative battery cable
   - Front wheel
   - Ball joint castle nut cotter pin, discard the cotter pin
   - Castle nut
   - Ball joint from the lower control arm assembly
   - Ball joint from the steering knuckle

To install:

3. Install or connect the following:
   - Ball joint to the steering knuckle. Torque the bolt to 36 ft. lbs. (49 Nm).
   - Ball joint to the lower control arm. Torque the castle nut to 29 ft. lbs. (39 Nm). Then, tighten the castle nut an additional 60 degrees until the slot in the castle nut is aligned with the cotter pin hole in the ball joint.
   - New cotter pin
   - Wheel
   - Negative battery cable

Lower Control Arm

REMOVAL & INSTALLATION

1. Before servicing the vehicle, refer to the precautions in the beginning of this section.
2. Remove or disconnect the following:
   - Wheel assembly
   - Stabilizer link
   - Ball joint from housing
   - Mounting bolts
   - Control arm

To install:

3. Install or connect the following:
   - Control arm to stabilizer. Torque the nut/bolt to 22 ft. lbs. (29 Nm).
   - Control arm to crossmember. Torque the nut/bolt to 72 ft. lbs. (98 Nm).
   - Control arm to rear mount. Torque the bolts to 181 ft. lbs. (245 Nm).
   - Ball joint to housing, Torque the nut to 36 ft. lbs. (49 Nm).
   - Wheel assembly

CONTROL ARM BUSHING REPLACEMENT

Front Bushing

1. Before servicing the vehicle, refer to the precautions in the beginning of this section.
2. Remove or disconnect the following:
   - Wheel
   - Control arm
3. Press the bushing out using Installer/Remover tool 927680000

To install:

4. Press the bushing in using Installer/Remover tool 927680000
5. Install or connect the following:
   - Control arm
   - Wheel

Rear Bushing

1. Before servicing the vehicle, refer to the precautions in the beginning of this section.
2. Remove or disconnect the following:
   - Wheel
   - Control arm and matchmark the bushing for reassembly
   - Nut and bushing

To install:

3. Install or connect the following:
   - Bushing into control arm and align the matchmark. Torque the nut to 140 ft. lbs. (190 Nm).
   - Control arm
   - Wheel

Wheel Bearings

ADJUSTMENT

The wheel bearings are not adjustable.

REMOVAL & INSTALLATION

Front

1. Before servicing the vehicle, refer to the precautions in the beginning of this section.
2. Remove the steering knuckle assembly.
3. Position the steering knuckle in a soft-jawed vise.
4. Press the hub from the steering knuckle. If the inner bearing race remains in the hub, press it out.
5. Remove or disconnect the following:
   - Rotor shield
   - Inner and outer seals
   - Snapring from the steering knuckle
6. Press the inner bearing race to remove the outer bearing.
7. Remove the Anti-lock Brakes (ABS) tone ring, if equipped
8. Press the wheel lugs from the hub.

➡To prevent deforming the hub, do not hammer the lugs out.

To install:

9. Press new wheel lugs into the hub.
10. If equipped, clean all foreign material from the hub and tone ring. Install the tone ring.
11. Clean the inside of the steering knuckle.
12. Remove the plastic lock from the inner race and press a new greased bearing into the hub by pressing the outer race.
13. Install the snapring into its groove.
14. Press a new outer oil seal until it contacts the bottom of the housing.
15. Press a new inner oil seal until it contacts the circlip.
16. Apply grease to the oil seal lips.
17. Install the rotor shield and tighten the bolts to 10 ft. lbs. (14 Nm).
18. Attach the hub to the steering knuckle.
19. Press a new bearing into the hub by driving the inner race.
20. Install the steering knuckle on the vehicle.

Rear

1. Before servicing the vehicle, refer to the precautions in the beginning of this section.
2. Disconnect the negative battery cable.
3. Loosen the parking brake adjustment.
4. Remove or disconnect the following:
   - Wheel
   - Axle nut
   - Caliper, leaving the line connected if equipped with disc brakes and suspend it aside, then remove the rotor
   - Drum and brake line, if equipped with drum brakes
   - Parking brake cable
   - Rear stabilizer from lateral link
   - Trailing link to the housing
   - Lateral link to the housing
   - Halfshaft
   - Anti-lock Brakes (ABS), speed sensor from the backing plate, if equipped
   - Strut from the housing
   - Housing assembly
5. Using Hub Stand 92708000 and Hub Remover 927420000
6. Remove or disconnect the following:

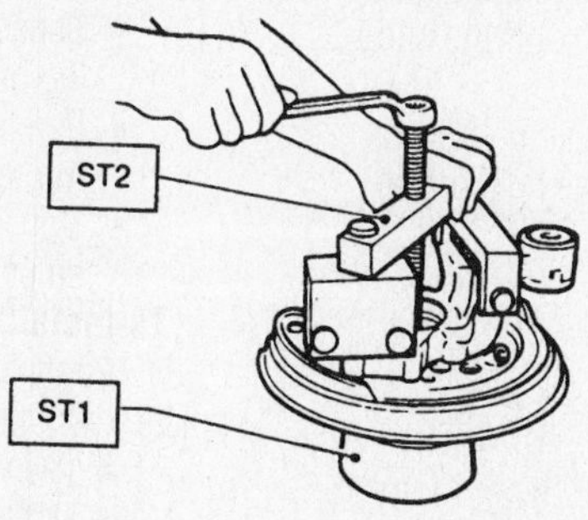

Use the proper tools to separate the hub from the housing to prevent damage

- Hub from the rear housing
- Backing plate from the housing.
- Outer, inner and sub oil seals.
- Snapring

7. Remove the bearing by pressing the inner race.

To install:

8. Clean the housing thoroughly.

➡Do not remove the plastic lock from the inner race when installing the bearing.

9. Install the new bearing into the housing by pressing the outer race.
10. Pack the bearing with grease.
11. Install the snapring.
12. Using Installer 927460000 seal driver, press in a new outer seal until it comes in contact with the snapring.
13. Using Installer 927450000 seal driver, press in a new inner seal until it contacts the bottom.
14. Install or connect the following:
    - New sub oil seal, apply grease to the oil seal lip
    - Backing plate. Torque the bolts to 38 ft. lbs. (52 Nm).
15. Using Installer 927450000 bearing driver, press in the hub into the housing.
16. Install or connect the following:
    - Housing to the strut. Torque the bolts to 108 ft. lbs. (147 Nm).
    - Halfshaft
    - Lateral link to the housing. Torque the bolt and new nut to 101 ft. lbs. (137 Nm).
    - Trailing link to the housing. Torque the bolt and new nut to 94 ft. lbs. (127 Nm).
    - Stabilizer to rear lateral link
    - Parking brake cable and brake
    - Brake line, if equipped with drum brakes
    - Rotor and caliper, if equipped with disc brakes
    - ABS speed sensor, if equipped
    - New axle nut and tighten it to 137 ft. lbs. (186 Nm). Stake the nut.
    - Wheel
    - Negative battery cable
17. Adjust the parking brake cable.

BRAKES

Brake Caliper

REMOVAL & INSTALLATION

Front

1. Before servicing the vehicle, refer to the precautions in the beginning of this section.
2. Remove or disconnect the following:
    - Front wheels
    - Brake hose from the caliper body
    - Caliper retainer bolts and the caliper
    - Cliper bracket, if necessary

To install:

3. Compress the piston assembly into the cylinder bore.
4. Install or connect the following:
    - Caliper bracket to the spindle assembly, if removed and secure in place with the retainer bolts. Tighten the retainer bolts to 59 ft. lbs. (80 Nm).
    - Caliper and tighten the retainers to 28 ft. lbs. (37 Nm).
    - Brake hose using new sealing washers, and tighten the fitting to 13 ft. lbs. (18 Nm)
5. Bleed the brake system.
6. Install the wheels and check the fluid level in the master cylinder.

Rear

1. Before servicing the vehicle, refer to the precautions in the beginning of this section.
2. Remove or disconnect the following:
    - Rear wheels
    - Brake hose from the caliper body
    - Caliper retainer bolts and the caliper
    - Caliper bracket, if necessary

To install:

3. Compress the piston assembly into the cylinder bore.
4. Install or connect the following:
    - Caliper bracket, if removed and secure in place with the retainer bolts. Tighten the retainer bolts to 38 ft. lbs. (52 Nm).
    - Caliper and tighten the retainers to 28 ft. lbs. (37 Nm).
    - Brake hose using new sealing washers, and tighten the fitting to 13 ft. lbs. (18 Nm)
5. Bleed the brake system.
6. Install the wheels and check the fluid level in the master cylinder.

Disc Brake Pads

REMOVAL & INSTALLATION

Front

1. Before servicing the vehicle, refer to the precautions in the beginning of this section.
2. Remove or disconnect the following:
    - Wheels
    - Lower caliper bolt
    - Swing the caliper upward to access the pads
    - Disc brake pads

To install:

3. Compress the caliper piston.
4. Install or connect the following:
    - New pads into the caliper brackets, being sure all shims and clips are in their original positions.
5. Swing the calipers down into position and install the caliper bolt. Tighten the bolt to 28 ft. lbs. (37 Nm).
6. Fill the master cylinder reservoir.
7. Install the wheels.

Rear

1. Before servicing the vehicle, refer to the precautions in the beginning of this section.

2. Remove or disconnect the following:
   - Wheels
   - Lower caliper bolt
   - Swing the caliper upward to access the pads
   - Disc brake pads

To install:

3. Compress the caliper piston.
4. Install or connect the following:
   - New pads into the caliper brackets, being sure all shims and clips are in their original positions.

5. Swing the calipers down into position and install the caliper bolt. Tighten the bolt to 28 ft. lbs. (37 Nm).
6. Fill the master cylinder reservoir.
7. Install the wheels.

Brake Drums

REMOVAL & INSTALLATION

1. Before servicing the vehicle, refer to the precautions in the beginning of this section.
2. Remove the rear wheels.
3. Release the parking brake.
4. If necessary, remove the adjusting hole cover from the backing plate and using a suitable tool, back off the shoe adjuster.
5. If the drum is difficult to remove, insert an 8mm bolt into the hole on the drum to push it off.
6. Remove the drum.

To install:

7. Install the drum.
8. Adjust the brake shoes.
9. Install the rear wheels.

Brake Shoes

REMOVAL & INSTALLATION

1. Before servicing the vehicle, refer to the precautions in the beginning of this section.
2. Remove or disconnect the following:
   - Wheels
   - Brake drum
   - Hold–down pins, springs and cups from the shoes
   - Lower return spring from both shoes
   - Shoes and adjuster from the backing plate
   - Parking brake cable from the parking lever on the trailing brake shoe
   - Upper shoe return spring and adjusting spring from the shoe

To install:

3. Apply brake grease to the backing plate where the brake shoes contact it.
4. Install or connect the following:
   - Upper return spring to the shoes
   - Parking brake cable to the lever
   - Shoes on the backing plate
   - Hold–down pins, springs and cups to the shoes
   - Lower return spring to both shoes

5. Set the outside diameter of the shoes less than 0.020–0.031 in. (0.5–0.8mm) compared to the inside diameter of the drum.
6. Install the drum and adjust the brake shoes.
7. Install the wheels.

SUZUKI

14

Esteem • Swift

SPECIFICATION AND MAINTENANCE CHARTS

VEHICLE AND ENGINE IDENTIFICATION

Engine							Model Year	
Code	Liters (cc)	Cu. in.	Cyl.	Fuel Sys.	Engine Type	Eng. Mfg.	Code	Year
2	1.3 (1298)	79.3	4	TFI	SOHC	Suzuki	Y	2000
3	1.6 (1590)	97.7	4	MFI	SOHC	Suzuki	1	2001
							2	2002
							3	2003
							4	2004

MFI: Multi-port Fuel Injection

TFI: Throttle body Fuel Injection

SOHC: Single Overhead Camshaft

42356-SUZC-C01

GENERAL ENGINE SPECIFICATIONS

Year	Engine ID/VIN	Engine Displacement Liters (cc)	Fuel System Type	Net Horsepower @ rpm	Net Torque @ rpm (ft. lbs.)	Bore x Stroke (in.)	Com-pression Ratio	Oil Pressure @ rpm
2000	2	1.3 (1298)	TFI	70@5500	74@3000	2.91x2.97	9.5:1	47-61@3000
	3	1.6 (1590)	MFI	98@6000	94@3200	2.95x3.54	9.5:1	47-61@4000
2001	2	1.3 (1298)	TFI	70@5500	74@3000	2.91x2.97	9.5:1	47-61@3000
	3	1.6 (1590)	MFI	98@6000	94@3200	2.95x3.54	9.5:1	47-61@4000
2002	2	1.3 (1298)	TFI	70@5500	74@3000	2.91x2.97	9.5:1	47-61@3000
	3	1.6 (1590)	MFI	98@6000	94@3200	2.95x3.54	9.5:1	47-61@4000
2003	2	1.3 (1298)	TFI	70@5500	74@3000	2.91x2.97	9.5:1	47-61@3000
	3	1.6 (1590)	MFI	98@6000	94@3200	2.95x3.54	9.5:1	47-61@4000

MFI: Multiport Fuel Injection

TFI: Throttle body Fuel Injection

42356-SUZC-C02

GASOLINE ENGINE TUNE-UP SPECIFICATIONS

Year	Engine Displacement Liters (cc)	Engine ID/VIN	Spark Plugs Gap (in.)	Ignition Timing (deg.)		Fuel Pump (psi)	Idle Speed (rpm)		Valve Clearance	
				MT	AT		MT	AT	In.	Ex.
2000	1.3 (1298)	2	0.029	5B	5B	13-20 ①	750	850	HYD	HYD
	1.6 (1590)	3	0.029	5B	5B	28-34 ①	750-800	750-800	②	②
2001	1.3 (1298)	2	0.029	5B	5B	13-20 ①	750	850	HYD	HYD
	1.6 (1590)	3	0.029	5B	5B	28-34 ①	750-800	750-800	②	②
2002	1.6 (1590)	3	0.029	5B	5B	28-34 ①	750-800	750-800	②	②
2003	1.6 (1590)	3	0.029	5B	5B	28-34 ①	750-800	750-800	②	②

Note: The Vehicle Emission Control Information label often reflects specification changes made during production. The label figures must be used if they differ from those in this chart.

HYD: Hydraulic

B: Before top dead center

① At idle

② When cold: 0.005-0.007
When hot: 0.007-0.008

42356-SUZC-C03

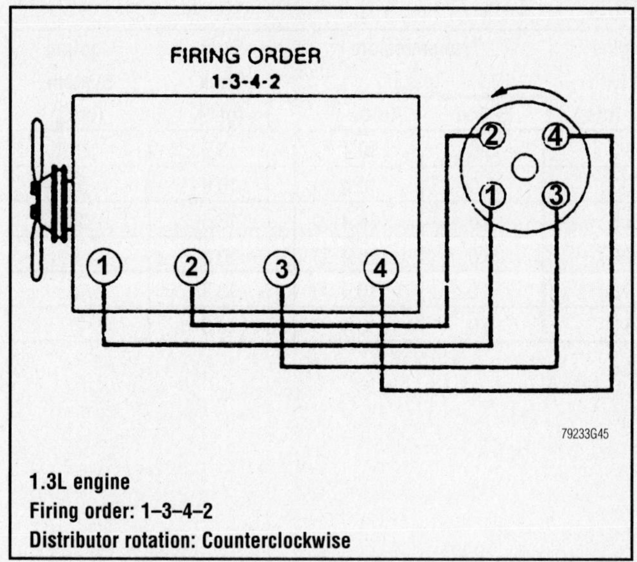

1.3L engine
Firing order: 1–3–4–2
Distributor rotation: Counterclockwise

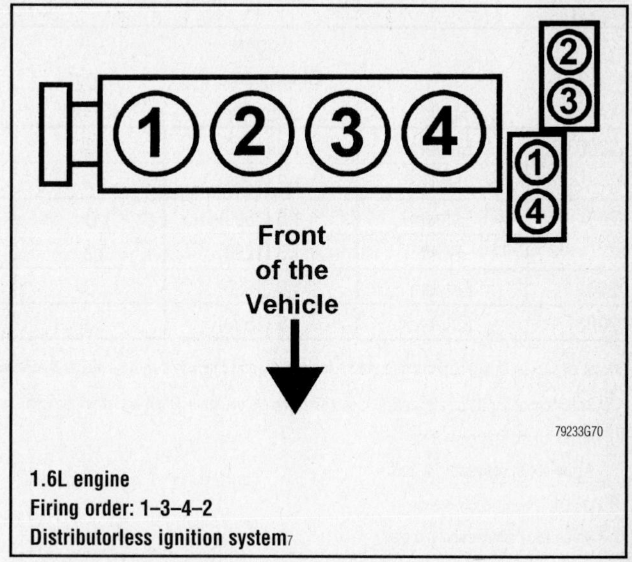

1.6L engine
Firing order: 1–3–4–2
Distributorless ignition system

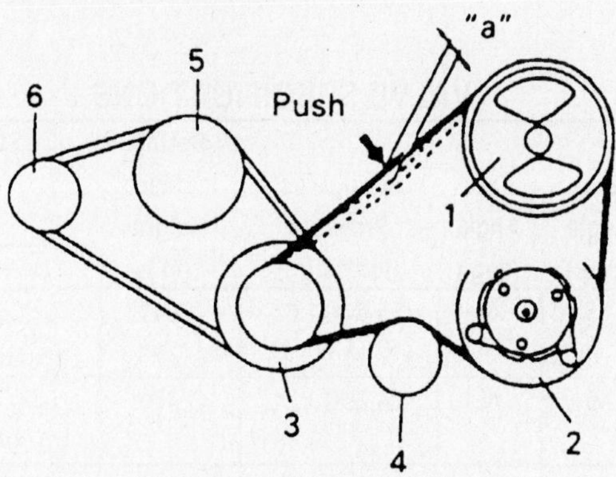

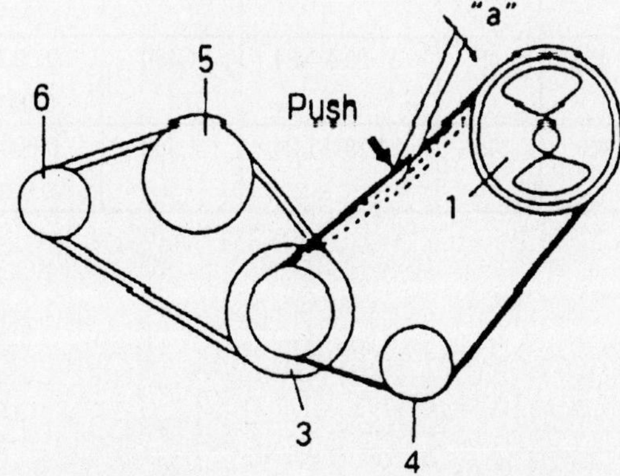

Accessory drive belt routing—1.3L and 1.6L engines

CAPACITIES

Year	Model	Engine Displacement Liters (cc)	Engine ID/VIN	Engine Oil with Filter (qts.)	Transmission (pts.) 5-Spd	Transmission (pts.) Auto.	Fuel Tank (gal.)	Cooling System (qts.)
2000	Esteem	1.6 (1590)	3	3.3	5.0	10.4 ①	13.5	②
	Swift	1.3 (1298)	2	3.3	5.0	10.4 ①	10.6	③
2001	Esteem	1.6 (1590)	3	3.3	5.0	10.4 ①	13.5	②
	Swift	1.3 (1298)	2	3.3	5.0	10.4 ①	10.6	③
2002	Esteem	1.6 (1590)	3	3.3	5.0	10.4 ①	13.5	②
2003	Esteem	1.6 (1590)	3	3.3	5.0	10.4 ①	13.5	②

Note: All capacities are approximate. Add fluid gradualy and check to be sure a proper fluid level is obtained.

① Specification for automatic transaxle is after complete overhaul. Drain and fill will be less

② Manual transmission: 4.8 qts.
Automatic transmission: 4.7 qts.

③ Manual transmission: 4.8 qts.
Automatic transmission: 4.9 qts.

42356-SUZC-C04

VALVE SPECIFICATIONS

Year	Engine ID/VIN	Engine Displacement Liters (cc)	Seat Angle (deg.)	Face Angle (deg.)	Spring Test Pressure (lbs. @ in.)	Spring Installed Height (in.)	Stem-to-Guide Clearance (in.) Intake	Stem-to-Guide Clearance (in.) Exhaust	Stem Diameter (in.) Intake	Stem Diameter (in.) Exhaust
2000	2	1.3 (1298)	45	45	55-64@1.63	1.941	0.0008-0.0019	0.0014-0.0025	0.2742-0.2748	0.2737-0.2742
	3	1.6 (1590)	45	45	24-28@1.24	1.450	0.0008-0.0018	0.0018-0.0028	0.2152-0.2157	0.2142-0.2148
2001	2	1.3 (1298)	45	45	55-64@1.63	1.941	0.0008-0.0019	0.0014-0.0025	0.2742-0.2748	0.2737-0.2742
	3	1.6 (1590)	45	45	24-28@1.24	1.450	0.0008-0.0018	0.0018-0.0028	0.2152-0.2157	0.2142-0.2148
2002	3	1.6 (1590)	45	45	24-28@1.24	1.450	0.0008-0.0018	0.0018-0.0028	0.2152-0.2157	0.2142-0.2148
2003	3	1.6 (1590)	45	45	24-28@1.24	1.450	0.0008-0.0018	0.0018-0.0028	0.2152-0.2157	0.2142-0.2148

42356-SUZC-C05

PISTON AND RING SPECIFICATIONS

All measurements are given in inches.

Year	Engine ID/VIN	Engine Displacement Liters (cc)	Piston Clearance	Ring Gap Top Compression	Ring Gap Bottom Compression	Ring Gap Oil Control	Ring Side Clearance Top Compression	Ring Side Clearance Bottom Compression	Ring Side Clearance Oil Control
2000	2	1.3 (1298)	0.0008-0.0015	0.0079-0.0118	0.0079-0.0118	0.0079-0.0275	0.0012-0.0027	0.0008-0.0023	snug
	3	1.6 (1590)	0.0008-0.0015	0.0079-0.0137	0.0079-0.0137	0.0079-0.0275	0.0012-0.0027	0.0008-0.0023	snug
2001	2	1.3 (1298)	0.0008-0.0015	0.0079-0.0118	0.0079-0.0118	0.0079-0.0275	0.0012-0.0027	0.0008-0.0023	snug
	3	1.6 (1590)	0.0008-0.0015	0.0079-0.0137	0.0079-0.0137	0.0079-0.0275	0.0012-0.0027	0.0008-0.0023	snug
2002	3	1.6 (1590)	0.0008-0.0015	0.0079-0.0137	0.0079-0.0137	0.0079-0.0275	0.0012-0.0027	0.0008-0.0023	snug
2003	3	1.6 (1590)	0.0008-0.0015	0.0079-0.0137	0.0079-0.0137	0.0079-0.0275	0.0012-0.0027	0.0008-0.0023	snug

42356-SUZC-C06

CRANKSHAFT AND CONNECTING ROD SPECIFICATIONS

All measurements are given in inches.

Year	Engine ID/VIN	Engine Displacement Liters (cc)	Crankshaft Main Brg. Journal Dia.	Crankshaft Main Brg. Oil Clearance	Crankshaft Shaft End-play	Crankshaft Thrust on No.	Connecting Rod Journal Diameter	Connecting Rod Oil Clearance	Connecting Rod Side Clearance
2000	3	1.3 (1298)	①	0.0008-0.0023	0.0044-0.0149	3	1.6529-1.6535	0.0008-0.0031	0.0039-0.0137
	6	1.6 (1590)	②	0.0008-0.0016	0.0044-0.0122	3	1.7316-1.7322	0.0008-0.0019	0.0039-0.0078
2001	3	1.3 (1298)	①	0.0008-0.0023	0.0044-0.0149	3	1.6529-1.6535	0.0008-0.0031	0.0039-0.0137
	6	1.6 (1590)	②	0.0008-0.0016	0.0044-0.0122	3	1.7316-1.7322	0.0008-0.0019	0.0039-0.0078
2002	6	1.6 (1590)	②	0.0008-0.0016	0.0044-0.0122	3	1.7316-1.7322	0.0008-0.0019	0.0039-0.0078
2003	6	1.6 (1590)	②	0.0008-0.0016	0.0044-0.0122	3	1.7316-1.7322	0.0008-0.0019	0.0039-0.0078

① No. 1: 1.7714-1.7716
No. 2: 1.7712-1.7714
No. 3: 1.7710-1.7712

② No. 1: 2.0470-2.0472
No. 2: 2.0468-2.0470
No. 3: 2.0465-2.0468

42356-SUZC-C07

TORQUE SPECIFICATIONS

All readings in ft. lbs.

Year	Engine ID/VIN	Engine Displacement Liters (cc)	Cylinder Head Bolts	Main Bearing Bolts	Rod Bearing Bolts	Crankshaft Damper Bolts	Flywheel Bolts	Manifold Intake	Manifold Exhaust	Spark Plugs	Lug Nut
2000	2	1.3 (1298)	①	36-41	24-26	76-83 ②	41-47	13-20	13-20	14-21	36-57
	3	1.6 (1590)	①	36-41	24-26	76-83 ②	57	13-20	13-20	14-21	58-80
2001	2	1.3 (1298)	①	36-41	24-26	76-83 ②	41-47	13-20	13-20	14-21	36-57
	3	1.6 (1590)	①	36-41	24-26	76-83 ②	57	13-20	13-20	14-21	58-80
2002	3	1.6 (1590)	①	36-41	24-26	76-83 ②	57	13-20	13-20	14-21	58-80
2003	3	1.6 (1590)	①	36-41	24-26	76-83 ②	57	13-20	13-20	14-21	58-80

① Step 1: 26 ft. lbs. (35 Nm)

Step 2: 41 ft. lbs. (55 Nm)

Step 3: 49 ft. lbs. (68 Nm)

② Specification shown is for crankshaft timing sprocket nut

42356-SUZC-C08

WHEEL ALIGNMENT

Year	Model		Caster Range (+/-Deg.)	Caster Preferred Setting (Deg.)	Camber Range (+/-Deg.)	Camber Preferred Setting (Deg.)	Toe-in (in.)	Steering Axis Inclination (Deg.)
2000	Swift	F	2.00	+3.00	1.00	+0.50	0.08+/-0.08	—
		R	—	—	—	—	0.18+/-0.06	—
	Esteem	F	2.00	+2.70	1.00	0	0+/-0.08	—
		R	—	—	—	0	0.08+/-0.08	—
2001	Swift	F	2.00	+3.00	1.00	+0.50	0.08+/-0.08	—
		R	—	—	—	—	0.18+/-0.06	—
	Esteem	F	2.00	+2.70	1.00	0	0+/-0.08	—
		R	—	—	—	0	0.08+/-0.08	—
2002	Esteem	F	2.00	+2.70	1.00	0	0+/-0.08	—
		R	—	—	—	0	0.08+/-0.08	—
2003	Esteem	F	2.00	+2.70	1.00	0	0+/-0.08	—
		R	—	—	—	0	0.08+/-0.08	—

42356-SUZC-C09

TIRE, WHEEL AND BALL JOINT SPECIFICATIONS

Year	Model	OEM Tires		Tire Pressures (psi)		Wheel	Ball Joint
		Standard	Optional	Front	Rear	Size	Inspection
2000	Swift	P155/80R13	None	32	32	4.5J	①
	Esteem GL	P175/70R13	None	30	30	5J	①
	Esteem GL Wagon	P185/60R14	None	30	30	5.5JJ	①
	Esteem GLX	P185/60R14	None	30	30	5.5JJ	①
	Esteem GLX Wagon	P195/55R15	None	29	29	5.5JJ	①
2001	Swift	P155/80R13	None	32	32	4.5J	①
	Esteem GL	P175/70R13	None	30	30	5J	①
	Esteem GL Wagon	P185/60R14	None	30	30	5.5JJ	①
	Esteem GLX	P185/60R14	None	30	30	5.5JJ	①
	Esteem GLX Wagon	P195/55R15	None	29	29	5.5JJ	①
2002	Esteem GL	P175/70R13	None	30	30	5J	①
	Esteem GL Wagon	P185/60R14	None	30	30	5.5JJ	①
	Esteem GLX	P185/60R14	None	30	30	5.5JJ	①
	Esteem GLX Wagon	P195/55R15	None	29	29	5.5JJ	①
2003	Esteem GL	P175/70R13	None	30	30	5J	①
	Esteem GL Wagon	P185/60R14	None	30	30	5.5JJ	①
	Esteem GLX	P185/60R14	None	30	30	5.5JJ	①
	Esteem GLX Wagon	P195/55R15	None	29	29	5.5JJ	①

OEM: Original Equipment Manufacturer

PSI: Pounds Per Square Inch

STD: Standard

OPT: Optional

① Replace if any measurable movement is found.

42356-SUZC-C10

BRAKE SPECIFICATIONS

All measurements in inches unless noted

Year	Model	Brake Disc			Brake Drum Diameter			Minimum Lining Thickness		Brake Caliper	
		Original Thickness	Minimum Thickness	Maximum Runout	Original Inside Diameter	Max. Wear Limit	Maximum Machine Diameter	Front	Rear	Bracket bolts ft lbs.	Mounting bolts ft lbs.
2000	Swift ①	0.670	0.620	0.004	7.09	7.87	7.87	0.236 ②	0.110 ②	36	27
	Swift	0.670	0.620	0.004	7.16	7.95	7.95	0.236 ②	0.110 ②	36	27
	Esteem	0.790	0.710	0.004	7.87	7.95	7.95	0.240 ②	0.110 ②	62	16
2001	Swift ①	0.670	0.620	0.004	7.09	7.87	7.87	0.236 ②	0.110 ②	36	27
	Swift	0.670	0.620	0.004	7.16	7.95	7.95	0.236 ②	0.110 ②	36	27
	Esteem	0.790	0.710	0.004	7.87	7.95	7.95	0.240 ②	0.110 ②	62	16
2002	Esteem	0.790	0.710	0.004	7.87	7.95	7.95	0.240 ②	0.110 ②	62	16
2003	Esteem	0.790	0.710	0.004	7.87	7.95	7.95	0.240 ②	0.110 ②	62	16

① Hatchback

② Measurement is for lining and backing together.

42356-SUZC-C11

SCHEDULED MAINTENANCE INTERVALS

SUZUKI—ESTEEM & SWIFT

TO BE SERVICED	TYPE OF SERVICE	VEHICLE MILEAGE INTERVAL (x1000)												
		7.5	15	22.5	30	37.5	45	52.5	60	67.5	75	82.5	90	97.5
Engine oil & filter	R	✓	✓	✓	✓	✓	✓	✓	✓	✓	✓	✓	✓	✓
Automatic transmission fluid & filter ①	S/I	✓	✓	✓	✓	✓	✓	✓	✓	✓	✓	✓	✓	✓
Clutch pedal free travel	S/I	✓	✓	✓	✓	✓	✓	✓	✓	✓	✓	✓	✓	✓
Drive axle boots	S/I	✓	✓	✓	✓	✓	✓	✓	✓	✓	✓	✓	✓	✓
Gear shift control lever/shift operation	S/I	✓	✓	✓	✓	✓	✓	✓	✓	✓	✓	✓	✓	✓
Inspect & rotate tires	S/I	✓	✓	✓	✓	✓	✓	✓	✓	✓	✓	✓	✓	✓
Manual transmission oil ②	S/I	✓	✓	✓	✓	✓	✓	✓	✓	✓	✓	✓	✓	✓
Power steering system	S/I	✓	✓	✓	✓	✓	✓	✓	✓	✓	✓	✓	✓	✓
Suspension system	S/I	✓	✓	✓	✓	✓	✓	✓	✓	✓	✓	✓	✓	✓
Brake discs, pads, drums & shoes	S/I	✓		✓		✓		✓		✓		✓		✓
Brake hoses, pipes, brake lever & cable	S/I	✓		✓		✓		✓		✓		✓		✓
Brake fluid ③	S/I		✓		✓		✓		✓		✓		✓	
Brake pedal	S/I		✓		✓		✓		✓		✓		✓	
Cooling system, hoses & connections	S/I		✓		✓		✓		✓		✓		✓	
Fuel tank, cap & lines	S/I		✓		✓		✓		✓		✓		✓	
Valve lash (clearance)	S/I		✓		✓		✓		✓		✓		✓	
Air cleaner filter element	R				✓				✓				✓	
Engine coolant	R				✓				✓				✓	
Spark plugs	R				✓				✓				✓	
Drive belts	S/I				✓				✓				✓	
Exhaust system	S/I				✓				✓				✓	

42356-SUZC-C12

SCHEDULED MAINTENANCE INTERVALS
SUZUKI—ESTEEM & SWIFT

TO BE SERVICED	TYPE OF SERVICE	VEHICLE MILEAGE INTERVAL (x1000)												
		7.5	15	22.5	30	37.5	45	52.5	60	67.5	75	82.5	90	97.5
Automatic transmission fluid hose	R						✓							
Camshaft timing belt	R								✓					
Ignition wiring	S/I								✓					

R: Replace S/I: Service or Inspect

① Replace every 100,000 miles.

② Replace every 15,000 miles.

③ Replace every 60,000 miles.

FREQUENT OPERATION MAINTENANCE (SEVERE SERVICE)

If a vehicle is operated under any of the following conditions it is considered severe service:

- Extremely dusty areas.

- 50% or more of the vehicle operation is in 32°C (90°F) or higher temperatures, or constant operation in temperatures below 0°C (32°F).

- Prolonged idling (vehicle operation in stop and go traffic).

- Frequent short running periods (engine does not warm to normal operating temperatures).

- Police, taxi, delivery usage or trailer towing usage.

Oil & oil filter: change every 3000 miles.

Brake discs, pads, drums & shoes: service or inspect initially at 3000 miles, 6000 miles, & every 12,000 miles thereafter.

Brake hoses & pipes: service or inspect initially at 3000 miles, 6000 miles & every 12,000 miles thereafter.

Air cleaner filter element: service or inspect ever 3000 miles & replace every 30,000 miles (if not replaced previously).

Automatic transmission fluid & filter: service or inspect every 6000 miles & replace every 15,000 miles (if not replaced previously).

Clutch pedal free travel: service or inspect every 6000 miles.

Inspect & rotate tires: service or inspect every 6000 miles.

Manual transmission oil: service or inspect every 6000 miles & replace every 12,000 miles (if not replaced previously).

Power steering system: service or inspect every 6000 miles.

Steering system: service or inspect every 6000 miles.

Suspension system: service or inspect every 6000 miles.

Drive belts: service or inspect every 15,000 miles.

Exhaust system: service or inspect every 15,000 miles.

42356-SUZC-C13

PRECAUTIONS

Before servicing any vehicle, please be sure to read all of the following precautions, which deal with personal safety, prevention of component damage, and important points to take into consideration when servicing a motor vehicle:

- Never open, service or drain the radiator or cooling system when the engine is hot; serious burns can occur from the steam and hot coolant.
- Observe all applicable safety precautions when working around fuel. Whenever servicing the fuel system, always work in a well-ventilated area. Do not allow fuel spray or vapors to come in contact with a spark, open flame, or excessive heat (a hot drop light, for example). Keep a dry chemical fire extinguisher near the work area. Always keep fuel in a container specifically designed for fuel storage; also, always properly seal fuel containers to avoid the possibility of fire or explosion. Refer to the additional fuel system precautions later in this section.
- Fuel injection systems often remain pressurized, even after the engine has been turned **OFF**. The fuel system pressure must be relieved before disconnecting any fuel lines. Failure to do so may result in fire and/or personal injury.
- Brake fluid often contains polyglycol ethers and polyglycols. Avoid contact with the eyes and wash your hands thoroughly after handling brake fluid. If you do get brake fluid in your eyes, flush your eyes with clean, running water for 15 minutes. If eye irritation persists, or if you have taken brake fluid internally, IMMEDIATELY seek medical assistance.
- The EPA warns that prolonged contact with used engine oil may cause a number of skin disorders, including cancer! You should make every effort to minimize your exposure to used engine oil. Protective gloves should be worn when changing oil. Wash your hands and any other exposed skin areas as soon as possible after exposure to used engine oil. Soap and water, or waterless hand cleaner should be used.
- All new vehicles are now equipped with an air bag system. The system must be disabled before performing service on or around system components, steering column, instrument panel components, wiring and sensors. Failure to follow safety and disabling procedures could result in accidental air bag deployment, possible personal injury and unnecessary system repairs.
- Always wear safety goggles when working with, or around, the air bag system. When carrying a non-deployed air bag, be sure the bag and trim cover are pointed away from your body. When placing a non-deployed air bag on a work surface, always face the bag and trim cover upward, away from the surface. This will reduce the motion of the module if it is accidentally deployed. Refer to the additional air bag system precautions later in this section.
- Clean, high quality brake fluid from a sealed container is essential to the safe and proper operation of the brake system. You should always buy the correct type of brake fluid for your vehicle. If the brake fluid becomes contaminated, completely flush the system with new fluid. Never reuse any brake fluid. Any brake fluid that is removed from the system should be discarded. Also, do not allow any brake fluid to come in contact with a painted surface; it will damage the paint.
- Never operate the engine without the proper amount and type of engine oil; doing so WILL result in severe engine damage.
- Timing belt maintenance is extremely important! Many models utilize an interference-type, non-freewheeling engine. If the timing belt breaks, the valves in the cylinder head may strike the pistons, causing potentially serious (also time-consuming and expensive) engine damage. Refer to the maintenance interval charts in the front of this section for the recommended replacement interval for the timing belt.
- Disconnecting the negative battery cable on some vehicles may interfere with the functions of the on-board computer system(s) and may require the computer to undergo a relearning process once the negative battery cable is reconnected.
- When servicing drum brakes, only dissemble and assemble one side at a time, leaving the remaining side intact for reference.
- Only an MVAC-trained, EPA-certified automotive technician should service the air conditioning system or its components.

ENGINE REPAIR

➡Disconnecting the negative battery cable on some vehicles may interfere with the functions of the on board computer systems and may require the computer to undergo a relearning process, once the negative battery cable is reconnected.

Distributor

REMOVAL & INSTALLATION

These models utilize a Distributorless Ignition System (DIS). With this system, the Electronic Control Module (ECM) determines proper ignition timing and time for the primary ignition coil circuit to turn **ON** and **OFF**.

Alternator

REMOVAL

Swift

1. Before servicing the vehicle, refer to the precautions in the beginning of this section.
2. Remove or disconnect the following:
   - Negative battery cable
   - **B** terminal wire and coupler from the alternator
   - Alternator drive belt adjusting bolt and loosen the adjuster arm bolt
   - Alternator cover
   - Alternator mounting bolts and nut
   - Alternator

Esteem

1. Before servicing the vehicle, refer to the precautions in the beginning of this section.
2. Remove or disconnect the following:
   - Negative battery cable
   - **B** terminal wire and coupler from the alternator
   - Alternator mounting bolts and loosen the drive belt adjusting bolt
   - Alternator cover
   - Alternator bracket bolts and the bracket
   - Alternator

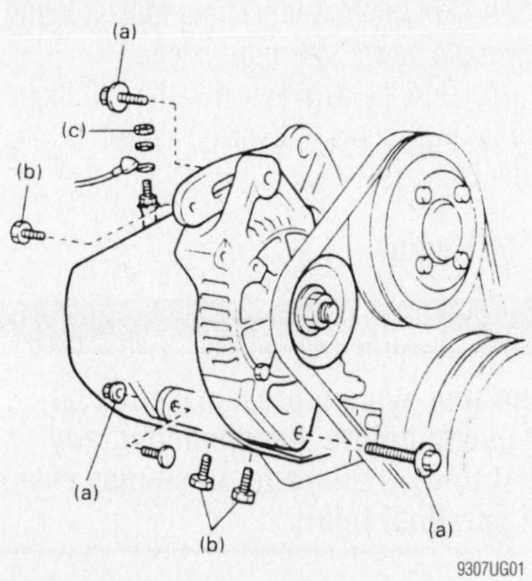

Alternator mounting—Swift models

INSTALLATION

Swift

1. Install or connect the following:
   - Alternator
   - Alternator mounting bolts and nut. Tighten to 13–20 ft. lbs. (18–28 Nm).
   - Alternator cover. Tighten the cover retainers to 36–60 inch lbs. (4–7 Nm).
   - Alternator drive belt adjusting bolt and tighten the adjuster arm bolt. Tighten to 13–20 ft. lbs. (18–28 Nm).
   - **B** terminal wire and coupler to the alternator
   - Negative battery cable

Esteem

1. Install or connect the following:
   - Alternator
   - Alternator bracket and bolts. Tighten the bolts to 16 ft. lbs. (23 Nm).
   - Alternator cover
   - Alternator drive belt adjusting bolt and mounting bolts. Tighten to 16 ft. lbs. (23 Nm).
   - **B** terminal wire and coupler to the alternator
   - Negative battery cable

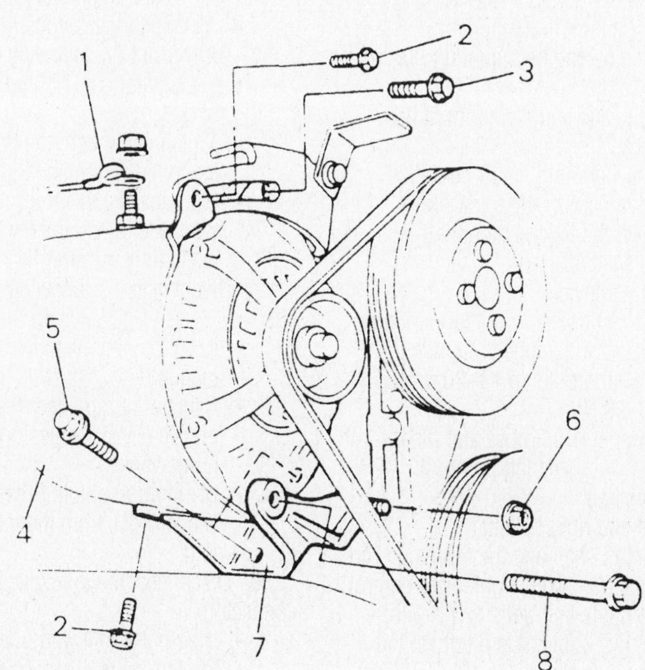

1. "B" terminal wire
2. Cover bolts
3. Drive belt adjusting bolt
4. Cover
5. Bracket bolts
6. Bracket nut
7. Bracket
8. Mounting bolt

9307UG12

Alternator mounting—Esteem models

Ignition Timing

ADJUSTMENT

Ignition timing is controlled by the Electronic Control Module (ECM). The ECM receives signals from various sensors mounted on the engine. No ignition timing adjustment is possible.

Engine Assembly

REMOVAL & INSTALLATION

1.3L Engine

✲✲ CAUTION

The fuel injection system remains under pressure after the engine has been turned OFF. Properly relieve fuel pressure before disconnecting any fuel lines. Failure to do so may result in fire or personal injury.

1. Before servicing the vehicle, refer to the precautions in the beginning of this section.
2. Properly relieve the fuel system pressure.
3. Remove or disconnect the following:
   - Battery and tray
   - Windshield washer hose from the hood
4. Using a grease pencil or marker, mark the hood hinge to hood outline. With the aid of an assistant, remove the hood.
5. Drain the cooling system.
6. Remove or disconnect the following:
   - Air cleaner assembly with the Mass Air Flow (MAF) sensor outlet hose
   - Radiator and cooling fan
7. Disconnect the following electrical wires and release the wiring harness from the clamps:
   - Ignition coil wire from the distributor cap, if equipped
   - Distributor electrical wires, if equipped
   - Ignition coil assembly, if equipped
   - Exhaust Gas Recirculation (EGR) solenoid vacuum valve
   - Radiator fan temperature switch
   - Engine coolant temperature gauge sensor
   - Engine Coolant Temperature (ECT) sensor
   - Idle Air Control (IAC) actuator
   - Ground wires from the intake manifold
   - Throttle Position (TP) sensor

- Fuel injector
- Camshaft Position (CMP) sensor, if equipped
- Oxygen (O_2S) sensor
- Oil pressure gauge sensor
- Alternator
- Starter
- Back-up light switch
- Negative battery cable from the transaxle
- Vehicle Speed Sensor (VSS)
- Noise filter ground wire
- Evaporative (EVAP) emission canister purge valve, if equipped

8. Disconnect the following vacuum hoses:
   - Brake booster hose from the intake manifold
   - Canister purge hose
   - Air conditioning valve hose
9. Remove or disconnect the following:
   - Fuel return hose and the fuel feed hose from the throttle body
   - Heater inlet and outlet hoses
10. Disconnect the following cables:
    - Accelerator cable from the throttle body
    - Clutch cable from the transaxle, if equipped
    - Speedometer cable from the transaxle, if equipped
    - Shift switch, if equipped with an automatic transaxle
    - VSS, if equipped
11. Remove or disconnect the following:
    - EVAP canister from the vehicle
    - Fender apron extensions
    - Exhaust pipe from the exhaust manifold
    - Control shaft and extension rod from the transaxle
12. Drain the engine and transaxle oil.
13. Remove or disconnect the following:
    - Left and right halfshafts

➡For engine and transaxle removal, it is not necessary to remove the halfshafts from the steering knuckles.

14. Remove or disconnect the following:
    - Air conditioning compressor from its mounting bracket with the hoses still attached, if equipped with air conditioning

➡Suspend the compressor where no damage will occur during engine removal and installation.

15. Remove or disconnect the following:
    - Power steering hoses from the power steering pump, if equipped

➡Plug the power steering hose, pipe and pump ports to minimize fluid loss.

- Rear torque rod bracket from the transaxle, if equipped with an automatic transaxle
- Rear mount from the body, if equipped with a manual transaxle

16. If equipped with an automatic transaxle, remove the rear mounting nut.
17. Install an engine lifting device.
18. Remove or disconnect the following:
    - Rear mount from the body
    - Left side engine mounting bracket bolts and bracket
    - Right side engine mount from its bracket
19. Before lifting the engine and assembly check to be sure that all the hoses, electric wires and cables are disconnected.
20. Remove the engine with the transaxle from the vehicle.

To install:

21. Lower the engine and transaxle into the engine compartment but do not remove the lifting device.
22. Install or connect the following:
    - Rear mount to the body
    - Left side engine mounting bracket and bolts
    - Right side engine mount to its bracket
    - Rear mounting nut, if equipped with an automatic transaxle
23. Tighten the engine mounting nuts and bolts to specification.
    - Lifting device
    - A/C compressor on its mounting bracket, if equipped. Tighten the mounting bolts to 13–20 ft. lbs. (18–28 Nm).
    - Power steering hose and pipe to the power steering pump, if equipped
    - Left and right halfshafts
    - Control shaft and the extension rod to the transaxle. Tighten the control shaft nuts and bolts to 11–14 ft. lbs. (15–20 Nm) and tighten the extension rod nut to 19–29 ft. lbs. (25–40 Nm).
    - Exhaust pipe to the exhaust manifold. Tighten the bolts to 29–36 ft. lbs. (40–50 Nm)
24. Fill the transaxle with gear oil.
25. Install the remaining components in the reverse order of removal.
26. Adjust the clutch pedal free-play.
27. Adjust the accelerator cable free-play.
28. Fill the engine with engine oil and the cooling system with coolant.
29. Fill the power steering reservoir and bleed the power steering system.
30. Run the engine and verify that there are no fuel, coolant, transmission or exhaust leaks.

1.6L Engine

⁂ CAUTION

The fuel system pressure must be relieved before disconnecting any fuel lines. Failure to do so may result in personal injury.

1. Before servicing the vehicle, refer to the precautions in the beginning of this section.
2. Relieve the fuel system pressure.
3. Mark the position of the hood on the hinges for installation reference, then remove the hood with the aid of an assistant.
4. Drain the cooling system.
5. Remove or disconnect the following:
   - Radiator and cooling fan
   - Air cleaner outlet hose
   - Air cleaner case bolts and case
6. Disconnect the following cables:
   - Accelerator cable from the throttle body
   - Clutch cable from the transaxle, if equipped with manual transmission
   - Gear select cable from the transaxle, if equipped with automatic transmission
7. Disconnect the following vacuum hoses:
   - Brake booster hose from the intake manifold
   - Canister purge hose from the Evaporative (EVAP) emission canister purge valve
   - Manifold Absolute Pressure (MAP) sensor hose from the intake manifold
8. Disconnect the following electrical connectors:
   - Distributor coil wire, if equipped
   - Ignition coils, if equipped
   - Camshaft Position (CMP) sensor
   - Engine oil pressure switch
   - Exhaust Gas Recirculation (EGR) solenoid vacuum valve
   - EVAP canister purge valve
   - Engine Coolant Temperature (ECT) sensor
   - Fuel injectors
   - Power steering pressure switch
   - Heated Oxygen Sensor (HO_2S)
   - Back-up light switch, manual transmission

- Shift switch, automatic transmission
- Forward clutch revolution sensor, automatic transmission
- Automatic transmission Vehicle Speed Sensor (VSS)
- Alternator
- Starter
- Battery negative cable from the transaxle
- Throttle Position (TP) sensor
- Idle Air Control (IAC) valve
- Manifold Absolute Pressure (MAP) sensor
- Crankshaft Position (CKP) sensor, if equipped
- All engine wires from the engine

9. Remove or disconnect the following:
   - Fuel feed hose from the feed pipe
   - Return hose from the fuel pressure regulator
   - Heater inlet and outlet hoses
   - Right and left engine undercovers
   - Front exhaust pipe from the exhaust manifold and center exhaust pipe
   - Gear shift control shaft from the transaxle and the extension rod, if equipped with a manual transaxle
10. Drain the engine and transaxle oil.
11. Remove or disconnect the following:
   - Left and right halfshafts
   - A/C compressor from the compressor bracket with the hoses still attached, if equipped. Position the air conditioning out of the way from the engine.
12. If equipped with power steering, drain the power steering pump of fluid.
13. Remove or disconnect the following:
   - Power steering hose from the power steering pump, if equipped
14. Install a lifting device on the engine.
15. Remove or disconnect the following:
   - Center member from the vehicle by unfastening the 7 nuts and 4 bolts
   - Left engine mount
   - Right engine mount and bracket
16. Check to be sure all cooling hoses, vacuum hoses and electrical wires are disconnected from the engine.
17. Lower the engine with the transaxle from the vehicle.

To install:

18. Raise the engine and transaxle into the engine compartment.
19. Install or connect the following:
   - Right engine mount with the bolts and nuts. Tighten the bolts and nuts to 40 ft. lbs. (55 Nm).
   - Left engine mount and install the bolts and nuts. Tighten the bolts and nuts to 40 ft. lbs. (55 Nm).
20. Install the center member using the 7 nuts and 4 bolts. Tighten the bolts and nuts as follows:
   a. Center member to the radiator support: 33 ft. lbs. (45 Nm)
   b. Center member to crossmember: 33 ft. lbs. (45 Nm)
   c. Engine mounts to center member nuts: 40 ft. lbs. (55 Nm)
21. Remove the lifting device.
22. Install or connect the following:
   - Power steering hose to the power steering pump, if equipped
   - A/C compressor on the air conditioning bracket on the engine, if equipped
   - Left and right halfshafts
   - Gear shift control shaft on the transaxle and the extension rod, if equipped with a manual transaxle
   - Front exhaust pipe on the center exhaust pipe and exhaust manifold
   - Remaining components in the reverse order of removal
23. Fill the cooling system, engine, transaxle and power steering pump.
24. Adjust all cables and check all connections.
25. Connect the negative battery cable to the battery.
26. Start the vehicle and check for leaks.

Water Pump

REMOVAL & INSTALLATION

1.3L Engines

1. Before servicing the vehicle, refer to the precautions in the beginning of this section.

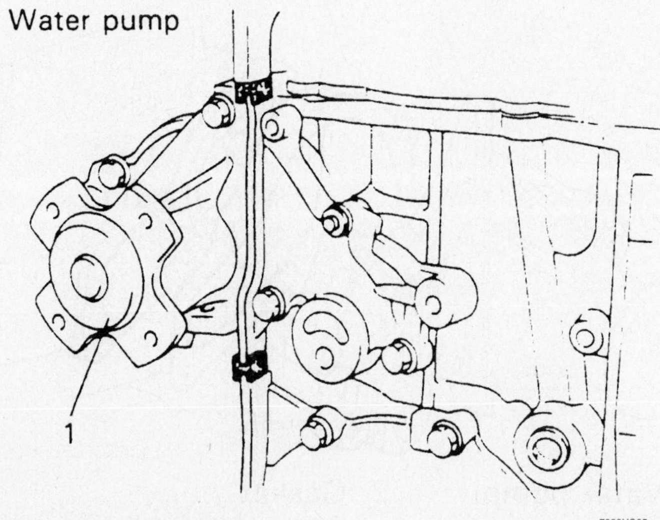

Water pump location—1.3L engine

2. Disconnect the negative battery cable.
3. Drain the cooling system into a suitable container and tighten the drain plug.
4. Remove or disconnect the following:
   - Air cleaner assembly and the Mass Air Flow (MAF) sensor and outlet hose
   - Air cleaner bracket
   - Right side fender apron clips by pushing the center pin

➡Do not push the center pin too far in, or it will fall off into the fender.

   - Power steering and air conditioning belt, if equipped
   - Loosen the water pump pulley bolts
   - Alternator drive belt
   - Water pump pulley
5. To remove the crankshaft pulley perform the following:
   a. If equipped with a manual transaxle, insert a suitable flat-bladed tool into the hole in the bell housing next to the exhaust pipe. This will lock the crankshaft in place.
   b. If equipped with an automatic transaxle, hold a suitable flat-bladed tool in line with the oil pan and insert into the teeth of the drive plate. This will lock the crankshaft in place.
   c. Loosen the crankshaft pulley bolts.
   d. Remove the crankshaft timing belt pulley bolt with a 17mm socket.
   e. Remove the pulley from the crankshaft.
   f. Install the crankshaft bolt.
   g. Remove the flat-bladed tool that was used to lock the crankshaft in place.

➡To remove the crankshaft pulley with the engine assembly mounted on the body, it is necessary to remove the crankshaft timing belt pulley bolt. If the engine assembly is dismounted, the bolt does not need to be removed.

6. Remove or disconnect the following:
   - Resonator and the timing belt outside cover
   - Loosen the right engine mounting bolt
   - Timing belt

**** WARNING**

After the timing belt is removed never turn the camshafts or the crankshaft. Interference may occur between the pistons and the valves causing component damage.

- Timing belt inside cover
- Water pump belt adjusting arm

7. Carefully remove the rubber seal between the water and oil pumps, and remove the seal between the water pump and the cylinder head.
8. Remove or disconnect the following:
   - Water pump bolts and the water pump

To install:

9. Clean the water pump mounting surface of old gasket material.
10. Install or connect the following:
    - New water pump gasket on the cylinder block
    - Water pump on the cylinder block and tighten the bolts to 84–108 inch lbs. (10–13 Nm)
    - Rubber seal between the water pump and the oil pump
    - Seal between the water pump and the cylinder head
    - Water pump belt adjusting arm
11. Install the timing belt inside cover.
12. With the crankshaft locked in position, remove the crankshaft bolt and install the crankshaft pulley. Tighten the crankshaft pulley bolts to 10–13 ft. lbs. (14–18 Nm). Using a 17mm socket, tighten the crankshaft timing belt pulley bolt to 76–83 ft. lbs. (105–115 Nm).
13. Install or connect the following:
    - Timing belt
    - Water pump pulley and drive belt. Tighten the water pump pulley bolts to 84–96 inch lbs. (9–12 Nm).
    - Remaining components in the reverse order of removal
14. Fill the cooling system.
15. Connect the negative battery cable.
16. Start the engine and top off the coolant as necessary.
17. Check the cooling system for leaks.
18. Check the ignition timing.

1.6L Engine

1. Before servicing the vehicle, refer to the precautions in the beginning of this section.
2. Disconnect the negative battery cable.
3. Drain the cooling system and tighten the drain plug.
4. Remove or disconnect the following:
   - Timing belt
   - Alternator adjusting arm
   - Oil dipstick guide and dipstick
   - Water pump bolts, gasket, water pump and rubber seal

To install:

5. Clean the water pump mounting surface of old gasket material.
6. Install or connect the following:
   - New water pump gasket on the cylinder block
   - Water pump on the cylinder block and tighten the bolts to 96 inch lbs. (11 Nm)
   - Rubber seal between the water pump and the oil pump
   - Seal between the water pump and the cylinder head
   - Timing belt
   - Alternator adjusting arm
   - Oil dipstick guide and dipstick using a new O-ring
7. Adjust drive belt tension.
8. Fill the cooling system with engine coolant.
9. Connect the negative battery cable.
10. Start the engine and top off the coolant as necessary.
11. Check the ignition timing.

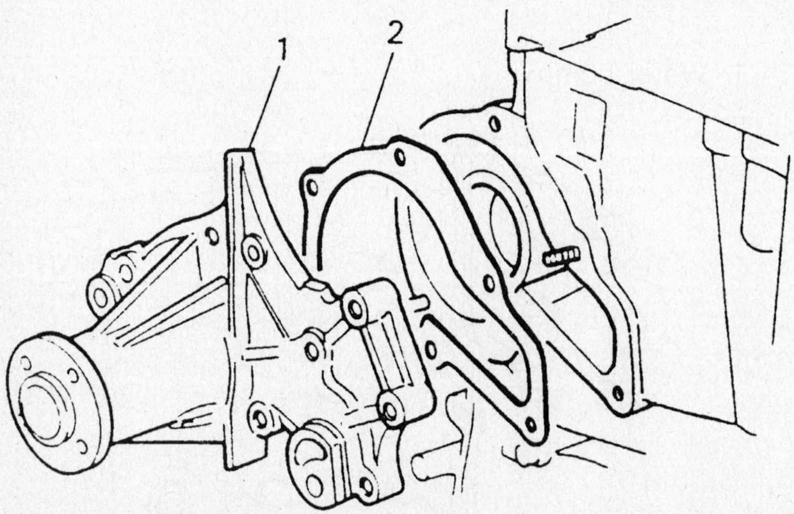

Exploded view of the water pump mounting—1.6L engine

Heater Core

REMOVAL & INSTALLATION

Esteem

1. Before servicing the vehicle, refer to the precautions in the beginning of this section.
2. Disconnect the negative battery cable.
3. Wait for a least 1 minute for the air bag system to deplete its energy before working on any part the instrument panel.
4. Drain the cooling system into a clean container for reuse.
5. Remove the instrument panel.
6. Remove the mode actuator by performing the following procedure:
   - Disconnect the mode actuator coupler.
   - At the heater housing, remove the mode actuator rod.
   - Remove the mode actuator from the heater housing.
7. Remove the rear air duct.
8. Remove the heater housing.
9. Remove the heater housing clips and screws; then, separate the heater housings.
10. Remove the heater core from the heater housing.

To install:

11. Install the heater core to the heater housing.

12. Assemble the heater housings and install the heater housing clips and screws.
13. Install the heater housing.
14. Install the rear air duct.
15. Install the mode actuator by performing the following procedure:
   - Install the mode actuator to the heater housing.
   - At the heater housing, install the mode actuator rod.
   - Connect the mode actuator coupler.
16. Install the instrument panel.
17. Refill the cooling system.
18. Connect the negative battery cable.
19. Operate the engine to normal operating temperatures; then, check the climate control operation and check for leaks.

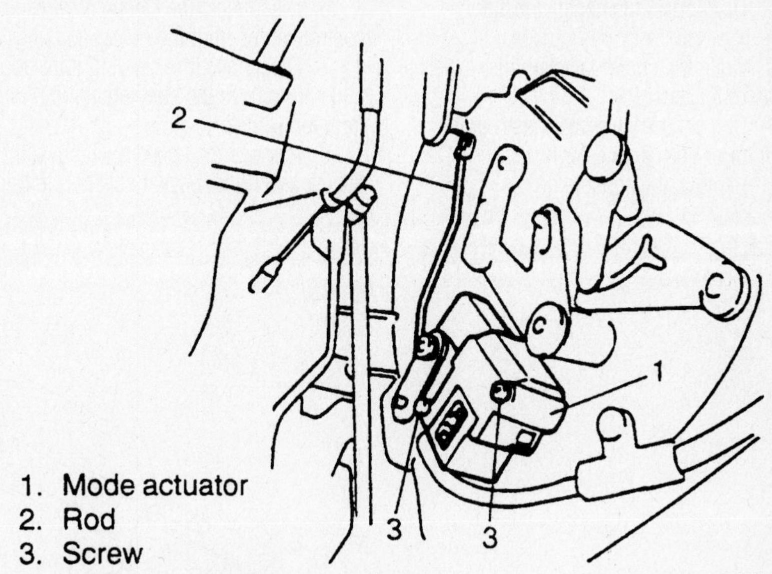

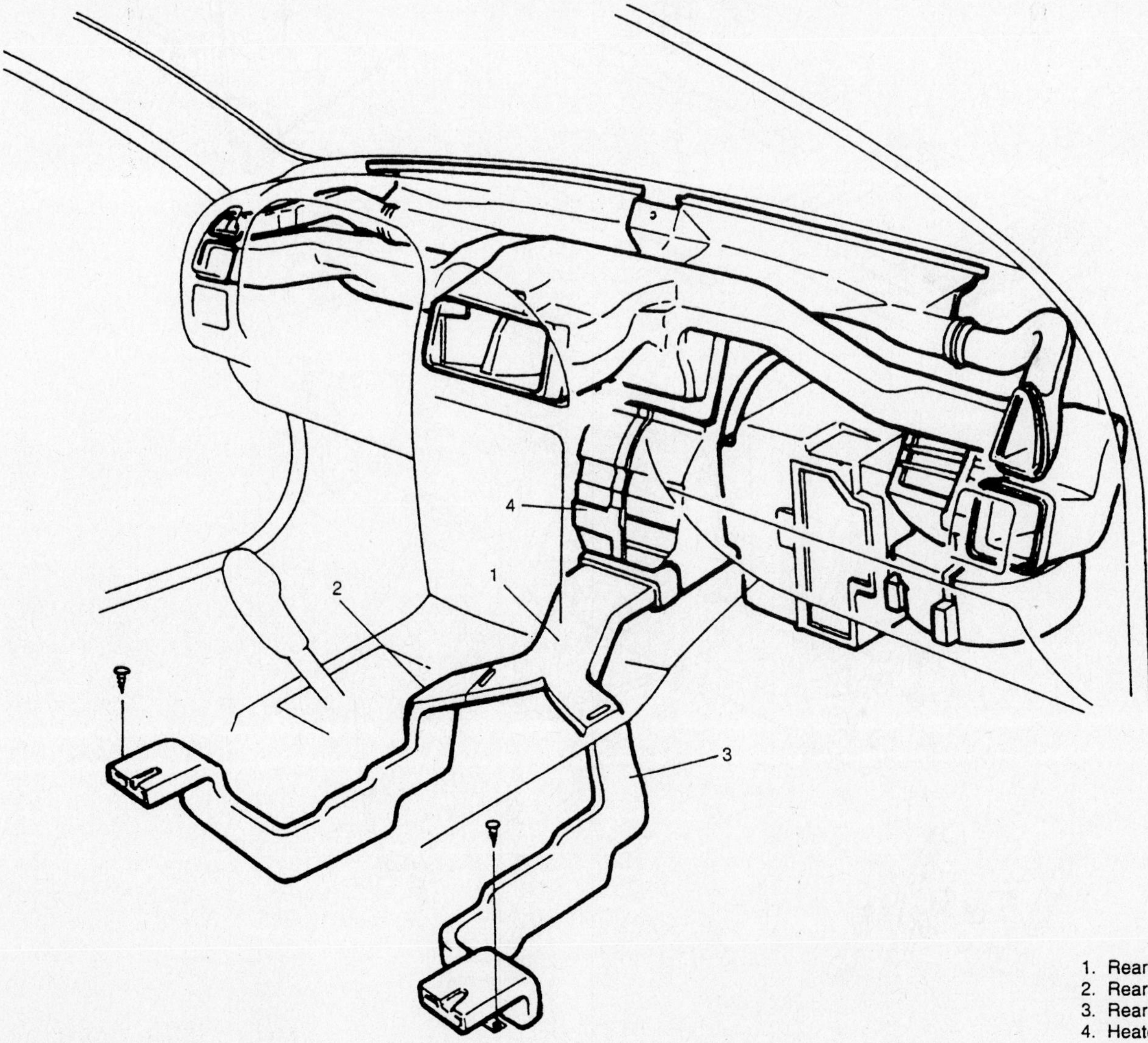

Swift

1. Before servicing the vehicle, refer to the precautions in the beginning of this section.

1
2
3
3
1. Mode actuator
2. Rod
3. Screw
93112GI5

View of the heater housing mode actuator—Esteem

1
2
3
4
1. Rear duct
2. Rear hose LH
3. Rear hose RH
4. Heater unit
93112GI4

Assembled view of the heater housing and related components—Esteem

2. Disconnect the negative battery cable.

3. Drain the cooling system into a clean container for reuse.

4. Disconnect the heater hoses at the heater core. Plug the heater hoses.

5. Remove the console box.

6. Disconnect the electrical wires and control cables from the heater housing.

7. Disconnect the steering joint upper bolt and remove the steering column unit from the vehicle.

8. Remove the speedometer assembly, the left and right speaker covers, the hood opening cable and the center garnish from the dash.

9. Remove the dashboard mounting bolts and the dashboard being careful to protect it from damage.

10. Remove the heater housing mounting bolts and remove it from the vehicle.

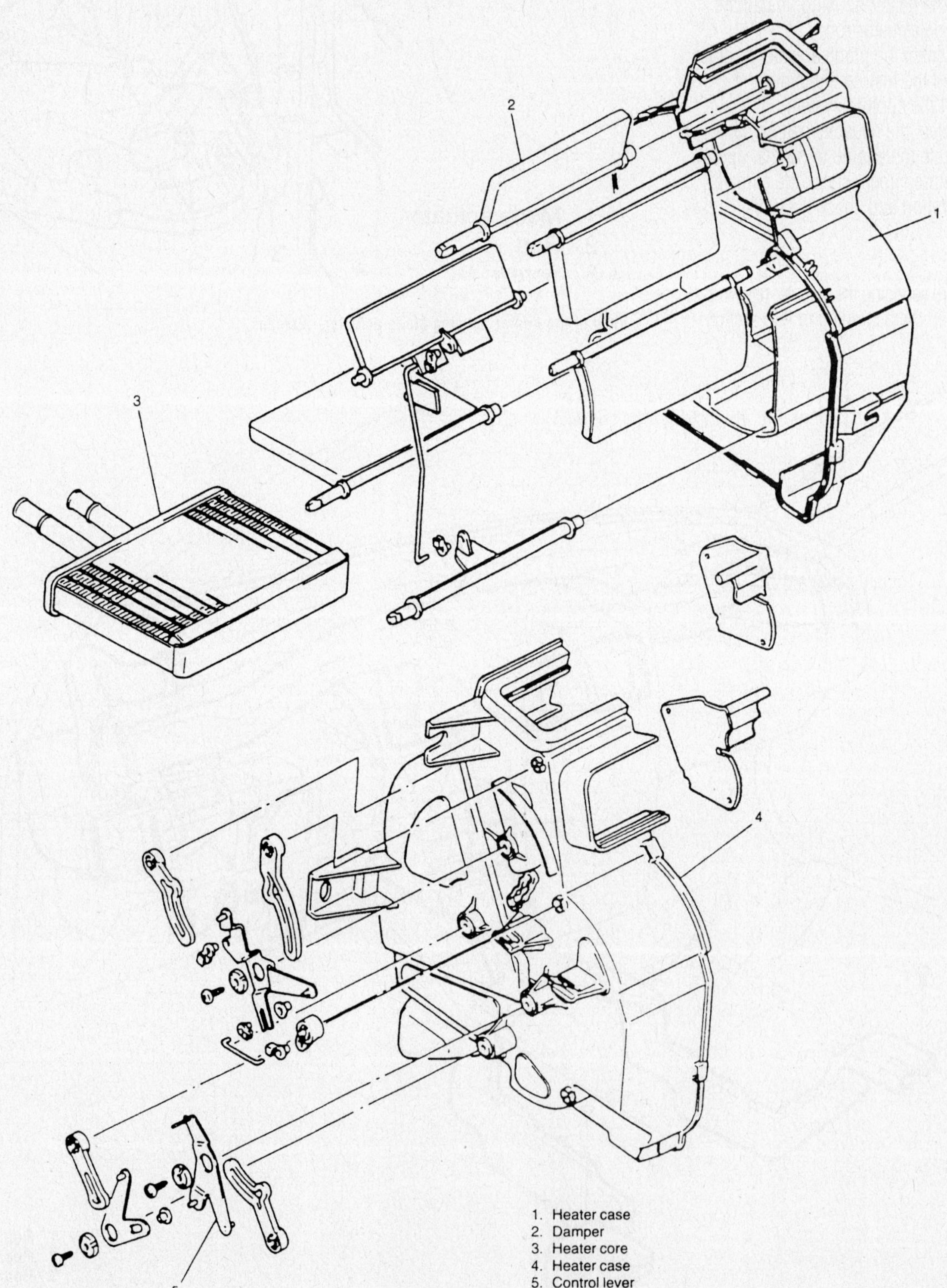

Exploded view of the heater core, heater housing and related components—Esteem

11. Remove the attaching screws and clips which hold the heater housing together and separate the housing to gain access to the heater core.

12. Remove the heater core from the case.

To install:

13. Install the heater core to the case.

14. Install the heater housing and the attaching screws and clips.

15. Install the heater housing and mounting bolts.

16. Install the dashboard and mounting bolts.

17. Install the speedometer assembly, the left and right speaker covers, the hood opening cable and the center garnish to the dash.

18. Install the steering column unit and connect the steering joint upper bolt.

19. Connect the electrical wires and control cables to the heater housing.

20. Install the console box.

21. Connect the heater hoses at the heater core.

22. Refill the cooling system.

23. Connect the negative battery cable.

24. Operate the engine to normal operating temperatures; then, check the climate control operation and check for leaks.

Cylinder Head

REMOVAL & INSTALLATION

1.3L Engines

✻✻ CAUTION

The fuel injection system remains under pressure after the engine has been turned OFF. Properly relieve fuel pressure before disconnecting any fuel lines. Failure to do so may result in fire or personal injury.

1. Before servicing the vehicle, refer to the precautions in the beginning of this section.
2. Relieve the fuel system pressure.
3. Drain the cooling system.
4. Remove or disconnect the following:
   - Air cleaner assembly
   - Ignition coil wire from the distributor cap, if equipped
   - Distributor electrical wires, if equipped
   - Ignition coil assembly, if equipped
   - Exhaust Gas Recirculation (EGR) solenoid vacuum valve
   - Radiator fan thermo switch
   - Engine Coolant Temperature (ECT) sensor
   - ECT gauge sensor
   - Idle Air Control (IAC) valve
   - Throttle Position (TP) sensor
   - Fuel injector
   - Ground wires from the intake manifold.
   - Oxygen (O_2S) sensor
   - Radiator hose from the thermostat housing
   - Heater hose from the intake manifold
   - Throttle body coolant outlet hose from the throttle body
   - Manifold Absolute Pressure (MAP) sensor hose from the intake manifold, if equipped
   - Canister hose from its pipe
   - Canister purge hose from the intake manifold
   - Brake booster from the intake manifold
   - Fuel return hose and the fuel feed hose from the throttle body
   - Throttle cable from the throttle body
   - Water pump and crankshaft pulleys
   - Timing belt
   - Rubber seal between the cylinder head and the water pump
   - Exhaust pipe from the exhaust manifold
   - Spark plug wire clamps from the cylinder head cover
   - Positive Crankcase Ventilation (PCV) hose
   - Cylinder head cover
5. Loosen all valve adjusting screws and allow the valves to close.
6. Remove or disconnect the following:
   - Cylinder head bolts in the reverse order of the tightening sequence
   - Cylinder head with the distributor, intake manifold and exhaust manifold
   - Distributor, intake manifold and exhaust manifold from the cylinder head
7. Clean the cylinder block mating surface of any old gasket material and clean any engine coolant from the cylinders.

To install:

8. Install or connect the following:
   - Intake and exhaust manifolds
   - Distributor on the cylinder head
   - New cylinder head gasket with the top mark facing up and toward the crankshaft pulley
   - Cylinder head on the engine block
9. Coat the cylinder head mounting bolts threads with clean engine oil.
10. Tighten the bolts as follows:
    a. Step 1: 26 ft. lbs. (35 Nm)
    b. Step 2: 41 ft. lbs. (55 Nm)
    c. Step 3: 49 ft. lbs. (68 Nm)
11. Install or connect the following:
    - Rubber seal between the cylinder head and the water pump
    - Timing belt
    - Exhaust pipe to the exhaust manifold, tighten the attaching bolts to 26–36 ft. lbs. (35–50 Nm)
    - Crankshaft and water pump pulleys
    - Throttle cable to the throttle body
    - Fuel return hose and the fuel feed hose to the throttle body
    - Remaining components in the reverse order of removal

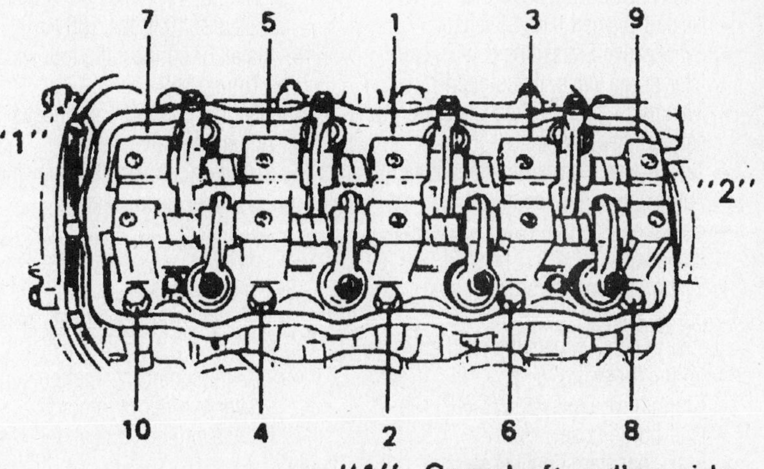

Cylinder head bolt torque sequence—1.3L engine

12. Refill the cooling system with coolant.
13. Connect the negative battery cable.
14. Adjust the ignition timing.
15. With the engine running, be sure that there are no fuel, coolant or exhaust leaks.

1.6L Engine

CAUTION

The fuel system pressure must be relieved before disconnecting any fuel lines. Failure to do so may result in personal injury.

1. Before servicing the vehicle, refer to the precautions in the beginning of this section.
2. Drain the cooling system.
3. Relieve the fuel system pressure.
4. Remove or disconnect the following:
   - Air cleaner outlet hose
   - Intake manifold rear stiffener bolt, alternator adjustment arm reinforcement bolt and right mounting bracket stiffener from the intake manifold
   - Heated Oxygen (HO_2S) sensor coupler and release its clamps
   - Exhaust from the manifold
5. Disconnect the electrical connectors from the following components:
   - Distributor, if equipped
   - Ignition coils, if equipped
   - Engine Coolant Temperature (ECT) sensor and gauge
   - Engine ground wire from intake manifold
   - Exhaust Gas Recirculation (EGR) solenoid vacuum valve
   - Fuel injectors
   - Throttle Position (TP) sensor
   - Idle Air Control (IAC) valve
   - Heated Oxygen (HO_2S) sensor
   - Evaporative Emissions (EVAP) canister purge valve, if equipped
   - Manifold Absolute Pressure (MAP) sensor, if equipped
   - Tank pressure control solenoid vacuum valve, if equipped
   - Camshaft Position (CMP) sensor, if equipped
   - Evaporative emissions solenoid purge valve
6. Label and disconnect the vacuum hoses from the following:
   - Evaporative Emissions (EVAP) canister purge hose
   - Brake booster supply hose
7. Remove or disconnect the following:
   - Fuel feed and return hoses from each pipe

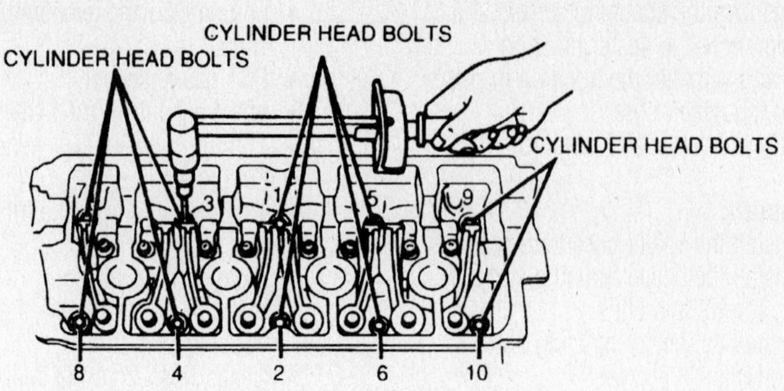

Cylinder head bolt torque sequence—1.6L engine

   - Cylinder head cover
   - Valve lash adjusting screws
8. Disconnect the following hoses:
   - Radiator inlet hose
   - Heater inlet hose
   - IAC valve outlet
9. Remove or disconnect the following:
   - Timing belt
   - Cylinder head bolts in reverse order of tightening. Once each bolt is loose, remove the bolts from the cylinder head.
10. Check to be sure all components are removed or disconnected before removing the cylinder head.
11. Remove the cylinder head with the intake manifold, exhaust manifold and distributor as an assembly.

To install:

12. Install a new cylinder head gasket and the cylinder head, with the distributor case, onto the cylinder block.
13. Tighten the cylinder head bolts, in sequence, using the following 3 Steps:
    a. Step 1: 26 ft. lbs. (35 Nm)
    b. Step 2: 41 ft. lbs. (55 Nm)
    c. Step 3: 49 ft. lbs. (68 Nm)
14. Install or connect the following:
    - Timing belt
    - Engine cooling water hoses
15. Adjust the valve lash.
16. Install or connect the following:
    - Cylinder head cover
    - Fuel feed and return hoses to each pipe
    - Vacuum hoses
    - All remaining electrical components
    - All remaining components in the reverse order of removal
17. Fill the engine coolant and check all fluids.
18. Connect the negative battery cable to the battery.
19. Start the engine and check for leaks.

Rocker Arms/Shafts

REMOVAL & INSTALLATION

Only the 1.3L engine uses rocker arms/shafts. On the 1.6L engine, the camshaft directly actuates the valves.

1.3L Engine

1. Before servicing the vehicle, refer to the precautions in the beginning of this section.
2. Drain the cooling system into a resealable container.
3. Remove or disconnect the following:
   - Negative battery cable
   - Timing belt
   - Cylinder head cover from the cylinder head
   - Camshaft Position (CMP) sensor coupler from the CMP sensor
   - High tension cords and couplers from the coil assemblies
   - Ignition coil from the exhaust manifold side
   - CMP case from the cylinder head
4. Loosen all valve adjusting screw lock nuts, turn the adjusting screws back all the way to allow the rocker arms to move freely.
5. Remove or disconnect the following:
   - Camshaft housing and the camshaft
   - Timing belt inside cover
   - Intake rocker arm with the clip from the rocker arm shaft. Make sure not to bend the clip when removing the intake rocker arm
   - Rocker arm shaft bolts
   - Exhaust rocker arms and spring by pulling out the shaft on the battery side. If necessary, remove the battery for clearance.
6. Inspect the rocker arms and shafts for wear and/or damage and replace parts as necessary.

a. If the tip of the valve lash adjuster is badly worn, replace the adjuster.

To install:

7. Apply engine oil to the rocker arms and the rocker arm shafts.
8. Install or connect the following:
   - O-ring on the shaft
   - Rocker arm shaft with its holes facing up
   - Rocker arm (exhaust side) and the rocker arm spring
   - Rocker arm shaft bolts and tighten the bolts to 96 inch lbs. (11 Nm)
9. Pour a small amount of engine oil into the arm pivot holding part of the shaft.
10. Install or connect the following:
    - Rocker arm (intake side), with the clips to the rocker arm shaft
    - Camshaft and housing
    - Camshaft oil seal
    - Timing belt inside cover
    - Camshaft timing belt pulley, while fitting the pin on the camshaft into the slot. Tighten the pulley bolt to 43 ft. lbs. (60 Nm).
    - Timing belt
    - Remaining components in the reverse order of removal.
11. Check the ignition timing.

1.6L Engine

1. Before servicing the vehicle, refer to the precautions in the beginning of this section.
2. Remove or disconnect the following:
   - Negative battery cable
   - Timing belt

WARNING

After the timing belt is removed, never turn the camshaft and crankshaft independently more than 90° in either direction. If turned, interference may occur between the piston and valves causing possible damage to the effected parts.

3. Using a camshaft sprocket holding tool, hold the sprocket stationary and remove the camshaft sprocket bolt.
4. Remove or disconnect the following:
   - Cylinder head cover
   - Distributor cap and distributor assembly from the engine, if equipped
   - Ignition coils and the Camshaft Position (CMP) sensor case from the cylinder head with the CMP sensor coupler disconnected
   - Loosen all of the valve adjusting screw locknuts until the rocker arms move freely
   - Camshaft housing and camshaft
   - Rocker arm shaft plug from the cylinder head
   - Timing belt inner cover-to-cylinder head bolts and the cover
   - All intake rocker arms and clips from the rocker shaft. Keep all parts in order so they can be reinstalled in their original locations.
   - 6 rocker arm shaft-to-cylinder head bolts. Push the rocker arm shaft through the rear of the cylinder head until the end of the rocker shaft appears.
   - O-ring from the rear of the rocker arm shaft
   - Exhaust rocker arms, rocker arm springs and rocker shaft by pulling the rocker arm shaft through the front of the cylinder head. Be sure to keep the parts in order for installation purposes.
5. Clean and inspect all parts for wear and/or damage. Replace parts as necessary.

To install:

6. Lubricate the rocker arms and shafts with clean engine oil before installation.
7. Install or connect the following:
   - Push the rocker shaft into the front of the cylinder head
   - Exhaust rocker arms and springs as the rocker arm shaft is being installed into the cylinder head
   - Push the rocker arm shaft through the rear of the cylinder head
   - New O-ring onto the rocker shaft
8. Rotate the rocker arm shaft so the flat machined surface is horizontal and facing downward, parallel with the cylinder head mating surface and slide the shaft back into the cylinder head.
9. Install or connect the following:
   - 6 rocker arm shaft bolts and tighten the rocker arm-to-cylinder head bolts to 96 inch lbs. (11 Nm). Fill the rocker arm shaft bolt holes with clean engine oil.
   - Intake rocker arms and clips onto the rocker arm shaft

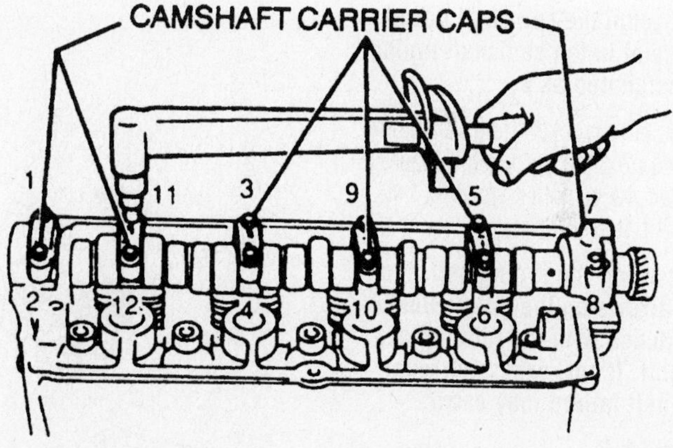

Camshaft carrier cap bolt torque sequence—1.6L engine

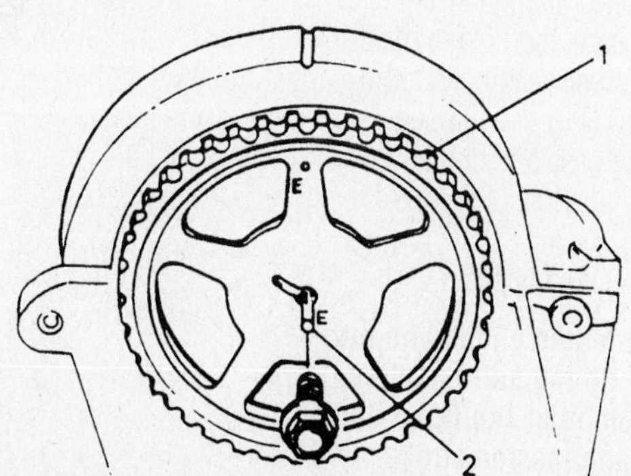

Align the dowel pin with the E-slot in the sprocket—1.6L engine

➡**The camshaft carrier caps are embossed with numbers and arrows to ensure correct assembly. The No. 1 camshaft carrier cap must be installed at the front of the cylinder head with the remaining carrier caps following in numerical order. The directional arrows must always point toward the front of the cylinder head.**

- Camshaft

✲✲ WARNING

If the camshaft carrier cap bolts are tightened at random, damage to the camshaft may occur.

10. Lubricate the new camshaft seal lip with clean engine oil
    - New camshaft seal into the cylinder head until it is flush with the camshaft carrier surface
    - Timing belt inner cover and tighten the cover-to-cylinder head bolts to 89 inch lbs. (10 Nm)
    - Rocker arm shaft plug into the cylinder head and tighten to 24 ft. lbs. (33 Nm)

➡**During camshaft timing belt sprocket installation, align the camshaft dowel pin with the slot in the camshaft timing belt gear designated as E.**

- Camshaft sprocket. Using a holding tool to hold the sprocket in place, tighten the camshaft sprocket bolt to 44 ft. lbs. (60 Nm).

➡**When installing the timing belt, the directional arrows on the timing belt must be matched with the rotation of the crankshaft. If not, excessive wear and timing belt failure may occur.**

- Timing belt

11. Apply RTV silicone rubber sealant to the surface of the distributor case that mates with the rear of the rocker arm shaft.
12. Install or connect the following:
    - Distributor case and tighten the 3 case-to-cylinder head bolts to 89 inch lbs. (10 Nm), if equipped
    - CMP sensor case and tighten the bolts to 96 inch lbs. (11 Nm), if equipped
    - Ignition coil, if equipped

➡**With the timing marks aligned on the sprockets and the timing belt installed, the number 4 piston is at Top Dead Center (TDC) of the compression stroke.**

- Distributor into the distributor case, if equipped. Be sure the rotor is aligned with the No. 4 tower on the distributor cap.
- Distributor cap, if equipped

13. Adjust the valve lash.
14. Install or connect the following:
    - Cylinder head cover onto the cylinder head, in the reverse order of removal. Clean all sealing surfaces and use a new gasket and O-rings. Tighten the cylinder head cover bolts to 89 inch lbs. (10 Nm).
    - Negative battery cable
15. Start the engine, allow it to reach normal operating temperature and check for leaks.
16. Check and adjust the ignition timing as necessary.

Intake Manifold

REMOVAL & INSTALLATION

✲✲ CAUTION

The fuel system pressure must be relieved before disconnecting any fuel lines. Failure to do so may result in personal injury.

1. Before servicing the vehicle, refer to the precautions in the beginning of this section.
2. Properly relieve the fuel system pressure.
3. Drain the coolant from the vehicle.

✲✲ CAUTION

To help avoid the danger of being burned, do not remove the drain plug and the radiator cap while the engine is still hot. Scalding fluid and steam can be blown out under pressure if the plug and cap are taken off too soon.

4. Remove or disconnect the following:
    - Negative battery cable
    - Air cleaner assembly
5. Disconnect the following electrical wires:
    - Exhaust Gas Recirculation (EGR) solenoid vacuum valve
    - Idle Speed Control (ISC) actuator

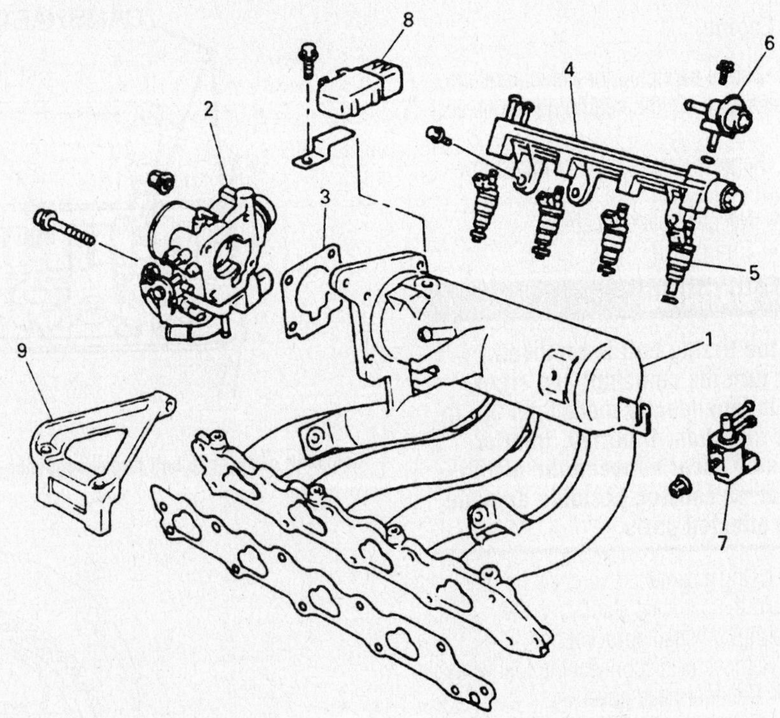

Intake manifold and related components—1.3L engine

- Ground wires from the intake manifold
- Fuel injectors
- Throttle Position (TP) sensor
- Idle Air Control (IAC) valve
- Tank pressure control solenoid vacuum valve, if equipped
- Manifold Absolute Pressure (MAP) sensor

6. Remove or disconnect the following:
   - Fuel return and feed hoses from the fuel pipes
   - Coolant hoses from the throttle body and the intake manifold
7. Disconnect the following vacuum hoses:
   - Canister purge hose from the intake manifold
   - Manifold Absolute Pressure (MAP) sensor hose from the intake manifold
   - Brake booster hose from the intake manifold
   - Positive Crankcase Ventilation (PCV) hose from the PCV valve
   - Accelerator cable from the throttle body
8. Remove or disconnect the following:
   - Intake manifold attaching nuts and bolts
   - Intake manifold and throttle body

To install:

9. Before installing the gasket, make sure that the mating surfaces of the intake manifold and the cylinder head are clean and undamaged.
10. Install or connect the following:
    - New intake manifold gasket on the cylinder head
    - Intake manifold and throttle body on the cylinder head
    - Intake manifold mounting nuts and bolts. Tighten the nuts and bolts to 13–20 ft. lbs. (18–28 Nm). Be sure that the clamps are properly installed on the lower intake manifold bolts.
    - Remaining components in the reverse order of removal
11. Refill the cooling system.
12. Connect the negative battery cable.
13. Start the engine and check for fuel and cooling system leaks.

Exhaust Manifold

REMOVAL & INSTALLATION

1.3L Engine

CAUTION

To avoid the danger of being burned, do not service the exhaust system while it is hot. Service should be performed only after the system cools down.

1. Before servicing the vehicle, refer to the precautions in the beginning of this section.
2. Remove or disconnect the following:
   - Negative battery cable
   - Heated Oxygen (HO_2S) sensor electrical coupler and release the wire from its clamps
   - Exhaust manifold cover
   - Exhaust manifold stiffener bolt
   - 2 bolts attaching the exhaust pipe to the exhaust manifold
   - Exhaust manifold mounting nuts and bolts
   - Exhaust manifold and the gasket

To install:

3. Before installing any components check the exhaust manifold and the engine for deterioration or damage and replace as necessary.
4. Install or connect the following:
   - Manifold gasket and the exhaust manifold on the engine. Tighten the nuts and bolts to 13–20 ft. lbs. (18–28 Nm).
   - Exhaust pipe gasket on the exhaust pipe and position the exhaust pipe
   - 2 bolts that attach the exhaust pipe on the exhaust manifold and

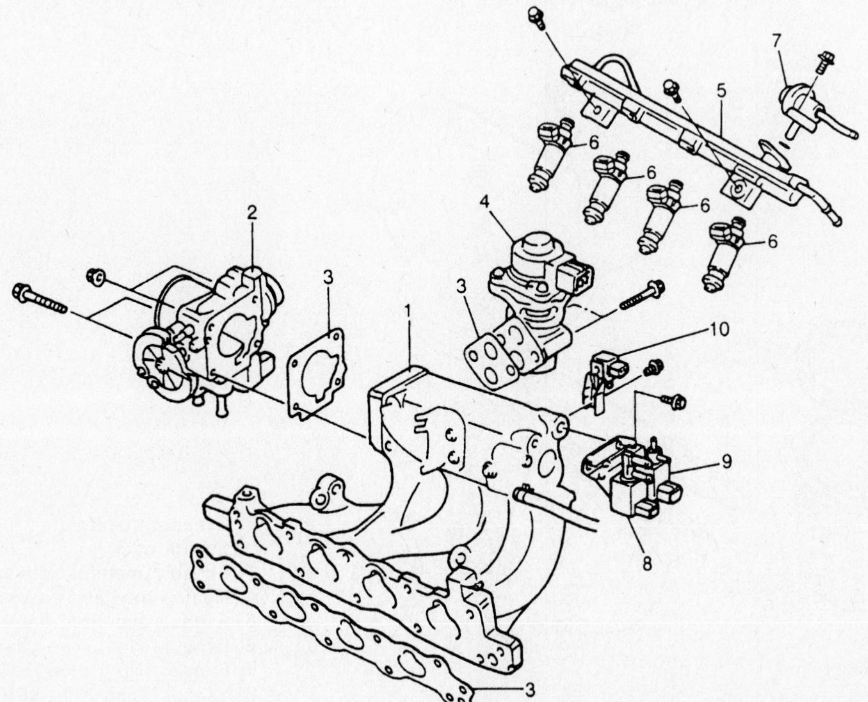

Intake manifold and related components—1.6L engine

tighten the bolts to 25–36 ft. lbs. (35–50 Nm)
- Exhaust manifold stiffener and tighten the bolt to 29–43 ft. lbs. (40–60 Nm)
- Remaining components in the reverse order of removal
- Negative battery cable

5. Run the engine and check for exhaust leaks.

1.6L Engine

1. Before servicing the vehicle, refer to the precautions in the beginning of this section.
2. Remove or disconnect the following:
   - Negative cable
   - 2 exhaust pipe bolts connecting the exhaust pipe to the exhaust manifold
   - Oxygen (O_2S) sensor lead wire at the coupler
   - Exhaust manifold heat shield from the manifold by unfastening the nut and bolt
   - Exhaust manifold mounting bolts and nuts
   - Exhaust manifold

To install:

3. Clean and inspect the sealing surfaces of the exhaust manifold and the cylinder head.
4. Install or connect the following:
   - New gaskets
   - Exhaust manifold and tighten the mounting bolts and nuts to 17 ft. lbs. (23 Nm)
   - O_2S lead wire at the coupler
   - 3 exhaust pipe bolts connecting the exhaust pipe to the exhaust manifold and tighten to 36 ft. lbs. (50 Nm)
   - Manifold heat shield to the exhaust manifold using the nuts and bolts
   - Negative battery cable
5. Check for exhaust leaks when finished.

Front Crankshaft Seal

REMOVAL & INSTALLATION

1.3L Engines

1. Remove or disconnect the following:
   - Negative battery cable
   - Timing belt
   - Crankshaft sprocket bolt using a suitable gear stopper to hold the flywheel
   - Sprocket bolt, sprocket and key
   - Seal from the oil pump housing, using a suitable tool

➡Be careful not to damage the crankshaft or the oil pump sealing surfaces when removing or installing the seal.

2. Clean and inspect the surfaces of the crankshaft and the oil pump assembly.

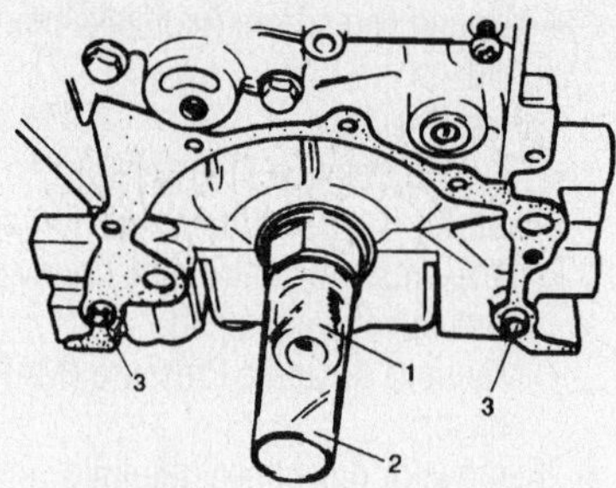

Oil seal guide tool—1.3L engine

To install:

3. Lubricate the new seal with clean engine oil.
4. Install or connect the following:
   - New seal over the crankshaft and into the oil pump, making sure the oil seal lip is not turned up. Use an oil seal guide tool.
   - Crankshaft sprocket and timing belt
   - Negative battery cable

1.6L Engine

➡The front oil seal can be removed from the engine without removing the oil pump.

1. Before servicing the vehicle, refer to the precautions in the beginning of this section.
2. Remove or disconnect the following:

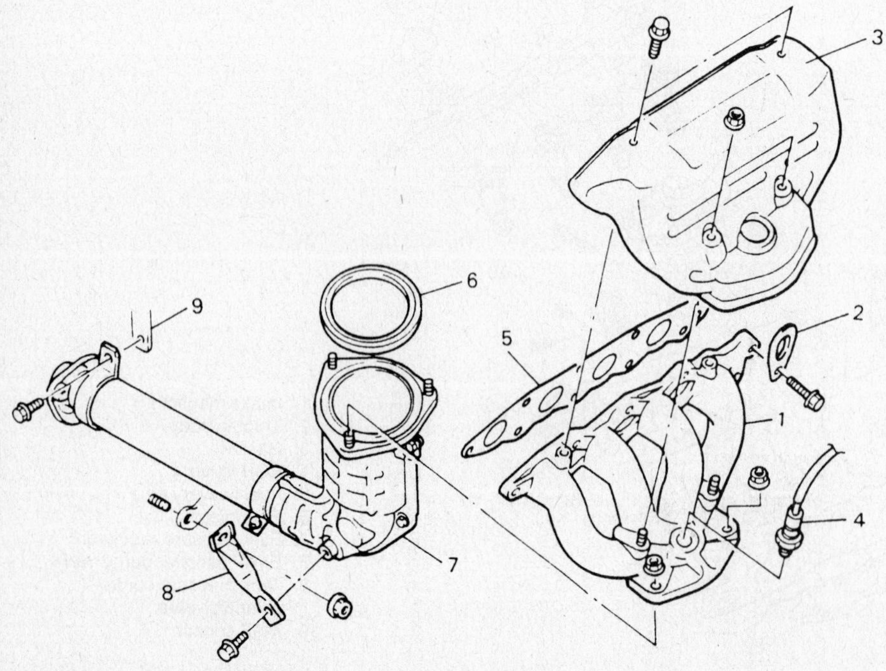

Exhaust manifold and related components—1.6L engine

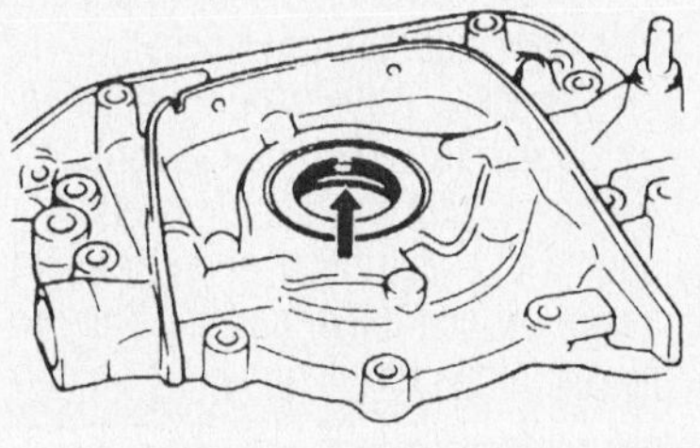

Front crankshaft oil seal location—1.6L engine

- Negative battery cable
- Timing belt
- Crankshaft timing belt sprocket

⁂ WARNING

When removing the front seal, be extremely careful not to damage the crankshaft.

3. Using a knife, cut off the oil seal lip.
4. Tape the end of a flat-bladed tool to avoid damaging the crankshaft. Pry out the oil seal using the taped end of the tool.
5. Inspect the oil seal contact surface on the crankshaft for signs of wear or damage.

To install:

6. Wipe the seal bore with a clean rag.
7. Apply multipurpose grease to the lip of a new oil seal.
8. Install or connect the following:
   - Oil seal into place using a seal installer tool. Be sure the seal surface is flush with the edge of the oil pump case. Work from the front of the cover. Be extremely careful not to damage the seal.
   - Crankshaft sprocket
   - Timing belt
   - Negative battery cable to the battery
9. Start the engine and check for leaks.

Camshaft and Valve Lifters

REMOVAL & INSTALLATION

1.3L Engine

1. Before servicing the vehicle, refer to the precautions in the beginning of this section.
2. Drain the cooling system into a resealable container.
3. Relieve the fuel system pressure.
4. Remove or disconnect the following:
   - Timing belt
   - Cylinder head cover from the cylinder head

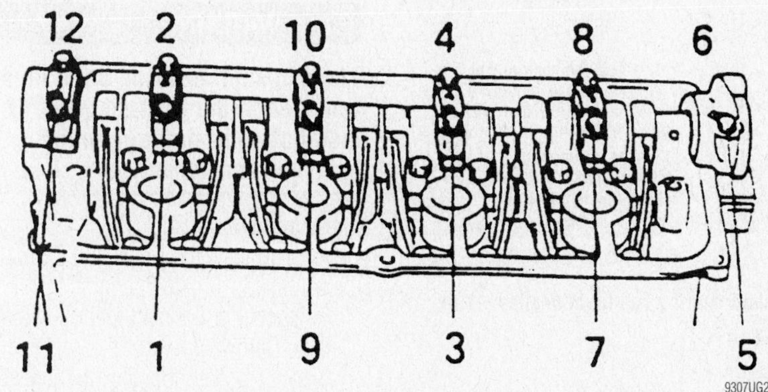

Loosen the camshaft bearing caps using several steps in the sequence shown—1.3L engines

   - Camshaft Position (CMP) sensor coupler from the CMP sensor
   - High tension cords and couplers from the coil assemblies
   - Ignition coil from the exhaust manifold side
   - CMP case from the cylinder head
5. Loosen all valve adjusting screw lock nuts, turn the adjusting screws back all the way to allow the rocker arms to move freely.
6. Remove or disconnect the following:
   - Camshaft bearing caps in the sequence illustrated in several steps
   - Camshaft
   - Timing belt inside cover
   - Intake rocker arm with the clip from the rocker arm shaft. Make sure not to bend the clip when removing the intake rocker arm
   - Rocker arm shaft bolts
   - Exhaust rocker arms and spring by pulling out the shaft on the battery side. If necessary, remove the battery for clearance.
7. Inspect the rocker arms and shafts for wear and/or damage and replace parts as necessary.
   a. If the tip of the valve lash adjuster is badly worn, replace the adjuster.

To install:

8. Apply engine oil to the rocker arms and the rocker arm shafts.
9. Install or connect the following:
   - O-ring on the shaft
   - Rocker arm shaft with its holes facing up
   - Rocker arm (exhaust side) and the rocker arm spring
   - Rocker arm shaft bolts and tighten the bolts to 96 inch lbs. (11 Nm)
10. Pour a small amount of engine oil into the arm pivot holding part of the shaft.
11. Install or connect the following:
   - Rocker arm (intake side), with the clips to the rocker arm shaft
   - Camshaft and the bearing caps. Tighten the bearing caps in several passes to reach the final torque of 96 inch lbs. (11 Nm). Refer to the accompanying illustration for the proper camshaft bearing cap torque sequence.
   - Camshaft oil seal
   - Timing belt inside cover
   - Camshaft timing belt pulley, while fitting the pin on the camshaft into

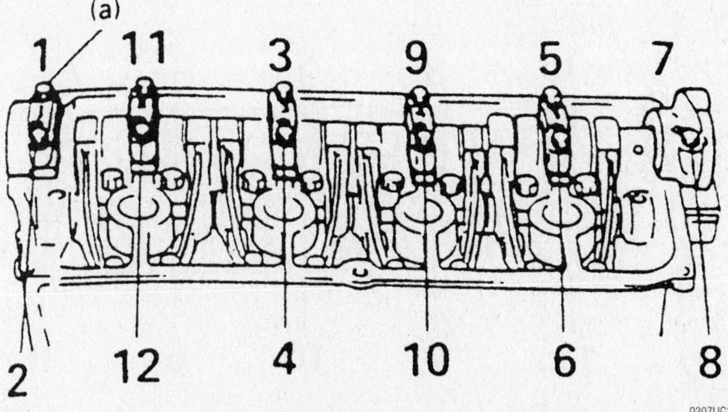

Tighten the camshaft bearing caps using several steps in the sequence shown—1.3L engines

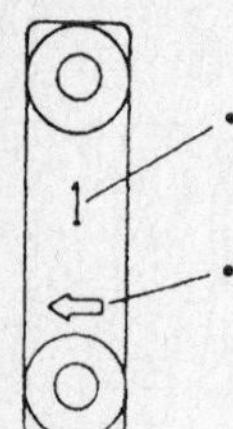
• Indicates position from timing belt side. Install in numerical order starting from timing belt side.
• Indicates direction of housing. Install so that arrow is directed toward timing belt side.
1
7923UG21

Camshaft bearing cap identification—1.6L engine

the slot. Tighten the pulley bolt to 43 ft. lbs. (60 Nm).
- Timing belt
- Remaining components in the reverse order of removal.

12. Check the ignition timing.

1.6L Engine

1. Before servicing the vehicle, refer to the precautions in the beginning of this section.
2. Relieve the fuel system pressure.
3. Remove or disconnect the following:
   - Water pump belt and pulley
   - Crankshaft pulley
   - Timing belt cover and the timing belt
   - Camshaft sprocket
   - Cylinder head cover
   - Distributor and distributor case, if equipped
   - Ignition coils and Camshaft Position (CMP) sensor case, with the CMP sensor housing coupler disconnected, if equipped
4. Loosen the valve adjusting screw locknuts and screws to allow all the valves to close.
5. Remove or disconnect the following:
   - Camshaft housing bolts
   - Housings
   - Camshaft

**** WARNING**

The camshaft housing bolts must be removed in the reverse order of installation or damage to the camshaft may occur.

To install:

6. Lubricate the lobes and journals of the camshaft with clean engine oil.
7. Install or connect the following:
   - Camshaft
   - Camshaft housing on the camshaft and cylinder head, starting with the number 1 housing

➡Embossed marks are provided on each camshaft housing, indicating position and direction for installation.

8. Apply engine oil to the camshaft journal sliding surface of each housing. Apply sealant to the mating surface of the number 6 housing which will mate with the cylinder head.
9. Apply engine oil to the housing bolts, and hand-tighten the bolts into the housing. Follow the tightening sequence in 3 to 4 even stages, finishing with a final torque of 96 inch lbs. (11 Nm).

**** WARNING**

The camshaft housing bolts must be tightened in the correct order or damage to the camshaft may occur.

10. Apply engine oil to the camshaft oil seal lip. Install the camshaft oil seal until the surface becomes flush.
11. Install or connect the following:
    - Camshaft sprocket
    - Timing belt
    - Timing belt cover
    - Crankshaft pulley
    - Water pump pulley
    - Water pump belt. Be sure the pin on the camshaft fits into the slot at the **E** mark on the camshaft sprocket. Tighten the sprocket bolt to 41–46 ft. lbs. (56–64 Nm).
12. Prior to installation on models equipped with a distributor, apply sealant to the area of the distributor housing that covers the rear of the rocker arm shaft on the cylinder head. Install the distributor and distributor housing to the cylinder head. Be sure the distributor is facing the correct firing position.
13. Prior to installation on models equipped with a Distributorless Ignition System (DIS), apply sealant to the area of the housing that covers the rear of the rocker arm shaft on the cylinder head. Install the CMP sensor and housing to the cylinder head and tighten the bolts to 96 inch lbs. (11 Nm).
    - Ignition coil assembly, if equipped
14. Adjust the valve lash.
15. Install or connect the following:
    - New cylinder head cover gasket and the cover
    - Negative battery cable
16. Start the engine and check for any water or oil leaks when finished.
17. Check and/or adjust the ignition timing as necessary.

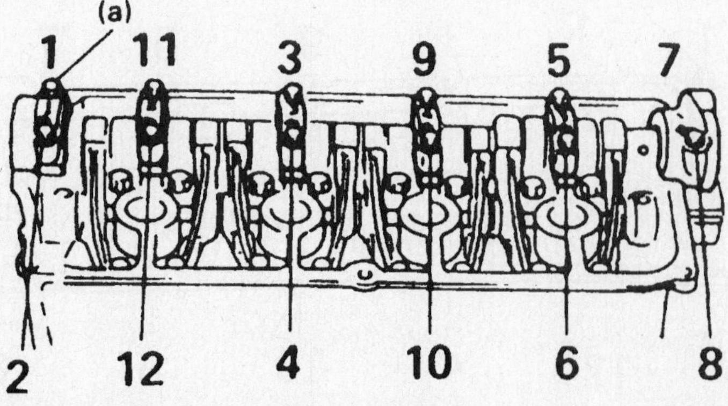
(a)
1
11
3
9
5
7
2
12
4
10
6
8
7923UG22

Camshaft housing bolt tightening sequence—1.6L engine

Valve Lash

ADJUSTMENT

1.3L Engines

Hydraulic valve lash adjusters are used to adjust the valve clearance to **0** lash automatically at all times. Adjustment is not required.

1.6L engine

1. Before servicing the vehicle, refer to the precautions in the beginning of this section.
2. Disconnect the negative battery cable.
3. Remove the cylinder head cover.
4. Remove the right front wheel and fender apron.
5. Turn the crankshaft pulley clockwise until the **V** mark on the pulley is aligned with the **0** calibration mark on the timing belt cover.
6. Remove the Camshaft Position (CMP) sensor and look in the hole for the notch on the camshaft rotor gear. If the notch is not visible in the hole, rotate the crankshaft 1 complete revolution until the notch is visible.

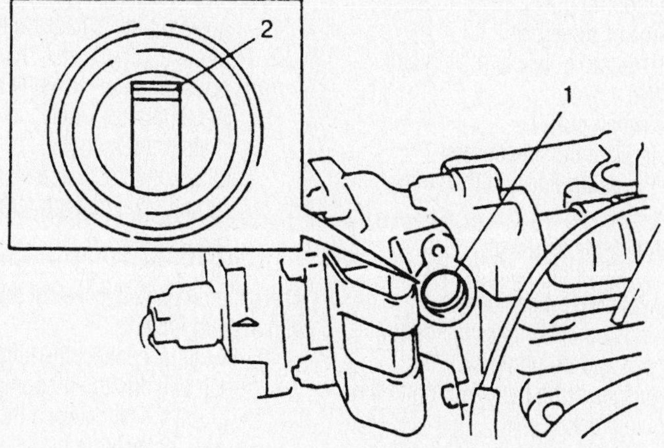

1. Hole for CMP sensor
2. Camshaft rotor gear

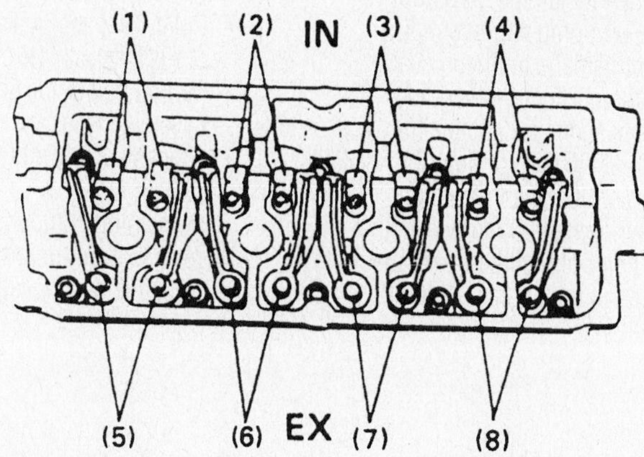

Valve numbered locations and camshaft rotor gear mark—1.6L engines

7. The valve lash is measured between the rocker arm adjusting screw and the valve stem. Use a thickness gauge to measure the gap.

8. Check the valve lash for the following valves:
- Intake valve of cylinder number 1 (ID 1)
- Intake valve of cylinder number 2 (ID 2)
- Exhaust valve of cylinder number 1 (ID 5)
- Exhaust valve of cylinder number 3 (ID 7)

9. If the valve lash is out of specification, adjust the specification after loosening the locknut and turning the adjusting screw. Hold the screw stationary while tightening the locknut. Recheck the specification after tightening the locknut.

10. Rotate the crankshaft 1 rotation clockwise and realign the timing marks.

11. Check the valve lash for the following valves:
- Intake valve of cylinder number 3 (ID 3)
- Intake valve of cylinder number 4 (ID 4)
- Exhaust valve of cylinder number 2 (ID 6)
- Exhaust valve of cylinder number 4 (ID 8)

12. Adjust the valves that are out of specification and recheck after tightening the locknut.

13. Install the remaining components.

14. Connect the negative battery cable.

Starter Motor

REMOVAL & INSTALLATION

1. Remove or disconnect the following:
- Negative battery cable
- Starter electrical connections
- 2 starter mounting bolts
- Starter motor

To install:

2. Install or connect the following:
- Starter
- 2 starter mounting bolts and tighten to 96 inch lbs. (10 Nm)
- Starter electrical connections
- Negative battery cable

Oil Pan

REMOVAL & INSTALLATION

1. Before servicing the vehicle, refer to the precautions in the beginning of this section.

2. Remove or disconnect the following:
- Negative battery cable
- Engine undercovers on 1.6L engines

3. Drain the engine oil into a suitable container.
- Lower plate on 1.3L models, from the clutch housing if equipped with a manual transaxle, or from the torque converter housing if equipped with a automatic transaxle.

4. On 1.6L models, remove or disconnect the following:
- Front exhaust pipe from the vehicle
- Transaxle stiffener plate from the engine and transaxle

5. Support the transmission and engine.

6. Remove or disconnect the following:
- Vehicle center member by removing the 7 nuts and 4 bolts from the center member, on 1.6L models
- Crankshaft Position (CKP) sensor from the oil pan, if equipped
- Oil pan retainer bolts
- Oil pan from the cylinder block
- Oil pump pick-up

To install:

7. Clean the mating surfaces of the oil pan and the engine block.

8. Install or connect the following:
- Oil pump strainer. Tighten the strainer bolt first at the bracket. Tighten the bolts to 96 inch lbs. (11 Nm).

9. Apply silicon sealant to the oil pan mating surface in one continuous bead.

10. Install or connect the following:
- Oil pan to the engine block and install the bolts. Start tightening the bolts at the center and move outward. Tighten the bolts to 96 inch lbs. (11 Nm).
- CKP sensor on the oil pan
- Drain plug and drain plug gasket on the oil pan
- Lower plate on the clutch housing, on 1.3L models; if equipped with a

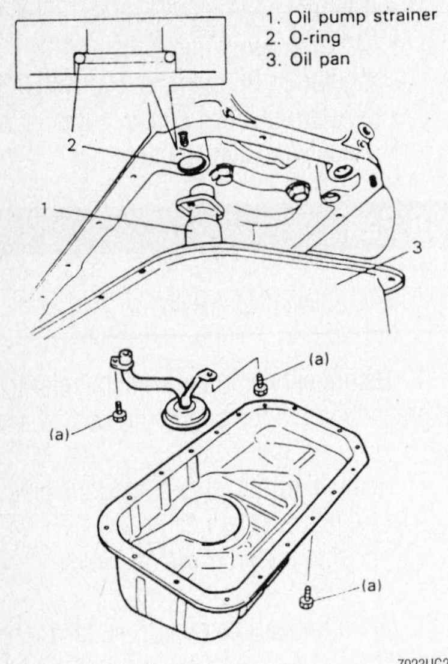

Oil pan mounting—1.3L and 1.6L engines

manual transaxle, or on the torque converter housing if equipped with an automatic transaxle

11. On 1.6L models, install or connect the following:
    - Center member on the vehicle. Tighten the bolts and nuts at the engine mounts to 40 ft. lbs. (55 Nm) and all other nuts to 33 ft. lbs. (45 Nm).
    - Transaxle stiffener plate and tighten the bolts to 37 ft. lbs. (50 Nm)
    - Exhaust pipe and tighten the nuts and bolts to 37 ft. lbs. (50 Nm)
    - Engine undercovers
12. Refill the engine with oil.
13. Connect the negative battery cable.
14. Start the engine and check for leaks.

Oil Pump

REMOVAL & INSTALLATION

1. Before servicing the vehicle, refer to the precautions in the beginning of this section.
2. Disconnect the negative battery cable.
3. Drain the engine oil.
4. Remove or disconnect the following:
    - Right side fender apron clips by pushing the center pin

**** WARNING**

Do not push the center pin too far in, or it will fall off into the fender.

- Power steering and air conditioning belt, if equipped
- Water pump pulley bolts, loosen
- Alternator drive belt
- Water pump pulley and alternator bracket
- Crankshaft pulley
- Timing belt outside covers
- Timing belt guide and the timing belt
- Engine oil level gauge
- Air conditioning compressor bracket bolts, if equipped
- Timing belt and crankshaft pulley
- Oil pan and oil pump pick-up
- 7 bolts securing the oil pump to the engine block
- Oil pump

To install:

5. Clean the engine block where the oil pump mounts, then install the oil pump gasket and the 2 oil pump alignment pins.
6. To prevent damage to the oil seal when installing the oil pump, fit a guide sleeve on the crankshaft and apply a thin coating of engine oil to the special tool.
7. Install or connect the following:
    - Oil pump. Install a long bolt in the lowest bolt hole on the intake manifold side of the engine. Install 2 long bolts in the 2 lowest bolt holes on the exhaust manifold side of the engine. Install the 4 short bolts in the other 4 bolt holes in the oil pump. Tighten all of the bolts to 96 inch lbs. (11 Nm). Check that the oil seal lip is not turned upward, then remove the special tool.
8. If the of the oil pump gasket bulges where the oil pan attaches cut the excess off with a sharp knife.
9. Install or connect the following:
    - Oil pan and the oil pump pick-up
    - Rubber seal between the oil pump and the water pump
    - Crankshaft key
    - Timing belt pulley
    - Crankshaft pulley pin
    - Pulley bolt and tighten to 80 ft. lbs. (110 Nm), with the crankshaft locked
    - Timing belt guide so that the concave side faces the oil pump
    - Crankshaft key and the timing belt pulley
    - Remaining components in the reverse order of removal
    - Negative battery cable

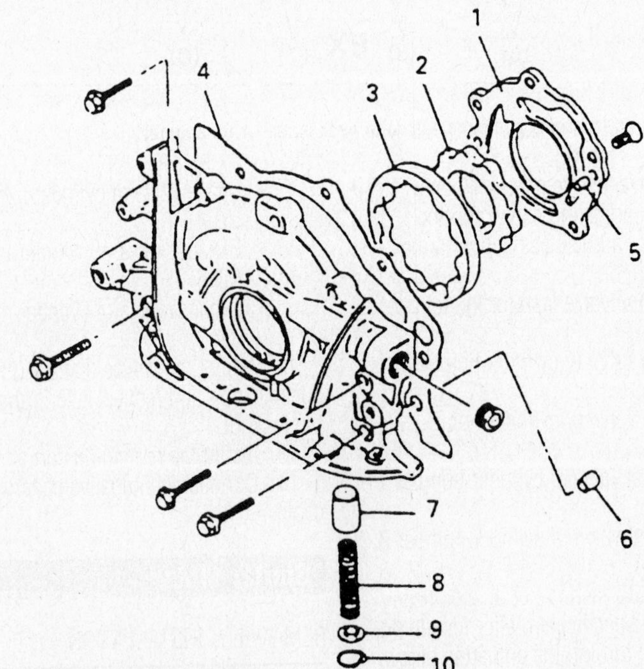

Exploded view of the oil pump—1.3L and 1.6L engine

10. Start the engine and check the engine oil pressure.

11. Check that no leaks are present.

Rear Main Seal

REMOVAL & INSTALLATION

1. Before servicing the vehicle, refer to the precautions in the beginning of this section.
2. Remove or disconnect the following:
   - Transaxle assembly
   - Flexplate/flywheel from the crankshaft
3. Carefully pry the oil seal out of the retainer without scratching the sealing surface of the crankshaft.

To install:

4. Apply engine oil the lip of the new seal.
5. Install or connect the following:
   - Seal in the retainer using a suitable seal driver
   - Flexplate/flywheel
   - Transaxle assembly

Timing Belt

REMOVAL & INSTALLATION

1.3L Engines

1. Before servicing the vehicle, refer to the precautions in the beginning of this section.
2. Remove all necessary components for access to the upper and lower timing belt outside covers, then remove the covers.
3. Align the camshaft timing belt pulley with its timing marks. The crankshaft and camshaft marks are straight up.
4. Remove the resonator and the timing belt outside cover.
5. Remove the tensioner stud and loosen the tensioner bolt.
6. Remove the tensioner spring and damper, then remove the timing belt.

⁂ WARNING

After the timing belt is removed never turn the camshaft or the crankshaft. Interference may occur between the pistons and the valves causing component damage.

7. Remove the tensioner and the tensioner plate.

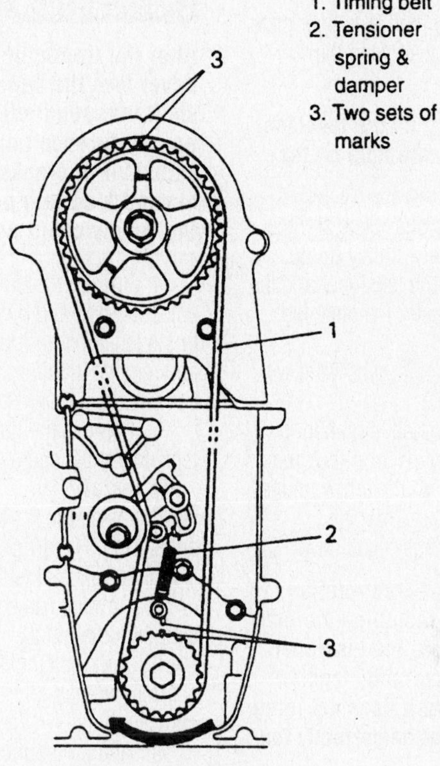

View of the timing belt and timing marks—Suzuki 1.0L engine

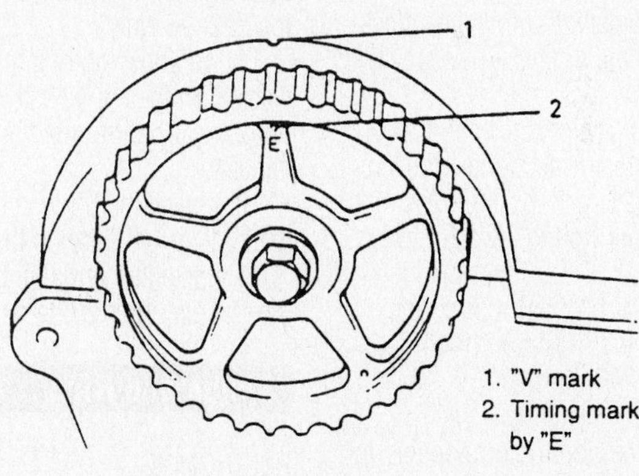

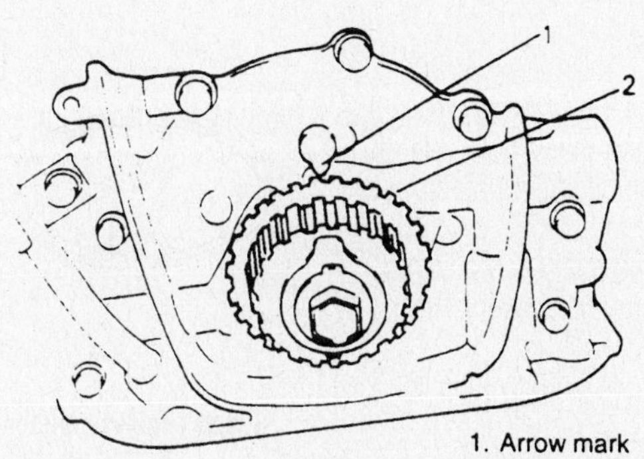

Match the "V" notch to the "E" mark on the camshaft, and the punch and arrow on the crankshaft to properly position the engine for belt service—Suzuki 1.3L and 1.6L engines

To install:

8. Install the timing belt tensioner plate and tensioner. Only hand-tighten the tensioner bolt.

➡Be sure that the lug on the tensioner plate is inserted into the hole on the tensioner.

9. Be sure the tensioner plate and the tensioner move uniformly. If they do not move together remove the tensioner and the tensioner plate and reinsert the plate lug into the tensioner hole.
10. Check the camshaft sprocket to verify that it has not moved.
11. Check the crankshaft alignment by verifying that the punch mark on the timing belt pulley is aligned with the arrow on the oil pump case.
12. Remove the cylinder head cover.

➡This is to permit the free rotation of the camshaft. When installing the timing belt on the pulleys, the tensioner spring force should correctly tension the belt. If the camshaft does not rotate freely the belt will not be correctly tensioned.

13. With the timing marks aligned, hold the tensioner plate up by hand and install the timing belt on the pulleys so there is no slack on the drive side of the belt.
14. Turn the crankshaft 2 rotations clockwise. Confirm that the timing marks are still properly aligned.
15. If the belt is free of slack and the alignment marks are correct tighten the tensioner stud to 84–96 inch lbs. (9–12 Nm). Tighten the tensioner bolt to 17–21 ft. lbs. (24–30 Nm).
16. Install the timing belt upper and lower outside covers. Tighten the timing cover bolts to 84–96 inch lbs. (9–12 Nm).
17. Install all remaining components in the reverse order of the removal procedure.

1.6L Engine

1. Before servicing the vehicle, refer to the precautions in the beginning of this section.
2. Remove all necessary components for access to the timing belt covers, then remove the covers.
3. Loosen but do not remove the tensioner bolt.

⁂ CAUTION

After the timing belt is removed, never turn the camshaft and crankshaft independently. This engine is an interference engine and if the camshaft or crankshaft is turned beyond a certain point, damage to the valves could occur.

4. Loosen the timing belt tensioner adjusting bolt and pivot nut. Apply pressure to the tensioner to loosen the timing belt, and remove the timing belt from the camshaft and crankshaft sprockets.
5. Remove the timing belt tensioner, tensioner plate and tensioner spring.

To install:

6. Install the timing belt tensioner, plate and spring. Hand-tighten the tensioner bolt and stud only at this time.
7. Turn the camshaft sprocket clockwise and align the timing marks.
8. Turn the crankshaft clockwise, using a 17mm wrench to crank the timing belt sprocket bolt.
9. Align the punch mark on the timing belt sprocket with the arrow mark on the oil pump.
10. With the timing marks aligned, remove any slack from the drive side of the belt. Tighten the tensioner bolt to 16–20 ft. lbs. (22–28 Nm).
11. To allow the belt to be free of any slack, turn the crankshaft clockwise 2 full rotations. Confirm that the timing marks are aligned.
12. Install the timing cover and tighten the bolts to 84–96 inch lbs. (9–12 Nm).
13. Install all remaining components in the reverse order of the removal procedure.

Piston and Ring

POSITIONING

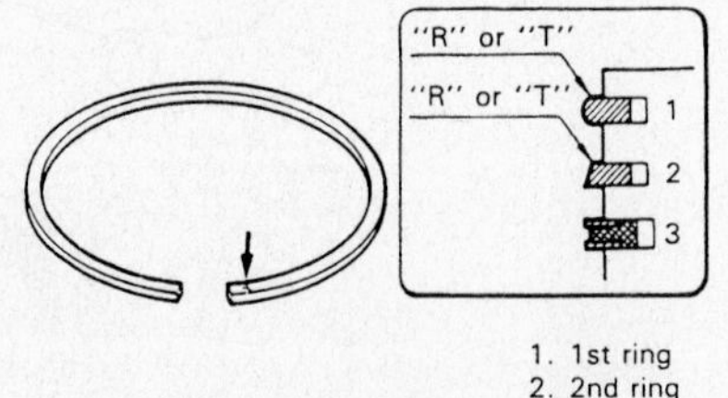

Suzuki engines—piston ring positioning

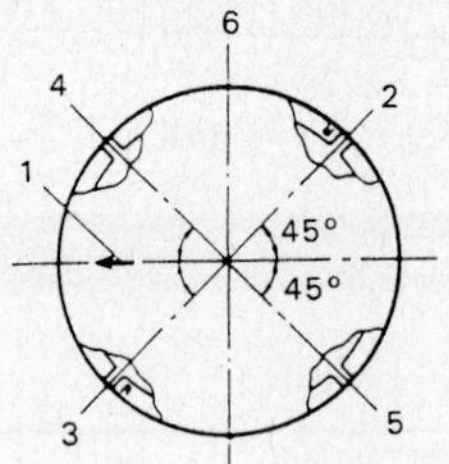

Suzuki engines—piston ring end-gap spacing

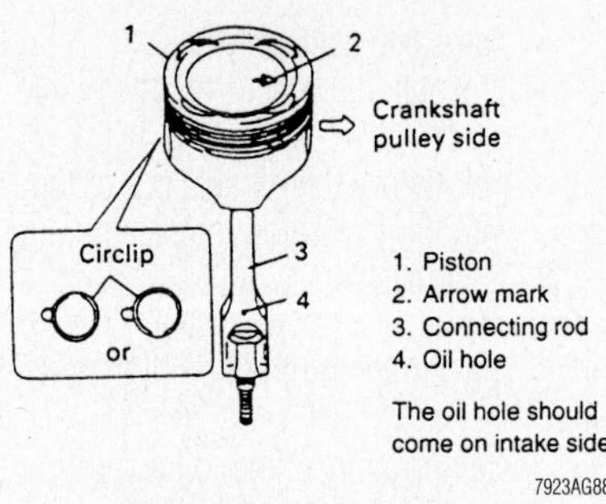

Suzuki engines—piston/connecting rod assembly-to-engine positioning

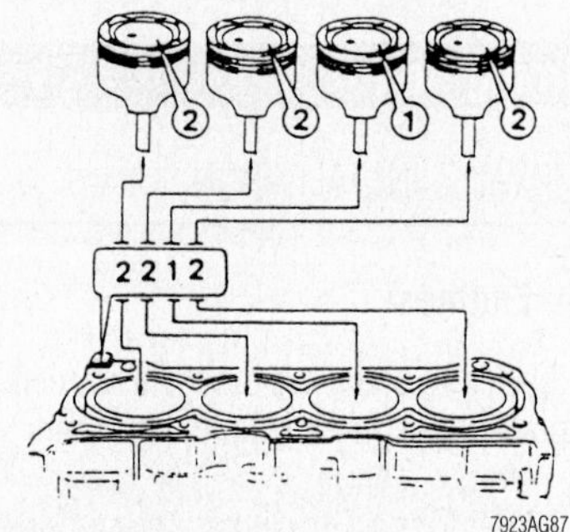

Suzuki engines—the piston ID number must match the number stamped in the engine block

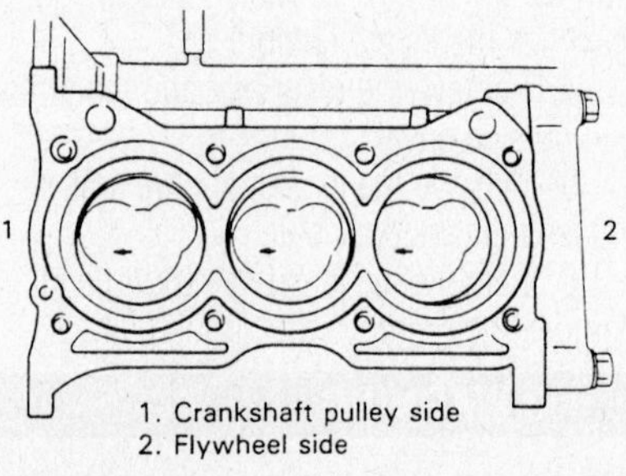

Suzuki engines—the directional arrow on the piston face must face the crankshaft pulley end of the engine

FUEL SYSTEM

Fuel System Service Precautions

Safety is the most important factor when performing not only fuel system maintenance, but any type of maintenance. Failure to conduct maintenance and repairs in a safe manner may result in serious personal injury or death. Maintenance and testing of the vehicle's fuel system components can be accomplished safely and effectively by adhering to the following rules and guidelines.

- To avoid the possibility of fire and personal injury, always disconnect the negative battery cable unless the repair or test procedure requires that battery voltage be applied.
- Always relieve the fuel system pressure prior to disconnecting any fuel system component (injector, fuel rail, pressure regulator, etc.), fitting or fuel line connection. Exercise extreme caution whenever relieving the fuel system pressure, to avoid exposing your skin, face and eyes to fuel spray. Please be advised that fuel under pressure may penetrate the skin or any part of the body that it contacts.
- Always place a shop towel or cloth around the fitting or connection prior to loosening to absorb any excess fuel due to spillage. Ensure that all fuel spillage (should it occur) is quickly removed from the engine surfaces. Ensure that all fuel soaked cloths or towels are deposited into a suitable waste container.
- Always keep a dry chemical (Class B) fire extinguisher near the work area.
- Do not allow fuel spray or fuel vapors to come into contact with a spark or open flame.
- Always use a back-up wrench when loosening and tightening fuel line connection fittings. This will prevent unnecessary stress and torsion on fuel line piping. Always follow the proper torque specifications.
- Always replace worn fuel fitting O-rings with new. Do not substitute fuel hose or equivalent, where fuel pipe is installed.

Fuel System Pressure

RELIEVING

**** CAUTION**

Care should be used when working around the fuel system. DO NOT smoke or expose the fuel system to any open flames. Keep a fire extinguisher handy.

1. Before servicing the vehicle, refer to the precautions in the beginning of this section.
2. Disconnect the negative battery cable from the battery.
3. Place the vehicle in **PARK** for automatic transmission or **NEUTRAL** for manual transmission.
4. Remove the relay box cover.
5. Disconnect the fuel pump relay from the relay box.
6. Remove the fuel filler cap from the filler neck to release the fuel vapor pressure in the fuel tank.
7. Start the vehicle and allow the engine to run until it stalls.
8. Crank the engine 3 more revolutions to eliminate any remaining pressure in the fuel lines.
9. Disconnect the negative battery cable.
10. Connect the fuel pump relay to the relay box.
11. After servicing the fuel system, connect the negative battery cable.
12. Start the engine and check for leaks in the system.

Fuel Filter

REMOVAL & INSTALLATION

**** CAUTION**

The fuel system pressure must be relieved before disconnecting any fuel lines. Failure to do so may result in personal injury.

1. Before servicing the vehicle, refer to the precautions in the beginning of this section.
2. Properly relieve the fuel system pressure.
3. Place a container under the fuel filter.
4. Remove or disconnect the following:
   - Fuel inlet hose from the fuel filter

**** CAUTION**

A small amount of fuel may be released after the fuel hose is disconnected. Cover the hose and pipe with a shop towel.

   - Outlet hose from the fuel feed pipe
   - 2 fuel filter mounting bracket bolts and remove the fuel filter from the frame with the outlet hose attached
   - Fuel filter from the bracket by unfastening the mounting bolt
   - Outlet hose from the fuel filter

To install:

5. Install or connect the following:
   - Fuel filter on the bracket
   - Mounting bolt
   - Remaining components in the reverse order of removal
   - Inlet and outlet hoses to the filter
   - Negative battery cable
6. With the ignition **ON** and the engine **OFF** check for leaks.

Fuel Pump

REMOVAL & INSTALLATION

1. Before servicing the vehicle, refer to the precautions in the beginning of this section.
2. Relieve the pressure from the fuel system.
3. Remove the rear seat cushion by removing or disconnecting the following:
   - Spare tire
   - Seat back by unfastening the 2 center mounting nuts and the 4 mounting screws
   - Fitting screws from the rear of the seat cushion
   - Lift the front of the seat cushion and remove the cushion
4. Remove or disconnect the following:
   - Fuel level gauge and the fuel pump lead wire couplers and detach the wire tape
   - Fuel filler hose from the fuel tank
   - Breather hose from the filler neck
5. Drain the fuel from the tank by pumping the fuel out through the fuel tank filler.

**** CAUTION**

Use a gasoline safe hand operated pump device to drain the fuel tank.

6. Remove or disconnect the following:
   - Fuel hoses from the fuel pipes, located near the fuel filter

**** CAUTION**

A small amount of fuel may be released after the fuel hose is disconnected. Cover the hose and pipe to be disconnected with a shop cloth.

7. Install a support (for example, a transmission jack) under the fuel tank.

8. Remove or disconnect the following:

- Fuel tank mounting hardware
- Tank from the vehicle
- Fuel lines from the fuel pump and sender assembly
- 12 screws that secure the fuel pump and fuel gauge assembly to the tank
- Pump and sender assembly
- Fuel pump electrical connectors
- Fuel strainer
- Fuel pump

To install:

9. Install or connect the following:

- Fuel pump on the fuel gauge assembly
- Fuel strainer on the fuel pump

➡Always install a new fuel pump strainer when replacing the fuel pump.

- Electrical connectors to the fuel pump
- Remaining components in the reverse order of removal
- Negative battery cable

10. Turn the ignition switch to the **ON** position, but leave the engine **OFF** and check for fuel leaks.

Fuel Injector

REMOVAL & INSTALLATION

Swift

1. Before servicing the vehicle, refer to the precautions in the beginning of this section.

2. Relieve the pressure from the fuel system.

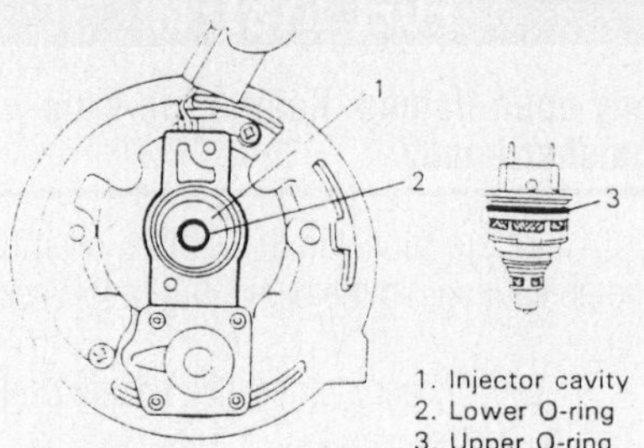

Install the lower injector O-ring (2) to the injector cavity (1) and the upper O-ring (3) to the injector

3. Remove or disconnect the following:

- Air cleaner assembly
- Air cleaner mounting stay from the throttle body
- Injector cover
- Injector from the throttle body
- Injector O-rings and discard

To install:

4. Inspect the injector filter for dirt or contamination. If present, clean and check for dirt in the fuel lines and tank.

5. Apply a tin coat of spindle oil or gasoline to the new O-rings.

6. Install or connect the following:

- Lower injector O-ring to the injector cavity and the upper O-ring to the injector
- Injector by pushing straight into the injector cavity. Do not turn the injector while pushing.

7. Make sure the injector cover O-ring is free of contamination and has not deteriorated. Apply a coat of spindle oil or gasoline to the O-ring.

8. Install or connect the following:

- Injector cover and tighten the retainers to 24–36 inch lbs. (3–4 Nm)
- Negative battery cable

9. Turn the ignition switch on and check for fuel leaks

- Air cleaner mounting stay to the throttle body
- Air cleaner assembly

Esteem

1. Before servicing the vehicle, refer to the precautions in the beginning of this section.

2. Relieve the pressure from the fuel system.

3. Remove or disconnect the following:

- Vacuum hose from the fuel pressure regulator
- Fuel injector electrical connections
- Fuel rail bolts
- Injector from the fuel rail
- Injector O-ring and discard

To install:

4. Coat the injector O-rings with gasoline.

5. Install or connect the following:

- Injector O-ring
- Injector to the fuel rail and intake manifold.

6. Make sure the injectors rotate smoothly. If not, the O-ring has been improperly installed and must be replaced.

- Fuel rail bolts. Once the bolts are tightened make sure the injectors rotate smoothly.
- Fuel injector electrical connections
- Vacuum hose to the fuel pressure regulator
- Negative battery cable

7. Start the vehicle and check for leaks.

DRIVE TRAIN

Transaxle Assembly

REMOVAL & INSTALLATION

Manual

1. Before servicing the vehicle, refer to the precautions in the beginning of this section.

2. Drain the transaxle oil.

3. Remove or disconnect the following:

- Battery cables
- Battery and tray
- Clutch cable adjusting nut, joint pin from the cable and cable from the bracket
- All the wiring harness clamps and connectors involved with the transaxle removal, tag if necessary for location to aid during installation
- Speedometer cable boot, case clip and cable from the case
- Radiator outlet pipe from the transmission side cover, on Swift models
- Transaxle retaining bolts
- Starter
- Fender apron extension on the left side
- Bolts connecting the exhaust pipe to the exhaust manifold and disconnect the joint

4. On Esteem models, with the engine supported, remove the vehicle center mounting member by removing the 7 nuts and 4 bolts.

- Gearshift control shaft nut and bolt
- Control shaft from the gearshift control shaft
- Extension rod nut and washers
- Clutch housing lower plate
- Sway bar
- Left and right front wheels
- Left and right ball joints
- Left and right halfshafts
- Transaxle stiffener
- Transmission to engine bolt and nut
- Engine rear mounting bracket bolts

5. Install an engine support.

6. Support the transaxle with a suitable jack.

7. Remove the left engine mounting bracket and stiffener.

8. Lower the transaxle with the engine attached. Pull the transaxle straight out toward the left side.

9. Lower and remove the transaxle.

To install:

10. Install or connect the following:
- Transaxle from the left side of the vehicle. Use care when inserting the pilot shaft into the clutch assembly. If the spline on the input shaft does not align with the clutch assembly spline, turn the crankshaft slightly to aid in spline alignment.

11. Raise the transaxle and engine.
- Left engine mounting bracket and stiffener. Tighten the bolts to 29–43 ft. lbs. (40–60 Nm).
- Rear engine mounting bracket bolts and tighten them to 29–43 ft. lbs. (40–60 Nm)

➡Before installing the bolts into the rear mounting bracket, apply sealant to the bolt threads.

- Transmission-to-engine bolt and nut. Tighten the nut and bolt to 29–43 ft. lbs. (40–60 Nm)

12. On Esteem models, install the center member on the vehicle and install the 7 nuts and 4 bolts. Tighten the bolts and nuts as follows:

a. Center member-to-radiator support: 33 ft. lbs. (45 Nm).

b. Center member-to-crossmember: 33 ft. lbs. (45 Nm).

c. Engine mounts-to-center member nuts: 40 ft. lbs. (55 Nm).

13. Install or connect the following:
- Transaxle stiffener

14. Lower the transaxle supporting jack.
- Left and right halfshafts
- Ball joints
- Sway bar
- Clutch housing lower plate
- Extension rod nut and washers. Tighten the rod nut to 18–28 ft. lbs. (25–40 Nm).
- Control shaft on gearshift
- Gearshift control shaft bolt and nut. Tighten the gearshift control shaft bolt and nut to 11–14 ft. lbs. (15–20 Nm).
- Exhaust pipe to the manifold and install the bolts. Tighten the bolts to 29–36 ft. lbs. (40–50 Nm).
- Left fender apron extension

15. Refill the transaxle with the recommended lubricant.

16. Remove the engine support fixture.

17. Install or connect the following:
- Starter
- Transaxle retaining bolts. Tighten the retaining bolts to 29–43 ft. lbs. (40–60 Nm).
- Remaining components in the reverse order of removal
- Negative battery cable and the ground strap on the transaxle

Automatic

1. Before servicing the vehicle, refer to the precautions in the beginning of this section.

2. Drain the cooling system.

3. Drain the transaxle fluid from the transaxle.

4. Remove or disconnect the following:
- Negative battery cable from the battery and transaxle
- Speedometer cable

5. Disconnect the following electrical connectors:
- Solenoids
- Vehicle Speed Sensor (VSS), if equipped
- Shift lever switch
- Forward clutch cylinder revolution sensor

6. Remove or disconnect the following:
- Wiring harness from the clamps on the transaxle
- Select cable from the transaxle
- Cooling system pipe from the transaxle
- Top transaxle-to-engine bolts
- Starter
- Exhaust manifold cover
- Front exhaust pipe from the exhaust manifold

7. Support the engine.
- Engine undercovers, if equipped.
- Transaxle cooler hoses

8. On Esteem models, with the engine supported, remove the vehicle center mounting member by removing the 7 nuts and 4 bolts.
- Front exhaust pipe from the vehicle
- Transaxle stiffener plate
- Transaxle housing lower plate
- Torque converter bolts. To lock the drive plate, engage a flat-bladed tool in the flywheel.
- Sway bar from the control arms
- Left and right front wheels
- Left and right ball joints from the steering knuckles
- Left and right halfshafts
- Engine rear mount and bracket

9. After removing the rear mount, remove the engine-to-transaxle bolt and nut located behind the rear bracket. Remove all bolts holding the engine to the transaxle.

10. Support the transaxle with a transaxle jack.

11. Remove or disconnect the following:
- Bolts from the engine left-hand mount
- Transaxle with the torque converter from the engine compartment

➡When removing the transaxle from the engine, move it parallel with the crankshaft and use care so not to apply excessive force to the drive plate and torque converter.

⁂ CAUTION

Be sure to keep the transaxle with the torque converter horizontal or facing up throughout the work. Should it be tilted with converter down, the converter may fall off and cause personal injury.

To install:

12. Install or connect the following:
- Transaxle to the engine assembly
- Transaxle attaching nuts and bolts
- Left-hand mounting bolts and tighten to 40 ft. lbs. (55 Nm)
- Engine-to-transaxle bolt and nut before installing the rear transaxle mount. Tighten the nut to 65 ft. lbs. (90 Nm).
- All bolts for the transaxle. Tighten the bolts to 65 ft. lbs. (90 Nm).
- Left halfshafts
- Ball joints
- Sway bar to the control arms
- Torque converter bolts and tighten the bolts to 14 ft. lbs. (19 Nm)
- Transaxle housing lower plate
- Stiffener plate with the 4 bolts. Tighten the bolts to 40 ft. lbs. (55 Nm).
- Exhaust pipe on the center pipe and tighten the bolts to 37 ft. lbs. (50 Nm)

13. On Esteem models, install the center member on the vehicle and install the 7 nuts and 4 bolts. Tighten the bolts and nuts as follows:

a. Center member-to-radiator support: 33 ft. lbs. (45 Nm).

b. Center member-to-crossmember: 33 ft. lbs. (45 Nm).

c. Engine mounts-to-center member nuts: 40 ft. lbs. (55 Nm).

14. Install or connect the following:
 - Oil hoses for the transaxle
 - Engine undercovers
15. Remove the engine support.
 - Exhaust pipe to the exhaust manifold and install the nuts
 - Exhaust manifold cover
 - Starter motor
 - Remaining components in the reverse order of removal

16. Fill the cooling system and the transaxle. Check all fluids.

17. Connect the negative battery cable to the transaxle and battery.

Clutch

ADJUSTMENT

1. Before servicing the vehicle, refer to the precautions in the beginning of this section.
2. Depress the clutch pedal lightly until tension on the clutch cable can be felt.
3. Measure the clutch pedal free-play. It should be 0.6–0.8 in. (15–20mm).
4. Adjust the clutch pedal free-play by tightening or loosening the clutch cable adjustment nut.
5. Measure the clutch lever free-play. It should be 0–0.08 in. (0–2mm). If the clutch release lever free-play exceeds specification, inspect the release shaft return spring for cracks or weakness.

➡Be sure the marks on the clutch release lever and release shaft are aligned. If they are not, remove the lever from the shaft, align the marks and repeat the free-play adjustment procedure.

REMOVAL & INSTALLATION

1. Before servicing the vehicle, refer to the precautions in the beginning of this section.
2. Remove the transaxle.

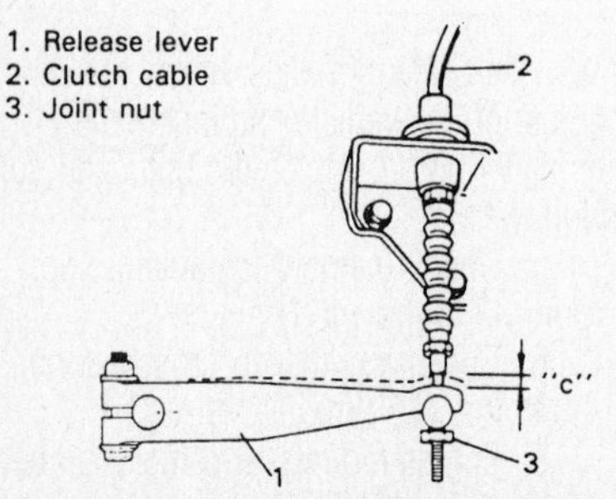

Clutch cable free-play adjustment

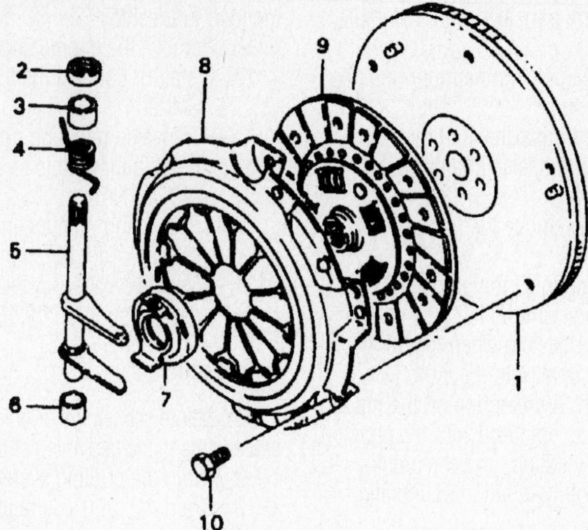

1. Flywheel
2. Release shaft seal
3. No. 2 bush
4. Return spring
5. Release shaft
6. No. 1 bush
7. Release bearing
8. Clutch cover
9. Clutch disc
10. Clutch cover bolt

7923UG27

Clutch component identification

3. Hold the flywheel stationary.
4. Matchmark the pressure plate and flywheel for installation reference.
5. Loosen the pressure plate attaching bolts 1 turn at a time (evenly) until the spring pressure is released.
6. Remove the clutch disc and pressure plate.

To install:

7. Clean the flywheel mating surfaces of all oil, grease and metal deposits. Inspect flywheel for cracks, heat checking or other defects and replace or resurface as necessary.
8. Check the wear on the facings of the clutch disc by measuring the depth of each rivet head depression. Replace clutch disc when rivet heads are 0.02 in. (0.5mm) below the surface of clutch surface.
9. Check the diaphragm spring and pressure plate for wear or damage. If the spring or plate is excessively worn, replace the pressure plate assembly.
10. Check the pilot bearing for smooth operation. If the bearing does not spin freely, replace it.
11. Position the clutch disc and pressure plate with the matchmarks aligned and install a clutch alignment tool.
12. Install the pressure plate bolts. Tighten the mounting bolts evenly and in a crisscross pattern to 13–20 ft. lbs. (18–28 Nm). Remove the alignment tool and the flywheel holding tool.
13. Lightly lubricate the transaxle input shaft splines, pilot bearing surface of the input shaft, and the release bearing with grease.
14. Install the transaxle.
15. Adjust the clutch cable.

Halfshaft

REMOVAL & INSTALLATION

1. Before servicing the vehicle, refer to the precautions in the beginning of this section.
2. Remove or disconnect the following:
 - Negative battery cable
 - Undo the sealer on the halfshaft nut
 - Halfshaft nut and washer
3. Drain the oil from the transmission.
4. On the left side shaft, use 2 large prybars to release the snapring fitting on the halfshaft inner joint from the differential.

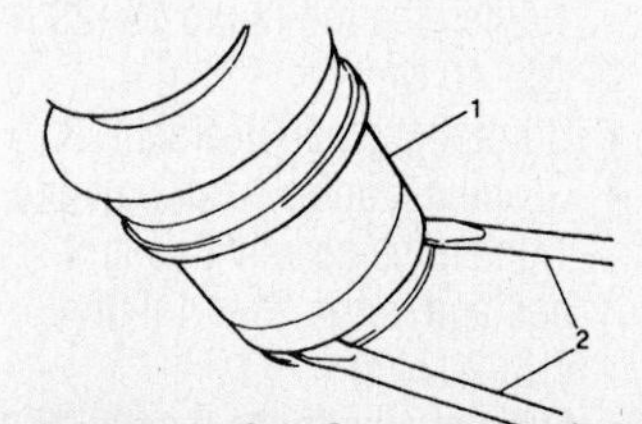

Disconnecting the left inboard joint

5. On the right side shaft, use a plastic hammer to drive the shaft joint to release the snapring fitting on the halfshaft inner joint from the differential.

6. Remove or disconnect the following:
- Sway bar attaching nut, washer and bushing from the suspension arm
- Ball joint stud bolt and nut
- Inboard joint from the differential by pulling it
- Outer joint from the steering knuckle
- Halfshaft from the vehicle

**** WARNING**

To prevent breakage of the boots, be careful not to bring the boots in contact with other components when removing the shaft assembly.

7. If the center shaft requires service, drain the transmission oil and remove the support bolts. Remove the center shaft from the differential, then from the vehicle.

To install:

8. If the center shaft was removed, install the shaft into the differential, then install the support bolts. Tighten the support bolts to 29–43 ft. lbs. (40–60 Nm).

9. Clean the grease seal on the steering knuckle and apply a small amount of fresh grease.

10. Install or connect the following:
- Wheel side joint on the steering knuckle, then the differential side joint on the differential. Seat the differential joint by hand, making sure that the snapring is seated.
- Halfshaft nut on the outer joint loosely, to hold it in position.

**** WARNING**

Do not hit the joints with a hammer to seat them. Use hand pressure only or component damage may occur.

- Suspension arm to the steering knuckle
- Ball joint stud
- Sway bar on the suspension arm
- Sway bar bushing, washer and nut. Tighten the nut to 17–20 ft. lbs. (23–28 Nm).

11. If equipped with a manual transaxle fill the transaxle with the specified gear oil.

12. Tighten the halfshaft nut to 109–145 ft. lbs. (150–200 Nm).

13. Connect the negative battery cable.

14. If equipped with an automatic transaxle, fill the transaxle with the specified transmission oil.

CV-Joints

OVERHAUL

Double Offset Joint (DOJ) Type

The Double Offset Joint (DOJ) type is identified by the outside shape of the differential side joint which has no dent.

1. Before servicing the vehicle, refer to the precautions in the beginning of this section.
2. Remove the driveshaft.
3. Remove the boot band from the differential side joint.
4. Slide the boot towards the center of the shaft and remove the snapring from the outer race.
5. Remove the shaft from the outer race.
6. Use a rag to clean off the grease, then remove the circlip that retains the cage using snapring pliers.
7. Using a suitable puller, draw the cage away and remove the boot from the shaft.
8. Inspect all components for wear and/or damage and replace as necessary.

To install:

9. Clean all components and allow them to completely dry.
10. Install the boot onto the shaft until its small diameter side fits to the shaft groove and attach it there with a boot band.
11. Using a pipe whose inner diameter is 0.906 in. (23mm) or more and outer diameter is 1.260 in. (32mm) or less, drive the cage into position. Install the cage using the smaller outside diameter side to the shaft end.
12. Install the circlip.
13. Apply grease to the entire surface of the cage.
14. Insert the cage into the outer race and fit the snapring into the groove of the outer race.

➡Position the opening of the snapring so that it will not be lined up with a ball.

15. Apply grease to the inside of the outer race and fit the boot to the outer race.
16. After fitting the boot, insert a screwdriver into the boot on the outer race side and allow air to enter the boot so that the air pressure in the boot equals atmospheric pressure
17. Fix the boot to the outer race with a boot band. Before tightening the band adjust the boot so that measurement D which is 8.10 in. (205.8mm) and measurement E which is 7.34 in. (186.4mm) are as shown in the accompanying illustration.
18. Install the driveshaft in the vehicle.

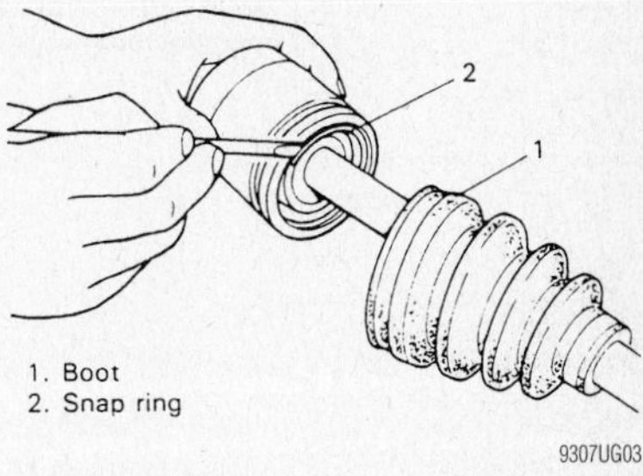

Remove the snapring from the outer race—DOJ type

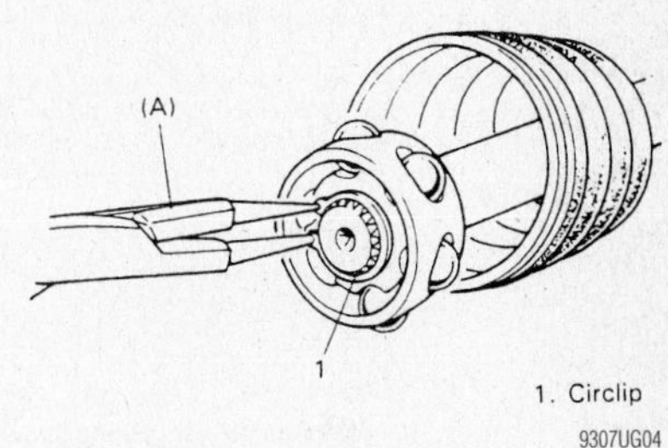

Remove the circlip that retains the cage—DOJ type

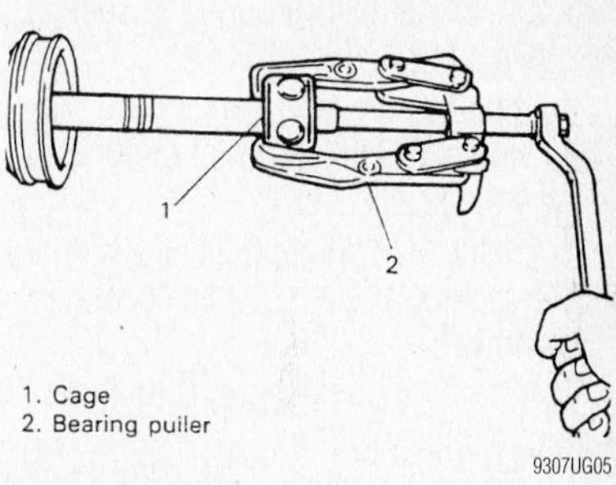

Draw the cage away using a suitable puller—DOJ type

Tripod Joint Type

The Tripod joint type can be identified by the 3 dent lines on the outside of the differential side joint.

1. Before servicing the vehicle, refer to the precautions in the beginning of this section.
2. Remove the driveshaft.
3. Remove the boot band and the Tripod joint housing.
4. Use a rag to clean off the grease, then remove the circlip using snapring pliers.
5. Remove the spider by using a suitable puller.
6. Remove the boot band and pull the differential side boot from the shaft.

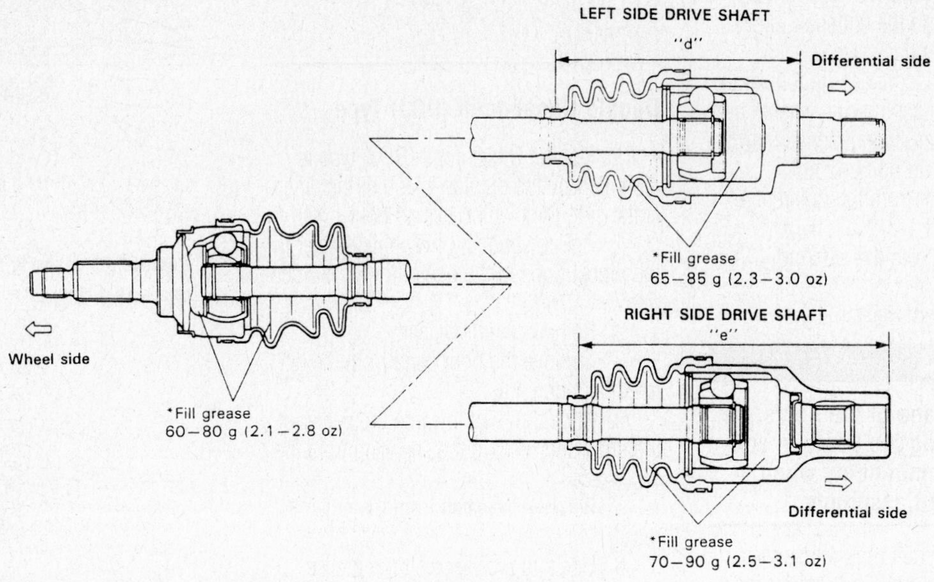

Adjust the boot so that measurements D and E are as illustrated—DOJ type

7. Remove the dynamic damper band, then pull the damper through the shaft, if equipped.

8. Remove the boot bands from the wheel side joint boot and pull the boot through the shaft.

9. Inspect all components for wear and/or damage and replace as necessary.

To install:

10. Clean all components and allow them to completely dry.

11. On Swift models, perform the following:

a. Apply grease to the wheel side joint. Use black grease in the tube included with the wheel side boot kit.

b. Install the wheel side boot on the shaft.

c. Fill the inside of the boot with grease and fasten the boot with a band.

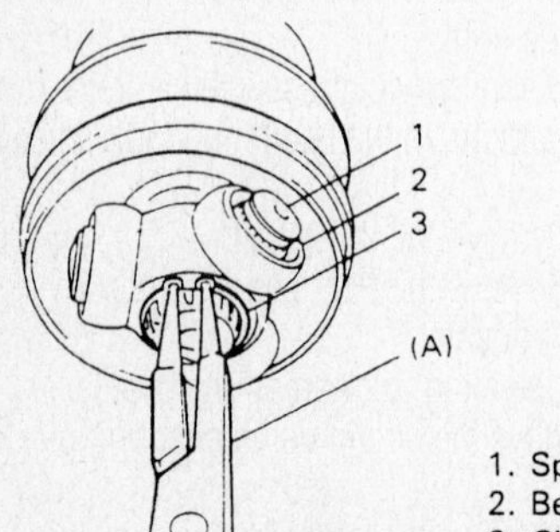

Remove the circlip that retains the spider assembly—Tripod type

d. Install the dynamic damper on the shaft, if equipped.

e. Install the differential side boot on the shaft.

f. Apply grease to the Tripod joint. Use the yellow grease in the tube included in the differential side joint kit.

g. Using a pipe whose inner diameter is 0.906 in. (23mm) or more and outer diameter is 1.260 in. (32mm) or less, drive the spider into position. Face the chamfered side inward and fasten it in place with the circlip.

h. Fill the boot with grease, then install the housing and joint it with the boot.

i. Fasten the boot bands.

12. On Esteem models, perform the following:

a. Apply grease to the wheel side joint. Use black grease in the tube included with the wheel side boot kit.

b. Install the wheel side boot on the shaft.

c. Fill the inside of the boot with grease.

d. Fit the boot band into the groove in the boot.

e. Tighten the boot band until its outer diameter is 3.05 in. (77.5mm).

f. Fold the boot band over the metal fixture.

g. Using a punch, caulk the center of the band folded over the fixture.

h. Cut the band about 0.28 in. (7mm) from the clip.

i. Clamp the band with the clip.

j. Fix the small diameter side of the boot band into the groove in the boot.

k. Tighten the boot band until its outer diameter is 1.14 in. (29mm).

l. Fold the boot band over the metal fixture.

m. Using a punch, caulk the center of the band folded over the fixture.

n. Install the dynamic damper on the right side driveshaft.

o. Install the differential side boot on the shaft.

p. Apply grease to the Tripod joint. Use the yellow grease in the tube included in the differential side joint kit.

q. Install the spider into position. Face the chamfered side inward and fasten it in place with the circlip.

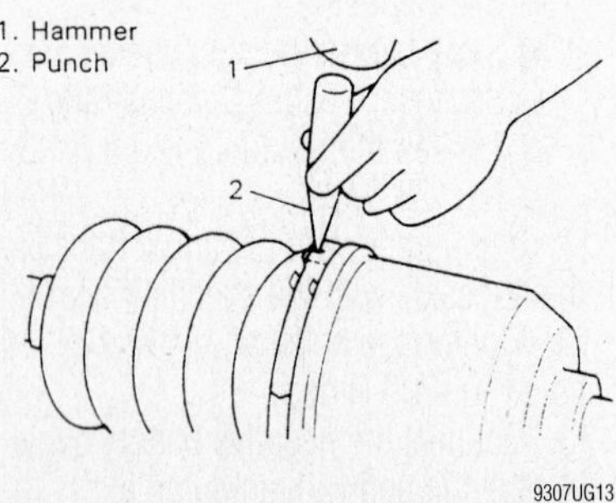

Using a punch, caulk the center of the band folded over the fixture—Esteem

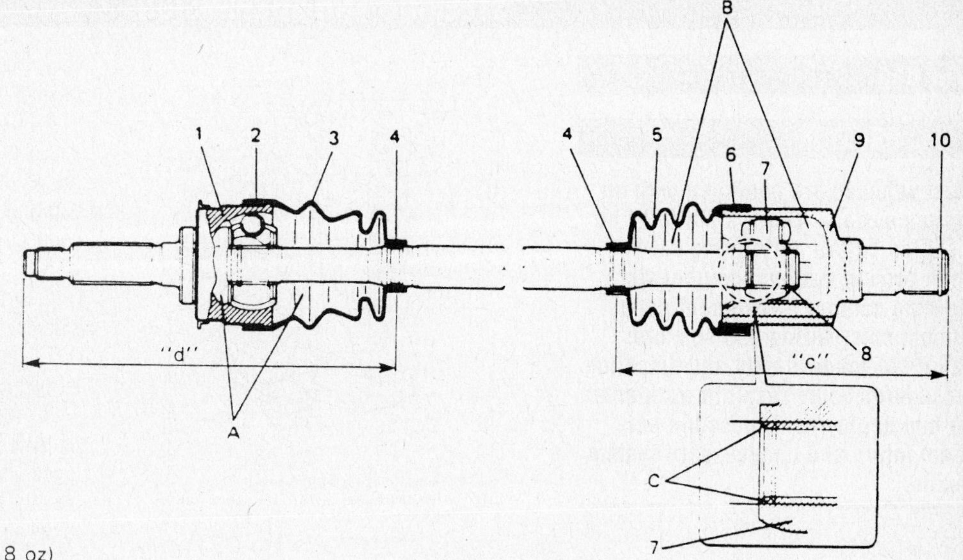

1. Wheel side joint assembly
2. Wheel side boot band
3. Wheel side boot
4. Boot band
5. Differential side boot
6. Differential side boot band
7. Tripod joint spider
8. Circlip
9. Tripod joint housing
10. Snap ring

A: Black grease (about 52 g/1.8 oz)
B: Yellow grease (about 100 g/3.5 oz)
C: Chamfered spline

Adjust the boot so that measurement D is correct—Esteem models equipped with a Tripod type joint

r. Fill the boot with grease, then install the housing and joint it with the boot.

s. Adjust the boot so that the when measured from the tip of the driveshaft (wheels side), to the inner (small) band; the measurement is 8 in. (204mm).

13. Insert a screwdriver into the boot on the outer race side and allow air to enter the boot so that the air pressure in the boot equals atmospheric pressure

a. Fasten the boot band so that diameter E which measures 3.11 in. (79mm) and diameter F which measures 1.14 in. (29mm) are correct. Refer to the accompanying illustration for the diameter locations.

14. Install the driveshaft.

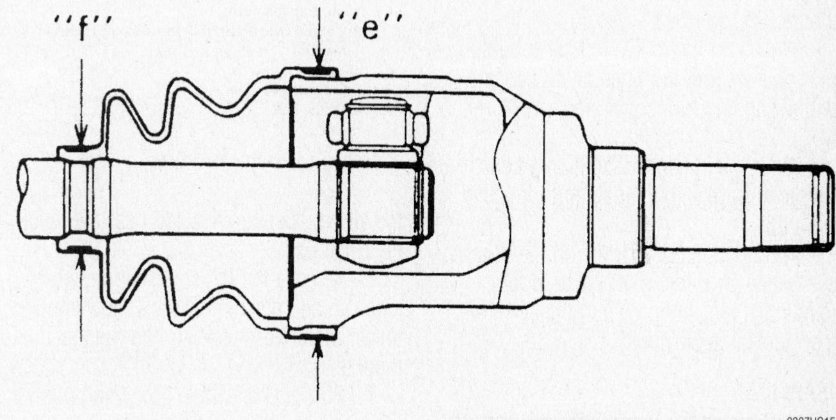

When tightening the bands, make sure that diameters E and F are correct—Esteem models

Air Bag

✻✻ CAUTION

Most vehicles are equipped with an air bag system. The system must be disabled before performing service on or around system components, steering column, instrument panel components, wiring and sensors. Failure to follow safety and disabling procedures could result in accidental air bag deployment, possible personal injury and unnecessary system repairs.

PRECAUTIONS

Several precautions must be observed when handling the inflator module to avoid accidental deployment and possible personal injury.

- Never carry the inflator module by the wires or connector on the underside of the module.
- When carrying a live inflator module, hold securely with both hands, and ensure that the bag and trim cover are pointed away.
- Place the inflator module on a bench or other surface with the bag and trim cover facing up.
- With the inflator module on the bench, never place anything on or close to the module that may be thrown in the event of an accidental deployment.

DISARMING

✻✻ WARNING

When performing service on or around the air bag system components or wiring, disable the air bag system. Failure to follow the procedures could result in possible deployment, personal injury or unneeded system repairs.

1. Before servicing the vehicle, refer to the precautions in the beginning of this section.
2. Disconnect the negative battery cable.
3. Turn the steering wheel so the wheels are pointing straight ahead.
4. Turn the ignition switch to the **LOCK** position and remove the key.
5. Remove the **AIR BAG-IG** fuse from

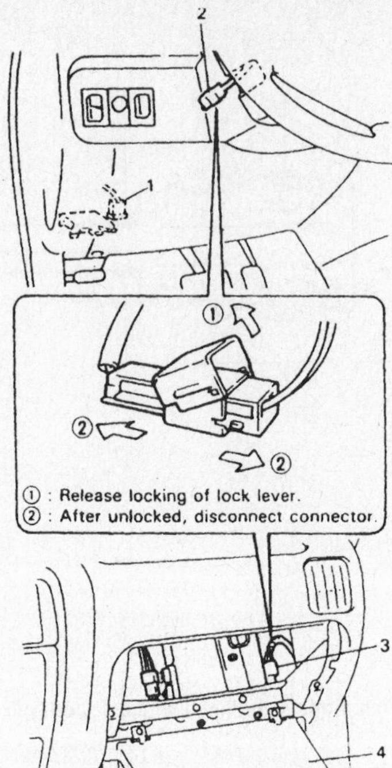

Air bag connector locations

the air bag fuse box located near the junction/fuse box.

6. Remove the left side steering wheel side cap and disconnect the yellow connector for the driver's side air bag (inflator) module.
7. Pull out the glove box while pushing in on the stoppers from the left and the right sides. Disconnect the yellow connector for the passenger air bag (inflator) module.

REARMING

✻✻ WARNING

When performing service on or around the air bag system components or wiring, disable the air bag system. Failure to follow the procedures could result in possible deployment, personal injury or unneeded system repairs.

1. Before servicing the vehicle, refer to the precautions in the beginning of this section.
2. Connect the negative battery cable.
3. Turn the ignition switch to the **LOCK** position and remove the key.
4. Connect the yellow connector for the passenger side air bag (inflator) module and the yellow connector for the driver's side air bag (inflator) module. Be sure to lock each connector with the lock lever.
5. Install the glove box assembly.
6. Install the left side steering wheel side cover.
7. Install the **AIR BAG-IG** fuse in the air bag fuse box.
8. Turn the ignition **ON** and verify that the **AIR BAG** warning lamp flashes 7 times, then turns off. If the system does not operate as described, diagnosis and repairs to the air bag system are necessary.

Rack and Pinion Steering Gear

REMOVAL & INSTALLATION

Manual

✻✻ WARNING

Be sure to set the front wheels straight ahead and remove the ignition key from the cylinder before starting repairs. If equipped with an air bag, the contact coil of the air bag system may be damaged if the key is not removed and the wheels are not straight ahead.

1. Before servicing the vehicle, refer to the precautions in the beginning of this section.
2. Disconnect the negative battery cable.
3. Slide the driver's seat back as far as possible.
4. Remove or disconnect the following:
   - Pull back the front part of the floor mat on the driver's side
   - Steering shaft joint cover
   - Steering shaft upper joint bolt, loosen but do not remove it
   - Steering shaft lower joint bolt
   - Lower joint from the pinion
   - Front wheels
   - Tie rod ends from the steering knuckles
   - Steering gear mounting bolts and the brackets
   - Steering gear case from the vehicle

To install:

5. Install or connect the following:
   - Steering gear, brackets and mounting bolts. Tighten bolts to 14–21 ft. lbs. (20–30 Nm).

- Tie rod ends to the steering knuckles

6. Be sure the steering wheel is in the straight-ahead position and the front wheels are pointing straight ahead.
7. Install or connect the following:
   - Steering shaft to the steering gear
   - Lower steering shaft-to-steering gear clinch bolt and tighten both steering joint bolts (upper and lower) to 14–21 ft. lbs. (20–30 Nm)
   - Remaining components in the reverse order of removal
   - Negative battery cable
8. Check and adjust the front wheel alignment.

Power

**** WARNING**

Be sure to set the front wheels straight ahead and remove the ignition key from the cylinder before starting repairs. If equipped with an air bag the contact coil of the air bag system may be damaged if the key is not removed and the wheels are not straight ahead.

1. Before servicing the vehicle, refer to the precautions in the beginning of this section.
2. Remove or disconnect the following:
   - Negative battery cable
   - Steering column joint covers
   - Steering shaft upper joint bolt, loosen but do not remove
   - Steering shaft lower joint bolt
   - Lower joint from the pinion
   - Front wheels
   - Tie rod ends from the steering knuckles
   - Front exhaust pipe, on Swift models
   - Engine rear torque rod with the torque rod and bracket, Swift models equipped with an automatic transaxle
   - Gearshift control shaft and extension rod from the transaxle, if equipped with a manual transaxle
   - Rear engine mount together with the bracket from the engine and suspension member, Esteem models
   - Mounting member from the suspension frame by removing the 2 bolts, Esteem models
   - High and low pressure lines from the rack and pinion

➡When the lines are disconnected plug the lines or place an oil pan under the vehicle.

   - Cylinder lines from the rack and pinion
   - Rack and pinion mounting bolts and brackets
   - Rack and pinion case from the vehicle

To install:

3. Install or connect the following:
   - Rack and pinion, brackets and mounting bolts. Tighten the bolts to 40 ft. lbs. (55 Nm) on Esteem models or 22–28 ft. lbs. (30–40 Nm) on Swift models.
   - Cylinder lines on the rack and pinion and tighten their fittings to 14–21 ft. lbs. (20–30 Nm)
   - High and low pressure lines to the rack and pinion. Tighten the fittings to 22–28 ft. lbs. (30–40 Nm).
   - Gearshift control shaft and extension rod to the transaxle, if equipped with a manual transaxle. Tighten the extension rod nut to 18–28 ft. lbs. (25–40 Nm) and tighten the control shaft nut and bolt to 11–14 ft. lbs. (15–20 Nm).
   - Engine rear torque rod with the torque rod and bracket, Swift models equipped with an automatic transaxle
   - Tie rod ends to the steering knuckles
4. Be sure the steering wheel is straight and the front wheels are pointing straight ahead.
   - Steering shaft to the rack and pinion
   - Lower steering shaft-to-rack and pinion clinch bolt and tighten both steering joint bolts (upper and lower) to 14–21 ft. lbs. (20–30 Nm)
   - Front wheels
   - Negative battery cable
5. Bleed the power steering system.
6. Lower the vehicle.
7. Check and adjust the front wheel alignment.

Strut

REMOVAL & INSTALLATION

Front

1. Before servicing the vehicle, refer to the precautions in the beginning of this section.
2. Remove or disconnect the following:

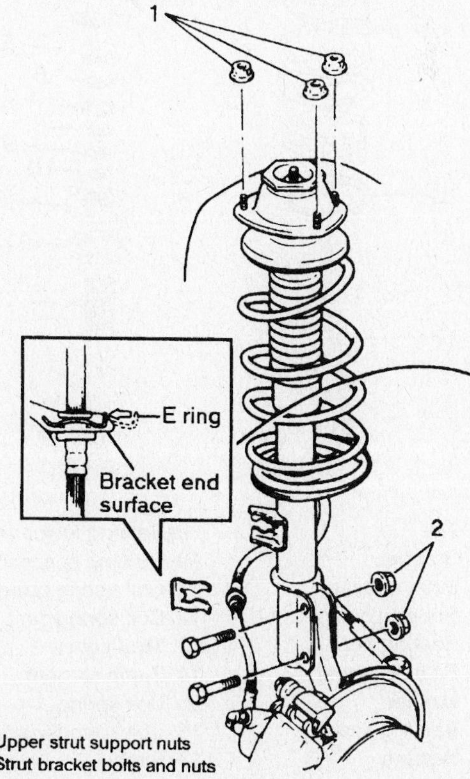

Strut assembly mounting

- Wheels
- Anti-Lock Brake (ABS) system wheel speed sensor, if equipped
- Brake hose clip, then the hose from the strut

3. Support the lower control arm with a floor jack.
4. Remove the strut bracket bolts.
5. Hold the strut by hand so it will not fall, and remove the upper strut support nuts from the engine compartment.

⁂ WARNING

Do not loosen the center nut at this time or serious injury or vehicle damage may result.

6. Remove the strut assembly from the vehicle.
7. Install a pair of coil spring compressors on the coil spring on the strut assembly. Turn the spring compressors alternately until the spring tension is released from the strut assembly. If the spring can be turned slightly, then it has been collapsed enough.

⁂ CAUTION

This procedure requires the use of a spring compressor. It cannot be performed without one. If you do not have access to this special tool, do not attempt to disassemble the strut. The coil spring is retained under considerable pressure. It can exert enough force to cause serious injury. Exercise extreme caution.

8. Keeping the spring collapsed, remove the strut center nut and remove the other components from the top of the strut assembly.
9. Remove the spring from the strut.

To install:

10. If you're installing a new spring, compress the spring with a pair of spring compressors. Be sure that the spring compresses to 9 inches (230mm) for installation.
11. Position the coil spring on the strut making sure that the end of the spring is mated to the stepped part of the lower seat.
12. Install or connect the following:
    - Bump stop on the strut rod
    - Strut cover
    - Spring seat
    - Upper spring seat
    - Bearing spacer. Align the strut bracket with the mark on the upper spring seat.
13. Clean the bearing lower washer and install it on the strut rod.
14. Clean the strut bearing and apply fresh grease to the bearing. Install the bearing on the lower washer.
15. Clean the bearing upper washer and install it.
16. Install these components in the following order:
    - Bearing upper seal
    - Bearing seat
    - Strut support
    - Inner spacer
    - Washer
    - Strut nut. Tighten the nut to 29–43 ft. lbs. (40–60 Nm). Apply a waterproof coating (paint or lacquer) to the nut and strut rod threads.
17. Loosen the spring compressors alternately, checking that the stepped part of the spring seat and spring are properly positioned.
18. Install or connect the following:
    - Strut assembly onto the vehicle
    - Upper support nuts loosely. Tighten the upper strut support nuts to 16–23 ft. lbs. (22–33 Nm) on Swift models or 20 ft. lbs. (28 Nm) on Esteem models.
    - Strut bracket nuts and bolts and tighten to 51–65 ft. lbs. (70–90 Nm)
    - ABS wheel speed sensor, if equipped
    - Brake hose clip
    - Wheels

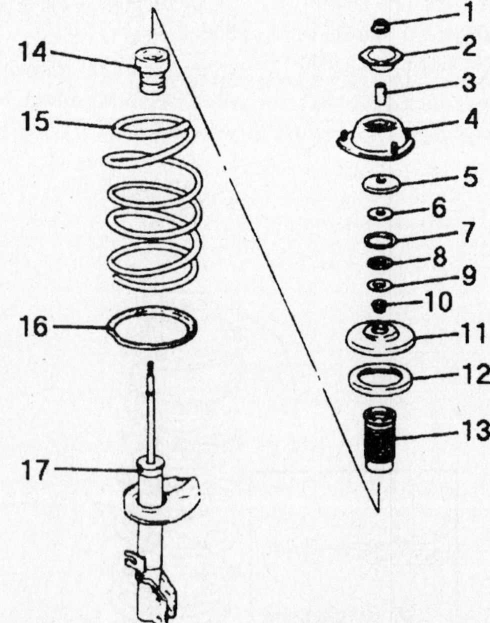

Exploded view of the strut assembly

Rear

ESTEEM

1. Before servicing the vehicle, refer to the precautions in the beginning of this section.
2. Remove wheel and tire assembly.
3. Place a jack under the lower control arm to support the suspension.
4. Remove or disconnect the following:
    - Brake line from the brake hose at the strut
    - E-ring securing the brake hose to the strut
    - Strut upper support nuts and push the strut down

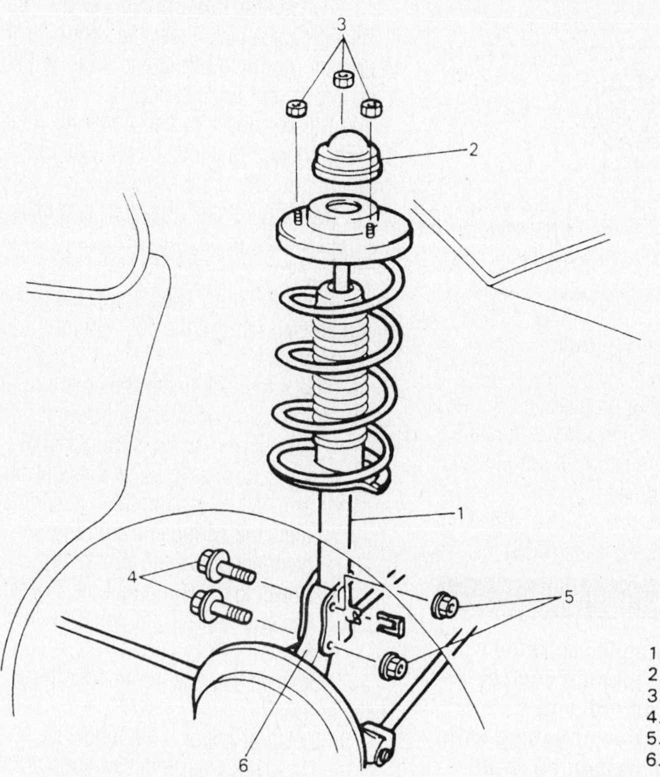

Rear strut assembly mounting—Esteem

- 2 bolts and nuts holding the strut to the rear knuckle
- Strut from the knuckle
- Strut from the vehicle

5. Install a pair of coil spring compressors on the coil spring on the strut assembly. Turn the spring compressors alternately until the spring tension is released from the strut assembly. If the spring can be turned slightly, then it has been collapsed enough.

✻✻ CAUTION

This procedure requires the use of a spring compressor. It cannot be performed without one. If you do not access to this special tool, do not attempt to disassemble the strut. The coil spring is retained under considerable pressure. It can exert enough force to cause serious injury. Exercise extreme caution.

6. Keeping the spring collapsed, remove the strut center nut and remove the other components from the top of the strut assembly.
7. Remove the spring from the strut.

To install:

8. If installing a new spring, compress the spring with a pair of spring compressors. Make sure that the spring compresses to 11½ inches (290mm) for installation.
9. Position the coil spring on the strut making sure that the end of the spring is mated to the stepped part of the lower seat.
10. Install the remaining components.
11. Install the strut support with the center nut. Tighten the nut to 40 ft. lbs. (55 Nm).
12. Loosen the spring compressors alternately, checking that the stepped part of the spring seat and spring are properly positioned.
13. Install or connect the following:
    - Strut in the vehicle
    - 2 bolts and nuts to hold the strut to the rear knuckle. Tighten the nuts 65 ft. lbs. (90 Nm).
14. Fully extend the strut and position the upper part of the strut into the vehicle's body. If the upper part of the strut does not reach the vehicle body, raise the jack under the control arm a little.
15. Install or connect the following:
    - Upper support nuts and tighten them to 20 ft. lbs. (28 Nm)
    - Brake hose to the strut and install the E-clip
    - Brake hose to the brake line
    - Wheels
16. Fill the master cylinder with brake fluid and bleed the brake system.

SWIFT

1. Before servicing the vehicle, refer to the precautions in the beginning of this section.
2. Remove the wheels.
3. Place a jack under the lower control arm to support the suspension.

✻✻ CAUTION

The coil spring is under extreme pressure. Be sure the lower control arm is firmly supported with a hydraulic jack before continuing with procedure. If this caution is not observed, serious bodily injury may result.

4. Remove or disconnect the following:
    - Strut support nuts and push the strut down

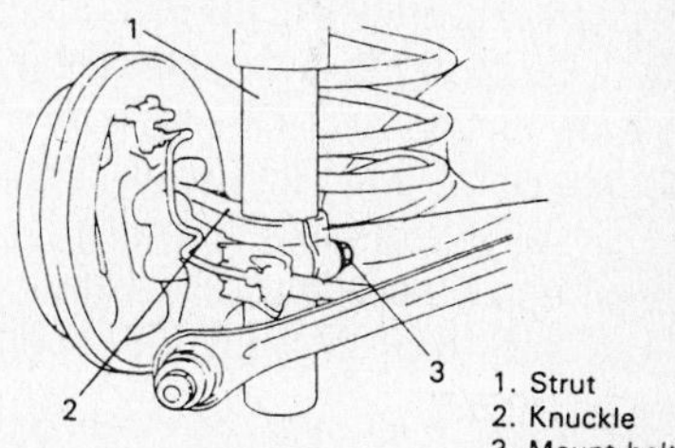

Rear strut upper mounting nuts—Swift

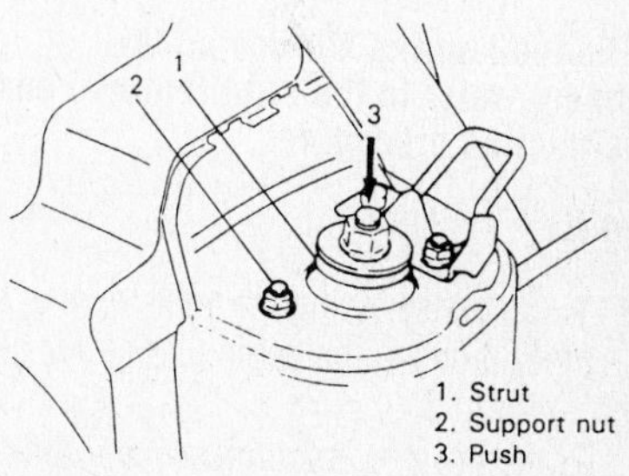

Rear strut lower mounting—Swift

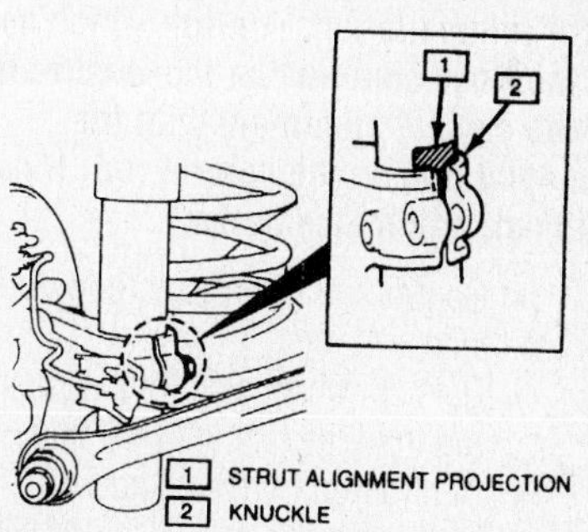

Shock-to-steering knuckle alignment—Swift

- Strut lower mounting bolt
- Strut from the knuckle. Compress the strut as short as possible for removal. If the strut is hard to remove, open the slit of the knuckle by inserting a wedge.

➡Do not open the knuckle slit wider than necessary. Do not lower the jack more than necessary during the strut removal to prevent the coil spring from coming off, or the brake flexible hose from stretching.

To install:

5. Install or connect the following:
   - Strut in the vehicle. Position the bottom of the alignment projection inside the knuckle opening.
   - Strut lower mounting bolt and tighten it to 36–50 ft. lbs. (50–70 Nm)
6. Fully extend the strut and position the upper part of the strut into the vehicle's body. If the upper part of the strut does not reach the vehicle's body, raise the jack under the control arm a little.
7. Install or connect the following:
   - Upper support nuts and tighten them to 20–27 ft. lbs. (28–38 Nm)
   - Wheels

Coil Spring

REMOVAL & INSTALLATION

➡For coil spring service on the esteem, refer to the strut removal and installation procedure.

Swift

1. Before servicing the vehicle, refer to the precautions in the beginning of this section.
2. Remove or disconnect the following:
   - Rear wheels

➡To facilitate the toe-in adjustment after reinstallation, confirm which one of the lines stamped on the washer is in the closest alignment with the stamped line on the control rod. If not marked, add matchmarks.

- Control rod inside bolt, body center side
- Outside (wheel side) of the control rod from the rear knuckle stud
- Control rod from the knuckle
- Nuts, washers, and bushings connecting the rear sway bar to the rear lower control arms

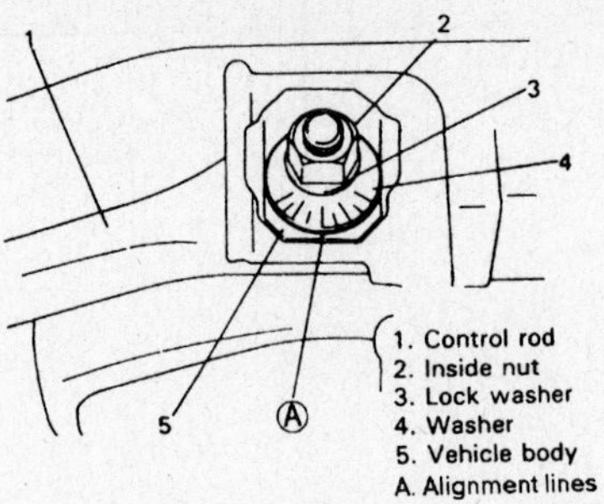

Control rod inside mount—Swift

- Rear mount nut on the control arm, loosen but do not remove the bolt
- Front nut of the control arm, loosen only at this time
- Wheel speed sensor, if equipped with Anti-Lock Brakes (ABS)

⁂ CAUTION

The coil spring is under extreme pressure. Be sure that the control arm is firmly supported with a hydraulic jack before continuing with procedure. If this precaution is not observed, serious bodily injury may result.

3. Loosen the lower mount nut on the knuckle. Place a jack under the control arm to prevent it from lowering and remove the lower mount nut on the knuckle.
4. Raise the jack placed under the control arm enough to allow the removal of the lower mount bolt of the knuckle.
5. Move the brake drum/backing plate toward the outside of the vehicle body so as to separate the lower mount of the knuckle from the control arm. Then, lower the jack gradually and remove the coil spring.

To install:

6. Place the jack under the control arm.
7. Install or connect the following:
   - Coil spring on the spring seat of the control arm. Raise the control arm. When seating the coil spring, mate the spring end with the stepped part of the control arm.
   - Lower knuckle mount bolt. Tighten the bolt to 29–33 ft. lbs. (40–45 Nm).
8. Remove the jack from under the suspension arm.
9. Install or connect the following:
   - Rear sway bar joints to the rear control arms and install the bushings, washers, and nuts. Tighten

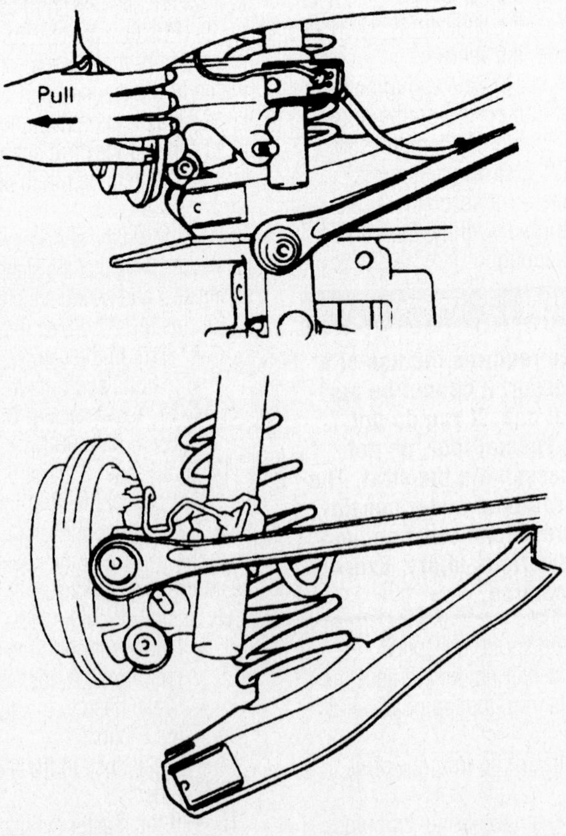

Disconnecting the knuckle and removing the coil spring

the nuts to 16–20 ft. lbs. (22–28 Nm).
- Control rod
- Inside control rod bolt and the outside control rod nut, but do not tighten them at this time
- Wheel speed sensor, if equipped
- Wheels and lower the vehicle

10. Tighten the control rod inside and outside nuts to 51–65 ft. lbs. (70–90 Nm).

➡When tightening the nuts, the vehicle should be off the hoist and in a non-loaded state. Also when tightening the inside nut, align the line stamped on the body with the line on the washer as confirmed before removal or align the matchmarks if marked.

11. Tighten the suspension arm front nuts to 36–50 ft. lbs. (50–70 Nm) and the rear nuts to 29–33 ft. lbs. (40–45 Nm). After tightening the suspension arm front nut, be sure that the washer is not tilted.
12. Check the rear wheel alignment.

Lower Ball Joint

REMOVAL & INSTALLATION

The lower ball joint is an integral part of the lower control arm assembly. If the ball joint is found to be defective the whole lower control arm assembly must be replaced.

Lower Control Arm

REMOVAL & INSTALLATION

The lower control arm and ball joint are a complete unit that will not separate.

1. Before servicing the vehicle, refer to the precautions in the beginning of this section.
2. Remove or disconnect the following:
   - Front wheels
   - Sway bar link nut, washer, and cushion
   - Ball joint stud bolt and nut
   - Lower control arm front bushing bolt
   - 2 bolts holding the lower control arm bracket to the vehicle
   - Lower control arm from the steering knuckle
   - Lower control arm from the vehicle

To install:

3. Install or connect the following:
   - Ball joint on the knuckle and secure it with the nut and bolt
   - Lower control arm rear bracket and bolts
   - Lower control arm front mounting bolt
   - Sway bar link to the control arm and install the cushion, washer and nut
4. Tighten the retainers to their proper torque specifications as follows:
   a. Control arm rear mounting bracket bolts: 27 ft. lbs. (37 Nm).
   b. Front mounting bolt: 65 ft. lbs. (90 Nm).
   c. Ball joint nut and bolt: 44 ft. lbs. (60 Nm).
   d. Sway bar link nut: 18 ft. lbs. (26 Nm).
5. Install the wheels.
6. Check the front wheel alignment.

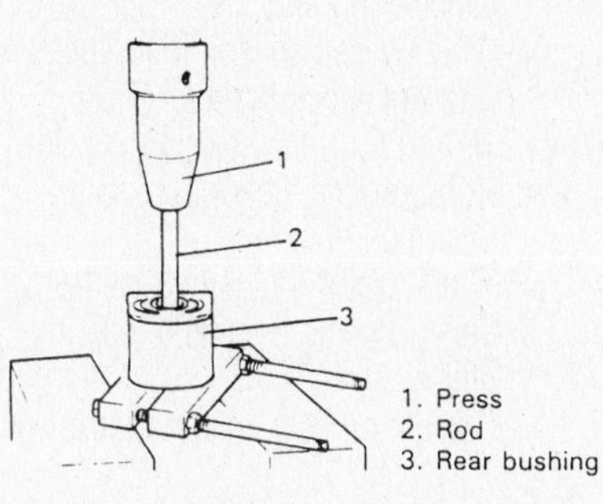

Use a suitable hydraulic press to remove the rear lower control arm bushing

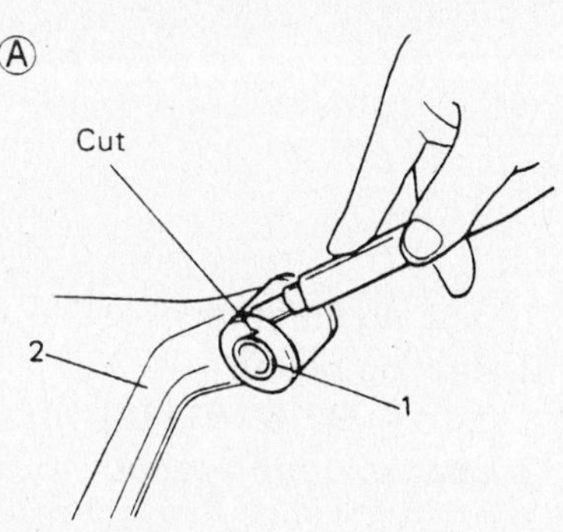

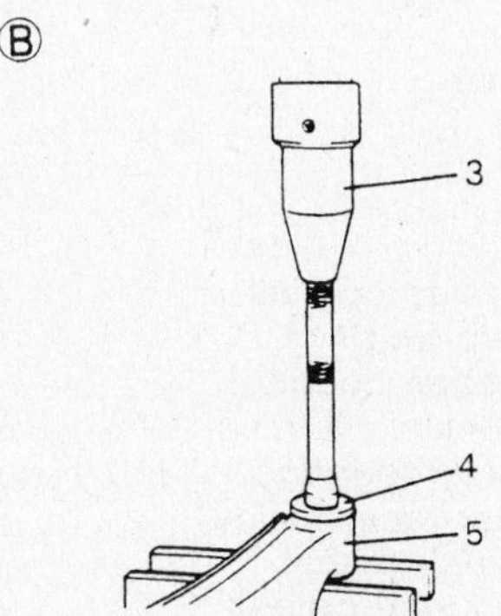

Cut the flange from the front bushing, then using a suitable hydraulic press; remove the rear lower control arm bushing

CONTROL ARM BUSHING REPLACEMENT

1. Before servicing the vehicle, refer to the precautions in the beginning of this section.
2. Remove the lower control arm.
3. Use a hydraulic press to remove the rear bushing.
4. Cut the flange from the front bushing.
5. Use a hydraulic press to remove the front bushing.

To install:

6. Apply a solution of soapy water to the outer diameter of the front bushing, this will aid in installation.

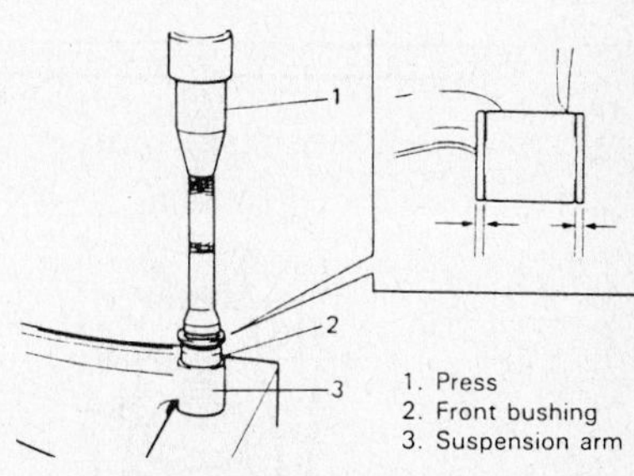

The front bushing should be positioned equally as shown after being pressed into position

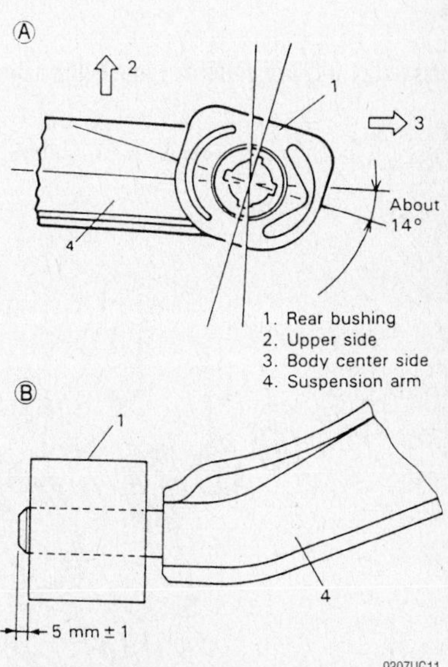

Install the rear bushing should be positioned as shown (A) before driving it into position (B)

7. Press the front bushing into its bore using a hydraulic press until the bushing is equal on the right and left of the arm as shown in the accompanying illustration.

8. Apply a solution of soapy water to the outer diameter of the rear bushing, this will aid in installation.

9. Install the rear bushing into the lower control arm in the direction and angle shown in figure A of the accompanying illustration.

10. Drive the bushing into position as shown in figure B of the accompanying illustration.

Wheel Bearings

ADJUSTMENT

The front and rear wheel bearings are a cartridge type design and cannot be adjusted.

REMOVAL & INSTALLATION

Front

➡Always replace bearing races as a complete set.

1. Before servicing the vehicle, refer to the precautions in the beginning of this section.
2. Remove or disconnect the following:
   - Front wheel
   - Brake caliper, carrier and disc from the steering knuckle
   - Wheel speed sensor, if equipped with Anti-Lock Brakes (ABS)
   - Tie rod from the steering knuckle
   - Hub from the steering knuckle
   - Steering knuckle
   - Wheel bearing outside race from the hub using a suitable bearing puller
   - Outside oil seal, snapring, outside bearing, inside oil seal and the inside bearing in that order

➡Once the bearing outer race is removed, the bearing set (outer race, bearings and inner races) should be replaced.

   - Bearing outer race from the knuckle by pressing the race out of the knuckle

To install

➡When installing the oil seals, be careful not to deform or tilt. Damage to the rubber part of the seal may occur.

3. Install or connect the following:
   - New bearing outer race into the knuckle using a press and the following tools: Bearing Installer Handle 09924–74510, Bearing And Oil Seal Installer 09944–68210 and a Bearing Installer Support 0994–78210
4. Apply lithium grease to the bearing races, bearings and oil seal lips.
5. Install or connect the following:
   - Outside bearing on the steering knuckle
   - Snapring to hold the outside bearing in place, then the outside oil seal
   - Inside bearing to the steering knuckle
   - Inside race and oil seal to the steering knuckle. When installing the inside oil seal, drive the oil seal in until it contacts the steering knuckle.

⁂ WARNING

If equipped with ABS use caution when installing the oil seal, because the seal has a hole that must align with the speed sensor position.

   - Outside race to the wheel hub using a bearing installer
6. Using a press and the proper tools, press the hub into the steering knuckle. After installation, be sure the hub is installed straight and turns freely.
7. Install or connect the following:
   - Steering knuckle in the vehicle
   - Tie rod end
   - Brake caliper, carrier and disc on the steering knuckle
   - Front wheel

Rear

WITH WHEEL HUBS

1. Before servicing the vehicle, refer to the precautions in the beginning of this section.

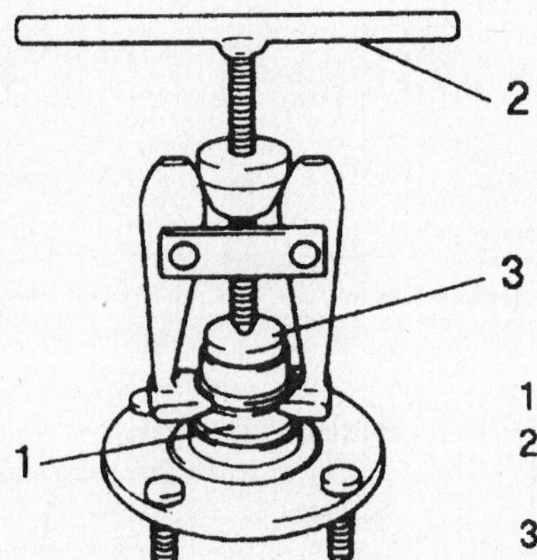

Use a 2- or 3-jaw puller to remove the bearing race from the hub

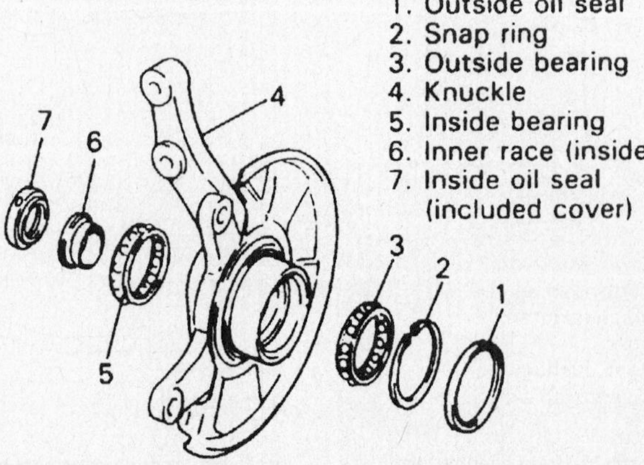

Steering knuckle bearing components—Swift

2. Set the parking brake.
3. Remove or disconnect the following:
   - Rear wheels
   - Rear brake drums
4. Use a brass drift and knock the wheel bearings from the drum assembly.

To install:

5. Position the inner wheel bearing on the drum with the sealed side facing out. Using a rear wheel Bearing Installer 09913–76010, install the rear wheel bearing. Install the wheel bearing spacer into the drum.
6. Install or connect the following:
   - Outer wheel bearing with the sealed side facing out, using a wheel bearing installer
7. Fill the space in the brake drum in between the wheel bearings to about 40% capacity with wheel bearing grease.
8. Install or connect the following:
   - Brake drum
   - Spindle washer and a new spindle nut. Tighten the spindle nut to 58–86 ft. lbs. (80–120 Nm).
9. Coat the spindle nut and the spindle dust cap with sealer.

➡When installing the spindle cap, hammer lightly several times on the collar of the cap until the collar comes closely into contact with the brake drum. If the fitting part of the cap is deformed or damaged or if it fits loose, replace the cap with a new one.

10. Depress the brake pedal with about 66 lbs. (30 kg) of force 3 to 5 times to obtain proper drum to shoe clearance.
11. Install the wheels.
12. Check to ensure that the brake drum is free from dragging and proper braking is obtained.

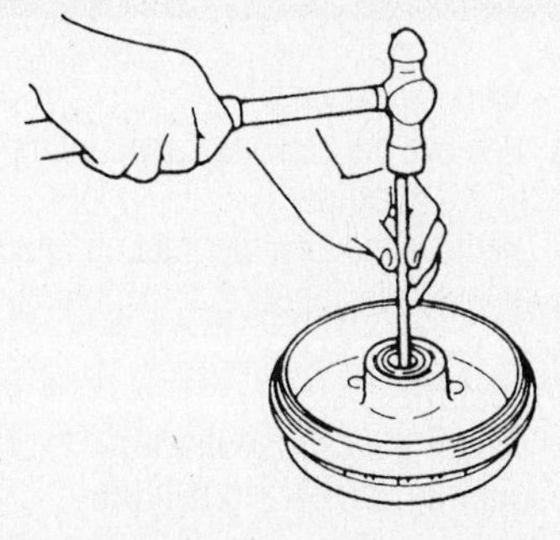

Use a hammer and drift to remove the wheel bearings from the hub—Swift

WITHOUT WHEEL HUBS

1. Before servicing the vehicle, refer to the precautions in the beginning of this section.
2. Set the parking brake.
3. Remove or disconnect the following:
   - Rear wheels
   - Brake drum, if equipped with drum brakes
   - Caliper, carrier and disc, if equipped with rear disc brakes
4. Release the parking brake.
5. Remove or disconnect the following:
   - Spindle cap without deforming it
   - Sealer from the spindle nut
   - Spindle nut and washer
6. Using a brake hub removal tool and a slide hammer remove the hub from the spindle.

➡The wheel bearing and hub are a solid unit. When the wheel bearing is found defective and it is necessary to replace it, replace the hub assembly.

To install:

7. Install or connect the following:

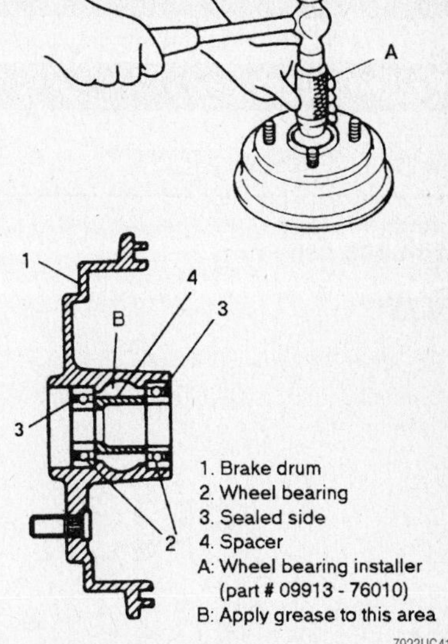

Wheel bearing installation—Swift

   - Wheel hub, washer and a new spindle nut. Tighten the spindle nut to 108–144 ft. lbs. (150–200 Nm).
8. Coat the spindle nut with sealer
9. Install or connect the following:
   - Spindle cap
   - Brake drums, if equipped with rear drum brakes
   - Brake caliper carrier and disc, if equipped with rear disc brakes
10. Depress the brake pedal with about 66 lbs. (30 kg) of force 3 to 5 times to obtain proper drum/rotor to shoe/pad clearance.
11. Install the wheel and tighten the lug nuts to 36–58 ft. lbs. (50–80 Nm).
12. Check to ensure that the brakes are free from dragging and that proper braking is obtained.

BRAKES

Brake Caliper

REMOVAL & INSTALLATION

Swift and Esteem

FRONT

1. Before servicing the vehicle, refer to the precautions in the beginning of this section.
2. Remove the wheels.
3. Remove the brake hose mounting bolt from the brake caliper.
4. Remove the 2 caliper mounting bolts, then remove the caliper from the mounting carrier.

To install:

5. Install the caliper assembly to the mounting carrier. Tighten the mounting bolts to 16–23 ft. lbs. (22–32 Nm).
6. Install the brake hose to the caliper. Always use new washers and tighten the mounting bolt to 14–18 ft. lbs. (20–25 Nm).
7. Bleed the brake system and install the front wheels.
8. Check the level of the brake fluid in the master cylinder reservoir.

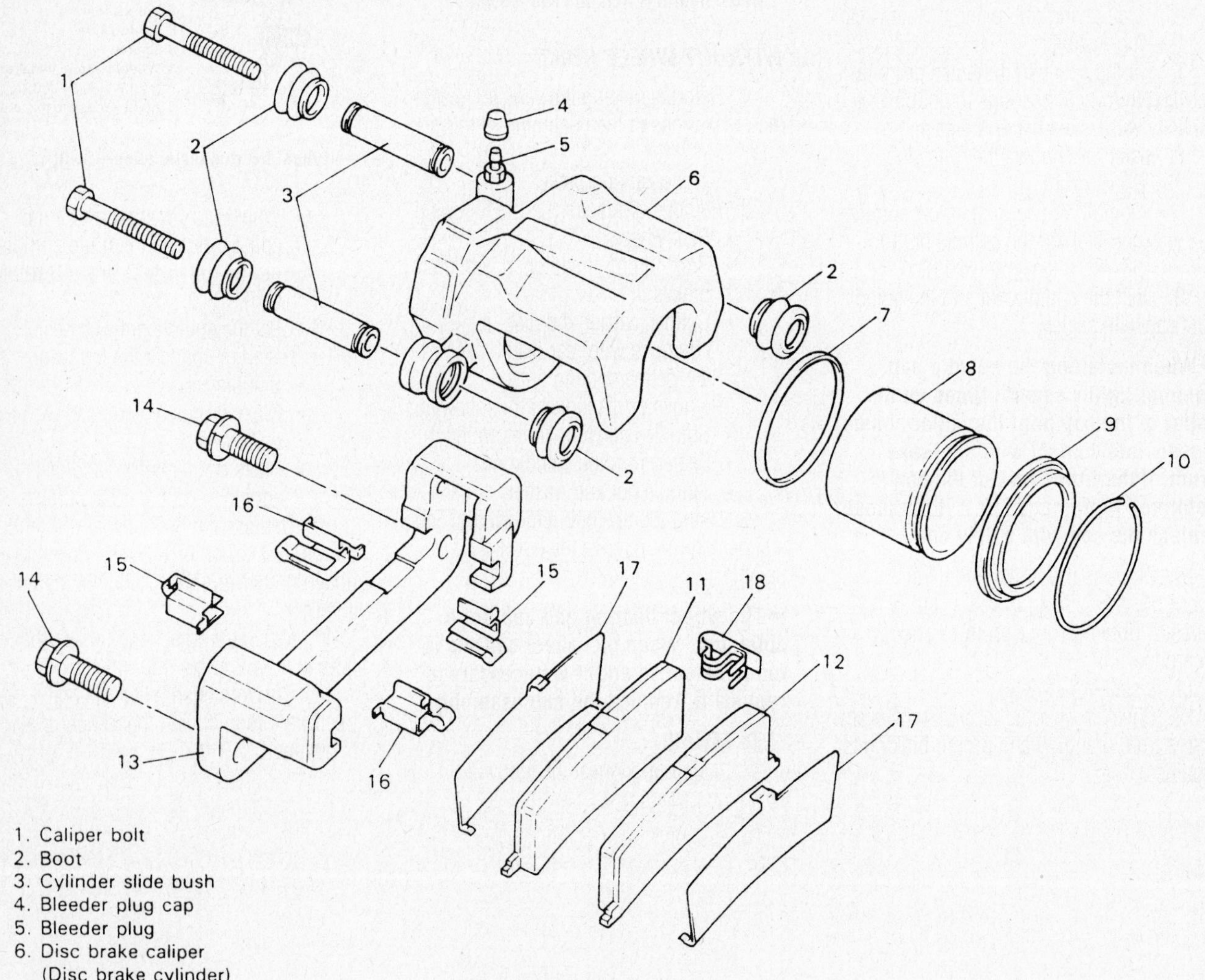

Front caliper—Esteem

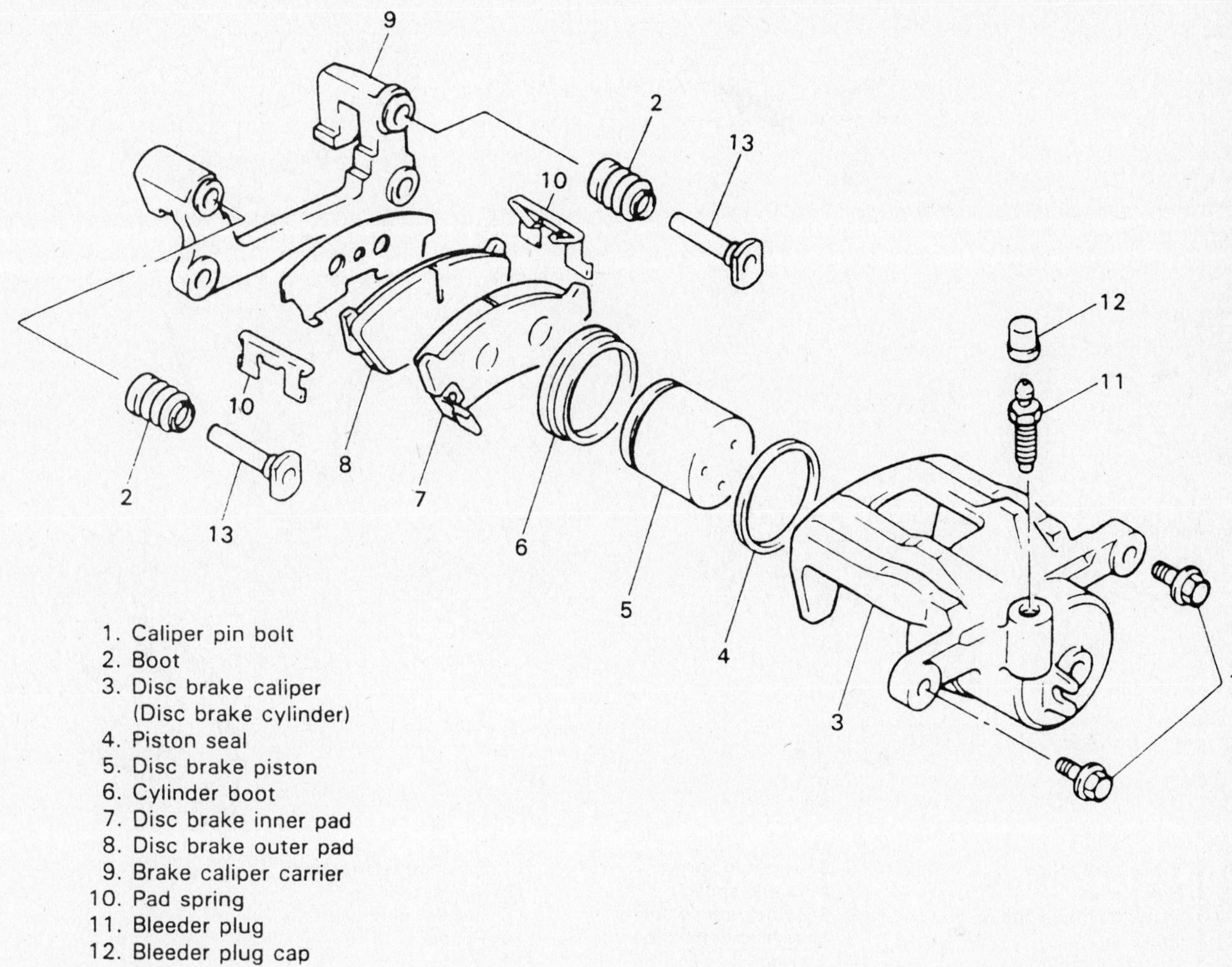

Front caliper—Swift

Disc Brake Pads

REMOVAL & INSTALLATION

Swift and Esteem

FRONT

1. Before servicing the vehicle, refer to the precautions in the beginning of this section.
2. Remove the wheels.
3. Remove the 2 caliper mounting bolts and then remove the caliper from the caliper carrier.
4. Remove the brake pads.

To install:

5. Install the pads into the caliper carrier.
6. On front calipers, press the caliper piston back into the caliper.
7. On rear calipers, use a piston installer 09945-16030, to rotate the caliper piston clockwise into the caliper bore. Lubricate the piston boot with silicone grease to avoid twisting the piston boot.
8. Install the caliper assembly to the caliper carrier. Tighten the bolts to 16–23 ft. lbs. (22–32 Nm).
9. Install the wheels.

Brake Drums

REMOVAL & INSTALLATION

Swift and Esteem

WITHOUT WHEEL HUBS

1. Before servicing the vehicle, refer to the precautions in the beginning of this section.
2. Remove the rear wheels.
3. Remove the spindle dust cap.
4. Remove the sealer from the spindle nut and remove the spindle nut and washer.
5. Install the brake drum removal tool 09943-17911, to the brake drum, then attach a slide hammer to the tool and remove the brake drum.

To install:

6. Install the brake drum.
7. Install the spindle washer and a new spindle nut. Torque the spindle nut to 58–86 ft. lbs. (80–120 Nm).
8. Apply sealer to the spindle nut.
9. Install the spindle dust cap.
10. Install the wheels.

WITH WHEEL HUBS

1. Before servicing the vehicle, refer to the precautions in the beginning of this section.
2. Remove the rear wheels.
3. Remove the 2 brake drum screws.
4. Pull the brake drum off using two 8mm bolts.

To install:

5. Install the brake drum.

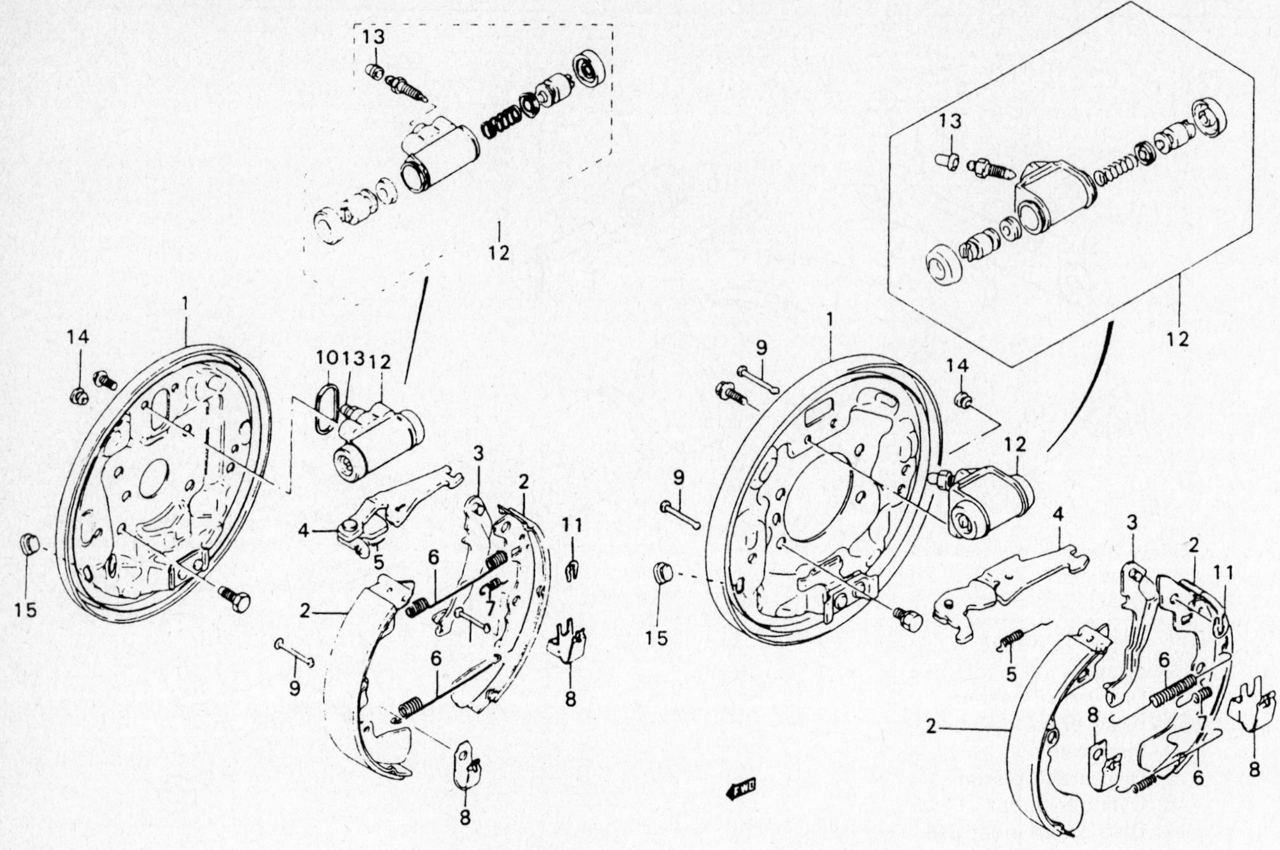

Rear drum brakes—Swift and Esteem

6. Tighten the brake drum screws.
7. Install the wheels.

Brake Shoes

REMOVAL & INSTALLATION

Swift and Esteem

1. Before servicing the vehicle, refer to the precautions in the beginning of this section.
2. Remove the rear wheels.
3. Remove the rear brake drums,
4. Remove the upper and lower springs from the brake shoes.
5. Remove the anti-rattle spring and brake shoe adjustment strut.
6. Remove the primary and secondary brake shoe hold-down springs and remove the shoes from vehicle.
7. Remove the clip securing the parking brake shoe lever to the secondary shoe.

To install:

8. Install the clip securing the parking brake shoe lever to the secondary shoe.
9. Install the primary and secondary brake shoes to the vehicle and secure with the hold-down springs.
10. Install the brake adjustment strut and anti-rattle spring.
11. Install the upper and lower return springs to the primary and secondary brake shoes.
12. Install the brake drum
13. Install the wheels.
14. Press the brake pedal 3–5 times to adjust the brake shoe clearance.
15. Adjust the parking brake cable.

SUZUKI

15

Grand Vitara • Vitara

SPECIFICATION CHARTS

ENGINE AND VEHICLE IDENTIFICATION CHART

Engine Code						
Code	Liters (cc)	Cu. In.	Cyl.	Fuel Sys.	Engine Type	Eng. Mfg.
0	1.6 (1590)	97	4	MFI	SOHC	Suzuki
5	2.0 (1997)	121.8	4	MFI	DOHC	Suzuki
6	2.5 (2494)	152	6	MFI	DOHC	Suzuki

MFI: Multiport Fuel Injection

DOHC: Dual Overhead Cam

SOHC: Single Overhead Cam

Model Year	
Code	Year
Y	2000
1	2001
2	2002
3	2003
4	2004

42356-SUZT-C01

GENERAL ENGINE SPECIFICATIONS

Year	Model	Engine Displacement Liters (cc)	Engine ID/VIN	Fuel System Type	Net Horsepower @ rpm	Net Torque @ rpm (ft. lbs.)	Bore x Stroke (in.)	Com-pression Ratio	Oil Pressure @ rpm
2000	Vitara	1.6 (1590)	0	MFI	95@5600	98@4000	2.95x3.54	9.5:1	47-61@3000
		2.0 (1997)	5	MFI	127@6000	134@3000	3.31x3.54	NA	55-67@4000
	Grand Vitara	2.5 (2494)	6	MFI	140@6500	151@4000	3.31x2.95	9.5:1	55-67@4000
2001	Vitara	2.0 (1997)	5	MFI	127@6000	134@3000	3.31x3.54	NA	55-67@4000
	Grand Vitara	2.5 (2494)	6	MFI	140@6500	151@4000	3.31x2.95	9.5:1	55-67@4000
2002	Vitara	2.0 (1997)	5	MFI	127@6000	134@3000	3.31x3.54	NA	55-67@4000
	Grand Vitara	2.5 (2494)	6	MFI	140@6500	151@4000	3.31x2.95	9.5:1	55-67@4000
2003	Vitara	2.0 (1997)	5	MFI	127@6000	134@3000	3.31x3.54	NA	55-67@4000
	Grand Vitara	2.5 (2494)	6	MFI	140@6500	151@4000	3.31x2.95	9.5:1	55-67@4000

MFI: Multi-port Fuel Injection

NA: Not available

42356-SUZT-C02

ENGINE TUNE-UP SPECIFICATIONS

Year	Engine Displacement Liters (cc)	Engine ID/VIN	Spark Plugs Gap (in.)	Ignition Timing (deg.) MT	Ignition Timing (deg.) AT	Fuel Pump (psi)	Idle Speed (rpm) MT	Idle Speed (rpm) AT	Valve Clearance In.	Valve Clearance Ex.
2000	1.6 (1590)	0	0.040	5B	5B	30-37	700-800	700-800	0.0050-0.0070	0.0050-0.0070
	2.0 (1997)	5	0.040	5B	5B	30-37	700-800	700-800	HYD	HYD
	2.5 (2494)	6	0.040	5B	5B	30-45	700-800	700-800	HYD	HYD
2001	2.0 (1997)	5	0.040	5B	5B	30-37	700-800	700-800	HYD	HYD
	2.5 (2494)	6	0.040	5B	5B	30-45	700-800	700-800	HYD	HYD
2002	2.0 (1997)	5	0.040	5B	5B	30-37	700-800	700-800	HYD	HYD
	2.5 (2494)	6	0.040	5B	5B	30-45	700-800	700-800	HYD	HYD
2003	2.0 (1997)	5	0.040	5B	5B	30-37	700-800	700-800	HYD	HYD
	2.5 (2494)	6	0.040	5B	5B	30-45	700-800	700-800	HYD	HYD

HYD: Hydraulic

42356-SUZT-C03

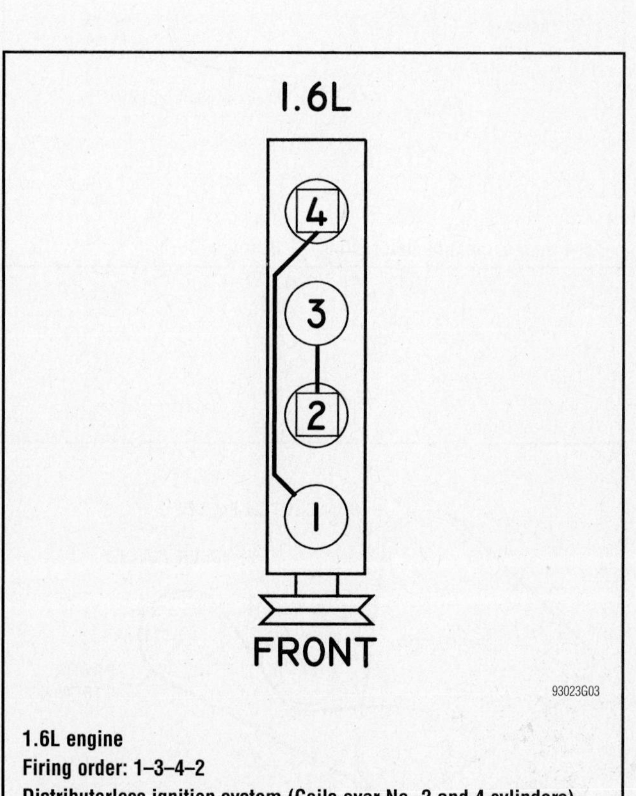

1.6L engine
Firing order: 1–3–4–2
Distributorless ignition system (Coils over No. 2 and 4 cylinders)

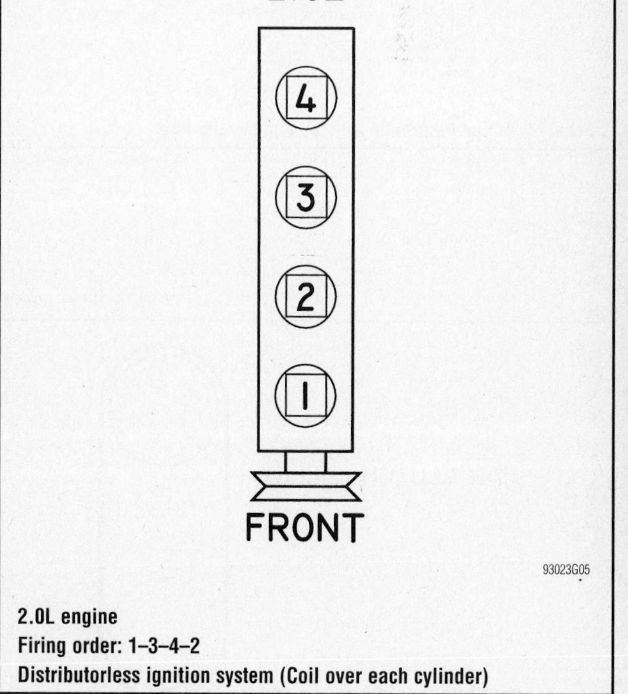

2.0L engine
Firing order: 1–3–4–2
Distributorless ignition system (Coil over each cylinder)

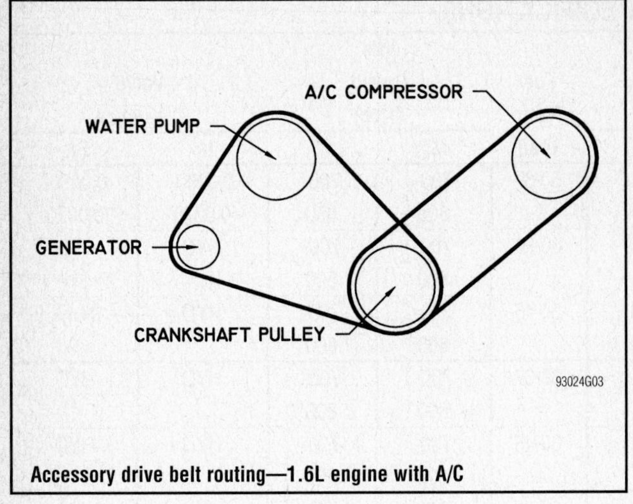

Accessory drive belt routing—1.6L engine with A/C

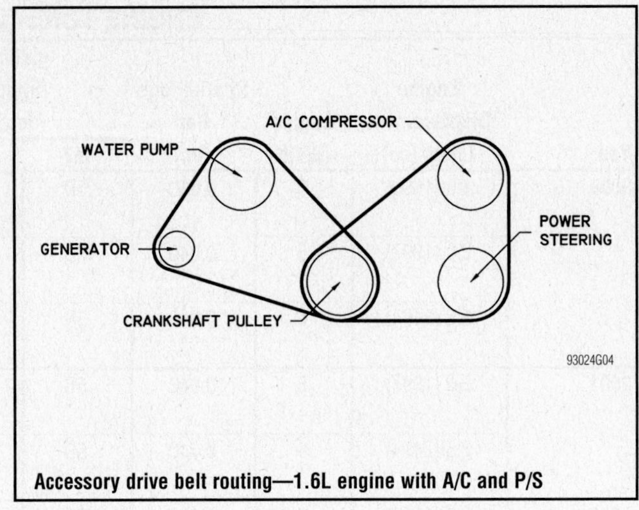

Accessory drive belt routing—1.6L engine with A/C and P/S

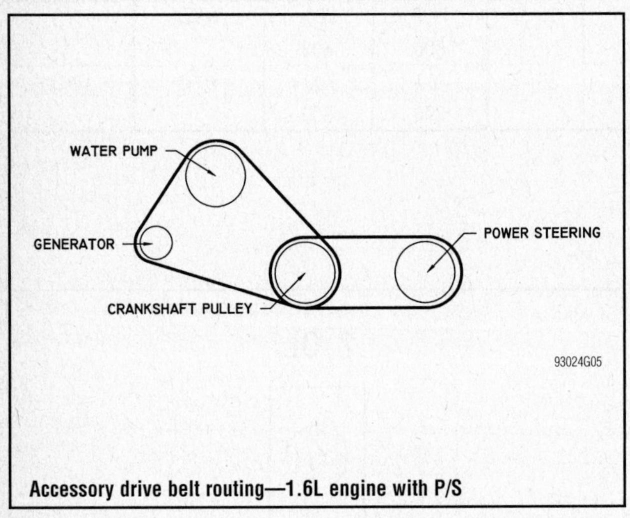

Accessory drive belt routing—1.6L engine with P/S

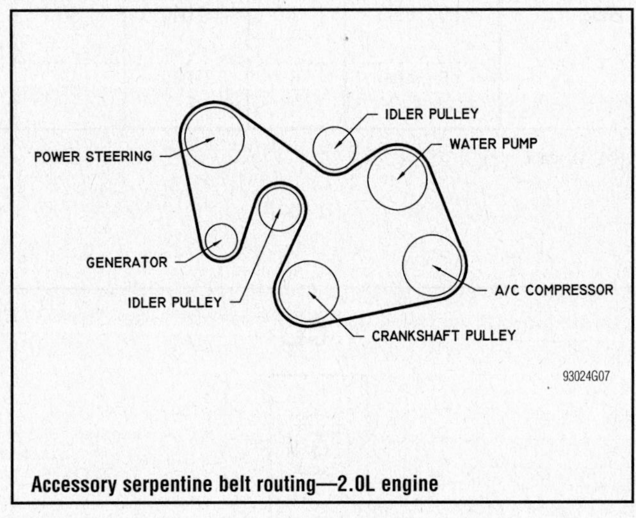

Accessory serpentine belt routing—2.0L engine

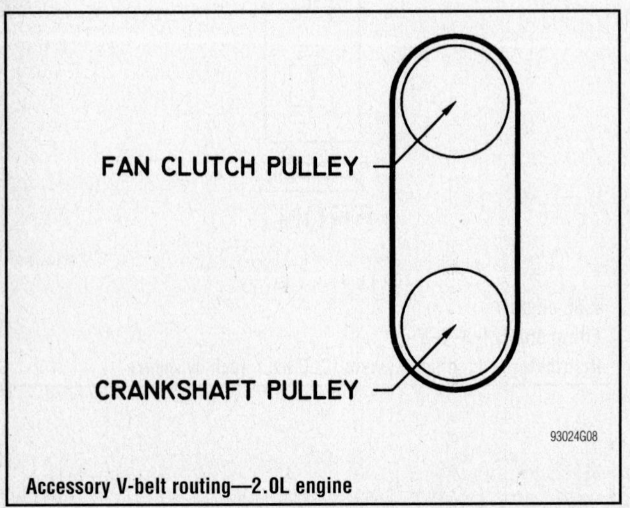

Accessory V-belt routing—2.0L engine

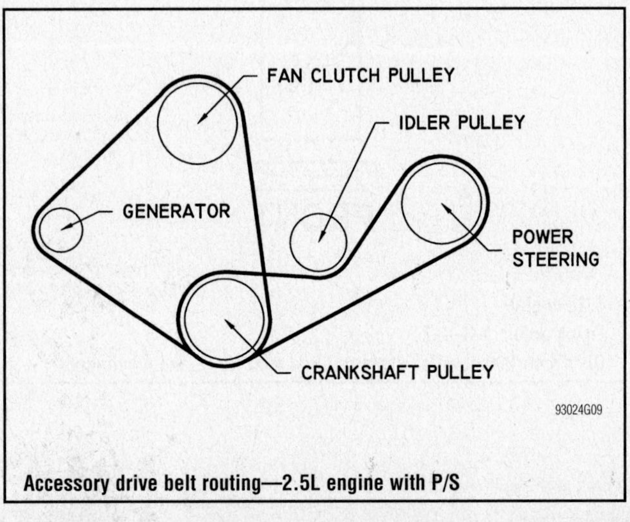

Accessory drive belt routing—2.5L engine with P/S

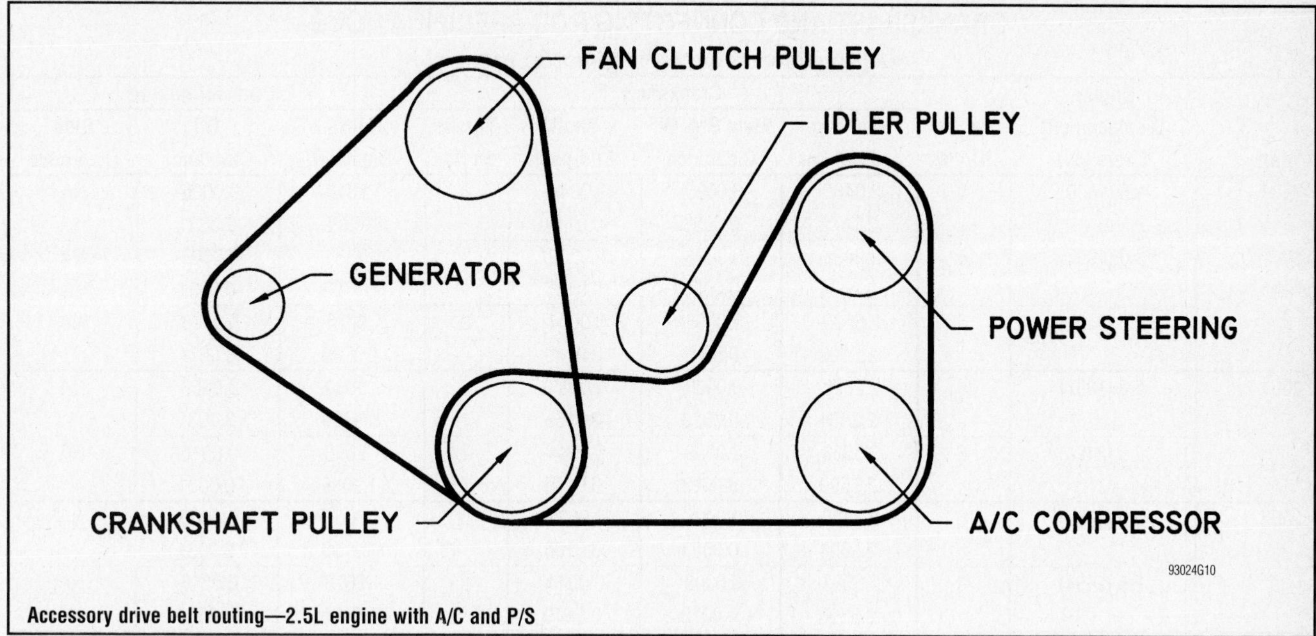

Accessory drive belt routing—2.5L engine with A/C and P/S

CAPACITIES

Year	Model	Engine Displacement Liters (cc)	Engine ID/VIN	Engine Oil with Filter (qts.)	Transmission (pts.) 5-Spd	Transmission (pts.) Auto.	Transfer Case (pts.)	Drive Axle Front (pts.)	Drive Axle Rear (pts.)	Fuel Tank (gal.)	Cooling System (qts.)
2000	Vitara	1.6 (1590)	0	4.75	①	②	3.6	2.1	4.6	11	5.5
		2.0 (1997)	5	5.9	①	②	3.6	2.1	4.6	③	5.5
	Grand Vitara	2.5 (2494)	6	6	5.5	②	3.6	2.1	4.6	18.5	5.5
2001	Vitara	2.0 (1997)	5	5.9	①	④	3.6	2.1	4.6	③	6.9
	Grand Vitara	2.5 (2494)	6	6	5.5	④	3.6	2.1	4.6	18.5	8.5
2002	Vitara	2.0 (1997)	5	5.9	①	④	3.6	2.1	4.6	③	6.9
	Grand Vitara	2.5 (2494)	6	6	5.5	④	3.6	2.1	4.6	18.5	8.5
2003	Vitara	2.0 (1997)	5	5.9	①	④	3.6	2.1	4.6	③	6.9
	Grand Vitara	2.5 (2494)	6	6	5.5	④	3.6	2.1	4.6	18.5	8.5

Note: All capacities are approximate. Add fluid gradually and check to be sure a proper fluid level is obtained.

① 2-wheel drive model: 4.0 pts.
4-wheel drive model: 3.2 pts.

② 3-speed transmission:
Fluid drain, and filter and pan removal only: 5.9 pts.
After complete transmission overhaul: 10.8 pts.
4-speed overdrive transmission:
Fluid drain, and filter and pan removal only: 5.3 pts.
After complete transmission overhaul: 14.6 pts.

③ 2-door model: 11 gals.
4-door model: 14.5 gals.

④ 2WD: 6.0 pts
4WD: 5.2 pts.

42356-SUZT-C04

CRANKSHAFT AND CONNECTING ROD SPECIFICATIONS

All measurements are given in inches.

			Crankshaft				Connecting Rod		
Year	Engine Displacement Liters (cc)	Engine ID/VIN	Main Brg. Journal Dia.	Main Brg. Oil Clearance	Shaft End-play	Thrust on No.	Journal Diameter	Oil Clearance	Side Clearance
2000	1.6 (1590)	0	2.0465-2.0472	0.0006-0.0023	0.0044-0.0149	3	1.7316-1.7322	0.0008-0.0031	NA
	2.0 (1997)	5	2.2828-2.2834	0.0008-0.0023	0.0039-0.0165	3	1.9660-1.9709	0.0008-0.0031	NA
	2.5 (2494)	6	2.5583-2.5590	0.0008-0.0023	0.0044-0.0149	2	1.9678-1.9685	0.0016-0.0031	NA
2001	2.0 (1997)	5	2.2828-2.2834	0.0008-0.0023	0.0039-0.0165	3	1.9660-1.9709	0.0008-0.0031	NA
	2.5 (2494)	6	2.5583-2.5590	0.0008-0.0023	0.0044-0.0149	2	1.9678-1.9685	0.0016-0.0031	NA
2002	2.0 (1997)	5	2.2828-2.2834	0.0008-0.0023	0.0039-0.0165	3	1.9660-1.9709	0.0008-0.0031	NA
	2.5 (2494)	6	2.5583-2.5590	0.0008-0.0023	0.0044-0.0149	2	1.9678-1.9685	0.0016-0.0031	NA
2003	2.0 (1997)	5	2.2828-2.2834	0.0008-0.0023	0.0039-0.0165	3	1.9660-1.9709	0.0008-0.0031	NA
	2.5 (2494)	6	2.5583-2.5590	0.0008-0.0023	0.0044-0.0149	2	1.9678-1.9685	0.0016-0.0031	NA

NA: Not Available

42356-SUZT-C05

VALVE SPECIFICATIONS

							Stem-to-Guide Clearance (in.)		Stem Diameter (in.)	
Year	Engine Displacement Liters (cc)	Engine ID/VIN	Seat Angle (deg.)	Face Angle (deg.)	Spring Test Pressure (lbs. @ in.)	Spring Installed Height (in.)	Intake	Exhaust	Intake	Exhaust
2000	1.6 (1590)	0	45	45	24-28@1.24	1.245	0.0008-0.0018	0.0018-0.0028	0.2152-0.2157	0.2142-0.2148
	2.0 (1997)	5	45	45	①	1.250	0.0008-0.0027	0.0018 0.0035	0.2348-0.2354	0.2339-0.2344
	2.5 (2494)	6	45	45	①	1.250	0.0008-0.0027	0.0018-0.0035	0.2348-0.2354	0.2339-0.2344
2001	2.0 (1997)	5	45	45	①	1.250	0.0008-0.0027	0.0018 0.0035	0.2348-0.2354	0.2339-0.2344
	2.5 (2494)	6	45	45	①	1.250	0.0008-0.0027	0.0018-0.0035	0.2348-0.2354	0.2339-0.2344
2002	2.0 (1997)	5	45	45	①	1.250	0.0008-0.0027	0.0018 0.0035	0.2348-0.2354	0.2339-0.2344
	2.5 (2494)	6	45	45	①	1.250	0.0008-0.0027	0.0018-0.0035	0.2348-0.2354	0.2339-0.2344
2003	2.0 (1997)	5	45	45	①	1.250	0.0008-0.0027	0.0018 0.0035	0.2348-0.2354	0.2339-0.2344
	2.5 (2494)	6	45	45	①	1.250	0.0008-0.0027	0.0018-0.0035	0.2348-0.2354	0.2339-0.2344

① Inner: 13.6-17.4@1.08
Outer: 30.4-39.2@1.25

42356-SUZT-C06

PISTON AND RING SPECIFICATIONS

All measurements are given in inches.

Year	Engine Displacement Liters (cc)	Engine ID/VIN	Piston Clearance	Ring Gap			Ring Side Clearance		
				Top Compression	Bottom Compression	Oil Control	Top Compression	Bottom Compression	Oil Control
2000	1.6 (1590)	0	0.0008-0.0015	0.0079-0.0275	0.0138-0.0275	0.0039-0.0669	0.0012-0.0027	0.0008-0.0023	NA
	2.0 (1997)	5	0.0008-0.0015	0.0079-0.0276	0.0138-0.0276	0.0079-0.0709	0.0012-0.0027	0.0008-0.0023	NA
	2.5 (2494)	6	0.0008-0.0015	0.0079-0.0276	0.0138-0.0276	0.0079-0.0709	0.0012-0.0027	0.0008-0.0023	NA
2001	2.0 (1997)	5	0.0008-0.0015	0.0079-0.0276	0.0138-0.0276	0.0079-0.0709	0.0012-0.0027	0.0008-0.0023	NA
	2.5 (2494)	6	0.0008-0.0015	0.0079-0.0276	0.0138-0.0276	0.0079-0.0709	0.0012-0.0027	0.0008-0.0023	NA
2002	2.0 (1997)	5	0.0008-0.0015	0.0079-0.0276	0.0138-0.0276	0.0079-0.0709	0.0012-0.0027	0.0008-0.0023	NA
	2.5 (2494)	6	0.0008-0.0015	0.0079-0.0276	0.0138-0.0276	0.0079-0.0709	0.0012-0.0027	0.0008-0.0023	NA
2003	2.0 (1997)	5	0.0008-0.0015	0.0079-0.0276	0.0138-0.0276	0.0079-0.0709	0.0012-0.0027	0.0008-0.0023	NA
	2.5 (2494)	6	0.0008-0.0015	0.0079-0.0276	0.0138-0.0276	0.0079-0.0709	0.0012-0.0027	0.0008-0.0023	NA

NA: Not Available

42356-SUZT-C07

TORQUE SPECIFICATIONS

All readings in ft. lbs.

Year	Engine Displacement Liters (cc)	Engine ID/VIN	Cylinder Head Bolts	Main Bearing Bolts	Rod Bearing Bolts	Crankshaft Damper Bolts	Flywheel Bolts	Manifold Intake	Manifold Exhaust	Spark Plugs	Lug Nut
2000	1.6 (1590)	0	①	36-41	24-26	94 ②	57	13-20	13-20	14-21	58-80
	2.0 (1997)	5	③	④	33	109	51	13-20	13-20	14-21	58-80
	2.5 (2494)	6	③	④	33	109	51	16.5	22.0	18	69
2001	2.0 (1997)	5	③	④	33	109	51	13-20	13-20	14-21	70
	2.5 (2494)	6	③	④	33	109	51	17	22	18	70
	2.0 (1997)	C	③	④	33	109	51	17	17	18	70
2002	2.0 (1997)	5	③	④	33	109	51	13-20	13-20	14-21	70
	2.5 (2494)	6	③	④	33	109	51	17	22	18	70
2003	2.0 (1997)	5	③	④	33	109	51	13-20	13-20	14-21	70
	2.5 (2494)	6	③	④	33	109	51	17	22	18	70

① Step 1: 26 ft. lbs.
Step 2: 41 ft. lbs.
Step 3: Loosen in reverse order to 0 ft. lbs.
Step 4: 26 ft. lbs.
Step 5: 52 ft. lbs.

② Value shown is for crankshaft timing belt sprocket

③ Step 1: 38 ft. lbs.
Step 2: 61 ft. lbs.
Step 3: Loosen in reverse order to 0 ft. lbs.
Step 4: 38 ft. lbs.
Step 5: 76 ft. lbs.
Step 6: Tighten 6mm bolt to 8 ft. lbs.

④ 10mm: 43.5 ft. lbs.
8mm: 19.5 ft. lbs.

⑤ Use multiple passes to arrive at final torque.

⑥ Value shown is for crankshaft timing belt sprocket

42356-SUZT-C08

WHEEL ALIGNMENT

Year	Model	Caster Range (+/-Deg.)	Caster Preferred Setting (Deg.)	Camber Range (+/-Deg.)	Camber Preferred Setting (Deg.)	Toe-in (in.)	Steering Axis Inclination (Deg.)
2000	Vitara	1.00	+2.66	1.00	0	0+/-0.08	—
	Grand Vitara	1.00	+2.66	1.00	0	0+/-0.08	—
2001	Vitara	1.00	+2.66	1.00	0	0+/-0.08	—
	Grand Vitara	1.00	+2.66	1.00	0	0+/-0.08	—
2002	Vitara	1.00	+2.66	1.00	0	0+/-0.08	—
	Grand Vitara	1.00	+2.66	1.00	0	0+/-0.08	—
2003	Vitara	1.00	+2.66	1.00	0	0+/-0.08	—
	Grand Vitara	1.00	+2.66	1.00	0	0+/-0.08	—

42356-SUZT-C09

TIRE, WHEEL AND BALL JOINT SPECIFICATIONS

		OEM Tires		Tire Pressures (psi)		Wheel	Ball Joint
Year	Model	Standard	Optional	Front	Rear	Size	Inspection
2000	Vitara 1.6L 2wd	P195/75SR15	none	26	26	5.5JJ	①
	Vitara 1.6L 4wd	P205/75R15	none	26	26	6.5JJ	①
	Vitara 2.0L	P215/65R16	none	26	26	6.5JJ	①
	Grand Vitara	P235/65SR16	none	26	26	7-JJ	①
2001	Vitara 2.0L	P215/65R16	none	26	26	6.5JJ	①
	Grand Vitara	P235/65SR16	none	26	26	7-JJ	①
2002	Vitara 2.0L	P215/65R16	none	26	26	6.5JJ	①
	Grand Vitara	P235/65SR16	none	26	26	7-JJ	①
2003	Vitara 2.0L	P215/65R16	none	26	26	6.5JJ	①
	Grand Vitara	P235/65SR16	none	26	26	7-JJ	①

OEM: Original Equipment Manufacturer

PSI: Pounds Per Square Inch

STD: Standard

OPT: Optional

① Replace if any measurable movement is found.

42356-SUZT-C10

BRAKE SPECIFICATIONS

All measurements in inches unless noted

		Brake Disc			Brake Drum Diameter			Minimum Lining Thickness		Brake Caliper	
Year	Model	Original Thickness	Minimum Thickness	Maximum Runout	Original Inside Diameter	Max. Wear Limit	Maximum Machine Diameter	Front	Rear	Bracket Bolts (ft. lbs.)	Mounting Bolts (ft. lbs.)
2000	Vitara	0.670	0.590	0.006	8.66	8.74	8.74	0.08	0.04	61.5	19-21
	Grand Vitara	0.866	0.787	0.006	8.66	8.74	8.74	0.08	0.04	61.5	①
2001	Vitara	0.670	0.590	0.006	8.66	8.74	8.74	0.08	0.04	61.5	19-21
	Grand Vitara	0.866	0.787	0.006	8.66	8.74	8.74	0.08	0.04	61.5	①
2002	Vitara	0.670	0.590	0.006	8.66	8.74	8.74	0.08	0.04	61.5	19-21
	Grand Vitara	0.866	0.787	0.006	8.66	8.74	8.74	0.08	0.04	61.5	①
2003	Vitara	0.670	0.590	0.006	8.66	8.74	8.74	0.08	0.04	61.5	19-21
	Grand Vitara	0.866	0.787	0.006	8.66	8.74	8.74	0.08	0.04	61.5	①

① 10mm: 37 ft. lbs.
12mm: 62 ft. lbs.

42356-SUZT-C11

SCHEDULED MAINTENANCE INTERVALS
SUZUKI—VITARA, GRAND VITARA

TO BE SERVICED	TYPE OF SERVICE	VEHICLE MILEAGE INTERVAL (x1000)												
		7.5	15	22.5	30	37.5	45	52.5	60	67.5	75	82.5	90	97.5
Engine oil & filter	R	✓	✓	✓	✓	✓	✓	✓	✓	✓	✓	✓	✓	✓
Automatic transmission fluid ①	S/I	✓	✓	✓	✓	✓	✓	✓	✓	✓	✓	✓	✓	✓
Manual transmission oil ②	S/I	✓	✓	✓	✓	✓	✓	✓	✓	✓	✓	✓	✓	✓
Steering system	S/I	✓	✓	✓	✓	✓	✓	✓	✓	✓	✓	✓	✓	✓
Transfer & differential oil ②	S/I	✓	✓	✓	✓	✓	✓	✓	✓	✓	✓	✓	✓	✓
Wheel discs & free wheeling hubs	S/I	✓	✓	✓	✓	✓	✓	✓	✓	✓	✓	✓	✓	✓
Suspension system	S/I	✓	✓	✓	✓	✓	✓	✓	✓	✓	✓	✓	✓	✓
Brake discs & pads (front)	S/I		✓		✓		✓		✓		✓		✓	
Brake drums & shoes (rear)	S/I		✓		✓		✓		✓		✓		✓	
Brake fluid ③	S/I		✓		✓		✓		✓		✓		✓	
Brake hoses & pipes	S/I		✓		✓		✓		✓		✓		✓	
Brake pedal	S/I		✓		✓		✓		✓		✓		✓	
Brake lever & cable	S/I		✓		✓		✓		✓		✓		✓	
Clutch	S/I		✓		✓		✓		✓		✓		✓	
Idle speed	S/I		✓		✓		✓		✓		✓		✓	
Propeller shafts	S/I		✓		✓		✓		✓		✓		✓	
Valve lash (clearance)	S/I		✓		✓		✓		✓		✓		✓	
Wheel bearings	S/I		✓		✓		✓		✓		✓		✓	
Air cleaner filter element	R				✓				✓				✓	
Engine coolant	R				✓				✓				✓	
Fuel filter	R				✓				✓				✓	
Spark plugs	R				✓				✓				✓	
Cooling system hoses	S/I				✓				✓				✓	
Drive belt(s)	S/I				✓				✓				✓	
Exhaust pipes & mountings	S/I				✓				✓				✓	
Fuel lines & connections	S/I				✓				✓				✓	
Camshaft timing belt	R								✓				✓	
Distributor cap & rotor	S/I								✓					
Emission-related hoses & tubes	S/I								✓					
Oxygen sensor	S/I											✓		
EVAP canister	R	every 100,000 miles												
PCV valve	R							✓						
EGR system	S/I							✓						

42356-SUZT-C12

SCHEDULED MAINTENANCE INTERVALS
SUZUKI—VITARA, GRAND VITARA

TO BE SERVICED	TYPE OF SERVICE	VEHICLE MILEAGE INTERVAL (x1000)												
		7.5	15	22.5	30	37.5	45	52.5	60	67.5	75	82.5	90	97.5
Fuel Injectors	S/I	every 100,000 miles												
TWC converter	S/I	every 100,000 miles												

R: Replace S/I: Service or Inspect

① Replace at 100,000 miles.

② Replace oil every 30,000 miles.

③ Replace every 60,000 miles.

FREQUENT OPERATION MAINTENANCE (SEVERE SERVICE)

If a vehicle is operated under any of the following conditions it is considered severe service:

- Extremely dusty areas.

- 50% or more of the vehicle operation is in 32°C (90°F) or higher temperatures, or constant operation in temperatures below 0°C (32°F).

- Prolonged idling (vehicle operation in stop and go traffic).

- Frequent short running periods (engine does not warm to normal operating temperatures).

- Police, taxi, delivery usage or trailer towing usage.

Oil & oil filter: replace every 3000 miles.

Air cleaner filter element: service or inspect every 3000 miles & replace every 15,000 miles.

Steering wheel free play, gear box oil & linkage: service or inspect every 3000 miles.

Brake & nuts on chassis: tighten every 6000 miles.

Brake discs & pads (front): service or inspect every 6000 miles.

Brake drums & shoes (rear): service or inspect every 6000 miles.

Exhaust pipes & mountings: tighten every 6000 miles.

Propeller shafts: service or inspect every 6000 miles.

Automatic transmission fluid & filter: replace every 15,000 miles.

Distributor cap & ignition wires: service or inspect every 15,000 miles.

Drive belt(s): service or inspect every 15,000 miles.

Manual transmission oil: replace every 15,000 miles.

Transfer & differential oil: replace every 15,000 miles.

42356-SUZT-C13

PRECAUTIONS

Before servicing any vehicle, please be sure to read all of the following precautions, which deal with personal safety, prevention of component damage and important points to take into consideration when servicing a motor vehicle:

- Never open, service or drain the radiator or cooling system when the engine is hot; serious burns can occur from the steam and hot coolant.
- Observe all applicable safety precautions when working around fuel. Whenever servicing the fuel system, always work in a well-ventilated area. Do not allow fuel spray or vapors to come in contact with a spark, open flame, or excessive heat (a hot drop light, for example). Keep a dry chemical fire extinguisher near the work area. Always keep fuel in a container specifically designed for fuel storage; also, always properly seal fuel containers to avoid the possibility of fire or explosion. Refer to the additional fuel system precautions later in this section.
- Fuel injection systems often remain pressurized, even after the engine has been turned **OFF**. The fuel system pressure must be relieved before disconnecting any fuel lines. Failure to do so may result in fire and/or personal injury.
- Brake fluid often contains polyglycol ethers and polyglycols. Avoid contact with the eyes and wash your hands thoroughly after handling brake fluid. If you do get brake fluid in your eyes, flush your eyes with clean, running water for 15 minutes. If eye irritation persists, or if you have taken brake fluid internally, IMMEDIATELY seek medical assistance.
- The EPA warns that prolonged contact with used engine oil may cause a number of skin disorders, including cancer. You should make every effort to minimize your exposure to used engine oil. Protective gloves should be worn when changing oil. Wash your hands and any other exposed skin areas as soon as possible after exposure to used engine oil. Soap and water, or waterless hand cleaner should be used.
- All new vehicles are now equipped with an air bag system, often referred to as a Supplemental Restraint System (SRS) or Supplemental Inflatable Restraint (SIR) system. The system must be disabled before performing service on or around system components, steering column, instrument panel components, wiring and sensors. Failure to follow safety and disabling procedures could result in accidental air bag deployment, possible personal injury, and unnecessary system repairs.
- Always wear safety goggles when working with, or around, the air bag system. When carrying a non-deployed air bag, be sure the bag and trim cover are pointed away from your body. When placing a non-deployed air bag on a work surface, always face the bag and trim cover upward, away from the surface. This will reduce the motion of the module if it is accidentally deployed. Refer to the additional air bag system precautions later in this section.
- Clean, high quality brake fluid from a sealed container is essential to the safe and proper operation of the brake system. You should always buy the correct type of brake fluid for your vehicle. If the brake fluid becomes contaminated, completely flush the system with new fluid. Never reuse any brake fluid. Any brake fluid that is removed from the system should be discarded. Also, do not allow any brake fluid to come in contact with a painted surface; it will damage the paint.
- Never operate the engine without the proper amount and type of engine oil; doing so WILL result in severe engine damage.
- Timing belt maintenance is extremely important. Many models utilize an interference-type, non-freewheeling engine. If the timing belt breaks, the valves in the cylinder head may strike the pistons, causing potentially serious (also time-consuming and expensive) engine damage. Refer to the maintenance interval charts in the front of this section for the recommended replacement interval for the timing belt.
- Disconnecting the negative battery cable on some vehicles may interfere with the functions of the on-board computer system(s) and may require the computer to undergo a relearning process once the negative battery cable is reconnected.
- When servicing drum brakes, only disassemble and assemble one side at a time, leaving the remaining side intact for reference.

ENGINE REPAIR

➡Disconnecting the negative battery cable on some vehicles may interfere with the functions of the on board computer system. The computer may undergo a relearning process once the negative battery cable is reconnected.

Distributor

REMOVAL

All engines are equipped with a Distributorless Ignition System (DIS).

Alternator

REMOVAL

1. Before servicing the vehicle, refer to the precautions in the beginning of this section.
2. Remove or disconnect the following:
   - Negative battery cable
   - Evaporative Emission (EVAP) canister
   - Accessory drive belt
   - Alternator harness connectors
   - Alternator mounting bracket
   - Alternator

INSTALLATION

Install or connect the following:

- Alternator
- Alternator mounting bracket. Tighten the bolts to 20 ft. lbs. (27 Nm).
- Alternator harness connectors
- Accessory drive belt. Tighten the alternator bolts to 24 ft. lbs. (33 Nm).
- EVAP canister
- Negative battery cable

Ignition Timing

ADJUSTMENT

1.6L Engine

This engine is equipped with a Distributorless Ignition System (DIS). All timing functions are controlled by the Powertrain Control Module (PCM). No adjustment is possible.

2.0L and 2.5L Engines

➡The 2.0L and 2.5L engines use a Camshaft Position (CMP) sensor that is rotated to set base timing.

➡Check and adjust the ignition timing with the engine at normal operating temperature, all electrical accessories OFF and transmission in P, N for auto-

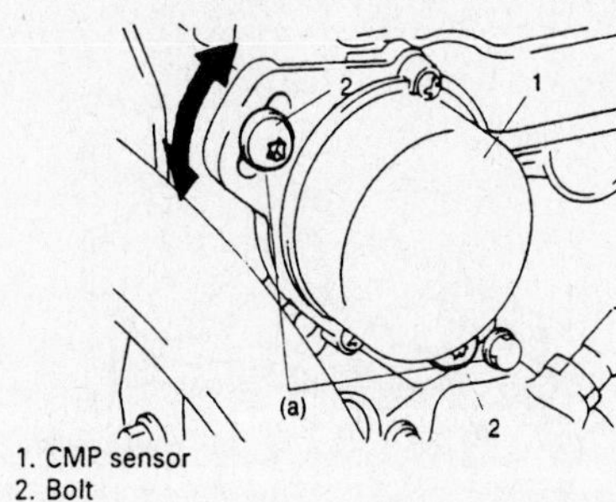

Camshaft position sensor

matic transmission or neutral for manual transmission.

1. Before servicing the vehicle, refer to the precautions in the beginning of this section.
2. With the engine **OFF**, connect a jumper wire between terminals **D** and **E** of the Data Link Connector (DLC).
3. Connect a timing light to the No. 1 spark plug wire and start the engine.
4. Ignition timing at idle should be 4–6 degrees Before Top Dead Center (BTDC).
5. Adjust the timing as necessary, then turn the engine **OFF**.
6. Remove the jumper wire from the DLC and remove the timing light.

Engine Assembly

REMOVAL & INSTALLATION

1.6L Engine

1. Before servicing the vehicle, refer to the precautions in the beginning of this section.
2. Relieve the fuel system pressure.
3. Drain the cooling system and engine oil.
4. Remove or disconnect the following:
   - Negative battery cable
   - Hood
   - Strut tower bar, if equipped
   - Cooling fan and shroud
   - Heater hoses
   - Radiator hoses
   - Bypass hose
   - Radiator
   - Air intake tube
   - Accelerator cable
   - Transmission cable, if equipped
   - Positive Crankcase Ventilation (PCV) valve and hose
   - Exhaust Gas Recirculation (EGR) valve and temperature sensor
   - EGR vacuum valve connector
   - EGR bypass valve connector
   - Idle Air Control (IAC) valve hoses and connector
   - Fuel lines
   - Main engine control wiring harness connectors at the firewall
   - Throttle Position (TP) sensor connector
   - Heated Oxygen (HO_2S) sensor connectors
   - Engine Coolant Temperature (ECT) sensor connector
   - Temperature gauge sender connector
   - A/C temperature switch, if equipped
   - Injector harness connectors
   - Alternator wiring connectors
   - Manifold Absolute Pressure (MAP) sensor connector and vacuum line
   - Brake booster vacuum line
   - Evaporative Emission (EVAP) canister and hoses
   - Distributor, if equipped
   - Front skidplate, if equipped
   - Power steering hoses
   - A/C compressor, if equipped
   - Flywheel access cover
   - Torque converter, if equipped
   - Clutch cable, if equipped
   - Transmission cooler lines, if equipped
   - Exhaust front pipe
   - Starter motor
   - Transmission flange fasteners and support the transmission
   - Left and right engine mounts
   - Engine

To install:

5. Install or connect the following:
   - Engine. Tighten the mount bolts to 40 ft. lbs. (54 Nm).
   - Transmission flange fasteners. Tighten them to 62 ft. lbs. (85 Nm).
   - Starter motor
   - Exhaust front pipe
   - Transmission cooler lines, if equipped
   - Clutch cable, if equipped
   - Torque converter, if equipped. Tighten the bolts to 40 ft. lbs. (54 Nm).
   - Flywheel access cover
   - A/C compressor, if equipped
   - Power steering hoses
   - Front skidplate, if equipped
   - Distributor, if equipped
   - EVAP canister and hoses
   - Brake booster vacuum line
   - MAP sensor connector and vacuum line
   - Alternator wiring connectors
   - Injector harness connectors
   - A/C temperature switch, if equipped
   - Temperature gauge sender connector
   - ECT sensor connector
   - HO_2S sensor connectors
   - TP sensor connector
   - Main engine control wiring harness connectors at the firewall
   - Fuel lines
   - IAC valve hoses and connector
   - EGR bypass valve connector
   - EGR vacuum valve connector
   - EGR valve and temperature sensor
   - PCV valve and hose
   - Transmission cable, if equipped
   - Accelerator cable
   - Air intake tube
   - Radiator
   - Bypass hose
   - Radiator hoses
   - Heater hoses
   - Cooling fan and shroud
   - Strut tower bar, if equipped
   - Hood
   - Negative battery cable
6. Fill the crankcase to the correct level.
7. Fill the cooling system.
8. Start the engine and check for leaks.

2.0L and 2.5L Engines

1. Before servicing the vehicle, refer to the precautions in the beginning of this section.
2. Relieve the fuel system pressure.
3. Drain the cooling system.
4. Drain the engine oil.
5. Remove or disconnect the following:
   - Negative battery cable
   - Hood
   - Heater hoses
   - Radiator hoses
   - Cooling fan and shroud
   - Radiator overflow tank
   - Radiator
   - Accelerator cable
   - Transmission cable, if equipped
   - Strut tower bar, if equipped
   - Air intake assembly
   - Engine oil dipstick tube
   - Transmission oil dipstick tube, if equipped
   - Ignition coil covers
   - Ignition coil connectors
   - Injector connectors
   - Camshaft Position (CMP) sensor connector
   - Crankshaft Position (CKP) sensor connector
   - Throttle Position (TP) sensor connector
   - Mass Air Flow (MAF) sensor connector

- Idle Air Control (IAC) valve
- Intake manifold ground cable
- Evaporative Emission (EVAP) canister purge valve
- Exhaust Gas Recirculation (EGR) valve connector
- Heated Oxygen (HO_2S) sensor connectors
- Engine Coolant Temperature (ECT) sensor connector
- Alternator wiring connectors
- Oil pressure gauge sender connector
- Power Steering Pressure (PSP) switch connector
- Alternator bracket ground cable
- Brake booster vacuum line
- Tank pressure control vacuum valve hose
- Fuel lines
- EVAP canister
- Power steering pump
- A/C compressor
- Steering shaft lower assembly
- Front differential housing, if equipped
- Exhaust front pipe and bracket
- Transmission oil cooler lines, if equipped
- Transmission stiffener brackets, if equipped
- Flywheel access cover
- Torque converter, if equipped
- Starter motor
- Transmission flange fasteners and support the transmission
- Left and right engine mounts
- Engine

To install:

6. Install or connect the following:
    - Engine
    - Left and right engine mounts. Tighten the nuts to 36 ft. lbs. (50 Nm).
    - Transmission flange fasteners. Tighten them to 58 ft. lbs. (80 Nm).
    - Starter motor
    - Torque converter. Tighten the bolts to 47 ft. lbs. (65 Nm).
    - Flywheel access cover
    - Transmission stiffener brackets. Tighten the bolts to 36 ft. lbs. (50 Nm).
    - Transmission oil cooler lines, if equipped
    - Exhaust front pipe and bracket
    - Front differential housing, if equipped
    - Steering shaft lower assembly
    - A/C compressor
    - Power steering pump
    - EVAP canister
    - Fuel lines
    - Tank pressure control vacuum valve hose
    - Brake booster vacuum line
    - Alternator bracket ground cable
    - PSP switch connector
    - Oil pressure gauge sender connector
    - Alternator wiring connectors
    - ECT sensor connector
    - HO_2S sensor connectors
    - EGR valve connector
    - EVAP canister purge valve
    - Intake manifold ground cable
    - IAC valve
    - MAF sensor connector
    - TP sensor connector
    - CKP sensor connector
    - CMP sensor connector
    - Injector connectors
    - Ignition coil connectors
    - Ignition coil covers
    - Transmission oil dipstick tube, if equipped
    - Engine oil dipstick tube
    - Air intake assembly
    - Strut tower bar, if equipped
    - Transmission cable, if equipped
    - Accelerator cable
    - Radiator
    - Radiator overflow tank
    - Cooling fan and shroud
    - Radiator hoses
    - Heater hoses
    - Hood
    - Negative battery cable
7. Fill the crankcase to the correct level.
8. Fill the cooling system.
9. Start the engine and check for leaks.

Water Pump

REMOVAL & INSTALLATION

1.6L Engines

1. Before servicing the vehicle, refer to the precautions in the beginning of this section.
2. Drain the cooling system.
3. Remove or disconnect the following:
    - Negative battery cable
    - Accessory drive belts
    - Cooling fan and shroud
    - Front cover
    - Timing belt
    - Oil dipstick tube
    - Alternator bracket
    - Timing belt tensioner
    - Water pump

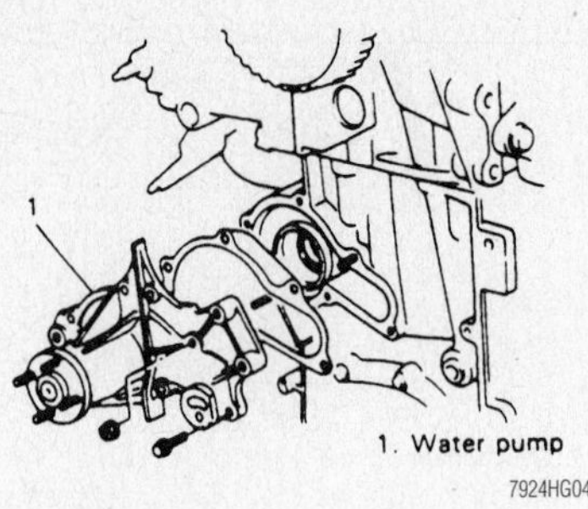

Exploded view of the water pump mounting—1.6L engines

To install:

4. Install or connect the following:
    - Water pump with a new gasket. Tighten the bolts to 106 inch lbs. (12 Nm).
    - Timing belt tensioner
    - Alternator bracket
    - Oil dipstick tube. Tighten the bolt to 97 inch lbs. (11 Nm).
    - Timing belt
    - Front cover
    - Cooling fan and shroud
    - Accessory drive belts
    - Negative battery cable
5. Fill the cooling system.
6. Start the engine and check for leaks.

2.0L Engines

1. Before servicing the vehicle, refer to the precautions in the beginning of this section.
2. Drain the cooling system.
3. Remove or disconnect the following:
    - Negative battery cable
    - Radiator hose at the thermostat housing
    - Heater outlet pipe bolt
    - Alternator belt
    - Water pump

To install:

➡Use new water pump bolts for assembly.

4. Install or connect the following:
    - Water pump with a new O-ring seal. Tighten the bolts to 19 ft. lbs. (27 Nm).
    - Alternator belt
    - Heater outlet pipe bolt
    - Radiator hose at the thermostat housing
    - Negative battery cable
5. Fill the cooling system.
6. Start the engine and check for leaks.

2.5L Engine

1. Before servicing the vehicle, refer to the precautions in the beginning of this section.

2. Drain the cooling system.
3. Remove or disconnect the following:
   - Negative battery cable
   - Accessory drive belts
   - Front cover
   - Water pump

To install:

4. Install or connect the following:
   - Water pump with a new O-ring seal. Tighten the bolts to 19 ft. lbs. (27 Nm).
   - Front cover
   - Accessory drive belts
   - Negative battery cable
5. Fill the cooling system.
6. Start the engine and check for leaks.

Heater Core

REMOVAL & INSTALLATION

1. Disconnect the negative battery cable.
2. To disable the air bag system, perform the following procedure:
   a. Position the front wheels so that they are pointing straight ahead.
   b. Turn the ignition switch to the LOCK position.
   c. In the fuse box, remove the AIR BAG fuse.
   d. Under the steering column, locate the contact coil/combination switch assembly's yellow connector; then, unlock and disconnect the connector.
   e. Pull outward on the glove box while pushing the stopper located at both sides and locate the passenger's side air bag module yellow connector; then, unlock and disconnect the connector.

➡With the AIR BAG fuse removed and the ignition switch turned ON; the air bag warning light may be ON; this is normal operation and does not indicate an air bag malfunction.

3. Drain the cooling system into a clean container for reuse.
4. Disconnect the heater hoses from the heater core.
5. Remove the instrument panel by performing the following procedure:
   a. Remove the console.
   b. Remove the glove box and the column hole cover.
   c. Disconnect the electrical connector and cables from the heater housing and blower motor assembly.
   d. Remove the steering column.
   e. Disconnect the speedometer connector and the speedometer assembly.
   f. Remove the hood opener.
   g. Disconnect the instrument panel electrical connectors.
   h. Remove the instrument panel-to-chassis screws and bolts.
   i. Using an assistant, remove the instrument panel.
6. If equipped with air conditioning, perform the following procedure:
   a. Discharge and recover the air conditioning refrigerant.
   b. If equipped with a G16 or a J20 engine, disconnect the suction pipe and liquid pipe from the air conditioning housing. Plug the openings to prevent contamination.
   c. If equipped with an H25 engine, disconnect the compressor suction pipe and receiver/drier outlet pipe from the air conditioning housing. Plug the openings to prevent contamination.
   d. Remove the blower motor assembly.
   e. Disconnect the thermistor wire coupler.
   f. Remove the air conditioning housing.
7. Disconnect the rear duct from the heater housing.
8. Disconnect the mode actuator electrical connectors.
9. If equipped, remove the air conditioning controller.
10. If equipped, disconnect and remove the Sensing and Diagnostic Module (SDM) or air bag controller module.
11. Remove the heater housing.
12. Remove the heater core pipe clamps and grommet.
13. Remove the heater core from the heater housing.

To install:

14. Install the heater core to the heater housing.
15. Install the heater core pipe clamps and grommet.
16. Install the heater housing.

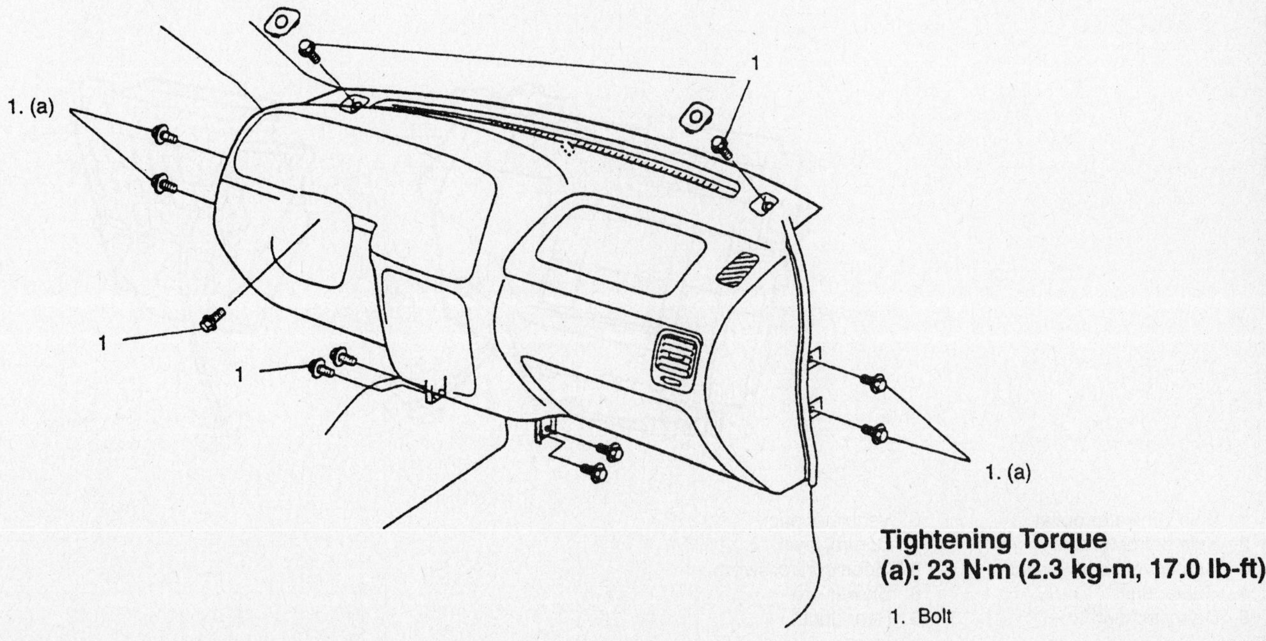

View of the instrument panel and fasteners—Suzuki Vitara

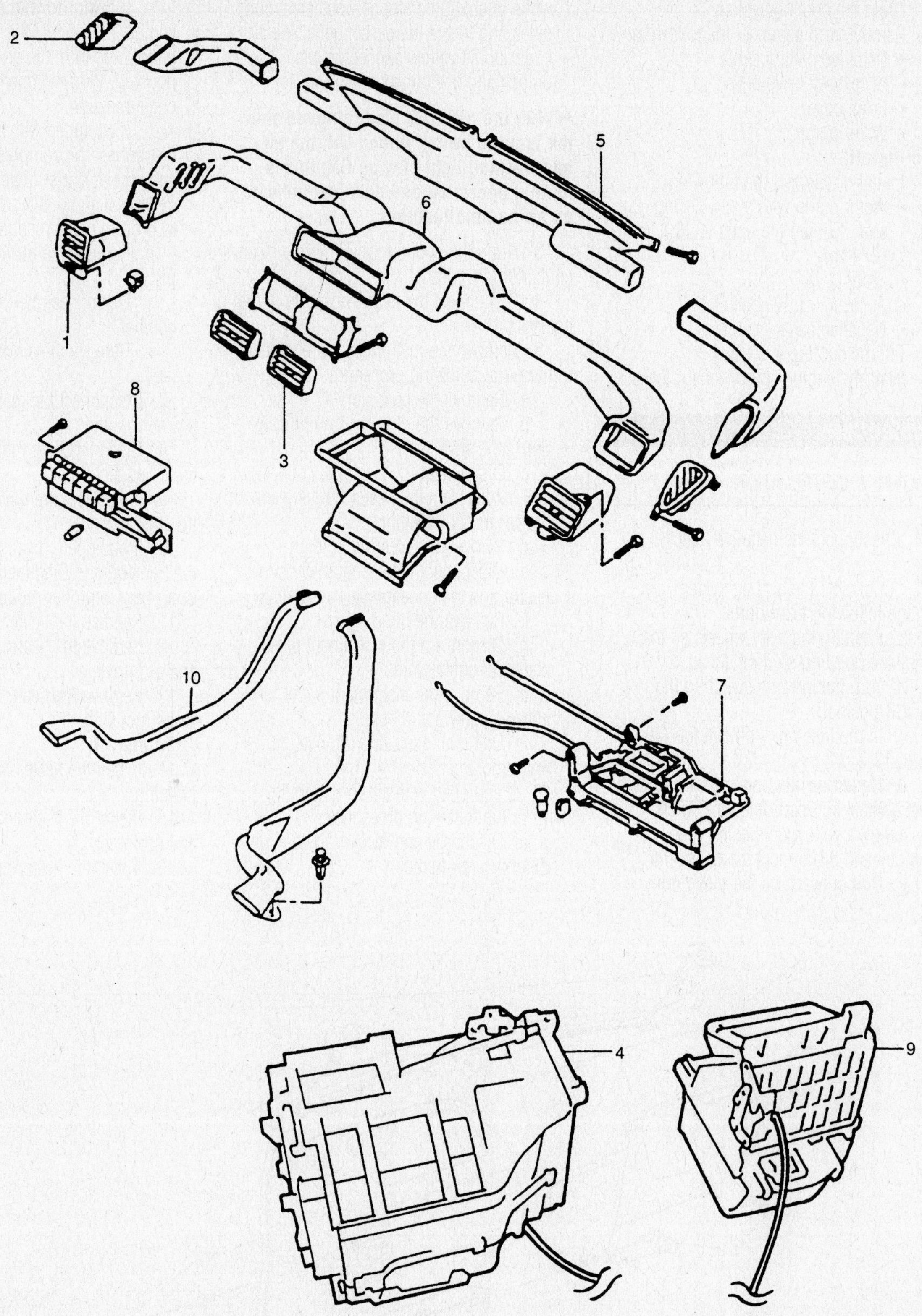

1. Side ventilator outlet
2. Side defroster outlet
3. Center ventilatior outlet
4. Heater unit
5. Defroster duct
6. Ventilator duct
7. Control lever
8. Mode control switch
9. Blower unit
10. Rear duct

93113GE9

Exploded view of the heater housing and ventilation ducts—Suzuki Vitara

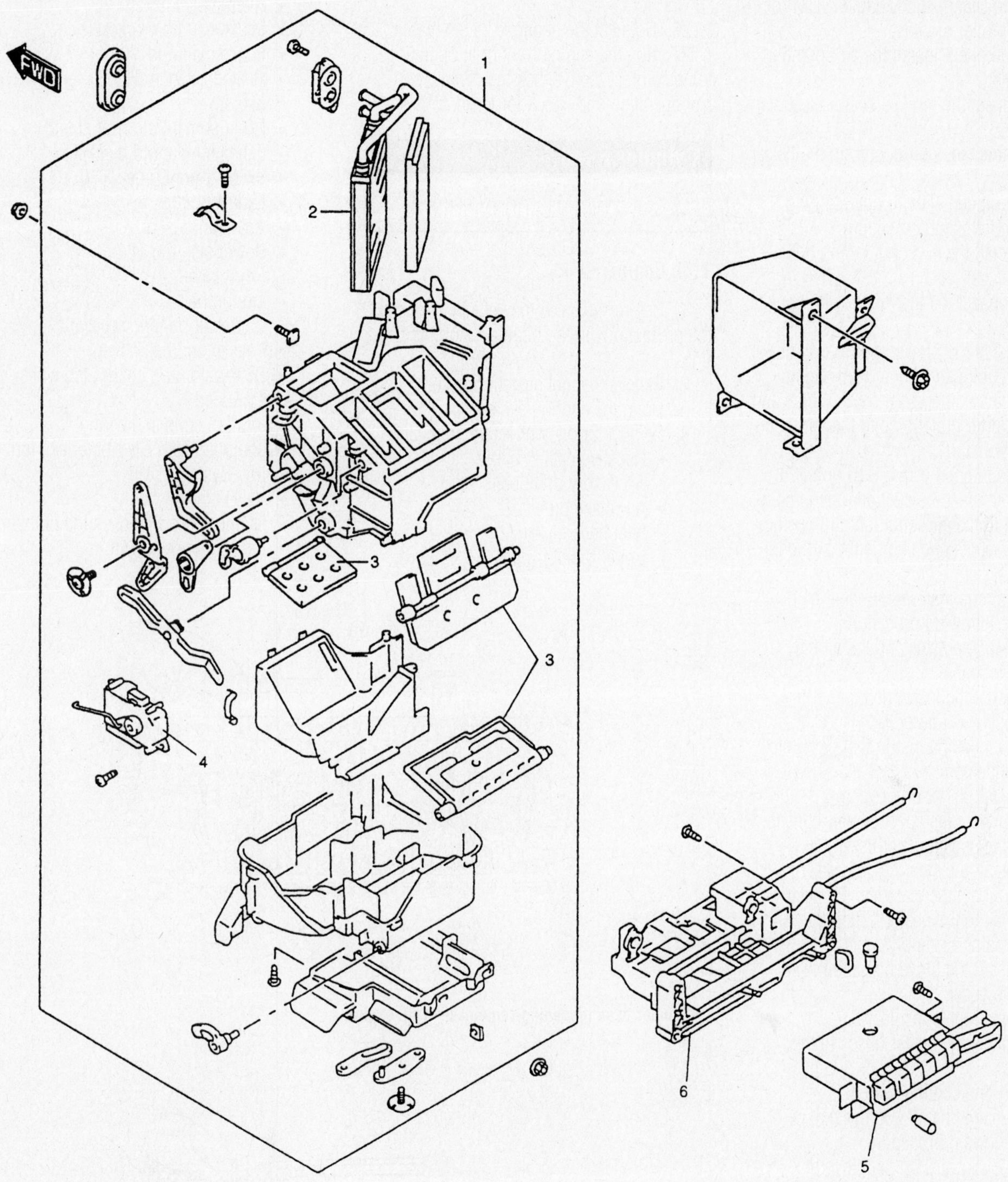

1. Heater assembly
2. Heater core
3. Damper
4. Mode actuator
5. Mode control switch
6. Control lever assembly

93113GE0

Exploded view of the heater core, heater housing and related components—Suzuki Vitara

17. If equipped, connect and install the Sensing and Diagnostic Module (SDM) or air bag controller module.

18. If equipped, install the air conditioning controller.

19. Connect the mode actuator electrical connectors.

20. Connect the rear duct from the heater housing.

21. If equipped with air conditioning, perform the following procedure:

a. Install the air conditioning housing.

b. Connect the thermistor wire coupler.

c. Install the blower motor assembly.

d. If equipped with an H25 engine, connect the compressor suction pipe and receiver/drier outlet pipe to the air conditioning housing.

e. If equipped with a G16 or a J20 engine, connect the suction pipe and liquid pipe to the air conditioning housing.

f. Evacuate and charge the air conditioning system.

22. Install the instrument panel by performing the following procedure:

a. Using an assistant, install the instrument panel.

b. Install the instrument panel-to-chassis screws and bolts.

c. Connect the instrument panel electrical connectors.

d. Install the hood opener.

e. Connect the speedometer connector and the speedometer assembly.

f. Install the steering column.

g. Connect the electrical connector and cables to the heater housing and blower motor assembly.

h. Install the glove box and the column hole cover.

i. Install the console.

23. Connect the heater hoses to the heater core.

24. Refill the cooling system.

25. To enable the air bag system, perform the following procedure:

a. Push inward on the glove box while pushing the stopper located at both sides and connect the passenger's side air bag module yellow connector and lock it.

b. Under the steering column, connect the contact coil/combination switch assembly's yellow connector.

c. In the fuse box, install the AIR BAG fuse.

d. Turn the ignition switch ON and verify that the AIR BAG warning light flashes 7 times and turns OFF; if the system does not operate as described, perform the Air Bad Diagnostic System Check.

26. Connect the negative battery cable.

27. Run the engine to normal operating temperatures; then, check the climate control operation and check for leaks.

Cylinder Head

REMOVAL & INSTALLATION

1.6L Engine

1. Before servicing the vehicle, refer to the precautions in the beginning of this section.
2. Relieve the fuel system pressure.
3. Drain the cooling system.
4. Remove or disconnect the following:
   - Negative battery cable
   - Accessory drive belts
   - Air intake pipe
   - Fuel lines
   - Upper radiator hose
   - Coolant bypass hose
   - Alternator bracket
   - Intake manifold brackets
   - Intake manifold
   - Heated Oxygen (HO_2S) sensor connectors
   - Exhaust manifold heat shield
   - Exhaust manifold bracket
   - Exhaust front pipe
   - Exhaust manifold
   - A/C compressor
   - Power steering pump
   - Front cover
   - Timing belt
   - Camshaft timing sprocket
   - Rear timing belt cover
   - Distributor and spark plug wires, if equipped
   - Ignition coils and wiring, if equipped with Distributorless Ignition System (DIS)
   - Valve cover
   - Cylinder head. Loosen the bolts in the sequence shown.

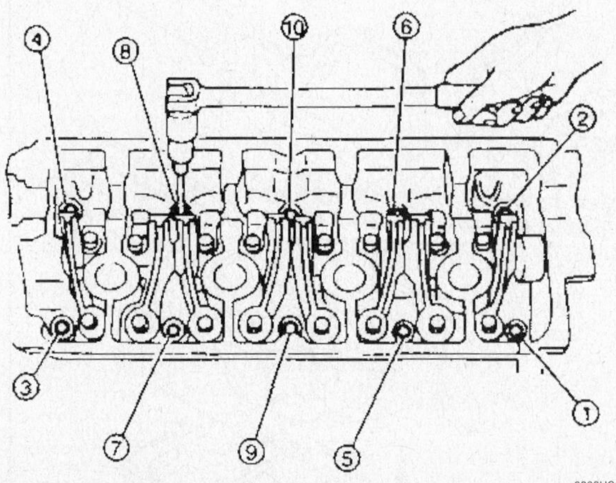

Cylinder head loosening sequence—1.6L engine

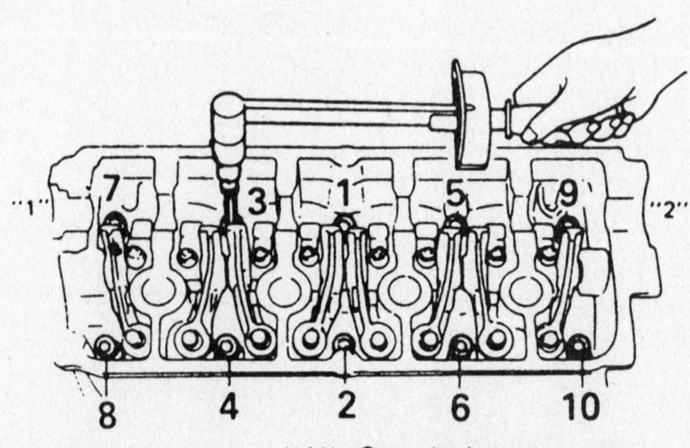

Cylinder head torque sequence—1.6L engine

To install:

5. Install the cylinder head with a new gasket.
6. Tighten the bolts in sequence as follows:
   a. Step 1: 26 ft. lbs. (35 Nm)
   b. Step 2: 41 ft. lbs. (55 Nm)
   c. Step 3: Loosen all bolts to 0 ft. lbs. (0 Nm)
   d. Step 4: 26 ft. lbs. (35 Nm)
   e. Step 5: 52 ft. lbs. (70 Nm)
7. Install or connect the following:
   - Valve cover
   - Ignition coils and wiring, if equipped with DIS
   - Distributor and spark plug wires, if equipped
   - Rear timing belt cover
   - Camshaft timing sprocket
   - Timing belt
   - Front cover
   - Power steering pump
   - A/C compressor
   - Exhaust manifold
   - Exhaust front pipe
   - Exhaust manifold bracket
   - Exhaust manifold heat shield
   - HO_2S sensor connectors
   - Intake manifold
   - Intake manifold brackets
   - Alternator bracket
   - Coolant bypass hose
   - Upper radiator hose
   - Fuel lines
   - Air intake pipe
   - Accessory drive belts
   - Negative battery cable
8. Fill the cooling system.
9. Start the engine and check for leaks.

2.0L Engines

1. Before servicing the vehicle, refer to the precautions in the beginning of this section.
2. Relieve the fuel system pressure.
3. Drain the cooling system.
4. Drain the engine oil.
5. Remove or disconnect the following:
   - Negative battery cable
   - Strut tower brace
   - Air intake tube
   - Exhaust Gas Recirculation (EGR) valve connector
   - Idle Air Control (IAC) valve connector
   - Throttle Position (TP) sensor connector
   - Evaporative Emission (EVAP) canister purge valve connector and hose
   - Intake manifold ground cable

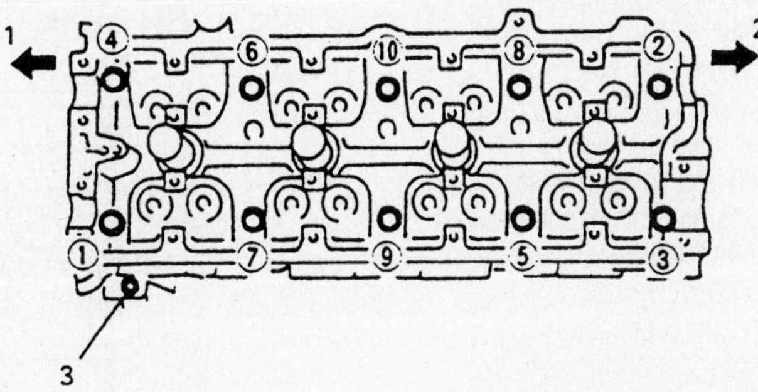

Cylinder head loosening sequence—2.0L engines

   - Heated Oxygen (HO_2S) sensor connectors
   - Camshaft Position (CMP) sensor connector
   - Engine Coolant Temperature (ECT) sensor connector
   - Fuel injector connectors
   - Ignition coils
   - Accelerator cable
   - Transmission cable, if equipped
   - Brake booster vacuum hose
   - Radiator hose
   - Bypass hose
   - Heater hose
   - Fuel lines
   - Intake manifold bracket
   - Water pipe
   - Valve cover
   - Accessory drive belts
   - Oil pan
   - Front cover
   - Timing chains
   - Camshafts
   - Exhaust front pipe
   - Exhaust manifold bracket
   - Cylinder head. Loosen the bolts in the sequence shown.

To install:

6. Install the cylinder head with a new gasket.
7. Tighten the bolts in sequence as follows:
   a. Step 1: 38 ft. lbs. (52 Nm)
   b. Step 2: 61 ft. lbs. (84 Nm)
   c. Step 3: Loosen all bolts to 0 ft. lbs. (0 Nm)
   d. Step 4: 38 ft. lbs. (52 Nm)
   e. Step 5: 76 ft. lbs. (105 Nm)
   f. Step 6: 6mm bolt to 96 inch lbs. (8 Nm)
8. Install or connect the following:
   - Exhaust manifold bracket
   - Exhaust front pipe
   - Camshafts
   - Timing chains
   - Front cover
   - Oil pan
   - Accessory drive belts
   - Valve cover
   - Water pipe

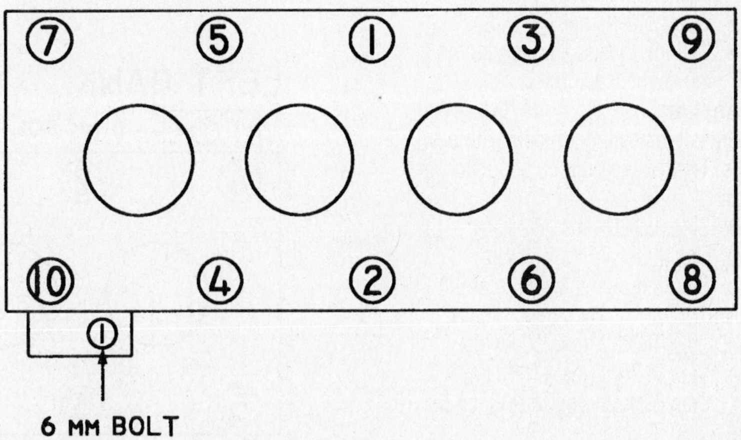

Cylinder head torque sequence—2.0L engines

- Intake manifold bracket
- Fuel lines
- Heater hose
- Bypass hose
- Radiator hose
- Brake booster vacuum hose
- Transmission cable, if equipped
- Accelerator cable
- Ignition coils
- Fuel injector connectors
- ECT sensor connector
- CMP sensor connector
- HO_2S sensor connectors
- Intake manifold ground cable
- EVAP canister purge valve connector and hose
- TP sensor connector
- IAC valve connector
- EGR valve connector
- Air intake tube
- Strut tower brace
- Negative battery cable

9. Fill the crankcase to the correct level.
10. Fill the cooling system.
11. Start the engine and check for leaks.

2.5L Engine

1. Before servicing the vehicle, refer to the precautions in the beginning of this section.
2. Relieve the fuel system pressure.
3. Drain the cooling system and engine oil.
4. Remove or disconnect the following:
   - Negative battery cable
   - Intake manifold
   - Ignition coil covers and ignition coils
   - Valve covers
   - Oil pan
   - Timing chain cover and timing chains
   - Camshaft Position (CMP) sensor
   - Camshafts
   - Exhaust manifolds
   - Water outlet caps
   - Cylinder heads. Loosen the bolts in the sequence shown.

To install:

5. Install the cylinder heads with new gaskets. Tighten the bolts in sequence as follows:
   a. Step 1: 38 ft. lbs. (52 Nm)
   b. Step 2: 61 ft. lbs. (84 Nm)
   c. Step 3: Loosen all bolts to 0 ft. lbs. (0 Nm)
   d. Step 4: 38 ft. lbs. (52 Nm)
   e. Step 5: 76 ft. lbs. (105 Nm)
   f. Step 6: 6mm bolt to 96 inch lbs. (8 Nm)
6. Install or connect the following:

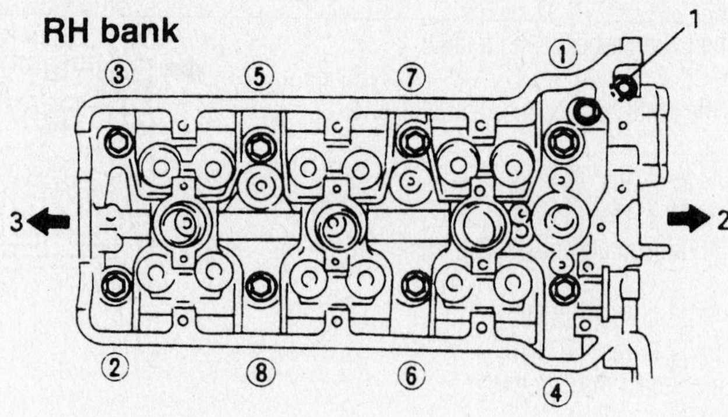

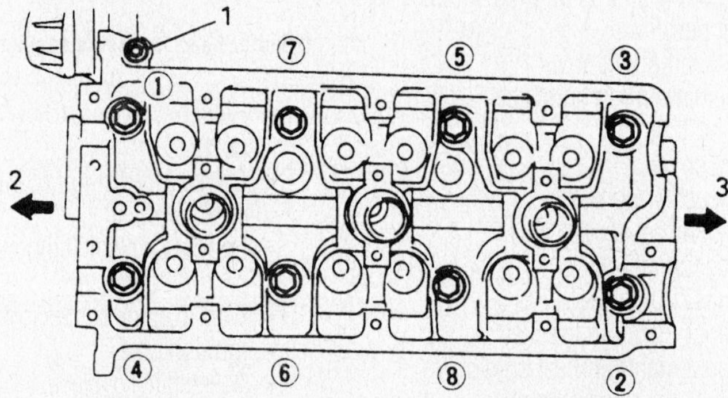

Cylinder head loosening sequence—2.5L engine

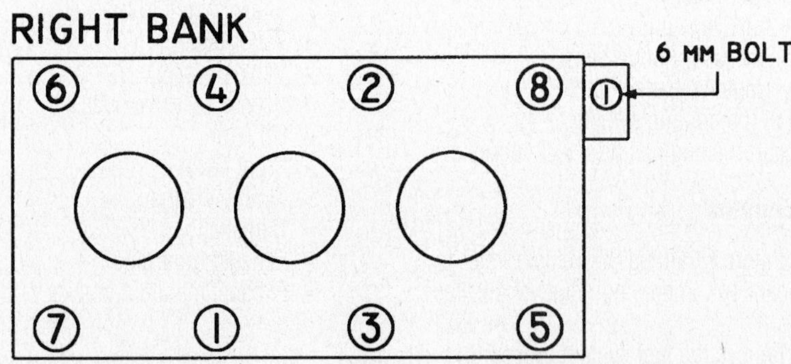

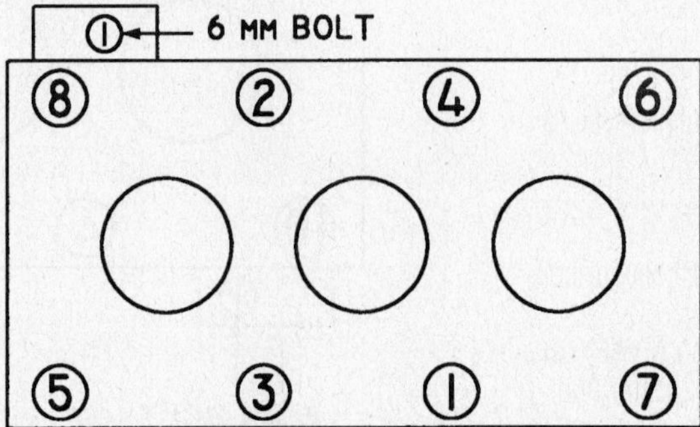

Cylinder head torque sequence—2.5L engine

- Water outlet caps
- Exhaust manifolds
- Camshafts
- CMP sensor
- Timing chain cover and timing chains
- Oil pan
- Valve covers
- Ignition coil covers and ignition coils
- Intake manifold
- Negative battery cable

7. Fill the crankcase to the correct level
8. Fill the cooling system.
9. Start the engine and check for leaks.

Rocker Arms/Shafts

REMOVAL & INSTALLATION

1.6L Engine

1. Before servicing the vehicle, refer to the precautions in the beginning of this section.
2. Drain the cooling system.
3. Remove or disconnect the following:
   - Negative battery cable
   - Accessory drive belts
   - Cooling fan and shroud
   - Radiator and hoses
   - Distributor, if equipped
   - Camshaft Position (CMP) sensor, if equipped
   - Valve cover
   - Front cover
   - Timing belt
   - Camshaft
   - Rocker arm shaft plug
   - Rear timing belt cover
4. Loosen the rocker arm locknuts and back the valve adjusters off until all rocker arms move freely with no tension.

➡Keep all valvetrain components in order for assembly.

5. Remove or disconnect the following:
   - Intake rocker arms and clips
   - Rocker arm shaft bolts
6. Push the rocker arm shaft towards the rear of the cylinder head and remove the rocker arm shaft O-ring.
7. Remove the exhaust rocker arms and springs by pulling the rocker arm shaft out of the front of the cylinder head.

To install:

8. Insert the rocker arm shaft into the front of the cylinder head, while installing the exhaust rocker arms and springs in their original positions.
9. Push the end of the rocker arm shaft out of the rear of the cylinder head and install a new O-ring.
10. Install or connect the following:
    - Rocker arm shaft bolts. Tighten them to 96 inch lbs. (8 Nm).
    - Intake rocker arms and clips in their original positions
    - Rear timing belt cover
    - Rocker arm shaft plug. Tighten it to 24 ft. lbs. (33 Nm).
    - Camshaft
    - Timing belt and adjust the valve clearance
    - Valve cover
    - Front cover
    - Distributor, if equipped
    - CMP sensor, if equipped
    - Radiator and hoses
    - Cooling fan and shroud
    - Accessory drive belts
    - Negative battery cable
11. Fill the cooling system.
12. Start the engine and check for leaks.

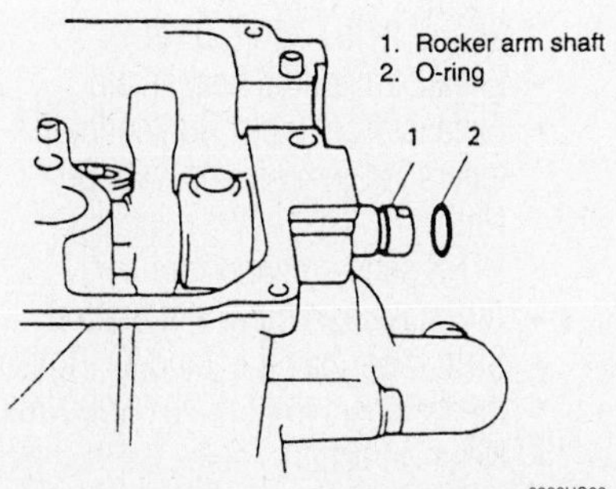

Rocker arm shaft and O-ring—1.6L engine

2.0L and 2.5L Engines

The 2.0L and 2.5L engines do not utilize rocker arms or rocker arm shafts.

Intake Manifold

REMOVAL & INSTALLATION

1.6L Engine

1. Before servicing the vehicle, refer to the precautions in the beginning of this section.
2. Relieve the fuel system pressure.
3. Drain the cooling system.
4. Remove or disconnect the following:
   - Negative battery cable
   - Air intake pipe
   - Accelerator cable and bracket
   - Transmission cable, if equipped
   - Throttle Position (TP) sensor connector
   - Idle Air Control (IAC) valve connector and hoses
   - Engine Coolant Temperature (ECT) sensor connector
   - Coolant temperature gauge sender connector
   - A/C coolant temperature switch connector, if equipped
   - Evaporative Emission (EVAP) canister purge valve connector and vacuum line
   - Exhaust Gas Recirculation (EGR) temperature sensor connector
   - EGR vacuum valve connector
   - EGR bypass valve connector
   - EGR vacuum lines
   - Fuel injector connectors
   - Intake manifold ground cable
   - Transmission vacuum line, if equipped
   - Brake booster vacuum line
   - Manifold Absolute Pressure (MAP) sensor vacuum line
   - Fuel lines
   - Upper radiator hose
   - Coolant bypass hose
   - Alternator bracket
   - Intake manifold brackets
   - Intake manifold

To install:

5. Install or connect the following:
   - Intake manifold with a new gasket. Tighten the nuts to 17 ft. lbs. (23 Nm).
   - Intake manifold brackets. Tighten the fasteners to 36 ft. lbs. (50 Nm).
   - Alternator bracket. Tighten the fasteners to 36 ft. lbs. (50 Nm).
   - Coolant bypass hose
   - Upper radiator hose
   - Fuel lines
   - MAP sensor vacuum line
   - Brake booster vacuum line
   - Transmission vacuum line, if equipped
   - Intake manifold ground cable
   - Fuel injector connectors
   - EGR vacuum lines
   - EGR bypass valve connector
   - EGR vacuum valve connector
   - EGR temperature sensor connector
   - EVAP canister purge valve connector and vacuum line
   - A/C coolant temperature switch connector, if equipped
   - Coolant temperature gauge sender connector
   - ECT sensor connector
   - IAC valve connector and hoses
   - TP sensor connector
   - Transmission cable, if equipped
   - Accelerator cable and bracket
   - Air intake pipe
   - Negative battery cable

6. Fill the cooling system.
7. Start the engine and check for leaks.

2.0L Engines

1. Before servicing the vehicle, refer to the precautions in the beginning of this section.
2. Relieve the fuel system pressure.
3. Drain the cooling system.
4. Remove or disconnect the following:
   - Negative battery cable
   - Air intake tube
   - Exhaust Gas Recirculation (EGR) valve connector
   - Idle Air Control (IAC) valve connector
   - Throttle Position (TP) sensor connector
   - Evaporative Emissions (EVAP) canister purge valve connector and hose
   - Intake manifold ground cable
   - Manifold Absolute Pressure (MAP) sensor connector
   - Accelerator cable
   - Transmission cable, if equipped
   - Brake booster vacuum hose
   - Positive Crankcase Ventilation (PCV) valve and hose
   - Fuel pressure regulator vacuum hose
   - Intake manifold vacuum hose
   - Throttle body coolant hoses
   - Water bypass pipe
   - Fuel lines
   - Fuel supply manifold with injectors attached
   - Intake manifold support brackets
   - Intake manifold water pipe
   - Intake manifold

To install:

5. Install or connect the following:
   - Intake manifold with a new gasket. Tighten the fasteners to 17 ft. lbs. (23 Nm).
   - Intake manifold water pipe
   - Intake manifold front support bracket. Tighten the bolts to 36 ft. lbs. (50 Nm).
   - Intake manifold rear support bracket. Tighten the bolts to 18 ft. lbs. (25 Nm).
   - Fuel supply manifold with injectors attached
   - Fuel lines
   - Water bypass pipe
   - Throttle body coolant hoses
   - Intake manifold vacuum hose
   - Fuel pressure regulator vacuum hose
   - PCV valve and hose
   - Brake booster vacuum hose
   - Transmission cable, if equipped
   - Accelerator cable
   - MAP sensor connector
   - Intake manifold ground cable
   - EVAP canister purge valve connector and hose
   - TP sensor connector
   - IAC valve connector
   - EGR valve connector
   - Air intake tube
   - Negative battery cable
6. Fill the cooling system.
7. Start the engine and check for leaks.

2.5L Engine

1. Before servicing the vehicle, refer to the precautions in the beginning of this section.
2. Relieve the fuel system pressure.
3. Drain the cooling system.
4. Remove or disconnect the following:
   - Negative battery cable
   - Strut tower bar
   - Intake Air Temperature (IAT) sensor connector
   - Surge tank cover
   - Air intake assembly
   - Accelerator cable
   - Transmission cable, if equipped
   - Throttle body coolant hoses
   - Fuel injector connectors
   - Throttle Position (TP) sensor connector
   - Mass Air Flow (MAF) sensor connector
   - Idle Air Control (IAC) valve connector
   - Intake manifold ground cables
   - Brake booster vacuum hose
   - Evaporative Emissions (EVAP) canister purge valve connector and hoses
   - Exhaust Gas Recirculation (EGR) valve connector
   - Positive Crankcase Ventilation (PCV) valve and hose
   - Heater hoses
   - EGR pipe
   - Fuel lines
   - Throttle body and intake collector
   - Intake manifold

To install:

5. Install or connect the following:
   - Intake manifold with new gaskets. Tighten the fasteners to 16 ft. lbs. (23 Nm).
   - Throttle body and intake collector with new gaskets. Tighten the fasteners to 102 inch lbs. (12 Nm).
   - Fuel lines
   - EGR pipe
   - Heater hoses
   - PCV valve and hose
   - EGR valve connector
   - EVAP canister purge valve connector and hoses
   - Brake booster vacuum hose
   - Intake manifold ground cables
   - IAC valve connector
   - MAF sensor connector
   - TP sensor connector
   - Fuel injector connectors
   - Throttle body coolant hoses
   - Transmission cable, if equipped
   - Accelerator cable
   - Air intake assembly
   - Surge tank cover
   - IAT sensor connector
   - Strut tower bar
   - Negative battery cable
6. Fill the cooling system.
7. Start the engine and check for leaks.

Exhaust Manifold

REMOVAL & INSTALLATION

1.6L and 2.0L Engines

1. Before servicing the vehicle, refer to the precautions in the beginning of this section.
2. Remove or disconnect the following:
   - Negative battery cable
   - Strut tower bar, if equipped
   - Air intake assembly and bracket
   - Heated Oxygen (HO_2S) sensor connector
   - Exhaust front pipe
   - Exhaust manifold heat shield
   - Exhaust manifold bracket, if equipped
   - Exhaust manifold

To install:

3. Install or connect the following:
   - Exhaust manifold with a new gasket. Tighten the fasteners to 13–20 ft. lbs. (18–28 Nm).
   - Exhaust manifold bracket, if equipped. Tighten the bolts to 36–43 ft. lbs. (50–60 Nm).
   - Exhaust manifold heat shield
   - Exhaust front pipe. Tighten the fasteners to 29–43 ft. lbs. (40–60 Nm).
   - HO_2S sensor connector
   - Air intake assembly and bracket
   - Strut tower bar, if equipped. Tighten the fasteners to 66 ft. lbs. (90 Nm).
   - Negative battery cable
4. Start the engine and check for leaks.

2.5L Engine

1. Before servicing the vehicle, refer to the precautions in the beginning of this section.
2. Remove or disconnect the following:
   - Negative battery cable
   - Strut tower bar
   - Air intake assembly
   - Heated Oxygen (HO_2S) sensor connectors
   - Oil dipstick tube
   - Exhaust Gas Recirculation (EGR) pipe
   - Exhaust manifold heat shields
   - Evaporative Emissions (EVAP) canister
   - Front driveshaft, if equipped
   - Exhaust front pipe
   - Exhaust manifold brace
   - Exhaust manifolds

To install:

3. Install or connect the following:
   - Exhaust manifolds with new gaskets. Tighten the nuts to 21 ft. lbs. (30 Nm).
   - Exhaust manifold brace
   - Exhaust front pipe. Tighten the fasteners to 37 ft. lbs. (50 Nm).
   - Front driveshaft, if equipped
   - EVAP canister
   - Exhaust manifold heat shields
   - EGR pipe
   - Oil dipstick tube
   - HO_2S sensor connectors
   - Air intake assembly
   - Strut tower bar
   - Negative battery cable
4. Start the engine and check for leaks.

Front Crankshaft Seal

REMOVAL & INSTALLATION

1.6L Engine

1. Before servicing the vehicle, refer to the precautions in the beginning of this section.
2. Drain the cooling system.
3. Remove or disconnect the following:
   - Negative battery cable
   - Accessory drive belts
   - Cooling fan and shroud
   - Water pump pulley
   - Crankshaft pulley
   - Front cover
   - Timing belt
   - Crankshaft timing sprocket
   - Front crankshaft seal

To install:

4. Install or connect the following:

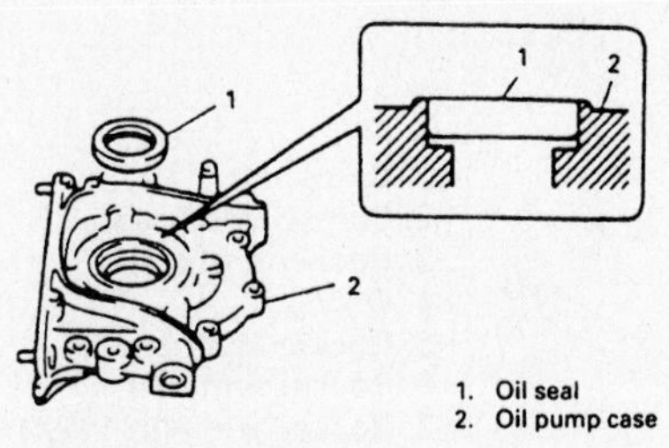

Install the new oil pump seal flush with the oil pump housing—1.6L engine

   - Front crankshaft seal flush with the oil pump housing
   - Crankshaft timing sprocket. Tighten the bolt to 94 ft. lbs. (128 Nm).
   - Timing belt
   - Front cover
   - Crankshaft pulley. Tighten the bolts to 12 ft. lbs. (16 Nm).
   - Water pump pulley
   - Cooling fan and shroud
   - Accessory drive belts
   - Negative battery cable
5. Start the engine and check for leaks.

Camshaft and Valve Lifters

REMOVAL & INSTALLATION

1.6L Engine

1. Before servicing the vehicle, refer to the precautions in the beginning of this section.
2. Drain the cooling system.
3. Remove or disconnect the following:
   - Negative battery cable
   - Radiator
   - Accessory drive belts
   - Crankshaft pulley
   - Front cover
   - Timing belt
   - Camshaft sprocket
   - Valve cover
   - Distributor and case, if equipped
   - Camshaft Position (CMP) sensor and case, if equipped
4. Loosen the rocker arm locknuts and back the valve adjusters off until all rocker arms move freely with no tension.
5. Remove or disconnect the following:
   - Camshaft bearing caps. Loosen the bolts in reverse of the tightening sequence.
   - Camshaft

To install:

6. Install or connect the following:
   - Camshaft
   - Camshaft bearing caps. Tighten the bolts in sequence to 96 inch lbs. (11 Nm).
   - CMP sensor and case, if equipped
   - Distributor and case, if equipped
   - Camshaft sprocket. Tighten the bolt to 44 ft. lbs. (60 Nm).
   - Timing belt and adjust the valve clearance
   - Valve cover
   - Front cover
   - Crankshaft pulley. Tighten the bolts to 12 ft. lbs. (16 Nm).
   - Accessory drive belts
   - Radiator
   - Negative battery cable
7. Fill the cooling system.
8. Start the engine and check for leaks.

2.0L Engines

1. Before servicing the vehicle, refer to the precautions in the beginning of this section.
2. Drain the engine oil.
3. Drain the cooling system.
4. Remove or disconnect the following:
   - Negative battery cable
   - Oil pan

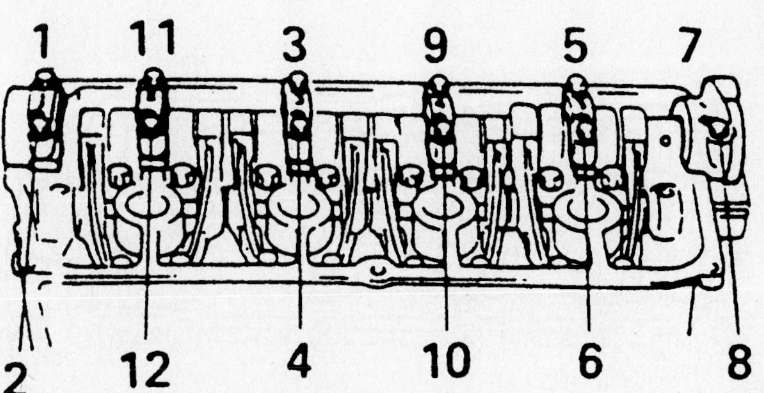

Camshaft housing torque sequence—1.6L engine

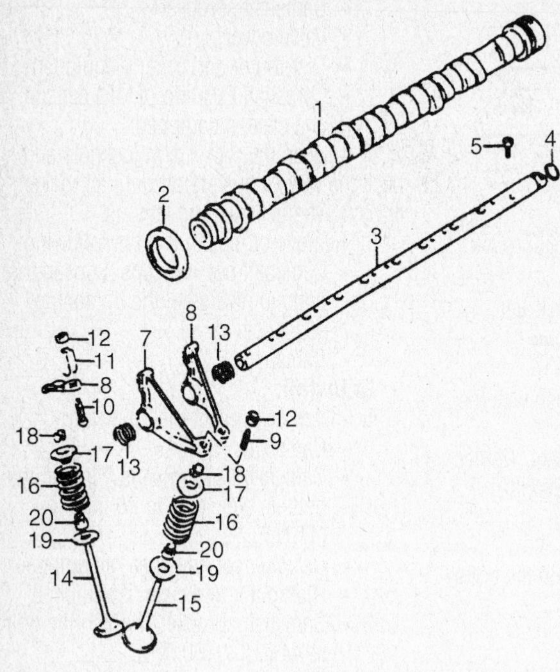

1. Camshaft
2. Camshaft oil seal
3. Rocker arm shaft
4. O ring
5. Rocker shaft bolt
6. Rocker arm (IN)
7. Rocker arm No. 1 (EX)
8. Rocker arm No. 2 (EX)
9. Valve adjusting screw
10. Valve adjusting screw
11. Clip
12. Lock nut
13. Rocker arm spring
14. Intake valve
15. Exhaust valve
16. Valve spring
17. Valve spring retainer
18. Valve cotter
19. Valve spring seat
20. Valve stem seal

7924HG24

Exploded view of the valve train components—1.6L engine

- Valve cover
- Accessory drive belts
- Crankshaft pulley
- Front cover
- Secondary timing chain
- Camshaft Position (CMP) sensor

➡Keep all valvetrain components in order for installation.

- Camshaft bearing caps. Loosen the bolts in several steps in reverse of the tightening sequence.
- Camshafts
- Hydraulic lash adjusters

To install:

5. Install or connect the following:
   - Hydraulic lash adjusters in their original positions
   - Camshafts
   - Camshaft bearing caps in their original positions. Tighten the bolts in several steps in sequence to 96 inch lbs. (11 Nm).
   - CMP sensor
   - Secondary timing chain
   - Front cover
   - Crankshaft pulley. Tighten the bolt to 109 ft. lbs. (148 Nm).
   - Accessory drive belts
   - Valve cover
   - Oil pan
   - Negative battery cable
6. Fill the crankcase to the correct level.
7. Fill the cooling system.
8. Start the engine and check for leaks.

⁂ WARNING

Wait ½ hour after installing the lash adjusters and camshafts before cranking or starting the engine to allow the lash adjusters to bleed down. Operating the engine before this time period may result in interference between the valves and pistons.

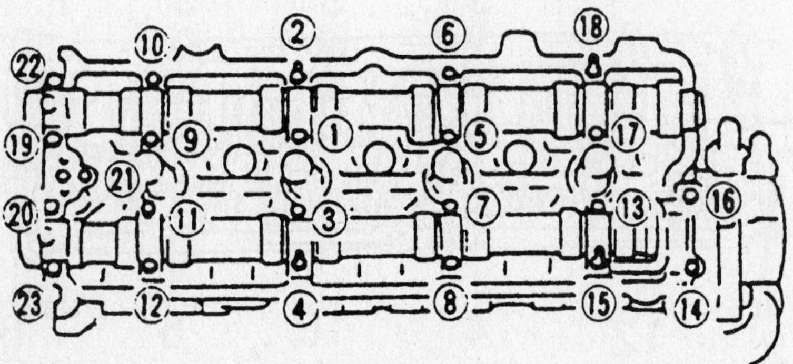

7924HG26

Camshaft housing torque sequence—2.0L engines

2.5L Engine

1. Before servicing the vehicle, refer to the precautions in the beginning of this section.
2. Drain the engine oil.
3. Drain the cooling system.
4. Remove or disconnect the following:
   - Negative battery cable
   - Intake manifold
   - Oil pan
   - Accessory drive belts
   - Water pump pulley
   - Crankshaft pulley
   - Timing chain cover
5. Align the timing marks as shown.

⁂ WARNING

Do not allow the crankshaft or camshafts to rotate once the timing chains have been removed. Valve or piston damage could result.

6. Remove or disconnect the following:
   - Left bank secondary timing chain
   - Primary timing chain
   - Valve covers

➡Keep all valvetrain components in order for assembly.

- Right bank camshaft bearing caps. Loosen the bolts in several steps and in the sequence shown.

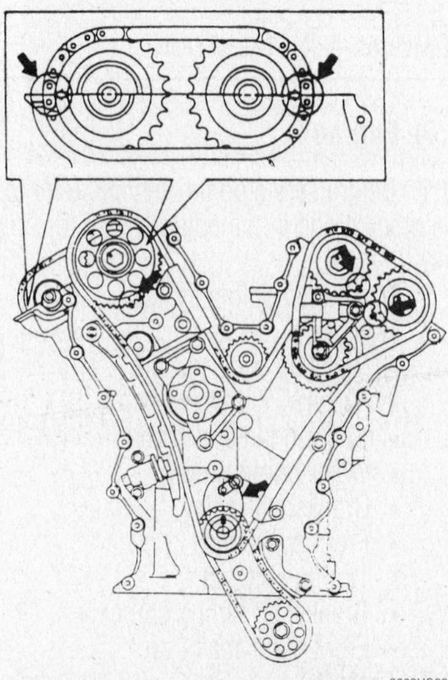

9302HG02

Timing mark alignment—2.5L engine

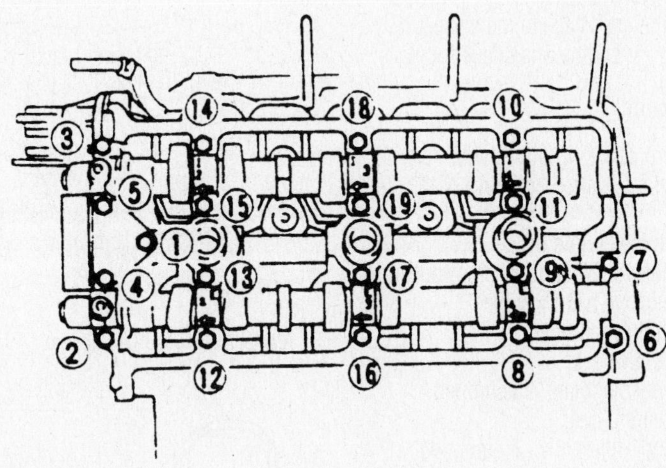

Left bank camshaft housing loosening sequence—2.5L engine

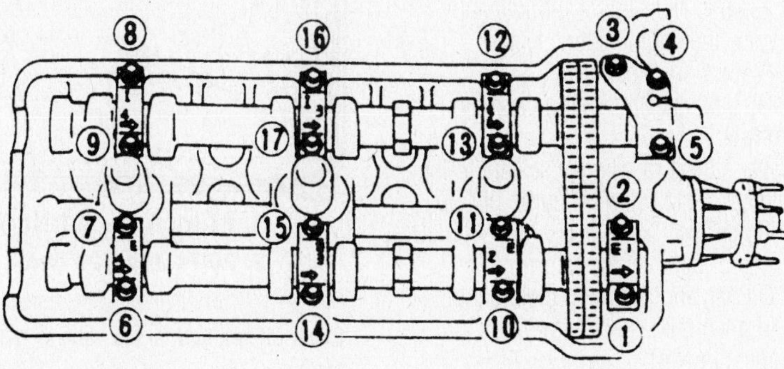

Right bank camshaft housing loosening sequence—2.5L engine

- Right bank secondary timing chain, exhaust and intake camshafts as an assembly
- Camshaft Position (CMP) sensor
- Left bank camshaft bearing caps. Loosen the bolts in several steps and in the sequence shown.
- Left bank camshafts
- Hydraulic lash adjusters

To install:

7. Install or connect the following:
   - Hydraulic lash adjusters in their original positions
   - Left bank camshafts
   - Left bank camshaft bearing caps. Tighten the bolts in several steps and in reverse of the loosening sequence to 102 inch lbs. (12 Nm).
   - CMP sensor
   - Right bank secondary timing chain, exhaust and intake camshafts as an assembly
   - Right bank camshaft bearing caps. Tighten the bolts in several steps and in reverse of the loosening sequence to 102 inch lbs. (12 Nm).

**** WARNING**

Wait ½ hour after installing the lash adjusters and camshafts before cranking or starting the engine to allow the lash adjusters to bleed down. Operating the engine before this time period may result in interference between the valves and pistons.

   - Valve covers. Tighten the bolts to 90 inch lbs. (10.5 Nm).
   - Primary timing chain
   - Left bank secondary timing chain
   - Timing chain cover
   - Crankshaft pulley. Tighten the bolt to 109 ft. lbs. (148 Nm).
   - Water pump pulley
   - Accessory drive belts
   - Oil pan
   - Intake manifold
   - Negative battery cable
8. Fill the crankcase to the correct level.
9. Fill the cooling system.
10. Start the engine and check for leaks.

Valve Lash

ADJUSTMENT

1.6L Engines

➡Measure valve clearance with the engine cold.

1. Before servicing the vehicle, refer to the precautions in the beginning of this section.
2. Remove the valve cover.
3. Set the engine to Top Dead Center (TDC) of the compression stroke for the cylinder to be adjusted.
4. Check the valve clearance. The valve clearance specifications are as follows:
   - Intake valves: 0.005–0.007 inches (0.13–0.17mm)
   - Exhaust valves: 0.005–0.007 inches (0.13–0.17mm)
5. After adjustment, tighten the locknuts to 11–14 ft. lbs. (15–19 Nm).
6. Repeat for each valve to be adjusted.

2.0L and 2.5L Engines

2.0L and 2.5L engines utilize automatic hydraulic lash adjusters to maintain proper valve lash at all times. Periodic valve lash inspection and adjustment is not necessary or possible.

Starter Motor

REMOVAL & INSTALLATION

1. Before servicing the vehicle, refer to the precautions in the beginning of this section.
2. Remove or disconnect the following:
   - Negative battery cable
   - Starter motor wiring connectors
   - Starter motor

To install:

3. Install or connect the following:
   - Starter motor. Tighten the bolts to 22 ft. lbs. (30 Nm).
   - Starter motor wiring connectors. Tighten the solenoid nut to 11 ft. lbs. (15 Nm).
   - Negative battery cable

Oil Pan

REMOVAL & INSTALLATION

1.6L Engines

1. Before servicing the vehicle, refer to the precautions in the beginning of this section.

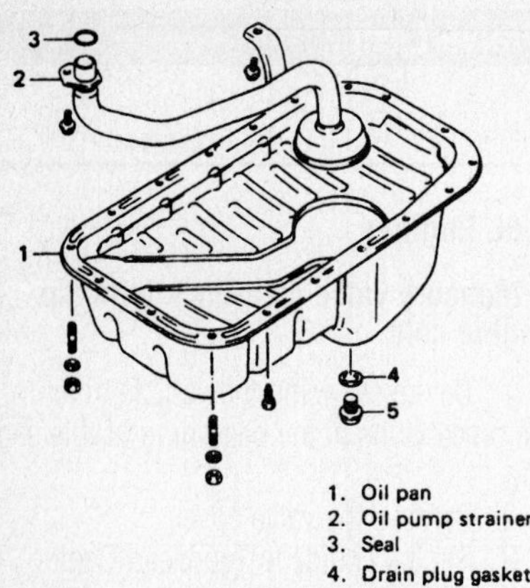

Exploded view of the oil pan and pump pickup mounting—1.6L engine

2. Drain the engine oil.
3. Remove or disconnect the following:
   - Negative battery cable
   - Front skidplate, if equipped
   - Front differential, if equipped
   - Crankshaft Position (CKP) sensor
   - Left transmission stiffener bracket, if equipped
   - Flywheel access panel
   - Oil pan and oil pump pickup tube

To install:

4. Apply a bead of silicone sealant to the oil pan flange.
5. Install or connect the following:
   - New oil pump pickup tube O-ring seal
   - Oil pan and oil pump pickup tube. Tighten the fasteners to 97 inch lbs. (11 Nm).
   - Flywheel access panel
   - Left transmission stiffener bracket, if equipped
   - CKP sensor
   - Front differential, if equipped
   - Front skidplate, if equipped. Tighten the bolts to 40 ft. lbs. (55 Nm).

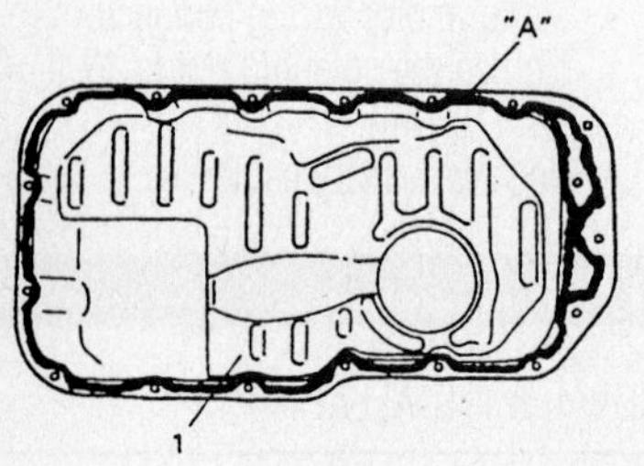

Before installing the oil pan, apply a continuous bead of silicone sealant to the oil pan mating flange—all engines

   - Negative battery cable
6. Fill the crankcase to the correct level.
7. Start the engine and check for leaks.

2.0L Engines

1. Before servicing the vehicle, refer to the precautions in the beginning of this section.
2. Drain the engine oil.
3. Remove or disconnect the following:
   - Negative battery cable
   - Oil dipstick tube
   - Front wheels
   - Front skidplate, if equipped
   - Steering gear
   - Front differential, if equipped
   - Left transmission stiffener bracket, if equipped
   - Flywheel access panel
   - Exhaust front pipe
   - Left and right motor mounts and raise the engine about 1 inch (25mm) for clearance
   - Oil pan and oil pump pickup tube

To install:

4. Apply a bead of silicone sealant to the oil pan flange. Install new oil pump pickup tube O-ring seals.
5. Install or connect the following:
   - Oil pan and oil pump pickup tube. Tighten the fasteners to 97 inch lbs. (11 Nm).
   - Left and right engine mounts. Tighten the nuts to 36 ft. lbs. (50 Nm).
   - Exhaust front pipe
   - Flywheel access panel
   - Left transmission stiffener bracket, if equipped
   - Front differential, if equipped
   - Steering gear
   - Front skidplate, if equipped
   - Front wheels
   - Oil dipstick tube
   - Negative battery cable
6. Fill the crankcase to the correct level.
7. Start the engine and check for leaks.

2.5L Engine

1. Before servicing the vehicle, refer to the precautions in the beginning of this section.
2. Drain the engine oil.
3. Remove or disconnect the following:
   - Negative battery cable
   - Oil dipstick tube
   - Front wheels
   - Front skidplate, if equipped
   - Steering gear
   - Front differential, if equipped
   - Lower oil pan

Lower crankcase O-ring seal

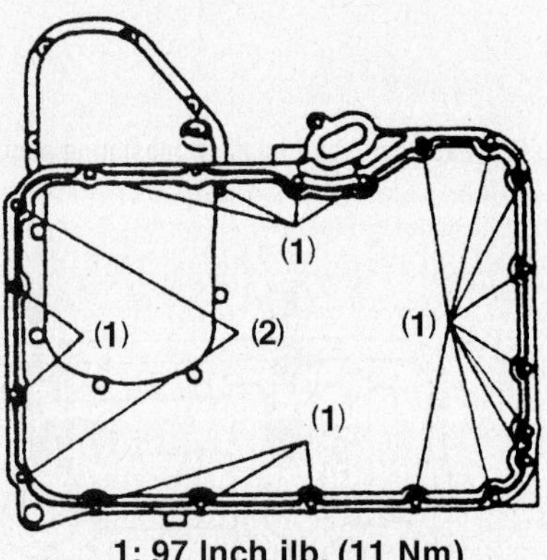

Upper oil pan bolt torque values—2.5L engine

   - Oil pickup tube bracket
   - Radiator outlet pipe
   - Upper oil pan and oil pickup tube

To install:

4. Install a new O-ring to the lower crankcase.
5. Apply a bead of silicone sealant to the upper oil pan flange.
6. Install or connect the following:
   - New oil pump pickup tube O-ring seals
   - Upper oil pan and oil pump pickup tube and tighten the fasteners as shown
   - Radiator outlet pipe
   - Oil pickup tube bracket
   - Lower oil pan. Tighten the bolts to 97 inch lbs. (11 Nm).
   - Front differential, if equipped
   - Steering gear
   - Front skidplate, if equipped
   - Front wheels
   - Oil dipstick tube
   - Negative battery cable
7. Fill the crankcase to the correct level.
8. Start the engine and check for leaks.

Oil Pump

REMOVAL & INSTALLATION

1.6L Engines

1. Before servicing the vehicle, refer to the precautions in the beginning of this section.
2. Drain the engine oil.
3. Drain the cooling system.
4. Remove or disconnect the following:
    - Negative battery cable
    - Accessory drive belts
    - Crankshaft pulley
    - Front cover
    - Timing belt
    - Alternator and bracket
    - A/C compressor and bracket, if equipped
    - Oil pan and oil pump pickup tube
    - Crankshaft timing sprocket
    - Oil pump

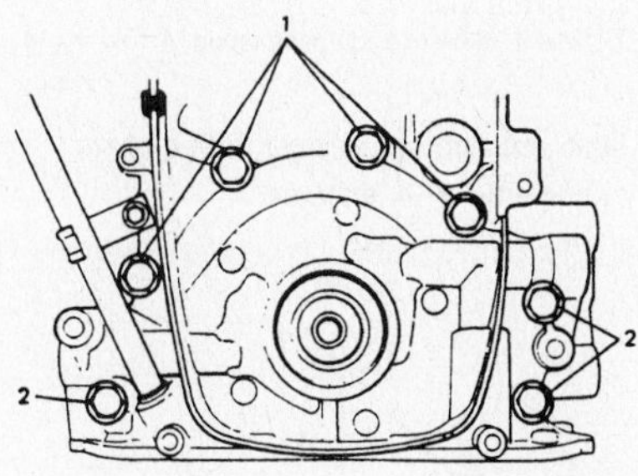

Oil pump housing short (1) and long bolt (2) locations—1.6L engines

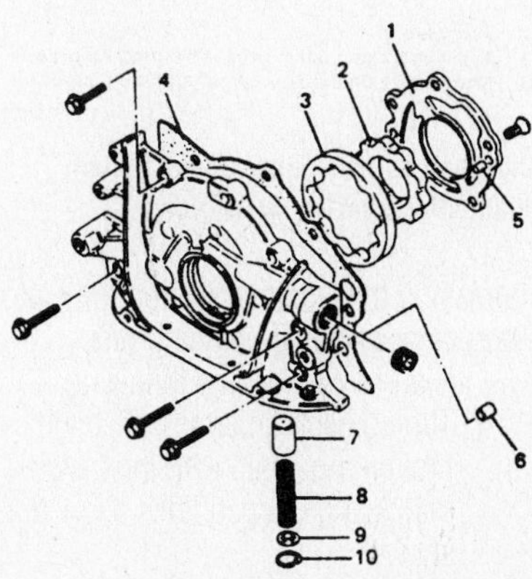

Exploded view of the oil pump housing—1.6L engines

➡The oil pump bolts are different lengths. Note their location for assembly.

To install:

5. Install or connect the following:
    - Oil pump with a new gasket. Tighten the bolts to 97 inch lbs. (11 Nm).
    - Crankshaft timing sprocket. Tighten the bolt to 94 ft. lbs. (130 Nm).
    - Oil pan and oil pump pickup tube
    - A/C compressor and bracket, if equipped
    - Alternator and bracket
    - Timing belt
    - Front cover
    - Crankshaft pulley. Tighten the bolts to 12 ft. lbs. (16 Nm).
    - Accessory drive belts
    - Negative battery cable
6. Fill the crankcase to the correct level.
7. Start the engine and check for leaks.

2.0L Engines

1. Before servicing the vehicle, refer to the precautions in the beginning of this section.
2. Drain the engine oil.
3. Remove or disconnect the following:
    - Negative battery cable
    - Oil pan and pickup tube
    - Oil pump sprocket cover
    - Oil pump

**** WARNING**

Do not remove the sprocket from the oil pump. Damage to the oil pump center shaft and abnormal pump operation may result.

To install:

4. Install or connect the following:
    - Oil pump. Tighten the bolts to 20 ft. lbs. (27 Nm).
    - Oil pump sprocket cover. Tighten the bolts to 108 inch lbs. (12 Nm).
    - Oil pan and pickup tube
    - Negative battery cable
5. Fill the crankcase to the correct level.
6. Start the engine and check for leaks.

2.5L Engine

1. Before servicing the vehicle, refer to the precautions in the beginning of this section.
2. Drain the cooling system.
3. Drain the engine oil.
4. Remove or disconnect the following:
    - Negative battery cable
    - Accessory drive belts

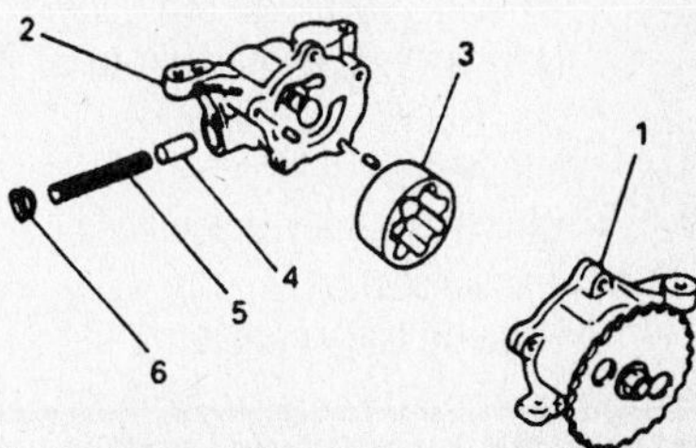

Exploded view of oil pump—2.0L and 2.5L engines

    - Intake manifold
    - Oil pan and oil pickup tube
    - Front cover
    - Oil pump chain guide
    - Oil pump

**** WARNING**

Do not remove the sprocket from the oil pump. Damage to the oil pump center shaft and abnormal pump operation may result.

To install:

5. Install or connect the following:
    - Oil pump. Tighten the bolts to 20 ft. lbs. (27 Nm).
    - Oil pump chain guide. Tighten the bolts to 97 inch lbs. (11 Nm).
    - Front cover
    - Oil pan and oil pickup tube
    - Intake manifold
    - Accessory drive belts
    - Negative battery cable
6. Fill the crankcase to the correct level.
7. Fill the cooling system.
8. Start the engine and check for leaks.

Rear Main Seal

REMOVAL & INSTALLATION

1. Before servicing the vehicle, refer to the precautions in the beginning of this section.
2. Remove or disconnect the following:
    - Negative battery cable
    - Transmission
    - Clutch assembly, if equipped
    - Flywheel
    - Rear main seal

To install:

3. Install or connect the following:
    - Rear main seal flush with the cylinder block
    - Flywheel. Tighten the bolts in a

crossing pattern to 58 ft. lbs. (79 Nm) for 1.6L engines or to 51 ft. lbs. (69 Nm) for all other engines.
- Clutch assembly, if equipped
- Transmission
- Negative battery cable

Timing Chain, Sprockets, Front Cover and Seal

REMOVAL & INSTALLATION

2.0L Engines

1. Before servicing the vehicle, refer to the precautions in the beginning of this section.
2. Drain the cooling system.
3. Drain the engine oil.
4. Remove or disconnect the following:
   - Negative battery cable
   - Oil pan and pickup tube
   - Valve cover
   - Bypass pipe and hose
   - Accessory drive belts
   - Cooling fan and shroud
   - Water pump pulley
   - Alternator belt tensioner and idler pulleys
   - Upper radiator hose
   - A/C compressor and bracket, if equipped
   - Crankshaft pulley
   - Front crankshaft seal
   - Front cover
5. Rotate the crankshaft to align the timing marks as shown.

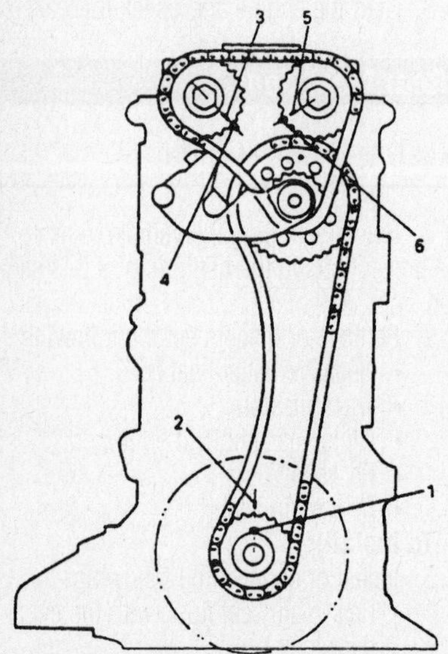

Timing mark alignment—2.0L engines

⁂ WARNING

Do not allow the crankshaft or camshafts to rotate once the timing chains have been removed. Valve or piston damage could result.

6. Remove or disconnect the following:
   - Second timing chain tensioner
   - Camshaft sprockets and second timing chain
   - First timing chain tensioner
   - Timing chain idler sprocket and first timing chain

To install:

7. Prepare the timing chain tensioners for installation by releasing the latches,

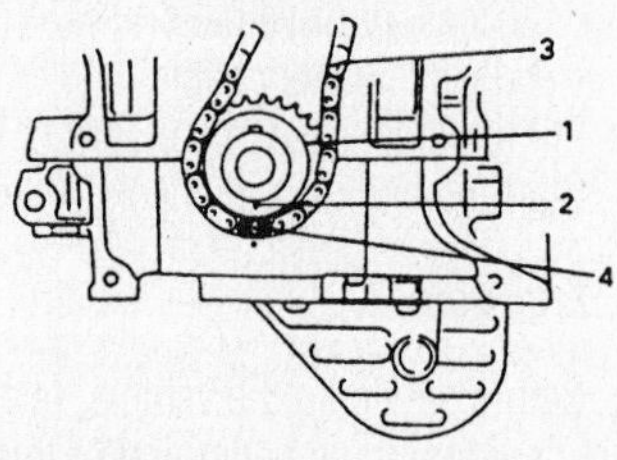

Crankshaft and first timing chain alignment—2.0L engines

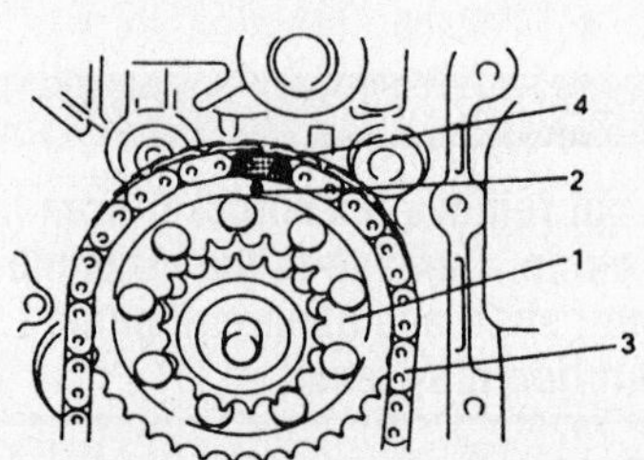

Idler sprocket and first timing chain alignment—2.0L engines

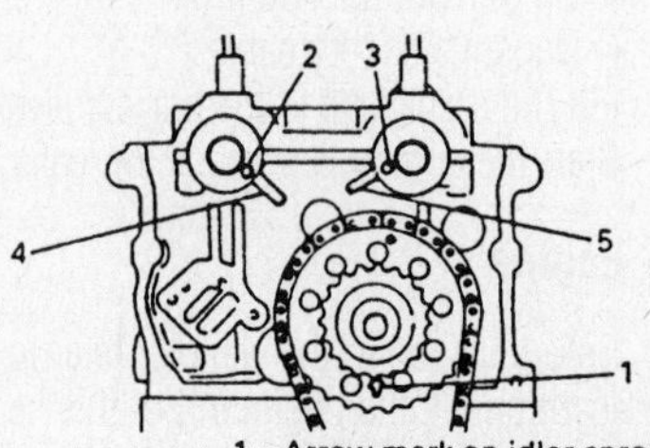

Idler sprocket and camshaft alignment—2.0L engines

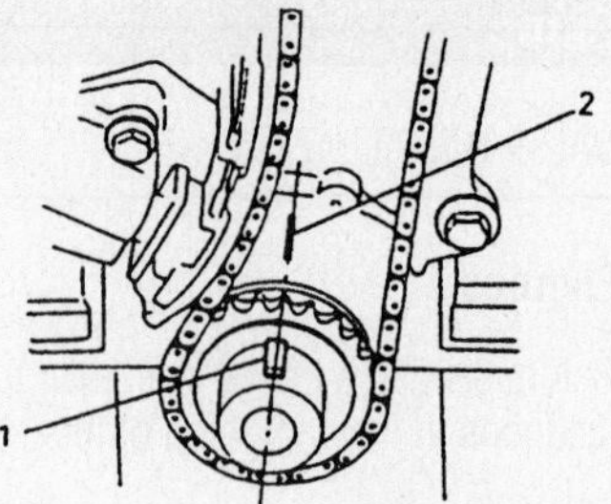

Crankshaft sprocket alignment—2.0L engines

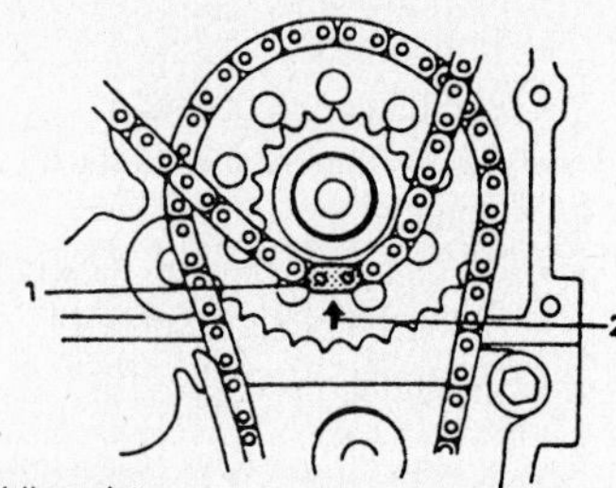

Idler sprocket and second timing chain alignment—2.0L engines

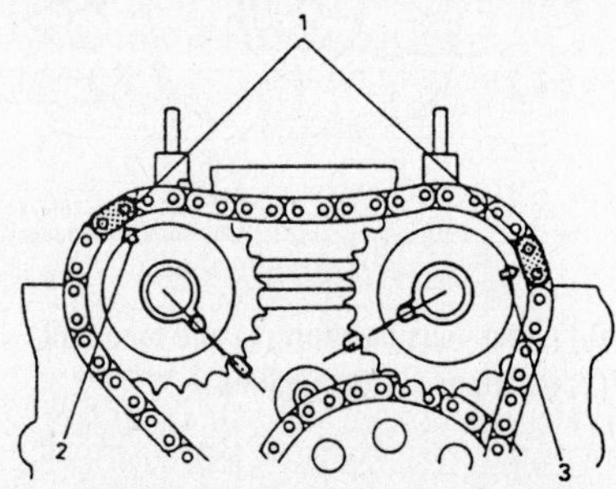

Camshaft sprocket and second timing chain alignment—2.0L engines

compressing the tensioner piston fully into the bore and installing retaining pins.

8. Install or connect the following:
   - Timing chain idler sprocket and first timing chain with the matchmarks and colored links aligned as shown
   - First timing chain tensioner. Tighten the bolts to 97 inch lbs. (11 Nm).
   - Camshaft sprockets and second timing chain with the matchmarks and colored links aligned as shown
   - Second timing chain tensioner.

Tighten the bolts to 97 inch lbs. (11 Nm) and the nut to 33 ft. lbs. (45 Nm).

9. Tighten the camshaft sprocket bolts to 59 ft. lbs. (80 Nm).

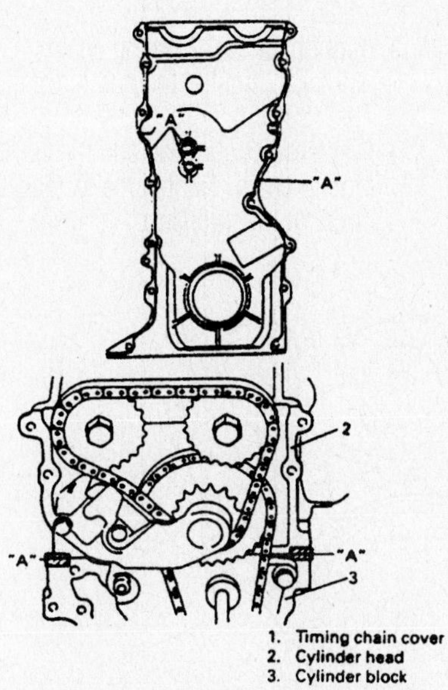

Prior to installing the timing chain cover on the engine block and cylinder head, apply silicone sealant to the cover as indicated (areas marked A)—2.0L and 2.5L engines

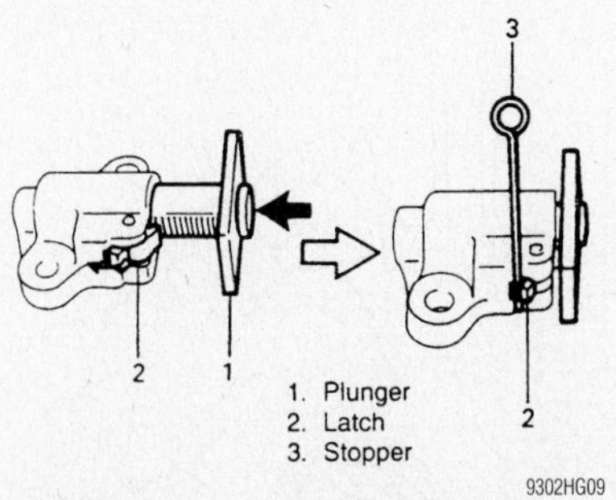

Preparing the No. 1 timing chain tensioner adjuster for installation—2.0L and 2.5L engines

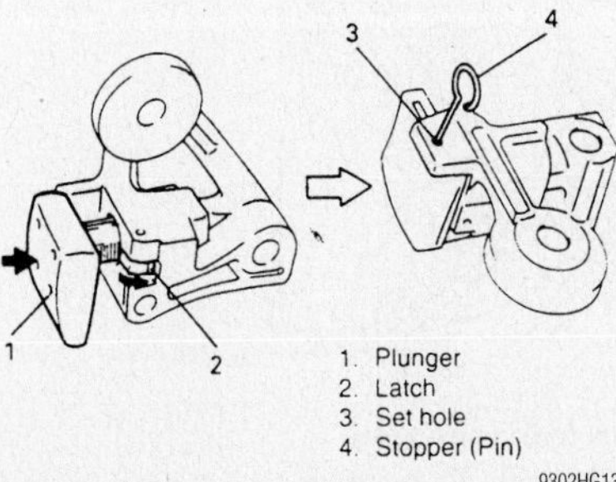

No. 2 timing chain tensioner—left bank tensioner shown—2.0L and 2.5L engines

10. Remove the timing chain tensioner retaining pins.

11. Rotate the crankshaft two complete turns and check that the timing marks align.

12. Install or connect the following:

- Front cover. Apply sealant as shown.
- Front crankshaft seal
- Crankshaft pulley. Tighten the bolt to 109 ft. lbs. (130 Nm).
- A/C compressor and bracket, if equipped
- Upper radiator hose
- Alternator belt tensioner and idler pulleys
- Water pump pulley
- Cooling fan and shroud
- Accessory drive belts
- Bypass pipe and hose
- Valve cover
- Oil pan and pickup tube
- Negative battery cable

13. Fill the crankcase to the correct level.

14. Fill the cooling system.

15. Start the engine and check for leaks.

2.5L Engine

1. Before servicing the vehicle, refer to the precautions in the beginning of this section.

2. Drain the cooling system.

3. Drain the engine oil.

4. Remove or disconnect the following:

- Negative battery cable
- Intake manifold
- Ignition coils
- Valve covers
- Accessory drive belts
- Cooling fan and shroud
- Water pump pulley
- Radiator
- Power steering pump and brackets
- Oil pan and pickup tube
- Crankshaft pulley
- Front crankshaft seal
- Crankshaft Position (CKP) sensor
- Front cover

5. Rotate the crankshaft so that the timing marks are aligned as shown.

⁂ WARNING

Do not allow the crankshaft or camshafts to rotate once the timing chains have been removed. Valve or piston damage could result.

6. Remove or disconnect the following:

- Left bank No. 2 timing chain tensioner
- Left bank intake and exhaust camshaft sprockets with the No. 2 timing chain
- No. 1 timing chain guides

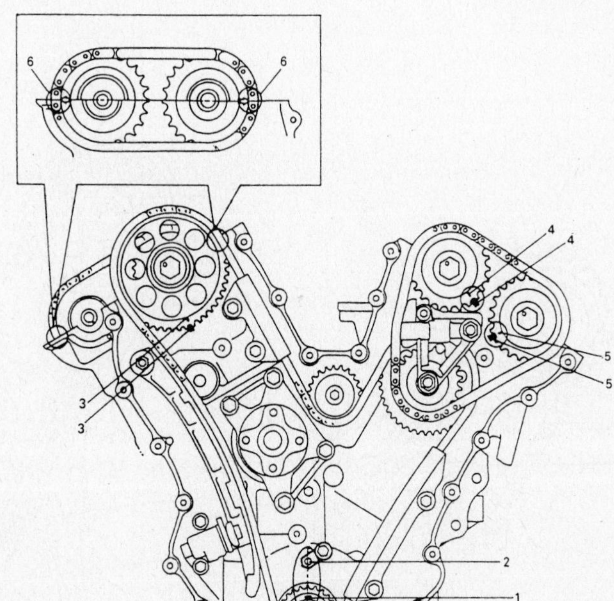

Timing chain alignment marks—2.5L engine

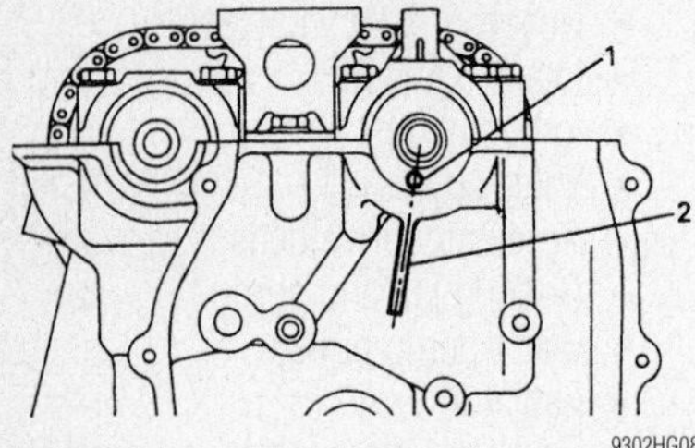

Right bank camshaft timing marks—2.5L engine

- No. 1 timing chain tensioner
- Center idler sprocket and the No. 1 timing chain
- Right bank No. 1 timing chain sprocket

7. The right bank No. 2 timing chain is removed with the intake and exhaust camshafts.

To install:

8. Prepare the timing chain tensioners for installation by releasing the latches, compressing the tensioner piston fully into the bore and installing retaining pins.

9. Align the timing chain sprocket matchmarks and colored chain links as shown during assembly.

10. Install or connect the following:
   - Right bank intake and exhaust camshafts with the No. 2 timing chain
   - Right bank No. 1 timing chain sprocket. Tighten the bolt to 58 ft. lbs. (80 Nm).
   - Center idler sprocket and the No. 1 timing chain. Tighten the fastener to 32 ft. lbs. (45 Nm).

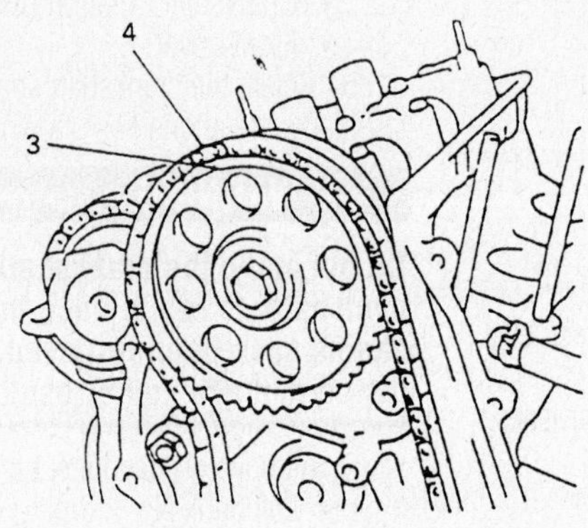

3. Match mark of RH bank 1st timing chain sprocket
4. Silver plate (LH) of 1st timing chain

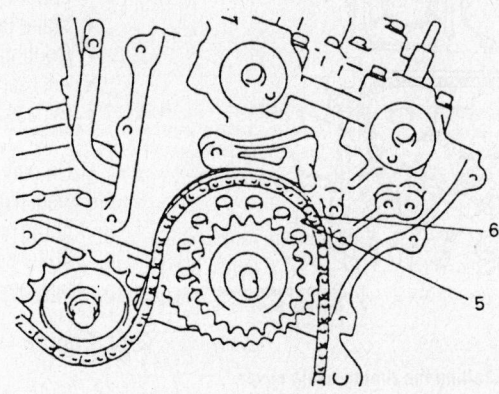

5. Match mark of idler sprocket No.2
6. Silver plate (RH) of 1st timing chain

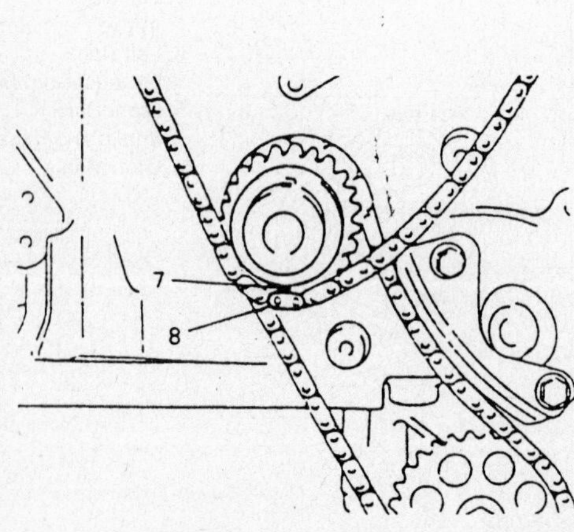

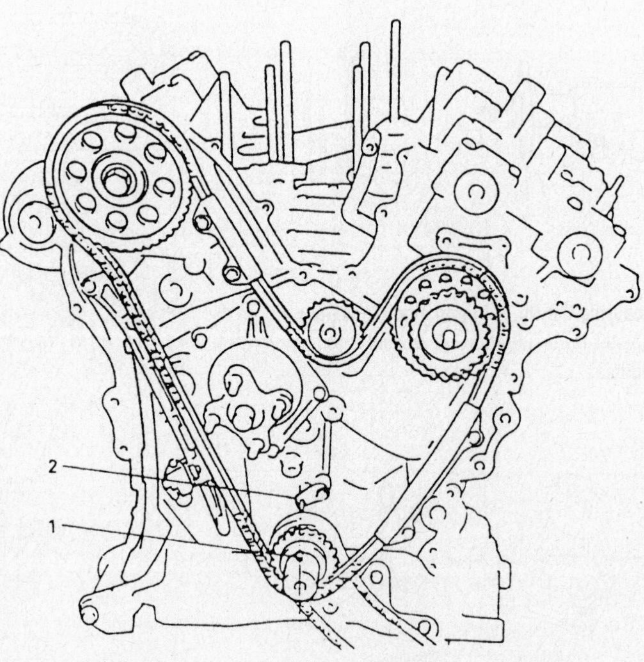

7. Match mark of crankshaft timing sprocket
8. Gold or Yellow plate of 1st timing chain

1. Crank timing pulley key
2. Oil jet

9302HG10

No. 1 timing chain alignment—2.5L engine

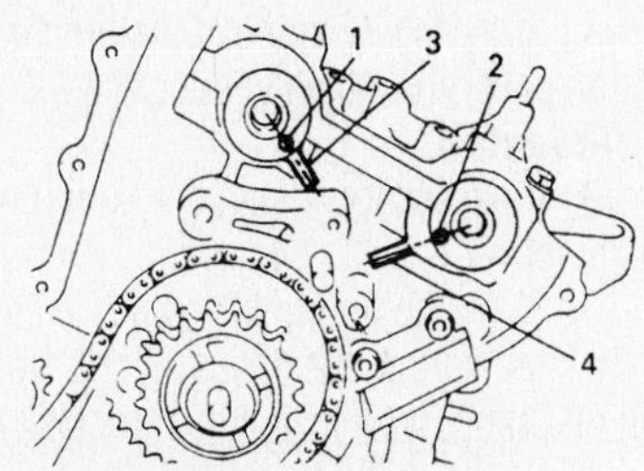

Left bank camshaft alignment—2.5L engine

- No. 1 timing chain tensioner and guides. Tighten the bolts to 97 inch lbs. (11 Nm).
- Left bank intake and exhaust camshaft sprockets with the No. 2 timing chain. Tighten the bolts to 57 ft. lbs. (80 Nm).
- Left bank No. 2 timing chain tensioner. Tighten the bolts to 97 inch lbs. (11 Nm).

11. Remove the retaining pins from the timing chain tensioners.
12. Rotate the crankshaft two complete turns and check that the timing marks align.
13. Install or connect the following:
    - Front cover. Tighten the bolts to 97 inch lbs. (11 Nm).
    - CKP sensor
    - Front crankshaft seal
    - Crankshaft pulley. Tighten the bolt to 109 ft. lbs. (148 Nm).
    - Oil pan and pickup tube
    - Power steering pump and brackets
    - Radiator
    - Water pump pulley
    - Cooling fan and shroud
    - Accessory drive belts
    - Valve covers
    - Ignition coils
    - Intake manifold
    - Negative battery cable

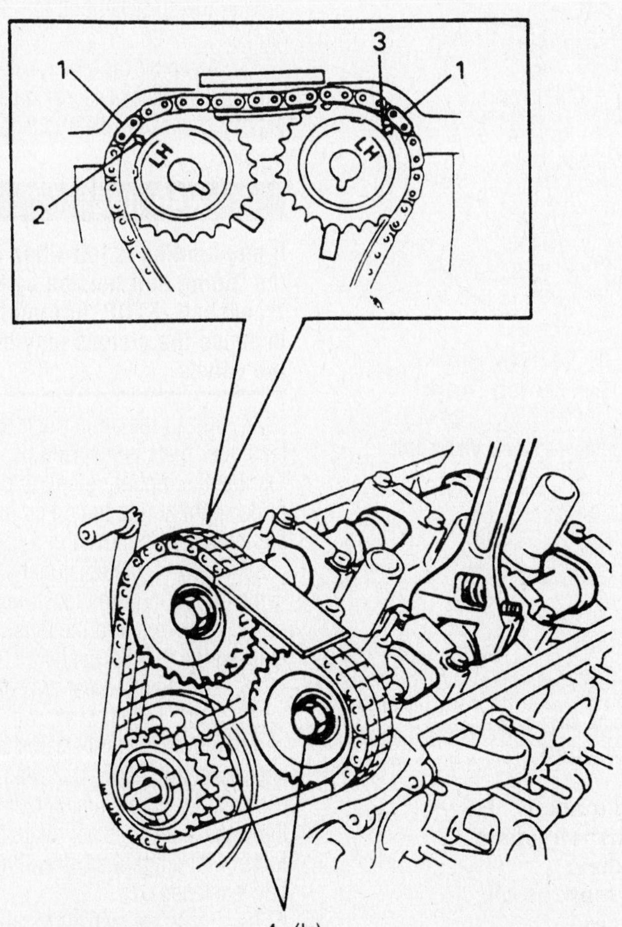

Align the left bank No. 2 chain silver links—2.5L engine

14. Fill the crankcase to the correct level.
15. Fill the cooling system.
16. Start the engine and check for leaks.

Timing Belt

REMOVAL & INSTALLATION

1.6L Engine

WARNING

Do not rotate the crankshaft counterclockwise or attempt to rotate the crankshaft by turning the camshaft sprocket.

1. Remove the timing belt cover.
2. If the timing belt is not already marked with a directional arrow, use white paint, a grease pencil or correction fluid to do so.
3. Rotate the crankshaft clockwise until the timing mark on the camshaft sprocket and the "V" mark on the timing belt inside cover are aligned, and the punch mark on the crankshaft sprocket is aligned with the mark on the engine.

WARNING

Do not rotate the crankshaft or camshaft once the timing belt is removed, because the valves and pistons can come into contact, which may cause internal engine damage.

4. Disconnect one end of the tensioner spring. Loosen the timing belt tensioner bolt and stud, then, using your finger, press the tensioner plate up and remove the timing belt from the crankshaft and camshaft sprockets.
5. Remove the timing belt tensioner, tensioner plate and spring from the engine.
6. Install Suzuki tool 09917-68220, or equivalent, onto the camshaft sprocket to hold the camshaft from rotating. Loosen the camshaft sprocket retaining bolt, then pull the camshaft sprocket off of the end of the camshaft.
7. Remove the crankshaft timing belt sprocket by loosening the center bolt, while preventing the crankshaft from rotating. To hold the crankshaft from turning, use Suzuki tool 09927-56010, or equivalent, or a large prybar inserted in the transmission housing slot and the flywheel teeth. Pull the sprocket off of the end of the crankshaft. Be sure to retain the crankshaft sprocket key and belt guide for assembly.

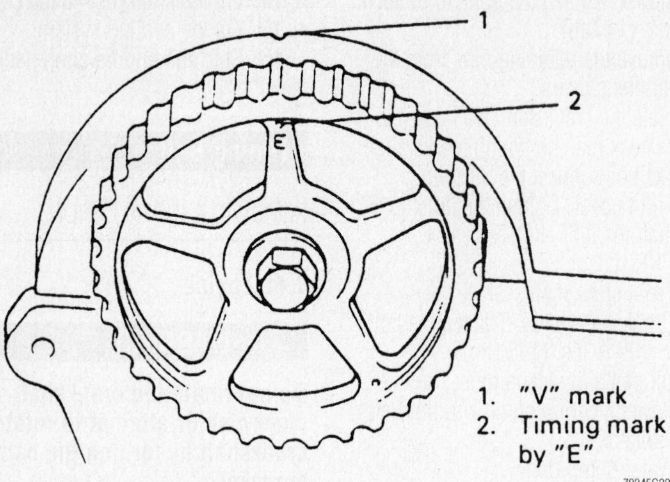

Camshaft timing marks — Suzuki Vitara 1.6L 16-valve engine

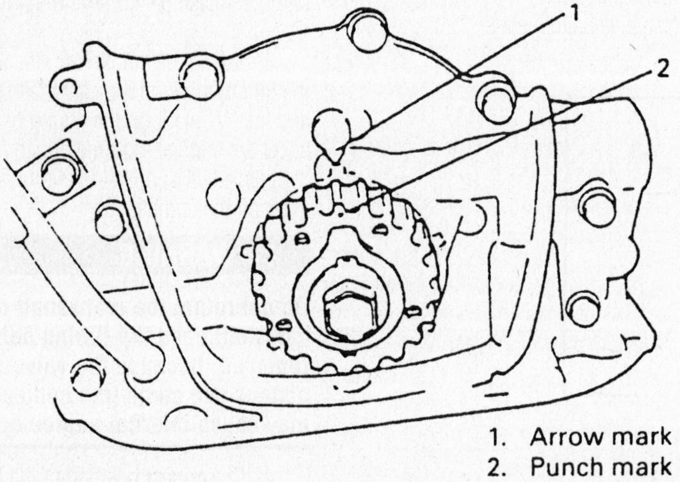

Align the punch mark with the arrow for proper timing belt installation — Suzuki Vitara 1.6L 16-valve engine

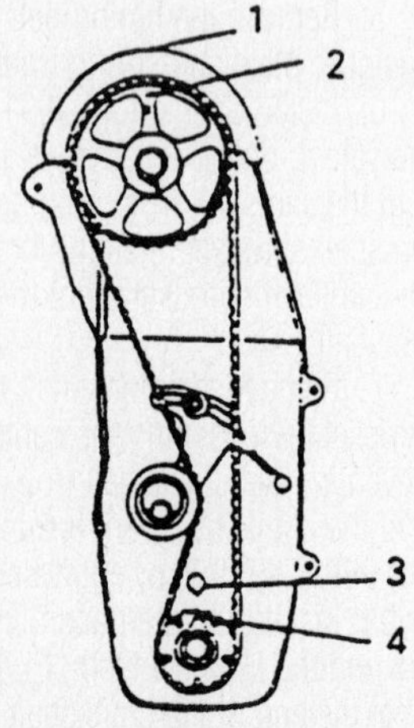

Rotate the crankshaft clockwise until the camshaft and crankshaft timing marks are aligned — Suzuki Vitara 1.6L 16-valve engine

8. If necessary, remove the timing belt inside cover from the cylinder head.

To install:

9. If necessary, install the timing belt inside cover.
10. Slide the timing belt guide on the crankshaft so that the concave side faces the oil pump, then install the sprocket key in the groove in the crankshaft.
11. Slide the pulley onto the crankshaft, and install the center retaining bolt. Tighten the center bolt to 80 ft. lbs. (110 Nm). To hold the crankshaft from turning, use Suzuki tool 09927-56010, or equivalent, or a large prybar inserted in the transmission housing slot and the flywheel teeth.
12. Install the timing belt camshaft sprocket, ensuring that the slot in the sprocket engages the camshaft (pulley) pin; this ensures that the sprocket is properly positioned on the end of the camshaft. Secure the camshaft with the holding tool used during removal, then tighten the sprocket bolt to 44 ft. lbs. (60 Nm).
13. Assemble the timing belt tensioner plate and the tensioner, making sure that the lug of the tensioner plate engages the tensioner.

**** WARNING**

If any binding is felt when adjusting the timing belt tension by turning the crankshaft, STOP turning the engine, because the pistons may be hitting the valves.

14. Install the timing belt tensioner, tensioner plate and spring on the engine. Tighten the mounting bolt and stud only finger-tight at this time. Ensure that when the tensioner is moved in a counterclockwise direction, the tensioner moves in the same direction. If the tensioner does not move, remove it and the tensioner plate to reassemble them properly.
15. Loosen all rocker arm valve lash locknuts and adjusting screws. This will permit movement of the camshaft without any rocker arm associated drag, which is essential for proper timing belt tensioning. If the camshaft does not rotate freely (free of rocker arm drag), the belt will not be properly tensioned.
16. Rotate the camshaft sprocket clockwise until the timing mark on the sprocket and the "V" mark on the timing belt inside cover are aligned.
17. Using a wrench, or socket and

breaker bar, on the crankshaft sprocket center bolt, turn the crankshaft clockwise until the punch mark on the sprocket is aligned with the arrow mark on the oil pump.

18. With the camshaft and crankshaft marks properly aligned, push the tensioner up with your finger and install the timing belt on the 2 sprockets, ensuring that the drive side of the belt is free of all slack. Release your finger from the tensioner. Be sure to install the timing belt so that the directional arrow is pointing in the appropriate direction.

➡In this position, the No. 4 cylinder is at Top Dead Center (TDC) on the compression stroke.

19. Rotate the crankshaft clockwise 2 full revolutions, then tighten the tensioner stud to 97 inch lbs. (11 Nm). Then, tighten the tensioner bolt to 18 ft. lbs. (24 Nm).

20. Ensure that all 4 timing marks are still aligned as before; if they are not, remove the timing belt, and install and tension it again.

21. Install the timing belt cover and all related components.

Piston and Ring

POSITIONING

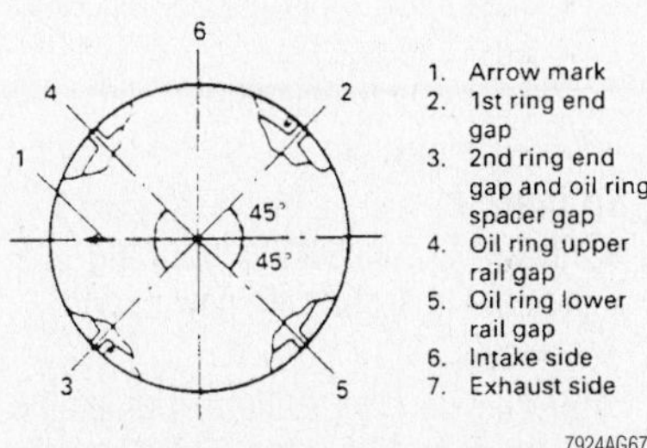

Piston ring end-gap spacing—All engines

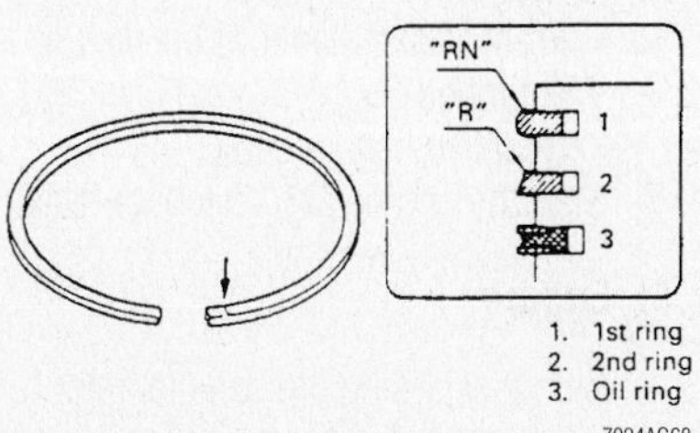

Compression ring identification marks—All engines

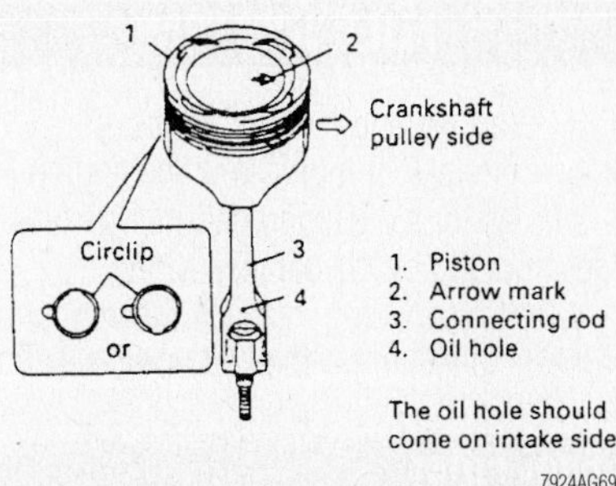

Piston and connecting rod positioning—1.6L engine

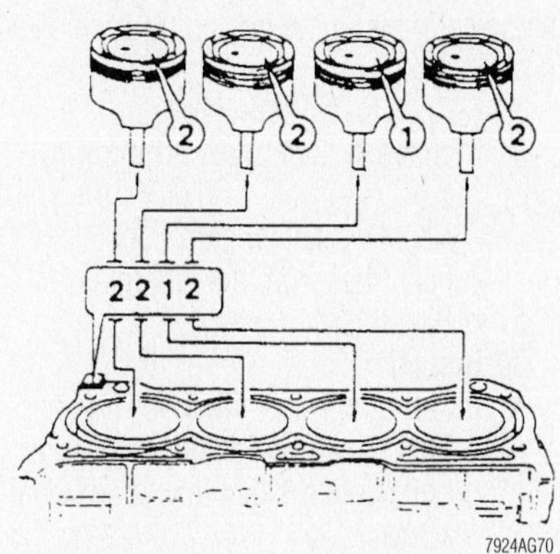

Piston installation—1.6L engine

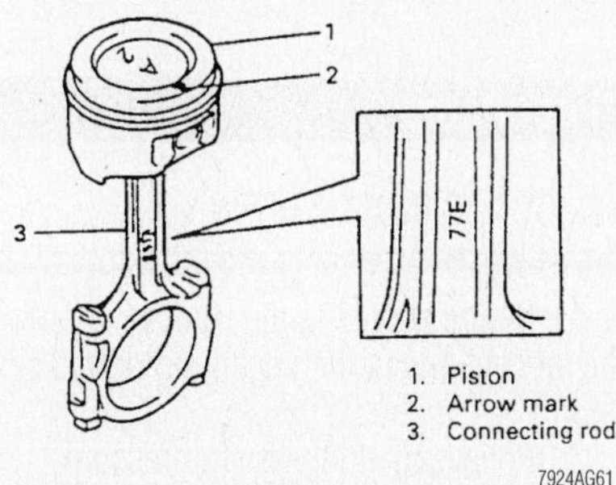

Piston and connecting rod positioning—2.0L and 2.5L engines

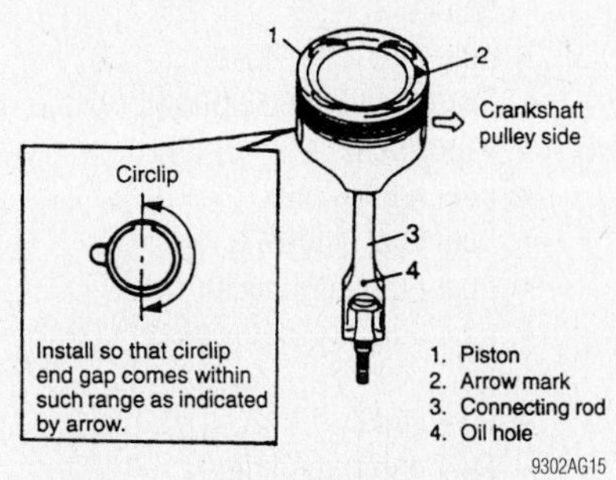

Piston pin circlip installation—2.5L engine

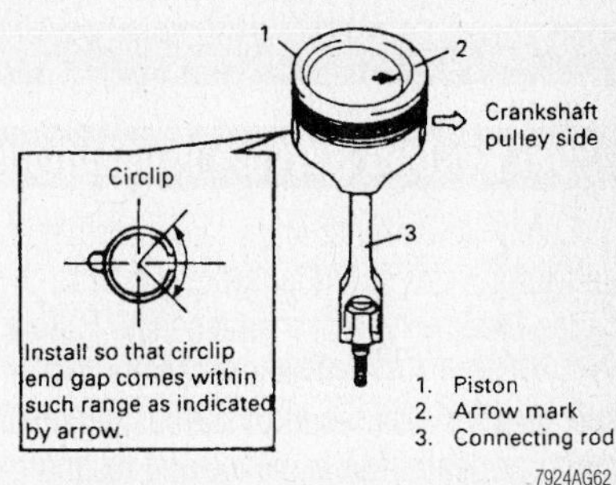

Piston pin circlip installation—2.0L engines

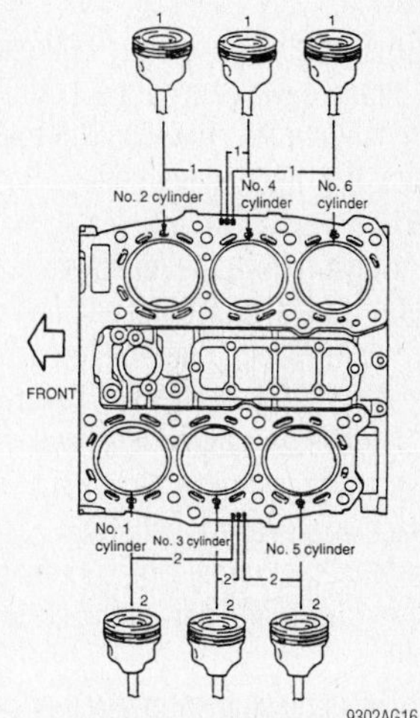

Piston identification—2.5L engine

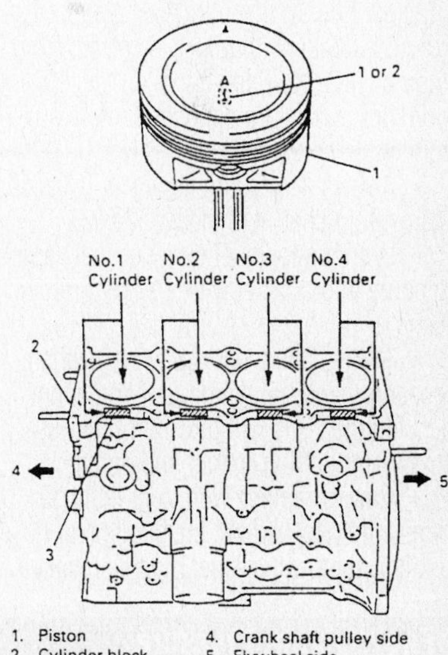

Piston identification—2.0L engines
Match pistons with "1" indicators to red cylinder paint marks
Match pistons with "2" indicators to blue cylinder paint marks

FUEL SYSTEM

Fuel System Service Precautions

Safety is the most important factor when performing not only fuel system maintenance but any type of maintenance. Failure to conduct maintenance and repairs in a safe manner may result in serious personal injury or death. Maintenance and testing of the vehicle fuel system components can be accomplished safely and effectively by adhering to the following rules and guidelines.

- To avoid the possibility of fire and personal injury, always disconnect the negative battery cable unless the repair or test procedure requires that battery voltage be applied.
- Always relieve the fuel system pressure prior to disconnecting any fuel system component (injector, fuel rail, pressure regulator, etc.), fitting or fuel line connection. Exercise extreme caution whenever relieving fuel system pressure to avoid exposing skin, face and eyes to fuel spray. Please be advised that fuel under pressure may penetrate the skin or any part of the body that it contacts.
- Always place a shop towel or cloth around the fitting or connection prior to loosening to absorb any excess fuel due to spillage. Ensure that all fuel spillage (should it occur) is quickly removed from engine surfaces. Ensure that all fuel soaked cloths or towels are deposited into a suitable waste container.
- Always keep a dry chemical (Class B) fire extinguisher near the work area.
- Do not allow fuel spray or fuel vapors to come into contact with a spark or open flame.
- Always use a backup wrench when loosening and tightening fuel line connection fittings. This will prevent unnecessary stress and torsion to fuel line piping.
- Always replace worn fuel fitting O-rings with new. Do not substitute fuel hose or equivalent, where fuel pipe is installed.

Fuel System Pressure

RELIEVING

1. Before servicing the vehicle, refer to the precautions in the beginning of this section.
2. Detach the wiring harness connector from the fuel pump relay, located under the left-hand side of the instrument panel near the ECM.
3. Start the engine and run it until it stops from lack of fuel. Crank the engine 2–3 times for a 3 second period. The fuel lines should now be depressurized.
4. After servicing, reattach the wiring harness connector to the fuel pump relay.

Fuel Filter

REMOVAL & INSTALLATION

1. Before servicing the vehicle, refer to the precautions in the beginning of this section.
2. Relieve fuel system pressure.
3. Remove or disconnect the following:
   - Negative battery cable
   - Fuel lines from the fuel filter
   - Fuel filter

To install:

4. Install or connect the following:
   - Fuel filter and tighten the bracket bolt. Note the fuel flow directional arrow.
   - Fuel lines to the fuel filter.
   - Negative battery cable
5. Start the engine and inspect the fuel filter connections for leaks.

Fuel Pump

REMOVAL & INSTALLATION

1. Before servicing the vehicle, refer to the precautions in the beginning of this section.
2. Relieve the fuel system pressure.
3. Remove or disconnect the following:
   - Negative battery cable
   - Fuel filler hose and vent hose
   - Fuel tank inlet valve and drain the fuel tank
   - Fuel filter inlet hose
   - Evaporative Emissions (EVAP) vapor hose
   - Fuel return line
   - Fuel tank skidplate
   - Fuel pump connector
   - Fuel tank pressure sensor connector
   - Fuel tank
   - Fuel pump module

To install:

4. Install or connect the following:
   - Fuel pump module with a new seal. Tighten the bolts to 44 inch lbs. (5 Nm).
   - Fuel tank. Tighten the strap bolts to 37 ft. lbs. (50 Nm).
   - Fuel tank pressure sensor connector
   - Fuel pump connector
   - Fuel tank skidplate
   - Fuel return line
   - EVAP vapor hose
   - Fuel filter inlet hose
   - Fuel tank inlet valve and drain the fuel tank
   - Fuel filler hose and vent hose
   - Negative battery cable
5. Fill the fuel tank.
6. Start the engine and check for leaks.

Fuel Injector

REMOVAL & INSTALLATION

1.6L and 2.0L Engines

1. Before servicing the vehicle, refer to the precautions in the beginning of this section.
2. Relieve the fuel system pressure.
3. Remove or disconnect the following:
   - Negative battery cable
   - Front intake manifold bracket, if equipped
   - Positive Crankcase Ventilation (PCV) valve and hose
   - Fuel injector harness connectors
   - Fuel line bracket
   - Fuel supply manifold
   - Fuel injectors

To install:

4. Install or connect the following:
   - Fuel injectors with new O-ring seals
   - Fuel supply manifold. Tighten the bolts to 17 ft. lbs. (23 Nm).
   - Fuel line bracket
   - Fuel injector harness connectors
   - PCV valve and hose
   - Front intake manifold bracket, if equipped
   - Negative battery cable
5. Start the engine and check for leaks.

2.5L Engine

1. Before servicing the vehicle, refer to the precautions in the beginning of this section.
2. Relieve the fuel system pressure.
3. Remove or disconnect the following:
   - Negative battery cable
   - Air intake tube

- Throttle body intake collector
- Fuel lines
- Fuel pressure regulator vacuum line
- Fuel injector harness connectors
- Fuel supply manifold connect pipe
- Fuel supply manifolds
- Fuel injectors

To install:

4. Install or connect the following:
   - Fuel injectors with new O-ring seals
   - Fuel supply manifolds. Tighten the bolts to 17 ft. lbs. (23 Nm).
   - Fuel supply manifold connect pipe. Tighten the bolts to 22 ft. lbs. (30 Nm).
   - Fuel injector harness connectors
   - Fuel pressure regulator vacuum line
   - Fuel lines
   - Throttle body intake collector
   - Air intake tube
   - Negative battery cable
5. Start the engine and check for leaks.

DRIVE TRAIN

Transmission Assembly

REMOVAL & INSTALLATION

Manual Transmission

1. Before servicing the vehicle, refer to the precautions in the beginning of this section.
2. Drain the transmission fluid.
3. Drain the transfer case fluid, if equipped.
4. Remove or disconnect the following:
   - Negative battery cable
   - Shift lever boots
   - Gear shift lever
   - Transfer case shift lever, if equipped
   - 4WD switch connector, if equipped
   - Reverse light switch connector
   - Starter motor
   - Front driveshaft, if equipped
   - Rear driveshaft
   - Speedometer cable, if equipped
   - Vehicle Speed (VSS) sensor, if equipped
   - Clutch slave cylinder or cable
   - Flywheel access cover
   - Transmission flange bolts and nuts
   - Transmission braces, if equipped
5. Support the transmission with a jack and remove the transmission mount and crossmember.

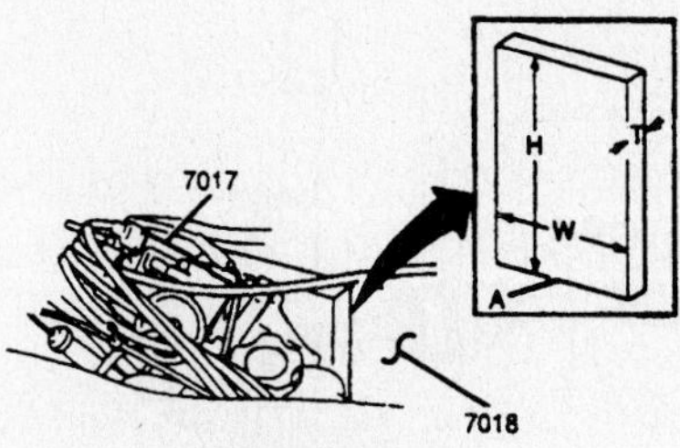

Support the engine with a wooden block between the cylinder head and the firewall—All models

6. Place a wooden block at the rear of the cylinder head as shown to support the engine when the transmission is removed.
7. Lower the transmission away from the vehicle.

To install:

8. Install or connect the following:
   - Transmission. Tighten the flange fasteners to 62–72 ft. lbs. (85–98 Nm).
   - Transmission mount and crossmember. Tighten the fasteners to 29–43 ft. lbs. (40–60 Nm).
   - Transmission braces, if equipped. Tighten the bolts to 62–72 ft. lbs. (85–98 Nm).
   - Flywheel access cover
   - Clutch slave cylinder or cable
   - VSS sensor, if equipped
   - Speedometer cable, if equipped
   - Rear driveshaft
   - Front driveshaft, if equipped
   - Starter motor
   - Reverse light switch connector
   - 4WD switch connector, if equipped
   - Transfer case shift lever, if equipped
   - Gear shift lever
   - Shift lever boots
   - Negative battery cable
9. Fill the transmission to the correct level.
10. Fill the transfer case, if equipped.

Automatic Transmission

1. Before servicing the vehicle, refer to the precautions in the beginning of this section.
2. Drain the transfer case oil, if equipped.
3. Remove or disconnect the following:
   - Negative battery cable
   - Center console and transfer case shift lever, if equipped
   - Transmission dipstick tube
   - Transmission wiring harness connectors
   - Starter motor
   - Front driveshaft, if equipped
   - Rear driveshaft
   - Gear select cable and bracket
   - Throttle Valve (TV) cable, if equipped
   - Exhaust front pipe
   - Transmission oil cooler lines
   - Transmission brace
   - Flywheel access cover
   - Torque converter
   - Speedometer cable, if equipped
   - Vehicle Speed (VSS) sensor connector, if equipped
   - Transmission flange bolts and nuts
   - Transmission braces, if equipped
4. Support the transmission with a jack and remove the transmission mount and crossmember.
5. Place a wooden block at the rear of the cylinder head as shown to support the engine when the transmission is removed.
6. Lower the transmission away from the vehicle.

To install:

7. Install or connect the following:
   - Transmission. Tighten the flange fasteners to 62–72 ft. lbs. (85–98 Nm).
   - Transmission mount and crossmember. Tighten the fasteners to 29–43 ft. lbs. (40–60 Nm).
   - Transmission braces, if equipped. Tighten the bolts to 62–72 ft. lbs. (85–98 Nm).
   - VSS sensor connector, if equipped
   - Speedometer cable, if equipped
   - Torque converter. Tighten the bolts to 47 ft. lbs. (65 Nm).
   - Flywheel access cover
   - Transmission brace
   - Transmission oil cooler lines
   - Exhaust front pipe
   - TV cable, if equipped
   - Gear select cable and bracket
   - Rear driveshaft
   - Front driveshaft, if equipped
   - Starter motor
   - Transmission wiring harness connectors
   - Transmission dipstick tube

- Center console and transfer case shift lever, if equipped
- Negative battery cable

8. Fill the transmission to the correct level.
9. Fill the transfer case, if equipped.

Clutch

ADJUSTMENTS

These vehicles are equipped with a hydraulic clutch system. No adjustment is necessary.

REMOVAL & INSTALLATION

1. Before servicing the vehicle, refer to the precautions in the beginning of this section.
2. Remove the transmission.
3. Loosen the pressure plate mounting bolts in a 2-step crisscross sequence until the spring tension is relieved.
4. Remove the pressure plate and the clutch disc.

To install:

5. Using a clutch alignment tool, assemble the clutch disc and pressure plate onto the flywheel.
6. Tighten the pressure plate bolts in multiple passes to 17 ft. lbs. (23 Nm).
7. Install the transmission.
8. Check for proper clutch operation.

Hydraulic Clutch System

BLEEDING

1. Before servicing the vehicle, refer to the precautions in the beginning of this section.
2. Fill the master cylinder reservoir to the MAX line with clean brake fluid and keep it at least half full throughout the bleeding procedure.
3. From beneath the vehicle, remove the bleeder plug cap, then attach a clear vinyl tube to the slave cylinder bleeder plug. Insert the open end of the hose into a container.
4. Have an assistant depress the clutch pedal. Open the bleeder after the pedal is depressed.
5. Close the bleeder before releasing the clutch pedal.
6. Repeat until all air bubbles are gone from the hydraulic fluid.
7. Install the bleeder plug cap.
8. Fill the clutch master cylinder fluid reservoir to the specified full level.

Transfer Case Assembly

REMOVAL & INSTALLATION

1. Before servicing the vehicle, refer to the precautions in the beginning of this section.
2. Drain the transfer case oil.
3. Remove or disconnect the following:
   - Negative battery cable
   - Distributor or Camshaft Position (CMP) sensor, if equipped
   - Center console
   - Transmission shift lever and case, if equipped with a manual transmission
   - Transfer case shift lever
   - Front and rear driveshafts
   - Exhaust center pipe
   - Speedometer cable or Vehicle Speed (VSS) sensor, as equipped
   - Vent hose
   - 4WD switch connector
4. Support the transmission with a jack and remove the transmission mount and crossmember.
5. Place a wooden block at the rear of the cylinder head as shown to support the engine when the transfer case is removed.
6. Lower the transfer case away from the vehicle.

To install:

7. Install or connect the following:
   - Transfer case. Tighten the bolts to 30 ft. lbs. (41 Nm).
   - 4WD switch connector
   - Vent hose
   - Speedometer cable or VSS sensor, as equipped
   - Exhaust center pipe
   - Front and rear driveshafts. Tighten the bolts to 36 ft. lbs. (50 Nm).
   - Transfer case shift lever
   - Transmission shift lever and case, if equipped with a manual transmission
   - Center console
   - Distributor or CMP sensor, if equipped
   - Negative battery cable
8. Fill the transfer case.

Halfshaft

REMOVAL & INSTALLATION

Left

1. Before servicing the vehicle, refer to the precautions in the beginning of this section.
2. Remove or disconnect the following:
   - Front wheel
   - Hub drive flange or locking hub, as equipped
   - Snapring
   - Thrust washer
   - Halfshaft flange fasteners
   - Halfshaft

To install:

3. Install or connect the following:
   - Halfshaft. Tighten the flange bolts to 37 ft. lbs. (50 Nm).
   - Thrust washer
   - Snapring
   - Locking hub, if equipped. Tighten the bolts to 24 ft. lbs. (33 Nm).

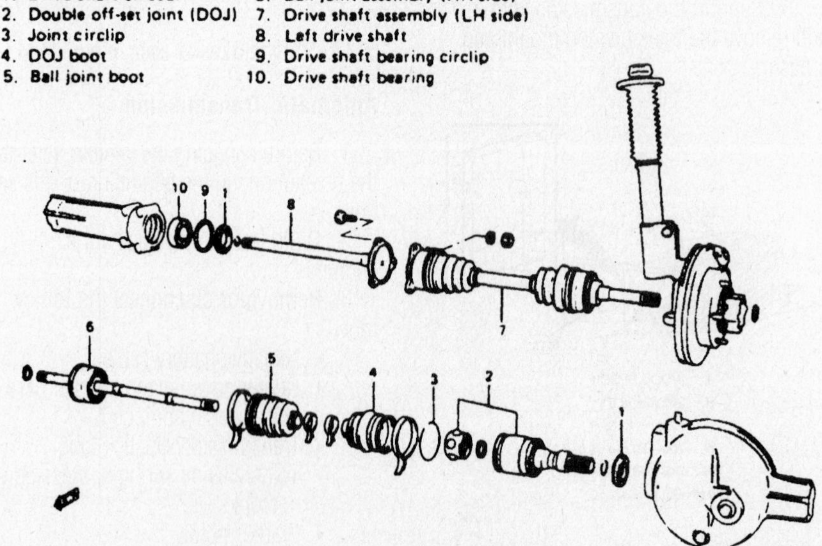

Exploded view of the left- and right-hand halfshaft assemblies

- Hub drive flange, if equipped. Tighten the bolts to 35 ft. lbs. (48 Nm).
- Front wheel

Right

1. Before servicing the vehicle, refer to the precautions in the beginning of this section.
2. Remove or disconnect the following:
   - Front wheel
   - Hub drive flange or locking hub, as equipped
   - Snapring
   - Thrust washer
   - Brake caliper
   - Wheel speed sensor, if equipped
   - Brake rotor
   - Stabilizer bar link
   - Outer tie rod end
   - Lower ball joint
   - Strut bracket bolts
   - Steering knuckle and wheel hub
3. Pry the inboard joint out of the differential and remove the halfshaft.

To install:

4. Insert the inboard joint into the differential until the circlip is felt to seat.
5. Install or connect the following:
   - Steering knuckle and wheel hub
   - Strut bracket bolts. Tighten them to 70 ft. lbs. (95 Nm).
   - Lower ball joint. Tighten the nut to 40 ft. lbs. (55 Nm).
   - Outer tie rod end. Tighten the nut to 35 ft. lbs. (48 Nm).
   - Stabilizer bar link. Tighten the nut to 21 ft. lbs. (29 Nm).
   - Brake rotor
   - Wheel speed sensor, if equipped
   - Brake caliper
   - Thrust washer
   - Snapring
   - Locking hub, if equipped. Tighten the bolts to 24 ft. lbs. (33 Nm).
   - Hub drive flange, if equipped. Tighten the bolts to 35 ft. lbs. (48 Nm).
   - Front wheel
6. Check the wheel alignment and adjust as necessary.

CV-Joints

OVERHAUL

Outer CV-Joint

The outer CV-joint is serviced with the axle shaft as an assembly. The outer CV-joint boot can be serviced by removing the inner CV-joint.

Inner CV-Joint

1. Before servicing the vehicle, refer to the precautions in the beginning of this section.
2. Remove or disconnect the following:
   - Halfshaft from the vehicle
   - Grease boot clamps
   - Outer race snapring
   - Outer race
   - Shaft snapring
   - Inner race, cage and balls

To install:

3. Install or connect the following:
   - Inner race, cage and balls
   - Shaft snapring
   - Outer race
   - Outer race snapring
4. Fill the outer race and the grease boot with CV-joint grease and tighten the boot clamps.
5. Install the axle halfshaft.

Manual Locking Hubs

REMOVAL & INSTALLATION

1. Before servicing the vehicle, refer to the precautions in the beginning of this section.
2. Set the selector knob to the **FREE** position.
3. Remove or disconnect the following:
   - Hub cover assembly
   - Hub body assembly

To install:

4. Install the hub body. Tighten the bolts to 18 ft. lbs. (25 Nm).

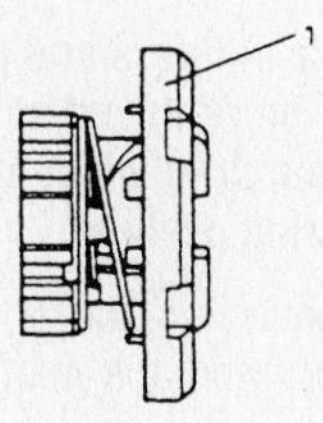

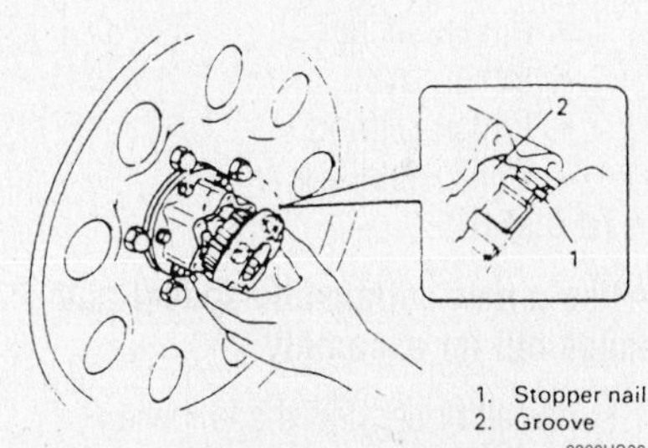

Manual hub alignment—All models

5. Align the hub cover stopper nail with the groove in the hub body and install the hub cover. Tighten the bolts to 90 inch lbs. (10 Nm).
6. Check for proper hub operation.

Automatic Locking Hubs

REMOVAL & INSTALLATION

1. Before servicing the vehicle, refer to the precautions in the beginning of this section.
2. Unlock the hub by setting the transfer case in the 2H position and driving backwards at least 6.5 feet (2 meters).
3. Remove or disconnect the following:
   - Hub sub assembly
   - Hub brake assembly

To install:

4. Align the brake assembly key with the slot in the spindle and install the brake assembly.
5. Align the matchmark on the sub assembly with the mark on the brake assembly and install the sub assembly. Tighten the hub bolts to 24 ft. lbs. (33 Nm).
6. Check for proper hub operation.

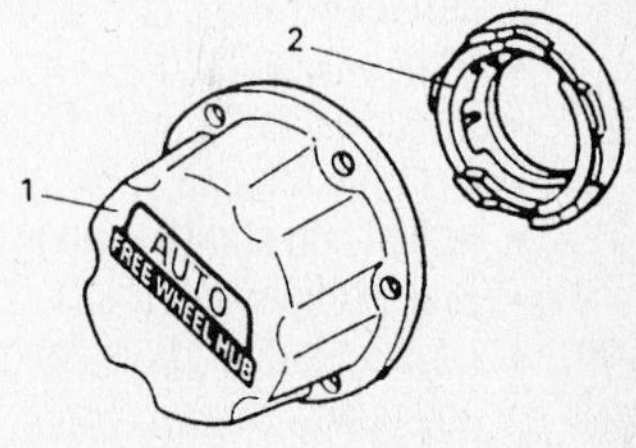

Automatic hub—All models

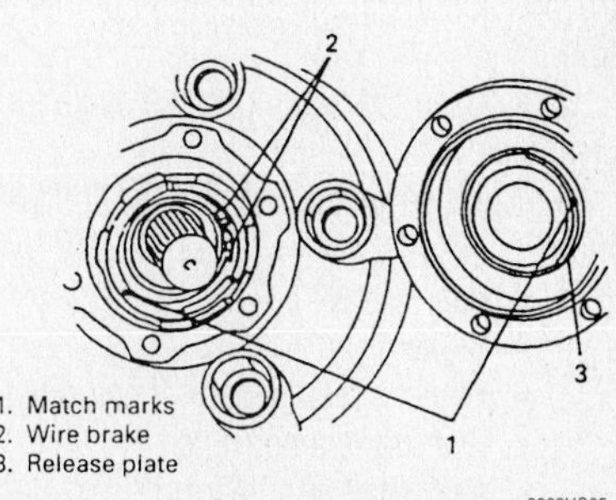

Automatic hub matchmarks—All models

Spindle Bearings

REMOVAL, PACKING & INSTALLATION

1. Before servicing the vehicle, refer to the precautions in the beginning of this section.
2. Support the control arm with a stand or floor jack.
3. Remove or disconnect the following:
   - Front wheel
   - Locking hub or drive flange, as equipped
   - Brake caliper and rotor
   - Wheel hub and bearing assembly
   - Outer tie rod end
   - Lower ball joint
   - Strut bracket bolts
   - Wheel spindle and steering knuckle assembly
   - Inner oil seal
   - Spindle bearing

To install:

4. Fill the recess in the wheel spindle with lithium grease.
5. Coat the spindle bearing and wheel spindle mating surfaces with sealant.
6. Press or drive the spindle bearing into the wheel spindle.
7. Install or connect the following:
   - Inner oil seal
   - Wheel spindle and steering knuckle assembly
   - Strut bracket bolts
   - Lower ball joint
   - Outer tie rod end
   - Wheel hub and bearing assembly
   - Brake caliper and rotor
   - Locking hub or drive flange, as equipped
   - Front wheel

Axle Shaft, Bearing and Seal

REMOVAL & INSTALLATION

1. Before servicing the vehicle, refer to the precautions in the beginning of this section.
2. Loosen the parking brake cable for clearance.
3. Remove or disconnect the following:
   - Rear wheel
   - Brake drum
   - Wheel speed sensor, if equipped
   - Bearing retainer nuts
   - Axle shaft and bearing
   - Axle shaft inner oil seal
4. If equipped with ABS, grind a flat spot on the wheel speed sensor tone ring, then split the ring with a chisel.
5. Grind flat spots on the bearing retainer and split it with a chisel.
6. Press the wheel bearing off the axle shaft.
7. Remove the bearing retainer and the outer oil seal.

To install:

8. Install or connect the following:
   - Outer oil seal to the bearing retainer
   - Bearing retainer to the axle shaft
   - Bearing and retainer ring pressed onto the axle shaft
   - Wheel speed sensor tone ring pressed onto the axle shaft, if equipped
   - Axle shaft inner oil seal
   - Axle shaft and bearing
   - Bearing retainer nuts. Tighten them to 17 ft. lbs. (23 Nm).
   - Wheel speed sensor, if equipped
   - Brake drum
   - Rear wheel
9. Fill the rear differential to the correct level.

Pinion Seal

REMOVAL & INSTALLATION

1. Before servicing the vehicle, refer to the precautions in the beginning of this section.
2. Remove or disconnect the following:
   - Driveshaft
   - Wheels
   - Brake calipers and pads or brake drum

➡The brake calipers and pads or brake drum must be removed so that there is no additional drag when measuring pinion bearing preload.

3. Use an inch lb. torque wrench and measure and record the amount of torque required to maintain pinion rotation through several revolutions.
4. Remove or disconnect the following:
   - Pinion flange
   - Pinion seal
   - Pinion bearing
   - Collapsible spacer

To install:

➡Use a new collapsible spacer and flange nut for assembly.

5. Install or connect the following:
   - Collapsible spacer
   - Pinion bearing
   - Pinion seal
   - Pinion flange
6. Rotate the pinion flange occasionally while tightening the flange nut to make sure the pinion bearings seat correctly.
7. Take frequent bearing preload torque readings. Tighten the flange nut to achieve the preload torque readings originally recorded.

✱✱ CAUTION

Never loosen the pinion nut to reduce bearing preload. If it is necessary to reduce bearing preload, install a new collapsible spacer and pinion nut.

8. Install or connect the following:
   - Driveshaft
   - Brake calipers and pads or brake drum
   - Wheels
9. Fill the differential with gear lubricant and check for leaks.

Axle Housing Assembly

REMOVAL & INSTALLATION

1. Before servicing the vehicle, refer to the precautions in the beginning of this section.
2. Drain the gear oil.
3. Support the vehicle at the frame with a hoist or jackstands.
4. Support the rear axle with a floor jack.
5. Remove or disconnect the following:
   - Rear wheels
   - Rear brake drums
   - Rear axle shafts
   - Load sensing proportioning valve linkage, if equipped
   - Brake fluid hose
   - Brake backing plates
   - Wheel speed sensor connector, if equipped
   - Axle vent tube
   - Rear driveshaft
   - Differential carrier assembly
   - Shock absorber lower bolts
   - Coil springs
   - Upper rods
   - Lower rods
   - Lateral rod
   - Axle housing

To install:

6. Install or connect the following:
   - Axle housing
   - Upper rods
   - Lower rods

- Coil springs
- Lateral rod
- Shock absorber lower bolts
- Differential carrier assembly. Tighten the nuts to 40 ft. lbs. (55 Nm).
- Rear driveshaft
- Axle vent tube
- Wheel speed sensor connector, if equipped
- Brake backing plates
- Brake fluid hose
- Load sensing proportioning valve linkage, if equipped
- Rear axle shafts
- Rear brake drums
- Rear wheels

7. Fill the rear axle to the correct level.
8. Lower the vehicle so that the rear suspension is at curb height.
9. Tighten the upper, lower and lateral rod fasteners to 65 ft. lbs. (90 Nm).
10. Tighten the lower shock absorber fasteners to 62 ft. lbs. (85 Nm).

STEERING AND SUSPENSION

Air Bag

CAUTION

Some vehicles are equipped with an air bag system. The system must be disarmed before performing service on, or around, system components, the steering column, instrument panel components, wiring and sensors. Failure to follow the safety precautions and the disarming procedure could result in accidental air bag deployment, possible injury and unnecessary system repairs.

PRECAUTIONS

Several precautions must be observed when handling the inflator module to avoid accidental deployment and possible personal injury.

- Never carry the inflator module by the wires or connector on the underside of the module.
- When carrying a live inflator module, hold securely with both hands and ensure that the bag/trim cover are pointed away.
- Place the inflator module on a bench or other surface with the bag and trim cover facing up.
- With the inflator module on the bench, never place anything on or close to the module which may be thrown in the event of an accidental deployment.
- Never use air bag component parts from another vehicle.
- If there is a chance of electrical shock to any of the air bag components, remove the air bag module before servicing the vehicle.

DISARMING

1. Before servicing the vehicle, refer to the precautions in the beginning of this section.
2. Remove or disconnect the following:
   - Negative battery cable
   - AIR BAG fuse

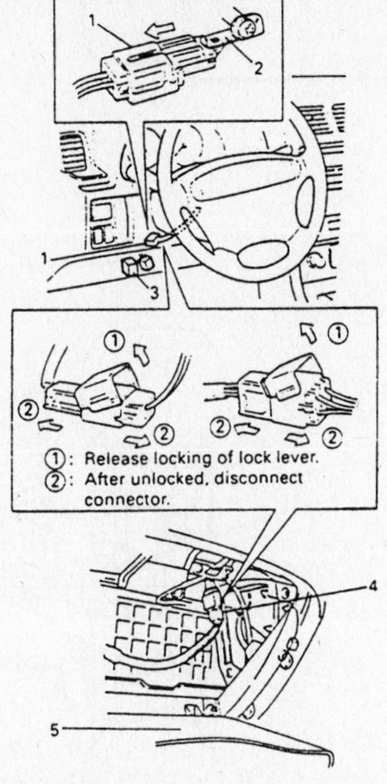
①: Release locking of lock lever.
②: After unlocked, disconnect connector.

1. Yellow connector of driver air bag (inflator) module
2. Connector stay
3. Air bag fuse box
4. Yellow connector of passenger air bag (inflator) module
5. Glove box

7924HG36

Air bag component location and identification—All models

- Driver air bag connector
- Glove box
- Passenger air bag connector

ARMING

When repairs are complete, install or connect the following:

- Passenger air bag connector
- Glove box
- Driver air bag connector
- AIR BAG fuse
- Negative battery cable

Recirculating Ball Steering Gear

REMOVAL & INSTALLATION

1. Before servicing the vehicle, refer to the precautions in the beginning of this section.
2. Remove or disconnect the following:
   - Skidplate, if equipped
   - Coolant overflow tank
   - Intermediate shaft pinch bolt
   - Power steering hoses
   - Pitman arm center link joint
   - Steering gearbox
   - Pitman arm

To install:

3. Install or connect the following:
   - Pitman arm. Tighten the nut to 102 ft. lbs. (140 Nm).
   - Steering gearbox. Tighten the bolts to 62 ft. lbs. (85 Nm).
   - Pitman arm center link joint. Tighten the nut to 37 ft. lbs. (50 Nm).
   - Power steering hoses
   - Intermediate shaft pinch bolt. Tighten it to 18 ft. lbs. (25 Nm).
   - Coolant overflow tank
   - Skidplate, if equipped
4. Fill the power steering system.
5. Start the engine and check for leaks.
6. Check the wheel alignment and adjust as necessary.

Power Rack and Pinion Steering Gear

REMOVAL & INSTALLATION

1. Before servicing the vehicle, refer to the precautions in the beginning of this section.
2. Remove or disconnect the following:
   - Power steering hoses
   - Intermediate shaft pinch bolt
   - Front wheels
   - Outer tie rod ends
   - Steering gear

To install:

3. Install or connect the following:
   - Steering gear. Tighten the bolts to 40 ft. lbs. (55 Nm).
   - Outer tie rod ends. Tighten the nuts to 32 ft. lbs. (43 Nm).
   - Front wheels
   - Intermediate shaft pinch bolt. Tighten it to 18 ft. lbs. (25 Nm).

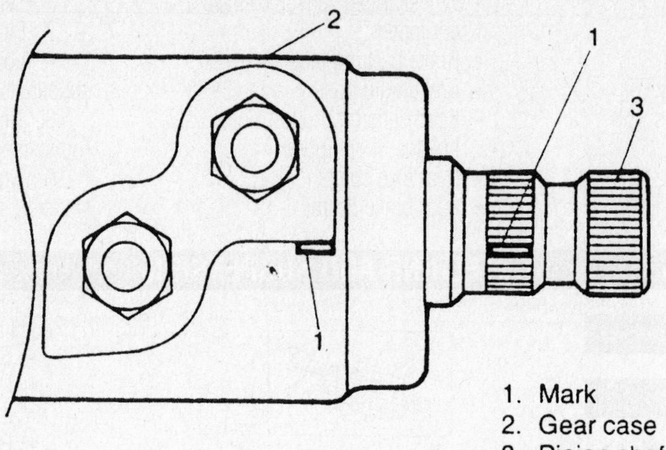

Steering gear centering marks

- Power steering hoses

4. Fill the power steering system.
5. Start the engine and check for leaks.
6. Check the wheel alignment and adjust as necessary.

Strut

REMOVAL & INSTALLATION

1. Before servicing the vehicle, refer to the precautions in the beginning of this section.
2. Support the control arm with a stand or floor jack.
3. Remove or disconnect the following:
   - Front wheel
   - Brake hose bracket
   - Strut bracket bolts
   - Upper strut mount nuts
   - Strut

To install:

4. Install or connect the following:
   - Strut. Tighten the upper mount nuts to 40 ft. lbs. (55 Nm) and the bracket bolts to 70 ft. lbs. (95 Nm).
   - Brake hose bracket
   - Front wheel
5. Check the wheel alignment and adjust as necessary.

Shock Absorber

REMOVAL & INSTALLATION

1. Before servicing the vehicle, refer to the precautions in the beginning of this section.
2. Support the rear axle housing with a hydraulic jack or stand.
3. Remove or disconnect the following:

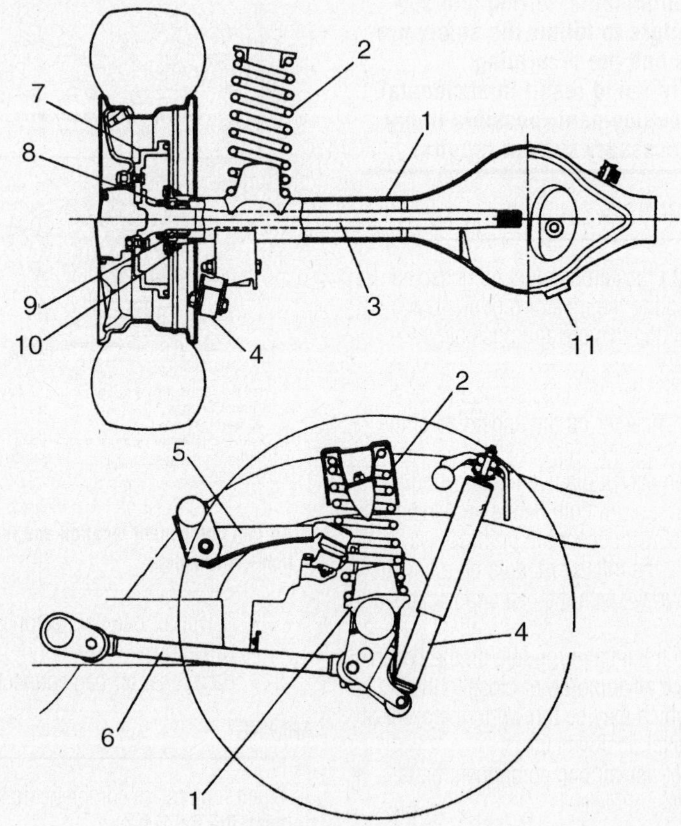

Rear suspension component identification

- Shock absorber upper locknut and retaining nut
- Lower shock absorber mounting nut and bolt
- Rear shock absorber

To install:

4. Install or connect the following:
   - Rear shock absorber
   - Lower mounting nut and bolt
   - Upper retaining nut and locknut
5. Torque the upper mounting nuts to 21 ft. lbs. (22–35 Nm) and the lower mounting nut/bolt to 62 ft. lbs. (85 Nm) for 2000–01 models; 74 ft. lbs. (100 Nm) for 2002–03 models.
6. Remove the jack or stand from the rear axle assembly.

Coil Spring

REMOVAL & INSTALLATION

Front

1. Before servicing the vehicle, refer to the precautions in the beginning of this section.
2. Support the vehicle at the frame with a hoist or jackstand.
3. Support the control arm with a floor jack.
4. Remove or disconnect the following:
   - Front wheel
   - Brake caliper and rotor
   - Locking hub or drive flange, if equipped
   - Axle shaft snapring and thrust washer, if equipped
   - Wheel speed sensor, if equipped
   - Stabilizer bar link
   - Lower ball joint
   - Strut bracket bolts
5. Lower the floor jack and remove the coil spring.

To install:

➡The bottom of the spring has a larger diameter than the top.

6. Install the coil spring onto the control arm and raise the floor jack.
7. Install or connect the following:
   - Strut bracket bolts. Tighten them to 70 ft. lbs. (95 Nm).
   - Lower ball joint. Tighten the nut to 40 ft. lbs. (55 Nm).
   - Stabilizer bar link. Tighten the nut to 21 ft. lbs. (29 Nm).
   - Wheel speed sensor, if equipped
   - Axle shaft snapring and thrust washer, if equipped
   - Locking hub or drive flange, if equipped
   - Brake caliper and rotor
   - Front wheel
8. Check the wheel alignment and adjust as necessary.

Rear

1. Before servicing the vehicle, refer to the precautions in the beginning of this section.
2. Support the vehicle at the frame with a hoist or jackstand.
3. Support the rear axle housing with a floor jack.
4. Remove or disconnect the following:
   - Rear wheels
   - Parking brake cable hanger
   - Shock absorber lower mounting bolts
   - Wheel speed sensor harness clamps, if equipped
   - Brake pipe E-ring
   - Axle vent hose
5. Lower the floor jack and remove the coil springs.

To install:

6. Install the coil springs onto the axle spring seats and raise the floor jack.
7. Install or connect the following:
   - Axle vent hose
   - Brake pipe E-ring
   - Wheel speed sensor harness clamps, if equipped
   - Shock absorber lower mounting bolts. Tighten them to 62 ft. lbs. (85 Nm) for 2000–01 models; 74 ft. lbs. (100 Nm) for 2002–03 models.
   - Parking brake cable hanger
   - Rear wheels

Lower Ball Joint

REMOVAL & INSTALLATION

The lower ball joint is serviced with the lower control arm as an assembly.

Lower Control Arm

REMOVAL & INSTALLATION

1. Before servicing the vehicle, refer to the precautions in the beginning of this section.
2. Support the vehicle at the frame with a hoist or jackstand.
3. Support the control arm with a floor jack.
4. Remove or disconnect the following:
   - Front wheel
   - Brake caliper and rotor
   - Locking hub or drive flange, if equipped
   - Axle shaft snapring and thrust washer, if equipped
   - Wheel speed sensor, if equipped
   - Stabilizer bar link
   - Lower ball joint
   - Strut bracket bolts
5. Lower the floor jack and remove the coil spring.
6. Remove the inner control arm bolts and remove the control arm.

To install:

7. Install the inner control arm bolts.
8. Install the coil spring onto the control arm and raise the floor jack.
9. Install or connect the following:
   - Strut bracket bolts. Tighten them to 70 ft. lbs. (95 Nm).
   - Lower ball joint. Tighten the nut to 40 ft. lbs. (55 Nm).
   - Stabilizer bar link. Tighten the nut to 21 ft. lbs. (29 Nm).
   - Wheel speed sensor, if equipped
   - Axle shaft snapring and thrust washer, if equipped
   - Locking hub or drive flange, if equipped
   - Brake caliper and rotor
   - Front wheel
10. Lower the vehicle so that the front suspension is at curb height.
11. Tighten the front inner bolt to 62 ft. lbs. (85 Nm) and the rear inner bolt to 92 ft. lbs. (127 Nm).
12. Check the wheel alignment and adjust as necessary.

CONTROL ARM BUSHING REPLACEMENT

1. Before servicing the vehicle, refer to the precautions in the beginning of this section.
2. Remove the control arm from the vehicle.
3. Remove the control arm bushings with a hydraulic press.

To install:

4. Lubricate the control arm bushings with liquid soap.
5. Press the bushings into the control arm until the bushing flange contacts the housing edge of the control arm.
6. Install the control arm to the vehicle.
7. Check the wheel alignment and adjust as necessary.

Wheel Bearings

ADJUSTMENT

The wheel bearings are not adjustable.

REMOVAL & INSTALLATION

1. Before servicing the vehicle, refer to the precautions in the beginning of this section.
2. Remove or disconnect the following:
   - Front wheel
   - Brake caliper and rotor
   - Locking hub or hub drive flange, if equipped
   - Hub grease cap, if equipped
   - Wheel speed sensor, if equipped
   - Wheel bearing lockwasher
   - Wheel bearing locknut and inner washer
   - Wheel hub and bearing assembly
   - Wheel hub oil seal
   - Wheel bearing oil seal
   - Snapring
3. Press the wheel bearing and race out of the hub.

To install:

4. Press the wheel bearing and race into the hub so that the race is fully seated in the hub bore.
5. Install or connect the following:
   - Snapring
   - Wheel bearing oil seal
   - Wheel hub oil seal
   - Wheel hub and bearing assembly
   - Wheel bearing locknut and inner washer. Tighten the nut to 157 ft. lbs. (216 Nm).
   - Wheel bearing lockwasher. Tighten the retaining screws to 13 inch lbs. (1.5 Nm).
   - Wheel speed sensor, if equipped
   - Hub grease cap, if equipped
   - Locking hub or hub drive flange, if equipped
   - Brake caliper and rotor
   - Front wheel

BRAKES

Brake Caliper

REMOVAL & INSTALLATION

1. Raise and safely support the vehicle.
2. Remove the wheels.
3. Disconnect and plug the brake line.
4. Remove the caliper mounting bolts (guide pins) and remove the caliper from the vehicle.

To install:

5. Install the caliper on the vehicle. Tighten the mounting bolts as follows:
   - Vitara: 20 ft. lbs. (27 Nm).
   - Grand Vitara: 10mm bolt to 37 ft. lbs. (50 Nm) and 12mm bolt to 62 ft. lbs. (84Nm)
6. Connect the hydraulic brake line, using 2 new washers. Torque the union bolt to 17 ft. lbs. (23 Nm).
7. Replace the front wheels.
8. Lower the vehicle.
9. Fill the brake reservoir and bleed the hydraulic brake system.

Disc Brake Pads

REMOVAL & INSTALLATION

1. Siphon about ⅔ of the fluid out of the master cylinder.
2. Raise and safely support the vehicle.
3. Remove the wheels.
4. Remove the brake caliper mounting bolts and remove the caliper from the mounting bracket.
5. Support the caliper with a wire.

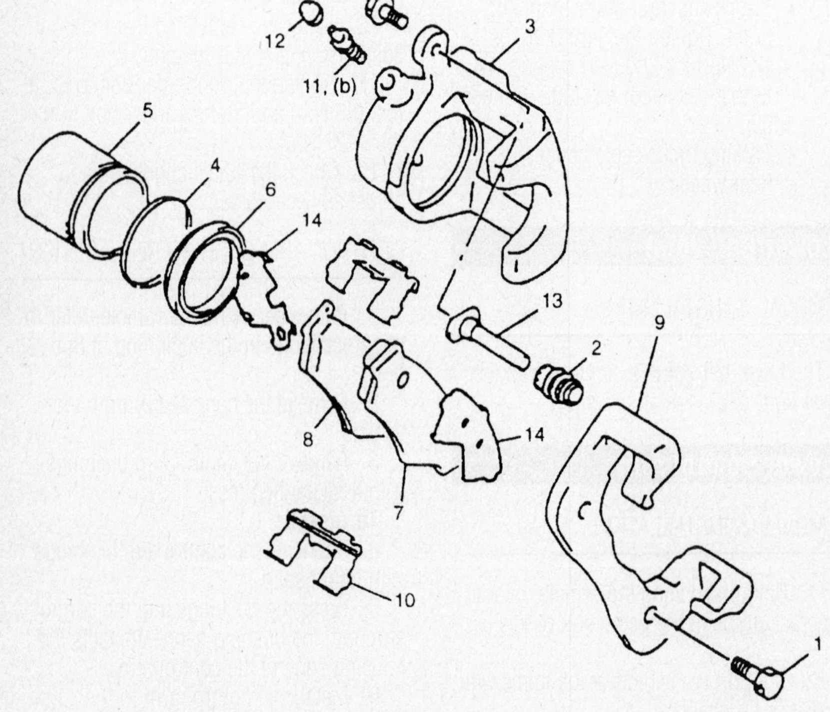

1. Caliper (slide) pin bolt
2. Boot
3. Disc brake caliper (disc brake cylinder)
4. Piston seal
5. Disc brake piston
6. Cylinder boot
7. Disc brake inner pad
8. Disc brake outer pad
9. Brake caliper carrier
10. Pad spring
11. Bleeder plug
12. Bleeder plug cap
13. Caliper pin
14. Anti noise shim
15. Inner shim

Tightening torque
(a): 8.0 N·m (0.80 kg-m, 6.0 lb-ft)
(b): 8.5 N·m (0.85 kg-m, 6.5 lb-ft)

93026G41

Front disc brake components—Vitara

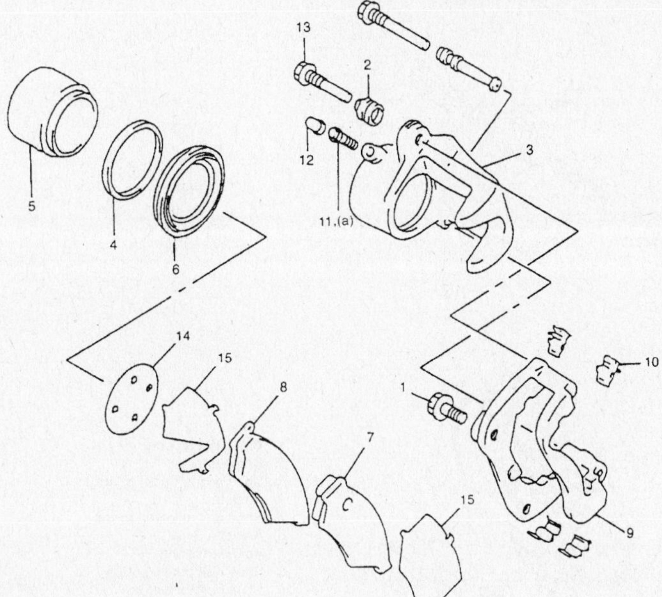

Front disc brake components—Grand Vitara

6. Using a large pair of plies or a C-clamp compress the caliper piston back into the bore.
7. Remove the disc brake pads and any shims from the caliper mounting bracket.

To install:

8. Install the brake pads and any shims removed from the caliper mounting bracket.
9. Install the caliper on the mounting bracket and install the mounting bolts.
10. Install the front wheels and lower the vehicle.

CAUTION

Do not attempt to drive the vehicle until after the following step is performed.

11. Depress the brake pedal repeatedly until a firm pedal is obtained. Do not attempt to drive the vehicle unless a firm pedal is obtained.
12. Check the fluid level in the master cylinder. Add fresh brake fluid, as necessary.
13. Road-test the vehicle.

Brake Drums

REMOVAL & INSTALLATION

1. Raise and safely support the vehicle.
2. Remove the rear wheel(s).
3. Release the parking brake.
4. Remove the parking brake lever cover screws and loosen the brake cable locking nut.
5. Install 2, 8mm bolts into the brake drum holes and uniformly tighten each bolt. Tighten each bolt until the brake drum is removed from the vehicle. If there is difficulty in removing the drum, insert a small tool through the hole in the rear of the backing plate, and hold the automatic adjusting lever away from the adjuster. Using another narrow, flat tool at the same time, reduce the brake shoe adjuster by turning the adjusting wheel.

To install:

6. Install the brake drum and pull the parking brake lever all the way up until a clicking sound can no longer be heard.

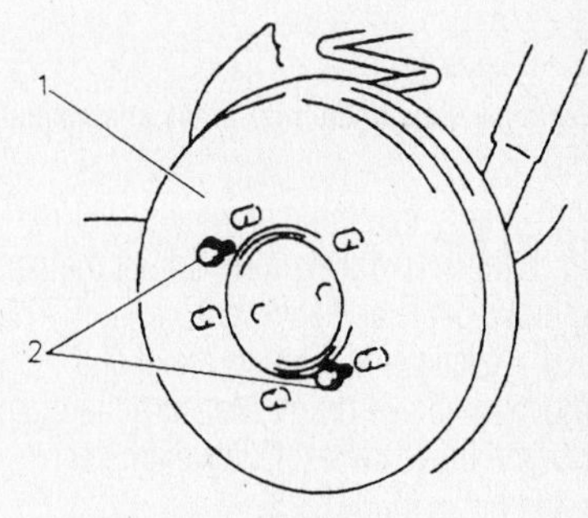

Removing the brake drum with the two 8mm bolts

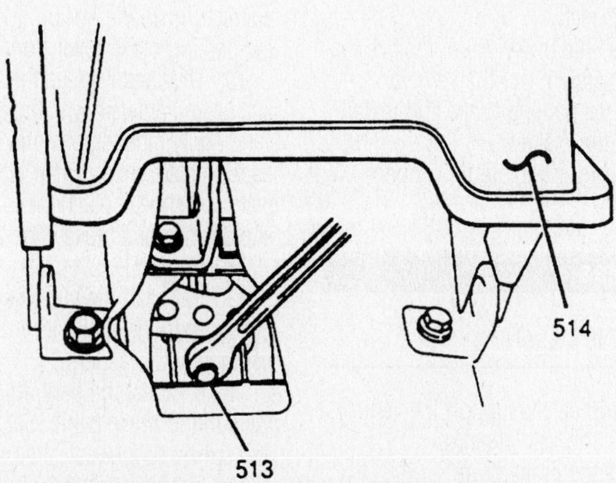

Reducing the adjuster to remove the brake drum

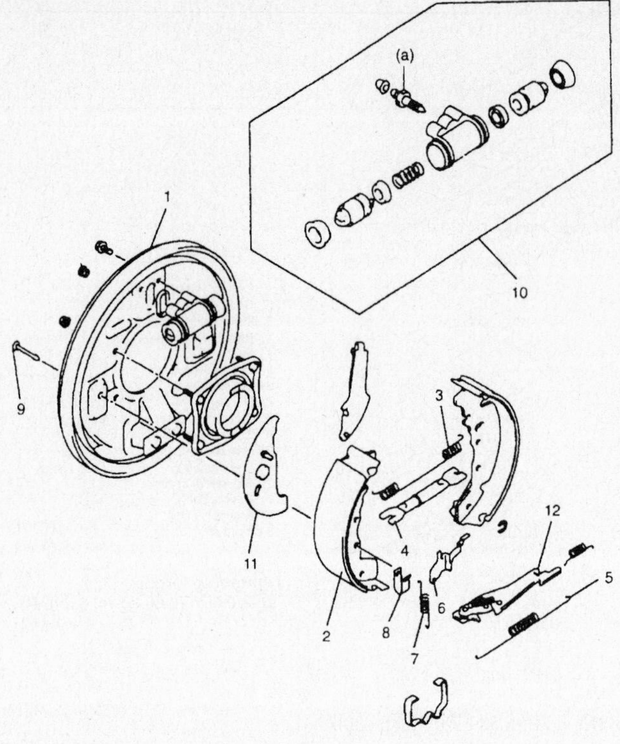

Exploded view of the rear brake components—Vitara and Grand Vitara

7. Verify that the rear wheels will not turn. If the rear wheels turn, adjust the parking brake cable as necessary.
8. Release the parking brake and remove the brake drum. Measure the diameter of the brake shoes. Outer diameter should be as follows:
   - For 2 door models: 8.638 (0.0012 inches (219 (0.3mm)
   - For 4 door models: 9.980 (0.0079 inches (253.5 (0.2mm)
9. If the brake shoe clearance is not correct, adjust the brake shoes until the clearance is correct.
10. Reinstall the brake drum, replace the wheel(s), and safely lower the vehicle.
11. Adjust the parking brake and install the cover with the 2 screws.
12. Road-test the vehicle for proper brake operation.

Brake Shoes

REMOVAL & INSTALLATION

1. Raise and safely support the vehicle.
2. Remove the rear wheel(s).
3. Remove the brake drum.
4. Using a suitable tool, remove the brake shoe return spring.
5. Using a brake spring hold-down tool, disengage the hold-down spring and retainers from the front shoe. Remove the hold-down retainer pinch
6. Disconnect the anchor spring from the front shoe and remove the front shoe.
7. Remove the anchor spring from the rear shoe. Using a brake spring hold-down tool, disengage the hold-down spring and retainers from the rear shoe. Remove the hold-down pinch
8. Disengage the parking brake lever from the parking brake cable and remove the rear shoe.
9. Remove the C-washer, the automatic adjuster lever and spring, the C-washer, and the parking brake lever from the rear shoe.
10. Thoroughly clean the backing plate and brake hardware with brake cleaning solvent. Apply high temperature grease to the backing plate shoe contact points, anchor plate and shoe contact points, adjusting bolt, and adjuster and brake shoe contact points.

To install:

11. Reinstall the automatic adjuster lever and the parking brake lever to the rear shoe using new C-washers.
12. Connect the parking brake lever to the parking brake cable. Set the adjuster and spring to the rear shoe.
13. Set the rear brake shoe in place, install the hold-down pin and install the hold-down spring and retainers. Make sure that the shoe is inserted in the wheel cylinder and that the other end is in the anchor plate.
14. Install the anchor spring to the rear shoe.
15. Install the front shoe to the other end of the anchor spring and set the front shoe in place. Make sure that the front shoe engages the wheel cylinder, adjuster mechanism and spring, and the anchor plate.
16. Reinstall the front brake shoe hold-down pin and secure with the hold-down spring and retainers using a suitable tool.
17. Install the return spring.
18. Install the brake drum and pull the parking brake lever all the way up until a clicking sound can no longer be heard.
19. Verify that the rear wheels will not turn. If the rear wheels turn, adjust the parking brake cable as necessary.
20. Release the parking brake and remove the brake drum. Measure the diameter of the brake shoes. Brake diameter should be as follows:
   - For 2 door models: 8.638 (0.0012 inches (219 (0.3mm)
   - For 4 door models: 9.980 (0.0079 inches (253.5 (0.2mm)
21. If the brake shoe clearance is not correct, adjust the brake shoes until the clearance is correct.
22. Reinstall the brake drum, replace the wheel(s), and safely lower the vehicle.
23. Road-test the vehicle for proper brake operation.

TOYOTA

4Runner

16

SPECIFICATION AND MAINTENANCE CHARTS

ENGINE AND VEHICLE IDENTIFICATION

	Engine					
Code ①	Liters (cc)	Cu. In.	Cyl.	Fuel Sys.	Engine Type	Eng. Mfg.
5VZ-FE	3.4 (3378)	206	6	MFI	DOHC	Toyota
1GR-FE	4.0 (3956)	241	6	SFI	DOHC	Toyota
2UZ-FE	4.7 (4664)	285	8	SFI	DOHC	Toyota

Model Year	
Code ②	Year
1	2001
2	2002
3	2003
4	2004
5	2005

SFI: Sequential Fuel Injection

MFI: Multi-port Fuel Injection

DOHC: Double Overhead Camshaft

① Stamped on the left side of the engine block

② 10th digit of the Vehicle Identification Number (VIN)

67170-4RUN-C01

GENERAL ENGINE SPECIFICATIONS

Year	Model	Engine Displacement Liters	Engine Series ID	Net Horsepower @ rpm	Net Torque @ rpm (ft. lbs.)	Bore x Stroke (in.)	Com-pression Ratio	Oil Pressure @ rpm
2001	4Runner	3.4	5VZ-FE	183@4800	217@3600	3.68x3.23	9.6:1	36-75@3000
2002	4Runner	3.4	5VZ-FE	183@4800	217@3600	3.68x3.23	9.6:1	36-75@3000
2003	4Runner	4.0	1GR-FE	245@5200	282@3200	3.70x3.74	NA	43-85@3000
		4.7	2UZ-FE	245@4800	315@3400	3.70x3.30	9.6:1	45-65@3000
2004	4Runner	4.0	1GR-FE	245@5200	282@3200	3.70x3.74	NA	43-85@3000
		4.7	2UZ-FE	245@4800	315@3400	3.70x3.30	9.6:1	45-65@3000

NA: Information not available

67170-4RUN-C02

ENGINE TUNE-UP SPECIFICATIONS

Year	Engine Displacement Liters	Engine ID	Spark Plug Gap (in.)	Ignition Timing (deg.)*	Fuel Pump (psi)	Idle Speed (rpm)	Valve Clearance	
							Intake	Exhaust
2001	3.4	5VZ-FE	0.031	8-12	38-44	650-750	0.006-0.010	0.010-0.014
2002	3.4	5VZ-FE	0.039-0.043	8-12	38-44	650-750	0.006-0.009	0.011-0.014
2003	4.0	1GR-FE	0.043	8-12	41-42	650-750	0.006-0.010	0.011-0.015
	4.7	2UZ-FE	0.043	8-12	38-44	650-750	0.006-0.009	0.011-0.014
2004	4.0	1GR-FE	0.043	8-12	41-42	650-750	0.006-0.010	0.011-0.015
	4.7	2UZ-FE	0.043	8-12	38-44	650-750	0.006-0.009	0.011-0.014

NOTE: The Vehicle Emission Control Information label often reflects specification changes made during production.

The label figures must be used if they differ from those in this chart.

* With terminals TC and CG connected to DLC3

67170-4RUN-C03

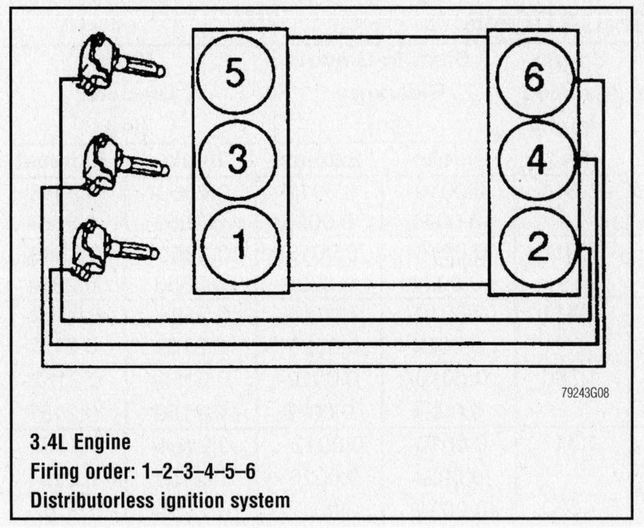

3.4L Engine
Firing order: 1–2–3–4–5–6
Distributorless ignition system

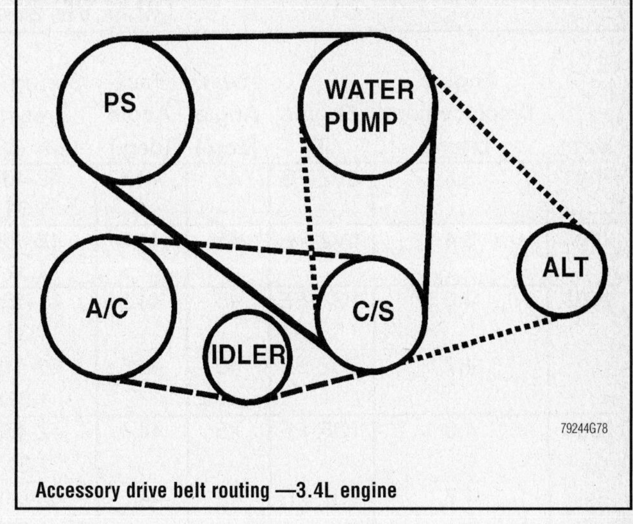

Accessory drive belt routing —3.4L engine

CAPACITIES

Year	Model	Engine Displacement Liters	Engine ID	Engine Oil with Filter (qts.)	Transmission (pts)*	Transfer Case (pts.)	Drive Axle Front (pts.)	Drive Axle Rear (pts.)	Fuel Tank (gal.)	Cooling System (qts.)
2001	4Runner	3.4	5VZ-FE	5.5	①	2.6	②	③	18.5	④
2002	4Runner	3.4	5VZ-FE	5.5	①	2.6	②	⑤	18.5	④
2003	4Runner	4.0	1GR-FE	5.5	⑥	3.0	3.0	6.5	23.0	10.4
		4.7	2UZ-FE	6.5	⑥	3.0	3.0	6.5	23.0	13.0
2004	4Runner	4.0	1GR-FE	5.5	⑥	3.0	3.0	6.5	23.0	10.4
		4.7	2UZ-FE	6.5	⑥	3.0	3.0	6.5	23.0	13.0

* Drain and refill. After draining, add the following amounts, then, fill to the cold full line.

① 2wd: 3.4
4wd: 4.2

② Without ADD: 2.32
With ADD: 2.44

③ 2WD: 5.8
4WD with differential locks: 5.8
4WD without differential locks: 5.2

④ With rear heater: 9.5
Without rear heater: 8.5

⑤ 2wd: 5.82
4wd w/o diff. Lock: 5.18
4wd w/diff. Lock: 5.82

⑥ A340E, A340F: 4.2
A750E, A750F: 6.4

67170-4RUN-C04

VALVE SPECIFICATIONS

Year	Engine Displacement Liters	Engine ID	Seat Angle (deg.)	Face Angle (deg.)	Spring Test Pressure (lbs. @ in.)	Spring Installed Height (in.)	Stem-to-Guide Clearance (in.) Intake	Stem-to-Guide Clearance (in.) Exhaust	Stem Diameter (in.) Intake	Stem Diameter (in.) Exhaust
2001	3.4	5VZ-FE	45	44.5	42-46@ 1.311	1.311	0.0010-0.0024	0.0012-0.0026	0.2350-0.2356	0.2348-0.2354
2002	3.4	5VZ-FE	45	44.5	42-46@ 1.311	1.311	0.0010-0.0024	0.0012-0.0026	0.2350-0.2356	0.2348-0.2354
2003	4.0	1GR-FE	45	44.5	42-46@ 1.311	1.311	0.0010-0.0024	0.0012-0.0026	0.2154-0.2159	0.2152-0.2157
	4.7	2UZ-FE	45	44.5	46-51@ 1.378	1.380	0.0010-0.0024	0.0012-0.0026	0.2154-0.2159	0.2152-0.2157
2004	4.0	1GR-FE	45	44.5	42-46@ 1.311	1.311	0.0010-0.0024	0.0012-0.0026	0.2154-0.2159	0.2152-0.2157
	4.7	2UZ-FE	45	44.5	46-51@ 1.378	1.380	0.0010-0.0024	0.0012-0.0026	0.2154-0.2159	0.2152-0.2157

67170-4RUN-C05

CRANKSHAFT AND CONNECTING ROD SPECIFICATIONS

All measurements are given in inches.

Year	Engine Displacement Liters	Engine ID	Crankshaft Main Brg. Journal Dia.	Crankshaft Main Brg. Oil Clearance	Crankshaft Shaft End-play	Crankshaft Thrust on No.	Connecting Rod Journal Diameter	Connecting Rod Oil Clearance	Connecting Rod Side Clearance
2001	3.4	5VZ-FE	2.5191-2.5197	0.0008-0.0015	0.0008-0.0087	2	2.1648-2.1654	0.0009-0.0021	0.0059-0.0130
2002	3.4	5VZ-FE	2.5191-2.5197	①	0.0008-0.0087	2	2.1648-2.1654	0.0009-0.0021	0.0059-0.0130
2003	4.0	1GR-FE	2.8342-2.8346	0.0007-0.0012	0.0016-0.0094	3	2.2044-2.2047	0.0002-0.0004	0.0059-0.0118
	4.7	2UZ-FE	2.6373-2.6378	0.0016-0.0023	0.0008-0.0087	3	2.0465-2.0472	0.0011-0.0021	0.0063-0.0138
2004	4.0	1GR-FE	2.8342-2.8346	0.0007-0.0012	0.0016-0.0094	3	2.2044-2.2047	0.0002-0.0004	0.0059-0.0118
	4.7	2UZ-FE	2.6373-2.6378	0.0016-0.0023	0.0008-0.0087	3	2.0465-2.0472	0.0011-0.0021	0.0063-0.0138

① No. 1: 0.0008-0.0015 in.
All others: 0.0009-0.0017 in.

67170-4RUN-C06

PISTON AND RING SPECIFICATIONS

All measurements are given in inches.

Year	Engine Displ. Liters	Engine ID	Piston Clearance	Ring Gap Top Comp.	Ring Gap Bottom Comp.	Ring Gap Oil Control	Ring Side Clearance Top Comp.	Ring Side Clearance Bottom Comp.	Ring Side Clearance Oil Control
2001	3.4	5VZ-FE	0.0053-0.0060	0.0118-0.0197	0.0157-0.0236	0.0059-0.0217	0.0016-0.0031	0.0012-0.0028	SNUG
2002	3.4	5VZ-FE	0.0053-0.0060	0.0118-0.0197	0.0157-0.0236	0.0059-0.0217	0.0016-0.0031	0.0012-0.0028	SNUG
2003	4.0	1GR-FE	0.0031-0.0040	0.0118-0.0157	0.0157-0.0197	0.0039-0.0157	0.0008-0.0028	0.0008-0.0024	0.0028-0.006
	4.7	2UZ-FE	0.0035-0.0044	0.0118-0.0197	0.0157-0.0256	0.0051-0.0189	0.0012-0.0031	0.0012-0.0028	SNUG
2004	4.0	1GR-FE	0.0031-0.0040	0.0118-0.0157	0.0157-0.0197	0.0039-0.0157	0.0008-0.0028	0.0008-0.0024	0.0028-0.006
	4.7	2UZ-FE	0.0035-0.0044	0.0118-0.0197	0.0157-0.0256	0.0051-0.0189	0.0012-0.0031	0.0012-0.0028	SNUG

67170-4RUN-C07

TORQUE SPECIFICATIONS

All readings in ft. lbs.

Year	Engine Displacement Liters	Engine ID	Cylinder Head Bolts	Main Bearing Bolts	Rod Bearing Bolts	Crankshaft Damper Bolts	Flywheel Bolts	Manifold Intake	Manifold Exhaust	Spark Plugs	Oil Pan Drain Plug
2001	3.4	5VZ-FE	①	②	③	184	63-67	13	30	13	28
2002	3.4	5VZ-FE	④	②	③	213	61	13	30	13	28
2003	4.0	1GR-FE	⑤	⑥	③	184	61	19	22	15	30
	4.7	2UZ-FE	⑦	⑧	③	181	⑨	13	33	13	29
2004	4.0	1GR-FE	⑤	⑥	③	184	61	19	22	15	30
	4.7	2UZ-FE	⑦	⑧	③	181	⑨	13	33	13	29

① Step 1: 25 ft. lbs.
Step 2: Plus 90 degrees
Recessed head: 13 ft. lbs.

② Step 1: 45 ft. lbs.
Step 2: Plus 90 degrees

③ Step 1: 18 ft. lbs.
Step 2: Plus 90 degrees

④ 12-pointed bolts
Step 1: 25 ft. lbs.
Step 2: +90 degrees
Step 3: +90 degrees
Recessed heads: 13 ft. lbs.

⑤ Right side: step 1, 27 ft. lbs.; step 2, plus 180 degrees
Left side: step 1, 27 ft. lbs.; step 2, plus 180 degrees; 14mm cap screw, 22 ft. lbs.

⑥ 12-point bolt heads: step 1, 45 ft. lbs.; step 2, plus 90 degrees; 12mm capscrew, 18 ft. lbs.

⑦ Step 1: 24 ft. lbs.
Step 2: plus 90 degrees
Step 3: plus 90 degrees

⑧ Step 1: 20 ft lbs.
Step 2: plus 90 degrees

⑨ Step 1: 22 ft. lbs.
Step 2: plus 90 degrees

67170-4RUN-C08

WHEEL ALIGNMENT

Year	Model	Caster Range (+/-Deg.)	Caster Preferred Setting (Deg.)	Camber Range (+/-Deg.)	Camber Preferred Setting (Deg.)	Toe-in (in.)	Steering Axis Inclination (Deg.)
2001	2WD	0.75	+3.25	0.75	-0.25	0.08+/-0.08	11+/-0.75
	4WD	0.75	+3.06	0.75	-0.25	0.08+/-0.08	11+/-0.75
2002	2WD	0.75	+3.25	0.75	-0.25	0.08+/-0.08	11+/-0.75
	4WD	0.75	+3.06	0.75	-0.25	0.08+/-0.08	11+/-0.75
2003	2WD exc. air suspension	0.75	+3.38	0.75	-0.47	0.08+/-0.16	12.97+/-0.75
	2WD w/air suspension	0.75	+3.55	0.75	-0.50	0.08+/-0.16	13.00+/-0.75
	4WD exc. air suspension	0.75	+3.22	0.75	-0.15	0.08+/-0.16	12.65+/-0/75
	4WD w/air suspension	0.75	+3.37	0.75	-0.17	0.08+/-0.16	12.67+/-0.75
2004	2WD exc. air suspension	0.75	+3.38	0.75	-0.47	0.08+/-0.16	12.97+/-0.75
	2WD w/air suspension	0.75	+3.55	0.75	-0.50	0.08+/-0.16	13.00+/-0.75
	4WD exc. air suspension	0.75	+3.22	0.75	-0.15	0.08+/-0.16	12.65+/-0/75
	4WD w/air suspension	0.75	+3.37	0.75	-0.17	0.08+/-0.16	12.67+/-0.75

67170-4RUN-C09

TIRE, WHEEL AND BALL JOINT SPECIFICATIONS

Year	Model	OEM Tires Standard	OEM Tires Optional	Tire Pressures (psi) Front	Tire Pressures (psi) Rear	Wheel Size	Ball Joint Inspection	Lugnut Torque (ft. lbs.)
2001	4Runner	P225/75R15	P265/70R16	Std: 29 Opt: 32	Std: 29 Opt: 32	7-JJ	6-39 ②	83
2002	4Runner SR5	P225/75R15	P265/70R16	Std: 29 Opt: 32	Std: 29 Opt: 32	7-JJ	③	83
	4Runner Limited	P265/70R16	None	32	32	7-JJ	③	83
2003	4Runner SR5	P225/75R15	P265/70R16	Std: 29 Opt: 32	Std: 29 Opt: 32	7-JJ	③	83
	4Runner Limited	P265/70R16	None	32 Opt: 32	32 Opt: 32	7-JJ	③	83
2004	4Runner	P265/70R16	P265/65R17	32	32	④	⑤	83

OEM: Original Equipment Manufacturer

PSI: Pounds Per Square Inch

STD: Standard

OPT: Optional

① Replace if any measurable movement is found.

② Torque required in inch lbs. to rotate ball joint when removed from the knuckle

③ Turning torque: upper 6-39 in. lbs.; lower 0.8-21.7 in. lbs.

④ 16 in. steel: 7J
16 inch aluminum: 7JJ
17 inch: 7.5JJ

⑤ Turning torque: upper less than 40 inch lbs.; lower, less than 27 inch lbs.

67170-4RUN-C10

BRAKE SPECIFICATIONS

All measurements in inches unless noted

			Brake Disc			Brake Drum Diameter				Brake Caliper	
						Original	Max.	Maximum	Minimum	Bracket	Mounting
			Original	Minimum	Maximum	Inside	Wear	Machine	Lining	Bolts	Bolts
Year	Model		Thickness	Thickness	Runout	Diameter	Limit	Diameter	Thickness	(ft. lbs.)	(ft. lbs.)
2001	4Runner		0.866	0.787	0.0028	11.61	—	11.69	0.039	—	90
2002	4Runner		0.866	0.787	0.0028	11.61	—	11.69	0.039	—	90
2003	4Runner	F	1.102	1.024	0.0020	—	—	—	0.039	—	91
		R	0.709	0.630	0.0079	—	—	—	0.039	77	65
2004	4Runner	F	1.102	1.024	0.0020	—	—	—	0.039	—	91
		R	0.709	0.630	0.0079	—	—	—	0.039	77	65

67170-4RUN-C11

SCHEDULED MAINTENANCE INTERVALS

TOYOTA—4RUNNER

		VEHICLE MILEAGE INTERVAL (x1000)																		
TO BE SERVICED	TYPE OF SERVICE	5	10	15	20	25	30	35	40	45	50	55	60	65	70	75	80	85	90	95
Automatic transmission and differential fluid	S/I			✓			✓			✓			✓			✓			✓	
Ball joints and boots	S/I			✓			✓			✓			✓			✓			✓	
Brake system	S/I			✓			✓			✓			✓			✓			✓	
Charcoal canister	S/I												✓							
Drive belts	S/I						✓						✓						✓	
Driveshaft bushing	L						✓						✓						✓	
Engine coolant	R						✓						✓						✓	
Engine oil & filter	R	✓	✓	✓	✓	✓	✓	✓	✓	✓	✓	✓	✓	✓	✓	✓	✓	✓	✓	✓
Exhaust system	S/I			✓			✓			✓			✓			✓			✓	
Fuel tank cap gasket	S/I						✓						✓						✓	
Halfshaft boots & flange bolts	S/I			✓			✓			✓			✓			✓			✓	
Limited slip differential fluid	R						✓						✓						✓	
Manual transmission and differential fluid	S/I						✓						✓						✓	
Platinum spark plugs	R												✓							
Propeller shaft (4WD)	L			✓			✓			✓			✓			✓			✓	
Propeller shaft bolts	S/I			✓			✓			✓			✓			✓			✓	
Rack and pinion assembly	S/I			✓			✓			✓			✓			✓			✓	
Rear wheel bearing	L						✓						✓						✓	
Steering linkage	S/I			✓			✓			✓			✓			✓			✓	
Valves	S/I												✓							

R: Replace   S/I: Service or Inspect   L: Lubricate

FREQUENT OPERATION MAINTENANCE (SEVERE SERVICE)

If a vehicle is operated under any of the following conditions it is considered severe service:

- Towing a trailer or using a camper or car-top carrier.

- Repeated short trips of less than 5 miles in temperatures below freezing.

- Excessive idling or low-speed driving for long distances as in heavy commercial use, such as delivery, taxi or police cars.

- Operating on rough, muddy or salt-covered roads.

- Operating on unpaved or dusty roads.

Oil filter: service or inspect every 5000 miles or 4 months, whichever occurs first.

Brake linings and discs or drums: service or inspect every 5000 miles or 4 months, whichever occurs first.

Steering linkage: service or inspect every 5000 miles or 4 months, whichever occurs first.

Ball joints and boots: service or inspect every 5000 miles or 4 months, whichever occurs first.

Brake discs & pads (front): service or inspect every 6000 miles.

Halfshaft boots: service or inspect every 5000 miles or 4 months. Retighten the flange bolts, whichever occurs first.

Body chassis bolts and nuts: service or inspect every 5000 miles or 4 months, whichever occurs first.

Transmission and differential fluid: replace every 15,000 miles or 12 months, whichever occurs first.

Transfer case and differential fluid: replace every 15,000 miles or 12 months, whichever occurs first.

Timing belt: replace every 60,000 miles or 48 months, whichever occurs first.

67170-4RUN-C12

PRECAUTIONS

Before servicing any vehicle, please be sure to read all of the following precautions, which deal with personal safety, prevention of component damage, and important points to take into consideration when servicing a motor vehicle:

- Never open, service or drain the radiator or cooling system when the engine is hot; serious burns can occur from the steam and hot coolant.
- Observe all applicable safety precautions when working around fuel. Whenever servicing the fuel system, always work in a well-ventilated area. Do not allow fuel spray or vapors to come in contact with a spark, open flame, or excessive heat (a hot drop light, for example). Keep a dry chemical fire extinguisher near the work area. Always keep fuel in a container specifically designed for fuel storage; also, always properly seal fuel containers to avoid the possibility of fire or explosion. Refer to the additional fuel system precautions later in this section.
- Fuel injection systems often remain pressurized, even after the engine has been turned **OFF**. The fuel system pressure must be relieved before disconnecting any fuel lines. Failure to do so may result in fire and/or personal injury.
- Brake fluid often contains polyglycol ethers and polyglycols. Avoid contact with the eyes and wash your hands thoroughly after handling brake fluid. If you do get brake fluid in your eyes, flush your eyes with clean, running water for 15 minutes. If eye irritation persists, or if you have taken brake fluid internally, IMMEDIATELY seek medical assistance.
- The EPA warns that prolonged contact with used engine oil may cause a number of skin disorders, including cancer. You should make every effort to minimize your exposure to used engine oil. Protective gloves should be worn when changing oil. Wash your hands and any other exposed skin areas as soon as possible after exposure to used engine oil. Soap and water, or waterless hand cleaner should be used.
- All new vehicles are now equipped with an air bag system. The system must be disabled before performing service on or around system components, steering column, instrument panel components, wiring and sensors. Failure to follow safety and disabling procedures could result in accidental air bag deployment, possible personal injury and unnecessary system repairs.
- Always wear safety goggles when working with, or around, the air bag system. When carrying a non-deployed air bag, be sure the bag and trim cover are pointed away from your body. When placing a non-deployed air bag on a work surface, always face the bag and trim cover upward, away from the surface. This will reduce the motion of the module if it is accidentally deployed. Refer to the additional air bag system precautions later in this section.
- NEVER disconnect the negative battery cable with the ignition **ON** or the engine running. Removing power from the computer control module with the ignition **ON** may destroy the module.
- Clean, high quality brake fluid from a sealed container is essential to the safe and proper operation of the brake system. You should always buy the correct type of brake fluid for your vehicle. If the brake fluid becomes contaminated, completely flush the system with new fluid. Never reuse any brake fluid. Any brake fluid that is removed from the system should be discarded. Also, do not allow any brake fluid to come in contact with a painted surface; it will damage the paint.
- Never operate the engine without the proper amount and type of engine oil; doing so WILL result in severe engine damage.
- Timing belt maintenance is extremely important. Many models utilize an interference-type, non-freewheeling engine. If the timing belt breaks, the valves in the cylinder head may strike the pistons, causing potentially serious (also time-consuming and expensive) engine damage. Refer to the maintenance interval charts in the front of this manual for the recommended replacement interval for the timing belt, and to the timing belt section for belt replacement and inspection.
- Disconnecting the negative battery cable on some vehicles may interfere with the functions of the on-board computer system(s) and may require the computer to undergo a relearning process once the negative battery cable is reconnected.
- When servicing drum brakes, only disassemble and assemble one side at a time, leaving the remaining side intact for reference.
- Only an MVAC-trained, EPA-certified automotive technician should service the air conditioning system or its components.

ENGINE REPAIR

➡Disconnecting the negative battery cable on some vehicles may interfere with the functions of the on board computer system. The computer may undergo a relearning process once the negative battery cable is reconnected.

Distributor

All models are equipped with a distibutorless ignition system.

Alternator

REMOVAL

3.4L Engine

1. Remove or disconnect the following:
   - Negative battery cable
   - Alternator wiring
   - Alternator locknut, pivot bolt, nut and adjusting bolt
   - Drive belt
   - Alternator

To install:

Install or connect the following:

- Alternator
- Drive belt. Tighten the locknut 25 ft. lbs. (33 Nm) and the pivot bolt 38 ft. lbs. (51 Nm).
- Alternator wiring
- Negative battery cable

4.0L Engine

1. Remove the V-bank cover.
2. Remove the engine under-cover.
3. Remove the accessory drive belt.
4. Remove the battery.
5. Disconnect the wiring harness from the alternator.
6. Remove the 2 mounting bolts and lift out the alternator.
7. Installation is the reverse of removal. Torque the mounting bolts to 32 ft. lbs. (43 Nm).

4.7L Engine

1. Disconnect the battery ground cable.
2. Remove the radiator upper seal (11 clips).
3. Remove the fan and shroud.
4. Remove the power steering pump and position it out of the way without disconnecting the hoses.
5. Disconnect the alternator wiring harness.

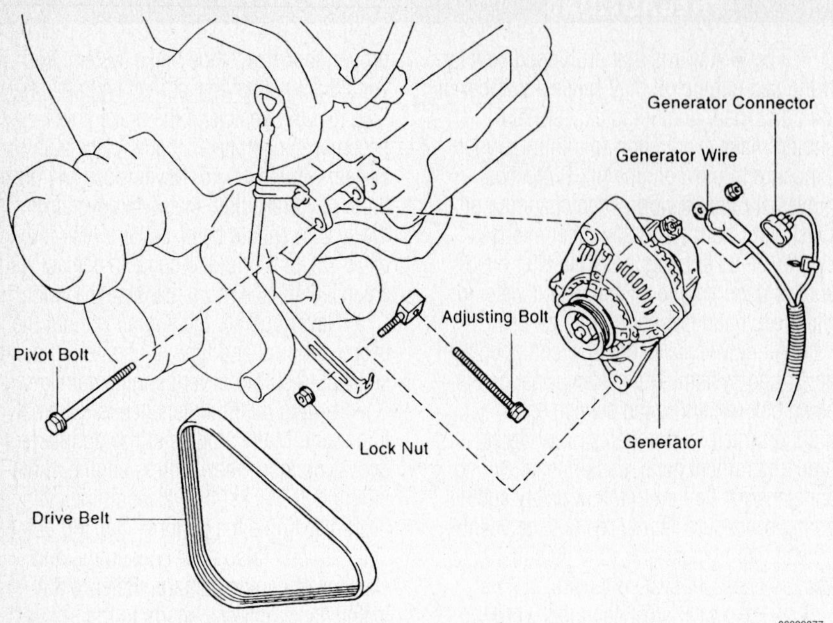

Exploded view of the alternator and drive belt—3.4L engine

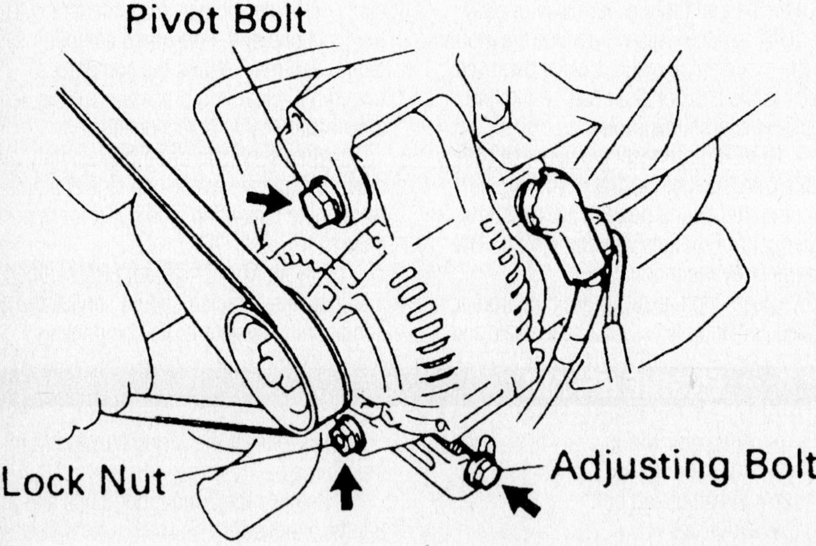

Locations of the adjusting and pivot bolts and the locknut—3.4L engine

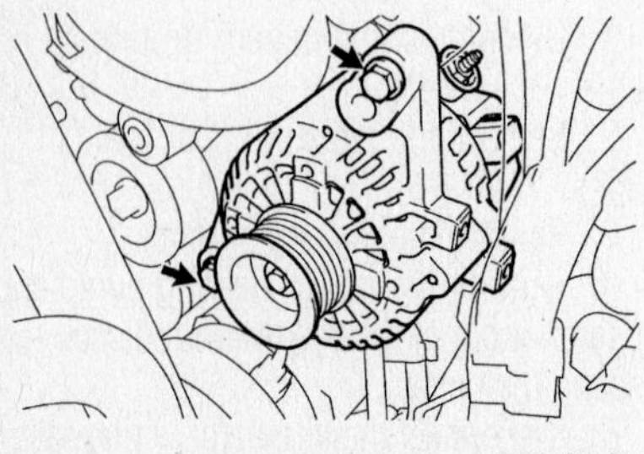

Alternator mounting—4.0L engine

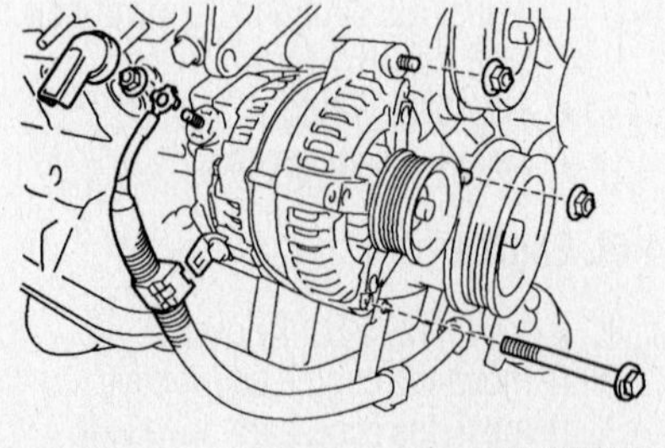

Alternator mounting—4.7L engine

6. Remove the bolt and 2 nuts, and lift out the alternator.

7. Installation is the reverse of removal. Observe the following torques:

- Alternator bolt: 29 ft. lbs. (39 Nm)
- Alternator M10 nut: 29 ft. lbs. (39 Nm)
- Alternator M8 nut: 11 ft. lbs. (16 Nm)
- Power steering pump fasteners: 32 ft. lbs. (43 Nm)
- Fan nuts: 21 ft. lbs. (29 Nm)

Ignition Timing

ADJUSTMENT

All engines are equipped with a Distributorless Ignition System (DIS). No timing adjustment is possible.

Engine Assembly

REMOVAL & INSTALLATION

3.4L

2WD

1. Before servicing the vehicle, refer to the precautions in the beginning of this section.
2. Properly relieve the fuel system pressure.
3. Remove or disconnect the following:
   - Hood
   - Battery
   - Engine under covers
4. Drain the engine coolant.
5. Drain the engine oil.
   - Radiator
   - Fan with the fluid coupling and fan pulleys
   - Air cleaner cap
   - Mass Air Flow (MAF) meter and the resonator
   - Air cleaner case and filter
6. Disconnect the following hoses:
   - Heater hoses
   - Brake booster vacuum hose
   - Evaporative Emissions (EVAP) hose
   - Fuel return hose
   - Fuel inlet hose
7. Detach the starter wire and connectors, as follows:
   - Ground strap, by removing the bolt
   - 3 starter wire clamps and connector
8. Detach the alternator connector and wire.
9. Disconnect the engine wiring harness, as follows:

- Glove box door
- Lower the finish No. 2 panel
- 4 ECM connectors
- 2 cassette connectors and the 2 wire clamps from the lower finish panel
- Engine wiring harness clamp

10. Remove or disconnect the following:
    - Igniter connector
    - Ground strap
    - Vacuum Switching Valve (VSV) connector for the Evaporative Emissions (EVAP)
    - Vapor pressure sensor connector and clamp
    - Vapor connector for the vapor pressure sensor and clamp
    - 2 engine wiring harness retainer-to-cowl panel nuts and pull out the engine wiring harness
    - Driveshaft from the transmission
    - Speedometer cable
    - Front exhaust pipe
    - Nut and the control cable
11. Place a jack under the transmission.
12. Remove or disconnect the following:
    - Transmission rear mounting bracket by removing the 8 bolts
    - Bolt and the air conditioning compressor wire clamp, if equipped with air conditioning
13. If necessary, install a No. 2 engine hanger with 2 bolts. Tighten the 2 bolts to 30 ft. lbs. (40 Nm).
14. Attach the engine hoist chain to the 2 engine hangers.
15. Remove or disconnect the following:
    - 4 engine front mounting insulators-to-frame bolts and nuts
    - Engine and transmission

To install:

16. Install or connect the following:
    - Engine
    - Engine mounts to the body mountings. Install the bolts and nuts but do not tighten at this time.
17. Remove the engine chain hoist the No. 2 engine hanger.
18. Install or connect the following:
    - Air conditioning wire with the bolt, if equipped with air conditioning
    - Transmission mounting bracket. Tighten the frame bolts to 43 ft. lbs. (58 Nm) and the mounting insulator bolts to 13 ft. lbs. (18 Nm).
    - Tighten the engine mounting nuts and bolts to 28 ft. lbs. (38 Nm).
    - Control cable
    - Front exhaust pipe
    - Speedometer cable
    - Driveshaft
    - All engine wiring harness, hoses and cables
    - Fan with the fluid coupling and fan pulleys. Tighten the nuts to 48 inch lbs. (5.4 Nm).
    - Air cleaner case and air filter
    - MAF meter, resonator and the air cleaner cap
    - Radiator
19. Fill the engine with oil.
20. Fill the engine and radiator with coolant.
21. Install or connect the following:
    - Engine undercover
    - Battery
    - Hood
22. Start the engine and check for leaks.
23. Make any necessary adjustments and road test the vehicle.

4WD

1. Before servicing the vehicle, refer to the precautions in the beginning of this section.
2. Remove or disconnect the following:
    - Transmission
    - Hood
3. Release the fuel system pressure.
4. Remove or disconnect the following:
    - Battery
    - Engine undercovers
5. Drain the engine coolant.
6. Drain the engine oil.
7. Remove or disconnect the following:
    - Radiator
    - Fan with the fluid coupling and fan pulleys
    - Air cleaner cap
    - Mass Air Flow (MAF) meter and the resonator
8. Disconnect the following hoses:
    - Heater hoses
    - Brake booster vacuum hose
    - Evaporative Emissions (EVAP) hose
    - Automatic Disconnecting Differential (ADD) vacuum hose
    - Fuel return hose
    - Fuel inlet hose
9. Detach the starter wire and connectors, as follows:
    - Ground strap, by removing the bolt
    - 3 starter wire clamps and connector
10. Detach the alternator connector and wire.
11. Disconnect the engine wiring harness, as follows:
    - Glove box door
    - Lower the finish No. 2 panel
    - 4 ECM connectors
    - 2 cassette connectors and the 2 wire clamps from the lower finish panel
    - Engine wiring harness clamp
12. Disconnect the following:
    - Igniter connector
    - Ground strap
    - Vacuum Switching Valve (VSV) connector for the EVAP
    - Vapor pressure sensor connector and clamp
    - Vapor connector for the vapor pressure sensor and clamp
13. Remove or disconnect the following:
    - 2 engine wiring harness retainer-to-cowl panel nuts and wiring harness
    - Air conditioning compressor wire clamp, if equipped with air conditioning
14. If necessary, install a No. 2 engine hanger with 2 bolts. Tighten the 2 bolts to 30 ft. lbs. (40 Nm).
15. Attach the engine hoist chain to the 2 engine hangers.
16. Remove or disconnect the following:
    - 4 engine front mounting insulators-to-frame bolts and nuts
    - Engine

To install:

17. Install or connect the following:
    - Engine
    - Engine mounts-to-body mountings. Install the bolts and nuts but do not tighten at this time.
18. Remove the engine chain hoist the No. 2 engine hanger.
19. Install or connect the following:
    - Air conditioning wire with the bolt, if equipped with air conditioning
    - Tighten the engine mounting nuts and bolts to 28 ft. lbs. (38 Nm).
    - Engine wiring harness
    - Engine wiring harness clamp
    - All wires, hoses and cables
    - Fan with the fluid coupling and fan pulleys. Tighten the nuts to 48 inch lbs. (5.4 Nm).
    - Air cleaner case and air filter
    - MAF meter, resonator and the air cleaner cap
    - Radiator
20. Fill the engine with oil.
21. Fill the engine and radiator with coolant.
22. Install or connect the following:
    - Transmission and refill it with transmission oil
    - Engine undercover
    - Battery
    - Hood
23. Start the engine, make any necessary adjustments and check for leaks.

4.0L Engine

1. Relieve fuel system pressure
2. Remove the transmission.
3. Drain the coolant.
4. Drain the engine oil.
5. Remove the battery.
6. Remove the hood.
7. Remove the V-bank cover.
8. Remove the air cleaner assembly.
9. Remove the radiator support upper seal (11 clips).
10. Remove the radiator.
11. Remove the fan shroud.
12. Remove the accessory drive belt.
13. Remove the fan.
14. Remove the fan pulley.
15. Remove the power steering pump.
16. Remove the alternator.
17. Remove the A/C compressor. Don't disconnect the lines.
18. Disconnect the fuel lines.
19. Disconnect the heater hoses.
20. Remove the upper intake manifold (air surge tank).
21. Remove the right front door scuff plate.
22. Remove the right cowl side trim plate.
23. Remove the glove compartment door.
24. Disconnect the connectors for the ECM, 4wd ECU, and instrument panel wiring. Pull the harness into the engine compartment.
25. Disconnect the front differential connector.
26. Remove the engine ground cable.
27. Attach lifting plates as shown.
28. Take up the weight of the engine with a crane.
29. Remove the engine mounting bracket bolts.
30. Remove the engine.
31. Installation is the reverse of removal. Observe the following torques:
    - Engine mount brackets: 28 ft. lbs. (38 Nm)
    - Hood: 10 ft. lbs. (13 Nm)
    - V-bank cover: 66 inch lbs. (7.5 Nm)

Engine lifting bracket positions—4.0L engine

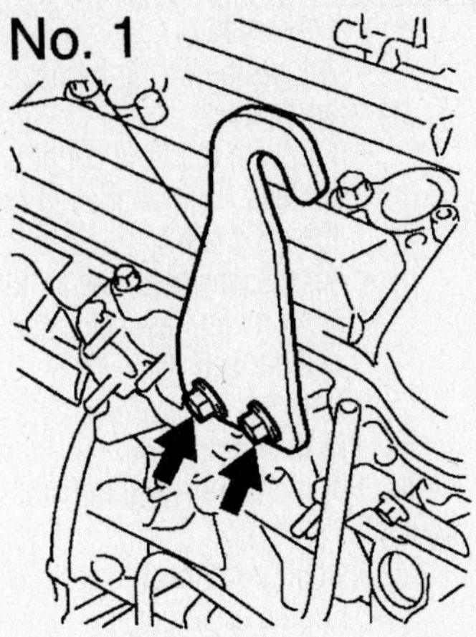

Engine mounts and related parts—4.7L engine

4.7L Engine

1. Relieve fuel system pressure.
2. Remove the transmission.
3. Remove the hood.
4. Remove the throttle body cover.
5. Remove the air cleaner assembly.
6. Remove the engine under-covers.
7. Drain the coolant and engine oil.
8. Disconnect the fuel lines.
9. Remove the fuel vapor line.
10. Remove the accessory drive belt.
11. Remove the power steering pump.
12. Remove the alternator.
13. Remove the A/C compressor. Don't disconnect the lines.
14. Remove the fan.
15. Remove the transmission filler tube.
16. Remove the oil level sending unit.
17. Remove the exhaust manifolds.
18. Disconnect the heater hoses.
19. Remove the right front door scuff plate.
20. Remove the right cowl side trim plate.
21. Remove the glove compartment door.
22. Disconnect the connectors for the ECM, 4wd ECU, and instrument panel wiring. Pull the harness into the engine compartment.
23. Disconnect the front differential connector.
24. Remove the engine ground cables.
25. Attach a crane to the lifting plates.
26. Take up the weight of the engine with a crane.
27. Remove the engine mounting bracket bolts.
28. Remove the engine.
29. Installation is the reverse of removal. Observe the following torques:
    - Engine mount brackets: 28 ft. lbs. (38 Nm)
    - Exhaust manifolds: 33 ft. lbs. (44 Nm)
    - Oil level sending unit: 11 ft. lbs. (15 Nm)
    - Fan: 21 ft. lbs. (29 Nm)
    - A/C compressor: bolt, 34 ft. lbs. (47 Nm); nut, 18 ft. lbs. (25 Nm)
    - Power steering pump: 32 ft. lbs. (43 Nm)
    - Hood: 10 ft. lbs. (13 Nm)
    - V-bank cover: 66 inch lbs. (7.5 Nm)

Heater Core

REMOVAL & INSTALLATION

2001–02 Models

FRONT HEATER

1. Disconnect the negative battery cable.

**** CAUTION**

After the negative battery cable has been disconnected, wait at least 1½ minutes for the air bag module to deplete its energy.

2. Drain the cooling system into a clean container for reuse.
3. Disconnect the heater hoses from the heater core.
4. Remove the steering wheel by performing the following procedure:
    a. Position the front wheels in the straight-ahead position.
    b. At both sides of the steering wheel, remove the side covers.
    c. Using a Torx® wrench, loosen the steering wheel screws until the screw's circumference ring catches on the screw case.

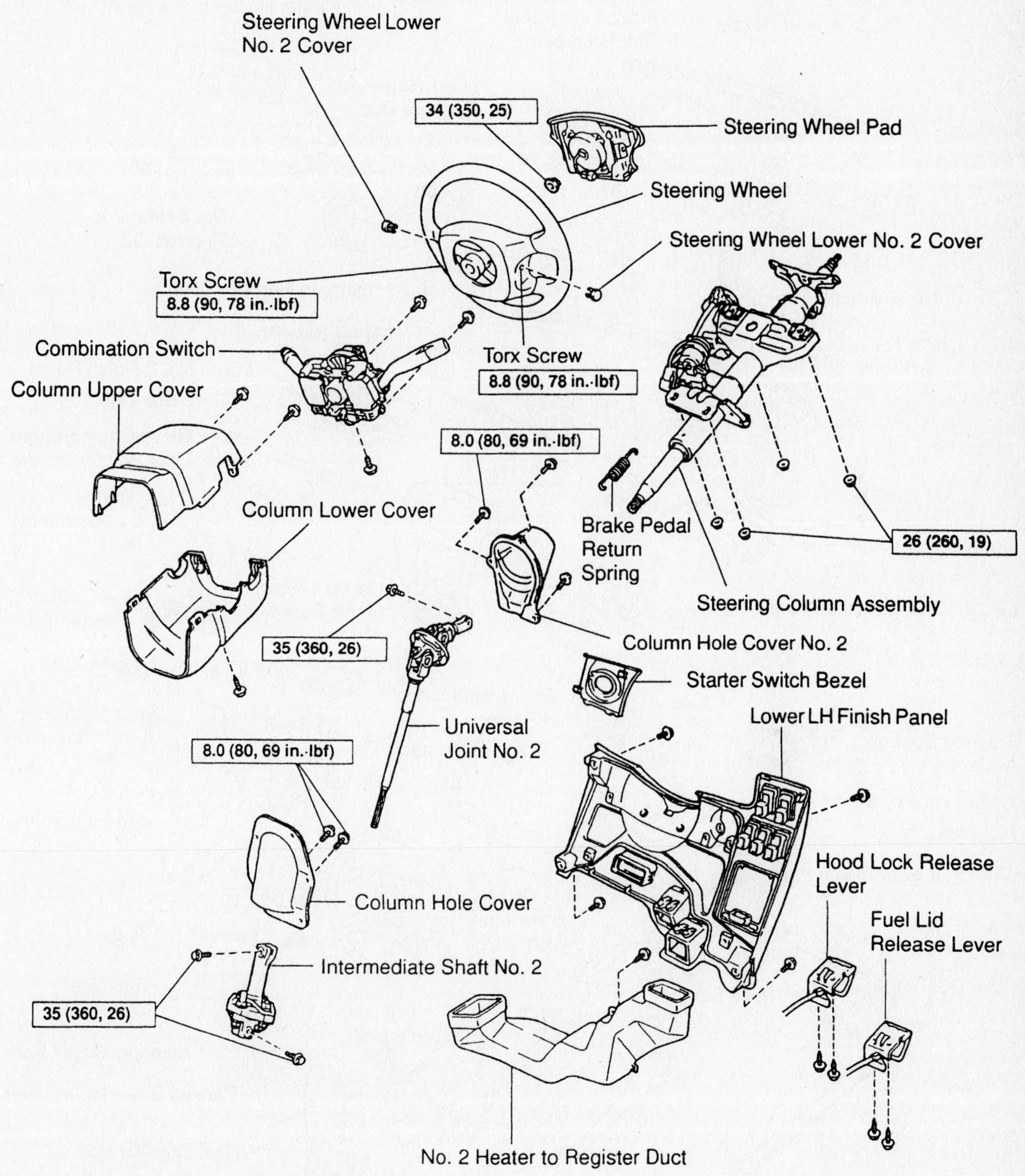

Exploded view of the steering wheel, air bag module, steering column and related components—2001–02

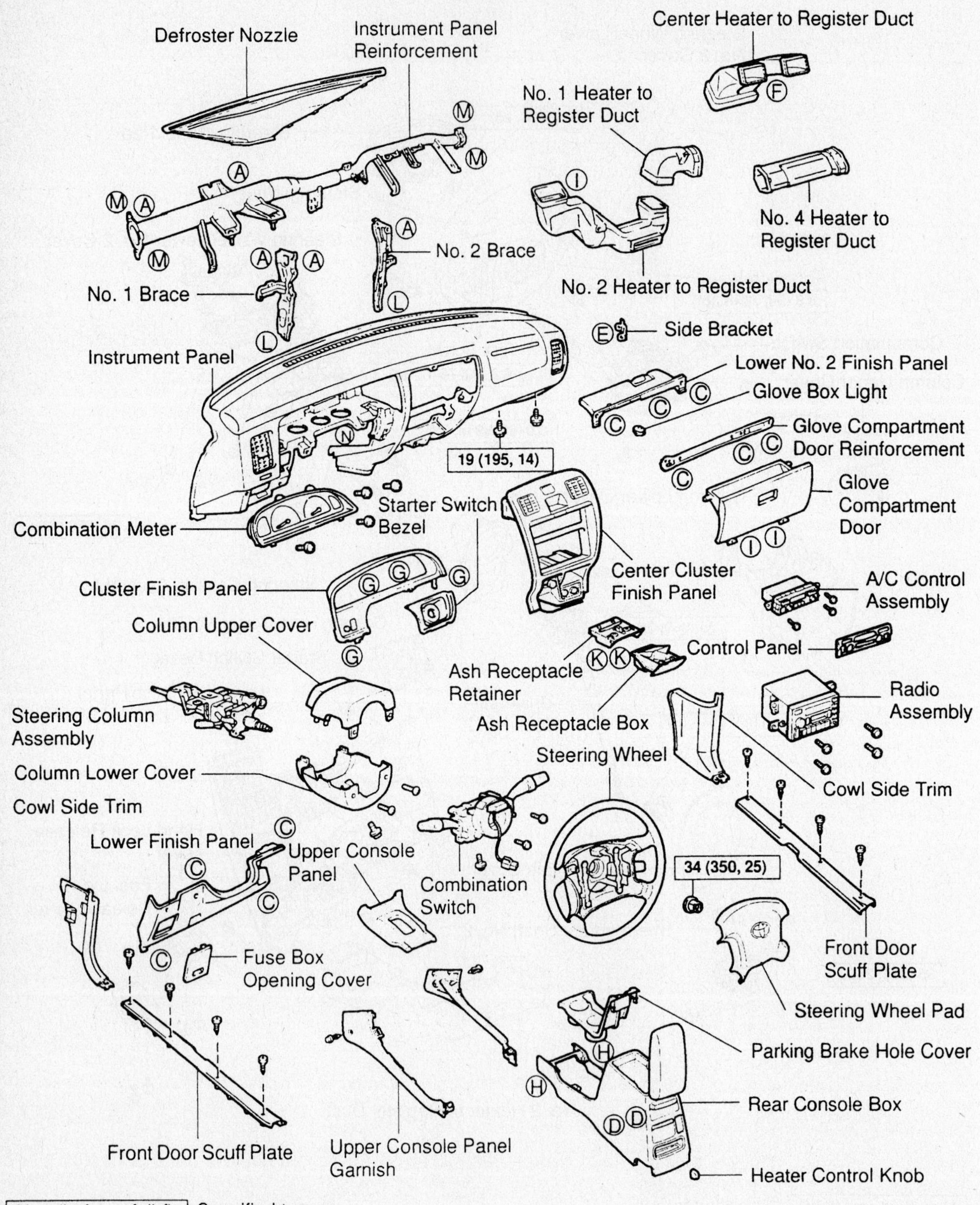
Defroster Nozzle
Instrument Panel Reinforcement
Center Heater to Register Duct
No. 1 Heater to Register Duct
No. 4 Heater to Register Duct
No. 2 Brace
No. 2 Heater to Register Duct
No. 1 Brace
Side Bracket
Instrument Panel
Lower No. 2 Finish Panel
Glove Box Light
Glove Compartment Door Reinforcement
19 (195, 14)
Glove Compartment Door
Starter Switch Bezel
Combination Meter
Center Cluster Finish Panel
A/C Control Assembly
Cluster Finish Panel
Column Upper Cover
Control Panel
Ash Receptacle Retainer
Radio Assembly
Steering Column Assembly
Ash Receptacle Box
Steering Wheel
Column Lower Cover
Cowl Side Trim
Cowl Side Trim
Lower Finish Panel
Upper Console Panel
Combination Switch
34 (350, 25)
Front Door Scuff Plate
Fuse Box Opening Cover
Steering Wheel Pad
Parking Brake Hole Cover
Rear Console Box
Front Door Scuff Plate
Upper Console Panel Garnish
Heater Control Knob
N·m (kgf·cm, ft·lbf) : Specified torque
93113GJ0

Exploded view of the instrument panel and related components—2001–02

d. Carefully, lift the air bag module, disconnect the electrical connector and remove the air bag.

⁂ CAUTION

Place the air bag module in a safe location with the front facing upward.

e. Remove the steering wheel nut.

f. Using a steering wheel puller, press the steering wheel from the steering column.

5. Remove the instrument panel and reinforcement by performing the following procedure:

a. Remove both front door scuff plates.

b. Remove both cowl side trims.

c. Remove the 2 hood lock release lever screws and the hood lock release lever.

d. Remove the 2 fuel lid release lever screws and the fuel lid release lever.

e. Remove the 4 lower finish panel bolts and the panel.

f. Remove the No. 1 and No. 2 heater-to-register duct screw and the ducts.

g. Pry out the starter switch bezel.

h. Remove the steering column cover screws and the covers.

i. Remove the combination switch-to-steering column screws, disconnect the electrical connector and the combination switch.

j. Remove the steering column-to-instrument panel nuts/bolts and the lower steering column bolt; then carefully, remove the steering column.

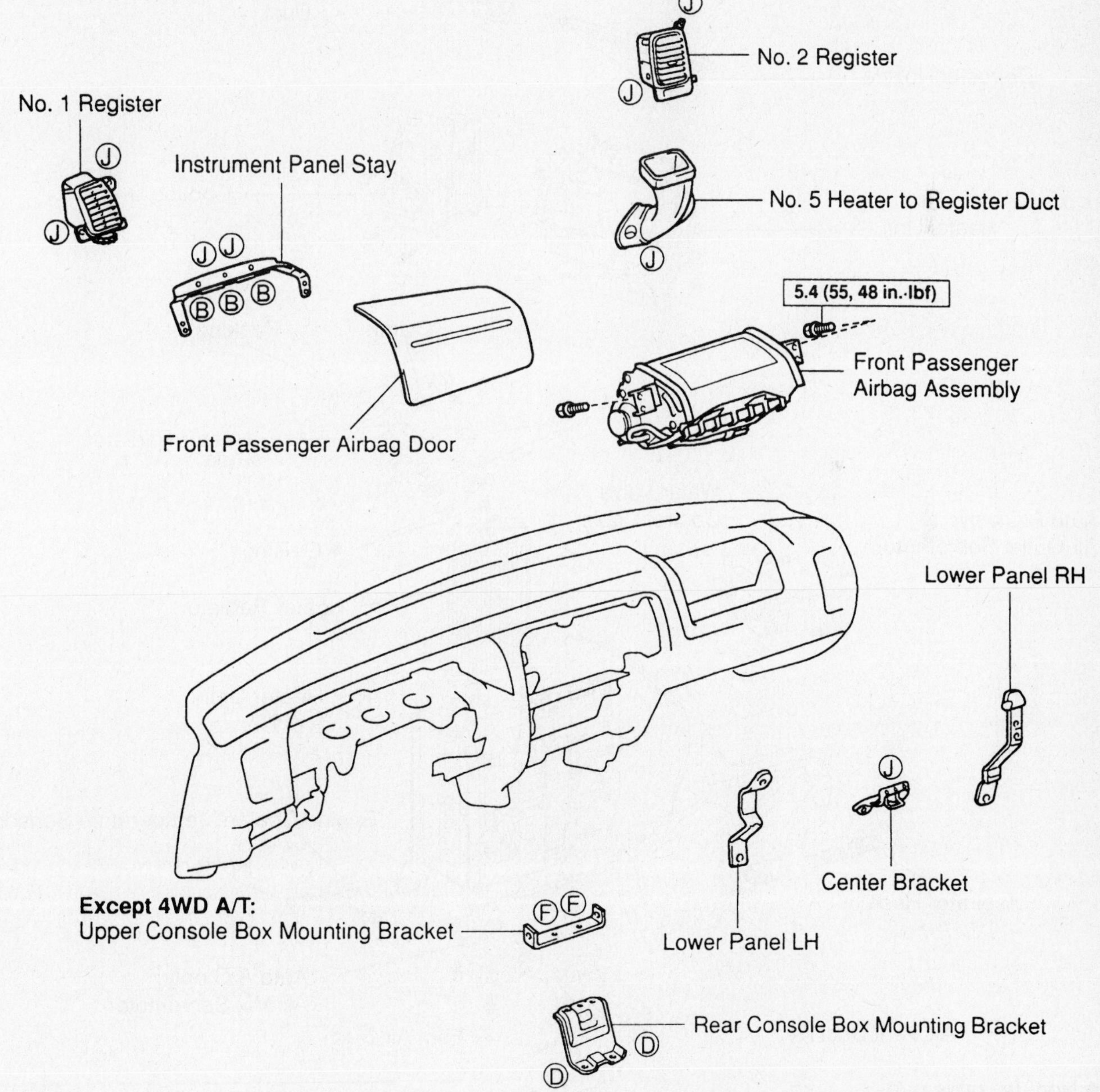

Exploded view of the instrument panel air bag module, ventilation components and brackets—2001–02

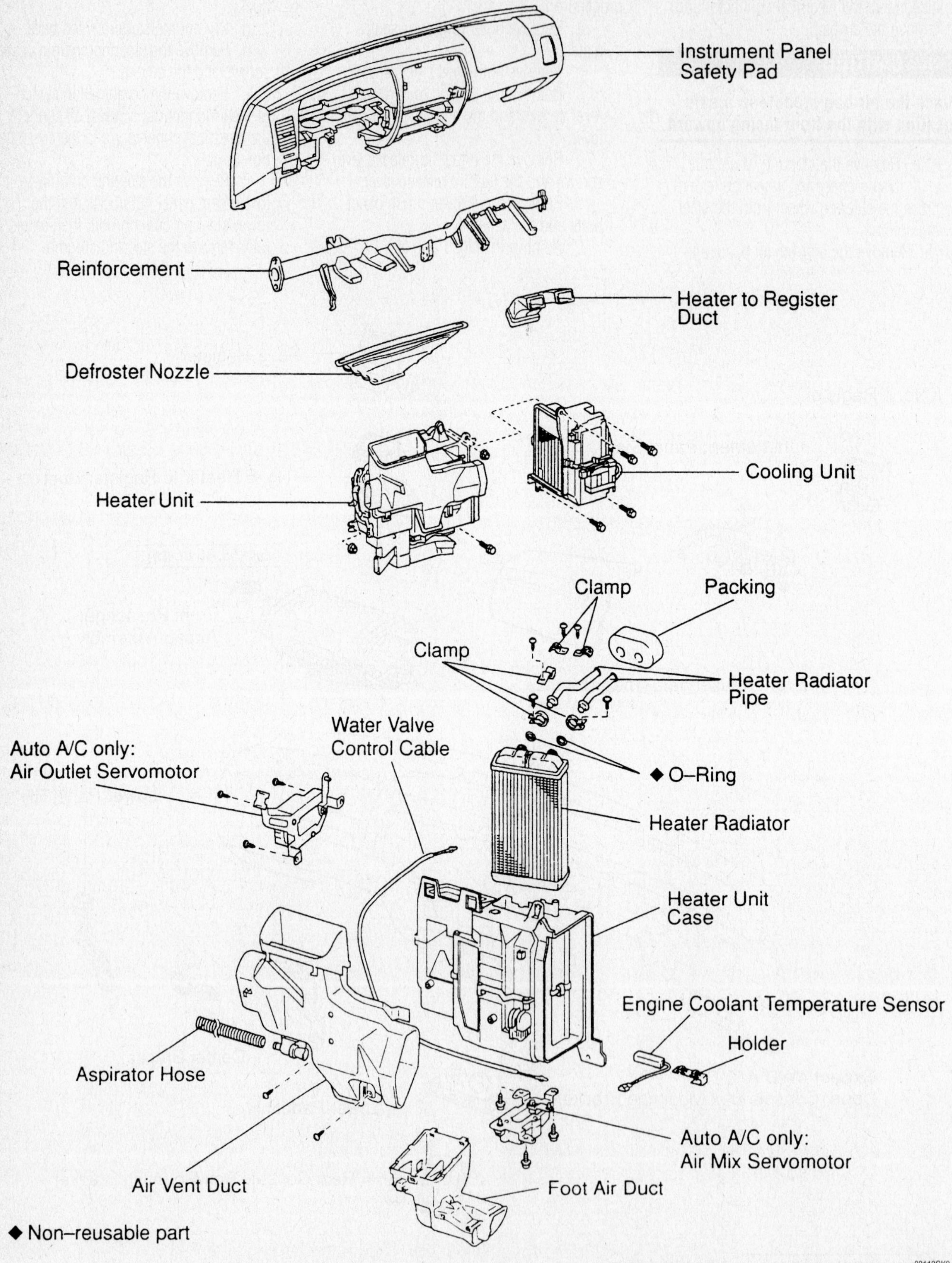

Exploded view of the front heater core, heater housing, evaporator housing and related components—2001–02

k. Remove the 4 cluster finish panel screws and the panel.

l. Remove the 4 combination meter screws, disconnect the electrical connectors and remove the combination meter.

m. Pry out the parking brake hole cover.

n. Pry out the upper console panel.

o. Disengage the 7 center cluster finish panel clips and remove the panel.

➡Remove the center cluster finish panel clips by starting at the bottom and working toward the top.

p. Remove the heater control knobs.

q. Remove the 2 rear console box bolts/screws and the rear console box.

r. Remove the upper console panel garnish.

s. Remove the 2 glove compartment door screws and the door.

t. Disconnect the passenger's side air bag module electrical connector.

u. Remove the glove box light.

v. Remove the 3 lower No. 2 finish panel bolts and the panel.

w. Remove the 3 glove compartment door reinforcement bolts and the reinforcement.

x. Remove the No. 4 heater-to-register duct.

y. Remove the radio assembly.

z. Remove the side bracket bolt and the bracket.

aa. If equipped with manual air conditioning, remove the heater control assembly.

bb. If equipped with automatic air conditioning, remove the air conditioning control assembly.

cc. Remove the instrument panel-to-chassis nut and 2 bolts; then, remove the instrument panel.

dd. Remove the instrument panel reinforcement-to-chassis nuts/bolts and the reinforcement.

6. Remove the defroster nozzle and heater-to-register duct.

7. Remove the evaporator housing by performing the following procedure:

a. Discharge and recover the air conditioning system refrigerant.

b. Disconnect the refrigerant lines from the evaporator core. Discard the O-rings and plug the openings to prevent contamination.

c. Disconnect the electrical connectors.

d. Remove the 3 evaporator housing-to-chassis screws and the housing.

8. Disconnect the mode control servomotor connector.

9. Disconnect the aspirator hose from the room temperature sensor.

10. Disconnect the heater valve control cable.

11. Remove the heater housing-to-chassis nuts and the heater housing.

12. Remove the 3 heater core-to-heater housing screws, the 2 clips and clamp.

13. Remove the heater core from the heater housing.

To install:

14. Install the 3 heater core-to-heater housing screws, the 2 clips and clamp.

15. Install the heater housing and the heater housing-to-chassis nuts.

16. Connect the heater valve control cable.

17. Connect the aspirator hose to the room temperature sensor.

18. Connect the mode control servomotor connector.

19. Install the defroster nozzle and heater-to-register duct.

20. Install the evaporator housing by performing the following procedure:

a. Install the evaporator housing and the 3 housing-to-chassis screws.

b. Connect the electrical connectors.

c. Using new O-rings, connect the refrigerant lines to the evaporator core.

21. Install the instrument panel and reinforcement by performing the following procedure:

a. Install the instrument panel reinforcement and the reinforcement-to-chassis nuts/bolts.

b. Install the instrument panel and the instrument panel-to-chassis nut and 2 bolts.

c. If equipped with automatic air conditioning, install the air conditioning control assembly.

d. If equipped with manual air conditioning, install the heater control assembly.

e. Install the side bracket and the bracket bolt.

f. Install the radio assembly.

g. Install the No. 4 heater-to-register duct.

h. Install the glove compartment door reinforcement and the 3 reinforcement bolts.

i. Install the lower No. 2 finish panel and the 3 panel bolts.

j. Install the glove box light.

k. Connect the passenger's side air bag module electrical connector.

l. Install the glove compartment door and the 2 door screws.

m. Install the upper console panel garnish.

n. Install the rear console box and the 2 rear console box bolts/screws.

o. Install the heater control knobs.

p. Install the center cluster finish panel and engage the 7 panel clips.

q. Install the upper console panel.

r. Install the parking brake hole cover.

s. Install the combination meter, connect the electrical connectors and install the 4 combination meter screws.

t. Install the cluster finish panel and the 4 panel screws.

u. Install the steering column and torque the steering column-to-instrument panel nuts to 19 ft. lbs. (26 Nm) and the lower steering column bolt to 26 ft. lbs. (35 Nm).

v. Install the combination switch-to-steering column, connect the electrical connector and the combination switch screws.

w. Install the steering column cover and the cover screws.

x. Pry out the starter switch bezel.

y. Install the No. 1 and No. 2 heater-to-register duct and the duct screws.

z. Install the lower finish panel and the 4 panel bolts.

aa. Install the fuel lid release lever and the 2 fuel lid release lever screws.

bb. Install the hood lock release lever and the 2 hood lock release lever screws.

cc. Install both cowl side trims.

dd. Install both front door scuff plates.

22. Install the steering wheel by performing the following procedure:

a. Install the steering wheel to the steering column.

b. Install the steering wheel nut and torque the nut to 25 ft. lbs. (34 Nm).

c. Carefully, install the air bag module and connect the electrical connector.

d. Using a Torx® wrench, torque the steering wheel screws to 78 inch lbs. (8.8 Nm).

e. At both sides of the steering wheel, install the side covers.

23. Connect the heater hoses to the heater core.

24. Refill the cooling system.

25. Connect the negative battery cable.

26. Evacuate and charge the air conditioning system.

27. Run the engine to normal operating temperatures; then, check the climate control operation and check for leaks.

REAR AUXILIARY HEATER

1. Disconnect the negative battery cable.

2. Drain the cooling system into a clean container for reuse.

3. Remove the front seats.
4. Remove the center console box.
5. Move the floor carpet backward.
6. Disconnect the rear heater hoses from the rear heater core.
7. Remove the rear heater duct bolt, screw and duct.
8. Remove the rear heater control assembly.
9. Disconnect the electrical connectors.
10. Remove the 3 rear heater housing-to-chassis screws and the housing.
11. Remove the blower resistor, the rear heater relay and with wiring harness from the rear heater housing.
12. Remove the rear heater housing case screws and separate the cases.
13. Remove the rear heater core.

To install:

14. Install the rear heater core.
15. Assemble the rear heater housing case and install the case screws.
16. Install the blower resistor, the rear heater relay and wiring harness to the rear heater housing.
17. Install the rear heater housing and the 3 housing-to-chassis screws.

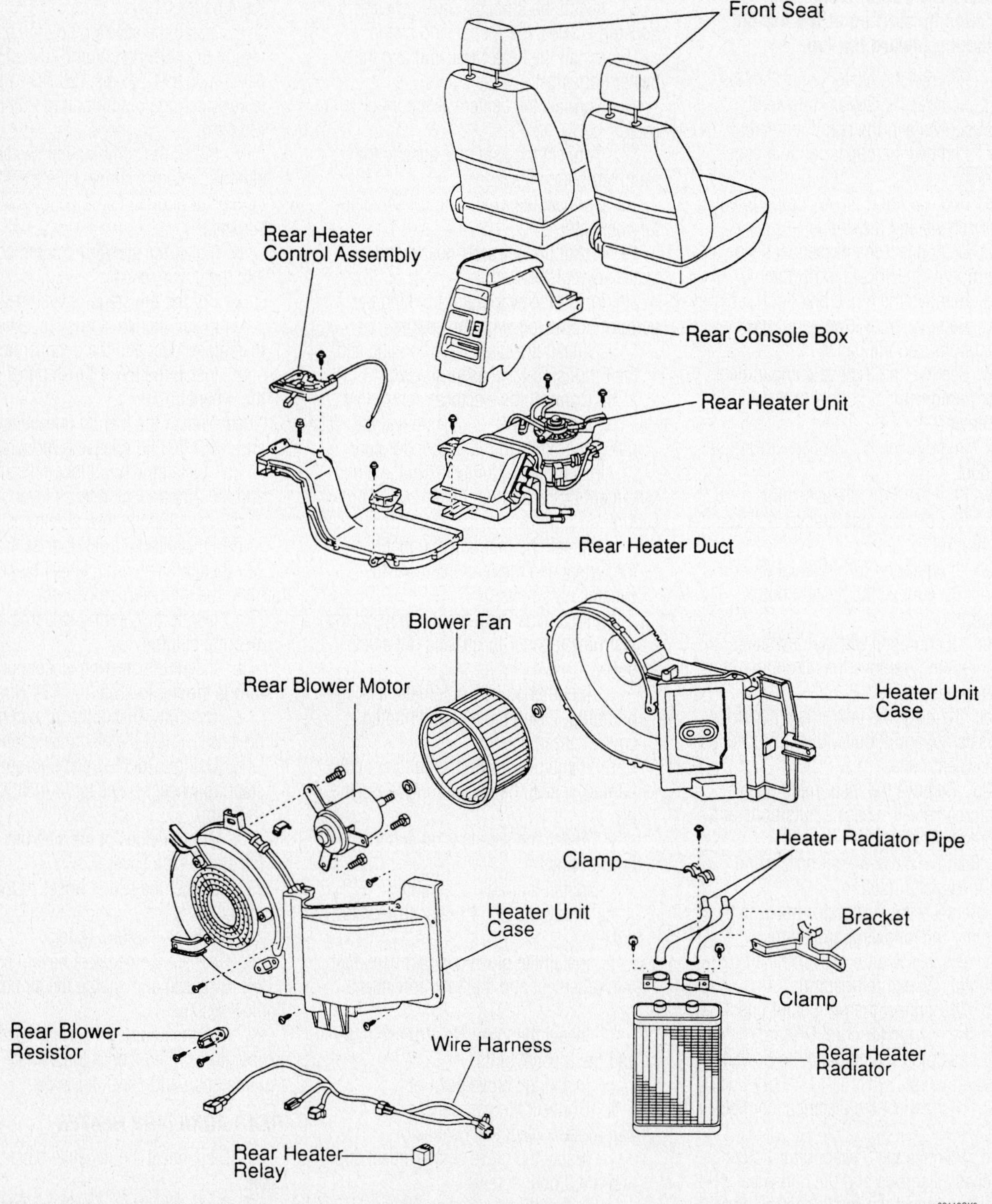

Exploded view of the rear heater housing and heater core—2001–02

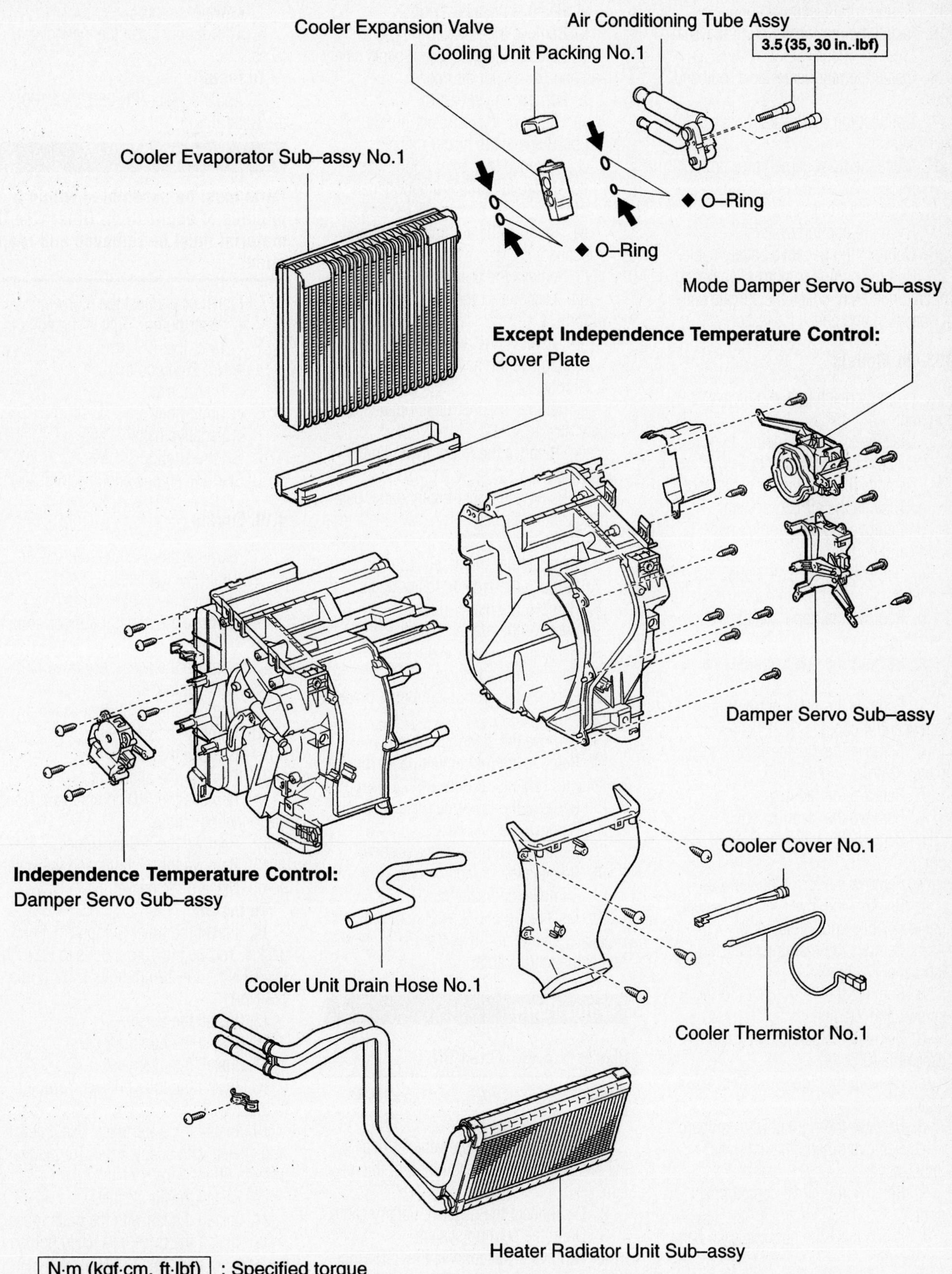
Cooler Expansion Valve
Air Conditioning Tube Assy
Cooling Unit Packing No.1
3.5 (35, 30 in.·lbf)
Cooler Evaporator Sub–assy No.1
◆ O–Ring
◆ O–Ring
Mode Damper Servo Sub–assy
Except Independence Temperature Control:
Cover Plate
Damper Servo Sub–assy
Cooler Cover No.1
Independence Temperature Control:
Damper Servo Sub–assy
Cooler Unit Drain Hose No.1
Cooler Thermistor No.1
Heater Radiator Unit Sub–assy
N·m (kgf·cm, ft·lbf) : Specified torque
◆ Non–reusable part
Compressor oil ND–OIL 8 or equivalent
67170-4RUN-G04

Heater core and related components—2003–04 models

18. Connect the electrical connectors.
19. Install the rear heater control assembly.
20. Install the rear heater duct, bolt and screw.
21. Connect the rear heater hoses to the rear heater core.
22. Move the floor carpet foreword.
23. Install the center console box.
24. Install the front seats.
25. Refill the cooling system.
26. Connect the negative battery cable.
27. Run the engine to normal operating temperatures; then, check the climate control operation and check for leaks.

2003–04 Models

1. Properly discharge and recover the refrigerant.
2. Disconnect the refrigerant lines.
3. Disconnect the heater hoses.
4. Remove the instrument panel pad:
   a. Disable the air bag system.
   b. Remove the lower finish panel (2 bolts and 4 clips).
   c. Remove the console upper rear panel (5 clips and 2 retainers)
   d. Remove the upper panel (6 clips and 2 retainers).
   e. Remove the console upper panel garnishes.
   f. Remove the A/C control unit (1 screw and 4 clips).
   g. Remove the cluster finish center panel (3 bolts and 8 clips).
   h. Remove the radio.
   i. Remove the steering wheel.
   j. Remove the steering column covers.
   k. Remove the turn signal switch.
   l. Remove the lower left instrument panel (2 bolts and 2 retainers).
   m. Remove the instrument cluster finish panel (7 clips).
   n. Remove the instrument cluster (3 screws and 2 bolts).
   o. Remove the console box rear assembly (6 bolts).
   p. Remove the right front door scuff plate.
   q. Remove the right cowl trim board.
   r. Remove the instrument panel under-cover.
   s. Remove the glove compartment door.
   t. Disconnect the passenger air bag connector.
   u. Remove the instrument panel lower finish panel.
   v. Remove the instrument panel register.
   w. Remove the assist grip.
   x. Remove the A-pillar trim panels.
   y. Remove the instrument panel safety pad and passenger air bag.
   z. Remove the speakers.
   aa. Remove the defroster nozzle.
   bb. Remove the heater ducts.
   cc. Remove the meter hood retainer.
   dd. Remove the instrument panel bracket.
   ee. Remove the automatic light control sensor.
   ff. Remove the thermistor.
   gg. Disconnect the instrument panel wiring.
   hh. Remove the antenna wire.
   ii. Remove the navigation antenna assembly.
   jj. Remove the instrument panel left door air bag.
   kk. Remove the instrument panel safety pad.
   ll. Remove the instrument panel passenger door air bag.
5. Remove the ECM.
6. Remove the A/C amplifier.
7. Remove the side defroster ducts.
8. Remove the heater ducts.
9. Remove the rear air ducts.
10. Remove the console duct.
11. Remove the air ducts.
12. Remove the instrument panel brace bracket.
13. Remove the steering column.
14. Remove the instrument panel reinforcement (18 bolts, 8 nuts and 27 clamps).
15. Remove the lower defroster nozzle.
16. Remove the A/C case.
17. Remove the mode damper servo.
18. Remove the damper servo.
19. Remove the heater core.
20. Installation is the reverse of removal. Evacuate, charge and leak test the system. Adjust the spiral cable.

Water Pump

REMOVAL & INSTALLATION

3.4L

1. Before servicing the vehicle, refer to the precautions in the beginning of this section.
2. Disconnect the negative battery cable.
3. Drain the cooling system.
4. Remove or disconnect the following:
   - Timing belt
   - Thermostat
   - No. 2 oil cooler hose from the water pump
   - Water pump
5. Thoroughly clean the mating surfaces.

To install:

6. Apply sealant (PN 08826-00100) to the water pump.

WARNING

Parts must be assembled within 5 minutes of application. Otherwise the material must be removed and reapplied.

7. Install or connect the following:
   - Water pump. Tighten the bolts to 14 ft. lbs. (20 Nm).
   - No. 2 oil cooler hose
   - Thermostat
   - Timing belt
   - Negative battery cable
8. Fill the cooling system.
9. Start the engine and check for leaks.

4.0L Engine

1. Remove the engine under-cover.
2. Drain the coolant.
3. Remove the V-bank cover.
4. Remove the radiator support upper seal.
5. Remove the accessory drive belt.
6. Remove the fan.
7. Remove the air cleaner assembly.
8. Remove the water inlet.
9. Remove the idler pulley.
10. Remove the alternator.
11. Remove the A/C compressor. Don't disconnect the lines.
12. Remove the belt tensioner.
13. Remove the 17 bolts and the water pump. Discard the gasket.

To install:

14. Install the water pump with a new gasket. Torque the 10mm bolts to 80 inch lbs. (9 Nm); the 12mm bolts to 17 ft. lbs. (23 Nm).
15. Install the tensioner.
16. Install the compressor.
17. Install the alternator.
18. Install the idler pulleys. Torque to 29 ft. lbs. (39 Nm).
19. Install the water inlet. Use a new O-ring coated with soapy water. Torque the 5 bolts to 80 inch lbs. (9 Nm).
20. Install the air cleaner.
21. Install the fan with the bolts loose.
22. Install the drive belt. Then tighten the fan bolts to 15 ft. lbs. (21 Nm).
23. Install the upper seal.
24. Fill the cooling system.
25. Install the V-bank cover. Torque to 66 inch lbs. (7.5 Nm).

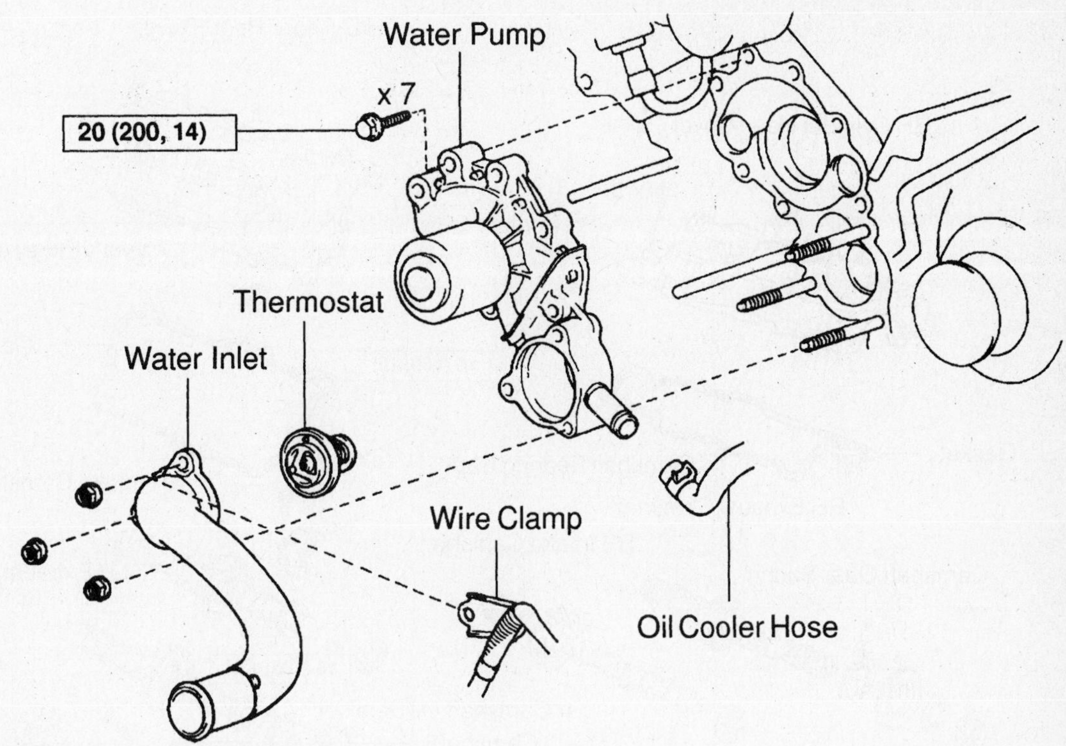

Water pump installation—3.4L engine

Water pump mounting—4.0L engine

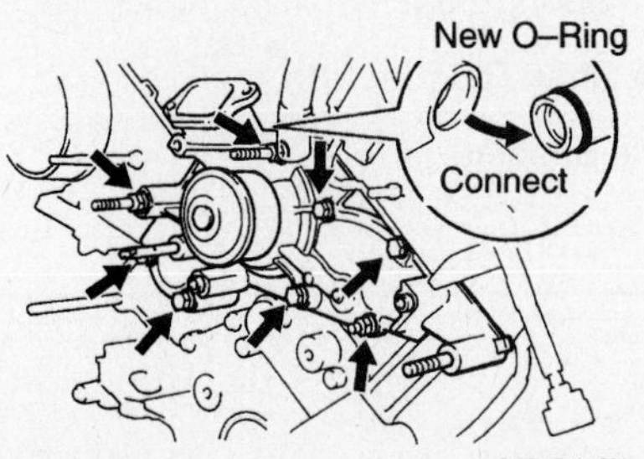

Water pump installation—4.7L engine

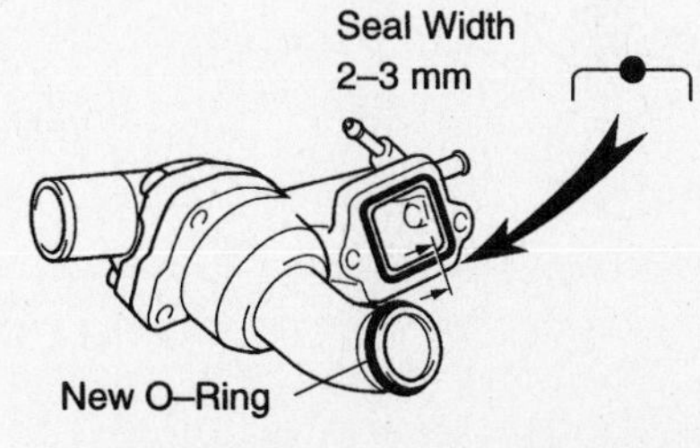

Water inlet installation—4.7L engine

26. Install the under-cover. Torque to 21 ft. lbs. (29 Nm).

4.7L Engine

1. Remove the timing belt.
2. Remove the water inlet housing. Remove the silicone gasket material.
3. Remove the 5 bolts, 2 studs and the nut. Remove the water pump. Discard the gasket.
4. Installation is the reverse of removal. Use soapy water on the new bypass pipe O-ring. Use a new water pump gasket and torque the bolts to 16 ft. lbs. (21 Nm); the studs and the nut to 13 ft. lbs. (18 Nm).
5. Apply a 2–3 mm wide bead of silicone gasket material to the inlet housing and install it within 5 minutes. Torque to 13 ft. lbs. (18 Nm).

Cylinder Head

REMOVAL & INSTALLATION

3.4L

1. Before servicing the vehicle, refer to the precautions in the beginning of this section.

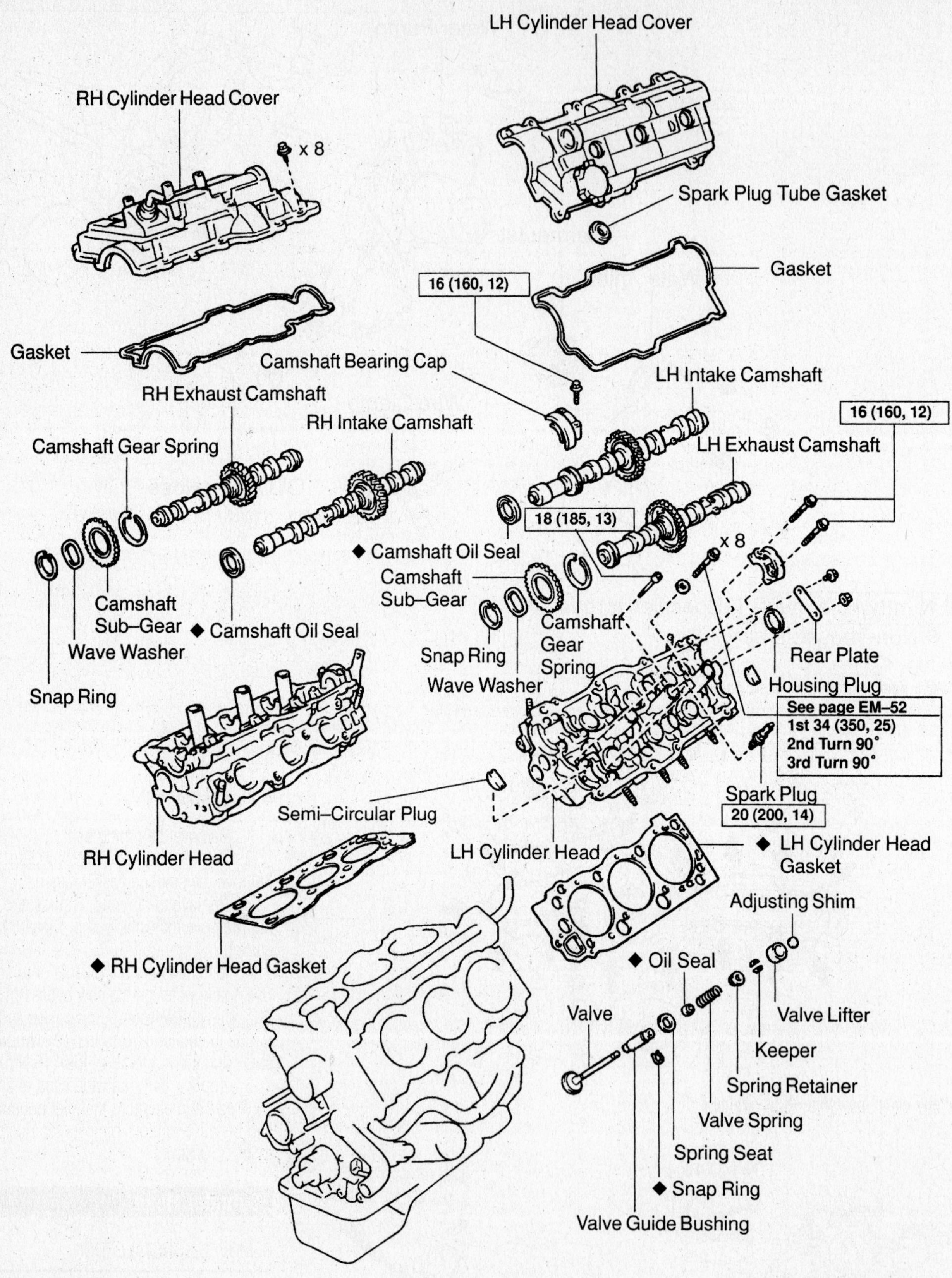

N·m (kgf·cm, ft·lbf) : Specified torque
◆ Non–reusable part

67170-4RUN-G133

Cylinder heads and related parts—3.4L engine

2. Disconnect the negative battery cable.
3. Relieve the fuel system pressure.
4. Remove the engine undercover.
5. Drain the cooling system.
6. Remove or disconnect the following:
   - Front exhaust pipe
   - Air cleaner cap
   - Mass Air Flow (MAF) meter and the resonator
7. Disconnect the following cables:
   - Actuator cable with the bracket, if equipped with cruise control
   - Accelerator cable
   - Throttle cable, if equipped with an automatic transmission
8. Disconnect the following hoses:
   - Heater hose
   - Brake booster vacuum hose
   - Evaporative Emissions (EVAP) hose
   - Automatic Disconnecting Differential (ADD) vacuum hose, for 4-wheel drive
   - Fuel inlet and fuel return hose
9. Remove or disconnect the following:
   - Spark plug wires with the ignition coils
   - Spark plugs
   - Intake chamber stay
   - No. 2 timing belt cover
   - Air intake chamber assembly
10. Remove the following connectors and hoses:
    - Throttle Position (TP) sensor connector
    - Idle Air Control (IAC) valve connector
    - Positive Crankcase Ventilation (PCV) hoses
    - Water bypass hoses
    - Air assist hose from the throttle body
    - Intake air connector
11. Disconnect the engine wiring harness protector, as follows:
    - 6 injector connectors
    - Engine Coolant Temperature (ECT) sensor and sender gauge connectors
    - Engine wiring harness protector from the cylinder head
12. Remove or disconnect the following:
    - Fuel pressure regulator
    - Intake manifold assembly
13. Set the No. 1 cylinder at Top Dead Center (TDC) of the compression stroke, as follows:
    a. Turn the crankshaft pulley and align its groove with the timing mark **0** of the No. 1 timing belt cover.
    b. Check that the timing marks of the camshaft timing pulleys and the No. 3 timing belt cover are aligned. If not, turn the crankshaft pulley 1 revolution (360 degrees).
14. Remove or disconnect the following:
    - Timing belt tensioner, by alternately loosening the 2 bolts
    - Timing belt
15. Remove the camshaft timing pulleys, as follows:
    a. Using Variable Pin Wrench Set 09960-10010, remove the pulley bolt, the timing pulley and the knock pin.
    b. Remove the 2 timing pulleys with the timing belt.
16. Remove or disconnect the following:
    - Bolt and the No. 2 idler pulley
    - Camshaft Position (CMP) sensor
    - No. 3 timing belt cover
    - Alternator from the engine
    - Alternator bracket
    - Power steering pump and move it aside without disconnecting the pump lines
    - Exhaust crossover pipe and gaskets, by removing the 6 nuts
    - Left-hand exhaust manifold, by removing the heat insulator and 6 nuts
    - Right-hand exhaust manifold, by removing the heat insulator and 6 nuts
    - 8 bolts, seal washers, cylinder head cover and gasket.

➡Remove both cylinder head covers

    - Semi-circular plugs
    - Right exhaust and intake camshafts
    - Left exhaust and intake camshafts
    - Valve lifters and shims from the cylinder head; arrange the valve lifters and shims in correct order
17. Remove the cylinder heads, as follows:
    - Bolt and ground strap
    - Cylinder head (recessed head) bolt on the cylinder head, using an 8mm hexagon wrench; then, repeat the procedure for the other side.
    - 8 cylinder head (12-pointed head) bolts, on each cylinder head.

➡Loosen the bolts in several passes, in the sequence shown.

    - 16 cylinder head bolts and plate washers
    - Cylinder head

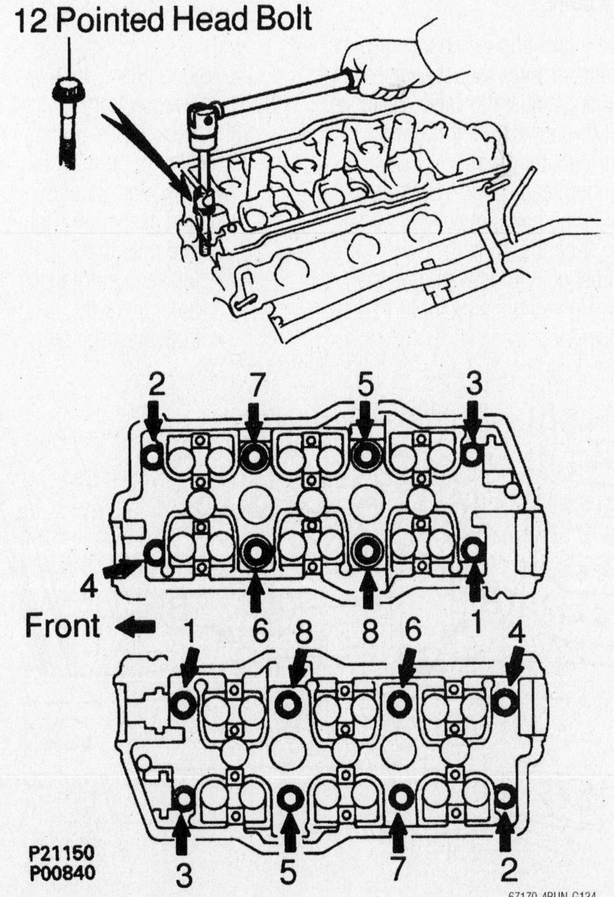

Head bolt loosening sequence—3.4L engine

To install:

18. Clean all surfaces.
19. Install or connect the following:
   - New cylinder head gaskets
   - Cylinder heads
20. Apply a light coat of engine oil on the threads and under the heads of the cylinder head bolts.
21. Tighten the cylinder head bolts using several passes, in sequence, as follows:
   - Step 1: 25 ft. lbs. (34 Nm)
   - Step 2: Mark the front of the cylinder head bolt with paint
   - Step 3: Turn 90 degrees
   - Step 4: Check that the painted mark is now at a 90 degrees angle to the front
22. Install the recessed head cylinder head bolts, as follows:
   - Step 1: Apply a light coat of engine oil on the threads and under the heads of the cylinder head bolts
   - Step 2: Tighten the cylinder head bolts, using a 8mm hexagon wrench, to 13 ft. lbs. (18 Nm).
   - Bolt and ground strap
23. Install or connect the following:
   - Valve lifters and shims

➡Check that the valve lifter rotates smoothly by hand.

   - Right intake and exhaust camshafts
   - Left intake and exhaust camshafts
24. Check and adjust the valve clearance.
25. Install or connect the following:
   - Semi-circular plugs
   - Cylinder head covers. Uniformly, tighten the bolts, in several passes, to 53 inch lbs. (6 Nm).
   - Exhaust manifolds with new gaskets. Tighten the nuts to 30 ft. lbs. (40 Nm).
   - Exhaust manifold heat insulators. Tighten the nuts to 71 inch lbs. (8 Nm).
   - Exhaust crossover pipe. Tighten the nuts to 33 ft. lbs. (45 Nm).
   - Power steering pump
   - Alternator bracket. Tighten the fasteners to 14 ft. lbs. (18 Nm).
   - Alternator
   - No. 3 timing belt cover. Tighten the bolts to 80 inch lbs. (9 Nm).
   - CMP sensor. Tighten it to 71 inch lbs. (8 Nm).
   - Timing belt
   - No. 2 timing belt idler bolt. Tighten the bolt to 30 ft. lbs. (40 Nm).

➡Check that the pulley bracket moves smoothly.

   - Left camshaft timing pulley
26. Set the No. 1 cylinder to TDC of the compression stroke, as follows:
   a. Connect the timing belt to the left camshaft timing pulley.
   b. Check that the installation mark on the timing belt is aligned with the end of the No. 1 timing belt cover.
   c. Install the right camshaft timing pulley and the timing belt.
   d. Set the timing belt tensioner. Alternately, tighten the bolts to 20 ft. lbs. (28 Nm).
27. Using pliers, remove the 1.5mm hexagon wrench from the belt tensioner.
28. Check the valve timing.
29. Install or connect the following:
   - New gaskets and the intake manifold assembly. Tighten the bolts and nuts to 13 ft. lbs. (18 Nm).
   - Intake manifold stay. Tighten the bolts to 14 ft. lbs. (18 Nm).
   - Fuel pressure regulator
30. Connect the engine wiring harness to the intake manifold, as follows:
   - Engine wiring harness to the cylinder head
   - 3 engine wiring harness clamps
31. Install or connect the following:
   - 6 injector connectors
   - ECT sender gauge connector
   - ECT sensor connector
   - Intake air connector
   - Air intake chamber assembly. Tighten the bolts and nuts to 13 ft. lbs. (18 Nm).
   - Intake chamber stay
   - No. 2 timing belt cover. Tighten the bolts to 80 inch lbs. (9 Nm).
   - PCV hoses
   - Water bypass hoses
   - Air assist hose to the throttle body
   - IAC valve connector
   - TP sensor connector
   - CMP sensor connector to the No. 2 timing belt cover
   - 3 spark plug wire clamps
32. Connect the following hoses:
   - Brake booster vacuum hose
   - EVAP hose
   - Automatic Disconnecting Differential (ADD) vacuum hose, for 4-wheel drive
   - Fuel inlet and fuel return hose
   - Heater hose
33. Install or connect the following:
   - Oil dipstick and guide, using a new O-ring
   - Spark plugs
   - Spark plug wires, with the ignition coils
   - Alternator drive belt
34. Connect the following cables:
   - Actuator cable with the bracket, if equipped with cruise control
   - Accelerator cable
   - Throttle cable, if equipped with an automatic transmission
   - MAF meter, resonator and air cleaner cap
   - Front exhaust pipe
   - Negative battery cable
35. Fill the radiator with engine coolant.
36. Start the engine and check for leaks.
37. Check the ignition timing.
38. Install the engine undercover.
39. Road test the vehicle.
40. Recheck all fluid levels.

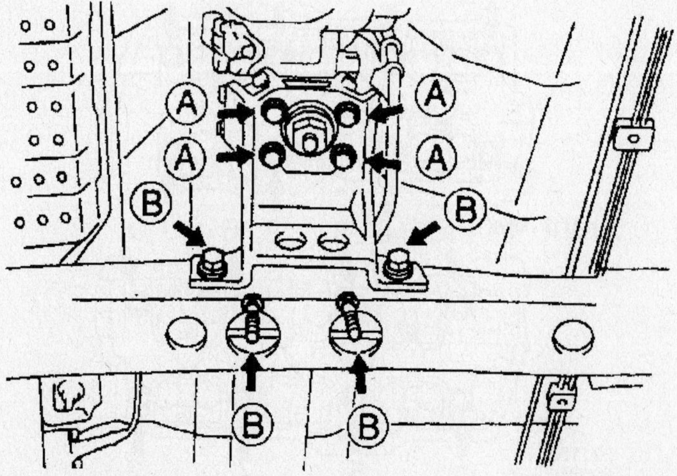

Cylinder head torque sequence—3.4L engine

4.0L Engine

RIGHT SIDE

1. Remove the timing chain.
2. Remove the air cleaner.
3. Remove the transmission oil filler tube.

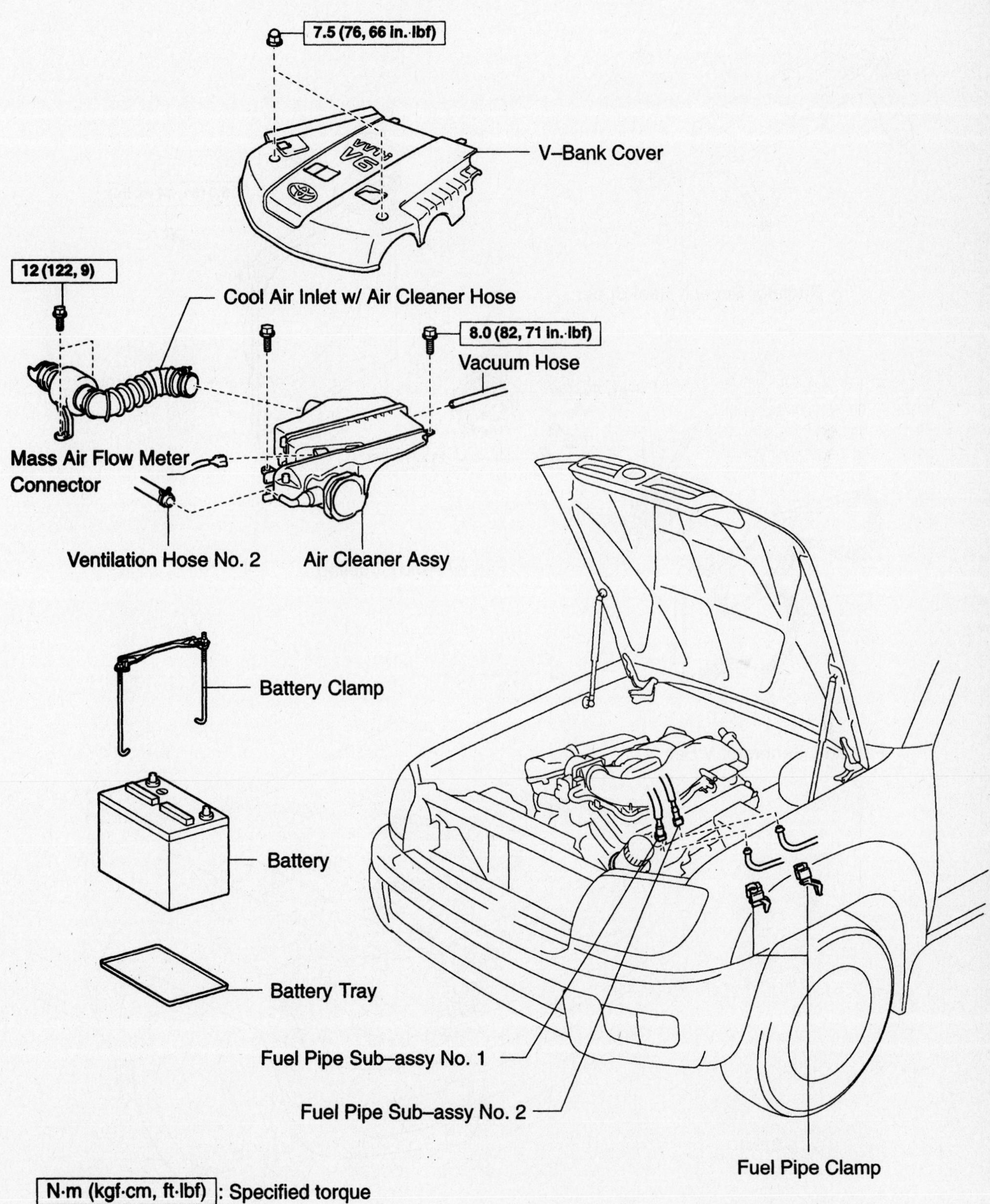

For right cylinder head removal, remove these parts—Part 1

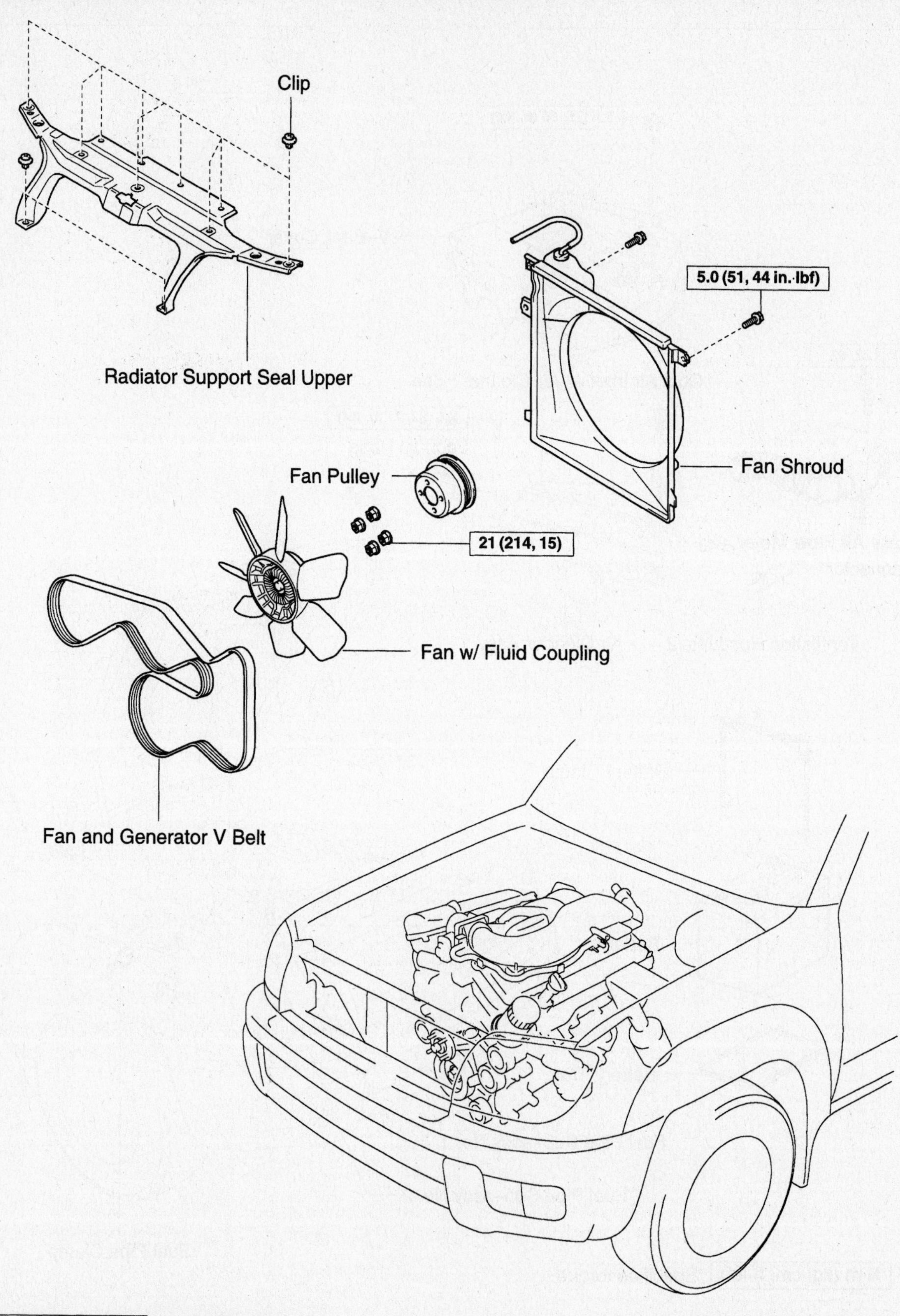

For right cylinder head removal, remove these parts—Part 2

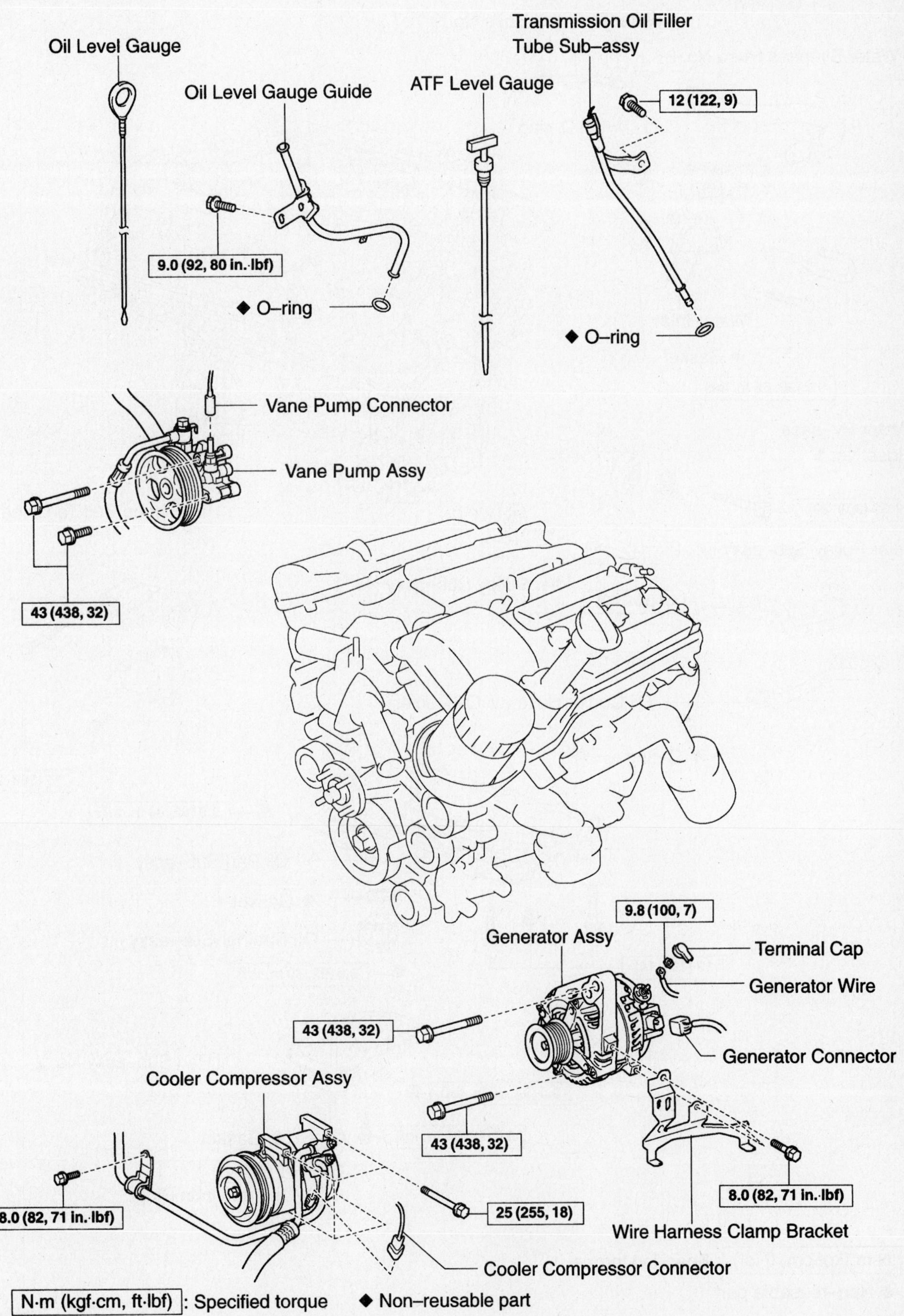

For right cylinder head removal, remove these parts—Part 3

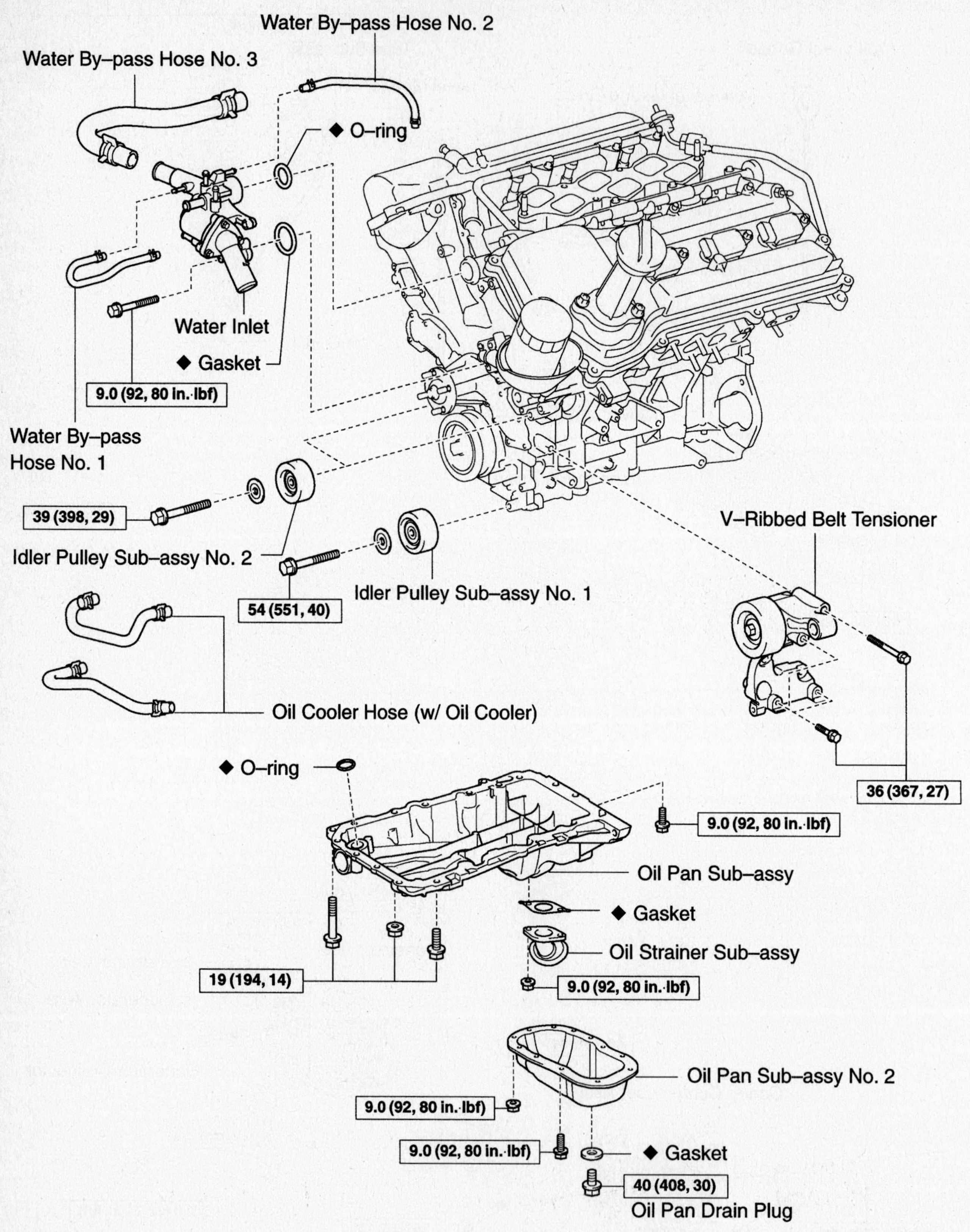

For right cylinder head removal, remove these parts—Part 4

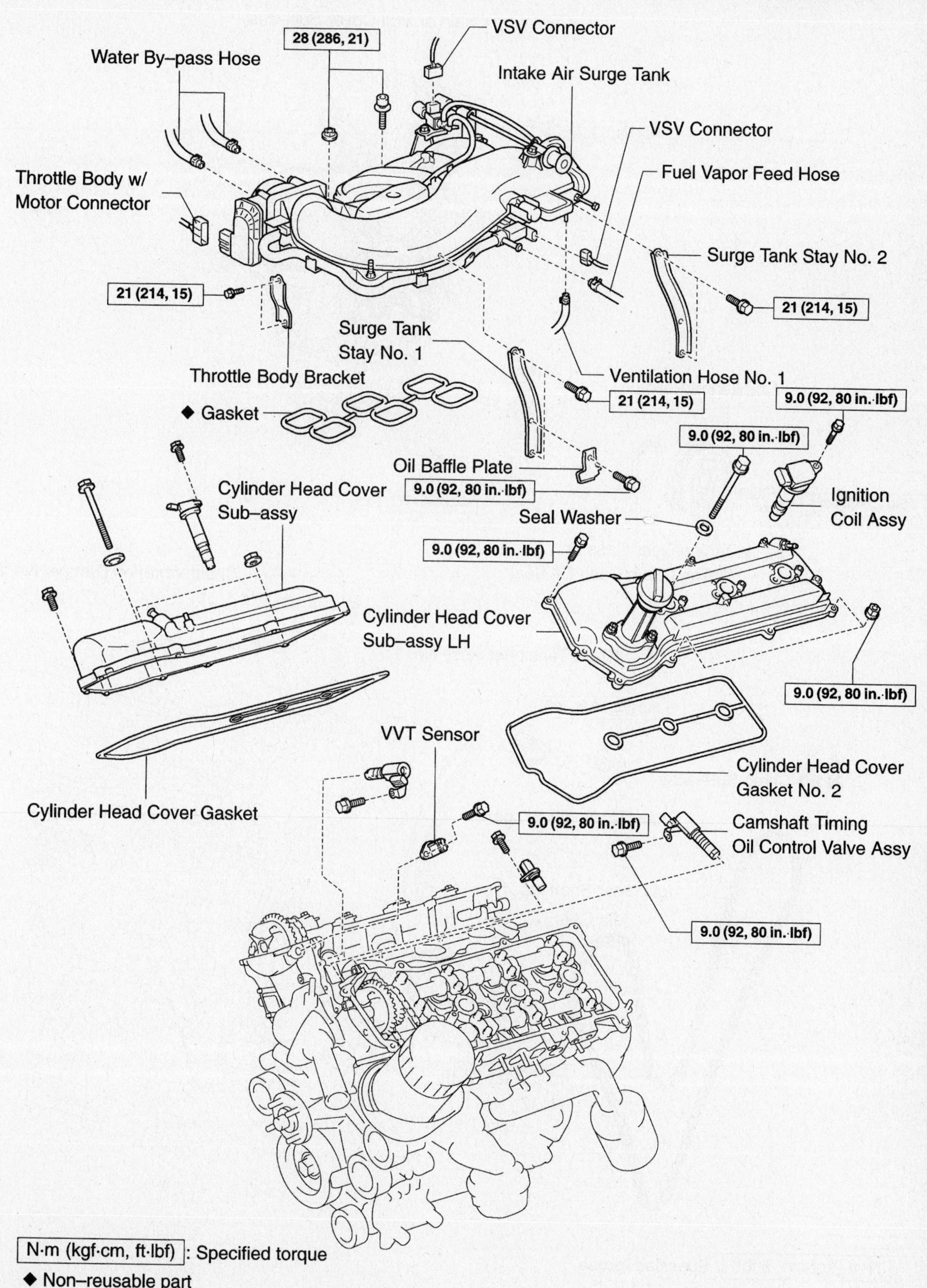
28 (286, 21)
VSV Connector
Water By–pass Hose
Intake Air Surge Tank
VSV Connector
Fuel Vapor Feed Hose
Throttle Body w/
Motor Connector
Surge Tank Stay No. 2
21 (214, 15)
21 (214, 15)
Surge Tank
Stay No. 1
Throttle Body Bracket
Ventilation Hose No. 1
21 (214, 15)
9.0 (92, 80 in.·lbf)
◆ Gasket
9.0 (92, 80 in.·lbf)
Oil Baffle Plate
Cylinder Head Cover
Sub–assy
9.0 (92, 80 in.·lbf)
Ignition
Coil Assy
Seal Washer
9.0 (92, 80 in.·lbf)
Cylinder Head Cover
Sub–assy LH
9.0 (92, 80 in.·lbf)
VVT Sensor
Cylinder Head Cover
Gasket No. 2
Cylinder Head Cover Gasket
9.0 (92, 80 in.·lbf)
Camshaft Timing
Oil Control Valve Assy
9.0 (92, 80 in.·lbf)
N·m (kgf·cm, ft·lbf) : Specified torque
◆ Non–reusable part
67170-4RUN-G11

For right cylinder head removal, remove these parts—Part 5

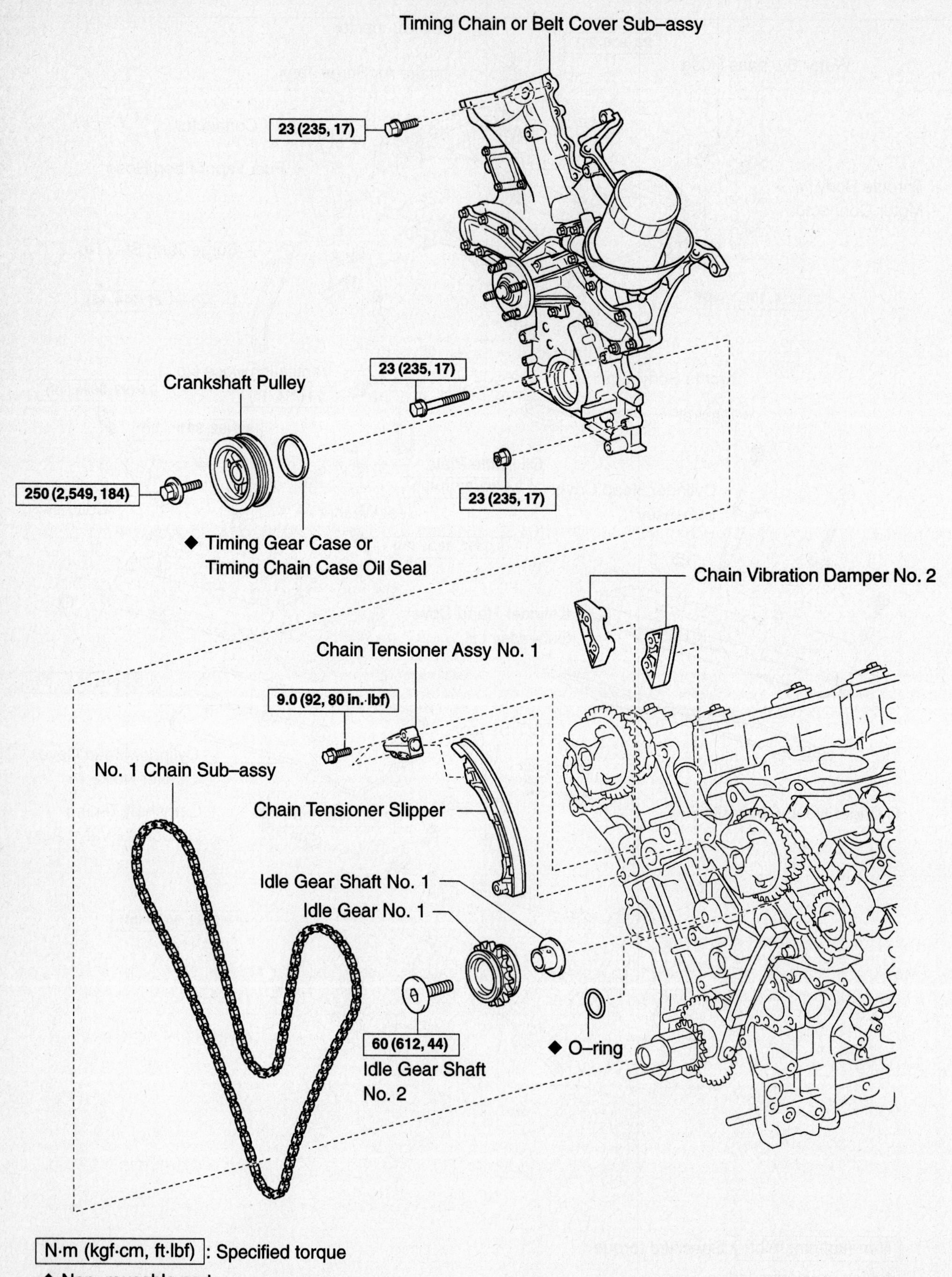

For right cylinder head removal, remove these parts—Part 6

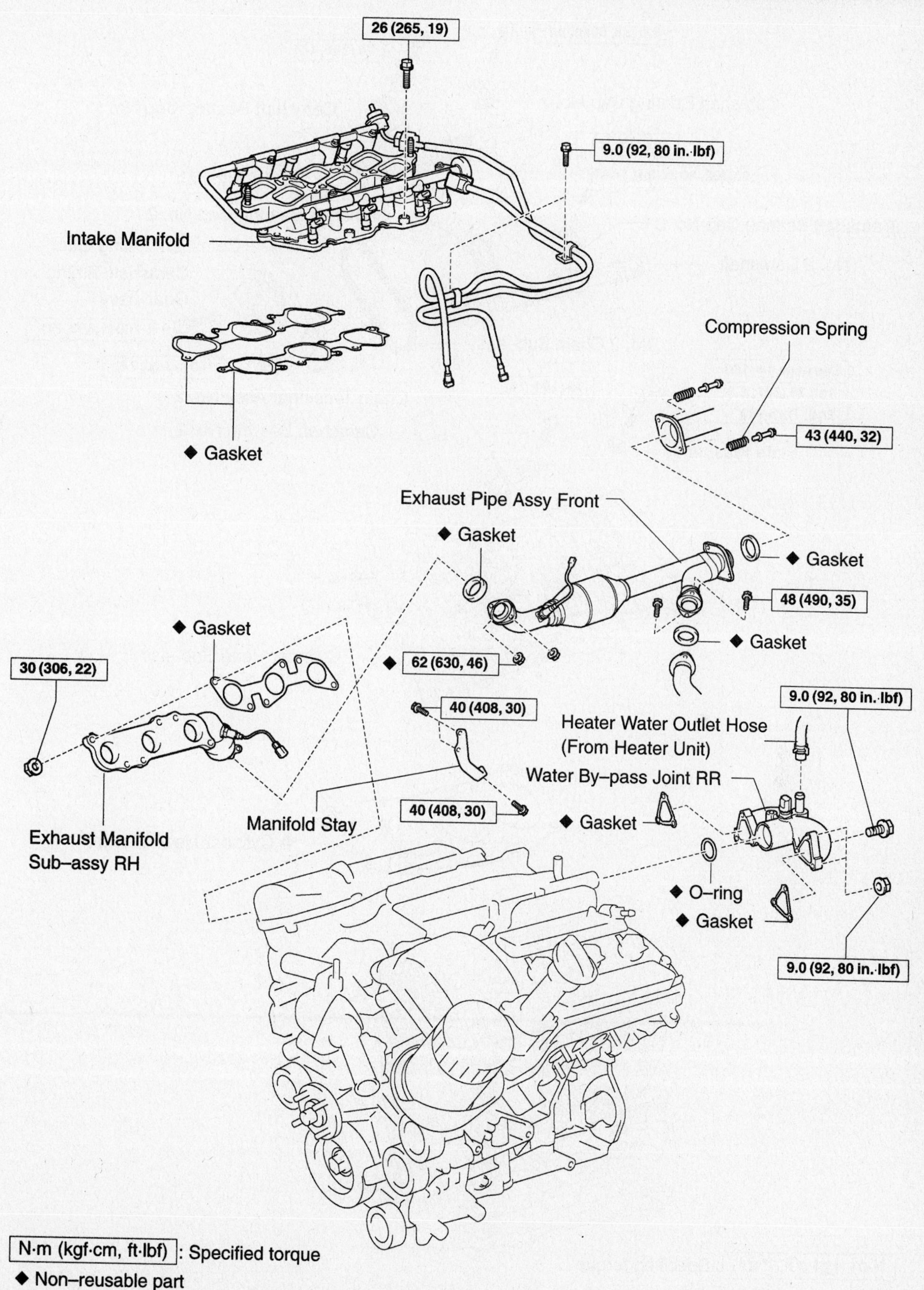

For right cylinder head removal, remove these parts—Part 7

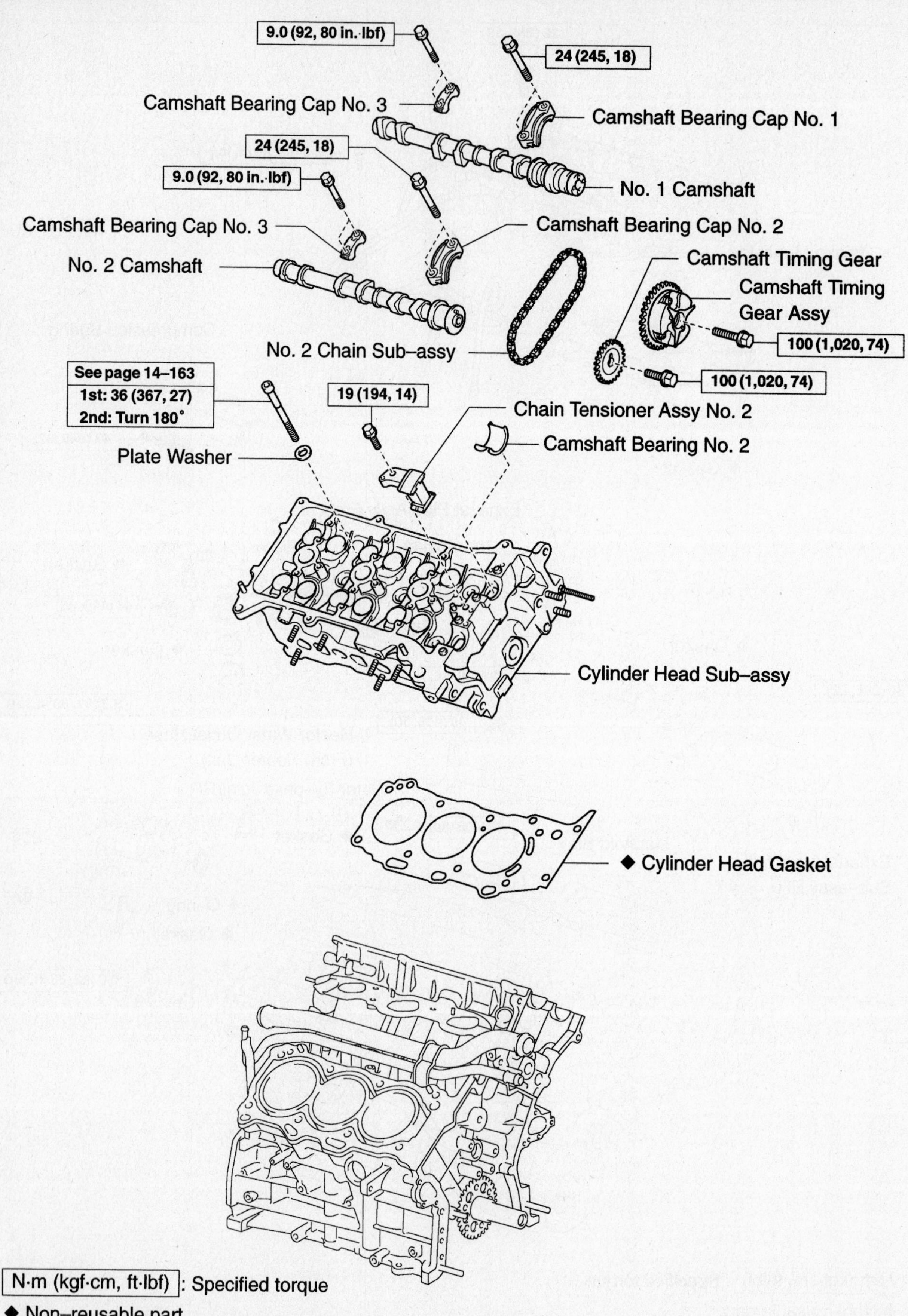

For right cylinder head removal, remove these parts—Part 8

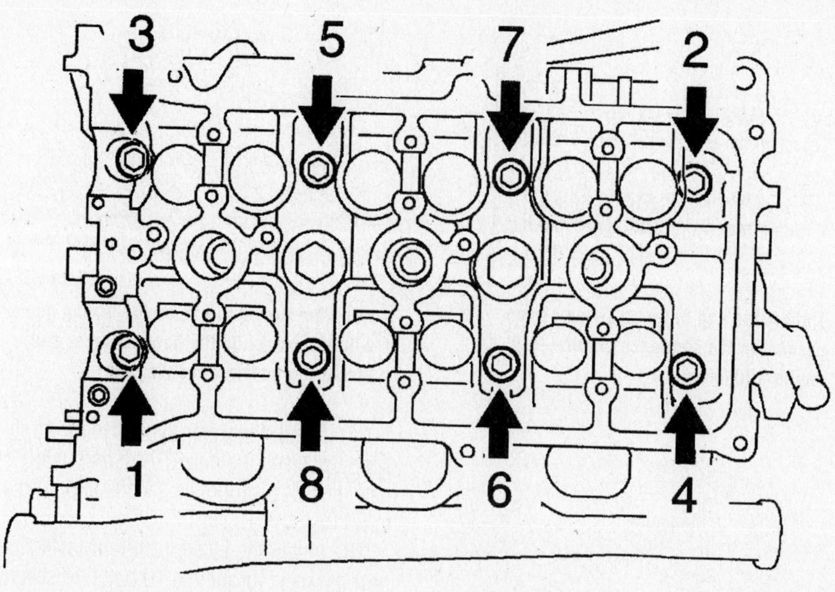

67170-4RUN-G16

Cylinder head bolt loosening sequence—4.0L engine right side

4. Remove the front exhaust pipe.
5. Remove the exhaust manifold.
6. Remove the fuel pipes.
7. Remove the intake manifold.
8. Remove the water by-pass joint.
9. Remove the right camshaft timing gears and chain. Insert a 1mm diameter pin into the tensioner hole to retain it.
10. Remove the tensioner.
11. Loosen the bolts in several steps, in the sequence shown and remove the camshafts. Keep the caps in order.

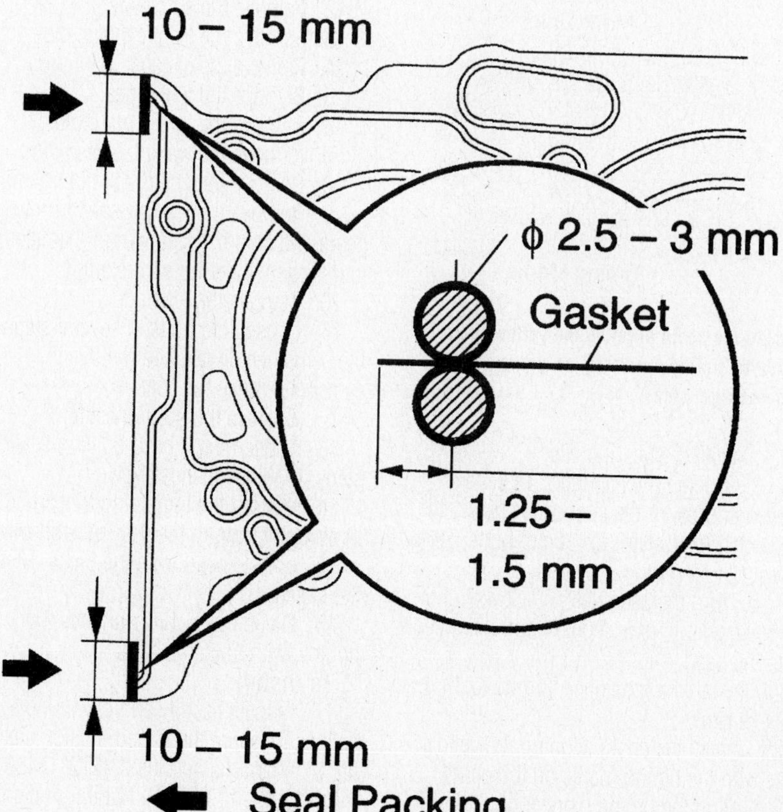

67170-4RUN-G18

Apply a continuous 3mm diameter bead of gasket material 08826-00080, or equivalent, to the gasket as shown—4.0L engine right side

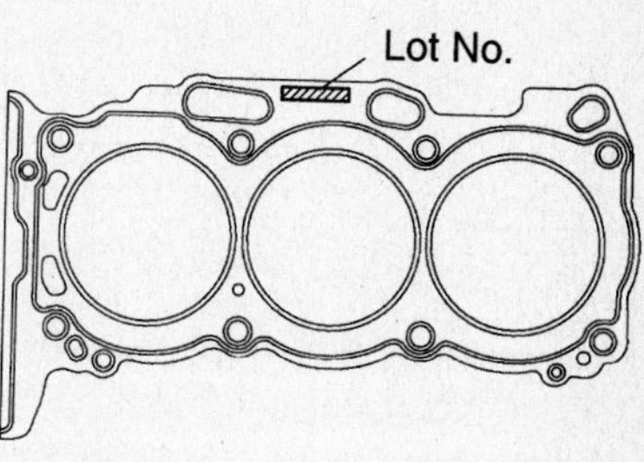

67170-4RUN-G19

Cylinder head gasket placement—4.0L right side

12. Remove the ground cables.
13. Loosen the head bolts in several steps, in the sequence shown.
14. Lift the head from the block. Discard the gasket.
15. Clean the mating surfaces thoroughly, without scratching.

To install:

16. Inspect each head bolt. Using a caliper, measure the thread outside diameter. Standard diameter is 0.4272–0.4331 inch (10.85–11.00mm). Minimum diameter is 0.421 inch (10.7mm).
17. Carefully remove any old silicone gasket material.
18. Apply a continuous 3mm diameter bead of gasket material 08826-00080, or equivalent, to the gasket as shown.

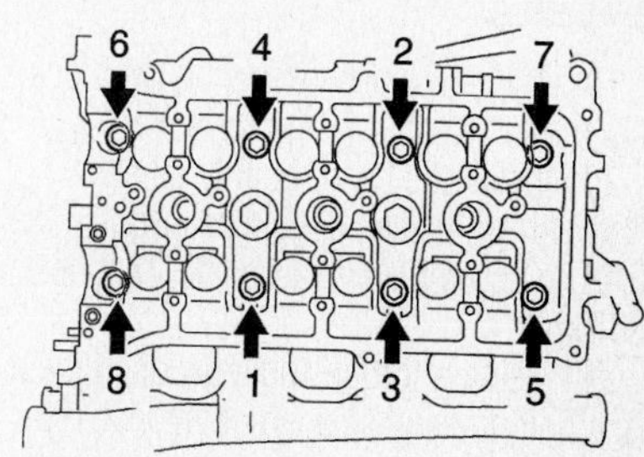

67170-4RUN-G20

Cylinder head bolt tightening sequence—4.0L engine right side

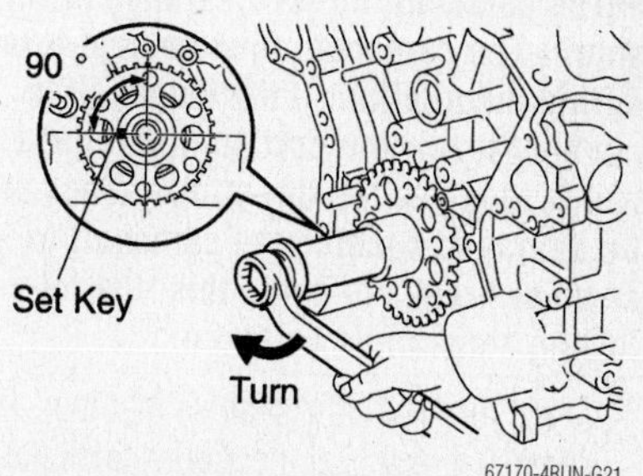

67170-4RUN-G21

Turn the crankshaft so that the crankshaft key is at the 9:00 o'clock position—4.0L engine right side

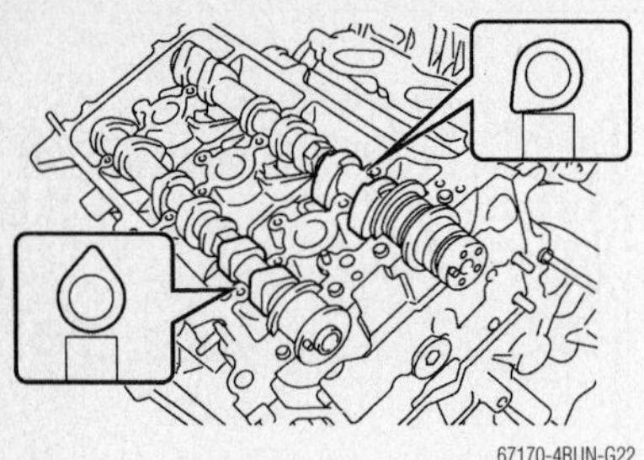

Place the camshafts on the head as shown—4.0L engine right side

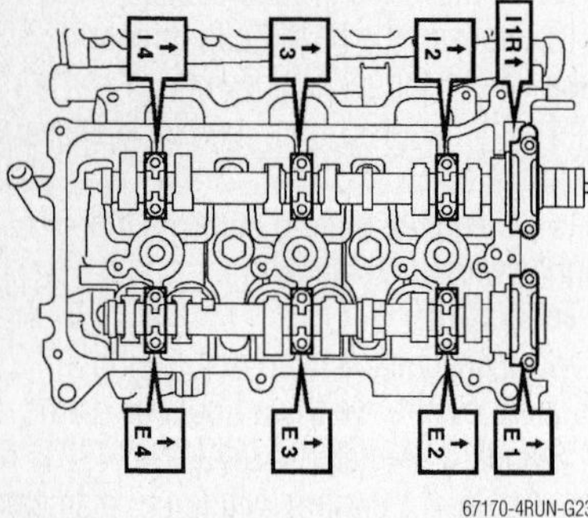

Install the bearing caps—4.0L engine right side

➡**The cylinder head must be installed within 3 minutes and the bolts tightened with 15 minutes. Otherwise, the silicone material must be removed and re-applied.**

19. Position a new gasket on the block with the lot number facing up.

20. Install the head. Lightly oil each head bolt with clean engine oil. Tighten the bolts, in several even passes, to 27 ft. lbs. (36 Nm).

21. Using a torque angle gauge, tighten each bolt, in sequence, an additional 180 degrees. If an angle gauge is not available, make a paint mark on each bolt head and the cylinder head, and tighten each bolt 180 degrees.

22. Install the camshafts.

➡**The camshaft thrust clearance is very small. The camshaft must be kept level during installation. If the camshaft is not kept level, the portion of the head which receives the thrust may crack or be damaged causing the camshaft to seize or break. To avoid this, the following steps must be taken:**

a. Turn the crankshaft so that the crankshaft key is at the 9:00 o'clock position.

b. Place the camshafts on the head as shown.

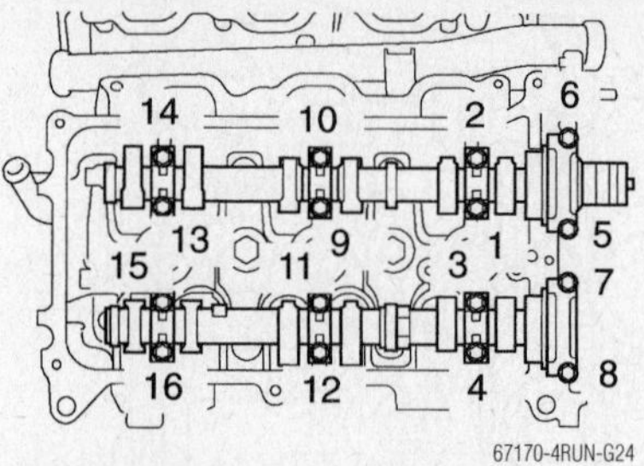

Tighten the cap bolts, in several even passes, in the sequence shown—4.0L engine right side

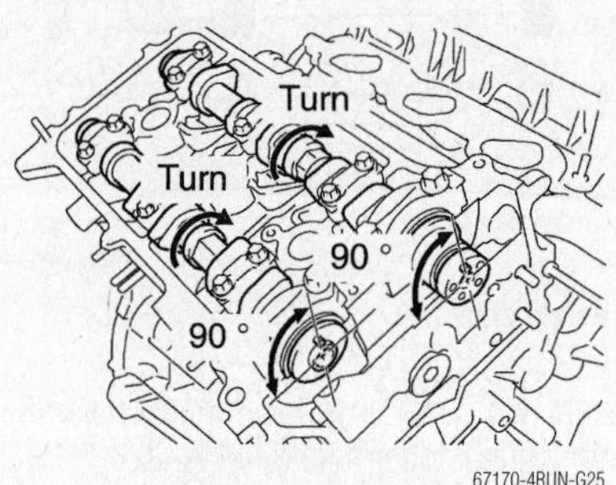

Turn the camshafts clockwise until the knock pin is in a position 90 degrees to the head—4.0L engine right side

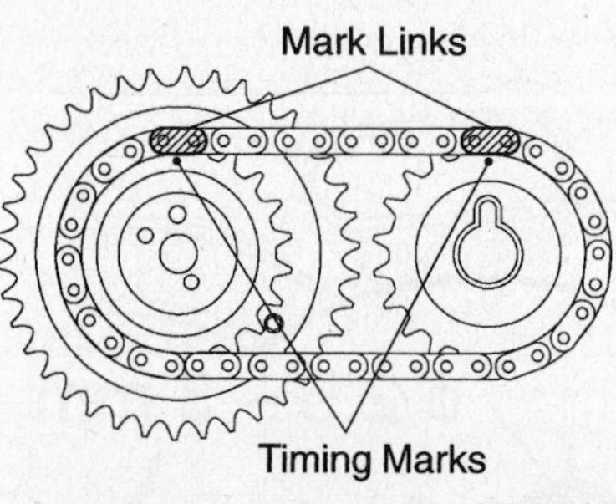

Install the chain aligning the yellow links with the timing marks on the gears—4.0L engine right side

23. Install the bearing caps.

a. Tighten the cap bolts, in several even passes, in the sequence shown, to 80 inch lbs (9 Nm) for 10mm bolts; 18 ft. lbs. (24 Nm) for 12mm bolts.

b. Turn the camshafts clockwise until the knock pin is in a position 90 degrees to the head.

24. Install the tensioner. Torque to 14 ft. lbs. (19 Nm).

25. Install the chain aligning the yellow links with the timing marks on the gears.

26. Align the timing marks on the timing gears with the marks on the bearing caps, and install the gear/chain assembly. Torque the bolts to 74 ft. lbs. (100 Nm).

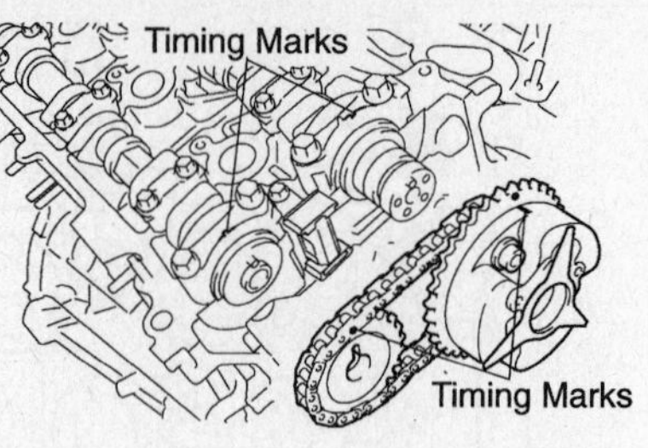

Align the timing marks on the timing gears with the marks on the bearing caps, and install the gear/chain assembly

27. Install the water bypass joint. Apply soapy water to the new O-ring and using 2 new gaskets. Torque the 2 bolts and 4 nuts to 80 inch lbs. (9 Nm).

28. Install the intake manifold, using new gaskets. Torque the 10 bolts, in several even passes, to 19 ft. lbs. (26 Nm).

29. Connect the fuel lines.

30. Install the exhaust manifold. Torque to 22 ft. lbs. (30 Nm). Torque the manifold stay to 30 ft. lbs. (40 Nm).

31. The remainder of installation is the reverse of removal.

LEFT SIDE

1. Remove the timing chain.
2. Remove the air cleaner.
3. Remove the front exhaust pipe.
4. Remove the exhaust manifold.
5. Remove the fuel pipes.
6. Remove the intake manifold.
7. Remove the water by-pass joint.
8. Remove the chain vibration damper.
9. Remove the left camshaft timing gears and chain. Insert a 1mm diameter pin into the tensioner hole to retain it.
10. Remove the tensioner.
11. Loosen the bolts in several steps, in the sequence shown and remove the camshafts. Keep the caps in order.
12. Remove the ground cable.
13. Loosen the 2 head bolts in several steps, in the sequence shown.
14. Loosen the remaining 8 head bolts in several steps, in the sequence shown.
15. Lift the head from the block. Discard the gasket.
16. Clean the mating surfaces thoroughly, without scratching.

To install:

17. Inspect each head bolt. Using a caliper, measure the thread outside diameter. Standard diameter is 0.4272–0.4331 inch (10.85–11.00mm). Minimum diameter is 0.421 inch (10.7mm).

18. Carefully remove any old silicone gasket material.

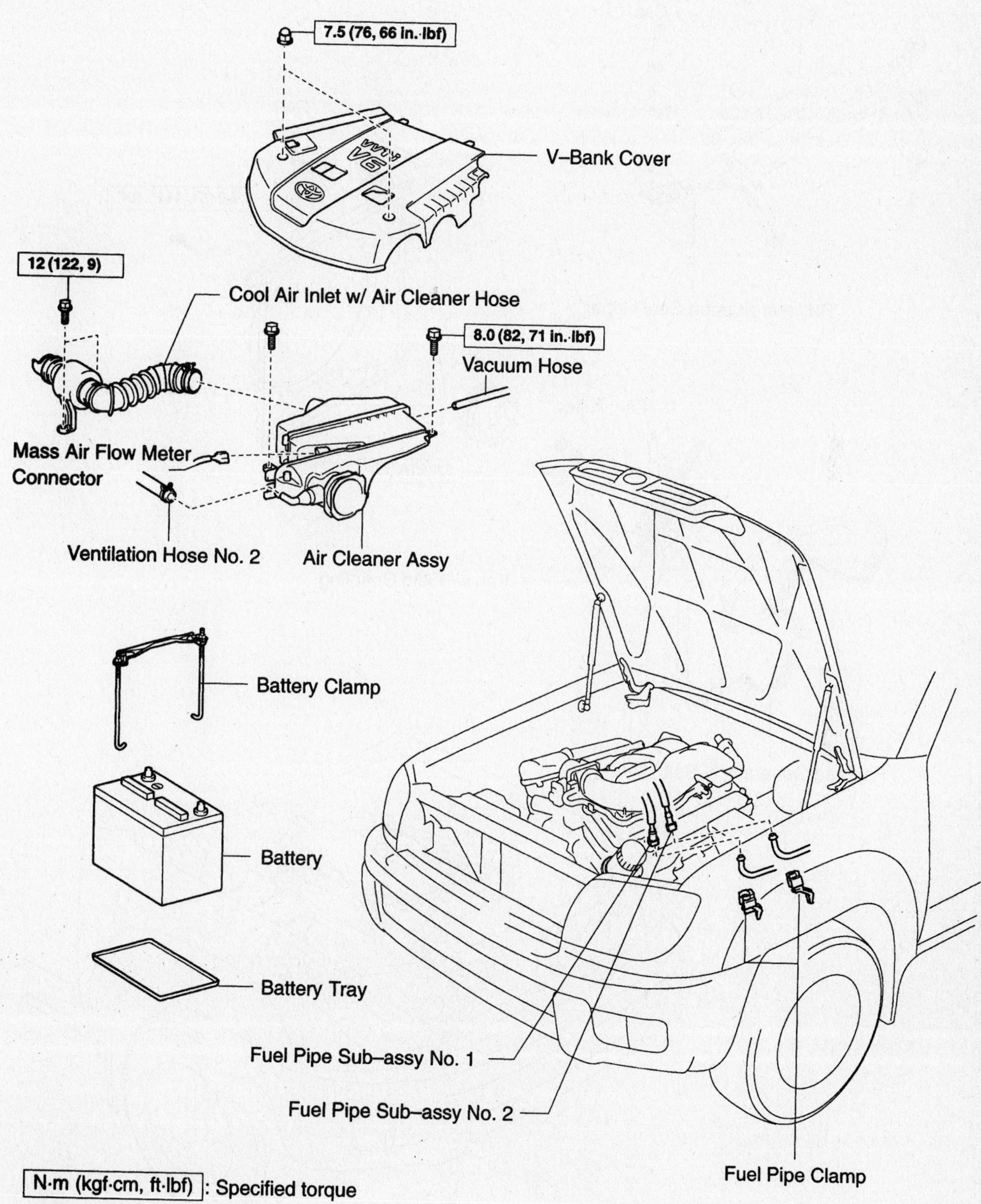

For left cylinder head removal, remove these parts—Part 1

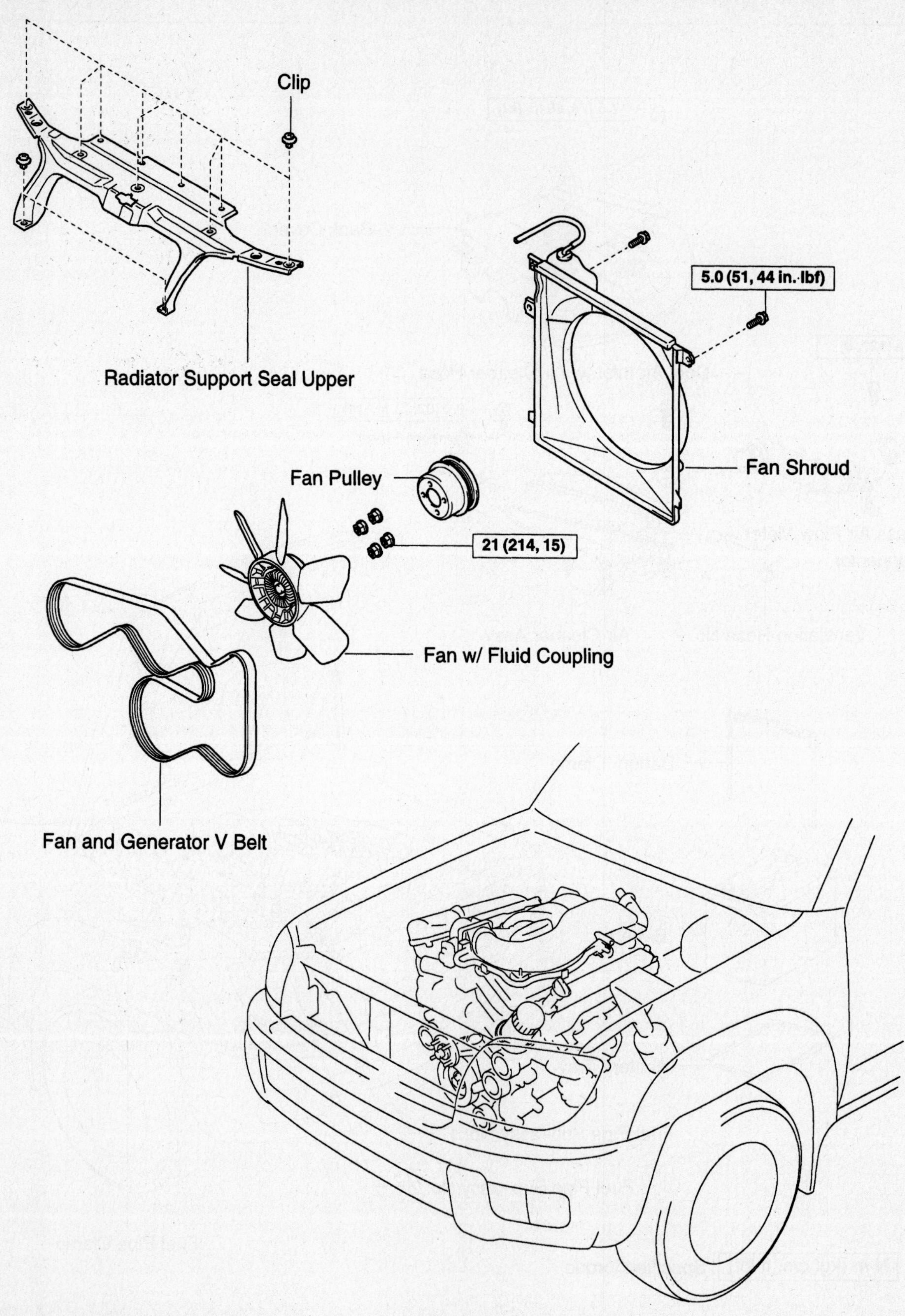

For left cylinder head removal, remove these parts—Part 2

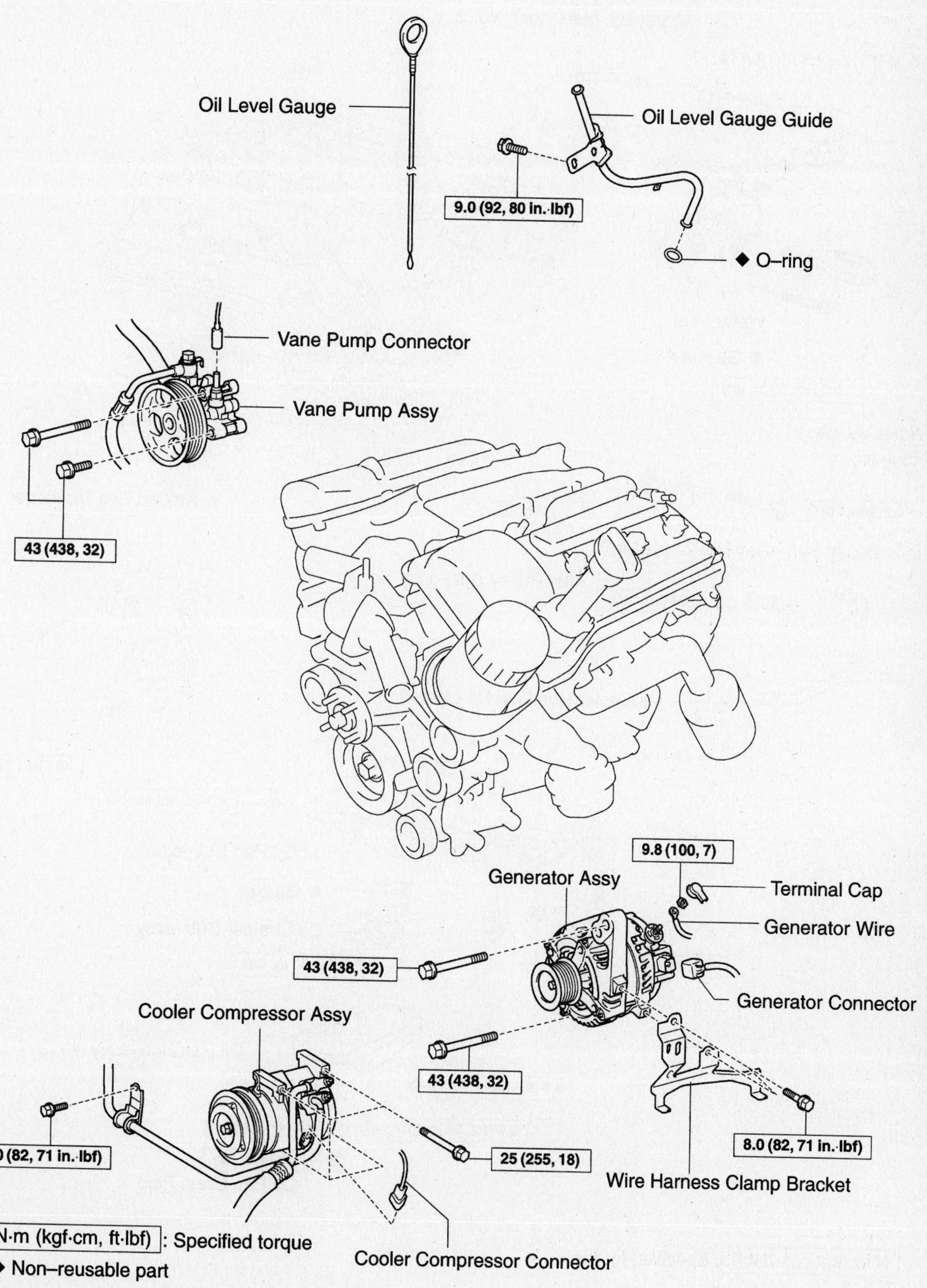

For left cylinder head removal, remove these parts—Part 3

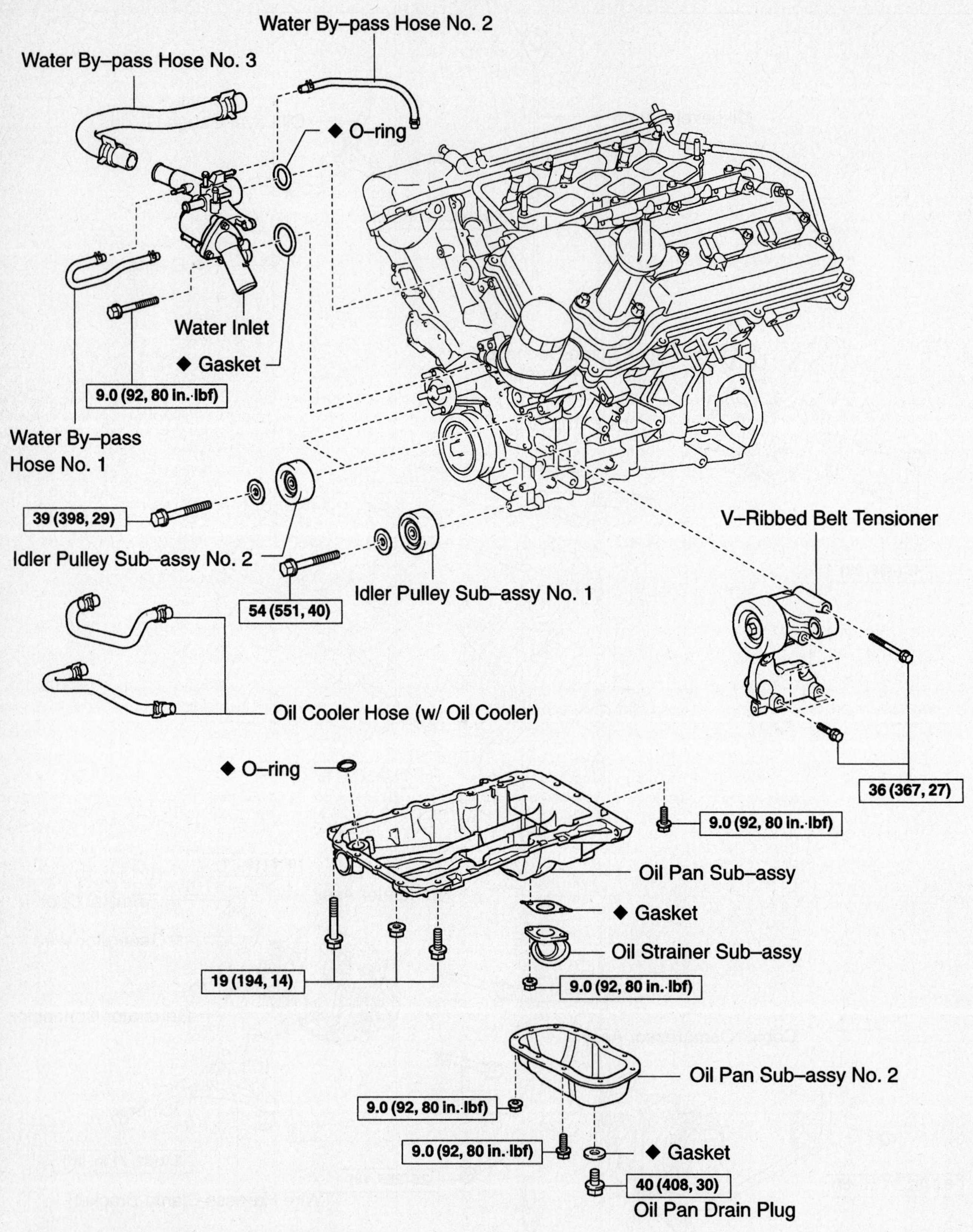

For left cylinder head removal, remove these parts—Part 4

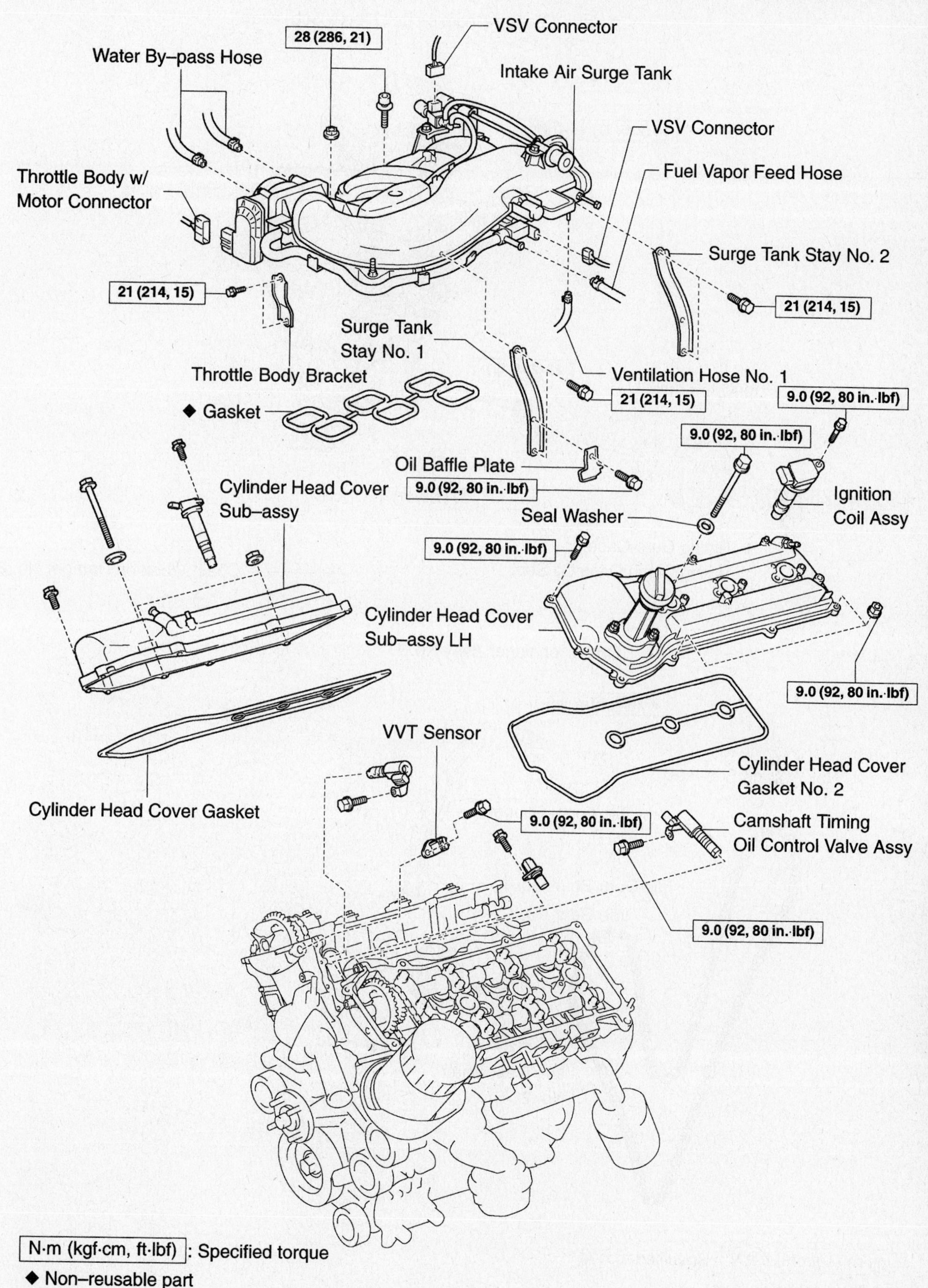

For left cylinder head removal, remove these parts—Part 5

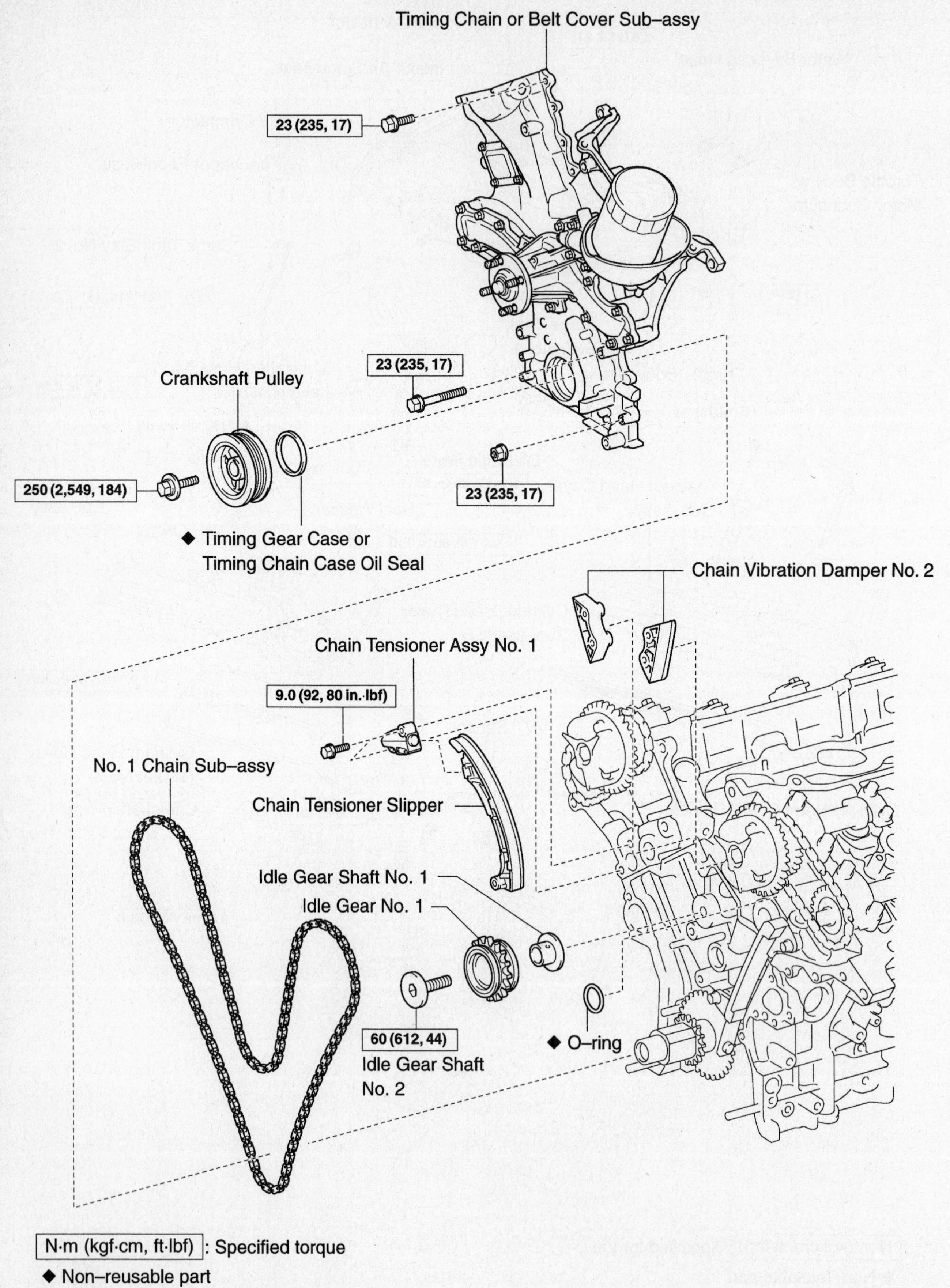

For left cylinder head removal, remove these parts—Part 6

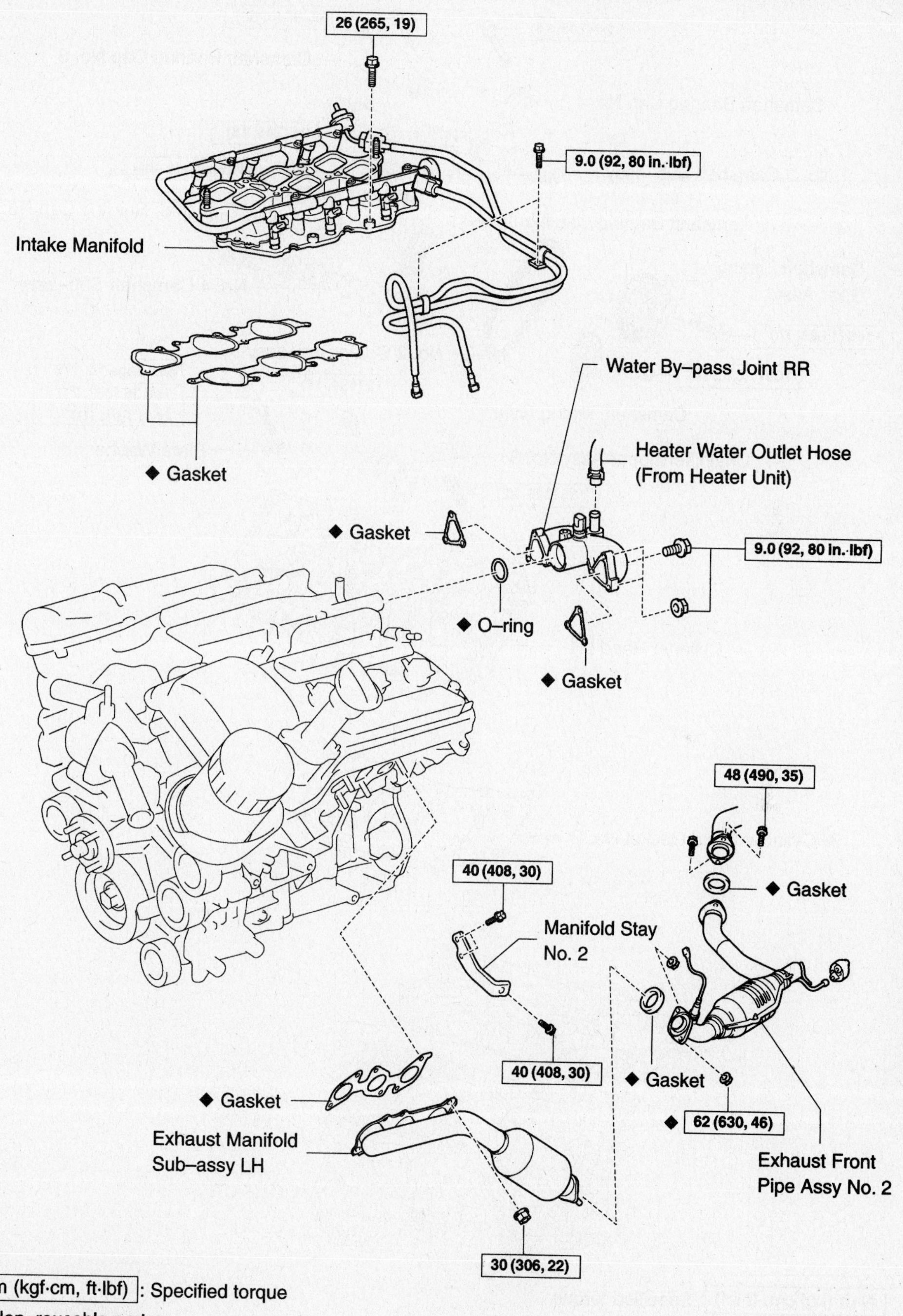

For left cylinder head removal, remove these parts—Part 7

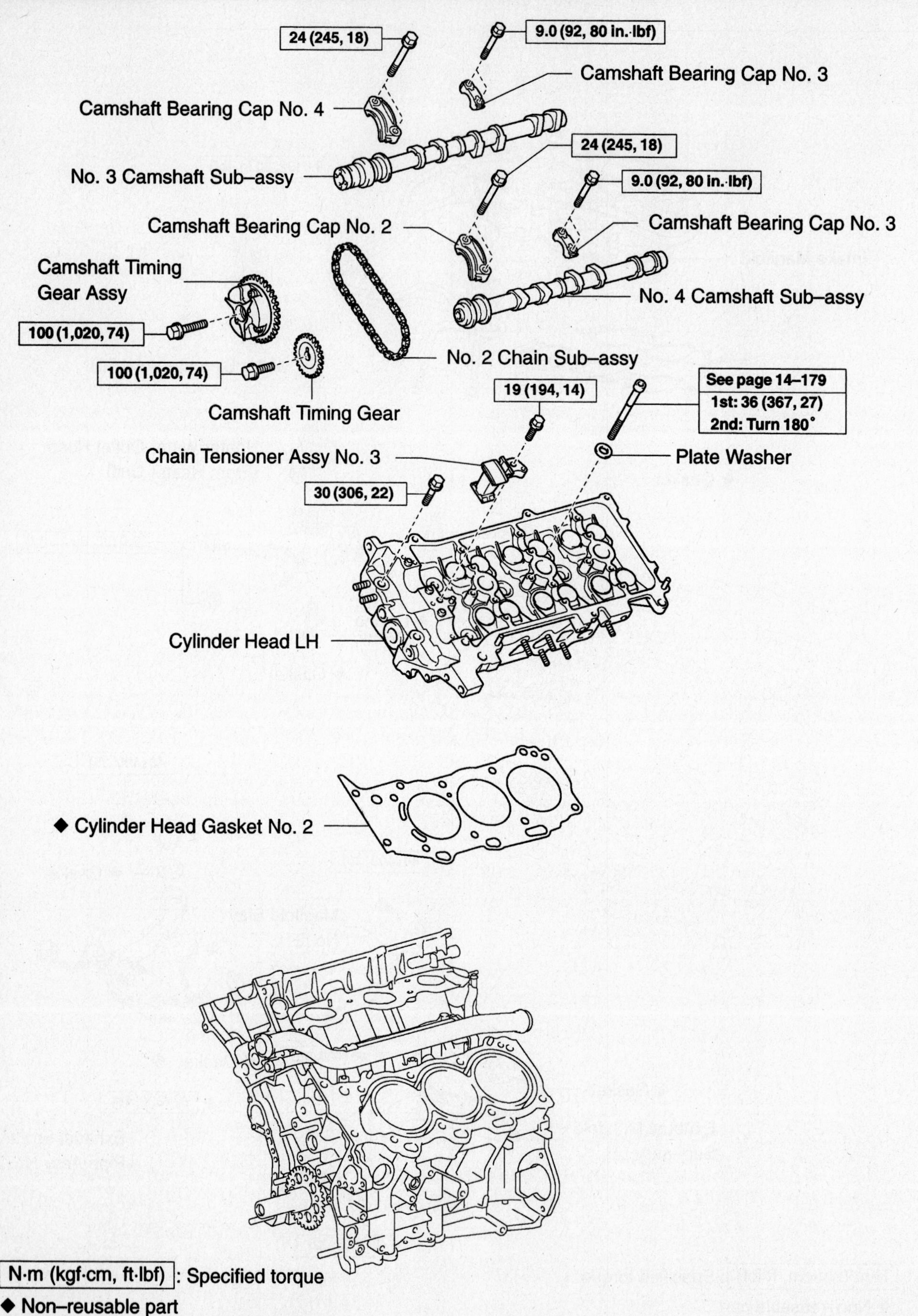

For left cylinder head removal, remove these parts—Part 8

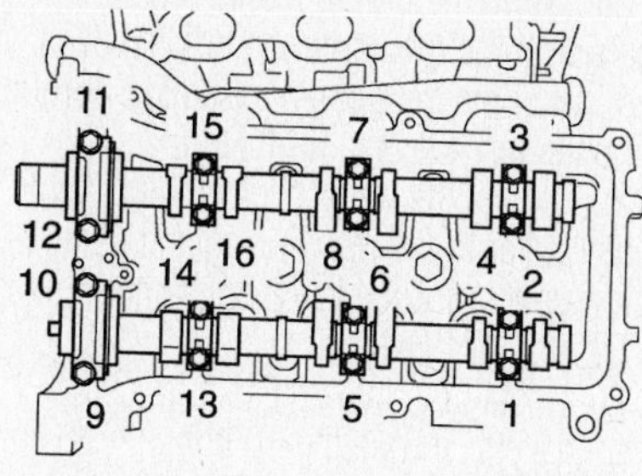

Camshaft bearing cap loosening sequence—4.0L engine, left side

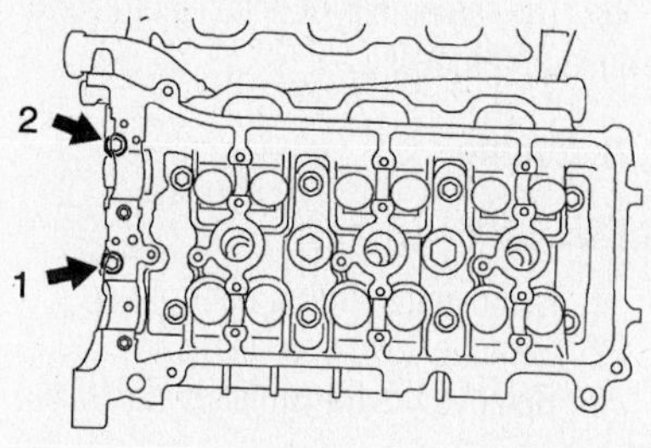

Loosen these 2 head bolts in several steps, in the sequence shown—4.0L engine, left side

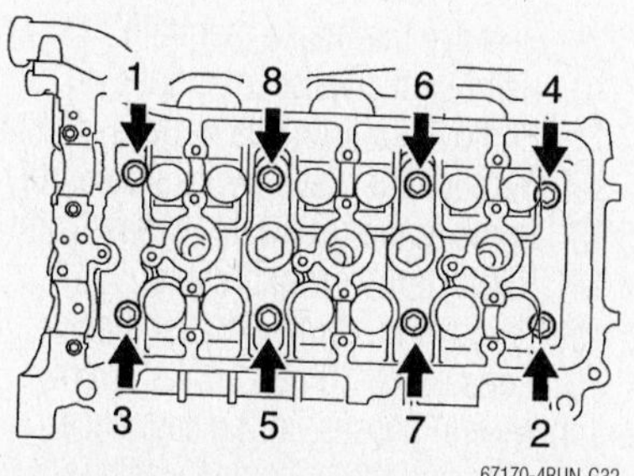

Cylinder head bolt loosening sequence—4.0L engine, left side

19. Apply a continuous 3mm diameter bead of gasket material 08826-00080, or equivalent, to the gasket as shown.

➡**The cylinder head must be installed within 3 minutes and the bolts tightened with 15 minutes. Otherwise, the silicone material must be removed and re-applied.**

20. Position a new gasket on the block with the lot number facing up.

21. Install the head. Lightly oil each head bolt with clean engine oil. Tighten the bolts, in several even passes, to 27 ft. lbs. (36 Nm).

22. Using a torque angle gauge, tighten each bolt, in sequence, an additional 180 degrees. If an angle gauge is not available, make a paint mark on each bolt head and

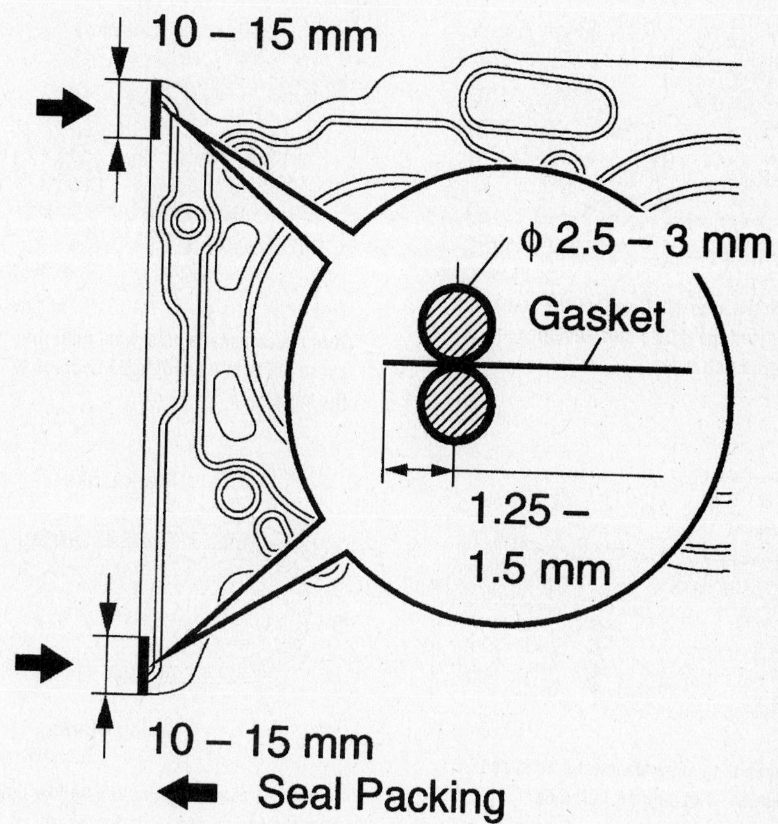

Apply a continuous 3mm diameter bead of gasket material 08826-00080, or equivalent, to the gasket as shown—4.0L engine

the cylinder head, and tighten each bolt 180 degrees.

23. Apply a light coat of clean engine oil to the threads and install the 2 front head bolts. Tighten the bolts in the sequence shown to 22 ft. lbs. (30 Nm).

24. Install the camshafts.

➡**The camshaft thrust clearance is very small. The camshaft must be kept level during installation. If the camshaft is not kept level, the portion of the head which receives the thrust may crack or be damaged causing the camshaft to seize or break. To avoid this, the following steps must be taken:**

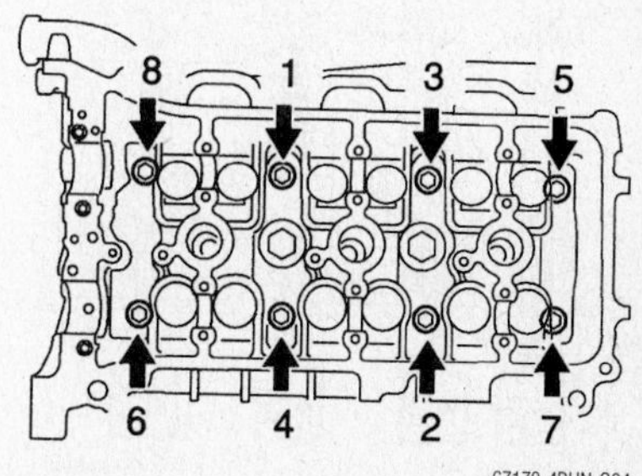

Cylinder head bolt tightening sequence—4.0L engine

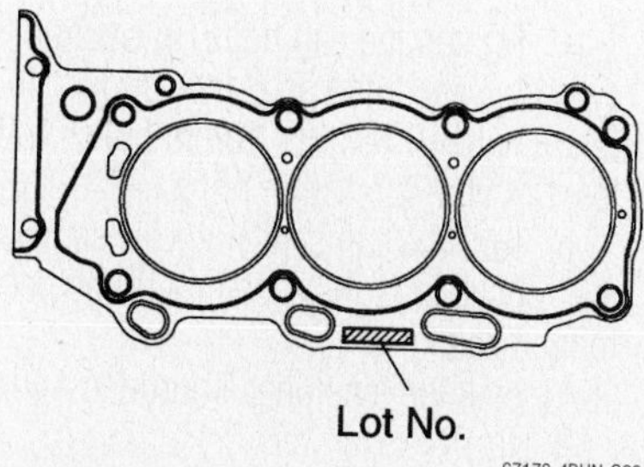

Cylinder head gasket placement—4.0L engine, left side

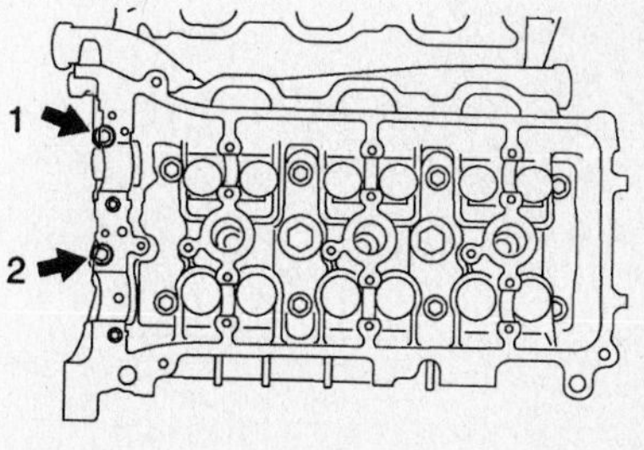

Front head bolt torque sequence—4.0L engine, left side

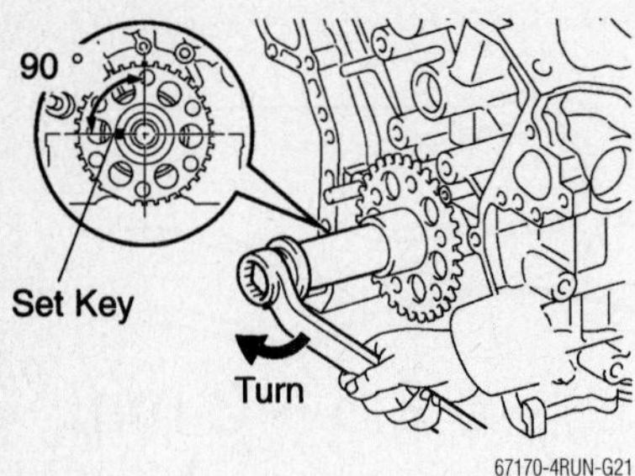

Turn the crankshaft so that the crankshaft key is at the 9:00 o'clock position—4.0L engine, left side

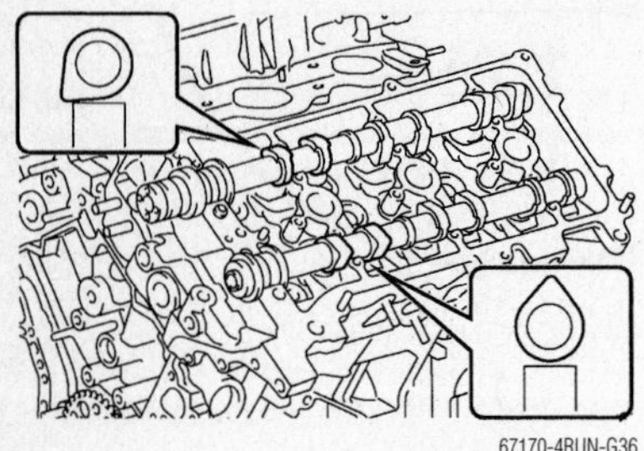

Place the camshafts on the head as shown—4.0L engine, left side

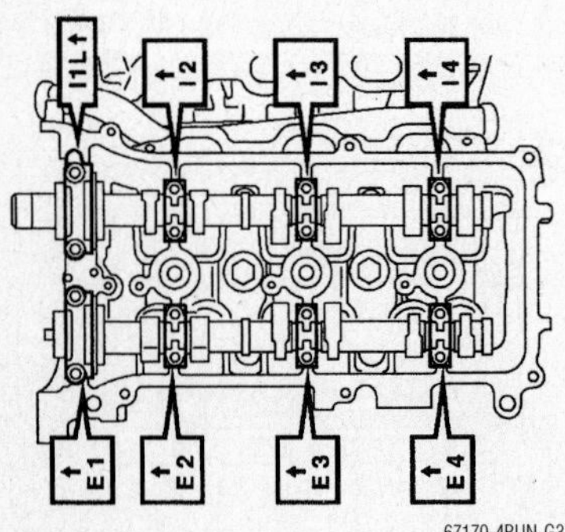

Install the bearing caps—4.0L engine, left side

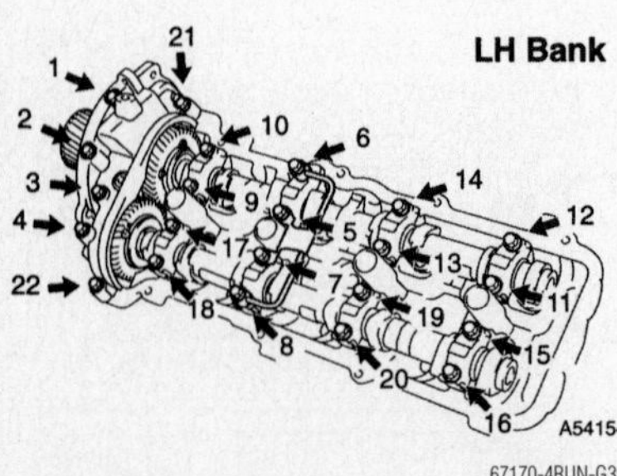

Tighten the cap bolts, in several even passes, in the sequence shown—4.0L engine, left side

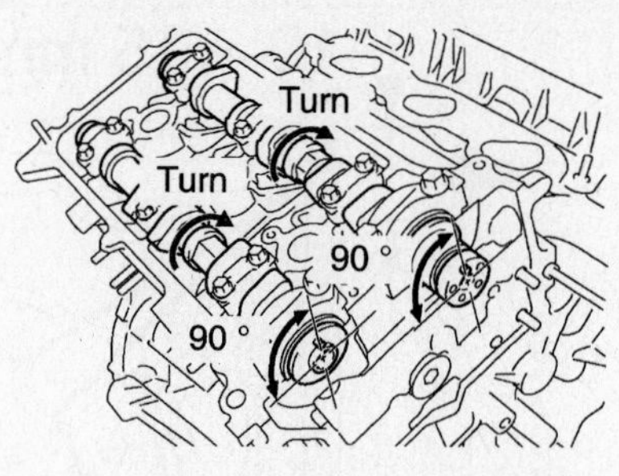

Turn the camshafts clockwise until the knock pin is in a position 90 degrees to the head

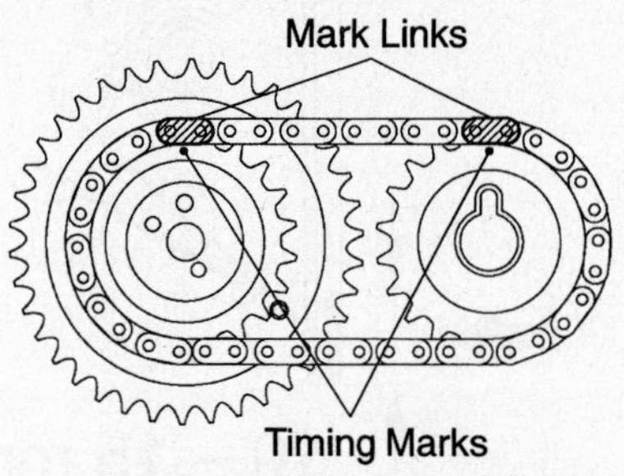

Install the chain aligning the yellow links with the timing marks on the gears

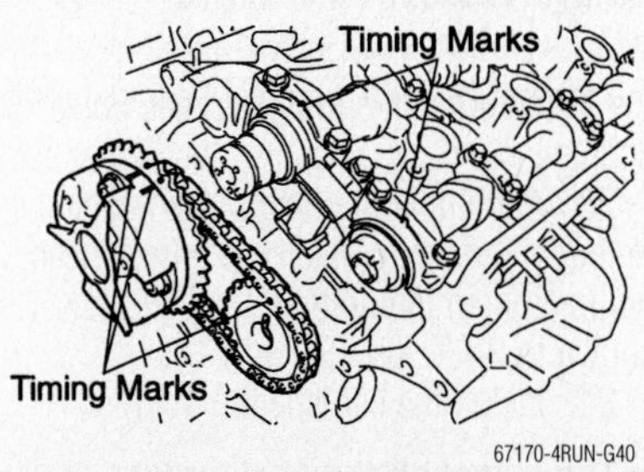

Align the timing marks on the timing gears with the marks on the bearing caps

a. Turn the crankshaft so that the crankshaft key is at the 9:00 o'clock position.

b. Place the camshafts on the head as shown.

25. Install the bearing caps.

a. Tighten the cap bolts, in several even passes, in the sequence shown, to 80 inch lbs (9 Nm) for 10mm bolts; 18 ft. lbs. (24 Nm) for 12mm bolts.

b. Turn the camshafts clockwise until the knock pin is in a position 90 degrees to the head.

26. Install the tensioner. Torque to 14 ft. lbs. (19 Nm).

27. Install the chain aligning the yellow links with the timing marks on the gears.

28. Align the timing marks on the timing gears with the marks on the bearing caps, and install the gear/chain assembly. Torque the bolts to 74 ft. lbs. (100 Nm).

29. Install the water bypass joint. Apply soapy water to the new O-ring and using 2 new gaskets. Torque the 2 bolts and 4 nuts to 80 inch lbs. (9 Nm).

30. Install the intake manifold, using new gaskets. Torque the 10 bolts, in several even passes, to 19 ft. lbs. (26 Nm).

31. Connect the fuel lines.

32. Install the exhaust manifold. Torque to 22 ft. lbs. (30 Nm). Torque the manifold stay to 30 ft. lbs. (40 Nm).

33. The remainder of installation is the reverse of removal.

4.7L Engine

RIGHT SIDE

1. Remove the timing belt.
2. Remove the camshaft.
3. Remove the transmission oil filler tube.
4. Remove the engine oil dipstick tube.
5. Disconnect the fuel lines.
6. Disconnect the vapor hose.
7. Disconnect the injector harness.
8. Disconnect the water bypass hoses.
9. Remove the intake manifold.
10. Remove the water inlet housing.
11. Remove the front water bypass joint.
12. Remove the water bypass assembly.
13. Remove the rear water bypass joint.
14. Remove the ground cable.
15. Remove the front exhaust pipe.
16. Loosen the 10 head bolts, evenly and in several passes, in the sequence shown. Remove the bolts and washers.
17. Clean the mating surfaces thoroughly, without scratching.

To install

18. Check the thread diameter of each head bolt. Outer diameter should be 0.3862–0.3921 inch (9.810–0.960mm). Minimum should be 0.3819 inch (9.700mm).

19. Install a new had gasket on the block with the ID mark up. For the right side, the mark is 2UR.

20. Install the cylinder head. Lightly oil the 10 bolts and tighten them, in several passes, to 24 ft. lbs. (32 Nm).

21. Using a torque angle gauge, tighten each bolt, in sequence, an additional 90 degrees, then, tighten each, in sequence another 90 degrees. If an angle gauge is not available, make a paint mark on each bolt head and the cylinder head, and tighten each bolt, in 2, 90 degrees stages, to 180 degrees.

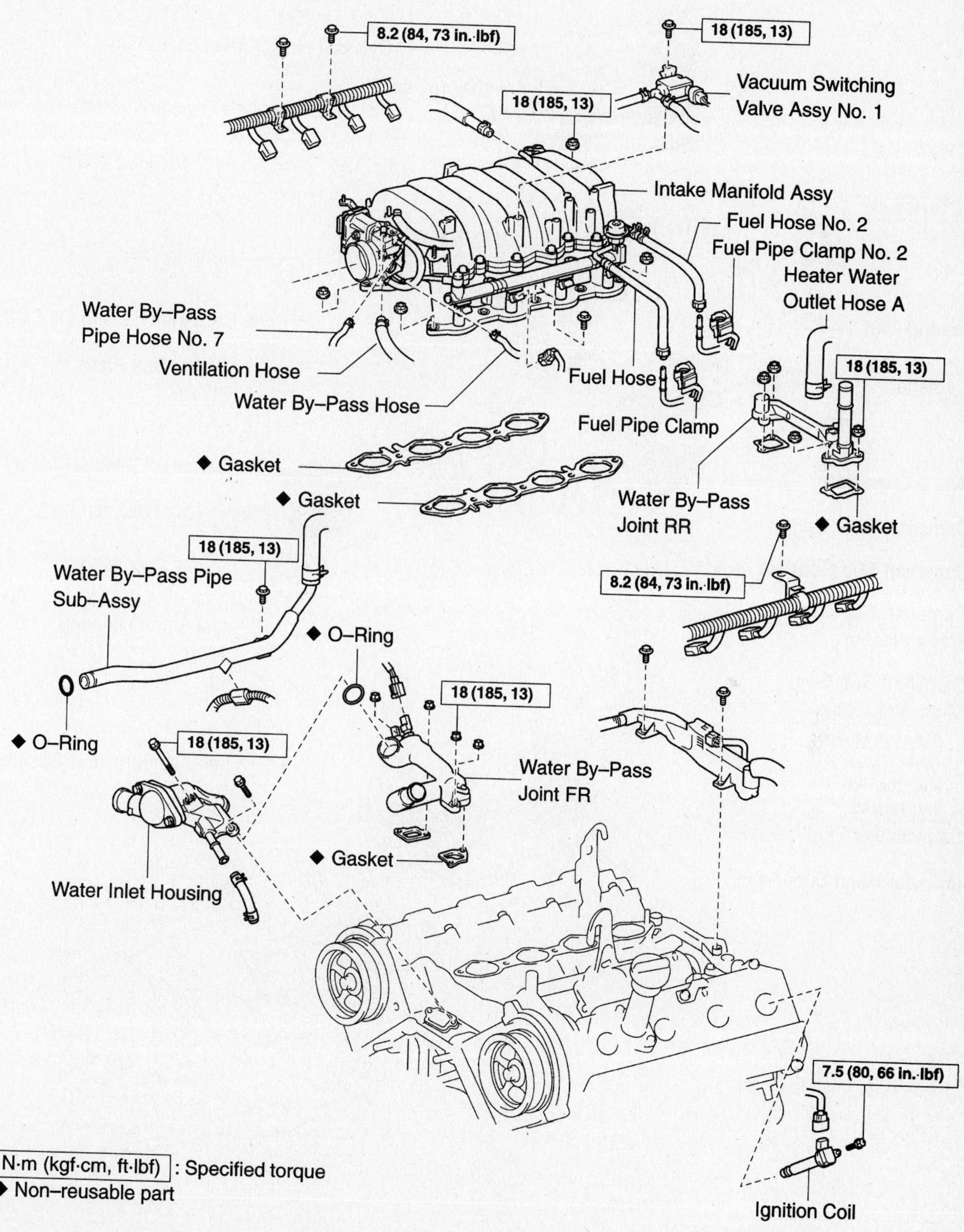

For 4.7L engine right cylinder head removal, remove these parts—Part 1

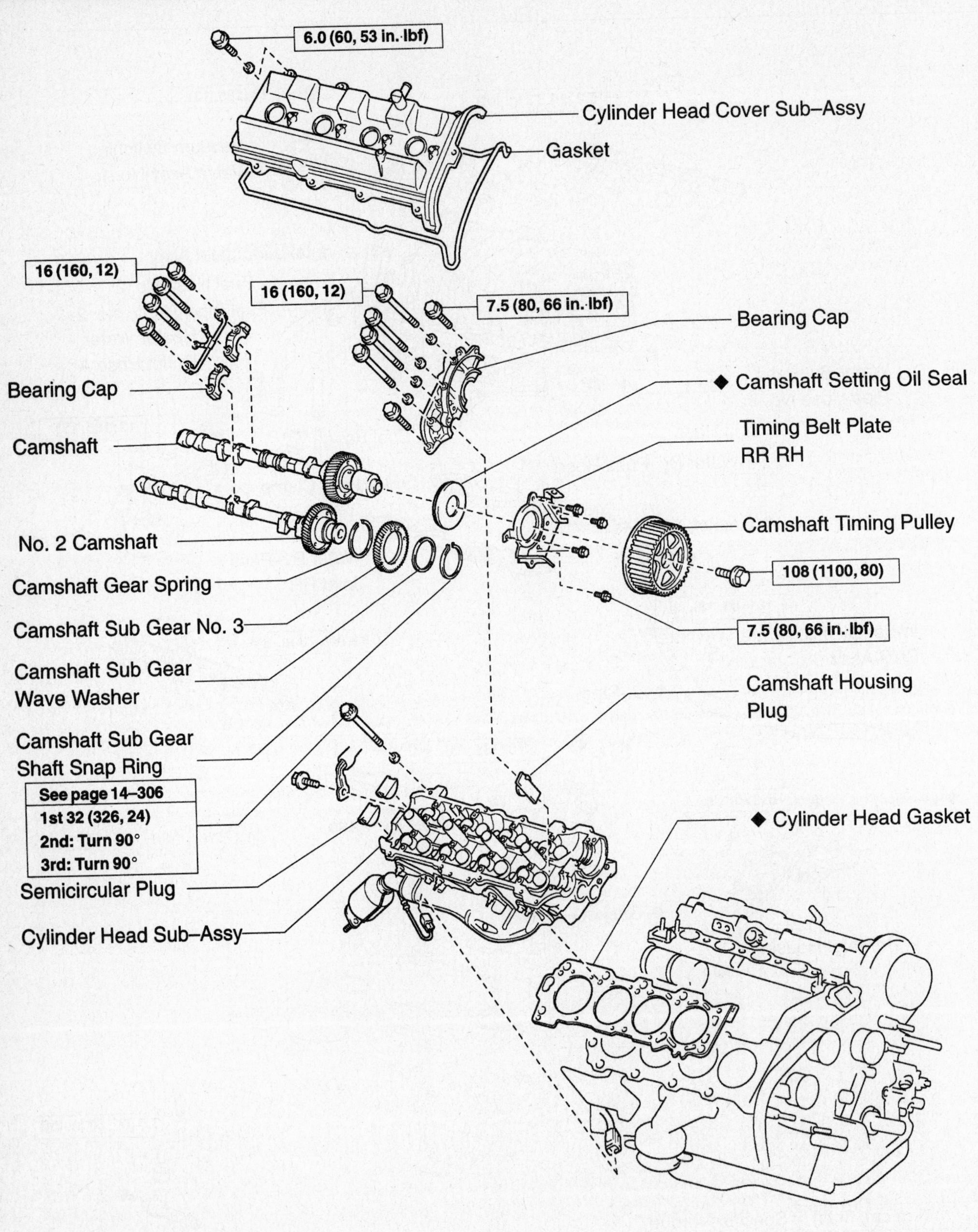

For 4.7L engine right cylinder head removal, remove these parts—Part 2

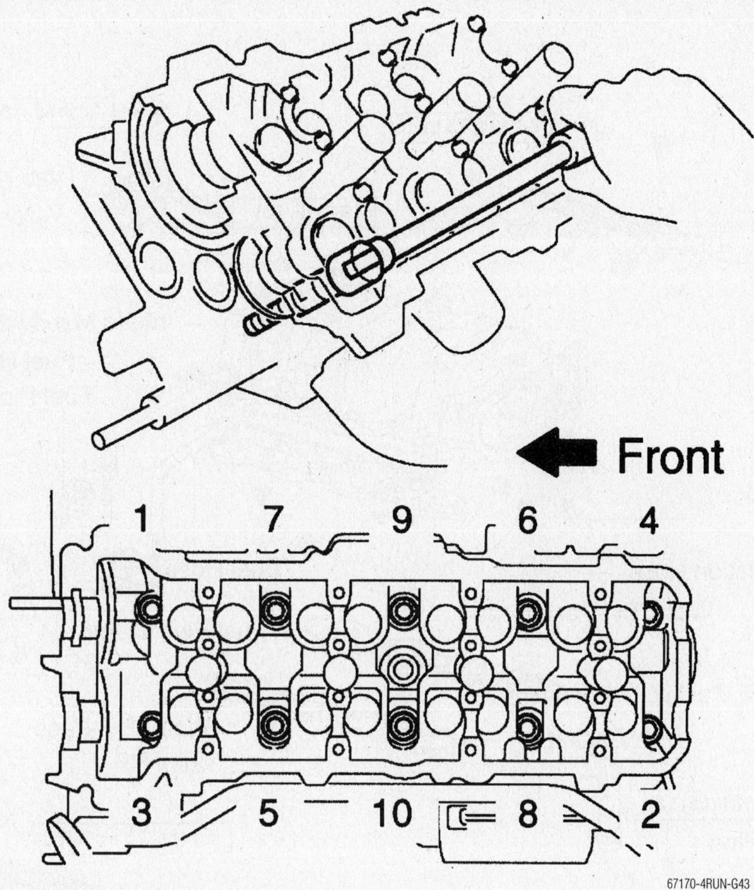

Cylinder head bolt loosening sequence—4.7L engine, right side

22. The remainder of installation is the reverse of removal. Observe the following torques:
    - Front exhaust pipe: 46 ft. lbs. (62 Nm)
    - Water inlet housing: 13 ft. lbs. (18 Nm).
    - Water bypass pipe: 13 ft. lbs. (18 Nm)
    - Intake manifold: 13 ft. lbs. (18 Nm)

LEFT SIDE

1. Remove the timing belt.
2. Remove the camshaft.
3. Remove the transmission oil filler tube.

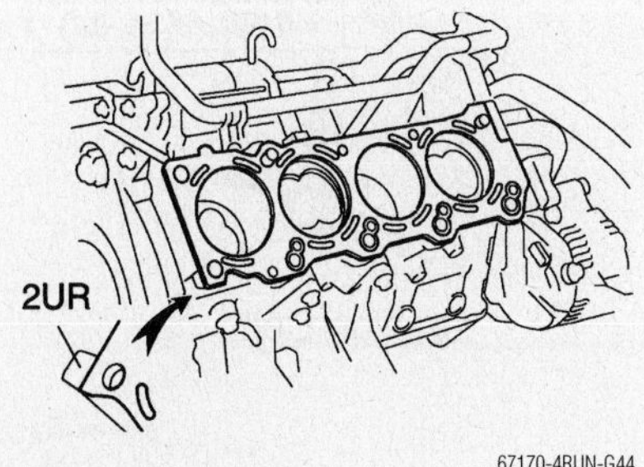

Cylinder head gasket positioning—4.7L engine, right side

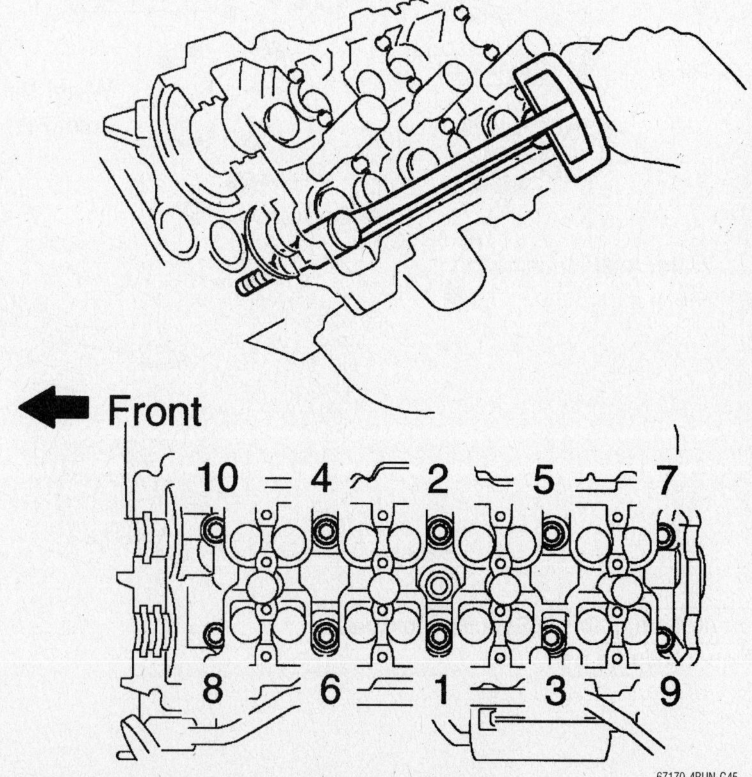

Cylinder head bolt torque sequence—4.7L engine, right side

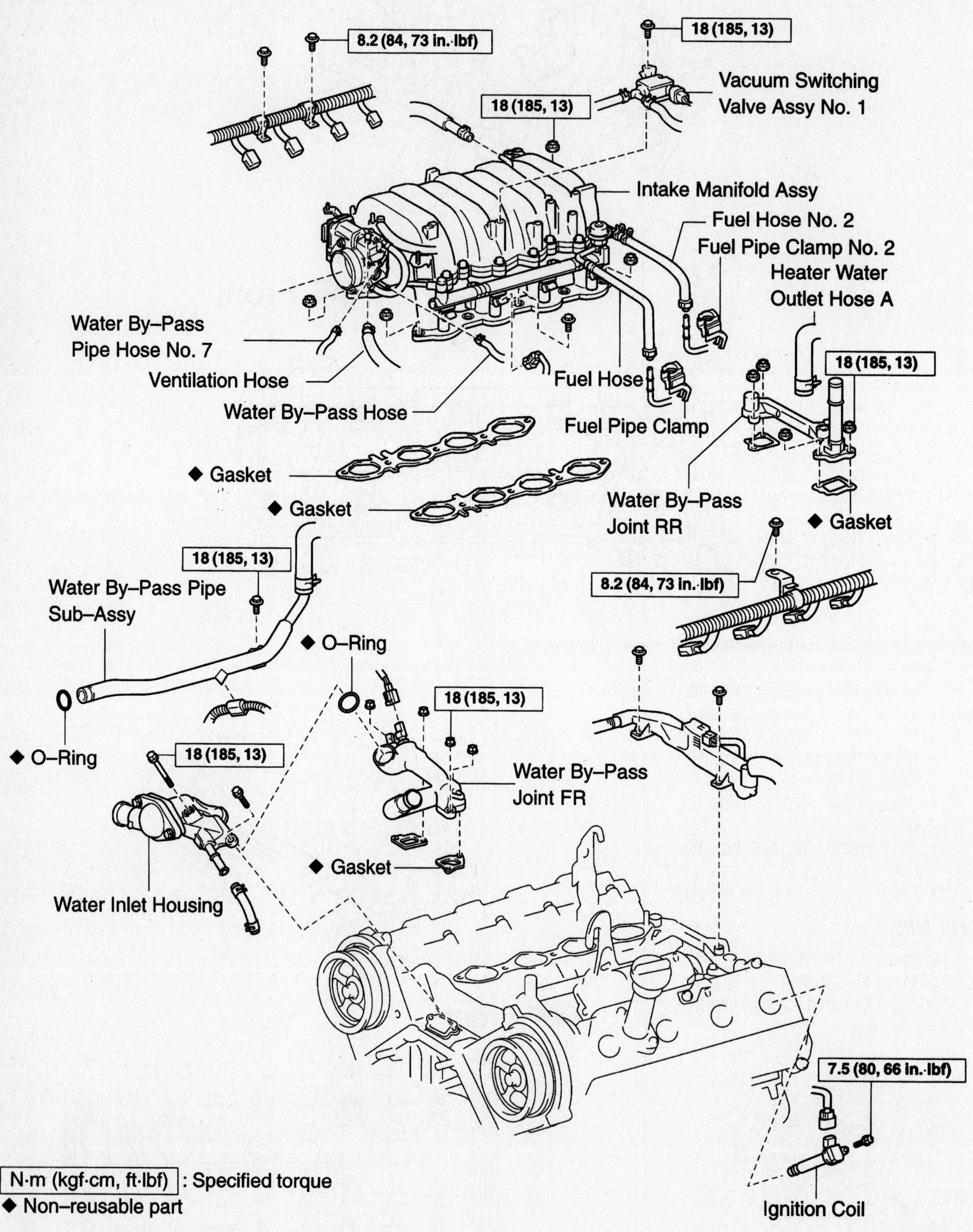

For 4.7L engine left cylinder head removal, remove these parts—Part 1

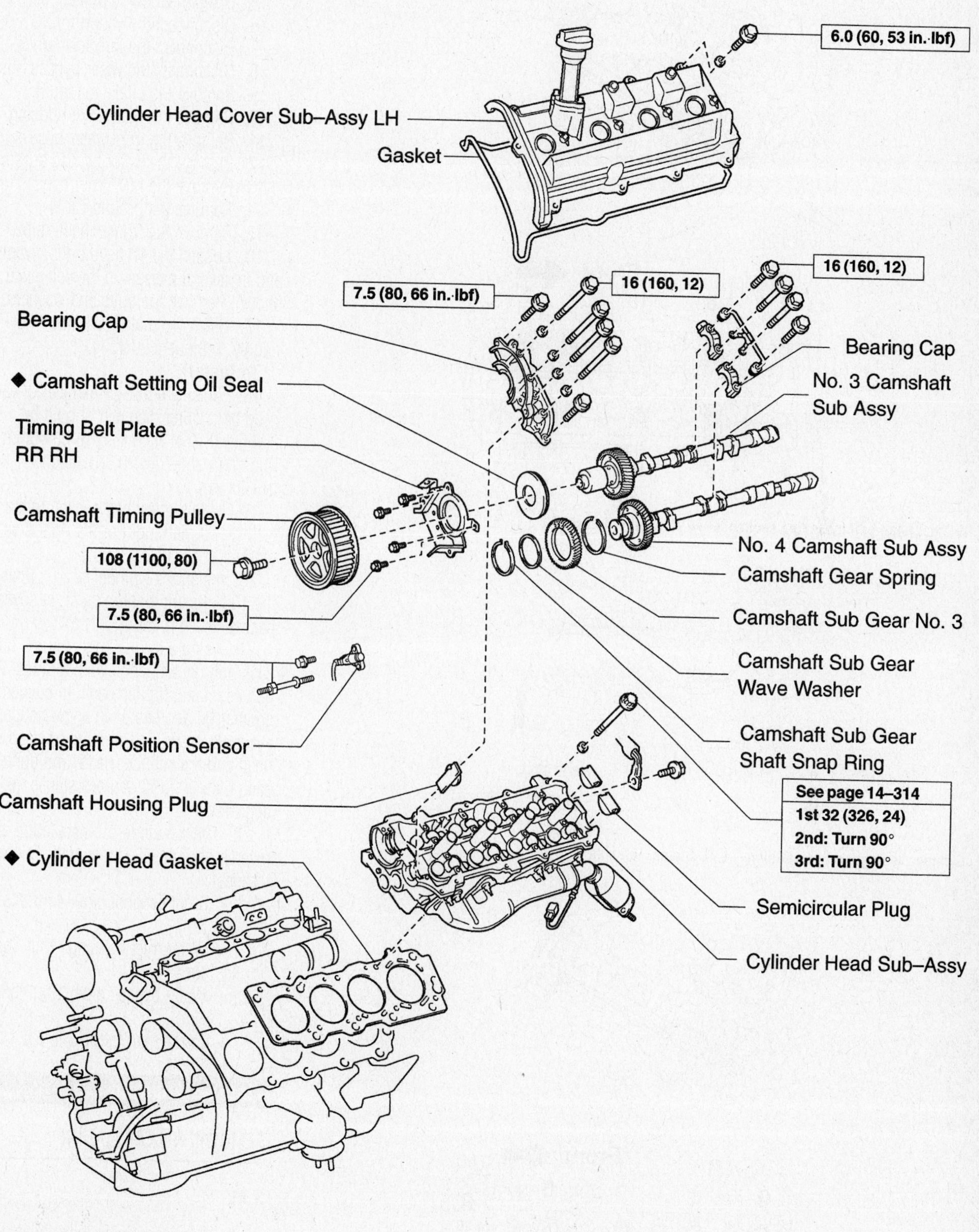

For 4.7L engine left cylinder head removal, remove these parts—Part 2

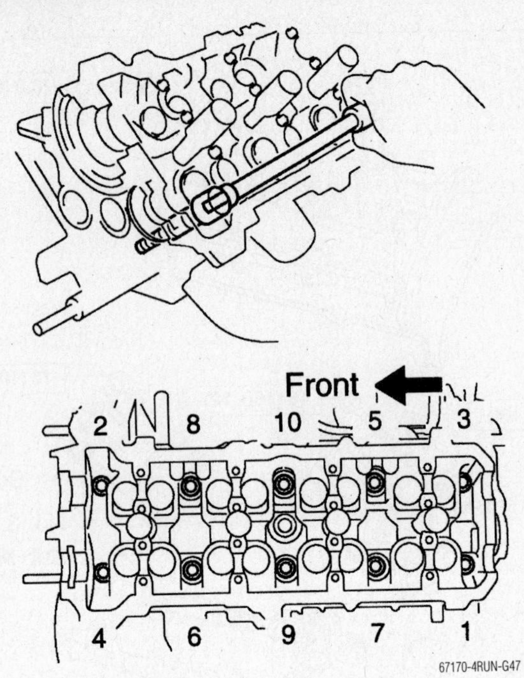

Cylinder head bolt loosening sequence—4.7L engine, left side

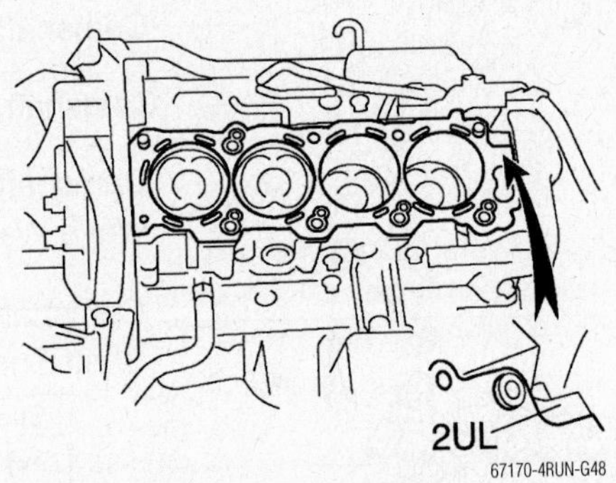

Cylinder head gasket positioning—4.7L engine, left side

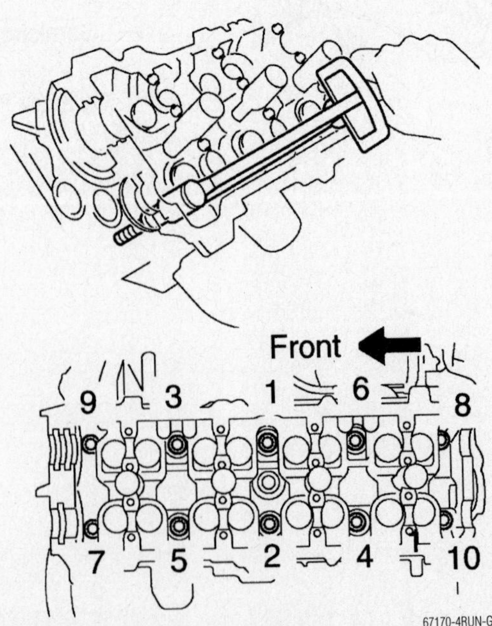

Cylinder head bolt torque sequence—4.7L engine, left side

4. Remove the engine oil dipstick tube.
5. Disconnect the fuel lines.
6. Disconnect the vapor hose.
7. Disconnect the injector harness.
8. Disconnect the water bypass hoses.
9. Remove the intake manifold.
10. Remove the water inlet housing.
11. Remove the front water bypass joint.
12. Remove the water bypass assembly.
13. Remove the rear water bypass joint.
14. Remove the ground cable.
15. Remove the front exhaust pipe.
16. Loosen the 10 head bolts, evenly and in several passes, in the sequence shown. Remove the bolts and washers.
17. Clean the mating surfaces thoroughly, without scratching.

To install

18. Check the thread diameter of each head bolt. Outer diameter should be 0.3862–0.3921 inch (9.810–0.960mm). Minimum should be 0.3819 inch (9.700mm).
19. Install a new had gasket on the block with the ID mark up. For the right side, the mark is 2UL.
20. Install the cylinder head. Lightly oil the 10 bolts and tighten them, in several passes, to 24 ft. lbs. (32 Nm).
21. Using a torque angle gauge, tighten each bolt, in sequence, an additional 90 degrees, then, tighten each, in sequence another 90 degrees. If an angle gauge is not available, make a paint mark on each bolt head and the cylinder head, and tighten each bolt, in 2, 90 degrees stages, to 180 degrees.
22. The remainder of installation is the reverse of removal. Observe the following torques:

- Front exhaust pipe: 46 ft. lbs. (62 Nm)
- Water inlet housing: 13 ft. lbs. (18 Nm).
- Water bypass pipe: 13 ft. lbs. (18 Nm)
- Intake manifold: 13 ft. lbs. (18 Nm)

Intake Manifold

REMOVAL & INSTALLATION

3.4L

1. Before servicing the vehicle, refer to the precautions in the beginning of this section.
2. Disconnect the negative battery cable.
3. Relieve the fuel system pressure.
4. Remove the engine undercover.

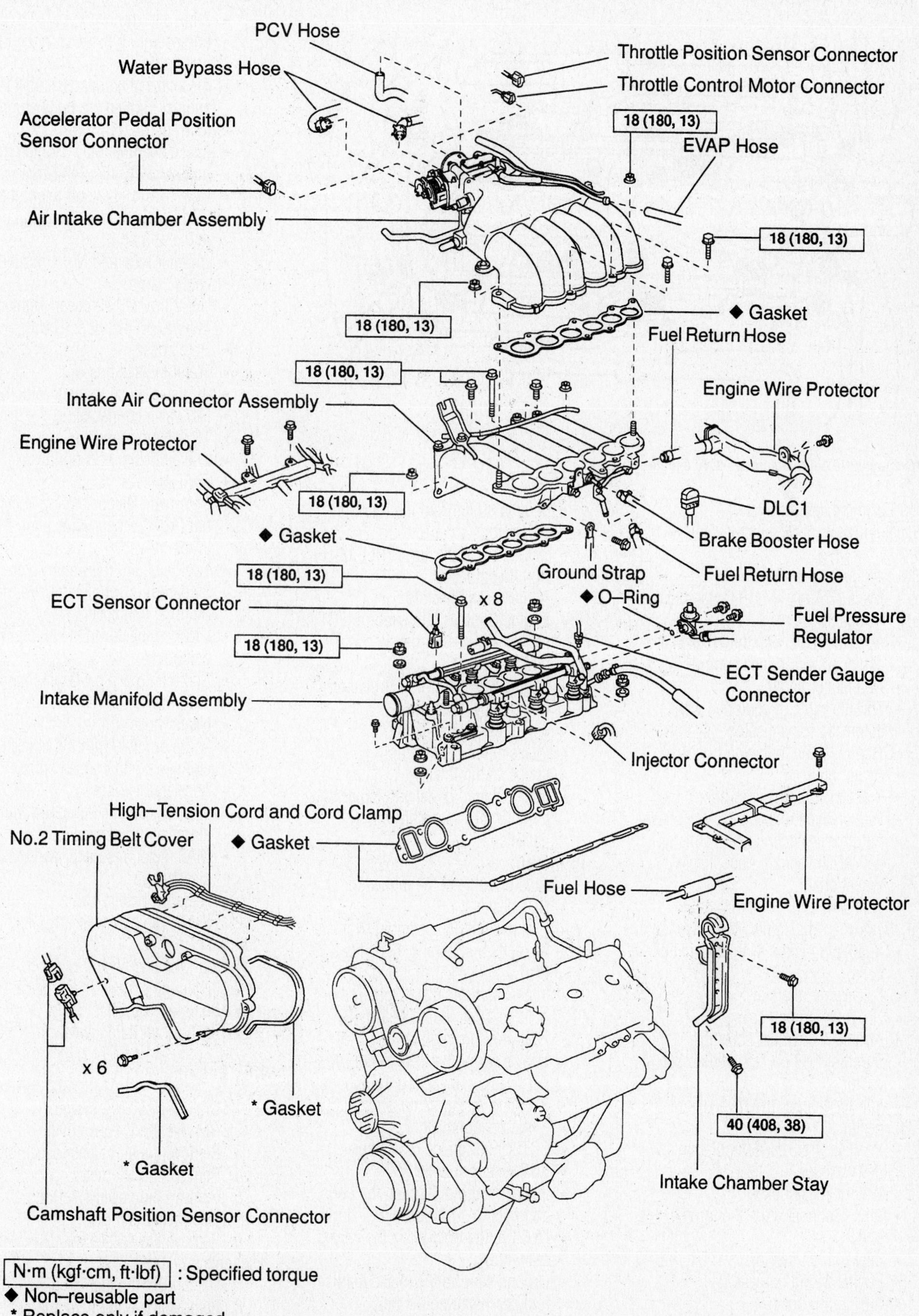
PCV Hose
Water Bypass Hose
Throttle Position Sensor Connector
Throttle Control Motor Connector
Accelerator Pedal Position Sensor Connector
18 (180, 13)
EVAP Hose
Air Intake Chamber Assembly
18 (180, 13)
18 (180, 13)
◆ Gasket
Fuel Return Hose
18 (180, 13)
Intake Air Connector Assembly
Engine Wire Protector
Engine Wire Protector
18 (180, 13)
DLC1
◆ Gasket
Brake Booster Hose
Ground Strap
Fuel Return Hose
18 (180, 13)
x 8
◆ O–Ring
ECT Sensor Connector
Fuel Pressure Regulator
18 (180, 13)
ECT Sender Gauge Connector
Intake Manifold Assembly
Injector Connector
High–Tension Cord and Cord Clamp
No.2 Timing Belt Cover
◆ Gasket
Fuel Hose
Engine Wire Protector
18 (180, 13)
x 6
* Gasket
40 (408, 38)
* Gasket
Intake Chamber Stay
Camshaft Position Sensor Connector
N·m (kgf·cm, ft·lbf) : Specified torque
◆ Non–reusable part
* Replace only if damaged
67170-4RUN-G143

Upper and lower intake manifold and related parts—3.4L engine

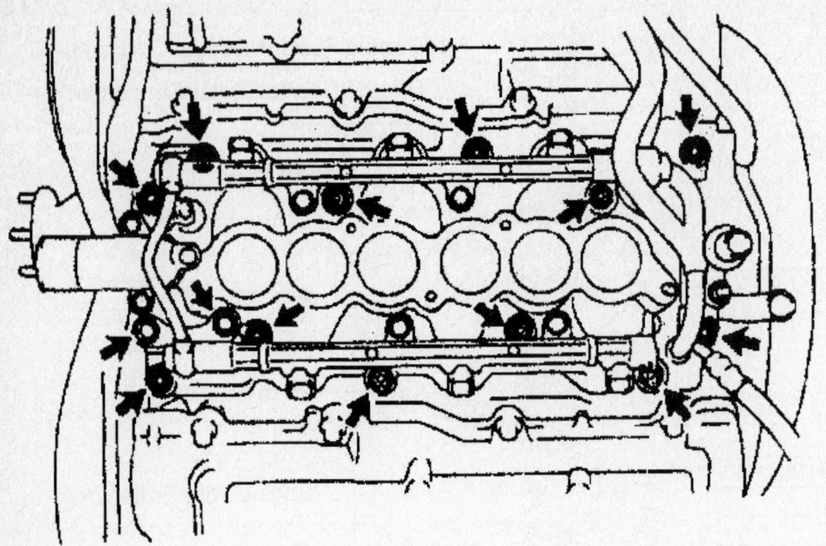

Intake manifold bolts and nuts—3.4L engine

5. Drain the cooling system.
6. Remove or disconnect the following:
    - Air cleaner cap
    - Mass Air Flow (MAF) meter and the resonator
    - Actuator cable with the bracket, if equipped with cruise control
    - Accelerator cable
    - Throttle cable, if equipped with automatic transmission
7. Disconnect the following hoses:
    - Heater hose
    - Brake booster vacuum hose
    - Evaporative Emissions (EVAP) hose
    - Automatic Disconnecting Differential (ADD) vacuum hose, for 4-Wheel drive
    - Fuel inlet and fuel return hose
8. Remove or disconnect the following:
    - Spark plug wires, with the ignition coils
    - Intake chamber stay
    - No. 2 timing belt cover
    - Air intake chamber assembly
    - Throttle Position (TP) sensor connector
    - Idle Air Control (IAC) valve connector
    - Positive Crankcase Ventilation (PCV) hoses
    - Water bypass hoses
    - Air assist hose from the throttle body
    - Intake air connector
    - Engine wiring harness
    - Fuel return hose
    - Vacuum hose, from the fuel pressure regulator
    - Ground strap, from the intake air connector
    - Data Link Connector 1 (DLC1), from the bracket
    - 6 injector connectors
    - Engine Coolant Temperature (ECT) sensor and sender gauge connectors
    - Engine wiring harness protector from the cylinder head
    - Fuel pressure regulator
    - Intake manifold assembly
    - Intake manifold stay
    - Intake manifold, delivery pipes and the injectors assembly with the gaskets

To install:

9. Install or connect the following:
    - New gaskets
    - Intake manifold assembly. Tighten the bolts and nuts to 13 ft. lbs. (18 Nm).
    - Intake manifold stay. Tighten the bolts to 14 ft. lbs. (18 Nm).
    - Fuel pressure regulator
    - Engine wiring harness to the cylinder head, by installing the 3 bolts
    - 3 engine wiring harness clamps
    - 6 injector connectors
    - ECT sender gauge connector
    - ECT sensor connector
    - Intake manifold. Tighten the bolts and nuts to 14 ft. lbs. (18.5 Nm).
    - DLC1 to the bracket on the intake manifold
    - Ground strap to the intake manifold, by installing the bolt
    - Brake booster vacuum hose, to the intake air connector
    - 2 fuel return hoses
    - Engine wiring harness to the intake manifold
    - Air intake chamber assembly to the engine. Tighten the bolts and nuts to 14 ft. lbs. (18.5 Nm).
    - Intake chamber stay. Tighten the bolts to 30 ft. lbs. (40 Nm).
    - New O-ring to the oil filler tube
    - Oil filler tube end into the tube hole in the oil pan
    - Oil filler tube and No. 1 throttle cable clamp
    - No. 2 timing belt cover. Tighten the bolts to 80 inch lbs. (9 Nm).
    - PCV hoses
    - Water bypass hoses
    - Air assist hose to the throttle body
    - IAC valve connector
    - TP sensor connector
    - Brake booster vacuum hose
    - EVAP hose
    - Automatic Disconnecting Differential (ADD) vacuum hose, for 4-Wheel drive
    - Fuel inlet and fuel return hose
    - 3 spark plug wire clamps to the No. 2 timing belt cover
    - CMP connector to the No. 2 timing belt cover
    - Spark plug wires with the ignition coils
    - Heater hose
    - Actuator cable with the bracket, if equipped with cruise control
    - Accelerator cable
    - Throttle cable, if equipped with automatic transmission
    - MAF meter, resonator and the air cleaner cap
    - Negative battery cable
10. Fill the radiator with engine coolant.
11. Start the engine and check for leaks.
12. Install the engine undercover.
13. Road test the vehicle.
14. Recheck all fluid levels.

4.0L Engine

1. Drain the cooling system.
2. Remove the V-bank cover.
3. Remove the air cleaner assembly.
4. Remove the alternator.
5. Remove the power steering pump. Don't disconnect the lines.
6. Remove the A/C compressor. Don't disconnect the lines.
7. Disconnect the fuel lines.
8. Remove the water inlet and outlet hoses.
9. Remove the water bypass hoses.

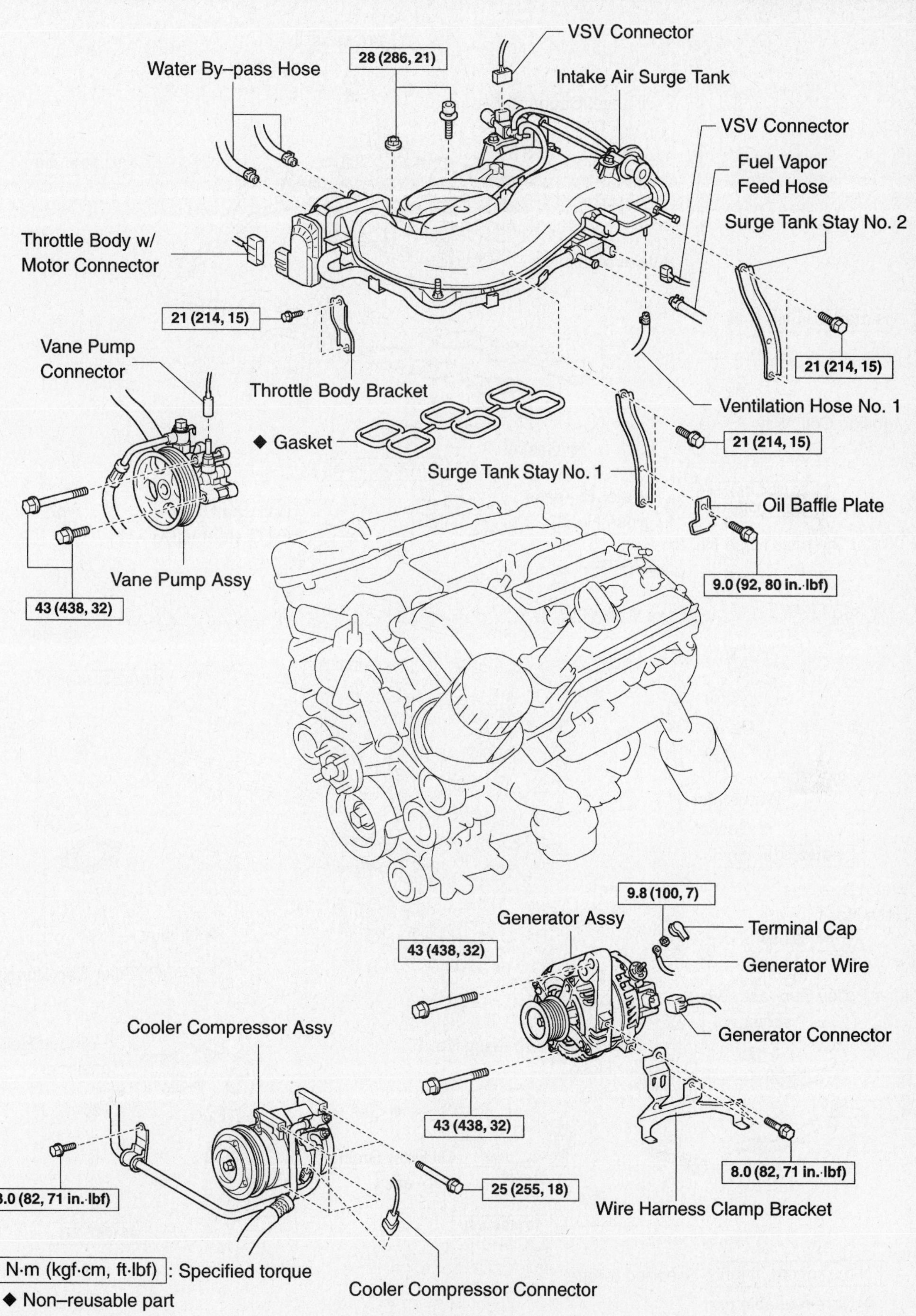

Upper intake manifold (intake air surge tank) and related parts—4.0L engine

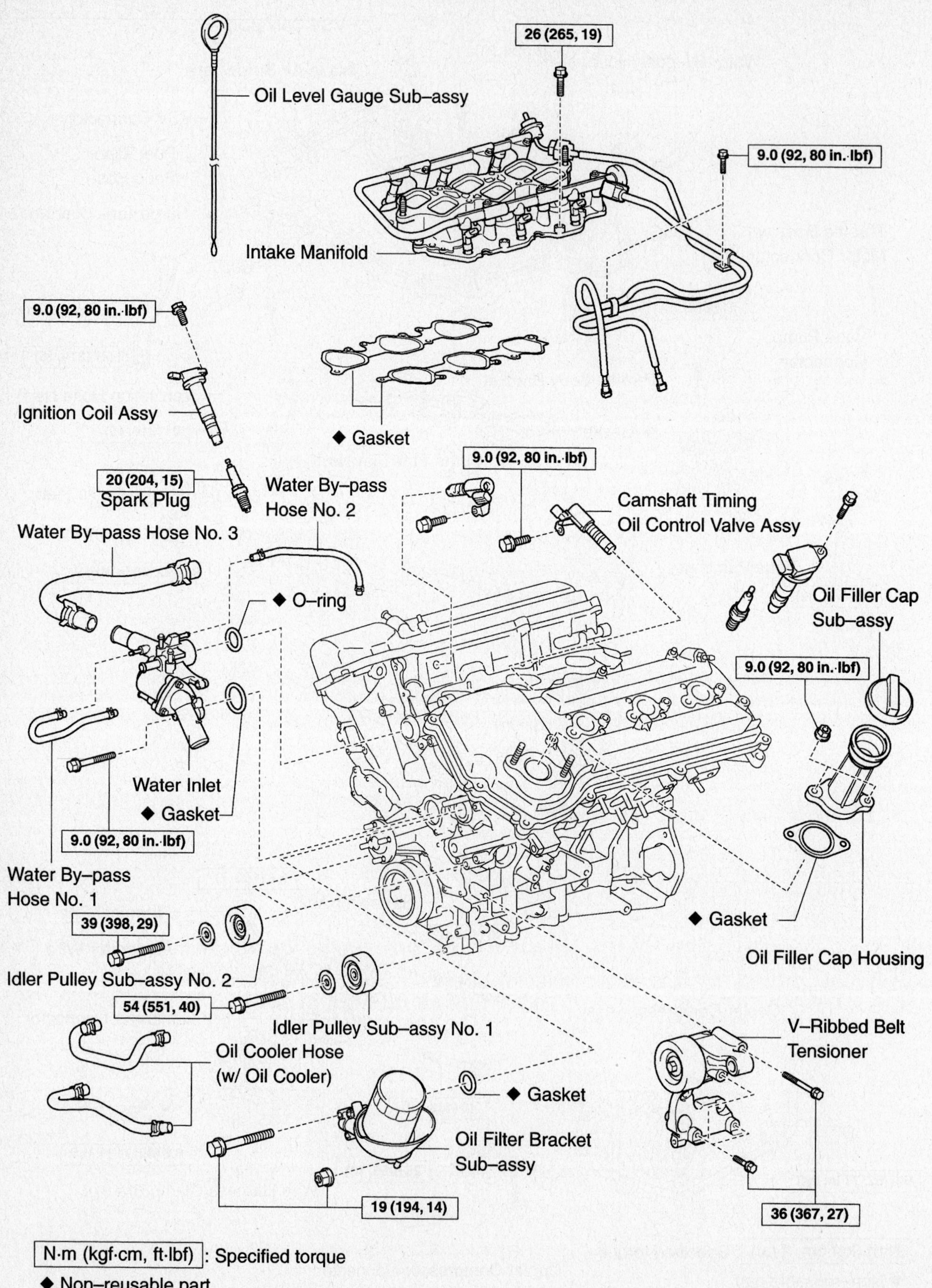

26 (265, 19)
Oil Level Gauge Sub–assy
9.0 (92, 80 in.·lbf)
Intake Manifold
9.0 (92, 80 in.·lbf)
Ignition Coil Assy
◆ Gasket
9.0 (92, 80 in.·lbf)
20 (204, 15)
Spark Plug
Water By–pass
Hose No. 2
Camshaft Timing
Oil Control Valve Assy
Water By–pass Hose No. 3
◆ O–ring
Oil Filler Cap
Sub–assy
9.0 (92, 80 in.·lbf)
Water Inlet
◆ Gasket
9.0 (92, 80 in.·lbf)
Water By–pass
Hose No. 1
39 (398, 29)
◆ Gasket
Oil Filler Cap Housing
Idler Pulley Sub–assy No. 2
54 (551, 40)
Idler Pulley Sub–assy No. 1
V–Ribbed Belt
Tensioner
Oil Cooler Hose
(w/ Oil Cooler)
◆ Gasket
Oil Filter Bracket
Sub–assy
19 (194, 14)
36 (367, 27)
N·m (kgf·cm, ft·lbf) : Specified torque
◆ Non–reusable part
67170-4RUN-G51

Lower intake manifold assembly and related parts—4.0L engine

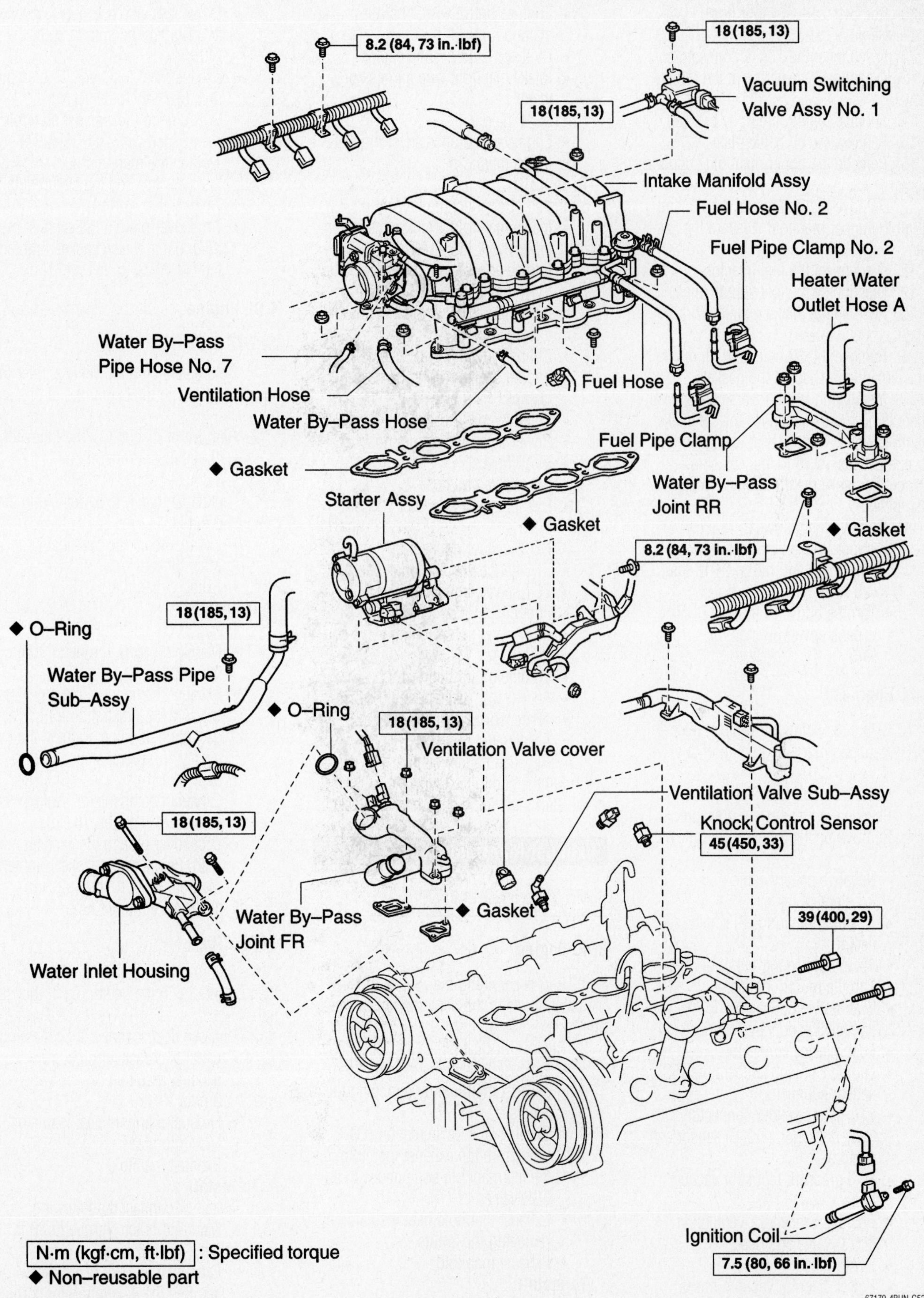
8.2 (84, 73 in.·lbf)
18 (185, 13)
Vacuum Switching Valve Assy No. 1
18 (185, 13)
Intake Manifold Assy
Fuel Hose No. 2
Fuel Pipe Clamp No. 2
Heater Water Outlet Hose A
Water By–Pass Pipe Hose No. 7
Ventilation Hose
Fuel Hose
Water By–Pass Hose
Fuel Pipe Clamp
◆ Gasket
Water By–Pass Joint RR
Starter Assy
◆ Gasket
◆ Gasket
8.2 (84, 73 in.·lbf)
◆ O–Ring
18 (185, 13)
Water By–Pass Pipe Sub–Assy
◆ O–Ring
18 (185, 13)
Ventilation Valve cover
Ventilation Valve Sub–Assy
18 (185, 13)
Knock Control Sensor
45 (450, 33)
Water By–Pass Joint FR
◆ Gasket
39 (400, 29)
Water Inlet Housing
Ignition Coil
7.5 (80, 66 in.·lbf)
N·m (kgf·cm, ft·lbf) : Specified torque
◆ Non–reusable part
67170-4RUN-G52

Intake manifold and related parts—4.7L engine

10. Remove the fuel vapor hose.
11. Remove the vent hose.
12. Disconnect the 2 VSV connectors.
13. Disconnect the wiring at the throttle body.
14. Remove the throttle body bracket.
15. Remove the oil baffle plate
16. Remove the upper manifold supports (stays).
17. Remove the 2 nuts and 4 bolts and lift off the upper manifold. Discard the gasket.
18. Disconnect the fuel injectors.
19. Disconnect the fuel supply lines.
20. Remove the intake manifold brace (stay).
21. Remove the 10 bolts and lift off the intake manifold. Discard the gaskets.
22. Installation is the reverse of removal. There is no torque sequence for the intake manifold bolts. Torque them evenly and in several passes, to 19 ft. lbs. (26 Nm). Torque the intake manifold brace to 30 ft. lbs. 40 Nm).
23. Observe the following torques:
    - Upper intake manifold (surge tank) nuts and bolts: 21 ft. lbs. (28 Nm)
    - Throttle body: 15 ft. lbs. (21 Nm)
    - V-bank cover: 66 inch lbs. (7.5 Nm)

4.7L Engine

1. Before servicing the vehicle, refer to the precautions in the beginning of this section.
2. Drain the cooling system.
3. Relieve the fuel system pressure.
4. Remove or disconnect the following:
    - Negative battery cable
    - Accelerator cable
    - Throttle Position (TP) sensor connector
    - Accelerator pedal position sensor
    - Throttle motor connector
    - Evaporative Emissions (EVAP) vacuum switching valve connector
    - Fuel injector connectors
    - Engine Coolant Temperature (ECT) sensor connector
    - ETC gauge sender connector
    - Heated Oxygen (HO_2S) sensor connectors
    - Fuel pressure regulator vacuum hose
    - Positive Crankcase Ventilation (PCV) valve and hose
    - EVAP hoses
    - Power steering vacuum hoses
    - Water bypass hose
    - Engine control wiring harness clamps
    - Cylinder head ground cables
    - Intake manifold wire harness protector
    - EVAP pipe
    - Engine appearance cover brackets
    - Intake manifold

To install:

5. Install or connect the following:
    - Intake manifold. Tighten the fasteners to 13 ft. lbs. (18 Nm).
    - Engine appearance cover brackets
    - EVAP pipe
    - Intake manifold wire harness protector
    - Cylinder head ground cables
    - Engine control wiring harness clamps
    - Water bypass hose
    - Power steering vacuum hoses
    - EVAP hoses
    - PCV valve and hose
    - Fuel pressure regulator vacuum hose
    - HO_2S sensor connectors
    - ETC gauge sender connector
    - ECT sensor connector
    - Fuel injector connectors
    - EVAP vacuum switching valve connector
    - Throttle motor connector
    - Accelerator pedal position sensor
    - TP sensor connector
    - Accelerator cable
    - Negative battery cable
6. Fill the cooling system.
7. Start the engine and check for leaks.

Exhaust Manifold

REMOVAL & INSTALLATION

3.4L Engine

1. Before servicing the vehicle, refer to the precautions in the beginning of this section.
2. Remove or disconnect the following:
    - Exhaust crossover pipe, from the exhaust manifold by removing the 3 nuts
    - Exhaust Gas Recirculation (EGR) pipe, from the exhaust manifold, on the left manifold equipped with an EGR valve
    - Exhaust manifold heat insulator, by removing the 3 nuts
    - Exhaust manifold

To install:

3. Install or connect the following:
    - Exhaust manifold, using a new gasket. Tighten the nuts to 30 ft. lbs. (40 Nm).
    - Exhaust heat insulator. Tighten the nuts to 71 inch lbs. (8 Nm).
    - EGR pipe to the exhaust manifold, if equipped with an EGR valve. Tighten the manifold nuts to 14 ft. lbs. (18 Nm) and the clamp nuts to 71 inch lbs. (8 Nm).
    - Crossover pipe to the exhaust manifold, using a new gasket. Tighten the nuts to 33 ft. lbs. (45 Nm).

4.0L Engine

RIGHT SIDE

1. Before servicing the vehicle, refer to the precautions in the beginning of this section.
2. Remove or disconnect the following:
    - Transmission filler tube
    - Exhaust crossover pipe, from the exhaust manifold by removing the 3 nuts
    - Exhaust manifold heat insulator
    - Manifold brace
    - Exhaust manifold

To install:

3. Install or connect the following:
    - Manifold brace. Torque to 30 ft. lbs. (40 Nm)
    - Exhaust manifold, using a new gasket. See the illustration for correct gasket positioning. Tighten the nuts evenly, in several steps, to 22 ft. lbs. (30 Nm).
    - Exhaust heat insulator. Tighten the nuts to 71 inch lbs. (8 Nm).
    - Crossover pipe to the exhaust manifold, using a new gasket. Tighten the nuts to 46 ft. lbs. (62 Nm).
    - Transmission filler tube

LEFT SIDE

1. Before servicing the vehicle, refer to the precautions in the beginning of this section.
2. Remove or disconnect the following:
    - Exhaust crossover pipe, from the exhaust manifold by removing the 3 nuts
    - Exhaust manifold heat insulator
    - Manifold brace
    - Exhaust manifold

To install:

3. Install or connect the following:
    - Manifold brace. Torque to 30 ft. lbs. (40 Nm)
    - Exhaust manifold, using a new gasket. See the illustration for correct gasket positioning. Tighten the nuts

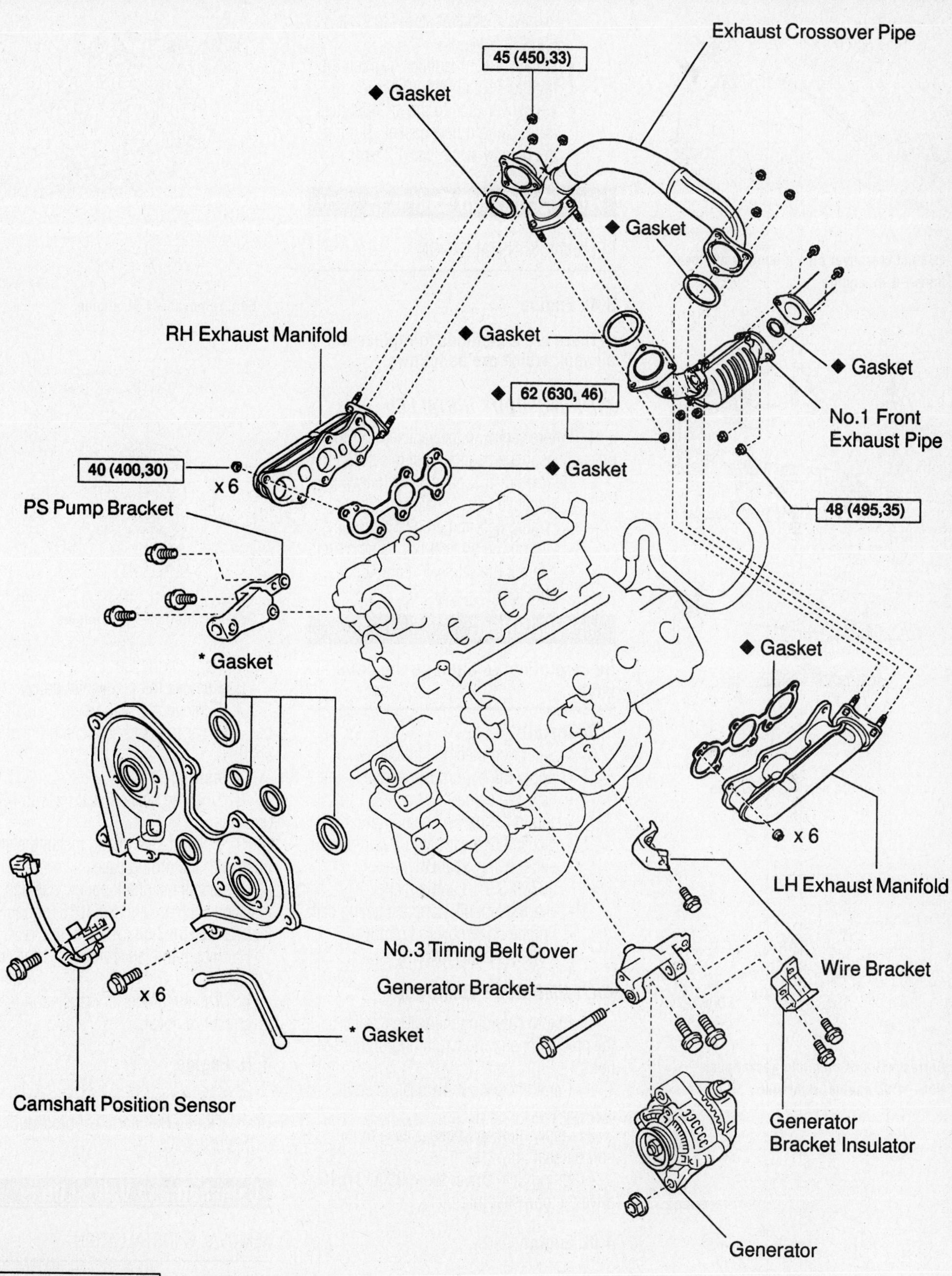

Exhaust manifolds and related parts—3.4L engine

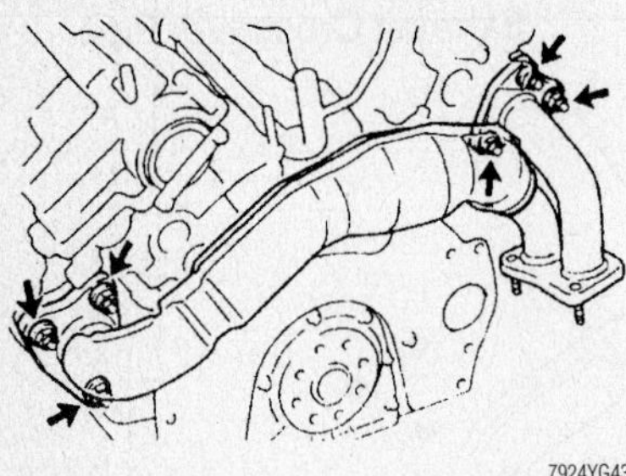

Exhaust crossover pipe mounting nut locations—3.4L engine

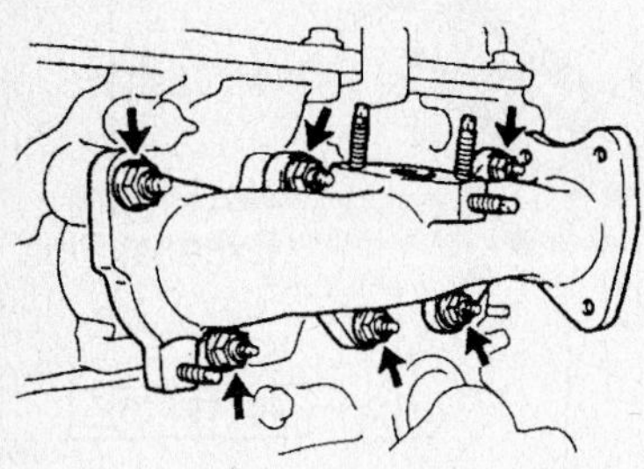

Exhaust manifold nuts—3.4L engine

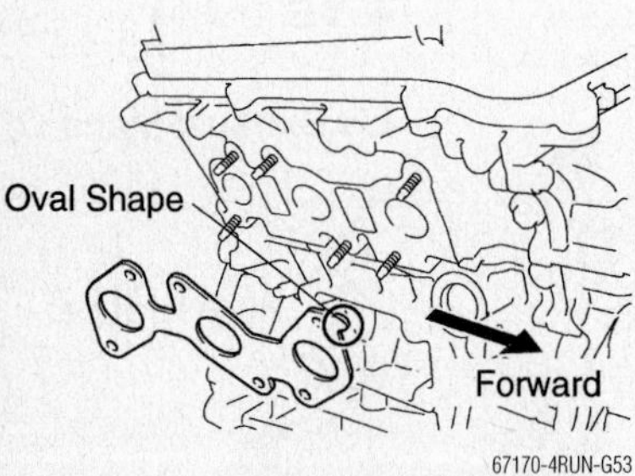

Correct exhaust manifold gasket position—4.0L engine, right side

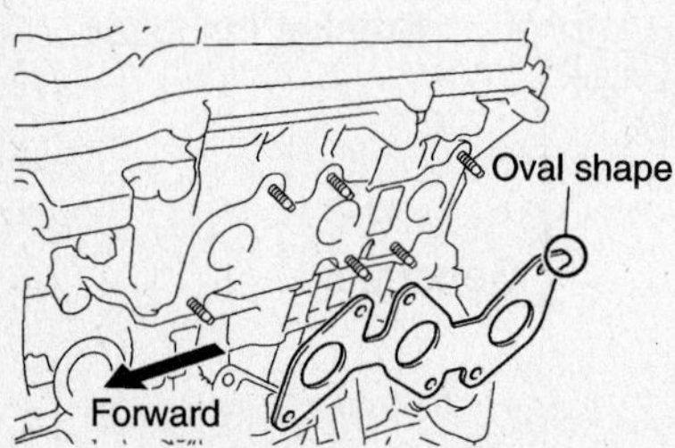

Correct exhaust manifold gasket position—4.0L engine, left side

evenly, in several steps, to 22 ft. lbs. (30 Nm).
- Exhaust heat insulator. Tighten the nuts to 71 inch lbs. (8 Nm).
- Crossover pipe to the exhaust manifold, using a new gasket. Tighten the nuts to 46 ft. lbs. (62 Nm).

Front Crankshaft Seal

REMOVAL & INSTALLATION

3.4L Engine

➡There are 2 methods to replace the oil seal, which are as follows:

OIL PUMP BODY INSTALLED

1. Before servicing the vehicle, refer to the precautions in the beginning of this section.
2. Remove or disconnect the following:
   - Negative battery cable
   - Timing belt and crankshaft pulley
   - Cut off the oil seal lip, using a knife
   - Pry out the oil seal, using a suitable tool

⁂ WARNING

Be careful not to damage the crankshaft.

To install:

3. Install or connect the following:
   - Apply multi-purpose grease to the new oil seal lip
   - Tap in the new oil seal until its surface is flush with the oil pump case edge, using Seal Driver tool 09309-37010 and a mallet
   - Crankshaft pulley and the timing belt
   - Engine undercover, if removed
   - Negative battery cable

OIL PUMP BODY REMOVED

1. Before servicing the vehicle, refer to the precautions in the beginning of this section.
2. Carefully pry out the seal using a suitable tool.
3. Apply multi-purpose grease to the new oil seal lip.
4. Using Seal Driver tool 09309-37010, drive the new seal into place.

4.0L Engine

1. Remove the V-bank cover.
2. Remove the radiator support upper seal.
3. Remove the engine under-cover.
4. Remove the accessory drive belt.

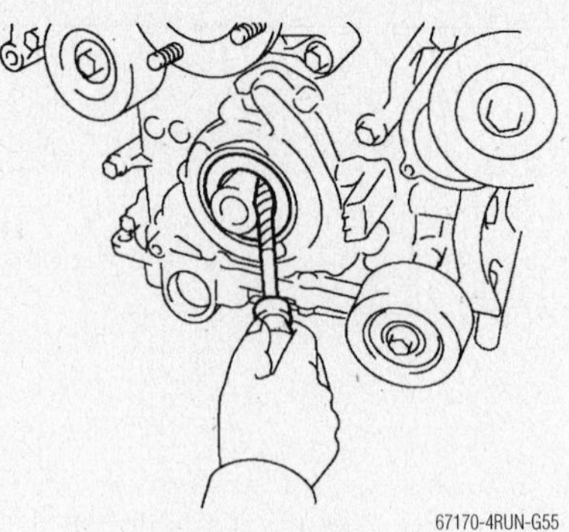

Seal removal—4.0L engine

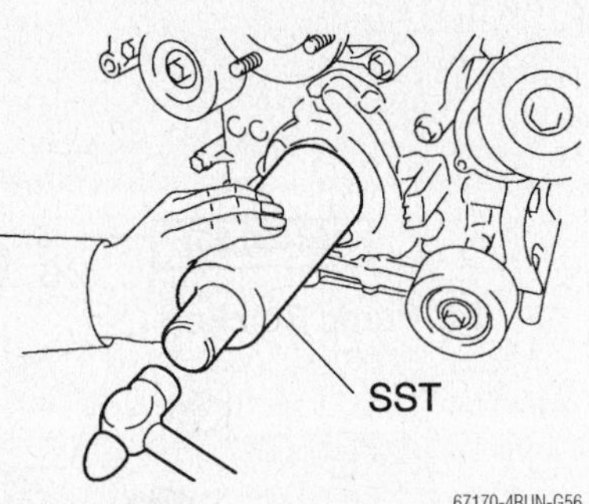

Seal installation—4.0L engine

5. Remove the fan.
6. Remove the crankshaft pulley.
7. Remove the seal, using a small pry-bar. Take car to avoid scratching the crankshaft.

To install:

8. Coat the new oil seal lip with MP grease.
9. Using a seal driver, install the seal flush with the timing case.
10. Install the crank pulley. Use and installer, such as 09213-54015, or equivalent. Torque the bolt to 184 ft. lbs. (250 Nm).
11. Install the fan. Torque to 15 ft. lbs. (21 Nm).
12. The remainder of installation is the reverse of removal.

4.7L Engine

For seal replacement, see the Front Cover, Timing Belt and Seal procedure, later in this chapter.

Camshaft and Valve Lifters

REMOVAL & INSTALLATION

3.4L Engine

1. Before servicing the vehicle, refer to the precautions in the beginning of this section.

2. Release the fuel pressure.
3. Remove or disconnect the following:
 - Negative battery cable
 - Engine undercover
4. Drain the cooling system.
5. Remove or disconnect the following:
 - Air cleaner cap
 - Mass Air Flow (MAF) meter and the resonator
 - Actuator cable with the bracket, if equipped with cruise control
 - Accelerator cable
 - Throttle cable, if equipped with an automatic transmission
 - Heater hose
 - Brake booster vacuum hose
 - Evaporative Emission (EVAP) hose
 - Automatic Disconnecting Differential (ADD) vacuum hose, for 4-wheel drive
 - Fuel inlet and fuel return hose
 - Spark plug wires, with the ignition coils
 - Intake chamber stay
 - Camshaft Position (CMP) sensor connector from the No. 2 timing belt cover
 - 3 spark plug wire clamps from the No. 2 timing belt cover
 - 6 bolts and the No. 2 timing belt cover
 - Air intake chamber assembly
 - Throttle Position (TP) sensor connector
 - Idle Air Control (IAC) valve connector
 - Positive Crankcase Ventilation (PCV) hoses
 - Water bypass hoses
 - Air assist hose from the throttle body
 - Intake air connector
 - 6 injector connectors
 - Engine Coolant Temperature (ECT) sensor and sender gauge connectors
 - Engine wiring harness protector, from the cylinder head
6. Set the No. 1 cylinder at Top Dead Center (TDC) of the compression stroke, as follows:

 a. Turn the crankshaft pulley and align its groove with the timing mark **0** on the No. 1 timing belt cover.

 b. Check that the timing marks of the camshaft timing pulleys and the No. 3 timing belt cover are aligned. If not, turn the crankshaft pulley 1 revolution (360 degrees).
7. Remove or disconnect the following:
 - Timing belt tensioner, by alternately loosening the 2 bolts
 - Timing belt
 - Camshaft timing pulley bolt, the timing pulley and the knock pin, using Variable Pin Wrench Set 09960-10010
 - Both timing pulleys
 - Bolt and the No. 2 idler pulley
 - Camshaft Position (CMP) sensor
 - Timing belt cover
 - 8 bolts, seal washers, cylinder head covers and gasket
 - Semi-circular plugs
8. Remove the right exhaust camshafts, as follows:

 a. Bring the service bolt hole of the driven sub-gear upward by turning the hexagon head portion of the exhaust camshaft with a wrench.

 b. Align the timing mark (2 dot marks) of the camshaft drive and driven gears by turning the camshaft with a wrench.

 c. Secure the exhaust camshaft sub-gear to the driven gear with a service bolt (6mm diameter, 16–20mm bolt length and 1.0mm in thread pitch).

➡When removing the camshaft, be sure that the torsional spring force of the sub-gear has been eliminated by the above operation.

 d. Uniformly loosen and remove the bearing cap bolts in several passes, in the sequence shown.

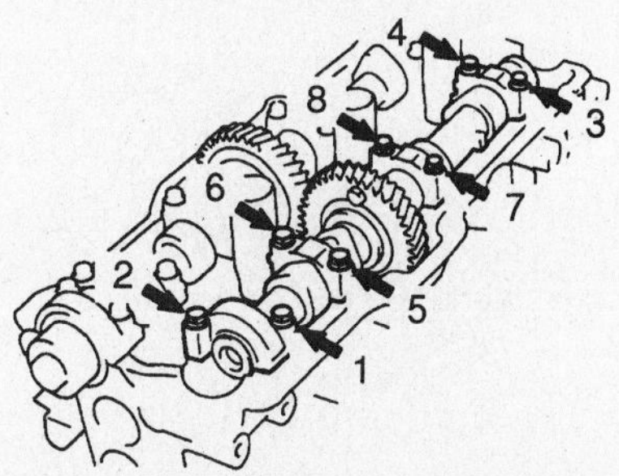

Intake camshaft bearing cap loosening sequence—3.4L engine, left side

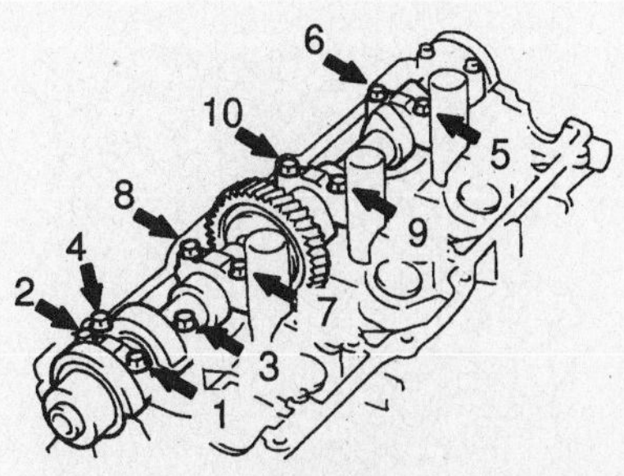

Exhaust camshaft bearing cap loosening sequence—3.4L engine, right side

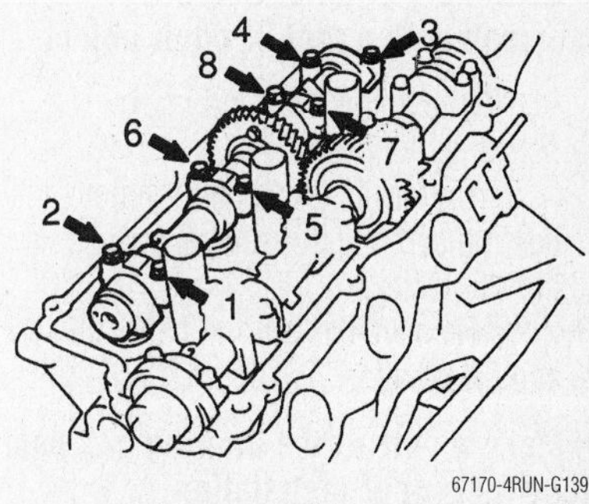

Intake camshaft bearing cap loosening sequence—3.4L engine, right side

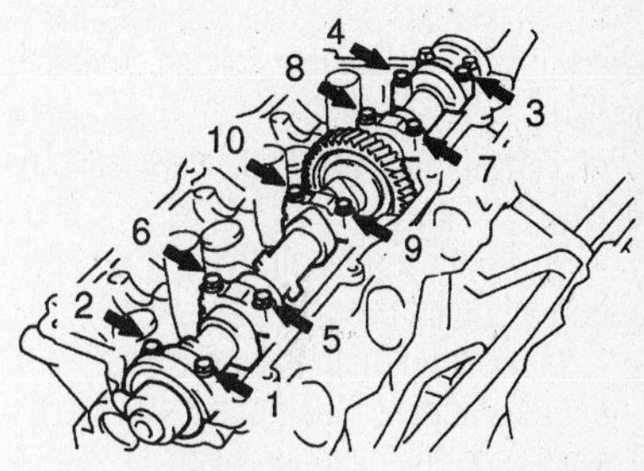

Exhaust camshaft bearing cap loosening sequence—3.4L engine, right side

 e. Remove the bearing caps and camshaft. Make a note of the bearing cap positions for proper installation.
9. Remove the right-hand intake camshaft, as follows:

 a. Uniformly loosen and remove the bearing cap bolts in several passes, in the sequence.

 b. Remove the bearing caps, oil seal and camshaft. Make a note of the bearing cap positions for proper installation.
10. Remove the left exhaust camshafts, as follows:

 a. Align the timing mark (1 dot mark) of the camshaft drive and driven gears by turning the camshaft with a wrench.

 b. Secure the exhaust camshaft sub-gear to the driven gear with a service bolt (6mm diameter, 16–20mm bolt length and 1.0mm in thread pitch).

➡When removing the camshaft, be sure the torsional spring force of the sub-gear has been eliminated by the above operation.

 c. Uniformly loosen and remove the bearing cap bolts in several passes, in the sequence.

 d. Remove the bearing caps and camshaft. Make a note of the bearing cap positions for proper installation.

➡Do not pry on or attempt to force the camshaft with a tool or other object.

11. Remove the left-hand intake camshaft, as follows:
   a. Uniformly loosen and remove the bearing cap bolts in several passes, in the sequence.
   b. Remove the bearing caps, oil seal and camshaft.

➡Make a note of the bearing cap positions for proper installation.

12. Remove the valve lifters and shims from the cylinder head. Arrange the valve lifters and shims in correct order.

To install:

13. Install the valve lifters and shims. Check that the valve lifter rotates smoothly by hand.
14. Install the right intake camshaft, as follows:
   a. Apply engine oil to the thrust portion of the intake camshaft.
   b. Position the intake camshaft at 90 degrees angle of the timing mark (2 dot marks) on the cylinder head.
   c. Install the bearing caps in their proper locations. Apply a light coat of engine oil to the threads and install the cap bolts.
   d. Apply a light coat of engine oil on the threads and under the heads of the bearing cap bolts.
   e. Uniformly tighten the cap bolts in the sequence shown to 12 ft. lbs. (16 Nm).
15. Install the right exhaust camshaft, as follows:
   a. Apply engine oil to the thrust portion of the intake camshaft.
   b. Align the timing marks (2 dot marks) of the camshaft drive and driven gears.
   c. Roll down the exhaust camshaft onto the bearing journals while engaging the gears with each other. Install the bearing caps in their proper locations.
   d. Apply a light coat of engine oil to the threads and install the cap bolts.
   e. Apply a light coat of engine oil on the threads and under the heads of the bearing cap bolts.
   f. Uniformly tighten the cap bolts in the sequence to 12 ft. lbs. (16 Nm).
   g. Remove the service bolt from the driven sub-gear. Check that the intake and exhaust camshafts turn smoothly.
   h. Align the timing marks (2 dot marks) of the camshaft drive and driven gears by turning the camshaft with a wrench.
16. Install the left intake camshaft, as follows:
   a. Apply engine oil to the thrust portion of the intake camshaft.
   b. Position the intake camshaft at 90 degrees angle of the timing mark (1 dot mark) on the cylinder head.
   c. Install the bearing caps in their proper locations. Apply a light coat of engine oil to the threads and install the cap bolts.
   d. Apply a light coat of engine oil on the threads and under the heads of the bearing cap bolts.
   e. Uniformly tighten the cap bolts in the sequence to 12 ft. lbs. (16 Nm).
17. Install the left exhaust camshaft, as follows:
   a. Apply engine oil to the thrust portion of the intake camshaft.
   b. Align the timing marks (1 dot mark) of the camshaft drive and driven gears.
   c. Roll down the exhaust camshaft onto the bearing journals while engaging the gears with each other. Install the bearing caps in their proper locations.
   d. Apply a light coat of engine oil to the threads and install the cap bolts.
   e. Apply a light coat of engine oil on the threads and under the heads of the bearing cap bolts.
   f. Uniformly tighten the cap bolts in the sequence to 12 ft. lbs. (16 Nm).
   g. Remove the service bolt.
18. Check and adjust the valve clearance.
19. Install or connect the following:
   - Semi-circular plugs
   - Cylinder head covers. Tighten the bolts, in several passes, to 53 inch lbs. (6 Nm).
   - No. 3 timing belt cover. Tighten the 6 bolts to 80 inch lbs. (9 Nm).
   - CMP sensor. Tighten it to 71 inch lbs. (8 Nm).
   - No. 2 timing belt idler. Tighten the bolt to 30 ft. lbs. (40 Nm).

➡Check that the pulley bracket moves smoothly.

20. Install the left camshaft timing pulley, as follows:
   a. Install the knock pin to the camshaft.
   b. Align the knock pin hose of the camshaft with the knock pin groove of the timing pulley.
   c. Slide the timing pulley on the camshaft with the flange side facing outward. Tighten the pulley bolt to 81 ft. lbs. (110 Nm).
21. Set the No. 1 cylinder to Top Dead Center (TDC) of the compression stroke, as follows:
   a. Turn the crankshaft pulley, and align its groove with the timing mark **0** on the No. 1 timing belt cover.

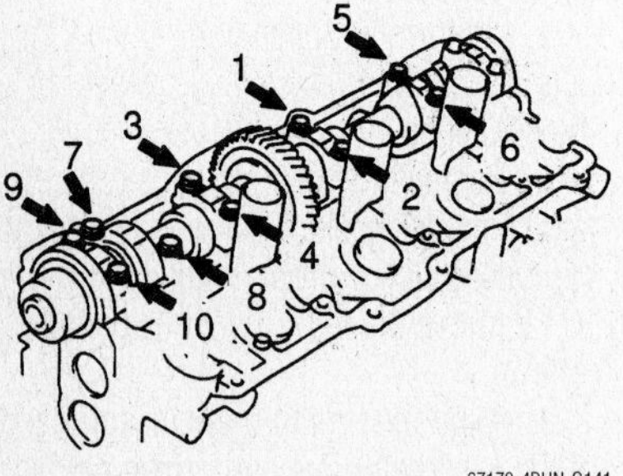

Intake camshaft bearing cap torque sequence—3.4L engine, left side

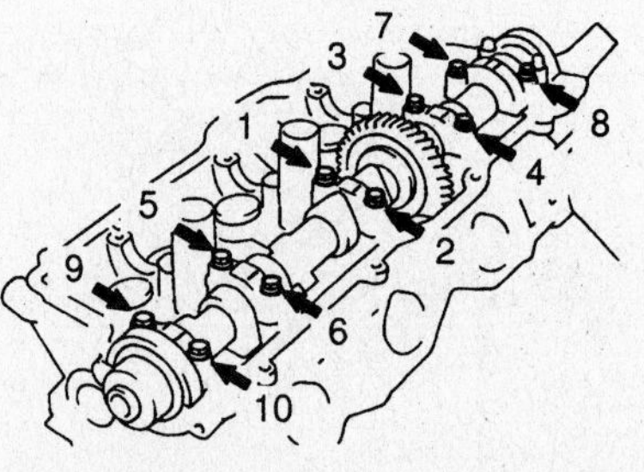

Intake camshaft bearing cap torque sequence—3.4L engine, right side

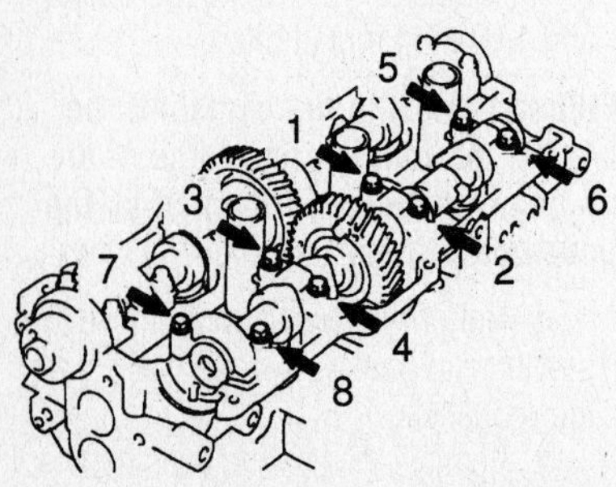

Exhaust camshaft bearing cap torque sequence—3.4L engine, left side

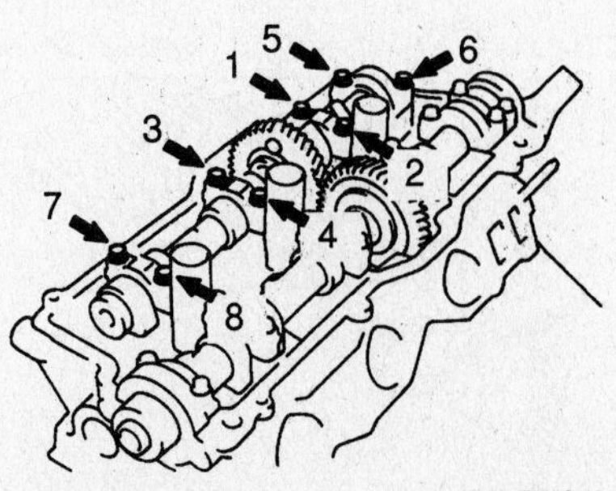

Exhaust camshaft bearing cap torque sequence—3.4L engine, right side

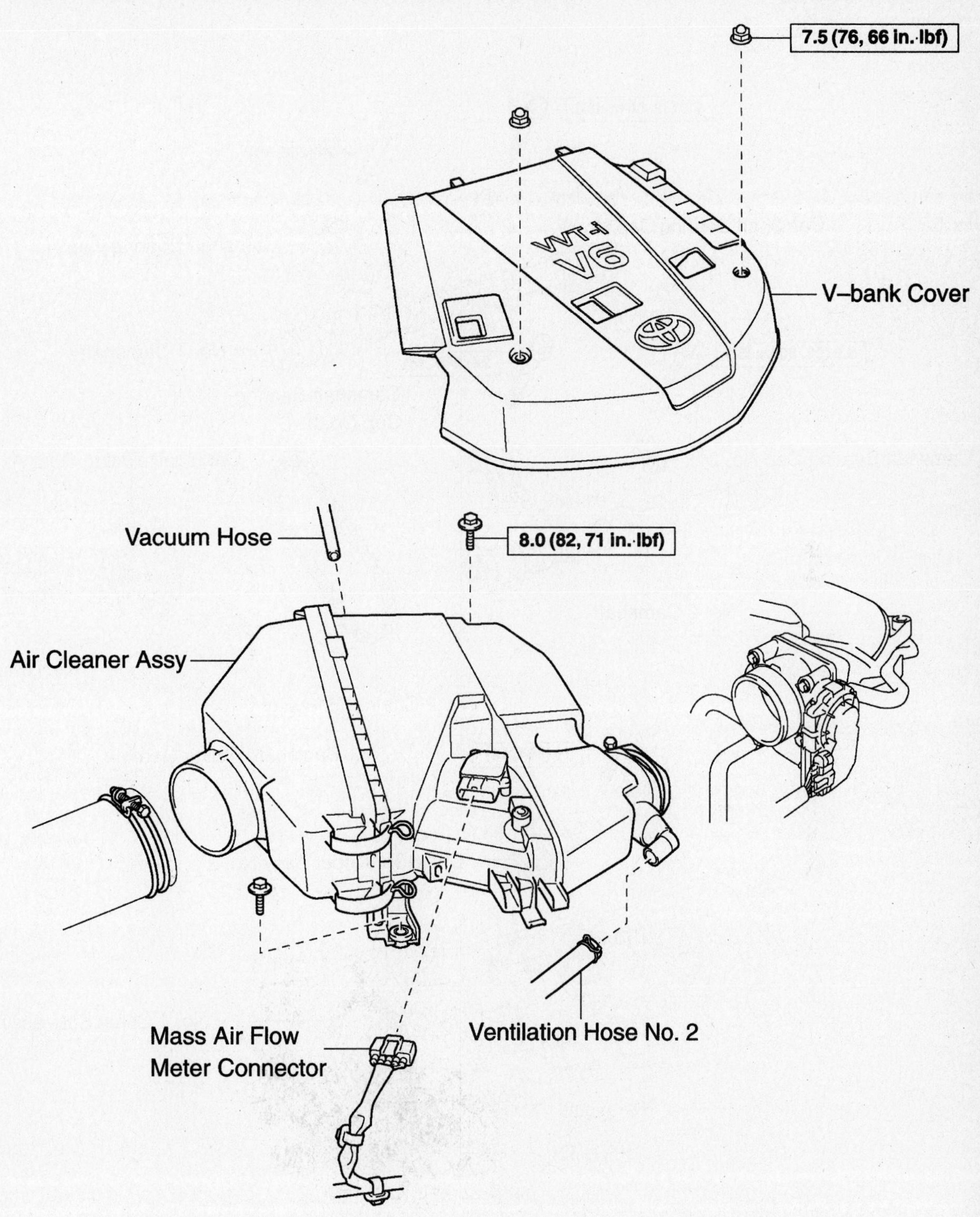

V-bank cover and air cleaner assembly—4.0L engine

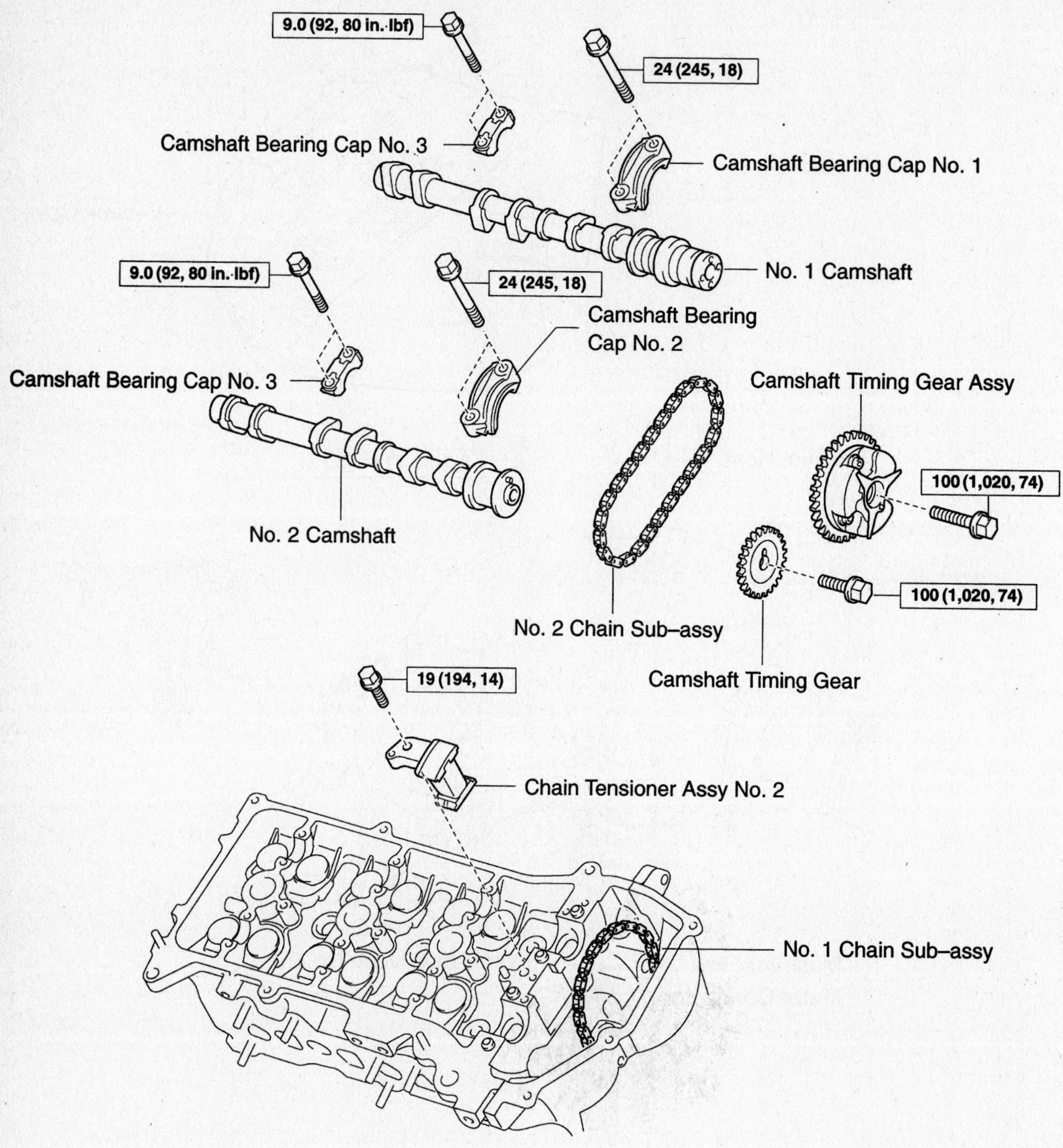

Right side camshaft and related parts—4.0L engine

b. Turn the camshaft, align the knock pin hole of the camshaft with the timing mark of the No. 3 timing belt cover.

c. Turn the camshaft timing pulley, align the timing marks of the camshaft timing pulley and the No. 3 timing belt cover.

22. Install or connect the following:
- Timing belt to the left camshaft timing pulley. Check that the installation mark on the timing belt is aligned with the end of the No. 1 timing belt cover.
- Right camshaft timing pulley
- Timing belt

23. Set the timing belt tensioner, as follows:

a. Using a press, slowly press in the pushrod using 220–2,205 lbs. (981–9,807 N) of force.

b. Align the holes of the pushrod and housing, pass a 1.5mm hexagon wrench through the holes to keep the setting position of the pushrod.

c. Release the press and install the dust boot to the tensioner.

24. Install the timing belt tensioner and alternately tighten the bolts to 20 ft. lbs. (28 Nm). Using pliers, remove the 1.5mm hexagon wrench from the belt tensioner.

25. Check the valve timing, as follows:

a. Slowly turn the crankshaft pulley 2 revolutions from the TDC-to-TDC. Always turn the crankshaft pulley clockwise.

b. Check that each pulley aligns with the timing marks. If the timing marks do not align, remove the timing belt and reinstall it.

26. Install or connect the following:
- Engine wiring harness to the cylinder head
- 3 engine wiring harness clamps.
- 6 injector connectors
- ECT sender gauge connector
- ECT sensor connector
- Intake air connector
- Air intake chamber assembly. Tighten the 4 bolts and 2 nuts to 13 ft. lbs. (18 Nm).
- Intake chamber stay. Tighten the 2 bolts to 30 ft. lbs. (40 Nm).
- New O-ring to the oil filler tube
- Oil filler tube end into the oil pan tube hole
- Oil filler tube and No. 1 throttle cable clamp
- No. 2 timing belt cover. Tighten the bolts to 80 inch lbs. (9 Nm).
- Remaining components
- Negative battery cable

27. Fill the cooling system.

28. Start the engine and check for leaks.

29. Check the ignition timing.

30. Install the engine undercover.

31. Road test the vehicle.

32. Recheck all fluid levels.

4.0L Engine

RIGHT SIDE

1. Drain the coolant.
2. Remove the V-bank cover.

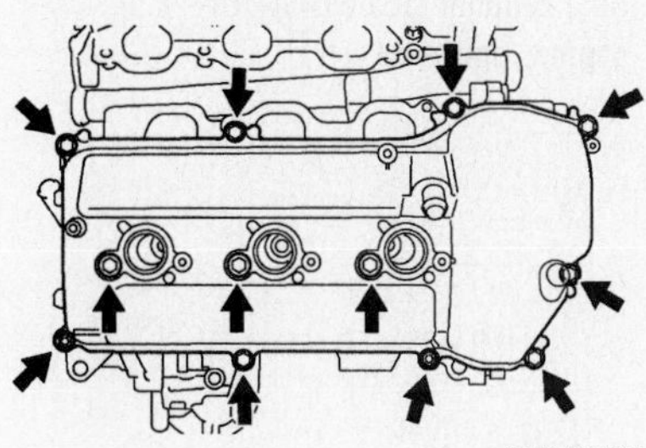

Right cylinder head cover—4.0L engine

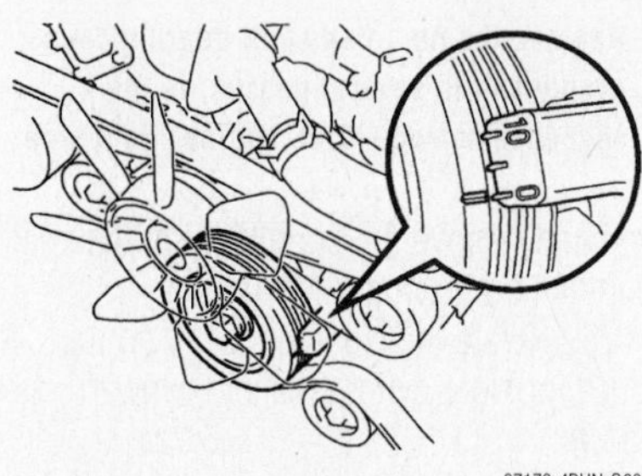

Turn the crankshaft pulley to align the notch with the timing mark "0" on the chain cover—4.0L engine, right side

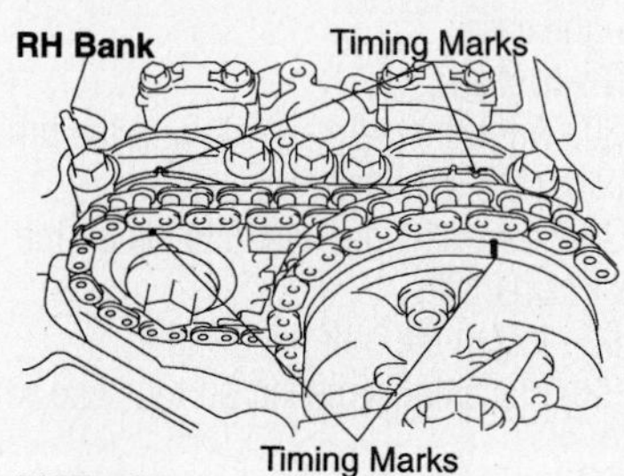

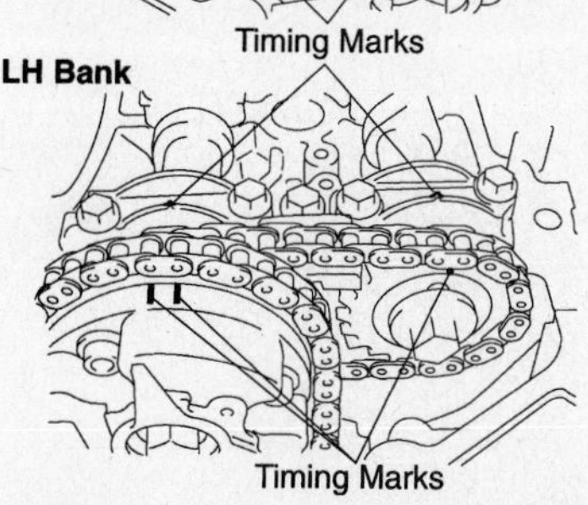

Check that the camshaft and chain timing marks are aligned—4.0L engine, right side

3. Remove the air cleaner assembly.
4. Remove the water bypass hoses.
5. Remove the fuel vapor hose.
6. Remove the vent hose.
7. Disconnect the 2 VSV connectors.
8. Disconnect the wiring at the throttle body.
9. Remove the throttle body bracket.
10. Remove the oil baffle plate
11. Remove the upper manifold supports (stays).
12. Remove the 2 nuts and 4 bolts and lift off the upper manifold. Discard the gasket.
13. Remove the cylinder head cover.
14. Turn the crankshaft pulley to align the notch with the timing mark "0" on the chain cover.
15. Check that the camshaft and chain timing marks are aligned as shown. If they're not, rotate the crankshaft one full turn (360 degrees) clockwise.
16. Using paint, matchmark the chain links that correspond to the timing marks on the camshaft sprockets.
17. Remove the 4 bolts and the timing chain cover plate and gasket.
18. While turning the stopper plate of the tensioner upward, push in on the plunger. Then, turn the stopper plate downward and insert a 3.5mm diameter metal pin into the holes in the stopper plate and tensioner.

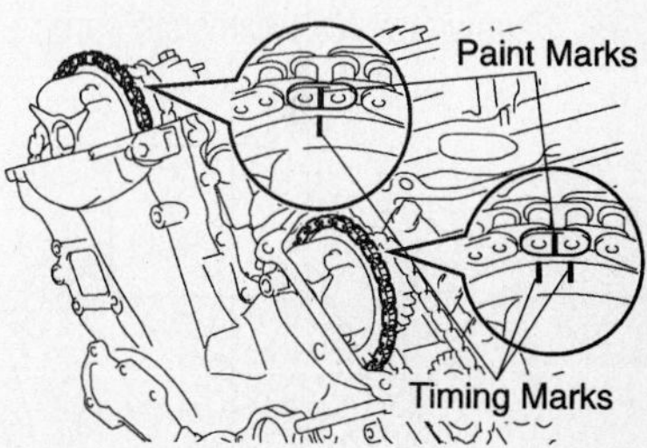

Using paint, matchmark the chain links that correspond to the timing marks on the camshaft sprockets—4.0L engine, right side

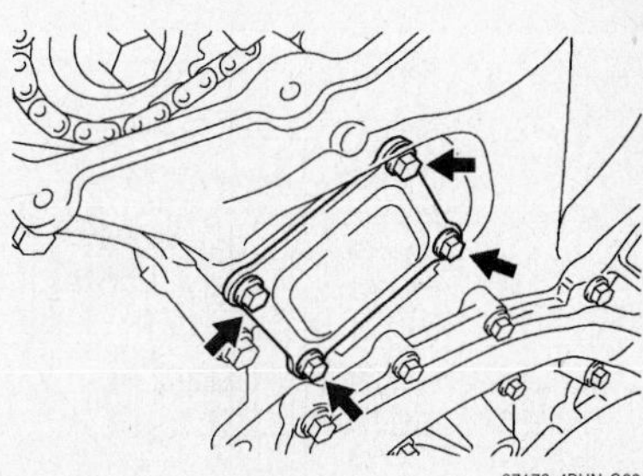

Remove the 4 bolts and the timing chain cover plate and gasket—4.0L engine, right side

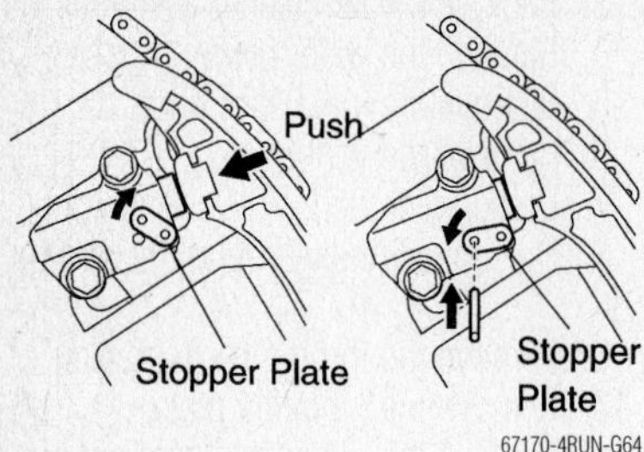

Locking the tensioner—4.0L engine—4.0L engine, right side

19. Remove the 2 bolts and the tensioner.

⁂ WARNING

The camshaft thrust clearances are very small. Keep the camshaft level while removing it to avoid damage to the cylinder head.

20. While lifting the No.2 tensioner, insert a 1mm dia. pin to hold it.
21. Remove the No.2 (outer) camshaft timing gear bolt. Remove the gear.
22. Rotate the camshaft counterclockwise so that the lobes for No.1 cylinder face upward.
23. Remove the bearing caps evenly and in several passes, in the sequence shown.
24. Remove the caps and the No.2 camshaft.
25. Remove the No.1 camshaft sprocket bolt and remove the sprocket. Lift the chain from the sprocket and secure it.

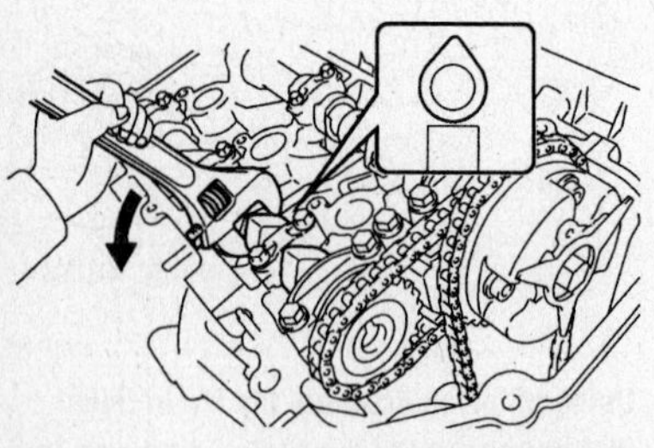

Rotate the No.2 camshaft counterclockwise so that the lobes for No.1 cylinder face upward—4.0L engine, right side

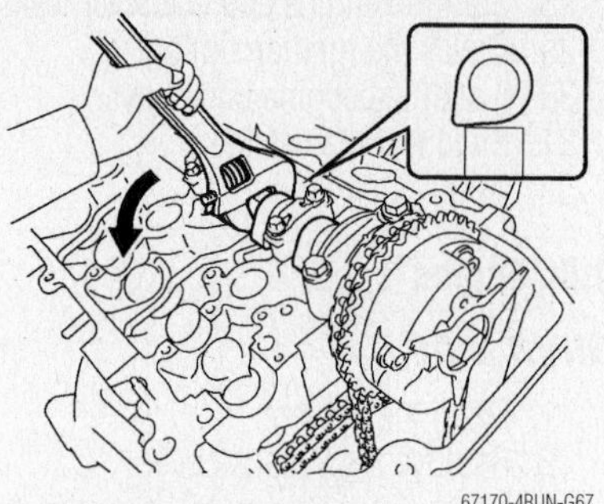

Rotate the No.1 camshaft so the lobes for No.1 cylinder face downward—4.0L engine, right side

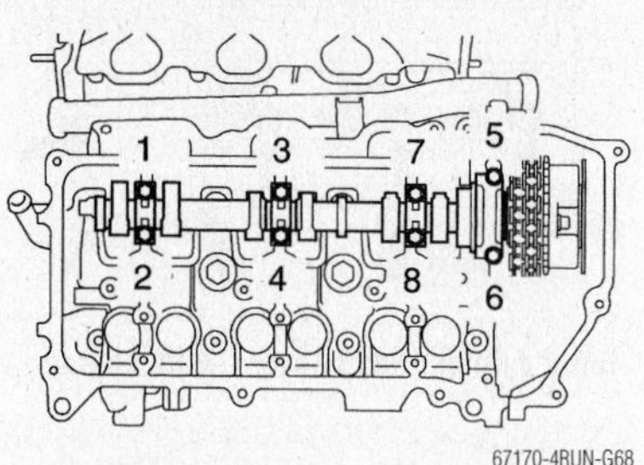

Remove the no.1 camshaft bearing caps evenly and in several passes, in the sequence shown—4.0L engine, right side

26. Rotate the No.1 camshaft so the lobes for No.1 cylinder face downward, as shown.
27. Remove the bearing caps evenly and in several passes, in the sequence shown.
28. Remove the caps and the No.1 camshaft. Keep the caps in order.
29. Tie the chain so that it doesn't fall into the case.

To install:

30. Align the painted link with the timing mark on the sprocket.
31. Install the gear and chains on the No.1 camshaft.
32. Install the bolt hand-tight.
33. Place the camshaft on the head with the No.1 cylinder lobes facing downward as before.

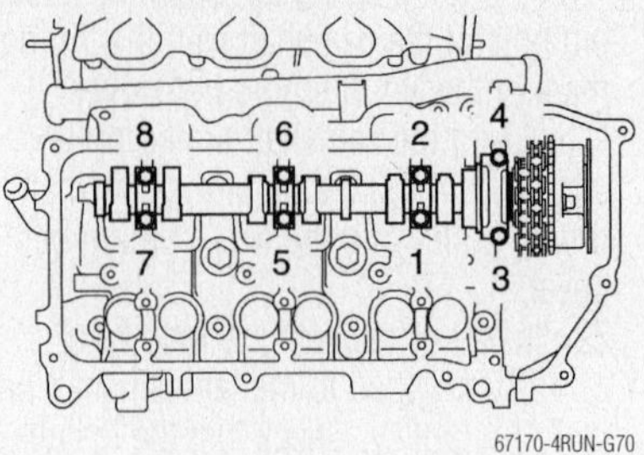

No.1 camshaft bolt torque sequence—4.0L engine, right side

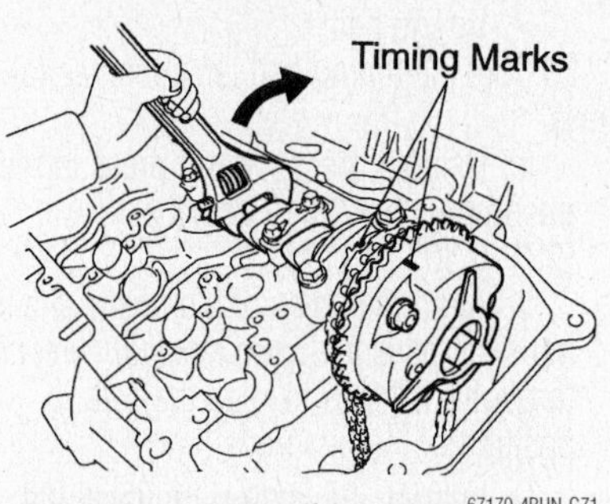

Rotate the No.1 camshaft clockwise so that the timing mark on the camshaft sprocket is aligned with the timing mark on the bearing cap—4.0L engine, right side

34. Install the caps. Lightly oil the bolts and install them.
35. Torque the bolts evenly, and in several passes, in the sequence shown, to 10mm bolts, 80 inch lbs. (9 Nm); 12m bolts 18 ft. lbs. (24 Nm).
36. Rotate the No.1 camshaft clockwise so that the timing mark on the camshaft sprocket is aligned with the timing mark on the bearing cap.
37. Align the paint mark on the chain with the timing mark on the sprocket.
38. Torque the sprocket bolt to 74 ft. lbs. (100 Nm).
39. Install the No.2 chain tensioner.

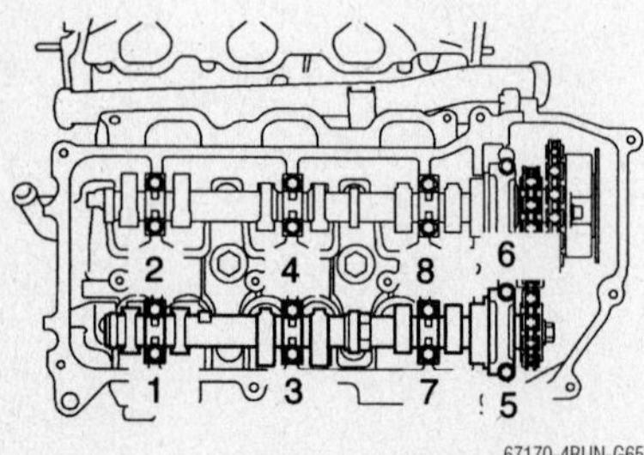

No.2 camshaft cap bolt loosening sequence—4.0L engine, right side

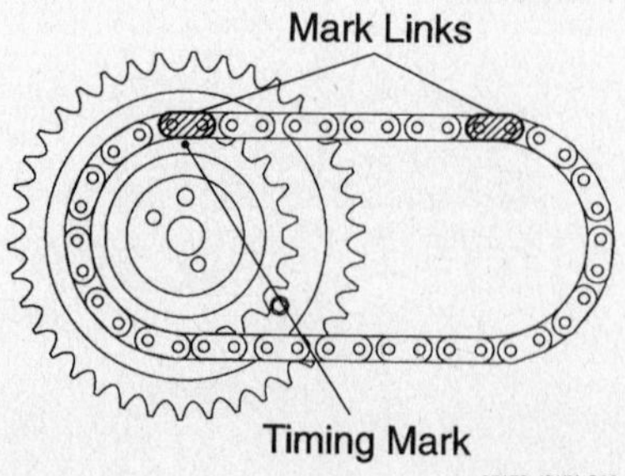

Painted link and timing mark alignment—4.0L engine, right side

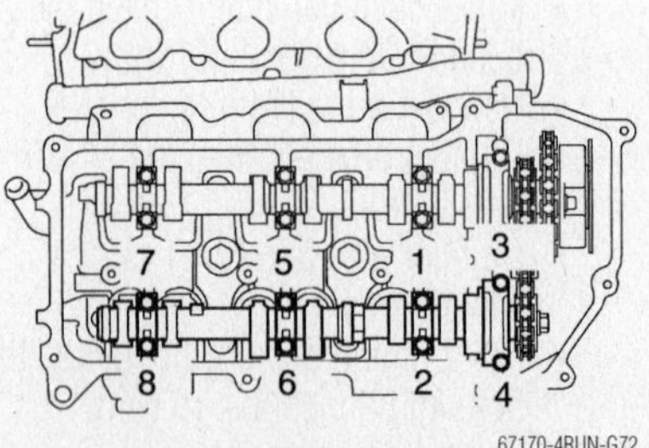

No.2 camshaft bolt torque sequence—4.0L engine, right side

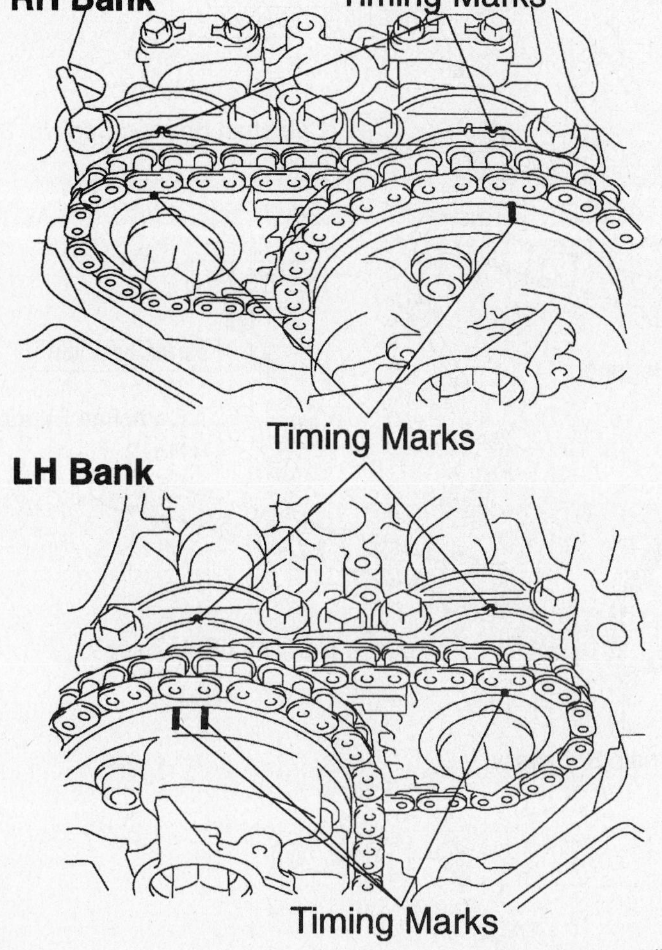

Check that the camshaft timing marks are aligned as shown—4.0L engine, right side

40. Temporarily, install the No.2 camshaft sprocket, aligning the timing marks. Torque the tensioner bolt to 14 ft. lbs. (19 Nm).
41. Place the No.2 camshaft on the head with the No.1 cylinder lobes facing upwards, as before.
42. Install the caps. Lightly oil the bolts and install them.
43. Torque the bolts evenly, and in several passes, in the sequence shown, to 10mm bolts, 80 inch lbs. (9 Nm); 12m bolts 18 ft. lbs. (24 Nm).
44. Rotate the No.2 camshaft clockwise so that the alignment pin is aligned with the hole in the sprocket.
45. Torque the sprocket bolt to 74 ft. lbs. (100 Nm).
46. Remove the pin from the No.2 tensioner.
47. Install the No.1 tensioner. Torque the bolts to 80 inch lbs. (9 Nm). Remove the pin.
48. Install the timing chain cover plate, using a new gasket. Torque the bolts to 80 inch lbs. (9 Nm).
49. Turn the crankshaft pulley two complete revolutions clockwise and align the "0" mark with the notch on the pulley. Check that the camshaft timing marks are aligned as shown.
50. Check the valve clearance as described below.
51. Clean all old gasket material from the head and head cover. Apply a 2–3mm dia. Bead of silicone gasket material to the head mating surface. Place the cover in position within 3 minutes.
52. Install the washers and bolts. Torque the bolts to 80 inch lbs. (9 Nm) within 15 minutes.
53. The remainder of installation is the reverse of removal.

LEFT SIDE

1. Drain the coolant.
2. Remove the V-bank cover.
3. Remove the air cleaner assembly.
4. Remove the water bypass hoses.
5. Remove the fuel vapor hose.
6. Remove the vent hose.
7. Disconnect the 2 VSV connectors.
8. Disconnect the wiring at the throttle body.
9. Remove the throttle body bracket.
10. Remove the oil baffle plate
11. Remove the upper manifold supports (stays).
12. Remove the 2 nuts and 4 bolts and lift off the upper manifold. Discard the gasket.
13. Remove the ignition coil assembly.
14. Remove the cylinder head cover.
15. Turn the crankshaft pulley to align the notch with the timing mark "0" on the chain cover.
16. Check that the camshaft and chain timing marks are aligned as shown. If they're not, rotate the crankshaft one full turn (360 degrees) clockwise.
17. Using paint, matchmark the chain links that correspond to the timing marks on the camshaft sprockets.
18. Remove the 4 bolts and the timing chain cover plate and gasket.
19. While turning the stopper plate of the tensioner upward, push in on the plunger. Then, turn the stopper plate downward and insert a 3.5mm diameter metal pin into the holes in the stopper plate and tensioner.
20. Remove the 2 bolts and the tensioner.

⁂ WARNING

The camshaft thrust clearances are very small. Keep the camshaft level while removing it to avoid damage to the cylinder head.

21. While lifting the No.4 tensioner, insert a 1mm dia. pin to hold it.
22. Remove the No.4 (outer) camshaft timing gear bolt. Remove the gear.
23. Remove the bearing caps evenly and in several passes, in the sequence shown.
24. Remove the caps and the No.4 camshaft.
25. Remove the No.3 tensioner.
26. Release the chain tension by turning the crank pulley slightly counterclockwise.
27. Remove the No.3 camshaft sprocket bolt and remove the sprocket. Lift the chain from the sprocket and secure it.
28. Remove the bearing caps evenly and in several passes, in the sequence shown.
29. Remove the caps and the No.3 camshaft. Keep the caps in order.
30. Tie the chain so that it doesn't fall into the case.

To install:

31. Align the painted link with the timing mark on the sprocket.

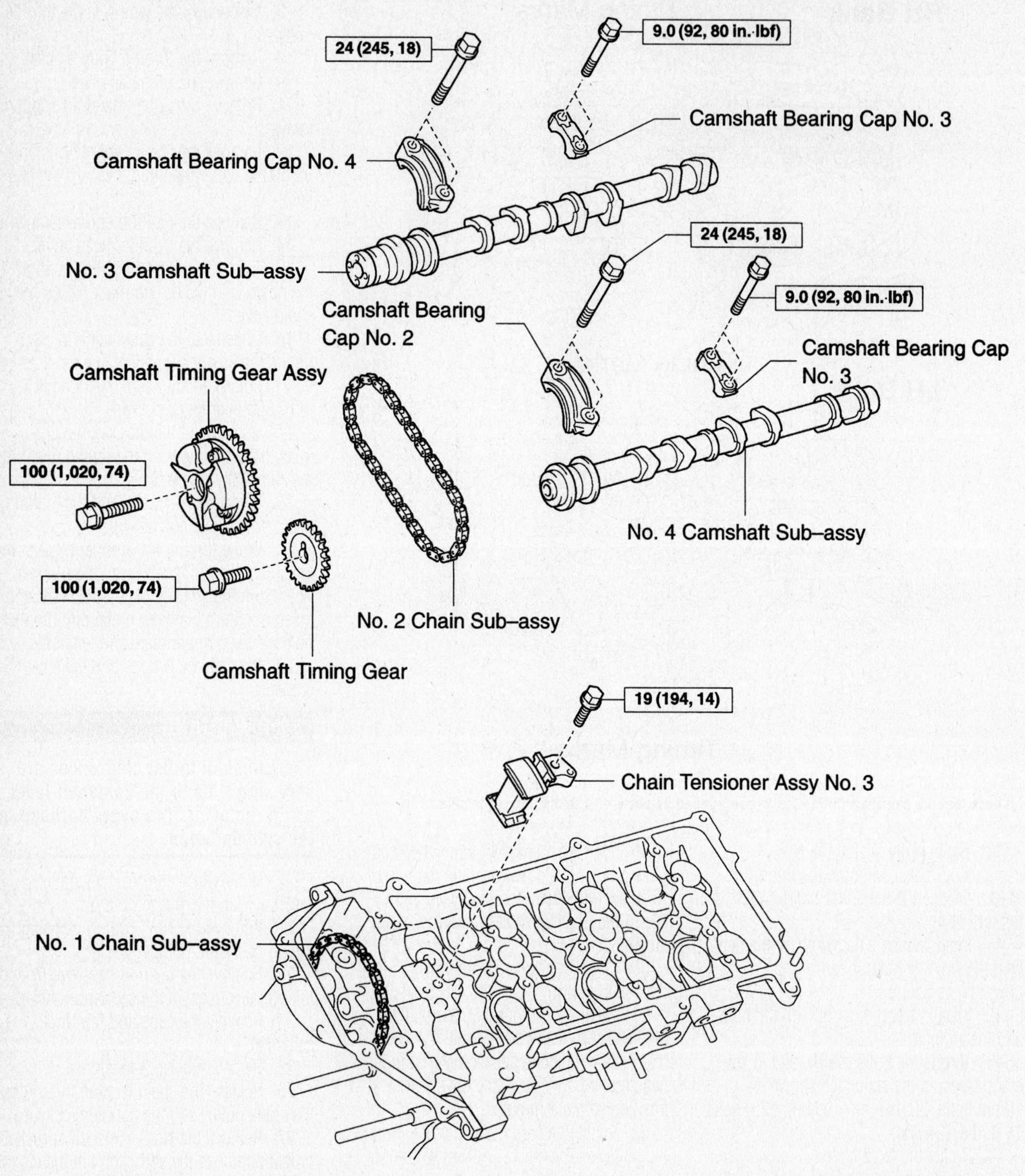

Left side camshaft and related parts—4.0L engine

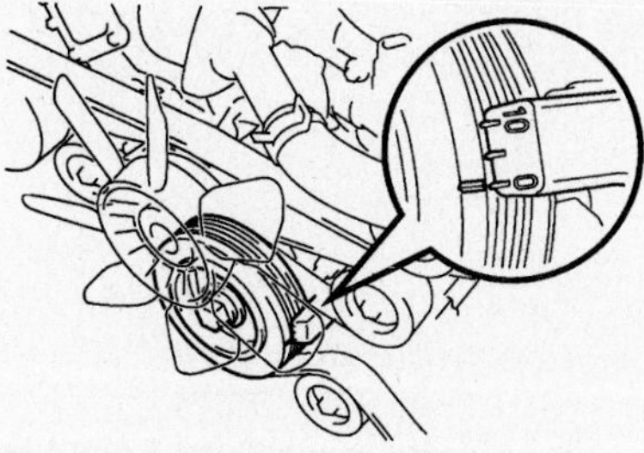

Turn the crankshaft pulley to align the notch with the timing mark "0" on the chain cover—4.0L engine

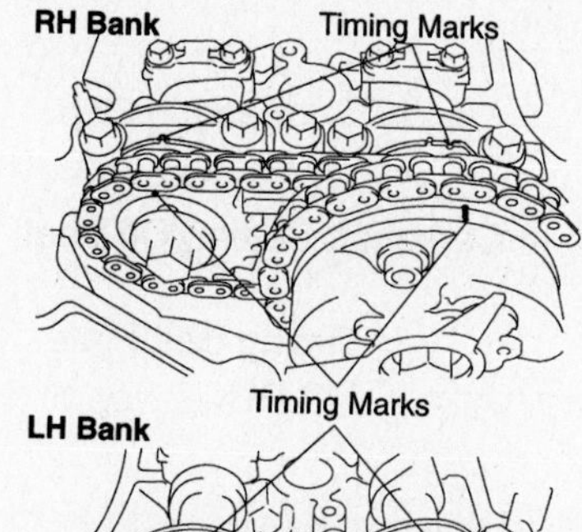

Check that the camshaft and chain timing marks are aligned—4.0L engine

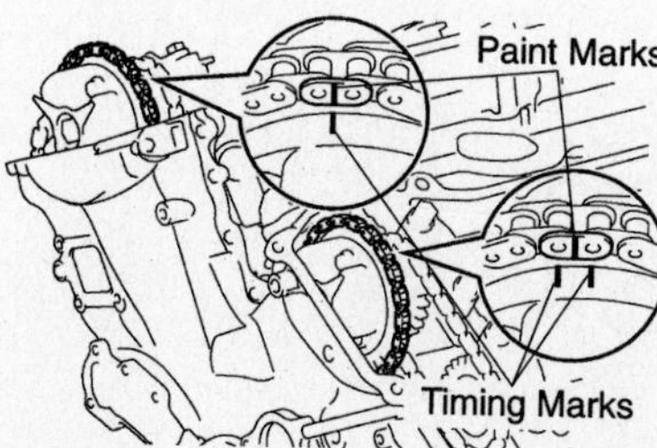

Using paint, matchmark the chain links that correspond to the timing marks on the camshaft sprockets—4.0L engine

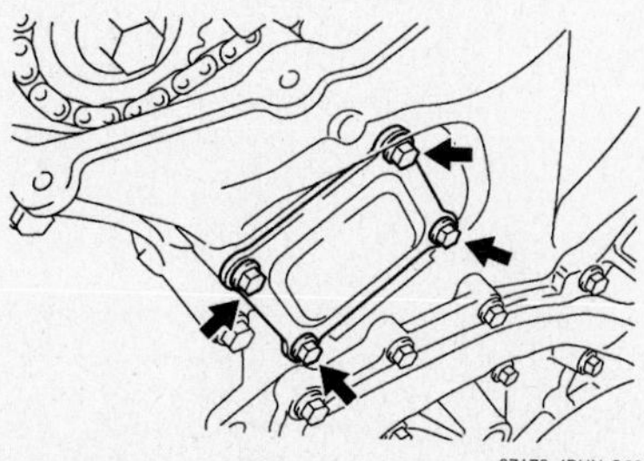

Remove the 4 bolts and the timing chain cover plate and gasket—4.0L engine

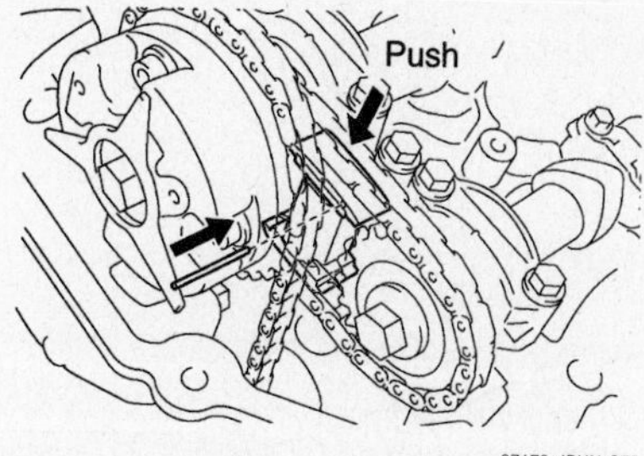

Locking the tensioner—4.0L engine left side

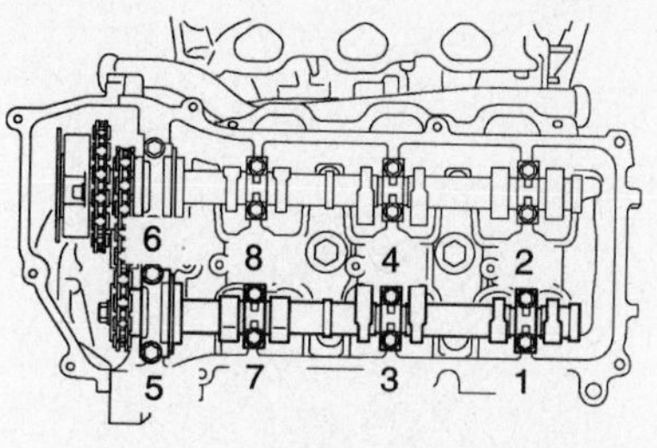

No.4 camshaft bearing cap bolt loosening sequence—4.0L engine, left side

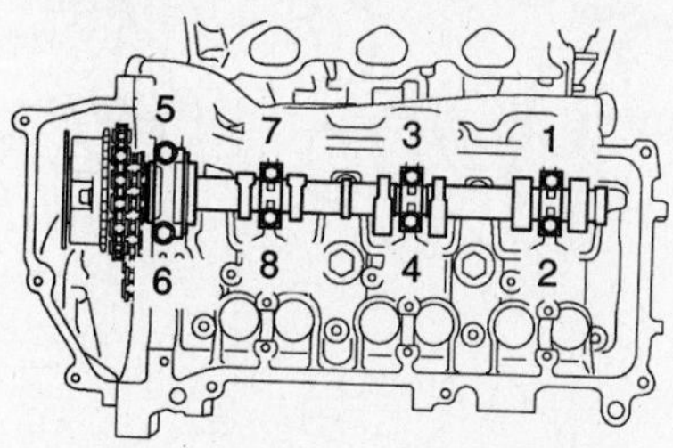

Remove the no.3 camshaft bearing caps evenly and in several passes, in the sequence shown—4.0L engine, left side

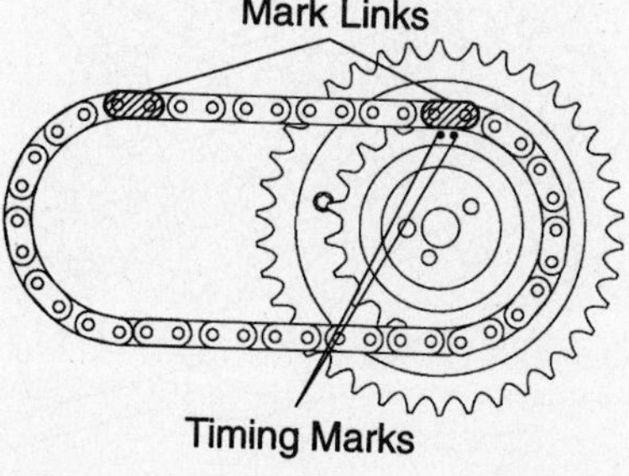

Painted link and timing mark alignment—4.0L engine, left side

32. Install the gear and chains on the No.3 camshaft.

33. Install the bolt hand-tight.

34. Place the camshaft on the head with the No.3 cylinder lobes facing downward as shown.

Place the camshaft on the head with the No.3 cylinder lobes facing downward as shown

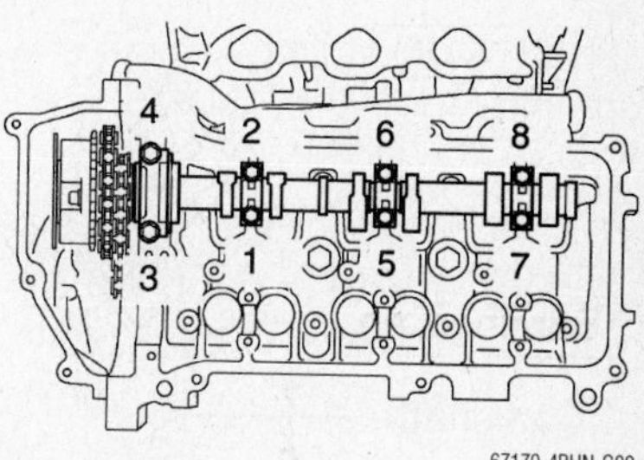

No.3 camshaft bolt torque sequence—4.0L engine, left side

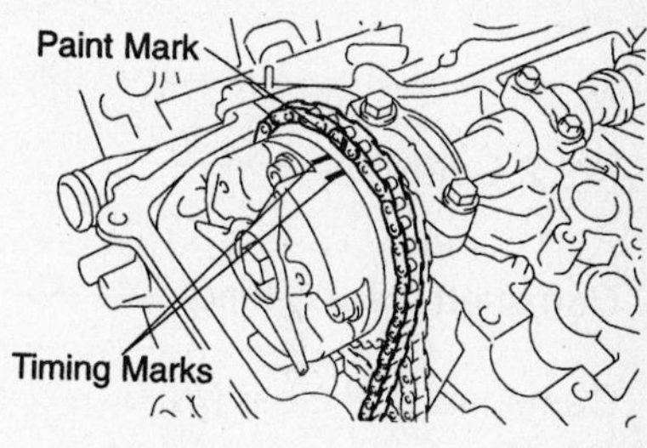

Rotate the No.3 camshaft clockwise so that the timing mark on the camshaft sprocket is aligned with the timing mark on the bearing cap—4.0L engine, left side

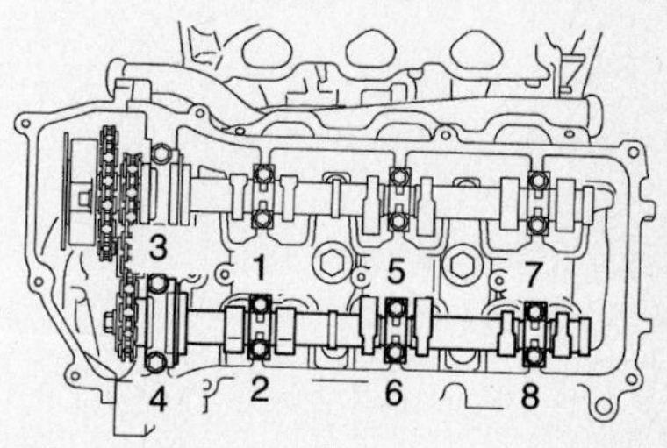

No.4 camshaft bolt torque sequence—4.0L engine, left side

35. Install the caps. Lightly oil the bolts and install them.

36. Torque the bolts evenly, and in several passes, in the sequence shown, to 10mm bolts, 80 inch lbs. (9 Nm); 12m bolts 18 ft. lbs. (24 Nm).

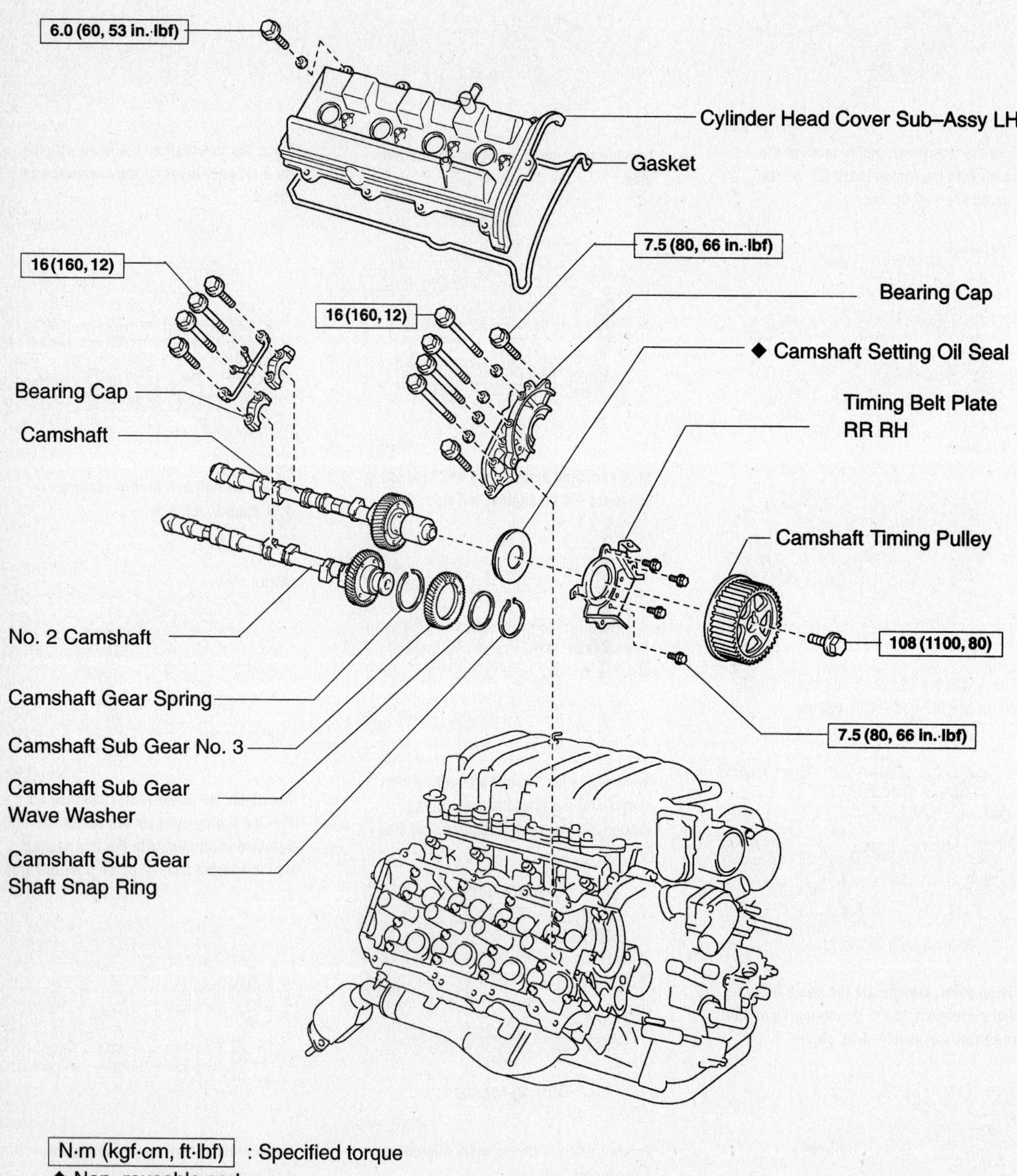

Right camshaft and related parts—4.7L engine

37. Rotate the No.3 camshaft clockwise so that the timing mark on the camshaft sprocket is aligned with the timing mark on the bearing cap.

38. Align the paint mark on the chain with the timing mark on the sprocket.

39. Torque the sprocket bolt to 74 ft. lbs. (100 Nm).

40. Install the No.4 chain tensioner.

41. Temporarily, install the No.4 camshaft sprocket, aligning the timing marks. Torque the tensioner bolt to 14 ft. lbs. (19 Nm).

42. Place the No.4 camshaft on the head with the No.1 cylinder lobes facing upwards.

43. Install the caps. Lightly oil the bolts and install them.

44. Torque the bolts evenly, and in several passes, in the sequence shown, to 10mm bolts, 80 inch lbs. (9 Nm); 12m bolts 18 ft. lbs. (24 Nm).

45. Rotate the No.2 camshaft clockwise so that the alignment pin is aligned with the hole in the sprocket.

46. Torque the sprocket bolt to 74 ft. lbs. (100 Nm).

47. Remove the pin from the No.3 tensioner.

48. Install the No.4 tensioner. Torque the bolts to 80 inch lbs. (9 Nm). Remove the pin.

49. Install the timing chain cover plate, using a new gasket. Torque the bolts to 80 inch lbs. (9 Nm).

50. Turn the crankshaft pulley two complete revolutions clockwise and align the "0" mark with the notch on the pulley. Check that the camshaft timing marks are aligned.

51. Check the valve clearance as described below.

52. Clean all old gasket material from the head and head cover. Apply a 2–3mm dia. Bead of silicone gasket material to the head mating surface. Place the cover in position within 3 minutes.

53. Install the washers and bolts. Torque the bolts to 80 inch lbs. (9 Nm) within 15 minutes.

54. The remainder of installation is the reverse of removal.

4.7L Engine

RIGHT SIDE

1. Remove the timing belt.
2. Remove the camshaft timing pulley.
3. Remove the right rear belt plate.
4. Remove the ignition coil.
5. Remove the cylinder head cover.

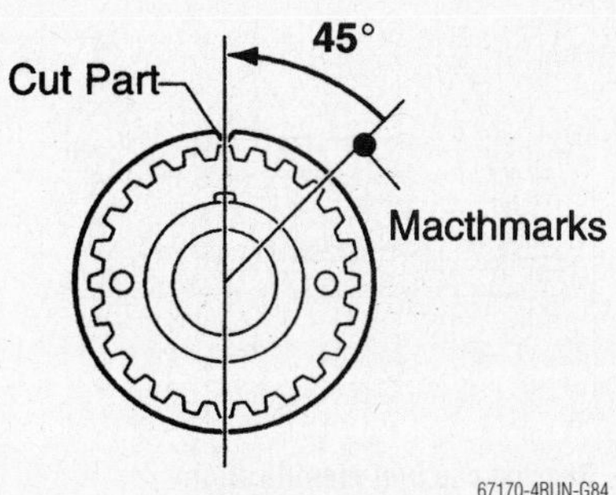

Set the crankshaft pulley at the correct angle as shown—4.7L engine

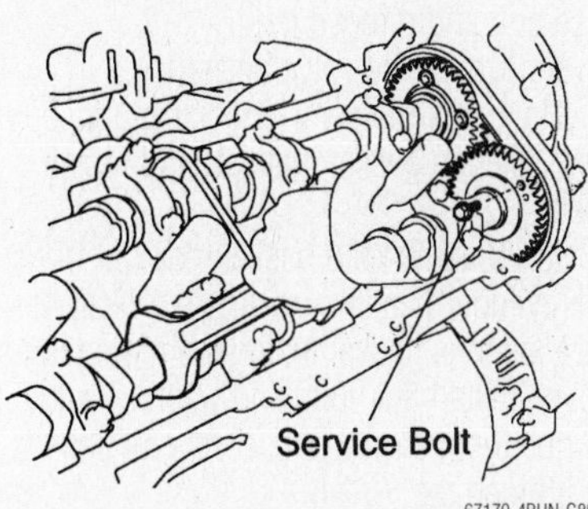

Secure the sub-gear to the main gear with a bolt 6mm x 1.0 x 16-20mm long—4.7L engine

6. Set the crankshaft pulley at the correct angle as shown.

7. Bring the service bolt hole of the sub-gear upward by turning the exhaust camshaft. Secure the sub-gear to the main gear with a bolt 6mm x 1.0 x 16-20mm long.

8. Set the timing mark (1 dot) of the camshaft main gear at approximately 10 degrees by turning the exhaust camshaft.

9. Loosen the 22 camshaft bearing cap bolts, evenly and in several passes, in the sequence shown.

10. Remove the oil feed pipe and bearing caps. Keep the caps in order.

11. Place the camshaft in a soft-jawed vise as shown.

12. Turn the sub-gear clockwise and remove the service bolt.

13. Remove the snapring, wave washer, sub-gear and camshaft gear spring.

14. Remove the camshaft housing plug.

15. Remove the semi-circular plug.

To install:

16. Remove all old silicone gasket material from the camshaft housing plug. Apply new gasket material as shown, and install the plug.

17. Assemble the camshaft gear spring, sub-gear and wave washer. Install the snapring.

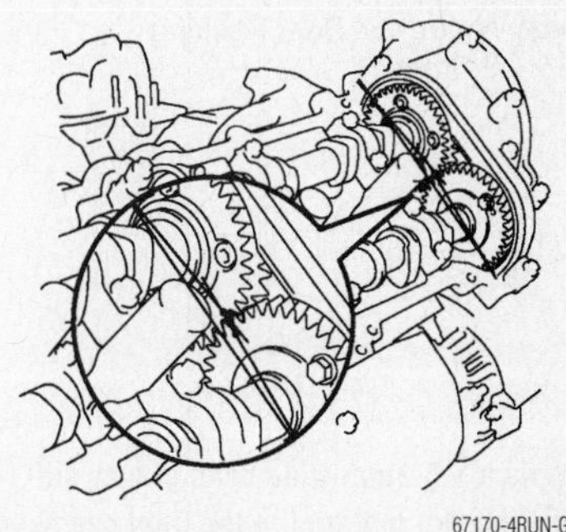

Set the timing mark of the camshaft main gear at approximately 10 degrees by turning the exhaust camshaft—4.7L engine

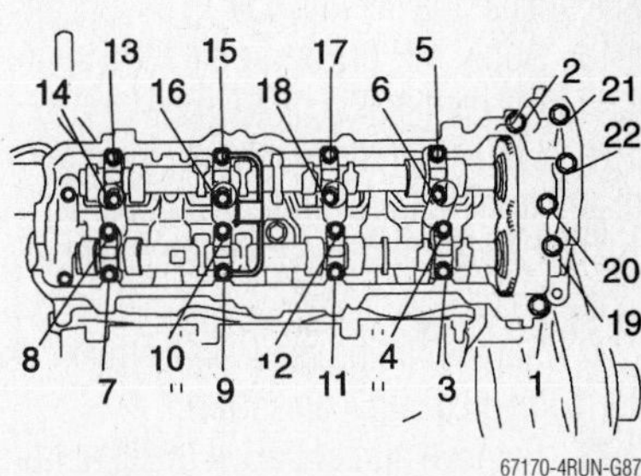

Loosen the 22 camshaft bearing cap bolts, evenly and in several passes, in the sequence shown—4.7L engine

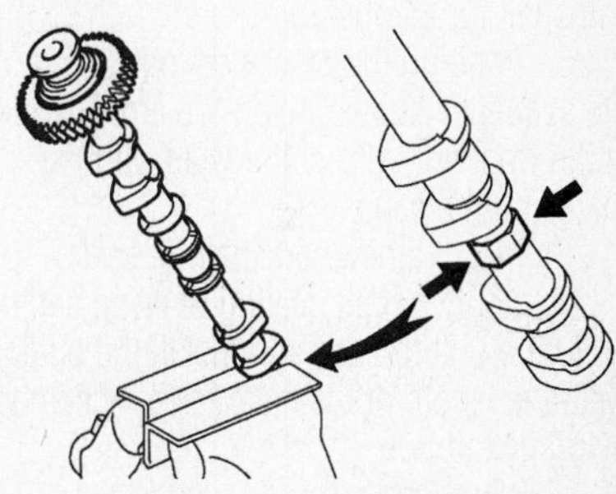

Place the camshaft in a soft-jawed vise as shown—4.7L engine

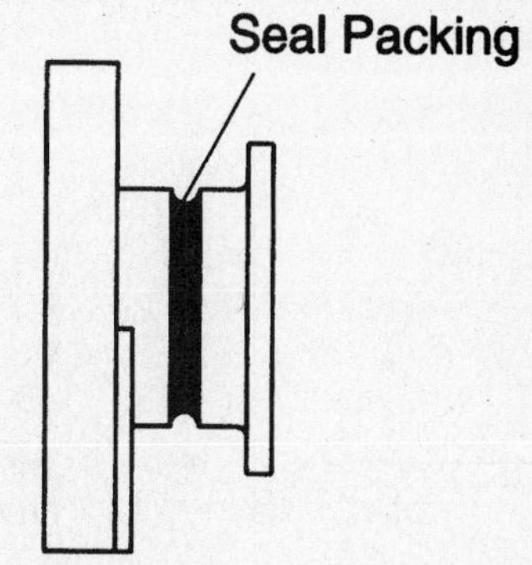

Apply new gasket material as shown, and install the plug—4.7L engine

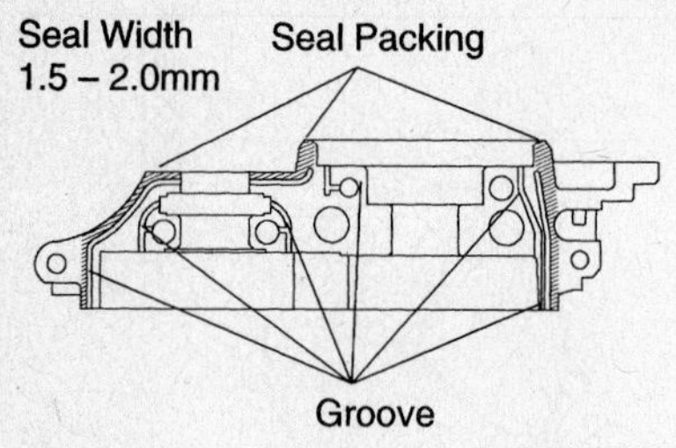

Apply a 1.5-2mm wide bead of new silicone gasket material to the front cap as shown

18. Align the holes in the main and sub-gears by turning the sub-gear counterclockwise. Install the service bolt.
19. Apply MP grease to the thrust portion of the camshafts.
20. Verify that the crankshaft hasn't rotated from the previously set position.
21. CAREFULLY position the camshafts on the head.
22. Set the timing mark on the main gear at a 10 degree angle as shown.
23. Remove all old gasket material from the front bearing cap. Apply a 1.5-2mm wide bead of new silicone gasket material to the front cap as shown.
24. Place the front cap on the head. Installing the front cap will determine the camshaft thrust position.
25. Position the other bearing caps in the sequence shown. Align the arrow marks at the front and rear of the head with the marks on the caps.
26. Install a new camshaft oil seal.
27. Apply a light coating of clean engine oil to the bolt threads and install the bolts. Note the length of the bolts in the accompanying illustration.
- A: 94mm
- B: 72mm
- C: 25mm
- D: 52mm
- E: 38mm

28. Tighten the bolts evenly, and in several passes to

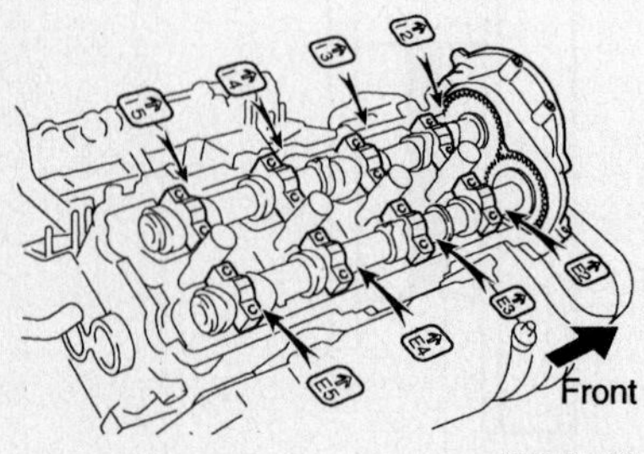

Camshaft bearing cap positioning sequence

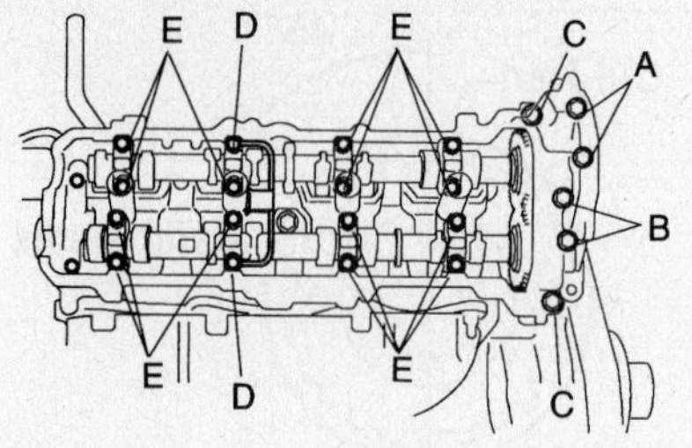

Bearing cap bolt identification

- Bolts C: 66 inch lbs. (7.5 Nm)
- All other bolts: 12 ft. lbs. (16 Nm)

29. Turn the camshafts to access the service bolt and remove it.
30. Check the valve clearance.
31. Clean the semi-circular plugs, apply new silicone gasket material and install them.
32. Clean all old gasket material from the cylinder head cover. Apply a bead of new silicone gasket material and install the cover. Install the bolts and washers and torque them evenly and in several passes, to 53 inch lbs. (6 Nm).
33. Install the coil. Torque the bolts to 66 inch lbs. (7.5 Nm).
34. Install the timing belt plate. Torque the bolts to 66 inch lbs. (7.5 Nm).
35. Install the camshaft pulley. Torque to 80 ft. lbs. (108 Nm).
36. Install the timing belt.

LEFT SIDE

1. Remove the timing belt.
2. Remove the camshaft timing pulley.
3. Remove the camshaft position sensor.
4. Remove the timing belt rear plate.
5. Remove the ignition coil.
6. Remove the cylinder head cover.
7. Set the crankshaft pulley at the correct angle as shown.
8. Bring the service bolt hole of the sub-gear upward by turning the exhaust camshaft. Secure the sub-gear to the main gear with a bolt 6mm x 1.0 x 16-20mm long.
9. Align the timing mark (2 dots) of the camshaft drive gears by turning the exhaust camshaft as shown.
10. Loosen the 22 camshaft bearing cap bolts, evenly and in several passes, in the sequence shown.
11. Remove the oil feed pipe and bearing caps. Keep the caps in order.
12. Place the camshaft in a soft-jawed vise as shown.
13. Turn the sub-gear clockwise and remove the service bolt.
14. Remove the snapring, wave washer, sub-gear and camshaft gear spring.
15. Remove the camshaft housing plug.
16. Remove the semi-circular plug.

To install:

17. Remove all old silicone gasket material from the camshaft housing plug. Apply new gasket material as shown, and install the plug.
18. Assemble the camshaft gear spring, sub-gear and wave washer. Install the snapring.
19. Align the holes in the main and sub-gears by turning the sub-gear counterclockwise. Install the service bolt.
20. Apply MP grease to the thrust portion of the camshafts.
21. Verify that the crankshaft hasn't rotated from the previously set position.
22. CAREFULLY position the camshafts on the head.

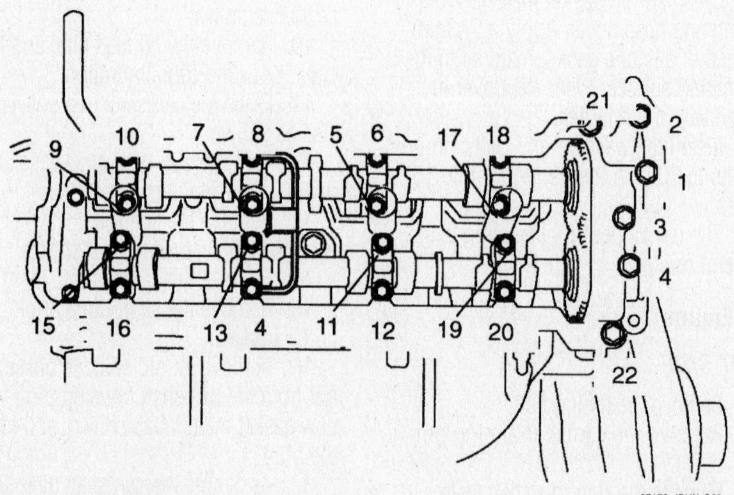

Tighten the bolts evenly, and in several passes in the sequence shown—4.7L engine, right side

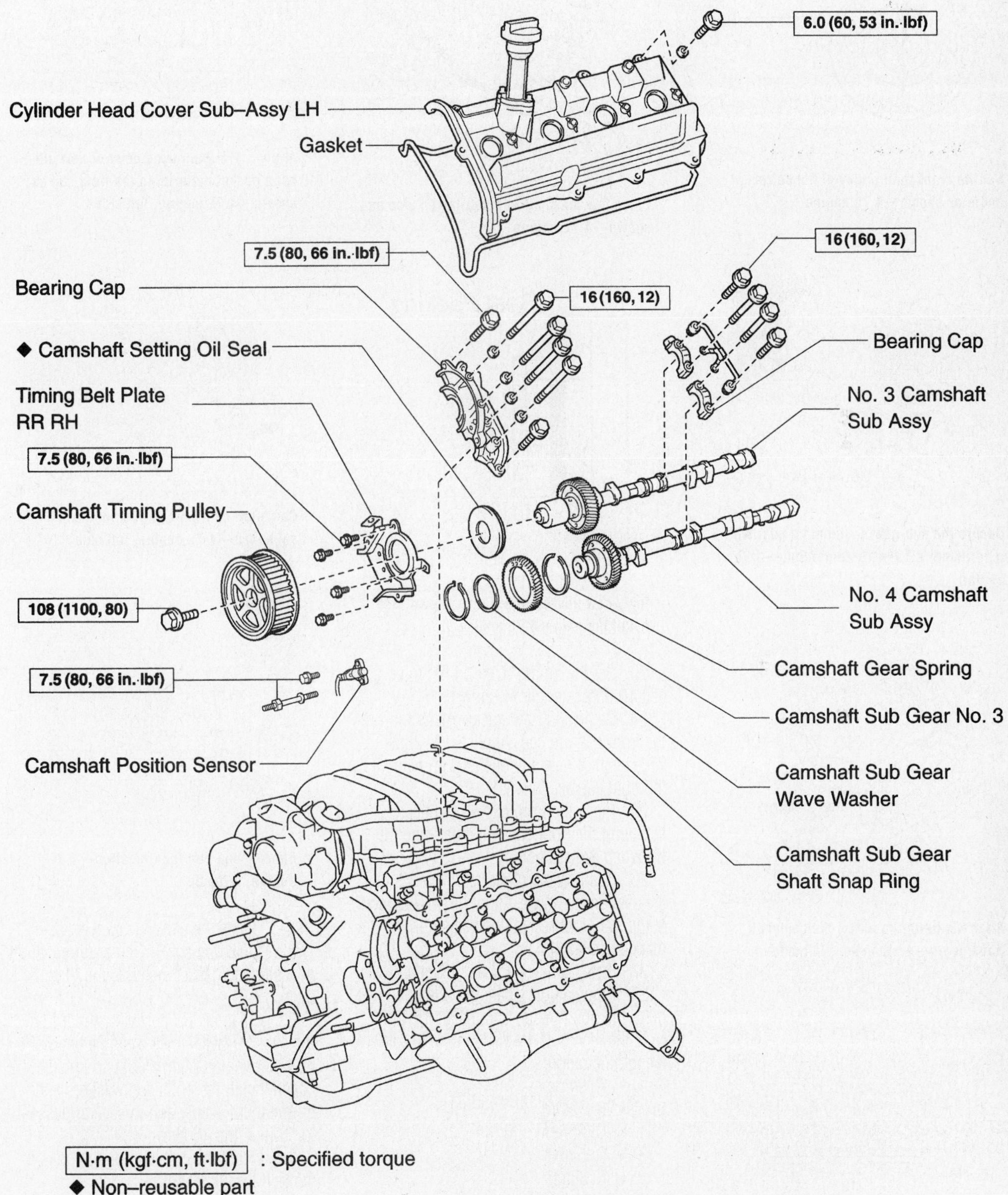
6.0 (60, 53 in.·lbf)
Cylinder Head Cover Sub–Assy LH
Gasket
7.5 (80, 66 in.·lbf)
16 (160, 12)
Bearing Cap
16 (160, 12)
◆ Camshaft Setting Oil Seal
Bearing Cap
Timing Belt Plate RR RH
No. 3 Camshaft Sub Assy
7.5 (80, 66 in.·lbf)
Camshaft Timing Pulley
108 (1100, 80)
No. 4 Camshaft Sub Assy
Camshaft Gear Spring
7.5 (80, 66 in.·lbf)
Camshaft Sub Gear No. 3
Camshaft Position Sensor
Camshaft Sub Gear Wave Washer
Camshaft Sub Gear Shaft Snap Ring
N·m (kgf·cm, ft·lbf) : Specified torque
◆ Non–reusable part
67170-4RUN-G94

Left camshaft and related parts—4.7L engine

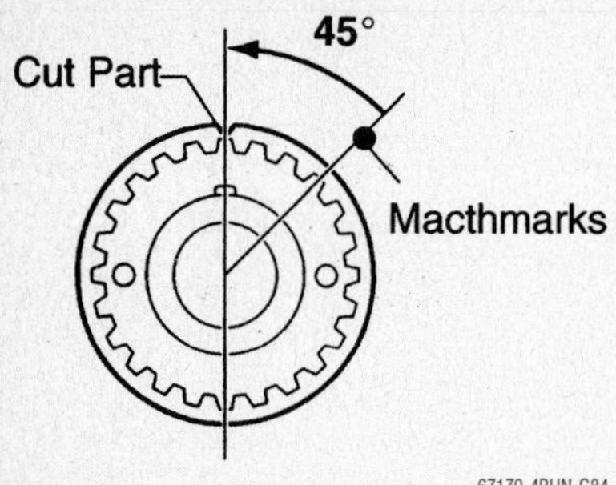

Set the crankshaft pulley at the correct angle as shown—4.7L engine

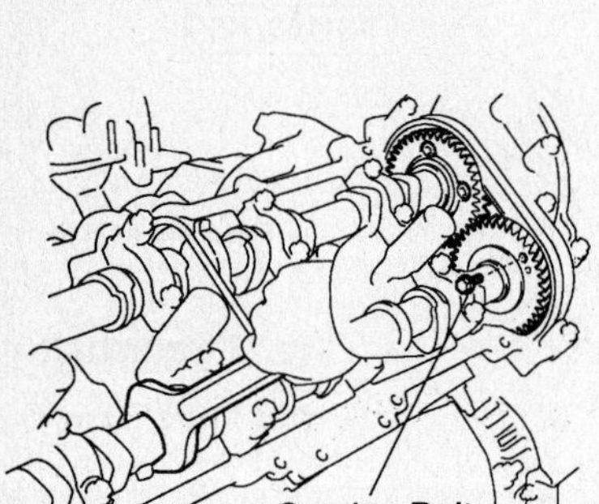

Secure the sub-gear to the main gear with a bolt 6mm x 1.0 x 16-20mm long—4.7L engine

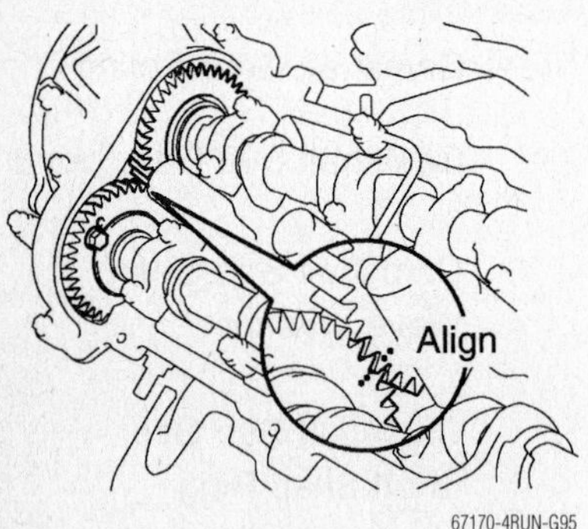

Align the timing marks of the camshaft drive gears—4.7L engine, left side

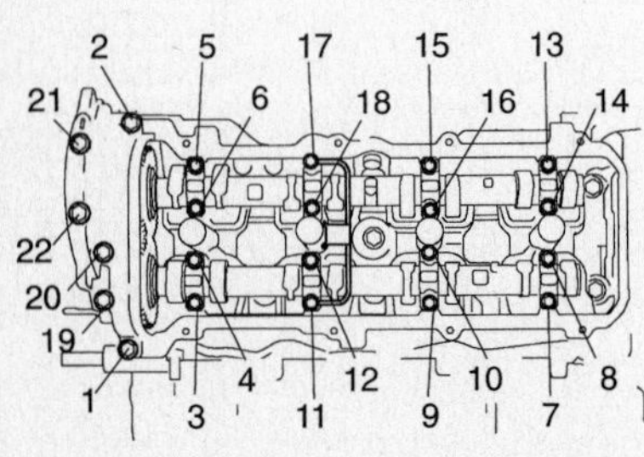

Loosen and remove the 22 camshaft bearing cap bolts, evenly and in several passes, in the sequence shown—4.7L engine, left side

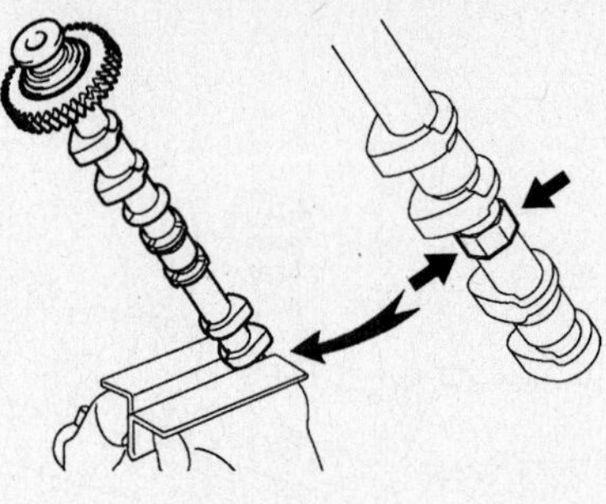

Place the camshaft in a soft-jawed vise as shown—4.7L engine

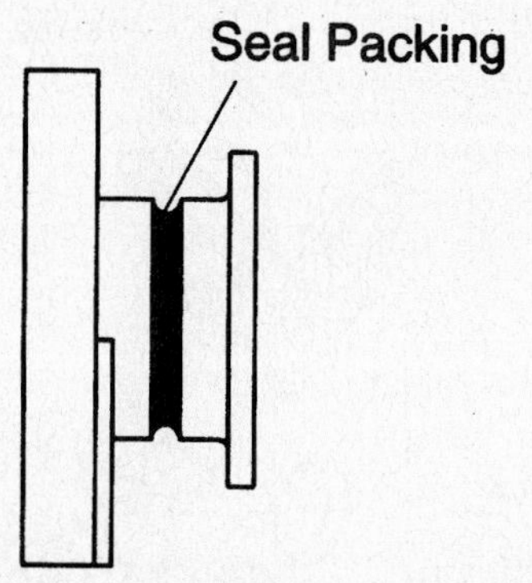

Apply new gasket material as shown, and install the plug—4.7L engine

23. Set the timing mark on the main gear at a 10 degree angle as shown.

24. Remove all old gasket material from the front bearing cap. Apply a 1.5-2mm wide bead of new silicone gasket material to the front cap as shown.

25. Place the front cap on the head. Installing the front cap will determine the camshaft thrust position.

26. Position the other bearing caps in the sequence shown. Align the arrow marks at the front and rear of the head with the marks on the caps.

27. Install a new camshaft oil seal.

28. Apply a light coating of clean engine oil to the bolt threads and install the bolts. Note the length of the bolts in the accompanying illustration.

- A: 94mm
- B: 72mm
- C: 25mm
- D: 52mm
- E: 38mm

29. Tighten the bolts evenly, and in several passes in the sequence shown, to:

- Bolts C: 66 inch lbs. (7.5 Nm)
- All other bolts: 12 ft. lbs. (16 Nm)

30. Turn the camshafts to access the service bolt and remove it.

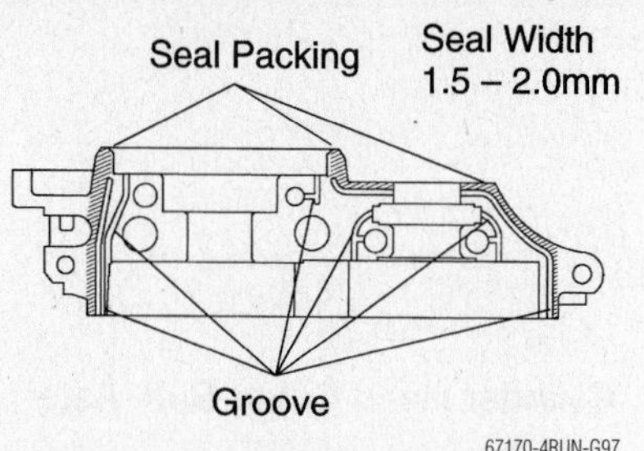

Apply a 1.5-2mm wide bead of new silicone gasket material to the front cap as shown—4.7L engine, left side

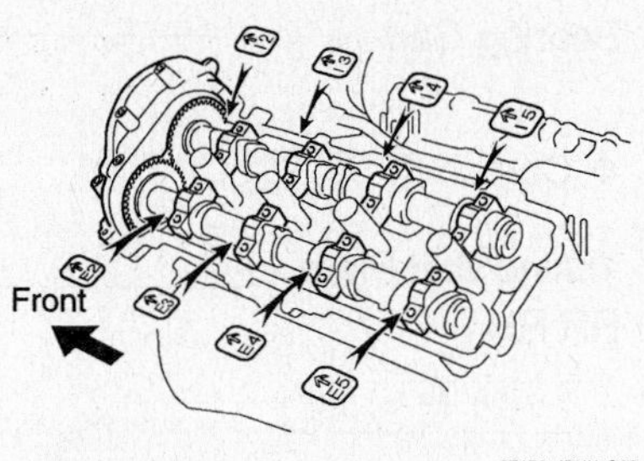

Camshaft bearing cap positioning sequence—4.7L engine, left side

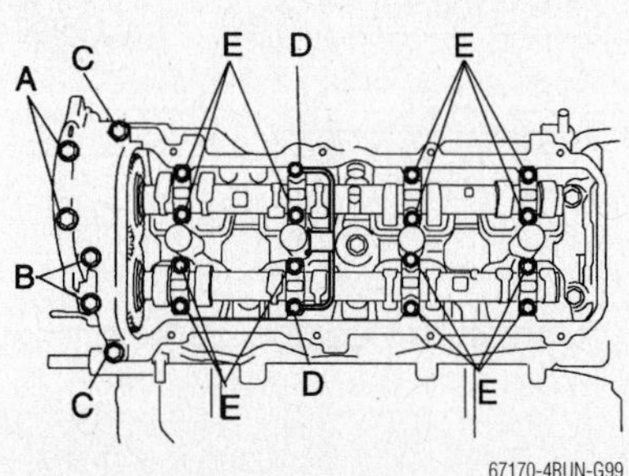

Bearing cap bolt identification—4.7L engine, left side

31. Check the valve clearance.

32. Clean the semi-circular plugs, apply new silicone gasket material and install them.

33. Clean all old gasket material from the cylinder head cover. Apply a bead of new silicone gasket material and install the cover. Install the bolts and washers and torque them evenly and in several passes, to 53 inch lbs. (6 Nm).

34. Install the coil. Torque the bolts to 66 inch lbs. (7.5 Nm).

35. Install the timing belt plate. Torque the bolts to 66 inch lbs. (7.5 Nm).

36. Install the camshaft pulley. Torque to 80 ft. lbs. (108 Nm).

37. Install the timing belt.

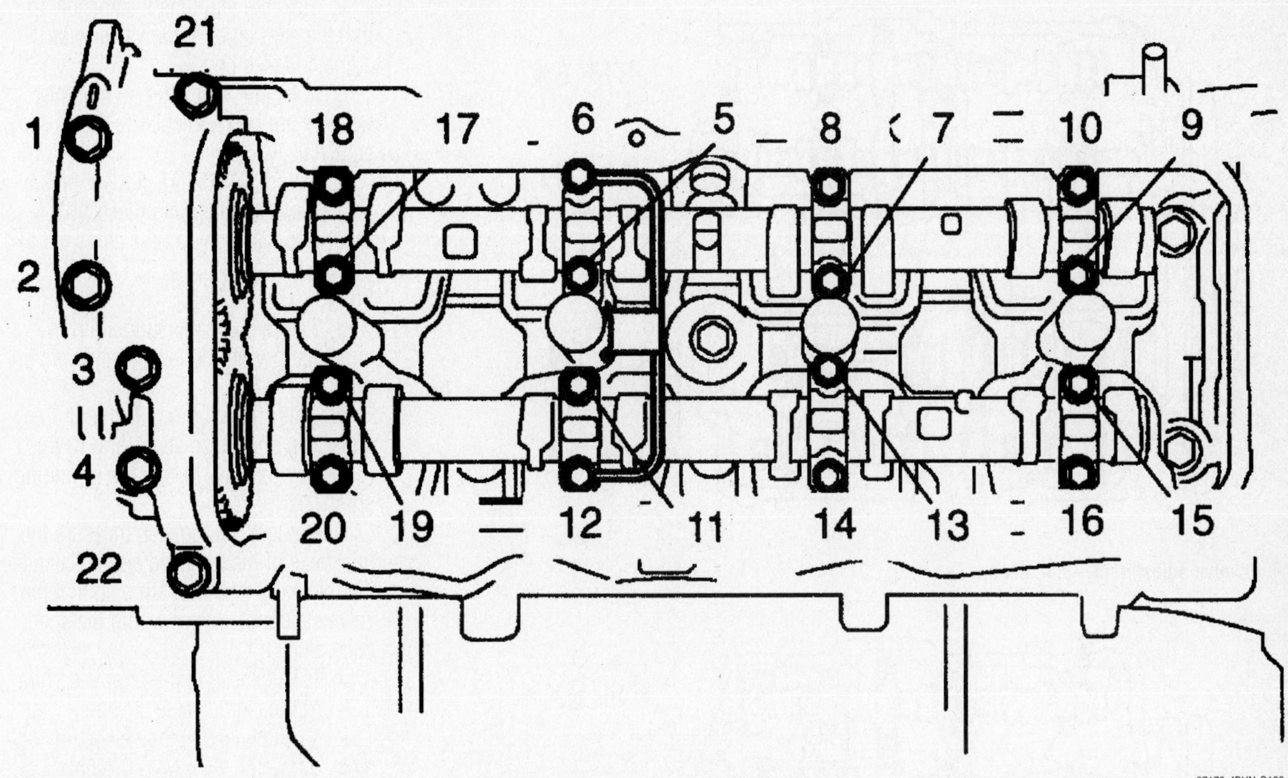

Tighten the bolts evenly, and in several passes in the sequence shown—4.7L engine, left side

Valve Lash

ADJUSTMENT

3.4L Engine

1. Before servicing the vehicle, refer to the precautions in the beginning of this section.
2. Disconnect the negative battery cable.
3. Drain the engine coolant.
4. Remove or disconnect the following:
   - Air intake connector
   - Cylinder head cover
5. Set the No. 1 cylinder to Top Dead Center (TDC) of the compression stroke, as follows:
   a. Turn the crankshaft pulley clockwise and align its groove with the **0** mark on the timing chain cover.
   b. Check that the timing marks (1 and 2 dots) of the camshaft drive and driven gears are in a straight line on the cylinder head surface. If not, turn the crankshaft 1 revolution (360 degrees) and align the marks.
6. Inspect the valve clearance, as follows:
   a. Measure the clearance between the valve lifter and the camshaft. Measure the 1st intake and the 3rd exhaust valves on the right head and the 6th intake and the 2nd exhaust valves on the left head.
   b. Turn the crankshaft ⅔ of a revolution (240 degrees) and adjust the 3rd intake and the 5th exhaust valves on the right head and the 2nd intake and the 4th exhaust valves on the left head.
   c. Turn the crankshaft ⅔ of a revolution (240 degrees) and adjust the 5th intake and the 1st exhaust valves on the right head and the 4th intake and the 6th exhaust valves on the left head.
7. Valve clearance cold should be:
   - Intake: 0.006–0.009 in. (0.13–0.23mm)
   - Exhaust: 0.011–0.014 in. (0.27–0.37mm)
8. Adjust the valve clearance by using adjusting shims, as follows:

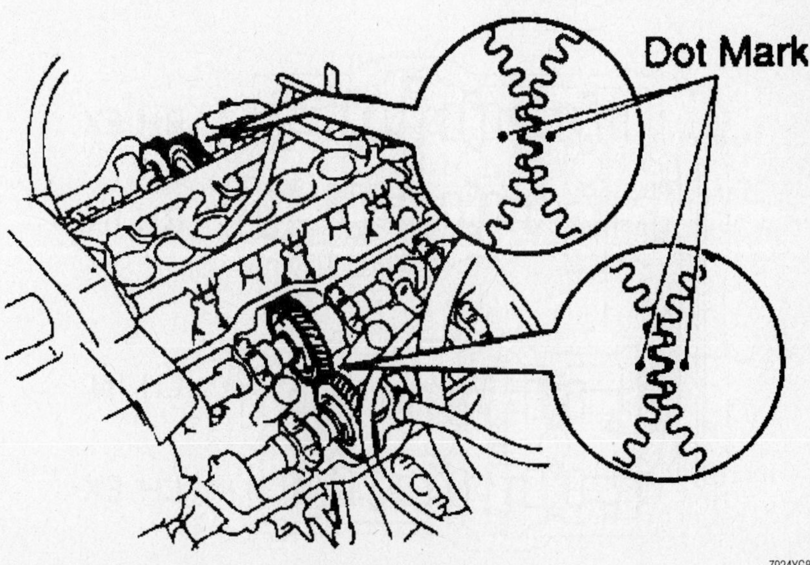

Aligning the timing marks—3.4L engine

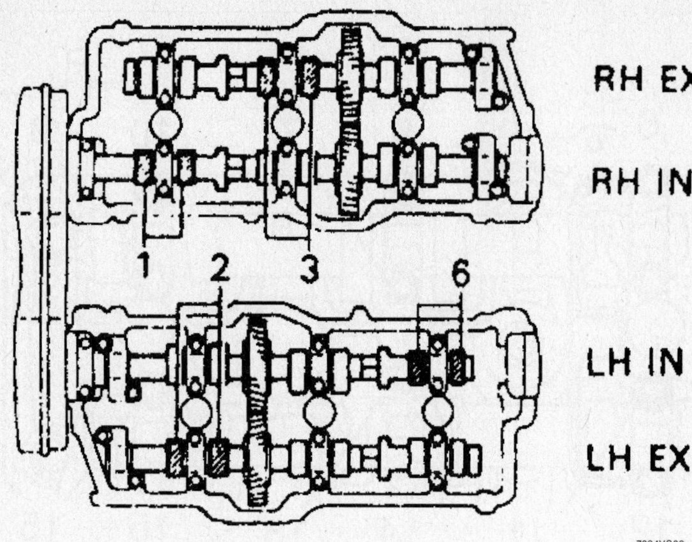

First valve adjustment—3.4L engine

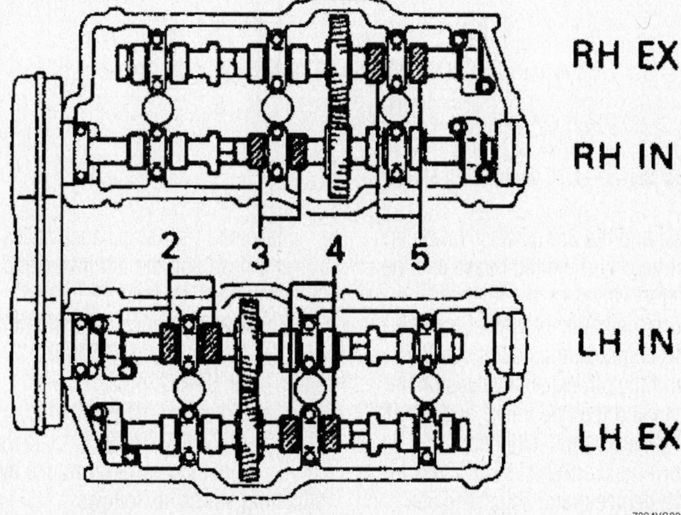

Second valve adjustment—3.4L engine

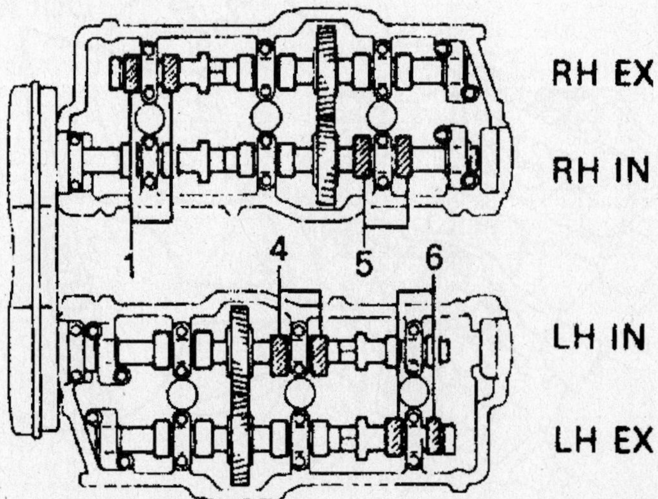

Third valve adjustment—3.4L engine

a. Turn the equipment camshaft so that the cam lobe for the valve to be adjusted faces up.

b. Turn the valve lifter so that the notches are perpendicular to the camshaft.

c. Using SST 09248-55040, press down the valve lifter and place SST 09248-05420, between the camshaft and the valve lifter. Remove SST 09248-55040.

d. Remove the adjusting shim with a small flat prying tool and a magnetic finger.

e. Determine the replacement adjusting shim size according to the following formula or use the adjusting shim charts.

f. Using a micrometer, measure the thickness of the removed shim. Calculate the thickness of a new shim so that the valve clearance comes within the specified value.

- T: Thickness of the removed shim
- A: Measured valve clearance
- N: Thickness of the new shim

g. Intake: N = T + (A – 0.007 in. (0.18mm))

h. Exhaust: N = T + (A – 0.013 in. (0.32mm))

i. Install a new adjusting shim. Place it on the valve lifter. Using the SST 09248-55040, press down the valve lifter and remove SST 09248-05420.

j. Recheck the valve clearance.

9. Install or connect the following:
   - Cylinder head cover
   - Intake air connector
   - Negative battery cable
10. Refill with engine coolant.
11. Start the engine and check for leaks.

4.0L Engine

1. Drain the coolant.
2. Remove the V-bank cover.
3. Remove the air cleaner assembly.
4. Remove the upper intake manifold (air surge tank).
5. Remove the ignition coil.
6. Remove the cylinder head cover.
7. Turn the crankshaft clockwise and align its groove with the "0" mark on the timing cover.
8. Check that the timing marks on the camshaft sprockets are aligned with the timing marks on the bearings caps, as shown. If not, rotate the crankshaft 360 degrees clockwise.
9. Check the valve clearance for the

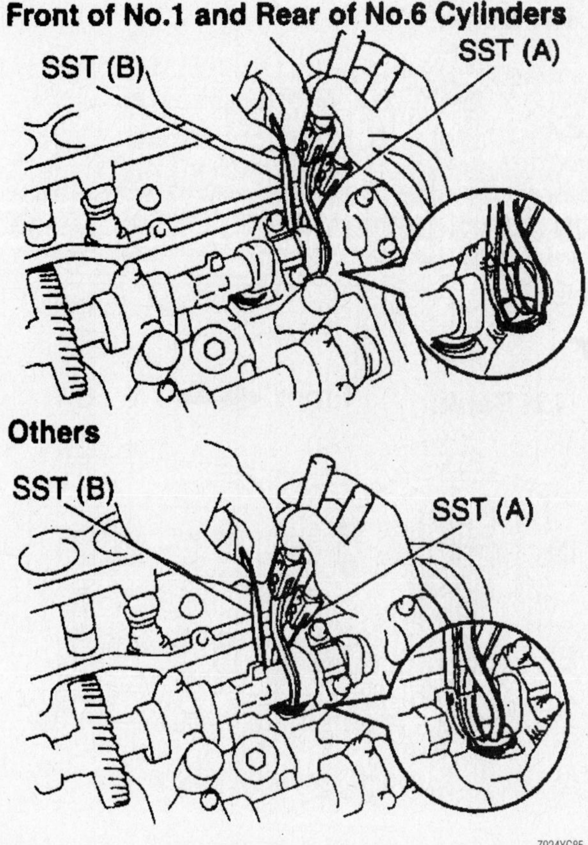

Removing the adjusting shim—3.4L engine

valves shown in Figure A. Clearance should be:

- Intake: 0.006–0.010 in. (0.15–0.25mm)
- Exhaust: 0.011–0.015 inch (0.29–0.39mm)

10. If any are not correct, record how far off they are. This will determine replacement lifter(s).

11. Turn the crankshaft ⅔ turn (240 degrees) clockwise. Check the valves shown in Figure B.

12. If any are not correct, record how far off they are. This will determine replacement lifter(s).

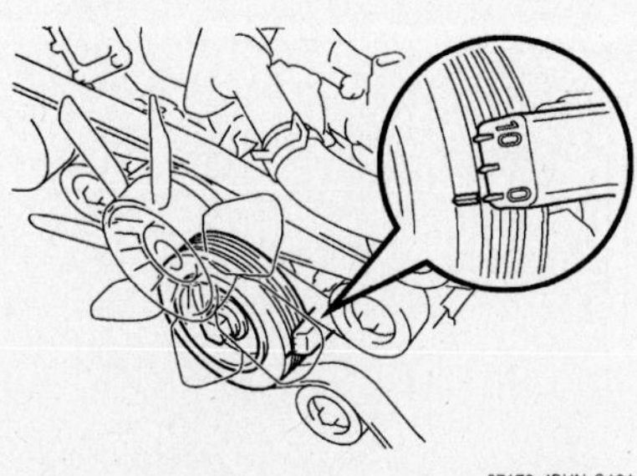

Turn the crankshaft clockwise and align its groove with the "0" mark on the timing cover—4.0L engine

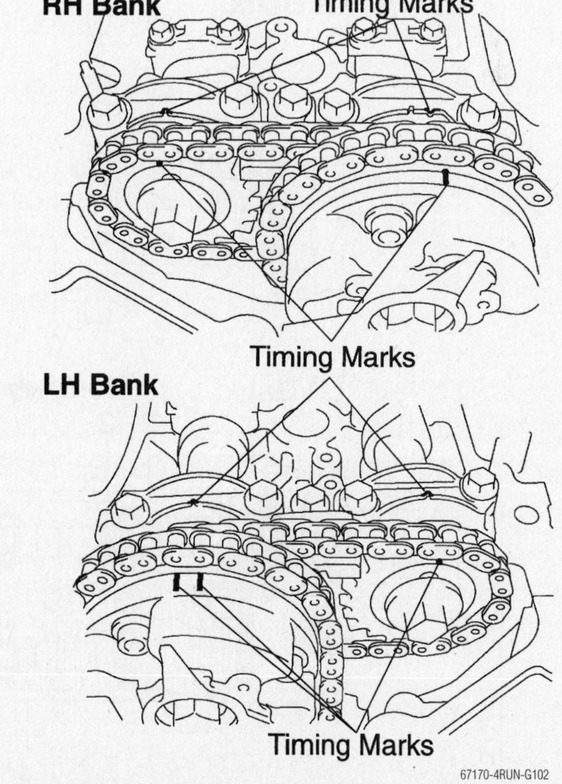

Check that the timing marks on the camshaft sprockets are aligned with the timing marks on the bearings caps—4.0L engine

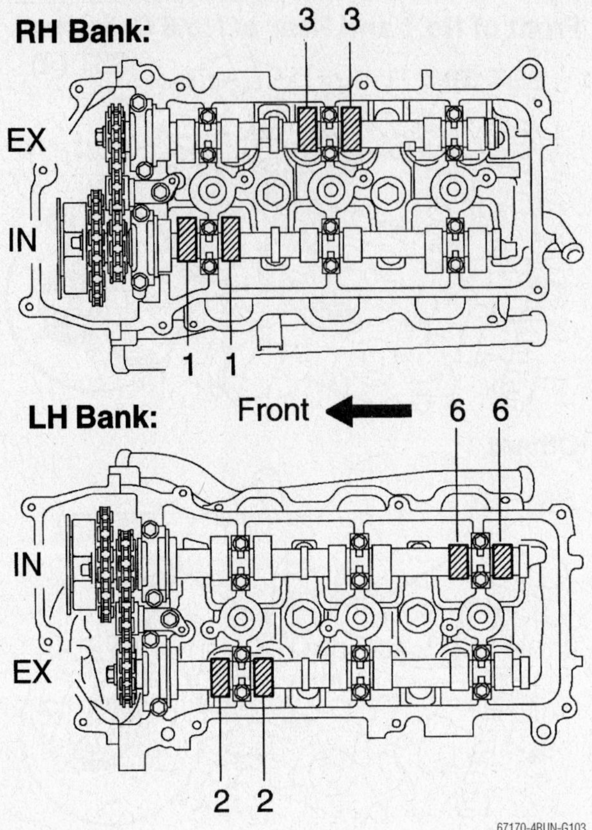

Valve clearance check, Figure A—4.0L engine

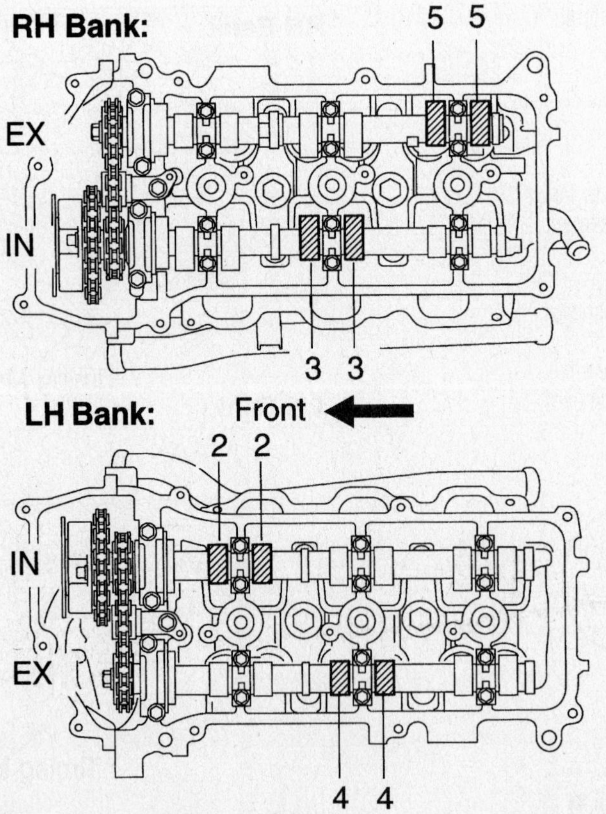

Valve clearance check, Figure B—4.0L engine

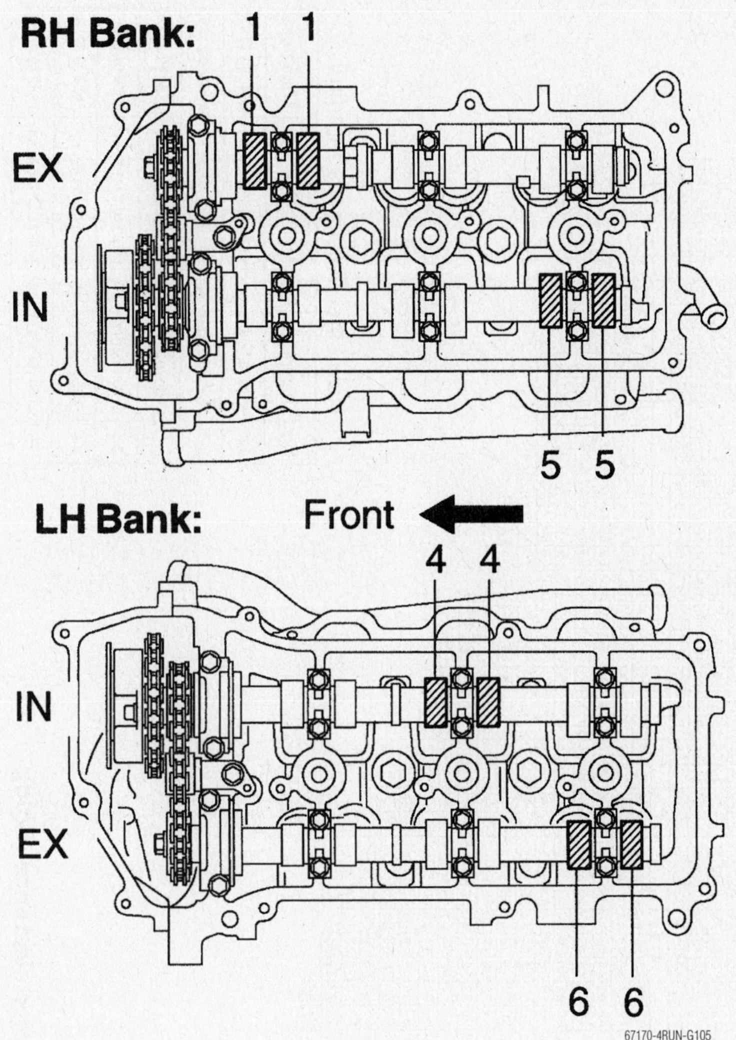

Valve clearance check, Figure C—4.0L engine

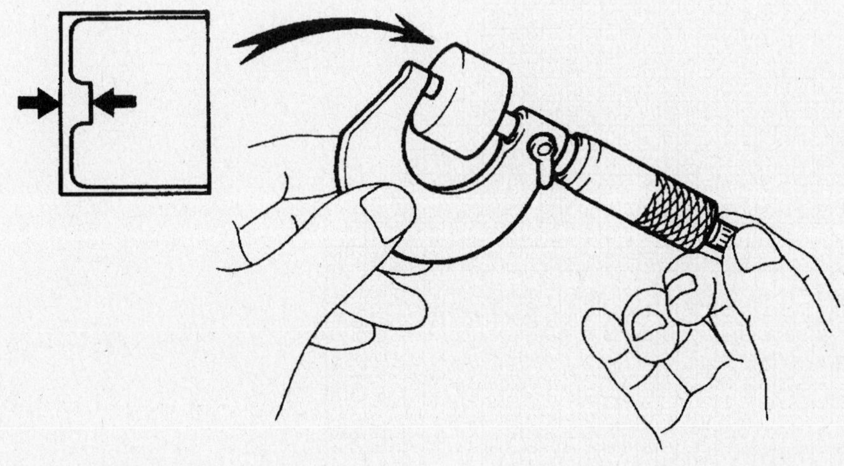

Measure the lifter at these points—4.0L engine

13. Turn the crankshaft ⅔ turn (240 degrees) clockwise. Check the valves shown in Figure C.
14. If any are not correct, record how far off they are. This will determine replacement lifter(s).
15. If adjustment is necessary, remove the camshaft(s).
16. Remove the lifters to be replaced. Measure the lifter at the points shown.
17. Determine the size of the replacement lifter using the accompanying charts. Lifters are available in 35 sizes, in increments of 0.020mm, from 5.060mm to 5.740mm.
18. Install the camshaft(s).
19. The remainder of installation is the reverse of removal.

4.7L Engine

1. Drain the coolant.
2. Remove the battery ground cable.
3. Remove the throttle body cover.

Valve Lifter Selection Chart (Intake)

Measure clearance mm (in.) \ Installed lifter thickness mm (in.)	5.060 (0.1992)	5.080 (0.2000)	5.100 (0.2008)	5.120 (0.2016)	5.140 (0.2024)	5.160 (0.2031)	5.180 (0.2039)	5.200 (0.2047)	5.210 (0.2051)	5.220 (0.2055)	5.230 (0.2059)	5.240 (0.2063)	5.250 (0.2067)	5.260 (0.2071)	5.270 (0.2075)	5.280 (0.2079)	5.290 (0.2083)	5.300 (0.2087)	5.310 (0.2091)	5.320 (0.2094)	5.330 (0.2098)	5.340 (0.2102)	5.350 (0.2106)	5.360 (0.2110)	5.370 (0.2114)	5.380 (0.2118)	5.390 (0.2122)	5.400 (0.2126)	5.410 (0.2130)	5.420 (0.2134)	5.430 (0.2138)	5.440 (0.2142)	5.450 (0.2146)	5.460 (0.2150)	5.470 (0.2154)	5.480 (0.2157)	5.490 (0.2161)	5.500 (0.2165)	5.510 (0.2169)	5.520 (0.2173)	5.530 (0.2177)	5.540 (0.2181)	5.550 (0.2185)	5.560 (0.2189)	5.570 (0.2193)	5.580 (0.2197)	5.590 (0.2201)	5.600 (0.2205)	5.620 (0.2213)	5.640 (0.2220)	5.660 (0.2228)	5.680 (0.2236)	5.700 (0.2244)	5.720 (0.2252)	5.740 (0.2260)
0.000 - 0.020 (0.0000 - 0.0008)													06	08	08	10	10	12	12	14	14	16	16	18	18	20	20	22	22	24	24	26	26	28	28	30	30	32	32	34	34	36	36	38	38	40	40	42	44	46	48	50	52	54	56
0.021 - 0.040 (0.0008 - 0.0016)											06	08	08	10	10	12	12	14	14	16	16	18	18	20	20	22	22	24	24	26	26	28	28	30	30	32	32	34	34	36	36	38	38	40	40	42	42	44	46	48	50	52	54	56	58
0.041 - 0.060 (0.0016 - 0.0024)									06	08	08	10	10	12	12	14	14	16	16	18	18	20	20	22	22	24	24	26	26	28	28	30	30	32	32	34	34	36	36	38	38	40	40	42	42	44	44	46	48	50	52	54	56	58	60
0.061 - 0.080 (0.0024 - 0.0031)								08	08	10	10	12	12	14	14	16	16	18	18	20	20	22	22	24	24	26	26	28	28	30	30	32	32	34	34	36	36	38	38	40	40	42	42	44	44	46	46	48	50	52	54	56	58	60	62
0.081 - 0.100 (0.0032 - 0.0039)							08	10	10	12	12	14	14	16	16	18	18	20	20	22	22	24	24	26	26	28	28	30	30	32	32	34	34	36	36	38	38	40	40	42	42	44	44	46	46	48	48	50	52	54	56	58	60	62	64
0.101 - 0.120 (0.0040 - 0.0047)						08	10	12	12	14	14	16	16	18	18	20	20	22	22	24	24	26	26	28	28	30	30	32	32	34	34	36	36	38	38	40	40	42	42	44	44	46	46	48	48	50	50	52	54	56	58	60	62	64	66
0.121 - 0.140 (0.0048 - 0.0055)					08	10	12	14	14	16	16	18	18	20	20	22	22	24	24	26	26	28	28	30	30	32	32	34	34	36	36	38	38	40	40	42	42	44	44	46	46	48	48	50	50	52	52	54	56	58	60	62	64	66	68
0.141 - 0.149 (0.0056 - 0.0059)				06	08	10	12	14	16	16	18	18	20	20	22	22	24	24	26	26	28	28	30	30	32	32	34	34	36	36	38	38	40	40	42	42	44	44	46	46	48	48	50	50	52	52	54	54	56	58	60	62	64	66	68
0.251 - 0.270 (0.0099 - 0.0106)	12	14	16	18	20	22	24	26	28	28	30	30	32	32	34	34	36	36	38	38	40	40	42	42	44	44	46	46	48	48	50	50	52	52	54	54	56	56	58	58	60	60	62	62	64	64	66	66	68	70	72				
0.271 - 0.290 (0.0107 - 0.0114)	14	16	18	20	22	24	26	28	30	30	32	32	34	34	36	36	38	38	40	40	42	42	44	44	46	46	48	48	50	50	52	52	54	54	56	56	58	58	60	60	62	62	64	64	66	66	68	68	70	72					
0.291 - 0.310 (0.0115 - 0.0122)	16	18	20	22	24	26	28	30	32	32	34	34	36	36	38	38	40	40	42	42	44	44	46	46	48	48	50	50	52	52	54	54	56	56	58	58	60	60	62	62	64	64	66	66	68	68	70	70	72						
0.311 - 0.330 (0.0122 - 0.0130)	18	20	22	24	26	28	30	32	34	34	36	36	38	38	40	40	42	42	44	44	46	46	48	48	50	50	52	52	54	54	56	56	58	58	60	60	62	62	64	64	66	66	68	68	70	70	72	72							
0.331 - 0.350 (0.0130 - 0.0138)	20	22	24	26	28	30	32	34	36	36	38	38	40	40	42	42	44	44	46	46	48	48	50	50	52	52	54	54	56	56	58	58	60	60	62	62	64	64	66	66	68	68	70	70	72	72	74								
0.351 - 0.370 (0.0138 - 0.0146)	22	24	26	28	30	32	34	36	38	38	40	40	42	42	44	44	46	46	48	48	50	50	52	52	54	54	56	56	58	58	60	60	62	62	64	64	66	66	68	68	70	70	72	72	74										
0.371 - 0.390 (0.0146 - 0.0154)	24	26	28	30	32	34	36	38	40	40	42	42	44	44	46	46	48	48	50	50	52	52	54	54	56	56	58	58	60	60	62	62	64	64	66	66	68	68	70	70	72	72	74												
0.391 - 0.410 (0.0154 - 0.0161)	26	28	30	32	34	36	38	40	42	42	44	44	46	46	48	48	50	50	52	52	54	54	56	56	58	58	60	60	62	62	64	64	66	66	68	68	70	70	72	72	74														
0.411 - 0.430 (0.0162 - 0.0169)	28	30	32	34	36	38	40	42	44	44	46	46	48	48	50	50	52	52	54	54	56	56	58	58	60	60	62	62	64	64	66	66	68	68	70	70	72	72	74																
0.431 - 0.450 (0.0170 - 0.0177)	30	32	34	36	38	40	42	44	46	46	48	48	50	50	52	52	54	54	56	56	58	58	60	60	62	62	64	64	66	66	68	68	70	70	72	72	74																		
0.451 - 0.470 (0.0178 - 0.0185)	32	34	36	38	40	42	44	46	48	48	50	50	52	52	54	54	56	56	58	58	60	60	62	62	64	64	66	66	68	68	70	70	72	72	74																				
0.471 - 0.490 (0.0185 - 0.0193)	34	36	38	40	42	44	46	48	50	50	52	52	54	54	56	56	58	58	60	60	62	62	64	64	66	66	68	68	70	70	72	72	74																						
0.491 - 0.510 (0.0193 - 0.0201)	36	38	40	42	44	46	48	50	52	52	54	54	56	56	58	58	60	60	62	62	64	64	66	66	68	68	70	70	72	72	74																								
0.511 - 0.530 (0.0201 - 0.0209)	38	40	42	44	46	48	50	52	54	54	56	56	58	58	60	60	62	62	64	64	66	66	68	68	70	70	72	72	74																										
0.531 - 0.550 (0.0209 - 0.0217)	40	42	44	46	48	50	52	54	56	56	58	58	60	60	62	62	64	64	66	66	68	68	70	70	72	72	74																												
0.551 - 0.570 (0.0217 - 0.0224)	42	44	46	48	50	52	54	56	58	58	60	60	62	62	64	64	66	66	68	68	70	70	72	72	74																														
0.571 - 0.590 (0.0225 - 0.0232)	44	46	48	50	52	54	56	58	60	60	62	62	64	64	66	66	68	68	70	70	72	72	74																																
0.591 - 0.610 (0.0233 - 0.0240)	46	48	50	52	54	56	58	60	62	62	64	64	66	66	68	68	70	70	72	72	74																																		
0.611 - 0.630 (0.0241 - 0.0248)	48	50	52	54	56	58	60	62	64	64	66	66	68	68	70	70	72	72	74																																				
0.631 - 0.650 (0.0248 - 0.0256)	50	52	54	56	58	60	62	64	66	66	68	68	70	70	72	72	74																																						
0.651 - 0.670 (0.0256 - 0.0264)	52	54	56	58	60	62	64	66	68	68	70	70	72	72	74																																								
0.671 - 0.690 (0.0264 - 0.0272)	54	56	58	60	62	64	66	68	70	70	72	72	74																																										
0.691 - 0.710 (0.0272 - 0.0280)	56	58	60	62	64	66	68	70	72	72	74																																												
0.711 - 0.730 (0.0280 - 0.0287)	58	60	62	64	66	68	70	72	74																																														
0.731 - 0.750 (0.0288 - 0.0295)	60	62	64	66	68	70	72																																																
0.751 - 0.770 (0.0296 - 0.0303)	62	64	66	68	70	72																																																	
0.771 - 0.790 (0.0304 - 0.0311)	64	66	68	70	72																																																		
0.791 - 0.810 (0.0311 - 0.0319)	66	68	70	72																																																			
0.811 - 0.830 (0.0319 - 0.0327)	68	70	72																																																				
0.831 - 0.850 (0.0327 - 0.0335)	70	72																																																					
0.851 - 0.870 (0.0335 - 0.0343)	72																																																							

Intake valve clearance (Cold):
0.15 – 0.25 mm (0.0059 – 0.0098 in.)

EXAMPLE:
The 5.250 mm (0.2067 in.) lifter is installed, and the measured clearance is 0.400 mm (0.0158 in.). Replace the 5.250 mm (0.2067 in.) shim with a new No. 46 lifter.

New Lifter Thickness mm (in.)

No.	Thickness	No.	Thickness	No.	Thickness
06	5.060 (0.1992)	30	5.300 (0.2087)	54	5.540 (0.2181)
08	5.080 (0.2000)	32	5.320 (0.2094)	56	5.560 (0.2189)
10	5.100 (0.2008)	34	5.340 (0.2102)	58	5.580 (0.2197)
12	5.120 (0.2016)	36	5.360 (0.2110)	60	5.600 (0.2205)
14	5.140 (0.2024)	38	5.380 (0.2118)	62	5.620 (0.2213)
16	5.160 (0.2031)	40	5.400 (0.2126)	64	5.640 (0.2220)
18	5.180 (0.2039)	42	5.420 (0.2134)	66	5.660 (0.2228)
20	5.200 (0.2047)	44	5.440 (0.2142)	68	5.680 (0.2236)
22	5.220 (0.2055)	46	5.460 (0.2150)	70	5.700 (0.2244)
24	5.240 (0.2063)	48	5.480 (0.2157)	72	5.720 (0.2252)
26	5.260 (0.2071)	50	5.500 (0.2165)	74	5.740 (0.2260)
28	5.280 (0.2079)	52	5.520 (0.2173)		

67170-4RUN-G107

Intake valve lifter selection chart—4.0L engine

Valve Lifter Selection Chart (Exhaust)

Measure clearance mm (in.) / Installed lifter thickness mm (in.)	5.060 (0.1992)	5.080 (0.2000)	5.100 (0.2008)	5.120 (0.2016)	5.140 (0.2024)	5.160 (0.2031)	5.180 (0.2039)	5.200 (0.2047)	5.210 (0.2051)	5.220 (0.2055)	5.230 (0.2059)	5.240 (0.2063)	5.250 (0.2067)	5.260 (0.2071)	5.270 (0.2075)	5.280 (0.2079)	5.290 (0.2083)	5.300 (0.2087)	5.310 (0.2091)	5.320 (0.2094)	5.330 (0.2098)	5.340 (0.2102)	5.350 (0.2106)	5.360 (0.2110)	5.370 (0.2114)	5.380 (0.2118)	5.390 (0.2122)	5.400 (0.2126)	5.410 (0.2130)	5.420 (0.2134)	5.430 (0.2138)	5.440 (0.2142)	5.450 (0.2146)	5.460 (0.2150)	5.470 (0.2154)	5.480 (0.2157)	5.490 (0.2161)	5.500 (0.2165)	5.510 (0.2169)	5.520 (0.2173)	5.530 (0.2177)	5.540 (0.2181)	5.550 (0.2185)	5.560 (0.2189)	5.570 (0.2193)	5.580 (0.2197)	5.590 (0.2201)	5.600 (0.2205)	5.620 (0.2213)	5.640 (0.2220)	5.660 (0.2228)	5.680 (0.2236)	5.700 (0.2244)	5.720 (0.2252)	5.740 (0.2260)
0.000 - 0.020 (0.0000 - 0.0008)																											06	08	08	10	10	12	12	14	14	16	16	18	18	20	20	22	22	24	24	26	26	28	30	32	34	36	38	40	42
0.021 - 0.040 (0.0008 - 0.0016)																									06	08	08	10	10	12	12	14	14	16	16	18	18	20	20	22	22	24	24	26	26	28	28	30	32	34	36	38	40	42	44
0.041 - 0.060 (0.0016 - 0.0024)																							06	08	08	10	10	12	12	14	14	16	16	18	18	20	20	22	22	24	24	26	26	28	28	30	30	32	34	36	38	40	42	44	46
0.061 - 0.080 (0.0024 - 0.0031)																					06	08	08	10	10	12	12	14	14	16	16	18	18	20	20	22	22	24	24	26	26	28	28	30	30	32	32	34	36	38	40	42	44	46	48
0.081 - 0.100 (0.0032 - 0.0039)																			06	08	08	10	10	12	12	14	14	16	16	18	18	20	20	22	22	24	24	26	26	28	28	30	30	32	32	34	34	36	38	40	42	44	46	48	50
0.101 - 0.120 (0.0040 - 0.0047)																	06	08	08	10	10	12	12	14	14	16	16	18	18	20	20	22	22	24	24	26	26	28	28	30	30	32	32	34	34	36	36	38	40	42	44	46	48	50	52
0.121 - 0.140 (0.0048 - 0.0055)															06	08	08	10	10	12	12	14	14	16	16	18	18	20	20	22	22	24	24	26	26	28	28	30	30	32	32	34	34	36	36	38	38	40	42	44	46	48	50	52	54
0.141 - 0.160 (0.0056 - 0.0063)													06	08	08	10	10	12	12	14	14	16	16	18	18	20	20	22	22	24	24	26	26	28	28	30	30	32	32	34	34	36	36	38	38	40	40	42	44	46	48	50	52	54	56
0.161 - 0.180 (0.0063 - 0.0071)											06	08	08	10	10	12	12	14	14	16	16	18	18	20	20	22	22	24	24	26	26	28	28	30	30	32	32	34	34	36	36	38	38	40	40	42	42	44	46	48	50	52	54	56	58
0.181 - 0.200 (0.0071 - 0.0079)									06	08	08	10	10	12	12	14	14	16	16	18	18	20	20	22	22	24	24	26	26	28	28	30	30	32	32	34	34	36	36	38	38	40	40	42	42	44	44	46	48	50	52	54	56	58	60
0.201 - 0.220 (0.0079 - 0.0087)								08	08	10	10	12	12	14	14	16	16	18	18	20	20	22	22	24	24	26	26	28	28	30	30	32	32	34	34	36	36	38	38	40	40	42	42	44	44	46	46	48	50	52	54	56	58	60	62
0.221 - 0.240 (0.0087 - 0.0094)							08	10	10	12	12	14	14	16	16	18	18	20	20	22	22	24	24	26	26	28	28	30	30	32	32	34	34	36	36	38	38	40	40	42	42	44	44	46	46	48	48	50	52	54	56	58	60	62	64
0.241 - 0.260 (0.0095 - 0.0102)						08	10	12	12	14	14	16	16	18	18	20	20	22	22	24	24	26	26	28	28	30	30	32	32	34	34	36	36	38	38	40	40	42	42	44	44	46	46	48	48	50	50	52	54	56	58	60	62	64	66
0.261 - 0.280 (0.0103 - 0.0110)					08	10	12	14	14	16	16	18	18	20	20	22	22	24	24	26	26	28	28	30	30	32	32	34	34	36	36	38	38	40	40	42	42	44	44	46	46	48	48	50	50	52	52	54	56	58	60	62	64	66	68
0.281 - 0.289 (0.0111 - 0.0114)				06	08	10	12	14	16	16	18	18	20	20	22	22	24	24	26	26	28	28	30	30	32	32	34	34	36	36	38	38	40	40	42	42	44	44	46	46	48	48	50	50	52	52	54	54	56	58	60	62	64	66	68
0.391 - 0.410 (0.0154 - 0.0161)	12	14	16	18	20	22	24	26	28	28	30	30	32	32	34	34	36	36	38	38	40	40	42	42	44	44	46	46	48	48	50	50	52	52	54	54	56	56	58	58	60	60	62	62	64	64	66	66	68	70	72				
0.411 - 0.430 (0.0162 - 0.0169)	14	16	18	20	22	24	26	28	30	30	32	32	34	34	36	36	38	38	40	40	42	42	44	44	46	46	48	48	50	50	52	52	54	54	56	56	58	58	60	60	62	62	64	64	66	66	68	68	70	72					
0.431 - 0.450 (0.0170 - 0.0177)	16	18	20	22	24	26	28	30	32	32	34	34	36	36	38	38	40	40	42	42	44	44	46	46	48	48	50	50	52	52	54	54	56	56	58	58	60	60	62	62	64	64	66	66	68	68	70	70	72						
0.451 - 0.470 (0.0178 - 0.0185)	18	20	22	24	26	28	30	32	34	34	36	36	38	38	40	40	42	42	44	44	46	46	48	48	50	50	52	52	54	54	56	56	58	58	60	60	62	62	64	64	66	66	68	68	70	70	72	72							
0.471 - 0.490 (0.0185 - 0.0193)	20	22	24	26	28	30	32	34	36	36	38	38	40	40	42	42	44	44	46	46	48	48	50	50	52	52	54	54	56	56	58	58	60	60	62	62	64	64	66	66	68	68	70	70	72	72	74								
0.491 - 0.510 (0.0193 - 0.0201)	22	24	26	28	30	32	34	36	38	38	40	40	42	42	44	44	46	46	48	48	50	50	52	52	54	54	56	56	58	58	60	60	62	62	64	64	66	66	68	68	70	70	72	72	74										
0.511 - 0.530 (0.0201 - 0.0209)	24	26	28	30	32	34	36	38	40	40	42	42	44	44	46	46	48	48	50	50	52	52	54	54	56	56	58	58	60	60	62	62	64	64	66	66	68	68	70	70	72	72	74												
0.531 - 0.550 (0.0209 - 0.0217)	26	28	30	32	34	36	38	40	42	42	44	44	46	46	48	48	50	50	52	52	54	54	56	56	58	58	60	60	62	62	64	64	66	66	68	68	70	70	72	72	74														
0.551 - 0.570 (0.0217 - 0.0224)	28	30	32	34	36	38	40	42	44	44	46	46	48	48	50	50	52	52	54	54	56	56	58	58	60	60	62	62	64	64	66	66	68	68	70	70	72	72	74																
0.571 - 0.590 (0.0225 - 0.0232)	30	32	34	36	38	40	42	44	46	46	48	48	50	50	52	52	54	54	56	56	58	58	60	60	62	62	64	64	66	66	68	68	70	70	72	72	74																		
0.591 - 0.610 (0.0233 - 0.0240)	32	34	36	38	40	42	44	46	48	48	50	50	52	52	54	54	56	56	58	58	60	60	62	62	64	64	66	66	68	68	70	70	72	72	74																				
0.611 - 0.630 (0.0241 - 0.0248)	34	36	38	40	42	44	46	48	50	50	52	52	54	54	56	56	58	58	60	60	62	62	64	64	66	66	68	68	70	70	72	72	74																						
0.631 - 0.650 (0.0248 - 0.0256)	36	38	40	42	44	46	48	50	52	52	54	54	56	56	58	58	60	60	62	62	64	64	66	66	68	68	70	70	72	72	74																								
0.651 - 0.670 (0.0256 - 0.0264)	38	40	42	44	46	48	50	52	54	54	56	56	58	58	60	60	62	62	64	64	66	66	68	68	70	70	72	72	74																										
0.671 - 0.690 (0.0264 - 0.0272)	40	42	44	46	48	50	52	54	56	56	58	58	60	60	62	62	64	64	66	66	68	68	70	70	72	72	74																												
0.691 - 0.710 (0.0272 - 0.0280)	42	44	46	48	50	52	54	56	58	58	60	60	62	62	64	64	66	66	68	68	70	70	72	72	74																														
0.711 - 0.730 (0.0280 - 0.0287)	44	46	48	50	52	54	56	58	60	60	62	62	64	64	66	66	68	68	70	70	72	72	74																																
0.731 - 0.750 (0.0288 - 0.0295)	46	48	50	52	54	56	58	60	62	62	64	64	66	66	68	68	70	70	72	72	74																																		
0.751 - 0.770 (0.0296 - 0.0303)	48	50	52	54	56	58	60	62	64	64	66	66	68	68	70	70	72	72	74																																				
0.771 - 0.790 (0.0304 - 0.0311)	50	52	54	56	58	60	62	64	66	66	68	68	70	70	72	72	74																																						
0.791 - 0.810 (0.0311 - 0.0319)	52	54	56	58	60	62	64	66	68	68	70	70	72	72	74																																								
0.811 - 0.830 (0.0319 - 0.0327)	54	56	58	60	62	64	66	68	70	70	72	72	74																																										
0.831 - 0.850 (0.0327 - 0.0335)	56	58	60	62	64	66	68	70	72	72	74																																												
0.851 - 0.870 (0.0335 - 0.0343)	58	60	62	64	66	68	70	72	74																																														
0.871 - 0.890 (0.0343 - 0.0350)	60	62	64	66	68	70	72																																																
0.891 - 0.910 (0.0351 - 0.0358)	62	64	66	68	70	72																																																	
0.911 - 0.930 (0.0359 - 0.0366)	64	66	68	70	72																																																		
0.931 - 0.950 (0.0367 - 0.0374)	66	68	70	72																																																			
0.951 - 0.970 (0.0374 - 0.0382)	68	70	72																																																				
0.971 - 0.990 (0.0382 - 0.0390)	70	72																																																					
0.991 - 1.010 (0.0390 - 0.0398)	72																																																						

Exhaust valve clearance (Cold):
0.29 – 0.39 mm (0.0114 – 0.0154 in.)

EXAMPLE:
The 5.340 mm (0.2102 in.) lifter is installed, and the measured clearance is 0.480 mm (0.0189 in.). Replace the 5.340 mm (0.2102 in.) shim with a new No. 48 lifter.

New Lifter Thickness mm (in.)

No.	Thickness	No.	Thickness	No.	Thickness
06	5.060 (0.1992)	30	5.300 (0.2087)	54	5.540 (0.2181)
08	5.080 (0.2000)	32	5.320 (0.2094)	56	5.560 (0.2189)
10	5.100 (0.2008)	34	5.340 (0.2102)	58	5.580 (0.2197)
12	5.120 (0.2016)	36	5.360 (0.2110)	60	5.600 (0.2205)
14	5.140 (0.2024)	38	5.380 (0.2118)	62	5.620 (0.2213)
16	5.160 (0.2031)	40	5.400 (0.2126)	64	5.640 (0.2220)
18	5.180 (0.2039)	42	5.420 (0.2134)	66	5.660 (0.2228)
20	5.200 (0.2047)	44	5.440 (0.2142)	68	5.680 (0.2236)
22	5.220 (0.2055)	46	5.460 (0.2150)	70	5.700 (0.2244)
24	5.240 (0.2063)	48	5.480 (0.2157)	72	5.720 (0.2252)
26	5.260 (0.2071)	50	5.500 (0.2165)	74	5.740 (0.2260)
28	5.280 (0.2079)	52	5.520 (0.2173)		

67170-4RUN-G108

Exhaust valve lifter selection chart—4.0L engine

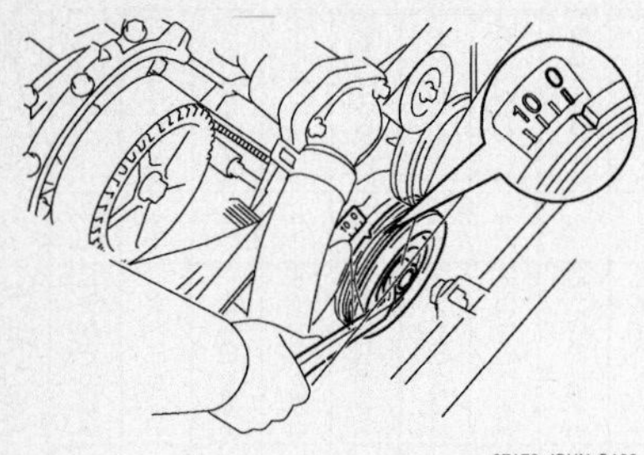

Turn the crankshaft clockwise and align its groove with the "0" mark on the timing cover—4.7L engine

4. Remove the air cleaner.
5. Remove the radiator upper support seal.
6. Remove the accessory drive belt.
7. Remove the oil cooler pipe.
8. Remove the timing belt covers.
9. Remove the cylinder head covers.
10. Turn the crankshaft clockwise and align its groove with the "0" mark on the timing cover.
11. Check that the timing marks on the camshaft sprockets are aligned with the timing marks on the bearings caps, as shown. If not, rotate the crankshaft 360 degrees clockwise.
12. Check the valve clearance on the valves indicated in Figure A. Valve clearance should be:
    - Intake: 0.006–0.010 inch (0.15–0.25mm)
    - Exhaust: 0.010–0.014 inch (0.25–0.35mm)
13. If any valve is not within specifications, record how far out it is. This will determine the replacement shim.
14. Turn the crankshaft 1 full turn (360 degrees) clockwise.
15. Check the valve clearance on the valves indicated in Figure B. Valve clearance should be:
    - Intake: 0.006–0.010 inch (0.15–0.25mm)
    - Exhaust: 0.010–0.014 inch (0.25–0.35mm)

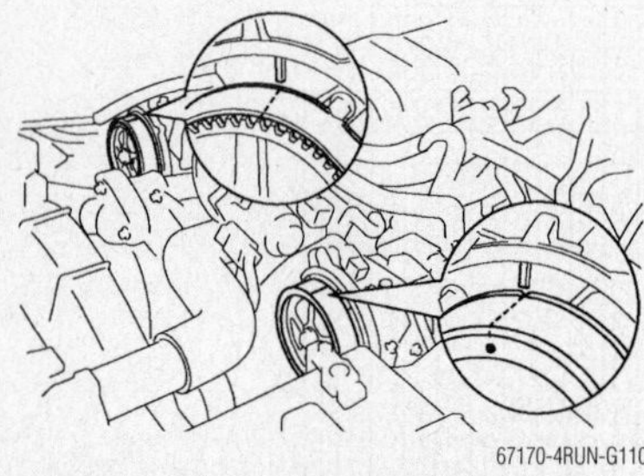

Check that the timing marks on the camshaft sprockets are aligned with the timing marks on the bearings caps—4.7L engine

RH Cylinder Head

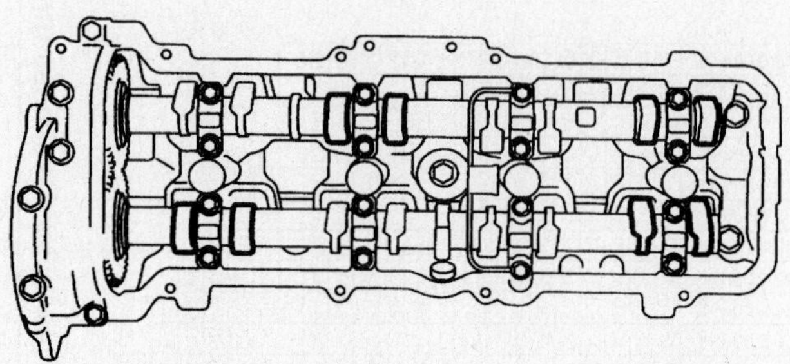

LH Cylinder Head

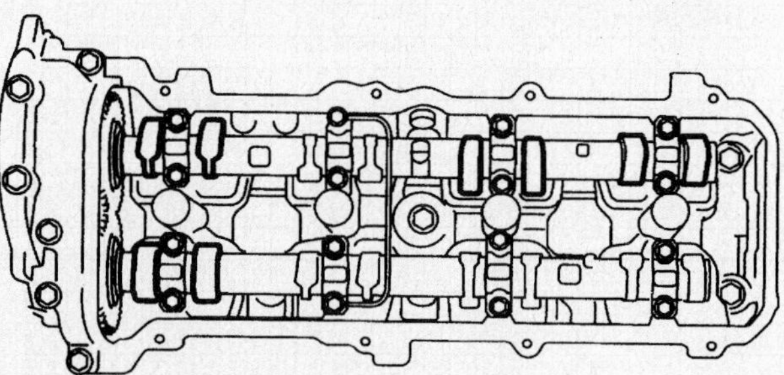

Valve clearance check, Figure A—4.7L engine

RH Cylinder Head

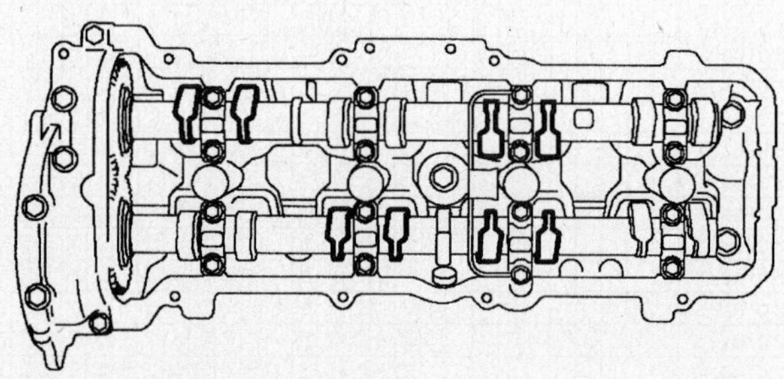

LH Cylinder Head

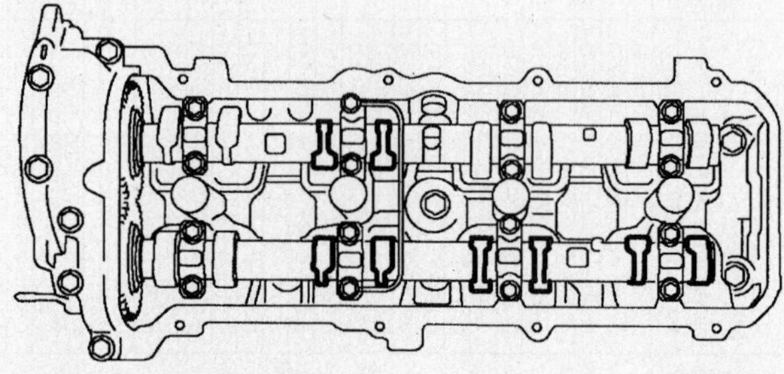

Valve clearance check, Figure B—4.7L engine

Adjusting Shim Selection Chart (Intake)

Measured clearance mm (in.) \ Installed shim thickness mm (in.)	2.000 (0.0787)	2.020 (0.0795)	2.040 (0.0803)	2.060 (0.0811)	2.080 (0.0819)	2.100 (0.0827)	2.120 (0.0835)	2.140 (0.0843)	2.160 (0.0850)	2.180 (0.0858)	2.200 (0.0866)	2.210 (0.0870)	2.220 (0.0874)	2.230 (0.0878)	2.240 (0.0882)	2.250 (0.0886)	2.260 (0.0890)	2.270 (0.0894)	2.280 (0.0898)	2.290 (0.0902)	2.300 (0.0906)	2.310 (0.0909)	2.320 (0.0913)	2.330 (0.0917)	2.340 (0.0921)	2.350 (0.0925)	2.360 (0.0929)	2.370 (0.0933)	2.380 (0.0937)	2.390 (0.0941)	2.400 (0.0945)	2.410 (0.0949)	2.420 (0.0953)	2.430 (0.0957)	2.440 (0.0961)	2.450 (0.0965)	2.460 (0.0969)	2.470 (0.0972)	2.480 (0.0976)	2.490 (0.0980)	2.500 (0.0984)	2.510 (0.0988)	2.520 (0.0992)	2.530 (0.0996)	2.540 (0.1000)	2.550 (0.1004)	2.560 (0.1008)	2.570 (0.1012)	2.580 (0.1016)	2.590 (0.1020)	2.600 (0.1024)	2.620 (0.1031)	2.640 (0.1039)	2.660 (0.1047)	2.680 (0.1055)	2.700 (0.1063)	2.720 (0.1071)	2.740 (0.1079)	2.760 (0.1087)	2.780 (0.1094)	2.800 (0.1102)
0.000-0.030 (0.0000-0.0012)							00	00	00	00	02	02	04	04	06	06	08	08	10	10	12	12	14	14	16	16	18	18	20	20	22	22	24	24	26	26	28	28	30	30	32	32	34	34	36	36	38	38	40	40	42	44	46	48	50	52	54	56	58	60	62
0.031-0.050 (0.0012-0.0020)						00	00	00	00	02	04	06	06	08	08	10	10	12	12	14	14	16	16	18	18	20	20	22	22	24	24	26	26	28	28	30	30	32	32	34	34	36	36	38	38	40	40	42	42	44	44	46	48	50	52	54	56	58	60	62	64
0.051-0.070 (0.0020-0.0028)					00	00	00	00	02	04	06	08	08	10	10	12	12	14	14	16	16	18	18	20	20	22	22	24	24	26	26	28	28	30	30	32	32	34	34	36	36	38	38	40	40	42	42	44	44	46	46	48	50	52	54	56	58	60	62	64	66
0.071-0.090 (0.0028-0.0035)				00	00	00	00	02	04	06	08	10	10	12	12	14	14	16	16	18	18	20	20	22	22	24	24	26	26	28	28	30	30	32	32	34	34	36	36	38	38	40	40	42	42	44	44	46	46	48	48	50	52	54	56	58	60	62	64	66	68
0.091-0.110 (0.0036-0.0043)			00	00	00	00	02	04	06	08	10	12	12	14	14	16	16	18	18	20	20	22	22	24	24	26	26	28	28	30	30	32	32	34	34	36	36	38	38	40	40	42	42	44	44	46	46	48	48	50	50	52	54	56	58	60	62	64	66	68	70
0.111-0.130 (0.0044-0.0051)		00	00	00	00	02	04	06	08	10	12	14	14	16	16	18	18	20	20	22	22	24	24	26	26	28	28	30	30	32	32	34	34	36	36	38	38	40	40	42	42	44	44	46	46	48	48	50	50	52	52	54	56	58	60	62	64	66	68	70	72
0.131-0.149 (0.0052-0.0059)		00	00	00	02	04	06	08	10	12	14	16	16	18	18	20	20	22	22	24	24	26	26	28	28	30	30	32	32	34	34	36	36	38	38	40	40	42	42	44	44	46	46	48	48	50	50	52	52	54	54	56	58	60	62	64	66	68	70	72	74
0.150-0.250 (0.0059-0.0098)																																																													
0.251-0.270 (0.0099-0.0106)	06	08	10	12	14	16	18	20	22	24	26	28	28	30	30	32	32	34	34	36	36	38	38	40	40	42	42	44	44	46	46	48	48	50	50	52	52	54	54	56	56	58	58	60	60	62	62	64	64	66	66	68	70	72	74	76	78	80	80	80	
0.271-0.290 (0.0107-0.0114)	08	10	12	14	16	18	20	22	24	26	28	30	30	32	32	34	34	36	36	38	38	40	40	42	42	44	44	46	46	48	48	50	50	52	52	54	54	56	56	58	58	60	60	62	62	64	64	66	66	68	68	70	72	74	76	78	80	80	80		
0.291-0.310 (0.0115-0.0122)	10	12	14	16	18	20	22	24	26	28	30	32	32	34	34	36	36	38	38	40	40	42	42	44	44	46	46	48	48	50	50	52	52	54	54	56	56	58	58	60	60	62	62	64	64	66	66	68	68	70	70	72	74	76	78	80	80	80			
0.311-0.330 (0.0122-0.0130)	12	14	16	18	20	22	24	26	28	30	32	34	34	36	36	38	38	40	40	42	42	44	44	46	46	48	48	50	50	52	52	54	54	56	56	58	58	60	60	62	62	64	64	66	66	68	68	70	70	72	72	74	76	78	80	80	80				
0.331-0.350 (0.0130-0.0138)	14	16	18	20	22	24	26	28	30	32	34	36	36	38	38	40	40	42	42	44	44	46	46	48	48	50	50	52	52	54	54	56	56	58	58	60	60	62	62	64	64	66	66	68	68	70	70	72	72	74	74	76	78	80	80	80					
0.351-0.370 (0.0138-0.0146)	16	18	20	22	24	26	28	30	32	34	36	38	38	40	40	42	42	44	44	46	46	48	48	50	50	52	52	54	54	56	56	58	58	60	60	62	62	64	64	66	66	68	68	70	70	72	72	74	74	76	76	78	80	80	80						
0.371-0.390 (0.0146-0.0154)	18	20	22	24	26	28	30	32	34	36	38	40	40	42	42	44	44	46	46	48	48	50	50	52	52	54	54	56	56	58	58	60	60	62	62	64	64	66	66	68	68	70	70	72	72	74	74	76	76	78	78	80	80	80							
0.391-0.410 (0.0154-0.0161)	20	22	24	26	28	30	32	34	36	38	40	42	42	44	44	46	46	48	48	50	50	52	52	54	54	56	56	58	58	60	60	62	62	64	64	66	66	68	68	70	70	72	72	74	74	76	76	78	78	80	80	80	80								
0.411-0.430 (0.0162-0.0169)	22	24	26	28	30	32	34	36	38	40	42	44	44	46	46	48	48	50	50	52	52	54	54	56	56	58	58	60	60	62	62	64	64	66	66	68	68	70	70	72	72	74	74	76	76	78	78	80	80	80	80	80									
0.431-0.450 (0.0170-0.0177)	24	26	28	30	32	34	36	38	40	42	44	46	46	48	48	50	50	52	52	54	54	56	56	58	58	60	60	62	62	64	64	66	66	68	68	70	70	72	72	74	74	76	76	78	78	80	80	80	80	80	80										
0.451-0.470 (0.0178-0.0185)	26	28	30	32	34	36	38	40	42	44	46	48	48	50	50	52	52	54	54	56	56	58	58	60	60	62	62	64	64	66	66	68	68	70	70	72	72	74	74	76	76	78	78	80	80	80	80	80	80	80											
0.471-0.490 (0.0185-0.0193)	28	30	32	34	36	38	40	42	44	46	48	50	50	52	52	54	54	56	56	58	58	60	60	62	62	64	64	66	66	68	68	70	70	72	72	74	74	76	76	78	78	80	80	80	80	80	80	80													
0.491-0.510 (0.0193-0.0201)	30	32	34	36	38	40	42	44	46	48	50	52	52	54	54	56	56	58	58	60	60	62	62	64	64	66	66	68	68	70	70	72	72	74	74	76	76	78	78	80	80	80	80	80	80	80															
0.511-0.530 (0.0201-0.0209)	32	34	36	38	40	42	44	46	48	50	52	54	54	56	56	58	58	60	60	62	62	64	64	66	66	68	68	70	70	72	72	74	74	76	76	78	78	80	80	80	80	80	80	80																	
0.531-0.550 (0.0209-0.0217)	34	36	38	40	42	44	46	48	50	52	54	56	56	58	58	60	60	62	62	64	64	66	66	68	68	70	70	72	72	74	74	76	76	78	78	80	80	80	80	80	80	80																			
0.551-0.570 (0.0217-0.0224)	36	38	40	42	44	46	48	50	52	54	56	58	58	60	60	62	62	64	64	66	66	68	68	70	70	72	72	74	74	76	76	78	78	80	80	80	80	80	80	80																					
0.571-0.590 (0.0225-0.0232)	38	40	42	44	46	48	50	52	54	56	58	60	60	62	62	64	64	66	66	68	68	70	70	72	72	74	74	76	76	78	78	80	80	80	80	80	80	80																							
0.591-0.610 (0.0233-0.0240)	40	42	44	46	48	50	52	54	56	58	60	62	62	64	64	66	66	68	68	70	70	72	72	74	74	76	76	78	78	80	80	80	80	80	80	80																									
0.611-0.630 (0.0241-0.0248)	42	44	46	48	50	52	54	56	58	60	62	64	64	66	66	68	68	70	70	72	72	74	74	76	76	78	78	80	80	80	80	80	80	80																											
0.631-0.650 (0.0248-0.0256)	44	46	48	50	52	54	56	58	60	62	64	66	66	68	68	70	70	72	72	74	74	76	76	78	78	80	80	80	80	80	80	80																													
0.651-0.670 (0.0256-0.0264)	46	48	50	52	54	56	58	60	62	64	66	68	68	70	70	72	72	74	74	76	76	78	78	80	80	80	80	80	80	80																															
0.671-0.690 (0.0264-0.0272)	48	50	52	54	56	58	60	62	64	66	68	70	70	72	72	74	74	76	76	78	78	80	80	80	80	80	80	80																																	
0.691-0.710 (0.0272-0.0280)	50	52	54	56	58	60	62	64	66	68	70	72	72	74	74	76	76	78	78	80	80	80	80	80	80	80																																			
0.711-0.730 (0.0280-0.0287)	52	54	56	58	60	62	64	66	68	70	72	74	74	76	76	78	78	80	80	80	80	80	80	80																																					
0.731-0.750 (0.0288-0.0295)	54	56	58	60	62	64	66	68	70	72	74	76	76	78	78	80	80	80	80	80	80	80																																							
0.751-0.770 (0.0296-0.0303)	56	58	60	62	64	66	68	70	72	74	76	78	78	80	80	80	80	80	80	80																																									
0.771-0.790 (0.0304-0.0311)	58	60	62	64	66	68	70	72	74	76	78	80	80	80	80	80	80	80																																											
0.791-0.810 (0.0311-0.0319)	60	62	64	66	68	70	72	74	76	78	80	80	80	80	80	80																																													
0.811-0.830 (0.0319-0.0327)	62	64	66	68	70	72	74	76	78	80	80	80	80	80																																															
0.831-0.850 (0.0327-0.0335)	64	66	68	70	72	74	76	78	80	80	80	80																																																	
0.851-0.870 (0.0335-0.0343)	66	68	70	72	74	76	78	80	80	80																																																			
0.871-0.890 (0.0343-0.0350)	68	70	72	74	76	78	80	80	80																																																				
0.891-0.910 (0.0351-0.0358)	70	72	74	76	78	80	80	80																																																					
0.911-0.930 (0.0359-0.0366)	72	74	76	78	80	80	80																																																							
0.931-0.950 (0.0367-0.0374)	74	76	78	80	80	80																																																								
0.951-0.970 (0.0374-0.0382)	76	78	80	80	80																																																									
0.971-0.990 (0.0382-0.0390)	78	80	80	80																																																										
0.991-1.010 (0.0390-0.0398)	80	80	80																																																											
1.011-1.030 (0.0398-0.0406)	80	80																																																												
1.031-1.050 (0.0406-0.0413)	80																																																													

Intake valve clearance (Cold):
0.15 – 0.25 mm (0.006 – 0.010 in.)

EXAMPLE:
The 2.300mm (0.0906 in.) shim is installed, and the measured clearance is 0.440 mm (0.0173 in.). Replace the 2.300 mm (0.0906 in.) shim with a No. 54 shim.

Shim No.	Thickness	Shim No.	Thickness	Shim No.	Thickness
00	2.000 (0.0787)	28	2.280 (0.0898)	56	2.560 (0.1008)
02	2.020 (0.0795)	30	2.300 (0.0906)	58	2.580 (0.1016)
04	2.040 (0.0803)	32	2.320 (0.0913)	60	2.600 (0.1024)
06	2.060 (0.0811)	34	2.340 (0.0921)	62	2.620 (0.1031)
08	2.080 (0.0819)	36	2.360 (0.0929)	64	2.640 (0.1039)
10	2.100 (0.0827)	38	2.380 (0.0937)	66	2.660 (0.1047)
12	2.120 (0.0835)	40	2.400 (0.0945)	68	2.680 (0.1055)
14	2.140 (0.0843)	42	2.420 (0.0953)	70	2.700 (0.1063)
16	2.160 (0.0850)	44	2.440 (0.0961)	72	2.720 (0.1071)
18	2.180 (0.0858)	46	2.460 (0.0969)	74	2.740 (0.1079)
20	2.200 (0.0866)	48	2.480 (0.0976)	76	2.760 (0.1087)
22	2.220 (0.0874)	50	2.500 (0.0984)	78	2.780 (0.1094)
24	2.240 (0.0882)	52	2.520 (0.0992)	80	2.800 (0.1102)
26	2.260 (0.0890)	54	2.540 (0.1000)		

67170-4RUN-G113

Intake valve shim selection chart—4.7L engine

Adjusting Shim Selection Chart (Exhaust)

Measured clearance mm (in.) \ Installed shim thickness mm (in.)	2.000 (0.0787)	2.020 (0.0795)	2.040 (0.0803)	2.060 (0.0811)	2.080 (0.0819)	2.100 (0.0827)	2.120 (0.0835)	2.140 (0.0843)	2.160 (0.0850)	2.180 (0.0858)	2.200 (0.0866)	2.210 (0.0870)	2.220 (0.0874)	2.230 (0.0878)	2.240 (0.0882)	2.250 (0.0886)	2.260 (0.0890)	2.270 (0.0894)	2.280 (0.0898)	2.290 (0.0902)	2.300 (0.0906)	2.310 (0.0909)	2.320 (0.0913)	2.330 (0.0917)	2.340 (0.0921)	2.350 (0.0925)	2.360 (0.0929)	2.370 (0.0933)	2.380 (0.0937)	2.390 (0.0941)	2.400 (0.0945)	2.410 (0.0949)	2.420 (0.0953)	2.430 (0.0957)	2.440 (0.0961)	2.450 (0.0965)	2.460 (0.0969)	2.470 (0.0972)	2.480 (0.0976)	2.490 (0.0980)	2.500 (0.0984)	2.510 (0.0988)	2.520 (0.0992)	2.530 (0.0996)	2.540 (0.1000)	2.550 (0.1004)	2.560 (0.1008)	2.570 (0.1012)	2.580 (0.1016)	2.590 (0.1020)	2.600 (0.1024)	2.620 (0.1031)	2.640 (0.1039)	2.660 (0.1047)	2.680 (0.1055)	2.700 (0.1063)	2.720 (0.1071)	2.740 (0.1079)	2.760 (0.1087)	2.780 (0.1094)	2.800 (0.1102)
0.000-0.030 (0.0000-0.0012)												00	00	00	00	00	00	00	00	00	02	02	04	04	06	06	08	08	10	10	12	12	14	14	16	16	18	18	20	20	22	22	24	24	26	26	28	28	30	30	32	34	36	38	40	42	44	46	48	50	52
0.031-0.050 (0.0012-0.0020)											00	00	00	00	00	00	00	02	02	04	04	06	06	08	08	10	10	12	12	14	14	16	16	18	18	20	20	22	22	24	24	26	26	28	28	30	30	32	32	34	34	36	38	40	42	44	46	48	50	52	54
0.051-0.070 (0.0020-0.0028)										00	00	00	00	00	00	02	02	04	04	06	06	08	08	10	10	12	12	14	14	16	16	18	18	20	20	22	22	24	24	26	26	28	28	30	30	32	32	34	34	36	36	38	40	42	44	46	48	50	52	54	56
0.071-0.090 (0.0028-0.0035)									00	00	00	00	00	02	02	04	04	06	06	08	08	10	10	12	12	14	14	16	16	18	18	20	20	22	22	24	24	26	26	28	28	30	30	32	32	34	34	36	36	38	38	40	42	44	46	48	50	52	54	56	58
0.091-0.110 (0.0036-0.0043)								00	00	00	00	02	02	04	04	06	06	08	08	10	10	12	12	14	14	16	16	18	18	20	20	22	22	24	24	26	26	28	28	30	30	32	32	34	34	36	36	38	38	40	40	42	44	46	48	50	52	54	56	58	60
0.111-0.130 (0.0044-0.0051)							00	00	00	00	02	04	04	06	06	08	08	10	10	12	12	14	14	16	16	18	18	20	20	22	22	24	24	26	26	28	28	30	30	32	32	34	34	36	36	38	38	40	40	42	42	44	46	48	50	52	54	56	58	60	62
0.131-0.150 (0.0052-0.0059)						00	00	00	00	02	04	06	06	08	08	10	10	12	12	14	14	16	16	18	18	20	20	22	22	24	24	26	26	28	28	30	30	32	32	34	34	36	36	38	38	40	40	42	42	44	44	46	48	50	52	54	56	58	60	62	64
0.151-0.170 (0.0059-0.0067)					00	00	00	00	02	04	06	08	08	10	10	12	12	14	14	16	16	18	18	20	20	22	22	24	24	26	26	28	28	30	30	32	32	34	34	36	36	38	38	40	40	42	42	44	44	46	46	48	50	52	54	56	58	60	62	64	66
0.171-0.190 (0.0067-0.0075)				00	00	00	00	02	04	06	08	10	10	12	12	14	14	16	16	18	18	20	20	22	22	24	24	26	26	28	28	30	30	32	32	34	34	36	36	38	38	40	40	42	42	44	44	46	46	48	48	50	52	54	56	58	60	62	64	66	68
0.191-0.210 (0.0075-0.0083)			00	00	00	00	02	04	06	08	10	12	12	14	14	16	16	18	18	20	20	22	22	24	24	26	26	28	28	30	30	32	32	34	34	36	36	38	38	40	40	42	42	44	44	46	46	48	48	50	50	52	54	56	58	60	62	64	66	68	70
0.211-0.230 (0.0083-0.0091)		00	00	00	00	02	04	06	08	10	12	14	14	16	16	18	18	20	20	22	22	24	24	26	26	28	28	30	30	32	32	34	34	36	36	38	38	40	40	42	42	44	44	46	46	48	48	50	50	52	52	54	56	58	60	62	64	66	68	70	72
0.231-0.249 (0.0091-0.0098)		00	00	00	02	04	06	08	10	12	14	16	16	18	18	20	20	22	22	24	24	26	26	28	28	30	30	32	32	34	34	36	36	38	38	40	40	42	42	44	44	46	46	48	48	50	50	52	52	54	54	56	58	60	62	64	66	68	70	72	74
0.250-0.350 (0.0098-0.0138)																																																													
0.351-0.370 (0.0138-0.0146)	06	08	10	12	14	16	18	20	22	24	26	28	28	30	30	32	32	34	34	36	36	38	38	40	40	42	42	44	44	46	46	48	48	50	50	52	52	54	54	56	56	58	58	60	60	62	62	64	64	66	66	68	70	72	74	76	78	80	80	80	
0.371-0.390 (0.0146-0.0154)	08	10	12	14	16	18	20	22	24	26	28	30	30	32	32	34	34	36	36	38	38	40	40	42	42	44	44	46	46	48	48	50	50	52	52	54	54	56	56	58	58	60	60	62	62	64	64	66	66	68	68	70	72	74	76	78	80	80	80		
0.391-0.410 (0.0154-0.0161)	10	12	14	16	18	20	22	24	26	28	30	32	32	34	34	36	36	38	38	40	40	42	42	44	44	46	46	48	48	50	50	52	52	54	54	56	56	58	58	60	60	62	62	64	64	66	66	68	68	70	70	72	74	76	78	80	80	80			
0.411-0.430 (0.0162-0.0169)	12	14	16	18	20	22	24	26	28	30	32	34	34	36	36	38	38	40	40	42	42	44	44	46	46	48	48	50	50	52	52	54	54	56	56	58	58	60	60	62	62	64	64	66	66	68	68	70	70	72	72	74	76	78	80	80	80				
0.431-0.450 (0.0170-0.0177)	14	16	18	20	22	24	26	28	30	32	34	36	36	38	38	40	40	42	42	44	44	46	46	48	48	50	50	52	52	54	54	56	56	58	58	60	60	62	62	64	64	66	66	68	68	70	70	72	72	74	74	76	78	80	80	80					
0.451-0.470 (0.0178-0.0185)	16	18	20	22	24	26	28	30	32	34	36	38	38	40	40	42	42	44	44	46	46	48	48	50	50	52	52	54	54	56	56	58	58	60	60	62	62	64	64	66	66	68	68	70	70	72	72	74	74	76	76	78	80	80	80						
0.471-0.490 (0.0185-0.0193)	18	20	22	24	26	28	30	32	34	36	38	40	40	42	42	44	44	46	46	48	48	50	50	52	52	54	54	56	56	58	58	60	60	62	62	64	64	66	66	68	68	70	70	72	72	74	74	76	76	78	78	80	80	80							
0.491-0.510 (0.0193-0.0201)	20	22	24	26	28	30	32	34	36	38	40	42	42	44	44	46	46	48	48	50	50	52	52	54	54	56	56	58	58	60	60	62	62	64	64	66	66	68	68	70	70	72	72	74	74	76	76	78	78	80	80	80	80								
0.511-0.530 (0.0201-0.0209)	22	24	26	28	30	32	34	36	38	40	42	44	44	46	46	48	48	50	50	52	52	54	54	56	56	58	58	60	60	62	62	64	64	66	66	68	68	70	70	72	72	74	74	76	76	78	78	80	80	80	80	80									
0.531-0.550 (0.0209-0.0217)	24	26	28	30	32	34	36	38	40	42	44	46	46	48	48	50	50	52	52	54	54	56	56	58	58	60	60	62	62	64	64	66	66	68	68	70	70	72	72	74	74	76	76	78	78	80	80	80	80	80	80										
0.551-0.570 (0.0217-0.0224)	26	28	30	32	34	36	38	40	42	44	46	48	48	50	50	52	52	54	54	56	56	58	58	60	60	62	62	64	64	66	66	68	68	70	70	72	72	74	74	76	76	78	78	80	80	80	80	80	80	80											
0.571-0.590 (0.0225-0.0232)	28	30	32	34	36	38	40	42	44	46	48	50	50	52	52	54	54	56	56	58	58	60	60	62	62	64	64	66	66	68	68	70	70	72	72	74	74	76	76	78	78	80	80	80	80	80	80	80
0.591-0.610 (0.0233-0.0240)	30	32	34	36	38	40	42	44	46	48	50	52	52	54	54	56	56	58	58	60	60	62	62	64	64	66	66	68	68	70	70	72	72	74	74	76	76	78	78	80	80	80	80	80	80	80
0.611-0.630 (0.0241-0.0248)	32	34	36	38	40	42	44	46	48	50	52	54	54	56	56	58	58	60	60	62	62	64	64	66	66	68	68	70	70	72	72	74	74	76	76	78	78	80	80	80	80	80	80	80
0.631-0.650 (0.0248-0.0256)	34	36	38	40	42	44	46	48	50	52	54	56	56	58	58	60	60	62	62	64	64	66	66	68	68	70	70	72	72	74	74	76	76	78	78	80	80	80	80	80	80	80
0.651-0.670 (0.0256-0.0264)	36	38	40	42	44	46	48	50	52	54	56	58	58	60	60	62	62	64	64	66	66	68	68	70	70	72	72	74	74	76	76	78	78	80	80	80	80	80	80	80
0.671-0.690 (0.0264-0.0272)	38	40	42	44	46	48	50	52	54	56	58	60	60	62	62	64	64	66	66	68	68	70	70	72	72	74	74	76	76	78	78	80	80	80	80	80	80	80
0.691-0.710 (0.0272-0.0280)	40	42	44	46	48	50	52	54	56	58	60	62	62	64	64	66	66	68	68	70	70	72	72	74	74	76	76	78	78	80	80	80	80	80	80	80
0.711-0.730 (0.0280-0.0287)	42	44	46	48	50	52	54	56	58	60	62	64	64	66	66	68	68	70	70	72	72	74	74	76	76	78	78	80	80	80	80	80	80	80
0.731-0.750 (0.0288-0.0295)	44	46	48	50	52	54	56	58	60	62	64	66	66	68	68	70	70	72	72	74	74	76	76	78	78	80	80	80	80	80	80	80
0.751-0.770 (0.0296-0.0303)	46	48	50	52	54	56	58	60	62	64	66	68	68	70	70	72	72	74	74	76	76	78	78	80	80	80	80	80	80	80
0.771-0.790 (0.0304-0.0311)	48	50	52	54	56	58	60	62	64	66	68	70	70	72	72	74	74	76	76	78	78	80	80	80	80	80	80	80
0.791-0.810 (0.0311-0.0319)	50	52	54	56	58	60	62	64	66	68	70	72	72	74	74	76	76	78	78	80	80	80	80	80	80	80
0.811-0.830 (0.0319-0.0327)	52	54	56	58	60	62	64	66	68	70	72	74	74	76	76	78	78	80	80	80	80	80	80	80
0.831-0.850 (0.0327-0.0335)	54	56	58	60	62	64	66	68	70	72	74	76	76	78	78	80	80	80	80	80	80	80
0.851-0.870 (0.0335-0.0343)	56	58	60	62	64	66	68	70	72	74	76	78	78	80	80	80	80	80	80	80
0.871-0.890 (0.0343-0.0350)	58	60	62	64	66	68	70	72	74	76	78	80	80	80	80	80	80	80
0.891-0.910 (0.0351-0.0358)	60	62	64	66	68	70	72	74	76	78	80	80	80	80	80	80
0.911-0.930 (0.0359-0.0366)	62	64	66	68	70	72	74	76	78	80	80	80	80	80
0.931-0.950 (0.0367-0.0374)	64	66	68	70	72	74	76	78	80	80	80	80
0.951-0.970 (0.0374-0.0382)	66	68	70	72	74	76	78	80	80	80
0.971-0.990 (0.0382-0.0390)	68	70	72	74	76	78	80	80	80
0.991-1.010 (0.0390-0.0398)	70	72	74	76	78	80	80	80
1.011-1.030 (0.0398-0.0406)	72	74	76	78	80	80	80
1.031-1.050 (0.0406-0.0413)	74	76	78	80	80	80
1.051-1.070 (0.0414-0.0421)	76	78	80	80	80
1.071-1.090 (0.0422-0.0429)	78	80	80	80
1.091-1.110 (0.0430-0.0437)	80	80	80
1.111-1.130 (0.0437-0.0445)	80	80
1.131-1.150 (0.0445-0.0453)	80

Shim No.	Thickness	Shim No.	Thickness	Shim No.	Thickness
00	2.000 (0.0787)	28	2.280 (0.0898)	56	2.560 (0.1008)
02	2.020 (0.0795)	30	2.300 (0.0906)	58	2.580 (0.1016)
04	2.040 (0.0803)	32	2.320 (0.0913)	60	2.600 (0.1024)
06	2.060 (0.0811)	34	2.340 (0.0921)	62	2.620 (0.1031)
08	2.080 (0.0819)	36	2.360 (0.0929)	64	2.640 (0.1039)
10	2.100 (0.0827)	38	2.380 (0.0937)	66	2.660 (0.1047)
12	2.120 (0.0835)	40	2.400 (0.0945)	68	2.680 (0.1055)
14	2.140 (0.0843)	42	2.420 (0.0953)	70	2.700 (0.1063)
16	2.160 (0.0850)	44	2.440 (0.0961)	72	2.720 (0.1071)
18	2.180 (0.0858)	46	2.460 (0.0969)	74	2.740 (0.1079)
20	2.200 (0.0866)	48	2.480 (0.0976)	76	2.760 (0.1087)
22	2.220 (0.0874)	50	2.500 (0.0984)	78	2.780 (0.1094)
24	2.240 (0.0882)	52	2.520 (0.0992)	80	2.800 (0.1102)
26	2.260 (0.0890)	54	2.540 (0.1000)		

Exhaust valve clearance (Cold):
0.25 – 0.35 mm (0.010 – 0.014 in.)

EXAMPLE:
The 2.300mm (0.0906 in.) shim is installed, and the measured clearance is 0.440 mm (0.0173 in.). Replace the 2.300 mm (0.0906 in.) shim with a No. 44 shim.

67170-4RUN-G114

Exhaust valve shim selection chart—4.7L engine

16. If any valve is not within specifications, record how far out it is. This will determine the replacement shim.

17. If any valve requires adjustment, remove the camshaft(s).

18. Remove the lifter(s) and shims to be replaced. Using a micrometer, measure the shim thickness. Use the accompanying charts to determine the size of the new shim. Shims are available in 41 sizes in increments of 0.020mm from 2.00mm to 2.80mm.

19. Once all out of specification shims are replaced, install the camshafts.

20. The remainder of installation is the reverse of removal.

Starter Motor

REMOVAL & INSTALLATION

3.4L Engine

1. Before servicing the vehicle, refer to the precautions in the beginning of this section.
2. Remove or disconnect the following:
   - Negative battery cable
   - Starter electrical connectors
   - Starter

To install:

3. Install or connect the following:
   - Starter. Tighten the fasteners to 29 ft. lbs. (39 Nm).
   - Electrical connections
   - Negative battery cable

4.0L Engine

WITH 2WD

1. Remove the rear under-cover.
2. Remove the No.2 manifold support.
3. Disconnect the starter wiring.
4. Remove the 2 bolts and lower the starter.
5. Installation is the reverse of removal. Torque the starter mounting bolts to 27 ft. lbs. (37 Nm). Torque the manifold brace to 30 ft. lbs. (40 Nm). Torque the under-cover to 21 ft. lbs. (29 Nm).

WITH 4WD

1. Remove the rear under-cover.
2. Remove the front exhaust pipe.
3. Remove the fender splash shield.
4. Disconnect the steering intermediate shaft.
5. Disconnect the starter wiring.
6. Remove the 2 bolts and lower the starter.
7. Installation is the reverse of removal. Torque the starter mounting bolts to 27 ft. lbs. (37 Nm). Torque the intermediate shaft bolts to 27 ft. lbs. (36 Nm). Torque the under-cover to 21 ft. lbs. (29 Nm).

Oil Pan

REMOVAL & INSTALLATION

3.4L Engine

1. Before servicing the vehicle, refer to the precautions in the beginning of this section.
2. Disconnect the negative battery cable.
3. Drain the engine oil.
4. Remove or disconnect the following:
   - Engine undercover
   - Front differential, if equipped with 4WD
   - Oil pan, separate it from the engine using SST 09032-00100 and a brass bar

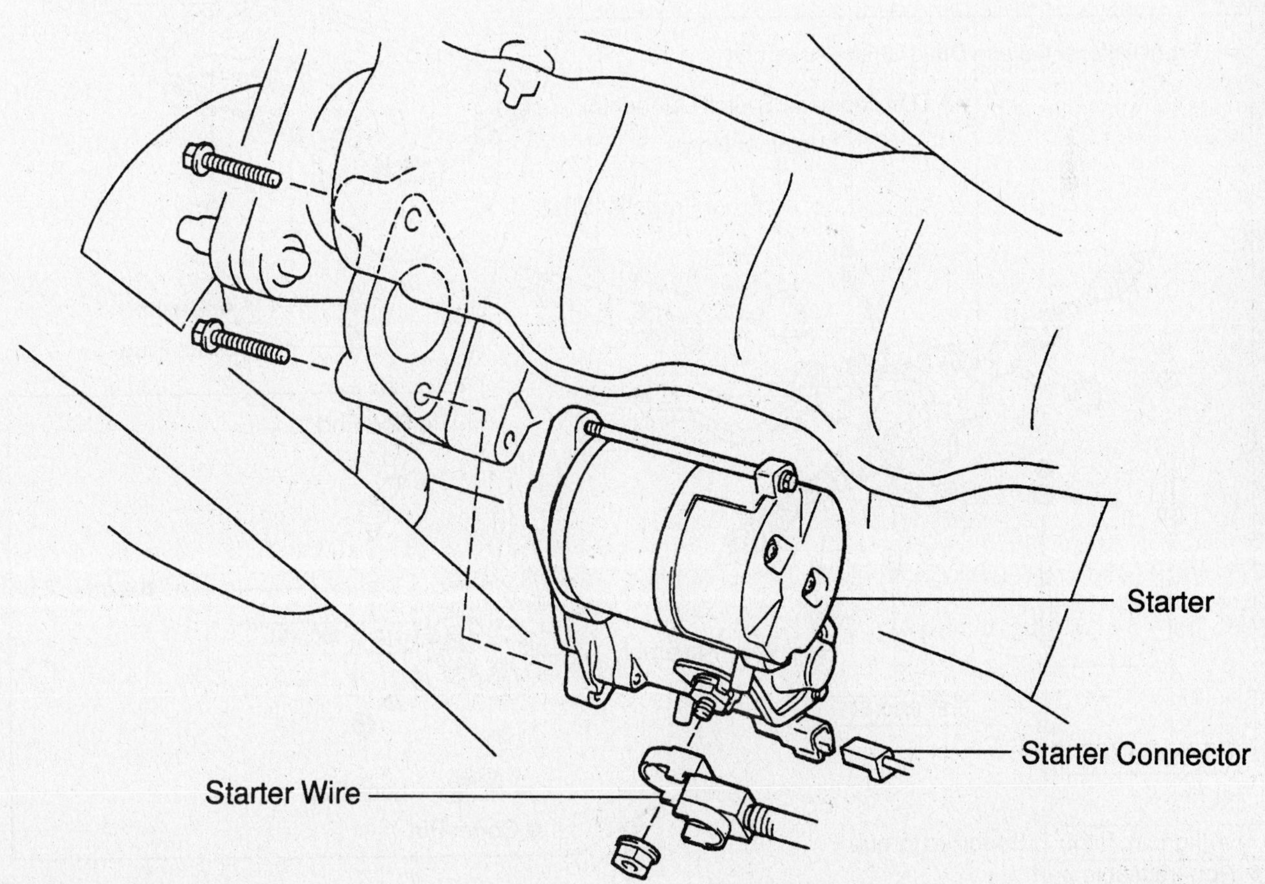

Starter mounting—3.4L engine

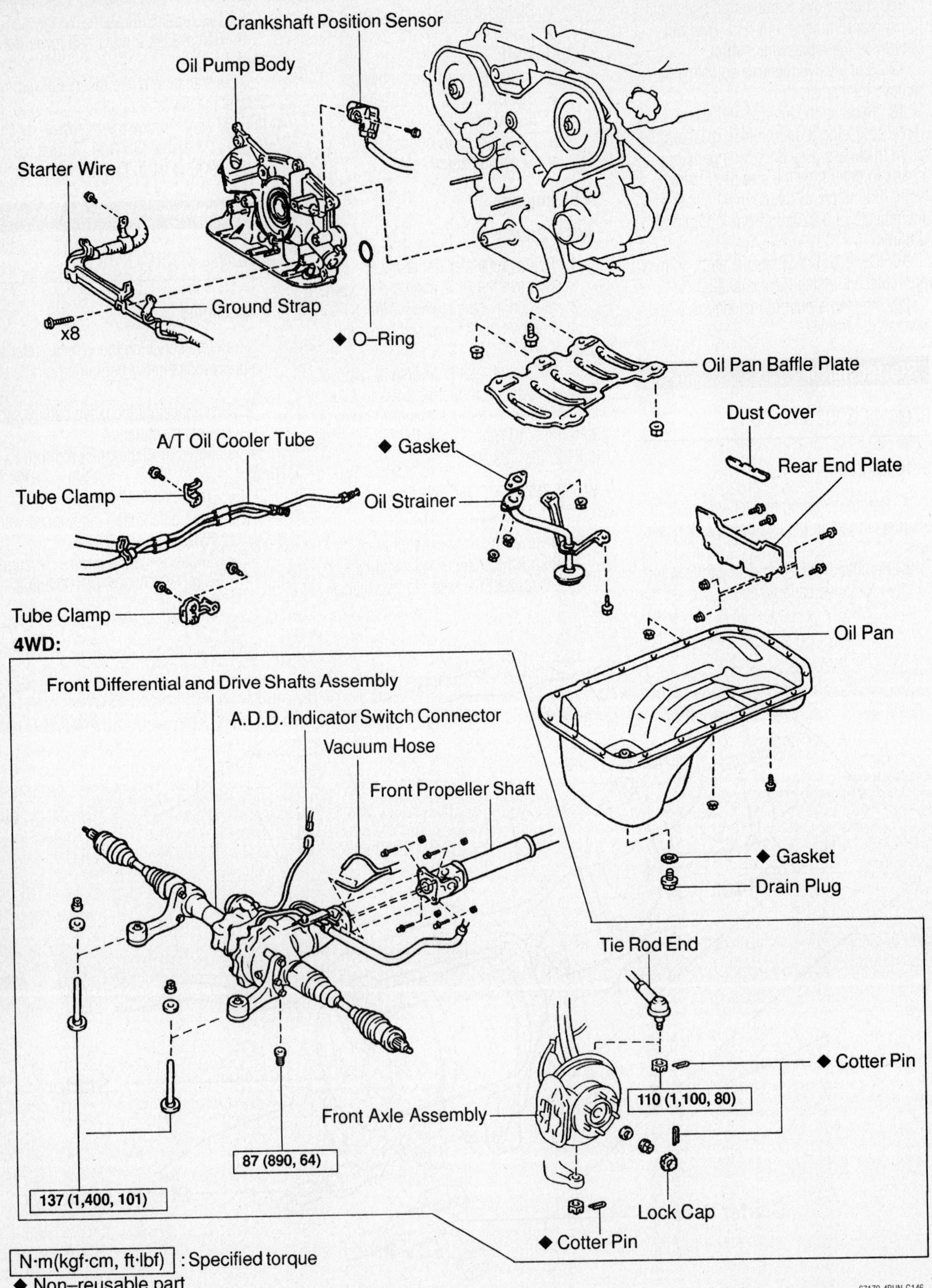
Crankshaft Position Sensor
Oil Pump Body
Starter Wire
Ground Strap
x8
◆ O–Ring
Oil Pan Baffle Plate
Dust Cover
Rear End Plate
A/T Oil Cooler Tube
◆ Gasket
Tube Clamp
Oil Strainer
Tube Clamp
Oil Pan
4WD:
Front Differential and Drive Shafts Assembly
A.D.D. Indicator Switch Connector
Vacuum Hose
Front Propeller Shaft
◆ Gasket
Drain Plug
Tie Rod End
◆ Cotter Pin
110 (1,100, 80)
Front Axle Assembly
87 (890, 64)
137 (1,400, 101)
Lock Cap
◆ Cotter Pin
N·m(kgf·cm, ft·lbf) : Specified torque
◆ Non–reusable part
67170-4RUN-G146

Oil pan and related parts—3.4L engine

To install:

5. Apply seal packing to the oil pan.
6. Install the oil pan to the cylinder block. Tighten the nuts and bolts to 108 inch lbs. (13 Nm)

⁂ WARNING

If parts are not assembled within 5 minutes of applying time, the effectiveness of the seal packing is lost and must be removed and reapplied.

7. Install or connect the following:
   - Front differential, if removed
   - Engine undercover
   - Negative battery cable
8. Fill with engine oil.
9. Start the engine and check for leaks.

4.0L Engine

LOWER PAN

1. Drain the oil.
2. Lift and support the front end with jackstands.
3. Remove the under-covers.
4. Lower the front axle (4wd) as far as possible.
5. Remove the 10 bolts and 2 nuts from the oil pan.
6. Insert a gasket separator tool to cut the sealer.
7. Break loose the oil pan.

To install:

8. Clean the mating surfaces of the lower and upper pan.
9. Apply a continuous 3–4mm dia. bead of silicone gasket material to the oil pan mating surface.
10. Install the pan within 3 minutes after applying the sealer.
11. Torque the nuts and bolts evenly and in several passes, to 80 inch lbs. (9 Nm).

UPPER PAN

1. Remove the lower pan.
2. Remove the strainer.

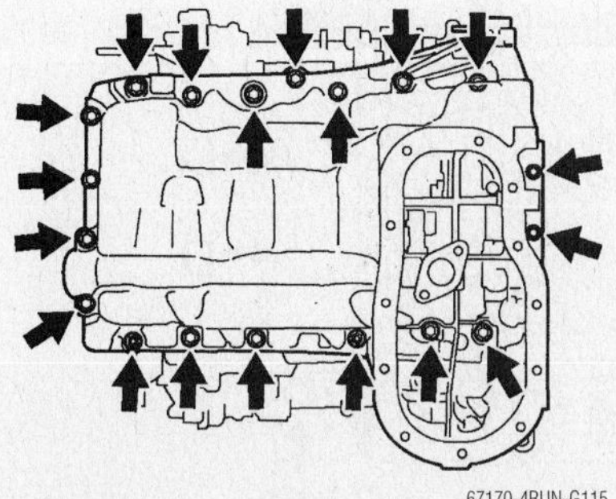

Upper oil pan fastener locations—4.0L engine

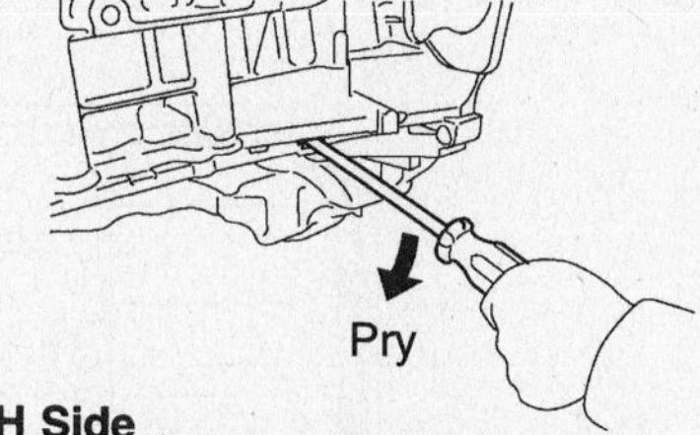

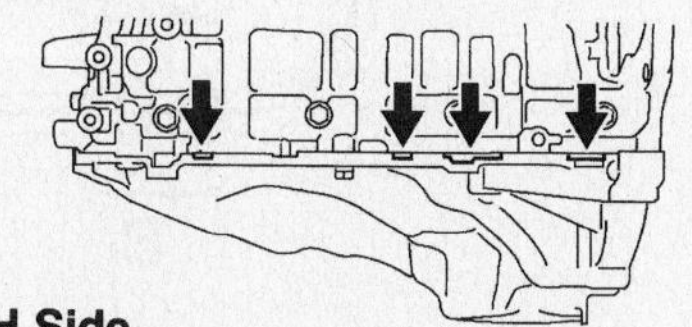

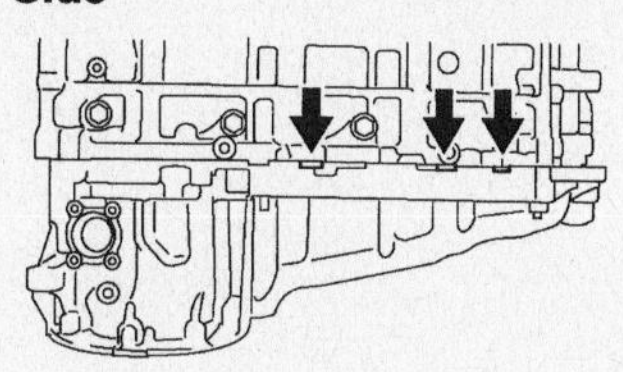

Upper oil pan prying locations—4.0L engine

3. Remove the 4 bellhousing cover bolts and remove the cover.
4. Remove the 17 bolts and 2 nuts from the upper pan.
5. Using a small prybar, pry loose the upper pan at the locations shown.

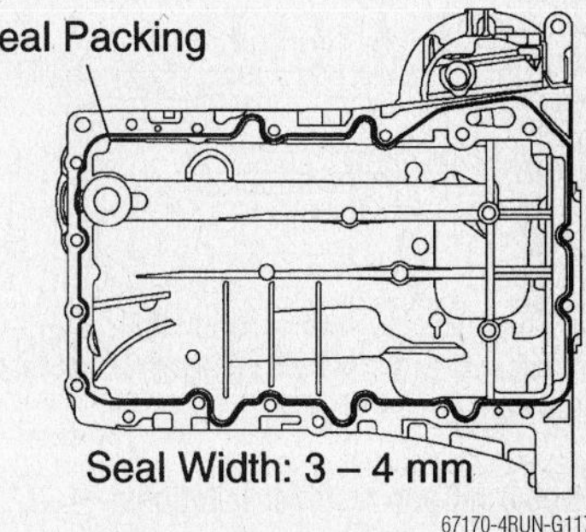

Apply a continuous 3–4mm dia. bead of silicone gasket material to the upper oil pan—4.0L engine

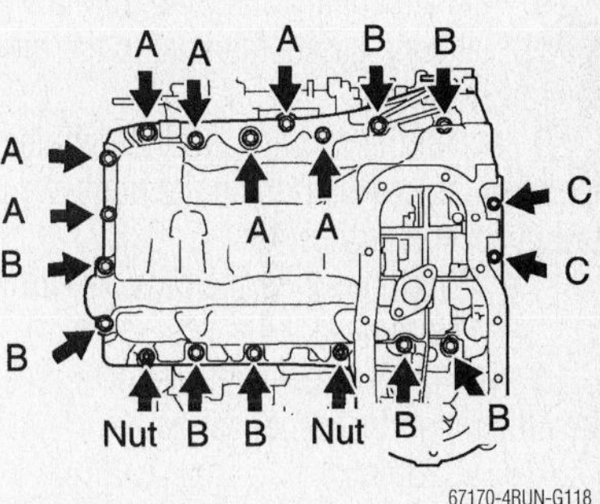

Upper oil pan bolt identification—4.0L engine

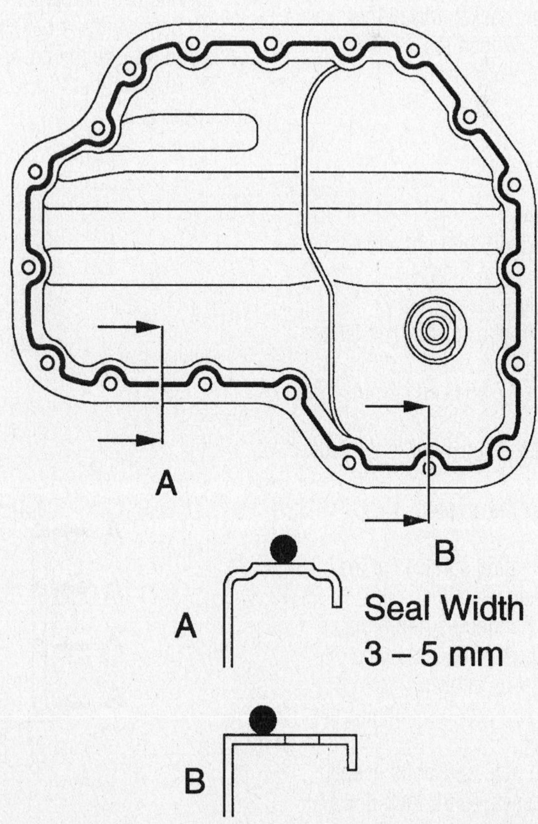

Apply a continuous 3–5mm dia. bead of silicone gasket material to the oil pan mating surface

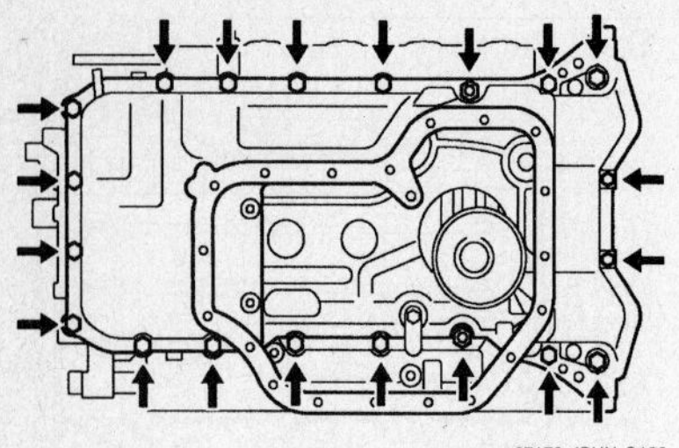

Upper oil pan fastener locations—4.7L engine

To install:

6. Clean the mating surfaces of the pan and block thoroughly.
7. Install a new O-ring on the pump.
8. Apply a continuous 3–4mm dia. bead of silicone gasket material to the oil pan, as shown.
9. Install the oil pan within 3 minutes of applying the sealer. Tighten the bolts evenly and in several passes, to:
   - 10mm head: 80 inch lbs. (9 Nm)
   - 12mm head: 14 ft. lbs. (19 Nm)
10. The bolts are different lengths. See the illustration for identification.
    - A: 25mm
    - B: 40mm
    - C: 14mm
11. Install the cover bolts. Torque to 27 ft. lbs. (37 Nm).
12. Using a new gasket, install the strainer. Torque to 80 inch lbs. (9 Nm).
13. Install the lower pan.

4.7L Engine

LOWER PAN

1. Drain the oil.
2. Lift and support the front end with jackstands.
3. Remove the under-covers.
4. Lower the front axle (4wd) as far as possible.
5. Remove the 17 bolts and 2 nuts from the oil pan.
6. Insert a gasket separator tool to cut the sealer.
7. Break loose the oil pan.

To install:

8. Clean the mating surfaces of the lower and upper pan.
9. Apply a continuous 3–5mm dia. bead of silicone gasket material to the oil pan mating surface, as shown.
10. Install the pan within 5 minutes after applying the sealer.
11. Torque the nuts and bolts evenly and in several passes, to 66 inch lbs. (7.5 Nm).

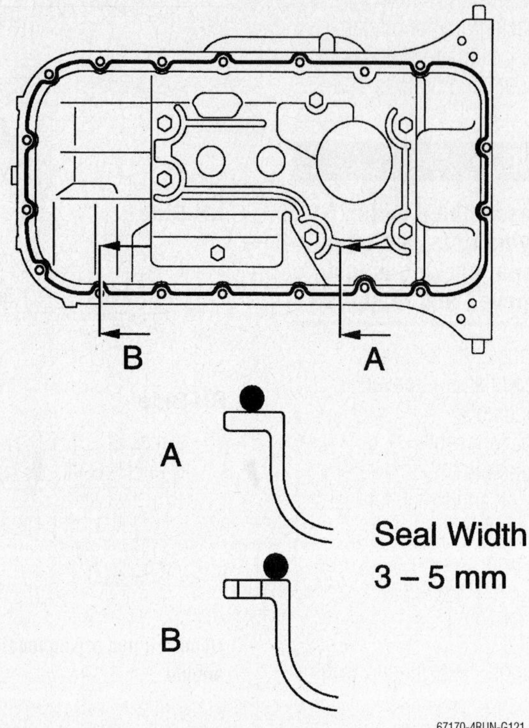

Apply a continuous 3–5mm dia. bead of silicone gasket material to the upper oil pan—4.7L engine

UPPER PAN

1. Remove the lower pan.
2. Remove the baffle.
3. Remove the 4 bellhousing cover bolts and remove the cover.
4. Remove the 18 bolts and 2 nuts from the upper pan.
5. Using a small prybar, pry loose the upper pan.

To install:

6. Clean the mating surfaces of the pan and block thoroughly.
7. Install a new O-ring on the pump.
8. Apply a continuous 3–5mm dia.

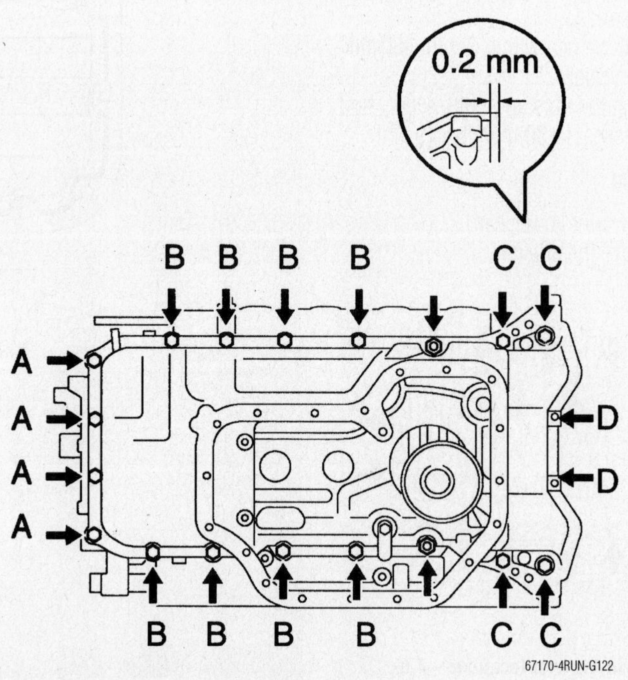

Upper oil pan bolt identification—4.7L engine

bead of silicone gasket material to the oil pan, as shown.

9. Install the oil pan within 5 minutes of applying the sealer. Tighten the bolts and nuts evenly and in several passes, to:

- Bolts A and D: 66 inch lbs. (7.5 Nm)
- Bolts B, C and the nuts: 21 ft. lbs. (28 Nm)

10. The bolts are different lengths. See the illustration for identification.

- A: 20mm
- B: 25mm
- C: 60mm
- D: 35mm

11. Install the cover bolts. Torque to 27 ft. lbs. (37 Nm).
12. Using a new gasket, install the strainer. Torque to 80 inch lbs. (9 Nm).
13. Install the lower pan.

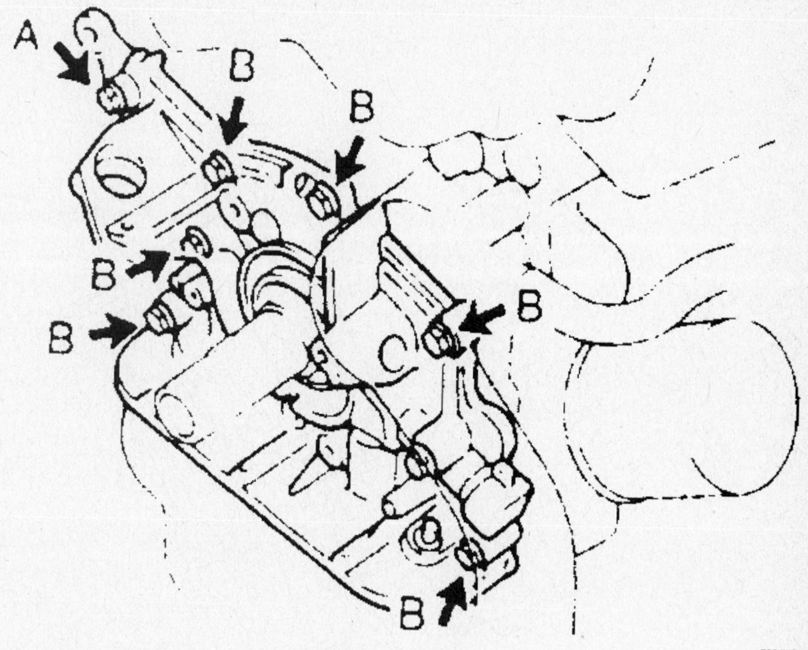

Oil pump bolt identification—3.4L engine

Oil Pump

REMOVAL & INSTALLATION

3.4L Engine

1. Before servicing the vehicle, refer to the precautions in the beginning of this section.
2. Remove or disconnect the following:

- Negative battery cable
- Engine undercover
- Crankshaft timing pulley
- Front differential, if equipped with 4WD

3. Drain the engine oil from the engine.
4. Remove or disconnect the following:

- Timing belt and crankshaft gear
- Oil cooler tube and clamp, if equipped with automatic transmission
- Stiffener plate
- Flywheel housing undercover and dust cover

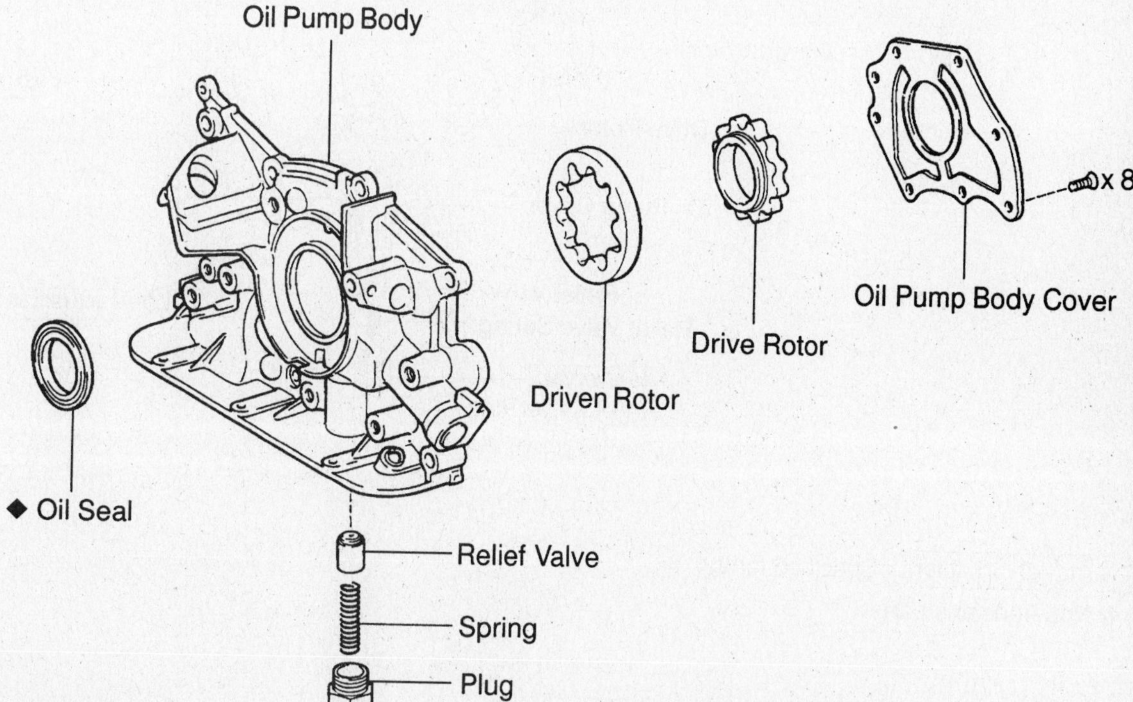

Oil pump—3.4L engine

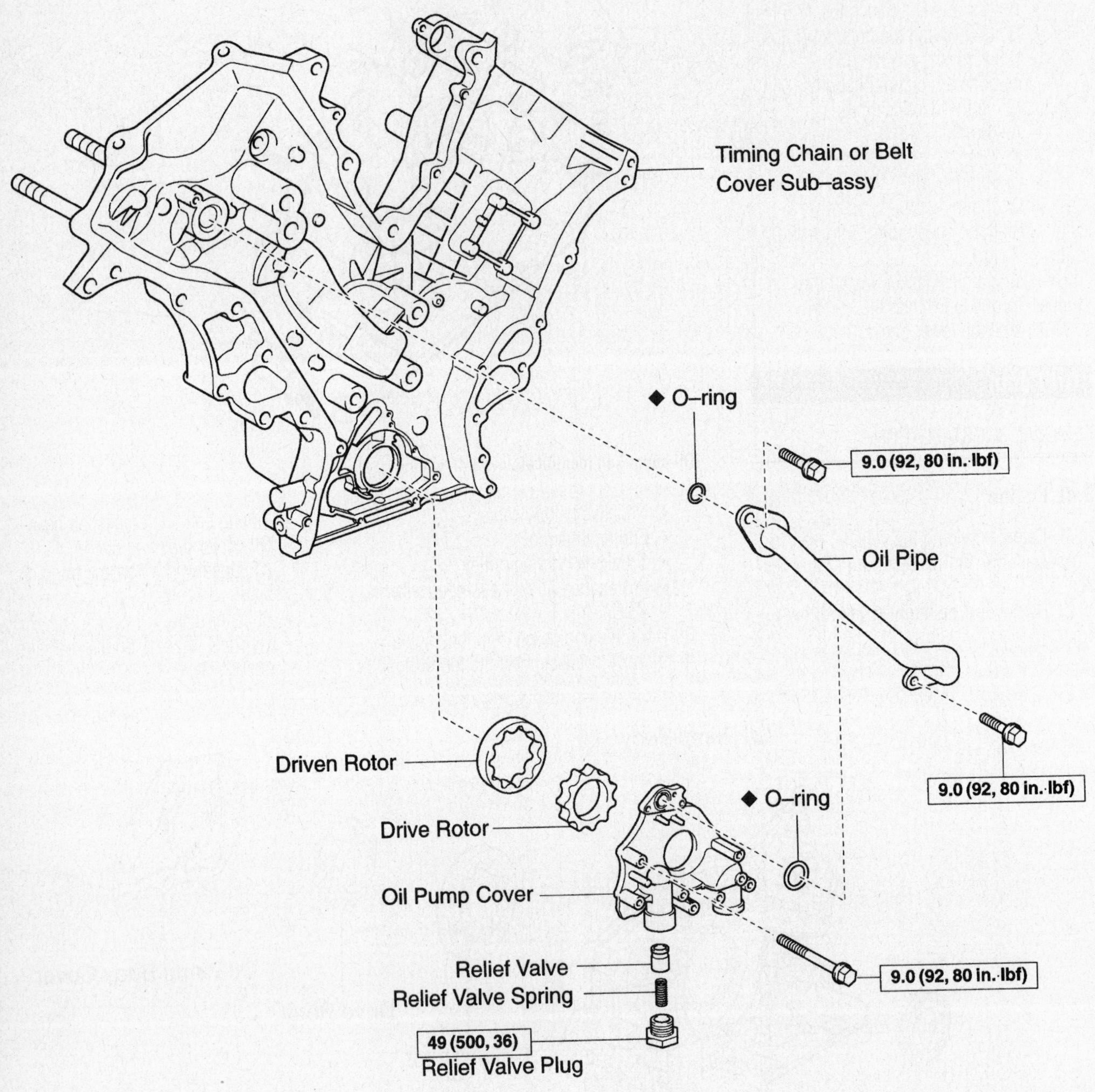

Oil pump and related parts—4.0L engine

- Rear end cover and dust cover
- Starter wire clamp
- Crankshaft Position (CKP) sensor
- Oil pan

➡**Be careful not to damage the baffle plate flange.**

- Oil strainer
- Oil baffle plate
- Oil pump body by removing the 8 bolts.
- O-ring from the cylinder block

To install:

5. Install or connect the following:
   - Apply Seal Packing PN 08826-00080 to the oil pump
   - New O-ring into the groove of the cylinder block
   - Oil pump to the crankshaft with the splined teeth of the drive rotor engaged with the large teeth of the crankshaft. Tighten the oil pump bolts "A" 15 ft. lbs. (20 Nm) and bolts "B" 31 ft. lbs. (42 Nm)
   - CKP
   - Oil pan baffle plate
   - Oil strainer with a new gasket. Tighten the bolts to 13 ft. lbs. (18 Nm).
   - Remaining components
   - Negative battery cable
6. Fill with engine oil.
7. Start the engine and check for leaks.

4.0L Engine

1. Remove the timing chain cover.
2. Remove the oil pipe, 3 bolts.
3. Remove the 2 O-rings.
4. Remove the 7 bolts, oil pump cover drive and driven rotors.
5. Check that the relief valve falls smoothly into the valve hole under its own weight.
6. Place the rotors into the timing chain cover with the marks facing upward.
7. Using a feeler gauge, check the clearance between the drive and driven rotor tips. Standard clearance is 0.06–0.16mm. If the clearance exceeds 0.16mm, replace the rotors as a set.
8. Inspect the rotor side clearance with a feeler gage and straight edge. Clearance should be 0.03–0.09mm. If the clearance exceeds 0.09mm, replace the rotors as a set. If necessary, replace the timing chain cover.
9. Using a feeler gauge, check the clearance between the driven rotor and the body. Clearance should be 0.250–0.325 mm. If clearance exceeds 0.325mm, replace the rotors as a set. If necessary, replace the timing chain cover.
10. Coat the relief valve with clean engine oil and install it. Torque to 36 ft. lbs. (49 Nm).
11. Coat the rotors with clean engine oil. Place the rotors in the timing chain cover with the timing marks aligned and facing the oil pump cover.
12. Install the cover. Torque the bolts to 80 inch lbs. (9 Nm).
13. Install new O-rings.
14. Install the oil pipe. Torque to 80 inch lbs. (9 Nm).

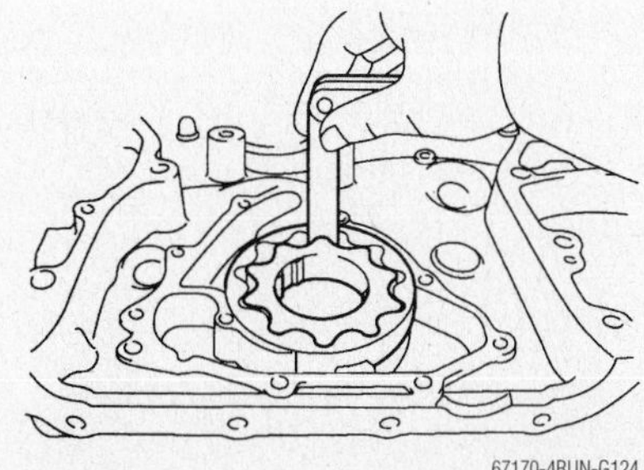

Using a feeler gauge, check the clearance between the drive and driven rotor tips—4.0L engine

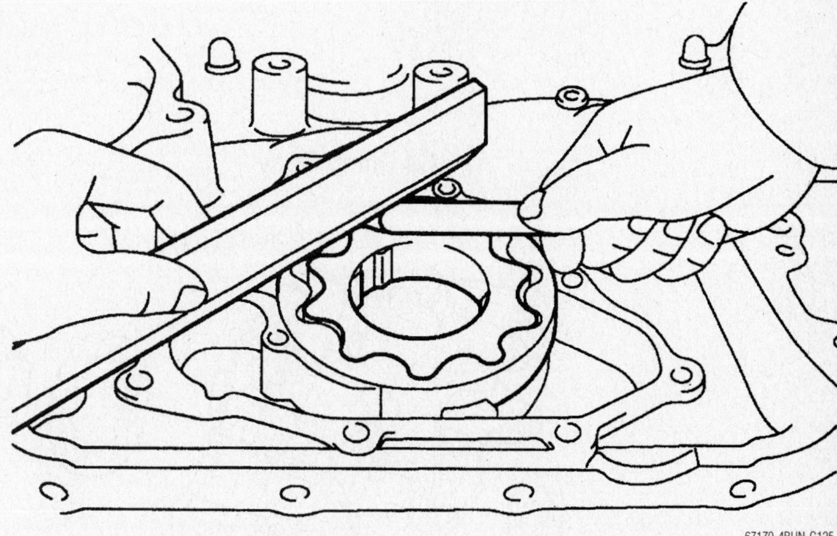

Inspect the rotor side clearance with a feeler gage and straight edge—4.0L engine

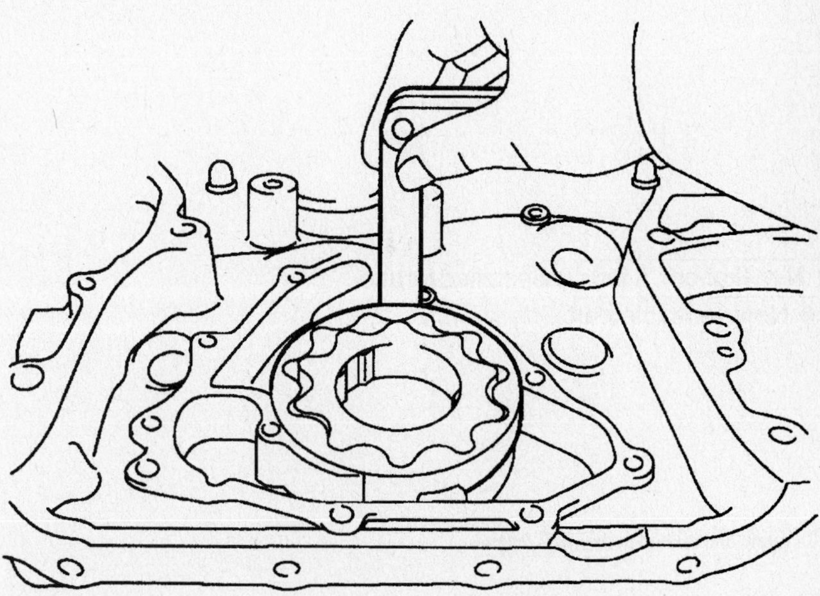

Using a feeler gauge, check the clearance between the driven rotor and the body—4.0L engine

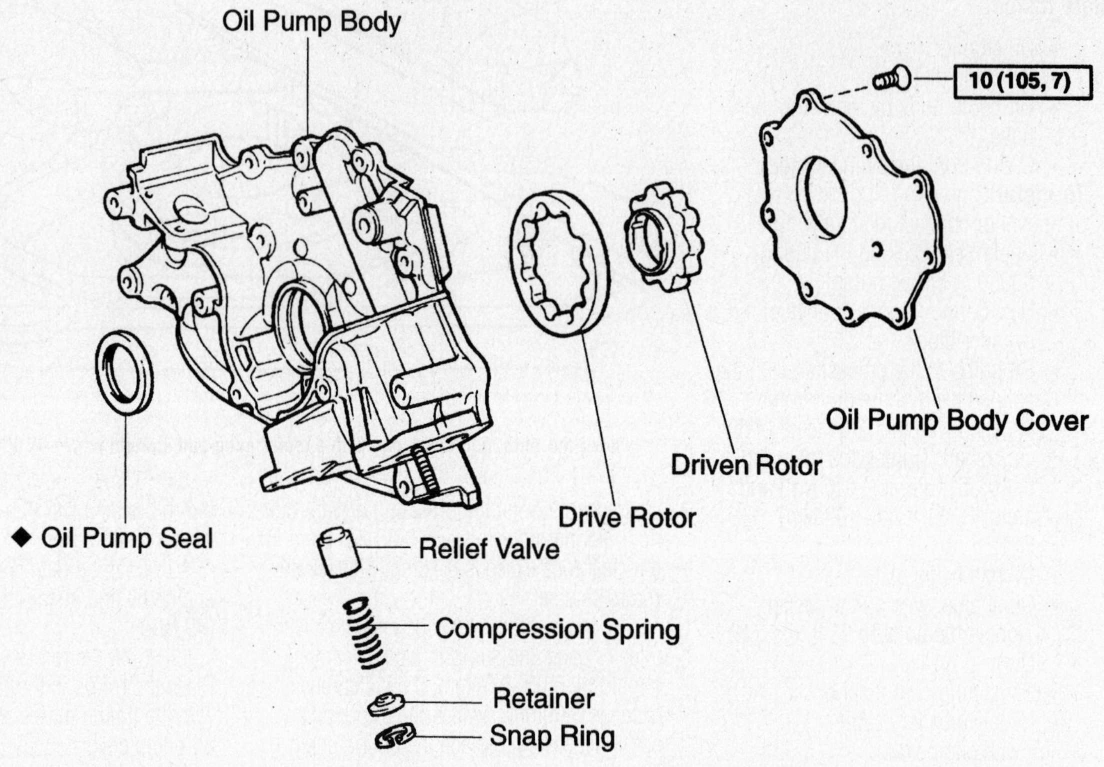

Oil pump and related parts—4.7L engine

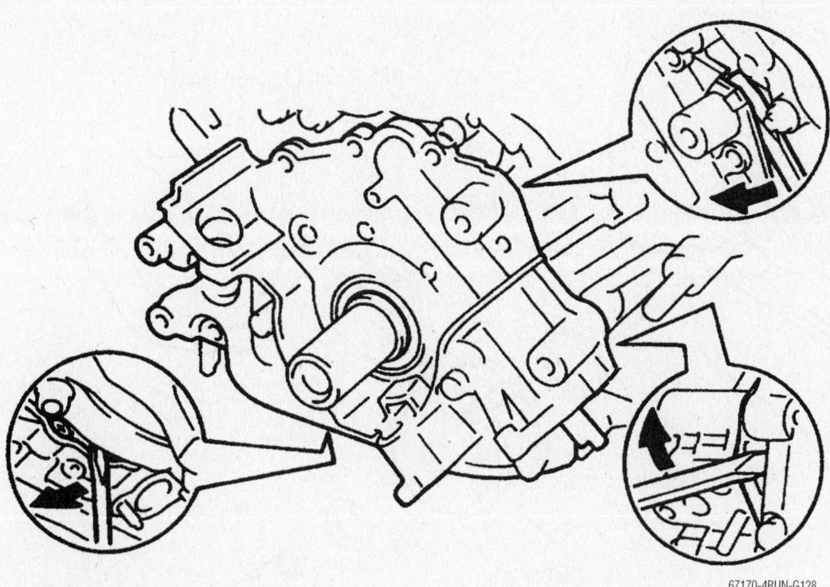

Oil pump prying positions—4.7L engine

4.7L Engine

1. Remove the engine.
2. Remove the timing belt.
3. Remove the crankshaft pulley.
4. Remove the crankshaft position sensor.
5. Remove the oil cooler.
6. Remove the lower oil pan.
7. Remove the oil strainer.
8. Remove the baffle plate.
9. Remove the upper oil pan.
10. Remove the 8 bolts and, prying at the positions shown, pry off the oil pump.
11. Remove the O-ring from the block.
12. Remove the timing belt cover seal.

To install:

13. Thoroughly clean the mating surfaces of the pump and block.
14. Install a new front cover seal.
15. Install a new O-ring on the block.
16. Apply new formed-in-place silicone gasket material to the oil pump, as shown. The pump must be installed within 5 minutes of sealer application.
17. Install the bolts. Torque the bolts evenly and in several passes, to:
    - Bolts with 6mm and 12mm heads: 11 ft. lbs. (16 Nm)
    - Bolts with 14mm heads: 23 ft. lbs. (31 Nm)
18. The remainder of installation is the reverse of removal.

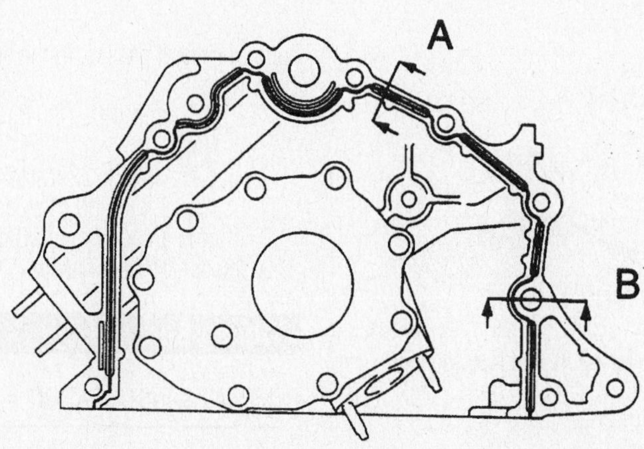

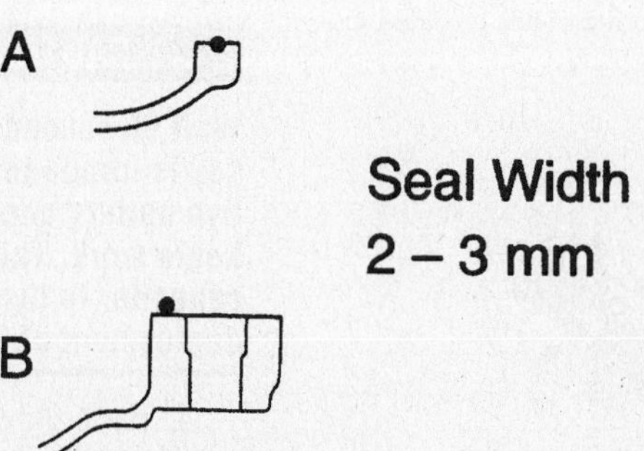

Gasket material application—4.7L engine oil pump

Oil Cooler

REMOVAL & INSTALLATION

3.4L Engine

1. Drain the coolant.
2. Disconnect the 2 hoses from the cooler.
3. Remove the bolt, gaskets, cooler and O-ring.
4. Installation is the reverse of removal. Use a new O-ring. Torque the cooler bolt to 43 ft. lbs. (59 Nm).

4.0L Engine

1. Drain the coolant.
2. Remove the oil filter.
3. Remove the V-bank cover.
4. Disconnect the 2 hoses from the cooler.
5. Remove the bolt, washer and cooler.
6. Installation is the reverse of removal. Use new O-rings. Torque the cooler bolt to 50 ft. lbs. (68 Nm). Torque the V-bank cover to 66 inch lbs. (7.5 Nm).

4.7L Engine

1. Remove the engine under-cover.
2. Drain the coolant.
3. Drain the engine oil.
4. Remove the oil filter.
5. Disconnect the 2 hoses from the cooler.
6. Remove the bolt, washer and cooler. Discard the O-ring.
7. Installation is the reverse of removal. Use new O-rings. Torque the cooler bolt to 51 ft. lbs. (69 Nm).

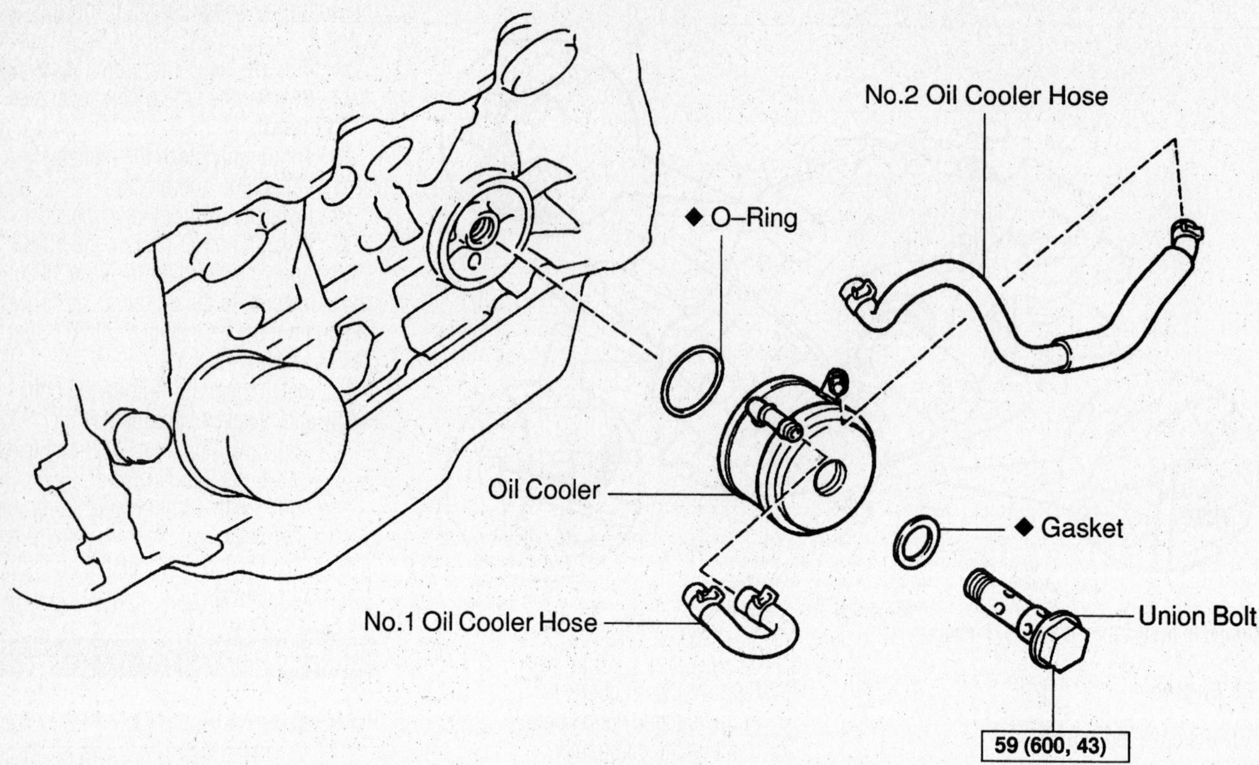

Oil Cooler—3.4L engine

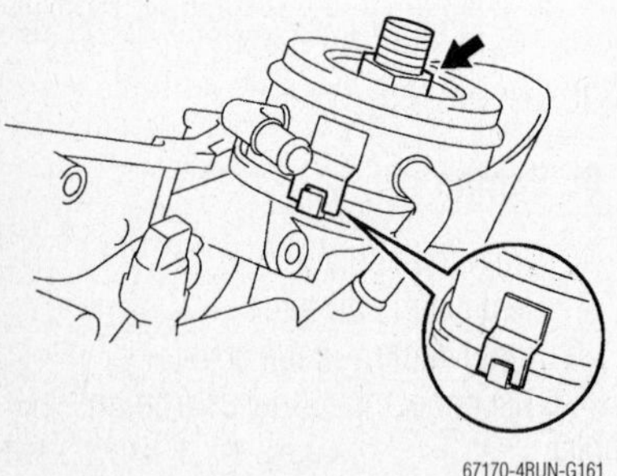

Oil Cooler—4.0L engine

Rear Main Seal

REMOVAL & INSTALLATION

3.4L, 4.0L and 4.7L Engines

1. Before servicing the vehicle, refer to the precautions in the beginning of this section.
2. Remove the transmission
3. Remove the driveplate and spacers.
4. Cut off the rubber lip portion of the seal with a sharp knife.
5. Pry out the oil seal.

To install:

6. Coat the lip of the new spacer with MP grease.

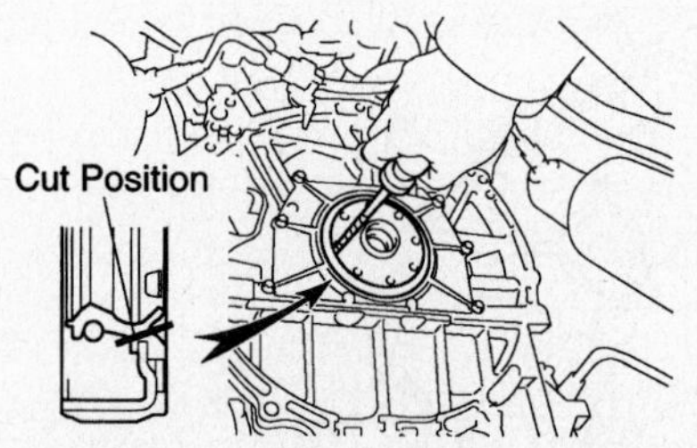

Cutting the seal—3.4L and 4.0L engines

7. Install the rear main seal so that it is flush with the seal retainer housing.
8. Install the driveplate and spacers. Apply thread lock to the bolts. Tighten the bolts in several passes, in the sequence shown, to:
    - 3.4L and 4.0L: 61 ft. lbs. (83 Nm)
    - 4.7L: step 1, 22 ft. lbs. (30 Nm); step 2, plus 90 degrees; step 3, plus an additional 90 degrees
9. Install the transmission.

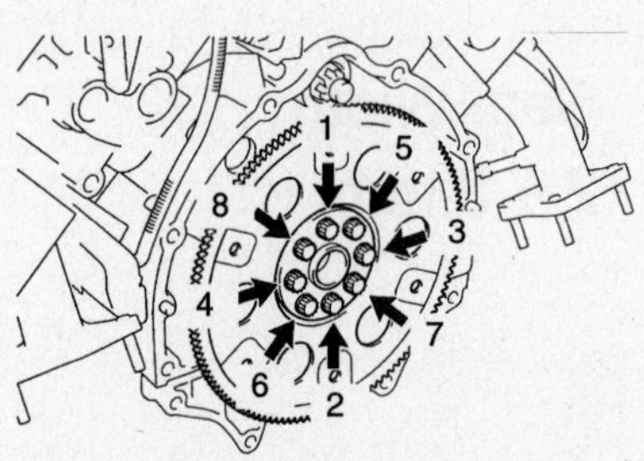

Driveplate bolt torque sequence

Timing Belt

REMOVAL & INSTALLATION

3.4L Engine

1. Disconnect the negative battery cable.

CAUTION

Wait 90 seconds from the time the key is turned to LOCK and the negative battery cable is disconnected to begin work. This allows the SRS capacitor to discharge and prevent deployment of the air bag(s).

2. Raise and safely support the vehicle.
3. Remove the engine undercover.
4. Drain the engine coolant.

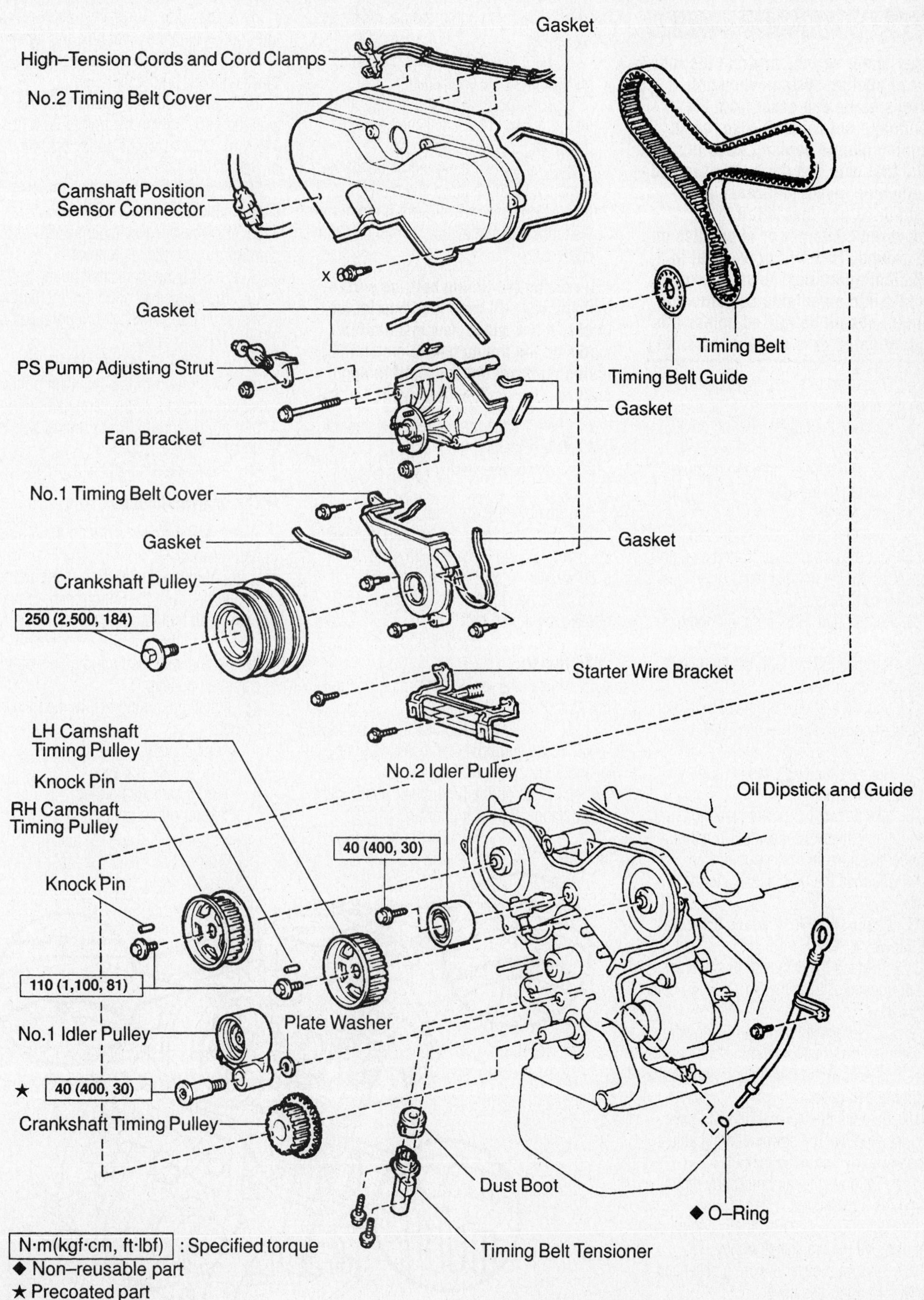

Gasket
High–Tension Cords and Cord Clamps
No.2 Timing Belt Cover
Camshaft Position Sensor Connector
x 6
Gasket
PS Pump Adjusting Strut
Fan Bracket
Timing Belt
Timing Belt Guide
Gasket
No.1 Timing Belt Cover
Gasket
Gasket
Crankshaft Pulley
250 (2,500, 184)
Starter Wire Bracket
LH Camshaft Timing Pulley
Knock Pin
RH Camshaft Timing Pulley
No.2 Idler Pulley
Oil Dipstick and Guide
40 (400, 30)
Knock Pin
110 (1,100, 81)
Plate Washer
No.1 Idler Pulley
★ 40 (400, 30)
Crankshaft Timing Pulley
Dust Boot
◆ O–Ring
Timing Belt Tensioner
N·m(kgf·cm, ft·lbf) : Specified torque
◆ Non–reusable part
★ Precoated part
67170-4RUN-G148

Timing belt and related parts—3.4L engine

✱✱ CAUTION

Never open, service or drain the radiator or cooling system when hot; serious burns can occur from the steam and hot coolant. Also, when draining engine coolant, keep in mind that cats and dogs are attracted to ethylene glycol antifreeze and could drink any that is left in an uncovered container or in puddles on the ground. This will prove fatal in sufficient quantities. Always drain coolant into a sealable container. Coolant should be reused unless it is contaminated or is several years old.

5. Disconnect the upper radiator hose from the engine.
6. Remove the power steering drive belt.
7. Remove the air conditioning drive belt by loosening the idler pulley nut and the adjusting bolt.
8. If equipped with air conditioning, disconnect the compressor from the engine and set aside. Do not disconnect the lines from the compressor.
9. If equipped with air conditioning, disconnect the air conditioning bracket.
10. Remove the fan with the fluid coupling and fan pulleys.
11. Loosen the lockbolt, pivot bolt, and the adjusting bolt and the alternator drive belt.
12. Remove the No. 2 fan shroud by removing the 2 clips.
13. Disconnect the power steering pump from the engine and set aside. Do not disconnect the lines from the pump.
14. Remove the oil dipstick and the guide.
15. Remove the No. 2 timing belt cover as follows:
   a. Detach the camshaft position sensor connector from the No. 2 timing belt cover.
   b. Disconnect the 4 spark plug wire clamps from the No. 2 timing belt cover.
   c. Remove the 6 bolts and remove the timing belt cover.
16. Remove the fan bracket as follows:
   a. Remove the power steering adjusting strut by removing the nut.
   b. Remove the fan bracket by removing the bolt and nut.
17. Using SST 09213-54015, or equivalent, remove the crankshaft pulley.
18. Remove the starter wire bracket and the No. 1 timing belt cover.
19. Remove the timing belt guide.
20. Set the No. 1 cylinder at Top Dead Center (TDC) of the compression stroke, as follows:
   a. Temporarily install the crankshaft pulley bolt to the crankshaft.
   b. Turn the crankshaft and align the timing marks of the crankshaft timing pulley and the oil pump body.
   c. Check that the timing marks of the camshaft timing pulleys and the No. 3 timing belt cover are aligned. If not, turn the crankshaft pulley one revolution (360 degrees).

➡If reusing the timing belt, be sure that you can still read the installation marks. If not, place new installation marks on the timing belt to match the timing marks of the camshaft timing pulleys.

21. Remove the timing belt tensioner by alternately loosening the 2 bolts.
22. Remove the right and left camshaft pulleys.
23. Remove the No. 2 idler pulley.
24. Using a 10mm hex wrench, remove the pivot bolt, No. 1 idler pulley and the plate washer.
25. Remove the timing belt guide and remove the timing belt.
26. Remove the crankshaft timing pulley.

To install:

27. Install the crankshaft timing belt pulley, as follows:
   a. Align the timing belt pulley set key with the key groove of the timing pulley and slide on the timing pulley.
   b. Slide on the timing belt pulley with the flange side facing inward.
28. Install the plate washer and the No. 1 idler pulley with the pivot bolt and tighten it to 26 ft. lbs. (35 Nm). Check that the pulley bracket moves smoothly.
29. Install the No. 2 timing belt idler with the bolt. Tighten the bolt to 30 ft. lbs. (40 Nm). Check that the pulley bracket moves smoothly.
30. Install the left and right camshaft timing pulleys.
31. Set the No. 1 cylinder to TDC of the compression stroke, as follows:
   a. Using the crankshaft pulley bolt, turn the crankshaft and align the timing marks of the crankshaft timing pulley and the oil pump body.
   b. Using SST 09960-10010, or equivalent, to turn the camshaft pulley to align the marks of the camshaft timing belt pulley and the No. 3 timing belt cover.
32. Install the timing belt, as follows:

➡The engine should be cold.

   a. Face the front mark on the timing belt forward.
   b. Align the installation mark on the timing belt with the timing mark of the crankshaft timing pulley.
   c. Align the installation marks on the timing belt with the timing marks of the camshaft pulleys.
33. Install the timing belt in the following order:
   - Left camshaft pulley
   - No. 2 idler pulley
   - Right camshaft pulley
   - Water pump pulley

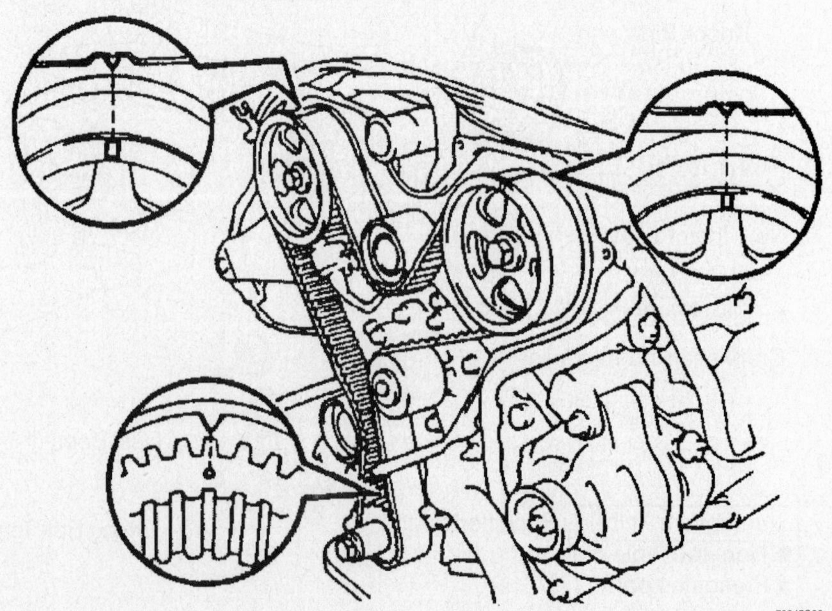

Crankshaft and camshaft timing mark locations—3.4L engine

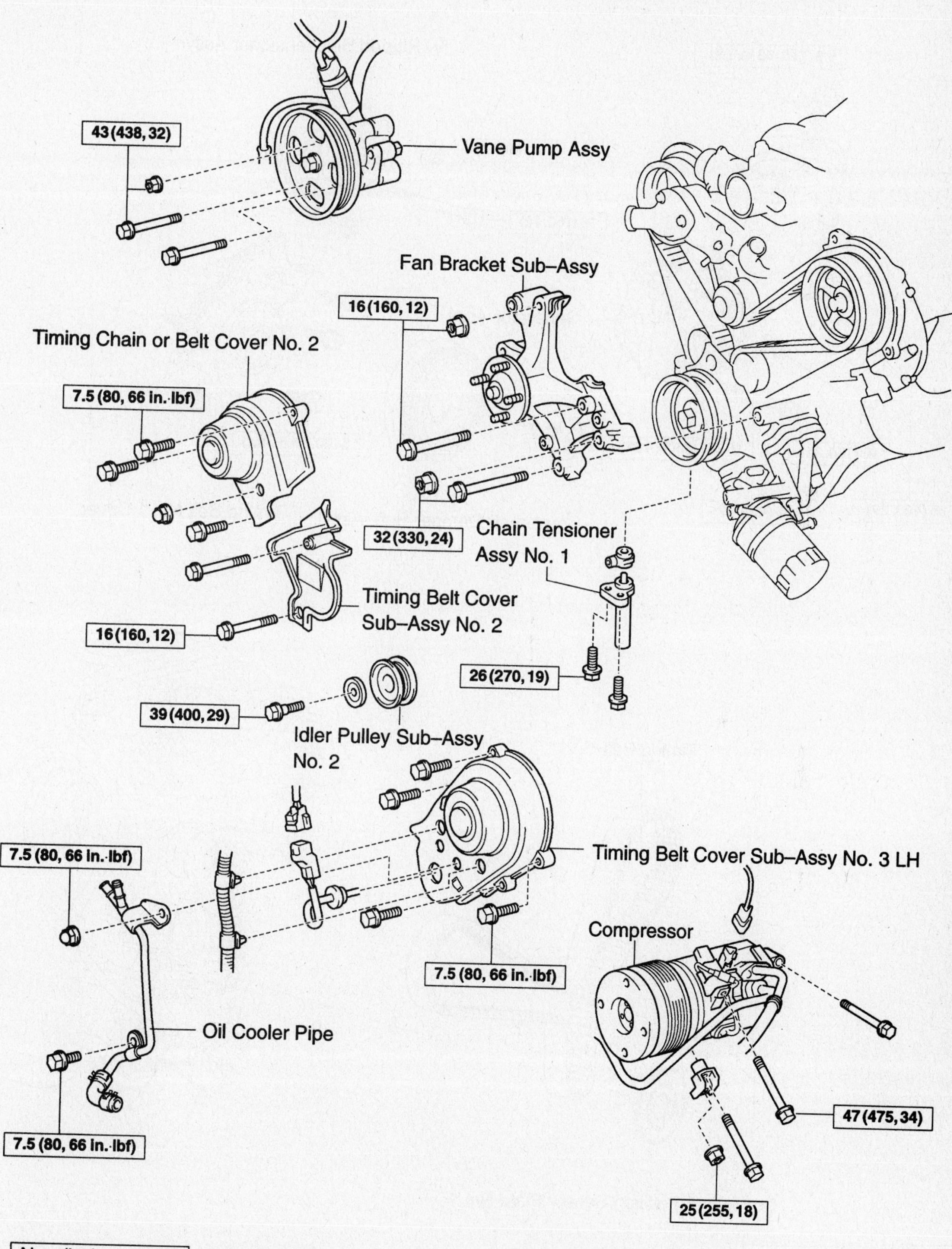

These parts must be removed prior to timing belt removal—4.7L engine

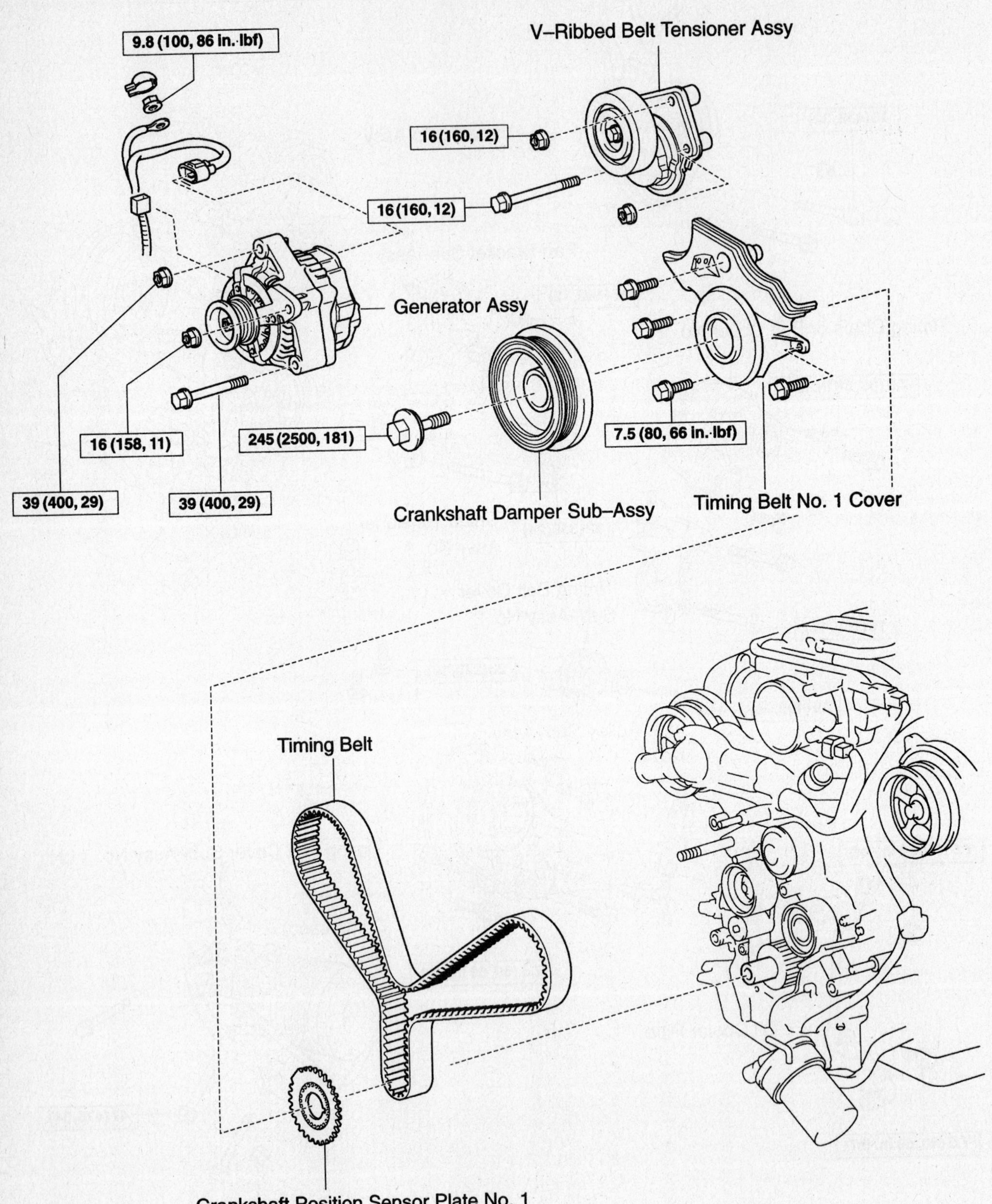

Timing belt and related parts—4.7L engine

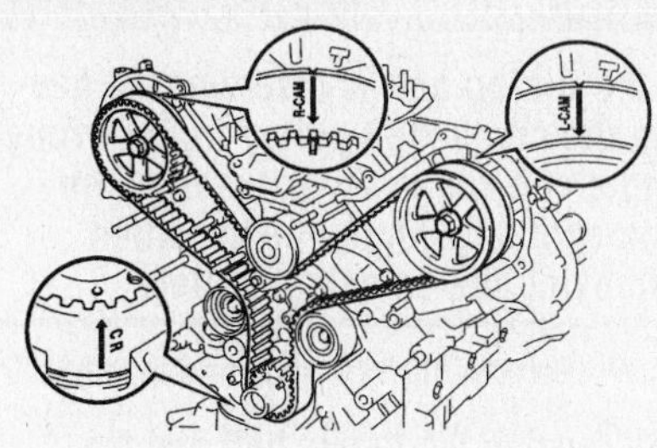

67170-4RUN-G151

Turn the crankshaft clockwise to align the marks

- Crankshaft pulley
- No. 1 idler pulley

⁂ WARNING

If any binding is felt when adjusting the timing belt tension by turning the crankshaft, STOP turning the engine, because the pistons may be hitting the valves.

34. Set the timing belt tensioner as follows:
    a. Using a press, slowly press in the pushrod using 220–2205 lbs. (981–9807 N) of force.
    b. Align the holes of the pushrod and housing, pass a 1.27mm wrench through the holes to keep the setting position of the pushrod.
    c. Release the press and install the dust boot to the tensioner.
35. Install the timing belt tensioner and alternately tighten the bolts to 20 ft. lbs. (27 Nm). Using pliers, remove the 1.27mm wrench from the belt tensioner.
36. Check the valve timing, as follows:
    a. Slowly turn the crankshaft and align the timing marks of the crankshaft timing pulley and the oil pump body. Always turn the crankshaft pulley clockwise.
    b. Check that the timing marks of the right and left timing pulleys align with the timing marks of the No. 3 timing belt cover. If the marks do not align, remove the timing belt and reinstall it.
37. Install the timing belt guide with the cup side facing outward.
38. Install the No. 1 timing belt cover and starter wire bracket. Tighten the timing belt cover fasteners to 80 inch lbs. (9 Nm).
39. Install the crankshaft pulley, as follows:
    a. Align the pulley set key with the key groove of the pulley and slide the pulley.
    b. Using SST 09213-54014, or equivalent, tighten the bolt to 184 ft. lbs. (250 Nm).
40. Install the fan bracket with the bolt and nut.
41. Install the No. 2 timing belt cover, and tighten the bolts to 80 inch lbs. (9 Nm). Install the remaining components.
42. Fill the cooling system with coolant.
43. Connect the negative battery cable.
44. Start the engine and check for leaks.
45. Check the ignition timing.

4.7L Engine

1. Drain the coolant.
2. Remove the battery ground cable.

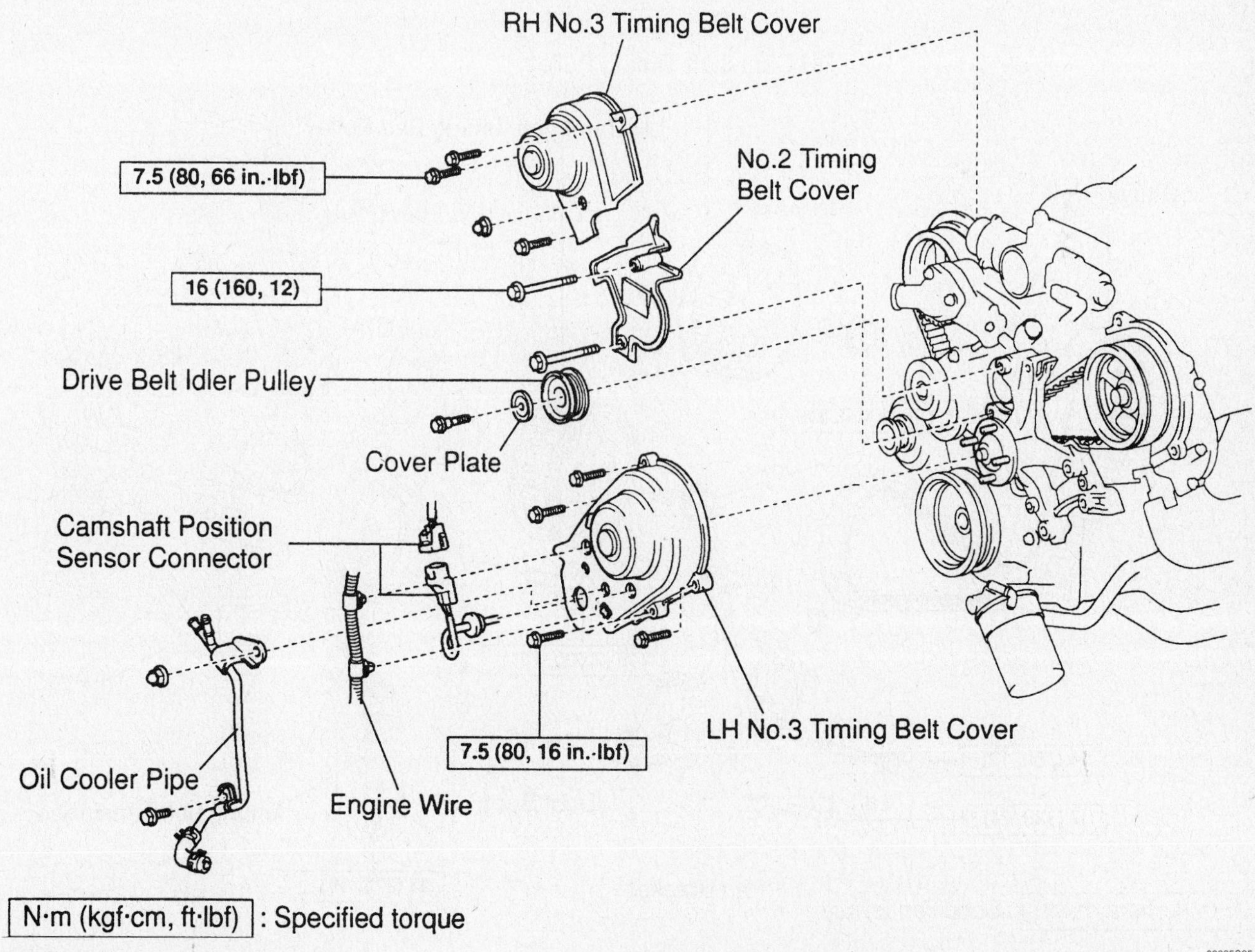

RH No.3 Timing Belt Cover
7.5 (80, 66 in.·lbf)
No.2 Timing Belt Cover
16 (160, 12)
Drive Belt Idler Pulley
Cover Plate
Camshaft Position Sensor Connector
LH No.3 Timing Belt Cover
7.5 (80, 16 in.·lbf)
Oil Cooler Pipe
Engine Wire
N·m (kgf·cm, ft·lbf) : Specified torque

93025G25

Exploded view of upper timing belt covers

3. Remove the throttle body cover.
4. Remove the air cleaner.
5. Remove the radiator support upper seal.
6. Remove the accessory drive belt.
7. Remove the fan.
8. Remove the power steering pump. Don't disconnect the hoses.
9. Remove the alternator.
10. Remove the A/C compressor. Don't disconnect the lines.
11. Remove the idler pulley.
12. Remove the oil cooler pipe.
13. Remove the upper timing belt covers.
14. Remove the accessory drive belt tensioner.
15. Remove the fan bracket.
16. Remove the crankshaft damper with a puller.
17. Remove the lower belt cover.
18. Remove the crankshaft position sensor plate.
19. If re-using the timing belt, check the installation marks on the belt. There are 3 marks. Turn the crankshaft clockwise to align the marks as shown. If the marks have disappeared, matchmark the belt and each pulley.
20. Using the Crankshaft Pulley Holding tool 09213-70010, Bolt tool 90105-08076 and Companion Flange Holding tool 09330-00021, or equivalent, loosen the crankshaft pulley bolt.
21. Position the No. 1 cylinder to approximately 50 degrees After Top Dead Center (ATDC) of the compression stroke by performing the following procedures:
   a. Rotate the crankshaft pulley (CLOCKWISE) to align its groove with the timing mark "0" on the lower (No. 1) timing belt cover.
   b. Check that the camshaft sprocket timing marks are aligned with the rear timing belt plate marks; if not, rotate the crankshaft 1 revolution (360 degrees).
   c. Rotate the crankshaft pulley approximately 50 degrees (CLOCKWISE) and align the crankshaft pulley timing mark between the centers of the crankshaft pulley bolt and the idler pulley bolt.

WARNING

If the timing belt is disengaged, having the crankshaft pulley in the wrong angle can cause the valve to come into contact with the piston when removing the camshaft pulley.

22. Remove the crankshaft pulley bolt.

➡If reusing the timing belt and the installation marks have disappeared, place new installation marks on the timing belt to match the camshaft timing sprocket marks.

➡To avoid meshing the timing sprocket and the timing belt, secure one with a string; then, place matchmarks on the timing belt and the right-side camshaft timing sprocket.

23. Remove the timing belt tensioner bolts and the tensioner.
24. Using the Camshaft Holding tool 09960-10010, or equivalent, slightly turn the left-side camshaft sprocket clockwise to loosen the tension spring. Then, disconnect

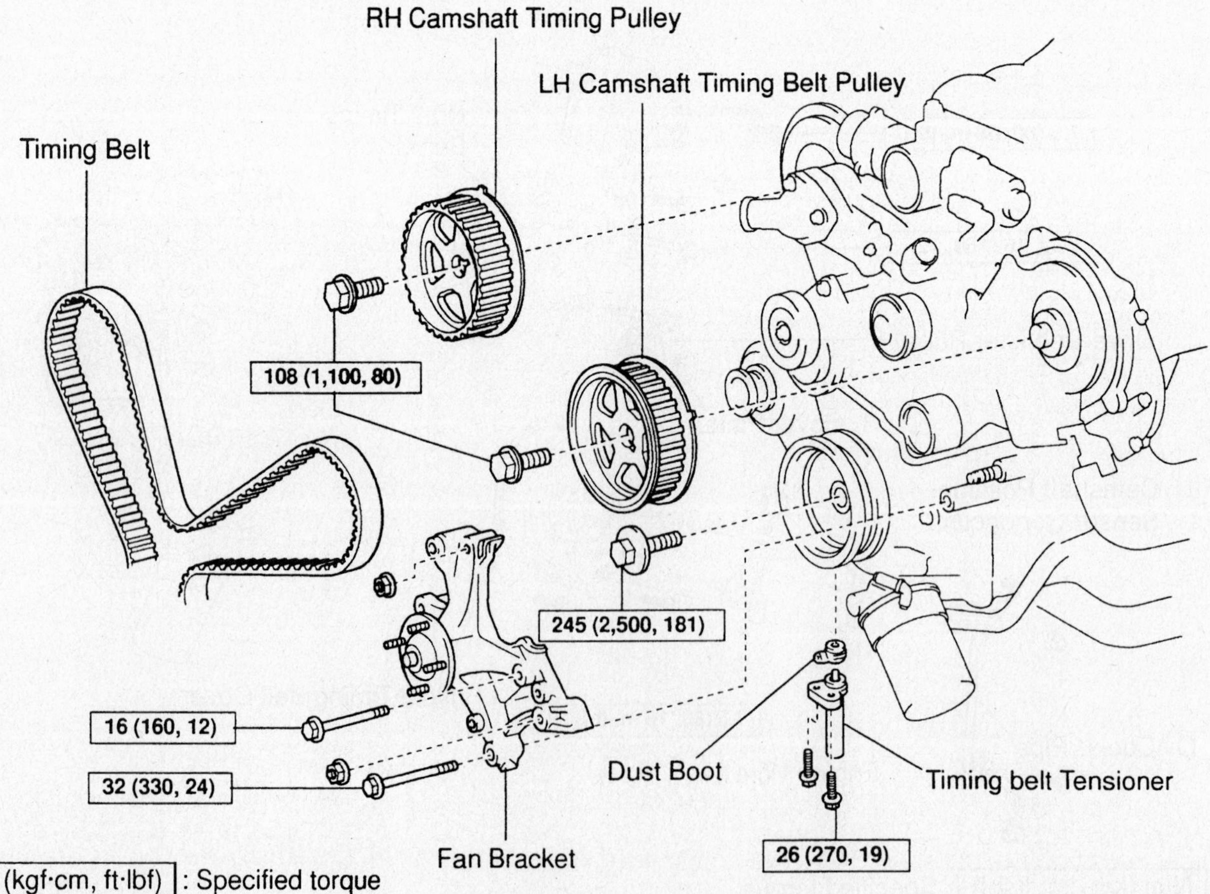

Exploded view of upper timing sprockets and components

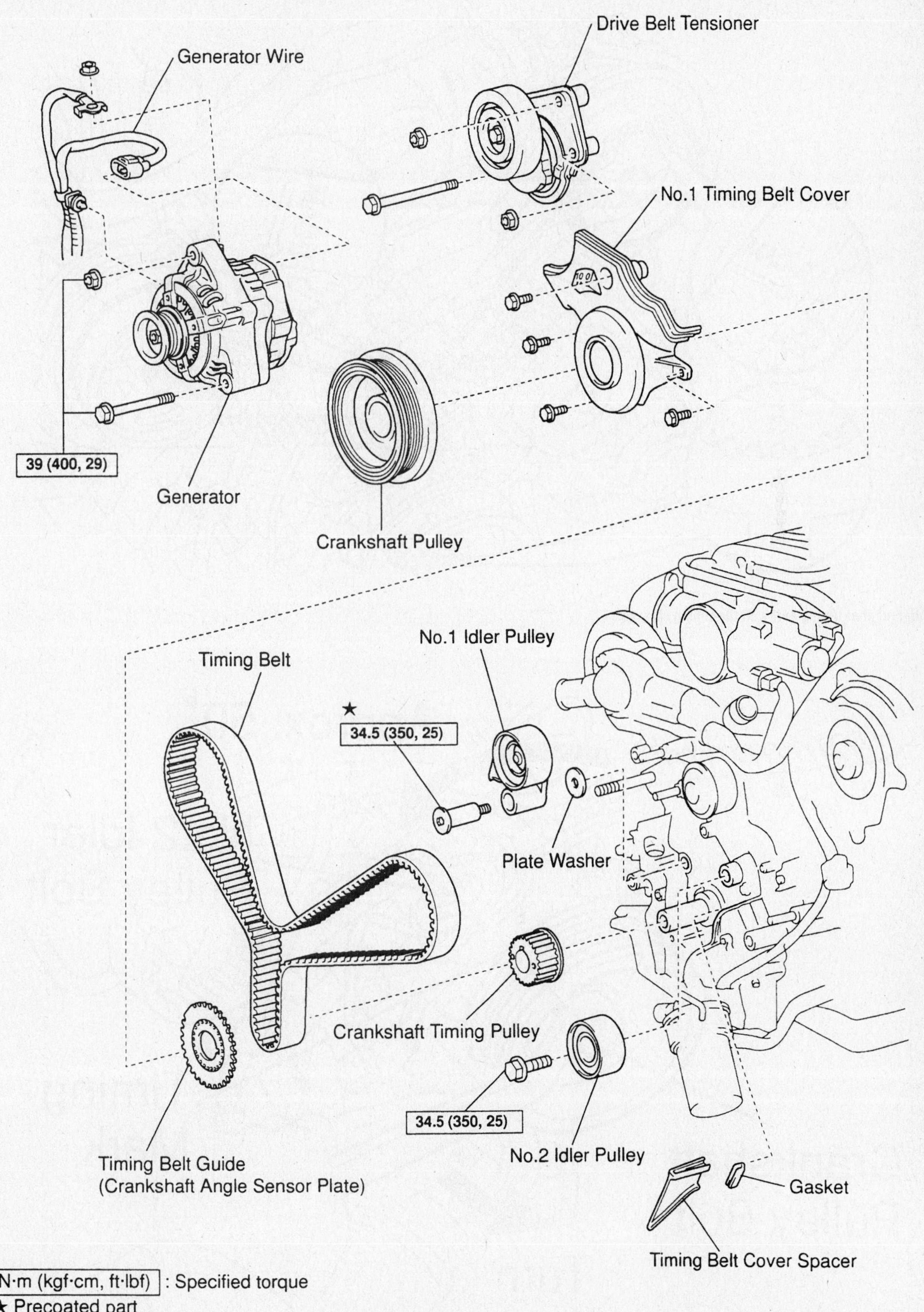

Exploded view of lower timing belt cover, sprockets and components

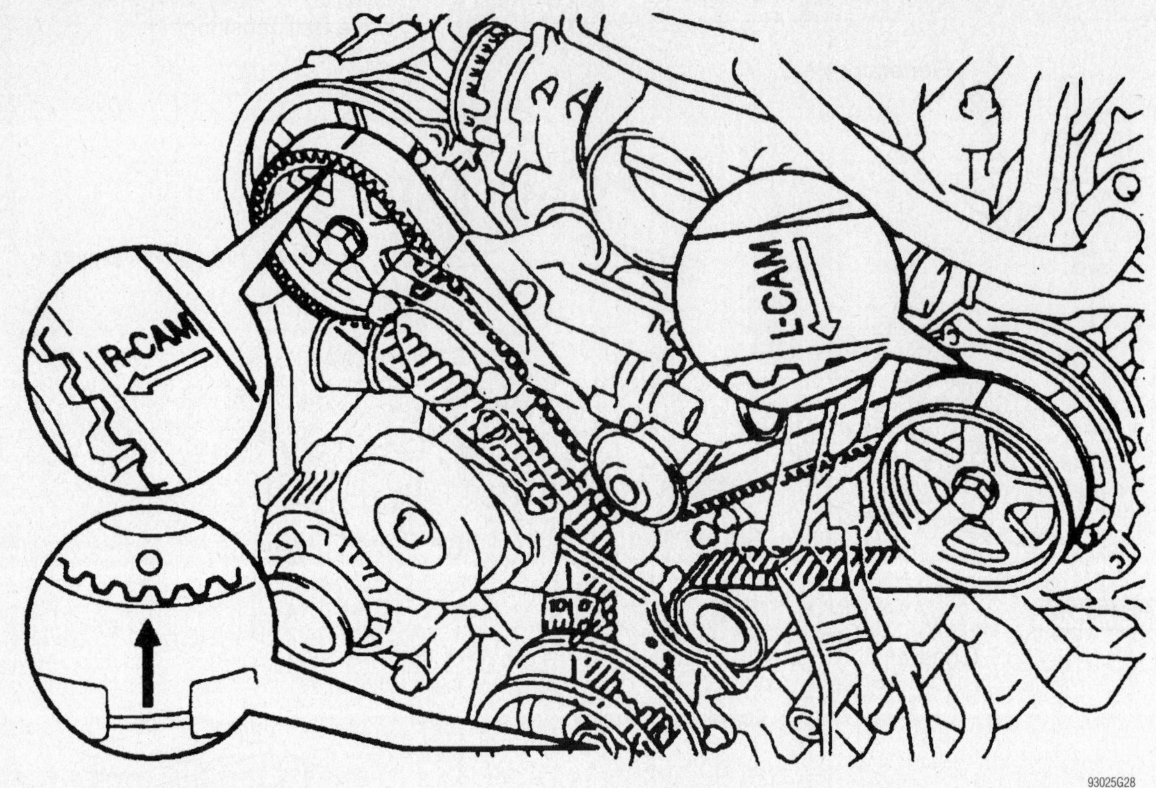

Alignment of timing belt with the timing sprockets

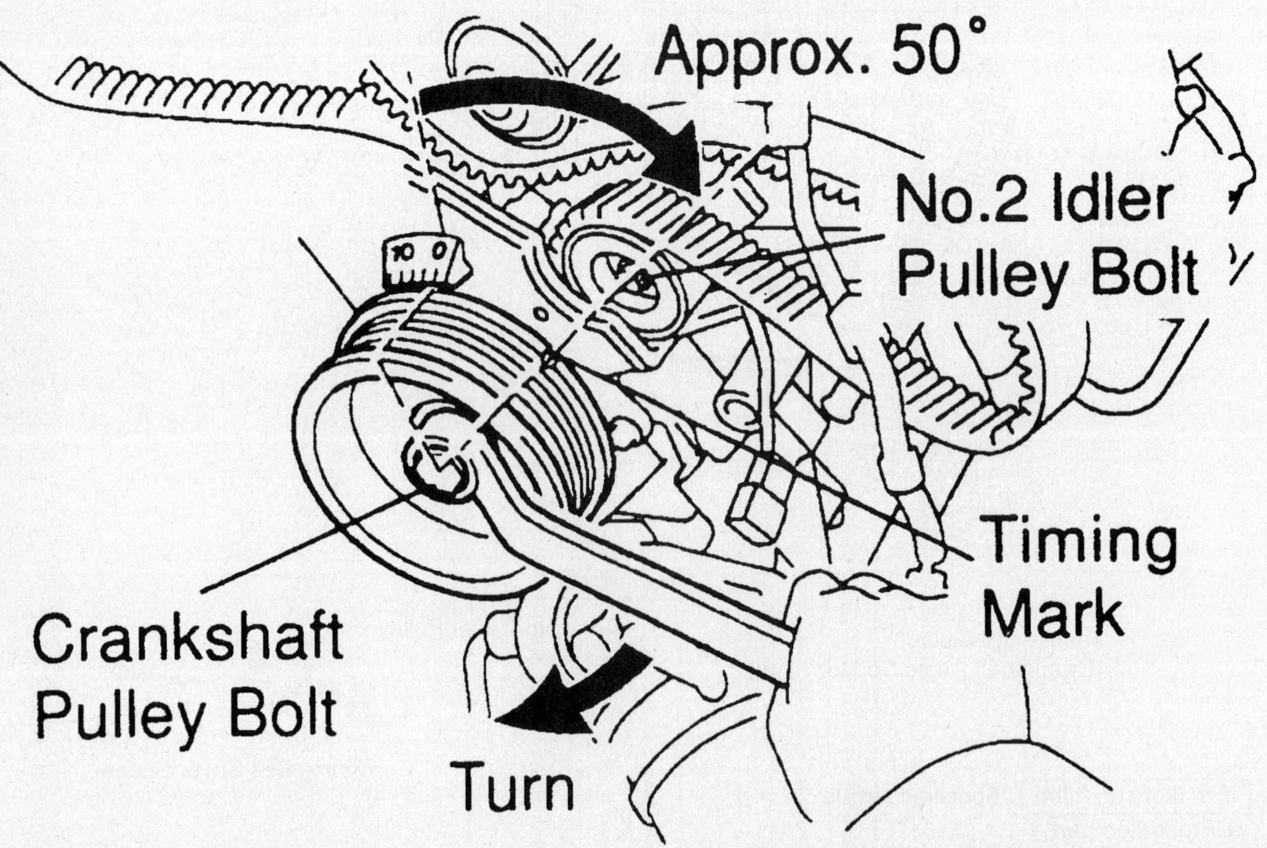

Aligning of crankshaft pulley timing mark with the center line of the crankshaft pulley bolt and the idler pulley bolt the timing belt from the camshaft sprockets.

25. Remove the alternator by performing the following procedures:

a. Disconnect the electrical connector from the alternator.

b. Remove the rubber cap/nut and disconnect the battery wire from the alternator.

c. Disconnect the wire clamp from the alternator cord clip.

d. Remove the alternator-to-engine nuts/bolts and the alternator.

26. Remove the serpentine drive belt tensioner nuts/bolts and the tensioner.

27. Using the Crankshaft Puller Assembly tool 09950-50012, or equivalent, press the crankshaft pulley from the crankshaft.

WARNING

DO NOT rotate the crankshaft pulley.

28. Remove the lower (No. 1) timing belt cover bolts and the cover.

29. Remove the timing belt guide, spacer and the timing belt.

To install:

➡With the timing belt removed, this is a perfect opportunity to inspect and/or replace the water pump.

30. Inspect the timing belt tensioner by performing the following procedures:

a. Inspect the seal for leakage; if leakage is suspected, replace the tensioner.

b. Using both hands to hold the tensioner facing upward, strongly press the pushrod against a solid surface. If the pushrod moves, replace the tensioner.

WARNING

Never hold the tensioner with the pushrod facing downward.

c. Measure the pushrod protrusion from the housing end, it should be 0.413–0.453 in. (10.5–11.5mm). If the protrusion is not as specified, replace the tensioner.

31. Temporarily install the timing belt by performing the following procedures:

a. Align the timing belt's installation mark with the crankshaft timing sprocket.

b. Install the timing belt on the crankshaft timing sprocket, the No. 1 idler pulley and the No. 2 idler pulley.

32. Install the gasket to the timing belt cover spacer and install the cover spacer.

33. Install the timing belt guide with the cup side facing outward.

34. Install the lower (No. 1) timing belt cover.

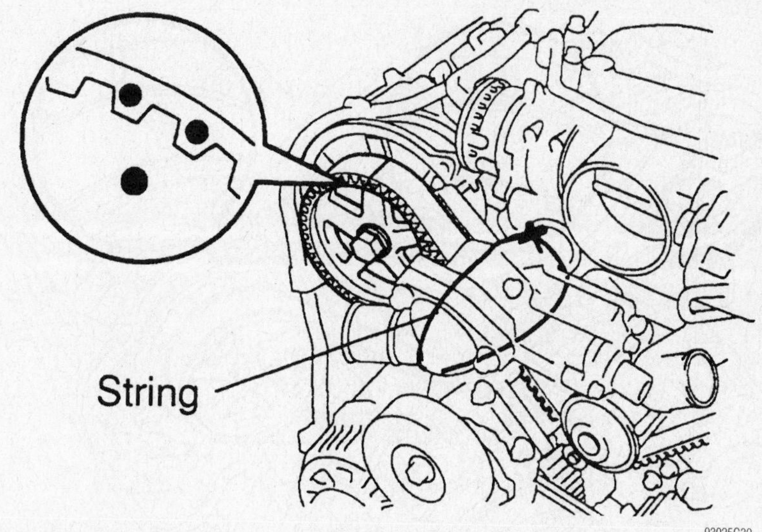

Securing the timing belt with string and matchmarking the camshaft with the timing belt

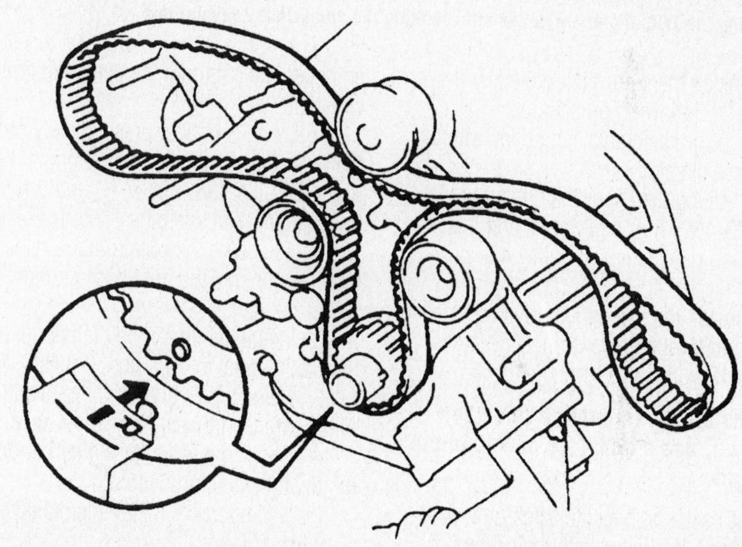

Installing the timing belt on the crankshaft sprocket

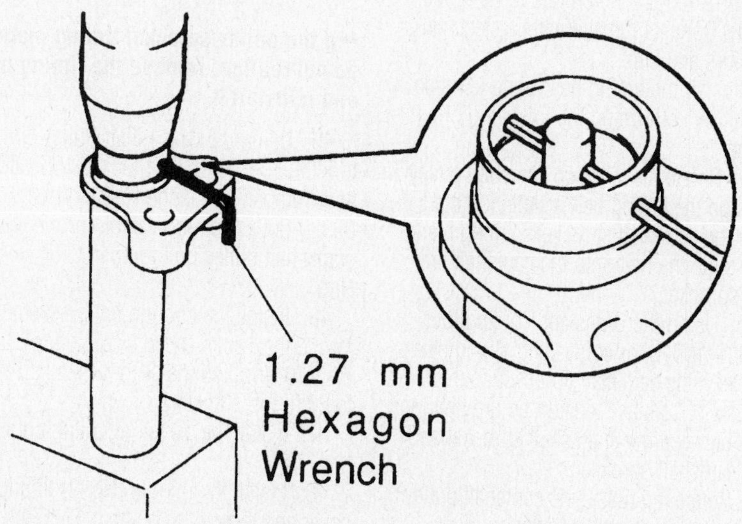

Securing the timing belt tensioner pushrod

Checking the TDC alignment marks after rotating the crankshaft 2 revolutions

35. Install the crankshaft pulley by performing the following procedures:
 a. Align the crankshaft pulley with the crankshaft key.
 b. Using the Crankshaft Installer tool 09223-46011, or equivalent, and a hammer, tap the crankshaft pulley into position.
36. Install the serpentine drive belt tensioner and torque the tensioner-to-engine bolts to 12 ft. lbs. (16 Nm).

➡To install the serpentine drive belt tensioner, use a bolt 4.18 in. (106mm) in length.

37. Check that the crankshaft pulley's timing mark is aligned with the centers of the idler pulley and crankshaft pulley bolts.
38. Install the alternator and torque the alternator-to-engine nuts/bolts to 29 ft. lbs. (39 Nm). Connect the alternator's electrical connectors and clip.
39. Install the timing belt to the left-side camshaft by performing the following procedures:
 a. Rotate the left-side camshaft pulley to align the timing belt installation mark with the camshaft sprocket's timing mark and slide the belt onto the camshaft timing sprocket.
 b. Using the Camshaft Holding tool 09960-10010, or equivalent, slightly turn the left-side camshaft sprocket counterclockwise to place tension on the timing belt between the crankshaft sprocket and the camshaft sprocket.
40. Rotate the right-side camshaft pulley to align the timing belt installation mark with the camshaft sprocket's timing mark and slide the belt onto the camshaft timing sprocket.
41. Using a vertical press, slowly press the pushrod into the housing using 200–2205 lbs. (981–9807 N) until the holes align, then, install a 1.27mm Allen® wrench to secure the pushrod and release the press. Install the dust boot on the tensioner housing.
42. Install the timing belt tensioner and torque the bolts to 19 ft. lbs. (26 Nm).
43. Using a pair of pliers, remove the Allen® wrench from the tensioner housing.
44. Check the valve timing by performing the following procedure:
 a. Temporarily install the crankshaft pulley bolt.
 b. Slowly, rotate the crankshaft pulley 2 revolutions (CLOCKWISE) and realign the TDC marks.

➡If the pulley/sprocket timing marks do not realign, remove the timing belt and reinstall it.

45. Using the Crankshaft Pulley Holding tool 09213-70010, Bolt tool 90105-08076 and Companion Flange Holding tool 09330-00021, or equivalent, torque the crankshaft pulley bolt to 181 ft. lbs. (245 Nm).
46. Install the cooling fan bracket and torque the 12mm (head size) bolt to 12 ft. lbs. (16 Nm) and the 14mm (head size) bolt to 24 ft. lbs. (32 Nm).
47. Install the air conditioning compressor.
48. Install the middle (No. 2) timing belt cover and torque the bolts to 12 ft. lbs. (16 Nm).
49. Install the upper right-side (No. 3) timing belt cover and torque the bolts to 66 inch lbs. (7.5 Nm).
50. Install the upper left-side (No. 3) timing belt cover by performing the following procedures:
 a. Install the oil cooler tube and bolt.
 b. Feed the Camshaft Position Sensor (CPS) through the left-side (No. 3) timing belt cover hole.
 c. Install the left-side (No. 3) timing belt cover and torque the bolts to 66 inch lbs. (7.5 Nm).
 d. Install the wire grommet to the left-side (No. 3) timing belt cover.
 e. Install the sensor connector to the connector bracket and connect the sensor connector.
 f. Install the sensor wire and the engine wire to the clamps on the left-side (No. 3) timing belt cover.
51. Install the drive belt idler pulley and cover plate; then, torque the pulley bolt to 27 ft. lbs. (37 Nm).
52. To complete the installation, reverse the removal procedures.
53. Refill the cooling system and connect the negative battery cable.

Timing Chain, Sprockets, Front Cover and Seal

REMOVAL & INSTALLATION

4.0L Engine

1. Remove the steering gear.
2. Remove the front differential carrier.
3. Drain the coolant.
4. Remove the battery.
5. Drain the engine oil.
6. Remove the V-bank cover.
7. Remove the radiator upper support seal.
8. Remove the accessory drive belt.
9. Remove the fan.
10. Remove the vent hose.
11. Remove the air cleaner.
12. Remove the oil dipstick tube.
13. Remove the water inlet.
14. Remove the power steering pump. Don't disconnect the lines.
15. Remove the alternator. Don't disconnect the lines.
16. Remove the A/C compressor. Don't disconnect the lines.
17. Remove the accessory drive belt tensioner.
18. Remove the idler pulleys.
19. Remove the crankshaft pulley.
20. Remove the lower oil pan.

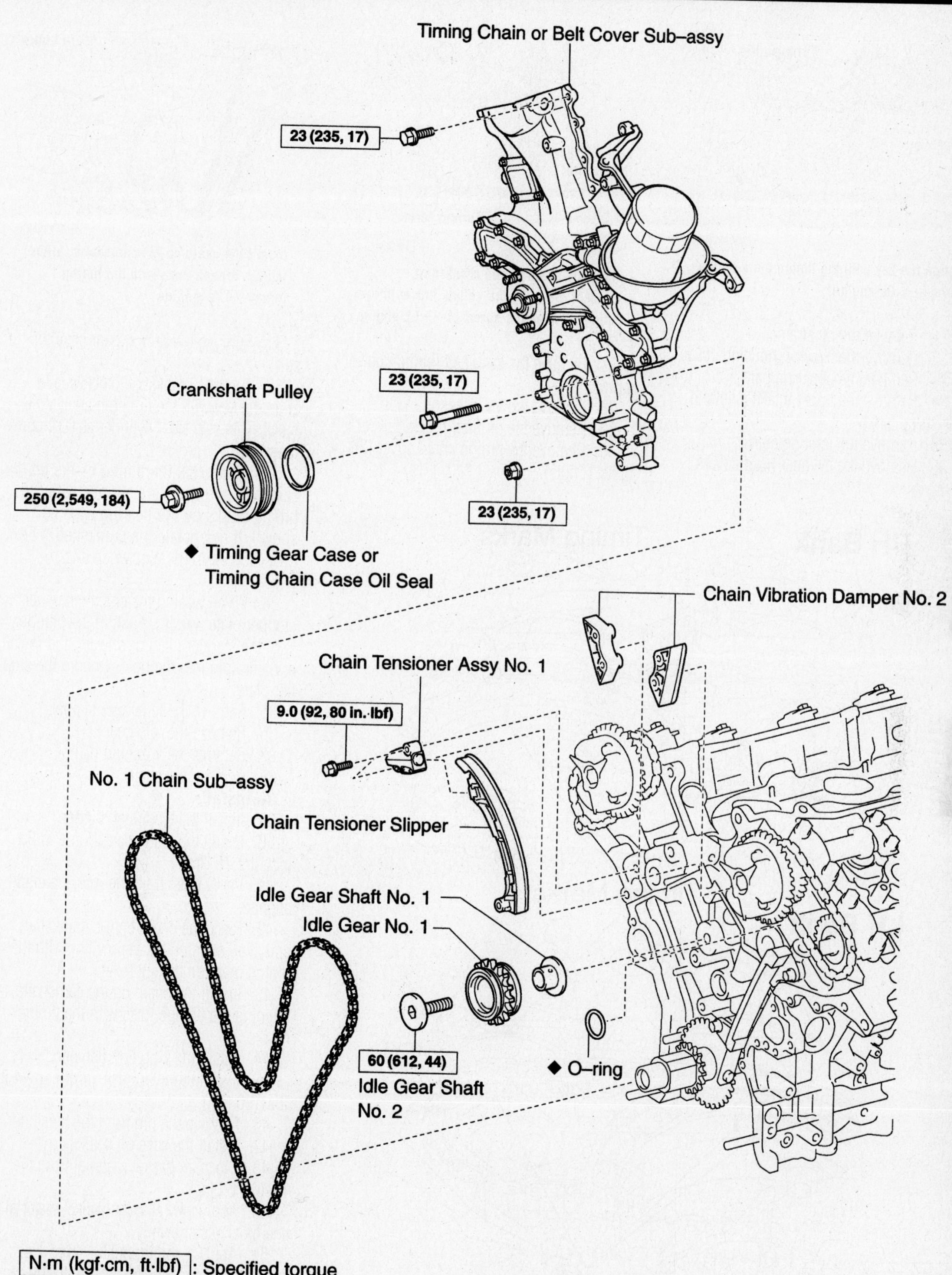
Timing Chain or Belt Cover Sub–assy
23 (235, 17)
Crankshaft Pulley
23 (235, 17)
250 (2,549, 184)
23 (235, 17)
◆ Timing Gear Case or
Timing Chain Case Oil Seal
Chain Vibration Damper No. 2
Chain Tensioner Assy No. 1
9.0 (92, 80 in.·lbf)
No. 1 Chain Sub–assy
Chain Tensioner Slipper
Idle Gear Shaft No. 1
Idle Gear No. 1
60 (612, 44)
Idle Gear Shaft
No. 2
◆ O–ring
N·m (kgf·cm, ft·lbf) : Specified torque
◆ Non–reusable part
67170-4RUN-G152

Timing chain and related parts—4.0L engine

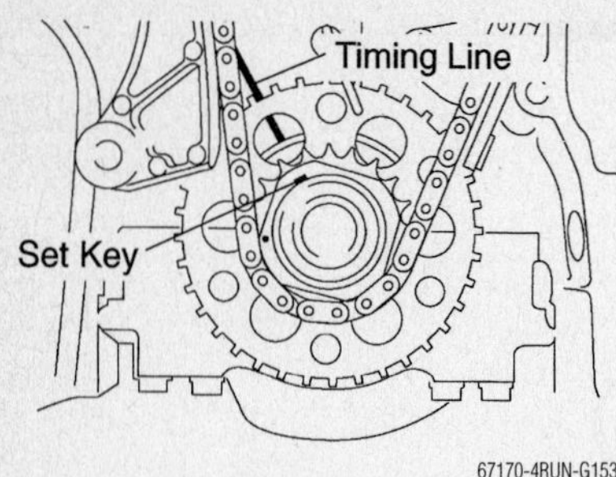

Align the key with the timing mark on the block—4.0L engine

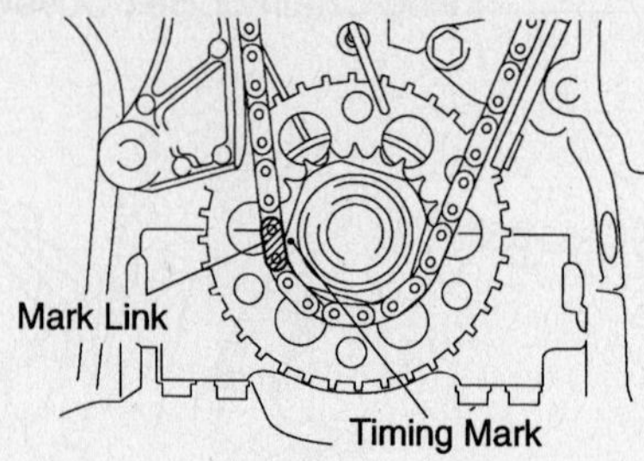

Place the chain on the crankshaft sprocket, aligning the yellow link with the timing mark on the sprocket—4.0L engine

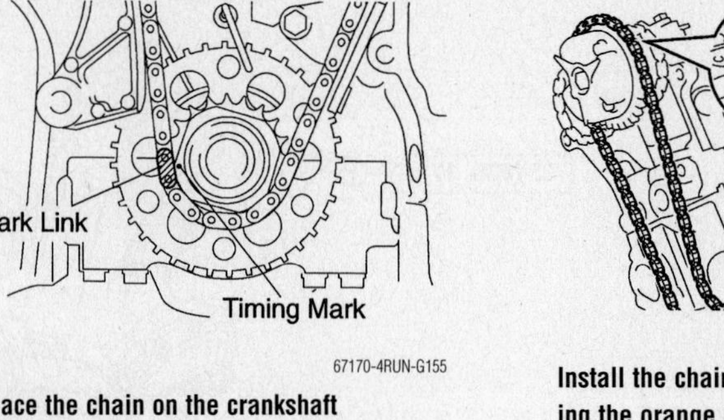

Install the chain on the camshafts, aligning the orange links with the timing marks—4.0L engine

21. Remove the oil strainer.
22. Remove the flywheel housing cover.
23. Remove the upper oil pan.
24. Remove the upper intake manifold (air surge tank).
25. Remove the ignition coil.
26. Remove the cylinder head covers.
27. Remove the camshaft timing control valve.
28. Remove the VVT sensor.
29. Remove the oil filter bracket.
30. Remove the timing chain cover (24 bolts, 2 nuts).

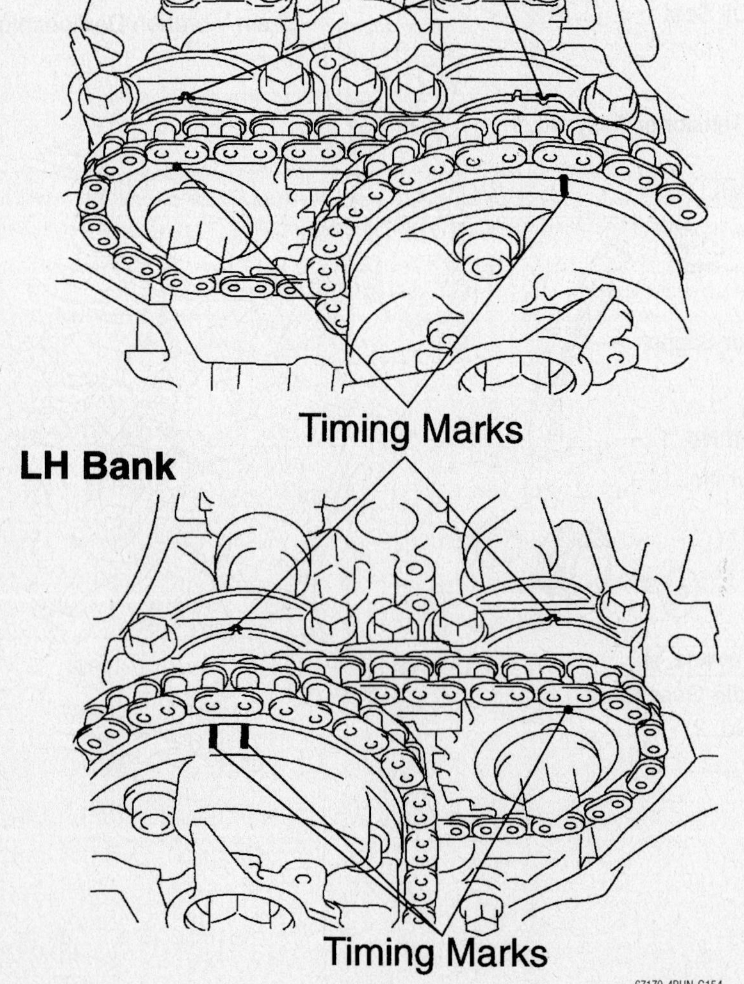

Check that the timing marks on the camshaft sprockets are aligned with the timing marks on the bearing caps—4.0L engine

31. Remove the timing gear case oil seal.
32. Turn the crankshaft clockwise to align the key with the timing mark on the block. This sets No.1 cylinder at TDC compression.
33. Check that the timing marks on the camshaft sprockets are aligned with the timing marks on the bearing caps as shown. If not, rotate the crankshaft 1 full turn (360 degrees) clockwise and recheck.
34. While turning the chain tensioner stopper plate upward, push in the plunger. Turn the stopper plate downward and install a 3.5mm pin into the holes to hold the stopper plate.
35. Remove the tensioner slipper.
36. Remove the idle gear.
37. Remove the vibration damper.
38. Remove the chain.

To install:

39. Install the tensioner slipper.
40. Install the tensioner. Torque to 80 inch lbs. (9 Nm).
41. Verify that all timing marks are still aligned.
42. Place the chain on the crankshaft sprocket, aligning the yellow link with the timing mark on the sprocket.
43. Install the chain on the camshafts, aligning the orange links with the timing marks.
44. Install the vibration damper.
45. Install the idler gear. Torque to 44 ft. lbs. (60 Nm).
46. Remove the pin from the tensioner.
47. Install the case oil seal.
48. Clean all gasket material from the case and block.
49. Apply new silicone gasket material to the block, as shown
50. Apply a continuous 3–4mm dia. bead of silicone gasket material to the case as shown.
51. Align the keyway of the oil pump drive rotor with the rectangular portion of

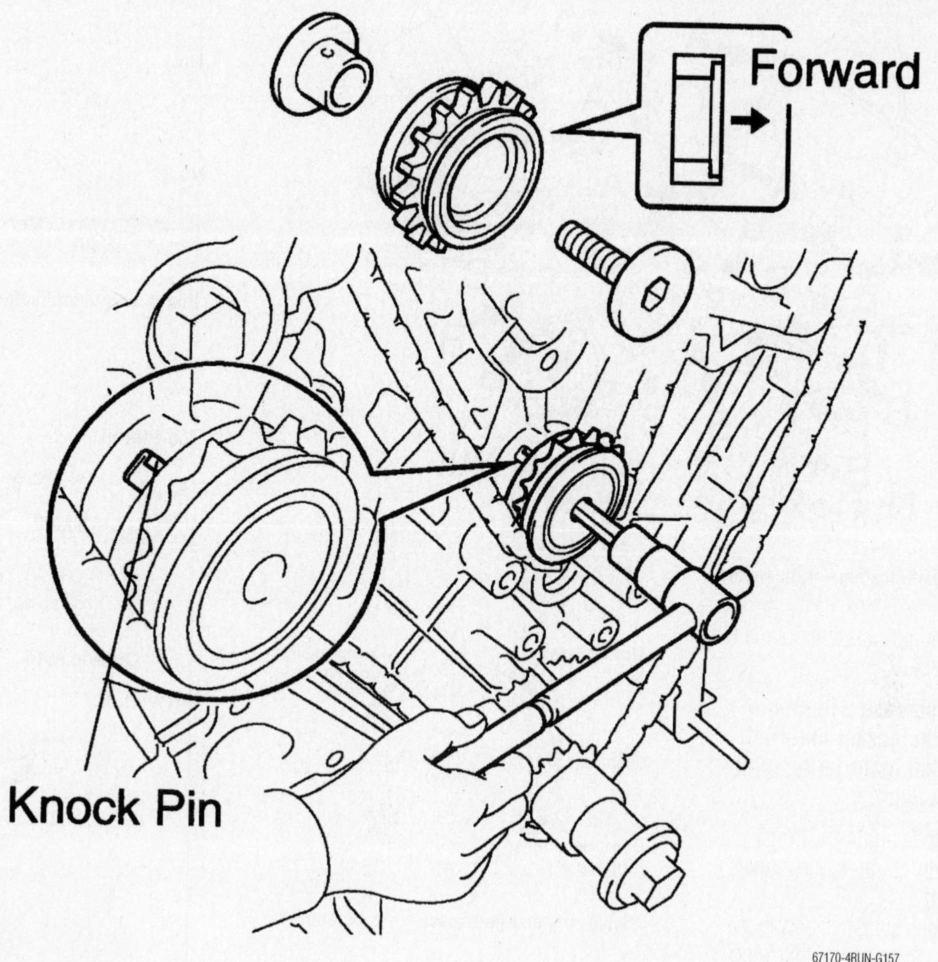

Idler gear installation—4.0L engine

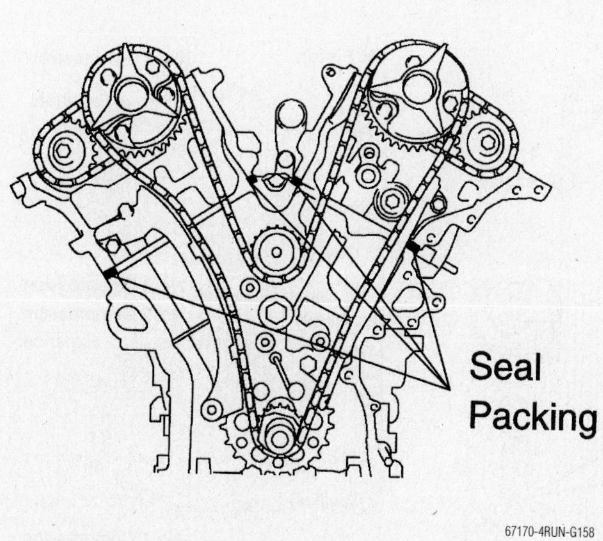

Gasket material application points on the block—4.0L engine

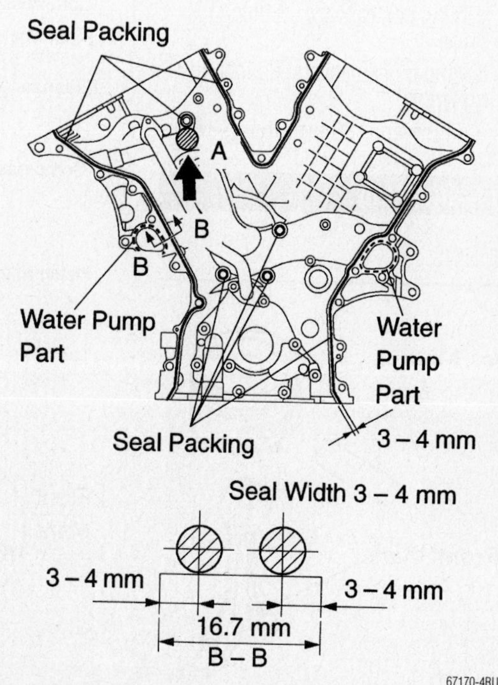

Apply a continuous 3–4mm dia. bead of silicone gasket material to the case—4.0L engine

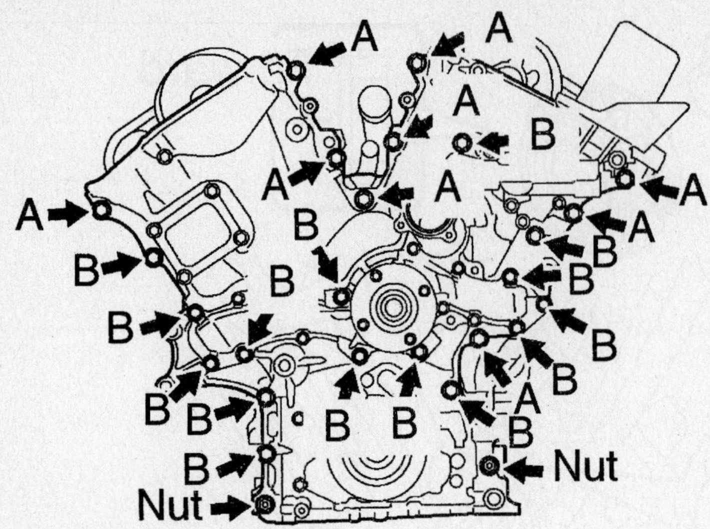

Timing case bolt identification—4.0L engine

the crankshaft timing sprocket, and slide the timing case into place.

➡Install the timing case with 3 minutes of applying the gasket material. Do not apply gasket material to point "A" in the illustration.

52. Install the bolts and nuts. Bolts marked "A" are 25mm long; those marked "B" are 55mm long.

53. The remainder of installation is the reverse of removal. Observe the following torques:

- Oil filter bracket: 14 ft. lbs. (19 Nm)
- VVT sensor: 71 inch lbs. (8 Nm)
- Timing control valve: 80 inch lbs. (9 Nm)
- Cylinder head covers: 80 inch lbs. (9 Nm)
- ignition coil: 80 inch lbs. (9 Nm)

Piston and Ring

POSITIONING

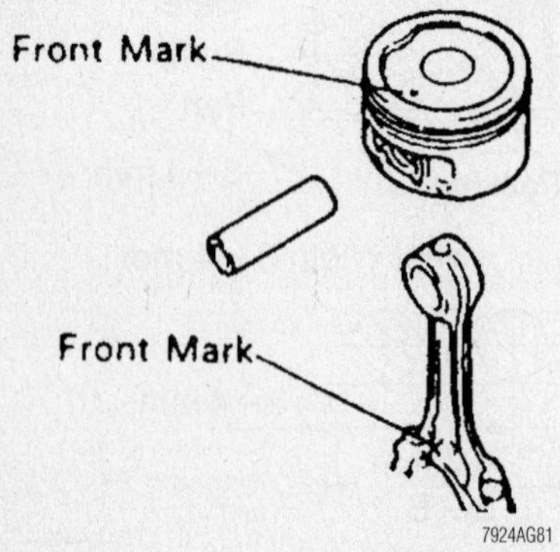

Piston to connecting rod assembly—3.4L engines

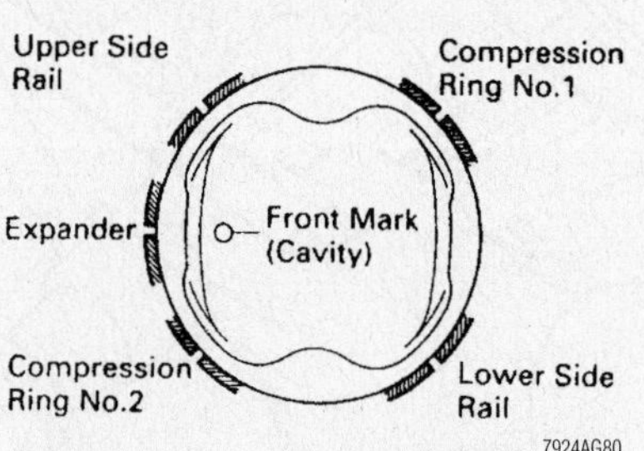

Piston ring end-gap spacing—3.4L engine

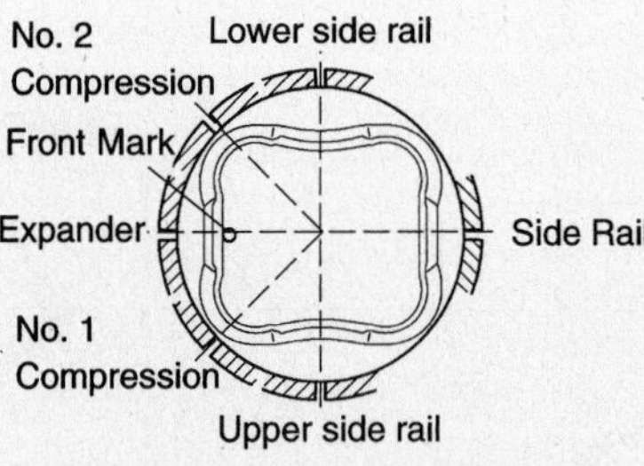

Piston ring end-gap spacing—4.0L engine

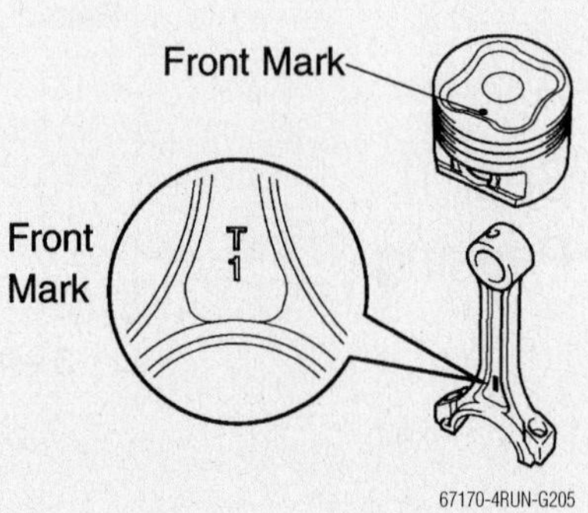

Piston to connecting rod assembly—4.0L engines

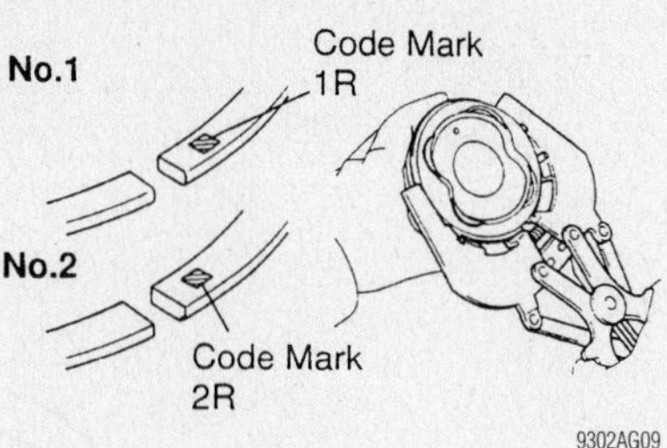

Piston ring identification—4.7L engine

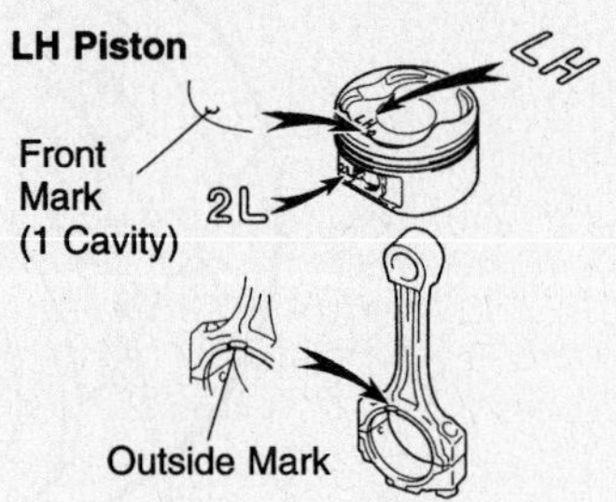

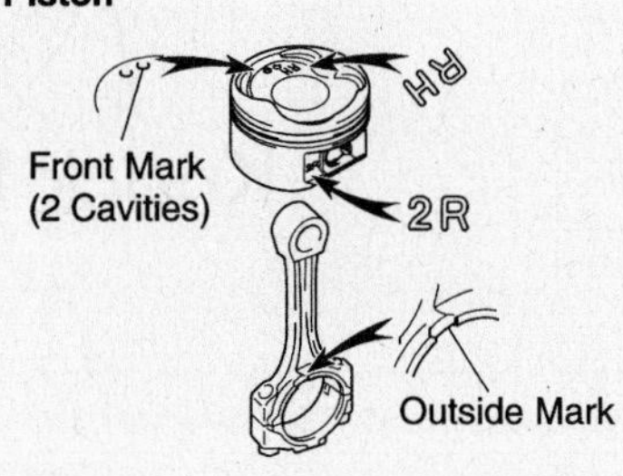

Piston to connecting rod assembly—4.7L engines

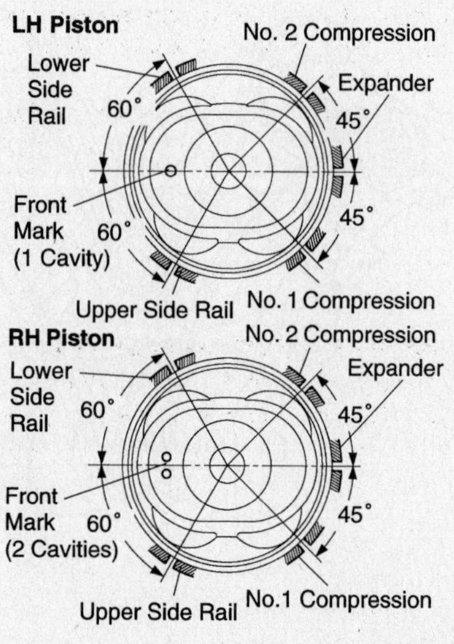

Piston ring end-gap spacing—4.7L engine

FUEL SYSTEM

Fuel System Service Precautions

Safety is the most important factor when performing not only fuel system maintenance but any type of maintenance. Failure to conduct maintenance and repairs in a safe manner may result in serious personal injury or death. Maintenance and testing of the vehicle's fuel system components can be accomplished safely and effectively by adhering to the following rules and guidelines.

- To avoid the possibility of fire and personal injury, always disconnect the negative battery cable unless the repair or test procedure requires that battery voltage be applied.
- Always relieve the fuel system pressure prior to disconnecting any fuel system component (injector, fuel rail, pressure regulator, etc.), fitting or fuel line connection. Exercise extreme caution whenever relieving fuel system pressure, to avoid exposing skin, face and eyes to fuel spray. Please be advised that fuel under pressure may penetrate the skin or any part of the body that it contacts.
- Always place a shop towel or cloth around the fitting or connection prior to loosening to absorb any excess fuel due to spillage. Ensure that all fuel spillage (should it occur) is quickly removed from engine surfaces. Ensure that all fuel soaked cloths or towels are deposited into a suitable waste container.
- Always keep a dry chemical (Class B) fire extinguisher near the work area.
- Do not allow fuel spray or fuel vapors to come into contact with a spark or open flame.
- Always use a back-up wrench when loosening and tightening fuel line connection fittings. This will prevent unnecessary stress and torsion to fuel line piping.
- Always replace worn fuel fitting O-rings with new. Do not substitute fuel hose or equivalent, where fuel pipe is installed.

Fuel System Pressure

RELIEVING

2001–02

1. Before servicing the vehicle, refer to the precautions in the beginning of this section.
2. Disconnect the negative battery terminal.
3. Place a catch-pan under the joint to be disconnected. A large quantity of fuel may be released when the joint is opened.
4. Wear eye or full face protection.
5. Place a shop towel over the area and slowly loosen the joint using a wrench of the correct size. Use a back-up wrench if needed.
6. Allow the fuel left in the line to bleed off slowly before fully disconnecting the joint.
7. Plug the opened lines immediately to prevent fuel spillage or the entry of dirt.
8. Dispose of the released fuel properly.
9. After connecting fuel lines, connect the negative battery cable and start the engine.
10. Check for leaks and repair as needed.

2003–04

1. Remove the fuel pump relay from the engine compartment relay block. Start the engine and allow it to run out of fuel. Turn the ignition off. Try to restart the engine to make sure there is no fuel pressure. Turn the ignition off, disconnect the battery ground and install the relay. Remove the fuel tnak cap.
2. Place a catch-pan under the joint to be disconnected. A large quantity of fuel may be released when the joint is opened.
3. Wear eye or full face protection.
4. Place a shop towel over the area and slowly loosen the joint using a wrench of the correct size. Use a back-up wrench if needed.
5. Allow the fuel left in the line to bleed off slowly before fully disconnecting the joint.
6. Plug the opened lines immediately to prevent fuel spillage or the entry of dirt.

Fuel Pump

REMOVAL & INSTALLATION

1. Before servicing the vehicle, refer to the precautions in the beginning of this section.
2. Relieve the fuel pressure.
3. Disconnect the negative battery cable from the battery.
4. Drain the fuel from the fuel tank.
5. Remove or disconnect the following:
   - Fuel tank
   - Fuel pump connector from the clamp
   - Access plate bolts, then pull out the fuel pump assembly from the fuel tank
   - Gasket(s) from the pump bracket
   - Fuel pump connector
   - Bracket from the lower side of the fuel pump
   - Fuel pump from the fuel hose
   - Rubber cushion, the clip and the fuel filter at the bottom of the fuel pump

To install:

6. Install or connect the following:
   - Fuel pump filter to the fuel pump with a new clip
   - Fuel pump to the fuel pump bracket
   - Fuel hose to the outlet port of the fuel pump
   - Fuel pump connector
   - Fuel pump assembly with a new gasket(s). Tighten the bolts to 31 inch lbs. (4 Nm).
   - Fuel pump connector to the clamp
   - Fuel tank
   - All electrical and fuel connections
   - Negative battery cable
7. Refill the fuel tank and check for leaks.

4.0L and 4.7L Engines

1. Before servicing the vehicle, refer to the precautions in the beginning of this section.
2. Relieve the fuel pressure.
3. Disconnect the negative battery cable from the battery.
4. Drain the fuel from the fuel tank.
5. Remove the rear floor mat.
6. Remove the floor access cover.
7. Disconnect the wiring from the pump.
8. Disconnect the filler hose.
9. Remove the skid plate.
10. Disconnect the supply and return lines.
11. Disconnect the breather hose.
12. Disconnect the vent hose.
13. Support the tank with a jack. Remove the bolts and tank bands and lower the tank from the vehicle.
14. Disconnect the supply and return lines from the pump and tank.
15. Using a special tool, such as 09808-14020, or equivalent, unscrew the retainer. Pull out the module.
16. Installation is the reverse of removal. Use a new gasket on the module. When installing the module, make sure that the tab on the plate fits into the slot in the tank. Use

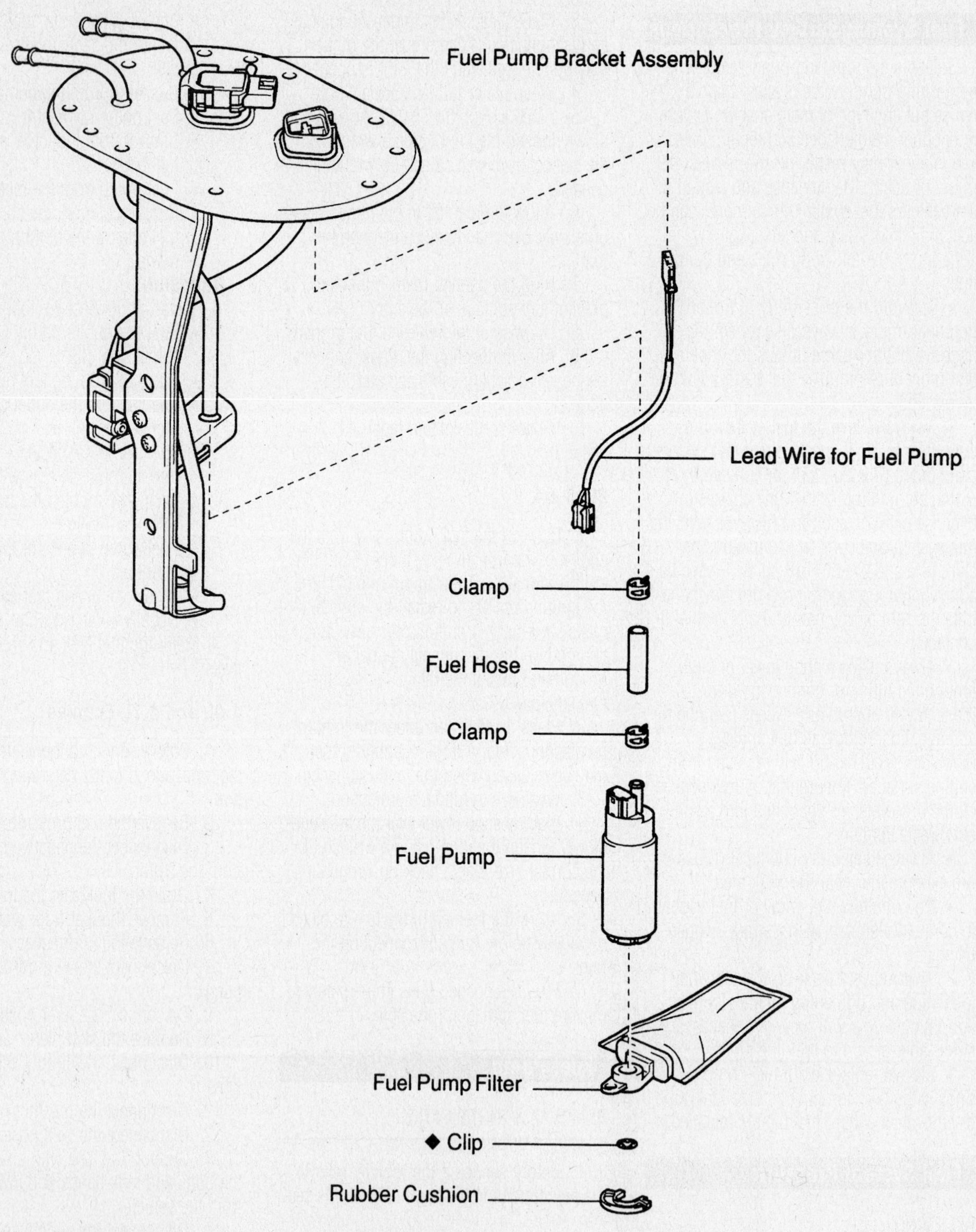

Fuel pump and related parts—3.4L engine

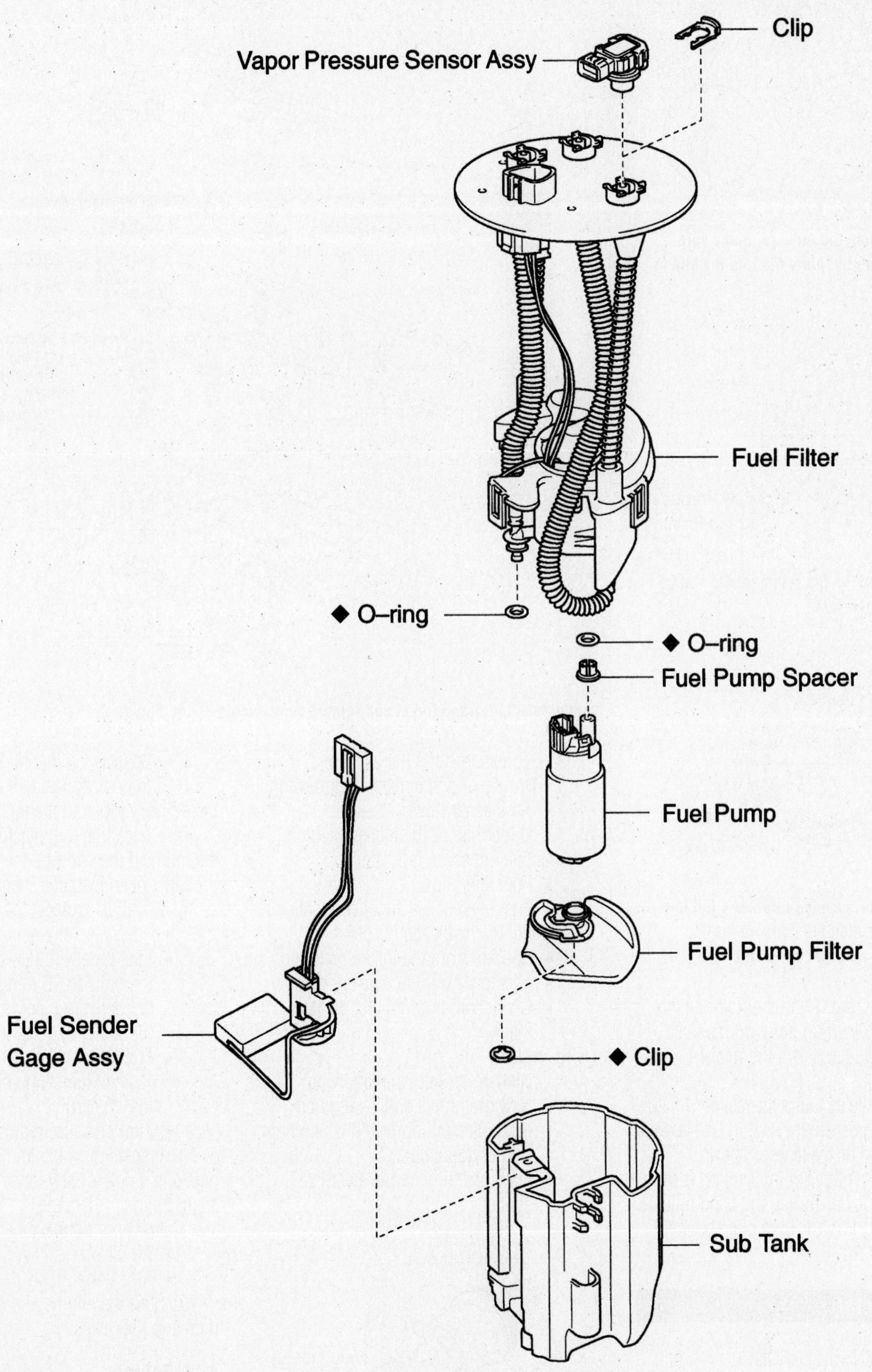

Fuel pump and related parts—4.0L and 4.7L engines

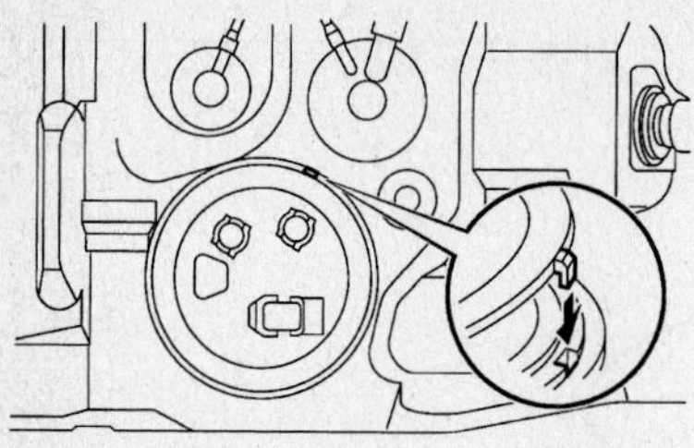

When installing the module, make sure that the tab on the plate fits into the slot in the tank

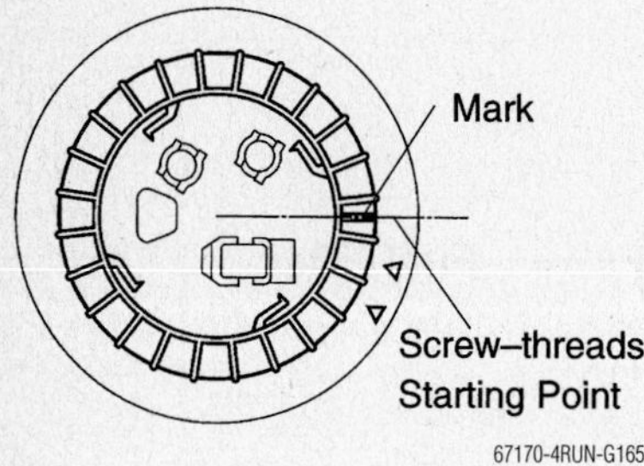

Align the mark on the retainer with a starting point on the tank

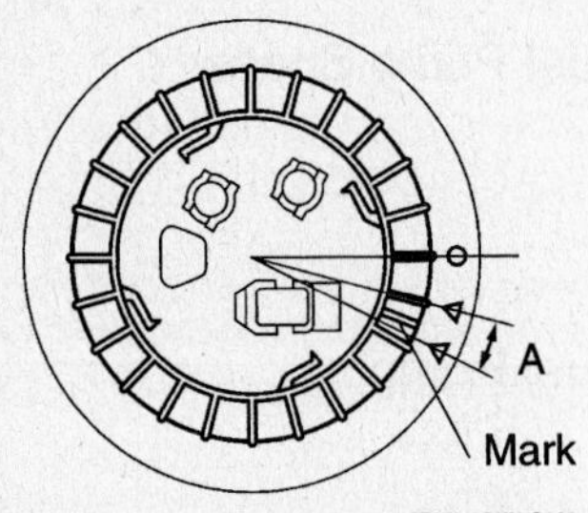

Using the tool, turn the retainer 1 full turn more and position the mark on the retainer into a range "A" on the tank

a new retainer ring. Align the mark on the retainer with a starting point on the tank. Hold the module and turn the retainer 1 full turn, by hand.

17. Using the tool, turn the retainer 1 full turn more and position the mark on the retainer into a range "A" on the tank, as shown.

18. The remainder of installation is the reverse of removal. Torque the band bolts to 30 ft. lbs. (40 Nm); the skid plate bolts to 15 ft. lbs. (20 Nm).

Fuel Injector

REMOVAL & INSTALLATION

3.4L Engine

1. Before servicing the vehicle, refer to the precautions in the beginning of this section.

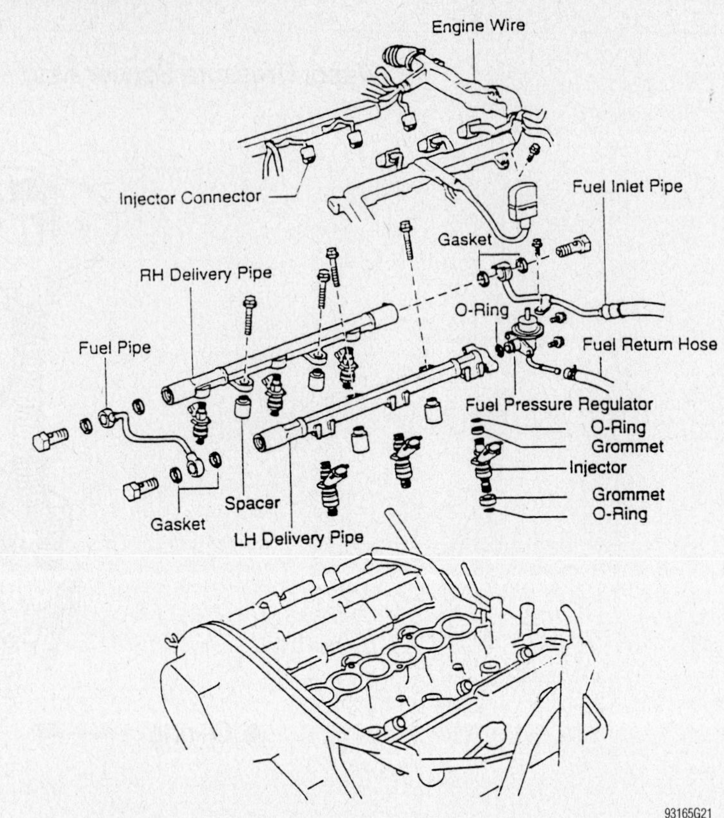

Fuel injector arrangement and related components—3.4L engine

2. Depressurize the fuel system.
3. Remove or disconnect the following:
   - Air cleaner hose
   - Upper half of the intake manifold
   - Fuel pressure regulator
   - Fuel inlet pipe
   - Fuel injector electrical connections
   - Fuel rail with the injectors
   - Spacers from the intake manifold
   - Injectors from the delivery pipes
   - O-rings and grommets, discard them

To install:

4. Install or connect the following:
   - New grommets and O-rings on each injector, lubricated with a light coat of gasoline
   - Fuel injector with the electrical connector facing outward

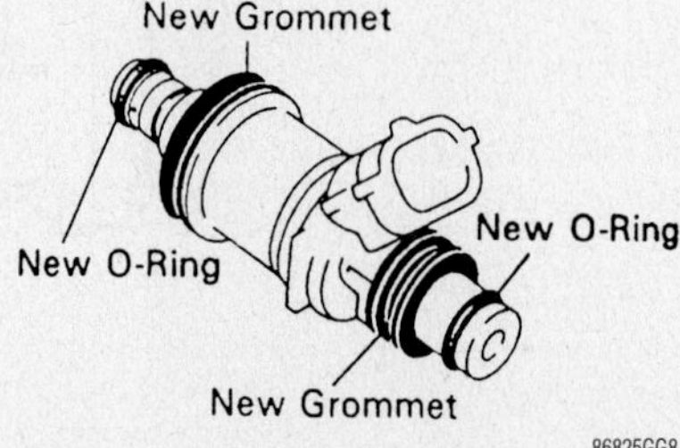

Install new O-rings and grommets on each injector—3.4L engine

   - Spacers on the intake manifold
5. Temporarily install the bolts to hold the delivery pipes to the intake manifold.
6. Check that the injectors rotate smoothly. If they do not, the O-rings have probably been installed incorrectly.
7. Install or connect the following:
   - Fuel injector electrical connectors
   - Fuel pipe with new gaskets. Tighten the bolts to 25 ft. lbs. (34 Nm) and the delivery pipes-to-intake manifold bolts to 10 ft. lbs. (13 Nm).
   - Fuel pipe union with new gaskets. Tighten the clamp bolt to 71 inch lbs. (8 Nm).
   - Fuel pressure regulator
8. Inspect the vacuum lines and connections. Look for any loose connections, sharp bends or damage.
9. Install or connect the following:
   - Air cleaner
   - Air cleaner hose
10. Start the engine and check for vacuum and fuel leaks.

4.0L Engine

1. Before servicing the vehicle, refer to the precautions in the beginning of this section.
2. Depressurize the fuel system.
3. Drain the coolant.

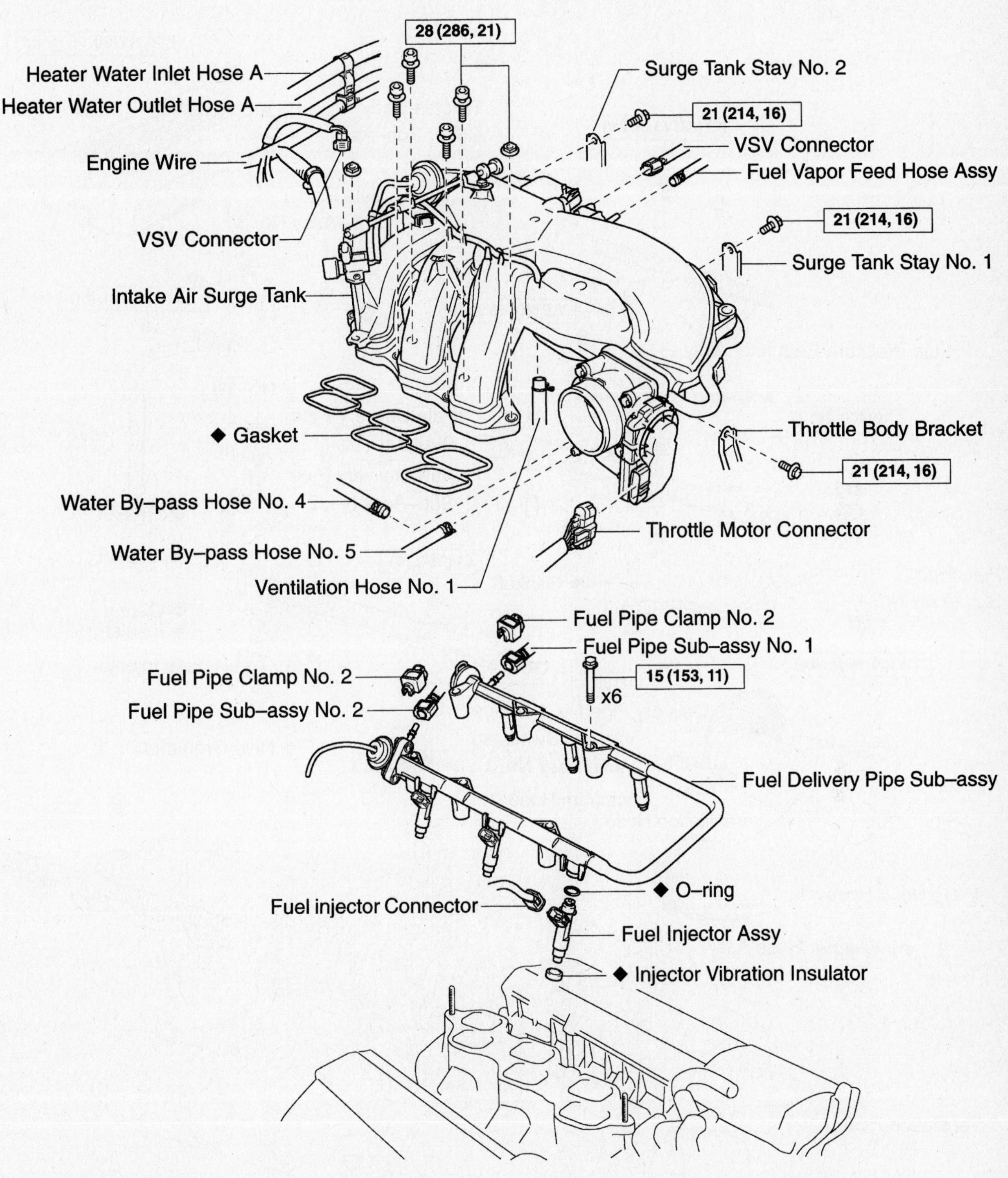
28 (286, 21)
Heater Water Inlet Hose A
Heater Water Outlet Hose A
Engine Wire
VSV Connector
Intake Air Surge Tank
Surge Tank Stay No. 2
21 (214, 16)
VSV Connector
Fuel Vapor Feed Hose Assy
21 (214, 16)
Surge Tank Stay No. 1
◆ Gasket
Water By–pass Hose No. 4
Water By–pass Hose No. 5
Ventilation Hose No. 1
Throttle Body Bracket
21 (214, 16)
Throttle Motor Connector
Fuel Pipe Clamp No. 2
Fuel Pipe Sub–assy No. 1
15 (153, 11)
x6
Fuel Pipe Clamp No. 2
Fuel Pipe Sub–assy No. 2
Fuel Delivery Pipe Sub–assy
◆ O–ring
Fuel injector Connector
Fuel Injector Assy
◆ Injector Vibration Insulator
N·m (kgf·cm, ft·lbf) : Specified torque
◆ Non–reusable part
67170-4RUN-G167

Fuel injectors and related parts—4.0L engine

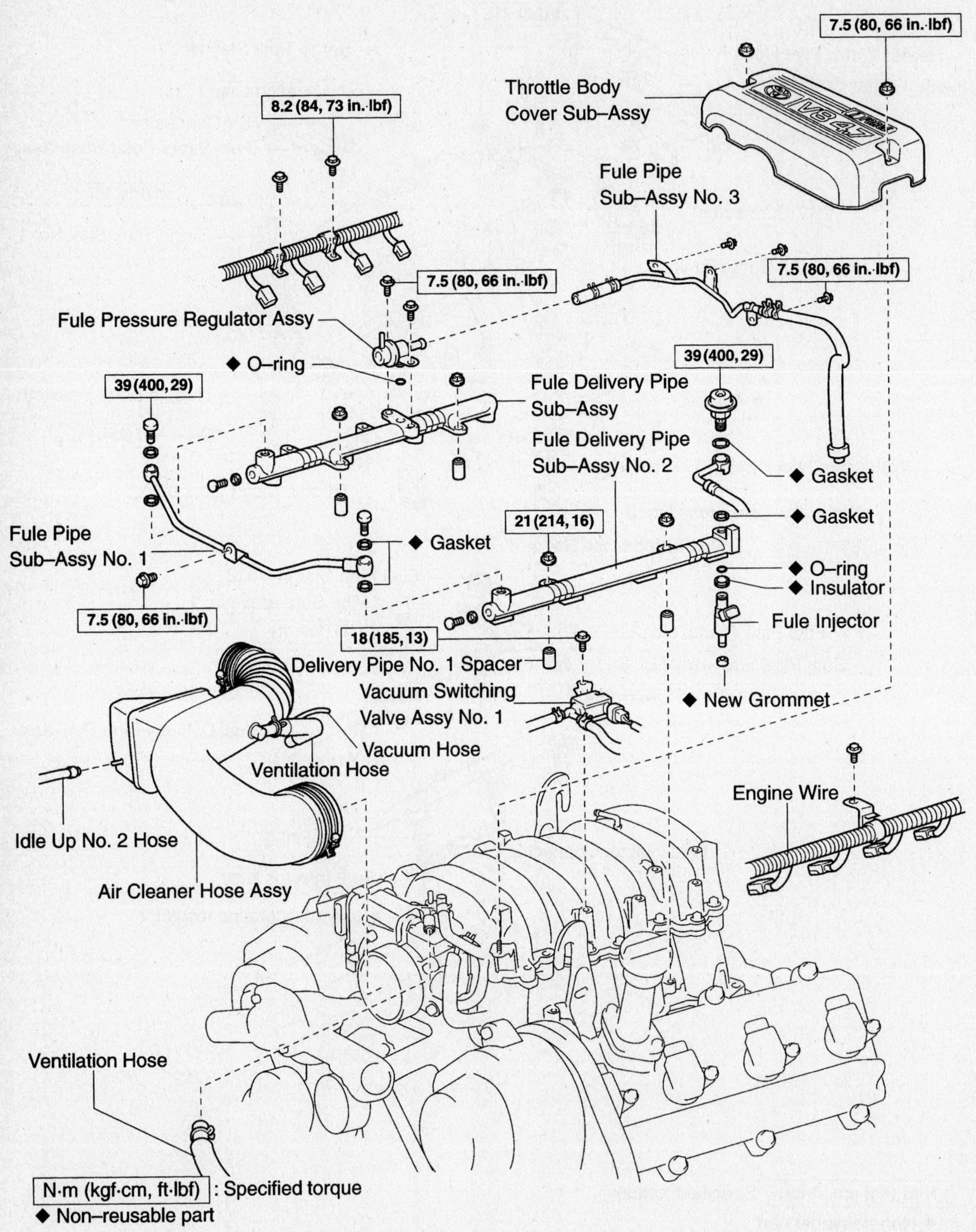

7.5 (80, 66 in.·lbf)
Throttle Body Cover Sub–Assy
8.2 (84, 73 in.·lbf)
Fule Pipe Sub–Assy No. 3
7.5 (80, 66 in.·lbf)
7.5 (80, 66 in.·lbf)
Fule Pressure Regulator Assy
◆ O–ring
39 (400, 29)
39 (400, 29)
Fule Delivery Pipe Sub–Assy
Fule Delivery Pipe Sub–Assy No. 2
◆ Gasket
◆ Gasket
◆ Gasket
21 (214, 16)
Fule Pipe Sub–Assy No. 1
◆ O–ring
◆ Insulator
7.5 (80, 66 in.·lbf)
Fule Injector
18 (185, 13)
Delivery Pipe No. 1 Spacer
Vacuum Switching Valve Assy No. 1
◆ New Grommet
Vacuum Hose
Ventilation Hose
Engine Wire
Idle Up No. 2 Hose
Air Cleaner Hose Assy
Ventilation Hose
N·m (kgf·cm, ft·lbf) : Specified torque
◆ Non–reusable part
67170-4RUN-G168

Fuel injectors and related parts—4.7L engine

4. Remove the V-bank cover.
5. Remove the air cleaner.
6. Disconnect the water bypass hoses.
7. Disconnect the fuel vapor hose.
8. Disconnect the vent hose.
9. Remove the upper intake manifold (air surge tank).
10. Disconnect the fuel supply lines.
11. Disconnect the injector wires.
12. Remove the injector bolts and pull the injector/fuel rail assembly from the heads.
13. Remove the injectors from the fuel rail.

To install:

14. Install a new insulator on each injector.
15. Coat new O-rings with gasoline and install them on the injectors.
16. Install the injectors on the fuel rail with a twisting motion. Position the connector outward.
17. Install the assembly onto the heads. Install the bolts finger-tight. Make sure each injector twists freely If any doesn't, replace the O-ring.
18. Tighten the bolts to 11 ft. lbs. (15 Nm).
19. Connect the wiring.
20. The remainder of installation is the reverse of removal.

4.7L Engine

1. Before servicing the vehicle, refer to the precautions in the beginning of this section.
2. Depressurize the fuel system.
3. Remove the air cleaner.
4. Disconnect the fuel pipes.
5. Disconnect the fuel hose from the pulsation damper.
6. Disconnect the vacuum switching valve.
7. Disconnect the wiring from the injectors.
8. Remove the nuts and remove the left and right fuel rails with the injectors.
9. Remove the injectors from the rails. Discard the O-rings, grommets and insulators from each injector.

To install:

10. Install a new grommet and insulator on each injector.
11. Coat each new O-ring with clean gasoline and install them on the injectors.
12. Install the injectors on the rails with a twisting motion. Make sure that they can rotate freely.
13. Install the rails/injectors on the intake manifold. Hand-tighten the nuts. Check that each injector can rotate freely. If not, replace the O-ring. Tighten the nuts to 16 ft. lbs. (21 Nm).
14. Using new gaskets, connect the fuel line to the pulsation damper. Torque to 29 ft. lbs. (39 Nm).
15. Install the vacuum switching valve. Torque to 13 ft. lbs. (18 Nm).
16. Connect fuel pipe No.3. Torque to 66 inch lbs.
17. Connect fuel pipe No.1 using new gaskets. Torque to 29 ft. lbs. (39 Nm).
18. Install the air cleaner.
19. Install the throttle body cover.

DRIVE TRAIN

Automatic Transmission

REMOVAL & INSTALLATION

2001–02

MODEL A340D, A340E AND A340H TRANSMISSIONS

➡The transfer case and the transmission should be removed as an assembly.

1. Before servicing the vehicle, refer to the precautions in the beginning of this section.
2. Remove or disconnect the following:
   - Negative battery cable
   - Air cleaner assembly, if necessary
   - Transmission throttle cable from the throttle body
   - Engine undercover
3. Drain the transmission and transfer case (if applicable) fluid.
4. Remove or disconnect the following:
   - Wiring connectors from the transmission and transfer case, if applicable.
   - Starter
5. Matchmarks on the front and rear driveshaft flanges and the differential pinion flanges. These marks must be aligned during installation.
6. Remove or disconnect the following:
   - Front and rear driveshaft flanges.
   - Center bearing bracket bolts, if equipped with a 2-piece driveshaft
   - Driveshaft
   - Speedometer cable
   - Front exhaust pipe and bracket
   - Transmission oil cooler lines, at the transmission
   - Oil cooler lines bracket and the transmission oil filler tube, as required
7. Support the transmission, using a jack with a wooden block placed between the jack and the transmission pan. Raise the transmission, just enough to take the weight off of the rear mount.
8. Remove or disconnect the following:
   - Rear engine mount with the bracket, the rear crossmember and the transfer case undercover, if applicable
   - Dynamic damper, for Regular Cab only
   - No. 2 cross-shaft bracket
9. Place a wooden block(s) between the engine oil pan and the front frame crossmember.
10. Slowly, lower the transmission until the engine rests on the wooden block(s).
11. Remove or disconnect the following:
    - Torque converter cover to gain access to the converter bolts
    - Torque converter bolts, by rotating the crankshaft to access the bolts through the service holes
    - Stiffener plates from the transmission
    - Shift control rod and the transfer case shift lever
12. For the A340H transmission remove or disconnect, perform the following:
    - Cross-shaft and the No. 2 shifting rod
    - Front stabilizer bar
    - Differential mount bolts, by supporting the front differential with a jack
    - Transmission and transfer case, by slowly lowering the front differential so there is enough clearance, if applicable
    - Differential, if enough clearance can't be obtained
13. Remove or disconnect the following:
    - Stabilizer bar
    - Auxiliary frame crossmember, if equipped
14. For A340D transmissions, obtain a bolt of the same dimensions as the torque converter bolts. Cut the head off of the bolt and hacksaw a slot in the bolt opposite the threaded end. Thread the guide pin into one of the torque converter bolt holes. The guide pin will help keep the converter with the transmission.

➡This modified bolt is used as a guide pin. 2 guide pins are needed to properly install the transmission.

15. Remove or disconnect the following:
- Transmission bolts, then move the transmission rearward by prying on the dowel pins through the service hole
- Transmission/transfer case assembly
- Transfer case from the transmission

To install:

16. Connect the transfer case to the transmission.

17. Apply a coat of multi-purpose grease to the torque converter stub shaft and the corresponding pilot hole in the flexplate.

18. Install or connect the following:
- Torque converter into the front of the transmission Push inward on the torque converter while rotating it to completely couple the torque converter to the transmission.

19. To be sure the converter is properly installed, measure the distance between the torque converter mounting lugs and the front mounting face of the transmission. The proper distance is 0.71 in. (18mm) for the A340H transmission or 0.79 in. (20mm) for the A340D, A340E and A340F transmissions.

20. For A340D transmissions, install guide pins into 2 opposite mounting lugs of the torque converter.

21. Install or connect the following:
- Transmission. Tighten the bolts to 47 ft. lbs. (63 Nm).
- Torque converter bolts, by rotating the crankshaft. Tighten the bolts evenly to 30 ft. lbs. (41 Nm) for the A340H, A3430D and A340E transmissions or to 20 ft. lbs. (27 Nm) for the A340F transmission.
- Torque converter access cover

22. Remove the wood block(s) from under the engine oil pan.

23. Install or connect the following:
- Transmission crossmember. Tighten the bolts to 70 ft. lbs. (95 Nm).
- Rear mount and bracket. Tighten the bracket bolts to 43 ft. lbs. (58 Nm) and the bracket-to-rear mount bolts to 108 inch lbs. (13 Nm).
- Transmission onto the crossmember. Tighten the transmission-to-mount bolts to 18 ft. lbs. (25 Nm).

24. Remove the wooden blocks from between the frame and the engine and the support from under the transmission.

25. Install or connect the following:
- Front differential, for the A340H transmission. Tighten the 2 rear mount bolts to 123 ft. lbs. (167 Nm) and the front mount through-bolt to 108 ft. lbs. (147 Nm).

➡If the differential oil was drained, refill it at this time.

- Shift control rod and the transfer case shift lever
- Front stabilizer bar, if applicable
- Cross-shaft and the No. 2 shifting rod, if applicable
- Stiffener plates. Tighten the bolts to 27 ft. lbs. (37 Nm).
- Transfer case undercover and the dynamic damper, if equipped. Tighten the dynamic damper mount bolts to 27 ft. lbs. (37 Nm).
- No. 2 cross-shaft bracket
- Oil filler tube and the oil cooler pipe bracket
- Oil cooler lines to the transmission. Tighten the fittings to 25 ft. lbs. (34 Nm).
- Front exhaust pipe and the support bracket
- Speedometer cable
- Front and rear driveshaft flanges with the differential pinion flanges, by aligning the matchmarks. Tighten the bolts to 54 ft. lbs. (74 Nm).
- Starter
- Wiring connectors to the transmission and the transfer case, if applicable
- Engine undercover
- Transmission throttle cable, by adjusting it
- Air cleaner assembly, if removed
- Negative battery cable

26. Refill the transmission and the transfer case, if applicable.

27. Start the engine and check for leaks.

28. Road test the vehicle for proper operation.

29. Recheck all fluid levels.

MODEL A340F TRANSMISSION

1. Before servicing the vehicle, refer to the precautions in the beginning of this section.

2. Remove or disconnect the following:
- Negative battery cable
- Throttle cable, from the engine compartment
- Automatic Transmission Fluid (ATF) dipstick
- Oil filler pipe

3. Remove the transmission shift lever assembly and transfer shift lever, as follows:
- Rear console upper panel, by disconnecting the connectors
- Heater control knobs
- Center cluster finish panel, by disconnecting the connectors
- Transfer shift lever knob, without the 2–4 selector
- Bolt and the transfer shift lever knob, with the 2–4 selector
- Front console upper panel
- 2–4 selector connector, if equipped
- Transfer shift lever knob
- Shift control rod
- Transmission shift lever assembly connector and the 8 screws
- Shift lever snapring, using pliers and pull out it from the transfer case
- Engine undercover
- Front and rear driveshafts
- Exhaust pipe

4. Disconnect the following connectors from the transmission:
- No. 2 Vehicle Speed Sensor (VSS) connector
- Solenoid connector
- Automatic Transmission Fluid (ATF) temperature sensor connector
- Park/neutral position switch connector

5. Detach the following connectors from the transfer case:
- Transfer neutral position switch connector
- Transfer L4 position switch connector
- Transfer 4WD position switch connector
- Actuator connector (2–4 selector only)

6. Remove or disconnect the following:
- Wiring harness from the transmission and the transfer case
- Both oil cooler pipes
- Rear end-plate and torque converter clutch mounting bolt

7. Support the transmission with a jackstand.

8. Remove or disconnect the following:
- Engine rear mount bolts
- 4 bolts and the crossmember
- Starter
- Transmission

To install:

9. Install or connect the following:
- Transmission. Tighten the bolts to 53 ft. lbs. (71 Nm).
- Starter. Tighten the bolts to 29 ft. lbs. (39 Nm).
- Crossmember. Tighten the 4 bolts to 48 ft. lbs. (65 Nm).
- Engine rear mount. Tighten the 4 bolts to 14 ft. lbs. (19 Nm).
- Clutch converter bolts, by installing the green colored bolt before the other 5. Tighten the bolts to 30 ft. lbs. (41 Nm).

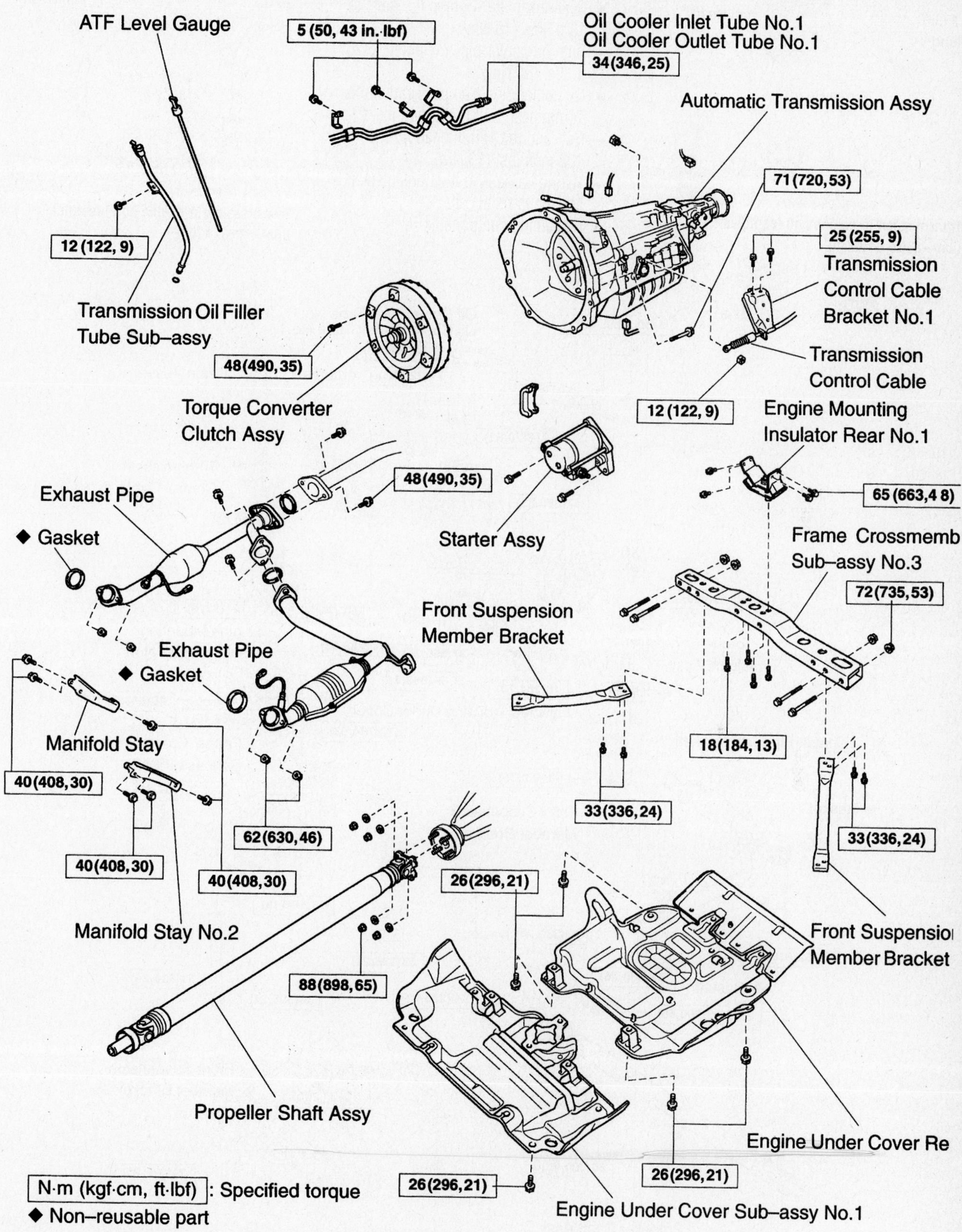
ATF Level Gauge
5 (50, 43 in.·lbf)
Oil Cooler Inlet Tube No.1
Oil Cooler Outlet Tube No.1
34(346,25)
Automatic Transmission Assy
71(720,53)
12(122, 9)
25(255, 9)
Transmission Control Cable Bracket No.1
Transmission Oil Filler Tube Sub–assy
48(490,35)
Transmission Control Cable
Torque Converter Clutch Assy
12(122, 9)
Engine Mounting Insulator Rear No.1
48(490,35)
Exhaust Pipe
65(663,4 8)
◆ Gasket
Starter Assy
Frame Crossmemb
Sub–assy No.3
72(735,53)
Front Suspension Member Bracket
Exhaust Pipe
◆ Gasket
Manifold Stay
18(184,13)
40(408,30)
33(336,24)
62(630,46)
33(336,24)
40(408,30)
40(408,30)
26(296,21)
Manifold Stay No.2
Front Suspensio
Member Bracket
88(898,65)
Propeller Shaft Assy
Engine Under Cover Re
26(296,21)
26(296,21)
Engine Under Cover Sub–assy No.1
N·m (kgf·cm, ft·lbf) : Specified torque
◆ Non–reusable part
67170-4RUN-G169

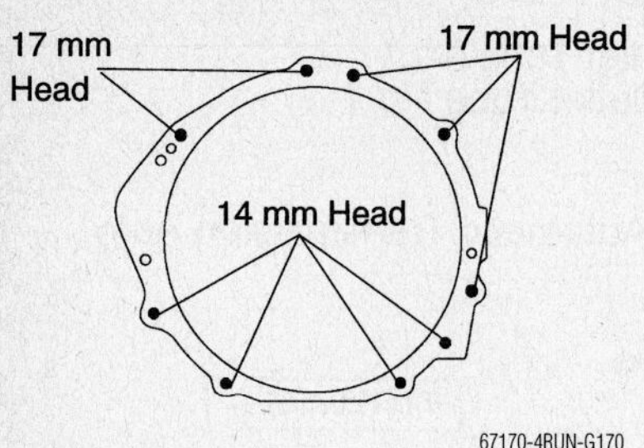

17 mm Head
17 mm Head
14 mm Head
67170-4RUN-G170

Transmission-to-engine bolt identification—2003-04 A340E

- Rear end-plate. Tighten the bolts to 13 ft. lbs. (18 Nm).
- Both oil cooler pipes. Tighten to 25 ft. lbs. (34 Nm).
- Oil cooler pipe clamps. Tighten the 10mm head bolt to 48 inch lbs. (5 Nm) and the 12mm head bolt to 108 inch lbs. (13 Nm).
- Wiring harness to the transmission and the transfer case
- Remaining components

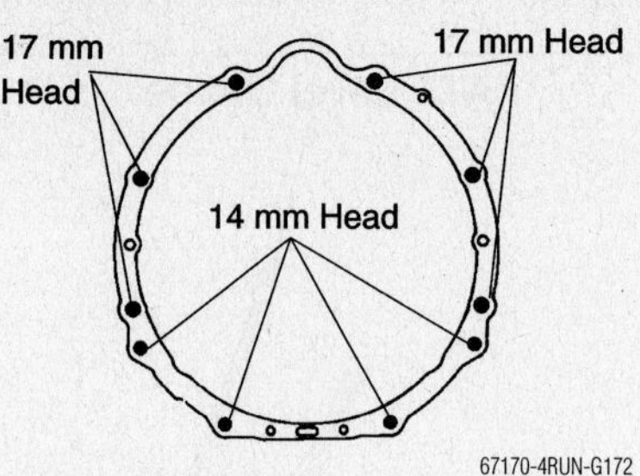

17 mm Head
17 mm Head
14 mm Head
67170-4RUN-G172

Transmission-to-engine bolt identification—2003–A750E

2WD:

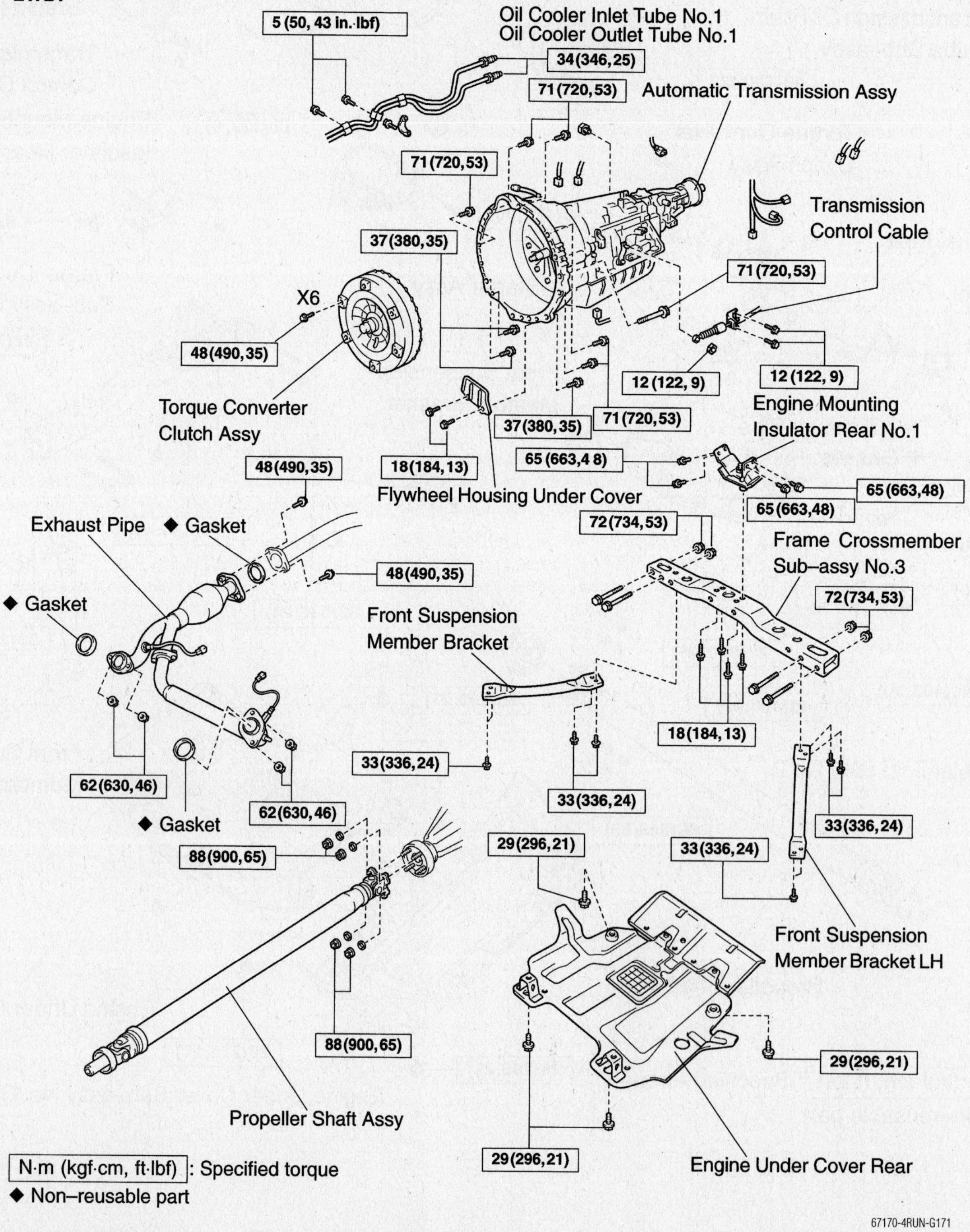

5 (50, 43 in.·lbf)
Oil Cooler Inlet Tube No.1
Oil Cooler Outlet Tube No.1
34(346,25)
71(720,53)
Automatic Transmission Assy
71(720,53)
Transmission Control Cable
37(380,35)
71(720,53)
X6
48(490,35)
12(122, 9)
12(122, 9)
Torque Converter Clutch Assy
37(380,35)
71(720,53)
Engine Mounting Insulator Rear No.1
48(490,35)
18(184,13)
65(663,4 8)
65(663,48)
65(663,48)
Flywheel Housing Under Cover
Exhaust Pipe
◆ Gasket
72(734,53)
Frame Crossmember Sub–assy No.3
48(490,35)
◆ Gasket
72(734,53)
Front Suspension Member Bracket
18(184,13)
33(336,24)
62(630,46)
33(336,24)
62(630,46)
◆ Gasket
33(336,24)
88(900,65)
29(296,21)
33(336,24)
Front Suspension Member Bracket LH
88(900,65)
29(296,21)
Propeller Shaft Assy
29(296,21)
Engine Under Cover Rear
N·m (kgf·cm, ft·lbf) : Specified torque
◆ Non–reusable part
67170-4RUN-G171

2003–04 A750E and related parts

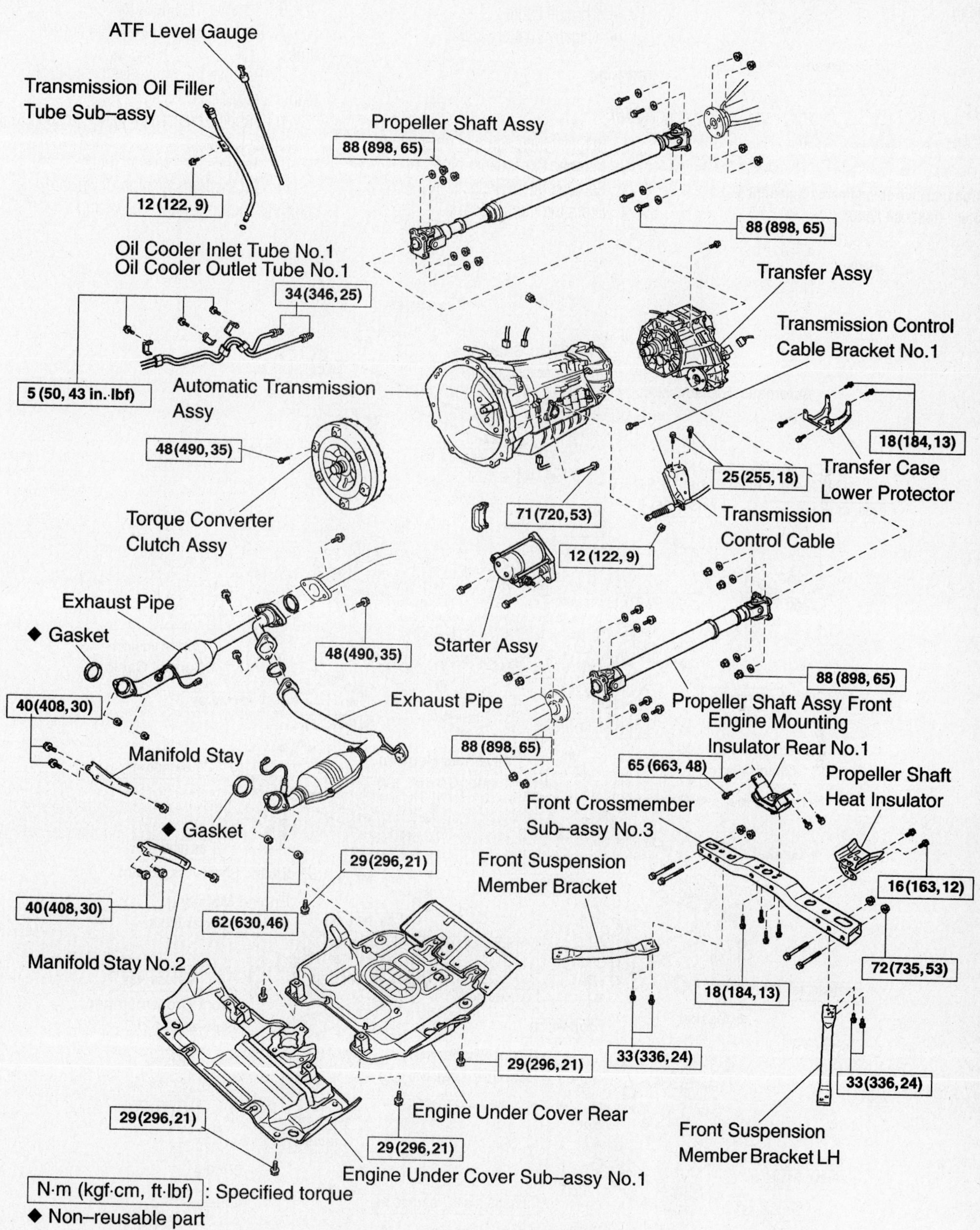
ATF Level Gauge
Transmission Oil Filler Tube Sub–assy
12 (122, 9)
Propeller Shaft Assy
88 (898, 65)
88 (898, 65)
Oil Cooler Inlet Tube No.1
Oil Cooler Outlet Tube No.1
34 (346, 25)
Transfer Assy
Transmission Control Cable Bracket No.1
5 (50, 43 in.·lbf)
Automatic Transmission Assy
48 (490, 35)
18 (184, 13)
Transfer Case Lower Protector
25 (255, 18)
71 (720, 53)
Torque Converter Clutch Assy
Transmission Control Cable
12 (122, 9)
Exhaust Pipe
◆ Gasket
48 (490, 35)
Starter Assy
88 (898, 65)
40 (408, 30)
Exhaust Pipe
Propeller Shaft Assy Front
Engine Mounting Insulator Rear No.1
Manifold Stay
88 (898, 65)
65 (663, 48)
Propeller Shaft Heat Insulator
Front Crossmember Sub–assy No.3
◆ Gasket
29 (296, 21)
Front Suspension Member Bracket
16 (163, 12)
40 (408, 30)
62 (630, 46)
72 (735, 53)
Manifold Stay No.2
18 (184, 13)
29 (296, 21)
33 (336, 24)
33 (336, 24)
Engine Under Cover Rear
29 (296, 21)
Front Suspension Member Bracket LH
29 (296, 21)
Engine Under Cover Sub–assy No.1
N·m (kgf·cm, ft·lbf) : Specified torque
◆ Non–reusable part
67170-4RUN-G173

17 mm Head
17 mm Head
14 mm Head
67170-4RUN-G174

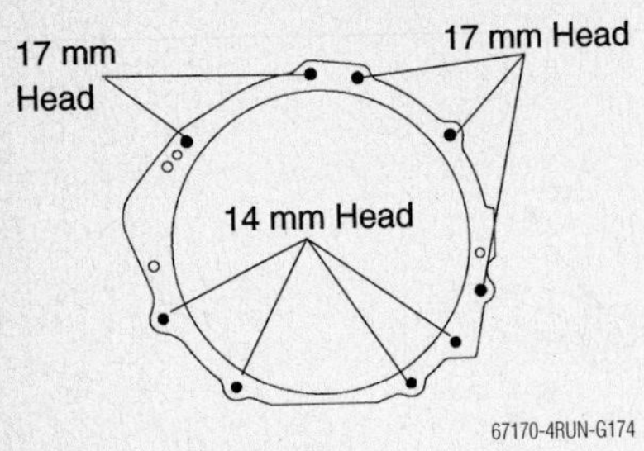

Transmission-to-engine bolt identification—2003–04 A340F

10. Fill the transmission and transfer case with transmission fluid.
    - Throttle cable
    - Negative battery cable

2003–04

A340E

1. Remove the engine under-covers.
2. Remove the exhaust pipe.
3. Remove the driveshaft.
4. Drain the transmission.
5. Remove the front crossmember brackets.
6. Remove the filler tube.
7. Remove the transmission cooler lines.
8. Remove the transmission control cable and bracket.
9. Support the transmission with a transmission jack.
10. Remove the crossmember.
11. Tilt the transmission downward, slightly and disconnect the wiring.

4WD:

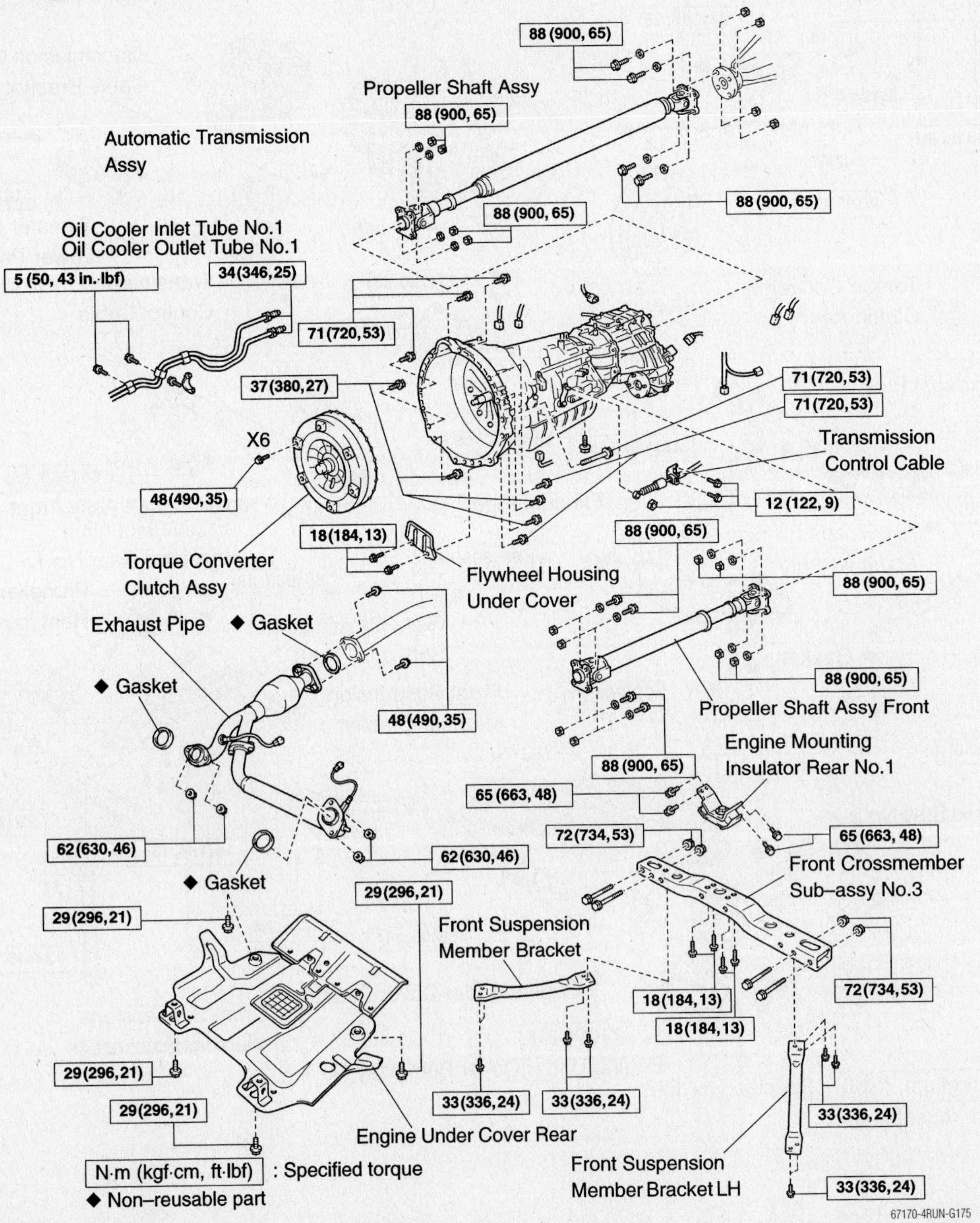

88 (900, 65)
Propeller Shaft Assy
88 (900, 65)
Automatic Transmission Assy
88 (900, 65)
88 (900, 65)
Oil Cooler Inlet Tube No.1
Oil Cooler Outlet Tube No.1
5 (50, 43 in.·lbf)
34(346,25)
71(720,53)
37(380,27)
71(720,53)
71(720,53)
X6
Transmission Control Cable
48(490,35)
12 (122, 9)
18(184,13)
88 (900, 65)
Torque Converter Clutch Assy
Flywheel Housing Under Cover
88 (900, 65)
Exhaust Pipe
◆ Gasket
◆ Gasket
88 (900, 65)
48(490,35)
Propeller Shaft Assy Front
Engine Mounting Insulator Rear No.1
88 (900, 65)
65 (663, 48)
72(734,53)
65 (663, 48)
62(630,46)
62(630,46)
Front Crossmember Sub–assy No.3
◆ Gasket
29(296,21)
29(296,21)
Front Suspension Member Bracket
72(734,53)
18(184,13)
18(184,13)
29(296,21)
29(296,21)
33(336,24)
33(336,24)
33(336,24)
Engine Under Cover Rear
N·m (kgf·cm, ft·lbf) : Specified torque
◆ Non–reusable part
Front Suspension Member Bracket LH
33(336,24)
67170-4RUN-G175

2003–04 A750F and related parts

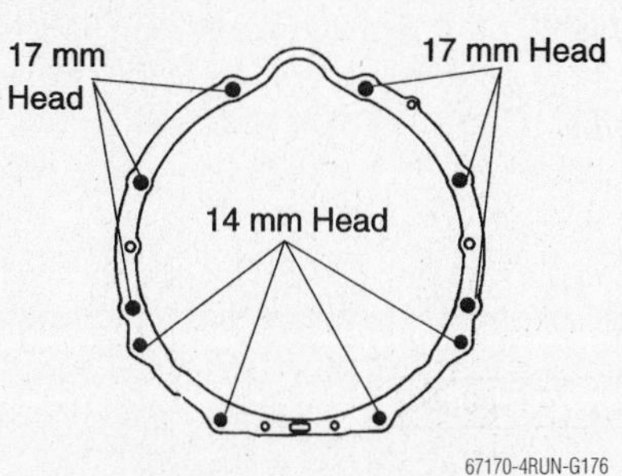

Transmission-to-engine bolt identification—2003–04 A750F

12. Remove the starter.
13. Remove the manifold stays.
14. Remove the bellhousing cover plate and turn the crankshaft to access and remove the 6 torque converter bolts.
15. Remove the 9 transmission-to-engine bolts and remove the transmission.
16. Remove the rear insulator.
17. Installation is the reverse of removal. Observe the following torques:
    - Rear insulator: 48 ft. lbs. (65 Nm)
    - Transmission-to-engine bolts (see illustration): 14mm head, 27 ft. lbs. (37 Nm); 17mm head, 53 ft. lbs. (71 Nm)
    - Torque converter bolts (install black bolt first): 35 ft. lbs. (48 Nm)
    - Manifold stays: 30 ft. lbs. (40 Nm)
    - Crossmember-to-frame: 53 ft. lbs. (72 Nm)
    - Crossmember-to-insulator: 13 ft. lbs. (18 Nm)
    - Crossmember brackets: 24 ft. lbs. (33 Nm)
    - Bellhousing cover: 13 ft. lbs. (18 Nm)
    - Engine under-covers: 21 ft. lbs. (26 Nm)
    - Transmission drain plug: 15 ft. lbs. (20 Nm)

A750E

1. Remove the engine under-covers.
2. Remove the exhaust pipe.
3. Remove the driveshaft.
4. Drain the transmission.
5. Remove the front crossmember brackets.
6. Remove the transmission cooler lines.
7. Remove the transmission control cable and bracket.
8. Support the transmission with a transmission jack.
9. Remove the crossmember.
10. Tilt the transmission downward, slightly and disconnect the wiring.
11. Remove the bellhousing cover plate and turn the crankshaft to access and remove the 6 torque converter bolts.
12. Remove the 10 transmission-to-engine bolts and remove the transmission.
13. Remove the rear insulator.
14. Installation is the reverse of removal. Observe the following torques:
    - Rear insulator: 48 ft. lbs. (65 Nm)
    - Transmission-to-engine bolts (see illustration): 14mm head, 27 ft. lbs. (37 Nm); 17mm head, 53 ft. lbs. (71 Nm)
    - Torque converter bolts (install black bolt first): 35 ft. lbs. (48 Nm)
    - Manifold stays: 30 ft. lbs. (40 Nm)
    - Bellhousing cover: 13 ft. lbs. (18 Nm)
    - Crossmember-to-frame: 53 ft. lbs. (72 Nm)
    - Crossmember-to-insulator: 13 ft. lbs. (18 Nm)
    - Crossmember brackets: 24 ft. lbs. (33 Nm)
    - Transmission drain plug: 21 ft. lbs. (28 Nm)
    - Engine under-covers: 21 ft. lbs. (26 Nm)

A340F

1. Remove the engine under-covers.
2. Remove the exhaust pipe.
3. Remove the front and rear driveshafts.
4. Drain the transmission.

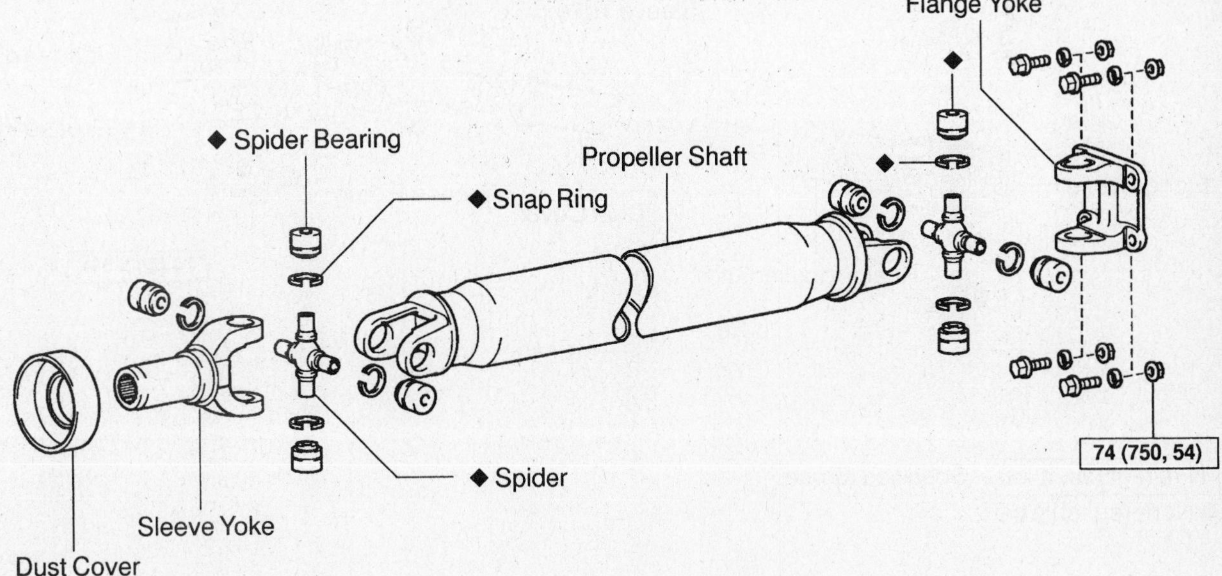

Rear driveshaft with 2WD—2001–02

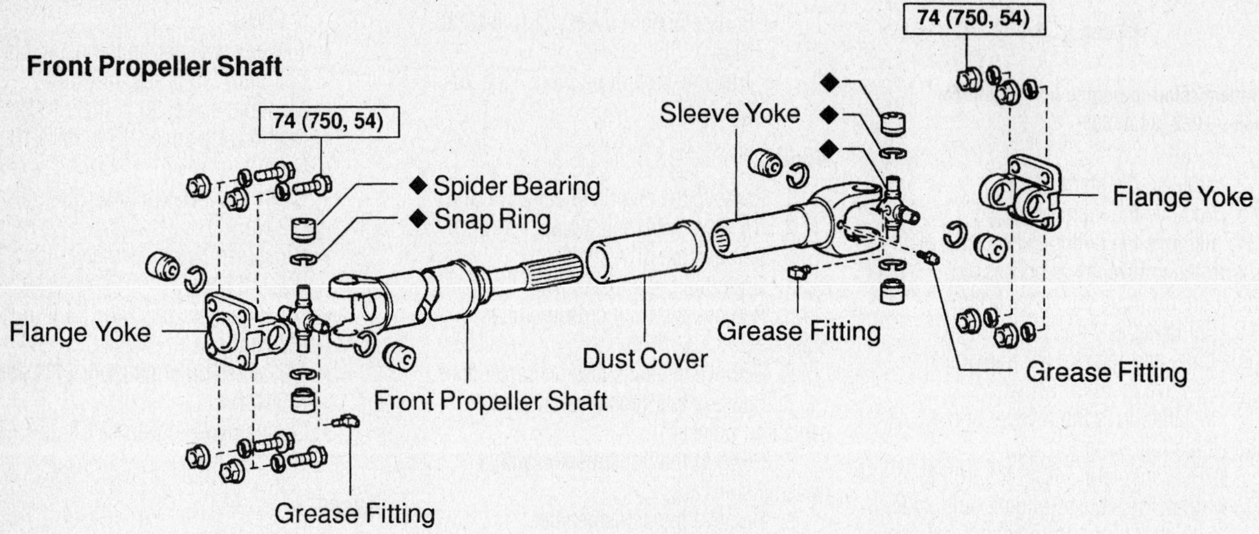

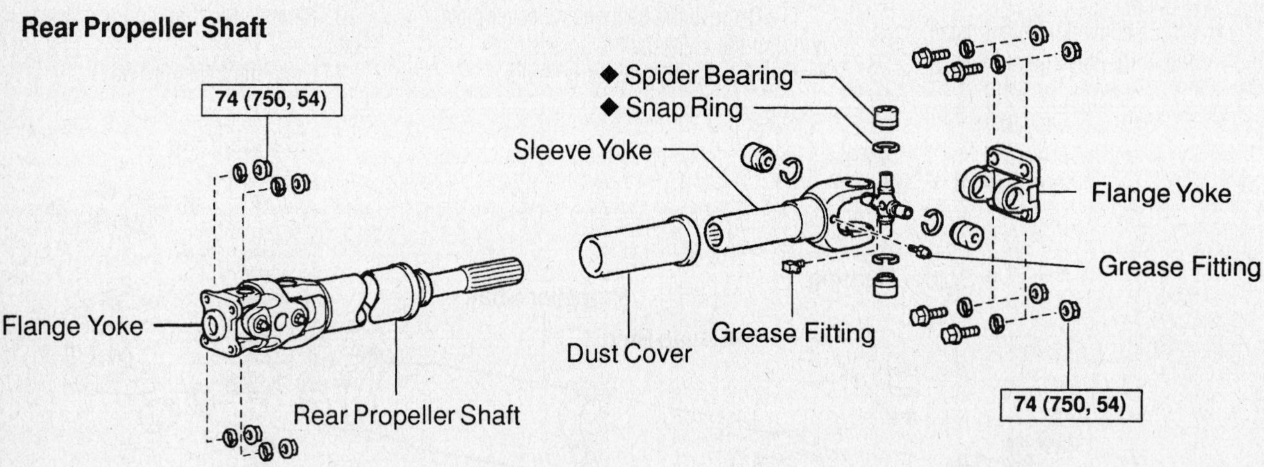

N·m (kgf·cm, ft·lbf) : Specified torque

◆ Non–reusable part

67170-4RUN-G180

5. Remove the driveshaft heat insulator.
6. Remove the front crossmember brackets.
7. Remove the filler tube.
8. Remove the transmission cooler lines.
9. Remove the transmission control cable and bracket.
10. Support the transmission with a transmission jack.
11. Remove the crossmember.
12. Remove the skid plate.
13. Tilt the transmission downward, slightly and disconnect the wiring.
14. Remove the starter.
15. Remove the manifold stays.
16. Remove the bellhousing cover plate and turn the crankshaft to access and remove the 6 torque converter bolts.
17. Remove the 9 transmission-to-engine bolts and remove the transmission.
18. Separate the transfer case from the transmission.
19. Remove the rear insulator.
20. Installation is the reverse of removal. Observe the following torques:

- Rear insulator: 48 ft. lbs. (65 Nm)
- Transmission-to-engine bolts (see illustration): 14mm head, 27 ft. lbs. (37 Nm); 17mm head, 53 ft. lbs. (71 Nm)
- Torque converter bolts (install black bolt first): 35 ft. lbs. (48 Nm)
- Manifold stays: 30 ft. lbs. (40 Nm)
- Bellhousing cover: 13 ft. lbs. (18 Nm)
- Crossmember-to-frame: 53 ft. lbs. (72 Nm)
- Crossmember-to-insulator: 13 ft. lbs. (18 Nm)
- Crossmember brackets: 24 ft. lbs. (33 Nm)
- Transmission drain plug: 15 ft. lbs. (20 Nm)
- Engine under-covers: 21 ft. lbs. (26 Nm)

A750F

1. Remove the engine under-covers.
2. Remove the exhaust pipe.
3. Remove the front and rear driveshafts.
4. Drain the transmission.
5. Remove the front crossmember brackets.
6. Remove the transmission cooler lines.
7. Remove the transmission control cable and bracket.
8. Support the transmission with a transmission jack.
9. Remove the crossmember.
10. Remove the skid plate.
11. Tilt the transmission downward, slightly and disconnect the wiring.
12. Remove the bellhousing cover plate and turn the crankshaft to access and remove the 6 torque converter bolts.
13. Remove the 10 transmission-to-engine bolts and remove the transmission.
14. Separate the transfer case from the transmission.
15. Remove the rear insulator.
16. Installation is the reverse of removal. Observe the following torques:

- Rear insulator: 48 ft. lbs. (65 Nm)
- Transmission-to-engine bolts (see illustration): 14mm head, 27 ft. lbs.

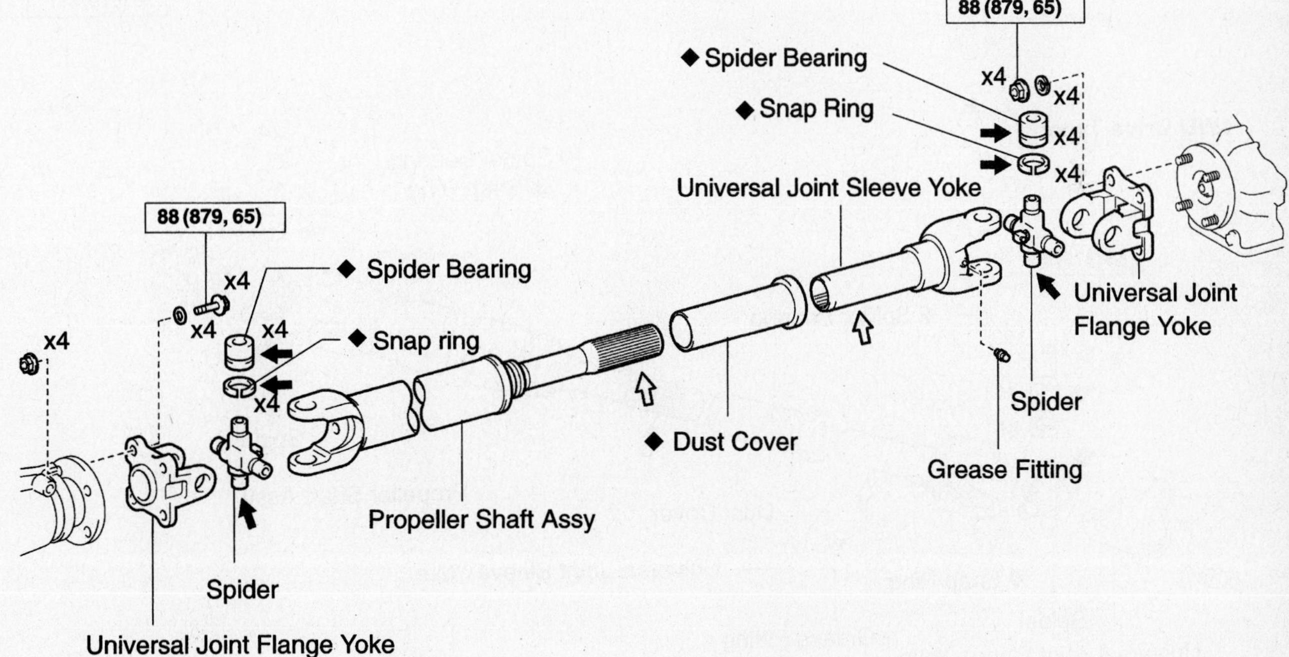

Front driveshaft—2003–04

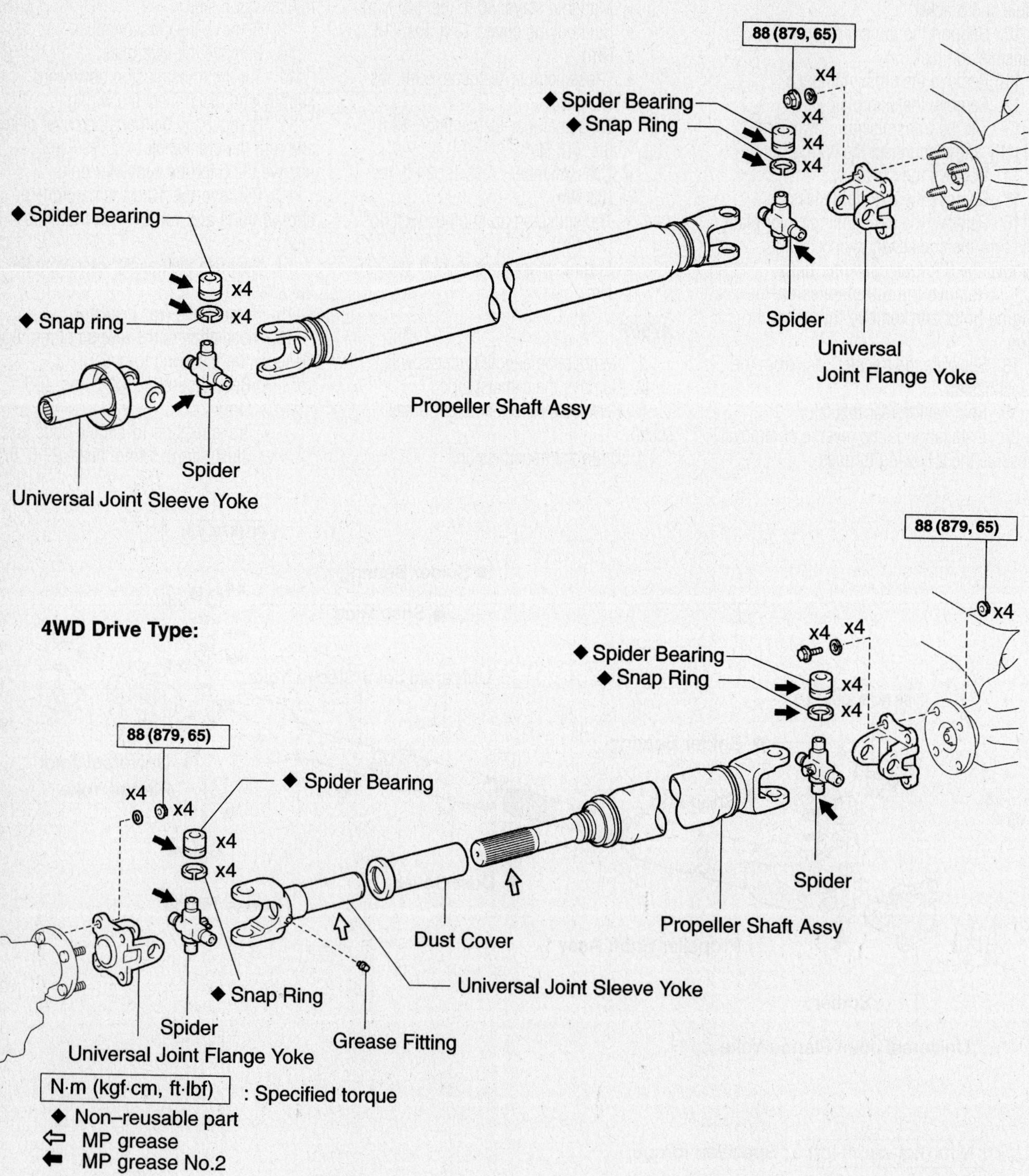
2WD Drive Type:
88 (879, 65)
x4
◆ Spider Bearing
◆ Snap Ring
x4
x4
x4
◆ Spider Bearing
x4
◆ Snap ring
x4
Spider
Universal Joint Flange Yoke
Propeller Shaft Assy
Spider
Universal Joint Sleeve Yoke
4WD Drive Type:
88 (879, 65)
x4
x4
x4
◆ Spider Bearing
◆ Snap Ring
x4
x4
88 (879, 65)
x4
x4
◆ Spider Bearing
x4
x4
Spider
Propeller Shaft Assy
Dust Cover
Universal Joint Sleeve Yoke
◆ Snap Ring
Spider
Grease Fitting
Universal Joint Flange Yoke
N·m (kgf·cm, ft·lbf) : Specified torque
◆ Non–reusable part
MP grease
MP grease No.2
67170-4RUN-G178

(37 Nm); 17mm head, 53 ft. lbs. (71 Nm)
- Torque converter bolts (install black bolt first): 35 ft. lbs. (48 Nm)
- Bellhousing cover: 13 ft. lbs. (18 Nm)
- Crossmember-to-frame: 53 ft. lbs. (72 Nm)
- Crossmember-to-insulator: 13 ft. lbs. (18 Nm)
- Crossmember brackets: 24 ft. lbs. (33 Nm)
- Transmission drain plug: 15 ft. lbs. (20 Nm)
- Engine under-covers: 21 ft. lbs. (26 Nm)

Transfer Case Assembly

REMOVAL & INSTALLATION

2001–02

1. Before servicing the vehicle, refer to the precautions in the beginning of this section.
2. Disconnect the negative battery cable.
3. Drain the transmission and the transfer case.
4. Remove or disconnect the following:
   - Transfer case with the transmission
   - Breather hose from the transfer upper cover and the transmission control retainer, if equipped with an automatic transmission
   - Rear engine mounting
   - Dynamic damper
5. Remove the driveshaft upper dust cover and the transfer from the transmissions, as follows:
   - Dust cover bolt from the bracket
   - Transfer case adapter rear mounting bolts
   - Transfer case, by pulling it straight up and away from the transmission.

✱✱ WARNING

Be careful not to damage the adapter rear oil seal with the transfer input gear spline.

To install:

6. Install the transfer case and the driveshaft upper dust cover to the transmission with a new gasket, as follows:
   - Shift the 2 shift fork shafts to the high 4 position
   - Apply MP grease to the adapter oil seal
   - New gasket to the transfer adapter
   - Transfer case to the transmission.

✱✱ WARNING

Take care not to damage the oil seal by the input gear spline.

   - Transfer case adapter. Tighten the rear bolts to 27 ft. lbs. (37 Nm).
   - Dust cover to the bracket. Tighten the bolt to 17 ft. lbs. (23 Nm).
7. Install or connect the following:
   - Engine rear mount. Tighten the bolts to 19 ft. lbs. (25 Nm).
   - Dynamic damper. Tighten the bolts to 27 ft. lbs. (37 Nm).
   - Breather hose, if equipped with an automatic transmission
   - Transfer case with the transmission to the engine
8. Fill the transmission and the transfer case with oil.
9. Test drive the vehicle and check the abnormal noise and smooth operation.
10. Recheck the fluid levels.

2003–04

1. Remove the transmission and transfer case as an assembly.
2. Remove the 8 bolts and 2 clamps, and separate the transfer case from the transmission.
3. Installation is the reverse of removal. Torque the bolts to 17 ft. lbs. (24 Nm).

Halfshaft

REMOVAL & INSTALLATION

2001–02

1. Before servicing the vehicle, refer to the precautions in the beginning of this section.
2. Remove the front wheel.
3. Drain the differential oil from the differential.
4. Remove the halfshaft locknut, as follows:
   - Grease cap
   - Cotter pin and the lockcap
   - Locknut, while applying the brakes
5. Remove or disconnect the following:
   - Halfshaft, using a brass bar and a hammer
   - Lower control arm
   - Halfshaft, by pushing the steering knuckle outward
   - Snapring from the inboard shaft

To install:

6. Install or connect the following:
   - Snapring to the inboard shaft
   - Halfshaft
   - Steering knuckle
   - Lower control arm. Tighten the nut to 105 ft. lbs. (142 Nm).
7. Connect the halfshaft, as follows:
   - Set the snapring opening side facing downward
   - Strike the inboard joint into the differential, using SST 09631-10030 and a hammer
   - Check that the halfshaft cannot be pulled out by hand
8. Install or connect the following:
   - Locknut, while applying the brakes. Tighten it to 174 ft. lbs. (235 Nm).
   - Adjusting cap and new cotter pin
   - Grease cap
   - Front wheel
9. Fill the differential with oil.

2003–04

1. Before servicing the vehicle, refer to the precautions in the beginning of this section.
2. Remove the front wheel.
3. Drain the differential oil from the differential.
4. Remove the speed sensor harness.
5. Remove the halfshaft locknut, as follows:
   - Grease cap
   - Cotter pin and the lockcap
   - Locknut, while applying the brakes
6. Remove or disconnect the following:
   - Tie rod end from the knuckle
   - Lower control arm from the ball joint
   - Halfshaft from the steering knuckle
   - Halfshaft from the differential with a slide hammer

To install:

7. Install or connect the following:
   - Halfshaft
   - Steering knuckle
   - Lower control arm. Tighten the nut to 166 ft. lbs. (225 Nm).
8. Connect the halfshaft, as follows:
   - Strike the inboard joint into the differential, using SST 09631-10030 and a hammer
   - Check that the halfshaft cannot be pulled out by hand
9. Install or connect the following:
   - Tie rod end. Torque to 67 ft. lbs. (91 Nm)
   - Speed sensor harness
   - Locknut, while applying the brakes. Tighten it to 217 ft. lbs. (294 Nm).

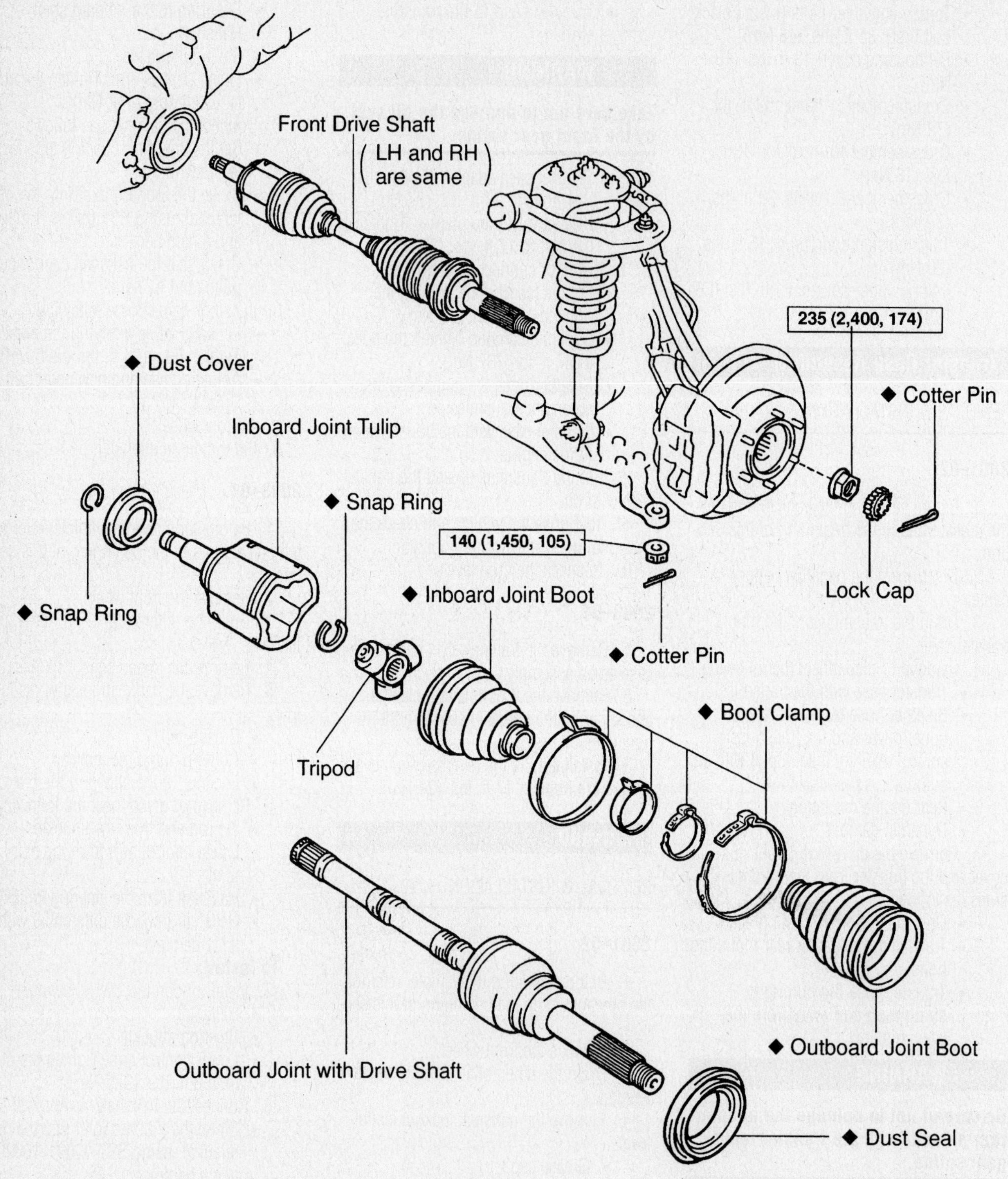

Front halfshaft and related parts—2001–02

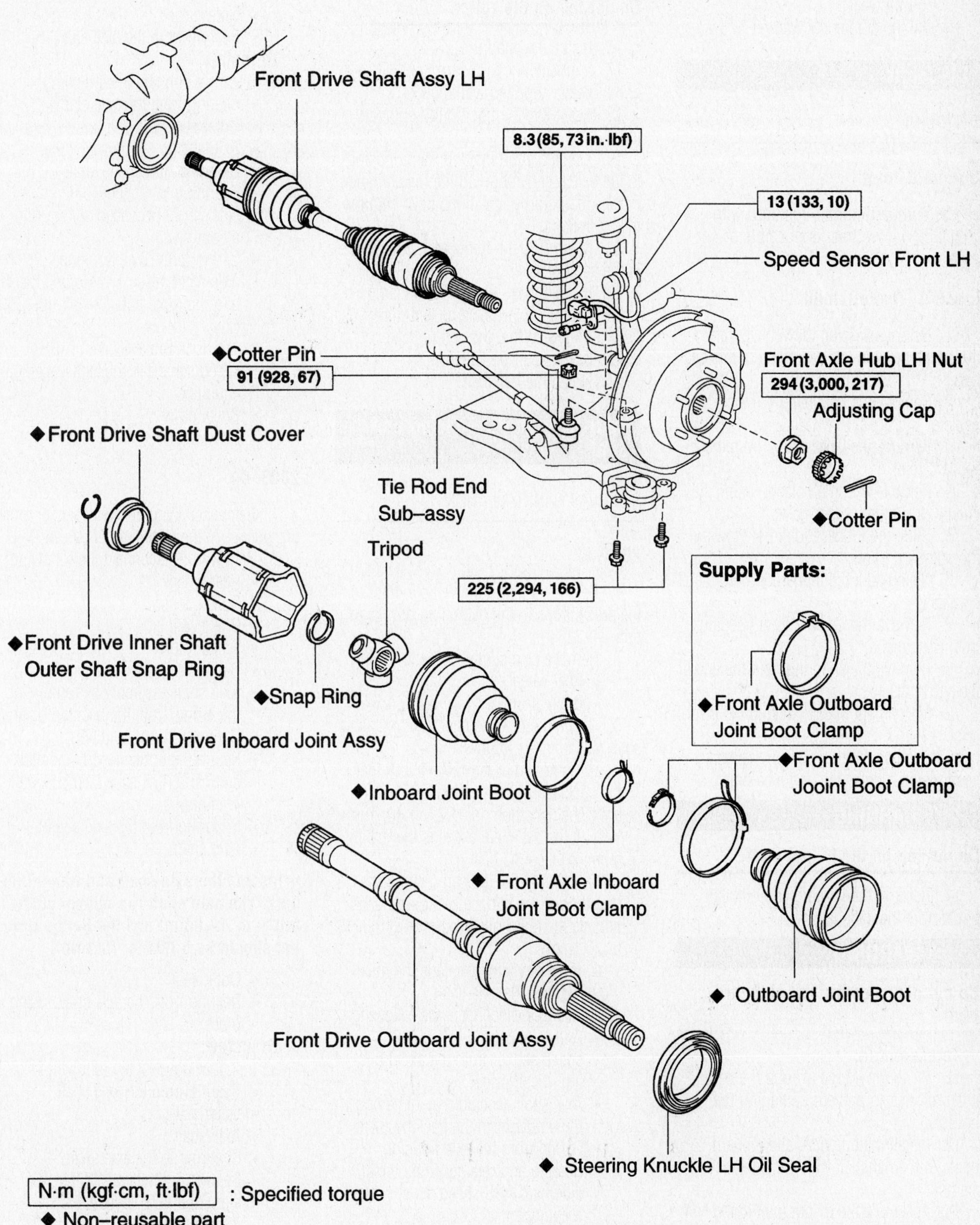
Front Drive Shaft Assy LH
8.3 (85, 73 in.·lbf)
13 (133, 10)
Speed Sensor Front LH
◆Cotter Pin
91 (928, 67)
Front Axle Hub LH Nut
294 (3,000, 217)
Adjusting Cap
◆Front Drive Shaft Dust Cover
Tie Rod End
Sub–assy
◆Cotter Pin
Tripod
225 (2,294, 166)
Supply Parts:
◆Front Drive Inner Shaft
Outer Shaft Snap Ring
◆Snap Ring
◆Front Axle Outboard
Joint Boot Clamp
Front Drive Inboard Joint Assy
◆Front Axle Outboard
Jooint Boot Clamp
◆Inboard Joint Boot
◆ Front Axle Inboard
Joint Boot Clamp
◆ Outboard Joint Boot
Front Drive Outboard Joint Assy
◆ Steering Knuckle LH Oil Seal
N·m (kgf·cm, ft·lbf) : Specified torque
◆ Non–reusable part
67170-4RUN-G182

Front halfshaft and related parts, left side shown—2003–04

- Adjusting cap and new cotter pin
- Grease cap
- Front wheel

10. Fill the differential with oil.

CV-Joints

OVERHAUL

Outboard Joint

The outboard joint is replaced with halfshaft; no overhaul is possible or necessary.

Inboard (Tri-Pot) Joint

1. Before servicing the vehicle, refer to the precautions in the beginning of this section.
2. Remove the halfshaft from the vehicle.
3. Remove the large clamp from the inboard joint.
4. Remove the small clamp, using side cutters, from the inboard joint.
5. Slide the inboard joint boot toward the outboard joint.
6. Matchmark the inboard joint to the halfshaft.
7. Remove the inboard joint housing from the halfshaft.
8. Remove the snapring from the end of the halfshaft.
9. Matchmark the halfshaft to the tri-pot joint.
10. Remove the tri-pot from the halfshaft, using a brass bar and a hammer.

⁂ WARNING

Do not tap on the tri-pot joint.

11. Remove the inboard and outboard boots from the halfshaft.

⁂ WARNING

Do not disassemble the outboard joint.

To assemble:

12. Wrap vinyl tape around the halfshaft splines to prevent damaging the boots.
13. Install the outboard and inboard boots to the halfshaft with the small end clamps.
14. Assemble the tri-pot joint to the halfshaft with the beveled side facing the outboard joint and align the matchmarks.
15. Install the tri-pot joint, using a brass bar and a hammer.

⁂ WARNING

Do not tap on the roller.

16. Install the snapring.
17. Lubricate the outboard joint with ½ of the grease supplied with the kit.
18. Assemble the boot to the outboard joint
19. Assemble the inboard joint housing to the halfshaft by aligning the matchmarks.
20. Temporarily install the boot onto the tri-pot housing.
21. Make sure the boots are positioned in the shaft grooves.
22. Install a new inboard joint clamp.
23. Crimp the large clamp with tool 09521-24010 so that the crimp clearance is 0.039–0.059 in. (1.0–1.5mm).
24. Install the halfshaft.

Rear Axle Shaft, Bearing and Seal

REMOVAL & INSTALLATION

2001–02

1. Before servicing the vehicle, refer to the precautions in the beginning of this section.
2. Remove or disconnect the following:
   - Rear wheel
   - Brake drum
3. Check the bearing backlash and axle shaft deviation, as follows:
   a. Using a dial indicator, check that the backlash in the bearing shaft direction. The maximum is 0.027 in. (0.7mm).
   b. If the backlash exceeds the maximum, replace the bearing.
   c. Using a dial indicator, check the deviation at the surface of the axle shaft outside the hub bolt. Maximum is 0.0039 in. (0.1mm).
   d. If the deviation exceeds the maximum, replace the axle shaft.
4. Remove or disconnect the following:
   - Anti-lock Brake System (ABS) speed sensor from the axle housing, if equipped
   - Axle shaft assembly by removing the 4 nuts from the backing plate
   - O-ring from the axle housing
   - Bearing and retainer (differential side) and ABS speed sensor rotor, if equipped
   - Snapring from the axle shaft
   - Axle shaft

➡Inspect the axle shaft and flange run-outs. The axle shaft run-out should be 0.079 in. (2.0mm) and the flange run-out should be 0.004 in. (0.1mm).

- Outer seal
- Bearing from the axle shaft

To install:

5. Install or connect the following:
   - Bearing from the axle shaft
   - Outer seal
   - Axle shaft
   - Snapring to the axle shaft
   - Bearing and retainer (differential side) and ABS speed sensor rotor, if equipped
   - O-ring to the axle housing
   - Axle shaft assembly. Tighten the 4 backing plate nuts to 48 ft. lbs. (66 Nm).
   - Anti-lock Brake System (ABS) speed sensor to the axle housing, if equipped
   - Brake drum
   - Rear wheel

2003–04

1. Before servicing the vehicle, refer to the precautions in the beginning of this section.
2. Remove or disconnect the following:
   - Rear wheel
   - Speed sensor
   - Brake caliper
   - Brake rotor
   - Parking brake assembly
   - Axle shaft assembly by removing the 4 nuts from the backing plate
   - O-ring from the axle housing
   - Bearing and retainer (differential side) and ABS speed sensor rotor, if equipped
   - Snapring from the axle shaft
   - Axle shaft

➡Inspect the axle shaft and flange run-outs. The axle shaft run-out should be 0.079 in. (2.0mm) and the flange run-out should be 0.004 in. (0.1mm).

- Outer seal
- Bearing from the axle shaft, using a press

To install:

3. Install or connect the following:
   - Bearing from the axle shaft
   - Outer seal
   - Axle shaft
   - Snapring to the axle shaft
   - Bearing and retainer (differential side) and ABS speed sensor rotor, if equipped
   - O-ring to the axle housing
   - Axle shaft assembly. Tighten the 4 backing plate nuts to 91 ft. lbs. (123 Nm).

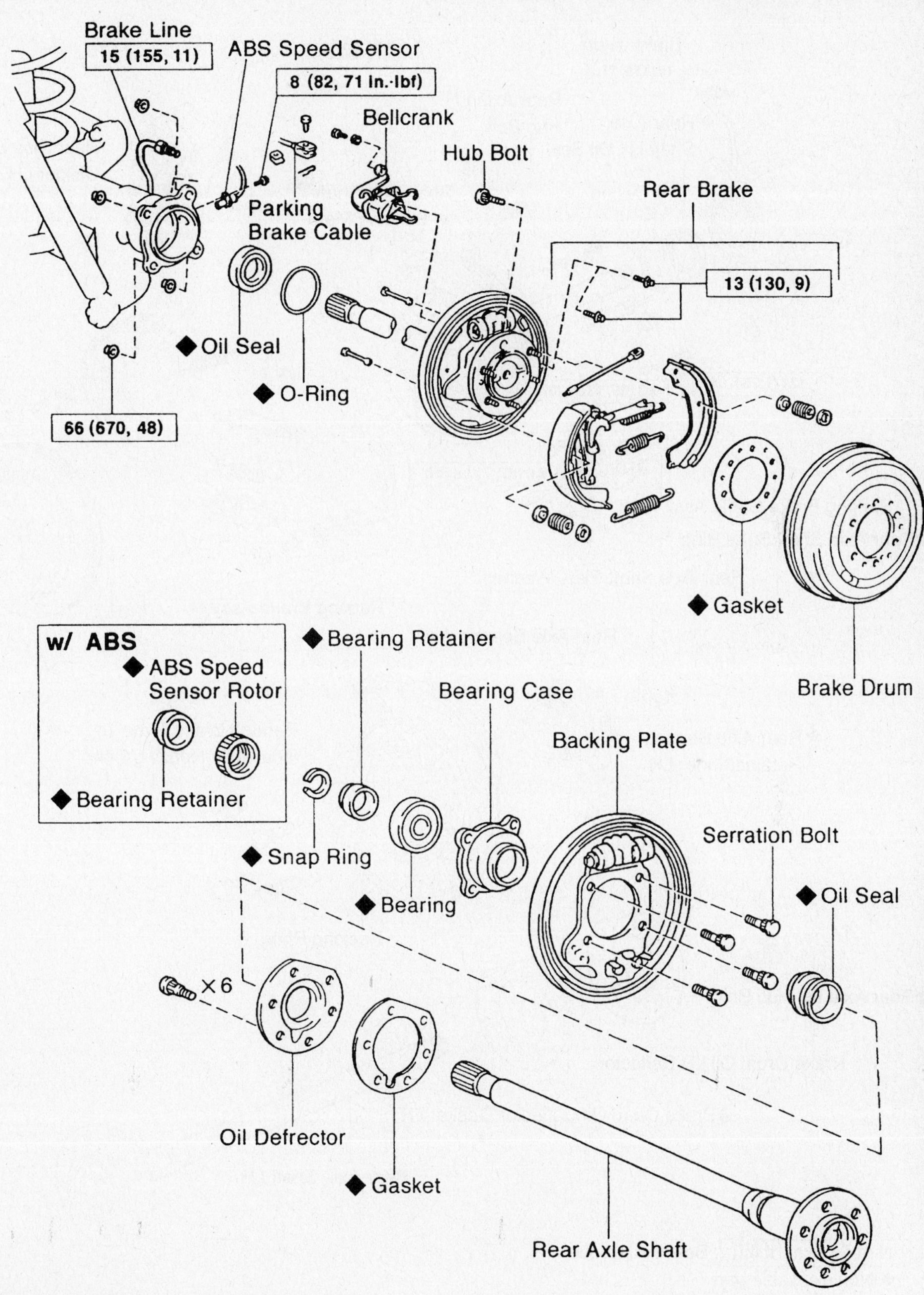

Exploded view of the rear axle shaft and components—2001–02

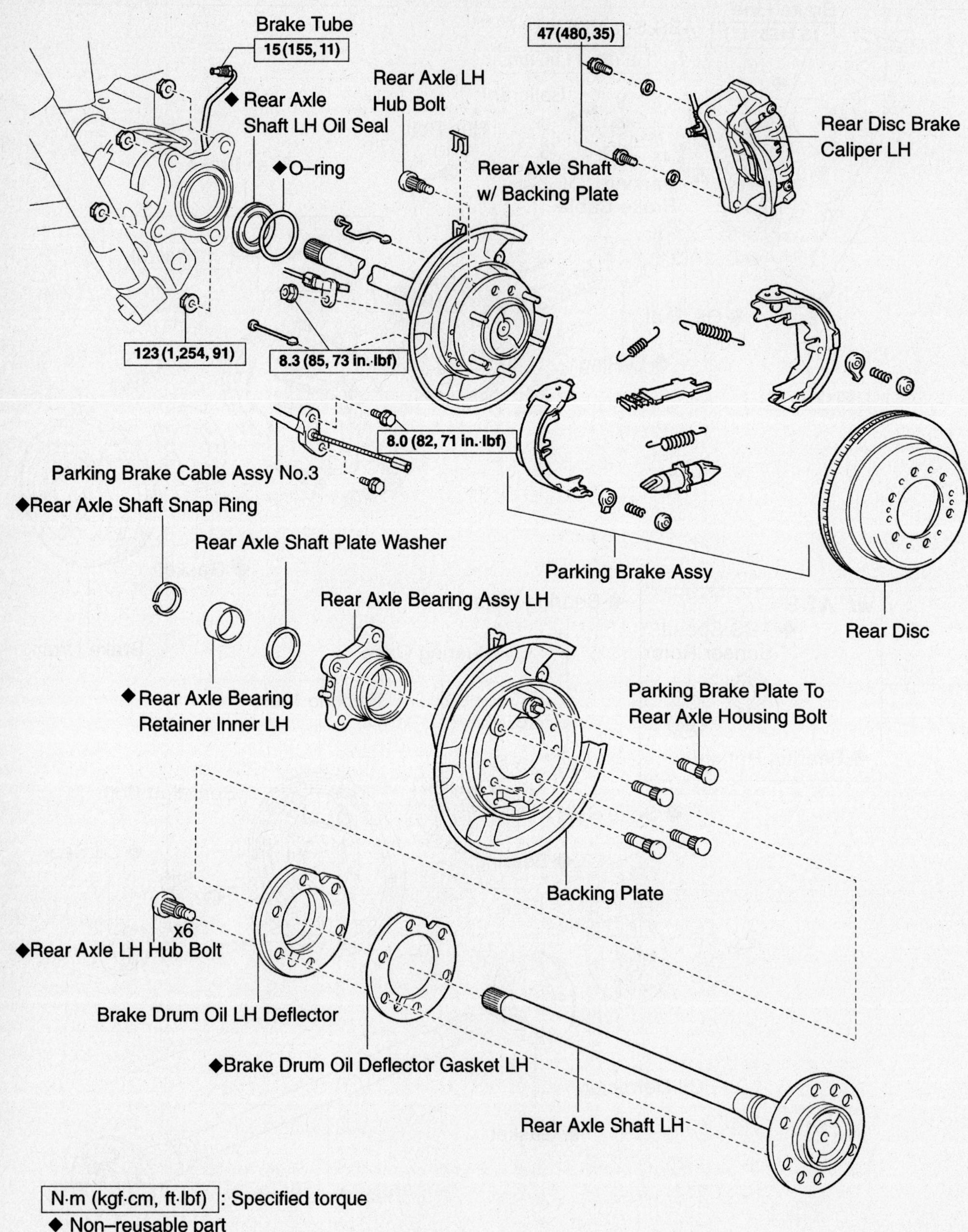

Brake Tube
15 (155, 11)
47 (480, 35)
Rear Axle LH Hub Bolt
◆ Rear Axle Shaft LH Oil Seal
Rear Disc Brake Caliper LH
◆O–ring
Rear Axle Shaft w/ Backing Plate
123 (1,254, 91)
8.3 (85, 73 in.·lbf)
8.0 (82, 71 in.·lbf)
Parking Brake Cable Assy No.3
◆Rear Axle Shaft Snap Ring
Rear Axle Shaft Plate Washer
Parking Brake Assy
Rear Axle Bearing Assy LH
Rear Disc
◆Rear Axle Bearing Retainer Inner LH
Parking Brake Plate To Rear Axle Housing Bolt
Backing Plate
x6
◆Rear Axle LH Hub Bolt
Brake Drum Oil LH Deflector
◆Brake Drum Oil Deflector Gasket LH
Rear Axle Shaft LH
N·m (kgf·cm, ft·lbf) : Specified torque
◆ Non–reusable part
67170-4RUN-G183

Exploded view of the rear axle shaft and components—2003–04

- Anti-lock Brake System (ABS) speed sensor to the axle housing, if equipped
- Brake rotor
- Brake caliper
- Rear wheel

Pinion Seal

REMOVAL & INSTALLATION

Front

2001–02

1. Before servicing the vehicle, refer to the precautions in the beginning of this section.
2. Drain the differential oil.
3. Remove or disconnect the following:
   - Front driveshaft by matchmarking it
   - Companion flange nut, by loosen the staked portion
   - Companion flange, using a screw-type extractor
   - Oil seal, using an extractor

To install:

4. Install a new oil seal, to a depth of 0.059 in. (1.5mm) below the lip, using a seal driver.
5. Lubricate the seal lip with multi-purpose grease.
6. Install or connect the following:
   - Companion flange, coat the threads with multi-purpose grease
   - New companion flange nut. Tighten it to 89 ft. lbs. (120 Nm).
7. Measure the bearing preload, using a torque wrench. The correct preload should be 5–9 inch lbs. (0.6–1.0 Nm) for a used bearing or 10–17 inch lbs. (1–2 Nm) for a new bearing.

➡If the preload is greater that specified, replace the bearing spacer. If the preload is less than specified, tighten the companion flange nut in 9 ft. lbs. (13 Nm) increments until the correct preload is achieved. Maximum torque for the nut is 165 ft. lbs. (223 Nm). If the value is exceeded, the bearing spacer must be replaced; do not back off the flange nut to lower the torque or preload.

8. Install the front driveshaft by aligning the matchmarks.
9. Check the companion flange run-out; maximum allowable run-out is 0.003 in. (0.10mm).
10. Stake the pinion flange nut.
11. Refill the differential with oil.

2003–04

1. Before servicing the vehicle, refer to the precautions in the beginning of this section.
2. Drain the differential oil.
3. Remove or disconnect the following:
   - Front driveshaft by matchmarking it
   - Companion flange nut, by loosen the staked portion
   - Companion flange, using a screw-type extractor
   - Oil seal, using an extractor

To install:

4. Install a new oil seal, to a depth of 0.171 in. (4.35mm) below the lip, using a seal driver.
5. Lubricate the seal lip with multi-purpose grease.
6. Install or connect the following:
   - Companion flange, coat the threads with multi-purpose grease
   - New companion flange nut. Tighten it to 89 ft. lbs. (120 Nm).
7. Measure the bearing preload, using a torque wrench. The correct preload should be 8.7–13.9 inch lbs. (0.98–1.57 Nm) for a new bearing or 4.3–6.9 inch lbs. (0.49–0.78 Nm) for a used bearing.

➡If the preload is greater that specified, replace the bearing spacer. If the preload is less than specified, tighten the companion flange nut in 9 ft. lbs. (13 Nm) increments until the correct preload is achieved. Maximum torque for the nut is 268 ft. lbs. (370 Nm). If the value is exceeded, the bearing spacer must be replaced; do not back off the flange nut to lower the torque or preload.

8. Install the front driveshaft by aligning the matchmarks.
9. Check the companion flange run-out; maximum allowable run-out is 0.003 in. (0.10mm).
10. Stake the pinion flange nut.
11. Refill the differential with oil.

Rear

2001–02

1. Before servicing the vehicle, refer to the precautions in the beginning of this section.
2. Drain the differential oil.
3. Remove or disconnect the following:
   - Driveshaft by matchmarking it
   - Companion flange nut, by loosen the staked portion
   - Companion flange, using a screw-type extractor
   - Oil seal, using an extractor

To install:

4. Install a new oil seal, to a depth of 0.059 in. (1.5mm) below the lip, using a seal driver.
5. Lubricate the seal lip with multi-purpose grease.
6. Install or connect the following:
   - Companion flange, coat the threads with multi-purpose grease
   - New companion flange nut. Tighten it to 89 ft. lbs. (120 Nm).
7. Measure the bearing preload, using a torque wrench. The correct preload should be 5–9 inch lbs. (0.6–1.0 Nm) for a used bearing or 10–17 inch lbs. (1–2 Nm) for a new bearing.

➡If the preload is greater that specified, replace the bearing spacer. If the preload is less than specified, tighten the companion flange nut in 9 ft. lbs. (13 Nm) increments until the correct preload is achieved. Maximum torque for the nut is 165 ft. lbs. (223 Nm). If the value is exceeded, the bearing spacer must be replaced; do not back off the flange nut to lower the torque or preload.

8. Install the driveshaft by aligning the matchmarks.
9. Check the companion flange run-out; maximum allowable run-out is 0.003 in. (0.10mm).
10. Stake the pinion flange nut.
11. Refill the differential with oil.

2003–04

1. Before servicing the vehicle, refer to the precautions in the beginning of this section.
2. Drain the differential oil.
3. Remove or disconnect the following:
   - Driveshaft by matchmarking it
   - Companion flange nut, by loosen the staked portion
   - Companion flange, using a screw-type extractor
   - Oil seal, using an extractor

To install:

4. Install a new oil seal, to a depth of 0.0276–0.0512 in. (0.7–1.3mm) below the lip, using a seal driver.
5. Lubricate the seal lip with multi-purpose grease.

6. Install or connect the following:
   - Companion flange, coat the threads with multi-purpose grease
   - New companion flange nut. Tighten it to 159 ft. lbs. (215 Nm).

7. Measure the bearing preload, using a torque wrench. The correct preload should be 5–7.5 inch lbs. (0.56–0.85 Nm) for a used bearing or 9.3–14.5 inch lbs. (1.05–1.64 Nm) for a new bearing.

➡**If the preload is not within that specified, replace the bearing spacer.**

8. Install the driveshaft by aligning the matchmarks.
9. Check the companion flange run-out; maximum allowable run-out is 0.003 in. (0.10mm).
10. Stake the pinion flange nut.
11. Refill the differential with oil.

STEERING AND SUSPENSION

Air Bag

⁂ CAUTION

Some vehicles are equipped with an air bag system. The system must be disarmed before performing service on, or around, system components, the steering column, instrument panel components, wiring and sensors. Failure to follow the safety precautions and the disarming procedure could result in accidental air bag deployment, possible injury and unnecessary system repairs.

PRECAUTIONS

Several precautions must be observed when handling the inflator module to avoid accidental deployment and possible personal injury.

- Never carry the inflator module by the wires or connector on the underside of the module.
- When carrying a live inflator module, hold securely with both hands and ensure that the bag and trim cover are pointed away.
- Place the inflator module on a bench or other surface with the bag and trim cover facing up.
- With the inflator module on the bench, never place anything on or close to the module which may be thrown in the event of an accidental deployment.

DISARMING

To avoid personal injury when working on vehicles equipped with an air bag, the negative battery cable must be disconnected and at least 90 seconds must elapse before working on the system. Failure to do so may result in deployment of the air bag.

Power Rack And Pinion Steering Gear

REMOVAL & INSTALLATION

2001–02

1. Before servicing the vehicle, refer to the precautions in the beginning of this section.
2. Disarm the airbag system.
3. Place the wheels in the straight-ahead position.
4. Remove or disconnect the following:
   - Negative battery cable
   - Steering wheel
   - Engine under-cover
   - Stabilizer bar
   - Right and left tie rod ends from the knuckle
   - Intermediate shaft from the steering rack, by matchmarking it
   - Pressure feed and the return tubes, using SST 09631-22020
   - Mount bracket and the grommet, from the power steering rack assembly
   - Power steering rack and pinion

To install:

5. Install or connect the following:
   - Power steering rack and pinion. Tighten the mounting bolts to (see the illustration):
   - A & C: 123 ft. lbs. (165 Nm)
   - B: 96 ft. lbs. (130 Nm).
   - Grommet and mount bracket to the gear assembly. Tighten the bolts to 65 ft. lbs. (88 Nm).
   - New O-ring
   - Pressure feed and return tubes. Tighten the line fittings to 30 ft. lbs. (41 Nm) for the pressure line; 33 ft. lbs. (45 Nm) for the return line.
   - Intermediate shaft to the steering rack, by aligning the matchmarks

➡**If installing a new rack assembly, be sure the steering wheel and the rack are centered.**

   - Right and left tie rod ends. Tighten nuts to 67 ft. lbs. (90 Nm).
   - New cotter pins
   - Negative battery cable

6. Center the spiral cable.
7. Install the steering wheel. Torque the nut to 37 ft. lbs. (50 Nm)
8. Check the steering wheel center point.
9. Check the fluid level and bleed the power steering system.
10. Install the under-cover. Torque to 22 ft. lbs. (30 Nm).
11. Check the front wheel alignment.

2003–04

1. Disarm the airbag system.
2. Remove the negative battery cable.
3. Place the wheels in the straight-ahead position.
4. Remove the steering wheel.
5. Remove the lower steering column cover.
6. Remove the turn signal switch.
7. Remove the spiral cable.
8. Remove the wheels.
9. Remove the engine under-cover.
10. Remove the stabilizer bar.
11. Remove the tie rod ends from the knuckles.
12. Disconnect the pressure and return lines from the gear.
13. Remove the 2 bolts and nuts. The nuts have detents. Never turn the nuts, just the bolts.
14. Remove the gear.
15. Installation is the reverse of removal. Center the spiral cable. Observe the following torques"
   - Pressure and return lines: 30 ft. lbs. (40 Nm)
   - Steering gear: 74 ft. lbs. (100 Nm)
   - Tie rod end-to-knuckle: 67 ft. lbs. (91 Nm)
   - Tie rod end-to-steering gear locknut: 65 ft. lbs. (88 Nm)
   - Steering wheel nut: 37 ft. lbs. (50 Nm)

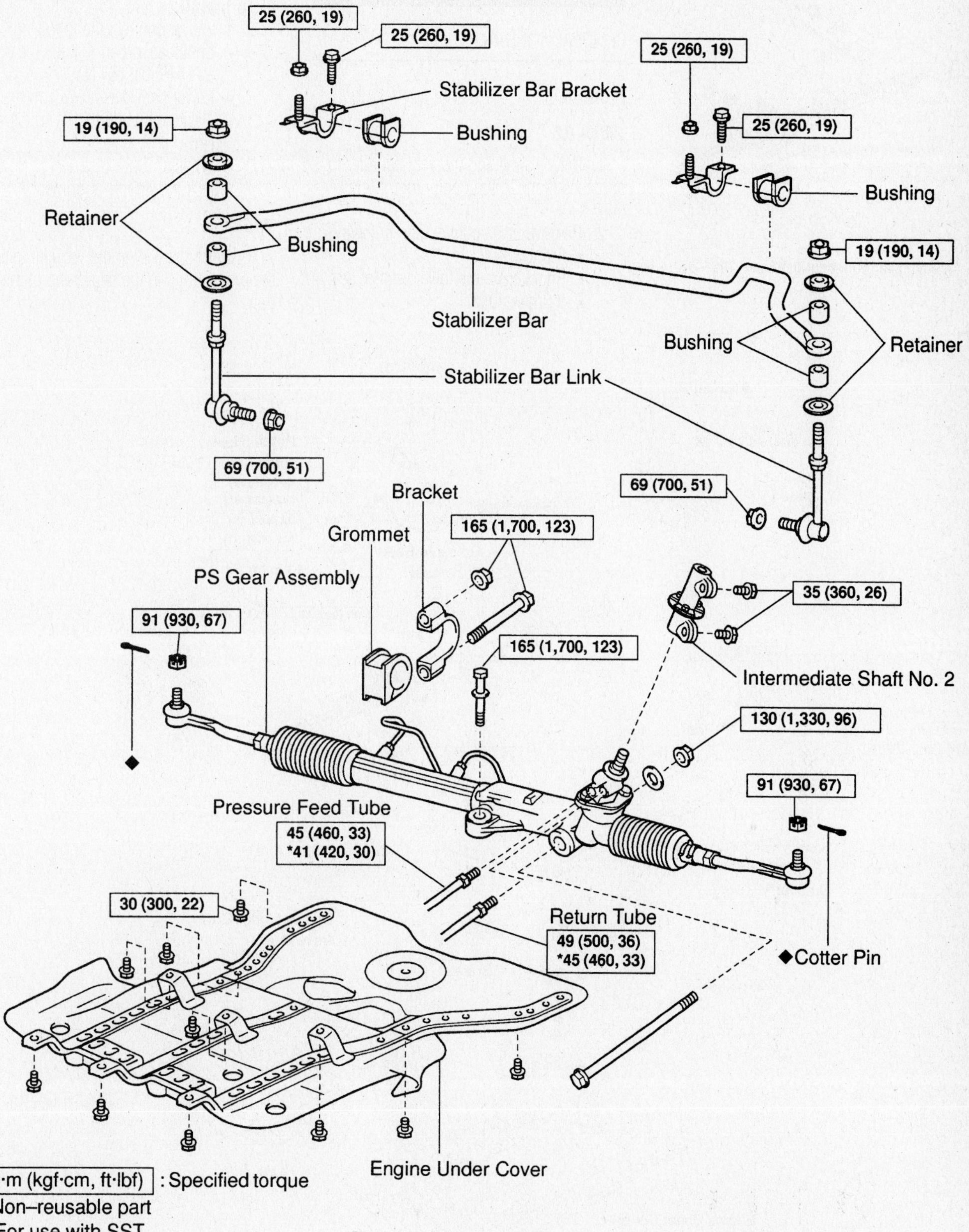
25 (260, 19)
25 (260, 19)
25 (260, 19)
Stabilizer Bar Bracket
19 (190, 14)
Bushing
25 (260, 19)
Bushing
Retainer
Bushing
19 (190, 14)
Stabilizer Bar
Bushing
Retainer
Stabilizer Bar Link
69 (700, 51)
69 (700, 51)
Bracket
Grommet
165 (1,700, 123)
PS Gear Assembly
35 (360, 26)
91 (930, 67)
165 (1,700, 123)
Intermediate Shaft No. 2
130 (1,330, 96)
91 (930, 67)
Pressure Feed Tube
45 (460, 33)
*41 (420, 30)
30 (300, 22)
Return Tube
49 (500, 36)
*45 (460, 33)
◆Cotter Pin
Engine Under Cover
N·m (kgf·cm, ft·lbf) : Specified torque
◆Non–reusable part
* For use with SST
67170-4RUN-G184

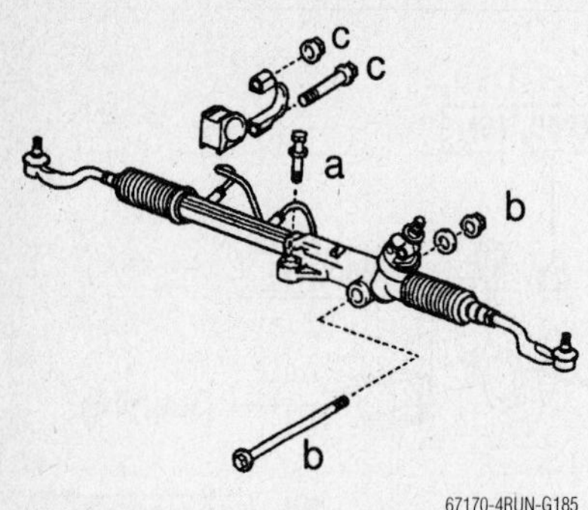

Steering gear bolt identification—2001–02

Shock Absorber

REMOVAL & INSTALLATION

Front

2001–02

1. Before servicing the vehicle, refer to the precautions in the beginning of this section.
2. Remove or disconnect the following:
   - Front wheel
   - Shock from the lower control arm
   - 3 upper nuts
   - Shock absorber

To install:

3. Install or connect the following:
   - Shock absorber. Tighten the 3 nuts to 47 ft. lbs. (64 Nm).
   - Lower shock-to-lower control arm. Tighten the bolt to 101 ft. lbs. (135 Nm).
   - Wheels
4. Check the vehicle alignment.

2003–04

1. Before servicing the vehicle, refer to the precautions in the beginning of this section.

N·m (kgf·cm, ft·lbf) : Specified torque
◆ Non–reusable part
* For use with SST

67170-4RUN-G186

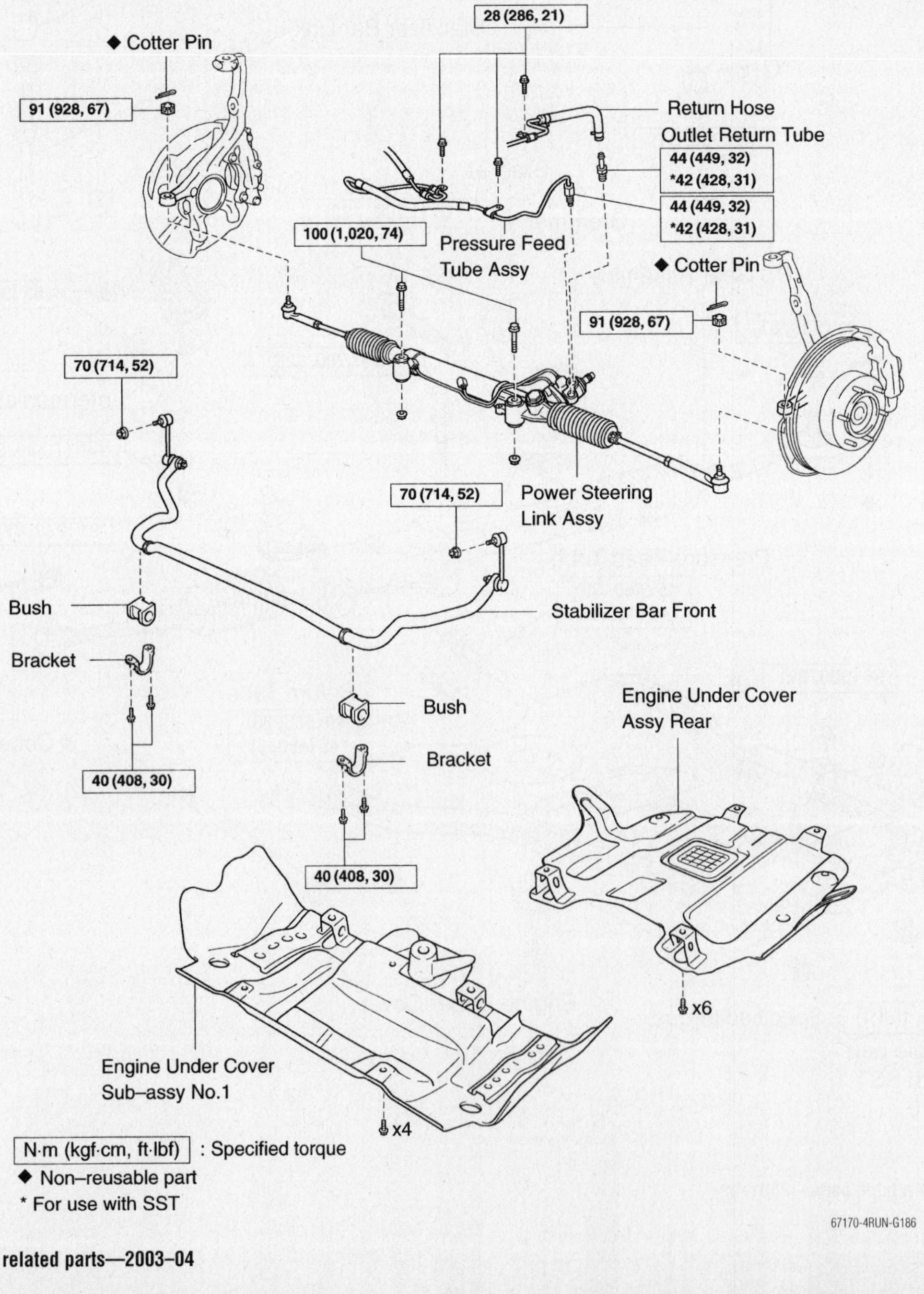

Steering gear and related parts—2003–04

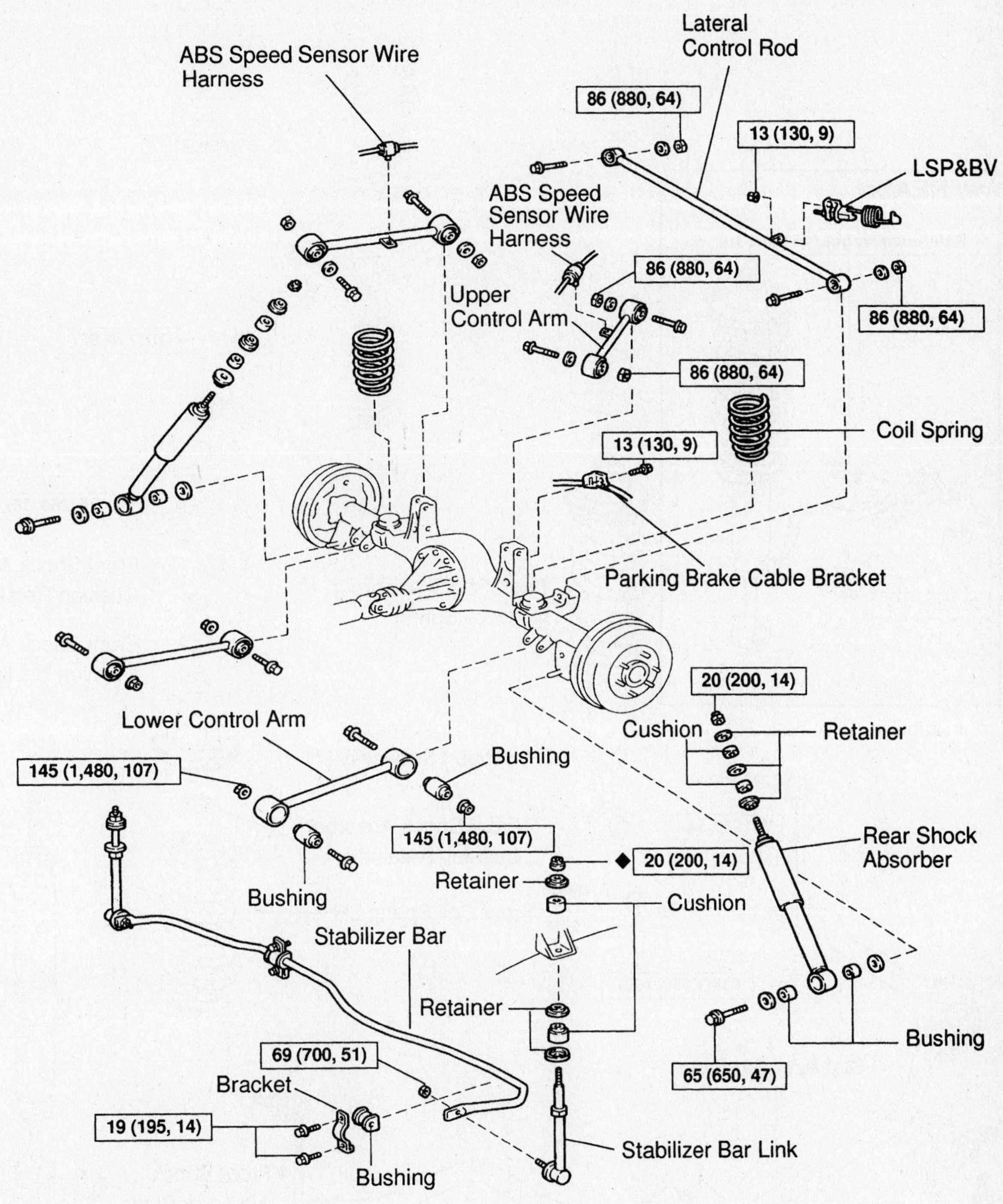

N·m (kgf·cm, ft·lbf) : Specified torque

◆ Non–reusable part

67170-4RUN-G187

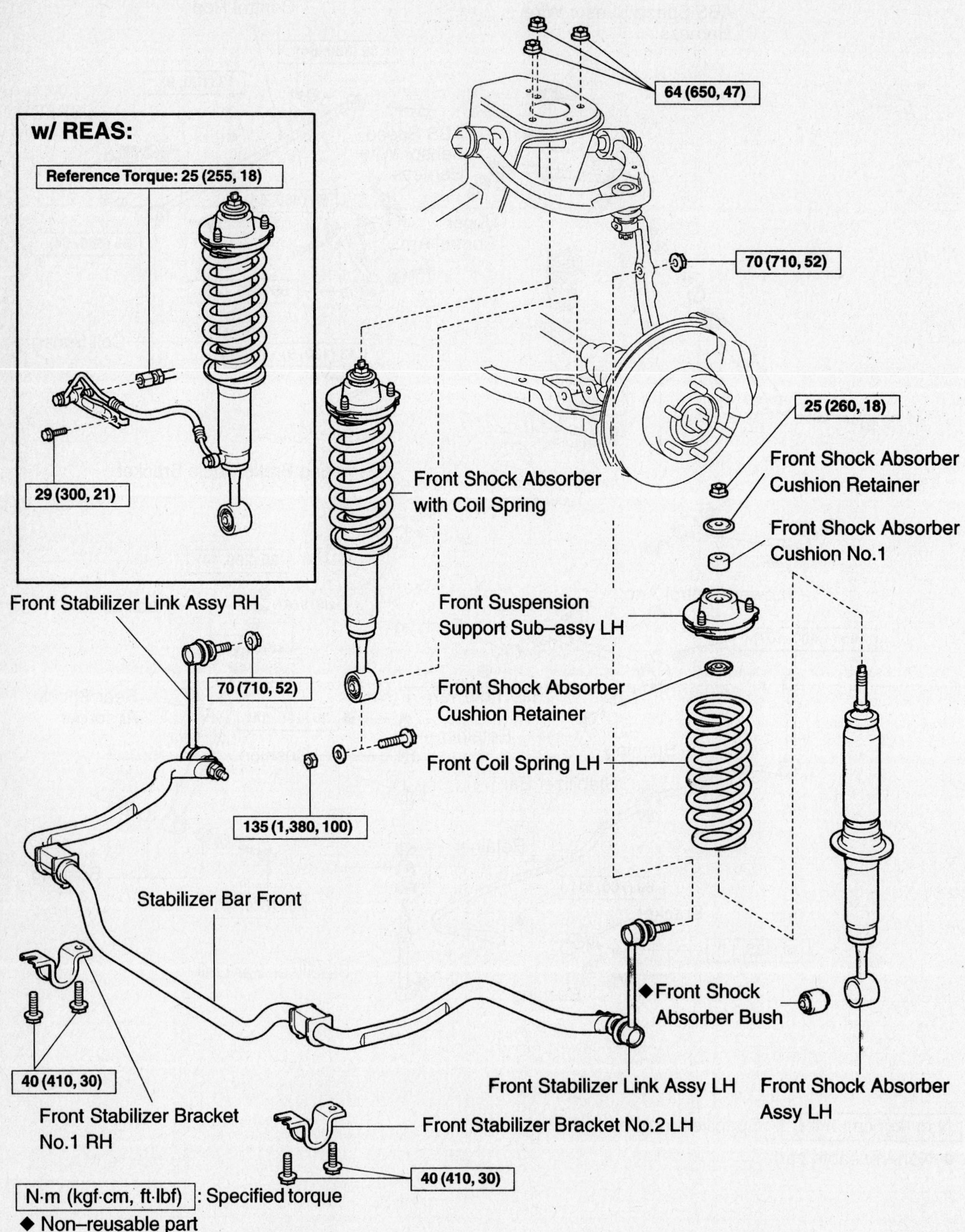

64 (650, 47)
w/ REAS:
Reference Torque: 25 (255, 18)
29 (300, 21)
70 (710, 52)
25 (260, 18)
Front Shock Absorber with Coil Spring
Front Shock Absorber Cushion Retainer
Front Shock Absorber Cushion No.1
Front Stabilizer Link Assy RH
70 (710, 52)
Front Suspension Support Sub–assy LH
Front Shock Absorber Cushion Retainer
Front Coil Spring LH
135 (1,380, 100)
Stabilizer Bar Front
◆Front Shock Absorber Bush
40 (410, 30)
Front Stabilizer Link Assy LH
Front Shock Absorber Assy LH
Front Stabilizer Bracket No.1 RH
Front Stabilizer Bracket No.2 LH
40 (410, 30)
N·m (kgf·cm, ft·lbf) : Specified torque
◆ Non–reusable part
67170-4RUN-G188

Front shock absorber and related parts—2003–04

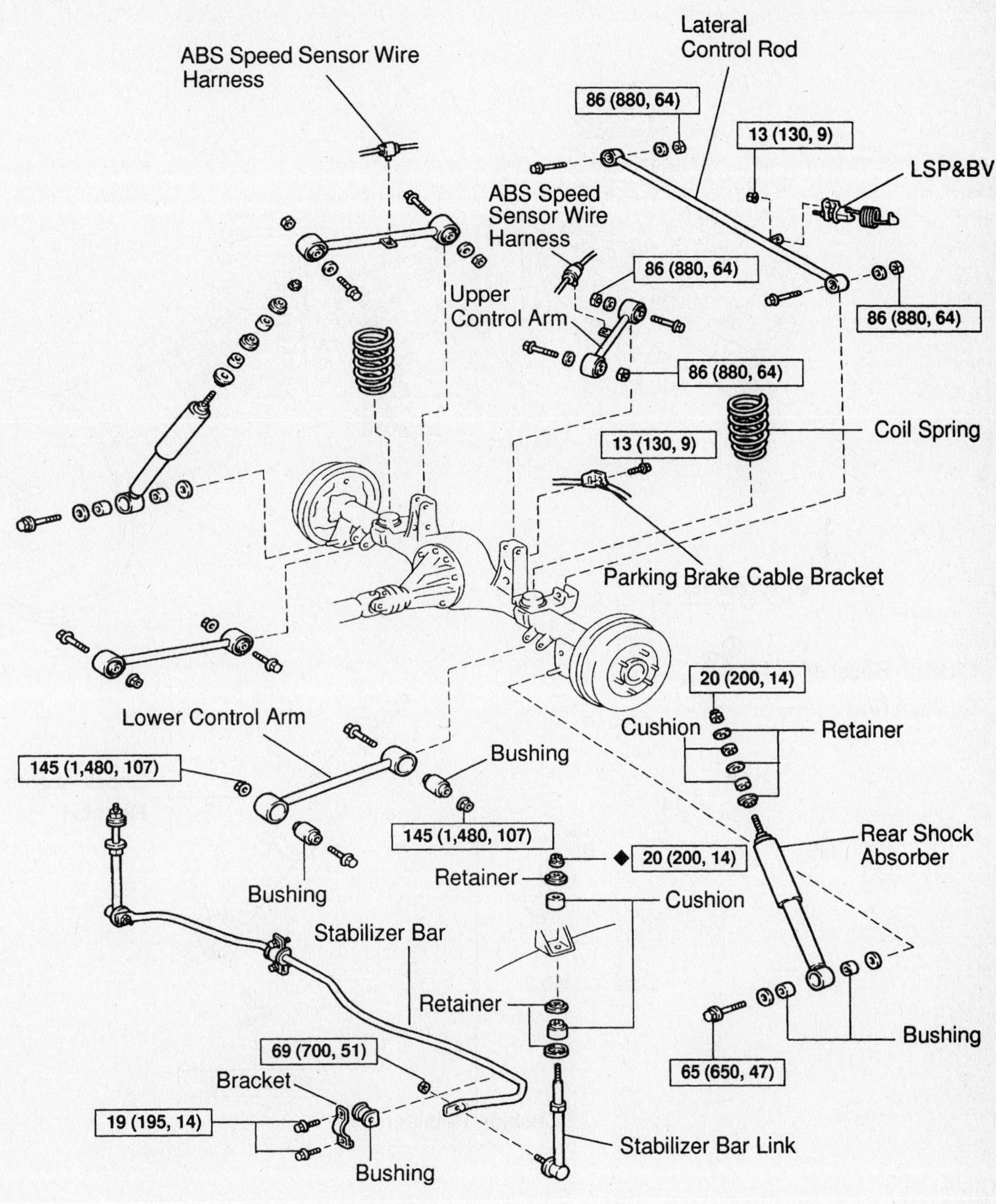

N·m (kgf·cm, ft·lbf) : Specified torque

◆ Non–reusable part

67170-4RUN-G190

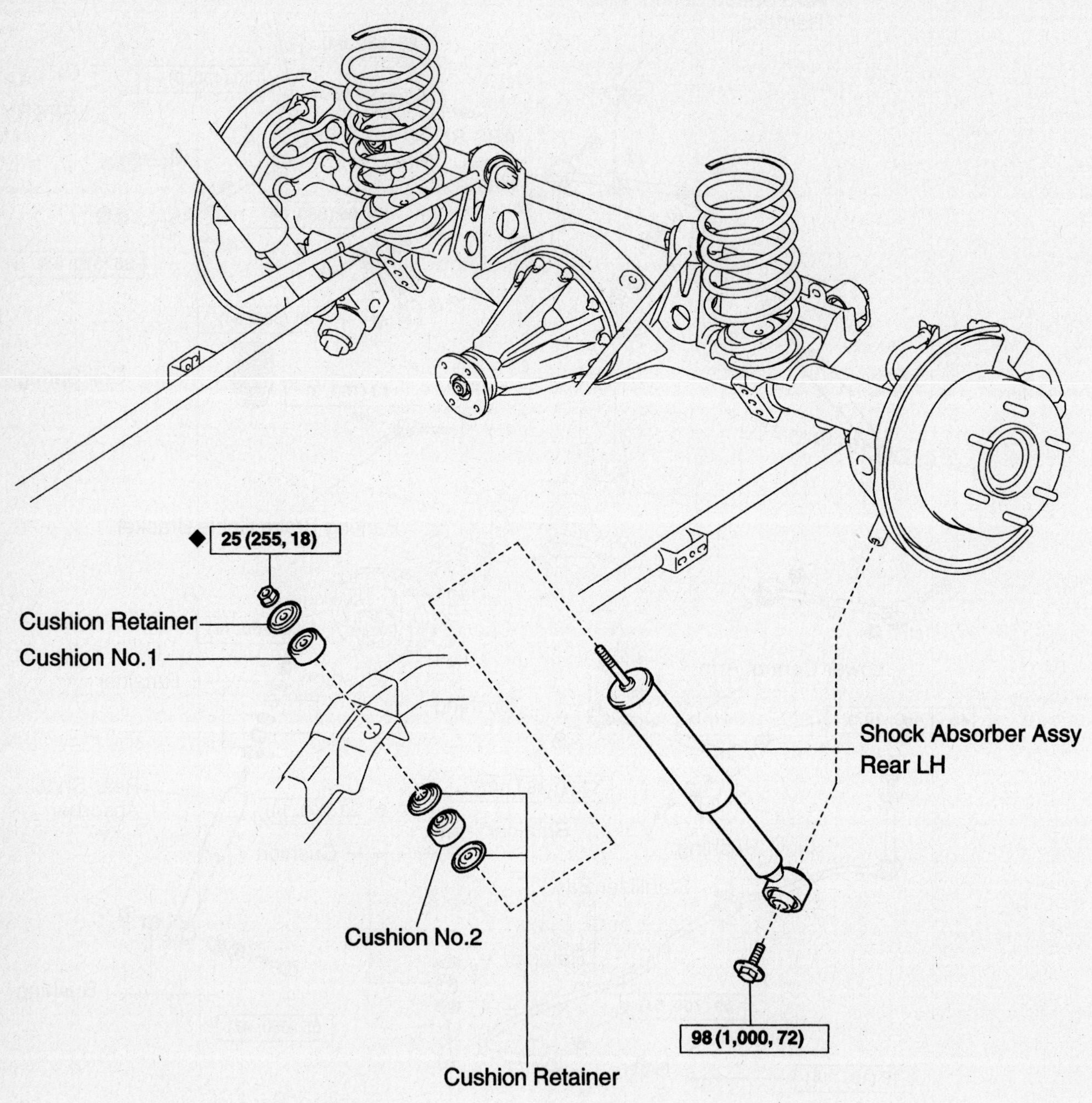

N·m (kgf·cm, ft·lbf) : Specified torque

◆ Non–reusable part

67170-4RUN-G191

Rear suspension—2003–04 without air suspension

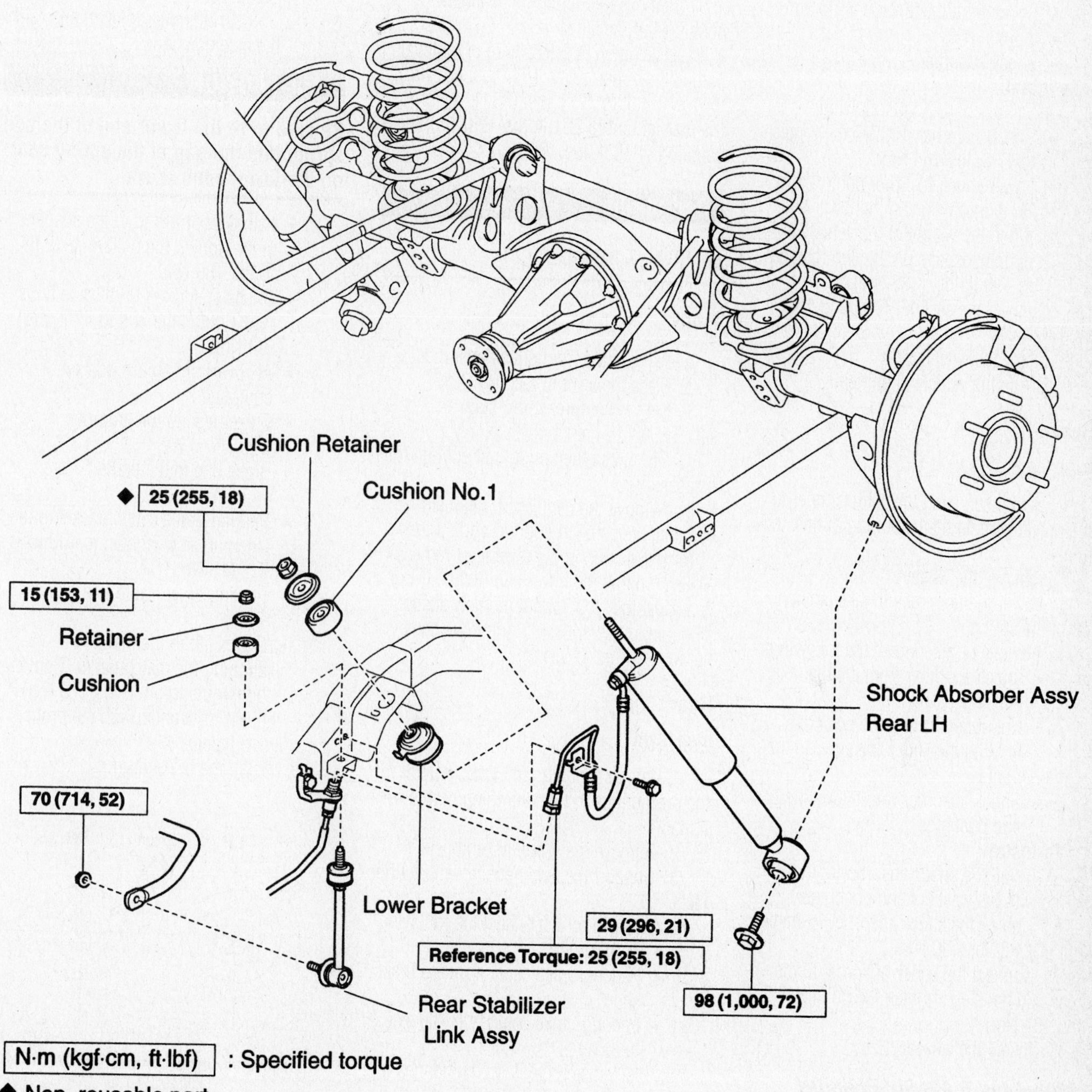

Rear suspension—2003–04 with air suspension

2. Remove or disconnect the following:
- Front wheel
- Stabilizer bar

3. With air shocks, lower the suspension as far as possible and disconnect the air line.
- Shock from the lower control arm
- 3 upper nuts
- Shock absorber

To install:

4. Install or connect the following:
- Shock absorber. Tighten the 3 nuts to 47 ft. lbs. (64 Nm).
- Lower shock-to-lower control arm. Temporarily tighten the lower bolt. When the weight of the vehicle is on the suspension, tighten the bolt to 100 ft. lbs. (135 Nm).

5. Connect the air line. Torque to 18 ft. lbs. (25 Nm)
6. Install the wheels.
7. Check the vehicle alignment.

Rear

2001–02

1. Before servicing the vehicle, refer to the precautions in the beginning of this section.
2. Remove the wheel.
3. Lower the floor jack to take tension off of the spring.
4. Remove or disconnect the following:
- Shock absorber from the rear axle housing
- Nut, retainers and the cushions holding the shock absorber to the frame
- Shock absorber with the washers and bushings

To install:

5. Install the shock absorber to the frame with the washers and bushings.
6. Tighten the shock absorber-to-frame nut to 14 ft. lbs. (20 Nm)
7. Connect the shock absorber to the rear axle housing. Tighten the bolt to 47 ft. lbs. (64 Nm)
8. Install the wheels.

2003–04 W/O AIR SUSPENSION

1. Remove the wheel.
2. Support the rear axle with a jack.
3. Remove the lower shock bolt.
4. Remove the upper nut, washers, bushings and retainer.
5. Installation is the reverse of removal. Torque the upper nut to 18 ft. lbs. (25 Nm); the lower bolt to 72 ft. lbs. (98 Nm).

2003–04 WITH AIR SUSPENSION

1. Remove the wheel.
2. Remove the stabilizer bar link.
3. Remove the air line bracket nut.
4. Lower the axle to fully extend the shocks.
5. Disconnect the air line coupling.
6. Remove the lower shock bolt.
7. Remove the upper nut, washers, bushings and retainer.
8. Installation is the reverse of removal. Torque the upper nut to 18 ft. lbs. (25 Nm); the lower bolt to 72 ft. lbs. (98 Nm). Torque the bracket bolt to 221 ft. lbs. (29 Nm); the coupling to 18 ft. lbs. (25 Nm).

Coil Spring

REMOVAL & INSTALLATION

Front

1. Remove the front shock absorber.
2. Place the strut in a compressor.
3. Set the compressor to span an 8 coil set.
4. Compress the spring just enough for disassembly.
5. Remove the center nut, retainers, cushion, suspension support and spring.
6. Installation is the reverse of removal. Make sure that the lower end of the spring is against the stop. Orient the new suspension support as shown. Tighten the center nut to 18 ft. lbs. (25 Nm).

Rear

2001–02

1. Before servicing the vehicle, refer to the precautions in the beginning of this section.
2. Remove the wheel assemblies.
3. Support the axle housing with a floor jack.
4. Remove or disconnect the following:
- Brake drum
- Parking brake cable from the brake shoe
- Parking brake cable from the axle housing.

5. Place matchmarks on the flanges for the driveshaft and differential.
6. Remove or disconnect the following:
- Driveshaft from the differential
- Brake hose line from the brake hose
- Brake hose-to-brake bracket clip
- Brake hose from the body
- Anti-lock Brake System (ABS) wiring harness bracket, if equipped with ABS
- Shock absorbers from the axle housing
- Lateral control rod nuts/bolts
- Control rod from the suspension
- Coil spring, by lower the rear axle housing

To install:

7. Install or connect the following:
- Coil springs into position and raise the axle housing.

⁂ CAUTION

Be sure to fit the lower end of the coil spring into the gap of the spring seat on the lower control arm.

- Lateral control rod to the suspension. Tighten the bolts/nuts to 64 ft. lbs. (86 Nm).
- Shock absorbers to the axle housing. Tighten the bolt to 47 ft. lbs. (64 Nm).
- ABS wiring harness bracket, if equipped
- Brake hose to the bracket
- Clip
- Brake line to the brake hose. Tighten the tube.
- Parking brake cable bracket to the axle housing. Tighten to 108 inch lbs. (13 Nm).
- Parking brake cable to the brake shoes
- Driveshaft to the differential, by aligning the matchmarks. Tighten bolts/nuts to 54 ft. lbs. (73 Nm).

8. Fill the differential with the proper amount and type of oil.
9. Install the brake drum.

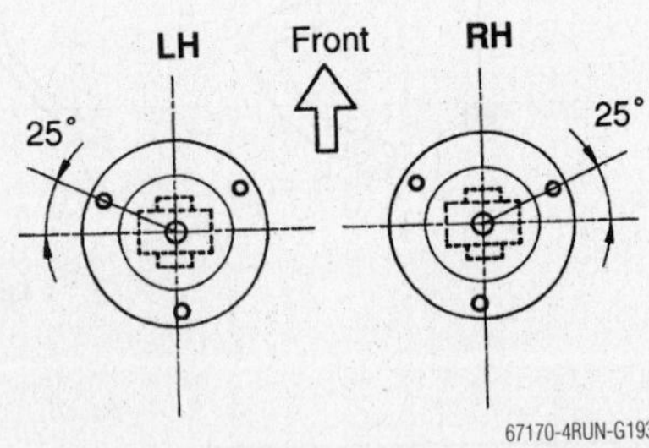

Suspension support orientation—2001–02

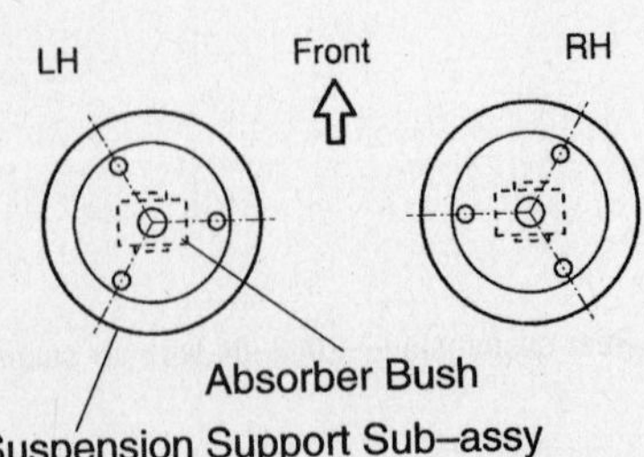

Suspension support orientation—2003–04

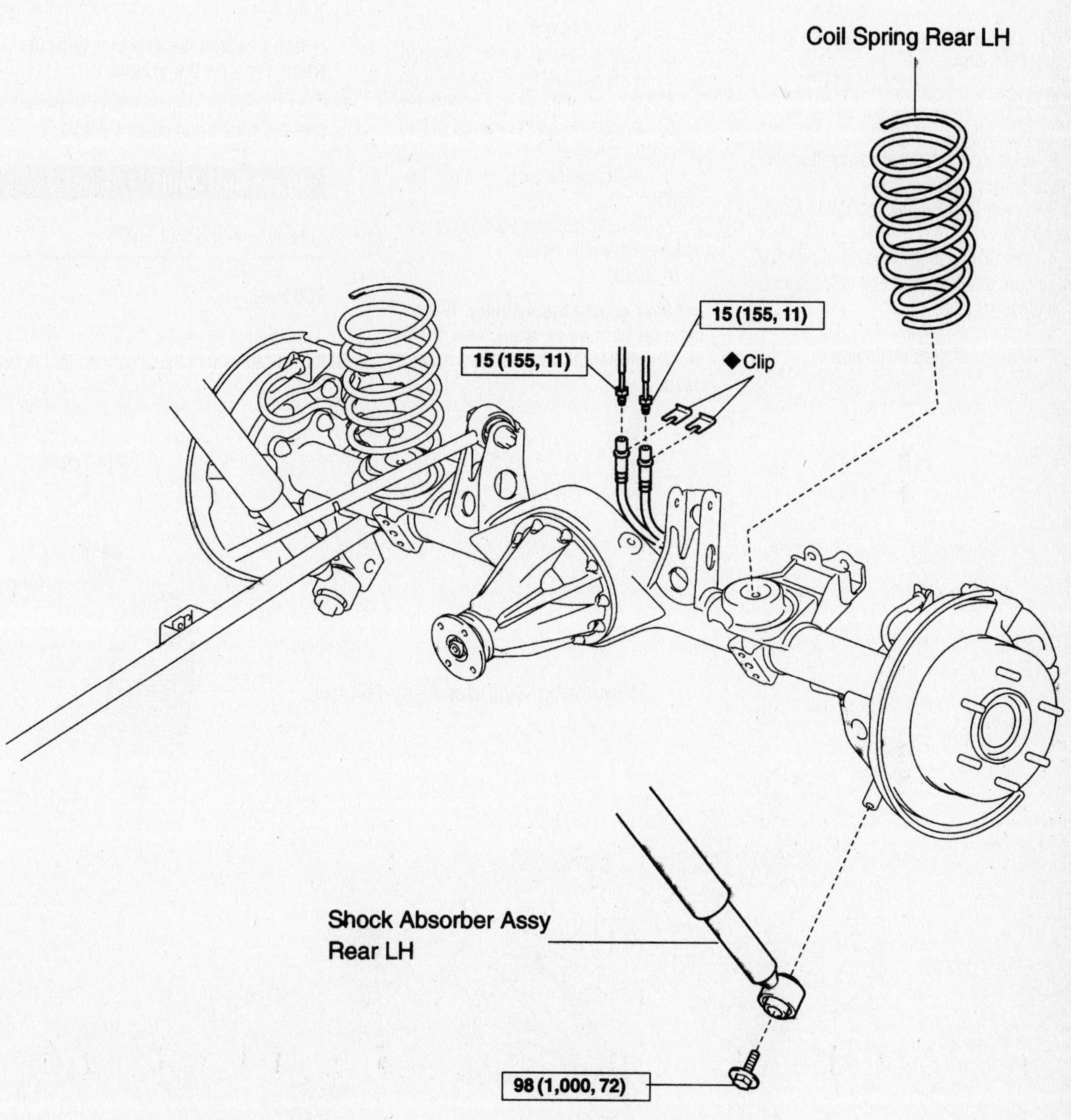

N·m (kgf·cm, ft·lbf) : Specified torque

◆ Non–reusable part

67170-4RUN-G195

10. Bleed the brake system.
11. Install the wheel assemblies.
12. Lower the vehicle and bounce the vehicle several times to stabilize the suspension.
13. Tighten the lower control arm to 107 ft. lbs. (145 Nm).

2003–04

1. Remove the rear shock.
2. Disconnect the brake pipes from the hoses at the bracket.
3. Lower the axle just enough to remove the spring.
4. Installation is the reverse of removal. Make sure that the lower end of the spring is against the stop.
5. Install the shock.
6. Refill and bleed the brakes.

Rear Air Spring

REMOVAL & INSTALLATION

1. Remove the wheels.
2. Support the frame with jackstands.
3. Lower the axle as far as it will go.
4. Disconnect the height control tube.
5. Remove the clip at the top of the pneumatic cylinder.
6. Discharge the air to collapse the cylinder.
7. Turn the cylinder 90 degrees and remove it from the axle.

To install

➡Never extend the cylinder. If the cylinder is to be re-used, new O-rings, plate and height control plug must be used.

8. Install the cylinder with the clip.
9. Connect the height control tube.
10. Raise the axle until it just contacts the bottom of the spring and install the clip at the lower end into the keyhole in the axle.

➡Don't extend the cylinder until the wheels are on the ground.

11. Once the vehicle is on the ground, start the engine and check for leaks.

Upper Ball Joint

REMOVAL & INSTALLATION

2001–02

1. Before servicing the vehicle, refer to the precautions in the beginning of this section.

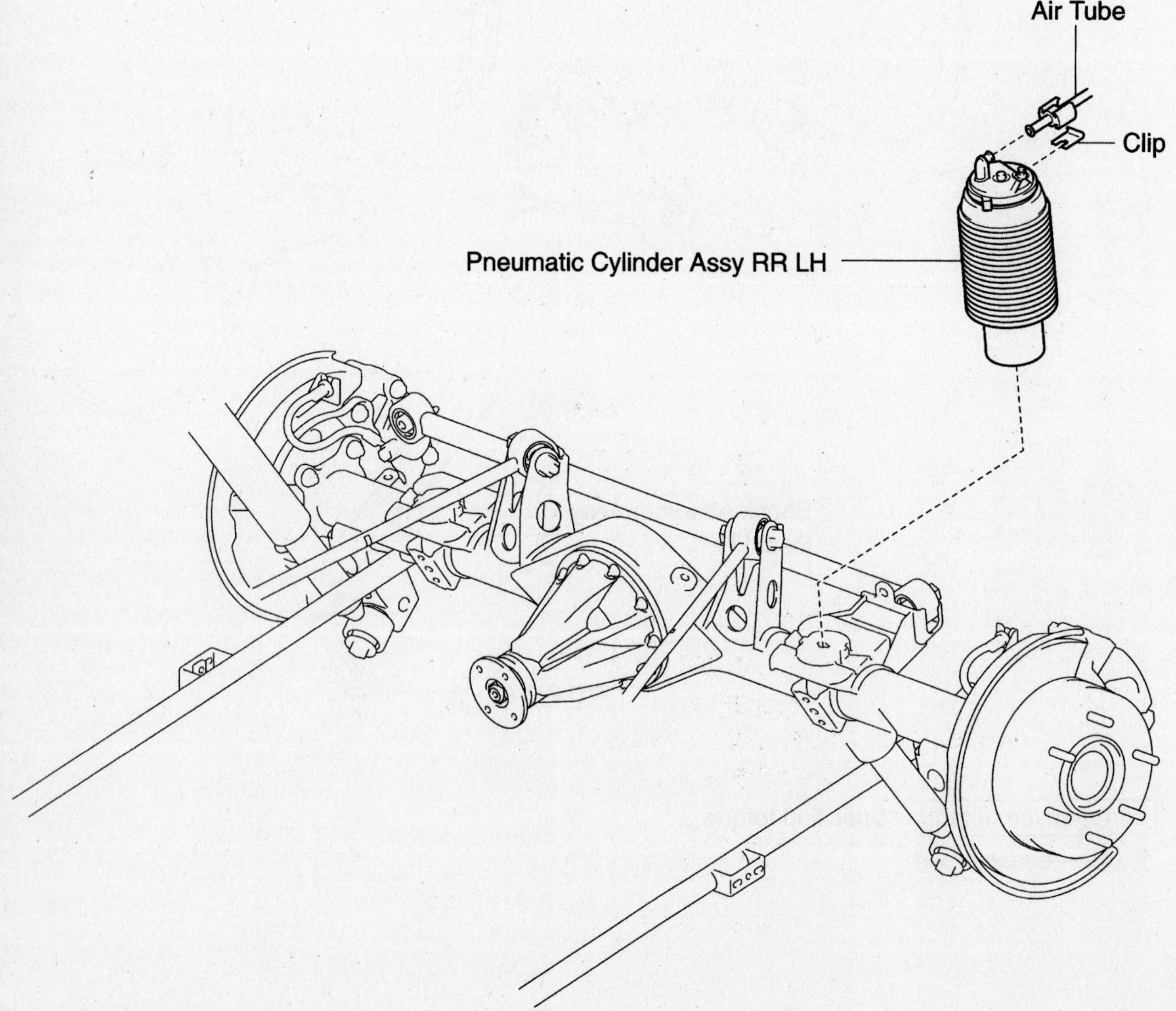

Rear air spring

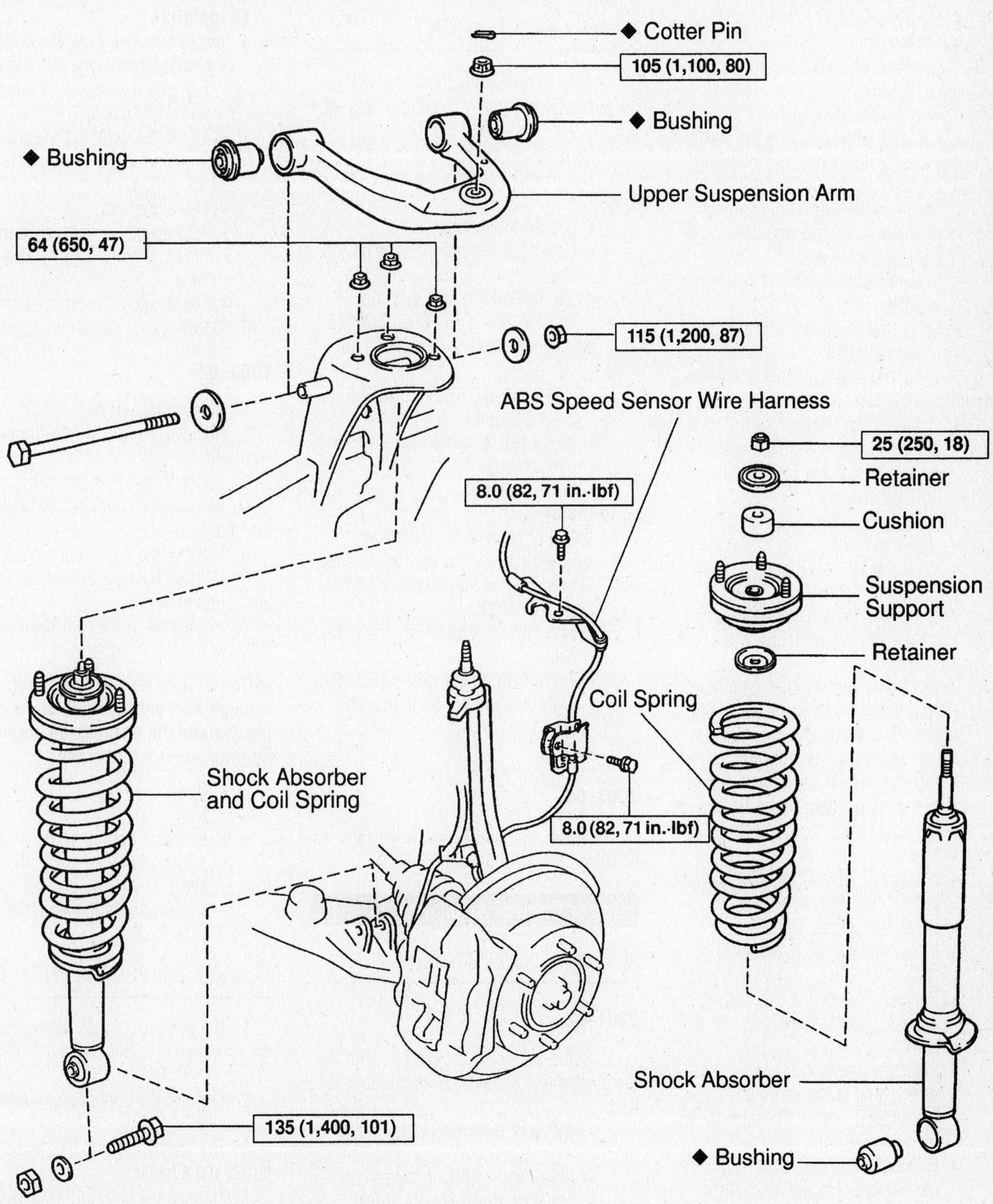

N·m (kgf·cm, ft·lbf) : Specified torque

◆ Non–reusable part

67170-4RUN-G197

Upper control arm and related parts—2001–02

2. Remove or disconnect the following:
   - Front wheels
   - Strut assembly
   - Grease cap

3. If equipped with 4WD, disconnect the halfshaft, as follows:
   - Cotter pin and lockcap
   - Locknut, while applying the brakes

4. Remove or disconnect the following:
   - Anti-lock Brake System (ABS) speed sensor and wiring harness clamp from the steering knuckle, if equipped with ABS
   - Brake line bracket from the steering knuckle
   - Front brake caliper and the rotor
   - Lower ball joint

5. Remove the steering knuckle with the axle hub, as follows:
   - Cotter pin and loosen the nut
   - Steering knuckle from the upper control arm, using SST 09950-40010
   - Steering knuckle
   - Upper ball joint
   - Wire and the boot
   - Snapring
   - Upper ball joint, using SST 09950-40010 and a deep socket wrench.

To install:

6. Install the upper ball joint, as follows:
   - New ball joint with a new snapring
   - New boot secured with a new wire

7. Install or connect the following:
   - Steering knuckle with the axle hub to the upper control arm. Tighten the nut to 80 ft. lbs. (108 Nm).
   - New cotter pin
   - Lower ball joint. Tighten the 4 bolts to 59 ft. lbs. (80 Nm).
   - Rotor and caliper. Tighten the caliper bolts to 90 ft. lbs. (123 Nm).
   - Brake line bracket to the steering knuckle. Tighten it to 21 ft. lbs. (28 Nm).
   - ABS speed sensor and wiring harness clamp to the steering knuckle, if equipped. Tighten the bolts to 72 inch lbs. (8 Nm).
   - Halfshaft, if disconnected. Tighten the nut to 174 ft. lbs. (235 Nm).
   - Grease cap
   - Strut
   - Front wheel

8. Check the alignment

2003–04

The upper ball joint is not replaceable. If defective, the arm must be replaced.

Lower Ball Joint

REMOVAL & INSTALLATION

1. Before servicing the vehicle, refer to the precautions in the beginning of this section.

2. Remove the wheel.

3. Disconnect the tie rod end, as follows:
   - Loosen the 4 bolts
   - Cotter pin and nut from the tie rod end
   - Tie rod end from the steering knuckle, using SST 09610-20012

4. Remove the lower ball joint, as follows:
   - Cotter pin and the nut from the lower ball joint
   - Lower ball joint from the lower suspension arm
   - Lower ball joint

To install:

5. Install or connect the following:
   - Lower ball joint to the lower control arm. Tighten the 4 bolts to 59 ft. lbs. (80 Nm).
   - Nut. Tighten it to 105 ft. lbs. (142 Nm).
   - Tie rod end to the steering knuckle. Tighten the nut to 66 ft. lbs. (90 Nm).
   - Wheel

2003–04

The lower ball joint is not replaceable. If defective, the arm must be replaced.

Upper Control Arm

REMOVAL & INSTALLATION

2001–02

1. Before servicing the vehicle, refer to the precautions in the beginning of this section.

2. Remove or disconnect the following:
   - Shock and coil spring assembly
   - Anti-lock Brake System (ABS) speed sensor wire harness clamp

3. Disconnect the upper ball joint, as follows:
   - Cotter pins and loosen the nut
   - Upper ball joint from the control arm, using a ball joint separator
   - Support the steering knuckle
   - Nut

4. Detach the control arm, by removing the nut, bolt, washers and lowering the arm.

To install:

5. Install or connect the following:
   - Upper control arm with the washer, bolt and nut. Tighten the nut to 87 ft. lbs. (115 Nm).
   - Upper ball joint to the control arm. Tighten the mounting nut to 80 ft. lbs. (105 Nm).
   - New cotter pin
   - ABS speed sensor wire harness clamp. Tighten it to 71 inch lbs. (8 Nm).
   - Shock and coil spring assembly

6. Check and/or adjust the alignment.

2003–04

1. Remove the wheel.
2. Remove the ABS wiring from the arm and strut.
3. Support the lower arm with a jack.
4. Remove the cotter pin and nut from the upper ball stud.
5. Remove the brake line bracket bolt.
6. Remove the bolt, 2 washers and nut and remove the arm.
7. Installation is the reverse of removal. Observe the following torques:

➡Do not fully tighten the upper arm through-bolt until the vehicle is on the ground and the suspension stabilized by jouncing it a few times.

   - Brake line bracket: 51 inch lbs. (5.8 Nm)
   - Ball stud nut: 81 ft. lbs. (110 Nm)
   - ABS wire brackets: 9 ft. lbs. (13 Nm)
   - Upper arm through-bolt: 85 ft. lbs. (115 Nm)

CONTROL ARM BUSHING REPLACEMENT

1. Before servicing the vehicle, refer to the precautions in the beginning of this section.
2. Remove the upper control arm from the vehicle.
3. Pry up the bushing flange, using a chisel and a hammer.
4. Using tools 09613-26010, 09613-20060 and 09950-00020 and a shop press, remove the bushing.

To install:

5. Using tools 09223-00010, 09506-35010 and a shop press, press the new bushing into the upper control arm.
6. Install the upper control arm to the chassis.
7. Check and/or adjust the alignment.

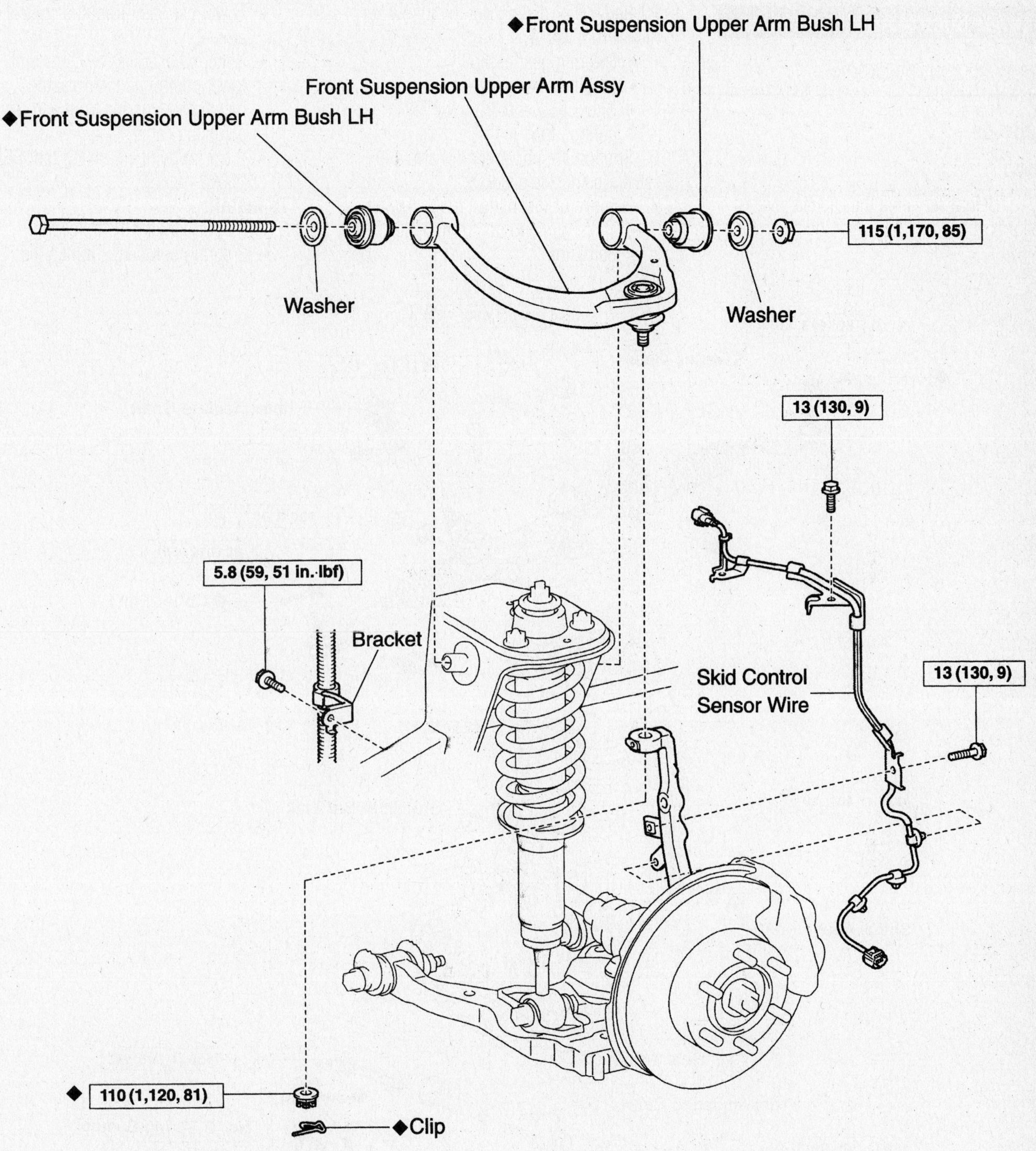

N·m (kgf·cm, ft·lbf) : Specified torque
◆ Non–reusable part

67170-4RUN-G198

Upper control arm and related parts—2003–04

Lower Control Arm

REMOVAL & INSTALLATION

2001–02

2WD

1. Before servicing the vehicle, refer to the precautions in the beginning of this section.
2. Remove or disconnect the following:
   - Front wheel
   - Steering gear assembly
   - Stabilizer bar link
   - Shock absorber from the lower control arm
3. Support the upper control arm and the steering knuckle securely.
4. Remove or disconnect the following:
   - Cotter pin and nut from the lower ball joint
   - Lower ball joint from the control arm
   - Bolts, nuts, adjusting cams and lower control arm, by placing matchmarks on the front and rear adjusting cams
   - 2 spring bumpers, using tool SST 09922-10010

To install:

5. Install or connect the following:
   - 2 spring bumpers. Tighten the

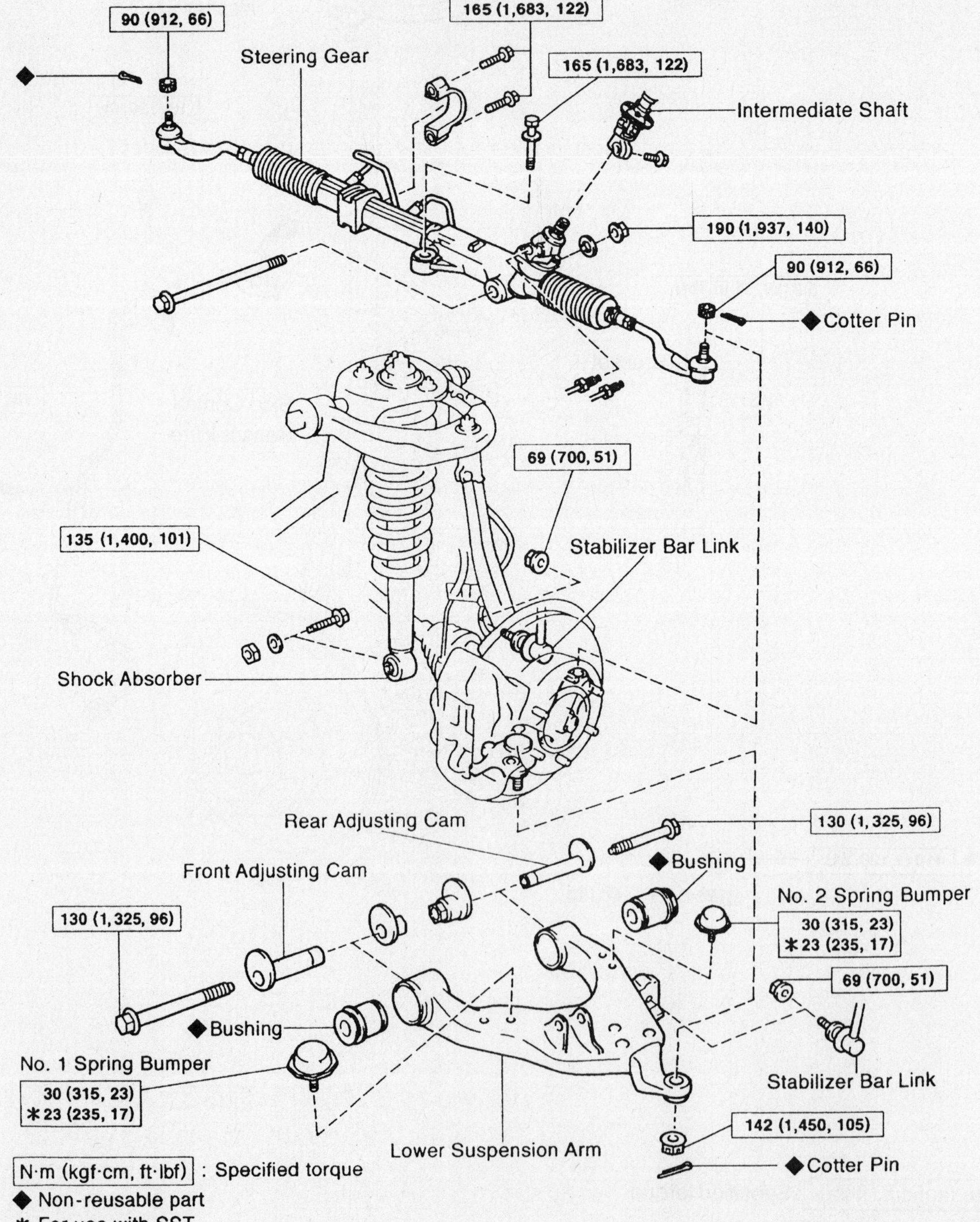

Exploded view of the lower control arm and associated components—2001–02

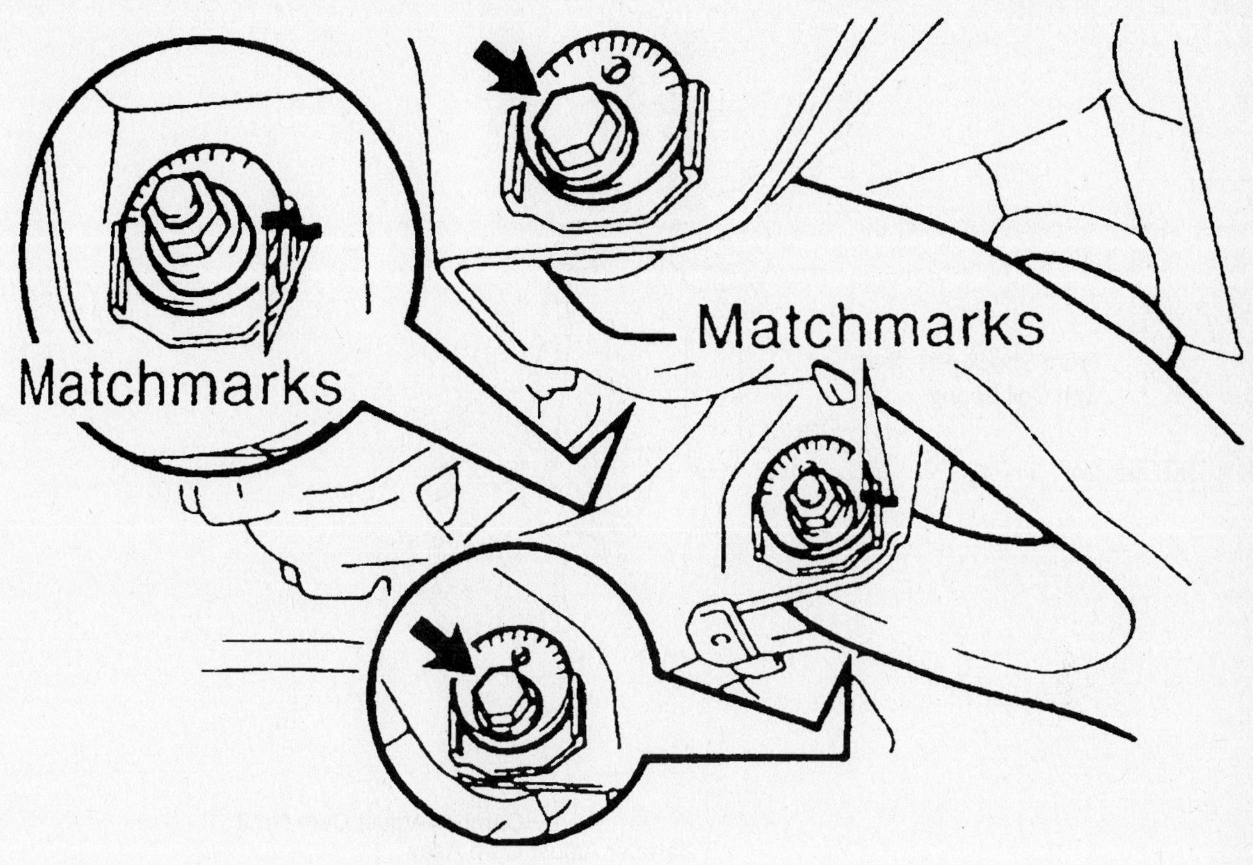

Place matchmarks on the front and rear adjusting cams—2001–02

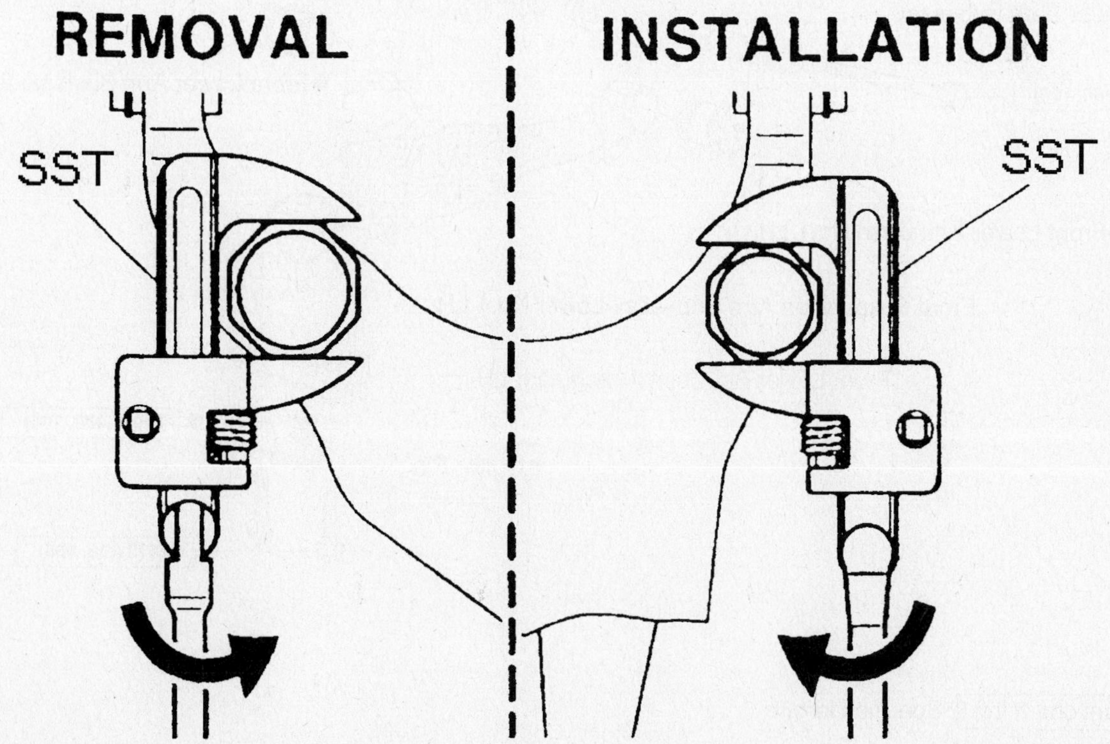

Remove the spring bumpers with special tool 09922-10010—2001–02

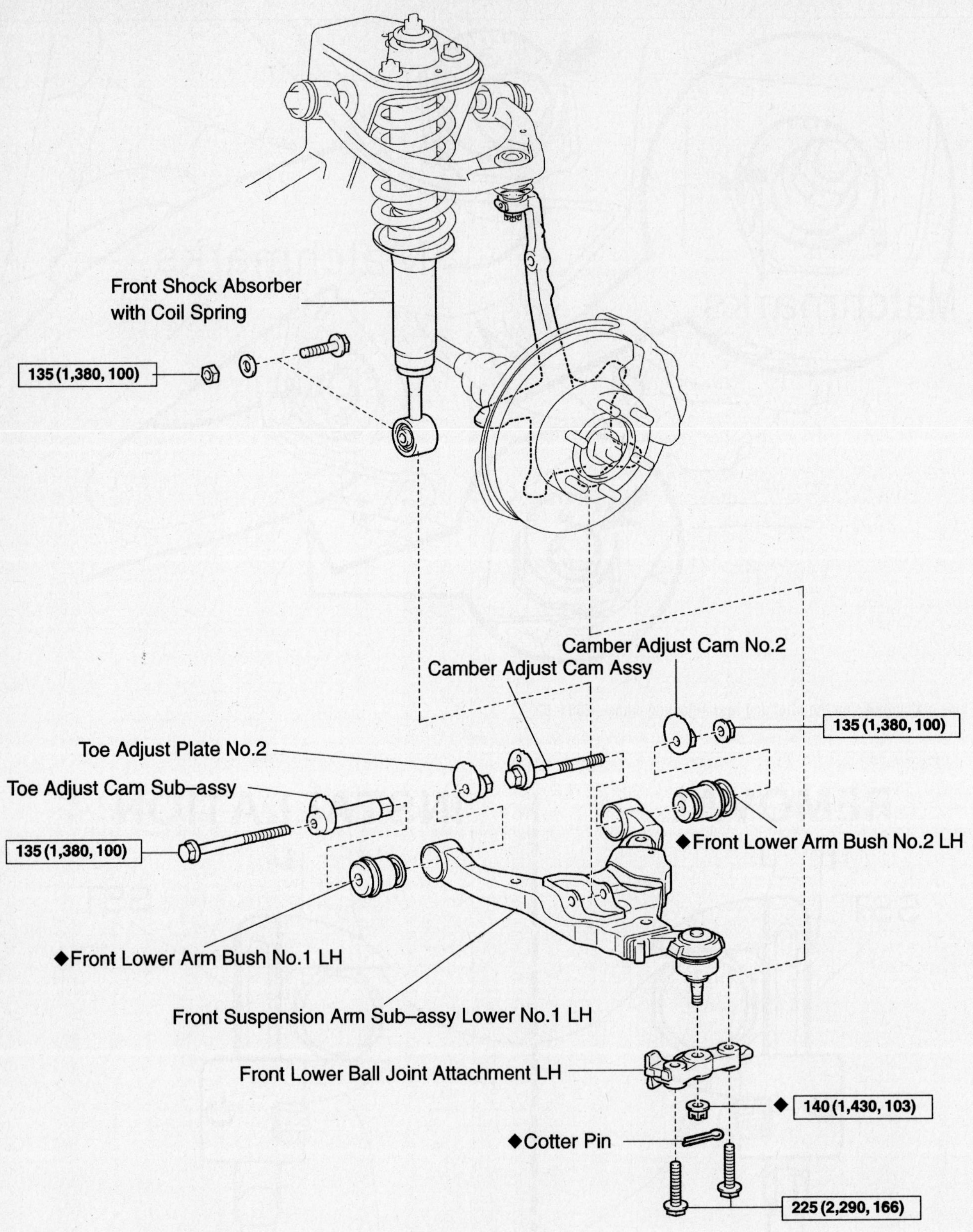

Front Shock Absorber with Coil Spring
135 (1,380, 100)
Camber Adjust Cam No.2
Camber Adjust Cam Assy
135 (1,380, 100)
Toe Adjust Plate No.2
Toe Adjust Cam Sub–assy
135 (1,380, 100)
◆Front Lower Arm Bush No.2 LH
◆Front Lower Arm Bush No.1 LH
Front Suspension Arm Sub–assy Lower No.1 LH
Front Lower Ball Joint Attachment LH
◆ 140 (1,430, 103)
◆Cotter Pin
225 (2,290, 166)
N·m (kgf·cm, ft·lbf) : Specified torque
◆ Non–reusable part
67170-4RUN-G199

Lower control arm and related parts—2003–04

spring bumpers to 17 ft. lbs. (23 Nm).
- Lower control arm and adjusting cams. Tighten the nuts and bolts to 96 ft. lbs. (130 Nm),
- Lower ball joint to the control arm
- Cotter pin and nut to the lower ball joint. Tighten the nut to 105 ft. lbs. (142 Nm).
- Shock absorber to the lower control arm
- Stabilizer bar link
- Steering gear assembly
- Front wheel

6. Check and/or adjust the alignment.

4WD

1. Before servicing the vehicle, refer to the precautions in the beginning of this section.
2. Remove or disconnect the following:
   - Front wheel
   - Steering gear assembly
   - Stabilizer bar link
   - Shock absorber from lower control arm
3. Support the upper control and steering knuckle securely.
4. Remove or disconnect the following:
   - Cotter pin and nut from the lower ball joint
   - Lower ball joint from the lower control arm
5. Place matchmarks on the front and rear adjusting cams.
6. Remove or disconnect the following:
   - 2 bolts, nuts, adjusting cams and lower control arm
   - Spring bumpers with a special tool 09922-10010.

To install:

7. Install or connect the following:
   - Spring bumpers. Tighten to 17 ft. lbs. (23 Nm).
   - Lower control arm, placing it in the appropriate position with the matchmarks. Tighten the arm to 96 ft. lbs. (130 Nm).
   - Lower ball joint. Tighten the nut to 105 ft. lbs. (142 Nm).
   - Shock absorber to the lower control arm
   - Stabilizer bar link
   - Steering gear assembly
   - Front wheel. Tighten the lug nuts.
8. Check and/or adjust the alignment.

2003–04

1. Remove the wheel.
2. Remove the lower shock absorber bolt.
3. Remove the ball joint attachment bolts.
4. Matchmark the camber and toe adjustment cams.
5. Remove the nuts, camber adjustment cams, toe adjustment cams, toe plate and lower arm.
6. Mount the arm in a vise. Remove the cotter pin and ball stud nut, then remove the ball joint attachment using a puller.

To install:

7. Position the lower arm and install the bolts, cams, plate and nuts.

➡Don't fully tighten these bolts until the vehicle is on the ground and the suspension stabilized by jouncing a few times.

8. Attach the lower ball joint attachment. Torque the ball stud nut to 103 ft. lbs. (140 Nm).
9. Connect the lower ball joint attachment. Torque the bolts to 166 ft. lbs. (225 Nm).
10. Connect the lower end of the shock absorber. Don't full tighten the bolt until the vehicle is on the ground.
11. Install the wheel. Lower the vehicle to the ground. Jounce the suspension a few times and tighten the upper arm bolts to 100 ft. lbs. (135 Nm); the lower shock bolt to 100 ft. lbs. (135 Nm).

CONTROL ARM BUSHING REPLACEMENT

1. Before servicing the vehicle, refer to the precautions in the beginning of this section.
2. Remove the lower control arm from the vehicle.
3. Pry up the bushing flange, using a chisel and a hammer.
4. Using tools 09613-26010, 09632-36010 and 09950-00020 and a shop press, remove the bushing.

To install:

5. Using tools 09316-20011, 09710-30021 and a shop press, press the new bushing into the lower control arm.
6. Install the lower control arm to the chassis. Torque the bolts to 96 ft. lbs. (130 Nm).
7. Check and/or adjust the alignment.

Front Wheel Bearing

ADJUSTMENT

The wheel bearings are sealed unit; no adjustment is possible.

REMOVAL & INSTALLATION

2001–02

1. Before servicing the vehicle, refer to the precautions in the beginning of this section.
2. Remove or disconnect the following:
   - Front wheels
   - Shock absorber
   - Grease cap
3. On 4WD, remove the halfshaft, as follows:
   - Cotter pin and lockcap
   - Locknut, while applying the brakes
4. Remove or disconnect the following:
   - Anti-lock Brake System (ABS) speed sensor and wiring harness clamp from the steering knuckle, if equipped with ABS
   - Brake line bracket from the steering knuckle
   - Front brake caliper and rotor
   - 4 bolts and the lower ball joint
5. Remove the steering knuckle with the axle hub, as follows:
   - Cotter pin and loosen the nut
   - Steering knuckle, using SST 09950-40010
6. Clamp the axle hub in a soft jaw vise.
7. Remove or disconnect the following:
   - Grease cap, for 2WD
   - Inside oil seal, for 4WD
   - 4 bolts and shift the brake dust cover towards the hub side
   - Axle hub from the steering knuckle, using SST 09710-30021
   - Bearing spacer and Anti-lock Brake System (ABS) speed sensor rotor/spacer
   - Oil seal (outside) from the steering knuckle, using a flat pry bar
8. Remove the bearing from the steering knuckle, as follows:
   - Snapring
   - Bearing from the steering knuckle, using SST 09950-60020 and 09950-70010 and a press

To install:

9. Install a new bearing, as follows:
   - New bearing to the steering knuckle, using SST 09527-17011 and 09950-60020 and a press
   - New snapring
10. Install or connect the following:
    - New outside oil seal, using SST 09223-15030 and a plastic hammer. Coat MP grease to the oil seal lip.
    - Brake dust cover to the steering knuckle. Tighten the 4 bolts to 13 ft. lbs. (18 Nm).

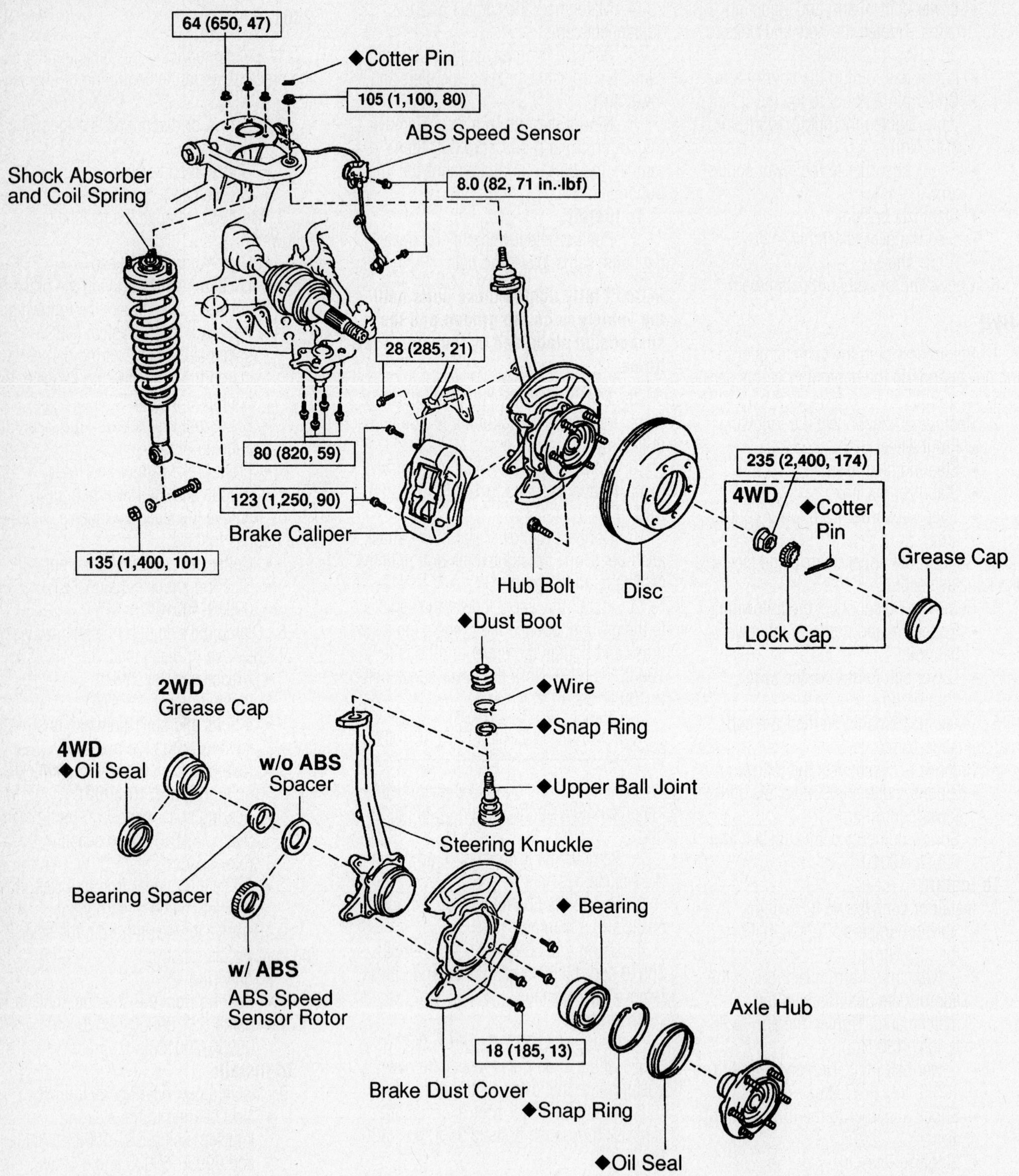
64 (650, 47)
◆Cotter Pin
105 (1,100, 80)
ABS Speed Sensor
Shock Absorber and Coil Spring
8.0 (82, 71 in.·lbf)
28 (285, 21)
80 (820, 59)
123 (1,250, 90)
Brake Caliper
135 (1,400, 101)
Hub Bolt
Disc
235 (2,400, 174)
4WD
◆Cotter Pin
Grease Cap
Lock Cap
◆Dust Boot
◆Wire
◆Snap Ring
◆Upper Ball Joint
2WD
Grease Cap
4WD
◆Oil Seal
w/o ABS
Spacer
Steering Knuckle
Bearing Spacer
◆ Bearing
w/ ABS
ABS Speed Sensor Rotor
Axle Hub
18 (185, 13)
Brake Dust Cover
◆Snap Ring
◆Oil Seal
N·m (kgf·cm, ft·lbf) : Specified torque
◆ Non–reusable part
67170-4RUN-G200

Front hub and related parts—2001–02

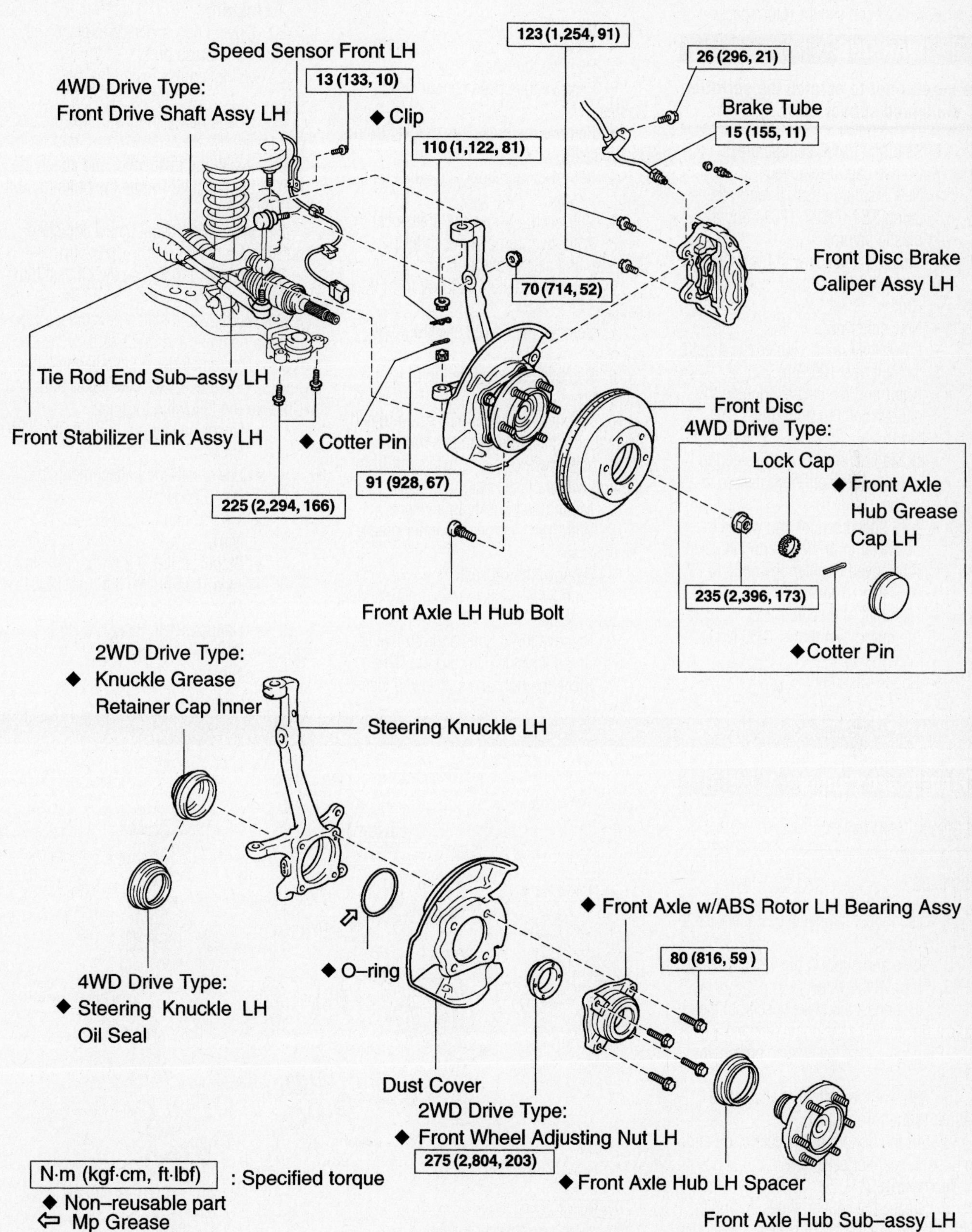
Speed Sensor Front LH
4WD Drive Type:
Front Drive Shaft Assy LH
13 (133, 10)
◆ Clip
110 (1,122, 81)
123 (1,254, 91)
26 (296, 21)
Brake Tube
15 (155, 11)
70 (714, 52)
Front Disc Brake
Caliper Assy LH
Tie Rod End Sub–assy LH
Front Stabilizer Link Assy LH
◆ Cotter Pin
91 (928, 67)
225 (2,294, 166)
Front Disc
4WD Drive Type:
Lock Cap
◆ Front Axle
Hub Grease
Cap LH
235 (2,396, 173)
◆Cotter Pin
Front Axle LH Hub Bolt
2WD Drive Type:
◆ Knuckle Grease
Retainer Cap Inner
Steering Knuckle LH
◆ Front Axle w/ABS Rotor LH Bearing Assy
80 (816, 59)
◆ O–ring
4WD Drive Type:
◆ Steering Knuckle LH
Oil Seal
Dust Cover
2WD Drive Type:
◆ Front Wheel Adjusting Nut LH
275 (2,804, 203)
◆ Front Axle Hub LH Spacer
Front Axle Hub Sub–assy LH
N·m (kgf·cm, ft·lbf) : Specified torque
◆ Non–reusable part
⇦ Mp Grease
67170-4RUN-G201

- Axle hub to the steering knuckle, using a press
- ABS speed sensor rotor/spacer.

✻✻ WARNING

Be careful not to scratch the serration of the speed sensor rotor.

- Bearing spacer, using a press
- Grease cap, if removed
- New inside oil seal, if removed, using SST 09527-17011 and a plastic hammer
- Steering knuckle with the axle hub. Tighten the nut to 80 ft. lbs. (108 Nm).
- New cotter pin
- Lower ball joint. Tighten the 4 bolts to 59 ft. lbs. (80 Nm).
- Rotor and the caliper. Tighten the caliper bolts to 90 ft. lbs. (123 Nm).
- Brake line bracket to the steering knuckle. Tighten the fasteners to 21 ft. lbs. (28 Nm).
- ABS speed sensor and wiring harness clamp to the steering knuckle, if removed. Tighten the bolts to 72 inch lbs. (8 Nm).
- Halfshaft, if disconnected. Tighten the nut to 174 ft. lbs. (235 Nm).
- Grease cap
- Shock absorber
- Front wheel
- Negative battery cable

2003–04

1. Remove the wheel.
2. Remove the speed sensor wiring harness.
3. Remove the brake line bracket from the knuckle.
4. Remove the caliper and suspend it out of the way.
5. With 4wd, remove the grease cap.
6. With 4wd, remove the cotter pin, lock cap and axle shaft nut.
7. Remove the stabilizer link from the knuckle.
8. Remove the tie rod end from the knuckle.
9. Remove the lower ball joint attachment nuts.
10. Remove the upper ball stud nut and separate the ball stud from the knuckle.
11. With 4wd, drive the axle shaft from the hub with a plastic hammer.
12. Mount the knuckle in a vise.
13. With 2wd, remove the inner grease retainer cap.
14. Remove the oil seal.
15. With 2wd, unstake the adjusting nut. Remove the nut.
16. Remove the 4 bolts and remove the hub from the knuckle. Discard the O-ring.
17. Place the hub in a soft-jawed vise.
18. Remove the bearing with a puller set.
19. Remove the spacer.

To install:

20. Drive a new spacer into place with a hammer and brass drift.
21. Using a press install the new bearing.
22. Coat a new O-ring with MP grease and install it.
23. Install the dust cover and attach the hub to the knuckle. Torque the bolts to 59 ft. lbs. (80 Nm).
24. With 2wd, install a new adjusting nut. Torque to 203 ft. lbs. (275 Nm).
25. With 4wd, press a new oil seal into place.
26. With 2wd, install a new grease cap, using a hammer and brass drift.
27. The remainder of installation is the reverse of removal Use new cotter pins. Observe the following torques:
    - Upper ball stud nut: 81 ft. lbs. (110 Nm)
    - Lower ball joint attachment: 166 ft. lbs. (225 Nm)
    - Tie rod end nut: 67 ft. lbs. (91 Nm)
    - Stabilizer link: 52 ft. lbs. (70 Nm)
    - Axle shaft nut: 173 ft. lbs. (235 Nm)
    - Caliper: 91 ft. lbs. (123 Nm)
    - Brake line bracket: 21 ft. lbs. (29 Nm)

BRAKES

Front Brake Caliper

REMOVAL AND INSTALLATION

2001–02

1. Disconnect the negative battery cable from the battery.
2. Raise and support the vehicle safely.
3. Remove the wheels.
4. Disconnect the brake hose from the caliper by removing the union bolt and 2 gaskets. Plug the end of the hose to prevent loss of fluid.
5. Remove the bolts that attach the caliper to the torque plate.
6. Lift the bottom of the caliper up and remove the caliper assembly.

To install:

7. Grease the caliper slides and bolts with lithium grease or equivalent. Install the caliper and secure with the bolts. Torque the bolts to 90 ft. lbs. (123 Nm).
8. Connect the brake hose to the caliper, using 2 new washers. Make sure the

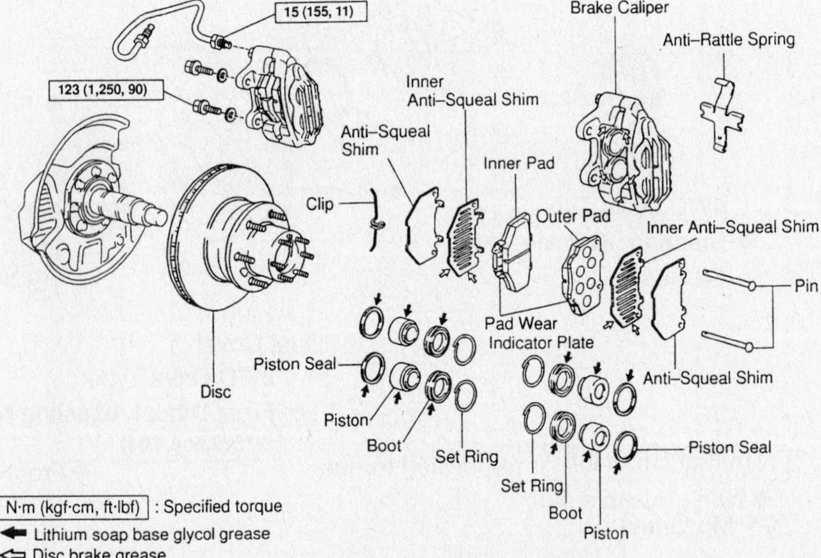

Caliper assembly—2001–02

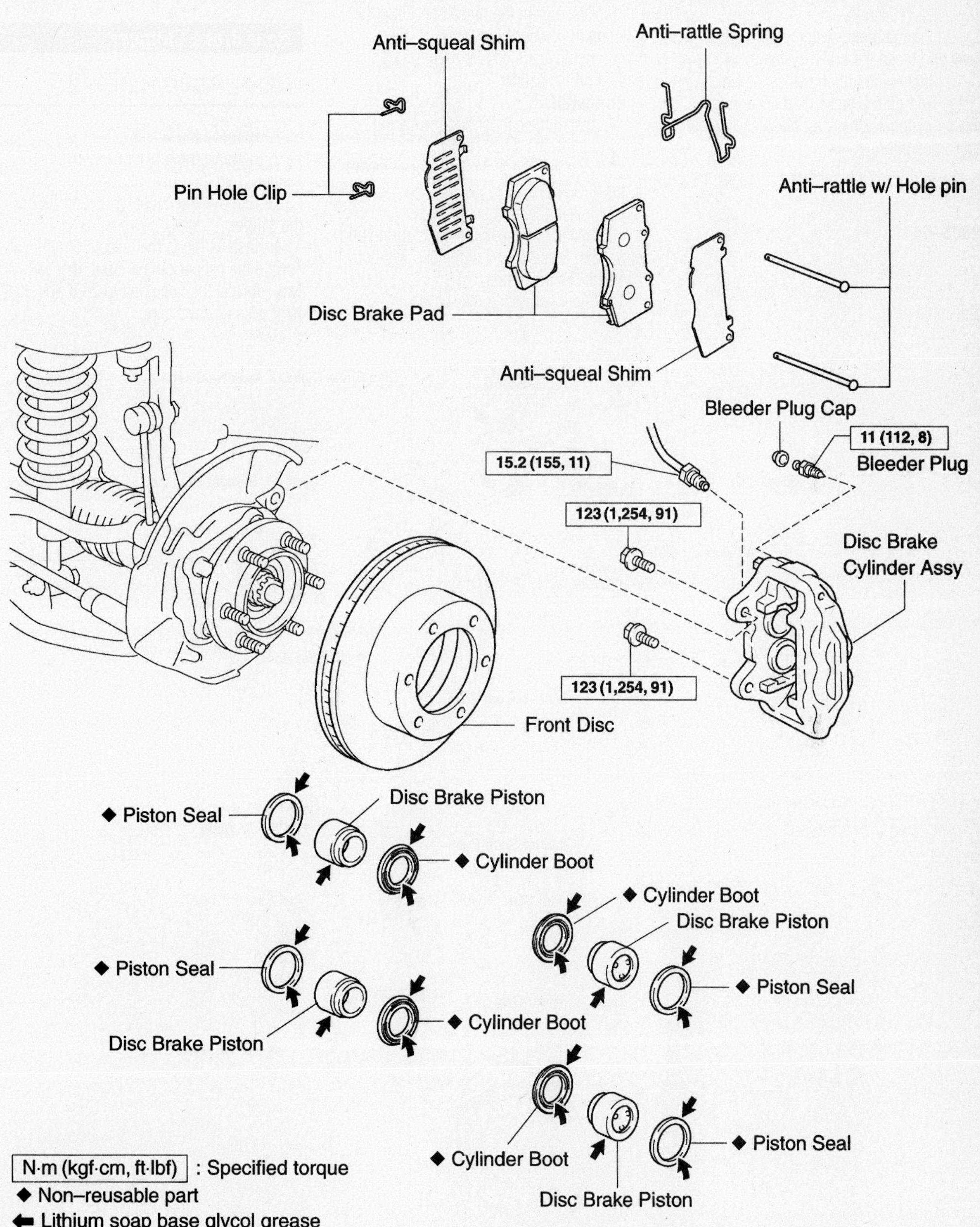

Front brake caliper—2003–04

flexible hose lock is securely in the lock hole of the caliper. Torque the union bolt to 11 ft. lbs. (15 Nm).

9. Fill the brake system to the proper level and bleed the brake system.

10. Install the tire and wheel assembly.

11. Top off the brake fluid level in the master cylinder. Check for leaks and proper brake operation.

12. Connect the negative battery cable to the battery.

2003–04

1. Remove the wheel.

2. Remove the clips and remove the anti-rattle pins.

3. Remove the anti-rattle spring.

4. Remove the pads and shims.

5. Disconnect the brake line. Plug the opening to prevent fluid loss.

6. Remove the caliper mounting bolts and lift off the caliper.

To install:

7. Install the caliper and torque the bolts to 91 ft. lbs. (123 Nm).

8. Connect the brake line. Torque to 11 ft. lbs. (15 Nm).

9. Install the shims and pads.

10. Install the anti-rattle pins, spring and new clips. Make sure that the clip with the handle faces the inside.

11. Refill and bleed the system.

12. Depress the brake pedal a few times to seat the pads.

Rear Brake Caliper

REMOVAL AND INSTALLATION

1. Remove the wheel.

2. Remove the brake hose from the caliper. Plug the hose to prevent fluid loss.

3. Remove the caliper pins and lift off the caliper.

4. Installation is the reverse of removal. Torque the caliper pins to 65 ft. lbs. (88 Nm). Torque the hose bolt to 23 ft. lbs. (31 Nm). Use new washers.

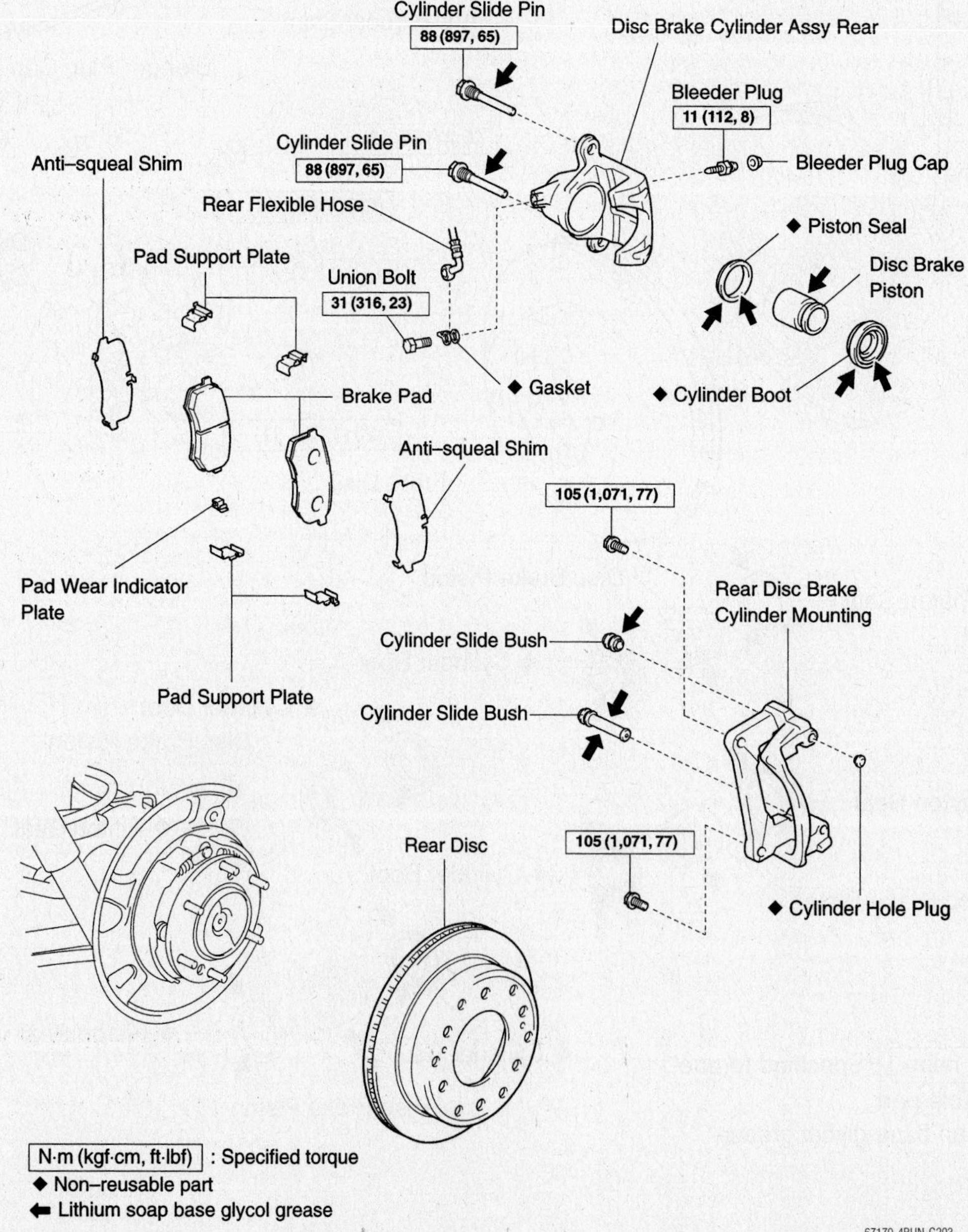

Rear brake caliper

Front Disc Brake Pads

REMOVAL AND INSTALLATION

2001–02

1. Raise the vehicle and support it safely.
2. Remove the wheel and tire assembly.
3. Remove the clip, pins, and the anti-rattle spring.
4. Remove the pads and the anti-squeal shims.
5. Remove the caliper, but do not disconnect the brake hose.

To install:

6. Before installing the new pads, check the disc thickness and disc runout.
7. Siphon out a small amount of brake fluid from the reservoir.
8. Temporarily install the old inner brake pad. Press in the pistons with a C-clamp or equivalent. Remove the old inner brake pad.
9. Apply disc brake grease to both sides of the inner anti-squeal shim. Install the anti-squeal shims to the new pads.
10. Install the pads.
11. Install the anti-rattle springs and pins. Install the clip.
12. Install the caliper and the mounting bolts. Torque the mounting bolts to 90 ft. lbs. (123 Nm).
13. Install the wheel and tire assembly.
14. Check and adjust the fluid level. Apply the brake pedal several times.
15. Road-test the vehicle for proper operation.

2003–04

1. Remove the wheel.
2. Remove the clips and remove the anti-rattle pins.
3. Remove the anti-rattle spring.
4. Remove the pads and shims.

To install:

5. Compress the piston with a C-clamp or large pliers.
6. Install the shims and pads.
7. Install the anti-rattle pins, spring and new clips. Make sure that the clip with the handle faces the inside.
8. Depress the brake pedal a few times to seat the pads.

Rear Disc Brake Pads

REMOVAL AND INSTALLATION

1. Raise the vehicle and support it safely.
2. Remove the wheel and tire assembly.
3. Remove the brake caliper and suspend it with a wire so the hose is not stretched or stressed.
4. Remove the brake pads, anti-squeal shim, pad support plates and wear indicators.

To install:

5. Before installing the new pads, check the disc thickness and disc runout.
6. Temporarily install the old inner brake pad. Press in the piston with a C-clamp or equivalent. Remove the old inner brake pad.
7. Install the pad support plates.
8. Install the pad wear indicator plate to each pads.
9. Install the anti-squeal shim to the outer pad. Install the pads so the wear indicator plate is facing upward.
10. Install the brake caliper. Torque the main sliding pin and the sub pin to 65 ft. lbs. (88 Nm).
11. Install the wheel and tire assembly.
12. Apply the brake pedal several times.
13. Road-test the vehicle for proper operation.

Brake Drums

REMOVAL AND INSTALLATION

1. Raise and safely support the vehicle.
2. Remove the rear wheel(s).

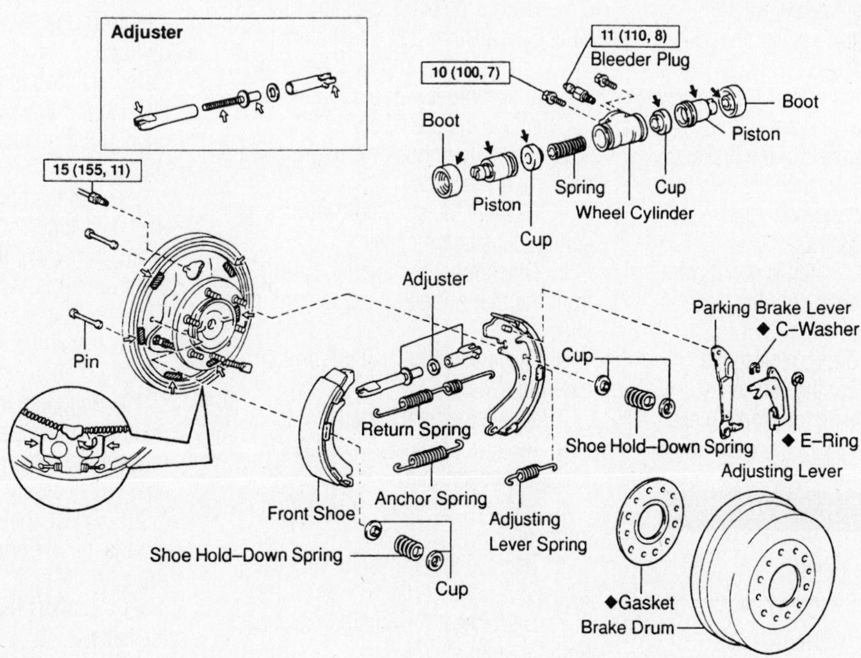

Exploded view of the rear brake drums components

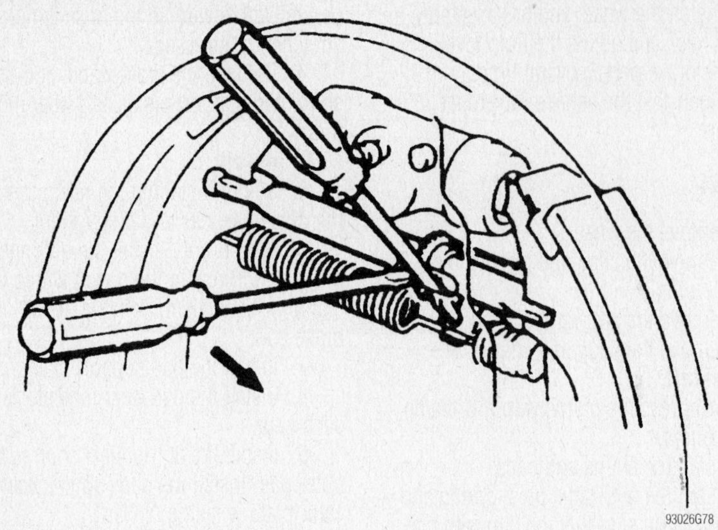

Use a brake adjusting tool (brake spoon) and a prytool to adjust the brake shoes through the adjusting hole

3. Remove the brake drum from the axle hub. If there is difficulty in removing the drum, insert a suitable tool through the hole in the rear of the backing plate, and hold the automatic adjusting lever away from the adjuster. Using another suitable tool at the same time, reduce the brake shoe adjuster by turning the adjusting wheel.

To install:

4. Install the brake drum and pull the parking brake lever all the way up until a clicking sound can no longer be heard.
5. Verify that the rear wheels will not turn. If the rear wheels turn, adjust the parking brake cable as necessary.
6. Release the parking brake and remove the brake drum. Measure the brake drum inside diameter and diameter of the brake shoes. Check that the difference between the diameters is the correct shoe clearance. Clearance is 0.024 in. (6mm).
7. If the brake shoe clearance is not correct, adjust the brake shoes until the clearance is correct.
8. Install the brake drum, replace the wheel(s), and safely lower the vehicle.
9. Road-test the vehicle for proper brake operation.

Brake Shoes

REMOVAL AND INSTALLATION

1. Loosen the rear wheel lug nuts slightly.
2. Block the front wheels, raise the rear of the vehicle, and safely support it with jackstands.
3. Remove the wheel lug nuts and the wheel.
4. Remove the brake drum.
5. If the drum is difficult to remove, perform the following:
   a. Insert a flat prying tool through the hole in the brake drum and hold the automatic adjusting lever away from the adjuster.
   b. Reduce the brake shoe adjustment by turning the adjuster bolt with a brake tool.
   c. The drum should now be loose enough to remove without much effort.
6. Remove the rear shoe.
   a. Carefully unhook the return spring from the brake shoe.
   b. Remove the shoe hold-down spring, cups and the pin.
   c. Disconnect the anchor spring from the rear shoe and remove the rear shoe.
   d. Disconnect the anchor spring from the front shoe.
7. Remove the front shoe.
   a. Remove the shoe hold-down spring, cups and pin.
   b. Remove the return spring from the front shoe.
   c. Remove the front shoe with the adjuster.
   d. Disconnect the parking brake cable from the front shoe.

To install:

8. Inspect the shoes for signs of unusual wear or scoring.
9. Check the wheel cylinder for any sign of fluid seepage or frozen pistons.
10. Clean and inspect the brake backing plate and all other components. Check that the brake drum inner diameter is within specified limits. Lubricate the backing plate at the positions the brakes come in contact with the backing plate. Also lubricate the anchor plate.
11. Mount the automatic adjuster assembly onto a new rear brake shoe.
12. Install the front shoe.
    a. Install the parking brake cable to the front shoe.
    b. Install the front shoe with the adjuster.
    c. Install the return spring to the front shoe.
    d. Install the shoe hold-down spring, cups and pin.
13. Install the rear shoe.
    a. Install the anchor spring to the front shoe.
    b. Install the anchor spring to the rear shoe and install the rear shoe.
    c. Install the shoe hold-down spring, cups and the pin.
    d. Hook the return spring to the brake shoe.
14. Install the brake drum.
15. Adjust the brake shoes until a slight drag is felt when the drum is spun by hand.
16. Remove the brake drum and check the clearance between brake shoes and brake drum. Adjust the clearance to specification.
17. Pull the parking lever all the way up until a clicking sound can no longer be heard. Verify that the drum doesn't turn. If the drum turns, adjust the parking brake cable.
18. Install the rear wheels, tighten the wheel lug nuts and lower the vehicle.
19. Retighten the wheel lug nuts and pump the brake pedal a few times before moving the vehicle. Adjust the rear brakes again if necessary.
20. Check the level of brake fluid in the master cylinder, then perform a test drive.
21. Connect the negative battery cable to the battery.

TOYOTA

17

Avalon • Camry • Camry Solara • Celica • Corolla • Echo • MR2

SPECIFICATION AND MAINTENANCE CHARTS

ENGINE AND VEHICLE IDENTIFICATION

Engine						
Code ①	Liters (cc)	Cu. In.	Cyl.	Fuel Sys.	Engine Type	Eng. Mfg.
1NZ-FE	1.5 (1496)	91	4	EFI	DOHC	Toyota
1ZZ-FE	1.8 (1794)	109	4	EFI	DOHC	Toyota
2ZZ-GE	1.8 (1796)	109	4	EFI	DOHC	Toyota
5S-FE	2.2 (2264)	138	4	EFI	DOHC	Toyota
2AZ-FE	2.4 (2398)	146	4	EFI	DOHC	Toyota
1MZ-FE	3.0 (2995)	183	6	EFI	DOHC	Toyota
3MZ-FE	3.3 (3311)	202	6	SFI	DOHC	Toyota

Model Year	
Code ②	Year
1	2001
2	2002
3	2003
4	2004
5	2005

EFI: Electronic Fuel Injection

SFI: Sequential Multi-port Fuel Injection

DOHC: Double Overhead Camshaft

① 8th digit of VIN

② 10th digit of VIN

67170-TOYO-C01

GENERAL ENGINE SPECIFICATIONS

Year	Model	Engine Displacement Liters (cc)	Engine Series (ID/VIN)	Fuel System	Net Horsepower @ rpm	Net Torque @ rpm (ft. lbs.)	Bore x Stroke (in.)	Com-pression Ratio	Oil Pressure psi @ idle
2001	Avalon	3.0 (2995)	1MZ-FE	EFI	200@5200	214@5200	3.44x3.27	10.5:1	4.3
	Camry	2.2 (2164)	5S-FE	EFI	133@5200	147@4400	3.43x3.58	9.5:1	4.3
	Camry	3.0 (2995)	1MZ-FE	EFI	194@5200	209@4400	3.44x3.27	10.5:1	4.3
	Camry Solara	2.2 (2164)	5S-FE	EFI	135@5200	147@4400	3.43x3.58	9.5:1	4.3
	Camry Solara	3.0 (2995)	1MZ-FE	EFI	200@5200	214@4400	3.44x3.27	10.5:1	4.3
	Corolla	1.8 (1794)	1ZZ-FE	EFI	120@5600	122@4400	3.11x3.60	10.0:1	4.3
	Celica	1.8 (1794)	1ZZ-FE	EFI	140@6400	125@4200	3.11x3.60	10.0:1	4.3
	Celica GT-S	1.8 (1796)	2ZZ-GE	EFI	180@7600	130@6800	3.23x3.35	11.5:1	4.3
	Echo	1.5 (1496)	1NZ-FE	EFI	108@5999	105@3999	2.95x3.32	10.5:0	3.3
	MR 2	1.8 (1794)	1ZZ-FE	EFI	140@6400	125@4200	3.11x3.60	10.0:1	4.3
2002	Avalon	3.0 (2995)	1MZ-FE	EFI	200@5200	214@5200	3.44x3.27	10.5:1	4.3
	Camry	2.4 (2398)	2AZ-FE	EFI	157@5600	162@4000	3.48x3.84	9.5:1	4.3
	Camry	3.0 (2995)	1MZ-FE	EFI	194@5200	209@4400	3.44x3.27	10.5:1	4.3
	Camry Solara	2.4 (2398)	2AZ-FE	EFI	157@5600	162@4000	3.48x3.84	9.5:1	4.3
	Camry Solara	3.0 (2995)	1MZ-FE	EFI	200@5200	214@4400	3.44x3.27	10.5:1	4.3
	Corolla	1.8 (1794)	1ZZ-FE	EFI	120@5600	122@4400	3.11x3.60	10.0:1	4.3
	Celica	1.8 (1794)	1ZZ-FE	EFI	140@6400	125@4200	3.11x3.60	10.0:1	4.3
	Celica GT-S	1.8 (1796)	2ZZ-GE	EFI	180@7600	130@6800	3.23x3.35	11.5:1	4.3
	Echo	1.5 (1496)	1NZ-FE	EFI	108@5999	105@3999	2.95x3.32	10.5:0	3.3
	MR 2	1.8 (1794)	1ZZ-FE	EFI	140@6400	125@4200	3.11x3.60	10.0:1	4.3
2003	Avalon	3.0 (2995)	1MZ-FE	EFI	200@5200	214@5200	3.44x3.27	10.5:1	4.3
	Camry	2.4 (2398)	2AZ-FE	EFI	157@5600	162@4000	3.48x3.84	9.5:1	4.3
	Camry	3.0 (2995)	1MZ-FE	EFI	194@5200	209@4400	3.44x3.27	10.5:1	4.3
	Camry Solara	2.4 (2398)	2AZ-FE	EFI	157@5600	162@4000	3.48x3.84	9.5:1	4.3
	Camry Solara	3.0 (2995)	1MZ-FE	EFI	200@5200	214@4400	3.44x3.27	10.5:1	4.3
	Corolla	1.8 (1794)	1ZZ-FE	EFI	120@5600	122@4400	3.11x3.60	10.0:1	4.3
	Celica	1.8 (1794)	1ZZ-FE	EFI	140@6400	125@4200	3.11x3.60	10.0:1	4.3
	Celica GT-S	1.8 (1796)	2ZZ-GE	EFI	180@7600	130@6800	3.23x3.35	11.5:1	4.3
	Echo	1.5 (1496)	1NZ-FE	EFI	108@5999	105@3999	2.95x3.32	10.5:0	3.3
	MR 2	1.8 (1794)	1ZZ-FE	EFI	140@6400	125@4200	3.11x3.60	10.0:1	4.3
2004	Avalon	3.0 (2995)	1MZ-FE	EFI	200@5200	214@5200	3.44x3.27	10.5:1	4.3
	Camry	2.4 (2398)	2AZ-FE	EFI	157@5600	162@4000	3.48x3.84	9.5:1	4.3
	Camry	3.0 (2995)	1MZ-FE	EFI	194@5200	209@4400	3.44x3.27	10.5:1	4.3
	Camry	3.3 (3311)	3MZ-FE	EFI	225@5600	240@3600	3.62x3.27	10.8:1	36-78@3000
	Camry Solara	2.4 (2398)	2AZ-FE	EFI	157@5600	162@4000	3.48x3.84	9.5:1	4.3
	Camry Solara	3.3 (3311)	3MZ-FE	EFI	225@5600	240@3600	3.62x3.27	10.8:1	36-78@3000
	Corolla	1.8 (1794)	1ZZ-FE	EFI	120@5600	122@4400	3.11x3.60	10.0:1	4.3
	Celica	1.8 (1794)	1ZZ-FE	EFI	140@6400	125@4200	3.11x3.60	10.0:1	4.3
	Celica GT-S	1.8 (1796)	2ZZ-GE	EFI	180@7600	130@6800	3.23x3.35	11.5:1	4.3
	Echo	1.5 (1496)	1NZ-FE	EFI	108@6000	105@4000	2.95x3.33	10.5:1	4.3
	MR 2	1.8 (1794)	1ZZ-FE	EFI	140@6400	125@4200	3.11x3.60	10.0:1	4.3

EFI: Electronic Fuel Injection

67170-TOYO-C02

ENGINE TUNE-UP SPECIFICATIONS

Year	Engine Displacement Liters (cc)	Engine ID/VIN	Spark Plug Gap (in.)	Ignition Timing (deg.) ①	Fuel Pump (psi)	Idle Speed (rpm)		Valve Clearance	
						MT	AT	In.	Ex.
2001	1.5 (1496)	1NZ-FE	0.043	8-12 BTDC	44-50	700-800	700-800	0.006-0.010	0.011-0.014
	1.8 (1794)	1ZZ-FE	0.043	10-18 BTDC	44-50	650-750	700-800	0.006-0.010	0.010-0.014
	1.8 (1796)	2ZZ-GE	0.043	8-12 BTDC	44-50	750-850	700-800	0.006-0.010	0.014-0.018
	2.2 (2164)	5S-FE	0.043	8-12 BTDC	44-50	700-800	700-800	0.007-0.011	0.011-0.015
	3.0 (2952)	1MZ-FE	0.043	8-12 BTDC	44-50	650-750	650-750	0.006-0.010	0.010-0.014
2002	1.5 (1496)	1NZ-FE	0.043	8-12 BTDC	44-50	700-800	700-800	0.006-0.010	0.011-0.014
	1.8 (1794)	1ZZ-FE	0.043	10-18 BTDC	44-50	650-750	700-800	0.006-0.010	0.010-0.014
	1.8 (1796)	2ZZ-GE	0.043	8-12 BTDC	44-50	750-850	700-800	0.006-0.010	0.014-0.018
	2.4 (2398)	2AZ-FE	0.043	8-12 BTDC	44-50	700-800	700-800	0.007-0.011	0.011-0.015
	3.0 (2952)	1MZ-FE	0.043	8-12 BTDC	44-50	650-750	650-750	0.006-0.010	0.010-0.014
2003	1.5 (1496)	1NZ-FE	0.043	8-12 BTDC	44-50	700-800	700-800	0.006-0.010	0.011-0.014
	1.8 (1794)	1ZZ-FE	0.043	10-18 BTDC	44-50	650-750	700-800	0.006-0.010	0.010-0.014
	1.8 (1796)	2ZZ-GE	0.043	8-12 BTDC	44-50	750-850	700-800	0.006-0.010	0.014-0.018
	2.4 (2398)	2AZ-FE	0.043	8-12 BTDC	44-50	700-800	700-800	0.007-0.011	0.011-0.015
	3.0 (2952)	1MZ-FE	0.043	8-12 BTDC	44-50	650-750	650-750	0.006-0.010	0.010-0.014
2004	1.5 (1496)	1NZ-FE	0.043	8-12 BTDC	44-50	700-800	700-800	0.006-0.010	0.011-0.014
	1.8 (1794)	1ZZ-FE	0.043	10-18 BTDC	44-50	650-750	700-800	0.006-0.010	0.010-0.014
	1.8 (1796)	2ZZ-GE	0.043	8-12 BTDC	44-50	750-850	700-800	0.006-0.010	0.014-0.018
	2.4 (2398)	2AZ-FE	0.043	8-12 BTDC	44-50	700-800	700-800	0.007-0.011	0.011-0.015
	3.0 (2952)	1MZ-FE	0.043	8-12 BTDC	44-50	650-750	650-750	0.006-0.010	0.010-0.014
	3.3 (3311)	3MZ-FE	0.043	8-12 BTDC	44-50	650-750	650-750	0.006-0.010	0.010-0.014

Note: The Vehicle Emission Control Information label often reflects specification changes made during production. The label figures must be used if they differ from those in this chart.

① With terminal TE1 and E1 connected of DLC1

67170-TOYO-C03

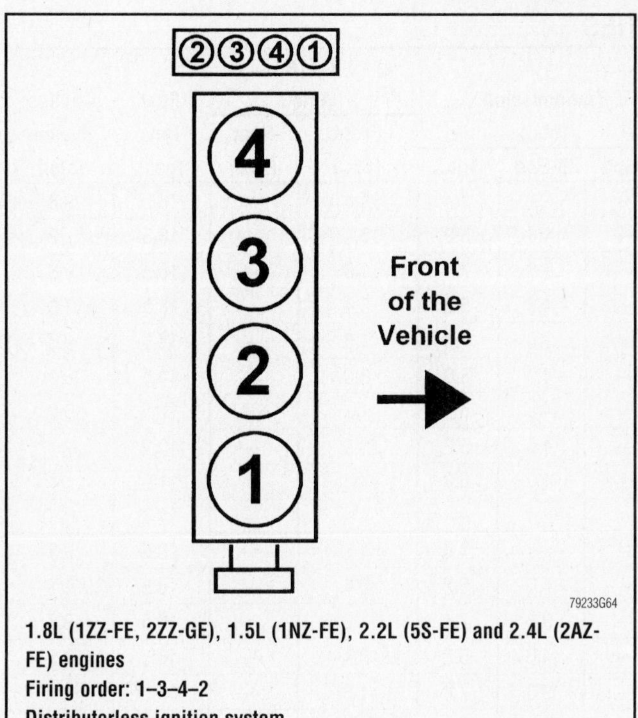

1.8L (1ZZ-FE, 2ZZ-GE), 1.5L (1NZ-FE), 2.2L (5S-FE) and 2.4L (2AZ-FE) engines
Firing order: 1–3–4–2
Distributorless ignition system

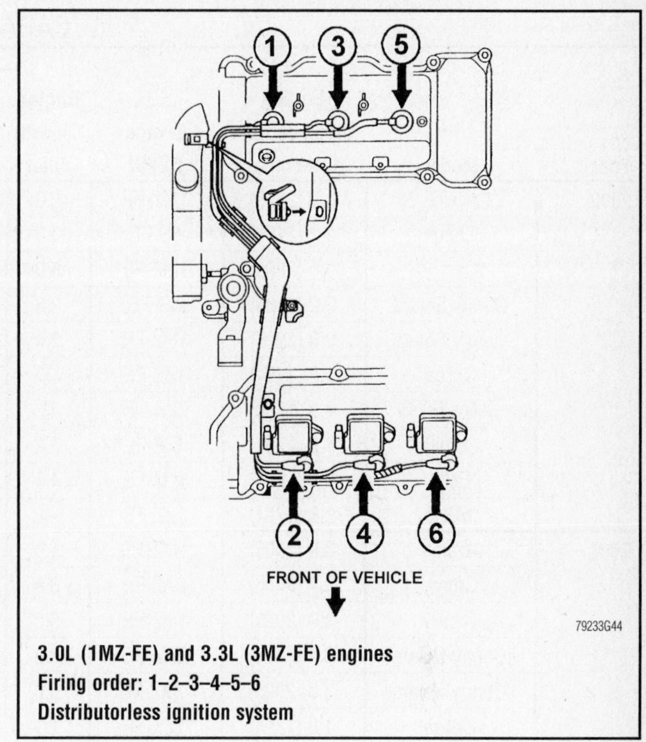

3.0L (1MZ-FE) and 3.3L (3MZ-FE) engines
Firing order: 1–2–3–4–5–6
Distributorless ignition system

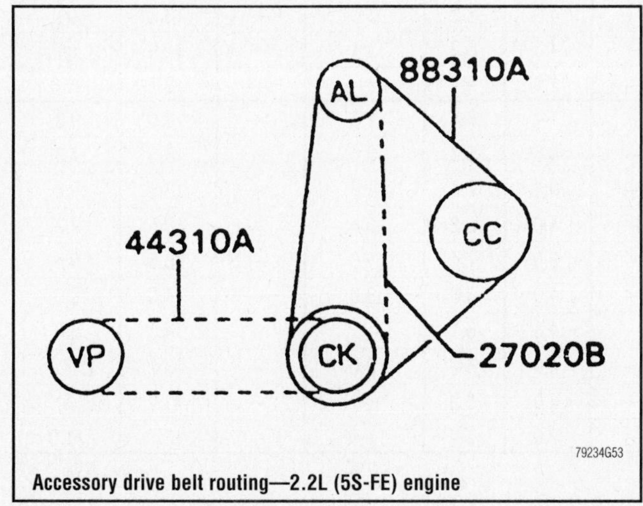

Accessory drive belt routing—2.2L (5S-FE) engine

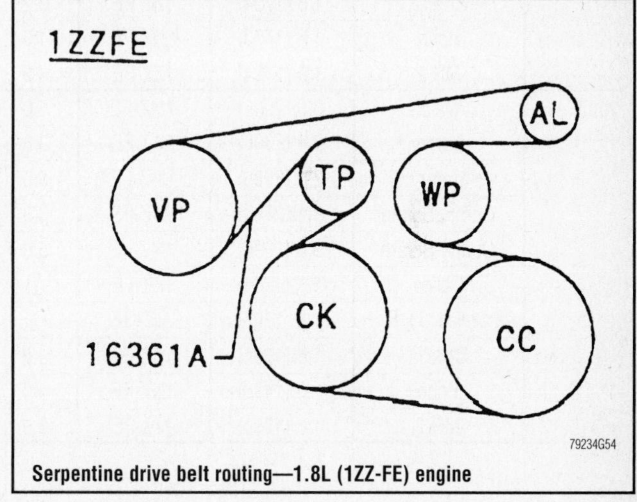

Serpentine drive belt routing—1.8L (1ZZ-FE) engine

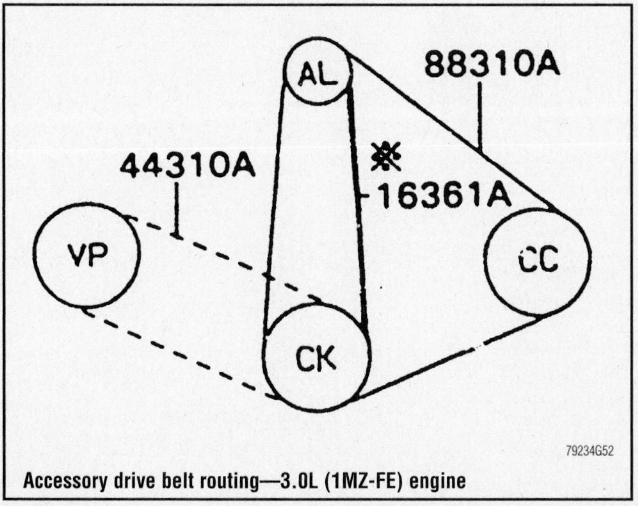

Accessory drive belt routing—3.0L (1MZ-FE) engine

CAPACITIES

		Engine Displacement	Engine	Engine Oil with	Transmission (pts.)			Drive Axle		Fuel Tank	Cooling System
Year	Model	Liters (cc)	ID/VIN	Filter	4-Spd	5-Spd	Auto.	Front (pts.)	Rear (pts.)	(gal.)	(qts.)
2000	Avalon	3.0 (2995)	1MZ-FE	5.0	—	—	7.4	1.8	—	18.5	9.8
	Camry	2.2 (2164)	5S-FE	3.8	—	4.6	5.2	3.4	—	18.5	7.3
	Camry	3.0 (2995)	1MZ-FE	5.0	—	9.8	7.4	1.8	—	18.5	9.6
	Camry Solara	2.2 (2164)	5S-FE	3.8	—	4.6	5.2	3.4	—	18.5	7.3
	Camry Solara	3.0 (2995)	1MZ-FE	5.0	—	9.8	7.4	1.8	—	18.5	9.6
	Celica	1.8 (1794)	1ZZ-FE	①	—	4.0	8.6	3.0	—	14.5	6.0
	Celica GT-S	1.8 (1796)	2ZZ-GE	②	—	4.8	6.1	③	—	14.5	6.0
	Corolla	1.8 (1794)	1ZZ-FE	3.2	—	4.0	5.2	3.0	—	13.2	④
	Echo	1.5 (1496)	1NZ-FE	4.0	—	4.0	6.2	③	—	11.9	⑤
	MR 2	1.8 (1794)	1ZZ-FE	3.9	—	2.0	—	—	—	12.7	11.0
2001	Avalon	3.0 (2995)	1MZ-FE	5.0	—	—	7.4	1.8	—	18.5	9.8
	Camry	2.4 (2398)	2AZ-FE	3.8	—	4.6	5.2	3.4	—	18.5	7.3
	Camry	3.0 (2995)	1MZ-FE	5.0	—	9.8	7.4	1.8	—	18.5	9.6
	Camry Solara	2.4 (2398)	2AZ-FE	3.8	—	4.6	5.2	3.4	—	18.5	7.3
	Camry Solara	3.0 (2995)	1MZ-FE	5.0	—	9.8	7.4	1.8	—	18.5	9.6
	Celica	1.8 (1794)	1ZZ-FE	①	—	4.0	8.6	3.0	—	14.5	6.0
	Celica GT-S	1.8 (1796)	2ZZ-GE	②	—	4.8	6.1	③	—	14.5	6.0
	Corolla	1.8 (1794)	1ZZ-FE	3.2	—	4.0	5.2	3.0	—	13.2	④
	Echo	1.5 (1496)	1NZ-FE	4.0	—	4.0	6.2	③	—	11.9	⑤
	MR 2	1.8 (1794)	1ZZ-FE	3.9	—	2.0	—	—	—	12.7	11.0
2002	Avalon	3.0 (2995)	1MZ-FE	5.0	—	—	7.4	1.8	—	18.5	9.8
	Camry	2.4 (2398)	2AZ-FE	3.8	—	4.6	5.2	3.4	—	18.5	7.3
	Camry	3.0 (2995)	1MZ-FE	5.0	—	9.8	7.4	1.8	—	18.5	9.6
	Camry Solara	2.4 (2398)	2AZ-FE	3.8	—	4.6	5.2	3.4	—	18.5	7.3
	Camry Solara	3.0 (2995)	1MZ-FE	5.0	—	9.8	7.4	1.8	—	18.5	9.6
	Celica	1.8 (1794)	1ZZ-FE	①	—	4.0	8.6	3.0	—	14.5	6.0
	Celica GT-S	1.8 (1796)	2ZZ-GE	②	—	4.8	6.1	③	—	14.5	6.0
	Corolla	1.8 (1794)	1ZZ-FE	3.2	—	4.0	5.2	3.0	—	13.2	④
	Echo	1.5 (1496)	1NZ-FE	4.0	—	4.0	6.2	③	—	11.9	⑤
	MR 2	1.8 (1794)	1ZZ-FE	3.9	—	2.0	—	—	—	12.7	11.0

67170-TOYO-C04

CAPACITIES

Year	Model	Engine Displacement Liters (cc)	Engine ID/VIN	Engine Oil with Filter	Transmission (pts.) 4-Spd	Transmission (pts.) 5-Spd	Transmission (pts.) Auto.	Drive Axle Front (pts.)	Drive Axle Rear (pts.)	Fuel Tank (gal.)	Cooling System (qts.)
2003	Avalon	3.0 (2995)	1MZ-FE	5.0	—	—	7.4	1.8	—	18.5	9.8
	Camry	2.2 (2164)	2AZ-FE	3.8	—	4.6	5.2	3.4	—	18.5	7.3
	Camry	3.0 (2995)	1MZ-FE	5.0	—	9.8	7.4	1.8	—	18.5	9.6
	Camry Solara	2.4 (2398)	2AZ-FE	3.8	—	4.6	5.2	3.4	—	18.5	7.3
	Camry Solara	3.0 (2995)	1MZ-FE	5.0	—	9.8	7.4	1.8	—	18.5	9.6
	Celica	1.8 (1794)	1ZZ-FE	①	—	4.0	8.6	3.0	—	14.5	6.0
	Celica GT-S	1.8 (1796)	2ZZ-GE	②	—	4.8	6.1	③	—	14.5	6.0
	Corolla	1.8 (1794)	1ZZ-FE	3.2	—	4.0	5.2	3.0	—	13.2	④
	Echo	1.5 (1496)	1NZ-FE	4.0	—	4.0	6.2	③	—	11.9	⑤
	MR 2	1.8 (1794)	1ZZ-FE	3.9	—	2.0	—	—	—	12.7	11.0
2004	Avalon	3.0 (2995)	1MZ-FE	5.0	—	—	7.4	1.8	—	18.5	9.8
	Camry	2.4 (2398)	2AZ-FE	3.8	—	4.6	5.2	3.4	—	18.5	7.3
	Camry	3.0 (2995)	1MZ-FE	5.0	—	9.8	7.4	1.8	—	18.5	9.6
	Camry	3.3 (3311)	3MZ-FE	5.0	—	—	—	—	—	18.5	—
	Camry Solara	2.4 (2398)	2AZ-FE	3.8	—	4.6	5.2	3.4	—	18.5	7.3
	Camry Solara	3.3 (3311)	3MZ-FE	5.0	—	—	—	—	—	18.5	—
	Celica	1.8 (1794)	1ZZ-FE	①	—	4.0	8.6	3.0	—	14.5	6.0
	Celica GT-S	1.8 (1796)	2ZZ-GE	②	—	4.8	6.1	③	—	14.5	6.0
	Corolla	1.8 (1794)	1ZZ-FE	3.2	—	4.0	5.2	3.0	—	13.2	④
	Echo	1.5 (1496)	1NZ-FE	4.0	—	4.0	6.2	③	—	11.9	⑤
	MR 2	1.8 (1794)	1ZZ-FE	3.9	—	2.0	—	—	—	12.7	11.0

Note: All capacities are approximate. Add fluid gradually and check to be sure a proper fluid level is obtained.

① w/oil cooler: 4.0
wo/oil cooler: 3.7

② w/oil cooler: 4.8
wo/oil cooler: 4.6

③ Included in transaxle capacity

④ M/T with Nippodenso radiator: 5.6
A/T with Nippodenso radiator: 6.2
M/T with Harrison radiator: 6.3
A/T with Harrison radiator: 6.2

⑤ w/MT: 4.7
w/AT: 4.5

67170-TOYO-C05

VALVE SPECIFICATIONS

Year	Engine Displacement Liters (cc)	Engine ID/VIN	Seat Angle (deg.)	Face Angle (deg.)	Spring Test Pressure (lbs. @ in.)	Spring Installed Height (in.)	Stem-to-Guide Clearance (in.) Intake	Stem-to-Guide Clearance (in.) Exhaust	Stem Diameter (in.) Intake	Stem Diameter (in.) Exhaust
2001	1.5 (1496)	1NZ-FE	45	44.5	33.5-37@ 1.280	1.280	0.0010-0.0024	0.0012-0.0026	0.1957-0.1963	0.1955-0.1961
	1.8 (1794)	1ZZ-FE	45	44.5	31.3-34.8@ 1.252	1.323	0.0010-0.0024	0.0012-0.0025	0.2154-0.2352	0.2152-0.2350
	1.8 (1796)	2ZZ-GE	45	44.5	①	1.516	0.0010-0.0023	0.0012-0.0025	0.2150-0.2156	0.2144-0.2154
	2.2 (2164)	5S-FE	45	44.5	36.8-42.5@ 1.366	1.366	0.0010-0.0024	0.0012-0.0026	0.2350-0.2356	0.2348-0.2354
	3.0 (2995)	1MZ-FE	45	44.5	41.9-46.3@ 1.331	1.331	0.0010-0.0024	0.0012-0.0026	0.2154-0.2159	0.2152-0.2157
2002	1.5 (1496)	1NZ-FE	45	44.5	33.5-37@ 1.280	1.280	0.0010-0.0024	0.0012-0.0026	0.1957-0.1963	0.1955-0.1961
	1.8 (1794)	1ZZ-FE	45	44.5	31.3-34.8@ 1.252	1.323	0.0010-0.0024	0.0012-0.0025	0.2154-0.2352	0.2152-0.2350
	1.8 (1796)	2ZZ-GE	45	44.5	①	1.516	0.0010-0.0023	0.0012-0.0025	0.2150-0.2156	0.2144-0.2154
	2.4 (2398)	2AZ-FE	45	45	②	—	0.0010-0.0024	0.0012-0.0026	0.2154-0.2159	0.2152-0.2157
	3.0 (2995)	1MZ-FE	45	44.5	41.9-46.3@ 1.331	1.331	0.0010-0.0024	0.0012-0.0026	0.2154-0.2159	0.2152-0.2157
2003	1.5 (1496)	1NZ-FE	45	44.5	33.5-37@ 1.280	1.280	0.0010-0.0024	0.0012-0.0026	0.1957-0.1963	0.1955-0.1961
	1.8 (1794)	1ZZ-FE	45	44.5	31.3-34.8@ 1.252	1.323	0.0010-0.0024	0.0012-0.0025	0.2154-0.2352	0.2152-0.2350
	1.8 (1796)	2ZZ-GE	45	44.5	①	1.516	0.0010-0.0023	0.0012-0.0025	0.2150-0.2156	0.2144-0.2154
	2.4 (2398)	2AZ-FE	45	45	②	—	0.0010-0.0024	0.0012-0.0026	0.2154-0.2159	0.2152-0.2157
	3.0 (2995)	1MZ-FE	45	44.5	41.9-46.3@ 1.331	1.331	0.0010-0.0024	0.0012-0.0026	0.2154-0.2159	0.2152-0.2157
2004	1.5 (1496)	1NZ-FE	45	44.5	33.5-37@ 1.280	1.280	0.0010-0.0024	0.0012-0.0026	0.1957-0.1963	0.1955-0.1961
	1.8 (1794)	1ZZ-FE	45	44.5	31.3-34.8@ 1.252	1.323	0.0010-0.0024	0.0012-0.0025	0.2154-0.2352	0.2152-0.2350
	1.8 (1796)	2ZZ-GE	45	44.5	①	1.516	0.0010-0.0023	0.0012-0.0025	0.2150-0.2156	0.2144-0.2154
	2.4 (2398)	2AZ-FE	45	45	②	—	0.0010-0.0024	0.0012-0.0026	0.2154-0.2159	0.2152-0.2157
	3.0 (2995)	1MZ-FE	45	44.5	41.9-46.3@ 1.331	1.331	0.0010-0.0024	0.0012-0.0026	0.2154-0.2159	0.2152-0.2157
	3.3 (3311)	3MZ-FE	NA	44.5	41.9-46.3@ 1.331	NA	0.0010-0.0024	0.0012-0.0026	0.2154-0.2159	0.2152-0.2157

① Intake: 49.6-55.5@1.516
Exhaust: 47.6-52.6@1.516

② Inner spring free length: 1.799 in. (45.7 mm)

67170-TOYO-C06

TORQUE SPECIFICATIONS

All readings in ft. lbs.

Year	Engine Displacement Liters (cc)	Engine ID/VIN	Cylinder Head Bolts	Main Bearing Bolts	Rod Bearing Bolts	Crankshaft Damper Bolts	Flywheel Bolts	Manifold Intake	Manifold Exhaust	Spark Plugs	Lug Nuts
2001	1.5 (1496)	1NZ-FE	①	②	③	94	④	22	20	13	76
	1.8 (1794)	1ZZ-FE	④	⑤	⑥	102	④	14	27	13	76
	1.8 (1796)	2ZZ-GE	⑦	⑤	⑧	87	④	⑨	37	13	76
	2.2 (2164)	5S-FE	④	43	⑩	80	⑥	14	36	13	76
	3.0 (2995)	1MZ-FE	⑪	⑫	⑩	159	61	11	36	13	76
2002	1.5 (1496)	1NZ-FE	①	②	③	94	④	22	20	13	76
	1.8 (1794)	1ZZ-FE	④	⑤	⑥	102	④	14	27	13	76
	1.8 (1796)	2ZZ-GE	⑦	⑤	⑧	87	④	⑨	37	13	76
	2.4 (2398)	2AZ-FE	⑬	⑭	⑩	125	⑮	22	27	13	76
	3.0 (2995)	1MZ-FE	⑪	⑫	⑩	159	61	11	36	13	76
2003	1.5 (1496)	1NZ-FE	①	②	③	94	④	22	20	13	76
	1.8 (1794)	1ZZ-FE	④	⑤	⑥	102	④	14	27	13	76
	1.8 (1796)	2ZZ-GE	⑦	⑤	⑧	87	④	⑨	37	13	76
	2.4 (2398)	2AZ-FE	⑬	⑭	⑩	125	⑮	22	27	13	76
	3.0 (2995)	1MZ-FE	⑪	⑫	⑩	159	61	11	36	13	76
2004	1.5 (1496)	1NZ-FE	①	②	③	94	④	22	20	13	76
	1.8 (1794)	1ZZ-FE	④	⑤	⑥	102	④	14	27	13	76
	1.8 (1796)	2ZZ-GE	⑦	⑤	⑧	87	④	⑨	37	13	76
	2.4 (2398)	2AZ-FE	⑬	⑭	⑩	125	⑮	22	27	13	76
	3.0 (2995)	1MZ-FE	⑪	⑫	⑩	159	61	11	36	13	76
	3.3 (3311)	3MZ-FE	⑪	⑫	⑩	162	61	11	36	18	76

① Step 1: 22 ft. lbs.
Step 2: 90 degree turn
Step 3: 90 degree turn

② Step 1: 16 ft. lbs.
Step 2: 90 degree turn

③ Step 1: 11 ft. lbs.
Step 2: 90 degree turn

④ Step 1: 36 ft. lbs.
Step 2: 90 degree turn

⑤ 12 pointed bolts:
Step 1: 16 ft. lbs.
Step 2: 32 ft. lbs.
Step 3: 45 degree turn
Step 4: 45 degree turn
Hex head bolts: 14 ft. lbs.

⑥ Step 1: 15 ft. lbs.
Step 2: 90 degree turn

⑦ Step 1: 26 ft. lbs.
Step 2: 180 degree turn

⑧ Step 1: 22 ft. lbs.
Step 2: 90 degree turn
Recessed head bolt: 13 ft. lbs.

⑨ 4 upper bolts and 1 nut: 20 ft. lbs.
1 lower bolt: 34 ft. lbs.

⑩ Step 1: 18 ft. lbs.
Step 2: Plus 90 degrees

⑪ Head bolt:
Step 1: 40 ft. lbs.
Step 2: Plus 90 degrees
Recessed head bolt: 13 ft. lbs.

⑫ 6-point bolts: 20 ft. lbs.
12-point bolts:
Step 1: 16 ft. lbs.
Step 2: Plus an additional 90 degrees

⑬ Step 1: Several passes in sequence to 58 ft. lbs
Step 2: Plus 90 degrees

⑭ Step 1: 15 ft. lbs.
Step 2: 29 ft. lbs.
Step 3: Plus 90 degrees

⑮ Manual Transmission: 96 ft. lbs.
Automatic Transmission: 72 ft. lbs.

67170-TOYO-C07

PISTON AND RING SPECIFICATIONS

All measurements are given in inches.

	Engine			Ring Gap			Ring Side Clearance		
Year	Displacement Liters (cc)	Engine ID/VIN	Piston Clearance	Top Compression	Bottom Compression	Oil Control	Top Compression	Bottom Compression	Oil Control
2001	1.5 (1496)	1NZ-FE	0.0022-0.0023	0.0098-0.0138	0.0138-0.0197	0.0039-0.0138	0.0012-0.0028	0.0012-0.0028	SNUG
	1.8 (1762)	1ZZ-FE	0.0033-0.0041	0.0098-0.0138	0.0138-0.0197	0.0039-0.0157	0.0012-0.0028	0.0012-0.0028	SNUG
	1.8 (1796)	2ZZ-GE	0.0003-0.0015	0.0098-0.0138	0.0138-0.0197	0.0059-0.0157	0.0012-0.0028	0.0012-0.0028	SNUG
	2.2 (2164)	5S-FE	0.0055-0.0063	0.0106-0.0197	0.0138-0.0236	0.0079-0.0217	0.0016-0.0031	0.0012-0.0028	SNUG
	3.0 (2995)	1MZ-FE	0.0033-0.0042	0.0098-0.0138	0.0138-0.0177	0.0059-0.0157	0.0008-0.0028	0.0008-0.0024	SNUG
2002	1.5 (1496)	1NZ-FE	0.0022-0.0023	0.0098-0.0138	0.0138-0.0197	0.0039-0.0138	0.0012-0.0028	0.0012-0.0028	SNUG
	1.8 (1762)	1ZZ-FE	0.0033-0.0041	0.0098-0.0138	0.0138-0.0197	0.0039-0.0157	0.0012-0.0028	0.0012-0.0028	SNUG
	1.8 (1796)	2ZZ-GE	0.0003-0.0015	0.0098-0.0138	0.0138-0.0197	0.0059-0.0157	0.0012-0.0028	0.0012-0.0028	SNUG
	2.4 (2398)	2AZ-FE	0.0020-0.0029	0.0087-0.0126	0.0197-0.0236	0.0039-0.0138	0.0012-0.0028	0.0012-0.0028	SNUG
	3.0 (2995)	1MZ-FE	0.0033-0.0042	0.0098-0.0138	0.0138-0.0177	0.0059-0.0157	0.0008-0.0028	0.0008-0.0024	SNUG
2003	1.5 (1496)	1NZ-FE	0.0022-0.0023	0.0098-0.0138	0.0138-0.0197	0.0039-0.0138	0.0012-0.0028	0.0012-0.0028	SNUG
	1.8 (1762)	1ZZ-FE	0.0033-0.0041	0.0098-0.0138	0.0138-0.0197	0.0039-0.0157	0.0012-0.0028	0.0012-0.0028	SNUG
	1.8 (1796)	2ZZ-GE	0.0003-0.0015	0.0098-0.0138	0.0138-0.0197	0.0059-0.0157	0.0012-0.0028	0.0012-0.0028	SNUG
	2.4 (2398)	2AZ-FE	0.0020-0.0029	0.0087-0.0126	0.0197-0.0236	0.0039-0.0138	0.0012-0.0028	0.0012-0.0028	SNUG
	3.0 (2995)	1MZ-FE	0.0033-0.0042	0.0098-0.0138	0.0138-0.0177	0.0059-0.0157	0.0008-0.0028	0.0008-0.0024	SNUG
2004	1.5 (1496)	1NZ-FE	0.0022-0.0023	0.0098-0.0138	0.0138-0.0197	0.0039-0.0138	0.0012-0.0028	0.0012-0.0028	SNUG
	1.8 (1762)	1ZZ-FE	0.0033-0.0041	0.0098-0.0138	0.0138-0.0197	0.0039-0.0157	0.0012-0.0028	0.0012-0.0028	SNUG
	1.8 (1796)	2ZZ-GE	0.0003-0.0015	0.0098-0.0138	0.0138-0.0197	0.0059-0.0157	0.0012-0.0028	0.0012-0.0028	SNUG
	2.4 (2398)	2AZ-FE	0.0020-0.0029	0.0087-0.0126	0.0197-0.0236	0.0039-0.0138	0.0012-0.0028	0.0012-0.0028	SNUG
	3.0 (2995)	1MZ-FE	0.0033-0.0042	0.0098-0.0138	0.0138-0.0177	0.0059-0.0157	0.0008-0.0028	0.0008-0.0024	SNUG
	3.3 (3311)	3MZ-FE	0.0013-0.0023	0.0118-0.0157	0.0197-0.0236	0.0059-0.0157	0.0012-0.0031	0.0008-0.0024	SNUG

67170-TOYO-C08

CRANKSHAFT AND CONNECTING ROD SPECIFICATIONS

All measurements are given in inches.

	Engine		Crankshaft				Connecting Rod		
	Displacement	Engine	Main Brg.	Main Brg. Oil	Shaft	Thrust	Journal	Oil	Side
Year	Liters (cc)	ID/VIN	Journal Dia.	Clearance	End-play	on No.	Diameter	Clearance	Clearance
2001	1.5 (1496)	1NZ-FE	①	0.0004-0.0009	0.0035-0.0075	3	1.5745-1.5748	0.0006-0.0016	0.0063-0.0142
	1.8 (1762)	1ZZ-FE	②	0.0006-0.0013	0.0008-0.0087	3	1.7320-1.7323	0.0011-0.0024	0.0063-0.0135
	1.8 (1796)	2ZZ-GE	②	0.0006-0.0013	0.0016-0.0094	3	1.7713-1.7717	0.0011-0.0020	0.0063-0.0135
	2.2 (2164)	5S-FE	2.1653-2.6550	0.0010-0.0017	0.0008-0.0087	3	2.0466-2.0472	0.0009-0.0022	0.0063-0.0123
	3.0 (2995)	1MZ-FE	2.4011-2.4016	0.0010-0.0018	0.0016-0.0095	2	2.0863-2.8660	0.0015-0.0025	0.0059-0.0118
2002	1.5 (1496)	1NZ-FE	①	0.0004-0.0009	0.0035-0.0075	3	1.5745-1.5748	0.0006-0.0016	0.0063-0.0142
	1.8 (1762)	1ZZ-FE	②	0.0006-0.0013	0.0008-0.0087	3	1.7320-1.7323	0.0011-0.0024	0.0063-0.0135
	1.8 (1796)	2ZZ-GE	②	0.0006-0.0013	0.0016-0.0094	3	1.7713-1.7717	0.0011-0.0020	0.0063-0.0135
	2.4 (2398)	2AZ-FE	2.1648-2.1654	0.0007-0.0016	0.0063-0.0143	3	1.8894-1.8898	0.0009-0.0022	0.0063-0.0123
	3.0 (2995)	1MZ-FE	2.4011-2.4016	0.0010-0.0018	0.0016-0.0095	2	2.0863-2.8660	0.0015-0.0025	0.0059-0.0118
2003	1.5 (1496)	1NZ-FE	①	0.0004-0.0009	0.0035-0.0075	3	1.5745-1.5748	0.0006-0.0016	0.0063-0.0142
	1.8 (1762)	1ZZ-FE	②	0.0006-0.0013	0.0008-0.0087	3	1.7320-1.7323	0.0011-0.0024	0.0063-0.0135
	1.8 (1796)	2ZZ-GE	②	0.0006-0.0013	0.0016-0.0094	3	1.7713-1.7717	0.0011-0.0020	0.0063-0.0135
	2.4 (2398)	2AZ-FE	2.1648-2.1654	0.0007-0.0016	0.0063-0.0143	3	1.8894-1.8898	0.0009-0.0022	0.0063-0.0123
	3.0 (2995)	1MZ-FE	2.4011-2.4016	0.0010-0.0018	0.0016-0.0095	2	2.0863-2.8660	0.0015-0.0025	0.0059-0.0118
2004	1.5 (1496)	1NZ-FE	①	0.0004-0.0009	0.0035-0.0075	3	1.5745-1.5748	0.0006-0.0016	0.0063-0.0142
	1.8 (1762)	1ZZ-FE	②	0.0006-0.0013	0.0008-0.0087	3	1.7320-1.7323	0.0011-0.0024	0.0063-0.0135
	1.8 (1796)	2ZZ-GE	②	0.0006-0.0013	0.0016-0.0094	3	1.7713-1.7717	0.0011-0.0020	0.0063-0.0135
	2.4 (2398)	2AZ-FE	2.1648-2.1654	0.0007-0.0016	0.0063-0.0143	3	1.8894-1.8898	0.0009-0.0022	0.0063-0.0123
	3.0 (2995)	1MZ-FE	2.4011-2.4016	0.0010-0.0018	0.0016-0.0095	2	2.0863-2.8660	0.0015-0.0025	0.0059-0.0118
	3.3 (3311)	3MZ-FE	2.4403-2.4409	③	0.0016-0.0094	4	2.0863-2.0866	0.0015-0.0026	0.0059-0.0118

① Reference mark:
0: 1.81102-1.81110
1: 1.81110-1.81118
2: 1.81118-1.81126
3: 1.81126-1.81133
4: 1.81133-1.81141
5: 1.81141-1.81149

② Reference mark:
0: 1.8897-1.8898
1: 1.8896-1.8897
2: 1.8895-1.8896
3: 1.8894-1.8895
4. 1.8893-1.8894
5: 1.8892-1.8893

③ Journal No. 1 and 4: 0.0006 - 0.0013 inch
Journal No. 2 and 3: 0.0010 - 0.0018 inch

67170-TOYO-C09

WHEEL ALIGNMENT

			Caster		Camber			
Year	Model		Range (+/-Deg.)	Preferred Setting (Deg.)	Range (+/-Deg.)	Preferred Setting (Deg.)	Toe-in (in.)	Steering Axis Inclination (Deg.)
2001	Avalon	F	0.75	+2.17	0.75	-0.62	0+/-0.08	13.07+/-0.75
		R	—	—	0.75	-0.72	0.16+/-0.08	—
	Camry 4-cyl.	F	0.75	+2.18	0.75	-0.60	0+/-0.08	13.08+/-0.75
		R	—	—	0.75	-0.70	0.16+/-0.08	—
	Camry 6-cyl.	F	0.75	+2.09	0.75	-0.60	0+/-0.08	13.08+/-0.75
		R	—	—	0.75	-0.75	0.16+/-0.08	—
	Camry Solara	F	0.75	+2.08	0.75	-0.52	0+/-0.08	12.09+/-0.75
		R	—	—	0.75	-0.65	0.16+/-0.08	—
	Celica	F	0.75	+2.05	0.75	-0.77	0+/-0.08	14.97+/-0.75
		R	—	—	0.75	-1.17	0.14+/-0.08	—
	Corolla	F	0.75	+1.19	0.75	-0.18	0.04+/-0.08	12.38+/-0.75
		R	—	—	0.75	-0.92	0.16+/-0.08	—
	Echo	F	0.75	+1.60	0.75	-0.58	0+/-0.08	10.08+/-0.75
		R	—	—	0.75	-0.93	0.11+/-0.12	—
	MR 2	F	0.75	3.08	0.75	-0.47	0.06+/-0.08	14.52+/-0.75
		R	—	—	0.75	-1.05	0.12+/-0.08	—
2002	Avalon	F	0.75	+2.17	0.75	-0.62	0+/-0.08	13.07+/-0.75
		R	—	—	0.75	-0.72	0.16+/-0.08	—
	Camry 4-cyl.	F	0.75	+2.18	0.75	-0.60	0+/-0.08	13.08+/-0.75
		R	—	—	0.75	-0.70	0.16+/-0.08	—
	Camry 6-cyl.	F	0.75	+2.09	0.75	-0.60	0+/-0.08	13.08+/-0.75
		R	—	—	0.75	-0.75	0.16+/-0.08	—
	Camry Solara	F	0.75	+2.08	0.75	-0.52	0+/-0.08	12.09+/-0.75
		R	—	—	0.75	-0.65	0.16+/-0.08	—
	Celica	F	0.75	+2.05	0.75	-0.77	0+/-0.08	14.97+/-0.75
		R	—	—	0.75	-1.17	0.14+/-0.08	—
	Corolla	F	0.75	+1.19	0.75	-0.18	0.04+/-0.08	12.38+/-0.75
		R	—	—	0.75	-0.92	0.16+/-0.08	—
	Echo	F	0.75	+1.60	0.75	-0.58	0+/-0.08	10.08+/-0.75
		R	—	—	0.75	-0.93	0.11+/-0.12	—
	MR 2	F	0.75	3.08	0.75	-0.47	0.06+/-0.08	14.52+/-0.75
		R	—	—	0.75	-1.05	0.12+/-0.08	—
2003	Avalon	F	0.75	+2.17	0.75	-0.62	0+/-0.08	13.07+/-0.75
		R	—	—	0.75	-0.72	0.16+/-0.08	—
	Camry 4-cyl.	F	0.75	+2.18	0.75	-0.60	0+/-0.08	11.45+/-0.75
		R	—	—	0.75	-1.27	0.16+/-0.08	—
	Camry 6-cyl.	F	0.75	+2.09	0.75	-0.60	0+/-0.08	13.08+/-0.75
		R	—	—	0.75	-0.75	0.16+/-0.08	—
	Camry Solara	F	0.75	+2.08	0.75	-0.52	0+/-0.08	12.09+/-0.75
		R	—	—	0.75	-0.65	0.16+/-0.08	—
	Celica	F	0.75	+2.05	0.75	-0.77	0+/-0.08	14.97+/-0.75
		R	—	—	0.75	-1.17	0.14+/-0.08	—
	Corolla	F	0.75	+1.19	0.75	-0.18	0.04+/-0.08	12.38+/-0.75
		R	—	—	0.75	-0.92	0.16+/-0.08	—
	Echo	F	0.75	+1.60	0.75	-0.58	0+/-0.08	10.08+/-0.75
		R	—	—	0.75	-0.93	0.11+/-0.12	—
	MR 2	F	0.75	3.08	0.75	-0.47	0.06+/-0.08	14.52+/-0.75
		R	—	—	0.75	-1.05	0.12+/-0.08	—

67170-TOYO-C10

WHEEL ALIGNMENT

Year	Model		Caster Range (+/-Deg.)	Caster Preferred Setting (Deg.)	Camber Range (+/-Deg.)	Camber Preferred Setting (Deg.)	Toe-in (in.)	Steering Axis Inclination (Deg.)
2004	Avalon	F	0.75	+2.17	0.75	-0.62	0+/-0.08	13.07+/-0.75
		R	—	—	0.75	-0.72	0.16+/-0.08	—
	Camry 4-cyl.	F	0.75	+2.65 ①	0.75	-0.72	0+/-0.08	11.45+/-0.75
		R	—	—	0.75	-1.27	0.16+/-0.08	—
	Camry 6-cyl.	F	0.75	+2.62 ②	0.75	-0.72	0+/-0.08	11.45+/-0.75
		R	—	—	0.75	-1.27	0.16+/-0.08	—
	Camry Solara 4-cyl.	F	0.75	+2.62 ③	0.75	-0.73 ④	0+/-0.08	11.47+/-0.75
		R	—	—	0.75	-1.35	0.16+/-0.08	—
	Camry Solara 6-cyl.	F	0.75	+2.83 ⑤	0.75	-0.75 ④	0+/-0.08	11.52+/-0.75
		R	—	—	0.75	-1.37	0.16+/-0.08	—
	Celica	F	0.75	+2.12	0.75	-0.47	0+/-0.08	13.15+/-0.75
		R	—	—	0.75	-1.18	0.12+/-0.08	—
	Celica GTS	F	0.75	+2.02	0.75	-0.42	0+/-0.08	13.07+/-0.75
		R	—	—	0.75	-1.18	0.12+/-0.08	—
	Corolla	F	0.75	+2.83	0.75	-0.53	0+/-0.08	11.35+/-0.75
		R	—	—	0.5	-1.45	0.10+/-0.10	—
	Echo	F	0.75	⑥	0.75	-0.58	0+/-0.08	10.08+/-0.75
		R	—	—	0.75	-0.93	0.13+/-0.12	—
	MR 2	F	0.75	3.08	0.75	-0.47	0.06+/-0.08	14.52+/-0.75
		R	—	—	0.75	-1.05	0.12+/-0.08	—

① Sport: 2.72
② 3MZ-FE: 2.65
③ Sport: 2.93
④ Sport: 0.77
⑤ Sport: 2.88
⑥ Sedan M/S: 0.97
Sedan P/S: 1.93
Hatchback M/S: 0.88
Hatchback P/S: 1.85

67170-TOYO-C11

TIRE, WHEEL AND BALL JOINT SPECIFICATIONS

		OEM Tires		Tire Pressures (psi)		Wheel	Ball Joint
Year	Model	Standard	Optional	Front	Rear	Size	Inspection
2001	Avalon	P205/65HR15	None	32	32	6-JJ	9-30 in. ①
	Camry	P205/65HR15	P205/60HR16	32	35	6-JJ	9-26 in. ①
	Camry Solara	P215/60R16	P205/60R16	—	—	—	9-30 in. ①
	Celica	205/55VR15	P205/55VR15	33	33	6.5-JJ	9-26 in. ①
	Corolla CE	P175/65R14	None	30	30	5.5-JJ	9-26 in. ①
	Corolla S, LE	P185/65R14	None	30	30	5.5-JJ	9-26 in. ①
	Echo	P175/65R14	None	32	32	5.5-JJ	9-26 in. ①
	MR 2	F: 205/50R16 R: 215/50R16	None	33	33	6-JJ	9-26 in. ①
2002	Avalon	P205/65HR15	None	32	32	6-JJ	9-30 in. ①
	Camry, LE	P205/65HR15	None	29	29	6-JJ	9-30 in. ①
	Camry SE, XLE	P215/60HR16	None	29	29	6-JJ	9-30 in. ①
	Camry Solara	P215/60R16	P205/60R16	—	—	—	9-30 in. ①
	Celica	205/55VR15	P205/55VR15	33	33	6.5-JJ	9-26 in. ①
	Corolla CE	P175/65R14	None	30	30	5.5-JJ	9-26 in. ①
	Corolla S, LE	P185/65R14	None	30	30	5.5-JJ	9-26 in. ①
	Echo	P175/65R14	None	32	32	5.5-JJ	9-26 in. ①
	MR 2	F: 205/50R16 R: 215/50R16	None	33	33	6-JJ	9-26 in. ①
2003	Avalon	P205/65HR15	None	32	32	6-JJ	9-30 in. ①
	Camry, LE	P205/65HR15	None	29	29	6-JJ	9-30 in. ①
	Camry SE, XLE	P215/60HR16	None	29	29	6-JJ	9-30 in. ①
	Camry Solara	P205/65R15	P205/60R16	—	—	—	9-30 in. ①
	Celica	205/55VR15	P205/55VR15	33	33	6.5-JJ	9-26 in. ①
	Corolla CE	P185/65R15	None	30	30	5.5-JJ	9-43 in. ①
	Corolla S, LE	P195/65R15	None	30	30	5.5-JJ	9-43 in. ①
	Echo	P175/65R14	None	32	32	5.5-JJ	9-26 in. ①
	MR 2	F: 205/50R16 R: 215/50R16	None	33	33	6-JJ	9-26 in. ①
2004	Avalon	P205/65R15	P205/60R16	31	31	6-JJ	9-30 in. ①
	Camry, LE	P205/65R15	None	29	29	6-JJ	9-30 in. ①
	Camry SE, XLE	P215/60R16	None	29	29	6-JJ	9-30 in. ①
	Camry Solara	P215/60R16	②	—	—	—	9-30 in. ①
	Celica	P195/60R15	P205/55R15	29 ③	29 ③	6.5-JJ	9-26 in. ①
	Corolla CE	P185/65R15	None	30	30	5.5-JJ	9-43 in. ①
	Corolla S, LE	P195/65R15	None	30	30	5.5-JJ	9-43 in. ①
	Echo	P175/65R14	P185/60R15	32	32	5.5-JJ	5.2-30 in. ①
	MR 2	F: 205/50R16 R: 215/50R16	None	33	33	6-JJ	9-26 in. ①

OEM: Original Equipment Manufacturer

PSI: Pounds Per Square Inch

① Torque required in inch lbs. to rotate ball joint when removed from the knuckle

② Sport and convertible models, 215/55R17

67170-TOYO-C12

BRAKE SPECIFICATIONS

All measurements in inches unless noted

			Brake Disc			Brake Drum Diameter				Brake Caliper	
Year	Model		Original Thickness	Minimum Thickness	Maximum Runout	Original Inside Diameter	Max. Wear Limit	Maximum Machine Diameter	Minimum Lining Thickness	Bracket Bolts (ft. lbs.)	Mounting Bolts (ft. lbs.)
2001	Avalon	F	1.102	1.024	0.0020	—	—	—	0.039	25	79
		R	0.354	0.315	0.0059	—	—	—	0.039	25	34
	Camry	F	1.102	1.024	0.0020	—	—	—	0.039	25	79
		R	0.394	0.354	0.0059	9.00	—	9.08	0.039	14	20
	Camry Solara	F	1.102	1.024	0.0020	—	—	—	0.039	25	79
		R	0.394	0.354	0.0059	9.00	—	9.08	0.039	14	20
	Celica	F	1.102	1.024	0.0020	—	—	—	0.039	25	79
		R	0.354	0.314	0.0059	7.87	—	7.91	0.039	—	34
	Celica GT-S	F	0.984	0.906	0.0020	—	—	—	0.039	25	79
		R	0.354	0.295	0.0059	—	—	—	0.039	—	34
	Corolla		0.866	0.787	0.0020	7.87	—	7.91	0.039	25	65
	Echo		0.709	0.630	0.0020	7.09	—	7.13	0.039	25	65
	MR 2	F	0.787	0.709	0.0020	—	—	—	0.039	80	25
		R	0.630	0.591	0.0039	—	—	—	0.039	44	15
2002	Avalon	F	1.102	1.024	0.0020	—	—	—	0.039	25	79
		R	0.354	0.315	0.0059	—	—	—	0.039	25	34
	Camry	F	1.102	1.024	0.0020	—	—	—	0.039	25	79
		R	0.394	0.354	0.0059	9.00	—	9.08	0.039	14	20
	Camry Solara	F	1.102	1.024	0.0020	—	—	—	0.039	25	79
		R	0.394	0.354	0.0059	9.00	—	9.08	0.039	14	20
	Celica	F	1.102	1.024	0.0020	—	—	—	0.039	25	79
		R	0.354	0.314	0.0059	7.87	—	7.91	0.039	—	34
	Celica GT-S	F	0.984	0.906	0.0020	—	—	—	0.039	25	79
		R	0.354	0.295	0.0059	—	—	—	0.039	—	34
	Corolla		0.866	0.787	0.0020	7.87	—	7.91	0.039	25	65
	Echo		0.709	0.630	0.0020	7.09	—	7.13	0.039	25	65
	MR 2	F	0.787	0.709	0.0020	—	—	—	0.039	80	25
		R	0.630	0.591	0.0039	—	—	—	0.039	44	15
2003	Avalon	F	1.102	1.024	0.0020	—	—	—	0.039	25	79
		R	0.354	0.315	0.0059	—	—	—	0.039	25	34
	Camry	F	1.102	1.024	0.0020	—	—	—	0.039	25	79
		R	0.394	0.354	0.0059	9.00	—	9.08	0.039	14	20
	Camry Solara	F	1.102	1.024	0.0020	—	—	—	0.039	25	79
		R	0.394	0.354	0.0059	9.00	—	9.08	0.039	14	20
	Celica	F	1.102	1.024	0.0020	—	—	—	0.039	25	79
		R	0.354	0.314	0.0059	7.87	—	7.91	0.039	—	34
	Celica GT-S	F	0.984	0.906	0.002	—	—	—	0.039	25	79
		R	0.354	0.295	0.006	—	—	—	0.039	—	34
	Corolla		0.866	0.787	0.0020	7.87	—	7.91	0.039	25	65
	Echo		0.709	0.630	0.0020	7.09	—	7.13	0.039	25	65
	MR 2	F	0.787	0.709	0.0020	—	—	—	0.039	80	25
		R	0.630	0.591	0.0039	—	—	—	0.039	44	15

67170-TOYO-C13

BRAKE SPECIFICATIONS

All measurements in inches unless noted

			Brake Disc			Brake Drum Diameter				Brake Caliper	
Year	Model		Original Thickness	Minimum Thickness	Maximum Runout	Original Inside Diameter	Max. Wear Limit	Maximum Machine Diameter	Minimum Lining Thickness	Bracket Bolts (ft. lbs.)	Mounting Bolts (ft. lbs.)
2004	Avalon	F	1.102	1.024	0.0020	—	—	—	0.039	25	79
		R	0.354	0.315	0.0059	—	—	—	0.039	25	34
	Camry	F	1.102	1.024	0.0020	—	—	—	0.039	25	79
		R	0.472	0.413	0.0059	9.00	—	9.08	0.039	29	35
	Camry Solara	F	1.102	1.024	0.0020	—	—	—	0.039	25	79
		R	0.472	0.413	0.0059	—	—	—	0.039	32	46
	Celica	F	0.984	0.906	0.002	—	—	—	0.039	25	79
		R	0.354	0.295	0.006	7.87	—	7.91	0.039	—	34
	Corolla		0.984	0.906	0.0020	7.87	—	7.91	0.039	25	79
	Echo		0.709	0.630	0.0020	7.09	—	7.13	0.039	25	65
	MR 2	F	0.787	0.709	0.0020	—	—	—	0.039	80	25
		R	0.630	0.591	0.0039	—	—	—	0.039	44	15

F: Front

R: Rear

67170-TOYO-C14

SCHEDULED MAINTENANCE INTERVALS

TOYOTA—AVALON, CAMRY, CELICA, COROLLA, ECHO, MR 2 & SOLARA

TO BE SERVICED	TYPE OF SERVICE	VEHICLE MILEAGE INTERVAL (x1000)												
		7.5	15	22.5	30	37.5	45	52.5	60	67.5	75	82.5	90	97.5
Engine oil & filter	R	✓	✓	✓	✓	✓	✓	✓	✓	✓	✓	✓	✓	✓
Drive belts	S/I								✓	✓	✓	✓	✓	✓
Automatic transaxle fluid & filter	S/I		✓		✓		✓		✓		✓		✓	
Ball joints & dust covers	S/I		✓		✓		✓		✓		✓		✓	
Bolts & nuts on body & chassis	S/I		✓		✓		✓		✓		✓		✓	
Brake line pipes & hoses	S/I		✓		✓		✓		✓		✓		✓	
Brake linings & drums	S/I		✓		✓		✓		✓		✓		✓	
Brake pads & discs (front & rear if equipped)	S/I		✓		✓		✓		✓		✓		✓	
Differential oil	S/I		✓		✓		✓		✓		✓		✓	
Drive shaft boots	S/I		✓		✓		✓		✓		✓		✓	
Manual transaxle oil	S/I		✓		✓		✓		✓		✓		✓	
Steering gear housing oil	S/I		✓		✓		✓		✓		✓		✓	
Steering linkage	S/I		✓		✓		✓		✓		✓		✓	
Air filter	R				✓				✓				✓	
Spark plugs	R				✓				✓				✓	
Spark plugs (platinum tip)	R								✓					
Exhaust system	S/I				✓				✓				✓	
Fuel lines & connections	S/I				✓				✓				✓	
Valve clearance	S/I				✓				✓				✓	
Engine coolant	R						✓				✓			
Fuel tank cap gasket	R								✓					
Charcoal canister	S/I								✓					

R: Replace S/I: Service or Inspect

FREQUENT OPERATION MAINTENANCE (SEVERE SERVICE)

If a vehicle is operated under any of the following conditions it is considered severe service:

- Extremely dusty areas.

- 50% or more of the vehicle operation is in 32°C (90°F) or higher temperatures, or constant operation in temperatures below 0°C (32°F).

- Prolonged idling (vehicle operation in stop and go traffic).

- Frequent short running periods (engine does not warm to normal operating temperatures).

- Police, taxi, delivery usage or trailer towing usage.

Oil & oil filter change: change every 2500 miles.

Bolts & nuts on chassis & body: tighten every 7500 miles.

Ball joints & dust covers: service or inspect every 12,000 miles.

Brake linings & drums: service or inspect ever 12,000 miles.

Brake pads & discs (front & rear if equipped): service or inspect every 12,000 miles.

Drive shaft boots: service or inspect every 12,000 miles.

Steering linkage: service or inspect every 12,000 miles.

Air filter: service or inspect every 15,000 miles.

Exhaust system: service or inspect every 15,000 miles.

Timing belt: replace every 60,000 miles.

67170-TOYO-C15

PRECAUTIONS

Before servicing any vehicle, please be sure to read all of the following precautions, which deal with personal safety, prevention of component damage, and important points to take into consideration when servicing a motor vehicle:

- Never open, service or drain the radiator or cooling system when the engine is hot; serious burns can occur from the steam and hot coolant.
- Observe all applicable safety precautions when working around fuel. Whenever servicing the fuel system, always work in a well-ventilated area. Do not allow fuel spray or vapors to come in contact with a spark, open flame, or excessive heat (a hot drop light, for example). Keep a dry chemical fire extinguisher near the work area. Always keep fuel in a container specifically designed for fuel storage; also, always properly seal fuel containers to avoid the possibility of fire or explosion. Refer to the additional fuel system precautions later in this section.
- Fuel injection systems often remain pressurized, even after the engine has been turned **OFF**. The fuel system pressure must be relieved before disconnecting any fuel lines. Failure to do so may result in fire and/or personal injury.
- Brake fluid often contains polyglycol ethers and polyglycols. Avoid contact with the eyes and wash your hands thoroughly after handling brake fluid. If you do get brake fluid in your eyes, flush your eyes with clean, running water for 15 minutes. If eye irritation persists, or if you have taken brake fluid internally, IMMEDIATELY seek medical assistance.
- The EPA warns that prolonged contact with used engine oil may cause a number of skin disorders, including cancer! You should make every effort to minimize your exposure to used engine oil. Protective gloves should be worn when changing oil. Wash your hands and any other exposed skin areas as soon as possible after exposure to used engine oil. Soap and water, or waterless hand cleaner should be used.
- All new vehicles are now equipped with an air bag system. The system must be disabled before performing service on or around system components, steering column, instrument panel components, wiring and sensors. Failure to follow safety and disabling procedures could result in accidental air bag deployment, possible personal injury and unnecessary system repairs.
- Always wear safety goggles when working with, or around, the air bag system. When carrying a non-deployed air bag, be sure the bag and trim cover are pointed away from your body. When placing a non-deployed air bag on a work surface, always face the bag and trim cover upward, away from the surface. This will reduce the motion of the module if it is accidentally deployed. Refer to the additional air bag system precautions later in this section.
- Clean, high quality brake fluid from a sealed container is essential to the safe and proper operation of the brake system. You should always buy the correct type of brake fluid for your vehicle. If the brake fluid becomes contaminated, completely flush the system with new fluid. Never reuse any brake fluid. Any brake fluid that is removed from the system should be discarded. Also, do not allow any brake fluid to come in contact with a painted surface; it will damage the paint.
- Never operate the engine without the proper amount and type of engine oil; doing so WILL result in severe engine damage.
- Timing belt maintenance is extremely important! Many models utilize an interference-type, non-freewheeling engine. If the timing belt breaks, the valves in the cylinder head may strike the pistons, causing potentially serious (also time-consuming and expensive) engine damage. Refer to the maintenance interval charts in the front of this section for the recommended replacement interval for the timing belt.
- Disconnecting the negative battery cable on some vehicles may interfere with the functions of the on-board computer system(s) and may require the computer to undergo a relearning process once the negative battery cable is reconnected.
- When servicing drum brakes, only disassemble and assemble one side at a time, leaving the remaining side intact for reference.

ENGINE REPAIR

Alternator

REMOVAL & INSTALLATION

Avalon, Camry and Solara

1. Before servicing the vehicle, refer to the precautions in the beginning of this section.
2. Remove or disconnect the following:
   - Negative battery cable. On models with an airbag, wait at least 90 seconds from the time that the ignition switch is turned to the LOCK position and the battery is disconnected before performing any further work.
   - Air cleaner, if necessary
   - Drive belt from the pulley
   - Harness and wire (and nut) from the alternator
   - The two bolts and engine mounting stay if necessary
   - Alternator

To install:

3. Install or connect the following:
   - Alternator
   - Drive belt. Torque the bolts. 2AZ-FE: M8 bolt—15 ft. lbs. (21 Nm); M10 bolt—38 ft. lbs. (52 Nm). Except 2AZ-FE: Pivot bolt—40 ft. lbs. (54 Nm); Lock bolt—14 ft. lbs. (19 Nm).
   - Wiring
   - Negative and starter battery cables

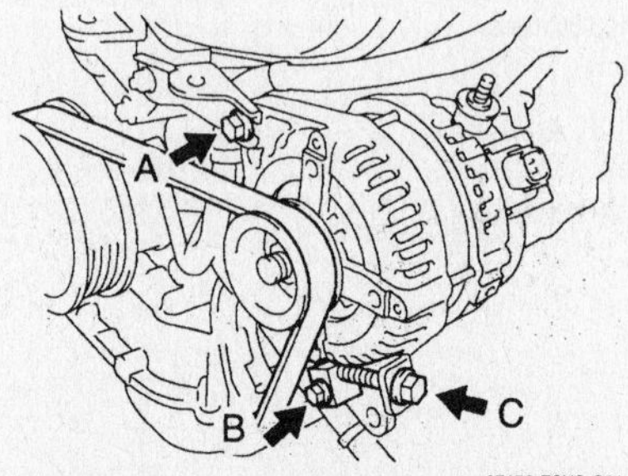

(A) pivot bolt, (B) lock bolt and (C) adjusting bolt

Celica

1. Before servicing the vehicle, refer to the precautions in the beginning of this section.
2. Remove or disconnect the following:
   - Negative battery cable
   - Drive belt
   - Wiring
   - On 2ZZ-GE, 2 bolts and bracket
   - Alternator

To install:

3. Install or connect the following:
   - Alternator
   - Torque the larger bolt—40 ft. lbs.

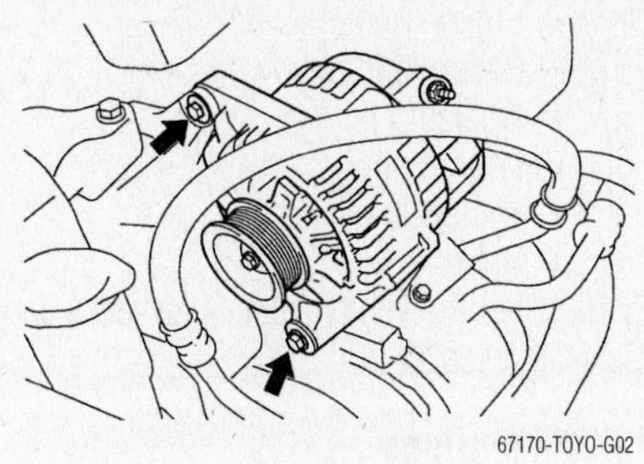

Alternator with 2 mounting bolts

(54 Nm); smaller bolt—18 ft. lbs. (25 Nm)

- On 2ZZ-GE, 2 bolts and bracket. Torque to 21 ft. lbs (29 Nm).
- Drive belt

- Wiring
- Negative battery cable

Corolla

1. Before servicing the vehicle, refer to the precautions in the beginning of this section.

➡It may be necessary to remove the gravel shield and work from underneath the car in order to gain access to the alternator retaining bolts.

2. Remove or disconnect the following:
   - Negative battery cable
   - Wiring from the alternator
   - Drive belt
   - Alternator

To install:

3. Install or connect the following:
   - Alternator. Torque the smaller bolt to 18 ft. lbs. (25 Nm) and the larger bolt to 40 ft. lbs. (54 Nm).
   - Alternator connector and wiring
   - Drive belt
   - Negative battery cable

Echo

1. Before servicing the vehicle, refer to the precautions in the beginning of this section.
2. Remove or disconnect the following:
   - Negative battery cable
   - Wire clamp from the rectifier
   - Alternator harness

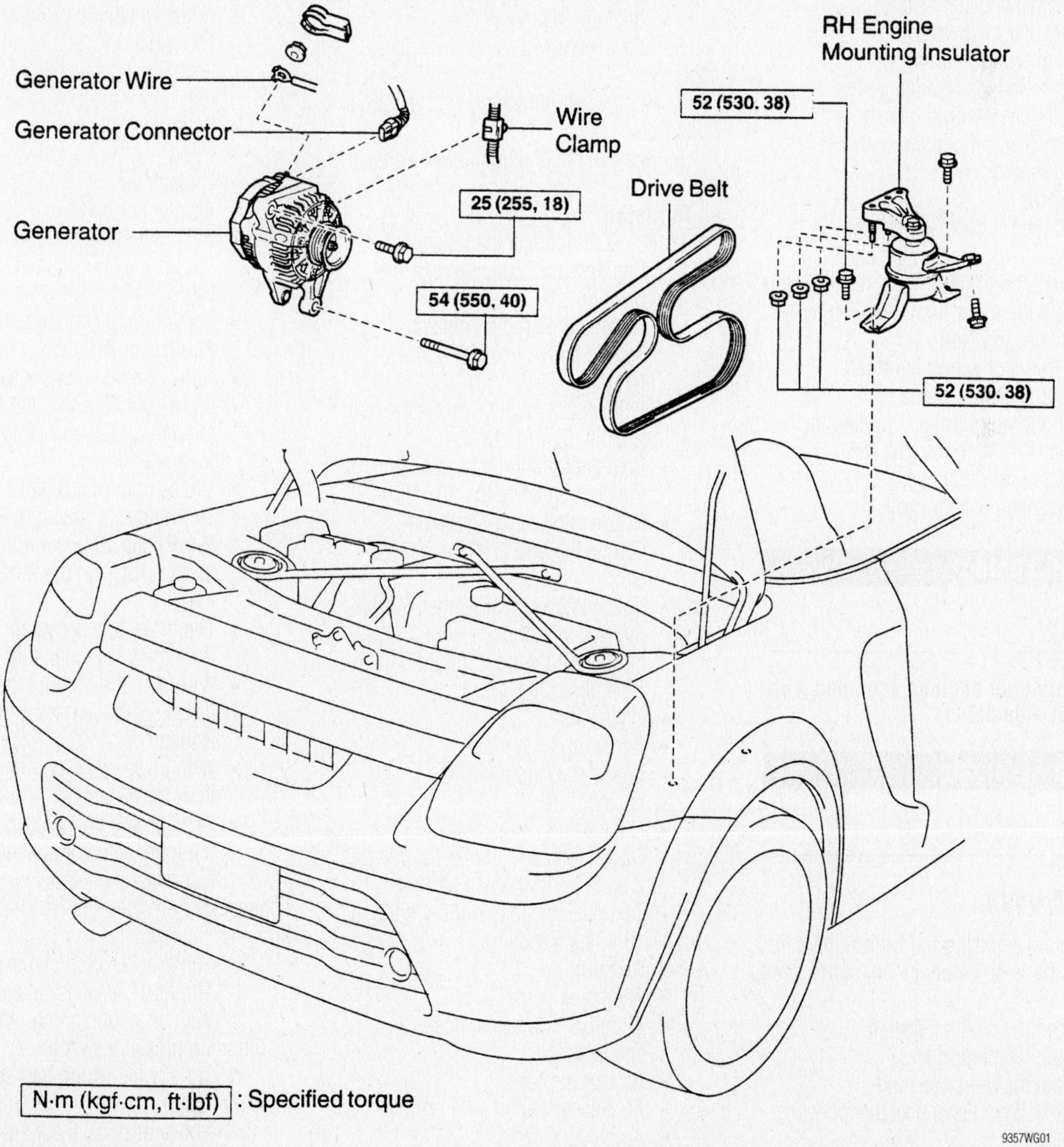

Generator Wire
Generator Connector
Generator
Wire Clamp
25 (255, 18)
54 (550, 40)
Drive Belt
RH Engine Mounting Insulator
52 (530. 38)
52 (530. 38)
N·m (kgf·cm, ft·lbf) : Specified torque
9357WG01

MR2 alternator mounting exploded view

- 2 bolts
- Alternator

To install:

3. Install or connect the following:
   - Alternator and hand tighten the bolts
   - Drive belt and adjust if necessary. Torque the 14 mm bolt to 14 ft. lbs. (18 Nm) and the 17 mm bolt to 40 ft. lbs. (54 Nm).
   - Alternator connector and clamp
   - Alternator electrical wiring
   - Negative battery cable

MR2

1. Before servicing the vehicle, refer to the precautions in the beginning of this section.
2. Remove or disconnect the following:
   - Negative battery cable
   - Accessory drive belt
   - Right motor mount
   - Wire harness clamp
   - Alternator wiring harness
   - Alternator mounting bolts
   - Alternator

To install:

3. Install or connect the following:
   - Alternator. Tighten the upper mounting bolt to 18 ft. lbs. (25 Nm) and the lower mounting bolt to 40 ft. lbs. (54 Nm).
   - Alternator wiring harness
   - Wire harness clamp
   - Right motor mount. Tighten the bolts to 38 ft. lbs. (52 Nm).
   - Accessory drive belt
   - Negative battery cable

Ignition Timing

ADJUSTMENT

➡The timing on engines equipped with DIS is not adjustable.

Engine Assembly

REMOVAL & INSTALLATION

1NZ-FE Engines

1. Before servicing the vehicle, refer to the precautions in the beginning of this section.
2. Drain the cooling system.
3. Drain the engine oil.
4. Drain the transaxle fluid.
5. Remove or disconnect the following:
   - Hood
   - Battery and tray
   - Outer front cowl top panel
   - Engine under covers
   - Accelerator cable
   - Air cleaner cap and related parts
   - Air cleaner case and related parts
   - All tubes, hoses and connectors attached to the engine
   - Accessory drive belts
   - Alternator
   - Radiator
   - With A/C, position the compressor out of the way
   - With MT, the clutch release cylinder
   - Transaxle control cables
   - Fusible link
   - Power steering pump
   - Center exhaust pipe
   - Halfshafts
   - Suspension crossmember
   - Rear engine mount
6. Attach a shop crane to the engine hangers.
7. Remove all remaining engine mount bolts/nuts.
8. Remove the engine/transaxle assembly.

To install:

9. Install or connect the following:
   - Engine/transaxle assembly into position
   - Right engine mount insulator. Torque the bolts to 35 ft. lbs. (47 Nm).
   - Left engine mount. Torque the bolts to 35 ft. lbs. (47 Nm).
   - Rear engine mount bracket. Torque the bolts to 35 ft. lbs. (47 Nm).
   - Rear engine mount insulator. Torque the bolt to 47 ft. lbs. (64 Nm).
   - Suspension crossmember. Torque the rear mount bolts to 86 ft. lbs. (116 Nm); the front mount bolts to 52 ft. lbs. (70 Nm).
   - Halfshafts
   - Center exhaust pipe
   - Power steering pump
   - Fusible link
   - Transaxle control cables
   - Clutch release cylinder. Torque the 2 bolts to 10 ft. lbs. (13 Nm).
   - Compressor. Torque the 4 bolts to 18 ft. lbs. (25 Nm).
   - Radiator
   - Alternator
   - All remaining tubes, hoses and connectors
   - Air cleaner case
   - Air cleaner cap
   - Accelerator cable
   - Outer front cowl top panel
   - Engine under covers
   - Battery and tray
   - Coolant
   - Engine oil
   - Transaxle oil
   - Hood
10. Start the vehicle, check for leaks and repair if necessary.

1ZZ-FE Engines

COROLLA

1. Before servicing the vehicle, refer to the precautions in the beginning of this section.
2. Relieve the fuel system pressure.
3. Drain the cooling system.
4. Drain the engine oil.
5. Drain the transaxle fluid.
6. Remove or disconnect the following:
   - Negative battery cable. On vehicles equipped with an air bag, wait at least 90 seconds before proceeding.
   - Battery
   - Hood
   - Undercover
   - Accelerator cable
   - With automatic transmission, throttle cable from the accelerator cable.
   - Radiator and cooling fan
   - Air cleaner assembly
   - Coolant reservoir tank stay
   - Electrical connector, the hose, the mounting bolt, and remove the washer tank
   - Cruise control actuator
   - The Manifold Absolute Pressure (MAP) sensor vacuum hose from the gas filter on the intake manifold
   - The brake booster vacuum hose from the intake manifold
   - With air conditioning: the air conditioning vacuum hose from the actuator
   - With power steering: the air hose from the air pipe
   - With air conditioning: the air conditioning actuator connector
7. Disconnect the following wires and connectors from the right-hand fender apron as follows:
   a. The ground strap connector.
   b. The MAP sensor connector.
   c. With air conditioning: the air conditioning pressure switch.
   d. The engine wiring harness from the fender apron.
8. Remove or disconnect the following:
   - Data Link Connector 1 (DLC1) con-

nector and ground strap from the left-hand fender apron.
- Engine relay box and 4 connectors.
- Charcoal canister
- Heater hoses from water inlet housing
- Fuel inlet and return hoses
- With manual transmission, clutch release cylinder without disconnecting the pipe
- Transaxle control cable(s)

9. To disconnect the engine wiring harness, disconnect or remove the following components:
- Left-hand and right-hand front door scuff plate
- Lower finish panel
- Lower panel with the glove compartment
- Radio and center cluster finish panel
- Rear console box
- On manual transmission, shift lever knob
- On automatic transmission, shifting hole bezel
- Lower center finish panel
- Floor carpet bracket
- The 3 ECM connectors and cowl wire connector

10. Remove or disconnect the remaining components:
- Air conditioning compressor
- Front exhaust pipe
- Halfshafts
- Power steering pump
- Engine mounting center member
- Through-bolt and nut holding the mounting insulator to the mounting bracket
- Engine and transaxle assembly
- Front and rear engine mounting bracket
- Starter
- Separate the transaxle from the engine

To install:

11. Install or connect the following:
- Engine to the transaxle
- Starter
- Rear engine mounting bracket bolts: 57 ft. lbs. (77 Nm).
- Front engine mounting bracket. bolts: 57 ft. lbs. (77 Nm).
- Engine and transaxle assembly into the vehicle
- Engine mounting center member.
- Front engine mounting insulator through-bolt and nut. Torque the bolt to 64 ft. lbs. (87 Nm).
- Halfshafts
- Front exhaust pipe
- Power steering pump. Torque the bolts to 29 ft. lbs. (39 Nm).
- Drive belt
- Air conditioner compressor. Torque the bolts to 18 ft. lbs. (25 Nm).
- Drive belt and reconnect the connector.

12. To install and connect the engine wiring harness, perform the following:
- Push the wire through the cowl
- Connect the 3 ECM connectors
- Attach the cowl wire connector
- Floor carpet bracket
- Center lower finish panel
- With automatic transmission, install the shifting hole bezel, with manual transmission, install the shift lever knob
- Rear console box
- Center cluster finish panel and the radio
- Lower panel with the glove compartment door
- Right and left-hand door scuff plates
- Lower finish panel

13. Install or connect the following:
- With manual transmission, clutch release cylinder
- Transaxle control cable(s)
- Fuel return and inlet hose. Torque the bolt to 22 ft. lbs. (29 Nm).
- Heater hoses to the water inlet housing
- Charcoal canister

14. Connect the following wires and connectors on the left-hand fender apron:
- The 4 connectors to the engine relay box
- Engine relay box
- The DLC1 connector
- The connector on the fender apron
- The ground strap on the fender apron

15. Install or connect on the right-hand fender apron:
- The ground strap connector
- The MAP sensor connector
- With air conditioning, the air conditioning pressure switch
- The engine wire from the fender apron

16. Install or connect the following:
- With A/C, the actuator connector
- With power steering, the air hoses to the air pipe
- The vacuum hose from the MAP sensor to the gas filter to the intake chamber
- The brake booster vacuum hose to the air intake chamber
- With A/C, the vacuum hose from the actuator
- With cruise control, actuator, actuator cable and cover
- Electrical connector and vinyl hose
- Washer tank with the bolt
- Coolant reservoir tank stay
- Air cleaner
- Radiator and cooling fan
- With automatic transmission, connect the throttle cable.
- Accelerator cable
- All fluids
- Negative battery cable
- Undercovers and hood

17. Start the vehicle, check for leaks and repair if necessary.

CELICA

1. Before servicing the vehicle, refer to the precautions in the beginning of this section.
2. Release the fuel system pressure.
3. Drain the engine oil.
4. Drain the cooling system.
5. Drain the transaxle fluid.
6. Remove or disconnect the following:
- Negative battery cable. On vehicles equipped with an air bag, wait at least 90 seconds before proceeding.
- Battery
- Hood
- Undercover
- Accelerator cable, cable bracket, and clamps.
- Air cleaner
- Cruise control actuator cable
- Radiator
- MAP sensor vacuum hose from the gas filter on the intake manifold
- Power steering air hose from the intake manifold
- Power steering hose from the air the air pipe
- Brake booster vacuum hose from the intake manifold
- Air conditioning idle-up valve
- Air conditioning idle-up valve hose from the intake manifold
- Air conditioning idle-up valve hose from the air pipe
- DLC1 from the bracket
- Engine wiring harness from the bracket
- Ground cable from the body and the ground strap from the body
- Heater hoses from the water outlet
- Heater hose from the water bypass pipe
- Fuel inlet hose from the fuel filter and the fuel return hose from the return pipe
- EVAP hose from the charcoal canister

- Engine wiring harness from the engine compartment relay box

7. To remove the engine wiring harness from the passenger's compartment, remove or disconnect the following:
    - The scuff plate
    - The cowl side trim
    - The finish panel from the lower instrument panel
    - Front side of the floor carpet
    - The wiring harness from the clamp of the ECM bracket
    - The 3 ECM connectors
    - The circuit opening relay connector
    - The 3 connectors from the connectors on the bracket
    - The A/C amplifier connector
    - The MAP sensor connector
    - The MAP sensor wire from the clamp on the bracket
    - The wire clamp from the bracket
    - The 2 nuts holding the engine wiring harness to the cowl
8. Remove or disconnect the following:
    - Front exhaust pipe
    - Halfshafts
    - Alternator drive belt
    - Air conditioning drive belt, compressor connector, and compressor. Do not disconnect the air conditioning lines
    - Remove the drive belt and remove the 4 bolts that secure the power steering pump. Without disconnecting the lines, securely hang the pump out of the way
    - A/C relay box
    - On manual transmission, clutch release cylinder from the transaxle
    - Transaxle control cable(s)
    - On automatic transmission, transaxle control cable from the engine mounting center member.
    - Exhaust pipe support bracket.
9. To remove the engine mounting center member, remove the following components:
    - The 2 dust covers from the rear side of the member
    - The A/C pipe from the bracket
    - The bolt and nut holding the front engine mounting bracket to the mounting insulator
    - The bolt holding the rear engine mounting bracket to the insulator
    - The bolt and 2 nuts holding the rear engine mounting insulator to the front suspension member
    - The 2 bolts and the rear engine mounting bracket, and the center member with the rear mounting insulator
10. Remove or disconnect the remaining components:
    - Attach an engine chain hoist to the engine hangers
    - Left-hand engine mounting bracket from the mounting insulator
    - Through-bolt and the left-hand mounting insulator
    - Ground strap connector
    - Right-hand engine mounting bracket from the mounting insulator
    - Lift the engine and transaxle assembly from the vehicle
    - Transaxle from the engine assembly

To install:

11. Install or connect the following:
    - Transaxle to the engine assembly.
    - Engine into the engine compartment
    - Right-hand engine mounting bracket to the mounting insulator. Temporarily install the 3 nuts.
    - Left-hand engine mounting insulator to the body with the through-bolt
    - Left-hand engine mounting bracket to the mounting insulator and install the 2 bolts and nut. Bolts and nut: 47 ft. lbs. (64 Nm).
    - Left-hand engine mounting through-bolt to the body. Bolt: 54 ft. lbs. (73 Nm).
    - Right-hand mounting bracket to the insulator. 12mm nut: 21 ft. lbs. (28 Nm), 14mm nut: 38 ft. lbs. (52 Nm).
    - Engine ground strap connector and remove the engine hoist.
12. To install the engine mounting center member, perform the following:
    - Attach the center member together with the rear engine mounting insulator to the front suspension member
    - Temporarily install the 2 bolts and nut holding the center member to the body
    - Install the rear engine mounting bracket. Bolts: 58 ft. lbs. (78 Nm)
    - Temporarily install the bolt and 2 nuts holding the rear engine mounting insulator to the front suspension member
    - Temporarily install the bolt holding the rear engine mounting bracket to the insulator
    - Temporarily install the bolt and nut holding the front engine mounting bracket to the insulator
    - Tighten the 2 bolts holding the center member to the body to 26 ft. lbs. (35 Nm)
    - Tighten the bolt and 2 nuts holding the rear mounting insulator to the front suspension member to 59 ft. lbs. (80 Nm)
    - Tighten the bolt holding the rear engine mounting bracket to the insulator to 65 ft. lbs. (88 Nm)
    - Tighten the bolt and nut holding the front engine mounting bracket to the insulator to 65 ft. lbs. (88 Nm
    - Install the air conditioning pipe to the bracket and install the 2 dust covers to the center member
13. Install or connect the following:
    - Exhaust pipe support bracket. Bolts to 14 ft. lbs. (19 Nm).
    - Transaxle control cable(s)
    - On automatic transmission, transaxle control cable to the engine mounting center member.
    - On manual transmission, clutch release cylinder. Bolts: 108 inch lbs. (12 Nm), then attach the bracket with the bolt.
    - A/C relay box to the body
    - Power steering pump. 12mm bolts: 14 ft. lbs. (19 Nm). 14mm bolts: 29 ft. lbs. (39 Nm). Install the drive belt. Adjusting bolt: 29 ft. lbs. (39 Nm).
    - A/C compressor. Bolts: 18 ft. lbs. (25 Nm).
    - A/C drive belt with the adjusting bolt. Torque the locknut to 29 ft. lbs. (39 Nm). Connect the connector.
    - Alternator drive belt
    - Halfshafts
    - Front exhaust pipe
14. To install the engine wiring harness in the passenger compartment, perform the following:
    - Push the harness through the cowl panel, install the retainer to the cowl with the 2 nuts and install the wire clamp to the bracket
    - Connect the harness to the clamp on the ECM
    - Connect the 3 ECM connectors and the circuit opening relay connector
    - Connect the 3 connectors to the connectors on the bracket
    - Connect the A/C amplifier connector
    - Install the floor carpet, the lower instrument panel finish panel, the cowl side trim panel, and the scuff plate
15. Install or connect the following:
    - Engine wiring harness with the 2 connectors to the engine compart-

ment relay box and install the relay box covers
- MAP sensor connector
- MAP sensor wire to the clamp on the bracket
- MAP sensor vacuum hose to the gas filter on the intake manifold
- Brake booster vacuum hose to the intake manifold
- A/C idle-up valve connector
- A/C idle-up valve hose to the intake manifold
- A/C idle-up valve hose to the air pipe
- DLC1 to the bracket
- Engine harness protector to the bracket
- Ground cable and the ground strap
- Heater hose to the water outlet and the heater hose to the water bypass pipe
- Fuel inlet hose to the fuel filter
- Fuel inlet hose with 2 new gaskets and the union bolt. Bolt: 22 ft. lbs. (30 Nm).
- Fuel return hose to the return pipe and connect the EVAP hose to the charcoal canister
- Power steering air hoses to the intake manifold and the air pipe
- Radiator
- Actuator cable to the clamps, on models equipped with cruise control
- Accelerator cable to the throttle body, cable bracket, and the clamps
- Air cleaner
- Battery tray and battery
- Hood
- Negative battery cable
- Engine undercover

16. Refill the transaxle assembly, engine oil and coolant.
17. Start the vehicle, check for leaks and repair if necessary.

MR2

1. Before servicing the vehicle, refer to the precautions in the beginning of this section.
2. Drain the cooling system.
3. Relieve the fuel system pressure.
4. Drain the engine oil
5. Drain the transaxle fluid
6. Remove or disconnect the following:
   - Engine hood
   - Rear suspension upper brace
   - Engine undercovers
   - Battery and tray
   - Air cleaner assembly and intake air hose
   - Accelerator cable
   - Rear bumper cover
   - Heated Oxygen (HO_2S) sensor connectors
   - Front exhaust pipe
   - Accessory drive belt
   - A/C compressor
   - Engine wiring harness connectors in the luggage compartment. Pull the harness through the firewall.
   - Transaxle control ECU harness connector
   - Engine wire at Junction Box No. 1
   - Engine ground wires
   - Wire brackets
   - Heater hoses
   - Radiator hoses
   - Fuel line
   - Clutch hose
   - Shift cables
   - Axle halfshafts
7. Support the drivetrain from below.
   - Left and right motor mounts
   - Front and rear motor mounts
   - Rear subframe
8. Lower the drivetrain assembly from the vehicle.
9. Remove the starter.
10. Remove the transaxle flange bolts and separate the engine and transaxle.

To install:

11. Installation is the reverse of the removal procedure, while using the following torque values:
    - Rear subframe. Tighten the bolts to 59 ft. lbs. (80 Nm).
    - Front and rear motor mounts. Tighten the through bolts to 69 ft. lbs. (93 Nm).
    - Left and right motor mounts. Tighten the mounting fasteners to 38 ft. lbs. (52 Nm) and the through bolts to 64 ft. lbs. (87 Nm).
    - A/C compressor. Tighten the bolts to 18 ft. lbs. (25 Nm).
    - Front exhaust pipe. Tighten the bolts to 32 ft. lbs. (43 Nm) and the nut to 46 ft. lbs. (62 Nm).

2ZZ-GE Engines

1. Before servicing the vehicle, refer to the precautions in the beginning of this section.
2. Release the fuel system pressure.
3. Drain the cooling system.
4. Drain the engine oil.
5. Drain the transaxle fluid.
6. Remove or disconnect the following:
   - Negative battery cable. On vehicles equipped with an air bag, wait at least 90 seconds before proceeding.
   - Battery
   - Hood
   - Undercover
   - Accelerator cable, cable bracket, and clamps.
   - Air cleaner
   - ECM box
   - Cruise control actuator cable
   - Radiator
   - MAP sensor vacuum hose from the gas filter on the intake manifold
   - Power steering air hose from the intake manifold
   - Power steering hose from the air the air pipe
   - Brake booster vacuum hose from the intake manifold
   - A/C idle-up valve
   - A/C idle-up valve hose from the intake manifold
   - A/C idle-up valve hose from the air pipe
   - DLC1 from the bracket
   - Engine wiring harness from the bracket
   - Ground cable from the body and the ground strap from the body
   - Heater hoses from the water outlet
   - Heater hose from the water bypass pipe
   - Fuel inlet hose from the fuel filter and the fuel return hose from the return pipe
   - EVAP hose from the charcoal canister
   - Engine wiring harness from the engine compartment relay box
7. To remove the engine wiring harness from the passenger's compartment, remove or disconnect the following:
   - The scuff plate
   - The cowl side trim
   - The finish panel from the lower instrument panel
   - Remove the front side of the floor carpet
   - The wiring harness from the clamp of the ECM bracket
   - The 3 ECM connectors
   - The circuit opening relay connector
   - The 3 connectors from the connectors on the bracket
   - The A/C amplifier connector
   - The MAP sensor connector
   - The MAP sensor wire from the clamp on the bracket
   - The wire clamp from the bracket
   - The 2 nuts holding the engine wiring harness to the cowl
8. Remove or disconnect the following:
   - Front exhaust pipe
   - Exhaust manifold

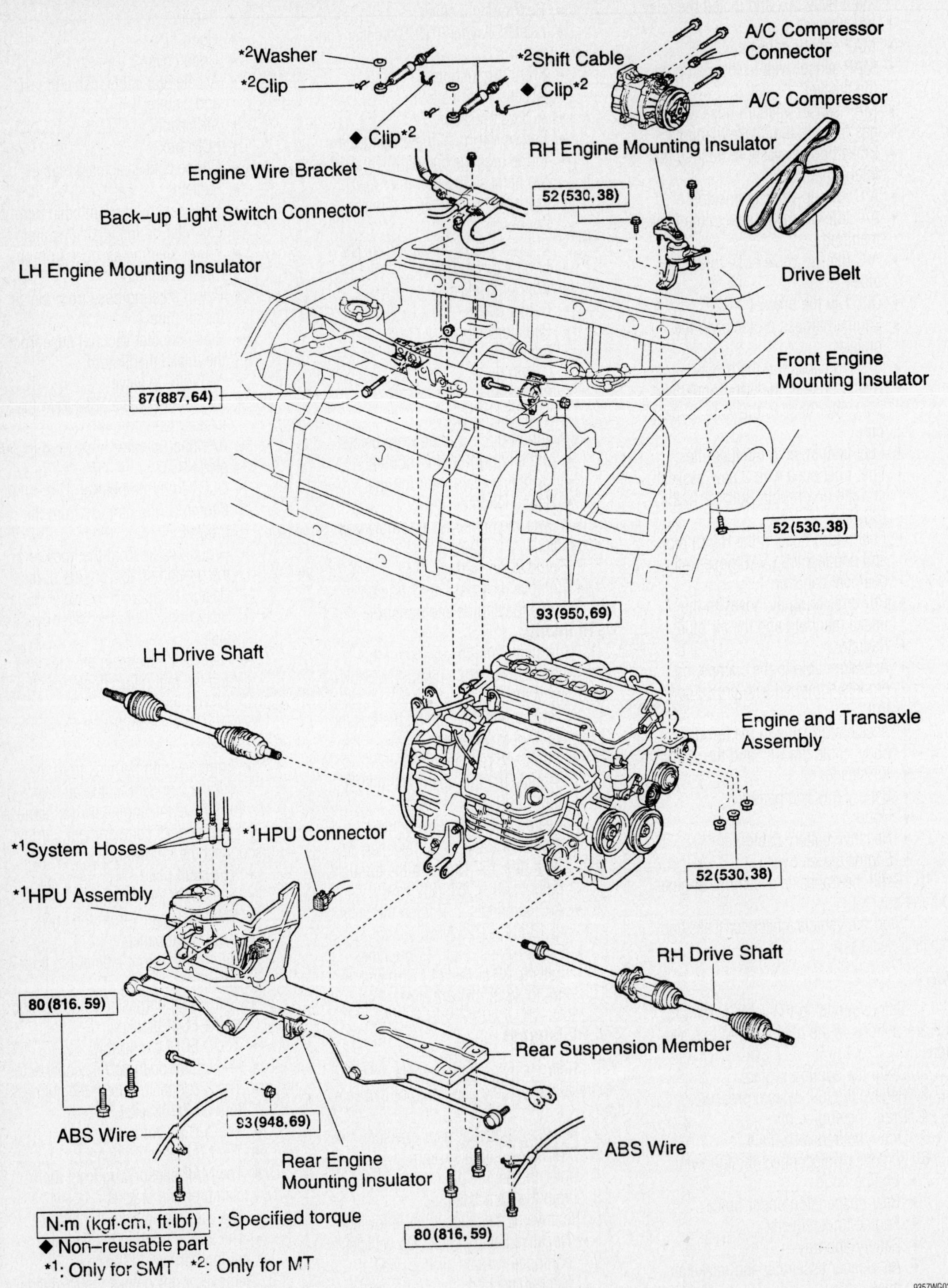
*2Washer
*2Clip
◆ Clip*2
*2Shift Cable
◆ Clip*2
A/C Compressor Connector
A/C Compressor
Engine Wire Bracket
Back-up Light Switch Connector
RH Engine Mounting Insulator
52 (530, 38)
Drive Belt
LH Engine Mounting Insulator
Front Engine Mounting Insulator
87 (887, 64)
52 (530, 38)
93 (950, 69)
LH Drive Shaft
Engine and Transaxle Assembly
*1System Hoses
*1HPU Connector
*1HPU Assembly
52 (530, 38)
RH Drive Shaft
80 (816, 59)
Rear Suspension Member
93 (948, 69)
ABS Wire
ABS Wire
Rear Engine Mounting Insulator
N·m (kgf·cm, ft·lbf) : Specified torque
◆ Non-reusable part
*1: Only for SMT *2: Only for MT
80 (816, 59)
9357WG02

MR2 engine mounting exploded view

- Halfshafts
- Alternator drive belt
- A/C drive belt, compressor connector, and compressor. Do not disconnect the A/C lines
- Remove the drive belt and remove the 4 bolts that secure the power steering pump. Without disconnecting the lines, securely hang the pump out of the way.
- A/C relay box
- On manual transmission, clutch release cylinder from the transaxle
- Transaxle control cable(s)
- On automatic transmission, transaxle control cable from the engine mounting center member.
- Exhaust pipe support bracket

9. To remove the engine mounting center member, remove the following components:
   - The 2 dust covers from the rear side of the member
   - The air conditioning pipe from the bracket
   - The bolt and nut holding the front engine mounting bracket to the mounting insulator
   - The bolt holding the rear engine mounting bracket to the insulator
   - The bolt and 2 nuts holding the rear engine mounting insulator to the front suspension member
   - The 2 bolts and the rear engine mounting bracket, and the center member with the rear mounting insulator
10. Remove or disconnect the remaining components:
    - Attach an engine chain hoist to the engine hangers
    - Left-hand engine mounting bracket from the mounting insulator
    - Through-bolt and the left-hand mounting insulator
    - Ground strap connector
    - Right-hand engine mounting bracket from the mounting insulator
    - Lift the engine and transaxle assembly from the vehicle
    - Transaxle from the engine assembly

To install:

11. Install or connect the following:
    - Transaxle to the engine assembly
    - Engine into the engine compartment
    - Right-hand engine mounting bracket to the mounting insulator. Temporarily install the 3 nuts.
    - Left-hand engine mounting insulator to the body with the through-bolt
    - Left-hand engine mounting bracket to the mounting insulator and install the 2 bolts and nut. Bolts and nut: 47 ft. lbs. (64 Nm).
    - Left-hand engine mounting through-bolt to the body. Bolt: 54 ft. lbs. (73 Nm).
    - Right-hand mounting bracket to the insulator. 12mm nut: 21 ft. lbs. (28 Nm), 14mm nut: 38 ft. lbs. (52 Nm).
    - Engine ground strap connector and remove the engine hoist
12. To install the engine mounting center member, perform the following:
    - Attach the center member together with the rear engine mounting insulator to the front suspension member
    - Temporarily install the 2 bolts and nut holding the center member to the body
    - Install the rear engine mounting bracket. Bolts: 58 ft. lbs. (78 Nm)
    - Temporarily install the bolt and 2 nuts holding the rear engine mounting insulator to the front suspension member
    - Temporarily install the bolt holding the rear engine mounting bracket to the insulator
    - Temporarily install the bolt and nut holding the front engine mounting bracket to the insulator
    - Tighten the 2 bolts holding the center member to the body to 26 ft. lbs. (35 Nm)
    - Tighten the bolt and 2 nuts holding the rear mounting insulator to the front suspension member to 59 ft. lbs. (80 Nm)
    - Tighten the bolt holding the rear engine mounting bracket to the insulator to 65 ft. lbs. (88 Nm)
    - Tighten the bolt and nut holding the front engine mounting bracket to the insulator to 65 ft. lbs. (88 Nm
    - Install the A/C pipe to the bracket and install the 2 dust covers to the center member
    - Exhaust manifold
13. Install or connect the following:
    - Exhaust pipe support bracket. Bolts to 14 ft. lbs. (19 Nm).
    - Transaxle control cable(s)
    - On automatic transmission, transaxle control cable to the engine mounting center member.
    - On manual transmission, clutch release cylinder. Bolts: 108 inch lbs. (12 Nm), then attach the bracket with the bolt
    - A/C relay box to the body
    - Power steering pump. 12mm bolts: 14 ft. lbs. (19 Nm). 14mm bolts: 29 ft. lbs. (39 Nm). Install the drive belt. Adjusting bolt: 29 ft. lbs. (39 Nm).
    - A/C compressor. Bolts: 18 ft. lbs. (25 Nm).
    - A/C drive belt with the adjusting bolt and tighten the idler pulley locknut to 29 ft. lbs. (39 Nm). Connect the connector.
    - Alternator drive belt
    - Halfshafts
    - Front exhaust pipe
14. To install the engine wiring harness in the passenger compartment, perform the following:
    - Push the harness through the cowl panel, install the retainer to the cowl with the 2 nuts and install the wire clamp to the bracket
    - Connect the harness to the clamp on the ECM
    - Connect the 3 ECM connectors and the circuit opening relay connector
    - Connect the 3 connectors to the connectors on the bracket
    - A/C amplifier connector
    - Install the floor carpet, the lower instrument panel finish panel, the cowl side trim panel, and the scuff plate
15. Install or connect the following:
    - Engine wiring harness with the 2 connectors to the engine compartment relay box and install the relay box covers.
    - MAP sensor connector
    - MAP sensor wire to the clamp on the bracket
    - MAP sensor vacuum hose to the gas filter on the intake manifold
    - Brake booster vacuum hose to the intake manifold
    - A/C idle-up valve connector
    - A/C idle-up valve hose to the intake manifold
    - A/C idle-up valve hose to the air pipe
    - DLC1 to the bracket
    - Engine harness protector to the bracket
    - Ground cable and the ground strap
    - Heater hose to the water outlet and the heater hose to the water bypass pipe
    - Fuel inlet hose to the fuel filter
    - Fuel inlet hose with 2 new gaskets and the union bolt. Bolt: 22 ft. lbs. (30 Nm).
    - Fuel return hose to the return pipe

and connect the EVAP hose to the charcoal canister
- Power steering air hoses to the intake manifold and the air pipe.
- Radiator
- On models equipped with cruise control, install the actuator cable to the clamps.
- Accelerator cable to the throttle body, cable bracket, and the clamps.
- Air cleaner
- Battery tray and battery
- Hood
- Refill the transaxle assembly, engine oil and coolant.
- Negative battery cable
- Engine undercover

16. Start the vehicle, check for leaks and repair if necessary.

5S-FE Engines

1. Before servicing the vehicle, refer to the precautions in the beginning of this section.
2. Drain the cooling system.
3. Drain the engine oil.
4. Drain the transaxle fluid.
5. Remove or disconnect the following:
   - Negative battery cable.

➡On vehicles equipped with an air bag, wait at least 90 seconds before proceeding after disconnecting the negative battery cable.

6. Remove or disconnect the following:
   - On the Camry Solara, strut tower brace
   - Battery and battery tray
   - Hood
   - Engine undercover
   - Accelerator cable from the throttle body. With automatic transmission, throttle cable
   - Air cleaner, resonator, and air intake hose
   - On models with cruise control, actuator cover, actuator with the bracket.
   - Ground strap at the battery carrier
   - Radiator and coolant reservoir hose
   - Washer tank, electrical lead and hose
7. To disconnect the wiring harness, tag and disconnect the following:
   - The 5 connectors to the engine relay box
   - The igniter connector
   - The noise filter connector
   - The connector at the left-hand fender apron
   - The 2 ground straps from the left-hand and right-hand fender aprons
   - The DLC1
   - Disconnect the MAP sensor connector
8. Remove or disconnect the following:
   - Dash panel undercover, glove compartment door, glove compartment, cowl harness connectors and the 2 ECM connectors
   - Heater hoses, fuel return hose, and fuel inlet hose
   - With manual transmission, starter and the clutch release cylinder. Don't disconnect the hydraulic line, simply hang the cylinder out of the way
   - Transaxle control cables at the transaxle
   - Tag and disconnect all remaining vacuum hoses and connectors
   - Engine wire from the cowl panel
   - A/C compressor, without disconnecting the refrigerant lines
   - Front exhaust pipe bracket and the front pipe from the exhaust manifold
   - Halfshafts
   - Without disconnecting the hydraulic lines, power steering pump
   - Left engine mounting insulator
   - Right rear engine mounting insulator
   - Front right engine mounting insulator
9. Attach an engine lifting device to the lift hooks. Remove the 3 bolts and disconnect the control rod. Slowly and carefully, lift the engine/transaxle assembly out of the engine compartment.
10. If equipped with automatic transmission, remove the starter. Separate the engine assembly from the transaxle.

To install:

11. Install or connect the following:
    - Engine assembly to the transaxle
    - With automatic transmission, starter
    - Engine control rod. Bolts: 47 ft. lbs. (64 Nm).
    - Front right engine mount. Bolts: 59 ft. lbs. (80 Nm).
    - Rear engine mount. Nuts: 48 ft. lbs. (66 Nm).
    - Left mount. Bolts (3 or 4): 47 ft. lbs. (64 Nm).
    - Power steering pump. Bolts: 31 ft. lbs. (43 Nm). Install the drive belt and connect the 2 air hoses to the air pipe.
    - Halfshafts
    - Front pipe to the manifold. Nuts: 46 ft. lbs. (62 Nm).
    - A/C compressor. Bolts: 20 ft. lbs. (27 Nm).
12. Feed the engine harness through the cowl and reattach the clamp to the cowl. Make the following connections:
    - The 2 ECM connectors
    - The 2 cowl wire connectors
    - Install the glove compartment and door
    - Install the lower instrument panel and the undercover
13. Install or connect the following:
    - Vacuum hoses and the transaxle control cables
    - On manual transmission vehicles, release cylinder and the starter
    - Fuel inlet hose and tighten it to 22 ft. lbs. (29 Nm). Connect the return hose and the 2 heater hoses.
    - Attach the 5 connectors to the relay box
    - The connectors from the left-hand fender apron
    - Install the engine relay box

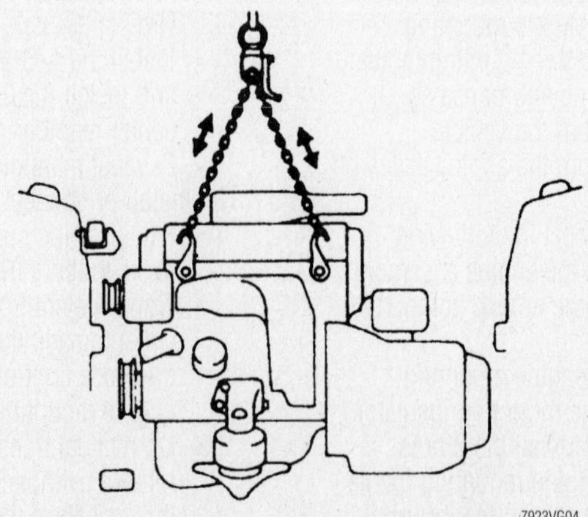

Use a hoist to remove the engine assembly—Camry 5S-FE engine

- The igniter connector
- On California models, the ignition coil connector
- The noise filter connector
- The 2 ground straps from the left-hand and right-hand fender apron
- The DLC1
- The MAP sensor connector
- Washer tank and connect the electrical lead and hose
- Coolant reservoir hose and the radiator
- With cruise control, actuator and bracket. Connect the actuator connector and install the cover.
- Ground strap to the battery carrier.
- Air cleaner assembly
- On California models, air hose to the air cleaner assembly and connect the air intake temperature sensor connector
- With automatic transmission, throttle cable
- Accelerator cable
- Battery tray and battery
- On Camry Solara models, strut tower brace
- Hood
- Oil and coolant
- Negative battery cable
- Undercover

14. Start the vehicle, check for leaks and repair if necessary.

2AZ-FE Engines

1. Before servicing the vehicle, refer to the precautions in the beginning of this section.
2. Relieve the fuel pressure from the fuel lines.
3. Drain the engine coolant from the cooling system.
4. Remove or disconnect the following:
   - Front wheels
   - Engine under covers
   - Front fender apron seal
   - Drain engine oil
   - Drain transmission fluid
   - Engine cover sub-assembly
   - Battery cables and remove battery
   - Air cleaner assembly, brackets and inlets
   - Engine stabilizing control rod
   - Oil cooler inlet and outlet hoses
   - Engine mounting stay and bracket
   - V-belts
   - Steering gear outlet return tube
   - Union to connector tube hose
   - Transmission control cable assembly
   - Heater inlet and outlet hoses
   - Radiator inlet and outlet hoses
   - Fuel pipe sub-assembly
   - Engine wire from ECU and junction box
   - Engine harness from engine compartment junction block
   - Alternator wiring and alternator
   - A/C compressor (Do NOT disconnect hoses)
   - Front exhaust pipe support bracket
   - Front exhaust pipe assembly
   - Front stabilizer link assembly
   - Both front axle hub nuts
   - Both front speed sensors
   - Separate both outer tie rod ends
   - Separate front lower suspension arm sub assembly, both sides
   - Drive plate and torque converter clutch setting bolts (6)
   - Separate steering intermediate shaft assembly
   - Attach engine hoist
   - 4 bolts and 2 nuts from RH & LH frame side rail plate
   - 4 bolts and 2 nuts from RH & LH front suspension member
   - Carefully remove engine assembly

To install:

5. Lower engine and transmission assembly onto the engine compartment
6. Install or connect the following:
   - RH & LH side rail plates; torque large bolt 63 ft. lbs. (85 Nm) and small bolt and nuts 24 ft. lbs. (32 Nm)
   - RH & LH front suspension member brace. Torque large bolt to 63 ft. lbs (85 Nm) and small bolt and nuts 24 ft. lbs. (32 Nm)
   - Steering intermediate shaft assembly
   - Drive plate and torque converter clutch setting bolts. Torque to 30 ft. lbs. (41 Nm).
   - LH & RH lower suspension arm sub-assemblies
   - LH & RH tie rod assemblies
   - LH & RH speed sensors
   - LH & RH front axle hub nuts. Torque to 217 ft. lbs. (294 Nm)
   - LH & RH stabilizer link assemblies
   - Front exhaust pipe assembly
   - Front exhaust support bracket
   - Fuel pipe sub-assembly
   - Transmission control cable assembly
   - A/C compressor assembly
   - Alternator belt adjusting bar and bracket
   - RH engine mounting stay
   - Engine stabilizing control rod
   - Alternator assembly
   - A/C compressor V-belt
   - Inspect drive belt deflector and tensioner
   - Air cleaner assembly, brackets and inlets
   - Verify vacuum hose connections
   - Add automatic transmission fluid
   - Add engine oil
   - Add power steering fluid
   - Inspect automatic transaxle fluid
   - Check for oil, coolant, fuel and exhaust leaks
   - Adjust front wheel alignment
   - Adjust ignition timing and engine idle speed
   - Inspect CO/HC
   - Check ABS sensor signal

1MZ-FE Engines

AVALON AND 2001 CAMRY/SOLARA

1. Before servicing the vehicle, refer to the precautions in the beginning of this section.
2. Properly relieve the fuel system pressure.
3. Drain the cooling system.
4. Drain the engine oil.
5. Drain the transaxle fluid.
6. Remove or disconnect the following:
   - Negative battery cable.

➡On vehicles equipped with an air bag, wait at least 90 seconds before proceeding after disconnecting the negative battery cable.

   - Hood
   - Battery and battery tray
   - Accelerator and throttle cables
   - Cruise control actuator, if equipped
   - Air cleaner assembly, mass air flow meter and air cleaner hose
   - Radiator
   - Engine relay box
   - 2 igniter connectors
   - Noise filter connector
   - Connector from the left-hand fender apron
   - 2 ground straps and any other electrical connections keeping them from being removed.
   - Vacuum hoses from the engine.
   - Fuel inlet and return hoses
   - Heater hoses
   - Transaxle control cable from the transaxle
   - Instrument panel undercover, the lower instrument panel and glove box assembly
   - 3 ECM connectors, the 5 cowl wire connectors, and the cooling fan

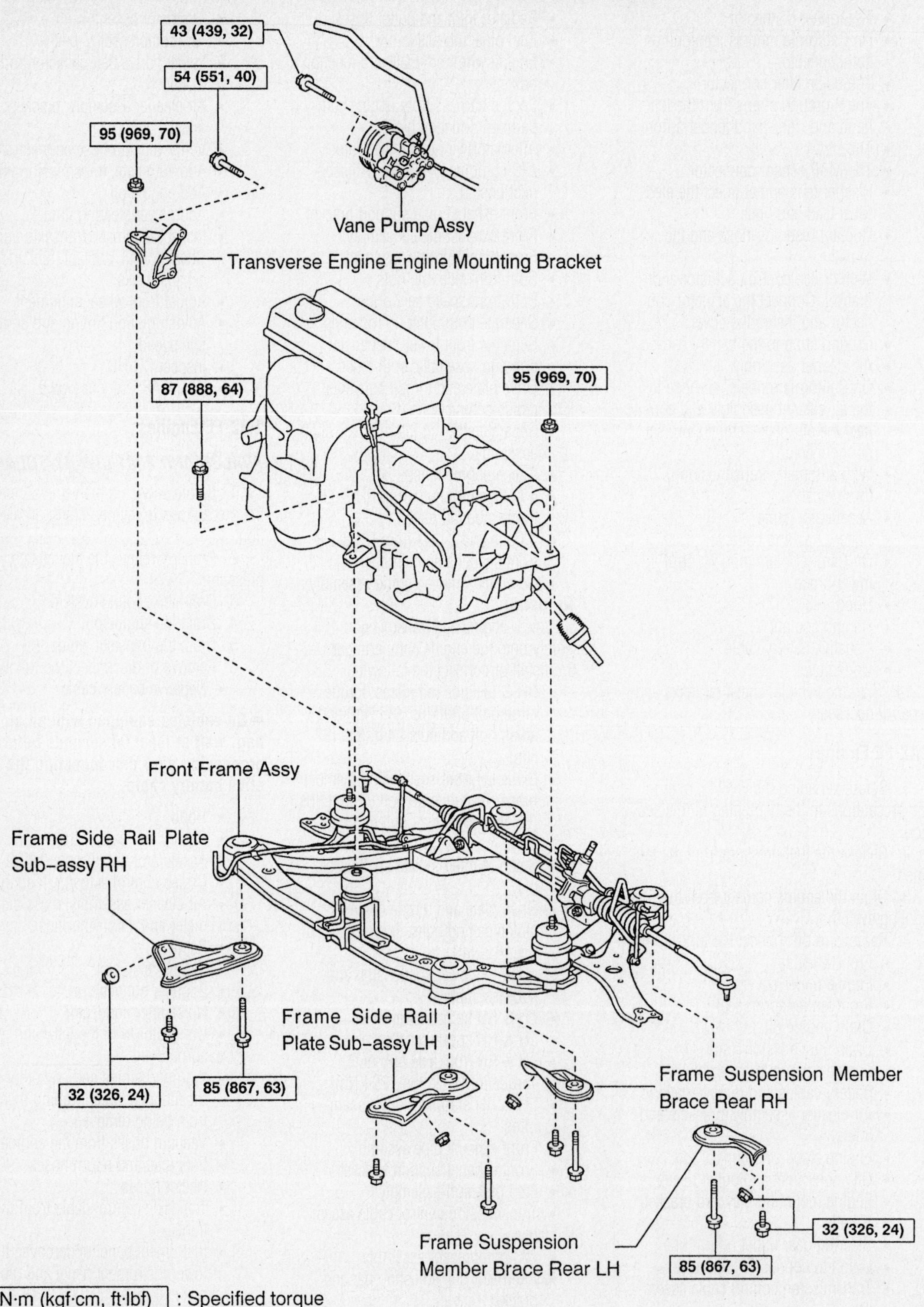
43 (439, 32)
54 (551, 40)
95 (969, 70)
Vane Pump Assy
Transverse Engine Engine Mounting Bracket
87 (888, 64)
95 (969, 70)
Front Frame Assy
Frame Side Rail Plate
Sub-assy RH
Frame Side Rail
Plate Sub-assy LH
Frame Suspension Member
Brace Rear RH
32 (326, 24)
85 (867, 63)
32 (326, 24)
85 (867, 63)
Frame Suspension
Member Brace Rear LH
N·m (kgf·cm, ft·lbf) : Specified torque
67170-TOYO-G03

2AZ-FE engine mounting

ECM connector. Push the engine wire through the cowl panel.
- Front exhaust pipe
- Halfshafts
- Power steering pressure tube
- Power steering pump
- A/C compressor without disconnecting the hoses
- Left-hand engine mounting insulator
- Right-hand engine mounting insulator
- Engine mounting shock absorber
- Front right engine mounting insulator

7. Attach a hoist chain to the engine hangers.
8. Remove or disconnect the following:
- Coolant reservoir hose and reservoir tank
- Right-side engine mounting stay bracket
- Engine control rod and bracket assembly

➡Make certain all wires, connectors and hoses are cleared from the engine.

- Engine/transaxle assembly from the vehicle

To install:

9. Carefully lower the engine position. Keep the engine level while aligning the engine mounts.
10. Install or connect the following:
- Engine control rod and bracket. Tighten to 47 ft. lbs. (64 Nm).
- Right engine mount stay bracket. Tighten to 23 ft. lbs. (31 Nm).
- Engine ground straps.
- Coolant reservoir tank
- Front engine insulator. Tighten to 48 ft. lbs. (66 Nm).
- Engine mounting shock absorber. Tighten to 35 ft. lbs. (48 Nm).
- Left and right engine mounts. Tighten to 48 ft. lbs. (66 Nm).
- Power steering pump and A/C compressor
- Power steering pressure tube
- Halfshafts and front exhaust pipe
- Engine wires and connectors
- Transaxle control cable to the transaxle
- Fuel hoses and heater hoses
- All vacuum hoses, wiring and connectors
- Radiator
- Cruise control actuator
- Throttle cable and accelerator cable
- MAF meter, the air cleaner assembly, and air cleaner hose
- Coolant and engine oil
- Battery tray and battery
- Transaxle fluid to the proper level
- Hood
- Negative battery cable

11. Start the vehicle, check for leaks and repair if necessary.

EXCEPT AVALON AND 2001 CAMRY/SOLARA

1. Before servicing the vehicle, refer to the precautions in the beginning of this section.
2. Relieve the fuel pressure from the fuel lines.
3. Drain the engine coolant from the cooling system.
4. Remove or disconnect the following:
- Front wheels
- Engine under covers
- Front fender apron seal
- Drain engine oil
- Drain automatic transmission fluid
- V-bank cover sub-assembly
- Radiator lower air deflector
- Battery cables and remove battery
- Battery tray
- Air cleaner assembly, brackets and inlets
- Intake air resonator sub-assembly
- A/C compressor V-belt
- Alternator wiring and alternator
- Engine stabilizing control rod
- Engine mounting stay
- Alternator bracket
- Alternator adjusting bar
- A/C compressor (Do NOT disconnect hoses)
- Transmission control cable assembly
- Union to check valve hose
- Fuel vapor feed hose
- Fuel pipe sub-assembly
- Heater inlet and outlet hoses
- Radiator inlet and outlet hoses
- Oil cooler inlet and outlet hoses
- Steering gear outlet return tube
- Glove compartment door assembly
- Engine wire from ECU and junction box
- Engine harness from engine compartment junction block
- Front exhaust pipe support bracket
- Rear exhaust pipe support bracket
- Front exhaust pipe assembly
- Front stabilizer link assembly
- Rear stabilizer link assembly
- Both front axle hub nuts
- Both front speed sensors
- Separate both outer tie rod ends
- Separate front lower suspension arm sub assembly, both sides
- Left and Right drive axles
- Separate steering intermediate shaft assembly
- Attach engine hoist
- 4 bolts and 2 nuts from RH & LH frame side rail plate
- 4 bolts and 2 nuts from RH & LH front suspension member
- Carefully remove engine assembly

To install:

5. Lower engine and transmission assembly onto
the engine compartment
6. Install or connect the following:
- RH & LH side rail plates; torque large bolt 63 ft. lbs. (85 Nm) and small bolt and nuts 24 ft. lbs. (32 Nm)
- RH & LH front suspension member brace. Torque large bolt to 63 ft. lbs (85 Nm) and small bolt and nuts 24 ft. lbs. (32 Nm).
- Steering intermediate shaft assembly
- LH & RH axel shaft assemblies
- LH & RH lower suspension arm sub-assemblies
- LH & RH tie rod assemblies
- LH & RH speed sensors
- LH & RH front axle hub nuts. Torque to 217 ft. lbs. (294 Nm).
- LH & RH stabilizer link assemblies
- Front exhaust pipe assembly
- Rear exhaust support bracket
- Front exhaust support bracket
- Fuel pipe sub-assembly
- Transmission control cable assembly
- A/C compressor assembly
- Alternator belt adjusting bar and bracket
- RH engine mounting stay
- Engine stabilizing control rod
- Alternator assembly
- A/C compressor V-belt
- Inspect drive belt deflector and tensioner
- Intake air resonator. Torque to 44 inch lbs. (5 Nm)
- Air cleaner assembly, brackets and inlets
- Verify vacuum hose connections
- Add automatic transmission fluid
- Add engine oil
- Add power steering fluid
- Inspect automatic transaxle fluid
- Check for oil, coolant, fuel and exhaust leaks
- Adjust front wheel alignment
- Adjust ignition timing and engine idle speed
- Inspect CO/HC
- Check ABS sensor signal

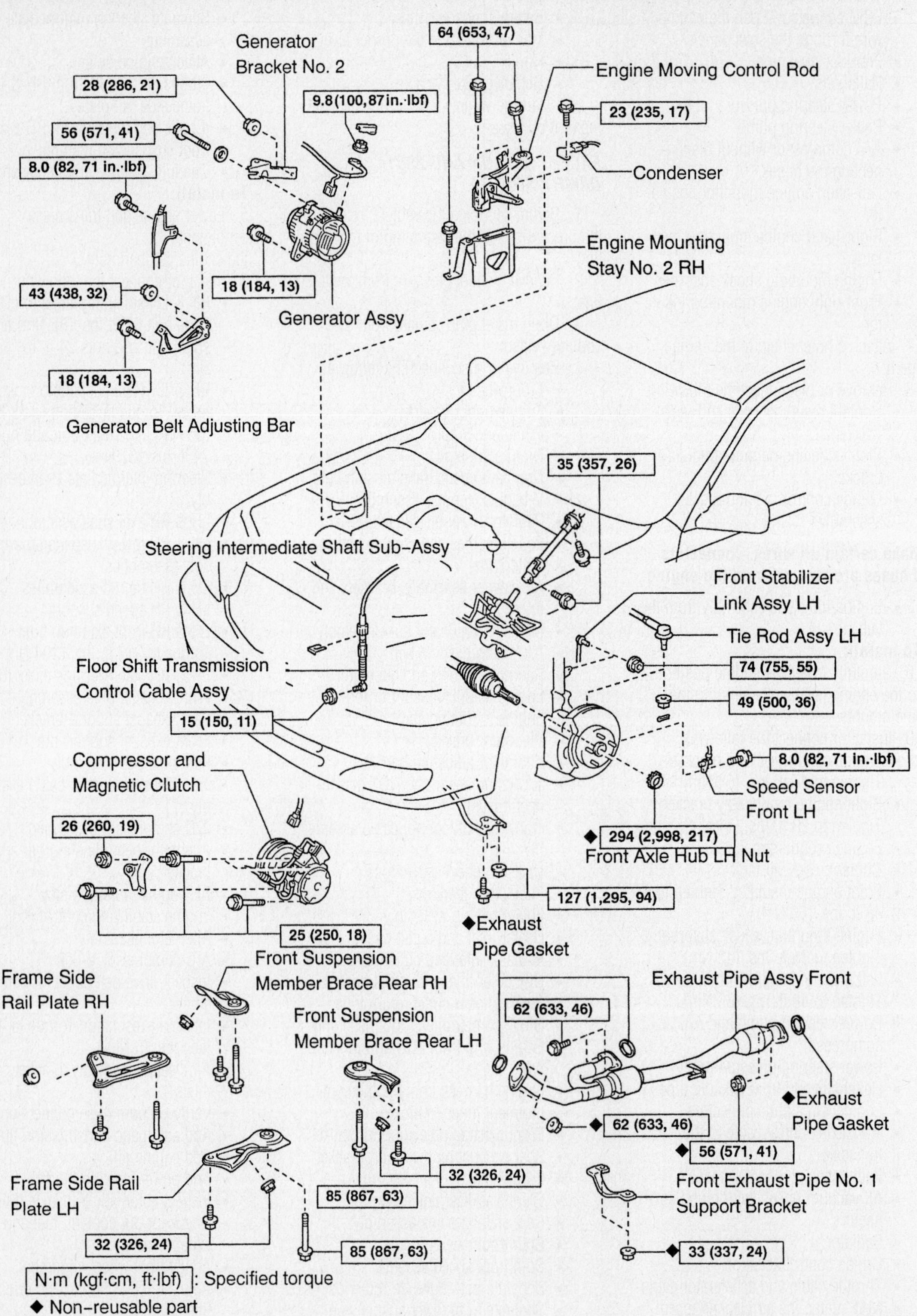
Generator Bracket No. 2
64 (653, 47)
28 (286, 21)
9.8(100,87in.·lbf)
Engine Moving Control Rod
23 (235, 17)
56 (571, 41)
8.0 (82, 71 in.·lbf)
Condenser
Engine Mounting Stay No. 2 RH
18 (184, 13)
43 (438, 32)
Generator Assy
18 (184, 13)
Generator Belt Adjusting Bar
35 (357, 26)
Steering Intermediate Shaft Sub-Assy
Front Stabilizer Link Assy LH
Tie Rod Assy LH
74 (755, 55)
Floor Shift Transmission Control Cable Assy
49 (500, 36)
15 (150, 11)
8.0 (82, 71 in.·lbf)
Compressor and Magnetic Clutch
Speed Sensor Front LH
26 (260, 19)
◆ 294 (2,998, 217)
Front Axle Hub LH Nut
127 (1,295, 94)
◆Exhaust Pipe Gasket
25 (250, 18)
Front Suspension Member Brace Rear RH
Frame Side Rail Plate RH
Exhaust Pipe Assy Front
Front Suspension Member Brace Rear LH
62 (633, 46)
◆Exhaust Pipe Gasket
◆ 62 (633, 46)
◆ 56 (571, 41)
32 (326, 24)
Frame Side Rail Plate LH
85 (867, 63)
Front Exhaust Pipe No. 1 Support Bracket
32 (326, 24)
85 (867, 63)
◆ 33 (337, 24)
N·m (kgf·cm, ft·lbf) : Specified torque
◆ Non-reusable part
67170-TOYO-G04

2004 1MZ-FE and 3MZ-FE engine mounting accessories

3MZ-FE Engines

1. Before servicing the vehicle, refer to the precautions in the beginning of this section.
2. Relieve the fuel pressure from the fuel lines.
3. Drain the engine coolant from the cooling system.
4. Remove or disconnect the following:
   - Front wheels
   - Engine under covers
   - Front fender apron seal
   - Drain engine oil
   - Drain automatic transmission fluid
   - V-bank cover sub-assembly
   - Radiator lower air deflector
   - Battery cables and remove battery
   - Battery tray
   - Air cleaner assembly, brackets and inlets
   - Intake air resonator sub-assembly
   - A/C compressor V-belt
   - Alternator wiring and alternator
   - Engine stabilizing control rod

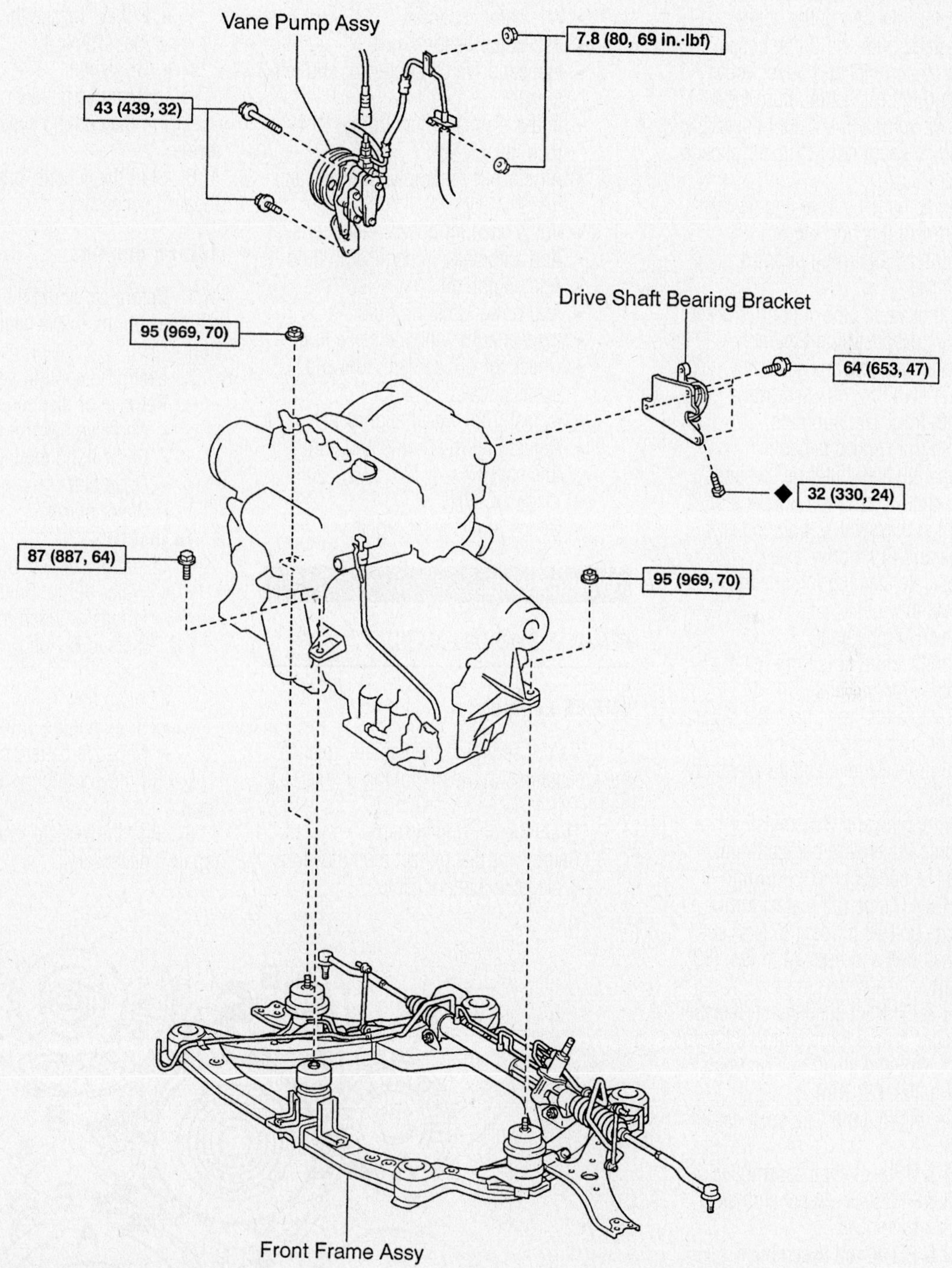

2004 1MZ-FE and 3MZ-FE engine mounting

- Engine mounting stay
- Alternator bracket
- Alternator adjusting bar
- A/C compressor (Do NOT disconnect hoses)
- Transmission control cable assembly
- Union to check valve hose
- Fuel vapor feed hose
- Fuel pipe sub-assembly
- Heater inlet and outlet hoses
- Radiator inlet and outlet hoses
- Oil cooler inlet and outlet hoses
- Steering gear outlet return tube
- Glove compartment door assembly
- Engine wire from ECU and junction box
- Engine harness from engine compartment junction block
- Front exhaust pipe support bracket
- Rear exhaust pipe support bracket
- Front exhaust pipe assembly
- Front stabilizer link assembly
- Rear stabilizer link assembly
- Both front axle hub nuts
- Both front speed sensors
- Separate both outer tie rod ends
- Separate front lower suspension arm sub assembly, both sides
- Left and Right drive axles
- Separate steering intermediate shaft assembly
- Attach engine hoist
- 4 bolts and 2 nuts from RH & LH frame side rail plate
- 4 bolts and 2 nuts from RH & LH front suspension member
- Carefully remove engine assembly

To install:

5. Lower engine and transmission assembly onto the engine compartment
6. Install or connect the following:
   - RH & LH side rail plates; torque large bolt 63 ft. lbs. (85 Nm) and small bolt and nuts 24 ft. lbs. (32 Nm)
   - RH & LH front suspension member brace. Torque large bolt to 63 ft. lbs (85 Nm) and small bolt and nuts 24 ft. lbs. (32 Nm).
   - Steering intermediate shaft assembly
   - LH & RH axel shaft assemblies
   - LH & RH lower suspension arm sub-assemblies
   - LH & RH tie rod assemblies
   - LH & RH speed sensors
   - LH & RH front axle hub nuts. Torque to 217 ft. lbs. (294 Nm).
   - LH & RH stabilizer link assemblies
   - Front exhaust pipe assembly
   - Rear exhaust support bracket
   - Front exhaust support bracket
   - Fuel pipe sub-assembly
   - Transmission control cable assembly
   - A/C compressor assembly
   - Alternator belt adjusting bar and bracket
   - RH engine mounting stay
   - Engine stabilizing control rod
   - Alternator assembly
   - A/C compressor V-belt
   - Inspect drive belt deflector and tensioner
   - Intake air resonator. Torque to 44 inch lbs. (5 Nm).
   - Air cleaner assembly, brackets and inlets
   - Verify vacuum hose connections
   - Add automatic transmission fluid
   - Add engine oil
   - Add power steering fluid
   - Inspect automatic transaxle fluid
   - Check for oil, coolant, fuel and exhaust leaks
   - Adjust front wheel alignment
   - Adjust ignition timing and engine idle speed
   - Inspect CO/HC
   - Check ABS sensor signal

Water Pump

REMOVAL & INSTALLATION

1NZ-FE Engines

1. Before servicing the vehicle, refer to the precautions in the beginning of this section.
2. Drain the cooling system.
3. Remove or disconnect the following:
   - Negative battery cable
   - Accessory drive belt
   - Water pump pulley, using a holding tool
   - Water pump and gasket (3 bolts; 2 nuts)

To install:

4. Install or connect:
   - Water pump, with a new gasket. Torque the nuts and bolts to 96 inch lbs. (11 Nm).
   - Pulley. Torque the bolts to 11 ft. lbs. (15 Nm).
   - Drive belt
   - Negative battery cable
5. Fill the cooling system to the proper level.
6. Start the vehicle, check for leaks and repair if necessary.

1ZZ-FE Engines

1. Before servicing the vehicle, refer to the precautions in the beginning of this section.
2. Drain the cooling system.
3. Remove or disconnect the following:
   - Negative battery cable
   - Right-hand engine under cover
   - Drive belt
   - Water pump

To install:

4. Install or connect the following:
   - Water pump. Bolts marked **A** (short): 80 inch lbs. (9 Nm). Bolts marked **B** (long): 96 inch lbs. (11 Nm).
   - Drive belt
   - Right engine under cover
   - Negative battery cable
5. Fill the cooling system to the proper level.
6. Start the vehicle, check for leaks and repair if necessary.

Water pump bolt identification—1.8L (1ZZ-FE) engine

2ZZ-GE Engines

1. Before servicing the vehicle, refer to the precautions in the beginning of this section.
2. Drain the cooling system.
3. Remove or disconnect the following:
   - Negative battery cable
   - Right-hand engine under cover
   - Drive belt
   - Water pump pulley
   - Water pump and O-ring

To install:

4. Install or connect the following:
   - Water pump with new O-ring. Bolts: 80 inch lbs. (9 Nm).
   - Water pump pulley. Bolts: 11 ft. lbs. (15 Nm).
   - Drive belt
   - Right engine under cover
   - Negative battery cable
5. Fill the cooling system to the proper level.
6. Start the vehicle, check for leaks and repair if necessary.

5S-FE Engines

1. Before servicing the vehicle, refer to the precautions in the beginning of this section.
2. Drain the cooling system.
3. Remove or disconnect the following:
   - Negative battery cable.

➡On vehicles equipped with an air bag, wait at least 90 seconds before proceeding after disconnecting the negative battery cable.

   - Right engine undercover
   - Lower radiator hose from the water outlet
   - Timing belt, timing belt tension spring, and the No. 2 idler pulley
   - Alternator, drive belt and the adjusting bar if necessary
   - 2 nuts holding the water pump to the water bypass pipe and remove the 3 bolts in sequence.
   - Water pump cover assembly
   - Gasket and 2 O-rings from the water pump and the bypass pipe.
   - Water pump from the water pump cover by removing the 3 bolts in sequence.

To install:

4. Install or connect the following:
   - Water pump to the water pump cover. Bolts: 78 inch lbs. (9 Nm) in proper sequence.
   - Water pump cover to the water bypass pipe, but do not install the nuts yet.
   - Water pump and tighten the 3 bolts in sequence. Bolts: 78 inch lbs. (9 Nm). Nuts: 82 inch lbs. (9 Nm).
   - Alternator drive belt adjusting bar. Bolt: 13 ft. lbs. (18 Nm).
   - No. 2 idler pulley and the timing belt tension spring
   - Lower radiator hose
   - Timing belt
   - Right engine undercover
   - Negative battery cable
5. Fill the cooling system to the proper level.
6. Start the vehicle, check for leaks and repair if necessary.

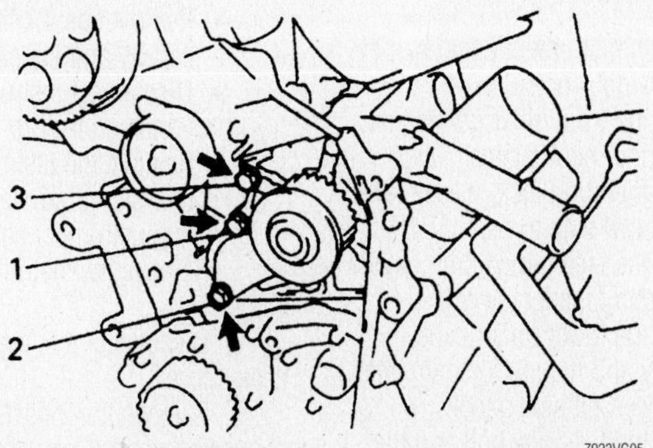

Install the 3 water pump bolts in this sequence—2.2L (5S-FE) engine

2AZ-FE Engines

1. Before servicing the vehicle, refer to the precautions in the beginning of this section.
2. Disconnect the negative battery cable.
3. Drain the cooling system.
4. Remove or disconnect the following:
   - RH front wheel
   - RH fender apron seal
   - Engine stabilizer control rod
   - RH engine stay and bracket
   - Alternator
   - Water pump pulley
   - Water pump assembly

2AZ-FE water pump mounting bolts

To install:

5. Install or connect the following:
   - Water pump assembly with new gasket. Torque to 80 inch lbs. (9.0 Nm)
   - Water pump pulley
   - Alternator
   - Engine mounting bracket
   - Engine stabilizing control rod
   - Engine mounting stay
   - Right front apron seal and wheel
   - Add coolant and check for leaks

1MZ-FE and 3MZ-FE Engines

AVALON AND 2001 CAMRY/SOLARA

1. Before servicing the vehicle, refer to the precautions in the beginning of this section.
2. Drain the cooling system.
3. Remove or disconnect the following:
   - Negative battery cable.

➡On vehicles equipped with an air bag, wait at least 90 seconds before proceeding after disconnecting the negative battery cable.

   - Timing belt
   - No. 2 idler pulley
   - 3 clamps and engine wire from the rear timing belt cover
   - Rear timing belt cover
   - Water pump

To install:

4. Install or connect the following:
   - Liquid sealer to the gasket, water pump and engine block.
   - Water pump. Bolts and nuts: 53 inch lbs. (6 Nm).
   - Rear timing belt cover. Bolts: 74 inch lbs. (9 Nm).
   - Engine wire with the 3 clamps to the rear timing belt cover.
   - No. 2 idler pulley. Bolt: 32 ft. lbs. (43 Nm).
   - With the flange side **outward**, right-hand camshaft pulley. Align the knock pin hole on the camshaft pulley with the knock pin on the camshaft. Bolt: 65 ft. lbs. (88 Nm).

- With the flange side **inward**, left-hand camshaft pulley. Align the knock pin hole on the camshaft pulley with the knock pin on the camshaft. Bolt: 94 ft. lbs. (125 Nm).
- Timing belt
- Negative battery cable

5. Fill the cooling system to the proper level.
6. Start the vehicle, check for leaks and repair if necessary.

EXCEPT AVALON AND 2001 CAMRY/SOLARA

1. Before servicing the vehicle, refer to the precautions in the beginning of this section.
2. Disconnect the negative battery cable.
3. Drain the cooling system.
4. Remove or disconnect the following:
   - RH front wheel
   - RH fender apron seal
   - A/C drive belt
   - PS drive belt
   - Engine stabilizer control rod
   - RH engine stay
   - Alternator bracket #2
   - Crankshaft pulley
   - Both timing belt covers
   - RH engine mounting bracket
   - Timing belt cover 1 and 2
   - Timing belt, guide and idler pulley sub-assembly #1
   - Camshaft timing pulleys and idler pulley sub-assembly #2
   - Timing belt cover #3
   - Water pump assembly

To install:

5. Install or connect the following:
   - Water pump assembly with new gasket. Torque to 71 inch lbs. (8 Nm).
   - Timing belt idler #1. Torque to 25 ft. lbs (24 Nm).
   - Timing belt cover #3
   - Camshaft timing pulleys
   - Timing belt idler sub-assembly
   - Timing belt, tensioner assembly and guide
   - Engine mounting bracket
   - Upper and lower timing belts covers
   - Crankshaft pulley
   - Alternator bracket
   - Engine mounting stay
   - Engine stabilizing control rod
   - PS pump
   - A/C drive belt
   - Inspect drive belt tension
   - Right front wheel
   - Add coolant and check for leaks

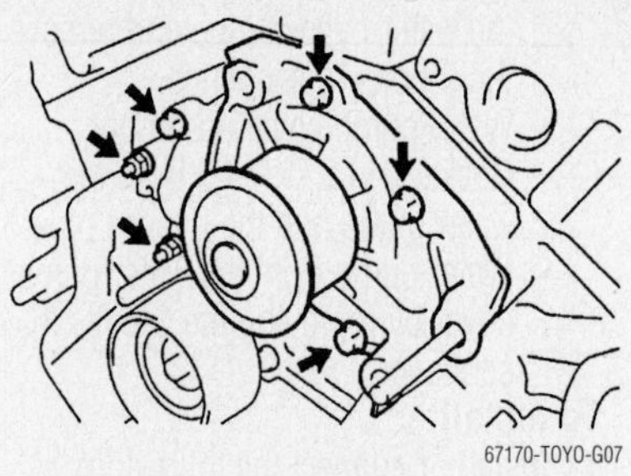

1MZ-FE and 3MZ-FE water pump mounting bolts

Heater Core

REMOVAL & INSTALLATION

Avalon

1. Disconnect the negative battery cable.

⁂ CAUTION

After disconnecting the negative battery cable, wait for at least 1½ minutes for the SRS to deplete its energy.

2. Drain the cooling system into a clean container for reuse.
3. Remove the air bag module and the steering wheel by performing the following procedure:
   - Place the front wheels in the straight-ahead position.
   - At both sides of the steering wheel, remove the screw covers.
   - Using a Torx socket, loosen the 2 air bag module-to-steering wheel Torx screws until the circumference ring catches on the screw case.
   - Carefully, remove the air bag module and disconnect the electrical connector.
   - Remove the steering wheel nut.
   - Using a steering wheel puller, press the steering wheel from the steering column.
4. Remove the instrument panel by performing the following procedure:
   - Remove the front pillar garnishes and the door scuff plates.
   - Remove the hood lock release lever and the cowl side trims.
   - Remove the steering column covers and the combination switch.
   - Remove the lower finish panel assembly and the instrument panel finish lower left side panel.
   - Remove the fuse box bolt and the No. 2 heater-to-register duct.
   - Remove the parking brake release lever and the No. 2 undercover.
   - Remove the lower No. 2 finish panel.
   - If equipped with a column shifter, disconnect the shift control cable from the shift lever housing; then, disconnect the shift control cable from the steering column cable bracket.
   - Matchmark the steering column shaft and the control valve shaft.
   - Remove the steering column shaft-to-intermediate shaft bolt.
   - Remove the steering column-to-instrument panel nuts and remove the steering column assembly.
   - Inside the glove compartment, pry out the glove compartment door finish plate.
   - Pull out the air bag electrical connector and disconnect it.
   - Remove the 3 glove compartment door-to-instrument panel nuts and the door.
   - Remove the 4 glove compartment-to-instrument panel screws and the glove compartment; then, disconnect the glove box light connector.
   - Remove the passenger's side air bag module-to-instrument panel 2 bolts and 4 nuts. Carefully, remove the air bag module from the instrument panel.
   - Remove the center cluster finish panel and the radio.
   - Remove the heater control assembly.
   - If equipped with a floor shifter, remove the upper console panel, the rear console box and the front console box.
   - If equipped with a column shifter, remove the finish panel.
   - Pry out the cluster finish panel.
   - Remove the 6 cluster finish panel screws, the cluster finish panel assembly.
   - Remove the 4 combination meter screws and the combination meter.
   - Disconnect instrument panel electrical connectors.
   - Remove the instrument panel-to-chassis nuts/bolts and the remove the instrument panel.
5. Remove the instrument panels No. 2 brace.
6. Disconnect the heater hoses from the heater core.
7. Remove the 2 heater pipes-to-heater core screws and clips; then, disconnect the heater pipes from the heater core.

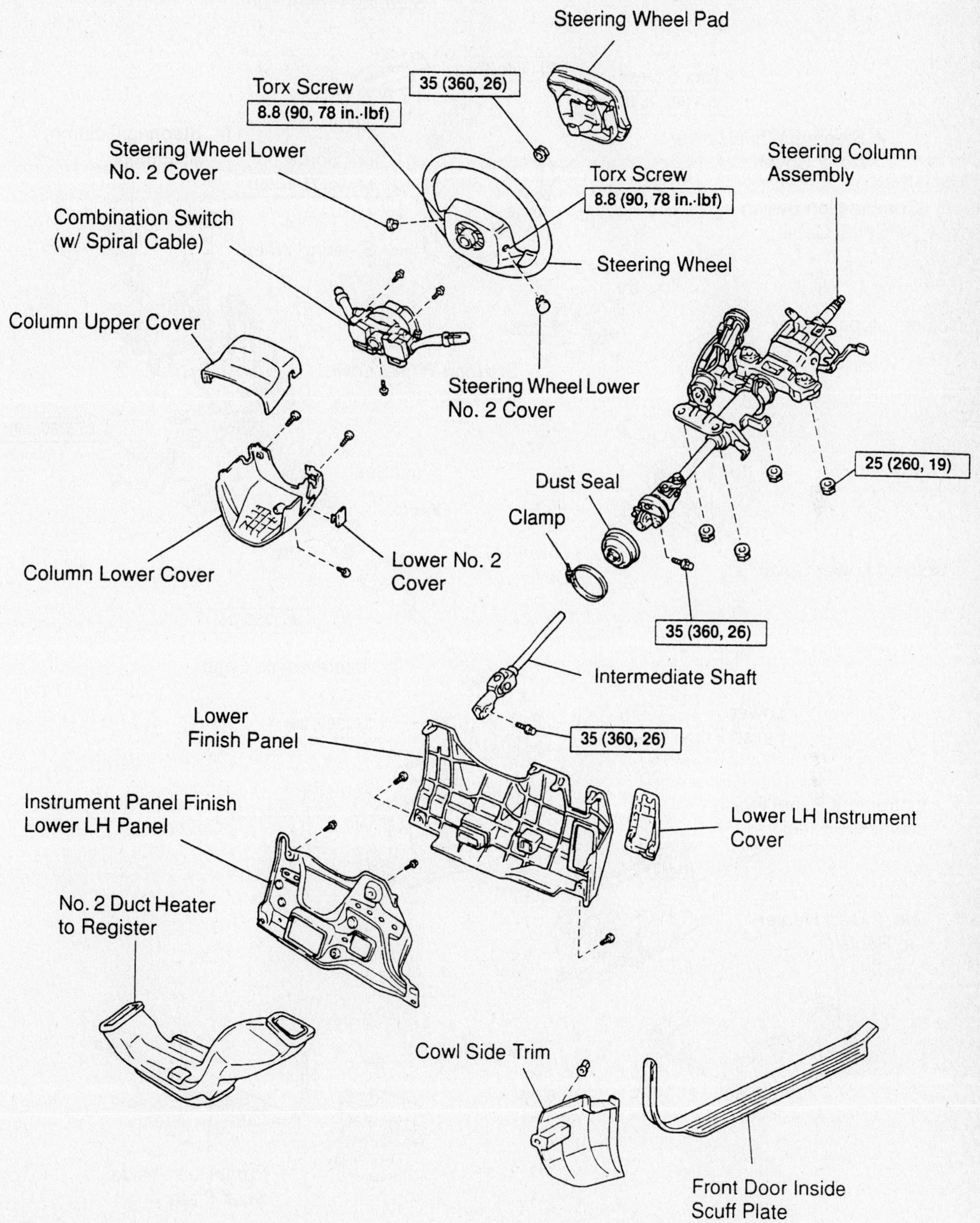

Exploded view of the air bag module, the steering wheel, the floor shift steering column and related components—Avalon

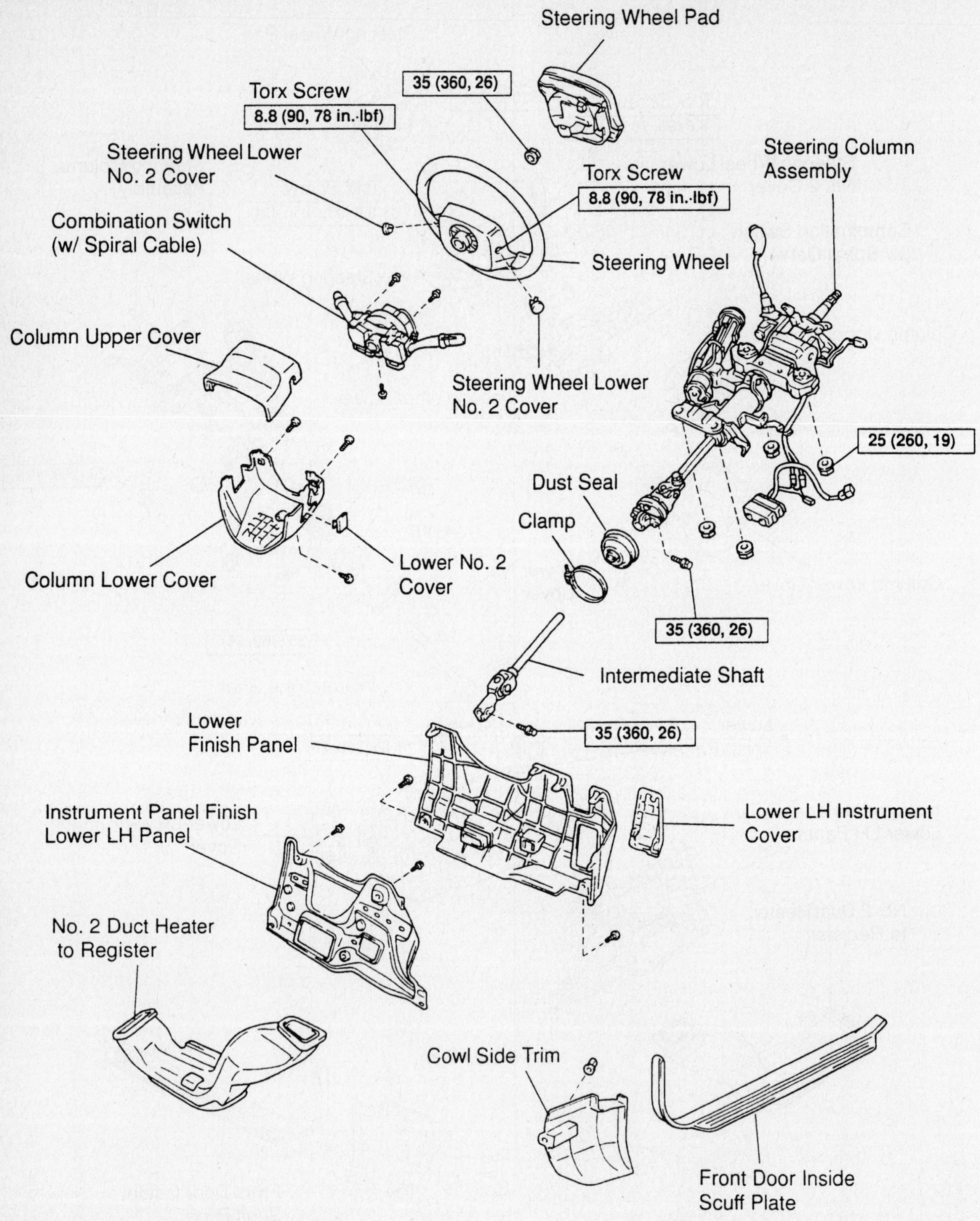

Exploded view of the air bag module, the steering wheel, the column shift steering column and related components—Avalon

8. Remove the heater core O-rings and discard them.

9. Remove the heater core from the heater housing.

To install:

10. Install the heater core to the heater housing.

11. Install new heater core O-rings.

12. Install the heater pipes to the heater core; then, the 2 heater pipes-to-heater core screws and clips.

13. Install the heater hoses to the heater core.

14. Install the instrument panels No. 2 brace.

15. Install the instrument panel by performing the following procedure:

- Install the instrument panel and the instrument panel-to-chassis nuts/bolts.
- Install instrument panel electrical connectors.
- Install the combination meter and the 4 combination meter screws.
- Install the cluster finish panel, the 6 cluster finish panel assembly screws.
- Install the cluster finish panel.
- If equipped with a column shifter, install the finish panel.
- If equipped with a floor shifter, install the upper console panel, the rear console box and the front console box.
- Install the heater control assembly.
- Install the center cluster finish panel and the radio.
- Carefully, install the air bag module to the instrument panel; then, install the passenger's side air bag module-to-instrument panel 2 bolts and 4 nuts.
- Install the glove compartment and the 4 glove compartment-to-instrument panel screws; then, install the glove box light connector.
- Install the glove compartment door and the 3 door-to-instrument panel nuts.
- Connect the air bag electrical connector.
- Inside the glove compartment, install the glove compartment door finish plate.
- Install the steering column assembly and the steering column-to-instrument panel nuts; then, torque the nuts to 19 ft. lbs. (25 Nm).
- Align the matchmarks and install the steering column shaft-to-intermediate shaft bolt.
- If equipped with a column shifter, install the shift control cable to the shift lever housing; then, install the shift control cable to the steering column cable bracket.
- Install the lower No. 2 finish panel.
- Install the parking brake release lever and the No. 2 undercover.
- Install the fuse box bolt and the No. 2 heater-to-register duct.
- Install the lower finish panel assembly and the instrument panel finish lower left side panel.
- Install the steering column covers and the combination switch.
- Install the hood lock release lever and the cowl side trims.
- Install the front pillar garnishes and the door scuff plates.

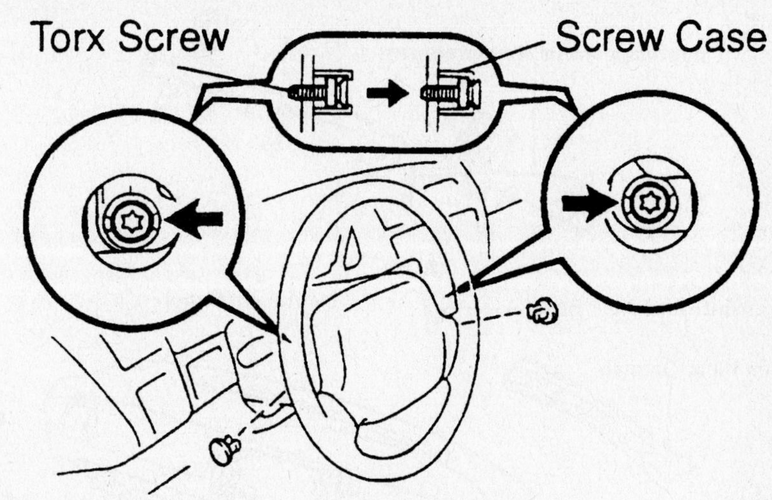

Releasing the air bag module-to-steering wheel screws—Avalon

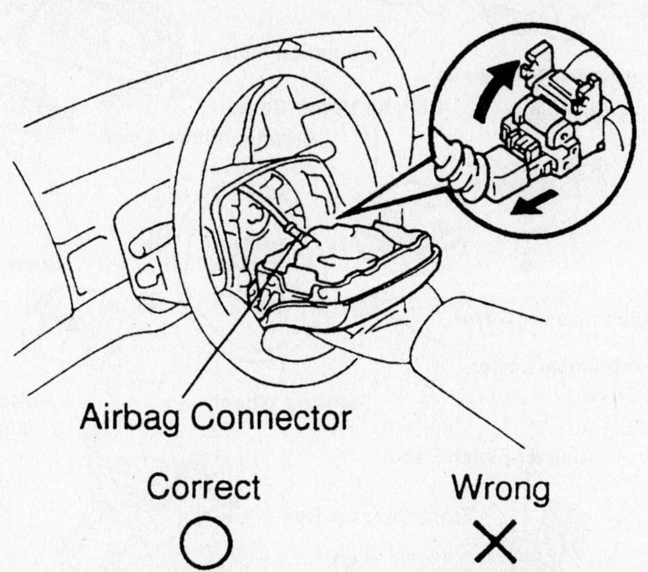

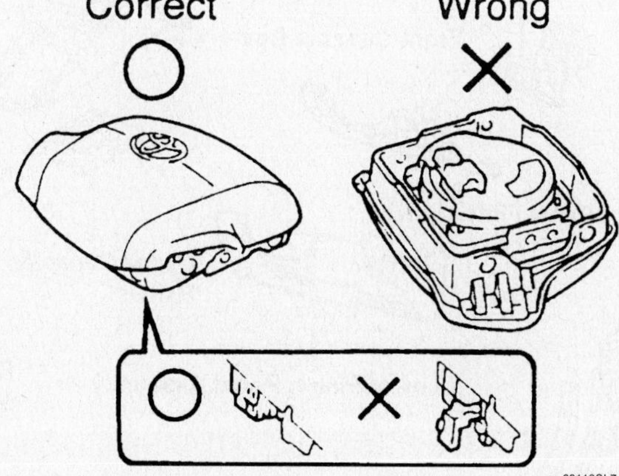

Disconnecting and positioning the air bag module—Avalon

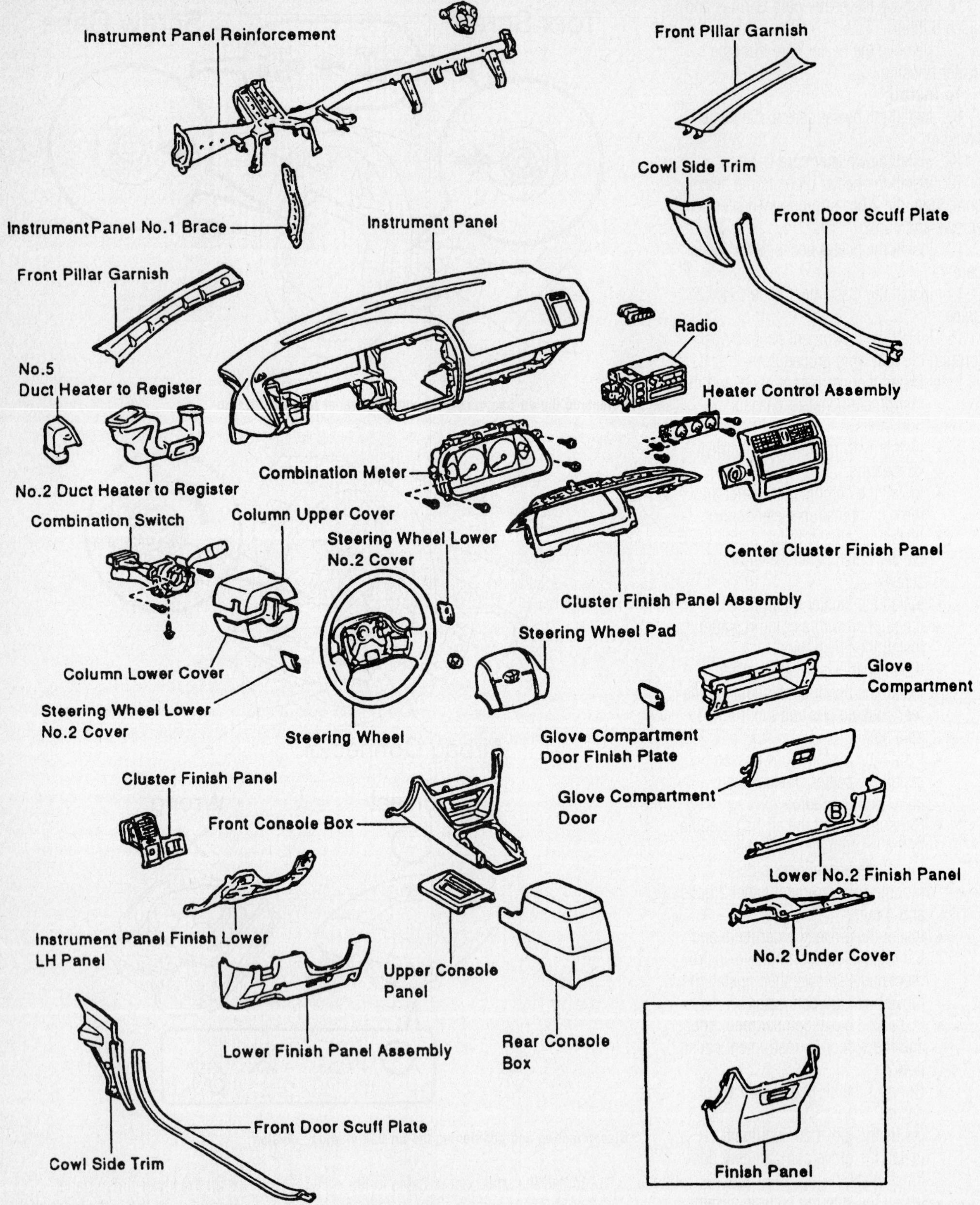
Instrument Panel Reinforcement
Front Pillar Garnish
Cowl Side Trim
InstrumentPanel No.1 Brace
Instrument Panel
Front Door Scuff Plate
Front Pillar Garnish
Radio
No.5
Duct Heater to Register
Heater Control Assembly
Combination Meter
No.2 Duct Heater to Register
Combination Switch
Column Upper Cover
Steering Wheel Lower
No.2 Cover
Center Cluster Finish Panel
Cluster Finish Panel Assembly
Steering Wheel Pad
Column Lower Cover
Glove
Compartment
Steering Wheel Lower
No.2 Cover
Steering Wheel
Glove Compartment
Door Finish Plate
Cluster Finish Panel
Glove Compartment
Door
Front Console Box
B
Lower No.2 Finish Panel
Instrument Panel Finish Lower
LH Panel
No.2 Under Cover
Upper Console
Panel
Lower Finish Panel Assembly
Rear Console
Box
Front Door Scuff Plate
Cowl Side Trim
Finish Panel
93112GA5

Exploded view of the instrument panel and related components—Avalon

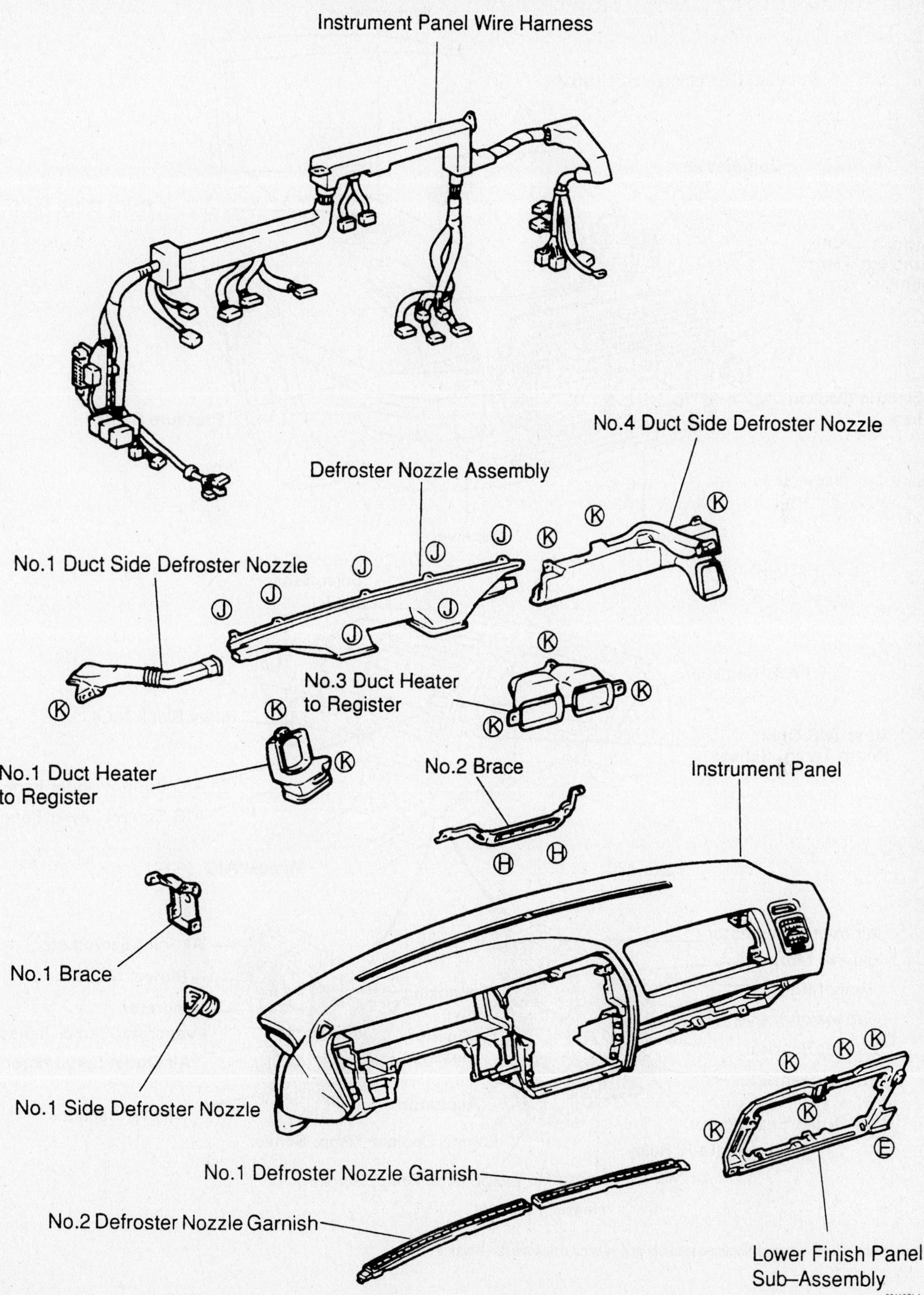

Exploded view of the wiring harness, ventilation system and related components—Avalon

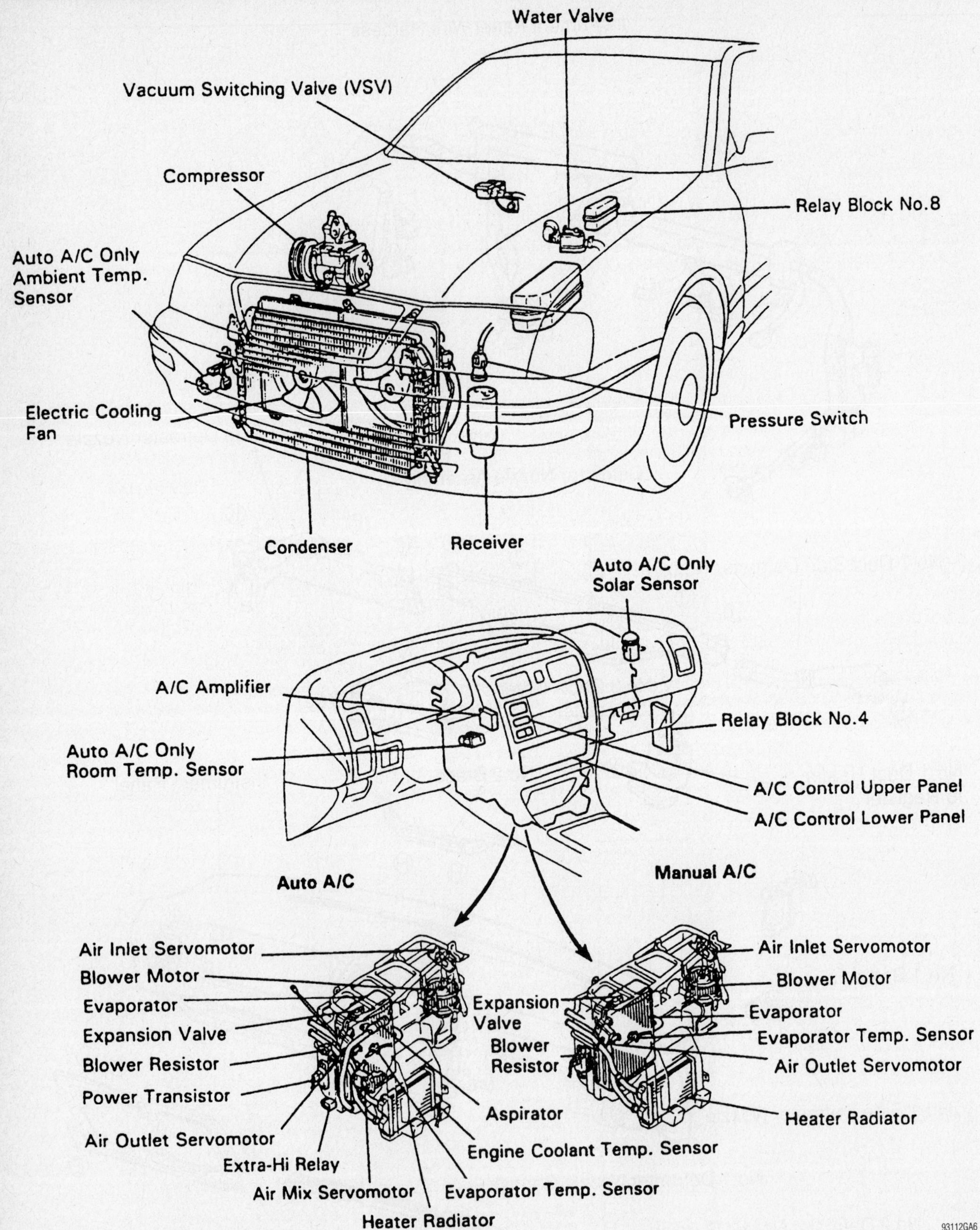
Water Valve
Vacuum Switching Valve (VSV)
Compressor
Relay Block No.8
Auto A/C Only
Ambient Temp.
Sensor
Electric Cooling
Fan
Pressure Switch
Condenser
Receiver
Auto A/C Only
Solar Sensor
A/C Amplifier
Relay Block No.4
Auto A/C Only
Room Temp. Sensor
A/C Control Upper Panel
A/C Control Lower Panel
Auto A/C
Manual A/C
Air Inlet Servomotor
Blower Motor
Evaporator
Expansion Valve
Blower Resistor
Power Transistor
Air Outlet Servomotor
Extra-Hi Relay
Air Mix Servomotor
Heater Radiator
Evaporator Temp. Sensor
Engine Coolant Temp. Sensor
Aspirator
Expansion
Valve
Blower
Resistor
Air Inlet Servomotor
Blower Motor
Evaporator
Evaporator Temp. Sensor
Air Outlet Servomotor
Heater Radiator
93112GA6

View of the heater/air conditioning assembly and related components—Avalon

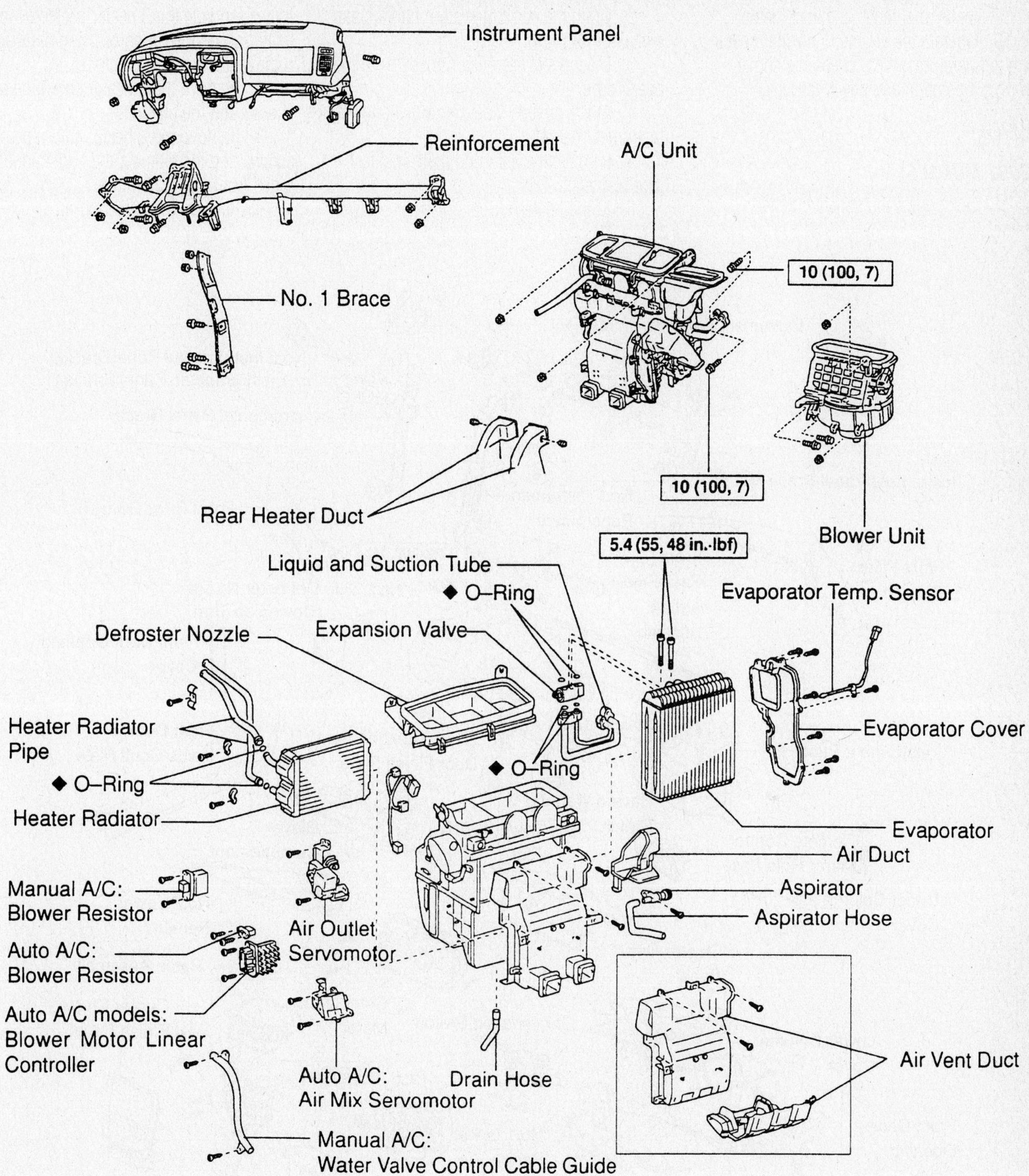

Exploded view of the evaporator housing, heater housing, heater core and related components—Avalon

16. Install the air bag module and the steering wheel by performing the following procedure:
    - Align the matchmarks and install the steering wheel to the steering column.
    - Install the steering wheel nut and torque to 26 ft. lbs. (35 Nm).
    - Carefully, connect the electrical connector and install the air bag module.
    - Using a Torx socket, tighten the 2 air bag module-to-steering wheel Torx screws to 78 inch lbs. (8.8 Nm).
    - At both sides of the steering wheel, install the screw covers.
17. Refill the cooling system.

18. Install the negative battery cable.
19. Operate the engine to normal operating temperatures; then, check the climate control operation and check for leaks.

Camry

2001 MODELS

1. Disconnect the negative battery cable.
2. Drain the cooling system into a clean container for reuse.
3. Disconnect the heater hoses from the heater core.
4. At the driver's side, remove the lower instrument panel.
5. Remove the left hand instrument lower panel.
6. At the heater/air conditioning housing, disengage the 3 heater protector-to-heater/air conditioning housing clips and remove the heater protector.
7. Remove the 3 heater core pipe clamp screws and the clamps.
8. Remove the 2 heater core pipe clamp screws and the clamps; then, disconnect the pipes from the heater core.
9. Remove the heater core.

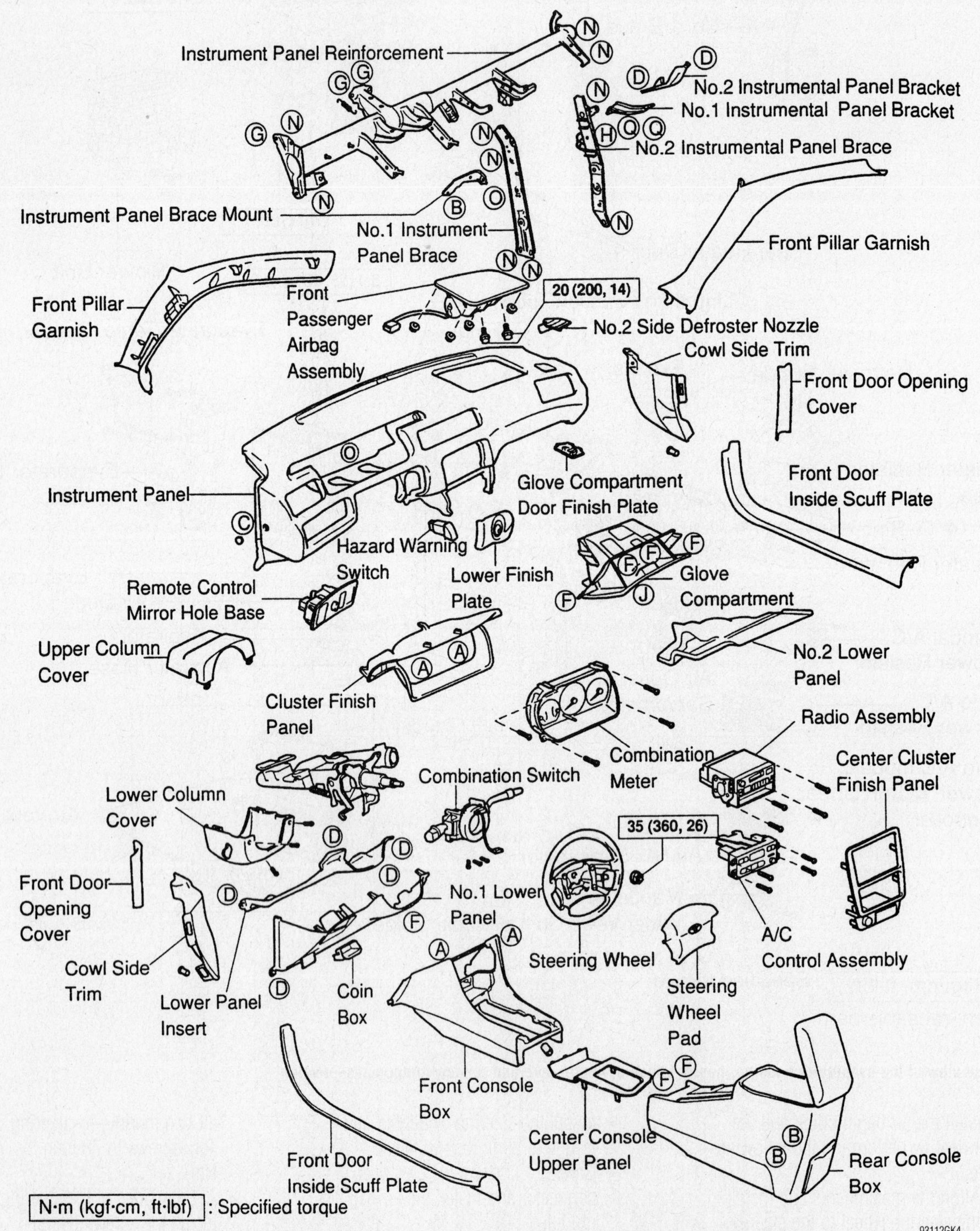

Exploded view of the instrument panel, heater/air conditioning housing and related components—2001 Camry

To install:

10. Install the heater core.

11. Connect the pipes to the heater core and install the heater core pipe clamp and the 2 clamp screws.

12. Install the heater core pipe clamps and the 3 clamp screws.

13. At the heater/air conditioning housing, install the heater protector and engage the 3 heater protector-to-heater/air conditioning housing clips.

14. Install the left hand instrument lower panel.

15. At the driver's side, install the lower instrument panel.

16. Connect the heater hoses to the heater core.

17. Refill the cooling system.

18. Connect the negative battery cable.

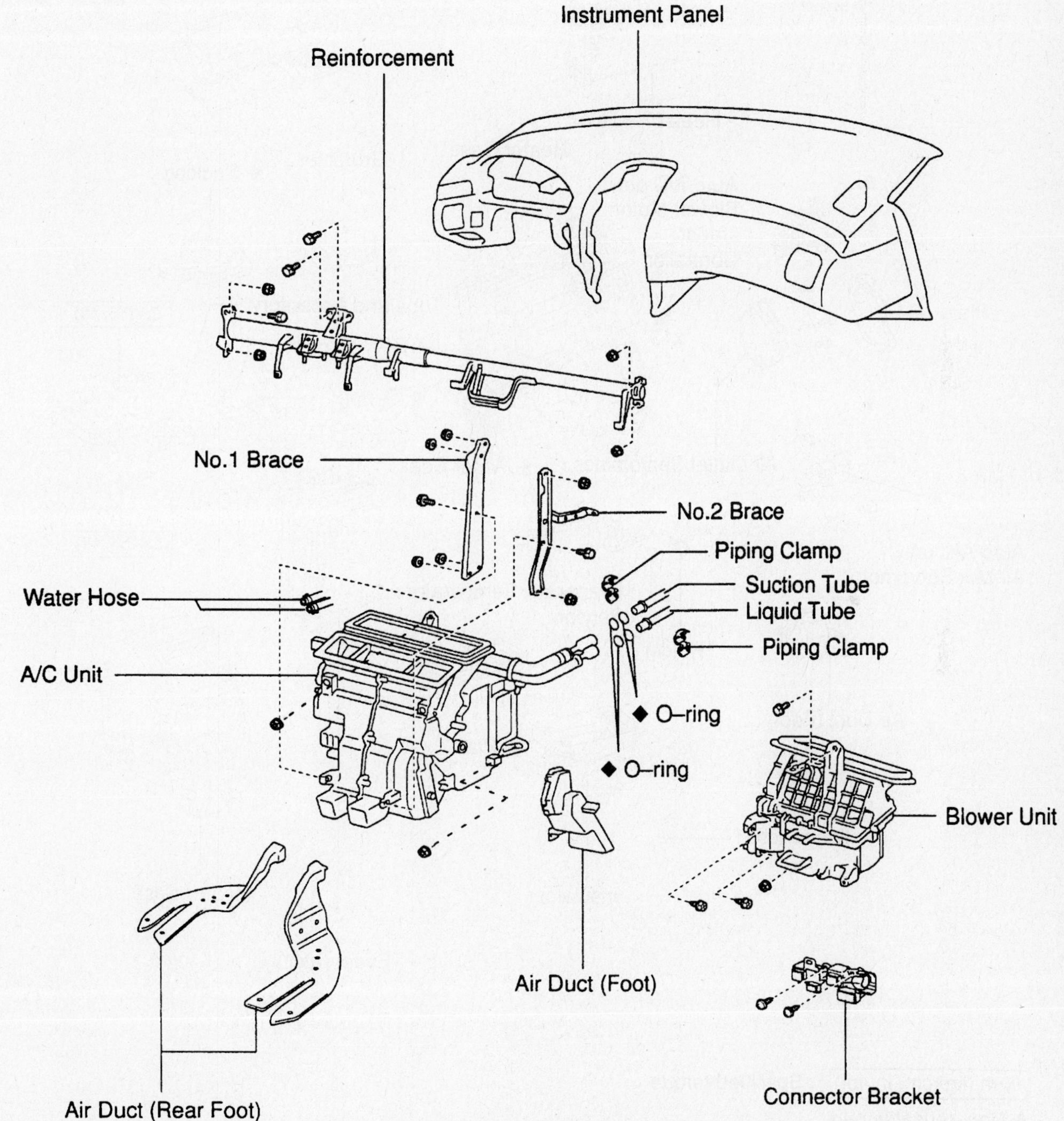

Exploded view of the instrument panel, heater/air conditioning housing and related components—2001 Camry

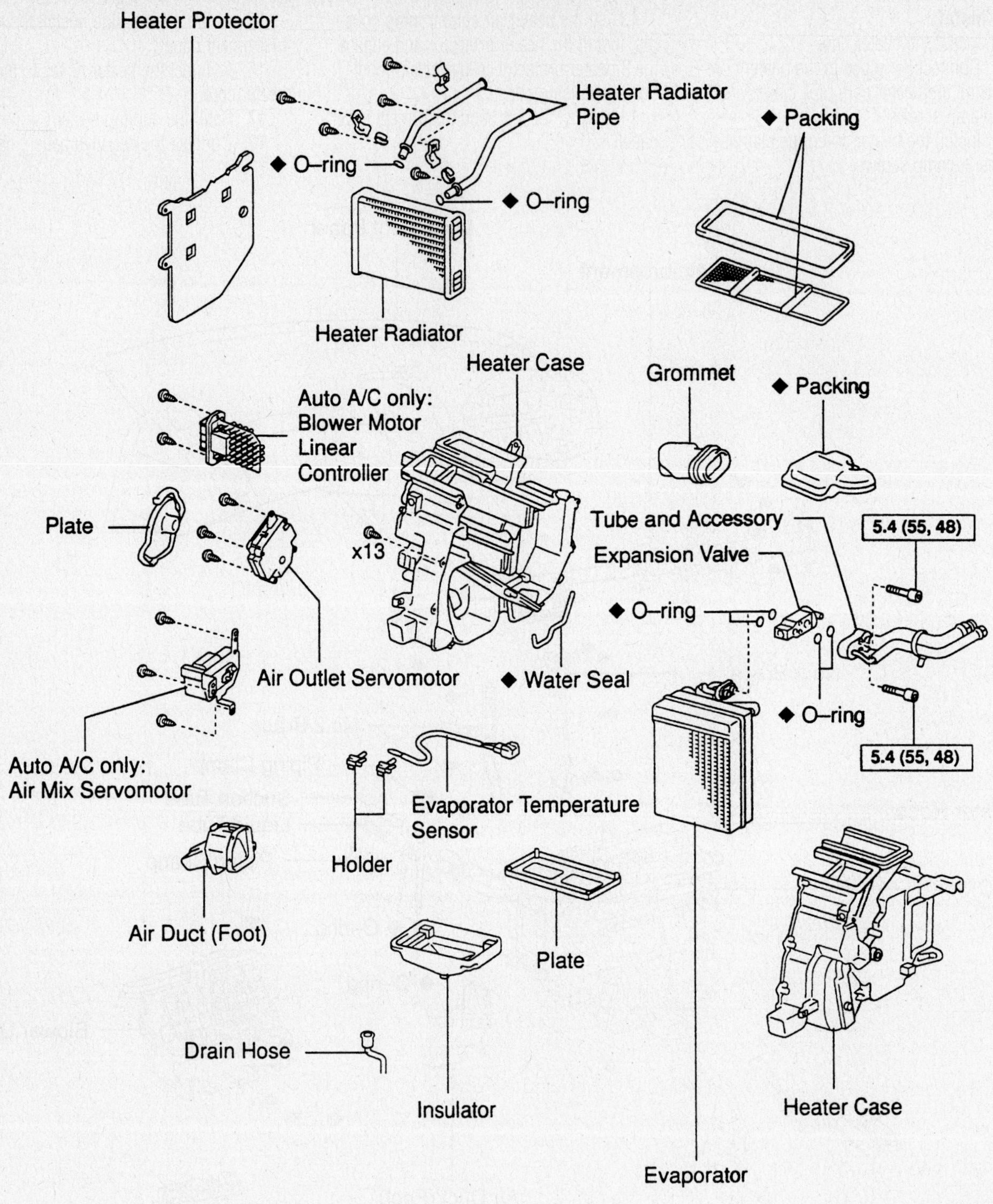

Exploded view of the heater/air conditioning housing, heater core, evaporator core and related components—2001 Camry

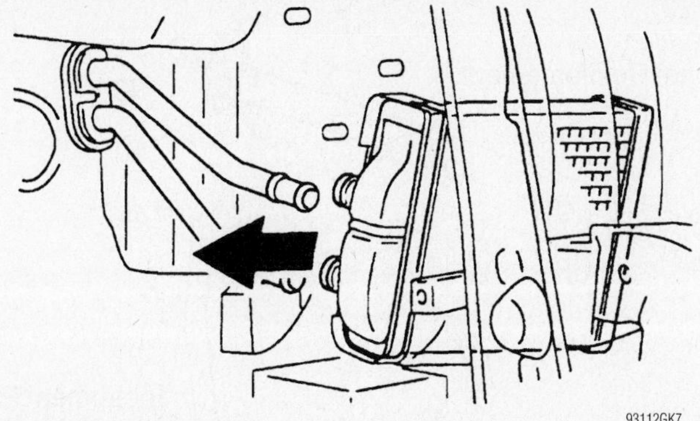

View of the heater core—2001Camry

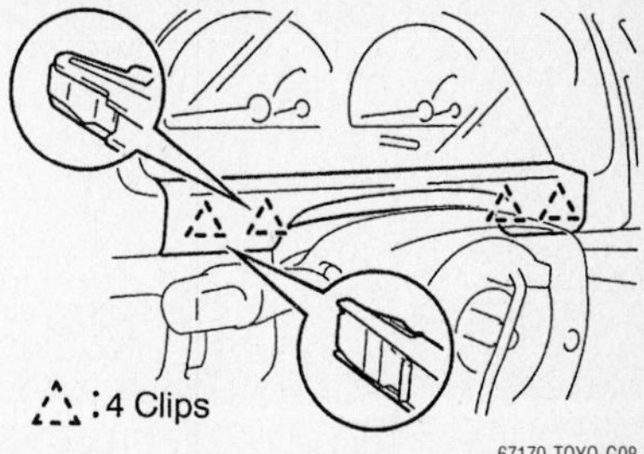

Instrument cluster finish panel clip locations—except 2001 Camry

EXCEPT 2001 MODELS

➡Removal of the heater core requires removal of the entire heater air conditioning assembly.

1. Before servicing the vehicle, refer to the precautions in the beginning of this section.
2. Drain the cooling system into a clean container for reuse.
3. Disconnect the negative battery cable. Wait 90 seconds before doing any further work while the airbag system de-energizes.
4. Discharge and recover the air conditioning system refrigerant.
5. Disconnect A/C suction hose (No. 1) and Liquid pipe (A)
   a. Install SST to piping clamp
   b. Push down SST and release the clamp lock

⁂ WARNING

Be careful not to deform the tube, when pushing the SST

➡Cap the open fittings immediately to prevent system contamination

6. Remove or disconnect the following:
   - Heater core hoses
7. Disassemble the dash components as follows
   - Lower steering column cover
   - Steering wheel
   - Instrument cluster finish panel sub-assembly
   - Steering column cover
   - Headlamp dimmer switch assembly
   - Windshield wiper switch assembly
   - Combination meter assembly
   - Door scuff plates
   - LH instrument panel under cover sub-assembly
   - Cowl side trim
   - Coin box
   - LH upper instrument panel sub assembly
   - LH lower instrument panel brace
   - Air conditioning control assembly
   - Instrument panel center cluster finish panel
   - Glove box door
   - RH lower instrument panel sub-assembly
   - Shift lever knob on M/T
   - Remove the console panel
   - Rear console box
   - Cup holder and ashtray
   - Upper console panel
   - Front console box
   - Instrument panel finish plate
   - LH & RH front pillar garnish
   - Instrument panel speakers
   - Passenger air bag connector
   - Instrument panel safety pad cap
   - Instrument panel safety pad sub-assembly

⁂ WARNING

Follow air bag removal procedures

8. Remove or disconnect the following:
   - Rear air ducts
   - Console box duct
   - Floor shift parking lock cable assembly

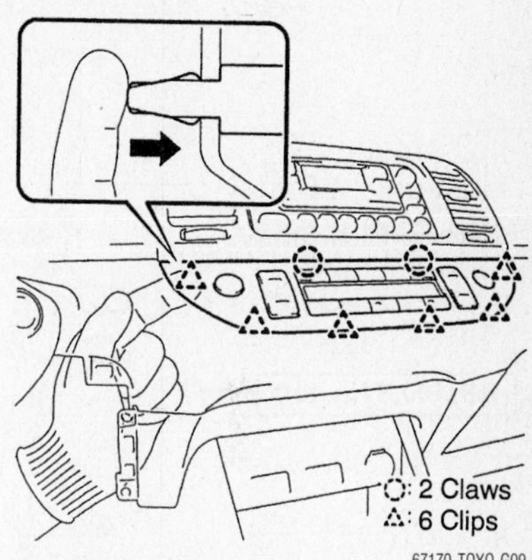

Air conditioning control panel clips—except 2001 Camry

Radio panel clips—except 2001 Camry

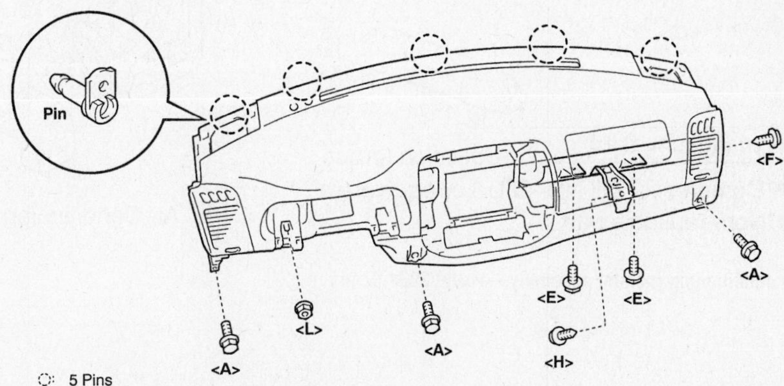

Instrument panel safety pad fastener locations—except 2001 Camry

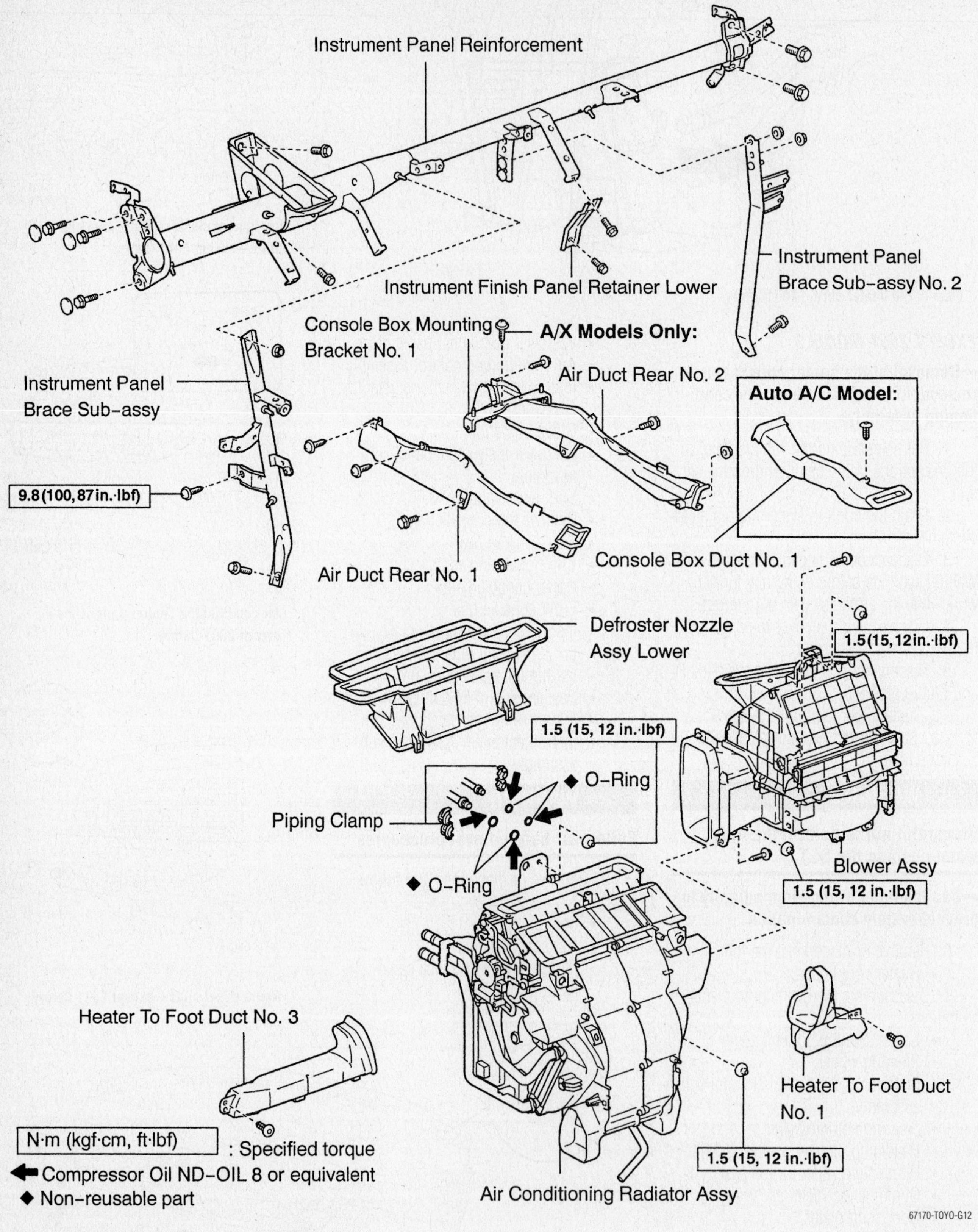
Instrument Panel Reinforcement
Instrument Panel
Brace Sub-assy No. 2
Instrument Finish Panel Retainer Lower
Console Box Mounting
Bracket No. 1
A/X Models Only:
Air Duct Rear No. 2
Instrument Panel
Brace Sub-assy
Auto A/C Model:
9.8 (100, 87 in.·lbf)
Air Duct Rear No. 1
Console Box Duct No. 1
Defroster Nozzle
Assy Lower
1.5 (15, 12 in.·lbf)
1.5 (15, 12 in.·lbf)
◆ O-Ring
Piping Clamp
◆ O-Ring
Blower Assy
1.5 (15, 12 in.·lbf)
Heater To Foot Duct No. 3
Heater To Foot Duct
No. 1
N·m (kgf·cm, ft·lbf) : Specified torque
Compressor Oil ND-OIL 8 or equivalent
◆ Non-reusable part
1.5 (15, 12 in.·lbf)
Air Conditioning Radiator Assy
67170-TOYO-G12

Air conditioning radiator assembly—except 2001 Camry

- Windshield wiper relay assembly
- Instrument panel brace assemblies
- Lower instrument finish panel retainer
- Heater to foot ducts
- Steering column assembly
- Instrument panel reinforcements
- Heater blower assembly
- Lower defroster nozzles
- Air conditioning radiator assembly
- Mode damper servo sub-assembly
- Airmix damper servo sub-assembly
- Heater radiator (core) sub-assembly

To install:

Install or connect the following:

9. Heater core unit sub assembly
   - Heater core into A/C assembly
   - Screw and clamp
   - Air conditioning radiator assembly
   - Lower defroster nozzles
   - Heater blower assembly
   - Instrument panel reinforcements
   - Steering column assembly
   - Heater to foot ducts
   - Lower instrument finish panel retainer
   - Instrument panel brace assemblies
   - Windshield wiper relay assembly
   - Floor shift parking lock cable assembly
   - Console box duct
   - Rear air ducts

10. Reassemble the dash components in the reverse of removal.

11. Install or connect the following:
    - Heater core hoses
    - A/C suction and pressure hoses, attach with bolt and plate

➡Lubricate O-rings with compressor oil

- Negative battery cable
- Fill cooling system with coolant
- Evacuate and recharge A/C system
- Warm up engine and inspect for coolant leaks

Celica

1. Disconnect the negative battery cable.
2. Drain the cooling system into a clean container for reuse.
3. Discharge and recover the air conditioning system refrigerant.
4. Remove the evaporator housing by performing the following procedure:
   - Disconnect the refrigerant lines from the evaporator core. Discard the O-rings. Plug the openings to prevent contamination.
   - Remove the 2 grommets and the drain pipe grommet.
   - At the passenger's side, remove the lower finish panel.
   - Disconnect the evaporator housing's electrical connector.
   - Remove the evaporator housing-to-chassis 3 nuts and 4 bolts.
   - Remove the evaporator housing.
5. Remove the heater valve-to-cowl bolt.
6. Remove the heater hoses from the heater core pipes.
7. Remove the heater pipe grommets.
8. Remove the air bag module and the steering wheel by performing the following procedure:
   - Place the front wheels in the straight-ahead position.
   - At both sides of the steering wheel, remove the screw covers.
   - Using a Torx socket, loosen the 2 air bag module-to-steering wheel Torx screws until the circumference ring catches on the screw case.
   - Carefully, remove the air bag module and disconnect the electrical connector.
   - Remove the steering wheel nut.
   - Using a steering wheel puller, press the steering wheel from the steering column.
9. Remove the instrument panel and reinforcement by performing the following procedure:
   - Remove the front pillar lower garnishes and the pillar garnishes.
   - Remove the front door scuff plates.
   - Remove the cowl side trim boards.
   - Remove the steering column covers.
   - Pry out the upper console panel by disengaging the 4 clips.
   - Remove the console box.
   - Remove the No. 1 finish panel.
   - Remove the finish panel and the combination switch.
   - Remove the lower cluster finish panel and the cluster finish panel.
   - Remove the No. 1 register and the combination meter.
   - Remove the 2 center cluster finish panel screws; then, pry out the finish panel.
   - Remove the air conditioning control assembly.
   - Remove the lower finish panel.
   - Remove the glove compartment door finish plate located inside the lower finish panel.
   - Pull up the air bag electrical connector and disconnect it.
   - Remove the 5 lower finish panel screws and the panel.
   - Remove the 4 bolts and lower pad inserts.
   - Remove the 4 lower center finish panel screws and the panel.
   - Remove the No. 2 side defroster nozzle.
   - Disconnect the instrument panel electrical connectors.
   - Remove the instrument panel 9 bolts and 2 nuts.
   - Remove the instrument panel.
   - Remove the No. 1 center console bracket bolt and 2 nuts.
   - Remove the No. 2 center console bracket nut and bolt.
   - Remove the brake spring.
   - Remove the center console bracket support 6 bolts and 2 nuts.
   - Remove the center console bracket.
10. Disconnect the connector and the electrical connector from the heater housing.
11. Remove the 4 heater housing-to-chassis nuts and heater housing.
12. Remove the 2 air outlet damper control servo motor screws and the servo motor.
13. Remove the air vent duct.
14. Remove the 2 heater air duct screws, the 2 clips and the duct.
15. Remove the 3 heater pipe clamp screws and the clamps.
16. Remove the heater core from the heater housing.

To install:

17. Install the heater core to the heater housing.
18. Install the heater pipe clamp and the 3 clamp screws.
19. Install the heater air duct, the 2 clips and the 2 duct screws.
20. Install the air vent duct.
21. Install the air outlet damper control servo motor and the 2 servo motor screws.
22. Install the heater housing and 4 heater housing-to-chassis nuts.
23. Connect the connector and the electrical connector to the heater housing.
24. Install the air bag module and the steering wheel by performing the following procedure:
    - Align the matchmarks and install the steering wheel to the steering column.
    - Install the steering wheel nut and torque the nut to 26 ft. lbs. (35 Nm).
    - Carefully, connect the electrical connector and install the air bag module.

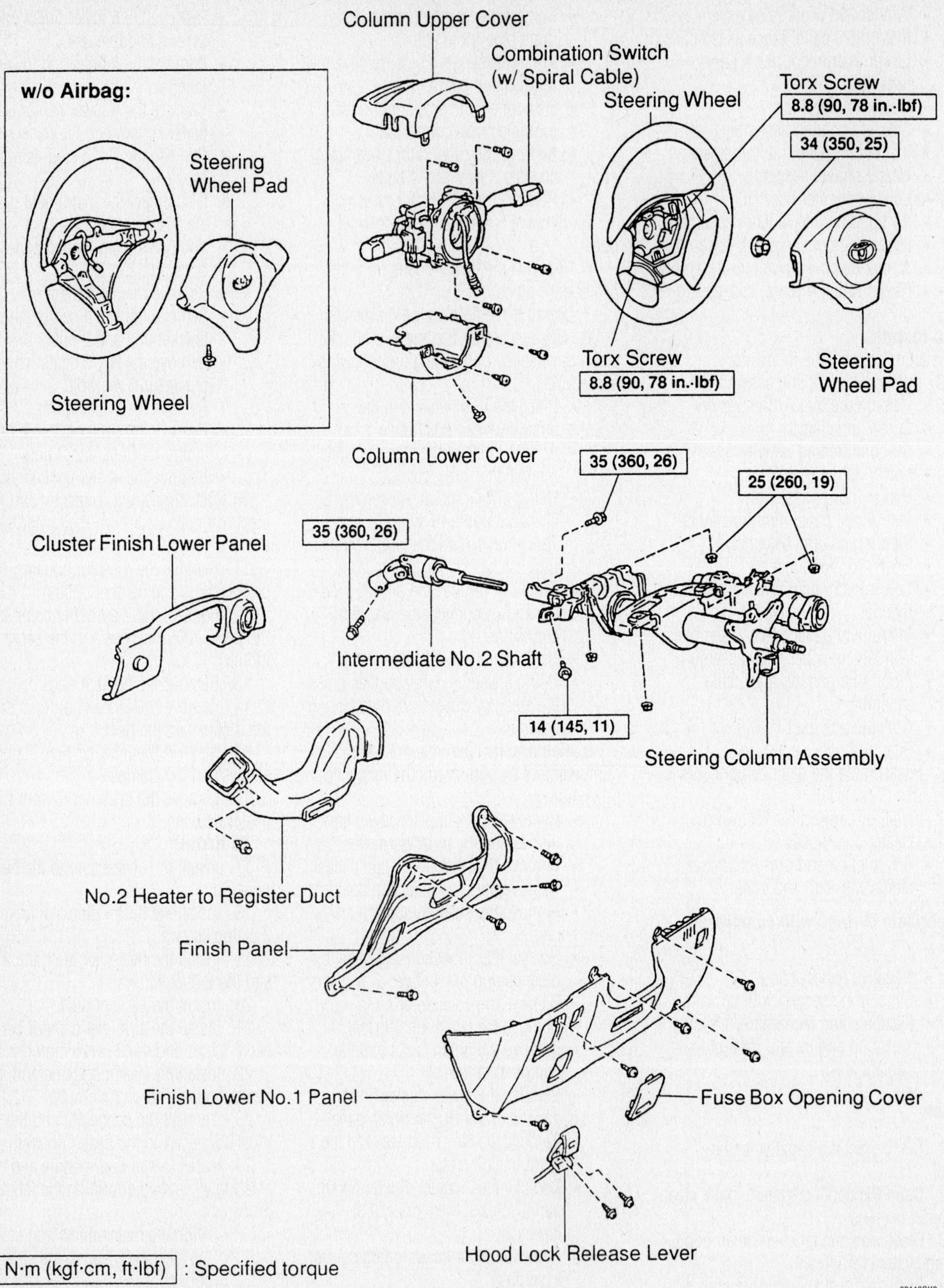

Exploded view of the air bag module, steering wheel and related components (typical)—Celica

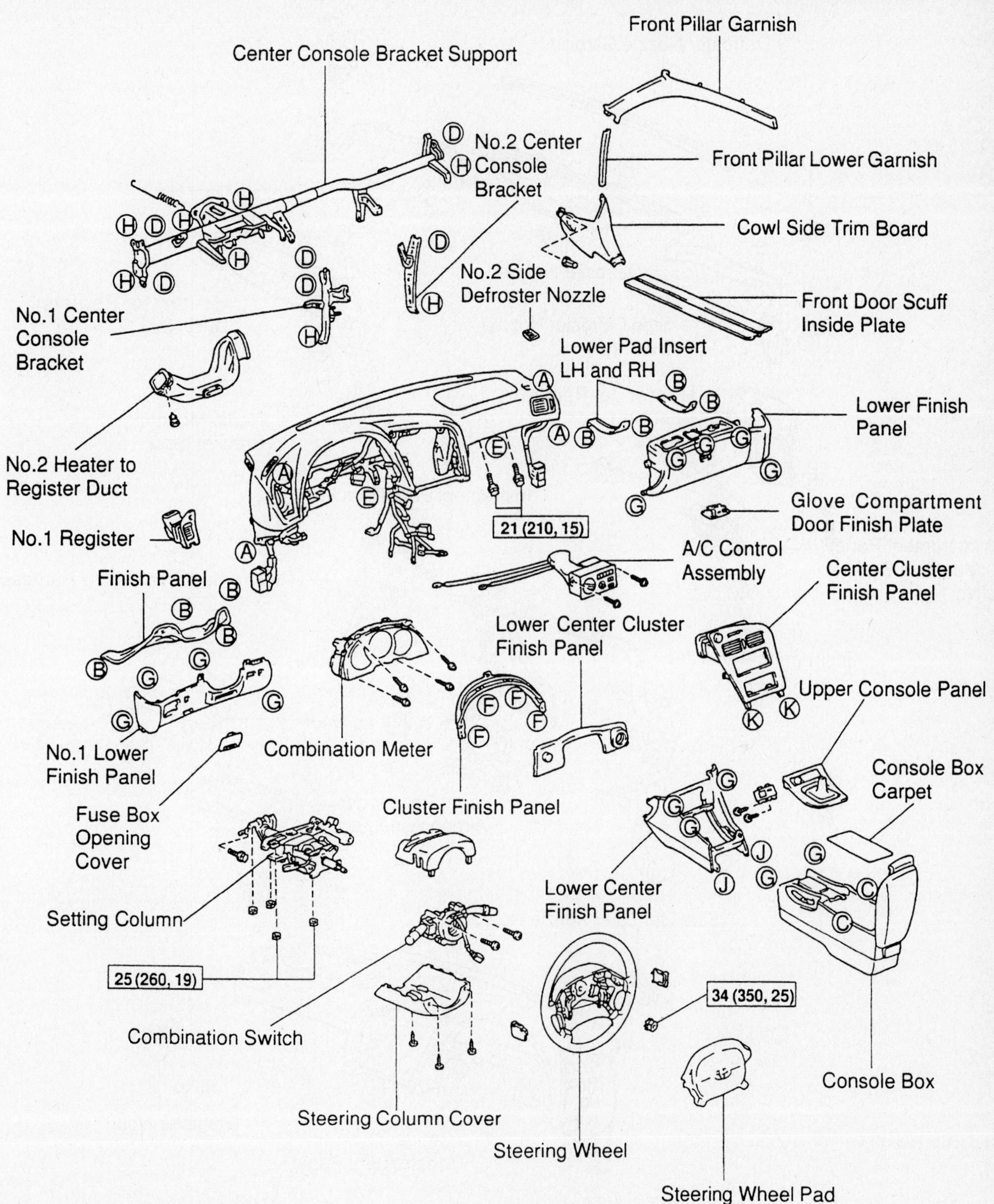
Front Pillar Garnish
Center Console Bracket Support
No.2 Center Console Bracket
Front Pillar Lower Garnish
Cowl Side Trim Board
No.2 Side Defroster Nozzle
Front Door Scuff Inside Plate
No.1 Center Console Bracket
Lower Pad Insert LH and RH
Lower Finish Panel
No.2 Heater to Register Duct
21 (210, 15)
Glove Compartment Door Finish Plate
No.1 Register
A/C Control Assembly
Center Cluster Finish Panel
Finish Panel
Lower Center Cluster Finish Panel
Upper Console Panel
No.1 Lower Finish Panel
Combination Meter
Console Box Carpet
Fuse Box Opening Cover
Cluster Finish Panel
Lower Center Finish Panel
Setting Column
25 (260, 19)
34 (350, 25)
Combination Switch
Console Box
Steering Column Cover
Steering Wheel
Steering Wheel Pad
N·m (kgf·cm, ft·lbf) : Specified torque
93112GK9

Exploded view of the instrument panel and related components—Celica

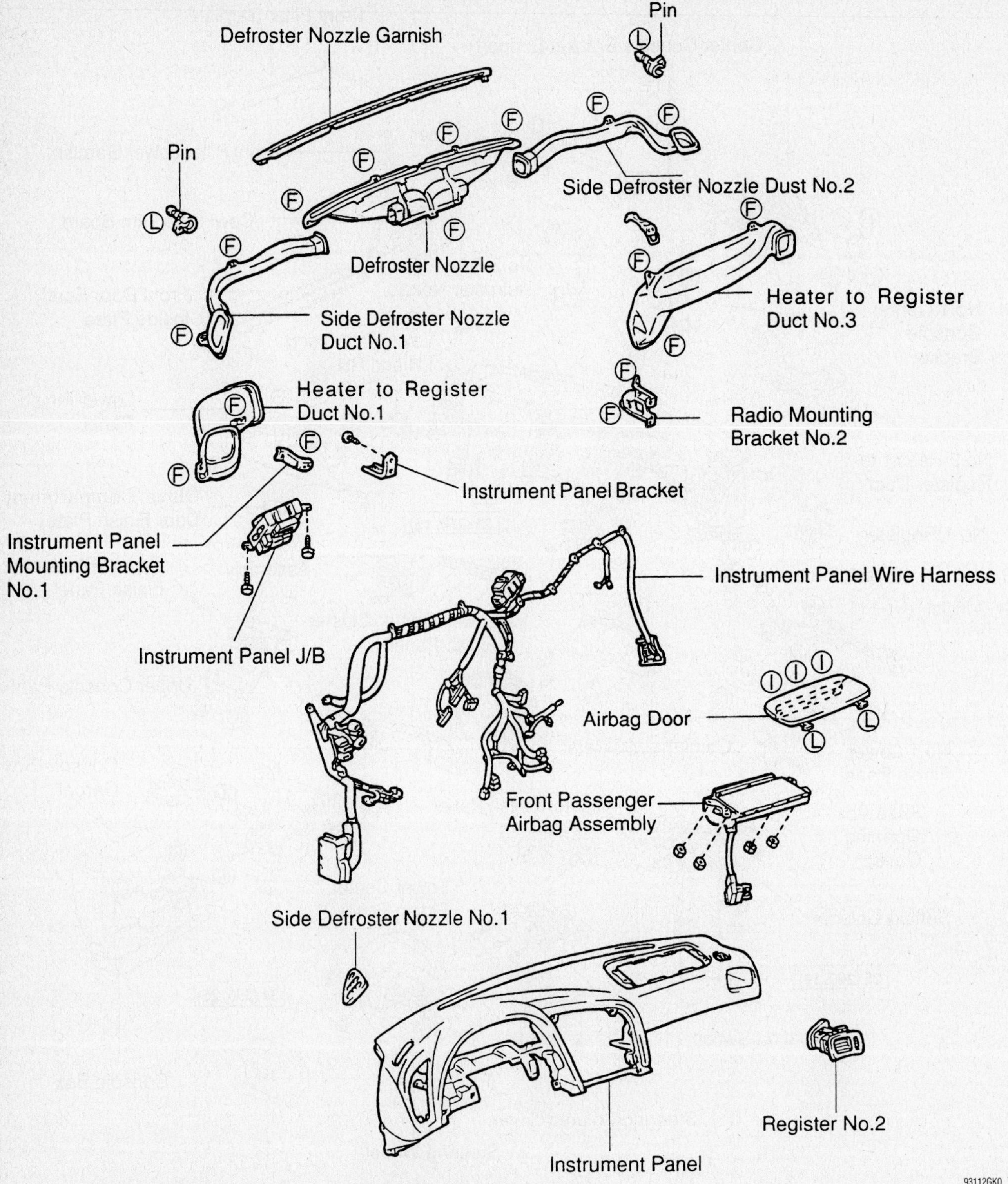

Exploded view of the ventilation system, wiring harness and related components—Celica

- Using a Torx socket, tighten the 2 air bag module-to-steering wheel Torx screws to 78 inch lbs. (8.8 Nm).
- At both sides of the steering wheel, install the screw covers.

25. Install the instrument panel and reinforcement by performing the following procedure:
    - Install the center console bracket.
    - Install the center console bracket support 6 bolts and 2 nuts.
    - Install the brake spring.
    - Install the No. 2 center console bracket nut and bolt.
    - Install the No. 1 center console bracket bolt and 2 nuts.
    - Install the instrument panel.

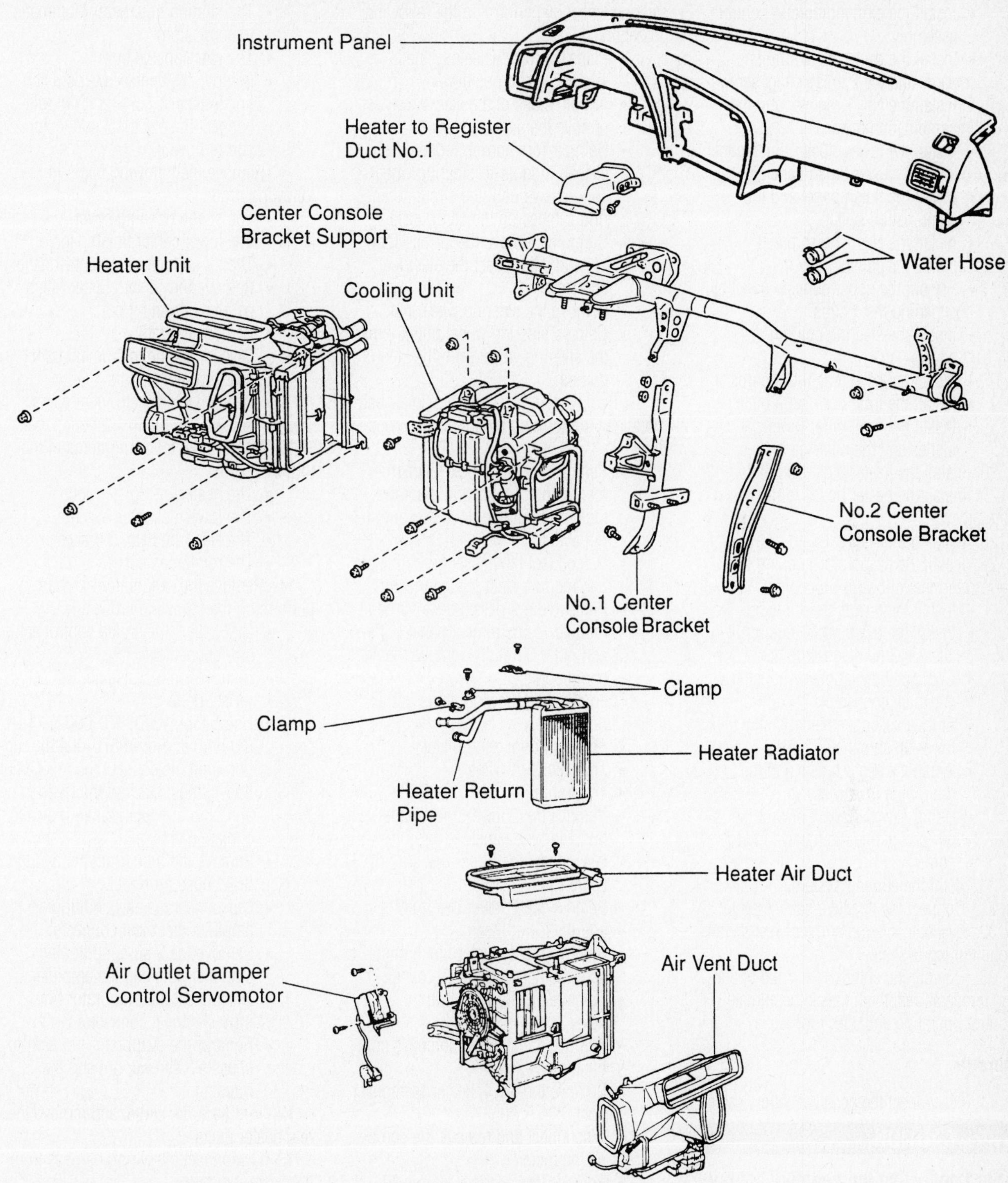

Exploded view of the heater core, heater housing, evaporator housing and related components—Celica

- Install the instrument panel 9 bolts and 2 nuts.
- Connect the instrument panel electrical connectors.
- Install the No. 2 side defroster nozzle.
- Install the lower center finish panel and the 4 panel screws.
- Install the 4 bolts and lower pad inserts.
- Install the lower finish panel and the 5 panel screws.
- Connect the air bag electrical connector.
- Install the glove compartment door finish plate located inside the lower finish panel.
- Install the lower finish panel.

- Install the air conditioning control assembly.
- Install the center cluster finish panel and the 2 finish panel screws.
- Install the No. 1 register and the combination meter.
- Install the lower cluster finish panel and the cluster finish panel.
- Install the finish panel and the combination switch.
- Install the No. 1 finish panel.
- Install the console box.
- Pry out the upper console panel by engaging the 4 clips.
- Install the steering column covers.
- Install the cowl side trim boards.
- Install the front door scuff plates.
- Install the front pillar lower garnishes and the pillar garnishes.

26. Install the heater pipe grommets.

27. Install the heater hoses to the heater core pipes.

28. Install the heater valve-to-cowl bolt.

29. Install the evaporator housing by performing the following procedure:

- Install the evaporator housing.
- Install the evaporator housing-to-chassis 3 nuts and 4 bolts.
- Connect the evaporator housing's electrical connector.
- At the passenger's side, install the lower finish panel.
- Install the 2 grommets and the drain pipe grommet.
- Using new O-rings, connect the refrigerant lines to the evaporator core.

30. Refill the cooling system.

31. Connect the negative battery cable.

32. Evacuate, charge and leak test the air conditioning system.

33. Operate the engine to normal operating temperatures; then, check the climate control operation and check for leaks.

Corolla

1. Disconnect the negative battery cable.

CAUTION

After Connecting the negative battery cable, wait for at least 1½ minutes for the SRS to deplete its energy.

2. Drain the cooling system into a clean container for reuse.

3. Disconnect the heater hoses from the heater core.

4. Discharge and recover the air conditioning system refrigerant.

5. Remove the air bag module and the steering wheel by performing the following procedure:

- Place the front wheels in the straight-ahead position.
- At both sides of the steering wheel, remove the screw covers.
- Using a Torx socket, loosen the 2 air bag module-to-steering wheel Torx screws until the circumference ring catches on the screw case.
- Carefully, remove the air bag module and disconnect the electrical connector.
- Remove the steering wheel nut.
- Using a steering wheel puller, press the steering wheel from the steering column.

6. Remove the passenger's side air bag module by performing the following procedure:

- Remove the glove box-to-instrument panel 2 bolts and 3 screws; then, pull out the glove box and remove it.
- Disconnect the passenger's side air bag module electrical connector.
- Remove the 2 passenger's air bag module-to-instrument panel bolts and nuts. Carefully, remove the air bag module.

7. Remove the following items in the following order:

- The front door scuff plates
- The cowl side trims
- The front pillar garnishes
- Remove the 2 lower finish panel bolts and the panel.
- Disconnect the hood lock control cable.
- Remove the 2 lower insert bolts and the lower insert.
- Remove the 2 cluster finish panel screws; then, pry out the panel.
- Remove the steering column cover.
- Remove the steering column's combination switch.
- Remove the No. 2 heater-to-register duct.
- Disconnect and remove the combination meter.
- Remove the steering column-to-instrument panel fasteners, the steering column shaft pinch bolt and the steering column.

8. Remove the following items in the following order:

- The lower left side panel
- The lower right side panel
- The shifting hole knob (manual transmission)
- The shifting hole bezel (automatic transmission)
- The rear console box
- Remove the center lower cluster finish panel by prying it out, disconnect the electrical connector and remove it.

9. Remove the following items in the following order:

- The stereo opening cover
- The lower center finish panel
- The air conditioning control panel
- Remove the center cluster finish panel by prying it out.
- Remove the radio.
- Disconnect the instrument panel electrical connectors.
- Remove the 4 instrument panel screws and the instrument panel.

10. Remove the following items in the following order:

- The heater-to-register duct
- The lower defroster nozzle
- The No. 1 and No. 2 braces
- The reinforcement

11. Remove the evaporator housing by performing the following procedure:

- Disconnect the cruise control actuator connector.
- Remove the 3 cruise control actuator set bolts.
- Using tool No. 09870-00025 (liquid line) and/or 09870-00015 (suction line), disconnect the refrigerant line clamps. Discard the O-rings and plug the openings to prevent contamination.
- Remove the tube grommet and the drain hose grommet.
- Disconnect the air conditioning amplifier electrical connector.
- Remove the 2 air conditioning amplifier nuts and the amplifier.
- Disconnect the evaporator housing's electrical connectors.
- Remove the evaporator housing-to-chassis 3 screws, nut and the housing.

12. Pull back the carpet and remove the rear heater ducts.

13. Disconnect the wiring harness from the heater housing.

14. Remove the 3 heater housing-to-chassis nuts and the heater housing.

15. Remove the air duct-to-heater housing screw, the 5 clips and the duct.

16. Release the 2 heater core cover clips and the cover.

17. Remove the heater core-to-heater housing screw, clamp and the heater core.

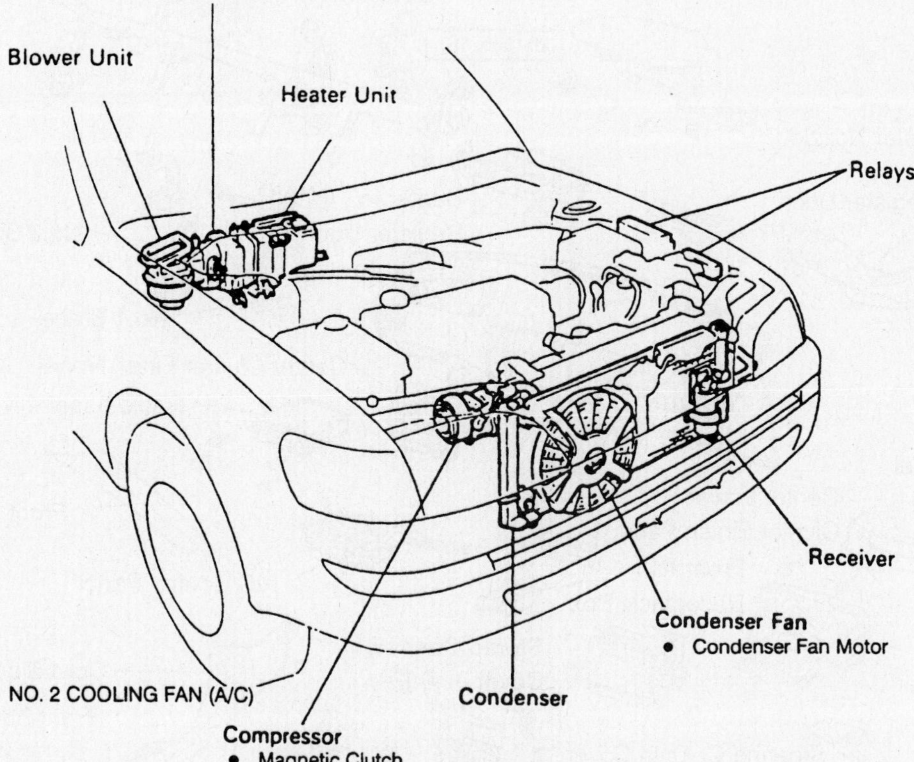

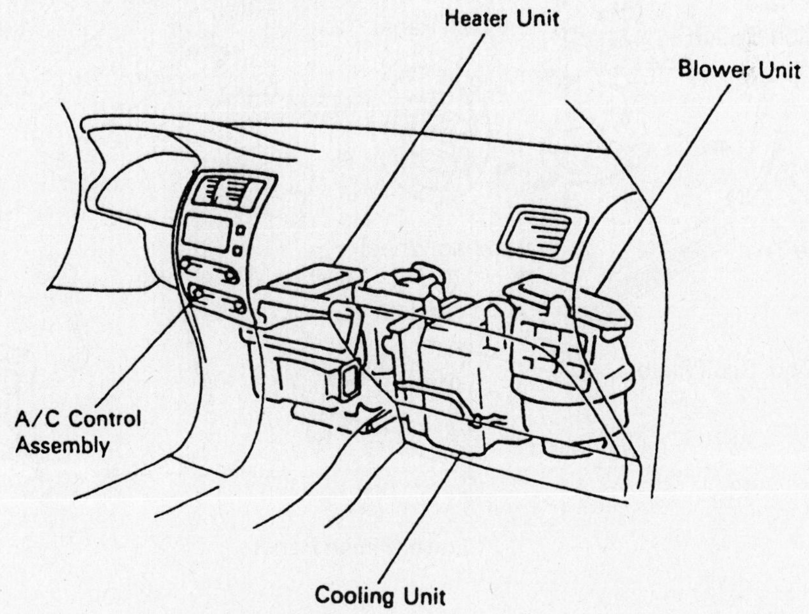

93112GA4

View of the heater/air conditioning assembly and related components—Corolla

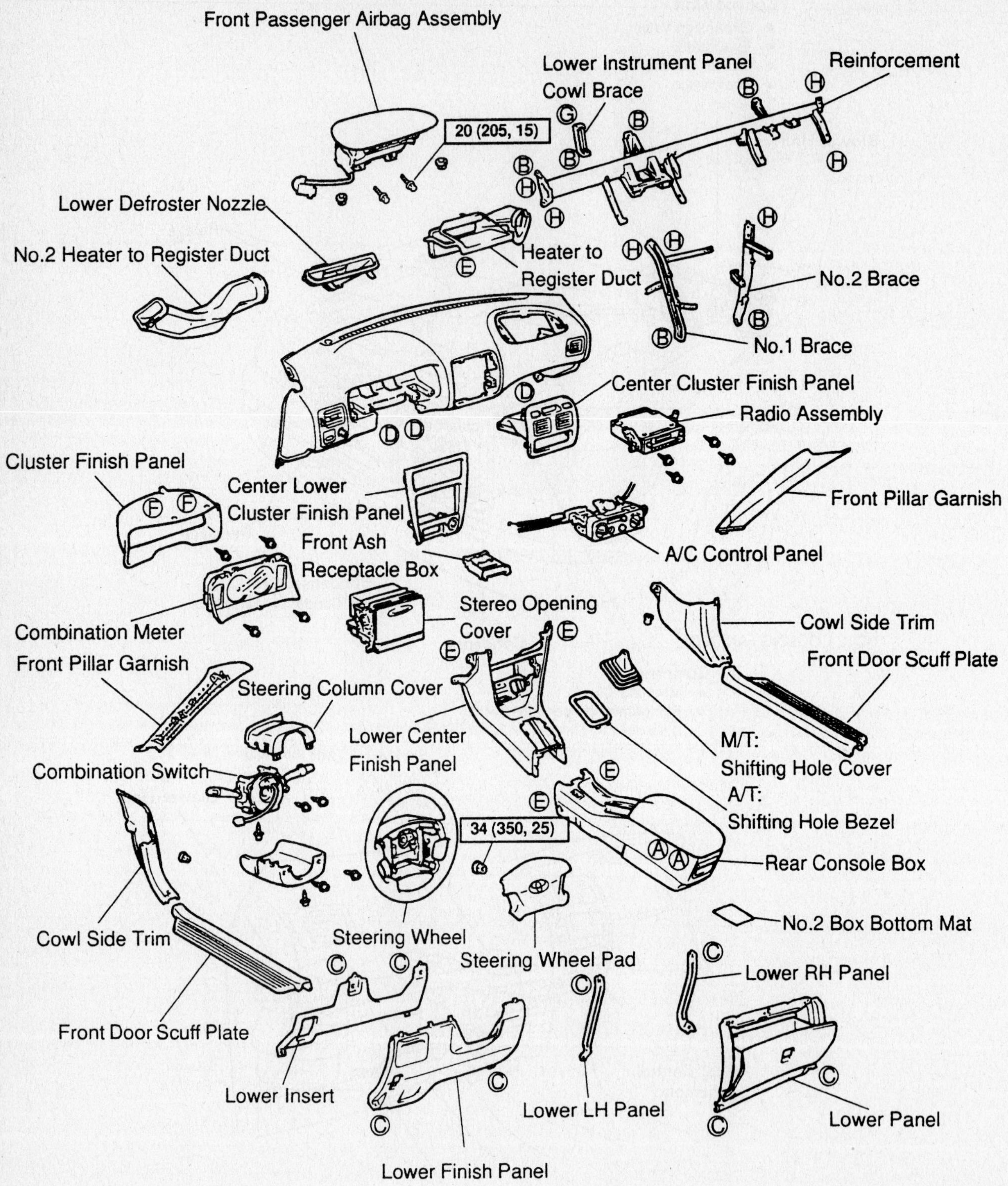
Front Passenger Airbag Assembly
Lower Instrument Panel
Cowl Brace
Reinforcement
20 (205, 15)
Lower Defroster Nozzle
No.2 Heater to Register Duct
Heater to
Register Duct
No.2 Brace
No.1 Brace
Center Cluster Finish Panel
Radio Assembly
Cluster Finish Panel
Center Lower
Cluster Finish Panel
Front Pillar Garnish
Front Ash
Receptacle Box
A/C Control Panel
Stereo Opening
Cover
Cowl Side Trim
Combination Meter
Front Pillar Garnish
Front Door Scuff Plate
Steering Column Cover
Lower Center
Finish Panel
M/T:
Shifting Hole Cover
A/T:
Shifting Hole Bezel
Combination Switch
34 (350, 25)
Rear Console Box
No.2 Box Bottom Mat
Cowl Side Trim
Steering Wheel
Steering Wheel Pad
Lower RH Panel
Front Door Scuff Plate
Lower Insert
Lower LH Panel
Lower Panel
Lower Finish Panel
N·m (kgf·cm, ft·lbf) : Specified torque
93112GM4

Exploded view of the instrument panel and related components—Corolla

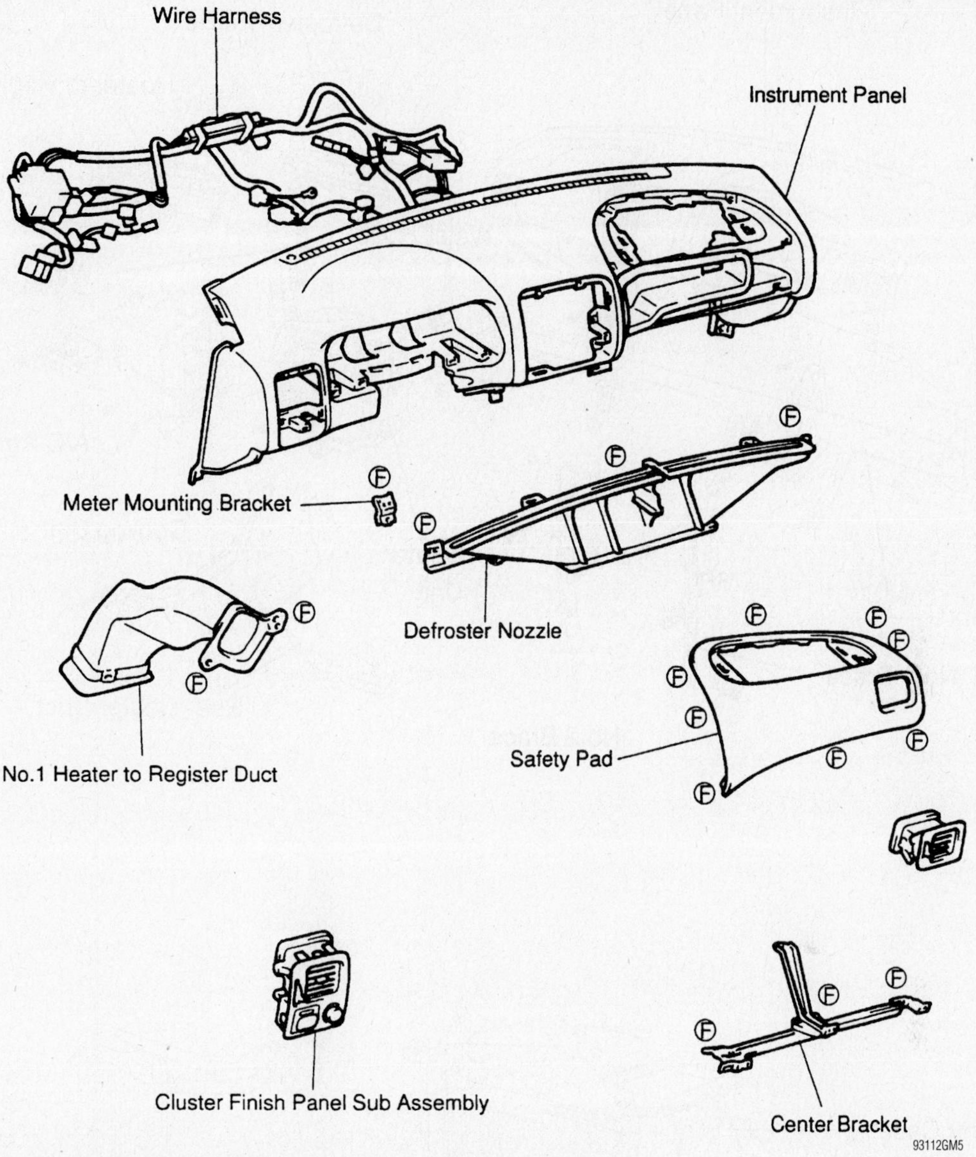

Wire Harness
Instrument Panel
Meter Mounting Bracket
Defroster Nozzle
No.1 Heater to Register Duct
Safety Pad
Cluster Finish Panel Sub Assembly
Center Bracket
93112GM5

Exploded view of the ventilation system, wiring harness and related components—Corolla

To install:

18. Install the heater core, clamp and the heater core-to-heater housing screw.

19. Install the heater core cover and the 2 cover clips.

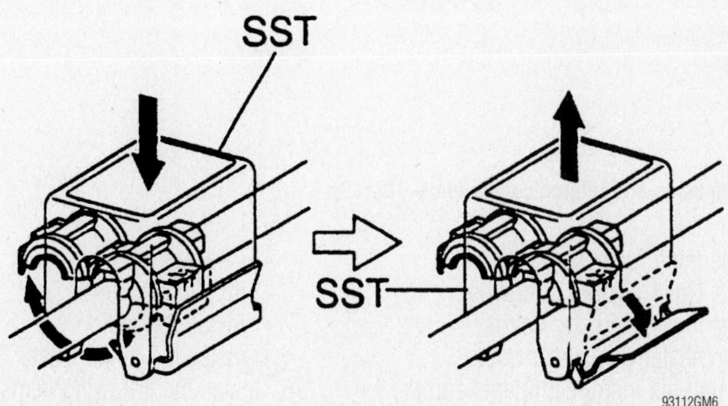

Using the special tool to remove the air conditioning refrigerant line clamps—Corolla

20. Install the air duct, the 5 clips and the duct-to-heater housing screw.

21. Install the heater housing and the 3 heater housing-to-chassis nuts.

22. Connect the wiring harness to the heater housing.

23. Install the rear heater ducts and the carpet.

24. Install the evaporator housing by performing the following procedure:

- Install the evaporator housing-to-chassis 3 screws, nut and the housing.
- Connect the evaporator housing's electrical connectors.
- Install the 2 air conditioning amplifier nuts and the amplifier.
- Connect the air conditioning amplifier electrical connector.
- Install the tube grommet and the drain hose grommet.
- Using new O-rings, connect the refrigerant line clamps.

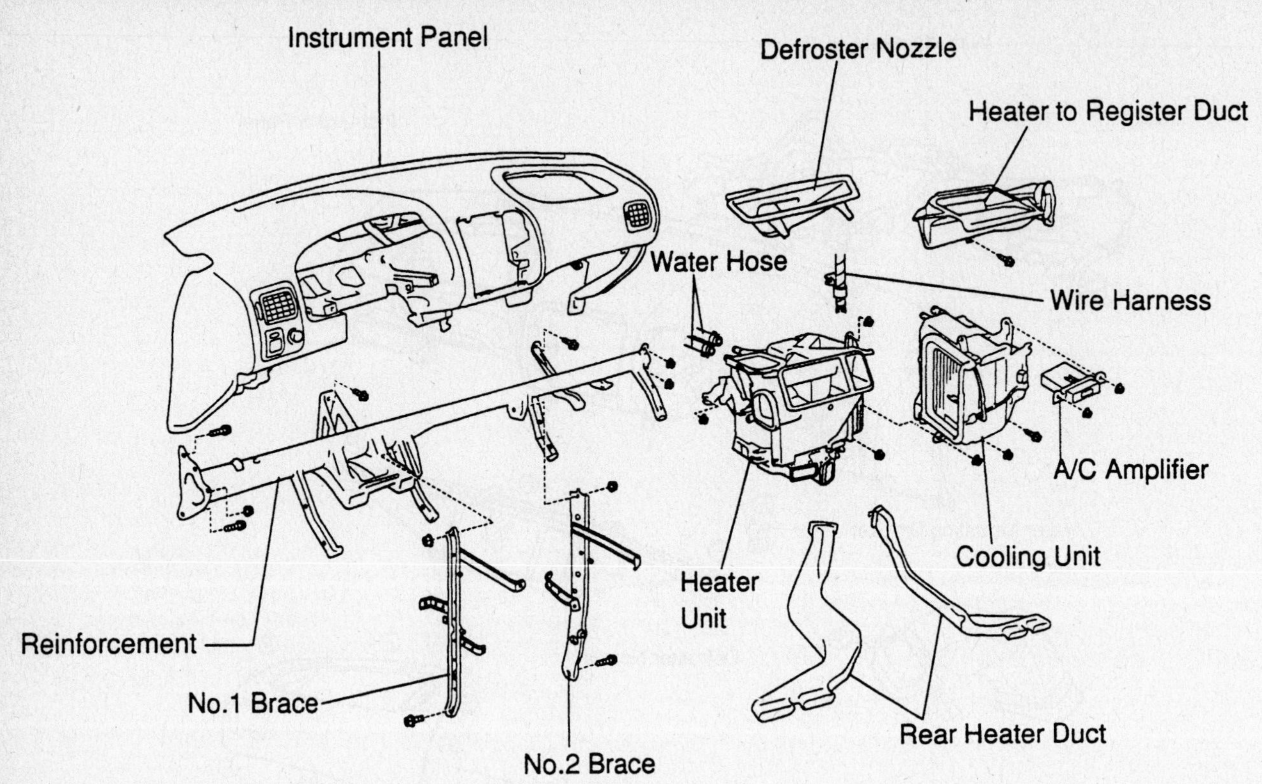

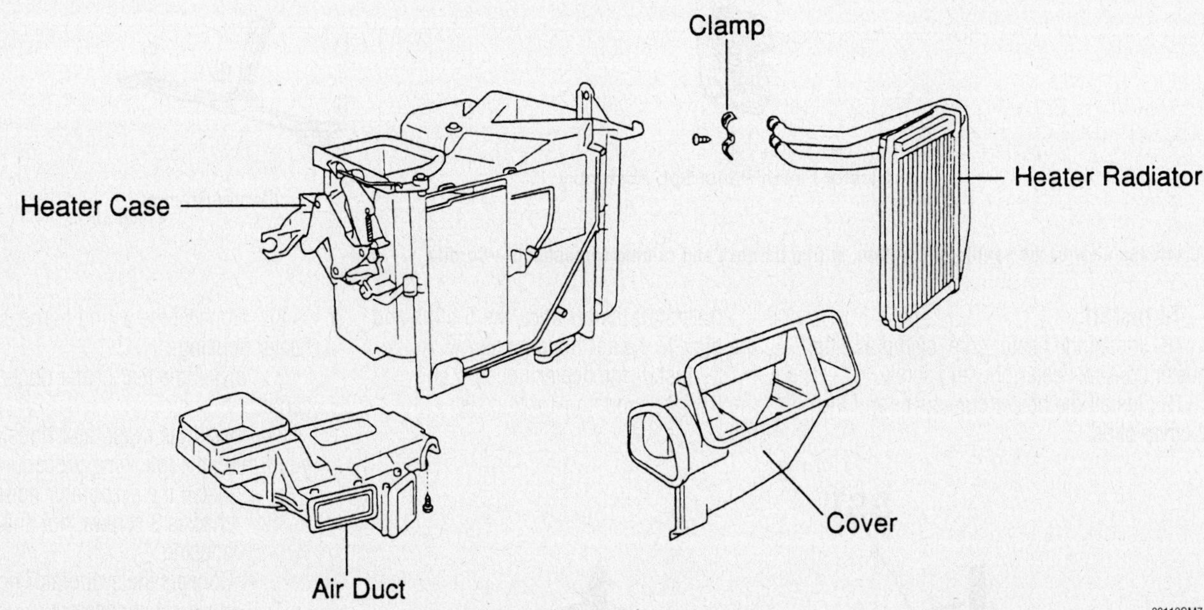

Exploded view of the heater core, heater housing, evaporator housing and related components—Corolla

- Install the 3 cruise control actuator set bolts.
- Connect the cruise control actuator connector.

25. Install the following items in the following order:

- The reinforcement
- The No. 1 and No. 2 braces
- The lower defroster nozzle
- The heater-to-register duct
- Install the instrument panel and the 4 instrument panel screws.
- Connect the instrument panel electrical connectors.
- Install the radio.
- Install the center cluster finish panel.

26. Install the following items in the following order:

- The air conditioning control panel
- The lower center finish panel
- The stereo opening cover
- Connect the electrical connector and install the center lower cluster finish panel.

27. Install the following items in the following order:
    - The rear console box
    - The shifting hole bezel (automatic transmission)
    - The shifting hole knob (manual transmission)
    - The lower right side panel
    - The lower left side panel
    - Install the steering column, the steering column shaft pinch bolt and the steering column-to-instrument panel fasteners.
    - Connect and install the combination meter.
    - Install the No. 2 heater-to-register duct.
    - Install the steering column's combination switch.
    - Install the steering column cover.
    - Install the cluster finish panel and the 2 panel screws.
    - Install the lower insert and the 2 lower insert bolts.
    - Connect the hood lock control cable.
    - Install the lower finish panel and the 2 panel bolts.
28. Install the following items in the following order:
    - The front pillar garnishes
    - The cowl side trims
    - The front door scuff plates
29. Install the passenger's side air bag module by performing the following procedure:
    - Carefully, Install the air bag module and the 2 passenger's air bag module-to-instrument panel bolts and nuts.
    - Connect the passenger's side air bag module electrical connector.
    - Install the glove box and the glove box-to-instrument panel 2 bolts and 3 screws.
30. Install the air bag module and the steering wheel by performing the following procedure:
    - Align the matchmarks and install the steering wheel to the steering column.
    - Install the steering wheel nut and torque the nut to 26 ft. lbs. (35 Nm).
    - Carefully, connect the electrical connector and install the air bag module.
    - Using a Torx socket, tighten the 2 air bag module-to-steering wheel Torx screws to 78 inch lbs. (8.8 Nm).
    - At both sides of the steering wheel, install the screw covers.
31. Connect the heater hoses to the heater core.
32. Refill the cooling system.
33. Connect the negative battery cable.
34. Evacuate, charge and leak test the air conditioning system refrigerant.

MR2

1. Before servicing the vehicle, refer to the precautions in the beginning of this section.
2. Drain the cooling system.
3. Discharge the A/C system.
4. Remove or disconnect the following:
    - Negative battery cable
    - Luggage compartment trim box cover
    - Heater hoses
    - Pipe grommets
    - Steering wheel
    - Door scuff plates
    - Cowl side trims
    - No. 1 lower finish panel
    - Instrument panel lower pad insert
    - Steering column covers
    - Spiral cable
    - Wiper/washer switch
    - Light control /headlight dimmer switch
    - Instrument cluster finish panel
    - Combination meter
    - Glove compartment door
    - Passenger airbag assembly
    - Passenger airbag manual switch
    - Hood release lever
    - Center instrument cluster finish panel
    - Ash tray outer case
    - Stereo finish panel
    - Instrument panel brace inner and outer covers
    - Heater control cables
    - Heater control assembly
    - Steering column
    - Front pillar garnishes
    - Instrument panel lower mount bracket
    - Center instrument panel bracket
    - Instrument panel braces
    - Instrument panel
    - Instrument panel reinforcement
    - A/C system suction tube and liquid line
    - Blower resistor connector
    - Thermistor connector
    - Cooling unit
    - Heater unit
    - Heater core

To install:

5. Install or connect the following:
    - Heater core
    - Heater unit
    - Cooling unit
    - Thermistor connector
    - Blower resistor connector
    - A/C system suction tube and liquid line
    - Instrument panel reinforcement
    - Instrument panel
    - Instrument panel braces
    - Center instrument panel bracket
    - Instrument panel lower mount bracket
    - Front pillar garnishes
    - Steering column
    - Heater control assembly
    - Heater control cables
    - Instrument panel brace inner and outer covers
    - Stereo finish panel
    - Ash tray outer case
    - Center instrument cluster finish panel
    - Hood release lever
    - Passenger airbag manual switch
    - Passenger airbag assembly
    - Glove compartment door
    - Combination meter
    - Instrument cluster finish panel
    - Light control /headlight dimmer switch
    - Wiper/washer switch
    - Spiral cable
    - Steering column covers
    - Instrument panel lower pad insert
    - No. 1 lower finish panel
    - Cowl side trims
    - Door scuff plates
    - Steering wheel
    - Pipe grommets
    - Heater hoses
    - Luggage compartment trim box cover
    - Negative battery cable
6. Fill the cooling system.
7. Recharge the A/C system.
8. Start the engine and check for leaks.

Solara

1. Disconnect the negative battery cable.
2. Drain the cooling system into a clean container for reuse.
3. Disconnect the heater hoses from the heater core.
4. At the driver's side, remove the lower instrument panel.

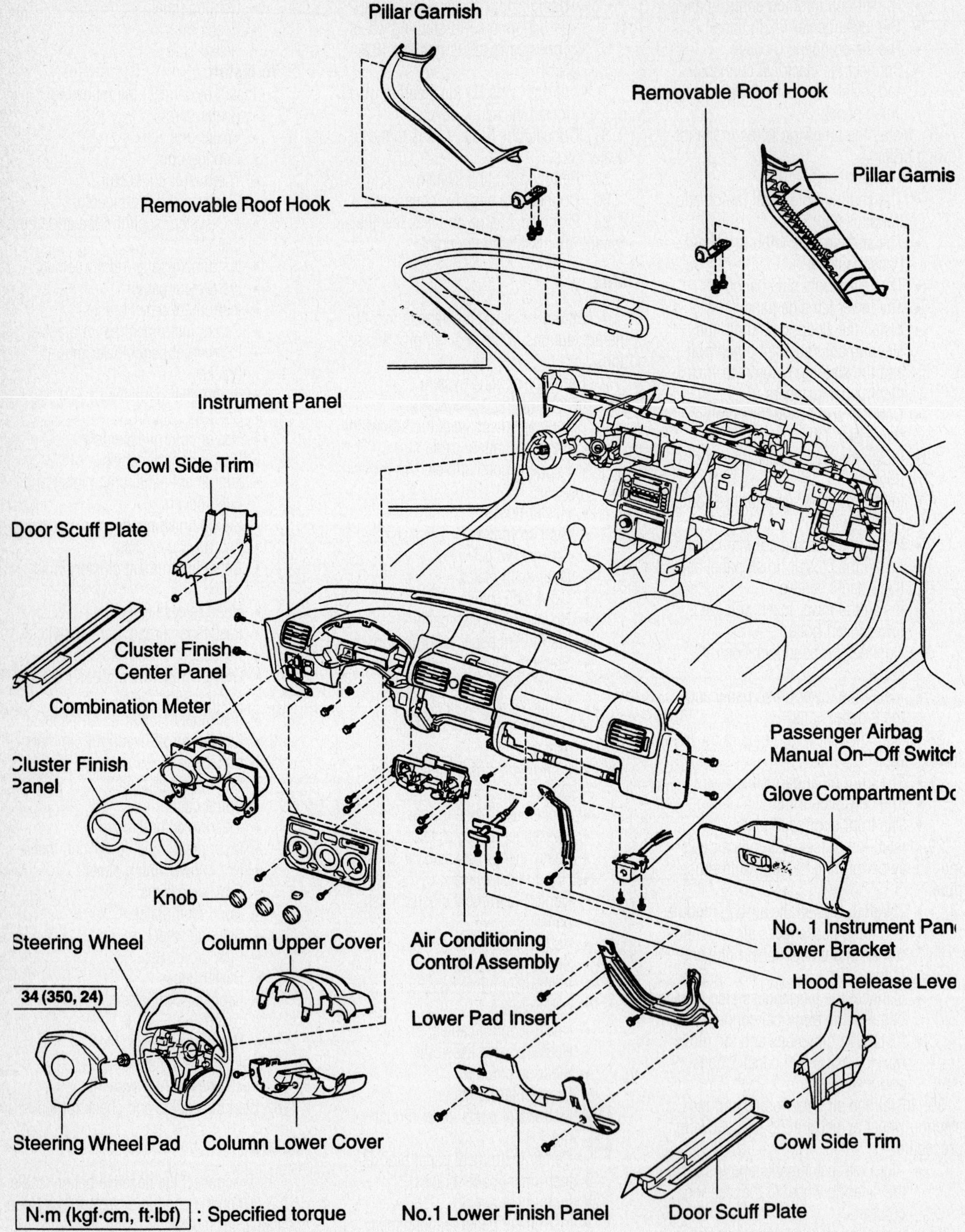
Pillar Garnish
Removable Roof Hook
Pillar Garnis
Removable Roof Hook
Instrument Panel
Cowl Side Trim
Door Scuff Plate
Cluster Finish
Center Panel
Combination Meter
Cluster Finish
Panel
Passenger Airbag
Manual On–Off Switch
Glove Compartment Dc
Knob
Steering Wheel
Column Upper Cover
Air Conditioning
Control Assembly
No. 1 Instrument Pan
Lower Bracket
Hood Release Leve
34 (350, 24)
Lower Pad Insert
Steering Wheel Pad
Column Lower Cover
Cowl Side Trim
N·m (kgf·cm, ft·lbf) : Specified torque
No.1 Lower Finish Panel
Door Scuff Plate
9357WG15

Instrument panel removal exploded view—MR2

x 18
Clip
Luggage Compartment
Trim Box Cover
Piping Clamp
(Quick Joint)
Cooling Unit
Thermistor Connector
Blower Resister Connector
9357WG16

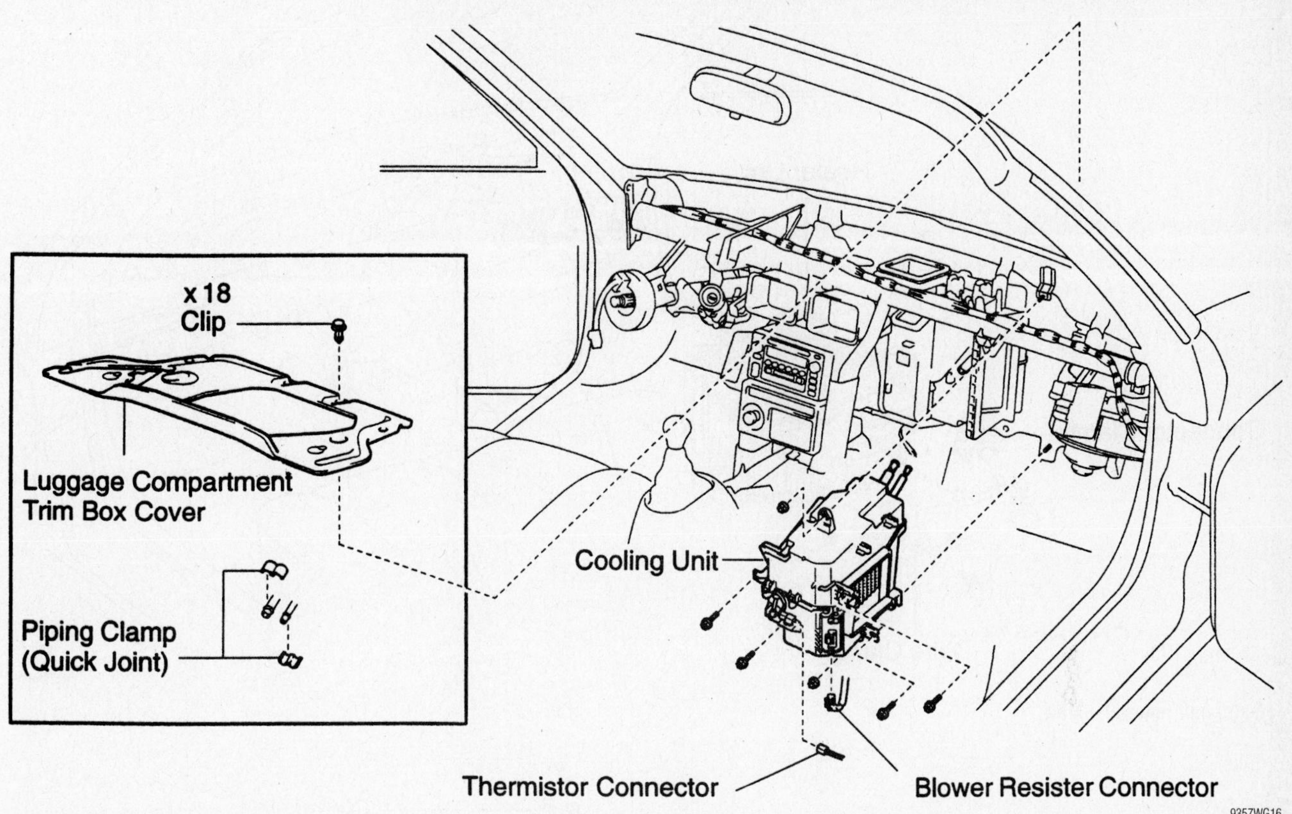

Evaporator case removal—MR2

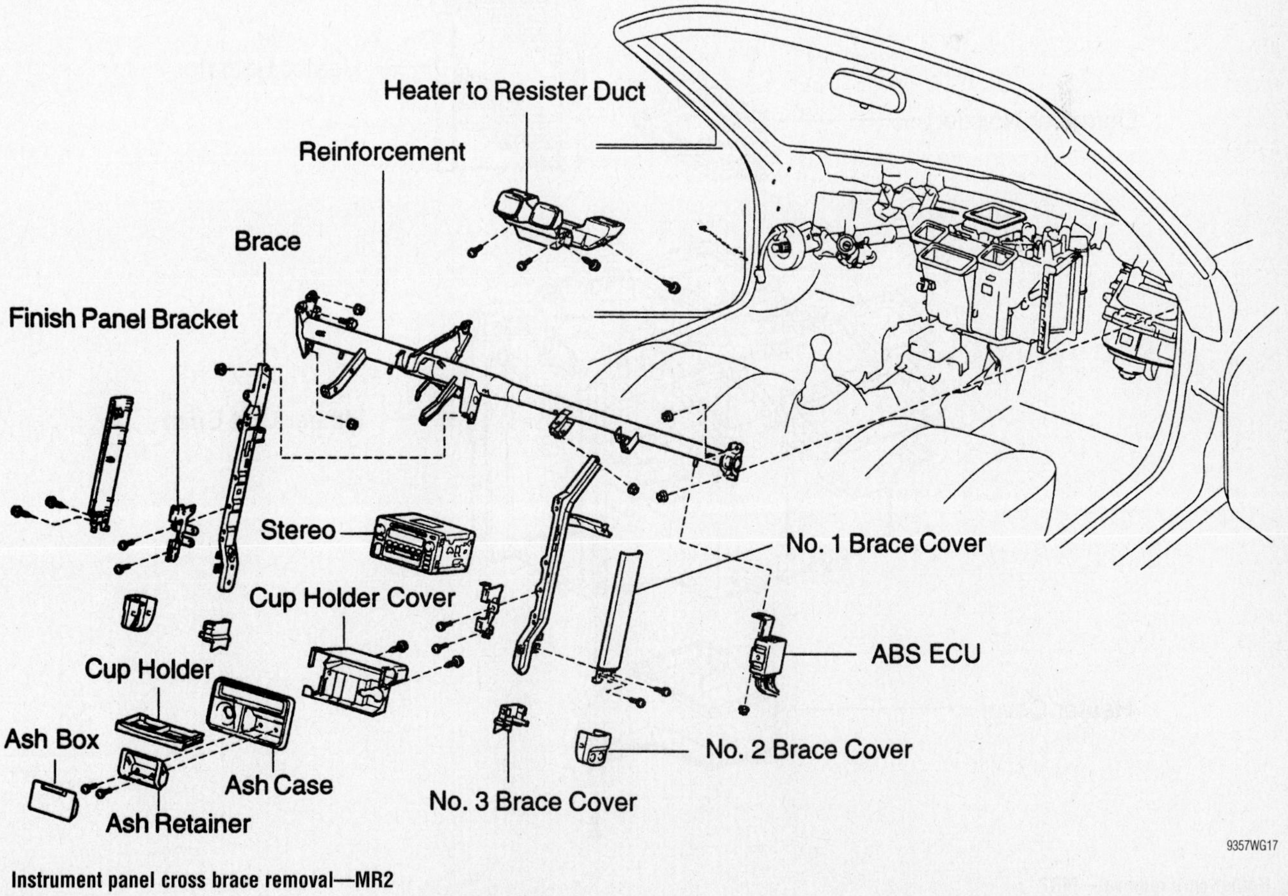

Heater to Resister Duct
Reinforcement
Brace
Finish Panel Bracket
Stereo
Cup Holder Cover
Cup Holder
Ash Box
Ash Case
Ash Retainer
No. 3 Brace Cover
No. 2 Brace Cover
ABS ECU
No. 1 Brace Cover
9357WG17

Instrument panel cross brace removal—MR2

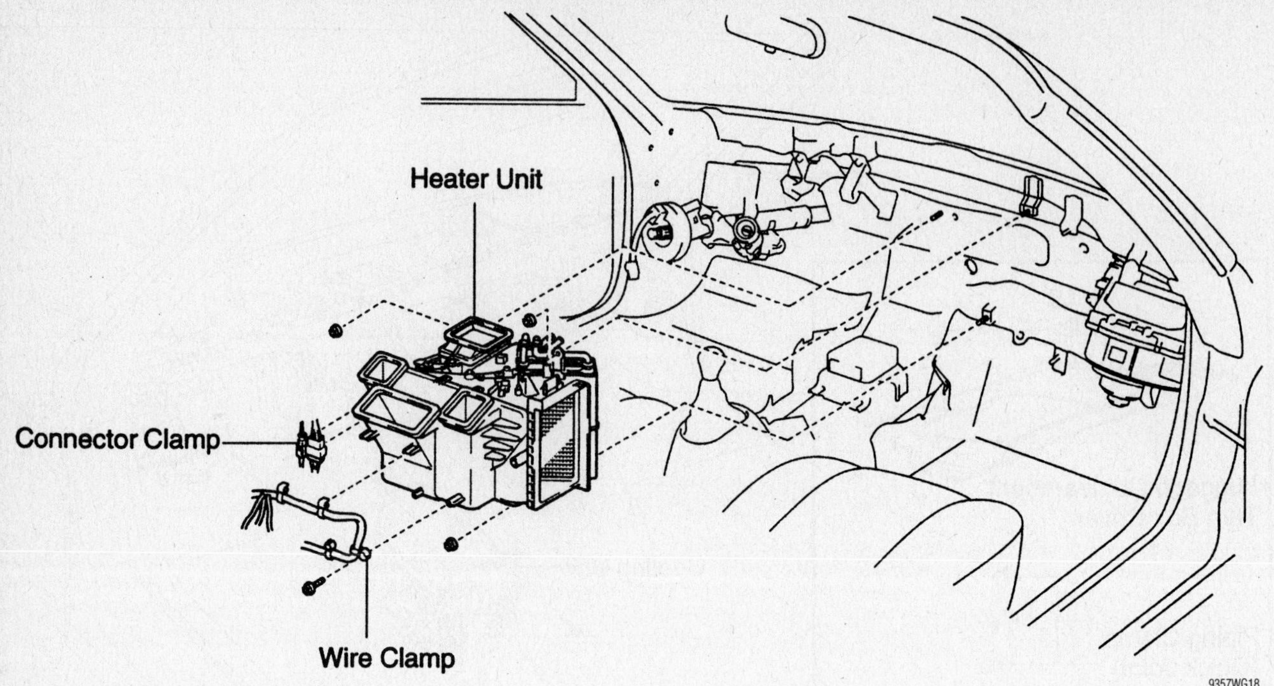

Heater unit removal—MR2

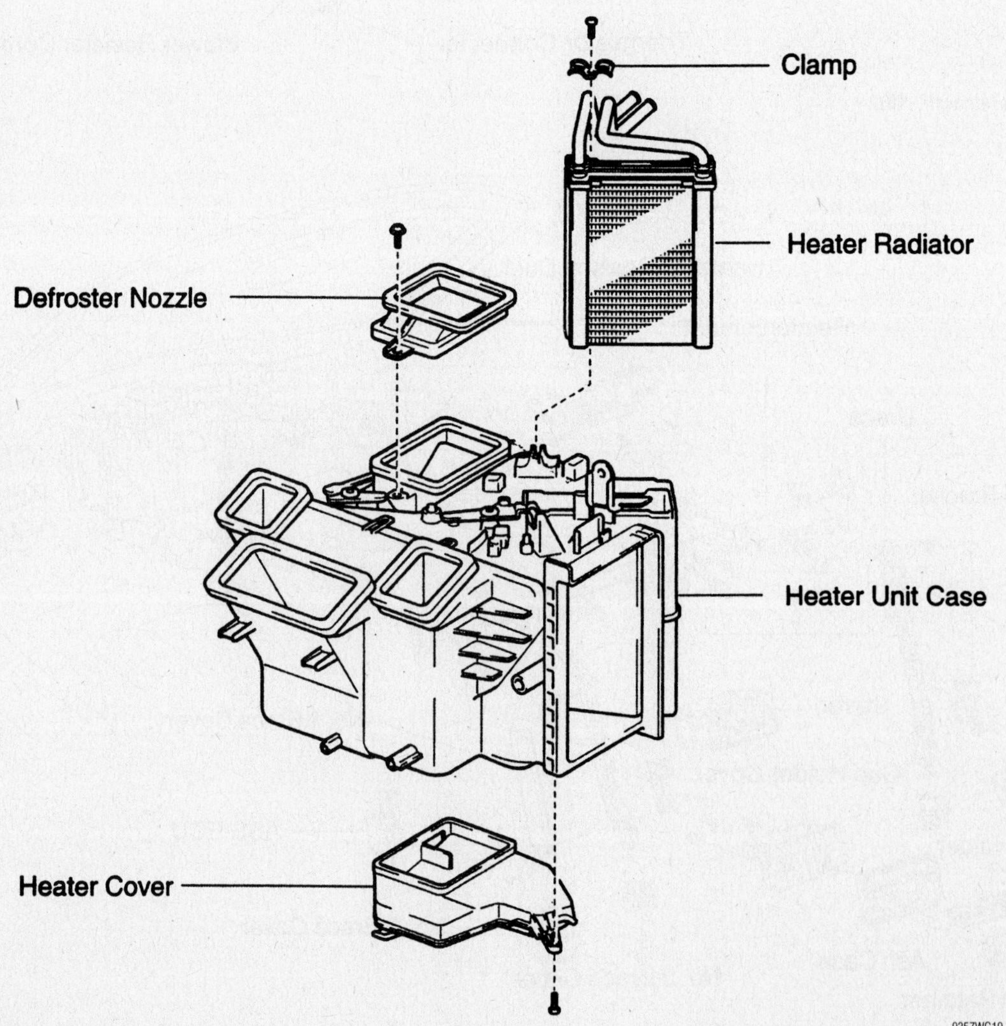

Heater core removal—MR2

5. Remove the left hand instrument lower panel.

6. At the heater/air conditioning housing, disengage the 3 heater protector-to-heater/air conditioning housing clips and remove the heater protector.

7. Remove the 3 heater core pipe clamp screws and the clamps.

8. Remove the 2 heater core pipe clamp screws and the clamps; then, disconnect the pipes from the heater core.

9. Remove the heater core.

To install:

10. Install the heater core.

11. Connect the pipes to the heater core and install the heater core pipe clamp and the 2 clamp screws.

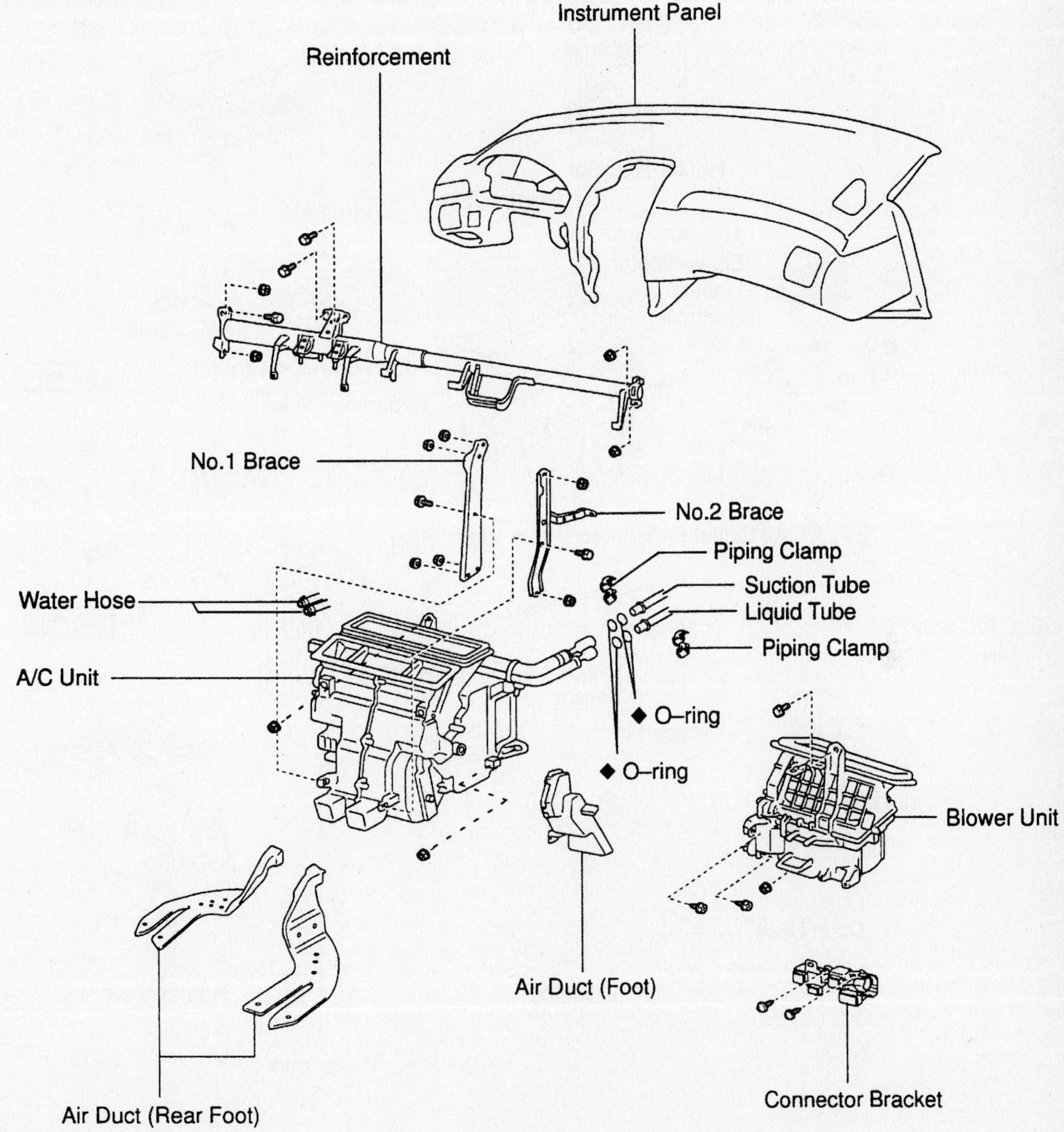

Exploded view of the instrument panel, heater/air conditioning housing and related components—Solara

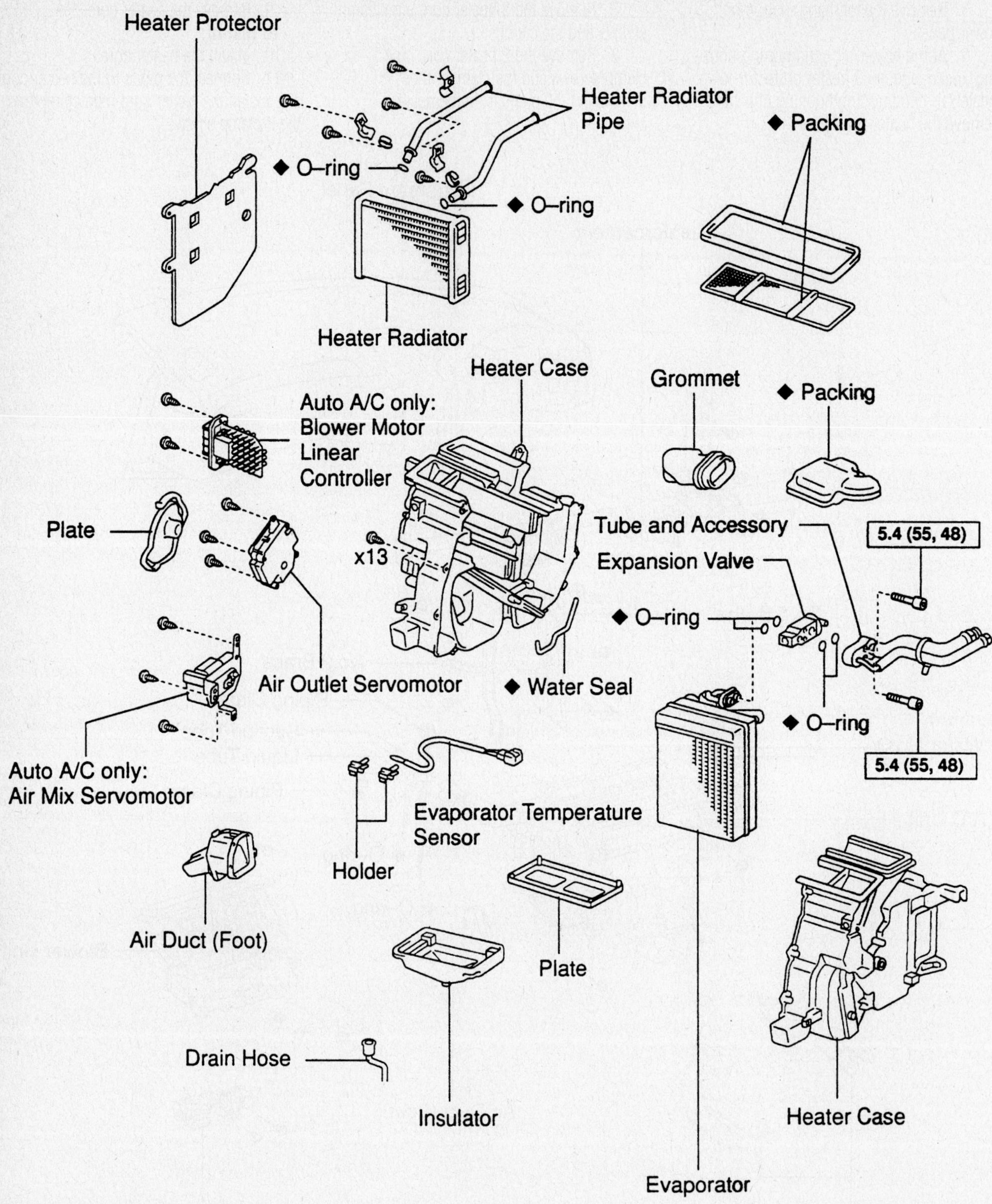

Exploded view of the heater/air conditioning housing, heater core, evaporator core and related components—Solara

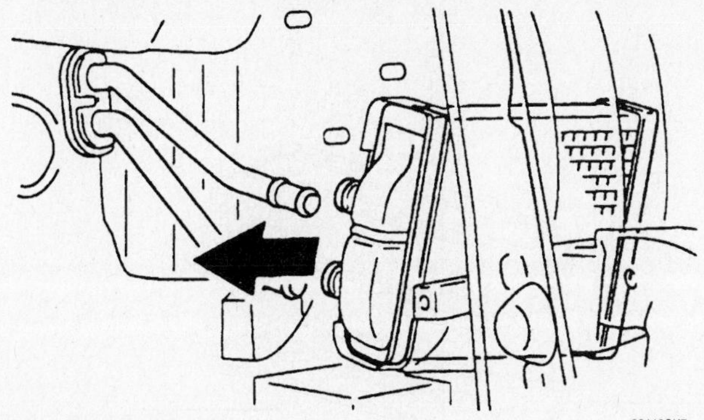
93112GK7

View of the heater core—Solara

12. Install the heater core pipe clamps and the 3 clamp screws.
13. At the heater/air conditioning housing, install the heater protector and engage the 3 heater protector-to-heater/air conditioning housing clips.
14. Install the left hand instrument lower panel.
15. At the driver's side, install the lower instrument panel.
16. Connect the heater hoses to the heater core.
17. Refill the cooling system.
18. Connect the negative battery cable.

Cylinder Head

REMOVAL & INSTALLATION

1NZ-FE Engines

1. Before servicing the vehicle, refer to the precautions in the beginning of this section.
2. Drain the cooling system.
3. Remove or disconnect the following:
   - Negative battery cable
   - Water filler
   - Outer front cowl top panel
   - Alternator
   - Air cleaner
   - Accelerator cable
   - Center exhaust pipe
   - Exhaust manifold support
   - Exhaust manifold
   - Ignition coil
   - Spark plugs
   - PCV hoses
   - Throttle body
   - Engine wiring harness at the head
   - Intake manifold
   - Camshaft position sensor
   - ECT sensor
   - Oil control valve
   - PCV valve
   - Oil filer cap
   - Cylinder head cover
   - Fuel injectors
   - Timing chain cover
   - Camshaft sprockets and valve timing control assembly
   - Camshafts
   - Cylinder head. Remove the bolts in a circular pattern, in several stages, starting from the ends and working towards the center

To install:

4. Install or connect the following:
   - Cylinder head, using a new gasket. The Lod. No. on the gasket faces UP
5. Torque the cylinder head bolts, in sequence, in 3 steps:
   a. Step 1: 22 ft. lbs. (29 Nm).
   b. Step 2: Plus a 90 degree turn.
   c. Step 3: Plus a 90 degree turn.
6. Install or connect the following:
   - Water bypass pipe. Torque the bolt to 80 inch lbs. (9 Nm).
   - Camshafts
7. Camshaft bearing caps in 2 stages:
   a. Step: 10 ft. lbs. (13 Nm).
   b. Step 2: 17 ft. lbs. (23 Nm).
8. Install or connect the following:
   - Sprockets and valve timing controller assembly, aligning the knock pin and hole. Torque the bolts to 47 ft. lbs. (64 Nm).
   - Check and adjust the valves
   - Cylinder head cover
   - Oil filler cap
   - PCV valve
   - ECT sensor
   - Camshaft position sensor
   - Timing chain cover
   - Intake manifold
   - Engine wiring harness
   - Throttle body
   - PCV hoses
   - Spark plugs
   - Ignition coils
   - Exhaust manifold
   - Exhaust manifold support. Torque the bolts to 27 ft. lbs. (37 Nm).
   - Front exhaust pipe. Torque the nuts to 46 ft. lbs. (62 Nm).
   - Accelerator cable
   - Air cleaner
   - Alternator
   - Water filler
   - Negative battery cable
9. Fill the cooling system to the proper level.
10. Start the vehicle, check for leaks and repair if necessary.

1ZZ-FE Engines

1. Before servicing the vehicle, refer to the precautions in the beginning of this section.
2. Drain the cooling system.
3. Remove or disconnect the following:
   - Battery

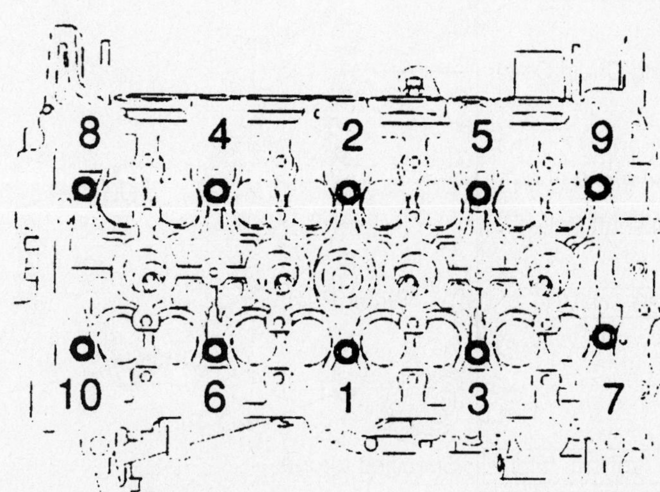

9307WG76

Fig. 1 Cylinder head bolt tightening sequence—1NZ-FE engine

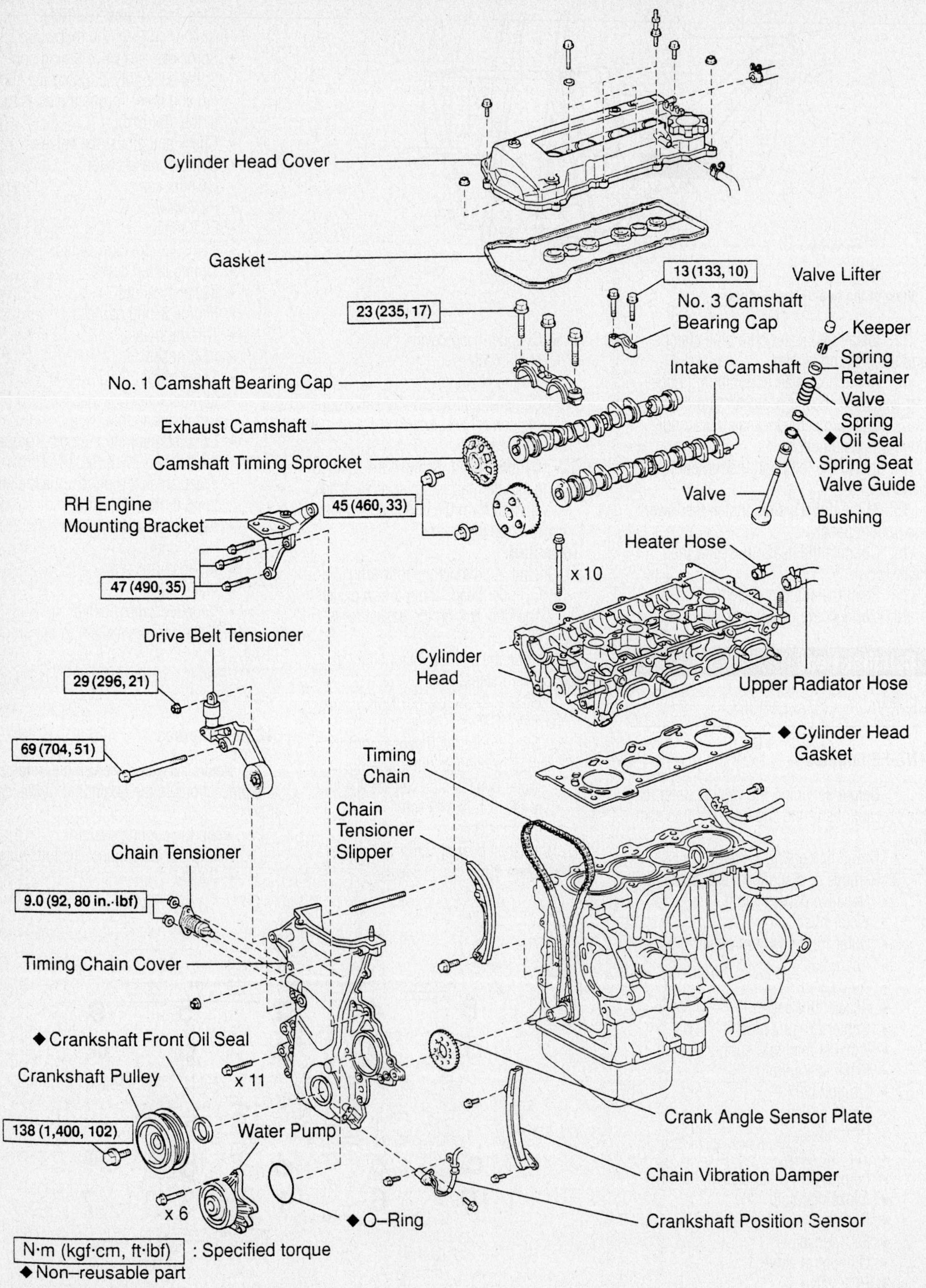
Cylinder Head Cover
Gasket
13 (133, 10)
Valve Lifter
23 (235, 17)
No. 3 Camshaft Bearing Cap
Keeper
Spring Retainer
No. 1 Camshaft Bearing Cap
Intake Camshaft
Valve Spring
Exhaust Camshaft
◆ Oil Seal
Camshaft Timing Sprocket
Spring Seat
Valve Guide
RH Engine Mounting Bracket
45 (460, 33)
Valve
Bushing
Heater Hose
47 (490, 35)
x 10
Drive Belt Tensioner
Cylinder Head
29 (296, 21)
Upper Radiator Hose
◆ Cylinder Head Gasket
69 (704, 51)
Timing Chain
Chain Tensioner Slipper
Chain Tensioner
9.0 (92, 80 in.·lbf)
Timing Chain Cover
◆ Crankshaft Front Oil Seal
Crankshaft Pulley
x 11
Crank Angle Sensor Plate
138 (1,400, 102)
Water Pump
Chain Vibration Damper
x 6
◆ O–Ring
Crankshaft Position Sensor
N·m (kgf·cm, ft·lbf) : Specified torque
◆ Non–reusable part
9307WG92

Cylinder head component removal—1ZZ-FE engine

- ECU box
- Coolant reservoir
- Air cleaner assembly
- Accelerator cable
- Alternator
- Exhaust pipe
- Exhaust manifold
- Coils
- Spark plugs
- PCV hoses
- Throttle body
- Injectors
- Wiring harness
- Intake manifold
- Camshaft position sensor
- ECT sensor
- PCV valve
- Oil filler cap
- Camshaft sprockets
- Camshafts
- Hoses
- Cylinder head bolts in sequence. To prevent damage to the cylinder head, loosen each bolt about ¼ of a turn during each pass until the bolts are loose.
- Cylinder head

To install:

4. Clean and degrease the surface of the cylinder head and engine block.
5. Install or connect the following:
   - New gasket on the engine block with the Lod No. stamp facing up.
   - Cylinder head
   - Apply a light coat of oil to cylinder head bolt threads and tighten in sequence. Replace any bolt that appears deformed. Bolts: 36 ft. lbs. (49 Nm).
   - Tighten each bolt in sequence an additional 90 degree turn.
   - Camshafts
   - Sprockets
   - Oil filler cap
   - PCV valve
   - ECT sensor
   - Intake manifold
   - Wiring harness
   - Exhaust manifold
   - Exhaust pipe
   - Alternator
   - accelerator cable
   - Air cleaner
   - ECM box
   - Battery
6. Fill the cooling system to the proper level.
7. Start the vehicle, check for leaks and repair if necessary.

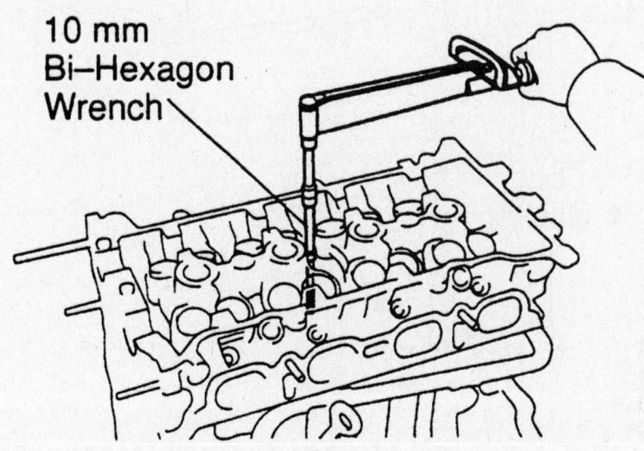

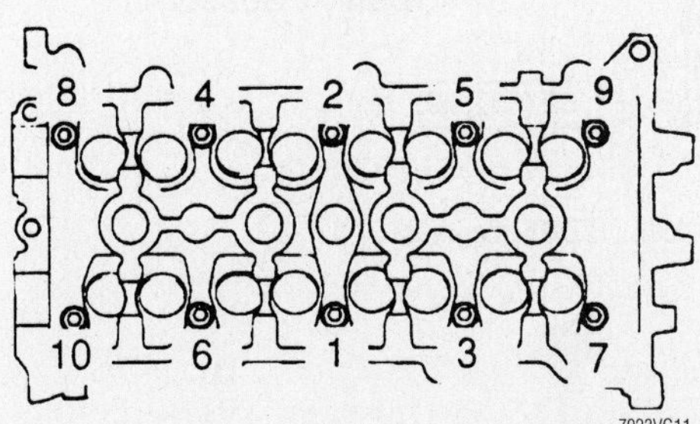

Cylinder head bolt tightening sequence—1ZZ-FE and 2ZZ-GE engines

2ZZ-GE Engines

1. Before servicing the vehicle, refer to the precautions in the beginning of this section.
2. Drain the cooling system.
3. Remove or disconnect the following:
   - Battery
   - ECU box
   - Coolant reservoir
   - Air cleaner assembly
   - Accelerator cable
   - Alternator
   - Exhaust pipe
   - Exhaust manifold
   - Coils
   - Spark plugs
   - PCV hoses
   - Throttle body
   - Injectors
   - Wiring harness
   - Intake manifold
   - Camshaft position sensor
   - ECT sensor
   - PCV valve
   - Oil filler cap
   - Camshaft sprockets
   - Camshafts
   - Hoses
   - Cylinder head bolts in sequence. To prevent damage to the cylinder head, loosen each bolt about ¼ of a turn during each pass until the bolts are loose.
   - Cylinder head

To install:

4. Clean and degrease the surface of the cylinder head and engine block.
5. Install or connect the following:
   - New gasket on the engine block with the Lod No. stamp facing up.
   - Cylinder head
   - Apply a light coat of oil to cylinder head bolt threads and tighten in sequence. Replace any bolt that appears deformed. Bolts: 26 ft. lbs. (49 Nm).
   - Torque each bolt in sequence an additional 180 degree turn.
   - Camshafts
   - Sprockets
   - Oil filler cap
   - PCV valve
   - ECT sensor
   - Intake manifold
   - Wiring harness
   - Exhaust manifold
   - Exhaust pipe
   - Alternator
   - accelerator cable
   - Air cleaner
   - ECM box
   - Battery
6. Fill the cooling system to the proper level.
7. Start the vehicle, check for leaks and repair if necessary.

5S-FE Engines

1. Before servicing the vehicle, refer to the precautions in the beginning of this section.
2. Drain the cooling system.
3. Remove or disconnect the following:
   - Negative battery cable.

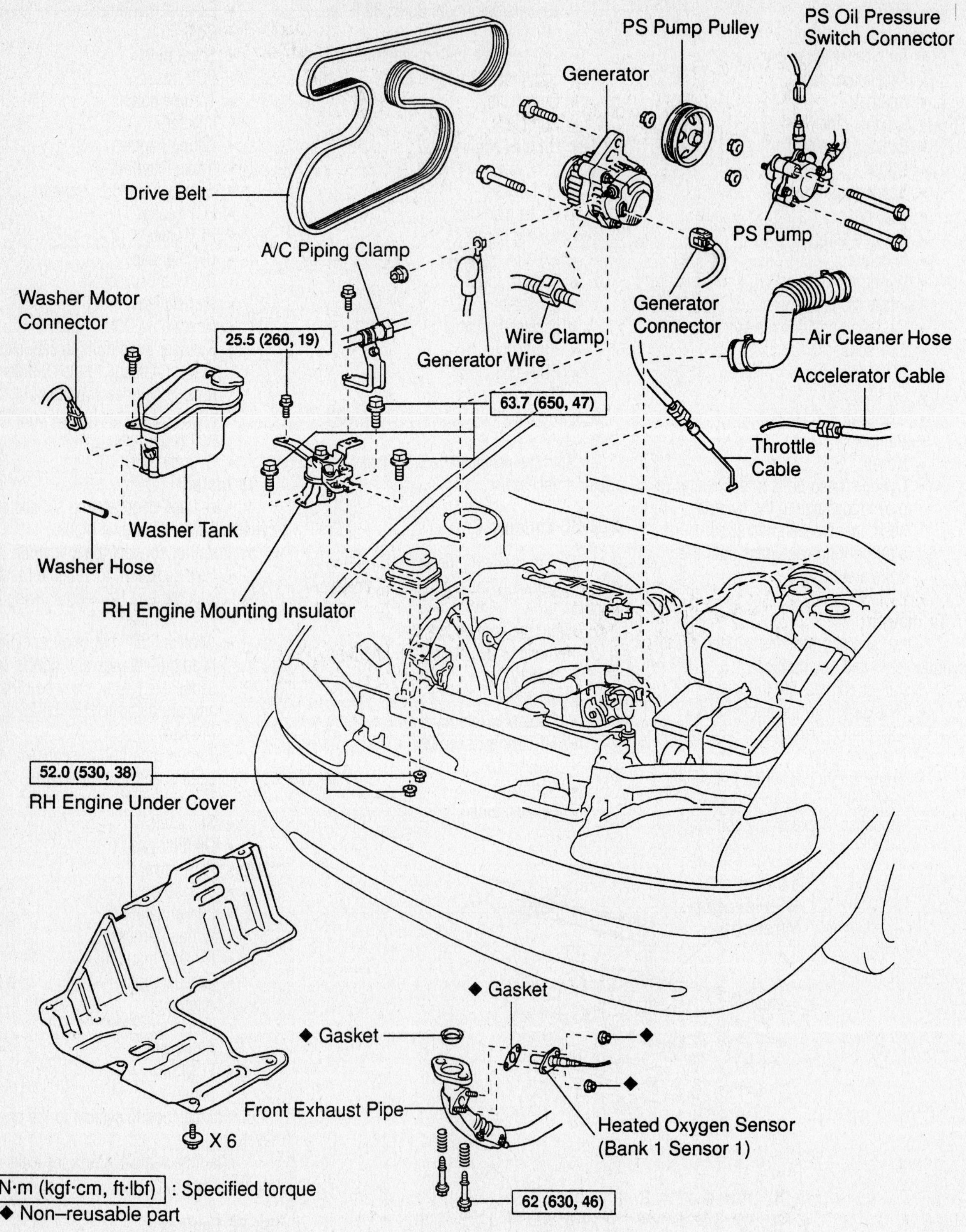

Exploded view of engine accessories and right-hand engine under cover—1.8L (1ZZ-FE) Engines.

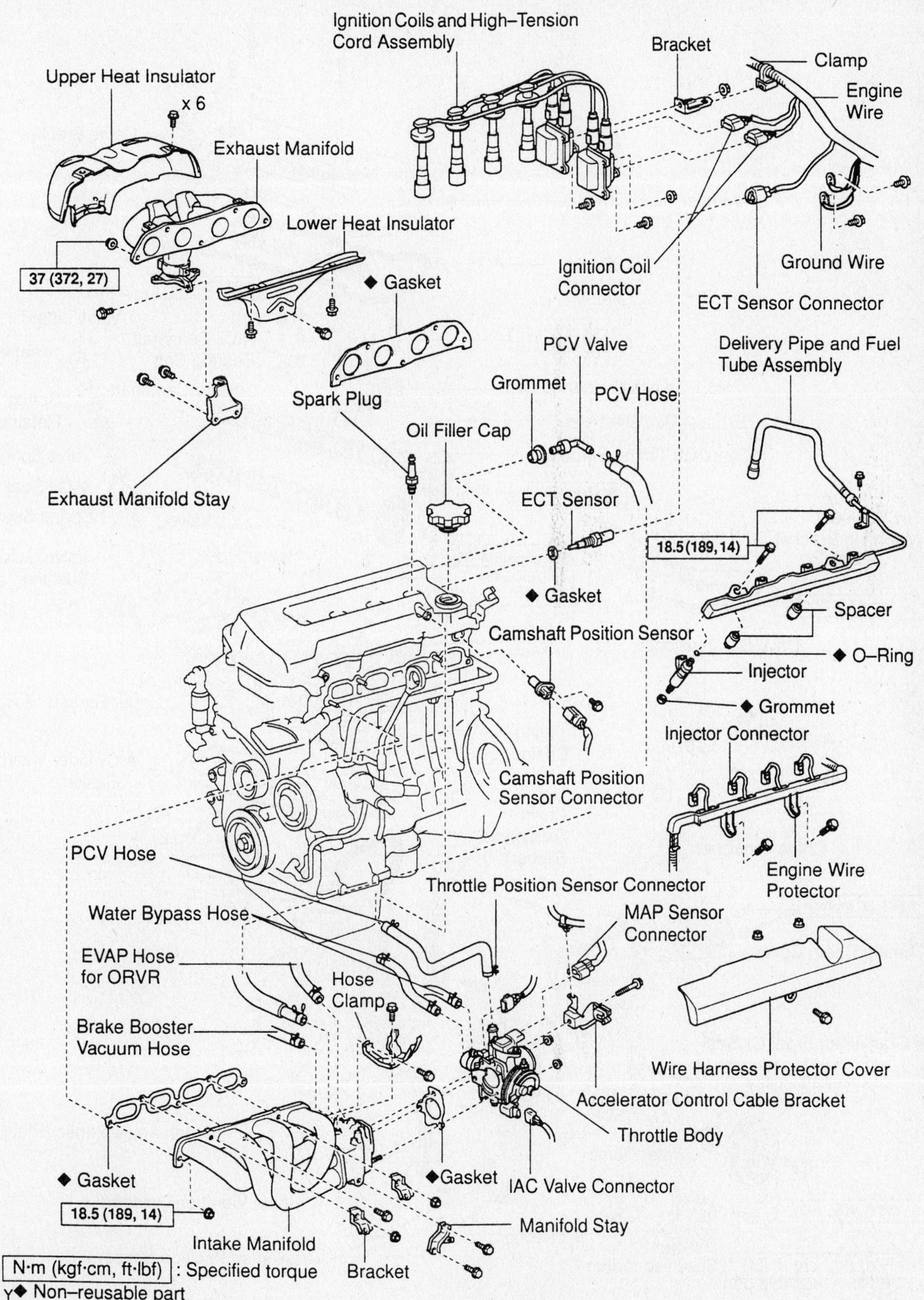
Ignition Coils and High–Tension Cord Assembly
Bracket
Clamp
Engine Wire
Upper Heat Insulator
x 6
Exhaust Manifold
Lower Heat Insulator
37 (372, 27)
◆ Gasket
Ignition Coil Connector
Ground Wire
ECT Sensor Connector
PCV Valve
Delivery Pipe and Fuel Tube Assembly
Grommet
PCV Hose
Spark Plug
Oil Filler Cap
Exhaust Manifold Stay
ECT Sensor
18.5 (189, 14)
◆ Gasket
Spacer
Camshaft Position Sensor
◆ O–Ring
Injector
◆ Grommet
Injector Connector
Camshaft Position Sensor Connector
PCV Hose
Engine Wire Protector
Throttle Position Sensor Connector
Water Bypass Hose
MAP Sensor Connector
EVAP Hose for ORVR
Hose Clamp
Brake Booster Vacuum Hose
Wire Harness Protector Cover
Accelerator Control Cable Bracket
Throttle Body
◆ Gasket
◆Gasket
IAC Valve Connector
18.5 (189, 14)
Manifold Stay
Intake Manifold
N·m (kgf·cm, ft·lbf) : Specified torque
Bracket
◆ Non–reusable part
9301WG04

View of engine intake, exhaust, ignition, and fuel system location—1.8L (1ZZ-FE) engines.

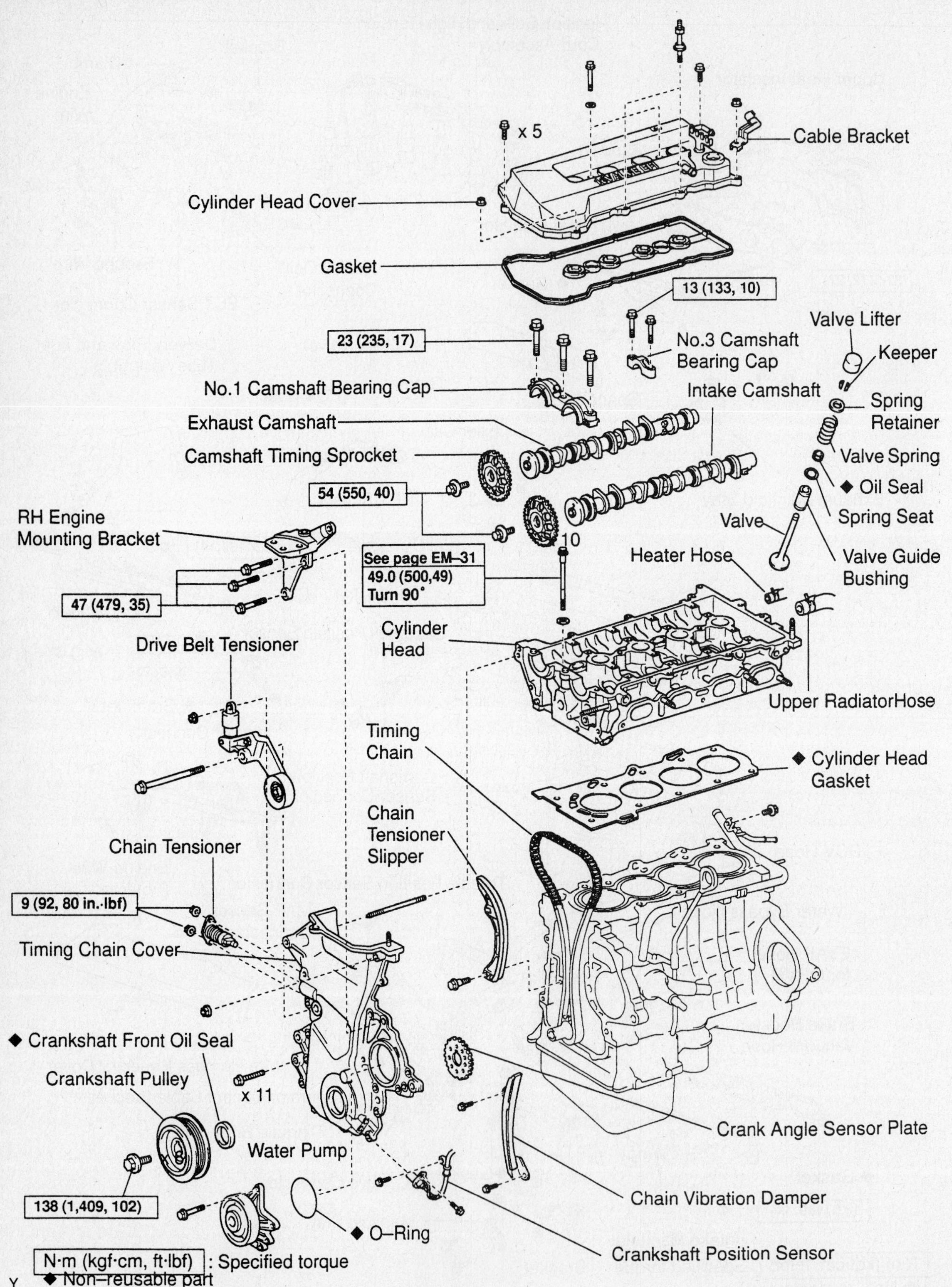
x 5
Cable Bracket
Cylinder Head Cover
Gasket
13 (133, 10)
Valve Lifter
23 (235, 17)
No.3 Camshaft Bearing Cap
Keeper
No.1 Camshaft Bearing Cap
Intake Camshaft
Spring Retainer
Exhaust Camshaft
Camshaft Timing Sprocket
Valve Spring
◆ Oil Seal
54 (550, 40)
Valve
Spring Seat
RH Engine Mounting Bracket
10
Heater Hose
Valve Guide Bushing
See page EM–31
49.0 (500,49)
Turn 90°
47 (479, 35)
Cylinder Head
Drive Belt Tensioner
Upper RadiatorHose
Timing Chain
◆ Cylinder Head Gasket
Chain Tensioner Slipper
Chain Tensioner
9 (92, 80 in.·lbf)
Timing Chain Cover
◆ Crankshaft Front Oil Seal
Crankshaft Pulley
x 11
Crank Angle Sensor Plate
Water Pump
138 (1,409, 102)
Chain Vibration Damper
◆ O–Ring
Crankshaft Position Sensor
N·m (kgf·cm, ft·lbf) : Specified torque
◆ Non–reusable part
Y
9301WG05

Illustration of disassembled cylinder head assembly—1.8L (1ZZ-FE) Engines.

Position the head gasket correctly on the cylinder head—1.8L (1ZZ-FE) engine

➡On vehicles equipped with an air bag, wait at least 90 seconds before proceeding after disconnecting the negative battery cable.

- All necessary components to gain access to the cylinder head using the diagram above.

➡On California vehicles, remove the Vacuum Switching Valve (VSV) for fuel pressure control and EGR. On all vehicles (except California), remove the VSV for EGR.

- Timing belt

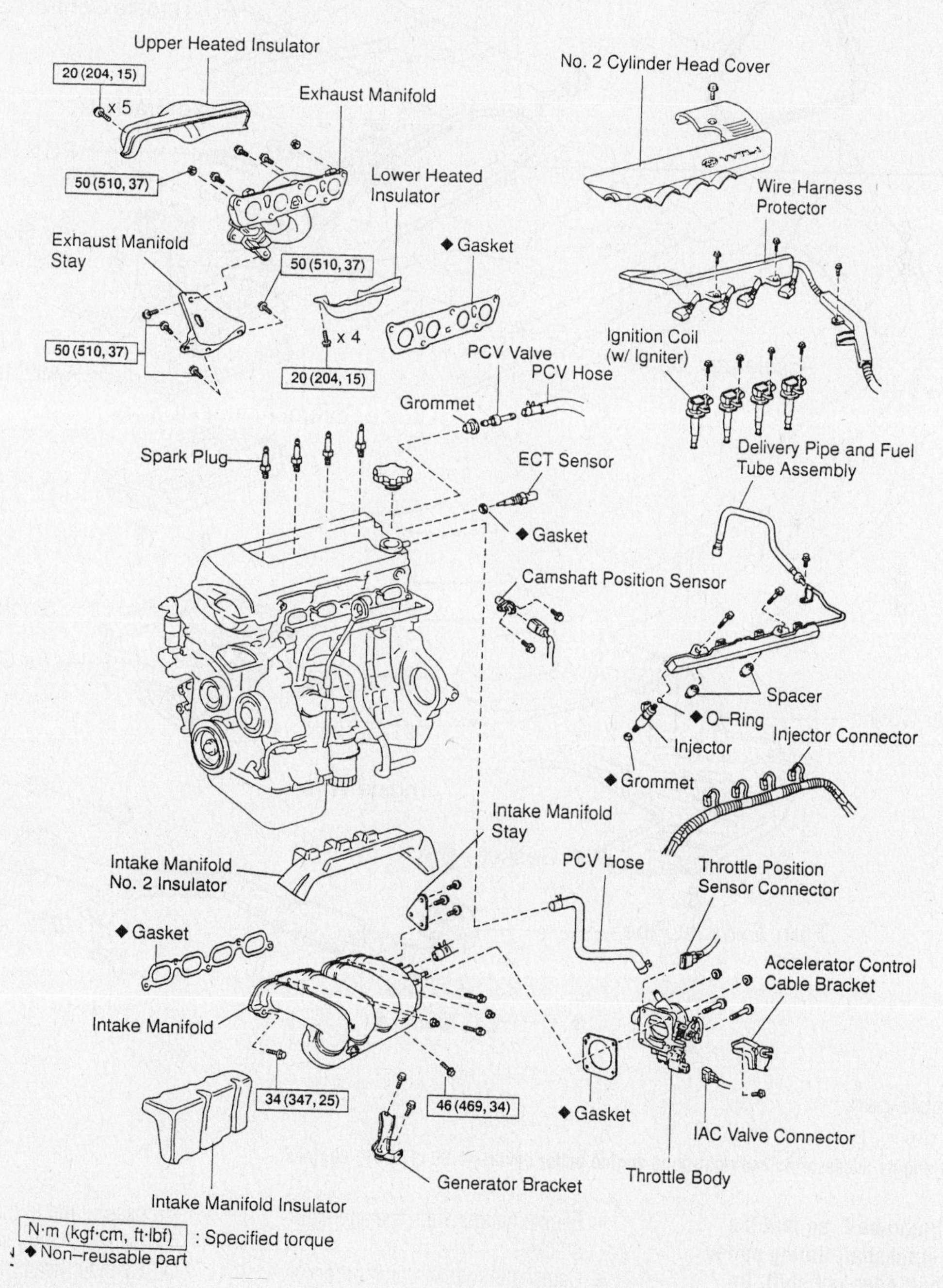

Cylinder head component exploded view—2ZZ-GE engine

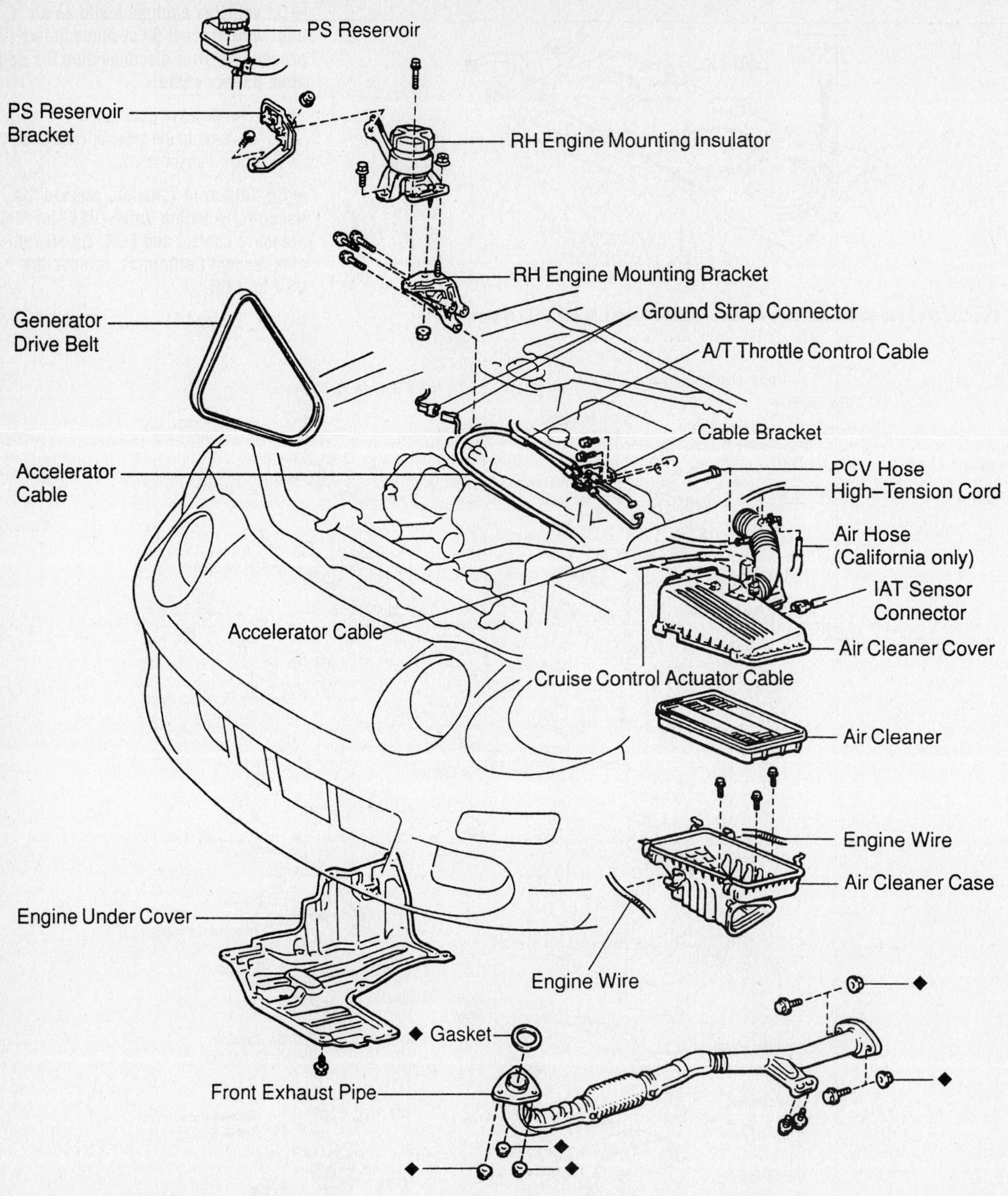

Exploded view of engine accessories and right-hand engine under cover—1.8L (1ZZ-FE) Engines.

➡Support the timing belt, so that the meshing of the crankshaft timing pulley and the timing belt does not shift. Be careful not to drop anything inside the timing belt cover.

- Engine hangers and the alternator bracket.
- Camshafts following the proper sequences and procedures.
- Cylinder head bolts in several passes and in the reverse order of the installation sequence.
- Cylinder head from the cylinder block, disengaging the cylinder head from the block dowel pins.

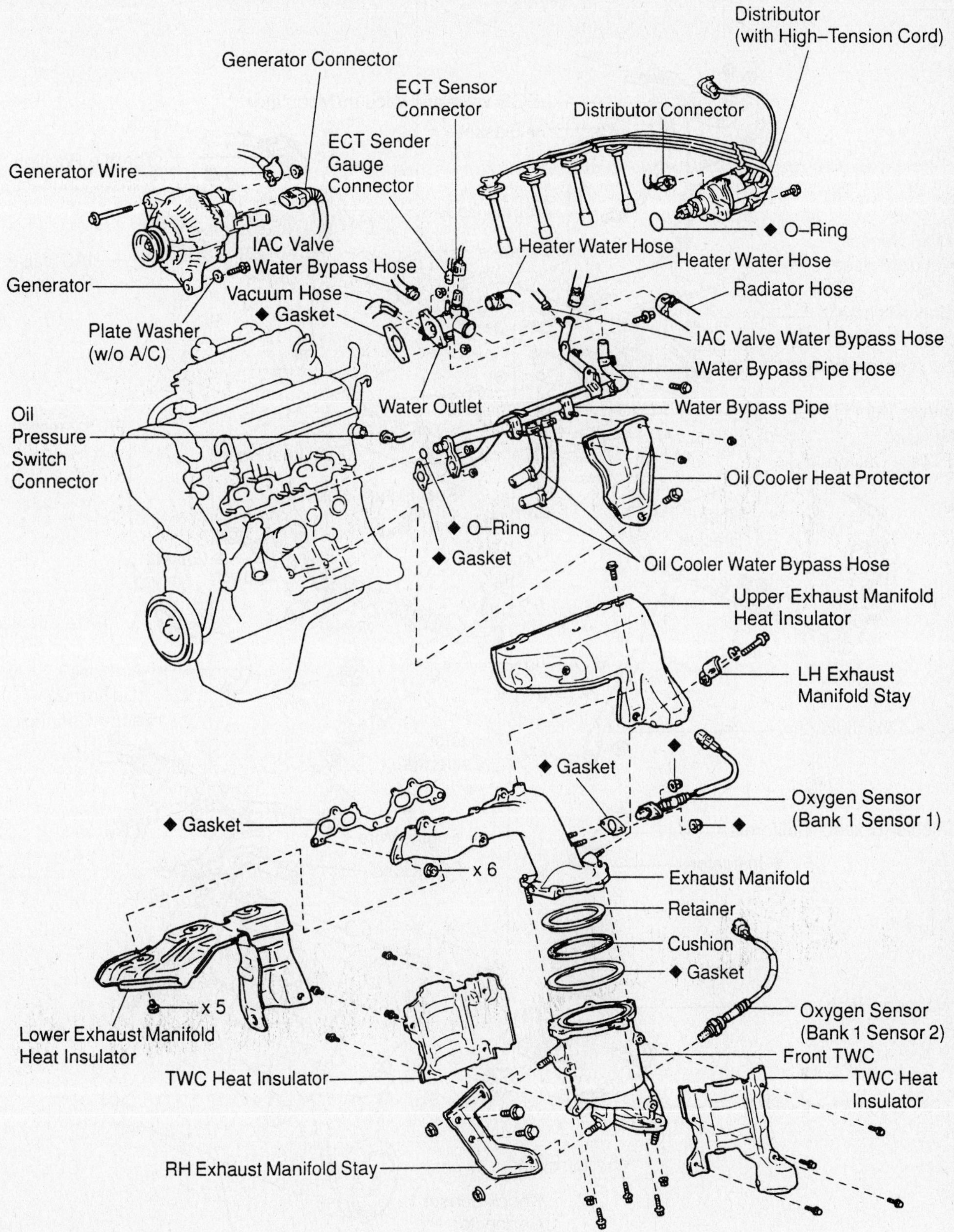
Distributor
(with High–Tension Cord)
Generator Connector
ECT Sensor
Connector
Distributor Connector
ECT Sender
Gauge
Connector
Generator Wire
◆ O–Ring
IAC Valve
Heater Water Hose
Water Bypass Hose
Heater Water Hose
Generator
Vacuum Hose
Radiator Hose
◆ Gasket
Plate Washer
(w/o A/C)
IAC Valve Water Bypass Hose
Water Bypass Pipe Hose
Oil
Pressure
Switch
Connector
Water Outlet
Water Bypass Pipe
Oil Cooler Heat Protector
◆ O–Ring
◆ Gasket
Oil Cooler Water Bypass Hose
Upper Exhaust Manifold
Heat Insulator
LH Exhaust
Manifold Stay
◆ Gasket
Oxygen Sensor
(Bank 1 Sensor 1)
◆ Gasket
x 6
Exhaust Manifold
Retainer
Cushion
◆ Gasket
x 5
Oxygen Sensor
(Bank 1 Sensor 2)
Lower Exhaust Manifold
Heat Insulator
Front TWC
TWC Heat Insulator
TWC Heat
Insulator
RH Exhaust Manifold Stay
◆ Non–reusable part
9301WG10

View of engine, exhaust, ignition and fuel system location—1.8L (1ZZ-FE) Engines.

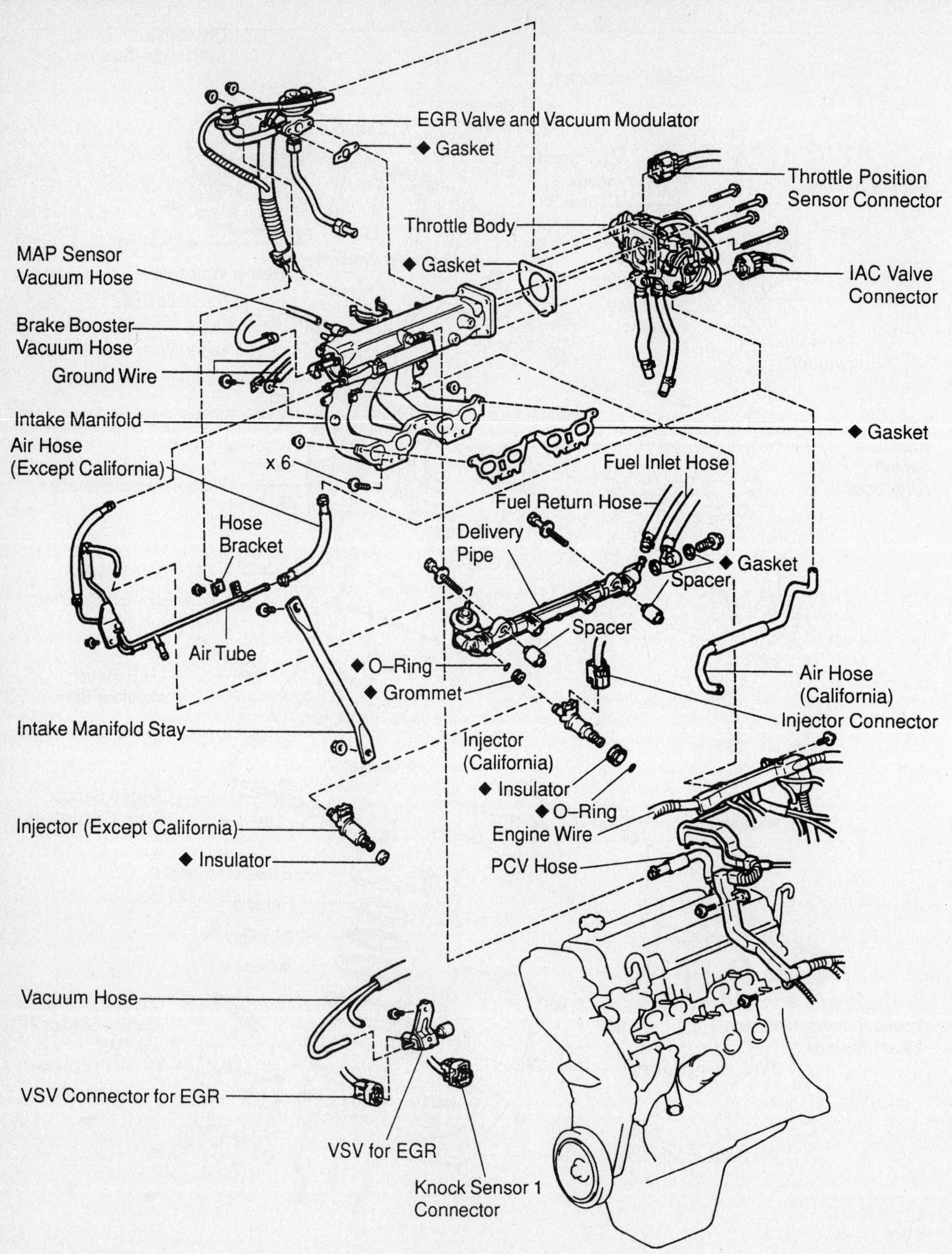

View of engine, intake, ignition and fuel system location—1.8L (1ZZ-FE) Engines.

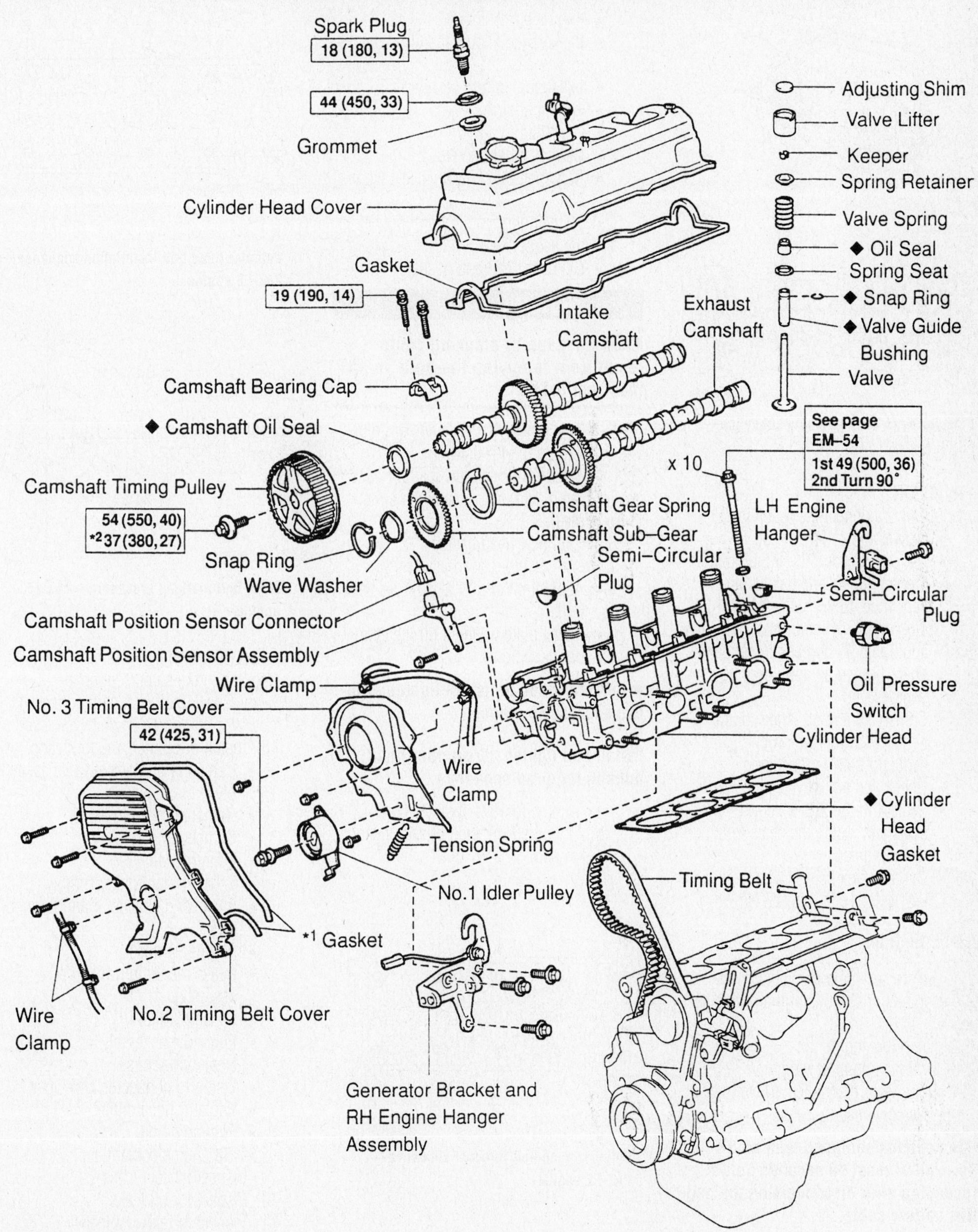
Spark Plug
18 (180, 13)
44 (450, 33)
Grommet
Cylinder Head Cover
Gasket
19 (190, 14)
Intake Camshaft
Exhaust Camshaft
Camshaft Bearing Cap
◆ Camshaft Oil Seal
Camshaft Timing Pulley
54 (550, 40)
*2 37 (380, 27)
Snap Ring
Wave Washer
Camshaft Position Sensor Connector
Camshaft Position Sensor Assembly
Wire Clamp
No. 3 Timing Belt Cover
42 (425, 31)
Wire Clamp
Tension Spring
No.1 Idler Pulley
*1 Gasket
Wire Clamp
No.2 Timing Belt Cover
Generator Bracket and RH Engine Hanger Assembly
Camshaft Gear Spring
Camshaft Sub–Gear
Semi–Circular Plug
Adjusting Shim
Valve Lifter
Keeper
Spring Retainer
Valve Spring
◆ Oil Seal
Spring Seat
◆ Snap Ring
◆ Valve Guide Bushing
Valve
See page EM–54
x 10
1st 49 (500, 36)
2nd Turn 90°
LH Engine Hanger
Semi–Circular Plug
Oil Pressure Switch
Cylinder Head
◆ Cylinder Head Gasket
Timing Belt
N·m (kgf·cm, ft·lbf) : Specified torque
*1 Replace only if damaged
*2 For use with SST
◆ Non–reusable part
9301WG12

Illustration of disassembled cylinder head assembly—1.8L (1ZZ-FE) Engines.

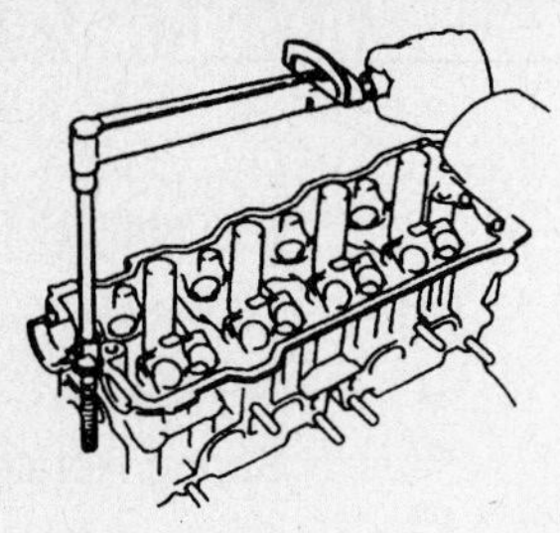

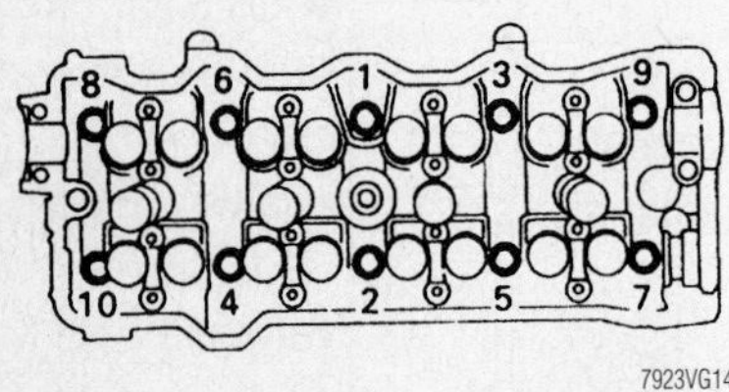

Cylinder head bolt tightening sequence—5S-FE engine

To install:

4. Install or connect the following:
   - Gasket and cylinder head
   - Cylinder head bolts, lightly oiled. Tighten, in several passes and in sequence, to 36 ft. lbs. (49 Nm). Tighten the cylinder head bolts an additional 90 degrees in sequence. Then, tighten an additional 90 degrees.
   - Camshafts and all other components following the proper sequences and procedures.
   - Check and adjust valve clearance.
   - Sealant to the 2 new semi-circular seals and install the seals to the cylinder head.
   - Cylinder head cover with new gasket. Nuts: 17 ft. lbs. (23 Nm).

2AZ-FE Engines

1. Before servicing the vehicle, refer to the precautions in the beginning of this section.
2. Drain the engine oil.
3. Drain the cooling system.
4. Remove or disconnect the following:
   - Negative battery cable.

➡On vehicles equipped with an air bag, wait at least 90 seconds before proceeding after disconnecting the negative battery cable.

   - Radiator hose outlet
   - Union to connector tube hose
   - Heater inlet hose
   - Fuel tube assembly
   - Intake manifold
   - Intake manifold runner valve assembly
   - Engine harness
   - Intake and exhaust manifold insulators
   - Exhaust manifold assembly
   - Timing chain
   - Camshafts
   - Camshaft bearing No. 2
   - Camshaft oil control valve assembly
   - 10 cylinder head bolts uniformly in the sequence
   - Cylinder head and gasket

⁂ WARNING

Head warpage or cracking could result from removing the bolts in an incorrect order.

5. Inspect the cylinder head set bolts. They should be 6.350–6.465 in. (161.3–164.2mm) in length. If length is greater than maximum, replace the bolts.

To install:

6. Install new head gasket
7. Install cylinder head assembly
   - Apply light oil to the cylinder head bolts
   - Install plate washers on the cylinder head bolts

➡Cylinder head bolts are tightened in two successive steps.

Install and tighten 10 cylinder head bolts in required sequence.

a. Tighten to 58 ft. lbs. (79 Nm)

b. Mark the front side of each head bolt with paint.

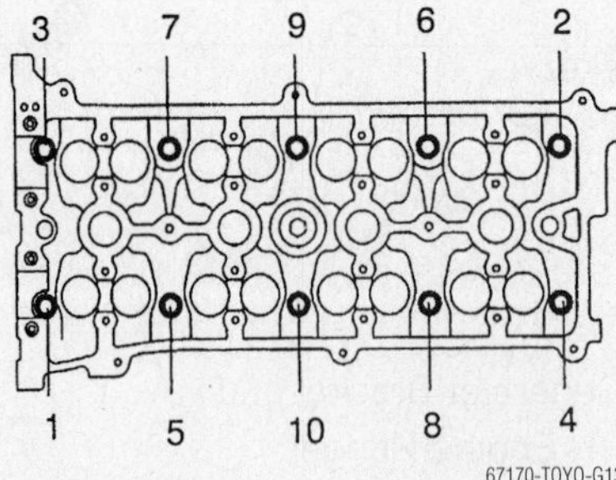

Cylinder head bolt removal sequence—2AZ-FE engine

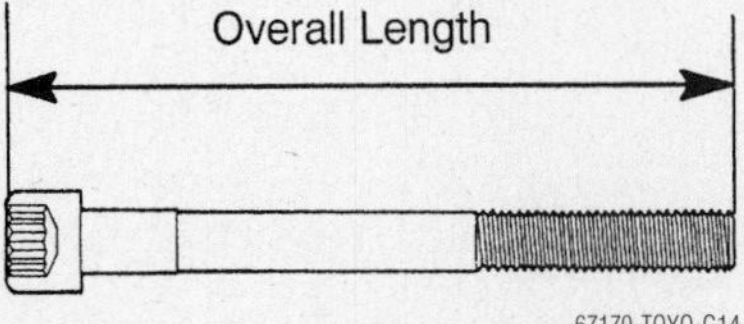

Measuring bolt length—2AZ-FE engine

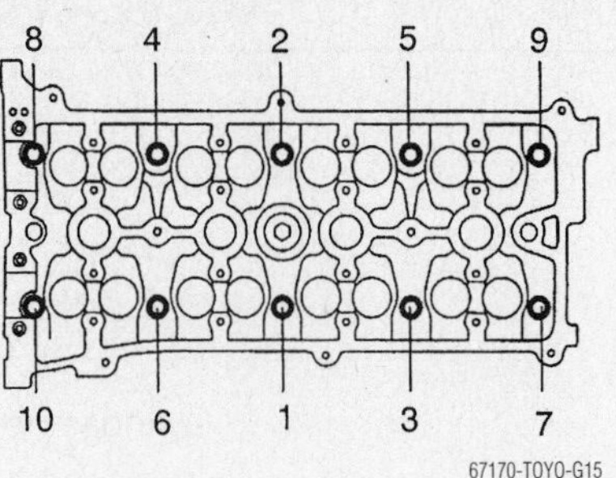

Cylinder head bolt installation sequence—2AZ-FE engine

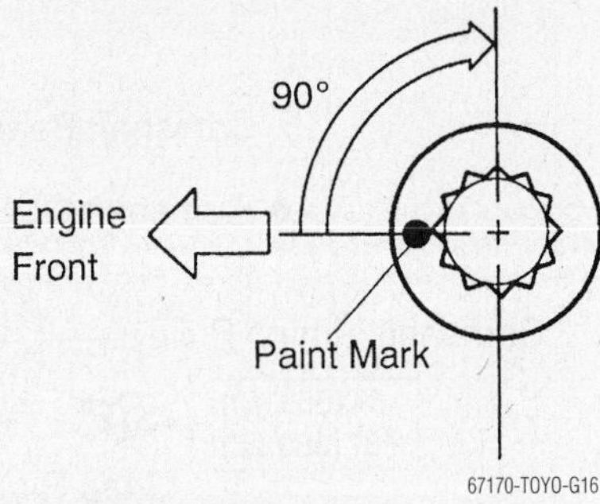

Head bolt marking procedure—2AZ-FE engine

c. Retighten cylinder head bolts 90 degrees in the same sequence

d. Check that each painted mark is now at a 90 degree angle to the front

8. Install or connect the following:
   - Camshaft oil control valve assembly
   - Camshaft bearing No. 2
   - Camshafts
   - Timing chain
   - Exhaust manifold assembly
   - Intake and exhaust manifold insulators
   - Engine harness
   - Intake manifold runner valve assembly
   - Intake manifold
   - Fuel tube assembly
   - Heater inlet hose
   - Union to connector tube hose
   - Radiator hose outlet
   - Fill engine oil
   - Fill engine coolant
9. Inspect for fuel leaks
10. Inspect for oil leaks
11. Inspect for exhaust leaks

1MZ-FE and 3MZ-FE Engines

2001 MODELS

1. Before servicing the vehicle, refer to the precautions in the beginning of this section.
2. Drain the engine oil.

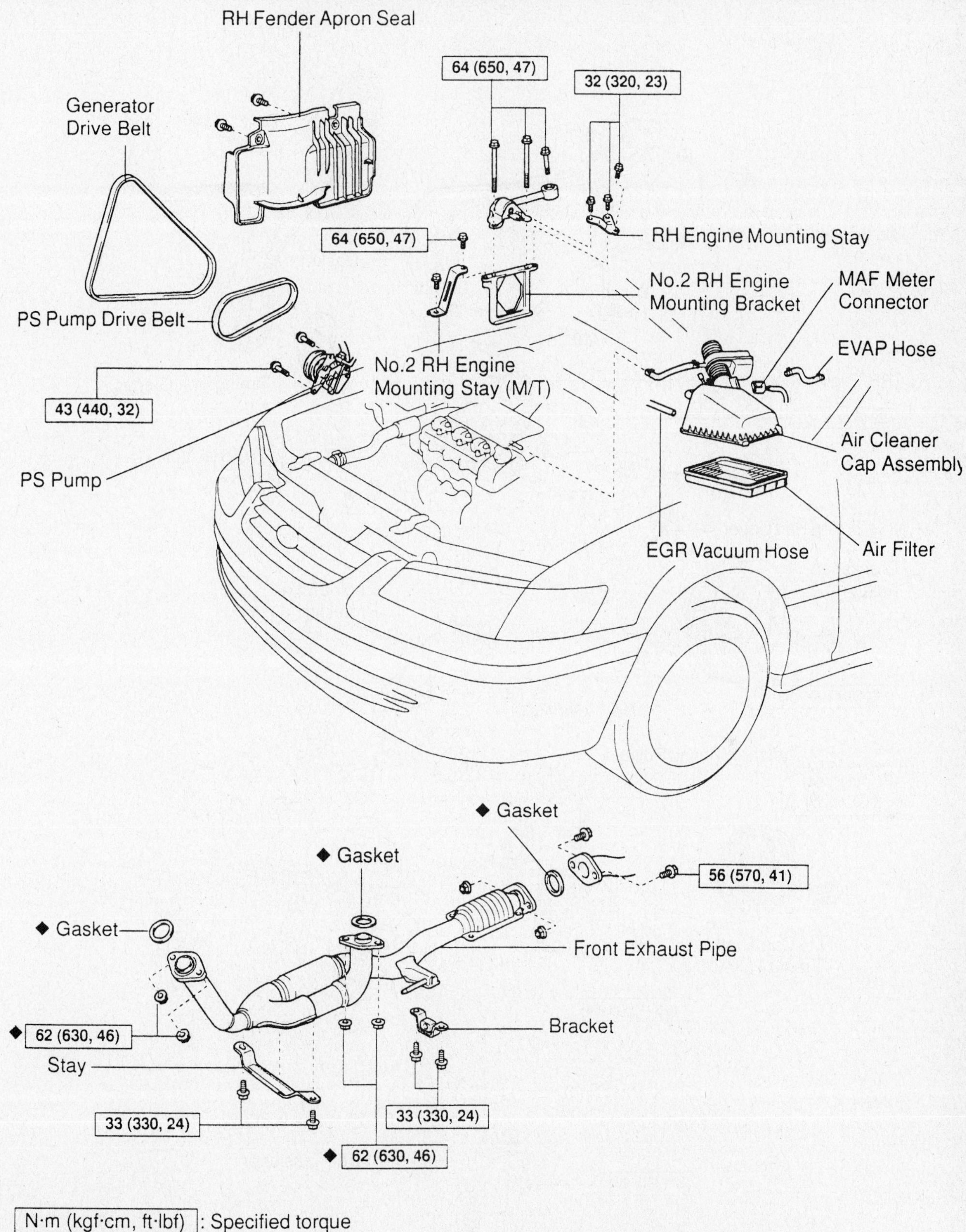

Removal of the right-hand fender apron, front exhaust pipe, and air cleaner assembly—2001 1MZ-FE engine

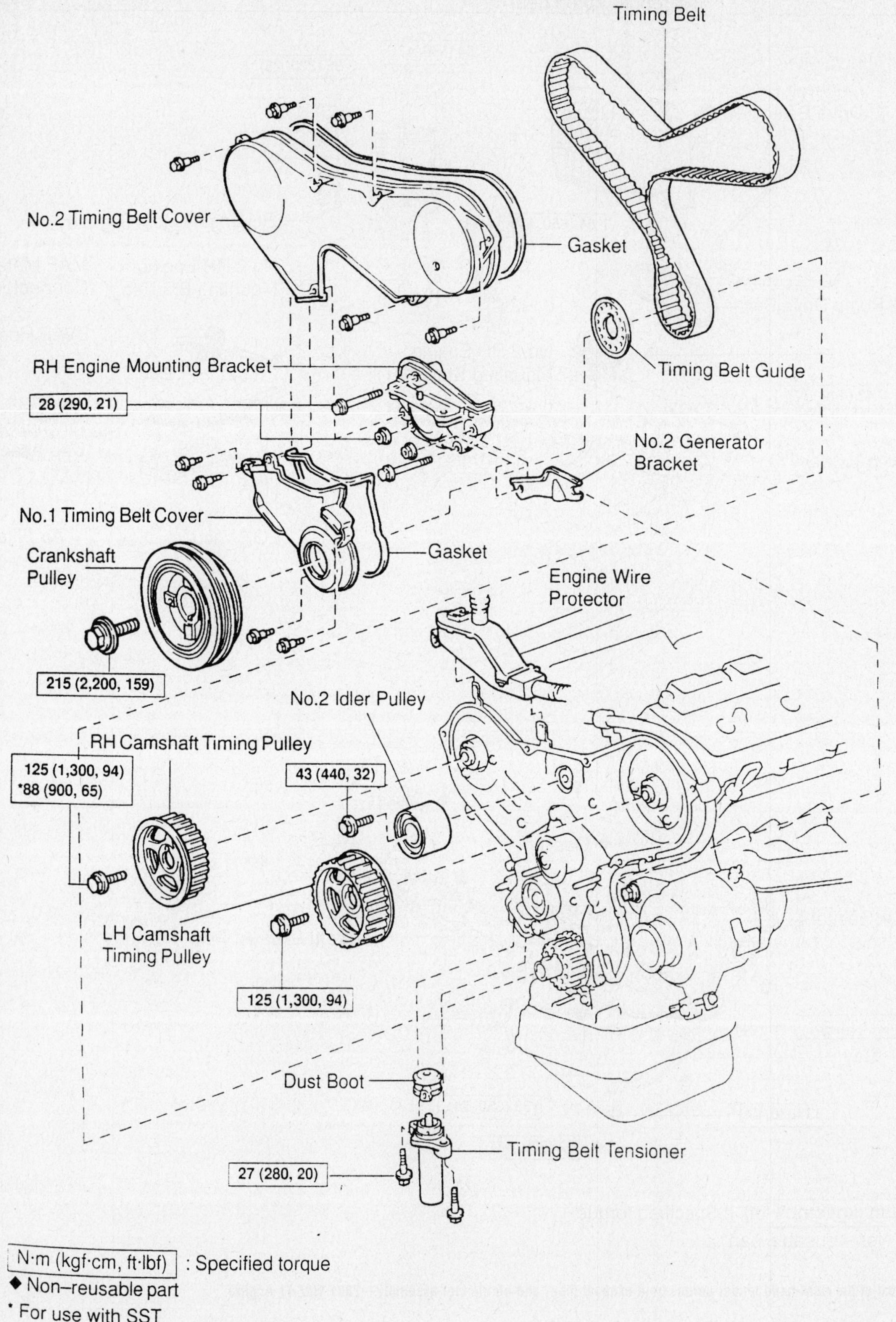
Timing Belt
No.2 Timing Belt Cover
Gasket
Timing Belt Guide
RH Engine Mounting Bracket
28 (290, 21)
No.2 Generator Bracket
No.1 Timing Belt Cover
Crankshaft Pulley
Gasket
Engine Wire Protector
215 (2,200, 159)
No.2 Idler Pulley
RH Camshaft Timing Pulley
125 (1,300, 94)
*88 (900, 65)
43 (440, 32)
LH Camshaft Timing Pulley
125 (1,300, 94)
Dust Boot
Timing Belt Tensioner
27 (280, 20)
N·m (kgf·cm, ft·lbf) : Specified torque
◆ Non–reusable part
* For use with SST
9301WG23

Timing belt and pulleys—2001 1MZ-FE engine

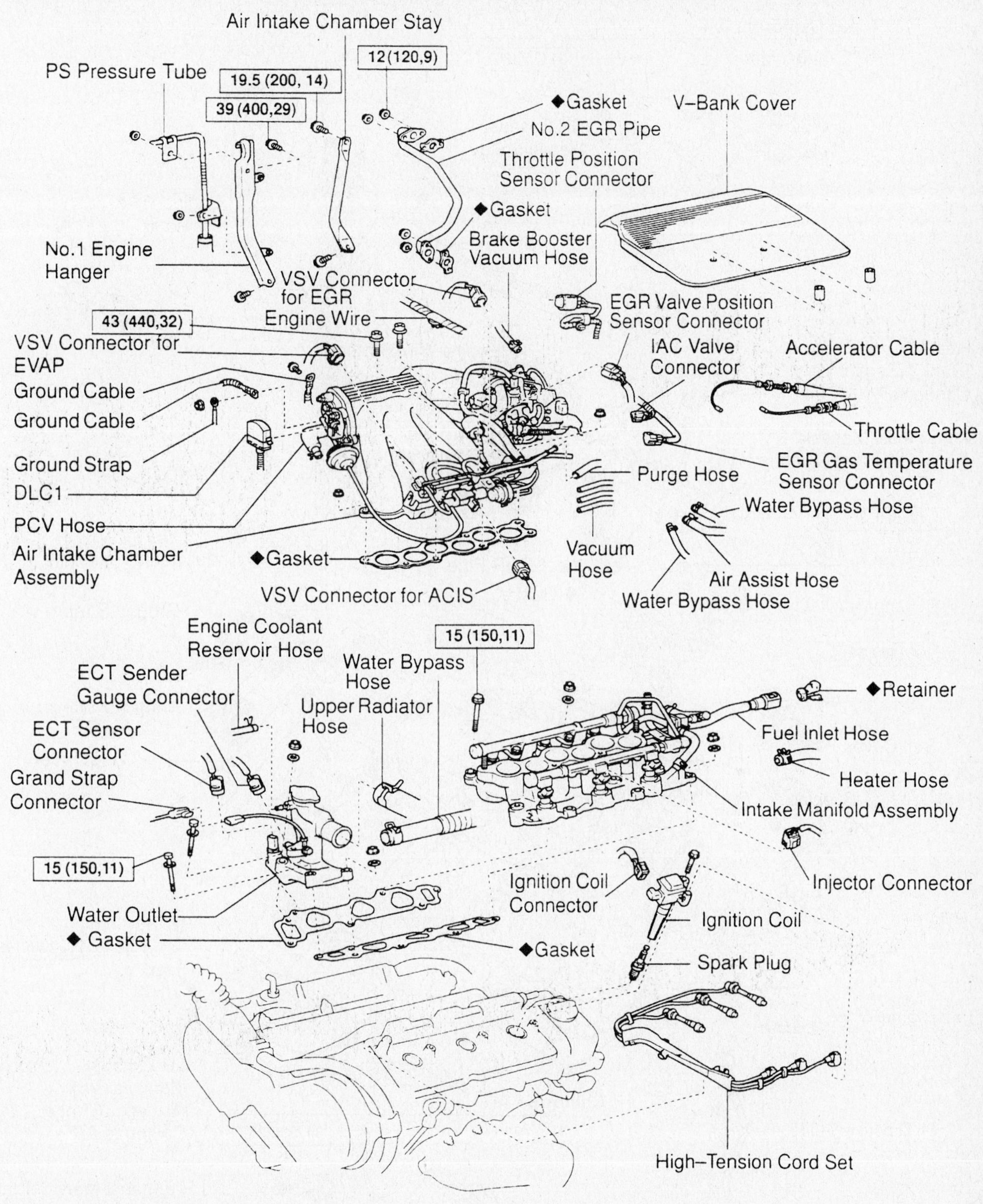
Air Intake Chamber Stay
PS Pressure Tube
19.5 (200, 14)
12(120,9)
39 (400,29)
◆Gasket
V–Bank Cover
No.2 EGR Pipe
Throttle Position
Sensor Connector
◆Gasket
Brake Booster
Vacuum Hose
No.1 Engine
Hanger
VSV Connector
for EGR
Engine Wire
43 (440,32)
EGR Valve Position
Sensor Connector
VSV Connector for
EVAP
IAC Valve
Connector
Accelerator Cable
Ground Cable
Ground Cable
Throttle Cable
Ground Strap
Purge Hose
EGR Gas Temperature
Sensor Connector
DLC1
Water Bypass Hose
PCV Hose
Air Intake Chamber
Assembly
◆Gasket
Vacuum
Hose
Air Assist Hose
VSV Connector for ACIS
Water Bypass Hose
Engine Coolant
Reservoir Hose
15 (150,11)
Water Bypass
Hose
ECT Sender
Gauge Connector
◆Retainer
Upper Radiator
Hose
ECT Sensor
Connector
Fuel Inlet Hose
Grand Strap
Connector
Heater Hose
Intake Manifold Assembly
15 (150,11)
Injector Connector
Ignition Coil
Connector
Water Outlet
Ignition Coil
◆ Gasket
◆Gasket
Spark Plug
High–Tension Cord Set
N·m (kgf·cm, ft·lbf) : Specified torque
◆ Non–reusable part
9301WG22

Intake manifold removal—2001 1MZ-FE engine

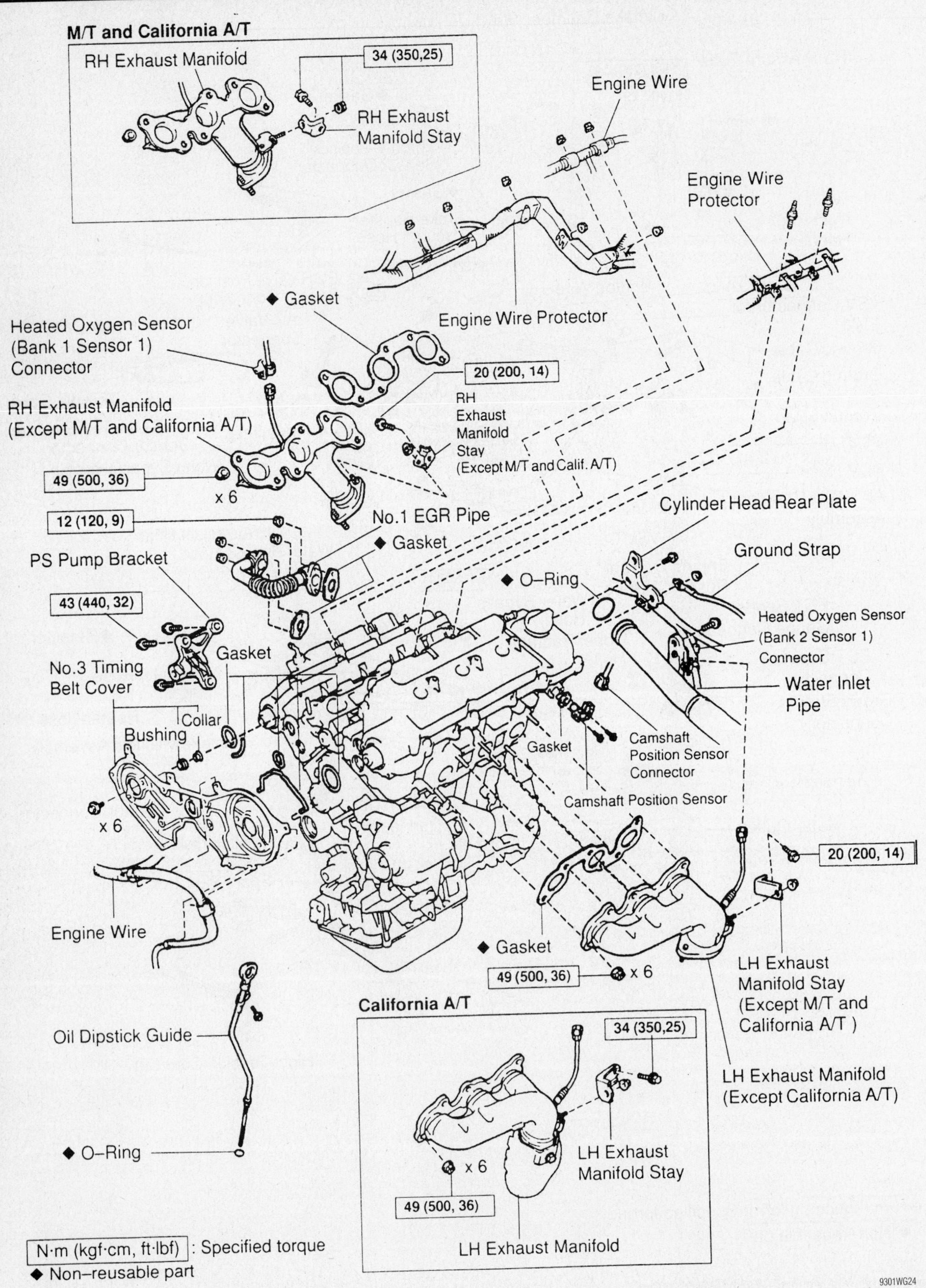
M/T and California A/T
RH Exhaust Manifold
34 (350,25)
RH Exhaust Manifold Stay
Engine Wire
Engine Wire Protector
◆ Gasket
Engine Wire Protector
Heated Oxygen Sensor (Bank 1 Sensor 1) Connector
20 (200, 14)
RH Exhaust Manifold Stay (Except M/T and Calif. A/T)
RH Exhaust Manifold (Except M/T and California A/T)
49 (500, 36)
x 6
12 (120, 9)
No.1 EGR Pipe
◆ Gasket
Cylinder Head Rear Plate
Ground Strap
PS Pump Bracket
◆ O–Ring
43 (440, 32)
Heated Oxygen Sensor (Bank 2 Sensor 1) Connector
Gasket
No.3 Timing Belt Cover
Water Inlet Pipe
Collar
Bushing
Camshaft Position Sensor Connector
Gasket
x 6
Camshaft Position Sensor
20 (200, 14)
Engine Wire
◆ Gasket
49 (500, 36)
x 6
LH Exhaust Manifold Stay (Except M/T and California A/T)
California A/T
34 (350,25)
Oil Dipstick Guide
LH Exhaust Manifold (Except California A/T)
◆ O–Ring
LH Exhaust Manifold Stay
x 6
49 (500, 36)
LH Exhaust Manifold
N·m (kgf·cm, ft·lbf) : Specified torque
◆ Non–reusable part
9301WG24

Exhaust manifold removal—2001 1MZ-FE engine

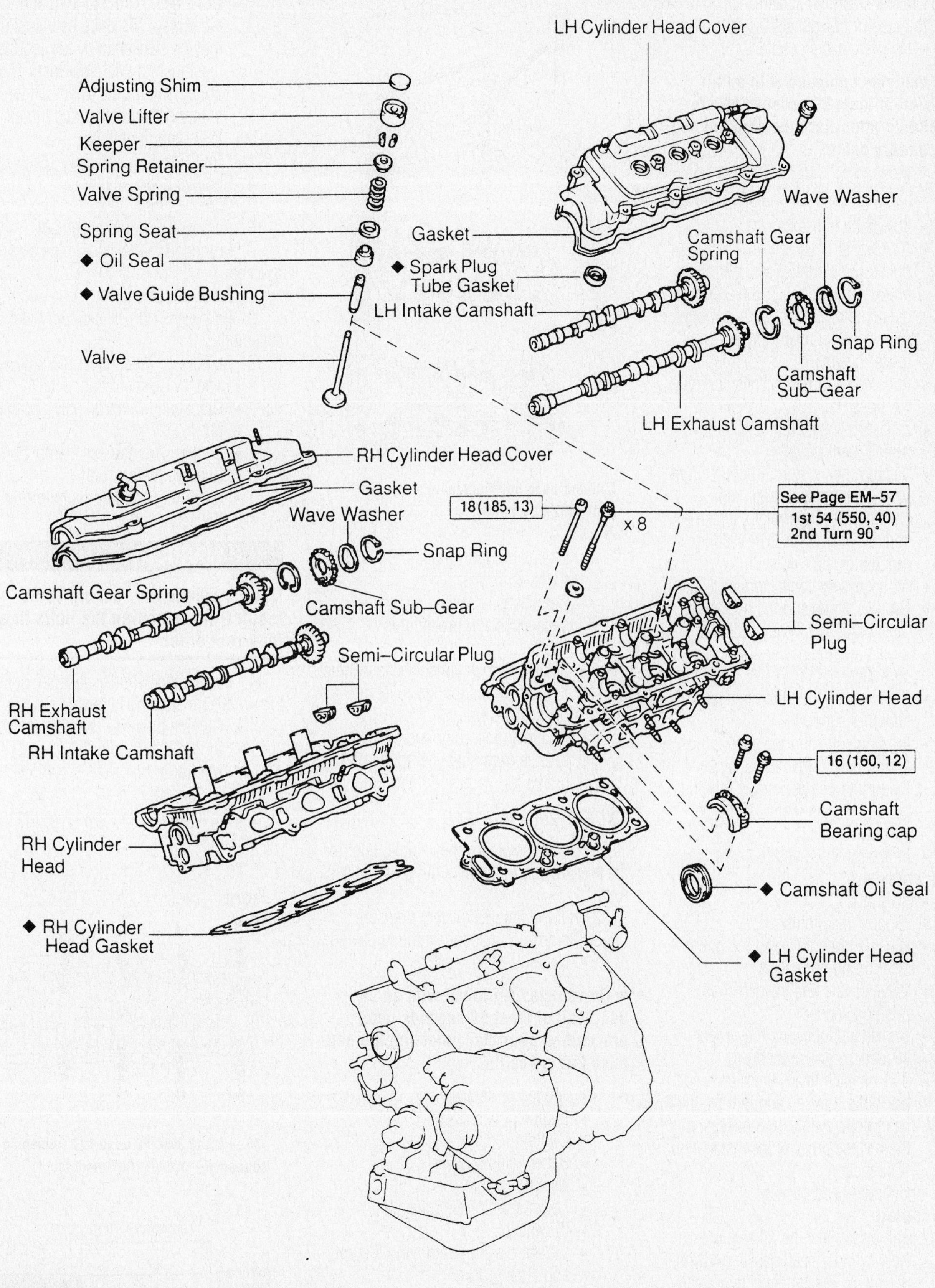

Cylinder head gasket position and torque specification—2001 1MZ-FE engine

3. Drain the cooling system.
4. Remove or disconnect the following:
   - Negative battery cable.

➡On vehicles equipped with an air bag, wait at least 90 seconds before proceeding after disconnecting the negative battery cable.

- Accelerator cable and the throttle cable on vehicles equipped with an automatic transaxle.
- Air cleaner cover, air flow meter, air duct, cruise control actuator and bracket (if equipped), the 2 engine ground straps, right engine mounting support, radiator hoses, and 2 heater hoses
- Fuel feed and return lines and plug the lines
- Pressure hose, plug, and remove from V-bank cover
- Vacuum hoses from: Fuel pressure VSV; Fuel pressure regulator; Cylinder head rear plate; Intake air control valve VSV; EGR vacuum modulator; EGR valve
- All necessary connectors and hoses
- Ground straps and the hydraulic motor pressure hose
- 2 nuts and the power steering pressure tube
- Engine hanger and the intake chamber support
- Ignition coils and the spark plugs, timing belt, camshaft pulleys and the timing belt rear cover, cylinder head rear plate, water inlet pipe, and water outlet
- Intake manifold and fuel rail assembly
- EGR pipe
- Exhaust manifolds
- Dipstick assembly and the power steering pump bracket
- Valve covers and the camshaft position sensor
- Camshafts following the proper sequences and procedures
- 2 (1 on each head) 8mm recessed hex bolts. Loosen and remove the 8 head bolts evenly, in 3 passes, in the reverse order of the installation sequence
- Cylinder head gasket

To install:

5. Install or connect the following:
   - New cylinder head gasket on cylinder block. Place the cylinder head onto the gasket.
   - Cylinder head bolts into the cylinder head. Bolts: 40 ft. lbs. (54 Nm), then an additional 90 degrees

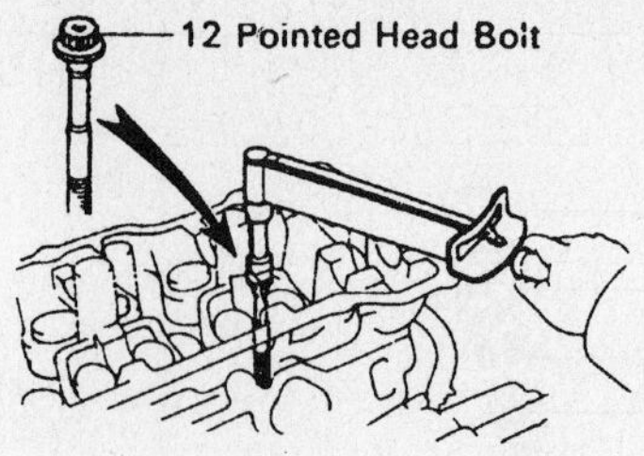

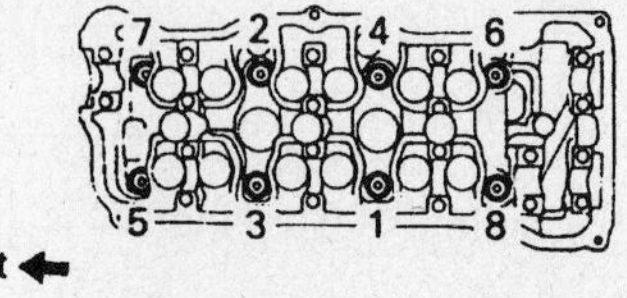

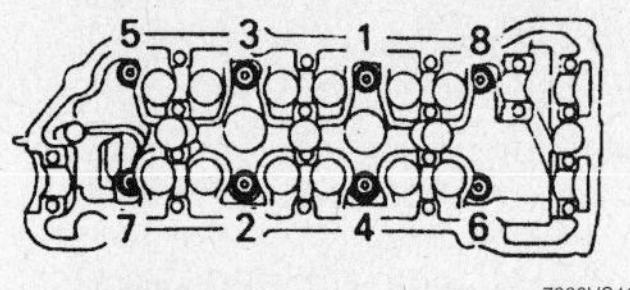

Cylinder head bolt tightening sequence—2001 1MZ-FE engine

   - 2 remaining 8mm bolts. Bolts: 13 ft. lbs. (18 Nm).
   - Camshafts following the proper sequences and procedures
   - Check and adjust the valves
   - Sealant to the cylinder heads where the camshaft supports meet the cylinder heads
   - All remaining components in reverse of removal using the diagrams for the proper torque figure.

EXCEPT 2001 MODELS

1. Before servicing the vehicle, refer to the precautions in the beginning of this section.
2. Relieve the fuel system pressure.
3. Remove or disconnect the following:
   - Negative battery cable.

➡On vehicles equipped with an air bag, wait at least 90 seconds before proceeding after disconnecting the negative battery cable.

- Oil pan protector
- Engine undercover
- Coolant
- Battery clamp cover
- Air cleaner inlet
- Lower radiator air deflector
- RF wheel
- V bank cover by removing the bolt and 2 cap nuts
- Air cleaner and intake air connector assembly
- Emission control valve
- Air intake surge tank
- Drive belt, fluid coupling and the fan pulley. The drive belt tension may be slackened by turning the tensioner counterclockwise. The pulley bolt for the drive belt tensioner has a left-handed thread.
- PS pump drive belt
- Radiator

4. Remove intake manifold
5. Remove timing belt
6. Remove PS pump assembly
7. Front and rear exhaust pipe and brackets
8. Remove camshafts
9. Remove LH or RH cylinder head assemblies
10. Remove or disconnect the following:
    - The VVT Sensor connector
    - Camshaft timing oil valve connector
    - Engine wire harness clamp
    - Remove the hex bolt
    - 8 cylinder head bolts uniformly in the sequence

WARNING

Head warpage or cracking could result from removing the bolts in an incorrect order.

11. Inspect the cylinder head set bolts. Ensure they match the following:
    - Outside diameter is .3524 to .3563 in. (8.95 to 9.05mm)
    - Minimum diameter is .3445 in. (8.75mm)
12. If diameter is lees than minimum, replace the bolts.

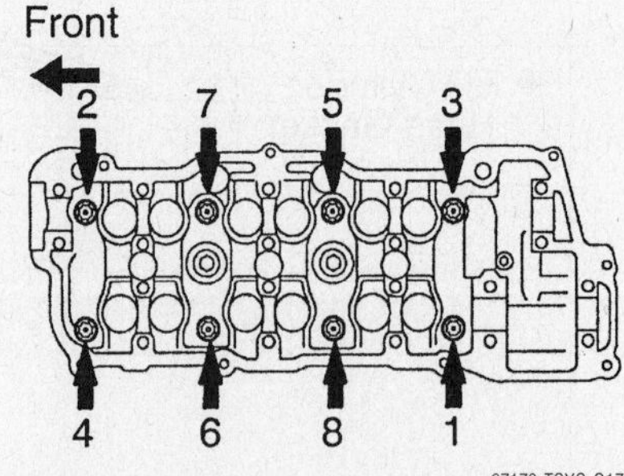

1MZ-FE and 3MZ-FE head bolt loosening sequence—except 2001 models

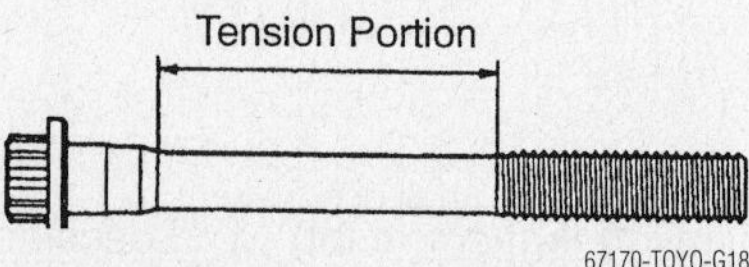

Check the diameter of the bolt in this area—except 2001 models

To install:

13. Install new head gasket with R mark upward
14. Install cylinder head assembly
    - Apply light oil to the cylinder head bolts
    - Install plate washers on the cylinder head bolts

➡**Cylinder head bolts are tightened in two successive steps.**

Install and tighten 8 cylinder head bolts in required sequence.

a. Tighten to 40 ft. lbs (54 Nm)
b. Mark the front side of each head bolt with paint.
c. Retighten cylinder head bolts 90 degrees in the same sequence

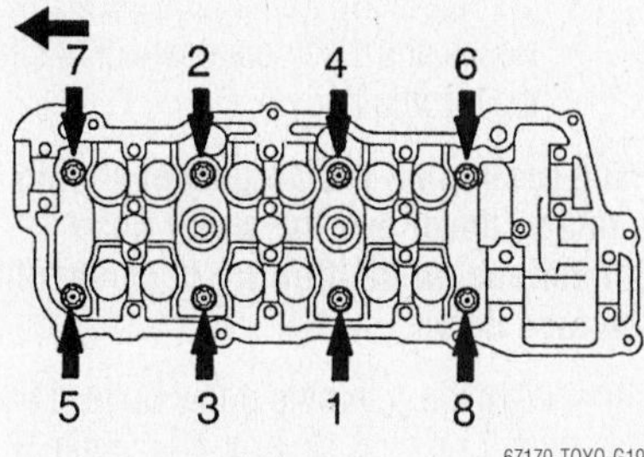

1MZ-FE and 3MZ-FE head bolt tightening sequence—except 2001 models

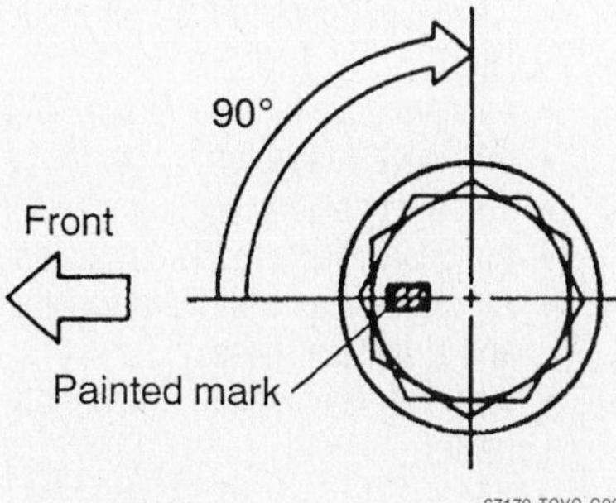

Mark the bolts as shown—except 2001 models

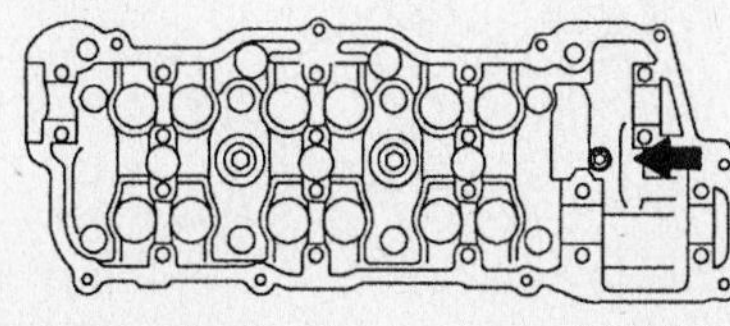

Hex head bolt location—except 2001 models

d. Check that each painted mark is now at a 90 degree angle to the front

15. Install or connect the following:
    - Tighten the hex bolt to 14 ft. lbs. (19 Nm)
    - Wiring harness clamp
    - Camshaft timing oil valve connector
    - Camshaft assemblies
    - Valve cover assemblies
    - Exhaust manifold assemblies and support brackets. Torque to 36 ft. lbs. (49 Nm)
    - Exhaust manifold heat insulators
    - PS pump assembly
    - Timing belt inner cover
    - Camshaft timing pulleys
    - Timing belt and idler assemblies
    - RH engine mounting bracket
    - Timing belt covers
    - Alternator bracket
    - Engine mounting stay No 2
    - Engine stabilizer rod
    - PS drive belt
    - A/C compressor drive belt
    - Water outlet
    - Intake manifold assembly
    - Intake air surge tank
    - Emission control valve set
    - Air cleaner assembly
    - Vacuum hoses
    - V-bank cover sub-assembly
    - Front suspension upper brace
    - RF wheel
    - Install engine oil
    - Installed engine coolant
16. Inspect for fuel leaks
17. Inspect for oil leaks
18. Inspect for exhaust leaks

Intake Manifold

REMOVAL & INSTALLATION

1NZ-FE Engines

1. Before servicing the vehicle, refer to the precautions in the beginning of this section.
2. Drain the cooling system.
3. Remove or disconnect the following:
    - Negative battery cable
    - Water filler
    - Outer front cowl top panel
    - Alternator
    - Air cleaner
    - Accelerator cable
    - Ignition coil
    - PCV hoses
    - Throttle body
    - Intake manifold and discard the gasket

To install:

4. Install or connect the following:
    - Intake manifold with a new gasket. Uniformly tighten the bolts and nuts, in several passes, from the ends, working towards the center, to 22 ft. lbs. (30 Nm).
    - Engine wiring harness
    - Throttle body
    - PCV hoses
    - Ignition coils
    - Accelerator cable
    - Air cleaner
    - Alternator
    - Water filler
    - Negative battery cable
5. Fill the cooling system.
6. Start the vehicle, check for leaks and repair if necessary.

1ZZ-FE Engines

1. Before servicing the vehicle, refer to the precautions in the beginning of this section.
2. Drain the cooling system.
3. Remove or disconnect the following:
    - Negative battery cable
    - Drive belt and alternator
    - Air intake duct
    - Accelerator cable
    - Exhaust pipe from the manifold.
    - Exhaust manifold support bracket
    - Spark plug wires, then ignition coils
    - Spark plugs
    - PCV hoses
    - Throttle body assembly
    - 2 bolts securing the wiring harness protector
    - Wiring connectors and ground wires
    - Intake manifold support bracket
    - Intake manifold and gasket

To install:

4. Install or connect the following:
    - Intake manifold with a new gasket. Torque the bolts to 14 ft. lbs. (18.5 Nm).
    - Harness wiring to the cylinder head and harness protector
    - Fuel injectors, throttle body and the PCV hoses
    - Spark plugs and ignition coils. Bolts and nuts: 80 inch lbs. (9 Nm).
    - Exhaust manifold and support bracket. Bolts: 37 ft. lbs. (49 Nm).
    - Front exhaust pipe to the manifold. Bolts: 46 ft. lbs. (62 Nm).
    - Oxygen sensor. Nuts: 14 ft. lbs. (20 Nm).

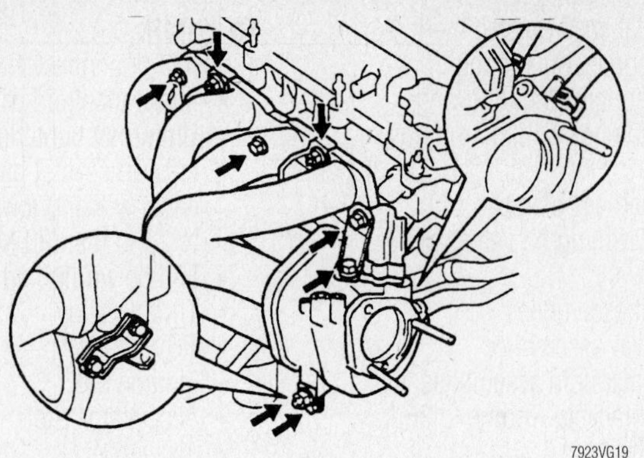

Intake manifold mounting fastener locations—1.8L (1ZZ-FE) engine

- Accelerator cable and air intake duct
- Alternator and drive belt
- Negative battery cable

5. Fill the cooling system.
6. Start the vehicle, check for leaks and repair if necessary.

2ZZ-GE Engines

1. Before servicing the vehicle, refer to the precautions in the beginning of this section.
2. Drain the cooling system.
3. Remove or disconnect the following:
   - Negative battery cable
   - Drive belt and alternator
   - Air intake duct
   - Accelerator cable
   - Spark plug wires, then ignition coils
   - Spark plugs
   - PCV hoses
   - Throttle body assembly
   - Wiring harness
   - Hoses and tubes connected to the head
   - Intake manifold support bracket
   - Intake manifold and gasket

To install:

4. Install or connect the following:
   - Intake manifold with a new gasket. Bolts A: 20 ft. lbs. (27 Nm); bolt B: 34 ft. lbs. (46 Nm)
   - Harness wiring to the cylinder head and harness protector
   - Fuel injectors, throttle body and the PCV hoses
   - Spark plugs and ignition coils. Bolts and nuts: 80 inch lbs. (9 Nm)
   - Oxygen sensor. Nuts: 14 ft. lbs. (20 Nm).
   - Accelerator cable and air intake duct
   - Alternator and drive belt
   - Negative battery cable
5. Fill the cooling system.

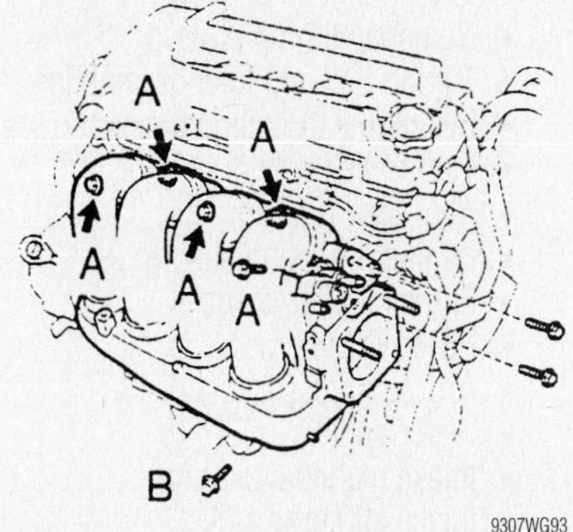

Intake manifold bolt installation—2ZZ-GE engine

6. Start the vehicle, check for leaks and repair if necessary.

5S-FE Engines

1. Before servicing the vehicle, refer to the precautions in the beginning of this section.
2. Drain the cooling system.
3. Remove or disconnect the following:
   - Negative battery cable.

➡On vehicles equipped with an air bag, wait at least 90 seconds before proceeding after disconnecting the negative battery cable.

   - Accelerator cable. With automatic transmission, throttle cable from the throttle body
   - Intake air temperature sensor connector
   - On California models, air cleaner hose
   - Air cleaner hose clamp bolt, air cleaner cap clips, air hose from the throttle body, and the air cleaner cap together with the resonator and air cleaner hose
   - Electrical connections and hoses from the throttle body
   - Throttle body. Type A throttle bodies are secured with 4 bolts and Type B throttle bodies are secured with 2 bolts and 2 nuts
   - Vacuum hose bracket and the engine wiring harness
   - EGR valve
   - No. 1 air intake chamber, and manifold stays
   - Intake manifold and gasket

To install:

4. Install or connect the following:
   - Intake manifold with a new gasket. Torque the bolts to 14 ft. lbs. (19 Nm)
   - Wire clamps to the wire brackets on the intake manifold
   - Vacuum hose bracket and engine wiring harness
   - No. 1 air intake chamber and manifold stays. 14mm bolts: 31 ft. lbs. (42 Nm). 12mm bolts: 16 ft. lbs. (22 Nm)
   - EGR valve
   - Throttle body with a new gasket
   - Hoses and electrical connections to the throttle body

➡The protrusion on the gasket should be facing down and the water hose connections on the throttle body should also face down.

   - On type A throttle body, Bolts: 14 ft. lbs. (19 Nm). Bolt A is 45mm in length and bolt B is 55mm.
   - On type B throttle body, bolts and nuts: 14 ft. lbs. (19 Nm)
   - PCV hose
   - 2 vacuum hoses to the EGR modulator
   - Vacuum hose to the TVV for EVAP
   - IAC valve connector
   - Throttle position sensor connector
   - Air cleaner hose to the throttle body
   - Air cleaner cap with the resonator and air cleaner hose
   - Intake air temperature sensor connector

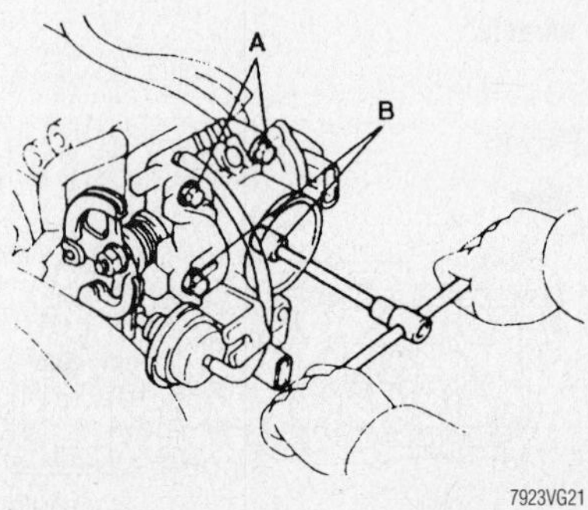

Mounting bolt length identification for the throttle body—2.2L (5S-FE) engine

- On California models, air hose to the air cleaner hose
- With automatic transmission, throttle cable
- Accelerator cable
- Negative battery cable

5. Fill the cooling system.
6. Start the vehicle, check for leaks and repair if necessary

2AZ-FE Engines

1. Before servicing the vehicle, refer to the precautions in the beginning of this section.
2. Relieve the fuel pressure from the fuel lines.
3. Drain the engine oil and coolant
4. Remove or disconnect the following:
   - Negative battery cable.

➡On vehicles equipped with an air bag, wait at least 90 seconds before proceeding after disconnecting the negative battery cable.

- Strut tower brace
- Radiator hose outlet
- Union to connector tube hose
- Water inlet water hose
- Fuel tube assembly
- Water bypass hoses from the throttle body
- Intake manifold and gasket

To install:

- Intake manifold and gasket. Tighten to 22 ft. lbs. (30 Nm)
- Water bypass hoses from the throttle body
- Fuel tube assembly
- Union to connector tube hose

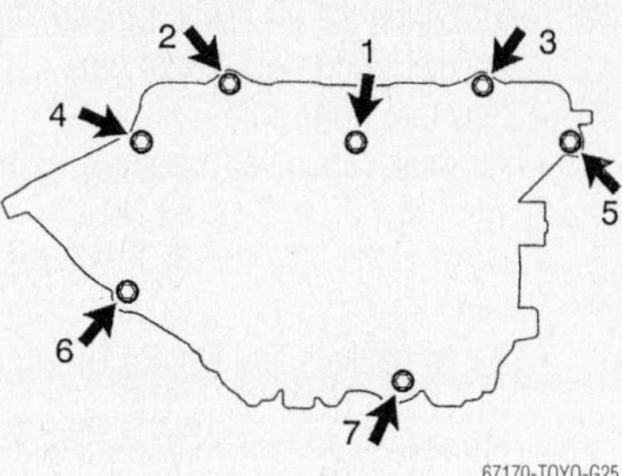

Intake manifold fastener tightening sequence—2AZ-FE engine

- Radiator hose outlet
- Strut tower brace. Tighten to 59 ft. lbs. (80 Nm)
- Add engine oil and coolant

5. Inspect for fuel leaks
6. Inspect for oil and coolant leaks

1MZ-FE and 3MZ-FE Engines

2001 MODELS

1. Before servicing the vehicle, refer to the precautions in the beginning of this section.
2. Relieve the fuel pressure from the fuel lines.
3. Remove or disconnect the following:
   - Negative battery cable.

➡On vehicles equipped with an air bag, wait at least 90 seconds before proceeding after disconnecting the negative battery cable.

- Air cleaner hose
- V-bank cover
- Air cleaner chamber assembly
- Accelerator cable

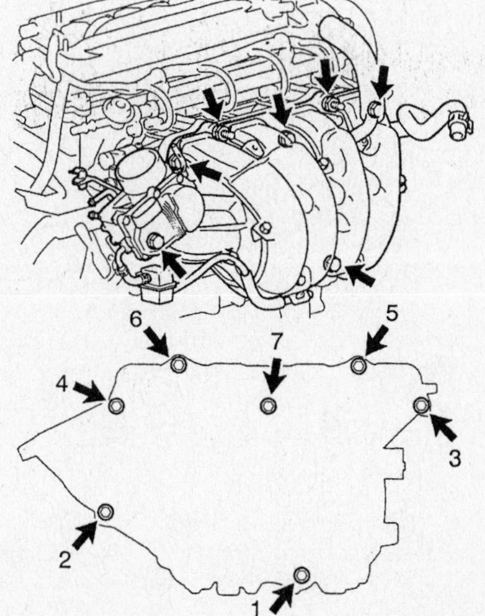

Intake manifold fastener location and loosening sequence—2AZ-FE engine

- Automatic transmission throttle cable
- TPS connector
- IAC valve connector
- EGR gas temperature sensor connector
- A/C idle up valve connector
- VSV connector for the Acoustic Control Induction System (ACIS)
- VSV connector for the fuel pressure control
- Disconnect the VSV for the EVAP
- VSV connector for the EGR
- DLC1 from the bracket on the intake air control valve
- Power steering pressure tube from the No. 1 engine hanger by removing the 2 bolts
- Brake booster vacuum hose from the intake air control valve for the ACIS
- PCV hose from the PCV valve on the right-hand cylinder head
- Ground strap and cable from the air intake air control valve from the ACIS
- Ground cable from the air intake chamber
- Vacuum hose clamp from fuel pipe
- 2 bypass hoses from the throttle body
- 2 power steering air hoses to the air intake chamber
- Air assist hose from the throttle body
- Remove the EVAP hose from the pipe on emission control valve set
- 2 vacuum hoses from the pipes on the cylinder head rear plate
- Vacuum sensing hose from the fuel pressure regulator
- Engine wire clamp from emission control valve set
- 2 bolts and the No. 1 engine hanger
- 2 bolts and the air intake chamber stay
- No. 2 EGR pipe and 2 gaskets by removing the 4 nuts.
- Hose from the VSV from the EVAP
- Air intake chamber assembly and gasket
- Fuel injector connectors
- Air assist hoses and pipe
- Fuel return hose from the No. 1 fuel pipe
- Fuel inlet hose from the fuel filter
- Delivery pipes and injectors from the engine
- Heater hoses
- Intake manifold and gasket

To install:

4. Thoroughly clean the intake manifold and cylinder head surfaces.
5. Install or connect the following:
   - Intake manifold with a new gasket. Torque the bolts to 11 ft. lbs. (15 Nm).
   - Heater hoses to the intake manifold
   - 2 new grommets to each injector
   - 2 new O-rings, lightly oiled, on each injector
   - Fuel injectors

➡Be sure to position the injector electrical connector outward

   - 4 spacers in position on the intake manifold
   - Right-hand delivery pipe and the No. 1 fuel pipe together with the 3 injectors in position on the intake manifold
   - 2 bolts, temporarily, holding the right-hand delivery pipe to the intake manifold
   - Bolt, temporarily, holding the No. 1 fuel pipe to the intake manifold
   - Left-hand delivery pipe and the No. 2 fuel pipe together with the 3 injectors in position
   - Fuel return hose to the fuel pressure regulator
   - 2 bolts, temporarily, holding the left-hand delivery pipe to the intake manifold
   - No. 2 fuel pipe, temporarily, to the left and right-hand delivery pipes with the union bolts and 2 new gaskets

➡Check that the injectors rotate smoothly. If the injectors do not rotate smoothly, the probable cause is incorrect installation of the O-rings. Replace the O-rings.

6. Tighten the 4 bolts holding the delivery pipes to the intake manifold. Bolts: 84 inch lbs. (10 Nm).
7. Tighten the bolt holding the No. 1 fuel pipe to the intake manifold. Bolts: 14 ft. lbs. (20 Nm).
8. Tighten the 2 union bolts holding the No. 2 fuel pipe to the delivery pipes. Bolts: 24 ft. lbs. (33 Nm).
9. Install or connect the following:
   - Fuel inlet hose to the fuel filter. Use 2 new gaskets when installing the union bolt
   - Fuel return hose to the No. 1 fuel pipe. When routing the fuel return hose, pass the hose under the heater hoses
   - Air assist hoses to the intake manifold, then the air assist pipe to the bracket on the No. 1 fuel pipe
   - Injector connectors
   - Air intake chamber assembly. Bolts and nuts: 32 ft. lbs. (43 Nm).
   - Hose to the VSV for the EVAP system
   - 2 new gaskets and No. 2 EGR pipe with the 4 nuts. Nuts: 108 inch lbs. (12 Nm).
   - No. 1 engine hanger with the 2 bolts. Bolts: 19 ft. lbs. (39 Nm).
   - Air intake chamber stay with the 2 bolts. Bolts: 14 ft. lbs. (20 Nm).
   - Brake booster vacuum hose to the intake air control valve for the ACIS
   - PCV hose to the PCV valve on the right-hand cylinder head
   - Ground strap and cable to the intake air control valve for the ACIS
   - Connect the ground cable and strap with the nut. Nut: 10 ft. lbs. (15 Nm).
   - Ground cable to air intake chamber
   - Vacuum hose clamp to fuel pipe
   - 2 water bypass hoses to the throttle body
   - Air assist hose to the throttle body
   - 2 power steering air hoses to the air intake chamber
   - Connect the EVAP hose to the pipe on the emission control valve set
   - 2 vacuum hoses to the pipes on the cylinder head rear plate
   - Vacuum sensing hose to the fuel pressure regulator
   - Engine wire clamp to the emission control valve set
   - Power steering pressure tube with the 2 nuts
   - TPS sensor connector
   - IAC valve connector
   - EGR gas temperature sensor connector
   - A/C idle up valve connector
   - VSV connector for the ACIS
   - VSV connector for the fuel pressure control
   - For California vehicles, install the VSV connector for the EVAP
   - VSV connector for the EGR
   - DLC1 to the bracket on the intake air control valve
   - Accelerator cable
   - Automatic transmission throttle cable
   - V-bank cover
   - Air cleaner hose
   - Negative battery cable
10. Fill the cooling system.
11. Start the vehicle, check for leaks and repair if necessary.

EXCEPT 2001 MODELS

1. Before servicing the vehicle, refer to the precautions in the beginning of this section.
2. Relieve the fuel pressure from the fuel lines.
3. Drain the engine oil and coolant
4. Remove or disconnect the following:
   - Negative battery cable.

➡On vehicles equipped with an air bag, wait at least 90 seconds before proceeding after disconnecting the negative battery cable.

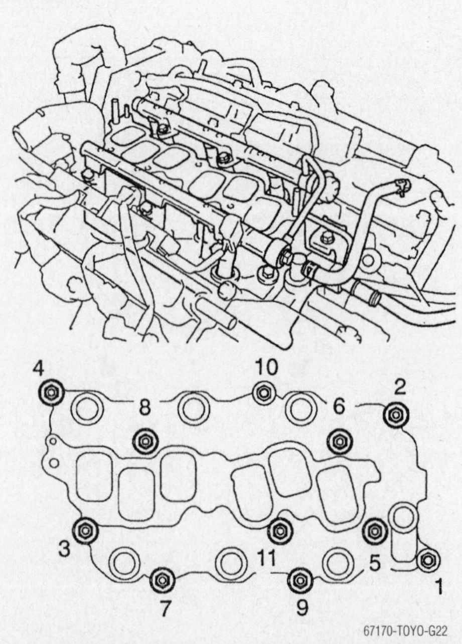

Manifold bolt locations and removal sequence—1MZ-FE and 3MZ-FE engines, except 2001

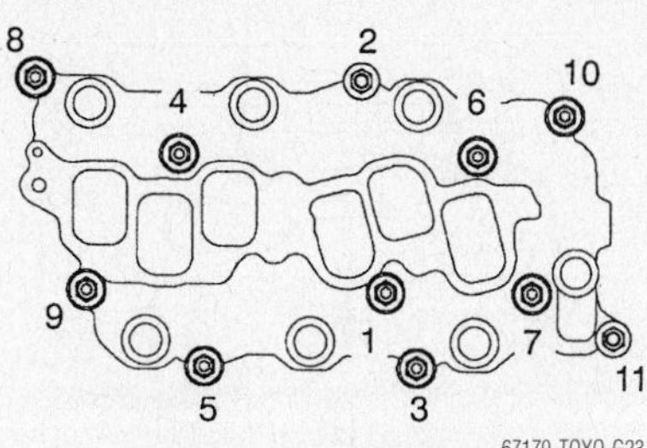

Manifold bolt tightening sequence—1MZ-FE and 3MZ-FE engines, except 2001

- V-bank cover
- Strut tower brace
- Air cleaner assembly and hose
- Fuel pipe assembly
- Heater inlet hose
- Manifold ground cable
- Injector plugs
- Manifold bolts and nuts in the correct sequence
- Intake manifold and gasket

To install:

5. Install or connect the following:
   - Intake manifold and gasket. Tighten the bolts and nuts in the correct sequence to 11 ft. lbs. (15 Nm)
   - Tighten the water outlet fasteners to 11 ft. lbs. (15 Nm)
   - Injector plugs
   - Manifold ground cable
   - Heater inlet hose
   - Fuel pipe assembly
   - Air cleaner assembly and hose
   - Strut tower brace. Tighten to 59 ft. lbs. (80 Nm)
   - V-bank cover
   - Add engine oil and coolant
6. Inspect for fuel leaks
7. Inspect for oil and coolant leaks

Exhaust Manifold

REMOVAL & INSTALLATION

1NZ-FE Engines

1. Before servicing the vehicle, refer to the precautions in the beginning of this section.
2. Remove or disconnect the following:
   - Negative battery cable.

➡On vehicles equipped with an air bag, wait at least 90 seconds before proceeding after disconnecting the negative battery cable.

   - All electrical wires and vacuum hoses that interfere with removal of the exhaust manifold.
   - Exhaust heat insulator
   - Exhaust pipe stay
   - Exhaust pipe from the manifold by removing the 2 bolts and 2 compression springs.
   - Exhaust manifold and gasket

To install:

3. Clean the gasket surfaces
4. Install or connect the following:
   - Exhaust manifold with a new gasket. Bolts: 20 ft. lbs. (27 Nm).
   - Exhaust stay to the engine and exhaust manifold. Bolt and nuts: 29 ft. lbs. (40 Nm).
   - Exhaust manifold heat insulator. Bolts: 71 inch lbs. (8 Nm).
   - Exhaust pipe to the exhaust manifold with the 2 compression springs and 2 bolts. Bolts: 46 ft. lbs. (62 Nm).
   - All electrical wires and vacuum hoses that were disconnected for removal of the exhaust manifold.
   - Negative battery cable

1ZZ-FE Engines

1. Before servicing the vehicle, refer to the precautions in the beginning of this section.
2. Drain the cooling system.
3. Remove or disconnect the following:
   - Negative battery cable
   - Drive belt and alternator
   - Air intake duct
   - Accelerator cable
   - Exhaust pipe from the manifold
   - Exhaust manifold support bracket
   - Heat insulator from the dash panel
   - Upper heat insulator
   - Exhaust manifold and gasket
   - If necessary, the lower heat insulator from the exhaust manifold.

To install:

4. Install or connect the following:
   - Lower heat insulator on the exhaust manifold. Bolts: 108 inch lbs. (12 Nm).
   - Exhaust manifold using a new gasket. Nuts, tightened several passes: 27 ft. lbs. (37 Nm).
   - Upper heat insulator. Bolts: 108 inch lbs. (12 Nm).
   - Heat insulator on the dash panel
   - Exhaust manifold support bracket. Bolts in an alternating pattern: 37 ft. lbs. (49 Nm).
   - Front exhaust pipe to the manifold. Bolts: 46 ft. lbs. (62 Nm).
   - Oxygen sensor, using new gasket and nuts. Nuts: 14 ft. lbs. (20 Nm).
   - Accelerator cable and air intake duct
   - Alternator and drive belt
   - Negative battery cable
5. Fill the cooling system.
6. Start the vehicle, check for leaks and repair if necessary.

2ZZ-GE Engines

1. Before servicing the vehicle, refer to the precautions in the beginning of this section.
2. Drain the cooling system.
3. Remove or disconnect the following:
   - Negative battery cable
   - Drive belt and alternator
   - Air intake duct
   - Accelerator cable
   - Exhaust pipe from the manifold
   - Exhaust manifold support bracket
   - Heat insulator from the dash panel.
   - Upper heat insulator
   - Exhaust manifold and gasket
   - If necessary, the lower heat insulator from the exhaust manifold.

To install:

4. Install or connect the following:
   - Lower heat insulator on the exhaust manifold. Bolts: 15 ft. lbs. (20 Nm).

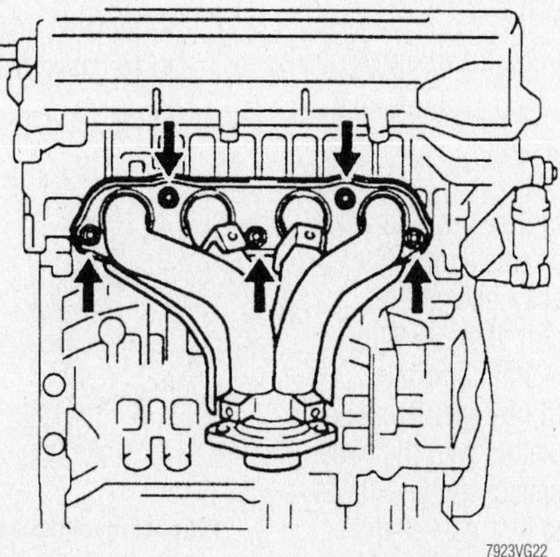

Exhaust manifold mounting nut locations—1.8L (1ZZ-FE) engine

- Exhaust manifold using a new gasket. Nuts, tightened several passes: 37 ft. lbs. (50 Nm).
- Upper heat insulator. Bolts: 15 ft. lbs. (20 Nm).
- Heat insulator on the dash panel.
- Exhaust manifold support bracket. Bolts: 37 ft. lbs. (49 Nm).
- Front exhaust pipe to the manifold. Bolts: 46 ft. lbs. (62 Nm).
- Oxygen sensor, using new gasket and nuts. Nuts: 14 ft. lbs. (20 Nm).
- Accelerator cable and air intake duct.
- Alternator and drive belt.
- Negative battery cable

5. Fill the cooling system.
6. Start the vehicle, check for leaks and repair if necessary.

5S-FE Engines

1. Before servicing the vehicle, refer to the precautions in the beginning of this section.
2. Remove or disconnect the following:
   - Negative battery cable. On vehicles equipped with an air bag, wait at least 90 seconds before proceeding
   - Bolts holding the front exhaust pipe to the mounting bracket
   - Front exhaust pipe from the manifold
   - Main oxygen sensor connector and the sub oxygen sensor connector
   - Left-hand exhaust manifold stay
   - Upper manifold heat insulator
   - Right-hand exhaust manifold stay
   - Exhaust manifold, and the 3-way catalytic converter assembly
   - Oxygen sensor and gasket from the exhaust manifold
   - Sub oxygen sensor from the 3-way catalytic converter
   - Lower heat insulator from the exhaust manifold
   - 3-way catalytic converter heat insulators
   - 3-way Catalytic Converter (TWC), gasket, retainer and cushion from the exhaust manifold

To install:

3. Place the cushion, retainer, and a new gasket on the TWC and reinstall it to the exhaust manifold with the bolts and the nuts. Bolts and nuts: 22 ft. lbs. (30 Nm).
4. Install or connect the following:
   - Lower manifold heat insulator and TWC heat insulators with the bolts.
   - Main oxygen sensor to the exhaust manifold with a new gasket and new nuts. Nuts: 14 ft. lbs. (19 Nm).
   - Sub oxygen sensor to the front TWC and tighten to 33 ft. lbs. (45 Nm).
   - Exhaust manifold, using new gasket and front TWC assembly to the engine with the nuts. Nuts, tighten uniformly in several passes: 36 ft. lbs. (49 Nm).
   - Right-hand exhaust manifold stay. Bolts and nuts: 31 ft. lbs. (42 Nm).
   - Upper heat insulator
   - Left-hand exhaust manifold stay. Bolt: 29 ft. lbs. (39 Nm). Nut: 31 ft. lbs. (42 Nm).
   - Main and the sub oxygen sensors connectors.
   - Front exhaust pipe with a new gasket to the TWC. Nuts: 46 ft. lbs. (62 Nm).
   - Front exhaust pipe to the exhaust pipe bracket. Bolts: 14 ft. lbs. (19 Nm).
   - Negative battery cable

2AZ-FE Engines

1. Before servicing the vehicle, refer to the precautions in the beginning of this section.
2. Remove or disconnect the following:
   - Negative battery terminal. If equipped with an air bag system, wait at least 90 seconds or longer before performing any other work
   - Exhaust manifold heat insulator
   - Oxygen sensor
   - Exhaust manifold converter stays
   - Exhaust pipe-to-manifold fasteners
   - Manifold nuts, manifold and gasket

To install:

3. Place a new gasket and reinstall the exhaust manifold with the nuts. Tighten to 27 ft. lbs. (33 Nm).
4. Install or connect the following:
   - Exhaust manifold converter stays. Tighten to 32 ft. lbs. (44 Nm)
   - Oxygen sensor
   - Exhaust manifold heat insulator
   - Negative battery cable

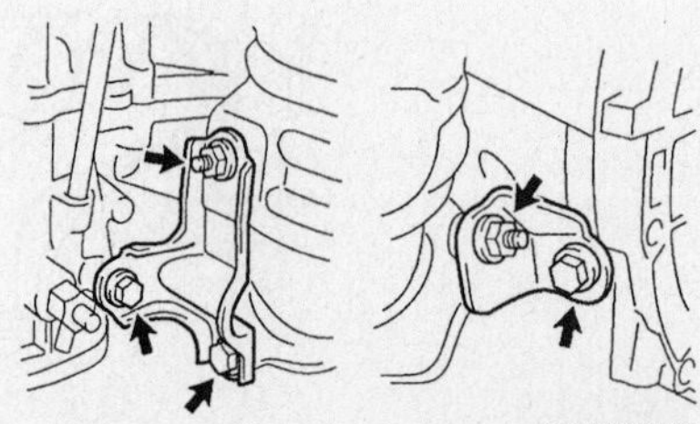

Exhaust manifold converter stays—2AZ-FE engine

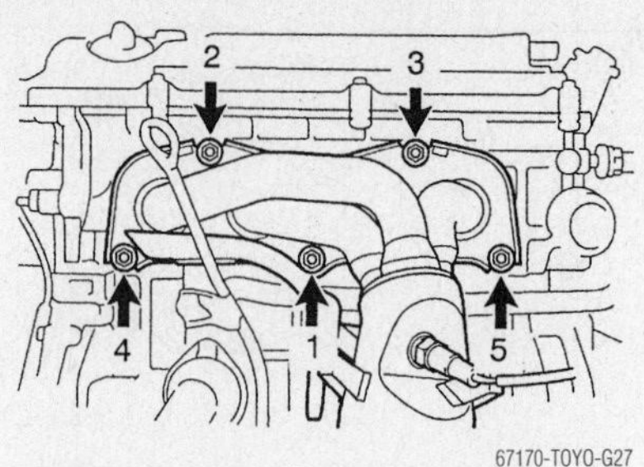

Exhaust manifold fastener tightening sequence—2AZ-FE engine

1MZ-FE Engines

1. Before servicing the vehicle, refer to the precautions in the beginning of this section.
2. Remove or disconnect the following:
   - Negative battery cable
   - Undercovers
   - Front exhaust pipe from the exhaust manifold
   - EGR pipe from the exhaust manifold
   - Heated oxygen sensor connector from the right exhaust manifold
   - Exhaust manifold stay
   - Exhaust manifold and gasket

To install:

3. Install or connect the following:
   - Exhaust manifold using new gasket. Bolts, uniformly tightened: 36 ft. lbs. (49 Nm).
   - Exhaust manifold stay. Bolt and nut: 15 ft. lbs. (20 Nm).
   - Heated oxygen sensor connector to the right exhaust manifold
   - EGR pipe, using new gaskets, to the exhaust manifold and the engine. Nuts: 108 inch lbs. (12 Nm).
   - Front exhaust pipe, using new gasket, to the exhaust manifold. Nuts: 46 ft. lbs. (62 Nm).
   - Undercovers
   - Negative battery cable

3MZ-FE Engines

1. Before servicing the vehicle, refer to the precautions in the beginning of this section.
2. Remove or disconnect the following:
   - Negative battery cable
   - Undercovers
   - Front exhaust pipes from the exhaust manifold
   - Heated oxygen sensor connector from the exhaust manifold
   - Heat shield insulator

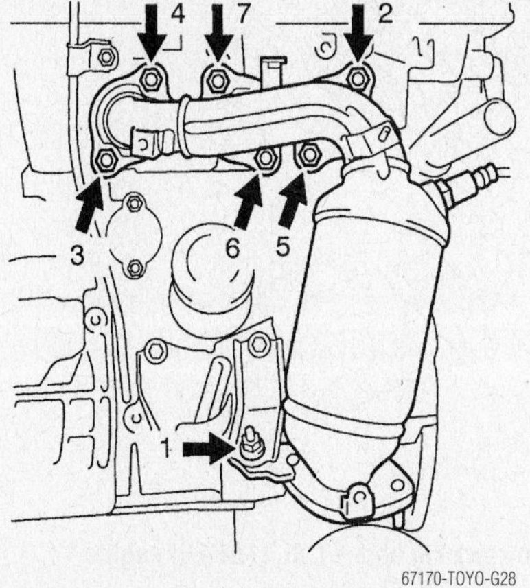

Front exhaust manifold fastener loosening sequence—3MZ-FE engine

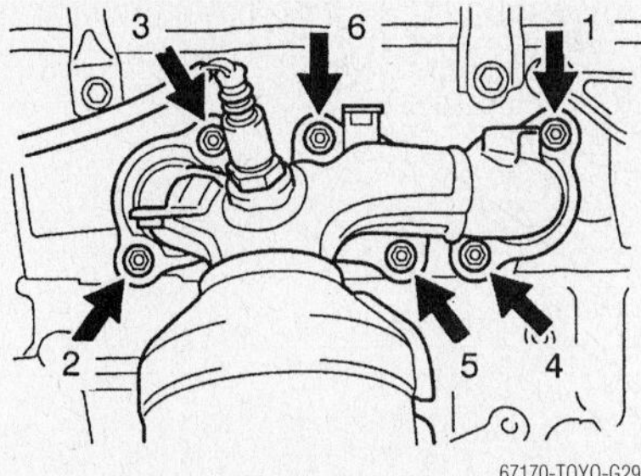

Rear exhaust manifold fastener loosening sequence—3MZ-FE engine

- For the front manifold, the exhaust manifold stay
- Exhaust manifold fasteners in the proper sequence
- Exhaust manifold and gasket

To install:

3. Install or connect the following:
   - Exhaust manifold using new gasket. Bolts, uniformly tightened in sequence: 36 ft. lbs. (49 Nm).

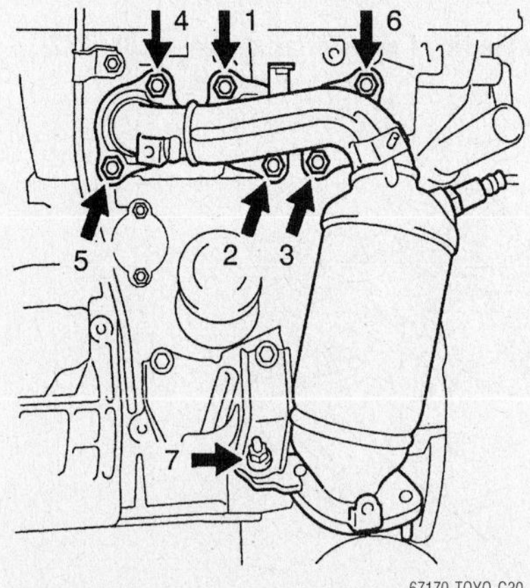

Front exhaust manifold fastener tightening sequence—3MZ-FE engine

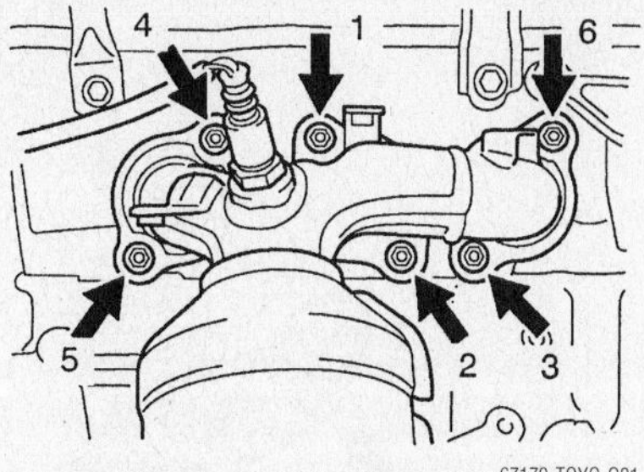

Rear exhaust manifold fastener tightening sequence—3MZ-FE engine

Retighten bolts 1 and 2 to specification again.
   - Exhaust manifold stay. Bolt and nut: 25 ft. lbs. (34 Nm).
   - Heat shield insulator
   - Heated oxygen sensor connector
   - Front exhaust pipe, using new gasket, to the exhaust manifold. Nuts: 46 ft. lbs. (62 Nm).
   - Undercovers
   - Negative battery cable

Camshaft(s)

REMOVAL & INSTALLATION

1NZ-FE Engines

1. Before servicing the vehicle, refer to the precautions in the beginning of this section.
2. Drain the cooling system.
3. Remove or disconnect the following:
   - Negative battery cable
   - Water filler
   - Outer front cowl top panel
   - Alternator
   - Air cleaner
   - Accelerator cable
   - Center exhaust pipe
   - Exhaust manifold support
   - Exhaust manifold
   - Ignition coil
   - Spark plugs
   - PCV hoses
   - Throttle body
   - Engine wiring harness at the head
   - Intake manifold
   - Camshaft position sensor
   - ECT sensor
   - Oil control valve
   - PCV valve
   - Oil filer cap
   - Cylinder head cover
   - Fuel injectors
   - Timing chain cover
   - Camshaft sprockets and valve timing control assembly
   - Camshafts

To install:

4. Install or connect the following:
   - Camshafts. Camshaft bearing caps in 2 stages: 1st10 ft. lbs. (13 Nm); 2nd17 ft. lbs. (23 Nm).
   - Sprockets and valve timing controller assembly, aligning the knock pin and hole. Torque the bolts to 47 ft. lbs. (64 Nm).
   - Check and adjust the valves.
   - Cylinder head cover
   - Oil filler cap
   - PCV valve
   - ECT sensor
   - Camshaft position sensor
   - Timing chain cover
   - Intake manifold
   - Engine wiring harness
   - Throttle body
   - PCV hoses
   - Spark plugs
   - Ignition coils
   - Exhaust manifold
   - Exhaust manifold support. Torque the bolts to 27 ft. lbs. (37 Nm).
   - Front exhaust pipe. Torque the nuts to 46 ft. lbs. (62 Nm).
   - Accelerator cable
   - Air cleaner

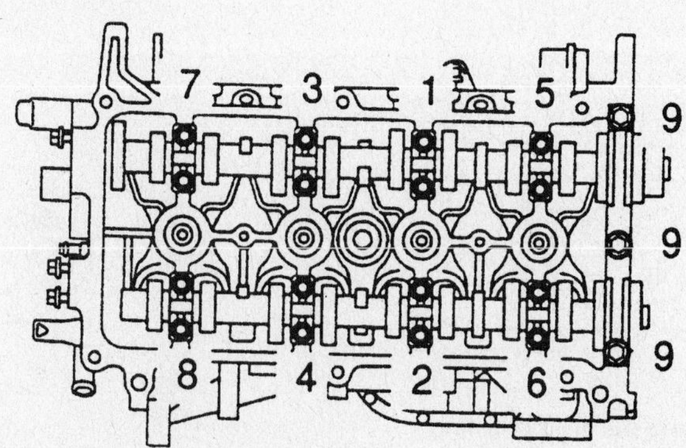

Camshaft bolt torque sequence—1NZ-FE engine

- Alternator
- Water filler
- Negative battery cable

5. Fill the cooling system to the proper level.
6. Start the vehicle, check for leaks and repair if necessary.

1ZZ-FE Engines

1. Before servicing the vehicle, refer to the precautions in the beginning of this section.
2. Remove or disconnect the following:
   - Negative battery cable. On vehicles equipped with an air bag, wait at least 90 seconds before proceeding.
   - Cylinder head cover
3. Turn the crankshaft so that the No. 1 piston is at TDC on the compression stroke. Check to see that the point marks on the camshaft sprockets are facing each other, if not, rotate the crankshaft 1 full revolution.
4. Tie the timing chain to each sprocket with string or wire to maintain correct valve timing.
5. Hold the camshafts with a wrench and remove the bolts securing the sprockets to the camshafts.
6. Using several passes, gradually remove the bearing cap bolts in the proper sequence. Then, remove the camshafts

To install:

7. Lubricate the camshafts with clean engine oil and place them on the cylinder head. Be sure to position the lobes for the No. 1 cylinder as shown in the illustration.
8. Install the bearing caps in their original positions. Apply clean engine oil to the threads and under the heads of the bearing cap bolts. After tightening the bolts on the No. 1 bearing cap to 17 ft. lbs. (23 Nm), tighten the remaining bolts in sequence using several passes to 10 ft. lbs. (13 Nm).
9. Check the valve clearance and make adjustments as needed.

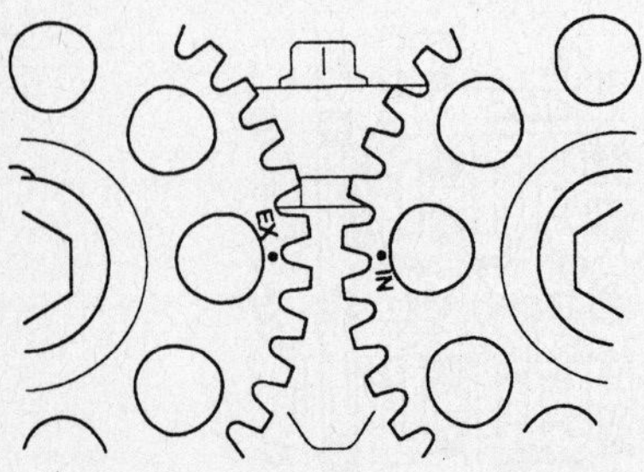

The sprocket marks will align when the No. 1 piston is at TDC on the compression stroke—1.8L (1ZZ-FE) engine

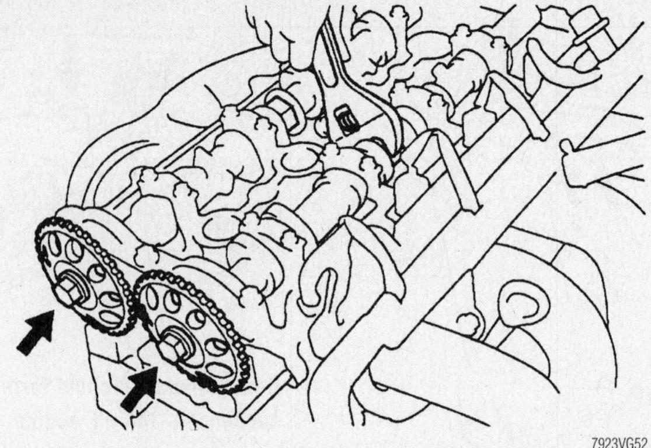

Hold the camshaft with a wrench while removing the sprocket bolt—1.8L (1ZZ-FE) engine

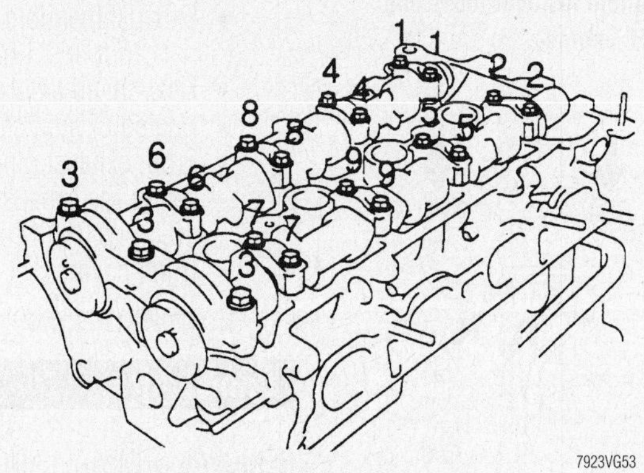

Camshaft bearing cap bolt removal sequence—1.8L (1ZZ-FE) engine

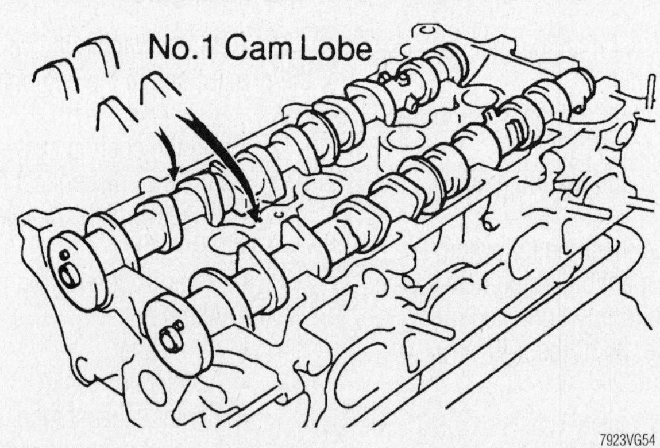

When installing the camshafts, position the lobes for the No. 1 cylinder as shown—1.8L (1ZZ-FE) engine

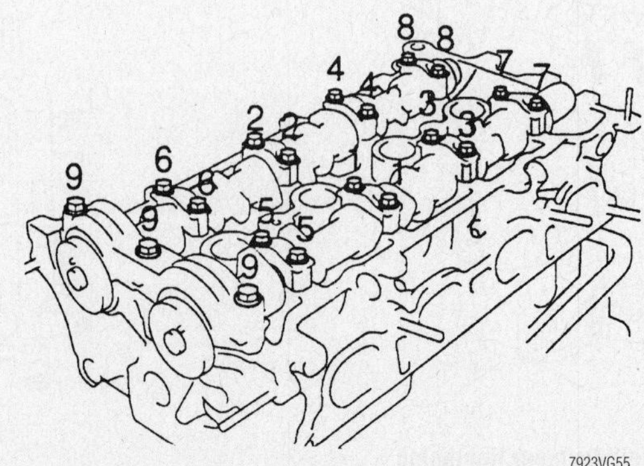

Camshaft bearing cap bolt tightening sequence—1.8L (1ZZ-FE) engine

10. Install or connect the following:
    - Camshaft sprockets and the chain
    - Cylinder head cover
    - Negative battery cable

2ZZ-GE Engines

1. Before servicing the vehicle, refer to the precautions in the beginning of this section.
2. Remove or disconnect the following:
    - Negative battery cable. On vehicles equipped with an air bag, wait at least 90 seconds before proceeding.
    - Cylinder head cover
3. Turn the crankshaft so that the No. 1 piston is at TDC on the compression stroke. Check to see that the point marks on the camshaft sprockets are facing each other, if not, rotate the crankshaft 1 full revolution.
4. Tie the timing chain to each sprocket with string or wire to maintain correct valve timing.
5. Hold the camshafts with a wrench and remove the bolts securing the sprockets to the camshafts.
6. Using several passes, gradually remove the bearing cap bolts in the proper sequence. Then, remove the camshafts

To install:

7. Lubricate the camshafts with clean engine oil and place them on the cylinder head. Be sure to position the lobes for the No. 1 cylinder as shown in the illustration.
8. Install the bearing caps in their original positions. Apply clean engine oil to the threads and under the heads of the bearing cap bolts. After tightening the bolts on the No. 1 bearing cap to 14 ft. lbs. (18 Nm), tighten the remaining bolts in sequence using several passes to 14 ft. lbs. (18 Nm).
9. Check the valve clearance and make adjustments as needed.
10. Install or connect the following:
    - Camshaft sprockets and the chain
    - Cylinder head cover
    - Negative battery cable

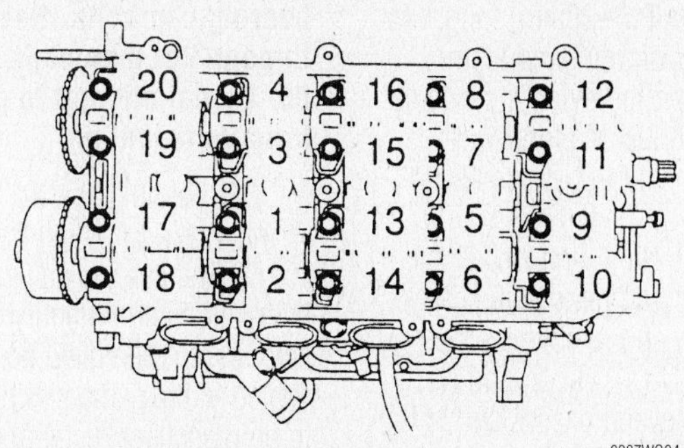

Camshaft bearing cap torque sequence—2ZZ-GE engines

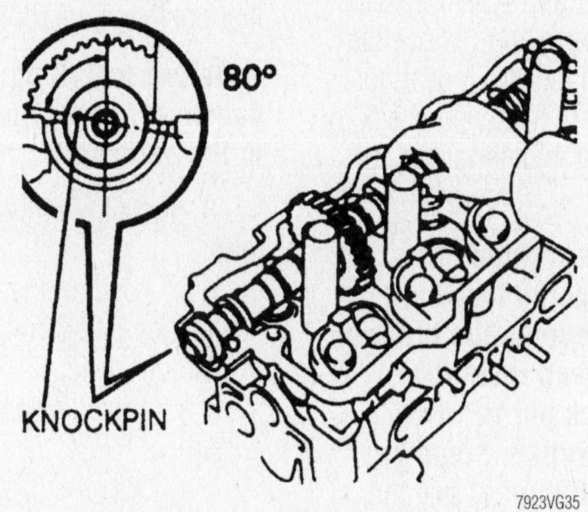

Intake camshaft removal and installation positioning—2.2L (5S-FE) engine

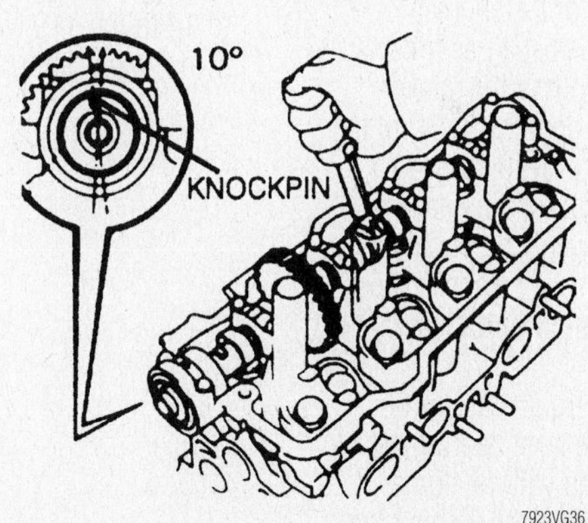

Exhaust camshaft removal and installation positioning—2.2L (5S-FE) engine

5S-FE Engines

1. Before servicing the vehicle, refer to the precautions in the beginning of this section.
2. Remove or disconnect the following:
    - Negative battery cable. On vehicles equipped with an air bag, wait at least 90 seconds before proceeding.
    - Spark plug wires
    - Timing belt, gears, and the covers
    - Any wire connectors, clamps, cables, or components necessary in order to remove the cylinder head cover.
    - Cylinder head cover and gasket
3. Set the No. 1 cylinder to TDC. Turn the crankshaft pulley and align its groove with the timing mark **0** of the No. 1 timing belt cover. Check that the valve lifters on the No. 1 cylinder are loose and valve lifters on the No. 4 cylinder are tight. If not, rotate the crankshaft 360 degrees.

➡Since the thrust clearance on both the intake and exhaust camshafts is small, the camshafts must be kept level during removal. If the camshafts are removed without being kept level, the camshaft may damage the bearing surface, causing the camshaft to seize during engine operation.

4. Remove exhaust camshaft as follows:
    a. Set the knock pin of the intake camshaft at 10–45 degrees BTDC of camshaft angle on the cylinder head. This angle will help to lift the exhaust camshaft level and evenly by pushing the No. 2 and No. 4 cylinder camshaft lobes of the exhaust camshaft toward their valve lifters.
    b. Secure the exhaust camshaft sub-gear to the main gear using a service bolt. The manufacturer recommends a

bolt 0.63–0.79 in. (16–20mm) long with a thread diameter of 6mm and a 1mm thread pitch. When removing the exhaust camshaft, be sure that the torsional spring force of the sub-gear has been eliminated.

c. Remove the No. 1 and No. 2 rear bearing cap bolts and remove the cap. Uniformly loosen and remove the bearing cap bolts on the No. 1, No. 2, and No. 4 bearing caps in several passes and in the reverse order of the installation sequence. Do not remove bearing cap bolts to No. 3 bearing cap at this time. Remove the No. 1, 2, and 4 bearing caps.

d. Alternately loosen and remove the bearing cap bolts on the No. 3 bearing cap. As these bolts are loosened check to see that the camshaft is being lifted out straight and level.

➡If the camshaft is not lifted out straight and level, tighten the No. 3 bearing cap bolts. Reverse the order of Steps 7c through 7a and reset the intake camshaft knock pin to 10–45 degrees BTDC, then repeat Steps 7a through 7c. Do not attempt to pry the camshaft from its mounting.

e. Remove the No. 3 bearing cap and exhaust camshaft from the engine.

5. Remove intake camshaft as follows:

a. Set the knock pin of the intake camshaft at 80–115 degrees BTDC of the camshaft angle on the cylinder head. This angle will help to lift the intake camshaft level and evenly by pushing No. 1 and No. 3 cylinder camshaft lobes of the intake camshaft toward their valve lifters.

b. Remove the 2 front bearing cap bolts and remove the front bearing cap and oil seal. If the cap will not come apart easily, leave it in place without the bolts.

c. Uniformly loosen and remove the bearing cap bolts to No. 1, No. 3, and the No. 4 bearing caps in several phases and in the reverse order of the installation sequence. Do not remove bearing cap bolts to the No. 2 bearing cap at this time. Remove No. 1, 3, and 4 bearing caps.

d. Alternately loosen and remove bearing cap bolts to the No. 2 bearing cap. As these bolts are loosened and after breaking the adhesion on the front bearing cap, check to see that the camshaft is being lifted out straight and level.

➡If the camshaft is not lifting out straight and level tighten the No. 2 bearing cap bolts. Reverse Steps 8b through 8d, then start over from Step 8b. Do not attempt to pry the camshaft from its mounting.

e. Remove the No. 2 bearing cap with the intake camshaft from the engine.

6. Remove the valve lifter shims and hydraulic lifters. Identify each lifter and shim as it is removed so it can be reinstalled in the same position. If the lifters are to be reused, store them upside down in a sealed container.

To install:

7. Install the valve lifters into their original positions and shims.

➡Before installing the intake camshaft, apply multi-purpose grease to the camshaft.

8. Install the intake camshaft, as follows:

a. Position the camshaft at 80–115 degrees BTDC of camshaft angle on the cylinder head.

b. Apply sealant to the front bearing cap.

c. Coat the bearing cap bolts with clean engine oil.

d. Tighten the camshaft bearing caps evenly in sequence and in several passes to 14 ft. lbs. (19 Nm).

e. Apply MP grease to a new oil seal lip, and by using a suitable tool, tap a new oil seal into place.

- Exhaust camshaft as follows:

f. Set the knock pin of the camshaft at 10–45 degrees BTDC of camshaft angle on the cylinder head.

g. Apply multipurpose grease to the camshaft.

h. Position the exhaust camshaft gear with the intake camshaft gear so that the timing marks are in alignment with one another. Be sure to use the proper alignment marks on the gears. Do not use the assembly reference marks.

i. Turn the intake camshaft clockwise or counterclockwise little by little until the exhaust camshaft sits in the bearing journals evenly without rocking the camshaft on the bearing journals.

j. Coat the bearing cap bolts with clean engine oil.

k. Tighten the camshaft bearing caps evenly in sequence and in several passes to 14 ft. lbs. (19 Nm). Remove the service bolt from the assembly.

9. Check and adjust valve clearance.

10. Install or connect the following:

- Cylinder head cover
- Timing belt and related components.
- Electrical connectors, cables, brackets, and components attached to the cylinder head cover.

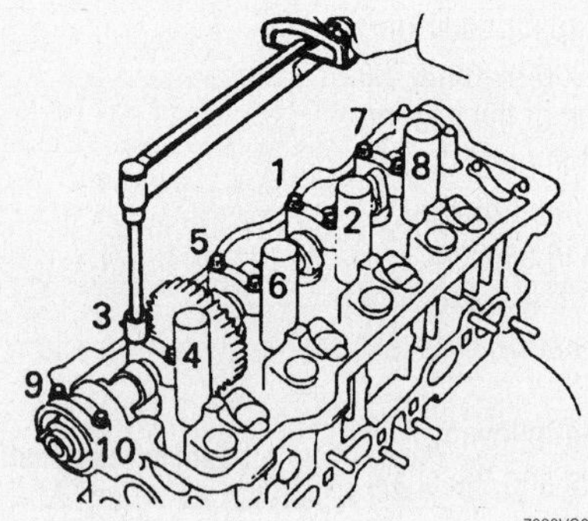

Intake camshaft bearing cap bolt tightening sequence—2.2L (5S-FE) engine

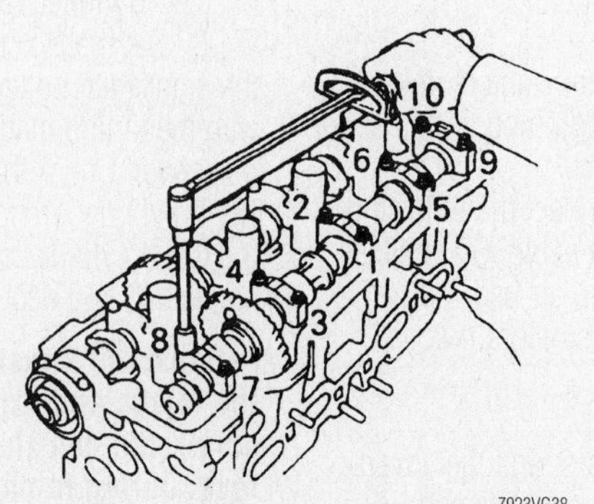

Exhaust camshaft bearing cap bolt tightening sequence—2.2L (5S-FE) engine

- Spark plug wires
- Negative battery cable

1MZ-FE Engines

1. Before servicing the vehicle, refer to the precautions in the beginning of this section.
2. Remove or disconnect the following:
   - Timing belt and idler pulley
   - Camshaft timing pulleys
   - Cylinder head covers

➡The thrust clearance on both the intake and exhaust camshafts is very small; the camshafts must be kept level during removal. If the camshafts are removed without being kept level, the camshaft may be caught in the cylinder head, causing the head to break or the camshaft to seize.

3. Remove the exhaust and intake camshafts from the right side cylinder head, as follows:
   a. Turn the camshaft with a wrench until the 2 pointed marks on the drive and driven gears are aligned. (The right camshaft gears have 2 marks apiece; the left side camshaft gears have 1 mark each.)
   b. Secure the exhaust camshaft sub-gear to the main gear using a service bolt. A bolt 0.63–0.79 in. (16–20mm) long with a 6mm thread diameter and a 1mm pitch is recommended. When removing the exhaust camshaft be sure the sub-gear is not loaded; all the force must be eliminated.
   c. Uniformly loosen and remove the exhaust camshaft bearing cap bolts in several passes and in the proper sequence. Remove the 8 bearing cap bolts and remove the caps, keeping them in the correct order.
   d. Remove the exhaust camshaft from the engine.
   e. Uniformly loosen and remove the 10 bearing cap bolts in several passes, in the proper sequence. Remove the bearing caps, keeping them in order, remove the oil seal, then, lift out the intake camshaft.
4. Remove the exhaust and intake camshafts from the left side cylinder head, as follows:
   a. Turn the camshaft with a wrench until the pointed marks on the drive and driven gears are aligned. (The right camshaft gears have 2 marks apiece; the left side camshaft gears have 1 mark each.)
   b. Secure the exhaust camshaft sub-gear to the main gear using a service bolt. A bolt 0.63–0.79 in. (16–20mm) long with a 6mm thread diameter and a 1mm pitch is recommended. When removing the exhaust camshaft be sure the sub-gear is not loaded; all the force must be eliminated.
   c. Uniformly loosen and remove the exhaust camshaft bearing cap bolts in several passes and in the proper sequence. Remove the 8 bearing cap bolts and remove the caps. Keep the caps in the correct order.
   d. Remove the exhaust camshaft from the engine.
   e. Uniformly loosen and remove the 10 bearing cap bolts in several passes, in the reverse order of the installation sequence. Remove the bearing caps, keeping them in order, remove the oil seal, then lift out the intake camshaft.
5. Remove the valve lifter shims and hydraulic lifters. If the lifters are to be reused, store them upside down in a sealed container.

To install:

6. Install the valve lifters into their original positions and shims. Check the valve clearance and replace the shims as necessary.

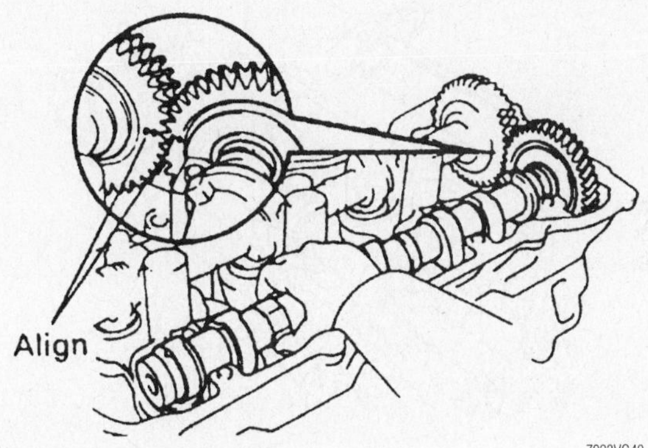

Aligning the camshaft gear timing marks for the right camshafts—3.0L (1MZ-FE) engine

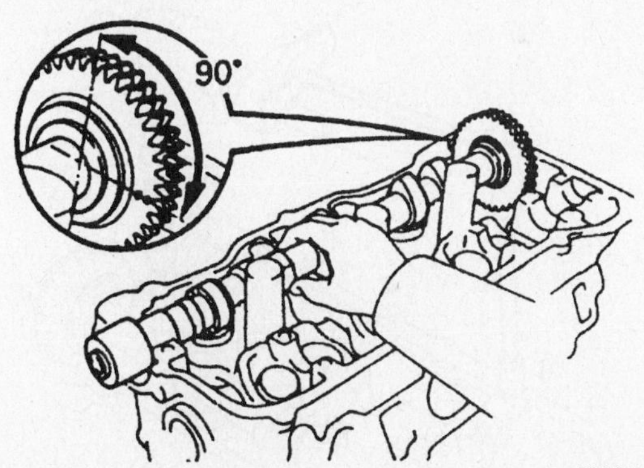

Camshaft installation for the right exhaust camshaft—3.0L (1MZ-FE) engine

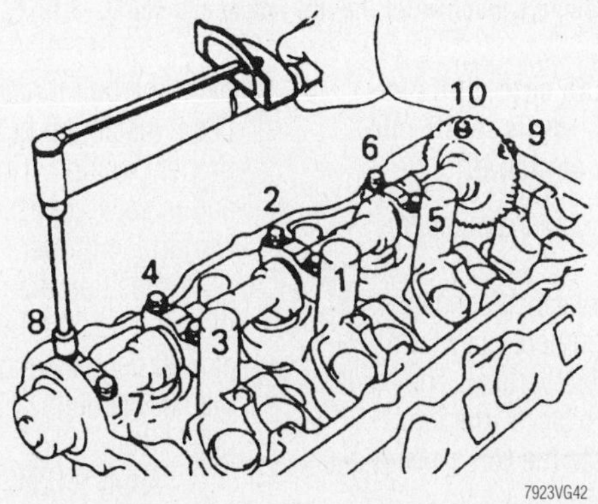

Bearing cap bolt tightening sequence for the right exhaust camshaft—3.0L (1MZ-FE) engine

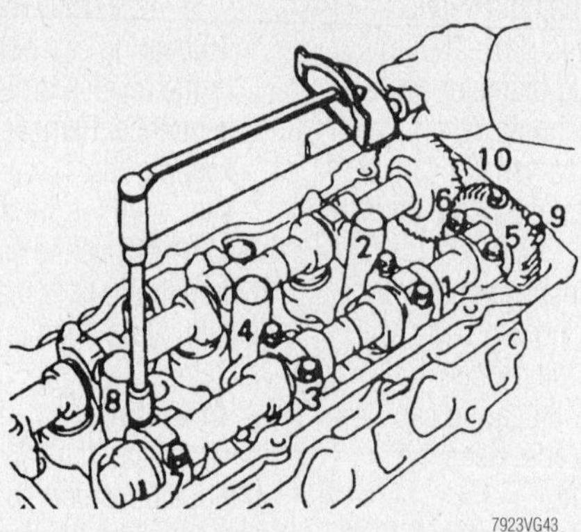

Bearing cap bolt tightening sequence for the right intake camshaft—3.0L (1MZ-FE) engine

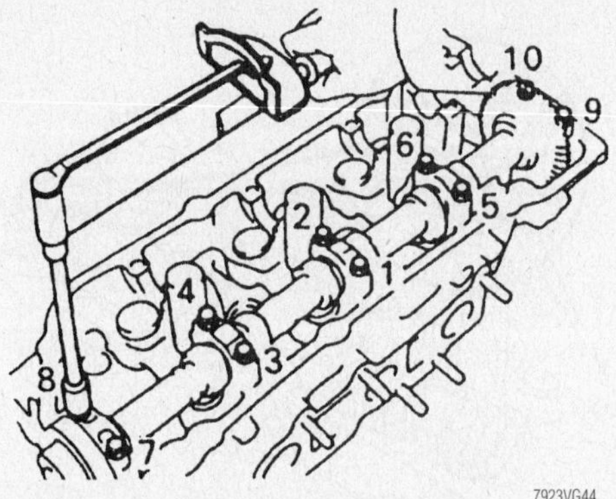

Bearing cap bolt tightening sequence for the left exhaust camshaft—3.0L (1MZ-FE) engine

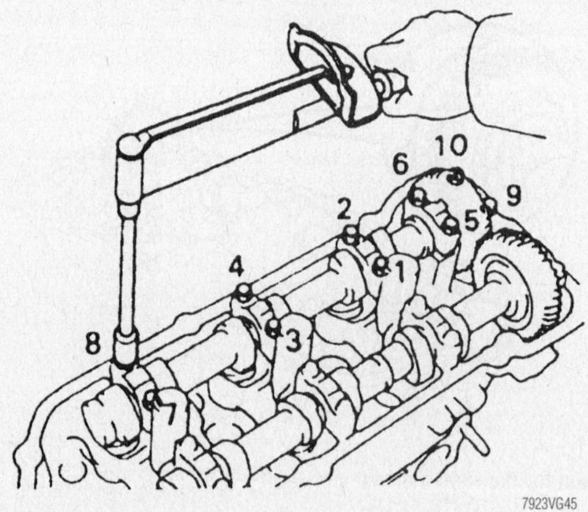

Bearing cap bolt tightening sequence for the left intake camshaft—3.0L (1MZ-FE) engine

➡Before installing the camshafts in either cylinder head, apply multi-purpose grease to each camshaft.

7. Install the right camshafts, as follows:
   a. Position the intake camshaft on the head so that the alignment marks are at a 90 degrees angle from vertical. The mark should be at the "3 o'clock" position.
   b. Apply sealant to the No. 1 bearing cap.
   c. Apply a light coat of clean engine oil to the bolt threads and under the bolt head. Install the bearing caps to their proper position. Tighten the bolts evenly and in several passes to 12 ft. lbs. (16 Nm) in the proper sequence.
   d. Position the exhaust camshaft on the head so that the alignment marks are at a 90 degrees angle from vertical. The mark must align with the marks on the other gear.
   e. Apply a light coat of clean engine oil to the bolt threads and under the bolt head. Install the bearing caps to their proper position. Tighten the bolts evenly and in several passes to 12 ft. lbs. (16 Nm) in the proper sequence.
   f. Remove the service bolt.
8. Install the left camshaft, as follows:
   a. Position the intake camshaft on the head so that the alignment mark is at a 90 degrees angle from vertical. The mark should be at the "9 o'clock" position.
   b. Apply sealant to the No. 1 bearing cap.
   c. Apply a light coat of clean engine oil to the bolt threads and under the bolt head. Install the bearing caps to their proper position. Tighten the bolts evenly and in several passes to 12 ft. lbs. (16 Nm) in the proper sequence.
   d. Position the exhaust camshaft on the head so that the alignment marks are at a 90 degree angle from vertical. The mark should be at the "3 o'clock" position and must align with the marks on the other gear.
   e. Apply a light coat of clean engine oil to the bolt threads and under the bolt head. Install the bearing caps to their proper position. Tighten the bolts evenly and in several passes to 12 ft. lbs. (16 Nm) in the proper sequence.
   f. Remove the service bolt.
9. Apply multi-purpose grease to new camshaft oil seals. Install the seals.
10. Install or connect the following:
    - No. 3 (rear) timing belt cover
    - Camshaft timing gears
    - Idler pulley, timing belt and covers
    - Cylinder head (valve) covers

3MZ-FE Engines

1. Before servicing the vehicle, refer to the precautions in the beginning of this section.
2. Remove or disconnect the following:
   - Negative battery cable. Wait at least 90 seconds before performing any other work
   - RH front wheel
   - Front suspension upper brace center
   - V-bank cover sub-assembly
   - Air cleaner assembly
   - Emission control valve set
   - Intake air surge tank
   - Ignition coil assembly
   - Cylinder head valve covers
   - Front fender apron seal
   - A/C Compressor and PS drive belts
   - Engine stabilizing rod
   - Engine mounting stay
   - Alternator bracket

- Crankshaft pulley
- Timing belt covers
- Timing belt, tensioners, idlers and guides
- Camshaft timing pulleys

➡Align the camshaft pulleys so that they can be returned to the original locations when reassembling.

- Inner timing belt cover

⁂ WARNING

Since the thrust clearance of the camshaft is small, the camshaft must be kept level while it is being removed. If the camshaft is not kept level, the portion of the cylinder head receiving the shaft thrust may crack or be damaged, causing the camshaft to seize or break.

3. Remove the camshaft using the following procedures
4. RH bank camshaft No.1 & LH bank camshaft No.2
   a. Align the (2 dot marks) of the camshaft drive and driven gear by turning the camshaft with a wrench.
   b. Secure the exhaust camshaft sub gear to the main gear with service bolt. Torque to 48 inch lbs. (5.4 Nm)

➡When removing the camshaft, make certain that the torsional spring force of the sub gear has been eliminated by installation of the service bolt.

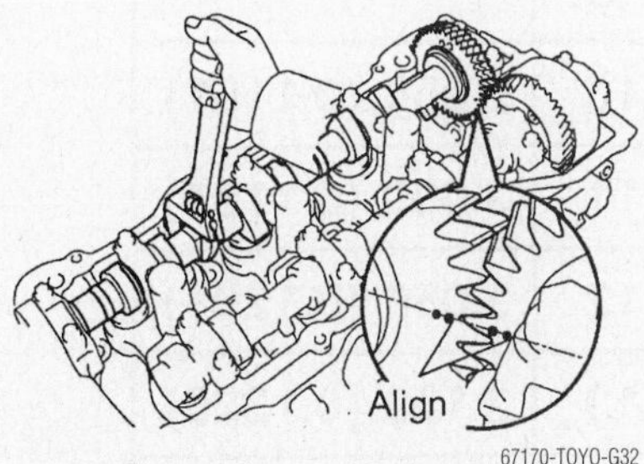

Aligning timing marks—3MZ-FE engine

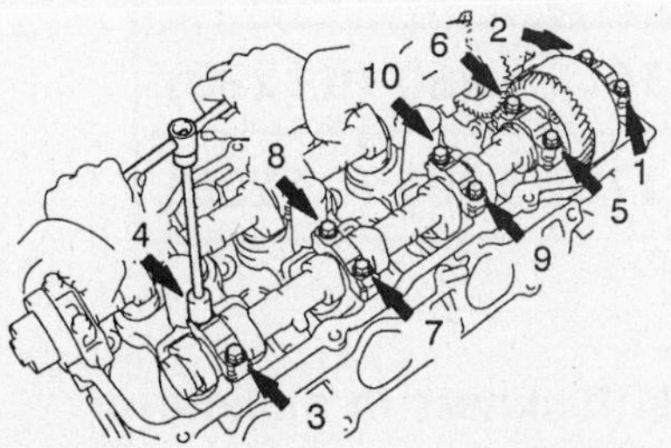

Camshaft bolt removal sequence Camshaft No.1—3MZ-FE engine

c. Using several steps, loosen the 10 bearing cap bolts uniformly in the sequence shown in the illustration. Remove the 5 bearing caps and camshaft

⁂ WARNING

Do Not pry out camshaft

⁂ WARNING

Do Not damage contact surface of the cylinder head that receives the shaft thrust.

5. LH bank camshaft No. 3 & No. 4
   a. Using several steps, loosen the 10 bearing cap bolts uniformly in the sequence shown in the illustration. Remove the 5 bearing caps and camshaft

⁂ WARNING

Do Not pry out camshaft

⁂ WARNING

Do Not damage contact surface of the cylinder head that receives the shaft thrust.

   b. Remove the oil seal from camshaft

To install:

6. Install RH No 2 Camshaft then RH Camshaft No. 1 using same procedure

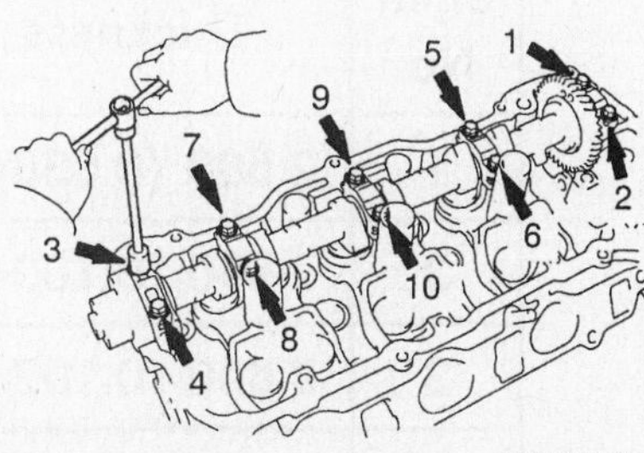

Camshaft bolt removal sequence Camshaft No.2 —3MZ-FE engine

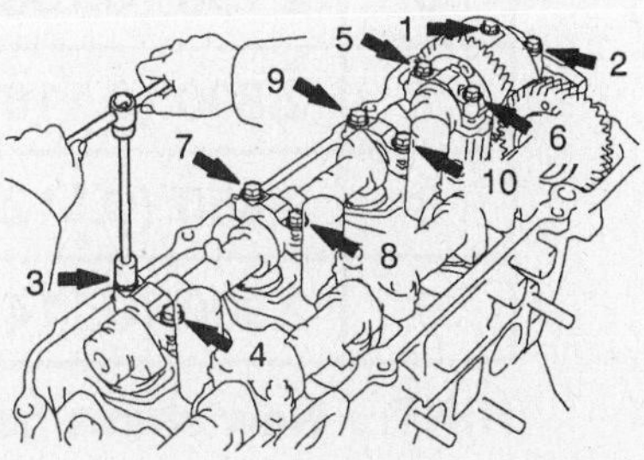

Camshaft bolt removal sequence camshaft No. 3—3MZ-FE engine

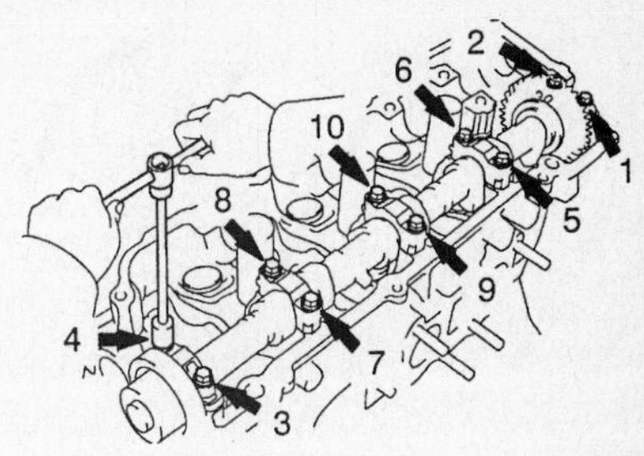

Camshaft bolt removal sequence camshaft No. 4—3MZ-FE engine

⁂ WARNING

Since the clearance of the camshaft is small, the camshaft must be kept level while bring installed. If the camshaft is not kept level, the cylinder head or camshaft may be damaged.

   a. Apply engine oil to the thrust portion and journal of camshaft
   b. No. 2 camshaft at 90 degree angel to the timing mark (2 dot marks) on the head.
   c. Multi-purpose grease to new oil seal
   d. Oil seal to camshaft

➡Do NOT turn over the oil seal lip

➡Insert oil seal until it stops

   e. Seal packing to bearing cap No. 1

➡Install bearing cap No. 1 within 5 minutes after applying seal packing

➡Do NOT expose seal packing to engine oil within 2 hours after installation

   f. 5 bearing caps in their proper locations
   g. Apply light coat of oil to the threads of the bearing caps
   h. Tighten the 10 bearing cap bolts in required sequence
   Torque to: 12 ft. lbs. (16 Nm)
7. Install LH camshaft No.3 and LH camshaft No. 4 using the above procedure
8. Install or connect the following:
   - Inner timing belt cover
   - RH then LH camshaft timing pulleys
   Torque to: 92 ft. lbs. (125 Nm)
   - Timing belt idlers, tensioners, guide and belt
   - RH engine mounting bracket
   - Timing belt covers
   - Crankshaft pulley
   - Alternator bracket

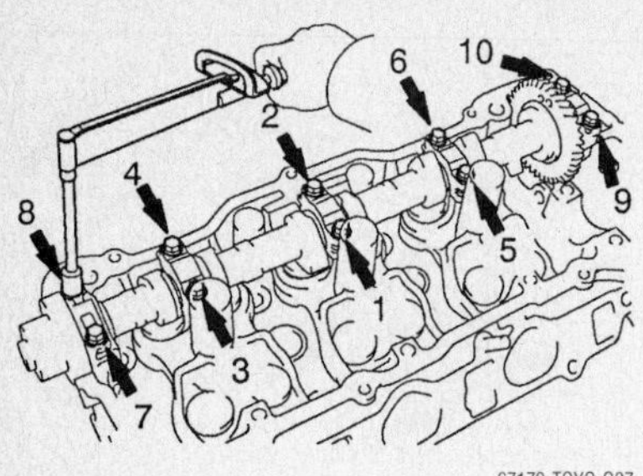

Bolt torque procedure RH camshaft No. 2—3MZ-FE engine

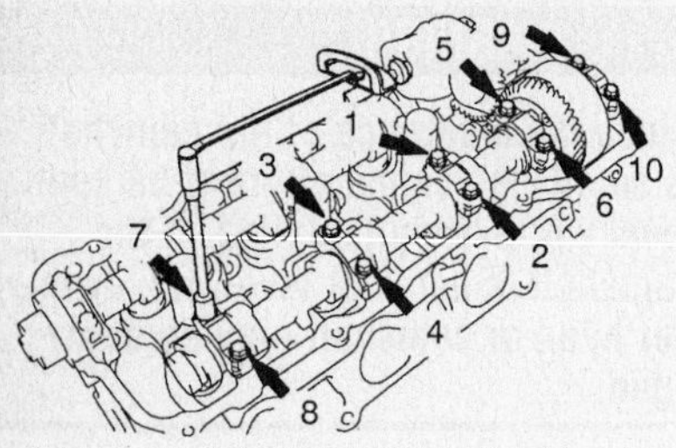

Bolt torque procedure RH camshaft No. 1—3MZ-FE engine

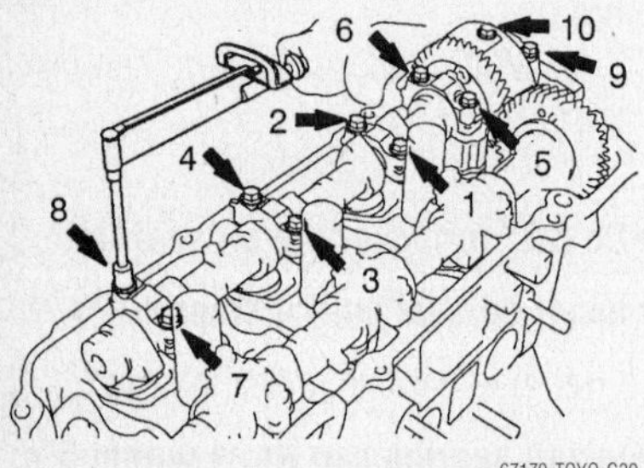

Bolt torque procedure LH camshaft No. 3—3MZ-FE engine

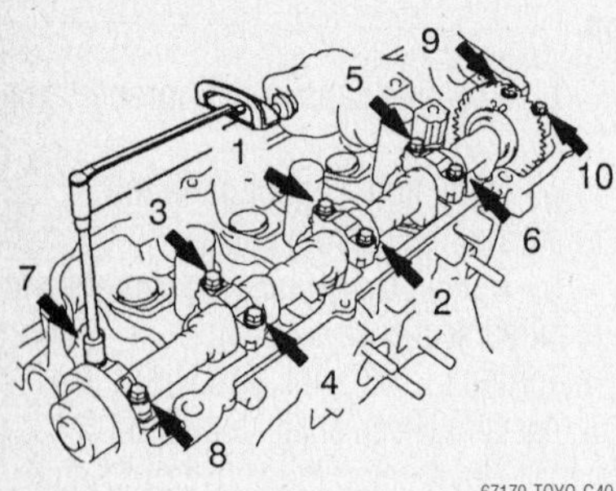

Bolt torque procedure LH camshaft No. 4—3MZ-FE engine

- RH engine mounting stay
- Engine stabilizing rod
- Inspect or adjust valve lash
- A/C compressor and PS drive belts
- Cylinder head valve covers
- Ignition coil assembly
- Intake air surge tank
- Emission control valve set
- Air cleaner assembly
- Vacuum hoses
- V-bank cover sub-assembly
- Front suspension upper center brace
- RF wheel
- Engine coolant. Check for leaks

Valve Lash

ADJUSTMENT

1NZ-FE Engines

➡Adjust the valve clearance when the engine is cold.

1. Before servicing the vehicle, refer to the precautions in the beginning of this section.
2. Remove or disconnect the following:
   - Negative battery cable. On vehicles equipped with an air bag, wait at least 90 seconds before proceeding.
   - Cylinder head cover
3. Turn the crankshaft pulley and align its groove with the timing mark **0** of the No. 1 timing cover.
4. Check that the timing marks on the camshaft sprockets and valve timing controller are facing up (12 o'clock). If not, turn the crankshaft 1 complete revolution (360 degrees).
5. Measure the clearance between the valve lifter and the camshaft. Record the measurements on the intake valves No. 1 and 2. Measure the exhaust valves at No. 1 and 3.
   a. The intake valve clearance cold is 0.006–0.010 in. (0.15–0.25mm).
   b. The exhaust valve clearance cold is 0.010–0.014 in. (0.25–0.35mm).
6. Turn the crankshaft pulley 1 revolution (360 degrees) and align the timing mark as before.
7. Measure the clearance between the valve lifter and the camshaft. Record the measurements on the intake valves No. 3 and 4. Measure the exhaust valves at No. 2 and 4.
   a. The intake valve clearance cold is 0.006–0.010 in. (0.15–0.25mm).
   b. The exhaust valve clearance cold is 0.010–0.014 in. (0.25–0.35mm).

Adjusting Shim Selection Chart

New shim thickness mm (in.)

Shim No.	Thickness	Shim No.	Thickness
1	2.500 (0.0984)	10	2.950 (0.1161)
2	2.550 (0.1004)	11	3.000 (0.1181)
3	2.600 (0.1024)	12	3.050 (0.1201)
4	2.650 (0.1043)	13	3.100 (0.1220)
5	2.700 (0.1063)	14	3.150 (0.1240)
6	2.750 (0.1083)	15	3.200 (0.1260)
7	2.800 (0.1102)	16	3.250 (0.1280)
8	2.850 (0.1122)	17	3.300 (0.1299)
9	2.900 (0.1142)		

HINT: New shims have the thickness in millimeters imprinted on the face.

7923VG57

Adjusting shim chart (intake and exhaust)—5S-FE and 1MZ-FE engines

1ZZ–FE: Valve Lifter Selection Chart (Intake)

Installed lifter thickness mm (in.) (columns 5.060–5.310)

Measured clearance mm (in.)	5.060 (0.1992)	5.080 (0.2000)	5.100 (0.2008)	5.120 (0.2016)	5.140 (0.2024)	5.160 (0.2031)	5.180 (0.2039)	5.200 (0.2047)	5.210 (0.2051)	5.220 (0.2055)	5.230 (0.2059)	5.240 (0.2063)	5.250 (0.2067)	5.260 (0.2071)	5.270 (0.2075)	5.280 (0.2079)	5.290 (0.2083)	5.300 (0.2087)	5.310 (0.2091)
0.000 – 0.030 (0.0000 – 0.0012)									06	06	06	06	06	08	10	10	12	12	14
0.031 – 0.050 (0.0012 – 0.0020)							06	06	06	06	08	08	10	10	12	12	14	14	16
0.051 – 0.070 (0.0020 – 0.0028)						06	06	06	08	08	10	10	12	12	14	14	16	16	18
0.071 – 0.090 (0.0028 – 0.0035)					06	06	06	08	10	10	12	12	14	14	16	16	18	18	20
0.091 – 0.110 (0.0036 – 0.0043)				06	06	06	08	10	12	12	14	14	16	16	18	18	20	20	22
0.111 – 0.130 (0.0044 – 0.0051)			06	06	06	08	10	12	14	14	16	16	18	18	20	20	22	22	24
0.131 – 0.149 (0.0052 – 0.0059)		06	06	06	08	10	12	14	16	16	18	18	20	20	22	22	24	24	26
0.150 – 0.250 (0.0059 – 0.0098)																			
0.251 – 0.270 (0.0099 – 0.0106)	12	14	16	18	20	22	24	26	28	28	30	30	32	32	34	34	36	36	38
0.271 – 0.290 (0.0107 – 0.0114)	14	16	18	20	22	24	26	28	30	30	32	32	34	34	36	36	38	38	40
0.291 – 0.310 (0.0115 – 0.0122)	16	18	20	22	24	26	28	30	32	32	34	34	36	36	38	38	40	40	42
0.311 – 0.330 (0.0122 – 0.0130)	18	20	22	24	26	28	30	32	34	34	36	36	38	38	40	40	42	42	44
0.331 – 0.350 (0.0130 – 0.0138)	20	22	24	26	28	30	32	34	36	36	38	38	40	40	42	42	44	44	46
0.351 – 0.370 (0.0138 – 0.0146)	22	24	26	28	30	32	34	36	38	38	40	40	42	42	44	44	46	46	48
0.371 – 0.390 (0.0146 – 0.0154)	24	26	28	30	32	34	36	38	40	40	42	42	44	44	46	46	48	48	50
0.391 – 0.410 (0.0154 – 0.0161)	26	28	30	32	34	36	38	40	42	42	44	44	46	46	48	48	50	50	52
0.411 – 0.430 (0.0162 – 0.0169)	28	30	32	34	36	38	40	42	44	44	46	46	48	48	50	50	52	52	54
0.431 – 0.450 (0.0170 – 0.0177)	30	32	34	36	38	40	42	44	46	46	48	48	50	50	52	52	54	54	56
0.451 – 0.470 (0.0178 – 0.0185)	32	34	36	38	40	42	44	46	48	48	50	50	52	52	54	54	56	56	58
0.471 – 0.490 (0.0185 – 0.0193)	34	36	38	40	42	44	46	48	50	50	52	52	54	54	56	56	58	58	60
0.491 – 0.510 (0.0193 – 0.0201)	36	38	40	42	44	46	48	50	52	52	54	54	56	56	58	58	60	60	62
0.511 – 0.530 (0.0201 – 0.0209)	38	40	42	44	46	48	50	52	54	54	56	56	58	58	60	60	62	62	64
0.531 – 0.550 (0.0209 – 0.0217)	40	42	44	46	48	50	52	54	56	56	58	58	60	60	62	62	64	64	66
0.551 – 0.570 (0.0217 – 0.0224)	42	44	46	48	50	52	54	56	58	58	60	60	62	62	64	64	66	66	68
0.571 – 0.590 (0.0225 – 0.0232)	44	46	48	50	52	54	56	58	60	60	62	62	64	64	66	66	68	68	70
0.591 – 0.610 (0.0233 – 0.0240)	46	48	50	52	54	56	58	60	62	62	64	64	66	66	68	68	70	70	72
0.611 – 0.630 (0.0241 – 0.0248)	48	50	52	54	56	58	60	62	64	64	66	66	68	68	70	70	72	72	74
0.631 – 0.650 (0.0248 – 0.0256)	50	52	54	56	58	60	62	64	66	66	68	68	70	70	72	72	74	74	74
0.651 – 0.670 (0.0256 – 0.0264)	52	54	56	58	60	62	64	66	68	68	70	70	72	72	74	74	74	74	74
0.671 – 0.690 (0.0264 – 0.0272)	54	56	58	60	62	64	66	68	70	70	72	72	74	74	74	74	74	74	
0.691 – 0.710 (0.0272 – 0.0280)	56	58	60	62	64	66	68	70	72	72	74	74	74	74	74	74			
0.711 – 0.730 (0.0280 – 0.0287)	58	60	62	64	66	68	70	72	74	74	74	74	74	74					
0.731 – 0.750 (0.0288 – 0.0295)	60	62	64	66	68	70	72	74	74	74	74	74							
0.751 – 0.770 (0.0296 – 0.0303)	62	64	66	68	70	72	74	74	74	74									
0.771 – 0.790 (0.0304 – 0.0311)	64	66	68	70	72	74	74	74											
0.791 – 0.810 (0.0311 – 0.0319)	66	68	70	72	74	74	74												
0.811 – 0.830 (0.0319 – 0.0327)	68	70	72	74	74	74													
0.831 – 0.850 (0.0327 – 0.0335)	70	72	74	74	74														
0.851 – 0.870 (0.0335 – 0.0343)	72	74	74	74															
0.871 – 0.890 (0.0343 – 0.0350)	74	74	74																
0.891 – 0.910 (0.0351 – 0.0358)	74	74																	
0.911 – 0.930 (0.0359 – 0.0366)	74																		

Installed lifter thickness mm (in.) (columns 5.320–5.500)

Measured clearance mm (in.)	5.320 (0.2094)	5.330 (0.2098)	5.340 (0.2102)	5.350 (0.2106)	5.360 (0.2110)	5.370 (0.2114)	5.380 (0.2118)	5.390 (0.2122)	5.400 (0.2126)	5.410 (0.2130)	5.420 (0.2134)	5.430 (0.2138)	5.440 (0.2142)	5.450 (0.2146)	5.460 (0.2150)	5.470 (0.2154)	5.480 (0.2157)	5.490 (0.2161)	5.500 (0.2165)
0.000 – 0.030 (0.0000 – 0.0012)	14	16	16	18	18	20	20	22	22	24	24	26	26	28	28	30	30	32	32
0.031 – 0.050 (0.0012 – 0.0020)	16	18	18	20	20	22	22	24	24	26	26	28	28	30	30	32	32	34	34
0.051 – 0.070 (0.0020 – 0.0028)	18	20	20	22	22	24	24	26	26	28	28	30	30	32	32	34	34	36	36
0.071 – 0.090 (0.0028 – 0.0035)	20	22	22	24	24	26	26	28	28	30	30	32	32	34	34	36	36	38	38
0.091 – 0.110 (0.0036 – 0.0043)	22	24	24	26	26	28	28	30	30	32	32	34	34	36	36	38	38	40	40
0.111 – 0.130 (0.0044 – 0.0051)	24	26	26	28	28	30	30	32	32	34	34	36	36	38	38	40	40	42	42
0.131 – 0.149 (0.0052 – 0.0059)	26	28	28	30	30	32	32	34	34	36	36	38	38	40	40	42	42	44	44
0.150 – 0.250 (0.0059 – 0.0098)																			
0.251 – 0.270 (0.0099 – 0.0106)	38	40	40	42	42	44	44	46	46	48	48	50	50	52	52	54	54	56	56
0.271 – 0.290 (0.0107 – 0.0114)	40	42	42	44	44	46	46	48	48	50	50	52	52	54	54	56	56	58	58
0.291 – 0.310 (0.0115 – 0.0122)	42	44	44	46	46	48	48	50	50	52	52	54	54	56	56	58	58	60	60
0.311 – 0.330 (0.0122 – 0.0130)	44	46	46	48	48	50	50	52	52	54	54	56	56	58	58	60	60	62	62
0.331 – 0.350 (0.0130 – 0.0138)	46	48	48	50	50	52	52	54	54	56	56	58	58	60	60	62	62	64	64
0.351 – 0.370 (0.0138 – 0.0146)	48	50	50	52	52	54	54	56	56	58	58	60	60	62	62	64	64	66	66
0.371 – 0.390 (0.0146 – 0.0154)	50	52	52	54	54	56	56	58	58	60	60	62	62	64	64	66	66	68	68
0.391 – 0.410 (0.0154 – 0.0161)	52	54	54	56	56	58	58	60	60	62	62	64	64	66	66	68	68	70	70
0.411 – 0.430 (0.0162 – 0.0169)	54	56	56	58	58	60	60	62	62	64	64	66	66	68	68	70	70	72	72
0.431 – 0.450 (0.0170 – 0.0177)	56	58	58	60	60	62	62	64	64	66	66	68	68	70	70	72	72	74	74
0.451 – 0.470 (0.0178 – 0.0185)	58	60	60	62	62	64	64	66	66	68	68	70	70	72	72	74	74	74	74
0.471 – 0.490 (0.0185 – 0.0193)	60	62	62	64	64	66	66	68	68	70	70	72	72	74	74	74	74	74	74
0.491 – 0.510 (0.0193 – 0.0201)	62	64	64	66	66	68	68	70	70	72	72	74	74	74	74	74	74		
0.511 – 0.530 (0.0201 – 0.0209)	64	66	66	68	68	70	70	72	72	74	74	74	74	74	74				
0.531 – 0.550 (0.0209 – 0.0217)	66	68	68	70	70	72	72	74	74	74	74	74	74						
0.551 – 0.570 (0.0217 – 0.0224)	68	70	70	72	72	74	74	74	74	74	74								
0.571 – 0.590 (0.0225 – 0.0232)	70	72	72	74	74	74	74	74	74										
0.591 – 0.610 (0.0233 – 0.0240)	72	74	74	74	74	74	74												
0.611 – 0.630 (0.0241 – 0.0248)	74	74	74	74	74														
0.631 – 0.650 (0.0248 – 0.0256)	74	74	74																
0.651 – 0.670 (0.0256 – 0.0264)	74																		
0.671 – 0.690 (0.0264 – 0.0272)																			
0.691 – 0.710 (0.0272 – 0.0280)																			
0.711 – 0.730 (0.0280 – 0.0287)																			
0.731 – 0.750 (0.0288 – 0.0295)																			
0.751 – 0.770 (0.0296 – 0.0303)																			
0.771 – 0.790 (0.0304 – 0.0311)																			
0.791 – 0.810 (0.0311 – 0.0319)																			
0.811 – 0.830 (0.0319 – 0.0327)																			
0.831 – 0.850 (0.0327 – 0.0335)																			
0.851 – 0.870 (0.0335 – 0.0343)																			
0.871 – 0.890 (0.0343 – 0.0350)																			
0.891 – 0.910 (0.0351 – 0.0358)																			
0.911 – 0.930 (0.0359 – 0.0366)																			

Installed lifter thickness mm (in.) (columns 5.510–5.740)

Measured clearance mm (in.)	5.510 (0.2169)	5.520 (0.2173)	5.530 (0.2177)	5.540 (0.2181)	5.550 (0.2185)	5.560 (0.2189)	5.570 (0.2193)	5.580 (0.2197)	5.590 (0.2201)	5.600 (0.2205)	5.620 (0.2213)	5.640 (0.2220)	5.660 (0.2228)	5.680 (0.2236)	5.700 (0.2244)	5.720 (0.2252)	5.740 (0.2260)
0.000 – 0.030 (0.0000 – 0.0012)	34	34	36	36	38	38	40	40	42	42	44	46	48	50	52	54	56
0.031 – 0.050 (0.0012 – 0.0020)	36	36	38	38	40	40	42	42	44	44	46	48	50	52	54	56	58
0.051 – 0.070 (0.0020 – 0.0028)	38	38	40	40	42	42	44	44	46	46	48	50	52	54	56	58	60
0.071 – 0.090 (0.0028 – 0.0035)	40	40	42	42	44	44	46	46	48	48	50	52	54	56	58	60	62
0.091 – 0.110 (0.0036 – 0.0043)	42	42	44	44	46	46	48	48	50	50	52	54	56	58	60	62	64
0.111 – 0.130 (0.0044 – 0.0051)	44	44	46	46	48	48	50	50	52	52	54	56	58	60	62	64	66
0.131 – 0.149 (0.0052 – 0.0059)	46	46	48	48	50	50	52	52	54	54	56	58	60	62	64	66	68
0.150 – 0.250 (0.0059 – 0.0098)																	
0.251 – 0.270 (0.0099 – 0.0106)	58	58	60	60	62	62	64	64	66	66	68	70	72	74	74	74	
0.271 – 0.290 (0.0107 – 0.0114)	60	60	62	62	64	64	66	66	68	68	70	72	74	74	74		
0.291 – 0.310 (0.0115 – 0.0122)	62	62	64	64	66	66	68	68	70	70	72	74	74	74			
0.311 – 0.330 (0.0122 – 0.0130)	64	64	66	66	68	68	70	70	72	72	74	74	74				
0.331 – 0.350 (0.0130 – 0.0138)	66	66	68	68	70	70	72	72	74	74	74	74					
0.351 – 0.370 (0.0138 – 0.0146)	68	68	70	70	72	72	74	74	74	74	74						
0.371 – 0.390 (0.0146 – 0.0154)	70	70	72	72	74	74	74	74	74	74							
0.391 – 0.410 (0.0154 – 0.0161)	72	72	74	74	74	74	74	74	74								
0.411 – 0.430 (0.0162 – 0.0169)	74	74	74	74	74	74											
0.431 – 0.450 (0.0170 – 0.0177)	74	74	74	74													
0.451 – 0.470 (0.0178 – 0.0185)	74	74															
0.471 – 0.490 (0.0185 – 0.0193)																	
0.491 – 0.510 (0.0193 – 0.0201)																	
0.511 – 0.530 (0.0201 – 0.0209)																	
0.531 – 0.550 (0.0209 – 0.0217)																	
0.551 – 0.570 (0.0217 – 0.0224)																	
0.571 – 0.590 (0.0225 – 0.0232)																	
0.591 – 0.610 (0.0233 – 0.0240)																	
0.611 – 0.630 (0.0241 – 0.0248)																	
0.631 – 0.650 (0.0248 – 0.0256)																	
0.651 – 0.670 (0.0256 – 0.0264)																	
0.671 – 0.690 (0.0264 – 0.0272)																	
0.691 – 0.710 (0.0272 – 0.0280)																	
0.711 – 0.730 (0.0280 – 0.0287)																	
0.731 – 0.750 (0.0288 – 0.0295)																	
0.751 – 0.770 (0.0296 – 0.0303)																	
0.771 – 0.790 (0.0304 – 0.0311)																	
0.791 – 0.810 (0.0311 – 0.0319)																	
0.811 – 0.830 (0.0319 – 0.0327)																	
0.831 – 0.850 (0.0327 – 0.0335)																	
0.851 – 0.870 (0.0335 – 0.0343)																	
0.871 – 0.890 (0.0343 – 0.0350)																	
0.891 – 0.910 (0.0351 – 0.0358)																	
0.911 – 0.930 (0.0359 – 0.0366)																	

Intake valve clearance (Cold):
0.15 – 0.25 mm (0.006 – 0.010 in.)
EXAMPLE: The 5.250 mm (0.2067 in.) lifter is installed, and the measured clearance is 0.400 mm (0.0157 in.).
Replace the 5.250 mm (0.2067 in.) lifter with a new No. 48 lifter.

New lifter thickness mm (in.)

Lifter No.	Thickness	Lifter No.	Thickness	Lifter No.	Thickness
06	5.060 (0.1992)	30	5.300 (0.2087)	54	5.540 (0.2181)
08	5.080 (0.2000)	32	5.320 (0.2094)	56	5.560 (0.2189)
10	5.100 (0.2008)	34	5.340 (0.2102)	58	5.580 (0.2197)
12	5.120 (0.2016)	36	5.360 (0.2110)	60	5.600 (0.2205)
14	5.140 (0.2024)	38	5.380 (0.2118)	62	5.620 (0.2213)
16	5.160 (0.2031)	40	5.400 (0.2126)	64	5.640 (0.2220)
18	5.180 (0.2039)	42	5.420 (0.2134)	66	5.660 (0.2228)
20	5.200 (0.2047)	44	5.440 (0.2142)	68	5.680 (0.2236)
22	5.220 (0.2055)	46	5.460 (0.2150)	70	5.700 (0.2244)
24	5.240 (0.2063)	48	5.480 (0.2157)	72	5.720 (0.2252)
26	5.260 (0.2071)	50	5.500 (0.2165)	74	5.740 (0.2260)
28	5.280 (0.2079)	52	5.520 (0.2173)		

9307WG70

Adjusting shim chart (intake)—1ZZ-FE engine

1ZZ-FE: Valve Lifter Selection Chart (Exhaust)

Measured clearance mm (in.) \ Installed lifter thickness mm (in.)	5.060 (0.1992)	5.080 (0.2000)	5.100 (0.2008)	5.120 (0.2016)	5.140 (0.2024)	5.160 (0.2031)	5.180 (0.2039)	5.200 (0.2047)	5.210 (0.2051)	5.220 (0.2055)	5.230 (0.2059)	5.240 (0.2063)	5.250 (0.2067)	5.260 (0.2071)	5.270 (0.2075)	5.280 (0.2079)	5.290 (0.2083)	5.300 (0.2087)	5.310 (0.2091)	5.320 (0.2094)	5.330 (0.2098)	5.340 (0.2102)	5.350 (0.2106)	5.360 (0.2110)	5.370 (0.2114)	5.380 (0.2118)	5.390 (0.2122)	5.400 (0.2126)	5.410 (0.2130)	5.420 (0.2134)	5.430 (0.2138)	5.440 (0.2142)	5.450 (0.2146)	5.460 (0.2150)	5.470 (0.2154)	5.480 (0.2157)	5.490 (0.2161)	5.500 (0.2165)	5.510 (0.2169)	5.520 (0.2173)	5.530 (0.2177)	5.540 (0.2181)	5.550 (0.2185)	5.560 (0.2189)	5.570 (0.2193)	5.580 (0.2197)	5.590 (0.2201)	5.600 (0.2205)	5.620 (0.2213)	5.640 (0.2220)	5.660 (0.2228)	5.680 (0.2236)	5.700 (0.2244)	5.720 (0.2252)	5.740 (0.2260)
0.000 – 0.030 (0.0000 – 0.0012)																			06	06	06	06	08	08	10	10	12	12	14	14	16	16	18	18	20	20	22	22	24	24	26	26	28	28	30	30	32	32	34	36	38	40	42	44	46
0.031 – 0.050 (0.0012 – 0.0020)																06	06	06	06	06	08	08	10	10	12	12	14	14	16	16	18	18	20	20	22	22	24	24	26	26	28	28	30	30	32	32	34	34	36	38	40	42	44	46	48
0.051 – 0.070 (0.0020 – 0.0028)														06	06	06	06	06	08	08	10	10	12	12	14	14	16	16	18	18	20	20	22	22	24	24	26	26	28	28	30	30	32	32	34	34	36	36	38	40	42	44	46	48	50
0.071 – 0.090 (0.0028 – 0.0035)												06	06	06	06	06	08	08	10	10	12	12	14	14	16	16	18	18	20	20	22	22	24	24	26	26	28	28	30	30	32	32	34	34	36	36	38	38	40	42	44	46	48	50	52
0.091 – 0.110 (0.0036 – 0.0043)											06	06	06	06	08	08	10	10	12	12	14	14	16	16	18	18	20	20	22	22	24	24	26	26	28	28	30	30	32	32	34	34	36	36	38	38	40	40	42	44	46	48	50	52	54
0.111 – 0.130 (0.0044 – 0.0051)									06	06	06	06	08	08	10	10	12	12	14	14	16	16	18	18	20	20	22	22	24	24	26	26	28	28	30	30	32	32	34	34	36	36	38	38	40	40	42	42	44	46	48	50	52	54	56
0.131 – 0.150 (0.0052 – 0.0059)								06	06	06	08	08	10	10	12	12	14	14	16	16	18	18	20	20	22	22	24	24	26	26	28	28	30	30	32	32	34	34	36	36	38	38	40	40	42	42	44	44	46	48	50	52	54	56	58
0.151 – 0.170 (0.0059 – 0.0067)						06	06	06	08	08	10	10	12	12	14	14	16	16	18	18	20	20	22	22	24	24	26	26	28	28	30	30	32	32	34	34	36	36	38	38	40	40	42	42	44	44	46	46	48	50	52	54	56	58	60
0.171 – 0.190 (0.0067 – 0.0075)					06	06	06	08	10	10	12	12	14	14	16	16	18	18	20	20	22	22	24	24	26	26	28	28	30	30	32	32	34	34	36	36	38	38	40	40	42	42	44	44	46	46	48	48	50	52	54	56	58	60	62
0.191 – 0.210 (0.0075 – 0.0083)				06	06	06	08	10	12	12	14	14	16	16	18	18	20	20	22	22	24	24	26	26	28	28	30	30	32	32	34	34	36	36	38	38	40	40	42	42	44	44	46	46	48	48	50	50	52	54	56	58	60	62	64
0.211 – 0.230 (0.0083 – 0.0091)			06	06	06	08	10	12	14	14	16	16	18	18	20	20	22	22	24	24	26	26	28	28	30	30	32	32	34	34	36	36	38	38	40	40	42	42	44	44	46	46	48	48	50	50	52	52	54	56	58	60	62	64	66
0.231 – 0.249 (0.0091 – 0.0098)		06	06	06	08	10	12	14	16	16	18	18	20	20	22	22	24	24	26	26	28	28	30	30	32	32	34	34	36	36	38	38	40	40	42	42	44	44	46	46	48	48	50	50	52	52	54	54	56	58	60	62	64	66	68
0.250 – 0.350 (0.0098 – 0.0138)																																																								
0.351 – 0.370 (0.0138 – 0.0146)	12	14	16	18	20	22	24	26	28	28	30	30	32	32	34	34	36	36	38	38	40	40	42	42	44	44	46	46	48	48	50	50	52	52	54	54	56	56	58	58	60	60	62	62	64	64	66	66	68	70	72	74	74	74	
0.371 – 0.390 (0.0146 – 0.0154)	14	16	18	20	22	24	26	28	30	30	32	32	34	34	36	36	38	38	40	40	42	42	44	44	46	46	48	48	50	50	52	52	54	54	56	56	58	58	60	60	62	62	64	64	66	66	68	68	70	72	74	74	74		
0.391 – 0.410 (0.0154 – 0.0161)	16	18	20	22	24	26	28	30	32	32	34	34	36	36	38	38	40	40	42	42	44	44	46	46	48	48	50	50	52	52	54	54	56	56	58	58	60	60	62	62	64	64	66	66	68	68	70	70	72	74	74	74			
0.411 – 0.430 (0.0162 – 0.0169)	18	20	22	24	26	28	30	32	34	34	36	36	38	38	40	40	42	42	44	44	46	46	48	48	50	50	52	52	54	54	56	56	58	58	60	60	62	62	64	64	66	66	68	68	70	70	72	72	74	74	74				
0.431 – 0.450 (0.0170 – 0.0177)	20	22	24	26	28	30	32	34	36	36	38	38	40	40	42	42	44	44	46	46	48	48	50	50	52	52	54	54	56	56	58	58	60	60	62	62	64	64	66	68	68	70	70	72	72	74	74	74	74	74					
0.451 – 0.470 (0.0178 – 0.0185)	22	24	26	28	30	32	34	36	38	38	40	40	42	42	44	44	46	46	48	48	50	50	52	52	54	54	56	56	58	58	60	60	62	62	64	64	66	66	68	68	70	70	72	72	74	74	74	74	74						
0.471 – 0.490 (0.0185 – 0.0193)	24	26	28	30	32	34	36	38	40	40	42	42	44	44	46	46	48	48	50	50	52	52	54	54	56	56	58	58	60	60	62	62	64	64	66	66	68	68	70	70	72	72	74	74	74	74	74	74							
0.491 – 0.510 (0.0193 – 0.0201)	26	28	30	32	34	36	38	40	42	42	44	44	46	46	48	48	50	50	52	52	54	54	56	56	58	58	60	60	62	62	64	64	66	66	68	68	70	70	72	72	74	74	74	74	74	74									
0.511 – 0.530 (0.0201 – 0.0209)	28	30	32	34	36	38	40	42	44	44	46	46	48	48	50	50	52	52	54	54	56	56	58	58	60	60	62	62	64	64	66	66	68	68	70	70	72	72	74	74	74	74	74	74											
0.531 – 0.550 (0.0209 – 0.0217)	30	32	34	36	38	40	42	44	46	46	48	48	50	50	52	52	54	54	56	56	58	58	60	60	62	62	64	64	66	66	68	68	70	70	72	72	74	74	74	74	74	74													
0.551 – 0.570 (0.0217 – 0.0224)	32	34	36	38	40	42	44	46	48	48	50	50	52	52	54	54	56	56	58	58	60	60	62	62	64	64	66	66	68	68	70	70	72	72	74	74	74	74	74	74															
0.571 – 0.590 (0.0225 – 0.0232)	34	36	38	40	42	44	46	48	50	50	52	52	54	54	56	56	58	58	60	60	62	62	64	64	66	66	68	68	70	70	72	72	74	74	74	74	74	74																	
0.591 – 0.610 (0.0233 – 0.0240)	36	38	40	42	44	46	48	50	52	52	54	54	56	56	58	58	60	60	62	62	64	64	66	66	68	68	70	70	72	72	74	74	74	74	74	74																			
0.611 – 0.630 (0.0241 – 0.0248)	38	40	42	44	46	48	50	52	54	54	56	56	58	58	60	60	62	62	64	64	66	66	68	68	70	70	72	72	74	74	74	74	74	74																					
0.631 – 0.650 (0.0248 – 0.0256)	40	42	44	46	48	50	52	54	56	56	58	58	60	60	62	62	64	64	66	66	68	68	70	70	72	72	74	74	74	74	74	74	74																						
0.651 – 0.670 (0.0256 – 0.0264)	42	44	46	48	50	52	54	56	58	58	60	60	62	62	64	64	66	66	68	68	70	70	72	72	74	74	74	74	74	74																									
0.671 – 0.690 (0.0264 – 0.0272)	44	46	48	50	52	54	56	58	60	60	62	62	64	64	66	66	68	68	70	70	72	72	74	74	74	74	74	74																											
0.691 – 0.710 (0.0272 – 0.0280)	46	48	50	52	54	56	58	60	62	62	64	64	66	66	68	68	70	70	72	72	74	74	74	74	74	74																													
0.711 – 0.730 (0.0280 – 0.0287)	48	50	52	54	56	58	60	62	64	64	66	66	68	68	70	70	72	72	74	74	74	74	74	74																															
0.731 – 0.750 (0.0288 – 0.0295)	50	52	54	56	58	60	62	64	66	66	68	68	70	70	72	72	74	74	74	74	74	74																																	
0.751 – 0.770 (0.0296 – 0.0303)	52	54	56	58	60	62	64	66	68	68	70	70	72	72	74	74	74	74	74	74																																			
0.771 – 0.790 (0.0304 – 0.0311)	54	56	58	60	62	64	66	68	70	70	72	72	74	74	74	74	74	74																																					
0.791 – 0.810 (0.0311 – 0.0319)	56	58	60	62	64	66	68	70	72	72	74	74	74	74	74	74																																							
0.811 – 0.830 (0.0319 – 0.0327)	58	60	62	64	66	68	70	72	74	74	74	74	74	74																																									
0.831 – 0.850 (0.0327 – 0.0335)	60	62	64	66	68	70	72	74	74	74	74	74																																											
0.851 – 0.870 (0.0335 – 0.0343)	62	64	66	68	70	72	74	74	74	74																																													
0.871 – 0.890 (0.0343 – 0.0350)	64	66	68	70	72	74	74	74																																															
0.891 – 0.910 (0.0351 – 0.0358)	66	68	70	72	74	74	74																																																
0.911 – 0.930 (0.0359 – 0.0366)	68	70	72	74	74	74																																																	
0.931 – 0.950 (0.0367 – 0.0374)	70	72	74	74	74																																																		
0.951 – 0.970 (0.0374 – 0.0382)	72	74	74	74																																																			
0.971 – 0.990 (0.0382 – 0.0390)	74	74	74																																																				
0.991 – 1.010 (0.0390 – 0.0398)	74	74																																																					
1.011 – 1.030 (0.0398 – 0.0406)	74																																																						

Exhaust valve clearance (Cold):
0.25 – 0.35 mm (0.010 – 0.014 in.)

EXAMPLE: The 5.340 mm (0.2102 in.) lifter is installed, and the measured clearance is 0.440 mm (0.0173 in.).
Replace the 5.340 mm (0.2102 in.) lifter with a new No. 48 lifter.

New lifter thickness mm (in.)

Lifter No.	Thickness	Lifter No.	Thickness	Lifter No.	Thickness
06	5.060 (0.1992)	30	5.300 (0.2087)	54	5.540 (0.2181)
08	5.080 (0.2000)	32	5.320 (0.2094)	56	5.560 (0.2189)
10	5.100 (0.2008)	34	5.340 (0.2102)	58	5.580 (0.2197)
12	5.120 (0.2016)	36	5.360 (0.2110)	60	5.600 (0.2205)
14	5.140 (0.2024)	38	5.380 (0.2118)	62	5.620 (0.2213)
16	5.160 (0.2031)	40	5.400 (0.2126)	64	5.640 (0.2220)
18	5.180 (0.2039)	42	5.420 (0.2134)	66	5.660 (0.2228)
20	5.200 (0.2047)	44	5.440 (0.2142)	68	5.680 (0.2236)
22	5.220 (0.2055)	46	5.460 (0.2150)	70	5.700 (0.2244)
24	5.240 (0.2063)	48	5.480 (0.2157)	72	5.720 (0.2252)
26	5.260 (0.2071)	50	5.500 (0.2165)	74	5.740 (0.2260)
28	5.280 (0.2079)	52	5.520 (0.2173)		

9307WG71

Adjusting shim chart (exhaust)—1ZZ-FE engine

2ZZ-GE: Valve Shim Selection Chart (Intake)

Measure clearance mm (in.) \ Installed lifter thickness mm (in.)	2.000 (0.0787)	2.020 (0.0795)	2.040 (0.0803)	2.060 (0.0811)	2.080 (0.0819)	2.100 (0.0827)	2.120 (0.0835)	2.140 (0.0843)	2.160 (0.0850)	2.180 (0.0858)	2.190 (0.0858)	2.200 (0.0862)	2.210 (0.0866)	2.220 (0.0870)	2.230 (0.0878)	2.240 (0.0882)	2.250 (0.0886)	2.260 (0.0890)	2.270 (0.0894)	2.280 (0.0898)	2.290 (0.0902)	2.300 (0.0906)	2.310 (0.0909)	2.320 (0.0913)	2.330 (0.0917)	2.340 (0.0921)	2.350 (0.0925)	2.360 (0.0929)	2.370 (0.0933)	2.380 (0.0937)	2.390 (0.0941)	2.400 (0.0945)	2.410 (0.0949)	2.420 (0.0953)	2.430 (0.0957)	2.440 (0.0961)	2.450 (0.0965)	2.460 (0.0969)	2.470 (0.0972)	2.480 (0.0976)	2.490 (0.0980)	2.500 (0.0984)	2.510 (0.0988)	2.520 (0.0992)	2.530 (0.0996)	2.540 (0.1000)	2.550 (0.1004)	2.560 (0.1008)	2.580 (0.1016)	2.600 (0.1024)	2.620 (0.1031)	2.640 (0.1039)	2.660 (0.1047)	2.680 (0.1055)	2.700 (0.1063)	2.720 (0.1071)	2.740 (0.1079)	2.760 (0.1087)	2.780 (0.1094)	2.800 (0.1102)
0.000-0.030 (0.0000-0.0012)									00	00	00	00	02	02	04	04	06	06	08	08	10	10	12	12	14	14	16	16	18	18	20	20	22	22	24	24	26	26	28	28	30	30	32	32	34	34	36	36	38	40	42	44	46	48	50	52	54	56	58	60
0.031-0.050 (0.0012-0.0020)							00	00	00	00	02	02	04	04	06	06	08	08	10	10	12	12	14	14	16	16	18	18	20	20	22	22	24	24	26	26	28	28	30	30	32	32	34	34	36	36	38	38	40	42	44	46	48	50	52	54	56	58	60	62
0.051-0.070 (0.0020-0.0028)						00	00	00	00	02	04	04	06	06	08	08	10	10	12	12	14	14	16	16	18	18	20	20	22	22	24	24	26	26	28	28	30	30	32	32	34	34	36	36	38	38	40	40	42	44	46	48	50	52	54	56	58	60	62	64
0.071-0.090 (0.0028-0.0035)					00	00	00	00	02	04	06	06	08	08	10	10	12	12	14	14	16	16	18	18	20	20	22	22	24	24	26	26	28	28	30	30	32	32	34	34	36	36	38	38	40	40	42	42	44	46	48	50	52	54	56	58	60	62	64	66
0.091-0.110 (0.0035-0.0043)				00	00	00	00	02	04	06	08	08	10	10	12	12	14	14	16	16	18	18	20	20	22	22	24	24	26	26	28	28	30	30	32	32	34	34	36	36	38	38	40	40	42	42	44	44	46	48	50	52	54	56	58	60	62	64	66	68
0.111-0.130 (0.0043-0.0051)			00	00	00	00	02	04	06	08	10	10	12	12	14	14	16	16	18	18	20	20	22	22	24	24	26	26	28	28	30	30	32	32	34	34	36	36	38	38	40	40	42	42	44	44	46	46	48	50	52	54	56	58	60	62	64	66	68	70
0.131-0.149 (0.0051-0.0059)		00	00	00	00	02	04	06	08	10	12	12	14	14	16	16	18	18	20	20	22	22	24	24	26	26	28	28	30	30	32	32	34	34	36	36	38	38	40	40	42	42	44	44	46	46	48	48	50	52	54	56	58	60	62	64	66	68	70	72
0.150-0.250 (0.0059-0.0098)																																																												
0.251-0.270 (0.0099-0.0106)	06	08	10	12	14	16	18	20	22	24	26	26	28	28	30	30	32	32	34	34	36	36	38	38	40	40	42	42	44	44	46	46	48	48	50	50	52	52	54	54	56	56	58	58	60	60	62	64	66	68	70	72	74	76	78	80	80	80	80	
0.271-0.290 (0.0107-0.0114)	08	10	12	14	16	18	20	22	24	26	28	28	30	30	32	32	34	34	36	36	38	38	40	40	42	42	44	44	46	46	48	48	50	50	52	52	54	54	56	56	58	58	60	60	62	62	64	64	66	68	70	72	74	76	78	80	80	80		
0.291-0.310 (0.0115-0.0122)	10	12	14	16	18	20	22	24	26	28	30	30	32	32	34	34	36	36	38	38	40	40	42	42	44	44	46	46	48	48	50	50	52	52	54	54	56	56	58	58	60	60	62	62	64	64	66	66	68	70	72	74	76	78	80	80	80			
0.311-0.330 (0.0122-0.0130)	12	14	16	18	20	22	24	26	28	30	32	32	34	34	36	36	38	38	40	40	42	42	44	44	46	46	48	48	50	50	52	52	54	54	56	56	58	58	60	60	62	62	64	64	66	66	68	68	70	72	74	76	78	80	80	80				
0.331-0.350 (0.0130-0.0138)	14	16	18	20	22	24	26	28	30	32	34	34	36	36	38	38	40	40	42	42	44	44	46	46	48	48	50	50	52	52	54	54	56	56	58	58	60	60	62	62	64	64	66	66	68	68	70	70	72	74	76	78	80	80	80					
0.351-0.370 (0.0138-0.0146)	16	18	20	22	24	26	28	30	32	34	36	36	38	38	40	40	42	42	44	44	46	46	48	48	50	50	52	52	54	54	56	56	58	58	60	60	62	62	64	64	66	66	68	68	70	70	72	72	74	76	78	80	80	80						
0.371-0.390 (0.0146-0.0154)	18	20	22	24	26	28	30	32	34	36	38	38	40	40	42	42	44	44	46	46	48	48	50	50	52	52	54	54	56	56	58	58	60	60	62	62	64	64	66	66	68	68	70	70	72	72	74	74	76	78	80	80	80							
0.391-0.410 (0.0154-0.0161)	20	22	24	26	28	30	32	34	36	38	40	40	42	42	44	44	46	46	48	58	50	50	52	52	54	54	56	56	58	58	60	60	62	62	64	64	66	66	68	68	70	70	72	72	74	74	76	76	78	80	80	80								
0.411-0.430 (0.0162-0.0169)	22	24	26	28	30	32	34	36	38	40	42	42	44	44	46	46	48	48	50	50	52	52	54	54	56	56	58	58	60	60	62	62	64	64	66	66	68	68	70	70	72	72	74	74	76	76	78	78	80	80	80									
0.431-0.450 (0.0170-0.0177)	24	26	28	30	32	34	36	38	40	42	44	44	46	46	48	48	50	50	52	52	54	54	56	56	58	58	60	60	62	62	64	64	66	66	68	68	70	70	72	72	74	74	76	76	78	78	80	80	80	80										
0.451-0.470 (0.0178-0.0185)	26	28	30	32	34	36	38	40	42	44	46	46	48	48	50	50	52	52	54	54	56	56	58	58	60	60	62	62	64	64	66	66	68	68	70	70	72	72	74	74	76	76	78	78	80	80	80	80	80											
0.471-0.490 (0.0185-0.0193)	28	30	32	34	36	38	40	42	44	46	48	48	50	50	52	52	54	54	56	56	58	58	60	60	62	62	64	64	66	66	68	68	70	70	72	72	74	74	76	76	78	78	80	80	80	80	80	80												
0.491-0.510 (0.0193-0.0201)	30	32	34	36	38	40	42	44	46	48	50	50	52	52	54	54	56	56	58	58	60	60	62	62	64	64	66	66	68	68	70	70	72	72	74	74	76	76	78	78	80	80	80	80	80	80														
0.511-0.530 (0.0201-0.0209)	32	34	36	38	40	42	44	46	48	50	52	52	54	54	56	56	58	58	60	60	62	62	64	64	66	66	68	68	70	70	72	72	74	74	76	76	78	78	80	80	80	80	80	80																
0.531-0.550 (0.0209-0.0217)	34	36	38	40	42	44	46	48	50	52	54	54	56	56	58	58	60	60	62	62	64	64	66	66	68	68	70	70	72	72	74	74	76	76	78	78	80	80	80	80	80	80																		
0.551-0.570 (0.0217-0.0224)	36	38	40	42	44	46	48	50	52	54	56	56	58	58	60	60	62	62	64	64	66	66	68	68	70	70	72	72	74	74	76	76	78	78	80	80	80	80	80	80																				
0.571-0.590 (0.0225-0.0232)	38	40	42	44	46	48	50	52	54	56	58	58	60	60	62	62	64	64	66	66	68	68	70	70	72	72	74	74	76	76	78	78	80	80	80	80	80	80																						
0.591-0.610 (0.0233-0.0240)	40	42	44	46	48	50	52	54	56	58	60	60	62	62	64	64	66	66	68	68	70	70	72	72	74	74	76	76	78	78	80	80	80	80	80	80																								
0.611-0.630 (0.0241-0.0248)	42	44	46	48	50	52	54	56	58	60	62	62	64	64	66	66	68	68	70	70	72	72	74	74	76	76	78	78	80	80	80	80	80	80																										
0.631-0.650 (0.0248-0.0256)	44	46	48	50	52	54	56	58	60	62	64	64	66	66	68	68	70	70	72	72	74	74	76	76	78	78	80	80	80	80	80	80																												
0.651-0.670 (0.0256-0.0264)	46	48	50	52	54	56	58	60	62	64	66	66	68	68	70	70	72	72	74	74	76	76	78	78	80	80	80	80	80	80																														
0.671-0.690 (0.0264-0.0272)	48	50	52	54	56	58	60	62	64	66	68	68	70	70	72	72	74	74	76	76	78	78	80	80	80	80	80	80																																
0.691-0.710 (0.0272-0.0280)	50	52	54	56	58	60	62	64	66	68	70	70	72	72	74	74	76	76	78	78	80	80	80	80	80	80																																		
0.711-0.730 (0.0280-0.0287)	52	54	56	58	60	62	64	66	68	70	72	72	74	74	76	76	78	78	80	80	80	80	80	80																																				
0.731-0.750 (0.0288-0.0295)	54	56	58	60	62	64	66	68	70	72	74	74	76	76	78	78	80	80	80	80	80	80																																						
0.751-0.770 (0.0296-0.0303)	56	58	60	62	64	66	68	70	72	74	76	76	78	78	80	80	80	80	80	80																																								
0.771-0.790 (0.0304-0.0311)	58	60	62	64	66	68	70	72	74	76	78	78	80	80	80	80	80	80																																										
0.791-0.810 (0.0311-0.0319)	60	62	64	66	68	70	72	74	76	78	80	80	80	80	80	80																																												
0.811-0.830 (0.0319-0.0327)	62	64	66	68	70	72	74	76	78	80	80	80	80	80																																														
0.831-0.850 (0.0327-0.0335)	64	66	68	70	72	74	76	78	80	80	80	80																																																
0.851-0.870 (0.0335-0.0343)	66	68	70	72	74	76	78	80	80	80																																																		
0.871-0.890 (0.0343-0.0350)	68	70	72	74	76	78	80	80	80																																																			
0.891-0.910 (0.0351-0.0358)	70	72	74	76	78	80	80	80																																																				
0.911-0.930 (0.0359-0.0366)	72	74	76	78	80	80	80																																																					
0.931-0.950 (0.0367-0.0374)	74	76	78	80	80	80																																																						
0.951-0.970 (0.0374-0.0382)	76	78	80	80	80																																																							
0.971-0.990 (0.0382-0.0390)	78	80	80	80																																																								
0.991-1.010 (0.0390-0.0398)	80	80	80																																																									
1.011-1.030 (0.0398-0.0406)	80	80																																																										
1.031-1.050 (0.0406-0.0413)	80																																																											

New Shim thickness mm (in.)

Shim No.	Thickness	Shim No.	Thickness	Shim No.	Thickness
00	2.000 (0.0787)	28	2.280 (0.0898)	56	2.560 (0.1008)
02	2.020 (0.0795)	30	2.300 (0.0906)	58	2.580 (0.1016)
04	2.040 (0.0803)	32	2.320 (0.0913)	60	2.600 (0.1024)
06	2.060 (0.0811)	34	2.340 (0.0921)	62	2.620 (0.1031)
08	2.080 (0.0819)	36	2.360 (0.0929)	64	2.640 (0.1039)
10	2.100 (0.0827)	38	2.380 (.0.0937)	66	2.660 (0.1047)
12	2.120 (0.0835)	40	2.400 (0.0945)	68	2.680 (0.1055)
14	2.140 (0.0843)	42	2.420 (0.0953)	70	2.700 (0.1063)
16	2.160 (0.0850)	44	2.440 (0.0961)	72	2.720 (0.1071)
18	2.180 (0.0858)	46	2.460 (0.0969)	74	2.740 (0.1079)
20	2.200 (0.0866)	48	2.480 (0.0976)	76	2.760 (0.1087)
22	2.220 (0.0874)	50	2.500 (0.0984)	78	2.780 (0.1094)
24	2.240 (0.0882)	52	2.520 (0.0992)	80	2.800 (0.1102)
26	2.260 (0.0890)	54	2.540 (0.1000)		

Intake valve clearance (Cold):
0.15 – 0.25 mm (0.006 – 0.010 in.)

EXAMPLE: The 2.200 mm (0.0826 in.) shim is installed, and the measured clearance is 0.400 mm (0.0157 in.). Replace the 2.400 mm (0.0945 in.) shim with a new No. 40 shim.

9307WG72

Adjusting shim chart (intake)—2ZZ-GE engine

2ZZ–GE: Valve Shim Selection Chart (Exhaust)

Measure clearance mm (in.) \ Installed lifter thickness mm (in.)	2.000 (0.0787)	2.020 (0.0795)	2.040 (0.0803)	2.060 (0.0811)	2.080 (0.0819)	2.100 (0.0827)	2.120 (0.0835)	2.140 (0.0843)	2.160 (0.0850)	2.180 (0.0858)	2.190 (0.0862)	2.200 (0.0866)	2.210 (0.0870)	2.220 (0.0874)	2.230 (0.0878)	2.240 (0.0882)	2.250 (0.0886)	2.260 (0.0890)	2.270 (0.0894)	2.280 (0.0898)	2.290 (0.0902)	2.300 (0.0906)	2.310 (0.0909)	2.320 (0.0913)	2.330 (0.0917)	2.340 (0.0921)	2.350 (0.0925)	2.360 (0.0929)	2.370 (0.0933)	2.380 (0.0937)	2.390 (0.0941)	2.400 (0.0945)	2.410 (0.0949)	2.420 (0.0953)	2.430 (0.0957)	2.440 (0.0961)	2.450 (0.0965)	2.460 (0.0969)	2.470 (0.0972)	2.480 (0.0976)	2.490 (0.0980)	2.500 (0.0984)	2.510 (0.0988)	2.520 (0.0992)	2.530 (0.0996)	2.540 (0.1000)	2.550 (0.1004)	2.560 (0.1008)	2.580 (0.1016)	2.600 (0.1024)	2.620 (0.1031)	2.640 (0.1039)	2.660 (0.1047)	2.680 (0.1055)	2.700 (0.1063)	2.720 (0.1071)	2.740 (0.1079)	2.760 (0.1087)	2.780 (0.1094)	2.800 (0.1102)
0.000–0.030 (0.0000–0.0012)																											00	00	00	00	00	00	02	02	04	04	06	06	08	08	10	10	12	12	14	14	16	16	18	20	22	24	26	28	30	32	34	36	38	40
0.031–0.050 (0.0012–0.0020)																								00	00	00	00	00	00	00	02	02	04	04	06	06	08	08	10	10	12	12	14	14	16	16	18	18	20	22	24	26	28	30	32	34	36	38	40	42
0.051–0.070 (0.0020–0.0028)																						00	00	00	00	00	00	00	02	02	04	04	06	06	08	08	10	10	12	12	14	14	16	16	18	18	20	20	22	24	26	28	30	32	34	36	38	40	42	44
0.071–0.090 (0.0028–0.0035)																				00	00	00	00	00	00	00	02	02	04	04	06	06	08	08	10	10	12	12	14	14	16	16	18	18	20	20	22	22	24	26	28	30	32	34	36	38	40	42	44	46
0.091–0.110 (0.0036–0.0043)																		00	00	00	00	00	00	00	02	02	04	04	06	06	08	08	10	10	12	12	14	14	16	16	18	18	20	20	22	22	24	24	26	28	30	32	34	36	38	40	42	44	46	48
0.111–0.130 (0.0044–0.0051)																00	00	00	00	00	00	00	02	02	04	04	06	06	08	08	10	10	12	12	14	14	16	16	18	18	20	20	22	22	24	24	26	36	28	30	32	34	36	38	40	42	44	46	48	50
0.131–0.150 (0.0052–0.0059)														00	00	00	00	00	00	00	02	02	04	04	06	06	08	08	10	10	12	12	14	14	16	16	18	18	20	20	22	22	24	24	26	26	28	28	30	32	34	36	38	40	42	44	46	48	50	52
0.151–0.170 (0.0059–0.0067)												00	00	00	00	00	00	00	02	02	04	04	06	06	08	08	10	10	12	12	14	14	16	16	18	18	20	20	22	22	24	24	26	26	28	28	30	30	32	34	36	38	40	42	44	46	48	50	52	54
0.171–0.190 (0.0067–0.0075)										00	00	00	00	00	00	00	02	02	04	04	06	06	08	08	10	10	12	12	14	14	16	16	18	18	20	20	22	22	24	24	26	26	28	28	30	30	32	32	34	36	38	40	42	44	46	48	50	52	54	56
0.191–0.210 (0.0075–0.0083)									00	00	00	00	00	00	02	02	04	04	06	06	08	08	10	10	12	12	14	14	16	16	18	18	20	20	22	22	24	24	26	26	28	28	30	30	32	32	34	34	36	38	40	42	44	46	48	50	52	54	56	58
0.211–0.230 (0.0083–0.0091)								00	00	00	00	00	02	02	04	04	06	06	08	08	10	10	12	12	14	14	16	16	18	18	20	20	22	22	24	24	26	26	28	28	30	30	32	32	34	34	36	36	38	40	42	44	46	48	50	52	54	56	58	60
0.231–0.250 (0.0091–0.0098)							00	00	00	00	02	02	04	04	06	06	08	08	10	10	12	12	14	14	16	16	18	18	20	20	22	22	24	24	26	26	28	28	30	30	32	32	34	34	36	36	38	38	40	42	44	46	48	50	52	54	56	58	60	62
0.251–0.270 (0.0099–0.0106)						00	00	00	00	02	04	04	06	06	08	08	10	10	12	12	14	14	16	16	18	18	20	20	22	22	24	24	26	26	28	28	30	30	32	32	34	34	36	36	38	38	40	40	42	44	46	48	50	52	54	56	58	60	62	64
0.271–0.290 (0.0107–0.0114)					00	00	00	00	02	04	06	06	08	08	10	10	12	12	14	14	16	16	18	18	20	20	22	22	24	24	26	26	28	28	30	30	32	32	34	34	36	36	38	38	40	40	42	42	44	46	48	50	52	54	56	58	60	62	64	66
0.291–0.310 (0.0115–0.0122)				00	00	00	00	02	04	06	08	08	10	10	12	12	14	14	16	16	18	18	20	20	22	22	24	24	26	26	28	28	30	30	32	32	34	34	36	36	38	38	40	40	42	42	44	44	46	48	50	52	54	56	58	60	62	64	66	68
0.311–0.330 (0.0122–0.0130)			00	00	00	00	02	04	06	08	10	10	12	12	14	14	16	16	18	18	20	20	22	22	24	24	26	26	28	28	30	30	32	32	34	34	36	36	38	38	40	40	42	42	44	44	46	46	48	50	52	54	56	58	60	62	64	66	68	70
0.331–0.349 (0.0130–0.0137)		00	00	00	00	02	04	06	08	10	12	12	14	14	16	16	18	18	20	20	22	22	24	24	26	26	28	28	30	30	32	32	34	34	36	36	38	38	40	40	42	42	44	44	46	46	48	48	50	52	54	56	58	60	62	64	66	68	70	72
0.350–0.450 (0.0138–0.0177)																																																												
0.451–0.470 (0.0178–0.0185)	06	08	10	12	14	16	18	20	22	24	26	26	28	28	30	30	32	32	34	34	36	36	38	38	40	42	42	44	44	46	46	48	48	50	50	52	52	54	54	56	56	58	58	60	60	62	62	64	66	68	70	72	74	76	78	80	80	80	80	
0.471–0.490 (0.0185–0.0193)	08	10	12	14	16	18	20	22	24	26	28	28	30	30	32	32	34	34	36	36	38	38	40	42	42	44	44	46	46	48	48	50	50	52	52	54	54	56	56	56	60	60	60	62	62	64	64	66	68	70	62	74	76	78	80	80	80	80		
0.491–0.510 (0.0193–0.0201)	10	12	14	16	18	20	22	24	26	28	30	30	32	32	34	34	36	36	38	38	40	42	42	44	44	46	46	48	48	50	50	52	52	54	54	56	56	58	58	60	60	62	62	64	64	66	66	68	70	72	74	76	78	80	80	80	80			
0.511–0.530 (0.0201–0.0209)	12	14	16	18	20	22	24	26	28	30	32	32	34	34	36	36	38	38	40	42	42	44	44	46	46	48	48	50	50	52	52	54	54	56	56	58	68	60	60	60	64	64	64	66	66	68	68	70	72	74	76	78	80	80	80	80				
0.531–0.550 (0.0209–0.0217)	14	16	18	20	22	24	26	28	30	32	34	34	36	36	38	38	40	42	42	44	44	46	46	48	48	50	50	52	52	54	54	56	56	58	58	60	60	62	62	64	64	66	66	68	68	70	70	72	74	76	78	80	80	80	80					
0.551–0.570 (0.0217–0.0224)	16	18	20	22	24	26	28	30	32	34	36	36	38	38	40	42	42	44	44	46	46	48	48	50	50	52	52	54	54	56	56	58	58	60	60	62	62	64	64	64	68	68	68	70	70	72	72	74	76	78	80	80	80	80						
0.571–0.590 (0.0225–0.0232)	18	20	22	24	26	28	30	32	34	36	38	38	40	42	42	44	44	46	46	48	48	50	50	52	52	54	54	56	56	58	68	60	60	62	62	64	64	66	66	68	68	70	70	72	72	74	74	76	78	80	80	80	80							
0.591–0.610 (0.0223–0.0240)	20	22	24	26	28	30	32	34	36	38	40	42	42	44	44	46	46	48	48	50	50	52	52	54	54	56	56	58	58	60	60	62	62	64	64	66	66	68	68	68	72	72	72	74	74	76	76	78	80	80	80	80								
0.611–0.630 (0.0241–0.0248)	22	24	26	28	30	32	34	36	38	40	42	44	44	46	46	48	48	50	50	52	52	54	54	56	56	58	58	60	60	62	62	64	64	66	66	68	68	70	70	72	72	74	74	76	76	78	78	80	80	80	80									
0.631–0.650 (0.0248–0.0256)	24	26	28	30	32	34	36	38	40	42	44	46	46	48	48	50	50	52	52	54	54	56	56	58	58	60	60	62	62	64	64	66	66	68	68	70	70	72	72	72	76	76	76	78	78	80	80	80	80	80										
0.651–0.670 (0.0256–0.0264)	25	28	30	32	34	36	38	40	42	44	46	48	48	50	50	52	52	54	54	56	56	58	58	60	60	62	62	64	64	66	66	68	68	70	70	72	72	74	74	76	76	78	78	80	80	80	80	80	80											
0.671–0.690 (0.0264–0.0272)	28	30	32	34	36	38	40	42	44	46	48	50	50	52	52	54	54	56	56	58	58	60	60	62	62	64	64	66	66	68	68	70	70	72	72	74	74	76	76	76	80	80	80	80	80	80	80	80												
0.691–0.710 (0.0272–0.0280)	30	32	34	36	38	40	42	44	46	48	50	52	52	54	54	56	56	58	58	60	60	62	62	64	64	66	66	68	68	70	70	72	72	74	74	76	76	78	78	80	80	80	80	80	80	80														
0.711–0.730 (0.0280–0.0287)	32	34	36	38	40	42	44	46	48	50	52	54	54	56	56	58	58	60	60	62	62	64	64	66	66	68	68	70	70	72	72	74	74	76	76	78	78	80	80	80	80	80	80	80																
0.731–0.750 (0.0288–0.0295)	34	36	38	40	42	44	46	48	50	52	54	56	56	58	58	60	60	62	62	64	64	66	66	68	68	70	70	72	72	74	74	76	76	78	78	80	80	80	80	80	80	80																		
0.751–0.770 (0.0296–0.0303)	36	38	40	42	44	46	48	50	52	54	56	58	58	60	60	62	62	64	64	66	66	68	68	70	70	72	72	74	74	76	76	78	78	80	80	80	80	80	80	80																				
0.771–0.790 (0.0304–0.0311)	38	40	42	44	46	48	50	52	54	56	58	60	60	62	62	64	64	66	66	68	68	70	70	72	72	74	74	76	76	78	78	80	80	80	80	80	80	80																						
0.791–0.810 (0.0311–0.0319)	40	42	44	46	48	50	52	54	56	58	60	62	62	64	64	66	66	68	68	70	70	72	72	74	74	76	76	78	78	80	80	80	80	80	80	80																								
0.811–0.830 (0.0319–0.0327)	42	44	46	48	50	52	54	56	58	60	62	64	64	66	66	68	68	70	70	72	72	74	74	76	76	78	78	80	80	80	80	80	80	80																										
0.831–0.850 (0.0327–0.0335)	44	46	48	50	52	54	56	58	60	62	64	66	66	68	68	70	70	72	72	74	74	76	76	78	78	80	80	80	80	80	80	80																												
0.851–0.870 (0.0335–0.0343)	46	48	50	52	54	56	58	60	62	64	66	68	68	70	70	72	72	74	74	76	76	78	78	80	80	80	80	80	80	80																														
0.871–0.890 (0.0343–0.0350)	48	50	52	54	56	58	60	62	64	66	68	70	70	72	72	74	74	76	76	78	78	80	80	80	80	80	80	80																																
0.891–0.910 (0.0351–0.0358)	50	52	54	56	58	60	62	64	66	68	70	72	72	74	74	76	76	78	78	80	80	80	80	80	80	80																																		
0.911–0.930 (0.0359–0.0366)	52	54	56	58	60	62	64	66	68	70	72	74	74	76	76	78	78	70	80	80	80	80	80	80																																				
0.931–0.950 (0.0367–0.0374)	54	56	58	60	62	64	66	68	70	72	74	76	76	78	78	80	80	80	80	80	80	80																																						
0.951–0.970 (0.0374–0.0382)	56	58	60	62	64	66	68	70	72	74	76	78	78	80	80	80	80	80	80	80																																								
0.971–0.990 (0.0382–0.0390)	58	60	62	64	66	68	70	72	74	76	78	80	80	80	80	80	80	80																																										
0.991–1.010 (0.0390–0.0398)	60	62	64	66	68	70	72	74	76	78	80	80	80	80	80	80																																												
1.011–1.030 (0.0398–0.0406)	62	64	66	68	[illegible]	72	74	76	78	80	80	80	80	80																																														
1.031–1.050 (0.0406–0.0413)	64	66	68	70	72	74	76	78	80	80	80	80																																																
1.051–1.070 (0.0414–0.0421)	66	68	70	72	74	76	78	80	80	80																																																		
1.071–1.090 (0.0422–0.0429)	68	70	72	74	76	78	80	80	80																																																			
1.091–1.110 (0.0430–0.0437)	70	72	74	76	78	80	80	80																																																				
1.111–1.130 (0.0437–0.0445)	72	74	76	78	80	80	80																																																					
1.131–1.150 (0.0445–0.0453)	74	76	78	80	80	80																																																						
1.151–1.170 (0.0453–0.0461)	76	78	80	80	80																																																							
1.171–1.190 (0.0461–0.0469)	78	80	80	80																																																								
1.191–1.210 (0.0469–0.0476)	80	80	80																																																									
1.211–1.230 (0.0477–0.0484)	80	80																																																										
1.231–1.250 (0.0485–0.0492)	80																																																												

Exhaust valve clearance (Cold):
0.35 – 0.45 mm (0.014 – 0.018 in.)

EXAMPLE: The 2.200 mm (0.0862 in.) shim is installed, and the measured clearance is 0.500 mm (0.0197 in.). Replace the 2.300 mm (0.0906 in.) shim with a new No. 30 shim.

New Shim thickness mm (in.)

Shim No.	Thickness	Shim No.	Thickness	Shim No.	Thickness
00	2.000 (0.0787)	28	2.280 (0.0898)	56	2.560 (0.1008)
02	2.020 (0.0795)	30	2.300 (0.0906)	58	2.580 (0.1016)
04	2.040 (0.0803)	32	2.320 (0.0913)	60	2.600 (0.1024)
06	2.060 (0.0811)	34	2.340 (0.0921)	62	2.620 (0.1031)
08	2.080 (0.0819)	36	2.360 (0.0929)	64	2.640 (0.1039)
10	2.100 (0.0827)	38	2.380 (.0.0937)	66	2.660 (0.1047)
12	2.120 (0.0835)	40	2.400 (0.0945)	68	2.680 (0.1055)
14	2.140 (0.0843)	42	2.420 (0.0953)	70	2.700 (0.1063)
16	2.160 (0.0850)	44	2.440 (0.0961)	72	2.720 (0.1071)
18	2.180 (0.0858)	46	2.460 (0.0969)	74	2.740 (0.1079)
20	2.200 (0.0866)	48	2.480 (0.0976)	76	2.760 (0.1087)
22	2.220 (0.0874)	50	2.500 (0.0984)	78	2.780 (0.1094)
24	2.240 (0.0882)	52	2.520 (0.0992)	80	2.800 (0.1102)
26	2.260 (0.0890)	54	2.540 (0.1000)		

9307WG73

Adjusting shim chart (exhaust)—2ZZ-GE engine

Adjusting Shim Selection Chart

New shim thickness mm (in.)

Shim No.	Thickness	Shim No.	Thickness
02	2.500 (0.0984)	20	2.950 (0.1161)
04	2.550 (0.1004)	22	3.000 (0.1181)
06	2.600 (0.1024)	24	3.050 (0.1201)
08	2.650 (0.1043)	26	3.100 (0.1220)
10	2.700 (0.1063)	28	3.150 (0.1240)
12	2.750 (0.1083)	30	3.200 (0.1260)
14	2.800 (0.1102)	32	3.250 (0.1280)
16	2.850 (0.1122)	34	3.300 (0.1299)
18	2.900 (0.1142)		

7923VG56

Adjusting shim chart (exhaust)—1NZ-FE engine

Adjusting Shim Selection Chart

New shim thickness mm (in.)

Shim No.	Thickness	Shim No.	Thickness
02	2.500 (0.0984)	20	2.950 (0.1161)
04	2.550 (0.1004)	22	3.000 (0.1181)
06	2.600 (0.1024)	24	3.050 (0.1201)
08	2.650 (0.1043)	26	3.100 (0.1220)
10	2.700 (0.1063)	28	3.150 (0.1240)
12	2.750 (0.1083)	30	3.200 (0.1260)
14	2.800 (0.1102)	32	3.250 (0.1280)
16	2.850 (0.1122)	34	3.300 (0.1299)
18	2.900 (0.1142)		

7923VG56

Adjusting shim chart (intake)—1NZ-FE engine

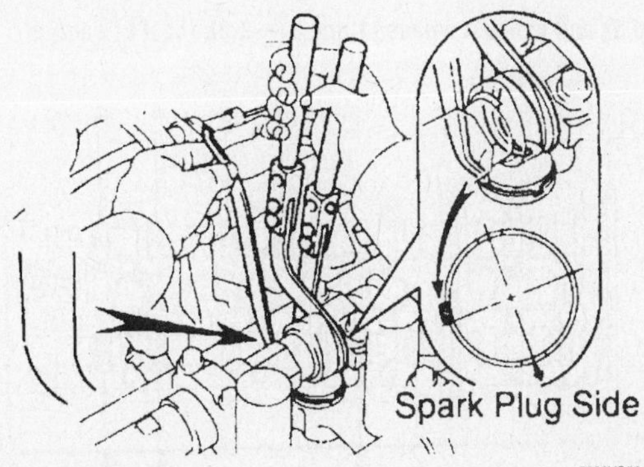

Common method of removing valve shims

8. To adjust the valve clearance:
   a. Set the No.1 cylinder at TDC compression. Place matchmarks on the timing chain and sprockets.
   b. Remove the 2 plugs from the timing chain cover.
   c. Turn the exhaust camshaft clockwise slightly while rotating the stopper plate on the tensioner downward. Push in on the tension plunger. When the stopper plate cannot be easily lowered, rotate the exhaust camshaft clockwise and counterclockwise slightly. Insert a 3mm bar into the holes in the stopper plate and tensioner to lock the tensioner. Remove the timing chain.
   d. Remove the valve timing controller assembly.
   e. Remove the lifters.
9. Determine the replacement adjusting shim size by either using the chart or the following formula:
   - Intake: N = T + A – 0.008 in. (0.20mm)
   - Exhaust: N = T + A – 0.012 in. (0.30mm)
   - T = Thickness of removed shim
   - A = Measured valve clearance
   - N = Thickness of new shim
10. Install a new shim.
11. Recheck the valve clearance.
12. Install the cylinder head covers.
13. Connect the negative battery cable.

2ZZ-GE Engines

➡Adjust the valve clearance when the engine is cold.

1. Before servicing the vehicle, refer to the precautions in the beginning of this section.
2. Remove or disconnect the following:
   - Negative battery cable.
   - Cylinder head covers
3. Turn the crankshaft pulley and align its groove with the timing mark **0** of the No. 1 timing cover.
4. Check that the timing marks on the camshaft sprockets are in alignment with the upper edge of the timing cover. If not, turn the crankshaft 1 complete revolution (360 degrees).
5. Measure the clearance between the valve lifter and the camshaft. Record the measurements on the intake valves No. 1 and 2. Measure the exhaust valves at No. 1 and 3.
   a. The intake valve clearance cold is 0.006–0.010 in. (0.15–0.25mm).
   b. The exhaust valve clearance cold is 0.014–0.018 in. (0.36–0.45mm).
6. Turn the crankshaft pulley 1 revolu-

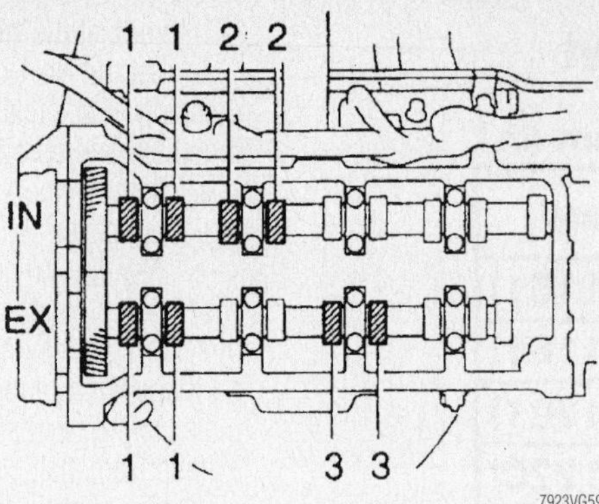

Intake valves (1 and 2) and exhaust valves (1 and 3)—1NZ-FE engine

tion (360 degrees) and align the groove with the timing mark **0** of the No.1 timing belt cover.

7. Measure the clearance between the valve lifter and the camshaft. Record the measurements on the intake valves No. 3 and 4. Measure the exhaust valves at No. 2 and 4.
   a. The intake valve clearance cold is 0.006–0.010 in. (0.15–0.25mm).
   b. The exhaust valve clearance cold is 0.014–0.018 in. (0.36–0.45mm).
8. To adjust the intake valve clearance:
   a. Remove the intake camshaft.
   b. Using a small screwdriver and a magnetic finger, remove the adjusting shim.
   c. Determine the replacement adjusting shim size by either using the chart or the following formula:
      - Intake: N = T + A – 0.008 in. (0.20mm)
      - T = Thickness of removed shim
      - A = Measured valve clearance
      - N = Thickness of new shim
   d. Install a new shim.
   e. Install intake camshaft.
   f. Recheck the valve clearance.
9. To adjust the exhaust valve clearance:
   a. Turn the crankshaft to position the cam lobe of the camshaft on the valve to be adjusted, upward.
   b. Turn the valve lifter so that the notch is perpendicular to the camshaft and facing the spark plug side.
   c. Using SST 09248–55040 (valve lifter press), or equivalent, hold the camshaft in place.
   d. Using SST 09248–55040 (valve lifter press), or equivalent, press down the valve lifter and place SST 09248–05420 (valve lifter stopper), or equivalent between the camshaft and valve lifter.
   e. Remove the SST 09248–44040 tool.
   f. Using a small screwdriver and a magnetic finger, remove the adjusting shim.
10. Determine the replacement adjusting shim size by either using the chart or the following formula:
    - Exhaust: N = T + A – 0.014 in. (0.36mm)
    - T = Thickness of removed shim
    - A = Measured valve clearance
    - N = Thickness of new shim
11. Install a new shim.
12. Recheck the valve clearance.
13. Install or connect the following:
    - Cylinder head covers
    - Negative battery cable

5S-FE Engines

➡Adjust the valve clearance when the engine is cold.

1. Before servicing the vehicle, refer to the precautions in the beginning of this section.
2. Remove or disconnect the following:
   - Negative battery cable. On vehicles equipped with an air bag, wait at least 90 seconds before proceeding.
   - Cylinder head covers
3. Turn the crankshaft pulley and align its groove with the timing mark **0** of the No. 1 timing cover.
4. Check that the timing marks on the camshaft sprockets are in alignment with the marks on the No. 4 timing cover. If not, turn the crankshaft 1 complete revolution (360 degrees).
5. Measure the clearance between the valve lifter and the camshaft. Record the measurements on the intake valves No. 1 and 2. Measure the exhaust valves at No. 1 and 3.
   a. The intake valve clearance cold is 0.007–0.011 in. (0.19–0.29mm).
   b. The exhaust valve clearance cold is 0.011–0.015 in. (0.28–0.38mm).
6. Turn the crankshaft pulley 1 revolution (360 degrees) and align the groove with the timing mark **0** of the No.1 timing belt cover.
7. Measure the clearance between the valve lifter and the camshaft. Record the measurements on the intake valves No. 3 and 4. Measure the exhaust valves at 2 and 4.

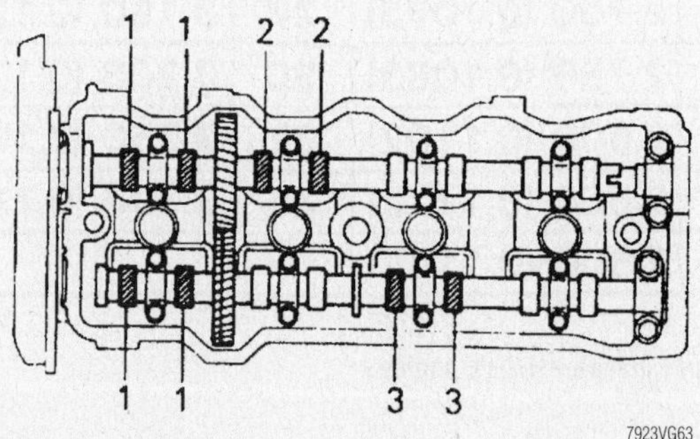

Intake valves (1 and 2) and exhaust valves (1 and 3)—2.2L (5S-FE) engine

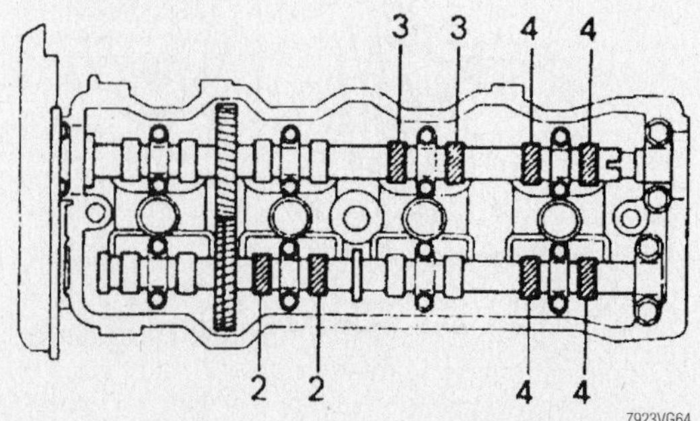

Intake valves (3 and 4) and exhaust valves (2 and 4)—2.2L (5S-FE) engine

a. The intake valve clearance cold is 0.007–0.011 in. (0.19–0.29mm).

b. The exhaust valve clearance cold is 0.011–0.015 in. (0.29–0.38mm).

8. To adjust the valve clearance:

a. Turn the crankshaft to position the cam lobe of the camshaft on the valve to be adjusted, upward.

b. Turn the valve lifter so that the notch is perpendicular to the camshaft and facing the spark plug side.

c. Using SST 09248–55040 (valve lifter press), or equivalent, hold the camshaft in place.

d. Using SST 09248–55040 (valve lifter press), or equivalent, press down the valve lifter and place SST 09248–05420 (valve lifter stopper), or equivalent between the camshaft and valve lifter.

e. Remove the SST 09248–44040 tool.

f. Using a small screwdriver and a magnetic finger, remove the adjusting shim.

9. Determine the replacement adjusting shim size by either using the chart or the following formula:

- Intake: N = T + A – 0.009 in. (24mm)
- Exhaust: N = T + A – 0.013 in. (0.33mm)
- T = Thickness of removed shim
- A = Measured valve clearance
- N = Thickness of new shim

10. Install a new shim.
11. Recheck the valve clearance.
12. Install or connect the following:
- Cylinder head covers
- Negative battery cable

1MZ-FE Engines

➡Adjust the valve clearance when the engine is cold.

1. Before servicing the vehicle, refer to the precautions in the beginning of this section.
2. Remove or disconnect the following:
- Negative battery cable. On vehicles equipped with an air bag, wait at least 90 seconds before proceeding.
- Accelerator/throttle cable from the throttle linkage
- Air cleaner cover, air flow meter, and air duct assembly
- V-bank cover
- Emission control valve set
- Air intake chamber
- Engine harness from the injectors and the ignition coils

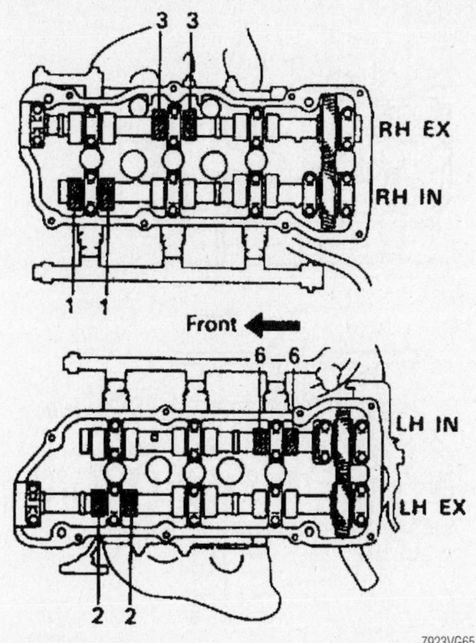

Adjust these valves during the 1st step—3.0L (1MZ-FE) engine

- Ignition coils and keep them in order for reassembly
- Spark plugs
- Cylinder head covers

3. Turn the crankshaft pulley and align its groove with the timing mark **0** of the No. 1 timing cover.
4. Check that the valve lifters on the No. 1 intake are loose and the No. 1 exhaust are tight. If not, turn the crankshaft 1 complete revolution (360 degrees).

➡All measurements should be written down. These recorded measurements will need to be used in conjunction with a mathematical formula to determine the thickness of the replacement shims.

5. Measure the clearance between the valve lifters and the camshaft. Record the measurements on valves No. 1 and 6 intake; No. 2 and 3 exhaust.

a. The intake valve clearance cold is 0.006–0.010 in. (0.15–0.25mm).

b. The exhaust valve clearance cold is 0.010–0.014 in. (0.25–0.35mm).

6. Turn the crankshaft ⅔ of a revolution (240 degrees). Record the measurements on valves No. 2 and 3 intake; No. 4 and 5 exhaust.
7. Turn the crankshaft another ⅔ of a revolution. Record the measurements on valves No. 4 and 5 intake; No. 1 and 6 exhaust.
8. Remove the adjusting shim by turn-

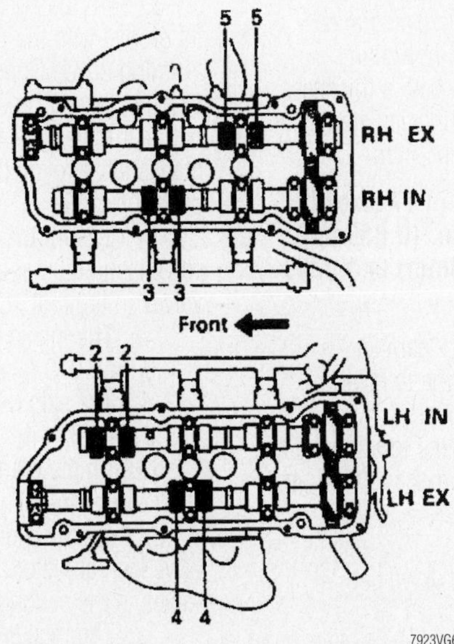

Adjust these valves during the 2nd step—3.0L (1MZ-FE) engine

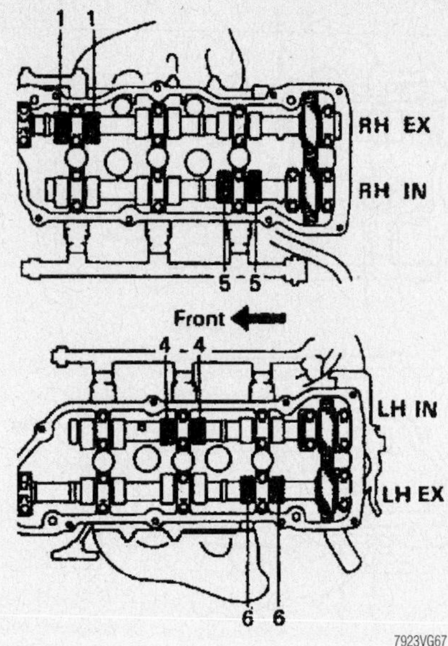

Adjust these valves during the 3rd step—3.0L (1MZ-FE) engine

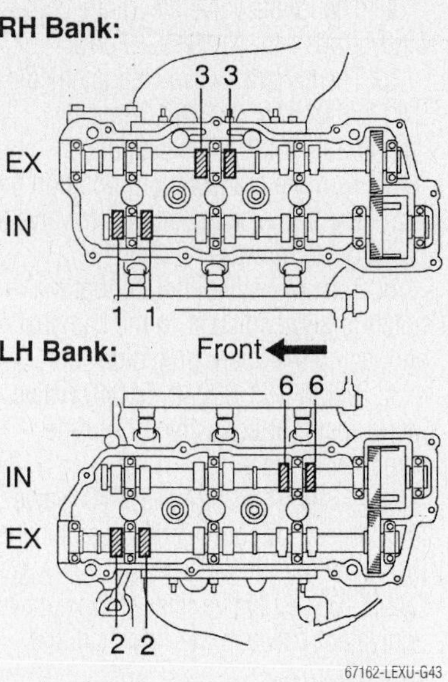

Measuring and adjust these valves first—3MZ-FE engine

ing the crankshaft to position the cam lobe of the camshaft in the up position on the valve to be adjusted. Using a small thin flat bladed tool, turn the valve lifter so that the notches are perpendicular to the camshaft. Press down the valve lifter with SST 09248–55010 part A, or equivalent. Place SST 09248–55010 part B between the camshaft and the valve lifter; remove part A.

9. Remove the adjusting shim with a magnet and a small screwdriver.

10. Determine the replacement adjusting shim size by either using the charts or the following formulas:

- Intake: N = T + (A – 0.008 in./0.020mm)
- Exhaust: N = T + (A – 0.012 in./0.30mm)
- T = Thickness of removed shim
- A = Measured valve clearance
- N = Thickness of new shim

11. Select a new shim with a thickness as close as possible to the calculated value. Install the new replacement shim.

➡Shims are available in 17 sizes in increments of 0.0020 in. (0.050mm), from 0.0984 in. (2.500mm) to 0.1299 in. (3.300mm).

12. Recheck the valve clearance.

13. Install or connect the following:

- Cylinder head covers
- Spark plugs and the ignition coils
- Engine wiring harness to the injectors and the coils
- Intake chamber
- Emission control valve set
- V-bank cover
- Air flow meter, air duct, and air cleaner cover
- Negative battery cable

3MZ-FE Engines

1. Before servicing the vehicle, refer to the precautions in the beginning of this section.

2. Remove or disconnect the following:

- Negative battery cable. Wait at least 90 seconds before performing any other work
- Front suspension upper brace center
- V-bank cover sub-assembly
- Air cleaner assembly
- Emission control valve set
- Intake air surge tank
- Ignition coil assembly
- Cylinder head valve covers

3. Turn the crankshaft pulley and align its groove with the timing mark **0** of the No. 1 timing cover. Check that the timing marks of the camshaft timing pulleys and timing belt rear plates are aligned. If not, turn the crankshaft 1 revolution (360 degrees) and align the mark.

4. Measure the clearance between the valve lash adjuster and the camshaft on the valves in the first sequence and record.

a. The intake valve clearance cold is 0.006–0.010 in. (0.15–0.25mm).

b. The exhaust valve clearance cold is 0.010–0.014 in. (0.25–0.35mm).

5. Turn the crankshaft 2/3 revolution and (240 degrees).

6. Measure the clearance between the valve lash adjuster and the camshaft on the valves in the second sequence and record.

a. The intake valve clearance cold is 0.006–0.010 in. (0.15–0.25mm).

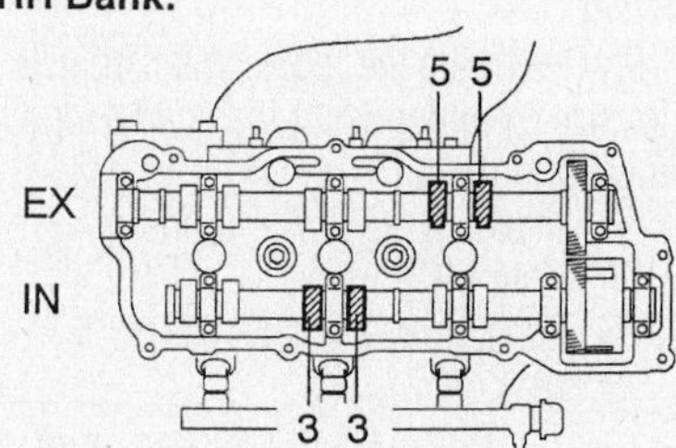

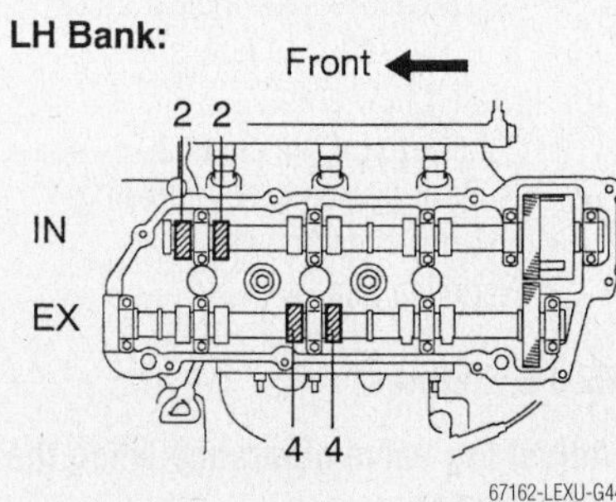

Measuring and adjust these valves second—3MZ-FE engine

b. The exhaust valve clearance cold is 0.010–0.014 in. (0.25–0.35mm).

7. Turn the crankshaft 2/3 revolution and (240 degrees).

8. Measure the clearance between the valve lash adjuster and the camshaft on the valves in the third sequence and record.

a. The intake valve clearance cold is 0.006–0.010 in. (0.15–0.25mm).

b. The exhaust valve clearance cold is 0.010–0.014 in. (0.25–0.35mm).

9. Adjust the valve lash using the following procedure

10. Remove the adjusting shim and turn

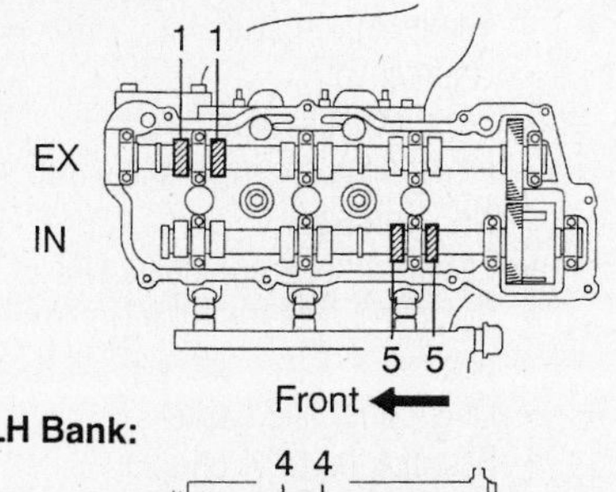

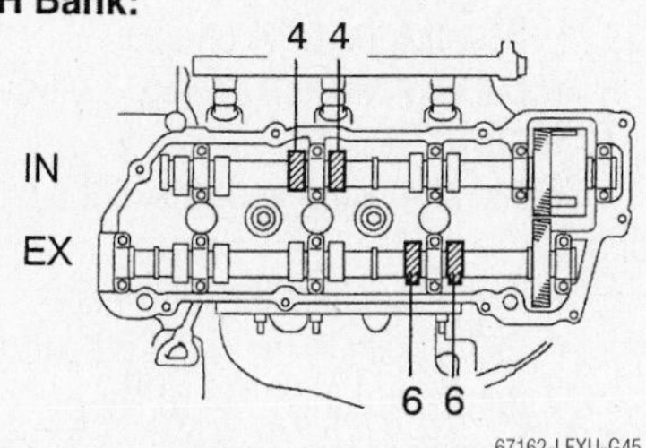

Measuring and adjust these valves third—3MZ-FE engine

the crankshaft to position the cam lobe of the camshaft on the adjusting valve upward. Position the hole in the shim toward the outside of the cylinder head. Press down the valve lash adjuster with the proper tool and place the proper tool between the camshaft and the valve lash adjuster. Remove the tool.

11. Remove the adjusting shim with the proper tool.

12. Determine the thickness of the replacement shim as follows:

a. T = Thickness of the used shim
b. A = Measured valve clearance
c. N = Thickness of new shim
d. Intake: N = T + (A – 0.006–0.010 in. (0.15–0.25mm))
e. Exhaust: N = T + (A – 0.010–0.014 in. (0.25–0.35mm))

➡Place the adjusting shim on the valve lifter with the imprinted number facing down.

13. Reinstall the following
- Cylinder head valve covers
- Ignition coil assembly
- Intake air surge tank
- Emission control valve set
- Air cleaner assembly
- Vacuum hoses
- Front suspension upper center brace
- Negitive battery cable

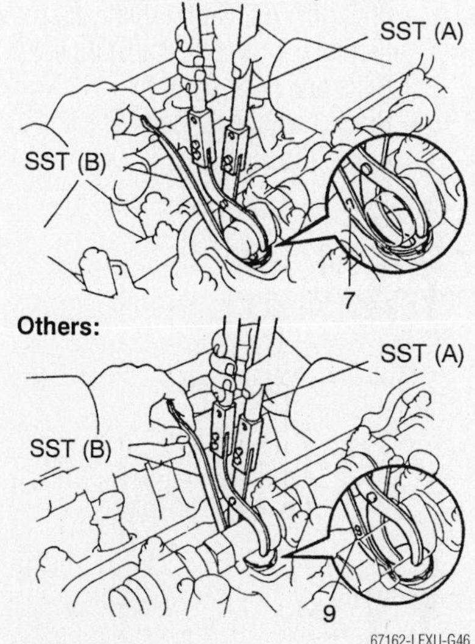

Using special tool to remove the adjusting shim—3MZ-FE engine

Oil Pan

REMOVAL & INSTALLATION

1NZ-FE Engines

1. Before servicing the vehicle, refer to the precautions in the beginning of this section.
2. Drain the engine oil.
3. Remove or disconnect the following:
   - Negative battery cable
   - Oil filter
   - Front exhaust pipe
   - Engine under covers
   - Oil pan bolts
4. Using a thin blade, cut the sealer holding the oil pan and lower the pan.
5. Installation is the reverse of removal. Use RTV sealer. Torque the bolts, in a criss-cross pattern, to 80 inch lbs. (9 Nm).

1ZZ-FE Engines

1. Before servicing the vehicle, refer to the precautions in the beginning of this section.
2. Drain the engine oil.
3. Remove or disconnect the following:
   - Negative battery cable. On vehicles equipped with an air bag, wait at least 90 seconds before proceeding.
   - Undercovers
   - Front exhaust pipe
   - Oil pan mounting bolts and nuts
   - Oil pan, cutting off the applied sealer.

To install:

4. Remove any old sealant from the oil pan flange and thoroughly clean the sealing surface.
5. Install or connect the following:
   - Oil pan. Tighten the bolts and nuts in several passes. Bolts and nuts: 80 inch lbs. (9 Nm).
   - Front exhaust pipe
   - Negative battery cable
   - Undercovers
6. Fill the engine with clean oil.
7. Start the vehicle, check for leaks and repair if necessary.

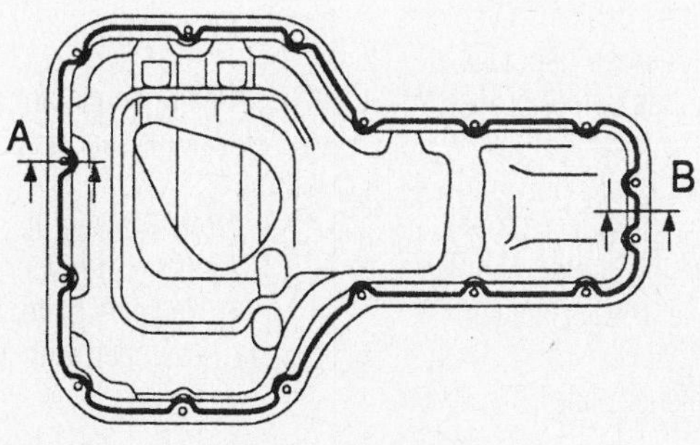

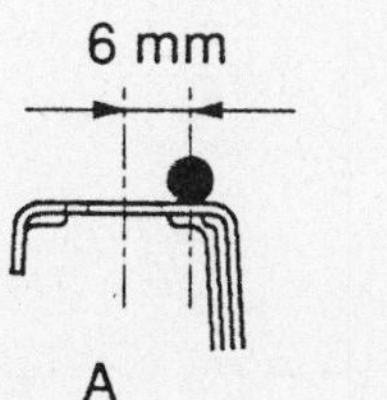

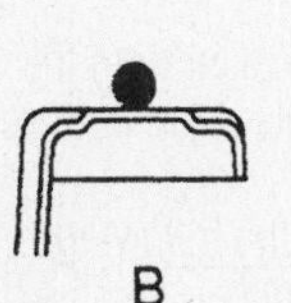

Apply sealant to the oil pan as shown—1.8L (1ZZ-FE) engine

2ZZ-GE Engines

1. Before servicing the vehicle, refer to the precautions in the beginning of this section.
2. Drain the engine oil.
3. Remove or disconnect the following:
   - Negative battery cable. On vehicles equipped with an air bag, wait at least 90 seconds before proceeding.
   - Undercovers
   - Front exhaust pipe
   - Oil pan mounting bolts and nuts
   - Oil pan, cutting off the applied sealer.

To install:

4. Remove any old sealant from the oil pan flange and thoroughly clean the sealing surface.
5. Install or connect the following:
   - Oil pan. Tighten the bolts and nuts in several passes. Bolts and nuts: 80 inch lbs. (9 Nm).
   - Front exhaust pipe
   - Negative battery cable
   - Undercovers
6. Fill the engine with clean oil.
7. Start the vehicle, check for leaks and repair if necessary.

5S-FE Engines

1. Before servicing the vehicle, refer to the precautions in the beginning of this section.
2. Drain the engine oil.
3. Remove or disconnect the following:
   - Negative battery cable. On vehicles equipped with an air bag, wait at least 90 seconds before proceeding.
   - Undercovers
   - Front exhaust pipe
   - Engine mounting center member
   - On Celica, the 3-way Catalytic Converter (TWC)
   - Rear end stiffener plate
   - Oil dipstick
   - Oil pan bolts and nuts. Cut off the applied sealer.

➡Do not use the cutting tool for the oil pump body side and rear oil seal retainer.

To install:

4. Remove any old sealant from the oil pan flange and thoroughly clean both sealing surfaces.
5. Install or connect the following:
   - Oil pan. Uniformly tighten the bolts and nuts in several passes. Bolts and nuts: 48 inch lbs. (5.4 Nm)

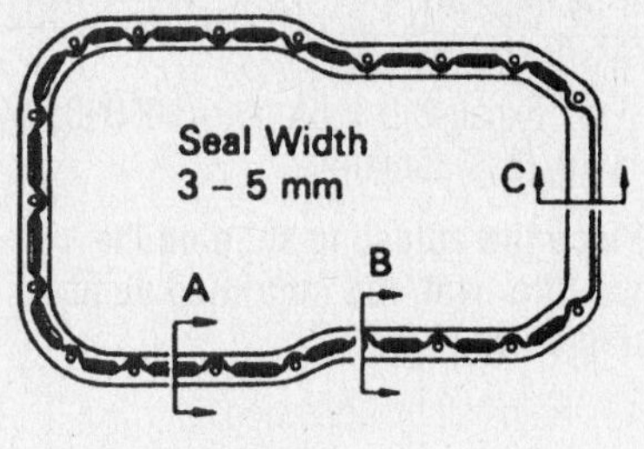

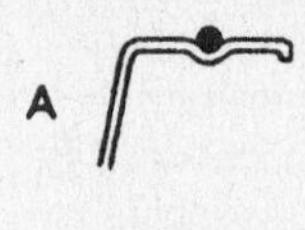

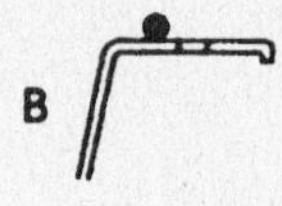

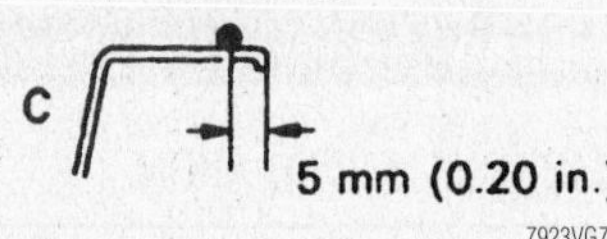

Oil pan sealing diagram—2.2L (5S-FE) engine

   - Oil dipstick
   - Rear end stiffener plate
   - On Celica, the TWC.
   - Engine mounting center member
   - Front exhaust pipe
   - Negative battery cable
   - Undercovers
6. Fill the engine with clean oil.
7. Start the vehicle, check for leaks and repair if necessary.

1MZ-FE Engines

1. Before servicing the vehicle, refer to the precautions in the beginning of this section.
2. Drain the engine oil.
3. Remove or disconnect the following:
   - Negative battery cable. On vehicles equipped with an air bag, wait at least 90 seconds before proceeding.
   - Fender apron seal
   - Undercover
   - Front exhaust pipe
   - Front exhaust pipe bracket from the No. 1 oil pan.
   - Flywheel housing undercover
   - Bolts and nuts from the No. 2 oil pan.
   - Oil strainer and gasket
   - Remove the No.1 oil pan
   - Baffle plate from the No. 1 oil pan

To install:

4. Clean all mating surfaces of the oil pans.
5. Install or connect the following:
   - Baffle plate to the No. 1 oil pan and tighten: 69 inch lbs. (8 Nm).
   - Install the No. 1 oil pan. with liquid sealant. Uniformly tighten the bolts and nuts in several passes. 10mm head bolt: 69 inch lbs. (8 Nm). 12mm head bolt: 14 ft. lbs. (19.5 Nm). 14mm head bolt-27 ft. lbs. (37.2 Nm)
   - Flywheel housing undercover. Bolts: 69 inch lbs. (7.8 Nm).
   - Oil strainer. Nuts: 69 inch lbs. (7.8 Nm).
   - No. 2 oil pan. Apply liquid sealant to the oil pan and engine block. Uniformly tighten the bolts and nuts in several passes. Bolts: 69 inch lbs. (7.8 Nm).
   - Flywheel housing undercover
   - Front exhaust pipe bracket to the No. 1 oil pan. Bolts: 15 ft. lbs. (21 Nm).
   - Front exhaust pipe. Exhaust manifolds to the front exhaust pipe nuts: 46 ft. lbs. (62 Nm). Front exhaust pipe to the center exhaust pipe. Bolts and nuts: 41 ft. lbs. (56 Nm).
   - Bracket with the 2 bolts. Bolts: 14 ft. lbs. (19 Nm).

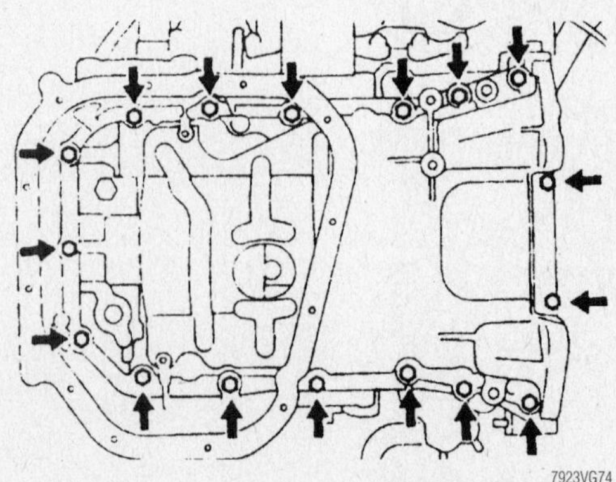

No. 1 oil pan mounting bolt locations—3.0L (1MZ-FE) engine

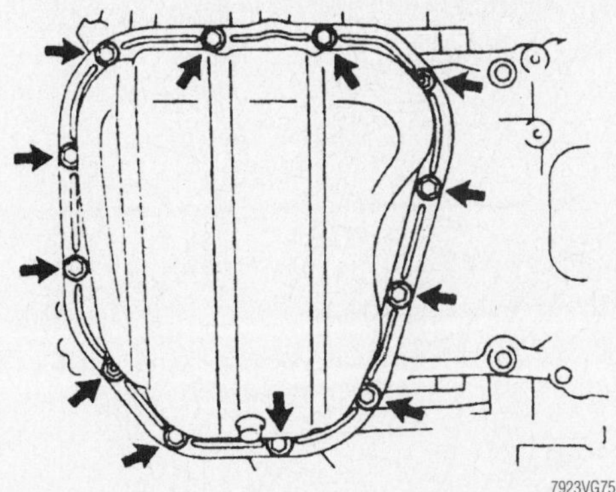

No. 2 oil pan mounting bolt locations—3.0L (1MZ-FE) engine

- Support stay with the 2 bolts. Bolts: 22 ft. lbs. (29 Nm).
- Undercover
- Right fender apron seal
- Negative battery cable

6. Fill the engine with clean oil.
7. Start the engine, check for leaks and repair if necessary.

3MZ-FE Engines

1. Before servicing the vehicle, refer to the precautions in the beginning of this section.

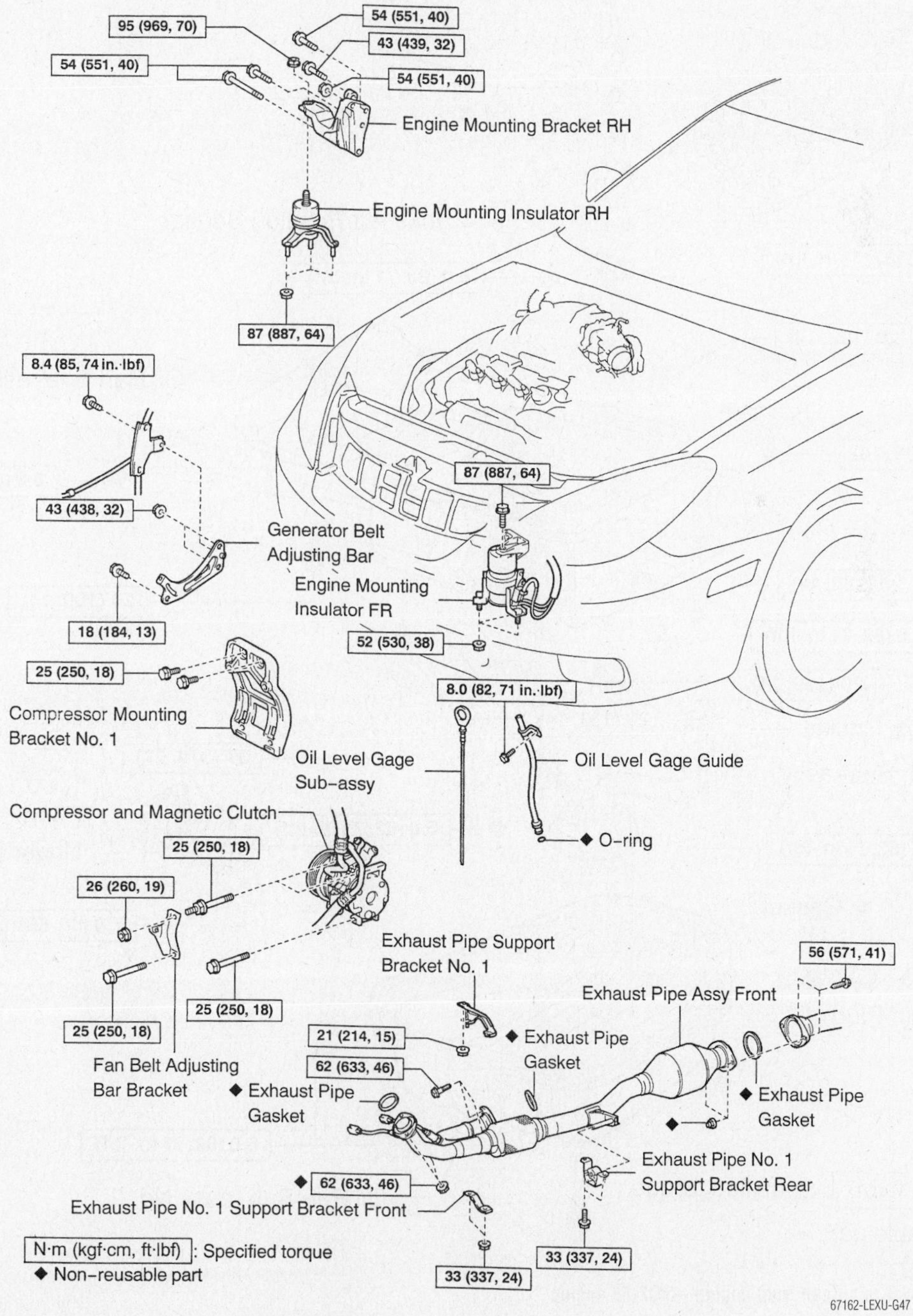

95 (969, 70)
54 (551, 40)
43 (439, 32)
54 (551, 40)
54 (551, 40)
Engine Mounting Bracket RH
Engine Mounting Insulator RH
87 (887, 64)
8.4 (85, 74 in.·lbf)
43 (438, 32)
Generator Belt Adjusting Bar
87 (887, 64)
Engine Mounting Insulator FR
18 (184, 13)
52 (530, 38)
25 (250, 18)
8.0 (82, 71 in.·lbf)
Compressor Mounting Bracket No. 1
Oil Level Gage Sub-assy
Oil Level Gage Guide
Compressor and Magnetic Clutch
◆ O-ring
25 (250, 18)
26 (260, 19)
Exhaust Pipe Support Bracket No. 1
56 (571, 41)
Exhaust Pipe Assy Front
25 (250, 18)
25 (250, 18)
21 (214, 15)
◆ Exhaust Pipe Gasket
62 (633, 46)
Fan Belt Adjusting Bar Bracket
◆ Exhaust Pipe Gasket
◆ Exhaust Pipe Gasket
Exhaust Pipe No. 1 Support Bracket Rear
◆ 62 (633, 46)
Exhaust Pipe No. 1 Support Bracket Front
N·m (kgf·cm, ft·lbf) : Specified torque
◆ Non-reusable part
33 (337, 24)
33 (337, 24)
67162-LEXU-G47

Exploded view, component removal for oil pan—3MZ-FE engine

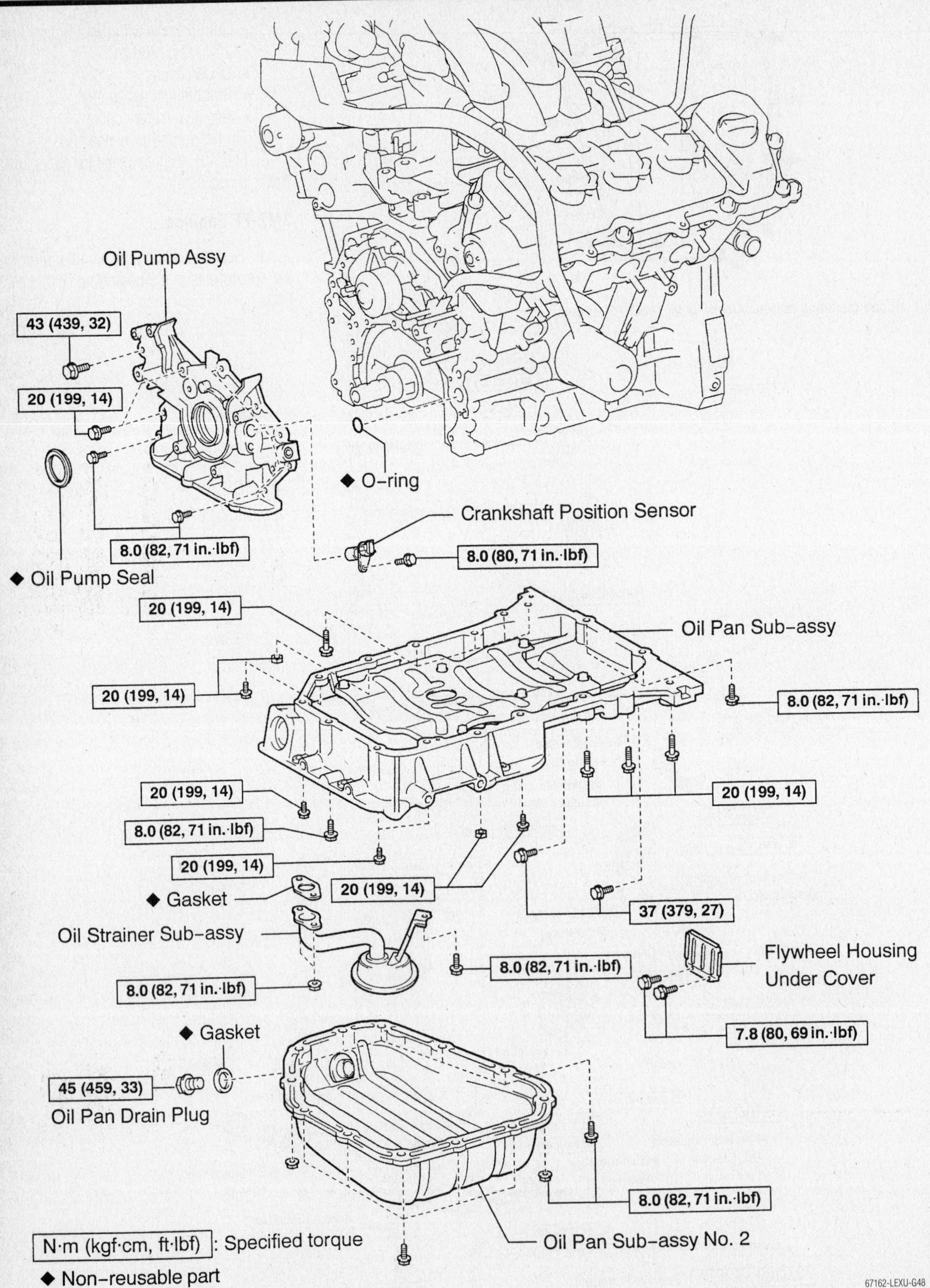

Exploded view, removal of oil pan from engine—3MZ-FE engine

2. Remove or disconnect the following:
- Battery negative cable
- RF wheel
- Engine under covers
- RH front fender apron seal
- A/C compressor, alternator and PS drive belts
- Engine stabilizer rod
- Engine mounting stay No2.
- Alternator bracket
- Crankshaft pulley
- Timing belt
- Crankshaft timing pulley
- Exhaust pipe and support brackets
- Oil gauge guide
- Alternator belt adjusting bar
- A/C compressor/clutch assembly
- A/C compressor mounting bracket

3. Separate FR engine mounting insulator

➡Do NOT remove the FR engine mounting at this time

a. Remove bolt and disconnect the power steering return hose
b. Remove the 4 nuts
c. Place a wooden block underneath the engine
d. Jack up the engine and remove the engine mounting insulator

⁂ WARNING

Be careful not to damage the oil pan

4. Remove or disconnect the following:
- RH engine mounting bracket
- 10 bolts and 2 nuts, gently pry off the oil pan sub assembly No.2.

⁂ WARNING

Be careful not to damage the oil pan flange area or the contact surface of the engine block.

- Oil strainer sub-assembly
- Flywheel housing under cover.

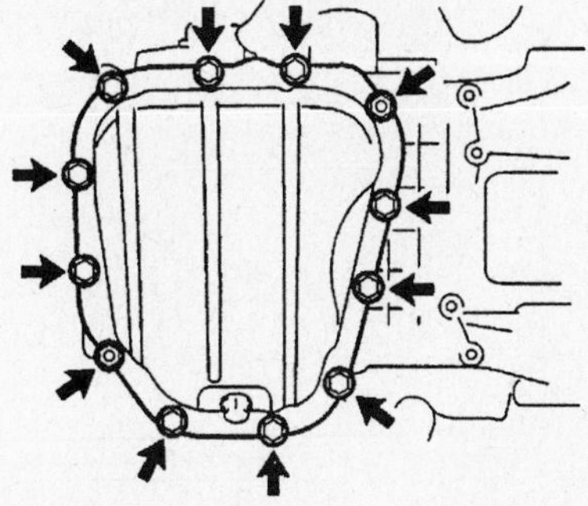

Oil pan sub-assembly No. 2—3MZ-FE engine

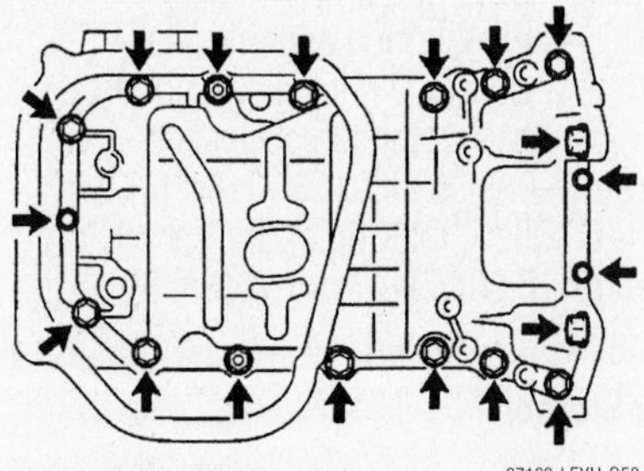

Oil pan sub-assembly No. 1—3MZ-FE engine

- Oil pan sub-assembly No. 1
- Engine oil level sensor

To install:

5. Install the oil pan sub-assembly No. 1.
6. Remove any old oil sealant from contact surface. Clean the surface thoroughly.
7. Apply a continuous bead of sealant 0.12 to 0.16 in (3 to 4 mm) around the block surface, making certain to surround the bolt holes.
8. Install the oil pan within 3 minutes after applying the sealant

➡Do NOT expose sealant to engine oil within 2 hours after installing

9. Install the oil pan using the 17 bolts and 2 nuts. Tighten uniformly in several steps as follows:
   a. 10 mm head 71 in. lbs (8.0 Nm).
   b. 12 mm head 14 ft. lbs.(20 Nm).
   c. 14 mm head 27 ft. lbs (37 Nm).
10. Install or connect the following:
   - Engine oil level sensor
   - Oil strainer assembly
   - Oil pan sub assembly No. 2
11. Remove any old oil sealant from contact surface. Clean the surface thoroughly.
12. Apply a continuous bead of sealant 0.16 to 0.20 in. (3 to 4 mm) around the block surface, making certain to surround the bolt holes.
13. Install the oil pan within 3 minutes after applying the sealant.

➡Do NOT expose sealant to engine oil within 2 hours after installing

14. Install the oil pan using the 10 bolts and 2 nuts. Torque to 71 inch lbs. (8.0 Nm).
15. Install or connect the following
   - RH engine mounting bracket; torque bolts "A" & "B" to 40 ft. lbs (54 Nm) and bolt "C" to 32 ft. lbs (43 Nm)
   - RH engine mounting insulator; torque nut "A" to 70 ft. lbs. (95 Nm) and nut "B" to 64 ft. lbs (87 Nm)

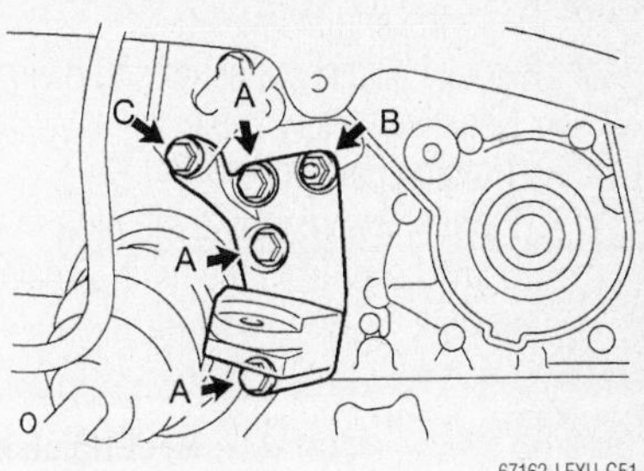

Engine mounting bracket tightening procedure—3MZ-FE engine

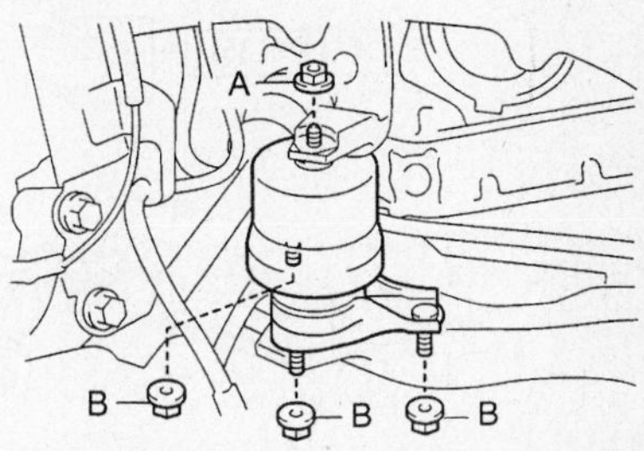

Engine mounting insulator tightening procedure—3MZ-FE engine

   - FR engine mounting insulator; torque bolt to 64 ft. lbs. (87 Nm) and nut to 38 ft. lbs (52 Nm)
   - A/C compressor mounting bracket; torque to 18 ft. lbs (25 Nm)
   - A/C compressor/clutch assembly; torque to 18 ft. lbs. (25 Nm)
   - Alternator belt adjusting bar; torque bolt to 18 ft. lbs (25 Nm) and nut to 19 ft. lbs. (26 Nm)
   - Oil level gage guide
   - Exhaust pipes and support brackets
   - Crankshaft timing pulley
   - Timing belt assembly
   - Crankshaft pulley
   - Alternator bracket
   - Engine mounting stay
   - Engine stabilizer rod
   - Alternator, PS pump and A/C drive belts
   - RH fender apron seal
   - RF wheel
   - Engine under covers
   - Negative battery cable
16. System initialization

Oil Pump

REMOVAL & INSTALLATION

1NZ-FE Engines

1. Before servicing the vehicle, refer to the precautions in the beginning of this section.

2. Drain the engine oil.
3. Remove or disconnect the following:
   - Negative battery cable
   - Timing chain cover
   - 2 bolts, 3 screws and the oil pump cover from the timing chain cover
   - Drive and driven rotors
   - Plug, spring and relief valve
4. Inspect the relief valve motion
5. Check rotor side clearance: 0.0012–0.0035 in. (0.03–0.09mm).
6. Check rotor tip clearance: 0.0024–0.0071 in. (0.06–0.18mm).
7. Check rotor-to-body clearance: 0.0098–0.0128 in. (0.250–0.325mm).

To install:

8. Install or connect the following:
   - Relief valve and spring. Torque the plug to 18 ft. lbs. (24 Nm).

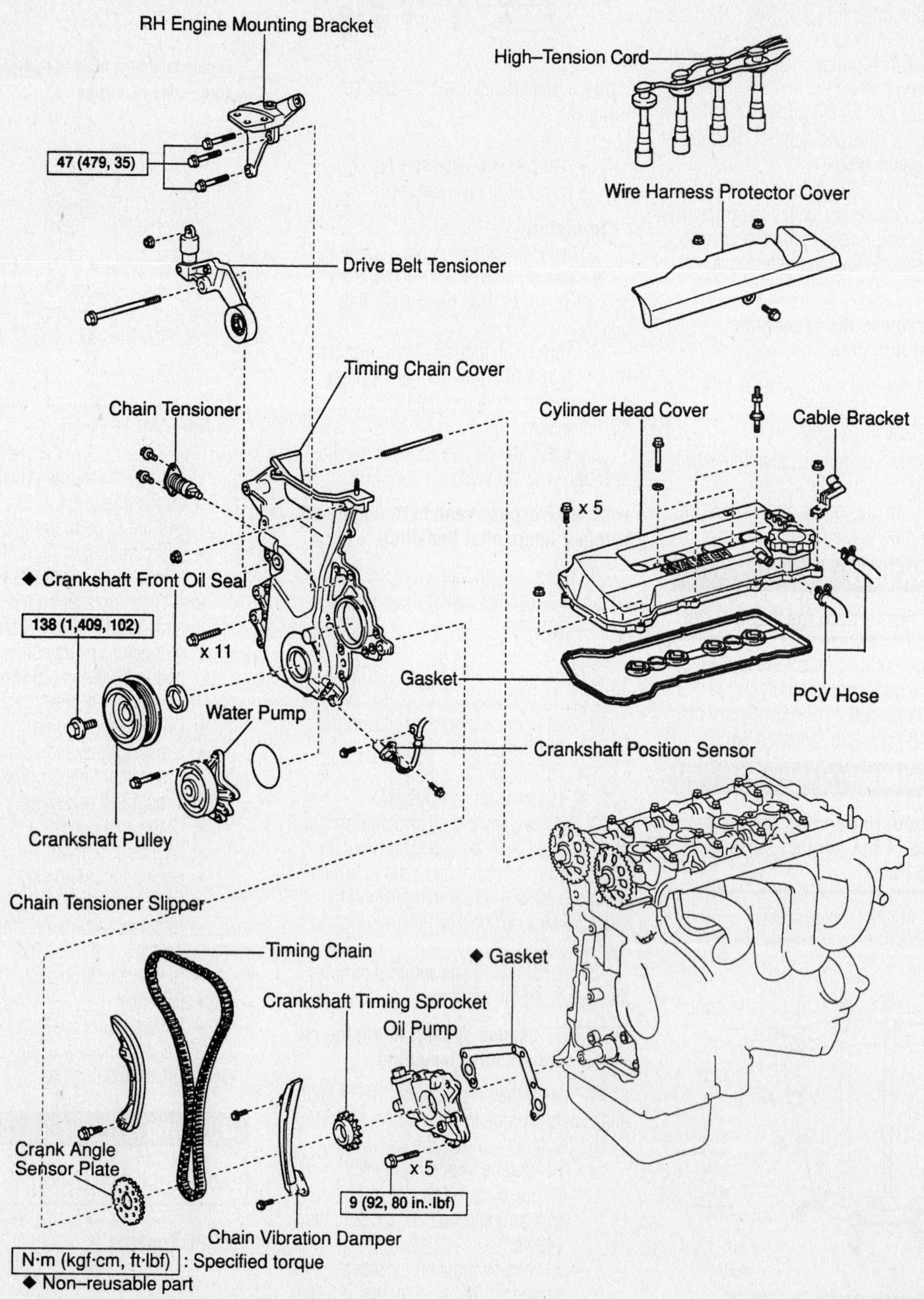

Exploded view of the oil pump mounting—1.8L (1ZZ-FE) engine

- Drive and driven rotors with the marks on the cover side
- Cover. Torque the bolts to 80 inch lbs. (9 Nm); the screws to 96 inch lbs. (11 Nm).
- Timing chain cover
- Negative battery cable

9. Fill the engine with clean oil.
10. Start the vehicle, check for leaks and repair if necessary.

1ZZ-FE Engines

1. Before servicing the vehicle, refer to the precautions in the beginning of this section.
2. Drain the engine oil.
3. Remove or disconnect the following:
   - Negative battery cable
   - Timing chain and crankshaft sprocket
   - Oil pump and gasket

To install:

4. Clean the mounting surface.
5. Install or connect the following:
   - Oil pump, with new gasket. Bolts: 80 inch lbs. (9 Nm).
   - Crankshaft sprocket and timing chain
   - Negative battery cable
6. Fill the engine with clean oil.
7. Start the vehicle, check for leaks and repair if necessary.

2ZZ-GE Engines

1. Before servicing the vehicle, refer to the precautions in the beginning of this section.
2. Drain the engine oil.
3. Remove or disconnect the following:
   - Negative battery cable
   - Timing chain and crankshaft sprocket
   - Oil pump and gasket

To install:

4. Clean the mounting surface.
5. Install or connect the following:
   - Oil pump, with new gasket. Bolts: 80 inch lbs. (9 Nm).
   - Crankshaft sprocket and timing chain
   - Negative battery cable
6. Fill the engine with clean oil.
7. Start the vehicle, check for leaks and repair if necessary.

5S-FE Engines

1. Before servicing the vehicle, refer to the precautions in the beginning of this section.
2. Drain the engine oil.
3. Remove or disconnect the following:
   - Negative battery cable. On vehicles equipped with an air bag, wait at least 90 seconds before proceeding.
   - Hood
   - Front exhaust pipe
   - Rear end stiffener plate
   - Oil dipstick and oil pan
   - Oil strainer and gasket
   - Timing belt and pulleys
   - Crankshaft position sensor
   - Oil pump and gasket

To install:

4. Install or connect the following:
   - Oil pump with new gasket. Bolts: 82 inch lbs. (9 Nm).

➡The long bolts are 1.38 in. (35mm) and all the others are 0.98 in. (25mm).

- Crankshaft position sensor
- Timing belt and pulleys
- Oil strainer with new gasket. Tighten to 48 inch lbs. (5 Nm).
- Oil pan and dipstick

➡The pan must be installed within 5 minutes of sealant application or the procedure will have to be repeated.

- Rear end stiffener plate. Bolts: 27 ft. lbs. (37 Nm).
- Front exhaust pipe
- Negative battery cable
- Hood

5. Fill the engine with clean oil.
6. Start the vehicle, check for leaks and repair if necessary.

1MZ-FE Engines

1. Before servicing the vehicle, refer to the precautions in the beginning of this section.
2. Drain the engine oil.
3. Remove or disconnect the following:
   - Negative battery cable. On vehicles equipped with an air bag, wait at least 90 seconds before proceeding.
   - Fender apron seal
   - Undercover
   - Front exhaust pipe
   - Front exhaust pipe bracket from the No. 1 oil pan
   - Alternator drive belt
   - A/C compressor
   - Power steering pump drive belt and adjusting strut
   - Timing belt and belt pulleys
   - Rear timing belt cover
   - A/C compressor housing bracket
   - No. 2 oil pan, oil strainer, No.1 oil pan and baffle plate
   - Crankshaft position sensor
   - 9 oil pump bolts. Make a note of the position of the each bolt.
   - Oil pump body by prying between the oil pump and main bearing cap
   - O-ring from the cylinder block
   - Plug, gasket, spring, and relief valve from the oil pump body
   - 9 screws, pump body cover, drive, and driven rotors

To install:

4. Install or connect the following:
   - Driven rotors, drive, pump body cover
   - 9 screws
   - Oil pump relief valve, spring, gasket, and the plug to the oil pump body
   - New O-ring on the cylinder block
   - Liquid sealant to the oil pump and engine block
   - Oil pump to the engine block
   - Bolts, uniformly tightened in several passes: 10mm head: 69 inch lbs. (8 Nm). 12mm head: 14 ft. lbs. (20 Nm)
   - Crankshaft position sensor. Bolt: 69 inch lbs. (8 Nm).
   - Baffle plate to the No. oil pan. Tighten to 69 inch lbs. (8 Nm).
   - No. 1 oil pan, oil strainer and No. 2 oil pan.
   - A/C compressor housing bracket. Bolts: 18 ft. lbs. (25 Nm).
   - Rear timing belt cover. Bolts: 74 inch lbs. (9 Nm).
   - Timing belt pulleys
   - Timing belt
   - Adjusting strut and power steering drive belt. Bolt and nut: 32 ft. lbs. (43 Nm).
   - A/C compressor
   - Alternator drive belt
   - Front exhaust pipe bracket to the No. 1 oil pan. Bolts: 15 ft. lbs. (21 Nm).
   - Front exhaust pipe
   - Undercover
   - Right fender apron seal
   - Negative battery cable
5. Fill the engine with clean oil.
6. Start the vehicle, check for leaks and repair if necessary.

3MZ-FE Engines

1. Follow engine oil pan removal procedure then perform the following steps.
2. Remove or disconnect the following:
   - Crankshaft position sensor
3. Oil pump assembly
   a. 9 bolts

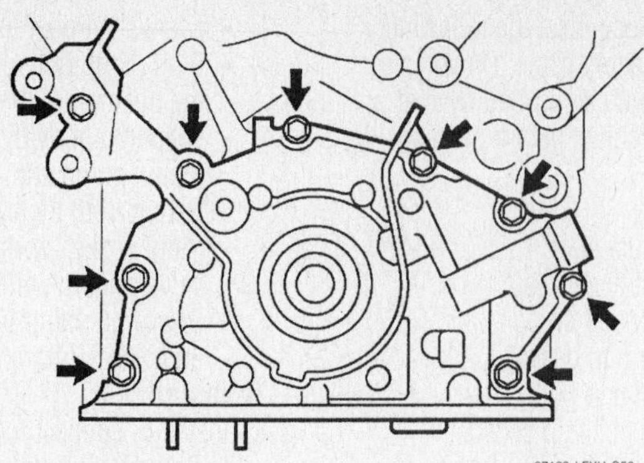

Oil pump removal bolt procedure—3MZ-FE engine

b. Remove oil pump by prying between the oil pump and bearing cap

c. Remove the oil ring

To install:

4. Install the oil pump seal using proper driver.

a. Tap in the seal until it is flush with the oil pump body

b. Apply multi-purpose grease to the seal lip.

5. Install oil pump assembly

a. Remove any old sealant from the mating surfaces

b. Apply a light coat of clean engine oil to the O-ring, then place it on the engine block.

c. Thoroughly clean the mating surface of any oil or old sealant

d. Apply a continuous bead of sealant on the oil pump body, making certain to surround the bolt holes.

e. Install the oil pump within 3 minutes after applying the sealant.

➡Do NOT expose the sealant to engine oil within 2 hours after installing

f. Align the key of the oil pump drive gear with the keyway located on the crankshaft, then slide the oil pump into place.

g. Install the oil pump with the 9 bolts. Tighten the bolts uniformly in several steps.

Torque to: Bolt A 71 in. lbs (8.0 Nm), Bolt B 14 ft lbs. (20 Nm), Bolt C 32 ft. lbs. (43 Nm)

6. Install crankshaft position sensor

7. Install oil pans using oil pan installation procedure

Rear Main Seal

REMOVAL & INSTALLATION

5S-FE, 1ZZ-FE, 2ZZ-GE and 1NZ-FE Engines

1. Remove or disconnect the following:
   - Transaxle
   - Clutch assembly
   - Flywheel or flexplate

2. Use a small sharp knife to cut off the lip of the oil seal. Take great care not to score any metal with the knife.

3. Use a small prytool to pry the old seal from the retaining plate. Be careful not to damage the plate. Protect the tip of the tool with tape and pad the fulcrum point with cloth.

4. Inspect the crankshaft and seal lip contact surfaces for any sign of damage.

To install:

5. Apply a light coat of multi-purpose grease to the lip of a new oil seal. Loosely fit the seal into place by hand, making sure it is not crooked.

6. Use a seal driver of the correct size to install the seal. Tap it into place until the surface of the seal is flush with the edge of the housing.

1MZ-FE Engine

1. Remove or disconnect the following:
   - Transaxle
   - Clutch cover assembly and flywheel or driveplate
   - Remove the rear end plate.
   - Oil seal retainer and gasket. Discard the gasket or sealant.

2. Use a small prybar to pry the oil seal from the retaining plate. Be careful not to damage the plate.

To install:

3. Clean the retainer contact surfaces thoroughly and lubricate the new oil seal with multi-purpose grease.

4. Drive the oil seal into the retainer until its surface is flush with the edge of the retainer. Make sure that the seal is installed evenly in the retainer to ensure proper sealing.

5. Apply a ⅛ inch bead of sealant to the oil seal retainer. Install the retainer and install the dust seal. Tighten the bolts to 69 inch lbs. (8 Nm).

6. Install the rear end plate.

7. On automatic transaxle equipped vehicles, install the driveplate.

8. On manual transaxle equipped vehicles, install the clutch disc and clutch cover.

9. Install the transaxle.

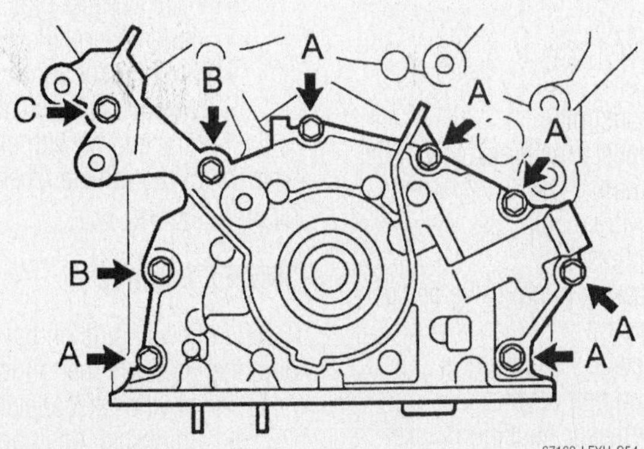

Oil pump bolt installation procedure—3MZ-FE engine

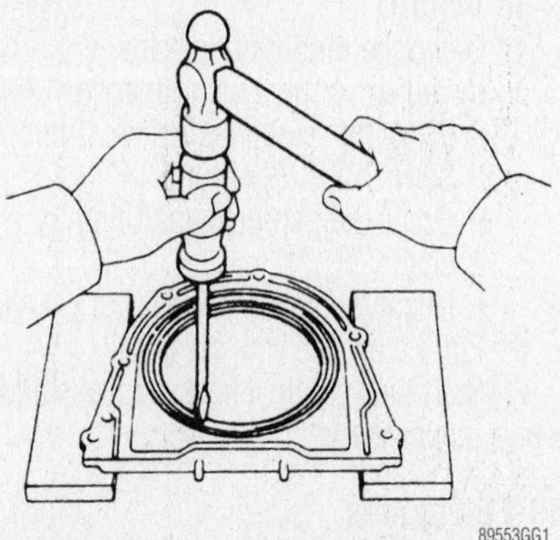

Always place the seal on blocks of wood, then tap the seal from the retainer—1MZ-FE engine

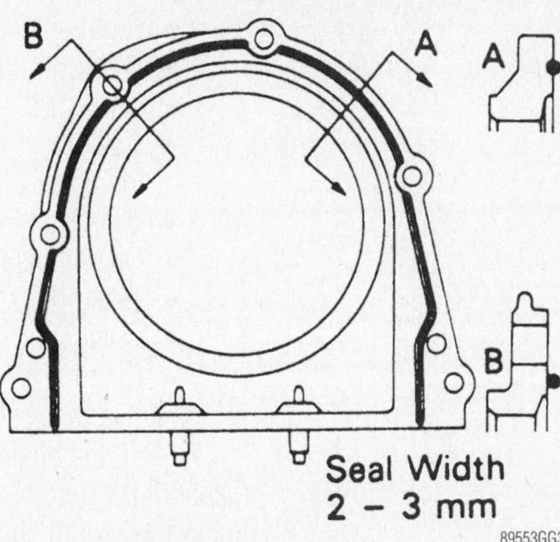

Apply sealant to the rear oil seal retainer—1MZ-FE engine

Front Crankshaft Seal

REMOVAL & INSTALLATION

➡The following procedures apply to engines using a timing belt. The procedures for front cover seals can be found later in this section.

5S-FE Engines

1. Before servicing the vehicle, refer to the precautions in the beginning of this section.
2. Remove or disconnect the following:
   - Negative battery cable. On vehicles equipped with an air bag, wait at least 90 seconds before proceeding

➡The front oil seal can be removed from the engine without removing the oil pump.

   - Timing belt covers and the timing belt
   - Front crankshaft gear from the crankshaft. Be sure not to damage any part of the crankshaft
3. Using a knife, cut off the oil seal lip.
4. Pry out the oil seal. Wrap the edge of the tool with a rag or tape to prevent damaging the crankshaft. Be careful not to damage the crankshaft.

To install:

5. Using a new seal, apply a thin layer of liquid sealer to the outside of the seal.
6. Apply multi purpose grease to the new oil seal lip.
7. Oil seal until its surface is flush with the oil pump body edge.
8. Install or connect the following:
   - Timing belt and the timing belt covers
   - All other components
   - Negative battery cable

1MZ-FE Engines

1. Before servicing the vehicle, refer to the precautions in the beginning of this section.
2. Remove or disconnect the following:
   - Negative battery cable. On vehicles equipped with an air bag, wait at least 90 seconds before proceeding.
   - Timing belt
   - Crankshaft timing gear
3. Cut out the lip portion of the oil seal.
4. Tape the end of a suitable prybar to protect the crankshaft and carefully remove the oil seal.

⁂ WARNING

Be careful not to damage the crankshaft sealing surface.

To install:

5. Apply multi-purpose grease to the lip of a new oil seal. Also apply a light coating of liquid sealant to the outside of the oil seal.
6. Oil seal, until its surface is flush with the oil pump case edge.
7. Install or connect the following:
   - Crankshaft timing gear
   - Timing belt
   - Negative battery cable

Timing Chain, Sprockets, Front Cover and Seal

REMOVAL & INSTALLATION

1ZZ-FE Engines

1. Before servicing the vehicle, refer to the precautions in the beginning of this section.
2. Drain the cooling system.
3. Remove or disconnect the following:
   - Negative battery cable
   - Right engine cover
   - Accessory drive belt and generator
   - Power steering pump, without disconnecting the hoses.
   - Right engine mount
   - Cylinder head cover
   - Turn the crankshaft so the No. 1 piston is at TDC on the compression stroke.
   - Crankshaft pulley
   - Crankshaft position sensor from the timing chain cover.
   - Accessory drive belt tensioner.
   - Right engine mounting bracket
   - Chain tensioner
   - Water pump
   - Timing chain cover
   - Crankshaft angle sensor plate
   - Timing chain tensioner slipper
   - Timing chain and crankshaft timing sprocket.
   - Timing chain vibration damper
   - Valve timing control assembly and camshaft timing sprocket
4. Drive the seal from the cover.
5. Pull the chain to its full length and measure the length of any 16 consecutive links. The length should not exceed 4.827 inches (122.6mm).
6. Check the slipper and damper wear. Maximum wear should not exceed 0.039 in. (1mm).
7. The tensioner plunger should move smoothly and lock into place with finger pressure.

To install:

8. Apply engine oil from the tip of the intake camshaft, back to 16mm.
9. Align the timing mark on the valve timing controller with the knock pin and gently push the valve timing controller onto the camshaft.
10. Set the No.1 piston to TDC compression. The key on the crankshaft should be at 12 o'clock.
11. Install or connect the following:
    - Sprockets. Torque the bolt to 33 ft. lbs. (45 Nm). Turn the camshafts to align the point marks on the sprockets.
    - Chain damper. Bolts: 96 inch lbs. (11 Nm).
    - Timing chain and crankshaft sprocket. Be sure to align the yellow chain link with the mark on the crankshaft sprocket.
    - Timing chain on the camshaft sprockets. Align the yellow links with the marks on the camshaft sprockets.

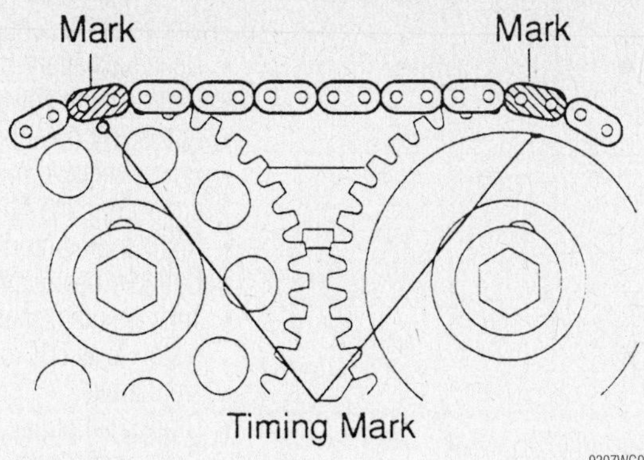

Timing chain link marks—1ZZ-FE and 2ZZ-GE engines

- Chain tensioner slipper. Bolt: 14 ft. lbs. (18.5 Nm).
- Crankshaft angle sensor plate with the **F** mark forward
- New seal in the front cover
- Silicone sealant to the timing chain cover as illustrated
- Timing chain cover

12. Water pump. Tighten the 10mm bolts marked "C" to 80 inch lbs. (9 Nm), those marked "A" to 10 ft. lbs. (13 Nm), and the remaining 10mm bolts to 96 inch lbs. (11 Nm). Tighten the 12mm bolts to 14 ft. lbs. (18.5 Nm). Be sure to install the bolts in their original locations. Bolt lengths:
    a. A: 1.77 in. (45mm).
    b. B: 1.38 in. (35mm).
    c. C: 1.18 in. (30mm).
    d. D: 0.98 in. (25mm).
13. With a Torx wrench, tighten the stud bolt to 82 inch lbs. (9.3 Nm).
14. Install or connect the following:
    - Right engine mounting bracket. Bolts, with sealant applied: 35 ft. lbs. (47 Nm).
    - Drive belt tensioner. Bolt: 51 ft. lbs. (69 Nm). Nut: 21 ft. lbs. (29 Nm).
    - Crankshaft position sensor. Tighten to 80 inch lbs. (9 Nm).
    - Crankshaft pulley. Bolt: 102 ft. lbs. (138 Nm).
15. Release the ratchet pawl and compress the chain tensioner. Place the hook on the pin to keep the tensioner compressed.
16. Install the tensioner, using a new O-ring. Torque the bolts to 80 inch lbs. (9 Nm).
17. Turn the crankshaft counterclockwise and remove the hook from the pin. Turn the crankshaft clockwise and be sure the slipper is pushed by the plunger.
18. Check the valve timing by turning the crankshaft clockwise until the mark of the pulley is aligned with the mark on the timing chain cover. The marks on the camshaft sprockets should be facing each other as shown.
19. Install or connect the following:
    - Silicone sealant to the 2 areas where the timing chain cover meets the cylinder head.
    - Cylinder head cover. Bolts with washers in the sequence shown: 80 inch lbs. (9 Nm). Bolts without washers: 96 inch lbs. (11 Nm).
    - Right engine mount. Bolts and nuts: 38 ft. lbs. (52 Nm).
    - Power steering pump
    - Alternator and drive belt
    - Right engine undercover
    - Negative battery cable
    - Washer tank
20. Fill the cooling system to the proper level.
21. Start the vehicle, check for leaks and repair if necessary.

1NZ-FE Engines

1. Before servicing the vehicle, refer to the precautions in the beginning of this section.
2. Drain the cooling system.
3. Remove or disconnect the following:
    - Negative battery cable
    - Right front wheel
    - Alternator
    - Power steering pump
    - Right engine mount insulator. Use a jack and wood block for support.
    - With A/C, the bolt holding the liquid tube to the insulator
    - Ignition coils
    - Cylinder head cover
4. Place No.1 cylinder on TDC compression. Make sure that the timing marks on the camshaft sprockets and valve timing controller assembly are facing UP (12 o'clock). If not, turn the crankshaft 360 degrees to align the marks.
5. Remove or disconnect the following:
    - Crankshaft pulley bolt
    - Pulley and pin
    - Crankshaft position sensor
    - Right engine mount bracket
    - Water pump
    - Oil control valve
    - 13 bolts, 1 nut and 1 stud bolt. Pry the cover off.
    - 2 O-rings from the block and pan
    - Chain tensioner
    - Tensioner slipper
    - Chain vibration damper
    - Chain
6. Drive the seal from the cover.
7. Pull the chain to its full length and measure the length of any 16 consecutive links. The length should not exceed 4.85 inches (123.2mm).
8. Check the slipper and damper wear. Maximum wear should not exceed 0.039 in. (1mm).
9. The tensioner plunger should move smoothly and lock into place with finger pressure.

To install:

10. Set the crankshaft at 140 degrees ATDC. Set the camshaft sprockets at 20 degrees ATDC; then, reset the crankshaft to 20 degrees ATDC.
11. Install or connect the following:
    - New seal, driven into place until flush with the cover edge
    - Vibration damper. Torque the bolts to 80 inch lbs. (9 Nm).
    - Timing chain

➡A new chain will have 3 marked links to align with the 3 sprockets, as shown in the accompanying illustration.

- Slipper
- Tensioner. Torque the bolts to 80 inch lbs. (9 Nm).
- Cover, using a 4–5mm bead of RTV sealer and new O-rings to the block and pan

12. Uniformly tighten the bolts in several passes, using the accompanying illustration, to:
    - A, C, E and G: 96 inch lbs. (11 Nm)
    - B, D and F: 18 ft. lbs. (24 Nm)
13. Bolt lengths are as follows:
    - A: 20mm
    - B: 30mm
    - C: 35mm
    - D: 20mm
    - E: 35mm
14. Install or connect the following:
    - Right engine mount bracket. Coat all but the end 2 threads of the bolt with RTV sealer. Torque the bolt to 41 ft. lbs. (55 Nm).
    - Crankshaft position sensor. Torque bolt A to 66 inch lbs. (7.5 Nm); bolts B to 96 inch lbs. (11 Nm).

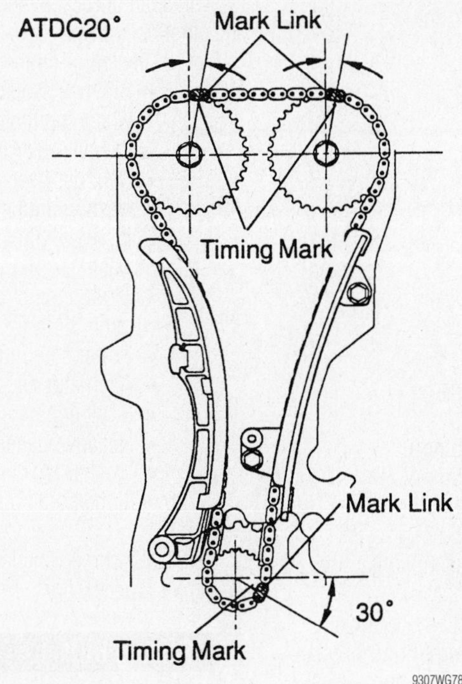

Timing chain installation—1NZ-FE engine

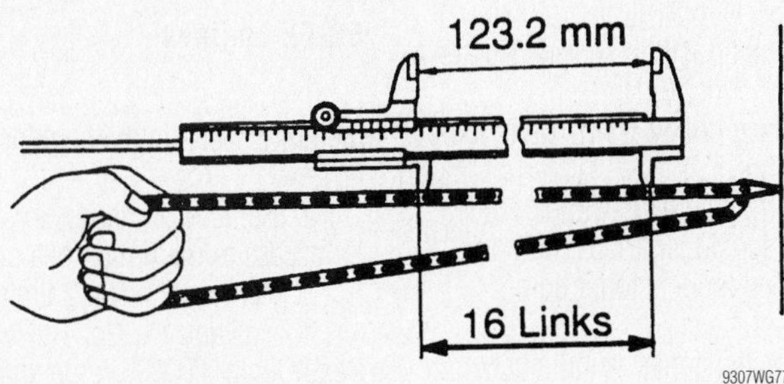

Measuring the timing chain—1NZ-FE engine

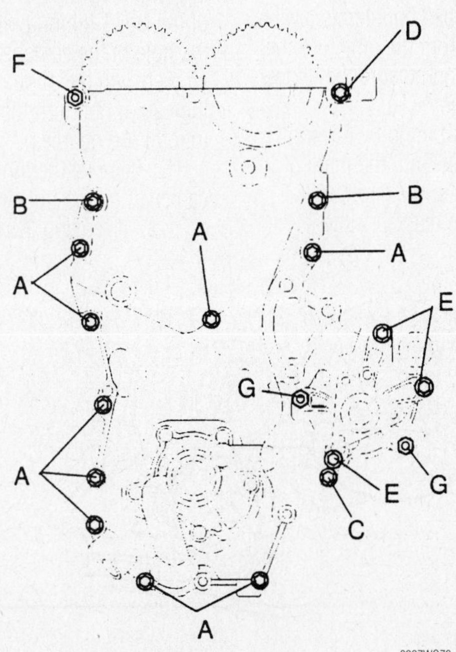

Timing cover bolt installation—1NZ-FE engine

- Oil control valve. Torque: 71 inch lbs. (8 Nm).
- Crankshaft pulley and pin. Torque the bolt to 94 ft. lbs. (128 Nm).
- Cylinder head cover with RTV gasket material at the 2 locations shown. Uniformly tighten the bolts and nuts, in several passes, to 84 inch lbs. (10 Nm).
- PCV hoses
- Ignition coils
- Right engine mount insulator. Torque the bolts and nuts to 35 ft. lbs. (47 Nm).
- Power steering pump
- Alternator
- Right under cover
- Wheel
- Negative battery cable

15. Fill the cooling system to the proper level.
16. Start the vehicle, check for leaks and repair if necessary.

2ZZ-GE Engines

1. Before servicing the vehicle, refer to the precautions in the beginning of this section.
2. Drain the cooling system.
3. Remove or disconnect the following:
   - Negative battery cable
   - Right engine cover
   - Accessory drive belt and generator
   - Power steering pump, without disconnecting the hoses.
   - Right engine mount
   - Cylinder head cover
   - Turn the crankshaft so the No. 1 piston is at TDC on the compression stroke.
   - Crankshaft pulley
   - Crankshaft position sensor from the timing chain cover.
   - Accessory drive belt tensioner.
   - Right engine mounting bracket
   - Chain tensioner
   - Water pump
   - Timing chain cover
   - Crankshaft angle sensor plate
   - Timing chain tensioner slipper
   - Timing chain and crankshaft timing sprocket.
   - Timing chain vibration damper
   - Valve timing control assembly and camshaft timing sprocket
4. Drive the seal from the cover.
5. Pull the chain to its full length and measure the length of any 16 consecutive links. The length should not exceed 4.827 inches (122.6mm).
6. Check the slipper and damper wear.

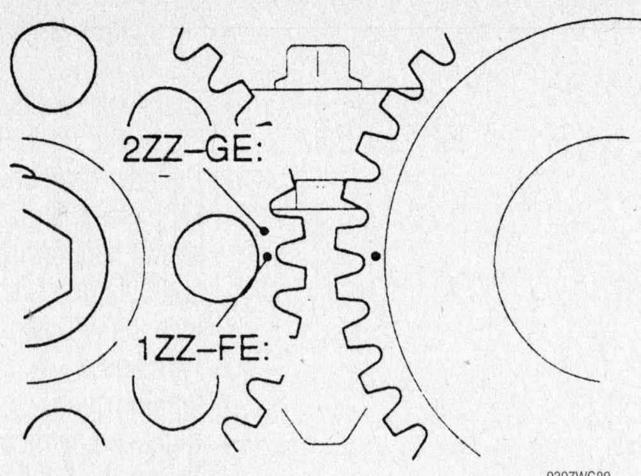

Timing mark identification at TDC compression—1ZZ-FE and 2ZZ-GE engines

Maximum wear should not exceed 0.039 in. (1mm).

7. The tensioner plunger should move smoothly and lock into place with finger pressure.

To install:

8. Apply engine oil from the tip of the intake camshaft, back to 16mm.

9. Align the timing mark on the valve timing controller with the knock pin and gently push the valve timing controller onto the camshaft.

10. Set the No.1 piston to TDC compression. The key on the crankshaft should be at 12 o'clock.

11. Install or connect the following:
- Sprockets. Torque the bolt to 33 ft. lbs. (45 Nm). Turn the camshafts to align the point marks on the sprockets.
- Chain damper. Bolts: 96 inch lbs. (11 Nm).
- Timing chain and crankshaft sprocket. Be sure to align the yellow chain link with the mark on the crankshaft sprocket.
- Timing chain on the camshaft sprockets. Align the yellow links with the marks on the camshaft sprockets.
- Chain tensioner slipper. Bolt: 14 ft. lbs. (18.5 Nm).
- Crankshaft angle sensor plate with the **F** mark forward
- New seal in the front cover
- Silicone sealant to the timing chain cover as illustrated
- Timing chain cover

12. Install the water pump. Tighten the 10mm bolts marked "C" to 80 inch lbs. (9 Nm), those marked "A" to 10 ft. lbs. (13 Nm), and the remaining 10mm bolts to 96 inch lbs. (11 Nm). Tighten the 12mm bolts to 14 ft. lbs. (18.5 Nm). Be sure to install the bolts in their original locations. Bolt lengths:

a. A: 1.77 in. (45mm).
b. B: 1.38 in. (35mm).
c. C: 1.18 in. (30mm).
d. D: 0.98 in. (25mm).

13. With a Torx wrench, tighten the stud bolt to 82 inch lbs. (9.3 Nm).

14. Install or connect the following:
- Right engine mounting bracket. Bolts, with sealant applied: 35 ft. lbs. (47 Nm).
- Drive belt tensioner. Bolt: 51 ft. lbs. (69 Nm). Nut: 21 ft. lbs. (29 Nm).
- Crankshaft position sensor. Tighten to 80 inch lbs. (9 Nm).
- Crankshaft pulley. Bolt: 102 ft. lbs. (138 Nm).

15. Release the ratchet pawl and compress the chain tensioner. Place the hook on the pin to keep the tensioner compressed.

16. Install the tensioner, using a new O-ring. Torque the bolts to 80 inch lbs. (9 Nm).

17. Turn the crankshaft counterclockwise and remove the hook from the pin. Turn the crankshaft clockwise and be sure the slipper is pushed by the plunger.

18. Check the valve timing by turning the crankshaft clockwise until the mark of the pulley is aligned with the mark on the timing chain cover. The marks on the camshaft sprockets should be facing each other as shown.

19. Install or connect the following:
- Silicone sealant to the 2 areas where the timing chain cover meets the cylinder head.
- Cylinder head cover. Bolts with washers in the sequence shown: 80 inch lbs. (9 Nm). Bolts without washers: 96 inch lbs. (11 Nm).
- Right engine mount. Bolts and nuts: 38 ft. lbs. (52 Nm).
- Power steering pump
- Alternator and drive belt
- Right engine undercover
- Negative battery cable
- Washer tank

20. Fill the cooling system to the proper level.

21. Start the vehicle, check for leaks and repair if necessary.

Timing Belt

REMOVAL & INSTALLATION

5S-FE Engines

1. Before servicing the vehicle, refer to the precautions in the beginning of this section.

2. Remove all necessary components for access to the timing belt covers.

3. Remove the No. 2 timing cover.

4. Position the No. 1 cylinder to Top Dead Center (TDC) on the compression stroke by turning the crankshaft pulley and aligning its groove with the timing mark **0** of the No. 1 timing belt cover. Check that the hole of the camshaft timing pulley is aligned with the alignment mark of the bearing cap. If not, turn the crankshaft one revolution (360 degrees).

5. Remove the timing belt from the camshaft timing pulley, as follows:

a. If reusing the belt, place match-

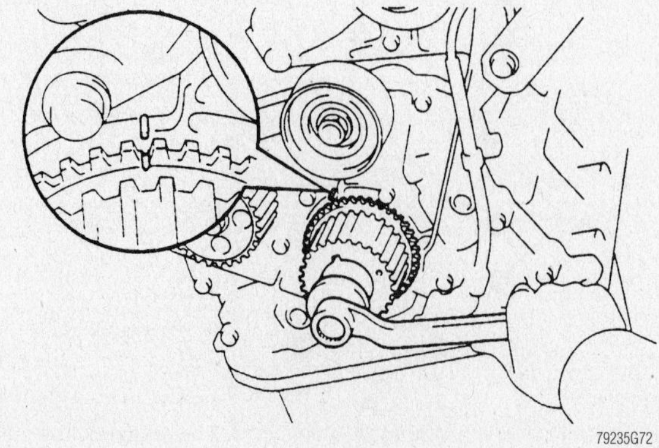

Crankshaft positioning for timing belt removal and installation—Toyota 2.2L (5S-FE) engine

marks on the timing belt and the camshaft pulley. Loosen the mount bolt of the No. 1 idler pulley and position the pulley toward the left as far as it will go. Tighten the bolt. Remove the belt from the camshaft pulley.

6. Remove the camshaft timing pulley as follows:

a. Using Toyota tools Nos. 09249-63010 and 09960-10010, remove the bolt and the camshaft pulley.

7. Remove the crankshaft pulley as follows:

a. Use Toyota tools Nos. 09213-54015 and 09330-00021 to hold the crankshaft pulley. Remove the pulley set bolt and remove the pulley using a puller.

8. Remove the No. 1 timing belt cover.

9. Remove the timing belt and the belt guide. If reusing the belt mark the belt and the crankshaft pulley in the direction of engine rotation and matchmark for correct installation.

To install:

10. Install the crankshaft timing pulley, as follows:

a. Align the timing pulley set key with the key groove of the pulley.

b. Slide on the timing pulley with the flange side facing inward.

11. Install the No. 2 idler pulley and tighten the bolt to 31 ft. lbs. (42 Nm). Be sure that the pulley moves freely.

12. Temporarily install the No. 1 idler pulley and tension spring. Pry the pulley toward the left as far as it will go. Tighten the bolt. Be sure that the pulley rotates freely.

13. Temporarily install the timing belt, as follows:

a. Using the crankshaft pulley bolt, turn the crankshaft and align the timing marks of the crankshaft timing pulley and the oil pump body.

b. If reusing the old belt, align the marks made during removal, and install the belt with the arrow pointing in the direction of the engine revolution.

14. Install the timing belt guide with the cup side facing outward.

15. Install the No. 1 timing belt cover.

16. Install the crankshaft pulley. Align the pulley set key with the key groove of the pulley and slide on the pulley. Tighten the bolt to 80 ft. lbs. (108 Nm).

17. Install the camshaft timing pulley as follows:

a. Align the camshaft knock pin with the knock pin groove of the pulley and slide on the timing pulley. Tighten the bolt to 40 ft. lbs. (54 Nm).

18. With the No. 1 cylinder set at TDC on the compression stroke, install the timing belt (all timing marks aligned). If reusing the belt, align with the marks made during the removal procedure:

a. Turn the crankshaft pulley and align its groove with the timing mark **0** of the No. 1 timing belt cover. Be sure the camshaft sprocket hole is aligned with the mark on the bearing cap.

19. Connect the timing belt to the camshaft timing pulley.

20. Check that the matchmark on the timing belt matches the end of the No. 1 timing belt cover.

21. Once the belt is installed, be sure that there is tension between the crankshaft timing pulley and the camshaft pulley.

22. Check the valve timing as follows:

a. Loosen the No. 1 idler pulley mount bolt ½ turn. Turn the crankshaft pulley two revolutions from TDC in the clockwise direction. Always turn the crankshaft pulley clockwise.

b. Be sure that the all the timing marks are aligned.

c. Slowly turn the crankshaft pulley 1 ⅞ revolutions. Align its groove with the mark at 45° Before Top Dead Center (BTDC) on the No. 1 timing belt cover for the No. 1 cylinder.

d. Tighten the No. 1 idler pulley mount bolt to 31 ft. lbs. (42 Nm).

23. Install the No. 2 timing belt cover as follows:

a. Install the upper gasket to the No. 1 timing belt cover.

b. Disconnect the engine wire protector between the cylinder head cover and the No. 3 timing belt cover.

c. Install the gasket to the timing belt cover.

d. Install the belt covers and all remaining components. During assembly, tighten the right engine mount bracket bolts to 38 ft. lbs. (52 Nm), the engine mount insulator bolt to 47 ft. lbs. (64 Nm), the through-bolt to 54 ft. lbs. (78 Nm), the power steering reservoir bracket bolt to 21 ft. lbs. (28 Nm), the power steering reservoir-to-bracket bolt to 27 ft. lbs. (37 Nm) and the nut to 38 ft. lbs. (52 Nm).

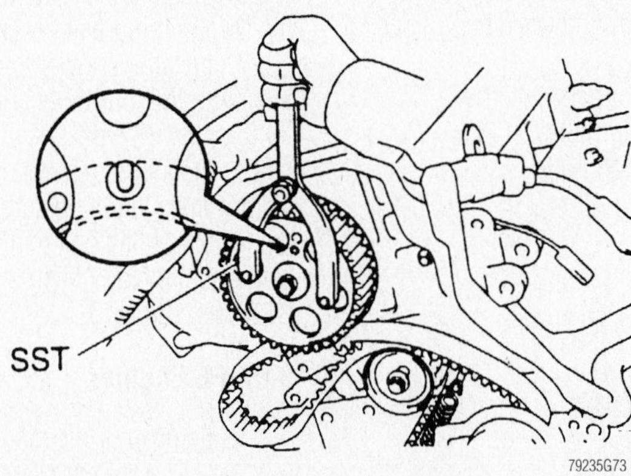

Using a spanner wrench, turn the camshaft into position so that the alignment mark is visible through the hole in the sprocket—Toyota 2.2L (5S-FE) engine

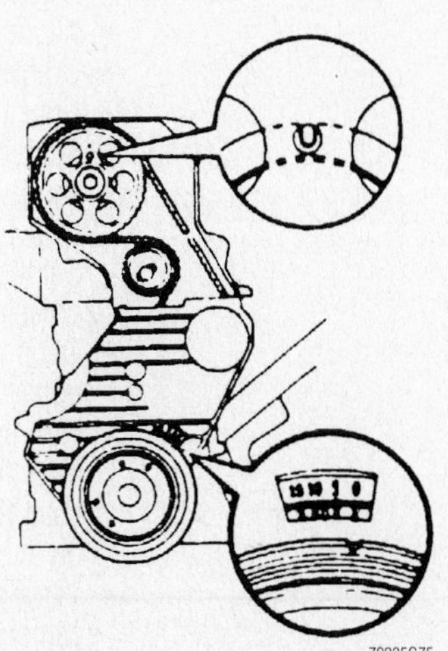

Sprocket alignment for timing belt replacement—Toyota 2.2L (5S-FE) engine

1MZ-FE Engines

1. Before servicing the vehicle, refer to the precautions in the beginning of this section.
2. Remove all necessary components for access to the timing belt covers.

CAUTION

If equipped with an air bag, be sure to disconnect the negative battery cable and wait at least 90 seconds before proceeding.

3. Remove the lower timing belt cover by removing the four bolts.
4. Remove the No. 2 timing belt cover as follows:
   a. Remove the bolt and disconnect the engine wire protector from the No. 3 (rear) timing belt cover.
   b. Disconnect the engine wire protector clamp from the No. 3 timing belt cover.
   c. Remove the five bolts from the No. 2 timing belt cover.
   d. Remove the No. 2 cover from the engine.
5. Remove the right engine mounting bracket by removing the nut and two bolts.
6. Remove the crankshaft timing belt guide.
7. Temporarily install the crankshaft pulley bolt.
8. Turn the crankshaft and align the crankshaft timing pulley groove with the oil pump alignment mark. Always turn the engine clockwise.
9. Ensure the timing mark of the camshaft timing pulleys and rear timing belt covers are aligned. If not, turn the engine over an additional 360 degrees (one revolution).
10. Remove the crankshaft pulley bolt.

➡If the belt is to be reused, align the installation marks on the belt to the marks on the pulleys. If the marks have worn off, make new ones.

11. Alternately loosen the two timing belt tensioner bolts. Remove the tensioner and dust boot.
12. Remove the timing belt.

To install:

13. Remove any oil or water from the pulleys.
14. Align the front mark of the timing belt with the dot mark of the crankshaft timing pulley.
15. Align the installation marks on the timing belt with the timing marks of the camshaft pulleys.
16. Install the timing belt in the following order:
   a. Crankshaft pulley.
   b. Water pump pulley.
   c. Left camshaft pulley.
   d. No. 2 idler pulley.
   e. Right camshaft pulley.
   f. No. 1 idler pulley.
17. Using a press, slowly press the timing belt tensioner until the holes of the pushrod and housing align. Insert a 0.05 in. (1.27mm) hexagonal Allen wrench through the holes to preserve the setting position.
18. Install the dust boot to the tensioner.
19. Install the tensioner with the two bolts. Alternately tighten the bolts to 20 ft. lbs. (27 Nm). Remove the Allen wrench.
20. Turn the crankshaft clockwise and align the crankshaft timing pulley groove with the oil pump alignment mark.
21. Ensure the camshaft timing marks align with the timing marks on the rear timing belt cover.
22. Install the timing belt guide.
23. Install the right engine mounting bracket and tighten the bolts to 21 ft. lbs. (28 Nm).
24. Install the upper timing belt cover with the five bolts. Tighten the bolts to 74 inch lbs. (8 Nm).
25. Install the engine wire protector clamp to the No. 3 timing belt cover.
26. Install the engine wire protector to the No. 3 timing belt cover with the bolt.
27. Install the lower timing belt cover by installing the four bolts. Tighten the bolts to 74 inch lbs. (8 Nm).
28. Install the remaining components. During installation be sure to tighten the crankshaft pulley bolt to 159 ft. lbs. (215 Nm) and the No. 2 alternator bracket nut to 21 ft. lbs. (28 Nm).

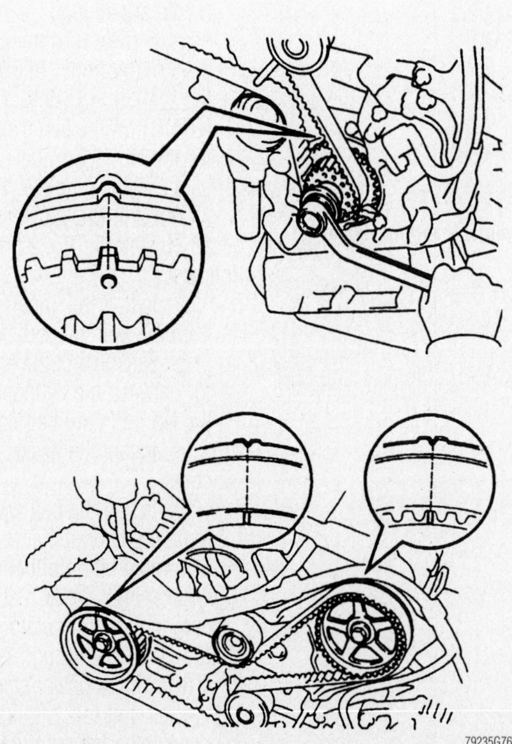

Camshaft and crankshaft timing belt sprocket alignment mark positioning for belt service—Toyota 3.0L (1MZ-FE) engine

3MZ-FE Engines

1. Remove or disconnect the following:
   - RH front fender apron seal
   - A/C drive belt
   - Engine stabilizing rod
   - Engine mounting stay
   - Alternator bracket
   - Crankshaft pulley
   - Upper and lower timing belt covers
   - Engine mounting bracket (If necessary)
2. Remove the timing belt
   - Set the No 1 cylinder to TDC/compression
   - Temporarily install the crankshaft pulley bolt
   - Turn the crankshaft clockwise, the align the timing mark on the crankshaft timing pulley with the mark on the oil pump body
   - Verify that the timing marks on the camshaft timing pulleys align with the timing marks on the inside tim-

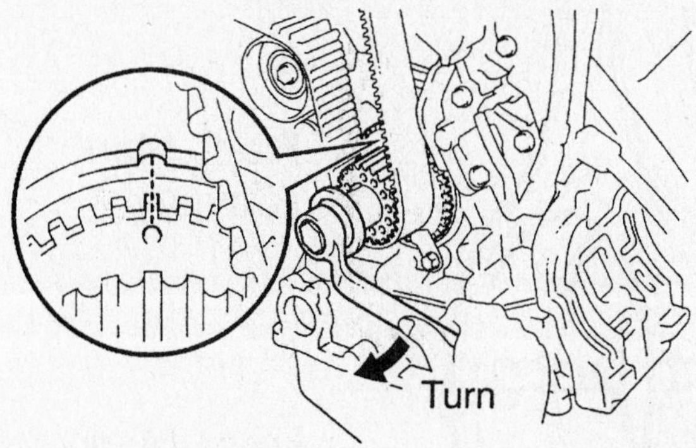

Aligning crankshaft pulley—3MZ-FE engine

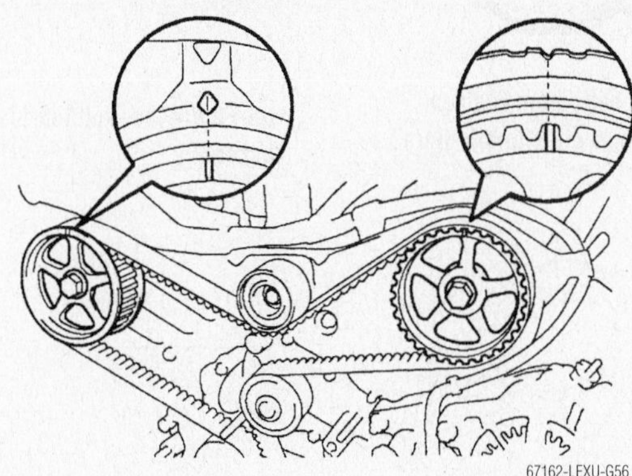

Aligning camshaft pulleys—3MZ-FE engine

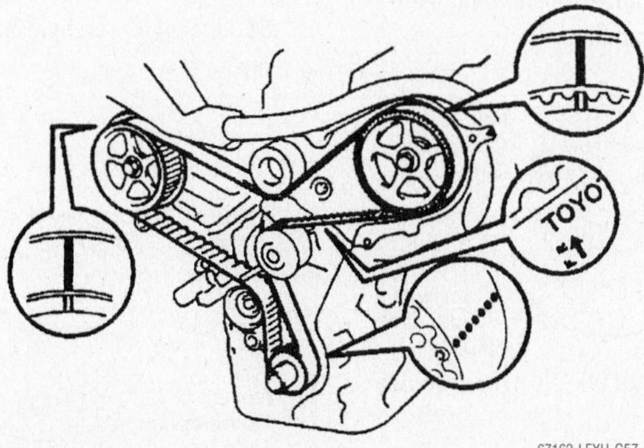

Marking the timing belt for reuse—3MZ-FE engine

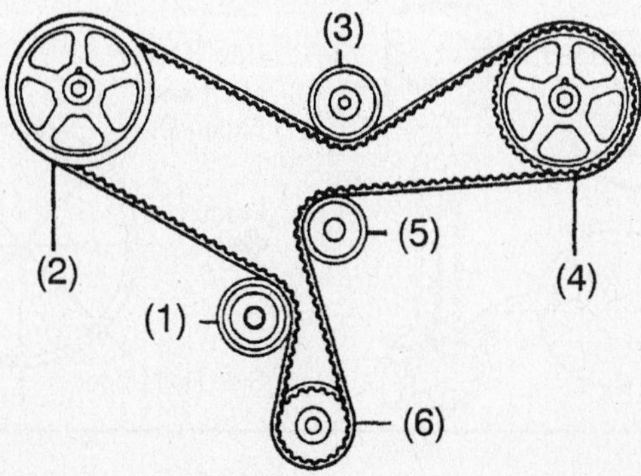

Removal of timing belt from pulleys—3MZ-FE engine

ing belt cover. If not, turn the crankshaft on revolution (360 degrees).

- If reusing the timing belt make certain there are 3 location marks on the timing belt corresponding with the marks on the timing gears.
- Set the No 1 cylinder to approximately 60 degrees BTDC/compression. Turn the crankshaft counterclockwise by approximately 60 degrees.

WARNING

If the timing belt is disengaged, having the crankshaft pulley set at the wrong angle can cause contact of the piston with the valves, causing damage to the valves. Always set the crankshaft pulley at the correct angle.

- Remove timing belt tensioner

➡Do NOT reinstall the timing belt tensioner with the plunger extended.

3. Remove the timing belt in the following order.
   - No 1 idler pulley
   - RH camshaft timing pulley
   - No. 2 idler pulley
   - LH camshaft timing pulley
   - Water pump pulley
   - Crankshaft timing pulley
4. Inspect the timing belt

➡do not reuse the timing belt if there is evidence of fraying, oil contamination, cracking, tooth damage or wear, or belt distortion. If there is any doubt in the belt condition, replace the belt.

To install:

5. After inspecting the pulleys for wear and checking for oil leaks install the timing belt using the following procedure.
   - Temporarily install the crankshaft pulley bolt
   - Make sure the crankshaft pulley is set at 60° BTDC
   - Align the timing marks on the camshaft pulley with the respective marks on the inside timing cover.
   - If reusing the old belt, align the marks previously made on the belt with the marks on the timing gears.
   - Install the belt in the reverse order used when removing the belt
6. Install the timing belt tensioner using the following procedure

➡Keep the tensioner in an upright position

- Slowly depress the push rod and align the hole with the hole in the housing
- Insert a 1.5 mm hexagon wrench through the holes to maintain the setting position of the push rod.
- Install the tensioner with the 2 bolts and torque to 20 ft. lbs (27 Nm).
- Remove the hexagon wrench

7. Slowly turn the crankshaft 2 revolutions clockwise.
8. Check to see that all timing marks are in alignment.
9. Remove the crankshaft bolt
10. Install the timing belt guide
11. Install all remaining components in the reverse order of removal.

Piston and Ring Positioning

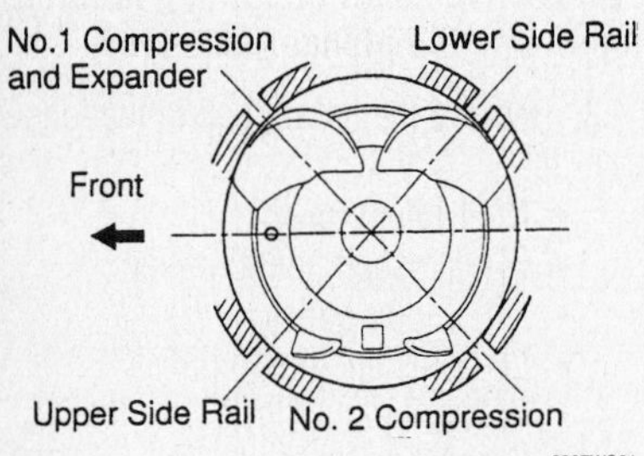

Piston ring positioning—1NZ-FE engine

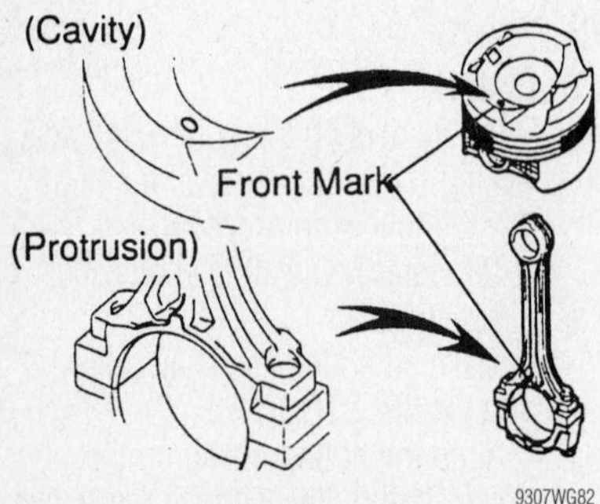

Piston and connecting rod positioning—1NZ-FE engine

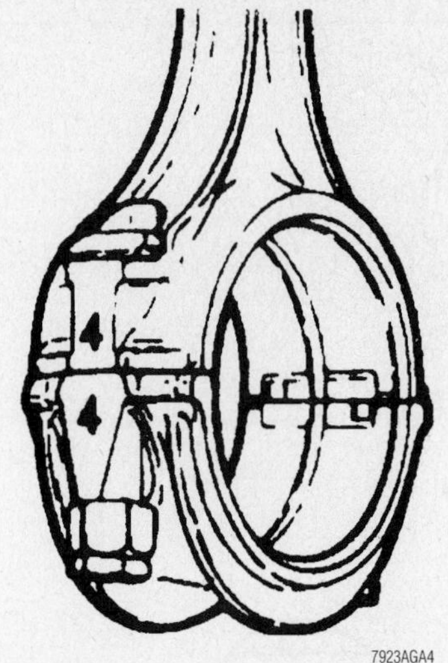

Before removing the caps from the connecting rods, be sure to matchmark them as shown

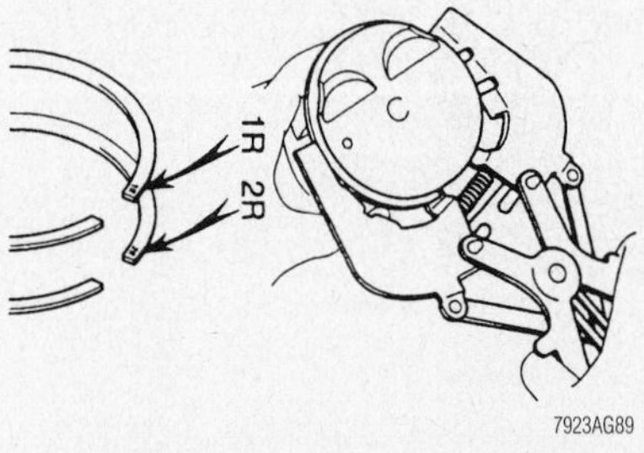

1ZZ-FE, 2ZZ-GE engine—piston ring identification mark locations

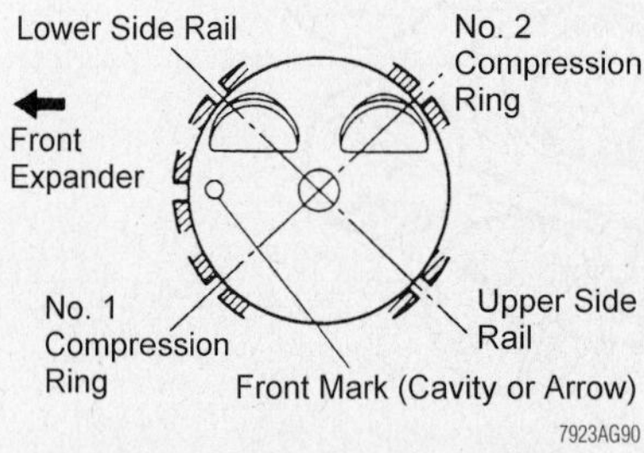

1ZZ-FE, 2ZZ-GE engine—piston ring end-gap spacing

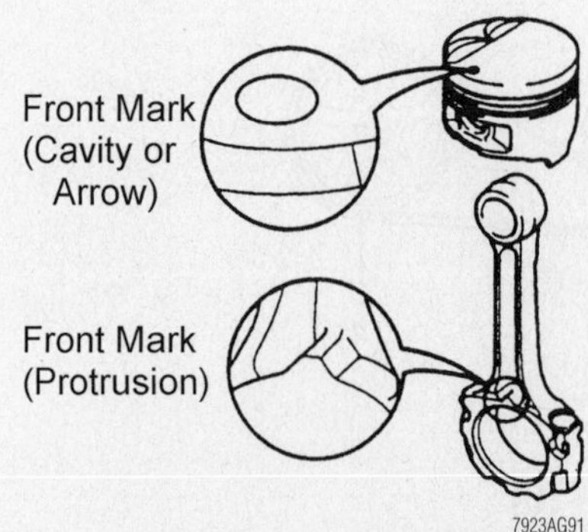

1ZZ-FE, 2ZZ-GE engine—piston-to-connecting rod assembly

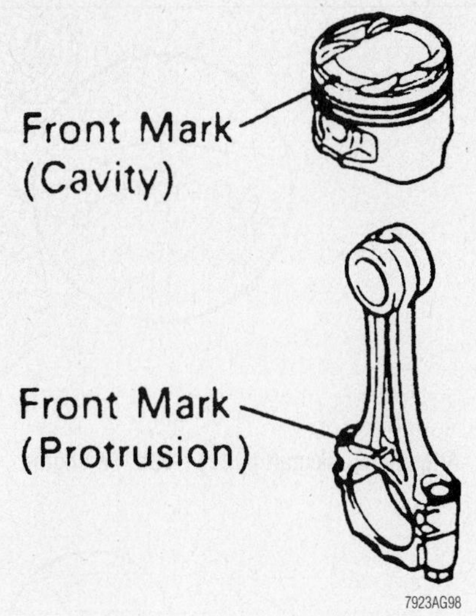

5S-FE engine—piston-to-connecting rod assembly

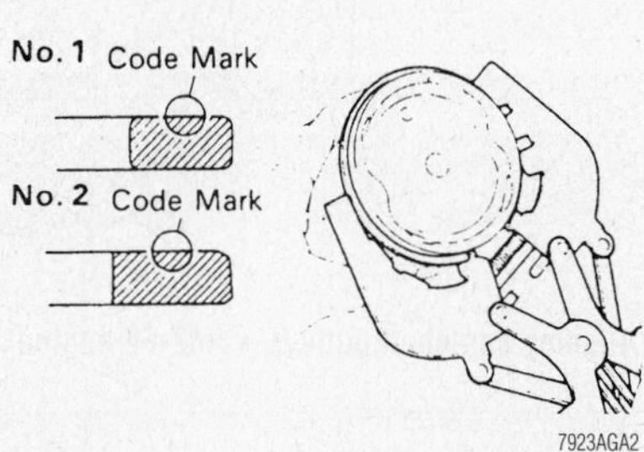

5S-FE engine—compression ring positioning

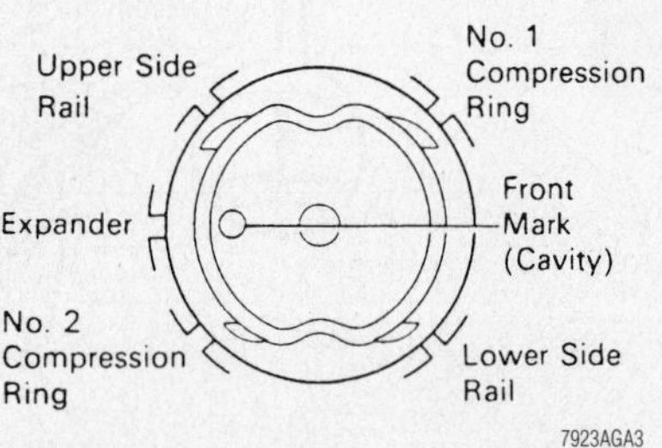

5S-FE engine—piston ring end-gap spacing

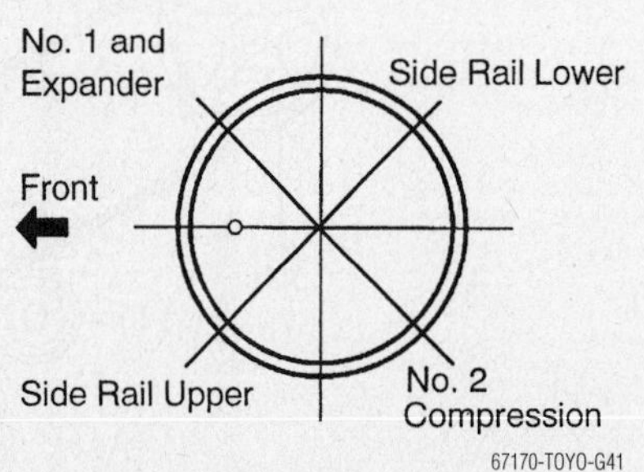

2AZ-FE engine—piston ring end-gap spacing

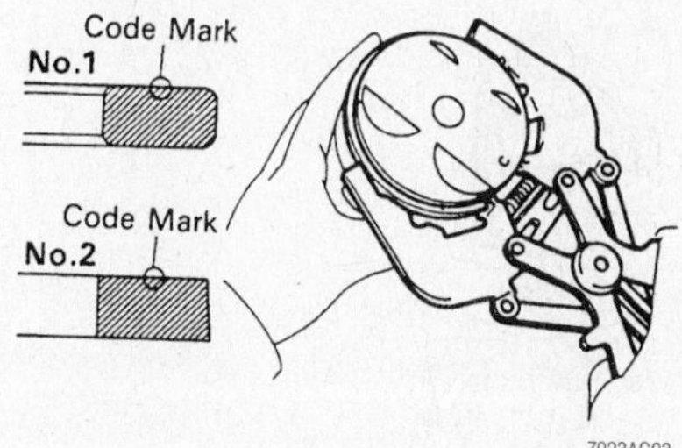

1MZ-FE engine—compression ring positioning

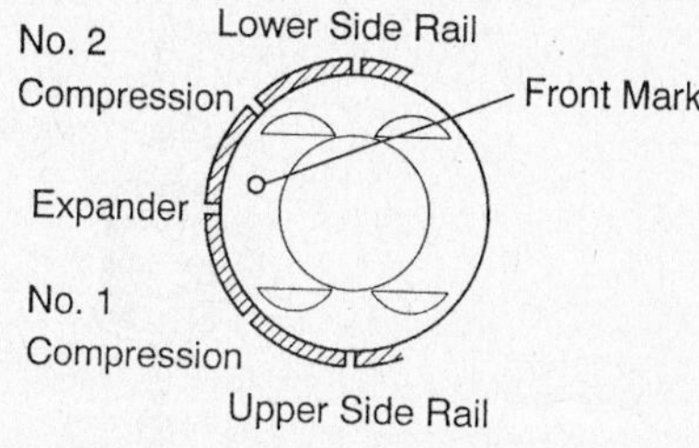

3MZ-FE engine—piston ring end-gap spacing

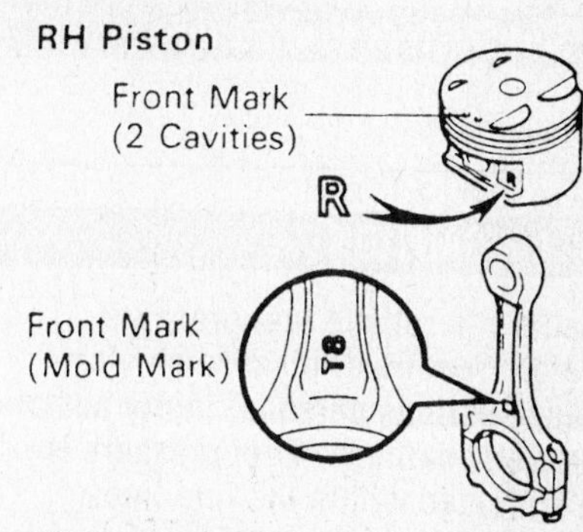

Camry 1MZ-FE engine—piston-to-connecting rod assembly

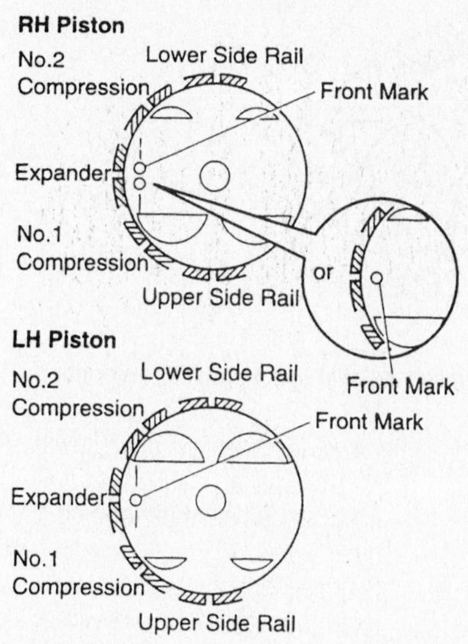

1MZ-FE engine—piston ring end-gap spacing

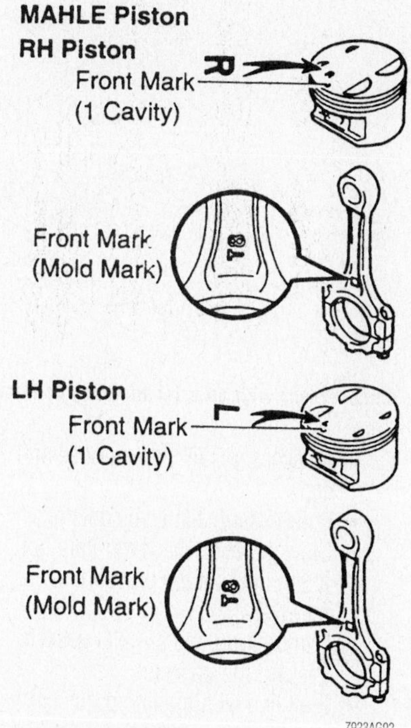

Avalon 1MZ-FE engine—piston-to-connecting rod assembly

FUEL SYSTEM

Fuel System Service Precautions

Safety is the most important factor when performing not only fuel system maintenance, but any type of maintenance. Failure to conduct maintenance and repairs in a safe manner may result in serious personal injury or death. Work on a vehicle's fuel system components can be accomplished safely and effectively by adhering to the following rules and guidelines.

• To avoid the possibility of fire and personal injury, always disconnect the negative battery cable unless the repair or test procedure requires that battery voltage by applied.

• Always relieve the fuel system pressure prior to disconnecting any fuel system component (injector, fuel rail, pressure regulator, etc.) fitting or fuel line connection. Exercise extreme caution whenever relieving fuel system pressure, to avoid exposing skin, face and eyes to fuel spray. Please be advised that fuel under pressure may penetrate the skin or any part of the body that it contacts.

• Always place a shop towel or rag around the fitting or connection prior to loosening to absorb any excess fuel due to spillage. Ensure that all fuel spillage is quickly remove from engine surfaces. Ensure that all fuel-soaked cloths or towels are deposited into a flame-proof waste container with a lid.

• Always keep a dry chemical (Class B) fire extinguisher near the work area.

• Do not allow fuel spray or fuel vapors to come into contact with a light bulb, spark or open flame.

• Always use a second wrench when loosening or tightening fuel line connections fittings. This will prevent unnecessary stress and torsion to fuel piping. Always follow the proper torque specifications.

• Always replace worn fuel fitting O-rings with new ones. Do not substitute fuel hose where rigid pipe is installed.

Fuel System Pressure

RELIEVING

CAUTION

Failure to relieve fuel pressure before repairs or disassembly can cause serious personal injury and/or property damage. Fuel pressure is maintained within the fuel lines, even if the engine is OFF or has not been run in a period of time. This pressure must be safely relieved before any fuel-bearing line or component is loosened or removed. On vehicles equipped with inflatable restraints or air bag systems, wait at least 90 seconds after disconnecting the battery cable before performing any other work. The back-up power will keep the restraint system energized for a period of time after the battery is disconnected.

1. Before servicing the vehicle, refer to the precautions in the beginning of this section.
2. Perform the following:
   - Remove the fuse for the fuel pump
   - Start the engine until the engine stalls
   - Disconnect the negative battery cable
   - Place a catch-pan under the joint to be disconnected. A large quantity of fuel may be released when the joint is opened
   - Wear eye or full face protection
   - Place a shop towel over the area and slowly release the joint using a wrench of the correct size.
   - Allow the any fuel left in the line to bleed off slowly before fully disconnecting the joint.
   - Plug the opened lines
3. After connecting fuel lines, install the fuse for the fuel pump and start the engine.

Fuel Filter

REMOVAL & INSTALLATION

1. Before servicing the vehicle, refer to the precautions in the beginning of this section.
2. Remove or disconnect the following:
   - Negative battery cable. On vehicles equipped with an air bag, wait at least 90 seconds before proceeding
   - Protective shield for the fuel filter

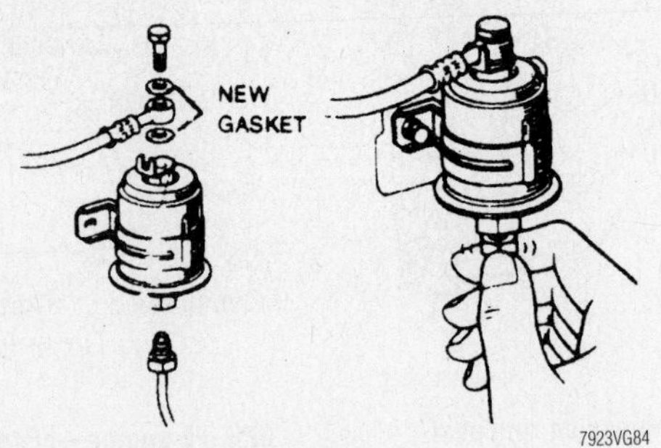

Always use new gaskets when replacing a fuel filter

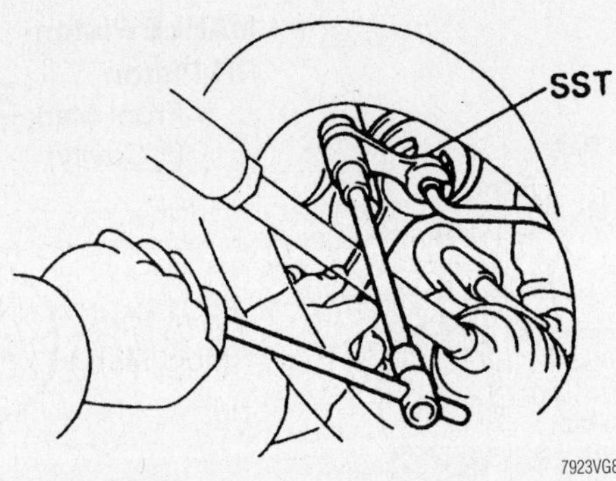

A line wrench with an extension may be needed to loosen the inlet line at the filter—Corolla

   - If necessary, air cleaner hose and cap
   - If necessary, charcoal canister.
   - Slowly loosen the lower flare nut fitting until all the pressure is relieved
   - Banjo fitting and 2 metal gaskets. Discard the gaskets
   - Fuel line with the flared nut from the filter
   - Filter from the mounting bracket

To install:

3. Install or connect the following:
   - New fuel filter
   - Banjo fitting with a new metal gasket on each side and install the union bolt. Bolt: 22 ft. lbs. (30 Nm).
   - Flare nut to the lower connection. Nut: 22 ft. lbs. (30 Nm).
   - Charcoal canister
   - Air cleaner hose and cap
   - Protective shield
   - Negative battery cable

Fuel Pump

REMOVAL & INSTALLATION

Avalon, Camry and Camry Solara

1. Before servicing the vehicle, refer to the precautions in the beginning of this section.
2. Remove or disconnect the following:
   - Negative battery cable. Wait at least 90 seconds before performing any other work.
   - Rear seat bottom and seat back
   - Rear floor service hole cover
   - Fuel suction w/pump and gauge tube assembly
   - 8 bolts and fuel tank vent tube set plate
   - Vapor pressure sensor assembly
   - Fuel suction hose and support
   - Fuel pump cushion rubber
   - Fuel sender gauge assembly
   - Fuel suction plate w/sender gauge
   - Fuel pump harness
   - Fuel pump
   - Fuel pump filter. Pry out clip and remove from pump
   - Fuel pressure regulator assembly

To install:

3. To install, reverse the removal procedure. Install a new O-ring on the fuel suction w/pump and gauge assembly. Torque the 8 bolts securing the fuel tank set plate to 52 inch lbs. (6.0 Nm).
4. Inspect fuel pump operation and check for fuel leaks.

Echo

1. Before servicing the vehicle, refer to the precautions in the beginning of this section.

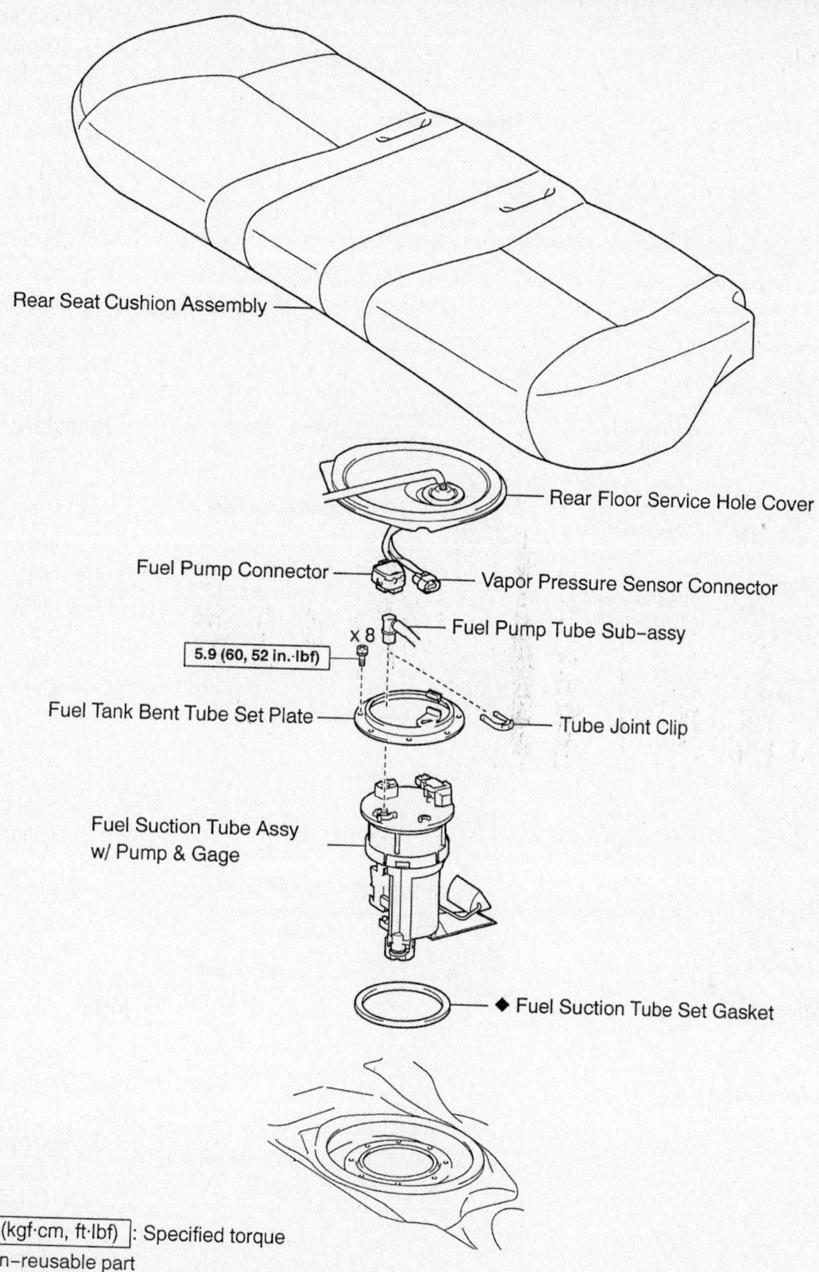

Exploded view for fuel pump removal—Camry shown, others similar

2. Relieve the fuel system pressure.
3. Remove or disconnect the following:
   - Negative battery cable. On vehicles equipped with an air bag, wait at least 90 seconds before proceeding.
   - Rear seat cushion
   - Floor service hole cover
   - Electrical connector at the fuel pump assembly
   - Fuel outlet pipe from the fuel pump bracket
   - Return hose from the fuel pump bracket.
   - Fuel pump set plate from the fuel tank by removing the 8 bolts
   - Fuel pump from the fuel bracket

To install:

4. Install or connect the following:
   - Fuel pump to the fuel tank. Bolts: 35 inch lbs. (4 Nm).
   - Return hose to the fuel pump bracket
   - Outlet pipe to the fuel pump bracket. Tighten to 21 ft. lbs. (28 Nm).
   - Service hole cover to the fuel tank
   - Fuel pump connector
   - Rear seat cushion
   - Negative battery cable

Celica and Corolla

1. Before servicing the vehicle, refer to the precautions in the beginning of this section.
2. Relieve the fuel system pressure.
3. Remove or disconnect the following:
   - Negative battery cable. On vehicles equipped with an air bag, wait at least 90 seconds before proceeding
   - Rear seat cushion and floor service hole cover
   - Access plate-to-fuel tank bolts, then pull out the plate/fuel pump assembly

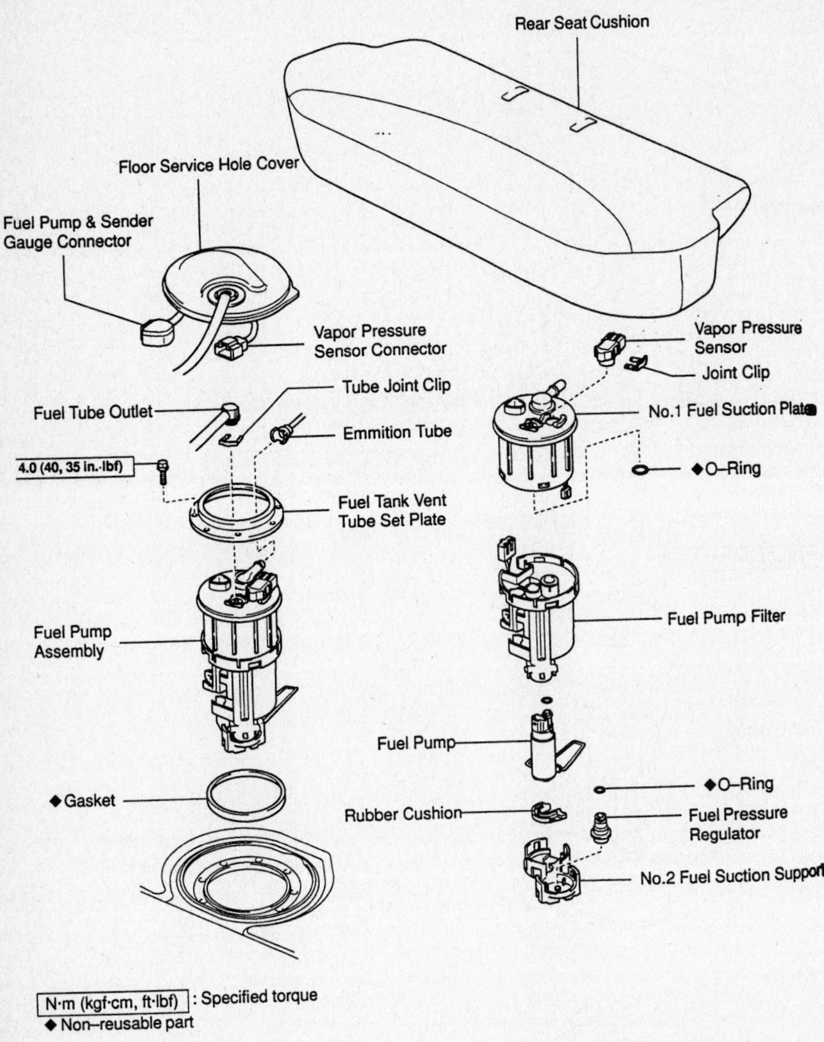

Fuel pump removal—Echo

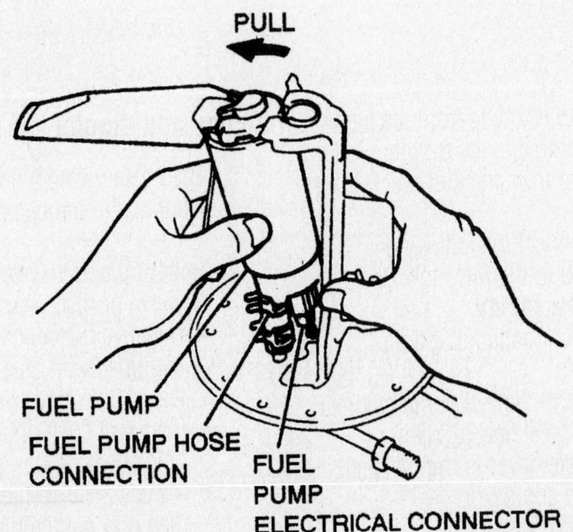

Pull the pump off the sender unit; the filter is still attached to the pump

- Fuel pump sender and fuel pump connector
- Outlet pipe from the fuel pump bracket
- Return hose from the pump bracket
- Fuel pump bracket assembly from the fuel tank
- Lower side of the fuel pump from the pump bracket
- Fuel pump connector
- Fuel hose from the fuel pump
- Rubber cushion from the pump
- Fuel filter from the pump by removing the small clip

To install:

4. Install or connect the following:
   - New cushion to the fuel pump
   - Fuel filter and new clip to the fuel pump
   - Fuel hose to the fuel pump, fuel pump connector and fuel pump to the bracket
   - Fuel pump bracket assembly to the fuel tank using a new gasket. Bolts: 30 inch lbs. (3 Nm).
   - Fuel return hose and the fuel outlet pipe to the fuel pump bracket
   - Fuel pump and fuel pump sender connector
   - Fuel tank
   - Negative battery cable
   - Floor service hole cover and rear seat cushion

MR2

1. Before servicing the vehicle, refer to the precautions in the beginning of this section.
2. Remove or disconnect the following:
   - Negative battery cable
   - Luggage compartment box
   - Service access panel
   - Fuel pump and gauge harness connector
   - Vapor pressure sensor harness connector
   - Fuel main tube
   - Emissions tube
   - Fuel pump module

To install:

3. Install or connect the following:
   - Fuel pump module. Tighten the bolts to 30 inch lbs. (3.4 Nm).
   - Emissions tube
   - Fuel main tube
   - Vapor pressure sensor harness connector
   - Fuel pump and gauge harness connector
   - Service access panel
   - Luggage compartment box

- Negative battery cable

4. Start the engine and check for leaks.

Fuel Injectors

REMOVAL & INSTALLATION

Avalon, Camry and Solara

5S-FE ENGINES

1. Before servicing the vehicle, refer to the precautions in the beginning of this section.
2. Drain the cooling system.
3. Remove or disconnect the following:
   - Negative battery cable
   - Accelerator cable from the throttle body
   - If equipped with automatic transmission, the throttle cable
   - Air intake temperature sensor connector
   - Cruise control actuator cable from the clamp on the resonator
   - Air cleaner
   - Wiring from the throttle position sensor and the ISC valve
   - Hoses from the PCV, EGR vacuum modulator and EVAP VSV
   - Throttle body
   - PS vacuum hoses
   - Hoses from the EVAP Bi-metal Vacuum Switching Valve (BVSV)
   - EGR valve and the vacuum modulator
   - Vacuum sensor hose at the air intake chamber
   - Brake booster vacuum hose and the vacuum sensing hose
   - With air conditioning, the magnet switch VSV wiring
   - Ground straps from the intake manifold
   - Knock sensor and EGR Vacuum Switching Valve (VSV) wiring
   - Engine wire harness
   - Stays or supports holding the air intake chamber and the intake manifold
   - Intake manifold
   - Wiring from each injector
   - Fuel inlet pipe
   - Fuel return hose
   - Delivery pipe or fuel rail along with the injectors
   - Insulators and rail spacers from the head
   - Injectors from the fuel rail
   - O-ring and grommet from each injector

Loosening and removing the pulsation damper—5S-FE engine

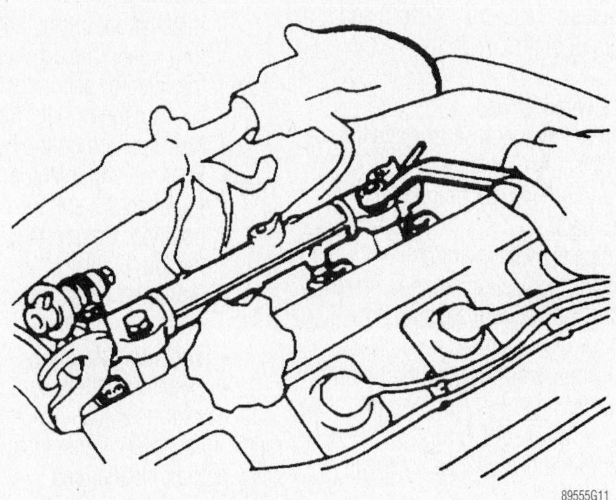

Unbolt and remove the fuel delivery pipe and injectors as an assembly

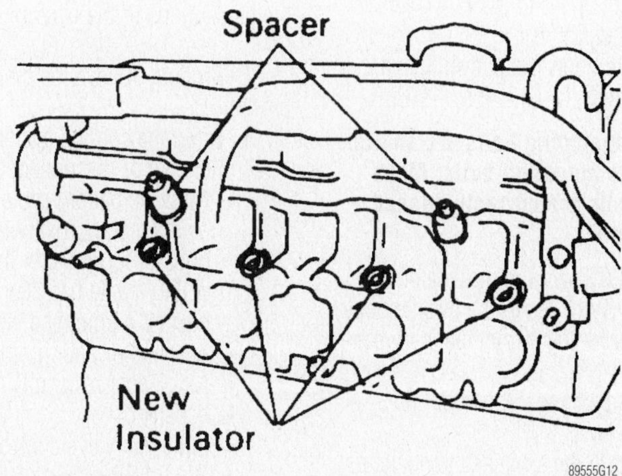

Place new injector seals and spacers in position on the cylinder head—5S-FE engine shown

To install:

4. Install or connect the following:
   - New grommet on each injector
   - New O-rings, coated with gasoline, on each injector
   - Injectors into the fuel rail while turning each left and right. After installation, check that the injectors turn freely in place; if not, remove the injector and inspect the O-ring for damage or deformation.

- New insulators and spacers on the head
- Fuel rail and injectors; check that the injectors still turn freely in position. Position the injector connectors upward
- Retaining bolts, tightening them to 9 ft. lbs. (13 Nm)
- Fuel return hose
- Fuel inlet pipe and pulsation damper to the delivery pipe. Use new gaskets; tighten the union bolt to 25 ft. lbs. (34 Nm).
- Wiring to each injector
- Intake manifold. Nuts and bolts evenly in several passes to 14 ft. lbs. (19 Nm).
- Air chamber and manifold stays. 14mm bolt: 31 ft. lbs. (42 Nm) and the 12mm bolt to 16 ft. lbs. (22 Nm).
- Engine wire harness
- Wiring to the knock sensor and the EGR VSV
- Both engine ground straps to the intake manifold
- A/C magnet switch wiring
- Hoses for the vacuum sensor, brake booster and vacuum sensing hose
- EGR valve and vacuum modulator. Use new gaskets. Tighten the union nut to 43 ft. lbs. (59 Nm) and the bolt to 9 ft. lbs. (13 Nm).
- Hoses to the charcoal canister and EGR VSV
- Wiring to the EGR temperature sensor if it was removed
- Vacuum hoses to the EVAP BVSV
- PS vacuum hoses
- Throttle body. Bolts, evenly and alternately, to 14 ft. lbs. (19 Nm).

➡The upper mounting bolts are shorter than the lower mounting bolts. Make certain the bolts are correctly placed before tightening.

- PCV, EGR vacuum modulator and EGR VSV hoses to the throttle body
- Wiring for the throttle position sensor
- Air cleaner cap, resonator and intake hose
- Wiring to the air intake temperature sensor
- Cruise control actuator cable
- Throttle control cable
- Accelerator cable
- Negative battery cable

5. Fill the cooling system to the proper level.
6. Start the vehicle, check for leaks and repair if necessary.

2AZ-FE ENGINES

1. Before servicing the vehicle, refer to the precautions in the beginning of this section.
2. Properly relieve the fuel system pressure.
3. Remove or disconnect the following:
   - Negative battery cable
   - Air cleaner cap and hose
   - Engine cover
   - Fuel tube assembly
   - Fuel delivery pipe with the injectors

To install:

4. Install or connect the following:
   - New O-rings, coated with gasoline, on each injector
   - Injectors into the fuel rail while turning each left and right. After installation, check that the injectors turn freely in place; if not, remove the injector and inspect the O-ring for damage or deformation.
   - New spacers on the head
   - Fuel rail and injectors; check that the injectors still turn freely in position. Position the injector connectors outward
   - Retaining bolts, tightening them to 15 ft. lbs. (20 Nm)
   - Fuel tube assembly
   - Negative battery cable
5. Start the vehicle, check for leaks and repair if necessary. Then install the engine cover and air cleaner cap.

1MZ-FE ENGINES

1. Before servicing the vehicle, refer to the precautions in the beginning of this section.
2. Properly relieve the fuel system pressure.
3. Drain the cooling system.
4. Remove or disconnect the following:
   - Negative battery cable. Work must be started approximately 90 seconds or longer after the negative battery cable has been disconnected, if equipped with an air bag
   - Accelerator and throttle cables
   - Air cleaner assembly
   - V-bank cover
   - Emission valve control set
   - No. 2 EGR pipe
   - Hydraulic motor pressure pipe from the water inlet and air inlet chamber
   - Air intake chamber assembly
   - Injector wiring
   - Air assist pipe from the bracket on the No. 1 fuel pipe
   - Air assist hoses from the intake manifold
   - Fuel return hose from the No. 1 fuel pipe
   - Fuel inlet hose for the fuel filter
   - 2 union bolts holding the No. 2 fuel pipe to the delivery pipes
   - Fuel return hose from the fuel pressure regulator
   - Union bolt for the right hand delivery pipe, 2 gaskets, 2 bolts, left hand delivery pipe together with the 3 injectors and the No. 2 fuel pipe
   - Union bolt for the delivery pipe and 2 gaskets from the No. 2 fuel pipe
   - The 3 bolts, right hand delivery pipe together with the 3 injectors and the No. 1 fuel pipe
   - The 4 spacers from the intake manifold
   - The 6 injectors from the delivery pipes
   - The two O-rings and two grommets from each injector

To install:

5. Install or connect the following:
   - 2 new grommets to each injector
   - New O-rings, with a light coat of fuel, to each injector
   - Injectors
   - The 4 spacers on the intake manifold
   - Right hand delivery pipe and the No. 1 fuel pipe together with the 3 injectors in position on the intake manifold
   - Bolt holding the right side delivery pipe, temporarily, to the intake manifold
   - Left hand delivery pipe and the No. 2 fuel pipe together with the 3 injectors in position on the intake manifold
   - Fuel return hose to the fuel pressure regulator
6. Temporarily install the 2 bolts holding the left hand delivery pipe to the intake manifold.
7. Temporarily install the No. 2 fuel pipe to the left side delivery pipe with the union bolt and 2 new gaskets.
8. Check that the injectors rotate smoothly. If they do not, Replace the O-rings.
9. Position the injector connector outward. Tighten the 4 bolts holding the delivery pipes to the intake manifold and tighten to 7 ft. lbs. (10 Nm). Tighten the bolt holding the No. 1 fuel pipe to the intake manifold to 14 ft. lbs. (20 Nm). Tighten the 2 union bolts holding the no. 2 fuel pipe to the delivery pipes to 24 ft. lbs. (32 Nm).
10. Install or connect the following:
    - Fuel inlet and return hoses. Union bolt: 22 ft. lbs. (30 Nm).

- Fuel return hose to the No. 1 fuel pipe. Pass the fuel return hose under the heater hoses
- Air assist hoses to the intake manifold
- Air assist pipe to the bracket on the No. 1 fuel pipe
- Fuel injector wiring connectors
- Air intake chamber assembly
- Hydraulic motor pressure pipe to the intake chamber. Bolts: 69 inch lbs. (8 Nm)
- No. 2 EGR pipe with new gaskets, tighten to 9 ft. lbs. (12 Nm)
- Emission control valve set
- V-bank cover
- Air cleaner hose
- Throttle and accelerator cables
- Negative battery cable

11. Fill the cooling system to the proper level.

12. Start the vehicle, check for leaks and repair if necessary.

3MZ-FE ENGINES

1. Before servicing the vehicle, refer to the precautions in the beginning of this section.
2. Remove or disconnect the following:
   - Negative battery cable
   - Suspension upper center brace
   - V-bank cover sub assembly
   - Air cleaner assembly
   - Emission control valve set
3. Remove intake air surge tank
4. Remove or disconnect the following:
   - Throttle motor connector
   - Water by-pass hoses
   - Union check valve hose
   - Ventilation hose
   - Pressure feed hose
   - Engine hangers
   - Surge tank stays
   - Bond cable connector
   - Emission control valve bracket
   - Intake air surge tank
   - Gasket from intake air surge tank
5. Remove fuel pipe assembly
6. Remove or disconnect the following:
   - Fuel pulsation damper and gasket
   - Fuel pipe union bolt and 2 gaskets
   - Bolt and separate the fuel pipe
7. Remove the fuel injector assembly
8. Remove or disconnect the following:
   - 6 fuel injector connectors
   - 4 bolts, then remove the fuel injector delivery pipes

⁂ WARNING

Be careful not to drop the fuel injectors when removing the fuel delivery pipes.

- 4 delivery pipe spacers from intake manifold
- 6 insulators from the intake manifold
- fuel injector from the fuel delivery pipes.

To install:

9. Install new fuel injector assembly
10. Install or connect the following
    - A new insulator and grommet to each injector
    - A light coat of gasoline to new O-rings and install them to each injector
    - A light coat of gasoline on the place where a delivery pipe touches an O-ring of the injector
    - Injector, while turning the clockwise and counterclockwise, into the delivery pipe
    - The 6 insulators and 4 spacers in position on the intake manifold
    - A light coat of gasoline on the place where an intake manifold touches an O-ring
    - The delivery pipes in position on the intake manifold
    - Temporarily, the 4 bolts holding the delivery pipe to the intake manifold

➡Check that the injectors rotate smoothly. If the injectors do not rotate smoothly, the probable cause is incorrect installation of the O-rings. Replace the O-rings.

- 4 bolts. Tighten bolts uniformly. Torque to: 7 ft. lbs. (10 Nm)
- 6 fuel injector connectors

11. Install fuel pipe assembly
12. Install or connect the following
    - 2 gaskets and fuel pipe union bolt
    - 2 gaskets and fuel pressure pulsation damper
    - Fuel pipe with bolt
13. Install intake air surge tank
14. Install or connect the following
    - New gaskets to intake air surge tank
    - Intake air surge tank and emission control valve bracket. Torque: 21 ft. lbs. (28 Nm).
    - 4 bolts. Torque: 21 ft. lbs. (28 Nm)
    - 2 bolts on surge tank stay. Torque: 21 ft. lbs. (28 Nm).
    - Engine hangers
    - Pressure feed tube

15. Remaining components in reverse order of removal procedure

16. Install engine coolant and check for leaks

Celica

1ZZ-FE ENGINES

1. Before servicing the vehicle, refer to the precautions in the beginning of this section.
2. Properly relieve the fuel system pressure.
3. Remove or disconnect the following:
   - No. 2 cylinder head cover
   - PCV hose
   - Fuel tube from the fuel pipe
   - Injector connectors
   - Delivery pipe and injectors
   - Spacers from the head
   - Injectors from the delivery pipe
   - O-ring and grommet from each injector

To install:

4. Install or connect the following:
   - New grommets
   - New O-rings coated with light machine oil
   - Injectors on the delivery pipe

➡Coat the contact point on the pipe with light machine oil and twist the injectors into place. The connector should face outward.

- Spacers

➡Coat the seats in the head where the injectors contact, with light machine oil.

- Delivery pipe and injectors

5. Loosely install the hold-down bolts and check that the injectors rotate smoothly. If they don't, the probable cause is incorrect O-ring installation.

6. Torque the hold-down bolts to 14 ft. lbs. (19 Nm).

7. Torque the fuel pipe bolt to 84 inch lbs. (9 Nm).

8. Connect the fuel line.
9. Install the PCV hose.
10. Install the No. 2 cover.

2ZZ-GE ENGINES

1. Before servicing the vehicle, refer to the precautions in the beginning of this section.
2. Properly relieve the fuel system pressure.
3. Remove or disconnect the following:
   - No. 2 cylinder head cover
   - Fuel tube from the fuel pipe
   - Injector connectors
   - Delivery pipe and injectors
   - Spacers from the head
   - Injectors from the delivery pipe
   - O-ring and grommet from each injector

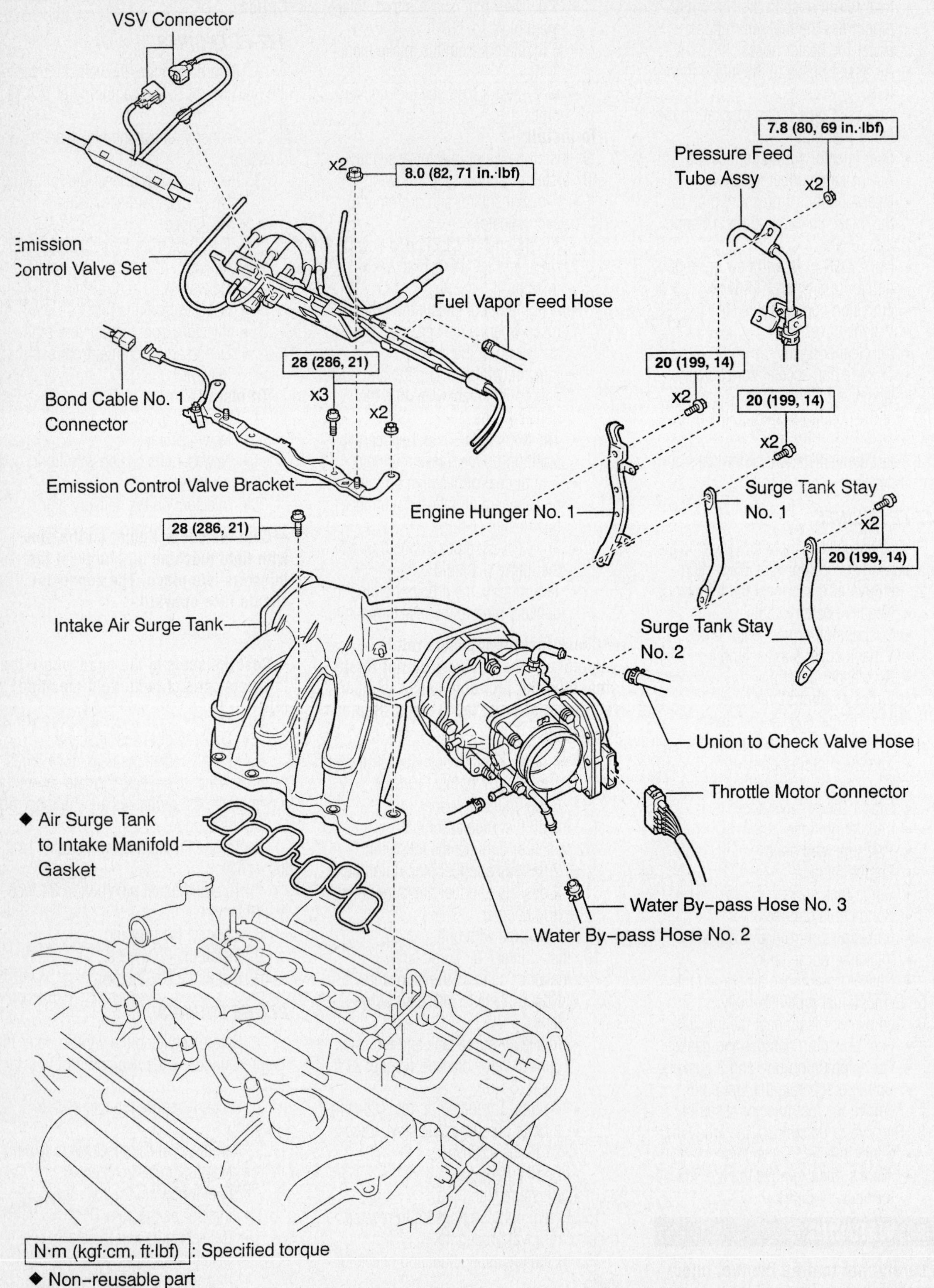
VSV Connector
x2
8.0 (82, 71 in.·lbf)
7.8 (80, 69 in.·lbf)
Pressure Feed
Tube Assy
x2
Emission
Control Valve Set
Fuel Vapor Feed Hose
28 (286, 21)
20 (199, 14)
Bond Cable No. 1
Connector
x3
x2
x2
20 (199, 14)
x2
Emission Control Valve Bracket
Surge Tank Stay
No. 1
Engine Hunger No. 1
x2
28 (286, 21)
20 (199, 14)
Intake Air Surge Tank
Surge Tank Stay
No. 2
Union to Check Valve Hose
Throttle Motor Connector
◆ Air Surge Tank
to Intake Manifold
Gasket
Water By-pass Hose No. 3
Water By-pass Hose No. 2
N·m (kgf·cm, ft·lbf) : Specified torque
◆ Non-reusable part
67162-LEXU-G61

Exploded view of Intake Air Surge tank—3MZ-FE engine

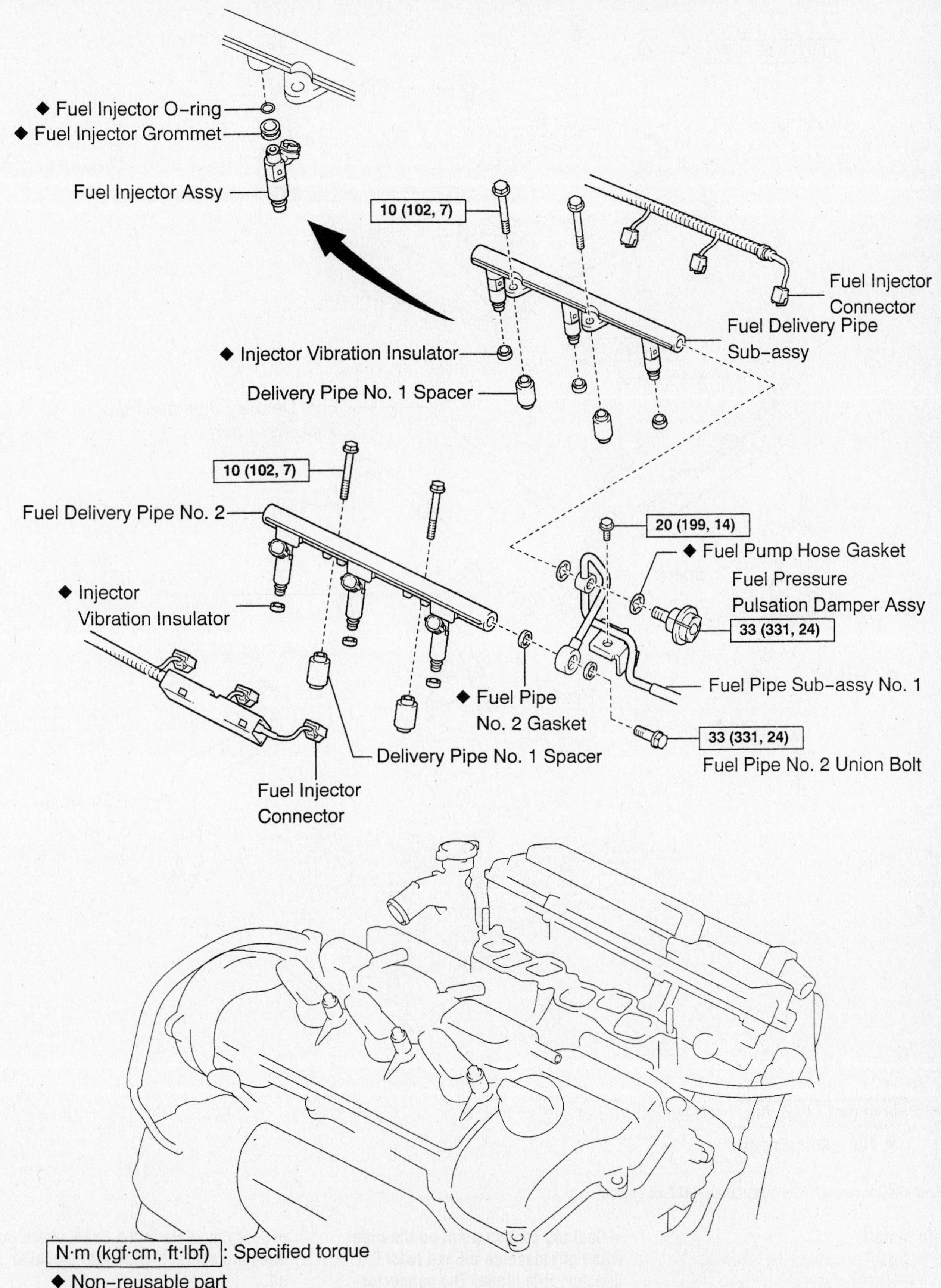

Exploded view of fuel injector delivery pipe and injectors—3MZ-FE engine

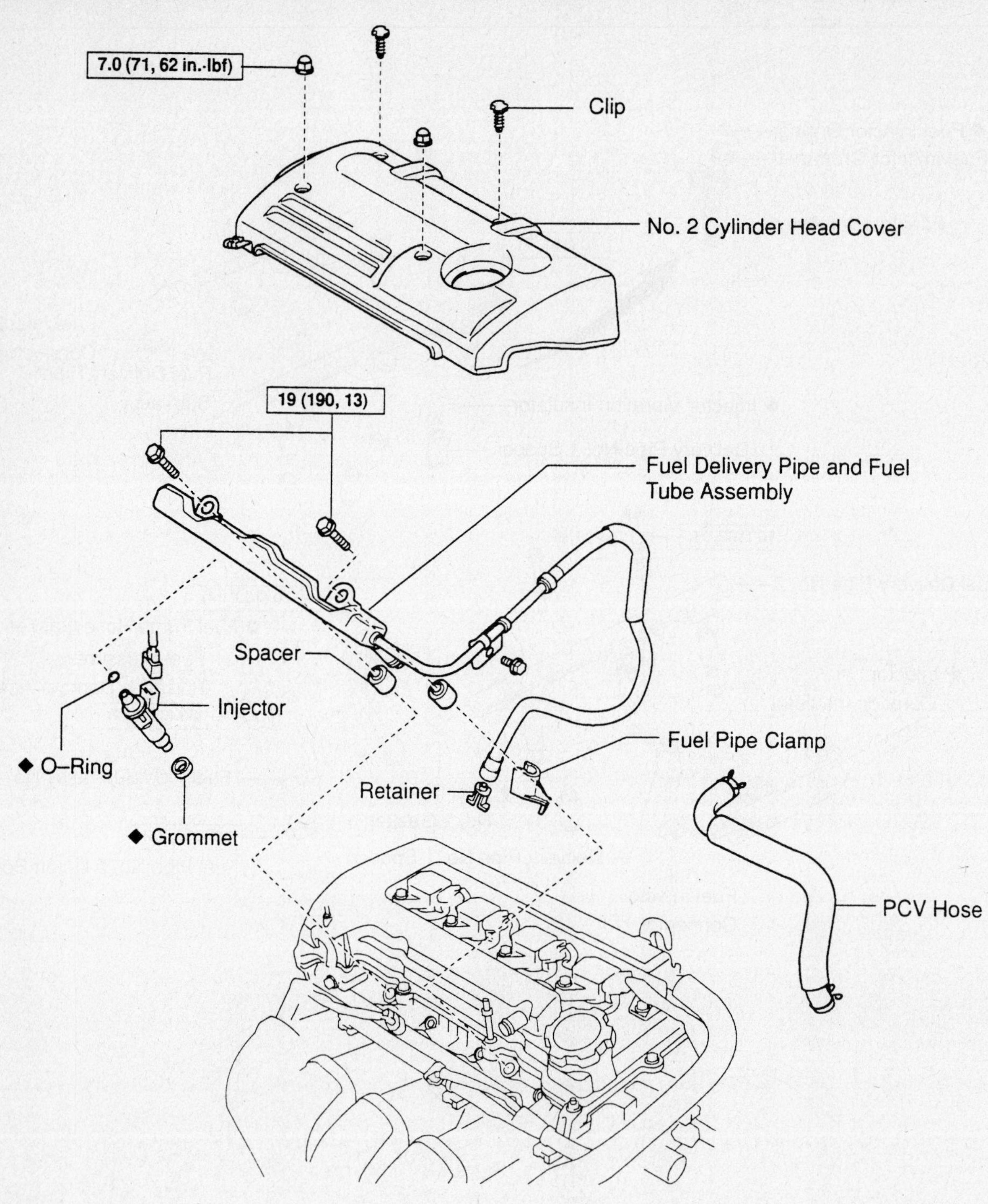

Fuel injector removal and installation—1ZZ-FE engine

To install:

4. Install or connect the following:
   - New grommets
   - New O-rings coated with light machine oil
   - Injectors on the delivery pipe

➡Coat the contact point on the pipe with light machine oil and twist the injectors into place. The connector should face outward.

- Spacers

➡Coat the seats in the head where the injectors contact, with light machine oil.

- Delivery pipe and injectors

5. Loosely install the hold-down bolts

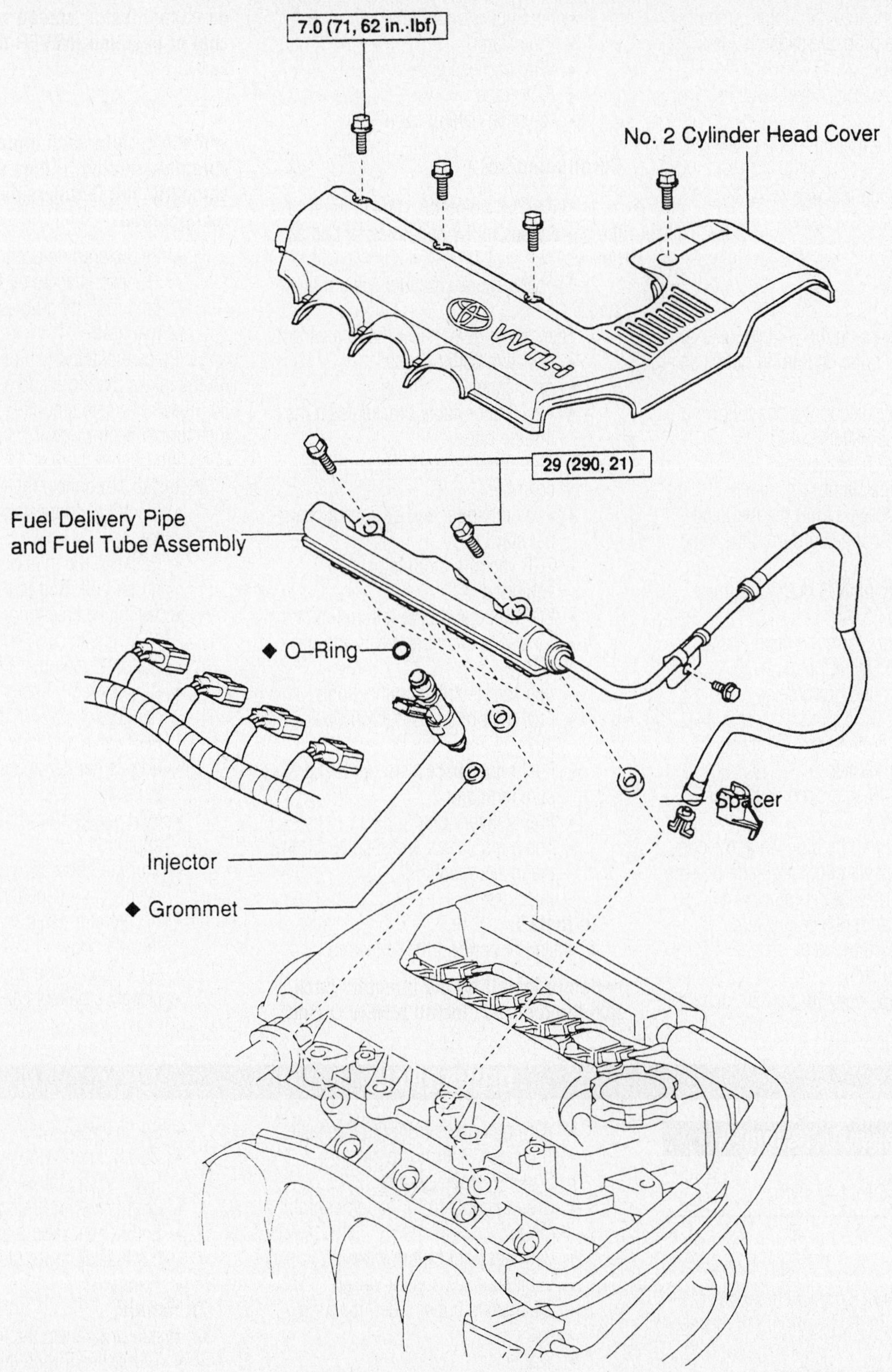

Fuel injector removal and installation—2ZZ-GE engine

and check that the injectors rotate smoothly. If they don't, the probable cause is incorrect O-ring installation.

6. Torque the hold-down bolts to 21 ft. lbs. (29 Nm).
7. Torque the fuel pipe bolt to 84 inch lbs. (9 Nm).
8. Connect the fuel line.
9. Install the PCV hose.
10. Install the No. 2 cover.

Echo

1. Before servicing the vehicle, refer to the precautions in the beginning of this section.
2. Remove or disconnect the following:
   - Negative battery cable
   - Cylinder head cover
   - Fuel pipe clamp
   - Fuel inlet line from the fuel pipe
   - Injector connectors from the injectors
   - Delivery pipe (3 bolts) and injectors
   - 2 spacers from the head
   - Injectors from the pipe
   - O-rings and grommets

To install:

3. Install or connect the following:
   - New grommets
   - New O-rings coated with clean engine oil
   - Injectors to the pipe. Coat the contact area with light machine oil. The injectors twist into place. The connector should face outward
   - Delivery pipe bolts. Torque: 14 ft. lbs. (19 Nm).
   - Fuel pipe bolt. Torque: 80 inch lbs. (9 Nm).
   - Fuel hose to fuel pipe
   - Pipe clamp
   - Wire harness cover
   - PCV hose
   - Negative battery cable

Corolla and MR2

1. Before servicing the vehicle, refer to the precautions in the beginning of this section.
2. Properly relieve the fuel system pressure.
3. Remove or disconnect the following:
   - Negative battery cable
   - Air cleaner
   - Accelerator cable bracket from the throttle body
   - Throttle body from the air intake chamber
   - Engine hanger and air intake chamber stay
   - EGR vacuum modulator if so equipped
   - EGR valve and pipe if so equipped
   - Air intake chamber cover and gasket
   - Injector electrical connections
   - Fuel inlet hose from the delivery pipe
   - Fuel return hose from the fuel pressure regulator
   - Fuel delivery pipe (rail)
   - The 4 insulators and 2 collars from the intake manifold
   - Injectors

To install:

4. Install or connect the following:

➡Before installing the injectors back into the fuel rail, install a NEW O-ring on each injector, coated with a light coat of gasoline (NEVER use oil of any sort).

   - Injectors

➡Make certain each injector can be smoothly rotated. If they do not rotate smoothly, the O-ring is not in its correct position.

   - Insulators into each injector hole
   - The two spacers on the delivery pipe mounting holes in the intake manifold

5. Place the delivery pipe and injectors on the intake manifold and again check that the injectors rotate smoothly. Position the injector connector upward. Install the two bolts and tighten them to 11 ft. lbs.
6. Install or connect the following:
   - Electrical connectors to each injector
   - Gaskets, the inlet pipe and fuel union bolt. Bolt to 22 ft. lbs.
   - Air intake chamber cover with a NEW gasket. Torque the retaining bolts in steps to 14 ft. lbs.
   - All necessary hoses and electrical connections
   - EGR valve and pipe if so equipped
   - Engine hanger and air intake chamber stay
   - EGR vacuum modulator if so equipped
   - Throttle body. Torque the bolts evenly (in a X-pattern) to 16 ft. lbs.
   - Accelerator cable bracket to the throttle body
   - Air cleaner hose and cap
   - Negative battery cable

DRIVE TRAIN

Manual Transaxle Assembly

REMOVAL & INSTALLATION

Echo

1. Before servicing the vehicle, refer to the precautions in the beginning of this section.
2. Drain the transaxle fluid.
3. Remove or disconnect the following:
   - Hood
   - Wiper arms
   - Right and left cowl top ventilator covers
   - No. 2 cylinder head cover
   - Battery
   - Air cleaner assembly
   - Wiring harness from the transaxle
   - Transaxle control cable
   - Clutch release cylinder
   - Ground cable from the left engine mount
   - Back-up light switch wires
   - Vehicle speed sensor wiring
   - 2 transaxle upper side mounting bolts
   - Starter
4. At this point, attach an engine crane to support the engine.
5. Remove or disconnect the following:
   - Left side engine under cover
   - Both halfshafts
   - 2 bolts and 1 nut securing the engine rear mount to the cross-member
   - Sliding yoke
   - Power steering hoses
   - Support the transaxle
   - Engine left mounting bracket
   - Engine rear mount and bracket
   - 5 transaxle lower side mount bolts
   - Transaxle

To install:

6. Install or connect the following:
   - Transaxle. Torque the 5 lower bolts to 25 ft. lbs. (33 Nm).
   - Engine rear mount and bracket. Torque the mount bolt and nut to 47 ft. lbs. (64 Nm); the bracket bolts to 36 ft. lbs. (49 Nm).
   - Engine left mounting bracket. Torque the bolts to 36 ft. lbs. (49 Nm).

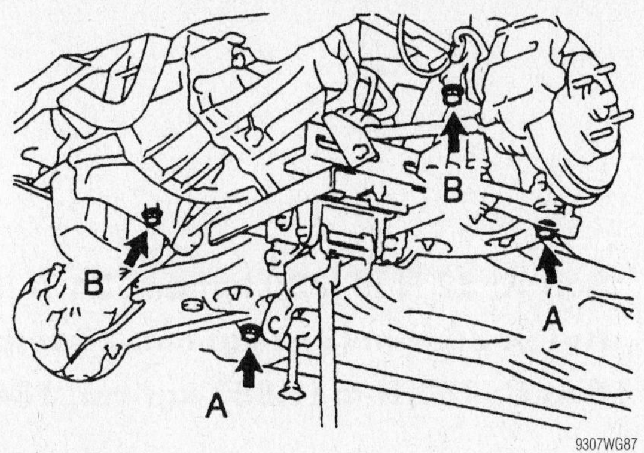

Sub-frame installation—Echo

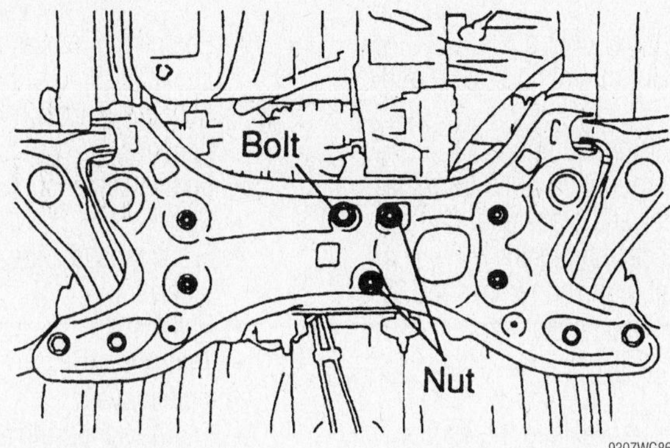

Crossmember installation—Echo

- Power steering hoses
- Sliding yoke
- 2 bolts and 1 nut securing the engine rear mount to the crossmember. Torque the bolts to 36 ft. lbs. (49 Nm).
- Both halfshafts
- Left side engine under cover
- Starter. Torque the bolts to 29 ft. lbs. (39 Nm).
- 2 transaxle upper side mounting bolts. Torque the bolts to 25 ft. lbs. (33 Nm).
- Vehicle speed sensor wiring
- Back-up light switch wires
- Ground cable from the left engine mount
- Clutch release cylinder
- Transaxle control cable
- Wiring harness from the transaxle
- Air cleaner assembly
- Battery
- No. 2 cylinder head cover
- Right and left cowl top ventilator covers
- Wiper arms
- Hood

7. Fill the transaxle to the proper level.
8. Start the vehicle, check for leaks and repair if necessary.

Camry and Solara

1. Before servicing the vehicle, refer to the precautions in the beginning of this section.
2. Drain the transaxle fluid.
3. Remove or disconnect the following:
   - Negative battery cable. On vehicles equipped with an air bag, wait at least 90 seconds before proceeding.
   - Air cleaner
   - With cruise control, cruise control actuator
   - Clutch release cylinder and tube clamp
   - Starter
   - Back-up light switch connector and ground strap
   - Wires clamp
   - Clips and washers that attach the transaxle control cables to the control levers
   - Transaxle control cables.
   - Speed sensor connector
   - Undercovers
   - Left and right halfshafts
   - 4 steering gear housing bolts.
   - Stabilizer bar bushing bracket
   - 2 set bolts and nuts
   - Steering gear box from the suspension member and suspend it securely
   - Exhaust pipe
   - Stiffener plate
   - Engine front mounting from the suspension member
   - Engine rear mounting from the front suspension member
   - Left engine mounting
   - Steering cooler pipe from the suspension member
   - 2 fender liner set screws
   - The 2 bolts and 4 nuts located on the outside of the suspension member brackets
   - The 4 larger bolts holding the suspension member to the vehicle body
   - The 2 front lower braces, rear braces, and the front suspension member.
   - Transaxle

To install:

4. Move the transaxle into position so that the input shaft spline is aligned with the clutch disc.
5. Install or connect the following:
   - Transaxle into the engine and secure with the lower mounting bolts. Bolts: 10mm mounting bolts: 47 ft. lbs. (63 Nm). 12mm bolts to 34 ft. lbs. (46 Nm).
   - Front suspension member and the 2 front lower braces and rear lower braces. 4 large bolts that hold the suspension member to the vehicle: 134 ft. lbs. (181 Nm); 2 outside bolts and 4 outside nuts: 24 ft. lbs. (32 Nm).
   - 2 fender liner set screws
   - Steering cooler pipe to the suspension member
   - Engine left mount. Bolts: 38 ft. lbs. (52 Nm); 2 nuts and 2 grommets. Nuts: 59 ft. lbs. (80 Nm).
   - Engine rear mounting to the front suspension member. Nuts: 59 ft. lbs. (80 Nm).
   - Engine front mounting to the suspension member. Bolt: 59 ft. lbs. (80 Nm).
   - Stiffener plate. Bolts: 27 ft. lbs. (37 Nm).
   - Exhaust pipe
   - Steering gear housing to the front suspension member. Bolts and nuts: 134 ft. lbs. (181 Nm).
   - Stabilizer bar bushing bracket. 4 bolts: 14 ft. lbs. (19 Nm).
   - Right and left halfshafts
   - Undercovers
   - Vehicle speed sensor

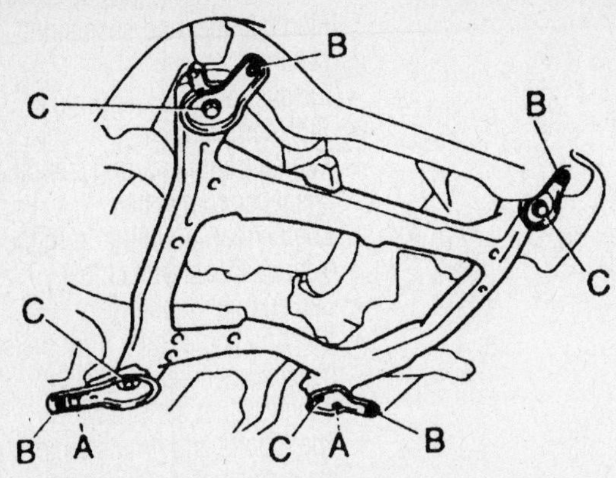

Front suspension member and fastener locations—Camry

- Control cables by installing the washers and clips
- Clamp that retains the wires to the transaxle
- Back-up light switch connector and ground cables
- Starter. Bolts: 29 ft. lbs. (39 Nm).
- Pipe clamp and clutch release cylinder to the transaxle. Bolts: 108 inch lbs. (13 Nm).
- Cruise control actuator
- Air cleaner
- Negative battery cable

6. Fill the transaxle fluid to the proper level.
7. Start the vehicle, check for leaks and repair if necessary.

Celica

1. Before servicing the vehicle, refer to the precautions in the beginning of this section.
2. Drain the transaxle fluid.
3. Remove or disconnect the following:
   - Negative battery cable. On vehicles equipped with an air bag, wait at least 90 seconds before proceeding
   - Air cleaner case assembly with hose
   - Release cylinder tube bracket
   - Clutch release cylinder
   - Back-up light switch connector
   - Ground cable on the transaxle
   - Shift cables from the transaxle
   - Vehicle speed sensor connector or the speedometer cable
   - Engine wire clamps
   - Starter set bolt from the transaxle upper side
   - Undercovers
   - Halfshafts
   - Front exhaust pipe and support bracket
   - Starter
   - Engine center support member
   - Engine rear mounting
   - Engine front mounting bracket and insulator
   - Engine left mounting bracket
   - Transaxle mounting bolts from the engine rear end plate side.
   - Transaxle case protector
   - Engine left side and remove the 3 upper transaxle bolts.
   - Transaxle

To install:

4. Position the transaxle to the engine and raise the engine right side. Align the input shaft with the clutch disc
5. Install or connect the following:
   - Transaxle to the engine. 3 upper transaxle bolts: 47 ft. lbs. (64 Nm).
   - Transaxle case protector. Bolts: 108 inch lbs. (13 Nm).
   - 4 transaxle lower bolts. Bolt A: 17 ft. lbs. (23 Nm); Bolt B: 34 ft. lbs. (46 Nm).
   - Left engine mounting bracket to the engine left mounting insulator. Bolts: 47 ft. lbs. (64 Nm).
   - Engine front mounting bracket and insulator. 2 bracket bolts: 57 ft. lbs. (77 Nm). Through-bolt: 64 ft. lbs. (87 Nm).
   - Engine rear mounting bracket and insulator. Bracket bolts: 57 ft. lbs. (77 Nm). Through-bolt: 64 ft. lbs. (87 Nm).
   - Engine center support member
   - Starter
   - Front exhaust pipe and support bracket
   - Halfshafts
   - Transaxle oil
   - Undercovers
   - Engine support fixture
   - Starter set bolt to the transaxle upper side. Bolt: 29 ft. lbs. (39 Nm).
   - Engine wire clamps
   - Vehicle speed sensor connector or the speedometer cable
   - Transaxle shift cables and ground cable
   - Back-up light switch connector
   - Release cylinder
   - Air cleaner case assembly
   - Negative battery cable
6. Fill the transaxle fluid to the proper level.
7. Start the vehicle, check for leaks and repair if necessary.

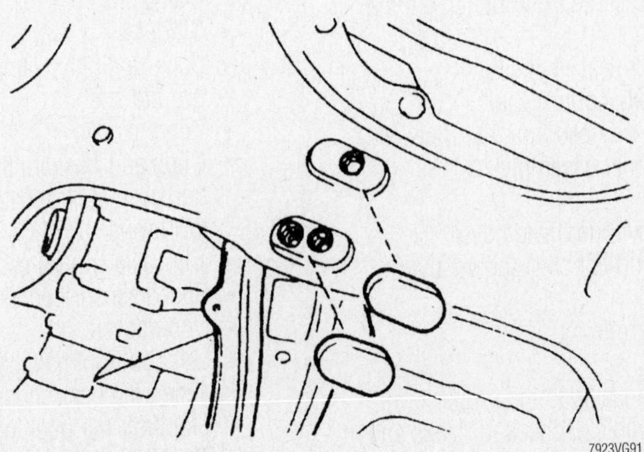

Rear mounting insulator set bolt locations—Celica

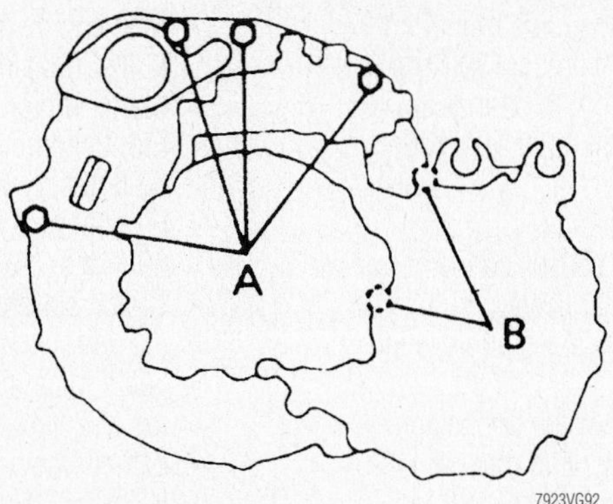

Upper transaxle mounting bolt locations—Celica

Celica GT-S

1. Before servicing the vehicle, refer to the precautions in the beginning of this section.
2. Drain the transaxle fluid.
3. Remove or disconnect the following:
   - Hood
   - No. 2 cylinder head cover
   - Radiator overflow bottle
   - Battery
   - Air cleaner assembly
   - ECM box
   - Wiring harness
   - Battery tray
   - Transaxle control cable
   - Engine ground cable
   - Speed sensor connector
   - Back-up light switch connector
   - Clutch release cylinder
   - Starter
   - Transaxle control cable bracket
   - 2 upper transaxle mounting bolts
4. Attach an engine crane to the engine.
5. Remove or disconnect the following:
   - Left engine mount bracket
   - Engine under covers
   - Both halfshafts
   - Front exhaust pipe
6. Unbolt the steering rack from the sub-frame and suspend it.
7. Remove or disconnect the following:
   - Stabilizer bar
   - Wheels
   - Ball joints from the lower arms
   - Lower arms and sub-frame
   - Engine rear mount and bracket
8. Raise the engine slightly.
9. Remove or disconnect the following:
   - 4 lower transaxle side bolts
   - Transaxle from the engine

To install:

10. Install or connect the following:
    - Transaxle to the engine. Install the 4 lower side bolts. Torque the 2 bottom bolts to 17 ft. lbs. (23 Nm); the 2 side bolts to 35 ft. lbs. (48 Nm).
    - Rear mount and bracket. Torque the bracket bolts to 47 ft. lbs. (64 Nm); the mount bolt to 64 ft. lbs. (87 Nm).
    - Lower control arms-to-frame. Bolts: 101 ft. lbs. (137 Nm); tie rod ball stud nuts: 36 ft. lbs. (49 Nm).
11. Install the sub-frame. Torque the bolts as illustrated:
    a. A and B: 38 ft. lbs. (52 Nm)
    b. C: 116 ft. lbs. (157 Nm)
    c. D: 29 ft. lbs. (39 Nm)
12. Install or connect the following:
    - Steering rack. Bolts: 43 ft. lbs. (58 Nm).
    - Front exhaust pipe. Bolts: 32 ft. lbs. (43 Nm).
    - Halfshafts
    - Under covers
    - Left mount bracket. Bolt: 44 ft. lbs. (60 Nm).
    - Left mount. Bolts: 44 ft. lbs. (60 Nm); nut: 59 ft. lbs. (80 Nm).
    - Transaxle upper side mount bolts. Torque: 47 ft. lbs. (64 Nm).
    - Control cable bracket. Bolts: 18 ft. lbs. (25 Nm).
    - Starter. Bolts: 28 ft. lbs. (37 Nm).
    - Clutch release cylinder
    - Wiring
    - Control cable
    - Battery tray
    - ECM
    - Air cleaner
    - Battery
    - Reservoir
    - No. 2 cylinder head cover
    - Hood
13. Fill the transaxle fluid to the proper level.
14. Start the vehicle, check for leaks and repair if necessary.

Corolla

1. Before servicing the vehicle, refer to the precautions in the beginning of this section.
2. Drain the transaxle fluid.
3. Remove or disconnect the following:
   - Negative battery cable. On vehicles equipped with an air bag, wait at least 90 seconds before proceeding.
   - Air cleaner case assembly with hose
   - Coolant reservoir tank
   - Release cylinder tube bracket
   - Clutch release cylinder

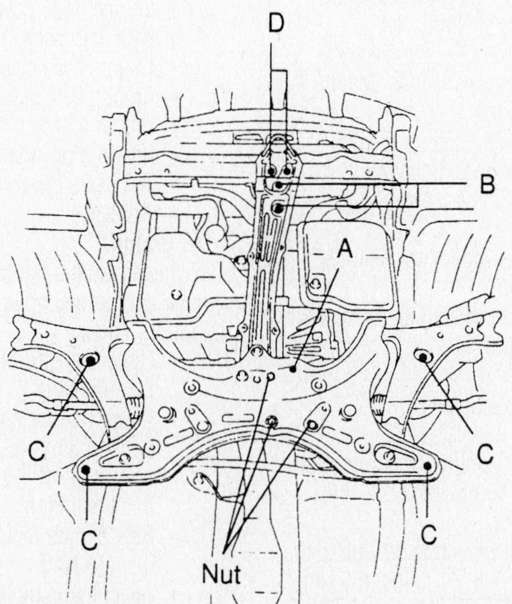

Celica sub-frame torques

- Back-up light switch connector
- Ground cable
- Shift cables from the transaxle
- Vehicle speed sensor connector or the speedometer cable
- Engine wire clamps
- Starter set bolt from the transaxle upper side
- 2 transaxle upper mounting bolts
- Engine left mounting stay
- Engine left mounting set bolt from the rear side
- Undercovers
- Lower ball joint from the lower arm.
- Halfshafts
- Front exhaust pipe
- Hole cover
- Engine front mounting set bolts
- Engine rear mounting
- Engine center support member
- Starter
- Transaxle mounting bolts from the engine rear end plate side
- Engine left mounting set bolts from the front side
- Transaxle mounting bolts from the engine front side, then engine rear side
- Transaxle

To install:

4. Align the input shaft with the clutch disc and install the transaxle to the engine. 12mm bolts: 47 ft. lbs. (64 Nm). 10mm bolts: 34 ft. lbs. (46 Nm).
5. Install or connect the following:
   - Left engine mount. Bolts: 41 ft. lbs. (56 Nm).
   - Transaxle mounting bolts to the engine rear end plate side. Bolts: 17 ft. lbs. (23 Nm).
   - Starter, lower bolt and electrical connector to the starter. Bolt: 29 ft. lbs. (39 Nm).
   - Engine center support member. Radiator support bolts: 45 ft. lbs. (61 Nm). Frame bolts: 152 ft. lbs. (206 Nm).
   - Engine rear mounting. Bolts: 35 ft. lbs. (48 Nm).
   - Engine front mounting. Bolts: 47 ft. lbs. (64 Nm).
   - Hole covers
   - Front exhaust pipe
   - Halfshafts
   - Lower ball joint to lower arm. Bolt and nuts: 105 ft. lbs. (142 Nm).
   - Undercovers
   - Engine left mounting set bolt to the rear side. Bolt: 41 ft. lbs. (56 Nm).
   - Engine left mounting stay. Bolt: 15 ft. lbs. (21 Nm).
   - 2 transaxle upper side mounting bolts. Bolts: 29 ft. lbs. (39 Nm).
   - Starter set bolt to the transaxle upper side. Bolt: 29 ft. lbs. (39 Nm).
   - Engine wire clamps
   - Vehicle speed sensor connector or the speedometer cable.
   - Transaxle shift cables and ground cable.
   - Back-up light switch connector.
   - Release cylinder and release cylinder tube bracket. Bolts: 108 inch lbs. (12 Nm).
   - Coolant reservoir tank
   - Air cleaner case assembly
   - Negative battery cable
6. Fill the transaxle fluid to the proper level.
7. Start the vehicle, check for leaks and repair if necessary.

MR2

1. Before servicing the vehicle, refer to the precautions in the beginning of this section.
2. Drain the transaxle oil.
3. Remove or disconnect the following:
   - Engine hood
   - Rear suspension upper brace
   - Air filter assembly
   - Battery and tray
   - Engine ground cable
   - Back-up light switch connector
   - Shift cables and bracket
   - Transaxle upper flange bolts
4. Attach an engine support fixture.
   - Left motor mount
   - Engine undercovers
   - Axle halfshafts
   - Clutch slave cylinder
   - Front motor mount and bracket
   - Starter motor
   - Rear motor mount and bracket
   - Transaxle lower flange bolts
   - Transaxle

To install:

5. Installation is the reverse of the removal procedure, while using the following torque values:
   - Upper transaxle flange bolts: 47 ft. lbs. (64 Nm)
   - Lower transaxle flange bolts: **A** bolts to 35 ft. lbs. (47 Nm) and **B** bolts to 17 ft. lbs. (23 Nm). Refer to the illustration.
   - Rear engine mount bracket: 58 ft. lbs. (78 Nm)
   - Rear engine mount bolts: 66 ft. lbs. (89 Nm)
   - Rear engine mount through bolt: 69 ft. lbs. (93 Nm)
   - Starter motor: 28 ft. lbs. (37 Nm)
   - Front engine mount bracket bolts: 56 ft. lbs. (78 Nm)
   - Front engine mount through bolt: 69 ft. lbs. (93 Nm)
   - Clutch slave cylinder: 108 Inch lbs. (12 Nm)
   - Left engine mount bracket: 38 ft. lbs. (52 Nm)
   - Left engine mount through bolt: 64 ft. lbs. (87 Nm)
   - Suspension upper brace bolts: 59 ft. lbs. (80 Nm)
   - Engine hood bolts: 15 ft. lbs. (20 Nm)
6. Fill the transaxle with the correct oil.

Upper transaxle flange bolts—MR2

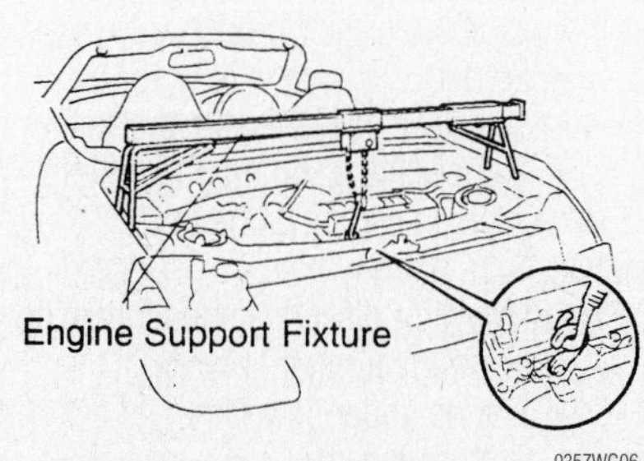

Engine support fixture—MR2

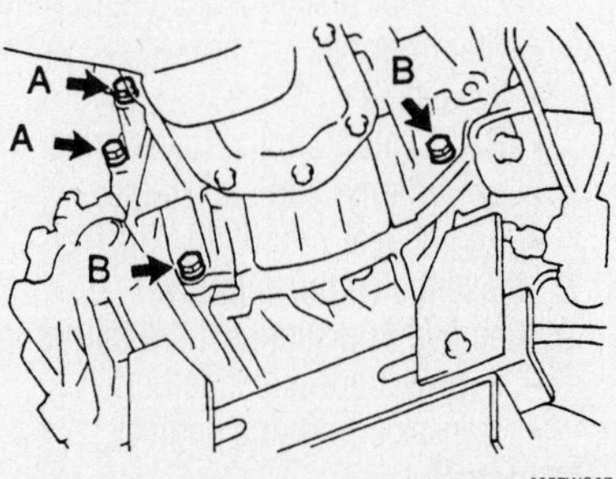

Lower transaxle flange bolts—MR2

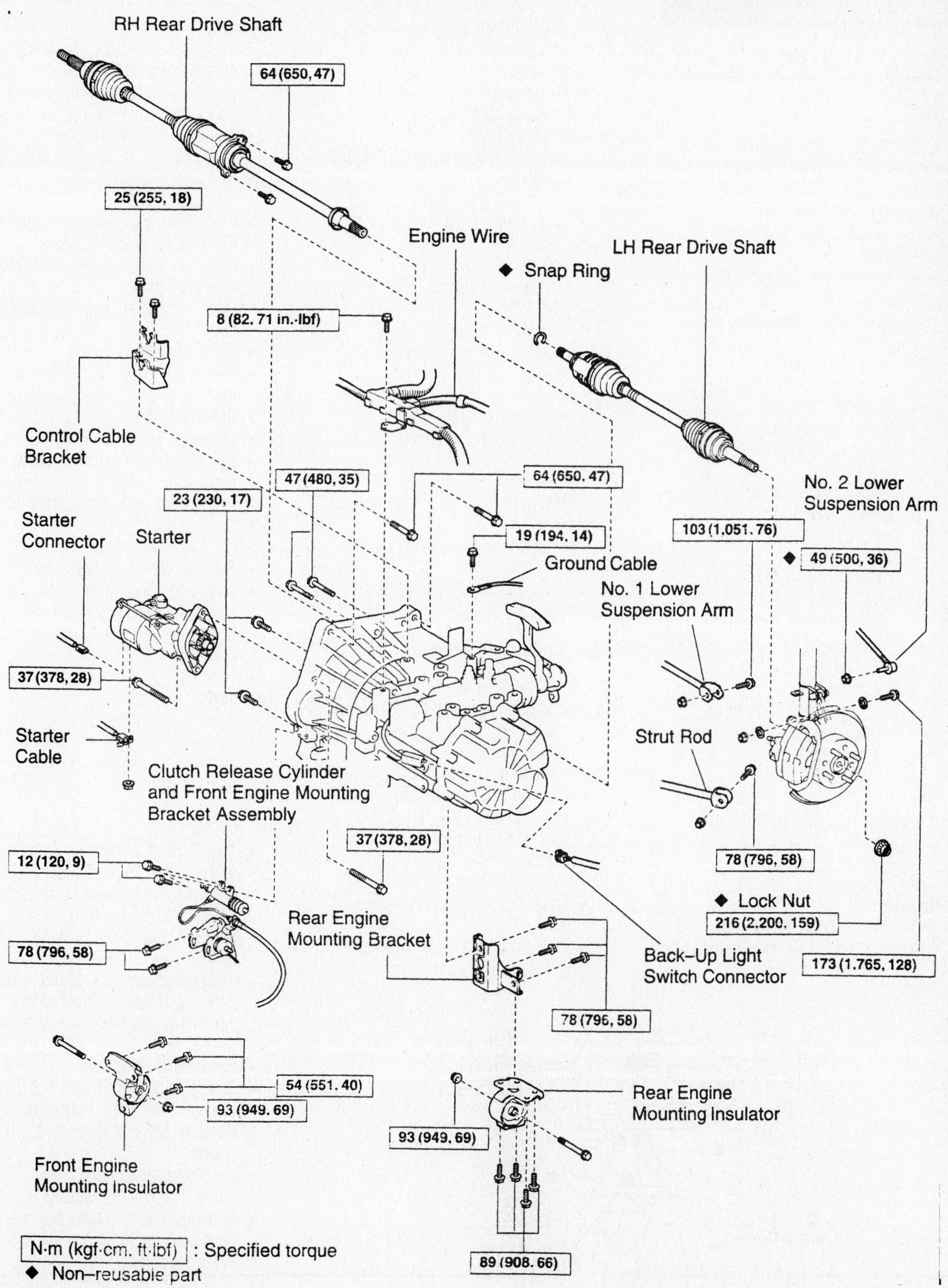
RH Rear Drive Shaft
64 (650, 47)
25 (255, 18)
Engine Wire
LH Rear Drive Shaft
◆ Snap Ring
8 (82.71 in.·lbf)
Control Cable Bracket
47 (480, 35)
64 (650, 47)
23 (230, 17)
No. 2 Lower Suspension Arm
Starter Connector
Starter
19 (194, 14)
103 (1,051.76)
Ground Cable
◆ 49 (500, 36)
No. 1 Lower Suspension Arm
37 (378, 28)
Strut Rod
Starter Cable
Clutch Release Cylinder and Front Engine Mounting Bracket Assembly
37 (378, 28)
12 (120, 9)
78 (796, 58)
◆ Lock Nut
216 (2,200, 159)
Rear Engine Mounting Bracket
78 (796, 58)
Back–Up Light Switch Connector
173 (1,765, 128)
78 (796, 58)
54 (551, 40)
93 (949, 69)
Rear Engine Mounting Insulator
93 (949, 69)
Front Engine Mounting Insulator
N·m (kgf·cm, ft·lbf) : Specified torque
◆ Non–reusable part
89 (908, 66)
9357WG04

Manual transaxle mounting exploded view—MR2

Automatic Transaxle

REMOVAL & INSTALLATION

Echo

1. Before servicing the vehicle, refer to the precautions in the beginning of this section.
2. Drain the transaxle fluid.
3. Remove or disconnect the following:
   - Hood
   - Right and left cowl top ventilator covers
   - No. 2 cylinder head cover
   - Battery
   - Air cleaner bracket
   - Wiring harness from the transaxle
   - Transaxle shift cable
   - Clutch release cylinder
   - Ground cable from the left engine mount
   - Park/Neutral switch wiring
   - Solenoid wiring
   - Direct Clutch Speed Sensor wiring
   - Filler pipe hose
   - 2 transaxle upper side mounting bolts
   - Engine under covers
   - Starter
   - Oil cooler hose
   - Both halfshafts
   - Torque converter access plug
   - Torque converter bolts and attach an engine crane to support the engine
   - Front suspension subframe (lower arms, steering gear and stabilizer)
   - 5 transaxle bolts
   - 2 transaxle mount bolts
   - Transaxle

To install:

4. Install or connect the following:
   - Transaxle. Torque the 5 bolts to 22 ft. lbs. (30 Nm).
   - 2 transaxle mount bolts. Torque the mount bolts to 36 ft. lbs. (49 Nm).
   - Torque converter bolts: 20 ft. lbs. (27 Nm).
   - Access plug
   - Subframe assembly
   - Oil cooler hoses
   - Starter. Torque the bolts to 29 ft. lbs. (39 Nm).
   - 2 transaxle upper side mounting bolts. Torque the bolts to 22 ft. lbs. (30 Nm).
   - Both halfshafts
   - Engine under covers
   - Filler hose
   - Transaxle control cable
   - Ground cable from the left engine mount
   - Wiring
   - Air cleaner bracket
   - No. 2 cylinder head cover
   - Right and left cowl top ventilator covers
   - Battery
   - Hood
5. Fill the transaxle fluid to the proper level.
6. Start the vehicle, check for leaks and repair if necessary.

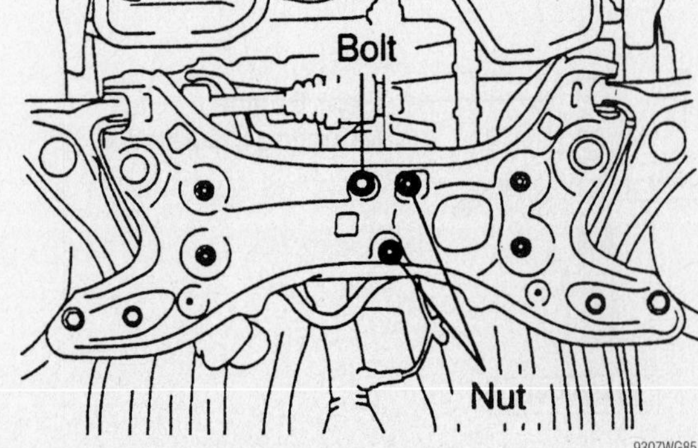

Rear mount installation—Echo

Avalon, Camry and Solara

1. Before servicing the vehicle, refer to the precautions in the beginning of this section.
2. Drain the transaxle fluid.
3. Remove or disconnect the following:
   - Negative battery cable. On vehicles equipped with an air bag, wait at least 90 seconds before proceeding
   - Battery
   - Air cleaner assembly
   - Throttle cable from the throttle body
   - Cruise control actuator cover and connector, if equipped
   - Ground wire
   - Starter
   - Speed sensor connectors, direct clutch speed sensor, and the park/neutral position switch connector on the transaxle
   - Solenoid connector on the transaxle
   - Shift control cable
   - Oil cooler hoses
   - Front side transaxle mounting bolts
   - Front engine mounting bolts
   - Oil cooler line mounting bolts from the front frame
   - Upper transaxle to engine mounting bolts
   - Front exhaust pipe
   - Engine side covers and undercovers
   - Both halfshafts
   - Front side engine mounting nut
   - Rear side engine mounting bolts (remove hole plugs)
   - Left side transaxle mounting bolts
   - Steering gear housing
   - Front frame assembly
   - Rear end plate mounting bolts
   - Torque converter cover
   - Torque converter retaining bolts
   - Remaining transaxle mounting bolts
   - Transaxle

To install:

4. Install or connect the following:
   - Transaxle aligning the 2 dowel pins on the block with the converter housing. 10mm bolts: 34 ft. lbs. (46 Nm); 12mm bolts: 47 ft. lbs. (64 Nm)
   - Torque converter bolts coated with sealer. Install the bolts starting with the green bolt followed by the rest. Bolts: 20 ft. lbs. (27 Nm).
   - Rear end plate. Bolts: 27 ft. lbs. (37 Nm).
   - Front frame assembly. 12mm bolts: 24 ft. lbs. (32 Nm); 19mm bolts: 134 ft. lbs. (181 Nm); Nut: 27 ft. lbs. (36 Nm).
   - Fender liner set screws
   - Steering gear to the frame. Bolts and nuts: 134 ft. lbs. (181 Nm).
   - Sway bar brackets. Bolts: 14 ft. lbs. (19 Nm).
   - Left transaxle mounting bolts. Bolts: 38 ft. lbs. (52 Nm).
   - Rear side mounting bolts and nuts. Bolts and nuts: 48 ft. lbs. (66 Nm). Install the plugs.
   - Front engine mounting nut. Nut: 59 ft. lbs. (80 Nm).
   - Halfshafts

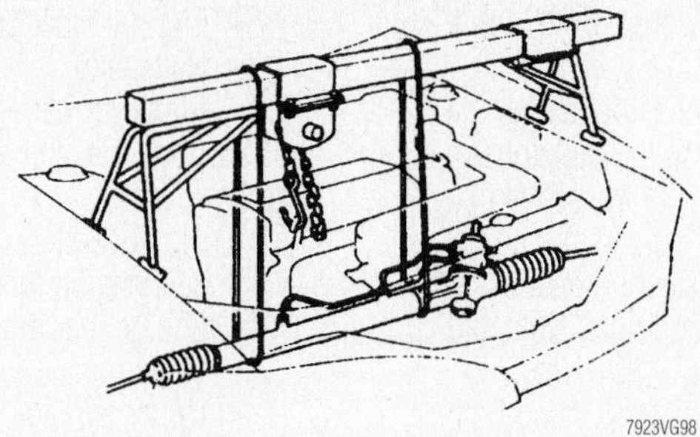

Tie the steering rack to the engine support fixture components, as shown—Avalon and Camry

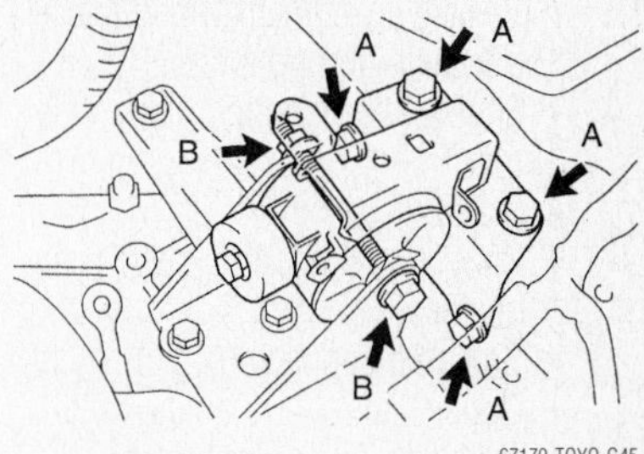

Mounting insulator fastener identification—Corolla

- Right and left engine side covers
- Lower engine cover
- Exhaust pipe. Nuts: 46 ft. lbs. (62 Nm).
- Exhaust pipe to the converter. Nuts and bolts: 32 ft. lbs. (43 Nm).
- Upper transaxle mounting bolts. Bolts: 47 ft. lbs. (64 Nm).
- Oil cooler clamping bolts to the front frame.
- Front side engine mounting bolts. Bolts: 59 ft. lbs. (80 Nm).
- Front side transaxle mounting bolts. Bolts: 59 ft. lbs. (80 Nm).
- Oil cooler hoses
- Shift control cable
- Solenoid electrical connector
- Park/neutral switch electrical connector
- Speed sensor and the direct clutch speed sensor connectors.
- Starter
- Ground strap
- Cruise control actuator and cover
- Throttle cable to the engine. Nuts: 11 ft. lbs. (15 Nm).
- Air cleaner
- Battery and battery cables.

5. Fill the transaxle to the proper level.
6. Start the vehicle, check for leaks and repair if necessary.

Corolla

1. Before servicing the vehicle, refer to the precautions in the beginning of this section.
2. Drain the transaxle fluid.
3. Remove or disconnect the following:
   - Negative battery cable. On vehicles equipped with an air bag, wait at least 90 seconds before proceeding
   - Negative battery cable from the transaxle
   - Transaxle level gauge
   - Reservoir tank and air cleaner assembly
   - Throttle cable from the bracket
   - Engine left mounting upper side bolts
   - Engine left mounting stay
   - Ground cable from the transaxle
   - Wiring harness clamp and throttle cable clamp
   - Undercovers
   - Left and right halfshafts
   - Front exhaust pipe
   - Engine support fixture
   - Mounting insulator
   - Starter
   - Vehicle speed sensor connector
   - Solenoid connector and park/neutral position switch connector. Remove the wiring harness clamps
   - Nut from the manual shift lever, then the control cable from the bracket by removing the clip
   - Oil cooler hoses
   - Transaxle filler tube
   - Converter cover
   - Torque converter bolts
   - Transaxle mounting bolts
   - Transaxle

To install:

4. Install or connect the following:
   - Transaxle. Tighten the bolts as follows: Bolt A: 47 ft. lbs. (64 Nm); Bolt B: 34 ft. lbs. (46 Nm); Bolt C: 17 ft. lbs. (23 Nm)
   - Torque converter bolts to the transaxle. Bolts: 20 ft. lbs. (28 Nm).
   - Torque converter cover.
   - Transaxle filler pipe
   - Oil cooler hoses and replace the clips to their original positions

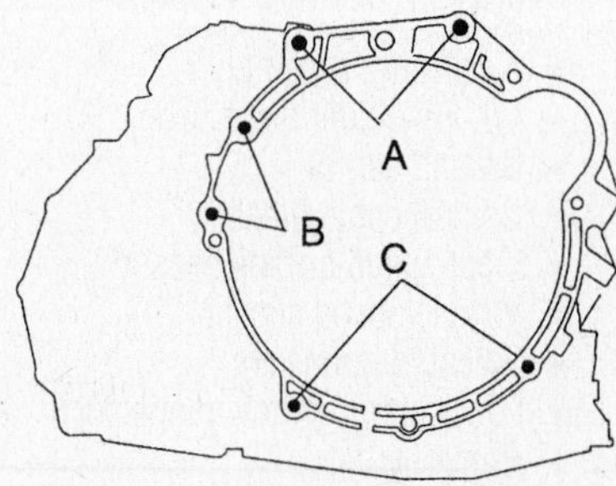

Transaxle mounting bolts—Corolla

   - Control cable for the transaxle to the bracket and install the clip
   - Control cable to the manual shaft lever by installing the nut
   - Solenoid connector and park/neutral position switch connector. Connect the wiring to the clamps
   - Vehicle speed sensor wiring
   - Starter. Bolt: 29 ft. lbs. (39 Nm).
   - Mounting insulator. Bolt A: 38 ft. lbs. (52 Nm); Bolt and nut B: 59 ft. lbs. (80 Nm)
   - Front exhaust pipe
   - Left and right halfshafts
   - Undercovers
   - Transaxle mounting bolts to the transaxle side
   - Wiring harness clamp and throttle cable clamp
   - Ground cable. Bolt: 7 ft. lbs. (10 Nm).
   - Engine left mounting upper side bolts and stay. Bolts: 38 ft. lbs. (52 Nm).
   - Throttle cable
   - Air cleaner and the reservoir tank
   - Transaxle level gauge
   - Negative battery cable

5. Fill the transaxle fluid to the proper level.
6. Start the vehicle, check for leaks and repair if necessary.

Celica

1. Before servicing the vehicle, refer to the precautions in the beginning of this section.
2. Drain the transaxle fluid.
3. Remove or disconnect the following:
   - Negative battery cable. On vehicles equipped with an air bag, wait at least 90 seconds before proceeding.
   - Throttle cable from the engine
   - Cruise control actuator. The cruise control actuator and bracket should be removed as an assembly.
   - Air cleaner assembly and battery.

- Vehicle speed sensor and the transaxle ground strap.
- Engine left mounting upper side bolt
- Starter
- Park/neutral position switch connector
- Solenoid connectors
- Upper transaxle retaining bolts
- Transaxle oil cooler hoses
- Undercover
- Both halfshafts
- Shift control cable from the control shaft lever and body bracket
- Engine rear mounting through-bolt
- Front exhaust pipe
- Air conditioner pipe bracket by removing the bolt
- Shift cable from the suspension member
- The 2 power steering gear assembly set bolts and nuts
- The 3 grommets from the center crossmember
- The 13 bolts and 2 nuts holding the suspension and center crossmembers
- Crossmembers from the vehicle
- No. 1 manifold stay
- Stiffener plate
- Torque converter bolts
- Transaxle

To install:

4. Install or connect the following:
   - Transaxle. 10mm bolt: 34 ft. lbs. (46 Nm); 12mm bolt: 47 ft. lbs. (64 Nm).
   - Torque converter bolts, with silicone applied to threads. Bolts: 18 ft. lbs. (25 Nm).
   - Stiffener plate. Bolts, alternately tightening: 12mm bolts: 15 ft. lbs. (21 Nm); 14mm bolts: 32 ft. lbs. (43 Nm).
   - No. 1 manifold stay. Bolt: 15 ft. lbs. (21 Nm). Nut: 32 ft. lbs. (43 Nm).
   - Raise the suspension member into position and install the 2 bolts to hold the suspension to the body. Bolts to 94 ft. lbs. (127 Nm).
   - The 3 bolts to hold the rear of the lower control arms to the subframe and body. Torque the bolt that goes through the lower control arm to 123 ft. lbs. (167 Nm) and the other 2 bolts to 130 ft. lbs. (175 Nm).
   - Side bolts
   - Center member
   - Engine rear mount and bracket. Nuts: 59 ft. lbs. (80 Nm); Bolt: 65 ft. lbs. (88 Nm).
   - Engine front mount. Bolts: 59 ft. lbs. (80 Nm).
   - The 2 front bolts connecting the center mount to the radiator support. Bolts: 26 ft. lbs. (35 Nm).
   - Grommets to the center member
   - The 2 power steering gear assembly set bolts and nuts. Nuts and bolts: 94 ft. lbs. (127 Nm).
   - The 2 shift cable mounting bolts
   - Air conditioner pipe bracket
   - Front exhaust pipe
   - Engine rear mounting bolt. Bolt: 64 ft. lbs. (88 Nm).
   - Shift control cable to the control shaft lever and body bracket. Install the clips
   - Left and right halfshafts
   - Undercovers
   - Oil cooler hoses with the 2 clips
   - Upper transaxle mounting bolts. Bolts: 47 ft. lbs. (64 Nm).
   - Solenoid connectors
   - Park/neutral position switch connector
   - Vehicle speed sensor connector and the ground strap to the transaxle
   - Starter. Bolts: 29 ft. lbs. (39 Nm).
   - Left mounting upper side bolt. Bolt: 47 ft. lbs. (64 Nm).
   - Air cleaner assembly
   - Cruise control actuator
   - Battery and cables
   - Throttle cable
5. Fill the transaxle fluid to the proper level.
6. Start the vehicle, check for leaks and repair if necessary.

Celica GT-S

1. Before servicing the vehicle, refer to the precautions in the beginning of this section.
2. Drain the cooling system.
3. Drain the transaxle fluid.
4. Remove or disconnect the following:
   - Hood
   - Battery
   - ECM
   - Air cleaner assembly
   - No. 2 cylinder head cover
   - Ground cables
   - Control cable bracket
   - Input speed turbine sensor
   - Vehicle speed sensor
   - Solenoid connector
   - Park/neutral switch connector
   - Control cable
   - Coolant reservoir
   - Starter
   - Engine under covers
   - Halfshafts
   - Stabilizer bar end links
5. Unbolt the steering rack and suspend it
   - 9 bolts and 3 nuts and lower the suspension member
   - Engine rear mount and attach and engine crane to support the engine.
   - Upper left side engine
   - Fluid cooler hoses and support the transaxle with a jack.
   - Torque converter access plug
   - Torque converter bolts
   - Transaxle bolts
   - Transaxle

To install:

6. Install or connect the following:
   - Transaxle. 2 lower bolts: 17 ft. lbs. (23 Nm); 2 lower side bolts: 34 ft. lbs. (46 Nm).
   - Torque converter bolts. Torque: 25 ft. lbs. (41 Nm).
   - Oil cooler hoses
   - Upper left side mount bolt and nut. Torque: 59 ft. lbs. (80 Nm).
   - Engine rear mount insulator. Bolt: 64 ft. lbs. (87 Nm).
7. Install the suspension member. Torque the bolts as illustrated:
   a. A: 116 ft. lbs. (157 Nm).
   b. B: 38 ft. lbs. (52 Nm).
   c. C: 29 ft. lbs. (39 Nm).
   d. Nut: 38 ft. lbs. (52 Nm).
8. Install or connect the following:
   - Steering rack. Bolts: 33 ft. lbs. (45 Nm).
   - Stabilizer bar end links. Torque: 32 ft. lbs. (44 Nm).
   - Halfshafts
   - Under covers
   - Starter. Torque: 28 ft. lbs. (37 Nm).
   - Coolant reservoir
   - Control cable
   - Wiring
   - Upper transaxle bolts. Torque: 47 ft. lbs. (64 Nm).
   - Control cable clamp
   - No. 2 cylinder head cover. Torque: 62 inch lbs. (7 Nm).
   - Air cleaner
   - ECM
   - Battery
   - Hood
9. Fill the cooling system.
10. Fill the transaxle to the proper level.
11. Start the vehicle, check for leaks and repair if necessary.

Clutch

REMOVAL & INSTALLATION

Camry, Solara and Echo

1. Before servicing the vehicle, refer to the precautions in the beginning of this section.
2. Remove or disconnect the following:
   - Negative battery cable. On vehicles equipped with an air bag, wait at least 90 seconds before proceeding.
   - Remove the transaxle assembly from the vehicle
   - Clutch pressure plate retaining bolts
   - Clutch cover
   - Clutch disc
   - Retaining clip and bearing from the transaxle
   - Release fork and boot assembly

To install:

3. Install or connect the following:
   - Clutch disc onto the flywheel
   - Clutch cover, aligning the matchmarks
   - Clutch cover retaining bolts. Bolts, tightened in a crisscross pattern: 14 ft. lbs. (19 Nm).
   - Boot, release fork, hub and bearing assemblies
   - Transaxle
   - Negative battery cable

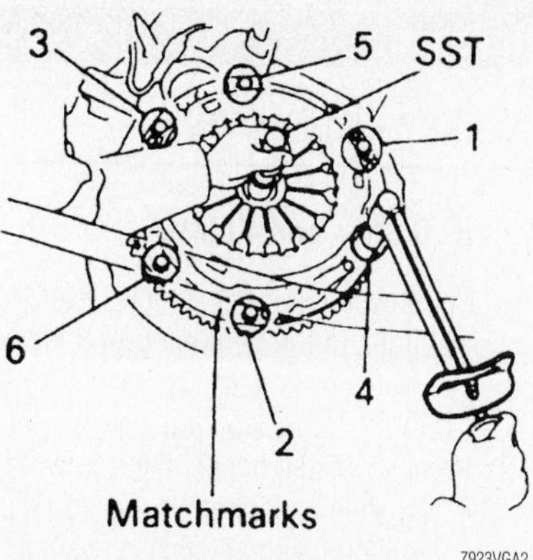

Tighten the pressure plate bolts according to the sequence shown—Camry

Corolla, Celica, and MR2

➡Do not allow grease or oil to get on any part of the disc, pressure plate, or flywheel surfaces.

1. Before servicing the vehicle, refer to the precautions in the beginning of this section.
2. Remove or disconnect the following:
   - Negative battery cable. On vehicles equipped with an air bag, wait at least 90 seconds before proceeding
   - Transaxle assembly
3. Make matchmarks on the clutch cover (pressure plate) and flywheel so that the pressure plate can be returned to its original position during installation.
4. Remove or disconnect the following:
   - Release fork bearing clips
   - Release bearing hub, complete with the release bearing
   - Release fork and support

✱✱ CAUTION

Slowly unfasten the bolts which attach the pressure plate. Loosen each bolt 1 turn at a time until the spring tension is released. If the bolts are released improperly the clutch assembly could fly apart, causing possible injury.

   - Pressure plate from the clutch cover/spring assembly
5. Inspect the disc, pressure plate and flywheel for damage and wear using a caliper to measure depth and width and a dial indicator to measure runout.
   a. The minimum clutch disc rivet head depth is 0.012 in. (0.3mm).
   b. The maximum clutch disc runout is 0.031 in. (0.8mm).
   c. The maximum pressure plate spring depth is 0.024 in. (0.6mm).
   d. The maximum pressure plate spring width is 0.197 in. (5.0mm).
   e. The maximum flywheel runout is 0.004 in. (0.1mm).
6. Replace or machine parts as necessary.

To install:

7. When reassembling, apply a thin coating of multipurpose grease to the release bearing hub and release fork contact

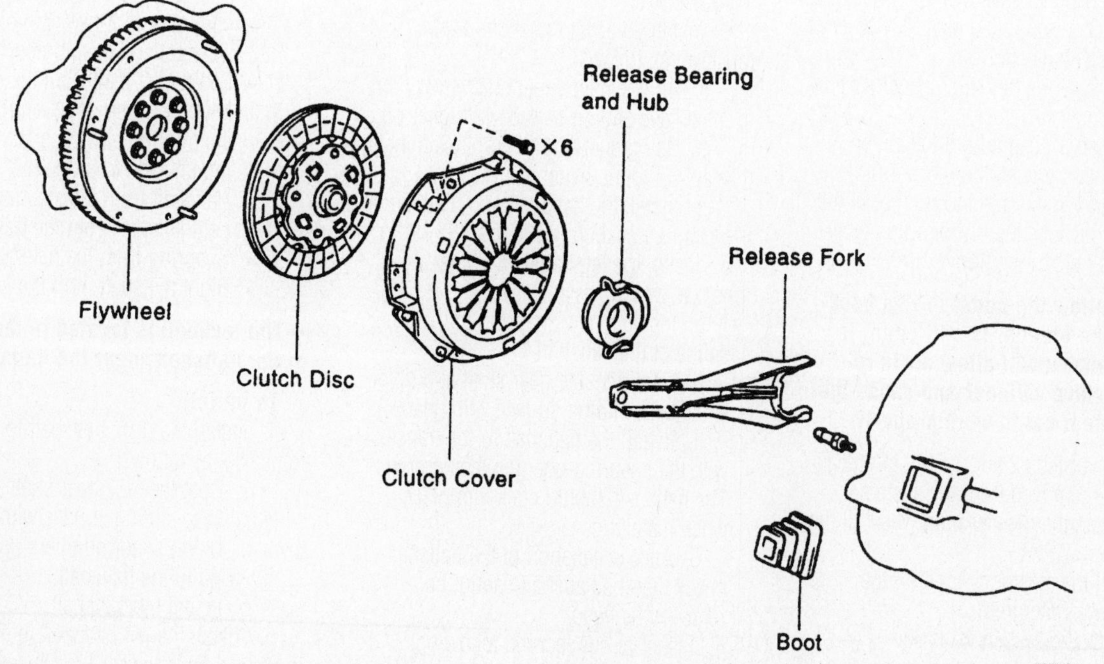

Clutch component assembly—Camry shown, others similar

points. Also, pack the groove inside the clutch hub with multipurpose grease and lubricate the pivot points of the release fork.

8. Install or connect the following:
   - Clutch disc and pressure plate. The bolts should be tightened in 2 or 3 steps, gradually and evenly. Final bolt torque is 14 ft. lbs. (19 Nm).
   - Release bearing, fork and boot
   - Transaxle assembly
   - Negative battery cable

ADJUSTMENTS

Hydraulic clutch actuating systems used in Toyota vehicles do not require adjustment.

Hydraulic Clutch System

BLEEDING

➡If any maintenance on the clutch system was performed or the system is suspected of containing air, bleed the system. Use care; brake fluid will remove the paint from any surface. If the brake fluid spills onto any painted surface, wash it off immediately with soap and water.

1. Before servicing the vehicle, refer to the precautions in the beginning of this section.
2. Fill the clutch reservoir with brake fluid. Check the reservoir level frequently and add fluid as needed.
3. Connect one end of a vinyl tube to the bleeder plug on the slave cylinder and submerge the other end into a clear container half-filled with brake fluid.
4. Slowly pump the clutch pedal several times.
5. Have an assistant hold the clutch pedal down and loosen the bleeder plug until fluid and/or air starts to run out of the bleeder plug. Close the bleeder plug while the pedal is held to the floor.

➡Do not allow the pedal to rise backup while the bleeder is still open. If this happens, it will allow air to reenter the slave cylinder and cause the clutch system not to work properly.

6. Repeat Steps 2 and 3 until all the air bubbles are removed from the system.
7. Tighten the bleeder plug when all the air is gone.
8. Refill the master cylinder to the proper level as required.
9. Check the system for leaks.

Halfshaft

REMOVAL & INSTALLATION

Avalon, Camry and Solara

1. Before servicing the vehicle, refer to the precautions in the beginning of this section.
2. Drain the transaxle fluid.
3. Remove or disconnect the following:
   - Negative battery cable. On vehicles equipped with an air bag, wait at least 90 seconds before proceeding.
   - Front fender apron seal
   - Tie rod end from the steering knuckle by removing the cotter pin and nut. Separate the tie rod from the steering knuckle
   - Stabilizer bar link from the lower control arm. Make note of the washers and cushions positions.
   - Lower ball joint from the steering knuckle. Push down on the lower control arm and separate the steering knuckle from the ball joint
   - Cotter pin, lock cap and locknut holding the halfshaft to the steering knuckle
   - Halfshaft from the steering knuckle
   - Left halfshaft from the transaxle
   - Snapring from the halfshaft.
   - Right halfshaft from the transaxle

➡The lockbolt is located in the center of the halfshaft, near the dampener.

To install:

4. Install the right halfshaft to the transaxle, as follows:
   a. Coat the side gear shaft and differential case sliding surface with gear oil.
   b. Using snapring pliers, install the snapring to the halfshaft.
   c. Install the halfshaft and the bearing lockbolt. Lockbolt: 24 ft. lbs. (32 Nm).
5. Install the left halfshaft to the transaxle, as follows:
   a. Install a new snapring to the inner spline of the halfshaft.
   b. Coat the side gear shaft and differential case sliding surface with gear oil.
   c. Install the halfshaft to the transaxle with the snapring opening facing down. The halfshaft should click into place when installing.
   d. After installation of the halfshaft, check that the halfshaft cannot be removed by hand.
6. Install or connect the following:
   - Halfshaft to the steering knuckle, then install the locknut. Locknut: 217 ft. lbs. (294 Nm).
   - Lock cap and new cotter pin to the halfshaft
   - Steering knuckle to the lower ball joint. Nuts and bolt: 94 ft. lbs. (127 Nm).
   - Stabilizer bar link to the lower control arm. Nut: 29 ft. lbs. (39 Nm).
   - Tie rod to the steering knuckle. Nut: 36 ft. lbs. (49 Nm).
   - New cotter pin to the tie rod end
   - Front fender apron seal
   - Front wheels
   - Negative battery cable
7. Fill the transaxle fluid to the proper level.
8. Start the vehicle, check for leaks and repair if necessary.

Echo

1. Before servicing the vehicle, refer to the precautions in the beginning of this section.
2. Drain the transaxle fluid.
3. Remove or disconnect the following:
   - Negative battery cable. On vehicles equipped with an air bag, wait at least 90 seconds before proceeding.
   - Both front wheels
   - Locknut holding the halfshaft to the steering knuckle
   - Tie rod end from the steering knuckle by removing the cotter pin and nut. Separate the tie rod from the steering knuckle
   - Stabilizer bar link from the lower control arm. Make note of the washers and cushions positions
   - Lower ball joint from the steering knuckle. Push down on the lower control arm and separate the steering knuckle from the ball joint
   - Halfshaft from the steering knuckle
   - Left halfshaft from the transaxle
   - Snapring from the halfshaft
   - Right halfshaft from the transaxle

➡The lockbolt is located in the center of the halfshaft, near the dampener.

To install:

4. Install the right halfshaft to the transaxle, as follows:
   a. Coat the side gear shaft and differential case sliding surface with gear oil.
   b. Using snapring pliers, install the snapring to the halfshaft.
   c. Install the halfshaft
5. Install the left halfshaft to the transaxle, as follows:

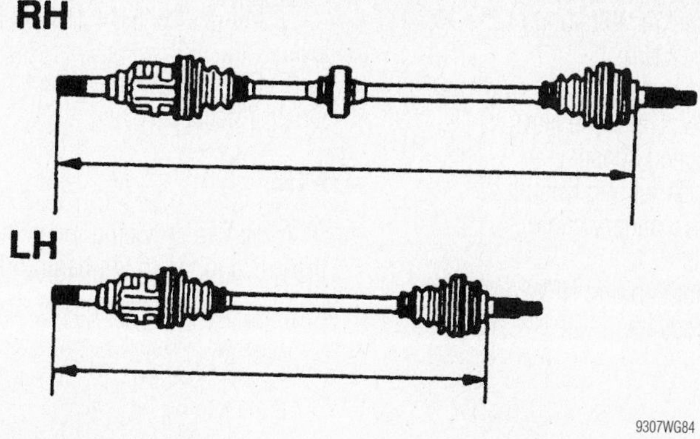

Measuring halfshaft length—Echo

a. Install a new snapring to the inner spline of the halfshaft.

b. Coat the side gear shaft and differential case sliding surface with gear oil.

c. Install the halfshaft to the transaxle with the snapring opening facing down. The halfshaft should click into place when installing.

d. After installation of the halfshaft, check that the halfshaft cannot be removed by hand.

6. Install or connect the following:
   - Halfshaft to the steering knuckle, then install the locknut. Locknut: 159 ft. lbs. (216 Nm).
   - Lock cap and new cotter pin to the halfshaft
   - Steering knuckle to the lower ball joint. Nut: 72 ft. lbs. (98 Nm).
   - Stabilizer bar link to the lower control arm. Nut: 29 ft. lbs. (39 Nm).
   - Tie rod to the steering knuckle. Nut: 36 ft. lbs. (49 Nm).
   - New cotter pin to the tie rod end
   - Both front wheels
   - Negative battery cable
7. Fill the transaxle fluid to the proper level.
8. Start the vehicle, check for leaks and repair if necessary.

Celica

➡The hub bearing could be damaged if subjected to the full weight of the vehicle, such as if the vehicle is moved without the halfshafts. If it is absolutely necessary to place the full vehicle weight on the hub bearing, first support the bearing with SST No. 09608–16041.

1. Before servicing the vehicle, refer to the precautions in the beginning of this section.
2. Drain the transaxle fluid.
3. Remove or disconnect the following:
   - Negative battery cable. On vehicles equipped with an air bag, wait at least 90 seconds before proceeding
   - Both front wheels
   - Cotter pin, locknut cap, and the bearing locknut
   - Undercovers
   - Tie rod ball joint from the steering knuckle
   - Stabilizer bar link from the lower suspension arm
   - Lower ball joint from the lower suspension arm
   - Halfshaft from the knuckle

➡Be careful not to damage the inner oil seal or the ABS sensor rotor on the halfshaft.

4. To remove the left side halfshaft, separate the halfshaft from the transaxle.
5. To remove the right side halfshaft perform the following steps:
   - Remove the 2 bolts of the center bearing bracket
   - Pull the halfshaft out together with the center bearing case and the center halfshaft.
   - Remove the center shaft with the right-hand halfshaft from the transaxle through the bearing bracket.

➡Do not damage the oil seal lip.

To install:

6. Install or connect the following:
   - Snapring opening side facing downward, on the oiled inboard joint tulip
   - Left side halfshaft into the transaxle
   - Right side halfshaft, with the bearing case and center shaft, into the transaxle
   - Center bearing case (right side). Bolts: 47 ft. lbs. (64 Nm)
7. After installing either halfshaft, check that there is 0.08–0.12 in. (2–3mm) of axial play. Check that the halfshaft is making contact with the pinion shaft and that the halfshaft cannot be pulled out.
8. Install or connect the following:
   - Halfshaft into the knuckle
   - Lower suspension arm to the lower ball joint. Bolt and nuts: 94 ft. lbs. (127 Nm).
   - Tie rod end to the steering knuckle. Nut: 36 ft. lbs. (49 Nm).
   - Stabilizer bar link to the lower suspension arm. Nuts: 33 ft. lbs. (44 Nm).
   - Front wheels
   - Locknut and washer. Locknut: 159 ft. lbs. (216 Nm).
   - Negative battery cable
   - Locknut cap and a new cotter pin.
   - Undercover
9. Fill the transaxle fluid to the proper level
10. Start the vehicle, check for leaks and repair if necessary.

Corolla

➡The hub bearing could be damaged if subjected to the full weight of the vehicle, such as if the vehicle is moved without the halfshafts. If it is absolutely necessary to place the full vehicle weight on the hub bearing, first support the bearing with SST No. 09608–16041.

1. Before servicing the vehicle, refer to the precautions in the beginning of this section.
2. Drain the transaxle fluid.
3. Remove or disconnect the following:
   - Negative battery cable. On vehicles equipped with an air bag, wait at least 90 seconds before proceeding
   - Cotter pin, locknut cap, and bearing locknut
   - Front wheels
   - Undercovers
   - With ABS, speed sensor
   - Tie rod ball joint from the steering knuckle
   - Lower ball joint from the lower suspension arm
4. Drive the halfshaft from the knuckle.

➡Most halfshafts can be separated from the knuckle using a brass or plastic hammer; some others may require the use of a puller.

5. Remove the halfshaft from the transaxle

To install:

6. Install or connect the following:
   - Snapring, opening side facing downward, to the inboard, oiled, joint tulip

- Halfshaft into the transaxle. After installing the halfshaft to the transaxle, check that there is 0.08–0.12 in. (2–3mm) of axial play. Check that the halfshaft is making contact with the pinion shaft and that the halfshaft cannot be pulled out
- Halfshaft into the knuckle
- Lower suspension arm to the steering knuckle. Nuts and bolts: 105 ft. lbs. (142 Nm).
- Tie rod end to the steering knuckle. Nut: 36 ft. lbs. (49 Nm).
- ABS speed sensor
- Hub locknut and washer
- Negative battery cable
- Wheels
- Locknut cap and NEW cotter pin.
- Undercovers

7. Fill the transaxle fluid to the proper level.
8. Start the vehicle, check for leaks and repair if necessary.

MR2

1. Before servicing the vehicle, refer to the precautions in the beginning of this section.

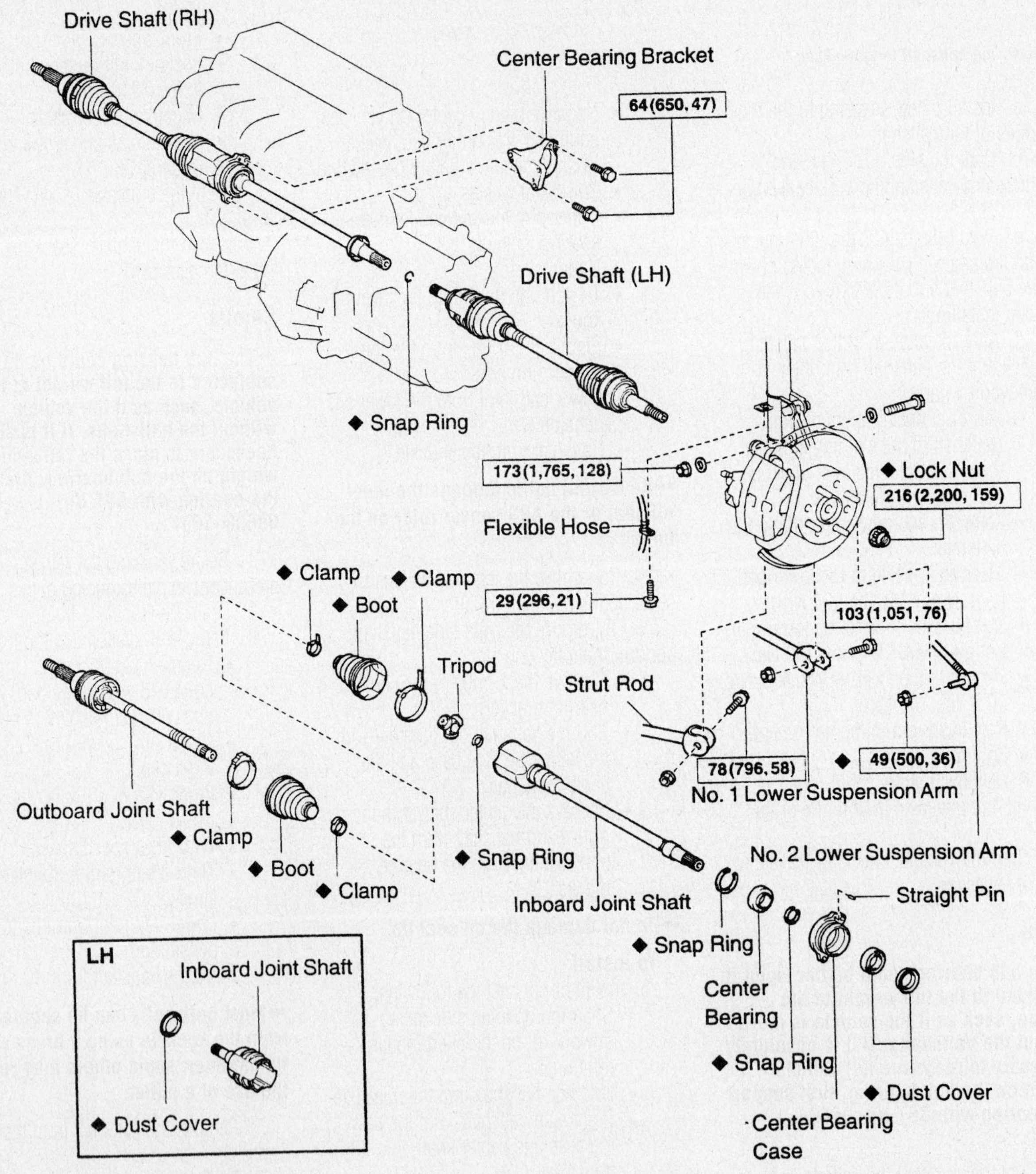

Axle halfshaft mounting exploded view—MR2

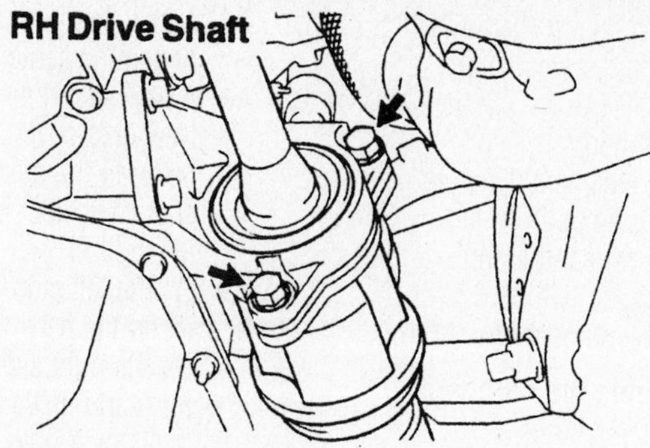

Axle halfshaft mounting detail—MR2

2. Remove or disconnect the following:
   - Rear wheel
   - Engine undercovers
   - Hub locknut
   - Brake fluid hose bracket
   - Strut rod
   - Lower suspension arms
   - Strut-to-spindle bolts
3. Drive the stub shaft from the axle hub with a plastic-faced hammer.
4. If removing the right halfshaft, remove the center bearing bracket bolts, then remove the axle halfshaft.
5. If removing the left halfshaft, drive the axle out of the transaxle with a hammer and brass punch.

To install:

➡Use new snap rings, circlips, and lock nuts for assembly.

➡Final tightening of suspension fasteners must take place with the suspension at curb height.

6. Install or connect the following:
   - Axle halfshaft. Tighten the center bearing bracket bolts to 47 ft. lbs. (64 Nm).
   - Strut-to-spindle-bolts and tighten them to 128 ft. lbs. (173 Nm).
   - Lower suspension arms. Tighten the No. 1 arm bolt to 76 ft. lbs. (103 Nm) and the No. 2 bolt to 36 ft. lbs. (49 Nm).
   - Strut rod. Tighten the bolt to 58 ft. lbs. (78 Nm).
   - Brake fluid hose bracket
   - Hub locknut: 159 ft. lbs. (216 Nm)
   - Engine undercovers
   - Rear wheel

CV-Joints

OVERHAUL

Avalon, Camry and Solara

1. Before servicing the vehicle, refer to the precautions in the beginning of this section.
2. Remove or disconnect the following:
   - Boot clamps
   - Inboard joint tulip
   - Snapring from the tri-pot
3. Place matchmarks on the driveshaft and tri-pot
4. Remove or disconnect the following:
   - Tri-pot joint off the driveshaft without hitting the joint roller
   - Inboard joint boot
   - On the right side, the clamp and dynamic damper
   - Clamps and the outboard drive boot. DO NOT disassemble the outboard joint
   - Dust cover from the inboard joint using a press

➡If equipped, be careful not to damage the ABS speed sensor rotor.

To assemble:

5. Install or connect the following:
   - Using a press, the inboard joint tulip into a new dust cover
   - On the right side, the clamp position in line with the groove of the halfshaft
   - Seal packing to the inboard joint cover
   - Inboard joint cover
   - Bolts and nuts and washers to keep the joint together. Tighten the bolts by hand
   - Outboard tulip joint and the outboard boot with about 4.8–5.5 ounces of grease
   - Boot onto the outboard joint
6. Pack the inboard tulip joint and boot with grease that was supplied with the boot kit.
7. Install or connect the following:
   - Inboard tulip joint onto the halfshaft
   - Boot onto the halfshaft
8. Before checking the standard length, bend the band and lock it.
9. Make sure that the boot is not stretched or squashed when the driveshaft is at standard length. Standard driveshaft length: 34.14 +/- 0.099 inch (867.3 +/- 2.5mm)
10. After making sure that the boots are in the shaft groove, secure the band
11. Pack in grease to the center driveshaft or side gear shaft. Use 5160 g (1.82.1 oz.)
12. Connect the driveshaft and the center driveshaft or the side gear shaft, placing a new gasket on the inboard joint without compressing the inboard boot.
13. Check to see that there is no play in the inboard and outboard joints and that the inboard joint slides smoothly in the thrust direction.

Echo

1. Before servicing the vehicle, refer to the precautions in the beginning of this section.

➡The outboard joint cannot be disassembled.

2. Remove or disconnect the following:
   - The 2 inboard boot clamps and slide the clamp down the shaft

➡Paint-mark the inboard joint shaft and tri-pot joint.

   - Inboard joint shaft from the outboard joint shaft

- Snapring from the tri-pot joint, and paint-mark the tri-pot and outboard joint shaft
- Tri-pot joint from the shaft with a brass hammer
- Inboard boot and clamps
- Damper (right shaft)
- Outboard joint boot
- Dust cover from the inboard joint shaft (press)

To assemble:

3. Install or connect the following:
   - Dust cover
   - Boots and clamps, loosely
   - Damper (right shaft)
   - Tri-pot joint, beveled edge toward the outboard shaft. Align the matchmarks and drive it into place
   - New snapring
   - Outboard boot. The grease capacity is 5.5–6 ounces (155–170 g).
   - Inboard shaft to outboard shaft
   - Inboard boot. The grease capacity is 4.4–4.8 ounces (125–135 g).

➡Toyota recommends different types of grease for the joints. OEM replacement boot kits have the grease color coded. The outboard grease is black; the inboard, yellow.

4. Check the boots at standard halfshaft length. The right shaft should be 32.02 inches +/- 0.197 in. (813.3mm +/- 5mm); the left shaft should be 22.61 inches +/- 0.197 in. (574.3mm +/- 5mm).

5. Check the position of the damper before installing the clamp. Distance from the outer face of the damper to the outer face of the outer joint should be 16.835 inches +/- 0.079 in. (427.6mm +/- 2mm).

Celica and MR2

1. Before servicing the vehicle, refer to the precautions in the beginning of this section.
2. Remove or disconnect the following:
   - Halfshaft
   - On the inboard joint tulip, the boot clamps. Slide the inboard joint boot toward the outboard joint.

➡Place matchmarks on the tri-pot and inboard joint tulip or center driveshaft. Do not punch the marks.

- Inboard joint tulip or center driveshaft from the driveshaft

3. On the tri-pot, use snapring expander, temporarily, slide the snapring toward the outboard joint side.

➡Place matchmarks on the driveshaft and tri-pot. Do not punch the marks.

4. Remove or disconnect the following:
   - Tri-pot from the driveshaft
   - Snapring
   - Inboard joint boot
   - On the right side with dynamic damper, the clamps from the damper and extract the unit
   - Clamps from the outboard joint
   - Boot from the outboard joint

➡Do not disassemble the outboard joint.

- On the left side, the dust cover from the inboard joint tulip
- On the right side, the dust cover from the center driveshaft
- On the right side, the snapring and bearing from the case
- The dust cover and snapring
- The bearing from the shaft and the other snapring
- The straight pin
- The No. 2 dust deflector

To assemble:

5. Install or connect the following:
   - On the right side, assemble the center driveshaft by inserting the straight pin into the bearing case
   - A new bearing into the case
   - A new snapring
   - A new bearing into the case assembly on the center driveshaft
   - A new snapring
   - New dust cover. The clearance between the dust cover and bearing should be kept in the ranges in the illustration.
   - On the left side, use a press and insert the new dust cover
   - On the right side, using a steel plate and press, press in a new dust cover until the distance from the tip of the center driveshaft to the cover reaches the specification as shown
6. Temporarily install boots and the damper. Before installing the outboard joint, wrap vinyl tape around the spline of the driveshaft to prevent damaging the boots and damper.
7. Install or connect the following:
   - Temporarily, the new outboard boot onto the driveshaft
   - On the dynamic damper, temporarily install the damper to the driveshaft.
   - Temporarily, a new inboard joint to the driveshaft
8. Install as follows:
   - A new snapring on the tri-pot joint

Place matchmarks on the tri-pot and outboard joints

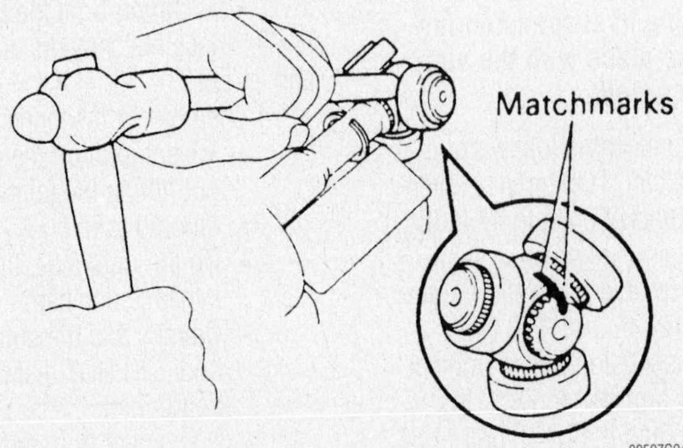

With matchmarks on the tri-pot, tap the joint for the driveshaft

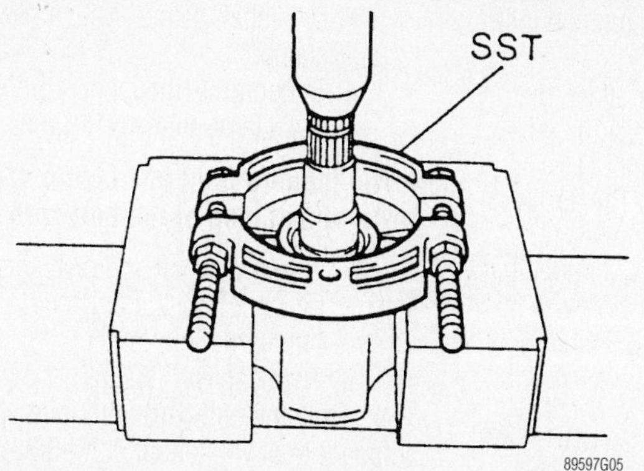

Using a press, remove the dust cover from the center driveshaft on the left side

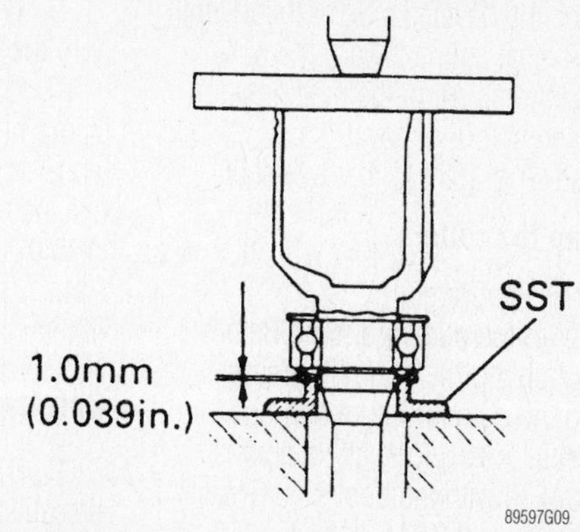

When installing the dust cover, the clearance should be within the ranges as specified

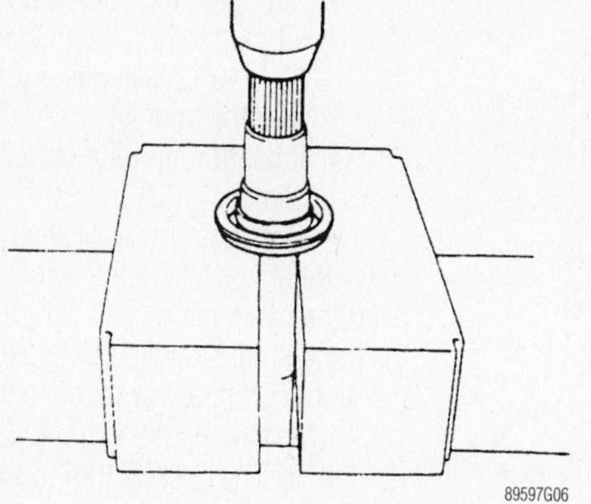

Use a press to drive the old dust cover out of the right side center driveshaft

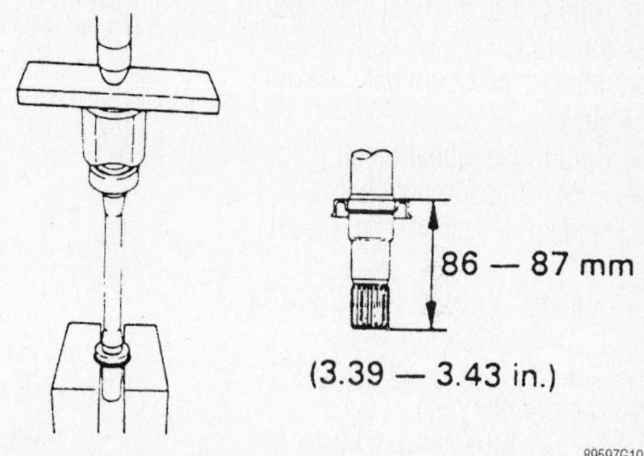

On the right side driveshaft, press the dust cover till the tip of the shaft reaches specifications

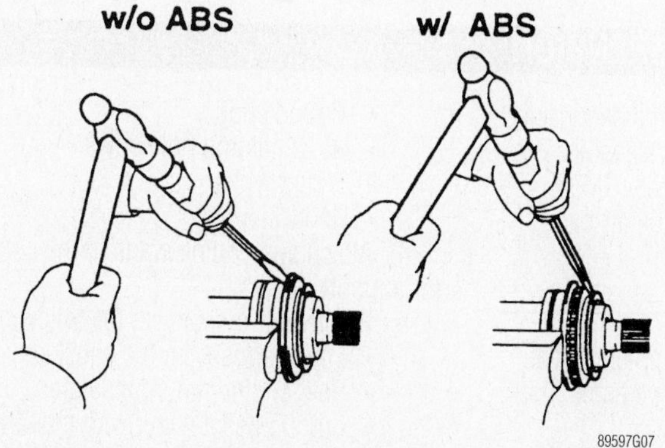

Removing the no. 2 dust deflector on models with and without ABS

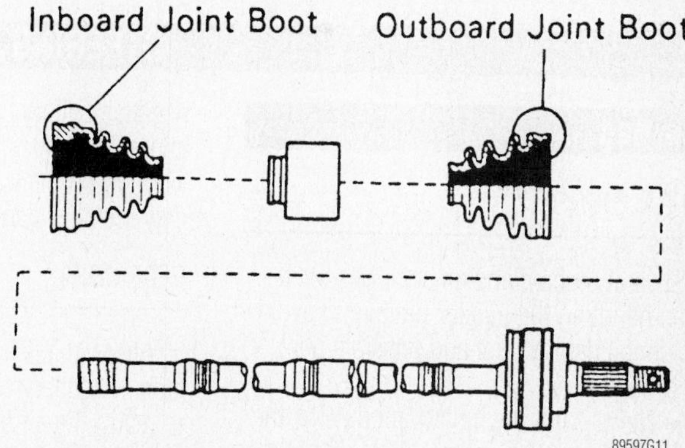

Temporarily install the driveshaft boots on the inboards and outboard and dynamic damper

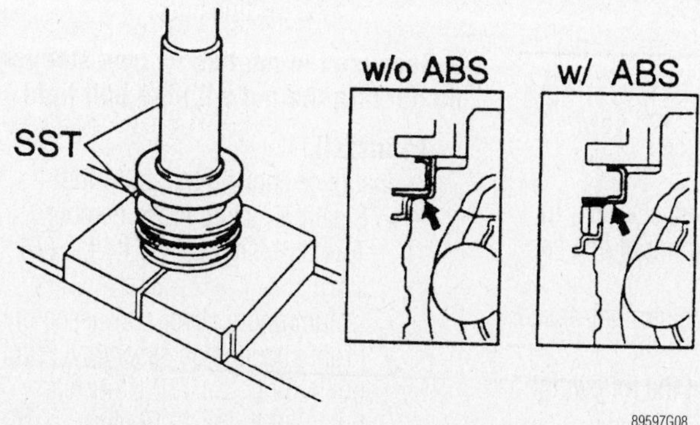

Press the new No. 2 deflector and seat it properly

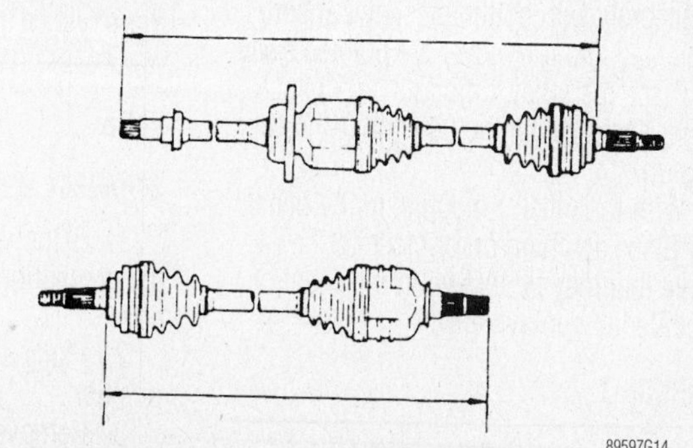

Measure the driveshaft length at these points and make sure it reaches specifications prior to installation

- Place the beveled side of the tri-pot joint axial spline toward the outboard joint. Align the matchmarks placed before removal
- The tri-pot joint to the driveshaft

➡Do not tap the roller.

- A new snapring
- Boot to the outboard joint. Before assembling the boot, fill grease into the outboard joint and boot. Special grease (usually black) is supplied with the boot kits. Grease capacity is 4.2–4.6 oz (120–130 g)
- Inboard joint tulip to the front driveshaft. Pack grease (usually yellow) into the boot and inboard joint tulip.
- Inboard joint tulip onto the driveshaft
- Boot to the inboard joint tulip
- Boot clamps to both boots

The standard driveshaft length is as follows:

- Automatic left: 22.516–22.910 inch (571.7–581.7mm)
- Automatic right: 33.74–34.13 inch (857.0–867.0mm
- Manual left: 22.34–22.73 inch (525.5–577.4mm)
- Manual right: 33.56–33.95 inch (852.5–862.5mm)

9. Bend the band of the clamp and lock onto the boot.
10. On the dynamic damper attach the clamp on the shaft groove. Check the distance as shown in the illustration. Correct distance should be 7.677–8.071 inch (195.0–205.0mm).
11. Using a screwdriver, tighten the band of the clamp and lock it.

Corolla

1. Before servicing the vehicle, refer to the precautions in the beginning of this section.
2. Remove or disconnect the following:
   - Inboard joint boot clips
   - Inboard joint tulip from the driveshaft
   - Snapring
   - Using a brass rod and hammer, the tri-pot joint off the driveshaft without hitting the joint roller
   - Inboard joint boot
   - Clamp and driveshaft damper
   - Clamps and the outboard drive boot. DO NOT disassemble the outboard joint.

To assemble:

3. Install or connect the following:

➡Before installing the boot, wrap the spline end of the shaft with masking tape to prevent damage to the boot.

- Driveshaft damper with a new clamp
- Temporarily, the inboard boot with new clamp to the drive joint

➡The inboard boot and clamp are larger than those of the outboard boot.

- The tri-pot onto the driveshaft with a brass rod and hammer without hitting the joint roller
- The snapring

4. Pack the outboard tulip joint and the outboard boot with about 0.26–0.33 lbs. ounces of grease that was supplied with the boot kit.
5. Install or connect the following:
   - Boot onto the outboard joint
   - Inboard tulip joint and boot with ½ lb. of grease that was supplied with the boot kit
   - Inboard tulip joint onto the driveshaft
   - Boot onto the driveshaft
6. Before checking the standard length, bend the band and lock it. Make sure that the boot is not stretched or squashed when the driveshaft is at standard length. Standard driveshaft length: LH: 540.2 mm (21.268 in.); RH: 857.4 mm (33.756 in.)

STEERING AND SUSPENSION

Air Bag

PRECAUTIONS

Several precautions must be observed when handling the inflator module to avoid accidental deployment and possible personal injury.

- Never carry the inflator module by the wires or connector on the underside of the module.
- When carrying a live inflator module, hold securely with both hands, and ensure that the bag and trim cover are pointed away.
- Place the inflator module on a bench or other surface with the bag and trim cover facing up.
- With the inflator module on the bench, never place anything on or close to the module that may be thrown in the event of an accidental deployment.

DISARMING

To avoid personal injury when working on vehicles equipped with an air bag, the negative battery cable must be disconnected and at least 90 seconds must elapse before working on the system. Failure to do so may result in deployment of the air bag.

REARMING

After vehicle service is completed, reattach the battery cables (positive cable first!) to rearm the air bag system.

Rack and Pinion Steering Gear

REMOVAL & INSTALLATION

Echo

MANUAL STEERING

1. Before servicing the vehicle, refer to the precautions in the beginning of this section.
2. Place the wheels in the straight-ahead position.
3. Remove or disconnect the following:
   - Steering wheel
   - Engine under covers
   - Tie rod ends
   - No. 2 column hole cover
   - Sliding yoke
   - Hood
4. Attach and engine crane to the engine for support.
5. Remove or disconnect the following:
   - Lower arms from the knuckles
   - Rear engine mount insulator
   - Front crossmember with the steering rack
   - Column hole cover sub-assembly
   - 4 bolts and nuts attaching the gear to the crossmember

➡Because the nut has its own stopper, do not turn the nut with the bolt tight.

To install:

6. Install or connect the following:
   - Steering rack to crossmember. Torque the nuts to 54 ft. lbs. (74 Nm).
   - Column hole cover sub-assembly
   - Front suspension assembly. Front bolts: 52 ft. lbs. (70 Nm); rear bolts: 86 ft. lbs. (116 Nm).

- Rear mount insulator. Torque the bolt and 2 nuts to 59 ft. lbs. (80 Nm).
- Lower arm to knuckle. Horizontal bolt: 65 ft. lbs. (88 Nm); vertical bolt: 97 ft. lbs. (132 Nm).
- Hod
- Sliding yoke
- No. 2 column hole cover
- Tie rods
- Under covers
- Steering wheel. Nut: 25 ft. lbs. (34 Nm).

POWER STEERING

1. Before servicing the vehicle, refer to the precautions in the beginning of this section.
2. Place the wheels in the straight-ahead position.
3. Remove or disconnect the following:
   - Steering wheel
   - Engine under covers
   - Tie rod ends
   - No. 2 column hole cover
   - Sliding yoke
   - Pressure and return lines
   - Hood
4. Attach and engine crane to the engine for support.
5. Remove or disconnect the following:
   - Lower arms from the knuckles
   - Rear engine mount insulator
   - Front crossmember with the steering rack
   - Stabilizer bar
   - Heat insulator
   - Damper
   - 4 bolts and nuts attaching the gear to the crossmember

➡Because the nut has its own stopper, do not turn the nut with the bolt tight.

To install:

6. Install or connect the following:
   - Steering rack to crossmember. Torque the nuts to 54 ft. lbs. (74 Nm).
   - Damper. Bolts: 13 ft. lbs. (18 Nm).
   - Heat insulator. Bolt: 26 ft. lbs. (35 Nm).
   - Stabilizer bar
   - Front suspension assembly. Front bolts: 52 ft. lbs. (70 Nm); rear bolts: 86 ft. lbs. (116 Nm).
   - Rear mount insulator. Torque the bolt and 2 nuts to 59 ft. lbs. (80 Nm).
   - Lower arm to knuckle. Horizontal bolt: 65 ft. lbs. (88 Nm); vertical bolt: 97 ft. lbs. (132 Nm).
   - Hood
   - Sliding yoke
   - No. 2 column hole cover
   - Tie rods
   - Pressure and return lines
   - Under covers
   - Steering wheel. Nut: 25 ft. lbs. (34 Nm).

Avalon, Camry and Solara

1. Before servicing the vehicle, refer to the precautions in the beginning of this section.
2. Position the front wheels straight ahead.
3. Remove or disconnect the following:
   - Negative battery cable. On vehicles equipped with an air bag, wait at least 90 seconds before proceeding.

➡If equipped with an air bag, disable the system and secure the steering wheel.

   - Front wheels
   - Left and right front fender apron seals.
   - Cotter pin and nut holding the steering knuckle to the tie rod end.
   - Tie rod end from the steering knuckle.

➡Place matchmarks on the intermediate shaft and the control valve shaft.

   - Lower bolt holding the control valve shaft to the intermediate shaft.
   - Intermediate shaft from steering rack housing.
   - Tube clamp
   - Return line and the pressure line from the control valve housing.
   - Stabilizer bar bolts and nuts. Do not remove the bar from the vehicle.
   - If necessary, rear engine mounting and bracket for additional clearance.
   - On the V6 engine, oxygen sensor.
   - Steering gear mounting bolts and nuts
   - Steering gear

To install:

4. Install or connect the following:
   - Steering gear. Nuts and bolts: 134 ft. lbs. (181 Nm).
   - On the V6, oxygen sensor.
   - Rear engine mounting bracket. Bolts: 38 ft. lbs. (52 Nm).
   - Stabilizer bar bolts and nuts.
   - Tube clamp. Nut: 84 inch lbs. (10 Nm).
   - Intermediate shaft to the steering rack. Bolts: 26 ft. lbs. (35 Nm).
   - Tie rods to the steering knuckles with the castellated nuts.
   - Front fender apron seals
   - Front wheels
   - Negative battery cable
   - Power steering fluid

Celica

1. Before servicing the vehicle, refer to the precautions in the beginning of this section.
2. Remove or disconnect the following:
   - Negative battery cable. On vehicles equipped with an air bag, wait at least 90 seconds before proceeding.
   - Right and left-hand engine under-covers.
   - Left and right-hand tie rod ends. Separate the tie rod using a puller.
   - Oxygen sensor
   - Front exhaust pipe and brackets.

➡Place matchmarks on the steering column intermediate shaft and the steering gear control valve shaft.

   - Lower bolt to the intermediate shaft
   - Intermediate shaft from the control valve shaft.
   - Pressure feed and return tubes from the steering rack.
   - Tube clamp bracket
3. Support the engine and transaxle with a support fixture.
4. Remove or disconnect the following:
   - Lower control arms from the lower ball joints.
   - Through-bolts to the front and rear mounting insulators.
   - Bolts holding the rear of the lower control arm to the sub-frame and body. Remove the bolts on both sides of the sub-frame.
5. Support the front sub-frame with a jack.
6. Remove or disconnect the following:
   - Bolts holding the sub-frame to the body. Lower the front sub-frame with the lower suspension arms and steering gear.
   - Power steering gear assembly by removing the set bolts and nuts from the sub-frame.

To install:

7. Install or connect the following:
   - Power steering gear assembly to the sub-frame. Bolts and nuts: 94 ft. lbs. (127 Nm).
   - Sub-frame to the body. Bolts: 94 ft. lbs. (127 Nm).

- Bolts to hold the rear of the lower control arms to the sub-frame and body. Bolt that goes through the lower control arm to 123 ft. lbs. (167 Nm) and the other 2 bolts to 130 ft. lbs. (175 Nm). Install the bolts to both sides
- Through-bolts to the front and rear mounting insulators. Bolts: 64 ft. lbs. (88 Nm).
- Lower control arms to the lower ball joints. Nuts and bolts: 94 ft. lbs. (127 Nm).
- Pressure feed and return tubes to the steering rack. Tubes: 26 ft. lbs. (36 Nm).
- Tube clamp bracket. Bolts: 108 inch lbs. (13 Nm).

8. Align the matchmarks on the intermediate shaft and control valve shaft.
9. Install or connect the following:
   - Lower bolt. Upper and lower bolts: 26 ft. lbs. (35 Nm).
   - Front exhaust pipe
   - Oxygen sensor
   - Tie rod ends to the steering knuckles
   - Right and left-hand engine undercovers
   - Negative battery cable
   - Power steering fluid

Corolla

1. Before servicing the vehicle, refer to the precautions in the beginning of this section.
2. Position the front wheels straight ahead.
3. Remove or disconnect the following:
   - Negative battery cable. On vehicles equipped with an air bag, wait at least 90 seconds before proceeding

➡If equipped with an air bag, disable the system and secure the steering wheel.

   - Steering column hole cover and loosen the upper pinch bolt on the sliding yoke
   - Lower pinch bolt at the pinion shaft
   - Front wheels
   - Left and right engine undercovers.
   - Left and right tie rod ends.
4. Install an engine support and tension it to support the engine without raising it.

⁂ CAUTION

The engine hoist is now in place and under tension. Use care when repositioning the vehicle and make necessary adjustments to the engine support.

5. Remove or disconnect the following:
   - Lower control arms from the ball joints
   - If equipped with a stabilizer bar, stabilizer bar links from both lower control arms
   - Right rear control arm bushing retaining bracket. Do this for both lower control arms
   - Stabilizer bar
   - Grommet in the crossmember
   - Bolt and nuts holding in the middle of the crossmember and support the crossmember with a jack
   - Bolts from the outer side of the suspension crossmember
   - Suspension crossmember with the lower suspension arms
   - Exhaust front pipe support
   - Engine rear mount insulator
   - Engine rear mount bracket
   - Pressure feed and return tubes
   - Brackets and grommets to the power steering rack
6. Slide the power steering gear assembly to the right side of the vehicle.

To install:

7. Install or connect the following:
   - Power steering assembly
   - Grommets and brackets. Nuts and bolts: 43 ft. lbs. (59 Nm).
   - Pressure feed and return tubes. Nuts: 26 ft. lbs. (36 Nm).
   - Engine rear mount bracket. Bolts: 57 ft. lbs. (77 Nm).
   - Engine rear mount insulator. Bolt: 64 ft. lbs. (87 Nm).
   - Exhaust front pipe support. Bolts: 14 ft. lbs. (19 Nm).
   - Outer 6 bolts to hold the crossmember to the vehicle. Bolts: 152 ft. lbs. (206 Nm).
   - Center crossmember-to-radiator support bolts: 45 ft. lbs. (61 Nm).
   - Lower A frame-to-center bolts: 161 ft. lbs. (218 Nm).
   - Lower A frame-to-outer bolts: 109 ft. lbs. (147 Nm).
   - Front, center and rear mount bolts: 45 ft. lbs. (61 Nm).
   - Grommet to the crossmember
   - Stabilizer bar
   - Lower control arm bushing retaining bracket. Do not tighten the bolts or nut at this time
   - Lower control arm to the lower ball joint. Bolt and nuts: 105 ft. lbs. (142 Nm). Connect both lower control arms to the ball joints
   - Stabilizer bar links to the lower control arms. Nuts: 33 ft. lbs. (44 Nm).
   - Sliding yoke to the pinion shaft. Lower bolt: 26 ft. lbs. (35 Nm). Tighten upper bolt: 20 ft. lbs. (27 Nm).
   - Steering column hole cover. Bolts: 43 inch lbs. (5 Nm).
   - Left and right-hand tie rod ends. Nuts: 36 ft. lbs. (49 Nm).
   - Front wheels
   - Control arm bracket bolts: 108 ft. lbs. (147 Nm).
   - Stabilizer bar bracket bolt: 37 ft. lbs. (50 Nm).
   - Bracket nut: 14 ft. lbs. (19 Nm).
   - Negative battery cable
8. Check and top off the power steering fluid.
9. Check and adjust the alignment, if needed.

MR2

➡Do not allow the steering wheel to rotate when disconnected from the steering gear. Damage to the air bag spiral cable may result.

1. Before servicing the vehicle, refer to the precautions in the beginning of this section.
2. Place the wheels pointing straight ahead and lock the steering wheel in place.
3. Remove or disconnect the following:
   - Luggage compartment trim box cover
   - Tool box
   - Power steering pressure and return lines from the power steering vane pump
   - Front luggage undercover
   - Outer tie rod ends
   - Steering intermediate shaft
   - Power steering gear mounting bolts
   - Power steering gear

To install:

4. Installation is the reverse of the removal procedure, while using the following torwue values:
   - Power steering gear mounting bolts: 42 ft. .lbs. (57 Nm)
   - Steering intermediate shaft pinch bolt: 26 ft. lbs. (35 Nm)
   - Outer tie rod ends: 36 ft. lbs. (49 Nm)
5. Fill and bleed the power steering system.

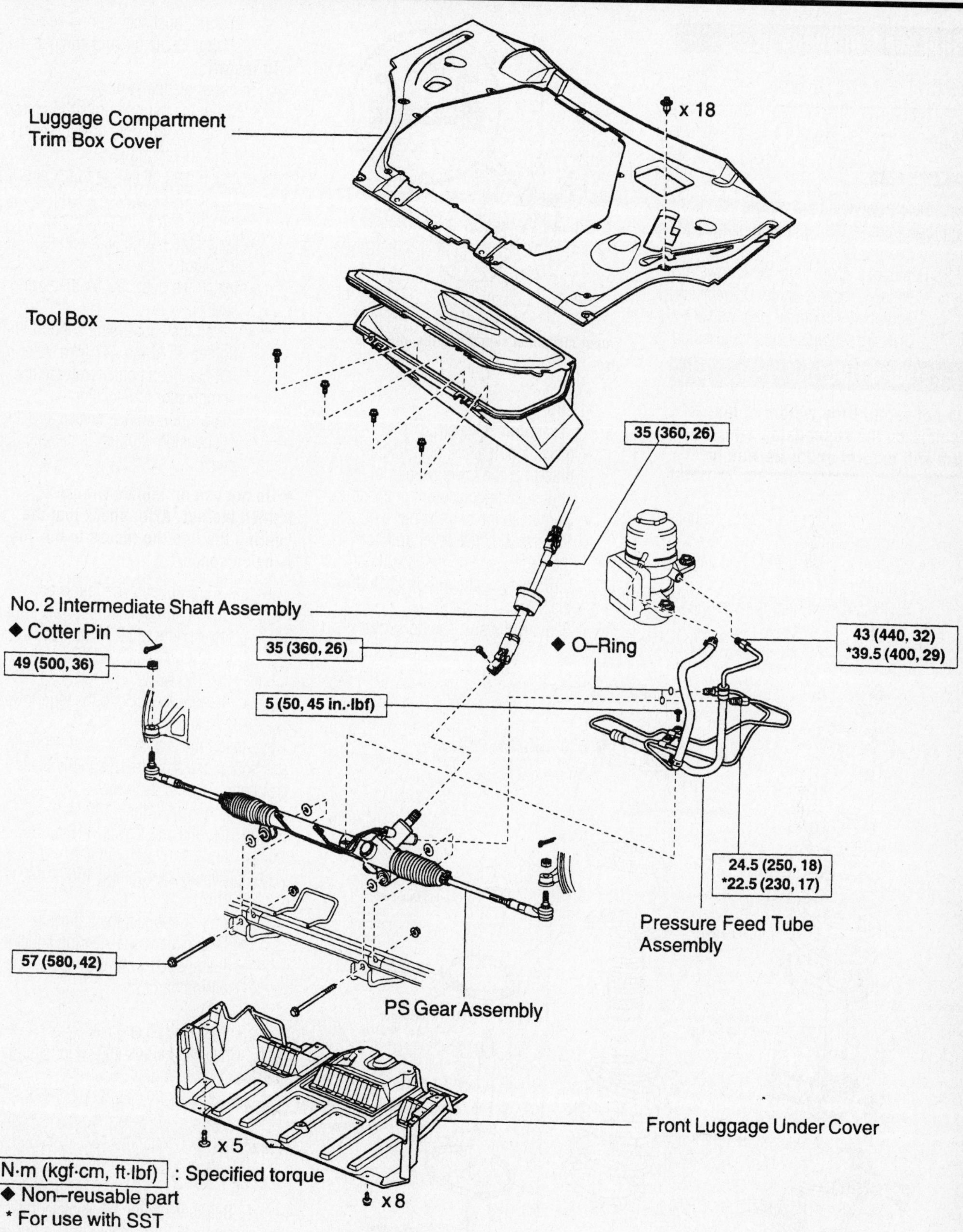

Power rack and pinion steering gear mounting exploded view—MR2

Strut and Coil Spring

REMOVAL & INSTALLATION

Front

EXCEPT MR2

1. Before servicing the vehicle, refer to the precautions in the beginning of this section.
2. Remove or disconnect the following:
   - Negative battery cable. On vehicles equipped with an air bag, wait at least 90 seconds before proceeding

⁂ WARNING

Do not support the weight of the vehicle on the suspension arm; the arm will deform under its weight.

   - Wheel
   - Bolt, and disconnect the brake hose from the strut
   - With ABS brakes, wiring harness from the strut
   - Bolts and strut from the steering knuckle

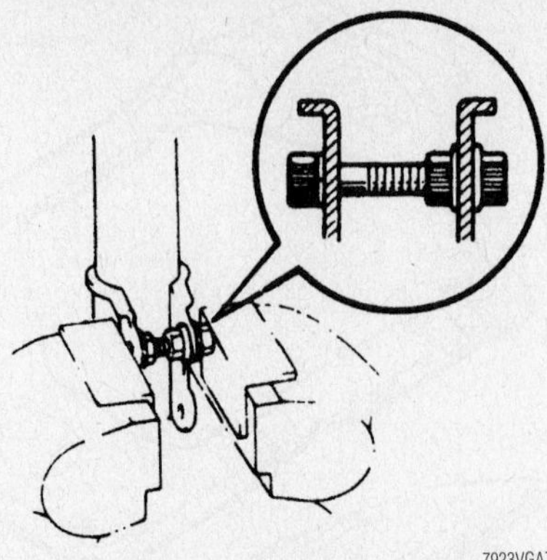

Proper method of supporting the strut in a vise

   - Strut
3. To disassemble the strut:
   - Install a bolt and 2 nuts to the bracket at the lower portion of the strut shell and secure it in a vise
   - Compress the coil spring
   - Remove the dust cover and hold the spring seat so that it will not turn. Remove the nut on the top of the strut
   - Remove the suspension support, bearing, dust seal, spring seat, spring, insulators and bumper

To install:

4. To assemble the strut:
   - Install the spring bumper to piston
   - Using a spring compressor, compress the spring
   - Install the coil spring to the strut. Fit the lower end of the coil spring into the gap of the lower seat
   - Install the spring seat with the insulator
   - Install the dust seal on the spring seat
   - Install the suspension support and tighten 35 ft. lbs. (47 Nm). After the nut has been tighten, release the compressor tool tension
   - Pack multipurpose grease into the suspension support. Install the dust cover

➡Do not use an impact wrench to tighten the nut. Also, check that the bearing fits into the recess in the suspension support.

5. Install or connect the following:
   - Nuts holding the strut to the strut tower. Nuts to 29 ft. lbs. (39 Nm), except on Avalon and Camry: 59 ft. lbs. (80 Nm)
   - Steering knuckle to the strut lower bracket
6. Insert the 2 bolts from the rear side and tighten the strut-to-steering knuckle arm bolts. Tighten as follows:
   - Echo: 97 ft. lbs. (132 Nm).
   - Corolla and Celica: 113 ft. lbs. (153 Nm).
   - Avalon and Camry: 155 ft. lbs. (210 Nm).
7. Install or connect the following:
   - Brake line to the steering knuckle
   - If equipped with ABS, secure the wiring harness
   - Wheel
   - Negative battery cable
8. Check and adjust the alignment, if needed.

MR2

Rear

EXCEPT MR 2

1. Before servicing the vehicle, refer to the precautions in the beginning of this section.
2. Remove or disconnect the following:
   - Negative battery cable from the battery. On vehicles equipped with an air bag, wait at least 90 seconds before proceeding

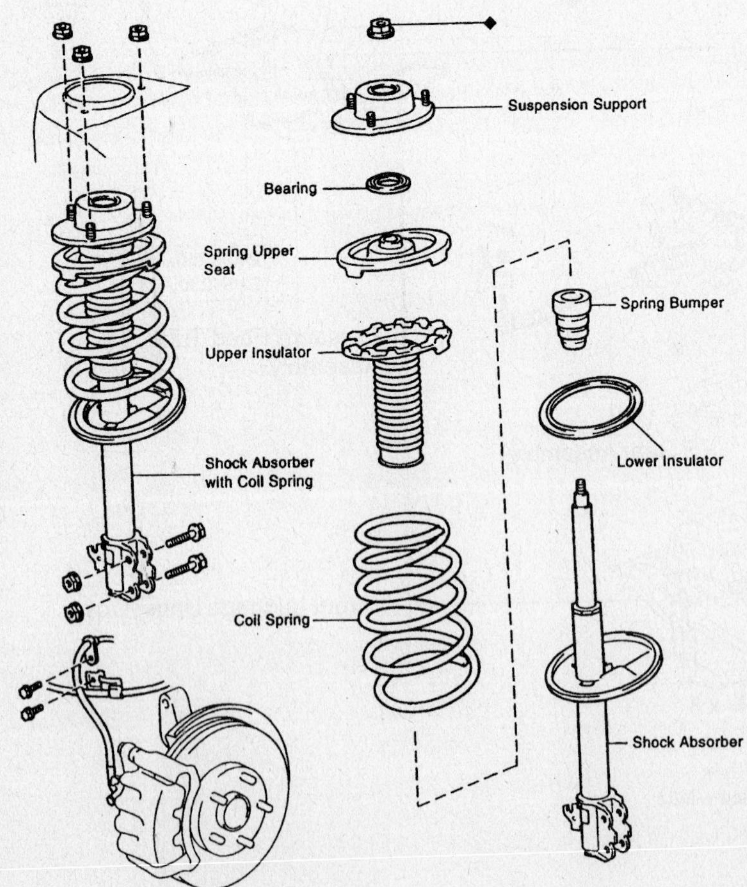

Common coil spring and strut component assembly

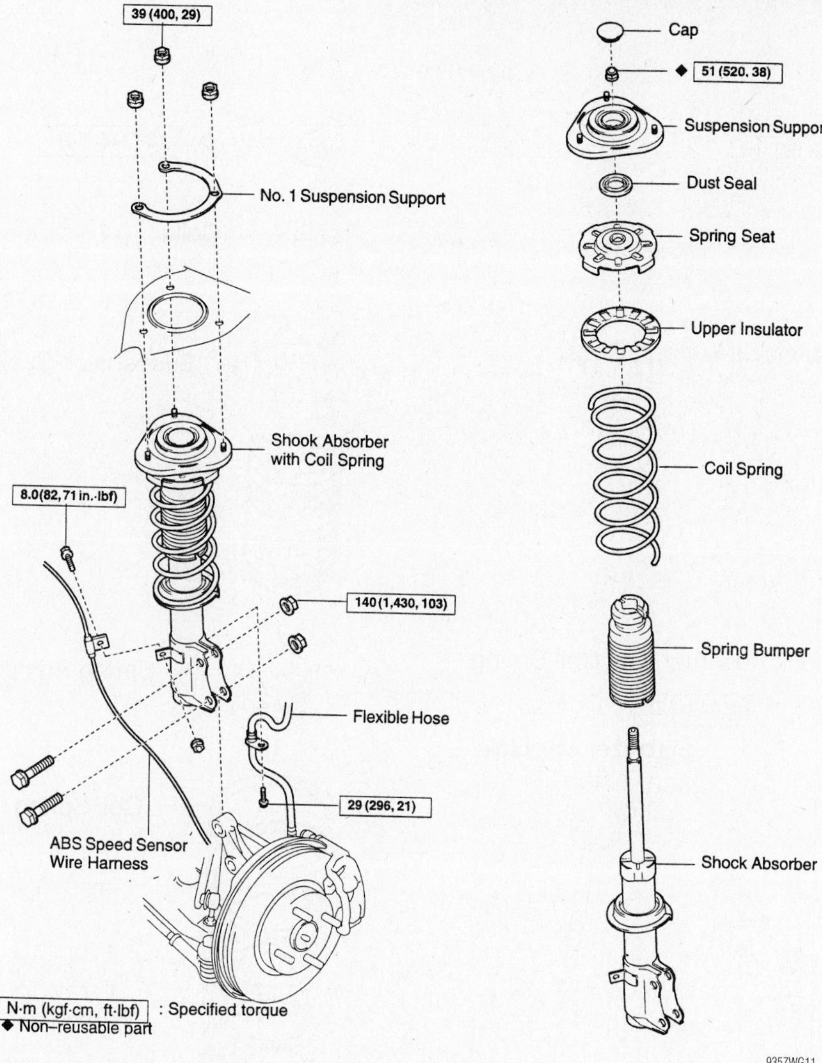

Front strut and coil spring mounting exploded view—MR2

- Rear seat cushion and any trim necessary to access the strut towers

3. Support the axle beam with a jack.
4. Remove or disconnect the following:
   - Wheel
   - With ABS, sensor wire from the strut
   - Stabilizer bar
5. Loosen the fasteners securing the strut to the axle carrier. Do not remove the bolts at this time.
6. Support the axle carrier with a jack.
7. Remove or disconnect the following:
   - Strut-to-strut tower nuts

CAUTION

Do not loosen the center nut on the top of the strut piston.

- Strut

8. To disassemble the strut:
   a. Place the strut assembly in a pipe vise or strut vise.

➡Do not attempt to clamp the strut assembly in a flat jaw vise as this will result in damage to the strut tube.

   b. Compress the spring until the upper suspension support is free of any spring tension. Do not over-compress the spring.
   c. Hold the upper support, then remove the nut on the end of the shock piston rod.
   d. Remove the support, coil spring, insulator, and bumper.
9. Inspect the strut as follows:
   a. Check the shock absorber by moving the piston shaft through its full range of travel. It should move smoothly and evenly throughout its entire travel without any trace of binding or notching.
   b. Use a small straightedge to check the piston shaft for any bending or deformation.
   c. Inspect the spring for any sign of deterioration or cracking. The waterproof coating on the coils should be intact to prevent rusting.

To install:

➡Never reuse a self-locking nut. Always replace self-locking nuts and cotter pins as applicable.

10. Assemble the strut as follows:
    a. Loosely assemble all components onto the strut assembly. Be sure the spring end aligns with the hollow in the lower seat.
    b. Align the upper suspension support with the piston rod and install the support.
    c. Align the suspension support with the strut lower bracket. This assures the spring will be properly seated top and bottom.
    d. Compress the spring to expose the strut piston rod threads.
    e. Install a new strut piston nut and tighten to the following:
       - Camry and Avalon: 36 ft. lbs. (49 Nm). Celica: 41 ft. lbs. (56 Nm). Corolla: 41 ft. lbs. (56 Nm).
    f. Remove the spring compressor. Be sure the paint mark on the upper support faces the outside of the strut.
11. Place the strut on the vehicle and install the nuts to hold the strut to the strut tower. Except Celica and Corolla: 29 ft. lbs. (39 Nm). Celica and Corolla: 59 ft. lbs. (80 Nm).
12. Install or connect the following:
    - Strut to the axle carrier and install the bolt and nut. Do not tighten at this time
    - Stabilizer link to the strut
    - Wheel. Bounce the vehicle up and down to stabilize the suspension
13. With the vehicle weight on the suspension, tighten the bolt holding the strut to the axle carrier as follows:
    - Corolla: 59 ft. lbs. (80 Nm)
    - Celica: 105 ft. lbs. (142 Nm)
    - Camry and Avalon: 188 ft. lbs. (255 Nm)
14. Install or connect the following:
    - Rear seat cushion and any trim
    - Negative battery cable

MR2

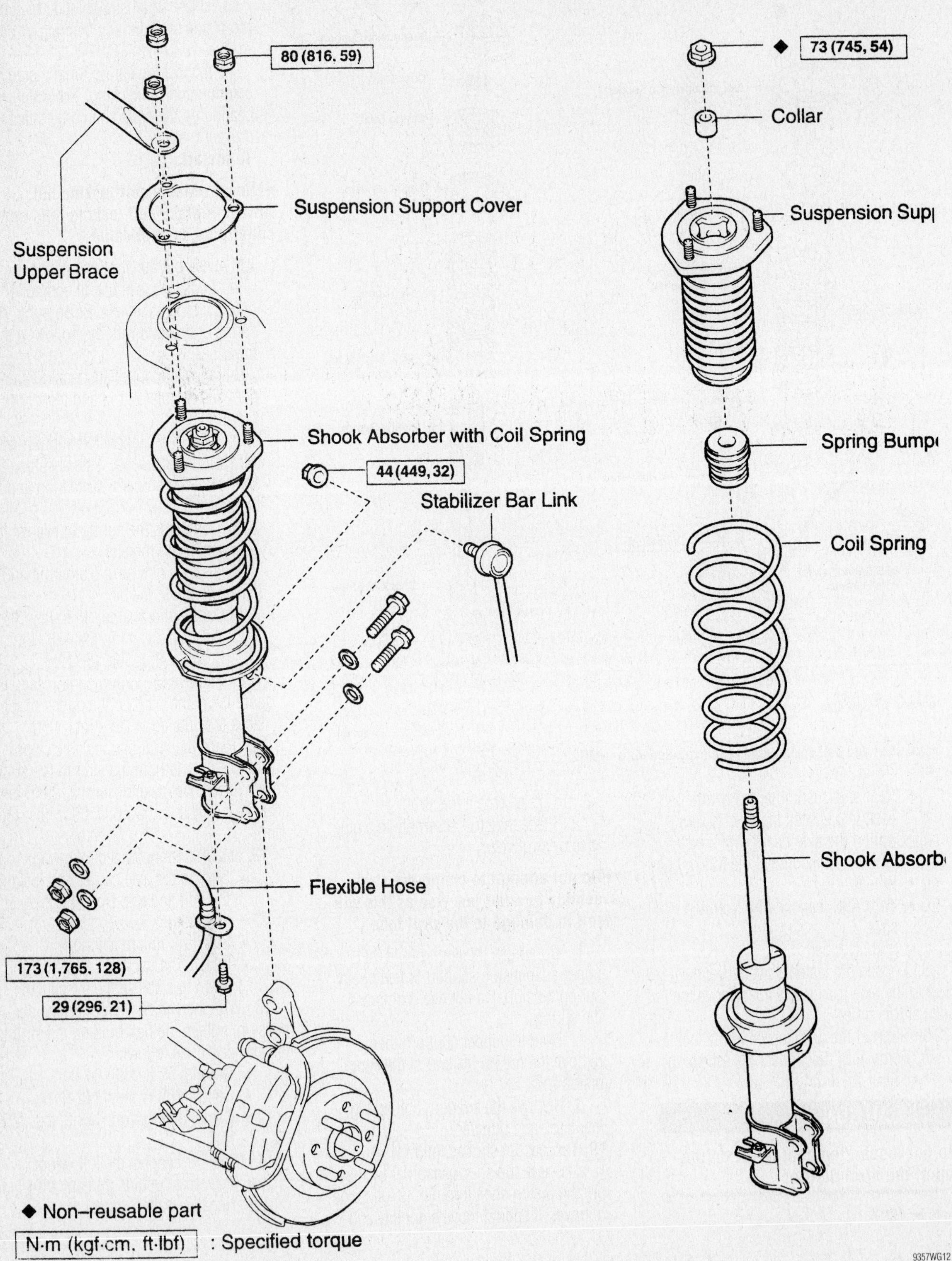

Rear strut and coil spring assembly exploded view—MR2

Rear Shock Absorber and Spring

REMOVAL AND INSTALLATION

Echo

1. Before servicing the vehicle, refer to the precautions in the beginning of this section.
2. Support the rear axle beam.
3. Remove or disconnect the following:
   - Wheels
   - Package tray trim
   - Rear seat
   - Door scuff plate and door opening trim
   - Quarter panel trim
   - Roof side inner garnish
   - Partition board
   - Shock absorber upper nuts and lower bolt
4. Lower the axle slowly and remove the spring

To install:

5. Position the upper insulator so that its gap fits into the end of the spring.
6. Place the lower insulator on the axle.
7. Raise the axle while positioning the shock. Torque the lower bolt to 36 ft. lbs. (49 Nm).
8. Position the lower nut on the rod so that the rod protrudes 15–18mm above the nut.
9. Install or connect the following:
   - Upper nut. Torque the nut to 18 ft. lbs. (25 Nm).
   - Partition board
   - Roof side inner garnish
   - Quarter panel
   - Door scuff plate and trim
   - Rear seat
   - Package tray trim
   - Wheels

Lower Ball Joint

REMOVAL & INSTALLATION

Except Echo and MR2

1. Before servicing the vehicle, refer to the precautions in the beginning of this section.
2. Remove or disconnect the following:
   - Negative battery cable. On vehicles equipped with an air bag, wait at least 90 seconds before proceeding
   - Front wheels
   - Cotter pin from the bearing locknut cap, then remove the cap
3. Depress the brake pedal and loosen the axle nut
4. Remove or disconnect the following:
   - Brake caliper attaching hardware, position the caliper aside with the hydraulic line still attached and suspend it with a wire
   - ABS speed sensor, if equipped
   - Rotor
5. Loosen the 2 nuts holding the strut to the steering knuckle assembly. Do not remove at this time.
6. Remove or disconnect the following:
   - Cotter pin and nut from the tie rod end. Using a tie rod end removal tool, separate the tie rod end from the steering knuckle
   - Steering knuckle from the strut assembly
   - Axle nut and grasp the hub and knuckle assembly. With a plastic hammer tap the axle shaft to remove knuckle and hub

➡Cover the halfshaft boot with a shop rag to protect it from any damage.

7. Clamp the steering knuckle in a vise and remove the dust deflector. Remove the nut holding the steering knuckle to the ball joint. Press the ball joint out of the steering knuckle.
8. Remove the ball joint from the arm.

To install:

9. Install the Lower ball joint to the lower arm. Tighten the fasteners to:
   a. Corolla: 66 ft. lbs. (89 Nm).
   b. Camry: 55 ft. lbs. (75 Nm).
   c. Avalon: 94 ft. lbs. (127 Nm).
   d. Celica: 105 ft. lbs. (142 Nm).
10. Install the ball joint to the steering knuckle. Tighten the ball joint-to-steering knuckle nut to:
   a. Celica and Corolla: 76 ft. lbs. (103 Nm).
   b. Avalon and Camry: 90 ft. lbs. (123 Nm).
11. Install or connect the following:
   - New cotter pin. Drive the deflector shield onto the knuckle
   - Knuckle and hub assembly to the axle and temporarily tighten the axle nut
   - Knuckle assembly to the lower strut bracket. Temporarily insert the mounting bolts from the rear and install the nuts
   - Tie rod end to the knuckle
12. Tighten the bolts on the lower side of the strut assembly.
13. Install or connect the following:
   - ABS speed sensor
   - Brake disc and the caliper
14. Tighten the axle nut.
15. Connect the negative battery cable.
16. Check and adjust the alignment, if needed.

MR2

The lower ball joint is not serviceable. If the lower ball joint is defective, replace the lower arm and ball joint as an assembly.

Echo

1. Before servicing the vehicle, refer to the precautions in the beginning of this section.

➡The ball joint is not replaceable.

2. Remove or disconnect the following:
   - Wheels
   - Stabilizer bar end link
   - Lower arm from the knuckle
3. For the right arm, jack up the engine slightly and disconnect the rear engine mount.
4. On either side, remove the 2 bolts and 1 nut and remove the arm.
5. Flip the ball stud back and forth a few times. Install the nut. Using a torque wrench, rotate the ball stud at the rate of 1 turn in 2–4 seconds. Take a torque reading on the 5th revolution. Turning torque should be at least 5 inch lbs. (0.59 Nm), and not more than 30 inch lbs. (3.4 Nm).

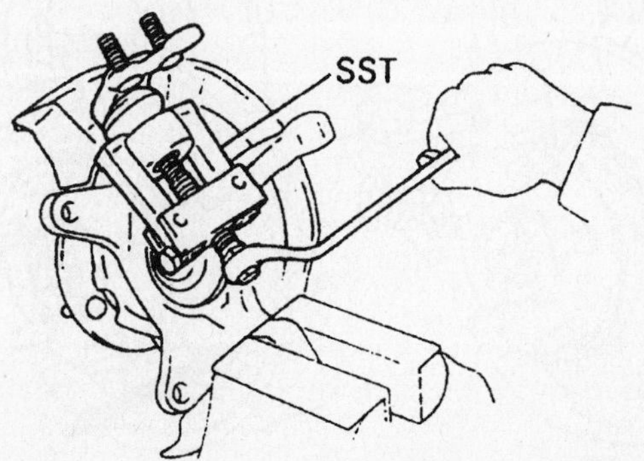

Removing the ball joint from the knuckle

To install:

6. Install or connect the following:
   - Lower control arm. Torque the bolts to A: 65 ft. lbs. (88 Nm); B: 97 ft. lbs. (132 Nm).

➡**DO NOT turn the nut.**

   - On the right side, the engine rear mount. Bolt and nuts: 59 ft. lbs. (80 Nm).
   - Lower arm to knuckle. Torque the nut to 72 ft. lbs. (98 Nm).
   - Stabilizer bar
   - Wheel

Wheel Bearings

ADJUSTMENT

Front

1. Before servicing the vehicle, refer to the precautions in the beginning of this section.
2. All models use a non-adjustable wheel bearing. To determine the condition of the wheel bearing, check the backlash in bearing shaft direction and the axle hub deviation. Maximum for backlash should be as follows:
   - Corolla, Echo, Avalon, Camry and Camry Solara: 0.0020 in. (0.05mm)
   - Celica: 0.0031 in. (0.08mm)
3. Maximum axle hub deviation is:
   - Corolla: 0.0028 in. (0.07mm)
   - Celica: 0.0028 in. (0.07mm)
   - Echo, Avalon and Camry: 0.0020 in. (0.05mm)
4. If the wheel bearing is out of specifications, replace the wheel bearing.

Rear

Check the backlash in bearing shaft direction and the axle hub deviation. Maximum for backlash should be 0.0020 in. (0.05mm). Maximum axle hub deviation is 0.0028 in. (0.07mm).

➡**The wheel bearing is non-adjustable. If the wheel bearing is out of specifications, replace the wheel bearing.**

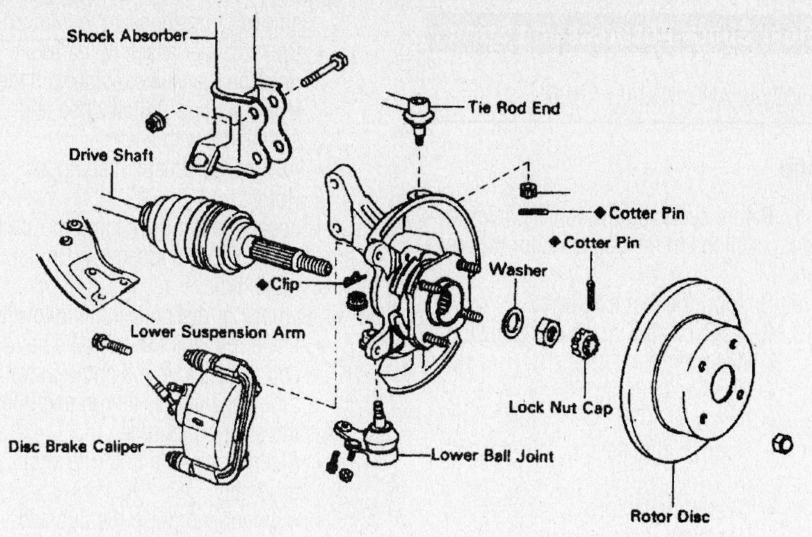

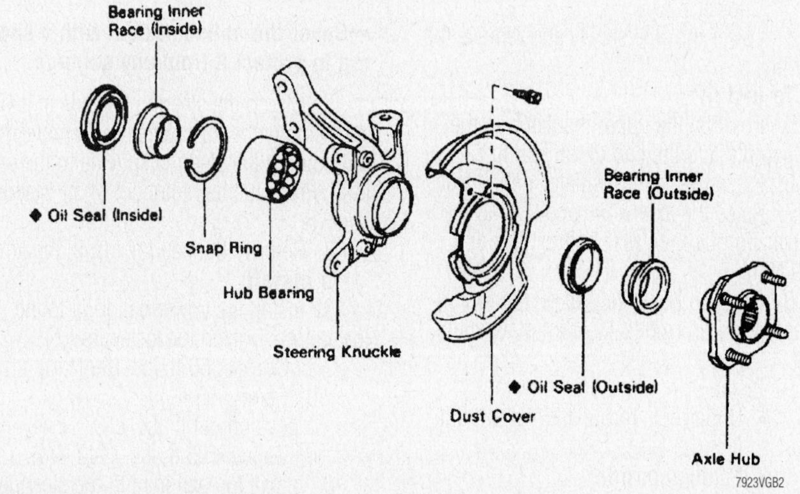

Steering knuckle and hub assembly

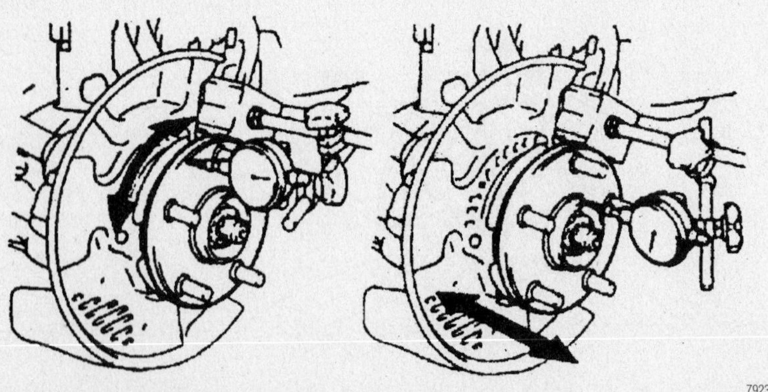

Checking the wheel bearings for deviation and free-play

REMOVAL & INSTALLATION

Front

1. Before servicing the vehicle, refer to the precautions in the beginning of this section.
2. Remove or disconnect the following:
   - Negative battery cable. On vehicles equipped with an air bag, wait at least 90 seconds before proceeding
   - Wheels
   - Axle nut cap
   - Axle nut
   - Caliper. Position the caliper aside with the hydraulic line still attached and suspend it with a wire.
   - ABS speed sensor
   - Rotor
3. Loosen the nuts on the lower side of the strut assembly. Do not remove at this time.

4. Remove or disconnect the following:
   - Tie rod end from the steering knuckle
   - Steering knuckle from the lower control arm
   - Knuckle from the strut assembly
   - Hub

➡**Cover the halfshaft boot with a shop rag to protect it from any damage.**

5. Clamp the steering knuckle in a vise and remove the dust deflector. Remove the nut holding the steering knuckle to the ball joint. Press the ball joint out of the steering knuckle.
6. Remove the inner axle seal.
7. Using a Torx®wrench, remove the bolts securing the dust cover.
8. Using hub puller, remove the hub and backing plate from the steering knuckle.
9. Using a proper sized driver and a press, remove the inner hub race from the axle hub.
10. Using seal removal tool, remove the outer axle seal.
11. Using snapring pliers, remove the snapring from the inner side of the steering knuckle.
12. Using a proper sized driver and a press, remove the bearing from the steering knuckle. The bearing is pressed from the front of the steering knuckle and is removed through the back of the steering knuckle.

To install:

13. Perform the following:
14. Using a proper sized driver and a press, install a new bearing to the steering knuckle.
15. Install the snapring to the steering knuckle using snapring pliers.
16. Using a seal driver and a hammer, install a new outer oil seal. Apply multipurpose grease to the oil seal lip.
17. Place the dust cover on the steering knuckle. Bolts: 78 inch lbs. (9 Nm).
18. Using a press and a proper sized driver, install the axle hub to the steering knuckle.
19. Attach the ball joint to the steering knuckle. Install a new cotter pin.
20. Using a seal driver and a hammer, install a new inner oil seal. Apply multipurpose grease to the oil seal lip.
21. Install the knuckle and hub assembly to the axle and temporarily tighten the axle nut.
22. Connect the knuckle assembly to the lower strut bracket. Temporarily insert the mounting bolts from the rear and install the nuts making sure the matchmarks made earlier are in alignment.

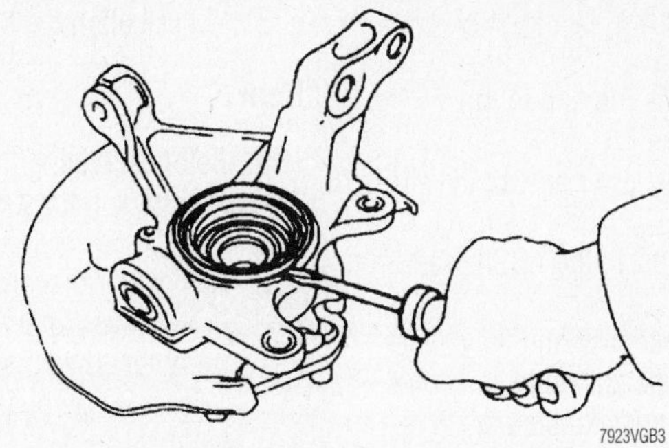

Removing the inner axle seal from the hub assembly

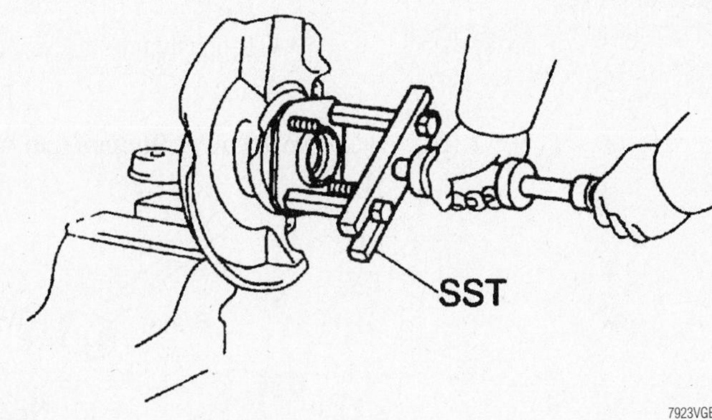

Removing the axle hub from the knuckle

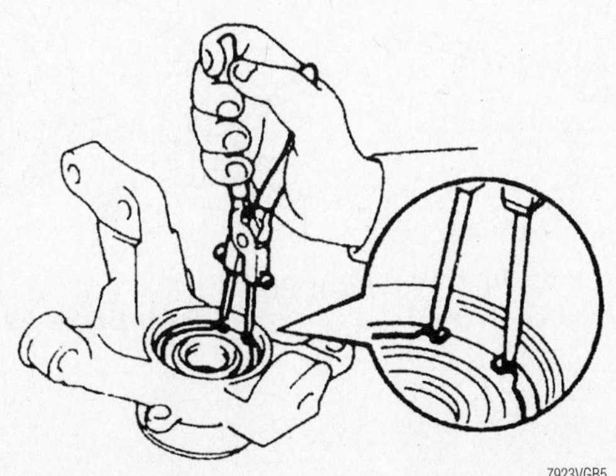

Removing the snapring from the knuckle before pressing out the bearing

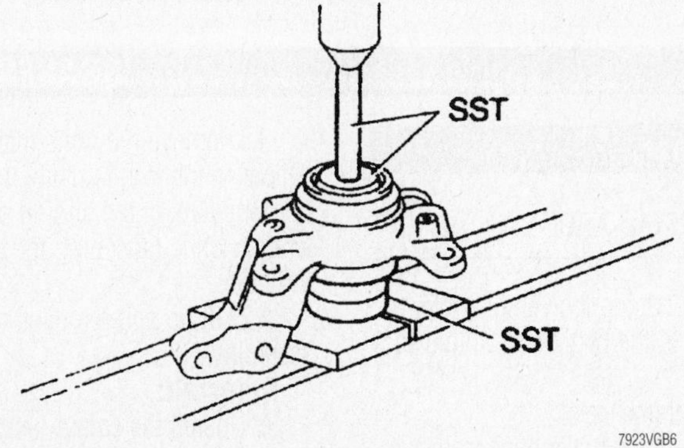

Removing the bearing from the steering knuckle using a press

23. Connect the lower ball joint to lower arm.
24. Connect the tie rod end to the knuckle.
25. Tighten on the lower side of the strut assembly.
26. If equipped, install the ABS speed sensor.
27. Install the brake disc and the caliper.
28. Tighten the axle nut while someone depresses the brake pedal. Install the adjusting nut cap and insert a new cotter pin.
29. Install the wheels to the vehicle. Verify that the wheel turns freely.
30. Connect the negative battery cable to the battery.
31. Check alignment.

Rear

1. Before servicing the vehicle, refer to the precautions in the beginning of this section.
2. Remove or disconnect the following:
   - Negative battery cable. On vehicles equipped with an air bag, wait at least 90 seconds before proceeding
   - Wheel
   - Brake drum or rotor
   - With ABS brakes, ABS wheel speed sensor
   - Hub
   - O-ring from the backing plate

To install:

3. Install or connect the following:
   - New O-ring onto the backing plate. Coat the O-ring with multipurpose grease
   - Hub to the knuckle. Bolts: Except Echo, 59 ft. lbs. (80 Nm); Echo, 38 ft. lbs. (52 Nm).
   - With ABS brakes, ABS wheel speed sensor
   - Brake drum or rotor
   - Wheel
   - Negative battery cable
4. Check and adjust the alignment, if needed.

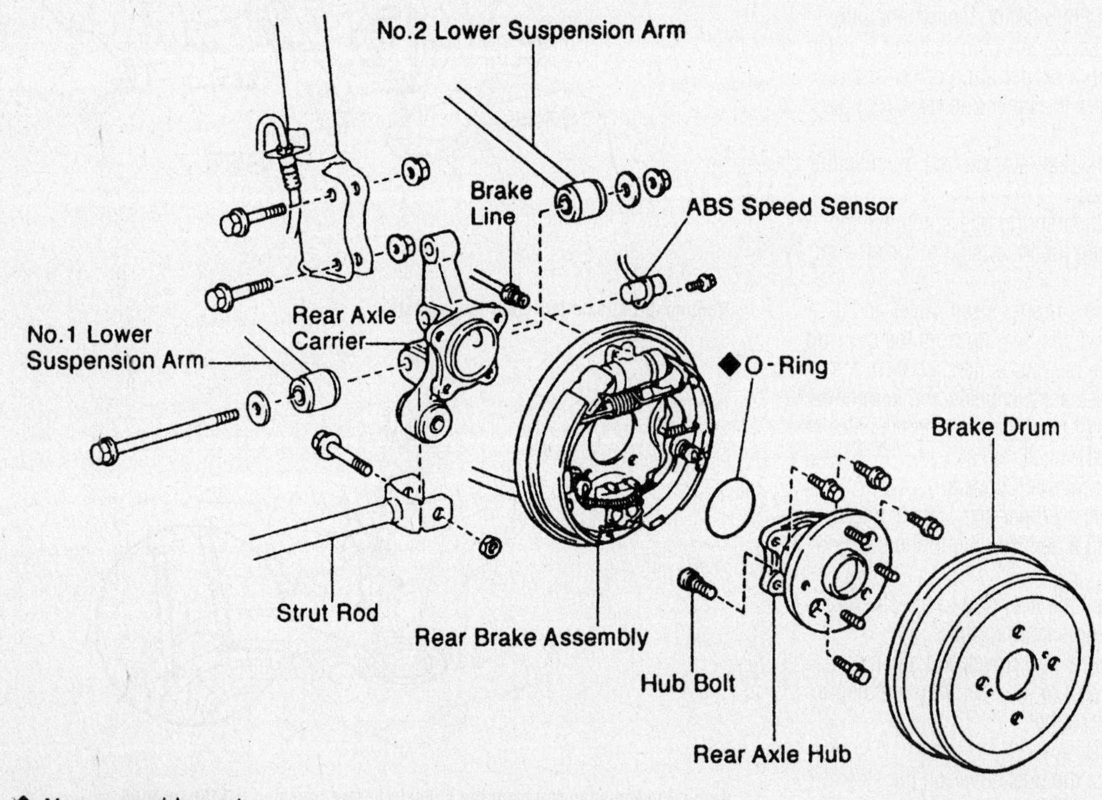

Exploded view of the hub and wheel bearing assembly—Corolla shown, Celica similar

BRAKES

Brake Caliper

REMOVAL & INSTALLATION

1. Before servicing the vehicle, refer to the precautions in the beginning of this section.
2. Remove the wheels.
3. Disconnect the brake hose from the caliper.
4. Remove the bolts that attach the caliper to the torque plate. If applicable, hold the flats of the sliding pin with a wrench while loosening the caliper attaching bolts.
5. Lift up and remove the caliper assembly.

To install:

6. Install the caliper and loosely install the bolts.
7. Hold the flats of the sliding pin with a wrench, then tighten the bolts. Tighten to 25 ft. lbs. (34 Nm) except on Celica rear calipers. Tighten the bolts on Celica rear calipers to 34 ft. lbs. (47 Nm).
8. Connect the brake hose to the caliper, using 2 new washers.
9. Fill the brake system to the proper level and bleed the brake system.

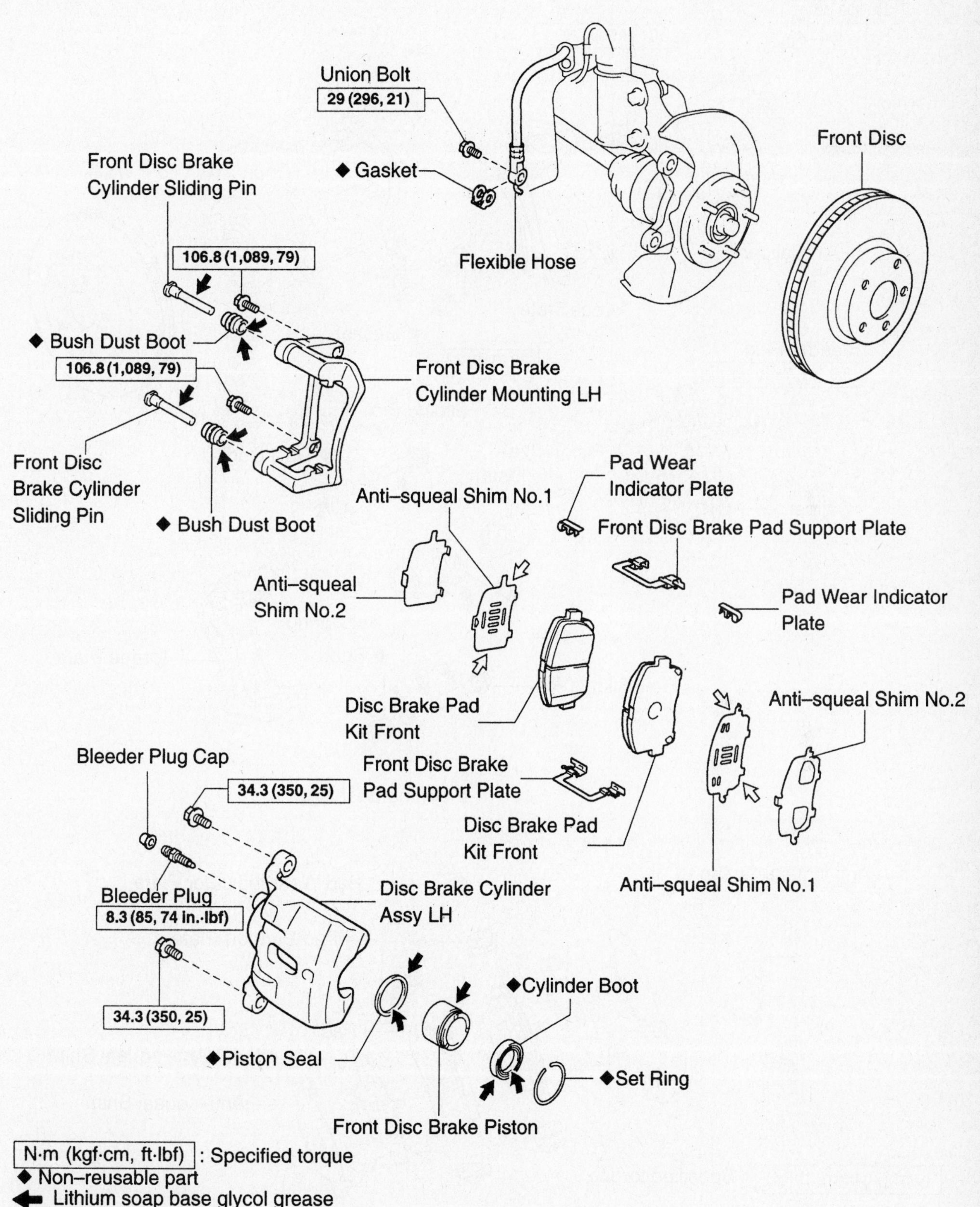
Union Bolt
29 (296, 21)
Front Disc
Front Disc Brake
Cylinder Sliding Pin
◆ Gasket
106.8 (1,089, 79)
Flexible Hose
◆ Bush Dust Boot
106.8 (1,089, 79)
Front Disc Brake
Cylinder Mounting LH
Front Disc
Brake Cylinder
Sliding Pin
◆ Bush Dust Boot
Pad Wear
Indicator Plate
Anti–squeal Shim No.1
Front Disc Brake Pad Support Plate
Anti–squeal
Shim No.2
Pad Wear Indicator
Plate
Disc Brake Pad
Kit Front
Anti–squeal Shim No.2
Bleeder Plug Cap
Front Disc Brake
Pad Support Plate
34.3 (350, 25)
Disc Brake Pad
Kit Front
Bleeder Plug
8.3 (85, 74 in.·lbf)
Disc Brake Cylinder
Assy LH
Anti–squeal Shim No.1
◆Cylinder Boot
34.3 (350, 25)
◆Piston Seal
◆Set Ring
Front Disc Brake Piston
N·m (kgf·cm, ft·lbf) : Specified torque
◆ Non–reusable part
Lithium soap base glycol grease
Disc brake grease
67170-TOYO-G46

Front caliper—Corolla

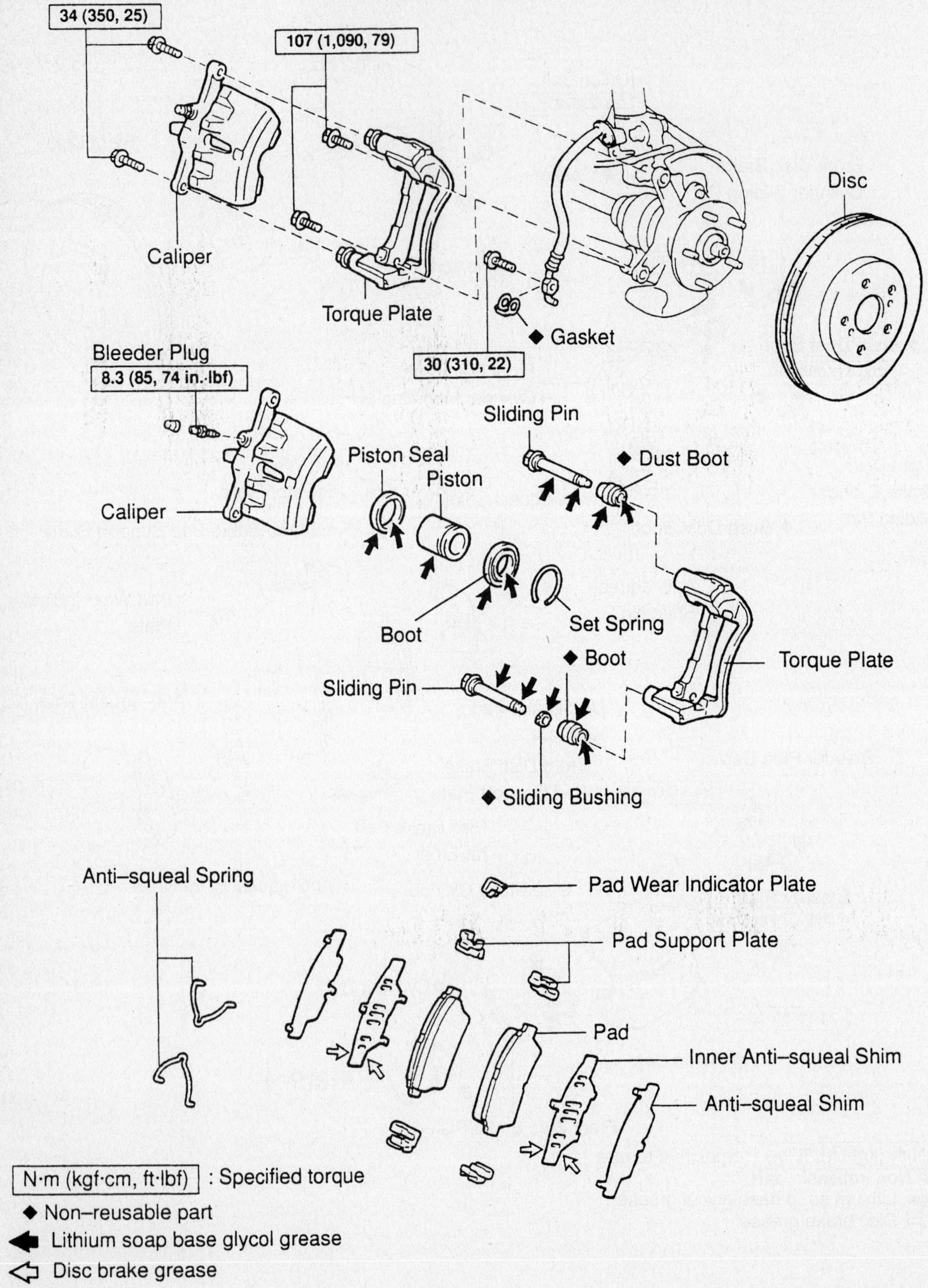

34 (350, 25)
107 (1,090, 79)
Disc
Caliper
Torque Plate
◆ Gasket
Bleeder Plug
8.3 (85, 74 in.·lbf)
30 (310, 22)
Sliding Pin
◆ Dust Boot
Piston Seal
Piston
Caliper
Boot
Set Spring
◆ Boot
Torque Plate
Sliding Pin
◆ Sliding Bushing
Anti–squeal Spring
Pad Wear Indicator Plate
Pad Support Plate
Pad
Inner Anti–squeal Shim
Anti–squeal Shim
N·m (kgf·cm, ft·lbf) : Specified torque
◆ Non–reusable part
Lithium soap base glycol grease
Disc brake grease
93016G67

Front caliper—Celica, Camry, Avalon

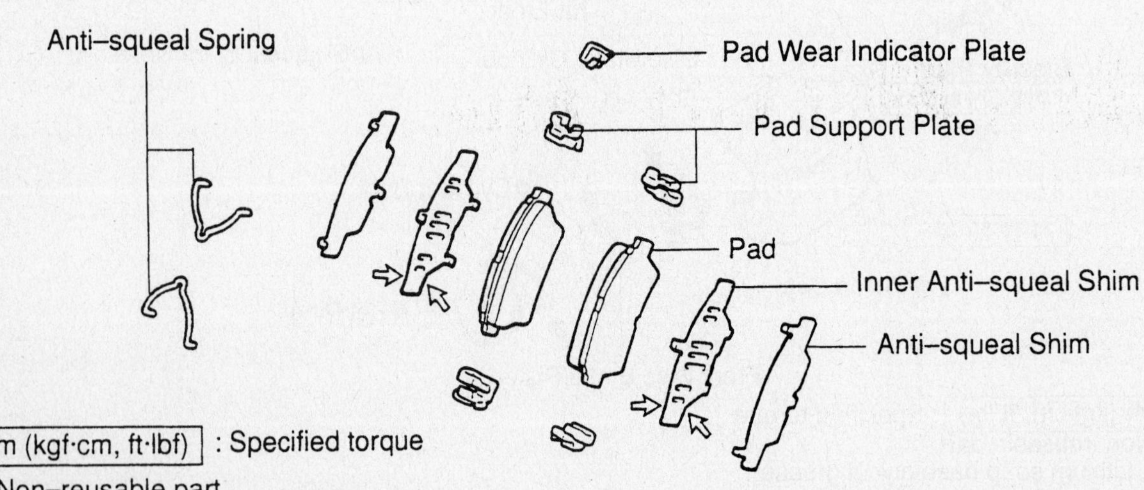

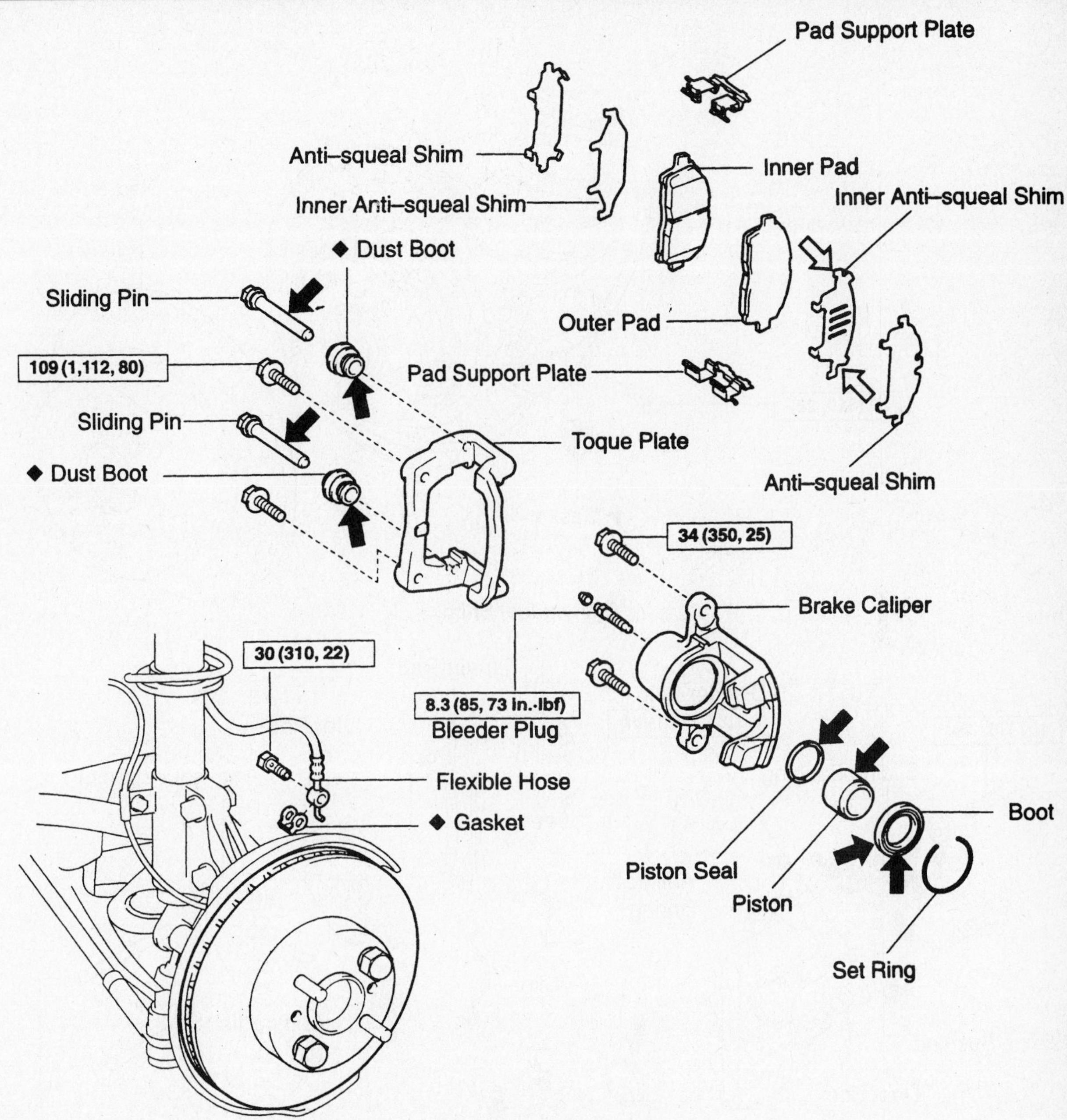

N·m (kgf·cm, ft·lbf) : Specified torque
◆ Non–reusable part
➡ Lithium soap base glycol grease
⇨ Disc brake grease

9357WG13

Front caliper—MR2

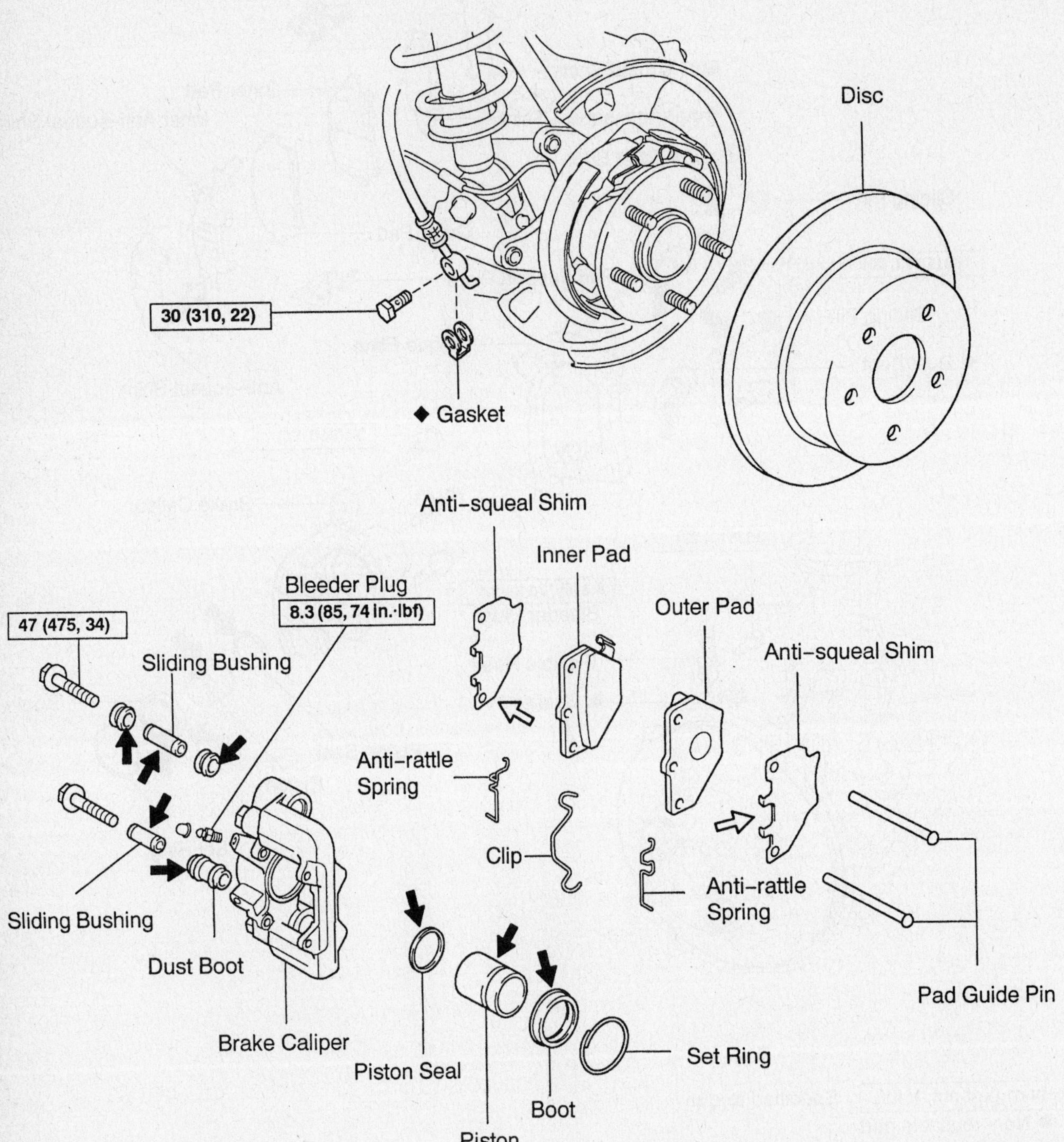
Disc
30 (310, 22)
◆ Gasket
Anti-squeal Shim
Inner Pad
Bleeder Plug
8.3 (85, 74 in.·lbf)
Outer Pad
47 (475, 34)
Sliding Bushing
Anti-squeal Shim
Anti-rattle Spring
Clip
Anti-rattle Spring
Sliding Bushing
Dust Boot
Pad Guide Pin
Brake Caliper
Piston Seal
Set Ring
Boot
Piston
N·m (kgf·cm, ft·lbf) : Specified torque
◆ Non-reusable part
Lithium soap base glycol grease
Disc brake grease

67170-TOYO-G47

Rear caliper—Celica

TMC made:

Union Bolt
29.4 (300, 22)
Rear LH Flexible Hose
◆Gasket
Rear Disc Brake Cylinder Slide Pin
34.3 (350, 25)
◆Rear Disc Brake Cylinder Slide Bush
Rear Disc Brake Bleeder Plug Cap
Rear Disc Brake Bleeder Plug
8.3 (85, 74 in.·lbf)
Disc Brake Cylinder Assy Rear LH
◆Piston Seal
Rear Disc Brake Disc Brake Piston
◆Set Ring
◆Cylinder Boot
Anti Squeal Shim No.1
◆Rear Disc Brake Bush Dust Boot
Anti Squeal Shim No.2
Rear Disc Brake Cylinder Slide Pin
39.2 (400, 29)
Rear Disc Brake Pad
Rear Disc Brake Pad Support Plate No.1
Rear Disc Brake Pad Support Plate No.2
Rear Disc Brake Pad
Pad Wear Indicator Plate
61.8 (630, 46)
Rear Disc
Rear Disc Brake Cylinder Mounting LH
61.8 (630, 46)
Pad Wear Indicator Plate
Parking Brake Shoe Adjusting Hole Plug
Anti Squeal Shim No.2
Anti Squeal Shim No.1
N·m (kgf·cm, ft·lbf) : Specified torque
◆ Non-reusable part
Lithium soap base glycol grease
Disc brake grease
67170-TOYO-G48

Rear caliper—Camry, TMC caliper

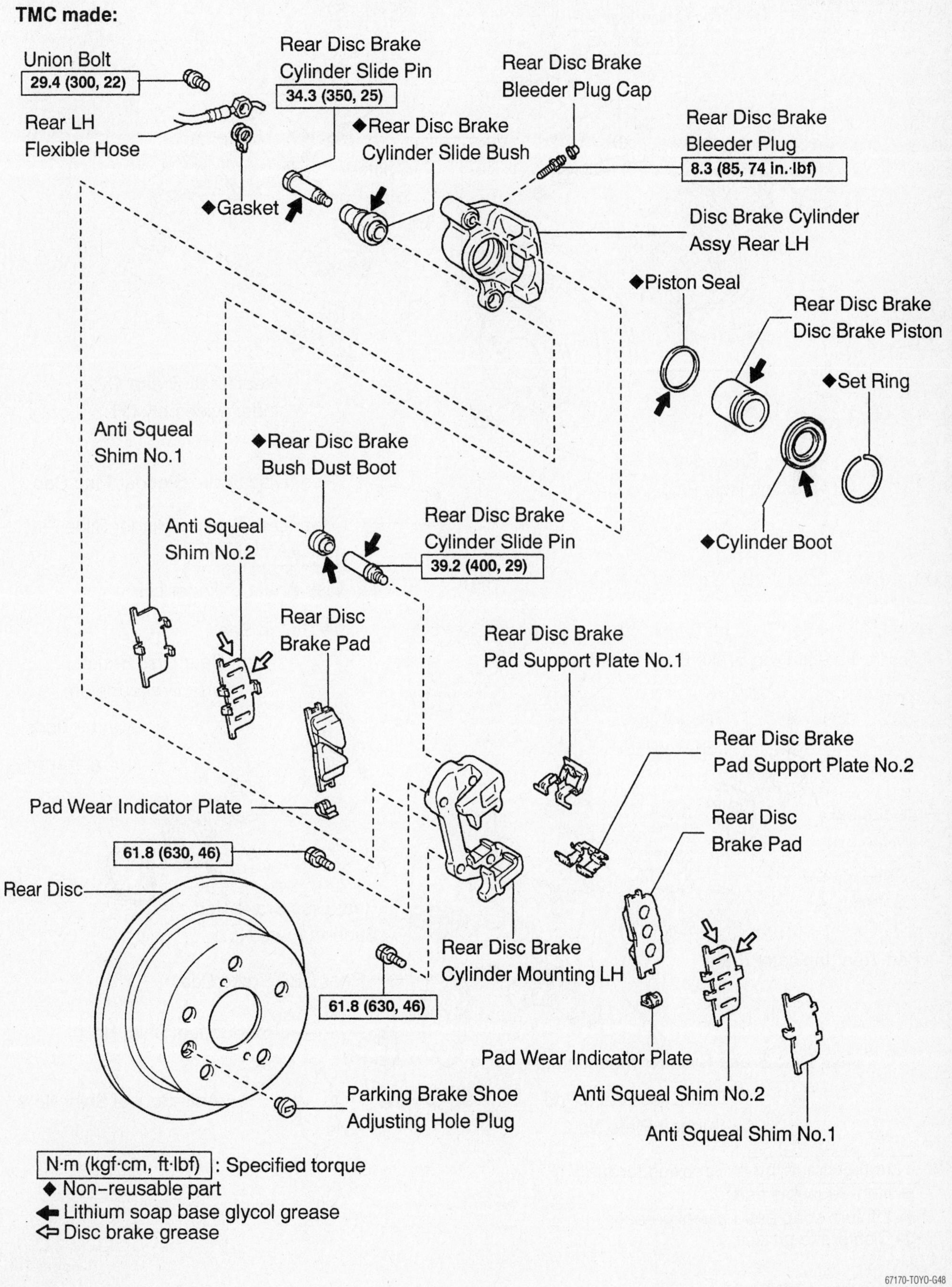

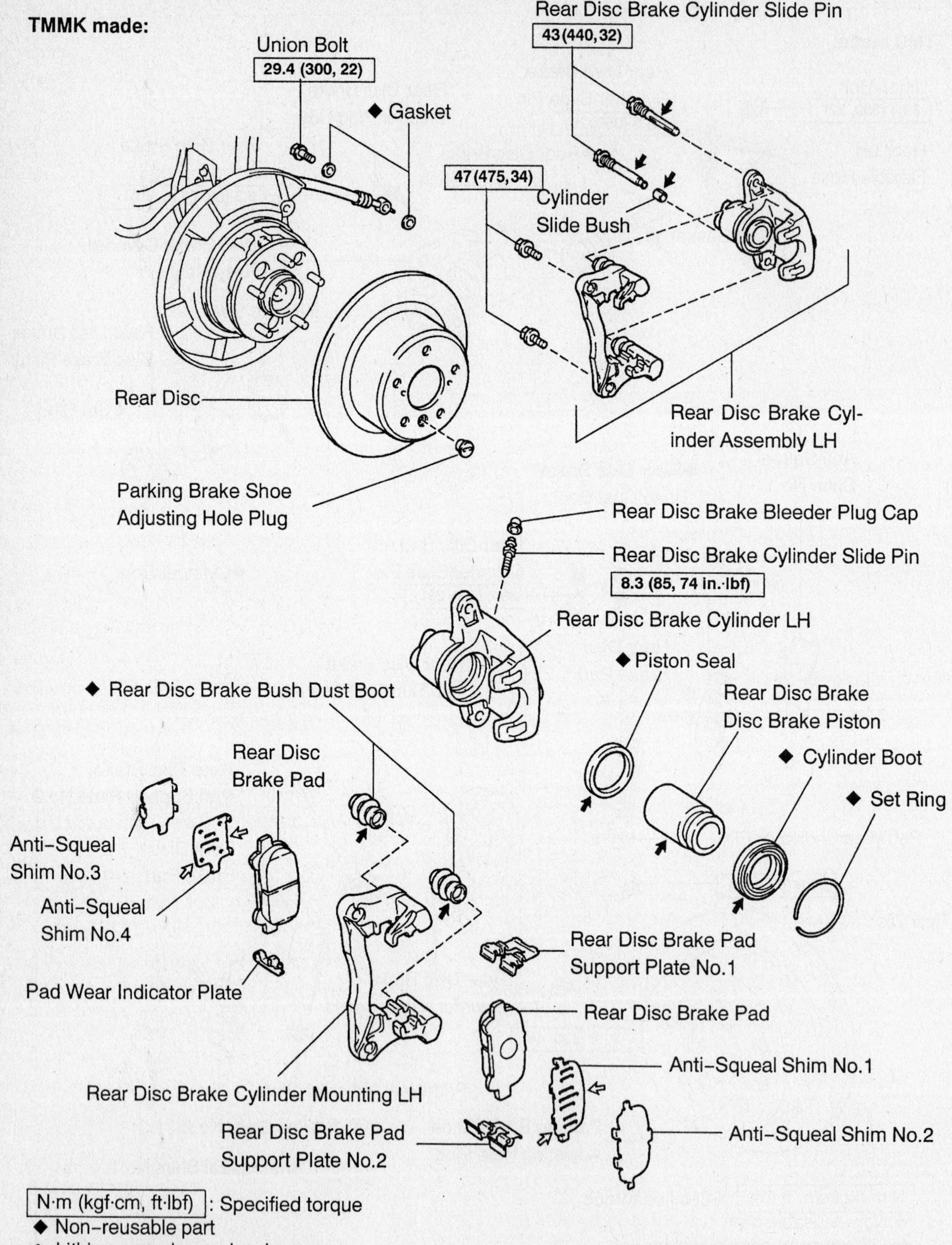
TMMK made:
Union Bolt
29.4 (300, 22)
◆ Gasket
Rear Disc Brake Cylinder Slide Pin
43 (440, 32)
47 (475, 34)
Cylinder Slide Bush
Rear Disc
Rear Disc Brake Cylinder Assembly LH
Parking Brake Shoe Adjusting Hole Plug
Rear Disc Brake Bleeder Plug Cap
Rear Disc Brake Cylinder Slide Pin
8.3 (85, 74 in.·lbf)
Rear Disc Brake Cylinder LH
◆ Piston Seal
◆ Rear Disc Brake Bush Dust Boot
Rear Disc Brake Disc Brake Piston
Rear Disc Brake Pad
◆ Cylinder Boot
◆ Set Ring
Anti-Squeal Shim No.3
Anti-Squeal Shim No.4
Pad Wear Indicator Plate
Rear Disc Brake Pad Support Plate No.1
Rear Disc Brake Pad
Anti-Squeal Shim No.1
Rear Disc Brake Cylinder Mounting LH
Rear Disc Brake Pad Support Plate No.2
Anti-Squeal Shim No.2
N·m (kgf·cm, ft·lbf) : Specified torque
◆ Non-reusable part
Lithium soap base glycol grease
Disc brake grease
67170-TOYO-G49

Rear caliper—Camry, TMMK caliper

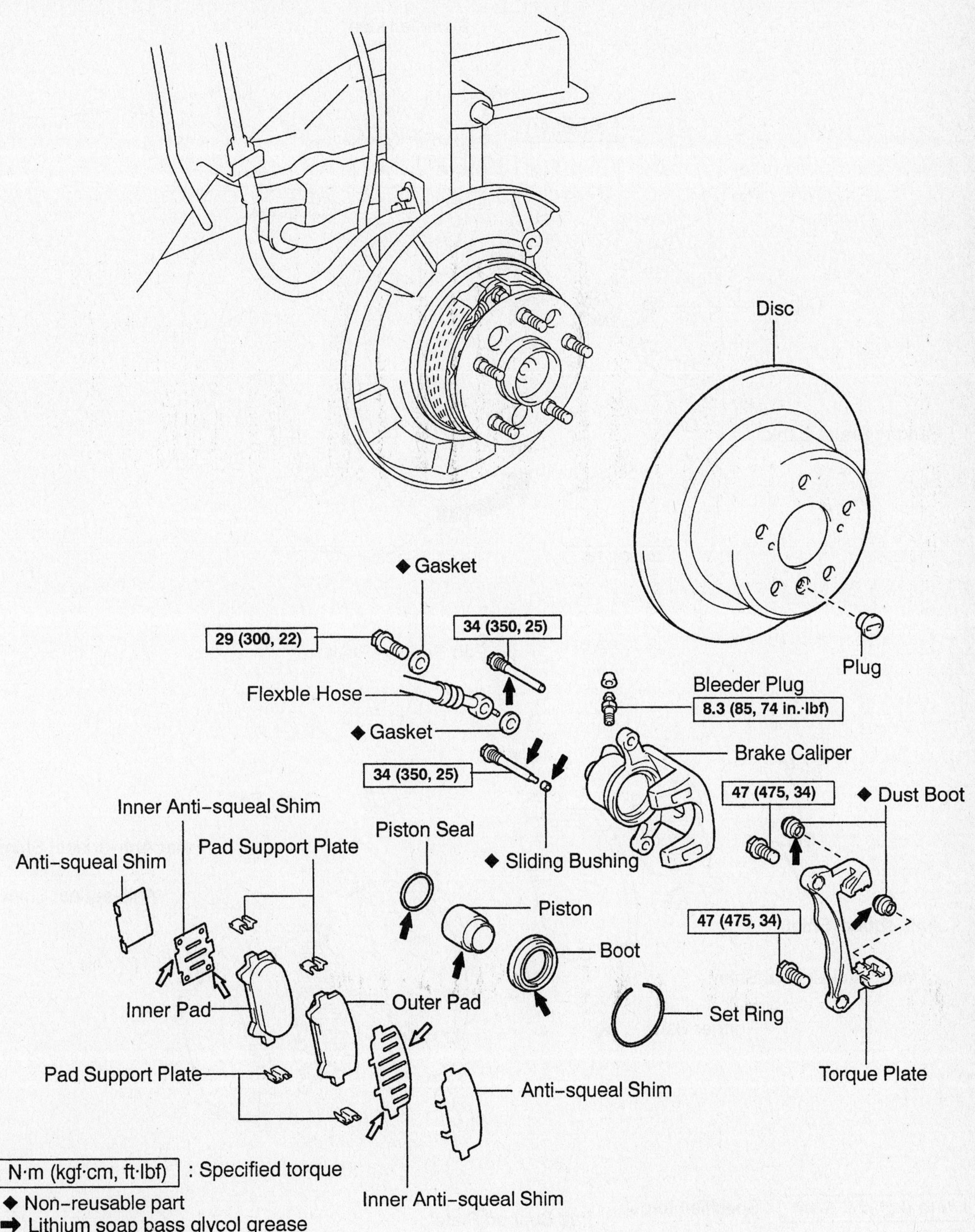
Disc
◆ Gasket
29 (300, 22)
34 (350, 25)
Plug
Flexble Hose
Bleeder Plug
8.3 (85, 74 in.·lbf)
◆ Gasket
Brake Caliper
34 (350, 25)
47 (475, 34)
◆ Dust Boot
Inner Anti-squeal Shim
Piston Seal
Pad Support Plate
Anti-squeal Shim
◆ Sliding Bushing
Piston
47 (475, 34)
Boot
Inner Pad
Outer Pad
Set Ring
Pad Support Plate
Torque Plate
Anti-squeal Shim
N·m (kgf·cm, ft·lbf) : Specified torque
◆ Non-reusable part
Lithium soap bass glycol grease
Disc brake grease
Inner Anti-squeal Shim
67170-TOYO-G50

Rear caliper—Avalon

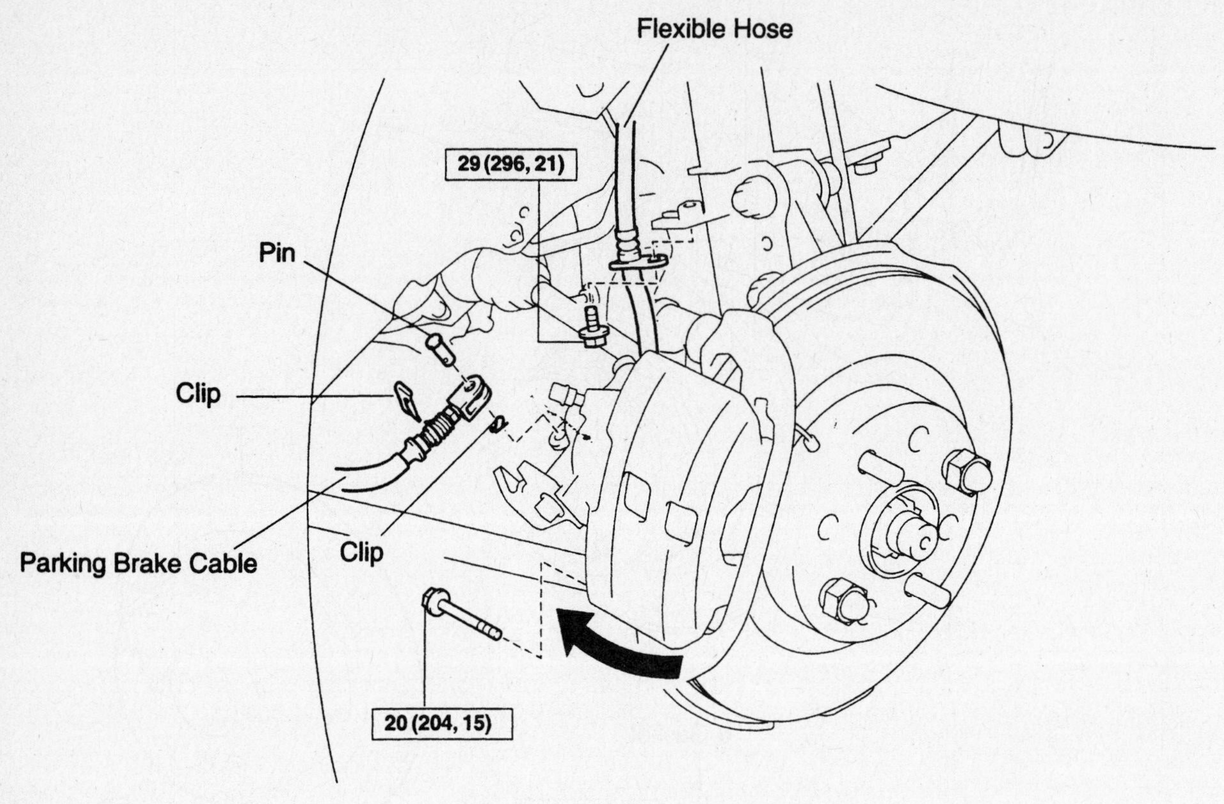
Flexible Hose
29 (296, 21)
Pin
Clip
Parking Brake Cable
Clip
20 (204, 15)

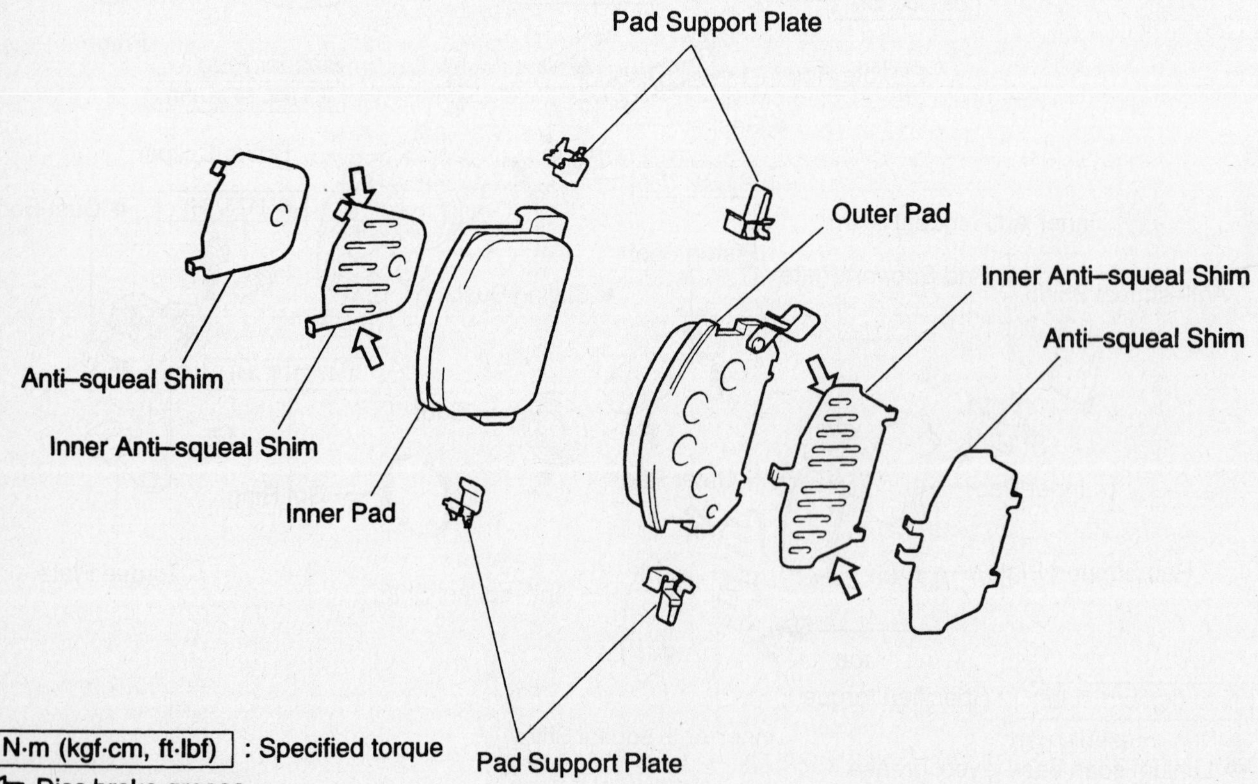
Pad Support Plate
Outer Pad
Inner Anti–squeal Shim
Anti–squeal Shim
Anti–squeal Shim
Inner Anti–squeal Shim
Inner Pad
Pad Support Plate
N·m (kgf·cm, ft·lbf) : Specified torque
⇦ Disc brake grease
9357WG14

Rear caliper—MR2

10. Add brake fluid to the reservoir to fill to the correct level.

11. Lower the vehicle to the ground.

Disc Brake Pads

REMOVAL & INSTALLATION

1. Before servicing the vehicle, refer to the precautions in the beginning of this section.
2. Remove the wheels.
3. Loosen and remove the caliper mounting bolts, then remove the caliper assembly, without disconnecting the brake line. Position it aside.
4. Slide out the old brake pads along with any anti-squeal shims, springs, pad wear indicators and pad support plates.

To install:

5. Install the pad support plates into the torque plate.
6. Install the pad wear indicators onto the pads. Be sure the arrow on the indicator plate is pointing in the direction of rotation.
7. Install the anti-squeal shims on the outside of each pad and then install the pad assemblies into the torque plate.
8. Compress the caliper piston into the bore. For Corolla rear calipers, use tool SST 09719-14020, to rotate the piston clockwise while pressing it into the bore until it locks.
9. Position the caliper back down over the pads.
10. Install and tighten the caliper mounting bolts.
11. Install the wheels. Check the brake fluid level.

Brake Drums

REMOVAL & INSTALLATION

1. Before servicing the vehicle, refer to the precautions in the beginning of this section.
2. Remove the wheels.
3. Remove the brake drum from the axle hub.

To install:

4. Install the brake drum.
5. Install the rear wheels, tighten the wheel lug nuts.

Brake Shoes

REMOVAL & INSTALLATION

All Models

1. Before servicing the vehicle, refer to the precautions in the beginning of this section.
2. Remove the wheels.
3. Remove the brake drum.
4. Unhook the return spring from the leading (front) brake shoe. Remove the hold-down spring and the pin. Pull out the brake shoe and unhook the anchor spring from the lower edge.
5. Remove the hold-down spring from the trailing (rear) shoe. Pull the shoe out with the adjuster strut, automatic adjuster assembly and springs attached and disconnect the parking brake cable. Unhook the return spring and then remove the adjusting strut. Remove the anchor spring.
6. Remove the adjusting strut. Unhook the adjusting lever spring from the rear shoe and then remove the automatic

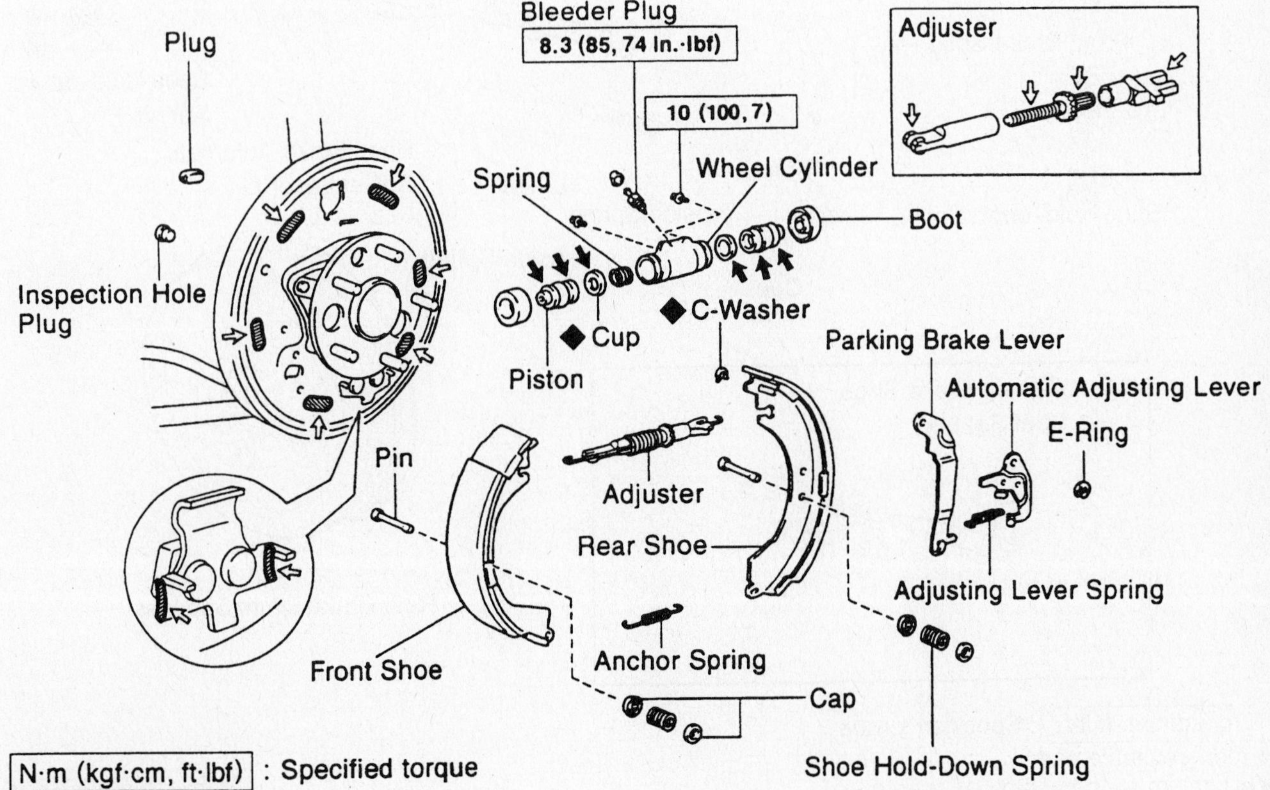

Rear drum brakes—Camry

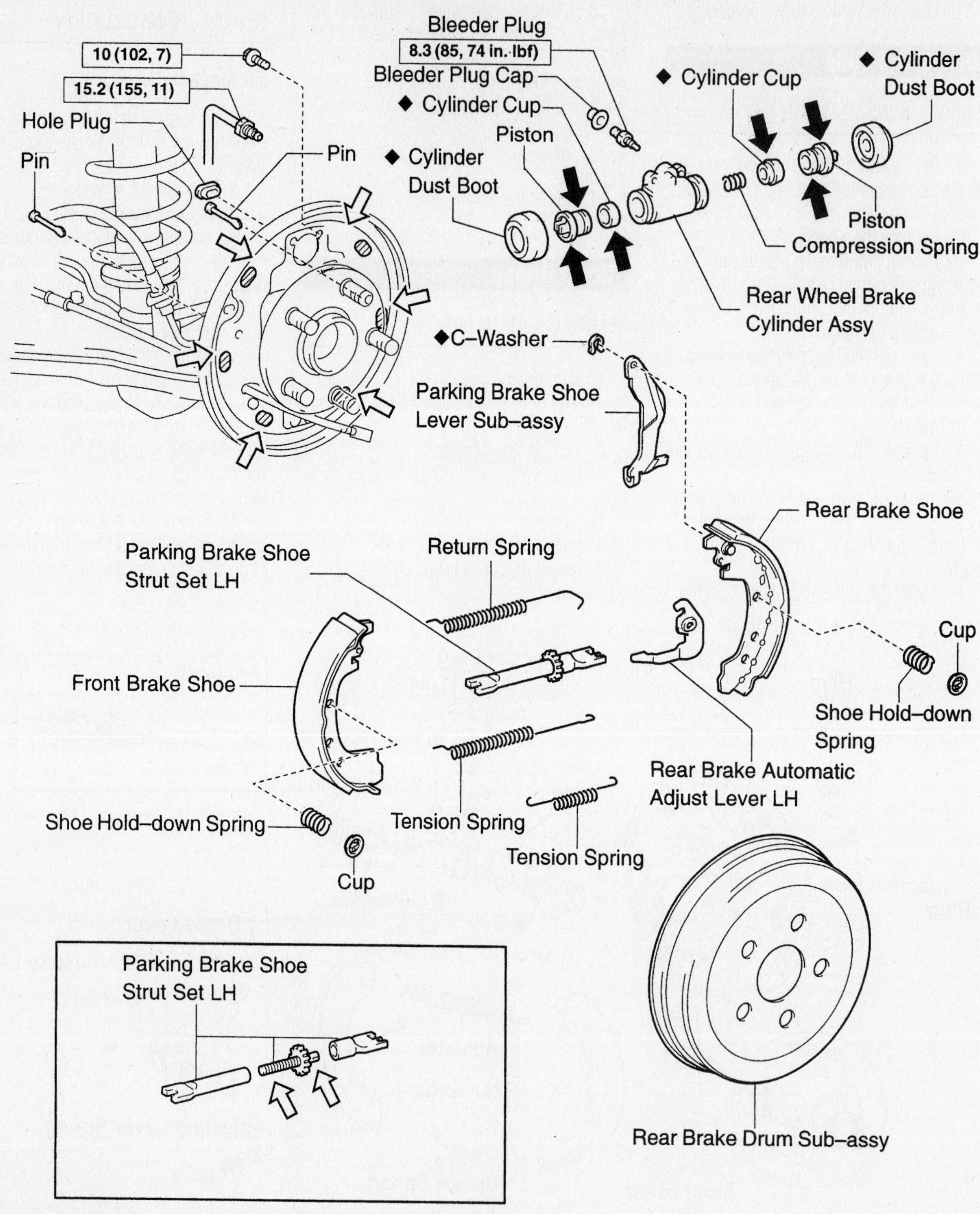
Bleeder Plug
8.3 (85, 74 in.·lbf)
Bleeder Plug Cap
10 (102, 7)
15.2 (155, 11)
Hole Plug
Pin
Pin
◆ Cylinder Cup
Piston
◆ Cylinder Dust Boot
◆ Cylinder Cup
◆ Cylinder Dust Boot
Piston
Compression Spring
Rear Wheel Brake Cylinder Assy
◆C–Washer
Parking Brake Shoe Lever Sub–assy
Rear Brake Shoe
Parking Brake Shoe Strut Set LH
Return Spring
Cup
Front Brake Shoe
Shoe Hold–down Spring
Rear Brake Automatic Adjust Lever LH
Shoe Hold–down Spring
Tension Spring
Tension Spring
Cup
Parking Brake Shoe Strut Set LH
Rear Brake Drum Sub–assy
N·m (kgf·cm, ft·lbf) : Specified torque
◆ Non–reusable part
Lithium soap base glycol grease
High temperature grease
67170-TOYO-G51

Rear drum brakes—Corolla

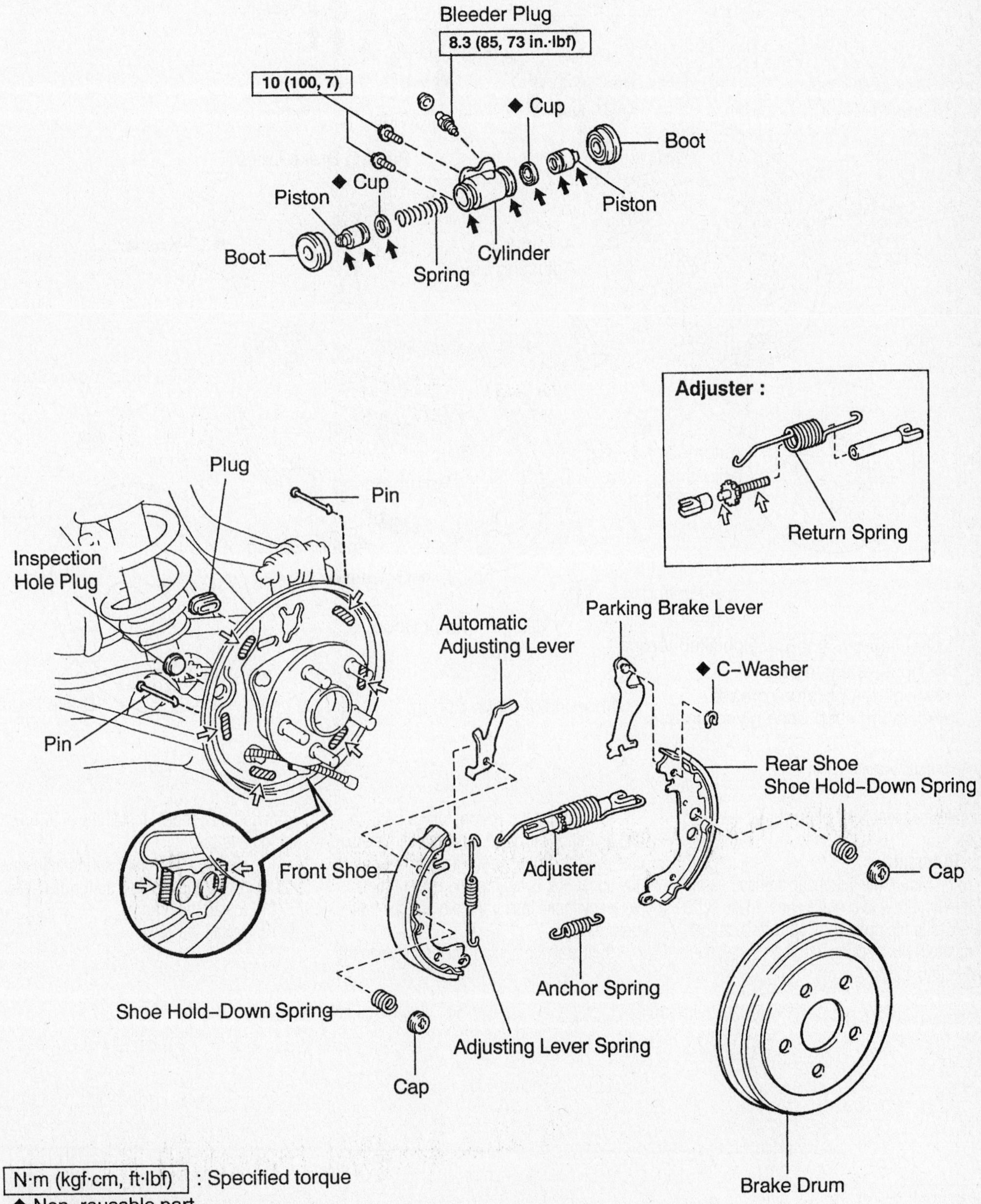

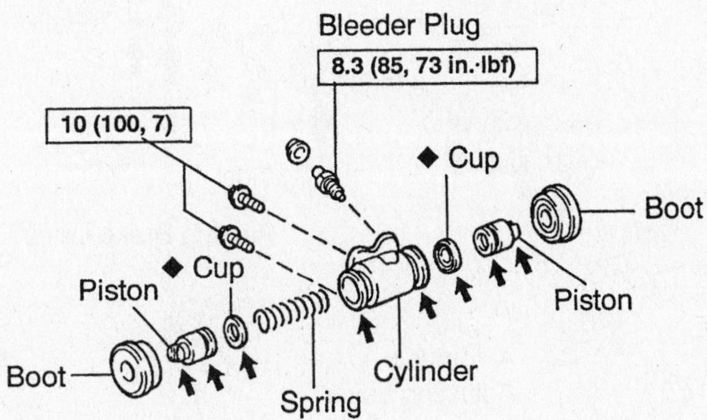

Bleeder Plug
8.3 (85, 73 in.·lbf)
10 (100, 7)
◆ Cup
Boot
◆ Cup
Piston
Piston
Boot
Cylinder
Spring
Adjuster :
Return Spring
Plug
Pin
Inspection Hole Plug
Pin
Parking Brake Lever
Automatic Adjusting Lever
◆ C-Washer
Rear Shoe
Shoe Hold-Down Spring
Front Shoe
Adjuster
Cap
Anchor Spring
Shoe Hold-Down Spring
Adjusting Lever Spring
Cap
Brake Drum

N·m (kgf·cm, ft·lbf) : Specified torque
◆ Non-reusable part
⇦ High temperature grease
⬅ Lithium soap base glycol grease

67170-TOYO-G52

Rear drum brakes—Celica

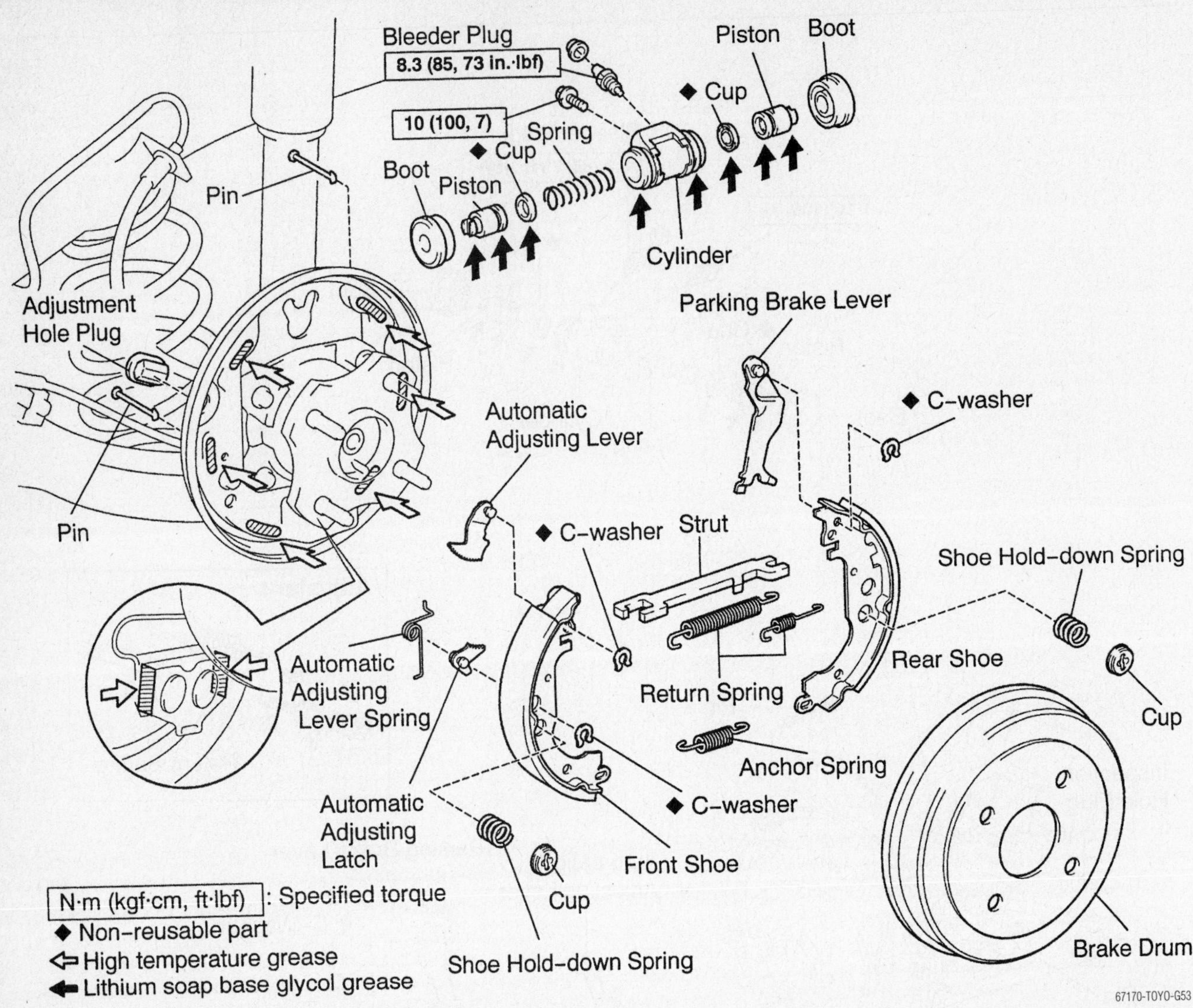

Rear drum brakes—Echo

adjuster assembly by popping out the C-clip.

To install:

7. Mount the automatic adjuster assembly onto a new rear brake shoe. Make sure the C-clip fits properly. Connect the adjusting strut/return spring and then install the adjusting spring.

8. Connect the parking brake cable to the rear shoe and then position the shoe so the lower end rides in the anchor plate and the upper end is against the boot in the wheel cylinder. Install the pin and the hold-down spring.

9. Install the anchor spring between the front and rear shoes. Install the hold-down spring and pin.

10. Connect the return spring/adjusting strut between the 2 shoes so it rides freely.

11. Install the drum.

12. Install the wheel.

TOYOTA

Highlander

18

SPECIFICATION AND MAINTENANCE CHARTS

ENGINE AND VEHICLE IDENTIFICATION

Engine						
Code ①	Liters (cc)	Cu. In.	Cyl.	Fuel Sys.	Engine Type	Eng. Mfg.
1MZ-FE	3.0 (2995)	183	6	SFI	DOHC	Toyota
2AZ-FE	2.4 (2362)	144	4	SFI	DOHC	Toyota
3MZ-FE	3.3 (NA)	NA	6	SFI	DOHC	Toyota

Model Year	
Code ②	Year
1	2001
2	2002
3	2003
4	2004
5	2005

SFI: Sequential Fuel Injection

DOHC: Double Overhead Camshaft

NA: Information not available

① Stamped on the left side of the engine block

② 10th digit of the Vehicle Identification Number (VIN)

67170-HIGH-C01

GENERAL ENGINE SPECIFICATIONS

Year	Model	Engine Displacement Liters	Engine Series ID	Net Horsepower @ rpm	Net Torque @ rpm (ft. lbs.)	Bore x Stroke (in.)	Com-pression Ratio	Oil Pressure @ rpm
2001	Highlander	2.4	2AZ-FE	155@5600	163@4000	3.48x3.78	NA	36@3000
		3.0	1MZ-FE	220@5800	222@4400	3.44x3.27	10.5:1	43-78@3000
2002	Highlander	2.4	2AZ-FE	155@5600	163@4000	3.48x3.78	NA	36@3000
		3.0	1MZ-FE	220@5800	222@4400	3.44x3.27	10.5:1	43-78@3000
2003	Highlander	2.4	2AZ-FE	155@5600	163@4000	3.48x3.78	NA	36@3000
		3.0	1MZ-FE	220@5800	222@4400	3.44x3.27	10.5:1	43-78@3000
2004	Highlander	2.4	2AZ-FE	155@5600	163@4000	3.48x3.78	NA	36@3000
		3.3	3MZ-FE	230@5600	242@3600	NA	NA	36-78@3000

NA: Information not available

67170-HIGH-C02

ENGINE TUNE-UP SPECIFICATIONS

Year	Engine Displacement Liters	Engine ID	Spark Plug Gap (in.)	Ignition Timing (deg.)*	Fuel Pump (psi)	Idle Speed (rpm)	Intake	Exhaust
2001	2.4	2AZ-FE	0.041	8-12B	44-50	600-700	0.007-0.011	0.012-0.016
	3.0	1MZ-FE	0.043	8-12B	44-50	650-750	0.006-0.010	0.010-0.014
2002	2.4	2AZ-FE	0.041	8-12B	44-50	600-700	0.007-0.011	0.012-0.016
	3.0	1MZ-FE	0.039-0.043	8-12B	44-50	650-750	0.006-0.010	0.010-0.014
2003	2.4	2AZ-FE	0.041	8-12B	44-50	600-700	0.007-0.011	0.012-0.016
	3.0	1MZ-FE	0.039-0.043	8-12B	44-50	650-750	0.006-0.010	0.010-0.014
2004	2.4	2AZ-FE	0.041	8-12B	44-50	600-700	0.007-0.011	0.012-0.016
	3.3	3MZ-FE	0.039-0.043	8-12B	44-50	650-750	0.006-0.010	0.010-0.014

NOTE: The Vehicle Emission Control Information label often reflects specification changes made during production.

The label figures must be used if they differ from those in this chart.

B: Before top dead center

* With terminals TC and CG connected to DLC3

67170-HIGH-C03

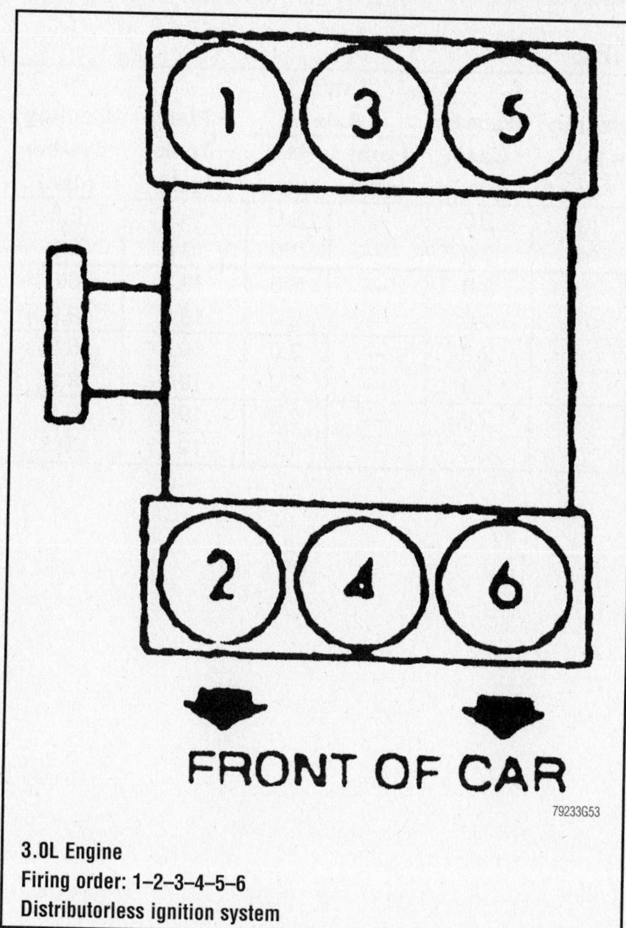

3.0L Engine
Firing order: 1–2–3–4–5–6
Distributorless ignition system

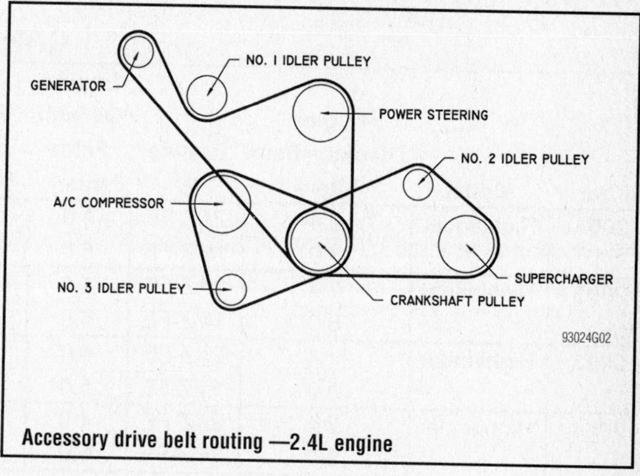

Accessory drive belt routing —2.4L engine

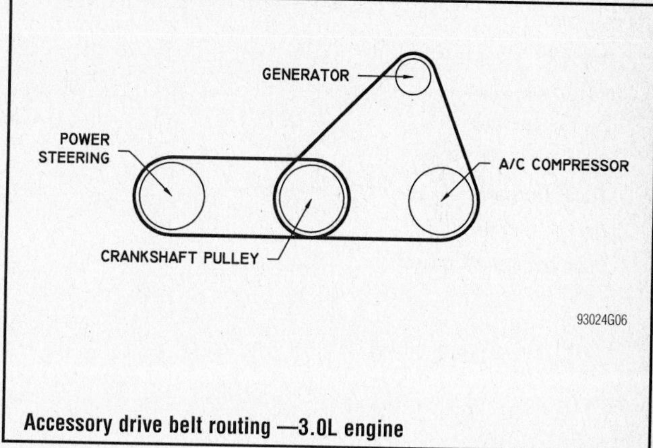

Accessory drive belt routing —3.0L engine

CAPACITIES

Year	Model	Engine Displacement Liters	Engine ID	Engine Oil with Filter (qts.)	Transmission (pts.) 5-Spd	Transmission (pts.) Auto.*	Transfer Case (pts.)	Drive Axle Front (pts.)	Drive Axle Rear (pts.)	Fuel Tank (gal.)	Cooling System (qts.)
2001	Highlander	2.4	2AZ-FE	4.0	—	①	2.0	—	2.0	19.8	6.8
		3.0	1MZ-FE	5.0	—	①	2.0	—	2.0	19.8	9.9
2002	Highlander	2.4	2AZ-FE	4.0	—	①	2.0	—	2.0	19.8	6.8
		3.0	1MZ-FE	5.0	—	①	2.0	—	2.0	19.8	9.9
2003	Highlander	2.4	2AZ-FE	4.0	—	①	2.0	—	2.0	19.8	6.8
		3.0	1MZ-FE	5.0	—	①	2.0	—	2.0	19.8	9.9
2004	Highlander	2.4	2AZ-FE	4.0	—	①	2.0	—	2.0	19.2	6.8
		3.3	3MZ-FE	5.0	—	②	2.0	—	1.9	19.2	10.3

*After draining, add the following amounts, then, fill to the cold full line.

① 2wd: 7.0
4wd: 8.2

② U151E Transaxle
Dry Fill: 18.6 pts.
Drain and Refill: 7.4 pts.
U151F Transaxle
Dry Fill: 19.0 pts.
Drain and Refill: 7.6 pts.

67170-HIGH-C04

VALVE SPECIFICATIONS

Year	Engine Displacement Liters	Engine ID	Seat Angle (deg.)	Face Angle (deg.)	Spring Test Pressure (lbs. @ in.)	Spring Installed Height (in.)	Stem-to-Guide Clearance (in.) Intake	Stem-to-Guide Clearance (in.) Exhaust	Stem Diameter (in.) Intake	Stem Diameter (in.) Exhaust
2001	2.4	2AZ-FE	45	44.5	NA	NA	0.0010-0.0024	0.0012-0.0026	0.2154-0.2159	0.2152-0.2157
	3.0	1MZ-FE	45	40.5	41.9-46.3@1.437	1.331	0.0010-0.0024	0.0012-0.0026	0.2154-0.2159	0.2152 0.2156
2002	2.4	2AZ-FE	45	44.5	NA	NA	0.0010-0.0024	0.0012-0.0026	0.2154-0.2159	0.2152-0.2157
	3.0	1MZ-FE	45	40.5	41.9-46.3@1.437	1.331	0.0010-0.0024	0.0012-0.0026	0.2154-0.2159	0.2152 0.2156
2003	2.4	2AZ-FE	45	44.5	NA	NA	0.0010-0.0024	0.0012-0.0026	0.2154-0.2159	0.2152-0.2157
	3.0	1MZ-FE	45	40.5	41.9-46.3@1.437	1.331	0.0010-0.0024	0.0012-0.0026	0.2154-0.2159	0.2152 0.2156
2004	2.4	2AZ-FE	45	44.5	NA	NA	0.0010-0.0024	0.0012-0.0026	0.2154-0.2159	0.2152-0.2157
	3.3	3MZ-FE	45	40.5	41.9-46.3@1.437	1.331	0.0010-0.0024	0.0012-0.0026	0.2154-0.2159	0.2152 0.2156

NA: Information not available

67170-HIGH-C05

CRANKSHAFT AND CONNECTING ROD SPECIFICATIONS

All measurements are given in inches.

Year	Engine Displacement Liters	Engine ID	Crankshaft				Connecting Rod		
			Main Brg. Journal Dia.	Main Brg. Oil Clearance	Shaft End-play	Thrust on No.	Journal Diameter	Oil Clearance	Side Clearance
2001	2.4	2AZ-FE	2.0654-2.1648	0.0009-0.0019	0.0016-0.0094	2	1.8894-1.8898	0.0009-0.0019	0.0063-0.0143
	3.0	1MZ-FE	2.4011-2.4016	①	0.0016-0.0095	2	2.0863-2.0866	0.0015-0.0025	0.0059-0.0188
2002	2.4	2AZ-FE	2.0654-2.1648	0.0009-0.0019	0.0016-0.0094	2	1.8894-1.8898	0.0009-0.0019	0.0063-0.0143
	3.0	1MZ-FE	2.4011-2.4016	①	0.0016-0.0095	2	2.0863-2.0866	0.0015-0.0025	0.0059-0.0188
2003	2.4	2AZ-FE	2.0654-2.1648	0.0009-0.0019	0.0016-0.0094	2	1.8894-1.8898	0.0009-0.0019	0.0063-0.0143
	3.0	1MZ-FE	2.4011-2.4016	①	0.0016-0.0095	2	2.0863-2.0866	0.0015-0.0025	0.0059-0.0188
2004	2.4	2AZ-FE	2.0654-2.1648	0.0009-0.0019	0.0016-0.0094	2	1.8894-1.8898	0.0009-0.0019	0.0063-0.0143
	3.3	3MZ-FE	2.4011-2.4016	①	0.0016-0.0095	2	2.0863-2.0866	0.0015-0.0026	0.0059-0.0118

① Journals 1 and 4: 0.0006 - 0.0013 in.
Journals 2 and 3: 0.0010 - 0.0018 in.

67170-HIGH-C06

PISTON AND RING SPECIFICATIONS

All measurements are given in inches.

Year	Engine Displ. Liters	Engine ID	Piston Clearance	Ring Gap			Ring Side Clearance		
				Top Comp.	Bottom Comp.	Oil Control	Top Comp.	Bottom Comp.	Oil Control
2001	2.4	2AZ-FE	0.0020-0.0029	0.0087-0.0126	0.0197-0.0236	0.0039-0.0138	0.0012-0.0028	0.0012-0.0028	SNUG
	3.0	1MZ-FE	0.0033-0.0042	0.0098-0.0138	0.0138-0.0177	0.0059-0.0157	0.0008-0.0028	0.0008-0.0024	SNUG
2002	2.4	2AZ-FE	0.0020-0.0029	0.0087-0.0126	0.0197-0.0236	0.0039-0.0138	0.0012-0.0028	0.0012-0.0028	SNUG
	3.0	1MZ-FE	0.0033-0.0042	0.0098-0.0138	0.0138-0.0177	0.0059-0.0157	0.0008-0.0028	0.0008-0.0024	SNUG
2003	2.4	2AZ-FE	0.0020-0.0029	0.0087-0.0126	0.0197-0.0236	0.0039-0.0138	0.0012-0.0028	0.0012-0.0028	SNUG
	3.0	1MZ-FE	0.0033-0.0042	0.0098-0.0138	0.0138-0.0177	0.0059-0.0157	0.0008-0.0028	0.0008-0.0024	SNUG
2004	2.4	2AZ-FE	0.0020-0.0029	0.0087-0.0126	0.0197-0.0236	0.0039-0.0138	0.0012-0.0028	0.0012-0.0028	SNUG
	3.3	3MZ-FE	0.0013-0.0023	0.0118-0.0138	0.0197-0.0236	0.0059-0.0157	0.0012-0.0031	0.0008-0.0024	0.0012-0.0043

67170-HIGH-C07

TORQUE SPECIFICATIONS

All readings in ft. lbs.

Year	Engine Displacement Liters	Engine ID	Cylinder Head Bolts	Main Bearing Bolts	Rod Bearing Bolts	Crankshaft Damper Bolts	Flywheel Bolts	Manifold		Spark Plugs	Oil Pan Drain Plug
								Intake	Exhaust		
2001	2.4	2AZ-FE	①	29	②	125	72	22	27	14	18
	3.0	1MZ-FE	③	④	②	159	61	32	36	13	33
2002	2.4	2AZ-FE	①	29	②	125	72	22	27	14	18
	3.0	1MZ-FE	④	④	②	159	61	32	36	13	33
2003	2.4	2AZ-FE	①	29	②	125	72	22	27	14	18
	3.0	1MZ-FE	③	④	②	159	61	32	36	13	33
2004	2.4	2AZ-FE	①	29	②	125	72	22	27	14	18
	3.3	3MZ-FE	⑤	④	②	162	61	11	36	18	33

① Step 1: 58
Step 2: plus 90 degrees

② Step 1: 18 ft. lbs.
Step 2: Plus 90 degrees

③ Step 1: 12 point bolts to 40 ft. lbs.
Step 2: 12 point bolts plus 90 degrees
Step 3: Hex head recessed bolt to 13 ft. lbs.

④ Step 1: 12 point cap bolts to 16 ft. lbs.
Step 2: 12 point cap bolts plus 90 degrees
Step 3: Hex head side bolts to 20 ft. lbs.

⑤ Step 1: 12 point bolts to 40 ft. lbs.
Step 2: 12 point bolts plus 90 degrees
Step 3: Hex head recessed bolt to 13 ft. lbs.

67170-HIGH-C08

WHEEL ALIGNMENT

Year	Model		Caster Range (+/-Deg.)	Caster Preferred Setting (Deg.)	Camber Range (+/-Deg.)	Camber Preferred Setting (Deg.)	Toe-in (in.)	Steering Axis Inclination (Deg.)
2001	Highlander	2WD F	0.75	+2.75	0.75	-0.67	0+/-0.08	10.75+/-0.75
		4WD F	0.75	+2.75	0.75	-0.58	0+/-0.08	10.58+/-0.75
		2WD R	—	—	0.75	-1.33	0.12+/-0.08	—
		4WD R	—	—	0.75	-0.75	0.12+/-0.08	—
2002	Highlander	2WD F	0.75	+2.75	0.75	-0.67	0+/-0.08	10.75+/-0.75
		4WD F	0.75	+2.75	0.75	-0.58	0+/-0.08	10.58+/-0.75
		2WD R	—	—	0.75	-1.33	0.12+/-0.08	—
		4WD R	—	—	0.75	-0.75	0.12+/-0.08	—
2003	Highlander	2WD F	0.75	+2.75	0.75	-0.67	0+/-0.08	10.75+/-0.75
		4WD F	0.75	+2.75	0.75	-0.58	0+/-0.08	10.58+/-0.75
		2WD R	—	—	0.75	-1.33	0.12+/-0.08	—
		4WD R	—	—	0.75	-0.75	0.12+/-0.08	—
2004	Highlander	2WD F	0.75	+2.75	0.75	-0.67	0+/-0.08	10.75+/-0.75
		4WD F	0.75	+2.75	0.75	-0.58	0+/-0.08	10.58+/-0.75
		2WD R	—	—	0.75	-1.33	0.12+/-0.08	—
		4WD R	—	—	0.75	-0.75	0.12+/-0.08	—

67170-HIGH-C09

TIRE, WHEEL AND BALL JOINT SPECIFICATIONS

Year	Model	OEM Tires		Tire Pressures (psi)		Wheel	Ball Joint	Lugnut Torque
		Standard	Optional	Front	Rear	Size	Inspection	(ft. lbs.)
2001	Highlander	P225/70R16	None	30	30	6.5-JJ	①	70
2002	Highlander	P225/70R16	None	30	30	6.5-JJ	①	70
2003	Highlander	P225/70R16	None	30	30	6.5-JJ	①	70
2004	Highlander	P225/70R16	P225/65R17	30	30	6.5-JJ	①	76

OEM: Original Equipment Manufacturer

PSI: Pounds Per Square Inch

STD: Standard

OPT: Optional

① Replace if any measurable movement is found.

67170-HIGH-C10

BRAKE SPECIFICATIONS

All measurements in inches unless noted

Year	Model		Brake Disc			Minimum	Brake Caliper	
			Original Thickness	Minimum Thickness	Maximum Runout	Lining Thickness	Bracket Bolts (ft. lbs.)	Mounting Bolts (ft. lbs.)
2001	Highlander	F	1.102	1.024	0.0020	0.039	79	25
		R	0.394	0.354	0.0059	0.039	43	25
2002	Highlander	F	1.102	1.024	0.0020	0.039	79	25
		R	0.394	0.354	0.0059	0.039	43	25
2003	Highlander	F	1.102	1.024	0.0020	0.039	79	25
		R	0.394	0.354	0.0059	0.039	43	25
2004	Highlander	F	1.102	1.024	0.0020	0.039	77	25
		R	0.394	0.335	0.0059	0.039	58	32

F: Front

R: Rear

67170-HIGH-C11

SCHEDULED MAINTENANCE INTERVALS

TOYOTA—HIGHLANDER

TO BE SERVICED	TYPE OF SERVICE	VEHICLE MILEAGE INTERVAL (x1000)												
		7.5	15	22.5	30	37.5	45	52.5	60	67.5	75	82.5	90	97.5
Engine oil & filter	R	✓	✓	✓	✓	✓	✓	✓	✓	✓	✓	✓	✓	✓
Automatic transmission fluid	S/I		✓		✓		✓		✓		✓		✓	
Ball joints & dust covers	S/I		✓		✓		✓		✓		✓		✓	
Bolts & nuts on chassis & body	S/I		✓		✓		✓		✓		✓		✓	
Brake linings & drums	S/I		✓		✓		✓		✓		✓		✓	
Brake line pipes & hoses	S/I		✓		✓		✓		✓		✓		✓	
Brake pads & discs (front & rear)	S/I		✓		✓		✓		✓		✓		✓	
Propeller shaft grease	S/I		✓		✓		✓		✓		✓		✓	
Steering knuckle & chassis grease	S/I		✓		✓		✓		✓		✓		✓	
Steering linkage	S/I		✓		✓		✓		✓		✓		✓	
Air cleaner filter	R				✓				✓				✓	
Spark plugs ①	R				✓				✓				✓	
Drive belts	S/I				✓				✓				✓	
Exhaust pipes & mountings	S/I				✓				✓				✓	
Fuel lines & connections	S/I				✓				✓				✓	
Engine coolant	R						✓				✓			
Charcoal canister	R								✓					
Fuel tank cap gasket	R								✓					
Heated oxygen sensors (except Calif.) ②	R													

R: Replace S/I: Service or Inspect

① Platinum plugs are replaced at 100,000 mile intervals

② Heated oxygen sensors (except Calif.): replace every 80,000 miles.

FREQUENT OPERATION MAINTENANCE (SEVERE SERVICE)

If a vehicle is operated under any of the following conditions it is considered severe service:

- Extremely dusty areas.

- 50% or more of the vehicle operation is in 32°C (90°F) or higher temperatures, or constant temperatures below 0°C (32°F).

- Prolonged idling (vehicle operation in stop and go traffic).

- Frequent short running periods (engine does not warm to normal operating temperatures).

- Police, taxi, delivery usage or trailer towing usage.

Air cleaner filter: service or inspect every 3750 miles

Engine oil & filter: replace every 3750 miles.

Ball joints & dust covers: service or inspect every 7500 miles.

Bolts & nuts on chassis & body: service or inspect every 7500 miles.

Brake pads & discs (front & rear): service or inspect every 7500 miles.

Steering knuckle & chassis grease: service or inspect every 7500 miles.

Steering linkage: service or inspect every 7500 miles.

Exhaust pipes & mountings: service or inspect every 15,000 miles.

67170-HIGH-C12

PRECAUTIONS

Before servicing any vehicle, please be sure to read all of the following precautions, which deal with personal safety, prevention of component damage, and important points to take into consideration when servicing a motor vehicle:

- Never open, service or drain the radiator or cooling system when the engine is hot; serious burns can occur from the steam and hot coolant.
- Observe all applicable safety precautions when working around fuel. Whenever servicing the fuel system, always work in a well-ventilated area. Do not allow fuel spray or vapors to come in contact with a spark, open flame, or excessive heat (a hot drop light, for example). Keep a dry chemical fire extinguisher near the work area. Always keep fuel in a container specifically designed for fuel storage; also, always properly seal fuel containers to avoid the possibility of fire or explosion. Refer to the additional fuel system precautions later in this section.
- Fuel injection systems often remain pressurized, even after the engine has been turned **OFF**. The fuel system pressure must be relieved before disconnecting any fuel lines. Failure to do so may result in fire and/or personal injury.
- Brake fluid often contains polyglycol ethers and polyglycols. Avoid contact with the eyes and wash your hands thoroughly after handling brake fluid. If you do get brake fluid in your eyes, flush your eyes with clean, running water for 15 minutes. If eye irritation persists, or if you have taken brake fluid internally, IMMEDIATELY seek medical assistance.
- The EPA warns that prolonged contact with used engine oil may cause a number of skin disorders, including cancer. You should make every effort to minimize your exposure to used engine oil. Protective gloves should be worn when changing oil. Wash your hands and any other exposed skin areas as soon as possible after exposure to used engine oil. Soap and water, or waterless hand cleaner should be used.
- All new vehicles are now equipped with an air bag system. The system must be disabled before performing service on or around system components, steering column, instrument panel components, wiring and sensors. Failure to follow safety and disabling procedures could result in accidental air bag deployment, possible personal injury and unnecessary system repairs.
- Always wear safety goggles when working with, or around, the air bag system. When carrying a non-deployed air bag, be sure the bag and trim cover are pointed away from your body. When placing a non-deployed air bag on a work surface, always face the bag and trim cover upward, away from the surface. This will reduce the motion of the module if it is accidentally deployed. Refer to the additional air bag system precautions later in this section.
- NEVER disconnect the negative battery cable with the ignition **ON** or the engine running. Removing power from the computer control module with the ignition **ON** may destroy the module.
- Clean, high quality brake fluid from a sealed container is essential to the safe and proper operation of the brake system. You should always buy the correct type of brake fluid for your vehicle. If the brake fluid becomes contaminated, completely flush the system with new fluid. Never reuse any brake fluid. Any brake fluid that is removed from the system should be discarded. Also, do not allow any brake fluid to come in contact with a painted surface; it will damage the paint.
- Never operate the engine without the proper amount and type of engine oil; doing so WILL result in severe engine damage.
- Timing belt maintenance is extremely important. Many models utilize an interference-type, non-freewheeling engine. If the timing belt breaks, the valves in the cylinder head may strike the pistons, causing potentially serious (also time-consuming and expensive) engine damage. Refer to the maintenance interval charts in the front of this manual for the recommended replacement interval for the timing belt, and to the timing belt section for belt replacement and inspection.
- Disconnecting the negative battery cable on some vehicles may interfere with the functions of the on-board computer system(s) and may require the computer to undergo a relearning process once the negative battery cable is reconnected.
- When servicing drum brakes, only disassemble and assemble one side at a time, leaving the remaining side intact for reference.
- Only an MVAC-trained, EPA-certified automotive technician should service the air conditioning system or its components.

ENGINE REPAIR

➡Disconnecting the negative battery cable on some vehicles may interfere with the functions of the on board computer system. The computer may undergo a relearning process once the negative battery cable is reconnected.

Alternator

REMOVAL

2.4L Engine

1. Before servicing the vehicle, refer to the precautions in the beginning of this section.
2. Remove or disconnect the following:
   - Electrical wiring from the alternator
   - Drive belt

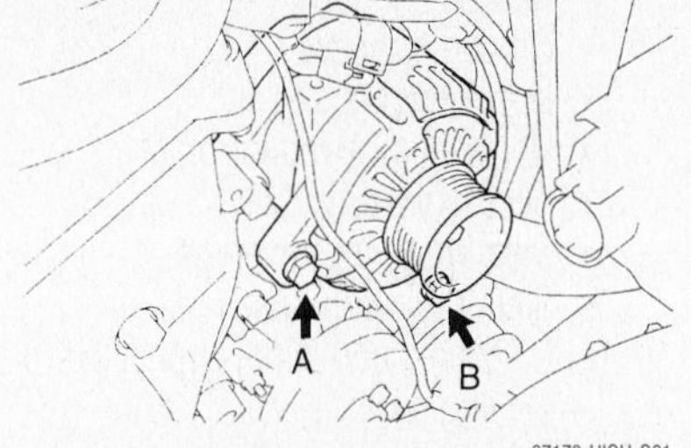

Alternator bolt locations—2.4L engine

   - 1 adjusting and 2 mounting bolts
   - Alternator
3. Installation is the reverse of removal. Observe the following torques:
   - M8 bolts: 15 ft. lbs. (21Nm)
   - M10 bolts: 38 ft. lbs. (52Nm)

3.0L Engine

1. Before servicing the vehicle, refer to the precautions in the beginning of this section.
2. Remove or disconnect the following:
   - Alternator electrical connectors
   - Wiring harness from the clip
   - Pivot bolt
   - Adjuster lockbolt
   - Drive belt
   - Alternator

To install:

3. Install or connect the following:
   - Alternator
   - Drive belt
   - Adjusting lockbolt. Tighten the bolt to 13 ft. lbs. (18 Nm).

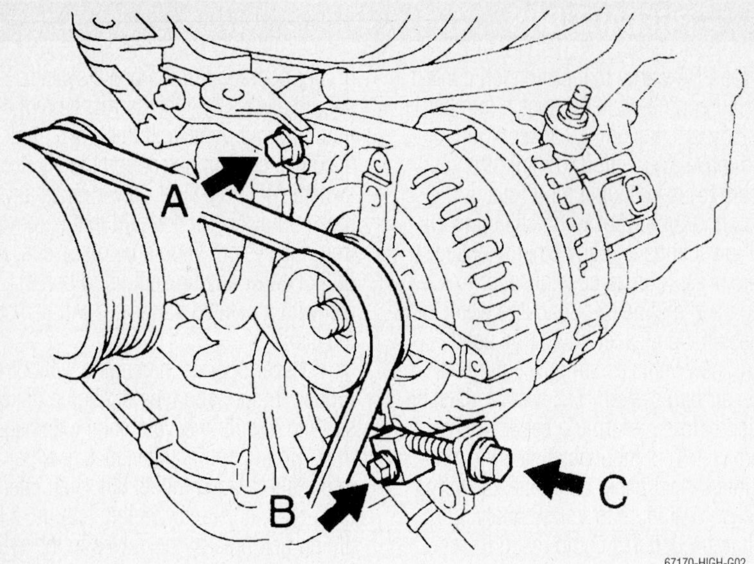

Alternator bolt locations—3.3L engine

- Pivot bolt. Tighten the bolt to 43 ft. lbs. (58 Nm).
- Wiring harness from the clip
- Alternator electrical connectors

3.3L Engine

1. Before servicing the vehicle, refer to the precautions in the beginning of this section.
2. Remove or disconnect the following:
   - Alternator electrical connectors
   - Wiring harness from the clip
   - Pivot bolt
   - Plate washer
   - Adjusting lockbolt
   - Drive belt
   - Alternator

To install:

3. Install or connect the following:
   - Alternator
   - Drive belt. Tension the belt to 170–180 lbs. for a new belt or 95–135 lbs. for a used belt.
   - Adjusting lockbolt. Tighten the bolt to 13 ft. lbs. (18 Nm).
   - Plate washer
   - Pivot bolt. Tighten the bolt to 41 ft. lbs. (56 Nm) for the 3.0L and 43 ft. lbs. (58 Nm) for the 3.3L.
   - Wiring harness from the clip
   - Alternator electrical connectors

Ignition Timing

ADJUSTMENT

All engines are equipped with a Distributorless Ignition System (DIS). No timing adjustment is possible.

Engine Assembly

REMOVAL & INSTALLATION

2.4L Engine

1. Before servicing the vehicle, refer to the precautions in the beginning of this section.
2. Matchmark the hood position.
3. Remove or disconnect the following:
   - Front wheels
   - No.1 engine undercover
   - Right and left fender splash shields
   - Right fender apron seal
   - Engine oil
   - Coolant
   - Transaxle fluid
   - Transfer case oil
   - Battery
   - Air cleaner
   - Radiator hoses
   - Oil cooler hoses
   - Upper engine stay
   - Upper engine mount bracket
   - Accessory drive belts
   - Steering pump reservoir
   - Steering pump hoses
   - All cables and wires connected to the engine
   - Exhaust pipe
   - Front drive shaft
   - Stabilizer links
   - Left and right axle hub nuts
   - Left and right speed sensors
   - Left and right tie rods
   - Left and right lower control arms
   - Torque converter-to-drive plate bolts
   - Intermediate steering shaft
   - AC compressor
4. Attach a crane, remove the 6 side rail plate subassembly bolts (3 each side) and the front suspension member rear brace.
5. Lift the engine out of the vehicle.
6. Installation is the reverse of removal. Observe the following torques:
   - Frame side plate bolts: Large 63 ft. lbs. (85Nm); small 24 ft. lbs. (32Nm)
   - Suspension member rear brace: Large 63 ft. lbs. (85Nm); small 24 ft. lbs. (32Nm)
   - Intermediate shaft bolt: 26 ft. lbs. (35Nm)
   - Torque converter bolts: 30 ft. lbs. (41Nm)
   - Lower control arms, bolts and nuts: 94 ft. lbs. (127Nm)
   - Tie rod nuts: 36 ft. lbs. (49Nm)
   - Speed sensors: 71 inch lbs. (8Nm)
   - Hub nuts: 217 ft. lbs. (294Nm)
   - Stabilizer link nuts: 55 ft. lbs. (74Nm)
   - Driveshaft nuts: 55 ft. lbs. (74Nm)
   - Engine mount bracket: 15 ft. lbs. (20Nm)
   - Engine mount stay: 47 ft. lbs. (64Nm)

3.0L Engine

1. Before servicing the vehicle, refer to the precautions in the beginning of this section.
2. Matchmark the hood position.
3. Remove or disconnect the following:
   - Front wheels
   - No.1 engine undercover
   - Right and left fender splash shields
   - Right fender apron seal
   - Coolant
   - Engine oil
   - Transaxle fluid
   - Transfer case oil
   - Battery
   - Air cleaner
   - Radiator and heater hoses
   - Oil cooler hoses
   - Upper engine stay
   - Upper engine mount bracket
   - Accessory drive belts
   - Steering pump reservoir
   - Steering pump hoses
   - All cables and wires connected to the engine
   - Exhaust pipes
   - Front drive shaft with center bearing
   - Stabilizer links
   - Starter

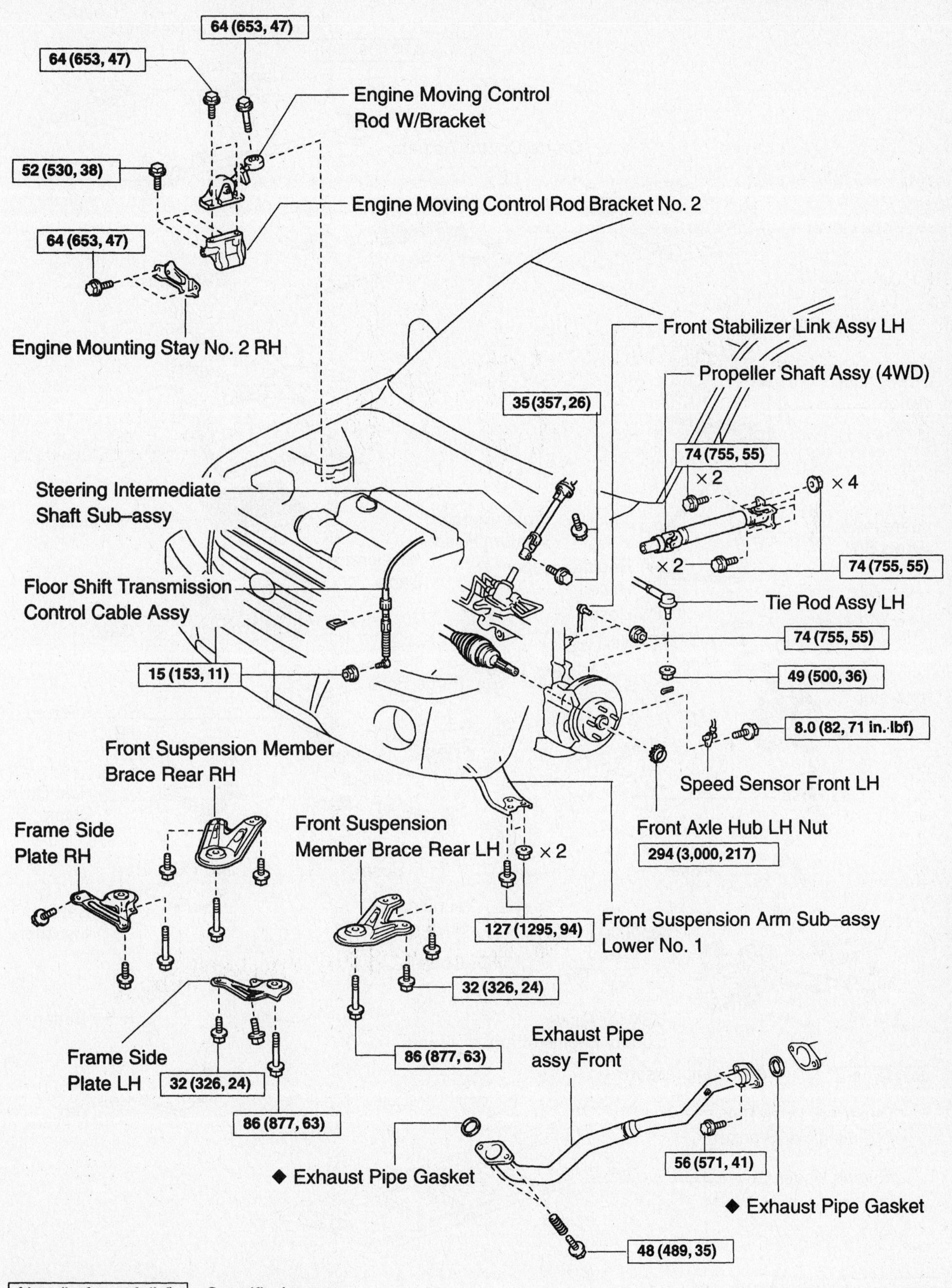
64 (653, 47)
64 (653, 47)
Engine Moving Control Rod W/Bracket
52 (530, 38)
Engine Moving Control Rod Bracket No. 2
64 (653, 47)
Engine Mounting Stay No. 2 RH
Front Stabilizer Link Assy LH
Propeller Shaft Assy (4WD)
35 (357, 26)
74 (755, 55)
× 2
× 4
Steering Intermediate Shaft Sub–assy
× 2
74 (755, 55)
Floor Shift Transmission Control Cable Assy
Tie Rod Assy LH
74 (755, 55)
15 (153, 11)
49 (500, 36)
8.0 (82, 71 in.·lbf)
Front Suspension Member Brace Rear RH
Speed Sensor Front LH
Frame Side Plate RH
Front Suspension Member Brace Rear LH
× 2
Front Axle Hub LH Nut
294 (3,000, 217)
127 (1295, 94)
Front Suspension Arm Sub–assy Lower No. 1
32 (326, 24)
Exhaust Pipe assy Front
86 (877, 63)
Frame Side Plate LH
32 (326, 24)
86 (877, 63)
56 (571, 41)
◆ Exhaust Pipe Gasket
◆ Exhaust Pipe Gasket
48 (489, 35)
N·m (kgf·cm, ft·lbf) : Specified torque
◆ Non–reusable part
67170-HIGH-G03

Engine mounts and related parts—2.4L engine

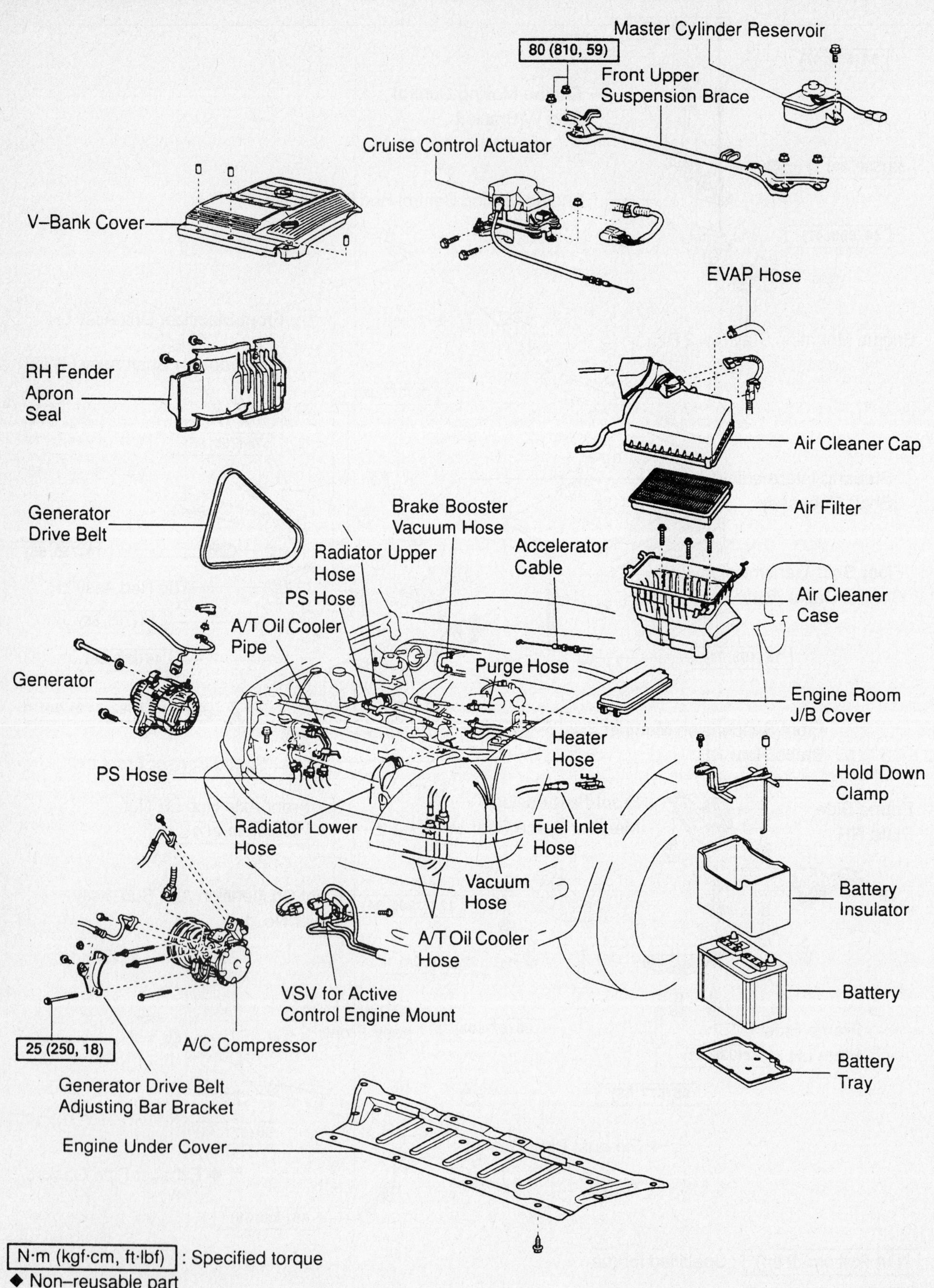
Master Cylinder Reservoir
80 (810, 59)
Front Upper
Suspension Brace
Cruise Control Actuator
V–Bank Cover
EVAP Hose
RH Fender
Apron
Seal
Air Cleaner Cap
Generator
Drive Belt
Air Filter
Brake Booster
Vacuum Hose
Radiator Upper
Hose
Accelerator
Cable
PS Hose
A/T Oil Cooler
Pipe
Air Cleaner
Case
Generator
Purge Hose
Engine Room
J/B Cover
Heater
Hose
PS Hose
Hold Down
Clamp
Radiator Lower
Hose
Fuel Inlet
Hose
Vacuum
Hose
Battery
Insulator
A/T Oil Cooler
Hose
Battery
VSV for Active
Control Engine Mount
25 (250, 18)
A/C Compressor
Battery
Tray
Generator Drive Belt
Adjusting Bar Bracket
Engine Under Cover
N·m (kgf·cm, ft·lbf) : Specified torque
◆ Non–reusable part
7924ZG84

Exploded view of engine pre-removal components—3.0L engine

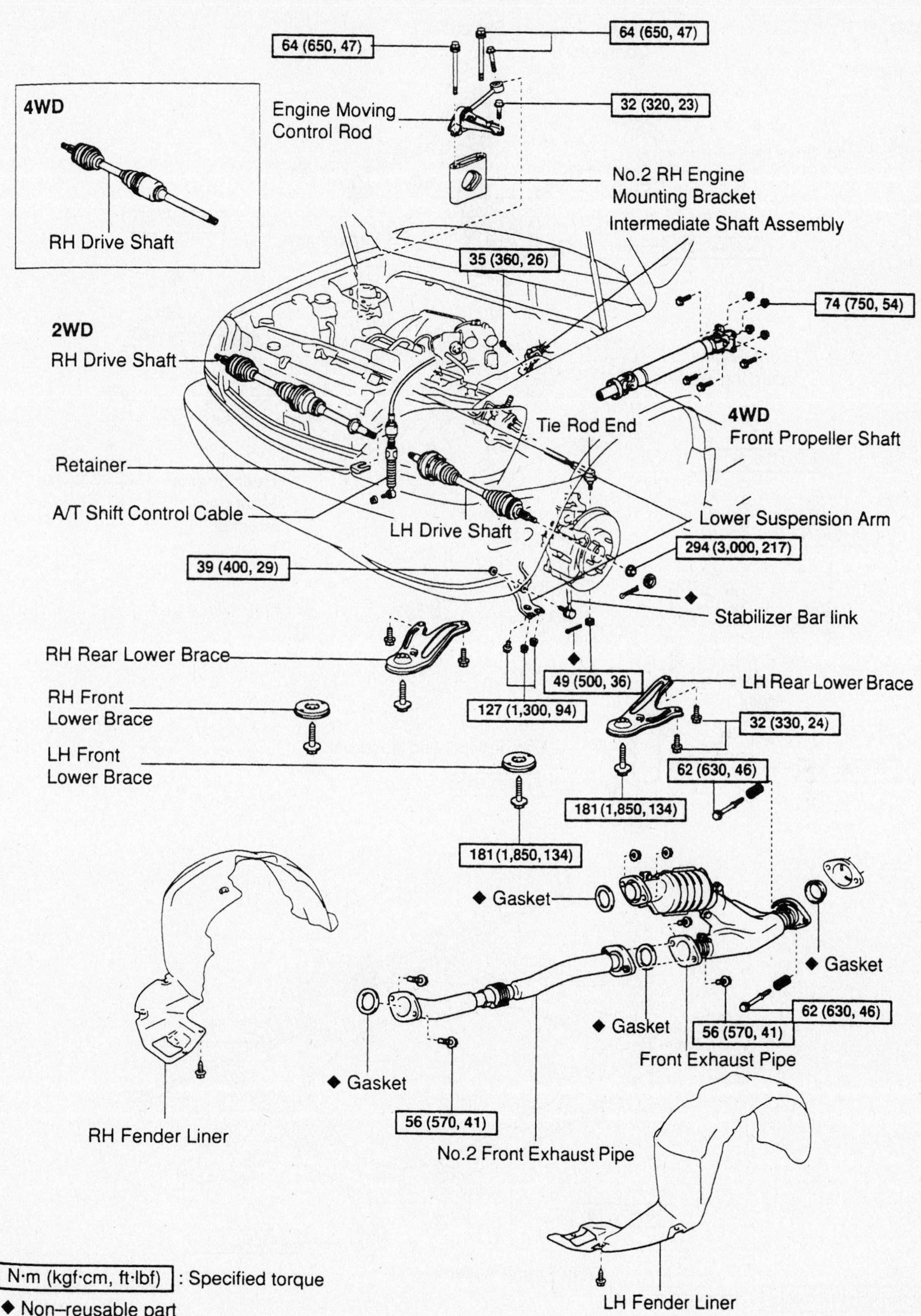

Exploded view of engine removal and installation tightening specifications of the related components—3.0L engine

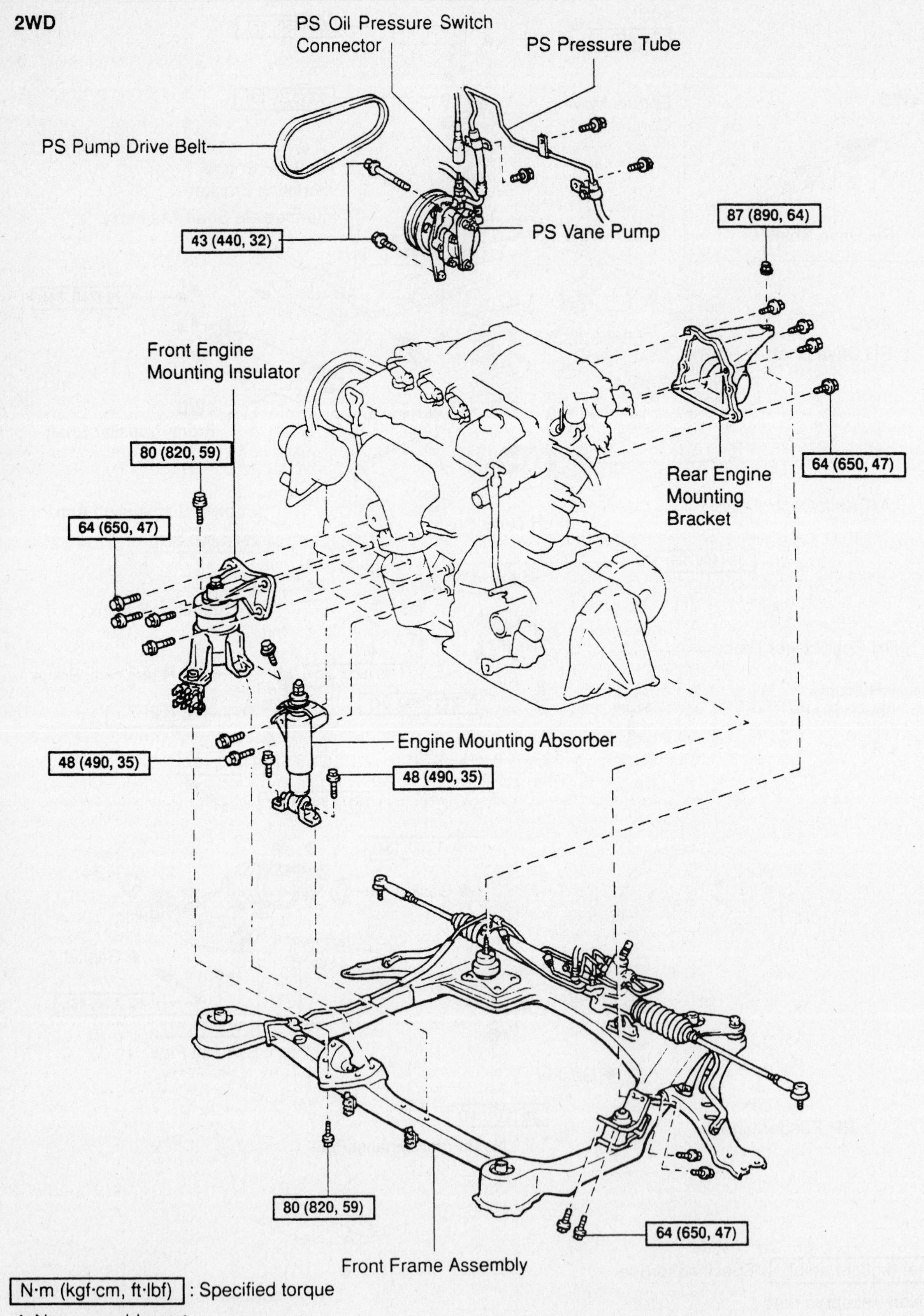

Exploded view of the suspension component removal and installation for engine removal—2WD

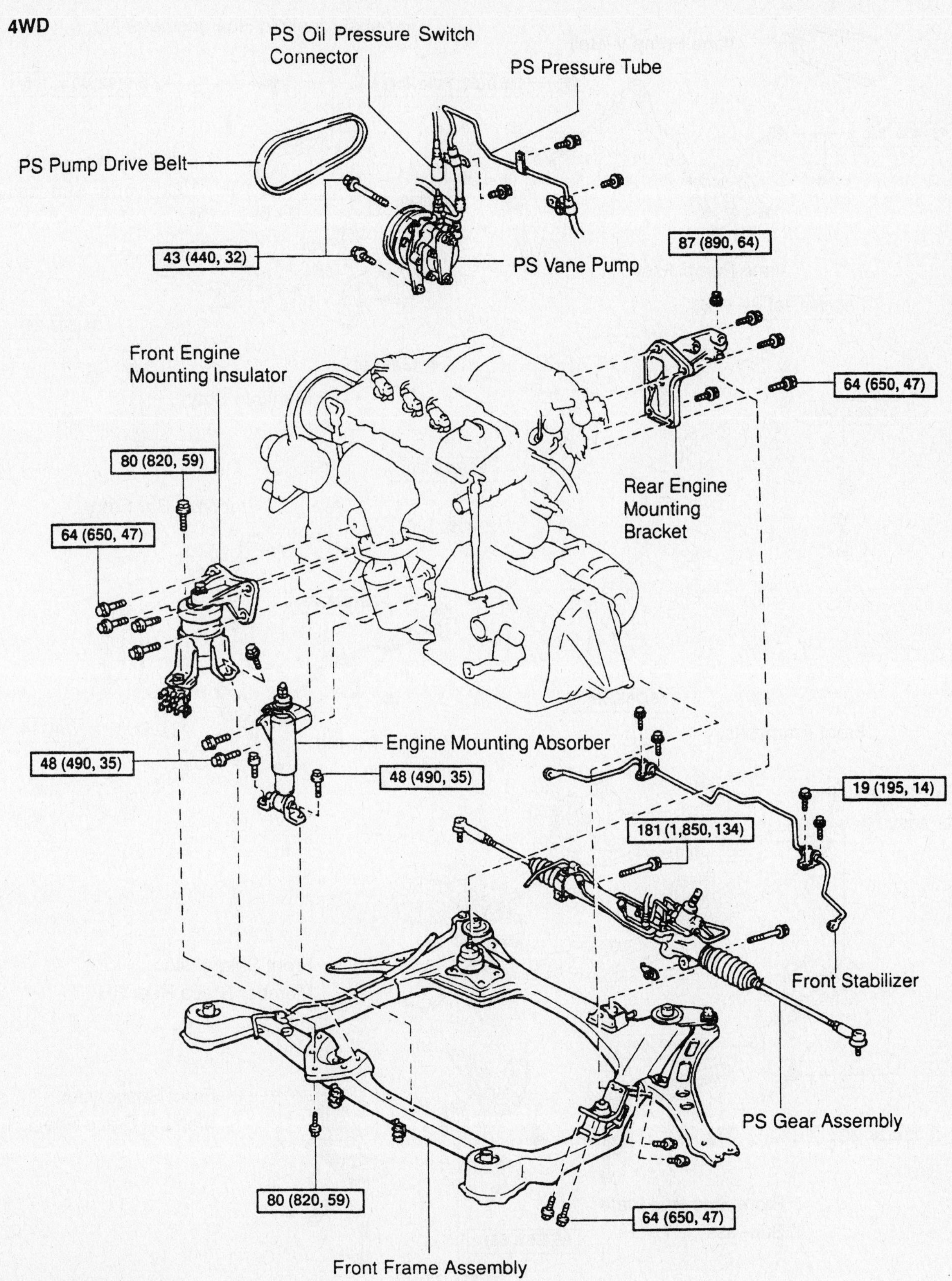

Exploded view of the suspension component removal and installation for engine removal—4WD

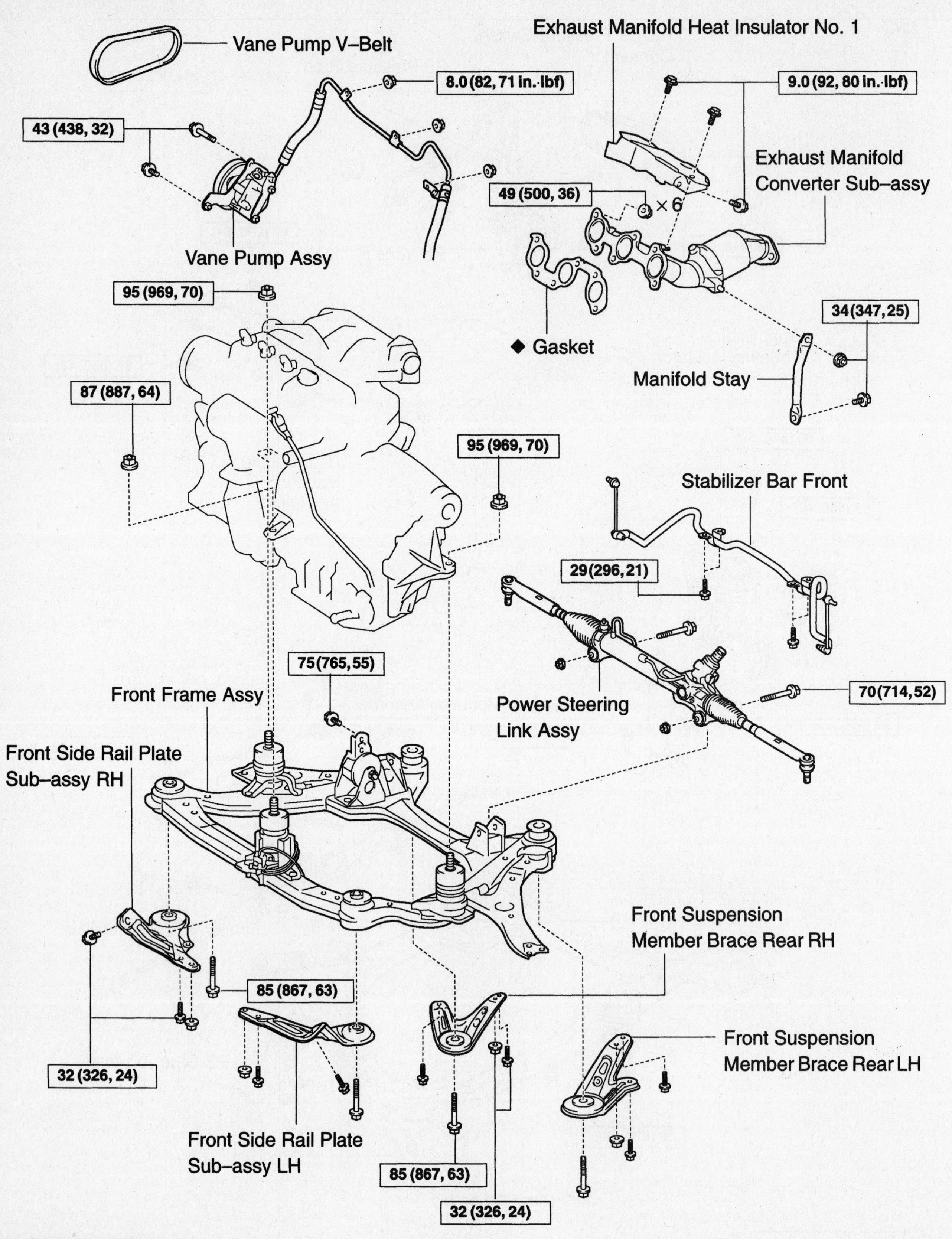

Vane Pump V–Belt
8.0 (82, 71 in.·lbf)
Exhaust Manifold Heat Insulator No. 1
9.0 (92, 80 in.·lbf)
43 (438, 32)
Exhaust Manifold
Converter Sub–assy
49 (500, 36)
× 6
Vane Pump Assy
95 (969, 70)
34 (347, 25)
◆ Gasket
Manifold Stay
87 (887, 64)
95 (969, 70)
Stabilizer Bar Front
29 (296, 21)
75 (765, 55)
Front Frame Assy
Power Steering
Link Assy
70 (714, 52)
Front Side Rail Plate
Sub–assy RH
Front Suspension
Member Brace Rear RH
85 (867, 63)
Front Suspension
Member Brace Rear LH
32 (326, 24)
Front Side Rail Plate
Sub–assy LH
85 (867, 63)
32 (326, 24)
N·m (kgf·cm, ft·lbf) : Specified torque
◆ Non–reusable part
67170-HIGH-G04

Engine mounting points and related parts—3.3L engine

- Alternator and brackets
- Left and right axle hub nuts
- Left and right speed sensors
- Left and right tie rods
- Left and right lower control arms
- Torque converter-to-drive plate bolts
- Intermediate steering shaft
- AC compressor

4. Attach a crane, remove the 6 side rail plate subassembly bolts (3 each side) and the front suspension member rear brace.
5. Lift the engine out of the vehicle.
6. Installation is the reverse of removal. Observe the following torques:
   - Frame side plate bolts: Large 63 ft. lbs. (85Nm); small 24 ft. lbs. (32Nm)
   - Suspension member rear brace: Large 63 ft. lbs. (85Nm); small 24 ft. lbs. (32Nm)
   - Intermediate shaft bolt: 26 ft. lbs. (35Nm)
   - Torque converter bolts: 30 ft. lbs. (41Nm)
   - Lower control arms, bolts and nuts: 94 ft. lbs. (127Nm)
   - Tie rod nuts: 36 ft. lbs. (49Nm)
   - Speed sensors: 71 inch lbs. (8Nm)
   - Hub nuts: 217 ft. lbs. (294Nm)
   - Stabilizer link nuts: 55 ft. lbs. (74Nm)
   - Driveshaft nuts: 55 ft. lbs. (74Nm)
   - Engine mount bracket: 15 ft. lbs. (20Nm)
   - Engine mount stay: 47 ft. lbs. (64Nm)

3.3L Engine

1. Before servicing the vehicle, refer to the precautions in the beginning of this section.
2. Drain the coolant, engine oil, transfer case fluid and transmission fluid.
3. Remove the front wheels.
4. Remove the engine undercover assembly.
5. Remove the left and right fender splash shields.
6. Remove the left and right fender apron seals.
7. Remove the wiper arms.
8. Remove the cowl top ventilator louver.
9. Remove the wiper linkage.
10. Remove the cowl top panel outer sub-assembly.
11. Remove the V-bank cover.
12. Remove the battery.
13. Remove the air cleaner assembly.
14. Remove the A/C compressor drive belt.
15. Remove the alternator.
16. Remove the engine roll brace.
17. Remove the front engine mount bracket.
18. Remove the alternator bracket.
19. Remove the alternator belt adjusting bar.
20. Remove the magnetic clutch from the A/C compressor.
21. Remove the transmission control cable.
22. Disconnect the check valve hose.
23. Disconnect the fuel vapor feed hose.
24. Disconnect the fuel pipes.
25. Disconnect the heater hoses.
26. Disconnect the radiator hoses.
27. Disconnect the oil cooler hoses.
28. Disconnect the power steering hoses.
29. Remove the glove compartment door.
30. Disconnect the engine wiring harness.
31. Remove the driveshaft (4wd).
32. Remove the exhaust pipes and brackets.
33. Remove the left and right stabilizer bar links.
34. Remove the left and right front axle hub nuts.
35. Remove the left and right speed sensors.
36. Remove the left and right tie rod ends.
37. Disconnect the left and right lower control arms.
38. Remove the left and right halfshafts.
39. Disconnect the steering intermediate shaft.
40. Disconnect the height control sensor link (air suspension).
41. Attach a lifting crane.
42. Remove the 6 bolts and 2 nuts, then, remove the left and right frame side rail plates.
43. Remove the 6 bolts and 2 nuts, then, remove the left and right front suspension rear braces.
44. Lift the engine/transaxle from the vehicle.
45. Installation is the reverse of removal. Observe the following torques:
    - Engine hanger: 14 ft. lbs. (20 Nm)
    - Alternator bracket: 43 ft. lbs. (58 Nm)
    - Right engine mount bracket: 40 ft. lbs. (54 Nm)
    - Manifold stay: 36 ft. lbs. (49 Nm)
    - Right rear engine mount bracket: 47 ft. lbs. (64 Nm)
    - Front frame nuts: 70 ft. lbs. (95 Nm)
    - Front right engine mount insulator nut: 64 ft. lbs. (87 Nm)
    - Right rear engine mount insulator bolts: 55 ft. lbs. (75 Nm)
    - Steering link: 52 ft. lbs. (70 Nm)
    - Stabilizer bar: 21 ft. lbs. (29 Nm)
    - Power steering pump adjusting bar: 32 ft. lbs. (43 Nm)
    - Power steering pressure tube nuts: 69 inch lbs. (8 Nm)
    - Left and right frame side rail plates: single end bolts 63 ft. lbs. (85 Nm); double end bolts and nuts: 24 ft. lbs. (32 Nm)
    - Left and right front suspension member rear braces: single end bolts 63 ft. lbs. (85 Nm); double end bolts and nuts: 24 ft. lbs. (32 Nm)
    - Height control sensor link: 48 inch lbs. (5 Nm)

Heater Core

REMOVAL & INSTALLATION

Front Heater

1. Disconnect the negative battery cable.
2. Drain the cooling system into a clean container for reuse.
3. Disconnect the heater hoses from the heater core.
4. Remove the steering wheel by performing the following procedure:
   a. Position the front wheels facing straight-ahead.
   b. Remove the steering wheel side covers.
   c. Using a Torx® wrench, loosen the 2 screws located at each side of the steering wheel until the screw's circumference groove catches on the screw case.
   d. Pull the air bag module from the steering wheel and disconnect the electrical connector.

CAUTION

Place the air bag module in a safe place with the front side facing upward.

   e. Remove the steering wheel nut.
   f. Place alignment marks on the steering wheel and the main shaft.
   g. Using a steering wheel puller, press the steering wheel from the steering column.
5. Remove the instrument panel and reinforcement by performing the following procedure:

a. Remove the front door scuff plates.

b. Remove the cowl side boards.

c. Remove the front door trim covers.

d. Remove the front pillar garnish by disengaging the 5 clips. If equipped with a tweeter speaker, disconnect the electrical connector.

e. Remove the steering column covers-to-steering column screws and the covers.

f. Remove the combination switch-to-steering column screws, disconnect the electrical connector(s) and remove the combination switch.

g. Remove the 2 hood open lever screws and the hood open lever.

h. Remove the 2 lower finish panel bolts and disengage the panel from the 3 clips.

i. Remove the 2 No. 1 safety pad insert bolts and the insert.

j. Remove the 2 No. 2 finish panel bolts and disengage the panel from the 4 clips.

k. In the left side of the glove compartment, pry out the glove box door finish plate and disconnect the air bag module connector.

l. Remove the glove box 3 nuts and 2 screws and the glove box.

m. Remove the center cluster finish panel by disengaging the claw (bottom center) and 4 clips (1 at each corner).

n. Remove the ashtray, the 2 ashtray receptacle box screws.

o. Remove the 4 lower center cluster finish panel screws and disconnect the connector.

p. Remove the clock, the No. 1 and No. 2 registers from the panel.

q. Remove the 3 cluster finish panel screws, disengage the 8 clips and remove the panel.

r. Remove the combination meter.

s. Remove the radio assembly.

t. Remove the heater control assembly.

u. Remove 2 passenger's side air bag module bolts; then, disconnect and remove the air bag module.

✲✲ CAUTION

Place the air bag module in a safe place with the front side facing upward.

v. Remove the instrument panel-to-chassis 5 bolts and nut.

w. Remove the audio amplifier.

x. Remove the No. 1 and No. 2 braces.

y. Remove the No. 2 cowl brace.

z. Remove the instrument panel reinforcement.

6. Remove the evaporator housing by performing the following procedure:

a. Discharge and recover the air conditioning system refrigerant.

b. In the engine compartment, remove the refrigerant lines-to-cowl connector bolts; then, disconnect the lines and discard the O-rings.

c. Disconnect the electrical connector at the evaporator housing.

d. Disconnect the wiring harness clamp.

e. Remove the evaporator housing-to-chassis 2 rivets, 3 bolts and nut.

f. Remove the evaporator housing.

7. Remove the 4 defroster nozzle nuts and the nozzle.

8. Disconnect and remove the theft deterrent and the wireless door lock ECUs.

9. Release the 2 air duct claws and the air duct.

10. Remove the 2 heater housing-to-chassis rivets and the heater housing.

➡When installing the heater housing, use new screws in place of the rivets.

11. Remove the heater core-to-heater housing cover.

12. Remove both heater core screws and clamps; then, remove the heater core.

To install:

13. Install the heater core and both heater core screws and clamps.

14. Install the heater core-to-heater housing cover.

➡When installing the heater housing, use new screws in place of the rivets.

15. Install the heater housing-to-chassis and the 2 heater housing screws.

16. Release the air duct and the air duct claws.

17. Connect and install the theft deterrent and the wireless door lock ECUs.

18. Install the defroster nozzle and the 4 nozzle nuts.

19. Install the evaporator housing by performing the following procedure:

a. Install the evaporator housing.

b. Install the evaporator housing-to-chassis 2 rivets, 3 bolts and nut.

c. Connect the wiring harness clamp.

d. Connect the electrical connector at the evaporator housing.

e. In the engine compartment, use new O-rings and install the refrigerant lines-to-cowl connector and install the bolts.

20. Install the instrument panel and reinforcement by performing the following procedure:

a. Install the instrument panel reinforcement.

b. Install the No. 2 cowl brace.

c. Install the No. 1 and No. 2 braces.

d. Install the audio amplifier.

e. Install the instrument panel-to-chassis 5 bolts and nut.

f. Connect and install the air bag module and the 2 passenger's side air bag module bolts.

g. Install the heater control assembly.

h. Install the radio assembly.

i. Install the combination meter.

j. Install the cluster finish panel, engage the 8 clips and install the panel screws.

k. Install the No. 1 and No. 2 registers and the clock to the panel.

l. Connect the lower center cluster finish panel connector and install the 4 lower center cluster finish panel screws.

m. Install the 2 ashtray receptacle box screws and the ashtray.

n. Install the center cluster finish panel by engaging the 4 clips (1 at each corner) and the claw (bottom center).

o. Install the glove box and the glove box 3 nuts and 2 screws.

p. In the left side of the glove compartment, connect the air bag module connector and install the glove box door finish plate.

q. Install the No. 2 finish panel, engage the 4 panel clips and install the 3 panel bolts.

r. Install the No. 1 safety pad insert and the 2 insert bolts.

s. Install the finish panel, engage the 3 finish panel clips and install 2 lower finish panel bolts.

t. Install the hood open lever and the 2 hood open lever screws.

u. Install the combination switch, connect the electrical connector(s) and install the combination switch-to-steering column screws.

v. Install the steering column covers and the covers-to-steering column screws.

w. Install the front pillar garnish by engaging the 5 clips. If equipped with a tweeter speaker, connect the electrical connector.

x. Install the front door trim covers.

y. Install the cowl side boards.

z. Install the front door scuff plates.

21. Install the steering wheel by performing the following procedure:

a. Install the steering wheel to the steering column.

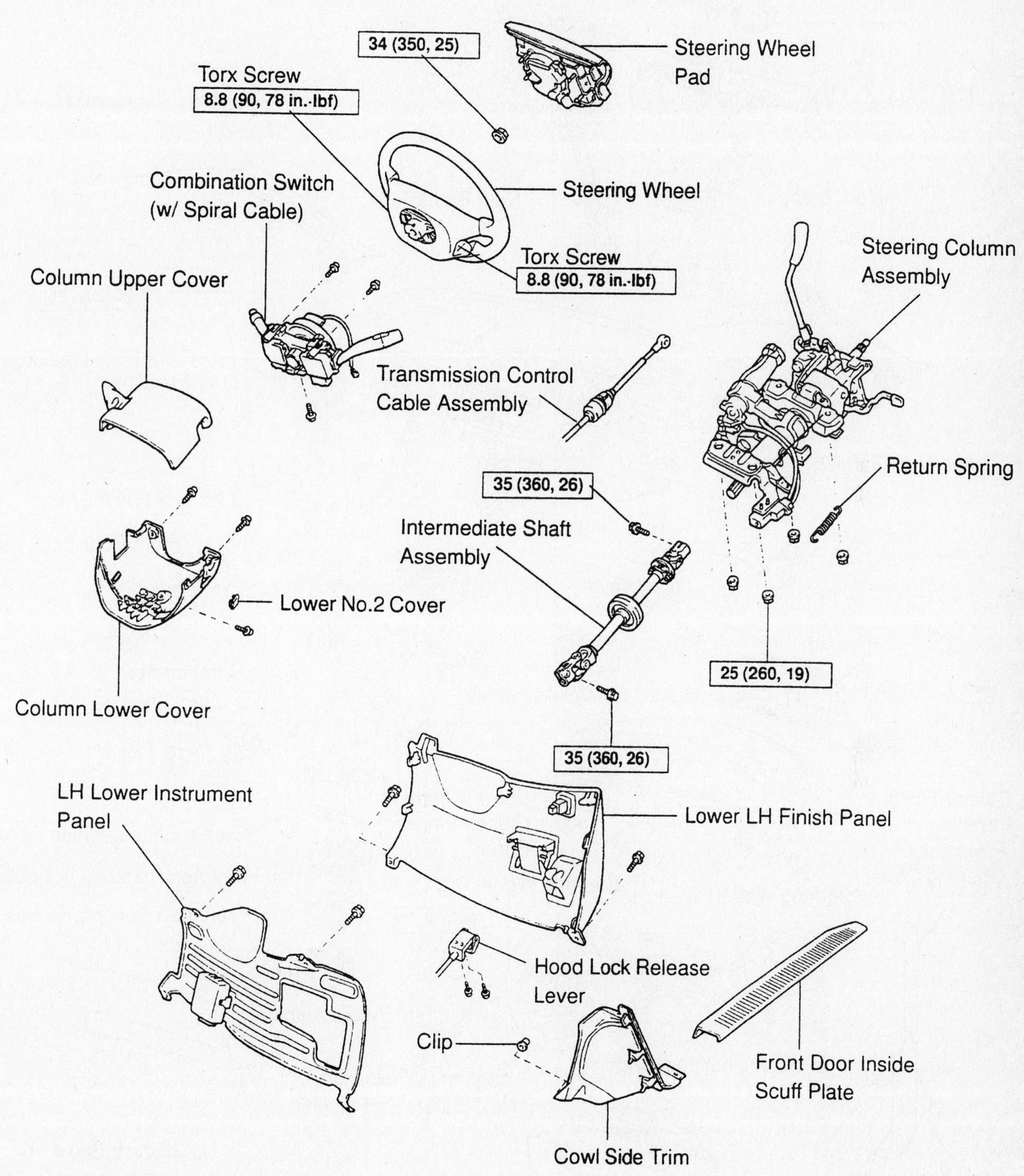

Exploded view of the steering wheel, steering column and related components

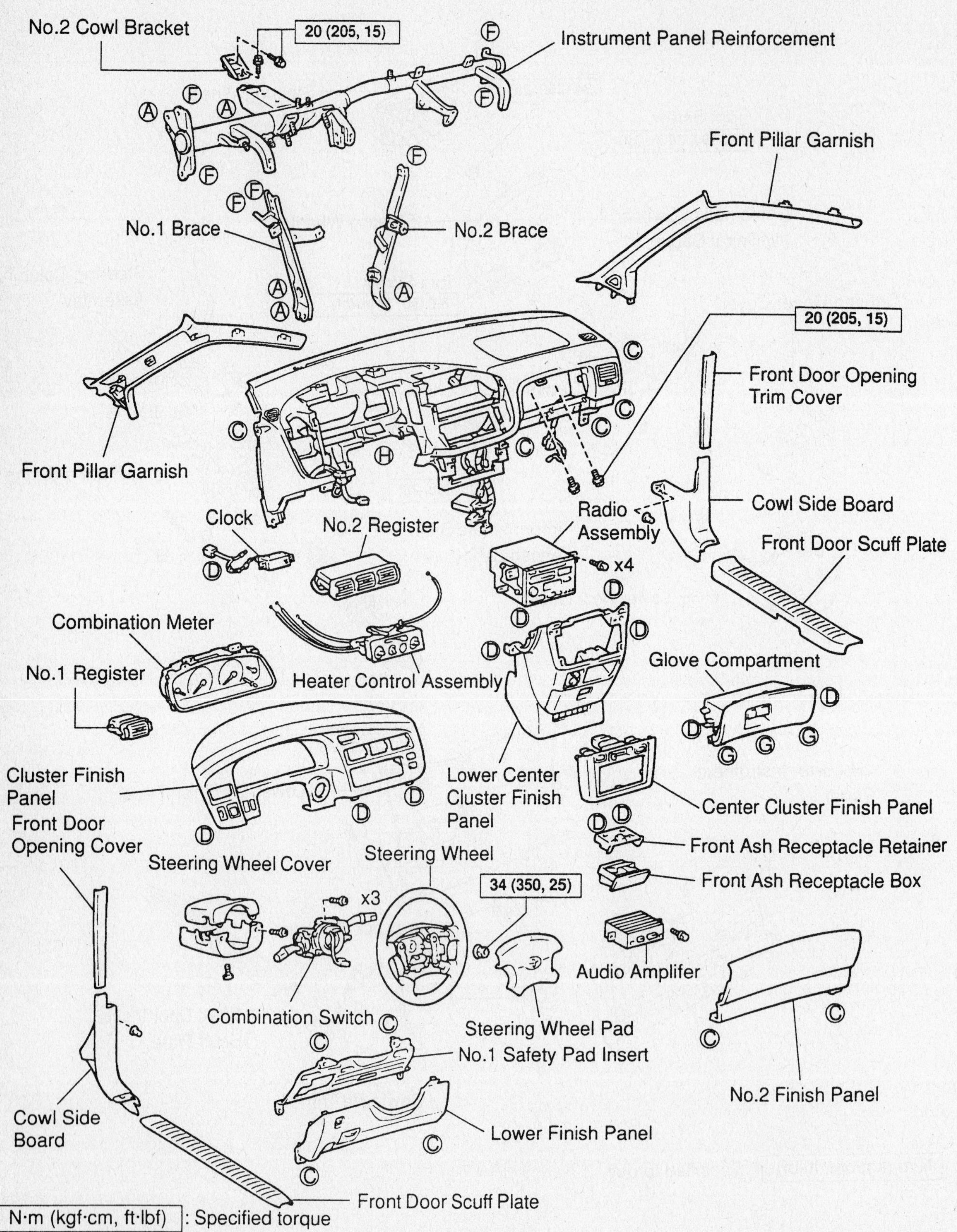

No.2 Cowl Bracket
20 (205, 15)
Instrument Panel Reinforcement
Front Pillar Garnish
No.1 Brace
No.2 Brace
20 (205, 15)
Front Door Opening Trim Cover
Front Pillar Garnish
Cowl Side Board
Clock
No.2 Register
Radio Assembly
x4
Front Door Scuff Plate
Combination Meter
No.1 Register
Heater Control Assembly
Glove Compartment
Cluster Finish Panel
Lower Center Cluster Finish Panel
Center Cluster Finish Panel
Front Door Opening Cover
Steering Wheel Cover
Steering Wheel
Front Ash Receptacle Retainer
34 (350, 25)
Front Ash Receptacle Box
x3
Audio Amplifer
Combination Switch
Steering Wheel Pad
No.1 Safety Pad Insert
No.2 Finish Panel
Cowl Side Board
Lower Finish Panel
Front Door Scuff Plate
N·m (kgf·cm, ft·lbf) : Specified torque
93113GH4

Exploded view of the instrument panel and related components

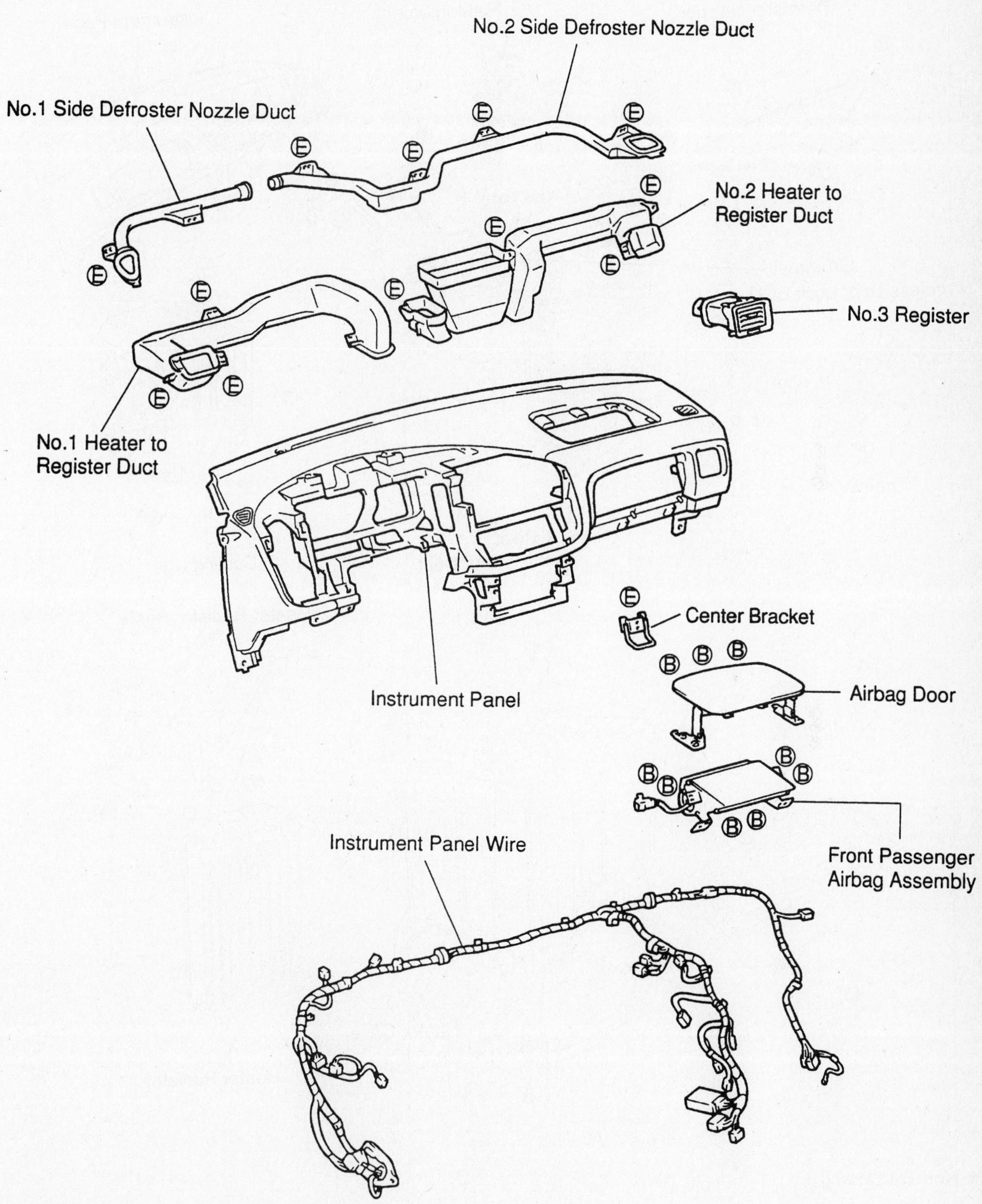

Exploded view of the ventilation system and related components

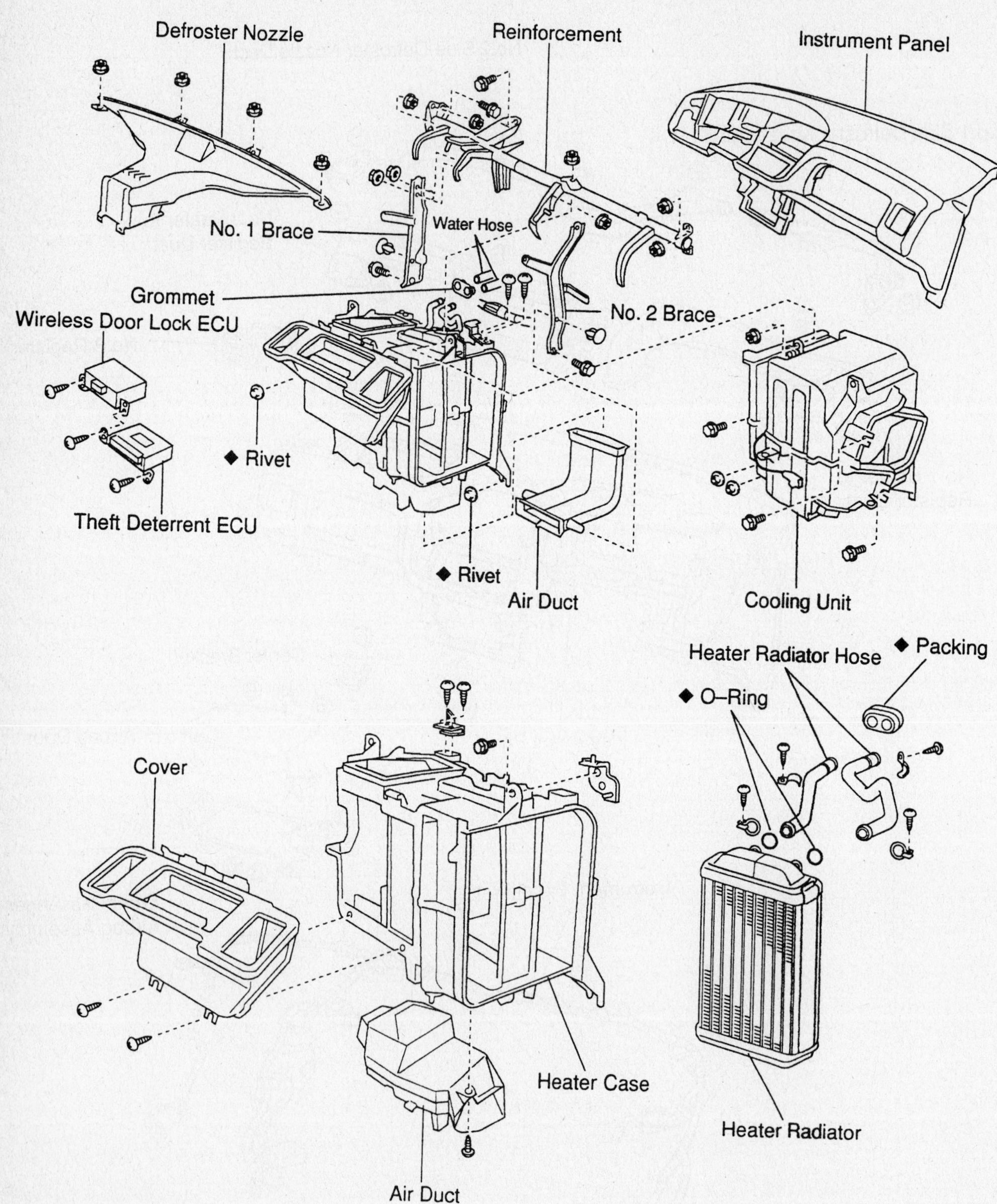

Exploded view of the heater core, heater housing, evaporator housing and related components

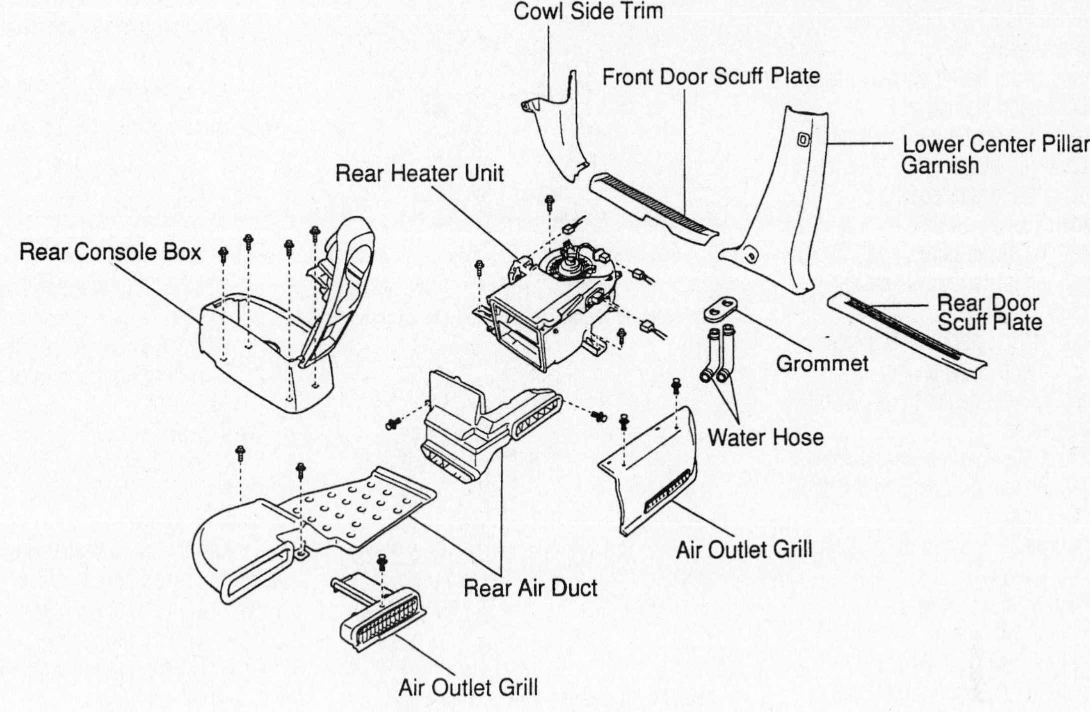

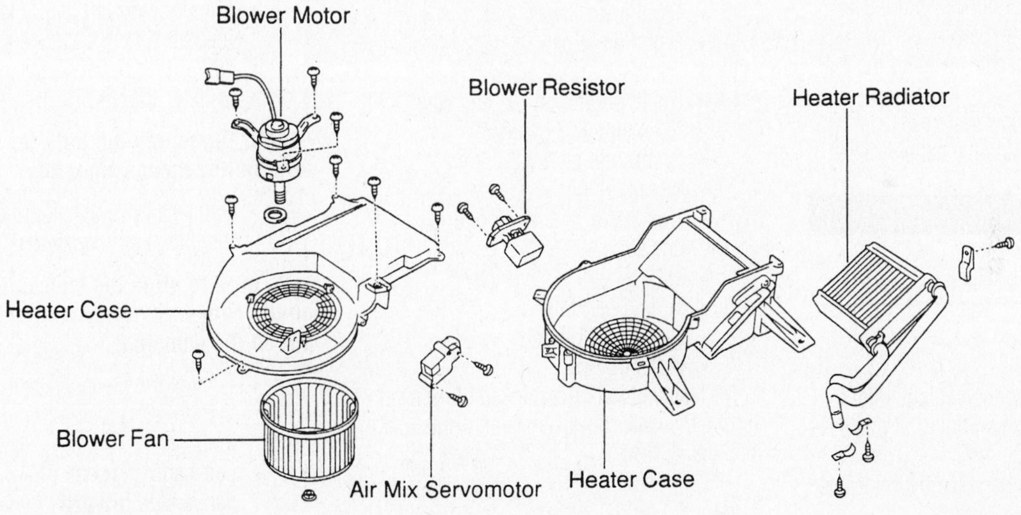

Exploded view of the rear heater core, the rear heater housing and related components

b. Align the steering wheel-to-main shaft marks.

c. Install the steering wheel nut and torque the nut to 25 ft. lbs. (34 Nm).

d. Install the air bag module to the steering wheel and connect the electrical connector.

e. Using a Torx® wrench, tighten the steering wheel screws to 78 inch lbs. (8.8 Nm).

f. Install the steering wheel side covers.

22. Connect the heater hoses to the heater core.
23. Refill the cooling system.
24. Connect the negative battery cable.
25. Evacuate and charge the air conditioning system.
26. Run the engine to normal operating temperatures; then, check the climate control operation and check for leaks.

Rear Auxiliary Heater

1. Disconnect the negative battery cable.
2. Drain the cooling system into a clean container for reuse.
3. Disconnect the heater hoses from the rear heater core.
4. Remove the front seats.
5. Remove the front door scuff plates.
6. Remove the cowl side trim.
7. Remove the rear door scuff plates.
8. Remove the lower door scuff plates.
9. Remove the rear console box.
10. Remove the left side air outlet grille.
11. Pull the carpet rearward.
12. Remove the 3 clips and the air outlet grille.
13. Remove the rear air duct 2 bolts, 2 clips and the duct.

14. Disconnect the electrical connectors.
15. Remove the 3 rear heater housing bolts and the housing.
16. Remove both heater core-to-heater housing screws and clamps.
17. Remove the heater core-to-heater housing screw and plate.
18. Remove the heater core.

To install:

19. Install the heater core.
20. Install the heater core-to-heater housing screw and plate.
21. Install both heater core-to-heater housing screws and clamps.
22. Install the rear heater housing and the 3 housing bolts.
23. Connect the electrical connectors.
24. Install the rear air duct and the duct 2 bolts and 2 clips.
25. Install the 3 clips and the air outlet grille.
26. Move the carpet forward.
27. Install the left side air outlet grille.
28. Install the rear console box.
29. Install the lower door scuff plates.
30. Install the rear door scuff plates.
31. Install the cowl side trim.
32. Install the front door scuff plates.
33. Install the front seats.
34. Connect the heater hoses to the rear heater core.
35. Refill the cooling system.
36. Connect the negative battery cable.

Water Pump

REMOVAL & INSTALLATION

2.4L Engine

1. Before servicing the vehicle, refer to the precautions in the beginning of this section.
2. Disconnect the negative battery cable.
3. Drain the engine coolant.
4. Remove or disconnect the following:
   - Alternator
   - Water pump pulley
   - Water pump
5. Installation is the reverse of removal. Torque the pump bolts and nuts to 80 inch lbs. (9Nm) and the pulley bolts to 19 ft. lbs. (26Nm).

3.0L Engine

1. Before servicing the vehicle, refer to the precautions in the beginning of this section.
2. Disconnect the negative battery cable.
3. Drain the engine coolant.
4. Remove or disconnect the following:

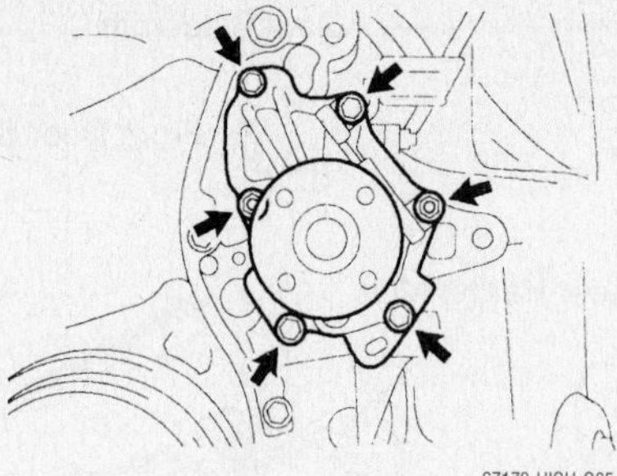

Water pump mounting bolts—2.4L engine

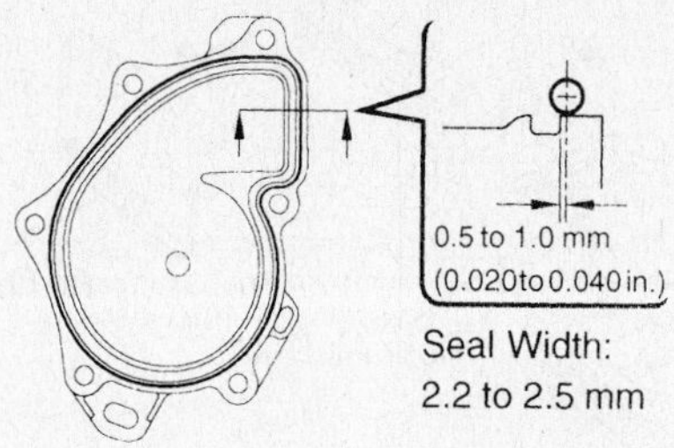

Sealer application—2.4L engine water pump

   - Right front wheel
   - Right fender apron seal
   - Accessory drive belts
   - Upper engine mount and stay
   - Alternator and bracket
   - Crankshaft pulley
   - Timing belt covers
   - Transverse engine mounting bracket
   - Timing belt
   - Timing belt idler
   - Camshaft pulley
   - Water pump
5. Installation is the reverse of removal. Torque the water pump bolts and nuts to 71 inch lbs. (8Nm).

3.3L

1. Before servicing the vehicle, refer to the precautions in the beginning of this section.
2. Disconnect the negative battery cable.
3. Drain the engine coolant.
4. Remove or disconnect the following:
   - Wiper and blade assembly
   - Top cowl seal and panel
   - Window washer hoses, from the ventilator louvers
   - Left and right ventilator louvers
   - Heater air duct
   - Front upper suspension brace
   - Timing belt
5. Mark the left and right camshaft pulleys with a touch of paint.
6. Remove or disconnect the following:
   - Right and left camshaft pulleys bolts
   - Pulleys from the engine

➡Be sure not to mix up the pulleys.

   - No. 2 idler pulley by removing the bolt
   - 3 clamps and engine wire from the rear timing belt cover
   - 6 No. 3 timing belt cover-to-engine bolts
   - Water pump nuts/bolts
   - Water pump and gasket from the engine

To install:

7. Check that the water pump turns smoothly. Also check the air hole for coolant leakage.
8. Apply liquid sealer to the gasket, water pump and engine block.
9. Install or connect the following:
   - Water pump, using a new gasket. Tighten the nuts/bolts to 53 inch lbs. (6 Nm).
   - Rear timing belt cover. Tighten the 6 bolts to 74 inch lbs. (9 Nm).
   - Engine wire with the 3 clamps to the rear timing belt cover
   - No. 2 idler pulley. Tighten the bolt to 32 ft. lbs. (43 Nm).

➡After tightening the bolt, be sure the idler pulley moves smoothly.

   - Right-hand camshaft pulley, with the flange side **outward**.

➡Be sure to align the knock pin hole on the camshaft pulley with the knock pin on the camshaft.

   - Tighten the camshaft bolt to 65 ft. lbs. (88 Nm), using the removal tools
   - Left-hand camshaft pulley, with the flange side **inward**.

➡Be sure to align the knock pin hole on the camshaft pulley with the knock pin on the camshaft.

   - Tighten the camshaft bolt to 94 ft. lbs. (125 Nm), using the removal tools
   - Timing belt
   - Front upper suspension brace. Tighten the nuts to 59 ft. lbs. (80 Nm).
10. Fill the engine coolant.
11. Install or connect the following:
   - Heater air duct
   - Left and right ventilator louvers
   - Window washer hoses to the ventilator louvers

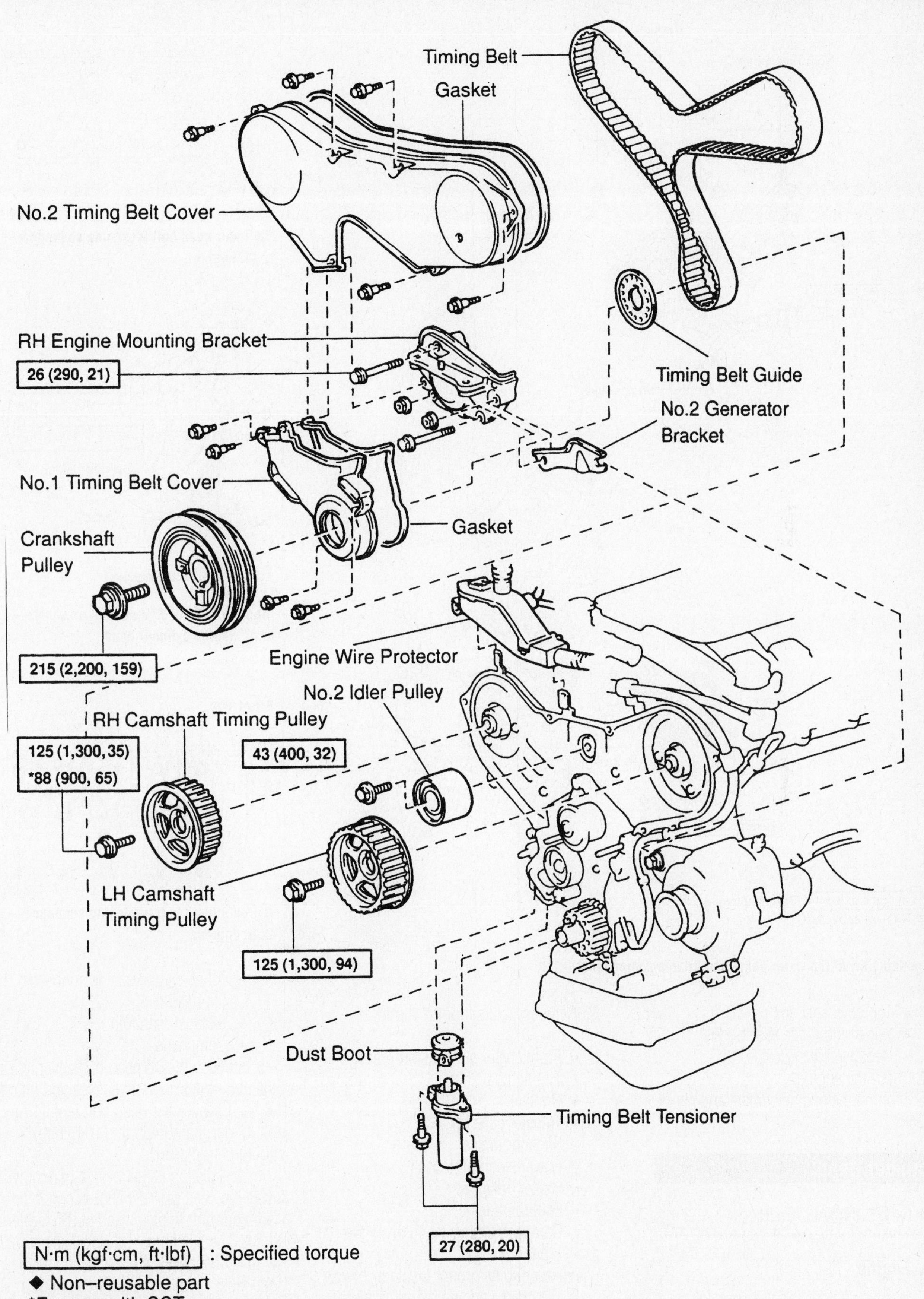

Exploded view of the components to gain access to the water pump—3.3L

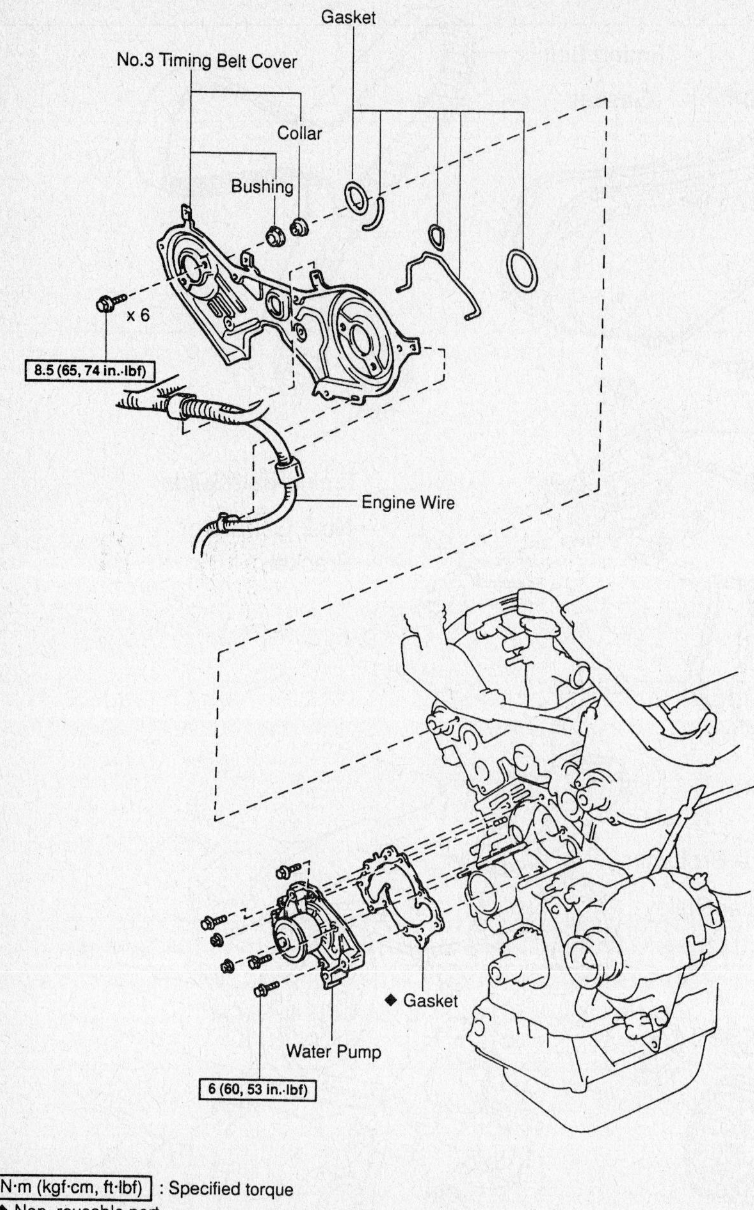

Exploded view of the water pump and related components—3.3L

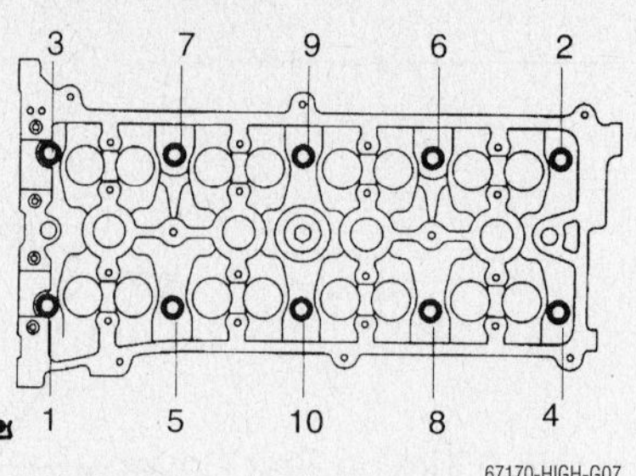

Cylinder head bolt loosening sequence—2.4L engine

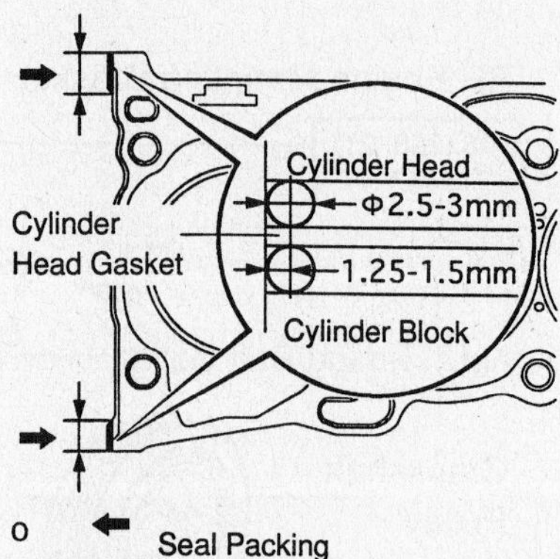

Apply a bead of RTV sealant as shown—2.4L engine cylinder head

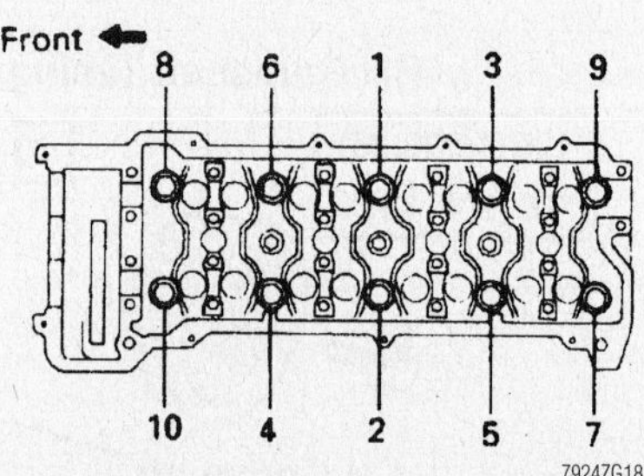

Cylinder head bolt tightening sequence—2.4L engines

- Top cowl seal and panel
- Wiper and blade assembly
- Negative battery cable

12. Start the engine.

13. Top off the engine coolant and check for leaks.

Cylinder Head

REMOVAL & INSTALLATION

2.4L Engine

1. Before servicing the vehicle, refer to the precautions in the beginning of this section.

2. Remove or disconnect the following:

- Front center suspension brace
- Timing chain
- Coolant
- Transfer case oil
- Radiator hoses
- Power steering hoses
- Heater hoses
- Fuel rail lines
- Camshaft timing oil control valve
- Front driveshaft
- Rear engine mount insulator (4wd)
- Transverse engine mount bracket (4wd)
- Intake manifold
- All wires and cables connected to the head
- Exhaust manifold
- Camshafts

3. Loosen the 10 head bolts evenly, a little at a time in several passes and lift off the head. Check the head bolt length. Any bolt longer than 6.465 in. (164.2mm) should be replaced.

4. Installation is the reverse of removal. Install the head gasket with the lot number stamp upward. Apply a bead of RTV sealer as shown. The head must be installed within 3 minutes of applying the sealer, and the head bolts must be tightened with 15 minutes. The head bolts must be tightened in sequence, in several passes, to 58 ft. lbs. (79Nm), then, an additional 90 degree turn each.

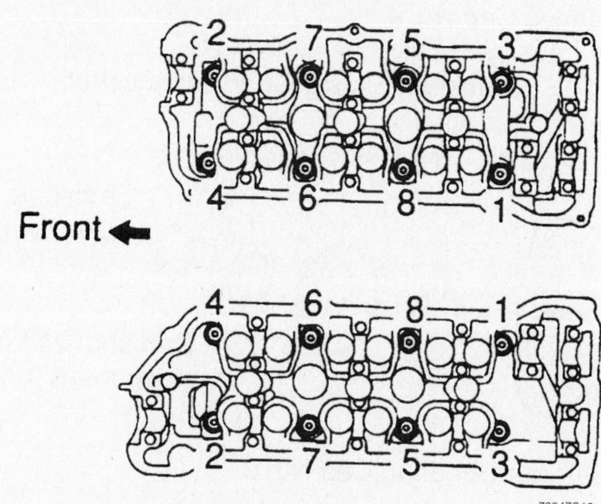

Cylinder head bolt loosening sequence—3.0L engine

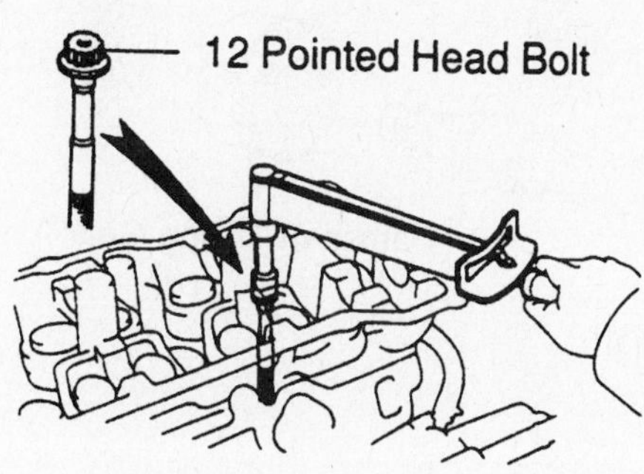

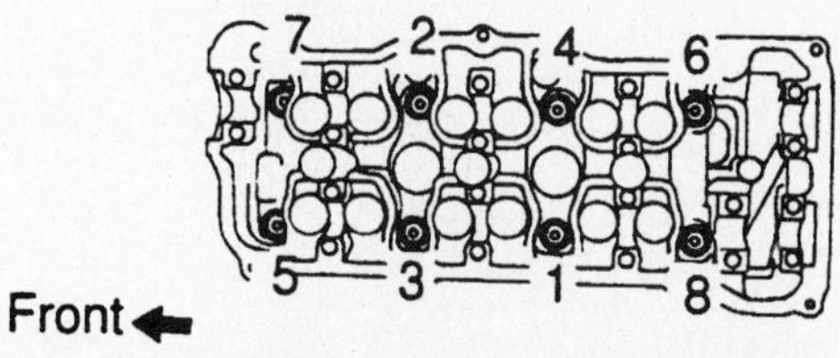

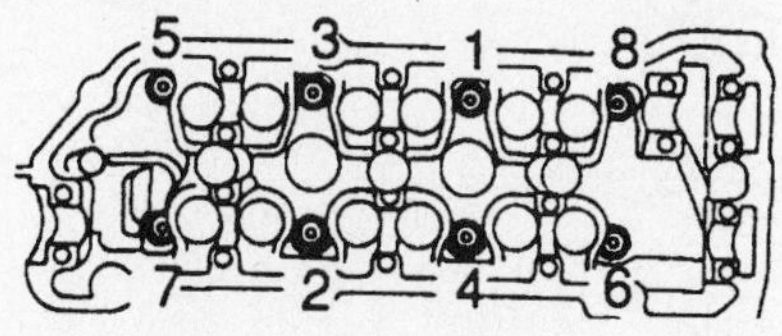

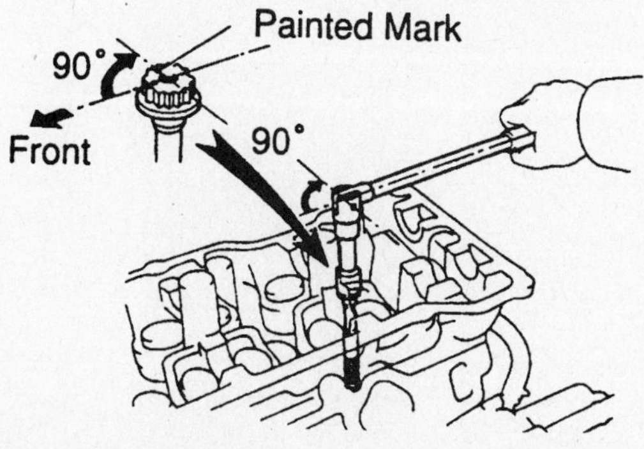

Cylinder head bolt tightening sequence—3.0L engine

3.0L Engine

1. Before servicing the vehicle, refer to the precautions in the beginning of this section.
2. Remove or disconnect the following:
   - Coolant
   - Engine oil
   - Exhaust pipes
   - Exhaust manifold
   - Camshaft cover
   - Upper center front suspension brace
   - Air cleaner
   - Intake air surge tank
   - Fuel rail
   - Heater hoses
   - Intake manifold
   - Radiator hose
   - Water outlet
   - Right front wheel
   - Right fender apron seal
   - Accessory drive belts
   - Engine roll stopper rod
   - Right engine mount
   - Alternator and brackets
   - Crankshaft pulley
   - Timing belt covers
   - Transverse engine mounting bracket
   - Timing belt
   - Timing belt tensioner
   - Power steering pump
   - Ignition coil pack
   - Camshafts
   - The hexagonal bolt, using an 8mm hex wrench, then the 8 head bolts
   - Cylinder head
3. Installation is the reverse of removal. Measure the head bolts. The minimum diameter of the stretch portion of each bolt should be at least 8.75mm (0.3775 in.). Replace any bolt that does not measure up. The head gasket is installed with the **R** mark upwards. Install the 8 head bolts first, tightened in sequence, in 2 equal steps, to 40 ft. lbs. (54Nm). Then tighten each an additional 90 degrees. Finally, install the hex bolt, torqued to 14 ft. lbs. (19 Nm).

3.3L Engines

1. Before servicing the vehicle, refer to the precautions in the beginning of this section.
2. Remove or disconnect the following:
   - Wiper and blade assembly
   - Top cowl seal and panel
   - Window washer hoses from the ventilator louvers
   - Left and right ventilator louvers
   - Heater air duct

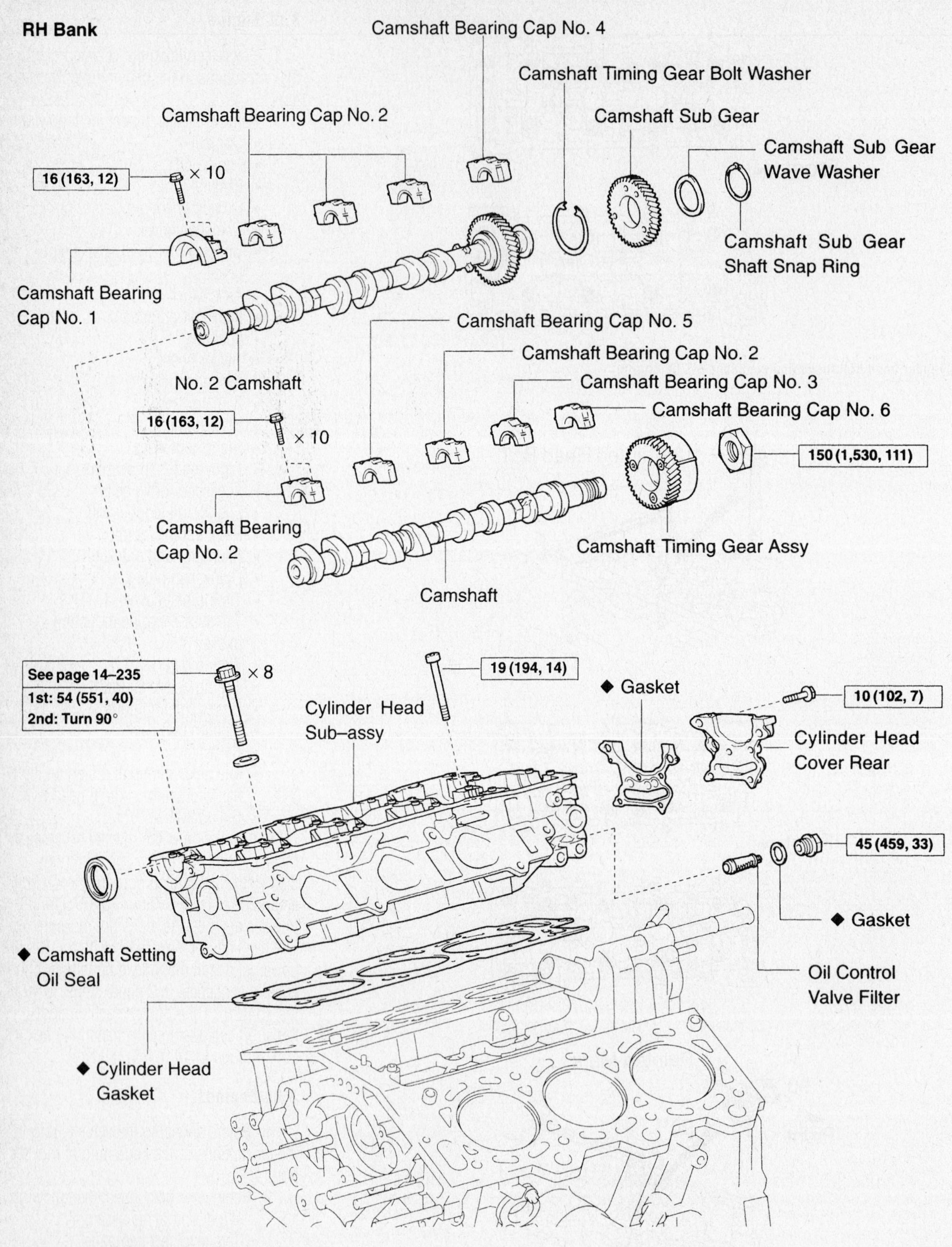
RH Bank
Camshaft Bearing Cap No. 4
Camshaft Timing Gear Bolt Washer
Camshaft Bearing Cap No. 2
Camshaft Sub Gear
Camshaft Sub Gear Wave Washer
16 (163, 12)
× 10
Camshaft Sub Gear Shaft Snap Ring
Camshaft Bearing Cap No. 1
Camshaft Bearing Cap No. 5
Camshaft Bearing Cap No. 2
No. 2 Camshaft
Camshaft Bearing Cap No. 3
Camshaft Bearing Cap No. 6
16 (163, 12)
× 10
150 (1,530, 111)
Camshaft Bearing Cap No. 2
Camshaft Timing Gear Assy
Camshaft
See page 14–235
1st: 54 (551, 40)
2nd: Turn 90°
× 8
19 (194, 14)
◆ Gasket
10 (102, 7)
Cylinder Head Sub–assy
Cylinder Head Cover Rear
45 (459, 33)
◆ Gasket
◆ Camshaft Setting Oil Seal
Oil Control Valve Filter
◆ Cylinder Head Gasket
N·m (kgf·cm, ft·lbf) : Specified torque
◆ Non–reusable part
67170-HIGH-G08

Right cylinder head and related parts—3.3L engine

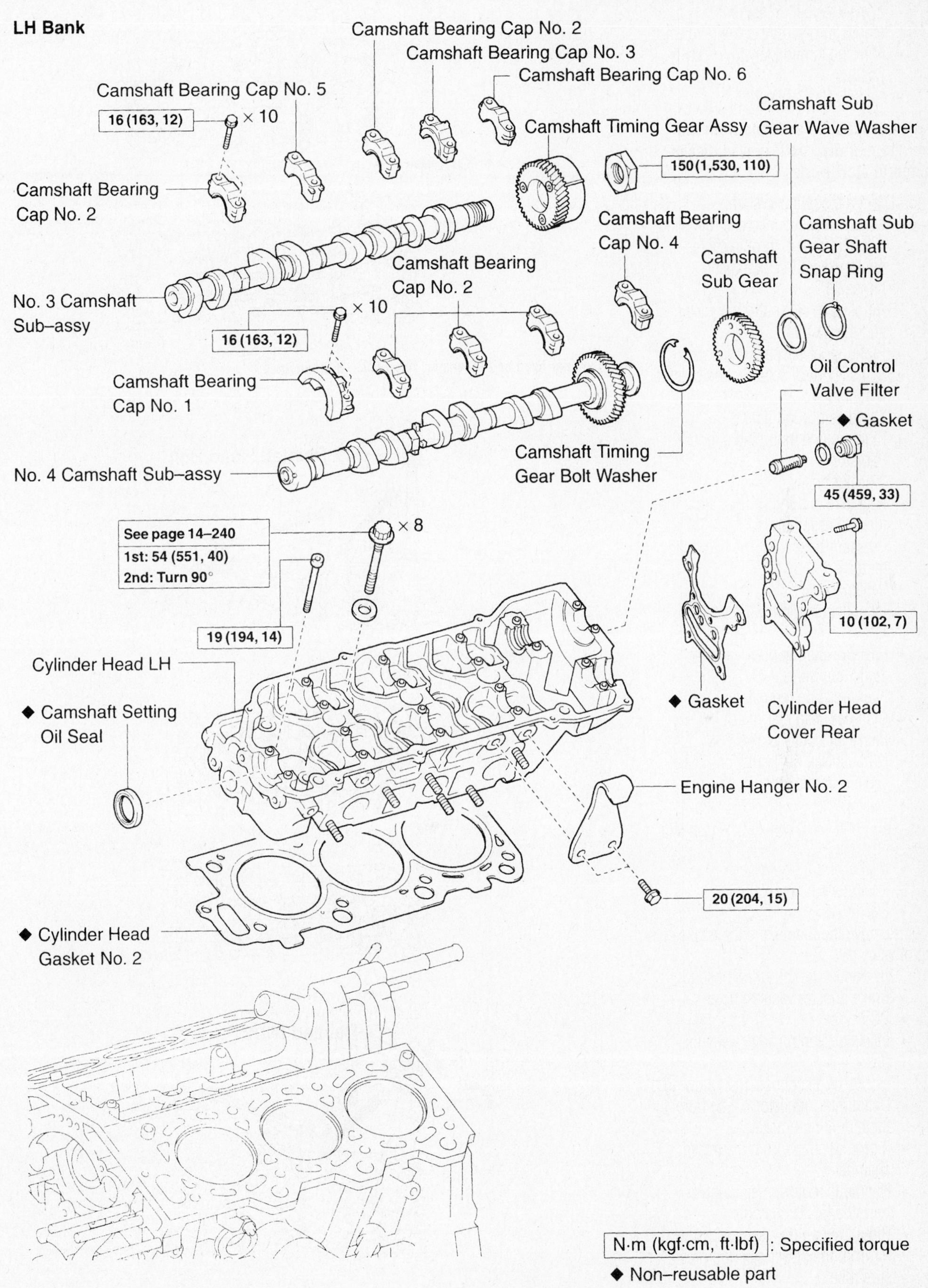
LH Bank
Camshaft Bearing Cap No. 2
Camshaft Bearing Cap No. 3
Camshaft Bearing Cap No. 6
Camshaft Bearing Cap No. 5
16 (163, 12)
× 10
Camshaft Sub Gear Wave Washer
Camshaft Timing Gear Assy
150 (1,530, 110)
Camshaft Bearing Cap No. 2
Camshaft Bearing Cap No. 4
Camshaft Sub Gear Shaft Snap Ring
Camshaft Sub Gear
Camshaft Bearing Cap No. 2
No. 3 Camshaft Sub–assy
× 10
16 (163, 12)
Camshaft Bearing Cap No. 1
Oil Control Valve Filter
◆ Gasket
Camshaft Timing Gear Bolt Washer
No. 4 Camshaft Sub–assy
45 (459, 33)
See page 14–240
1st: 54 (551, 40)
2nd: Turn 90°
× 8
19 (194, 14)
10 (102, 7)
Cylinder Head LH
◆ Camshaft Setting Oil Seal
◆ Gasket
Cylinder Head Cover Rear
Engine Hanger No. 2
20 (204, 15)
◆ Cylinder Head Gasket No. 2
N·m (kgf·cm, ft·lbf) : Specified torque
◆ Non–reusable part
67170-HIGH-G09

Left cylinder head and related parts—3.3L engine

3. Relieve the fuel pressure.
4. Remove or disconnect the following:
   - Turn the ignition key to the **OFF** position
   - Negative battery cable

➡Wait at least 90 seconds from the time the negative battery was disconnected to start work.

5. Drain the cooling system.
6. Remove or disconnect the following:
   - Accelerator and throttle cables, if equipped with an automatic transaxle
   - Air cleaner cover, air flow meter and the air duct
   - Front upper suspension brace
   - Cruise control actuator and bracket, if equipped
   - 2 engine ground straps
   - Right engine mounting support
   - Radiator hoses
   - 2 heater hoses
   - Fuel feed and return lines from the fuel rail assembly
   - Pressure hose from the hydraulic motor
   - V-bank cover
7. Disconnect the following vacuum hoses:
   - Fuel pressure control Vacuum Switching Valve (VSV)
   - Fuel pressure regulator
   - Cylinder head rear plate
   - Intake air control valve VSV
   - Exhaust Gas Recirculation (EGR) vacuum modulator
   - EGR valve
8. Disconnect the following wiring and hoses:
   - Intake air control valve
   - Fuel pressure regulator
   - EGR VSV
9. Remove the 2 nuts and the emission control valve set.
10. Disconnect the following hoses;
    - Brake booster vacuum hose
    - PCV hose
    - Intake air control valve vacuum hose
11. Remove or disconnect the following:
    - Data Link Connector (DLC) from the mounting bracket
    - 2 ground straps from the intake chamber
    - Hydraulic motor pressure hose from the intake chamber
    - Right Oxygen (O_2) sensor connector from the power steering pressure tube
    - 2 nuts and the power steering pressure tube from the intake chamber

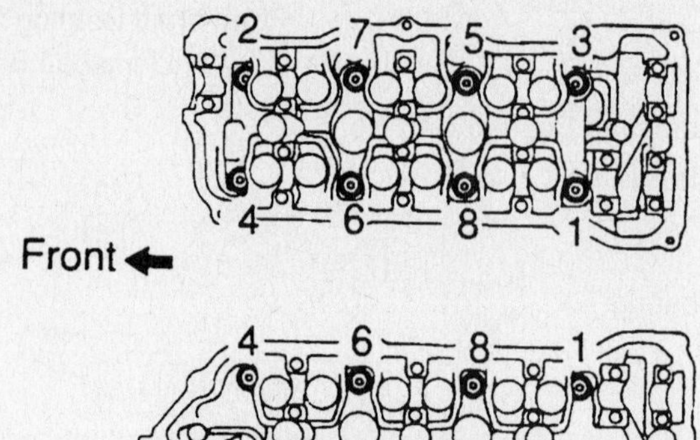

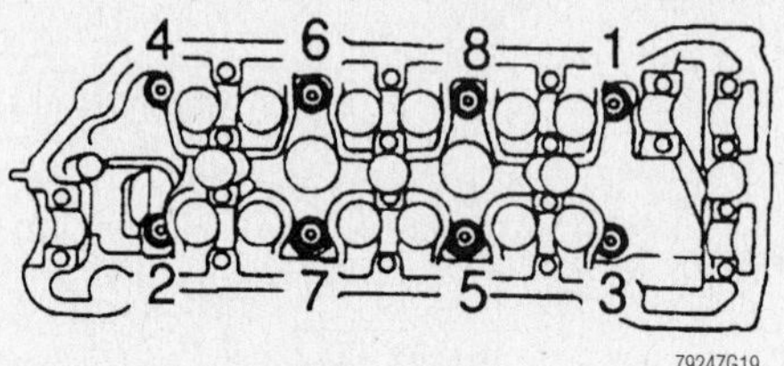

Cylinder head bolt loosening sequence—3.3L engines

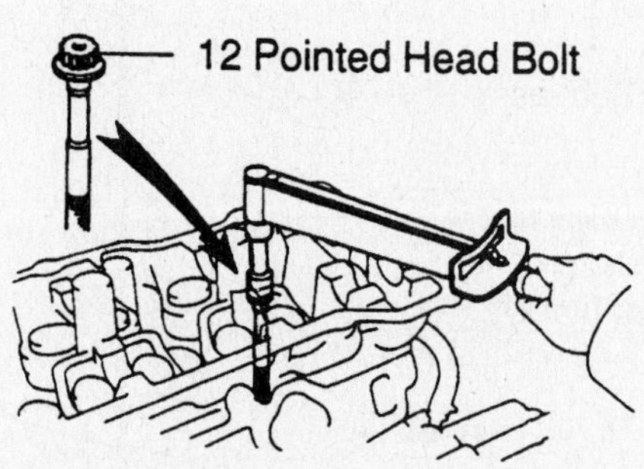

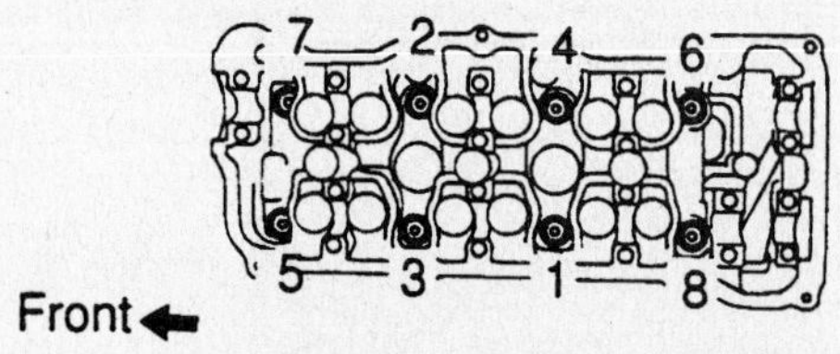

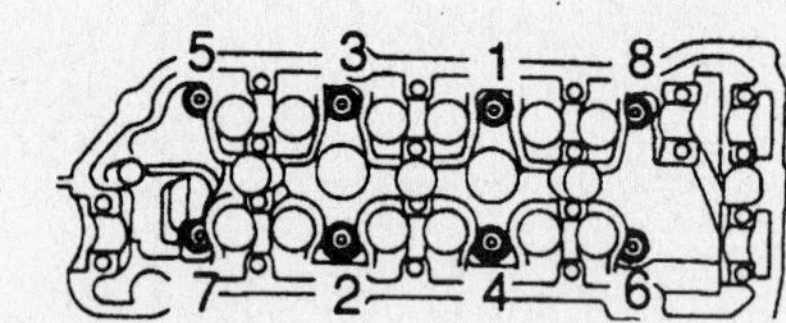

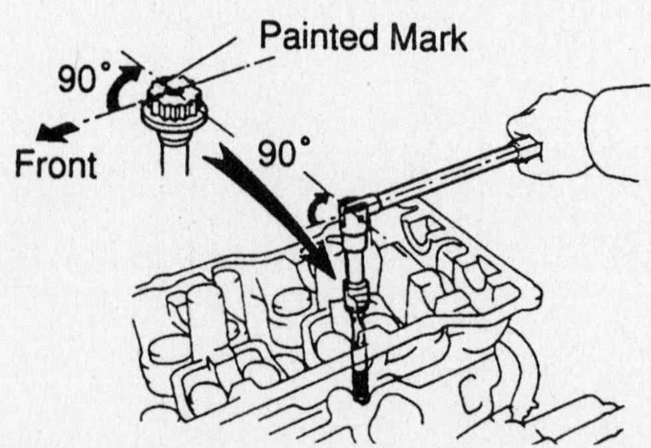

Cylinder head bolt tightening sequence—3.3L engine

- Both power steering air hoses
- Engine hanger and the intake chamber support
- EGR pipe and gaskets

12. Disconnect the following wiring:
    - Throttle Position (TP) sensor connector
    - Idle Air Control (IAC) valve connector
    - EGR gas temperature connector
    - Air conditioning idle up connector
13. Disconnect the following vacuum hoses:
    - 2 vacuum hoses from the Thermal Vacuum Valve (TVV)
    - Vacuum hose from the cylinder head rear plate
    - Vacuum hose from the charcoal canister
14. Remove or disconnect the following:
    - Air assist hose and the 2 water bypass hoses
    - Air intake chamber
    - Left engine wiring harness and move it aside
    - Wiring harness from the rear of the engine
    - Right engine wiring harness and move it aside
    - Ignition coils and move them aside
    - Timing belt
    - Camshaft pulleys and the timing belt rear cover
    - Cylinder head rear plate
    - Water inlet pipe
    - Air assist hose and vacuum hose
    - Intake manifold and fuel rail assembly
    - Water outlet
    - EGR pipe from the right exhaust manifold
    - Front exhaust pipe and exhaust manifolds
    - Dipstick assembly and the power steering pump bracket
    - Valve covers and the Camshaft Position (CMP) sensor
    - Camshafts
15. Be sure the engine is at/or near ambient temperature and remove the 2 (1 on each head) 8mm recessed hex bolts. Loosen and remove the 8 head bolts evenly, in 3 passes, in the reverse order of the installation sequence. Carefully lift the head from the engine; if necessary to pry the head loose, take great care not to damage the mating surfaces. Place the head on wood blocks in a clean work area.

➡If the cylinder head bolts are loosened out of sequence, warpage or cracking could result.

16. Remove the cylinder head gasket. With a gasket scraper, carefully remove all the old gasket material from the cylinder head and engine block surfaces.

To install:

17. Place the new cylinder head gasket onto the cylinder block.
18. Install the cylinder head, in sequence, using several steps, as follows:
    - Cylinder head onto the gasket
    - Cylinder head bolts lubricated with clean engine oil
    - Tighten the bolts in sequence in 3 steps to 40 ft. lbs. (54 Nm).

➡If any bolt does not meet the torque, replace it.

- Mark the forward edge of each bolt with paint, then tighten each bolt, in proper sequence, an additional 90 degrees.
- Check that each painted mark is now at a 90 degrees angle to the front

➡The paint mark applied to the bolt in the 9 o'clock position and should now be in the 12 o'clock position.

- Remaining 8mm bolts, lubricated with engine oil. Tighten both bolts to 13 ft. lbs. (18 Nm).

19. Install the camshafts.
20. Check and adjust the valves.
21. Apply sealant to the cylinder heads where the camshaft supports meet the cylinder heads.
22. Install or connect the following:
    - Cylinder head covers, using new gaskets
    - Dipstick and power steering pump bracket
    - Exhaust manifolds. Tighten the nuts to 36 ft. lbs. (49 Nm).
    - EGR pipe to the right exhaust manifold
    - Water outlet
    - Intake manifold and the fuel rail assembly. Tighten the intake manifold nuts/bolts to 11 ft. lbs. (15 Nm).
    - Air assist hose and the 2 water bypass hoses
    - Water inlet pipe and cylinder head rear plate
    - Timing belt rear cover and camshaft pulleys
    - Timing belt
    - Spark plugs and ignition coils
    - Right engine wiring harness
    - Wiring harness to the rear of the engine
    - Left engine wiring harness
    - Air intake chamber
    - EGR pipe, using new gaskets
23. Connect the following vacuum hoses:
    - The 2 TVV vacuum hoses
    - The vacuum hose to the rear cylinder head plate
    - Charcoal canister vacuum hose
24. Connect the following electrical wiring:
    - TP sensor connector
    - IAC valve connector
    - EGR gas temperature connector
    - Air conditioning idle up connector
25. Install or connect the following:
    - Engine hanger and the intake chamber support
    - Both power steering air hoses
    - Power steering pressure tube to the intake chamber
    - O_2 sensor connector to the pressure tube.
    - Both ground straps, to the intake chamber
    - DLC to the bracket
26. Connect the following hoses:
    - Power brake booster vacuum hose
    - PCV hose
    - IAC valve vacuum hose
27. Install or connect the following:
    - Emission control valve set and related vacuum hoses and connectors
    - V-bank cover
    - Pressure hose to the hydraulic motor
    - Fuel lines to the fuel rail assembly
    - Heater and radiator hoses
    - Right engine mounting support
    - both engine ground straps
    - Upper front suspension brace, if removed. Tighten the nuts to 59 ft. lbs. (80 Nm).
    - Cruise control actuator and bracket
    - Air cleaner, air flow meter and air duct assembly
    - Accelerator and throttle cables
28. Fill the cooling system.
29. Install or connect the following:
    - Negative battery cable
    - Heater air duct
    - Left and right ventilator louvers
    - Window washer hoses from the ventilator louvers
    - Top cowl seal and panel
    - Wiper and blade assembly
30. Start the engine and check for leaks.
31. Bleed the air from the cooling system.
32. Road test the vehicle and check for unusual noise, shock, slippage, correct shift points and smooth operation.
33. Recheck the coolant and engine oil levels.

Intake Manifold

REMOVAL & INSTALLATION

2.4L Engine

1. Before servicing the vehicle, refer to the precautions in the beginning of this section.
2. Disconnect the negative battery cable.
3. Release the fuel system pressure.
4. Drain the engine coolant.
5. Remove or disconnect the following:
   - Air cleaner cap
   - Mass Air Flow (MAF) meter and the resonator
   - Accelerator cable from the throttle body, if equipped with a manual transmission
   - Accelerator and throttle cables from the throttle body, if equipped with an automatic transmission
   - Cruise control cable from the actuator, if equipped with cruise control
   - Intake air connector
   - Air hose for Idle Air Control (IAC)
   - Vacuum sensing hose
   - Wire clamp for the engine wiring harness
   - Positive Crankcase Ventilation (PCV) hoses.
   - Engine wiring harness
   - Air conditioning compressor connector, if equipped with air conditioning
   - Oil pressure sensor connector
   - Engine Coolant Temperature (ECT) sensor connector
   - ECT sender gauge connector
   - Exhaust Gas Recirculation (EGR) gas temperature sensor connector
   - Vacuum Switching Valve (VSV) connector, for the EGR
   - 2 vacuum hoses, from the VSV for the EGR
   - Ground strap, from the cowl top panel
   - Engine wiring harness, from the air intake chamber
   - Throttle Position (TP) sensor connector
   - IAC valve connector
   - Crankshaft Position (CKP) sensor connector
   - Knock (KS) sensor connector
   - Data Link Connector 1 (DLC1), from the bracket
   - Engine wiring harness clamp
   - EGR pipe
   - Intake chamber stay
   - Air intake chamber assembly
6. Disconnect the following hoses:
   - Evaporative Emission (EVAP) hose, from the throttle body
   - Brake booster vacuum hose, from the union
   - Water bypass hose, from the water bypass pipe
   - Water bypass hose, from the cylinder head rear cover
   - Injector connectors
   - Fuel inlet pipe
   - Hoses and the fuel return pipe.
7. Remove the delivery pipe and injectors, as follows:
   - Delivery pipe, together with the 4 injectors

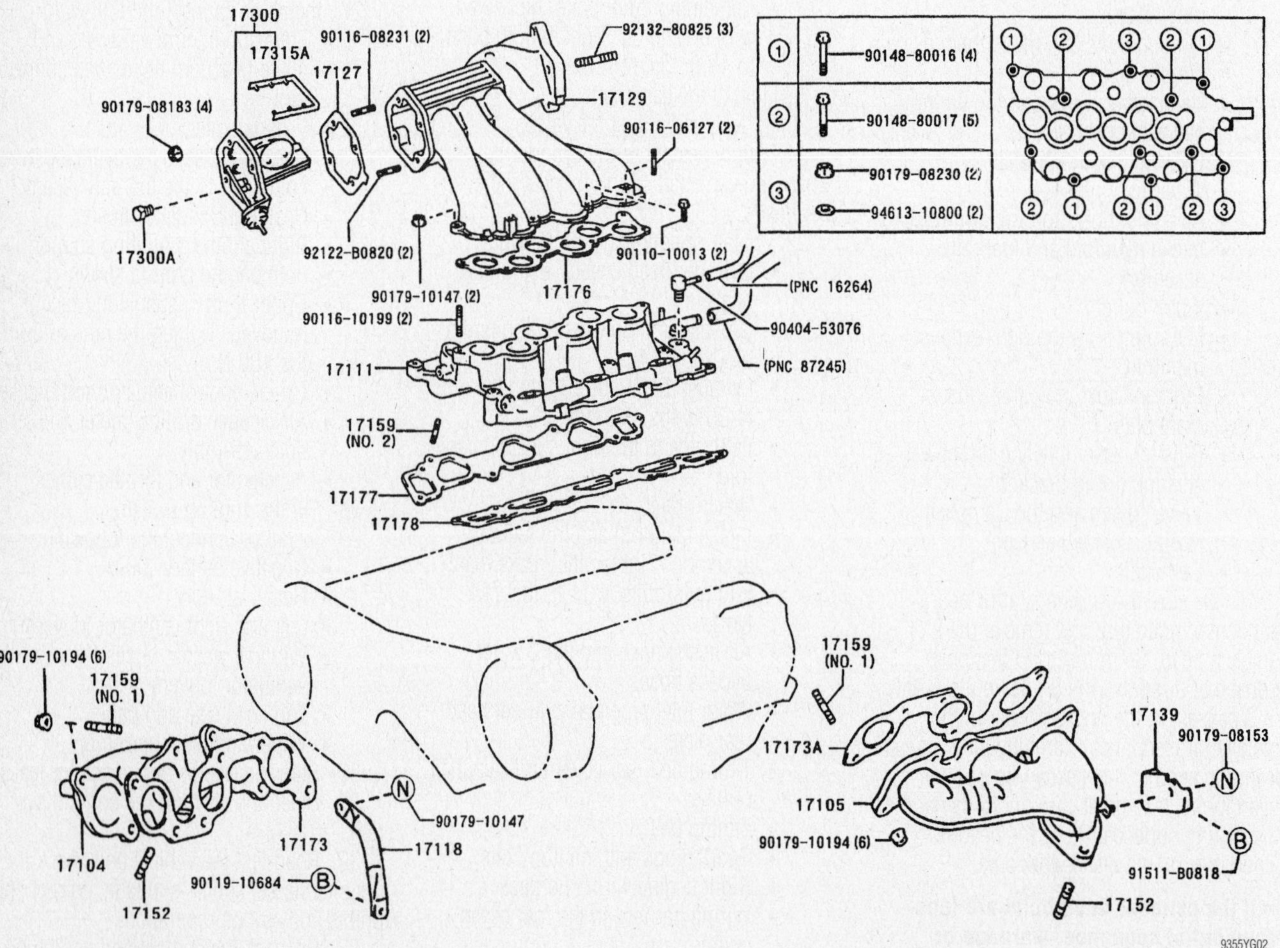

Intake manifold and related components—2001 2.4L engine

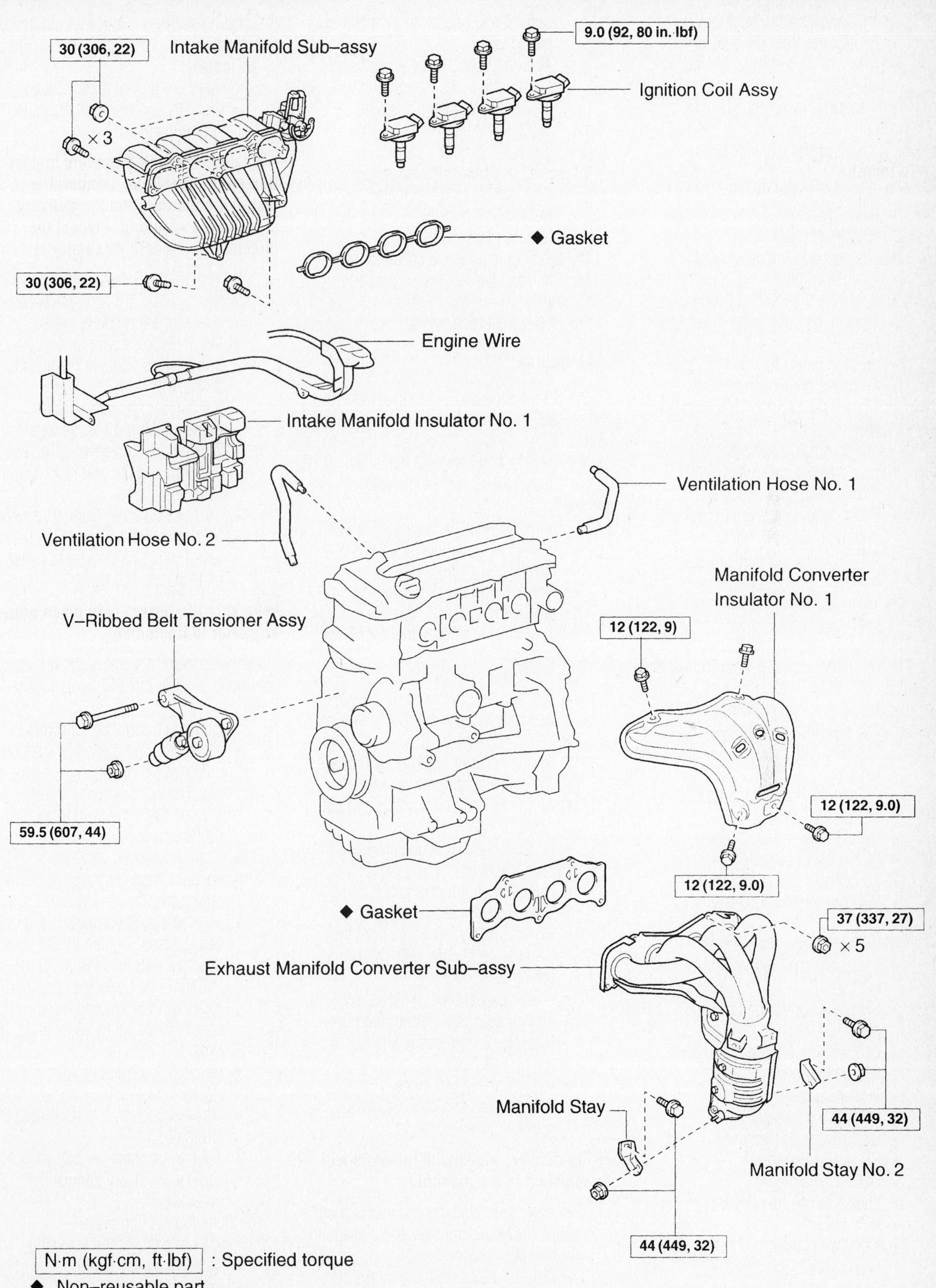

Intake manifold and related components—2004 2.4L engine

- 4 insulators from the 4 spacers
- 4 injectors, from the delivery pipe
- O-ring and grommets, from each injector
- 4 spacers, by carefully prying them out

8. Remove the intake manifold.

To install:

9. Install or connect the following:
   - Intake manifold. Tighten the bolts to 22 ft. lbs. (29 Nm).
   - Injectors and the delivery pipe
   - Fuel return pipe
   - Fuel inlet pipe, with a new gasket. Tighten the bolts to 22 ft. lbs. (29 Nm).
   - Injector connectors
   - Air intake chamber assembly. Tighten the bolts to 15 ft. lbs. (21 Nm).
10. Connect the following hoses:
    - Evaporative Emissions (EVAP) hose, to the throttle body
    - Brake booster vacuum hose, to the union
    - Water bypass hose, to water bypass pipe
    - Water bypass hose, to cylinder head rear cover
11. Install or connect the following:
    - Air intake chamber stay. Tighten the bolts to 15 ft. lbs. (20 Nm).
    - EGR pipe. Tighten bolts to 13 ft. lbs. (18 Nm), nut "A" to 14 ft. lbs. (19 Nm) and nut B to 15 ft. lbs. (20 Nm).
    - Air conditioning compressor connector
    - Oil pressure sensor connector
    - ECT sensor connector
    - ECT sender gauge connector
    - EGR gas temperature sensor connector
    - VSV connector for the EGR
    - 2 vacuum hose to the VSV for the EGR
    - Ground strap to the cowl top panel
    - Engine wiring harness to the air intake chamber
    - TP sensor connector
    - IAC valve connector
    - CKP sensor connector
    - KS sensor connector
    - DLC1 to the bracket
    - Engine wiring harness clamp
    - PCV hoses
    - Intake air connector. Tighten the bolts to 13 ft. lbs. (18 Nm).
    - Cruise control cable to the actuator, if equipped with cruise control
    - Accelerator cable to the throttle body, if equipped with a manual transmission
    - Accelerator and throttle cables to the throttle body, if equipped with an automatic transmission
12. Fill the engine and radiator with engine coolant.
13. Install or connect the following:
    - Air cleaner cap, MAF meter and resonator assembly
    - Negative battery cable
14. Start the engine and check for leaks.
15. Road test the vehicle for proper operation.
16. Recheck all fluid levels.

3.0L Engine

1. Before servicing the vehicle, refer to the precautions in the beginning of this section.
2. Remove or disconnect the following:
3. Properly relieve the fuel system pressure.
4. Drain and recycle the engine coolant.
5. Remove or disconnect the following:
   - Accelerator cable
   - Throttle cable
   - Air cleaner
   - Any wiring or hoses interfering with removal
   - Right side engine mount stay
   - Radiator and heater hoses in the way of the intake manifold removal
   - V-bank cover
   - All the vacuum hose and wiring for the emission control valve set
   - Air intake chamber and discard the gasket
   - Exhaust Gas Recirculation (EGR) pipe and discard the gaskets
   - Hydraulic motor pressure hose from the air intake chamber
   - Engine wiring harnesses from the left side, right side, rear and No. 3 timing belt cover
   - Front exhaust pipe, if necessary
   - Timing belt, camshaft timing pulleys, No. 2 idler pulley and No. 3 timing belt cover
   - Cylinder head rear plate
   - 2 bolts, nuts and plate washers with the intake manifold.

➡The delivery pipes with injectors will be attached to the manifold.

- Other fuel related components such as the No. 2 fuel pipe and pulsation damper, if needed
- Delivery pipes from the intake manifold

6. Clean and inspect the intake manifold mating surfaces. Scrape all old gasket material off.

To install:

7. Install or connect the following:
   - Delivery pipes with injectors to the intake manifold.

➡Be sure to place 4 spacers in position on the manifold. Temporarily install 4 bolts to retain the delivery pipes to the manifold. Inspect the injectors for smooth rotation.

- Tighten the delivery pipes bolts to 84 inch lbs. (10 Nm), once the injectors are properly seated
- No. 2 fuel pipe with union bolts and gaskets. Tighten the bolts to 24 ft. lbs. (32 Nm).
- No. 1 fuel pipe with pulsation damper, using 4 new gaskets. Tighten the damper to 35 ft. lbs. (32 Nm) and the bolt to 11 ft. lbs. (15 Nm).
- Fuel pressure regulator, if removed
- Intake manifold. Tighten the 9 bolts and 2 nuts in a crisscross pattern to 11 ft. lbs. (15 Nm).

➡Be sure the gasket is in place properly prior to tightening.

8. Retighten the water outlet mounting nuts/bolts to 11 ft. lbs. (15 Nm), if loosened.
9. Install or connect the following:
   - Air assist hose and water inlet pipe, using a new O-ring, by applying a small amount of soapy water. Tighten the fastener(s) to 14 ft. lbs. (20 Nm).
   - Ground strap
   - Vacuum hoses removed to the air intake chamber and vacuum tank
   - Any remaining components, using new gaskets. Tighten the air intake chamber nuts/bolts to 32 ft. lbs. (43 Nm), the EGR pipe nuts to 108 inch lbs. (12 Nm) and the emission control valve set to 69 inch lbs. (8 Nm).
   - Air cleaner assembly
   - Heater hoses
   - Throttle cable with bracket onto the throttle body
   - Accelerator cable, by adjusting it, if equipped with an automatic transaxle
10. Refill the cooling system
11. Install or connect the following:
    - Negative battery cable
    - Heater air duct
12. Start the engine and inspect for leaks.

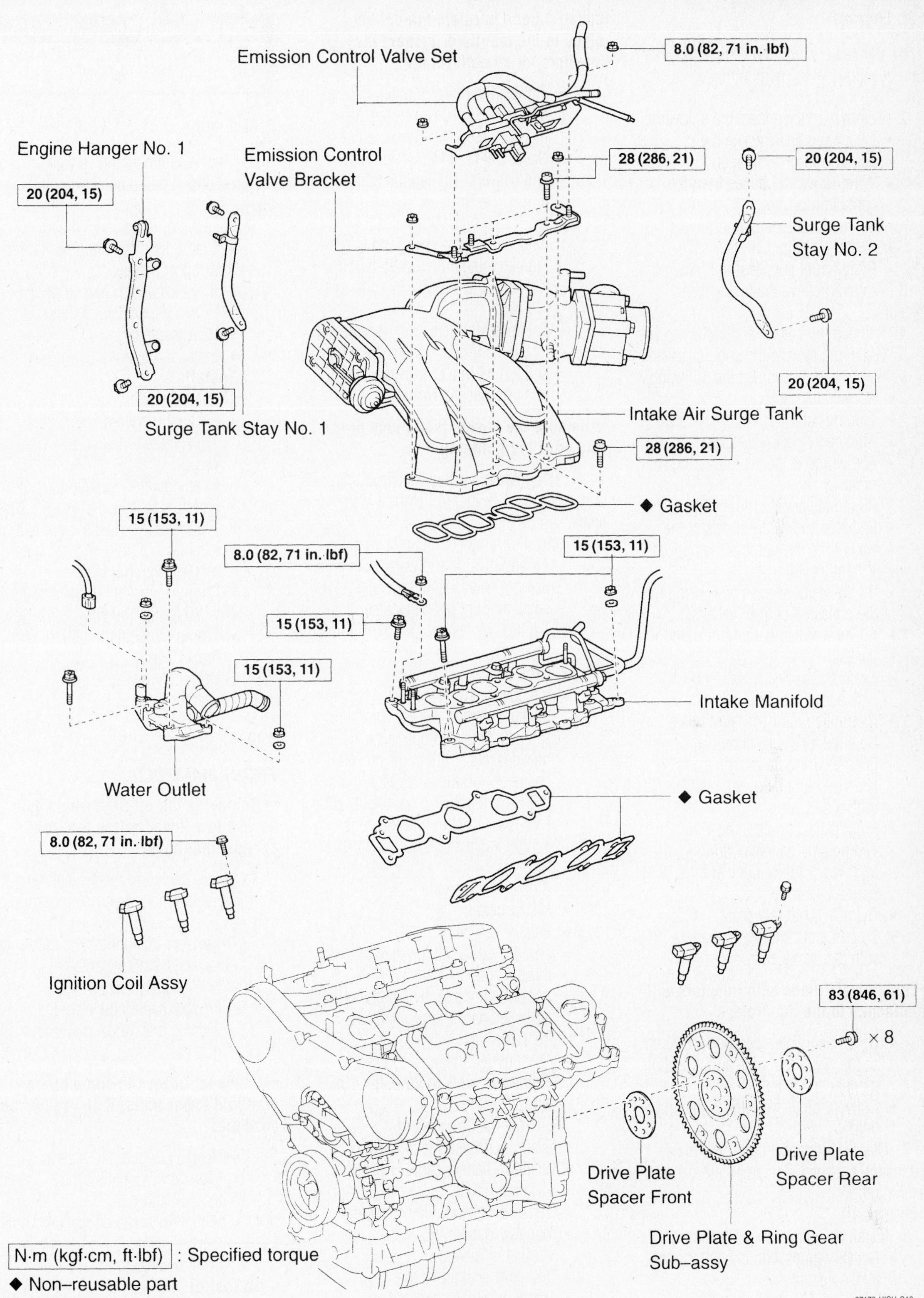
Emission Control Valve Set
8.0 (82, 71 in. lbf)
Engine Hanger No. 1
Emission Control Valve Bracket
28 (286, 21)
20 (204, 15)
20 (204, 15)
Surge Tank Stay No. 2
20 (204, 15)
20 (204, 15)
Surge Tank Stay No. 1
Intake Air Surge Tank
28 (286, 21)
◆ Gasket
15 (153, 11)
8.0 (82, 71 in. lbf)
15 (153, 11)
15 (153, 11)
15 (153, 11)
Intake Manifold
Water Outlet
◆ Gasket
8.0 (82, 71 in. lbf)
Ignition Coil Assy
83 (846, 61)
× 8
Drive Plate Spacer Rear
Drive Plate Spacer Front
Drive Plate & Ring Gear Sub–assy
N·m (kgf·cm, ft·lbf) : Specified torque
◆ Non–reusable part
67170-HIGH-G10

Intake manifold and related parts—3.3L engine

3.3L Engines

1. Before servicing the vehicle, refer to the precautions in the beginning of this section.
2. Remove or disconnect the following:
   - Wiper and blade assembly
   - Top cowl seal and panel
   - Window washer hoses from the ventilator louvers
   - Left and right ventilator louvers
   - Heater air duct
   - Front upper suspension brace
3. Properly relieve the fuel system pressure.
4. Remove the battery and battery tray.
5. Drain and recycle the engine coolant.
6. Remove or disconnect the following:
   - Accelerator cable
   - Throttle cable
   - Air cleaner cap assembly
   - Any wiring or hoses interfering with removal
   - Right side engine mount stay
   - Radiator and heater hoses in the way of the intake manifold removal
   - V-bank cover
   - All the vacuum hose and wiring for the emission control valve set
   - Air intake chamber and discard the gasket
   - Exhaust Gas Recirculation (EGR) pipe and discard the gaskets
   - Hydraulic motor pressure hose from the air intake chamber
   - Engine wiring harnesses from the left side, right side, rear and No. 3 timing belt cover
   - Front exhaust pipe, if necessary
   - Timing belt, camshaft timing pulleys, No. 2 idler pulley and No. 3 timing belt cover
   - Cylinder head rear plate
   - 2 bolts, nuts and plate washers with the intake manifold.

➡The delivery pipes with injectors will be attached to the manifold.

   - Other fuel related components such as the No. 2 fuel pipe and pulsation damper, if needed
   - Delivery pipes from the intake manifold
7. Clean and inspect the intake manifold mating surfaces. Scrape all old gasket material off.

To install:

8. Install or connect the following:
   - Delivery pipes with injectors to the intake manifold.

➡Be sure to place 4 spacers in position on the manifold. Temporarily install 4 bolts to retain the delivery pipes to the manifold. Inspect the injectors for smooth rotation.

   - Tighten the delivery pipes bolts to 84 inch lbs. (10 Nm), once the injectors are properly seated
   - No. 2 fuel pipe with union bolts and gaskets. Tighten the bolts to 24 ft. lbs. (32 Nm).
   - No. 1 fuel pipe with pulsation damper, using 4 new gaskets. Tighten the damper to 35 ft. lbs. (32 Nm) and the bolt to 11 ft. lbs. (15 Nm).
   - Fuel pressure regulator, if removed
   - Intake manifold. Tighten the 9 bolts and 2 nuts in a crisscross pattern to 11 ft. lbs. (15 Nm).

➡Be sure the gasket is in place properly prior to tightening.

9. Retighten the water outlet mounting nuts/bolts to 11 ft. lbs. (15 Nm), if loosened.
10. Install or connect the following:
    - Air assist hose and water inlet pipe, using a new O-ring, by applying a small amount of soapy water. Tighten the fastener(s) to 14 ft. lbs. (20 Nm).
    - Ground strap
    - Vacuum hoses removed to the air intake chamber and vacuum tank
    - Any remaining components, using new gaskets. Tighten the air intake chamber nuts/bolts to 32 ft. lbs. (43 Nm), the EGR pipe nuts to 108 inch lbs. (12 Nm) and the emission control valve set to 69 inch lbs. (8 Nm).
    - Air cleaner assembly
    - Heater hoses
    - Battery and tray
    - Throttle cable with bracket onto the throttle body
    - Accelerator cable, by adjusting it, if equipped with an automatic transaxle
    - Front upper suspension brace. Tighten the nuts to 59 ft. lbs. (80 Nm).
11. Refill the cooling system
12. Install or connect the following:
    - Negative battery cable
    - Heater air duct
    - Left and right ventilator louvers
    - Window washer hoses from the ventilator louvers
    - Top cowl seal and panel
    - Wiper and blade assembly
13. Start the engine and inspect for leaks.

Exhaust Manifold

REMOVAL & INSTALLATION

2.4L Engine

1. Before servicing the vehicle, refer to the precautions in the beginning of this section.
2. Remove or disconnect the following:
   - Clamp from the support bracket
   - Support bracket
   - Front exhaust pipe and gaskets from the exhaust manifold
   - Heat insulator
   - Exhaust manifold and gasket

To install:

3. Install or connect the following:
   - Exhaust manifold and gasket. Tighten the nuts to 36 ft. lbs. (49 Nm).
   - Heat insulator. Tighten the bolts and nuts to 48 inch lbs. (5.5 Nm).
   - Front exhaust pipe assembly to the exhaust manifold. Tighten the nuts to 46 ft. lbs. (62 Nm).
   - Support bracket. Tighten the bolts to 29 ft. lbs. (39 Nm).
   - Clamp. Tighten the bolt to 14 ft. lbs. (19 Nm).
4. Start the engine.
5. Check for exhaust leaks.

3.0L Engine

FRONT MANIFOLD

➡Removing the oil filter helps gain access to a lower bolt in the front exhaust manifold.

1. Before servicing the vehicle, refer to the precautions in the beginning of this section.
2. Remove or disconnect the following:
   - Negative battery cable
   - Engine undercovers
   - Front exhaust pipe from the exhaust manifolds, by removing the nuts

➡Check for access to some of the manifold lower bolts, if so remove any possible.

   - Heated Oxygen (HO_2) sensor
   - Exhaust manifold stay, by removing the bolt and nut
   - Remaining exhaust manifold nuts; then, separate the exhaust manifold from the engine

To install:

3. Install or connect the following:
   - Exhaust manifold, using a new gas-

ket. Uniformly, tighten the bolts to 36 ft. lbs. (49 Nm).
- Exhaust manifold stay. Tighten the nut/bolt to 15 ft. lbs. (20 Nm).
- Heated Oxygen (HO_2) sensor to the exhaust manifold
- Front exhaust pipe to the exhaust manifold, using a new gasket. Tighten both nuts to 46 ft. lbs. (62 Nm).
- Engine undercovers
- Negative battery cable

REAR MANIFOLD

1. Before servicing the vehicle, refer to the precautions in the beginning of this section.
2. Remove or disconnect the following:
   - Negative battery cable
   - Engine undercovers
   - Front exhaust pipe from both exhaust manifolds, from below the engine
   - Exhaust Gas Recirculation (EGR) pipe from the rear exhaust manifold, by removing the 4 nuts
   - Heated Oxygen (HO_2) sensor wiring, from the right exhaust manifold
   - Exhaust manifold stay
   - 6 exhaust manifold nuts and the exhaust manifold

To install:

3. Install or connect the following:
   - Exhaust manifold to the engine, using a new gasket. Tighten the 6 nuts to 36 ft. lbs. (49 Nm).
   - Exhaust manifold stay. Tighten the nut/bolt to 15 ft. lbs. (20 Nm).
   - HO_2sensor wiring to the exhaust manifold
   - EGR pipe to the exhaust manifold and the engine, using new gaskets. Tighten the 4 nuts to 108 inch lbs. (12 Nm).
   - Front exhaust pipe to the exhaust manifold, use a new gasket. Tighten both nuts to 46 ft. lbs. (62 Nm).
   - Engine undercovers
   - Negative battery cable

3.3L Engines

FRONT MANIFOLD

➡Removing the oil filter helps gain access to a lower bolt in the front exhaust manifold.

1. Before servicing the vehicle, refer to the precautions in the beginning of this section.
2. Remove or disconnect the following:
   - Negative battery cable
   - Engine undercovers
   - Front exhaust pipe from the exhaust manifolds, by removing the nuts

➡Check for access to some of the manifold lower bolts, if so remove any possible.

   - Heated Oxygen (HO_2) sensor
   - Exhaust manifold stay, by removing the bolt and nut
   - Remaining exhaust manifold nuts; then, separate the exhaust manifold from the engine

To install:

3. Install or connect the following:
   - Exhaust manifold, using a new gasket. Uniformly, tighten the bolts to 36 ft. lbs. (49 Nm).
   - Exhaust manifold stay. Tighten the nut/bolt to 15 ft. lbs. (20 Nm).
   - Heated Oxygen (HO_2) sensor to the exhaust manifold
   - Front exhaust pipe to the exhaust manifold, using a new gasket. Tighten both nuts to 46 ft. lbs. (62 Nm).
   - Engine undercovers
   - Negative battery cable

REAR MANIFOLD

1. Before servicing the vehicle, refer to the precautions in the beginning of this section.
2. Remove or disconnect the following:
   - Negative battery cable
   - Engine undercovers
   - Front exhaust pipe from both exhaust manifolds, from below the engine
   - Exhaust Gas Recirculation (EGR) pipe from the rear exhaust manifold, by removing the 4 nuts
   - Heated Oxygen (HO_2) sensor wiring, from the right exhaust manifold
   - Exhaust manifold stay
   - 6 exhaust manifold nuts and the exhaust manifold

To install:

3. Install or connect the following:
   - Exhaust manifold to the engine, using a new gasket. Tighten the 6 nuts to 36 ft. lbs. (49 Nm).
   - Exhaust manifold stay. Tighten the nut/bolt to 15 ft. lbs. (20 Nm).
   - HO_2 sensor wiring to the exhaust manifold
   - EGR pipe to the exhaust manifold and the engine, using new gaskets. Tighten the 4 nuts to 108 inch lbs. (12 Nm).
   - Front exhaust pipe to the exhaust manifold, use a new gasket. Tighten both nuts to 46 ft. lbs. (62 Nm).
   - Engine undercovers
   - Negative battery cable

Front Crankshaft Seal

For 2.4L engines, see the Timing Chain procedure later in this chapter.

REMOVAL & INSTALLATION

3.0L Engine

1. Before servicing the vehicle, refer to the precautions in the beginning of this section.
2. Remove or disconnect the following:
   - Engine coolant reservoir tank and the alternator belt
   - Right front wheel and the splash shield
   - Power steering pump drive belt, by loosening both bolts
   - Both ground wire connectors
   - Right engine mounting stay
   - Engine moving control rod and the No. 2 right engine mount bracket

➡To extract the engine bracket and control rod, raise the engine slightly.

   - No. 2 alternator bracket
   - Crankshaft pulley bolt, using a pry-bar and wrench or Crankshaft Pulley Holding tool 09213-54015 and Flange Holding tool 09330-00021
   - Crankshaft pulley, using a puller
   - No. 1 timing belt cover
3. Remove the No. 2 timing belt cover, as follows:
   - Engine wire protector from the No. 3 (rear) timing belt cover
   - Engine wire protector clamp from the No. 3 timing belt cover
   - 5 bolts from the No. 2 timing belt cover
   - No. 2 cover

To install:

4. Install or connect the following:
   - No. 2 timing belt cover, using a new gasket

➡Install it evenly to the part of the belt cover shaded black. After installation, press down on it so that the adhesive sticks to the belt cover firmly.

   - No. 2 timing belt cover. Tighten the 5 bolts to 74 inch lbs. (8 Nm).

- Engine wire protector clamp to the No. 3 timing belt cover
- Engine wire protector to the No. 3 timing belt cover with the bolt
- No. 3 timing belt cover, using a new gasket
- Tighten the 4 No. 1 timing belt cover bolts to 74 inch lbs. (8 Nm).
- Crankshaft pulley. Tighten the bolt to 159 ft. lbs. (215 Nm).
- No. 2 alternator bracket. Tighten the nut to 21 ft. lbs. (28 Nm). Do not tighten the pivot bolt at this time.
- No. 2 right engine mounting bracket and the moving control rod
- Right engine mount stay
- Both ground wire connectors
- Drive belts by adjusting them
- Coolant reservoir
- Right front splash shield and wheel
- Negative battery cable

5. Start the vehicle and check for any leaks.
6. Recheck the ignition timing.

3.3L Engines

1. Before servicing the vehicle, refer to the precautions in the beginning of this section.
2. Remove or disconnect the following:
   - Engine coolant reservoir tank and the alternator belt
   - Right front wheel and the splash shield
   - Power steering pump drive belt, by loosening both bolts
   - Both ground wire connectors
   - Right engine mounting stay
   - Engine moving control rod and the No. 2 right engine mount bracket

➡To extract the engine bracket and control rod, raise the engine slightly.

   - No. 2 alternator bracket
   - Crankshaft pulley bolt, using a pry-bar and wrench or Crankshaft Pulley Holding tool 09213-54015 and Flange Holding tool 09330-00021
   - Crankshaft pulley, using a puller
   - No. 1 timing belt cover
3. Remove the No. 2 timing belt cover, as follows:
   - Engine wire protector from the No. 3 (rear) timing belt cover
   - Engine wire protector clamp from the No. 3 timing belt cover
   - 5 bolts from the No. 2 timing belt cover
   - No. 2 cover

To install:

4. Install or connect the following:
   - No. 2 timing belt cover, using a new gasket

➡Install it evenly to the part of the belt cover shaded black. After installation, press down on it so that the adhesive sticks to the belt cover firmly.

   - No. 2 timing belt cover. Tighten the 5 bolts to 74 inch lbs. (8 Nm).
   - Engine wire protector clamp to the No. 3 timing belt cover
   - Engine wire protector to the No. 3 timing belt cover with the bolt
   - No. 3 timing belt cover, using a new gasket
   - Tighten the 4 No. 1 timing belt cover bolts to 74 inch lbs. (8 Nm).
   - Crankshaft pulley. Tighten the bolt to 159 ft. lbs. (215 Nm).
   - No. 2 alternator bracket. Tighten the nut to 21 ft. lbs. (28 Nm). Do not tighten the pivot bolt at this time.
   - No. 2 right engine mounting bracket and the moving control rod
   - Right engine mount stay
   - Both ground wire connectors
   - Drive belts by adjusting them
   - Coolant reservoir
   - Right front splash shield and wheel
   - Negative battery cable
5. Start the vehicle and check for any leaks.
6. Recheck the ignition timing.

Camshaft and Valve Lifters

REMOVAL & INSTALLATION

2.4L Engine

1. Before servicing the vehicle, refer to the precautions in the beginning of this section.
2. Disconnect the negative battery cable.
3. Drain the engine coolant.
4. Remove or disconnect the following:
   - Right front wheel
   - Right fender splash shield
   - Right fender apron seal
   - No.1 engine undercover
   - Coil pack
   - Cylinder head cover
5. Set the No.1 piston at TDC compression.
6. Remove or disconnect the following:
   - Timing chain tensioner No.1
7. Loosen the camshaft timing gear set bolt.
8. Raise the camshaft and remove the set bolt.
9. Remove or disconnect the following:
   - Timing gear and chain
   - Exhaust camshaft
10. Loosen the intake camshaft cap bolts in several passes, in the sequence shown. Remove the caps. Remove the camshaft.
11. Installation is the reverse of removal. Tighten the cap bolt, in several passes, in the sequences shown, to:
    - Front caps: 22 ft. lbs. (30Nm)
    - All other caps: 80 inch lbs. (9Nm)
12. See the Timing Chain Removal and Installation procedure.

3.0L Engine

1. Before servicing the vehicle, refer to the precautions in the beginning of this section.
2. Remove or disconnect the following:
   - Timing belt and idler pulley
   - Camshaft timing pulleys
   - Cylinder head covers

➡The thrust clearance on both the intake and exhaust camshafts is very small; the camshafts must be kept level during removal. If the camshafts are removed without being kept level, the camshaft may be caught in the cylinder head, causing the head to break or the camshaft to seize.

3. Remove the exhaust and intake camshafts from the right side cylinder head, as follows:
   a. Turn the camshaft with a wrench until the 2 pointed marks drive and driven gears are aligned. (The right camshaft gears have 2 marks apiece; the left side camshaft gears have 1 mark each.)
   b. Secure the exhaust camshaft sub-gear to the main gear using a service bolt. A bolt 0.63–0.79 in. (16–20mm) long with a 6mm thread diameter and a 1mm pitch is recommended. When removing the exhaust camshaft be sure the sub-gear is not loaded; all the force must be eliminated.
   c. Uniformly loosen and remove the exhaust camshaft bearing cap bolts in several passes and in the proper sequence. Remove the 8 bearing cap bolts and remove the caps, keeping them in the correct order.
   d. Remove the exhaust camshaft from the engine.
   e. Uniformly loosen and remove the 10 bearing cap bolts in several passes, in the proper sequence. Remove the bearing caps, keeping them in order, remove the oil seal, then lift out the intake camshaft.

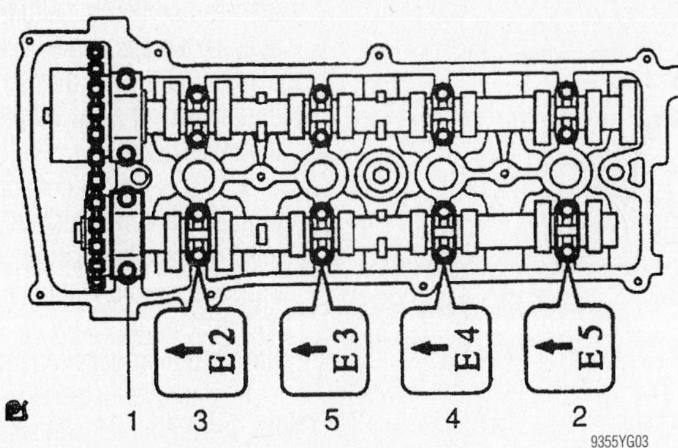

Exhaust camshaft cap bolt loosening sequence—2.4L engine

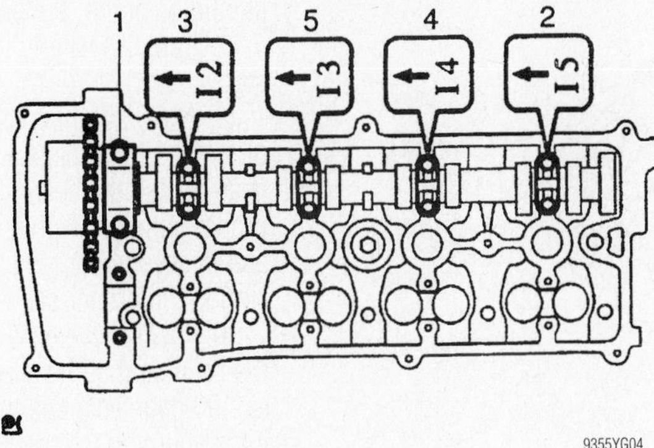

Intake camshaft cap bolt loosening sequence—2.4L Engine

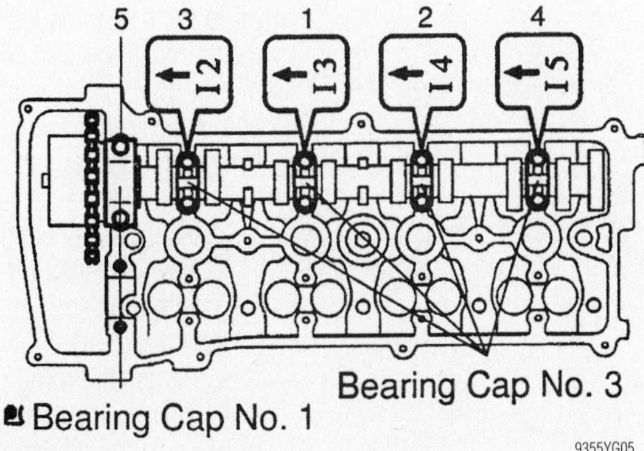

Intake camshaft cap bolt tightening sequence—2.4L engine

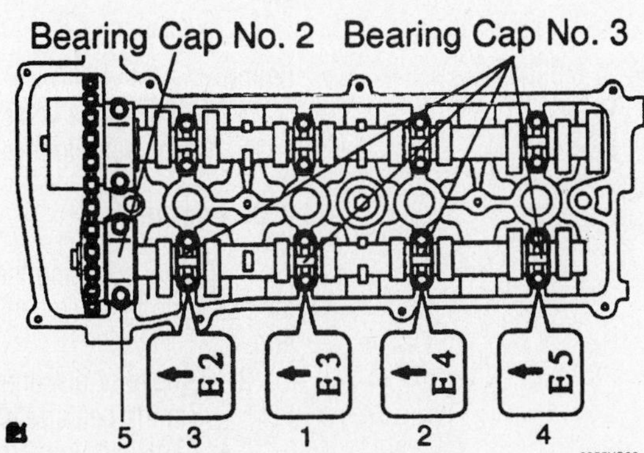

Exhaust camshaft cap bolt tightening sequence—2.4L engine

4. Remove the exhaust and intake camshafts from the left side cylinder head, as follows:

a. Turn the camshaft with a wrench until the pointed marks on the drive and driven gears are aligned. (The right camshaft gears have 2 marks apiece; the left side camshaft gears have 1 mark each.)

b. Secure the exhaust camshaft sub-gear to the main gear using a service bolt. A bolt 16–20mm long with a 6mm thread diameter and a 1mm pitch is recommended. When removing the exhaust camshaft be sure the sub-gear is not loaded; all the force must be eliminated.

c. Uniformly loosen and remove the exhaust camshaft bearing cap bolts in several passes and in the proper sequence. Remove the 8 bearing cap bolts and remove the caps. Keep the caps in the correct order.

d. Remove the exhaust camshaft from the engine.

e. Uniformly loosen and remove the 10 bearing cap bolts in several passes, in the reverse order of the installation sequence. Remove the bearing caps, keeping them in order, remove the oil seal, then lift out the intake camshaft.

5. Remove the valve lifter shims and hydraulic lifters. Identify each lifter and shim as it is removed so it can be reinstalled in the same position. If the lifters are to be reused, store them upside down in a sealed container.

To install:

6. Install the valve lifters into their original positions and install the shims. Check valve clearance and replace the shims as necessary.

7. When reinstalling, remember that the camshafts must be handled carefully and kept straight and level to avoid damage.

8. Before installing the camshafts in either cylinder head, apply multi-purpose grease to each camshaft.

9. Install the right camshafts, as follows:

a. Position the intake camshaft on the head so that the alignment marks are at a 90 degrees angle from vertical. The mark should be at the "3 o'clock" position.

b. Apply sealant to the No. 1 bearing cap.

c. Apply a light coat of clean engine oil to the bolt threads and under the bolt head. Install the bearing caps to their proper position. Tighten the bolts evenly and in several passes to 12 ft. lbs. (16 Nm) in the proper sequence.

d. Position the exhaust camshaft on

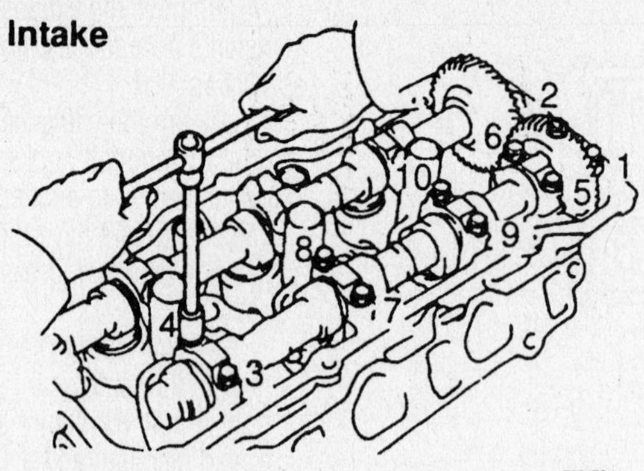

Right intake camshaft bearing cap bolt loosening sequence—3.0L engine

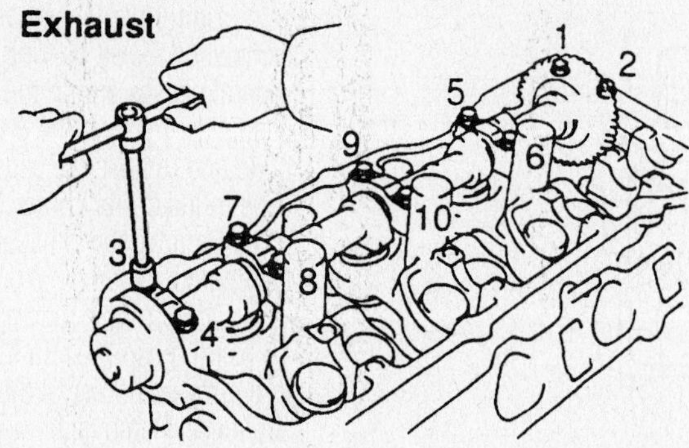

Right side exhaust camshaft bearing cap bolt loosening sequence—3.0L engine

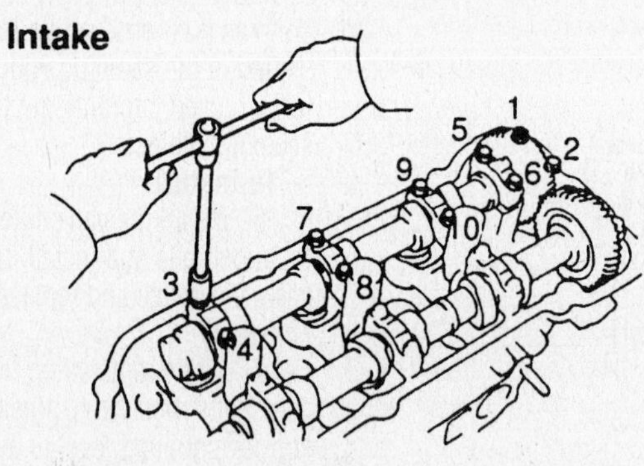

Left intake camshaft bearing cap bolt loosening sequence—3.0L engine

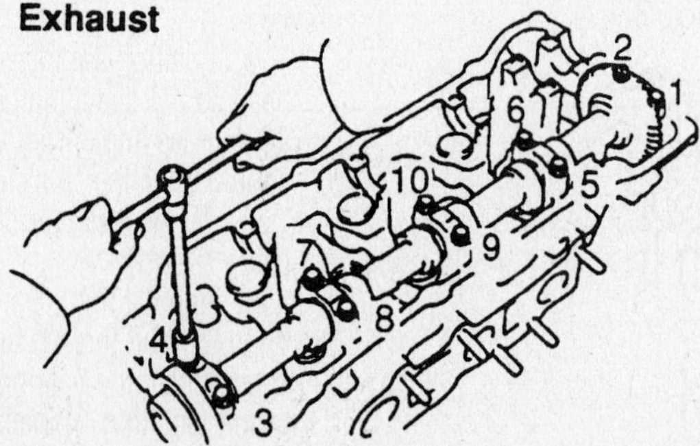

Left side exhaust camshaft bearing cap bolt loosening sequence—3.0L engine

the head so that the alignment marks are at a 90 degrees angle from vertical. The mark should be at the "9 o'clock" position and must align with the marks on the other gear.

e. Apply a light coat of clean engine oil to the bolt threads and under the bolt head. Install the bearing caps to their proper position. Tighten the bolts evenly and in several passes to 12 ft. lbs. (16 Nm) in the proper sequence.

f. Remove the service bolt.

10. Install the left camshafts, as follows:

a. Position the intake camshaft on the head so that the alignment mark is at a 90 degrees angle from vertical. The mark should be at the "9 o'clock" position.

b. Apply sealant to the No. 1 bearing cap.

c. Apply a light coat of clean engine oil to the bolt threads and under the bolt head. Install the bearing caps to their proper position. Tighten the bolts evenly and in several passes to 12 ft. lbs. (16 Nm) in the proper sequence.

d. Position the exhaust camshaft on the head so that the alignment marks are at a 90 degrees angle from vertical. The mark should be at the "3 o'clock" position and must align with the marks on the other gear.

e. Apply a light coat of clean engine oil to the bolt threads and under the bolt head. Install the bearing caps to their proper position. Tighten the bolts evenly and in several passes to 12 ft. lbs. (16 Nm) in the proper sequence.

f. Remove the service bolt.

11. Install or connect the following:
- New camshaft oil seals, lubricated with multi-purpose grease
- No. 3 (rear) timing belt cover
- Camshaft timing gears
- Idler pulley, timing belt and covers

12. Check and adjust the valve clearance.

13. Install the cylinder head (valve) covers.

14. Start the engine. Check the ignition timing.

15. Test drive the vehicle.

16. Check all fluid levels.

3.3L Engines

1. Before servicing the vehicle, refer to the precautions in the beginning of this section.

2. Remove or disconnect the following:
- Timing belt and idler pulley
- Camshaft timing pulleys
- Cylinder head covers

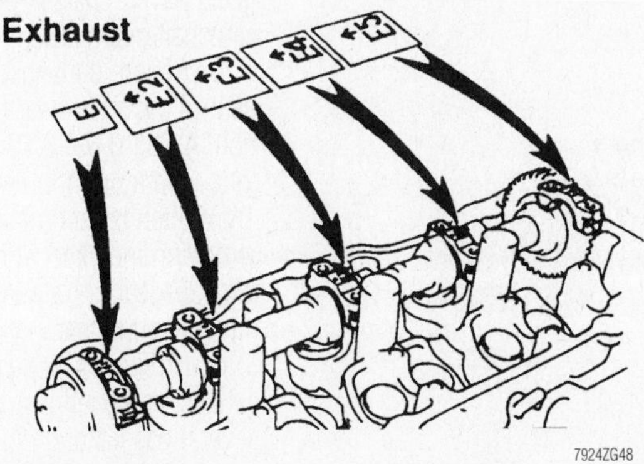

Right exhaust bearing caps must be placed in their proper locations—3.0L engine

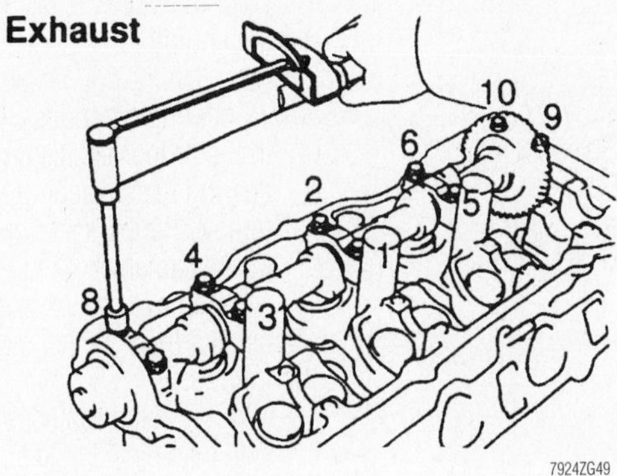

Right exhaust camshaft bearing cap bolt tightening sequence—3.0L engine

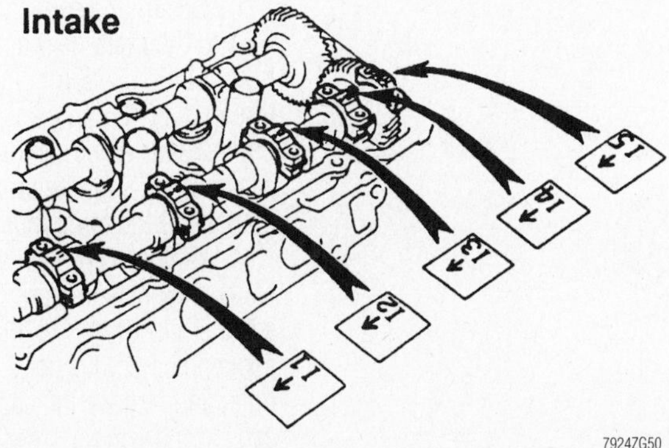

Right intake bearing caps must be placed in their proper locations—3.0L engine

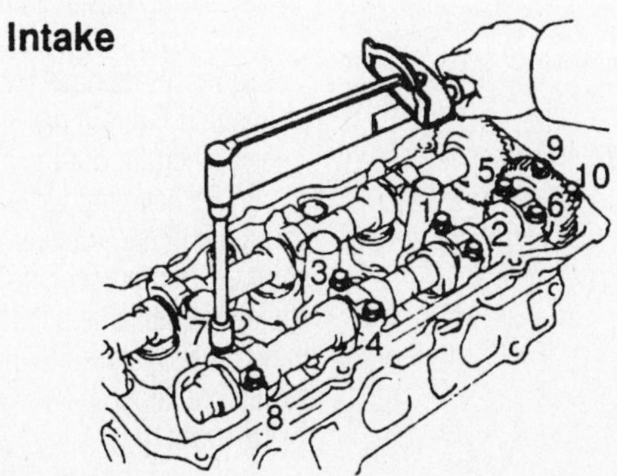

Right intake camshaft bearing cap bolt tightening sequence—3.0L engine

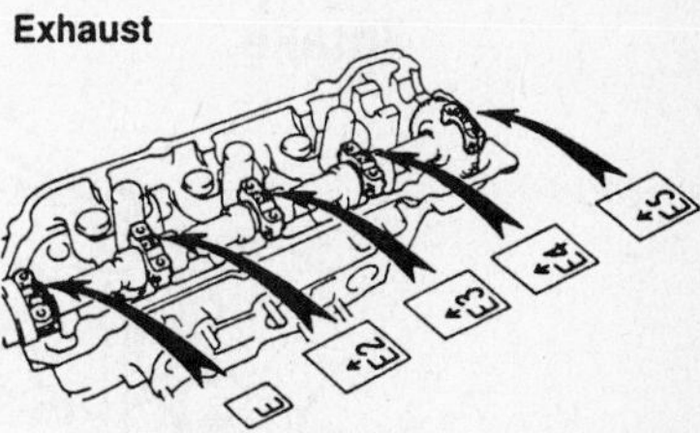

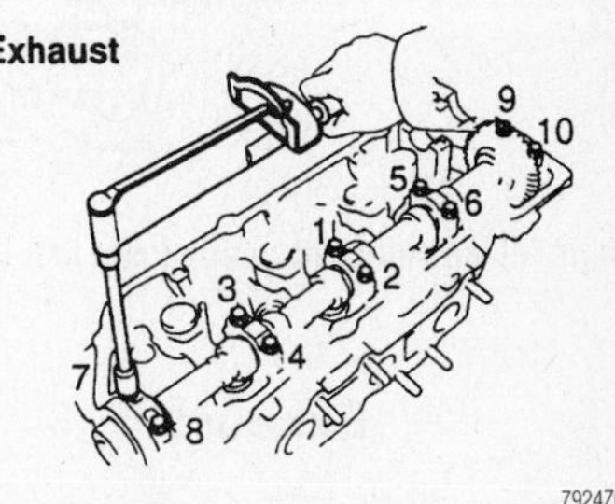

Left exhaust bearing caps locations and bolt tightening sequence—3.0L engine

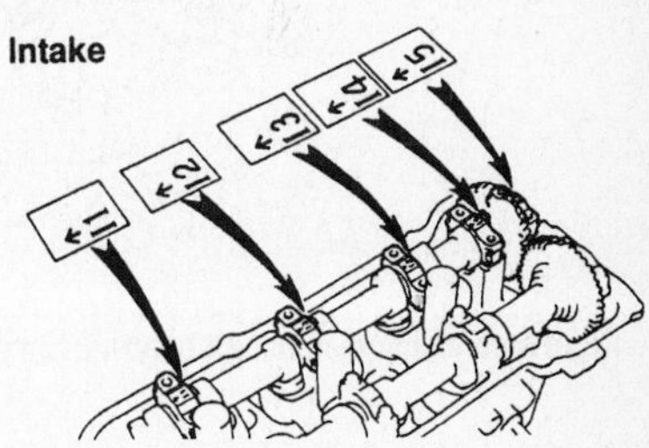

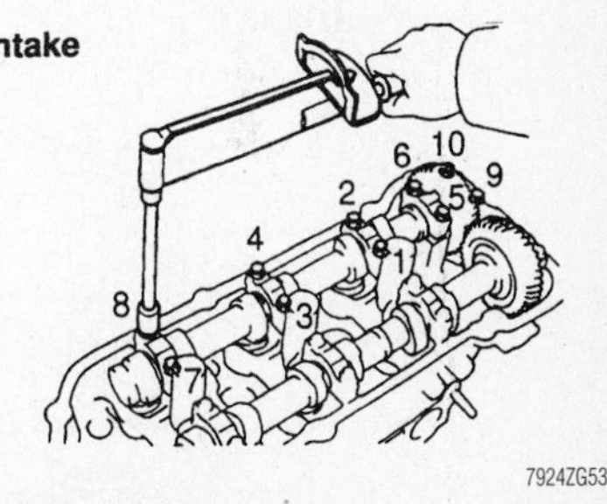

Left intake camshaft bearing cap locations and bolt tightening sequence—3.0L engine

➡The thrust clearance on both the intake and exhaust camshafts is very small; the camshafts must be kept level during removal. If the camshafts are removed without being kept level, the camshaft may be caught in the cylinder head, causing the head to break or the camshaft to seize.

3. Remove the exhaust and intake camshafts from the right side cylinder head, as follows:

a. Turn the camshaft with a wrench until the 2 pointed marks drive and driven gears are aligned. (The right camshaft

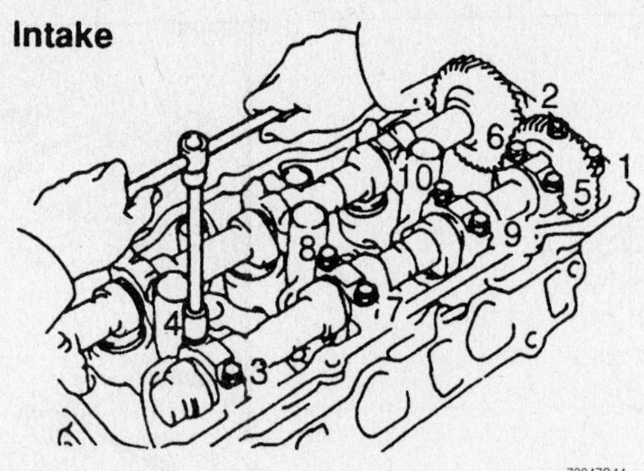

Right intake camshaft bearing cap bolt loosening sequence—3.3L engines

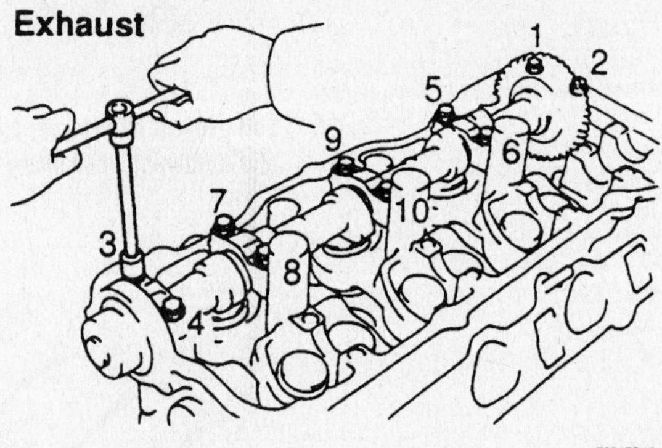

Right side exhaust camshaft bearing cap bolt loosening sequence—3.3L engines

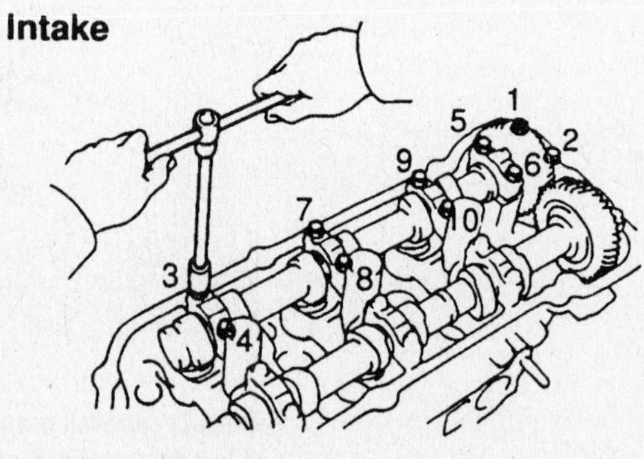

Left intake camshaft bearing cap bolt loosening sequence—3.3L engines

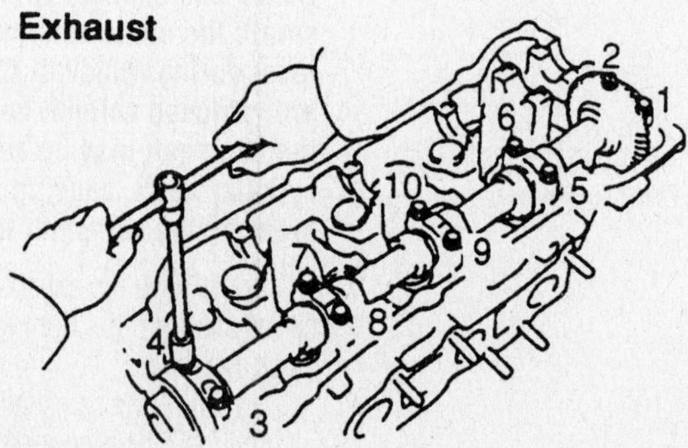

Left side exhaust camshaft bearing cap bolt loosening sequence—3.3L engines

gears have 2 marks apiece; the left side camshaft gears have 1 mark each.)

b. Secure the exhaust camshaft sub-gear to the main gear using a service bolt. A bolt 0.63–0.79 in. (16–20mm) long with a 6mm thread diameter and a 1mm pitch is recommended. When removing the exhaust camshaft be sure the sub-gear is not loaded; all the force must be eliminated.

c. Uniformly loosen and remove the exhaust camshaft bearing cap bolts in several passes and in the proper sequence. Remove the 8 bearing cap bolts and remove the caps, keeping them in the correct order.

d. Remove the exhaust camshaft from the engine.

e. Uniformly loosen and remove the 10 bearing cap bolts in several passes, in the proper sequence. Remove the bearing caps, keeping them in order, remove the oil seal, then lift out the intake camshaft.

4. Remove the exhaust and intake camshafts from the left side cylinder head, as follows:

a. Turn the camshaft with a wrench until the pointed marks on the drive and driven gears are aligned. (The right camshaft gears have 2 marks apiece; the left side camshaft gears have 1 mark each.)

b. Secure the exhaust camshaft sub-gear to the main gear using a service bolt. A bolt 16–20mm long with a 6mm thread diameter and a 1mm pitch is recommended. When removing the exhaust camshaft be sure the sub-gear is not loaded; all the force must be eliminated.

c. Uniformly loosen and remove the exhaust camshaft bearing cap bolts in several passes and in the proper sequence. Remove the 8 bearing cap bolts and remove the caps. Keep the caps in the correct order.

d. Remove the exhaust camshaft from the engine.

e. Uniformly loosen and remove the 10 bearing cap bolts in several passes, in the reverse order of the installation sequence. Remove the bearing caps, keeping them in order, remove the oil seal, then lift out the intake camshaft.

5. Remove the valve lifter shims and hydraulic lifters. Identify each lifter and shim as it is removed so it can be reinstalled in the same position. If the lifters are to be reused, store them upside down in a sealed container.

To install:

6. Install the valve lifters into their original positions and install the shims. Check

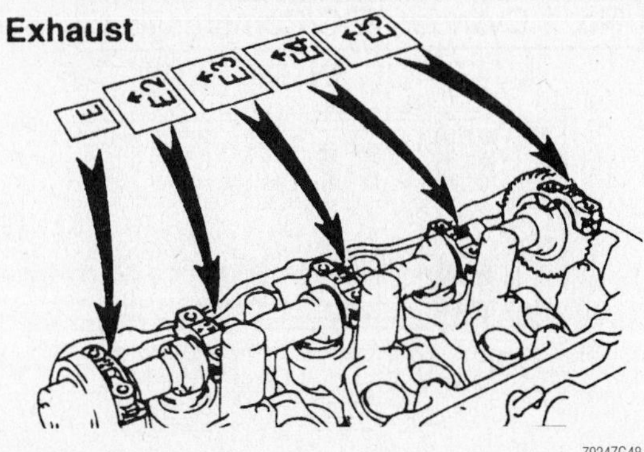

Right exhaust bearing caps must be placed in their proper locations—3.3L engines

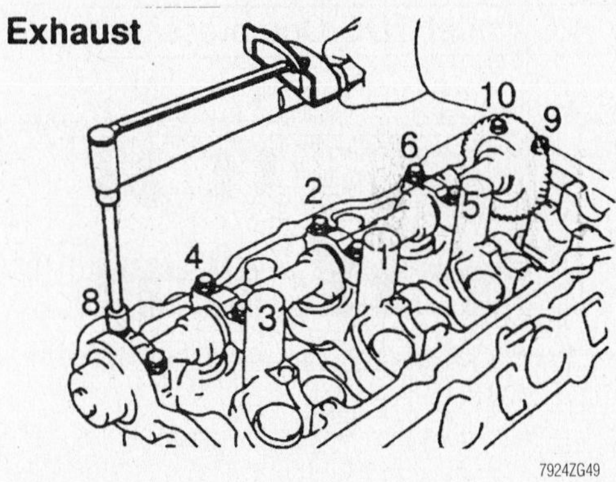

Right exhaust camshaft bearing cap bolt tightening sequence—3.3L engines

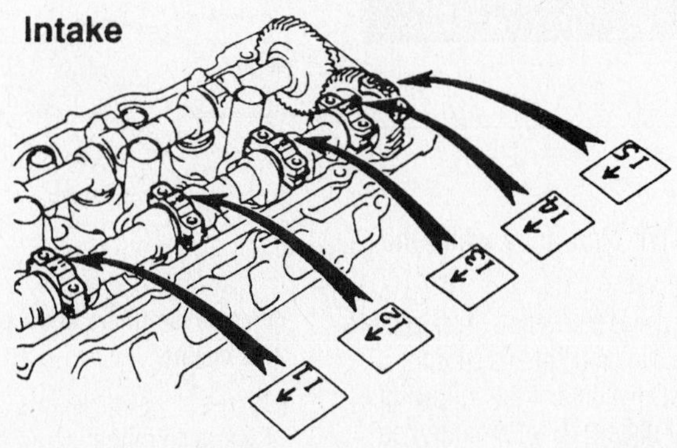

Right intake bearing caps must be placed in their proper locations—3.3L engines

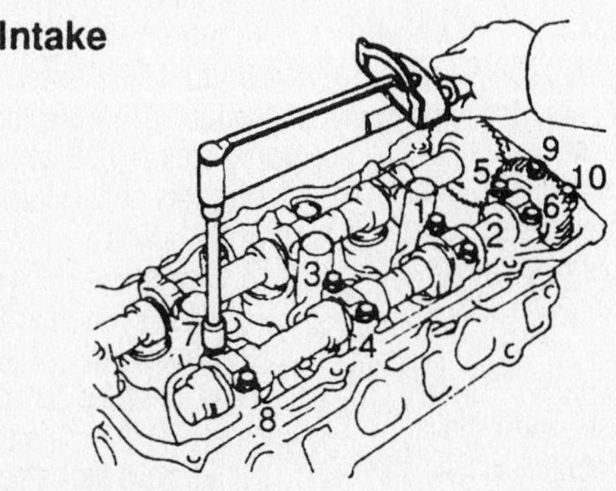

Right intake camshaft bearing cap bolt tightening sequence—3.3L engines

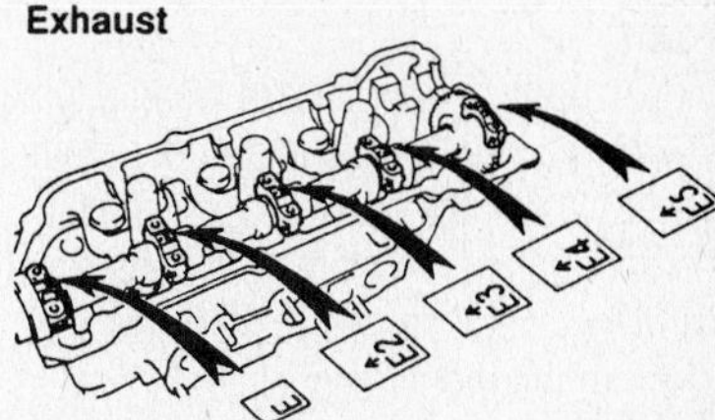

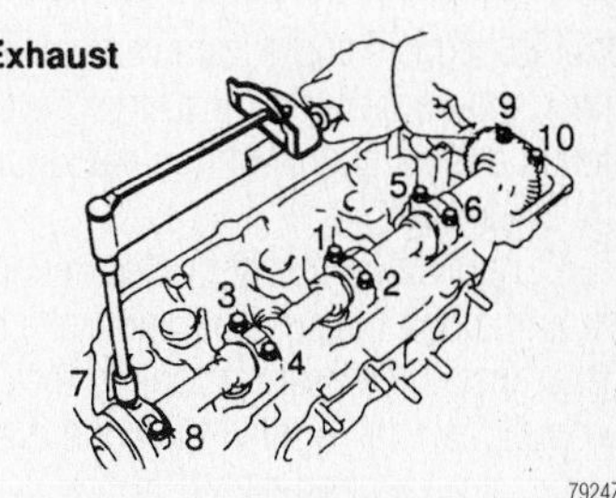

Left exhaust bearing caps locations and bolt tightening sequence—3.3L engines

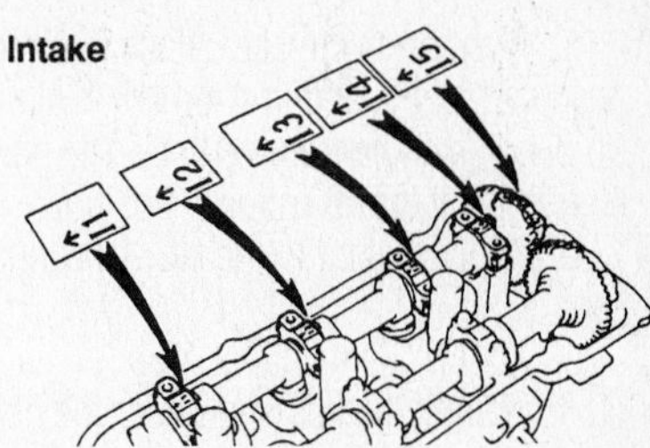

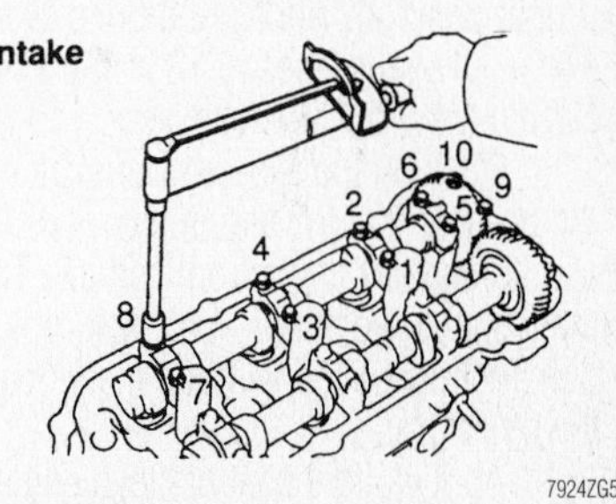

Left intake camshaft bearing cap locations and bolt tightening sequence—3.3L engines

valve clearance and replace the shims as necessary.

7. When reinstalling, remember that the camshafts must be handled carefully and kept straight and level to avoid damage.

8. Before installing the camshafts in either cylinder head, apply multi-purpose grease to each camshaft.

9. Install the right camshafts, as follows:

a. Position the intake camshaft on the head so that the alignment marks are at a 90 degrees angle from vertical. The mark should be at the "3 o'clock" position.

b. Apply sealant to the No. 1 bearing cap.

c. Apply a light coat of clean engine oil to the bolt threads and under the bolt head. Install the bearing caps to their proper position. Tighten the bolts evenly and in several passes to 12 ft. lbs. (16 Nm) in the proper sequence.

d. Position the exhaust camshaft on the head so that the alignment marks are at a 90 degrees angle from vertical. The mark should be at the "9 o'clock" position and must align with the marks on the other gear.

e. Apply a light coat of clean engine oil to the bolt threads and under the bolt head. Install the bearing caps to their proper position. Tighten the bolts evenly and in several passes to 12 ft. lbs. (16 Nm) in the proper sequence.

f. Remove the service bolt.

10. Install the left camshafts, as follows:

a. Position the intake camshaft on the head so that the alignment mark is at a 90 degrees angle from vertical. The mark should be at the "9 o'clock" position.

b. Apply sealant to the No. 1 bearing cap.

c. Apply a light coat of clean engine oil to the bolt threads and under the bolt head. Install the bearing caps to their proper position. Tighten the bolts evenly and in several passes to 12 ft. lbs. (16 Nm) in the proper sequence.

d. Position the exhaust camshaft on the head so that the alignment marks are at a 90 degrees angle from vertical. The mark should be at the "3 o'clock" position and must align with the marks on the other gear.

e. Apply a light coat of clean engine oil to the bolt threads and under the bolt head. Install the bearing caps to their proper position. Tighten the bolts evenly and in several passes to 12 ft. lbs. (16 Nm) in the proper sequence.

f. Remove the service bolt.

11. Install or connect the following:

- New camshaft oil seals, lubricated with multi-purpose grease
- No. 3 (rear) timing belt cover
- Camshaft timing gears
- Idler pulley, timing belt and covers

12. Check and adjust the valve clearance.

13. Install the cylinder head (valve) covers.

14. Start the engine. Check the ignition timing.

15. Test drive the vehicle.

16. Check all fluid levels.

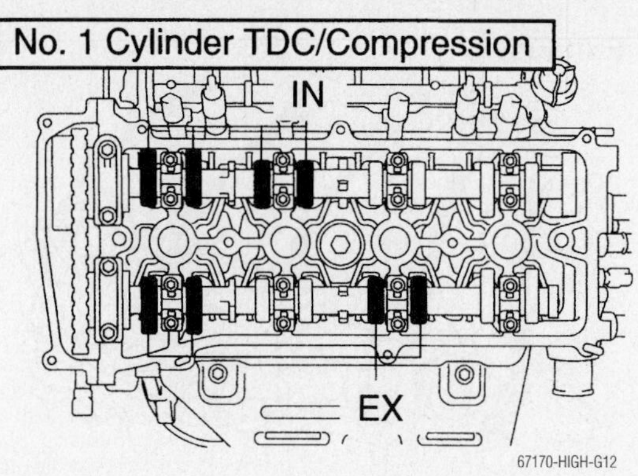

Check the clearance on these valves with the engine at No.1 TDC compression—2.4L engine

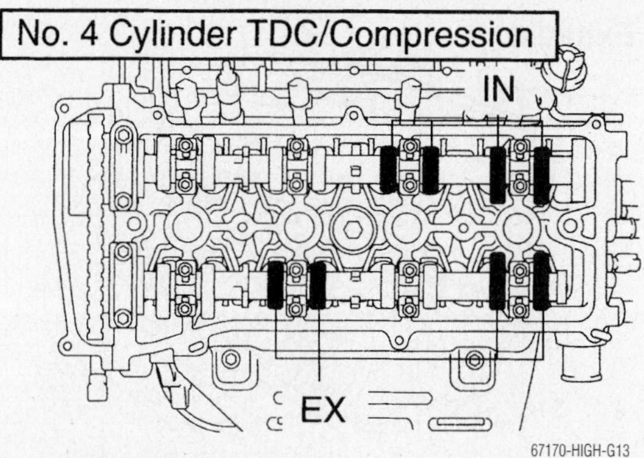

Check the clearance on these valves with the engine at No.4 TDC compression—2.4L engine

Valve Lash

ADJUSTMENT

2.4L Engine

➡Adjust the valve clearance when the engine is cold.

1. Before servicing the vehicle, refer to the precautions in the beginning of this section.

2. Remove or disconnect the following:

- Negative battery cable. If equipped with an air bag, wait at least 90 seconds before proceeding.
- Right front wheel, splash shield and apron seal
- engine undercover
- Coil pack
- Air intake hoses
- Cylinder head cover

3. Place the No.1 piston at TDC compression. Check only those valves shown. Record the clearance. If out of clearance, the measurement will be used to calculate the adjusting shims.

4. Place the No.4 piston at TDC compression. Check only those valves shown. Record the clearance. If out of clearance, the measurement will be used to calculate the adjusting shims.

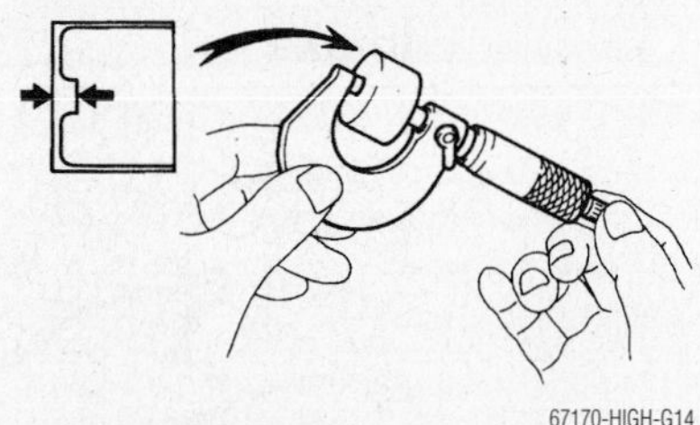

Measure the lifters at the point shown—2.4L engine

Clearance range is:

- Intake: 0.19–0.29mm (0.008–0.011 in.)
- Exhaust: 0.30–0.40mm (0.012–0.016 in.)

To adjust the valves:

5. Turn the crankshaft 1 complete revolution (360 degrees) clockwise and set the No.1 piston at TDC compression. Place matchmarks on the chain and camshaft sprocket.

6. Remove the tensioner.

7. Loosen the camshaft sprocket bolt.

8. Remove the exhaust camshaft bearing caps, raise the camshaft and remove the sprocket. Tie the chain out of the way.

9. Remove the camshaft.

10. Remove the lifters and keep them in order.

Valve Lifter Selection Chart (Intake)

Measure clearance mm (in.) \ Installed lifter thickness mm (in.)	5.060 (0.1992)	5.080 (0.2000)	5.100 (0.2008)	5.120 (0.2016)	5.140 (0.2024)	5.160 (0.2031)	5.180 (0.2039)	5.200 (0.2047)	5.210 (0.2051)	5.220 (0.2055)	5.230 (0.2059)	5.240 (0.2063)	5.250 (0.2067)	5.260 (0.2071)	5.270 (0.2075)	5.280 (0.2079)	5.290 (0.2083)	5.300 (0.2087)	5.310 (0.2091)	5.320 (0.2094)	5.330 (0.2098)	5.340 (0.2102)	5.350 (0.2106)	5.360 (0.2110)	5.370 (0.2114)	5.380 (0.2118)	5.390 (0.2122)	5.400 (0.2126)	5.410 (0.2130)	5.420 (0.2134)	5.430 (0.2138)	5.440 (0.2142)	5.450 (0.2146)	5.460 (0.2150)	5.470 (0.2154)	5.480 (0.2157)	5.490 (0.2161)	5.500 (0.2165)	5.510 (0.2169)	5.520 (0.2173)	5.530 (0.2177)	5.540 (0.2181)	5.550 (0.2185)	5.560 (0.2189)	5.570 (0.2193)	5.580 (0.2197)	5.590 (0.2201)	5.600 (0.2205)	5.620 (0.2213)	5.640 (0.2220)	5.660 (0.2228)	5.680 (0.2236)	5.700 (0.2244)	5.720 (0.2252)	5.740 (0.2260)
0.000-0.030 (0.0000-0.0012)													06	06	08	08	10	10	12	12	14	14	16	16	18	18	20	20	22	22	24	24	26	26	28	28	30	30	32	32	34	34	36	36	38	38	40	40	42	44	46	48	50	52	54
0.031-0.050 (0.0012-0.0020)									06	06	06	08	08	10	10	12	12	14	14	16	16	18	18	20	20	22	22	24	24	26	26	28	28	30	30	32	32	34	34	36	36	38	38	40	40	42	42	44	46	48	50	52	54	56	58
0.051-0.070 (0.0020-0.0028)								06	08	08	08	10	10	12	12	14	14	16	16	18	18	20	20	22	22	24	24	26	26	28	28	30	30	32	32	34	34	36	36	38	38	40	40	42	42	44	44	46	48	50	52	54	56	58	60
0.071-0.090 (0.0028-0.0035)							06	08	10	10	10	12	12	14	14	16	16	18	18	20	20	22	22	24	24	26	26	28	28	30	30	32	32	34	34	36	36	38	38	40	40	42	42	44	44	46	46	48	50	52	54	56	58	60	62
0.091-0.110 (0.0036-0.0043)						06	08	10	12	12	12	14	14	16	16	18	18	20	20	22	22	24	24	26	26	28	28	30	30	32	32	34	34	36	36	38	38	40	40	42	42	44	44	46	46	48	48	50	52	54	56	58	60	62	64
0.111-0.130 (0.0044-0.0051)					06	08	10	12	14	14	14	16	16	18	18	20	20	22	22	24	24	26	26	28	28	30	30	32	32	34	34	36	36	38	38	40	40	42	42	44	44	46	46	48	48	50	50	52	54	56	58	60	62	64	66
0.131-0.150 (0.0052-0.0059)				06	08	10	12	14	16	16	16	18	18	20	20	22	22	24	24	26	26	28	28	30	30	32	32	34	34	36	36	38	38	40	40	42	42	44	44	46	46	48	48	50	50	52	52	54	56	58	60	62	64	66	68
0.151-0.170 (0.0059-0.0067)			06	08	10	12	14	16	18	18	18	20	20	22	22	24	24	26	26	28	28	30	30	32	32	34	34	36	36	38	38	40	40	42	42	44	44	46	46	48	48	50	50	52	52	54	54	56	58	60	62	64	66	68	70
0.171-0.189 (0.0067-0.0074)		06	08	10	12	14	16	18	20	20	20	22	22	24	24	26	26	28	28	30	30	32	32	34	34	36	36	38	38	40	40	42	42	44	44	46	46	48	48	50	50	52	52	54	54	56	56	58	60	62	64	66	68	70	72
0.190-0.290 (0.0075-0.0114)																																																							
0.291-0.310 (0.0115-0.0122)	12	14	16	18	20	22	24	26	28	28	30	30	32	32	34	34	36	36	38	38	40	40	42	42	44	44	46	46	48	48	50	50	52	52	54	54	56	56	58	58	60	60	62	62	64	64	66	66	68	70	72	74	74	74	
0.311-0.330 (0.0122-0.0130)	14	16	18	20	22	24	26	28	30	30	32	32	34	34	36	36	38	38	40	40	42	42	44	44	46	46	48	48	50	50	52	52	54	54	56	56	58	58	60	60	62	62	64	64	66	66	68	68	70	72	74	74	74		
0.331-0.350 (0.0130-0.0138)	16	18	20	22	24	26	28	30	32	32	34	34	36	36	38	38	40	40	42	42	44	44	46	46	48	48	50	50	52	52	54	54	56	56	58	58	60	60	62	62	64	64	66	66	68	68	70	70	72	74	74	74			
0.351-0.370 (0.0138-0.0146)	18	20	22	24	26	28	30	32	34	34	36	36	38	38	40	40	42	42	44	44	46	46	48	48	50	50	52	52	54	54	56	56	58	58	60	60	62	62	64	64	66	66	68	68	70	70	72	72	74	74	74				
0.371-0.390 (0.0146-0.0154)	20	22	24	26	28	30	32	34	36	36	38	38	40	40	42	42	44	44	46	46	48	48	50	50	52	52	54	54	56	56	58	58	60	60	62	62	64	64	66	66	68	68	70	70	72	72	74	74	74	74					
0.391-0.410 (0.0154-0.0161)	22	24	26	28	30	32	34	36	38	38	40	40	42	42	44	44	46	46	48	48	50	50	52	52	54	54	56	56	58	58	60	60	62	62	64	64	66	66	68	68	70	70	72	72	74	74	74	74							
0.411-0.430 (0.0162-0.0169)	24	26	28	30	32	34	36	38	40	40	42	42	44	44	46	46	48	48	50	50	52	52	54	54	56	56	58	58	60	60	62	62	64	64	66	66	68	68	70	70	72	72	74	74	74	74									
0.431-0.450 (0.0170-0.0177)	26	28	30	32	34	36	38	40	42	42	44	44	46	46	48	48	50	50	52	52	54	54	56	56	58	58	60	60	62	62	64	64	66	66	68	68	70	70	72	72	74	74	74	74											
0.451-0.470 (0.0178-0.0185)	28	30	32	34	36	38	40	42	44	44	46	46	48	48	50	50	52	52	54	54	56	56	58	58	60	60	62	62	64	64	66	66	68	68	70	70	72	72	74	74	74	74													
0.471-0.490 (0.0185-0.0193)	30	32	34	36	38	40	42	44	46	46	48	48	50	50	52	52	54	54	56	56	58	58	60	60	62	62	64	64	66	66	68	68	70	70	72	72	74	74	74	74															
0.491-0.510 (0.0193-0.0201)	32	34	36	38	40	42	44	46	48	48	50	50	52	52	54	54	56	56	58	58	60	60	62	62	64	64	66	66	68	68	70	70	72	72	74	74	74	74																	
0.511-0.530 (0.0201-0.0209)	34	36	38	40	42	44	46	48	50	50	52	52	54	54	56	56	58	58	60	60	62	62	64	64	66	66	68	68	70	70	72	72	74	74	74	74																			
0.531-0.550 (0.0209-0.0217)	36	38	40	42	44	46	48	50	52	52	54	54	56	56	58	58	60	60	62	62	64	64	66	66	68	68	70	70	72	72	74	74	74	74																					
0.551-0.570 (0.0217-0.0224)	38	40	42	44	46	48	50	52	54	54	56	56	58	58	60	60	62	62	64	64	66	66	68	68	70	70	72	72	74	74	74	74																							
0.571-0.590 (0.0225-0.0232)	40	42	44	46	48	50	52	54	56	56	58	58	60	60	62	62	64	64	66	66	68	68	70	70	72	72	74	74	74	74																									
0.591-0.610 (0.0233-0.0240)	42	44	46	48	50	52	54	56	58	58	60	60	62	62	64	64	66	66	68	68	70	70	72	72	74	74	74	74																											
0.611-0.630 (0.0241-0.0248)	44	46	48	50	52	54	56	58	60	60	62	62	64	64	66	66	68	68	70	70	72	72	74	74	74	74																													
0.631-0.650 (0.0248-0.0256)	46	48	50	52	54	56	58	60	62	62	64	64	66	66	68	68	70	70	72	72	74	74	74	74																															
0.651-0.670 (0.0256-0.0264)	48	50	52	54	56	58	60	62	64	64	66	66	68	68	70	70	72	72	74	74	74	74																																	
0.671-0.690 (0.0264-0.0272)	50	52	54	56	58	60	62	64	66	66	68	68	70	70	72	72	74	74	74	74																																			
0.691-0.710 (0.0272-0.0280)	52	54	56	58	60	62	64	66	68	68	70	70	72	72	74	74	74	74																																					
0.711-0.730 (0.0280-0.0287)	54	56	58	60	62	64	66	68	70	70	72	72	74	74	74	74																																							
0.731-0.750 (0.0288-0.0295)	56	58	60	62	64	66	68	70	72	72	74	74	74	74																																									
0.751-0.770 (0.0296-0.0303)	58	60	62	64	66	68	70	72	74	74	74	74																																											
0.771-0.790 (0.0304-0.0311)	60	62	64	66	68	70	72	74	74	74																																													
0.791-0.810 (0.0311-0.0319)	62	64	66	68	70	72	74																																																
0.811-0.830 (0.0319-0.0327)	64	66	68	70	72	74																																																	
0.831-0.850 (0.0327-0.0335)	66	68	70	72	74																																																		
0.851-0.870 (0.0335-0.0343)	68	70	72	74																																																			
0.871-0.890 (0.0343-0.0350)	70	72	74																																																				
0.891-0.910 (0.0351-0.0358)	72	74																																																					
0.911-0.930 (0.0359-0.0366)	74																																																						

Intake valve clearance (Cold):
0.19 to 0.29 mm (0.008 to 0.011 in.)

EXAMPLE:

The 5.250 mm (0.2067 in.) lifter is installed, and the measured clearance is 0.400 mm (0.0157 in.). Replace the 5.250 mm (0.2067 in.) lifter with a new No. 42 lifter.

New Lifter Thickness mm (in.)

Lifter No.	Thickness	Lifter No.	Thickness	Lifter No.	Thickness
06	5.060 (0.1992)	30	5.300 (0.2087)	54	5.540 (0.2181)
08	5.080 (0.2000)	32	5.320 (0.2094)	56	5.560 (0.2189)
10	5.100 (0.2008)	34	5.340 (0.2102)	58	5.580 (0.2197)
12	5.120 (0.2016)	36	5.360 (0.2110)	60	5.600 (0.2205)
14	5.140 (0.2024)	38	5.380 (0.2118)	62	5.620 (0.2213)
16	5.160 (0.2031)	40	5.400 (0.2126)	64	5.640 (0.2220)
18	5.180 (0.2039)	42	5.420 (0.2134)	66	5.660 (0.2228)
20	5.200 (0.2047)	44	5.440 (0.2142)	68	5.680 (0.2236)
22	5.220 (0.2055)	46	5.460 (0.2150)	70	5.700 (0.2244)
24	5.240 (0.2063)	48	5.480 (0.2157)	72	5.720 (0.2252)
26	5.260 (0.2071)	50	5.500 (0.2165)	74	5.740 (0.2260)
28	5.280 (0.2079)	52	5.520 (0.2173)		

67170-HIGH-G15

Intake valve lifter selection chart—2.4L engine

Valve Lifter Selection Chart (Exhaust)

Measure clearance mm (in.) / Installed lifter thickness mm (in.)	5.060 (0.1992)	5.080 (0.2000)	5.100 (0.2008)	5.120 (0.2016)	5.140 (0.2024)	5.160 (0.2031)	5.180 (0.2039)	5.200 (0.2047)	5.210 (0.2051)	5.220 (0.2055)	5.230 (0.2059)	5.240 (0.2063)	5.250 (0.2067)	5.260 (0.2071)	5.270 (0.2075)	5.280 (0.2079)	5.290 (0.2083)	5.300 (0.2087)	5.310 (0.2091)	5.320 (0.2094)	5.330 (0.2098)	5.340 (0.2102)	5.350 (0.2106)	5.360 (0.2110)	5.370 (0.2114)	5.380 (0.2118)	5.390 (0.2122)	5.400 (0.2126)	5.410 (0.2130)	5.420 (0.2134)	5.430 (0.2138)	5.440 (0.2142)	5.450 (0.2146)	5.460 (0.2150)	5.470 (0.2154)	5.480 (0.2157)	5.490 (0.2161)	5.500 (0.2165)	5.510 (0.2169)	5.520 (0.2173)	5.530 (0.2177)	5.540 (0.2181)	5.550 (0.2185)	5.560 (0.2189)	5.570 (0.2193)	5.580 (0.2197)	5.590 (0.2201)	5.600 (0.2205)	5.620 (0.2213)	5.640 (0.2220)	5.660 (0.2228)	5.680 (0.2236)	5.700 (0.2244)	5.720 (0.2252)	5.740 (0.2260)
0.000-0.030 (0.0000-0.0012)																							06	06	08	08	10	10	12	12	14	14	16	16	18	18	20	20	22	22	24	24	26	26	28	28	30	30	32	34	36	38	40	42	44
0.031-0.050 (0.0012-0.0020)																					06	06	08	08	10	10	12	12	14	14	16	16	18	18	20	20	22	22	24	24	26	26	28	28	30	30	32	32	34	36	38	40	42	44	46
0.051-0.070 (0.0020-0.0028)																			06	06	08	08	10	10	12	12	14	14	16	16	18	18	20	20	22	22	24	24	26	26	28	28	30	30	32	32	34	34	36	38	40	42	44	46	48
0.071-0.090 (0.0028-0.0035)																	06	06	08	08	10	10	12	12	14	14	16	16	18	18	20	20	22	22	24	24	26	26	28	28	30	30	32	32	34	34	36	36	38	40	42	44	46	48	50
0.091-0.110 (0.0036-0.0043)															06	06	08	08	10	10	12	12	14	14	16	16	18	18	20	20	22	22	24	24	26	26	28	28	30	30	32	32	34	34	36	36	38	38	40	42	44	46	48	50	52
0.111-0.130 (0.0044-0.0051)													06	06	08	08	10	10	12	12	14	14	16	16	18	18	20	20	22	22	24	24	26	26	28	28	30	30	32	32	34	34	36	36	38	38	40	40	42	44	46	48	50	52	54
0.131-0.150 (0.0052-0.0059)											06	06	08	08	10	10	12	12	14	14	16	16	18	18	20	20	22	22	24	24	26	26	28	28	30	30	32	32	34	34	36	36	38	38	40	40	42	42	44	46	48	50	52	54	56
0.151-0.170 (0.0059-0.0067)									06	06	08	08	10	10	12	12	14	14	16	16	18	18	20	20	22	22	24	24	26	26	28	28	30	30	32	32	34	34	36	36	38	38	40	40	42	42	44	44	46	48	50	52	54	56	58
0.171-0.190 (0.0067-0.0075)								06	08	08	10	10	12	12	14	14	16	16	18	18	20	20	22	22	24	24	26	26	28	28	30	30	32	32	34	34	36	36	38	38	40	40	42	42	44	44	46	46	48	50	52	54	56	58	60
0.191-0.210 (0.0075-0.0083)							06	08	10	10	12	12	14	14	16	16	18	18	20	20	22	22	24	24	26	26	28	28	30	30	32	32	34	34	36	36	38	38	40	40	42	42	44	44	46	46	48	48	50	52	54	56	58	60	62
0.211-0.230 (0.0083-0.0091)						06	08	10	12	12	14	14	16	16	18	18	20	20	22	22	24	24	26	26	28	28	30	30	32	32	34	34	36	36	38	38	40	40	42	42	44	44	46	46	48	48	50	50	52	54	56	58	60	62	64
0.231-0.250 (0.0091-0.0098)					06	08	10	12	14	14	16	16	18	18	20	20	22	22	24	24	26	26	28	28	30	30	32	32	34	34	36	36	38	38	40	40	42	42	44	44	46	46	48	48	50	50	52	52	54	56	58	60	62	64	66
0.251-0.270 (0.0099-0.0106)				06	08	10	12	14	16	16	18	18	20	20	22	22	24	24	26	26	28	28	30	30	32	32	34	34	36	36	38	38	40	40	42	42	44	44	46	46	48	48	50	50	52	52	54	54	56	58	60	62	64	66	68
0.271-0.290 (0.0107-0.0114)			06	08	10	12	14	16	18	18	20	20	22	22	24	24	26	26	28	28	30	30	32	32	34	34	36	36	38	38	40	40	42	42	44	44	46	46	48	48	50	50	52	52	54	54	56	56	58	60	62	64	66	68	70
0.291-0.299 (0.0115-0.0118)		06	08	10	12	14	16	18	20	20	22	22	24	24	26	26	28	28	30	30	32	32	34	34	36	36	38	38	40	40	42	42	44	44	46	46	48	48	50	50	52	52	54	54	56	56	58	58	60	62	64	66	68	70	72
0.300-0.400 (0.0118-0.0157)																																																							
0.401-0.420 (0.0158-0.0165)	12	14	16	18	20	22	24	26	26	28	28	30	30	32	32	34	34	36	36	38	38	40	40	42	42	44	44	46	46	48	48	50	50	52	52	54	54	56	56	58	58	60	60	62	62	64	64	66	68	70	72	74	74	74	
0.421-0.440 (0.0166-0.0173)	14	16	18	20	22	24	26	28	28	30	30	32	32	34	34	36	36	38	38	40	40	42	42	44	44	46	46	48	48	50	50	52	52	54	54	56	56	58	58	60	60	62	62	64	64	66	66	68	70	72	74	74	74		
0.441-0.460 (0.0174-0.0181)	16	18	20	22	24	26	28	30	30	32	32	34	34	36	36	38	38	40	40	42	42	44	44	46	46	48	48	50	50	52	52	54	54	56	56	58	58	60	60	62	62	64	64	66	66	68	68	70	72	74	74	74			
0.461-0.480 (0.0181-0.0189)	18	20	22	24	26	28	30	32	32	34	34	36	36	38	38	40	40	42	42	44	44	46	46	48	48	50	50	52	52	54	54	56	56	58	58	60	60	62	62	64	64	66	66	68	68	70	70	72	74	74	74				
0.481-0.500 (0.0189-0.0197)	20	22	24	26	28	30	32	34	34	36	36	38	38	40	40	42	42	44	44	46	46	48	48	50	50	52	52	54	54	56	56	58	58	60	60	62	62	64	64	66	66	68	68	70	70	72	72	74	74	74					
0.501-0.520 (0.0197-0.0205)	22	24	26	28	30	32	34	36	36	38	38	40	40	42	42	44	44	46	46	48	48	50	50	52	52	54	54	56	56	58	58	60	60	62	62	64	64	66	66	68	68	70	70	72	72	74	74	74	74						
0.521-0.540 (0.0205-0.0213)	24	26	28	30	32	34	36	38	38	40	40	42	42	44	44	46	46	48	48	50	50	52	52	54	54	56	56	58	58	60	60	62	62	64	64	66	66	68	68	70	70	72	72	74	74	76	76	74							
0.541-0.560 (0.0213-0.0220)	26	28	30	32	34	36	38	40	40	42	42	44	44	46	46	48	48	50	50	52	52	54	54	56	56	58	58	60	60	62	62	64	64	66	66	68	68	70	70	72	72	74	74	74	74	74	74								
0.561-0.580 (0.0221-0.0228)	28	30	32	34	36	38	40	42	42	44	44	46	46	48	48	50	50	52	52	54	54	56	56	58	58	60	60	62	62	64	64	66	66	68	68	70	70	72	72	74	74	74	74	74	74										
0.581-0.600 (0.0229-0.0236)	30	32	34	36	38	40	42	44	44	46	46	48	48	50	50	52	52	54	54	56	56	58	58	60	60	62	62	64	64	66	66	68	68	70	70	72	72	74	74	74	74	74	74												
0.601-0.620 (0.0237-0.0244)	32	34	36	38	40	42	44	46	46	48	48	50	50	52	52	54	54	56	56	58	58	60	60	62	62	64	64	66	66	68	68	70	70	72	72	74	74	74	74	74	74														
0.621-0.640 (0.0244-0.0252)	34	36	38	40	42	44	46	48	48	50	50	52	52	54	54	56	56	58	58	60	60	62	62	64	64	66	66	68	68	70	70	72	72	74	74	74	74	74	74																
0.641-0.660 (0.0252-0.0260)	36	38	40	42	44	46	48	50	50	52	52	54	54	56	56	58	58	60	60	62	62	64	64	66	66	68	68	70	70	72	72	74	74	74	74	74	74																		
0.661-0.680 (0.0260-0.0268)	38	40	42	44	46	48	50	52	52	54	54	56	56	58	58	60	60	62	62	64	64	66	66	68	68	70	70	72	72	74	74	74	74	74	74																				
0.681-0.700 (0.0268-0.0276)	40	42	44	46	48	50	52	54	54	56	56	58	58	60	60	62	62	64	64	66	66	68	68	70	70	72	72	74	74	74	74	74	74																						
0.701-0.720 (0.0276-0.0283)	42	44	46	48	50	52	54	56	56	58	58	60	60	62	62	64	64	66	66	68	68	70	70	72	72	74	74	74	74	74	74																								
0.721-0.740 (0.0284-0.0291)	44	46	48	50	52	54	56	58	58	60	60	62	62	64	64	66	66	68	68	70	70	72	72	74	74	74	74	74	74																										
0.741-0.760 (0.0292-0.0299)	46	48	50	52	54	56	58	60	60	62	62	64	64	66	66	68	68	70	70	72	72	74	74	74	74	74	74																												
0.761-0.780 (0.0300-0.0307)	48	50	52	54	56	58	60	62	62	64	64	66	66	68	68	70	70	72	72	74	74	74	74	74	74																														
0.781-0.800 (0.0307-0.0315)	50	52	54	56	58	60	62	64	64	66	66	68	68	70	70	72	72	74	74	74	74	74	74																																
0.801-0.820 (0.0315-0.0323)	52	54	56	58	60	62	64	66	66	68	68	70	70	72	72	74	74	74	74	74	74																																		
0.821-0.840 (0.0323-0.0331)	54	56	58	60	62	64	66	68	68	70	70	72	72	74	74	74	74	74	74																																				
0.841-0.860 (0.0331-0.0339)	56	58	60	62	64	66	68	70	70	72	72	74	74	74	74	74	74																																						
0.861-0.880 (0.0339-0.0346)	58	60	62	64	66	68	70	72	72	74	74	74	74	74	74																																								
0.881-0.900 (0.0347-0.0354)	60	62	64	66	68	70	72	74	74	74	74	74	74																																										
0.901-0.920 (0.0355-0.0362)	62	64	66	68	70	72	74	74	74	74	74																																												
0.921-0.940 (0.0363-0.0370)	64	66	68	70	72	74	74	74	74																																														
0.941-0.960 (0.0370-0.0378)	66	68	70	72	74	74	74																																																
0.961-0.980 (0.0378-0.0386)	68	70	72	74	74	74																																																	
0.981-1.000 (0.0386-0.0394)	70	72	74	74	74																																																		
1.001-1.020 (0.0394-0.0402)	72	74	74	74																																																			
1.021-1.040 (0.0402-0.0409)	74	74	74																																																				
1.041-1.060 (0.0410-0.0417)	74	74																																																					
1.061-1.080 (0.0418-0.0425)	74																																																						

New lifter thickness mm (in.)

Lifter No.	Thickness	Lifter No.	Thickness	Lifter No.	Thickness
06	5.060 (0.1992)	30	5.300 (0.2087)	54	5.540 (0.2181)
08	5.080 (0.2000)	32	5.320 (0.2094)	56	5.560 (0.2189)
10	5.100 (0.2008)	34	5.340 (0.2102)	58	5.580 (0.2197)
12	5.120 (0.2016)	36	5.360 (0.2110)	60	5.600 (0.2205)
14	5.140 (0.2024)	38	5.380 (0.2118)	62	5.620 (0.2213)
16	5.160 (0.2031)	40	5.400 (0.2126)	64	5.640 (0.2220)
18	5.180 (0.2039)	42	5.420 (0.2134)	66	5.660 (0.2228)
20	5.200 (0.2047)	44	5.440 (0.2142)	68	5.680 (0.2236)
22	5.220 (0.2055)	46	5.460 (0.2150)	70	5.700 (0.2244)
24	5.240 (0.2063)	48	5.480 (0.2157)	72	5.720 (0.2252)
26	5.260 (0.2071)	50	5.500 (0.2165)	74	5.740 (0.2260)
28	5.280 (0.2079)	52	5.520 (0.2173)		

Exhaust valve clearance (Cold):
0.30 to 0.40 mm (0.012 to 0.016 in.)

EXAMPLE:

The 5.340 mm (0.2102 in.) lifter is installed, and the measured clearance is 0.440 mm (0.0173 in.). Replace the 5.340 mm (0.2102 in.) lifter with a new No. 44 lifter.

67170-HIGH-G16

Exhaust valve lifter selection chart—2.4L engine

11. Measure the thickness of any lifter on which the clearance was out of range. Calculate the thickness of the necessary replacement lifter. Lifters are available in 0.020mm increments from 5.060mm to 5.740mm.

12. For Camshaft and Timing Chain installation, see the respective procedures in this section.

3.0L Engine

➡Adjust the valve clearance when the engine is cold.

1. Before servicing the vehicle, refer to the precautions in the beginning of this section.
2. Remove or disconnect the following:
   - Negative battery cable. If equipped with an air bag, wait at least 90 seconds before proceeding.
   - Accelerator/throttle cable from the throttle linkage
   - Air cleaner cover, air flow meter and air duct assembly
   - V-bank cover
   - Emission control valve set
   - Air intake chamber
   - Engine harness from the injectors and the ignition coils
   - Ignition coils and keep them in order for reassembly
   - Spark plugs
   - Cylinder head covers
3. Turn the crankshaft pulley and align its groove with the timing mark **0** of the No. 1 timing cover.
4. Check that the valve lifters on the No. 1 intake are loose and the No. 1 exhaust are tight. If not, turn the crankshaft 1 complete revolution (360 degrees).

➡All measurements should be written down. These recorded measurements will need to be used in conjunction with a mathematical formula to determine the thickness of the replacement shims.

5. Measure the clearance between the valve lifters and the camshaft. Record the measurements on valves No. 1 and 6 intake; No. 2 and 3 exhaust.
   a. The intake valve clearance cold is 0.006–0.010 in. (0.15–0.25mm).
   b. The exhaust valve clearance cold is 0.010–0.014 in. (0.25–0.35mm).
6. Turn the crankshaft ⅔ of a revolution (240 degrees). Record the measurements on valves No. 2 and 3 intake; No. 4 and 5 exhaust.
7. Turn the crankshaft another ⅔ of a

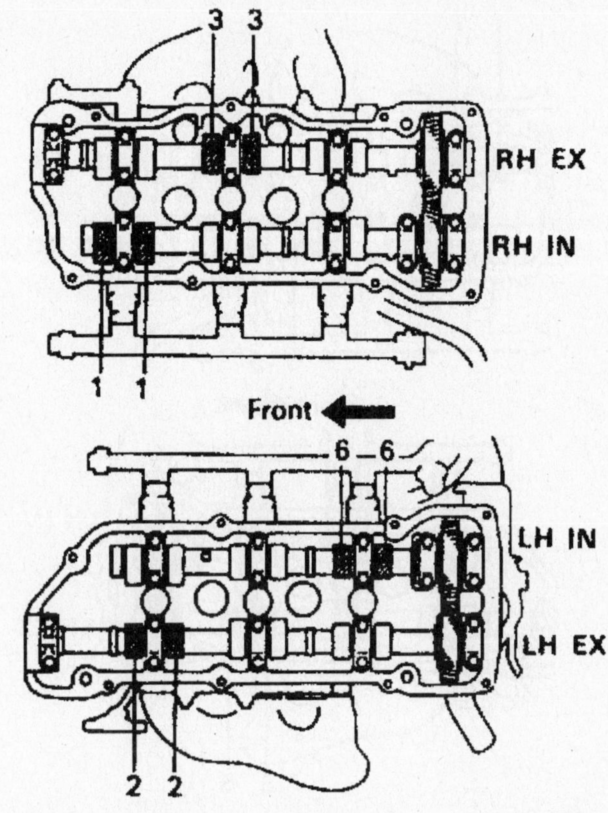

Adjust these valves during the 1st step—3.0L engine

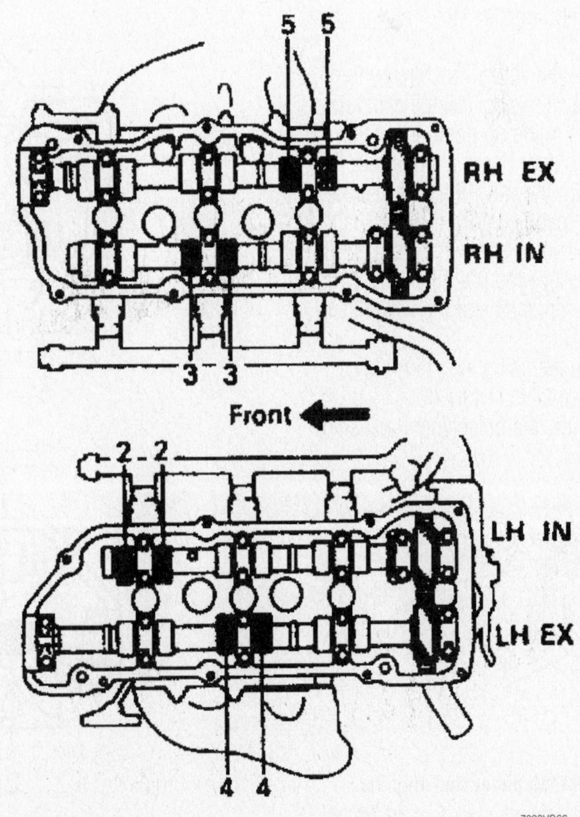

Adjust these valves during the 2nd step—3.0L engine

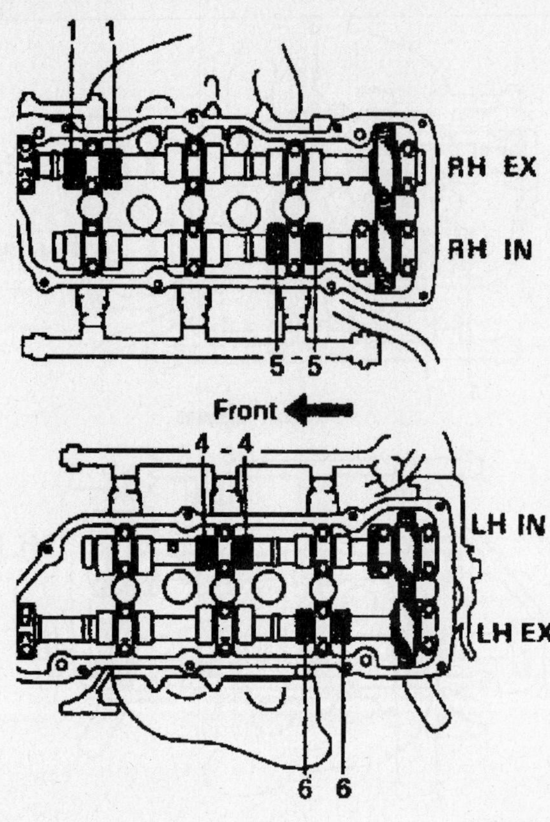

Adjust these valves during the 3rd step—3.0L engine

revolution. Record the measurements on valves No. 4 and 5 intake; No. 1 and 6 exhaust.

8. Remove the adjusting shim by turning the crankshaft to position the cam lobe of the camshaft in the up position on the valve to be adjusted. Using a small thin flat bladed tool, turn the valve lifter so that the notches are perpendicular to the camshaft. Press down the valve lifter with tool 09248-55010 part A. Place too 09248-55010 part B between the camshaft and the valve lifter; remove part A.

9. Remove the adjusting shim with a magnet and a small screwdriver.

10. Determine the replacement adjusting shim size by either using the charts or the following formulas:

- Intake: N = T + (A–0.008 in./0.020mm)
- Exhaust: N = T + (A–0.012 in./0.30mm)
- T = Thickness of removed shim
- A = Measured valve clearance
- N = Thickness of new shim

11. Select a new shim with a thickness as close as possible to the calculated value. Install the new replacement shim.

➡Shims are available in 17 sizes in increments of 0.0020 in. (0.050mm), from 0.0984 in. (2.500mm) to 0.1299 in. (3.300mm).

12. Recheck the valve clearance.

13. Install or connect the following:

- Cylinder head covers
- Spark plugs and the ignition coils
- Engine wiring harness to the injectors and the coils
- Intake chamber
- Emission control valve set
- V-bank cover
- Air flow meter, air duct and air cleaner cover
- Negative battery cable

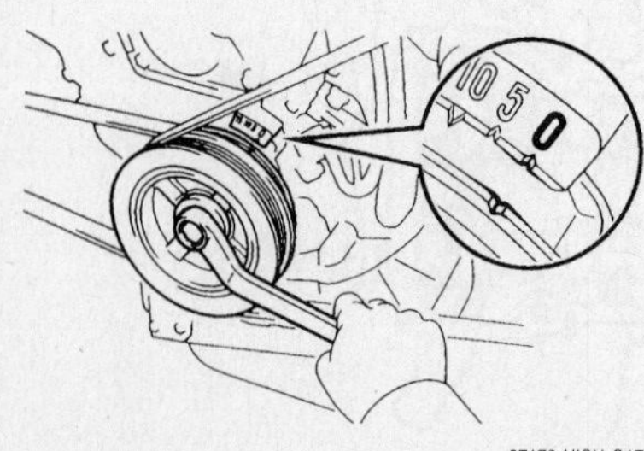

Turn the crankshaft pulley and align its groove with the timing mark 0 of the No. 1 timing cover

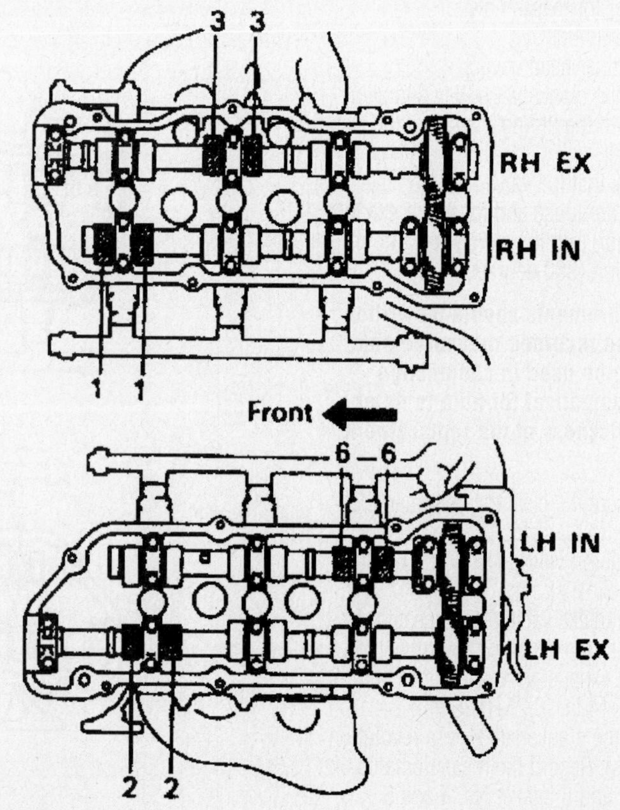

Adjust these valves during the 1st step—3.3L engine

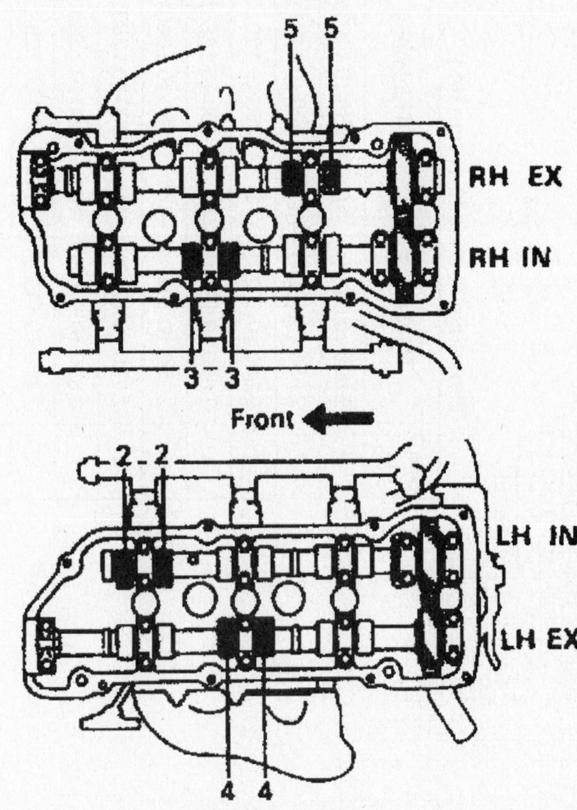

Adjust these valves during the 2nd step—3.3L engine

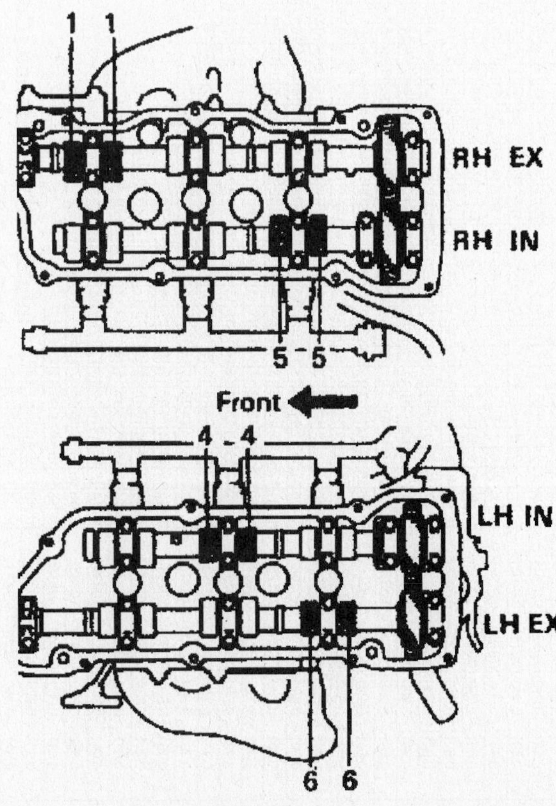

Adjust these valves during the 3rd step—3.3L engine

3.3L Engine

➡Adjust the valve clearance when the engine is cold.

1. Before servicing the vehicle, refer to the precautions in the beginning of this section.
2. Remove or disconnect the following:
   - Negative battery cable. If equipped with an air bag, wait at least 90 seconds before proceeding.
   - Accelerator/throttle cable from the throttle linkage
   - Air cleaner cover, air flow meter and air duct assembly
   - V-bank cover
   - Emission control valve set
   - Air intake chamber
   - Engine harness from the injectors and the ignition coils
   - Ignition coils and keep them in order for reassembly
   - Spark plugs
   - Cylinder head covers
3. Turn the crankshaft pulley and align its groove with the timing mark **0** of the No. 1 timing cover.
4. Check that the valve lifters on the No. 1 intake are loose and the No. 1 exhaust are tight. If not, turn the crankshaft 1 complete revolution (360 degrees).

➡All measurements should be written down. These recorded measurements will need to be used in conjunction with a mathematical formula to determine the thickness of the replacement shims.

5. Measure the clearance between the valve lifters and the camshaft. Record the measurements on valves No. 1 and 6 intake; No. 2 and 3 exhaust.
   a. The intake valve clearance cold is 0.006–0.010 in. (0.15–0.25mm).
   b. The exhaust valve clearance cold is 0.010–0.014 in. (0.25–0.35mm).
6. Turn the crankshaft ⅔ of a revolution (240 degrees). Record the measurements on valves No. 2 and 3 intake; No. 4 and 5 exhaust.
7. Turn the crankshaft another ⅔ of a revolution. Record the measurements on valves No. 4 and 5 intake; No. 1 and 6 exhaust.
8. Remove the adjusting shim by turning the crankshaft to position the cam lobe of the camshaft in the up position on the valve to be adjusted. Using a small thin flat bladed tool, turn the valve lifter so that the notches are perpendicular to the camshaft. Press down the valve lifter with tool 09248-55010 part A. Place too 09248-55010 part

Adjusting Shim Selection Chart (Intake)

Measured clearance mm (in.) / Installed shim thickness mm (in.)	2.500 (0.0984)	2.520 (0.0992)	2.540 (0.1000)	2.550 (0.1004)	2.560 (0.1008)	2.580 (0.1016)	2.600 (0.1024)	2.620 (0.1031)	2.640 (0.1039)	2.650 (0.1043)	2.660 (0.1047)	2.670 (0.1051)	2.680 (0.1055)	2.690 (0.1059)	2.700 (0.1063)	2.710 (0.1067)	2.720 (0.1071)	2.730 (0.1075)	2.740 (0.1079)	2.750 (0.1083)	2.760 (0.1087)	2.770 (0.1091)	2.780 (0.1094)	2.790 (0.1098)	2.800 (0.1102)	2.810 (0.1106)	2.820 (0.1110)	2.830 (0.1114)	2.840 (0.1118)	2.850 (0.1122)	2.860 (0.1126)	2.870 (0.1130)	2.880 (0.1134)	2.890 (0.1138)	2.900 (0.1142)	2.910 (0.1146)	2.920 (0.1150)	2.930 (0.1154)	2.940 (0.1157)	2.950 (0.1161)	2.960 (0.1165)	2.970 (0.1169)	2.980 (0.1173)	2.990 (0.1177)	3.000 (0.1181)	3.010 (0.1185)	3.020 (0.1189)	3.030 (0.1193)	3.040 (0.1197)	3.050 (0.1201)	3.060 (0.1205)	3.080 (0.1213)	3.100 (0.1220)	3.120 (0.1228)	3.140 (0.1236)	3.150 (0.1240)	3.160 (0.1244)	3.180 (0.1252)	3.200 (0.1260)	3.220 (0.1268)	3.240 (0.1276)	3.250 (0.1280)	3.260 (0.1283)	3.280 (0.1291)	3.300 (0.1299)
0.000 – 0.020 (0.0000 – 0.0008)									1	1	1	1	1	1	1	1	2	2	2	2	2	3	3	3	3	3	4	4	4	4	4	5	5	5	5	5	6	6	6	6	6	7	7	7	7	7	8	8	8	8	8	9	9	10	10	10	10	11	11	12	12	12	12	13	13
0.021 – 0.040 (0.0008 – 0.0016)								1	1	1	1	1	1	1	2	2	2	2	2	3	3	3	3	3	4	4	4	4	4	5	5	5	5	5	6	6	6	6	6	7	7	7	7	7	8	8	8	8	8	9	9	9	10	10	10	11	11	11	12	12	12	13	13	13	14
0.041 – 0.060 (0.0016 – 0.0024)							1	1	1	1	1	1	2	2	2	2	2	3	3	3	3	3	4	4	4	4	4	5	5	5	5	5	6	6	6	6	6	7	7	7	7	7	8	8	8	8	8	9	9	9	9	10	10	10	11	11	11	12	12	12	13	13	13	14	14
0.061 – 0.080 (0.0024 – 0.0031)						1	1	1	1	1	2	2	2	2	2	3	3	3	3	3	4	4	4	4	4	5	5	5	5	5	6	6	6	6	6	7	7	7	7	7	8	8	8	8	8	9	9	9	9	9	10	10	10	11	11	11	12	12	12	13	13	13	14	14	14
0.081 – 0.100 (0.0032 – 0.0039)				1	1	1	1	1	2	2	2	2	2	3	3	3	3	3	4	4	4	4	4	5	5	5	5	5	6	6	6	6	6	7	7	7	7	7	8	8	8	8	8	9	9	9	9	9	10	10	10	10	11	11	12	12	12	12	13	13	14	14	14	14	15
0.101 – 0.120 (0.0040 – 0.0047)			1	1	1	1	1	2	2	2	2	3	3	3	3	3	4	4	4	4	4	5	5	5	5	5	6	6	6	6	6	7	7	7	7	7	8	8	8	8	8	9	9	9	9	9	10	10	10	10	10	11	11	12	12	12	12	13	13	14	14	14	14	15	15
0.121 – 0.140 (0.0048 – 0.0055)		1	1	1	1	1	2	2	2	3	3	3	3	3	4	4	4	4	4	5	5	5	5	5	6	6	6	6	6	7	7	7	7	7	8	8	8	8	8	9	9	9	9	9	10	10	10	10	10	11	11	11	12	12	12	13	13	13	14	14	14	15	15	15	16
0.141 – 0.149 (0.0056 – 0.0059)		1	1	1	1	2	2	2	3	3	3	3	4	4	4	4	4	5	5	5	5	5	6	6	6	6	6	7	7	7	7	7	8	8	8	8	8	9	9	9	9	9	10	10	10	10	10	11	11	11	11	12	12	12	13	13	13	14	14	14	15	15	15	16	16
0.150 – 0.250 (0.0059 – 0.0098)																																																																	
0.251 – 0.260 (0.0099 – 0.0102)	2	3	3	3	3	4	4	5	5	5	5	6	6	6	6	6	7	7	7	7	7	8	8	8	8	8	9	9	9	9	9	10	10	10	10	10	11	11	11	11	11	12	12	12	12	12	13	13	13	13	13	14	14	15	15	15	15	16	16	17	17	17	17	17	
0.261 – 0.280 (0.0103 – 0.0110)	2	3	3	3	4	4	4	5	5	5	6	6	6	6	6	7	7	7	7	7	8	8	8	8	8	9	9	9	9	9	10	10	10	10	10	11	11	11	11	11	12	12	12	12	12	13	13	13	13	13	14	14	14	15	15	15	16	16	16	17	17	17	17	17	
0.281 – 0.300 (0.0111 – 0.0118)	3	3	4	4	4	4	5	5	6	6	6	6	6	7	7	7	7	7	8	8	8	8	8	9	9	9	9	9	10	10	10	10	10	11	11	11	11	11	12	12	12	12	12	13	13	13	13	13	14	14	14	14	15	15	16	16	16	16	17	17	17	17	17		
0.301 – 0.320 (0.0119 – 0.0126)	3	4	4	4	4	5	5	6	6	6	6	7	7	7	7	7	8	8	8	8	8	9	9	9	9	9	10	10	10	10	10	11	11	11	11	11	12	12	12	12	12	13	13	13	13	13	14	14	14	14	14	15	15	16	16	16	16	17	17	17	17				
0.321 – 0.340 (0.0126 – 0.0134)	4	4	4	5	5	5	6	6	6	7	7	7	7	7	8	8	8	8	8	9	9	9	9	9	10	10	10	10	10	11	11	11	11	11	12	12	12	12	12	13	13	13	13	13	14	14	14	14	14	15	15	15	16	16	16	17	17	17	17	17					
0.341 – 0.360 (0.0134 – 0.0142)	4	4	5	5	5	6	6	6	7	7	7	7	8	8	8	8	8	9	9	9	9	9	10	10	10	10	10	11	11	11	11	11	12	12	12	12	12	13	13	13	13	13	14	14	14	14	14	15	15	15	15	16	16	16	17	17	17	17	17						
0.361 – 0.380 (0.0142 – 0.0150)	4	5	5	5	6	6	6	7	7	7	8	8	8	8	8	9	9	9	9	9	10	10	10	10	10	11	11	11	11	11	12	12	12	12	12	13	13	13	13	13	14	14	14	14	14	15	15	15	15	15	16	16	16	17	17	17	17	17							
0.381 – 0.400 (0.0150 – 0.0157)	5	5	6	6	6	6	7	7	8	8	8	8	8	9	9	9	9	9	10	10	10	10	10	11	11	11	11	11	12	12	12	12	12	13	13	13	13	13	14	14	14	14	14	15	15	15	15	16	16	16	16	16	17	17	17	17	17								
0.401 – 0.420 (0.0158 – 0.0165)	5	6	6	6	6	7	7	8	8	8	8	9	9	9	9	9	10	10	10	10	10	11	11	11	11	11	12	12	12	12	12	13	13	13	13	13	14	14	14	14	14	15	15	15	15	15	16	16	16	16	16	17	17	17	17										
0.421 – 0.440 (0.0166 – 0.0173)	6	6	6	7	7	7	8	8	8	9	9	9	9	9	10	10	10	10	10	11	11	11	11	11	12	12	12	12	12	13	13	13	13	13	14	14	14	14	14	15	15	15	15	15	16	16	16	16	16	17	17	17	17	17											
0.441 – 0.460 (0.0174 – 0.0181)	6	6	7	7	7	8	8	8	9	9	9	9	10	10	10	10	10	11	11	11	11	11	12	12	12	12	12	13	13	13	13	13	14	14	14	14	14	15	15	15	15	15	16	16	16	16	16	17	17	17	17	17	17												
0.461 – 0.480 (0.0181 – 0.0189)	6	7	7	7	8	8	8	9	9	9	10	10	10	10	10	11	11	11	11	11	12	12	12	12	12	13	13	13	13	13	14	14	14	14	14	15	15	15	15	15	16	16	16	16	16	17	17	17	17	17	17	17													
0.481 – 0.500 (0.0189 – 0.0197)	7	7	8	8	8	8	9	9	10	10	10	10	10	11	11	11	11	11	12	12	12	12	12	13	13	13	13	13	14	14	14	14	14	15	15	15	15	15	16	16	16	16	16	17	17	17	17	17	17	17	17														
0.501 – 0.520 (0.0197 – 0.0205)	7	8	8	8	8	9	9	10	10	10	10	11	11	11	11	11	12	12	12	12	12	13	13	13	13	13	14	14	14	14	14	15	15	15	15	15	16	16	16	16	16	17	17	17	17	17	17	17	17																
0.521 – 0.540 (0.0205 – 0.0213)	8	8	8	9	9	9	10	10	10	11	11	11	11	11	12	12	12	12	12	13	13	13	13	13	14	14	14	14	14	15	15	15	15	15	16	16	16	16	16	17	17	17	17	17	17	17	17																		
0.541 – 0.560 (0.0213 – 0.0220)	8	8	9	9	9	10	10	10	11	11	11	11	12	12	12	12	12	13	13	13	13	13	14	14	14	14	14	15	15	15	15	15	16	16	16	16	16	17	17	17	17	17	17	17	17																				
0.561 – 0.580 (0.0221 – 0.0228)	8	9	9	9	10	10	10	11	11	11	12	12	12	12	12	13	13	13	13	13	14	14	14	14	14	15	15	15	15	15	16	16	16	16	16	17	17	17	17	17	17	17	17																						
0.581 – 0.600 (0.0229 – 0.0236)	9	9	10	10	10	10	11	11	12	12	12	12	12	13	13	13	13	13	14	14	14	14	14	15	15	15	15	15	16	16	16	16	16	17	17	17	17	17	17	17	17																								
0.601 – 0.620 (0.0237 – 0.0244)	9	10	10	10	10	11	11	12	12	12	12	13	13	13	13	13	14	14	14	14	14	15	15	15	15	15	16	16	16	16	16	17	17	17	17	17	17	17	17																										
0.621 – 0.640 (0.0244 – 0.0252)	10	10	10	11	11	11	12	12	12	13	13	13	13	13	14	14	14	14	14	15	15	15	15	15	16	16	16	16	16	17	17	17	17	17	17	17	17																												
0.641 – 0.660 (0.0252 – 0.0260)	10	10	11	11	11	12	12	12	13	13	13	13	14	14	14	14	14	15	15	15	15	15	16	16	16	16	16	17	17	17	17	17	17	17	17																														
0.661 – 0.680 (0.0260 – 0.0268)	10	11	11	11	12	12	12	13	13	13	14	14	14	14	14	15	15	15	15	15	16	16	16	16	16	17	17	17	17	17	17	17	17																																
0.681 – 0.700 (0.0268 – 0.0276)	11	11	12	12	12	12	13	13	14	14	14	14	14	15	15	15	15	15	16	16	16	16	16	17	17	17	17	17	17	17	17																																		
0.701 – 0.720 (0.0276 – 0.0283)	11	12	12	12	12	13	13	14	14	14	14	15	15	15	15	15	16	16	16	16	16	17	17	17	17	17	17	17	17																																				
0.721 – 0.740 (0.0284 – 0.0291)	12	12	12	13	13	13	14	14	14	15	15	15	15	15	16	16	16	16	16	17	17	17	17	17	17	17	17																																						
0.741 – 0.760 (0.0292 – 0.0299)	12	12	13	13	13	14	14	14	15	15	15	15	16	16	16	16	16	17	17	17	17	17	17	17	17																																								
0.761 – 0.780 (0.0300 – 0.0307)	12	13	13	13	14	14	14	15	15	15	16	16	16	16	16	17	17	17	17	17	17	17	17																																										
0.781 – 0.800 (0.0307 – 0.0315)	13	13	14	14	14	14	15	15	16	16	16	16	16	17	17	17	17	17	17	17	17																																												
0.801 – 0.820 (0.0315 – 0.0323)	13	14	14	14	14	15	15	16	16	16	16	17	17	17	17	17	17	17	17																																														
0.821 – 0.840 (0.0323 – 0.0331)	14	14	14	15	15	15	16	16	16	17	17	17	17	17	17	17	17																																																
0.841 – 0.860 (0.0331 – 0.0339)	14	14	15	15	15	16	16	16	17	17	17	17	17	17	17																																																		
0.861 – 0.880 (0.0339 – 0.0346)	14	15	15	15	16	16	16	17	17	17	17	17	17																																																				
0.881 – 0.900 (0.0347 – 0.0354)	15	15	16	16	16	16	17	17	17	17	17																																																								
0.901 – 0.920 (0.0355 – 0.0362)	15	16	16	16	16	17	17	17	17																																																										
0.921 – 0.940 (0.0363 – 0.0370)	16	16	16	17	17	17	17	17																																																											
0.941 – 0.960 (0.0370 – 0.0378)	16	16	17	17	17	17	17																																																													
0.961 – 0.980 (0.0378 – 0.0386)	16	17	17	17	17	17																																																														
0.981 – 1.000 (0.0386 – 0.0394)	17	17	17	17	17																																																															
1.001 – 1.020 (0.0394 – 0.0402)	17	17	17																																																																		
1.021 – 1.040 (0.0402 – 0.0409)	17	17																																																																			
1.041 – 1.050 (0.0410 – 0.0413)	17																																																																				

Intake valve clearance (Cold):
0.15 to 0.25 mm (0.006 to 0.010 In.)

EXAMPLE:
The 2.800 mm (0.1102 in.) shim is installed, and the measured clearance is 0.450 mm (0.0177 in.). Replace the 2.800 mm (0.1102 in.) shim with a new No. 12 shim.

New shim thickness mm (in.)

Shim No.	Thickness	Shim No.	Thickness
1	2.500 (0.0984)	10	2.950 (0.1161)
2	2.550 (0.1004)	11	3.000 (0.1181)
3	2.600 (0.1024)	12	3.050 (0.1201)
4	2.650 (0.1043)	13	3.100 (0.1220)
5	2.700 (0.1063)	14	3.150 (0.1240)
6	2.750 (0.1083)	15	3.200 (0.1260)
7	2.800 (0.1102)	16	3.250 (0.1280)
8	2.850 (0.1122)	17	3.300 (0.1299)
9	2.900 (0.1142)		

HINT:
A shim's thickness is written on its face in millimeters.

67170-HIGH-G18

Intake valve shim selection chart—3.3L engine

Adjusting Shim Selection Chart (Exhaust)

Measured clearance mm (in.) \ Installed shim thickness mm (in.)	2.500 (0.0984)	2.520 (0.0992)	2.540 (0.1000)	2.550 (0.1004)	2.560 (0.1008)	2.580 (0.1016)	2.600 (0.1024)	2.620 (0.1031)	2.640 (0.1039)	2.650 (0.1043)	2.660 (0.1047)	2.670 (0.1051)	2.680 (0.1055)	2.690 (0.1059)	2.700 (0.1063)	2.710 (0.1067)	2.720 (0.1071)	2.730 (0.1075)	2.740 (0.1079)	2.750 (0.1083)	2.760 (0.1087)	2.770 (0.1091)	2.780 (0.1094)	2.790 (0.1098)	2.800 (0.1102)	2.810 (0.1106)	2.820 (0.1110)	2.830 (0.1114)	2.840 (0.1118)	2.850 (0.1122)	2.860 (0.1126)	2.870 (0.1130)	2.880 (0.1134)	2.890 (0.1138)	2.900 (0.1142)	2.910 (0.1146)	2.920 (0.1150)	2.930 (0.1154)	2.940 (0.1157)	2.950 (0.1161)	2.960 (0.1165)	2.970 (0.1169)	2.980 (0.1173)	2.990 (0.1177)	3.000 (0.1181)	3.010 (0.1185)	3.020 (0.1189)	3.030 (0.1193)	3.040 (0.1197)	3.050 (0.1201)	3.060 (0.1205)	3.080 (0.1213)	3.100 (0.1220)	3.120 (0.1228)	3.140 (0.1236)	3.150 (0.1240)	3.160 (0.1244)	3.180 (0.1252)	3.200 (0.1260)	3.220 (0.1268)	3.240 (0.1276)	3.250 (0.1280)	3.260 (0.1283)	3.280 (0.1291)	3.300 (0.1299)
0.000 – 0.020 (0.0000 – 0.0008)																		1	1	1	1	1	1	1	1	1	2	2	2	2	2	3	3	3	3	3	4	4	4	4	4	5	5	5	5	5	6	6	6	6	6	7	7	8	8	8	8	9	9	10	10	10	10	11	11
0.021 – 0.040 (0.0008 – 0.0016)																1	1	1	1	1	1	1	1	1	2	2	2	2	2	3	3	3	3	3	4	4	4	4	4	5	5	5	5	5	6	6	6	6	6	7	7	7	8	8	8	9	9	9	10	10	10	11	11	11	12
0.041 – 0.060 (0.0016 – 0.0024)														1	1	1	1	1	1	1	1	1	2	2	2	2	2	3	3	3	3	3	4	4	4	4	4	5	5	5	5	5	6	6	6	6	6	7	7	7	7	8	8	8	9	9	9	10	10	10	11	11	11	12	12
0.061 – 0.080 (0.0024 – 0.0031)												1	1	1	1	1	1	1	1	1	2	2	2	2	2	3	3	3	3	3	4	4	4	4	4	5	5	5	5	5	6	6	6	6	6	7	7	7	7	7	8	8	8	9	9	9	10	10	10	11	11	11	12	12	12
0.081 – 0.100 (0.0032 – 0.0039)										1	1	1	1	1	1	1	1	1	2	2	2	2	2	3	3	3	3	3	4	4	4	4	4	5	5	5	5	5	6	6	6	6	6	7	7	7	7	7	8	8	8	8	9	9	10	10	10	10	11	11	12	12	12	12	13
0.101 – 0.120 (0.0040 – 0.0047)									1	1	1	1	1	1	1	1	2	2	2	2	2	3	3	3	3	3	4	4	4	4	4	5	5	5	5	5	6	6	6	6	6	7	7	7	7	7	8	8	8	8	8	9	9	10	10	10	10	11	11	12	12	12	12	13	13
0.121 – 0.140 (0.0048 – 0.0055)								1	1	1	1	1	1	1	2	2	2	2	2	3	3	3	3	3	4	4	4	4	4	5	5	5	5	5	6	6	6	6	6	7	7	7	7	7	8	8	8	8	8	9	9	9	10	10	10	11	11	11	12	12	12	13	13	13	14
0.141 – 0.160 (0.0056 – 0.0063)							1	1	1	1	1	1	2	2	2	2	2	3	3	3	3	3	4	4	4	4	4	5	5	5	5	5	6	6	6	6	6	7	7	7	7	7	8	8	8	8	8	9	9	9	9	10	10	10	11	11	11	12	12	12	13	13	13	14	14
0.161 – 0.180 (0.0063 – 0.0071)						1	1	1	1	1	2	2	2	2	2	3	3	3	3	3	4	4	4	4	4	5	5	5	5	5	6	6	6	6	6	7	7	7	7	7	8	8	8	8	8	9	9	9	9	9	10	10	10	11	11	11	12	12	12	13	13	13	14	14	14
0.181 – 0.200 (0.0071 – 0.0079)				1	1	1	1	1	2	2	2	2	2	3	3	3	3	3	4	4	4	4	4	5	5	5	5	5	6	6	6	6	6	7	7	7	7	7	8	8	8	8	8	9	9	9	9	9	10	10	10	10	11	11	12	12	12	12	13	13	14	14	14	14	15
0.201 – 0.220 (0.0079 – 0.0087)			1	1	1	1	1	2	2	2	2	3	3	3	3	3	4	4	4	4	4	5	5	5	5	5	6	6	6	6	6	7	7	7	7	7	8	8	8	8	8	9	9	9	9	9	10	10	10	10	10	11	11	12	12	12	12	13	13	14	14	14	14	15	15
0.221 – 0.240 (0.0087 – 0.0094)		1	1	1	1	1	2	2	2	3	3	3	3	3	4	4	4	4	4	5	5	5	5	5	6	6	6	6	6	7	7	7	7	7	8	8	8	8	8	9	9	9	9	9	10	10	10	10	10	11	11	11	12	12	12	13	13	13	14	14	14	15	15	15	16
0.241 – 0.249 (0.0095 – 0.0098)		1	1	1	1	2	2	2	3	3	3	3	4	4	4	4	4	5	5	5	5	5	6	6	6	6	6	7	7	7	7	7	8	8	8	8	8	9	9	9	9	9	10	10	10	10	10	11	11	11	11	12	12	12	13	13	13	14	14	14	15	15	15	16	16
0.250 – 0.350 (0.0098 – 0.0138)																																																																	
0.351 – 0.360 (0.0138 – 0.0142)	2	3	3	3	3	4	4	5	5	5	5	6	6	6	6	6	7	7	7	7	7	8	8	8	8	8	9	9	9	9	9	10	10	10	10	10	11	11	11	11	11	12	12	12	12	12	13	13	13	13	13	14	14	15	15	15	15	16	16	17	17	17	17	17	
0.361 – 0.380 (0.0142 – 0.0150)	2	3	3	3	4	4	4	5	5	5	6	6	6	6	6	7	7	7	7	7	8	8	8	8	8	9	9	9	9	9	10	10	10	10	10	11	11	11	11	11	12	12	12	12	12	13	13	13	13	13	14	14	14	15	15	15	16	16	16	17	17	17	17	17	
0.381 – 0.400 (0.0150 – 0.0157)	3	3	4	4	4	4	5	5	6	6	6	6	6	7	7	7	7	7	8	8	8	8	8	9	9	9	9	9	10	10	10	10	10	11	11	11	11	11	12	12	12	12	12	13	13	13	13	13	14	14	14	14	15	15	16	16	16	16	17	17	17	17	17		
0.401 – 0.420 (0.0158 – 0.0165)	3	4	4	4	4	5	5	6	6	6	6	7	7	7	7	7	8	8	8	8	8	9	9	9	9	9	10	10	10	10	10	11	11	11	11	11	12	12	12	12	12	13	13	13	13	13	14	14	14	14	14	15	15	16	16	16	16	17	17	17	17				
0.421 – 0.440 (0.0166 – 0.0173)	4	4	4	5	5	5	6	6	6	7	7	7	7	7	8	8	8	8	8	9	9	9	9	9	10	10	10	10	10	11	11	11	11	11	12	12	12	12	12	13	13	13	13	13	14	14	14	14	14	15	15	15	16	16	16	17	17	17	17	17					
0.441 – 0.460 (0.0174 – 0.0181)	4	4	5	5	5	6	6	6	7	7	7	7	8	8	8	8	8	9	9	9	9	9	10	10	10	10	10	11	11	11	11	11	12	12	12	12	12	13	13	13	13	13	14	14	14	14	14	15	15	15	15	16	16	16	17	17	17	17	17						
0.461 – 0.480 (0.0181 – 0.0189)	4	5	5	5	6	6	6	7	7	7	8	8	8	8	8	9	9	9	9	9	10	10	10	10	10	11	11	11	11	11	12	12	12	12	12	13	13	13	13	13	14	14	14	14	14	15	15	15	15	15	16	16	16	17	17	17	17	17							
0.481 – 0.500 (0.0189 – 0.0197)	5	5	6	6	6	6	7	7	8	8	8	8	8	9	9	9	9	9	10	10	10	10	10	11	11	11	11	11	12	12	12	12	12	13	13	13	13	13	14	14	14	14	14	15	15	15	15	15	16	16	16	16	17	17	17	17	17								
0.501 – 0.520 (0.0197 – 0.0205)	5	6	6	6	6	7	7	8	8	8	8	9	9	9	9	9	10	10	10	10	10	11	11	11	11	11	12	12	12	12	12	13	13	13	13	13	14	14	14	14	14	15	15	15	15	15	16	16	16	16	16	17	17	17	17										
0.521 – 0.540 (0.0205 – 0.0213)	6	6	6	7	7	7	8	8	8	9	9	9	9	9	10	10	10	10	10	11	11	11	11	11	12	12	12	12	12	13	13	13	13	13	14	14	14	14	14	15	15	15	15	15	16	16	16	16	16	17	17	17	17	17											
0.541 – 0.560 (0.0213 – 0.0220)	6	6	7	7	7	8	8	8	9	9	9	9	10	10	10	10	10	11	11	11	11	11	12	12	12	12	12	13	13	13	13	13	14	14	14	14	14	15	15	15	15	15	16	16	16	16	16	17	17	17	17	17	17												
0.561 – 0.580 (0.0221 – 0.0228)	6	7	7	7	8	8	8	9	9	9	10	10	10	10	10	11	11	11	11	11	12	12	12	12	12	13	13	13	13	13	14	14	14	14	14	15	15	15	15	15	16	16	16	16	16	17	17	17	17	17	17	17													
0.581 – 0.600 (0.0229 – 0.0236)	7	7	8	8	8	8	9	9	10	10	10	10	10	11	11	11	11	11	12	12	12	12	12	13	13	13	13	13	14	14	14	14	14	15	15	15	15	15	16	16	16	16	16	17	17	17	17	17	17	17	17														
0.601 – 0.620 (0.0237 – 0.0244)	7	8	8	8	8	9	9	10	10	10	10	11	11	11	11	11	12	12	12	12	12	13	13	13	13	13	14	14	14	14	14	15	15	15	15	15	16	16	16	16	16	17	17	17	17	17	17	17	17																
0.621 – 0.640 (0.0244 – 0.0252)	8	8	8	9	9	9	10	10	10	11	11	11	11	11	12	12	12	12	12	13	13	13	13	13	14	14	14	14	14	15	15	15	15	15	16	16	16	16	16	17	17	17	17	17	17	17	17																		
0.641 – 0.660 (0.0252 – 0.0260)	8	8	9	9	9	10	10	10	11	11	11	11	12	12	12	12	12	13	13	13	13	13	14	14	14	14	14	15	15	15	15	15	16	16	16	16	16	17	17	17	17	17	17	17	17																				
0.661 – 0.680 (0.0260 – 0.0268)	8	9	9	9	10	10	10	11	11	11	12	12	12	12	12	13	13	13	13	13	14	14	14	14	14	15	15	15	15	15	16	16	16	16	16	17	17	17	17	17	17	17	17																						
0.681 – 0.700 (0.0268 – 0.0276)	9	9	10	10	10	10	11	11	12	12	12	12	12	13	13	13	13	13	14	14	14	14	14	15	15	15	15	15	16	16	16	16	16	17	17	17	17	17	17	17	17																								
0.701 – 0.720 (0.0276 – 0.0283)	9	10	10	10	10	11	11	12	12	12	12	13	13	13	13	13	14	14	14	14	14	15	15	15	15	15	16	16	16	16	16	17	17	17	17	17	17	17	17																										
0.721 – 0.740 (0.0284 – 0.0291)	10	10	10	11	11	11	12	12	12	13	13	13	13	13	14	14	14	14	14	15	15	15	15	15	16	16	16	16	16	17	17	17	17	17	17	17	17																												
0.741 – 0.760 (0.0292 – 0.0299)	10	10	11	11	11	12	12	12	13	13	13	13	14	14	14	14	14	15	15	15	15	15	16	16	16	16	16	17	17	17	17	17	17	17	17																														
0.761 – 0.780 (0.0300 – 0.0307)	10	11	11	11	12	12	12	13	13	13	14	14	14	14	14	15	15	15	15	15	16	16	16	16	16	17	17	17	17	17	17	17	17																																
0.781 – 0.800 (0.0307 – 0.0315)	11	11	12	12	12	12	13	13	14	14	14	14	14	15	15	15	15	15	16	16	16	16	16	17	17	17	17	17	17	17	17																																		
0.801 – 0.820 (0.0315 – 0.0323)	11	12	12	12	12	13	13	14	14	14	14	15	15	15	15	15	16	16	16	16	16	17	17	17	17	17	17	17	17																																				
0.821 – 0.840 (0.0323 – 0.0331)	12	12	12	13	13	13	14	14	14	15	15	15	15	15	16	16	16	16	16	17	17	17	17	17	17	17	17																																						
0.841 – 0.860 (0.0331 – 0.0339)	12	12	13	13	13	14	14	14	15	15	15	15	16	16	16	16	16	17	17	17	17	17	17	17	17																																								
0.861 – 0.880 (0.0339 – 0.0346)	12	13	13	13	14	14	14	15	15	15	16	16	16	16	16	17	17	17	17	17	17	17	17																																										
0.881 – 0.900 (0.0347 – 0.0354)	13	13	14	14	14	14	15	15	16	16	16	16	16	17	17	17	17	17	17	17	17																																												
0.901 – 0.920 (0.0355 – 0.0362)	13	14	14	14	14	15	15	16	16	16	16	17	17	17	17	17	17	17	17																																														
0.921 – 0.940 (0.0363 – 0.0370)	14	14	14	15	15	15	16	16	16	17	17	17	17	17	17	17	17																																																
0.941 – 0.960 (0.0370 – 0.0378)	14	14	15	15	15	16	16	16	17	17	17	17	17	17	17																																																		
0.961 – 0.980 (0.0378 – 0.0386)	14	15	15	15	16	16	16	17	17	17	17	17	17																																																				
0.981 – 1.000 (0.0386 – 0.0394)	15	15	16	16	16	16	17	17	17	17	17																																																						
1.001 – 1.020 (0.0394 – 0.0402)	15	16	16	16	16	17	17	17	17																																																										
1.021 – 1.040 (0.0402 – 0.0409)	16	16	16	17	17	17	17	17																																																											
1.041 – 1.060 (0.0410 – 0.0417)	16	16	17	17	17	17	17																																																													
1.061 – 1.080 (0.0418 – 0.0425)	16	17	17	17	17	17																																																													
1.081 – 1.100 (0.0426 – 0.0433)	17	17	17	17	17																																																														
1.101 – 1.120 (0.0433 – 0.0441)	17	17	17																																																															
1.121 – 1.140 (0.0441 – 0.0449)	17	17																																																																	
1.141 – 1.150 (0.0449 – 0.0453)	17																																																																		

Exhaust valve clearance (Cold):
0.25 to 0.35 mm (0.010 to 0.014 in.)

EXAMPLE:
The 2.800 mm (0.1102 in.) shim is installed, and the measured clearance is 0.450 mm (0.0177 in.). Replace the 2.800 mm (0.1102 in.) shim with a new No. 10 shim.

New shim thickness mm (in.)

Shim No.	Thickness	Shim No.	Thickness
1	2.500 (0.0984)	10	2.950 (0.1161)
2	2.550 (0.1004)	11	3.000 (0.1181)
3	2.600 (0.1024)	12	3.050 (0.1201)
4	2.650 (0.1043)	13	3.100 (0.1220)
5	2.700 (0.1063)	14	3.150 (0.1240)
6	2.750 (0.1083)	15	3.200 (0.1260)
7	2.800 (0.1102)	16	3.250 (0.1280)
8	2.850 (0.1122)	17	3.300 (0.1299)
9	2.900 (0.1142)		

HINT:
A shim's thickness is written on its face in millimeters.

67170-HIGH-G19

Exhaust valve shim selection chart—3.3L engine

B between the camshaft and the valve lifter; remove part A.

9. Remove the adjusting shim with a magnet and a small screwdriver.

10. Determine the replacement adjusting shim size by either using the charts or the following formulas:

- Intake: N = T + (A–0.008 in./0.020mm)
- Exhaust: N = T + (A–0.012 in./0.30mm)
- T = Thickness of removed shim
- A = Measured valve clearance
- N = Thickness of new shim

11. Select a new shim with a thickness as close as possible to the calculated value. Install the new replacement shim.

➡Shims are available in 17 sizes in increments of 0.0020 in. (0.050mm), from 0.0984 in. (2.500mm) to 0.1299 in. (3.300mm).

12. Recheck the valve clearance.

13. Install or connect the following:

- Cylinder head covers
- Spark plugs and the ignition coils
- Engine wiring harness to the injectors and the coils
- Intake chamber
- Emission control valve set
- V-bank cover
- Air flow meter, air duct and air cleaner cover
- Negative battery cable

Starter Motor

REMOVAL & INSTALLATION

2.4L Engine

2001–02

1. Before servicing the vehicle, refer to the precautions in the beginning of this section.

2. Remove or disconnect the following:

- Air cleaner
- starter wiring
- Starter

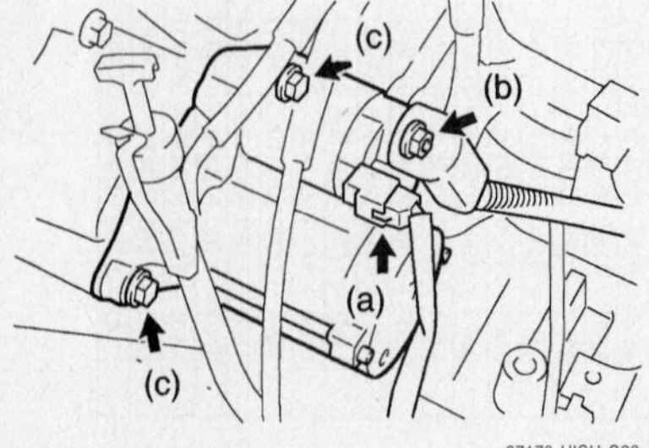

Starter mounting—2.4L engine

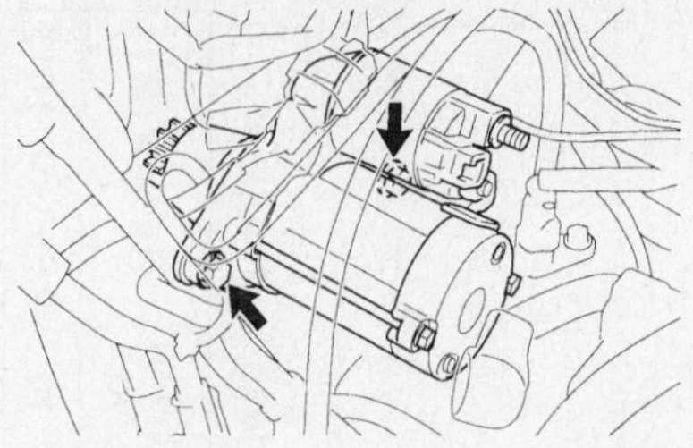

Starter mounting bolt locations—3.3L engine

3. Installation is the reverse of removal. Torque the starter bolts to 29 ft. lbs. (39Nm).

2003–04

1. Remove the battery.
2. Remove the battery tray.
3. Disconnect the starter wiring.
4. Remove the mounting bolts.
5. Lift out the starter.
6. Installation is the reverse of removal. Torque the mounting bolts to 27 ft. lbs. (37 Nm).

3.0L Engine

1. Before servicing the vehicle, refer to the precautions in the beginning of this section.

2. Remove or disconnect the following:

- Battery
- Air cleaner
- Starter

3. Installation is the reverse of removal. Torque the starter bolts to 31 ft. lbs. (42Nm).

3.3L Engine

1. Before servicing the vehicle, refer to the precautions in the beginning of this section.
2. Remove the battery.
3. Remove the battery tray.
4. Remove the wiring from the starter.
5. Remove the 2 bolts and lower the starter.
6. Installation is the reverse of removal. Torque the starter bolts to 27 ft. lbs. (37 Nm).

Oil Pan

REMOVAL & INSTALLATION

2.4L Engine

1. Before servicing the vehicle, refer to the precautions in the beginning of this section.

2. Remove or disconnect the following:

- Engine undercover
- Engine oil
- Oil pan bolts, nuts and pan

3. Installation is the reverse of removal. Torque the 12 bolts and 2 nuts to 80 inch lbs. (9Nm).

3.0L Engine

1. Before servicing the vehicle, refer to the precautions in the beginning of this section.

2. Remove or disconnect the following:

- Right front wheel
- Fender apron seal
- Engine undercover

3. Drain the engine oil from the engine.

4. Remove or disconnect the following:

- Front exhaust pipe
- Front exhaust pipe bracket from the No. 1 oil pan
- Flywheel housing undercover
- 10 bolts and 2 nuts to the No. 2 oil pan

5. Insert the blade of the Oil Pan Seal Cutting tool 09032-00100 between the No. 1 and No. 2 oil pans. Clean the surfaces of the oil pans.

6. Remove or disconnect the following:

- 3 oil strainer nuts and gasket

7. Remove the No. 1 oil pan, as follows:

- 2 bolts and the flywheel housing undercover
- 17 bolts and 2 nuts to the No. 1 oil pan

➡Make a note of the position of the each bolt. When replacing the bolts into the oil pan, place each bolt in the position from which it was removed.

- Oil pan, by prying the portions between the cylinder block and the oil pan

➡Be careful not to damage the contact surfaces.

- Baffle plate from the No. 1 oil pan

To install:

8. Clean all mating surfaces of the oil pans.

9. Install the baffle plate to the No. 1 oil pan and tighten to 69 inch lbs. (8 Nm).

10. Install the No. 1 oil pan, as follows:

a. Using a non residue solvent, clean both sealing surfaces to the oil pan.

b. Apply liquid sealant to the oil pan and engine block.

c. Install the oil pan with the 17 bolts and 2 nuts. Uniformly tighten the bolts and nuts in several passes.

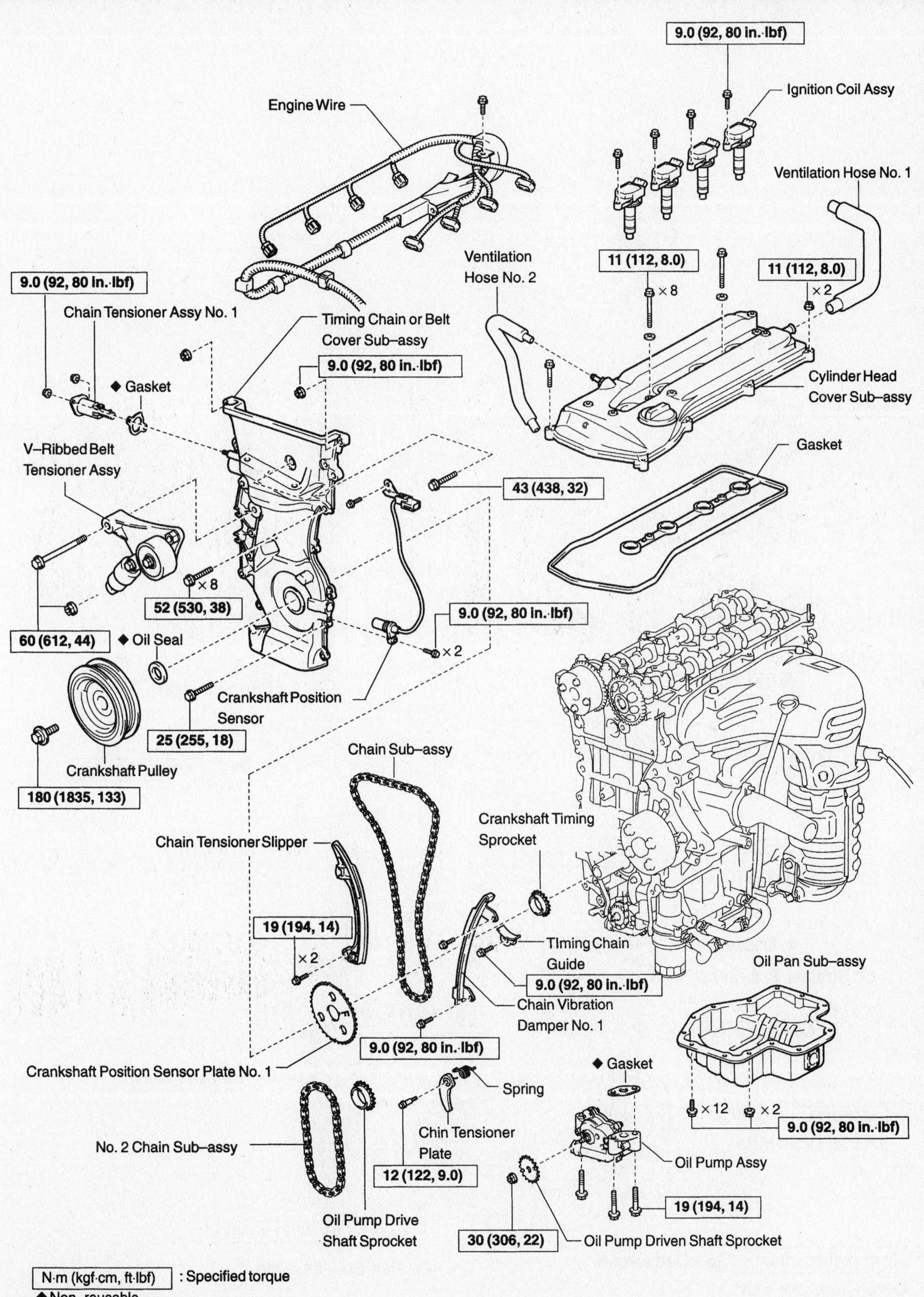
9.0 (92, 80 in.·lbf)
Ignition Coil Assy
Engine Wire
Ventilation Hose No. 1
Ventilation Hose No. 2
11 (112, 8.0)
×8
11 (112, 8.0)
×2
9.0 (92, 80 in.·lbf)
Chain Tensioner Assy No. 1
Timing Chain or Belt Cover Sub–assy
9.0 (92, 80 in.·lbf)
◆ Gasket
Cylinder Head Cover Sub–assy
V–Ribbed Belt Tensioner Assy
Gasket
43 (438, 32)
×8
52 (530, 38)
9.0 (92, 80 in.·lbf)
60 (612, 44)
◆ Oil Seal
×2
Crankshaft Position Sensor
25 (255, 18)
Chain Sub–assy
Crankshaft Pulley
180 (1835, 133)
Crankshaft Timing Sprocket
Chain Tensioner Slipper
19 (194, 14)
×2
TIming Chain Guide
Oil Pan Sub–assy
9.0 (92, 80 in.·lbf)
Chain Vibration Damper No. 1
9.0 (92, 80 in.·lbf)
◆ Gasket
Crankshaft Position Sensor Plate No. 1
Spring
×12
×2
9.0 (92, 80 in.·lbf)
Chin Tensioner Plate
No. 2 Chain Sub–assy
Oil Pump Assy
12 (122, 9.0)
19 (194, 14)
Oil Pump Drive Shaft Sprocket
30 (306, 22)
Oil Pump Driven Shaft Sprocket
N·m (kgf·cm, ft·lbf) : Specified torque
◆ Non–reusable
67170-HIGH-G22

Oil pan, pump and related parts—2.4L engine

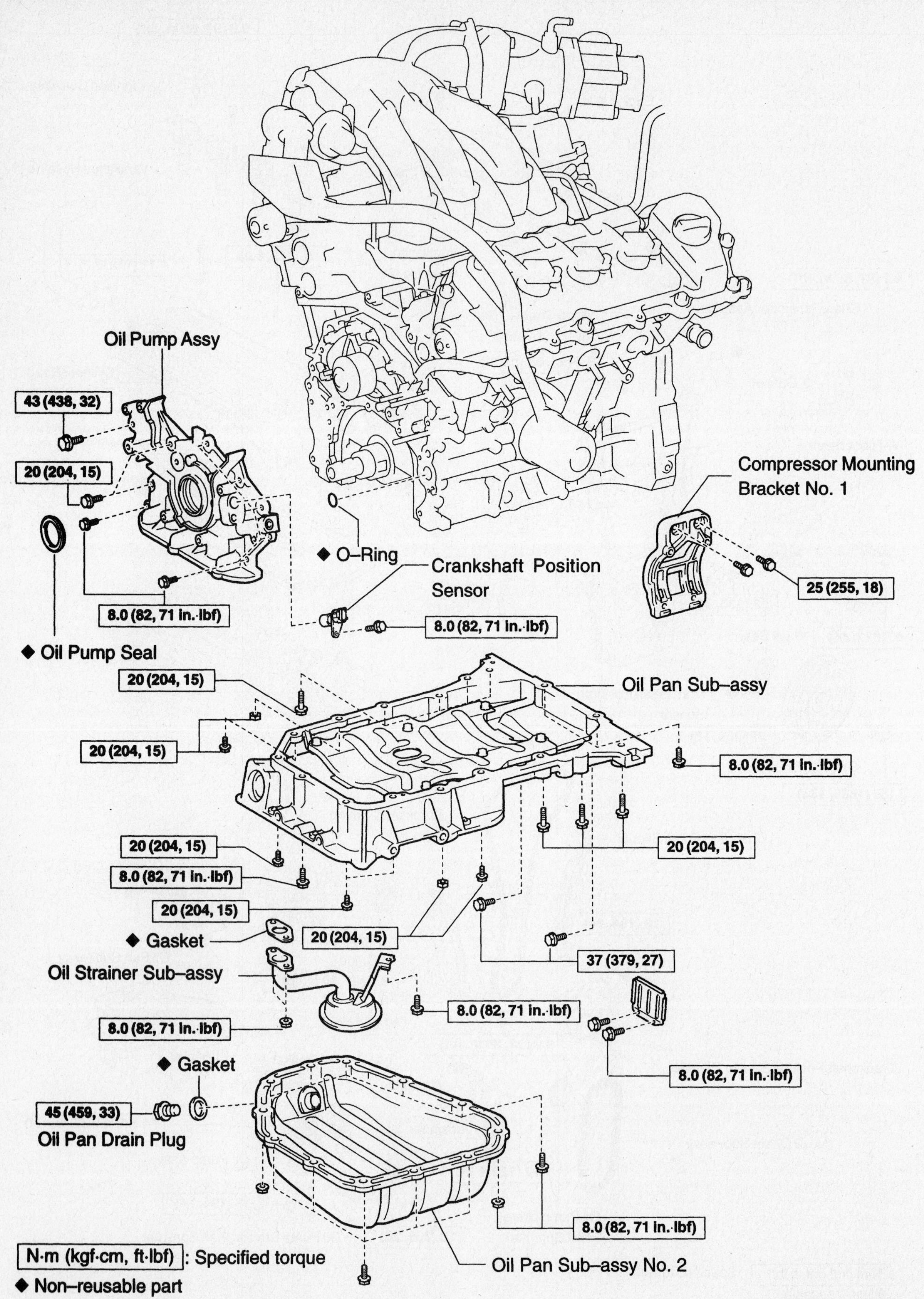

Oil pan, pump and related parts—3.3L engine

d. Tighten the No. 1 oil pan bolts, as follows:

- 10mm head bolt: 69 inch lbs. (8 Nm)
- 12mm head bolt: 14 ft. lbs. (20 Nm)
- 14mm head bolt: 27 ft. lbs. (37 Nm)

e. Install the flywheel housing undercover with the 2 bolts. Tighten the bolts to 69 inch lbs. (8 Nm).

11. Install the oil strainer with the 3 nuts. Tighten the nuts to 69 inch lbs. (8 Nm).
12. Install the No. 2 oil pan, as follows:

a. Using a non residue solvent, clean both sealing surfaces to the oil pan.

b. Apply liquid sealant to the oil pan and engine block.

c. Install the No. 2 oil pan with the 10 bolts and 2 nuts. Uniformly tighten the bolts and nuts in several passes. Tighten the bolts to 69 inch lbs. (8 Nm).

13. Install or connect the following:
    - Flywheel housing undercover
    - Front exhaust pipe bracket to the No. 1 oil pan. Tighten the bolts to 15 ft. lbs. (21 Nm).
14. Install the front exhaust pipe, as follows:
    - Temporarily install the 3 new gaskets and the front exhaust pipe with the 2 bolts and 6 nuts
    - Tighten the 4 exhaust manifolds-to-front exhaust pipe nuts to 46 ft. lbs. (62 Nm).
    - Tighten the both front exhaust pipe-to-center exhaust pipe nuts/bolts to 41 ft. lbs. (56 Nm).
    - Bracket. Tighten both bolts to 14 ft. lbs. (19 Nm).
    - Support stay. Tighten both bolts to 22 ft. lbs. (29 Nm).
15. Install or connect the following:
    - Engine undercover
    - Right fender apron seal
    - Right front wheel
16. Fill the engine with oil.
17. Start the engine and check for leaks.

3.3L Engines

1. Before servicing the vehicle, refer to the precautions in the beginning of this section.
2. Remove or disconnect the following:
    - Right front wheel
    - Fender apron seal
    - Engine undercover
3. Drain the engine oil from the engine.
4. Remove or disconnect the following:
    - Front exhaust pipe
    - Front exhaust pipe bracket from the No. 1 oil pan
    - Flywheel housing undercover
    - 10 bolts and 2 nuts to the No. 2 oil pan
5. Insert the blade of the Oil Pan Seal Cutting tool 09032-00100 between the No. 1 and No. 2 oil pans. Clean the surfaces of the oil pans.
6. Remove or disconnect the following:
    - 3 oil strainer nuts and gasket
7. Remove the No. 1 oil pan, as follows:
    - 2 bolts and the flywheel housing undercover
    - 17 bolts and 2 nuts to the No. 1 oil pan

➡Make a note of the position of the each bolt. When replacing the bolts into the oil pan, place each bolt in the position from which it was removed.

- Oil pan, by prying the portions between the cylinder block and the oil pan

➡Be careful not to damage the contact surfaces.

- Baffle plate from the No. 1 oil pan

To install:

8. Clean all mating surfaces of the oil pans.
9. Install the baffle plate to the No. 1 oil pan and tighten to 69 inch lbs. (8 Nm).
10. Install the No. 1 oil pan, as follows:

a. Using a non residue solvent, clean both sealing surfaces to the oil pan.

b. Apply liquid sealant to the oil pan and engine block.

c. Install the oil pan with the 17 bolts and 2 nuts. Uniformly tighten the bolts and nuts in several passes.

d. Tighten the No. 1 oil pan bolts, as follows:

- 10mm head bolt: 69 inch lbs. (8 Nm)
- 12mm head bolt: 14 ft. lbs. (20 Nm)
- 14mm head bolt: 27 ft. lbs. (37 Nm)

e. Install the flywheel housing undercover with the 2 bolts. Tighten the bolts to 69 inch lbs. (8 Nm).

11. Install the oil strainer with the 3 nuts. Tighten the nuts to 69 inch lbs. (8 Nm).
12. Install the No. 2 oil pan, as follows:

a. Using a non residue solvent, clean both sealing surfaces to the oil pan.

b. Apply liquid sealant to the oil pan and engine block.

c. Install the No. 2 oil pan with the 10 bolts and 2 nuts. Uniformly tighten the bolts and nuts in several passes. Tighten the bolts to 69 inch lbs. (8 Nm).

13. Install or connect the following:
    - Flywheel housing undercover
    - Front exhaust pipe bracket to the No. 1 oil pan. Tighten the bolts to 15 ft. lbs. (21 Nm).
14. Install the front exhaust pipe, as follows:
    - Temporarily install the 3 new gaskets and the front exhaust pipe with the 2 bolts and 6 nuts
    - Tighten the 4 exhaust manifolds-to-front exhaust pipe nuts to 46 ft. lbs. (62 Nm).
    - Tighten the both front exhaust pipe-to-center exhaust pipe nuts/bolts to 41 ft. lbs. (56 Nm).
    - Bracket. Tighten both bolts to 14 ft. lbs. (19 Nm).
    - Support stay. Tighten both bolts to 22 ft. lbs. (29 Nm).
15. Install or connect the following:
    - Engine undercover
    - Right fender apron seal
    - Right front wheel
16. Fill the engine with oil.
17. Start the engine and check for leaks.

Oil Pump

REMOVAL & INSTALLATION

2.4L Engine

1. Before servicing the vehicle, refer to the precautions in the beginning of this section.
2. Remove or disconnect the following:
    - Timing chain
    - Oil pump
3. Installation is the reverse of removal. Torque the pump bolts to 14 ft. lbs. (19Nm).

3.0L Engine

1. Before servicing the vehicle, refer to the precautions in the beginning of this section.
2. Remove or disconnect the following:
    - Oil pan
    - Crankshaft Position (CKP) sensor
    - 9 oil pump bolts

➡Make a note of the position of the each bolt. When replacing the bolts into the oil pump body, place each bolt in the position from which it was removed.

- Oil pump body, by prying between the oil pump and main bearing cap
- O-ring from the cylinder block
- Plug, gasket, spring and relief valve from the oil pump body

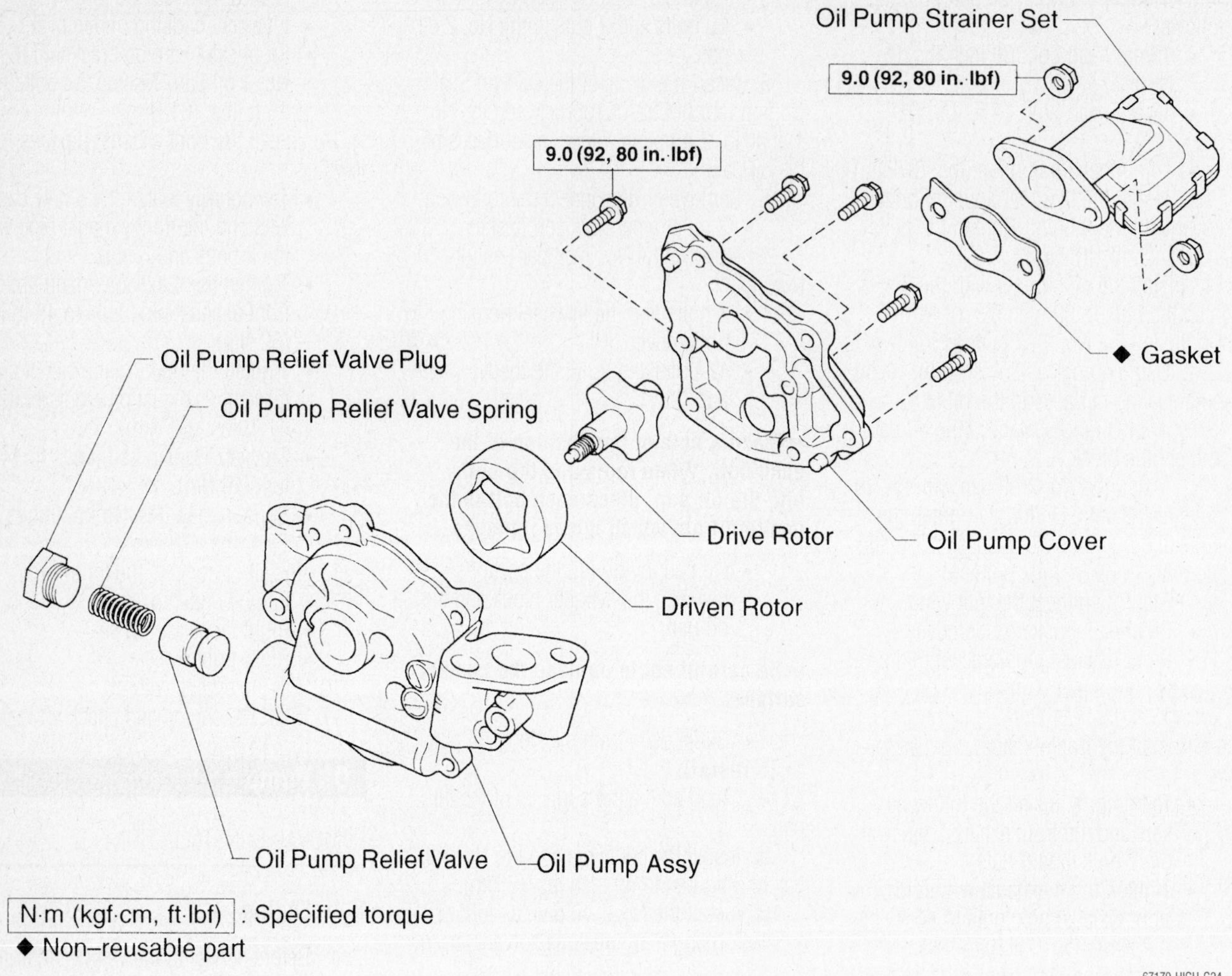

Oil pump exploded view—2.4L engine

- 9 screws, pump body cover, drive and driven rotors

To install:

3. Install or connect the following:
   - Driven rotors, drive, pump body cover, using the 9 screws
   - Oil pump relief valve, spring, gasket and the plug to the oil pump body
   - New O-ring on the cylinder block
4. Using a non residue solvent, clean both sealing surfaces to the oil pump.
5. Apply liquid sealant to the oil pump and engine block.
6. Install or connect the following:
   - Oil pump

➡Be sure to engage the splined teeth of the oil pump drive gear with the large teeth of the crankshaft.

- 9 oil pump bolts. Tighten the bolts in several passes to 69 inch lbs. (8 Nm), for 10mm or to 14 ft. lbs. (20 Nm), for 12mm.
- CKP sensor. Tighten the bolt to 69 inch lbs. (8 Nm).
- Baffle plate to the No. oil pan. Tighten to 69 inch lbs. (8 Nm).
- No. 1 oil pan, oil strainer and No. 2 oil pan

7. Refill the engine with oil.
8. Start the engine and inspect for leaks.
9. Recheck the engine oil level.

3.3L Engines

1. Before servicing the vehicle, refer to the precautions in the beginning of this section.
2. Remove or disconnect the following:
   - Oil pan
   - Crankshaft Position (CKP) sensor
   - 9 oil pump bolts

➡Make a note of the position of the each bolt. When replacing the bolts into the oil pump body, place each bolt in the position from which it was removed.

- Oil pump body, by prying between the oil pump and main bearing cap
- O-ring from the cylinder block
- Plug, gasket, spring and relief valve from the oil pump body
- 9 screws, pump body cover, drive and driven rotors

To install:

3. Install or connect the following:
   - Driven rotors, drive, pump body cover, using the 9 screws
   - Oil pump relief valve, spring, gasket and the plug to the oil pump body
   - New O-ring on the cylinder block
4. Using a non residue solvent, clean both sealing surfaces to the oil pump.
5. Apply liquid sealant to the oil pump and engine block.
6. Install or connect the following:
   - Oil pump

➡Be sure to engage the splined teeth of the oil pump drive gear with the large teeth of the crankshaft.

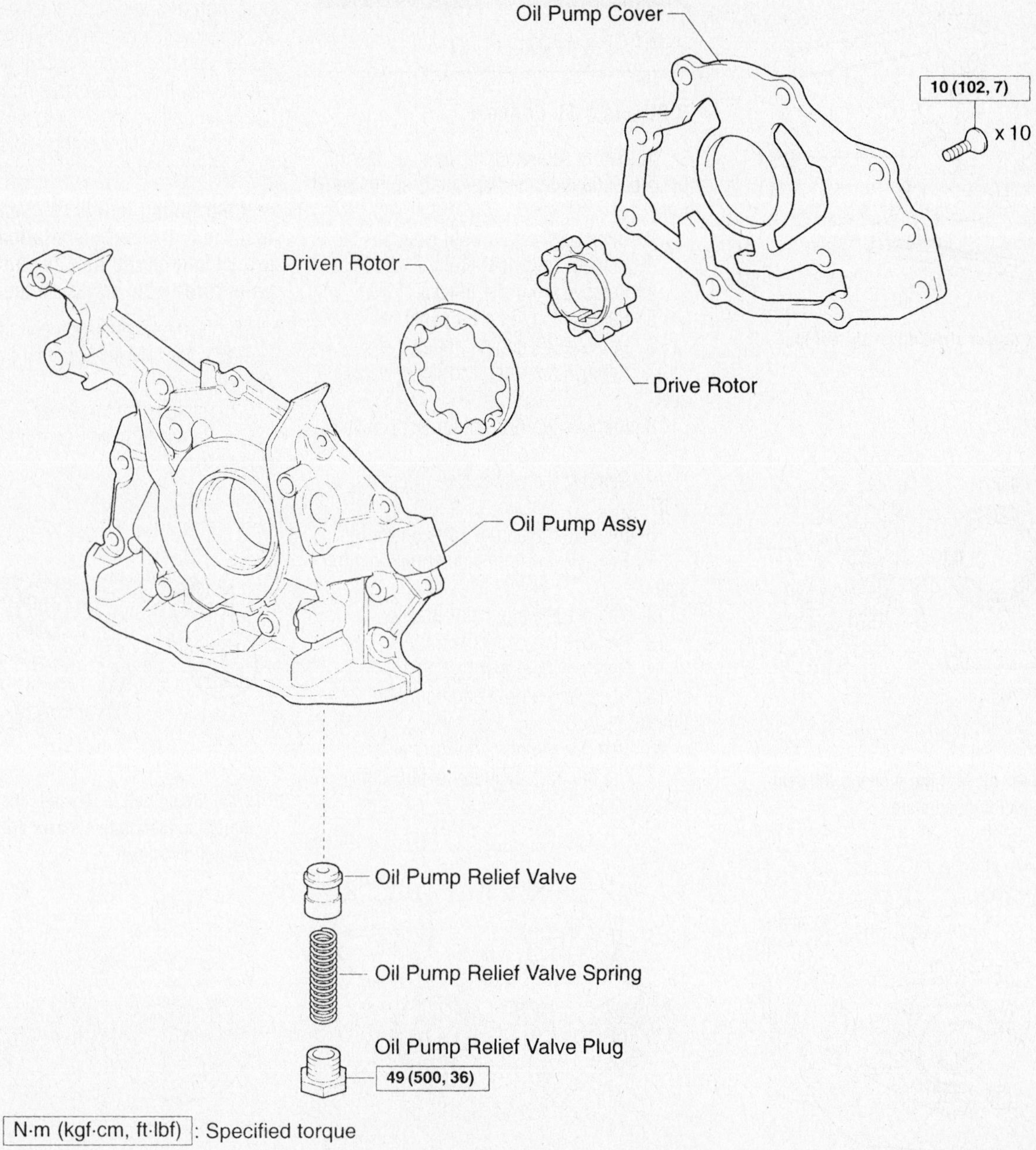

Oil Pump Cover

10 (102, 7)

x 10

Driven Rotor

Drive Rotor

Oil Pump Assy

Oil Pump Relief Valve

Oil Pump Relief Valve Spring

Oil Pump Relief Valve Plug

49 (500, 36)

N·m (kgf·cm, ft·lbf) : Specified torque

67170-HIGH-G25

Oil pump exploded view—3.3L engine

- 9 oil pump bolts. Tighten the bolts in several passes to 69 inch lbs. (8 Nm), for 10mm; 14 ft. lbs. (20 Nm), for 12mm; 32 ft. lbs. (43 Nm) for 14mm
- CKP sensor. Tighten the bolt to 69 inch lbs. (8 Nm).
- Baffle plate to the No. oil pan. Tighten to 69 inch lbs. (8 Nm).
- No. 1 oil pan, oil strainer and No. 2 oil pan

7. Refill the engine with oil.
8. Start the engine and inspect for leaks.
9. Recheck the engine oil level.

Rear Main Seal

REMOVAL & INSTALLATION

All Engines

If the rear oil seal retainer is not installed to the block, use a tapered ended screwdriver and hammer to remove the oil seal. Apply multi-purpose grease to the new oil seal lip. Using a seal driver, tap the seal into place. Be careful not to install it slantwise.

1. Before servicing the vehicle, refer to

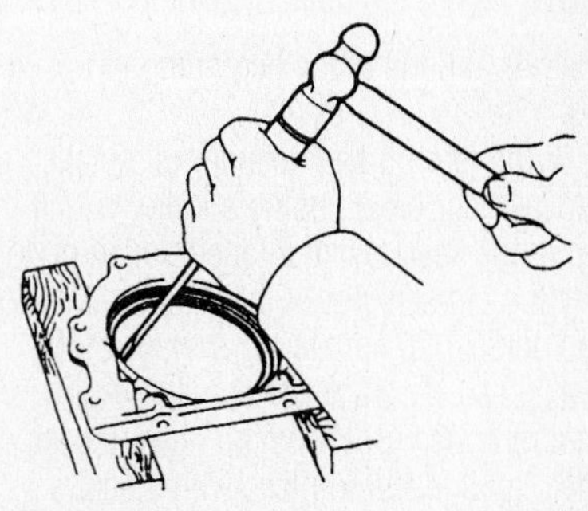

Carefully tap the old seal from the retainer

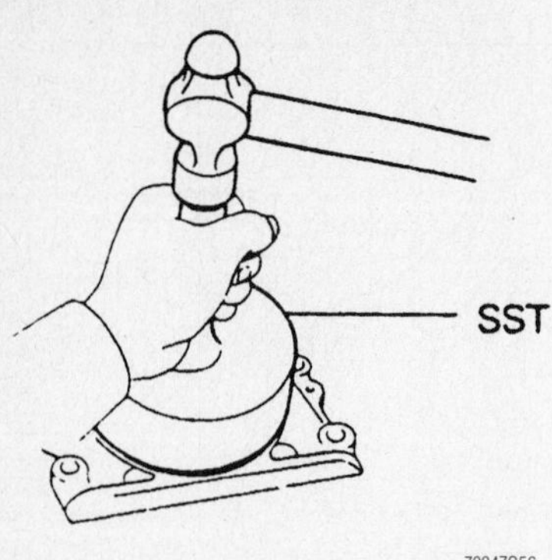

Use the proper sized driver to seat the seal

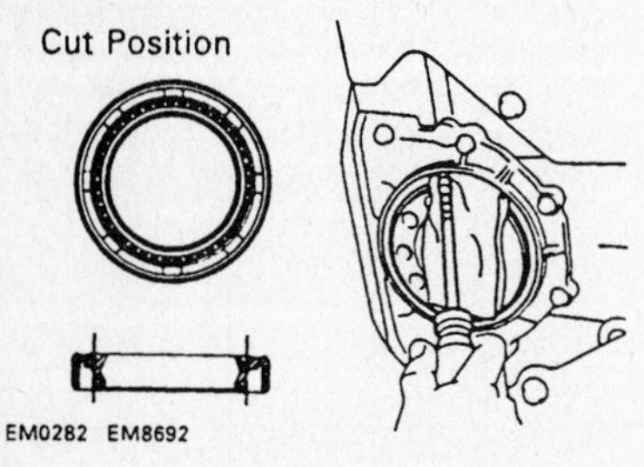

Cut off the oil seal lip, then pry the seal out of the retaining plate

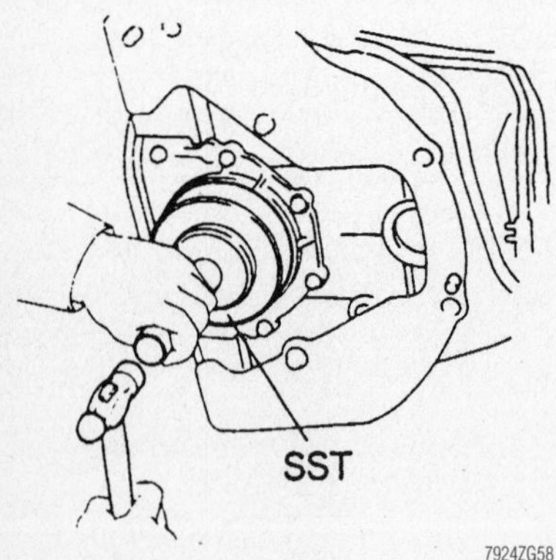

Tap a new seal into place

the precautions in the beginning of this section.

If the rear oil seal retainer is installed on the cylinder block, using a knife, cut off the lip of the seal. Using a taped ended prytool, pry the old seal out of the retainer. Inspect the oil seal lip contacting surface of the crankshaft for cracks or damage. Apply multipurpose grease to the new oil seal, then tap the seal in place with a seal installer. Be careful not to install the seal slantwise.

Timing Belt

REMOVAL & INSTALLATION

3.0L and 3.3L Engines

1. Before servicing the vehicle, refer to the precautions in the beginning of this section.
2. Remove the right front wheel.
3. Remove the wiper arms.
4. Remove the wiper linkage.
5. Remove the top cowl panel.
6. Remove the engine undercovers.
7. Remove the right front fender apron seal.
8. Remove the A/C compressor drive belt.
9. Remove the power steering pump belt.
10. Remove the engine roll control rod.
11. Remove the right side engine mount stay.
12. Remove the alternator bracket.
13. Remove the crankshaft pulley.
14. Remove the upper belt cover.
15. Remove the right engine mount bracket.
16. Remove the no. 2 timing belt guide.
17. Set the no.1 cylinder to TDC compression.

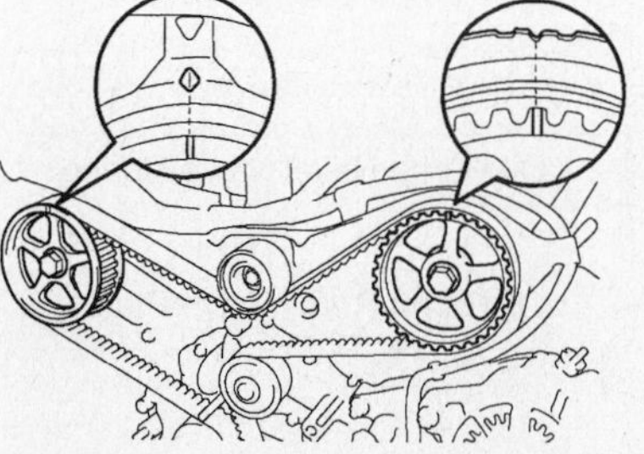

Check that the timing marks on the camshaft pulleys are aligned with the notches on the inner belt cover

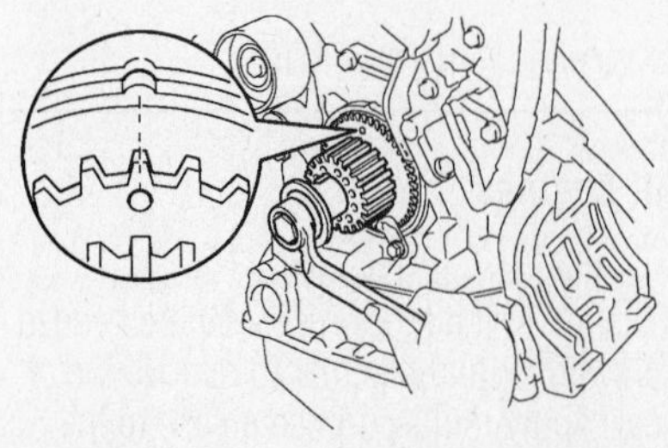

Turn the crankshaft clockwise to align the timing mark on the crankshaft timing pulley with the notch in the oil pump body

18. Temporarily install the crank pulley bolt. Turn the crankshaft clockwise to align the timing mark on the crankshaft timing pulley with the notch in the oil pump body. Check that the timing marks on the camshaft pulleys are aligned with the notches on the inner belt cover. If not, rotate the crankshaft 360 degrees clockwise.

➡If the timing belt is re-used, check that the 3 original installation marks are visible on the belt as shown. If not, paint three new marks on the belt.

19. Turn the crankshaft counterclockwise by 60 degrees. Make sure that the belt is still engaged.
20. Remove the tensioner.
21. Remove the belt from the pulleys in this order:

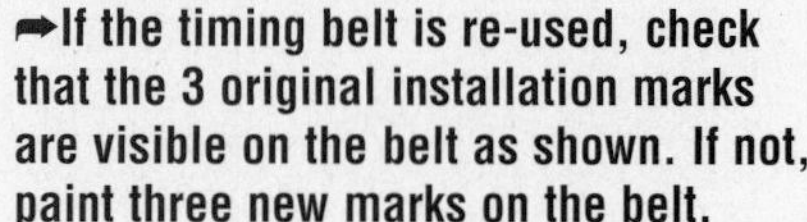
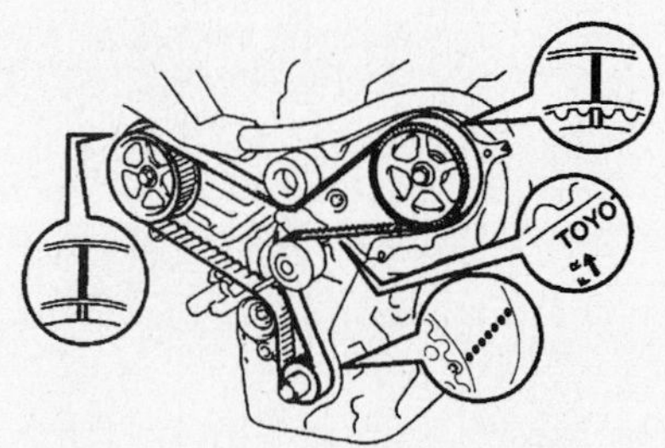

If the timing belt is re-used, check that the 3 original installation marks are visible on the belt as shown

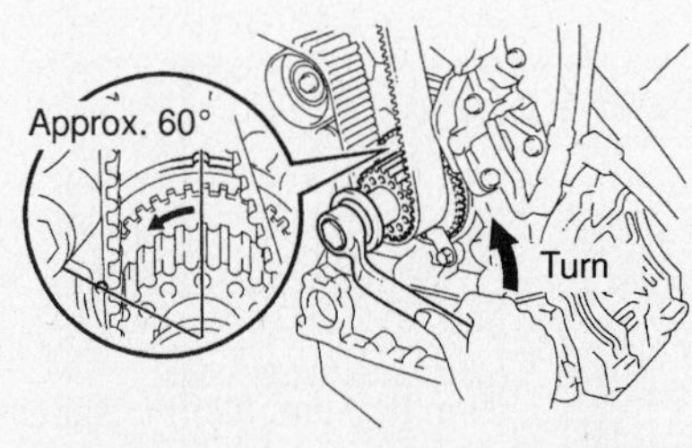

Turn the crankshaft counterclockwise by 60 degrees

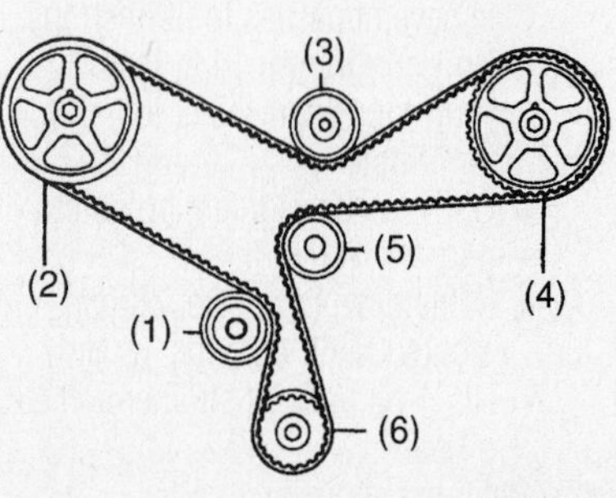

Remove the belt from the pulleys in this order

- Lower idler pulley
- Right camshaft pulley
- Upper idler pulley
- Left camshaft pulley
- Water pump pulley
- Crankshaft timing pulley

22. If the belt is being re-used, check it for wear or damage; don't twist it or turn it inside-out. If there is any doubt as to its condition, replace it.

To install:

23. Clean all the pulleys.
24. Turn the crankshaft another 60 degrees counterclockwise.
25. Turn the camshaft pulleys back into alignment so the marks align with the notches on the inner cover.
26. Turn the crankshaft back so that the timing mark aligns with the notch on the oil pump.
27. Align the installation marks on the belt with the timing marks on the pulleys.
28. Install the belt in this order:
    - Crankshaft
    - Water pump
    - Left camshaft
    - Upper idler
    - Right camshaft
    - Lower idler
29. Set the tensioner in a press and collapse the plunger. Do not apply more that 2,205 lbs (9.8 kN) of force. Insert a suitable metal rod through the holes to hold the plunger in position.
30. Install the tensioner and torque the 2 bolts alternately to 20 ft. lbs. (27 Nm).

⁂ WARNING

Be sure to tighten to bolts alternately and evenly so the tensioner seats flat.

31. Remove the metal rod from the tensioner.
32. Turn the crankshaft 2 full revolutions clockwise (720 degrees), and align the timing mark on the crank pulley with the notch on the oil pump.
33. Check the timing marks on the camshaft pulleys for alignment with the notches on the inner cover. If they do not align, remove the belt and align the mismatched mark(s).
34. The remainder of installation is the reverse of removal. Observe the following torques:
    - Right engine mount bracket: 21 ft. lbs. (28 Nm)
    - Right engine mount insulator: 70 ft. lbs. (95 Nm)
    - Timing belt covers: 75 inch lbs. (8.5 Nm)
    - Crankshaft pulley: 162 ft. lbs. (220 Nm)
    - Alternator bracket: 21 ft. lbs. (28 Nm)
    - Right engine mount stay: 47 ft. lbs. (64 Nm)
    - Engine roll control rod: tighten first A, then B, then C to 47 ft. lbs. (64 Nm). Torque D to 17 ft. lbs. (23 Nm)

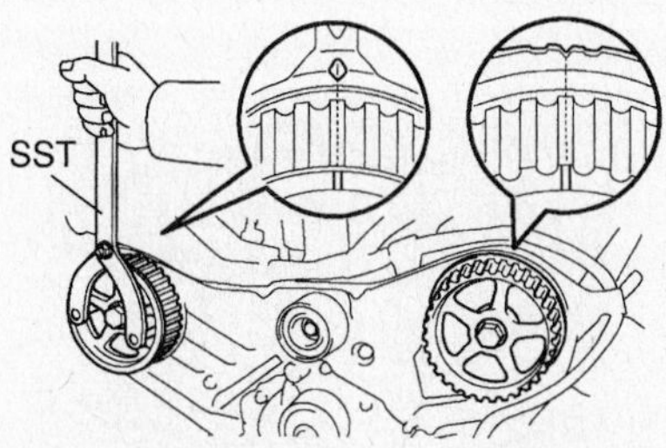

Turn the camshaft pulleys back into alignment so the marks align with the notches on the inner cover

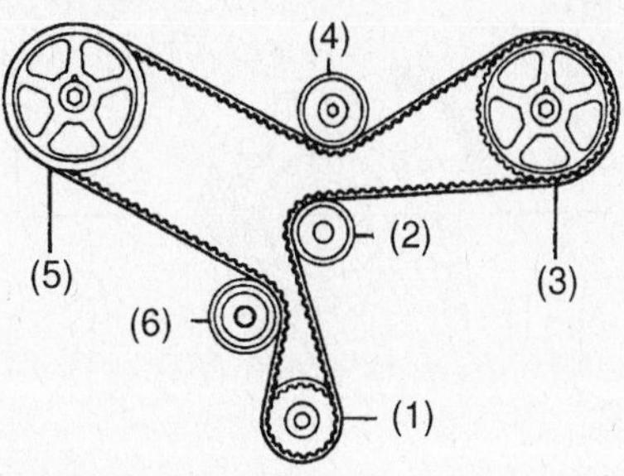

Install the belt in this order

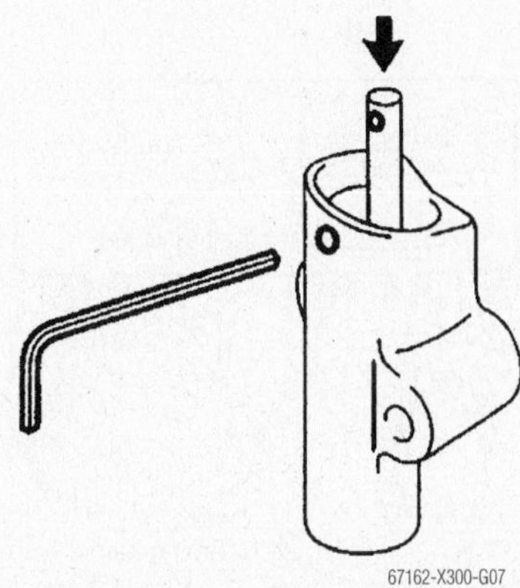

Set the tensioner in a press and collapse the plunger. Do not apply more that 2,205 lbs (9.8 kN) of force. Insert a suitable metal rod through the holes to hold the plunger in position

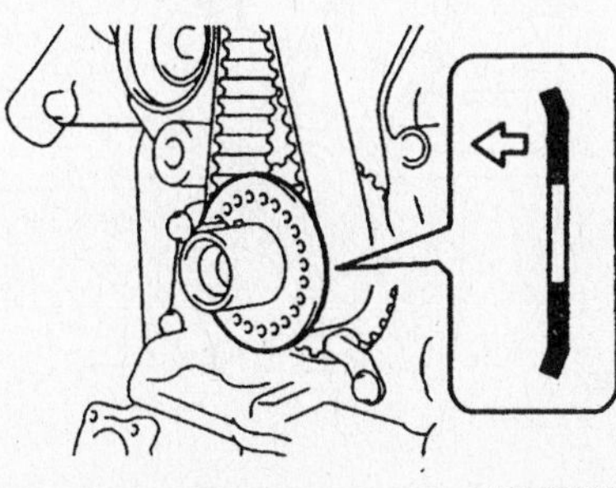

Install the timing belt guide with the cupped side facing front

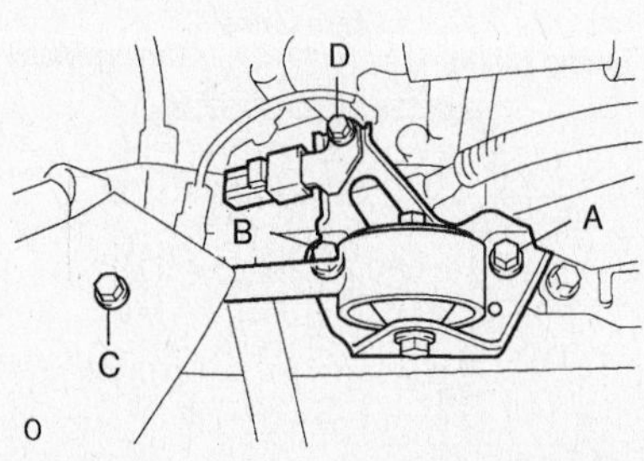

Tighten the engine roll control rod bolts in this order

Timing Chain, Sprockets, Front Cover and Seal

REMOVAL & INSTALLATION

2.4L Engine

1. Before servicing the vehicle, refer to the precautions in the beginning of this section.
2. Disconnect the negative battery cable.
3. Drain the engine coolant.
4. Remove or disconnect the following:
    - Hood
    - Engine oil
    - Right front wheel
    - Right fender splash shield
    - Right fender apron seal
    - No.1 engine undercover
    - Engine roll stopper and bracket
    - Exhaust pipe
    - Upper engine mount, right side
    - Accessory drive belts
    - Alternator
    - Power steering pump
5. Set the No.1 piston at TDC compression.
    - Crankshaft pulley
    - Oil pan
    - CKP sensor
    - Chain tensioner assembly No.1
    - V-belt tensioner
6. Take up the weight of the engine with a crane.

7. Remove or disconnect the following:
   - Transverse engine mount insulator
   - Transverse engine mount bracket
   - Timing chain cover (14 bolts and 2 nuts
   - CKP sensor plate
   - Chain tensioner slipper
   - Primary chain vibration damper
   - Primary chain and crankshaft sprocket

8. Turn the crankshaft 90 degrees counterclockwise and align the adjusting hole of the oil pump drive shaft gear with the groove in the pump.

9. Insert a 4mm diameter bar into the hole to lock the gear in position and remove the nut.

10. Remove the bolt, tensioner plate, spring, tensioner oil pump driveshaft gear and the chain.

To install:

11. Turn the crankshaft so that the key is in the left horizontal position.

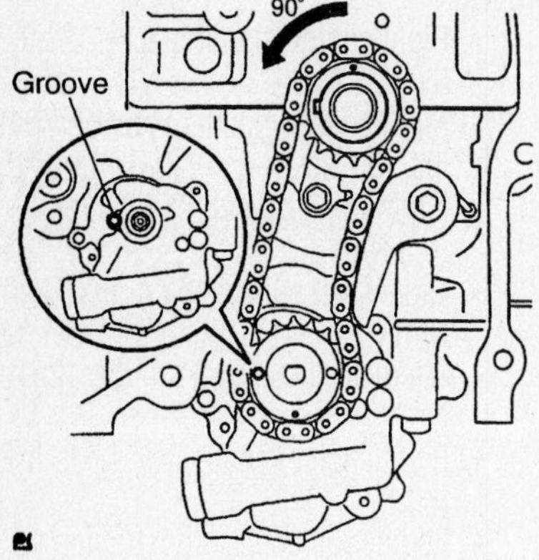

Aligning the adjusting hole and groove—2.4L engine

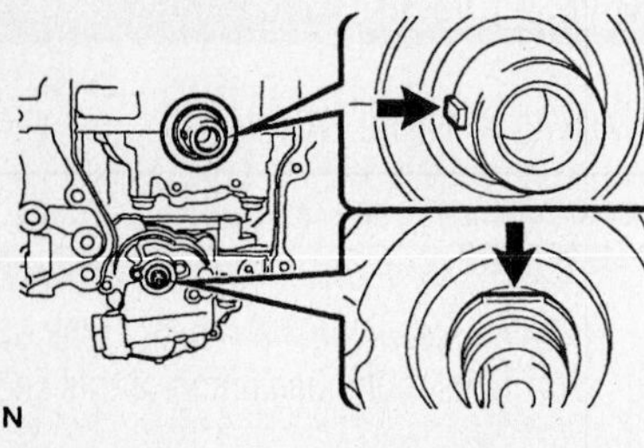

Aligning the crankshaft with the key in the left horizontal position—2.4L engine

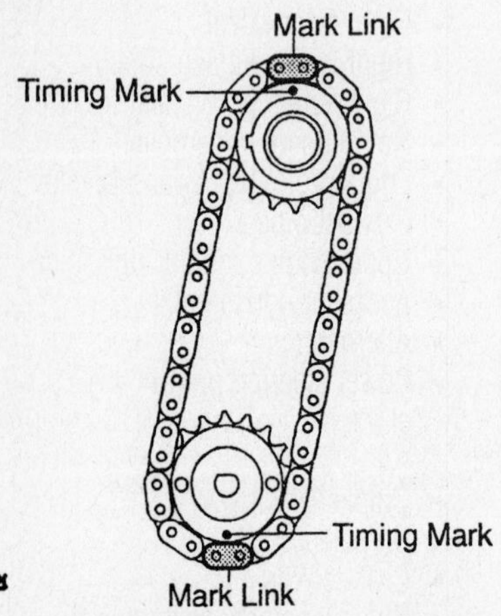

Install the secondary chain and gears with the timing marks aligned as shown—2.4L engine

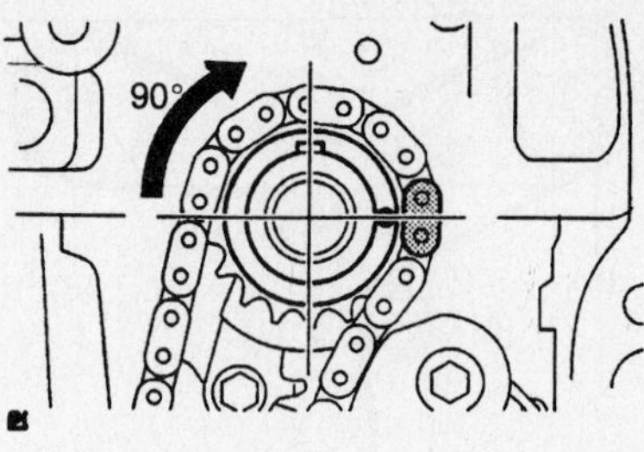

Aligning the crankshaft with the key in the 12 o'clock position—2.4L engine

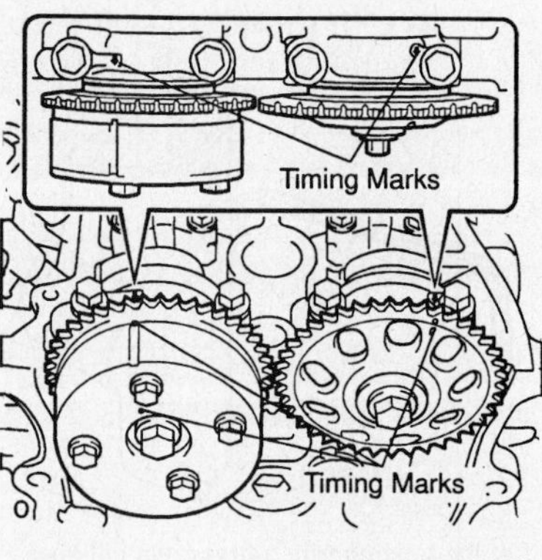

Aligning the timing marks with the No.1 piston at TDC compression—2.4L engine

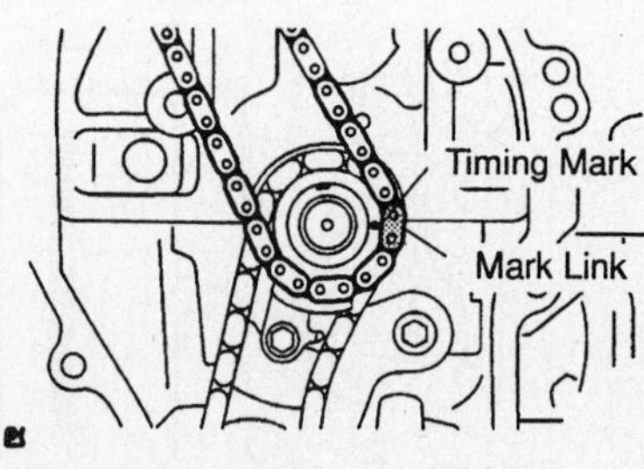

Aligning the timing chain bottom end marks—2.4L engine

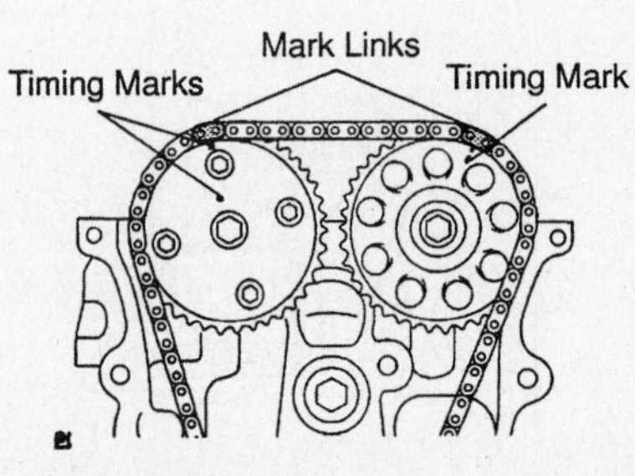

Aligning the timing chain upper end marks—2.4L engine

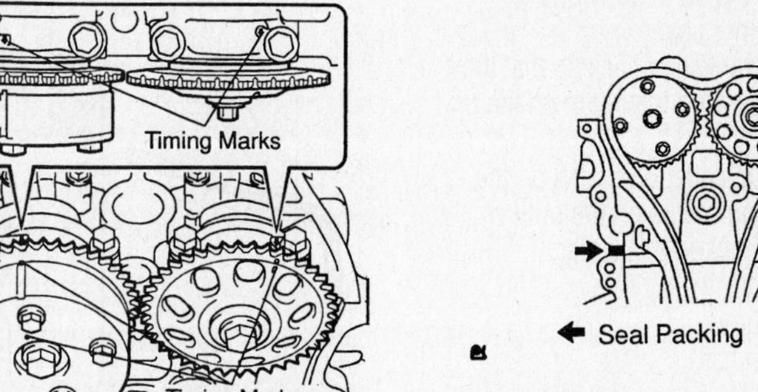

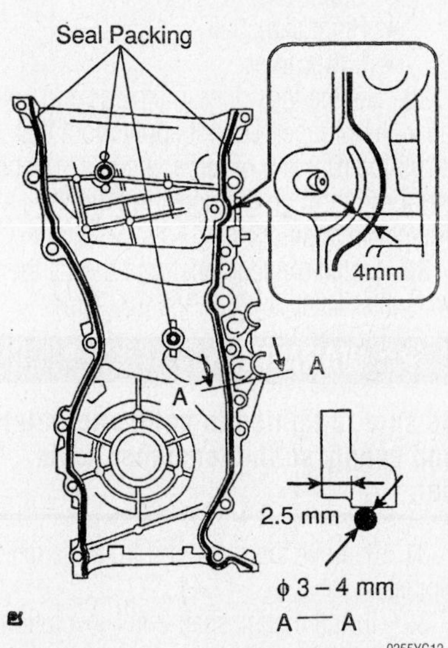

Timing cover sealant application—2.4L engine

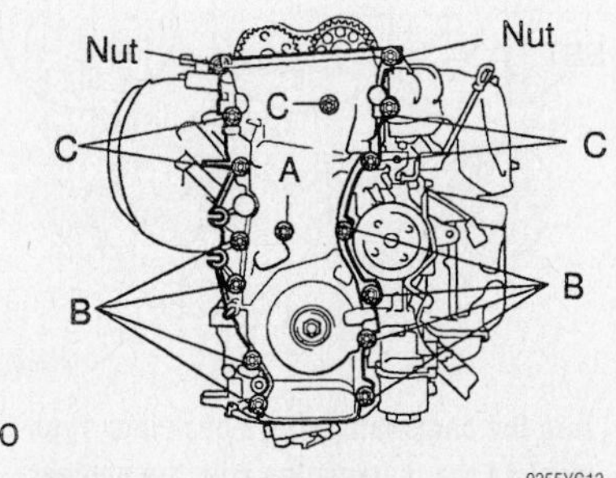

Timing cover bolt positions—2.4L engine

12. Install the secondary chain and gears with the timing marks aligned as shown.

13. Install the damper spring and tensioner plate. Torque the nut to 10 ft. lbs. (13Nm).

14. Align the oil pump adjusting hole and groove, lock it with the bar and torque the nut to 22 ft. lbs. (30Nm).

15. Rotate the crankshaft counterclockwise 90 degrees so the crankshaft key is at the 12 o'clock position and shown.

16. Install the primary chain damper. Torque the bolts to 80 inch lbs. (9Nm).

17. Set the No.1 piston at TDC compression with the timing marks aligned as shown.

18. Turn the crankshaft, using the pulley bolt, until the key is at the 12 o'clock position.

19. Install the bottom end of the chain, with sprocket, so that the colored links are aligned as shown.

20. Align the upper end timing marks as shown and install the chain.

21. Install the tensioner slipper. Torque the bolt to 14 ft. lbs. (19Nm).

22. Install the CKP sensor plate with the **F** mark outwards.

➡When installing the cover, use RTV sealant in the positions shown. The cover must be installed within 3 minutes of seal application. Do not start the engine within 2 hours of seal application.

23. Apply the sealant and install the cover. Torque the cover bolts as follows:

- Bolt A: 80 inch lbs. (9Nm)
- Bolts B: 15 ft. lbs. (21 Nm)
- Bolts C: 32 ft. lbs. (43Nm)
- Nuts: 80 inch lbs. (9Nm)

24. The remainder of installation is the reverse of removal.

Piston and Ring

POSITIONING

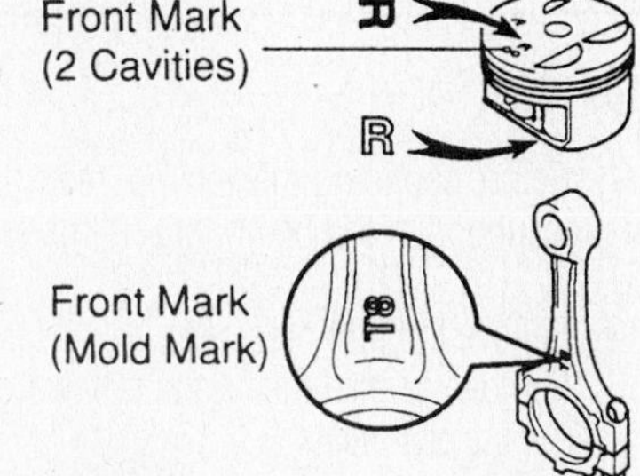

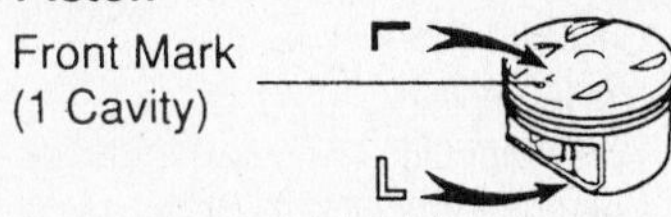

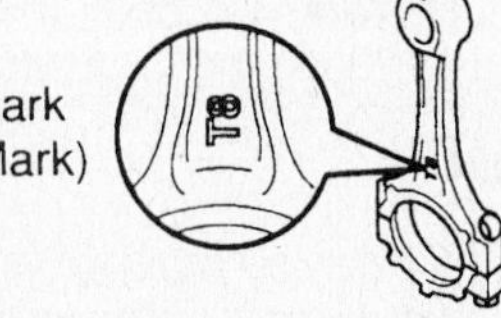

Piston/connecting rod-to-engine positioning–3.0L and 3.3L engines

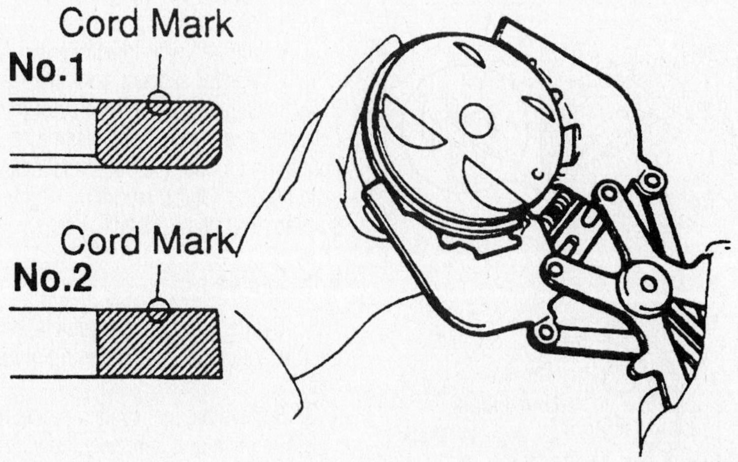

Piston ring positioning–3.0L and 3.3L engines

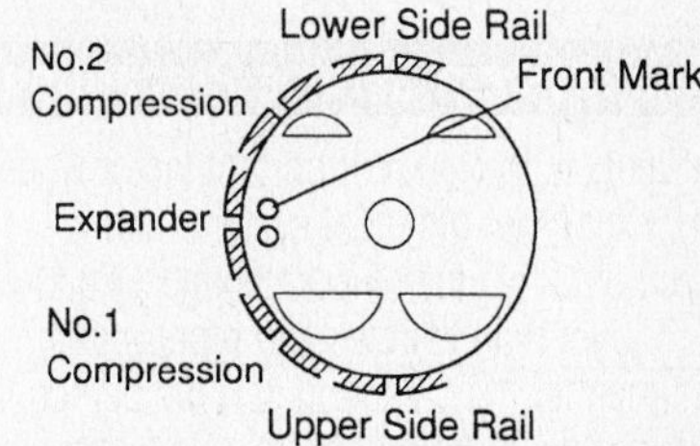

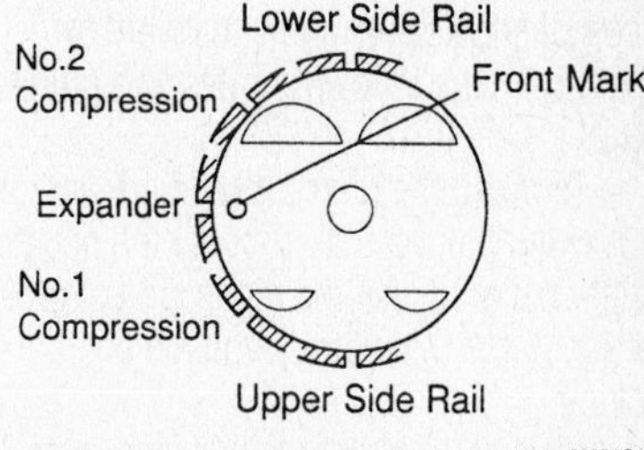

Piston ring identification–3.0L and 3.3L engines

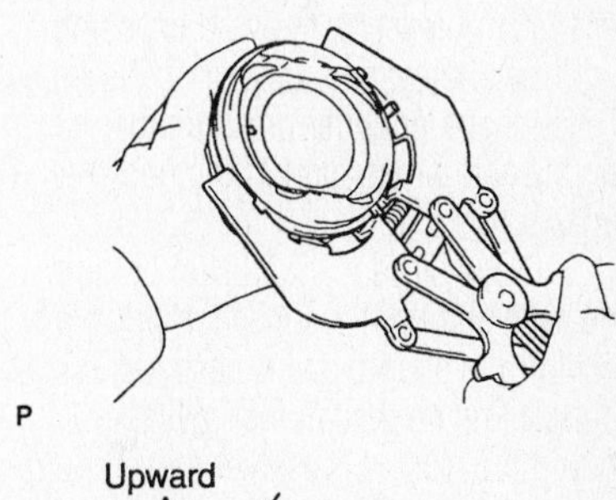

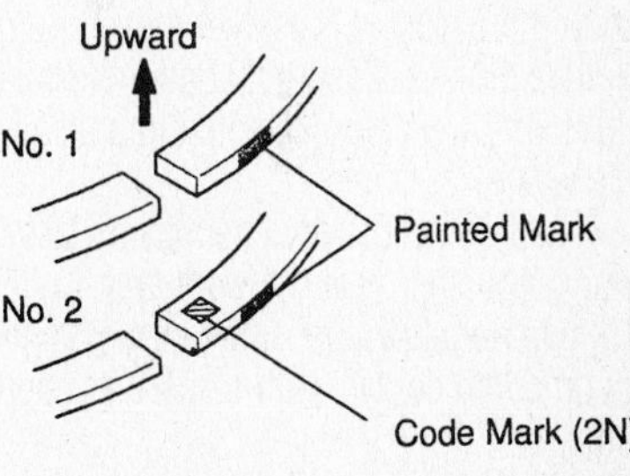

Piston ring identification—2.4L engine

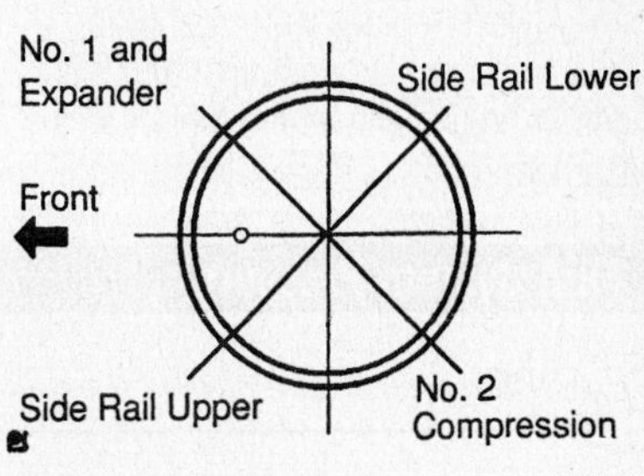

Piston ring installation—2.4L engine

FUEL SYSTEM

Fuel System Service Precautions

Safety is the most important factor when performing not only fuel system maintenance but any type of maintenance. Failure to conduct maintenance and repairs in a safe manner may result in serious personal injury or death. Maintenance and testing of the vehicle's fuel system components can be accomplished safely and effectively by adhering to the following rules and guidelines.

- To avoid the possibility of fire and personal injury, always disconnect the negative battery cable unless the repair or test procedure requires that battery voltage be applied.
- Always relieve the fuel system pressure prior to disconnecting any fuel system component (injector, fuel rail, pressure regulator, etc.), fitting or fuel line connection. Exercise extreme caution whenever relieving fuel system pressure, to avoid exposing skin, face and eyes to fuel spray. Please be advised that fuel under pressure may penetrate the skin or any part of the body that it contacts.
- Always place a shop towel or cloth around the fitting or connection prior to loosening to absorb any excess fuel due to spillage. Ensure that all fuel spillage (should it occur) is quickly removed from engine surfaces. Ensure that all fuel soaked cloths or towels are deposited into a suitable waste container.
- Always keep a dry chemical (Class B) fire extinguisher near the work area.
- Do not allow fuel spray or fuel vapors to come into contact with a spark or open flame.
- Always use a back-up wrench when loosening and tightening fuel line connection fittings. This will prevent unnecessary stress and torsion to fuel line piping.
- Always replace worn fuel fitting O-rings with new. Do not substitute fuel hose or equivalent, where fuel pipe is installed.

Fuel System Pressure

RELIEVING

Disconnect the fuel pump connector at the tank. Start the engine and let it run out of fuel. Turn the switch off. Disconnect the battery ground. Reconnect the fuel pump connector.

Fuel Filter

REMOVAL & INSTALLATION

The fuel filter is in the tank as part of the pump assembly and is not a normal maintenance item.

Fuel Pump

REMOVAL & INSTALLATION

2001–03

1. Before servicing the vehicle, refer to the precautions in the beginning of this section.
2. Relieve the fuel pressure.
3. Remove or disconnect the following:
   - Both rear seats
   - Carpet
   - Service hole cover
   - Connector
   - Joint clip and pull out the fuel tube
   - Vent tube set plate
   - Pump and gauge assembly
4. Installation is the reverse of removal.

2004

1. Before servicing the vehicle, refer to the precautions in the beginning of this section.
2. Relieve the fuel system pressure.
3. Remove the deck board assembly.
4. Remove the rear seats.

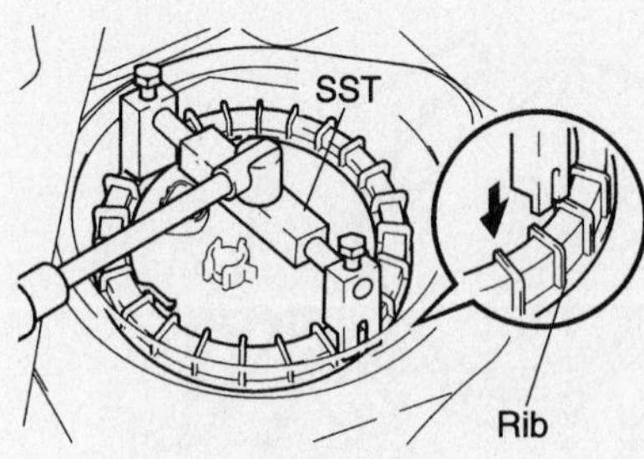

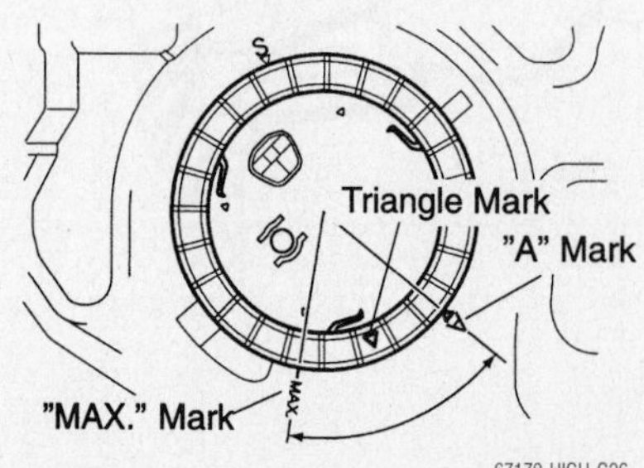

Lock ring positioning during installation

5. Remove the left side door scuff plate.
6. Remove the left side trim cover.
7. Remove the left rear seat side cover.
8. Remove the carpet.
9. Remove the rear floor service cover.
10. Disconnect the fuel pump wiring.
11. Disconnect the fuel hose.
12. Remove the fuel pump locking ring. Special tool 09808-14020 is available for this job.
13. Remove the pump assembly from the tank.
14. Installation is the reverse of removal. When installing the locking ring, tighten it first by hand, then, using the special tool turn the ring 1½ turn. The triangle mark on the lockring must be positioned between the **A** and **MAX** marks on the fuel tank.

**** WARNING**

No other type of tool should be used for this operation.

Fuel Injector

REMOVAL & INSTALLATION

2.4L Engine

1. Before servicing the vehicle, refer to the precautions in the beginning of this section.
2. Relieve the fuel system pressure.
3. Remove or disconnect the following:
   - Air cleaner and hoses
   - Fuel line from the rail
   - Injector connectors
   - Injector wiring harness
   - Fuel rail with injectors
   - Injector spacers from the head
4. Installation is the reverse of removal. Coat the new o-rings with clean fuel. Before tightening the fuel rail bolts, make sure that each injector rotates smoothly. Tighten the bolts to 15 ft. lbs. (20Nm).

3.0L Engine

1. Before servicing the vehicle, refer to the precautions in the beginning of this section.
2. Relieve the fuel system pressure.
3. Remove or disconnect the following:
   - Coolant
   - V-bank cover
   - Battery
   - Air cleaner assembly
   - Upper front suspension brace
   - Intake air surge tank

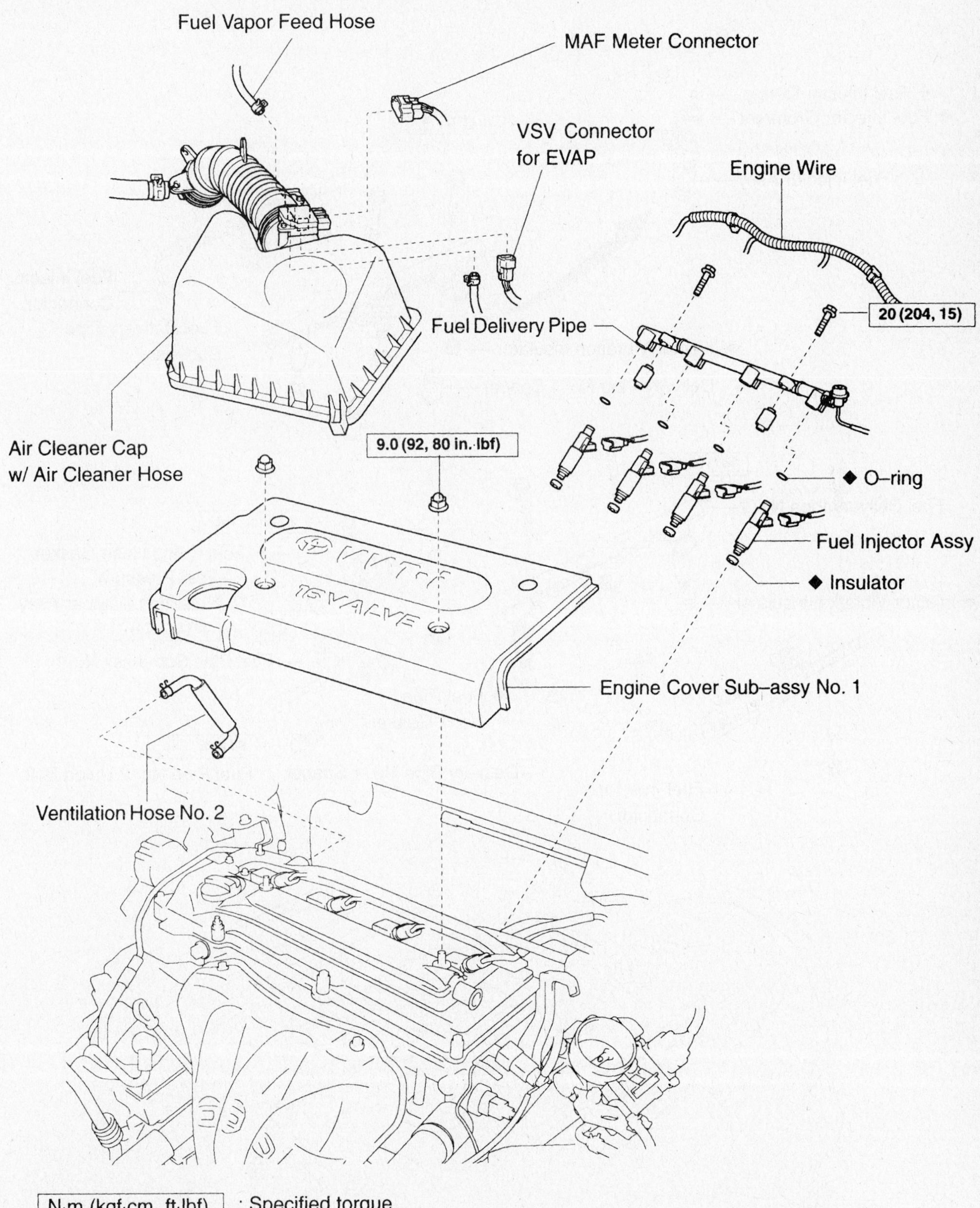

Fuel injectors and related parts—2.4L engine, 2004 model shown

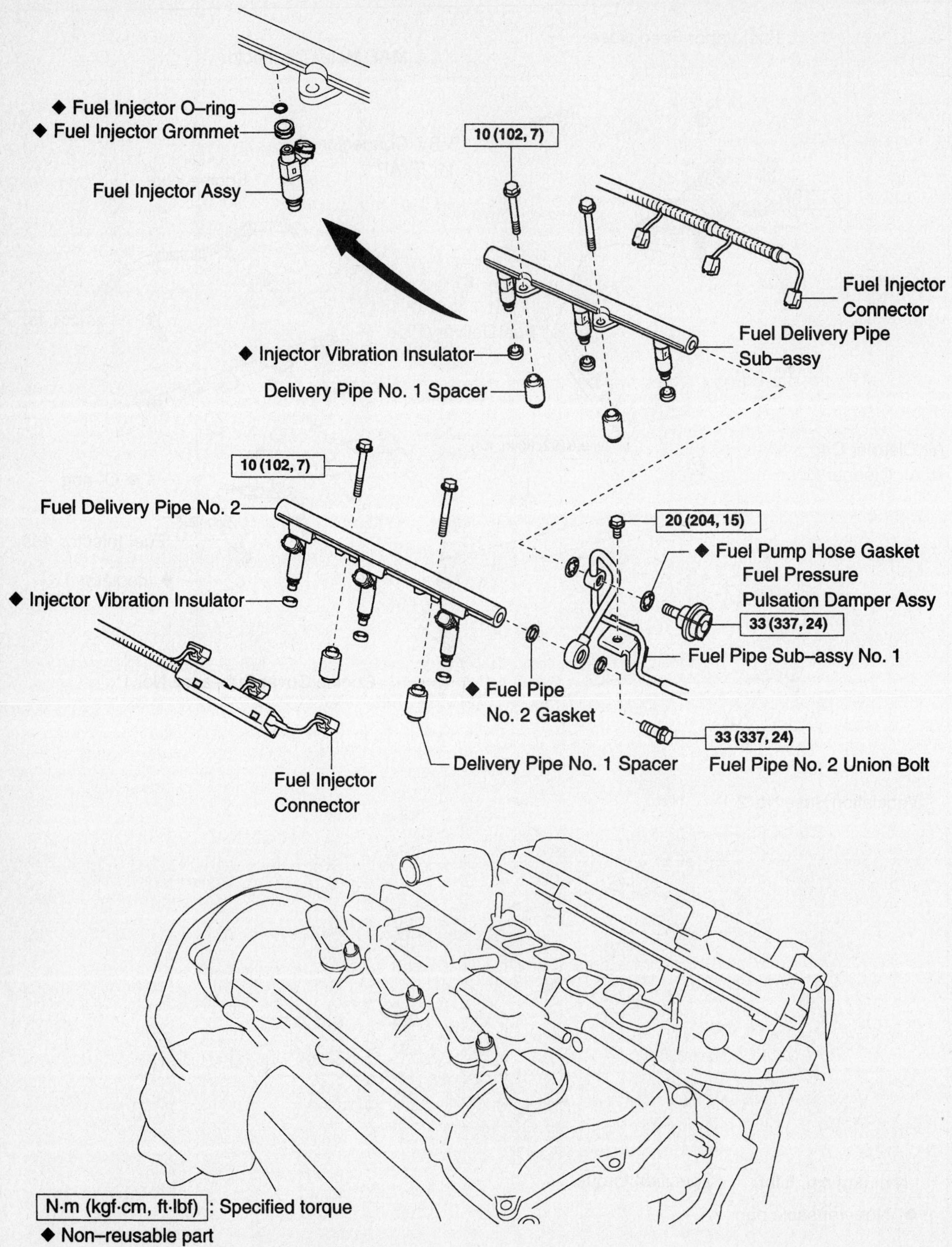

◆ Fuel Injector O–ring
◆ Fuel Injector Grommet
Fuel Injector Assy
10 (102, 7)
Fuel Injector Connector
Fuel Delivery Pipe Sub–assy
◆ Injector Vibration Insulator
Delivery Pipe No. 1 Spacer
10 (102, 7)
Fuel Delivery Pipe No. 2
20 (204, 15)
◆ Fuel Pump Hose Gasket
Fuel Pressure Pulsation Damper Assy
◆ Injector Vibration Insulator
33 (337, 24)
Fuel Pipe Sub–assy No. 1
◆ Fuel Pipe No. 2 Gasket
33 (337, 24)
Delivery Pipe No. 1 Spacer
Fuel Pipe No. 2 Union Bolt
Fuel Injector Connector
N·m (kgf·cm, ft·lbf) : Specified torque
◆ Non–reusable part
67170-HIGH-G28

Fuel injectors and related parts—3.3L engine

- Fuel supply line
- Injector connectors
- Fuel rail with injectors

4. Installation is the reverse of removal. Coat the new o-rings with clean fuel. Before tightening the fuel rail bolts, make sure that each injector rotates smoothly. Tighten the bolts to 84 inch lbs. (10Nm).

3.3L Engine

1. Before servicing the vehicle, refer to the precautions in the beginning of this section.
2. Relieve the fuel system pressure.
3. Drain the coolant.
4. Remove the wiper arms.
5. Remove the wiper linkage.
6. Remove the fender-to-cowl side seals.
7. Remove the rain sensor.
8. Remove the front shock absorber caps (air suspension).
9. Remove the 4 set nuts from the strut (w/o air suspension).
10. Remove the cowl top outer panel.
11. Remove the 6 set nuts from the shock absorber.
12. Remove the V-bank cover.
13. Remove the air cleaner assembly and inlet tubes.
14. Remove the emission control valve set.
15. Remove the upper intake manifold (intake air surge tank). Discard the gasket.
16. Remove the fuel pipe sub-assembly.
17. Disconnect the wiring at the injectors.
18. Remove the 4 bolts and 2 delivery pipe along with the injectors.
19. Remove the delivery pipe spacers and insulators from the manifold.
20. Pull each injector from the pipe.

To install:

21. Install new O-rings on each injector. Apply a light coating of gasoline to the O-rings and mating points on the pipes.
22. Using a twisting motion, install the injectors on the pipes.

➡Be careful to avoid twisting the O-rings. After installation, check that the injectors turn smoothly. If not, use new O-rings.

23. Install the pipes and injectors.
24. Loosely install the bolts and make sure that the injectors still turn freely. If not, replace the O-rings.
25. Torque the bolts to 84 inch lbs. (10 Nm).
26. The remainder of installation is the reverse of removal. Observe the following torques:
    - Fuel line union bolt: 24 ft. lbs. (33 Nm)
    - Pulsation damper: 24 ft. lbs. (33 Nm)
    - Fuel feed pipe: 14 ft. lbs. (20 Nm)
    - Upper intake manifold (air surge tank): 21 ft. lbs. (28 Nm)
    - Upper intake manifold stays: 14 ft. lbs. (20 Nm)

DRIVE TRAIN

Automatic Transaxle Assembly

REMOVAL & INSTALLATION

2.4L Engine

1. Remove the engine/transaxle assembly.
2. Remove the halfshafts.
3. Remove the engine mounting bracket (4wd).
4. Remove the transfer case (4wd).
5. Disconnect the wiring.
6. Remove the starter.
7. Remove the cables and hoses.
8. Remove the filler tube.
9. Remove the front engine mount bracket.
10. Remove the flywheel housing undercover.
11. Turn the crankshaft to gain access to the torque converter bolts. There is one green bolt.
12. Remove the 9 engine-to-transaxle bolts. Separate the transaxle from the engine.
13. Installation is the reverse of removal. Observe the following torques:
    - Transaxle-to-engine: bolts A 47 ft. lbs (64 Nm); bolts B 34 ft. lbs. (46 Nm); bolts C 27 ft. lbs. (37 Nm)
    - Torque converter bolts (use a thread locking compound such as Three Bond 1324): 30 ft. lbs. (41 Nm)

➡Install the green bolt first.

- Undercover: 69 inch lbs. (8 Nm)
- Engine mount bracket: 47 ft. lbs. (64 Nm)
- Transfer case-to-transaxle: 51 ft. lbs. (69 Nm)

3.0L Engine

1. Before servicing the vehicle, refer to the precautions in the beginning of this section.
2. Remove or disconnect the following:
    - Hood
    - Wiper and blade assembly
    - Top cowl seal and panel
    - Window washer hoses, from the ventilator louvers
    - Left and right ventilator louvers
    - Heater air duct
    - Battery and tray
    - Throttle cable
    - Front upper suspension brace
    - Cruise control actuator with its bracket, if equipped
    - Starter
    - Shift control cable
    - Driveshaft, for 4WD
    - Body-to-engine ground strap
    - Park/Neutral Position (PNP) switch, solenoid and ATF temperature connectors
    - 5 upper transaxle-to-engine mounting bolts
    - Front wheel
    - Engine undercover
    - Halfshafts
    - Front exhaust pipe
    - Stabilizer bar
    - Both steering gear mounting bolts and support it in the vehicle
    - Shift control cable from its bracket

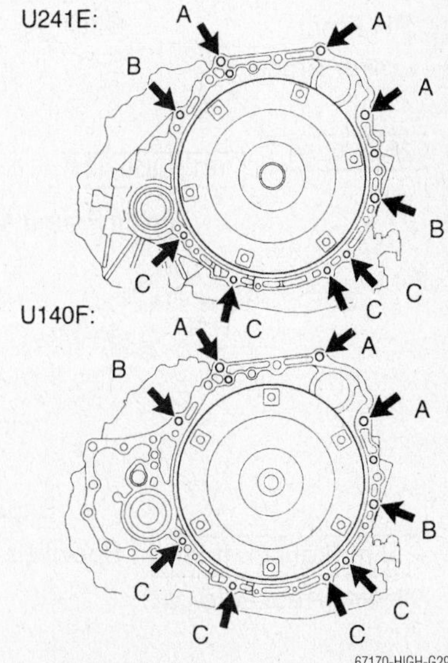

Engine-to-transaxle bolt identification—2.4L engine

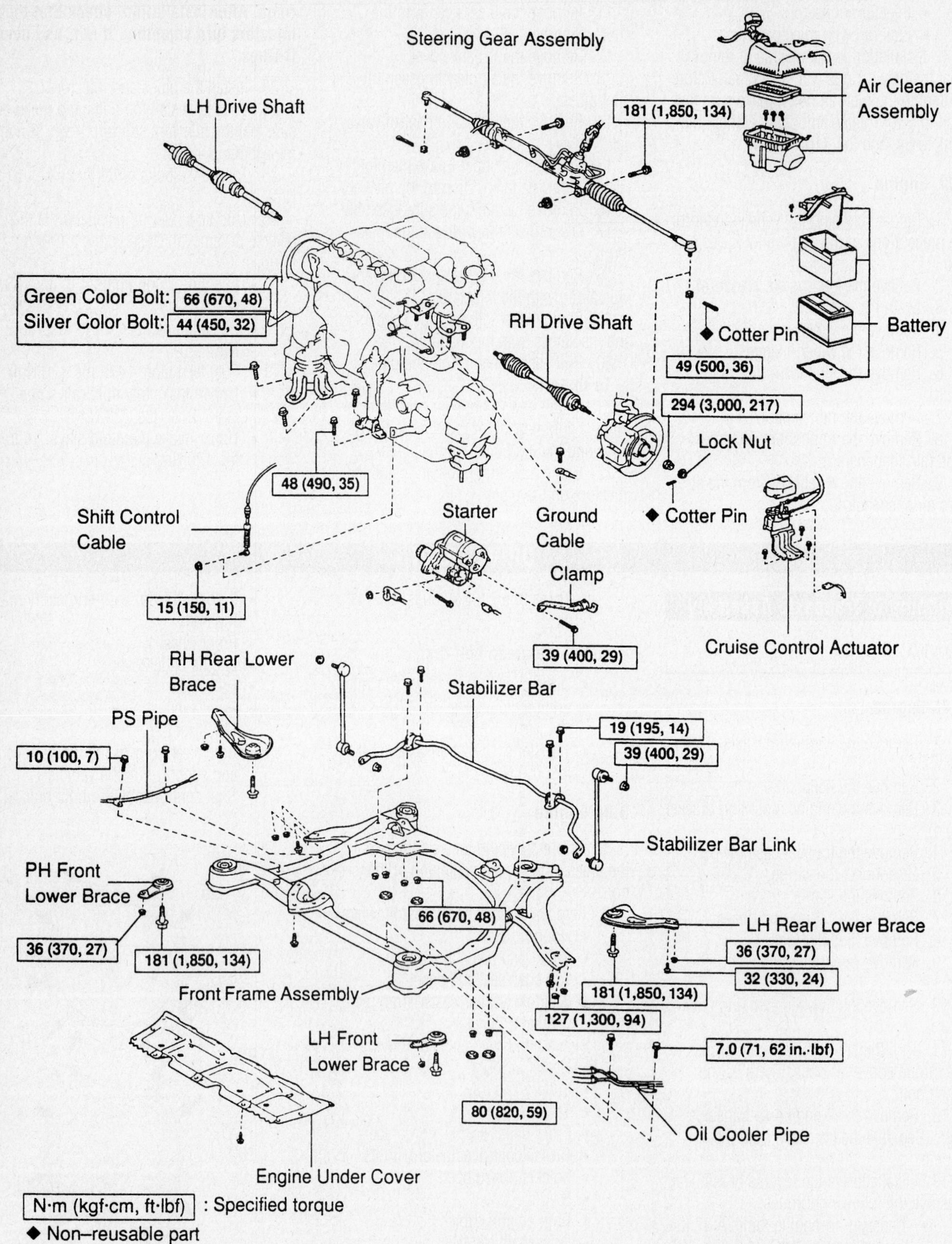

Exploded view of the transaxle removal and installation components—3.0L engine

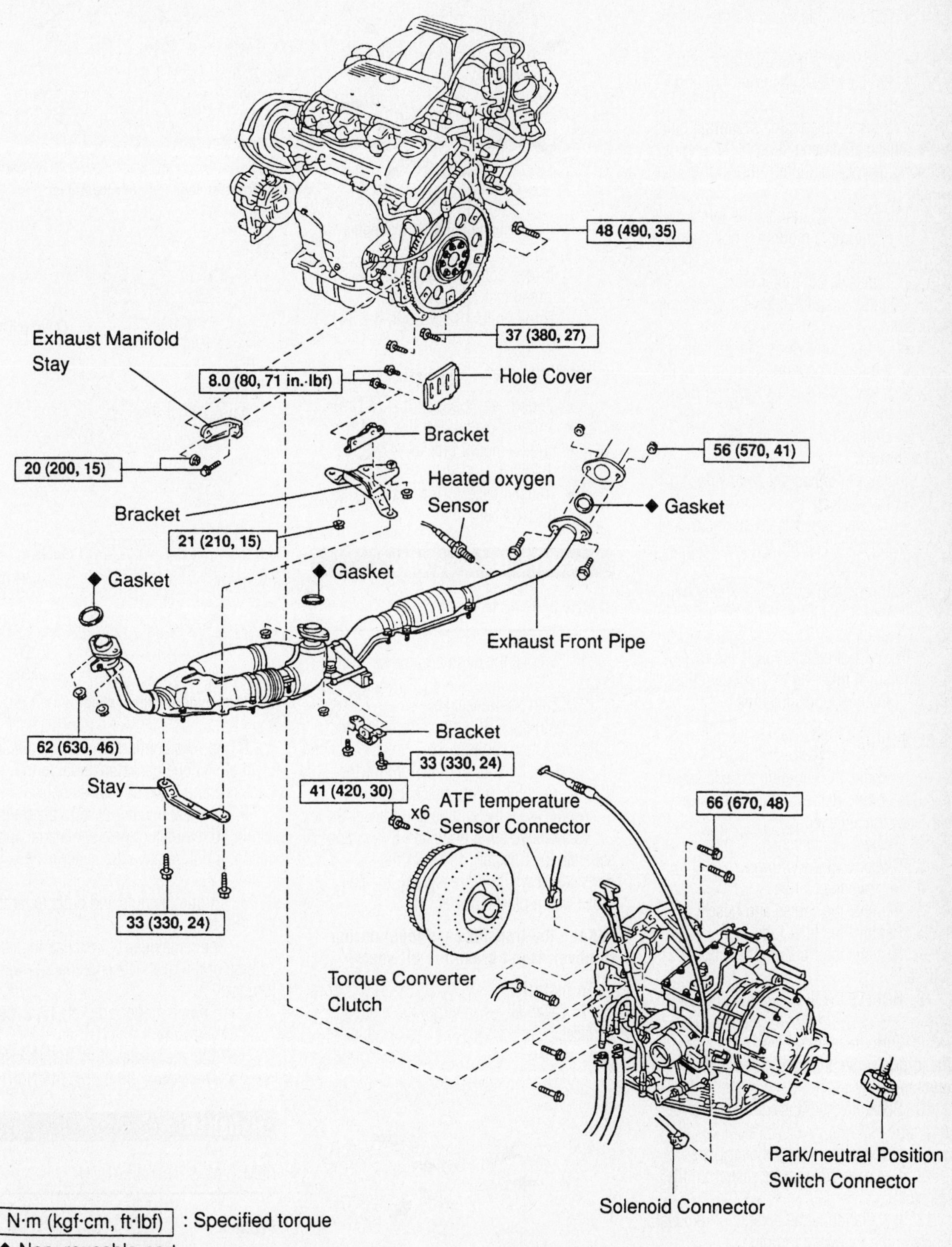

Exploded view of the transaxle removal and installation components—3.0L engine, Cont.

- Power steering pipe and the oil cooler clamps from the frame
- Both left-side transaxle mounting nuts
- Rear-side engine mounting nuts
- Engine shock absorber mounting bolts
- 3 front-side engine mounting bolts

3. Attach an engine sling to the engine hangers in order to support the engine weight.
4. Remove or disconnect the following:
   - Front frame mounting bolts and the frame
   - Transaxle oil cooler lines
5. Support the transaxle with a transmission jack.
   - Torque converter access cover
   - 6 torque converter mounting bolts
   - 3 lower transaxle-to-engine mounting bolts
   - Engine from the transaxle

To install:

6. Install or connect the following:
   - Transaxle
   - 3 lower transaxle-to-engine mounting bolts and tighten to the illustrated value.
   - Torque converter-to-flexplate bolts, starting with the black bolt, then the other 5.
7. The rest of installation is the reverse of the removal referring to the illustrations for the tightening specifications.

3.3L Engine

1. Remove the engine/transaxle assembly. See Engine Removal and Installation, earlier in this chapter.
2. Remove the halfshafts.
3. Disconnect the wiring.
4. Remove the starter.
5. Remove the cables and hoses.
6. Remove the filler tube.
7. Remove the front engine mount bracket.
8. Remove the flywheel housing undercover.
9. Turn the crankshaft to gain access to the torque converter bolts. There is one green bolt.
10. Separate the transaxle from the engine.
11. Remove the right stiffener plate.
12. Separate the transfer case from the transaxle.
13. Installation is the reverse of removal. Observe the following torques:
    - Transaxle-to-engine: bolts A 47 ft. lbs (64 Nm); bolts B 34 ft. lbs. (46 Nm); bolts C 27 ft. lbs. (37 Nm)

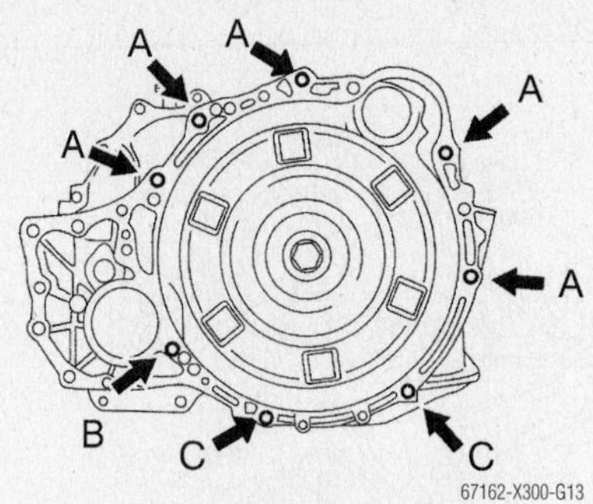

Transaxle-to-engine bolts—3.3L engine

- Torque converter bolts (use a thread locking compound such as Three Bond 1324): 30 ft. lbs. (41 Nm)

➡Install the green bolt first.

- Undercover: 69 inch lbs. (8 Nm)
- Stiffener plate: 25 ft. lbs. (34 Nm)
- Engine mount bracket: 47 ft. lbs. (64 Nm)
- Transfer case-to-transaxle: 51 ft. lbs. (69 Nm)

Transfer Case Assembly

REMOVAL & INSTALLATION

1. Remove the engine/transaxle assembly.
2. Drain the transaxle.
3. With the 2.4L engine, remove the stiffener plate (5 bolts).
4. With the 2.4L engine, remove the right rear engine mount bracket (3 bolts).
5. Separate the engine and transaxle.
6. Remove the 2 bolts and 6 nuts and separate the transfer case from the transaxle. It will be necessary to break it loose with a plastic mallet.

➡Keep the transfer case level during removal. Don't grasp the oil seals.

To install:

7. Clean all grease from the mating surfaces.

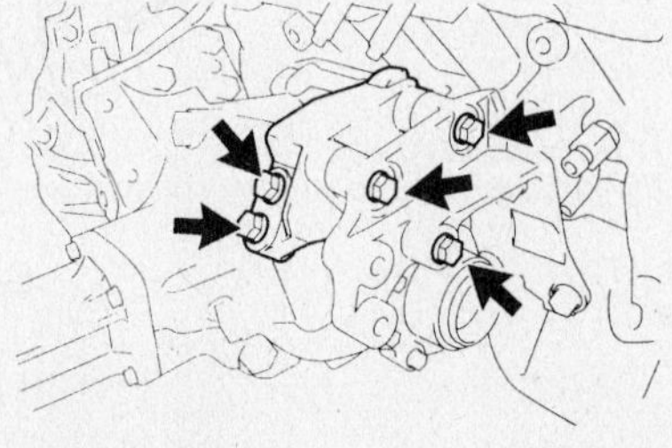

Stiffener plate

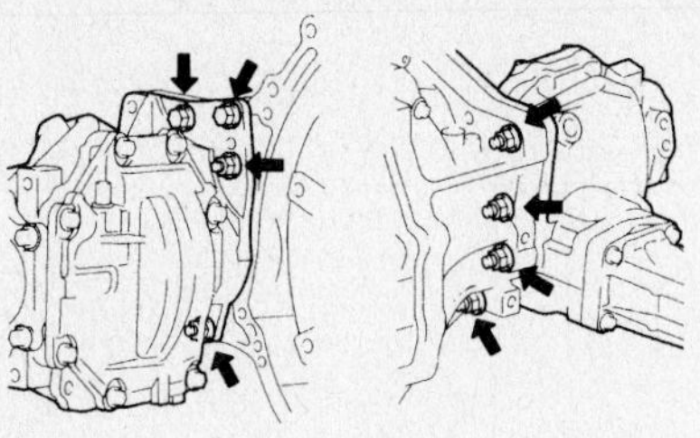

Transfer case fastener locations

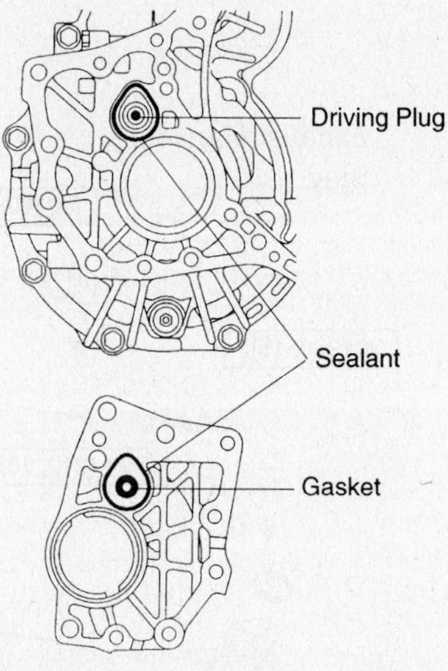

Gasket material application

8. Apply a continuous 1.2mm diameter bead of silicone gasket material to the transaxle and transfer case as shown.
9. Join the transfer case to the transaxle within 10 minutes of gasket material application. If not, remove the material and start again.
10. Torque the nuts and bolts to 51 ft. lbs. (69 Nm).
11. The remainder of installation is the reverse of removal. Observe the following torques:
    - Engine mount bracket: 47 ft. lbs. (64 Nm)
    - Stiffener plate: 25 ft. lbs. (34 Nm)
    - Drain plug: 36 ft. lbs. (49 Nm)

Halfshaft

REMOVAL & INSTALLATION

Front

1. Before servicing the vehicle, refer to the precautions in the beginning of this section.

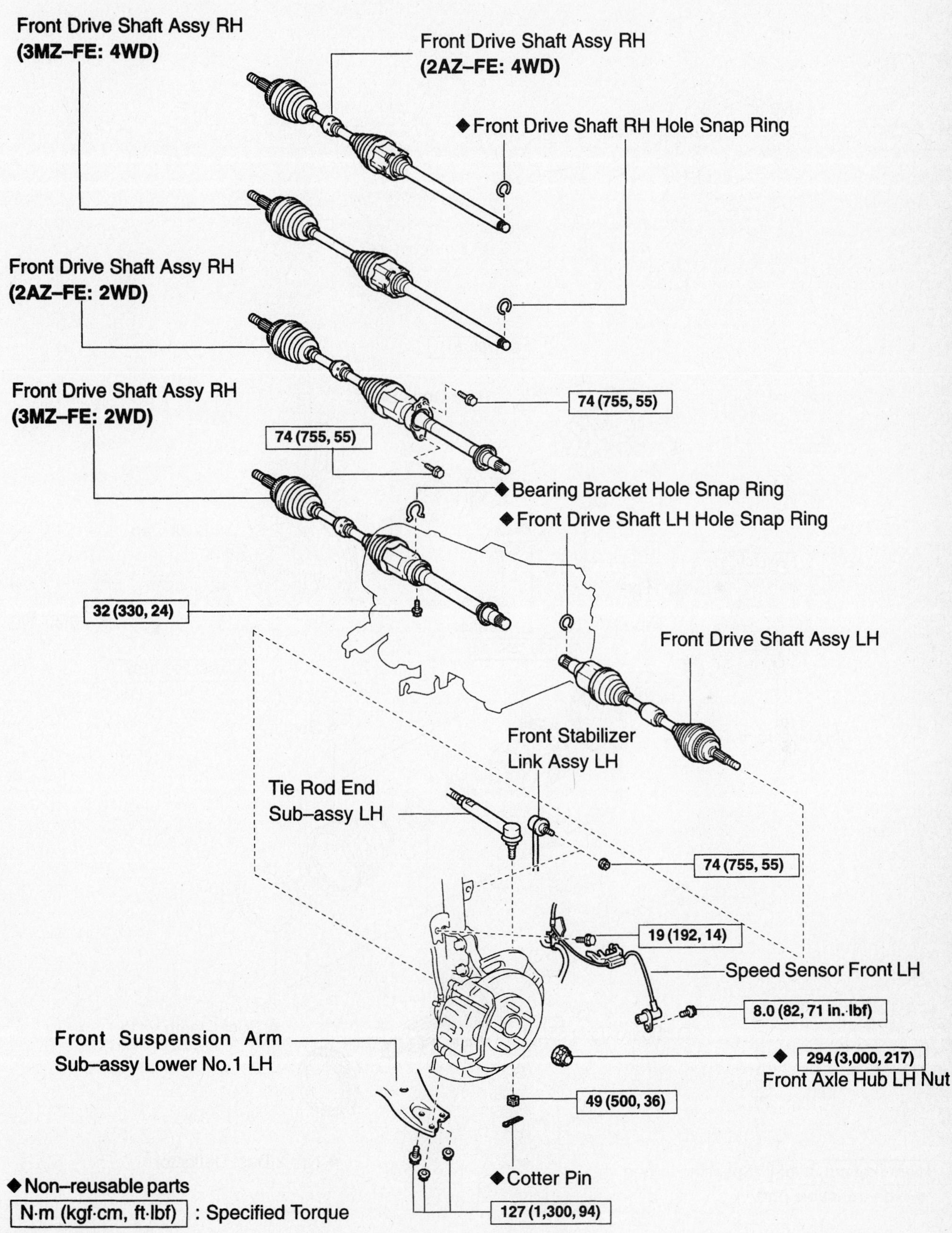
Front Drive Shaft Assy RH
(3MZ–FE: 4WD)
Front Drive Shaft Assy RH
(2AZ–FE: 4WD)
◆Front Drive Shaft RH Hole Snap Ring
Front Drive Shaft Assy RH
(2AZ–FE: 2WD)
Front Drive Shaft Assy RH
(3MZ–FE: 2WD)
74 (755, 55)
74 (755, 55)
◆Bearing Bracket Hole Snap Ring
◆Front Drive Shaft LH Hole Snap Ring
32 (330, 24)
Front Drive Shaft Assy LH
Front Stabilizer
Link Assy LH
Tie Rod End
Sub–assy LH
74 (755, 55)
19 (192, 14)
Speed Sensor Front LH
8.0 (82, 71 in.·lbf)
Front Suspension Arm
Sub–assy Lower No.1 LH
◆ 294 (3,000, 217)
Front Axle Hub LH Nut
49 (500, 36)
◆Cotter Pin
127 (1,300, 94)
◆Non–reusable parts
N·m (kgf·cm, ft·lbf) : Specified Torque
67170-HIGH-G33

Front halfshaft and related parts

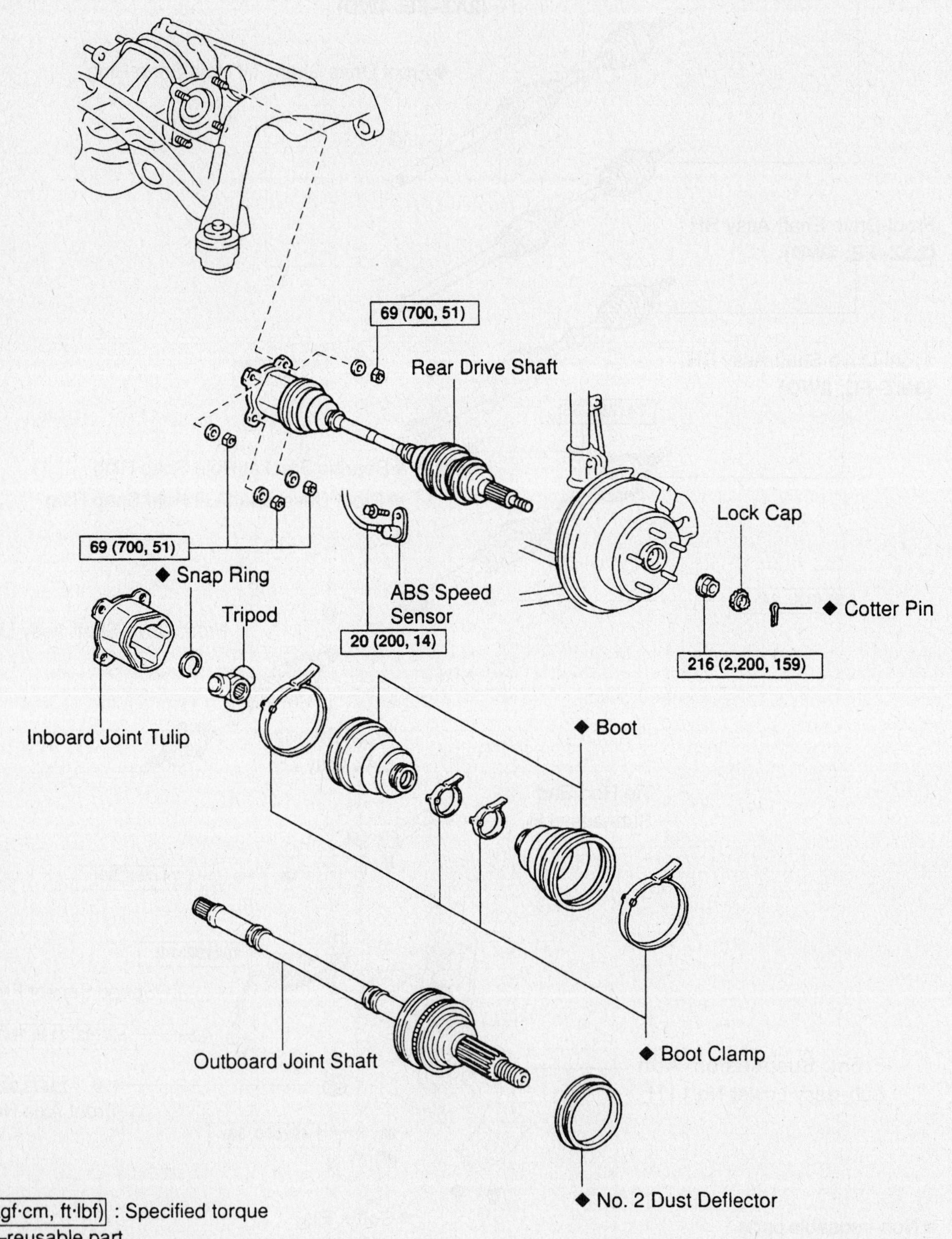

Exploded view of the rear halfshaft—2001–03

4WD:

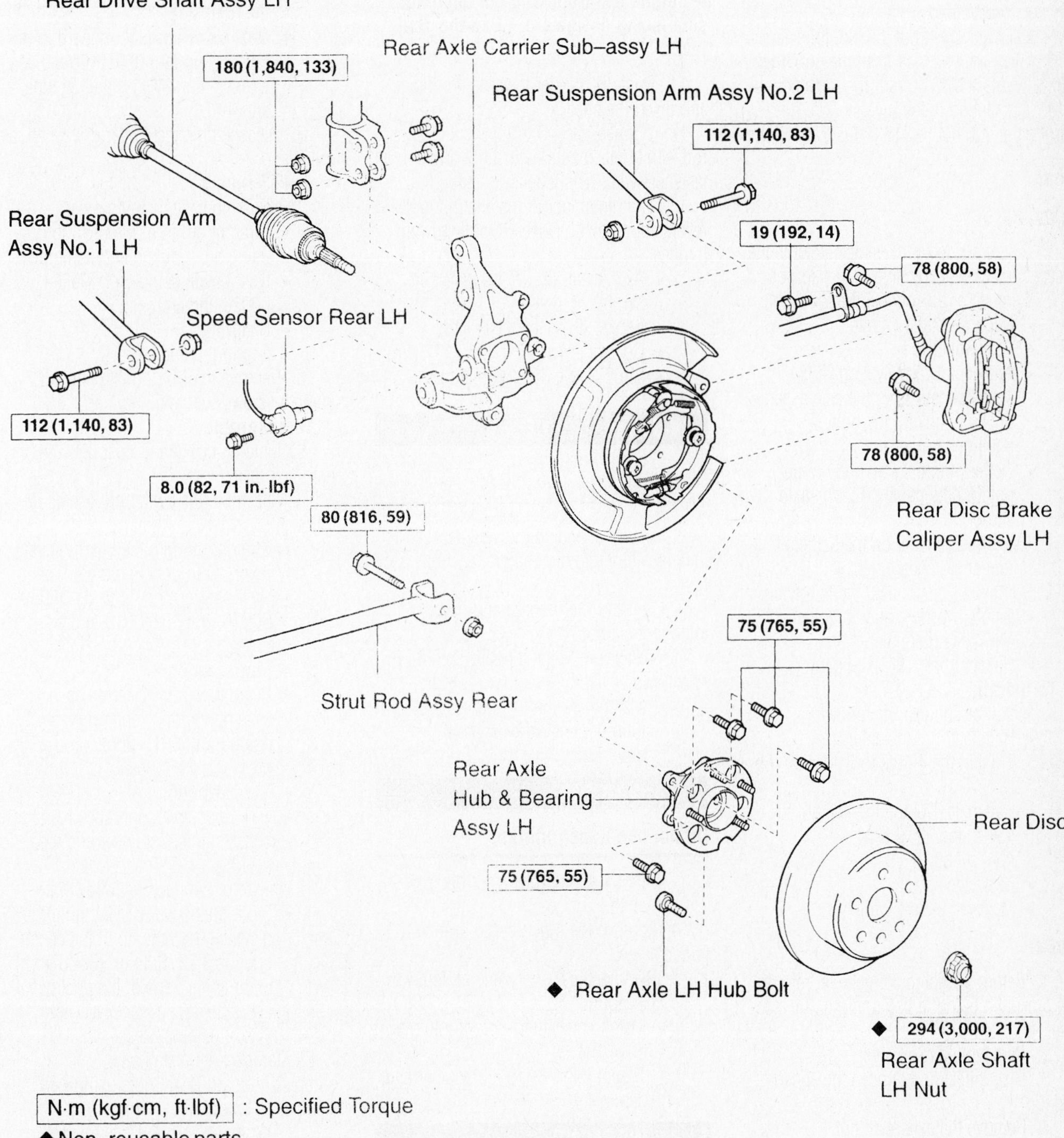
Rear Drive Shaft Assy LH
180 (1,840, 133)
Rear Axle Carrier Sub–assy LH
Rear Suspension Arm Assy No.2 LH
112 (1,140, 83)
Rear Suspension Arm
Assy No.1 LH
19 (192, 14)
78 (800, 58)
Speed Sensor Rear LH
112 (1,140, 83)
78 (800, 58)
8.0 (82, 71 in. lbf)
80 (816, 59)
Rear Disc Brake
Caliper Assy LH
75 (765, 55)
Strut Rod Assy Rear
Rear Axle
Hub & Bearing
Assy LH
Rear Disc
75 (765, 55)
◆ Rear Axle LH Hub Bolt
◆ 294 (3,000, 217)
Rear Axle Shaft
LH Nut
N·m (kgf·cm, ft·lbf) : Specified Torque
◆ Non–reusable parts
67170-HIGH-G34

Exploded view of the rear halfshaft—2004

2. Remove or disconnect the following:
   - Front wheels
   - Fender apron seal
   - Transaxle fluid
   - Transfer case oil (4wd)
   - Hub nut
   - Stabilizer bar link
   - Speed sensor
   - Tie rod end
   - Lower arm from the ball joint
3. Slide the halfshaft from the hub, then, carefully, pry the shaft from the transaxle.
4. Installation is the reverse of removal. Torque the hub nut to 217 ft. lbs. (294Nm).

Rear

2001–03

1. Before servicing the vehicle, refer to the precautions in the beginning of this section.
2. Remove or disconnect the following:
   - Negative battery cable
   - Rear wheels
   - Anti-lock Brake System (ABS) speed sensor from the axle assembly by removing the bolt, if equipped
   - Cotter pin, lock cap and the nut holding the halfshaft to the axle carrier
3. Place matchmarks on the halfshaft and differential side gear shaft.
4. Remove or disconnect the following:
   - 4 nuts, washers and the halfshaft from the differential
   - Halfshaft from the axle carrier

To install:

5. Install or connect the following:
   - Halfshaft into the axle carrier. Tighten the 4 nuts to 51 ft. lbs. (69 Nm).
   - Halfshaft. Tighten the locknut to 159 ft. lbs. (216 Nm).
   - ABS sensor
   - Rear wheels
   - Negative battery cable

2004

1. Before servicing the vehicle, refer to the precautions in the beginning of this section.
2. Remove the rear wheel.
3. Disconnect and remove the speed sensor.
4. Remove the axle shaft nut.
5. Disconnect the height control sensor.
6. Disconnect the rear control arms.
7. Disconnect the strut rod.
8. Push the rear axle carrier subassembly towards the outside and separate the shaft from the carrier.
9. Remove the shaft, keeping it level.

To install

10. Align the shaft splines and install the shaft with a brass bar and hammer.

➡Set the snapring with the opening side facing downward. Keep the shaft level.

11. Push the carrier towards the inside and insert the shaft.
12. Connect the control arms and strut rod with the fasteners hand-tight. Tighten all fasteners with the suspension loaded.
13. The remainder of installation is the reverse of removal. Observe the following torques:
   - Axle shaft nut: 217 ft. lbs. (294 Nm)
   - Wheel: 76 ft. lbs. (103 Nm)
   - Control arms: 83 ft. lbs. (112 Nm)
   - Strut: 133 ft. lbs. (180 Nm)

CV-Joints

OVERHAUL

1. Before servicing the vehicle, refer to the precautions in the beginning of this section.
2. Remove the inboard and outboard joint boot clamps.
3. Disassemble the inboard joint tulip, as follows:
   - Matchmark the tri-pot, inboard joint tulip or center driveshaft to the driveshaft

⁂ WARNING

Do not use punch marks.

   - Inboard joint tulip from the driveshaft
4. Remove the inboard and outboard joint clamps.
5. Remove the tri-pot joint, as follows:
   - Snapring
   - Matchmark the tri-pot joint to the driveshaft
   - Tri-pot joint, using a brass bar and hammer

⁂ WARNING

Do not tap the roller.

6. Remove or disconnect the following:
   - Inboard and outboard joint boots

➡Do not disassemble the outboard joint.

   - Dust cover from the center driveshaft, using a press, for 2WD on the right side
   - Dust cover from the inboard joint tulip, using tool 09950-00020 and a press, for 2WD on the left side and 4WD
7. Disassemble the center driveshaft, as follows:
   - Snapring
   - Bearing case, using a press
   - Straight pin from the bearing case, using a pin punch and hammer
   - Dust cover, using tool 09950-00020 and a press
   - Snapring
   - Bearing, using a press
8. Remove the No. 2 dust deflector, using a screwdriver and hammer.

To assemble:

9. Install a new No. 2 dust deflector, using a press.
10. Assemble the center driveshaft, as follows:
   - Straight pin into the bearing case, using a pin punch and hammer
   - New bearing, using tools 09959-60010, 09950-70010 and a press
   - New snapring
   - Bearing with the bearing case assembly to the center driveshaft, using tool 09710-30021 and a press
   - New snapring
   - New dust cover, until the clearance between the dust cover and the bearing is 0.039 in. (1.0mm)
11. Install or connect the following:
   - Right dust cover (2WD), until the distance from the tip of the center drive is 3.39–3.34 in. (86–87mm) to the inner edge of the dust cover
   - Left side dust cover (2WD and 4WD), using a press
12. Temporarily install new outboard/inboard joint boots using new clamps, as follows:
   a. Wrap tape around the driveshaft splines.
   b. Install the new outboard joint boot onto the driveshaft.

c. Install the new inboard joint boot onto the driveshaft.

13. Install the tri-pot joint, as follows:
- Tri-pot joint, face the beveled side toward the outboard joint and align the matchmarks
- Tri-pot joint onto the driveshaft, using a press

⁂ WARNING

Be careful not to tap the roller.

- New snapring

14. Install the outboard joint boot packed with grease from the boot kit.

15. Install the inboard joint tulip, as follows:
- Pack the inboard joint boot with grease from the boot kit
- Inboard joint tulip, by aligning the matchmarks
- Temporarily, install the inboard joint boot packed with grease from the kit

16. Install the boot clamps to both boots, as follows:
- Both boots to the shaft grooves
- Halfshaft length should be 33.055–33.449 in. (839.6–849.6mm) for the right side on 2WD with A/T, 21.397–21.791 in. (543.5–553.5mm) for the left side on 2WD with A/T, 19.929–20.323 in. (506.2–516.2mm) for the right side on 4WD or 19.803–20.197 in. (503–511mm) for the left side on 4WD
- Both new boot clamps boot
- Bend the band and lock it using a screwdriver

Pinion Seal

REMOVAL & INSTALLATION

Rear

1. Before servicing the vehicle, refer to the precautions in the beginning of this section.
2. Drain the differential oil.
3. Remove or disconnect the following:
   - Exhaust pipe
   - Driveshaft by matchmarking it
   - Companion flange nut, by loosen the staked portion
   - Companion flange, using a screw-type extractor
   - Oil seal, using an extractor
   - Slinger
   - Front bearing
   - Spacer

To install:

4. Install or connect the following
   - New spacer
   - Bearing
   - Slinger
   - New seal

➡Seal installation depth: 2.0mm +/- 0.3mm

   - Companion flange
   - New nut. Coat the threads with clean differential oil. Torque the nut to 80 ft. lbs. (108Nm).
5. The remainder of installation is the reverse of removal.

STEERING AND SUSPENSION

Air Bag

⁂ CAUTION

Some vehicles are equipped with an air bag system. The system must be disarmed before performing service on, or around, system components, the steering column, instrument panel components, wiring and sensors. Failure to follow the safety precautions and the disarming procedure could result in accidental air bag deployment, possible injury and unnecessary system repairs.

PRECAUTIONS

Several precautions must be observed when handling the inflator module to avoid accidental deployment and possible personal injury.

- Never carry the inflator module by the wires or connector on the underside of the module.
- When carrying a live inflator module, hold securely with both hands and ensure that the bag and trim cover are pointed away.
- Place the inflator module on a bench or other surface with the bag and trim cover facing up.
- With the inflator module on the bench, never place anything on or close to the module which may be thrown in the event of an accidental deployment.

DISARMING

To avoid personal injury when working on vehicles equipped with an air bag, the negative battery cable must be disconnected and at least 90 seconds must elapse before working on the system. Failure to do so may result in deployment of the air bag.

Power Rack And Pinion Steering Gear

REMOVAL & INSTALLATION

1. Before servicing the vehicle, refer to the precautions in the beginning of this section.
2. Remove or disconnect the following:
   - Negative battery cable

➡Wait at least 90 seconds before working on the vehicle to allow the Supplemental Restraint System (SRS) system to disarm.

   - Steering wheel
   - Front wheels
   - Tie rod ends
   - Intermediate shaft

➡Matchmark the shaft and gear.

   - Stabilizer bar end links
   - Pressure and return lines
   - Steering gear
   - Installation is the reverse of removal. Observe the following torques:
   - Rack mounting bolts: 52 ft. lbs. (70Nm)
   - Stabilizer bar end links: 55 ft. lbs. (74Nm)
   - Intermediate shaft bolt: 26 ft. lbs. (35Nm)
   - Tie rod end nuts: 36 ft. lbs. (49Nm)

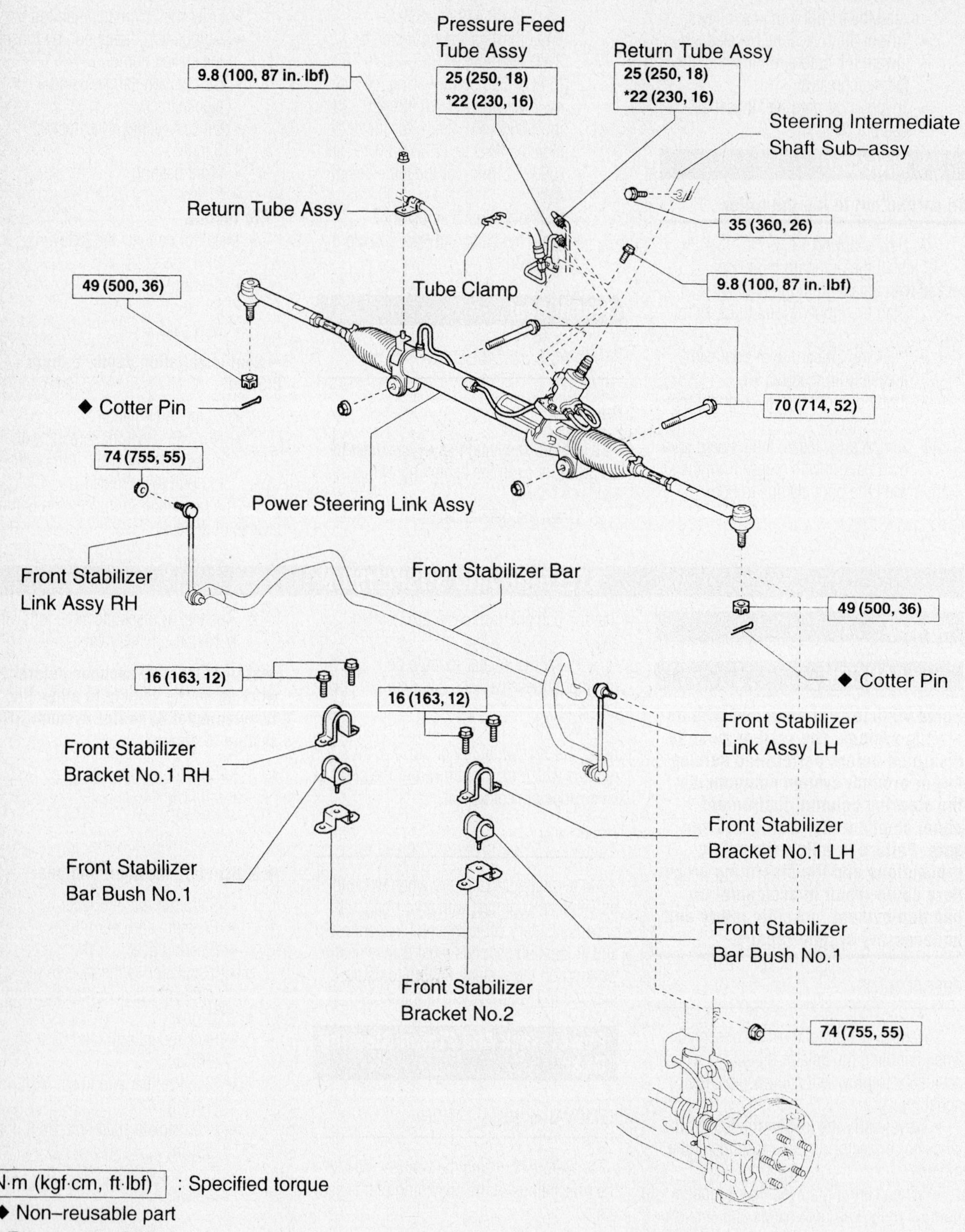
Pressure Feed Tube Assy
25 (250, 18)
*22 (230, 16)
Return Tube Assy
25 (250, 18)
*22 (230, 16)
9.8 (100, 87 in.·lbf)
Steering Intermediate Shaft Sub–assy
Return Tube Assy
35 (360, 26)
49 (500, 36)
Tube Clamp
9.8 (100, 87 in. lbf)
70 (714, 52)
◆ Cotter Pin
74 (755, 55)
Power Steering Link Assy
Front Stabilizer Link Assy RH
Front Stabilizer Bar
49 (500, 36)
◆ Cotter Pin
16 (163, 12)
16 (163, 12)
Front Stabilizer Link Assy LH
Front Stabilizer Bracket No.1 RH
Front Stabilizer Bracket No.1 LH
Front Stabilizer Bar Bush No.1
Front Stabilizer Bar Bush No.1
Front Stabilizer Bracket No.2
74 (755, 55)
N·m (kgf·cm, ft·lbf) : Specified torque
◆ Non–reusable part
* For use with SST
67170-HIGH-G35

Steering rack and related parts

Strut

REMOVAL & INSTALLATION

Front

2001–03

1. Before servicing the vehicle, refer to the precautions in the beginning of this section.

➡Do not support the weight of the vehicle on the suspension arm; the arm will deform under its weight.

2. Remove or disconnect the following:
   - Wheel
   - Stabilizer bar link
   - Brake hose and the Anti-lock Brake System (ABS) speed sensor wire from the strut
   - Strut lower end from the steering knuckle lower arm
   - 3 upper strut mounting plate-to-upper wheel arch nuts
   - Strut

To install:

3. Install or connect the following:
   - Tighten the 3 suspension support-to-wheel arch nuts to 59 ft. lbs. (80 Nm).
   - Tighten the strut-to-steering knuckle arm bolts to 155 ft. lbs. (210 Nm).
   - Sway bar link to the strut. Tighten the nut to 55 ft. lbs. (74 Nm).
   - ABS speed sensor and the brake hose to the strut, if equipped
   - Wheel
4. Check and/or adjust the front wheel alignment.

2004

1. Remove the wheel.
2. Disconnect the stabilizer bar link.
3. Loosen, don't remove, the strut locknut.
4. Disconnect the brake hose from the strut.
5. Remove the lower mounting bolts.
6. Remove the upper retaining nuts.

To install:

7. Position the strut and install the upper nuts. Torque to 59 ft. lbs. (80 Nm).
8. Install the lower bolts and torque to 170 ft. lbs. (230 Nm).
9. Connect the brake line.
10. Tighten the strut locknut to 36 ft. lbs. (49 Nm).
11. Connect the stabilizer links and torque to 55 ft. lbs. (74 Nm).
12. Install the wheel. Torque to 76 ft. lbs. (103 Nm).

Rear

2001–03

1. Before servicing the vehicle, refer to the precautions in the beginning of this section.

➡Do not support the weight of the vehicle on the suspension arm; the arm will deform under its weight.

2. Remove or disconnect the following:
   - Wheel
   - Brake hose and the Anti-lock Brake System (ABS) speed sensor wire from the strut
   - Sway bar link from the strut
3. Loosen, but do not remove the 2 lower bolts.
4. Support the axle carrier with a jack and remove cap.
5. If the strut is being disassembled, loosen the center nut.
6. Remove the 3 mounting nuts.
7. Lower the carrier and remove the 2 lower nuts and bolts.
8. Installation is the reverse of removal. Observe the following torques:
   - 3 mounting nuts: 29 ft. lbs. (39Nm)
   - 2 lower nuts: 188 ft. lbs. (255Nm)
   - Center nut: 36 ft. lbs. (49Nm)
   - Stabilizer link: 29 ft. lbs. (39Nm)

2004

1. Remove the tonneau cover.
2. Remove the left side deck trim cover.
3. Remove the rear wheels.
4. Remove the stabilizer link from the strut.
5. On 2-wheel drive models, disconnect the skid control sensor wire and brake hose from the strut and carrier.
6. On 4-wheel drive models, disconnect the brake hose and speed sensor from the strut and carrier.
7. Loosen the 2 nuts at the lower end of the strut, but don't remove them.
8. Support the rear axle carrier with a jack.
9. Remove the 3 upper strut nuts and lower the axle.
10. Remove the lower strut bolts and nuts. Lift out the strut.

To install:

11. Position the strut and install the 3 upper nuts. Torque to 43 ft. lbs. (58 Nm).
12. Lift the axle and install the 2 lower bolts and nuts. Torque to 133 ft. lbs. (180 Nm).
13. The remainder of installation is the reverse of removal. Observe the following torques:
   - Brake hose clamp: 14 ft. lbs. (19 Nm)
   - Wire-to-strut clamp: 44 inch lbs. (5 Nm)
   - Sensor clamp: 71 inch lbs. (8 Nm)
   - Stabilizer link: 29 ft. lbs. (39 Nm)

STRUT OVERHAUL

1. Before servicing the vehicle, refer to the precautions in the beginning of this section.
2. Remove or disconnect the following:
   - Wheel

➡If equipped, be careful not to damage the oil seal, driveshaft boot and/or speed sensor rotor when removing the steering knuckle.

   - Shock absorber (strut assembly)
3. Install a nut/bolt to the bracket at the lower portion of the strut assembly and secure it in a vise.
4. Compress the coil spring with a spring compressor.

⁂ CAUTION

The proper tools must be used for this procedure. The spring on the strut is under high pressure and can cause serious injury if not properly removed and installed.

5. Remove or disconnect the following:
   - Center retaining nut, by holding the spring seat
   - Support, dust seal, spring seat, insulator and spring from the strut assembly

To install:

6. Install the spring bumper and lower insulator to the strut assembly.
7. Compress the coil spring and fit the lower end of the spring into the spring seat gap.
8. Install or connect the following:
   - Upper insulator, spring seat, dust seal, support and spring seat. Tighten the new retaining nut to 36 ft. lbs. (49Nm).
   - Strut
   - Wheel
9. If required, bleed the brake system and check for leaks.
10. Check and/or adjust the front wheel alignment.

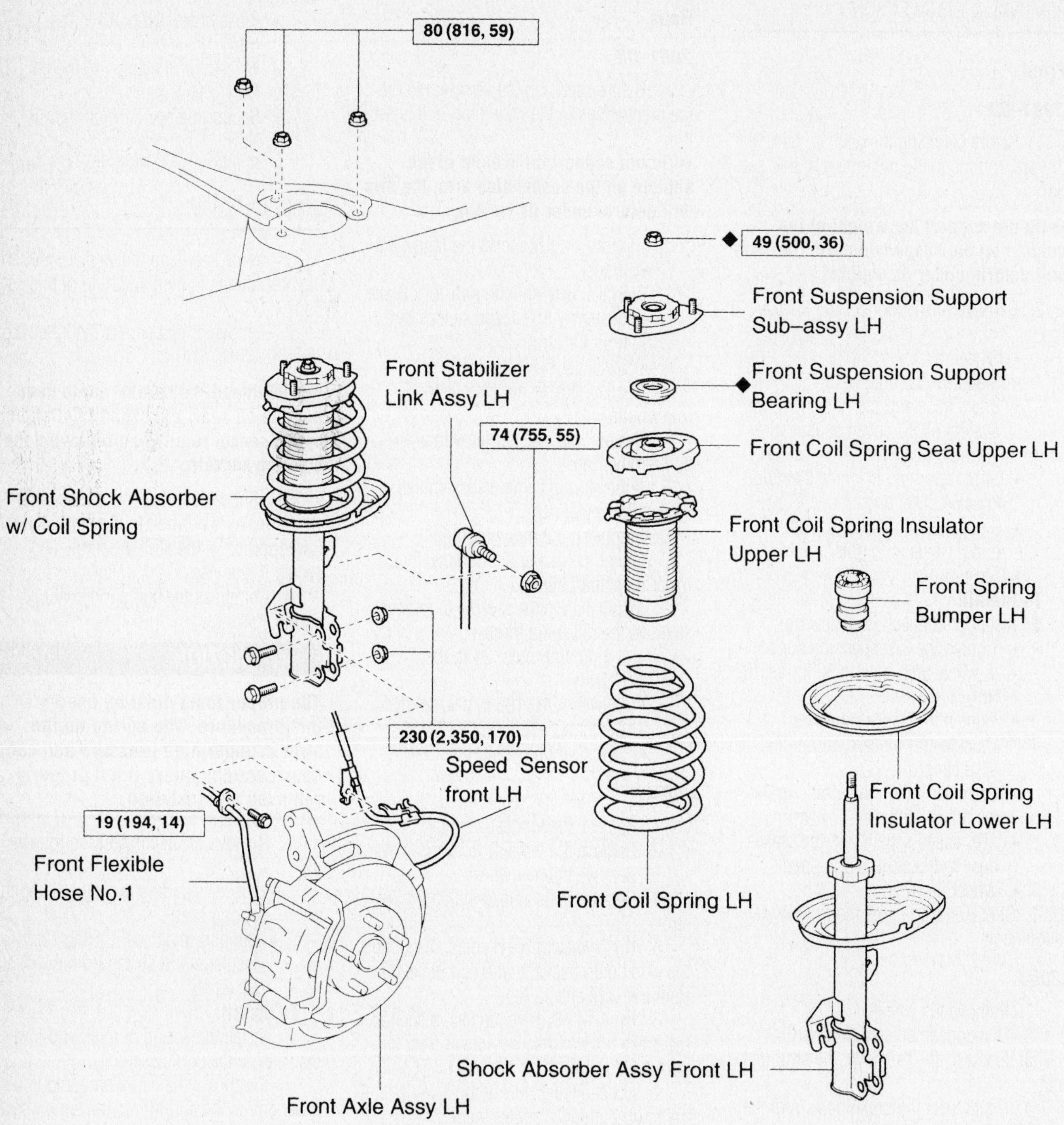

N·m (kgf·cm, ft·lbf) : Specified torque

◆ Non–reusable part

67170-HIGH-G36

Front strut and related parts—2004 models

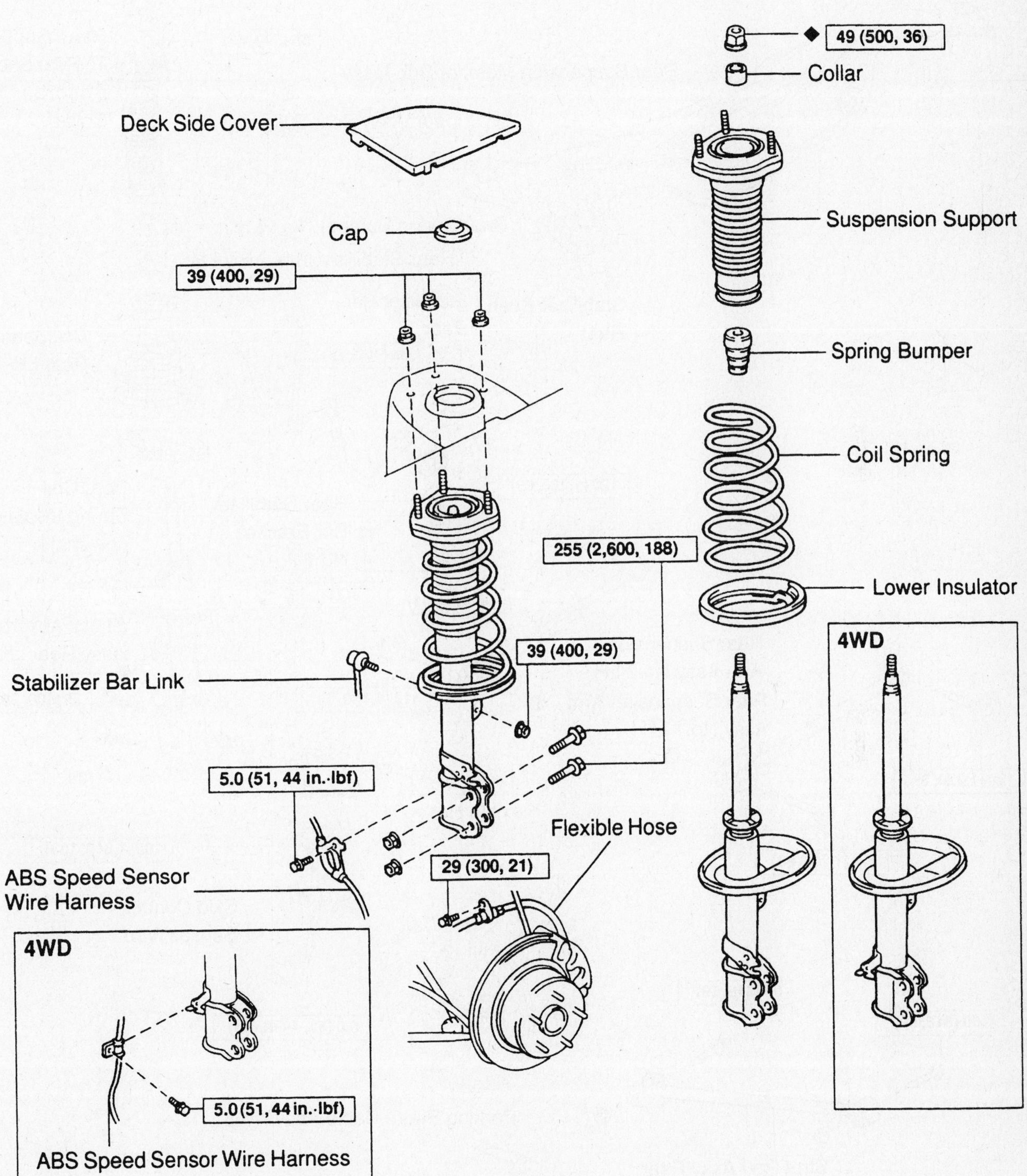

N·m (kgf·cm, ft·lbf) : Specified torque

◆ Non–reusable part

7924ZG89

Rear strut and related parts—2001–03 models

FF:

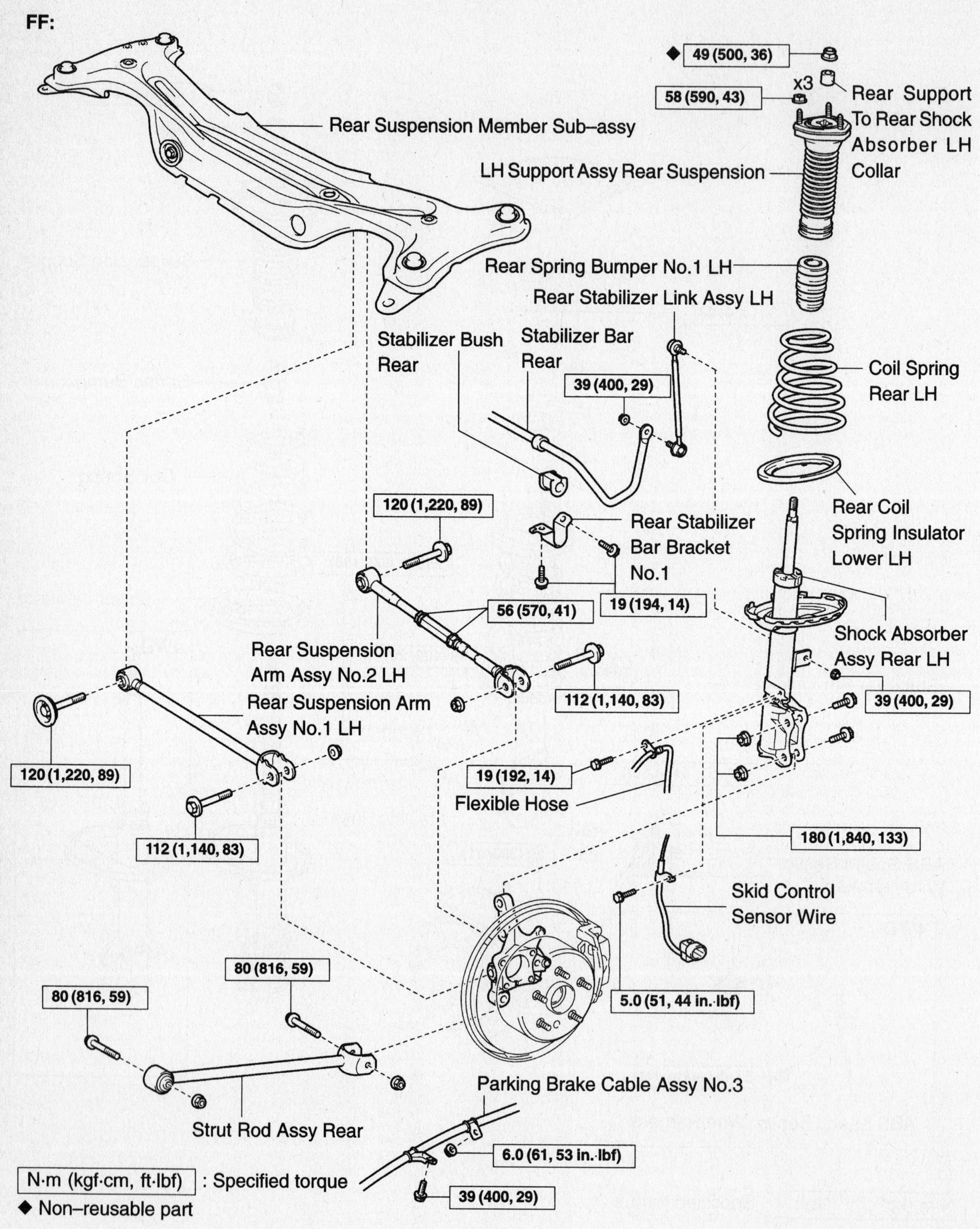
◆ 49 (500, 36)
58 (590, 43)
x3
Rear Support To Rear Shock Absorber LH Collar
Rear Suspension Member Sub–assy
LH Support Assy Rear Suspension
Rear Spring Bumper No.1 LH
Rear Stabilizer Link Assy LH
Stabilizer Bush Rear
Stabilizer Bar Rear
39 (400, 29)
Coil Spring Rear LH
120 (1,220, 89)
Rear Coil Spring Insulator Lower LH
Rear Stabilizer Bar Bracket No.1
56 (570, 41)
19 (194, 14)
Shock Absorber Assy Rear LH
Rear Suspension Arm Assy No.2 LH
Rear Suspension Arm Assy No.1 LH
112 (1,140, 83)
39 (400, 29)
120 (1,220, 89)
19 (192, 14)
Flexible Hose
112 (1,140, 83)
180 (1,840, 133)
Skid Control Sensor Wire
80 (816, 59)
80 (816, 59)
5.0 (51, 44 in.·lbf)
Parking Brake Cable Assy No.3
Strut Rod Assy Rear
6.0 (61, 53 in.·lbf)
N·m (kgf·cm, ft·lbf) : Specified torque
39 (400, 29)
◆ Non–reusable part
67170-HIGH-G37

Rear strut and related parts—2004 models

Coil Spring

REMOVAL & INSTALLATION

➡See Strut Overhaul in this section.

Lower Ball Joint

REMOVAL & INSTALLATION

1. Before servicing the vehicle, refer to the precautions in the beginning of this section.
2. Remove or disconnect the following:
   - Wheel
   - Hub nut
   - Caliper and rotor
   - Lower control arm from the ball joint
   - Tie rod end
   - Halfshaft
   - Ball joint from the knuckle
3. Installation is the reverse of removal. Observe the following torques:
   - Ball stud nut: 90 ft. lbs. (123Nm)
   - Lower arm-to-ball joint: 94 ft. lbs. (127Nm)
   - Tie rod end: 36 ft. lbs. (49Nm)

Lower Control Arm

REMOVAL & INSTALLATION

2001–03

1. Before servicing the vehicle, refer to the precautions in the beginning of this section.

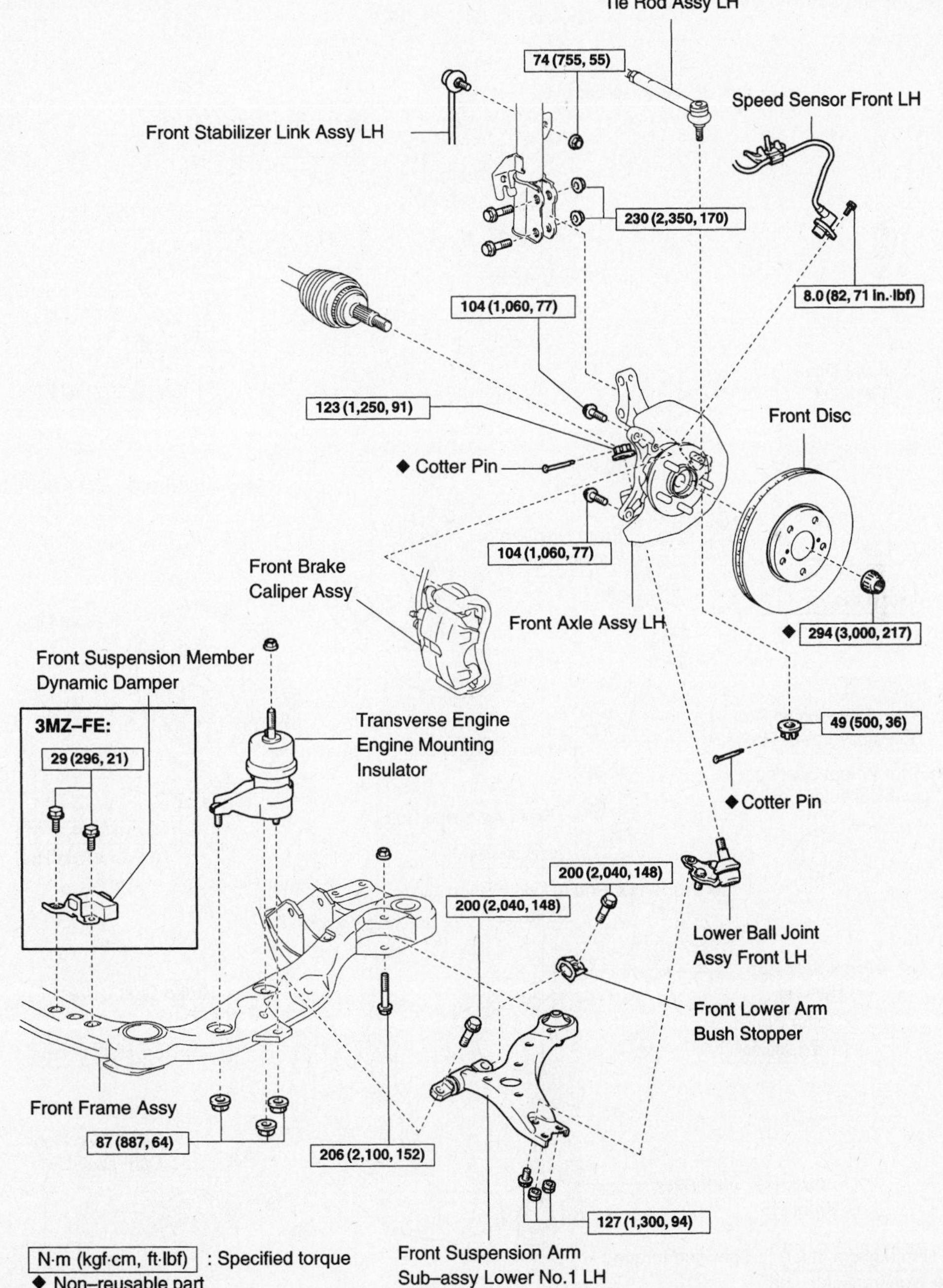

Lower control arm and related parts

2. Remove or disconnect the following:
   - Engine/transaxle assembly
   - Transverse engine mount insulator
   - 2 front and 1 rear lower arm mount bolts
   - Lower arm
3. Installation is the reverse of removal. Observe the following torques:
   - Front side bolts: 148 ft. lbs. (200Nm)
   - Rear arm bolt/nut: 152 ft. (206Nm)
   - Insulator: 64 ft. lbs. (87Nm)

2004

1. Remove the engine/transaxle assembly.
2. Remove the transverse engine mounting insulator.
3. Remove the 3 bolts securing the arm to the engine support member.
4. Remove the front lower arm bush stopper.
5. Remove the ball joint-to-arm bolts.
6. Installation is the reverse of removal. Observe the following torques:
   - 2 short arm-to-support bolts: 148 ft. lbs. (200 Nm)
   - 1 long arm-to-support bolt: 152 ft. lbs. (206 Nm)
   - Ball joint-to-arm: 94 ft. lbs. (127 Nm)
   - Transverse engine mounting insulator: 64 ft. lbs. (87 Nm)

CONTROL ARM BUSHING REPLACEMENT

➡The Highlander does not have replaceable bushings.

Front Wheel Bearing

REMOVAL

1. Remove the wheel.
2. Remove the hub nut.
3. Remove the caliper and suspend it out of the way.
4. Remove the brake rotor.
5. Disconnect the tie rod end from the knuckle.

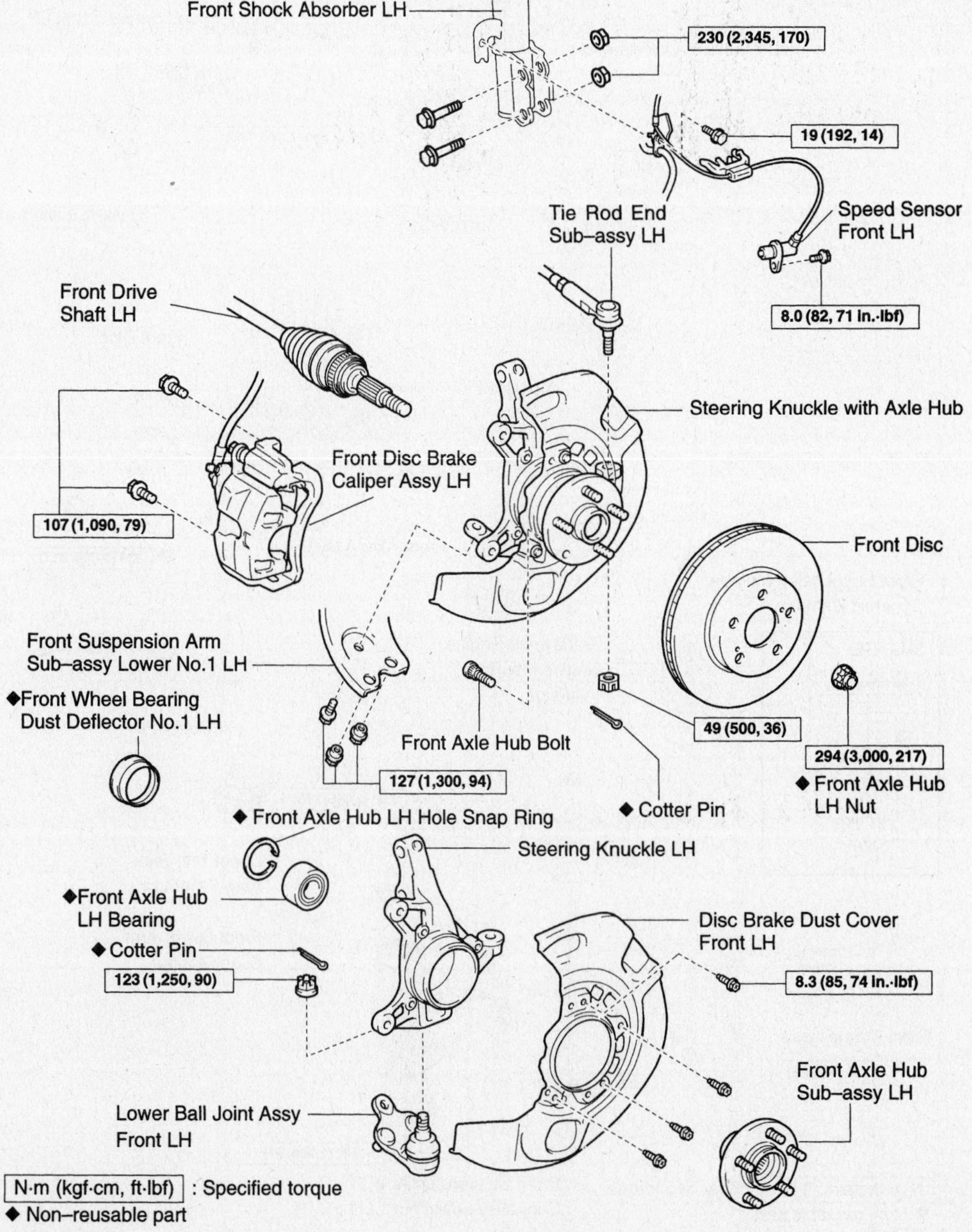

Front hub and related parts

6. Disconnect the control arm from the ball joint.
7. Remove the halfshaft from the knuckle.
8. Remove the lower strut-to-knuckle bolts and remove the hub/knuckle assembly.
9. Remove the ball joint.

DISASSEMBLY AND ASSEMBLY

1. Remove the inner seal from the hub.
2. Remove the snap ring from the hub.
3. Mount the knuckle in a vise and, using a slidehammer, remove the hub from the knuckle.
4. Remove the outer seal.
5. Using a press, the bearings and races can now be replaced.
6. Replace the snap ring and seals.
7. Press the knuckle onto the hub assembly.

INSTALLATION

1. Install the ball joint. Torque to 91 ft. lbs. (123 Nm). Advance the nut as much as 60 degrees to align the cotter pin hole. Use a new cotter pin.
2. Attach the knuckle assembly to the strut. Torque to 170 ft. lbs. (230 Nm).
3. The remainder of installation is the reverse of removal. Observe the following torques:
   - Caliper support: 77 ft. lbs. (104 Nm)
   - Hub nut: 217 ft. lbs. (294 Nm)
   - Caliper pins: 25 ft. lbs. (34 Nm)

Rear Wheel Bearings

REMOVAL AND INSTALLATION

2001–03

The 4wd rear wheel bearings are serviced as a unit with the hub and are not replaceable. See the Halfshaft Removal and Installation procedure.

WITH 2WD

1. Before servicing the vehicle, refer to the precautions in the beginning of this section.
2. Remove or disconnect the following:
   - Rear wheel
   - Flexible brake hose from the rear strut assembly
   - Brake caliper and support it on using wire
   - Brake rotor
3. Remove or disconnect the following:
   - Anti-lock Brake System (ABS) speed sensor connector
   - 4 rear axle hub assembly nuts
   - Hub assembly

To install:

4. Install or connect the following:
   - New hub assembly. Tighten the nuts to 59 ft. lbs. (80 Nm).
   - ABS speed sensor
   - Brake rotor
   - Brake caliper. Tighten the mounting bolts to 34 ft. lbs. (47 Nm).
   - Flexible brake hose to the rear strut assembly. Tighten the mounting bolt to 21 ft. lbs. (29 Nm).
   - Wheels
5. Test drive the vehicle.

2004

1. Remove the wheel.
2. Remove the speed sensor from the axle carrier.
3. Remove the axle shaft nut.
4. Remove the caliper and support assembly and suspend it out of the way.
5. Check the bearing backlash. It should not exceed 0.0020 in. (0.05mm). If it does, it must be replaced.
6. Remove the 4 bolts and remove the hub/bearing assembly.
7. Installation is the reverse of removal. Observe the following torques:
   - Hub/bearing assembly bolts: 55 ft. lbs. (75 Nm)
   - Caliper support: 58 ft. lbs. (78 Nm)
   - Brake hose clamp: 14 ft. lbs. (19 Nm)
   - Hub nut: 217 ft. lbs. (294 Nm)

BRAKES

Front Brake Caliper

REMOVAL AND INSTALLATION

2001–03

1. Before servicing the vehicle, refer to the precautions in the beginning of this section.
2. Remove or disconnect the following:
   - Front wheel
   - 2 mounting bolts and caliper
3. If the caliper is being replaced, disconnect the brake line and plug both openings.

➡Depending on the brake type, there may be either 1 or 2 sealing washers.

4. Installation is the reverse of removal. Torque the mounting bolts to 25 ft. lbs. (34Nm). If the brake hose was removed, torque the union bolt to 21 ft. lbs. (29Nm).

2004

1. Disconnect the brake line from the caliper and plug it.
2. Hold the caliper slide pins and remove the mounting bolts.
3. Lift off the caliper.
4. Remove the pads and anti-squeal shims.
5. Remove the wear indicator from the inner pad.
6. Installation is the reverse of removal. Grease the caliper slides and bolts with lithium grease or equivalent. Apply disc brake grease to the anti-squeal shims. Torque the caliper bolts to 25 ft. lbs. (34 Nm); the brake line union bolt to 21 ft. lbs. (29 Nm).

Rear Brake Caliper

REMOVAL AND INSTALLATION

2001–03

1. Before servicing the vehicle, refer to the precautions in the beginning of this section.
2. Remove or disconnect the following:
   - Rear wheel
   - Slide pin and caliper
3. If the caliper is being replaced, disconnect the brake line and plug both openings.

➡Depending on the brake type, there may be either 1 or 2 sealing washers.

4. Installation is the reverse of removal. Torque the slide pin to 25 ft. lbs. (34Nm). If

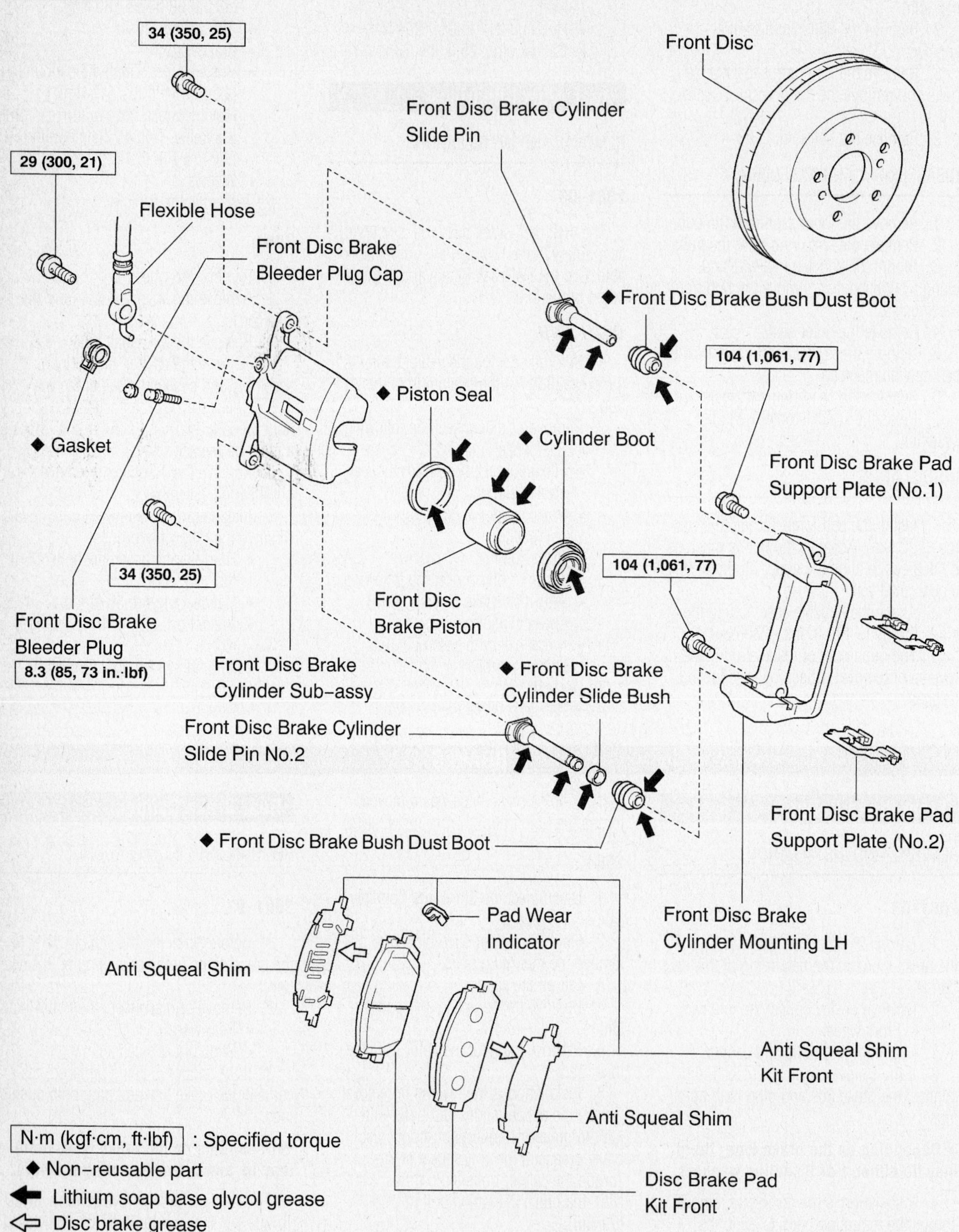
34 (350, 25)
Front Disc
Front Disc Brake Cylinder Slide Pin
29 (300, 21)
Flexible Hose
Front Disc Brake Bleeder Plug Cap
◆ Front Disc Brake Bush Dust Boot
104 (1,061, 77)
◆ Piston Seal
◆ Cylinder Boot
◆ Gasket
Front Disc Brake Pad Support Plate (No.1)
104 (1,061, 77)
34 (350, 25)
Front Disc Brake Piston
Front Disc Brake Bleeder Plug
8.3 (85, 73 in.·lbf)
Front Disc Brake Cylinder Sub-assy
◆ Front Disc Brake Cylinder Slide Bush
Front Disc Brake Cylinder Slide Pin No.2
◆ Front Disc Brake Bush Dust Boot
Front Disc Brake Pad Support Plate (No.2)
Pad Wear Indicator
Front Disc Brake Cylinder Mounting LH
Anti Squeal Shim
Anti Squeal Shim Kit Front
Anti Squeal Shim
N·m (kgf·cm, ft·lbf) : Specified torque
◆ Non-reusable part
Lithium soap base glycol grease
Disc brake grease
Disc Brake Pad Kit Front
67162-X300-G11

Front disc brake components—2004

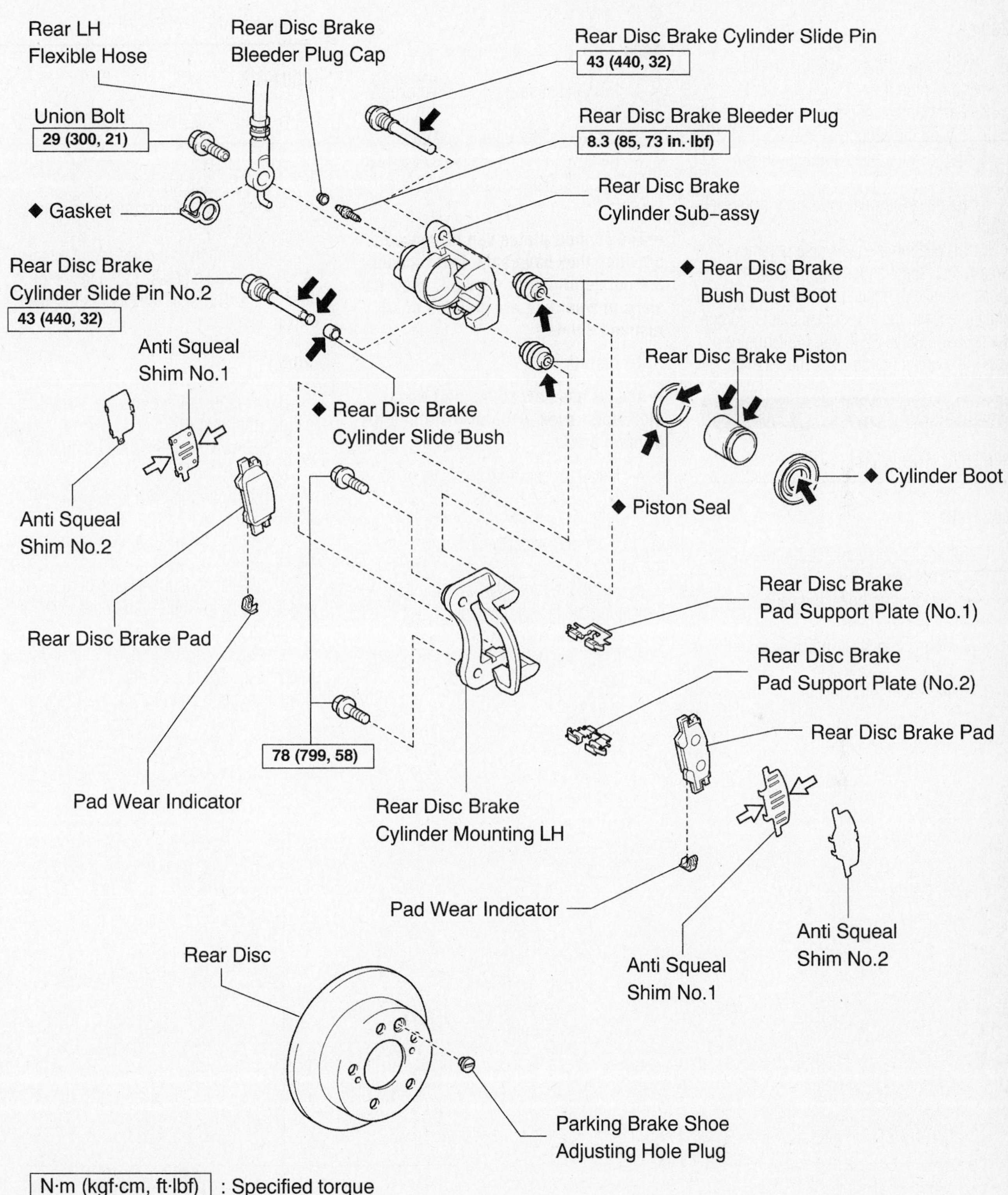
Rear LH
Flexible Hose
Rear Disc Brake
Bleeder Plug Cap
Rear Disc Brake Cylinder Slide Pin
43 (440, 32)
Union Bolt
29 (300, 21)
Rear Disc Brake Bleeder Plug
8.3 (85, 73 in.·lbf)
◆ Gasket
Rear Disc Brake
Cylinder Sub-assy
Rear Disc Brake
Cylinder Slide Pin No.2
43 (440, 32)
◆ Rear Disc Brake
Bush Dust Boot
Anti Squeal
Shim No.1
Rear Disc Brake Piston
◆ Rear Disc Brake
Cylinder Slide Bush
◆ Cylinder Boot
◆ Piston Seal
Anti Squeal
Shim No.2
Rear Disc Brake
Pad Support Plate (No.1)
Rear Disc Brake Pad
Rear Disc Brake
Pad Support Plate (No.2)
Rear Disc Brake Pad
78 (799, 58)
Pad Wear Indicator
Rear Disc Brake
Cylinder Mounting LH
Pad Wear Indicator
Anti Squeal
Shim No.2
Rear Disc
Anti Squeal
Shim No.1
Parking Brake Shoe
Adjusting Hole Plug
N·m (kgf·cm, ft·lbf) : Specified torque
◆ Non-reusable part
Lithium soap base glycol grease
Disc brake grease
67162-X300-G12

Rear disc brake components—2004

the brake hose was removed, torque the union bolt to 21 ft. lbs. (29Nm).

2004

1. Disconnect the brake line from the caliper and plug it.
2. Remove the caliper mounting bolts.
3. Lift off the caliper.
4. Remove the pads and anti-squeal shims.
5. Remove the wear indicators from each pad.
6. Installation is the reverse of removal. Grease the caliper slides and bolts with lithium grease or equivalent. Apply disc brake grease to the anti-squeal shims. Torque the caliper bolts to 32 ft. lbs. (43 Nm); the brake line union bolt to 21 ft. lbs. (29 Nm).

Front Disc Brake Pads

REMOVAL AND INSTALLATION

2001–03

1. Remove or disconnect the following:
   - Front wheel
   - 2 mounting bolts and caliper
   - Pads and shims
   - Shims and wear indicator from the pads
2. Installation is the reverse of removal. Apply disc brake grease to the inside of each shim. The wear indicator is installed on the inner pad.

2004

1. Hold the sliding pin and remove the lower bolt.
2. Lift the caliper up and secure it.
3. Remove the pads, 4 shims and wear indicator plate. Remove the 2 pad support plates.

➡The support plates can be reused, provided they have sufficient rebound, are not deformed or cracked, show no signs of wear and are cleaned of all rust and debris.

To install:

➡Always use new shims and wear indicators, even when re-installing the original pads.

4. Install a wear indicator plate on the inner pad.
5. Apply disc brake grease to both sides of the inner anti-squeal shims and install the shims.
6. Install the inner pad with the wear indicator plate facing upwards.
7. Install the outer pad.
8. Install the caliper. Torque the bolt to 25 ft. lbs.

Rear Disc Brake Pads

REMOVAL AND INSTALLATION

2001–03

1. Remove or disconnect the following:
   - Rear wheel
   - Caliper
   - Pads and shims
   - Shims and wear indicator from the pads
2. Installation is the reverse of removal. Apply disc brake grease to the inside of each shim. The wear indicator is installed on the inner pad.

2004

1. Disconnect the brake line from the caliper and plug it.
2. Remove the caliper mounting bolts.
3. Lift off the caliper.
4. Remove the pads and anti-squeal shims.
5. Remove the wear indicators from each pad.
6. Installation is the reverse of removal. Grease the caliper slides and bolts with lithium grease or equivalent. Apply disc brake grease to the anti-squeal shims. Torque the caliper bolts to 32 ft. lbs. (43 Nm); the brake line union bolt to 21 ft. lbs. (29 Nm).

TOYOTA

Land Cruiser • Sequoia

19

SPECIFICATION AND MAINTENANCE CHARTS

ENGINE AND VEHICLE IDENTIFICATION

Engine						
Code ①	Liters (cc)	Cu. In.	Cyl.	Fuel Sys.	Engine Type	Eng. Mfg.
2UZ-FE	4.7 (4664)	285	8	SFI	DOHC	Toyota

Model Year	
Code ②	Year
1	2001
2	2002
3	2003
4	2004
5	2005

SFI: Sequential Fuel Injection

DOHC: Double Overhead Camshaft

① Stamped on the left side of the engine block

② 10th digit of the Vehicle Identification Number (VIN)

67170-LCSQ-C01

GENERAL ENGINE SPECIFICATIONS

Year	Model	Engine Displacement Liters	Engine Series ID	Net Horsepower @ rpm	Net Torque @ rpm (ft. lbs.)	Bore x Stroke (in.)	Compression Ratio	Oil Pressure @ rpm
2001	Land Cruiser	4.7	2UZ-FE	245@4800	315@3400	3.70x3.30	9.6:1	45-65@3000
	Sequoia	4.7	2UZ-FE	245@4800	315@3400	3.70x3.30	9.6:1	45-65@3000
2002	Land Cruiser	4.7	2UZ-FE	245@4800	315@3400	3.70x3.30	9.6:1	45-65@3000
	Sequoia	4.7	2UZ-FE	245@4800	315@3400	3.70x3.30	9.6:1	45-65@3000
2003	Land Cruiser	4.7	2UZ-FE	245@4800	315@3400	3.70x3.30	9.6:1	45-65@3000
	Sequoia	4.7	2UZ-FE	245@4800	315@3400	3.70x3.30	9.6:1	45-65@3000
2004	Land Cruiser	4.7	2UZ-FE	245@4800	315@3400	3.70x3.31	9.6:1	45-65@3000
	Sequoia	4.7	2UZ-FE	245@4800	315@3400	3.70x3.31	9.6:1	45-65@3000

67170-LCSQ-C02

ENGINE TUNE-UP SPECIFICATIONS

Year	Engine Displacement Liters	Engine ID	Spark Plug Gap (in.)	Ignition Timing (deg.)*	Fuel Pump (psi)	Idle Speed (rpm)		Valve Clearance	
						MT	AT	Intake	Exhaust
2001	4.7	2UZ-FE	0.031	5-15B	38-44	—	650-750	0.006-0.010	0.010-0.014
2002	4.7	2UZ-FE	0.031	5-15B	38-44	—	650-750	0.006-0.010	0.010-0.014
2003	4.7	2UZ-FE	0.031	5-15B	38-44	—	650-750	0.006-0.010	0.010-0.014
2004	4.7	2UZ-FE	0.043	5-15B	38-44	—	650-750	0.006-0.010	0.010-0.014

NOTE: The Vehicle Emission Control Information label often reflects specification changes made during production.
The label figures must be used if they differ from those in this chart.

B: Before top dead center

* With terminals TC and E1 connected to DLC1

67170-LCSQ-C03

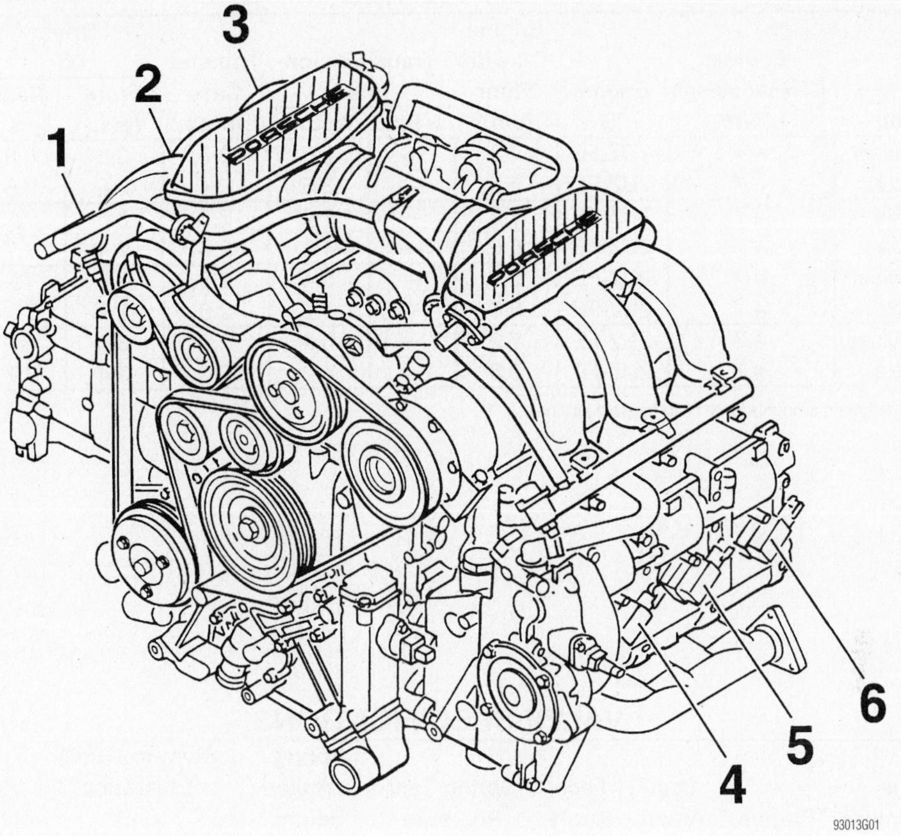

4.7L Engine
Firing order: 1–8–4–3–6–5–7–2
Distributorless ignition system

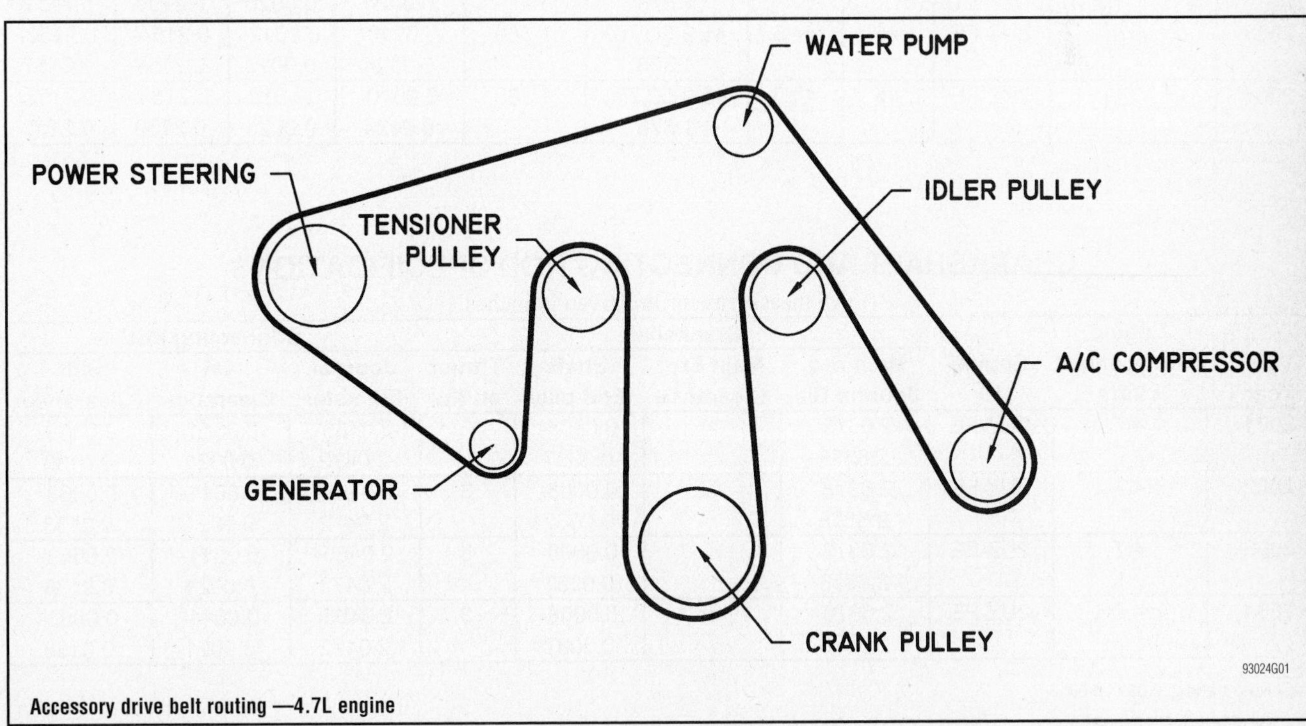

Accessory drive belt routing —4.7L engine

CAPACITIES

Year	Model	Engine Displacement Liters	Engine ID	Engine Oil with Filter (qts.)	Transmission (pts.) 5-Spd	Transmission (pts.) Auto.*	Transfer Case (pts.)	Drive Axle Front (pts.)	Drive Axle Rear (pts.)	Fuel Tank (gal.)	Cooling System (qts.)
2001	Land Cruiser	4.7	2UZ-FE	7.2	—	3.6	3.6	3.6	6.8	25.1	①
	Sequoia	4.7	2UZ-FE	5.5	—	3.6	2.4	3.6	6.6	21.6	②
2002	Land Cruiser	4.7	2UZ-FE	7.2	—	7.4	2.8	3.6	③	25.4	②
	Sequoia	4.7	2UZ-FE	6.6	—	4.2	2.6	2.4	7.7	26.1	12.3
2003	Land Cruiser	4.7	2UZ-FE	7.2	—	6.4	2.8	3.6	③	25.4	②
	Sequoia	4.7	2UZ-FE	6.6	—	4.2	2.6	2.4	7.7	26.1	12.3
2004	Land Cruiser	4.7	2UZ-FE	7.2	—	6.4	2.8	3.6	③	25.4	②
	Sequoia	4.7	2UZ-FE	6.6	—	4.2	2.6	2.4	7.7	26.1	12.3

*After draining, add the following amounts, then fill to the cold full line

① With rear heater: 9.5
Without rear heater: 8.5

② With rear heater: 16.2
Without rear heater: 15.6

③ w/o diff. Lock: 6.98
w/diff lock: 6.76

67170-LCSQ-C04

VALVE SPECIFICATIONS

Year	Engine Displacement Liters	Engine ID	Seat Angle (deg.)	Face Angle (deg.)	Spring Test Pressure (lbs. @ in.)	Spring Installed Height (in.)	Stem-to-Guide Clearance (in.) Intake	Stem-to-Guide Clearance (in.) Exhaust	Stem Diameter (in.) Intake	Stem Diameter (in.) Exhaust
2001	4.7	2UZ-FE	45	44.5	45.9-50.7@ 1.378	1.380	0.0010-0.0024	0.0012-0.0026	0.2154-0.2159	0.2152-0.2157
2002	4.7	2UZ-FE	45	44.5	45.9-50.7@ 1.378	1.380	0.0010-0.0024	0.0012-0.0026	0.2154-0.2159	0.2152-0.2157
2003	4.7	2UZ-FE	45	44.5	45.9-50.7@ 1.378	1.380	0.0010-0.0024	0.0012-0.0026	0.2154-0.2159	0.2152-0.2157
2004	4.7	2UZ-FE	45	44.5	45.9-50.7@ 1.378	1.380	0.0010-0.0024	0.0012-0.0026	0.2154-0.2159	0.2152-0.2157

67170-LCSQ-C05

CRANKSHAFT AND CONNECTING ROD SPECIFICATIONS

All measurements are given in inches.

Year	Engine Displacement Liters	Engine ID	Crankshaft Main Brg. Journal Dia.	Crankshaft Main Brg. Clearance	Crankshaft Shaft End-play	Crankshaft Thrust on No.	Connecting Rod Journal Diameter	Connecting Rod Oil Clearance	Connecting Rod Side Clearance
2001	4.7	2UZ-FE	2.6373-2.6378	①	0.0008-0.0087	3	2.0465-2.0472	0.0011-0.0021	0.0063-0.0138
2002	4.7	2UZ-FE	2.6373-2.6378	①	0.0008-0.0087	3	2.0465-2.0472	0.0011-0.0021	0.0063-0.0138
2003	4.7	2UZ-FE	2.6373-2.6378	①	0.0008-0.0087	3	2.0465-2.0472	0.0011-0.0021	0.0063-0.0138
2004	4.7	2UZ-FE	2.6373-2.6378	①	0.0008-0.0087	3	2.0465-2.0472	0.0011-0.0021	0.0063-0.0138

① Nos. 1 and 2: 0.0011-0.0018
All others: 0.0016-0.0023

67170-LCSQ-C06

PISTON AND RING SPECIFICATIONS

All measurements are given in inches.

Year	Engine Displacement Liters	Engine ID	Piston Clearance	Ring Gap Top Comp.	Ring Gap Bottom Comp.	Ring Gap Oil Control	Ring Side Clearance Top Comp.	Ring Side Clearance Bottom Comp.	Ring Side Clearance Oil Control
2000	4.7	2UZ-FE	0.0035-0.0044	0.0118-0.0197	0.0157-0.0256	0.0051-0.0189	0.0012-0.0031	0.0012-0.0028	SNUG
2001	4.7	2UZ-FE	0.0035-0.0044	0.0118-0.0157	0.0157-0.0217	0.0051-0.0150	0.0012-0.0031	0.0012-0.0028	SNUG
2002	4.7	2UZ-FE	0.0035-0.0044	0.0118-0.0157	0.0157-0.0217	0.0051-0.0150	0.0012-0.0031	0.0012-0.0028	SNUG
2003	4.7	2UZ-FE	0.0035-0.0044	0.0118-0.0157	0.0157-0.0217	0.0051-0.0150	0.0012-0.0031	0.0012-0.0028	SNUG
2004	4.7	2UZ-FE	0.0035-0.0044	0.0118-0.0157	0.0157-0.0217	0.0051-0.0150	0.0012-0.0031	0.0012-0.0028	SNUG

67170-LCSQ-C07

TORQUE SPECIFICATIONS

All readings in ft. lbs.

Year	Engine Displacement Liters	Engine ID	Cylinder Head Bolts	Main Bearing Bolts	Rod Bearing Bolts	Crankshaft Damper Bolts	Flywheel Bolts	Manifold Intake	Manifold Exhaust	Spark Plugs	Oil Pan Drain Plug
2001	4.7	2UZ-FE	①	②	③	181	④	13	33	13	29
2002	4.7	2UZ-FE	①	②	③	181	④	13	33	13	29
2003	4.7	2UZ-FE	①	②	③	181	④	13	33	13	29
2004	4.7	2UZ-FE	①	②	③	181	④	13	33	13	29

① Step 1: 24
Step 2: Plus 90 degrees
Step 3: Plus 90 degrees

② Step 1: 20 ft. lbs.
Step 2: Plus 90 degrees

③ Step 1: 18 ft. lbs.
Step 2: Plus 90 degrees

④ Step 1: 35 ft. lbs.
Step 2: Plus 90 degrees

67170-LCSQ-C08

WHEEL ALIGNMENT

Year	Model	Caster Range (+/-Deg.)	Caster Preferred Setting (Deg.)	Camber Range (+/-Deg.)	Camber Preferred Setting (Deg.)	Toe-in (in.)	Steering Axis Inclination (Deg.)
2001	Land Cruiser	0.75	+2.50	0.75	+0.08	0.04+/-0.08	12.17+/-0.75
	Sequoia	0.75	+2.95	0.75	+0.13	0.05+/-0.08	10.65+/-0.75
2002	Land Cruiser	0.75	+2.50	0.75	+0.08	0.04+/-0.08	12.17+/-0.75
	Sequoia	0.75	①	0.75	+0.13	0.05+/-0.08	10.635+/-0.75
2003	Land Cruiser	0.75	+2.50	0.75	+0.08	0.04+/-0.08	12.17+/-0.75
	Sequoia	0.75	①	0.75	+0.13	0.05+/-0.08	10.635+/-0.75
2004	Land Cruiser	0.75	+2.50	0.75	+0.08	0.04+/-0.08	12.17+/-0.75
	Sequoia	0.75	①	0.75	+0.13	0.05+/-0.08	10.635+/-0.75

Note: All alignment specifications are based on nominal ride height and standard tires

① P245/70R16: +2.95
P265/70R16: +3.00

67170-LCSQ-C09

TIRE, WHEEL AND BALL JOINT SPECIFICATIONS

Year	Model	OEM Tires		Tire Pressures (psi)		Wheel	Ball Joint	Lugnut Torque
		Standard	Optional	Front	Rear	Size	Inspection	(ft. lbs.)
2001	Land Cruiser	P275/70R15	None	32	32	8-JJ	①	97
	Sequoia	P245/70R16	P265/70R16	32	Std: 35; Opt: 32	7-JJ	②	97
2002	Land Cruiser	P275/70R16	None	29③	32③	8-JJ	①	97
	Sequoia	P245/70R16	P265/70R16	32	Std: 35; Opt: 32	7-JJ	②	97
2003	Land Cruiser	P275/65R17	P275/60R18	④	④	8-JJ	①	97
	Sequoia	P265/70R16	P265/6R17	32	Std: 35; Opt: 32	7-JJ	②	97
2004	Land Cruiser	P275/65R17	P275/60R18	④	④	8-JJ	①	97
	Sequoia	P265/70R16	P265/65R17	④	④	std: 7; opt: 7.5	②	97

OEM: Original Equipment Manufacturer
PSI: Pounds Per Square Inch
STD: Standard
OPT: Optional
NA: Not Available

① Replace if any measurable movement is found.
② Upper ball joint turning torque: 6-39 inch lbs.
Lower ball joint turning torque: 1-22 inch lbs.
Lower ball joint excessive play: 0.020 in.
③ Trailer towing: front 32; rear 35
④ See placard on vehicle

67170-LCSQ-C10

BRAKE SPECIFICATIONS

All measurements in inches unless noted

Year	Model		Brake Disc Original Thickness	Brake Disc Minimum Thickness	Brake Disc Maximum Runout	Minimum Lining Thickness	Brake Caliper Torque Plate (ft. lbs.)	Brake Caliper Mounting Bolts (ft. lbs.)
2001	Land Cruiser	F	1.260	1.181	0.0059	0.039	—	90
		R	0.709	0.611	—	0.039	65	76
	Sequoia	F	1.102	1.024	0.0028	0.039	—	90
		R	0.709	0.611	0.0039	0.039	77	65
2002	Land Cruiser	F	1.260	1.181	0.0028	0.039	—	90
		R	0.709	0.611	—	0.039	76	20
	Sequoia	F	1.102	1.024	0.0028	0.039	—	90
		R	0.709	0.611	0.0039	0.039	77	65
2003	Land Cruiser	F	1.260	1.181	0.0028	0.039	—	90
		R	0.709	0.611	—	0.039	76	20
	Sequoia	F	1.102	1.024	0.0028	0.039	—	90
		R	0.709	0.611	0.0039	0.039	77	65
2004	Land Cruiser	F	1.260	1.181	0.0028	0.039	—	90
		R	0.709	0.611	—	0.039	76	20
	Sequoia	F	1.102	1.024	0.0028	0.039	—	90
		R	0.709	0.611	0.0039	0.039	77	65

F: Front
R: Rear

67170-LCSQ-C11

SCHEDULED MAINTENANCE INTERVALS

Toyota Sequoia

TO BE SERVICED	TYPE OF SERVICE	VEHICLE MILEAGE INTERVAL (x1000)												
		7.5	15	22.5	30	37.5	45	52.5	60	67.5	75	82.5	90	97.5
Engine oil & filter	R	✓	✓	✓	✓	✓	✓	✓	✓	✓	✓	✓	✓	✓
Automatic transmission fluid & filter	S/I		✓		✓		✓		✓		✓		✓	
Ball joints & dust covers	S/I		✓		✓		✓		✓		✓		✓	
Bolts & nuts on chassis & body	S/I		✓		✓		✓		✓		✓		✓	
Brake line pipes & hoses	S/I		✓		✓		✓		✓		✓		✓	
Brake pads & discs	S/I		✓		✓		✓		✓		✓		✓	
Propeller shaft grease	S/I		✓		✓		✓		✓		✓		✓	
Steering knuckle & chassis grease	S/I		✓		✓		✓		✓		✓		✓	
Steering linkage	S/I		✓		✓		✓		✓		✓		✓	
Transfer and differential oil	S/I		✓		✓		✓		✓		✓		✓	
Air cleaner filter	R				✓				✓				✓	
Spark plugs ①	R				✓				✓				✓	
Drive belts	S/I				✓				✓				✓	
Exhaust pipes & mountings	S/I				✓				✓				✓	
Fuel lines & connections	S/I				✓				✓				✓	
Engine coolant	R						✓				✓			
Charcoal canister	R								✓					
Fuel tank cap gasket	R								✓					
Heated oxygen sensors (except Calif.)	R										✓			

R: Replace S/I: Service or Inspect

① Platinum plugs, replace every 100,000 miles

② Heated oxygen sensors (except Calif.): replace every 80,000 miles.

FREQUENT OPERATION MAINTENANCE (SEVERE SERVICE)

If a vehicle is operated under any of the following conditions it is considered severe service:

- Extremely dusty areas.

- 50% or more of the vehicle operation is in 32°C (90°F) or higher temperatures, or constant operation in temperatures below 0°C (32°F).

- Prolonged idling (vehicle operation in stop and go traffic).

- Frequent short running periods (engine does not warm to normal operating temperatures).

- Police, taxi, delivery usage or trailer towing usage.

Air cleaner filter: service or inspect every 3750 miles

Engine oil & filter: replace every 3750 miles.

Ball joints & dust covers: service or inspect every 7500 miles.

Bolts & nuts on chassis & body: service or inspect every 7500 miles.

Brake pads & discs (front & rear): service or inspect every 7500 miles.

Steering knuckle & chassis grease: service or inspect every 7500 miles.

Steering linkage: service or inspect every 7500 miles.

Propeller shaft grease: service or inspect every 7500 miles.

Exhaust pipes & mountings: service or inspect every 15,000 miles.

67170-LCSQ-C12

SCHEDULED MAINTENANCE INTERVALS

Toyota Land Cruiser

TO BE SERVICED	TYPE OF SERVICE	VEHICLE MILEAGE INTERVAL (x1000)																		
		5	10	15	20	25	30	35	40	45	50	55	60	65	70	75	80	85	90	95
Automatic transmission and differential fluid	S/I			✓			✓			✓			✓			✓			✓	
Ball joints and boots	S/I			✓			✓			✓			✓			✓			✓	
Brake system	S/I			✓			✓			✓			✓			✓			✓	
Charcoal canister	S/I												✓							
Drive belts	S/I						✓						✓						✓	
Driveshaft bushing (4WD)	L						✓						✓						✓	
Engine coolant	R						✓						✓						✓	
Engine oil & filter	R	✓	✓	✓	✓	✓	✓	✓	✓	✓	✓	✓	✓	✓	✓	✓	✓	✓	✓	✓
Exhaust pipes & mounts	S/I			✓			✓			✓			✓			✓			✓	
Fuel lines	S/I						✓						✓						✓	
Fuel tank cap gasket	S/I						✓						✓						✓	
Halfshaft boots & flange bolts	S/I			✓			✓			✓			✓			✓			✓	
Limited slip differential fluid	R						✓						✓						✓	
Non-platinum spark plugs	R						✓						✓						✓	
Platinum spark plugs	R												✓							
Propeller shaft (4WD)	L			✓			✓			✓			✓			✓			✓	
Propeller shaft bolts	S/I			✓			✓			✓			✓			✓			✓	
Rack and pinion assembly	S/I			✓			✓			✓			✓			✓			✓	
Rear wheel bearing	L						✓						✓						✓	
Steering Knuckle	L			✓			✓			✓			✓			✓			✓	
Steering linkage	S/I			✓			✓			✓			✓			✓			✓	
Valves	S/I												✓							

R: Replace S/I: Service or Inspect L: Lubricate

FREQUENT OPERATION MAINTENANCE (SEVERE SERVICE)

If a vehicle is operated under any of the following conditions it is considered severe service:

- Towing a trailer or using a camper or car-top carrier.

- Repeated short trips of less than 5 miles in temperatures below freezing.

- Excessive idling or low-speed driving for long distances as in heavy commercial use, such as delivery, taxi or police cars.

- Operating on rough, muddy or salt-covered roads.

- Operating on unpaved or dusty roads.

Oil filter: service or inspect every 5000 miles or 4 months, whichever occurs first.

Brake linings and discs or drums: service or inspect every 5000 miles or 4 months, whichever occurs first.

Steering linkage: service or inspect every 5000 miles or 4 months, whichever occurs first.

Ball joints and boots: service or inspect every 5000 miles or 4 months, whichever occurs first.

Brake discs & pads (front): service or inspect every 6000 miles.

Halfshaft boots: service or inspect every 5000 miles or 4 months. Retighten the flange bolts, whichever occurs first.

Body chassis bolts and nuts: service or inspect every 5000 miles or 4 months, whichever occurs first.

Transmission and differential fluid: replace every 15,000 miles or 12 months, whichever occurs first.

Transfer case and differential fluid: replace every 15,000 miles or 12 months, whichever occurs first.

Timing belt: replace every 60,000 miles or 48 months, whichever occurs first.

67170-LCSQ-C13

PRECAUTIONS

Before servicing any vehicle, please be sure to read all of the following precautions, which deal with personal safety, prevention of component damage, and important points to take into consideration when servicing a motor vehicle:

- Never open, service or drain the radiator or cooling system when the engine is hot; serious burns can occur from the steam and hot coolant.
- Observe all applicable safety precautions when working around fuel. Whenever servicing the fuel system, always work in a well-ventilated area. Do not allow fuel spray or vapors to come in contact with a spark, open flame, or excessive heat (a hot drop light, for example). Keep a dry chemical fire extinguisher near the work area. Always keep fuel in a container specifically designed for fuel storage; also, always properly seal fuel containers to avoid the possibility of fire or explosion. Refer to the additional fuel system precautions later in this section.
- Fuel injection systems often remain pressurized, even after the engine has been turned **OFF**. The fuel system pressure must be relieved before disconnecting any fuel lines. Failure to do so may result in fire and/or personal injury.
- Brake fluid often contains polyglycol ethers and polyglycols. Avoid contact with the eyes and wash your hands thoroughly after handling brake fluid. If you do get brake fluid in your eyes, flush your eyes with clean, running water for 15 minutes. If eye irritation persists, or if you have taken brake fluid internally, IMMEDIATELY seek medical assistance.
- The EPA warns that prolonged contact with used engine oil may cause a number of skin disorders, including cancer. You should make every effort to minimize your exposure to used engine oil. Protective gloves should be worn when changing oil. Wash your hands and any other exposed skin areas as soon as possible after exposure to used engine oil. Soap and water, or waterless hand cleaner should be used.
- All new vehicles are now equipped with an air bag system. The system must be disabled before performing service on or around system components, steering column, instrument panel components, wiring and sensors. Failure to follow safety and disabling procedures could result in accidental air bag deployment, possible personal injury and unnecessary system repairs.
- Always wear safety goggles when working with, or around, the air bag system. When carrying a non-deployed air bag, be sure the bag and trim cover are pointed away from your body. When placing a non-deployed air bag on a work surface, always face the bag and trim cover upward, away from the surface. This will reduce the motion of the module if it is accidentally deployed. Refer to the additional air bag system precautions later in this section.
- NEVER disconnect the negative battery cable with the ignition **ON** or the engine running. Removing power from the computer control module with the ignition **ON** may destroy the module.
- Clean, high quality brake fluid from a sealed container is essential to the safe and proper operation of the brake system. You should always buy the correct type of brake fluid for your vehicle. If the brake fluid becomes contaminated, completely flush the system with new fluid. Never reuse any brake fluid. Any brake fluid that is removed from the system should be discarded. Also, do not allow any brake fluid to come in contact with a painted surface; it will damage the paint.
- Never operate the engine without the proper amount and type of engine oil; doing so WILL result in severe engine damage.
- Timing belt maintenance is extremely important. Many models utilize an interference-type, non-freewheeling engine. If the timing belt breaks, the valves in the cylinder head may strike the pistons, causing potentially serious (also time-consuming and expensive) engine damage. Refer to the maintenance interval charts in the front of this manual for the recommended replacement interval for the timing belt, and to the timing belt section for belt replacement and inspection.
- Disconnecting the negative battery cable on some vehicles may interfere with the functions of the on-board computer system(s) and may require the computer to undergo a relearning process once the negative battery cable is reconnected.
- When servicing drum brakes, only disassemble and assemble one side at a time, leaving the remaining side intact for reference.
- Only an MVAC-trained, EPA-certified automotive technician should service the air conditioning system or its components.

ENGINE REPAIR

➡Disconnecting the negative battery cable on some vehicles may interfere with the functions of the on board computer system. The computer may undergo a relearning process once the negative battery cable is reconnected.

Distributor

All models are equipped with a distibutorless ignition system.

Alternator

REMOVAL

Land Cruiser

1. Before servicing the vehicle, refer to the precautions in the beginning of this section.
2. Drain the cooling system.
3. Remove or disconnect the following:
   - Negative battery cable
   - Accessory drive belt
   - Engine under cover
   - Radiator
   - Power steering pump pulley
   - Alternator harness connectors
   - Alternator

To install:

4. Install or connect the following:
   - Alternator. Tighten the fasteners to 29 ft. lbs. (39 Nm).
   - Alternator harness connectors
   - Power steering pump pulley
   - Radiator
   - Engine under cover
   - Accessory drive belt
   - Negative battery cable
5. Fill the cooling system.
6. Start the engine and check for leaks.

Sequoia

1. Before servicing the vehicle, refer to the precautions in the beginning of this section.
2. Drain the cooling system.
3. Remove or disconnect the following:
   - Negative battery cable
   - Accessory drive belt
   - Engine under cover
   - Power steering pump
   - Alternator harness connectors
   - Alternator

To install:

4. Install or connect the following:
   - Alternator. Tighten the fasteners to 29 ft. lbs. (39 Nm).

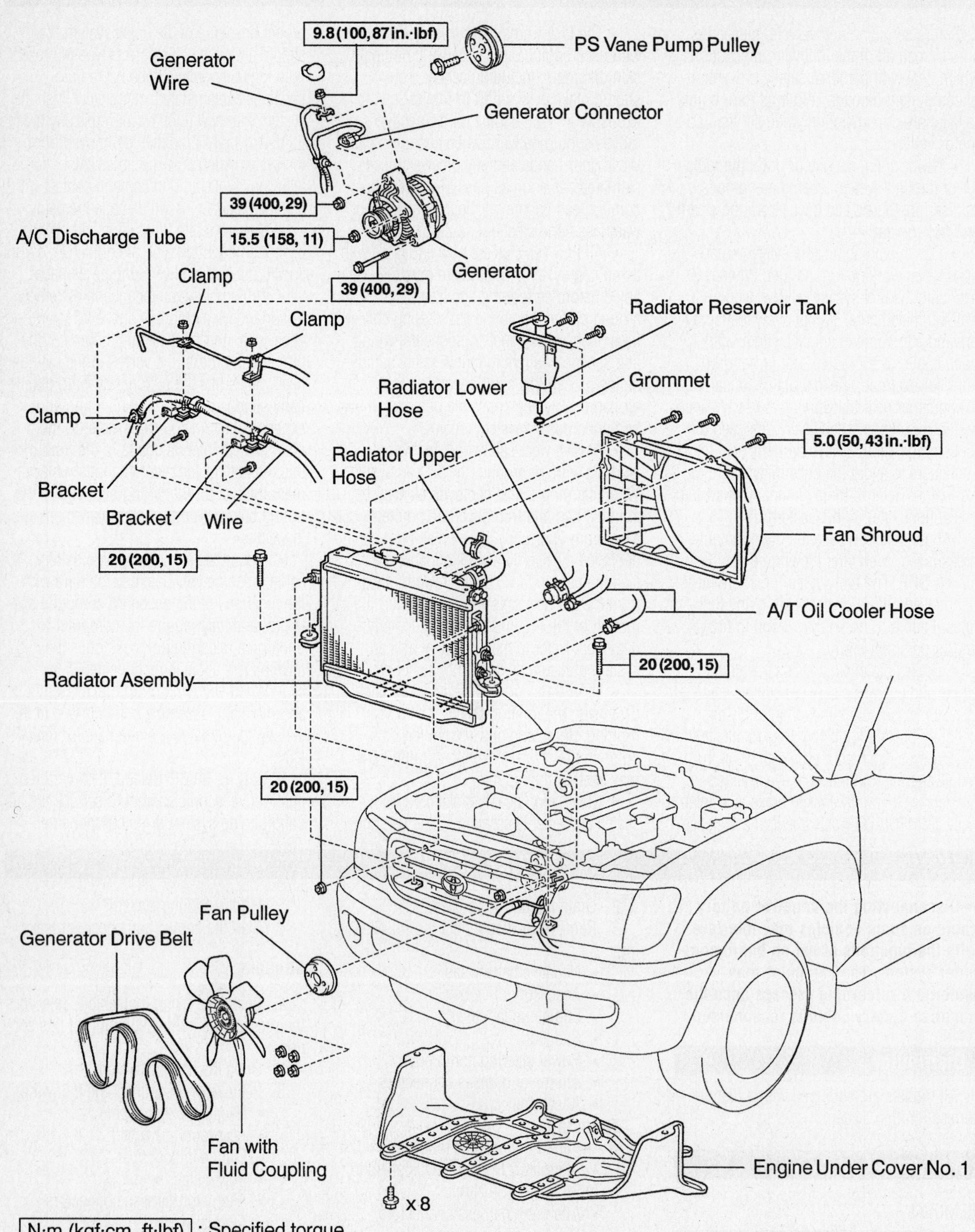
9.8 (100, 87 in.·lbf)
PS Vane Pump Pulley
Generator
Wire
Generator Connector
39 (400, 29)
A/C Discharge Tube
15.5 (158, 11)
Generator
Clamp
39 (400, 29)
Clamp
Radiator Reservoir Tank
Radiator Lower
Hose
Grommet
Clamp
5.0 (50, 43 in.·lbf)
Radiator Upper
Hose
Bracket
Bracket
Wire
Fan Shroud
20 (200, 15)
A/T Oil Cooler Hose
20 (200, 15)
Radiator Asembly
20 (200, 15)
Fan Pulley
Generator Drive Belt
Fan with
Fluid Coupling
Engine Under Cover No. 1
x 8
N·m (kgf·cm, ft·lbf) : Specified torque
67170-LCSQ-G01

Alternator and related parts—Land Cruiser

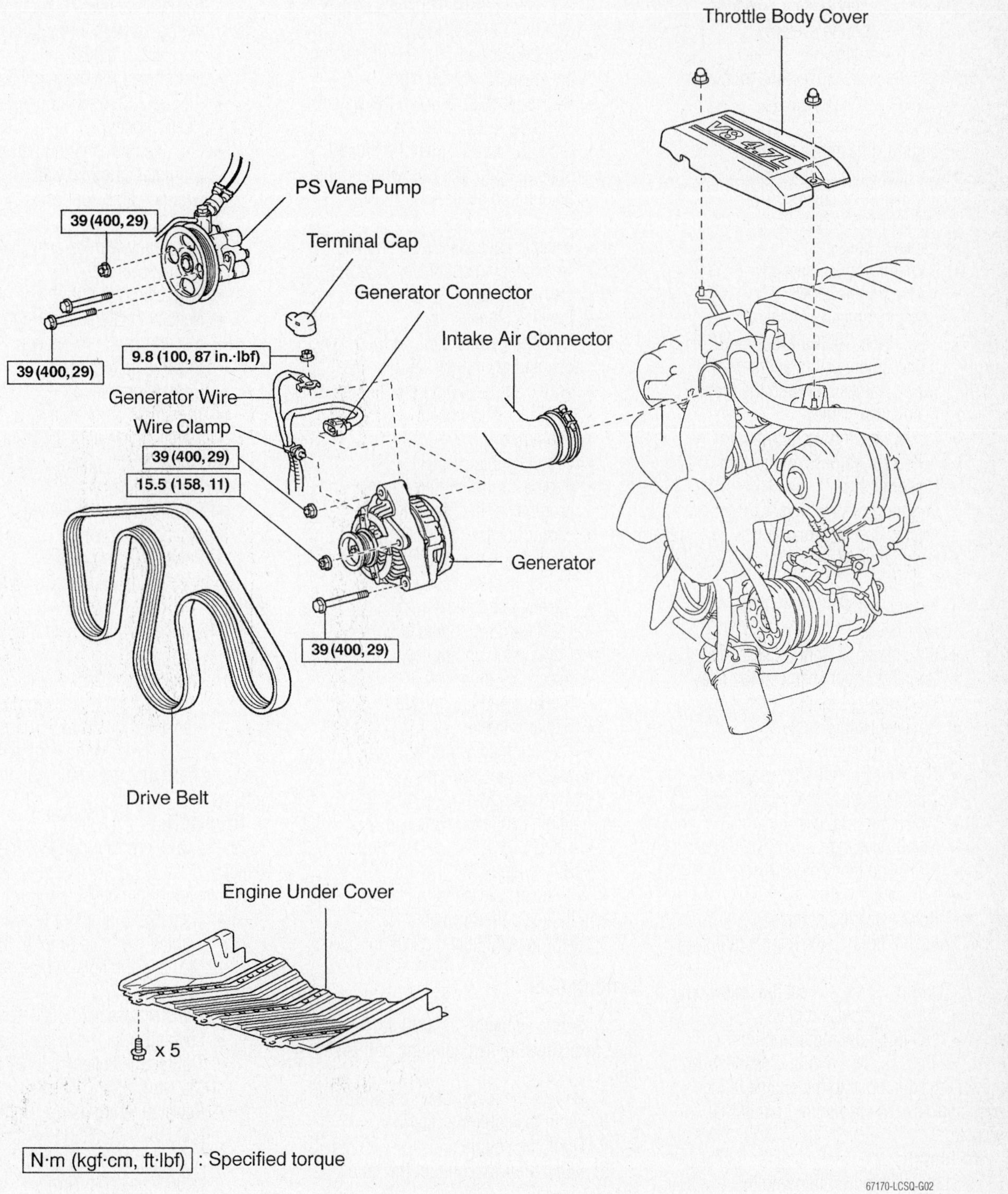

Throttle Body Cover
PS Vane Pump
39 (400, 29)
Terminal Cap
Generator Connector
Intake Air Connector
9.8 (100, 87 in.·lbf)
39 (400, 29)
Generator Wire
Wire Clamp
39 (400, 29)
15.5 (158, 11)
Generator
39 (400, 29)
Drive Belt

Engine Under Cover

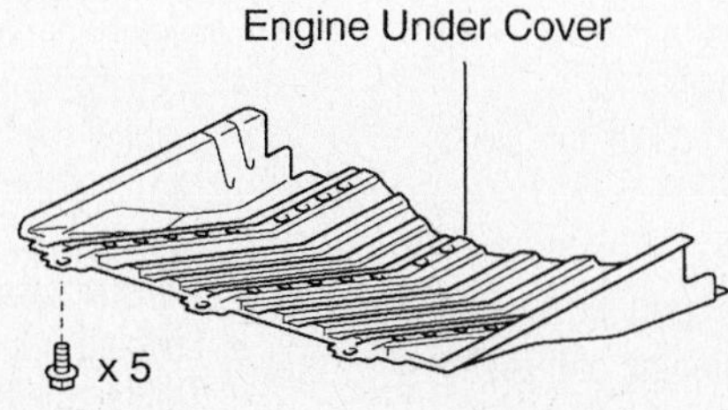

N·m (kgf·cm, ft·lbf) : Specified torque

67170-LCSQ-G02

Alternator and related parts—Sequoia

- Alternator harness connectors
- Power steering pump
- Engine under cover
- Accessory drive belt
- Negative battery cable

5. Fill the cooling system.
6. Start the engine and check for leaks.

Ignition Timing

ADJUSTMENT

All engines are equipped with a Distributorless Ignition System (DIS). No timing adjustment is possible.

Engine Assembly

REMOVAL & INSTALLATION

Sequoia

1. Before servicing the vehicle, refer to the precautions in the beginning of this section.

2. Relieve the fuel system pressure.
3. Drain the cooling system.
4. Drain the engine oil.
5. Remove or disconnect the following:
   - Battery and tray
   - Hood
   - Engine appearance cover
   - Air intake pipe
   - Engine under covers
   - Coolant recovery tank
   - Radiator hoses
   - Radiator and fan shroud
   - Accessory drive belt
   - Cooling fan and pulley
   - Powertrain Control Module (PCM) harness connectors and pass the wiring harness through the firewall
   - Accelerator cable
   - Power steering vacuum hoses
   - Alternator harness connectors
   - Heater hoses
   - Engine control wiring harness and grommet at the firewall
   - Ground cable connector
   - Fuel lines
   - Evaporative Emissions (EVAP) canister hoses
   - Wire clamp at right inner fender
   - Negative battery cable at the relay box and right inner fender
   - Positive battery cable
   - Center console
   - Transmission shift lever assembly
   - Transfer case shift lever and rod
   - Exhaust front pipes
   - Stabilizer bar
   - Front and rear driveshafts
   - A/C compressor
   - Power steering pump
6. Attach a hoist to the engine lifting eyes.
7. Remove or disconnect the following:
   - Transfer case skid plate
   - Left and right motor mounts
   - Transmission mount crossmember
8. Attach a hoist to the engine lifting eyes and raise the powertrain out of the vehicle.

To install:

9. Lower the powertrain into the vehicle.
10. Install or connect the following:
    - Transmission mount crossmember. Tighten the bolts to 37 ft. lbs. (50 Nm) and the nuts to 55 ft. lbs. (74 Nm).
    - Transfer case skid plate
    - Left and right motor mounts. Tighten the fasteners to 22 ft. lbs. (30 Nm).
    - Power steering pump. Tighten the bolts to 13 ft. lbs. (17 Nm).
    - A/C compressor. Tighten the bolts to 36 ft. lbs. (49 Nm).
    - Front driveshaft. Tighten the fasteners to 59 ft. lbs. (80 Nm).
    - Rear driveshaft. Tighten the fasteners to 78 ft. lbs. (106 Nm).
    - Stabilizer bar. Tighten the bracket bolts to 13 ft. lbs. (18 Nm) and the link nuts to 18 ft. lbs. (25 Nm).
    - Exhaust front pipes
    - Transfer case shift lever and rod
    - Transmission shift lever assembly
    - Center console
    - Positive battery cable
    - Negative battery cable at the relay box and right inner fender
    - Wire clamp at right inner fender
    - EVAP canister hoses
    - Fuel lines
    - Ground cable connector
    - Engine control wiring harness and grommet at the firewall
    - Heater hoses
    - Alternator harness connectors
    - Power steering vacuum hoses
    - Accelerator cable
    - PCM harness connectors
    - Cooling fan and pulley
    - Accessory drive belt
    - Radiator and fan shroud
    - Radiator hoses
    - Coolant recovery tank
    - Engine under covers
    - Air intake pipe
    - Engine appearance cover
    - Hood
    - Battery and tray
11. Fill the crankcase to the correct level.
12. Fill the cooling system.
13. Start the engine and check for leaks.

Land Cruiser

1. Before servicing the vehicle, refer to the precautions in the beginning of this section.
2. Relieve the fuel system pressure.
3. Drain the cooling system.
4. Drain the engine oil.
5. Remove or disconnect the following:
   - Battery and tray
   - Hood
   - Engine appearance cover
   - Air intake pipe
   - Engine under covers
   - Coolant recovery tank
   - Radiator hoses
   - Radiator and fan shroud
   - Accessory drive belt
   - Cooling fan and pulley
   - Powertrain Control Module (PCM) harness connectors and pass the wiring harness through the firewall
   - Accelerator cable
   - Power steering vacuum hoses
   - Alternator harness connectors
   - Heater hoses
   - Engine control wiring harness and grommet at the firewall
   - Ground cable connector
   - Fuel lines
   - Evaporative Emissions (EVAP) canister hoses
   - Wire clamp at right inner fender
   - Negative battery cable at the relay box and right inner fender
   - Positive battery cable
   - Center console
   - Transmission shift lever assembly
   - Transfer case shift lever and rod
   - Exhaust front pipes
   - Stabilizer bar
   - Front and rear driveshafts
   - A/C compressor
   - Power steering pump
6. Attach a hoist to the engine lifting eyes.
7. Remove or disconnect the following:
   - Transfer case skid plate
   - Left and right motor mounts
   - Transmission mount crossmember
8. Attach a hoist to the engine lifting eyes and raise the powertrain out of the vehicle.

To install:

9. Lower the powertrain into the vehicle.
10. Install or connect the following:
    - Transmission mount crossmember. Tighten the bolts to 37 ft. lbs. (50 Nm) and the nuts to 55 ft. lbs. (74 Nm).
    - Transfer case skid plate
    - Left and right motor mounts. Tighten the fasteners to 22 ft. lbs. (30 Nm).
    - Power steering pump. Tighten the bolts to 13 ft. lbs. (17 Nm).
    - A/C compressor. Tighten the bolts to 36 ft. lbs. (49 Nm).
    - Front driveshaft. Tighten the fasteners to 59 ft. lbs. (80 Nm).
    - Rear driveshaft. Tighten the fasteners to 78 ft. lbs. (106 Nm).
    - Stabilizer bar. Tighten the bracket bolts to 13 ft. lbs. (18 Nm) and the link nuts to 18 ft. lbs. (25 Nm).
    - Exhaust front pipes
    - Transfer case shift lever and rod
    - Transmission shift lever assembly
    - Center console
    - Positive battery cable

- Negative battery cable at the relay box and right inner fender
- Wire clamp at right inner fender
- EVAP canister hoses
- Fuel lines
- Ground cable connector
- Engine control wiring harness and grommet at the firewall
- Heater hoses
- Alternator harness connectors
- Power steering vacuum hoses
- Accelerator cable
- PCM harness connectors
- Cooling fan and pulley
- Accessory drive belt
- Radiator and fan shroud
- Radiator hoses
- Coolant recovery tank
- Engine under covers
- Air intake pipe
- Engine appearance cover
- Hood
- Battery and tray

11. Fill the crankcase to the correct level.
12. Fill the cooling system.
13. Start the engine and check for leaks.

Heater Core

REMOVAL & INSTALLATION

Land Cruiser

FRONT HEATER

1. Disconnect the negative battery cable.
2. Drain the cooling system into a clean container for reuse.
3. Disconnect the heater hoses from the heater core.
4. Remove the steering wheel by performing the following procedure:
   a. Position the front wheels facing straight-ahead.
   b. Remove the steering wheel side covers.
   c. Using a Torx® wrench, loosen the 2 screws located at each side of the steering wheel until the screw's circumference groove catches on the screw case.
   d. Pull the air bag module from the steering wheel and disconnect the electrical connector.

**** CAUTION**

Place the air bag module in a safe place with the front side facing upward.

   e. Remove the steering wheel nut.
   f. Place alignment marks on the steering wheel and the main shaft.
   g. Using a steering wheel puller, press the steering wheel from the steering column.
5. Remove the instrument panel and reinforcement by performing the following procedure:
   a. Remove the front door scuff plates, the cowl side trim and the front door opening trim.
   b. At the driver's side, remove the 2 assist grip plugs, the 2 screws and assist grip and the front pillar garnish.
   c. At the passenger's side, remove the 4 assist grip plugs, the 4 screws, the 2 assist grips and the front pillar garnish.
   d. Remove the instrument cluster finish panel.
   e. Remove the 2 screws and the hood lock control cable.
   f. Remove the 2 screws and the fuel lid control cable lever.
   g. Remove the lower No. 1 panel screw and the panel.
   h. Remove the lower left side panel.
   i. Remove the 3 steering column cover screws and the covers.
   j. At the steering column, disconnect the electrical connectors; then, remove the clamp, the 3 screws and the combination switch.
   k. Remove the No. 2 heater-to-register duct screw and the duct.
   l. Remove the steering column-to-instrument panel bolts and the steering column.
   m. At the combination meter, disconnect the electrical connectors; then, remove the 4 screws and the combination meter.
   n. Remove the glove compartment door stoppers, the 2 screws and the glove box door.
   o. At the passenger's side air bag module, remove the No. 1 undercover, pull the air bag connector up from the undercover and disconnect it; then, remove the air bag.

**** CAUTION**

Place the air bag module in a safe place with the front side facing upward.

   p. Remove the 3 lower No. 2 panel screws and the panel.
   q. Remove the center cluster; then, pry the center cluster from the dash by prying the 8 clips in the following order:
   - Left side
   - Right side
   - Top left side
   - Top right side

   r. Remove the 4 radio screws, pull the radio outward, disconnect the electrical connectors and remove the radio.
   s. At the rear console panel, remove the transfer shift lever knob. Pry the panel upward disengaging the 4 clips (2 on each side) and remove the panel.
   t. At the rear of the console, remove the 2 rear end panel-to-console screws; then, pry the end panel rearward disengaging the 2 clips and remove the panel.
   u. If not equipped with a rear air conditioning system, disconnect the connector and control cable; then, remove the 3 rear heater control panel screws and the panel.
   v. Remove the 4 rear console box-to-chassis screws/bolts and the console box.
   w. Remove the center lower cluster finish panel by prying panel rearward disengaging the 5 clips; then, disconnect the electrical connector.
   x. Remove the 2 front console-to-chassis bolts/screws, disengage the 2 clips and remove the console.
   y. At the instrument panel, disconnect the junction connectors (the connectors can be disconnected by loosening the bolts), the instrument panel-to-chassis 8 bolts and 2 nuts.

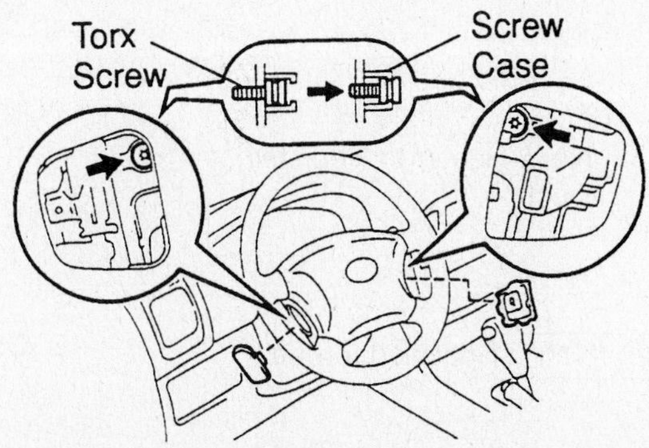

View the steering wheel's Torx® bolts—Toyota Land Cruiser

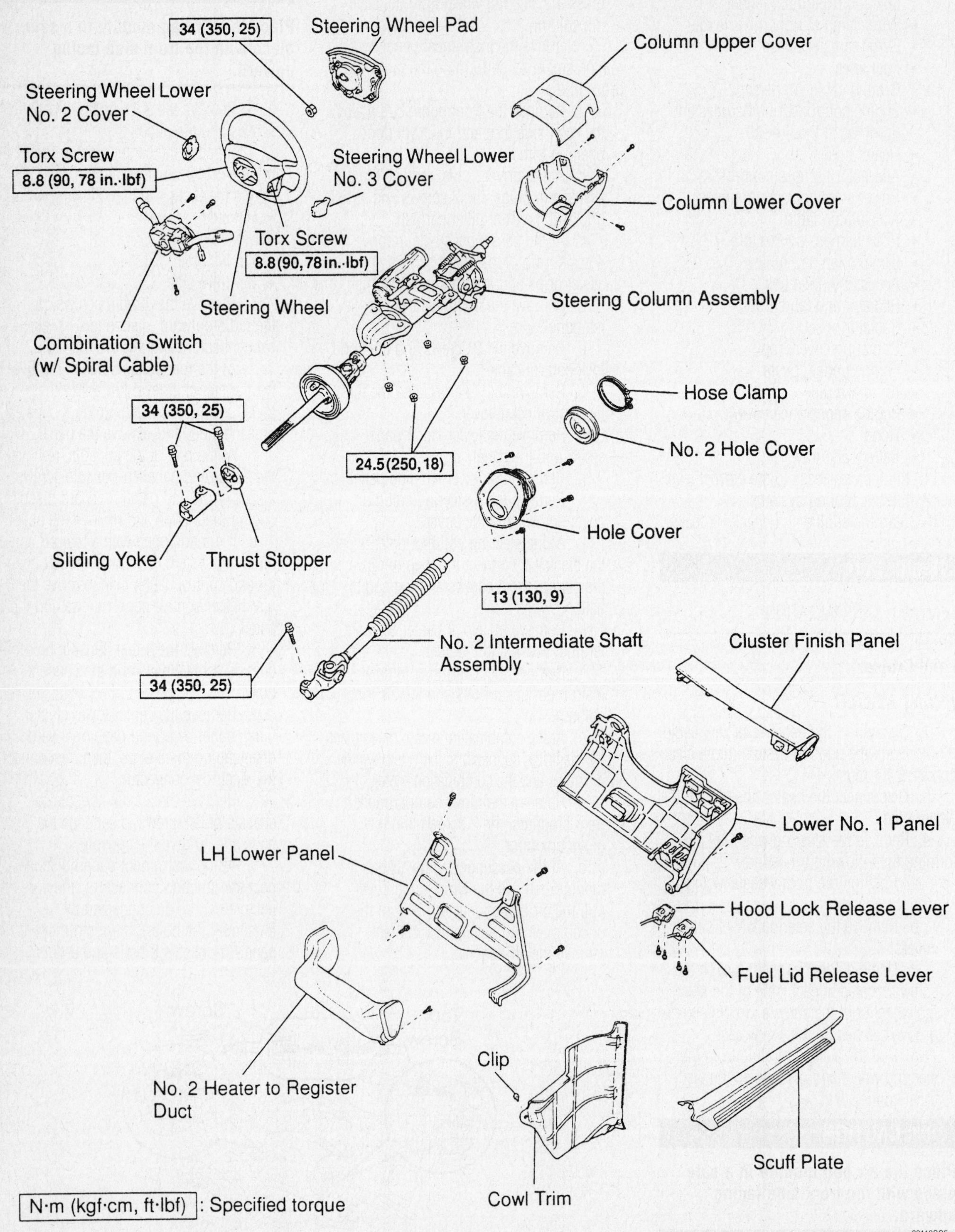

Exploded view the steering column—Toyota Land Cruiser (Part 1 of 2)

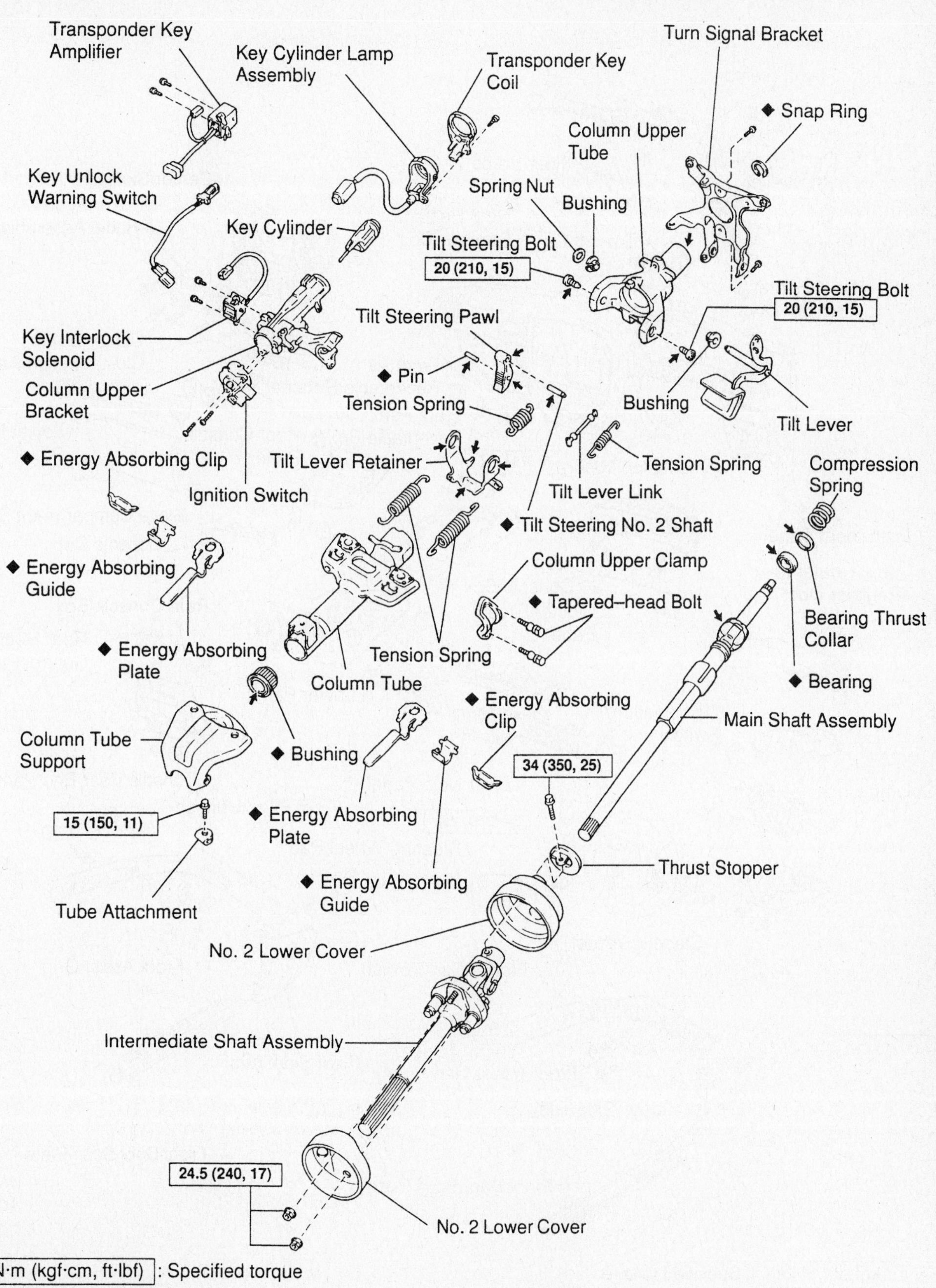
Transponder Key Amplifier
Key Cylinder Lamp Assembly
Transponder Key Coil
Turn Signal Bracket
◆ Snap Ring
Column Upper Tube
Key Unlock Warning Switch
Spring Nut
Bushing
Key Cylinder
Tilt Steering Bolt
20 (210, 15)
Tilt Steering Bolt
20 (210, 15)
Tilt Steering Pawl
Key Interlock Solenoid
◆ Pin
Column Upper Bracket
Tension Spring
Bushing
Tilt Lever
◆ Energy Absorbing Clip
Tilt Lever Retainer
Tension Spring
Compression Spring
Ignition Switch
Tilt Lever Link
◆ Tilt Steering No. 2 Shaft
◆ Energy Absorbing Guide
Column Upper Clamp
◆ Tapered–head Bolt
Bearing Thrust Collar
◆ Energy Absorbing Plate
Tension Spring
Column Tube
◆ Bearing
◆ Energy Absorbing Clip
Main Shaft Assembly
Column Tube Support
◆ Bushing
34 (350, 25)
15 (150, 11)
◆ Energy Absorbing Plate
Thrust Stopper
◆ Energy Absorbing Guide
Tube Attachment
No. 2 Lower Cover
Intermediate Shaft Assembly
24.5 (240, 17)
No. 2 Lower Cover
N·m (kgf·cm, ft·lbf) : Specified torque
◆ Non–reusable part
Molybdenum disulfide lithium base grease
93113GG6

Exploded view the steering column—Toyota Land Cruiser (Part 2 of 2)

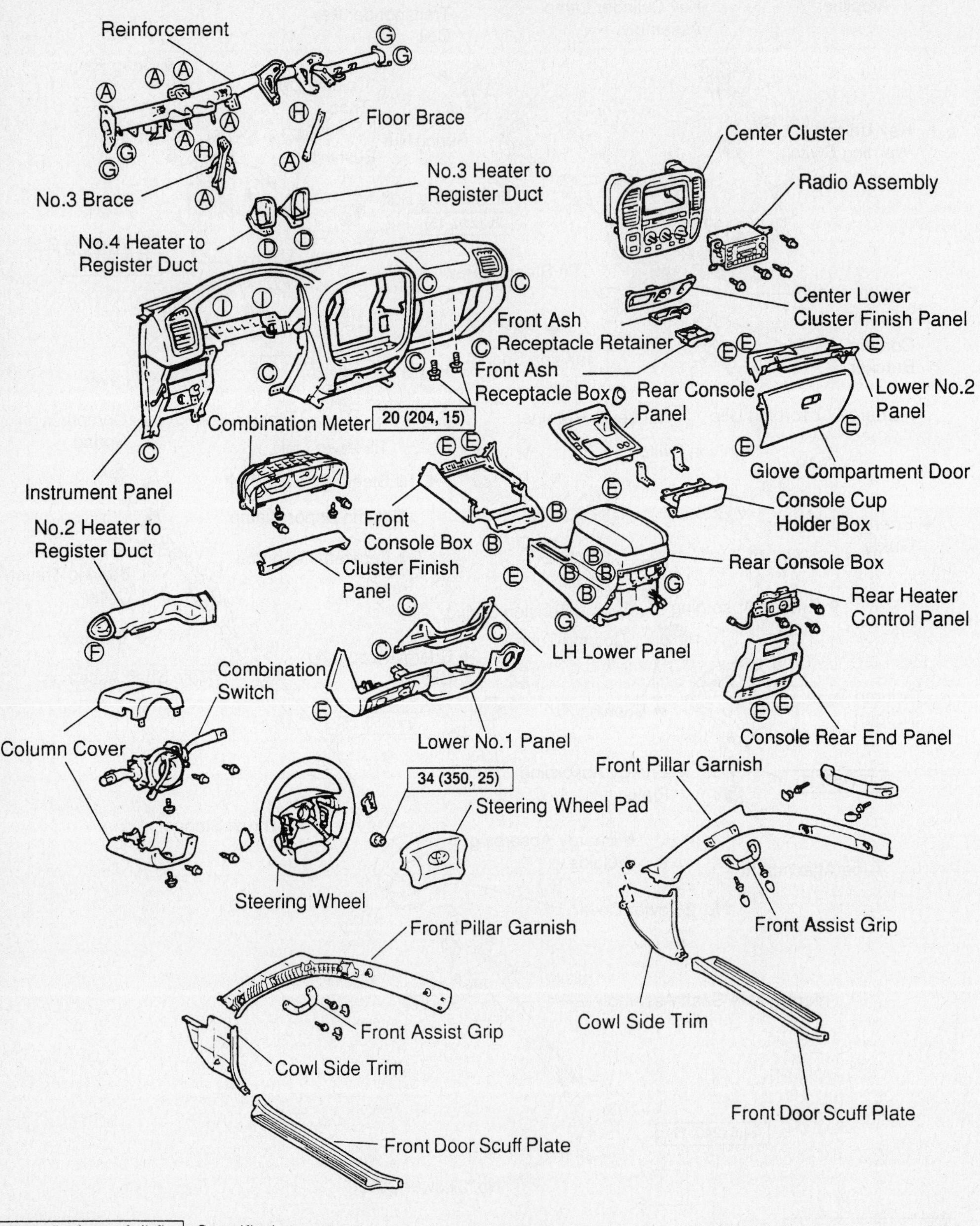
Reinforcement
Floor Brace
No.3 Brace
No.3 Heater to
Register Duct
No.4 Heater to
Register Duct
Center Cluster
Radio Assembly
Center Lower
Cluster Finish Panel
Front Ash
Receptacle Retainer
Front Ash
Receptacle Box
Rear Console
Panel
Lower No.2
Panel
Combination Meter
20 (204, 15)
Instrument Panel
Glove Compartment Door
No.2 Heater to
Register Duct
Front
Console Box
Cluster Finish
Panel
Console Cup
Holder Box
Rear Console Box
Rear Heater
Control Panel
LH Lower Panel
Combination
Switch
Column Cover
Lower No.1 Panel
Console Rear End Panel
Front Pillar Garnish
34 (350, 25)
Steering Wheel Pad
Steering Wheel
Front Pillar Garnish
Front Assist Grip
Front Assist Grip
Cowl Side Trim
Cowl Side Trim
Front Door Scuff Plate
Front Door Scuff Plate
N·m (kgf·cm, ft·lbf) : Specified torque
93113GG7

Exploded view the instrument panel and related components—Toyota Land Cruiser

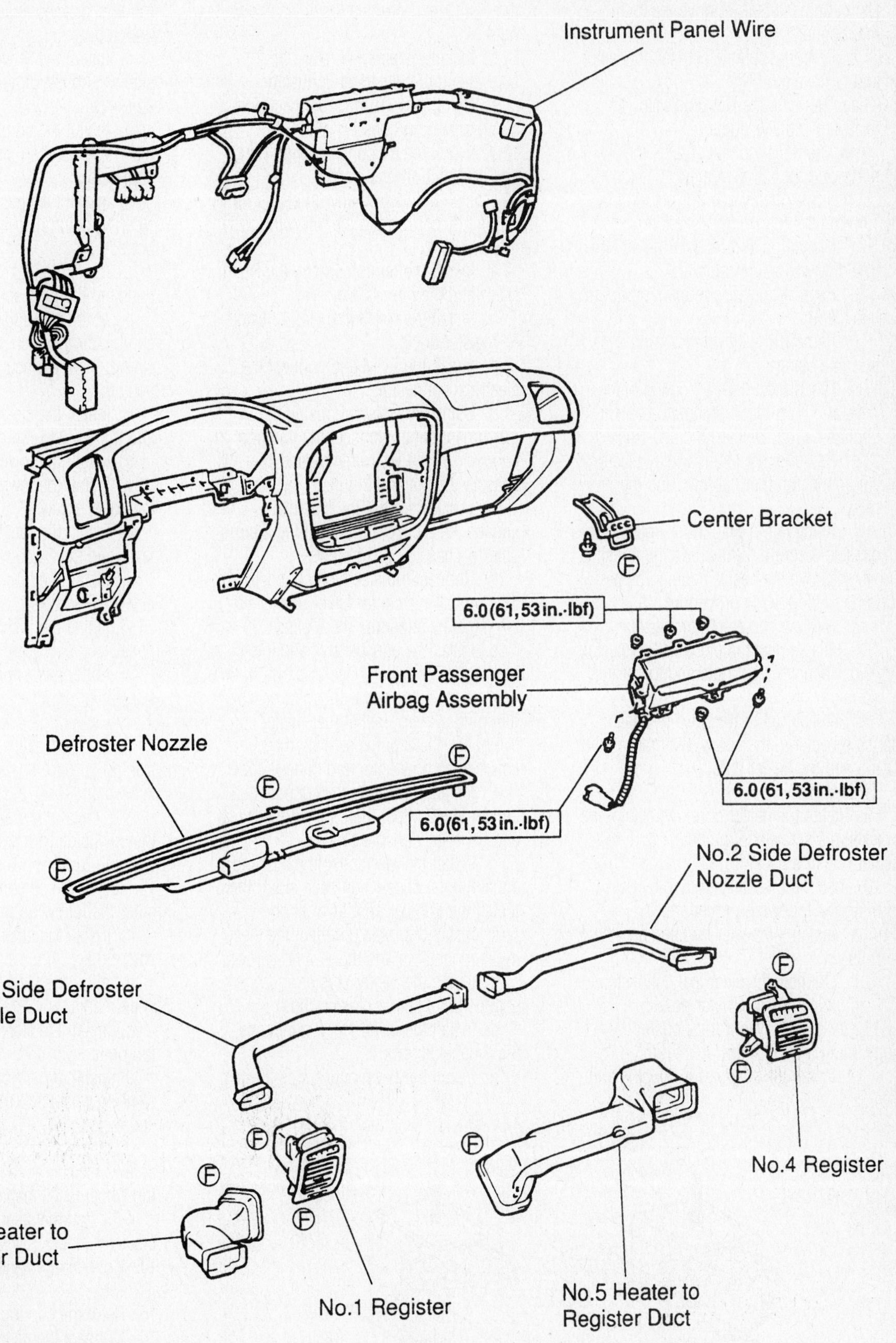

Exploded view the front ventilation ducts and related components—Toyota Land Cruiser

Using an assistant, remove the instrument panel.

z. Disconnect the electrical connector and remove the ECM.

aa. Remove the No. 3 and No. 4 heater-to-register ducts.

bb. Remove the floor brace, the No. 1 brace and the reinforcement.

6. Remove the evaporator housing by performing the following procedure:

a. Discharge and recover the air conditioning system refrigerant.

b. Remove the air conditioning liquid line clamp.

c. Remove the air conditioning suction line clamp.

d. Disconnect both air conditioning lines and plug the openings to prevent contamination. Discard the 4 O-rings.

e. Remove the antenna relay electrical connector, the 2 screws and the relay.

f. Remove the evaporator housing-to-chassis 4 screws/2 nuts and the housing.

7. Remove the heater housing by performing the following procedure:

a. Remove the defroster nozzle.

b. Disconnect the electrical connector.

c. Remove the 4 nuts and the heater housing.

8. Remove the heater core-to-heater housing packing, the screw, the bracket, the clamp and the heater core.

To install:

9. Install the heater core, the clamp, the bracket, the screw and the heater core-to-heater housing packing.

10. Install the heater housing by performing the following procedure:

a. Install the heater housing and the 4 nuts.

b. Connect the electrical connector.

c. Install the defroster nozzle.

11. Install the evaporator housing by performing the following procedure:

a. Install the evaporator housing and the housing-to-chassis 4 screws and 2 nuts.

b. Install the antenna relay, the 2 screws and the electrical connector.

c. Using new O-rings, connect both air conditioning lines.

d. Install the air conditioning liquid line and suction line clamp.

12. Install the instrument panel and reinforcement by performing the following procedure:

a. Install the reinforcement, the No. 1 brace and the floor brace.

b. Install the No. 3 and No. 4 heater-to-register ducts.

c. Install the ECM and connect the electrical connector.

d. Using an assistant, install the instrument panel, connect the junction connectors, the instrument panel-to-chassis 8 bolts and 2 nuts.

e. Install the front the console, engage the 2 clips and install the 2 console-to-chassis bolts/screws.

f. Connect the electrical connector; then, install the center lower cluster finish panel by engaging the 5 clips.

g. Install the console box and the 4 rear console box-to-chassis screws/bolts.

h. If not equipped with a rear air conditioning system, install rear heater control panel, the 3 panel screws; then, connect the connector and control cable.

i. Install the rear of the console and engage the 2 clips; then, install the 2 rear end panel-to-console screws.

j. Install the rear console panel and engage the 4 clips (2 on each side); then, install the transfer shift lever knob.

k. Install the radio, connect the electrical connectors and the 4 radio screws.

l. Install the center cluster and engage the 8 center cluster clips.

m. Install the lower No. 2 panel and the 3 panel screws.

n. Install the passenger's side air bag module, connect it and install the No. 1 undercover.

o. Install the glove box door, the 2 screws and the glove compartment door stoppers.

p. Install the combination meter and the 4 screws; then, connect the electrical connectors.

q. Install the steering column and the steering column-to-instrument panel bolts.

r. Install the No. 2 heater-to-register duct and the duct screw.

s. At the steering column, install the combination switch, the 3 screws and the clamp; then, connect the electrical connectors.

t. Install the steering column covers and the 3 covers screws.

u. Install the lower left side panel.

v. Install the lower No. 1 panel and the panel screw.

w. Install the fuel lid control cable lever and the 2 screws.

x. Install the hood lock control cable and the 2 screws.

y. Install the instrument cluster finish panel.

z. At the passenger's side, install the front pillar garnish, the 2 assist grips, the 4 screws and the 4 assist grip plugs.

aa. At the driver's side, install the front pillar garnish, assist grip, the 2 screws and the 2 assist grip plugs.

bb. Install the front door scuff plates, the cowl side trim and the front door opening trim.

13. Install the steering wheel by performing the following procedure:

a. Install the steering wheel to the steering column.

b. Align the steering wheel-to-main shaft marks.

c. Install the steering wheel nut and torque to 25 ft. lbs. (34 Nm).

d. Install the air bag module to the steering wheel and connect the electrical connector.

e. Using a Torx® wrench, tighten the 2 screws located at each side of the steering wheel to 78 inch lbs. (8.8 Nm).

f. Install the steering wheel side covers.

14. Connect the heater hoses to the heater core.

15. Refill the cooling system.

16. Connect the negative battery cable.

a. Evacuate and charge the air conditioning system refrigerant.

17. Run the engine to normal operating temperatures. Check the climate control operation and check for leaks.

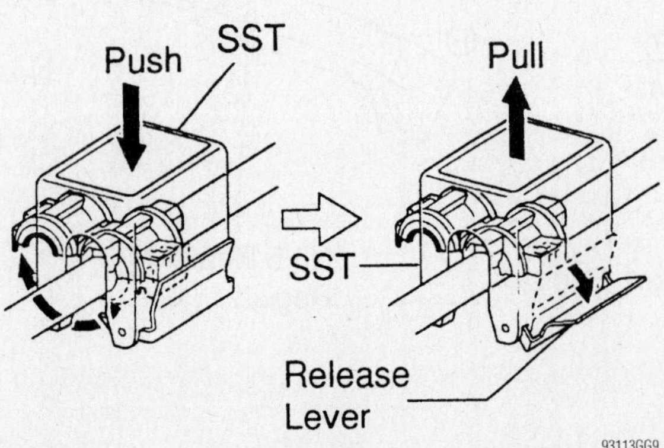

View the air conditioning line clamp removal tool—Toyota Land Cruiser

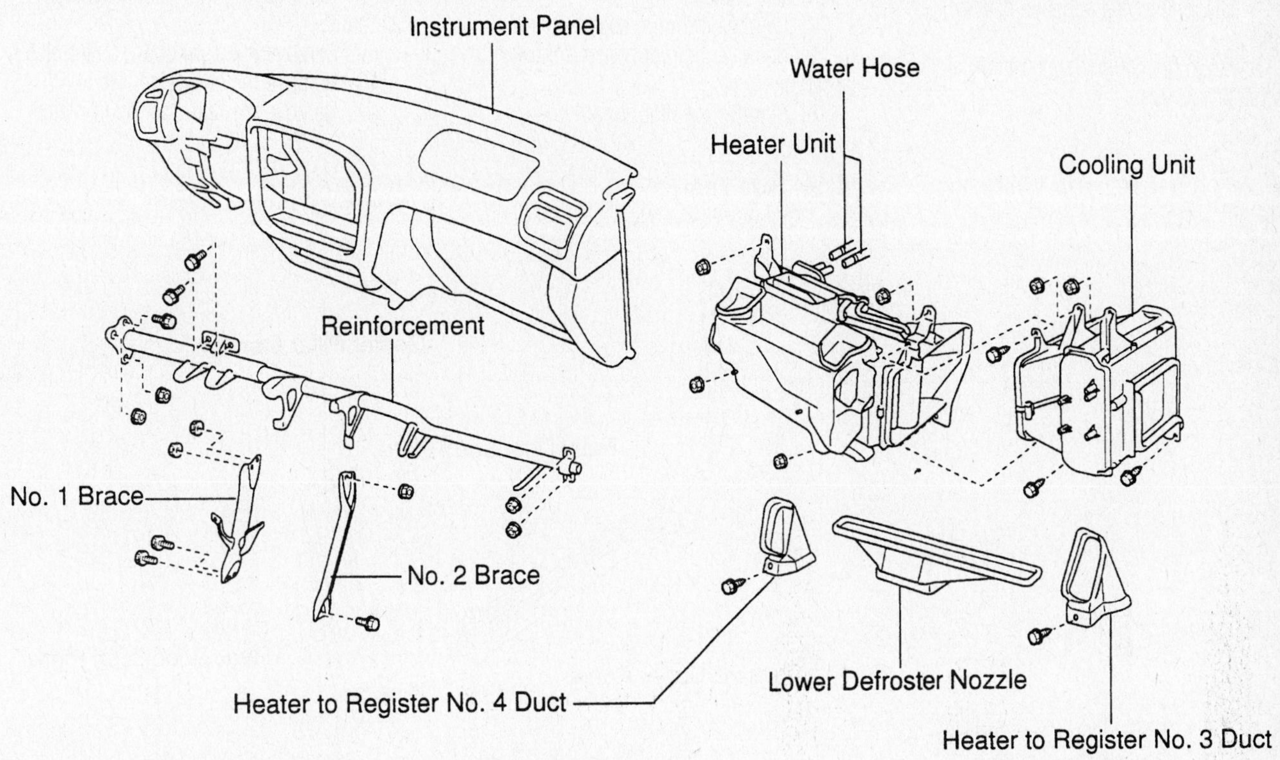

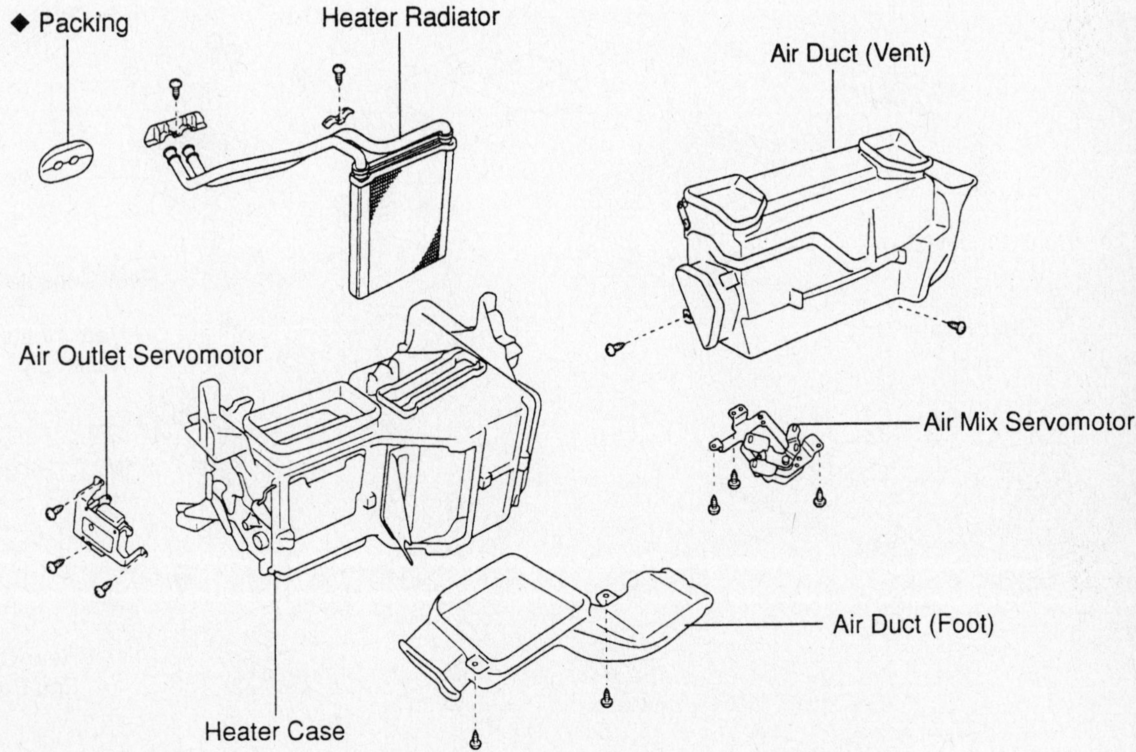

◆ Non–reusable part

93113GG0

Exploded view of the front heater core, heater housing, evaporator housing and related components—Toyota Land Cruiser

REAR AUXILIARY HEATER

1. Disconnect the negative battery cable.
2. Drain the cooling system into a clean container for reuse.
3. Disconnect the heater hoses from the rear heater core.
4. Remove the front seats.
5. Remove the rear heater control assembly.
6. Remove the rear console box.
7. Remove the front console box cover.
8. Remove the lower center cluster finish panel.
9. Remove the front door scuff plates.
10. Remove the cowl side trim.
11. Remove the rear door scuff plates.

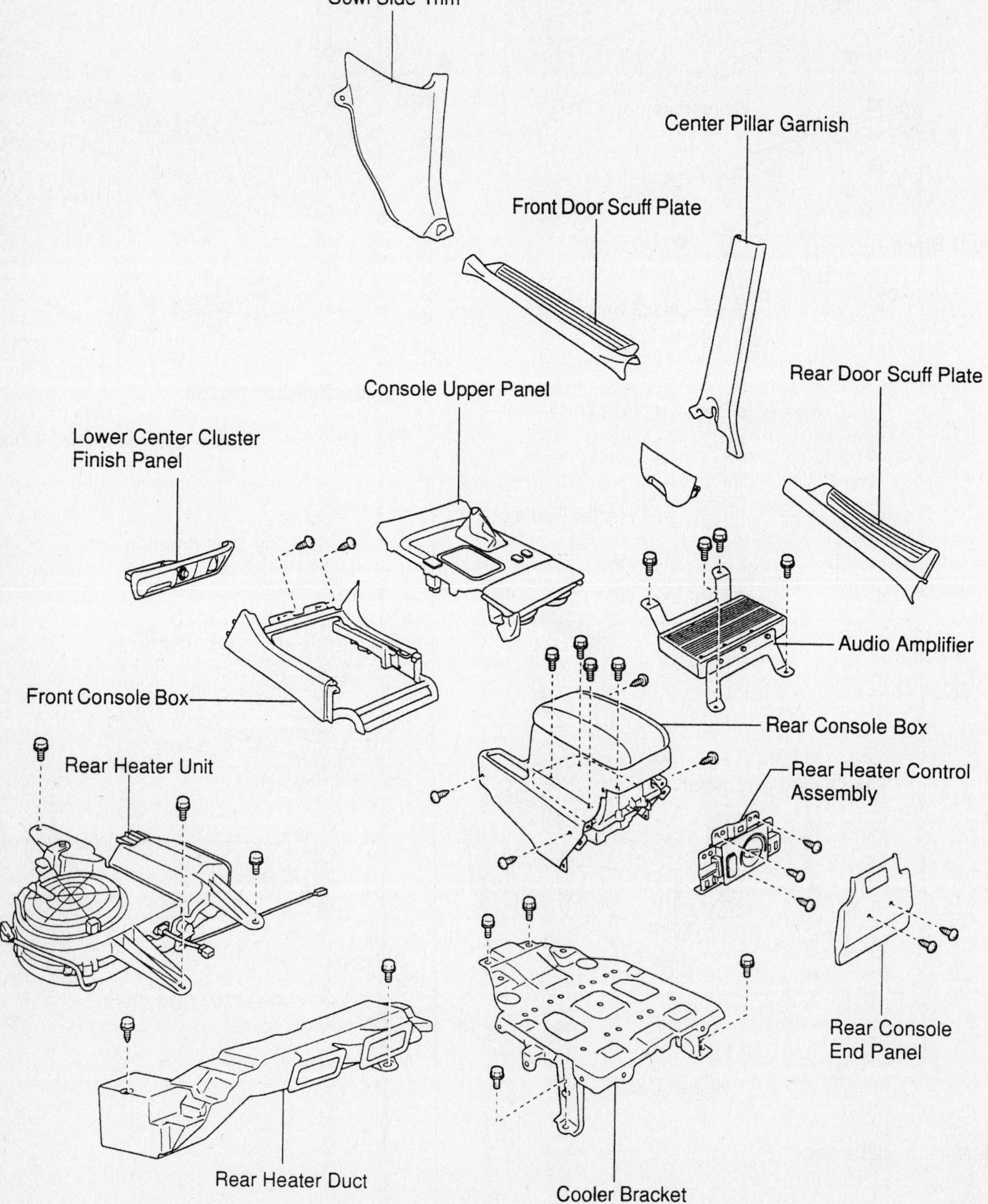

Exploded view of the rear heater housing and related components—Toyota Land Cruiser

12. Remove the center pillar garnishes.
13. Slide the carpet rearward.
14. Remove the cooler bracket bolts and the bracket.
15. Remove the rear heater duct bolt/screw and the duct.
16. Disconnect the rear heater housing electrical connector.
17. Remove the 3 rear heater housing-to-chassis bolts and the heater housing.
18. Remove the heater core-to-heater housing 3 screws and 2 clamps.
19. Remove the heater core from the heater housing.

To install:

20. Install the heater core to the heater housing.
21. Install the heater core-to-heater housing 3 screws and 2 clamps.
22. Install the heater housing and the 3 rear heater housing-to-chassis bolts.
23. Connect the rear heater housing electrical connector.
24. Install the rear heater duct and the duct bolt/screw.
25. Install the cooler bracket and the bracket bolts.
26. Slide the carpet rearward.
27. Install the center pillar garnishes.
28. Install the rear door scuff plates.
29. Install the cowl side trim.
30. Install the front door scuff plates.
31. Install the lower center cluster finish panel.
32. Install the front console box cover.
33. Install the rear console box.
34. Install the rear heater control assembly.
35. Install the front seats.
36. Connect the heater hoses to the rear heater core.
37. Refill the cooling system.
38. Connect the negative battery cable.

Sequoia

FRONT HEATER

1. Disconnect the negative battery cable.
2. Drain the cooling system into a clean container for reuse.

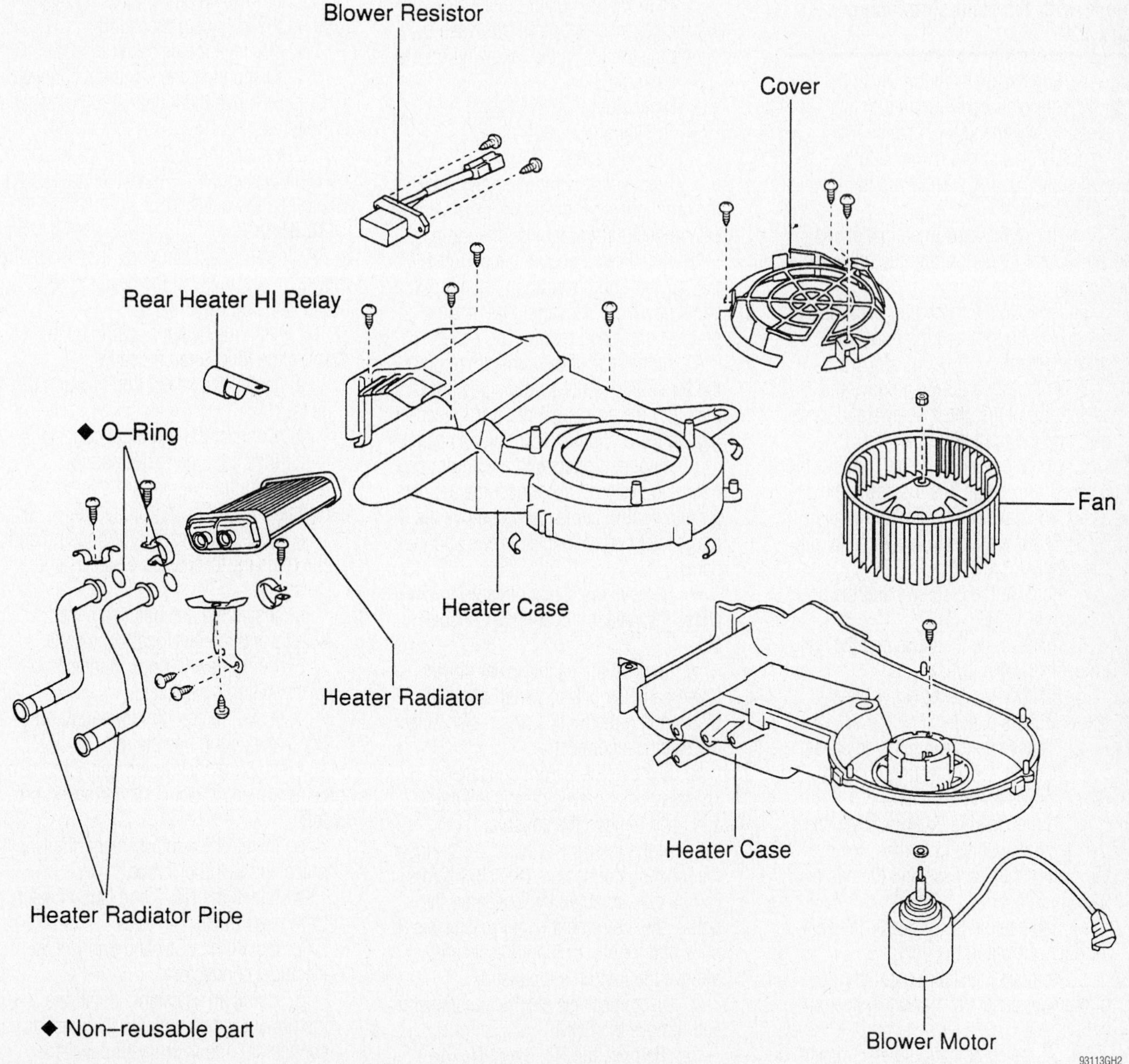

Exploded view of the rear heater core, heater housing and related components—Toyota Land Cruiser

3. Disconnect the heater hoses from the heater core.

4. Remove the steering wheel by performing the following procedure:

a. Position the front wheels facing straight-ahead.

b. Remove the steering wheel side covers.

c. Using a Torx® wrench, loosen the 2 screws located at each side of the steering wheel until the screw's circumference groove catches on the screw case.

d. Pull the air bag module from the steering wheel and disconnect the electrical connector.

⁂ CAUTION

Place the air bag module in a safe place with the front side facing upward.

e. Remove the steering wheel nut.

f. Place alignment marks on the steering wheel and the main shaft.

g. Using a steering wheel puller, press the steering wheel from the steering column.

5. Remove the instrument panel and reinforcement by performing the following procedure:

a. Remove the front door scuff plates, the cowl side trim and the front door opening trim.

b. At the driver's side, remove the 2 assist grip plugs, the 2 screws and assist grip and the front pillar garnish.

c. At the passenger's side, remove the 4 assist grip plugs, the 4 screws, the 2 assist grips and the front pillar garnish.

d. Remove the instrument cluster finish panel.

e. Remove the 2 screws and the hood lock control cable.

f. Remove the 2 screws and the fuel lid control cable lever.

g. Remove the lower No. 1 panel screw and the panel.

h. Remove the lower left side panel.

i. Remove the 3 steering column cover screws and the covers.

j. At the steering column, disconnect the electrical connectors; then, remove the clamp, the 3 screws and the combination switch.

k. Remove the No. 2 heater-to-register duct screw and the duct.

l. Remove the steering column-to-instrument panel bolts and the steering column.

m. At the combination meter, disconnect the electrical connectors; then, remove the 4 screws and the combination meter.

n. Remove the glove compartment door stoppers, the 2 screws and the glove box door.

o. At the passenger's side air bag module, remove the No. 1 undercover, pull the air bag connector up from the undercover and disconnect it; then, remove the air bag.

⁂ CAUTION

Place the air bag module in a safe place with the front side facing upward.

p. Remove the 3 lower No. 2 panel screws and the panel.

q. Remove the center cluster; then, pry the center cluster from the dash by prying the 8 clips in the following order:

- Left side
- Right side
- Top left side
- Top right side

r. Remove the 4 radio screws, pull the radio outward, disconnect the electrical connectors and remove the radio.

s. At the rear console panel, remove the transfer shift lever knob; then, pry the panel upward disengaging the 4 clips (2 on each side) and remove the panel.

t. At the rear of the console, remove the 2 rear end panel-to-console screws; then, pry the end panel rearward disengaging the 2 clips and remove the panel.

u. If not equipped with a rear air conditioning system, disconnect the connector and control cable; then, remove the 3 rear heater control panel screws and the panel.

v. Remove the 4 rear console box-to-chassis screws/bolts and the console box.

w. Remove the center lower cluster finish panel by prying panel rearward disengaging the 5 clips; then, disconnect the electrical connector.

x. Remove the 2 front console-to-chassis bolts/screws, disengage the 2 clips and remove the console.

y. At the instrument panel, disconnect the junction connectors (the connectors can be disconnected by loosening the bolts), the instrument panel-to-chassis 8 bolts and 2 nuts. Using an assistant, remove the instrument panel.

z. Disconnect the electrical connector and remove the ECM.

aa. Remove the No. 3 and No. 4 heater-to-register ducts.

bb. Remove the floor brace, the No. 1 brace and the reinforcement.

6. Remove the evaporator housing by performing the following procedure:

a. Discharge and recover the air conditioning system refrigerant.

b. Remove the air conditioning liquid line clamp.

c. Remove the air conditioning suction line clamp.

d. Disconnect both air conditioning lines and plug the openings to prevent contamination. Discard the 4 O-rings.

e. Remove the antenna relay electrical connector, the 2 screws and the relay.

f. Remove the evaporator housing-to-chassis 4 screws/2 nuts and the housing.

7. Remove the heater housing by performing the following procedure:

a. Remove the defroster nozzle.

b. Disconnect the electrical connector.

c. Remove the 4 nuts and the heater housing.

8. Remove the heater core-to-heater housing packing, the screw, the bracket, the clamp and the heater core.

To install:

9. Install the heater core, the clamp, the bracket, the screw and the heater core-to-heater housing packing.

10. Install the heater housing by performing the following procedure:

a. Install the heater housing and the 4 nuts.

b. Connect the electrical connector.

c. Install the defroster nozzle.

11. Install the evaporator housing by performing the following procedure:

a. Install the evaporator housing and the housing-to-chassis 4 screws and 2 nuts.

b. Install the antenna relay, the 2 screws and the electrical connector.

c. Using new O-rings, connect both air conditioning lines.

d. Install the air conditioning liquid line and suction line clamp.

12. Install the instrument panel and reinforcement by performing the following procedure:

a. Install the reinforcement, the No. 1 brace and the floor brace.

b. Install the No. 3 and No. 4 heater-to-register ducts.

c. Install the ECM and connect the electrical connector.

d. Using an assistant, install the instrument panel, connect the junction connectors, the instrument panel-to-chassis 8 bolts and 2 nuts.

e. Install the front the console,

engage the 2 clips and install the 2 console-to-chassis bolts/screws.

f. Connect the electrical connector; then, install the center lower cluster finish panel by engaging the 5 clips.

g. Install the console box and the 4 rear console box-to-chassis screws/bolts.

h. If not equipped with a rear air conditioning system, install rear heater control panel, the 3 panel screws; then, connect the connector and control cable.

i. Install the rear of the console and engage the 2 clips; then, install the 2 rear end panel-to-console screws.

j. Install the rear console panel and engage the 4 clips (2 on each side); then, install the transfer shift lever knob.

k. Install the radio, connect the electrical connectors and the 4 radio screws.

l. Install the center cluster and engage the 8 center cluster clips.

m. Install the lower No. 2 panel and the 3 panel screws.

n. Install the passenger's side air bag module, connect it and install the No. 1 undercover.

o. Install the glove box door, the 2 screws and the glove compartment door stoppers.

p. Install the combination meter and the 4 screws; then, connect the electrical connectors.

q. Install the steering column and the steering column-to-instrument panel bolts.

r. Install the No. 2 heater-to-register duct and the duct screw.

s. At the steering column, install the combination switch, the 3 screws and the clamp; then, connect the electrical connectors.

t. Install the steering column covers and the 3 covers screws.

u. Install the lower left side panel.

v. Install the lower No. 1 panel and the panel screw.

w. Install the fuel lid control cable lever and the 2 screws.

x. Install the hood lock control cable and the 2 screws.

y. Install the instrument cluster finish panel.

z. At the passenger's side, install the front pillar garnish, the 2 assist grips, the 4 screws and the 4 assist grip plugs.

aa. At the driver's side, install the front pillar garnish, assist grip, the 2 screws and the 2 assist grip plugs.

bb. Install the front door scuff plates, the cowl side trim and the front door opening trim.

13. Install the steering wheel by performing the following procedure:

a. Install the steering wheel to the steering column.

b. Align the steering wheel-to-main shaft marks.

c. Install the steering wheel nut and torque to 25 ft. lbs. (34 Nm).

d. Install the air bag module to the steering wheel and connect the electrical connector.

e. Using a Torx® wrench, tighten the 2 screws located at each side of the steering wheel to 78 inch lbs. (8.8 Nm).

f. Install the steering wheel side covers.

14. Connect the heater hoses to the heater core.

15. Refill the cooling system.

16. Connect the negative battery cable.

a. Evacuate and charge the air conditioning system refrigerant.

17. Run the engine to normal operating temperatures; then, check the climate control operation and check for leaks.

REAR AUXILIARY HEATER

1. Disconnect the negative battery cable.
2. Drain the cooling system into a clean container for reuse.
3. Disconnect the heater hoses from the rear heater core.
4. Remove the front seats.
5. Remove the rear heater control assembly.
6. Remove the rear console box.
7. Remove the front console box cover.
8. Remove the lower center cluster finish panel.
9. Remove the front door scuff plates.
10. Remove the cowl side trim.
11. Remove the rear door scuff plates.
12. Remove the center pillar garnishes.
13. Slide the carpet rearward.
14. Remove the cooler bracket bolts and the bracket.
15. Remove the rear heater duct bolt/screw and the duct.
16. Disconnect the rear heater housing electrical connector.
17. Remove the 3 rear heater housing-to-chassis bolts and the heater housing.
18. Remove the heater core-to-heater housing 3 screws and 2 clamps.
19. Remove the heater core from the heater housing.

To install:

20. Install the heater core to the heater housing.
21. Install the heater core-to-heater housing 3 screws and 2 clamps.
22. Install the heater housing and the 3 rear heater housing-to-chassis bolts.
23. Connect the rear heater housing electrical connector.
24. Install the rear heater duct and the duct bolt/screw.
25. Install the cooler bracket and the bracket bolts.
26. Slide the carpet rearward.
27. Install the center pillar garnishes.
28. Install the rear door scuff plates.
29. Install the cowl side trim.
30. Install the front door scuff plates.
31. Install the lower center cluster finish panel.
32. Install the front console box cover.
33. Install the rear console box.
34. Install the rear heater control assembly.
35. Install the front seats.
36. Connect the heater hoses to the rear heater core.
37. Refill the cooling system.
38. Connect the negative battery cable.

Water Pump

REMOVAL & INSTALLATION

1. Before servicing the vehicle, refer to the precautions in the beginning of this section.

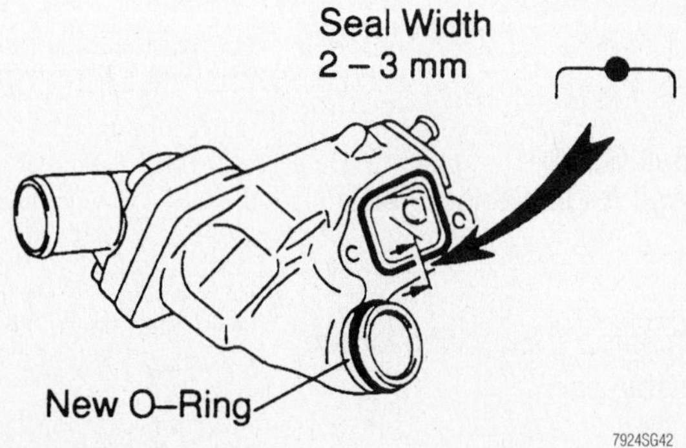

Water inlet housing sealant application

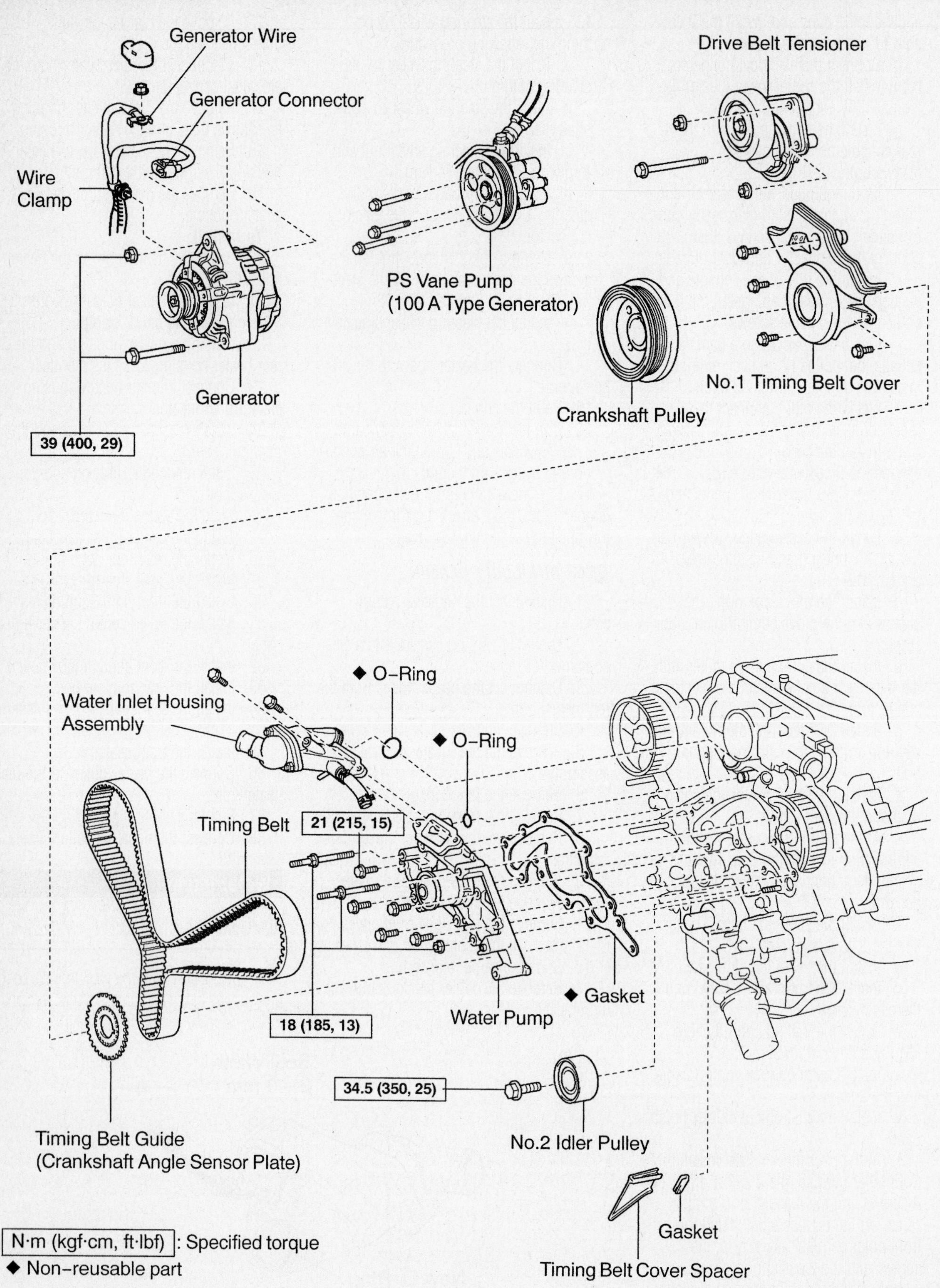
Generator Wire
Generator Connector
Wire Clamp
Generator
39 (400, 29)
PS Vane Pump
(100 A Type Generator)
Drive Belt Tensioner
Crankshaft Pulley
No.1 Timing Belt Cover
◆ O-Ring
Water Inlet Housing Assembly
◆ O-Ring
Timing Belt
21 (215, 15)
18 (185, 13)
Water Pump
◆ Gasket
34.5 (350, 25)
No.2 Idler Pulley
Timing Belt Guide
(Crankshaft Angle Sensor Plate)
Gasket
Timing Belt Cover Spacer
N·m (kgf·cm, ft·lbf) : Specified torque
◆ Non-reusable part
67170-LCSQ-G03

Water pump and related parts

2. Drain the cooling system.
3. Remove or disconnect the following:
    - Negative battery cable
    - Timing belt.
    - No. 2 idler pulley
    - Radiator hose
    - Bypass hose
    - Water inlet housing assembly
    - Water pump

To install:

4. Install or connect the following:
    - Water pump. Use a new gasket and tighten the bolts to 15 ft. lbs. (21 Nm). Tighten the stud bolt and nut to 13 ft. lbs. (18 Nm).
    - Water inlet housing assembly. Use a new O-ring and apply sealant as shown. Tighten the bolts to 13 ft. lbs. (18 Nm).
    - Bypass hose
    - Radiator hose
    - No. 2 idler pulley
    - Timing belt
    - Negative battery cable
5. Fill the cooling system.
6. Start the engine and check for leaks.

Cylinder Head

REMOVAL & INSTALLATION

1. Before servicing the vehicle, refer to the precautions in the beginning of this section.
2. Drain the cooling system.
3. Relieve the fuel system pressure.
4. Remove or disconnect the following:
    - Battery and tray
    - Engine appearance cover
    - Engine under covers
    - Air intake assembly
    - Accessory drive belt
    - A/C compressor and bracket
    - Cooling fan and bracket
    - Radiator
    - Idler pulley
    - Front covers
    - Timing belt.
    - Camshaft sprockets
    - Camshaft Position (CMP) sensor
    - Power steering pump
    - Exhaust front pipes
    - Transmission dipstick tube
    - Ignition coils
    - Rear timing belt covers
    - Fuel lines
    - Intake manifold
    - Water inlet housing assembly
    - Front and rear water bypass joints
    - Engine lifting eyes
    - Oil dipstick tube

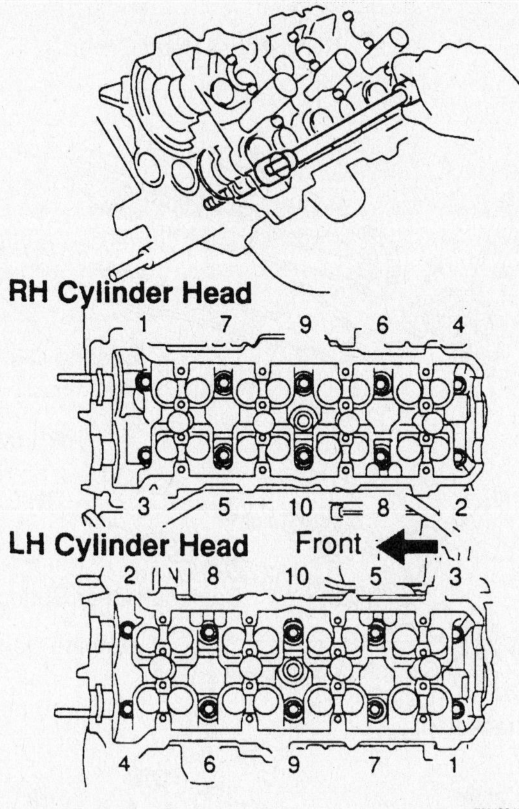

Cylinder head loosening sequence

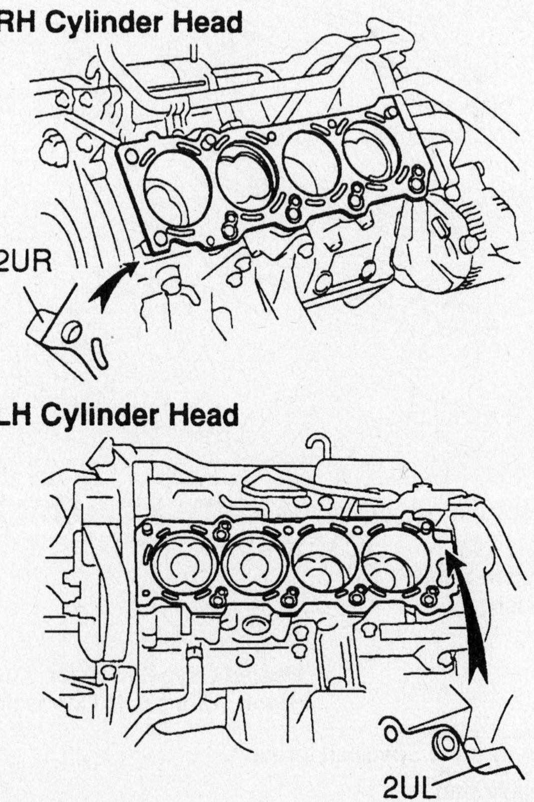

Cylinder head gasket identification

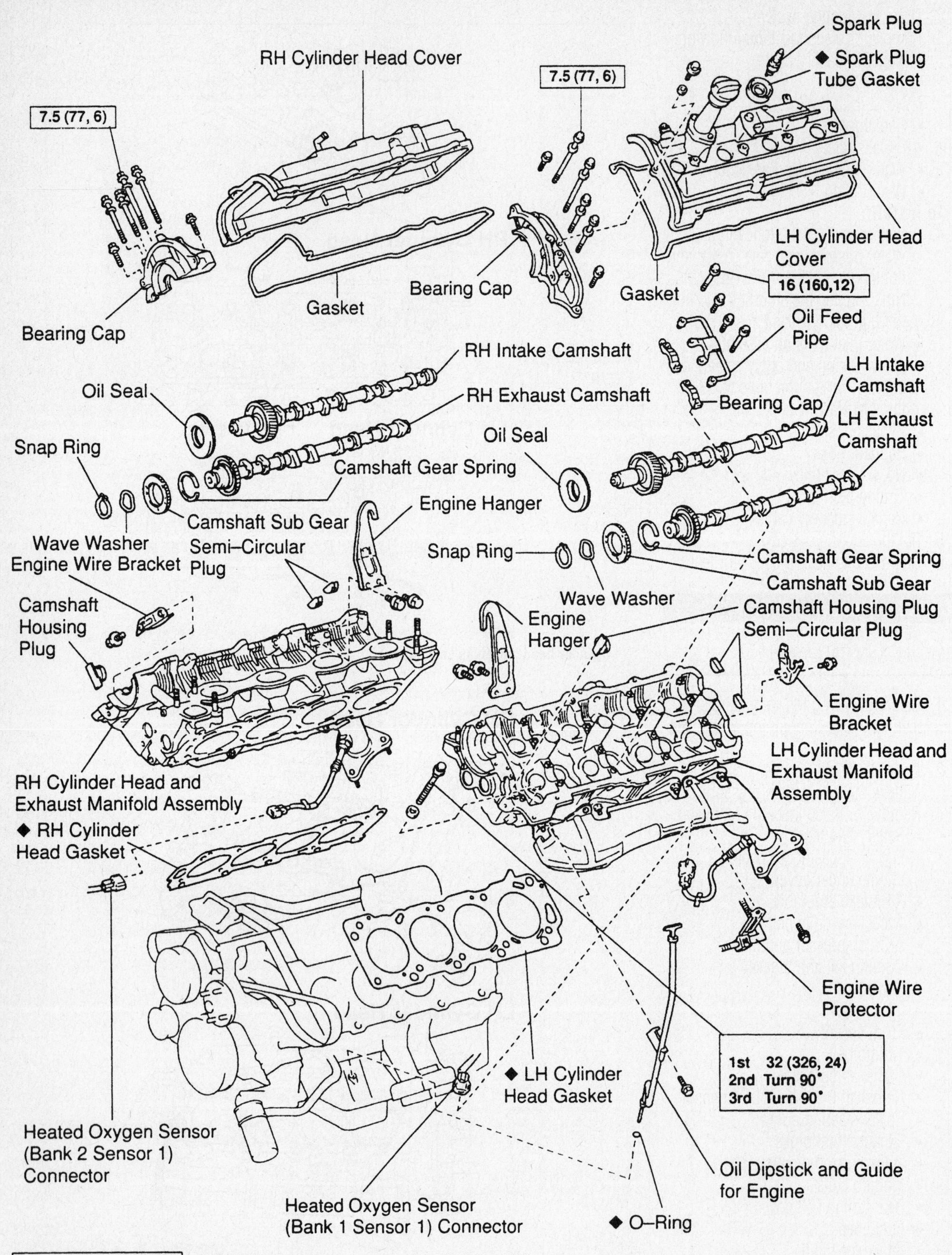
Spark Plug
◆ Spark Plug Tube Gasket
RH Cylinder Head Cover
7.5 (77, 6)
7.5 (77, 6)
LH Cylinder Head Cover
Gasket
Bearing Cap
Gasket
16 (160,12)
Bearing Cap
Oil Feed Pipe
RH Intake Camshaft
LH Intake Camshaft
Oil Seal
RH Exhaust Camshaft
Bearing Cap
LH Exhaust Camshaft
Oil Seal
Snap Ring
Camshaft Gear Spring
Engine Hanger
Camshaft Sub Gear
Wave Washer
Semi–Circular Plug
Engine Wire Bracket
Snap Ring
Camshaft Gear Spring
Camshaft Sub Gear
Camshaft Housing Plug
Wave Washer
Engine Hanger
Camshaft Housing Plug
Semi–Circular Plug
Engine Wire Bracket
LH Cylinder Head and Exhaust Manifold Assembly
RH Cylinder Head and Exhaust Manifold Assembly
◆ RH Cylinder Head Gasket
Engine Wire Protector
1st 32 (326, 24)
2nd Turn 90°
3rd Turn 90°
◆ LH Cylinder Head Gasket
Heated Oxygen Sensor (Bank 2 Sensor 1) Connector
Oil Dipstick and Guide for Engine
Heated Oxygen Sensor (Bank 1 Sensor 1) Connector
◆ O–Ring
N·m (kgf·cm, ft·lbf) : Specified torque
◆ Non–reusable part
7924SG49

Exploded view of the cylinder head mounting

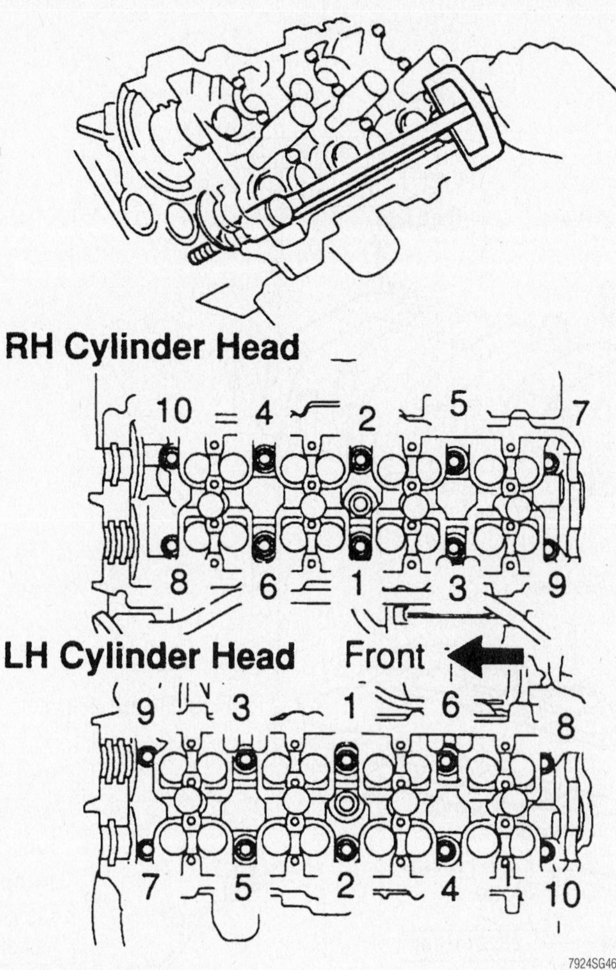

Cylinder head torque sequence

- Valve covers
- Camshafts
- Cylinder heads with the exhaust manifolds attached. Loosen the bolts in the sequence shown.

To install:

5. Install the cylinder heads with new gaskets. Tighten the bolts in sequence as follows:
   a. Step 1: 24 ft. lbs. (32 Nm)
   b. Step 2: Plus 180 degrees
6. Install or connect the following:
   - Camshafts
   - Valve covers
   - Oil dipstick tube
   - Engine lifting eyes
   - Front and rear water bypass joints
   - Water inlet housing assembly
   - Intake manifold
   - Fuel lines
   - Rear timing belt covers
   - Ignition coils
   - Transmission dipstick tube
   - Exhaust front pipes
   - Power steering pump
   - CMP sensor
   - Camshaft sprockets
   - Timing belt
   - Front covers
   - Idler pulley
   - Radiator
   - Cooling fan and bracket
   - A/C compressor and bracket
   - Accessory drive belt
   - Air intake assembly
   - Engine under covers
   - Engine appearance cover
   - Battery and tray
7. Fill the cooling system.
8. Start the engine and check for leaks.

Intake Manifold

REMOVAL & INSTALLATION

1. Before servicing the vehicle, refer to the precautions in the beginning of this section.
2. Drain the cooling system.
3. Relieve the fuel system pressure.
4. Remove or disconnect the following:
   - Negative battery cable
   - Engine appearance cover
   - Accelerator cable
   - Throttle Position (TP) sensor connector
   - Accelerator pedal position sensor
   - Throttle motor connector
   - Evaporative Emissions (EVAP) vacuum switching valve connector
   - Fuel injector connectors
   - Engine Coolant Temperature (ECT) sensor connector
   - ETC gauge sender connector
   - Heated Oxygen (HO_2S) sensor connectors
   - Fuel pressure regulator vacuum hose
   - Positive Crankcase Ventilation (PCV) valve and hose
   - EVAP hoses
   - Power steering vacuum hoses
   - Water bypass hose
   - Engine control wiring harness clamps
   - Cylinder head ground cables
   - Intake manifold wire harness protector
   - EVAP pipe
   - Engine appearance cover brackets
   - Intake manifold

To install:

5. Install or connect the following:
   - Intake manifold. Tighten the fasteners to 13 ft. lbs. (18 Nm).
   - Engine appearance cover brackets
   - EVAP pipe
   - Intake manifold wire harness protector
   - Cylinder head ground cables
   - Engine control wiring harness clamps
   - Water bypass hose
   - Power steering vacuum hoses
   - EVAP hoses
   - PCV valve and hose
   - Fuel pressure regulator vacuum hose
   - HO_2S sensor connectors
   - ETC gauge sender connector
   - ECT sensor connector
   - Fuel injector connectors
   - EVAP vacuum switching valve connector
   - Throttle motor connector
   - Accelerator pedal position sensor
   - TP sensor connector
   - Accelerator cable
   - Engine appearance cover
   - Negative battery cable
6. Fill the cooling system.
7. Start the engine and check for leaks.

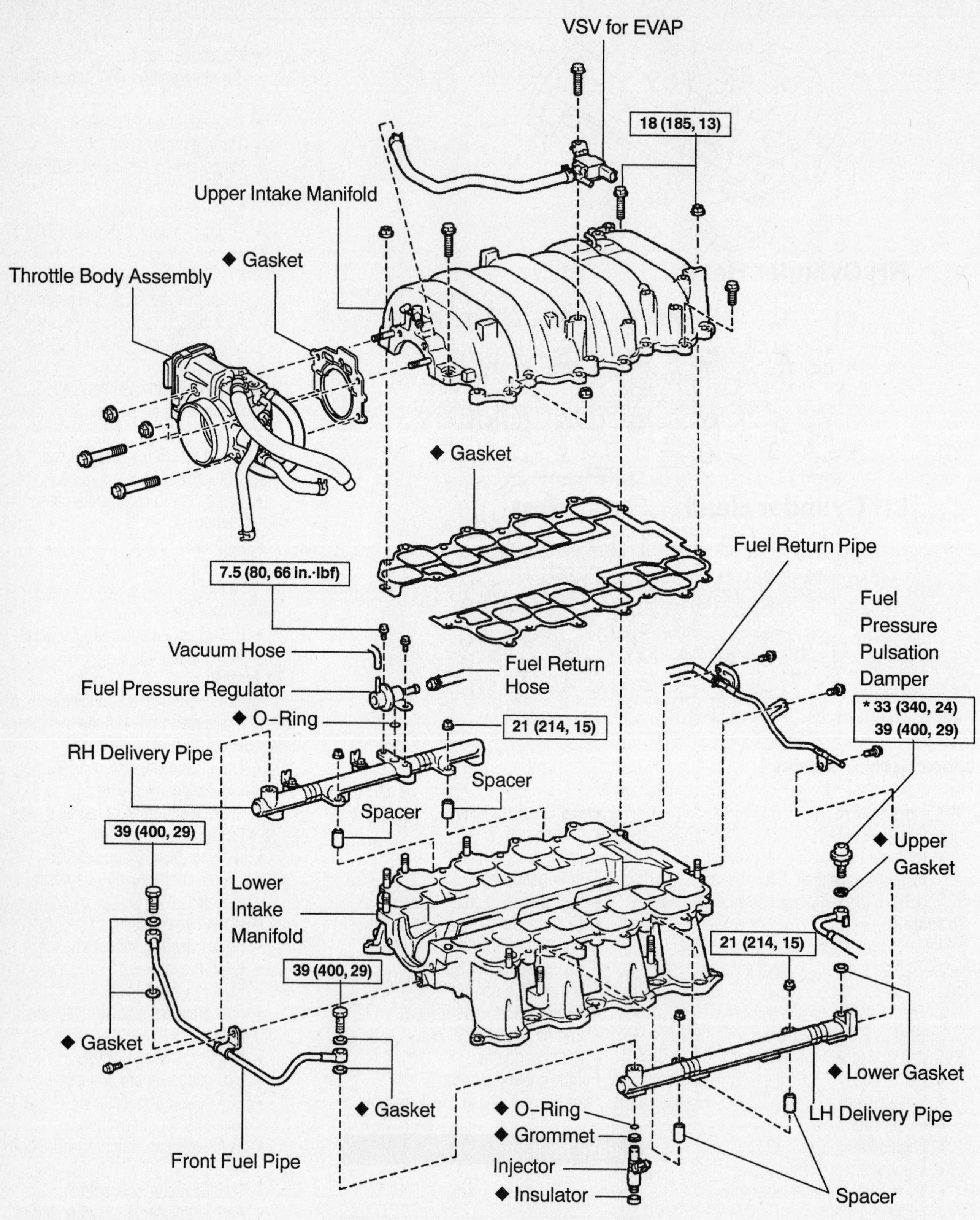
VSV for EVAP
18 (185, 13)
Upper Intake Manifold
◆ Gasket
Throttle Body Assembly
◆ Gasket
Fuel Return Pipe
7.5 (80, 66 in.·lbf)
Fuel
Pressure
Pulsation
Damper
Vacuum Hose
Fuel Return
Hose
Fuel Pressure Regulator
◆ O-Ring
21 (214, 15)
* 33 (340, 24)
39 (400, 29)
RH Delivery Pipe
Spacer
Spacer
39 (400, 29)
◆ Upper
Gasket
Lower
Intake
Manifold
21 (214, 15)
39 (400, 29)
◆ Gasket
◆ Lower Gasket
◆ Gasket
◆ O-Ring
LH Delivery Pipe
◆ Grommet
Front Fuel Pipe
Injector
◆ Insulator
Spacer
N·m (kgf·cm, ft·lbf) : Specified torque
◆ Non-reusable part
* For use with SST
67170-LCSQ-G04

Intake manifold and related parts—Land Cruiser

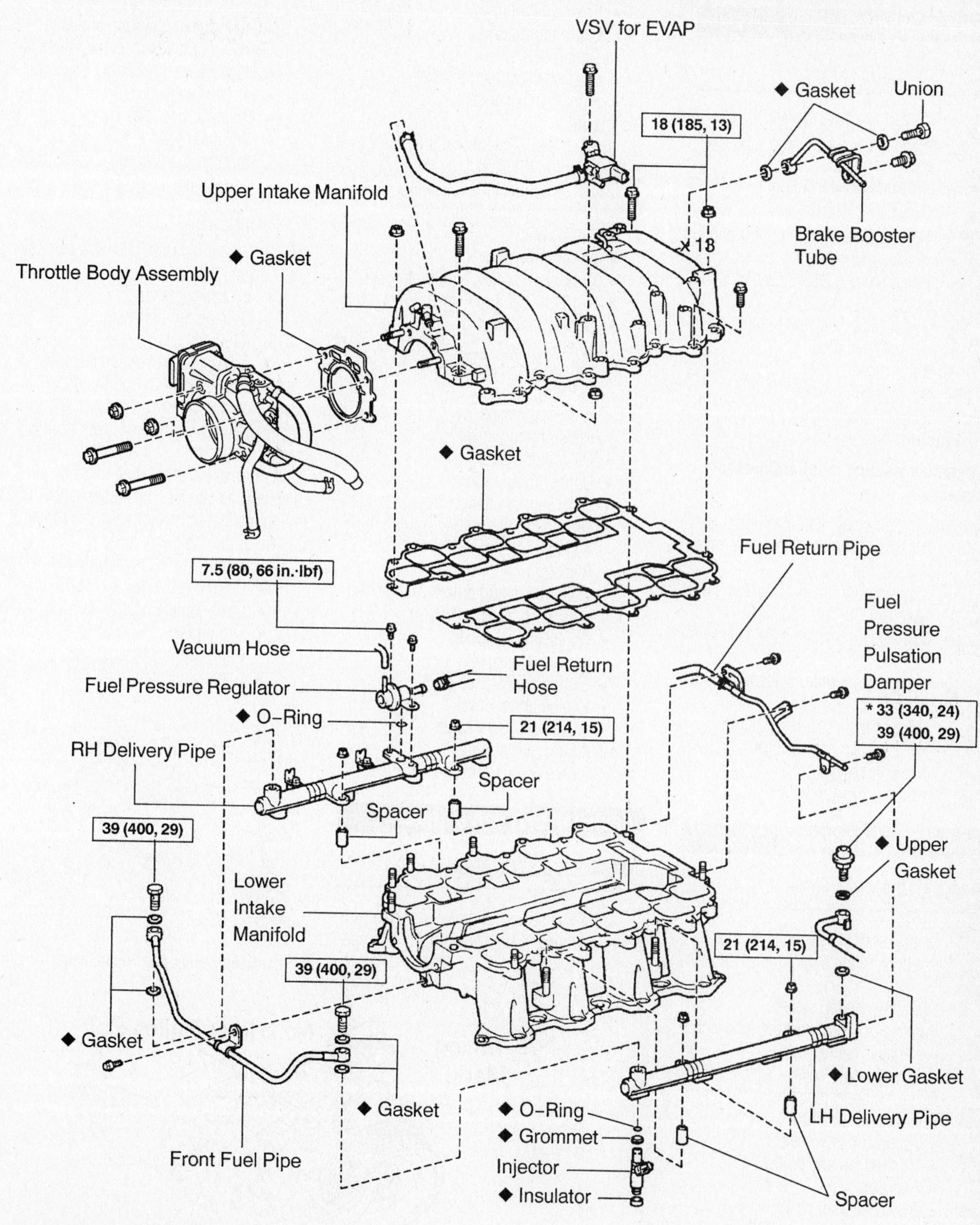

N·m (kgf·cm, ft·lbf) : Specified torque
◆ Non–reusable part
* For use with SST

67170-LCSQ-G05

Intake manifold and related parts—Sequoia

Exhaust Manifold

REMOVAL & INSTALLATION

1. Before servicing the vehicle, refer to the precautions in the beginning of this section.
2. Attach a hoist to the engine lifting eyes.
3. Remove or disconnect the following:
    - Negative battery cable
    - Heated Oxygen (HO_2S) sensor connectors
    - Exhaust manifold heat shield
    - Exhaust front pipe
    - Motor mount
    - Motor mount bracket
    - Exhaust manifold

To install:

➡Use new exhaust manifold nuts for assembly.

4. Install or connect the following:
    - Exhaust manifold. Tighten the nuts to 32 ft. lbs. (44 Nm).
    - Motor mount bracket. Tighten the bolts to 27 ft. lbs. (36 Nm).
    - Motor mount. Tighten the fasteners to 22 ft. lbs. (30 Nm).
    - Exhaust front pipe. Tighten the nuts to 46 ft. lbs. (62 Nm).
    - Exhaust manifold heat shield
    - HO_2S sensor connectors
    - Negative battery cable
5. Start the engine and check for leaks.

Front Crankshaft Seal

REMOVAL & INSTALLATION

1. Before servicing the vehicle, refer to the precautions in the beginning of this section.
2. Drain the cooling system.
3. Remove or disconnect the following:
    - Negative battery cable
    - Engine under cover
    - Engine appearance cover
    - Air intake assembly
    - Accessory drive belt
    - Cooling fan and pulley
    - Radiator
    - Drive belt idler pulley
    - Camshaft Position (CMP) sensor connector
    - Upper timing covers
    - Oil cooler pipe
    - Center timing cover
    - A/C compressor
    - Cooling fan bracket
    - Crankshaft pulley
    - Lower timing cover
    - Timing belt.
    - Crankshaft timing sprocket
    - Front crankshaft seal

To install:

4. Install the oil seal so that it is flush with the oil pump housing.
5. Install or connect the following:
    - Crankshaft timing sprocket
    - Timing belt
    - Lower timing cover
    - Crankshaft pulley. Tighten the bolt to 181 ft. lbs. (245 Nm).
    - Cooling fan bracket. Tighten the 12mm bolts to 12 ft. lbs. (16 Nm) and the 14mm bolts to 24 ft. lbs. (32 Nm).
    - A/C compressor
    - Center timing cover
    - Oil cooler pipe
    - Upper timing covers
    - CMP sensor connector
    - Drive belt idler pulley. Tighten the bolt to 27 ft. lbs. (37 Nm).
    - Radiator
    - Cooling fan and pulley. Tighten the nuts to 16 ft. lbs. (21 Nm).
    - Accessory drive belt
    - Air intake assembly
    - Engine appearance cover
    - Engine under cover
    - Negative battery cable
6. Fill the cooling system.
7. Start the engine and check for leaks.

Camshaft and Valve Lifters

REMOVAL & INSTALLATION

1. Before servicing the vehicle, refer to the precautions in the beginning of this section.
2. Drain the cooling system.
3. Relieve the fuel system pressure.
4. Remove or disconnect the following:
    - Negative battery cable
    - Engine under covers
    - Engine appearance cover
    - Air intake hose
    - Accessory drive belt
    - Cooling fan
    - Radiator
    - Idler pulley
    - Upper and middle timing belt covers
    - A/C compressor
    - Cooling fan bracket
    - Alternator
    - Accessory drive belt tensioner
5. Set the engine to Top Dead Center (TDC) with the camshaft sprocket timing marks aligned with the rear cover timing marks.
6. Rotate the crankshaft to 50 degrees After TDC as shown. The crankshaft pulley timing mark should align with the center of the No. 2 idler pulley bolt.
7. Remove or disconnect the following:
    - Crankshaft pulley
    - Lower timing cover
    - Timing belt.
    - Camshaft timing sprockets
    - Camshaft Position (CMP) sensor

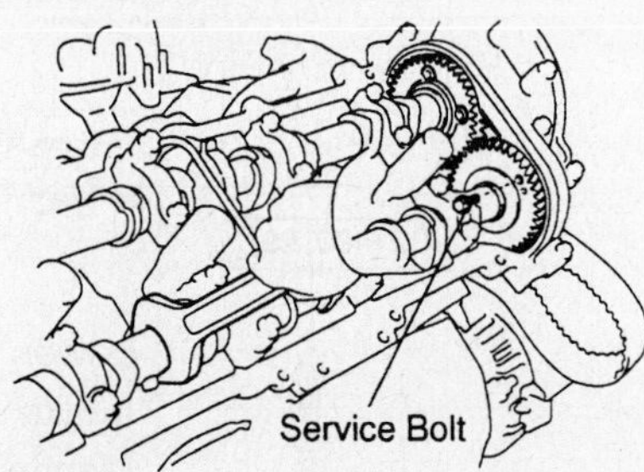

Camshaft service bolt installation

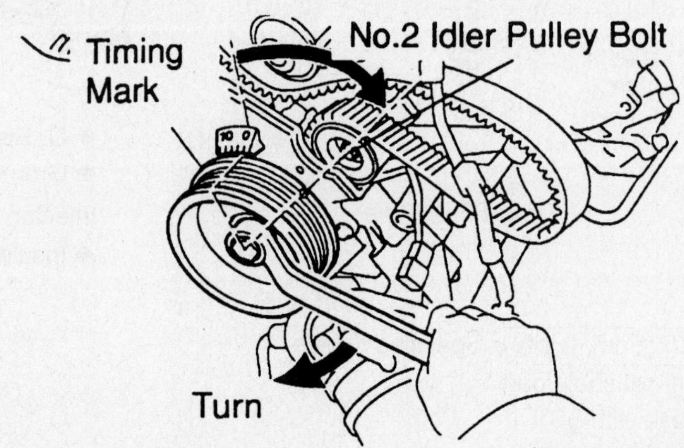

Setting the crankshaft to 50 degrees ATDC

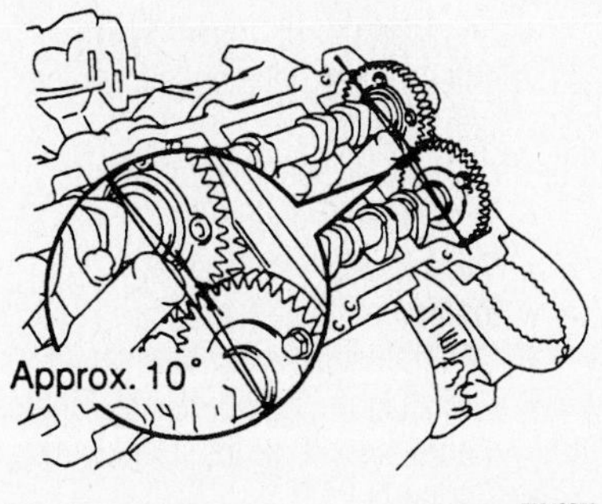

7924SG59

Right bank camshaft timing mark (1 dot marks) alignment

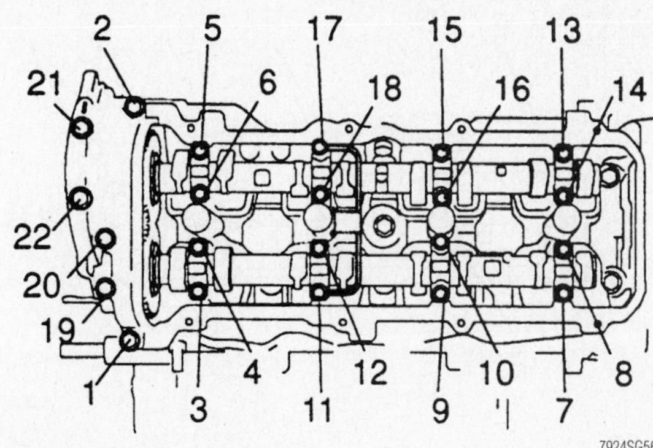

7924SG56

Left bank camshaft bearing cap loosening sequence

- Ignition coils
- Valve cover
- Timing belt rear covers

8. Rotate the right bank camshafts as necessary to access the exhaust camshaft sub-gear service bolt hole and install a 6mm x 1.0mm bolt.

➡**Keep all valvetrain components in order for assembly.**

9. Align the right bank camshaft 1 dot timing marks to a **10** degree angle as shown.

10. Loosen the bearing cap bolts in sequence and in several passes.

11. Remove the right bank camshafts.

12. Rotate the left bank camshafts as necessary to access the exhaust camshaft sub-gear service bolt hole and install a 6mm x 1.0mm bolt.

13. Align the left bank camshaft 2 dot timing marks as shown.

14. Loosen the bearing cap bolts in sequence and in several passes.

15. Remove the left bank camshafts.

16. Remove the valve lifters and shims.

To install:

17. Ensure that the crankshaft is at 50 degrees After TDC.

18. Install or connect the following:
- Valve lifters and shims in their original positions
- Right bank camshafts with the 1 dot timing marks at 10 degrees
- Left bank camshafts with the 2 dot timing marks aligned
- Left and right bank camshaft bearing caps in their original positions. Apply sealant to the front bearing caps as shown.
- Camshaft oil seals

19. The bearing cap bolts vary in length and are identified as follows:
- A: 3.70 inches (94mm)
- B: 2.83 inches (72mm)
- C: 0.98 inches (25mm)
- D: 2.05 inches (52mm)
- E: 1.50 inches (38mm)

20. Bolts in positions **A**, **B** and **C** are installed dry.

21. Lubricate the threads and under the contact flange for bolts in positions **D** and **E**.

22. Install oil feed pipes and the bearing cap bolts according to position in the illustrations.

23. Tighten the camshaft bearing bolts in sequence and in several passes to the following specifications:

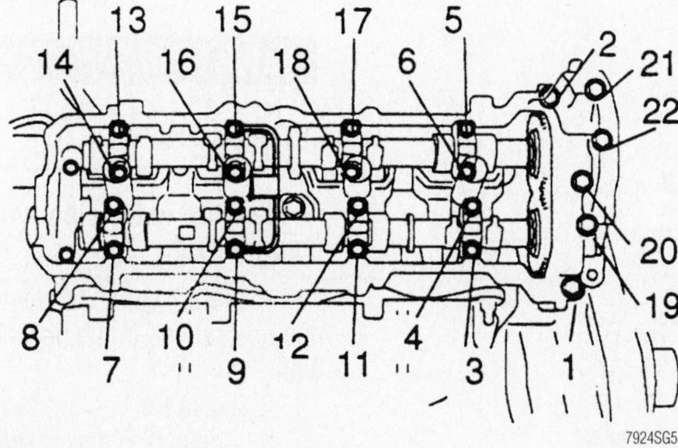

7924SG54

Right bank camshaft bearing cap loosening sequence

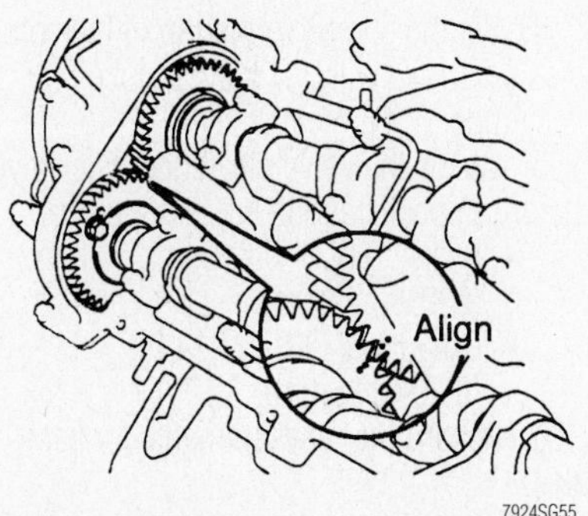

7924SG55

Left bank camshaft timing mark (2 dot marks) alignment

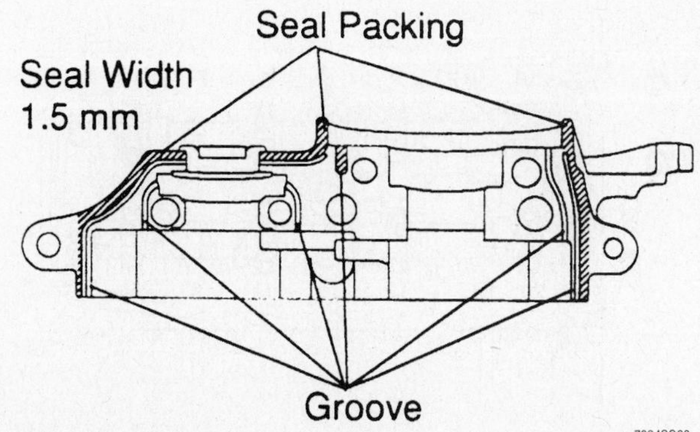

7924SG60

Apply a 1.5mm bead of sealant to the front bearing caps

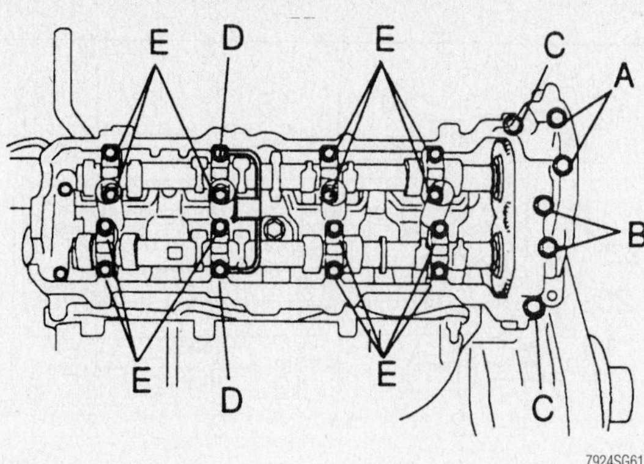

Right bank bearing cap bolt location

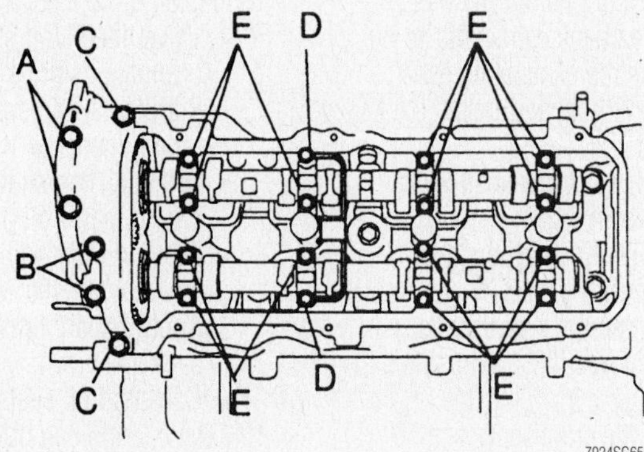

Left camshaft bearing cap bolt locations

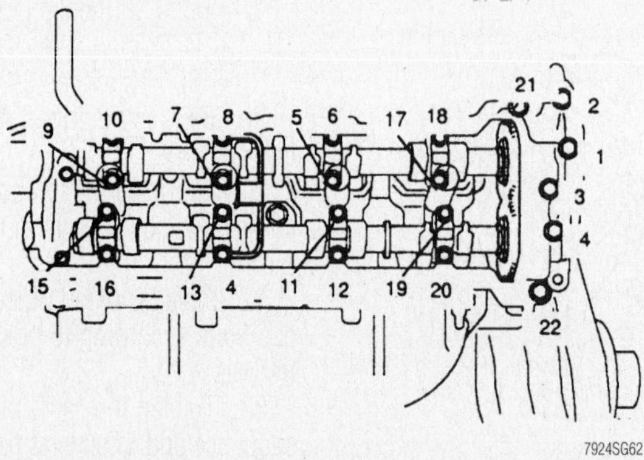

Right bank camshaft bearing cap bolt torque sequence

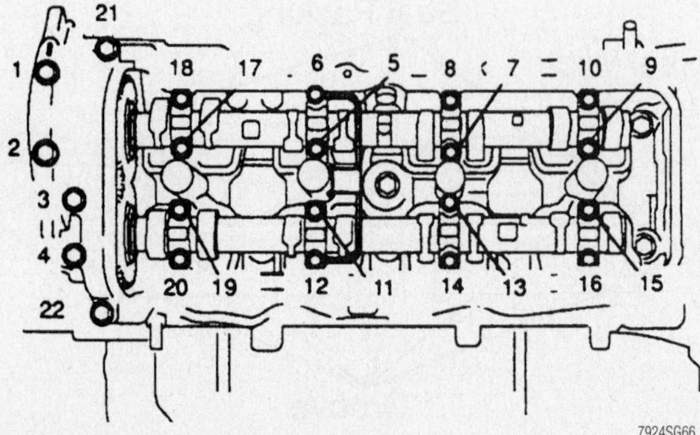

Left bank camshaft bearing cap bolt torque sequence

- Bolt C: 66 inch lbs. (7.5 Nm)
- All others: 12 ft. lbs. (16 Nm)

24. Remove the service bolts from the exhaust camshaft gears.
25. Install or connect the following:
    - Timing belt rear covers
    - Valve cover
    - Ignition coils
    - CMP sensor
    - Camshaft timing sprockets. Tighten the bolts to 80 ft. lbs. (108 Nm).
    - Timing belt
    - Lower timing cover
    - Crankshaft pulley. Tighten the bolt to 181 ft. lbs. (245 Nm).
    - Accessory drive belt tensioner
    - Alternator
    - Cooling fan bracket
    - A/C compressor
    - Upper and middle timing belt covers
    - Idler pulley. Tighten the bolt to 27 ft. lbs. (37 Nm).
    - Radiator
    - Cooling fan
    - Accessory drive belt
    - Air intake hose
    - Engine appearance cover
    - Engine under covers
    - Negative battery cable
26. Fill the cooling system.
27. Start the engine and check for leaks.

Valve Lash

ADJUSTMENT

➡Measure valve clearance with the engine cold.

1. Before servicing the vehicle, refer to the precautions in the beginning of this section.
2. Drain the cooling system.
3. Remove or disconnect the following:
    - Negative battery cable
    - Ignition coils
    - Valve covers
4. Set the engine to the top of the compression stroke with the valves closed for the cylinder to be measured.
5. Check the valve clearance. The valve clearance specifications are as follows:
    - Intake: 0.006–0.010 in. (0.15–0.25mm)
    - Exhaust: 0.010–0.014 in. (0.25–0.35mm)
6. Record the measurements for each valve.
7. When all valve clearances have been measured, remove the camshafts.
8. Remove the valve shims and mea-

Measured clearance mm (in.) \ Installed shim thickness mm (in.)	2.000 (0.0787)	2.020 (0.0795)	2.040 (0.0803)	2.060 (0.0811)	2.080 (0.0819)	2.100 (0.0827)	2.120 (0.0835)	2.140 (0.0843)	2.160 (0.0850)	2.180 (0.0858)	2.200 (0.0866)	2.210 (0.0870)	2.220 (0.0874)	2.230 (0.0878)	2.240 (0.0882)	2.250 (0.0886)	2.260 (0.0890)	2.270 (0.0894)	2.280 (0.0898)	2.290 (0.0902)	2.300 (0.0906)	2.310 (0.0909)	2.320 (0.0913)	2.330 (0.0917)	2.340 (0.0921)	2.350 (0.0925)	2.360 (0.0929)	2.370 (0.0933)	2.380 (0.0937)	2.390 (0.0941)	2.400 (0.0945)	2.410 (0.0949)	2.420 (0.0953)	2.430 (0.0957)	2.440 (0.0961)	2.450 (0.0965)	2.460 (0.0969)	2.470 (0.0972)	2.480 (0.0976)	2.490 (0.0980)	2.500 (0.0984)	2.510 (0.0988)	2.520 (0.0992)	2.530 (0.0996)	2.540 (0.1000)	2.550 (0.1004)	2.560 (0.1008)	2.570 (0.1012)	2.580 (0.1016)	2.590 (0.1020)	2.600 (0.1024)	2.620 (0.1031)	2.640 (0.1039)	2.660 (0.1047)	2.680 (0.1055)	2.700 (0.1063)	2.720 (0.1071)	2.740 (0.1079)	2.760 (0.1087)	2.780 (0.1094)	2.800 (0.1102)
0.000–0.030 (0.0000–0.0012)							00	00	00	00	02	02	04	04	06	06	08	08	10	10	12	12	14	14	16	16	18	18	20	20	22	22	24	24	26	26	28	28	30	30	32	32	34	34	36	36	38	38	40	40	42	44	46	48	50	52	54	56	58	60	62
0.031–0.050 (0.0012–0.0020)						00	00	00	00	02	04	06	06	08	08	10	10	12	12	14	14	16	16	18	18	20	20	22	22	24	24	26	26	28	28	30	30	32	32	34	34	36	36	38	38	40	40	42	42	44	44	46	48	50	52	54	56	58	60	62	64
0.051–0.070 (0.0020–0.0028)					00	00	00	00	02	04	06	08	08	10	10	12	12	14	14	16	16	18	18	20	20	22	22	24	24	26	26	28	28	30	30	32	32	34	34	36	36	38	38	40	40	42	42	44	44	46	46	48	50	52	54	56	58	60	62	64	66
0.071–0.090 (0.0028–0.0035)				00	00	00	00	02	04	06	08	10	10	12	12	14	14	16	16	18	18	20	20	22	22	24	24	26	26	28	28	30	30	32	32	34	34	36	36	38	38	40	40	42	42	44	44	46	46	48	48	50	52	54	56	58	60	62	64	66	68
0.091–0.110 (0.0036–0.0043)			00	00	00	00	02	04	06	08	10	12	12	14	14	16	16	18	18	20	20	22	22	24	24	26	26	28	28	30	30	32	32	34	34	36	36	38	38	40	40	42	42	44	44	46	46	48	48	50	50	52	54	56	58	60	62	64	66	68	70
0.111–0.130 (0.0044–0.0051)		00	00	00	00	02	04	06	08	10	12	14	14	16	16	18	18	20	20	22	22	24	24	26	26	28	28	30	30	32	32	34	34	36	36	38	38	40	40	42	42	44	44	46	46	48	48	50	50	52	52	54	56	58	60	62	64	66	68	70	72
0.131–0.149 (0.0052–0.0059)		00	00	00	02	04	06	08	10	12	14	16	16	18	18	20	20	22	22	24	24	26	26	28	28	30	30	32	32	34	34	36	36	38	38	40	40	42	42	44	44	46	46	48	48	50	50	52	52	54	54	56	58	60	62	64	66	68	70	72	74
0.150–0.250 (0.0059–0.0098)																																																													
0.251–0.270 (0.0099–0.0106)	06	08	10	12	14	16	18	20	22	24	26	28	28	30	30	32	32	34	34	36	36	38	38	40	40	42	42	44	44	46	46	48	48	50	50	52	52	54	54	56	56	58	58	60	60	62	62	64	64	66	66	68	70	72	74	76	78	80	80	80	
0.271–0.290 (0.0107–0.0114)	08	10	12	14	16	18	20	22	24	26	28	30	30	32	32	34	34	36	36	38	38	40	40	42	42	44	44	46	46	48	48	50	50	52	52	54	54	56	56	58	58	60	60	62	62	64	64	66	66	68	68	70	72	74	76	78	80	80	80		
0.291–0.310 (0.0115–0.0122)	10	12	14	16	18	20	22	24	26	28	30	32	32	34	34	36	36	38	38	40	40	42	42	44	44	46	46	48	48	50	50	52	52	54	54	56	56	58	58	60	60	62	62	64	64	66	66	68	68	70	70	72	74	76	78	80	80	80			
0.311–0.330 (0.0122–0.0130)	12	14	16	18	20	22	24	26	28	30	32	34	34	36	36	38	38	40	40	42	42	44	44	46	46	48	48	50	50	52	52	54	54	56	56	58	58	60	60	62	62	64	64	66	66	68	68	70	70	72	72	74	76	78	80	80	80				
0.331–0.350 (0.0130–0.0138)	14	16	18	20	22	24	26	28	30	32	34	36	36	38	38	40	40	42	42	44	44	46	46	48	48	50	50	52	52	54	54	56	56	58	58	60	60	62	62	64	64	66	66	68	68	70	70	72	72	74	74	76	78	80	80	80					
0.351–0.370 (0.0138–0.0146)	16	18	20	22	24	26	28	30	32	34	36	38	38	40	40	42	42	44	44	46	46	48	48	50	50	52	52	54	54	56	56	58	58	60	60	62	62	64	64	66	66	68	68	70	70	72	72	74	74	76	76	78	80	80	80						
0.371–0.390 (0.0146–0.0154)	18	20	22	24	26	28	30	32	34	36	38	40	40	42	42	44	44	46	46	48	48	50	50	52	52	54	54	56	56	58	58	60	60	62	62	64	64	66	66	68	68	70	70	72	72	74	74	76	76	78	78	80	80	80							
0.391–0.410 (0.0154–0.0161)	20	22	24	26	28	30	32	34	36	38	40	42	42	44	44	46	46	48	48	50	50	52	52	54	54	56	56	58	58	60	60	62	62	64	64	66	66	68	68	70	70	72	72	74	74	76	76	78	78	80	80	80	80								
0.411–0.430 (0.0162–0.0169)	22	24	26	28	30	32	34	36	38	40	42	44	44	46	46	48	48	50	50	52	52	54	54	56	56	58	58	60	60	62	62	64	64	66	66	68	68	70	70	72	72	74	74	76	76	78	78	80	80	80	80	80									
0.431–0.450 (0.0170–0.0177)	24	26	28	30	32	34	36	38	40	42	44	46	46	48	48	50	50	52	52	54	54	56	56	58	58	60	60	62	62	64	64	66	66	68	68	70	70	72	72	74	74	76	76	78	78	80	80	80	80	80	80										
0.451–0.470 (0.0178–0.0185)	26	28	30	32	34	36	38	40	42	44	46	48	48	50	50	52	52	54	54	56	56	58	58	60	60	62	62	64	64	66	66	68	68	70	70	72	72	74	74	76	76	78	78	80	80	80	80	80	80	80											
0.471–0.490 (0.0185–0.0193)	28	30	32	34	36	38	40	42	44	46	48	50	50	52	52	54	54	56	56	58	58	60	60	62	62	64	64	66	66	68	68	70	70	72	72	74	74	76	76	78	78	80	80	80	80	80	80	80													
0.491–0.510 (0.0193–0.0201)	30	32	34	36	38	40	42	44	46	48	50	52	52	54	54	56	56	58	58	60	60	62	62	64	64	66	66	68	68	70	70	72	72	74	74	76	76	78	78	80	80	80	80	80	80	80															
0.511–0.530 (0.0201–0.0209)	32	34	36	38	40	42	44	46	48	50	52	54	54	56	56	58	58	60	60	62	62	64	64	66	66	68	68	70	70	72	72	74	74	76	76	78	78	80	80	80	80	80	80	80																	
0.531–0.550 (0.0209–0.0217)	34	36	38	40	42	44	46	48	50	52	54	56	56	58	58	60	60	62	62	64	64	66	66	68	68	70	70	72	72	74	74	76	76	78	78	80	80	80	80	80	80	80																			
0.551–0.570 (0.0217–0.0224)	36	38	40	42	44	46	48	50	52	54	56	58	58	60	60	62	62	64	64	66	66	68	68	70	70	72	72	74	74	76	76	78	78	80	80	80	80	80	80	80																					
0.571–0.590 (0.0225–0.0232)	38	40	42	44	46	48	50	52	54	56	58	60	60	62	62	64	64	66	66	68	68	70	70	72	72	74	74	76	76	78	78	80	80	80	80	80	80	80																							
0.591–0.610 (0.0233–0.0240)	40	42	44	46	48	50	52	54	56	58	60	62	62	64	64	66	66	68	68	70	70	72	72	74	74	76	76	78	78	80	80	80	80	80	80	80																									
0.611–0.630 (0.0241–0.0248)	42	44	46	48	50	52	54	56	58	60	62	64	64	66	66	68	68	70	70	72	72	74	74	76	76	78	78	80	80	80	80	80	80	80																											
0.631–0.650 (0.0248–0.0256)	44	46	48	50	52	54	56	58	60	62	64	66	66	68	68	70	70	72	72	74	74	76	76	78	78	80	80	80	80	80	80	80																													
0.651–0.670 (0.0256–0.0264)	46	48	50	52	54	56	58	60	62	64	66	68	68	70	70	72	72	74	74	76	76	78	78	80	80	80	80	80	80	80																															
0.671–0.690 (0.0264–0.0272)	48	50	52	54	56	58	60	62	64	66	68	70	70	72	72	74	74	76	76	78	78	80	80	80	80	80	80	80																																	
0.691–0.710 (0.0272–0.0280)	50	52	54	56	58	60	62	64	66	68	70	72	72	74	74	76	76	78	78	80	80	80	80	80	80	80																																			
0.711–0.730 (0.0280–0.0287)	52	54	56	58	60	62	64	66	68	70	72	74	74	76	76	78	78	80	80	80	80	80	80	80																																					
0.731–0.750 (0.0288–0.0295)	54	56	58	60	62	64	66	68	70	72	74	76	76	78	78	80	80	80	80	80	80	80																																							
0.751–0.770 (0.0296–0.0303)	56	58	60	62	64	66	68	70	72	74	76	78	78	80	80	80	80	80	80	80																																									
0.771–0.790 (0.0304–0.0311)	58	60	62	64	66	68	70	72	74	76	78	80	80	80	80	80	80	80																																											
0.791–0.810 (0.0311–0.0319)	60	62	64	66	68	70	72	74	76	78	80	80	80	80	80	80																																													
0.811–0.830 (0.0319–0.0327)	62	64	66	68	70	72	74	76	78	80	80	80	80	80																																															
0.831–0.850 (0.0327–0.0335)	64	66	68	70	72	74	76	78	80	80	80	80																																																	
0.851–0.870 (0.0335–0.0343)	66	68	70	72	74	76	78	80	80	80																																																			
0.871–0.890 (0.0343–0.0350)	68	70	72	74	76	78	80	80	80																																																				
0.891–0.910 (0.0351–0.0358)	70	72	74	76	78	80	80	80																																																					
0.911–0.930 (0.0359–0.0366)	72	74	76	78	80	80	80																																																						
0.931–0.950 (0.0367–0.0374)	74	76	78	80	80	80																																																							
0.951–0.970 (0.0374–0.0382)	76	78	80	80	80																																																								
0.971–0.990 (0.0382–0.0390)	78	80	80	80																																																									
0.991–1.010 (0.0390–0.0398)	80	80	80																																																										
1.011–1.030 (0.0398–0.0406)	80	80																																																											
1.031–1.050 (0.0406–0.0413)	80																																																												

Intake valve clearance (Cold):
0.15 – 0.25 mm (0.006 – 0.010 in.)

EXAMPLE:
The 2.300 mm (0.0906 in.) shim is installed, and the measured clearance is 0.440 mm (0.0173 in.). Replace the 2.300 mm (0.0906 in.) shim with a No. 54 shim.

New shim thickness mm (in.)

Shim No.	Thickness	Shim No.	Thickness	Shim No.	Thickness
00	2.000 (0.0787)	28	2.280 (0.0898)	56	2.560 (0.1008)
02	2.020 (0.0795)	30	2.300 (0.0906)	58	2.580 (0.1016)
04	2.040 (0.0803)	32	2.320 (0.0913)	60	2.600 (0.1024)
06	2.060 (0.0811)	34	2.340 (0.0921)	62	2.620 (0.1031)
08	2.080 (0.0819)	36	2.360 (0.0929)	64	2.640 (0.1039)
10	2.100 (0.0827)	38	2.380 (0.0937)	66	2.660 (0.1047)
12	2.120 (0.0835)	40	2.400 (0.0945)	68	2.680 (0.1055)
14	2.140 (0.0843)	42	2.420 (0.0953)	70	2.700 (0.1063)
16	2.160 (0.0850)	44	2.440 (0.0961)	72	2.720 (0.1071)
18	2.180 (0.0858)	46	2.460 (0.0969)	74	2.740 (0.1079)
20	2.200 (0.0866)	48	2.480 (0.0976)	76	2.760 (0.1087)
22	2.220 (0.0874)	50	2.500 (0.0984)	78	2.780 (0.1094)
24	2.240 (0.0882)	52	2.520 (0.0992)	80	2.800 (0.1102)
26	2.260 (0.0890)	54	2.540 (0.1000)		

7924SG71

Intake valve clearance shim selection chart

Measured clearance mm (in.) / Installed shim thickness mm (in.)	2.000 (0.0787)	2.020 (0.0795)	2.040 (0.0803)	2.060 (0.0811)	2.080 (0.0819)	2.100 (0.0827)	2.120 (0.0835)	2.140 (0.0843)	2.160 (0.0850)	2.180 (0.0858)	2.200 (0.0866)	2.210 (0.0870)	2.220 (0.0874)	2.230 (0.0878)	2.240 (0.0882)	2.250 (0.0886)	2.260 (0.0890)	2.270 (0.0894)	2.280 (0.0898)	2.290 (0.0902)	2.300 (0.0906)	2.310 (0.0909)	2.320 (0.0913)	2.330 (0.0917)	2.340 (0.0921)	2.350 (0.0925)	2.360 (0.0929)	2.370 (0.0933)	2.380 (0.0937)	2.390 (0.0941)
0.000–0.030 (0.0000–0.0012)												00	00	00	00	00	00	00	00	00	02	02	04	04	06	06	08	08	10	10
0.031–0.050 (0.0012–0.0020)											00	00	00	00	00	00	00	02	02	04	04	06	06	08	08	10	10	12	12	14
0.051–0.070 (0.0020–0.0028)										00	00	00	00	00	00	02	02	04	04	06	06	08	08	10	10	12	12	14	14	16
0.071–0.090 (0.0028–0.0035)									00	00	00	00	00	02	02	04	04	06	06	08	08	10	10	12	12	14	14	16	16	18
0.091–0.110 (0.0036–0.0043)								00	00	00	00	02	02	04	04	06	06	08	08	10	10	12	12	14	14	16	16	18	18	20
0.111–0.130 (0.0044–0.0051)							00	00	00	00	02	04	04	06	06	08	08	10	10	12	12	14	14	16	16	18	18	20	20	22
0.131–0.150 (0.0052–0.0059)						00	00	00	00	02	04	06	06	08	08	10	10	12	12	14	14	16	16	18	18	20	20	22	22	24
0.151–0.170 (0.0059–0.0067)					00	00	00	00	02	04	06	08	08	10	10	12	12	14	14	16	16	18	18	20	20	22	22	24	24	26
0.171–0.190 (0.0067–0.0075)				00	00	00	00	02	04	06	08	10	10	12	12	14	14	16	16	18	18	20	20	22	22	24	24	26	26	28
0.191–0.210 (0.0075–0.0083)			00	00	00	00	02	04	06	08	10	12	12	14	14	16	16	18	18	20	20	22	22	24	24	26	26	28	28	30
0.211–0.230 (0.0083–0.0091)		00	00	00	00	02	04	06	08	10	12	14	14	16	16	18	18	20	20	22	22	24	24	26	26	28	28	30	30	32
0.231–0.249 (0.0091–0.0098)		00	00	00	02	04	06	08	10	12	14	16	16	18	18	20	20	22	22	24	24	26	26	28	28	30	30	32	32	34
0.250–0.350 (0.0098–0.0138)																														
0.351–0.370 (0.0138–0.0146)	06	08	10	12	14	16	18	20	22	24	26	28	28	30	30	32	32	34	34	36	36	38	38	40	40	42	42	44	44	46
0.371–0.390 (0.0146–0.0154)	08	10	12	14	16	18	20	22	24	26	28	30	30	32	32	34	34	36	36	38	38	40	40	42	42	44	44	46	46	48
0.391–0.410 (0.0154–0.0161)	10	12	14	16	18	20	22	24	26	28	30	32	32	34	34	36	36	38	38	40	40	42	42	44	44	46	46	48	48	50
0.411–0.430 (0.0162–0.0169)	12	14	16	18	20	22	24	26	28	30	32	34	34	36	36	38	38	40	40	42	42	44	44	46	46	48	48	50	50	52
0.431–0.450 (0.0170–0.0177)	14	16	18	20	22	24	26	28	30	32	34	36	36	38	38	40	40	42	42	44	44	46	46	48	48	50	50	52	52	54
0.451–0.470 (0.0178–0.0185)	16	18	20	22	24	26	28	30	32	34	36	38	38	40	40	42	42	44	44	46	46	48	48	50	50	52	52	54	54	56
0.471–0.490 (0.0185–0.0193)	18	20	22	24	26	28	30	32	34	36	38	40	40	42	42	44	44	46	46	48	48	50	50	52	52	54	54	56	56	58
0.491–0.510 (0.0193–0.0201)	20	22	24	26	28	30	32	34	36	38	40	42	42	44	44	46	46	48	48	50	50	52	52	54	54	56	56	58	58	60
0.511–0.530 (0.0201–0.0209)	22	24	26	28	30	32	34	36	38	40	42	44	44	46	46	48	48	50	50	52	52	54	54	56	56	58	58	60	60	62
0.531–0.550 (0.0209–0.0217)	24	26	28	30	32	34	36	38	40	42	44	46	46	48	48	50	50	52	52	54	54	56	56	58	58	60	60	62	62	64
0.551–0.570 (0.0217–0.0224)	26	28	30	32	34	36	38	40	42	44	46	48	48	50	50	52	52	54	54	56	56	58	58	60	60	62	62	64	64	66
0.571–0.590 (0.0225–0.0232)	28	30	32	34	36	38	40	42	44	46	48	50	50	52	52	54	54	56	56	58	58	60	60	62	62	64	64	66	66	68
0.591–0.610 (0.0233–0.0240)	30	32	34	36	38	40	42	44	46	48	50	52	52	54	54	56	56	58	58	60	60	62	62	64	64	66	66	68	68	70
0.611–0.630 (0.0241–0.0248)	32	34	36	38	40	42	44	46	48	50	52	54	54	56	56	58	58	60	60	62	62	64	64	66	66	68	68	70	70	72
0.631–0.650 (0.0248–0.0256)	34	36	38	40	42	44	46	48	50	52	54	56	56	58	58	60	60	62	62	64	64	66	66	68	68	70	70	72	72	74
0.651–0.670 (0.0256–0.0264)	36	38	40	42	44	46	48	50	52	54	56	58	58	60	60	62	62	64	64	66	66	68	68	70	70	72	72	74	74	76
0.671–0.690 (0.0264–0.0272)	38	40	42	44	46	48	50	52	54	56	58	60	60	62	62	64	64	66	66	68	68	70	70	72	72	74	74	76	76	78
0.691–0.710 (0.0272–0.0280)	40	42	44	46	48	50	52	54	56	58	60	62	62	64	64	66	66	68	68	70	70	72	72	74	74	76	76	78	78	80
0.711–0.730 (0.0280–0.0287)	42	44	46	48	50	52	54	56	58	60	62	64	64	66	66	68	68	70	70	72	72	74	74	76	76	78	78	80	80	80
0.731–0.750 (0.0288–0.0295)	44	46	48	50	52	54	56	58	60	62	64	66	66	68	68	70	70	72	72	74	74	76	76	78	78	80	80	80	80	80
0.751–0.770 (0.0296–0.0303)	46	48	50	52	54	56	58	60	62	64	66	68	68	70	70	72	72	74	74	76	76	78	78	80	80	80	80	80	80	80
0.771–0.790 (0.0304–0.0311)	48	50	52	54	56	58	60	62	64	66	68	70	70	72	72	74	74	76	76	78	78	80	80	80	80	80	80	80		
0.791–0.810 (0.0311–0.0319)	50	52	54	56	58	60	62	64	66	68	70	72	72	74	74	76	76	78	78	80	80	80	80	80	80	80				
0.811–0.830 (0.0319–0.0327)	52	54	56	58	60	62	64	66	68	70	72	74	74	76	76	78	78	80	80	80	80	80	80	80						
0.831–0.850 (0.0327–0.0335)	54	56	58	60	62	64	66	68	70	72	74	76	76	78	78	80	80	80	80	80	80	80								
0.851–0.870 (0.0335–0.0343)	56	58	60	62	64	66	68	70	72	74	76	78	78	80	80	80	80	80	80	80										
0.871–0.890 (0.0343–0.0350)	58	60	62	64	66	68	70	72	74	76	78	80	80	80	80	80	80	80												
0.891–0.910 (0.0351–0.0358)	60	62	64	66	68	70	72	74	76	78	80	80	80	80	80	80														
0.911–0.930 (0.0359–0.0366)	62	64	66	68	70	72	74	76	78	80	80	80	80	80																
0.931–0.950 (0.0367–0.0374)	64	66	68	70	72	74	76	78	80	80	80	80																		
0.951–0.970 (0.0374–0.0382)	66	68	70	72	74	76	78	80	80	80																				
0.971–0.990 (0.0382–0.0390)	68	70	72	74	76	78	80	80	80																					
0.991–1.010 (0.0390–0.0398)	70	72	74	76	78	80	80	80																						
1.011–1.030 (0.0398–0.0406)	72	74	76	78	80	80	80																							
1.031–1.050 (0.0406–0.0413)	74	76	78	80	80	80																								
1.051–1.070 (0.0414–0.0421)	76	78	80	80	80																									
1.071–1.090 (0.0422–0.0429)	78	80	80	80																										
1.091–1.110 (0.0430–0.0437)	80	80	80																											
1.111–1.130 (0.0437–0.0445)	80	80																												
1.131–1.150 (0.0445–0.0453)	80																													

Measured clearance mm (in.) / Installed shim thickness mm (in.)	2.400 (0.0945)	2.410 (0.0949)	2.420 (0.0953)	2.430 (0.0957)	2.440 (0.0961)	2.450 (0.0965)	2.460 (0.0969)	2.470 (0.0972)	2.480 (0.0976)	2.490 (0.0980)	2.500 (0.0984)	2.510 (0.0988)	2.520 (0.0992)	2.530 (0.0996)	2.540 (0.1000)	2.550 (0.1004)	2.560 (0.1008)	2.570 (0.1012)	2.580 (0.1016)	2.590 (0.1020)	2.600 (0.1024)	2.620 (0.1031)	2.640 (0.1039)	2.660 (0.1047)	2.680 (0.1055)	2.700 (0.1063)	2.720 (0.1071)	2.740 (0.1079)	2.760 (0.1087)	2.780 (0.1094)	2.800 (0.1102)
0.000–0.030 (0.0000–0.0012)	12	12	14	14	16	16	18	18	20	20	22	22	24	24	26	26	28	28	30	30	32	34	36	38	40	42	44	46	48	50	52
0.031–0.050 (0.0012–0.0020)	14	16	16	18	18	20	20	22	22	24	24	26	26	28	28	30	30	32	32	34	34	36	38	40	42	44	46	48	50	52	54
0.051–0.070 (0.0020–0.0028)	16	18	18	20	20	22	22	24	24	26	26	28	28	30	30	32	32	34	34	36	36	38	40	42	44	46	48	50	52	54	56
0.071–0.090 (0.0028–0.0035)	18	20	20	22	22	24	24	26	26	28	28	30	30	32	32	34	34	36	36	38	38	40	42	44	46	48	50	52	54	56	58
0.091–0.110 (0.0036–0.0043)	20	22	22	24	24	26	26	28	28	30	30	32	32	34	34	36	36	38	38	40	40	42	44	46	48	50	52	54	56	58	60
0.111–0.130 (0.0044–0.0051)	22	24	24	26	26	28	28	30	30	32	32	34	34	36	36	38	38	40	40	42	42	44	46	48	50	52	54	56	58	60	62
0.131–0.150 (0.0052–0.0059)	24	26	26	28	28	30	30	32	32	34	34	36	36	38	38	40	40	42	42	44	44	46	48	50	52	54	56	58	60	62	64
0.151–0.170 (0.0059–0.0067)	26	28	28	30	30	32	32	34	34	36	36	38	38	40	40	42	42	44	44	46	46	48	50	52	54	56	58	60	62	64	66
0.171–0.190 (0.0067–0.0075)	28	30	30	32	32	34	34	36	36	38	38	40	40	42	42	44	44	46	46	48	48	50	52	54	56	58	60	62	64	66	68
0.191–0.210 (0.0075–0.0083)	30	32	32	34	34	36	36	38	38	40	40	42	42	44	44	46	46	48	48	50	50	52	54	56	58	60	62	64	66	68	70
0.211–0.230 (0.0083–0.0091)	32	34	34	36	36	38	38	40	40	42	42	44	44	46	46	48	48	50	50	52	52	54	56	58	60	62	64	66	68	70	72
0.231–0.249 (0.0091–0.0098)	34	36	36	38	38	40	40	42	42	44	44	46	46	48	48	50	50	52	52	54	54	56	58	60	62	64	66	68	70	72	74
0.250–0.350 (0.0098–0.0138)																															
0.351–0.370 (0.0138–0.0146)	46	48	48	50	50	52	52	54	54	56	56	58	58	60	60	62	62	64	64	66	66	68	70	72	74	76	78	80	80	80	
0.371–0.390 (0.0146–0.0154)	48	50	50	52	52	54	54	56	56	58	58	60	60	62	62	64	64	66	66	68	68	70	72	74	76	78	80	80	80		
0.391–0.410 (0.0154–0.0161)	50	52	52	54	54	56	56	58	58	60	60	62	62	64	64	66	66	68	68	70	70	72	74	76	78	80	80	80			
0.411–0.430 (0.0162–0.0169)	52	54	54	56	56	58	58	60	60	62	62	64	64	66	66	68	68	70	70	72	72	74	76	78	80	80	80				
0.431–0.450 (0.0170–0.0177)	54	56	56	58	58	60	60	62	62	64	64	66	66	68	68	70	70	72	72	74	74	76	78	80	80	80					
0.451–0.470 (0.0178–0.0185)	56	58	58	60	60	62	62	64	64	66	66	68	68	70	70	72	72	74	74	76	76	78	80	80	80						
0.471–0.490 (0.0185–0.0193)	58	60	60	62	62	64	64	66	66	68	68	70	70	72	72	74	74	76	76	78	78	80	80	80							
0.491–0.510 (0.0193–0.0201)	60	62	62	64	64	66	66	68	68	70	70	72	72	74	74	76	76	78	78	80	80	80	80								
0.511–0.530 (0.0201–0.0209)	62	64	64	66	66	68	68	70	70	72	72	74	74	76	76	78	78	80	80	80	80	80									
0.531–0.550 (0.0209–0.0217)	64	66	66	68	68	70	70	72	72	74	74	76	76	78	78	80	80	80	80	80	80										
0.551–0.570 (0.0217–0.0224)	66	68	68	70	70	72	72	74	74	76	76	78	78	80	80	80	80	80	80	80											
0.571–0.590 (0.0225–0.0232)	68	70	70	72	72	74	74	76	76	78	78	80	80	80	80	80	80	80													
0.591–0.610 (0.0233–0.0240)	70	72	72	74	74	76	76	78	78	80	80	80	80	80	80	80															
0.611–0.630 (0.0241–0.0248)	72	74	74	76	76	78	78	80	80	80	80	80	80	80																	
0.631–0.650 (0.0248–0.0256)	74	76	76	78	78	80	80	80	80	80	80	80																			
0.651–0.670 (0.0256–0.0264)	76	78	78	80	80	80	80	80	80	80																					
0.671–0.690 (0.0264–0.0272)	78	80	80	80	80	80	80	80																							
0.691–0.710 (0.0272–0.0280)	80	80	80	80	80	80																									
0.711–0.730 (0.0280–0.0287)	80	80	80	80																											
0.731–0.750 (0.0288–0.0295)	80	80																													

Exhaust valve clearance (Cold):
0.25 – 0.35 mm (0.010 – 0.014 in.)

EXAMPLE:
The 2.300 mm (0.0906 in.) shim is installed, and the measured clearance is 0.440 mm (0.0173 in.). Replace the 2.300 mm (0.0906 in.) shim with a No. 44 shim.

New shim thickness mm (in.)

Shim No.	Thickness	Shim No.	Thickness	Shim No.	Thickness
00	2.000 (0.0787)	28	2.280 (0.0898)	56	2.560 (0.1008)
02	2.020 (0.0795)	30	2.300 (0.0906)	58	2.580 (0.1016)
04	2.040 (0.0803)	32	2.320 (0.0913)	60	2.600 (0.1024)
06	2.060 (0.0811)	34	2.340 (0.0921)	62	2.620 (0.1031)
08	2.080 (0.0819)	36	2.360 (0.0929)	64	2.640 (0.1039)
10	2.100 (0.0827)	38	2.380 (0.0937)	66	2.660 (0.1047)
12	2.120 (0.0835)	40	2.400 (0.0945)	68	2.680 (0.1055)
14	2.140 (0.0843)	42	2.420 (0.0953)	70	2.700 (0.1063)
16	2.160 (0.0850)	44	2.440 (0.0961)	72	2.720 (0.1071)
18	2.180 (0.0858)	46	2.460 (0.0969)	74	2.740 (0.1079)
20	2.200 (0.0866)	48	2.480 (0.0976)	76	2.760 (0.1087)
22	2.220 (0.0874)	50	2.500 (0.0984)	78	2.780 (0.1094)
24	2.240 (0.0882)	52	2.520 (0.0992)	80	2.800 (0.1102)
26	2.260 (0.0890)	54	2.540 (0.1000)		

7924SG72

Exhaust valve clearance shim selection chart

sure them. Note this measurement along with the clearance measurement recorded earlier.

9. Using the valve clearance and shim thickness measurements, find replacement shims in the Adjusting Shim Selection charts.

10. Install or connect the following:

- Replacement valve shims
- Camshafts
- Valve covers
- Ignition coils
- Negative battery cable

11. Fill the cooling system.

12. Start the engine and check for leaks.

Starter Motor

REMOVAL & INSTALLATION

1. Before servicing the vehicle, refer to the precautions in the beginning of this section.

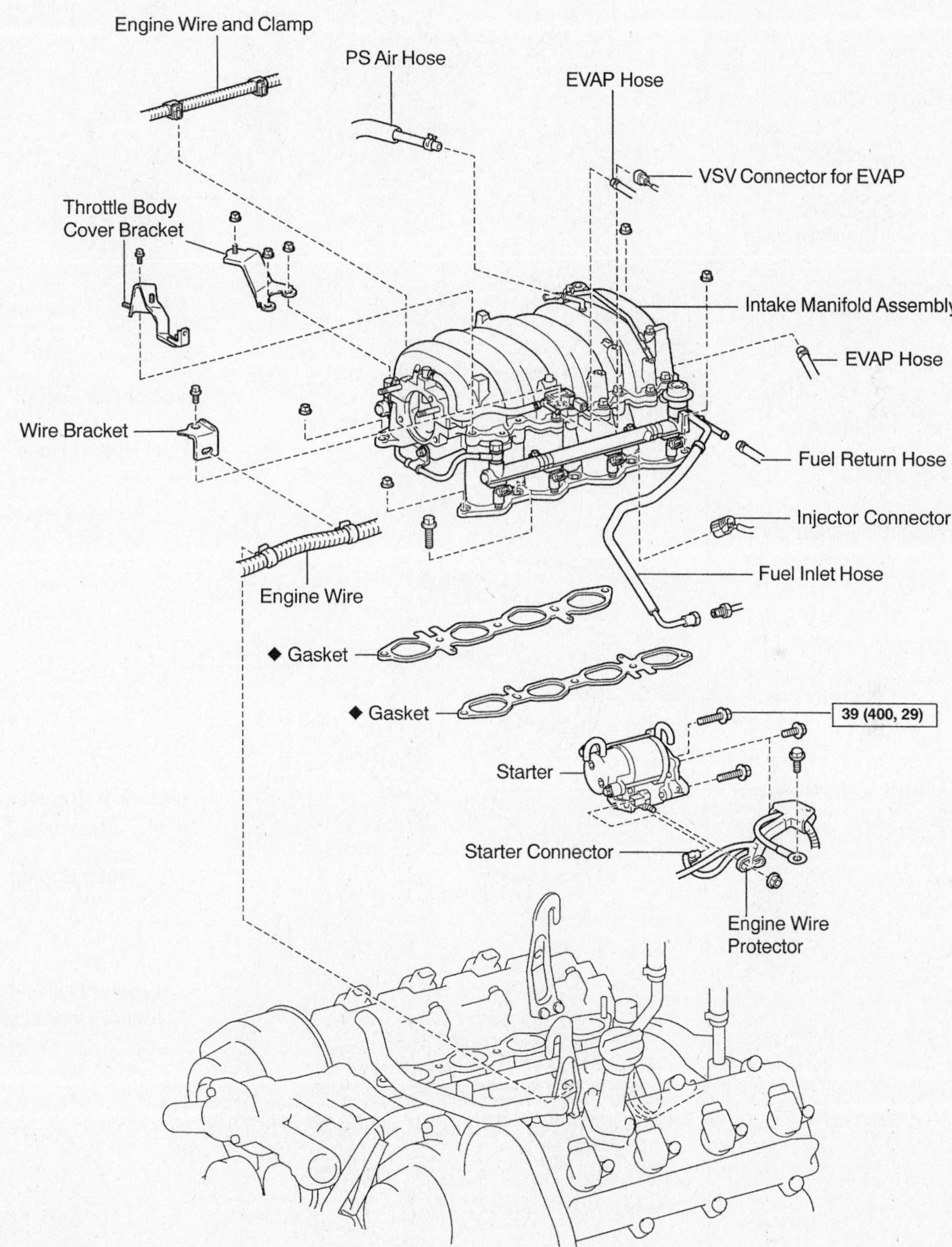

Starter removal—Sequoia

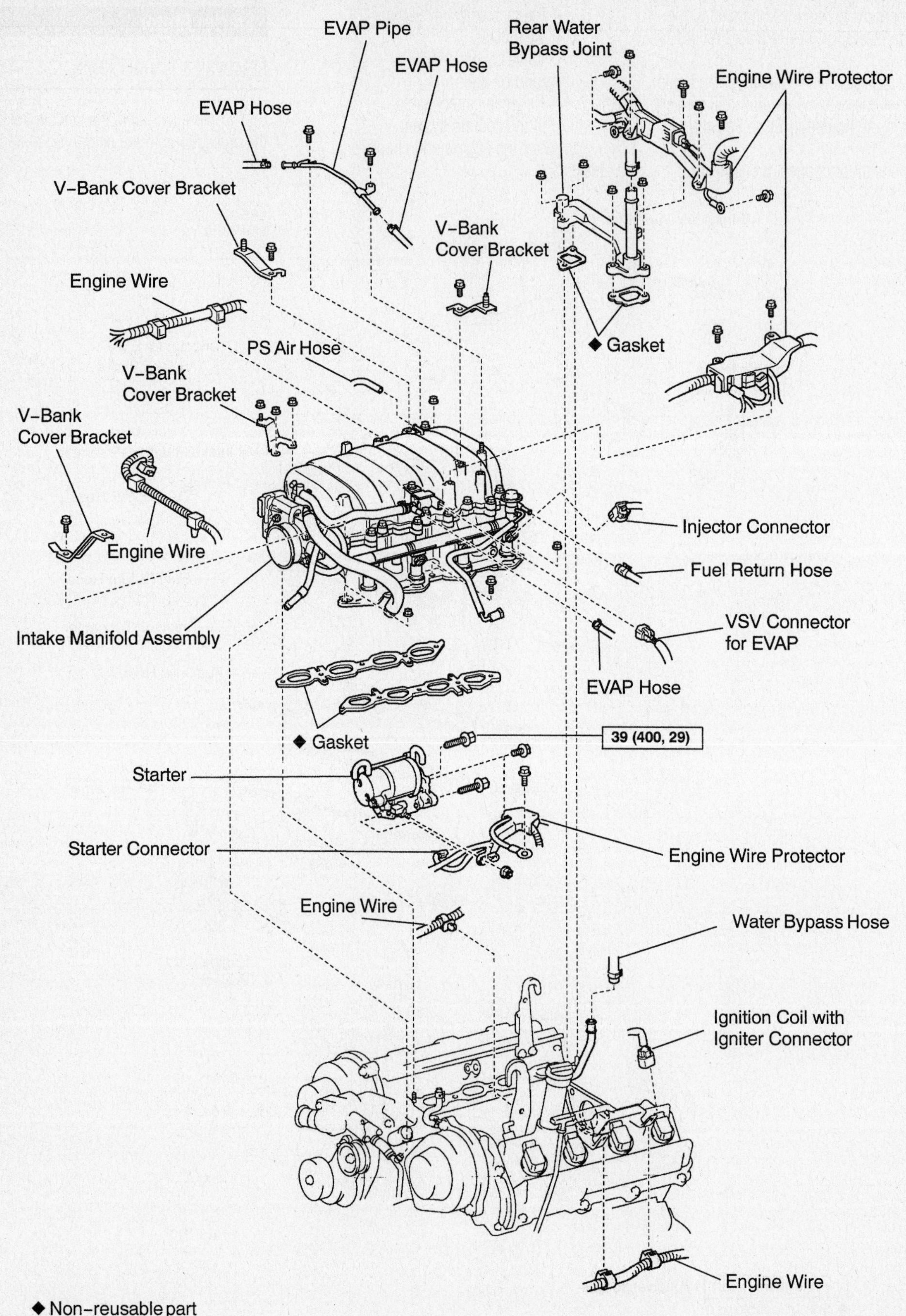
EVAP Pipe
Rear Water
Bypass Joint
EVAP Hose
Engine Wire Protector
EVAP Hose
V-Bank Cover Bracket
V-Bank
Cover Bracket
Engine Wire
◆ Gasket
PS Air Hose
V-Bank
Cover Bracket
V-Bank
Cover Bracket
Engine Wire
Injector Connector
Fuel Return Hose
Intake Manifold Assembly
VSV Connector
for EVAP
EVAP Hose
◆ Gasket
39 (400, 29)
Starter
Starter Connector
Engine Wire Protector
Engine Wire
Water Bypass Hose
Ignition Coil with
Igniter Connector
Engine Wire
◆ Non-reusable part
67170-LCSQ-G07

Starter removal—Land Cruiser

2. Drain the cooling system.
3. Relieve the fuel system pressure.
4. Remove or disconnect the following:
   - Negative battery cable
   - Engine appearance cover
   - Air intake tube
   - Intake manifold
   - Starter motor mounting bolts
   - Starter wiring connectors
   - Starter motor

To install:

5. Install or connect the following:
   - Starter motor
   - Starter wiring connectors. Tighten the cable nut to 86 inch lbs. (10 Nm).
   - Starter motor mounting bolts. Tighten the bolts to 29 ft. lbs. (39 Nm).
   - Intake manifold
   - Air intake tube
   - Engine appearance cover
   - Negative battery cable
6. Fill the cooling system.
7. Start the engine and check for leaks.

Timing Belt

REMOVAL & INSTALLATION

1. Disconnect the negative battery cable.
2. Raise and safely support the vehicle.
3. Remove the oil pan protector and the engine under cover.
4. Drain the cooling system and store the coolant for refilling purposes.
5. Lower the vehicle and remove the battery clamp cover.
6. From the top of the engine, remove the fuel return hose, the engine cover nuts/bolts and the cover.
7. Remove the air cleaner and the intake air connector assembly.
8. Remove the cooling fan pulley by performing the following procedures:
   a. Loosen the 4 fan clutch-to-fan pulley nuts.
   b. Using a box-end wrench on the serpentine drive belt tensioner bolt, rotate the tensioner counterclockwise and remove the drive belt.

➡The serpentine drive belt tensioner bolt is a left-hand thread.

   c. Remove the fan clutch-to-fan pulley nuts, the fan, the clutch assembly and the fan pulley.
9. Remove the radiator by performing the following procedures:
   a. Disconnect the upper, lower and reservoir hoses from the radiator.
   b. Disconnect and plug the automatic transmission oil cooler at the radiator. Disconnect the automatic transmission oil cooler hoses from the fan shroud clamp.
   c. Remove the radiator reservoir tank.
   d. Remove the fan shroud-to-radiator bolts and the shroud.
   e. Remove the 2 upper radiator-to-chassis nuts.
   f. Remove the middle radiator-to-chassis nut/bolts and brackets.
   g. Carefully, lift the radiator from the vehicle.
10. Remove the serpentine drive belt idler pulley bolt, cover plate and pulley.
11. Remove the right side (No. 3) timing belt cover.
12. Remove the left side (No. 3) timing belt cover by performing the following procedures:
   a. Disconnect the engine wire from both wire clamps.
   b. Disconnect the camshaft position sensor wire from the wire clamp on the left-side (No.3) timing belt cover.
   c. Disconnect the sensor connector from the connector bracket.
   d. Disconnect the sensor connector.
   e. Remove the wire grommet from the left-side (No. 3) timing belt cover.
   f. Remove the oil cooler tube bolts and tube.
13. Remove the middle (No. 2) timing belt cover bolts and cover.
14. Remove the cooling fan bracket nuts/bolts and bracket.

➡If reusing the timing belt, make sure that there are 3 installation marks on the belt; if there are none, install them.

15. Using the Crankshaft Pulley Holding tool 09213-70010, Bolt tool 90105-08076 and Companion Flange Holding tool 09330-00021, or equivalent, loosen the crankshaft pulley bolt.
16. Position the No. 1 cylinder to approximately 50 degrees After Top Dead Center (ATDC) of the compression stroke by performing the following procedures:
   a. Rotate the crankshaft pulley (CLOCKWISE) to align its groove with the timing mark "0" on the lower (No. 1) timing belt cover.
   b. Check that the camshaft sprocket timing marks are aligned with the rear timing belt plate marks; if not, rotate the crankshaft 1 revolution (360 degrees).
   c. Rotate the crankshaft pulley approximately 50 degrees (CLOCKWISE) and align the crankshaft pulley timing mark between the centers of the crankshaft pulley bolt and the idler pulley bolt.

⁂ WARNING

If the timing belt is disengaged, having the crankshaft pulley in the wrong angle can cause the valve to come into contact with the piston when removing the camshaft pulley.

17. Remove the crankshaft pulley bolt.

➡If reusing the timing belt and the installation marks have disappeared, place new installation marks on the timing belt to match the camshaft timing sprocket marks.

➡To avoid meshing the timing sprocket and the timing belt, secure one with a string; then, place matchmarks on the timing belt and the right-side camshaft timing sprocket.

18. Remove the timing belt tensioner bolts and the tensioner.
19. Using the Camshaft Holding tool 09960-10010, or equivalent, slightly turn the left-side camshaft sprocket clockwise to loosen the tension spring. Then, disconnect the timing belt from the camshaft sprockets.
20. Remove the alternator by performing the following procedures:
   a. Disconnect the electrical connector from the alternator.
   b. Remove the rubber cap/nut and disconnect the battery wire from the alternator.
   c. Disconnect the wire clamp from the alternator cord clip.
   d. Remove the alternator-to-engine nuts/bolts and the alternator.
21. Remove the serpentine drive belt tensioner nuts/bolts and the tensioner.
22. Using the Crankshaft Puller Assembly tool 09950-50012, or equivalent, press the crankshaft pulley from the crankshaft.

⁂ WARNING

DO NOT rotate the crankshaft pulley.

23. Remove the lower (No. 1) timing belt cover bolts and the cover.
24. Remove the timing belt guide, spacer and the timing belt.

To install:

➡With the timing belt removed, this is a perfect opportunity to inspect and/or replace the water pump.

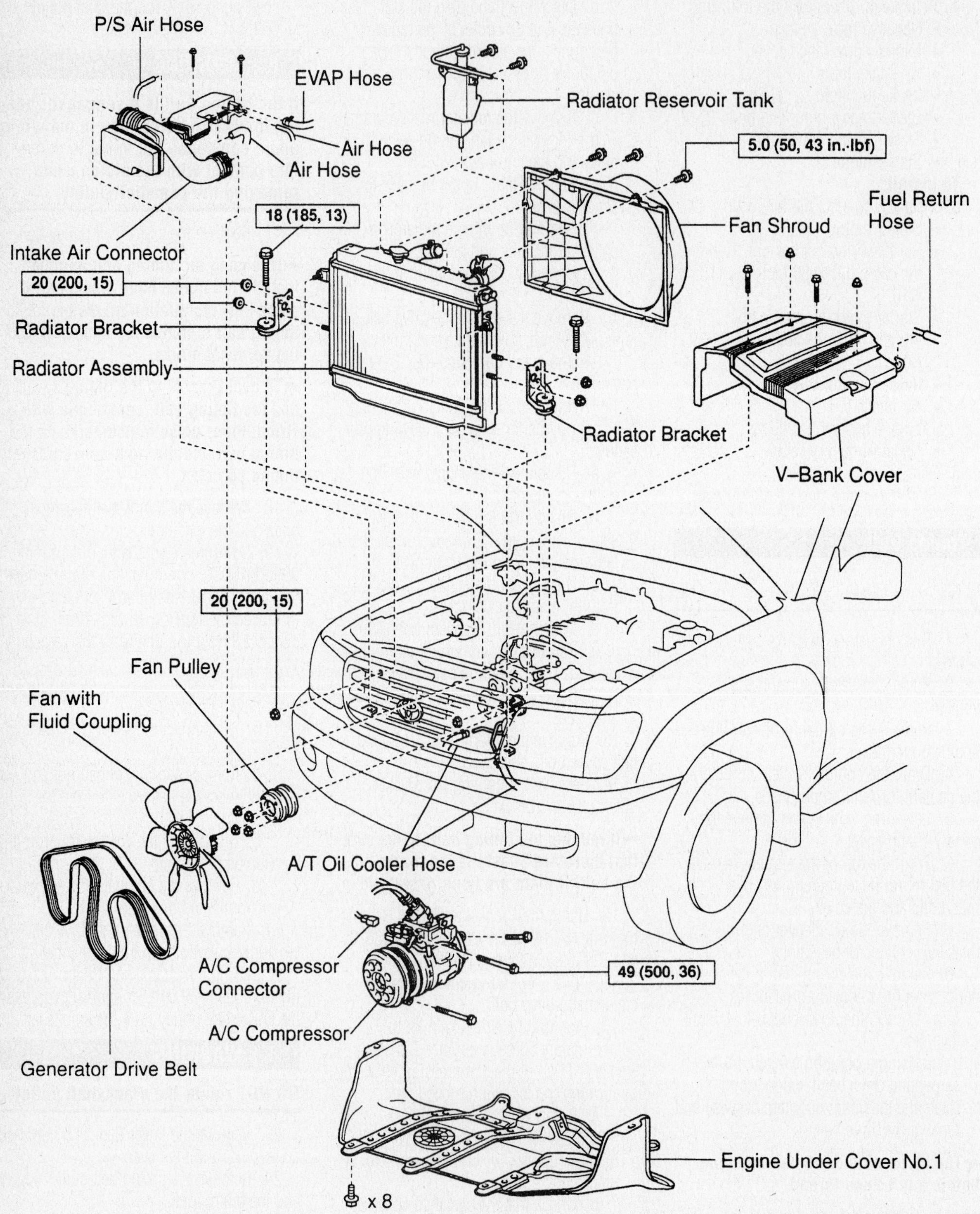

Exploded view of vehicle components for timing belt replacement—Land Cruiser

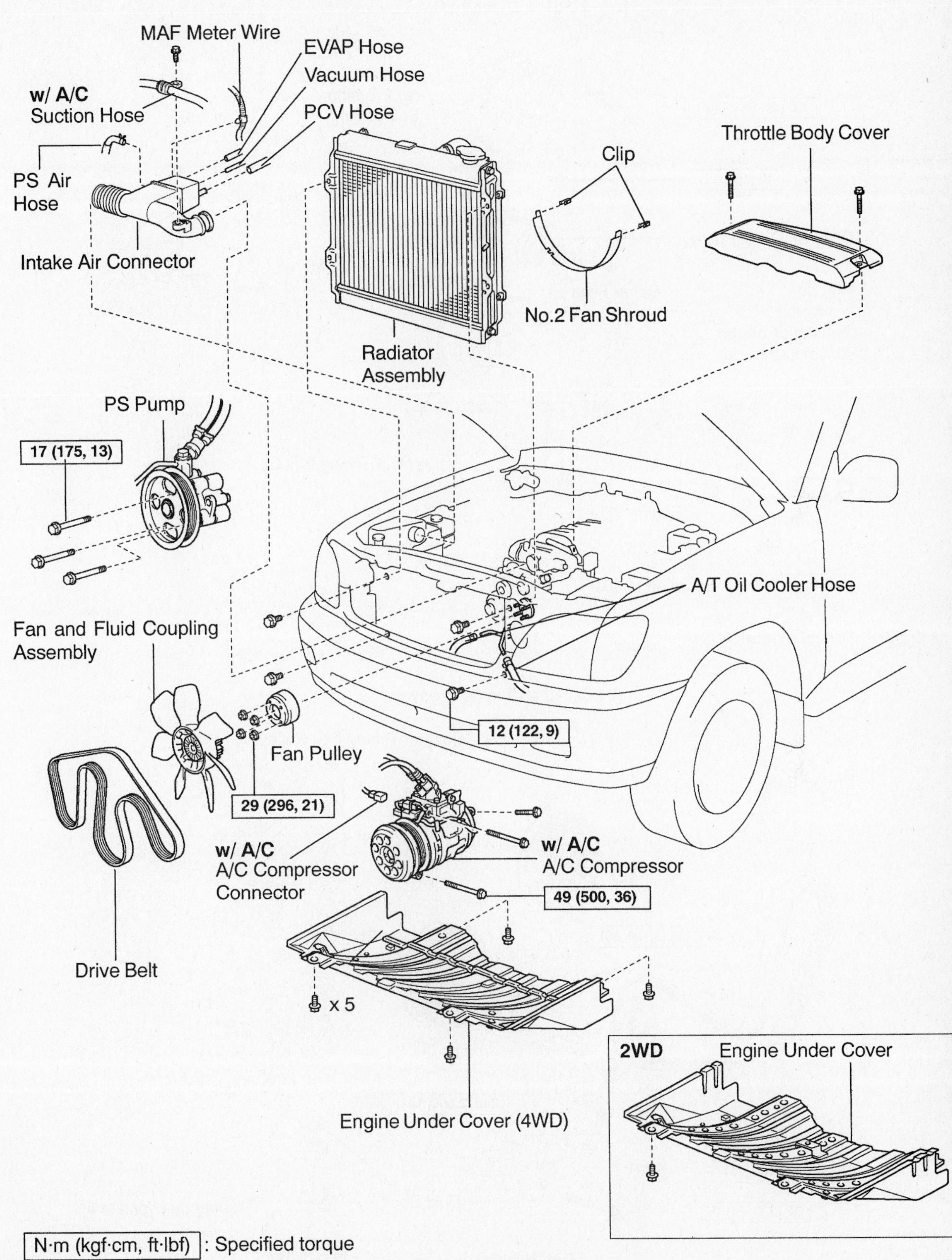

Exploded view of vehicle components for timing belt replacement—Sequoia

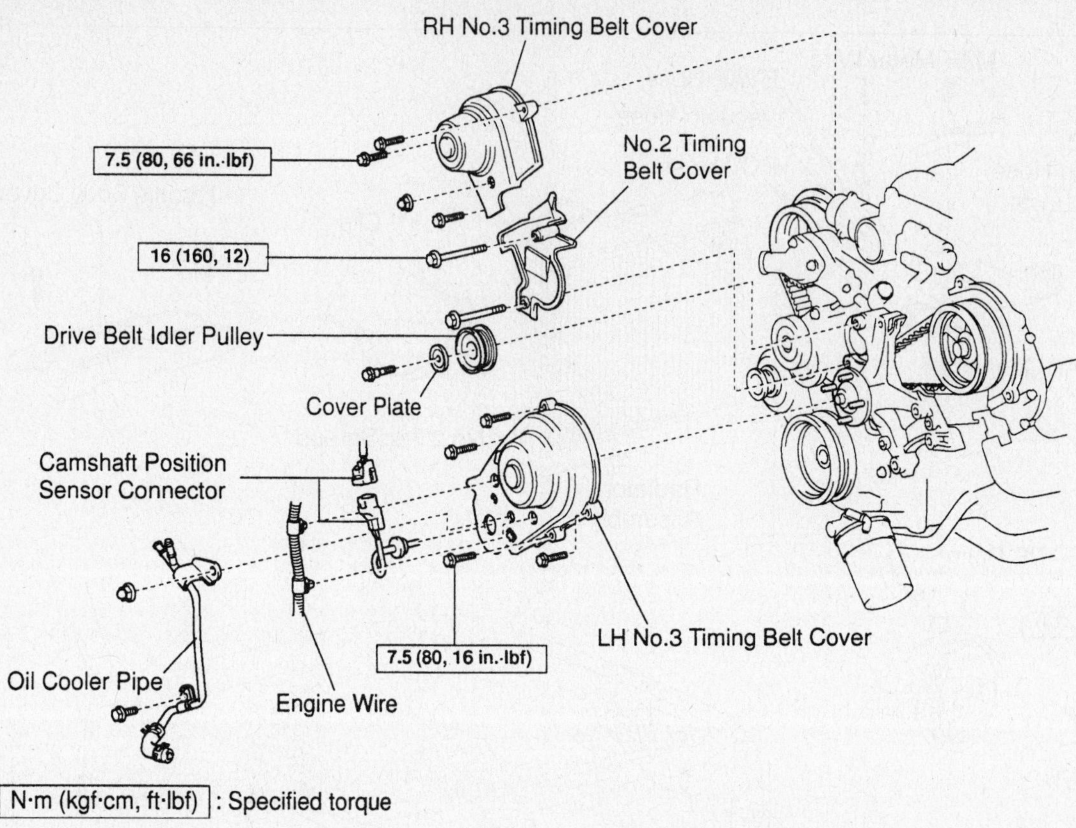

Exploded view of upper timing belt covers

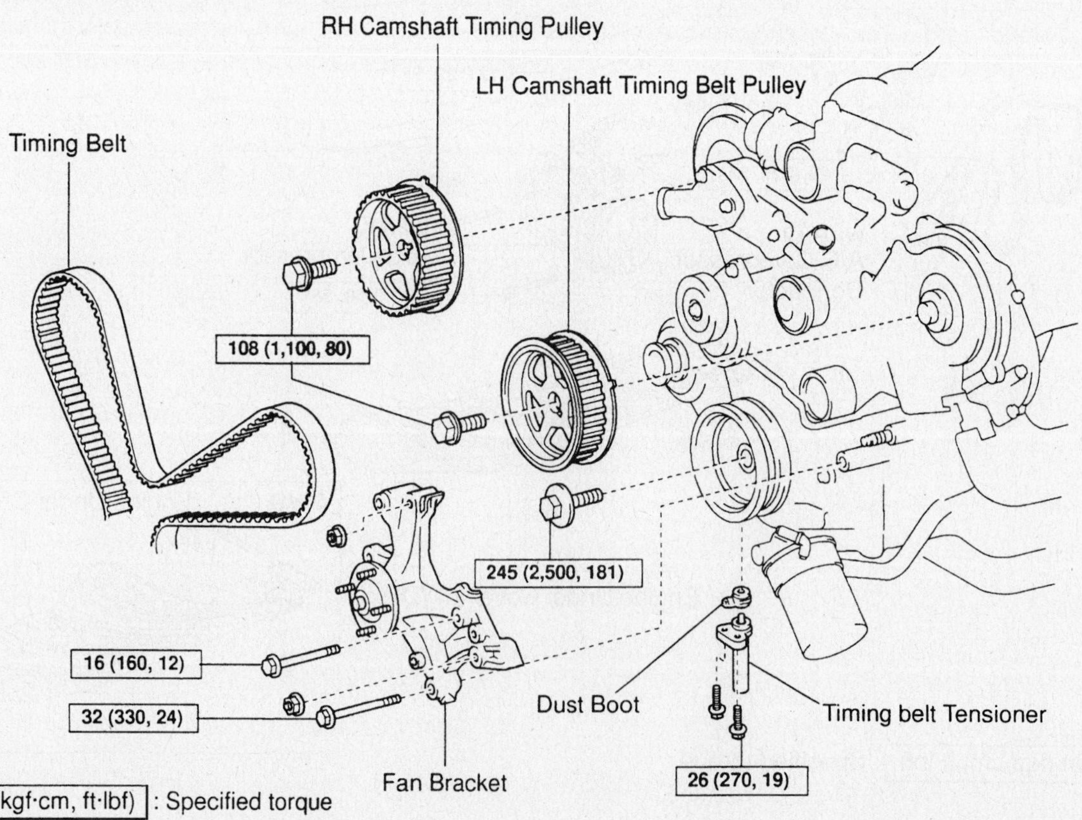

Exploded view of upper timing sprockets and components

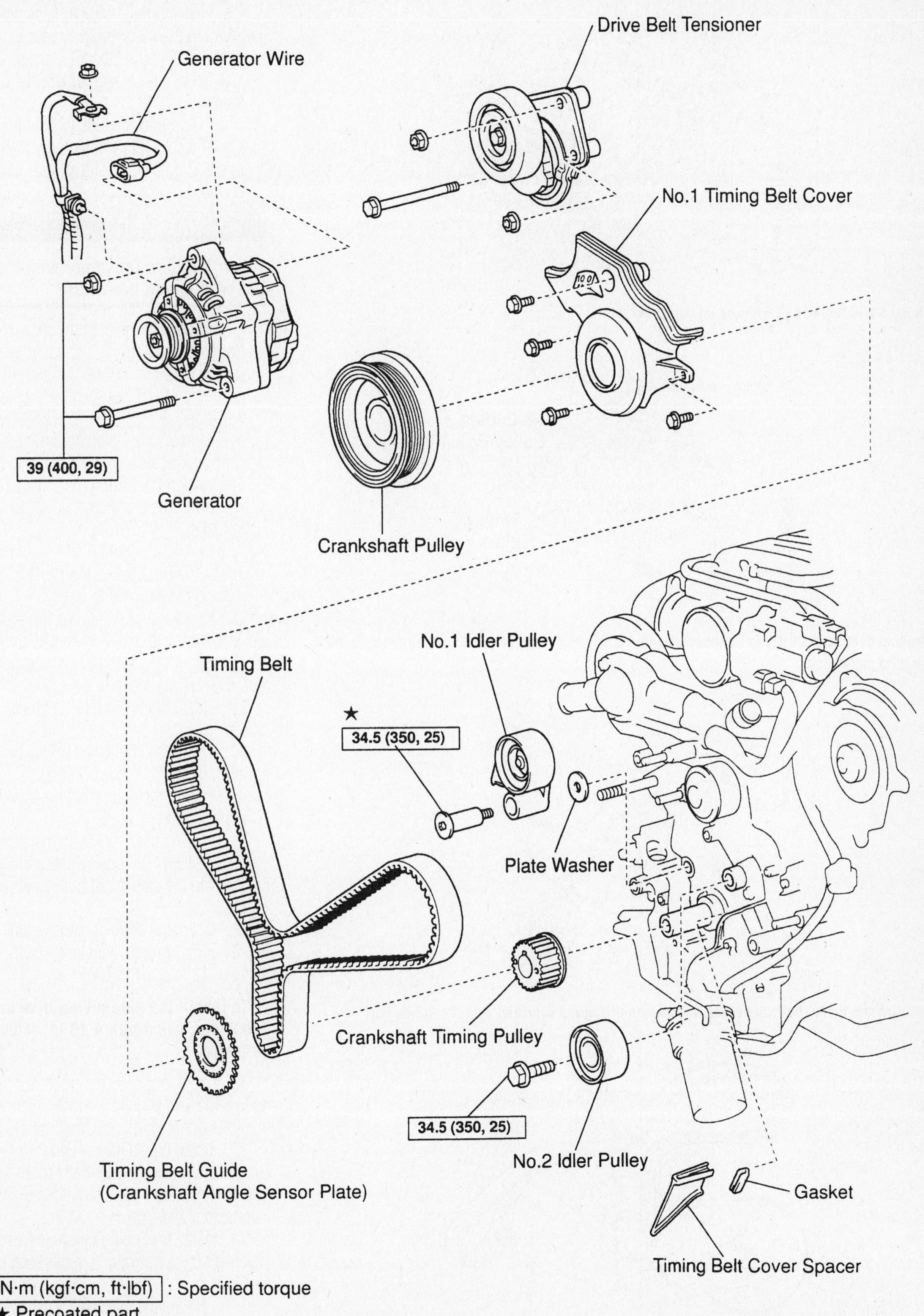

Exploded view of lower timing belt cover, sprockets and components

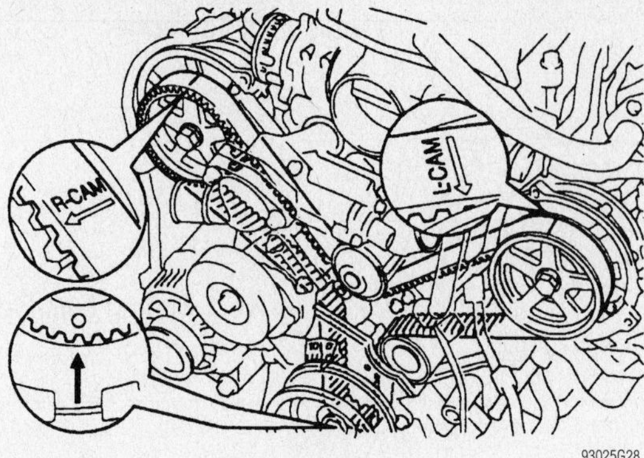

Alignment of timing belt with the timing sprockets

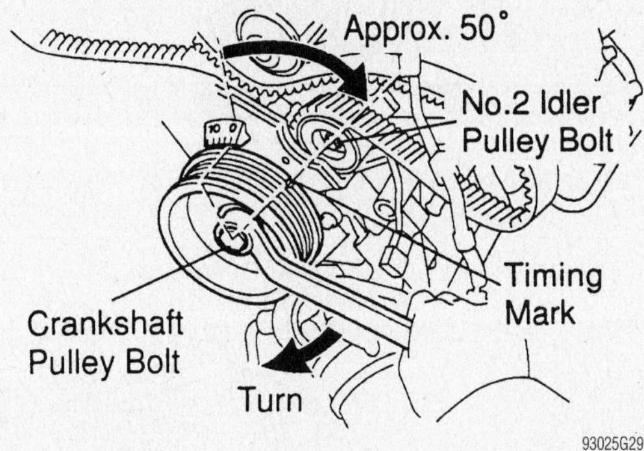

Aligning of crankshaft pulley timing mark with the center line of the crankshaft pulley bolt and the idler pulley bolt

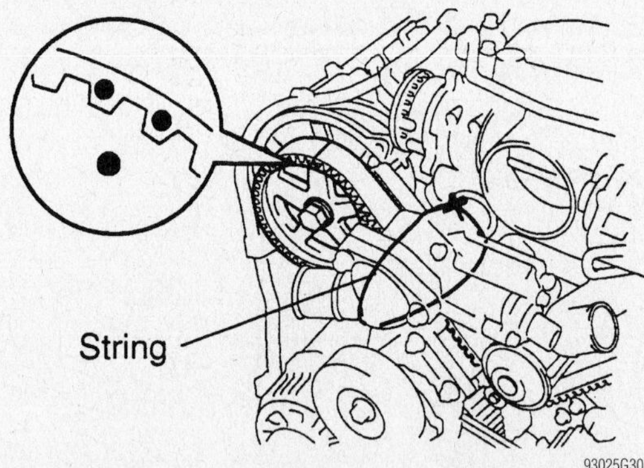

Securing the timing belt with string and matchmarking the camshaft with the timing belt

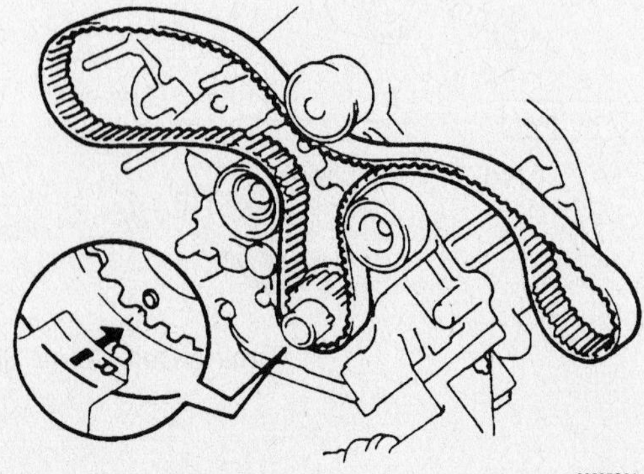

Installing the timing belt on the crankshaft sprocket

25. Inspect the timing belt tensioner by performing the following procedures:
   a. Inspect the seal for leakage; if leakage is suspected, replace the tensioner.
   b. Using both hands to hold the tensioner facing upward, strongly press the pushrod against a solid surface. If the pushrod moves, replace the tensioner.

⁂ WARNING

Never hold the tensioner with the pushrod facing downward.

   c. Measure the pushrod protrusion from the housing end, it should be 0.413–0.453 in. (10.5–11.5mm). If the protrusion is not as specified, replace the tensioner.
26. Temporarily install the timing belt by performing the following procedures:
   a. Align the timing belt's installation mark with the crankshaft timing sprocket.
   b. Install the timing belt on the crankshaft timing sprocket, the No. 1 idler pulley and the No. 2 idler pulley.
27. Install the gasket to the timing belt cover spacer and install the cover spacer.
28. Install the timing belt guide with the cup side facing outward.
29. Install the lower (No. 1) timing belt cover.
30. Install the crankshaft pulley by performing the following procedures:
   a. Align the crankshaft pulley with the crankshaft key.
   b. Using the Crankshaft Installer tool 09223-46011, or equivalent, and a hammer, tap the crankshaft pulley into position.
31. Install the serpentine drive belt tensioner and torque the tensioner-to-engine bolts to 12 ft. lbs. (16 Nm).

➡To install the serpentine drive belt tensioner, use a bolt 4.18 in. (106mm) in length.

32. Check that the crankshaft pulley's timing mark is aligned with the centers of the idler pulley and crankshaft pulley bolts.
33. Install the alternator and torque the alternator-to-engine nuts/bolts to 29 ft. lbs. (39 Nm). Connect the alternator's electrical connectors and clip.
34. Install the timing belt to the left-side camshaft by performing the following procedures:
   a. Rotate the left-side camshaft pulley to align the timing belt installation mark with the camshaft sprocket's timing mark

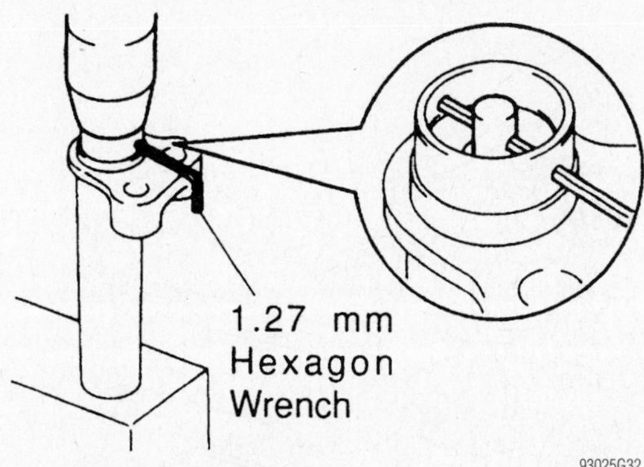

Securing the timing belt tensioner pushrod

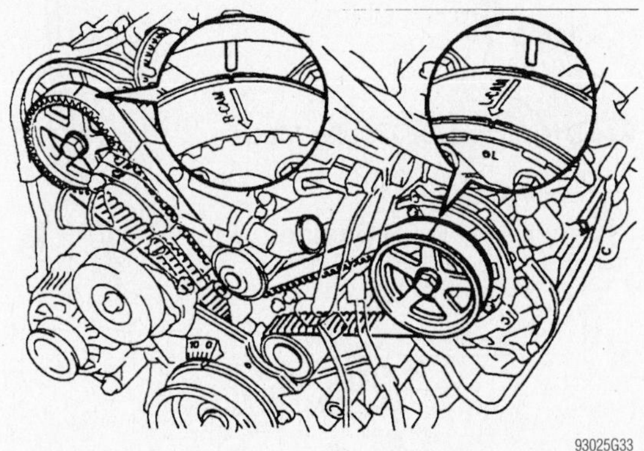

Checking the TDC alignment marks after rotating the crankshaft 2 revolutions

and slide the belt onto the camshaft timing sprocket.

b. Using the Camshaft Holding tool 09960-10010, or equivalent, slightly turn the left-side camshaft sprocket counterclockwise to place tension on the timing belt between the crankshaft sprocket and the camshaft sprocket.

35. Rotate the right-side camshaft pulley to align the timing belt installation mark with the camshaft sprocket's timing mark and slide the belt onto the camshaft timing sprocket.

36. Using a vertical press, slowly press the pushrod into the housing using 200–2205 lbs. (981–9807 N) until the holes align, then, install a 1.27mm Allen® wrench to secure the pushrod and release the press. Install the dust boot on the tensioner housing.

37. Install the timing belt tensioner and torque the bolts to 19 ft. lbs. (26 Nm).

38. Using a pair of pliers, remove the Allen® wrench from the tensioner housing.

39. Check the valve timing by performing the following procedure:

a. Temporarily install the crankshaft pulley bolt.

b. Slowly, rotate the crankshaft pulley 2 revolutions (CLOCKWISE) and realign the TDC marks.

➡If the pulley/sprocket timing marks do not realign, remove the timing belt and reinstall it.

40. Using the Crankshaft Pulley Holding tool 09213-70010, Bolt tool 90105-08076 and Companion Flange Holding tool 09330-00021, or equivalent, torque the crankshaft pulley bolt to 181 ft. lbs. (245 Nm).

41. Install the cooling fan bracket and torque the 12mm (head size) bolt to 12 ft. lbs. (16 Nm) and the 14mm (head size) bolt to 24 ft. lbs. (32 Nm).

42. Install the air conditioning compressor.

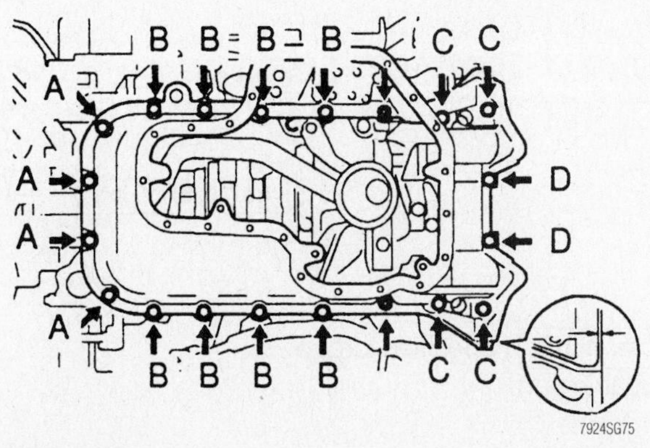

Upper oil pan bolt location

43. Install the middle (No. 2) timing belt cover and torque the bolts to 12 ft. lbs. (16 Nm).

44. Install the upper right-side (No. 3) timing belt cover and torque the bolts to 66 inch lbs. (7.5 Nm).

45. Install the upper left-side (No. 3) timing belt cover by performing the following procedures:

a. Install the oil cooler tube and bolt.

b. Feed the Camshaft Position Sensor (CPS) through the left-side (No. 3) timing belt cover hole.

c. Install the left-side (No. 3) timing belt cover and torque the bolts to 66 inch lbs. (7.5 Nm).

d. Install the wire grommet to the left-side (No. 3) timing belt cover.

e. Install the sensor connector to the connector bracket and connect the sensor connector.

f. Install the sensor wire and the engine wire to the clamps on the left-side (No. 3) timing belt cover.

46. Install the drive belt idler pulley and cover plate; then, torque the pulley bolt to 27 ft. lbs. (37 Nm).

47. To complete the installation, reverse the removal procedures.

48. Refill the cooling system and connect the negative battery cable.

Oil Pan

REMOVAL & INSTALLATION

1. Before servicing the vehicle, refer to the precautions in the beginning of this section.

2. Remove the engine from the vehicle and mount it on a stand.

3. Remove or disconnect the following:
   - Oil dipstick tube
   - Lower oil pan

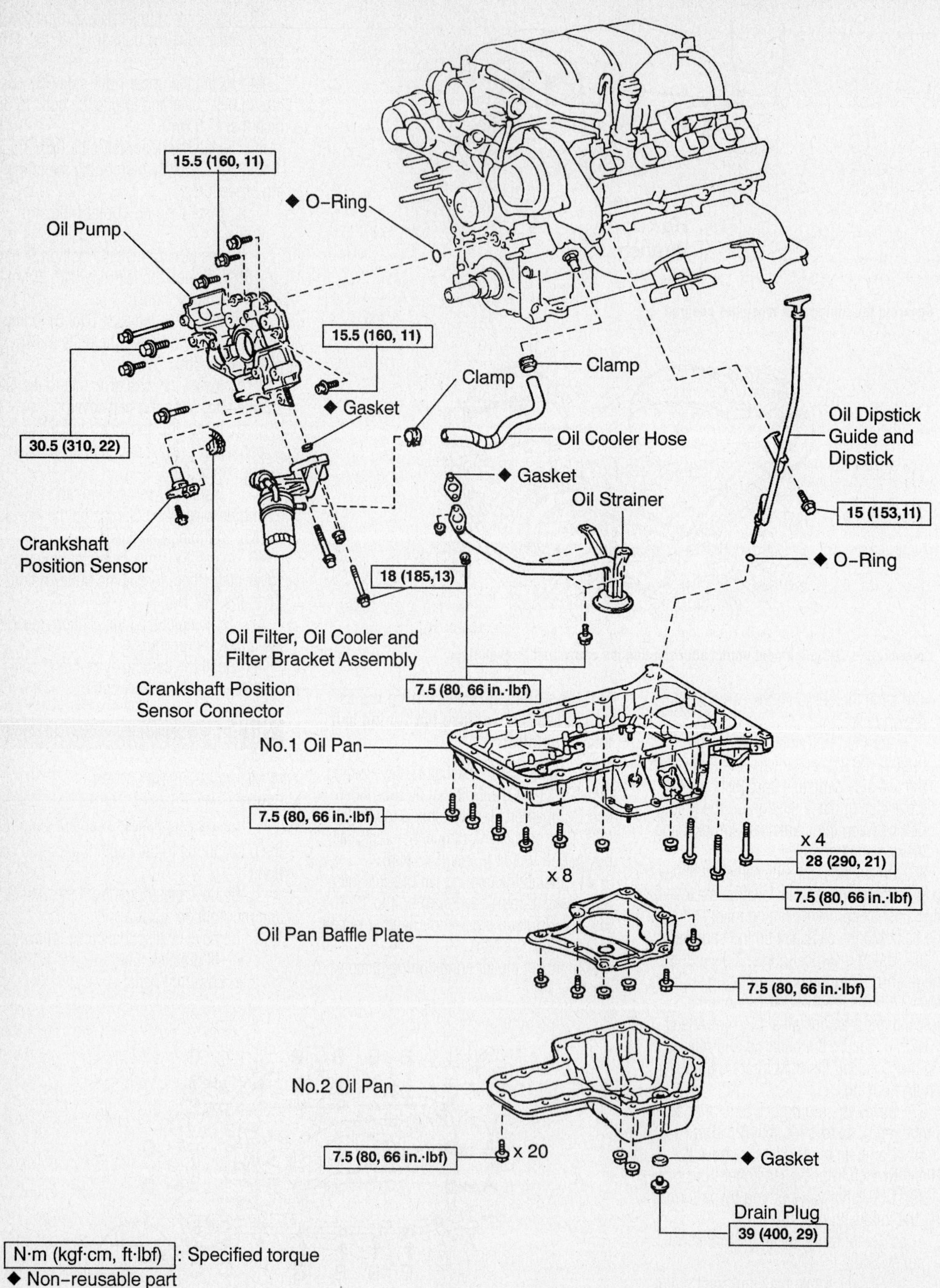
15.5 (160, 11)
◆ O–Ring
Oil Pump
15.5 (160, 11)
Clamp
Clamp
◆ Gasket
30.5 (310, 22)
Oil Cooler Hose
Oil Dipstick
Guide and
Dipstick
◆ Gasket
Oil Strainer
15 (153,11)
Crankshaft
Position Sensor
◆ O–Ring
18 (185,13)
Oil Filter, Oil Cooler and
Filter Bracket Assembly
Crankshaft Position
Sensor Connector
7.5 (80, 66 in.·lbf)
No.1 Oil Pan
7.5 (80, 66 in.·lbf)
x 4
28 (290, 21)
x 8
7.5 (80, 66 in.·lbf)
Oil Pan Baffle Plate
7.5 (80, 66 in.·lbf)
No.2 Oil Pan
7.5 (80, 66 in.·lbf)
x 20
◆ Gasket
Drain Plug
39 (400, 29)
N·m (kgf·cm, ft·lbf) : Specified torque
◆ Non–reusable part
67170-LCSQ-G09

Oil pan and pump—Land Cruiser

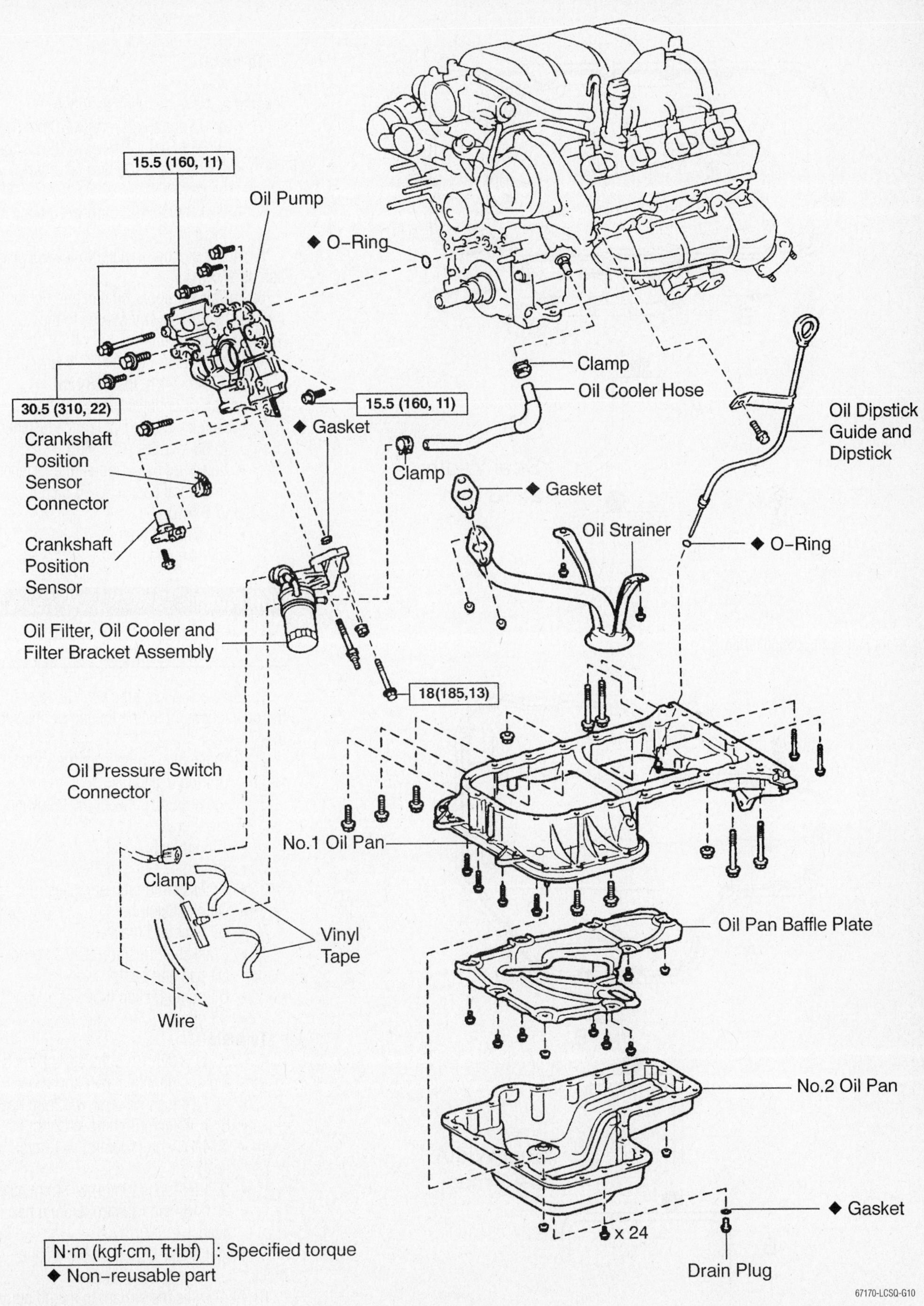
15.5 (160, 11)
Oil Pump
◆ O-Ring
30.5 (310, 22)
15.5 (160, 11)
◆ Gasket
Crankshaft Position Sensor Connector
Crankshaft Position Sensor
Oil Filter, Oil Cooler and Filter Bracket Assembly
18(185,13)
Clamp
Oil Cooler Hose
Clamp
◆ Gasket
Oil Strainer
Oil Dipstick Guide and Dipstick
◆ O-Ring
Oil Pressure Switch Connector
No.1 Oil Pan
Clamp
Vinyl Tape
Wire
Oil Pan Baffle Plate
No.2 Oil Pan
◆ Gasket
x 24
Drain Plug
N·m (kgf·cm, ft·lbf) : Specified torque
◆ Non-reusable part
67170-LCSQ-G10

Oil pan and pump—Sequoia

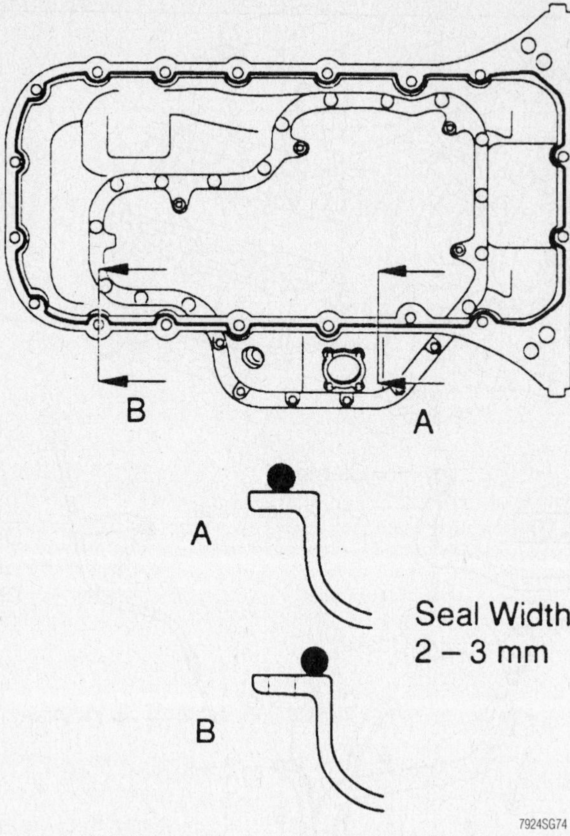

Upper oil pan sealant application

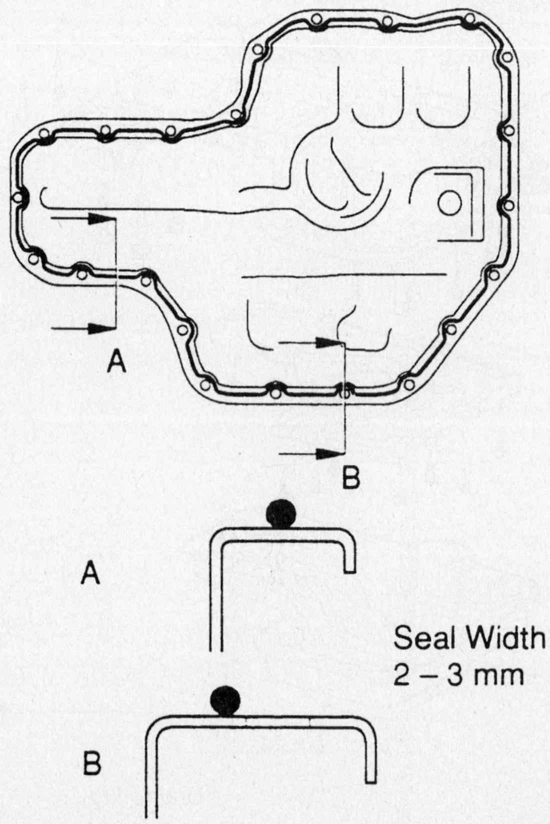

Lower oil pan sealant application

- Oil pan baffle
- Upper oil pan

To install:

4. The upper oil pan bolts are different lengths and are identified as follows:
   - A: 0.79 inch (20mm) w/10mm head
   - B: 0.98 inch (25mm) w/12mm head
   - C: 2.36 inch (60mm) w/12mm head
   - D: 1.38 inch (35mm) w/10mm head
5. Apply silicone sealant to the upper oil pan as shown.
6. Install the upper oil pan and tighten the fasteners in several passes to the following specifications:
   - 10mm: 66 inch lbs. (7.5 Nm)
   - 12mm: 21 ft. lbs. (28 Nm)
7. Install or connect the following:
   - Oil pan baffle. Tighten the fasteners to 66 inch lbs. (7.5 Nm).
   - Lower oil pan. Tighten the fasteners in several passes to 66 inch lbs. (7.5 Nm).
   - Oil dipstick tube
8. Install the engine.

Oil Pump

REMOVAL & INSTALLATION

1. Before servicing the vehicle, refer to the precautions in the beginning of this section.
2. Remove the engine from the vehicle and mount it on a stand.
3. Remove or disconnect the following:
   - Front cover
   - Timing belt.
   - Timing belt idler pulleys
   - Crankshaft timing sprocket
   - Oil dipstick tube
   - Oil filter and bracket
   - Crankshaft Position (CKP) sensor
   - Oil pan and baffle
   - Oil pump pickup tube
   - Oil pump

To install:

4. The upper oil pan bolts are different lengths and are identified as follows:
   - A: 1.38 inch (35mm) w/12mm head
   - B: 1.97 inch (50mm) w/12mm head
   - C: 4.17 inch (106mm) w/12mm head
   - D: 1.57 inch (40mm) w/14mm head
   - E: 1.18 inch (30mm) w/6mm hex head
5. Install a new O-ring on the engine block.
6. Apply silicone sealant to the oil pump housing as shown.

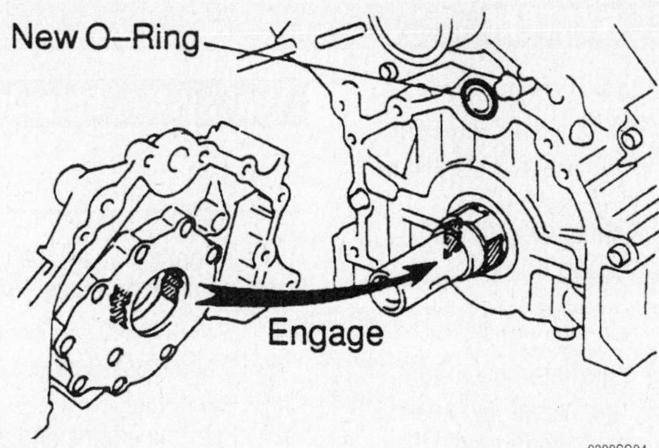

Location of the O-ring seal

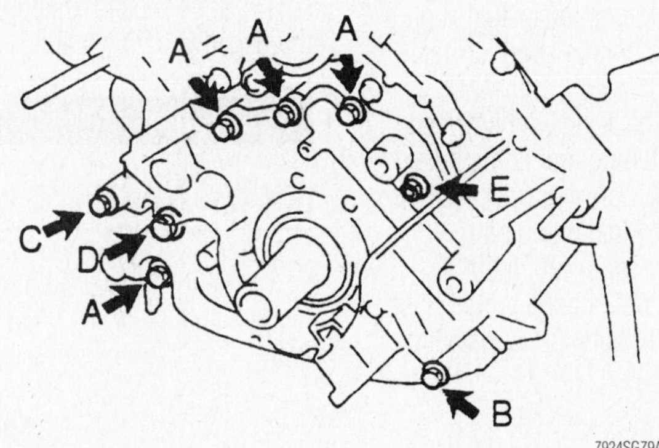

Oil pump bolt location

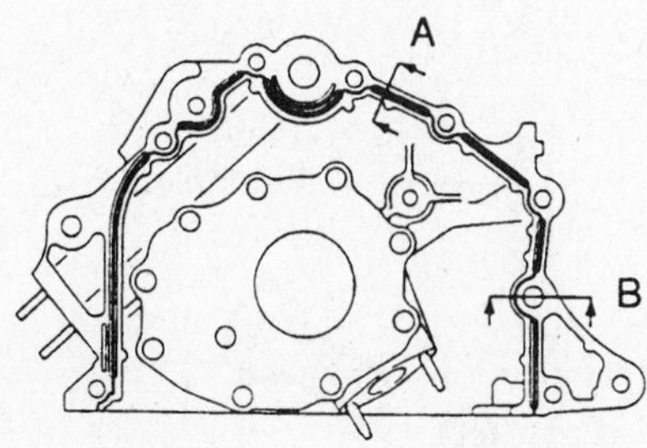

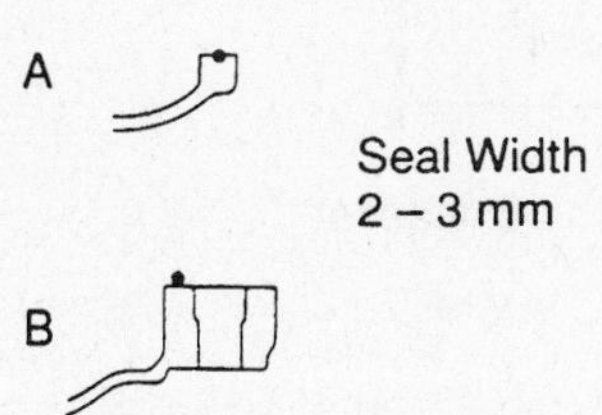

Oil pump housing sealant application

7. Install the oil pump. Tighten the bolts in several passes to the following specifications:
   - 12mm: 11 ft. lbs. (15.5 Nm)
   - 14mm: 22 ft. lbs. (30.5 Nm)
   - 6mm Hex: 11 ft. lbs. (15.5 Nm)
8. Install or connect the following:
   - Oil pump pickup tube. Tighten the bolts to 66 inch lbs. (7.5 Nm).
   - Oil pan and baffle
   - CKP sensor
   - Oil filter and bracket. Tighten the bolts to 13 ft. lbs. (18 Nm).
   - Oil dipstick tube
   - Crankshaft timing sprocket
   - Timing belt idler pulleys
   - Timing belt
   - Front cover
9. Install the engine.

Rear Main Seal

REMOVAL & INSTALLATION

1. Before servicing the vehicle, refer to the precautions in the beginning of this section.
2. Remove the transmission and flywheel from the vehicle.
3. Cut off the rubber lip portion of the seal with a sharp knife.
4. Pry out the oil seal.

To install:

5. Install the rear main seal so that it is flush with the seal retainer housing.
6. Install or connect the following:
   - Flywheel/driveplate. Tighten the bolts to 35 ft. lbs. (48 Nm) plus a 90 degree turn.
   - Transmission

Piston and Ring

POSITIONING

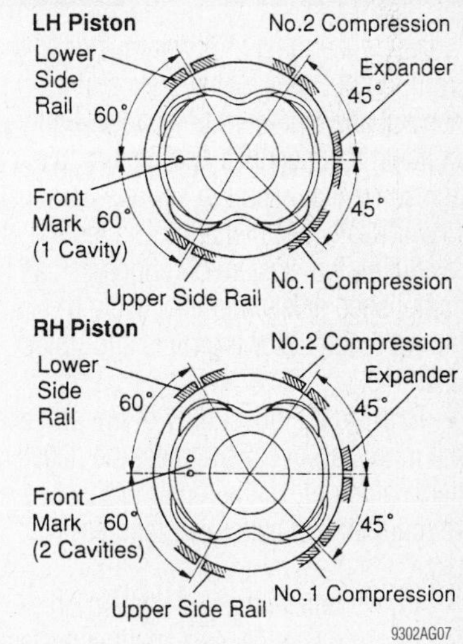

Piston ring positioning

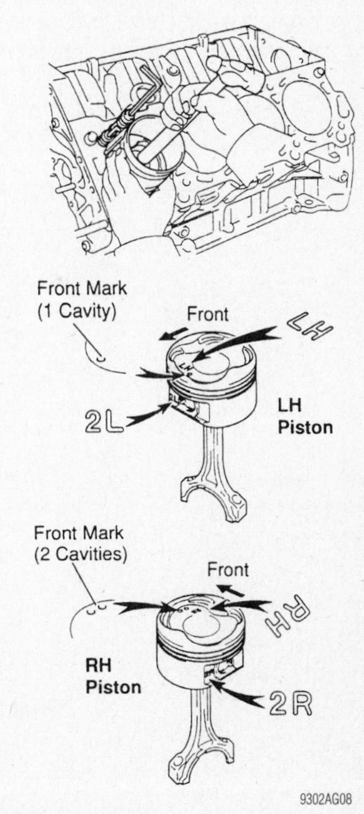

Piston positioning

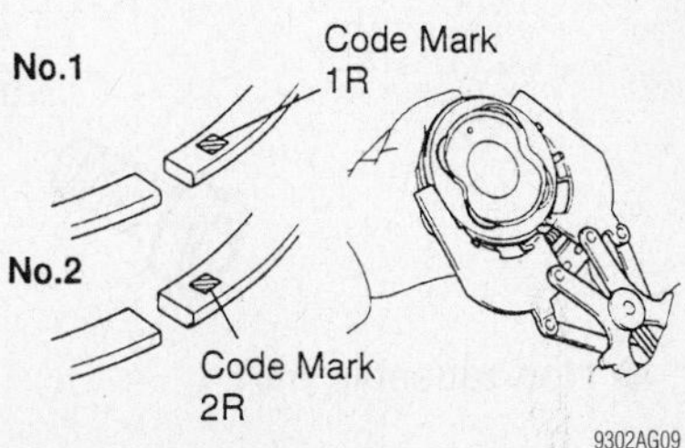

Piston ring identification

FUEL SYSTEM

Fuel System Service Precautions

Safety is the most important factor when performing not only fuel system maintenance but any type of maintenance. Failure to conduct maintenance and repairs in a safe manner may result in serious personal injury or death. Maintenance and testing of the vehicle's fuel system components can be accomplished safely and effectively by adhering to the following rules and guidelines.

- To avoid the possibility of fire and personal injury, always disconnect the negative battery cable unless the repair or test procedure requires that battery voltage be applied.
- Always relieve the fuel system pressure prior to disconnecting any fuel system component (injector, fuel rail, pressure regulator, etc.), fitting or fuel line connection. Exercise extreme caution whenever relieving fuel system pressure, to avoid exposing skin, face and eyes to fuel spray. Please be advised that fuel under pressure may penetrate the skin or any part of the body that it contacts.
- Always place a shop towel or cloth around the fitting or connection prior to loosening to absorb any excess fuel due to spillage. Ensure that all fuel spillage (should it occur) is quickly removed from engine surfaces. Ensure that all fuel soaked cloths or towels are deposited into a suitable waste container.
- Always keep a dry chemical (Class B) fire extinguisher near the work area.
- Do not allow fuel spray or fuel vapors to come into contact with a spark or open flame.
- Always use a back-up wrench when loosening and tightening fuel line connection fittings. This will prevent unnecessary stress and torsion to fuel line piping.
- Always replace worn fuel fitting O-rings with new. Do not substitute fuel hose or equivalent, where fuel pipe is installed.

Fuel System Pressure

RELIEVING

1. Before servicing the vehicle, refer to the precautions in the beginning of this section.
2. Disconnect the fuel pump connector near the fuel tank.
3. Start the engine and allow it to run until it stalls. Crank the engine for a few seconds to relieve additional fuel pressure.
4. Disconnect the negative battery cable.
5. When repairs are complete, connect the negative battery cable.

Fuel Filter

REMOVAL & INSTALLATION

1. Before servicing the vehicle, refer to the precautions in the beginning of this section.

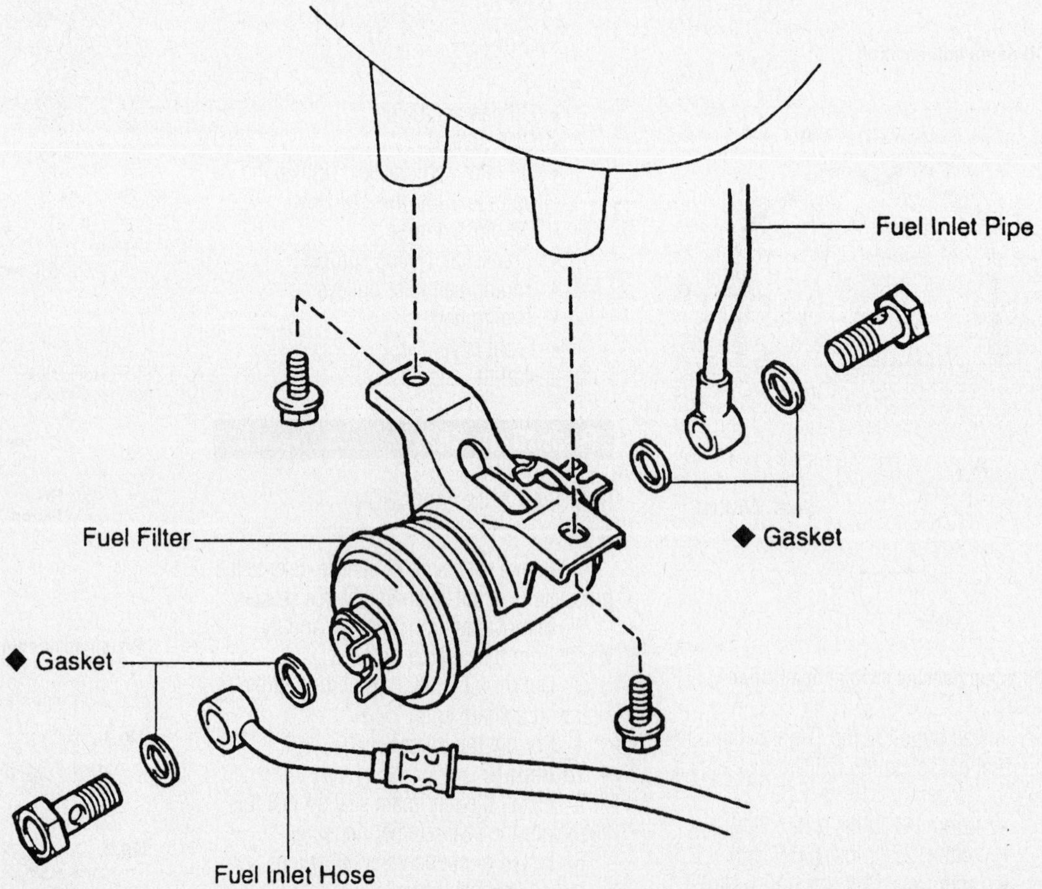

Always use new gaskets when replacing the fuel filter

2. Relieve the fuel system pressure.
3. Remove or disconnect the following:
   - Negative battery cable
   - Fuel lines
   - Fuel filter

To install:

4. Install the fuel filter.
5. Use new washers and tighten the fuel line bolts to the following specifications:
   - Banjo bolt fittings: 21 ft. lbs. (29 Nm)
   - Flare nut fitting: 28 ft. lbs. (38 Nm)
6. Connect the negative battery cable.
7. Start the engine and check for leaks.

Fuel Pump

REMOVAL & INSTALLATION

Sequoia

1. Before servicing the vehicle, refer to the precautions in the beginning of this section.

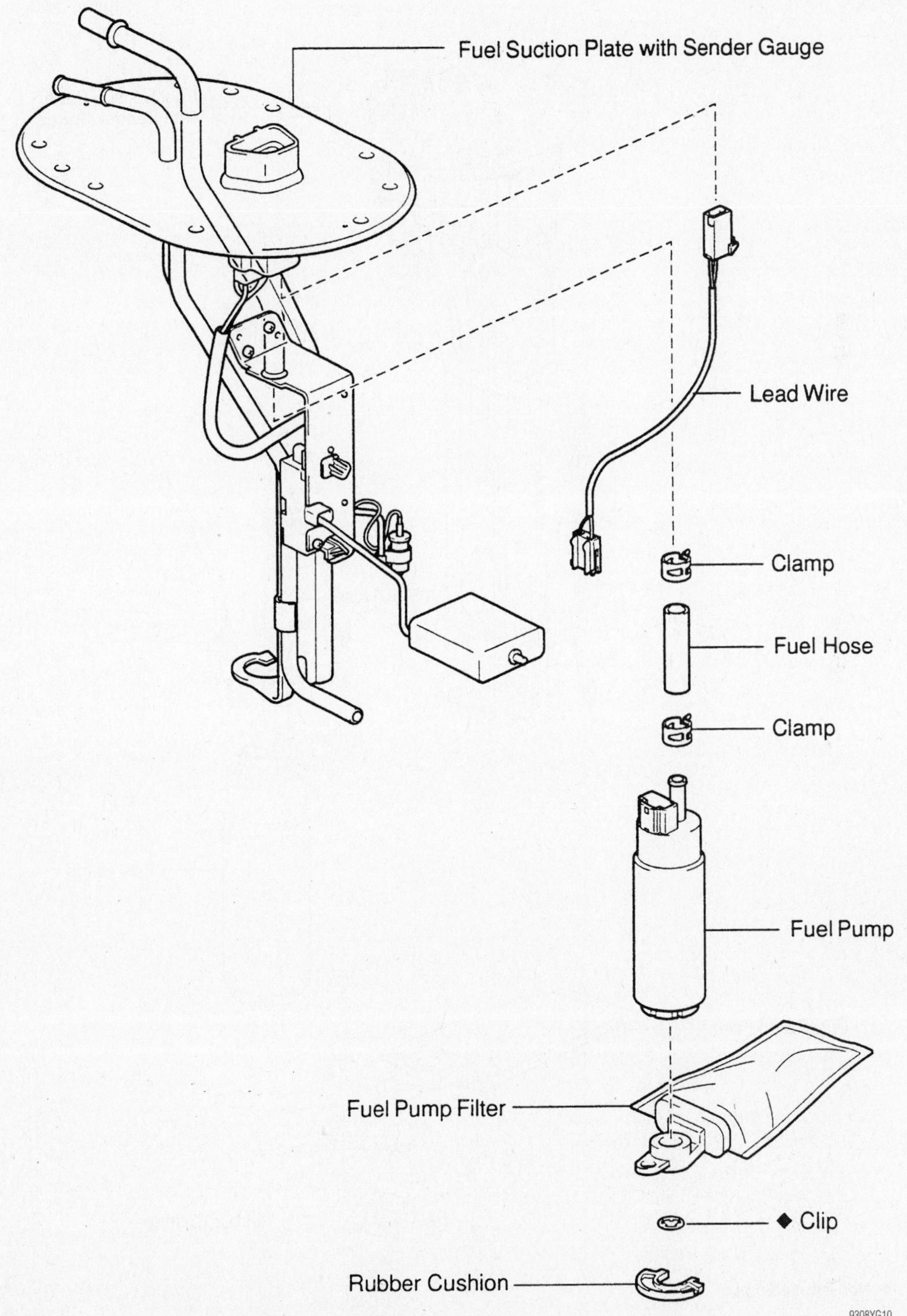

Exploded view of the fuel pump and related components—Sequoia

2. Relieve the fuel system pressure.
3. Remove or disconnect the following:
   - Negative battery cable
   - Fuel tank
   - Fuel pump harness connector
   - Fuel lines
   - Fuel pump module

To install:

4. Install or connect the following:
   - Fuel pump module. Tighten the bolts to 35 inch lbs. (4 Nm).
   - Fuel lines
   - Fuel pump harness connector
   - Fuel tank
   - Negative battery cable
5. Start the engine and check for leaks.

Land Cruiser

1. Before servicing the vehicle, refer to the precautions in the beginning of this section.
2. Relieve the fuel system pressure.
3. Remove or disconnect the following:
   - Negative battery cable
   - Rear seats
   - Door sill trim plates
   - Carpeting and floor mats
   - Access panel
   - Fuel pump harness connector
   - Fuel lines
   - Fuel pump module

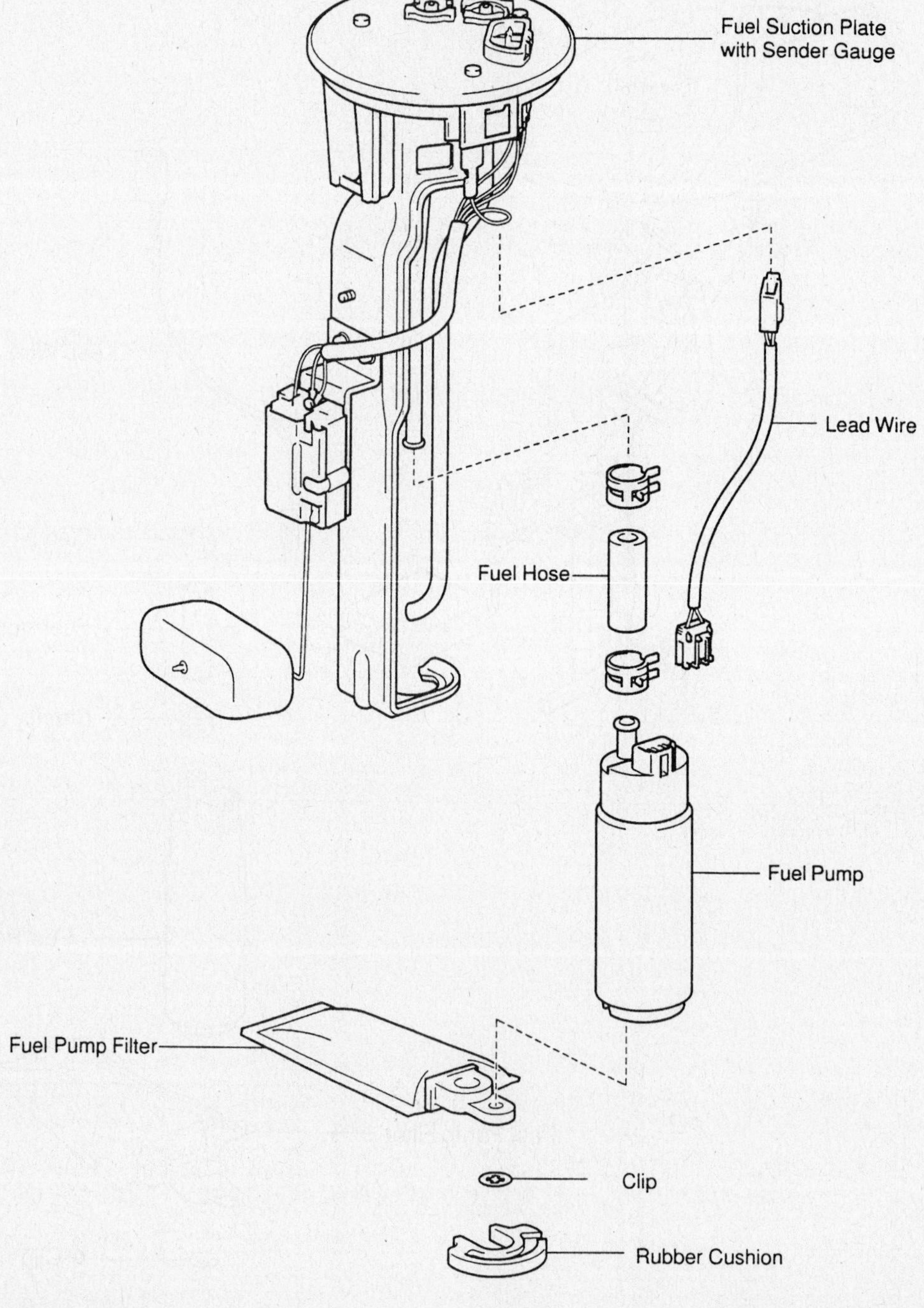

Exploded view of the fuel pump and related components— Land Cruiser

To install:

4. Install or connect the following:
   - Fuel pump module. Tighten the bolts to 35 inch lbs. (4 Nm).
   - Fuel lines
   - Fuel pump harness connector
   - Access panel
   - Carpeting and floor mats
   - Door sill trim plates
   - Rear seats
   - Negative battery cable
5. Start the engine and check for leaks.

Fuel Injector

REMOVAL & INSTALLATION

1. Before servicing the vehicle, refer to the precautions in the beginning of this section.

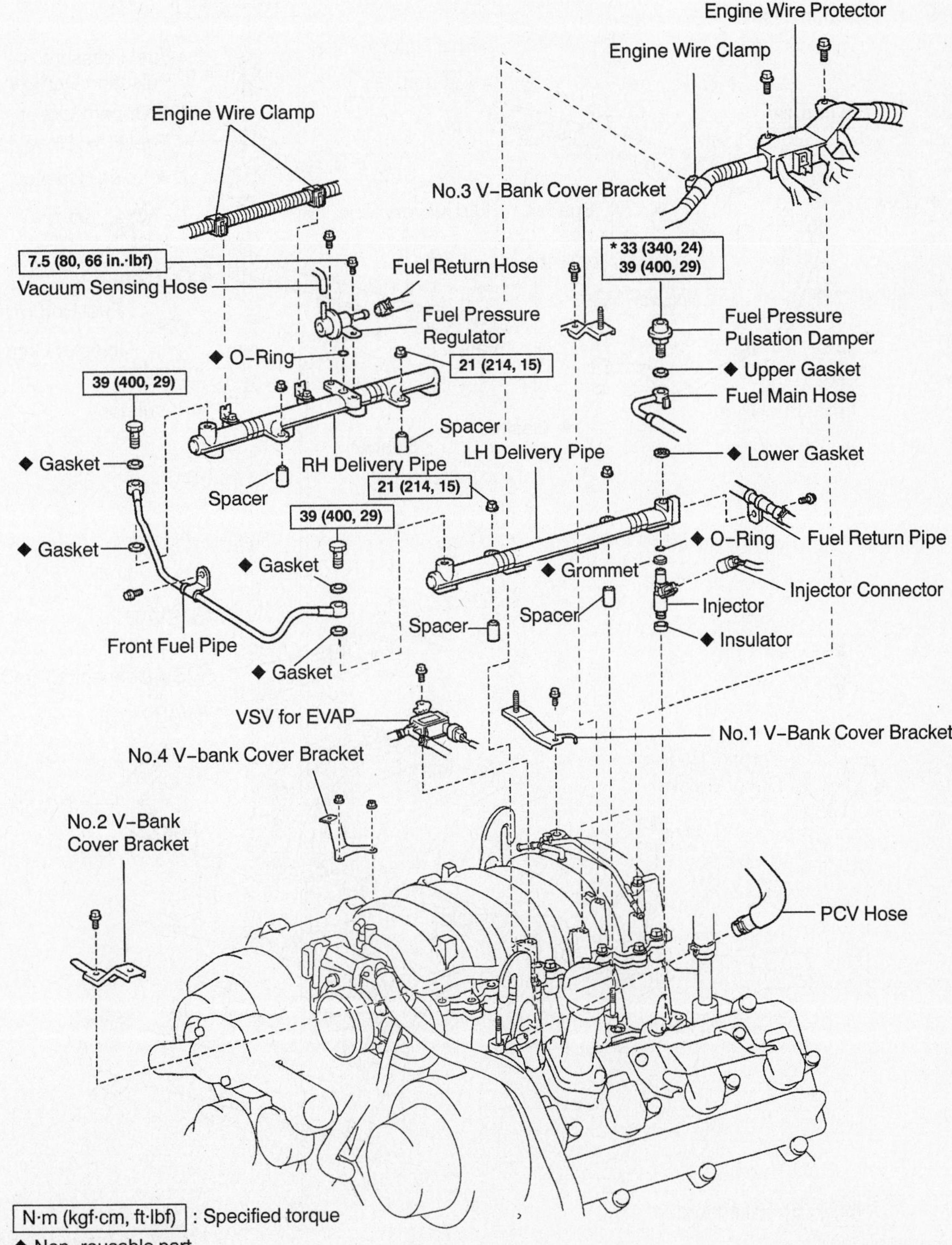

Fuel injectors and related parts—Land Cruiser

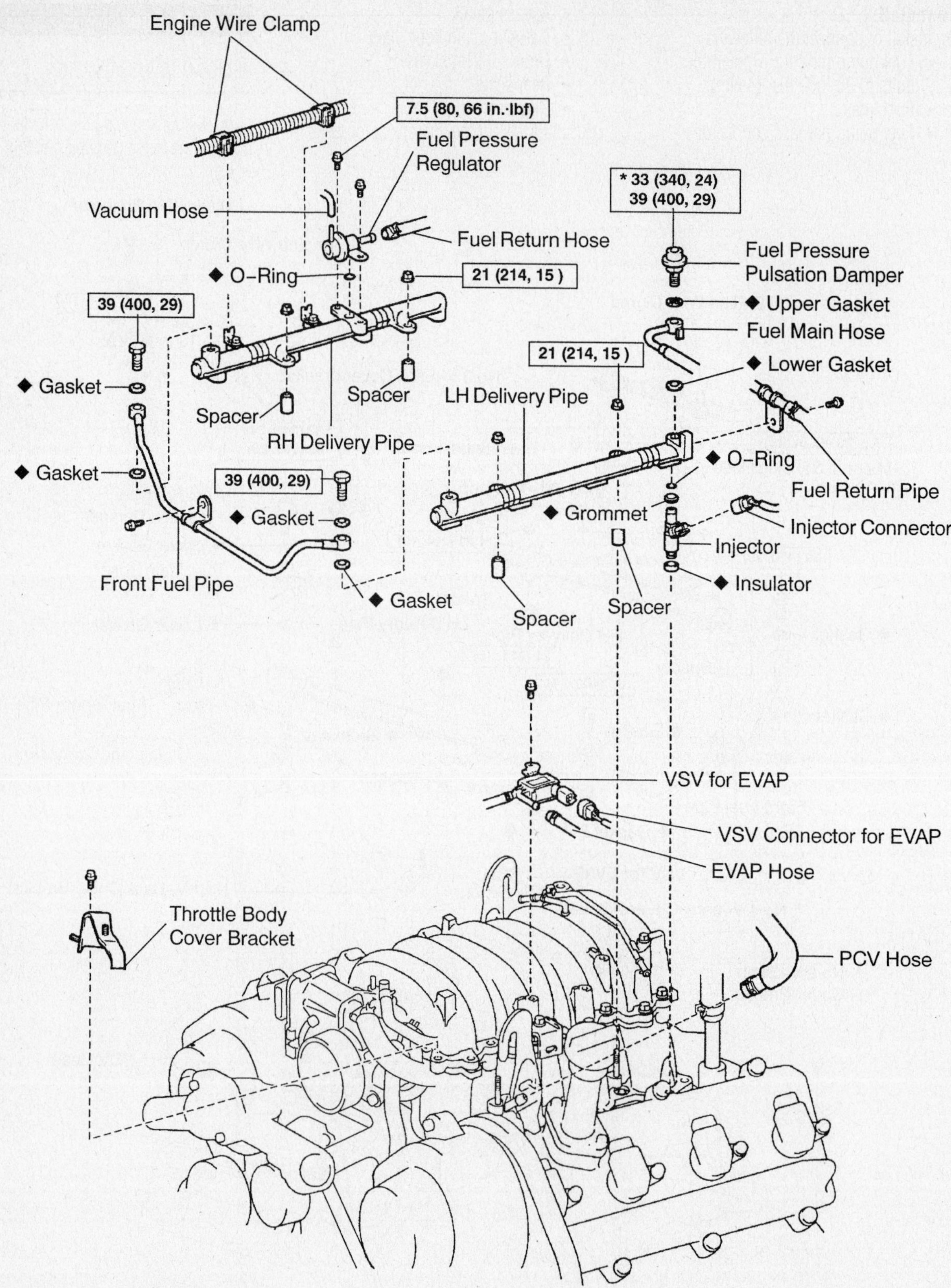
Engine Wire Clamp
7.5 (80, 66 in.·lbf)
Fuel Pressure Regulator
Vacuum Hose
Fuel Return Hose
* 33 (340, 24)
39 (400, 29)
Fuel Pressure Pulsation Damper
◆ O-Ring
21 (214, 15)
◆ Upper Gasket
39 (400, 29)
Fuel Main Hose
21 (214, 15)
◆ Lower Gasket
◆ Gasket
Spacer
LH Delivery Pipe
Spacer
RH Delivery Pipe
◆ O-Ring
◆ Gasket
39 (400, 29)
Fuel Return Pipe
◆ Grommet
◆ Gasket
Injector Connector
Injector
Front Fuel Pipe
◆ Insulator
◆ Gasket
Spacer
Spacer
VSV for EVAP
VSV Connector for EVAP
EVAP Hose
Throttle Body Cover Bracket
PCV Hose
N·m (kgf·cm, ft·lbf) : Specified torque
◆ Non-reusable part
* For use with SST
67170-LCSQ-G12

Fuel injectors and related parts—Sequoia

2. Relieve the fuel system pressure.
3. Remove or disconnect the following:
   - Negative battery cable
   - Engine appearance cover
   - Air intake tube
   - Fuel lines
   - Fuel pulsation damper
   - Fuel pressure regulator vacuum line
   - Accelerator cable and bracket
   - Positive Crankcase Ventilation (PCV) valve and hose
   - Evaporative Emissions (EVAP) vacuum switching valve
   - Engine appearance cover brackets
   - Fuel injector harness connectors
   - Engine harness protector
   - Fuel supply manifold crossover pipe
   - Fuel supply manifolds with injectors attached
   - Fuel injectors

To install:

4. Install the fuel injectors to the supply manifold with new O-ring seals and new grommets.
5. Install new injector insulators to the intake manifold.
6. Install or connect the following:
   - Fuel supply manifolds with injectors attached. Tighten the bolts to 66 inch lbs. (7.5 Nm).
   - Fuel supply manifold crossover pipe. Tighten the bolts to 29 ft. lbs. (39 Nm).
   - Engine harness protector
   - Fuel injector harness connectors
   - Engine appearance cover brackets
   - EVAP vacuum switching valve
   - PCV valve and hose
   - Accelerator cable and bracket
   - Fuel pressure regulator vacuum line
   - Fuel pulsation damper
   - Fuel lines
   - Air intake tube
   - Engine appearance cover
   - Negative battery cable
7. Start the engine and check for leaks.

DRIVE TRAIN

Automatic Transmission Assembly

REMOVAL & INSTALLATION

Sequoia

1. Before servicing the vehicle, refer to the precautions in the beginning of this section.
2. Remove or disconnect the following:
   - Oil filler pipe
   - No. 1 engine undercover
   - Exhaust pipes
   - Front and rear driveshafts
   - Nos. 1&2 vehicle speed sensors
   - Solenoid connector
   - Shift cable at the transmission
   - Overdrive sensor connector
   - Oil cooler lines
   - ATF temperature sensor
   - Park/Neutral switch
   - End plate and converter clutch mounting bolts
3. Raise the transmission slightly.
4. Remove or disconnect the following:
   - Crossmember
   - Rear mount insulator
   - Transmission and transfer case as a unit
5. Installation is the reverse of removal. Note the following torques:
   - Transmission-to-transfer case bolts: 53 ft. lbs. (71Nm)
   - Rear mount insulator-to-transmission: 48 ft. lbs. (65Nm)
   - Rear mount insulator-to-crossmember: 13 ft. lbs. (18Nm)
   - Crossmember-to-frame: 53 ft. lbs. (71Nm)
   - Torque converter clutch: 30 ft. lbs. (41Nm)

➡**Install the green bolt first.**

   - Rear end plate: 13 ft. lbs. (18Nm)
   - Oil cooler lines: 25 ft. lbs. (34Nm)
   - Shift control bracket: 13 ft. lbs. (18Nm)

Land Cruiser

1. Before servicing the vehicle, refer to the precautions in the beginning of this section.
2. Remove or disconnect the following:
   - Battery and tray
   - Air intake assembly
   - Cooling fan and shroud
   - Coolant recovery reservoir
   - Transmission dipstick tube
   - Center console
   - Transmission gear select lever and rod
   - Transfer case shift lever and rod
   - Engine under covers
   - Exhaust front pipes
   - Front and rear driveshafts
   - Vehicle Speed (VSS) sensor connectors
   - Overdrive clutch speed sensor connector
   - Solenoid harness connector
   - Transmission fluid temperature sensor connector
   - Park/Neutral Position (PNP) switch connector
   - Center differential lock indicator switch connector
   - L4 solenoid valve position switch connector
   - Motor actuator connector
   - Torque converter
   - Transmission oil cooler lines
   - Transmission mount crossmember. Support the transmission with a jack.
   - Transmission flange bolts
   - Transmission

To install:

3. Install or connect the following:
   - Transmission. Tighten the flange bolts to 53 ft. lbs. (72 Nm).
   - Transmission mount crossmember. Tighten the bolts to 37 ft. lbs. (50 Nm) and the nuts to 54 ft. lbs. (74 Nm).
   - Transmission oil cooler lines
   - Torque converter. Tighten the bolts to 35 ft. lbs. (48 Nm).
   - Motor actuator connector
   - L4 solenoid valve position switch connector
   - Center differential lock indicator switch connector
   - PNP switch connector
   - Transmission fluid temperature sensor connector
   - Solenoid harness connector
   - Overdrive clutch speed sensor connector
   - VSS sensor connectors
   - Front driveshaft. Tighten the fasteners to 59 ft. lbs. (80 Nm).
   - Rear driveshaft. Tighten the fasteners to 78 ft. lbs. (106 Nm).
   - Exhaust front pipes
   - Engine under covers
   - Transfer case shift lever and rod
   - Transmission gear select lever and rod
   - Center console
   - Transmission dipstick tube
   - Coolant recovery reservoir
   - Cooling fan and shroud
   - Air intake assembly
   - Battery and tray
4. Check the transmission and transfer case fluid levels and adjust as necessary.

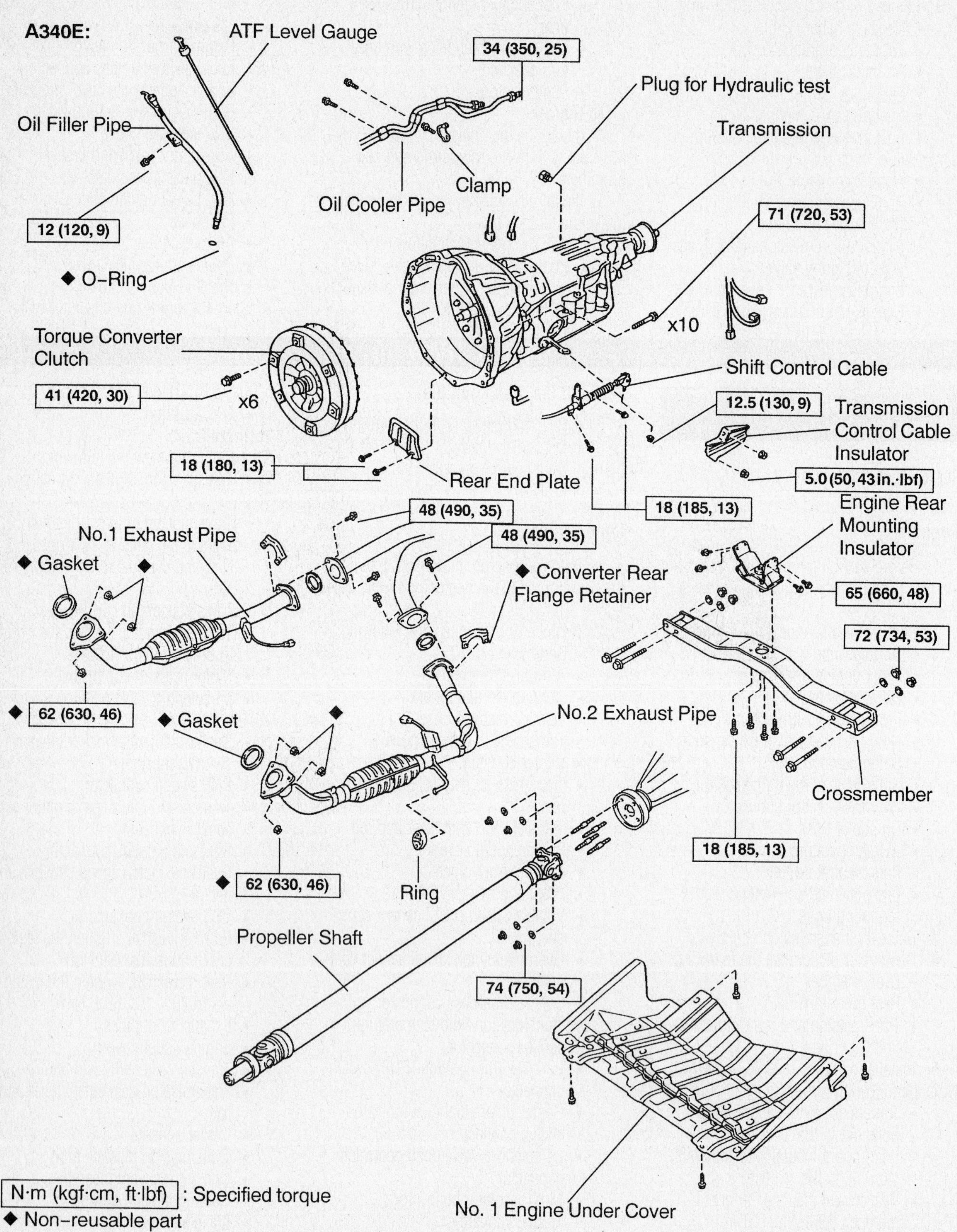
A340E:
ATF Level Gauge
34 (350, 25)
Plug for Hydraulic test
Oil Filler Pipe
Transmission
Clamp
Oil Cooler Pipe
71 (720, 53)
12 (120, 9)
◆ O-Ring
x10
Torque Converter
Clutch
Shift Control Cable
41 (420, 30)
x6
12.5 (130, 9)
Transmission
Control Cable
Insulator
18 (180, 13)
Rear End Plate
5.0 (50, 43 in.·lbf)
48 (490, 35)
18 (185, 13)
Engine Rear
Mounting
Insulator
No.1 Exhaust Pipe
48 (490, 35)
◆ Gasket
◆
◆ Converter Rear
Flange Retainer
65 (660, 48)
72 (734, 53)
◆ 62 (630, 46)
◆ Gasket
◆
No.2 Exhaust Pipe
Crossmember
18 (185, 13)
◆ 62 (630, 46)
Ring
Propeller Shaft
74 (750, 54)
N·m (kgf·cm, ft·lbf) : Specified torque
◆ Non-reusable part
No. 1 Engine Under Cover
67170-LCSQ-G13

Transmission and related parts—Sequoia

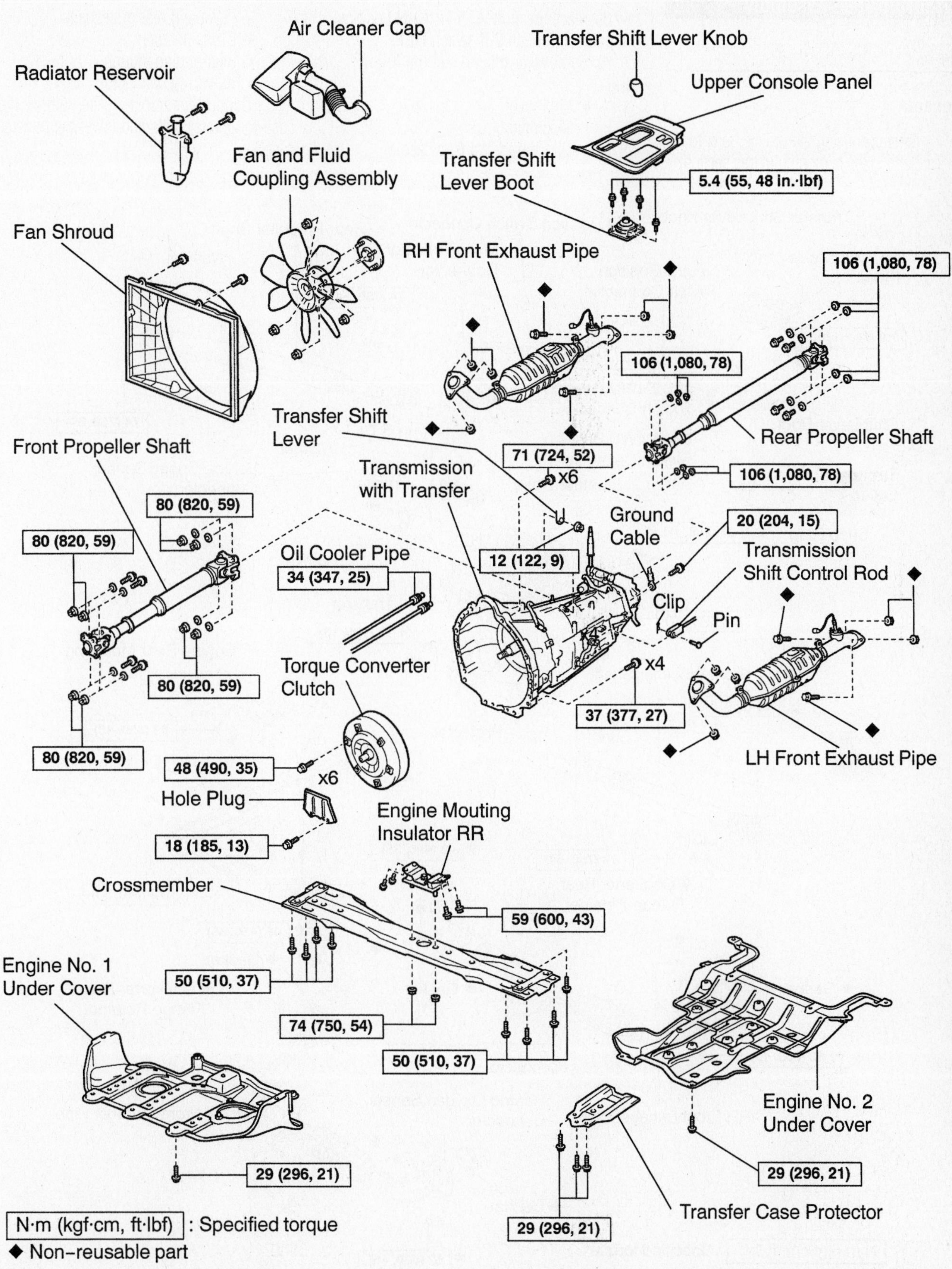
Air Cleaner Cap
Transfer Shift Lever Knob
Radiator Reservoir
Upper Console Panel
Fan and Fluid Coupling Assembly
Transfer Shift Lever Boot
5.4 (55, 48 in.·lbf)
Fan Shroud
RH Front Exhaust Pipe
106 (1,080, 78)
106 (1,080, 78)
Transfer Shift Lever
Rear Propeller Shaft
Front Propeller Shaft
71 (724, 52)
x6
Transmission with Transfer
106 (1,080, 78)
80 (820, 59)
80 (820, 59)
Ground Cable
20 (204, 15)
Oil Cooler Pipe
12 (122, 9)
Transmission Shift Control Rod
34 (347, 25)
Clip
Pin
x4
Torque Converter Clutch
80 (820, 59)
37 (377, 27)
80 (820, 59)
48 (490, 35)
x6
LH Front Exhaust Pipe
Hole Plug
Engine Mouting Insulator RR
18 (185, 13)
Crossmember
59 (600, 43)
Engine No. 1 Under Cover
50 (510, 37)
74 (750, 54)
50 (510, 37)
Engine No. 2 Under Cover
29 (296, 21)
29 (296, 21)
Transfer Case Protector
N·m (kgf·cm, ft·lbf) : Specified torque
◆ Non–reusable part
29 (296, 21)
67170-LCSQ-G14

Transmission and related parts—Land Cruiser

Transfer Case Assembly

REMOVAL & INSTALLATION

Sequoia

1. Before servicing the vehicle, refer to the precautions in the beginning of this section.
2. Drain the transfer case oil.
3. Place the shift lever in the **H** position and the one-touch 2-4 switch **OFF**.
4. Remove or disconnect the following:
   - Shift lever
   - Skid plate
   - Oil from the transfer case
   - Exhaust pipes
   - Front and rear driveshafts
   - Crossmember
   - Rear engine mount
   - All wiring connectors
5. Support the transfer case, remove the case-to-adapter bolts and lower the transfer case.

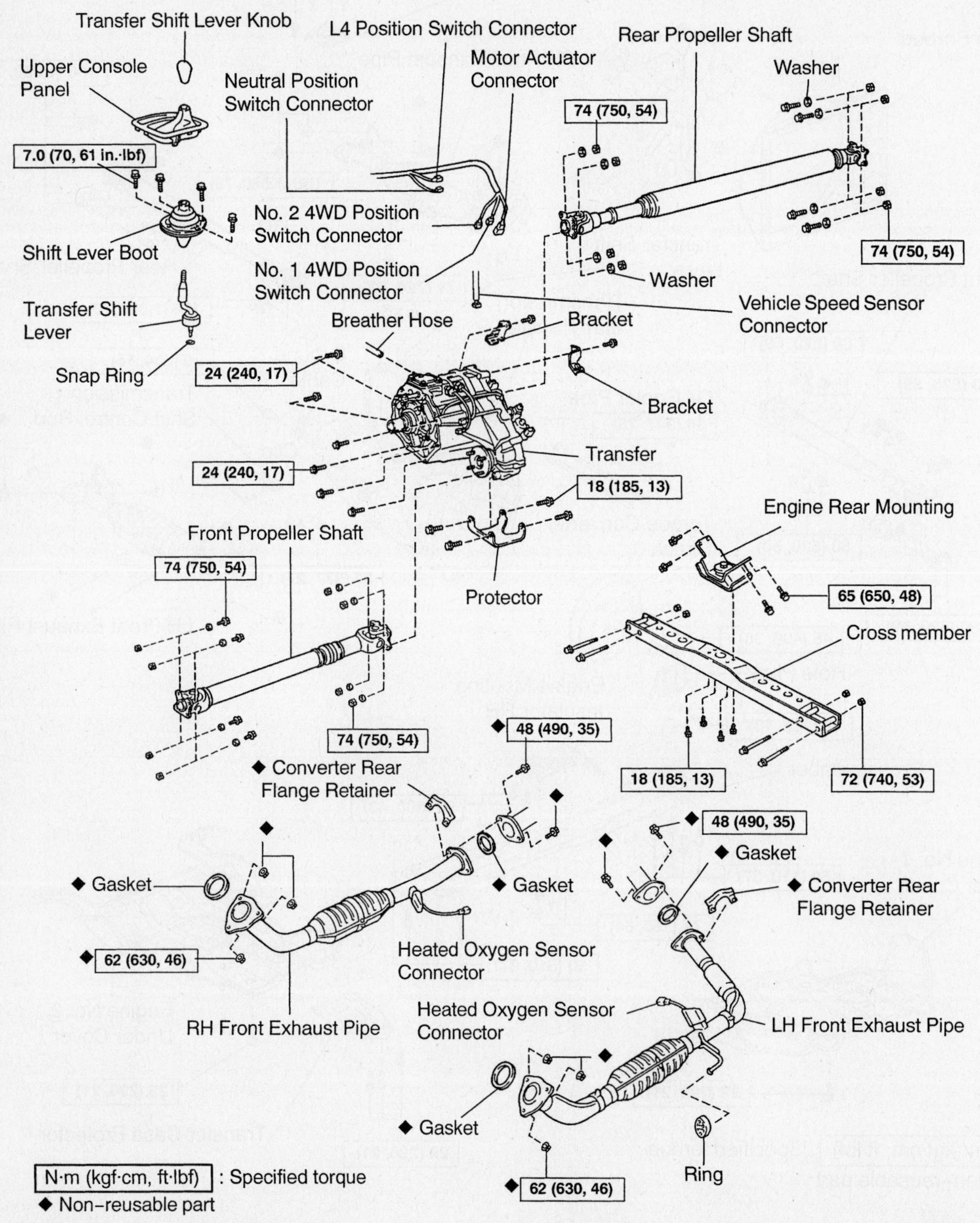

Transfer case and related parts—Sequoia

6. Installation is the reverse of removal. Note the following torques:
   - Transfer case-to-adapter bolts: 17 ft. lbs. (24Nm)
   - Engine rear mount-to-adapter: 48 ft. lbs. (65Nm)
   - Crossmember: 53 ft. lbs. (72Nm)
   - Engine rear mount set bolts: 13 ft. lbs. (18Nm)
   - Skid plate: 13 ft. lbs. (18Nm)

Land Cruiser

1. Before servicing the vehicle, refer to the precautions in the beginning of this section.
2. Drain the transfer case oil.
3. Remove or disconnect the following:
   - Transfer case protector
   - Front and rear driveshafts
   - Transfer case shift lever rod
   - Ground cable
   - Transmission mount crossmember. Support the transmission with a jack.
   - Transfer case vent hose
   - Vehicle Speed (VSS) sensor connector
   - Center differential lock indicator switch connector
   - Motor actuator connectors

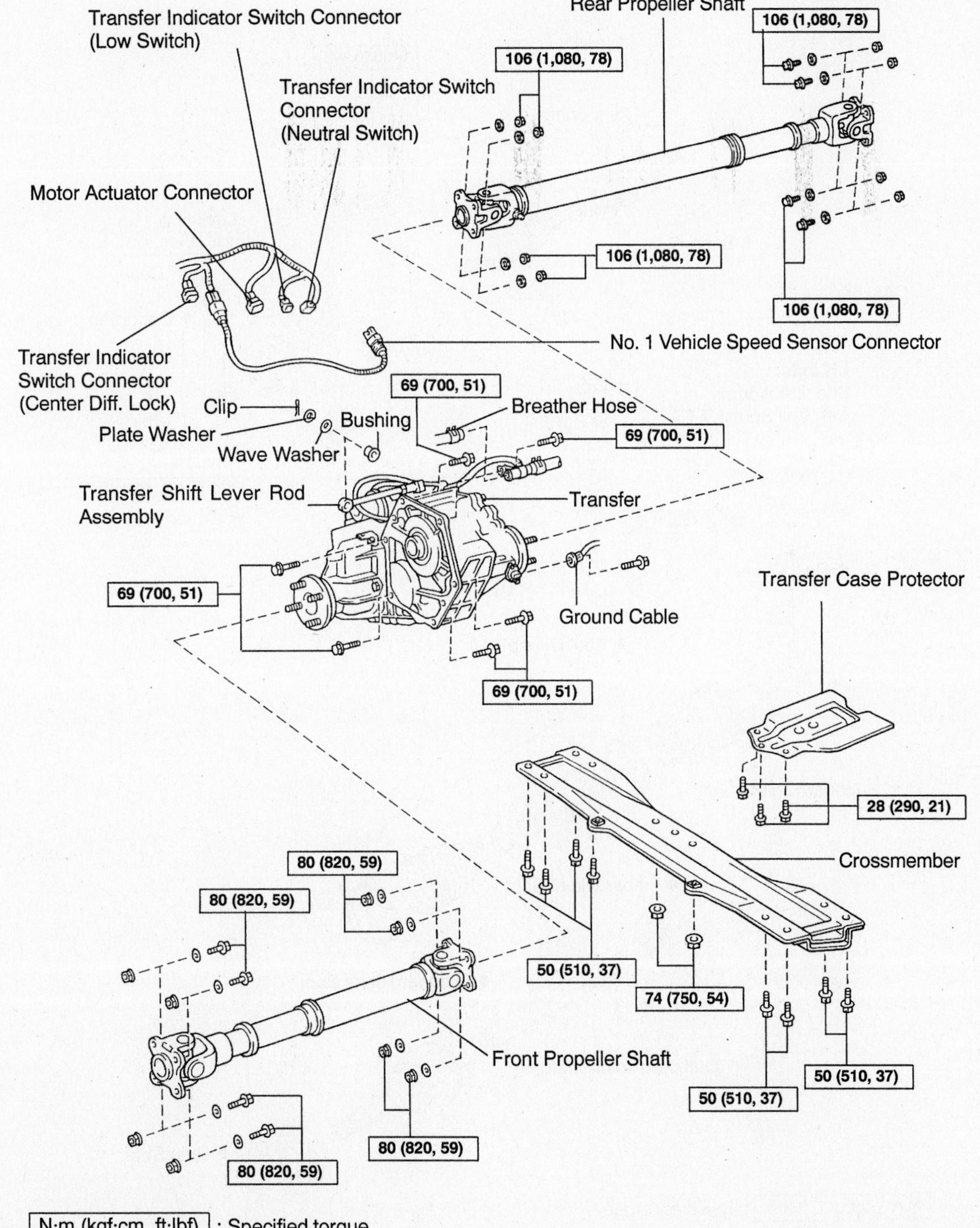

Transfer case and related parts—Land Cruiser

- Transfer case adapter bolts
- Transfer case

To install:

4. Install or connect the following:
   - Transfer case. Tighten the adapter bolts to 51 ft. lbs. (69 Nm).
   - Motor actuator connectors
   - Center differential lock indicator switch connector
   - VSS sensor connector
   - Transfer case vent hose
   - Transmission mount crossmember. Tighten the bolts to 37 ft. lbs. (50 Nm) and the nuts to 54 ft. lbs. (74 Nm).
   - Ground cable
   - Transfer case shift lever rod
   - Front driveshaft. Tighten the fasteners to 59 ft. lbs. (80 Nm).
   - Rear driveshaft. Tighten the fasteners to 78 ft. lbs. (106 Nm).
   - Transfer case protector
5. Fill the transfer case to the correct level.

Halfshaft

REMOVAL & INSTALLATION

Sequoia

1. Before servicing the vehicle, refer to the precautions in the beginning of this section.
2. Remove or disconnect the following:
   - Front wheel
   - Under cover

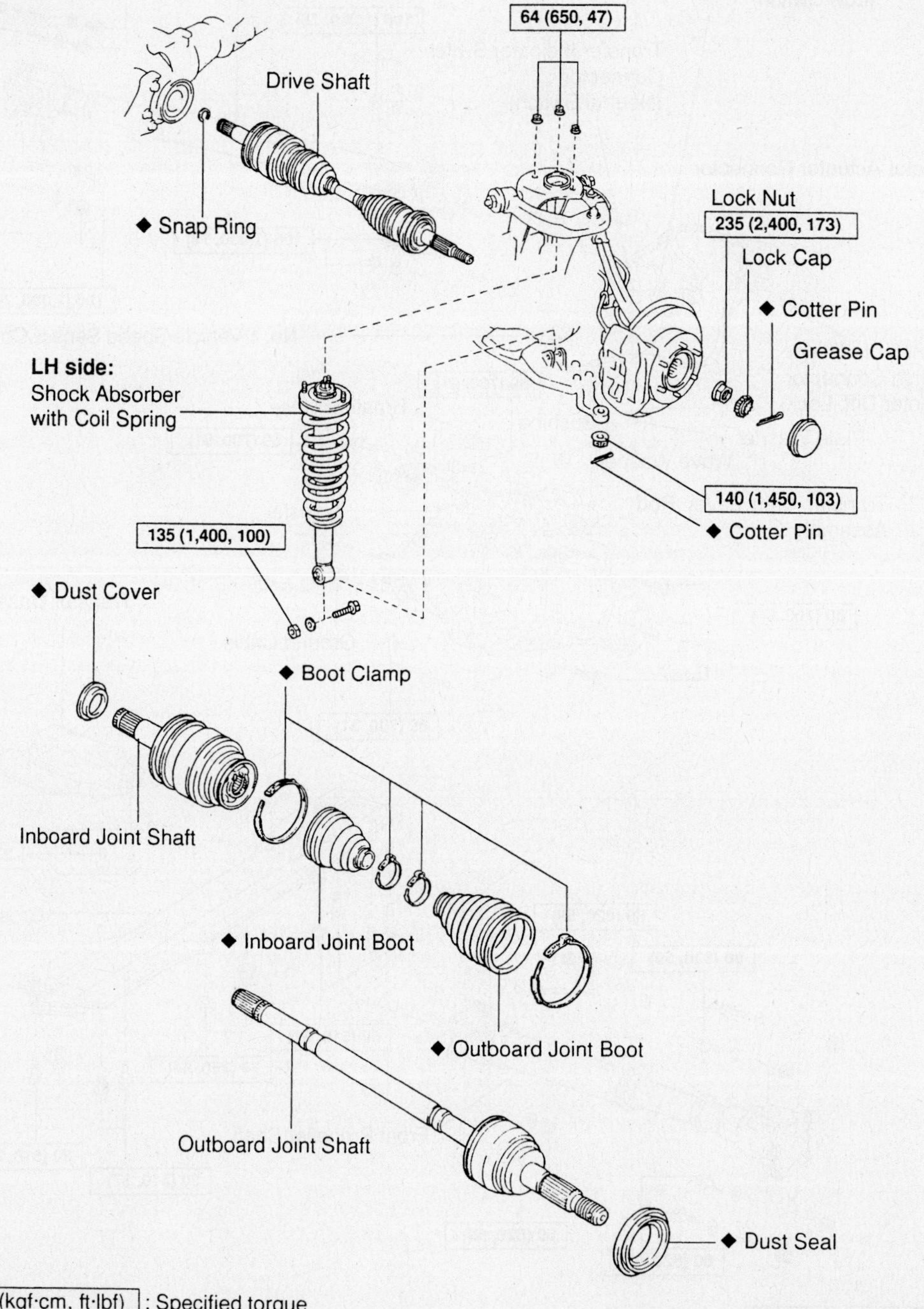

View of the halfshaft and related components—Sequoia

3. Drain the differential oil.
4. Remove or disconnect the following:
   - Grease cap
   - Cotter pin and lock cap
   - Halfshaft locknut by applying the brakes
   - Lower control arm from the lower ball joint
   - Halfshaft from the steering knuckle, using a plastic hammer
   - Left strut, for the left halfshaft
   - Right halfshaft, using a brass bar and a hammer
   - Left halfshaft, using tools 09520-01010 and 09520-24010
   - Snapring from the inboard joint shaft

To install:

5. Install or connect the following:
   - New snapring, onto the inboard joint shaft with the opening facing downward
   - Halfshafts to the differential using a brass bar and a hammer
   - Halfshafts to the steering knuckles

⁂ WARNING

Be careful not to damage the oil seal, boot or dust seal.

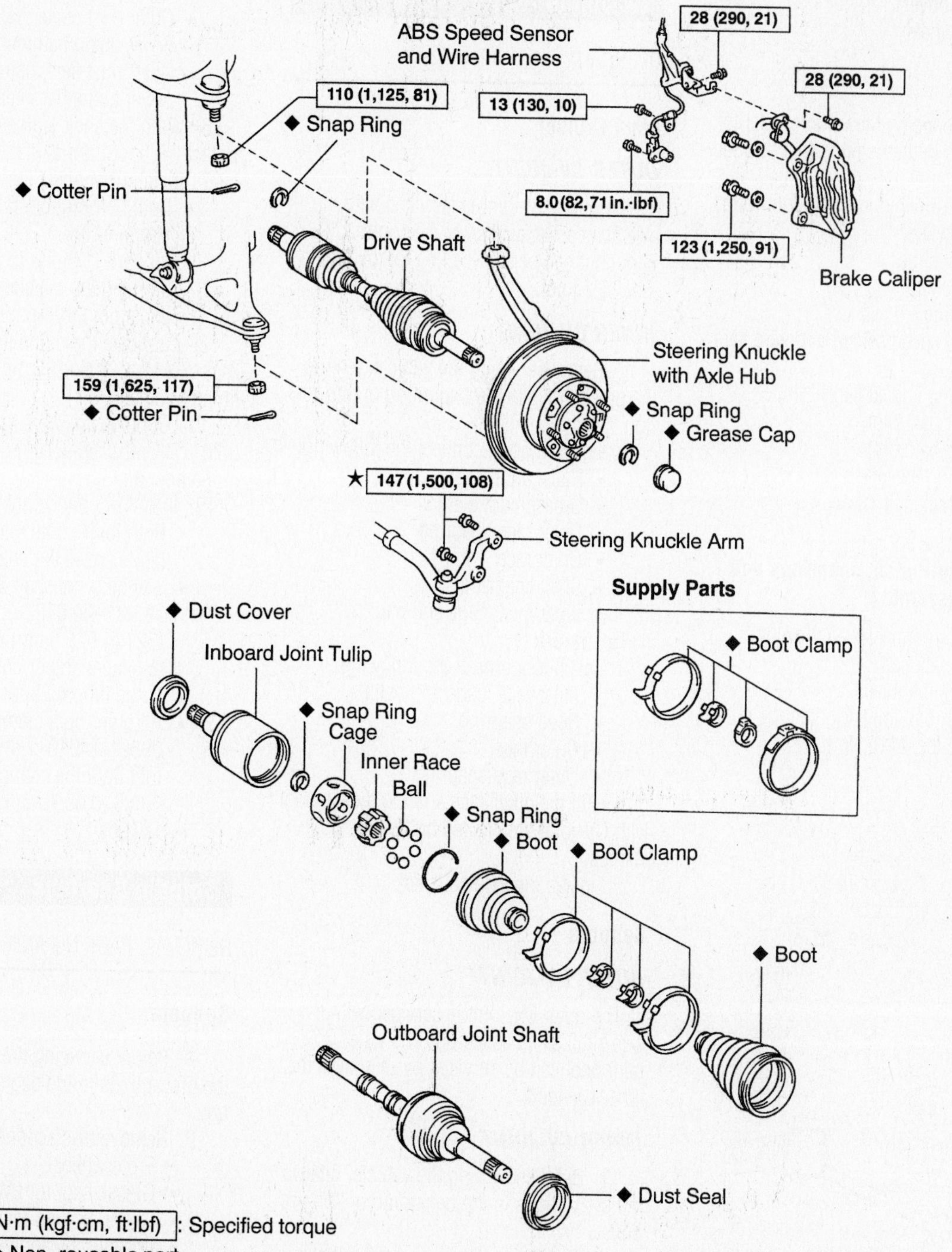

Halfshaft and related components—Land Cruiser

- Lower control arm to the lower ball joint using a new cotter pin. Torque the ball joint nut to 103 ft. lbs. (140 Nm).
- Halfshaft locknut by applying the brakes. Torque the nut to 173 ft. lbs. (235 Nm).
- Lock cap and a new cotter pin
- Grease cap

6. Refill the differential with oil.
7. Install or connect the following:
   - Under cover
   - Front wheel

Land Cruiser

1. Before servicing the vehicle, refer to the precautions in the beginning of this section.
2. Remove or disconnect the following:
   - Front wheel
   - Brake caliper
   - Grease cap
   - Snapring
   - Wheel speed sensor and wire harness
   - Steering knuckle arm
   - Lower ball joint
   - Upper ball joint
   - Steering knuckle
   - Axle halfshaft

To install:

➡Use new split pins, snaprings and circlips for assembly.

3. Install or connect the following:
   - Axle halfshaft
   - Steering knuckle
   - Upper ball joint. Tighten the nut to 81 ft. lbs. (110 Nm).

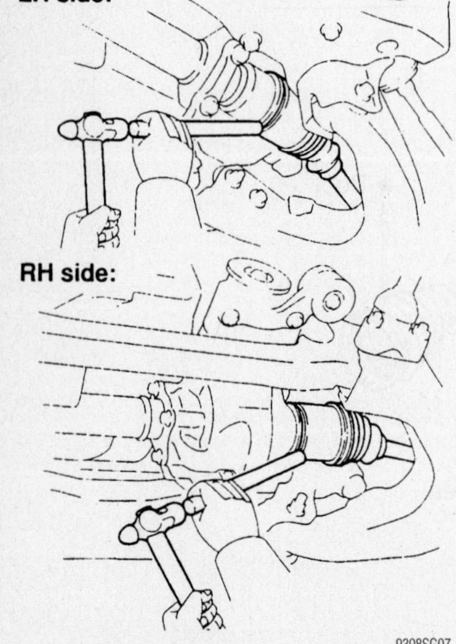

Axle halfshaft removal—Land Cruiser

- Lower ball joint. Tighten the nut to 117 ft. lbs. (159 Nm).
- Steering knuckle arm. Tighten the bolts to 108 ft. lbs. (147 Nm).
- Wheel speed sensor and wire harness
- Snapring
- Grease cap
- Brake caliper
- Front wheel

CV-Joints

OVERHAUL

Land Cruiser

OUTER CV-JOINT

The outer CV-joint is serviced with the axle shaft as an assembly. The outer CV-joint boot can be serviced by removing the inner CV-joint.

INNER CV-JOINT

1. Before servicing the vehicle, refer to the precautions in the beginning of this section.
2. Remove or disconnect the following:
   - Halfshaft from the vehicle
   - Grease boot clamps
   - Outer race snapring
   - Outer race
   - Shaft snapring
   - Inner race, cage and balls

To install:

3. Install or connect the following:
   - Inner race, cage and balls
   - Shaft snapring
   - Outer race
   - Outer race snapring
4. Fill the outer race and the grease boot with CV-joint grease and tighten the boot clamps.
5. Install the axle halfshaft.

Sequoia

OUTER CV-JOINT

The outer CV-joint is serviced with the axle shaft as an assembly. The outer CV-joint boot can be serviced by removing the inner CV-joint.

INNER CV-JOINT

1. Before servicing the vehicle, refer to the precautions in the beginning of this section.
2. Remove or disconnect the following:
   - Halfshaft from the vehicle
   - Large boot clamps
   - Small boot clamps
3. Matchmark inboard CV-joint to the shaft
4. Remove or disconnect the following:
   - Inboard CV-joint from the shaft by expanding the snapring
   - Both CV-joint boots
   - Outer dust seal, using a shop press and tool 09950-00020
   - Outer dust cover, using a shop press and tool 09950-00020

To install:

5. Install or connect the following:
   - Outer dust cover, using a screwdriver and a hammer
   - Outer dust seal, using a screwdriver and a hammer
6. Wrap the shaft splines with tape to protect the boot from damage.
7. Install or connect the following:
   - Both CV-joint boots with clamps, temporarily
   - Inboard CV-joint to the shaft by aligning the matchmarks and expanding the snapring
8. Lubricate the outboard joint with 7.23–7.94 oz. (205–225g) grease, provided in the boot kit.
9. Lubricate the inboard joint with 6.70–7.41 oz. (190–210g) grease, provided in the boot kit.
10. Install or connect the following:
    - Both joint boots making sure the boots are in the shaft groove
    - Standard halfshaft length is 20.531–20.689 in. (521.5–525.5mm) when the shaft is not expanded or contracted
    - Large inboard boot clamp
    - All other boot clamps using tool 09521-24010. Tighten the crimping tool until the clamp clearance is 0.039–0.059 in. (1.0–1.5mm)
    - Halfshaft

Spindle Bearings

REMOVAL, PACKING AND INSTALLATION

Sequoia

1. Before servicing the vehicle, refer to the precautions in the beginning of this section.
2. Remove or disconnect the following:
   - Front wheel
   - Grease cap, for 2WD
   - Cotter pin and lock cap, for 4WD
   - Locknut, by applying the brakes
   - Anti-lock Brake System (ABS) speed sensor and wiring harness clamp from the steering knuckle

- Brake line clamp from the steering knuckle

➡**Be careful not to damage the brake tube.**

- Brake caliper and disc

3. Support the steering knuckle
4. Remove or disconnect the following:
   - 4 lower ball joint-to-steering knuckle bolts
   - Upper ball joint cotter pin and loosen the nut
   - Upper ball joint from the steering knuckle using tool 09950-40011
   - Steering knuckle by placing it in a soft-jawed vise
   - Inside oil seal

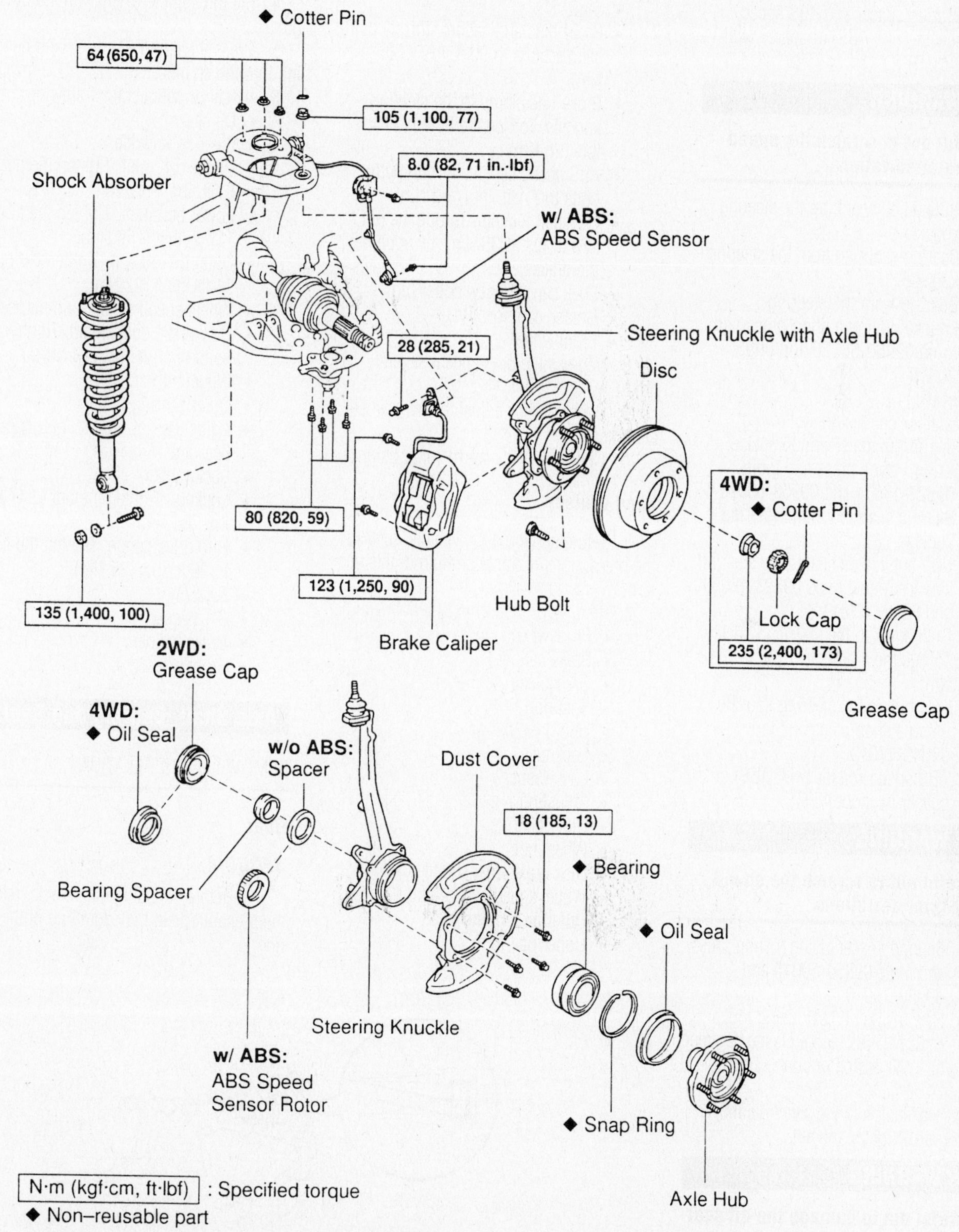

Exploded view of the front axle hub and related components—Sequoia

- 4 bolts and shift the dust cover towards the hub side (outside)
- Axle hub from the steering knuckle using tools 09710-30021 and 09950-40011
- Dust cover from the steering knuckle
- Bearing spacer and ABS speed sensor (with ABS) or spacer (without ABS)

⁂ WARNING

Be careful not to scratch the speed sensor rotor serrations

- Outside oil seal from the steering knuckle
- Bearing snapring from the steering knuckle
- Bearing from the steering knuckle, using a shop press and tools 09950-60020 and 09950-70010

To install:

5. Install or connect the following:
   - Bearing to the steering knuckle, using a shop press and tools 09527-17011 and 09950-60020
   - Bearing snapring to the steering knuckle
   - New outside oil seal to the steering knuckle, using tools 09223-15030 and 09527-17011
   - Dust cover to the steering knuckle. Torque the 4 bolts to 13 ft. lbs. (18 Nm).
   - Axle hub to the steering knuckle using a shop press and tool 09649-17010
   - ABS speed sensor (with ABS) or spacer (without ABS)

⁂ WARNING

Be careful not to scratch the speed sensor rotor serrations

- Bearing spacer, using a shop press and tools 09950-60010 and 09950-70010
- New inside oil seal, using tool 09527-17011 and a plastic hammer

6. For 4WD, install or connect the following:

a. Halfshaft into the axle hub and temporarily tighten the nut

⁂ WARNING

Be careful not to damage the oil seal or boot.

b. Steering knuckle to the upper control arm. Tighten the nut to 77 ft. lbs. (105 Nm).

c. New cotter pin.

7. Install or connect the following:
   - Lower ball joint to the steering knuckle. Torque the 4 bolts to 59 ft. lbs. (80 Nm).
   - Strut
   - Brake disc and caliper. Torque both caliper bolts to 90 ft. lbs. (123 Nm).
   - Brake line clamp to the steering knuckle. Torque the bolt to 21 ft. lbs. (28 Nm).
   - ABS speed sensor. Torque both bolts to 7.1 ft. lbs. (8.2 Nm).
   - Halfshaft locknut. Torque the nut to 173 ft. lbs. (235 Nm), by applying the brakes.
   - Lock cap and new cotter pin
   - Grease cap, for 2WD
   - Front wheel

8. Depress the brake pedal several times.

9. Check and/or adjust the front wheel alignment.

10. Check the ABS speed sensor signal.

Land Cruiser

1. Before servicing the vehicle, refer to the precautions in the beginning of this section.

2. Remove or disconnect the following:
   - Front wheel
   - Brake caliper
   - Grease cap
   - Snapring
   - Hub drive flange
   - Locknut
   - Lockwasher
   - Adjusting nut
   - Outer bearing
   - Wheel hub
   - Disc brake dust shield
   - Wheel speed sensor and harness
   - Outer tie rod end
   - Upper ball joint
   - Lower ball joint
   - Steering knuckle
   - Oil seal, bushing and spindle bearing

To install:

3. Coat the spindle bearing and bushing with lithium grease.

4. Fill the spindle cavity with lithium grease.

5. Press the spindle bearing and bushing into the spindle.

6. Install or connect the following:
   - Oil seal
   - Steering knuckle
   - Upper ball joint. Tighten the nut to 81 ft. lbs. (110 Nm).
   - Lower ball joint. Tighten the nut to 117 ft. lbs. (159 Nm).
   - Outer tie rod end. Tighten the nut to 91 ft. lbs. (122 Nm).
   - Wheel speed sensor and harness
   - Disc brake dust shield. Tighten the bolts to 13 ft. lbs. (18 Nm).
   - Wheel hub
   - Outer bearing
   - Adjusting nut. Adjust the wheel bearings.
   - Lockwasher
   - Locknut. Tighten the nut to 47 ft. lbs. (64 Nm).
   - Hub drive flange. Tighten the nuts to 26 ft. lbs. (35 Nm).
   - Snapring
   - Grease cap
   - Brake caliper
   - Front wheel

Axle Shaft, Bearing and Seal

REMOVAL & INSTALLATION

Rear

SEQUOIA

1. Before servicing the vehicle, refer to the precautions in the beginning of this section.

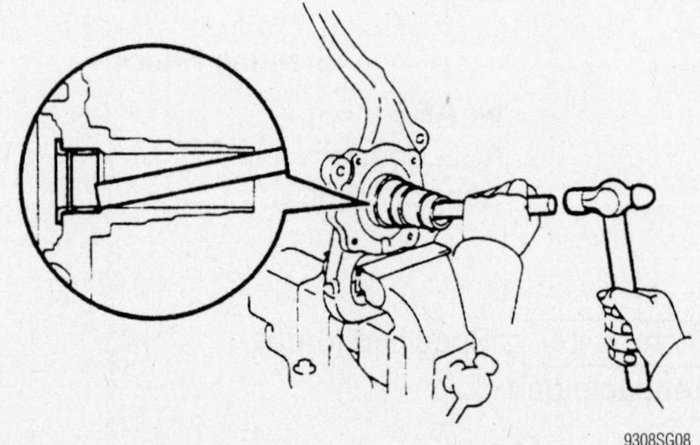

Removing the oil seal, bushing and spindle bearing—Land Cruiser

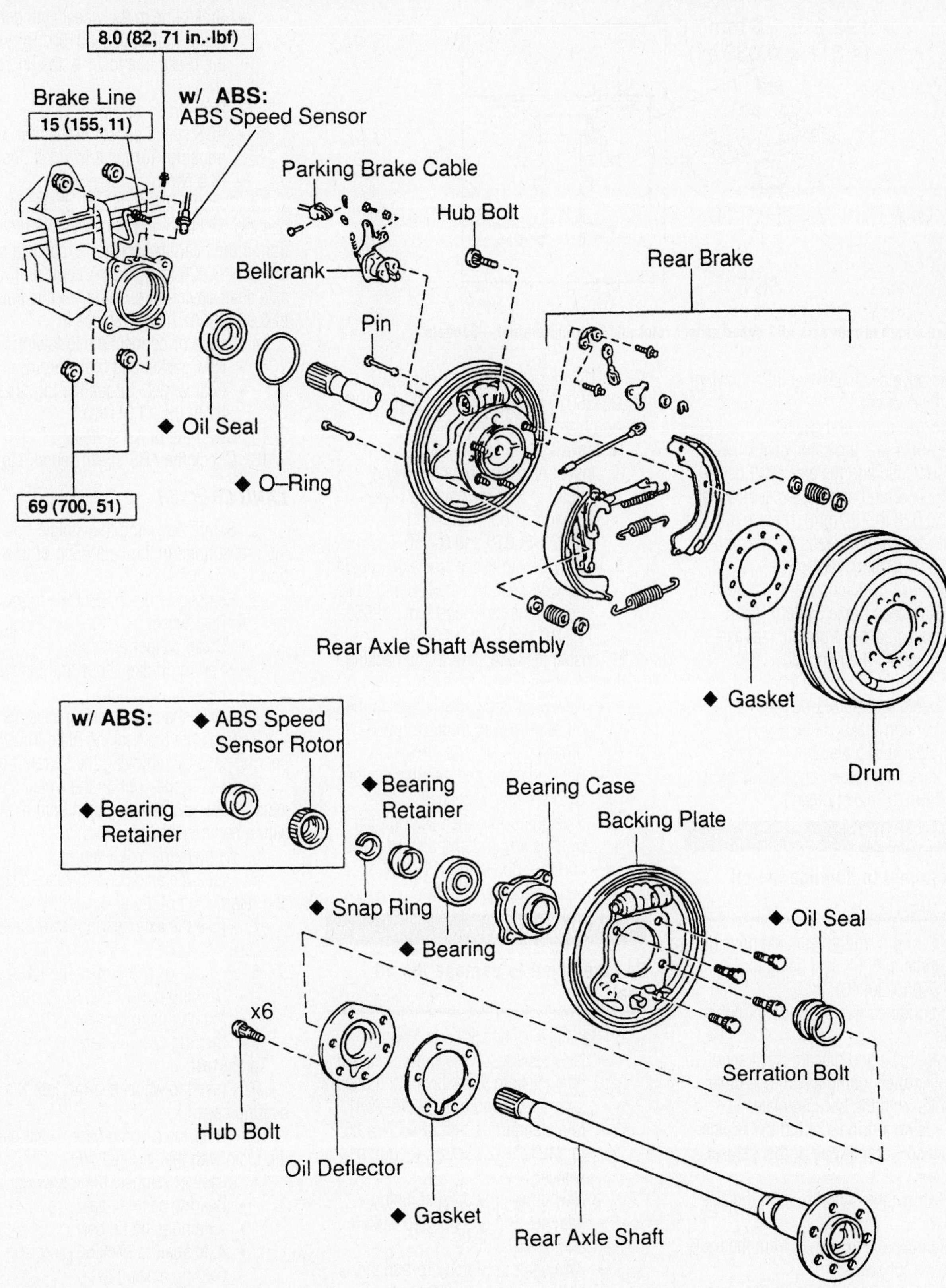
8.0 (82, 71 in.·lbf)
Brake Line
15 (155, 11)
w/ ABS:
ABS Speed Sensor
Parking Brake Cable
Hub Bolt
Bellcrank
Rear Brake
Pin
◆ Oil Seal
◆ O–Ring
69 (700, 51)
Rear Axle Shaft Assembly
◆ Gasket
Drum
w/ ABS:
◆ ABS Speed Sensor Rotor
◆ Bearing Retainer
◆ Bearing Retainer
Bearing Case
Backing Plate
◆ Snap Ring
◆ Bearing
◆ Oil Seal
x6
Serration Bolt
Hub Bolt
Oil Deflector
◆ Gasket
Rear Axle Shaft
N·m (kgf·cm, ft·lbf) : Specified torque
◆ Non–reusable part
9308YG14

Exploded view of the rear axle—Sequoia

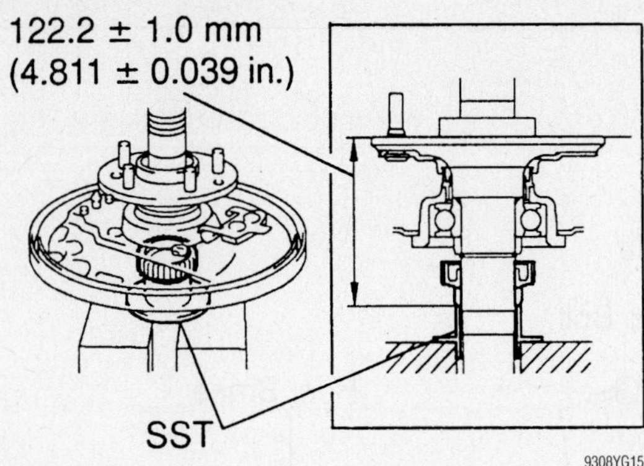

Standard length of rear axle ABS speed sensor rotor and bearing retainer—Sequoia

2. Remove or disconnect the following:
   - Rear wheel
   - Brake drum and gasket
3. Using a dial indicator, check the bearing backlash and the axle shaft deviation. If the bearing backlash exceeds a maximum or 0.028 in. (0.7mm), replace it. If the axle shaft deviation exceeds the maximum of 0.004 in. (0.1mm), replace it.
4. Remove or disconnect the following:
   - Anti-lock Brake System (ABS) speed sensor from the rear axle housing, if equipped
   - Brake line from the wheel cylinder, using tool 09023-00100
   - Parking brake cable
   - 4 backing plate nuts
   - Axle shaft assembly, by pulling it from the axle housing

⁂ WARNING

Be careful not to damage the oil seal.

   - O-ring from the rear axle housing
   - Inner side oil seal using tool 09308-00010
5. If equipped with ABS, perform the following:
   a. Remove and discard the 4 serration bolt nuts; then, using a hammer, drive the bolts from the backing plate.
   b. Using a grinder, grind the retainer and sensor rotor surfaces; then, chisel them out.
6. Remove the snapring from the axle shaft.
7. Remove the axle shaft from the backing plate, as follows:
   a. Position tool 09521-25011 onto the backing plate with the 4 nuts.
   b. Using a shop press, remove the axle shaft and bearing retainer from the backing plate.
8. Using tool 09308-00010, pull the oil seal from the backing plate.
9. Using a shop press and tools 09223-56010 and 09950-60010, press the bearing from the backing plate.

To install:

10. Install or connect the following:
    - Bearing into the backing plate, using a shop press and tools 09223-56010 and 09950-60010
    - New O-ring to the rear axle housing
    - New oil seal into the backing plate, using a hammer and tools 09950-70010 and 09950-60010
11. Install the axle shaft to the backing plate, as follows:
    - New outer side seal, lubricate the oil seal lip with multi-purpose grease
    - Backing plate and bearing retainer onto the rear axle shaft
    - Axle shaft onto the backing plate, by pressing it using a shop press and tool 09316-60011
    - New snapring

⁂ WARNING

Be careful not to damage the oil seal.

12. Install or connect the following:
    - New sensor rotor and new bearing retainer onto the axle shaft, using a shop press and tool 09316-60011 to a standard length of 4.77–4.85 in. (121.2–123.2mm), if equipped with ABS
    - New inner side oil seal, using a hammer and tools 09950-60020 and 09950-70010
    - Axle shaft assembly. Torque the bolts to 51 ft. lbs. (69 Nm).

⁂ WARNING

Be careful not to damage the oil seal.

    - Parking brake cable
    - Brake line to the wheel cylinder, using tool 09023-00100. Torque the brake line to 11 ft. lbs. (15 Nm).
    - Rear brake assembly
    - ABS speed sensor to the rear axle housing. Torque it to 7.1 ft. lbs. (8.0 Nm).
13. Using a dial indicator, check the bearing backlash and the axle shaft deviation. If the bearing backlash exceeds a maximum or 0.028 in. (0.7mm), replace it. If the axle shaft deviation exceeds the maximum of 0.004 in. (0.1mm), replace it.
14. Install or connect the following:
    - New gasket and brake drum
    - Rear wheel. Torque the lug nuts to 81 ft. lbs. (110 Nm).
15. Bleed the brake system.
16. Check the ABS speed sensor signal.

LAND CRUISER

1. Before servicing the vehicle, refer to the precautions in the beginning of this section.
2. Remove or disconnect the following:
   - Rear wheel
   - Brake caliper and rotor
   - Parking brake shoes and hardware
   - Bearing case nuts
   - Axle shaft and bearing assembly
3. Separate the backing plate from the bearing case by removing the serrated bolts.
4. Grind a flat spot on the wheel speed sensor rotor and retainer, then split them with a hammer and chisel.
5. Remove the axle snapring.
6. Press the axle bearing case, bearing and retainer off of the axle.
7. Press the axle bearing from the bearing case.
8. Remove or disconnect the following:
   - Backing plate
   - Axle housing oil seal
   - Bearing case oil seal

To install:

9. Press the wheel bearing into the bearing case.
10. Install the bearing case to the backing plate with the serrated bolts.
11. Install or connect the following:
    - Bearing case oil seal
    - Axle housing oil seal
    - Axle shaft to backing plate and bearing assembly
    - Bearing retainer
    - Axle snapring
    - Wheel speed sensor rotor and retainer
    - Axle shaft and bearing assembly to the axle housing. Tighten the nuts to 91 ft. lbs. (123 Nm).

- Parking brake shoes and hardware
- Brake caliper and rotor
- Rear wheel

Pinion Seal

REMOVAL & INSTALLATION

Front

SEQUOIA

1. Before servicing the vehicle, refer to the precautions in the beginning of this section.
2. Remove the under cover.
3. Drain the differential housing oil.
4. Remove the front driveshaft.
5. Remove the companion flange, as follows:
   - Loosen the staked part of the nut, using a chisel and a hammer
   - Companion flange nut, using tool 09330-00021
   - Companion flange, using tools 09950-30011 and 09954-03010
6. Remove the oil seal and slinger, as follows:
   - Oil seal, using tool 09308-10010
   - Oil slinger

To install:

7. Install or connect the following:
   - Oil slinger
   - New oil seal, using a hammer and tool 09554-22010 to a depth of 0.153–0.189 in. (4.2–4.8mm).
8. Install the companion flange, as follows:
   - Companion flange
   - New nut, lubricated with hypoid gear oil

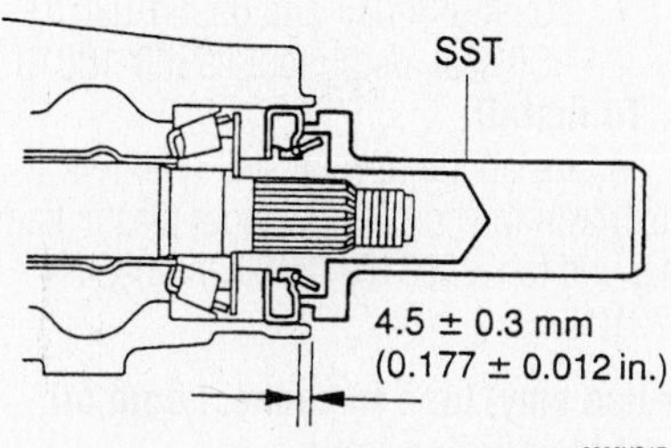

Positioning the Sequoia front pinion seal in the differential housing—Rear differential assembly is similar

   - Torque the nut to 80 ft. lbs. (108 Nm), using tool 09330-00021.
9. Adjust the drive pinion preload
10. Rotate the drive pinion, using a torque wrench while tightening the flange nut to make sure the bearing preload is 10.4–16.5 inch lbs. (1.2–1.9 Nm) for a new bearing or 5.2–8.7 inch lbs. (0.6–1.0 Nm) for a used bearing. Tighten the flange nut to achieve the preload torque readings originally recorded.

CAUTION

Never loosen the pinion nut to reduce bearing preload.

11. Install or connect the following:
   - Drive pinion nut, stake it
   - Front driveshaft. Tighten the fasteners to 54 ft. lbs. (74 Nm).
   - Under cover
12. Fill the differential with gear lubricant and check for leaks.

LAND CRUISER

1. Before servicing the vehicle, refer to the precautions in the beginning of this section.

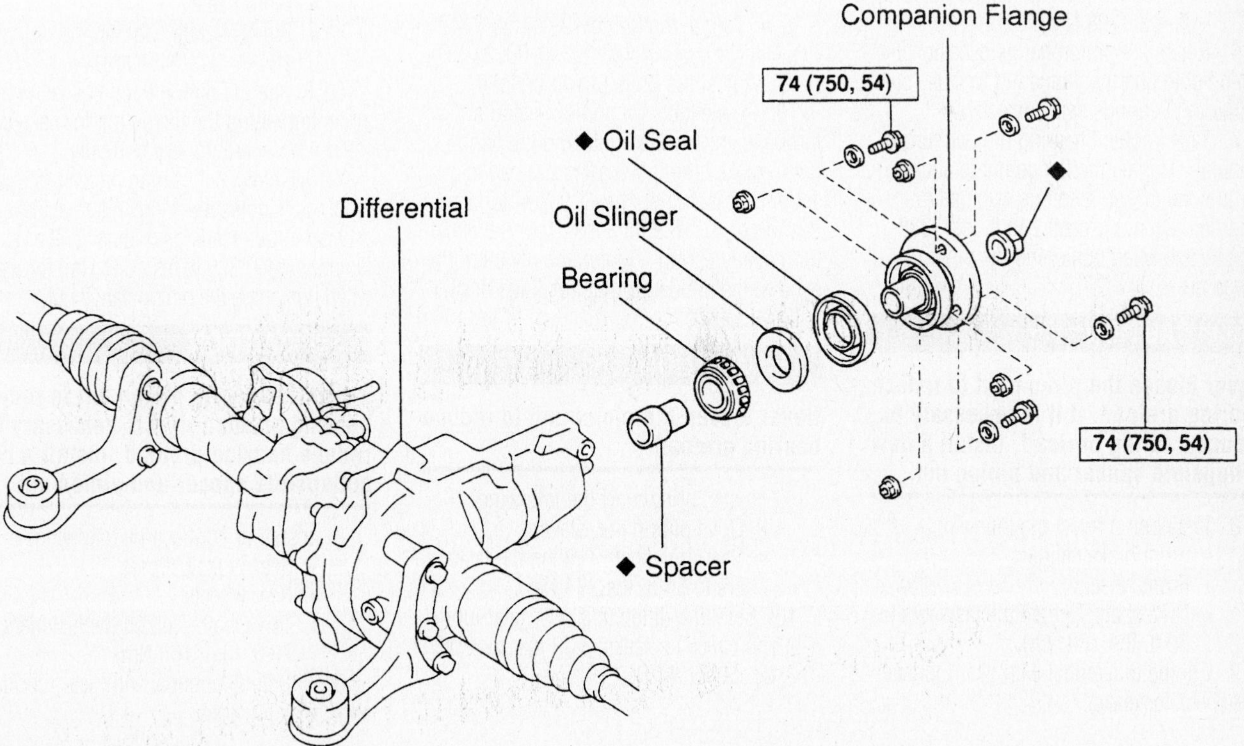

Exploded view of the Sequoia front differential assembly—Rear differential assembly is similar

2. Remove or disconnect the following:
   - Driveshaft
   - Front wheels
   - Front brake calipers

➡The front brake calipers must be removed so that there is no additional drag when measuring pinion bearing preload.

3. Use an inch lb. torque wrench and measure the amount of torque required to maintain pinion rotation through several revolutions.
4. Remove or disconnect the following:
   - Pinion flange
   - Oil seal
   - Oil slinger
   - Pinion bearing and race
   - Oil storage ring
   - Collapsible spacer

To install:

➡Use a new collapsible spacer and flange nut for assembly.

5. Install or connect the following:
   - Collapsible spacer
   - Oil storage ring
   - Pinion bearing and race
   - Pinion seal
   - Pinion flange. Tighten the nut to 80 ft. lbs. (108 Nm).
6. Rotate the pinion flange occasionally while tightening the flange nut to make sure the pinion bearings seat correctly.
7. Take frequent bearing preload torque readings. Tighten the flange nut to achieve the preload torque readings originally recorded. Do not exceed 249 ft. lbs. (338 Nm) torque when tightening the pinion flange nut.

⁂ CAUTION

Never loosen the pinion nut to reduce bearing preload. If it is necessary to reduce bearing preload, install a new collapsible spacer and pinion nut.

8. Install or connect the following:
   - Front brake calipers
   - Front wheels
   - Driveshaft. Tighten the fasteners to 59 ft. lbs. (80 Nm).
9. Fill the differential with gear lubricant and check for leaks.

Rear

SEQUOIA

1. Before servicing the vehicle, refer to the precautions in the beginning of this section.
2. Drain the differential housing oil.
3. Remove the rear driveshaft.
4. Remove the companion flange, as follows:
   - Loosen the staked part of the nut, using a chisel and a hammer
   - Companion flange nut, using tool 09330-00021
   - Companion flange, using tools 09950-30011 and 09954-03010
   - Oil seal, using tool 09308-10010

To install:

5. Install the new oil seal until it is flush with the housing, using a plastic hammer and tools 09316-12010 and 09649-17010

➡Use vinyl tape to connect both oil seal installation tools.

6. Install the companion flange, as follows:
   - Companion flange
   - New nut, lubricated with hypoid gear oil
   - Torque the nut to 109 ft. lbs. (147 Nm), using tool 09330-00021.
7. Adjust the drive pinion preload
8. Rotate the drive pinion, using a torque wrench while tightening the flange nut to make sure the bearing preload is 11.4–16.7 inch lbs. (1.3–1.9 Nm) for a new bearing or 4.3–6.9 inch lbs. (0.5–0.8 Nm) for a used bearing. Tighten the flange nut to achieve the preload torque readings originally recorded.

⁂ CAUTION

Never loosen the pinion nut to reduce bearing preload.

9. Install or connect the following:
   - Drive pinion nut, stake it
   - Rear driveshaft. Tighten the fasteners to 54 ft. lbs. (74 Nm).
10. Refill the differential with gear lubricant and check for leaks; 3.33 qts. for 2WD or 3.12 qts. for 4WD.

LAND CRUISER

1. Before servicing the vehicle, refer to the precautions in the beginning of this section.
2. Remove or disconnect the following:
   - Driveshaft
   - Rear wheels
   - Rear brake calipers

➡The rear brake calipers must be removed so that there is no additional drag when measuring pinion bearing preload.

3. Use an inch lb. torque wrench and measure the amount of torque required to maintain pinion rotation through several revolutions.
4. Remove or disconnect the following:
   - Pinion flange
   - Oil seal
   - Oil slinger
   - Pinion bearing and race
   - Collapsible spacer

To install:

➡Use a new collapsible spacer and flange nut for assembly.

5. Install or connect the following:
   - Collapsible spacer
   - Pinion bearing and race
   - Pinion seal
   - Pinion flange. Tighten the nut to 181 ft. lbs. (245 Nm).
6. Rotate the pinion flange occasionally while tightening the flange nut to make sure the pinion bearings seat correctly.
7. Take frequent bearing preload torque readings. Tighten the flange nut to achieve the preload torque readings originally recorded. Do not exceed 326 ft. lbs. (441 Nm) torque when tightening the pinion flange nut.

⁂ CAUTION

Never loosen the pinion nut to reduce bearing preload. If it is necessary to reduce bearing preload, install a new collapsible spacer and pinion nut.

8. Install or connect the following:
   - Rear brake calipers
   - Rear wheels
   - Driveshaft. Tighten the fasteners to 78 ft. lbs. (106 Nm).
9. Fill the differential with gear lubricant and check for leaks.

STEERING AND SUSPENSION

Air Bag

⁂ CAUTION

Some vehicles are equipped with an air bag system. The system must be disarmed before performing service on, or around, system components, the steering column, instrument panel components, wiring and sensors. Failure to follow the safety precautions and the disarming procedure could result in accidental air bag deployment, possible injury and unnecessary system repairs.

PRECAUTIONS

Several precautions must be observed when handling the inflator module to avoid accidental deployment and possible personal injury.

- Never carry the inflator module by the wires or connector on the underside of the module.
- When carrying a live inflator module, hold securely with both hands and ensure that the bag and trim cover are pointed away.
- Place the inflator module on a bench or other surface with the bag and trim cover facing up.
- With the inflator module on the bench, never place anything on or close to the module which may be thrown in the event of an accidental deployment.

DISARMING

To avoid personal injury when working on vehicles equipped with an air bag, the negative battery cable must be disconnected and at least 90 seconds must elapse before working on the system. Failure to do so may result in deployment of the air bag.

Power Rack and Pinion Steering Gear

REMOVAL & INSTALLATION

Sequoia

1. Before servicing the vehicle, refer to the precautions in the beginning of this section.
2. Position the front wheels in the straight-ahead position.
3. Remove or disconnect the following:
   - Engine under cover
   - Steering wheel pad
   - Steering wheel
   - Left and right outer tie-rod ends from the steering knuckles
4. Matchmark the No. 2 intermediate shaft to the steering gear input shaft.
5. Remove or disconnect the following:
   - Clamp plate
   - Pressure feed and return tubes from the power steering gear, using tool 09631-22020
   - Power steering gear assembly

To install:

6. Install or connect the following:
   - Power steering gear assembly. Torque the set bolt to 123 ft. lbs. (165 Nm) and the set nut/bolt to 96 ft. lbs. (91 Nm).
   - Pressure feed and return tubes to the power steering gear. Torque them to 27 ft. lbs. (32 Nm), using tool 09631-22020.
   - Clamp plate. Torque the bolt to 21 ft. lbs. (29 Nm).
   - No. 2 intermediate shaft to the steering gear input shaft
   - Left and right outer tie-rod ends to the steering knuckles. Torque the nuts to 67 ft. lbs. (91 Nm).
   - Steering wheel. Torque the nut to 26 ft. lbs. (35 Nm).
   - Steering wheel pad
   - Engine under cover
7. Fill and bleed the power steering system.
8. Check and/or adjust the wheel alignment, as necessary.

Land Cruiser

2001–03

1. Before servicing the vehicle, refer to the precautions in the beginning of this section.
2. Matchmark the intermediate shaft to the steering gear input shaft.
3. Remove or disconnect the following:
   - Negative battery cable
   - Engine under covers
   - Outer tie rod ends
   - Engine oil filter adapter
   - Intermediate steering shaft
   - Power steering hoses and bracket
   - Power steering gear

To install:

4. Install or connect the following:
   - Power steering gear. Tighten the fasteners to 74 ft. lbs. (100 Nm).
   - Power steering hoses and bracket
   - Intermediate steering shaft. Tighten the bolts to 25 ft. lbs. (34 Nm).
   - Engine oil filter adapter. Tighten the bolts to 13 ft. lbs. (18 Nm).
   - Outer tie rod ends. Tighten the nuts to 90 ft. lbs. (122 Nm).
   - Engine under covers
   - Negative battery cable
5. Fill the power steering fluid reservoir.
6. Check the wheel alignment and adjust as necessary.

2004

1. Place the front wheels in the straight-ahead position.
2. Remove the steering wheel.
3. Remove the engine under-covers.

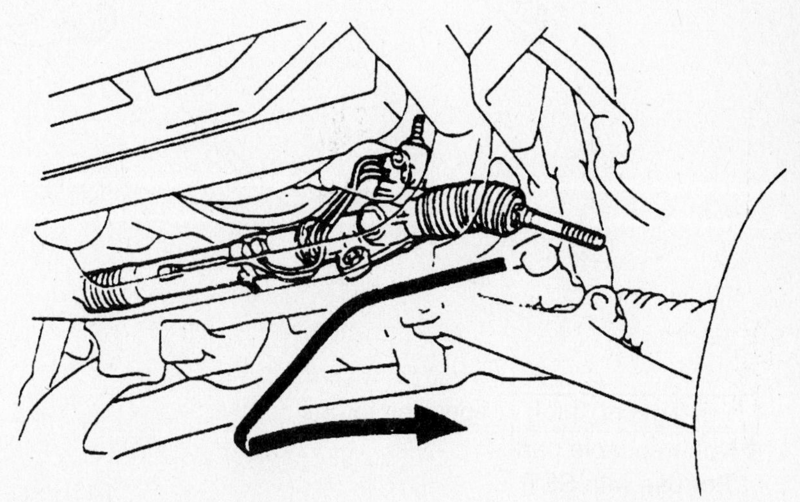

Power rack and pinion steering gear removal—Land Cruiser

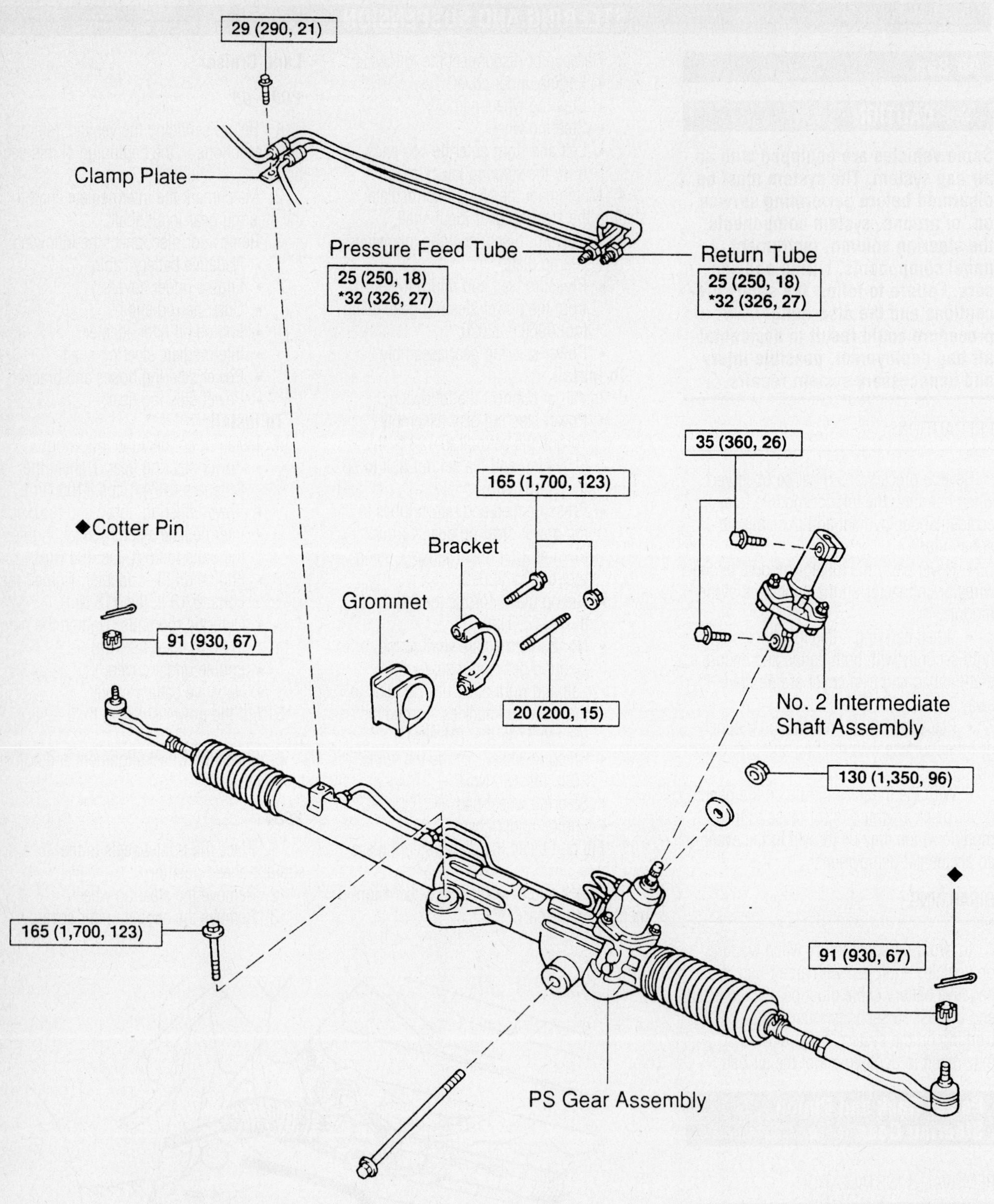

Exploded view of the power rack and pinion steering gear mounting—Sequoia

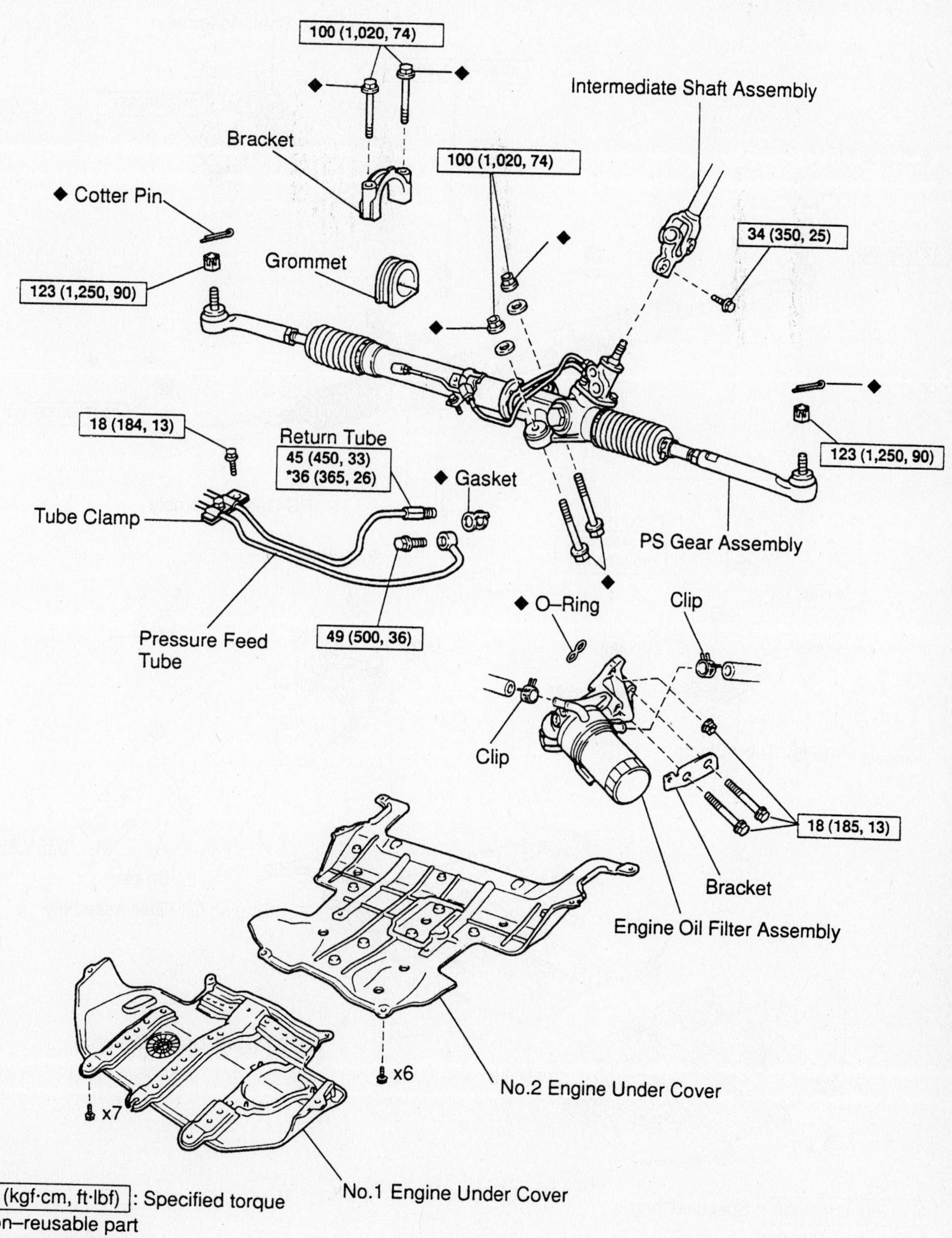

Exploded view of the rack and pinion steering gear mounting—2001-03 Land Cruiser

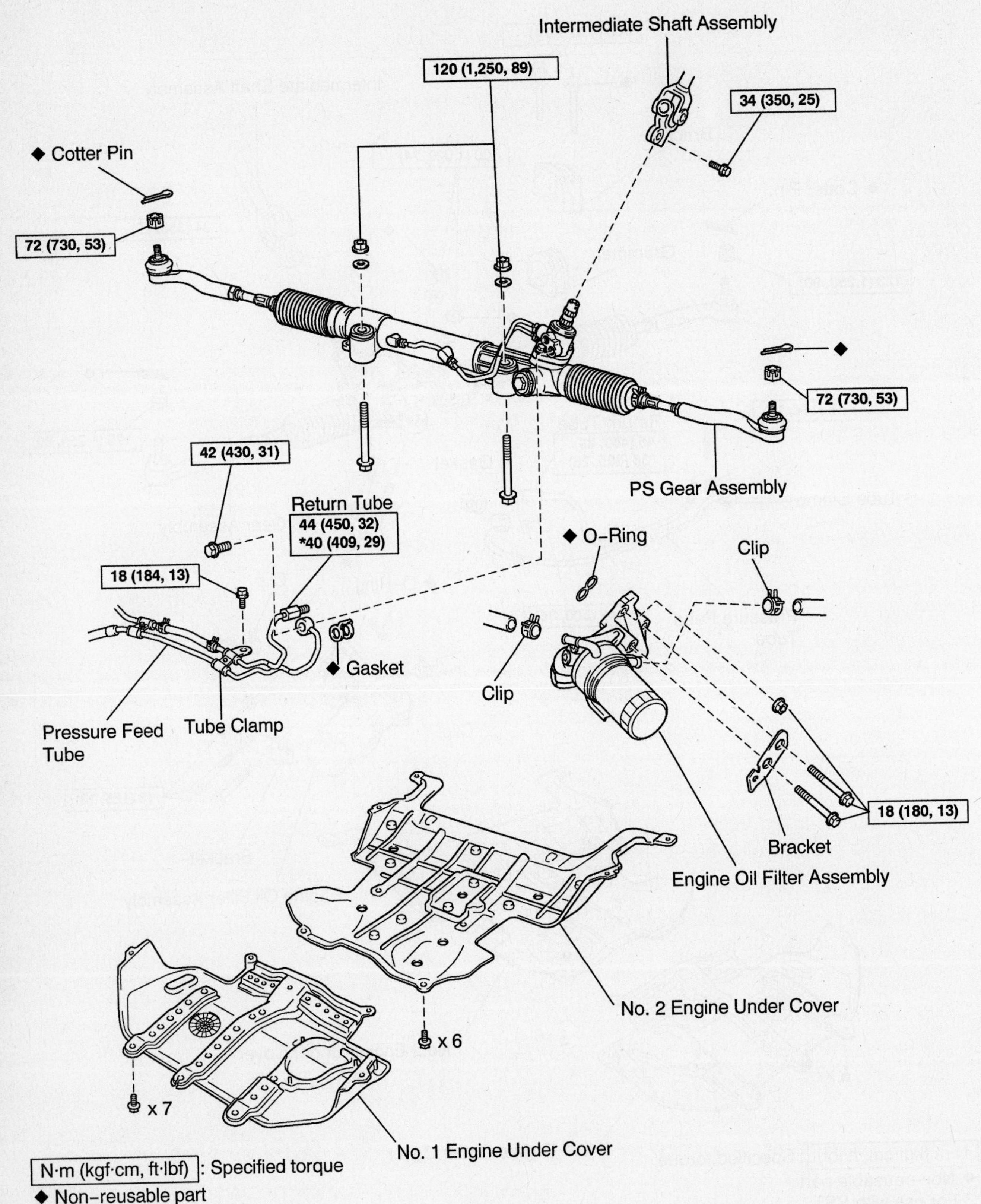

Steering gear and related components—2004 Land Cruiser

4. Disconnect the left and right tie rod ends.

5. Remove the oil filter.

6. Remove the oil filter adapter.

7. Turn the steering fully to the right, matchmark and disconnect the intermediate shaft from the gear.

8. Disconnect the pressure and return lines.

9. Remove the 2 mounting bolts and remove the gear, sliding it to the right and pulling it from the left side.

10. Installation is the reverse of removal. Observe the following torques:

- Steering gear mounting bolts: 89 ft. lbs. (120 Nm)
- Return line: 37 ft. lbs. (50 Nm). Use a torque wrench with a fulcrum length of 300mm.
- Pressure line: 31 ft. lbs. (42 Nm)
- Oil filter adapter: 13 ft. lbs. (18 Nm)
- Tie rod ends: 53 ft. lbs. (72 Nm)

Front Shock Absorber

REMOVAL & INSTALLATION

Land Cruiser

1. Before servicing the vehicle, refer to the precautions in the beginning of this section.
2. Support the axle with a jackstand.
3. Remove or disconnect the following:

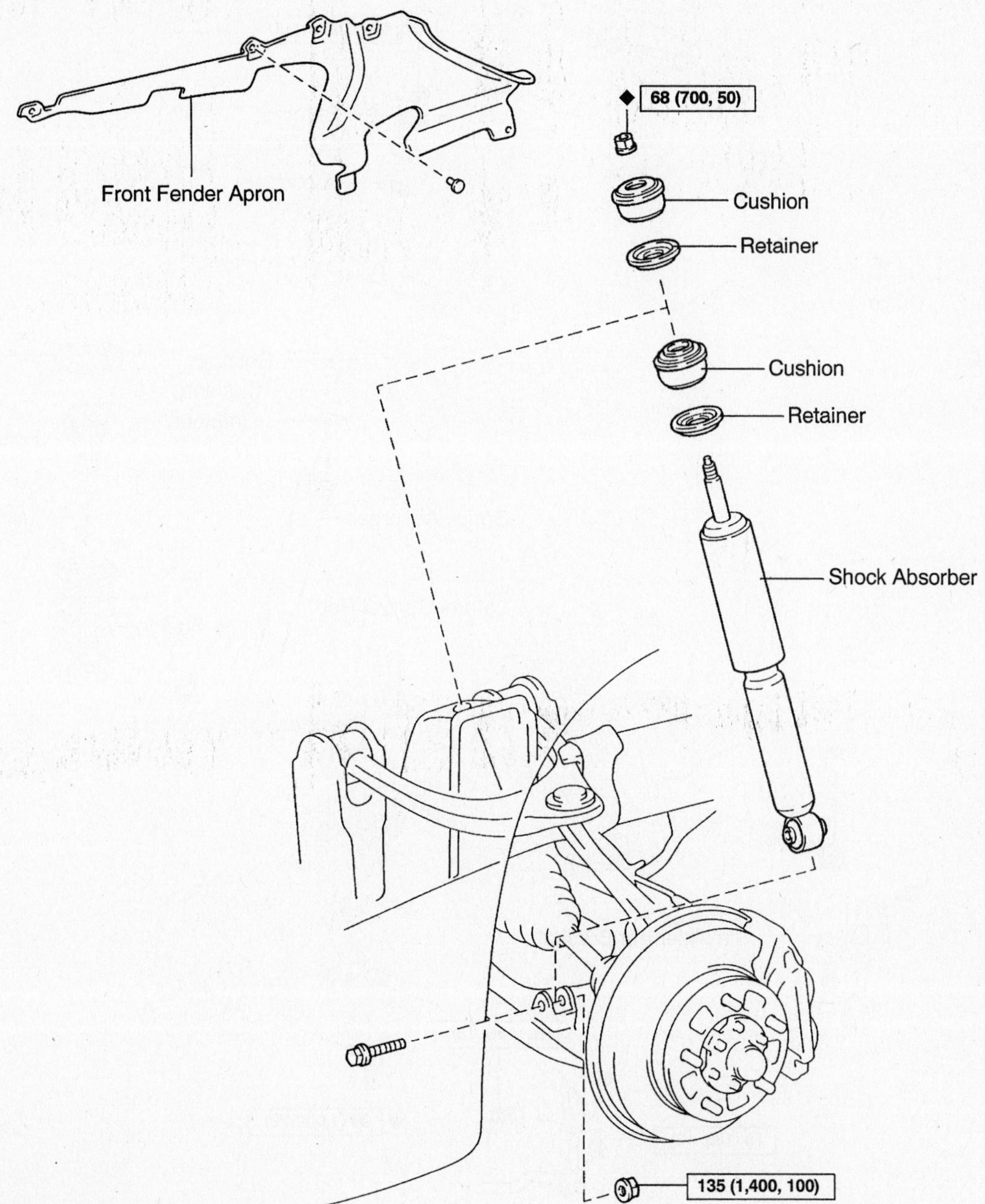

Front shock absorber—Land Cruiser

- Front wheel
- Shock absorber

To install:

4. Install or connect the following:
   - Shock absorber. Tighten the nut to 50 ft. lbs. (68 Nm) and the bolt to 100 ft. lbs. (135 Nm).
   - Front wheel

Rear Shock Absorber

REMOVAL & INSTALLATION

Land Cruiser

1. Before servicing the vehicle, refer to the precautions in the beginning of this section.
2. Support the axle with a jackstand.
3. Remove or disconnect the following:
   - Rear wheel
   - Shock absorber

To install:

4. Install or connect the following:
   - Shock absorber. Tighten the nut to 51 ft. lbs. (69 Nm) and the bolt to 72 ft. lbs. (98 Nm).
   - Rear wheel

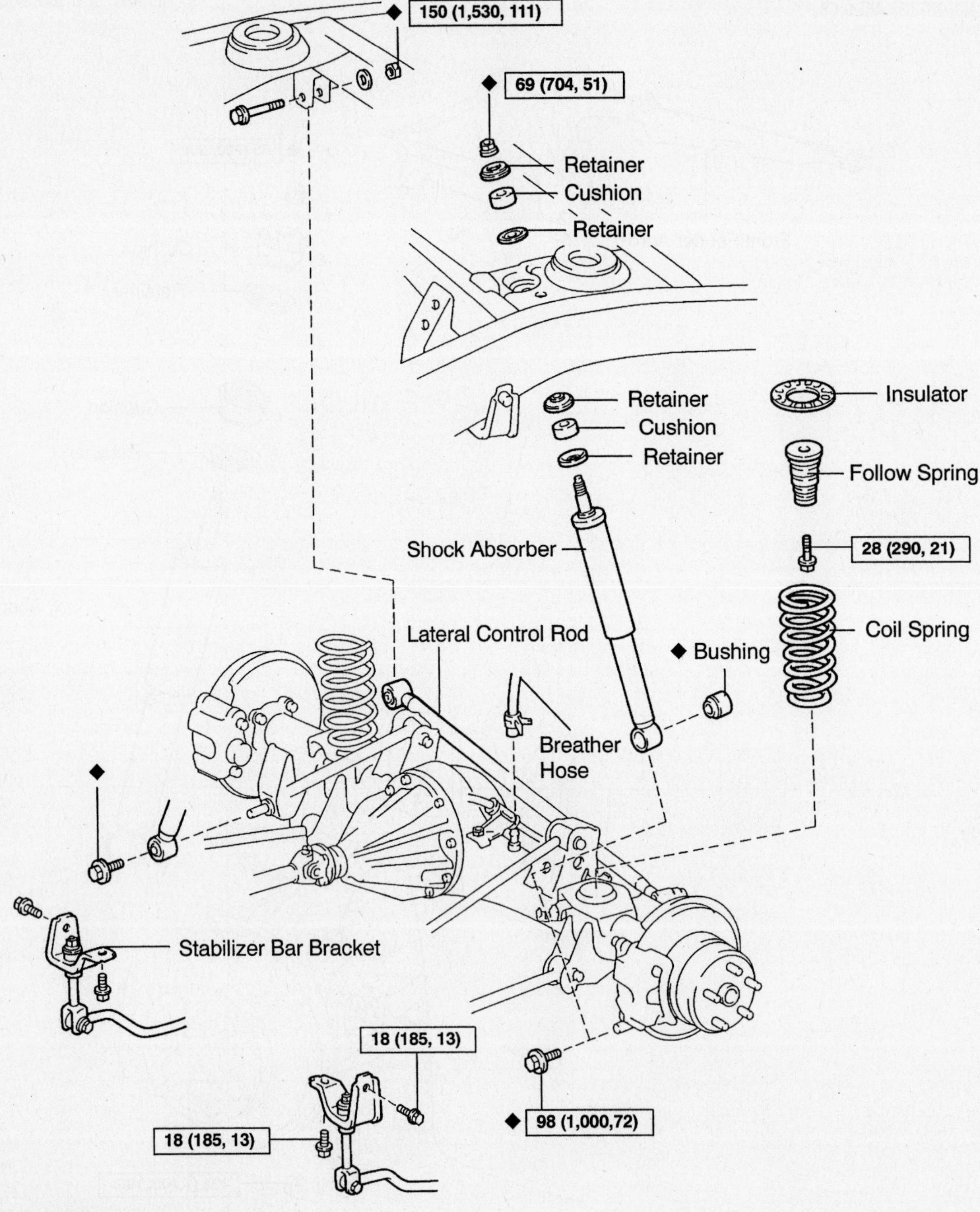

Rear suspension—Land Cruiser

Sequoia

2001–03

1. Before servicing the vehicle, refer to the precautions in the beginning of this section.
2. Remove or disconnect the following:
   - Rear wheel
   - Shock absorber

To install:

3. Install or connect the following:
   - Shock absorber. Tighten the upper nut to 15 ft. lbs. (20 Nm) and the lower nut/bolt to 64 ft. lbs. (87 Nm).
   - Rear wheel.

2004

1. Remove the rear wheels.
2. Support the axle with a jack.
3. Without Auto Leveler:
   a. Disconnect the lower end of the shock from the axle.
   b. Remove the upper nut, retainer and bushings.

Normal Type:

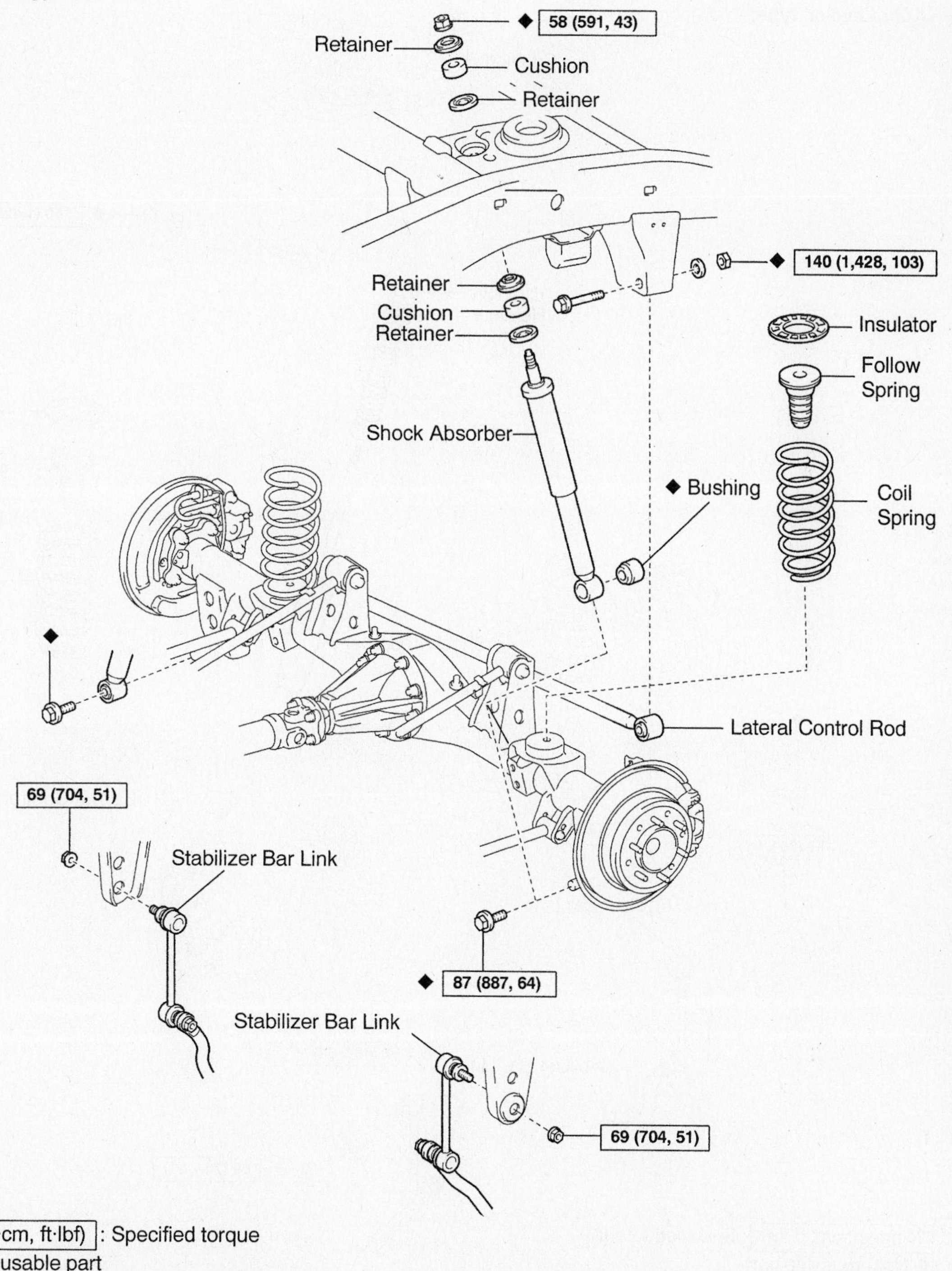

2004 Sequoia rear suspension, without Auto Leveler

4. With Auto Leveler:

CAUTION

Perform this procedure with the shock absorber stretched completely. The primary reaction force of the shock absorber is approximately 1,000 N.

a. Remove the bolt and disconnect the shock absorber from the axle.

b. Remove the upper nut, retainer and bushings.

5. Installation is the reverse of removal. Torque the lower end bolt to 64 ft. lbs. (87 Nm); the upper end nut to 43 ft. lbs. (58 Nm).

Front Strut

REMOVAL & INSTALLATION

Sequoia

1. Before servicing the vehicle, refer to the precautions in the beginning of this section.
2. Remove or disconnect the following:
   - Front wheel

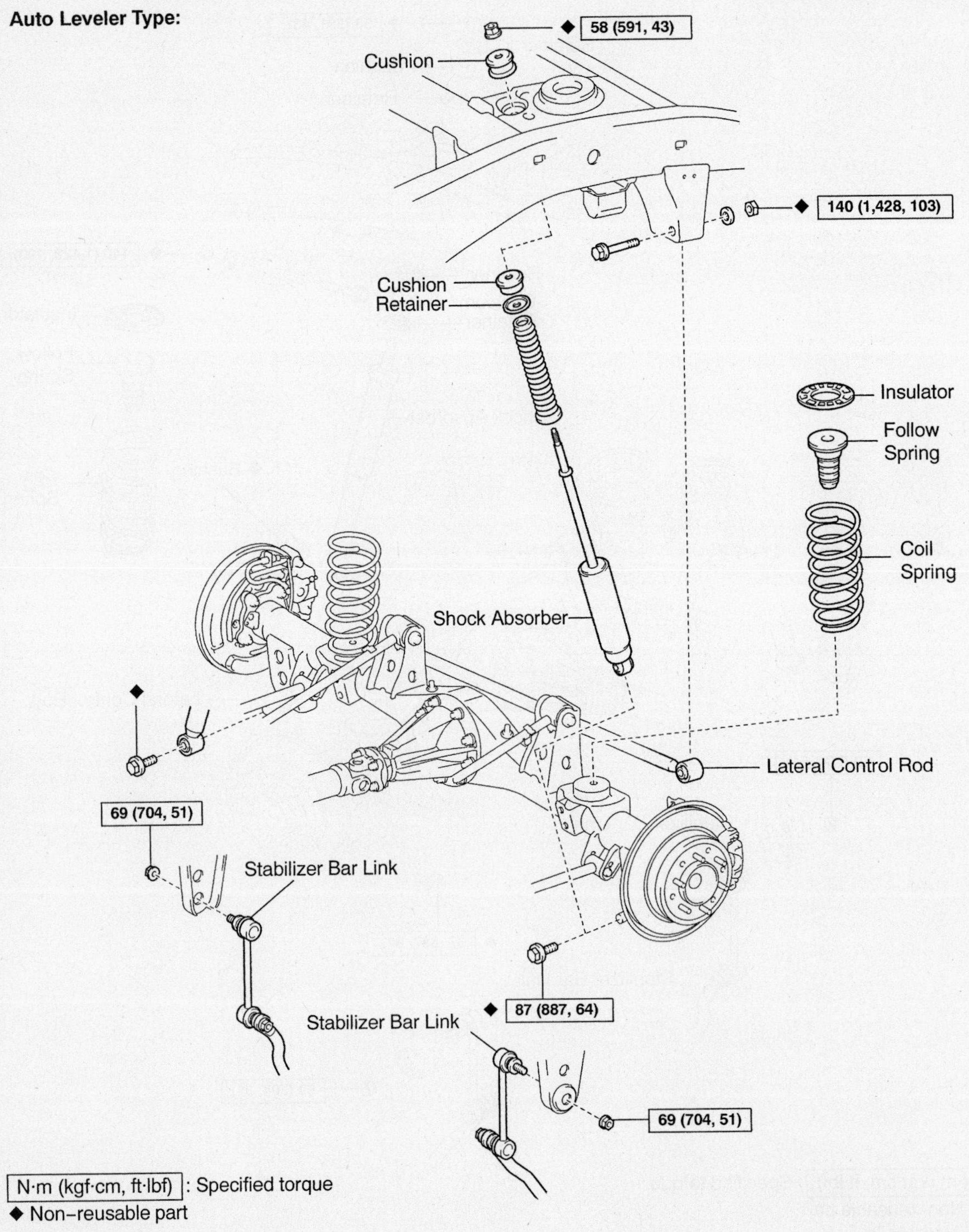

2004 Sequoia rear suspension, with Auto Leveler

- Strut-to-lower control arm nut/bolt and the strut
- Strut-to-chassis 3 nuts/bolts and the strut

To install:

3. Install or connect the following:

- Strut to the chassis. Torque the 3 nuts/bolts to 47 ft. lbs. (64 Nm).
- Strut to the lower control arm. Torque the nut/bolt to 100 ft. lbs. (135 Nm).
- Front wheel

DISASSEMBLY

1. Remove the strut.
2. Using a compressor, compress the coil spring.

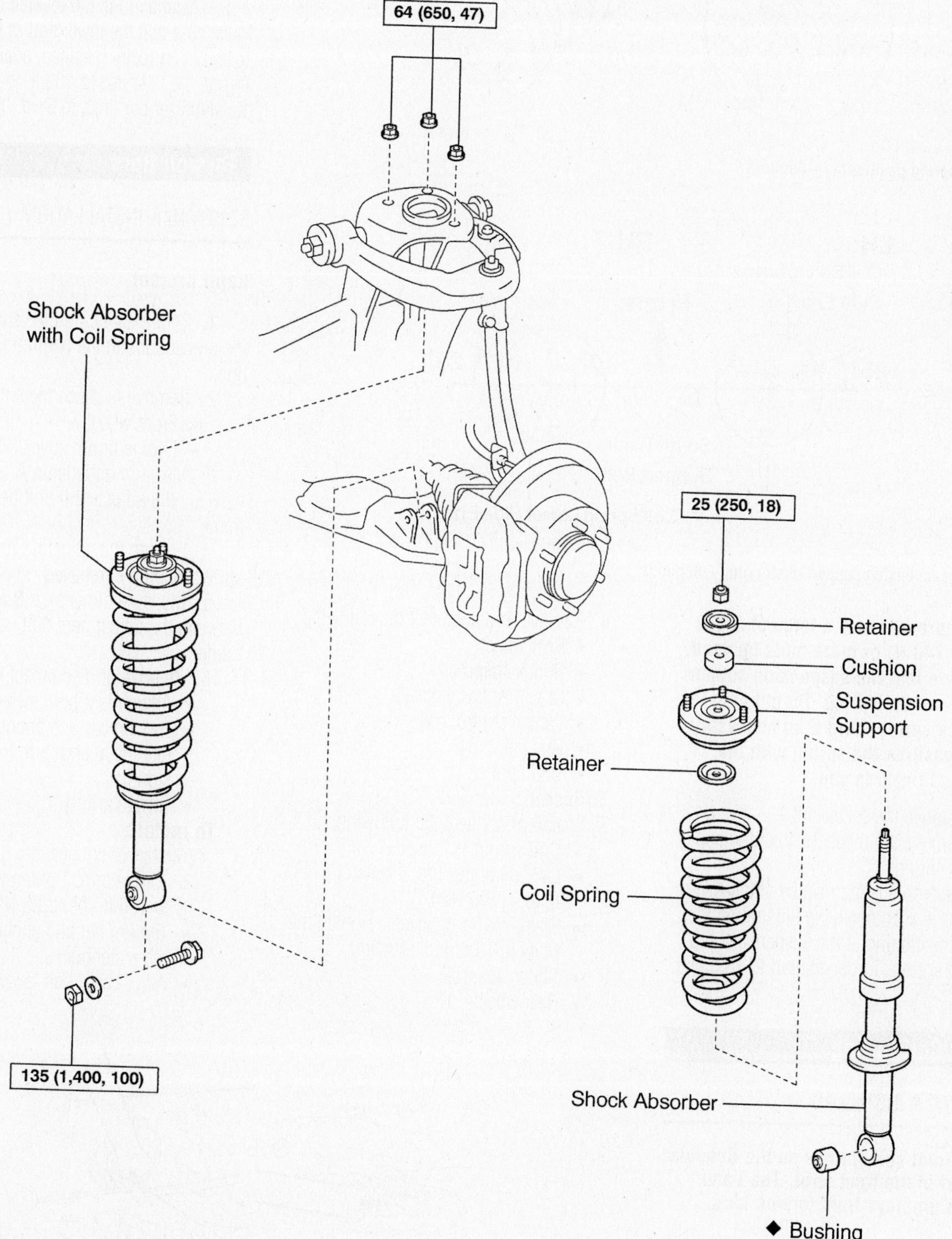

Front strut—Sequoia

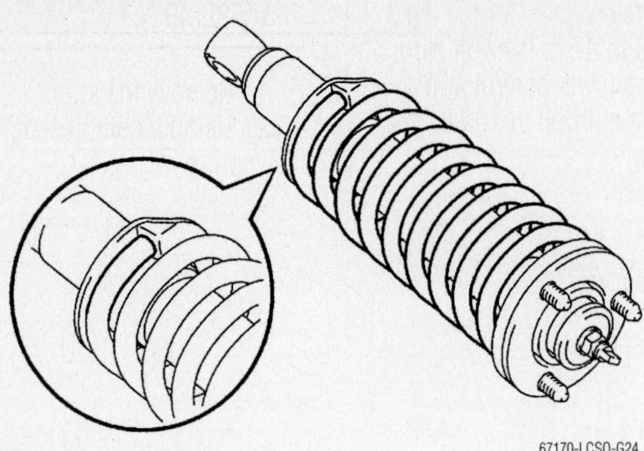

Correct spring positioning—Sequoia

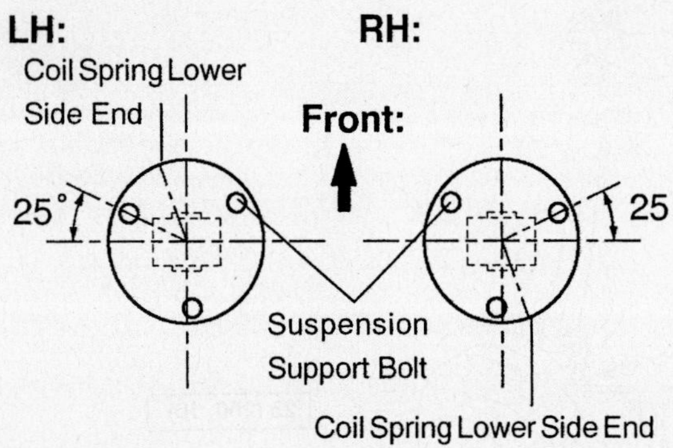

Correct suspension support positioning—Sequoia

➡A compressor with a force of 2,860 lbs. (12,740 N) or more must be used. Make sure that the suspension support is free from the spring. Do not compress the spring more than necessary. Do not position the spring with the upper end towards you.

3. Remove the center nut.
4. Remove the retainers, bushing support and spring.
5. Assembly is the reverse of disassembly. See the accompanying illustration for correct positioning of the suspension support and spring. Torque the nut to 18 ft. lbs. (25 Nm).

Rear Coil Spring

REMOVAL & INSTALLATION

➡The front coil springs on the Sequoia are part of the front strut. The Land Cruiser employs front torsion bars.

Land Cruiser

1. Before servicing the vehicle, refer to the precautions in the beginning of this section.
2. Support the vehicle at the frame.
3. Support the axle with a floor jack.
4. Remove or disconnect the following:
   - Rear wheel
   - Shock absorber
   - Stabilizer bar brackets
   - Lateral control rod
   - Breather hose
   - Coil spring

To install:

5. Install or connect the following:
   - Coil spring
   - Lateral control rod. Tighten to 111 ft. lbs. (150 Nm).
   - Stabilizer bar brackets. Tighten the bolts to 13 ft. lbs. (18 Nm)
   - Shock absorber
   - Rear wheel

Sequoia

1. Remove the shock absorber.
2. Disconnect the left and right stabilizer bar links.
3. Disconnect the lateral rod.
4. Lower the axle slowly and remove the spring.
5. Installation is the reverse of removal. Make sure that the lower end of the coil spring is correctly installed, against the stop. Torque the later rod to 103 ft. lbs. (140 Nm); the stabilizer bar links to 51 ft. lbs. (69 Nm).

Torsion Bars

REMOVAL & INSTALLATION

Land Cruiser

1. Before servicing the vehicle, refer to the precautions in the beginning of this section.
2. Remove or disconnect the following:
   - Front wheel
   - Engine under cover
3. Measure dimension **A** as shown between the adjustment bolt head and the frame.
4. Loosen the adjusting bolt until all spring tension is relieved.
5. Measure dimension **B** as shown between the adjustment bolt head and the frame.
6. Remove or disconnect the following:
   - Adjustment bolt, swivel and seat
   - Torsion bar and anchor arm. Separate the anchor arm from the torsion bar.
   - Torque arm

To install:

7. Install or connect the following:
   - Torque arm. Tighten the fasteners to 166 ft. lbs. (225 Nm).
   - Torsion bar and anchor arm. Align the matchmarks.
   - Adjustment bolt, swivel and seat

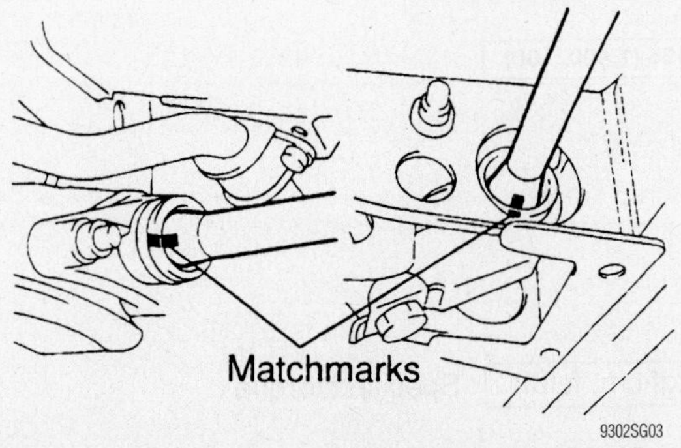

Matchmark the torsion bar to the anchor arm and torque arm—Land Cruiser

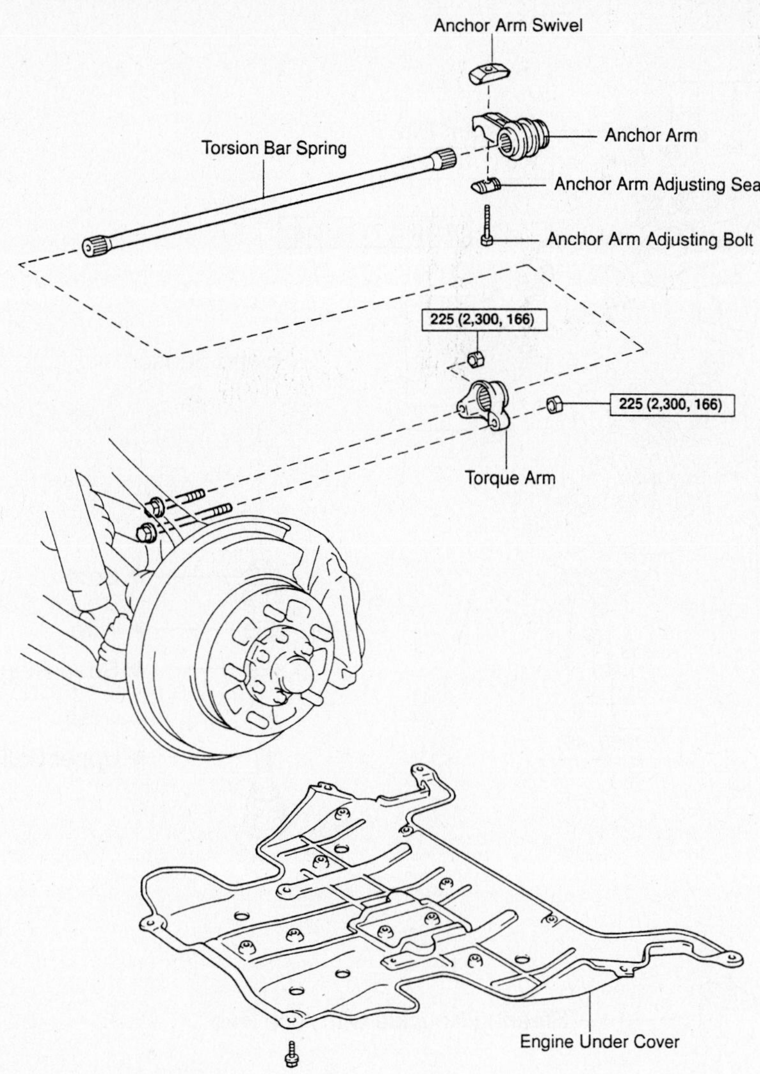

Torsion bar mounting exploded view—Land Cruiser

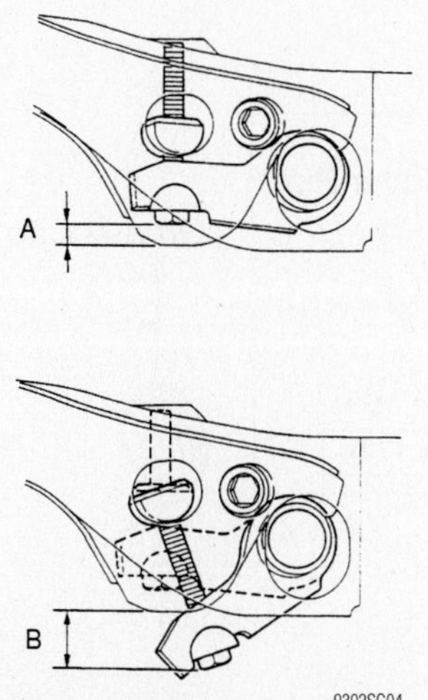

Reference measurements A and B—Land Cruiser

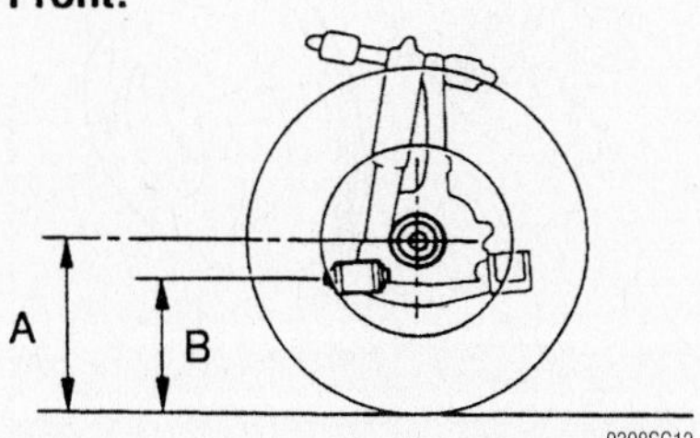

Ride height measurements A and B—Land Cruiser

8. Check that dimension **B** is close to the measurement made at disassembly.

9. If installing a new torsion bar, tighten the adjustment bolt until dimension **A** is as follows:

- Left torsion bar: 0.315–0.984 inches (8–25mm)
- Right torsion bar: 0.079–0.709 inches (2–18mm)

10. If installing the original torsion bar, tighten the adjustment bolt until dimension **A** is close to the measurement made at disassembly.

11. Install or connect the following:

- Engine under cover
- Front wheel

12. Place the vehicle on a flat, level surface and check the vehicle curb height as follows:

a. Step 1: Measure dimension **A** between the spindle center and the ground.

b. Step 2: Measure dimension **B** between the lower control arm front bolt center and the ground.

c. Step 3: Turn the adjusting bolt so that **A** minus **B** is equal to 2.795 inches (71mm).

Upper Ball Joint

REMOVAL & INSTALLATION

Land Cruiser

The upper ball joint is serviced with the upper control arm as an assembly.

Sequoia

1. Before servicing the vehicle, refer to the precautions in the beginning of this section.
2. Remove or disconnect the following:
   - Front wheel
   - Steering knuckle with the axle hub
   - Wire and boot
   - Snapring
   - Upper ball joint from the steering knuckle, using a deep socket wrench and tool 09050-40011, or equivalent press.

To install:

3. Install or connect the following:
   - New upper ball joint to the steering knuckle, using a deep socket and tool 09309-37010, or equivalent press.
   - New snapring
4. Using a torque wrench, inspect the upper ball joint rotation, as follows:

a. Flip the ball joint back-and-forth 5 times.

b. Using a torque wrench, continuously turn the nut 1 turn in 2–4 seconds.

c. Take the reading on the 5th turn; it should be 6–39 inch lbs. (0.7–4.4 Nm). If not, replace the upper ball joint.

5. Install or connect the following:
   - New boot secured with a wire
   - Front wheel.

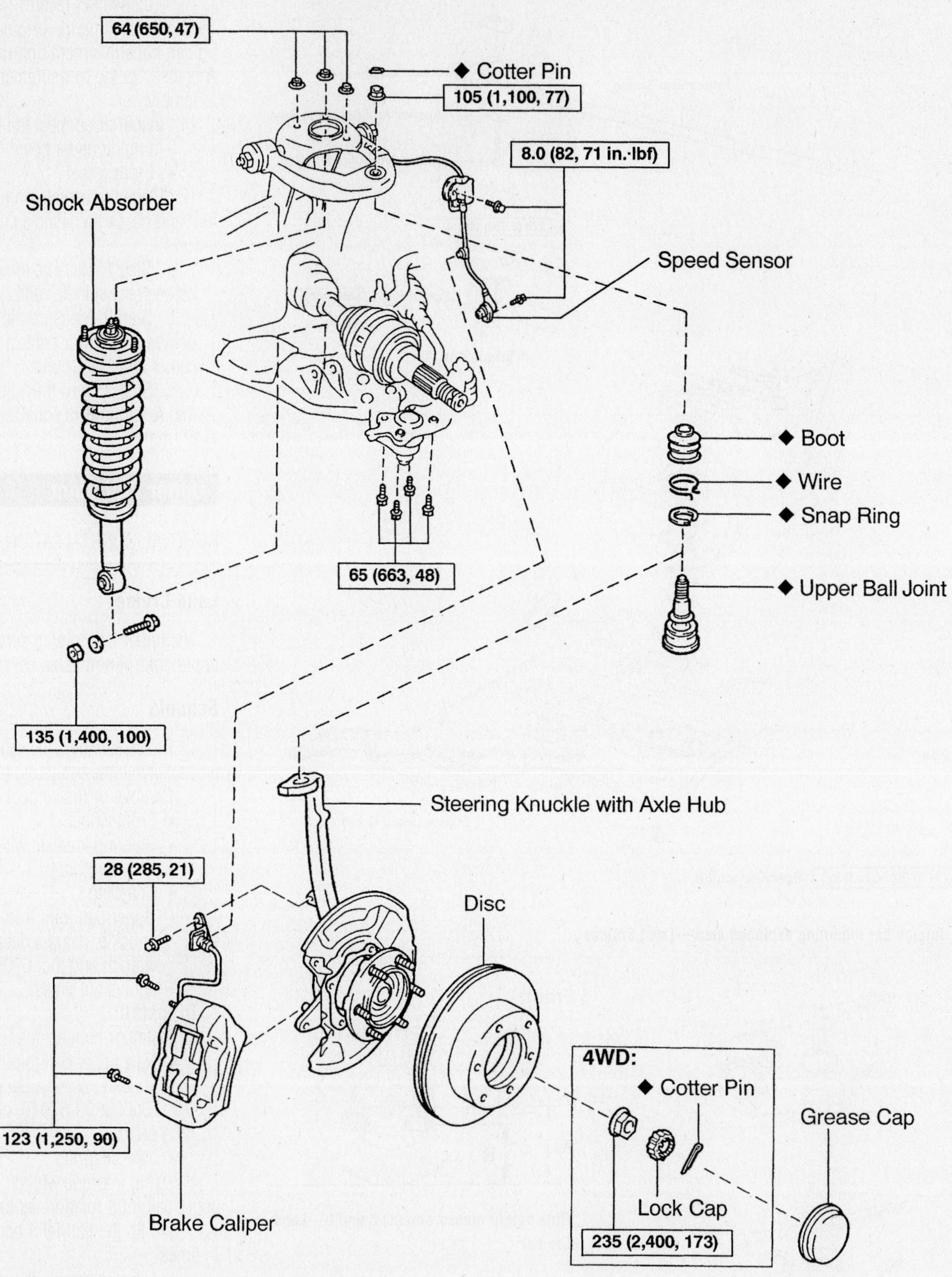

Upper ball joint and related parts—Sequoia

6. Check and/or adjust the front wheel alignment.

Lower Ball Joint

REMOVAL & INSTALLATION

Land Cruiser

The lower ball joint is serviced with the lower control arm as an assembly.

Sequoia

1. Before servicing the vehicle, refer to the precautions in the beginning of this section.
2. Remove or disconnect the following:
   - Front wheel
   - 4 lower ball joint set bolts
   - Tie-rod end from the lower ball joint, using tool 09610-20012
   - Lower ball joint nut.
   - Lower ball joint from the lower control arm, using tool 09628-62011

To install:

3. Install or connect the following:
   - New lower ball joint to the lower control arm stud. Torque the bolts to 117 ft. lbs. (159 Nm).
   - New cotter pin
   - Tie-rod end to the lower ball joint. Torque the nut to 67 ft. lbs. (91 Nm).
   - Lower ball joint set bolts. Torque the 4 bolts to 59 ft. lbs. (80 Nm) for 2001–03; 48 ft. lbs. (65 Nm) for 2004.
   - Front wheel.
4. Check and/or adjust the front wheel alignment.

Upper Control Arm

REMOVAL & INSTALLATION

Land Cruiser

1. Before servicing the vehicle, refer to the precautions in the beginning of this section.
2. Remove or disconnect the following:
   - Front wheel
   - Inner fender liner
   - Wheel speed sensor harness
   - Upper ball joint
   - Adjustment cam bolts
   - Upper control arm

To install:

3. Install or connect the following:
   - Upper control arm. Tighten the adjustment cam bolts to 72 ft. lbs. (98 Nm).
   - Upper ball joint. Tighten the nut to 81 ft. lbs. (110 Nm).
   - Wheel speed sensor harness. Tighten the bolts to 10 ft. lbs. (13 Nm).
   - Inner fender liner
   - Front wheel
4. Check the wheel alignment and adjust as necessary.

Sequoia

1. Before servicing the vehicle, refer to the precautions in the beginning of this section.
2. Remove or disconnect the following:
   - Front wheel
   - Strut
   - Wheel speed sensor harness, if equipped with Anti-lock Brake System (ABS)
3. Upper ball joint, as follows:
   - Cotter pin and loosen the nut

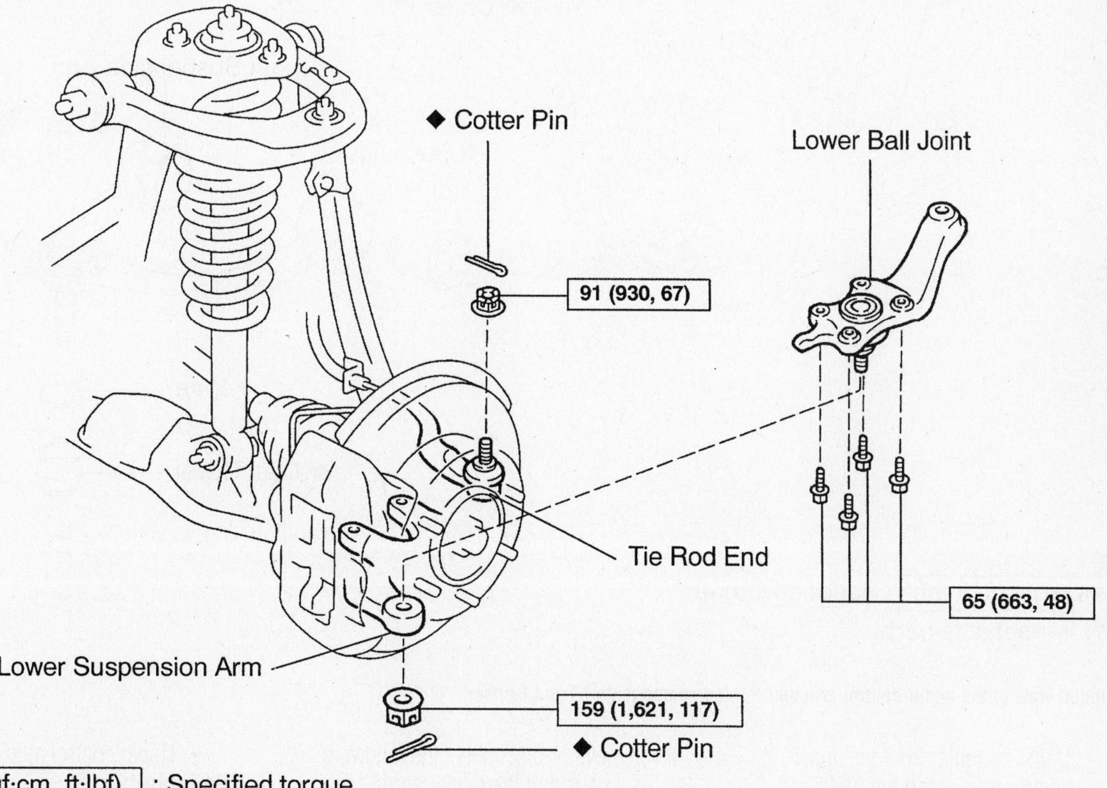

Lower ball joint installation—2004 Sequoia

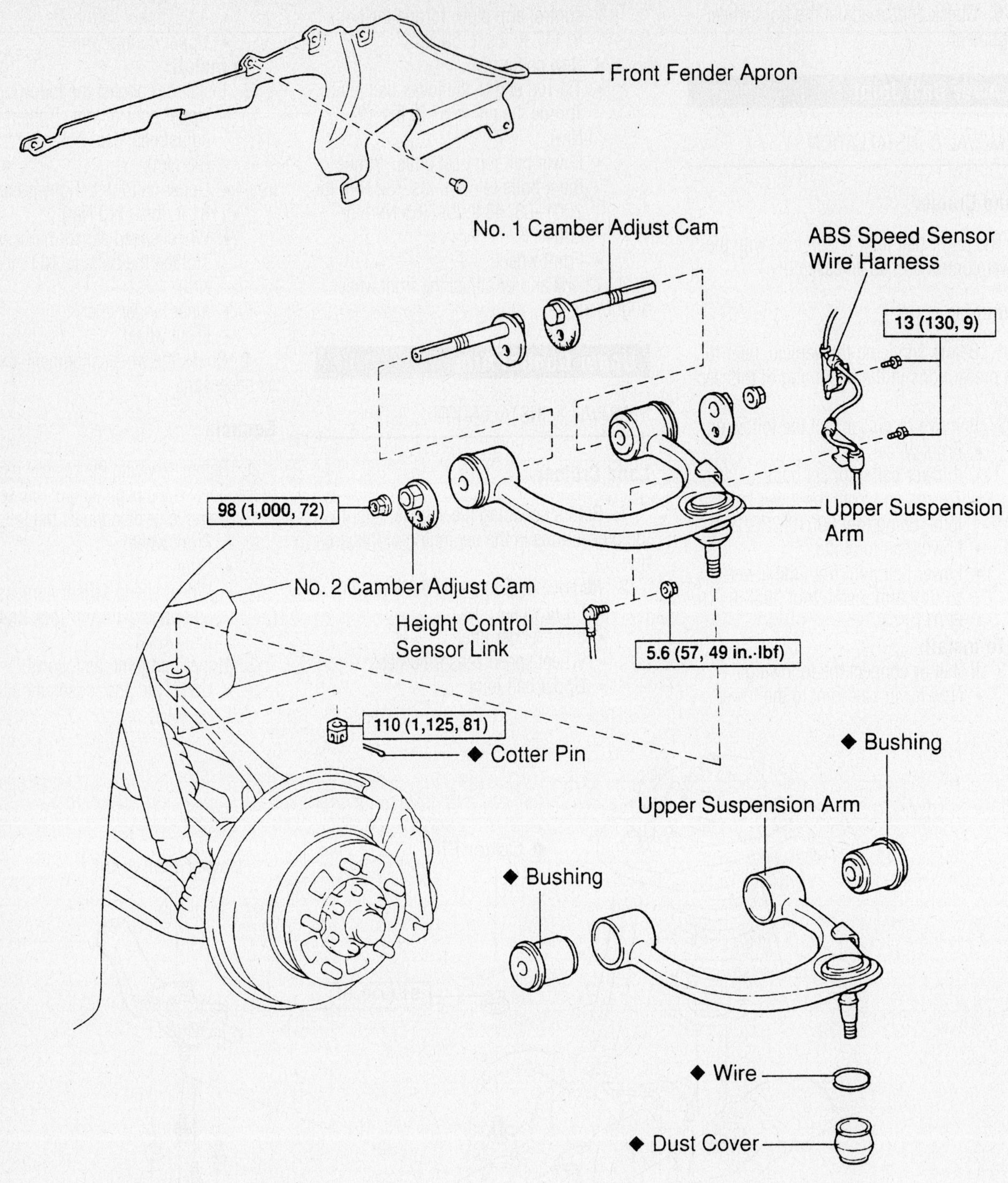

Exploded view of the upper control arm and related components—Land Cruiser

- Upper ball joint from the upper control arm, using tool 09950-40011
- Steering knuckle, support it securely
- Upper ball joint nut

4. Remove or disconnect the following:
   - 4 clips and the fender apron seal
   - Brake/fuel line clamp nut and clamp
   - Both upper control arm-to-chassis nuts/bolts
   - Upper control arm

To install:

5. Install or connect the following:
   - Upper control arm. Torque both upper control arm-to-chassis nuts/bolts to 72 ft. lbs. (98 Nm).

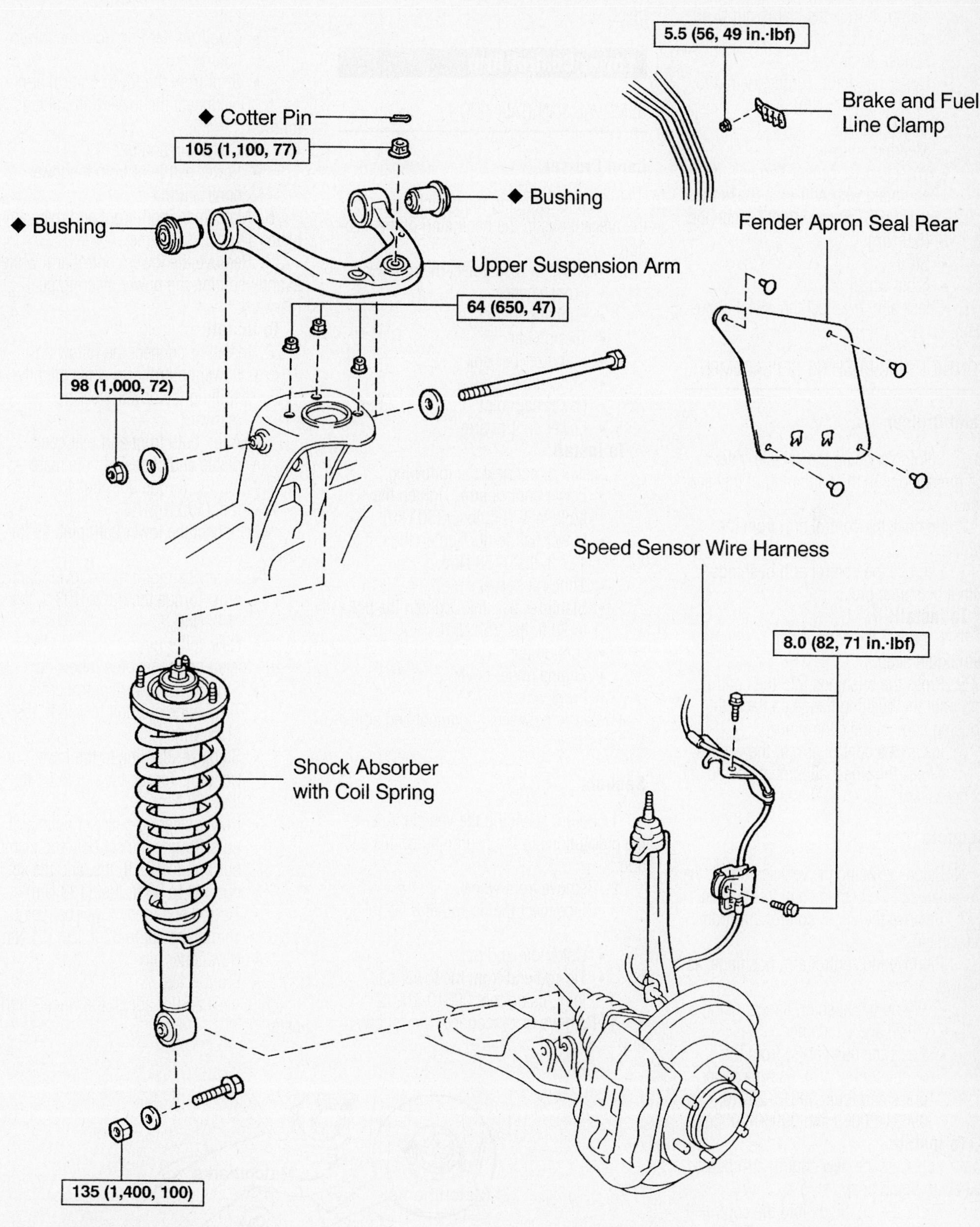

Upper control arm and related parts—Sequoia

- Brake/fuel line clamp nut and clamp. Torque the clamp nut to 49 inch lbs. (5.5 Nm).
- Fender apron seal
- Upper ball joint. Torque the nut to 77 ft. lbs. (105 Nm).
- New cotter pin
- Steering knuckle
- Wheel speed sensor harness, if equipped with Anti-lock Brake System (ABS). Torque it to 71 inch lbs. (8.0 Nm).
- Strut
- Front wheel

6. Check and/or adjust the wheel alignment.

CONTROL ARM BUSHING REPLACEMENT

Land Cruiser

1. Before servicing the vehicle, refer to the precautions in the beginning of this section.
2. Remove the control arm from the vehicle.
3. Remove the control arm bushings with a hydraulic press.

To install:

4. Lubricate the control arm bushings with liquid soap.
5. Press the bushings into the control arm until the bushing flange contacts the housing edge of the control arm.
6. Install the control arm to the vehicle.
7. Check the wheel alignment and adjust as necessary.

Sequoia

1. Before servicing the vehicle, refer to the precautions in the beginning of this section.
2. Remove the upper control arm from the vehicle.
3. Remove the control arm bushings, as follows:
   - Pry up the bushing flange, using a chisel and a hammer
   - Press the bushing(s) from the upper control arm, using a shop press and tools 09613-26010, 09631-20060 and 09950-00020

To install:

4. Lubricate the new control arm bushings with liquid soap.
5. Press the bushings into the control arm until the bushing flange contacts the housing edge of the control arm, using a shop press, a steel plate and tools 09631-12090 and 09710-30021
6. Install the upper control arm to the vehicle.
7. Check and/or adjust the wheel alignment.

Lower Control Arm

REMOVAL & INSTALLATION

Land Cruiser

1. Before servicing the vehicle, refer to the precautions in the beginning of this section.
2. Remove or disconnect the following:
   - Front wheel
   - Engine under cover
   - Torsion bar
   - Stabilizer bar link
   - Shock absorber
   - Lower ball joint
   - Lower control arm

To install:

3. Install or connect the following:
   - Lower control arm. Tighten the bolts to 170 ft. lbs. (230 Nm).
   - Lower ball joint. Tighten the nut to 117 ft. lbs. (159 Nm).
   - Shock absorber
   - Stabilizer bar link. Tighten the bolt to 38 ft. lbs. (52 Nm).
   - Torsion bar
   - Engine under cover
   - Front wheel
4. Check the wheel alignment and adjust as necessary.

Sequoia

1. Before servicing the vehicle, refer to the precautions in the beginning of this section.
2. Remove front wheel.
3. Disconnect the tie-rod end, as follows:
   - Cotter pin and nut
   - Tie-rod end from the lower ball joint, using tool 09610-20012
4. Remove or disconnect the following:
   - Power steering gear set bolts and nuts
   - Stabilizer bar link from the lower control arm
   - Strut from the lower control arm
5. Disconnect the lower ball joint, as follows:
   - Cotter pin and nut
   - Lower ball joint from the lower control arm
6. Matchmark both front and rear cam plates and chassis frame.
7. Remove the lower control arm while slightly shifting the power steering gear rearward.

To install:

8. Install or connect the following:
   - Lower control arm while slightly shifting the power steering gear rearward
   - Align both front and rear cam plates and chassis frame matchmarks. Torque both bolts to 96 ft. lbs. (130 Nm).
9. Connect the lower ball joint, as follows:
   - Lower ball joint to the lower control arm. Torque the nut to 103 ft. lbs. (140 Nm).
   - New cotter pin
10. Install or connect the following:
    - Strut to the lower control arm. Torque the nut/bolt to 100 ft. lbs. (135 Nm).
    - Stabilizer bar link to the lower control arm. Torque the nut to 51 ft. lbs. (69 Nm).
    - Power steering gear set bolts and nuts. Torque the set bolt and clamp nut/bolt to 122 ft. lbs. and the set nut/bolt to 96 ft. lbs. (130 Nm)
    - Tie-rod end to the lower ball joint. Torque the nut to 67 ft. lbs. (91 Nm).
    - New cotter pin
    - Front wheel.
11. Check and/or adjust the wheel alignment.

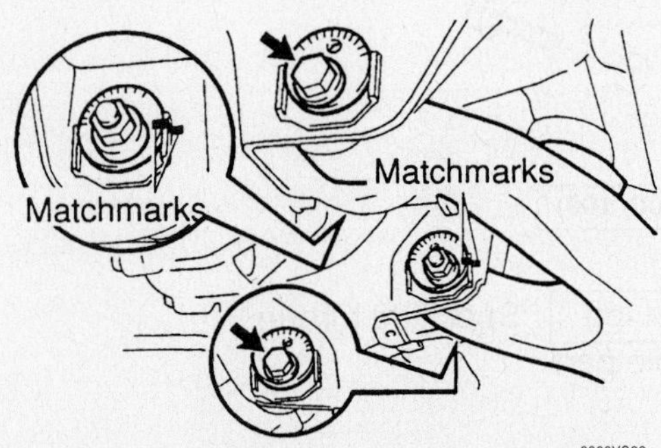

View of the lower control arm's cam plate alignment—Sequoia

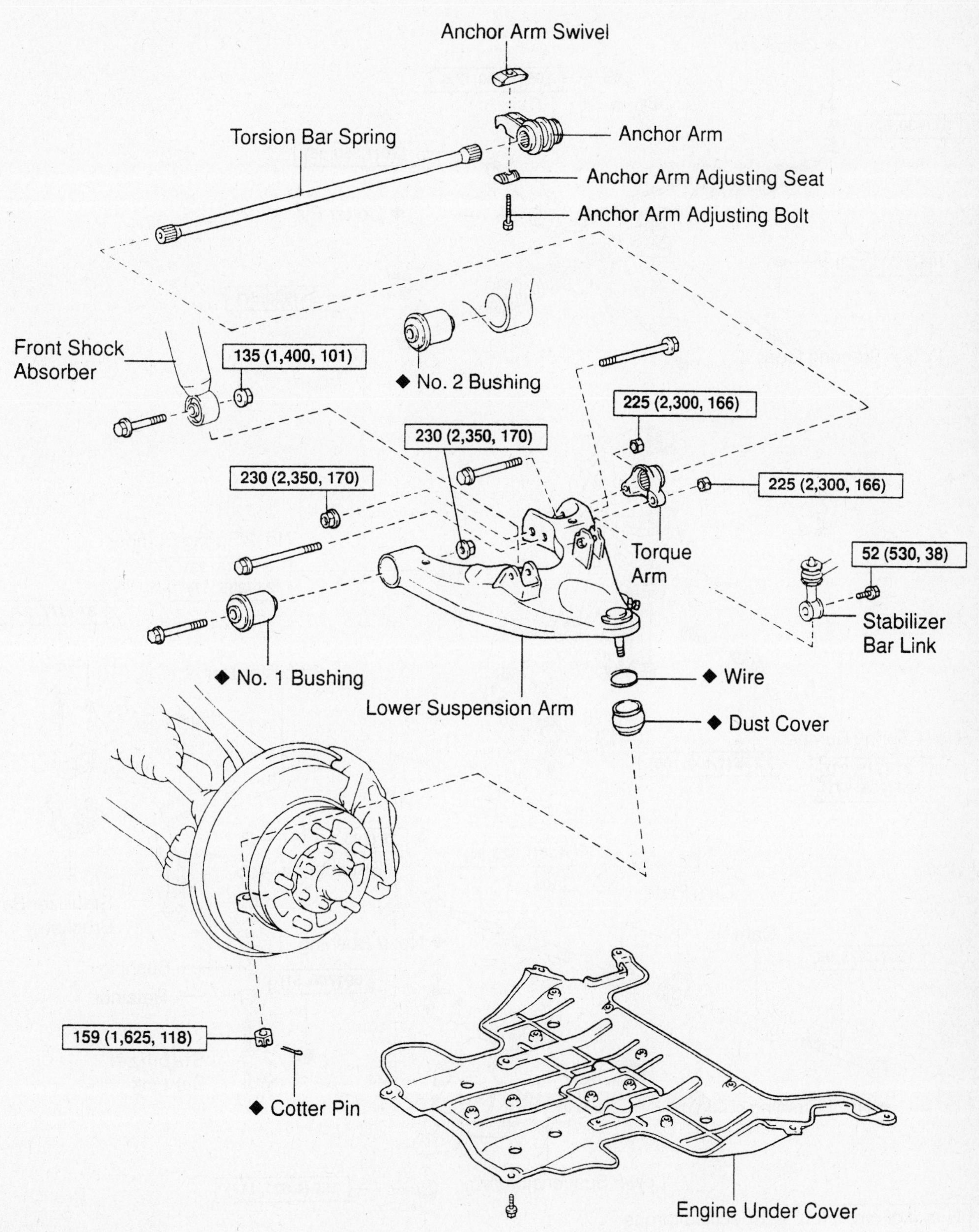

Exploded view of the lower control arm and related components—Land Cruiser

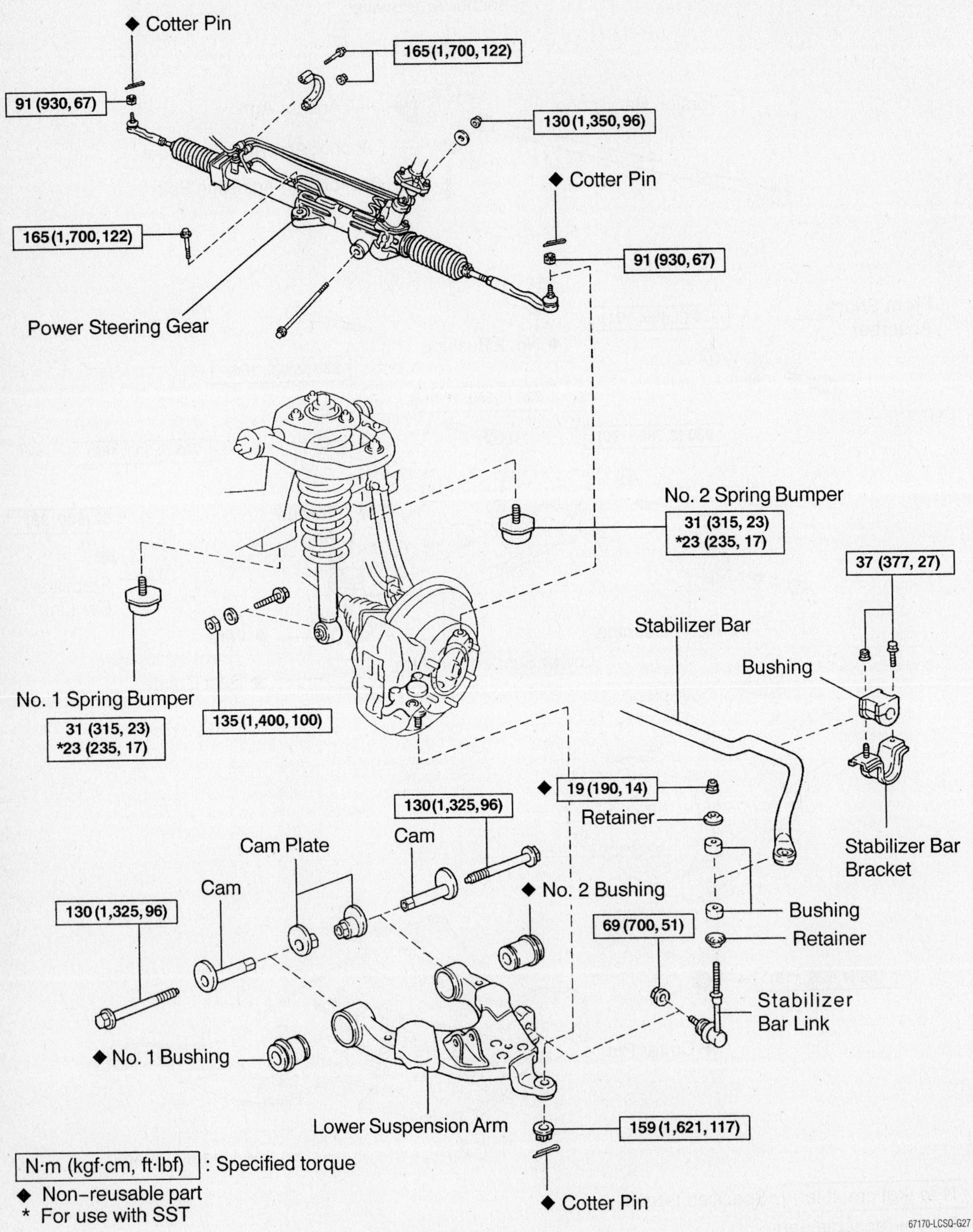
◆ Cotter Pin
165 (1,700, 122)
91 (930, 67)
130 (1,350, 96)
◆ Cotter Pin
165 (1,700, 122)
91 (930, 67)
Power Steering Gear
No. 2 Spring Bumper
31 (315, 23)
*23 (235, 17)
37 (377, 27)
Stabilizer Bar
Bushing
No. 1 Spring Bumper
31 (315, 23)
*23 (235, 17)
135 (1,400, 100)
◆ 19 (190, 14)
Retainer
130 (1,325, 96)
Cam
Cam Plate
Stabilizer Bar
Bracket
Cam
◆ No. 2 Bushing
Bushing
130 (1,325, 96)
69 (700, 51)
Retainer
Stabilizer
Bar Link
◆ No. 1 Bushing
Lower Suspension Arm
159 (1,621, 117)
N·m (kgf·cm, ft·lbf) : Specified torque
◆ Non-reusable part
* For use with SST
◆ Cotter Pin
67170-LCSQ-G27

Lower control arm and related parts—Sequoia

CONTROL ARM BUSHING REPLACEMENT

Land Cruiser

1. Before servicing the vehicle, refer to the precautions in the beginning of this section.
2. Remove the control arm from the vehicle.
3. Remove the control arm bushings with a hydraulic press.

To install:

4. Lubricate the control arm bushings with liquid soap.
5. Press the bushings into the control arm until the bushing flange contacts the housing edge of the control arm.
6. Install the control arm to the vehicle.
7. Check the wheel alignment and adjust as necessary.

Sequoia

1. Before servicing the vehicle, refer to the precautions in the beginning of this section.
2. Remove the lower control arm from the vehicle.
3. Remove the control arm bushings, as follows:
   - Pry up the bushing flange, using a chisel and a hammer
   - Press the bushing(s) from the upper control arm, using a shop press and tools 09613-26010, 09632-36010 and 09950-00020

To install:

4. Lubricate the new control arm bushings with liquid soap.
5. Press the No. 1 bushing into the control arm until the bushing flange contacts the housing edge of the control arm, using a shop press, a steel plate and tools 09631-12090 and 09502-12010, facing the correct direction.
6. Press the No. 2 bushing into the control arm until the bushing flange contacts the housing edge of the control arm, using a shop press, a steel plate and tools 09631-12090 and 09950-60020, facing the correct direction.
7. Install the lower control arm to the vehicle.
8. Check and/or adjust the wheel alignment.

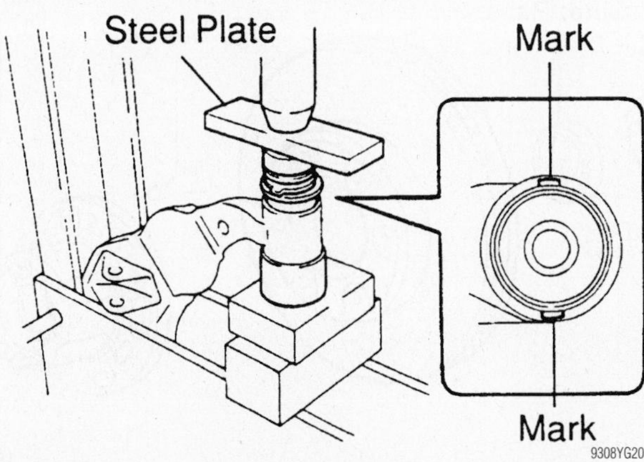

View of the No. 1 bushing's installed direction—Sequoia

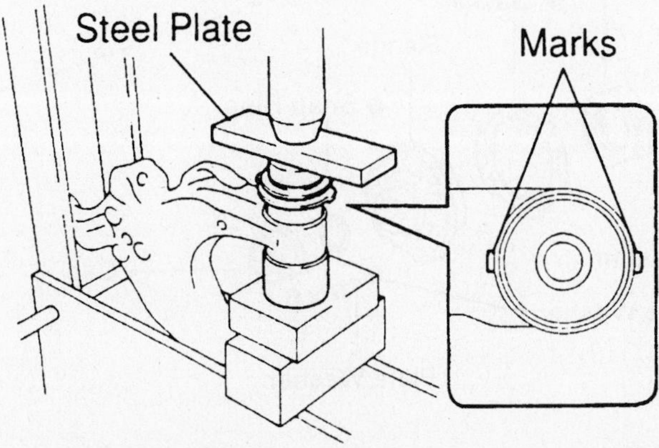

View of the No. 2 bushing's installed direction—Sequoia

Front Wheel Bearing

ADJUSTMENT

Land Cruiser

1. Before servicing the vehicle, refer to the precautions in the beginning of this section.
2. Remove or disconnect the following:
   - Front wheel
   - Brake caliper
   - Grease cap
   - Snapring
   - Hub drive flange
   - Locknut
   - Lockwasher
3. Tighten the adjusting nut to 43 ft. lbs. (59 Nm) while rotating the hub to seat the bearings.
4. Loosen the adjusting nut.
5. Tighten the adjusting nut to 48 inch lbs. (5.4 Nm) and check that the bearing has no play.
6. Check the bearing preload with a spring tension gauge. The preload should be 6.4–12.6 lbs. (28–56 N) for 2001–03; 9.5–15 lbs. (42–67 N) for 2004.
7. Install or connect the following:
   - Lockwasher
   - Locknut. Tighten the nut to 47 ft. lbs. (64 Nm). Recheck the preload.
   - Hub drive flange. Tighten the nuts to 26 ft. lbs. (35 Nm) for 2001–03; 24 ft. lbs. (33 Nm) for 2004.
   - Snapring. Pull out on the axle shaft and select a snapring that ensures that the clearance between the tip of the flange and the snapring is less than 0.008 inch (0.2mm).
   - Grease cap
   - Brake caliper
   - Front wheel

Sequoia

The wheel bearings are sealed unit; no adjustment is possible.

REMOVAL & INSTALLATION

Land Cruiser

1. Before servicing the vehicle, refer to the precautions in the beginning of this section.
2. Remove or disconnect the following:
   - Front wheel
   - Brake caliper
   - Grease cap
   - Snapring
   - Hub drive flange

- Locknut
- Lockwasher
- Adjusting nut
- Outer bearing
- Wheel hub
- Inner grease seal
- Inner bearing

To install:

3. Install or connect the following:
   - Inner bearing
   - Inner grease seal
   - Wheel hub
   - Outer bearing
   - Adjusting nut. Adjust the wheel bearings.
   - Lockwasher
   - Locknut. Tighten the nut to 47 ft. lbs. (64 Nm).
   - Hub drive flange. Tighten the nuts to 26 ft. lbs. (35 Nm) for 2001–03; 24 ft. lbs. (33 Nm) for 2004.
   - Snapring
   - Grease cap
   - Brake caliper
   - Front wheel

Sequoia

1. Before servicing the vehicle, refer to the precautions in the beginning of this section.
2. Remove or disconnect the following:
   - Front wheel
   - Grease cap
   - With 4wd, cotter pin, lock cap and halfshaft nut
   - Speed sensor wire from the knuckle
   - Caliper and rotor. Support the caliper out of the way.

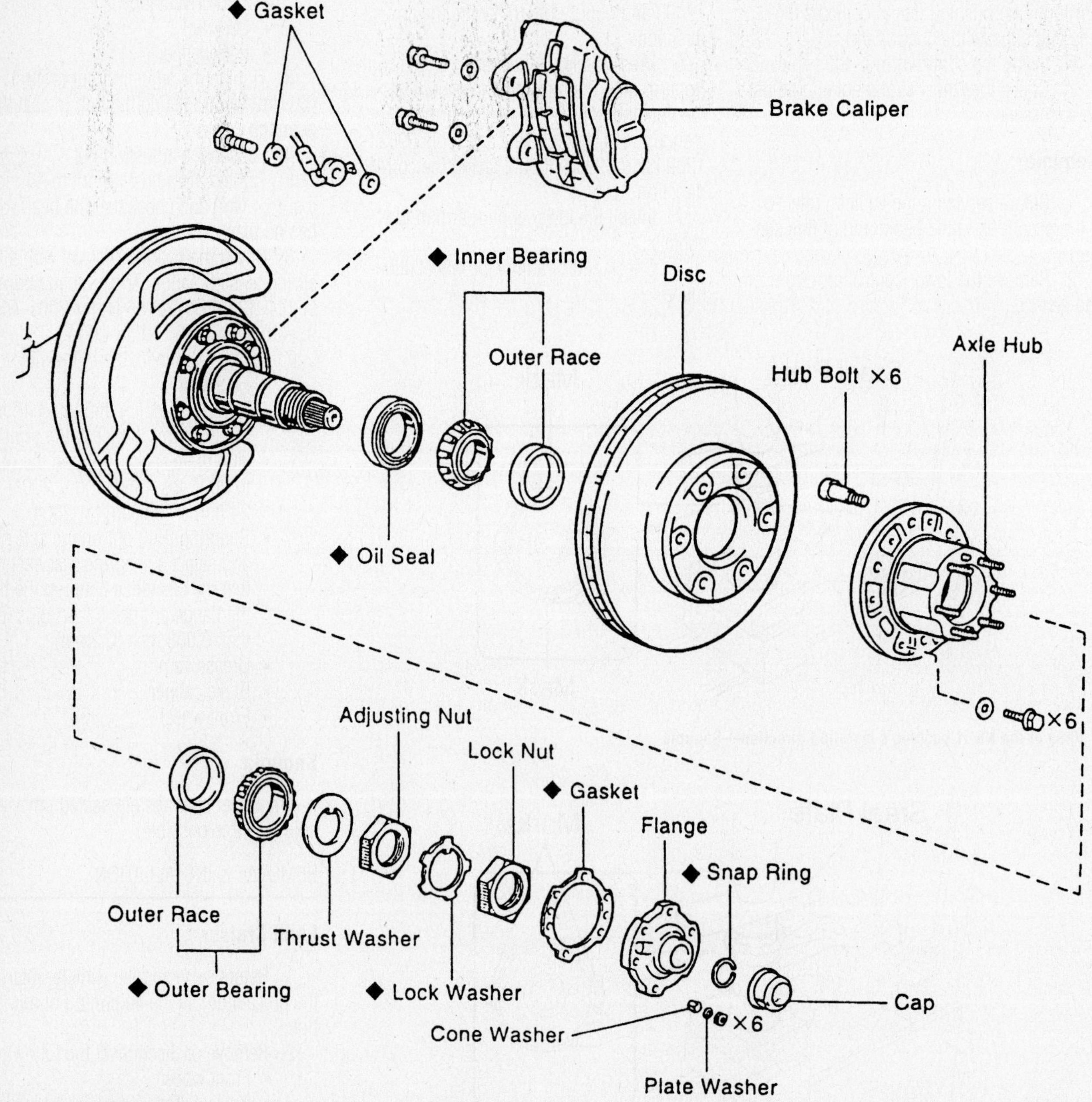

Exploded view of the front hub and related components—2001 Land Cruiser

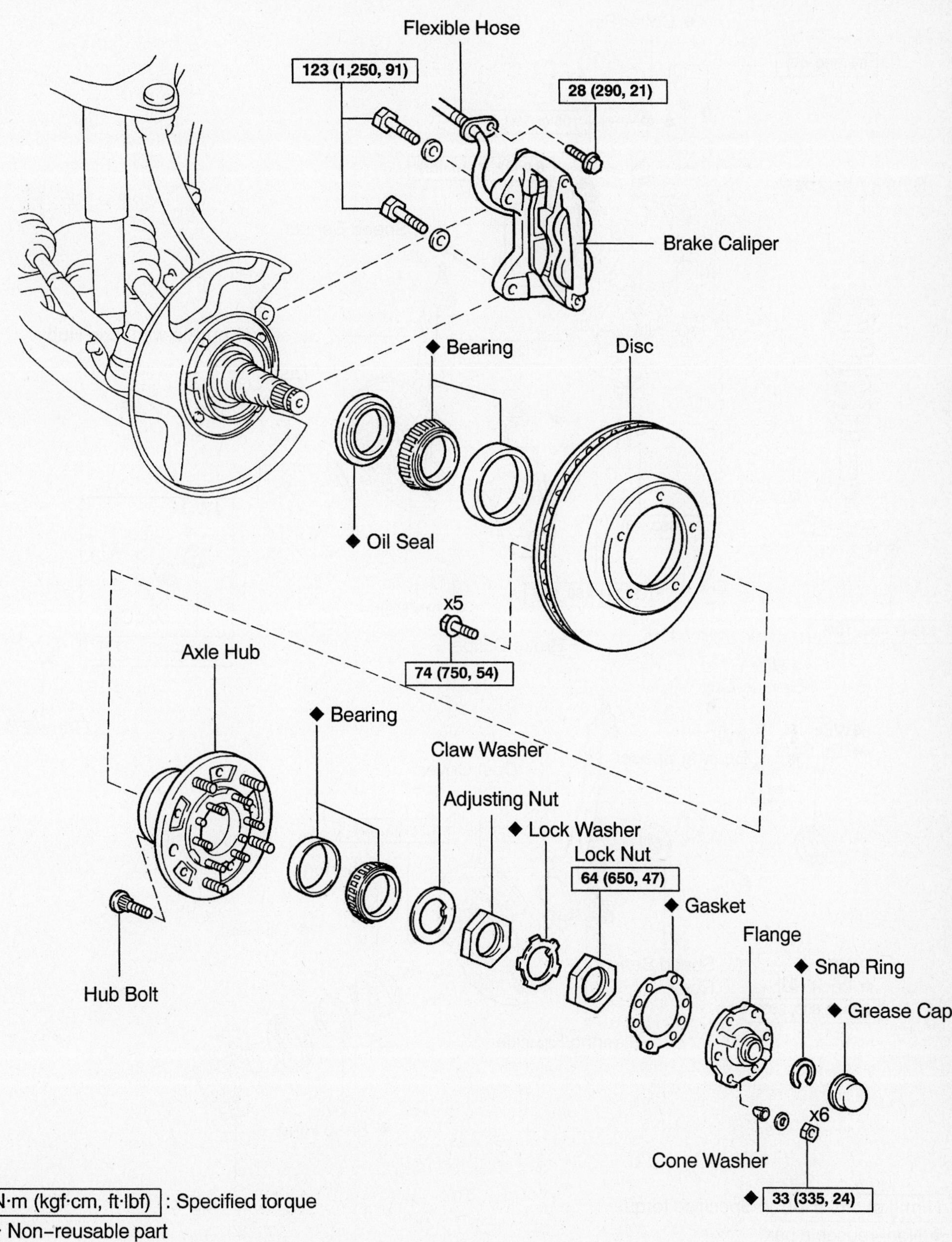

Exploded view of the front hub and related components—2004 Land Cruiser

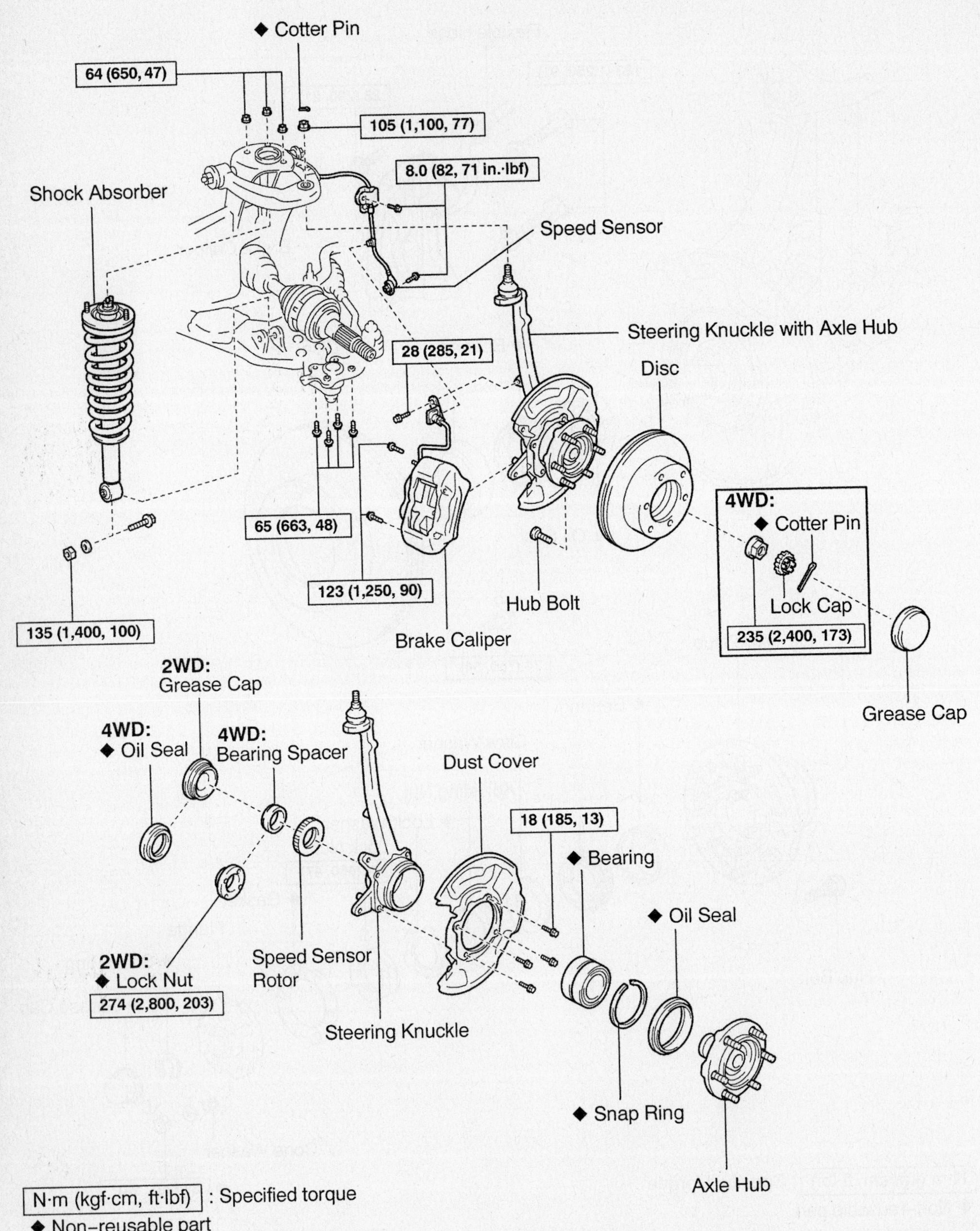
◆ Cotter Pin
64 (650, 47)
105 (1,100, 77)
8.0 (82, 71 in.·lbf)
Shock Absorber
Speed Sensor
Steering Knuckle with Axle Hub
28 (285, 21)
Disc
65 (663, 48)
4WD:
◆ Cotter Pin
123 (1,250, 90)
Hub Bolt
Lock Cap
135 (1,400, 100)
235 (2,400, 173)
Brake Caliper
2WD:
Grease Cap
Grease Cap
4WD:
◆ Oil Seal
4WD:
Bearing Spacer
Dust Cover
18 (185, 13)
◆ Bearing
◆ Oil Seal
2WD:
◆ Lock Nut
274 (2,800, 203)
Speed Sensor
Rotor
Steering Knuckle
◆ Snap Ring
Axle Hub
N·m (kgf·cm, ft·lbf) : Specified torque
◆ Non-reusable part
67170-LCSQ-G29

Front hub and related parts—Sequoia

- Lower ball joint
- Axle hub/steering knuckle assembly and place it in a vise
- With 2wd, the grease cap
- Inner grease seal, for 4WD

3. Remove the 4 bolts and shift the dust cover towards the hub side.
4. Using a suitable puller, remove the hub from the knuckle.
5. Remove the dust cover from the steering knuckle.
6. With 4wd, remove the bearing spacer and Anti-lock Brake System (ABS) speed sensor, if equipped with ABS

**** WARNING**

Be careful not to scratch the speed sensor rotor serrations.

7. Remove the outside oil seal from steering knuckle, using a small prybar
8. Remove the bearing from the steering knuckle, as follows:
   - Snapring
   - Bearing from the steering knuckle, using tools 09950-60020 and 09950-70010, or equivalent.

To install:

9. Install the bearing to the steering knuckle, as follows:
   - Bearing to the steering knuckle, using a press and tools 09950-60020 and 09527-17011, or equivalent
   - New snapring
10. Install the new outside oil seal to steering knuckle, using a plastic hammer and tools 09223-15030 and 09527-17011. Coat the seal lip with MP grease.
11. Install the axle hub to the steering knuckle, as follows:
   - Dust cover to the steering knuckle. Torque the 4 bolts to 13 ft. lbs. (18 Nm).
   - Axle hub to the steering knuckle, using a shop press and tool 09649-17010, or equivalent.
12. Install or connect the following:
   - Bearing spacer and Anti-lock Brake System (ABS) speed sensor, if equipped with ABS

**** WARNING**

Be careful not to scratch the speed sensor rotor serrations.

   - Bearing spacer, if not equipped with ABS, using a shop press and tools 09950-60010 and 09950-70010
   - With 2wd, install a new locknut. Torque to 203 ft. lbs. (274 Nm). Stake the nut.
   - With 4wd, the bearing spacer using a press
   - Grease cap, for 2WD
   - Inner grease seal, for 4WD, using a plastic hammer and tool 09527-17011
13. Install the hub/steering knuckle assembly. With 4wd, install the halfshaft and install the nut loosely.
14. Connect the knuckle to the upper arm. Torque the nut to 77 ft. lbs. (105 Nm). The nut can be tightened up to an additional 60 degrees to align the hole. Install a new cotter pin.
15. Connect the lower ball joint. Torque the 4 bolts to 48 ft. lbs. (65 Nm).
16. Install the shock absorber.
17. Install the rotor and caliper.
18. Attach the brake line to the knuckle.
19. Connect the speed sensor.
20. With 4wd, apply the brakes and tighten the shaft nut to 173 ft. lbs. (235 Nm). The nut can be tightened up to an additional 60 degrees to align the hole. Install a new cotter pin.
21. Install the grease cap.
22. Install the wheel.
23. Check the alignment.

BRAKES

Front Brake Caliper

REMOVAL & INSTALLATION

Land Cruiser

1. Disconnect the negative battery cable from the battery.
2. Raise and support the vehicle safely.
3. Remove the wheels.
4. Disconnect the brake hose from the caliper by removing the union bolt and 2 gaskets. Plug the end of the hose to prevent loss of fluid.
5. Remove the 2 caliper mounting bolts.
6. Lift the bottom of the caliper up and remove the caliper assembly.

To install:

7. Grease the caliper slides and bolts with lithium grease or equivalent. Install the caliper and secure with the bolts. Torque the bolts to 90 ft. lbs. (123 Nm).
8. Connect the brake hose to the caliper, using 2 new washers. Make sure the flexible hose lock is securely in the lock hole of the caliper. Torque the union bolt to 22 ft. lbs. (30 Nm).

◆ Gasket
◆ Gasket
Anti-Rattle Spring
Bleeder Plug
Brake Caliper
Anti-Squeal Shim
Inner Anti-Squeal Shim
Inner Pad
Clip
Outer Pad
Pin
Piston Seal
Piston
Boot
Set Ring
◆ Non-reusable part
Lithium soap base glycol grease
Disc Brake Grease

93026G75

Exploded view of the front disc brake components—2001 Land Cruiser

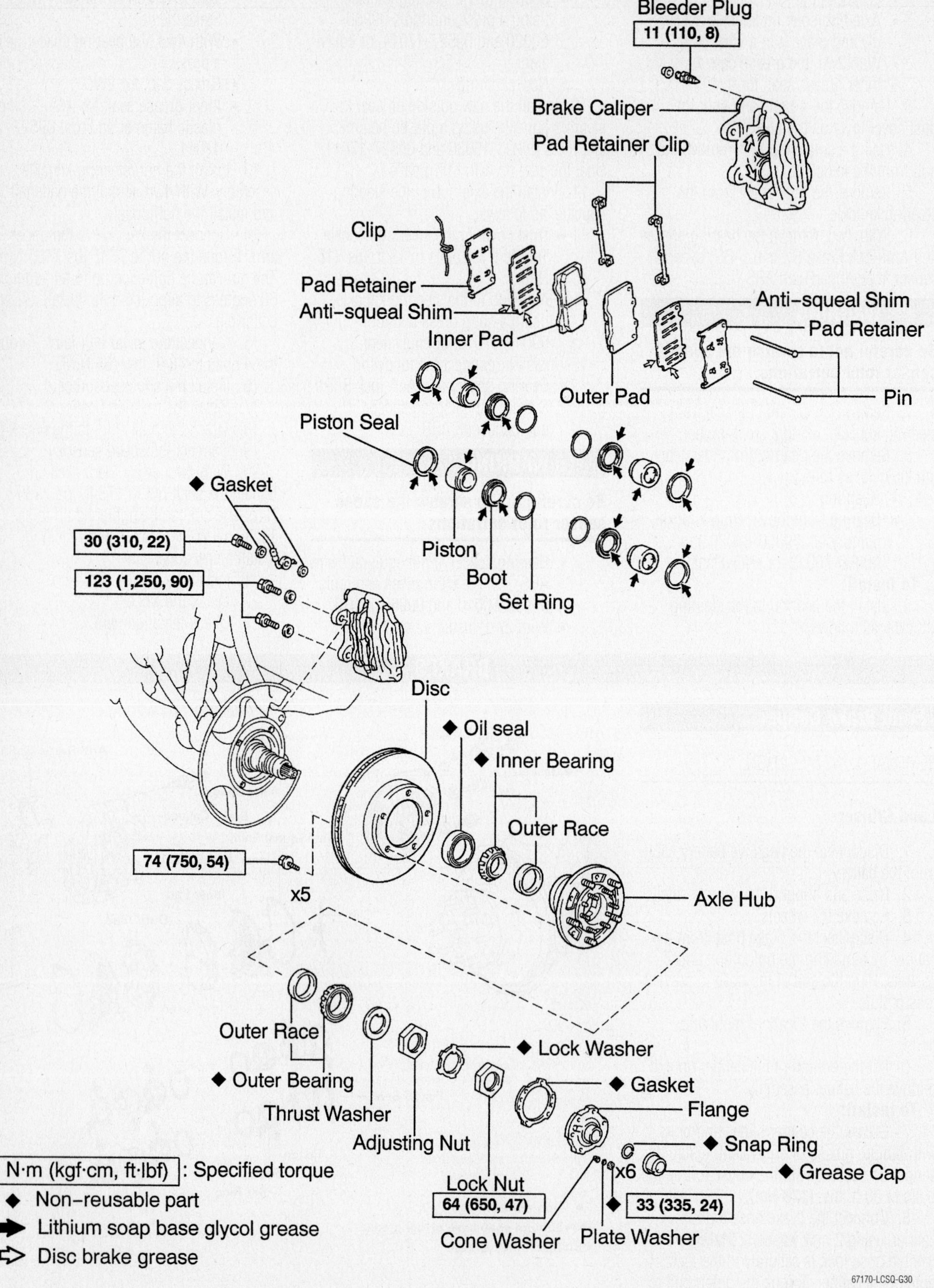

Exploded view of the front disc brake components—2004 Land Cruiser

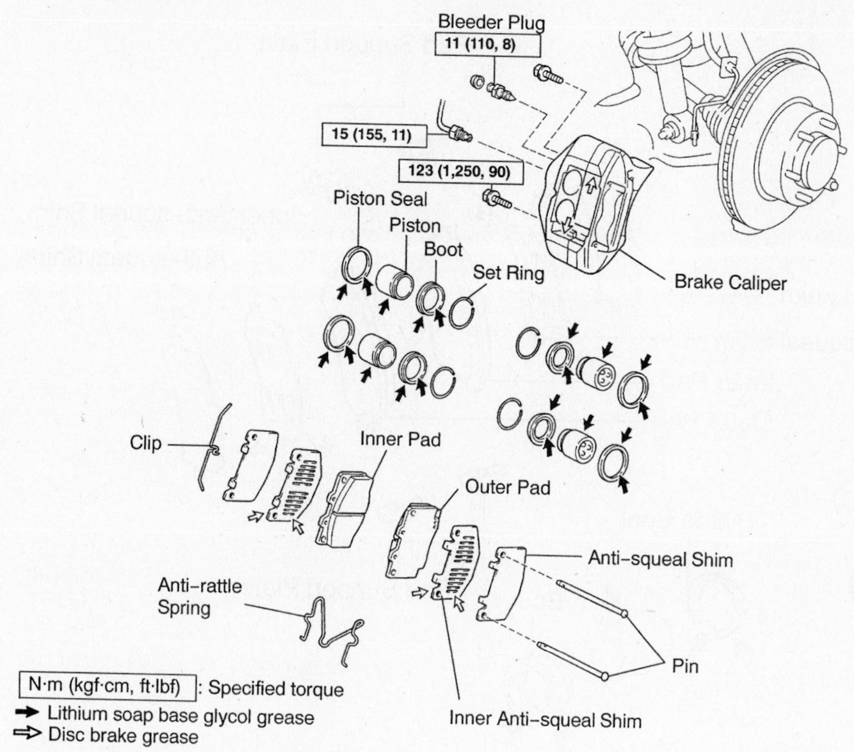

Front caliper and related parts—Sequoia

9. Fill the brake system to the proper level and bleed the brake system.
10. Install the tire and wheel assembly.
11. Top off the brake fluid level in the master cylinder. Check for leaks and proper brake operation.
12. Connect the negative battery cable to the battery.

Sequoia

1. Disconnect the negative battery cable from the battery.
2. Raise and support the vehicle safely.
3. Remove the wheels.
4. Disconnect the brake hose from the caliper. Plug the end of the hose to prevent loss of fluid.
5. Remove the bolts that attach the caliper to the torque plate.
6. Lift the bottom of the caliper up and remove the caliper assembly.

To install:

7. Grease the caliper slides and bolts with lithium grease or equivalent. Install the caliper and secure with the bolts. Torque the bolts to 90 ft. lbs. (123 Nm).
8. Connect the brake hose to the caliper. Torque 11 ft. lbs. (15 Nm).
9. Fill the brake system to the proper level and bleed the brake system.
10. Install the tire and wheel assembly.
11. Top off the brake fluid level in the master cylinder. Check for leaks and proper brake operation.
12. Connect the negative battery cable to the battery.

Rear Brake Caliper

REMOVAL & INSTALLATION

Land Cruiser

1. Remove the brake line from the caliper.
2. Hold the siding pin and remove the 2 bolts.

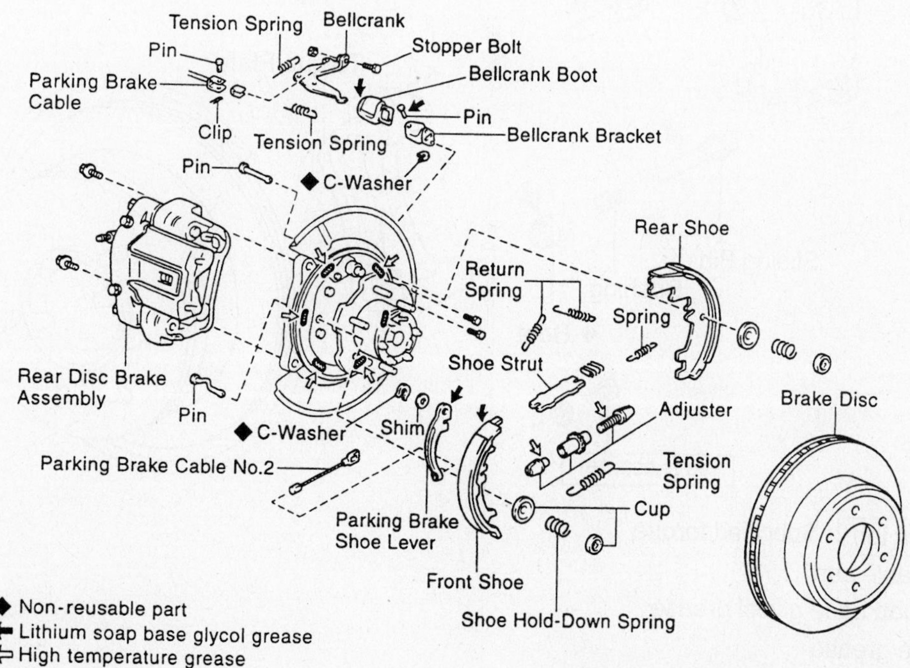

Rear disc brake components—2001 Land Cruiser

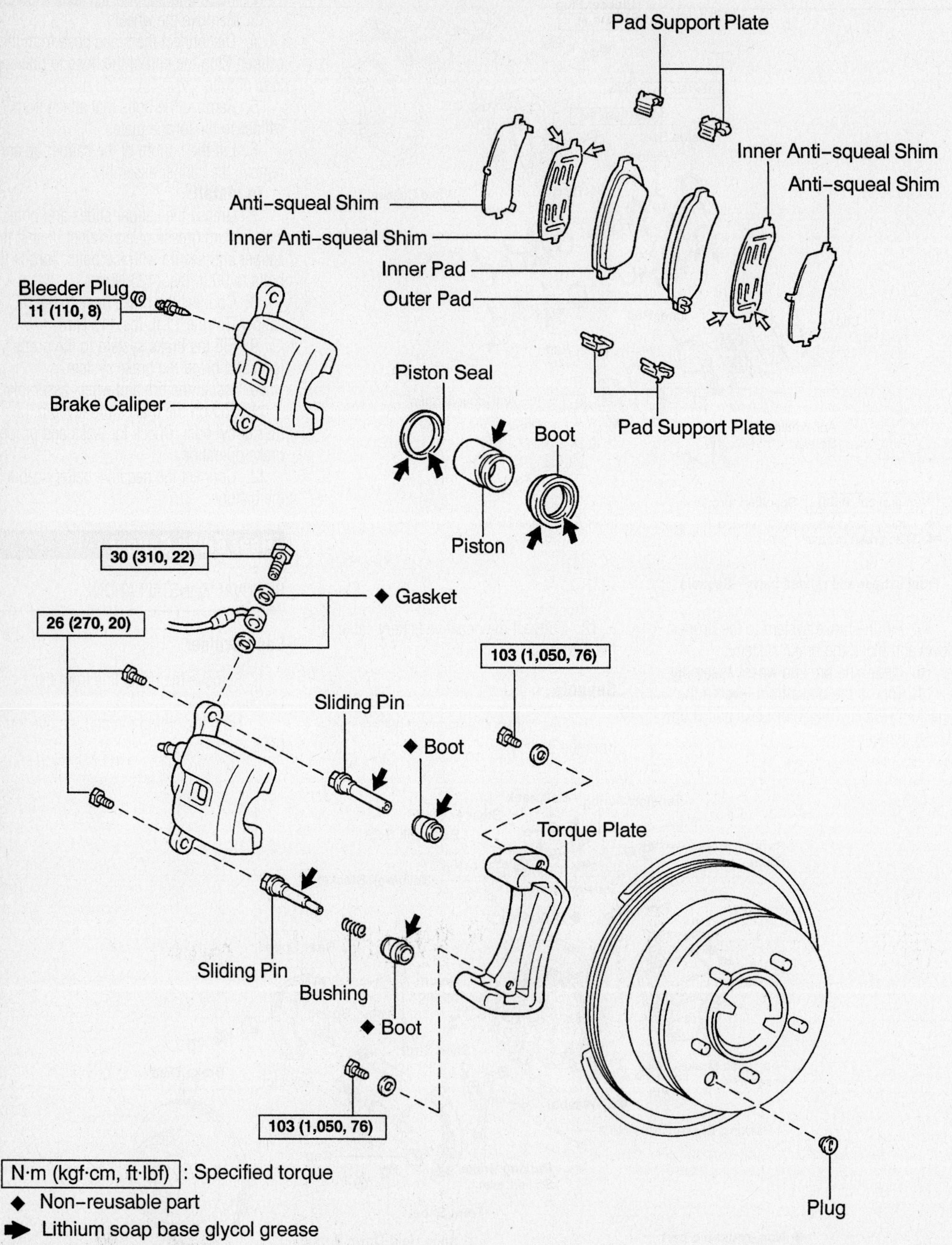

Rear brake caliper and related parts—2004 Land Cruiser

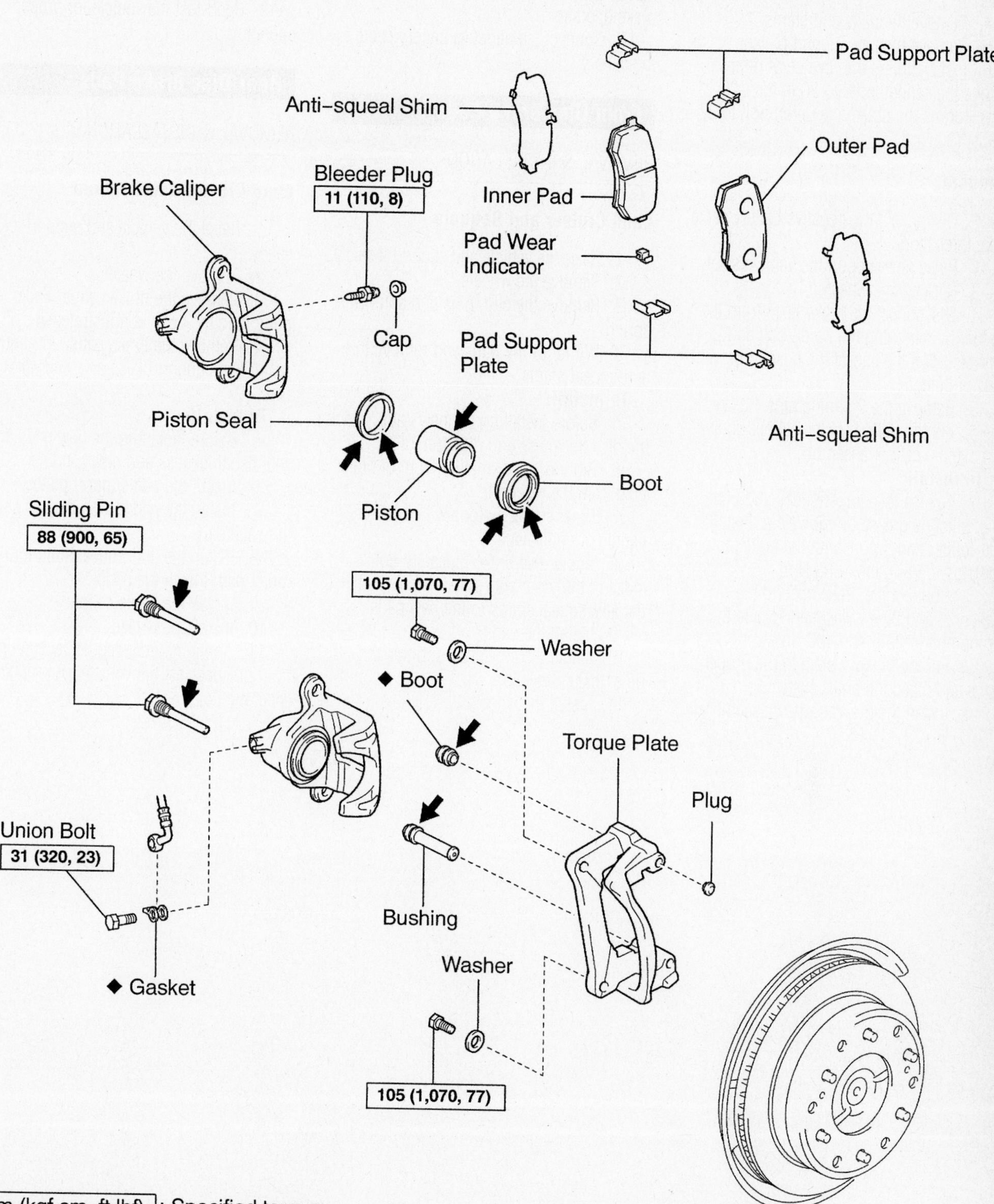
Pad Support Plate
Anti–squeal Shim
Outer Pad
Brake Caliper
Bleeder Plug
11 (110, 8)
Inner Pad
Pad Wear
Indicator
Cap
Pad Support
Plate
Piston Seal
Anti–squeal Shim
Boot
Sliding Pin
88 (900, 65)
Piston
105 (1,070, 77)
Washer
◆ Boot
Torque Plate
Plug
Union Bolt
31 (320, 23)
Bushing
Washer
◆ Gasket
105 (1,070, 77)
N·m (kgf·cm, ft·lbf) : Specified torque
◆ Non–reusable part
Lithium soap base glycol grease
67170-LCSQ-G33

Rear brake caliper—Sequoia

3. Remove the caliper from the torque plate.
4. Remove the pads and shims.
5. Remove the pad support plates.
6. Installation is the reverse of removal. Torque the caliper bolts to 20 ft. lbs. (26 Nm). Torque the brake line union bolt to 22 ft. lbs. (30 Nm).

Sequoia

1. Disconnect the negative battery cable from the battery.
2. Raise and support the vehicle safely.
3. Remove the wheels.
4. Disconnect the brake hose from the caliper by removing the union bolt and 2 gaskets. Plug the end of the hose to prevent loss of fluid.
5. Remove the 2 sliding pins.
6. Lift the bottom of the caliper up and remove the caliper assembly.

To install:

7. Grease the caliper slides and pins with silicone grease or equivalent. Install the caliper and secure with the bolts. Torque the pins to 65 ft. lbs. (883 Nm).
8. Connect the brake hose to the caliper, using 2 new washers. Torque the union bolt to 22 ft. lbs. (30 Nm).
9. Fill the brake system to the proper level and bleed the brake system.
10. Install the tire and wheel assembly.
11. Top off the brake fluid level in the master cylinder. Check for leaks and proper brake operation.
12. Connect the negative battery cable to the battery.

Front Disc Brake Pads

REMOVAL & INSTALLATION

Land Cruiser and Sequoia

1. Raise the vehicle and support it safely.
2. Remove the wheels.
3. Remove the clip, pins and anti-rattle spring.
4. Withdraw the pads and remove the anti-squeal shims.

To install:

5. Before installing the new pads, check the disc thickness and disc runout.
6. Siphon out a small amount of brake fluid from the reservoir.
7. Press in the pistons with a hammer handle or equivalent.
8. Apply disc brake grease to both sides of the inner anti-squeal shim. Install the anti-squeal shims to the new pads.
9. Install the pads.
10. Install the anti-rattle springs and pins. Install the clip.
11. Install the wheels.
12. Check and adjust the fluid level. Apply the brake pedal several times.
13. Road-test the vehicle for proper operation.

Rear Disc Brake Pads

REMOVAL & INSTALLATION

Land Cruiser and Sequoia

1. Raise the vehicle and support it safely.
2. Remove the wheels.
3. Remove the brake caliper and suspend it so the hose is not stretched.
4. Remove the brake pads, anti-squeal shim, pad support plates and wear indicators.

To install:

5. Before installing the new pads, check the disc thickness and disc runout.
6. Install the pad support plates.
7. Install the pad wear indicator plates on each pad.
8. Install the anti-squeal shim to the outer pad. Install the pads.
9. Install the brake caliper.
10. Install the wheels.
11. Apply the brake pedal several times.
12. Road-test the vehicle for proper operation.

TOYOTA

Matrix

20

SPECIFICATION AND MAINTENANCE CHARTS

ENGINE AND VEHICLE IDENTIFICATION

Engine						
Code ①	Liters (cc)	Cu. In.	Cyl.	Fuel Sys.	Engine Type	Eng. Mfg.
1ZZ-FE	1.8 (1794)	109	4	EFI	DOHC	Toyota
2ZZ-GE	1.8 (1796)	109.5	4	EFI	DOHC	Toyota

Model Year	
Code ②	Year
3	2003
4	2004
5	2005

EFI: Electronic Fuel Injection

DOHC: Double Overhead Camshaft

① 8th digit of VIN

② 10th digit of VIN

67170-TMAT-C01

GENERAL ENGINE SPECIFICATIONS

Year	Model	Engine Displacement Liters (VIN)	Net Horsepower @ rpm	Net Torque @ rpm (ft. lbs.)	Bore x Stroke (in.)	Com-pression Ratio	Oil Pressure @ idle
2003	Matrix	1.8 (1ZZ-FE)	①	②	3.11x3.60	10.0:1	27
	Matrix	1.8 (2ZZ-GE)	180@7600	130@6800	3.23x3.35	11.5:1	27
2004-05	Matrix	1.8 (1ZZ-FE)	①	②	3.11x3.60	10.0:1	27
	Matrix	1.8 (2ZZ-GE)	180@7600	130@6800	3.23x3.35	11.5:1	27

EFI: Electronic Fuel Injection

① 2WD models: 130@6000
4WD models: 123@6000

② 2WD models: 126@4200
4WD models: 119@4200

67170-TMAT-C02

ENGINE TUNE-UP SPECIFICATIONS

Year	Engine Displacement Liters (VIN)	Spark Plug Gap (in.)	Ignition Timing (deg.)	Fuel Pump (psi)	Idle Speed (rpm) MT	Idle Speed (rpm) AT	Valve Clearance In.	Valve Clearance Ex.
2003	1.8 (1ZZ-FE)	0.043	①	44-50	650-750	650-750	0.0059-0.0098	0.0098-0.0138
	1.8 (2ZZ-GE)	0.043	②	44-50	750-850	700-800	0.0031-0.0071	0.0087-0.0126
2004-05	1.8 (1ZZ-FE)	0.043	①	44-50	650-750	650-750	0.0059-0.0098	0.0098-0.0138
	1.8 (2ZZ-GE)	0.043	②	44-50	750-850	700-800	0.0031-0.0071	0.0087-0.0126

Note: The Vehicle Emission Control Information label often reflects specification changes made during production. The label figures must be used if they differ from those in this chart.

① With terminal TC and CG of DLC3 connected: 8-12 degrees BTDC
With terminal TC and CG of DLC3 disconnected: 10-18 degrees BTDC

② With terminal TC and CG of DLC3 connected: 8-12 degrees BTDC
With terminal TC and CG of DLC3 disconnected:
A/T: 10-18 degrees BTDC
M/T: 4-12 degrees BTDC

67170-TMAT-C03

CAPACITIES

Year	Model	Engine Displacement Liters (VIN)	Engine Oil with Filter	Transmission (pts.) 5-Spd	6-Spd	Auto.	Drive Axle Front (pts.)	Rear (pts.)	Fuel Tank (gal.)	Cooling System (qts.)
2003	Matrix	1.8 (1ZZ-FE)	3.9	4.0	4.0	5.2	NA	1.04	①	6.9
	Matrix	1.8 (2ZZ-GE)	4.7	4.8	4.8	6.1	NA	—	13.2	7.1
2004-05	Matrix	1.8 (1ZZ-FE)	3.9	4.0	4.0	5.2	NA	1.04	①	6.9
	Matrix	1.8 (2ZZ-GE)	4.7	4.8	4.8	6.1	NA	—	13.2	7.1

Note: All capacities are approximate. Add fluid gradually and check to be sure a proper fluid level is obtained.

NA - Not available

① 2WD: 13.2 gallons
4WD: 11.9 gallons

67170-TMAT-C04

VALVE SPECIFICATIONS

Year	Engine Displacement Liters (VIN)	Seat Angle (deg.)	Face Angle (deg.)	Spring Test Pressure (lbs. @ in.)	Spring Installed Height (in.)	Stem-to-Guide Clearance (in.) Intake	Exhaust	Stem Diameter (in.) Intake	Exhaust
2003	1.8 (1ZZ-FE)	45	44.5	31.3-34.8@ 1.252	1.323	0.0010-0.0024	0.0012-0.0026	0.2154-0.2159	0.2152-0.2158
	1.8 (2ZZ-GE)	45	44.5	①	1.516	0.0010-0.0023	0.0012-0.0025	0.2145-0.2156	0.2144-0.2154
2004-05	1.8 (1ZZ-FE)	45	44.5	31.3-34.8@ 1.252	1.323	0.0010-0.0024	0.0012-0.0026	0.2154-0.2159	0.2152-0.2158
	1.8 (2ZZ-GE)	45	44.5	①	1.516	0.0010-0.0023	0.0012-0.0025	0.2145-0.2156	0.2144-0.2154

① Intake: 49.6-55.5@1.516
Exhaust: 47.6-52.6@1.516

67170-TMAT-C05

CRANKSHAFT AND CONNECTING ROD SPECIFICATIONS

All measurements are given in inches.

Year	Engine Displacement Liters (cc)	Engine ID/VIN	Crankshaft Main Brg. Journal Dia.	Main Brg. Oil Clearance	Shaft End-play	Thrust on No.	Connecting Rod Journal Diameter	Oil Clearance	Side Clearance
2003	1.8 (1762)	1ZZ-FE	1.8893-1.8898	0.0006-0.0013	0.0008-0.0087	3	1.7320-1.7323	0.0011-0.0024	0.0063-0.0135
	1.8 (1796)	2ZZ-GE	1.8893-1.8898	0.0006-0.0013	0.0016-0.0094	3	1.7713-1.7717	0.0011-0.0020	0.0063-0.0135
2004-05	1.8 (1762)	1ZZ-FE	1.8893-1.8898	0.0006-0.0013	0.0008-0.0087	3	1.7320-1.7323	0.0011-0.0024	0.0063-0.0135
	1.8 (1796)	2ZZ-GE	1.8893-1.8898	0.0006-0.0013	0.0016-0.0094	3	1.7713-1.7717	0.0011-0.0020	0.0063-0.0135

67170-TMAT-C06

PISTON AND RING SPECIFICATIONS

All measurements are given in inches.

Year	Engine Displacement Liters (cc)	Engine ID/VIN	Piston Clearance	Ring Gap Top Compression	Ring Gap Bottom Compression	Ring Gap Oil Control	Ring Side Clearance Top Compression	Ring Side Clearance Bottom Compression	Ring Side Clearance Oil Control
2003	1.8 (1762)	1ZZ-FE	0.0026-0.0035	0.0098-0.0138	0.0138-0.0197	0.0059-0.0197	0.0008-0.0028	0.0012-0.0028	0.0012-0.0043
	1.8 (1796)	2ZZ-GE	0.0003-0.0015	0.0098-0.0138	0.0138-0.0197	NA	0.0009-0.0028	0.0012-0.0028	NA
2004-05	1.8 (1762)	1ZZ-FE	0.0026-0.0035	0.0098-0.0138	0.0138-0.0197	0.0059-0.0197	0.0008-0.0028	0.0012-0.0028	0.0012-0.0043
	1.8 (1796)	2ZZ-GE	0.0003-0.0015	0.0098-0.0138	0.0138-0.0197	NA	0.0009-0.0028	0.0012-0.0028	NA

NA - Not available

67170-TMAT-C07

TORQUE SPECIFICATIONS

All readings in ft. lbs.

Year	Engine Displacement Liters (VIN)	Cylinder Head Bolts	Main Bearing Bolts	Rod Bearing Bolts	Crankshaft Damper Bolts	Flywheel Bolts	Manifold Intake	Manifold Exhaust	Spark Plugs	Oil Pan Drain Plug
2003	1.8 (1ZZ-FE)	①	②	③	102	①	22	27	18	76
	1.8 (2ZZ-GE)	④	⑤	⑥	87	①	⑦	37	13	76
2004-05	1.8 (1ZZ-FE)	①	②	③	102	①	22	27	18	76
	1.8 (2ZZ-GE)	④	⑤	⑥	87	①	⑦	37	13	76

① Step 1: 36 ft. lbs.
Step 2: 90 degree turn

② 12 pointed bolts:
Step 1: 33 ft. lbs.
Step 2: 90 degree turn
Hex head bolts: 14 ft. lbs.

③ Step 1: 15 ft. lbs.
Step 2: 90 degree turn

④ Step 1: 26 ft. lbs.
Step 2: 180 degree turn

⑤ 12 pointed bolts:
Step 1: 16 ft. lbs.
Step 2: 32 ft. lbs.
Step 3: 45 degree turn
Step 4: 45 degree turn
Hex head bolts: 13 ft. lbs.

⑥ Step 1: 22 ft. lbs.
Step 2: 90 degree turn

⑦ Bolt A: 25 ft. lbs.
Bolt B: 34 ft. lbs.

67170-TMAT-C08

WHEEL ALIGNMENT

Year	Model		Caster Range (+/-Deg.)	Caster Preferred Setting (Deg.)	Camber Range (+/-Deg.)	Camber Preferred Setting (Deg.)	Toe-in (in.)
2003	Matrix - 2WD	F	0.75	+2.78	0.75	-0.77	0+/-0.08
		R	—	—	0.50	-1.45	0.11+/-0.11
	Matrix - 4WD	F	0.75	+2.77	0.75	-0.48	0+/-0.08
		R	—	—	0.75	-0.73	0.08+/-0.08
2004-05	Matrix - 2WD	F	0.75	+2.78	0.75	-0.77	0+/-0.08
		R	—	—	0.50	-1.45	0.11+/-0.11
	Matrix - 4WD	F	0.75	+2.77	0.75	-0.48	0+/-0.08
		R	—	—	0.75	-0.73	0.08+/-0.08

67170-TMAT-C09

TIRE, WHEEL AND BALL JOINT SPECIFICATIONS

Year	Model	OEM Tires Standard	OEM Tires Optional	Tire Pressures (psi) Front	Tire Pressures (psi) Rear	Wheel Size	Ball Joint Inspection	Lug Nuts
2003	Matrix	205/55R16	—	33	33	6.5-JJ	9-26 in. ①	76
2004-05	Matrix	205/55R16	—	33	33	6.5-JJ	9-26 in. ①	76

OEM: Original Equipment Manufacturer

PSI: Pounds Per Square Inch

STD: Standard

OPT: Optional

① Torque required in inch lbs. to rotate ball joint when removed from the knuckle

67170-TMAT-C10

BRAKE SPECIFICATIONS

All measurements in inches unless noted

Year	Model		Brake Disc Original Thickness	Brake Disc Minimum Thickness	Brake Disc Maximum Runout	Brake Drum Diameter Original Inside Diameter	Brake Drum Diameter Max. Wear Limit	Brake Drum Diameter Maximum Machine Diameter	Minimum Lining Thickness	Brake Caliper Bracket Bolts (ft. lbs.)	Brake Caliper Mounting Bolts (ft. lbs.)
2003	Matrix	F	0.984	0.906	0.0020	—	—	—	0.039	79	25
		R	0.354	0.295	0.0059	9.00	—	9.04	0.039	—	34
2004-05	Matrix	F	0.984	0.906	0.0020	—	—	—	0.039	79	25
		R	0.354	0.295	0.0059	9.00	—	9.04	0.039	—	34

F: Front

R: Rear

67170-TMAT-C11

SCHEDULED MAINTENANCE INTERVALS
TOYOTA—MATRIX

TO BE SERVICED	TYPE OF SERVICE	VEHICLE MILEAGE INTERVAL (x1000)												
		7.5	15	22.5	30	37.5	45	52.5	60	67.5	75	82.5	90	97.5
Engine oil & filter	R	✓	✓	✓	✓	✓	✓	✓	✓	✓	✓	✓	✓	✓
Drive belts	S/I								✓	✓	✓	✓	✓	✓
Automatic transaxle fluid & filter	S/I		✓		✓		✓		✓		✓		✓	
Ball joints & dust covers	S/I		✓		✓		✓		✓		✓		✓	
Bolts & nuts on body & chassis	S/I		✓		✓		✓		✓		✓		✓	
Brake line pipes & hoses	S/I		✓		✓		✓		✓		✓		✓	
Brake linings & drums	S/I		✓		✓		✓		✓		✓		✓	
Brake pads & discs (front & rear if equipped)	S/I		✓		✓		✓		✓		✓		✓	
Differential oil	S/I		✓		✓		✓		✓		✓		✓	
Drive shaft boots (except Supra)	S/I		✓		✓		✓		✓		✓		✓	
Manual transaxle oil	S/I		✓		✓		✓		✓		✓		✓	
Steering gear housing oil	S/I		✓		✓		✓		✓		✓		✓	
Steering linkage	S/I		✓		✓		✓		✓		✓		✓	
Air filter	R				✓				✓				✓	
Spark plugs	R				✓				✓				✓	
Spark plugs (platinum tip)	R								✓					
Exhaust system	S/I				✓				✓				✓	
Fuel lines & connections	S/I				✓				✓				✓	
Valve clearance	S/I				✓				✓				✓	
Engine coolant	R						✓				✓			
Fuel tank cap gasket	R								✓					
Charcoal canister	S/I								✓					

R: Replace I: Inspect A: Adjust

FREQUENT OPERATION MAINTENANCE (SEVERE SERVICE)

If a vehicle is operated under any of the following conditions it is considered severe service:

- Extremely dusty areas.

- 50% or more of the vehicle operation is in 32°C (90°F) or higher temperatures, or constant operation in temperatures below 0°C (32°F).

- Prolonged idling (vehicle operation in stop and go traffic).

- Frequent short running periods (engine does not warm to normal operating temperatures).

- Police, taxi, delivery usage or trailer towing usage.

Oil & oil filter: change every 6000 miles.

Bolts & nuts on chassis & body: tighten every 7500 miles.

Ball joints & dust covers: service or inspect every 12,000 miles.

Brake linings & drums: service or inspect ever 12,000 miles.

Brake pads & discs (front & rear if equipped): service or inspect every 12,000 miles.

Drive shaft boots & except Supra): service or inspect every 12,000 miles.

Steering linkage: service or inspect every 12,000 miles.

Air filter: service or inspect every 15,000 miles.

Exhaust system: service or inspect every 15,000 miles.

Timing belt: replace every 60,000 miles.

67170-TMAT-C12

PRECAUTIONS

Before servicing any vehicle, please be sure to read all of the following precautions, which deal with personal safety, prevention of component damage, and important points to take into consideration when servicing a motor vehicle:

- Never open, service or drain the radiator or cooling system when the engine is hot; serious burns can occur from the steam and hot coolant.
- Observe all applicable safety precautions when working around fuel. Whenever servicing the fuel system, always work in a well-ventilated area. Do not allow fuel spray or vapors to come in contact with a spark, open flame or excessive heat (a hot drop light, for example). Keep a dry chemical fire extinguisher near the work area. Always keep fuel in a container specifically designed for fuel storage; also, always properly seal fuel containers to avoid the possibility of fire or explosion. Refer to the additional fuel system precautions later in this section.
- Fuel injection systems often remain pressurized, even after the engine has been turned **OFF**. The fuel system pressure must be relieved before disconnecting any fuel lines. Failure to do so may result in fire and/or personal injury.
- Brake fluid often contains polyglycol ethers and polyglycols. Avoid contact with the eyes and wash your hands thoroughly after handling brake fluid. If you do get brake fluid in your eyes, flush your eyes with clean, running water for 15 minutes. If eye irritation persists, or if you have taken brake fluid internally, IMMEDIATELY seek medical assistance.
- The EPA warns that prolonged contact with used engine oil may cause a number of skin disorders, including cancer! You should make every effort to minimize your exposure to used engine oil. Protective gloves should be worn when changing oil. Wash your hands and any other exposed skin areas as soon as possible after exposure to used engine oil. Soap and water, or waterless hand cleaner should be used.
- All new vehicles are now equipped with an air bag system. The system must be disabled before performing service on or around system components, steering column, instrument panel components, wiring and sensors. Failure to follow safety and disabling procedures could result in accidental air bag deployment, possible personal injury and unnecessary system repairs.
- Always wear safety goggles when working with, or around, the air bag system. When carrying a non-deployed air bag, be sure the bag and trim cover are pointed away from your body. When placing a non-deployed air bag on a work surface, always face the bag and trim cover upward, away from the surface. This will reduce the motion of the module if it is accidentally deployed. Refer to the additional air bag system precautions later in this section.
- Clean, high quality brake fluid from a sealed container is essential to the safe and proper operation of the brake system. You should always buy the correct type of brake fluid for your vehicle. If the brake fluid becomes contaminated, completely flush the system with new fluid. Never reuse any brake fluid. Any brake fluid that is removed from the system should be discarded. Also, do not allow any brake fluid to come in contact with a painted surface; it will damage the paint.
- Never operate the engine without the proper amount and type of engine oil; doing so WILL result in severe engine damage.
- Timing belt maintenance is extremely important! Many models utilize an interference-type, non-freewheeling engine. If the timing belt breaks, the valves in the cylinder head may strike the pistons, causing potentially serious (also time-consuming and expensive) engine damage.
- Disconnecting the negative battery cable on some vehicles may interfere with the functions of the on-board computer system(s) and may require the computer to undergo a relearning process once the negative battery cable is reconnected.
- When servicing drum brakes, only disassemble and assemble one side at a time, leaving the remaining side intact for reference.
- Only an MVAC-trained, EPA-certified automotive technician should service the air conditioning system or its components.

ENGINE REPAIR

Alternator

REMOVAL & INSTALLATION

1. Before servicing the vehicle, refer to the precautions in the beginning of this section.
2. Remove or disconnect the following:
   - Negative battery cable
   - Drive belt
   - Wire clamp from the clip on the rectifier end frame
   - Rubber clamp and nut
   - Alternator wiring and connector
   - Alternator

To install:

3. Install or connect the following:
   - Alternator. Torque the 12mm bolt to 18 ft. lbs. (25 Nm) and the 14mm bolt to 39 ft. lbs. (54 Nm).
   - Alternator connector and wiring
   - Rubber clamp and nut
   - Wire clamp
   - Drive belt
   - Negative battery cable

Ignition Timing

ADJUSTMENT

➡The timing on engines equipped with Distributorless Ignition Systems (DIS) is not adjustable.

Engine Assembly

REMOVAL & INSTALLATION

1. Before servicing the vehicle, refer to the precautions in the beginning of this section.
2. Relieve the fuel system pressure.
3. Drain the cooling system.
4. Drain the engine oil.
5. Drain the transaxle fluid and transfer fluid, if equipped.
6. Remove or disconnect the following:
   - Negative battery cable. Wait at least 90 seconds before proceeding.
   - Battery
   - Hood
   - Undercovers
   - Radiator inlet and outlet hoses
   - Radiator hose outlet
   - Oil cooler inlet and outlet tubes
   - Upper radiator support and radiator, if equipped with A/C
   - Battery
   - Air cleaner assembly
   - Fuel pipe clamp
   - Fuel tube sub-assembly
   - Accelerator control cable

- Cruise control actuator, if equipped
- Union-to-connector tube hose
- Heater inlet and outlet hoses
- Transmission shift cable(s)
- Clutch release cylinder, on manual transaxle
- Glove compartment door
- Engine relay block cover
- 3 connectors from the relay block
- 2 ground cables
- Engine wire from the Engine Control Module (ECM) and junction block
- Engine wire from the cabin
- Drive belt
- Compressor and magnetic clutch, if equipped with A/C. Unbolt and position aside, DO NOT disconnect the lines.
- Vane pump oil reservoir from the bracket
- Return tube
- Right side front door scuff plate
- Right side cowl side trim plate
- Right side rear door scuff plate, AWD
- Lower right side center pillar garnish, AWD
- Right front seat, AWD
- Column hole cover silencer sheet
- Steering intermediate shaft
- Front floor panel brace, FWD
- Center exhaust pipe, AWD
- Propeller shaft with center bearing shaft, AWD
- Front exhaust pipe
- Front hub nuts
- Tie rod ends from the steering knuckles

7. Separate the front stabilizer links and lower control arm ball joints
- Front halfshafts

8. Remove the engine from the vehicle, as follows:

a. Set the engine lifter.

b. Remove the bolts and nuts, then remove the engine mounting insulator.

c. Remove the through bolt and nut, then detach the engine mounting insulator from the vehicle.

d. Remove the 6 bolts as shown.

e. Use a suitable tool to suspend the engine assembly, as shown in the figure.

f. No. 1 engine hanger: P/N 12281-15040 (1ZZ-FE), 12281-88600 (2ZZ-GE).

g. No. 2 engine hanger: P/N 12281-22021 (1ZZ-FE), 12281-88600 (2ZZ-GE).

h. Bolt: P/N 91512-B1016.

i. Torque the bolts to 28 ft. lbs. (38 Nm).

CAUTION

Do not try to suspend the engine by hooking the chain to any other part.

j. Attach an engine chain hoist to the hangers.

k. Using the chain block and sling device, suspend the engine.

l. Remove the engine and transaxle assembly from the vehicle.

9. Remove or disconnect the following components, as necessary:
- Vane pump
- Steering gear
- Crossmember
- Manifold stay
- Oxygen (O_2) sensor
- Exhaust manifold
- Starter
- Transaxle
- Transfer case
- Clutch
- Flywheel
- Alternator
- Ignition coil
- Fuel delivery pipe
- Intake manifold
- Oil level gauge
- Water inlet and bypass pipes
- Thermostat
- Oil pressure switch
- Crankshaft Position (CKP) sensor
- Knock Sensor (KS)
- Drive belt tensioner
- Engine mounts and brackets
- Coolant Temperature Sensor (CTS)

To install:

10. Install any removed components to the engine and transaxle assembly.

11. To install the engine:

a. Place the engine and transaxle on an engine lifter.

b. Install the engine with the transaxle to the vehicle.

c. Temporarily install the crossmember and 6 bolts.

d. Install the left engine mounting insulator. Tighten the bolts to 59 ft. lbs. (80 Nm).

e. Install the right engine mounting insulator. Tighten the bolts to 38 ft. lbs. (52 Nm).

f. Insert SST 09670-00010 to the positioning holes of the right handle crossmember and on the right handle of the vehicle. Temporarily tighten bolt A, then bolt B.

g. Insert SST 09670-00010 to the positioning holes of the left handle crossmember and on the left handle of the vehicle. Temporarily tighten bolt A, then bolt B.

h. Insert the SST to the positioning holes on the right-handle crossmember and right handle. Tighten bolt A to 116 ft.

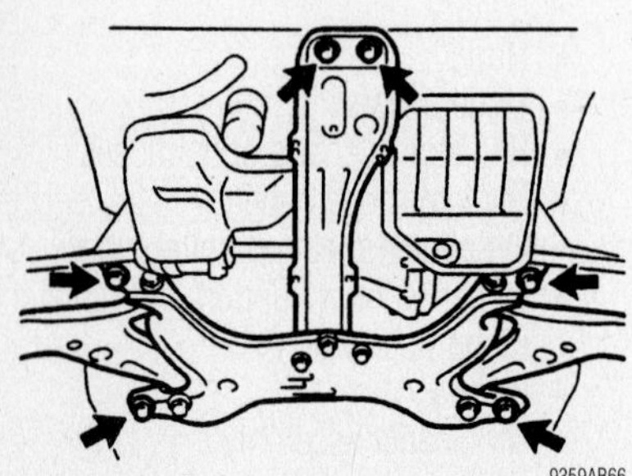

Remove the 6 bolts, as indicated by arrows

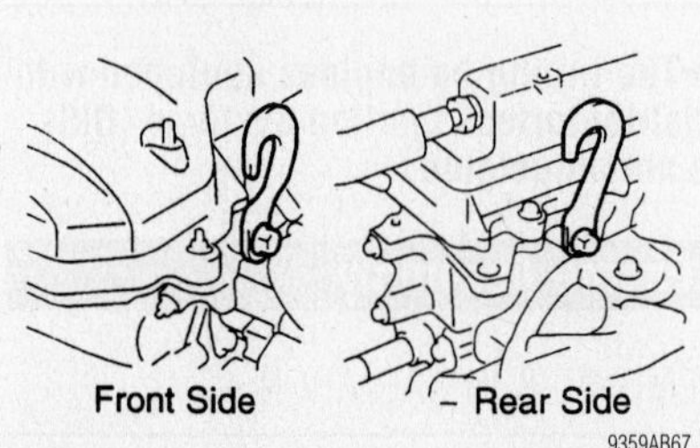

Install the engine hangers—1ZZ-FE shown, 2ZZ-GE similar

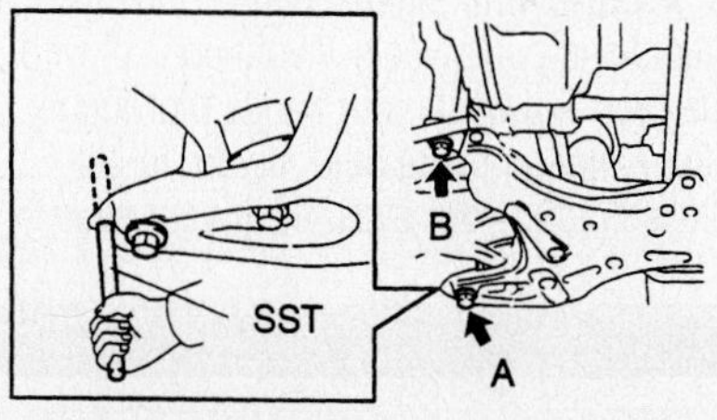

Insert the SST to the positioning holes of the right handle crossmember and on the right handle of the vehicle. Temporarily tighten bolt A, then bolt B

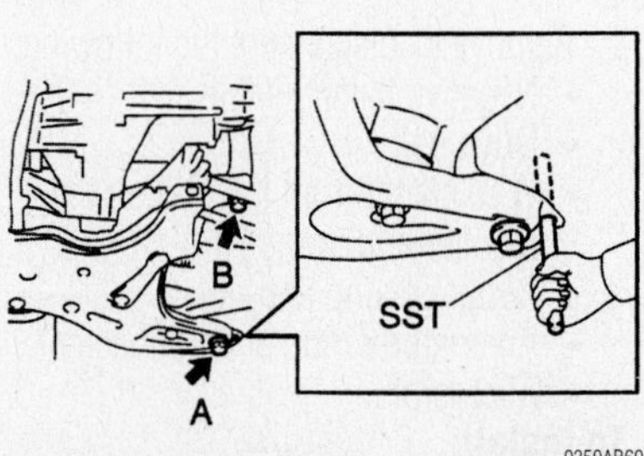

Insert the SST to the positioning holes of the left handle crossmember and on the left handle of the vehicle. Temporarily tighten bolt A, then bolt B

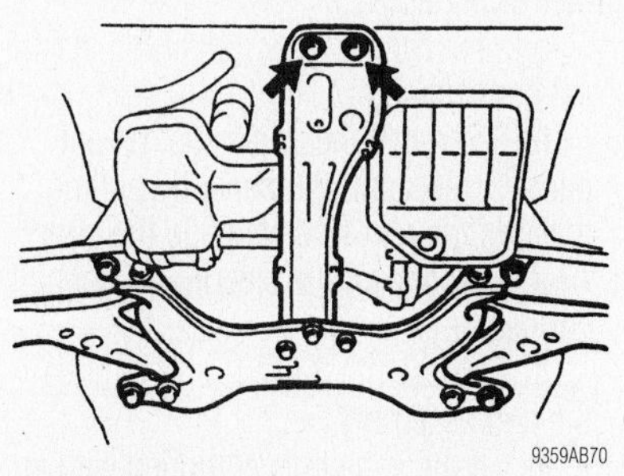
Tighten the 2 crossmember bolts, indicated by arrows

lbs. (157 Nm) and bolt B to 83 ft. lbs. (113 Nm).

i. Insert the SST to the positioning holes on the left-handle crossmember and left handle. Tighten bolt A to 116 ft. lbs. (157 Nm) and bolt B to 83 ft. lbs. (113 Nm).

j. Tighten the 2 crossmember bolts, shown in the figure, to 29 ft. lbs. (39 Nm).

12. Installation of the remaining components is the reverse of the removal procedure.

13. Make sure all fluid levels are accurate, then start the engine check for leaks.

Water Pump

REMOVAL & INSTALLATION

1ZZ-FE Engine

1. Before servicing the vehicle, refer to the precautions in the beginning of this section.
2. Drain the cooling system.
3. Remove or disconnect the following:
   - Negative battery cable
   - Right-hand engine under cover
   - Drive belt
   - Alternator
   - Water pump

To install:

4. Install or connect the following:
   - Water pump. Torque bolts marked **A** (short) to 80 inch lbs. (9 Nm) and bolts marked **B** (long) to 96 inch lbs. (11 Nm).
   - Alternator
   - Drive belt
   - Right engine under cover
   - Negative battery cable
5. Fill the cooling system to the proper level.
6. Start the vehicle, check for leaks and repair if necessary.

Water pump bolt identification—1.8L (1ZZ-FE) engine

2ZZ-GE Engine

1. Before servicing the vehicle, refer to the precautions in the beginning of this section.
2. Drain the cooling system.
3. Remove or disconnect the following:
   - Negative battery cable
   - Right-hand engine under cover
   - Drive belt
   - Alternator
   - Water pump pulley, using SST 09960-10010
   - Water pump and O-ring

To install:

4. Install or connect the following:
   - Water pump with new O-ring. Torque the bolts to 80 inch lbs. (9 Nm).
   - Water pump pulley, using SST 09960-10010. Torque the bolts to 11 ft. lbs. (15 Nm).
   - Alternator
   - Drive belt
   - Right engine under cover
   - Negative battery cable
5. Fill the cooling system to the proper level.
6. Start the vehicle, check for leaks and repair if necessary.

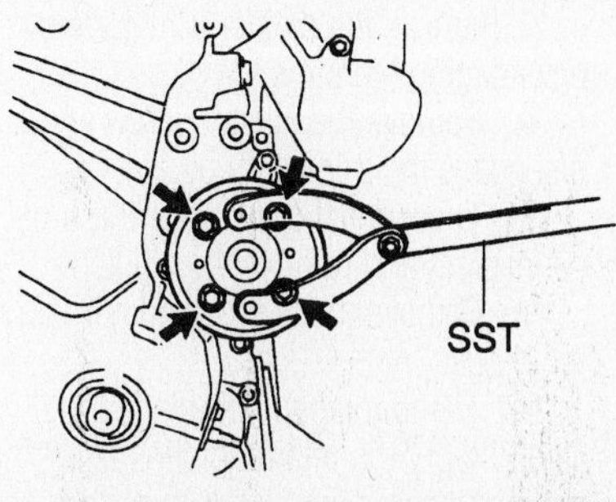

View of the special tool needed to remove and install the water pump pulley—2ZZ-GE engine

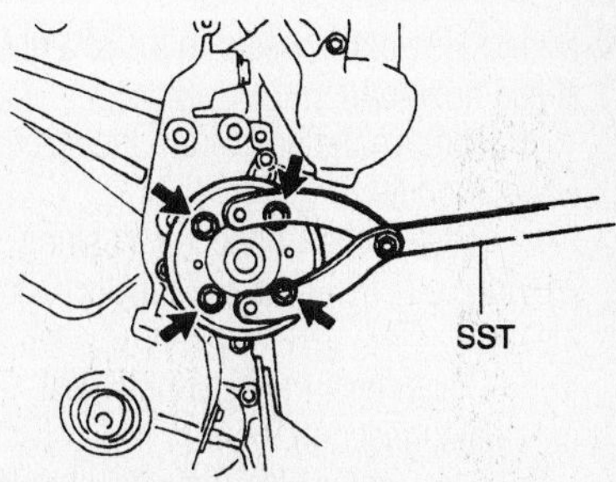

Water pump mounting and bolt locations—2ZZ-GE engine

Heater Core

REMOVAL & INSTALLATION

1. Before servicing the vehicle, refer to the precautions in the beginning of this section.
2. Drain the cooling system.
3. Discharge and recover the A/C system refrigerant using approved equipment.
4. Remove or disconnect the following:
   - Negative battery cable
   - Heater hoses from the core
   - Evaporator inlet and outlet tubes from the evaporator and cap the lines to avoid system contamination
5. Remove the instrument panel as follows:
   a. Disable the air bag system.
   b. Using a taped flat–bladed tool, carefully pry the retaining clips attaching the center trim plate to the instrument panel.
   c. Disconnect the A/C switch, hazard switch; rear defogger switch and passenger seat belt indicator switch electrical connections.
   d. Remove the radio retaining screws, clamp from the radio bracket, slide the radio forward to disconnect the power and antenna connections. Remove the radio.
   e. Remove the A/C switch and screw.
   f. Remove the hazard switch.
   g. Remove the rear defogger switch.
   h. Remove the manual transmission shift knob.
   i. Using a taped flat–bladed tool, carefully pry the retaining clips attaching the front floor console trim plate to the floor console assembly.
   j. Disconnect the 2 cigar lighter connectors.
   k. Disconnect the accessory power receptacle connectors.
   l. Remove the cigar lighters and power receptacle.
   m. Place both wheels in the straight ahead position.
   n. Remove the bolts from the steering wheel module.
   o. Release the Connector Position Assurance (CPA) from the inflator module.
   p. Disconnect the steering wheel module connectors.
   q. Remove the steering wheel module.
   r. Matchmark the steering wheel nut–to–shaft position, then remove the steering wheel nut and the wheel.

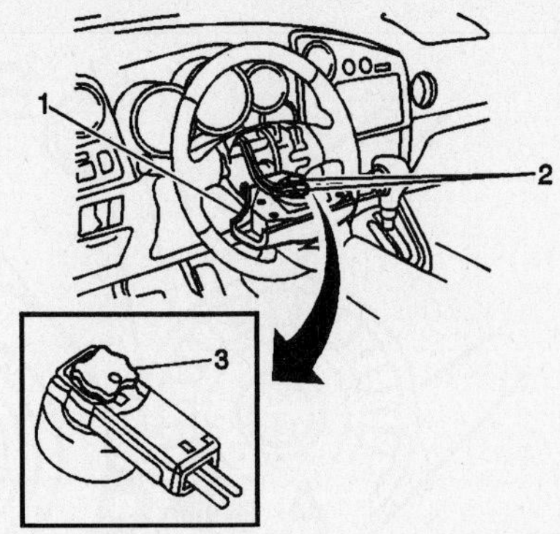

Exploded view of the CPA assembly

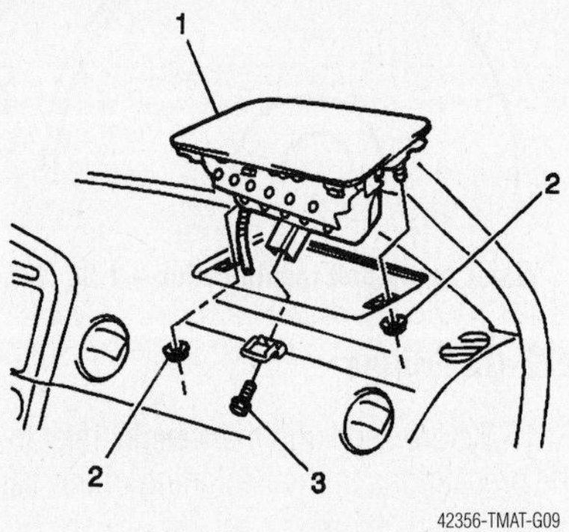

Remove the instrument panel module connectors and the passenger air bag assembly

   s. Remove the upper and lower steering column cover screws and the covers.
   t. Disconnect the turn signal/headlamp assembly connectors.
   u. Remove the turn signal/headlamp switch assembly
   v. Remove the wiper switch by depressing the tab.
   w. Remove the glove box.
   x. Disconnect the instrument panel connector.
   y. Remove the instrument panel module connectors and the passenger air bag assembly.
   z. Remove the cluster trim plate by disengaging the clips.
   aa. Remove the cluster screw and disengage the 2 lower clips.
   bb. Disconnect the cluster electrical connectors and remove the cluster.
   cc. Remove the windshield garnish moldings.
   dd. Using a taped flat–bladed tool, carefully pry the retaining clips attaching the instrument panel left trim plate to the instrument panel.
   ee. Disconnect the power mirror and dimmer switch connectors.
   ff. Remove the power mirror and dimmer switches.
   gg. Disconnect any remaining electrical connections.
   hh. Remove the upper instrument panel screws and the panel by pulling towards the rear to disengage the tabs.
   ii. Disconnect the steering wheel coil connector.
   jj. Release the 3 claws and remove the coil assembly.
   kk. If the vehicle is equipped with an automatic transmission, insert the key into the cylinder, turn to the ACC position, push in the release button, disconnect the park lock cable, remove the key from the cylinder and lock the steering wheel.
   ll. Move the silencer pad from the column.
   mm. Matchmark the steering shaft coupling to the shaft.
   nn. Loosen the upper bolt on the coupling.
   oo. Remove the lower bolt from the coupling.
   pp. Move the coupling onto the column shaft.
   qq. Disconnect the wiring harness clamps from the column.
   rr. Remove the 3 bolts and the column.
   ss. Remove the body hinge trim panels.
   tt. Remove the sill plates.
   uu. Remove the front floor console storage door.
   vv. Remove the screws attaching the console to the instrument panel, pull the console rewards and up and remove the front floor console.
   ww. Remove the HVAC retaining screw; disconnect the electrical connectors and module control, temperature control and A/C cables. Remove the unit.
   xx. Push in the clip and disconnect the cable from the manual selector shifter assembly.
   yy. Using a suitable prytool, disconnect the park lock cable from the bracket.
   zz. Disconnect the shift select cable from the manual selector lever.
   aa. Using a suitable prytool, disconnect the shift select cable from the shift lever plate.
   bb. Disconnect the electrical connectors and the wire harness clip.
   cc. Remove the nuts from the selector and remove the selector.
   dd. Disconnect the hood release cable from the release handle.
   ee. Remove the 8 bolts, 4 push

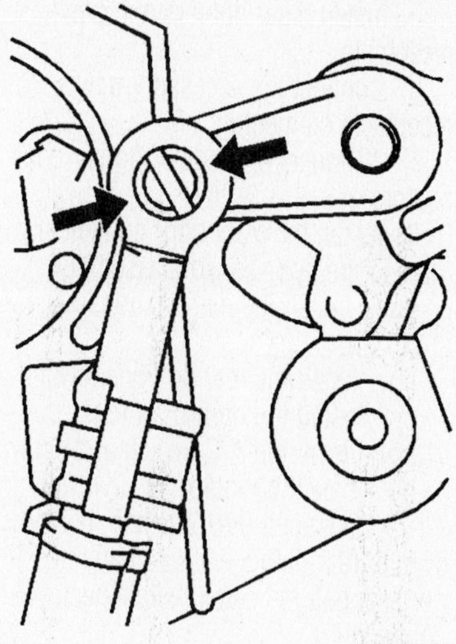

Push in the clip and disconnect the cable from the manual selector shifter assembly

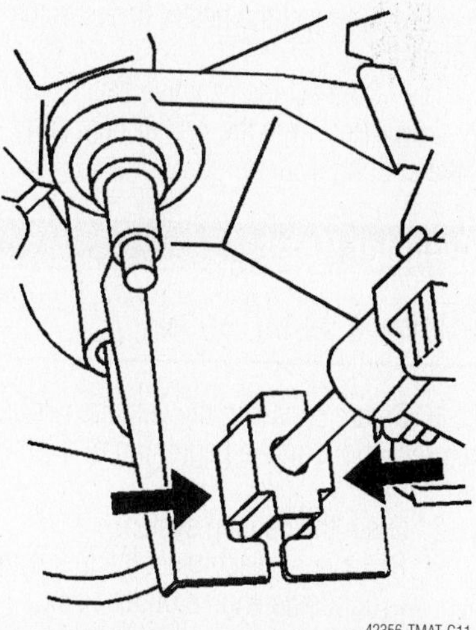

Using a suitable prytool, disconnect the park lock cable from the bracket

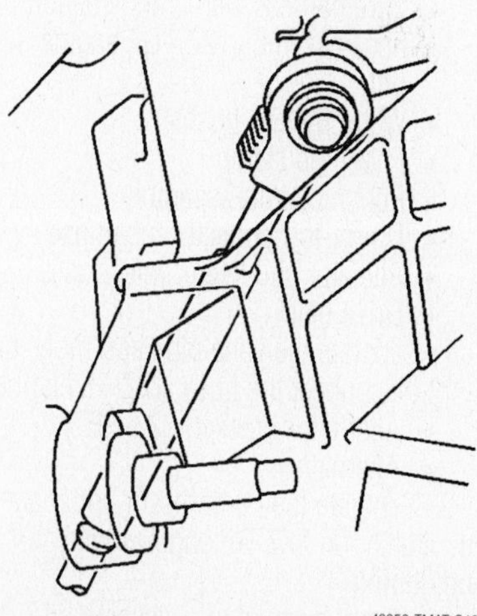

Disconnect the shift select cable from the manual selector lever

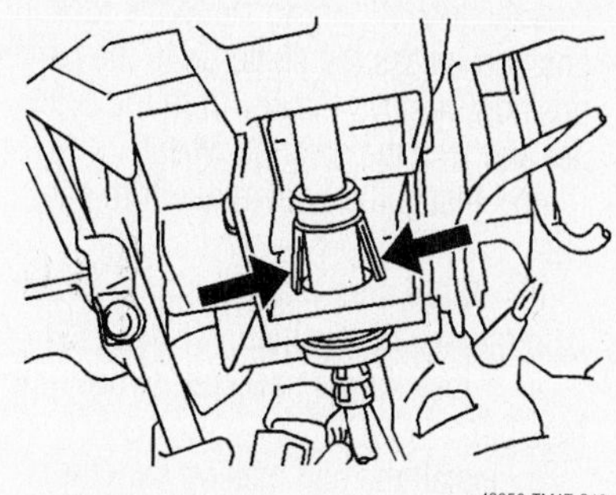

Using a suitable prytool, disconnect the shift select cable from the shift lever plate

retainers and the wire harness clamps from the lower instrument panel and remove the panel.

ff. Disengage the wiring harness clips, remove the bolts retaining the lower instrument panel pad and the pad.

gg. Remove the ground cable.

hh. Remove the connector housing bracket from the right instrument panel center support brace.

ii. Remove the left brace nut, right brace nut, left brace bolt, right brace bolt, left center support brace and right center support brace.

jj. Remove the windshield defroster nozzle duct from the heater case.

kk. Remove the 5 bolts and the nuts from the instrument panel reinforcement at the hinge pillars.

ll. Remove the instrument panel reinforcement.

mm. Disconnect the blower motor connector.

nn. Disconnect the rear ducts from the HVAC module.

oo. Remove the HVAC module.

pp. Remove the 12 bolts from the core case.

qq. Remove the heater core.

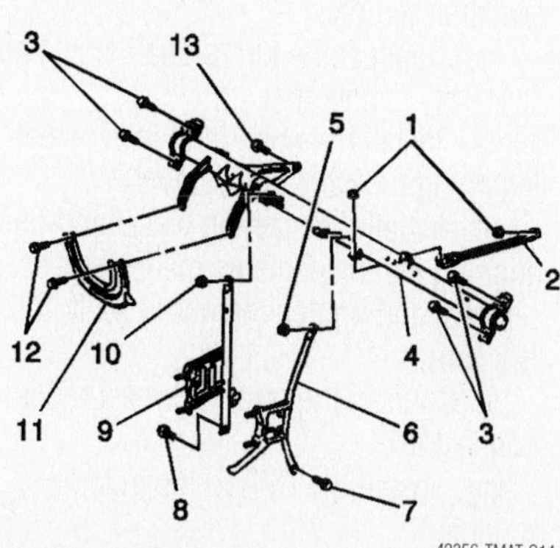

Exploded view of the instrument panel reinforcement

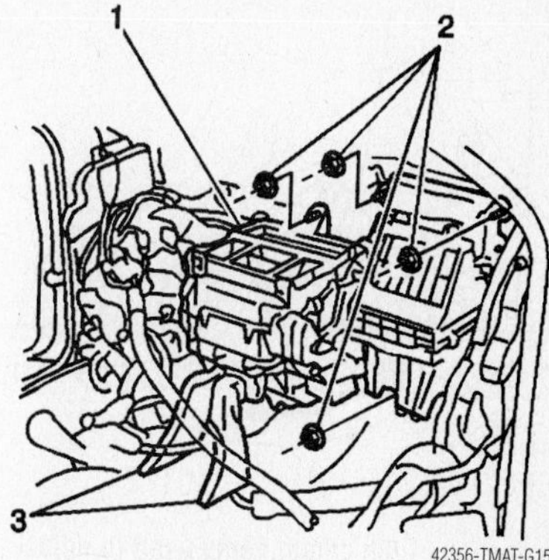

Remove the HVAC module

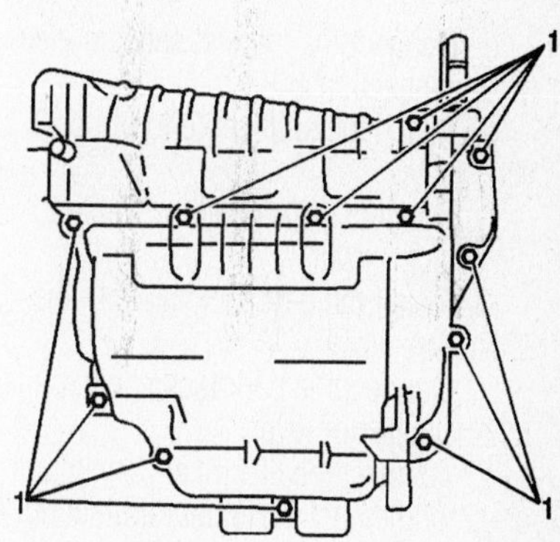

Remove the 12 bolts from the core case to access the heater core

To install:

a. Install the heater core.

b. Install the 12 bolts from the core case and tighten to 89 inch lbs. (10 Nm).

c. Install the HVAC module.

d. Connect the rear ducts to the HVAC module.

e. Connect the blower motor connector.

f. Install the instrument panel reinforcement.

g. Install the 5 bolts and the nuts to the instrument panel reinforcement at the hinge pillars. Tighten to 21 ft. lbs. (28 Nm).

h. Install the windshield defroster nozzle duct to the heater case.

i. Install left center support brace and right center support brace. Tighten the nuts and bolt to 15 ft. lbs. (20 Nm).

j. Install the connector housing bracket.

k. Connect the ground cable.

l. Install the lower instrument panel pad and the bolts and attach the wiring harness clips.

m. Connect the hood release cable to the release handle.

n. Install the lever and tighten the nuts to 12 ft. lbs. (18 Nm).

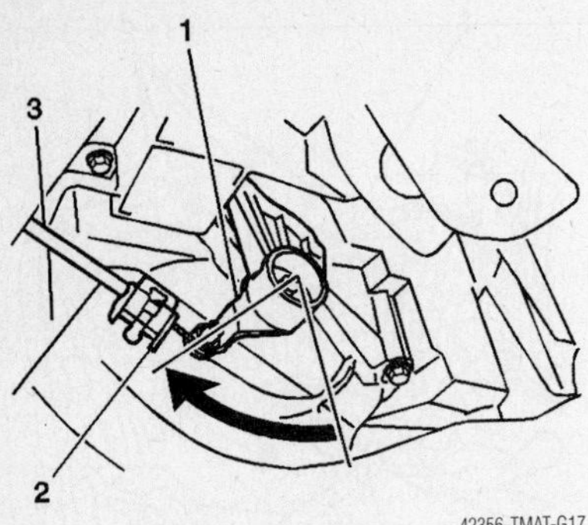

Location of the components used to adjust the temperature control cable (3) clip (2), door lever (1).

o. Connect the manual selector electrical connections.

p. Attach the shift cable to the shift lever plate.

q. Connect the shift select cable to the selector lever.

r. Install the park lock cable to the shift lever plate.

s. Connect the park lock cable to the manual selector lever.

t. Connect the electrical connectors and module control, temperature control and A/C cables. Install the HVAC unit.

u. Adjust the temperature control cable by setting the temperature control dial to coldest. Hold the door lever fully rearwards, clockwise. Attach the cable to the control clip.

v. Adjust the Mode linkage by setting the dial to defrost. Hold the door lever fully rearwards, clockwise. Attach the cable to the control clip.

w. Install the front floor console and tighten the screws.

x. Install the front floor console storage door.

y. Install the sill plates.

z. Install the column.

aa. Install the 3 bolts and tighten the lower bolt to 16 ft. lbs. (21 Nm) and the 2 upper bolts to 16 ft. lbs. (21 Nm).

bb. Align the matchmarks made prior to removal.

cc. Lower the coupling onto the shaft. Install the bolts and tighten to 26 ft. lbs. (35 Nm).

dd. Connect the wiring harness clamps to the column.

ee. Move the silencer pad to the column.

ff. If the vehicle is equipped with an automatic transmission, insert the key into the cylinder, turn to the ACC position, insert the park lock cable making sure the release button engages. Make sure the key will not rotate to the lock position unless the shifter is in the park position, remove the key from the cylinder and lock the steering wheel.

gg. Install the body hinge trim panels.

hh. Make sure the turn signal switch is in the neutral position.

ii. If installing a new coil, remove the lock pin.

jj. Install the coil making sure the 3 claws engage.

kk. While holding the coil casing, turn the coil center casing counterclockwise until the coil reaches its stop.

ll. Turn the coil center casing clockwise 2 ½ turns.

mm. Align the center casing with the arrow on the outer casing.

nn. Connect the coil electrical connector.

oo. Install the upper instrument panel and screws.

pp. Connect the electrical connections.

qq. Install the power mirror and dimmer switches.

rr. Connect the power mirror and dimmer switch connectors.

ss. Install the instrument panel left trim plate to the instrument panel.

tt. Install the windshield garnish moldings.

uu. Connect the cluster electrical connectors and install the cluster.

vv. Engage the cluster lower clips and install the screw.

ww. Install the cluster trim plate.

xx. Install the passenger air bag assembly and the instrument panel module connectors.

yy. Connect the instrument panel connector.

zz. Install the glove box.

aa. Install the wiper switch.

bb. Install the turn signal/headlamp switch assembly

cc. Connect the turn signal/headlamp assembly connectors.

dd. Install the upper and lower steering column covers and screws.

ee. Install the steering wheel and nut aligning the matchmarks made prior to removal and tighten the nut to 37 ft. lbs. (50 Nm).

ff. Connect the steering wheel module connectors.

gg. Install the CPA to the inflator module.

hh. Install the steering wheel module and tighten the retainers 78 inch lbs. (9 Nm)

ii. Install cigar lighters and power receptacle.

jj. Connect the accessory power receptacle connectors.

kk. Connect the 2 cigar lighter connectors.

ll. Install the front floor console trim plate to the floor console assembly.

mm. Install the manual transmission shift knob.

nn. Install the rear defogger switch.

oo. Install the hazard switch.

pp. Install the A/C switch and screw.

qq. Install the radio.

rr. Connect the A/C switch, hazard switch, rear defogger switch and passenger seat belt indicator switch electrical connections.

ss. Install the center trim plate to the instrument panel.

tt. Connect the evaporator inlet and outlet tubes.

uu. Connect the heater hoses to the core.

vv. Connect the negative battery cable.

ww. Recharge the A/C system and fill the cooling system.

Cylinder Head

REMOVAL & INSTALLATION

1. Before servicing the vehicle, refer to the precautions in the beginning of this section.
2. Drain the cooling system.
3. Remove or disconnect the following:
   - Right side engine under cover
   - Right front wheel and tire
   - Cylinder head cover
   - Air cleaner assembly with hose
   - Accelerator control cable
   - Wire harness clamp and suction hose assembly, 2ZZ-GE engine only
   - Water bypass hoses
   - Fuel pipe clamp
   - Fuel tube sub-assembly
   - Union-to-connector tube hose
   - Radiator and heater inlet hoses
   - Drive belt
4. Separate the vane pipe assembly, but do not disconnect the hose, 1ZZ-FE engine.
   - Alternator bracket, 2ZZ-GE
   - Alternator
5. Separate the compressor and magnetic clutch, on 2ZZ-GE engines with air conditioning.
   - Front exhaust pipe assembly
   - Power steering pump reservoir and position it aside, 1ZZ-FE engine

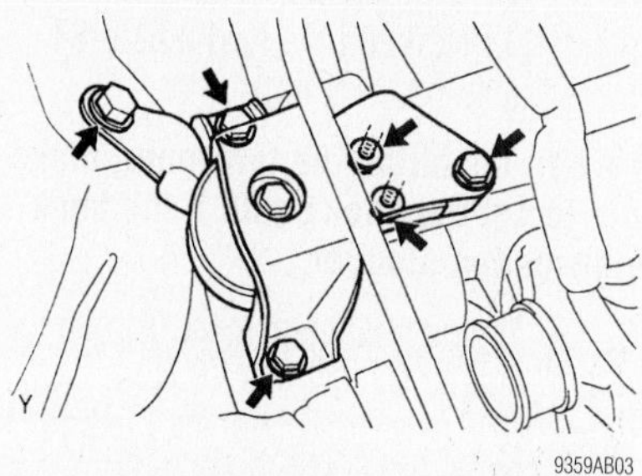

With the engine supported, remove the right side engine mount—1ZZ-FE engine shown, 2ZZ-GE similar

6. Place a jack with a wooden block under the vehicle for support, then remove the 4 bolts and 2 nuts and remove the right side engine mount.

7. Remove the engine wire, on 1ZZ-FE engines as follows:

a. Remove the 5 clamps from the brackets.

b. Detach the connectors.

c. Remove the ignition coil connectors.

d. Bolt and nut holding the engine wire.

8. Remove or disconnect the following:

- Ignition coil assembly
- Positive Crankcase Ventilation (PCV) hoses
- Valve (cylinder head) cover sub-assembly

9. Set the No. 1 cylinder to Top Dead Center (TDC) of the compressor stroke as follows:

a. Turn the crankshaft pulley, and align its groove with the "0" timing mark of the timing chain cover.

b. Make sure the point marks of the camshaft timing sprockets and Variable Valve Timing (VVT) timing sprockets are in a straight line as shown. If not, turn the crankshaft 1 complete revolution (360°) and align the marks.

10. Remove or disconnect the following:

- Crankshaft pulley, using SST 09960-10010
- Belt tensioner
- Exhaust manifold stay and head insulator, 2ZZ-GE engine
- Water pump pulley and pump
- Transverse engine mounting bracket
- Crankshaft Position (CKP) sensor
- No. 1 chain tensioner assembly, making sure not to revolve the crankshaft without the tensioner
- Timing chain or belt cover
- Timing gear cover oil seal

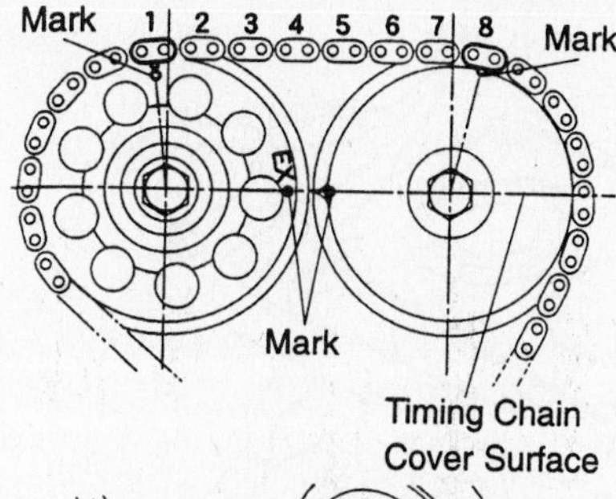

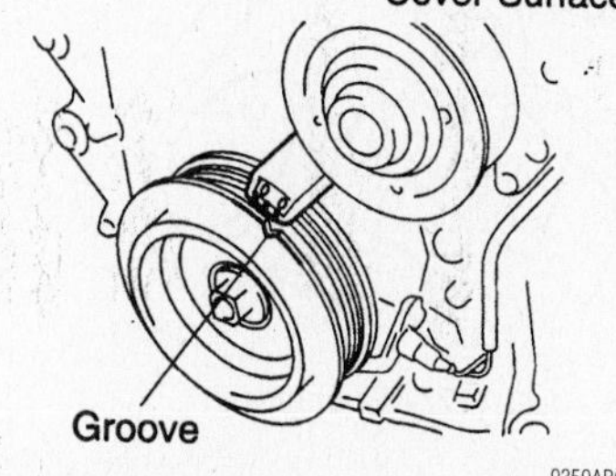

Proper timing mark alignment for TDC

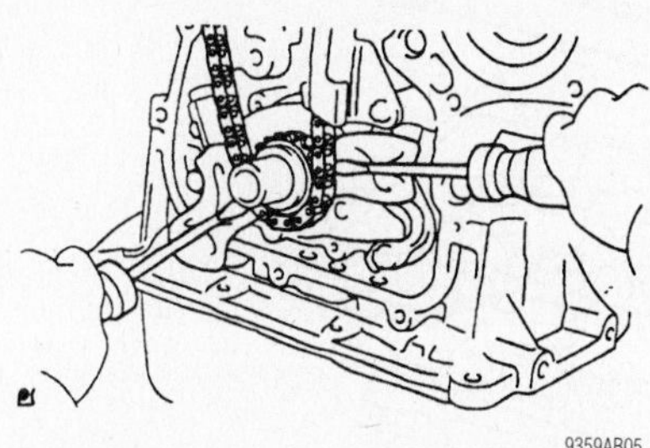

Remove the timing chain with the crankshaft gear

- CKP sensor plate No. 1
- Timing chain tensioner slipper
- Timing chain vibration damper No. 1

➡**In case you turn the camshafts with the timing chain removed, turn the crankshaft ¼ turn for the valve to avoid contact with the pistons.**

- Timing chain sub-assembly. Remove the chain with the crankshaft gear, using screwdrivers as shown.
- Surge tank stay, 2ZZ-GE engine
- Intake manifold
- Oil level gauge
- Water bypass pipe bolts and pipe, 1ZZ-FE engine
- Camshafts
- Camshaft timing oil control valve, 1ZZ-FE engine
- Manifold stay, 1ZZ-FE engine
- Cylinder head bolts in sequence. To prevent damage to the cylinder head, loosen each bolt about ¼ of a turn during each pass until the bolts are loose.
- Cylinder head

To install:

11. Clean and degrease the surface of the cylinder head and engine block.

12. Check the length of the cylinder head bolts. They should be 5.780–5.835 in. (146.8–148.2mm) long. If they are longer than 5.846 in. (148.5mm), they must be replaced.

13. Install or connect the following:

- New gasket on the engine block with the Lot No. stamp facing up.
- Cylinder head
- Apply a light coat of oil to cylinder head bolt threads and tighten in sequence, in several passes, to 36 ft. lbs. (49 Nm) for 1ZZ-FE engines or to 26 ft. lbs. (35 Nm) for 2ZZ-GE engines.
- Tighten each head bolt, in sequence, an additional 90 degree turn for 1ZZ-FE engines, or 180 degree turn for 2ZZ-GE engines.
- Manifold stay, 1ZZ-FE engine. Tighten the bolts to 36 ft. lbs. (49 Nm).
- Camshaft timing oil control valve, on 1ZZ-FE engines, and tighten to 80 inch lbs. (9 Nm)
- Camshaft

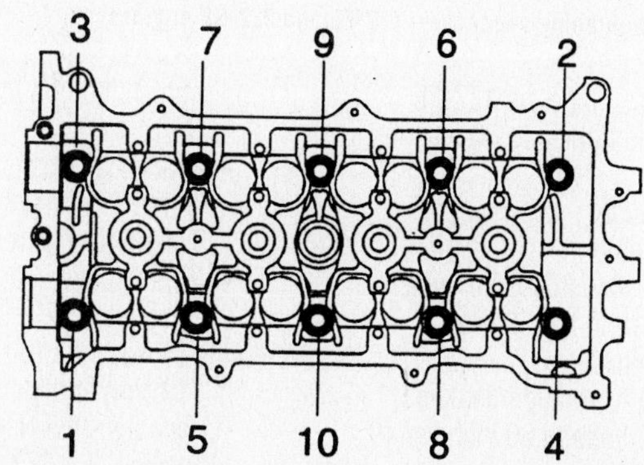

Cylinder head bolt loosening sequence

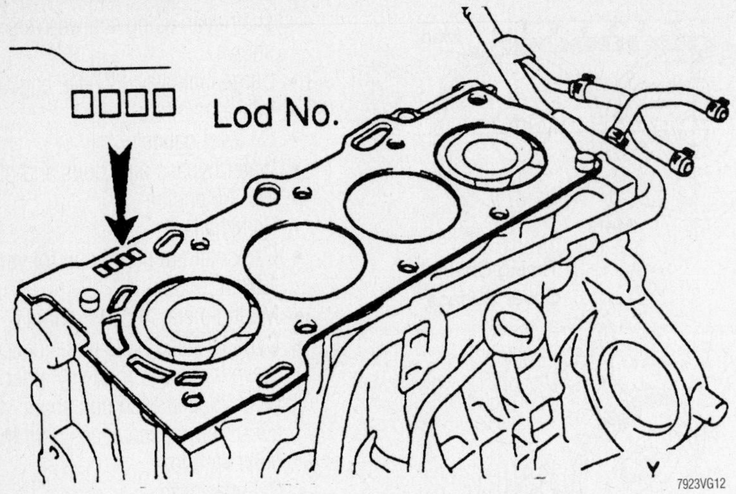

Position the head gasket correctly on the cylinder head—1.8L (1ZZ-FE) engine

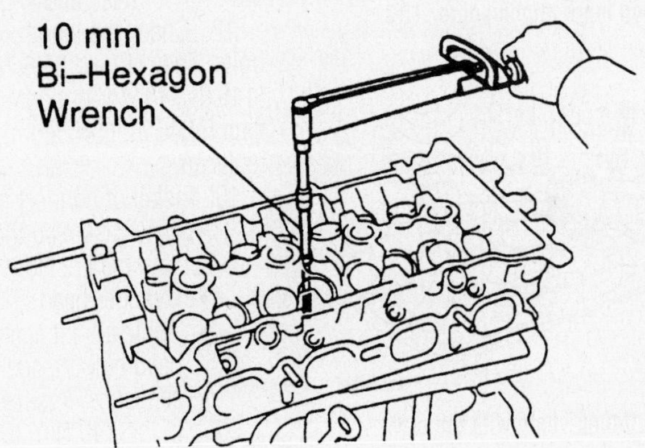

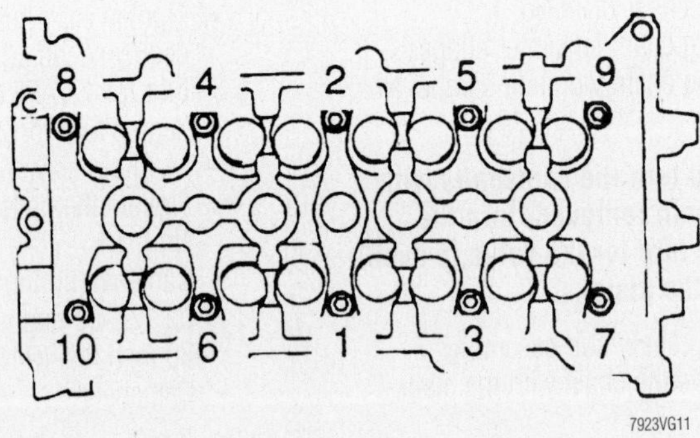

Cylinder head bolt tightening sequence—1ZZ-FE and 2ZZ-GE engines

- Water by-pass pipe, on 1ZZ-FE engines, and tighten to 80 inch lbs. (9 Nm)
- Oil level gauge
- Intake manifold
- Surge tank stay, 2ZZ-GE engine. Tighten to 18 ft. lbs. (24 Nm).
- Timing chain
- Timing chain vibration damper. Tighten the bolts to 80 inch lbs. (9 Nm).
- Timing chain tensioner slipper and tighten the bolt to 14 ft. lbs. (19 Nm).
- Crankshaft position sensor plate, with the "F" mark facing forward.
- Timing gear cover oil seal
- Timing cover. For 1ZZ-FE engine, tighten the "A" bolts to 10 ft. lbs. (13 Nm), the "B" bolts to 14 ft. lbs. (19 Nm) and the stud bolt to 84 inch lbs. (9.5 Nm), using a Torx® wrench. For 2ZZ-GE engines, tighten the M8 bolts to 15 ft. lbs. (21 Nm), the M6 bolts to 8 ft. lbs. (11 Nm) and the stud bolt to 84 inch lbs. (9.5 Nm).

➡When installing the tensioner, make sure to set the hook again if the hook releases the plunger.

- Timing chain tensioner. Torque the nuts to 80 inch lbs. (9 Nm).
- CKP sensor and tighten the bolts to 80 inch lbs. (9 Nm)
- Transverse engine mounting bracket. Tighten the bolts to 35 ft. lbs. (47 Nm).
- Water pump and pulley
- Exhaust manifold stay and heat insulator, 2ZZ-GE engine
- Belt tensioner. Tighten the nut to 21 ft. lbs. (29 Nm) and the bolt to 51 ft. lbs. (69 Nm) on 1ZZ-FE engines or to 74 ft. lbs. (100 Nm) on 2ZZ-GE engines.

14. Install the crankshaft pulley, as follows:

a. Align the pulley set key with the key groove of the pulley and slide on the pulley.

b. Use SST 09960-11010 to install the bolt and tighten to 102 ft. lbs. (138 Nm) for 1ZZ-FE engine or to 87 ft. lbs. (118 Nm) on 2ZZ-GE engines.

c. Turn the crankshaft counterclockwise and disconnect the plunger knock pin from the hook.

d. Turn the crankshaft clockwise and check that the slipper is pushed by the plunger. If the plunger does not spring out, press the slipper into the chain tensioner with a screwdriver so that the hook is released from the knock pin and the plunger springs out.

15. Install or connect the following:

- Cylinder head sub-assembly cover. Install seal packing into the locations shown and install within 3 minutes. Tighten the "A" bolts to 8 ft. lbs. (11 Nm) and the "B" bolts to 80 inch lbs. (9 Nm) for 1ZZ-FE engines. For 2ZZ-GE engines, tighten the bolts to 7 ft. lbs. (10 Nm).
- Ignition coil assembly. Torque the bolts to 80 inch lbs. (9 Nm).
- Engine wire and tighten to 80 inch lbs. (9 Nm), 1ZZ-FE
- Right side engine mount. Tighten to 38 ft. lbs. (52 Nm).
- Front exhaust pipe
- Vane pump, 1ZZ-FE
- Compressor and magnetic clutch, 2ZZ-GE
- Alternator bracket, 2ZZ-GE engine
- Alternator

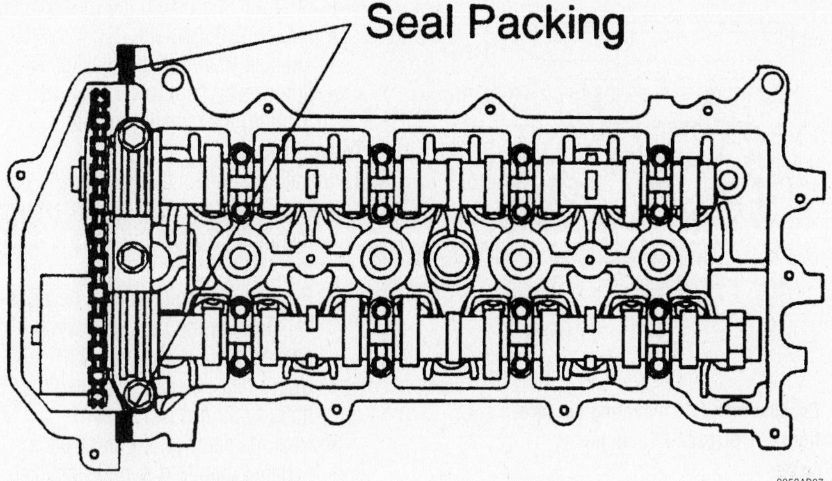

Seal packing installation locations

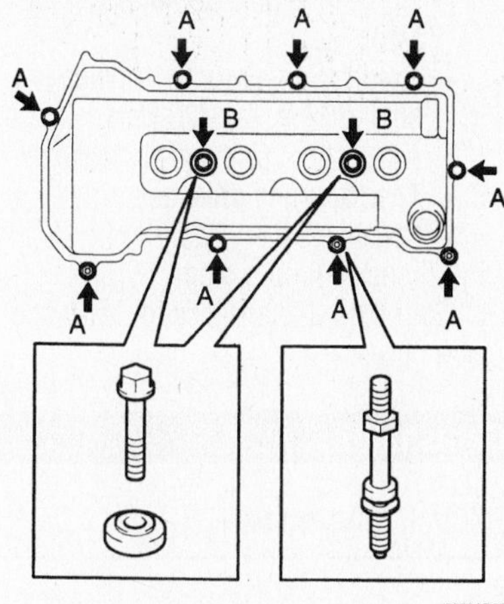

Cylinder head (valve) cover bolt locations—1ZZ-FE engine

- Suction hose and wire harness clamp, 2ZZ-GE engine
- Air cleaner and hose
- Main cylinder head cover and tighten to 62 inch lbs. (7 Nm)
- Right front wheel and tire. Tighten the lug nuts to 76 ft. lbs. (103 Nm).

16. Fill the cooling system to the proper level.

17. Start the vehicle, check for leaks and repair if necessary.

Intake Manifold

REMOVAL & INSTALLATION

1ZZ-FE Engine

1. Before servicing the vehicle, refer to the precautions in the beginning of this section.
2. Drain the cooling system.
3. Remove or disconnect the following:
   - Negative battery cable
   - Drive belt and alternator
   - Air intake duct
   - Accelerator cable
   - Exhaust pipe from the manifold.
   - Exhaust manifold support bracket
   - Spark plug wires, then ignition coils
   - Spark plugs
   - Positive Crankcase Ventilation (PCV) hoses
   - Throttle body assembly
   - 2 bolts securing the wiring harness protector
   - Wiring connectors and ground wires
   - Intake manifold support bracket
   - Intake manifold and gasket

To install:

4. Install or connect the following:
   - Intake manifold with a new gasket. Torque the bolts to 22 ft. lbs. (30 Nm).
   - Harness wiring to the cylinder head and harness protector
   - Fuel injectors, throttle body and the PCV hoses
   - Spark plugs and ignition coils. Tighten the bolts and nuts to 80 inch lbs. (9 Nm).
   - Exhaust manifold and support bracket. Tighten the bolts to 37 ft. lbs. (49 Nm).
   - Front exhaust pipe to the manifold. Tighten the bolts to 46 ft. lbs. (62 Nm).
   - Oxygen Sensor (O_2S). Tighten the nuts to 14 ft. lbs. (20 Nm).
   - Accelerator cable and air intake duct
   - Alternator and drive belt
   - Negative battery cable
5. Fill the cooling system.
6. Start the vehicle, check for leaks and repair if necessary.

2ZZ-GE Engine

1. Before servicing the vehicle, refer to the precautions in the beginning of this section.
2. Drain the cooling system.
3. Remove or disconnect the following:
   - Negative battery cable
   - Drive belt and alternator
   - Air intake duct
   - Accelerator cable
   - Spark plug wires, then ignition coils
   - Spark plugs
   - Positive Crankcase Ventilation (PCV) hoses
   - Throttle body assembly

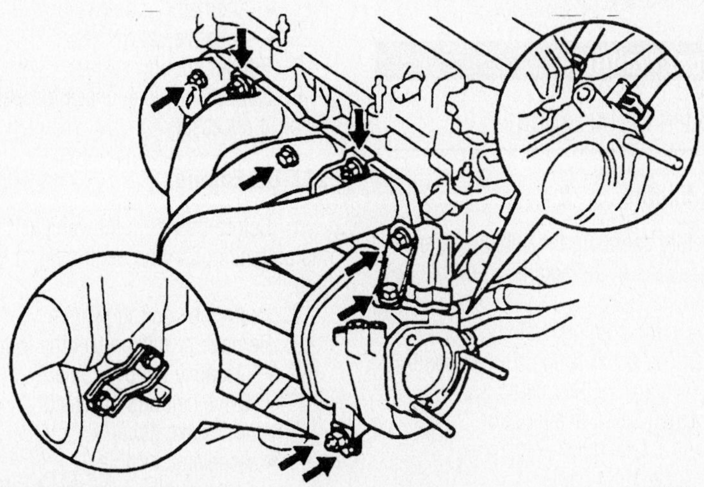

Intake manifold mounting fastener locations—1.8L (1ZZ-FE) engine

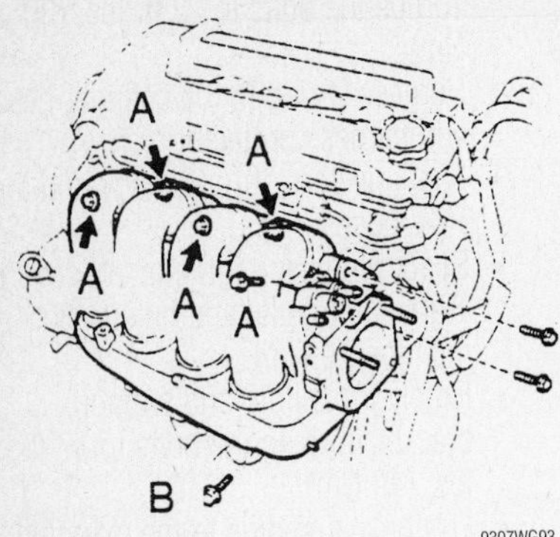

Intake manifold bolt installation—2ZZ-GE engine

- Wiring harness
- Hoses and tubes connected to the head
- Intake manifold support bracket
- Intake manifold and gasket

To install:

4. Install or connect the following:
   - Intake manifold with a new gasket. Tighten bolts A to 25 ft. lbs. (34 Nm) and bolt B to 34 ft. lbs. (46 Nm)
   - Harness wiring to the cylinder head and harness protector
   - Fuel injectors, throttle body and the PCV hoses
   - Spark plugs and ignition coils. Tighten the bolts and nuts to 80 inch lbs. (9 Nm).
   - Oxygen Sensor (O_2S). Tighten the nuts to 14 ft. lbs. (20 Nm).
   - Accelerator cable and air intake duct
   - Alternator and drive belt
   - Negative battery cable
5. Fill the cooling system.
6. Start the vehicle, check for leaks and repair if necessary.

Exhaust Manifold

REMOVAL & INSTALLATION

1ZZ-FE Engine

1. Before servicing the vehicle, refer to the precautions in the beginning of this section.
2. Drain the cooling system.
3. Remove or disconnect the following:
   - Negative battery cable
   - Drive belt and alternator
   - Air intake duct
   - Accelerator cable
   - Exhaust pipe from the manifold
   - Exhaust manifold support bracket

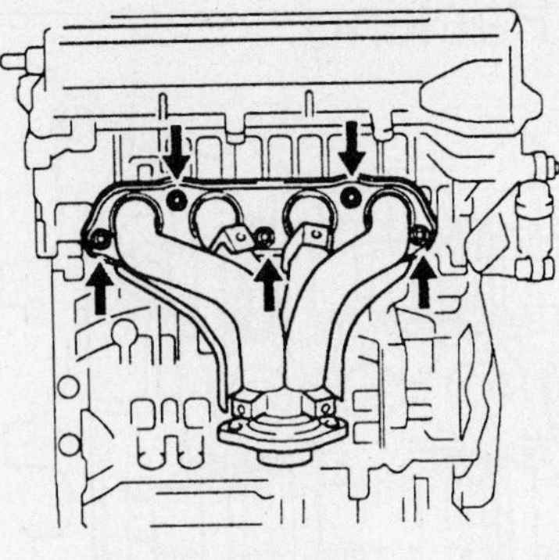

Exhaust manifold mounting nut locations—1.8L (1ZZ-FE) engine

   - Heat insulator from the dash panel
   - Upper heat insulator
   - Exhaust manifold and gasket
   - If necessary, the lower heat insulator from the exhaust manifold.

To install:

4. Install or connect the following:
   - Lower heat insulator on the exhaust manifold. Tighten the bolts to 108 inch lbs. (12 Nm).
   - Exhaust manifold using a new gasket. Tighten the nuts, in several passes, to 27 ft. lbs. (37 Nm).
   - Upper heat insulator. Tighten the bolts to 108 inch lbs. (12 Nm).
   - Heat insulator on the dash panel
   - Exhaust manifold support bracket. Tighten the bolts, in an alternating pattern, to 37 ft. lbs. (49 Nm).
   - Front exhaust pipe to the manifold. Tighten the bolts to 46 ft. lbs. (62 Nm).
   - Oxygen Sensor (O_2S). Tighten the nuts to 14 ft. lbs. (20 Nm).
   - Accelerator cable and air intake duct
   - Alternator and drive belt
   - Negative battery cable
5. Fill the cooling system.
6. Start the vehicle, check for leaks and repair if necessary.

2ZZ-GE Engine

1. Before servicing the vehicle, refer to the precautions in the beginning of this section.
2. Drain the cooling system.
3. Remove or disconnect the following:
   - Negative battery cable
   - Drive belt and alternator
   - Air intake duct
   - Accelerator cable
   - Exhaust pipe from the manifold
   - Exhaust manifold support bracket
   - Heat insulator from the dash panel.
   - Upper heat insulator
   - Exhaust manifold and gasket
   - If necessary, the lower heat insulator from the exhaust manifold.

To install:

4. Install or connect the following:
   - Lower heat insulator on the exhaust manifold. Tighten the bolts to 15 ft. lbs. (20 Nm).
   - Exhaust manifold using a new gasket. Tighten the nuts, in several passes to 37 ft. lbs. (50 Nm).
   - Upper heat insulator. Tighten the bolts to 15 ft. lbs. (20 Nm).
   - Heat insulator on the dash panel.
   - Exhaust manifold support bracket. Tighten the bolts to 37 ft. lbs. (49 Nm).
   - Front exhaust pipe to the manifold. Tighten the bolts to 46 ft. lbs. (62 Nm).
   - Oxygen Sensor (O_2S). Tighten the nuts to 14 ft. lbs. (20 Nm).
   - Accelerator cable and air intake duct.
   - Alternator and drive belt.
   - Negative battery cable
5. Fill the cooling system.
6. Start the vehicle, check for leaks and repair if necessary.

Camshaft(s)

REMOVAL & INSTALLATION

1ZZ-FE Engine

1. Before servicing the vehicle, refer to the precautions in the beginning of this section.
2. Remove or disconnect the following:
   - Negative battery cable
   - Right side engine under cover
   - Cylinder head cover
   - Suction hose sub-assembly, 2ZZ-GE engine
   - Drive belt

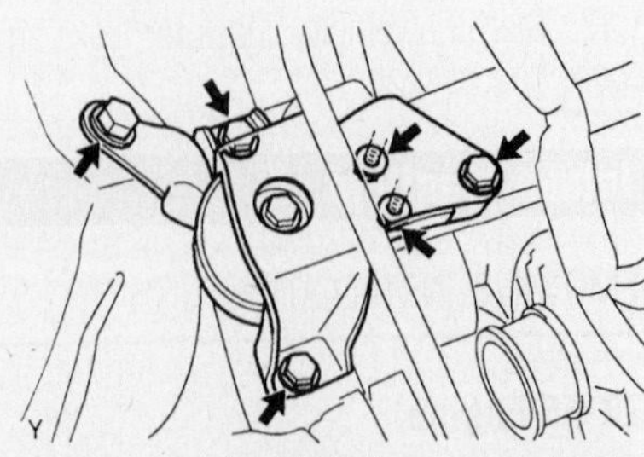

With the engine supported, remove the right side engine mount—1ZZ-FE engine shown, 2ZZ-GE similar

- Power steering pump reservoir and position it aside, 1ZZ-FE engine

3. Place a jack with a wooden block under the vehicle for support, then remove the 4 bolts and 2 nuts and remove the right side engine mount.

4. Remove the engine wire, on 1ZZ-FE engines:

   a. Remove the 5 clamps from the brackets.

   b. Detach the connectors.

   c. Remove the ignition coil connectors.

   d. Bolt and nut holding the engine wire.

5. Remove or disconnect the following:
   - Ignition coil assembly
   - Positive Crankcase Ventilation (PCV) hoses from the valve cover
   - Valve (cylinder head) cover sub-assembly

6. Set the No. 1 cylinder to Top Dead Center (TDC) of the compressor stroke as follows:

   a. Turn the crankshaft pulley, and align its groove with the "0" timing mark of the timing chain cover.

   b. Make sure the point marks of the camshaft timing sprockets and VVT timing sprockets are in a straight line as shown. If not, turn the crankshaft 1 complete revolution (360°) and align the marks.

7. Remove the drive belt tensioner.

⁂ WARNING

Do not turn the crankshaft without the tensioner installed.

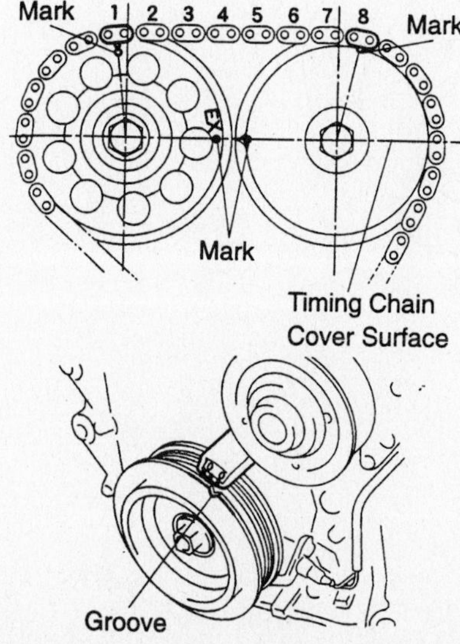

Proper timing mark alignment for TDC

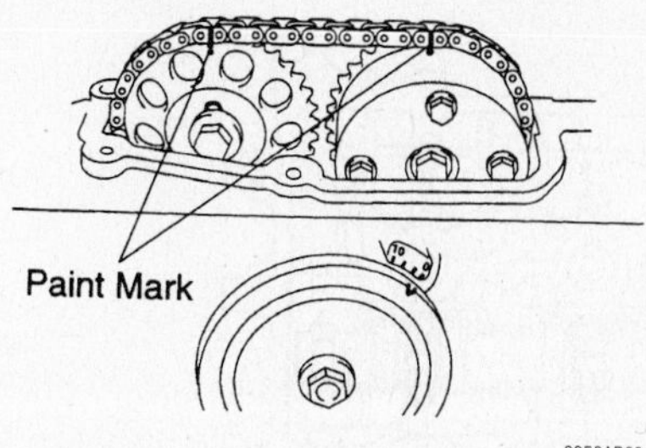

Matchmark the timing chain and cam sprockets

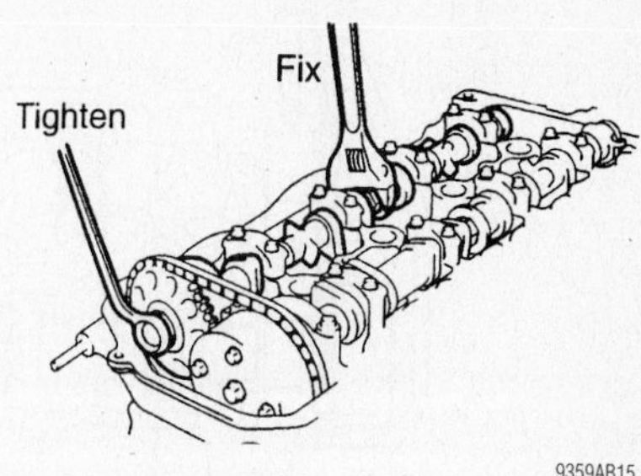

Hold the camshaft with a wrench while removing the set bolt

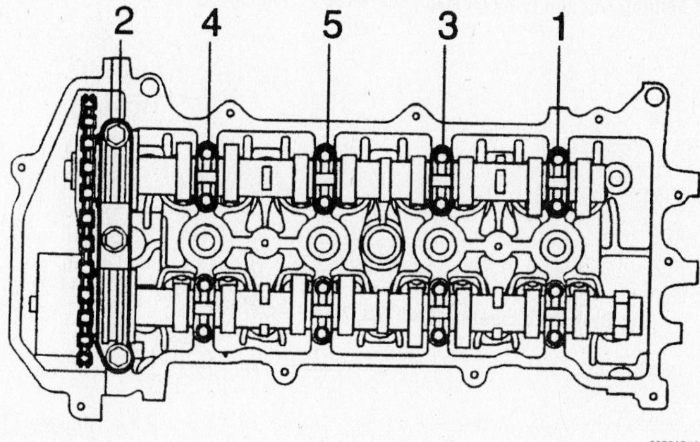

Camshaft bearing cap bolt removal sequence—1ZZ-FE engine

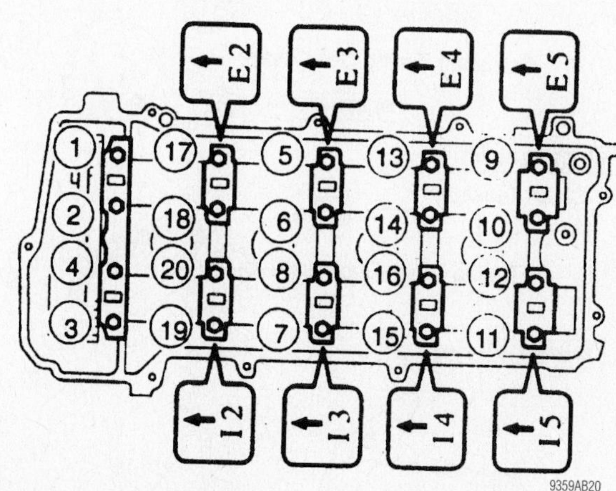

Camshaft bearing cap bolt removal sequence—2ZZ-GE engine

8. Make sure the No. 1 cylinder is at TDC of the compression stroke.

9. Matchmark the timing chain and camshaft sprockets

10. Remove the 2 nuts and chain tensioner.

11. Hold the camshafts with a wrench and loosen the camshaft set bolt.

12. Using several passes, gradually remove the bearing cap bolts from the No. 2 camshaft, in the proper sequence.

13. Remove the camshaft and timing gear as shown.

14. Using several passes, gradually

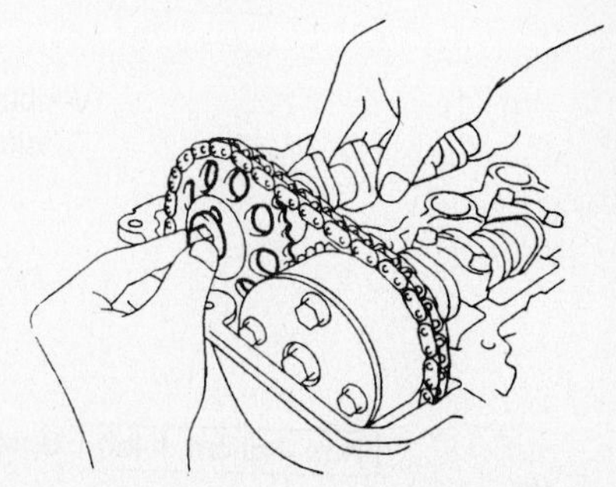

Carefully remove the cam and timing gear

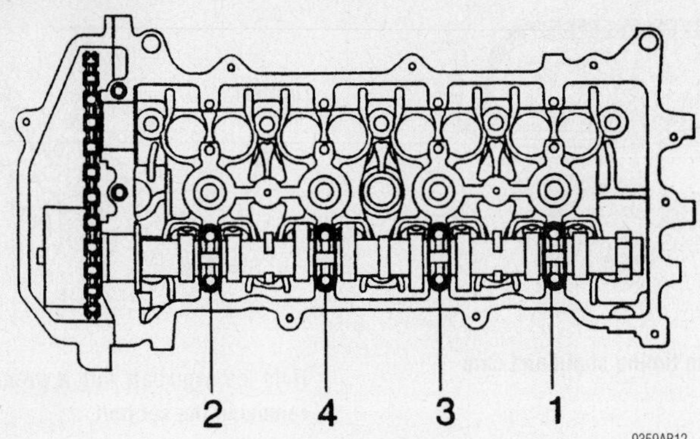

2
4
3
1
9359AB12

Camshaft bearing cap bolt removal sequence—1ZZ-FE engine

9359AB13

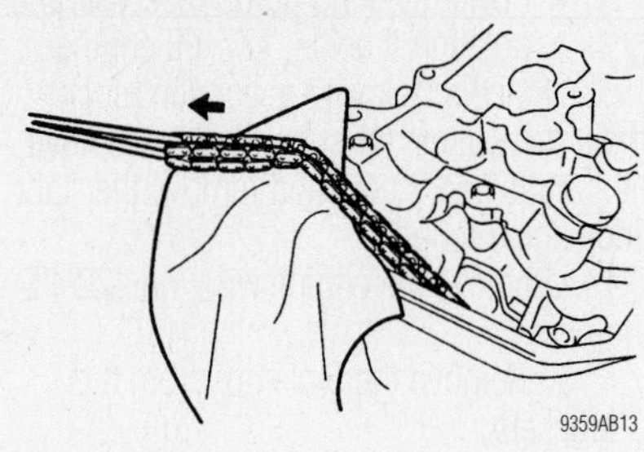

Secure the timing chain with string to prevent it from slipping down into the timing chain cover

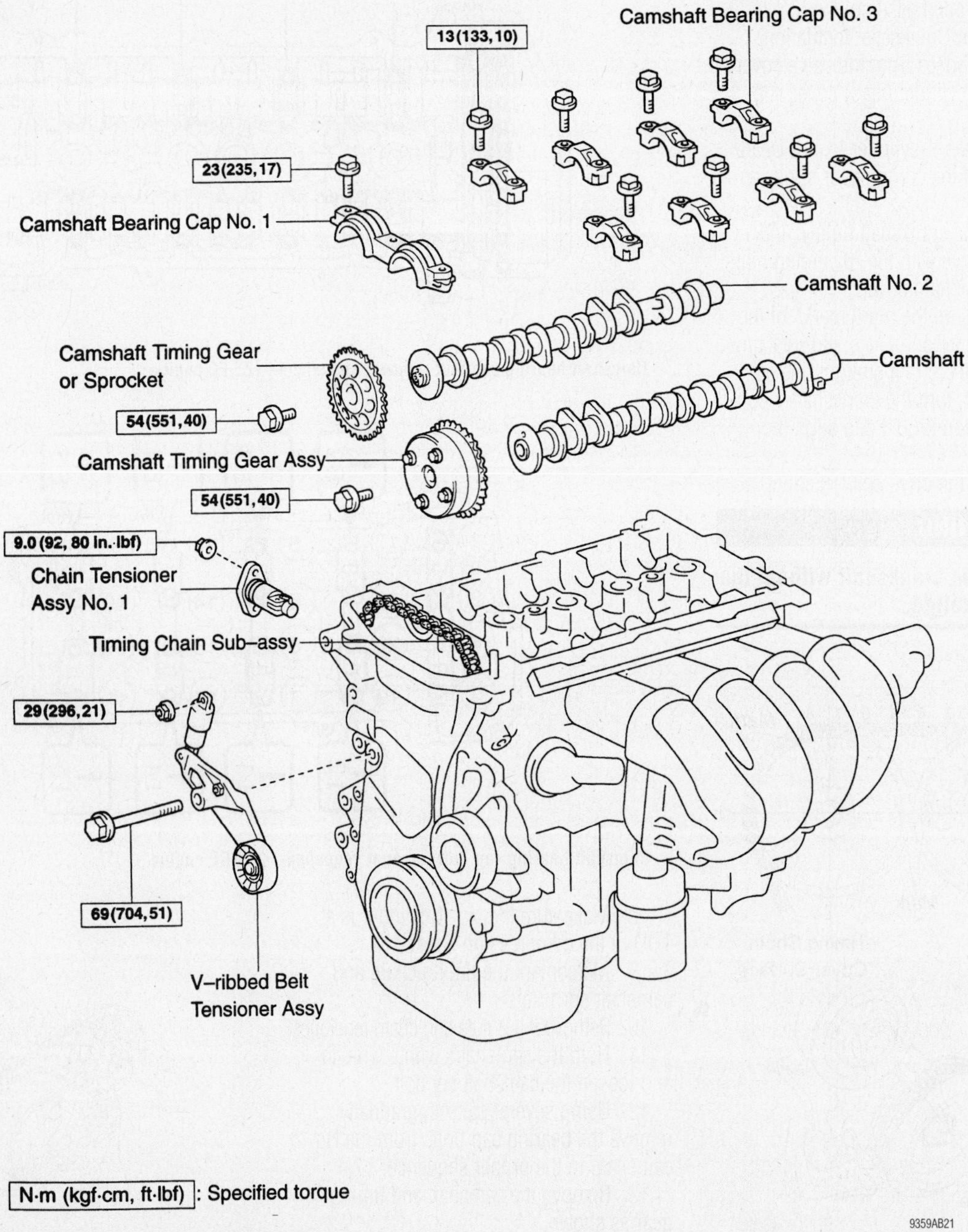

Camshaft Bearing Cap No. 3
13(133,10)
23(235,17)
Camshaft Bearing Cap No. 1
Camshaft No. 2
Camshaft Timing Gear or Sprocket
Camshaft
54(551,40)
Camshaft Timing Gear Assy
54(551,40)
9.0(92, 80 in.·lbf)
Chain Tensioner Assy No. 1
Timing Chain Sub–assy
29(296,21)
69(704,51)
V–ribbed Belt Tensioner Assy
N·m (kgf·cm, ft·lbf) : Specified torque
9359AB21

Exploded view of the camshafts and related components—1ZZ-FE engine

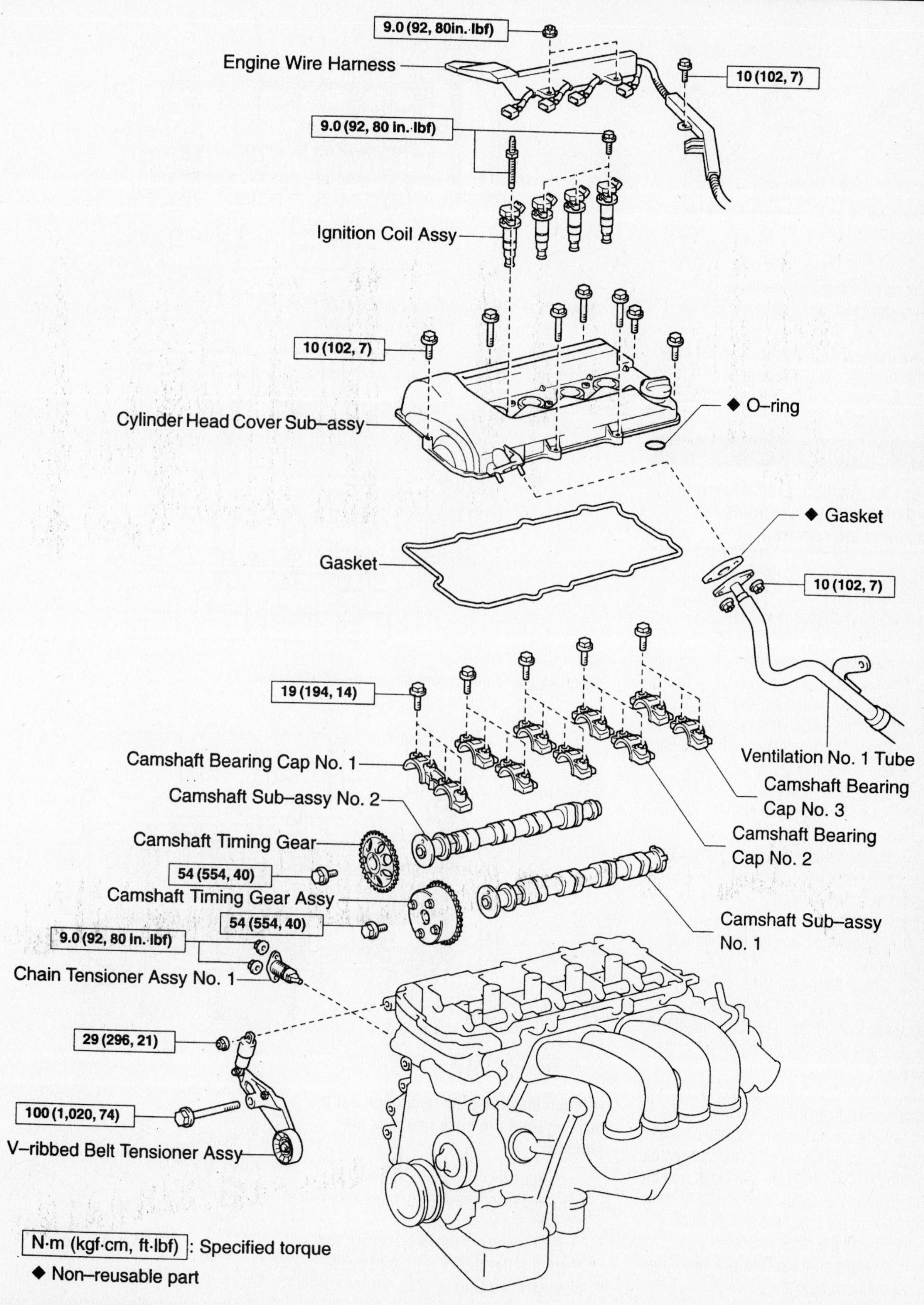

Exploded view of the camshafts and related components—2ZZ-GE engine

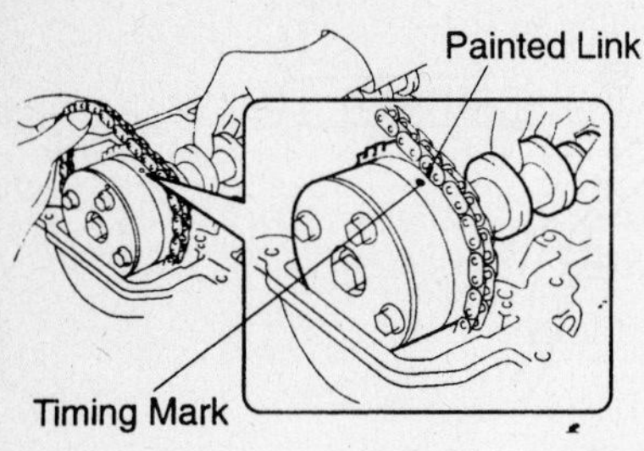

9359AB14

Make sure the alignment marks on the timing chain and camshaft gear match up

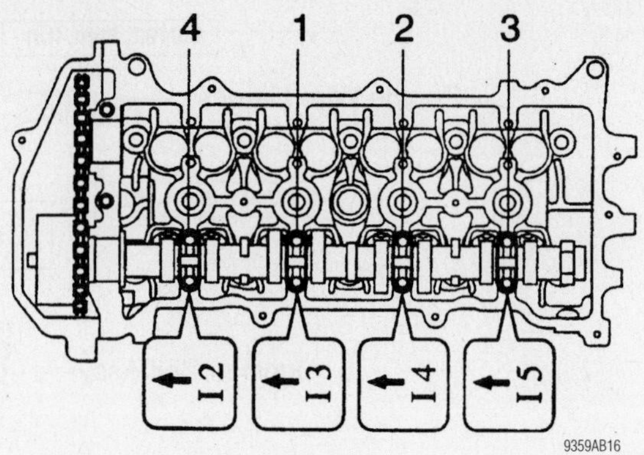

9359AB16

Camshaft cap bolt tightening sequence—1ZZ-FE engine

remove the bearing cap bolts from the other camshaft, in the proper sequence.

15. Remove the camshaft while holding the timing chain.

WARNING

Do not let anything drop down into the timing chain cover while the camshafts are removed.

16. Tie the timing chain with a string as shown, to prevent it from dropping down into the timing chain cover.

To install:

17. Position the camshaft on the cylinder head, then install the timing chain on the cam timing gear, with the painted links aligned with the marks on the timing gear.

18. Check the front marks and numbers and torque the camshaft cap bolts, in sequence, to 10 ft. lbs. (13 Nm) for 1ZZ-FE engine, or to 14 ft. lbs. (19 Nm) for 2ZZ-GE engines.

19. Put camshaft No. 2 on the cylinder head, with the painted links of the chain aligned with the mark on the timing gear.

20. Tighten the camshaft gear set bolt temporarily.

21. Check the front marks and numbers and torque the camshaft cap bolts, in sequence, to 10 ft. lbs. (13 Nm). Install the No. 1 bearing cap and tighten to 17 ft. lbs. (23 Nm).

22. Hold the camshaft secure with a wrench and tighten the set bolt to 40 ft. lbs. (54 Nm). Be careful not the damage the lifters.

23. Check to be sure the matchmarks on the timing chain and cam sprockets, and the alignment of the pulley groove with the timing mark on the cover are still aligned.

24. Install the chain tensioner:

a. Make sure the O-ring is clean, then set the hook as shown.

b. Oil the tensioner, then install and tighten to 80 inch lbs. (9 Nm).

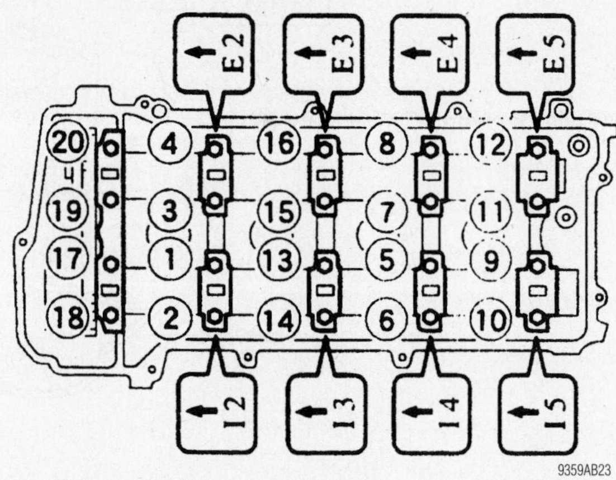

9359AB23

Camshaft cap bolt tightening sequence—2ZZ-GE engine

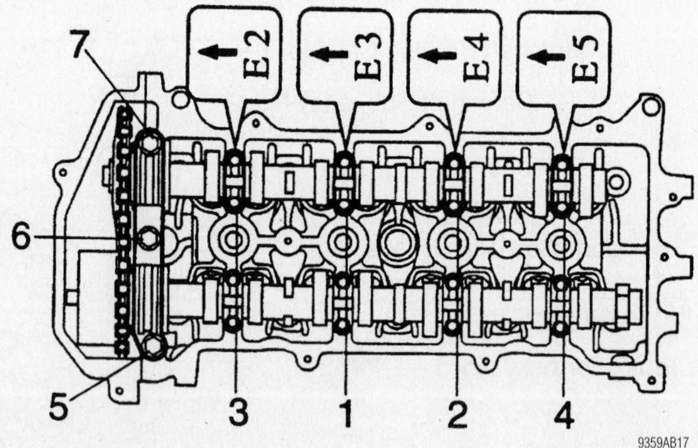

9359AB17

Camshaft cap bolt tightening sequence—1ZZ-FE

➡**When installing the tensioner, set the hook again if the hook releases the plunger.**

c. Turn the crankshaft counterclockwise, and disconnect the plunger knock pin from the hook.

d. Turn the crankshaft clockwise and check that the slipper is pushed by the plunger. If the plunger does not spring out, press the slipper into the chain tensioner with a screwdriver so that the

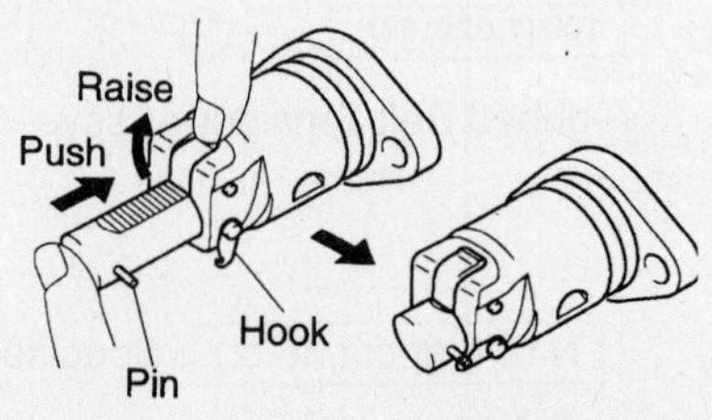

9359AB18

Set the timing chain tensioner hook properly

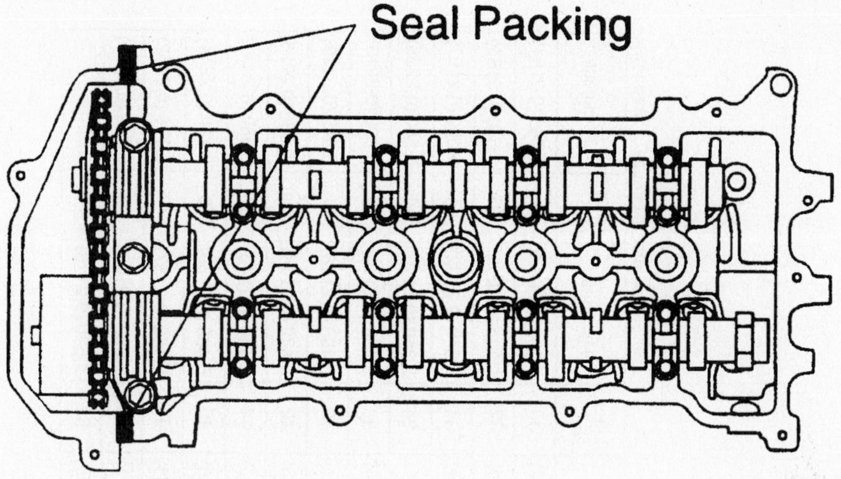

Seal packing installation locations

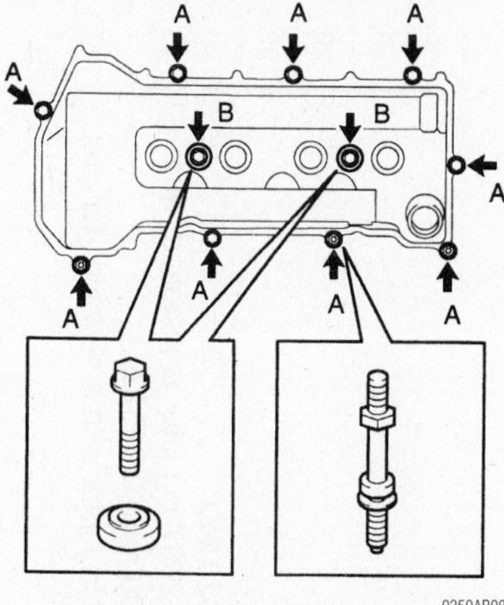

Cylinder head (valve) cover bolt locations—1ZZ-FE engine

hook is released from the knock pin and the plunger springs out.

25. Check the valve clearance and make adjustments as needed.
26. Install or connect the following:
    - Belt tensioner. Tighten the nut to 21 ft. lbs. (29 Nm) and the bolt to 51 ft. lbs. (69 Nm).
    - Cylinder head sub-assembly cover. Install seal packing into the locations shown and install within 3 minutes. Tighten the "A" bolts to 8 ft. lbs. (11 Nm) and the "B" bolts to 80 inch lbs. (9 Nm) for 1ZZ-FE engine and to 7 ft. lbs. (10 Nm) for 2ZZ-GE engines.
    - Ignition coil assembly. Torque the bolts to 80 inch lbs. (9 Nm).
    - Engine wire and tighten to 80 inch lbs. (9 Nm)
    - Right side engine mount. Tighten to 38 ft. lbs. (52 Nm).
    - Cylinder head (valve) cover
    - Negative battery cable

Valve Lash

ADJUSTMENT

1ZZ-FE Engine

➡Adjust the valve clearance when the engine is cold.

1. Before servicing the vehicle, refer to the precautions in the beginning of this section.
2. Remove or disconnect the following:
    - Negative battery cable.
    - Cylinder head covers
    - Engine wire
    - Ignition coil
    - Positive Crankcase Ventilation (PCV) hoses
    - Cylinder head cover sub-assembly
3. Set the No. 1 cylinder to Top Dead Center (TDC) of the compressor stroke as follows:
    a. Turn the crankshaft pulley, and align its groove with the "0" timing mark of the timing chain cover.
    b. Make sure the point marks of the camshaft timing sprockets and VVT timing sprockets are in a straight line as shown. If not, turn the crankshaft 1 complete revolution (360°) and align the marks.
4. Check the valve clearance of the first set of the valves shown:
    a. Use a feeler gauge to measure the clearance between the valve lifter and camshaft. The clearance of the intake valves should be 0.0059–0.0098 in. (0.15–0.25mm). The clearance of the exhaust valves should be 0.0098–0.0138 in. (0.25–0.35mm).
    b. Note the out-of-specification valve clearance measurements. You will need them later to determine the required replacement valve lifter.
    c. Turn the crankshaft 1 revolution (360°) to set the No. 4 cylinder to TDC.
5. Check the valve clearance of the second set of the valves shown:
    a. Use a feeler gauge to measure the clearance between the valve lifter and camshaft. The clearance of the intake valves should be 0.0059–0.0098 in. (0.15–0.25mm). The clearance of the exhaust valves should be 0.0098–0.0138 in. (0.25–0.35mm).
    b. Note the out-of-specification valve clearance measurements. You will need them later to determine the required replacement valve lifter.
6. Remove or disconnect the following:
    - Drive belt
    - Right side engine mount
    - Drive belt tensioner

⁂ WARNING

DO NOT turn the crankshaft while the tensioner is removed!

7. Set the No. 1 cylinder to TDC of the compression stroke.
    - Camshafts
    - Valve lifters.
8. Use a micrometer to measure the thickness of the used lifter. Calculate the thickness of a new lifter. so the valve clearance comes within the specified value:
    a. A: Thickness of new lifter.
    b. B: Thickness of used lifter.
    c. C: Measured valve clearance.
    d. Intake valve clearance: A = B + (C−0.0079 in. (0.20mm).
    e. Exhaust valve clearance: A = B + (C−0.0118 in. (0.30mm).
    f. Select a new lifter with a thickness as close as possible to the calculated values. Lifters come in 35 sizes in increments of 0.0008 in. (0.020mm) from 0.1992–0.2260 in (5.060–5.740mm).
9. Install or connect the following:
    - Camshafts
    - Drive belt tensioner
    - Right hand engine mount
    - Cylinder head (valve) cover sub-assembly
    - Ignition coil
    - Engine wire
    - Cylinder head (valve) cover
    - Negative battery cable

1ZZ–FE: Valve Lifter Selection Chart (Intake)

Measured clearance mm (in.) / Installed lifter thickness mm (in.)	5.060 (0.1992)	5.080 (0.2000)	5.100 (0.2008)	5.120 (0.2016)	5.140 (0.2024)	5.160 (0.2031)	5.180 (0.2039)	5.200 (0.2047)	5.210 (0.2051)	5.220 (0.2055)	5.230 (0.2059)	5.240 (0.2063)	5.250 (0.2067)	5.260 (0.2071)	5.270 (0.2075)	5.280 (0.2079)	5.290 (0.2083)	5.300 (0.2087)	5.310 (0.2091)	5.320 (0.2094)	5.330 (0.2098)	5.340 (0.2102)	5.350 (0.2106)	5.360 (0.2110)	5.370 (0.2114)	5.380 (0.2118)	5.390 (0.2122)	5.400 (0.2126)	5.410 (0.2130)	5.420 (0.2134)	5.430 (0.2138)	5.440 (0.2142)	5.450 (0.2146)	5.460 (0.2150)	5.470 (0.2154)	5.480 (0.2157)	5.490 (0.2161)	5.500 (0.2165)	5.510 (0.2169)	5.520 (0.2173)	5.530 (0.2177)	5.540 (0.2181)	5.550 (0.2185)	5.560 (0.2189)	5.570 (0.2193)	5.580 (0.2197)	5.590 (0.2201)	5.600 (0.2205)	5.620 (0.2213)	5.640 (0.2220)	5.660 (0.2228)	5.680 (0.2236)	5.700 (0.2244)	5.720 (0.2252)	5.740 (0.2260)
0.000 – 0.030 (0.0000 – 0.0012)									06	06	06	06	06	08	10	10	12	12	14	14	16	16	18	18	20	20	22	22	24	24	26	26	28	28	30	30	32	32	34	34	36	36	38	38	40	40	42	42	44	46	48	50	52	54	56
0.031 – 0.050 (0.0012 – 0.0020)							06	06	06	06	08	08	10	10	12	12	14	14	16	16	18	18	20	20	22	22	24	24	26	26	28	28	30	30	32	32	34	34	36	36	38	38	40	40	42	42	44	44	46	48	50	52	54	56	58
0.051 – 0.070 (0.0020 – 0.0028)						06	06	06	08	08	10	10	12	12	14	14	16	16	18	18	20	20	22	22	24	24	26	26	28	28	30	30	32	32	34	34	36	36	38	38	40	40	42	42	44	44	46	46	48	50	52	54	56	58	60
0.071 – 0.090 (0.0028 – 0.0035)					06	06	06	08	10	10	12	12	14	14	16	16	18	18	20	20	22	22	24	24	26	26	28	28	30	30	32	32	34	34	36	36	38	38	40	40	42	42	44	44	46	46	48	48	50	52	54	56	58	60	62
0.091 – 0.110 (0.0036 – 0.0043)				06	06	06	08	10	12	12	14	14	16	16	18	18	20	20	22	22	24	24	26	26	28	28	30	30	32	32	34	34	36	36	38	38	40	40	42	42	44	44	46	46	48	48	50	50	52	54	56	58	60	62	64
0.111 – 0.130 (0.0044 – 0.0051)			06	06	06	08	10	12	14	14	16	16	18	18	20	20	22	22	24	24	26	26	28	28	30	30	32	32	34	34	36	36	38	38	40	40	42	42	44	44	46	46	48	48	50	50	52	52	54	56	58	60	62	64	66
0.131 – 0.149 (0.0052 – 0.0059)		06	06	06	08	10	12	14	16	16	18	18	20	20	22	22	24	24	26	26	28	28	30	30	32	32	34	34	36	36	38	38	40	40	42	42	44	44	46	46	48	48	50	50	52	52	54	54	56	58	60	62	64	66	68
0.150 – 0.250 (0.0059 – 0.0098)																																																							
0.251 – 0.270 (0.0099 – 0.0106)	12	14	16	18	20	22	24	26	28	28	30	30	32	32	34	34	36	36	38	38	40	40	42	42	44	44	46	46	48	48	50	50	52	52	54	54	56	56	58	58	60	60	62	62	64	64	66	66	68	70	72	74	74	74	
0.271 – 0.290 (0.0107 – 0.0114)	14	16	18	20	22	24	26	28	30	30	32	32	34	34	36	36	38	38	40	40	42	42	44	44	46	46	48	48	50	50	52	52	54	54	56	56	58	58	60	60	62	62	64	64	66	66	68	68	70	72	74	74	74		
0.291 – 0.310 (0.0115 – 0.0122)	16	18	20	22	24	26	28	30	32	32	34	34	36	36	38	38	40	40	42	42	44	44	46	46	48	48	50	50	52	52	54	54	56	56	58	58	60	60	62	62	64	64	66	66	68	68	70	70	72	74	74	74			
0.311 – 0.330 (0.0122 – 0.0130)	18	20	22	24	26	28	30	32	34	34	36	36	38	38	40	40	42	42	44	44	46	46	48	48	50	50	52	52	54	54	56	56	58	58	60	60	62	62	64	64	66	66	68	68	70	70	72	72	74	74	74				
0.331 – 0.350 (0.0130 – 0.0138)	20	22	24	26	28	30	32	34	36	36	38	38	40	40	42	42	44	44	46	46	48	48	50	50	52	52	54	54	56	56	58	58	60	60	62	62	64	64	66	66	68	68	70	70	72	72	74	74	74	74					
0.351 – 0.370 (0.0138 – 0.0146)	22	24	26	28	30	32	34	36	38	38	40	40	42	42	44	44	46	46	48	48	50	50	52	52	54	54	56	56	58	58	60	60	62	62	64	64	66	66	68	68	70	70	72	72	74	74	74	74	74						
0.371 – 0.390 (0.0146 – 0.0154)	24	26	28	30	32	34	36	38	40	40	42	42	44	44	46	46	48	48	50	50	52	52	54	54	56	56	58	58	60	60	62	62	64	64	66	66	68	68	70	70	72	72	74	74	74	74	74	74							
0.391 – 0.410 (0.0154 – 0.0161)	26	28	30	32	34	36	38	40	42	42	44	44	46	46	48	48	50	50	52	52	54	54	56	56	58	58	60	60	62	62	64	64	66	66	68	68	70	70	72	72	74	74	74	74	74	74	74								
0.411 – 0.430 (0.0162 – 0.0169)	28	30	32	34	36	38	40	42	44	44	46	46	48	48	50	50	52	52	54	54	56	56	58	58	60	60	62	62	64	64	66	66	68	68	70	70	72	72	74	74	74	74	74	74											
0.431 – 0.450 (0.0170 – 0.0177)	30	32	34	36	38	40	42	44	46	46	48	48	50	50	52	52	54	54	56	56	58	58	60	60	62	62	64	64	66	66	68	68	70	70	72	72	74	74	74	74	74	74													
0.451 – 0.470 (0.0178 – 0.0185)	32	34	36	38	40	42	44	46	48	48	50	50	52	52	54	54	56	56	58	58	60	60	62	62	64	64	66	66	68	68	70	70	72	72	74	74	74	74	74	74															
0.471 – 0.490 (0.0185 – 0.0193)	34	36	38	40	42	44	46	48	50	50	52	52	54	54	56	56	58	58	60	60	62	62	64	64	66	66	68	68	70	70	72	72	74	74	74	74	74	74																	
0.491 – 0.510 (0.0193 – 0.0201)	36	38	40	42	44	46	48	50	52	52	54	54	56	56	58	58	60	60	62	62	64	64	66	66	68	68	70	70	72	72	74	74	74	74	74	74																			
0.511 – 0.530 (0.0201 – 0.0209)	38	40	42	44	46	48	50	52	54	54	56	56	58	58	60	60	62	62	64	64	66	66	68	68	70	70	72	72	74	74	74	74	74	74																					
0.531 – 0.550 (0.0209 – 0.0217)	40	42	44	46	48	50	52	54	56	56	58	58	60	60	62	62	64	64	66	66	68	68	70	70	72	72	74	74	74	74	74	74																							
0.551 – 0.570 (0.0217 – 0.0224)	42	44	46	48	50	52	54	56	58	58	60	60	62	62	64	64	66	66	68	68	70	70	72	72	74	74	74	74	74	74																									
0.571 – 0.590 (0.0225 – 0.0232)	44	46	48	50	52	54	56	58	60	60	62	62	64	64	66	66	68	68	70	70	72	72	74	74	74	74	74	74																											
0.591 – 0.610 (0.0233 – 0.0240)	46	48	50	52	54	56	58	60	62	62	64	64	66	66	68	68	70	70	72	72	74	74	74	74	74	74																													
0.611 – 0.630 (0.0241 – 0.0248)	48	50	52	54	56	58	60	62	64	64	66	66	68	68	70	70	72	72	74	74	74	74	74	74																															
0.631 – 0.650 (0.0248 – 0.0256)	50	52	54	56	58	60	62	64	66	66	68	68	70	70	72	72	74	74	74	74	74	74																																	
0.651 – 0.670 (0.0256 – 0.0264)	52	54	56	58	60	62	64	66	68	68	70	70	72	72	74	74	74	74	74	74																																			
0.671 – 0.690 (0.0264 – 0.0272)	54	56	58	60	62	64	66	68	70	70	72	72	74	74	74	74	74	74																																					
0.691 – 0.710 (0.0272 – 0.0280)	56	58	60	62	64	66	68	70	72	72	74	74	74	74	74	74																																							
0.711 – 0.730 (0.0280 – 0.0287)	58	60	62	64	66	68	70	72	74	74	74	74	74	74																																									
0.731 – 0.750 (0.0288 – 0.0295)	60	62	64	66	68	70	72	74	74	74	74	74																																											
0.751 – 0.770 (0.0296 – 0.0303)	62	64	66	68	70	72	74	74	74	74																																													
0.771 – 0.790 (0.0304 – 0.0311)	64	66	68	70	72	74	74	74																																															
0.791 – 0.810 (0.0311 – 0.0319)	66	68	70	72	74	74	74																																																
0.811 – 0.830 (0.0319 – 0.0327)	68	70	72	74	74	74																																																	
0.831 – 0.850 (0.0327 – 0.0335)	70	72	74	74	74																																																		
0.851 – 0.870 (0.0335 – 0.0343)	72	74	74	74																																																			
0.871 – 0.890 (0.0343 – 0.0350)	74	74	74																																																				
0.891 – 0.910 (0.0351 – 0.0358)	74	74																																																					
0.911 – 0.930 (0.0359 – 0.0366)	74																																																						

New lifter thickness mm (in.)

Lifter No.	Thickness	Lifter No.	Thickness	Lifter No.	Thickness
06	5.060 (0.1992)	30	5.300 (0.2087)	54	5.540 (0.2181)
08	5.080 (0.2000)	32	5.320 (0.2094)	56	5.560 (0.2189)
10	5.100 (0.2008)	34	5.340 (0.2102)	58	5.580 (0.2197)
12	5.120 (0.2016)	36	5.360 (0.2110)	60	5.600 (0.2205)
14	5.140 (0.2024)	38	5.380 (0.2118)	62	5.620 (0.2213)
16	5.160 (0.2031)	40	5.400 (0.2126)	64	5.640 (0.2220)
18	5.180 (0.2039)	42	5.420 (0.2134)	66	5.660 (0.2228)
20	5.200 (0.2047)	44	5.440 (0.2142)	68	5.680 (0.2236)
22	5.220 (0.2055)	46	5.460 (0.2150)	70	5.700 (0.2244)
24	5.240 (0.2063)	48	5.480 (0.2157)	72	5.720 (0.2252)
26	5.260 (0.2071)	50	5.500 (0.2165)	74	5.740 (0.2260)
28	5.280 (0.2079)	52	5.520 (0.2173)		

Intake valve clearance (Cold):
0.15 – 0.25 mm (0.006 – 0.010 in.)

EXAMPLE: The 5.250 mm (0.2067 in.) lifter is installed, and the measured clearance is 0.400 mm (0.0157 in.).
Replace the 5.250 mm (0.2067 in.) lifter with a new No. 48 lifter.

9307WG70

Adjusting shim chart (intake)—1ZZ-FE engine

1ZZ–FE: Valve Lifter Selection Chart (Exhaust)

Measured clearance mm (in.) / Installed lifter thickness mm (in.)	5.060 (0.1992)	5.080 (0.2000)	5.100 (0.2008)	5.120 (0.2016)	5.140 (0.2024)	5.160 (0.2031)	5.180 (0.2039)	5.200 (0.2047)	5.210 (0.2051)	5.220 (0.2055)	5.230 (0.2059)	5.240 (0.2063)	5.250 (0.2067)	5.260 (0.2071)	5.270 (0.2075)	5.280 (0.2079)	5.290 (0.2083)	5.300 (0.2087)	5.310 (0.2091)	5.320 (0.2094)	5.330 (0.2098)	5.340 (0.2102)	5.350 (0.2106)	5.360 (0.2110)	5.370 (0.2114)	5.380 (0.2118)	5.390 (0.2122)	5.400 (0.2126)	5.410 (0.2130)	5.420 (0.2134)	5.430 (0.2138)	5.440 (0.2142)	5.450 (0.2146)	5.460 (0.2150)	5.470 (0.2154)	5.480 (0.2157)	5.490 (0.2161)	5.500 (0.2165)	5.510 (0.2169)	5.520 (0.2173)	5.530 (0.2177)	5.540 (0.2181)	5.550 (0.2185)	5.560 (0.2189)	5.570 (0.2193)	5.580 (0.2197)	5.590 (0.2201)	5.600 (0.2205)	5.620 (0.2213)	5.640 (0.2220)	5.660 (0.2228)	5.680 (0.2236)	5.700 (0.2244)	5.720 (0.2252)	5.740 (0.2260)
0.000 – 0.030 (0.0000 – 0.0012)																			06	06	06	06	08	08	10	10	12	12	14	14	16	16	18	18	20	20	22	22	24	24	26	26	28	28	30	30	32	32	34	36	38	40	42	44	46
0.031 – 0.050 (0.0012 – 0.0020)																06	06	06	06	06	08	08	10	10	12	12	14	14	16	16	18	18	20	20	22	22	24	24	26	26	28	28	30	30	32	32	34	34	36	38	40	42	44	46	48
0.051 – 0.070 (0.0020 – 0.0028)														06	06	06	06	06	08	08	10	10	12	12	14	14	16	16	18	18	20	20	22	22	24	24	26	26	28	28	30	30	32	32	34	34	36	36	38	40	42	44	46	48	50
0.071 – 0.090 (0.0028 – 0.0035)												06	06	06	06	06	08	08	10	10	12	12	14	14	16	16	18	18	20	20	22	22	24	24	26	26	28	28	30	30	32	32	34	34	36	36	38	38	40	42	44	46	48	50	52
0.091 – 0.110 (0.0036 – 0.0043)											06	06	06	06	08	08	10	10	12	12	14	14	16	16	18	18	20	20	22	22	24	24	26	26	28	28	30	30	32	32	34	34	36	36	38	38	40	40	42	44	46	48	50	52	54
0.111 – 0.130 (0.0044 – 0.0051)									06	06	06	06	08	08	10	10	12	12	14	14	16	16	18	18	20	20	22	22	24	24	26	26	28	28	30	30	32	32	34	34	36	36	38	38	40	40	42	42	44	46	48	50	52	54	56
0.131 – 0.150 (0.0052 – 0.0059)								06	06	06	08	08	10	10	12	12	14	14	16	16	18	18	20	20	22	22	24	24	26	26	28	28	30	30	32	32	34	34	36	36	38	38	40	40	42	42	44	44	46	48	50	52	54	56	58
0.151 – 0.170 (0.0059 – 0.0067)						06	06	06	08	08	10	10	12	12	14	14	16	16	18	18	20	20	22	22	24	24	26	26	28	28	30	30	32	32	34	34	36	36	38	38	40	40	42	42	44	44	46	46	48	50	52	54	56	58	60
0.171 – 0.190 (0.0067 – 0.0075)					06	06	06	08	10	10	12	12	14	14	16	16	18	18	20	20	22	22	24	24	26	26	28	28	30	30	32	32	34	34	36	36	38	38	40	40	42	42	44	44	46	46	48	48	50	52	54	56	58	60	62
0.191 – 0.210 (0.0075 – 0.0083)				06	06	06	08	10	12	12	14	14	16	16	18	18	20	20	22	22	24	24	26	26	28	28	30	30	32	32	34	34	36	36	38	38	40	40	42	42	44	44	46	46	48	48	50	50	52	54	56	58	60	62	64
0.211 – 0.230 (0.0083 – 0.0091)			06	06	06	08	10	12	14	14	16	16	18	18	20	20	22	22	24	24	26	26	28	28	30	30	32	32	34	34	36	36	38	38	40	40	42	42	44	44	46	46	48	48	50	50	52	52	54	56	58	60	62	64	66
0.231 – 0.249 (0.0091 – 0.0098)		06	06	06	08	10	12	14	16	16	18	18	20	20	22	22	24	24	26	26	28	28	30	30	32	32	34	34	36	36	38	38	40	40	42	42	44	44	46	46	48	48	50	50	52	52	54	54	56	58	60	62	64	66	68
0.250 – 0.350 (0.0098 – 0.0138)																																																							
0.351 – 0.370 (0.0138 – 0.0146)	12	14	16	18	20	22	24	26	28	28	30	30	32	32	34	34	36	36	38	38	40	40	42	42	44	44	46	46	48	48	50	50	52	52	54	54	56	56	58	58	60	60	62	62	64	64	66	66	68	70	72	74	74	74	
0.371 – 0.390 (0.0146 – 0.0154)	14	16	18	20	22	24	26	28	30	30	32	32	34	34	36	36	38	38	40	40	42	42	44	44	46	46	48	48	50	50	52	52	54	54	56	56	58	58	60	60	62	62	64	64	66	66	68	68	70	72	74	74	74		
0.391 – 0.410 (0.0154 – 0.0161)	16	18	20	22	24	26	28	30	32	32	34	34	36	36	38	38	40	40	42	42	44	44	46	46	48	48	50	50	52	52	54	54	56	56	58	58	60	60	62	62	64	64	66	66	68	68	70	70	72	74	74	74			
0.411 – 0.430 (0.0162 – 0.0169)	18	20	22	24	26	28	30	32	34	34	36	36	38	38	40	40	42	42	44	44	46	46	48	48	50	50	52	52	54	54	56	56	58	58	60	60	62	62	64	64	66	66	68	68	70	70	72	72	74	74	74				
0.431 – 0.450 (0.0170 – 0.0177)	20	22	24	26	28	30	32	34	36	36	38	38	40	40	42	42	44	44	46	46	48	48	50	50	52	52	54	54	56	56	58	58	60	60	62	62	64	64	66	68	68	70	70	72	72	74	74	74	74	74					
0.451 – 0.470 (0.0178 – 0.0185)	22	24	26	28	30	32	34	36	38	38	40	40	42	42	44	44	46	46	48	48	50	50	52	52	54	54	56	56	58	58	60	60	62	62	64	64	66	66	68	68	70	70	72	72	74	74	74	74	74						
0.471 – 0.490 (0.0185 – 0.0193)	24	26	28	30	32	34	36	38	40	40	42	42	44	44	46	46	48	48	50	50	52	52	54	54	56	56	58	58	60	60	62	62	64	64	66	66	68	68	70	70	72	72	74	74	74	74	74	74							
0.491 – 0.510 (0.0193 – 0.0201)	26	28	30	32	34	36	38	40	42	42	44	44	46	46	48	48	50	50	52	52	54	54	56	56	58	58	60	60	62	62	64	64	66	66	68	68	70	70	72	72	74	74	74	74	74	74									
0.511 – 0.530 (0.0201 – 0.0209)	28	30	32	34	36	38	40	42	44	44	46	46	48	48	50	50	52	52	54	54	56	56	58	58	60	60	62	62	64	64	66	66	68	68	70	70	72	72	74	74	74	74	74	74											
0.531 – 0.550 (0.0209 – 0.0217)	30	32	34	36	38	40	42	44	46	46	48	48	50	50	52	52	54	54	56	56	58	58	60	60	62	62	64	64	66	66	68	68	70	70	72	72	74	74	74	74	74	74													
0.551 – 0.570 (0.0217 – 0.0224)	32	34	36	38	40	42	44	46	48	48	50	50	52	52	54	54	56	56	58	58	60	60	62	62	64	64	66	66	68	68	70	70	72	72	74	74	74	74	74	74															
0.571 – 0.590 (0.0225 – 0.0232)	34	36	38	40	42	44	46	48	50	50	52	52	54	54	56	56	58	58	60	60	62	62	64	64	66	66	68	68	70	70	72	72	74	74	74	74	74	74																	
0.591 – 0.610 (0.0233 – 0.0240)	36	38	40	42	44	46	48	50	52	52	54	54	56	56	58	58	60	60	62	62	64	64	66	66	68	68	70	70	72	72	74	74	74	74	74	74																			
0.611 – 0.630 (0.0241 – 0.0248)	38	40	42	44	46	48	50	52	54	54	56	56	58	58	60	60	62	62	64	64	66	66	68	68	70	70	72	72	74	74	74	74	74	74																					
0.631 – 0.650 (0.0248 – 0.0256)	40	42	44	46	48	50	52	54	56	56	58	58	60	60	62	62	64	64	66	66	68	68	70	70	72	72	74	74	74	74	74	74	74																						
0.651 – 0.670 (0.0256 – 0.0264)	42	44	46	48	50	52	54	56	58	58	60	60	62	62	64	64	66	66	68	68	70	70	72	72	74	74	74	74	74	74																									
0.671 – 0.690 (0.0264 – 0.0272)	44	46	48	50	52	54	56	58	60	60	62	62	64	64	66	66	68	68	70	70	72	72	74	74	74	74	74	74																											
0.691 – 0.710 (0.0272 – 0.0280)	46	48	50	52	54	56	58	60	62	62	64	64	66	66	68	68	70	70	72	72	74	74	74	74	74	74																													
0.711 – 0.730 (0.0280 – 0.0287)	48	50	52	54	56	58	60	62	64	64	66	66	68	68	70	70	72	72	74	74	74	74	74	74																															
0.731 – 0.750 (0.0288 – 0.0295)	50	52	54	56	58	60	62	64	66	66	68	68	70	70	72	72	74	74	74	74	74	74																																	
0.751 – 0.770 (0.0296 – 0.0303)	52	54	56	58	60	62	64	66	68	68	70	70	72	72	74	74	74	74	74	74																																			
0.771 – 0.790 (0.0304 – 0.0311)	54	56	58	60	62	64	66	68	70	70	72	72	74	74	74	74	74	74																																					
0.791 – 0.810 (0.0311 – 0.0319)	56	58	60	62	64	66	68	70	72	72	74	74	74	74	74	74																																							
0.811 – 0.830 (0.0319 – 0.0327)	58	60	62	64	66	68	70	72	74	74	74	74	74	74																																									
0.831 – 0.850 (0.0327 – 0.0335)	60	62	64	66	68	70	72	74	74	74	74	74																																											
0.851 – 0.870 (0.0335 – 0.0343)	62	64	66	68	70	72	74	74	74	74																																													
0.871 – 0.890 (0.0343 – 0.0350)	64	66	68	70	72	74	74	74																																															
0.891 – 0.910 (0.0351 – 0.0358)	66	68	70	72	74	74	74																																																
0.911 – 0.930 (0.0359 – 0.0366)	68	70	72	74	74	74																																																	
0.931 – 0.950 (0.0367 – 0.0374)	70	72	74	74	74																																																		
0.951 – 0.970 (0.0374 – 0.0382)	72	74	74	74																																																			
0.971 – 0.990 (0.0382 – 0.0390)	74	74	74																																																				
0.991 – 1.010 (0.0390 – 0.0398)	74	74																																																					
1.011 – 1.030 (0.0398 – 0.0406)	74																																																						

New lifter thickness mm (in.)

Lifter No.	Thickness	Lifter No.	Thickness	Lifter No.	Thickness
06	5.060 (0.1992)	30	5.300 (0.2087)	54	5.540 (0.2181)
08	5.080 (0.2000)	32	5.320 (0.2094)	56	5.560 (0.2189)
10	5.100 (0.2008)	34	5.340 (0.2102)	58	5.580 (0.2197)
12	5.120 (0.2016)	36	5.360 (0.2110)	60	5.600 (0.2205)
14	5.140 (0.2024)	38	5.380 (0.2118)	62	5.620 (0.2213)
16	5.160 (0.2031)	40	5.400 (0.2126)	64	5.640 (0.2220)
18	5.180 (0.2039)	42	5.420 (0.2134)	66	5.660 (0.2228)
20	5.200 (0.2047)	44	5.440 (0.2142)	68	5.680 (0.2236)
22	5.220 (0.2055)	46	5.460 (0.2150)	70	5.700 (0.2244)
24	5.240 (0.2063)	48	5.480 (0.2157)	72	5.720 (0.2252)
26	5.260 (0.2071)	50	5.500 (0.2165)	74	5.740 (0.2260)
28	5.280 (0.2079)	52	5.520 (0.2173)		

Exhaust valve clearance (Cold):
0.25 – 0.35 mm (0.010 – 0.014 in.)

EXAMPLE: The 5.340 mm (0.2102 in.) lifter is installed, and the measured clearance is 0.440 mm (0.0173 in.). Replace the 5.340 mm (0.2102 in.) lifter with a new No. 48 lifter.

9307WG71

Adjusting shim chart (exhaust)—1ZZ-FE engine

Measure clearance mm(in.) / Installed shim thickness mm(in.)	2.000 (0.0787)	2.020 (0.0795)	2.040 (0.0803)	2.060 (0.0811)	2.080 (0.0819)	2.100 (0.0827)	2.120 (0.0835)	2.140 (0.0843)	2.160 (0.0850)	2.180 (0.0858)	2.190 (0.0862)	2.200 (0.0866)	2.210 (0.0870)	2.220 (0.0874)	2.230 (0.0878)	2.240 (0.0882)	2.250 (0.0886)	2.260 (0.0890)	2.270 (0.0894)	2.280 (0.0898)	2.290 (0.0902)	2.300 (0.0906)	2.310 (0.0909)	2.320 (0.0913)	2.330 (0.0917)	2.340 (0.0921)	2.340 (0.0921)	2.350 (0.0925)	2.360 (0.0929)	2.370 (0.0933)
0.000 - 0.030 (0.0000 - 0.0012)										00	02	02	04	04	06	06	08	08	10	10	12	12	14	14	16	16	18	18	20	20
0.031 - 0.050 (0.0012 - 0.0020)								00	02	04	06	06	08	08	10	10	12	12	14	14	16	16	18	18	20	20	22	22	24	24
0.051 - 0.070 (0.0020 - 0.0028)							00	02	04	06	08	08	10	10	12	12	14	14	16	16	18	18	20	20	22	22	24	24	26	26
0.071 - 0.090 (0.0028 - 0.0035)					00	02	04	06	08	10	12	12	14	14	16	16	18	18	20	20	22	22	24	24	26	26	28	28	30	30
0.091 - 0.099 (0.0036 - 0.0039)				00	02	04	06	08	10	12	14	14	16	16	18	18	20	20	22	22	24	24	26	26	28	28	30	30	32	32
0.100 - 0.160 (0.0039 - 0.0063)																														
0.161 - 0.180 (0.0063 - 0.0071)	06	80	10	12	14	16	18	20	22	24	26	26	28	28	30	30	32	32	34	34	36	36	38	38	40	40	42	42	44	44
0.181 - 0.200 (0.0071 - 0.0079)	10	12	14	16	18	20	22	24	26	28	30	30	32	32	34	34	36	36	38	38	40	40	42	42	44	44	46	46	48	48
0.201 - 0.220 (0.0079 - 0.0087)	12	14	16	18	20	22	24	26	28	30	32	32	34	34	36	36	38	38	40	40	42	42	44	44	46	46	48	48	50	50
0.221 - 0.240 (0.0087 - 0.0094)	16	18	20	22	24	26	28	30	32	34	36	36	38	38	40	40	42	42	44	44	46	46	48	48	50	50	52	52	54	54
0.241 - 0.260 (0.0095 - 0.0102)	18	20	22	24	26	28	30	32	34	36	38	38	40	40	42	42	44	44	46	46	48	48	50	50	52	52	54	54	56	56
0.261 - 0.280 (0.0103 - 0.0110)	22	24	26	28	30	32	34	36	38	40	42	42	44	44	46	46	48	48	50	50	52	52	54	54	56	56	58	58	60	60
0.281 - 0.300 (0.0111 - 0.0118)	24	26	28	30	32	34	36	38	40	42	44	44	46	46	48	48	50	50	52	52	54	54	56	56	58	58	60	60	62	62
0.301 - 0.320 (0.0119 - 0.0126)	28	30	32	34	36	38	40	42	44	46	48	48	50	50	52	52	54	54	56	56	58	58	60	60	62	62	64	64	66	66
0.321 - 0.340 (0.0126 - 0.0134)	30	32	34	36	38	40	42	44	46	48	50	50	52	52	54	54	56	56	58	58	60	60	62	62	64	64	66	66	68	68
0.341 - 0.360 (0.0134 - 0.0142)	34	36	38	40	42	44	46	48	50	52	54	54	56	56	58	58	60	60	62	62	64	64	66	66	68	68	70	70	72	72
0.361 - 0.380 (0.0142 - 0.0150)	36	38	40	42	44	46	48	50	52	54	56	56	58	58	60	60	62	62	64	64	66	66	68	68	70	70	72	72	74	74
0.381 - 0.400 (0.0150 - 0.0157)	40	42	44	46	48	50	52	54	56	58	60	60	62	62	64	64	66	66	68	68	70	70	72	72	74	74	76	76	78	78
0.401 - 0.420 (0.0158 - 0.0165)	42	44	46	48	50	52	54	56	58	60	62	62	64	64	66	66	68	68	70	70	72	72	74	74	76	76	78	78	80	80
0.421 - 0.440 (0.0166 - 0.0173)	46	48	50	52	54	56	58	60	62	64	66	66	68	68	70	70	72	72	74	74	76	76	78	78	80	80				
0.441 - 0.460 (0.0174 - 0.0181)	48	50	52	54	56	58	60	62	64	66	68	68	70	70	72	72	74	74	76	76	78	78	80	80						
0.461 - 0.480 (0.0181 - 0.0189)	52	54	56	58	60	62	64	66	68	70	72	72	74	74	76	76	78	78	80	80										
0.481 - 0.500 (0.0189 - 0.0197)	54	56	58	60	62	64	66	68	70	72	74	74	76	76	78	78	80	80												
0.501 - 0.520 (0.0197 - 0.0205)	58	60	62	64	66	68	70	72	74	76	78	78	80	80																
0.521 - 0.540 (0.0205 - 0.0213)	60	62	64	66	68	70	72	74	76	78	80	80																		
0.541 - 0.560 (0.0213 - 0.0220)	64	66	68	70	72	74	76	78	80																					
0.561 - 0.580 (0.0221 - 0.0228)	66	68	70	72	74	76	78	80																						
0.581 - 0.600 (0.0229 - 0.0236)	70	72	74	76	78	80																								
0.601 - 0.620 (0.0237 - 0.0244)	72	74	76	78	80																									
0.621 - 0.640 (0.0244 - 0.0252)	76	78	80																											
0.641 - 0.660 (0.0252 - 0.0260)	78	80																												
0.661 - 0.680 (0.0260 - 0.0268)	80																													

Measure clearance mm(in.) / Installed shim thickness mm(in.)	2.380 (0.0937)	2.390 (0.0941)	2.400 (0.0945)	2.410 (0.0949)	2.420 (0.0953)	2.430 (0.0957)	2.440 (0.0961)	2.450 (0.0965)	2.460 (0.0969)	2.470 (0.0972)	2.480 (0.0976)	2.490 (0.0980)	2.500 (0.0984)	2.510 (0.0988)	2.520 (0.0992)	2.530 (0.0996)	2.540 (0.1000)	2.550 (0.1004)	2.560 (0.1008)	2.580 (0.1016)	2.600 (0.1024)	2.620 (0.1031)	2.640 (0.1039)	2.660 (0.1047)	2.680 (0.1055)	2.700 (0.1063)	2.720 (0.1071)	2.740 (0.1079)	2.760 (0.1087)	2.780 (0.1094)	2.800 (0.1102)
0.000 - 0.030 (0.0000 - 0.0012)	22	22	24	24	26	26	28	28	30	30	32	32	34	34	36	36	38	38	40	42	44	46	48	50	52	54	56	58	60	62	64
0.031 - 0.050 (0.0012 - 0.0020)	26	26	28	28	30	30	32	32	34	34	36	36	38	38	40	40	42	42	44	46	48	50	52	54	56	58	60	62	64	66	68
0.051 - 0.070 (0.0020 - 0.0028)	28	28	30	30	32	32	34	34	36	36	38	38	40	40	42	42	44	44	46	48	50	52	54	56	58	60	62	64	66	68	70
0.071 - 0.090 (0.0028 - 0.0035)	32	32	34	34	36	36	38	38	40	40	42	42	44	44	46	46	48	48	50	52	54	56	58	60	62	64	66	68	70	72	74
0.091 - 0.099 (0.0036 - 0.0039)	34	34	36	36	38	38	40	40	42	42	44	44	46	46	48	48	50	50	52	54	56	58	60	62	64	66	68	70	72	74	76
0.100 - 0.160 (0.0039 - 0.0063)																															
0.161 - 0.180 (0.0063 - 0.0071)	46	46	48	48	50	50	52	52	54	54	56	56	58	58	60	60	62	62	64	66	68	70	72	74	76	78	80				
0.181 - 0.200 (0.0071 - 0.0079)	50	50	52	52	54	54	56	56	58	58	60	60	62	62	64	64	66	66	68	70	72	74	76	78	80						
0.201 - 0.220 (0.0079 - 0.0087)	52	52	54	54	56	56	58	58	60	60	62	62	64	64	66	66	68	68	70	72	74	76	78	80							
0.221 - 0.240 (0.0087 - 0.0094)	56	56	58	58	60	60	62	62	64	64	66	66	68	68	70	70	72	72	74	76	78	80	82								
0.241 - 0.260 (0.0095 - 0.0102)	58	58	60	60	62	62	64	64	66	66	68	68	70	70	72	72	74	74	76	78	80										
0.261 - 0.280 (0.0103 - 0.0110)	62	62	64	64	66	66	68	68	70	70	72	72	74	74	76	76	78	78	80												
0.281 - 0.300 (0.0111 - 0.0118)	64	64	66	66	68	68	70	70	72	72	74	74	76	76	78	78	80	80													
0.301 - 0.320 (0.0119 - 0.0126)	68	68	70	70	72	72	74	74	76	76	78	78	80	80																	
0.321 - 0.340 (0.0126 - 0.0134)	70	70	72	72	74	74	76	76	78	78	80	80																			
0.341 - 0.360 (0.0134 - 0.0142)	74	74	76	76	78	78	80	80																							
0.361 - 0.380 (0.0142 - 0.0150)	76	76	78	78	80	80																									
0.381 - 0.400 (0.0150 - 0.0157)	80	80																													

New Shim thickness mm (in.)

Shim No.	Thickness	Shim No.	Thickness	Shim No.	Thickness
00	2.000 (0.0787)	28	2.280 (0.0898)	56	2.560 (0.1008)
02	2.020 (0.0795)	30	2.300 (0.0906)	58	2.580 (0.1016)
04	2.040 (0.0803)	32	2.320 (0.0913)	60	2.600 (0.1024)
06	2.060 (0.0811)	34	2.340 (0.0921)	62	2.620 (0.1031)
08	2.080 (0.0819)	36	2.360 (0.0929)	64	2.640 (0.1039)
10	2.100 (0.0827)	38	2.380 (0.0937)	66	2.660 (0.1047)
12	2.120 (0.0835)	40	2.400 (0.0945)	68	2.680 (0.1055)
14	2.140 (0.0843)	42	2.420 (0.0953)	70	2.700 (0.1063)
16	2.160 (0.0850)	44	2.440 (0.0961)	72	2.720 (0.1071)
18	2.180 (0.0858)	46	2.460 (0.0969)	74	2.740 (0.1079)
20	2.200 (0.0866)	48	2.480 (0.0976)	76	2.760 (0.1087)
22	2.220 (0.0874)	50	2.500 (0.0984)	78	2.780 (0.1094)
24	2.240 (0.0882)	52	2.520 (0.0992)	80	2.800 (0.1102)
26	2.260 (0.0890)	54	2.540 (0.1000)		

Intake valve clearance (Cold):
0.08 – 0.18 mm (0.0031 – 0.0071 in.)
EXAMPLE: The 2.200 mm (0.0826 in.) shim is installed, and the measured clearance is 0.400 mm (0.0157 in.).
Replace the 2.600 mm (0.1024 in.) shim with a new No. 60 shim.

9359AB26

Adjusting shim chart (intake)—2ZZ-GE engine

Measure clearance mm(in.) / Installed shim thickness mm(in.)	2.000 (0.0787)	2.020 (0.0795)	2.040 (0.0803)	2.060 (0.0811)	2.080 (0.0819)	2.100 (0.0827)	2.120 (0.0835)	2.140 (0.0843)	2.160 (0.0850)	2.180 (0.0858)	2.190 (0.0862)	2.200 (0.0866)	2.210 (0.0870)	2.220 (0.0874)	2.230 (0.0878)	2.240 (0.0882)	2.250 (0.0886)	2.260 (0.0890)	2.270 (0.0894)	2.280 (0.0898)	2.290 (0.0902)	2.300 (0.0906)	2.310 (0.0909)	2.320 (0.0913)	2.330 (0.0917)	2.340 (0.0921)	2.340 (0.0921)	2.350 (0.0925)	2.360 (0.0929)	2.370 (0.0933)	2.380 (0.0937)	2.390 (0.0941)	2.400 (0.0945)	2.410 (0.0949)	2.420 (0.0953)	2.430 (0.0957)	2.440 (0.0961)	2.450 (0.0965)	2.460 (0.0969)	2.470 (0.0972)	2.480 (0.0976)	2.490 (0.0980)	2.500 (0.0984)	2.510 (0.0988)	2.520 (0.0992)	2.530 (0.0996)	2.540 (0.1000)	2.550 (0.1004)	2.560 (0.1008)	2.580 (0.1016)	2.600 (0.1024)	2.620 (0.1031)	2.640 (0.1039)	2.660 (0.1047)	2.680 (0.1055)	2.700 (0.1063)	2.720 (0.1071)	2.740 (0.1079)	2.760 (0.1087)	2.780 (0.1094)	2.800 (0.1102)
0.000 - 0.030 (0.0000 - 0.0012)																															00	00	02	02	04	04	06	06	08	08	10	10	12	12	14	14	16	16	18	20	22	24	26	28	30	32	34	36	38	40	42
0.031 - 0.050 (0.0012 - 0.0020)																												00	04	04	06	06	08	08	10	10	12	12	14	14	16	16	18	18	20	20	22	22	24	26	28	30	32	34	36	38	40	42	44	46	48
0.051 - 0.070 (0.0020 - 0.0028)																								00	02	02	04	04	06	06	08	08	10	10	12	12	14	14	16	16	18	18	20	20	22	22	24	24	26	28	30	32	34	36	38	40	42	44	46	48	50
0.071 - 0.090 (0.0028 - 0.0035)																					00	00	02	02	04	04	06	06	08	08	10	10	12	12	14	14	16	16	18	18	20	20	22	22	24	24	26	26	28	30	32	34	36	38	40	42	44	46	48	50	52
0.091 - 0.110 (0.0036 - 0.0043)																		00	02	02	04	04	06	06	08	08	10	10	12	12	14	14	16	16	18	18	20	20	22	22	24	24	26	26	28	28	30	30	32	34	36	38	40	42	44	46	48	50	52	54	56
0.111 - 0.130 (0.0044 - 0.0051)															00	00	02	02	04	04	06	06	08	08	10	10	12	12	14	14	16	16	18	18	20	20	22	22	24	24	26	26	28	28	30	30	32	32	34	36	38	40	42	44	46	48	50	52	54	56	58
0.131 - 0.150 (0.0052 - 0.0059)												00	02	02	04	04	06	06	08	08	10	10	12	12	14	14	16	16	18	18	20	20	22	22	24	24	26	26	28	28	30	30	32	32	34	34	36	36	38	40	42	44	46	48	50	52	54	56	58	60	62
0.151 - 0.170 (0.0059 - 0.0067)										00	02	02	04	04	06	06	08	08	10	10	12	12	14	14	16	16	18	18	20	20	22	22	24	24	26	26	28	28	30	30	32	32	34	34	36	36	38	38	40	42	44	46	48	50	52	54	56	58	60	62	64
0.171 - 0.190 (0.0067 - 0.0075)								00	02	04	06	06	08	08	10	10	12	12	14	14	16	16	18	18	20	20	22	22	24	24	26	26	28	28	30	30	32	32	34	34	36	36	38	38	40	40	42	42	44	46	48	50	52	54	56	58	60	62	64	66	68
0.191 - 0.210 (0.0075 - 0.0083)							00	02	04	06	08	08	10	10	12	12	14	14	16	16	18	18	20	20	22	22	24	24	26	26	28	28	30	30	32	32	34	34	36	36	38	38	40	40	42	42	44	44	46	48	50	52	54	56	58	60	62	64	66	68	70
0.211 - 0.230 (0.0083 - 0.0091)					00	02	04	06	08	10	12	12	14	14	16	16	18	18	20	20	22	22	24	24	26	26	28	28	30	30	32	32	34	34	36	36	38	38	40	40	42	42	44	44	46	46	48	48	50	52	54	56	58	60	62	64	66	68	70	72	74
0.231 - 0.239 (0.0091 - 0.0094)				00	02	04	06	08	10	12	14	14	16	16	18	18	20	20	22	22	24	24	26	26	28	28	30	30	32	32	34	34	36	36	38	38	40	40	42	42	44	44	46	46	48	48	50	50	52	54	56	58	60	62	64	66	68	70	72	74	76
0.301 - 0.320 (0.0119 - 0.0126)	06	08	10	12	14	16	18	20	22	24	26	26	28	28	30	30	32	32	34	34	36	36	38	38	40	40	42	42	44	44	46	46	48	48	50	50	52	52	54	54	56	56	58	58	60	60	62	62	64	66	68	70	72	74	76	78	80	80			
0.321 - 0.340 (0.0126 - 0.0134)	08	10	12	14	16	18	20	22	24	26	28	28	30	30	32	32	34	34	36	36	38	38	40	40	42	42	44	44	46	46	48	48	50	50	52	52	54	54	56	56	58	58	60	60	62	62	64	64	66	68	70	72	74	76	78	80					
0.341 - 0.360 (0.0134 - 0.0142)	12	14	16	18	20	22	24	26	28	30	32	32	34	34	36	36	38	38	40	40	42	42	44	44	46	46	48	48	50	50	52	52	54	54	56	56	58	58	60	60	62	62	64	64	66	66	68	68	70	72	74	76	78	80							
0.361 - 0.380 (0.0142 - 0.0150)	16	18	20	22	24	26	28	30	32	34	36	36	38	38	40	40	42	42	44	44	46	46	48	48	50	50	52	52	54	54	56	56	58	58	60	60	62	62	64	64	66	66	68	68	70	70	72	72	74	76	78	80									
0.381 - 0.400 (0.0150 - 0.0157)	18	20	22	24	26	28	30	32	34	36	38	38	40	40	42	42	44	44	46	46	48	48	50	50	52	52	54	54	56	56	58	58	60	60	62	62	64	64	66	66	68	68	70	70	72	72	74	74	76	78	80										
0.401 - 0.420 (0.0158 - 0.0165)	22	24	26	28	30	32	34	36	38	40	42	42	44	44	46	46	48	48	50	50	52	52	54	54	56	56	58	58	60	60	62	62	64	64	66	66	68	68	70	70	72	72	74	74	76	76	78	78	80												
0.421 - 0.440 (0.0166 - 0.0173)	24	26	28	30	32	34	36	38	40	42	44	44	46	46	48	48	50	50	52	52	54	54	56	56	58	58	60	60	62	62	64	64	66	66	68	68	70	70	72	72	74	74	76	76	78	78	80	80													
0.441 - 0.460 (0.0174 - 0.0181)	28	30	32	34	36	38	40	42	44	46	48	48	50	50	52	52	54	54	56	56	58	58	60	60	62	62	64	64	66	66	68	68	70	70	72	72	74	74	76	76	78	78	80	80																	
0.461 - 0.480 (0.0181 - 0.0189)	30	32	34	36	38	40	42	44	46	48	50	50	52	52	54	54	56	56	58	58	60	60	62	62	64	64	66	66	68	68	70	70	72	72	74	74	76	76	78	78	80	80																			
0.481 - 0.500 (0.0189 - 0.0197)	34	36	38	40	42	44	46	48	50	52	54	54	56	56	58	58	60	60	62	62	64	64	66	66	68	68	70	70	72	72	74	74	76	76	78	78	80	80																							
0.501 - 0.520 (0.0197 - 0.0205)	36	38	40	42	44	46	48	50	52	54	56	56	58	58	60	60	62	62	64	64	66	66	68	68	70	70	72	72	74	74	76	76	78	78	80	80																									
0.521 - 0.540 (0.0205 - 0.0213)	40	42	44	46	48	50	52	54	56	58	60	60	62	62	64	64	66	66	68	68	70	70	72	72	74	74	76	76	78	78	80	80																													
0.541 - 0.560 (0.0213 - 0.0220)	42	44	46	48	50	52	54	56	58	60	62	62	64	64	66	66	68	68	70	70	72	72	74	74	76	76	78	78	80	80																															
0.561 - 0.580 (0.0221 - 0.0228)	46	48	50	52	54	56	58	60	62	64	66	66	68	68	70	70	72	72	74	74	76	76	78	78	80	80																																			
0.581 - 0.600 (0.0229 - 0.0236)	48	50	52	54	56	58	60	62	64	66	68	68	70	70	72	72	74	74	76	76	78	78	80	80																																					
0.601 - 0.620 (0.0237 - 0.0244)	52	54	56	58	60	62	64	66	68	70	72	72	74	74	76	76	78	78	80	80																																									
0.621 - 0.640 (0.0244 - 0.0252)	54	56	58	60	62	64	66	68	70	72	74	74	76	76	78	78	80	80																																											
0.641 - 0.660 (0.0252 - 0.0260)	58	60	62	64	66	68	70	72	74	76	78	78	80	80																																															
0.661 - 0.680 (0.0260 - 0.0268)	60	62	64	66	68	70	72	74	76	78	80	80																																																	
0.681 - 0.700 (0.0268 - 0.0276)	64	66	68	70	72	74	76	78	80																																																				
0.701 - 0.720 (0.0276 - 0.0283)	66	68	70	72	74	76	78	80																																																					
0.721 - 0.740 (0.0284 - 0.0291)	70	72	74	76	78	80																																																							
0.741 - 0.760 (0.0292 - 0.0299)	72	74	76	78	80																																																								
0.761 - 0.780 (0.0300 - 0.0307)	76	78	80																																																										
0.781 - 0.800 (0.0307 - 0.0315)	80																																																												
0.801 - 0.820 (0.0315 - 0.0323)	80																																																												

New Shim thickness mm (in.)

Shim No.	Thickness	Shim No.	Thickness	Shim No.	Thickness
00	2.000 (0.0787)	28	2.280 (0.0898)	56	2.560 (0.1008)
02	2.020 (0.0795)	30	2.300 (0.0906)	58	2.580 (0.1016)
04	2.040 (0.0803)	32	2.320 (0.0913)	60	2.600 (0.1024)
06	2.060 (0.0811)	34	2.340 (0.0921)	62	2.620 (0.1031)
08	2.080 (0.0819)	36	2.360 (0.0929)	64	2.640 (0.1039)
10	2.100 (0.0827)	38	2.380 (0.0937)	66	2.660 (0.1047)
12	2.120 (0.0835)	40	2.400 (0.0945)	68	2.680 (0.1055)
14	2.140 (0.0843)	42	2.420 (0.0953)	70	2.700 (0.1063)
16	2.160 (0.0850)	44	2.440 (0.0961)	72	2.720 (0.1071)
18	2.180 (0.0858)	46	2.460 (0.0969)	74	2.740 (0.1079)
20	2.200 (0.0866)	48	2.480 (0.0976)	76	2.760 (0.1087)
22	2.220 (0.0874)	50	2.500 (0.0984)	78	2.780 (0.1094)
24	2.240 (0.0882)	52	2.520 (0.0992)	80	2.800 (0.1102)
26	2.260 (0.0890)	54	2.540 (0.1000)		

Exhaust valve clearance (Cold):
0.22 – 0.32 mm (0.0087 – 0.0126 in.)

EXAMPLE: The 2.200 mm (0.0862 in.) shim is installed, and the measured clearance is 0.500 mm (0.0197 in.).
Replace the 2.540 mm (0.1000 in.) shim with a new No. 54 shim.

9359AB27

Adjusting shim chart (exhaust)—2ZZ-GE engine

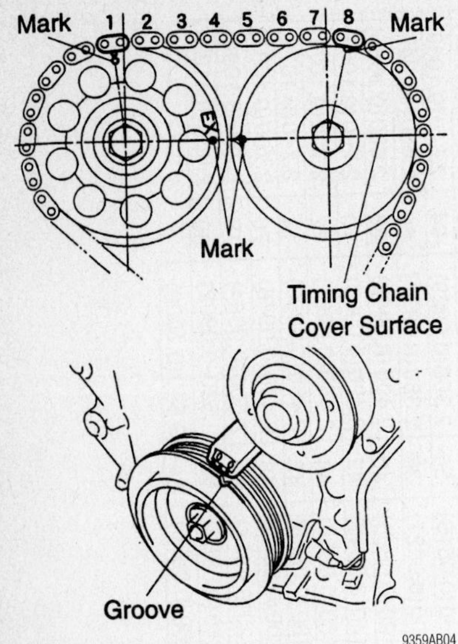

Proper timing mark alignment for TDC—1ZZ-FE and 2ZZ-GE engines

2ZZ-GE Engine

➡Adjust the valve clearance when the engine is cold.

1. Before servicing the vehicle, refer to the precautions in the beginning of this section.
2. Remove or disconnect the following:
   - Negative battery cable.
   - Right side engine under cover
   - Cylinder head cover
   - Ignition coil assembly
   - Wire harness clamp
   - Suction hose sub-assembly
   - Cylinder head cover sub-assembly
   - Drive belt
   - Right side engine mount
3. Set the No. 1 cylinder to Top Dead Center (TDC) of the compressor stroke as follows:
   a. Turn the crankshaft pulley, and align its groove with the "0" timing mark of the timing chain cover.
   b. Make sure the point marks of the camshaft timing sprockets and VVT timing sprockets are in a straight line as shown. If not, turn the crankshaft 1 complete revolution (360°) and align the marks.
4. Check the valve clearance of the first set of the valves shown:
   a. Use a feeler gauge to measure the clearance between the valve lifter and camshaft. The clearance of the intake valves should be 0.0031–0.0071 in. (0.08–0.18mm). The clearance of the exhaust valves should be 0.0087–0.0126 in. (0.22–0.32mm).

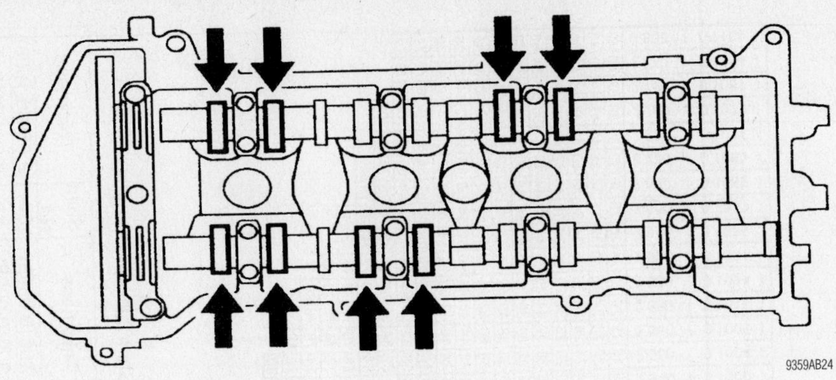

Check the clearance of the 1st set of valves–1ZZ-FE engine

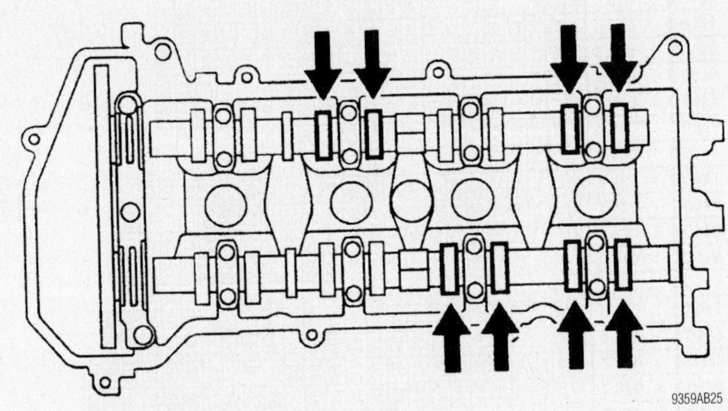

Check the clearance of the 2nd set of valves–1ZZ-FE engine

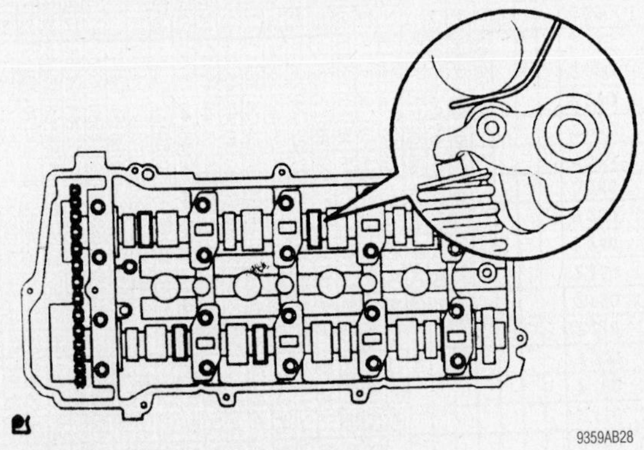

Check the clearance of the 1st set of valves–2ZZ-GE engine

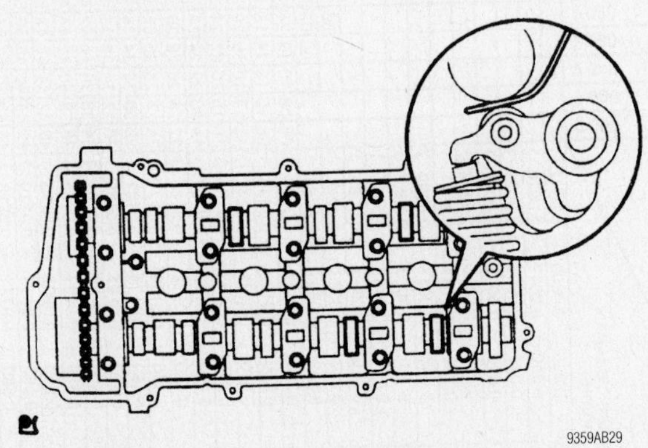

Check the clearance of the 2nd set of valves–2ZZ-GE engine

b. Note the out-of-specification valve clearance measurements. You will need them later to determine the required replacement valve lifter.

c. Turn the crankshaft 1 revolution (360°) to set the No. 4 cylinder to TDC.

5. Check the valve clearance of the second set of the valves shown:

a. Use a feeler gauge to measure the clearance between the valve lifter and camshaft. The clearance of the intake valves should be 0.0031–0.0071 in. (0.08–0.18mm). The clearance of the exhaust valves should be 0.0087–0.0126 in. (0.22–0.32mm).

b. Note the out-of-specification valve clearance measurements. You will need them later to determine the required replacement valve lifter.

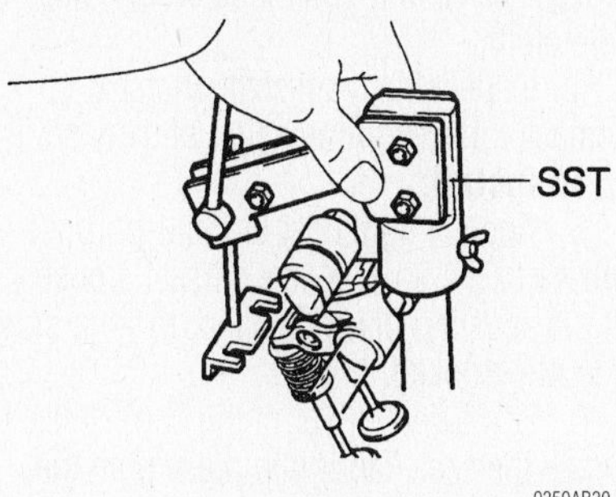

Insert the special tool into the plug tube—2ZZ-GE

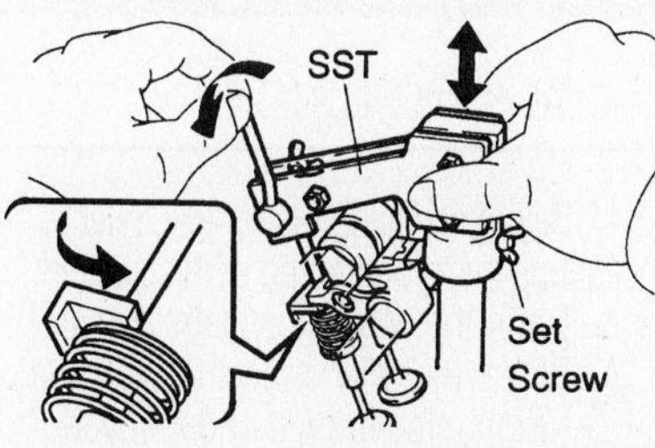

Operate the lever so that the SST's seat surface comes to contact with the valve retainer and lock them with the set screw

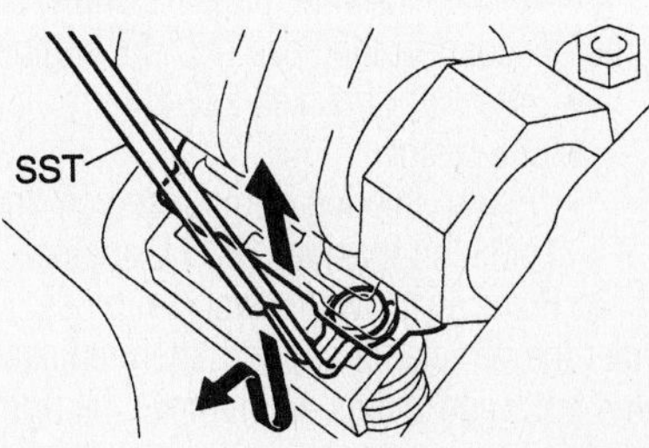

Setting the tool from the right side, makes shim removal easier—2ZZ-GE

6. To adjust the intake valve clearance:

a. Set the SST. Turn the crankshaft so the related rocker arm, where the valve clearance is adjusted, is fully pushed down.

➡Remove the spark plug and take off the compression.

b. Insert SST 09248-77010 into the plug tube. The tool cannot be inserted unless the set screw is loosened.

c. Operate the lever so that the SST's seat surface comes to contact with the valve retainer and lock them with the set screw. Clearance between the valve retainer and SST's set surface is not allowed. Be careful not to make clearance when inserting the SST, since clearance may unlock the keeper.

d. Lock the set screw on the tube side of the SST.

e. Rotate the crankshaft so that the camshaft is position as shown. During rotation, pay attention to the direction, to prevent the nose of the camshaft from interfering with the SST's shaft. Do not rotate the crankshaft excessively.

f. Lift the rocker arm to make room and remove the adjusting shim using SST 09248-77010.

7. Determine the size of the replaced shim according to the chart or the following formula:

a. Use a dial indicator to measure the thickness of the removed shim.

b. Calculate the thickness of a new shim so that the valve clearance comes within the specified value.

c. A: Thickness of new shim.

d. B: Thickness of used shim.

e. C: Measured valve clearance.

f. Intake: A = B + (C − 0.005 in. [0.13mm])

g. Exhaust: A = B + (C − 0.011 in. [0.27mm])

h. Select a new shim with a thickness as close as possible to the calculated values. Shims come in 41 sizes in increments of 0.0008 in. (0.020mm) from 0.0787–0.1102 in (2.0–2.8mm).

8. Lift the rocker arm to make room, then install the adjusting shim using the SST. To remove the tool from the shim, push down on the rocker arm.

9. Turn the crankshaft so the related rocker arm, where the valve clearance is adjusted, is fully pushed down.

10. Loosen the 2 set-screws, then remove the SST.

11. Install all components in the reverse of the removal procedure.

Starter

REMOVAL & INSTALLATION

1. Before servicing the vehicle, refer to the precautions in the beginning of this section.
2. Remove or disconnect the following:
   - Negative battery cable
   - Right side engine undercover
   - Starter wiring
   - Starter
3. Installation is the reverse of removal. Torque the bolts to 27 ft. lbs. (37 Nm) and the nut to 7 ft. lbs. (10 Nm).

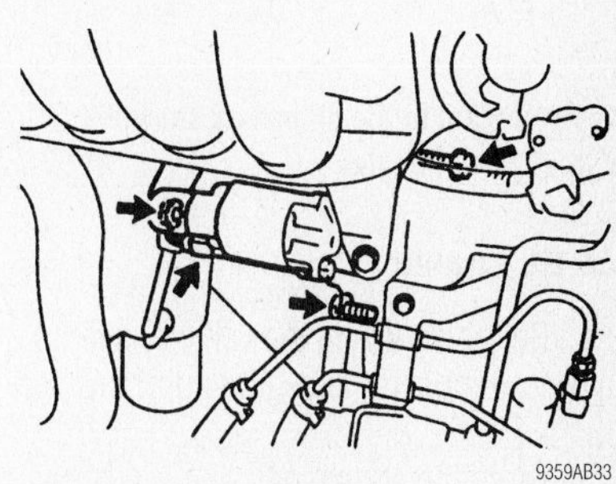

Starter mounting—Matrix

Oil Pan

REMOVAL & INSTALLATION

1ZZ-FE Engine

1. Before servicing the vehicle, refer to the precautions in the beginning of this section.
2. Drain the engine oil.
3. Remove or disconnect the following:
   - Negative battery cable
   - Undercovers
   - Front exhaust pipe
   - Oil pan mounting bolts and nuts
   - Oil pan, cutting off the applied sealer.

To install:

4. Remove any old sealant from the oil pan flange and thoroughly clean the sealing surface.
5. Install or connect the following:
   - Oil pan. Tighten the bolts and nuts in several passes to 80 inch lbs. (9 Nm).
   - Front exhaust pipe
   - Negative battery cable
   - Undercovers
6. Fill the engine with clean oil.
7. Start the vehicle, check for leaks and repair if necessary.

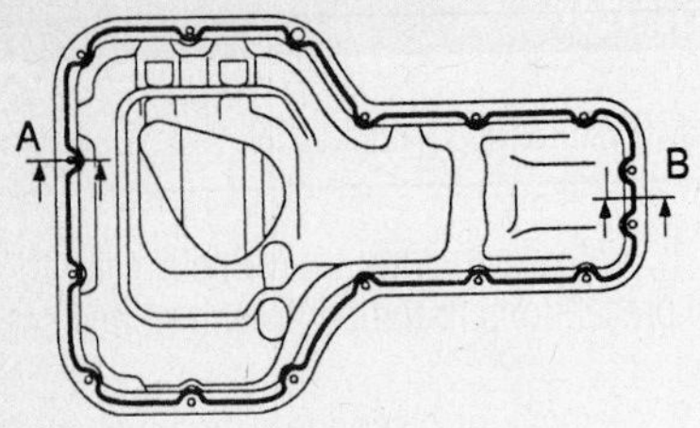

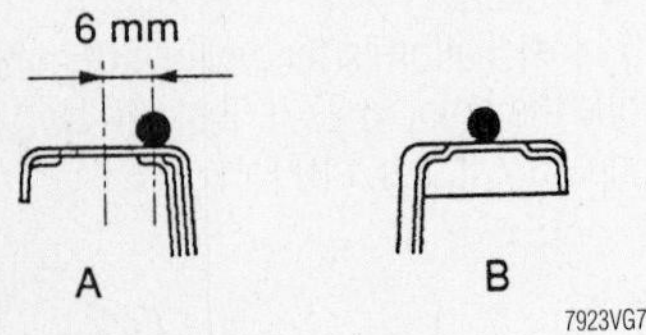

Apply sealant to the oil pan as shown—1.8L (1ZZ-FE) engine

2ZZ-GE Engine

1. Before servicing the vehicle, refer to the precautions in the beginning of this section.
2. Drain the engine oil.
3. Remove or disconnect the following:
   - Negative battery cable. On vehicles equipped with an air bag, wait at least 90 seconds before proceeding.
   - Undercovers
   - Front exhaust pipe
   - Oil pan mounting bolts and nuts
   - Oil pan, cutting off the applied sealer

To install:

4. Remove any old sealant from the oil pan flange and thoroughly clean the sealing surface.
5. Install or connect the following:
   - Oil pan. Tighten the bolts and nuts in several passes to 80 inch lbs. (9 Nm).
   - Front exhaust pipe
   - Negative battery cable
   - Undercovers
6. Fill the engine with clean oil.
7. Start the vehicle, check for leaks and repair if necessary.

Oil Pump

REMOVAL & INSTALLATION

1ZZ-FE Engine

1. Before servicing the vehicle, refer to the precautions in the beginning of this section.

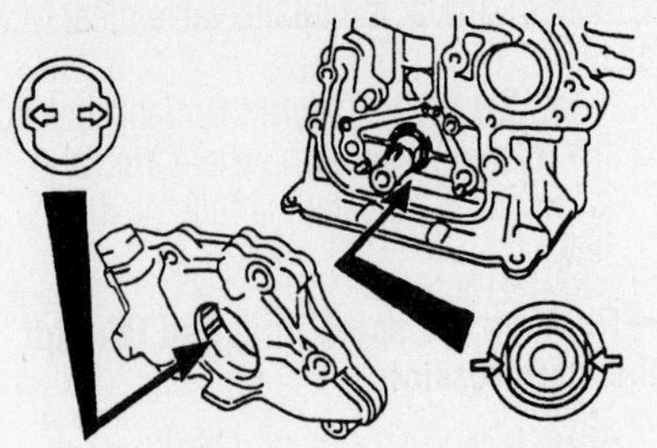

Oil pump mounting—1ZZ-FE and 2ZZ-GE engines

2. Drain the engine oil.
3. Remove or disconnect the following:
   - Negative battery cable
   - Timing chain and crankshaft sprocket
   - Timing chain vibration damper
   - Oil pump bolts, pump and gasket

To install:

4. Clean the mounting surface.
5. Install or connect the following:
   - Oil pump, with new gasket. Engage the spline teeth of the oil pump drive rotor with the larger teeth of the crankshaft, and slide the pump on.
   - Oil pump bolts and tighten to 80 inch lbs. (9 Nm)
   - Crankshaft vibration damper and tighten to 80 inch lbs. (9 Nm)
   - Crankshaft sprocket and timing chain
   - Negative battery cable
6. Fill the engine with clean oil.
7. Start the vehicle, check for leaks and repair if necessary.

2ZZ-GE Engine

1. Before servicing the vehicle, refer to the precautions in the beginning of this section.
2. Drain the engine oil.
3. Remove or disconnect the following:
   - Negative battery cable
   - Timing chain and crankshaft sprocket
   - Oil pump and gasket

To install:

4. Clean the mounting surface.
5. Install or connect the following:
   - Oil pump, with new gasket. Engage the spline teeth of the oil pump drive rotor with the larger teeth of the crankshaft, and slide the pump on.
   - Oil pump bolts and tighten to 80 inch lbs. (9 Nm)
   - Crankshaft sprocket and timing chain
   - Negative battery cable
6. Fill the engine with clean oil.
7. Start the vehicle, check for leaks and repair if necessary.

Rear Main Seal

REMOVAL & INSTALLATION

1. Remove or disconnect the following:
   - Transaxle
   - Clutch assembly
   - Flywheel or flexplate
2. Use a small sharp knife to cut off the lip of the oil seal. Take great care not to score any metal with the knife.
3. Use a small prytool to pry the old seal from the retaining plate. Be careful not to damage the plate. Protect the tip of the tool with tape and pad the fulcrum point with cloth.
4. Inspect the crankshaft and seal lip contact surfaces for any sign of damage.

To install:

5. Apply a light coat of multi-purpose grease to the lip of a new oil seal. Loosely fit the seal into place by hand, making sure it is not crooked.
6. Use a seal driver of the correct size to install the seal. Tap it into place until the surface of the seal is flush with the edge of the housing.

Timing Chain, Sprockets, Front Cover and Seal

REMOVAL & INSTALLATION

1. Before servicing the vehicle, refer to the precautions in the beginning of this section.
2. Drain the cooling system.
3. Remove or disconnect the following:
   - Right side engine under cover
   - Right front wheel and tire
   - Cylinder head cover
   - Wire harness clamp and suction hose assembly, 2ZZ-GE engine
   - Drive belt
4. Separate the vane pipe assembly, but do not disconnect the hose, 1ZZ-FE engine.
   - Alternator bracket, 2ZZ-GE
   - Alternator
   - Power steering pump reservoir and position it aside, 1ZZ-FE engine
5. Place a jack with a wooden block under the vehicle for support, then remove the 4 bolts and 2 nuts and remove the right side engine mount.
6. Remove the engine wire as follows, on 1ZZ-FE engines:

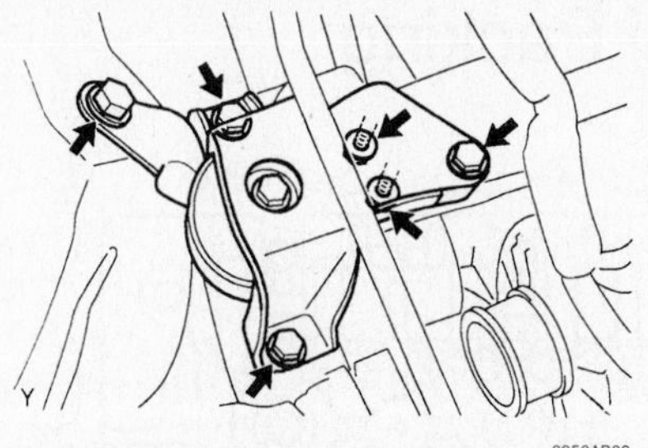
With the engine supported, remove the right side engine mount—1ZZ-FE engine shown, 2ZZ-GE similar

a. Remove the 5 clamps from the brackets.

b. Detach the connectors.

c. Remove the ignition coil connectors.

d. Bolt and nut holding the engine wire.

7. Remove the engine wire as follows, on 2ZZ-GE engines:

a. Detach the ignition coil, oil control valve and Crankshaft Position Sensor (CKP) sensor electrical connectors.

b. Bolt and nut for the engine ground, then position the engine wire aside

8. Remove or disconnect the following:

- Ignition coil assembly
- Positive Crankcase Ventilation (PCV) hoses from the cylinder head cover, if necessary
- Cylinder head (valve) cover sub-assembly

9. Set the No. 1 cylinder to Top Dead Center (TDC) of the compressor stroke as follows:

a. Turn the crankshaft pulley, and align its groove with the "0" timing mark of the timing chain cover.

b. Make sure the point marks of the camshaft timing sprockets and VVT timing sprockets are in a straight line as shown. If not, turn the crankshaft 1 complete revolution (360°) and align the marks.

- Crankshaft pulley, using SST 09960-10010
- Belt tensioner
- Water pump pulley, if equipped, and pump
- Transverse engine mounting bracket
- Crankshaft Position (CKP) sensor
- No. 1 chain tensioner assembly, making sure not to revolve the crankshaft without the tensioner
- Timing chain cover. The cover is retained with 11 bolts and nuts and a Torx® stud bolt. Pry the cover between the cylinder head and block to remove it.
- Timing gear cover oil seal
- CKP sensor plate No. 1
- Timing chain tensioner slipper

➡In case you turn the camshafts with the timing chain removed, turn the crankshaft ¼ turn for the valve to avoid contact with the pistons.

- Timing chain sub-assembly. Remove the chain with the crankshaft gear, using screwdrivers as shown

To install:

10. Set the No. 1 cylinder to TDC of the compression stroke:

a. Turn the hexagonal wrench head part of the camshafts, and align the point marks of the cam sprockets.

b. Using the crankshaft pulley bolt, turn the crankshaft and position the crankshaft set key upward.

11. Install or connect the following:

- Timing chain on the crank sprocket with the yellow link aligned with the mark on the crank sprocket. There are 3 yellow links on the timing chain.
- Crankshaft sprocket, using SST 09223-22010

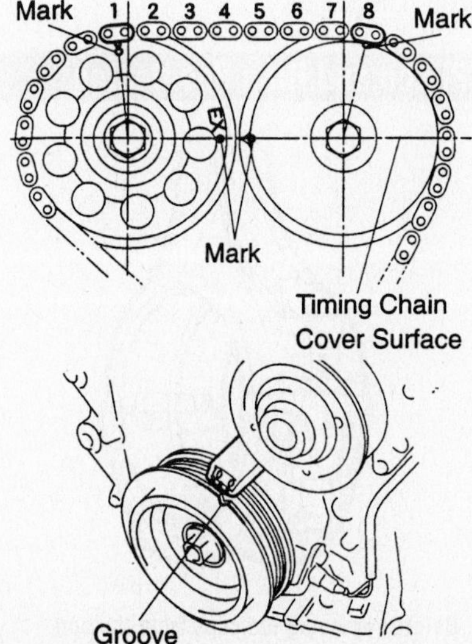

Proper timing mark alignment for TDC

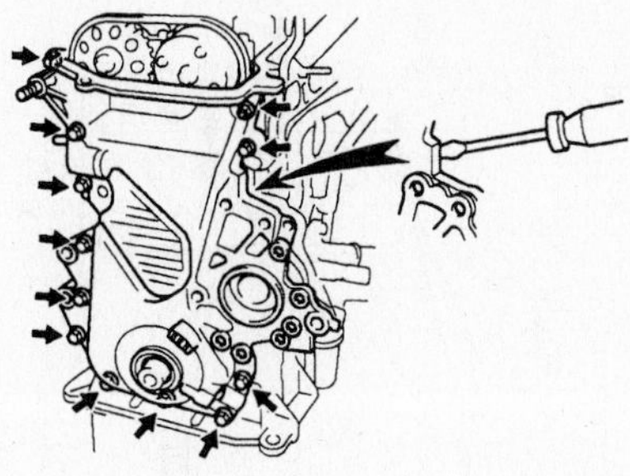
Timing chain cover mounting—1ZZ-FE engine shown, 2ZZ-GE similar

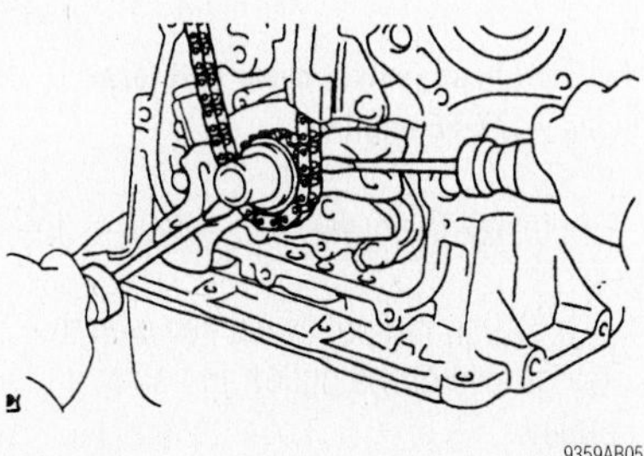
Remove the timing chain with the crankshaft gear

Proper alignment of the camshaft sprockets—1ZZ-FE engine

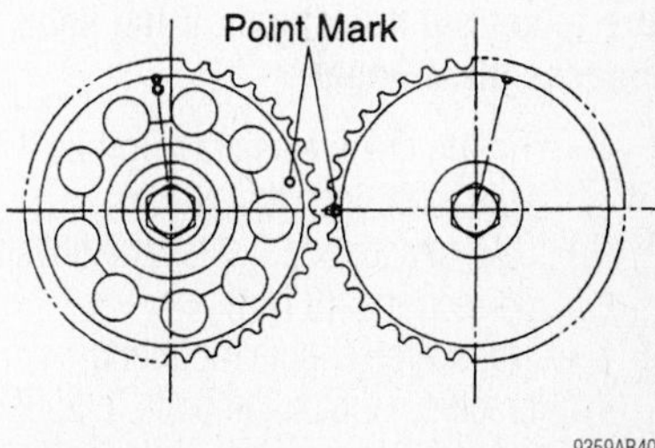

Proper alignment of the camshaft sprockets—2ZZ-GE engine

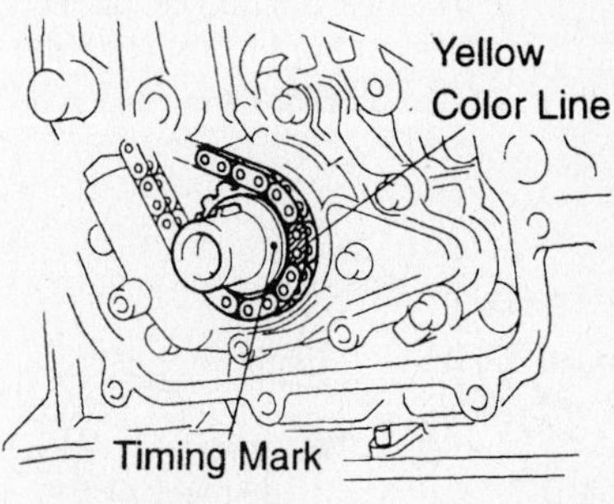

Make sure the yellow link is aligned with the crankshaft sprocket timing mark—1ZZ-FE and 2ZZ-GE engines

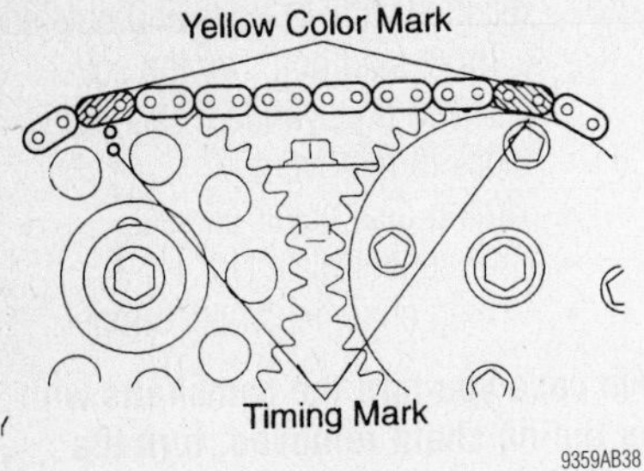

The yellow links of the timing chain must align with the camshaft sprocket timing marks—1ZZ-FE and 2ZZ-GE engines

- Timing chain on the camshaft sprockets with the yellow links aligned with the marks on the cam sprockets
- Timing chain tensioner slipper and tighten the bolt to 14 ft. lbs. (19 Nm)
- Crankshaft position sensor plate, with the "F" mark facing forward
- Timing gear cover oil seal
- Timing cover. For 1ZZ-FE engine, tighten the "A" bolts to 10 ft. lbs. (13 Nm), the "B" bolts to 14 ft. lbs. (19 Nm) and the stud bolt to 84 inch lbs. (9.5 Nm), using a Torx® wrench. For 2ZZ-GE engines, tighten the M8 bolts to 15 ft. lbs. (21 m), the M6 bolts to 8 ft. lbs. (11 Nm) and the stud bolt to 84 inch lbs. (9.5 Nm).

➡When installing the tensioner, make sure to set the hook again if the hook releases the plunger.

- Timing chain tensioner. Torque the nuts to 80 inch lbs. (9 Nm).
- CKP sensor and tighten the bolts to 80 inch lbs. (9 Nm)
- Transverse engine mounting bracket. Tighten the bolts to 35 ft. lbs. (47 Nm).
- Water pump and pulley
- Drive belt tensioner. Tighten the nut to 21 ft. lbs. (29 Nm) and the bolt to 51 ft. lbs. (69 Nm) on 1ZZ-FE engines or to 74 ft. lbs. (100 Nm) on 2ZZ-GE engines.

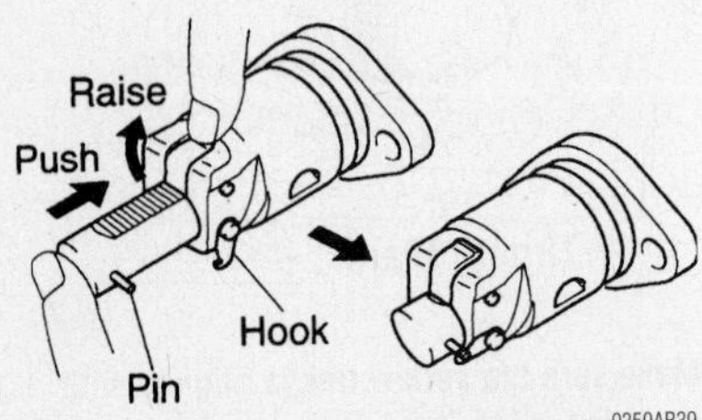

Timing chain tensioner—1ZZ-FE engine

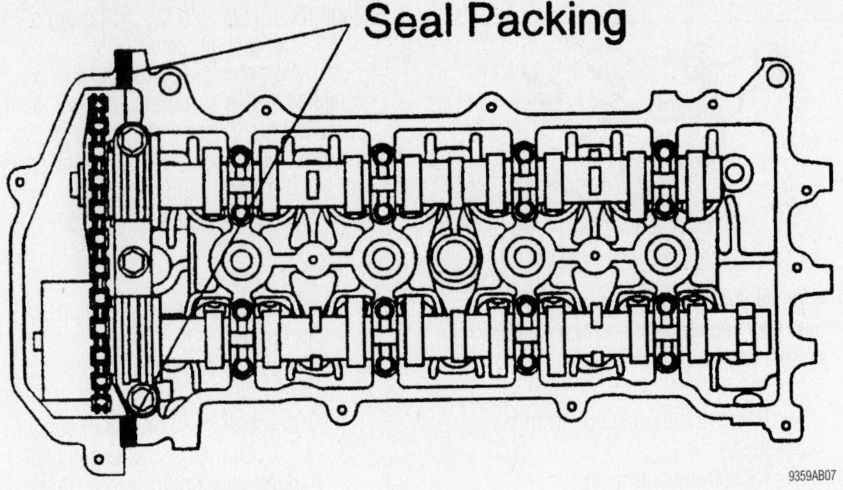

Seal packing installation locations

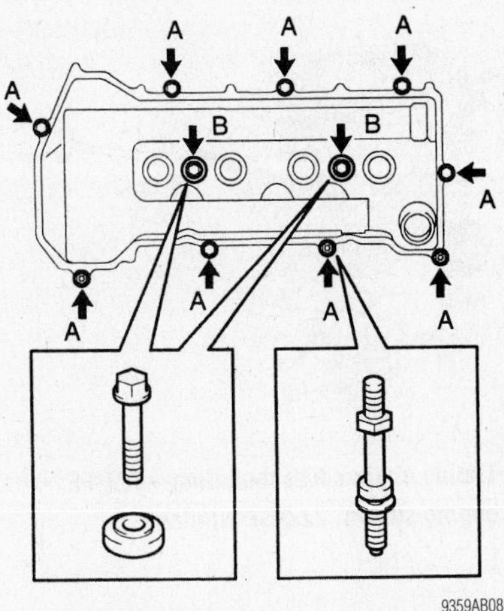

Cylinder head (valve) cover bolt locations—1ZZ-FE engine

12. Install the crankshaft pulley, as follows:
   a. Align the pulley set key with the key groove of the pulley and slide on the pulley.
   b. Use SST 09960-11010 to install the bolt and tighten to 102 ft. lbs. (138 Nm) for 1ZZ-FE engine or to 87 ft. lbs. (118 Nm) on 2ZZ-GE engines.
   c. Turn the crankshaft counterclockwise and disconnect the plunger knock pin from the hook.
   d. Turn the crankshaft clockwise and check that the slipper is pushed by the plunger. If the plunger does not spring out, press the slipper into the chain tensioner with a screwdriver so that the hook is released from the knock pin and the plunger springs out.
   - Cylinder head sub-assembly cover. Install seal packing into the locations shown and install within 3 minutes. Tighten the "A" bolts to 8 ft. lbs. (11 Nm) and the "B" bolts to 80 inch lbs. (9 Nm) for 1ZZ-FE engines. For 2ZZ-GE engines, tighten the bolts to 7 ft. lbs. (10 Nm).
   - Ignition coil assembly. Torque the bolts to 80 inch lbs. (9 Nm).
   - Engine wire and tighten to 80 inch lbs. (9 Nm)
   - Right side engine mount. Tighten to 38 ft. lbs. (52 Nm).
   - Alternator bracket, 2ZZ-GE engine
   - Alternator
   - Vane pump, 1ZZ-FE
   - Main cylinder head cover and tighten to 62 inch lbs. (7 Nm)
   - Right front wheel and tire. Tighten the lug nuts to 76 ft. lbs. (103 Nm).
13. Fill the cooling system to the proper level.
14. Start the vehicle, check for leaks and repair if necessary.

Piston and Ring Positioning

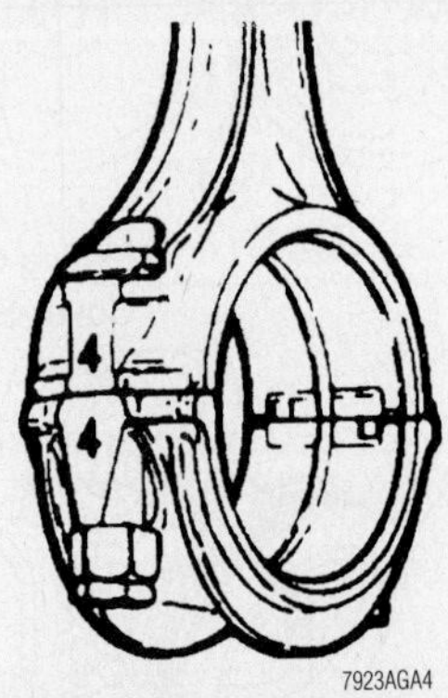

Before removing the caps from the connecting rods, be sure to matchmark them as shown

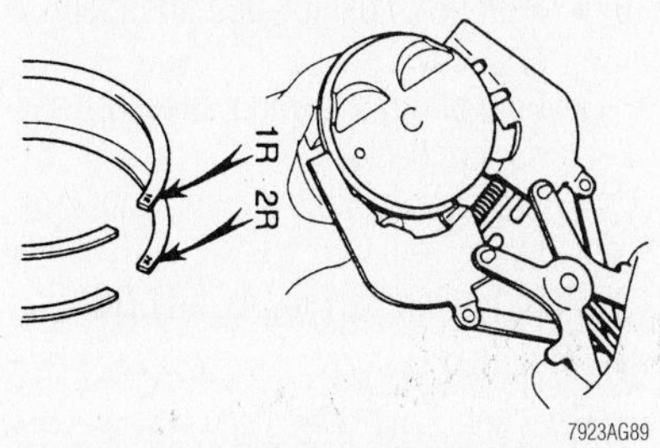

Piston ring identification mark locations—1ZZ-FE and 2ZZ-GE engines

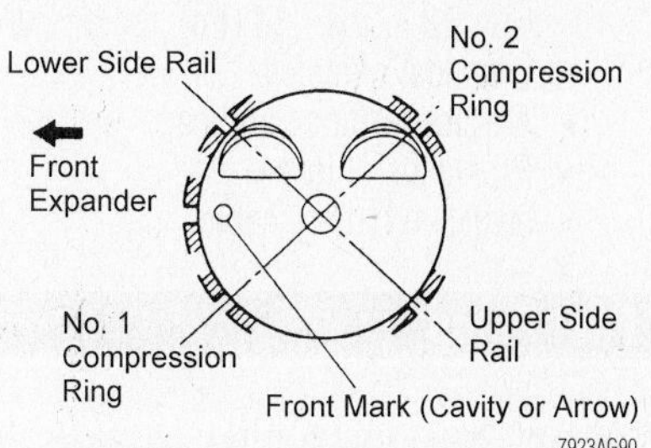

Piston ring end-gap spacing —1ZZ-FE and 2ZZ-GE engines

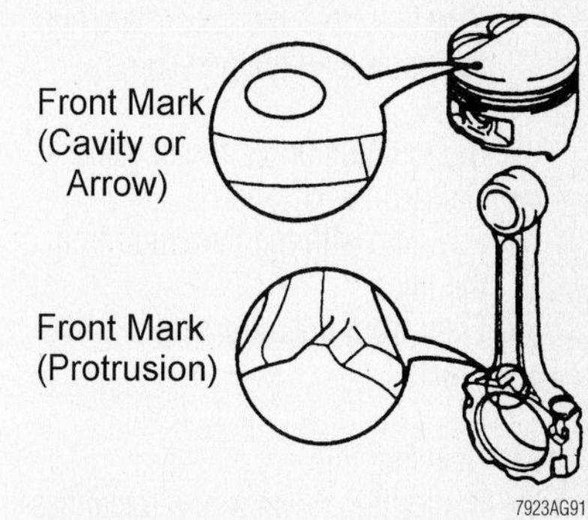

Piston-to-connecting rod assembly —1ZZ-FE and 2ZZ-GE engines

FUEL SYSTEM

Fuel System Service Precautions

Safety is the most important factor when performing not only fuel system maintenance, but any type of maintenance. Failure to conduct maintenance and repairs in a safe manner may result in serious personal injury or death. Work on a vehicle's fuel system components can be accomplished safely and effectively by adhering to the following rules and guidelines.

- To avoid the possibility of fire and personal injury, always disconnect the negative battery cable unless the repair or test procedure requires that battery voltage by applied.
- Always relieve the fuel system pressure prior to disconnecting any fuel system component (injector, fuel rail, pressure regulator, etc.) fitting or fuel line connection. Exercise extreme caution whenever relieving fuel system pressure, to avoid exposing skin, face and eyes to fuel spray. Please be advised that fuel under pressure may penetrate the skin or any part of the body that it contacts.
- Always place a shop towel or rag around the fitting or connection prior to loosening to absorb any excess fuel due to spillage. Ensure that all fuel spillage is quickly remove from engine surfaces. Ensure that all fuel-soaked cloths or towels are deposited into a flame-proof waste container with a lid.
- Always keep a dry chemical (Class B) fire extinguisher near the work area.
- Do not allow fuel spray or fuel vapors to come into contact with a light bulb, spark or open flame.
- Always use a second wrench when loosening or tightening fuel line connections fittings. This will prevent unnecessary stress and torsion to fuel piping. Always follow the proper torque specifications.
- Always replace worn fuel fitting O-rings with new ones. Do not substitute fuel hose where rigid pipe is installed.

Fuel System Pressure

RELIEVING

CAUTION

Failure to relieve fuel pressure before repairs or disassembly can cause serious personal injury and/or property damage. Fuel pressure is maintained within the fuel lines, even if the engine is OFF or has not been run in a period of time. This pressure must be safely relieved before any fuel-bearing line or component is loosened or removed. On vehicles equipped with inflatable restraints or air bag systems, wait at least 90 seconds after disconnecting the battery cable before performing any other work. The back-up power will keep the restraint system energized for a period of time after the battery is disconnected.

1. Before servicing the vehicle, refer to the precautions in the beginning of this section.
2. Perform the following:
   a. Remove the rear seat cushion.
   b. Remove the rear floor service hole cover.
   c. Disconnect the fuel pump connector.
   d. Start and run the engine, until it stalls.
   e. Turn the ignition key to the **LOCK** position.
   f. Disconnect the negative battery cable.
   g. Connect the fuel pump connector.
   h. Install the service hole cover and rear seat cushion.
   i. Place a catch-pan under the joint to be disconnected. A large quantity of fuel may be released when the joint is opened.
   j. Wear eye or full face protection.
   k. Place a shop towel over the area and slowly release the joint using a wrench of the correct size.
   l. Allow the any fuel left in the line to bleed off slowly before fully disconnecting the joint.
   m. Plug the opened lines.

Fuel Filter

REMOVAL & INSTALLATION

1. Before servicing the vehicle, refer to the precautions in the beginning of this section.
2. Relieve the fuel system pressure.
3. Remove or disconnect the following:
   - Negative battery cable
   - Protective shield for the fuel filter
   - Air cleaner hose and cap, if necessary
   - Charcoal canister, if necessary

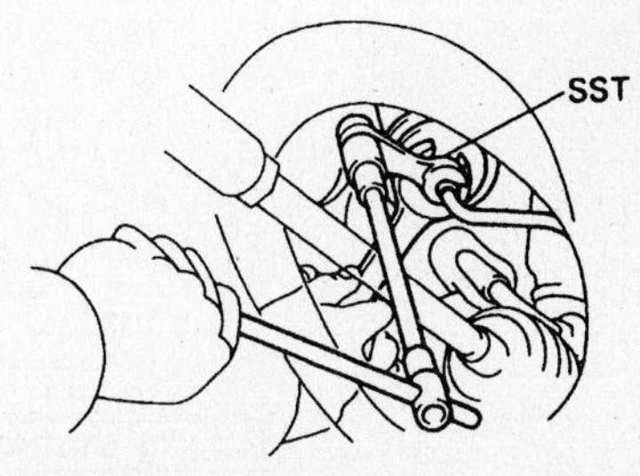

A line wrench with an extension may be needed to loosen the inlet line at the filter

- Slowly loosen the lower flare nut fitting until all the pressure is relieved
- Banjo fitting and 2 metal gaskets. Discard the gaskets.
- Fuel line with the flared nut from the filter
- Filter from the mounting bracket

To install:

4. Install or connect the following:
   - New fuel filter
   - Banjo fitting with a new metal gasket on each side and install the union bolt. Bolt: 22 ft. lbs. (30 Nm).
   - Flare nut to the lower connection. Nut: 22 ft. lbs. (30 Nm).
   - Charcoal canister
   - Air cleaner hose and cap
   - Protective shield
   - Negative battery cable

Fuel Pump

REMOVAL & INSTALLATION

1. Before servicing the vehicle, refer to the precautions in the beginning of this section.
2. Remove or disconnect the following:
   - Negative battery cable
   - Rear seat cushion and floor service hole cover
   - Fuel pump and vapor pressure sensor connectors
   - Start and run the engine, until it stalls
3. Turn the ignition key to the **LOCK** position.
   - Negative battery cable
4. Connect the fuel pump connector.
   - Fuel tank protector, AWD vehicles
   - Fuel tank main tube sub-assembly
   - Fuel emission tube sub-assembly No. 1, FWD vehicles

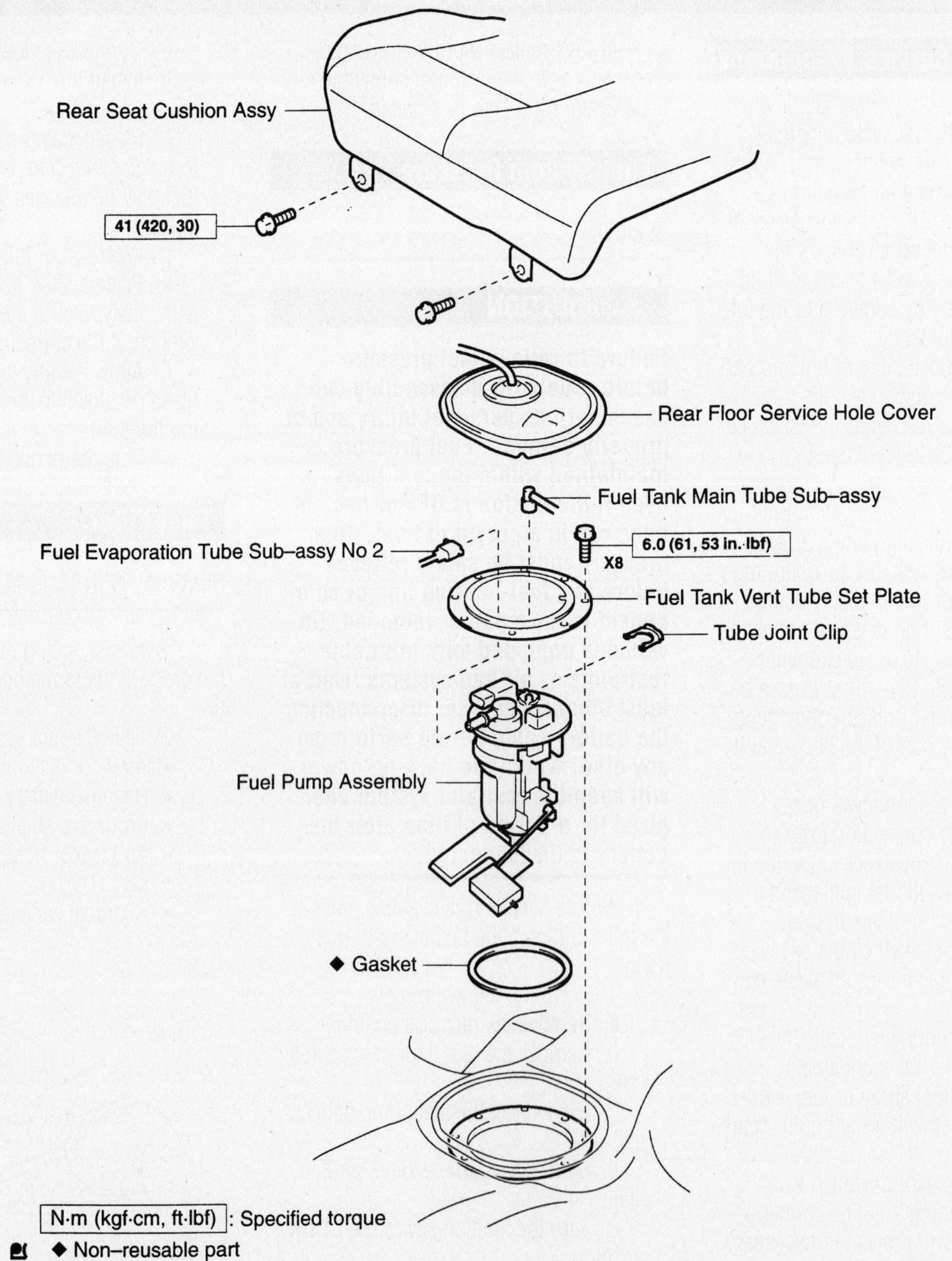

Exploded view of the fuel pump mounting—FWD shown, AWD similar

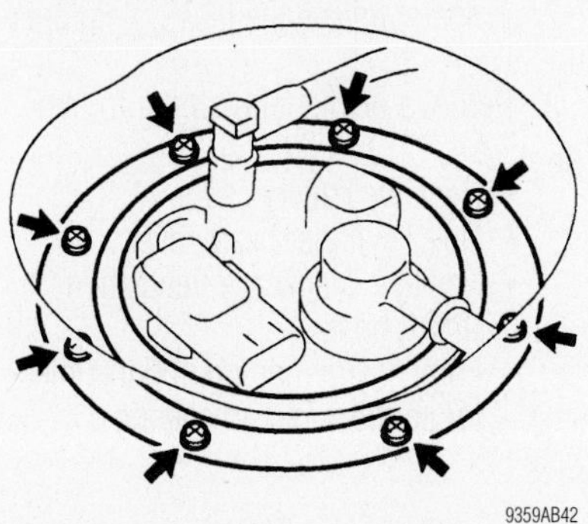

The fuel tank vent tube set plate is secured with 8 bolts on FWD vehicles

- Fuel tank vent tube set plate. The plate is secured with 8 bolts on FWD vehicles, or 5 bolts on AWD vehicles.
- Fuel pump assembly, being careful not to damage the filter or bend the arm of the fuel sender gauge
- Fuel suction tube set gasket
- Fuel suction support No. 2
- Fuel pump rubber cushion
- Fuel sender gauge assembly. Unplug the connector, then use a screwdriver to unlock the gauge and slide it to remove.
- Fuel section plate sub-assembly
- Vapor pressure sensor
- Fuel pump harness
- Fuel pump
- Fuel pump filter
- Fuel pressure regulator and O-ring

To install:

5. Install or connect the following:
   - New regulator O-ring and regulator
   - Fuel pump filter
   - Fuel pump
   - Vapor pressure sensor
   - Fuel suction tube set gasket
   - Fuel pump assembly
   - Fuel tank vent tube set plate. Tighten the bolts to 53 inch lbs. (6 Nm).

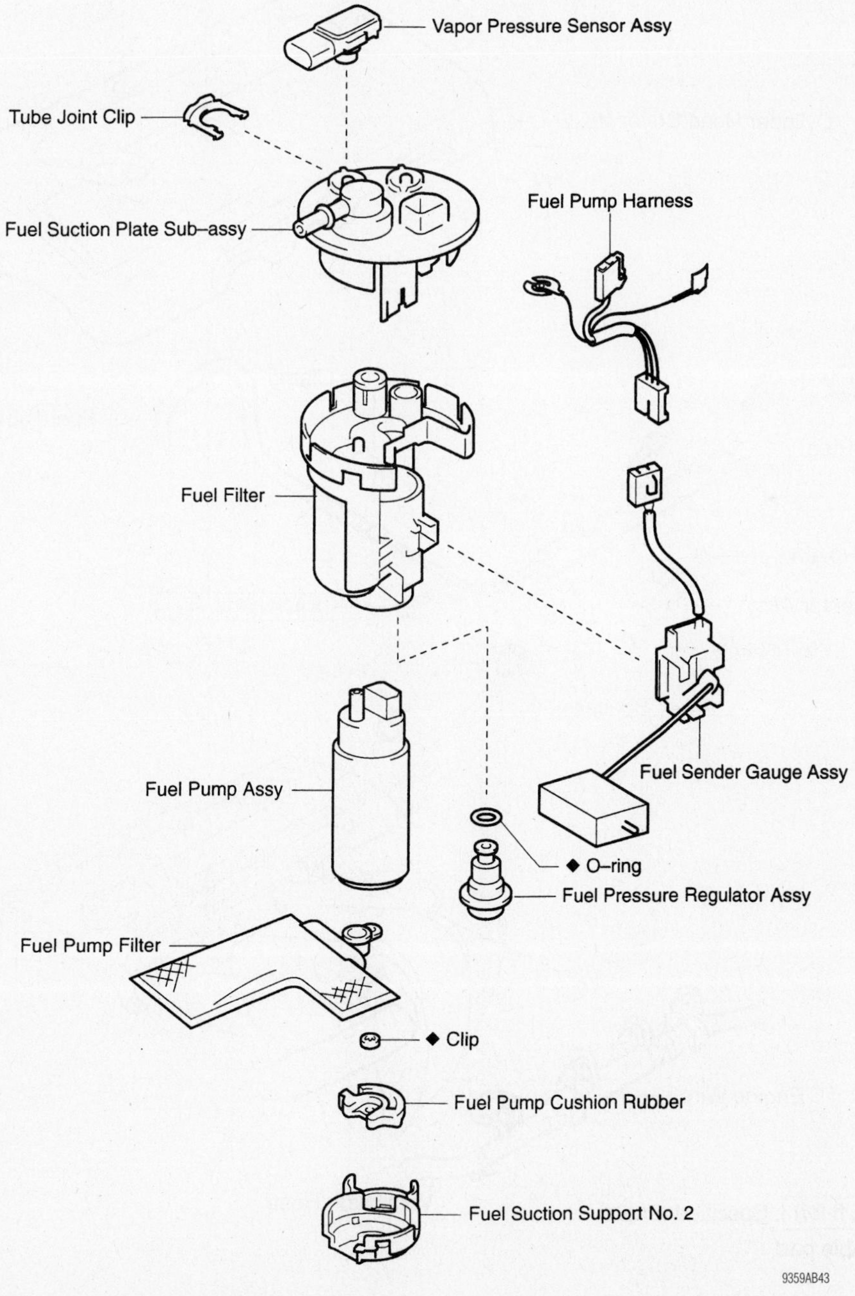

Vapor Pressure Sensor Assy
Tube Joint Clip
Fuel Suction Plate Sub–assy
Fuel Pump Harness
Fuel Filter
Fuel Sender Gauge Assy
Fuel Pump Assy
◆ O–ring
Fuel Pressure Regulator Assy
Fuel Pump Filter
◆ Clip
Fuel Pump Cushion Rubber
Fuel Suction Support No. 2
9359AB43

Fuel pump assembly components—FWD vehicles shown, AWD similar

- Connect the fuel emission tube sub-assembly
- Fuel tank main tube sub-assembly
- Fuel tank protector No. 2, AWD vehicles
- Negative battery cable. Check for fuel leaks.
- Floor service hole cover. Use butyl tape to seal the cover.
- Rear seat cushion

Fuel Injectors

REMOVAL & INSTALLATION

1ZZ-FE Engine

1. Before servicing the vehicle, refer to the precautions in the beginning of this section.

2. Properly relieve the fuel system pressure.

3. Remove or disconnect the following:

- Negative battery cable.
- No. 2 cylinder head cover
- Positive Crankcase Ventilation (PCV) hose
- Engine wire, unplugging the injector connectors and clamps

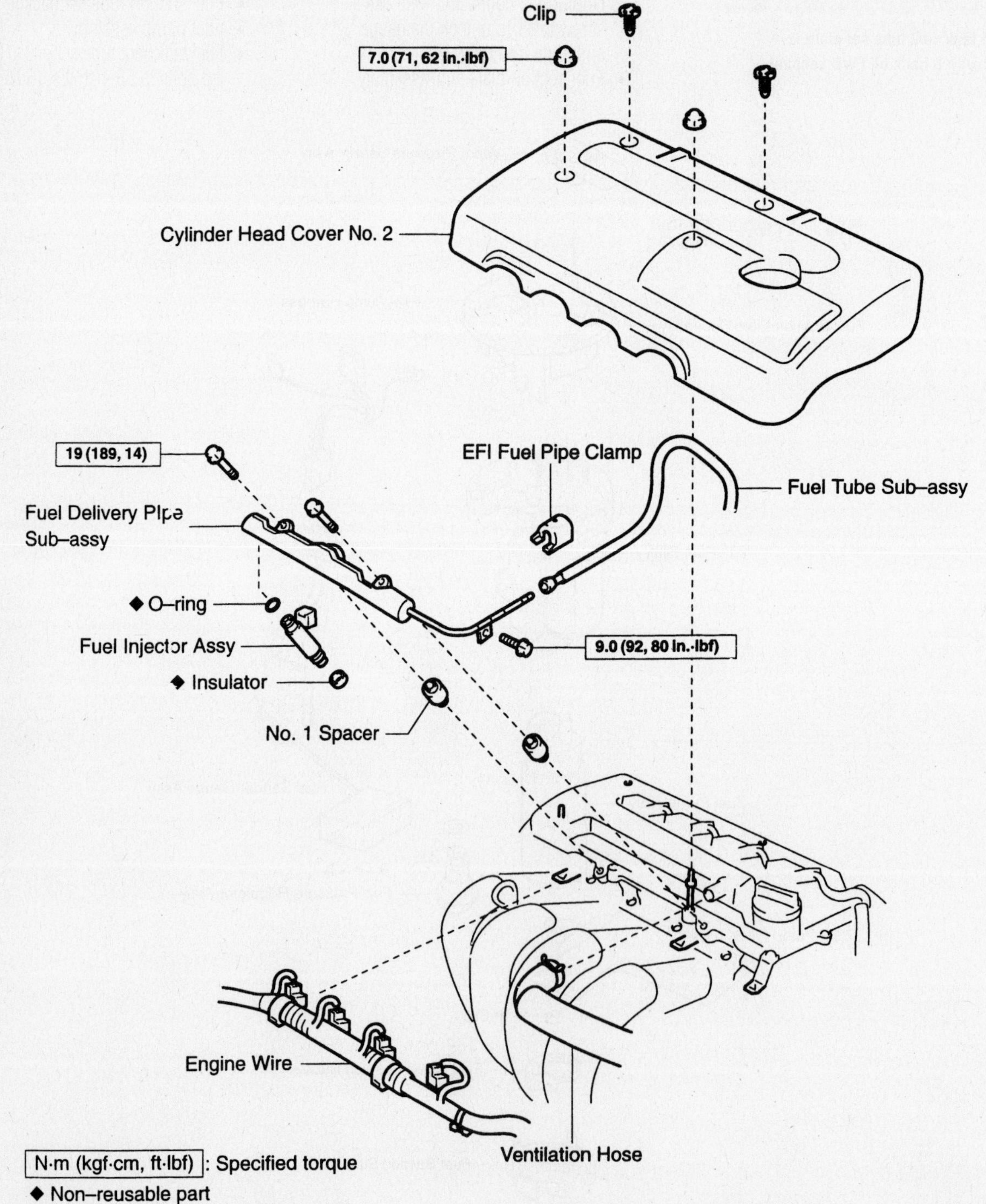

Fuel injector removal and installation—1ZZ-FE engine

- Fuel pipe clamp
- Fuel line/tube sub-assembly

⁂ WARNING

Be careful not to drop the fuel injectors when removing the delivery pipe.

- Fuel delivery pipe sub-assembly with the injectors attached
- Delivery pipe and injectors
- Spacers from the head
- Injectors from the delivery pipe
- O-ring and grommet from each injector

To install:

4. Install or connect the following:
   - New grommets
   - New O-rings coated with light machine oil
   - Injectors on the delivery pipe

➡Coat the contact point on the pipe with light machine oil and twist the injectors into place. The connector should face outward.

- Spacers

➡Coat the seats in the head where the injectors contact, with light machine oil.

- Delivery pipe and injectors

5. Loosely install the hold-down bolts and check that the injectors rotate smoothly. If they don't, the probable cause is incorrect O-ring installation. Torque the delivery pipe hold-down bolts to 14 ft. lbs. (19 Nm) and the fuel pipe bolt to 80 inch lbs. (9 Nm).

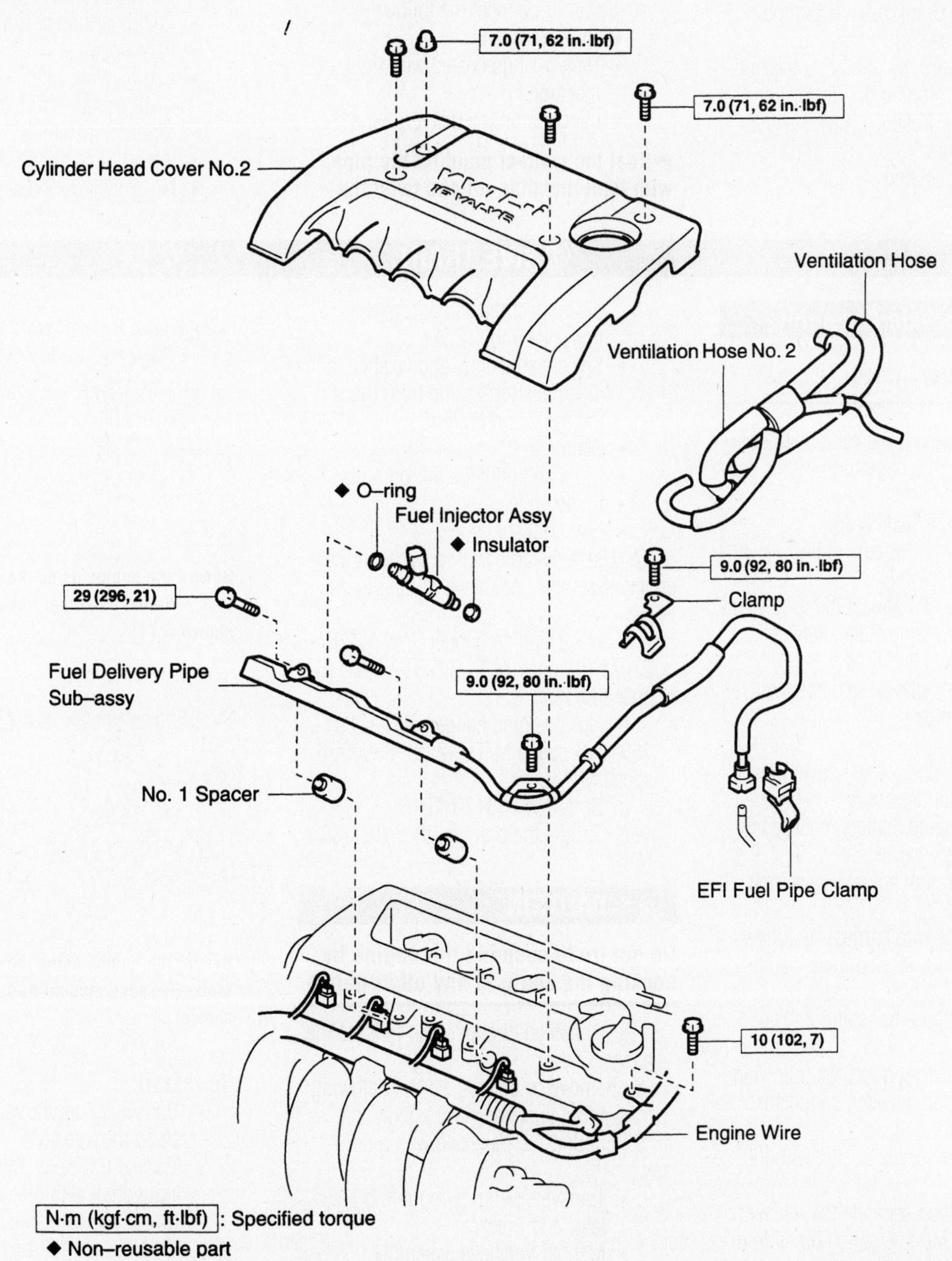

Fuel injector removal and installation—2ZZ-GE engine

- Engine wire, attaching the injector connectors and clamps
- Fuel line/tube sub-assembly
- PCV hose
- No. 2 cylinder head (valve) cover

2ZZ-GE ENGINE

1. Before servicing the vehicle, refer to the precautions in the beginning of this section.
2. Properly relieve the fuel system pressure.
3. Remove or disconnect the following:
   - Negative battery cable.
   - No. 2 cylinder head cover
   - Positive Crankcase Ventilation (PCV) hose
   - Engine wire, by removing the bolt, then unplugging the injector and Camshaft Position (CMP) sensor connectors
   - Fuel pipe clamp

⁂ WARNING

Be careful not to drop the fuel injectors when removing the delivery pipe.

   - Fuel delivery pipe sub-assembly with the injectors attached
   - Delivery pipe and injectors
   - Spacers from the head
   - Injectors from the delivery pipe
   - O-ring and grommet from each injector

To install:

4. Install or connect the following:
   - New grommets
   - New O-rings coated with light machine oil
   - Injectors on the delivery pipe

➡Coat the contact point on the pipe with light machine oil and twist the injectors into place. The connector should face outward.

   - Spacers

➡Coat the seats in the head where the injectors contact, with light machine oil.

   - Delivery pipe and injectors
5. Loosely install the hold-down bolts and check that the injectors rotate smoothly. If they don't, the probable cause is incorrect O-ring installation. Torque the delivery pipe hold-down bolts to 14 ft. lbs. (19 Nm) and the fuel pipe bolt to 80 inch lbs. (9 Nm).
   - Fuel line/tube sub-assembly
   - PCV hose
   - Engine wire, by connecting the CMP sensor and injector connectors and installing the bolt. Tighten the bolt to 7 ft. lbs. (10 Nm).
   - No. 2 cylinder head (valve) cover

DRIVE TRAIN

Manual Transaxle Assembly

REMOVAL & INSTALLATION

1. Before servicing the vehicle, refer to the precautions in the beginning of this section.
2. Drain the transaxle fluid.
3. Place the front wheels in the straight-ahead position.
4. Remove or disconnect the following:
   - Steering intermediate shaft
   - Front wheel and tires
   - Right and left side undercovers
   - Exhaust pipe
   - Hood
   - Cylinder head (valve) cover
   - Air cleaner assembly
   - Battery clamp, battery, battery tray and battery carrier
   - Cruise control actuator assembly, if equipped
5. Remove the wire harness as follows:
   a. Remove the wire harness clamp, 2 bolts and wire harness brackets.
   b. Remove the 2 bolts and 2 ground cables.
6. Remove or disconnect the following:
   - Back-up lamp switch connector, with ABS
   - Speed sensor connector, without ABS
   - 5 bolts, then separate the release cylinder with the clutch pipes from the transaxle
   - Shift cable clips and washer, then disconnect the cable from the transaxle and bracket
   - Select cable clips and washer, then disconnect the cable from the transaxle and bracket
   - Starter
   - Right and left side tie rod ends
   - Pressure feed tube
   - Front halfshafts
7. Use a suitable tool to suspend the engine assembly, as shown in the illustration:
   a. No. 1 engine hanger: P/N 12281-22021 (5-speed M/T), 12281-88600 (6-speed M/T).
   b. No. 2 engine hanger: P/N 12281-15040 (5-speed M/T), 12281-88600 (6-speed M/T).
   c. Bolt: P/N 91512-B1016.
   d. Torque the bolts to 28 ft. lbs. (38 Nm).

⁂ CAUTION

Do not try to suspend the engine by hooking the chain to any other part.

   e. Attach an engine chain hoist to the hangers.
8. Remove or disconnect the following:
   - Front suspension crossmember
9. Support the transaxle with a floor jack.
   - Transverse engine mounting insulator and brackets
   - Manual transaxle assembly
   - Transverse engine mounting brackets from the transaxle, if necessary

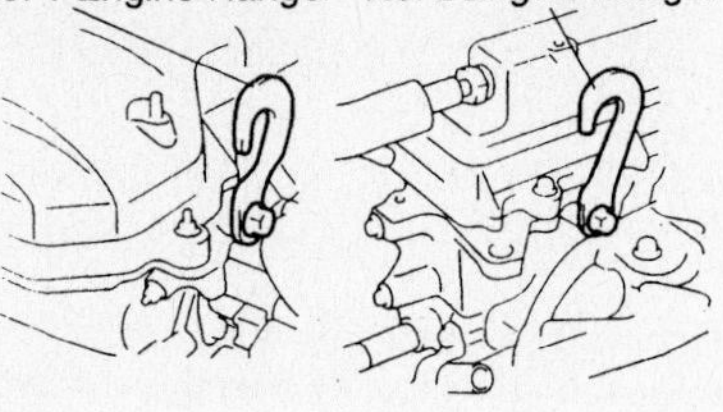

Secure the engine using the proper tools—5-speed manual transmission shown

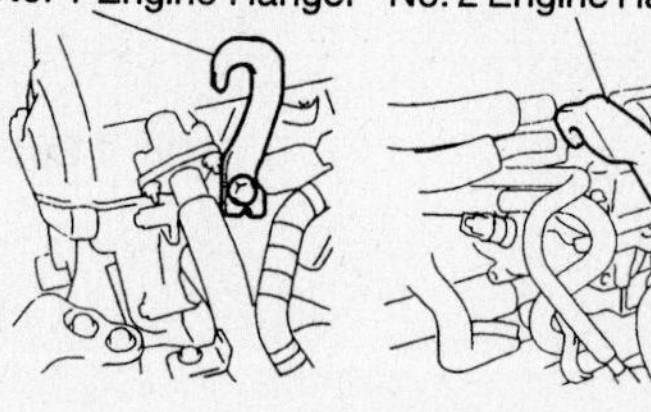

Secure the engine using the proper tools—6-speed manual transmission shown

To install:

- Transverse engine mounting brackets to the transaxle, if necessary
- Manual transaxle, by aligning the input shaft with the clutch disc. Torque the "A" bolts to 47 ft. lbs. (64 Nm), the "B" bolts to 35 ft. lbs. (47 Nm) and the "C" bolts to 17 ft. lbs. (23 Nm).

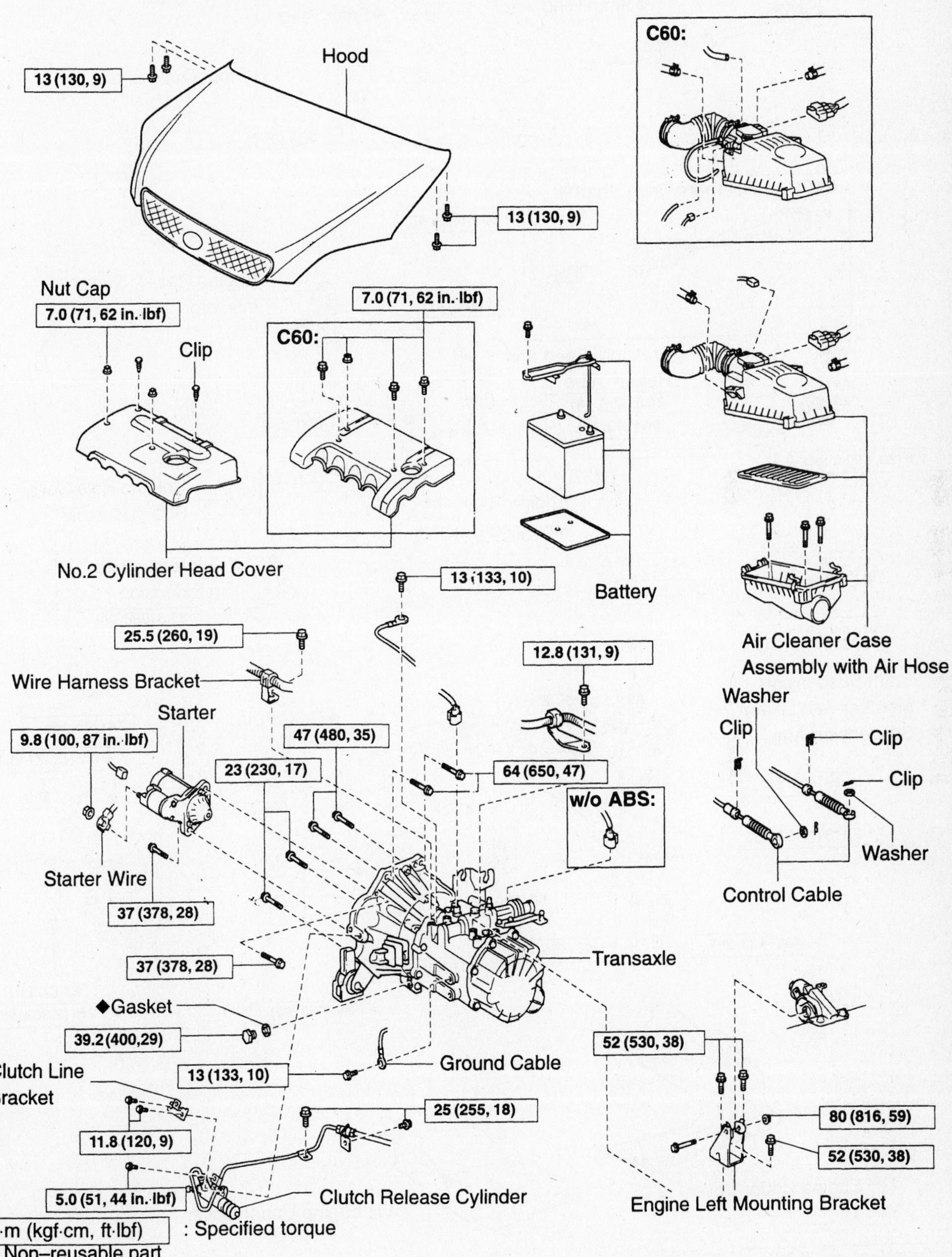

Hood
13 (130, 9)
C60:
13 (130, 9)
Nut Cap
7.0 (71, 62 in.·lbf)
Clip
C60:
7.0 (71, 62 in.·lbf)
No.2 Cylinder Head Cover
13 (133, 10)
Battery
Air Cleaner Case Assembly with Air Hose
25.5 (260, 19)
Wire Harness Bracket
Starter
12.8 (131, 9)
Washer
Clip
Clip
Clip
9.8 (100, 87 in.·lbf)
47 (480, 35)
23 (230, 17)
64 (650, 47)
w/o ABS:
Washer
Starter Wire
37 (378, 28)
Control Cable
37 (378, 28)
Transaxle
◆Gasket
39.2 (400,29)
52 (530, 38)
Clutch Line Bracket
13 (133, 10)
Ground Cable
25 (255, 18)
80 (816, 59)
11.8 (120, 9)
52 (530, 38)
5.0 (51, 44 in.·lbf)
Clutch Release Cylinder
Engine Left Mounting Bracket
N·m (kgf·cm, ft·lbf) : Specified torque
◆ Non–reusable part
9359AB47

Exploded view of the manual transaxle (1 of 2)

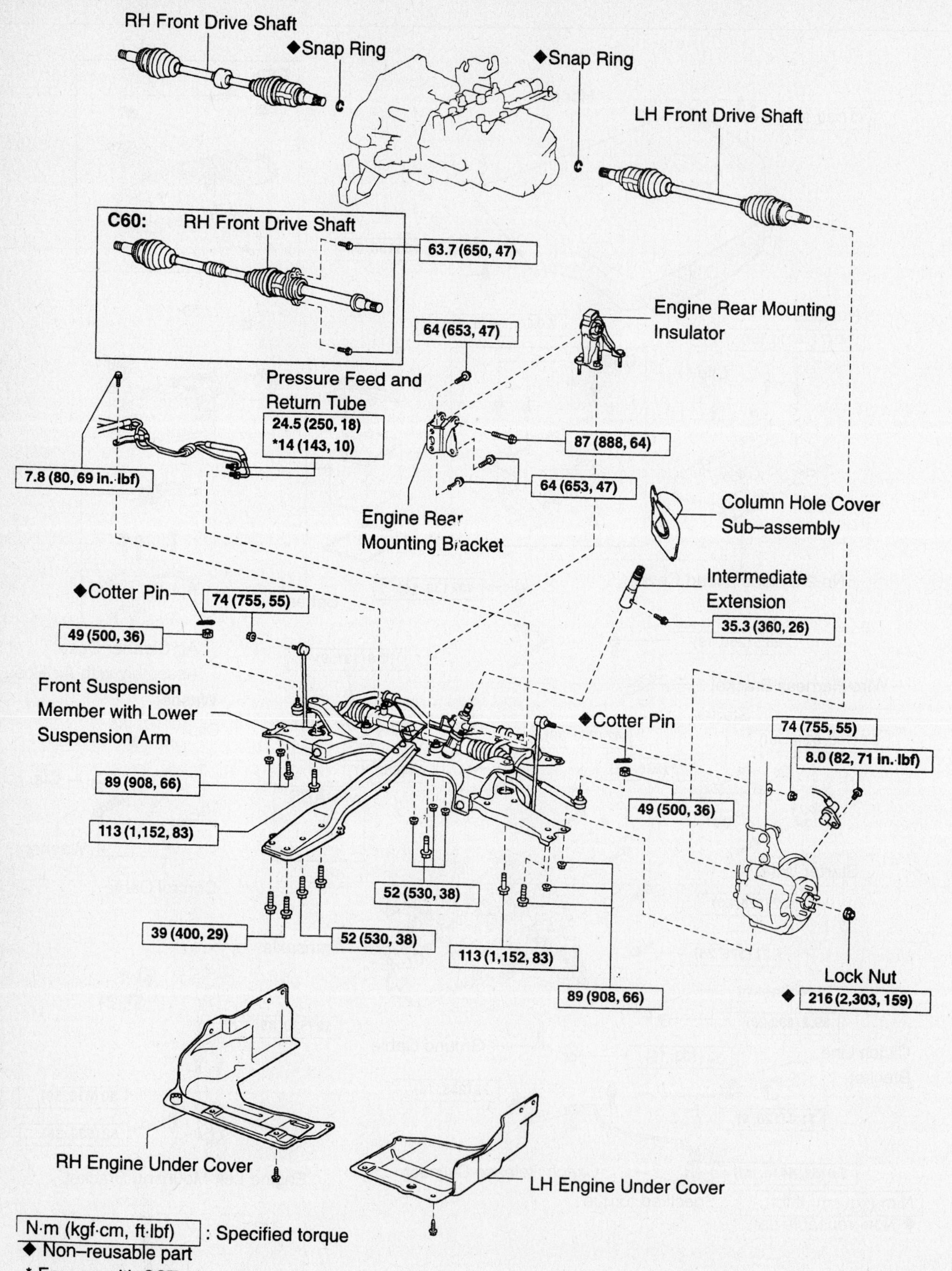
RH Front Drive Shaft
◆Snap Ring
◆Snap Ring
LH Front Drive Shaft
C60: RH Front Drive Shaft
63.7 (650, 47)
Engine Rear Mounting Insulator
64 (653, 47)
Pressure Feed and Return Tube
24.5 (250, 18)
*14 (143, 10)
87 (888, 64)
7.8 (80, 69 in.·lbf)
64 (653, 47)
Engine Rear Mounting Bracket
Column Hole Cover Sub–assembly
Intermediate Extension
◆Cotter Pin
74 (755, 55)
49 (500, 36)
35.3 (360, 26)
Front Suspension Member with Lower Suspension Arm
◆Cotter Pin
74 (755, 55)
8.0 (82, 71 in.·lbf)
89 (908, 66)
49 (500, 36)
113 (1,152, 83)
52 (530, 38)
39 (400, 29)
52 (530, 38)
113 (1,152, 83)
Lock Nut
89 (908, 66)
◆ 216 (2,303, 159)
RH Engine Under Cover
LH Engine Under Cover
N·m (kgf·cm, ft·lbf) : Specified torque
◆ Non–reusable part
* For use with SST
9359AB48

Exploded view of the manual transaxle (2 of 2)

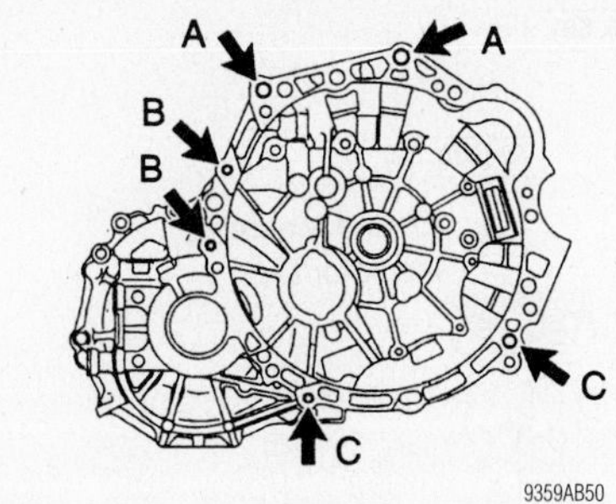

Manual transaxle bolt installation locations

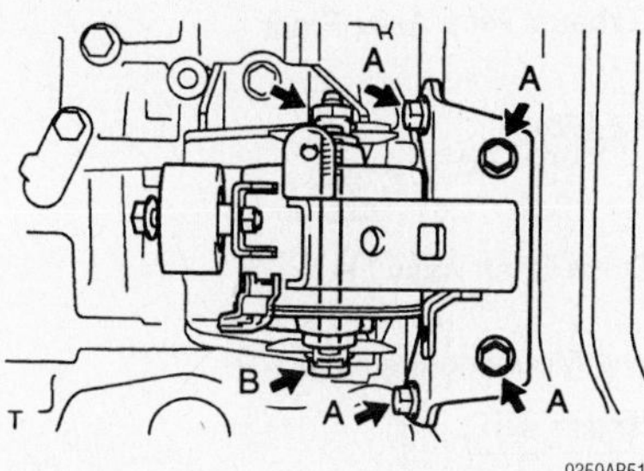

Transverse engine mounting insulator bolt locations

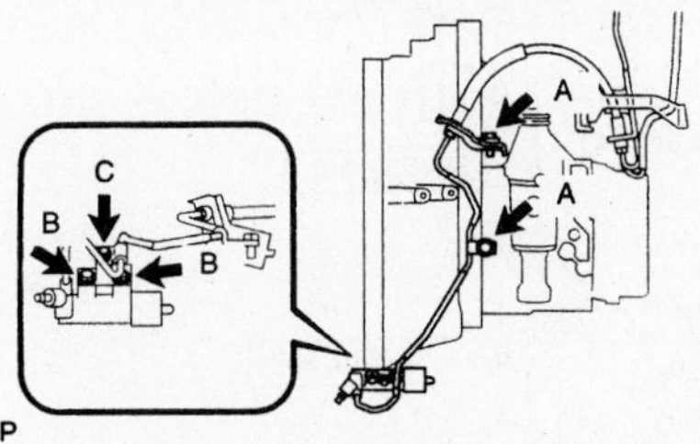

Clutch release cylinder bolt locations

- Transverse engine mounting bracket. Tighten to 38 ft. lbs. (52 Nm).
- Transverse engine mounting insulator. Tighten the "A" bolts to 38 ft. lbs. (52 Nm) and the "B" bolts to 59 ft. lbs. (80 Nm).

10. The remainder of installation is the reverse of the removal procedure, noting the following specifications:

a. Starter mounting bolts: 27 ft. lbs. (37 Nm).

b. Clutch release cylinder bolts: "A" bolts 19 ft. lbs. (25 Nm), "B" bolts 9 ft. lbs. (12 Nm) and "C" bolts 44 inch lbs. (5 Nm).

c. Battery carrier bolts: 10 ft. lbs. (13 Nm).

d. Battery clamp bolt: 44 inch lbs. (5 Nm).

e. Battery clamp nut: 31 inch lbs. (3.5 Nm).

f. Cylinder head cover bolts: 62 inch lbs. (7 Nm).

g. Hood bolts: 10 ft. lbs. (13 Nm).

h. Wheel lug nuts: 76 ft. lbs. (103 Nm).

11. Fill the transaxle fluid to the proper level.

12. Start the vehicle, check for leaks and repair if necessary.

Automatic Transaxle

REMOVAL & INSTALLATION

FWD—A246E & U240E Transaxles

1. Before servicing the vehicle, refer to the precautions in the beginning of this section.
2. Drain the transaxle fluid.
3. Remove or disconnect the following:
   - Negative battery cable
   - Hood
   - No. 2 cylinder head cover
   - Battery and battery carrier
   - Air cleaner assembly with hose
   - Floor shift cable transmission control shift
   - Transmission control cable support
   - No. 1 transmission control cable bracket
   - Wiring harness and brackets
   - Transmission wire connector
   - Park/neutral position switch connector, with Anti-lock Brake System (ABS)
   - Speedometer sensor connector, without ABS
   - Transmission revolution sensor connectors, if equipped
   - Transmission fluid filler tube
   - No. 1 oil cooler inlet and outlet tubes
   - Foot rest
   - Floor carpet
   - Oxygen (O_2) sensor connector
4. Suspend the engine as follows:

a. Disconnect the 2 Positive Crankcase Ventilation (PCV) hoses.

b. Install the No. 1 and No. 2 engine hangers in the correct direction.

c. No. 1 engine hanger: P/N 12281-22021 (A246E) or 12281-88600 (U240E).

d. No. 2 engine hanger: P/N 12281-15040 (A246E) or 12281-88600 (U240E)

e. Bolt: P/N 91512-B1016.

f. Torque the bolt to 28 ft. lbs. (38 Nm).

g. Attach an engine chain hoist to the engine hangers.

- Front wheels
- Right and left engine undercovers
- Front floor panel brace, U240E transaxle
- Front exhaust pipe
- Front halfshafts
- Automatic transmission case protector
- Starter

5. Support the transaxle with a floor jack
   - Left side transverse engine mounting insulator and bracket
   - Right side front and rear engine mount insulators
   - 4 bolts, dynamic damper and member sub-assembly
   - Front and rear right side transverse engine mounting brackets
   - Flywheel housing undercover
   - Automatic transaxle. Turn the crankshaft for access to the 6 bolts while holding the crankshaft pulley bolt with a wrench.
   - Torque converter clutch
6. Installation is the reverse of the removal procedure, noting the following specifications:

a. Automatic transaxle: Bolt "A" to 47 ft. lbs. (64 Nm), bolt "B" to 34 ft. lbs. (47 Nm) and bolt "C" to 17 ft. lbs. (23 Nm).

b. Torque converter bolts: 20 ft. lbs. (28 Nm).

c. Front and rear right transverse engine mounting bracket bolts: 47 ft. lbs. (64 Nm).

d. Member sub-assembly center bolts: "A" bolts to 29 ft. lbs. (39 Nm) and "B" bolts to 38 ft. lbs. (52 Nm).

e. Right rear engine mounting insulator-to-engine mounting bracket bolt: 64 ft. lbs. (87 Nm).

f. Right rear engine mount insulator nuts and bolt: 38 ft. lbs. (52 Nm).

g. Left side engine mounting bracket-to-transaxle bolts: 38 ft. lbs. (52 Nm).

h. Left side engine mounting insulator bolts and nut: Bolt "A" to 38 ft. lbs. (52 Nm), Bolt "B" and Nut "B" to 59 ft. lbs. (80 Nm).

i. Front right engine mount insulator-to-mounting bracket bolt and nut: 38 ft. lbs. (52 Nm).

j. Starter bolts: 29 ft. lbs. (39 Nm).

k. Automatic transmission case protector bolts: 14 ft. lbs. (18 Nm).

l. Wheel lug nuts: 76 ft. lbs. (103 Nm).

m. Oil cooler clamp bolts: 49 inch lbs. (5.5 Nm).

n. Oil cooler inlet and outlet tubes: 25 ft. lbs. (34 Nm).

o. Wire harness bracket bolt: 9 ft. lbs. (13 Nm).

p. Transmission control cable bracket bolts: 9 ft. lbs. (12 Nm).

q. Transmission control cable support: 9 ft. lbs. (12 Nm).

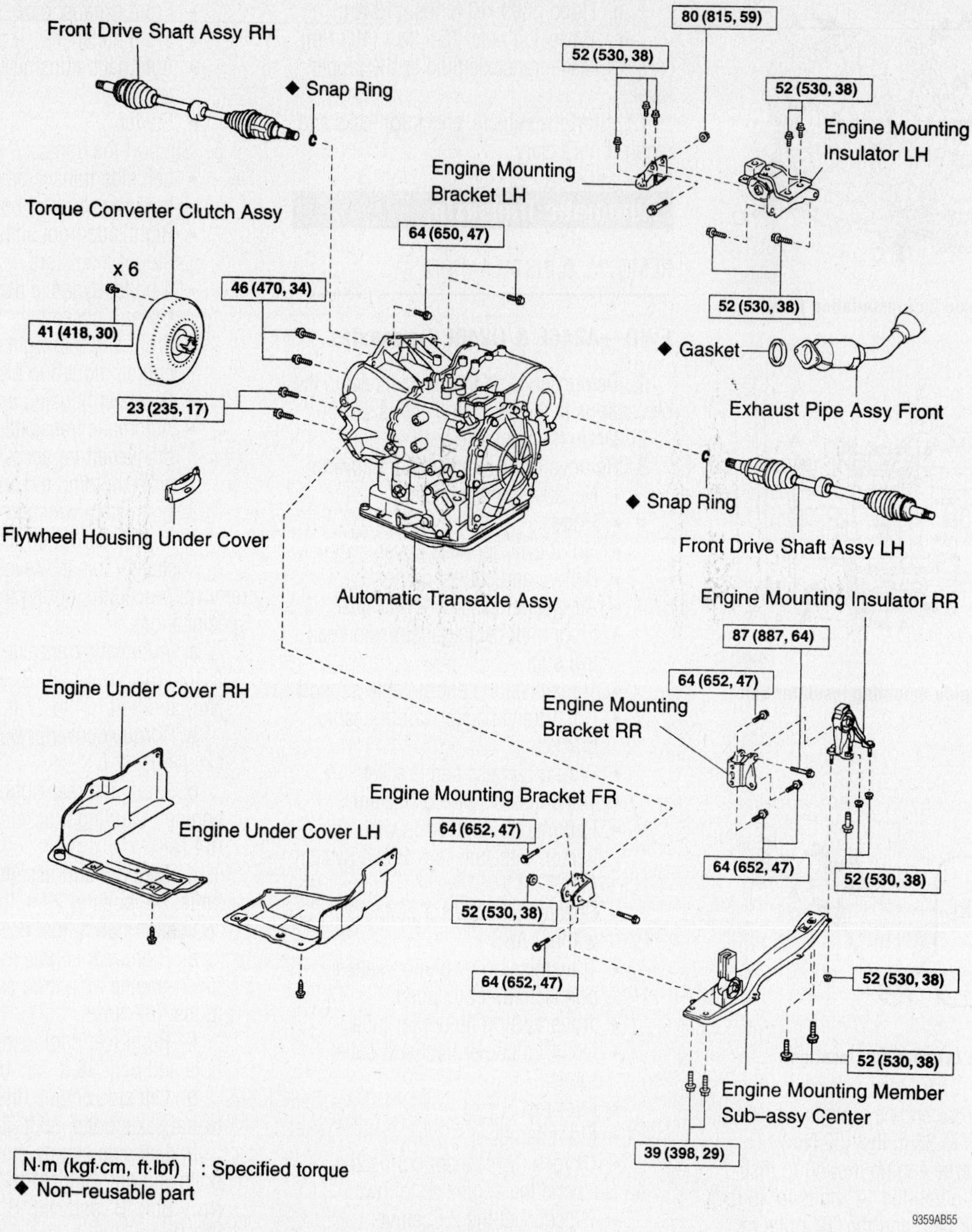

Front Drive Shaft Assy RH
◆ Snap Ring
80 (815, 59)
52 (530, 38)
52 (530, 38)
Engine Mounting Insulator LH
Engine Mounting Bracket LH
Torque Converter Clutch Assy
64 (650, 47)
x 6
46 (470, 34)
41 (418, 30)
52 (530, 38)
◆ Gasket
23 (235, 17)
Exhaust Pipe Assy Front
◆ Snap Ring
Flywheel Housing Under Cover
Front Drive Shaft Assy LH
Automatic Transaxle Assy
Engine Mounting Insulator RR
87 (887, 64)
64 (652, 47)
Engine Under Cover RH
Engine Mounting Bracket RR
Engine Mounting Bracket FR
Engine Under Cover LH
64 (652, 47)
64 (652, 47)
52 (530, 38)
52 (530, 38)
64 (652, 47)
52 (530, 38)
52 (530, 38)
Engine Mounting Member Sub–assy Center
39 (398, 29)
N·m (kgf·cm, ft·lbf) : Specified torque
◆ Non–reusable part

9359AB55

Automatic transaxle and related components—U240E transaxle shown, A246E similar

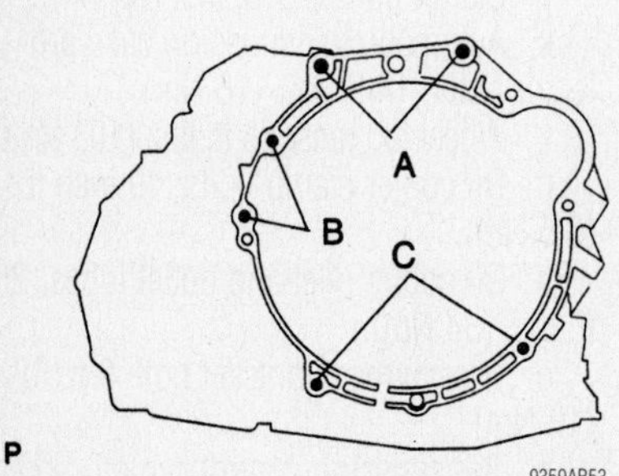

Automatic transaxle bolt locations

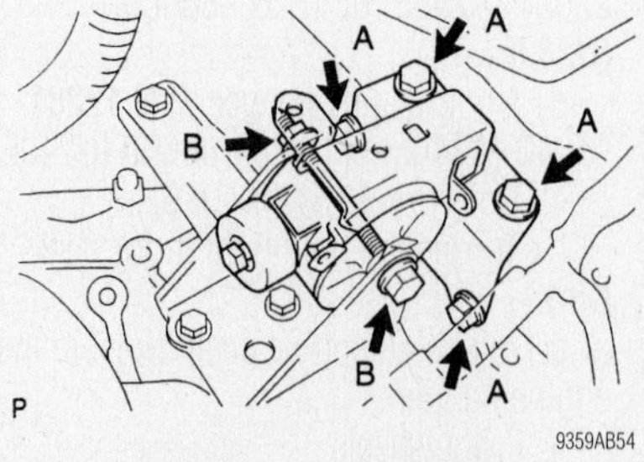

A
A
B
A
B
A
P
9359AB54

Left side engine mount insulator bolt and nut locations

r. Battery carrier: 10 ft. lbs. (13 Nm).

s. Air cleaner assembly: 62 inch lbs. (7 Nm).

t. Cylinder head cover bolts: 62 inch lbs. (7 Nm).

u. Hood bolts: 10 ft. lbs. (13 Nm).

7. Fill the transaxle fluid to the proper level.

8. Start the vehicle, check for leaks and repair if necessary.

AWD—U341F Transaxle

1. Before servicing the vehicle, refer to the precautions in the beginning of this section.

2. Drain the transaxle fluid.
3. Remove or disconnect the following:
   - Negative battery cable
   - Engine and transaxle assembly
   - Transfer case
   - Automatic transmission case protector
   - Front left side halfshaft
   - Transmission control cable support and bracket
   - Wire harness clamp bracket, bolts and 2 wire harnesses
   - Transmission wire connector
   - Park/neutral position switch connector
   - Transmission revolution sensor connectors, if equipped
   - Transmission fluid filler tube
   - Oil cooler inlet and outlet tubes
   - Transverse engine mounting brackets
   - Flywheel housing undercover
   - Automatic transaxle. Turn the crankshaft for access to the 6 bolts while holding the crankshaft pulley bolt with a wrench.
   - Torque converter clutch
4. Installation is the reverse of the removal procedure, noting the following specifications:
   a. Automatic transaxle: Bolt "A" to 47 ft. lbs. (64 Nm), bolt "B" to 34 ft. lbs. (47 Nm) and bolt "C" to 17 ft. lbs. (23 Nm).
   b. Oil cooler clamp bolts: 8 ft. lbs. (11 Nm) for the top bolt and 49 inch lbs. (5.5 Nm) for the bottom bolt
   c. Oil cooler inlet and outlet tube bolts: 25 ft. lbs. (34 Nm).

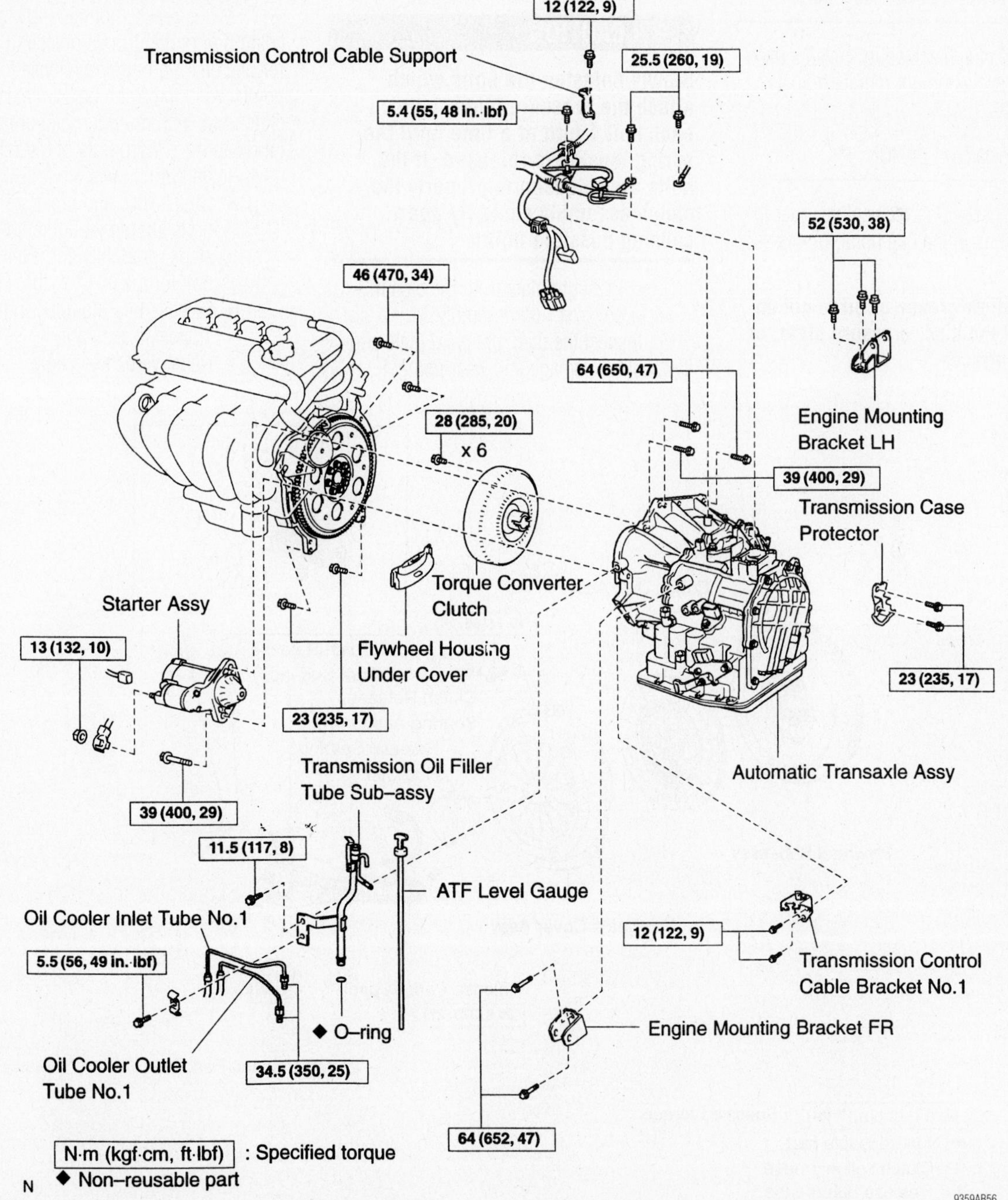

Automatic transaxle and related components—U341F transaxle

d. Wire harness clamp bracket bolt: 48 inch lbs. (5 Nm).

e. Transmission control cable bracket and support bolts: 9 ft. lbs. (12 Nm).

f. Automatic transmission case protector bolts: 17 ft. lbs. (23 Nm).

5. Fill the transaxle fluid to the proper level.

6. Start the vehicle, check for leaks and repair if necessary.

Clutch

ADJUSTMENTS

Hydraulic clutch actuating systems used in Toyota vehicles do not require adjustment.

REMOVAL & INSTALLATION

1. Before servicing the vehicle, refer to the precautions in the beginning of this section.

➡**Do not allow grease or oil to get on any part of the disc, pressure plate, or flywheel surfaces.**

2. Remove or disconnect the following:
   - Negative battery cable. On vehicles equipped with an air bag, wait at least 90 seconds before proceeding
   - Transaxle assembly

3. Make matchmarks on the clutch cover (pressure plate) and flywheel so that the pressure plate can be returned to its original position during installation.

4. Remove or disconnect the following:
   - Release fork bearing clips
   - Release bearing hub, complete with the release bearing
   - Release fork and support

✱✱ CAUTION

Slowly unfasten the bolts which attach the pressure plate. Loosen each bolt 1 turn at a time until the spring tension is released. If the bolts are released improperly the clutch assembly could fly apart, causing possible injury.

   - Pressure plate from the clutch cover/spring assembly

5. Inspect the disc, pressure plate and flywheel for damage and wear using a caliper to measure depth and width and a dial indicator to measure runout.

a. The minimum clutch disc rivet head depth is 0.012 in. (0.3mm).

b. The maximum clutch disc runout is 0.031 in. (0.8mm).

c. The maximum pressure plate spring depth is 0.024 in. (0.6mm).

d. The maximum pressure plate spring width is 0.197 in. (5.0mm).

e. The maximum flywheel runout is 0.004 in. (0.1mm).

6. Replace or machine parts as necessary.

To install:

7. When reassembling, apply a thin coating of multipurpose grease to the release bearing hub and release fork contact points. Also, pack the groove inside the clutch hub with multipurpose grease and lubricate the pivot points of the release fork.

8. Install or connect the following:
   - Clutch disc and pressure plate. The bolts should be tightened in 2 or 3 steps, gradually and evenly. Final bolt torque is 14 ft. lbs. (19 Nm).
   - Release bearing, fork and boot
   - Transaxle assembly
   - Negative battery cable

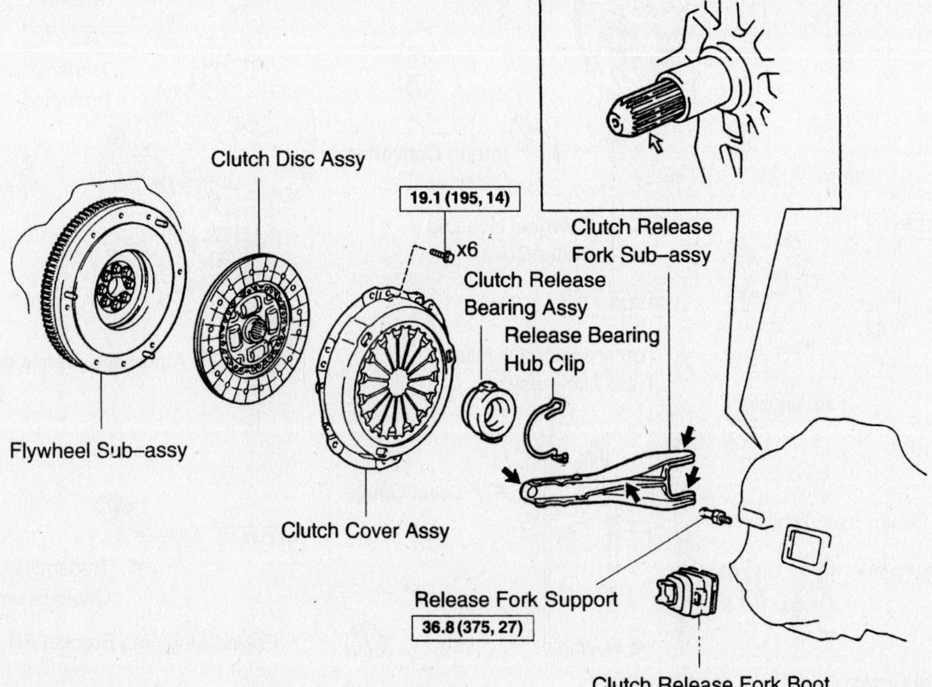

Exploded view of the clutch components

Hydraulic Clutch System

BLEEDING

➡If any maintenance on the clutch system was performed or the system is suspected of containing air, bleed the system. Use care; brake fluid will remove the paint from any surface. If the brake fluid spills onto any painted surface, wash it off immediately with soap and water.

1. Before servicing the vehicle, refer to the precautions in the beginning of this section.
2. Fill the clutch reservoir with brake fluid. Check the reservoir level frequently and add fluid as needed.
3. Connect one end of a vinyl tube to the bleeder plug on the slave cylinder and submerge the other end into a clear container half-filled with brake fluid.
4. Slowly pump the clutch pedal several times.
5. Have an assistant hold the clutch pedal down and loosen the bleeder plug until fluid and/or air starts to run out of the bleeder plug. Close the bleeder plug while the pedal is held to the floor.

➡Do not allow the pedal to rise backup while the bleeder is still open. If this happens, it will allow air to re-enter the slave cylinder and cause the clutch system not to work properly.

6. Repeat Steps 2 and 3 until all the air bubbles are removed from the system.
7. Tighten the bleeder plug when all the air is gone.
8. Refill the master cylinder to the proper level as required.
9. Check the system for leaks.

Transfer Case

REMOVAL & INSTALLATION

1. Before servicing the vehicle, refer to the precautions in the beginning of this section.
2. Drain the transfer case fluid.

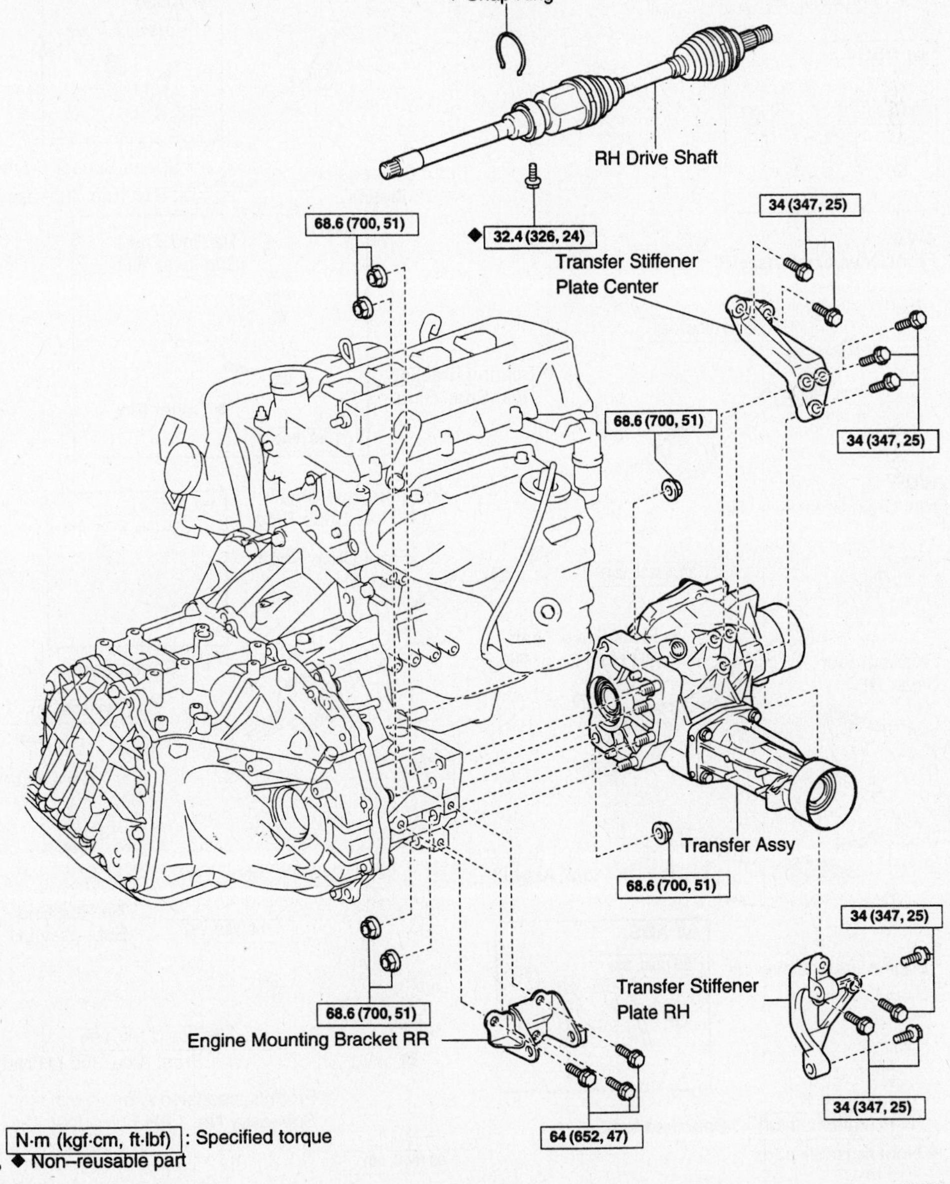

Exploded view of the transfer case mounting

3. Remove or disconnect the following:
   - Negative battery cable. Due to the air bag system, wait at least 90 seconds before proceeding
   - Engine and transaxle assembly
   - Separate vane pump
   - Steering gear
   - Crossmember
   - Manifold stay
   - Oxygen (O_2) sensor
   - Exhaust manifold heat shield
   - Exhaust manifold
   - Starter
   - Right side halfshaft
   - Transverse engine mounting bracket
   - Center and right side transfer stiffener plates

WARNING

When removing the transfer case, DO NOT touch the oil seal.

   - Transfer case bolts, and transfer assembly, using a mallet to dislodge it from the transaxle

4. Installation is the reverse of the removal procedure, noting the following specifications:
   a. Transfer case stiffener case bolts: 25 ft. lbs. (34 Nm).
   b. Engine mounting bracket bolts: 47 ft. lbs. (64 Nm).

5. Add fluid to the transfer case, and check for leaks.

Halfshaft

REMOVAL & INSTALLATION

➡The hub bearing could be damaged if subjected to the full weight of the vehicle, such as if the vehicle is moved without the halfshafts. If it is absolutely

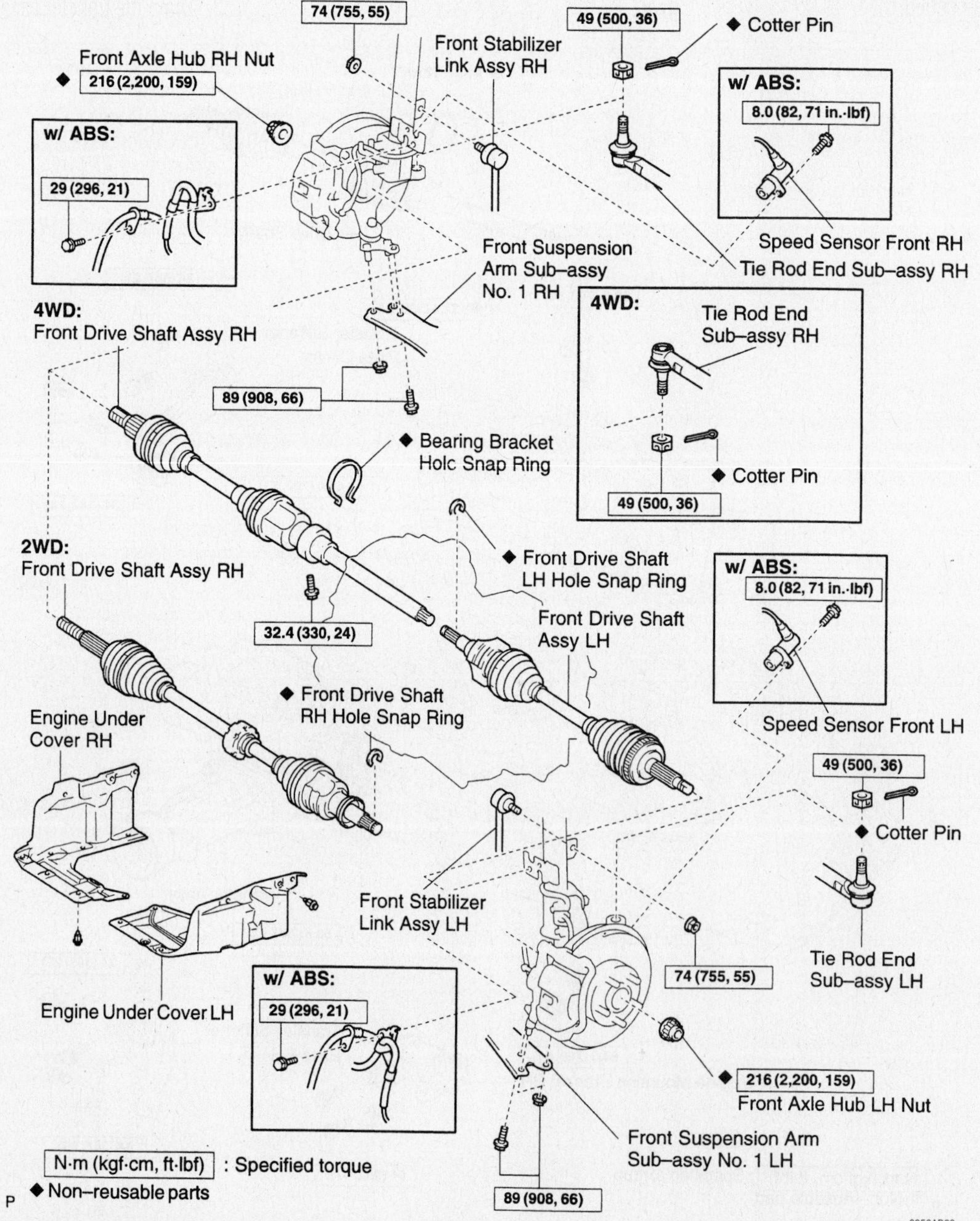

Halfshafts and related components

necessary to place the full vehicle weight on the hub bearing, first support the bearing with SST No. 09608–16041.

1. Before servicing the vehicle, refer to the precautions in the beginning of this section.
2. Drain the transaxle fluid.
3. Remove or disconnect the following:
   - Negative battery cable. Due to the air bag system, wait at least 90 seconds before proceeding.
   - Both front wheels
   - Cotter pin, locknut cap, and the hub nut
   - Undercovers
   - Speed sensors
   - Tie rod ball joint from the steering knuckle
   - Stabilizer bar link from the lower suspension arm
   - Lower ball joint from the lower suspension arm
   - Halfshaft from the knuckle

➡Be careful not to damage the inner oil seal or the ABS sensor rotor on the halfshaft.

4. To remove the left side halfshaft, separate the halfshaft from the transaxle.
5. To remove the right side halfshaft perform the following steps:
   - Remove the 2 bolts of the center bearing bracket
   - Pull the halfshaft out together with the center bearing case and the center halfshaft.
   - Remove the center shaft with the right-hand halfshaft from the transaxle through the bearing bracket.

➡Do not damage the oil seal lip.

To install:

6. Install or connect the following:
   - Snapring opening side facing downward, on the oiled inboard joint tulip
   - Left side halfshaft into the transaxle
   - Right side halfshaft, with the bearing case and center shaft, into the transaxle
   - Center bearing case (right side).
7. After installing either halfshaft, check that there is 0.08–0.12 in. (2–3mm) of axial play. Check that the halfshaft is making contact with the pinion shaft and that the halfshaft cannot be pulled out.
8. Install or connect the following:
   - Halfshaft into the knuckle
   - Lower suspension arm to the lower ball joint. Torque the bolt and nuts to 66 ft. lbs. (89 Nm).
   - Tie rod end to the steering knuckle. Tighten the nut to 36 ft. lbs. (49 Nm).
   - Stabilizer bar link to the lower suspension arm. Torque the nuts to 55 ft. lbs. (74 Nm).
   - Front wheels
   - Hub nut and washer and tighten to 159 ft. lbs. (216 Nm)
   - Negative battery cable
   - Locknut cap and a new cotter pin.
   - Speed sensors
   - Undercover
9. Fill the transaxle fluid to the proper level
10. Start the vehicle, check for leaks and repair if necessary.

CV-Joints

OVERHAUL

1. Before servicing the vehicle, refer to the precautions in the beginning of this section.

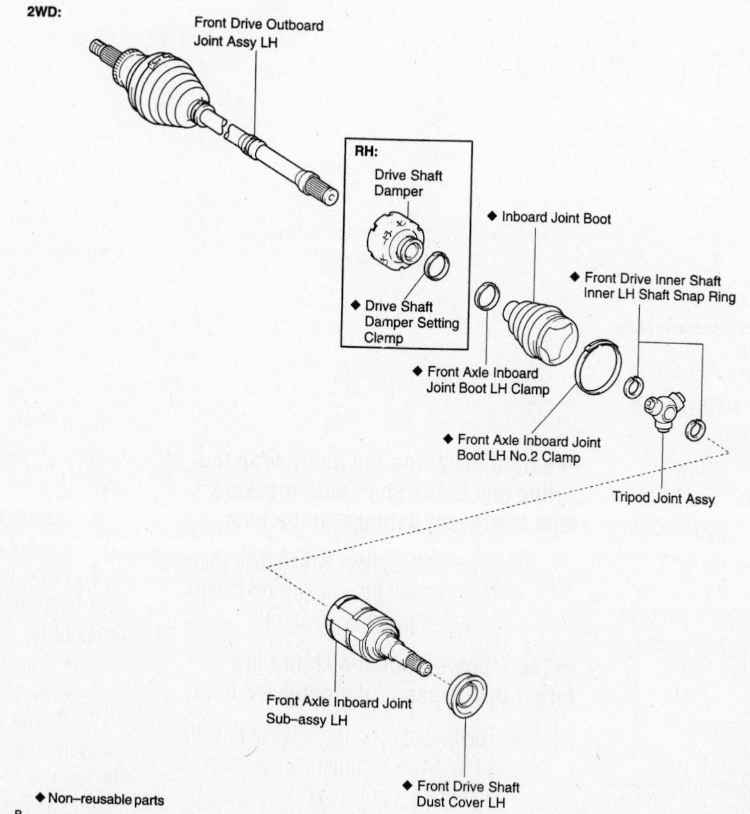

Exploded view of the CV-joint—FWD vehicles

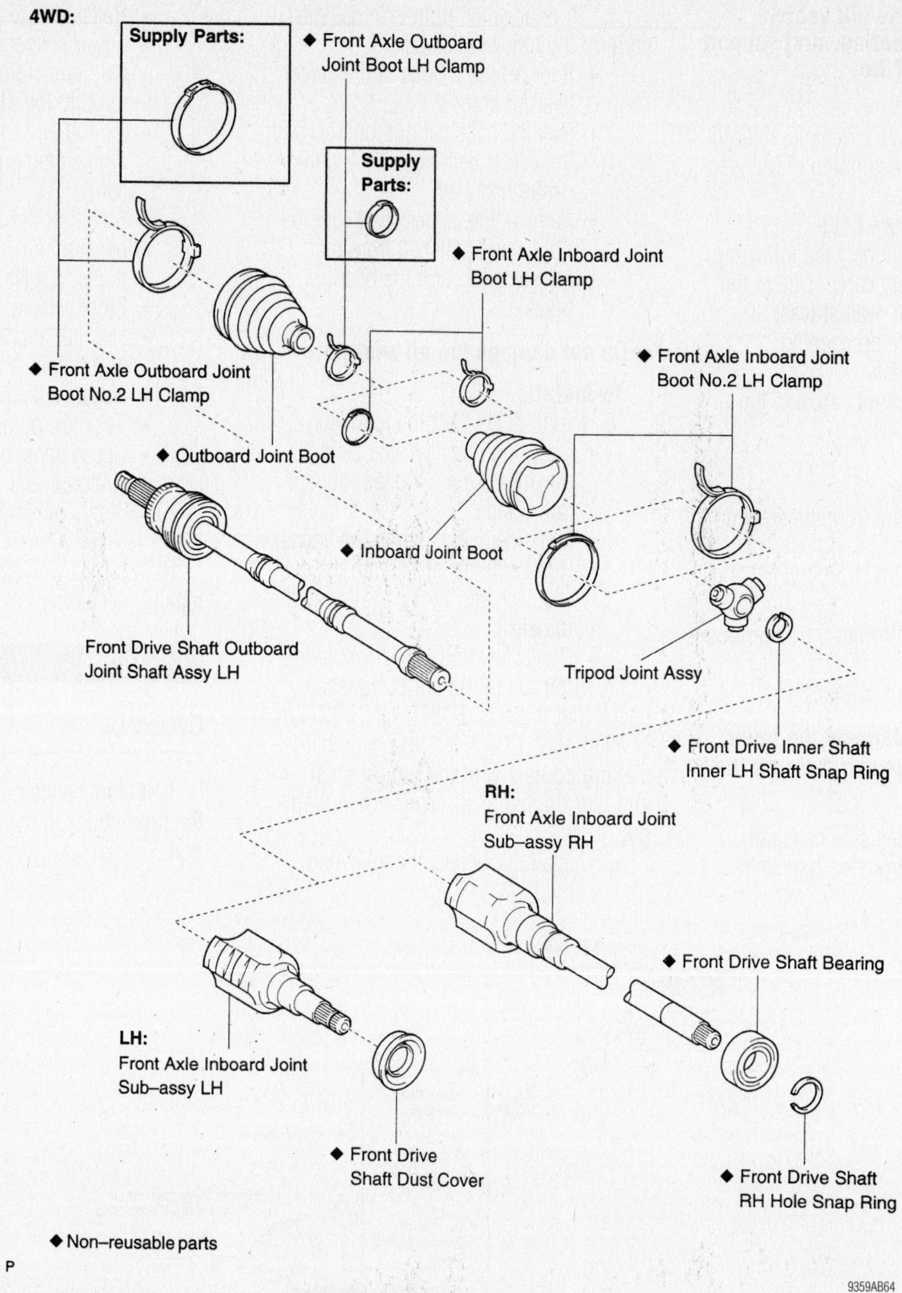

Exploded view of the CV-joint—AWD vehicles

2. Remove or disconnect the following:
   - Inboard joint boot clips
   - Inboard joint tulip from the driveshaft
   - Snapring
   - Using a brass rod and hammer, the tri-pot joint off the driveshaft without hitting the joint roller
   - Inboard joint boot
   - Clamp and driveshaft damper
   - Clamps and the outboard drive boot. DO NOT disassemble the outboard joint.

To assemble:

3. Install or connect the following:

➡Before installing the boot, wrap the spline end of the shaft with masking tape to prevent damage to the boot.

- Driveshaft damper with a new clamp
- Temporarily, the inboard boot with new clamp to the drive joint

➡The inboard boot and clamp are larger than those of the outboard boot.

- The tri-pot onto the driveshaft with a brass rod and hammer without hitting the joint roller
- The snapring

4. Pack the outboard tulip joint and the outboard boot with about 0.26–0.33 lbs. ounces of grease that was supplied with the boot kit.
5. Install or connect the following:
   - Boot onto the outboard joint
6. Pack the inboard tulip joint and boot with ½ lb. of grease that was supplied with the boot kit.
   - Inboard tulip joint onto the driveshaft
   - Boot onto the driveshaft
7. Before checking the standard length, bend the band and lock it. Make sure that the boot is not stretched or squashed when the driveshaft is at standard length. Standard driveshaft length: LH: 540.2 mm (21.268 in.); RH: 857.4 mm (33.756 in.)

STEERING AND SUSPENSION

Air Bag

PRECAUTIONS

Several precautions must be observed when handling the inflator module to avoid accidental deployment and possible personal injury.

- Never carry the inflator module by the wires or connector on the underside of the module.
- When carrying a live inflator module, hold securely with both hands, and ensure that the bag and trim cover are pointed away.
- Place the inflator module on a bench or other surface with the bag and trim cover facing up.
- With the inflator module on the bench, never place anything on or close to the module that may be thrown in the event of an accidental deployment.

DISARMING

To avoid personal injury when working on vehicles equipped with an air bag, the negative battery cable must be disconnected and at least 90 seconds must elapse before working on the system. Failure to do so may result in deployment of the air bag.

REARMING

After vehicle service is completed, reattach the battery cables (positive cable first!) to rearm the air bag system.

Rack and Pinion Steering Gear

REMOVAL & INSTALLATION

1. Before servicing the vehicle, refer to the precautions in the beginning of this section.
2. Position the front wheels straight ahead.
3. Remove or disconnect the following:
   - Negative battery cable. Because these vehicles are equipped with air bags, wait at least 90 seconds before proceeding.
   - Horn button
   - Steering wheel
   - Front wheels
   - Left and right engine undercovers
   - Left and right tie rod ends
   - Column hose cover silencer sheet
   - Steering intermediate shaft
   - Pressure feed and return tubes
   - Left and right side front stabilizer links
   - Right and left front lower control arms from the ball joints
   - Hood
   - No. 2 cylinder head (valve) cover
4. Install an engine support and tension it to support the engine without raising it.
   a. No. 1 engine hanger: P/N 12281-22021 1ZZ-FE, 12281-88600 2ZZ-GE.
   b. No. 2 engine hanger: P/N 12281-15040 1ZZ-FE, 12281-88600 2ZZ-GE.
   c. Bolt: P/N 91512-B1016.
   d. Torque the bolts to 28 ft. lbs. (38 Nm).

✱✱ CAUTION

Do not try to suspend the engine by hooking the chain to any other part.

   e. Attach an engine chain hoist to the hangers.

✱✱ CAUTION

The engine hoist is now in place and under tension. Use care when repositioning the vehicle and make necessary adjustments to the engine support.

5. Remove or disconnect the following:
   - Bolt and nuts holding in the middle of the crossmember and support the crossmember with a jack
   - Bolts from the outer side of the suspension crossmember
   - Suspension crossmember with the steering gear assembly
   - Steering intermediate shaft, after matchmarking it
   - Rack and pinion steering gear from the crossmember
6. Installation is the reverse of the removal procedure, noting the following specifications:
   a. Steering gear bolts and nuts: 43 ft. lbs. (58 Nm) FWD, 60 ft. lbs. (82 Nm) AWD.

1ZZ-FE:

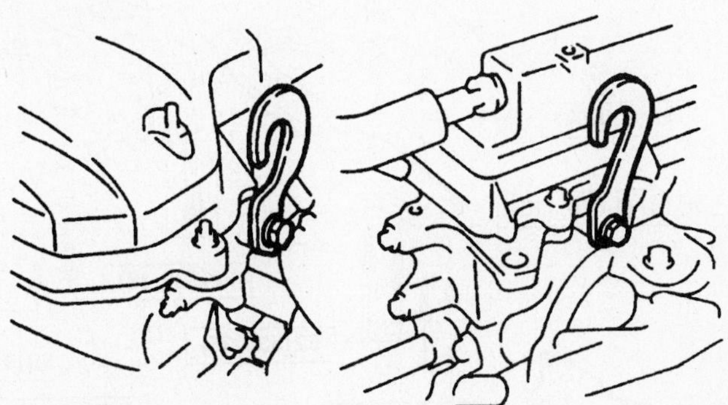

2ZZ-GE:

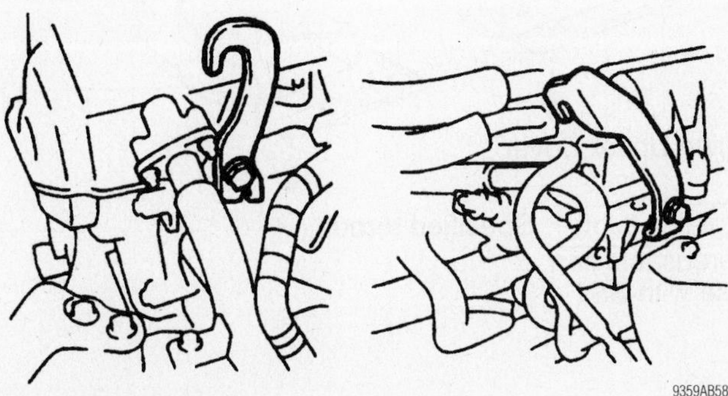

Proper installation of engine hangers

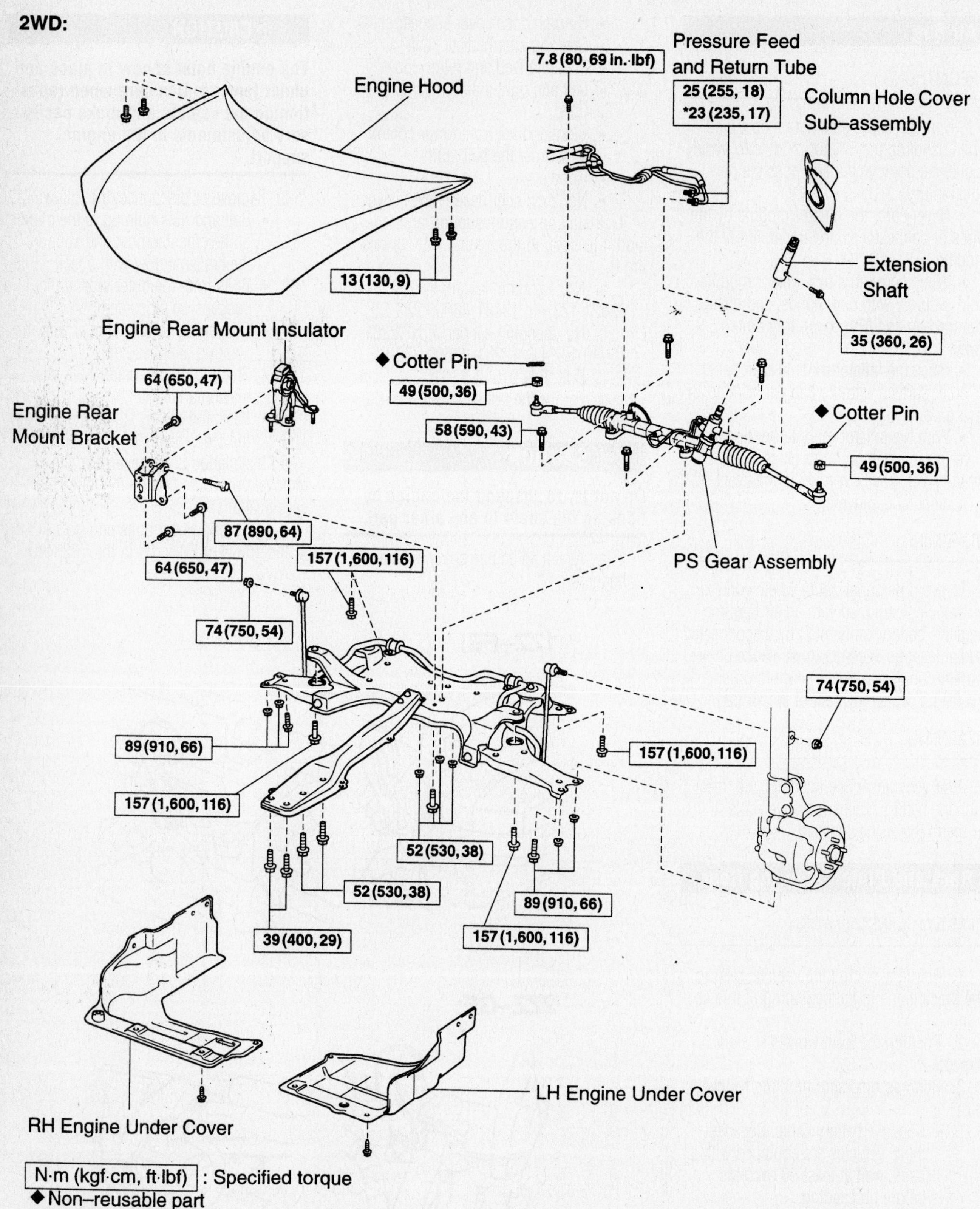

Exploded view of a typical power rack and pinion steering gear unit—FWD shown

b. Steering intermediate shaft: 26 ft. lbs. (35 Nm).

c. Suspension crossmember bolts: 116 ft. lbs. (157 Nm).

d. Engine mount insulator bolts: 38 ft. lbs. (52 Nm).

e. Center member-to-frame bolts: 29 ft. lbs. (39 Nm).

f. Stabilizer bar link-to-the lower control arms nuts: 55 ft. lbs. (74 Nm).

g. Fluid return and pressure tubes: 17 ft. lbs. (23 Nm).

h. Tie rod ends: 36 ft. lbs. (49 Nm).

i. Wheel lug nuts: 76 ft. lbs. (103 Nm).

7. Check and top off the power steering fluid.

8. Check and adjust the alignment, if needed.

Strut and Coil Spring

REMOVAL & INSTALLATION

Front

1. Before servicing the vehicle, refer to the precautions in the beginning of this section.

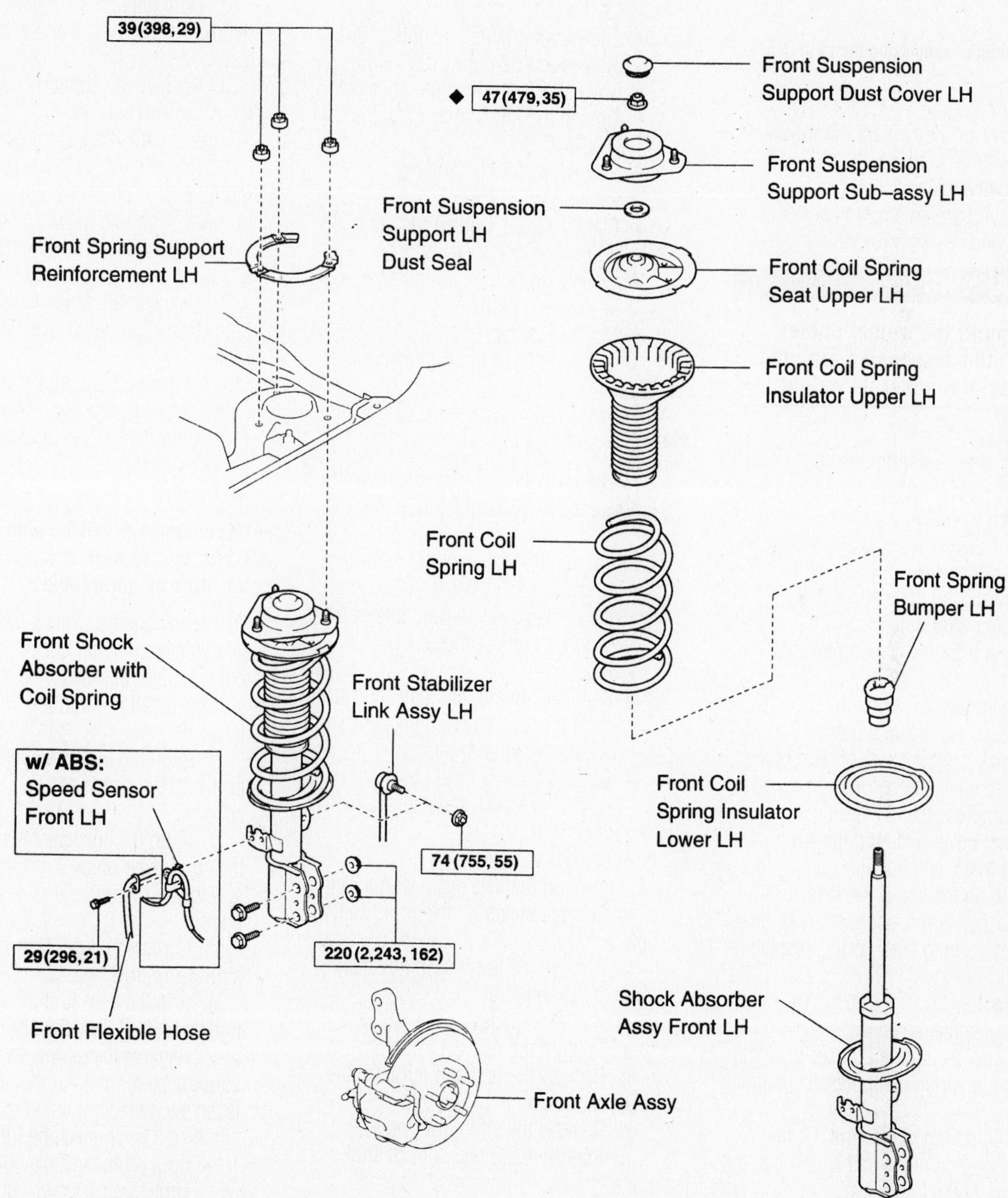

Common coil spring and strut component assembly

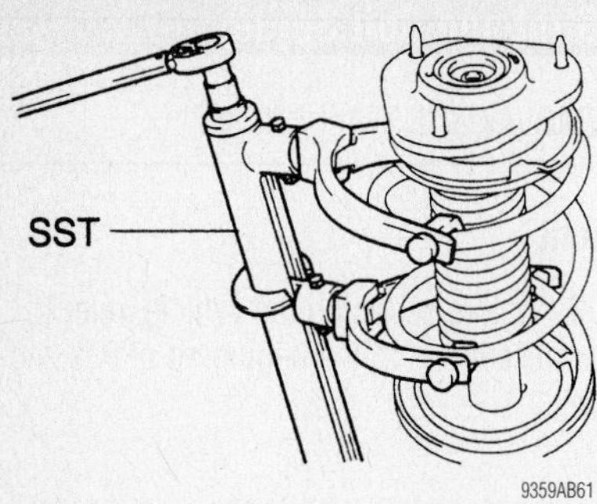

Proper method of supporting the strut in a vise

2. Remove or disconnect the following:
   - Negative battery cable. Because of the air bag system, wait at least 90 seconds before proceeding

⁂ WARNING

Do not support the weight of the vehicle on the suspension arm; the arm will deform under its weight.

   - Wheel
   - Stabilizer link from the strut
   - Bolt, and disconnect the brake hose from the strut
   - With ABS brakes, speed sensor wiring harness from the strut
   - Lower strut bolts and nuts
   - Upper strut nuts
   - Strut from the steering knuckle
   - Strut
3. To disassemble the strut:
   - Install a bolt and 2 nuts to the bracket at the lower portion of the strut shell and secure it in a vise
   - Compress the coil spring
   - Dust cover and hold the spring seat so that it will not turn
   - Nut on the top of the strut
   - Suspension support, bearing, dust seal, spring seat, spring, insulators and bumper

To install:

4. To assemble the strut:
   - Install the spring bumper to piston
5. Using a spring compressor, compress the spring.
   - Coil spring to the strut. Fit the lower end of the coil spring into the gap of the lower seat.
   - Spring seat with the insulator
   - Dust seal on the spring seat
   - Suspension support and tighten 35 ft. lbs. (47 Nm). After the nut has been tighten, release the compressor tool tension.
6. Pack multipurpose grease into the suspension support.
   - Dust cover.

➡Do not use an impact wrench to tighten the nut. Also, check that the bearing fits into the recess in the suspension support.

   - Strut
   - Nuts holding the strut to the strut tower. Tighten the nuts to 29 ft. lbs. (39 Nm).
   - 2 lower strut bolts and nuts. Tighten to 162 ft. lbs. (220 Nm).
   - Brake line to the steering knuckle. Tighten the line bolt to 21 ft. lbs. (29 Nm).
   - Secure the wiring harness, if equipped with ABS
   - Stabilizer link. Tighten the nut to 55 ft. lbs. (74 Nm).
   - Wheel. Tighten the lug nuts to 76 ft. lbs. (103 Nm).
   - Negative battery cable
7. Check and adjust the alignment, if needed.

Rear

1. Before servicing the vehicle, refer to the precautions in the beginning of this section.
2. Remove or disconnect the following:
   - Negative battery cable. Because of the air bag system, wait at least 90 seconds before proceeding.
   - Rear wheel
   - Rear deck board, luggage compartment tray and any trim necessary to access the strut towers
   - Shock absorber head cover
3. On AWD vehicles, separate the rear stabilizer link.
4. For FWD vehicles:
   a. Support the axle beam with a jack.
   b. Remove the strut tower nuts and bolt.
   c. Remove the lower strut nut, cushion retainer and strut .
5. For AWD vehicles:
   a. Support the rear control arm.
   b. Remove the bolt and nut from the rear control arm.
   c. Remove the strut tower nuts.
   d. Remove the 3 rear control arm bolts.
   e. Press the rear control arm down to the outside of the vehicle, then remove the strut.
6. To disassemble the strut:
   a. Place the strut assembly in a pipe vise or strut vise.

⁂ WARNING

Do not attempt to clamp the strut assembly in a flat jaw vise as this will result in damage to the strut tube.

   b. Compress the spring until the upper suspension support is free of any spring tension. Do not over-compress the spring.
   c. Hold the upper support, then remove the nut on the end of the shock piston rod.
   d. Remove the support, coil spring, insulator, and bumper.
7. Inspect the strut as follows:
   a. Check the shock absorber by moving the piston shaft through its full range of travel. It should move smoothly and evenly throughout its entire travel without any trace of binding or notching.
   b. Use a small straightedge to check the piston shaft for any bending or deformation.
   c. Inspect the spring for any sign of deterioration or cracking. The waterproof coating on the coils should be intact to prevent rusting.

To install:

➡Never reuse a self-locking nut. Always replace self-locking nuts and cotter pins as applicable.

8. Assemble the strut as follows:
   a. Loosely assemble all components onto the strut assembly. Be sure the spring end aligns with the hollow in the lower seat.
   b. Align the upper suspension support with the piston rod and install the support.
   c. Align the suspension support with the strut lower bracket. This assures the spring will be properly seated top and bottom.
   d. Compress the spring to expose the strut piston rod threads.
   e. Install a new strut piston nut and tighten to 41 ft. lbs. (56 Nm).
   f. Remove the spring compressor. Be sure the paint mark on the upper support faces the outside of the strut.
9. Install or connect the following:
   - Strut on the vehicle. Tighten the strut-to-strut tower nuts to 59 ft. lbs. (80 Nm).
   - Strut to the axle carrier and install the nut and cushion retainer/bolt snug. Do not fully tighten at this time.
   - Strut head cover

- Rear control arm (AWD). Tighten the bolts to 48 ft. lbs. (65 Nm).
- Rear stabilizer link (AWD)
- Trunk tray, deckboard and any other trim pieces removed
- Wheel

10. With the vehicle's weight on the suspension, tighten the bolt holding the strut to the axle carrier to 59 ft. lbs. (80 Nm) for FWD vehicles, or 103 ft. lbs. (140 Nm) for AWD vehicles.
    - Negative battery cable

11. Check and adjust the rear wheel alignment.

Lower Ball Joint

REMOVAL & INSTALLATION

1. Before servicing the vehicle, refer to the precautions in the beginning of this section.
2. Remove or disconnect the following:
    - Negative battery cable. Wait at least 90 seconds before proceeding.
    - Front wheel
3. Depress the brake pedal and loosen the hub nut
    - ABS speed sensor, if equipped
    - Cotter pin and nut from the tie rod end. Using a tie rod end removal tool, separate the tie rod end from the steering knuckle.
    - Lower control arm ball joint, using a suitable puller
    - Separate the front halfshaft
    - Lower ball joint cotter pin and castle nut
    - Lower ball joint from the steering knuckle using a puller

To install:

4. Install or connect the following:
    - Lower ball joint to the lower arm. Tighten the castle nut to 76 ft. lbs. (103 Nm).
    - New cotter pin
    - Front halfshaft
    - Lower control arm
    - Tie rod end to the knuckle
    - ABS speed sensor
    - Hub nut
    - Wheel
    - Negative battery cable
5. Check and adjust the alignment, if needed.

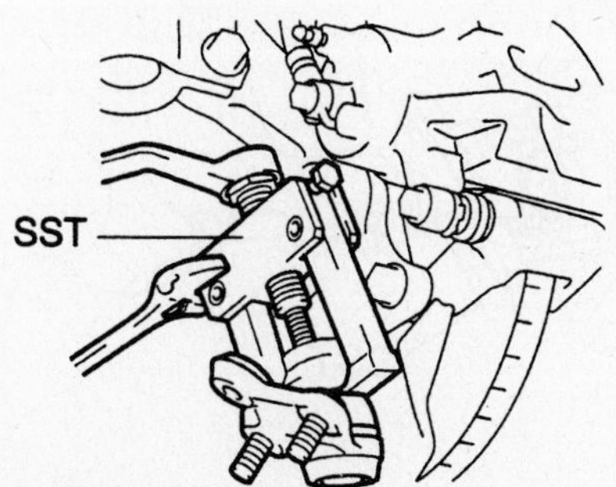

Removing the ball joint from the knuckle

Upper Control Arm

REMOVAL & INSTALLATION

Rear—AWD Only

1. Before servicing the vehicle, refer to the precautions in the beginning of this section.
2. Remove or disconnect the following:
    - Negative battery cable. Wait at least 90 seconds before proceeding
    - Rear wheel
    - Exhaust pipe
    - Propeller shaft with center bearing shaft
    - Rear stabilizer links
    - Rear hub nuts
    - Rear brake drum
    - Speed sensor
    - Front brake shoe
    - Parking brake shoe strut set
    - Rear brake shoe
    - Parking brake cables
    - Rear brake hoses
    - Separate the rear suspension arms
    - Separate the upper control arm
    - Rear drive axle assembly
    - Rear strut nut and bolt
    - Rear strut
    - Rear suspension arm
    - Rear suspension member
    - Upper control arm assembly. Matchmark the camber adjust cams and rear suspension member prior to removal.
3. Installation is the reverse of the removal procedure.

Lower Control Arm

REMOVAL & INSTALLATION

1. Before servicing the vehicle, refer to the precautions in the beginning of this section.
2. Remove or disconnect the following:
    - Negative battery cable. Wait at least 90 seconds before proceeding..
    - Front wheel
    - Stabilizer link
    - Bolt and nuts and separate the lower control arm from the lower ball joint
    - Bolts and nuts, then separate the steering gear. Loosen the bolt, since the nut cannot be rotated, then suspend the steering gear.
3. Support the engine, using the engine lifting hooks and the procedure under Engine Removal & Installation.
    - Crossmember
    - Lower control arm from the crossmember
4. Installation is the reverse of the removal procedure.

Wheel Bearings

REMOVAL & INSTALLATION

Front

1. Before servicing the vehicle, refer to the precautions in the beginning of this section.
2. Remove or disconnect the following:
    - Negative battery cable. On vehicles equipped with an air bag, wait at least 90 seconds before proceeding.
    - Wheels
    - Hub nut
    - Front stabilizer link
    - Anti-lock Brake System (ABS) speed sensor
    - Brake caliper
    - Rotor
    - Tie rod end from the steering knuckle
    - Lower control arm ball joint
    - Front halfshaft from the hub, using a mallet to tap it out. Be careful not to damage the boot or speed sensor.
3. Loosen the nuts on the lower side of the strut assembly. Do not remove at this time.
    - Lower ball joint using a puller
    - Tie rod end from the steering knuckle
    - Steering knuckle from the lower control arm
    - Knuckle from the strut assembly
    - Hub

➡Cover the halfshaft boot with a shop rag to protect it from any damage.

4. Clamp the steering knuckle in a vise and remove the dust deflector. Remove the nut holding the steering knuckle to the ball joint. Press the ball joint out of the steering knuckle.

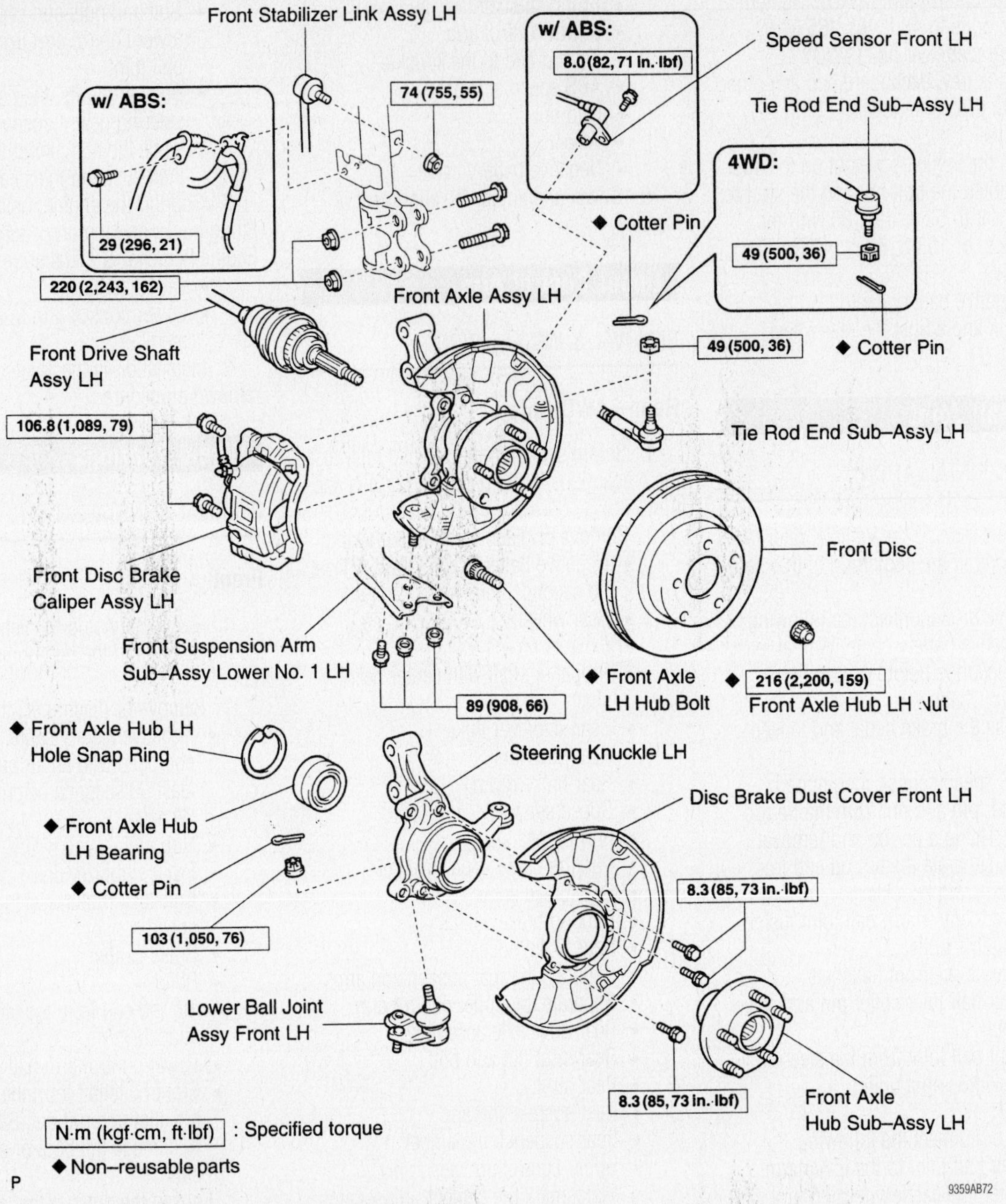

Front Stabilizer Link Assy LH

w/ ABS:

8.0 (82, 71 in.·lbf)

Speed Sensor Front LH

74 (755, 55)

w/ ABS:

Tie Rod End Sub–Assy LH

4WD:

29 (296, 21)

◆ Cotter Pin

49 (500, 36)

220 (2,243, 162)

Front Axle Assy LH

Front Drive Shaft Assy LH

49 (500, 36)

◆ Cotter Pin

106.8 (1,089, 79)

Tie Rod End Sub–Assy LH

Front Disc

Front Disc Brake Caliper Assy LH

Front Suspension Arm Sub–Assy Lower No. 1 LH

◆ Front Axle LH Hub Bolt

◆ 216 (2,200, 159)

Front Axle Hub LH Nut

89 (908, 66)

◆ Front Axle Hub LH Hole Snap Ring

Steering Knuckle LH

Disc Brake Dust Cover Front LH

◆ Front Axle Hub LH Bearing

◆ Cotter Pin

8.3 (85, 73 in.·lbf)

103 (1,050, 76)

Lower Ball Joint Assy Front LH

8.3 (85, 73 in.·lbf)

Front Axle Hub Sub–Assy LH

N·m (kgf·cm, ft·lbf) : Specified torque

◆ Non–reusable parts

P

9359AB72

Exploded view of the front hub and bearing, and related components

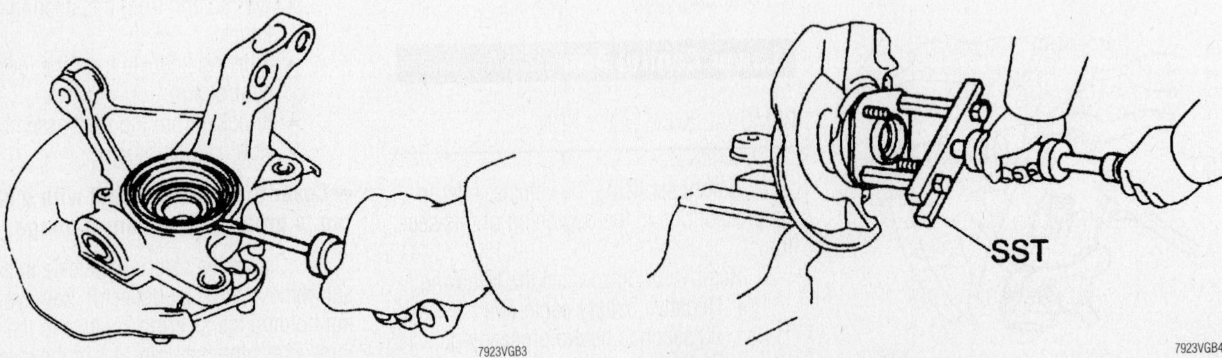

Removing the inner axle seal from the hub assembly

Removing the axle hub from the knuckle

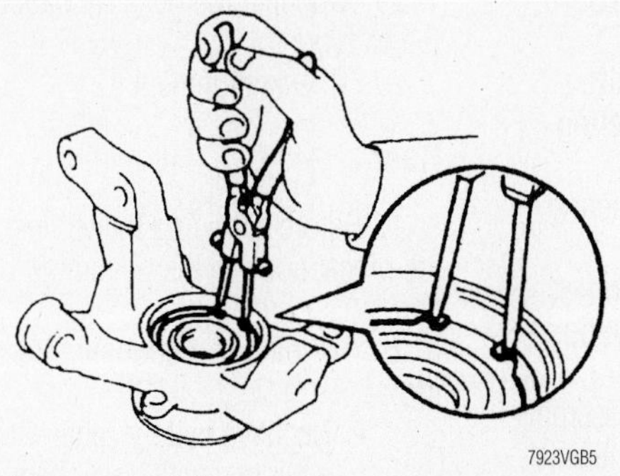

Removing the snapring from the knuckle before pressing out the bearing

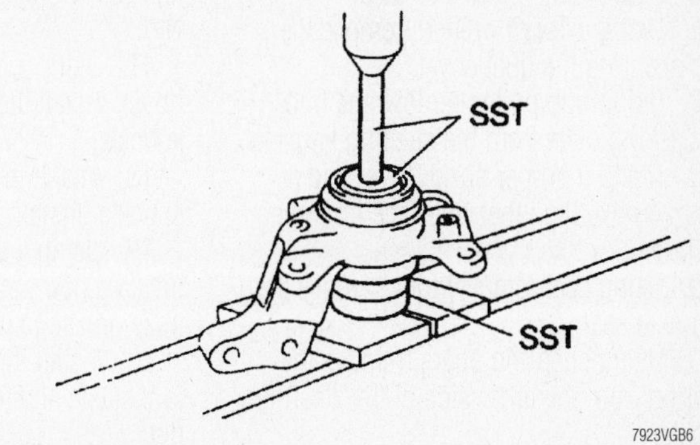

Removing the bearing from the steering knuckle using a press

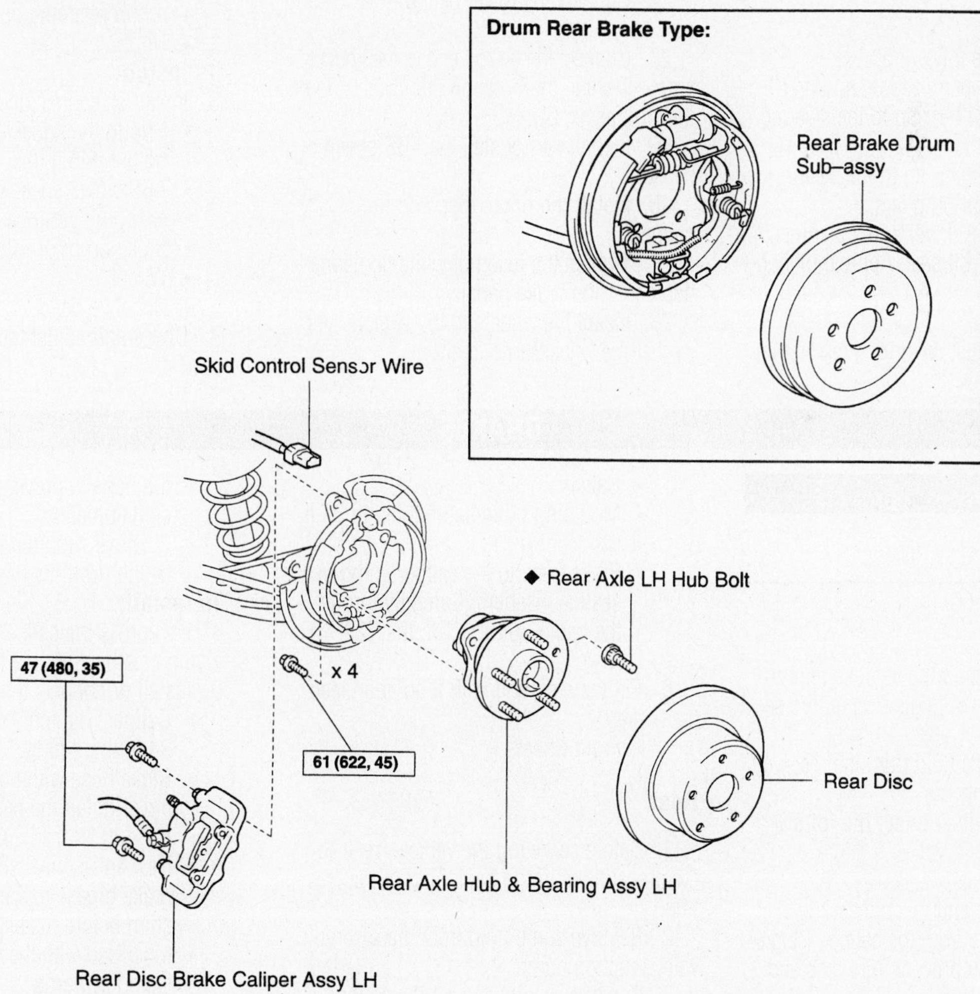

Exploded view of the hub and wheel bearing assembly

5. Remove the inner axle seal.
6. Using a Torx® wrench, remove the bolts securing the dust cover.
7. Using hub puller, remove the hub and backing plate from the steering knuckle.
8. Using a proper sized driver and a press, remove the inner hub race from the axle hub.
9. Using seal removal tool, remove the outer axle seal.
10. Using snapring pliers, remove the snapring from the inner side of the steering knuckle.
11. Using a proper sized driver and a press, remove the bearing from the steering knuckle. The bearing is pressed from the front of the steering knuckle and is removed through the back of the steering knuckle.

To install:

12. Perform the following:
13. Using a proper sized driver and a press, install a new bearing to the steering knuckle.
14. Install the snapring to the steering knuckle using snapring pliers.
15. Using a seal driver and a hammer, install a new outer oil seal. Apply multipurpose grease to the oil seal lip.
16. Place the dust cover on the steering knuckle. Tighten the bolts: 78 inch lbs. (9 Nm).
17. Using a press and a proper sized driver, install the axle hub to the steering knuckle.
18. Attach the ball joint to the steering knuckle. Install a new cotter pin.
19. Using a seal driver and a hammer, install a new inner oil seal. Apply multipurpose grease to the oil seal lip.
20. Install the knuckle and hub assembly to the axle and temporarily tighten the axle nut.
21. Connect the knuckle assembly to the lower strut bracket. Temporarily insert the mounting bolts from the rear and install the nuts making sure the matchmarks made earlier are in alignment.
22. Connect the lower ball joint to lower arm.
23. Connect the tie rod end to the knuckle.
24. Tighten the bolts on the lower side of the strut assembly.
25. If equipped, install the ABS speed sensor.
26. Install the brake disc and the caliper.
27. Tighten the axle nut while someone depresses the brake pedal.
28. Install the wheels to the vehicle. Verify that the wheel turns freely.
29. Connect the negative battery cable to the battery.
30. Check alignment.

Rear

1. Before servicing the vehicle, refer to the precautions in the beginning of this section.
2. Remove or disconnect the following:
    - Negative battery cable. On vehicles equipped with an air bag, wait at least 90 seconds before proceeding.
    - Wheel
    - Brake drum or rotor
    - With ABS brakes, ABS wheel speed sensor or skid control sensor, as applicable
    - 4 hub retaining bolts
    - Hub

To install:

3. Install or connect the following:
    - Hub to the knuckle. Tighten the bolts to 45 ft. lbs. (61 Nm).
    - ABS wheel speed or skid control sensor, if equipped
    - Brake drum or rotor
    - Wheel
    - Negative battery cable
4. Check and adjust the alignment, if needed.

BRAKES

Brake Caliper

REMOVAL & INSTALLATION

Front

1. Before servicing the vehicle, refer to the precautions in the beginning of this section.
2. Remove some fluid from the reservoir with a suction pump.
3. Remove or disconnect the following:
    - Front wheels
    - Banjo bolt and disconnect the brake hose from the caliper. Plug the hose to prevent fluid loss and contamination.
    - Mounting bolts while holding the slide pin
    - Caliper

To Install:

4. Compress the caliper piston using a C–clamp or other suitable tool.
5. Install or connect the following:
    - Caliper
    - Mounting bolts and tighten to 25 ft. lbs. (34 Nm)
    - Brake hose to the caliper using new sealing washers. Carefully torque the banjo bolt to 21 ft. lbs. (29 Nm).
6. Fill the reservoir with fluid and bleed the brakes.
    - Front wheels

Rear

1. Before servicing the vehicle, refer to the precautions in the beginning of this section.
2. Remove some fluid from the reservoir with a suction pump.
3. Remove or disconnect the following:
    - Rear wheels
    - Clip and both anti-rattle springs
    - Two pad guide pins
    - Pads with the shims
    - Banjo bolt and disconnect the brake hose from the caliper. Plug the hose to prevent fluid loss and contamination.
    - 2 caliper mounting bolts and the caliper from its mounting bracket

To Install:

4. Compress the caliper piston using a C–clamp or other suitable tool.
5. Install or connect the following:
    - Caliper. Tighten the caliper bolts to 34 ft. lbs. (46 Nm).
    - Brake hose with new sealing washers. Tighten the banjo bolt to 21 ft. lbs. (29 Nm).
    - New anti-squeal shims, apply disc brake grease to the inside of the shim before installation
    - Inner pad with the wear indicator facing upwards
    - Outer pad
    - Two pad guide pins
    - Anti-rattle springs and the clip
6. Fill the reservoir with fluid and bleed the brake system. Adjust the parking brake if necessary.
    - Rear wheels

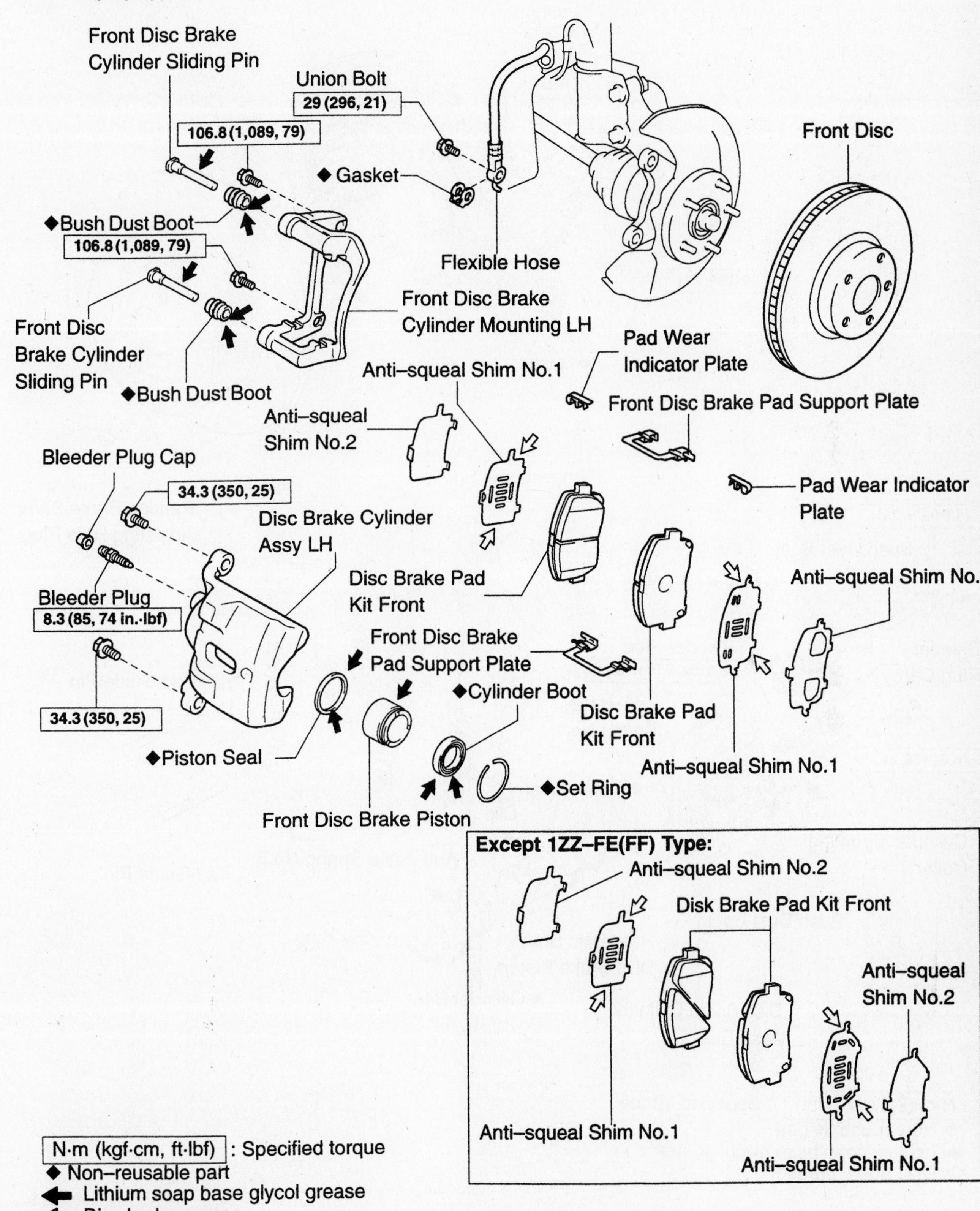
1ZZ-FE(FF) Type:
Front Disc Brake Cylinder Sliding Pin
Union Bolt
29 (296, 21)
106.8 (1,089, 79)
◆ Gasket
Front Disc
◆Bush Dust Boot
106.8 (1,089, 79)
Flexible Hose
Front Disc Brake Cylinder Mounting LH
Front Disc Brake Cylinder Sliding Pin
Pad Wear Indicator Plate
Anti-squeal Shim No.1
◆Bush Dust Boot
Front Disc Brake Pad Support Plate
Anti-squeal Shim No.2
Bleeder Plug Cap
Pad Wear Indicator Plate
34.3 (350, 25)
Disc Brake Cylinder Assy LH
Disc Brake Pad Kit Front
Anti-squeal Shim No.2
Bleeder Plug
8.3 (85, 74 in.·lbf)
Front Disc Brake Pad Support Plate
◆Cylinder Boot
34.3 (350, 25)
Disc Brake Pad Kit Front
◆Piston Seal
Anti-squeal Shim No.1
◆Set Ring
Front Disc Brake Piston
Except 1ZZ-FE(FF) Type:
Anti-squeal Shim No.2
Disk Brake Pad Kit Front
Anti-squeal Shim No.2
Anti-squeal Shim No.1
Anti-squeal Shim No.1
N·m (kgf·cm, ft·lbf) : Specified torque
◆ Non-reusable part
Lithium soap base glycol grease
Disc brake grease
N
42356-TMAT-G01

Exploded view of the front caliper components

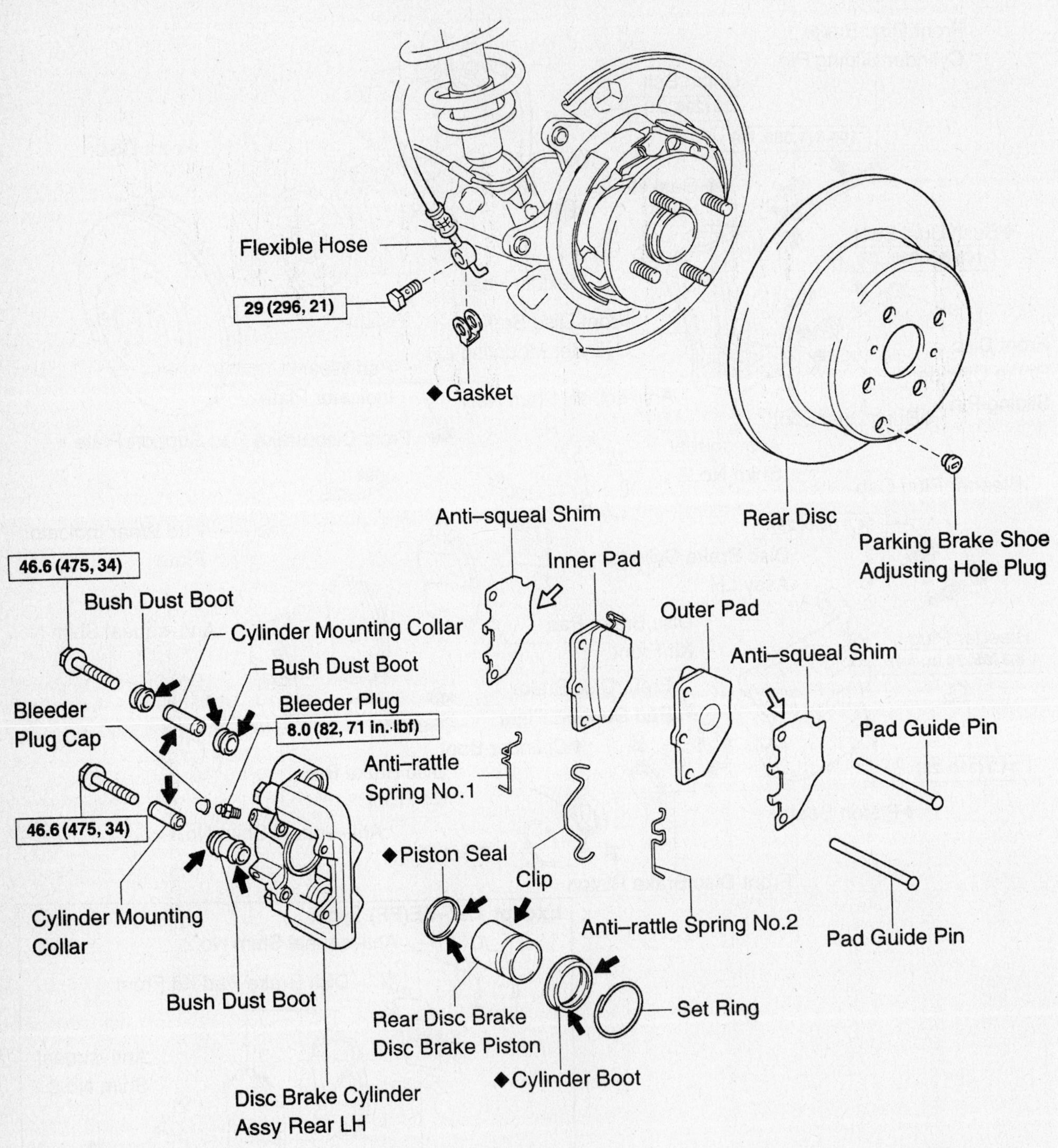

Flexible Hose
29 (296, 21)
◆Gasket
Rear Disc
Parking Brake Shoe Adjusting Hole Plug
Anti–squeal Shim
Inner Pad
Outer Pad
Anti–squeal Shim
46.6 (475, 34)
Bush Dust Boot
Cylinder Mounting Collar
Bush Dust Boot
Bleeder Plug
8.0 (82, 71 in.·lbf)
Bleeder Plug Cap
Anti–rattle Spring No.1
Pad Guide Pin
46.6 (475, 34)
◆Piston Seal
Clip
Cylinder Mounting Collar
Anti–rattle Spring No.2
Pad Guide Pin
Bush Dust Boot
Rear Disc Brake Disc Brake Piston
Set Ring
◆Cylinder Boot
Disc Brake Cylinder Assy Rear LH
N·m (kgf·cm, ft·lbf) : Specified torque
◆ Non–reusable part
Lithium soap base glycol grease
Disc brake grease
42356-TMAT-G02

Exploded view of the rear caliper components

Disc Brake Pads

REMOVAL & INSTALLATION

Front

1. Before servicing the vehicle, refer to the precautions in the beginning of this section.
2. Remove some fluid from the reservoir with a suction pump.
3. Remove or disconnect the following:
   - Front wheels
   - Mounting bolts while holding the slide pin
   - Caliper
   - Pads and shims
   - Both anti-squeal shims from the pads
   - Wear indicator plates from the pads

To Install:

4. Compress the caliper piston using a C–clamp or other suitable tool.
5. Install or connect the following:
   - Wear indicator plates
   - Both anti-squeal shims
   - Pads and shims
   - Caliper
   - Mounting bolts and tighten to 25 ft. lbs. (34 Nm)
6. Fill the reservoir with fluid and bleed the brakes, if necessary.
   - Front wheels

Rear

1. Before servicing the vehicle, refer to the precautions in the beginning of this section.
2. Remove some fluid from the reservoir with a suction pump.
3. Remove or disconnect the following:
   - Rear wheels
   - Clip and both anti-rattle springs
   - Two pad guide pins
   - Pads with the shims

To Install:

4. Compress the caliper piston using a C–clamp or other suitable tool.
5. Install or connect the following:
   - New anti-squeal shims, apply disc brake grease to the inside of the shim before installation
   - Inner pad with the wear indicator facing upwards
   - Outer pad
   - Two pad guide pins
   - Anti-rattle springs and the clip
6. Fill the reservoir with fluid and bleed the brake system. Adjust the parking brake if necessary.
   - Rear wheels

Brake Drums

REMOVAL & INSTALLATION

1. Before servicing the vehicle, refer to the precautions in the beginning of this section.
2. Remove or disconnect the following:
   - Wheel
   - Brake drum. If the drum will not pull of the axle, back off the automatic adjuster by turning the adjusting wheel.

To install:

3. Install or connect the following:
   - Drum on the axle
   - Wheel
4. Refill the master cylinder and pump pedal to attain full brake pedal before road-testing the vehicle.

Brake Shoes

REMOVAL & INSTALLATION

FWD Models

1. Before servicing the vehicle, refer to the precautions in the beginning of this section.
2. Remove or disconnect the following:
   - Wheel
   - Brake drum. If the drum will not pull of the axle, back off the automatic adjuster by turning the adjusting wheel.
   - Upper side tension spring with the spacer
   - Anchor side spring using needle nosed pliers
   - Hold-down springs and pins from the front shoe
   - Upper side return spring from the front shoe
   - Front shoe
   - Left hand parking brake shoe strut set from the front shoe
   - Automatic adjuster lever from the front shoe
   - Hold-down springs and pins from the rear shoe
   - Upper side return spring from the rear shoe
   - Parking brake cable from the rear shoe using needle nosed pliers
   - Rear brake shoe
   - C–washer using a suitable pry tool from the shoe
   - Parking brake lever from the shoe

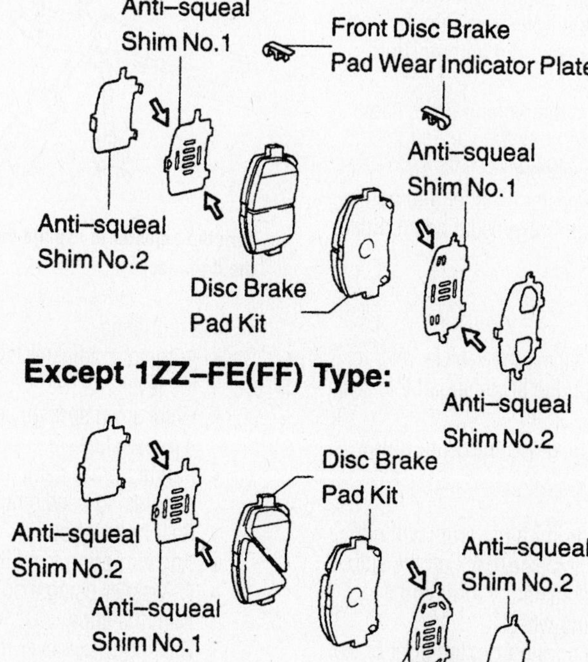

Exploded view of the front pads and related components

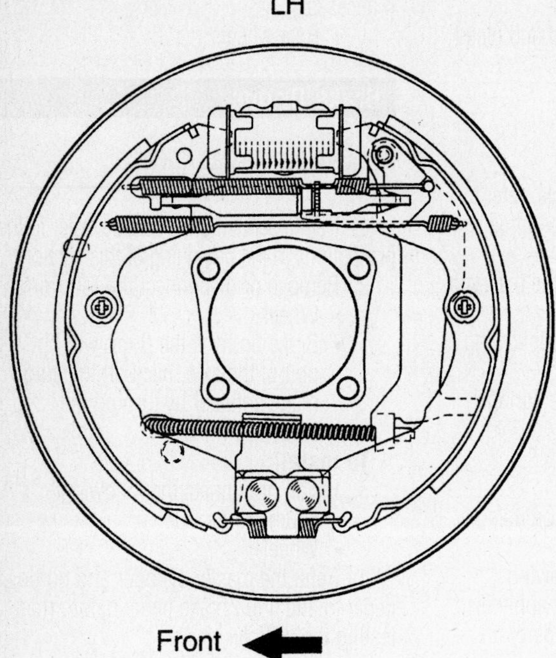

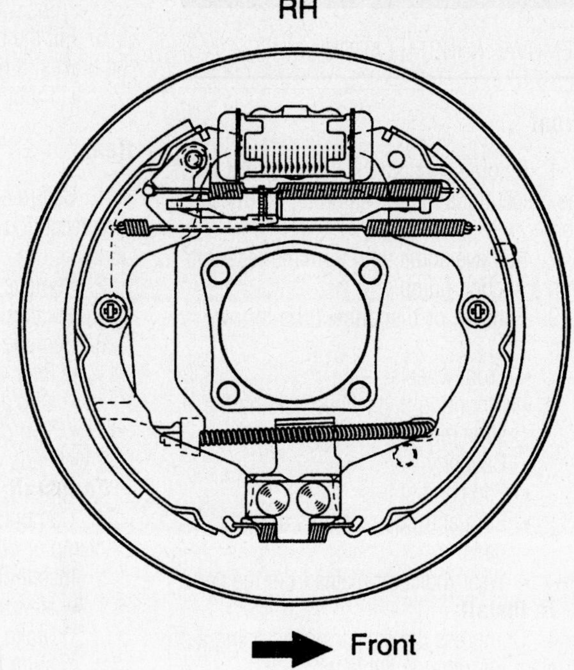

View of the properly installed rear drum brake assembly

To install:

3. Lubricate the contact points on the backing plate and the adjuster with lithium grease.

4. Install or connect the following:

- Parking brake lever, and attach using a new C–washer
- Parking brake lever to the shoe lever
- Upper return spring to the rear shoe
- Shoe assembly onto the backing plate
- Hold-down springs and pins to retain the rear shoe
- Automatic adjuster lever

5. Apply lithium grease to the adjuster bolt.

- Left hand parking brake shoe strut set
- Upper side return spring to the front shoe
- Hold-down springs and pins to the front shoe
- Anchor side spring using needle nosed pliers
- Upper side tension spring with the spacer

6. Adjust the rear brakes as follows:

a. Temporarily install the drum and hub nuts.

b. Remove the hole plug from the backing plate.

c. Turn the adjuster to expand the shoe until the drum locks.

d. Back off the adjuster eight notches using a suitable adjustment tool.

e. Install the hole plug. into the backing plate to prevent dirt and moisture from entering.

f. Readjust the parking brake cable as necessary.

7. Install the wheels.

8. Refill the master cylinder and pump pedal to attain full brake pedal before Road-testing the vehicle.

AWD Models

1. Before servicing the vehicle, refer to the precautions in the beginning of this section.

2. Remove or disconnect the following:

- Wheel
- Brake drum. If the drum will not pull of the axle, back off the automatic adjuster by turning the adjusting wheel.
- Return spring from the front brake shoe
- Anchor spring using needle nosed pliers
- Hold-down springs and pins from the front shoe

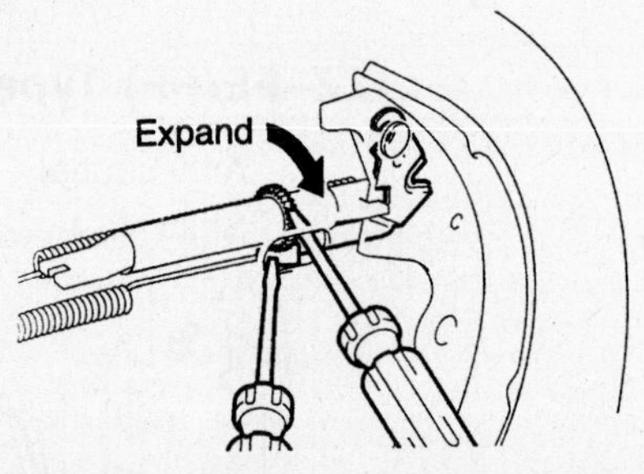

Turn the adjuster to expand the shoe until the drum locks

- Front shoe
- Automatic adjuster lever spring and the lever
- Hold-down springs and pins from the rear shoe
- Parking brake cable from the rear shoe using needle nosed pliers
- Rear brake shoe
- Anchor spring from the rear shoe
- C–washer using a suitable pry tool from the shoe
- Parking brake lever from the shoe

To install:

3. Lubricate the contact points on the backing plate and the adjuster with lithium grease.

4. Install or connect the following:

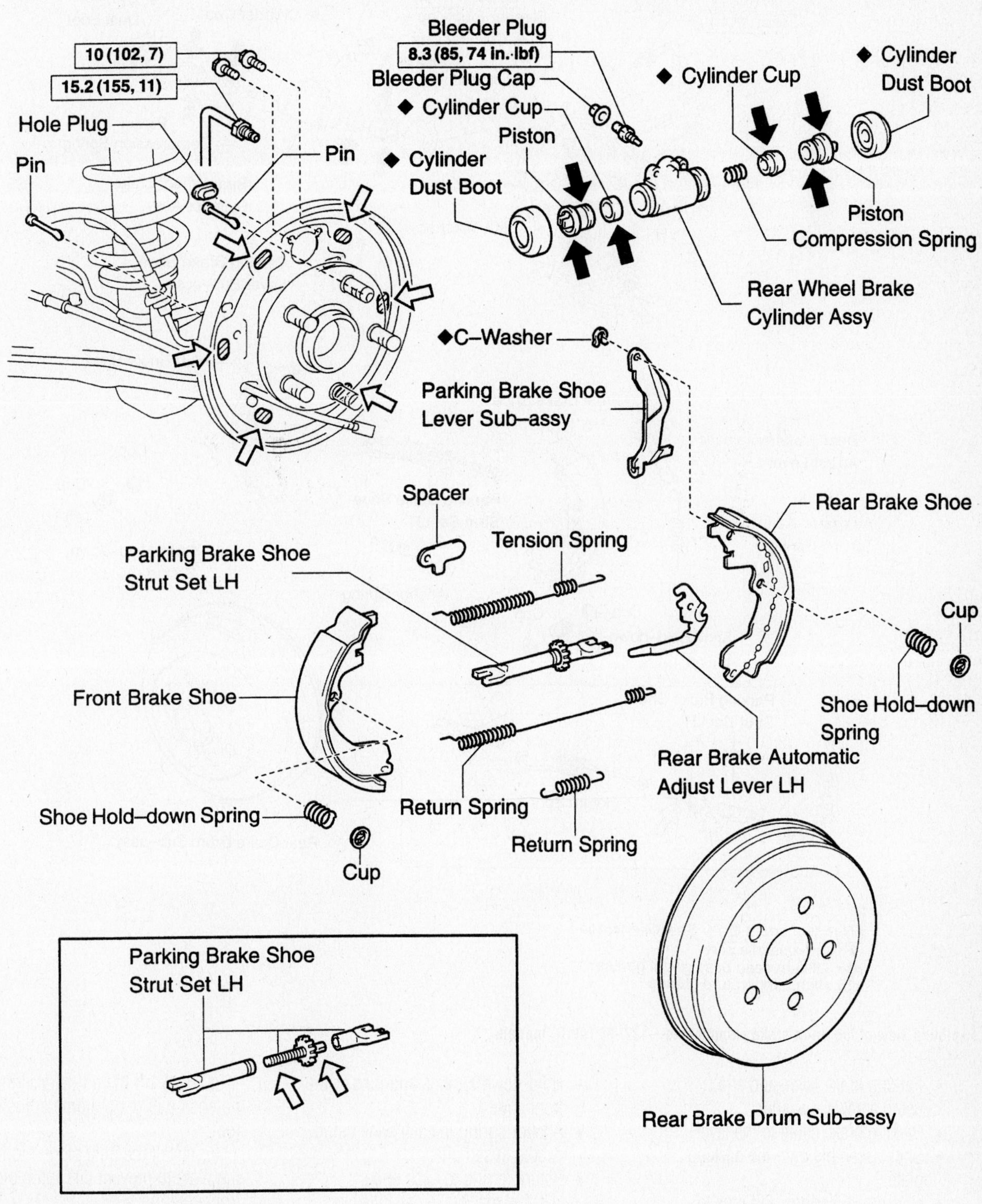
Bleeder Plug
10 (102, 7)
8.3 (85, 74 in.·lbf)
15.2 (155, 11)
Bleeder Plug Cap
◆ Cylinder Cup
◆ Cylinder Cup
◆ Cylinder Dust Boot
Hole Plug
Piston
Pin
Pin
◆ Cylinder Dust Boot
Piston
Compression Spring
Rear Wheel Brake Cylinder Assy
◆C–Washer
Parking Brake Shoe Lever Sub–assy
Spacer
Rear Brake Shoe
Tension Spring
Parking Brake Shoe Strut Set LH
Cup
Front Brake Shoe
Shoe Hold–down Spring
Rear Brake Automatic Adjust Lever LH
Return Spring
Shoe Hold–down Spring
Return Spring
Cup
Parking Brake Shoe Strut Set LH
Rear Brake Drum Sub–assy
N·m (kgf·cm, ft·lbf) : Specified torque
◆ Non–reusable part
Lithium soap base glycol grease
High temperature grease
P
42356-TMAT-G06

Exploded view of the drum brake components—1ZZ–FE (FWD) models

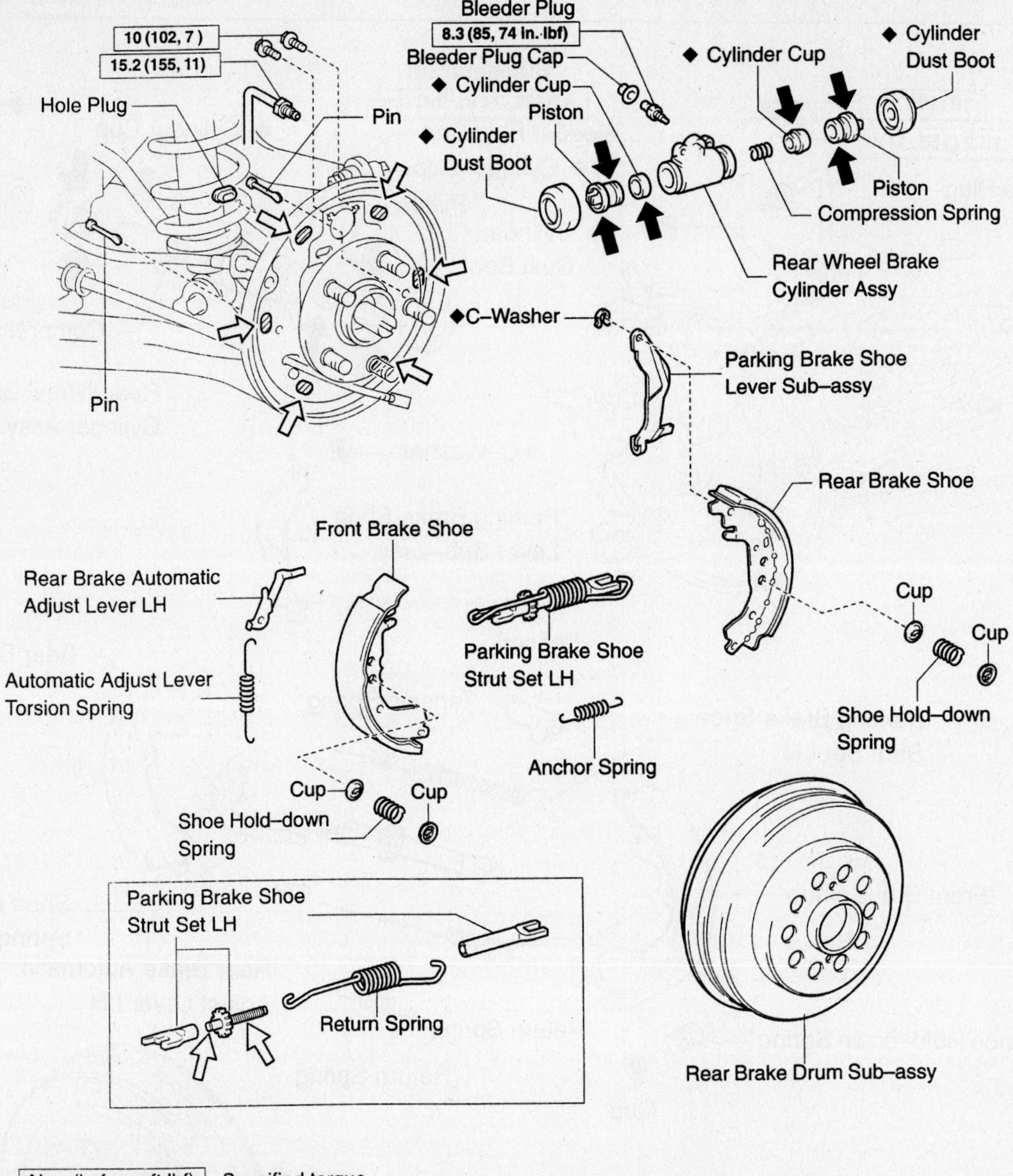

Exploded view of the drum brake components—1ZZ–FE (AWD) models

- Parking brake lever, and attach using a new C–washer
- Parking brake cable to the lever
- Shoe assembly onto the backing plate
- Hold-down springs and pins to retain the rear shoe
- Automatic adjuster lever

5. Apply lithium grease to the adjuster bolt.

- Left hand parking brake shoe strut set
- Hold-down springs and pins to the front shoe
- Anchor spring to each shoe using needle nosed pliers
- Return spring to each shoe

6. Adjust the rear brakes as follows:

a. Temporarily install the drum and hub nuts.

b. Remove the hole plug from the backing plate.

c. Turn the adjuster to expand the shoe until the drum locks.

d. Back off the adjuster eight notches using a suitable adjustment tool.

e. Install the hole plug. into the backing plate to prevent dirt and moisture from entering.

f. Readjust the parking brake cable as necessary.

7. Install the wheels.

8. Refill the master cylinder and pump pedal to attain full brake pedal before Road-testing the vehicle.

TOYOTA

RAV4

21

SPECIFICATION AND MAINTENANCE CHARTS

ENGINE AND VEHICLE IDENTIFICATION

Engine						
Code ①	Liters (cc)	Cu. In.	Cyl.	Fuel Sys.	Engine Type	Eng. Mfg.
1AZ-FE	2.0 (1998)	122	4	SFI	DOHC	Toyota
2AZ-FE	2.4 (2362)	144	4	SFI	DOHC	Toyota

Model Year	
Code ②	Year
1	2001
2	2002
3	2003
4	2004
5	2005

SFI: Sequential Fuel Injection

DOHC: Double Overhead Camshaft

① Stamped on the left side of the engine block

② 10th digit of the Vehicle Identification Number (VIN)

67170-RAV4-C01

GENERAL ENGINE SPECIFICATIONS

Year	Model	Engine Displacement Liters	Engine Series ID	Net Horsepower @ rpm	Net Torque @ rpm (ft. lbs.)	Bore x Stroke (in.)	Com-pression Ratio	Oil Pressure @ rpm
2001	RAV4	2.0	1AZ-FE	127@5400	132@4600	3.40x3.40	9.5:1	NA
2002	RAV4	2.0	1AZ-FE	127@5400	132@4600	3.40x3.40	9.5:1	NA
2003	RAV4	2.0	1AZ-FE	127@5400	132@4600	3.40x3.40	9.5:1	NA
2004	RAV4	2.4	2AZ-FE	155@5600	163@4000	3.48x3.78	NA	36@3000

NA: Not Available

67170-RAV4-C02

ENGINE TUNE-UP SPECIFICATIONS

Year	Engine Displacement Liters	Engine ID	Spark Plug Gap (in.)	Ignition Timing (deg.)*	Fuel Pump (psi)	Idle Speed (rpm)		Valve Clearance	
						MT	AT	Intake	Exhaust
2001	2.0	1AZ-FE	0.043	8-12B	44-50	700-800	700-800	0.007-0.011	0.011-0.015
2002	2.0	1AZ-FE	0.043	8-12B	44-50	700-800	700-800	0.008-0.011	0.012-0.016
2003	2.0	1AZ-FE	0.043	8-12B	44-50	700-800	700-800	0.008-0.011	0.012-0.016
2004	2.4	2AZ-FE	0.041	8-12B	44-50	600-700	600-700	0.007-0.011	0.012-0.016

NOTE: The Vehicle Emission Control Information label often reflects specification changes made during production.

The label figures must be used if they differ from those in this chart.

B: Before top dead center

* With terminals TC and CG connected to DLC3

67170-RAV4-C03

79243G67

2.0L Engine
Firing order: 1–3–4–2
Distributorless ignition system

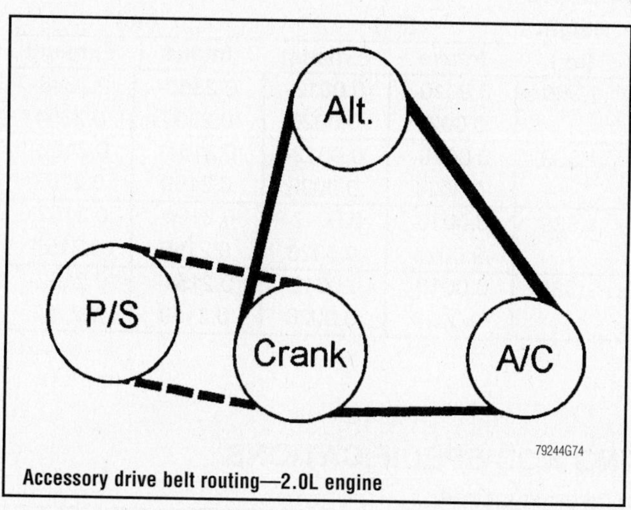

Accessory drive belt routing—2.0L engine

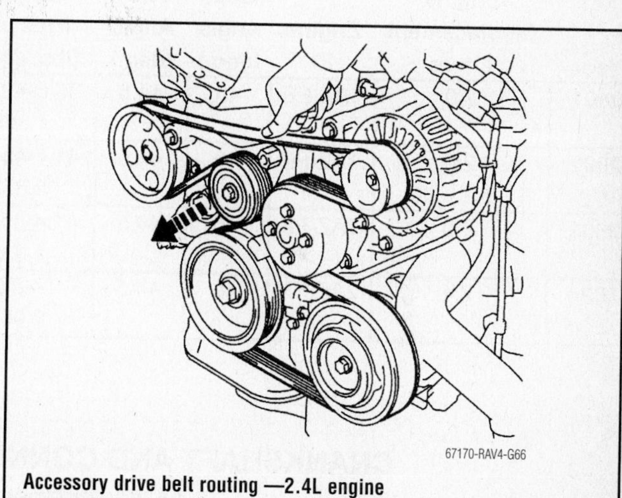

Accessory drive belt routing —2.4L engine

CAPACITIES

Year	Model	Engine Displacement Liters	Engine ID	Engine Oil with Filter (qts.)	Transmission (pts.) 5-Spd	Transmission (pts.) Auto.*	Transfer Case (pts.)	Drive Axle Front (pts.)	Drive Axle Rear (pts.)	Fuel Tank (gal.)	Cooling System (qts.)
2001	RAV4	2.0	1AZ-FE	4.4	①	②	2.0	—	2.0	14.8	③
2002	RAV4	2.0	1AZ-FE	4.6	①	②	2.0	—	2.0	14.8	③
2003	RAV4	2.0	1AZ-FE	4.6	①	②	2.0	—	2.0	14.8	③
2004	RAV4	2.4	2AZ-FE	4.0	①	②	2.0	—	2.0	14.8	③

*After draining, add the following amounts, then, fill to the cold full line.

① 2wd: 5.2
4wd: 7.2

② 2wd: 7.0
4wd: 8.2

③ MT: 6.7
AT: 6.6

67170-RAV4-C04

VALVE SPECIFICATIONS

Year	Engine Displacement Liters	Engine ID	Seat Angle (deg.)	Face Angle (deg.)	Spring Test Pressure (lbs. @ in.)	Spring Installed Height (in.)	Stem-to-Guide Clearance (in.) Intake	Stem-to-Guide Clearance (in.) Exhaust	Stem Diameter (in.) Intake	Stem Diameter (in.) Exhaust
2001	2.0	1AZ-FE	45	44.5	36.8-42.5@ 1.366	1.366	0.0010-0.0024	0.0012-0.0026	0.2350-0.2356	0.2348-0.2354
2002	2.0	1AZ-FE	45	44.5	41.4-45.9@ 1.339	1.339	0.0010-0.0024	0.0012-0.0026	0.2154-0.2159	0.2152-0.2157
2003	2.0	1AZ-FE	45	44.5	41.4-45.9@ 1.339	1.339	0.0010-0.0024	0.0012-0.0026	0.2154-0.2159	0.2152-0.2157
2004	2.4	2AZ-FE	45	44.5	77-86@ 0.969	1.339	0.0010-0.0024	0.0012-0.0026	0.2154-0.2159	0.2152-0.2157

67170-RAV4-C05

CRANKSHAFT AND CONNECTING ROD SPECIFICATIONS

All measurements are given in inches.

Year	Engine Displ. Liters	Engine ID	Crankshaft Main Brg. Journal Dia.	Crankshaft Main Brg. Oil Clearance	Crankshaft Shaft End-play	Crankshaft Thrust on No.	Connecting Rod Journal Diameter	Connecting Rod Oil Clearance	Connecting Rod Side Clearance
2001	2.0	1AZ-FE	2.1653-2.1655	0.0010-0.0017	0.0008-0.0087	3	2.0466-2.0472	0.0009-0.0022	0.0063-0.0123
2002	2.0	1AZ-FE	2.1649-2.1655	0.0010-0.0016	0.0008-0.0087	3	1.8894-1.8898	0.0009-0.0019	0.0063-0.0143
2003	2.0	1AZ-FE	2.1649-2.1655	0.0010-0.0016	0.0008-0.0087	3	1.8894-1.8898	0.0009-0.0019	0.0063-0.0143
2004	2.4	2AZ-FE	2.0654-2.1648	0.0009-0.0019	0.0016-0.0094	2	1.8894-1.8898	0.0009-0.0019	0.0063-0.0143

67170-RAV4-C06

PISTON AND RING SPECIFICATIONS

All measurements are given in inches.

Year	Engine Displ. Liters	Engine ID	Piston Clearance	Ring Gap Top Comp.	Ring Gap Bottom Comp.	Ring Gap Oil Control	Ring Side Clearance Top Comp.	Ring Side Clearance Bottom Comp.	Ring Side Clearance Oil Control
2001	2.0	1AZ-FE	0.0056-0.0064	0.0106-0.0185	0.0177-0.0256	0.0039-0.0177	0.0012-0.0028	0.0012-0.0028	SNUG
2002	2.0	1AZ-FE	0.0025-0.0034	0.0118-0.0157	0.0185-0.0244	0.0039-0.0138	0.0008-0.0028	0.0008-0.0024	0.0028-0.0059
2003	2.0	1AZ-FE	0.0025-0.0034	0.0118-0.0157	0.0185-0.0244	0.0039-0.0138	0.0008-0.0028	0.0008-0.0024	0.0028-0.0059
2004	2.4	2AZ-FE	0.0020-0.0029	0.0087-0.0126	0.0197-0.0236	0.0039-0.0138	0.0012-0.0028	0.0012-0.0028	SNUG

67170-RAV4-C07

TORQUE SPECIFICATIONS

All readings in ft. lbs.

Year	Engine Displacement Liters	Engine ID	Cylinder Head Bolts	Main Bearing Bolts	Rod Bearing Bolts	Crankshaft Damper Bolts	Flywheel Bolts	Manifold Intake	Manifold Exhaust	Spark Plugs	Oil Pan Drain Plug
2001	2.0	1AZ-FE	①	43	②	80	③	14	36	13	18
2002	2.0	1AZ-FE	④	⑤	②	125	⑥	22	25	15	18
2003	2.0	1AZ-FE	④	⑤	②	125	⑥	22	25	15	18
2004	2.4	2AZ-FE	④	⑤	②	132	⑥	22	27	14	18

① Step 1: 35 ft. lbs.
Step 2: Plus 90 degrees

② Step 1: 18 ft. lbs.
Step 2: Plus 90 degrees

③ Manual transmission: 65 ft. lbs.
Automatic transmission: 61 ft. lbs.

④ Step 1: 58
Step 2: plus 90 degrees

⑤ Step 1: 15 ft. lbs.
Step 2: 30 ft. lbs.
Step 3: +90 degrees

⑥ MT: 96 ft. lbs.
AT: 72 ft. lbs.

67170-RAV4-C08

WHEEL ALIGNMENT

			Caster		Camber			Steering
Year	Model		Range (+/-Deg.)	Preferred Setting (Deg.)	Range (+/-Deg.)	Preferred Setting (Deg.)	Toe-in (in.)	Axis Inclination (Deg.)
2001	RAV4	2WD	0.75	+1.42	0.75	-0.33	0+/-0.08	11+/-0.75
		4WD	0.75	+1.33	0.75	-0.25	0+/-0.08	11+/-0.75
2002	RAV4	2WD	0.75	+2.00	0.75	-0.42	0+/-0.08	11+/-0.75
		4WD	0.75	+1.92	0.75	-0.33	0.04+/-0.08	10.75+/-0.75
2003	RAV4	2WD	0.75	+2.00	0.75	-0.42	0+/-0.08	11+/-0.75
		4WD	0.75	+1.92	0.75	-0.33	0.04+/-0.08	10.75+/-0.75
2004	RAV4	2WD	0.75	+2.00	0.75	-0.42	0+/-0.08	11+/-0.75
		4WD	0.75	+1.92	0.75	-0.33	0.04+/-0.08	10.75+/-0.75

67170-RAV4-C09

TIRE, WHEEL AND BALL JOINT SPECIFICATIONS

		OEM Tires		Tire Pressures (psi)				Lugnut
Year	Model	Standard	Optional	Front	Rear	Wheel Size	Ball Joint Inspection	Torque (ft. lbs.)
2001	RAV4	P215/70R16	P235/60HR16	29	29	①	4-30 ②	76
2002	RAV4	P215/70R16	P235/60HR16	29	29	③	9-43 ②	76
2003	RAV4	P215/70R16	P235/60HR16	29	29	①	9-43 ②	76
2004	RAV4	P215/70R16	None	④	④	NA	9-43 ②	76

OEM: Original Equipment Manufacturer

PSI: Pounds Per Square Inch

STD: Standard

OPT: Optional

① Steel wheel: 6.5J; aluminum wheel: 7JJ

② Torque required in inch lbs. to rotate ball joint when removed from the knuckle

③ Replace if any measurable movement is found.

④ See placard on vehicle

67170-RAV4-C10

BRAKE SPECIFICATIONS

All measurements in inches unless noted

			Brake Disc			Brake Drum Diameter				Brake Caliper	
Year	Model		Original Thickness	Minimum Thickness	Maximum Runout	Original Inside Diameter	Max. Wear Limit	Maximum Machine Diameter	Minimum Lining Thickness	Bracket Bolts (ft. lbs.)	Mounting Bolts (ft. lbs.)
2001	RAV4		0.709	0.630	0.0020	9.000	—	9.079	0.039	78	20
2002	RAV4		0.984	0.906	0.0020	9.000	—	9.079	0.039	78	20
2003	RAV4		0.984	0.906	0.0020	9.000	—	9.079	0.039	78	20
2004	RAV4	F	0.984	0.906	0.0020	—	—	—	0.039	78	20
		R	0.354	0.317	0.004	9.000	—	9.079	0.039	—	33

F: Front

R: Rear

67170-RAV4-C11

SCHEDULED MAINTENANCE INTERVALS

TOYOTA—RAV4

TO BE SERVICED	TYPE OF SERVICE	VEHICLE MILEAGE INTERVAL (x1000)																		
		5	10	15	20	25	30	35	40	45	50	55	60	65	70	75	80	85	90	95
Automatic transmission and differential fluid	S/I			✓			✓			✓			✓			✓			✓	
Ball joints and boots	S/I			✓			✓			✓			✓			✓			✓	
Brake system	S/I			✓			✓			✓			✓			✓			✓	
Charcoal canister	S/I												✓							
Drive belts	S/I						✓						✓						✓	
Driveshaft bushing	L						✓						✓						✓	
Engine coolant	R						✓						✓						✓	
Engine oil & filter	R	✓	✓	✓	✓	✓	✓	✓	✓	✓	✓	✓	✓	✓	✓	✓	✓	✓	✓	✓
Exhaust pipes & mounts	S/I			✓			✓			✓			✓			✓			✓	
Fuel tank cap gasket	S/I						✓						✓						✓	
Halfshaft boots & flange bolts	S/I			✓			✓			✓			✓			✓			✓	
Limited slip differential fluid	R						✓						✓						✓	
Manual transmission and differential fluid	S/I						✓						✓						✓	
Platinum spark plugs	R												✓							
Propeller shaft bolts	S/I			✓			✓			✓			✓			✓			✓	
Steering linkage	S/I			✓			✓			✓			✓			✓			✓	
Transfer case and differential fluid	S/I			✓			✓			✓			✓			✓			✓	
Valves	S/I												✓							

R: Replace S/I: Service or Inspect L: Lubricate

FREQUENT OPERATION MAINTENANCE (SEVERE SERVICE)

If a vehicle is operated under any of the following conditions it is considered severe service:

- Towing a trailer or using a camper or car-top carrier.

- Repeated short trips of less than 5 miles in temperatures below freezing.

- Excessive idling or low-speed driving for long distances as in heavy commercial use, such as delivery, taxi or police cars.

- Operating on rough, muddy or salt-covered roads.

- Operating on unpaved or dusty roads.

Oil filter: service or inspect every 5000 miles or 4 months, whichever occurs first.

Brake linings and discs or drums: service or inspect every 5000 miles or 4 months, whichever occurs first.

Steering linkage: service or inspect every 5000 miles or 4 months, whichever occurs first.

Ball joints and boots: service or inspect every 5000 miles or 4 months, whichever occurs first.

Brake discs & pads (front): service or inspect every 6000 miles.

Halfshaft boots: service or inspect every 5000 miles or 4 months. Retighten the flange bolts, whichever occurs first.

Body chassis bolts and nuts: service or inspect every 5000 miles or 4 months, whichever occurs first.

Transmission and differential fluid: replace every 15,000 miles or 12 months, whichever occurs first.

Transfer case and differential fluid: replace every 15,000 miles or 12 months, whichever occurs first.

Timing belt: replace every 60,000 miles or 48 months, whichever occurs first.

67170-RAV4-C12

PRECAUTIONS

Before servicing any vehicle, please be sure to read all of the following precautions, which deal with personal safety, prevention of component damage, and important points to take into consideration when servicing a motor vehicle:

- Never open, service or drain the radiator or cooling system when the engine is hot; serious burns can occur from the steam and hot coolant.
- Observe all applicable safety precautions when working around fuel. Whenever servicing the fuel system, always work in a well-ventilated area. Do not allow fuel spray or vapors to come in contact with a spark, open flame, or excessive heat (a hot drop light, for example). Keep a dry chemical fire extinguisher near the work area. Always keep fuel in a container specifically designed for fuel storage; also, always properly seal fuel containers to avoid the possibility of fire or explosion. Refer to the additional fuel system precautions later in this section.
- Fuel injection systems often remain pressurized, even after the engine has been turned **OFF**. The fuel system pressure must be relieved before disconnecting any fuel lines. Failure to do so may result in fire and/or personal injury.
- Brake fluid often contains polyglycol ethers and polyglycols. Avoid contact with the eyes and wash your hands thoroughly after handling brake fluid. If you do get brake fluid in your eyes, flush your eyes with clean, running water for 15 minutes. If eye irritation persists, or if you have taken brake fluid internally, IMMEDIATELY seek medical assistance.
- The EPA warns that prolonged contact with used engine oil may cause a number of skin disorders, including cancer. You should make every effort to minimize your exposure to used engine oil. Protective gloves should be worn when changing oil. Wash your hands and any other exposed skin areas as soon as possible after exposure to used engine oil. Soap and water, or waterless hand cleaner should be used.
- All new vehicles are now equipped with an air bag system. The system must be disabled before performing service on or around system components, steering column, instrument panel components, wiring and sensors. Failure to follow safety and disabling procedures could result in accidental air bag deployment, possible personal injury and unnecessary system repairs.
- Always wear safety goggles when working with, or around, the air bag system. When carrying a non-deployed air bag, be sure the bag and trim cover are pointed away from your body. When placing a non-deployed air bag on a work surface, always face the bag and trim cover upward, away from the surface. This will reduce the motion of the module if it is accidentally deployed. Refer to the additional air bag system precautions later in this section.
- NEVER disconnect the negative battery cable with the ignition **ON** or the engine running. Removing power from the computer control module with the ignition **ON** may destroy the module.
- Clean, high quality brake fluid from a sealed container is essential to the safe and proper operation of the brake system. You should always buy the correct type of brake fluid for your vehicle. If the brake fluid becomes contaminated, completely flush the system with new fluid. Never reuse any brake fluid. Any brake fluid that is removed from the system should be discarded. Also, do not allow any brake fluid to come in contact with a painted surface; it will damage the paint.
- Never operate the engine without the proper amount and type of engine oil; doing so WILL result in severe engine damage.
- Timing belt maintenance is extremely important. Many models utilize an interference-type, non-freewheeling engine. If the timing belt breaks, the valves in the cylinder head may strike the pistons, causing potentially serious (also time-consuming and expensive) engine damage. Refer to the maintenance interval charts in the front of this manual for the recommended replacement interval for the timing belt, and to the timing belt section for belt replacement and inspection.
- Disconnecting the negative battery cable on some vehicles may interfere with the functions of the on-board computer system(s) and may require the computer to undergo a relearning process once the negative battery cable is reconnected.
- When servicing drum brakes, only disassemble and assemble one side at a time, leaving the remaining side intact for reference.
- Only an MVAC-trained, EPA-certified automotive technician should service the air conditioning system or its components.

ENGINE REPAIR

➡Disconnecting the negative battery cable on some vehicles may interfere with the functions of the on board computer system. The computer may undergo a relearning process once the negative battery cable is reconnected.

Distributor

All models are equipped with a distibutorless ignition system.

Alternator

REMOVAL

2.0L

1. Before servicing the vehicle, refer to the precautions in the beginning of this section.
2. Remove or disconnect the following:
   - Electrical wiring from the alternator
   - Loosen the adjusting lockbolt and the pivot bolt.
   - Loosen the adjusting bolt to relieve tension on the drive belt, if equipped with air conditioning
   - Drive belt

➡It may be necessary to remove other belts for access.

   - Pivot bolt and the adjusting lockbolt
   - Alternator

To install:

3. Install or connect the following:
   - Alternator
   - Adjusting lockbolt and the pivot bolt
   - Drive belt
4. Adjust the drive belt tension to:
   - New belt with A/C: 140–190 lbs.
   - Used belt with A/C: 100–120 lbs.
   - New belt without A/C: 100–150 lbs.
   - Used belt without A/C: 75–115 lbs.
5. Install or connect the following:
   - Tighten the pivot bolt to 38 ft. lbs. (52 Nm).
   - Tighten the adjusting lockbolt to 13 ft. lbs. (18 Nm).
   - Electrical wiring to the alternator

2.4L

1. Before servicing the vehicle, refer to the precautions in the beginning of this section.
2. Remove or disconnect the following:
   - Electrical wiring from the alternator
   - Drive belt
   - 1 adjusting and 2 mounting bolts
   - Alternator

To install:

3. Installation is the reverse of removal. Observe the following torques:
   - M8 bolts: 15 ft. lbs. (21Nm)
   - M10 bolts: 38 ft. lbs. (52Nm)

Ignition Timing

ADJUSTMENT

All engines are equipped with a Distributorless Ignition System (DIS). No timing adjustment is possible.

Engine Assembly

REMOVAL & INSTALLATION

1. Before servicing the vehicle, refer to the precautions in the beginning of this section.
2. Relieve the fuel system pressure.
3. Remove or disconnect the following:
   - Negative battery cable
   - Battery
   - Hood
   - Engine undercover
4. Drain the engine coolant and oil.
5. Drain the transaxle assembly.
6. Remove or disconnect the following:
   - Air cleaner and case
   - Accelerator cable from the throttle body, bracket and clamps
7. Disconnect and remove the engine wire from the No. 2 relay block, as follows:
   - No. 2 relay block from the body by removing the 2 bolts
   - Upper cover to the relay block
   - Electrical connectors
   - Engine wire, by removing the 2 nuts
   - Charcoal canister
   - Alternator
   - Upper and lower radiator hoses
   - Water inlet from the engine by removing the 2 nuts
   - Heater hoses
   - Fuel hose, by placing a rag under the fuel inlet hose
   - Starter by disconnecting the electrical connectors and 2 bolts, if equipped with manual transmission
   - Ground cable from the transaxle by removing the bolt
   - Clutch release cylinder from the transaxle, if equipped with manual transmission
   - Transaxle control cables (2 cable for manual transmission or 1 for automatic transmission) from the transaxle.
   - Transaxle cable from the front suspension crossmember and engine mounting centermember by removing the 2 bolts, if equipped with an automatic transmission
   - Transaxle oil cooler hoses, if equipped with an automatic transmission or 4WD with manual transmission
8. Detach the following:
   - Vapor pressure sensor connector
   - Igniter connector
   - Ignition coil connector
   - Noise filter connector
   - Ignition coil wire
   - Manifold Absolute Pressure (MAP) sensor connector
   - MAP sensor vacuum hose from the gas filter on the intake manifold
   - Brake booster hose from the intake manifold
   - Differential lock control solenoid connector, if equipped with a 4WD manual transmission
   - Ground strap from cowl
9. Detach the engine wire from the passenger compartment, as follows:
   - Right-hand scuff plate
   - Right-hand side trim
   - Right-hand carpet center cover
   - 2 ECM connectors
   - 2 connectors from the bracket connectors
   - No. 4 junction block connector
   - Wire clamp from the bracket
   - Engine wire from the passenger compartment
10. Remove the front exhaust pipe, as follows:
    - 3 nuts and the front exhaust pipe from the exhaust manifold, using a 14mm deep socket wrench; discard the gasket
    - 2 bolts and 2 nuts holding the front exhaust pipe to the catalytic converter
    - Front exhaust pipe and 2 gaskets
11. Remove the compressor from the engine and suspend the compressor securely.

➡It is not necessary to remove the air conditioning compressor lines in order to remove the engine.

12. Remove or disconnect the following:
    - Driveshaft, if equipped with 4WD
    - Halfshaft
    - Sway bar
13. Remove the front suspension crossmember assembly, as follows:
    - 2 centermember set nuts holding the centermember to the middle of the crossmember.
    - 2 rack and pinion assembly set bolts/nuts from the crossmember. Securely suspend the steering gear assembly.
    - Catalytic converter with pipe from the ring
    - Support the suspension crossmember with a jack
    - 6 bolts from the suspension crossmember
    - Suspension crossmember with the lower suspension arms
14. Remove the engine mounting centermember, as follows:
    - 2 bolts holding the centermember to the front engine mounting insulator
    - 2 bolts holding the centermember to the body
    - Centermember
15. Disconnect the power steering pump from the engine, as follows:
    - 2 vacuum hoses from the steering pump
    - Adjusting bolt for the power steering unit. Loosen the pivot bolt to the power steering pump and remove the drive belt. Use Torque Wrench Adapter tool 09249-63010 and a deep socket to loosen the pivot bolt.
    - Power steering pump from the engine by removing the 3 bracket bolts

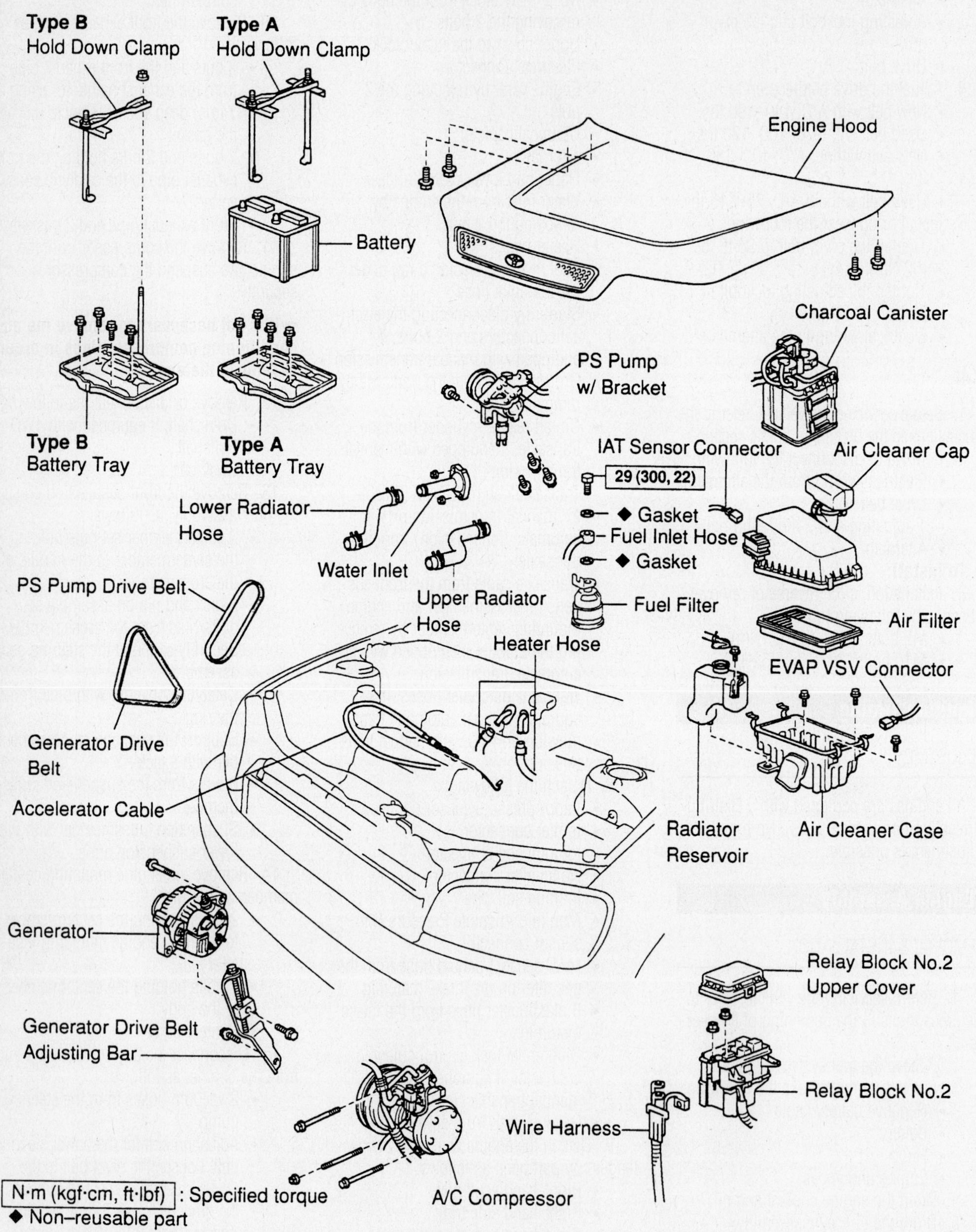
Type B
Hold Down Clamp
Type A
Hold Down Clamp
Engine Hood
Battery
Charcoal Canister
PS Pump
w/ Bracket
Type B
Battery Tray
Type A
Battery Tray
IAT Sensor Connector
Air Cleaner Cap
29 (300, 22)
◆ Gasket
Fuel Inlet Hose
◆ Gasket
Lower Radiator
Hose
Water Inlet
PS Pump Drive Belt
Upper Radiator
Hose
Fuel Filter
Air Filter
Heater Hose
EVAP VSV Connector
Generator Drive
Belt
Accelerator Cable
Radiator
Reservoir
Air Cleaner Case
Generator
Relay Block No.2
Upper Cover
Generator Drive Belt
Adjusting Bar
Relay Block No.2
Wire Harness
N·m (kgf·cm, ft·lbf) : Specified torque
◆ Non–reusable part
A/C Compressor
7924ZG04

Exploded view of the engine accessory removal components—2.0L

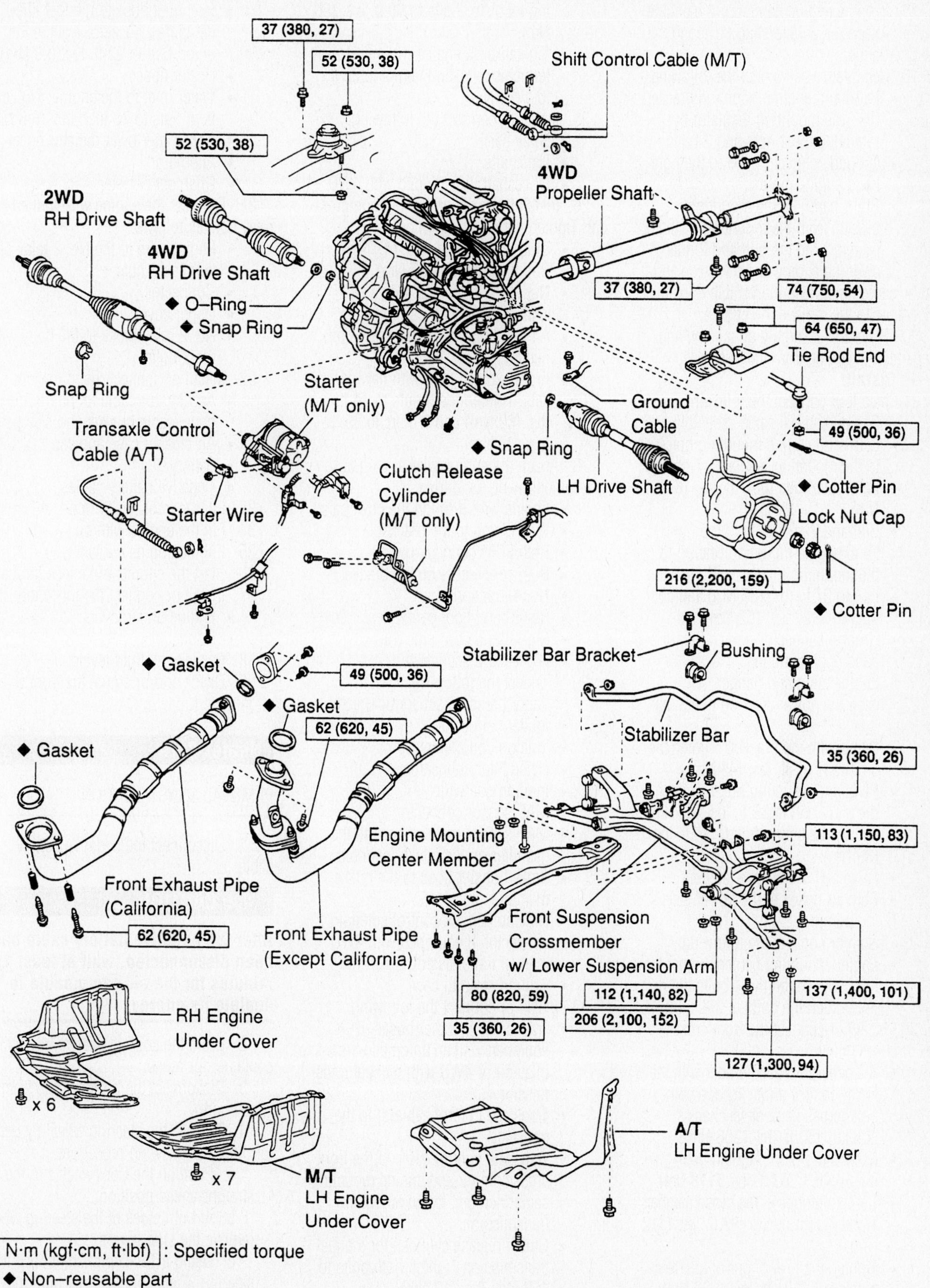
37 (380, 27)
52 (530, 38)
Shift Control Cable (M/T)
52 (530, 38)
4WD
Propeller Shaft
2WD
RH Drive Shaft
4WD
RH Drive Shaft
◆ O–Ring
◆ Snap Ring
37 (380, 27)
74 (750, 54)
64 (650, 47)
Tie Rod End
Snap Ring
Starter
(M/T only)
Ground
Cable
Transaxle Control
Cable (A/T)
◆ Snap Ring
49 (500, 36)
Clutch Release
Cylinder
(M/T only)
LH Drive Shaft
◆ Cotter Pin
Starter Wire
Lock Nut Cap
216 (2,200, 159)
◆ Cotter Pin
Stabilizer Bar Bracket
Bushing
◆ Gasket
49 (500, 36)
◆ Gasket
62 (620, 45)
Stabilizer Bar
◆ Gasket
35 (360, 26)
Engine Mounting
Center Member
113 (1,150, 83)
Front Exhaust Pipe
(California)
62 (620, 45)
Front Exhaust Pipe
(Except California)
Front Suspension
Crossmember
w/ Lower Suspension Arm
80 (820, 59)
112 (1,140, 82)
137 (1,400, 101)
35 (360, 26)
206 (2,100, 152)
127 (1,300, 94)
RH Engine
Under Cover
x 6
A/T
LH Engine Under Cover
x 7
M/T
LH Engine
Under Cover
N·m (kgf·cm, ft·lbf) : Specified torque
◆ Non–reusable part
7924ZG05

Exploded view of the engine removal—2.0L

16. Install a engine hanger to the engine.
17. Attach the engine sling device to the engine hangers.
18. Remove or disconnect the following:
   - Left-hand engine mounting bracket from the mounting insulator by removing the 2 nuts and 2 bolts
   - Ground connector next to the right-hand engine mount
   - Right-hand engine mounting bracket from the mounting insulator by removing the bolt and 2 nuts
19. Lower the engine and transaxle and at the same time, raise the vehicle to gain clearance to the remove the engine.
20. Place the assembly on a stand and separate the engine from the transaxle.

To install:

21. Install or connect the following:
   - Engine and transaxle assembly
   - Left-hand engine mounting bracket to the mounting insulator. Tighten both nuts/bolts to 47 ft. lbs. (64 Nm).
   - Bolt and 2 nuts to hold the right-hand engine mounting bracket to the mounting insulator. Tighten the bolt to 27 ft. lbs. (37 Nm) and both nuts to 38 ft. lbs. (52 Nm).
   - Ground connector next to the right-hand engine mount
   - Engine sling and hanger
22. Install the power steering pump, as follows:
   - Pump with the bracket. Tighten the 3 bolts to 32 ft. lbs. (43 Nm).
   - Pivot and adjusting bolts. Tighten the pivot bolt to 32 ft. lbs. (43 Nm) and the adjusting bolt to 29 ft. lbs. (39 Nm).
   - Drive belt. Adjust the tension.
   - Both air hoses to the power steering pump
23. Install or connect the following:
   - Engine mounting centermember to the body; install the 4 bolts but do not tighten the bolts at this time
24. Install or connect the front crossmember, as follows:
   - Suspension crossmember with the lower control arms. Torque both bolts crossmember-to-chassis bolts to 152 ft. lbs. (206 Nm).
   - Rack and pinion. Tighten both nuts/bolts to 83 ft. lbs. (113 Nm).
   - Centermember to the crossmember. Tighten both nuts to 82 ft. lbs. (112 Nm).
   - Tighten the lower control arm rear brackets to 101 ft. lbs. (137 Nm).
   - Tighten both engine mounting centermember-to-front engine mounting insulator bolts to 59 ft. lbs. (80 Nm).
   - Tighten both engine mounting centermember-to-body bolts to 26 ft. lbs. (35 Nm).
25. Install or connect the following:
   - Sway bar
   - Halfshafts
   - Driveshaft, if equipped with 4WD
26. Install the air conditioning compressor. Tighten nut/bolts, as follows:
   - Stud bolt to 34 ft. lbs. (47 Nm)
   - Bolt to 27 ft. lbs. (37 Nm)
   - Nut to 20 ft. lbs. (27 Nm)
27. Install or connect the following:
   - Air conditioning compressor connector
   - Front exhaust pipe with new gaskets. Tighten the 3 nuts to 46 ft. lbs. (62 Nm) and 2 bolts to 35 ft. lbs. (48 Nm).
28. Attach the engine wire to the passenger compartment, as follows:
   - Engine wire through the cowl panel
   - Wire clamp to the bracket
   - Both ECM connectors
   - Both connectors on the bracket
   - No. 4 junction block
   - Right-hand floor carpet center cover
   - Right-hand cowl side trim
   - Right-hand scuff plate
29. Connect the following:
   - Vapor pressure sensor connector
   - Igniter connector
   - Ignition coil connector
   - Noise filter connector
   - Ignition coil wire
   - MAP sensor connector
   - MAP sensor vacuum hose to the gas filter on the intake manifold
   - Brake booster hose to the intake manifold
   - Differential lock control solenoid connector, if equipped with 4WD manual transmission
   - Ground strap to cowl
30. Install or connect the following:
   - Transaxle oil cooler hoses, if equipped with an automatic transmission or 4WD with manual transmission
   - Transaxle control cable(s) to the transaxle
   - Transaxle control cable to the front crossmember and engine mounting centermember, for an automatic transmission
   - Clutch release cylinder, for manual transmission Tighten both bolts to 108 inch lbs. (12 Nm).
   - Ground cable to the transaxle
   - Starter, for a manual transmission
   - Fuel inlet hose to the fuel filter, using new gaskets. Tighten the union bolt to 22 ft. lbs. (29 Nm).
   - Heater hoses
   - Water inlet to the engine. Tighten both nuts to 78 inch lbs. (8.8 Nm).
   - Upper and lower radiator hoses
   - Alternator
   - Charcoal canister
31. Connect the engine wire to the No. 2 relay box, as follows:
   - Engine wire to the No. 2 relay block with the 2 nuts
   - Connector
   - Upper cover
   - No. 2 relay block to the body with the 2 bolts
32. Install or connect the following:
   - Accelerator cable to the throttle body, cable bracket and clamps
   - Air cleaner case and cap
   - Battery
   - Negative battery cables
33. Fill the transaxle with oil.
34. Fill the engine with oil.
35. Fill the engine coolant.
36. Start the engine and check for leaks.
37. Install or connect the following:
   - Engine undercovers
   - Hood
38. Recheck all fluid levels.
39. Check and/or adjust the front wheel alignment.

Heater Core

REMOVAL & INSTALLATION

1. Disconnect the negative battery cable.

CAUTION

After the negative battery cable has been disconnected, wait at least 1½ minutes for the air bag module to deplete its energy.

2. Drain the cooling system into a clean container for reuse.
3. Disconnect the heater hoses from the heater core.
4. Remove the steering wheel by performing the following procedure:
   a. Position the front wheels in the straight-ahead position.
   b. At both sides of the steering wheel, remove the side covers.
   c. Using a Torx® wrench, loosen the steering wheel Torx® screws until the screw's circumference ring catches on the screw case.

d. Carefully, lift the air bag module, disconnect the electrical connector and remove the air bag.

CAUTION

Place the air bag module in a safe location with the front facing upward.

e. Remove the steering wheel nut.

f. Using a steering wheel puller, press the steering wheel from the steering column.

5. Remove the instrument panel and reinforcement by performing the following procedure:

a. Disconnect the seat belt pre-tensioner connector.

b. Remove both front door scuff plates.

c. At both sides, remove the 2 cowl side trim board clips and the trim boards.

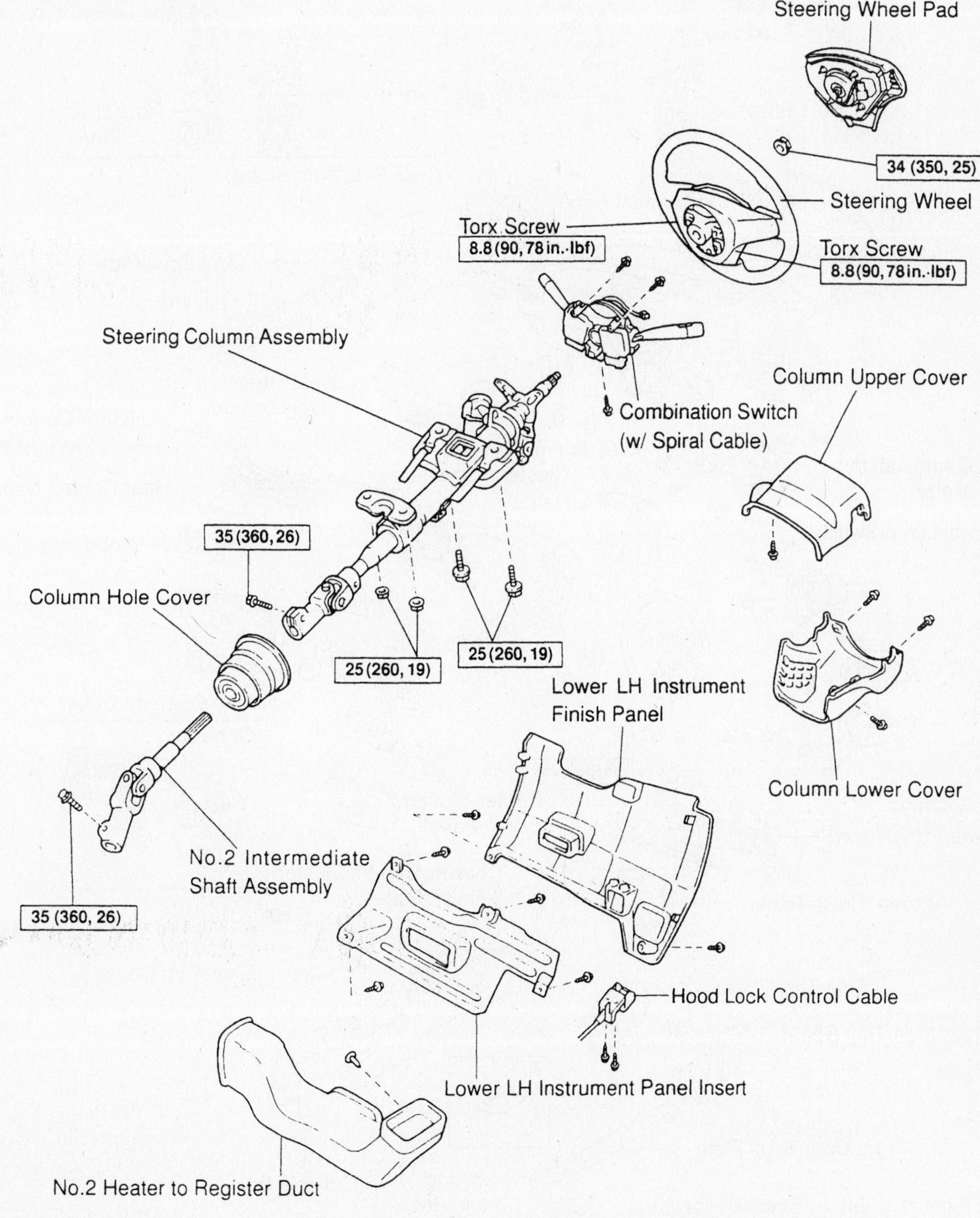

Exploded view of the steering wheel, air bag module, steering column and related components

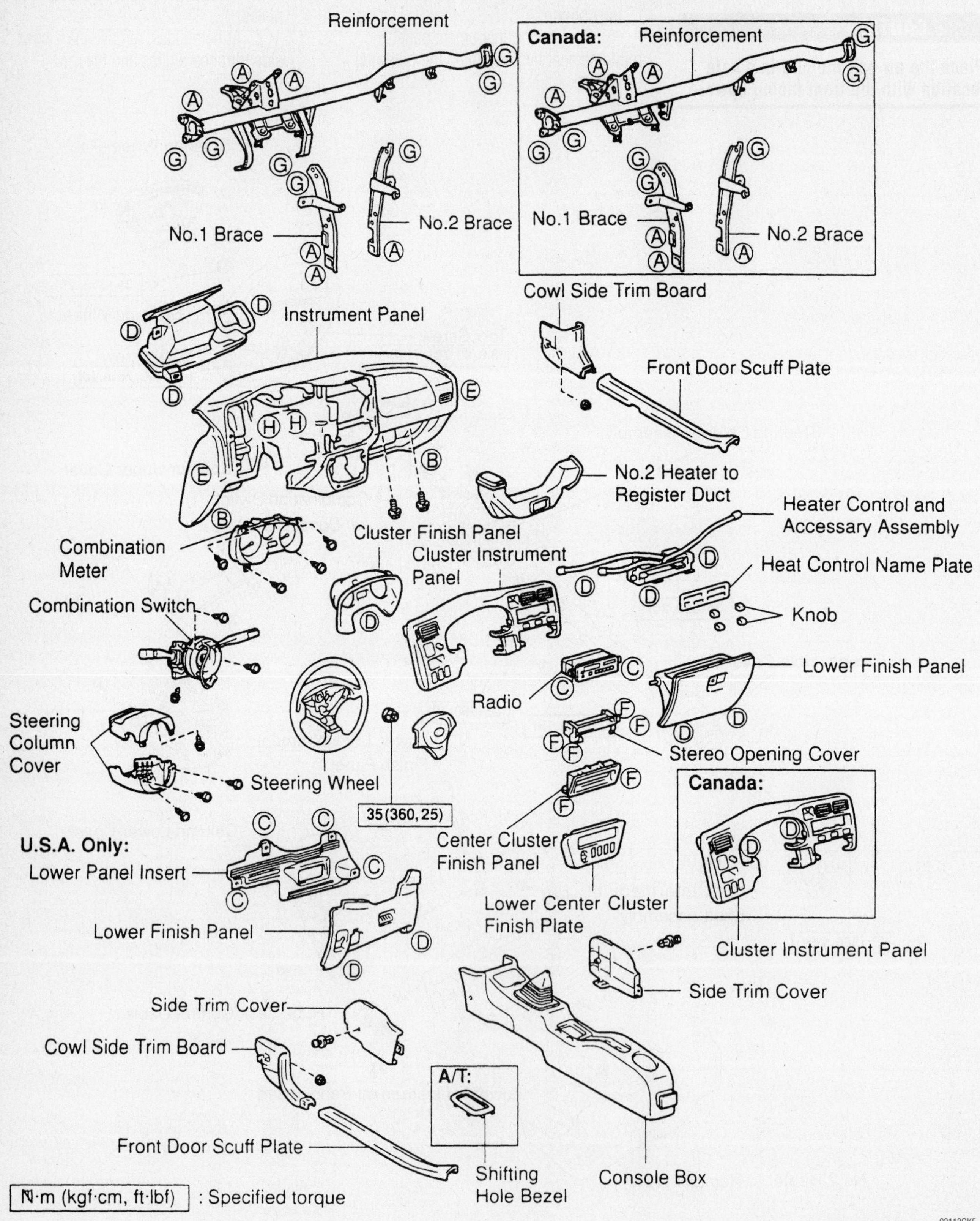

Reinforcement
Canada:
Reinforcement
No.1 Brace
No.2 Brace
No.1 Brace
No.2 Brace
Cowl Side Trim Board
Instrument Panel
Front Door Scuff Plate
No.2 Heater to Register Duct
Heater Control and Accessary Assembly
Combination Meter
Cluster Finish Panel
Cluster Instrument Panel
Heat Control Name Plate
Knob
Combination Switch
Lower Finish Panel
Radio
Steering Column Cover
Stereo Opening Cover
Steering Wheel
Canada:
35 (360, 25)
U.S.A. Only:
Lower Panel Insert
Center Cluster Finish Panel
Lower Center Cluster Finish Plate
Lower Finish Panel
Cluster Instrument Panel
Side Trim Cover
Side Trim Cover
Cowl Side Trim Board
A/T:
Front Door Scuff Plate
Shifting Hole Bezel
Console Box
N·m (kgf·cm, ft·lbf) : Specified torque
93113GK5

Exploded view of the instrument panel and related components

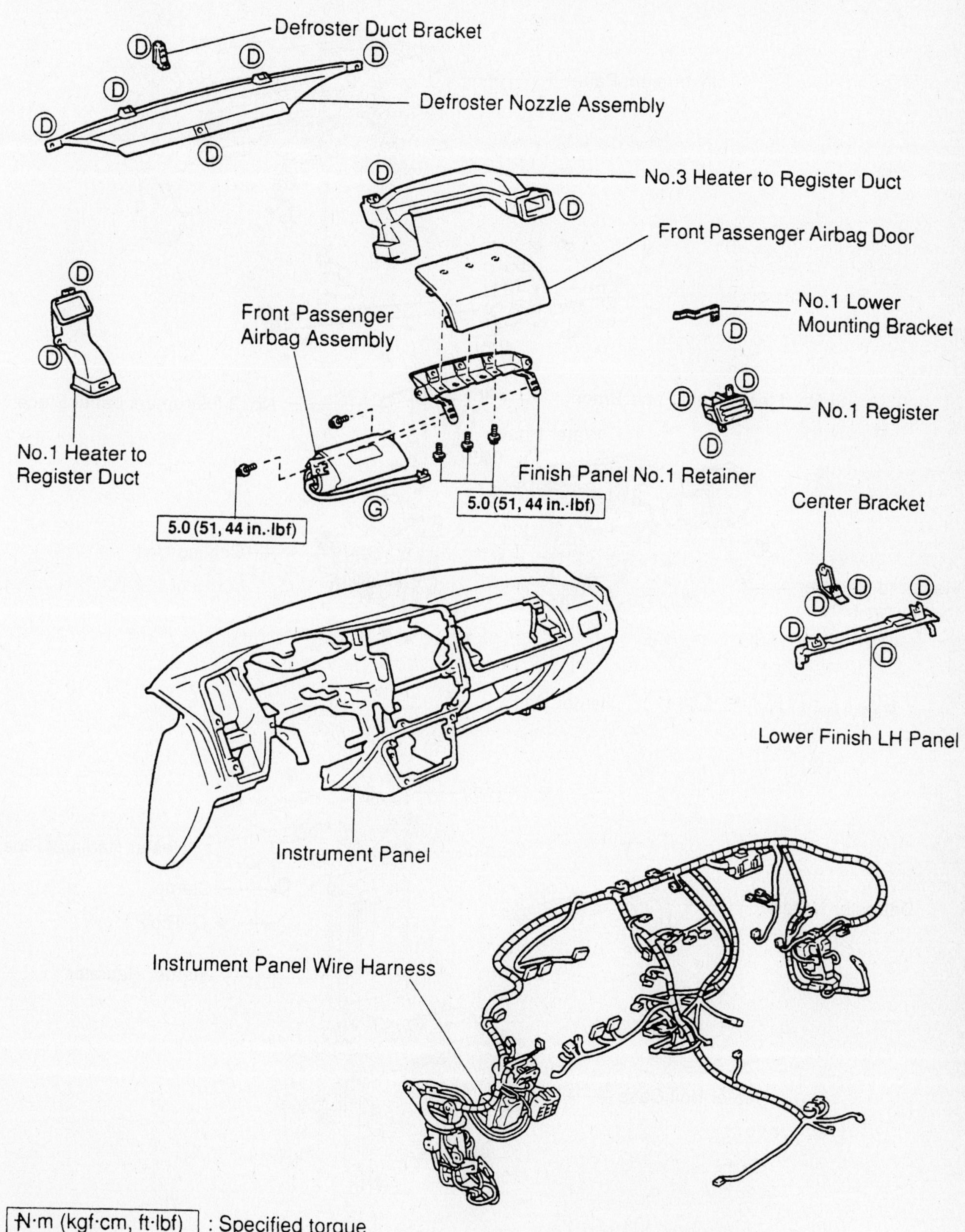

Exploded view of the instrument panel air bag module, ventilation components and wiring harness

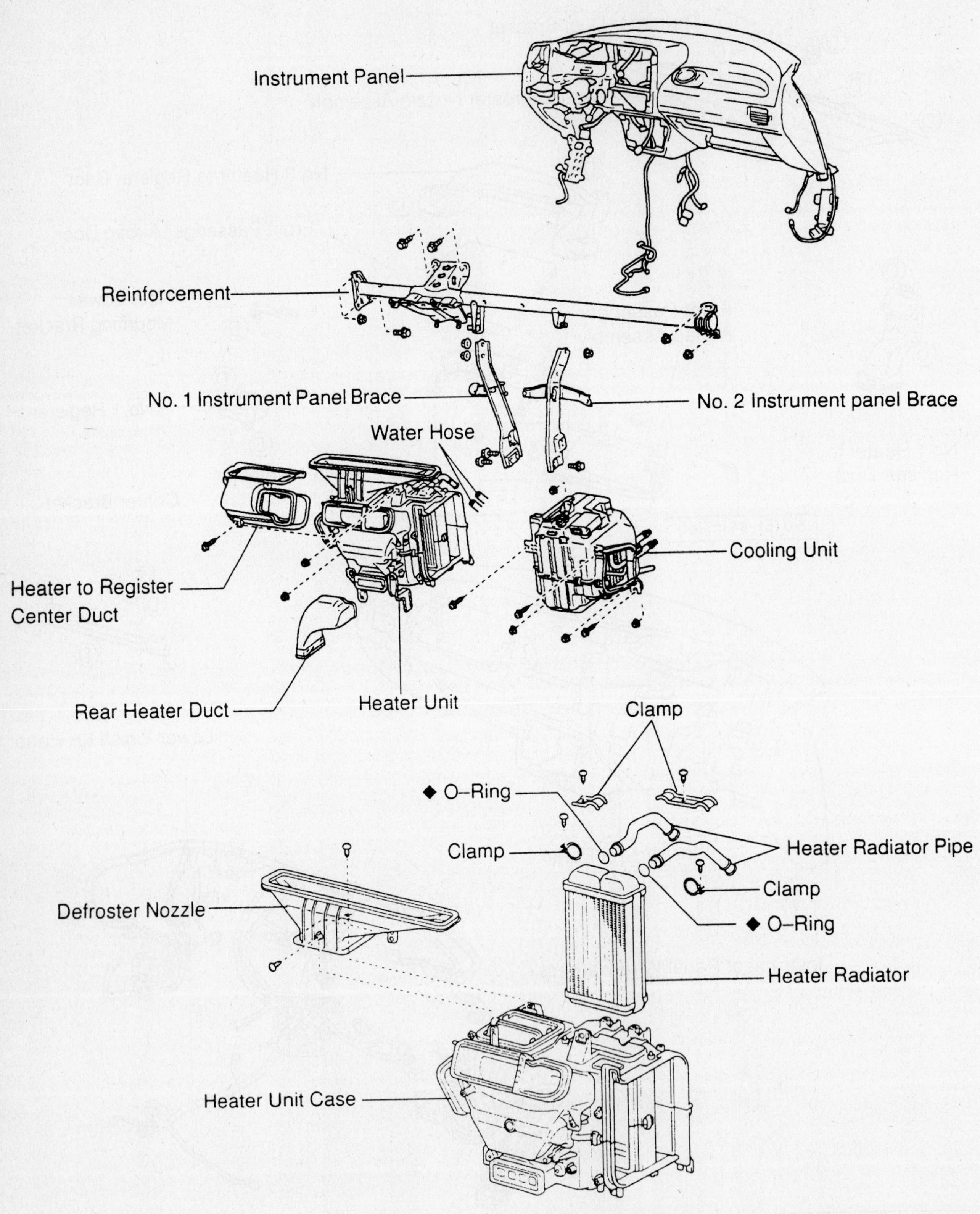

Exploded view of the heater core, heater housing, evaporator housing and related components

d. Remove the combination switch-to-steering column screws, disconnect the electrical connectors and remove the combination switch.

e. Remove the 4 steering column cover screws and the cover.

f. Remove the cluster finish panel screw and the panel.

g. Remove the 4 combination meter screws, disconnect the electrical connectors and the meter.

h. Remove the hood lock release lever.

i. Remove the 2 lower finish panel screws and the panel.

j. For USA models, remove the lower panel insert.

k. Remove the No. 2 heater-to-register duct.

l. Remove the steering column-to-instrument panel nuts/bolts and the lower steering column bolt; then carefully, remove the steering column.

m. Remove the 2 center cluster finish panel screws and the panel.

n. Pull off the heater control knobs.

➡For Canada models, remove the 2 screws.

o. Pry off the heater control name plate and the cluster instrument panel.

p. Remove the 3 heater control assembly screws and the assembly.

q. Disconnect the connectors and remove the cluster instrument panel.

r. Remove the heater control and accessory assembly.

s. Remove the radio.

t. Remove the side trim cover and the console box.

u. Remove the lower center cluster finish panel and disconnect the connectors.

v. Remove the stereo opening cover.

w. Remove the glove compartment door.

x. Remove the instrument panel-to-chassis fasteners.

y. Remove the instrument panel and disconnect the electrical connectors.

z. Remove the No. 1 and No. 2 brace nuts/bolts and the braces.

aa. Remove the instrument panel reinforcement-to-chassis nuts/bolts and the reinforcement.

6. Remove the evaporator housing by performing the following procedure:

a. Discharge and recover the air conditioning system refrigerant.

b. Disconnect the refrigerant lines from the evaporator core. Discard the O-rings and plug the openings to prevent contamination.

c. Disconnect the electrical connectors.

d. Remove the 3 evaporator housing-to-chassis nuts/bolts and the housing.

7. Remove the rear heater duct from the heater housing.

8. Remove the heater housing-to-chassis nuts and the housing.

9. Remove the 2 defroster nozzle-to-heater housing screws and the nozzle.

10. Remove the 2 heater core-to-heater housing screws, clamps and the heater core.

To install:

11. Install the heater core and the 2 heater core-to-heater housing screws and clamps.

12. Install the defroster nozzle and the 2 nozzle-to-heater housing screws.

13. Install the heater housing and the housing-to-chassis nuts.

14. Install the rear heater duct to the heater housing.

15. Install the evaporator housing by performing the following procedure:

a. Install the evaporator housing and the 3 housing-to-chassis nuts/bolts.

b. Connect the electrical connectors.

c. Using new O-rings, connect the refrigerant lines to the evaporator core.

16. Install the instrument panel and reinforcement by performing the following procedure:

a. Install the instrument panel reinforcement and the reinforcement-to-chassis nuts/bolts.

b. Install the No. 1 and No. 2 brace and the braces nuts/bolts.

c. Install the instrument panel and connect the electrical connectors.

d. Install the instrument panel-to-chassis fasteners.

e. Install the glove compartment door.

f. Install the stereo opening cover.

g. Install the lower center cluster finish panel and connect the connectors.

h. Install the side trim cover and the console box.

i. Install the radio.

j. Install the heater control and accessory assembly.

k. Connect the connectors and install the cluster instrument panel.

l. Install the heater control assembly and the 3 assembly screws.

m. Install the cluster instrument panel and the heater control name plate.

➡For Canada models, install the 2 screws.

n. Push on the heater control knobs.

o. Install the center cluster finish panel and the 2 panel screws.

p. Carefully, install the steering column. Then, install the steering column-to-instrument panel nuts/bolts and torque the nuts/bolts 19 ft. lbs. (25 Nm) and the lower steering column bolt to 26 ft. lbs. (5 Nm).

q. Install the No. 2 heater-to-register duct.

r. For USA models, install the lower panel insert.

s. Install the lower finish panel and the 2 panel screws.

t. Install the hood lock release lever.

u. Install the combination meter, connect the electrical connectors and the 4 meter screws.

v. Install the cluster finish panel and the panel screw.

w. Install the steering column cover and the 4 cover screws.

x. Install the combination switch, connect the electrical connectors and install the combination switch-to-steering column screws.

y. At both sides, install the cowl side trim board and the 2 trim boards clips.

z. Install both front door scuff plates.

aa. Connect the seat belt pre-tensioner connector.

17. Install the steering wheel by performing the following procedure:

a. Install the steering wheel to the steering column.

b. Install the steering wheel nut and torque the nut to 25 ft. lbs. (34 Nm).

c. Connect the electrical connector and carefully, install the air bag module.

d. Using a Torx® wrench, tighten the steering wheel screws to 78 inch lbs. (8.8 Nm).

e. At both sides of the steering wheel, install the side covers.

18. Connect the heater hoses to the heater core.

19. Refill the cooling system.

20. Connect the negative battery cable.

21. Evacuate and charge the air conditioning system.

22. Run the engine to normal operating temperatures. Check the climate control operation and check for leaks.

Water Pump

REMOVAL & INSTALLATION

2.0L Engine

1. Before servicing the vehicle, refer to the precautions in the beginning of this section.

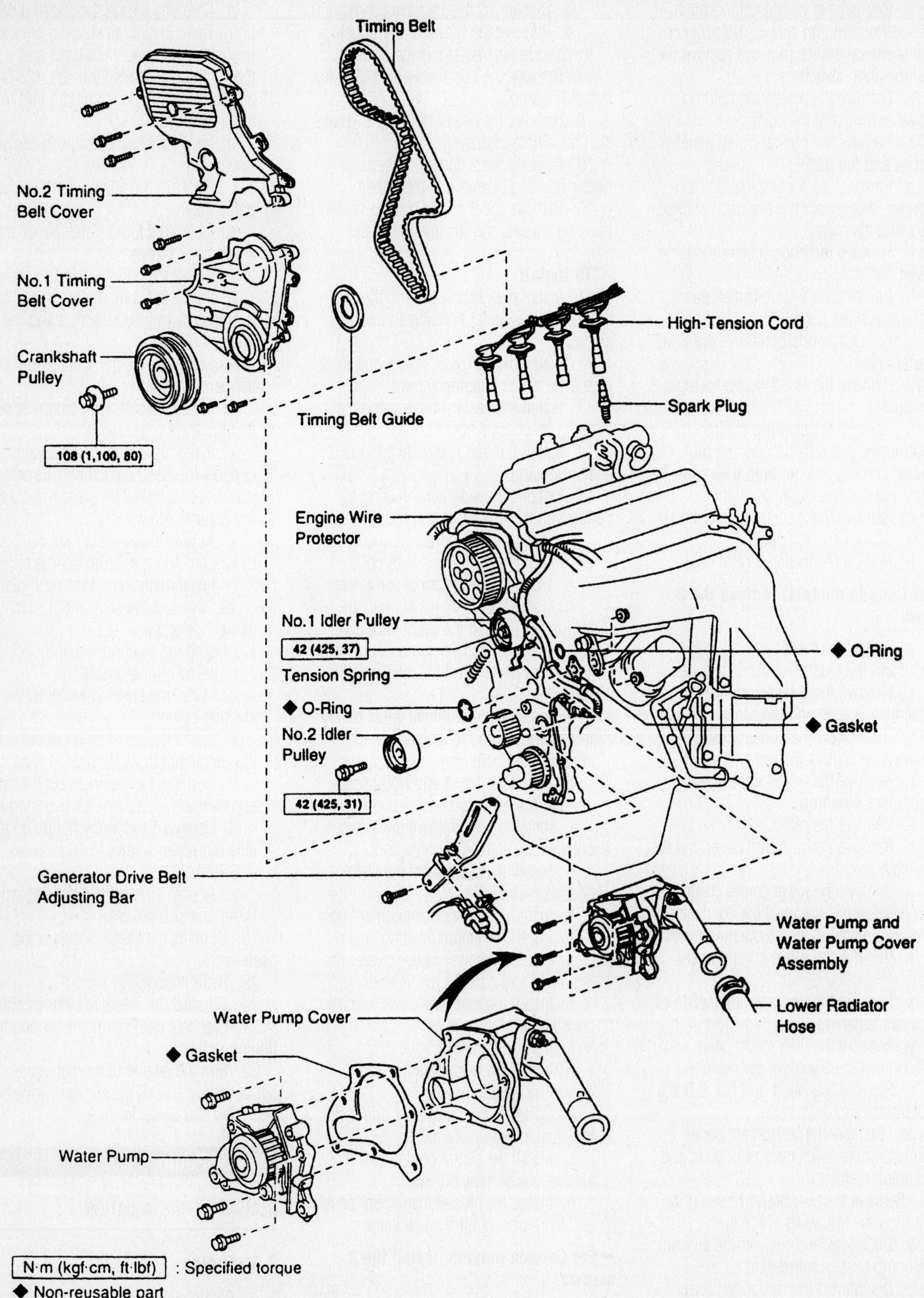

Exploded view of the water pump and related components—2.0L

2. Remove or disconnect the following:
   - Negative battery cable
   - Right-hand engine undercover

3. Drain the engine coolant from the radiator and engine.

4. Remove or disconnect the following:
   - Drive belt
   - Lower radiator hose from the water inlet
   - Drive belt tension spring and the No. 2 idler pulley
   - Crankshaft Position (CKP) sensor connector clamp
   - Alternator drive belt adjusting bar
   - 2 water pump-to-water bypass pipe nuts
   - 3 water pump bolts in the sequence
   - Water pump cover from the water bypass pipe
   - Water pump and water pump cover assembly
   - Gasket and 2 O-rings from the water pump and water bypass pipe
   - 3 bolts, water pump and gasket, from the water pump cover

To install:

5. Install or connect the following:
   - Water pump to the water pump cover, using a new gasket. Tighten the 3 bolts to 78 inch lbs. (9 Nm).
   - New O-ring and gasket to the water pump cover
   - New O-ring to the water bypass pipe, by applying soapy water to the O-ring
   - Water pump cover to the water bypass pipe; do not install the nuts at this time
   - Water pump. Tighten the 3 bolts, in sequence, to 78 inch lbs. (9 Nm).
   - Water pump cover to the water pump pipe. Tighten the 2 bolts to 82 inch lbs. (9 Nm).
   - Alternator drive belt adjusting bar. Tighten the bolt to 20 ft. lbs. (27 Nm).
   - CKP connector clamp
   - No. 2 idler pulley and drive belt tension spring
   - Lower radiator hose
   - Drive belt
   - Negative battery cable

6. Fill the engine and radiator with engine coolant.

7. Start the engine and check for leaks.

8. Install the right-hand engine undercover.

2.4L Engine

1. Before servicing the vehicle, refer to the precautions in the beginning of this section.

2. Drain the cooling system.

3. Remove the accessory drive belt.

4. Remove the water pump pulley.

5. Remove the crankshaft position sensor wire and clamp.

6. Remove the 4 bolts and 2 nuts from the pump.

7. Using a small prybar, remove the water pump.

8. Remove all traces of the gasket material from the pump and block.

To install:

9. Apply a 2.5mm wide bead of RTV gasket material to the pump sealing surface as shown.

➡Install the pump with 5 minutes of applying the sealer or the sealer will have to be removed and new sealer applied.

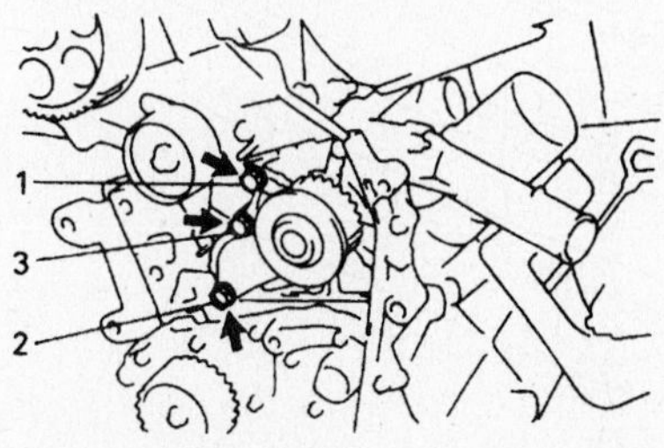

Loosening sequence for the water pump bolts—2.0L engine

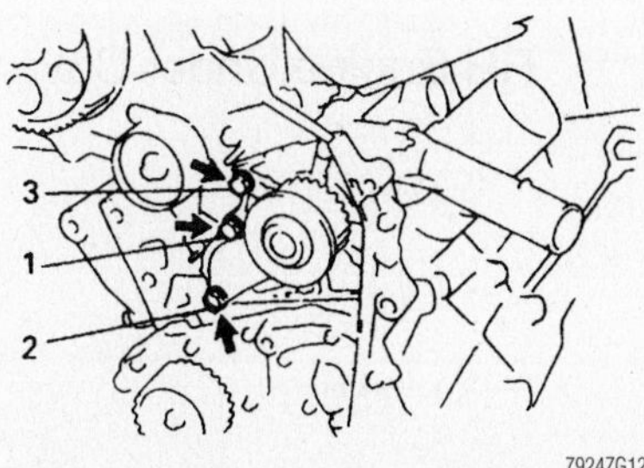

Tightening sequence for the water pump bolts—2.0L engine

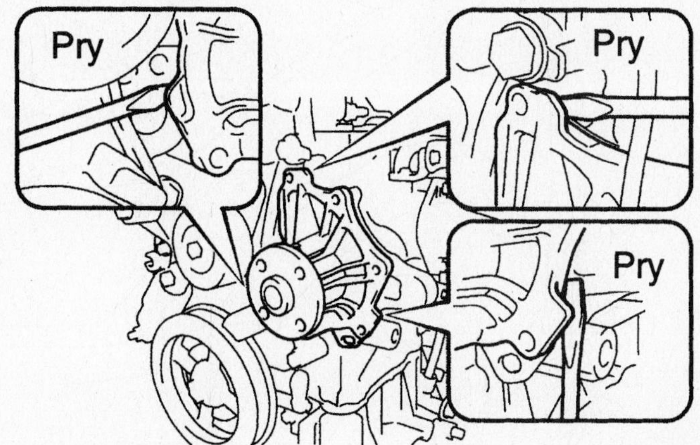

Pry at these points—2.4L engine

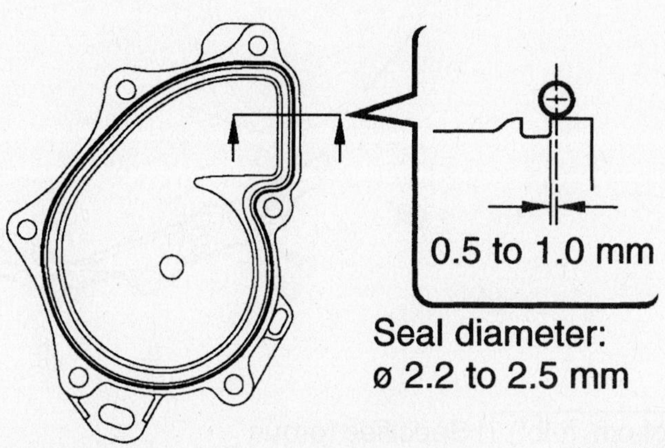

Apply a 2.5mm wide bead of RTV gasket material to the pump sealing surface as shown—2.4L engine

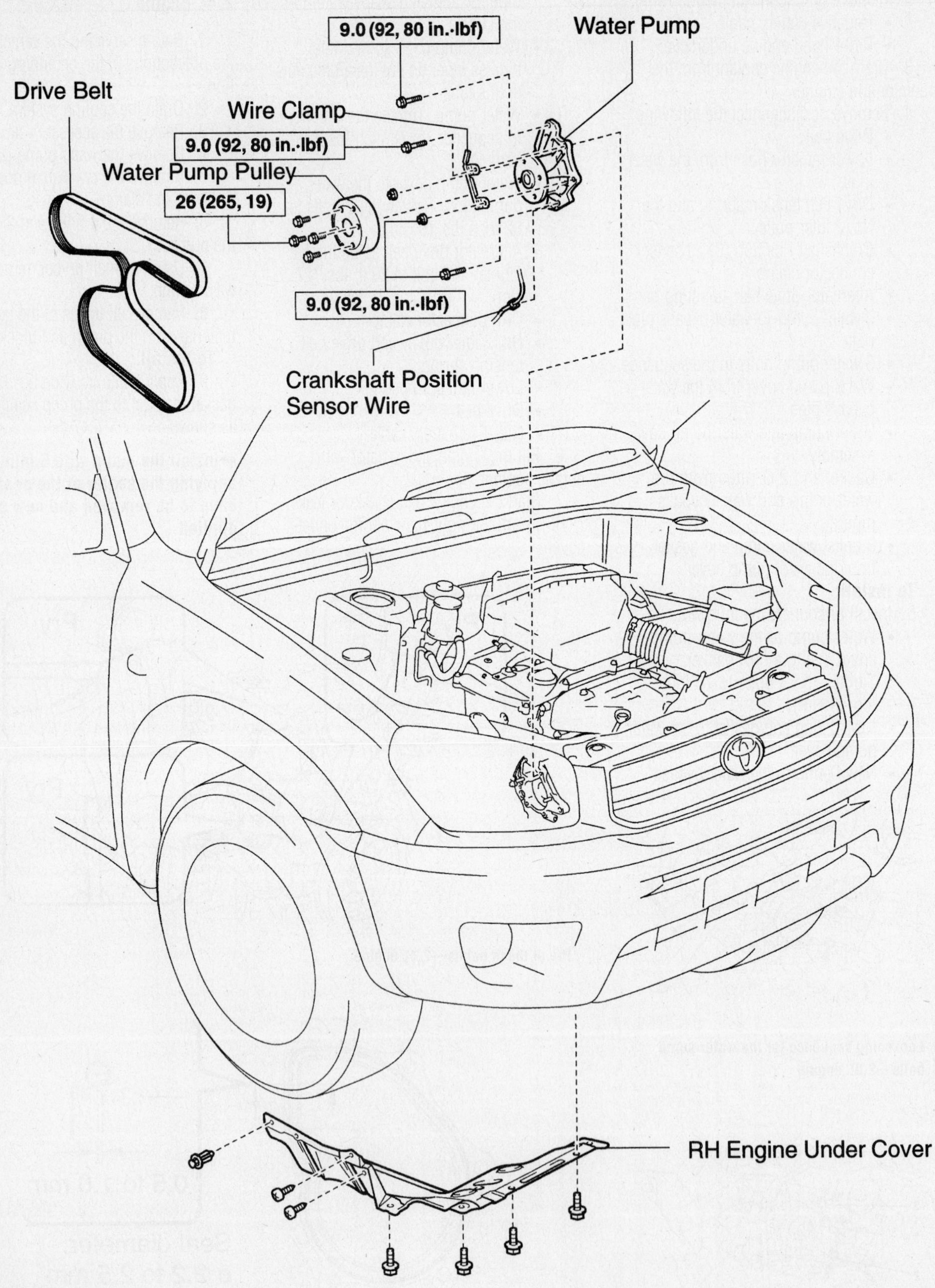

Water pump and related parts—2.4L engine

10. Install the pump and torque the nuts and bolts to 80 inch lbs. (9 Nm).
11. The remainder of installation is the reverse of removal. Refill the cooling system.

Cylinder Head

REMOVAL & INSTALLATION

2.0L Engine

1. Before servicing the vehicle, refer to the precautions in the beginning of this section.
2. Release the fuel system pressure.
3. Remove or disconnect the following:
   - Negative battery cable
   - Right-hand engine undercover
4. Drain the engine coolant.
5. Remove or disconnect the following:
   - Camshafts
   - Cylinder head bolts in several passes
   - Cylinder head with the intake manifold
   - Air hose from the intake manifold
   - 2 bolts and the air tube
   - Intake manifold and gasket
   - Air hose from the cylinder head port
   - Air hose
   - Fuel delivery pipe and the injectors
   - Oil pressure switch

To install:

6. Install or connect the following:
   - Oil pressure switch
   - Fuel injectors and the delivery pipe
   - Air hose to the cylinder head port
   - Intake manifold with new gaskets. Tighten the nut/bolts to 14 ft. lbs. (19 Nm).
   - Air tube with the 2 bolts
   - Air hose to the intake manifold
7. Clean the gasket mating surfaces using care not to damage the aluminum components, replace the gasket; then, lower the cylinder head onto the engine. Be sure the dowel pins are aligned and no hoses or wires are between the head and cylinder block.
8. For 2000–01, tighten the cylinder head bolts as follows:
   a. Apply a light coat of engine oil to the cylinder head bolts.
   b. Tighten the cylinder head bolts, in several passes, in sequence, to 36 ft. lbs. (49 Nm).
   c. Mark the front of the cylinder head bolt with paint.
   d. Retighten the cylinder head bolts by 90 degrees in sequence.

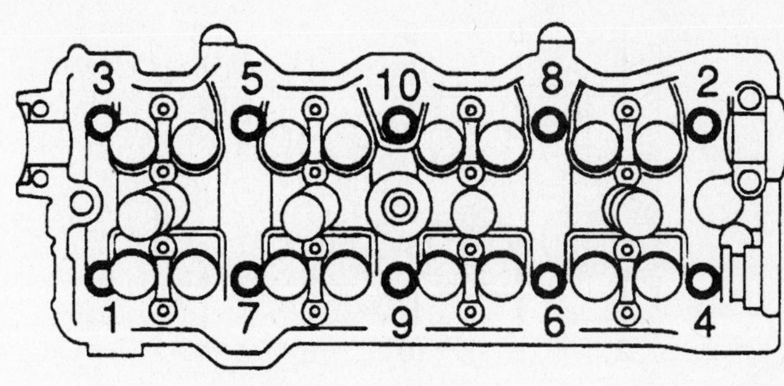

Cylinder head bolts installation sequence—2.0L engine

   e. Retighten an additional 90 degrees and be sure that the paint mark is now positioned toward the rear.
9. For 2002, tighten the cylinder head bolts in 3 progressive steps, as follows:
   a. Apply a light coat of engine oil to the cylinder head bolts.
   b. Tighten the cylinder head bolts, in sequence, to 15 ft. lbs. (26 Nm).
   c. Tighten the cylinder head bolts, in sequence, to 30 ft. lbs. (52 Nm).
   d. Mark the front of the cylinder head bolt with paint.
   e. Retighten the cylinder head bolts by 90 degrees in sequence.
10. Install or connect the following:
   - Intake and exhaust camshafts
   - Negative battery cable
11. Refill the engine with coolant, start the engine, warm up and check for leaks.
12. Bleed the cooling system and top off coolant as necessary.
13. Install the right-hand engine undercover.
14. Check ignition timing and road test the vehicle for proper operation.

2.4L Engine

1. Before servicing the vehicle, refer to the precautions in the beginning of this section.
2. Properly relieve the fuel system pressure.
3. Remove the right engine under-cover.
4. Drain the coolant.
5. Drain the oil.
6. Remove the air cleaner assembly.
7. Remove the drive belt.
8. Remove the alternator.
9. Remove the power steering pump.
10. Remove the ignition coils.
11. Remove the spark plugs.
12. Remove the injectors.
13. Remove the exhaust manifold.
14. Remove the oil filler cap.
15. Remove the PCV hoses and valve.
16. Remove the intake manifold.
17. Remove all wires and harnesses connected to the head.
18. Remove the timing chain.
19. Remove the camshaft sprocket.
20. Remove the VVT sprocket.

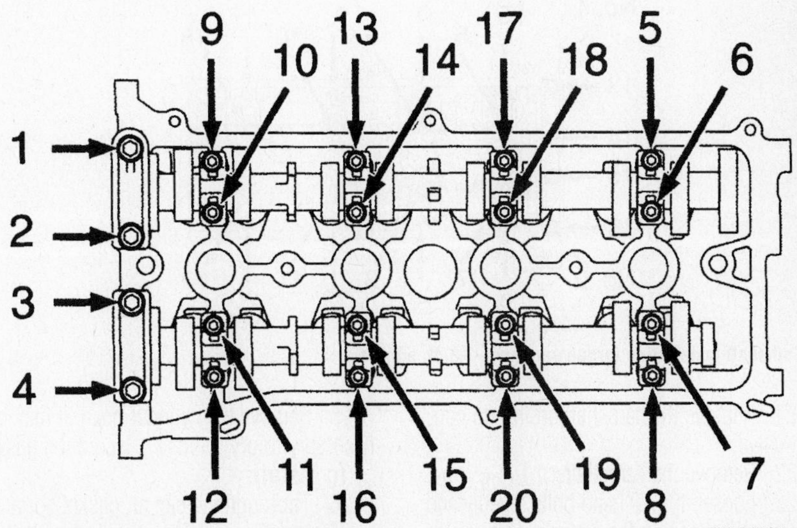

Camshaft removal sequence—2.4L engine

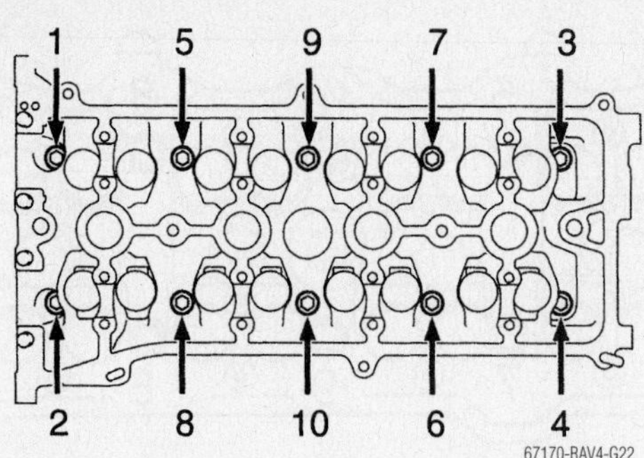

Head bolt removal sequence—2.4L engine

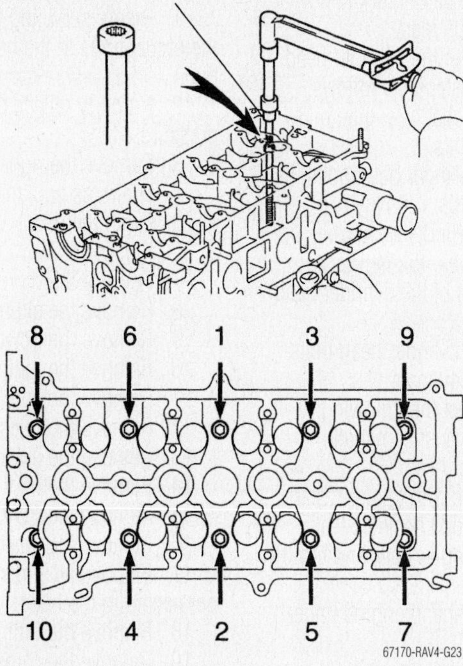

Cylinder head bolt torque sequence—2.4L engine

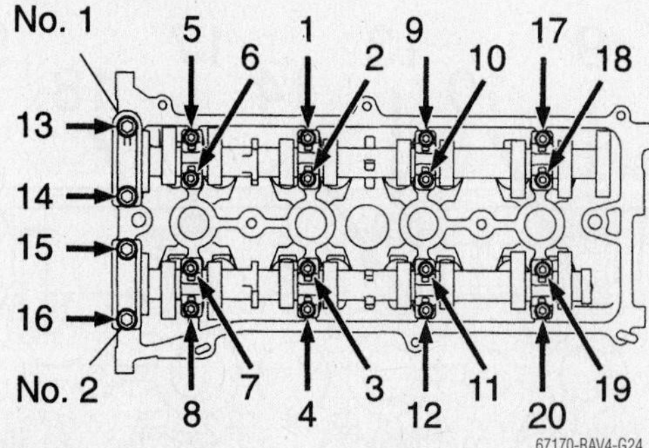

Camshaft installation torque sequence—2.4L engine

21. Remove the camshaft timing oil control valve.
22. Remove the camshafts.
23. Loosen the 10 head bolts, evenly and in several passes, in the sequence shown.
24. Remove the plate washers.
25. Remove the cylinder head. It may be necessary to pry it loose. Discard the gasket.

To install:

26. Thoroughly clean all gasket surfaces.
27. Install a new gasket with the identification number upwards.
28. Carefully install the head.
29. Apply a light coating of engine oil to the threads and under the head of each bolt. Install the bolts and tighten them evenly and in several passes, in the sequence shown, to 58 ft. lbs. (79 Nm).
30. Matchmark the head of each bolt and the cylinder head. Tighten each bolt in sequence an additional 90 degrees.
31. Install the camshafts.
32. Check and adjust the valve clearance.
33. Install the camshaft and VVT sprockets.
34. Install the timing chain.
35. Install the camshaft position sensor.
36. Install the oil control valve.
37. Install the oil filler cap.
38. Install the intake manifold.
39. Install the PCV valve and hoses.
40. Install the engine wiring.
41. Install the exhaust manifold.
42. Install the injectors.
43. Install the spark plugs.
44. Install the ignition coils.
45. Install the power steering pump.
46. Install the alternator.
47. Install the drive belt.
48. Install the air cleaner assembly.
49. Refill all fluids.
50. Install the under-cover.

Intake Manifold

REMOVAL & INSTALLATION

2.0L Engine

1. Before servicing the vehicle, refer to the precautions in the beginning of this section.
2. Properly relieve the fuel system pressure.
3. Remove or disconnect the following:
   - Negative battery cable
   - Air cleaner assembly
   - Throttle body from the intake manifold
4. Disconnect the engine wire from the intake manifold, as follows:
   - 4 injector connectors
   - 2 engine wire clamps from the intake manifold wire brackets
   - Engine wire protector from the right-hand side of the intake manifold by removing the bolt
   - Engine wire from the wire clamp
5. Remove the EGR valve, EGR pipe and modulator, as follows:
   - Both vacuum hoses from the Exhaust Gas Recirculation (EGR) Vacuum Switching Valve (VSV)

- Vacuum modulator from the clamp on the intake manifold
- Loosen the cylinder head side of the EGR pipe union nut
- Both nuts, the EGR valve, pipe assembly and gasket
- Vacuum modulator

6. Disconnect the following hoses:
   - Fuel filter vacuum sensor hose on the intake manifold
   - Brake booster vacuum hose from the intake manifold
   - Ground strap from the intake manifold
7. Remove or disconnect the following:
   - Intake manifold stay by removing the 2 bolts
   - Control cable from the clamp on the rear side of the intake manifold, if equipped with automatic transmission
   - Air hose from the intake manifold
   - Air tube from the intake manifold, by removing the 2 bolts
   - 6 bolts and 2 nuts from the intake manifold
   - Intake manifold

To install:

8. Install or connect the following:
   - Intake manifold. Tighten the 6 bolts and 2 nuts to 14 ft. lbs. (19 Nm).
   - Air tube with the 2 bolts
   - Air hose to the intake manifold
   - Control cable to the clamp on the rear side of the intake manifold, if equipped with an automatic transmission
   - Intake manifold stay. Tighten both bolts to 31 ft. lbs. (42 Nm).
9. Connect the following hoses:
   - Ground strap to the intake manifold
   - Brake booster vacuum hose to the intake manifold
   - Fuel filter vacuum sensor hose to the intake manifold
10. Install the EGR valve, EGR pipe and the vacuum modulator, as follows:
    - Vacuum modulator
    - EGR valve and pipe. Tighten both nuts to 108 inch lbs. (13 Nm) and the union nut to 43 ft. lbs. (59 Nm).
    - Vacuum hoses
11. Install or connect the following:
    - Engine wire and injectors

➡The No. 1 and No. 3 injector connectors are brown, and the No. 2 and No. 4 injector connectors are gray.

- Throttle body to the intake manifold
- Air cleaner assembly
- Negative battery cable

2.4L Engine

1. Remove the right engine under-cover.
2. Drain the coolant.
3. Drain the oil.
4. Remove the air cleaner assembly.
5. Remove the drive belt.
6. Remove the alternator.
7. Remove the power steering pump.
8. Remove the ignition coils.
9. Remove the spark plugs.
10. Remove the injectors.
11. Remove the oil filler cap.
12. Remove the PCV hoses and valve.
13. Disconnect the TPS.
14. Remove the 2 water hoses and 2 vacuum hoses.
15. Disconnect the wiring harness.
16. Remove the 5 bolts and 2 nuts.
17. Remove the intake manifold.

To install:

18. Thoroughly clean all gasket surfaces.
19. Install the intake manifold, using a new gasket.
20. Tighten the nuts and bolts evenly and in several passes, to 22 ft. lbs. (30 Nm).
21. Install the PCV valve and hoses.
22. Install the engine wiring.
23. Install the injectors.
24. Install the spark plugs.
25. Install the ignition coils.
26. Install the power steering pump.
27. Install the alternator.
28. Install the drive belt.
29. Install the air cleaner assembly.
30. Refill all fluids.
31. Install the under-cover.

Exhaust Manifold

REMOVAL & INSTALLATION

2.0L Engine

1. Before servicing the vehicle, refer to the precautions in the beginning of this section.

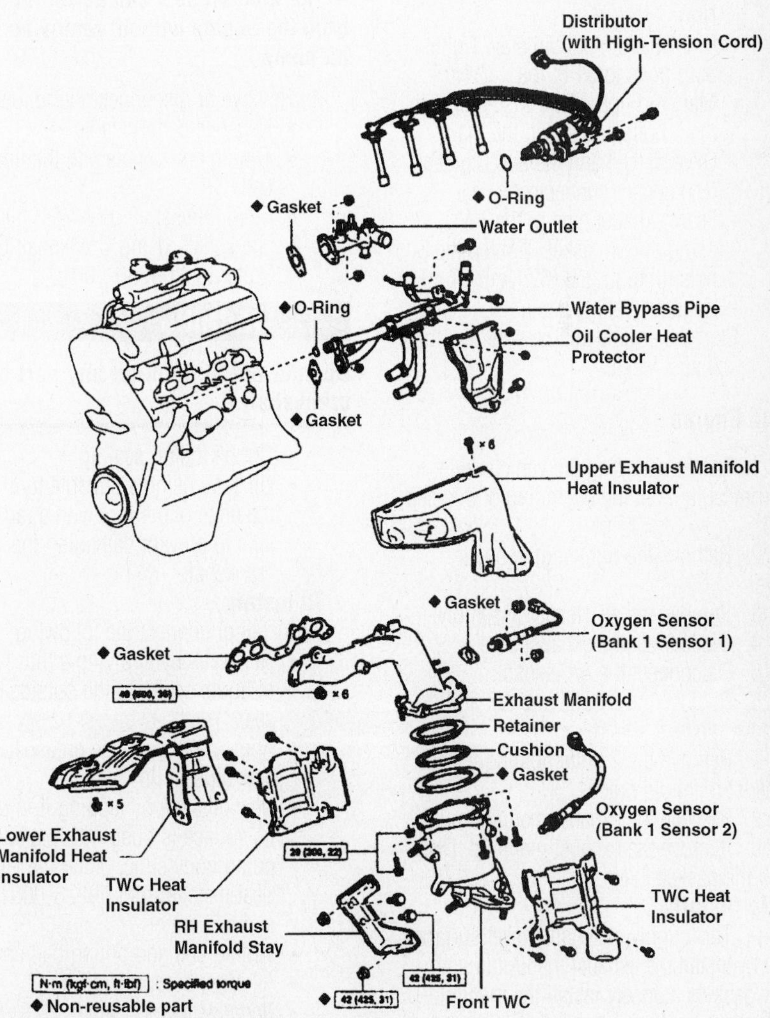

Exploded view of the exhaust manifold and components—2.0L engine

2. Remove or disconnect the following:
   - Negative battery cable
   - Front exhaust pipe from the exhaust manifold, using a 14mm deep socket wrench; discard the gasket
   - Main Oxygen (O_2) sensor and the sub Oxygen (O_2) sensor connectors
   - 6 bolts and the upper manifold heat insulator
   - 2 right-hand exhaust manifold stay-to-cylinder block bolts
   - 6 nuts, the exhaust manifold and the Three-Way Catalytic (TWC) converter assembly
   - Exhaust manifold and front catalytic converter

To install:

3. Install or connect the following:
   - Catalytic converter to the exhaust manifold. Tighten the nuts/bolts to 22 ft. lbs. (29 Nm).
   - Exhaust manifold and the front TWC assembly. Tighten the 6 nuts, in several passes to 36 ft. lbs. (49 Nm).
   - Right-hand manifold stay. Tighten both bolts to 31 ft. lbs. (42 Nm).
   - Manifold upper heat insulator with the 6 bolts and attach the main Oxygen (O_2) and the sub Oxygen (O_2) sensor connectors
   - Front exhaust pipe to the TWC, using a new gasket. Tighten the 3 nuts to 46 ft. lbs. (62 Nm).
   - Negative battery cable
4. Start the engine and be sure that there are no exhaust leaks.

2.4L Engine

1. Before servicing the vehicle, refer to the precautions in the beginning of this section.
2. Remove the right engine under-cover.
3. Remove the air cleaner assembly.
4. Disconnect the exhaust pipe.
5. Disconnect the A/F sensor.
6. Remove the 3 bolts and nut and remove the heat shield.
7. Remove the 2 bolts holding the exhaust manifold braces.
8. Remove the exhaust manifold nuts.
9. Remove the exhaust manifold. Discard the gaskets.

To install:

10. Thoroughly clean all gasket surfaces.
11. Install the exhaust manifold using new gaskets. Loosely install the manifold-to-brace nuts. Install the 5 manifold nuts and torque them to 27 ft. lbs. (37 Nm).
12. Tighten the brace nuts to 32 ft. lbs. (44 Nm).
13. Install and tighten the brace-to-crankcase nuts to 32 ft. lbs. (44 Nm).
14. Connect the exhaust pipe. Torque to 32 ft. lbs. (44 Nm).
15. Install the heat shield. Torque to 9 ft. lbs. (12 Nm).
16. Install the air cleaner assembly.
17. Install the under-cover.

Front Crankshaft Seal

REMOVAL & INSTALLATION

For front crankshaft seal on the 2.4L engine, see, "Timing Chain, Cover and Seal".

2.0L Engine

1. Before servicing the vehicle, refer to the precautions in the beginning of this section.

➡The front oil seal can be removed from the engine without removing the oil pump.

2. Remove or disconnect the following:
   - Negative battery cable
   - Timing belt covers and the timing belt
   - Front crankshaft gear from the crankshaft, using Crankshaft Gear Puller tool 09950-50010

⁂ WARNING

Be sure not to damage any part of the crankshaft.

   - Cut off the oil seal lip
   - Oil seal, using a suitable tool. Wrap the edge of the tool with a rag or tape to prevent damaging the crankshaft.

To install:

3. Install or connect the following:
   - New seal, by applying a thin layer of liquid sealer to the outside of the seal
   - Apply multi purpose grease to the new oil seal lip
   - New oil seal, by tapping it in until its surface is flush with the oil pump body edge, using the Oil Seal Installer tool 09226-00010 and a hammer
   - Timing belt and timing belt covers
   - All other components
   - Negative battery cable
4. Start the engine and check for leaks.

Camshaft and Valve Lifters

REMOVAL & INSTALLATION

2.0L Engine

1. Before servicing the vehicle, refer to the precautions in the beginning of this section.
2. Remove or disconnect the following:
   - Negative battery cable
   - Cylinder head cover and the upper timing belt cover
3. Rotate the crankshaft to set the engine at Top Dead Center (TDC)/compression for the No. 1 cylinder.

➡Due to the small thrust clearance on both the intake and exhaust camshafts, the camshafts must be kept level during removal. If the camshafts are removed without being kept level, the camshaft may be caught in the cylinder head causing the head to break or the camshaft to seize.

4. Remove the camshaft timing sprocket and the timing belt.
5. Set the knock pin of the intake camshaft at 10–45 degrees Before Top Dead Center (BTDC) of camshaft angle. This angle will help to lift the exhaust camshaft level and evenly by pushing No. 2 and No. 4 cylinder camshaft lobes of the exhaust camshaft toward their valve lifters.
6. Secure the exhaust camshaft sub-gear to the main gear using a service bolt. The manufacturer recommends a bolt 0.63–0.79 in. (16–20mm) long with a thread diameter of 6mm and a 1mm thread pitch. When removing the exhaust camshaft, be sure that the torsional spring force of the sub-gear has been eliminated.
7. Remove the No. 1 and No. 2 rear bearing cap bolts and remove the cap. Uniformly loosen and remove bearing cap bolts

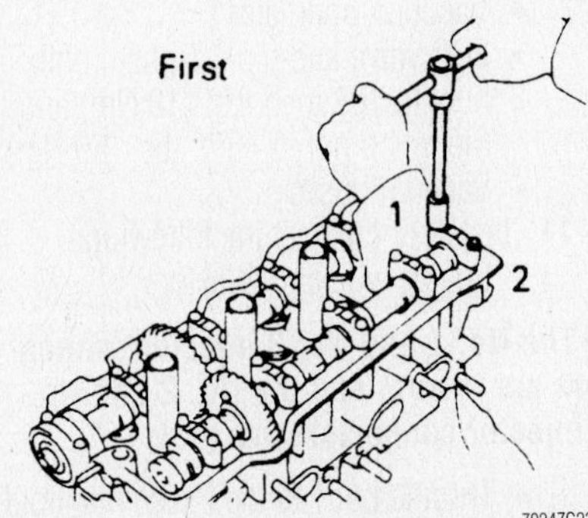

Exhaust camshaft bolt removal: step 1—2.0L

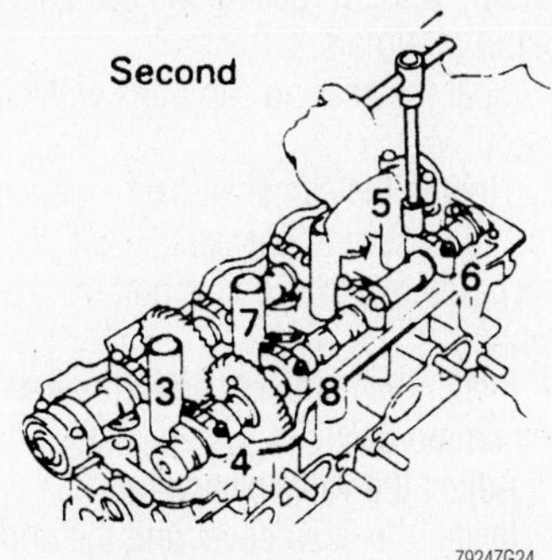

Exhaust camshaft bolt removal: step 2—2.0L

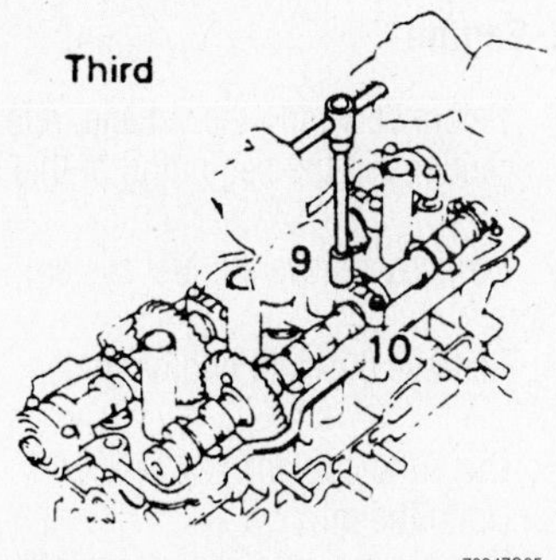

Exhaust camshaft bolt removal: step 3—2.0L

No. 3 to No. 8 in several passes and in the proper sequence. Do not remove bearing cap bolts No. 9 and 10 at this time. Remove the No. 1, 2 and 4 bearing caps.

8. Alternately loosen and remove bearing cap bolts No. 9 and 10. As these bolts are loosened check to see that the camshaft is being lifted out straight and level.

➡If the camshaft is not lifting out straight and level retighten No. 9 and 10 bearing cap bolts. Reverse the order of steps 5 through 7 and reset the intake camshaft knock pin to 10–45 degrees BTDC and repeat steps 5 through 7 again. Do not attempt to pry the camshaft from its mounting.

9. Remove the No. 3 bearing cap and exhaust camshaft from the engine.

10. Set the knock pin of the intake camshaft at 80–115 degrees BTDC of camshaft angle. This angle will help to lift the intake camshaft level and evenly by pushing No. 1 and No. 3 cylinder camshaft lobes of the intake camshaft toward their valve lifters.

11. Remove the No. 1 and No. 2 front bearing cap bolts and remove the front bearing cap and oil seal. If the cap will not come apart easily, leave it in place without the bolts.

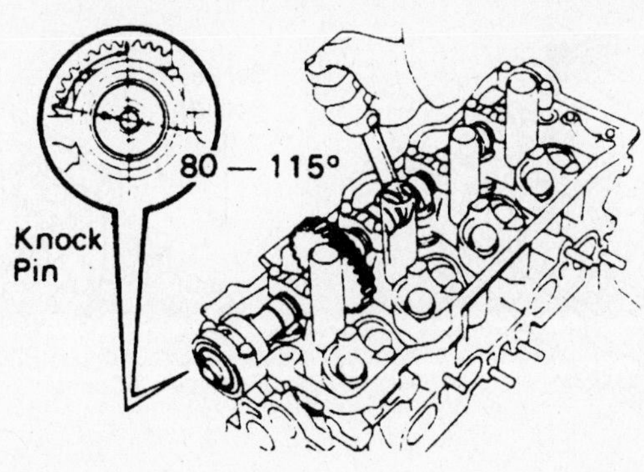

Intake camshaft knock pin alignment—2.0L

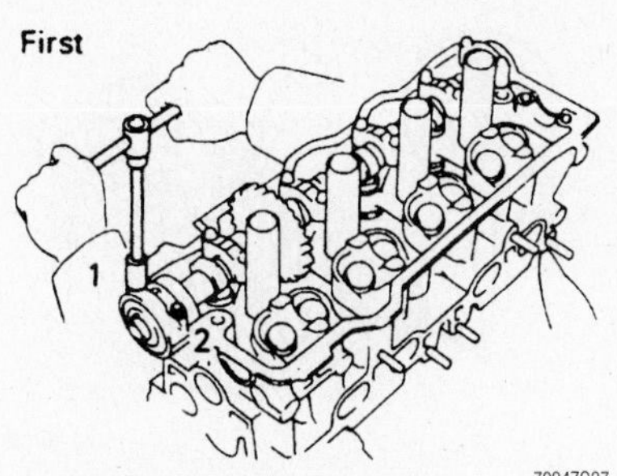

Intake camshaft bolt removal: step 1—2.0L

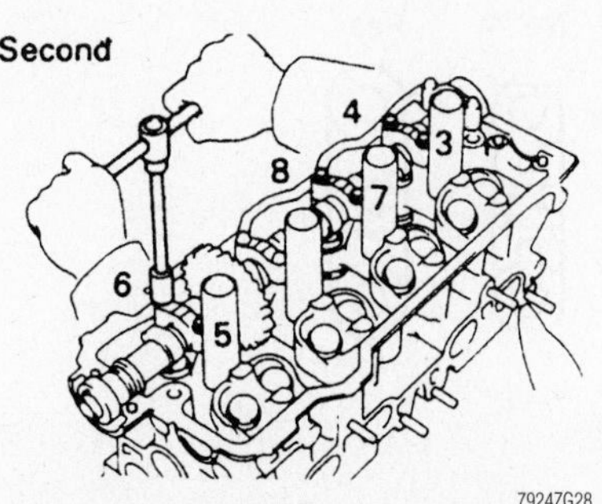

Intake camshaft bolt removal: step 2—2.0L

12. Uniformly loosen and remove bearing cap bolts No. 3 to No. 8 in several phases and in the proper sequence. Do not remove bearing cap bolts No. 9 and 10 at this time. Remove No. 1, 3 and 4 bearing caps.

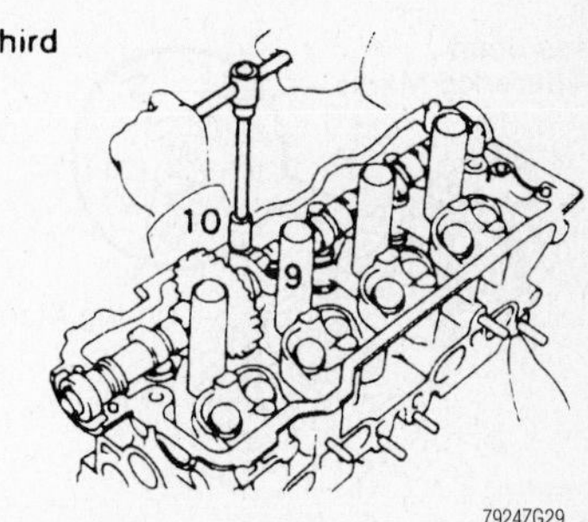

Intake camshaft bolt removal: step 3—2.0L

13. Alternately loosen and remove bearing cap bolts No. 9 and 10. As these bolts are loosened and after breaking the adhesion on the front bearing cap, check to see that the camshaft is being lifted out straight and level.

➡If the camshaft is not lifting out straight and level retighten No. 9 and 10 bearing cap bolts. Reverse steps 10 through 12, than start over from step 10. Do not attempt to pry the camshaft from its mounting.

14. Remove the No. 2 bearing cap with the intake camshaft from the engine.

15. Remove the valve adjusting shims from the engine. Be sure to replace the shims to their original location.

To install:

16. Install the valve adjusting shims to the engine.

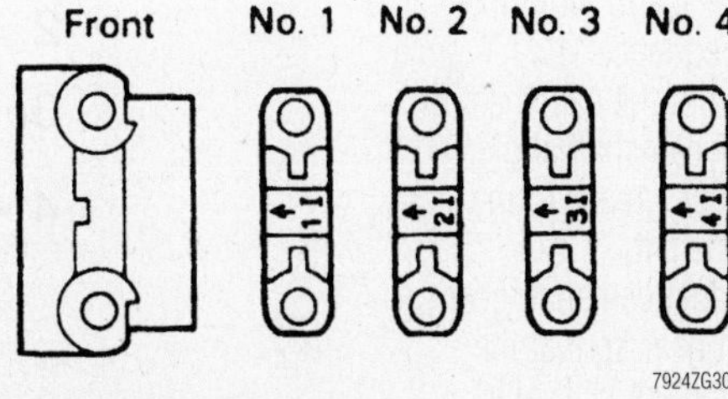

Intake camshaft bearing cap positioning—2.0L

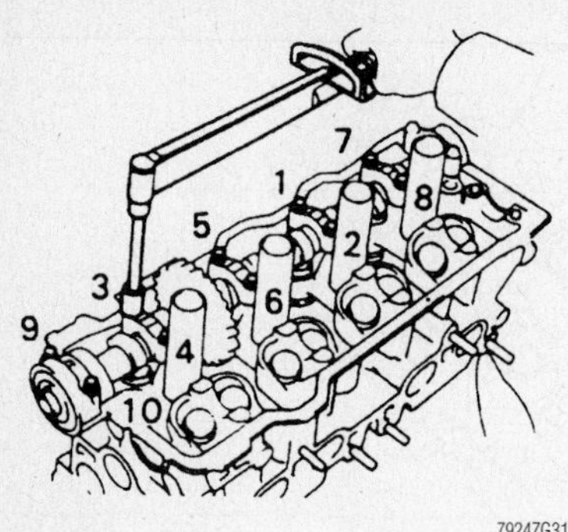

Intake camshaft bolt tightening sequence—2.0L

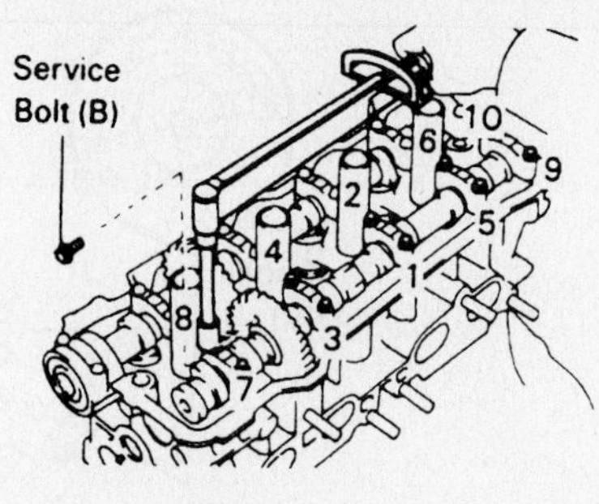

Exhaust camshaft bolt tightening sequence—2.0L

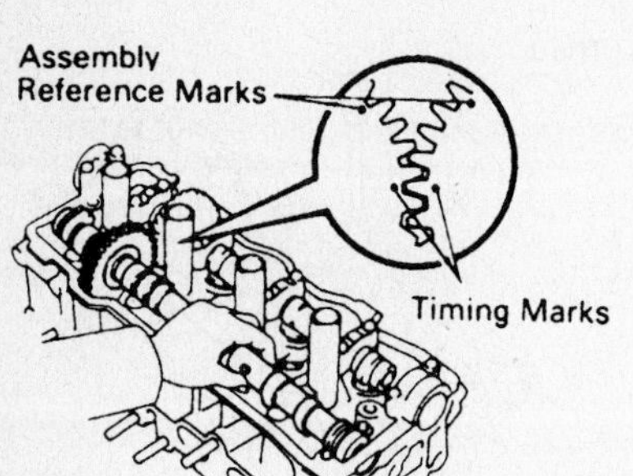

Camshaft timing mark alignment—2.0L

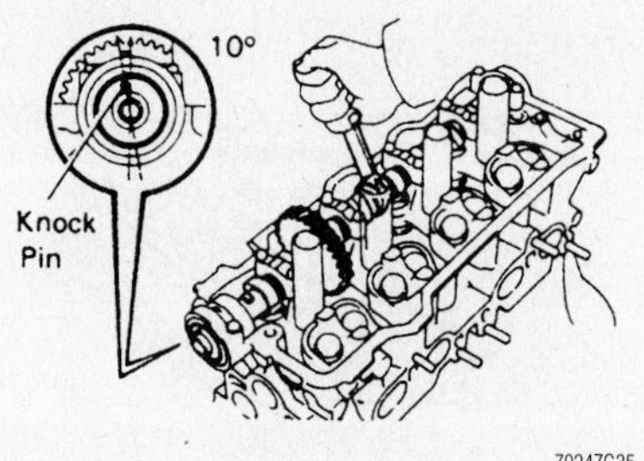

Exhaust camshaft knock pin alignment—2.0L

No. 1 No. 2 No. 3 No. 4 Rear

Exhaust camshaft bearing cap positioning—2.0L

17. Before installing the intake camshaft, apply multi-purpose grease to the thrust portion of the camshaft.
18. Position the camshaft at 80–115 degrees BTDC of camshaft angle on the cylinder head.
19. Apply sealant to the front bearing cap.
20. Coat the bearing cap bolts with clean engine oil.
21. Tighten the camshaft bearing caps evenly and in several passes to 14 ft. lbs. (19 Nm) in the proper sequence.
22. Set the knock pin of the camshaft at 10–45 degrees BTDC of camshaft angle.
23. Apply multipurpose grease to the thrust portion of the camshaft.
24. Position the exhaust camshaft gear with the intake camshaft gear so that the timing marks are in alignment with one another. Be sure to use the proper alignment marks on the gears. Do not use the assembly reference marks.
25. Turn the intake camshaft clockwise or counterclockwise little by little until the exhaust camshaft sits in the bearing journals evenly without rocking the camshaft on the bearing journals.
26. Coat the bearing cap bolts with clean engine oil.
27. Tighten the camshaft bearing caps evenly and in several passes to 14 ft. lbs. (19 Nm). Remove the service bolt from the assembly.
28. Install the camshaft timing pulleys and the timing belt.
29. Adjust the valve clearance.
30. Install the head cover and the upper timing cover. Reconnect the negative battery cable.
31. Start the engine and check for leaks.
32. Check and adjust the ignition timing.

2.4L Engine

1. Before servicing the vehicle, refer to the precautions in the beginning of this section.
2. Properly relieve the fuel system pressure.
3. Remove the right engine undercover.
4. Drain the coolant.
5. Drain the oil.
6. Remove the air cleaner assembly.
7. Remove the drive belt.
8. Remove the alternator.
9. Remove the power steering pump.
10. Remove the ignition coils.
11. Remove the spark plugs.
12. Remove the injectors.
13. Remove the exhaust manifold.
14. Remove the oil filler cap.
15. Remove the PCV hoses and valve.
16. Remove the intake manifold.
17. Remove all wires and harnesses connected to the head.
18. Remove the timing chain.
19. Remove the camshaft sprocket.
20. Remove the VVT sprocket.
21. Remove the camshaft timing oil control valve.

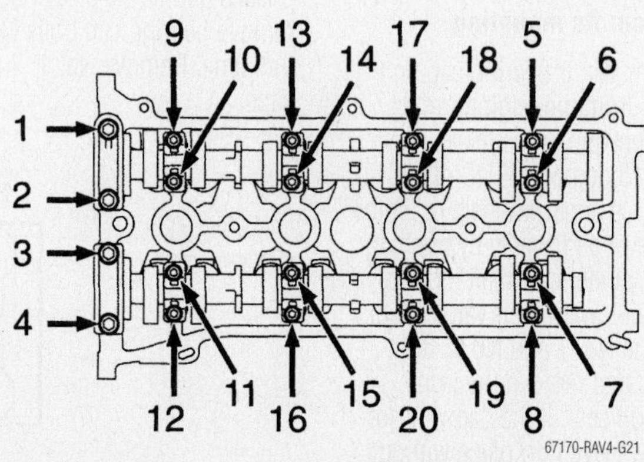

Loosen the camshaft bearing cap bolts evenly, in several passes, in the sequence shown

22. Loosen the camshaft bearing cap bolts evenly, in several passes, in the sequence shown, and remove them.

23. Mark and remove the bearing caps. Make sure they are identified for installation.

24. Remove the camshafts.

To install:

25. Coat the camshafts and bearings with clean engine oil.

26. Install the camshafts with the No.1 cam lobes facing as shown.

37. Install the engine wiring.
38. Install the exhaust manifold.
39. Install the injectors.
40. Install the spark plugs.
41. Install the ignition coils.
42. Install the power steering pump.
43. Install the alternator.
44. Install the drive belt.
45. Install the air cleaner assembly.
46. Refill all fluids.
47. Install the under-cover.

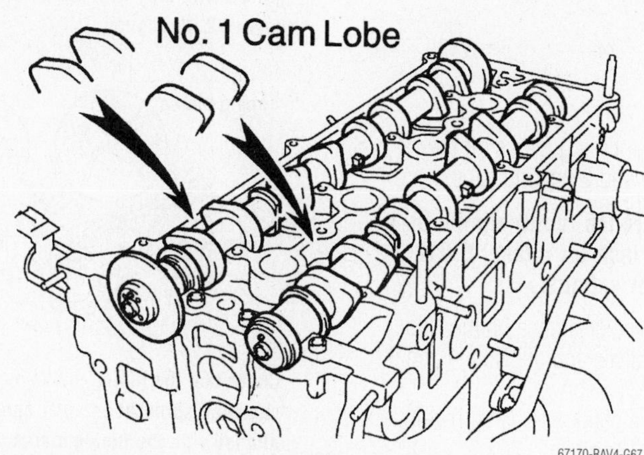

Install the camshafts with the No.1 cam lobes facing as shown

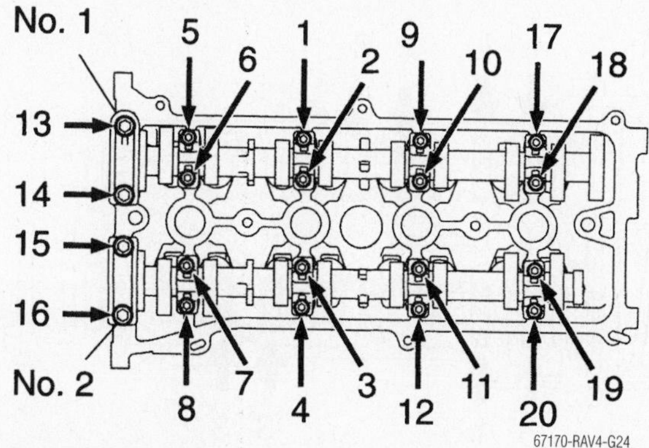

Torque the bolts evenly, in several passes, in the sequence shown

27. Coat the threads and heads of the bearing cap bolts with clean engine oil.

28. Torque the bolts evenly, in several passes, in the sequence shown, to 22 ft. lbs. (29.5 Nm) for Nos. 1 and 2; 80 inch lbs. (9 Nm) for all the others.

29. Check and adjust the valve clearance.

30. Install the camshaft and VVT sprockets.

31. Install the timing chain.
32. Install the camshaft position sensor.
33. Install the oil control valve.
34. Install the oil filler cap.
35. Install the intake manifold.
36. Install the PCV valve and hoses.

Valve Lash

ADJUSTMENT

2.0L Engine

1. Before servicing the vehicle, refer to the precautions in the beginning of this section.

2. Remove the cylinder head covers.

3. Use a wrench to turn the crankshaft until the notch in the pulley aligns with timing mark **0** of the No. 1 timing belt cover. This will ensure that the No. 1 piston is at Top Dead Center (TDC) of the compression stroke.

➡Check that the valve lifters on the No. 1 cylinder are loose and those on the No. 4 cylinder are tight. If not, rotate the crankshaft 1 complete revolution (360 degrees) and then realign the marks.

4. Using a flat feeler gauge measure the clearance between the camshaft lobe and the valve lifter on the first set of valves shown. This measurement should correspond to specifications.

➡If the measurement is within specifications, go on to the next step. If not, record the measurement taken for each individual valve.

5. Rotate the crankshaft 1 complete revolution and realign the timing marks.

6. Measure the clearance of the second set of valves.

➡If the measurement for this set of valves (and also the previous one) is within specifications, go no further, the procedure is finished. If not, record the measurements and proceed to the next step.

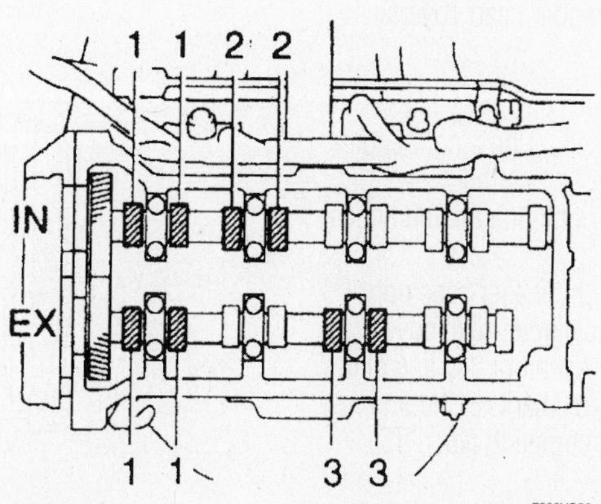

Adjust these valve first—2.0L

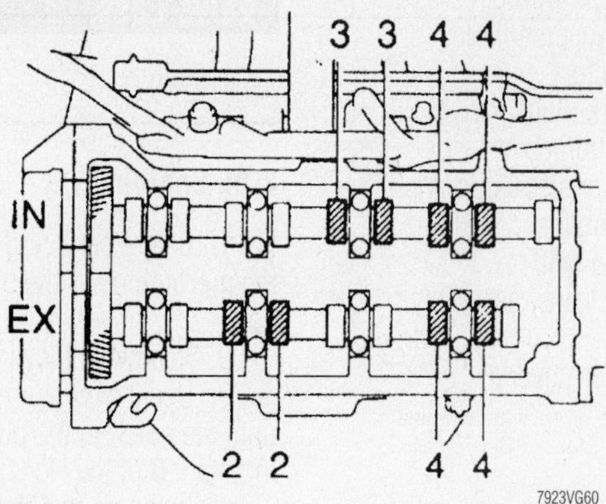

Adjust these valve second—2.0L

7. Rotate the crankshaft to position the intake camshaft lobe of the cylinder to be adjusted, facing upward.

➡Both intake and exhaust valve clearance may be adjusted at the same time, if required.

8. Using a suitable tool, turn the valve lifter so the notch is easily accessible.
9. Install tool 09248-55010 between both camshaft lobes and turn the handle so the tool presses down both intake and exhaust valve lifters evenly.
10. Using a suitable tool and a magnet, remove the valve shims.
11. Measure the thickness of the old shim with a micrometer. Using this measurement and the clearance made earlier (from Step 3 or 5), determine what size replacement shim will be required in order to bring the valve clearance into specification.

➡Replacement shims are available in 27 sizes, in increments of 0.0020 in. (0.05mm). Shim sizes are 0.0787–0.1299 in. (2.00–3.30mm).

12. Install the new shim, remove the special tool; then, recheck the valve clearances.
13. Install the cylinder head covers.

2.4L Engine

Perform this procedure on a cold engine only!

1. Remove the right side engine undercover.
2. Remove the air cleaner assembly.
3. Remove the cylinder head cover.
4. Turn the crankshaft pulley and align its groove with the "0" mark on the timing cover. This sets the engine to No. 1 TDC compression.
5. Check that the timing marks on the camshaft sprocket and VVT sprocket are aligned with the timing marks on the camshaft No.1 and 2 bearing caps.

➡Valve clearance (cold) should be 0.008-0.011 inch for intake; 0.012-0.016 inch for exhaust.

6. Check the clearance on the Nos. 1 and 2 intake valves and the Nos. 1 and 3 exhaust valves, as shown.
7. Turn the crankshaft 1 full revolution clockwise (360 degrees).
8. Check the clearance on Nos. 3 and 4 intake, and Nos. 2 and 4 exhaust.

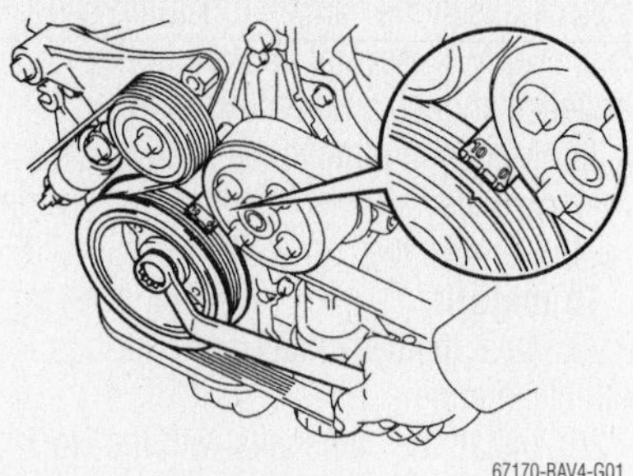

Turn the crankshaft pulley and align its groove with the "0" mark on the timing cover—2.4L engine

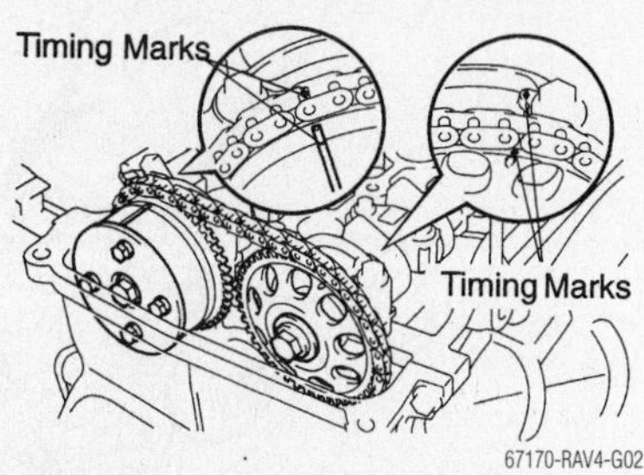

Check that the timing marks on the camshaft sprocket and VVT sprocket are aligned with the timing marks on the camshaft No.1 and 2 bearing caps—2.4L engine

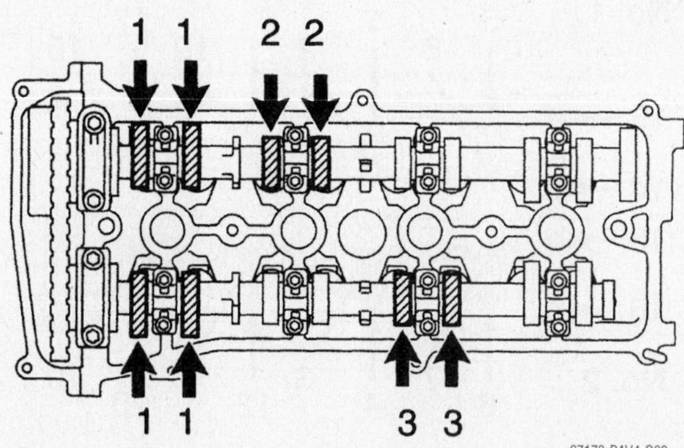

Check the clearance on the Nos. 1 and 2 intake valves and the Nos. 1 and 3 exhaust valves—2.4L engine

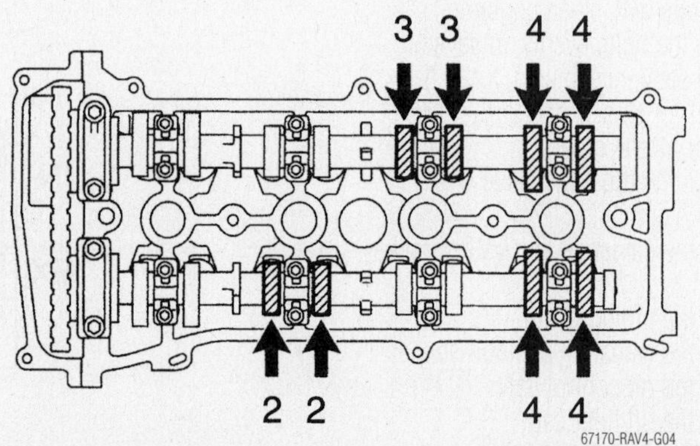

Check the clearance on Nos. 3 and 4 intake, and Nos. 2 and 4 exhaust—2.4L engine

If Adjustment Is Needed

9. Reset the crankshaft to No. 1 TDC compression.
10. Place matchmarks on the timing chain and camshaft sprockets.
11. Remove the chain tensioner and gaskets.
12. Loosen the exhaust camshaft sprocket bolt.
13. Remove the exhaust camshaft bearing caps, evenly, in several passes, in the sequence shown.
14. Lift the camshaft and remove the sprocket together with the timing chain.
15. Remove the intake camshaft bearing caps, evenly, in several passes, in the sequence shown.
16. Remove the intake camshaft. Tie the timing chain out of the way as shown.
17. For any valve needing adjustment, remove the lifter. Determine replacement lifter size.
    a. Using a micrometer, measure the lifter thickness.
    b. Calculate the thickness of a new liter to bring the valve clearance into the proper range.
    c. Select a lifter with a thickness as close as possible to correct the specified value, from the accompanying charts. Lifters are available in 35 sizes in 0.0008 inch (0.020mm) increments from 5.060mm to 5.740mm.

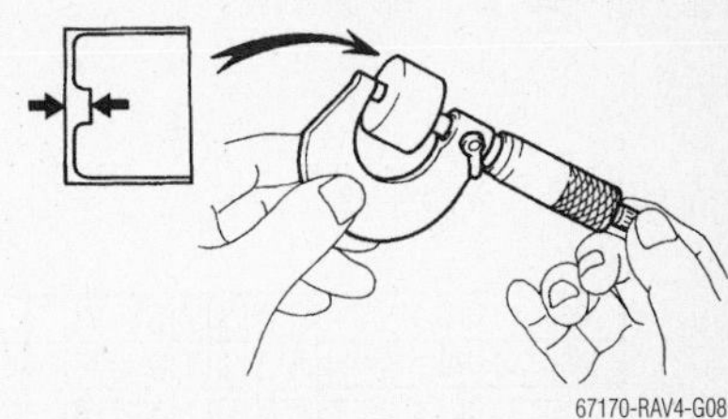

67170-RAV4-G08

Using a micrometer, measure the lifter thickness—2.4L engine

➡An ID number inside the lifter shows the 2 decimal place size. So, a 38 mark would indicate a lifter that is 5.38mm thick.

    d. Coat the replacement lifter with clean engine oil and install it.
18. When all new lifters are installed, align the crankshaft timing mark with the "0" mark on the timing cover.
19. Hold the chain and install the intake camshaft, aligning all marks.
20. Coat the threads and heads of the bearing cap bolts with clean engine oil.
21. Torque the bolts evenly, in several passes, in the sequence shown, to 22 ft. lbs. (29.5 Nm) for Nos. 1 and 2; 80 inch lbs. (9 Nm) for all the others.
22. Install the exhaust camshaft, aligning all marks and install the bearing caps in the same manner as you did with the intake caps.
23. Recheck all timing marks.
24. Install the tensioner. Torque to 80 inch lbs. (9 Nm).
25. Recheck the valve timing by setting the crankshaft timing mark to align with the "0" mark on the timing cover. All timing marks should align
26. The remainder of installation is the reverse of removal.

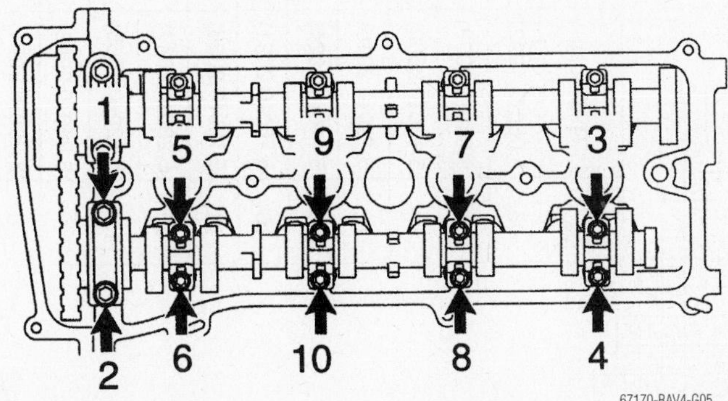

67170-RAV4-G05

Remove the exhaust camshaft bearing caps, evenly, in several passes, in the sequence shown—2.4L engine

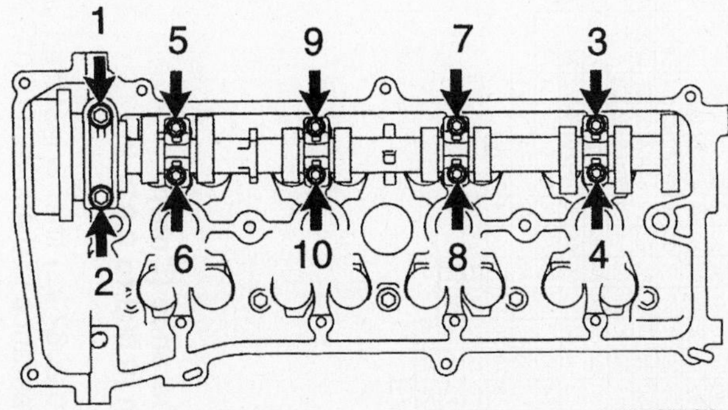

67170-RAV4-G06

Remove the intake camshaft bearing caps, evenly, in several passes, in the sequence shown—2.4L engine

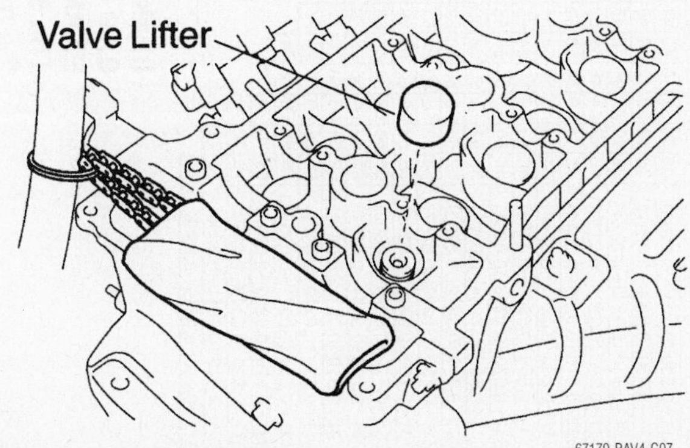

67170-RAV4-G07

Tie the timing chain out of the way—2.4L engine

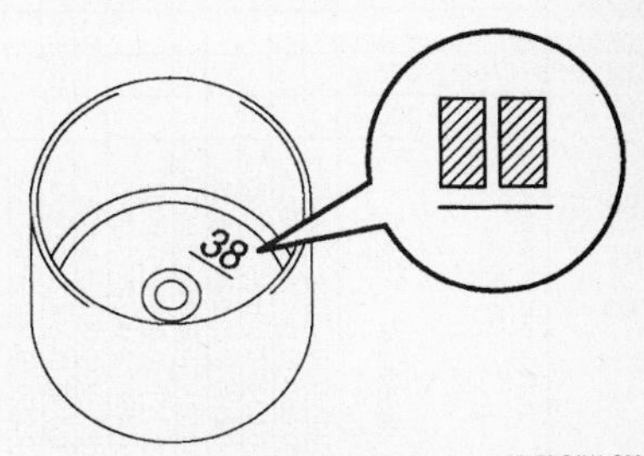

67170-RAV4-G09

An ID number inside the lifter shows the 2 decimal place size—2.4L engine

Valve Lifter Selection Chart (Intake)

Installed lifter thickness mm (in.) / Measure clearance mm (in.)	5.060 (0.1992)	5.080 (0.2000)	5.100 (0.2008)	5.120 (0.2016)	5.140 (0.2024)	5.160 (0.2031)	5.180 (0.2039)	5.200 (0.2047)	5.210 (0.2051)	5.220 (0.2055)	5.230 (0.2059)	5.240 (0.2063)	5.250 (0.2067)	5.260 (0.2071)	5.270 (0.2075)	5.280 (0.2079)	5.290 (0.2083)	5.300 (0.2087)	5.310 (0.2091)	5.320 (0.2094)
0.000-0.030 (0.0000-0.0012)													06	06	08	08	10	10	12	12
0.031-0.050 (0.0012-0.0020)									06	06	06	08	08	10	10	12	12	14	14	16
0.051-0.070 (0.0020-0.0028)								06	08	08	08	10	10	12	12	14	14	16	16	18
0.071-0.090 (0.0028-0.0035)							06	08	10	10	10	12	12	14	14	16	16	18	18	20
0.091-0.110 (0.0036-0.0043)						06	08	10	12	12	12	14	14	16	16	18	18	20	20	22
0.111-0.130 (0.0044-0.0051)					06	08	10	12	14	14	14	16	16	18	18	20	20	22	22	24
0.131-0.150 (0.0052-0.0059)				06	08	10	12	14	16	16	16	18	18	20	20	22	22	24	24	26
0.151-0.170 (0.0059-0.0067)			06	08	10	12	14	16	18	18	18	20	20	22	22	24	24	26	26	28
0.171-0.189 (0.0067-0.0074)		06	08	10	12	14	16	18	20	20	20	22	22	24	24	26	26	28	28	30
0.190-0.290 (0.0075-0.0114)																				
0.291-0.310 (0.0115-0.0122)	12	14	16	18	20	22	24	26	28	28	30	30	32	32	34	34	36	36	38	38
0.311-0.330 (0.0122-0.0130)	14	16	18	20	22	24	26	28	30	30	32	32	34	34	36	36	38	38	40	40
0.331-0.350 (0.0130-0.0138)	16	18	20	22	24	26	28	30	32	32	34	34	36	36	38	38	40	40	42	42
0.351-0.370 (0.0138-0.0146)	18	20	22	24	26	28	30	32	34	34	36	36	38	38	40	40	42	42	44	44
0.371-0.390 (0.0146-0.0154)	20	22	24	26	28	30	32	34	36	36	38	38	40	40	42	42	44	44	46	46
0.391-0.410 (0.0154-0.0161)	22	24	26	28	30	32	34	36	38	38	40	40	42	42	44	44	46	46	48	48
0.411-0.430 (0.0162-0.0169)	24	26	28	30	32	34	36	38	40	40	42	42	44	44	46	46	48	48	50	50
0.431-0.450 (0.0170-0.0177)	26	28	30	32	34	36	38	40	42	42	44	44	46	46	48	48	50	50	52	52
0.451-0.470 (0.0178-0.0185)	28	30	32	34	36	38	40	42	44	44	46	46	48	48	50	50	52	52	54	54
0.471-0.490 (0.0185-0.0193)	30	32	34	36	38	40	42	44	46	46	48	48	50	50	52	52	54	54	56	56
0.491-0.510 (0.0193-0.0201)	32	34	36	38	40	42	44	46	48	48	50	50	52	52	54	54	56	56	58	58
0.511-0.530 (0.0201-0.0209)	34	36	38	40	42	44	46	48	50	50	52	52	54	54	56	56	58	58	60	60
0.531-0.550 (0.0209-0.0217)	36	38	40	42	44	46	48	50	52	52	54	54	56	56	58	58	60	60	62	62
0.551-0.570 (0.0217-0.0224)	38	40	42	44	46	48	50	52	54	54	56	56	58	58	60	60	62	62	64	64
0.571-0.590 (0.0225-0.0232)	40	42	44	46	48	50	52	54	56	56	58	58	60	60	62	62	64	64	66	66
0.591-0.610 (0.0233-0.0240)	42	44	46	48	50	52	54	56	58	58	60	60	62	62	64	64	66	66	68	68
0.611-0.630 (0.0241-0.0248)	44	46	48	50	52	54	56	58	60	60	62	62	64	64	66	66	68	68	70	70
0.631-0.650 (0.0248-0.0256)	46	48	50	52	54	56	58	60	62	62	64	64	66	66	68	68	70	70	72	72
0.651-0.670 (0.0256-0.0264)	48	50	52	54	56	58	60	62	64	64	66	66	68	68	70	70	72	72	74	74
0.671-0.690 (0.0264-0.0272)	50	52	54	56	58	60	62	64	66	66	68	68	70	70	72	72	74	74	74	74
0.691-0.710 (0.0272-0.0280)	52	54	56	58	60	62	64	66	68	68	70	70	72	72	74	74	74	74		
0.711-0.730 (0.0280-0.0287)	54	56	58	60	62	64	66	68	70	70	72	72	74	74	74	74				
0.731-0.750 (0.0288-0.0295)	56	58	60	62	64	66	68	70	72	72	74	74	74	74						
0.751-0.770 (0.0296-0.0303)	58	60	62	64	66	68	70	72	74	74	74	74								
0.771-0.790 (0.0304-0.0311)	60	62	64	66	68	70	72	74	74	74										
0.791-0.810 (0.0311-0.0319)	62	64	66	68	70	72	74													
0.811-0.830 (0.0319-0.0327)	64	66	68	70	72	74														
0.831-0.850 (0.0327-0.0335)	66	68	70	72	74															
0.851-0.870 (0.0335-0.0343)	68	70	72	74																
0.871-0.890 (0.0343-0.0350)	70	72	74																	
0.891-0.910 (0.0351-0.0358)	72	74																		
0.911-0.930 (0.0359-0.0366)	74																			

Installed lifter thickness mm (in.) / Measure clearance mm (in.)	5.330 (0.2098)	5.340 (0.2102)	5.350 (0.2106)	5.360 (0.2110)	5.370 (0.2114)	5.380 (0.2118)	5.390 (0.2122)	5.400 (0.2126)	5.410 (0.2130)	5.420 (0.2134)	5.430 (0.2138)	5.440 (0.2142)	5.450 (0.2146)	5.460 (0.2150)	5.470 (0.2154)	5.480 (0.2157)	5.490 (0.2161)	5.500 (0.2165)	5.510 (0.2169)	5.520 (0.2173)	5.530 (0.2177)	5.540 (0.2181)	5.550 (0.2185)	5.560 (0.2189)	5.570 (0.2193)	5.580 (0.2197)	5.590 (0.2201)	5.600 (0.2205)	5.620 (0.2213)	5.640 (0.2220)	5.660 (0.2228)	5.680 (0.2236)	5.700 (0.2244)	5.720 (0.2252)	5.740 (0.2260)
0.000-0.030 (0.0000-0.0012)	14	14	16	16	18	18	20	20	22	22	24	24	26	26	28	28	30	30	32	32	34	34	36	36	38	38	40	40	42	44	46	48	50	52	54
0.031-0.050 (0.0012-0.0020)	16	18	18	20	20	22	22	24	24	26	26	28	28	30	30	32	32	34	34	36	36	38	38	40	40	42	42	44	46	48	50	52	54	56	58
0.051-0.070 (0.0020-0.0028)	18	20	20	22	22	24	24	26	26	28	28	30	30	32	32	34	34	36	36	38	38	40	40	42	42	44	44	46	48	50	52	54	56	58	60
0.071-0.090 (0.0028-0.0035)	20	22	22	24	24	26	26	28	28	30	30	32	32	34	34	36	36	38	38	40	40	42	42	44	44	46	46	48	50	52	54	56	58	60	62
0.091-0.110 (0.0036-0.0043)	22	24	24	26	26	28	28	30	30	32	32	34	34	36	36	38	38	40	40	42	42	44	44	46	46	48	48	50	52	54	56	58	60	62	64
0.111-0.130 (0.0044-0.0051)	24	26	26	28	28	30	30	32	32	34	34	36	36	38	38	40	40	42	42	44	44	46	46	48	48	50	50	52	54	56	58	60	62	64	66
0.131-0.150 (0.0052-0.0059)	26	28	28	30	30	32	32	34	34	36	36	38	38	40	40	42	42	44	44	46	46	48	48	50	50	52	52	54	56	58	60	62	64	66	68
0.151-0.170 (0.0059-0.0067)	28	30	30	32	32	34	34	36	36	38	38	40	40	42	42	44	44	46	46	48	48	50	50	52	52	54	54	56	58	60	62	64	66	68	70
0.171-0.189 (0.0067-0.0074)	30	32	32	34	34	36	36	38	38	40	40	42	42	44	44	46	46	48	48	50	50	52	52	54	54	56	56	58	60	62	64	66	68	70	72
0.190-0.290 (0.0075-0.0114)																																			
0.291-0.310 (0.0115-0.0122)	40	40	42	42	44	44	46	46	48	48	50	50	52	52	54	54	56	56	58	58	60	60	62	62	64	64	66	66	68	70	72	74	74	74	
0.311-0.330 (0.0122-0.0130)	42	42	44	44	46	46	48	48	50	50	52	52	54	54	56	56	58	58	60	60	62	62	64	64	66	66	68	68	70	72	74	74	74		
0.331-0.350 (0.0130-0.0138)	44	44	46	46	48	48	50	50	52	52	54	54	56	56	58	58	60	60	62	62	64	64	66	66	68	68	70	70	72	74	74	74			
0.351-0.370 (0.0138-0.0146)	46	46	48	48	50	50	52	52	54	54	56	56	58	58	60	60	62	62	64	64	66	66	68	68	70	70	72	72	74	74	74				
0.371-0.390 (0.0146-0.0154)	48	48	50	50	52	52	54	54	56	56	58	58	60	60	62	62	64	64	66	66	68	68	70	70	72	72	74	74	74	74					
0.391-0.410 (0.0154-0.0161)	50	50	52	52	54	54	56	56	58	58	60	60	62	62	64	64	66	66	68	68	70	70	72	72	74	74	74	74							
0.411-0.430 (0.0162-0.0169)	52	52	54	54	56	56	58	58	60	60	62	62	64	64	66	66	68	68	70	70	72	72	74	74	74	74									
0.431-0.450 (0.0170-0.0177)	54	54	56	56	58	58	60	60	62	62	64	64	66	66	68	68	70	70	72	72	74	74	74	74											
0.451-0.470 (0.0178-0.0185)	56	56	58	58	60	60	62	62	64	64	66	66	68	68	70	70	72	72	74	74	74	74													
0.471-0.490 (0.0185-0.0193)	58	58	60	60	62	62	64	64	66	66	68	68	70	70	72	72	74	74	74	74															
0.491-0.510 (0.0193-0.0201)	60	60	62	62	64	64	66	66	68	68	70	70	72	72	74	74	74	74																	
0.511-0.530 (0.0201-0.0209)	62	62	64	64	66	66	68	68	70	70	72	72	74	74	74	74																			
0.531-0.550 (0.0209-0.0217)	64	64	66	66	68	68	70	70	72	72	74	74	74	74																					
0.551-0.570 (0.0217-0.0224)	66	66	68	68	70	70	72	72	74	74	74	74																							
0.571-0.590 (0.0225-0.0232)	68	68	70	70	72	72	74	74	74	74																									
0.591-0.610 (0.0233-0.0240)	70	70	72	72	74	74	74	74																											
0.611-0.630 (0.0241-0.0248)	72	72	74	74	74	74																													
0.631-0.650 (0.0248-0.0256)	74	74	74	74																															
0.651-0.670 (0.0256-0.0264)	74	74																																	

New Lifter Thickness mm (in.)

Lifter No.	Thickness	Lifter No.	Thickness	Lifter No.	Thickness
06	5.060 (0.1992)	30	5.300 (0.2087)	54	5.540 (0.2181)
08	5.080 (0.2000)	32	5.320 (0.2094)	56	5.560 (0.2189)
10	5.100 (0.2008)	34	5.340 (0.2102)	58	5.580 (0.2197)
12	5.120 (0.2016)	36	5.360 (0.2110)	60	5.600 (0.2205)
14	5.140 (0.2024)	38	5.380 (0.2118)	62	5.620 (0.2213)
16	5.160 (0.2031)	40	5.400 (0.2126)	64	5.640 (0.2220)
18	5.180 (0.2039)	42	5.420 (0.2134)	66	5.660 (0.2228)
20	5.200 (0.2047)	44	5.440 (0.2142)	68	5.680 (0.2236)
22	5.220 (0.2055)	46	5.460 (0.2150)	70	5.700 (0.2244)
24	5.240 (0.2063)	48	5.480 (0.2157)	72	5.720 (0.2252)
26	5.260 (0.2071)	50	5.500 (0.2165)	74	5.740 (0.2260)
28	5.280 (0.2079)	52	5.520 (0.2173)		

Intake valve clearance (Cold):
0.19 to 0.29 mm (0.008 to 0.011 in.)
EXAMPLE:
The 5.250 mm (0.2067 in.) lifter is installed, and the measured clearance is 0.400 mm (0.0157 in.).
Replace the 5.250 mm (0.2067 in.) lifter with a new No. 42 lifter.

67170-RAV4-G10

Intake valve lifter size selection chart—2.4L engine

Valve Lifter Selection Chart (Exhaust)

Measure clearance mm (in.) \ Installed lifter thickness mm (in.)	5.060 (0.1992)	5.080 (0.2000)	5.100 (0.2008)	5.120 (0.2016)	5.140 (0.2024)	5.160 (0.2031)	5.180 (0.2039)	5.200 (0.2047)	5.210 (0.2051)	5.220 (0.2055)	5.230 (0.2059)	5.240 (0.2063)	5.250 (0.2067)	5.260 (0.2071)	5.270 (0.2075)	5.280 (0.2079)	5.290 (0.2083)	5.300 (0.2087)	5.310 (0.2091)	5.320 (0.2094)	5.330 (0.2098)	5.340 (0.2102)	5.350 (0.2106)	5.360 (0.2110)	5.370 (0.2114)	5.380 (0.2118)	5.390 (0.2122)	5.400 (0.2126)	5.410 (0.2130)	5.420 (0.2134)	5.430 (0.2138)	5.440 (0.2142)	5.450 (0.2146)	5.460 (0.2150)	5.470 (0.2154)	5.480 (0.2157)	5.490 (0.2161)	5.500 (0.2165)	5.510 (0.2169)	5.520 (0.2173)	5.530 (0.2177)	5.540 (0.2181)	5.550 (0.2185)	5.560 (0.2189)	5.570 (0.2193)	5.580 (0.2197)	5.590 (0.2201)	5.600 (0.2205)	5.620 (0.2213)	5.640 (0.2220)	5.660 (0.2228)	5.680 (0.2236)	5.700 (0.2244)	5.720 (0.2252)	5.740 (0.2260)
0.000–0.030 (0.0000–0.0012)																							06	06	08	08	10	10	12	12	14	14	16	16	18	18	20	20	22	22	24	24	26	26	28	28	30	30	32	34	36	38	40	42	44
0.031–0.050 (0.0012–0.0020)																					06	06	08	08	10	10	12	12	14	14	16	16	18	18	20	20	22	22	24	24	26	26	28	28	30	30	32	32	34	36	38	40	42	44	46
0.051–0.070 (0.0020–0.0028)																			06	06	08	08	10	10	12	12	14	14	16	16	18	18	20	20	22	22	24	24	26	26	28	28	30	30	32	32	34	34	36	38	40	42	44	46	48
0.071–0.090 (0.0028–0.0035)																	06	06	08	08	10	10	12	12	14	14	16	16	18	18	20	20	22	22	24	24	26	26	28	28	30	30	32	32	34	34	36	36	38	40	42	44	46	48	50
0.091–0.110 (0.0036–0.0043)															06	06	08	08	10	10	12	12	14	14	16	16	18	18	20	20	22	22	24	24	26	26	28	28	30	30	32	32	34	34	36	36	38	38	40	42	44	46	48	50	52
0.111–0.130 (0.0044–0.0051)													06	06	08	08	10	10	12	12	14	14	16	16	18	18	20	20	22	22	24	24	26	26	28	28	30	30	32	32	34	34	36	36	38	38	40	40	42	44	46	48	50	52	54
0.131–0.150 (0.0052–0.0059)											06	06	08	08	10	10	12	12	14	14	16	16	18	18	20	20	22	22	24	24	26	26	28	28	30	30	32	32	34	34	36	36	38	38	40	40	42	42	44	46	48	50	52	54	56
0.151–0.170 (0.0059–0.0067)									06	06	08	08	10	10	12	12	14	14	16	16	18	18	20	20	22	22	24	24	26	26	28	28	30	30	32	32	34	34	36	36	38	38	40	40	42	42	44	44	46	48	50	52	54	56	58
0.171–0.190 (0.0067–0.0075)								06	08	08	10	10	12	12	14	14	16	16	18	18	20	20	22	22	24	24	26	26	28	28	30	30	32	32	34	34	36	36	38	38	40	40	42	42	44	44	46	46	48	50	52	54	56	58	60
0.191–0.210 (0.0075–0.0083)							06	08	10	10	12	12	14	14	16	16	18	18	20	20	22	22	24	24	26	26	28	28	30	30	32	32	34	34	36	36	38	38	40	40	42	42	44	44	46	46	48	48	50	52	54	56	58	60	62
0.211–0.230 (0.0083–0.0091)						06	08	10	12	12	14	14	16	16	18	18	20	20	22	22	24	24	26	26	28	28	30	30	32	32	34	34	36	36	38	38	40	40	42	42	44	44	46	46	48	48	50	50	52	54	56	58	60	62	64
0.231–0.250 (0.0091–0.0098)					06	08	10	12	14	14	16	16	18	18	20	20	22	22	24	24	26	26	28	28	30	30	32	32	34	34	36	36	38	38	40	40	42	42	44	44	46	46	48	48	50	50	52	52	54	56	58	60	62	64	66
0.251–0.270 (0.0099–0.0106)				06	08	10	12	14	16	16	18	18	20	20	22	22	24	24	26	26	28	28	30	30	32	32	34	34	36	36	38	38	40	40	42	42	44	44	46	46	48	48	50	50	52	52	54	54	56	58	60	62	64	66	68
0.271–0.290 (0.0107–0.0114)			06	08	10	12	14	16	18	18	20	20	22	22	24	24	26	26	28	28	30	30	32	32	34	34	36	36	38	38	40	40	42	42	44	44	46	46	48	48	50	50	52	52	54	54	56	56	58	60	62	64	66	68	70
0.291–0.299 (0.0115–0.0118)		06	08	10	12	14	16	18	20	20	22	22	24	24	26	26	28	28	30	30	32	32	34	34	36	36	38	38	40	40	42	42	44	44	46	46	48	48	50	50	52	52	54	54	56	56	58	58	60	62	64	66	68	70	72
0.300–0.400 (0.0118–0.0157)																																																							
0.401–0.420 (0.0158–0.0165)	12	14	16	18	20	22	24	26	26	28	28	30	30	32	32	34	34	36	36	38	38	40	40	42	42	44	44	46	46	48	48	50	50	52	52	54	54	56	56	58	58	60	60	62	62	64	64	66	68	70	72	74	74	74	
0.421–0.440 (0.0166–0.0173)	14	16	18	20	22	24	26	28	28	30	30	32	32	34	34	36	36	38	38	40	40	42	42	44	44	46	46	48	48	50	50	52	52	54	54	56	56	58	58	60	60	62	62	64	64	66	66	68	70	72	74	74	74		
0.441–0.460 (0.0174–0.0181)	16	18	20	22	24	26	28	30	30	32	32	34	34	36	36	38	38	40	40	42	42	44	44	46	46	48	48	50	50	52	52	54	54	56	56	58	58	60	60	62	62	64	64	66	66	68	68	70	72	74	74	74			
0.461–0.480 (0.0181–0.0189)	18	20	22	24	26	28	30	32	32	34	34	36	36	38	38	40	40	42	42	44	44	46	46	48	48	50	50	52	52	54	54	56	56	58	58	60	60	62	62	64	64	66	66	68	68	70	70	72	74	74	74				
0.481–0.500 (0.0189–0.0197)	20	22	24	26	28	30	32	34	34	36	36	38	38	40	40	42	42	44	44	46	46	48	48	50	50	52	52	54	54	56	56	58	58	60	60	62	62	64	64	66	66	68	68	70	70	72	72	74	74	74					
0.501–0.520 (0.0197–0.0205)	22	24	26	28	30	32	34	36	36	38	38	40	40	42	42	44	44	46	46	48	48	50	50	52	52	54	54	56	56	58	58	60	60	62	62	64	64	66	66	68	68	70	70	72	72	74	74	74	74						
0.521–0.540 (0.0205–0.0213)	24	26	28	30	32	34	36	38	38	40	40	42	42	44	44	46	46	48	48	50	50	52	52	54	54	56	56	58	58	60	60	62	62	64	64	66	66	68	68	70	70	72	72	74	74	76	76	74							
0.541–0.560 (0.0213–0.0220)	26	28	30	32	34	36	38	40	40	42	42	44	44	46	46	48	48	50	50	52	52	54	54	56	56	58	58	60	60	62	62	64	64	66	66	68	68	70	70	72	72	74	74	74	74	74	74								
0.561–0.580 (0.0221–0.0228)	28	30	32	34	36	38	40	42	42	44	44	46	46	48	48	50	50	52	52	54	54	56	56	58	58	60	60	62	62	64	64	66	66	68	68	70	70	72	72	74	74	74	74	74	74										
0.581–0.600 (0.0229–0.0236)	30	32	34	36	38	40	42	44	44	46	46	48	48	50	50	52	52	54	54	56	56	58	58	60	60	62	62	64	64	66	66	68	68	70	70	72	72	74	74	74	74	74	74												
0.601–0.620 (0.0237–0.0244)	32	34	36	38	40	42	44	46	46	48	48	50	50	52	52	54	54	56	56	58	58	60	60	62	62	64	64	66	66	68	68	70	70	72	72	74	74	74	74	74	74														
0.621–0.640 (0.0244–0.0252)	34	36	38	40	42	44	46	48	48	50	50	52	52	54	54	56	56	58	58	60	60	62	62	64	64	66	66	68	68	70	70	72	72	74	74	74	74	74	74																
0.641–0.660 (0.0252–0.0260)	36	38	40	42	44	46	48	50	50	52	52	54	54	56	56	58	58	60	60	62	62	64	64	66	66	68	68	70	70	72	72	74	74	74	74	74	74																		
0.661–0.680 (0.0260–0.0268)	38	40	42	44	46	48	50	52	52	54	54	56	56	58	58	60	60	62	62	64	64	66	66	68	68	70	70	72	72	74	74	74	74	74	74																				
0.681–0.700 (0.0268–0.0276)	40	42	44	46	48	50	52	54	54	56	56	58	58	60	60	62	62	64	64	66	66	68	68	70	70	72	72	74	74	74	74	74	74																						
0.701–0.720 (0.0276–0.0283)	42	44	46	48	50	52	54	56	56	58	58	60	60	62	62	64	64	66	66	68	68	70	70	72	72	74	74	74	74	74	74																								
0.721–0.740 (0.0284–0.0291)	44	46	48	50	52	54	56	58	58	60	60	62	62	64	64	66	66	68	68	70	70	72	72	74	74	74	74	74	74																										
0.741–0.760 (0.0292–0.0299)	46	48	50	52	54	56	58	60	60	62	62	64	64	66	66	68	68	70	70	72	72	74	74	74	74	74	74																												
0.761–0.780 (0.0300–0.0307)	48	50	52	54	56	58	60	62	62	64	64	66	66	68	68	70	70	72	72	74	74	74	74	74	74																														
0.781–0.800 (0.0307–0.0315)	50	52	54	56	58	60	62	64	64	66	66	68	68	70	70	72	72	74	74	74	74	74	74																																
0.801–0.820 (0.0315–0.0323)	52	54	56	58	60	62	64	66	66	68	68	70	70	72	72	74	74	74	74	74	74																																		
0.821–0.840 (0.0323–0.0331)	54	56	58	60	62	64	66	68	68	70	70	72	72	74	74	74	74	74	74																																				
0.841–0.860 (0.0331–0.0339)	56	58	60	62	64	66	68	70	70	72	72	74	74	74	74	74	74																																						
0.861–0.880 (0.0339–0.0346)	58	60	62	64	66	68	70	72	72	74	74	74	74	74	74																																								
0.881–0.900 (0.0347–0.0354)	60	62	64	66	68	70	72	74	74	74	74	74	74																																										
0.901–0.920 (0.0355–0.0362)	62	64	66	68	70	72	74	74	74	74	74																																												
0.921–0.940 (0.0363–0.0370)	64	66	68	70	72	74	74	74	74																																														
0.941–0.960 (0.0370–0.0378)	66	68	70	72	74	74	74																																																
0.961–0.980 (0.0378–0.0386)	68	70	72	74	74	74																																																	
0.981–1.000 (0.0386–0.0394)	70	72	74	74	74																																																		
1.001–1.020 (0.0394–0.0402)	72	74	74	74																																																			
1.021–1.040 (0.0402–0.0409)	74	74	74																																																				
1.041–1.060 (0.0410–0.0417)	74	74																																																					
1.061–1.080 (0.0418–0.0425)	74																																																						

Exhaust valve clearance (Cold):
0.30 to 0.40 mm (0.012 to 0.016 in.)
EXAMPLE:
The 5.340 mm (0.2102 in.) lifter is installed, and the measured clearance is 0.430 mm (0.0169 in.).
Replace the 5.340 mm (0.2102 in.) lifter with a new No. 42 lifter.

New Lifter Thickness mm (in.)

Lifter No.	Thickness	Lifter No.	Thickness	Lifter No.	Thickness
06	5.060 (0.1992)	30	5.300 (0.2087)	54	5.540 (0.2181)
08	5.080 (0.2000)	32	5.320 (0.2094)	56	5.560 (0.2189)
10	5.100 (0.2008)	34	5.340 (0.2102)	58	5.580 (0.2197)
12	5.120 (0.2016)	36	5.360 (0.2110)	60	5.600 (0.2205)
14	5.140 (0.2024)	38	5.380 (0.2118)	62	5.620 (0.2213)
16	5.160 (0.2031)	40	5.400 (0.2126)	64	5.640 (0.2220)
18	5.180 (0.2039)	42	5.420 (0.2134)	66	5.660 (0.2228)
20	5.200 (0.2047)	44	5.440 (0.2142)	68	5.680 (0.2236)
22	5.220 (0.2055)	46	5.460 (0.2150)	70	5.700 (0.2244)
24	5.240 (0.2063)	48	5.480 (0.2157)	72	5.720 (0.2252)
26	5.260 (0.2071)	50	5.500 (0.2165)	74	5.740 (0.2260)
28	5.280 (0.2079)	52	5.520 (0.2173)		

67170-RAV4-G11

Exhaust valve lifter size selection chart—2.4L engine

Starter Motor

REMOVAL & INSTALLATION

2.0L RAV4

1. Before servicing the vehicle, refer to the precautions in the beginning of this section.
2. Remove the engine coolant reservoir.
3. Remove the air cleaner cap assembly, as follows:
 a. Disconnect the following:
 - Skid control relay connectors
 - High tension cord from the air cleaner hose and resonator
 - Intake Air Temperature (IAT) sensor connector
 - Positive Crankcase Ventilation (PCV) hose from the air cleaner hose
 - Air hose from the air cleaner cap assembly

 b. The air cleaner cap assembly from the air cleaner case assembly by removing the 4 clamps.
 c. Loosen the hose clamp and disconnect the air cleaner hose from the throttle body.
 d. Remove the air cleaner cap assembly.
4. Remove or disconnect the following:
 - Vacuum Switching Valve (VSV) from the air cleaner case assembly
 - 3 bolts and the air cleaner case assembly
 - Starter electrical connectors
 - Both bolts and the starter

To install:

5. Install or connect the following:
 - Starter. Tighten the bolts to 29 ft. lbs. (38 Nm).
 - Starter electrical connectors
 - Air cleaner case assembly
 - VSV to the air cleaner case assembly
6. Install or connect the air cleaner cap assembly, as follows:
 a. Install the air cleaner cap assembly.
 b. Connect the air cleaner hose to the throttle body and tighten the hose clamp.
 c. Install the air cleaner cap assembly to the air cleaner case assembly by installing the 4 clamps.
 d. Connect the following:
 - Air hose to the air cleaner cap assembly
 - PCV hose to the air cleaner hose
 - IAT sensor connector
 - High tension cord to the air cleaner hose and resonator
 - Skid control relay connectors
7. Install the engine coolant reservoir.

2.4L Engine

1. Before servicing the vehicle, refer to the precautions in the beginning of this section.
2. Disconnect the hose from the transmission level gauge.
3. Disconnect the wires from the starter.
4. Remove the 2 bolts and lift out the starter.
5. Installation is the reverse of removal. Torque the bolts to 27 ft. lbs. (37 Nm).

Oil Pan

REMOVAL & INSTALLATION

2.0L Engine

1. Before servicing the vehicle, refer to the precautions in the beginning of this section.
2. Remove the right-hand engine undercover.
3. Drain the crankcase oil.
4. Remove or disconnect the following:
 - Dipstick
 - Front exhaust pipe
 - Stiffener plate from the engine by removing the 2 (manual transmission) or 3 (automatic transmission) bolts.
 - 2 nuts and 17 bolts from the oil pan
 - Oil pan and discard the gasket

To install:

5. Clean all gasket surfaces completely.
6. Apply a thin bead of sealer to the oil pan mounting surfaces.
7. Install or connect the following:
 - Oil pan. Tighten the nuts/bolts to 48 inch lbs. (5 Nm).
 - Stiffener plate. Tighten the bolts to 27 ft. lbs. (37 Nm).
 - Front exhaust pipe
8. Fill the engine with oil to the proper level.
9. Start the engine and check for leaks. Recheck the engine oil level.
10. Install the right engine cover.

2.4L Engine

1. Before servicing the vehicle, refer to the precautions in the beginning of this section.
2. Remove the engine undercover.
3. Drain the engine oil.
4. Disconnect the exhaust pipe.
5. Remove the oil pan bolts, nuts and pan.
6. Installation is the reverse of removal. Clean the gasket mating surfaces. Always use new gasket material. The new RTV material should be a bead about 4mm in diameter. Parts must be assembled within 5 minutes. Torque the bolts and nuts to 80 inch lbs. (9Nm).

Oil Pump

REMOVAL & INSTALLATION

2.0L Engine

1. Before servicing the vehicle, refer to the precautions in the beginning of this section.
2. Remove or disconnect the following:
 - Negative battery cable
 - Hood
 - Right-hand engine undercover
3. Drain the engine oil.
4. Remove or disconnect the following:
 - Front exhaust pipe
 - Rear end stiffener plate
 - Oil dipstick
 - 17 bolts and 2 nuts from the oil pan
5. Insert the blade of the Oil Pan Seal Cutting tool 09032-00100 between the oil pan and the cylinder block; then, cut off the applied sealer and remove the oil pan

➡Do not use the tool for the oil pump body side and rear oil seal retainer.

6. Remove the bolts, nuts, oil strainer and gasket.
7. Carefully suspend the engine with a sling device.
8. Remove or disconnect the following:
 - Timing belt
 - No. 2 idler pulley and crankshaft timing pulley
 - Oil pump's pulley, using the Variable Pin Wrench Set 09960-10010
 - Crankshaft Position (CKP) sensor
 - Oil pump, by discarding the gasket

To install:

9. Install or connect the following:
 - Oil pump, using a new gasket. Tighten the 12 bolts to 82 inch lbs. (9 Nm).

➡The long bolts are 35mm and all the others are 25mm.

 - CKP sensor
 - Oil pump pulley. Tighten the nut to 18 ft. lbs. (24 Nm).

- Crankshaft timing pulley and No. 2 idler pulley
- Timing belt

10. Remove the engine sling.
11. Install the oil strainer with a new gasket. Tighten the nuts/bolts to 48 inch lbs. (5 Nm).
12. Remove any old sealant from the oil pan flange and thoroughly clean both sealing surfaces.
13. Apply a 3–5mm bead of sealant to the oil pan flange.

➡The pan must be installed within 5 minutes of sealant application or the procedure will have to be repeated.

14. Install or connect the following:
    - Oil pan. Tighten the 17 bolts and 2 nuts to 48 inch lbs. (5 Nm).
    - Dipstick
    - Rear end stiffener plate. Tighten the bolts to 27 ft. lbs. (37 Nm).
    - Front exhaust pipe
    - Negative battery cable
    - Hood
15. Refill the engine with oil.

⁂ WARNING

Be sure to prime the oil pump prior to initial engine start-up or engine damage may occur because of low oil pressure.

16. Start the engine and check for leaks.
17. Recheck the engine oil level.
18. Install the right-hand engine undercover.

2.4L Engine

REMOVAL

1. Drain the oil.
2. Remove the oil pan.
3. Remove the air cleaner assembly.
4. Remove the ABS actuator.
5. Remove the right engine mount insulator.
6. Remove the timing chain.
7. Remove the crankshaft sprocket.
8. Remove the oil pump drive chain and sprockets.
9. Remove the 3 bolts and the oil pump.

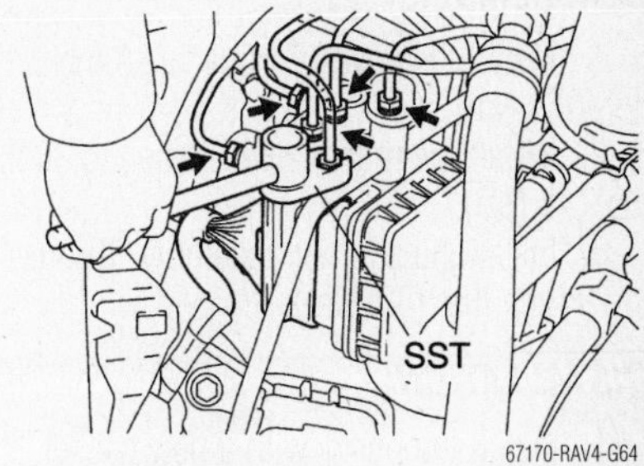

ABS actuator removal

INSPECTION

1. Remove the relief valve spring.
2. Remove the pump body cover.
3. Check the clearance between the tip of the drive rotor and the driven rotor. Clearance should be 0.0138 inch max.
4. Place a straight-edge across the rotors and pump body. Check the side clearance. Clearance should be 0.0063 inch max.
5. Check the clearance between the driven rotor and pump body. Clearance should be 0.0128 inch max.

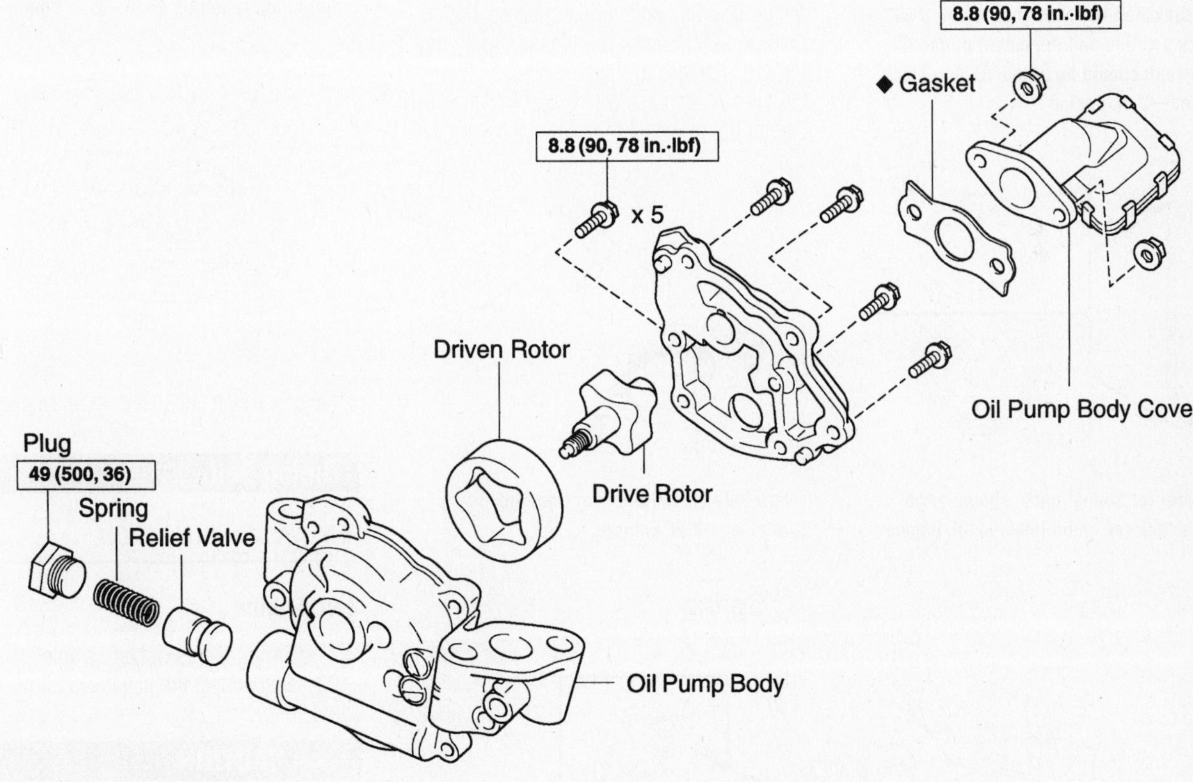

Oil pump—2.4L engine

REASSEMBLY

1. Install the cover. Torque to 78 inch lbs. (8.8 Nm).
2. Install the relief valve. Torque to 36 ft. lbs. (49 Nm).
3. Install the strainer assembly. Torque to 78 inch lbs. (8.8 Nm).

INSTALLATION

1. Install the pump with a new gasket. Torque to 14 ft. lbs. (19 Nm).
2. Install the drive chain and sprockets. The crankshaft key should be at the 9:00 o'clock position and the cutout on the oil pump shaft should be at the 12:00 o'clock position. The sprocket timing marks should align with the colored chain links. Torque the sprocket nut to 22 ft. lbs. (29 Nm).
3. Install the tensioner. Torque to 9 ft. lbs. (12 Nm).
4. Install the oil pan.
5. The remainder of installation is the reverse of removal. Torque the exhaust pipe-to-manifold nuts to 32 ft. lbs. (43 Nm); the pip-to-pipe nuts to 36 ft. lbs. (49 Nm).

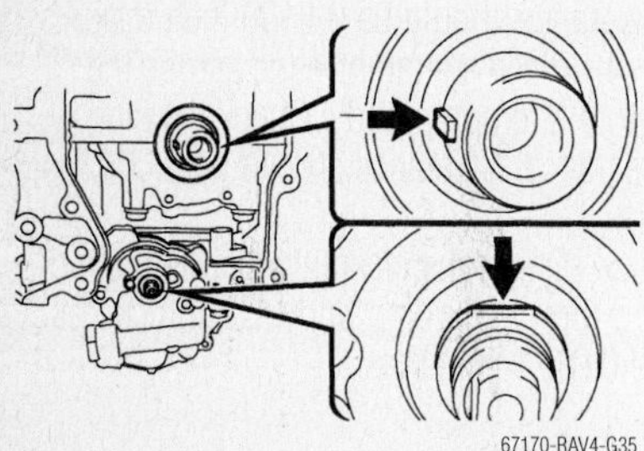

67170-RAV4-G35

The crankshaft key should be at the 9:00 o'clock position and the cutout on the oil pump shaft should be at the 12:00 o'clock position—2.4L engine

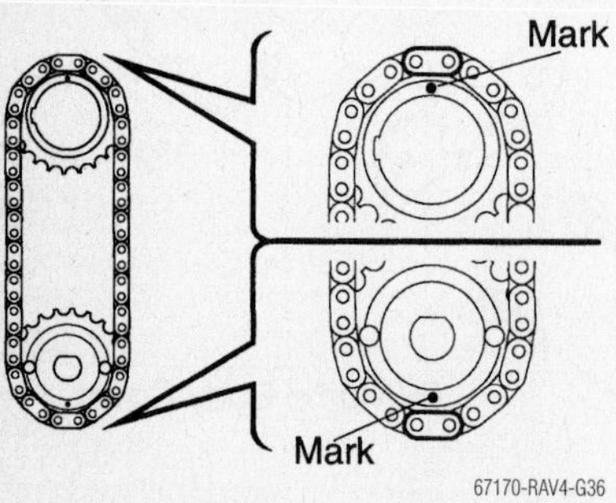

67170-RAV4-G36

The sprocket timing marks should align with the colored chain links—2.4L engine

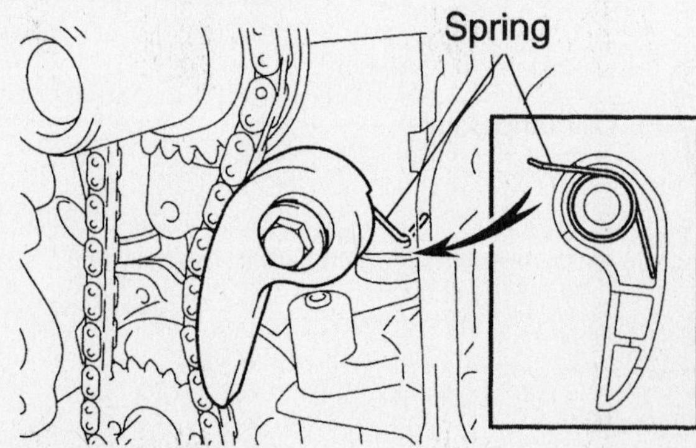

67170-RAV4-G37

Tension spring positioning—2.4L engine

Rear Main Seal

REMOVAL & INSTALLATION

If the rear oil seal retainer is not installed to the block, use a tapered ended screw-driver and hammer to remove the oil seal. Apply multi-purpose grease to the new oil seal lip. Using a seal driver, tap the seal into place. Be careful not to install it slantwise.

1. Before servicing the vehicle, refer to the precautions in the beginning of this section.

If the rear oil seal retainer is installed on the cylinder block, using a knife, cut off the lip of the seal. Using a taped ended prytool, pry the old seal out of the retainer. Inspect the oil seal lip contacting surface of the crankshaft for cracks or damage. Apply multipurpose grease to the new oil seal, then tap the seal in place with a seal installer. Be careful not to install the seal slantwise.

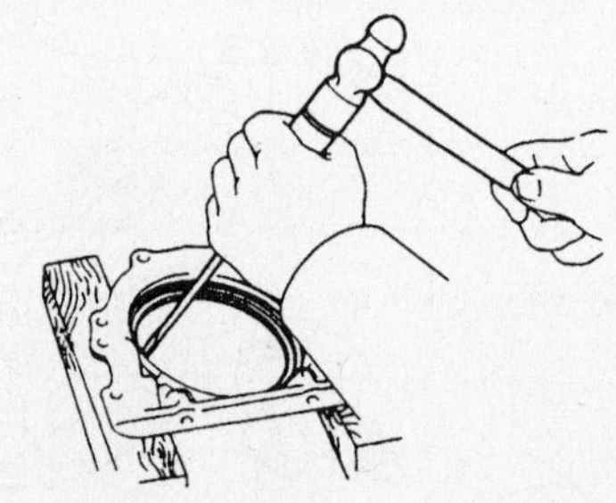

7924ZG55

Carefully tap the old seal from the retainer—2.0L Engines

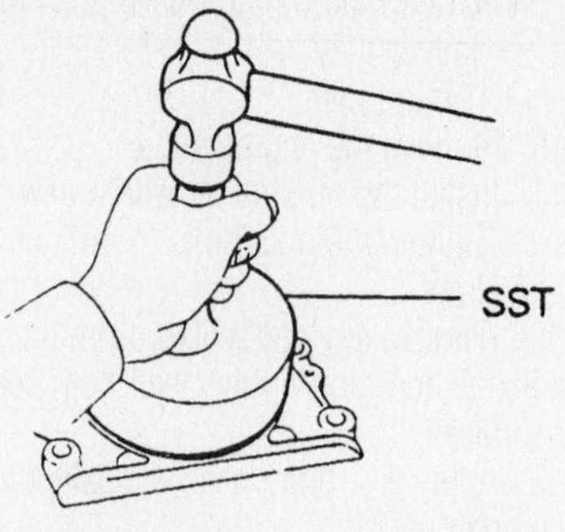

7924ZG56

Use the proper sized driver to seat the seal—2.0L Engines

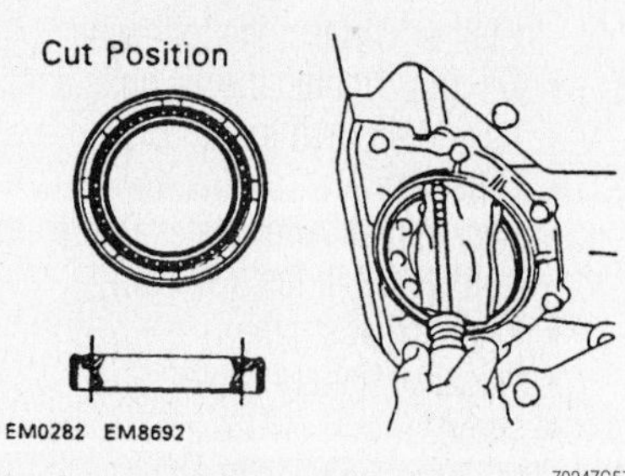

7924ZG57

Cut off the oil seal lip, then pry the seal out of the retaining plate—2.0L Engines

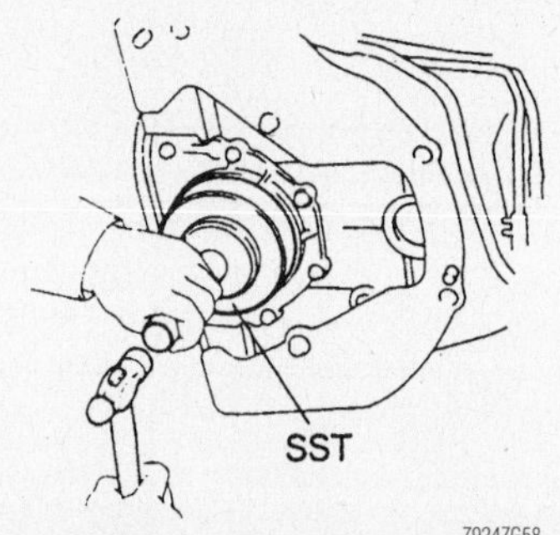

7924ZG58

Tap a new seal into place—2.0L Engines

Timing Belt

REMOVAL & INSTALLATION

2.0L Engine

The timing belt is not adjustable.

1. Disconnect the negative battery cable.

**** CAUTION**

To avoid air bag deployment, if equipped, work must be started after approximately 90 seconds or longer from the time the ignition switch is turned to the LOCK position and the negative battery cable is disconnected from the battery.

2. Disconnect the power steering reservoir tank and remove the reservoir bracket.

3. Detach the wiring harness bracket for the Data Link Connector 1 (DLC1).

4. Remove the alternator and alternator bracket.

5. If equipped with ABS brakes, remove the ABS actuator.

6. Remove the right front wheel and the fender apron seal.

7. Remove the power steering drive belt.

8. Slightly raise the engine using a block of wood and floor jack under the oil pan to prevent damage.

9. Remove the 4 bolts, 2 nuts, and right-hand mounting bracket.

10. Remove the spark plugs.

11. Using SST 09213-54015, or equivalent, loosen the crankshaft pulley bolt and remove it by pulling it straight off the crankshaft.

12. Using SST 09249-63010, or equivalent, loosen the retaining bolts and remove the right engine mounting bracket.

13. Remove the upper (No. 2) timing belt cover.

14. Install the crankshaft pulley to the crankshaft and temporarily install the retaining bolt.

15. Turn the crankshaft pulley and align its groove with the timing mark **0** of the No. 1 timing belt cover. Check that the hole of the camshaft timing pulley is aligned with the timing mark of the bearing cap. If not, turn the crankshaft 360 degrees and align the marks.

➡If the timing belt is to be reused, matchmark the timing belt to the timing pulleys and timing belt covers so the belt can be reinstalled in its original position. Also, be sure to mark an arrow on the belt to indicate which direction it was turning.

16. Remove the timing belt from the camshaft timing pulley.

17. Hold the camshaft sprocket with a spanner wrench and remove the mounting bolt. Remove the camshaft pulley.

18. Remove the crankshaft pulley bolt and remove the crankshaft pulley.

19. Remove the No. 1 timing belt cover.

20. Remove the timing belt guide and the timing belt.

21. Remove the No. 1 idler pulley and tension spring.

22. Remove the No. 2 idler pulley.

23. Remove the crankshaft timing pulley.

24. Support the oil pump sprocket with a spanner wrench, then remove the mounting bolt and remove the sprocket.

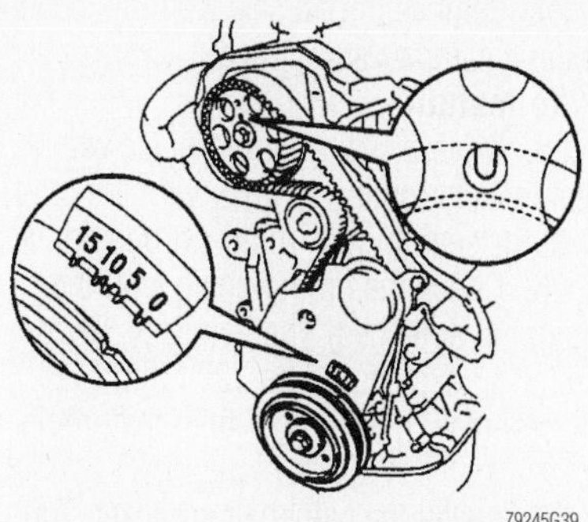

It is necessary to align the timing reference indicators prior to removing the timing belt—2.0L engine

To install:

25. Install the oil pump pulley. Tighten the nut to 18 ft. lbs. (24 Nm).

26. Install the crankshaft timing pulley. Align the pulley set key with the key groove of the pulley. Slide on the pulley facing the flange side inward.

27. Install the No. 2 idler pulley and tighten the mounting bolt to 31 ft. lbs. (42 Nm). Be sure that the pulley moves smoothly.

28. Install the No. 1 idler pulley with the bolt and the tension spring. Pry the pulley toward the left as far as it will go and tighten the bolt. Make sure that the pulley moves smoothly.

29. Temporarily install the timing belt. Using the crankshaft pulley bolt, turn the crankshaft and position the key groove of the crankshaft timing pulley upward. If reusing the timing belt, align the points marked during removal.

30. Install the timing belt on the crankshaft timing pulley, oil pump pulley, No. 1 idler pulley, water pump pulley and the No. 2 idler pulley.

31. Install the timing belt guide.

➡If the old timing belt is being reinstalled, be sure the directional arrow is facing in the original direction and that the belt and sprocket/cover matchmarks are properly aligned.

32. Install the lower (No. 1) timing belt cover and new gasket with the 4 bolts.

33. Align the crankshaft pulley set key with the pulley key groove. Temporarily install the crankshaft pulley and bolt.

34. Align the camshaft knock pin with the groove of the pulley, and slide the timing pulley onto the camshaft with the plate washer and set bolt.

35. Tighten the pulley set bolt to 40 ft. lbs. (54 Nm).

WARNING

If any binding is felt when adjusting the timing belt tension by turning the crankshaft, STOP turning the engine, because the pistons may be hitting the valves.

36. Turn the crankshaft pulley and align the **0** mark on the lower (No. 1) timing belt cover.

37. Finish installing the timing belt and check the valve timing, as follows:

a. If reusing the old timing belt, align the matchmarks made previously and install the timing belt onto the camshaft pulley.

b. Align the marks on the timing belt with the marks on the camshaft pulley.

c. Loosen the No. 1 idler pulley set bolt ½ turn.

d. Turn the crankshaft pulley 2 complete revolutions TDC to TDC. ALWAYS turn the crankshaft CLOCKWISE. Check that the pulleys are still in alignment with the timing marks.

e. If the No. 1 idler pulley uses a green tension spring, slowly turn the crankshaft pulley 1⅞ revolutions, and align its groove with the mark at 45 degrees BTDC (for the No. 1 cylinder) of the No. 1 timing belt cover.

f. Tighten the No. 1 idler pulley set bolt to 31 ft. lbs. (42 Nm).

g. Be sure there is belt tension between the crankshaft and camshaft timing pulleys.

38. Place the right-hand engine mounting bracket in position but do not install the bolts.

39. Install the upper (No. 2) timing cover with a new gasket(s).

40. Remove the engine crankshaft pulley bolt and pulley.

41. Using SST 09249-63010, or equivalent, install the mounting bolts for the right-hand mounting bracket. Tighten the mounting bolts to 38 ft. lbs. (52 Nm).

42. Align the crankshaft pulley set key with the pulley key groove. Install the pulley. Tighten the pulley bolt to 80 ft. lbs. (108 Nm).

43. Install the spark plugs.

44. Install the right-hand mounting insulator, as follows:

a. Attach the mounting insulator to the body and mounting bracket with the 4 bolts and 2 nuts.

b. Tighten the 3 bolts to hold the mounting insulator to the body. Tighten the bolts to 47 ft. lbs. (64 Nm).

c. Tighten the 2 nuts and bolt to hold

the mounting insulator to the mounting bracket. Tighten the bolt to 27 ft. lbs. (37 Nm) and the nut to 38 ft. lbs. (52 Nm).

45. Install and adjust the power steering pump drive belt.
46. Install the right-hand engine under-cover.
47. Install the right front wheel.
48. Lower the engine.
49. If equipped, install the ABS actuator.
50. Install the alternator and alternator bracket.
51. Install the wiring harness bracket for the DLC1.
52. Install the power steering reservoir bracket and reservoir.
53. Connect the negative battery cable.
54. Start the engine and check the timing.

Timing Cover, Chain and Seal

REMOVAL & INSTALLATION

2.4L Engine

1. Drain the oil.
2. Remove the right under-cover.
3. Remove the power steering pump.
4. Disconnect the brake lines and remove the ABS actuator. Plug the lines.
5. Remove the right engine mount insulator.
6. Remove the accessory drive belt.
7. Remove the alternator.
8. Remove the air cleaner assembly.
9. Remove the ignition coils.
10. Remove the cylinder head cover.
11. Turn the crankshaft clockwise and align the groove in the pulley with the "0" mark on the timing cover. Check that the timing marks on the cam sprockets align with the timing marks on the Nos. 1 and 2 bearing caps. If the marks don't align, turn the crankshaft one full turn (360 degrees) clockwise.
12. Remove the crank pulley.
13. Remove the chain tensioner.
14. Remove the drive belt tensioner.
15. Remove the crankshaft position sensor.
16. Remove the oil pan.
17. Remove the timing chain cover (14 bolts and 2 nuts)
18. Remove the crank angle sensor plate.
19. Remove the chain tension slipper.
20. Remove the chain vibration damper.
21. Remove the chain guide.
22. Remove the timing chain.
23. Remove the crankshaft timing sprocket.
24. Remove the oil seal from the cover, using a hammer and punch.

To install:

25. Drive a new seal into place with a seal installer until it is flush with the cover. Coat the seal lip with multi-purpose grease.
26. Check the timing chain with the chain full stretched. The length of 16 consecutive links should be 122.6mm (4.827 inches) max. If it's longer than that, replace it.
27. Install the camshaft sprocket. Torque to 40 ft. lbs. (54 Nm).
28. Position the VVT sprocket so that the sprocket pin groove is slight right of the camshaft pin. Press the VVT into place turning it counterclockwise.
29. Install the bolt and torque to 40 ft. lbs. (54 Nm). Check that the valve timing controller turns clockwise and that it is locked securely when the lock pin hole is at the locking point.
30. Turn the camshafts so that the timing marks on the cam sprockets align with the timing marks on the Nos. 1 and 2 bearing caps. If the marks don't align, turn the crankshaft one full turn (360 degrees) clockwise.
31. Turn the crankshaft so that the key is at the 12:00 o'clock position.

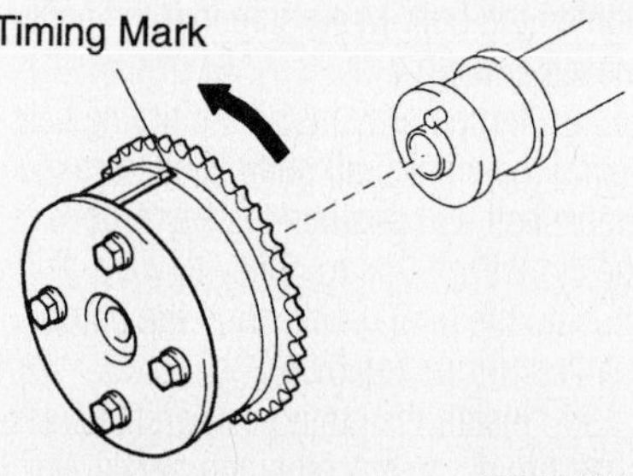

Position the VVT sprocket so that the sprocket pin groove is slight right of the camshaft pin—2.4L engine

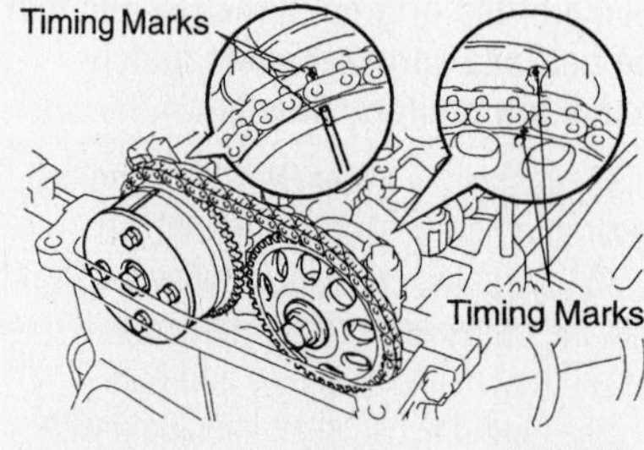

Turn the camshafts so that the timing marks on the cam sprockets align with the timing marks on the Nos. 1 and 2 bearing caps—2.4L engine

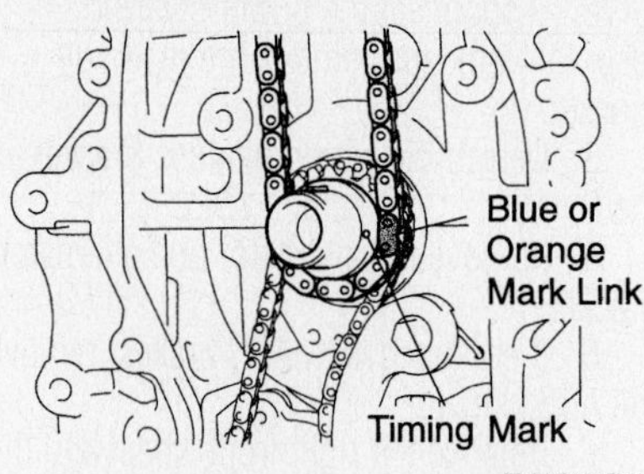

Install the timing chain on the crank sprocket with the blue or orange link is aligned with the timing mark on the crank sprocket—2.4L engine

32. Install the chain vibration damper. Torque to 80 inch lbs. (9 Nm).
33. Install the crank timing sprocket.
34. Install the timing chain on the crank sprocket with the blue or orange link is aligned with the timing mark on the crank sprocket.
35. Install the timing chain on the camshaft sprockets with the gold or yellow links aligned with the timing marks on the sprockets.
36. Install the timing chain guide. Torque to 80 inch lbs. (9 Nm).
37. Install the chain tension slipper. Torque to 14 ft. lbs. (19 Nm).

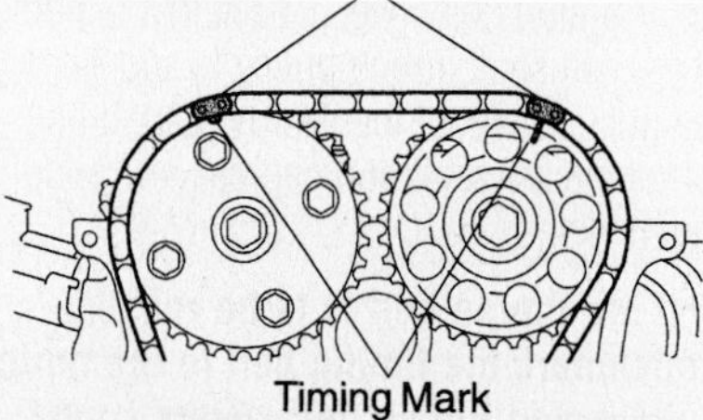

Install the timing chain on the camshaft sprockets with the gold or yellow links aligned with the timing marks on the sprockets—2.4L engine

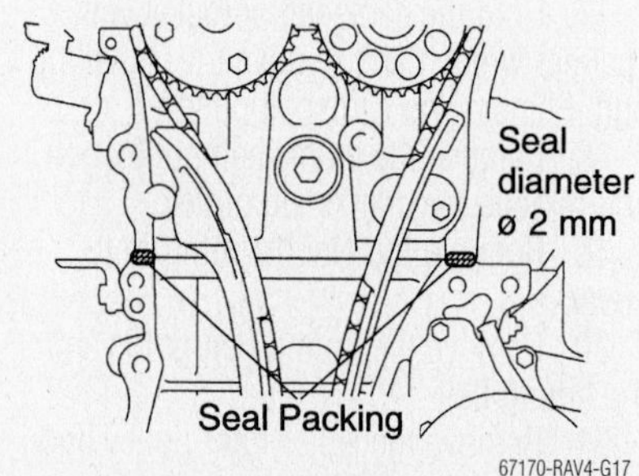

Apply sealer to the 2 areas where the crankcase and block meet—2.4L engine

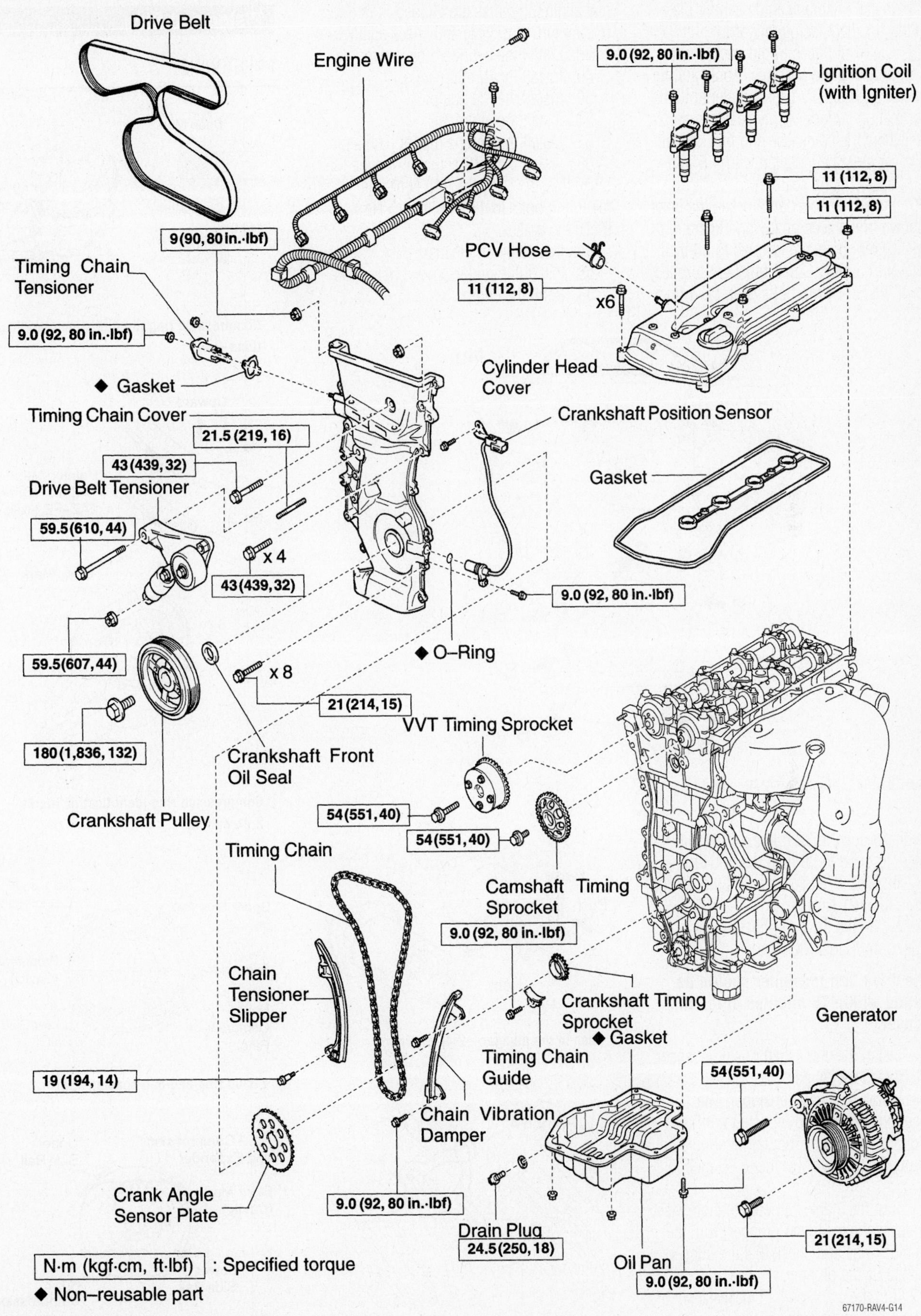
Drive Belt
Engine Wire
9.0 (92, 80 in.·lbf)
Ignition Coil (with Igniter)
11 (112, 8)
11 (112, 8)
9 (90, 80 in.·lbf)
PCV Hose
Timing Chain Tensioner
11 (112, 8)
x6
9.0 (92, 80 in.·lbf)
◆ Gasket
Cylinder Head Cover
Timing Chain Cover
Crankshaft Position Sensor
21.5 (219, 16)
43 (439, 32)
Gasket
Drive Belt Tensioner
59.5 (610, 44)
x 4
43 (439, 32)
9.0 (92, 80 in.·lbf)
59.5 (607, 44)
◆ O–Ring
x 8
21 (214, 15)
VVT Timing Sprocket
180 (1,836, 132)
Crankshaft Front Oil Seal
54 (551, 40)
Crankshaft Pulley
54 (551, 40)
Timing Chain
Camshaft Timing Sprocket
9.0 (92, 80 in.·lbf)
Chain Tensioner Slipper
Crankshaft Timing Sprocket
Generator
◆ Gasket
Timing Chain Guide
54 (551, 40)
19 (194, 14)
Chain Vibration Damper
Crank Angle Sensor Plate
9.0 (92, 80 in.·lbf)
Drain Plug
24.5 (250, 18)
21 (214, 15)
N·m (kgf·cm, ft·lbf) : Specified torque
Oil Pan
9.0 (92, 80 in.·lbf)
◆ Non–reusable part
67170-RAV4-G14

Timing chain and related parts—2.4L engine

38. Install the crank angle sensor plate with the "F" mark facing forward.

39. Clean all traces of old gasket material from the mating surfaces. Apply a 4mm diameter bead of RTV gasket material to the cover mating surface. Apply sealer to the 2 areas where the crankcase and block meet. The cover must be installed with 3-5 minutes of applying the sealer.

40. Install the cover and torque the bolts as shown in the accompanying illustration. Bolt A 10 mm bolts are torqued to 80 inch lbs. (9 Nm); bolt B 12mm bolts are torqued to 15 ft. lbs. (21 Nm); bolts C and D 14mm bolts are torqued to 32 ft. lbs. (43 Nm); the nuts are torqued to 80 inch lbs. (9 Nm).

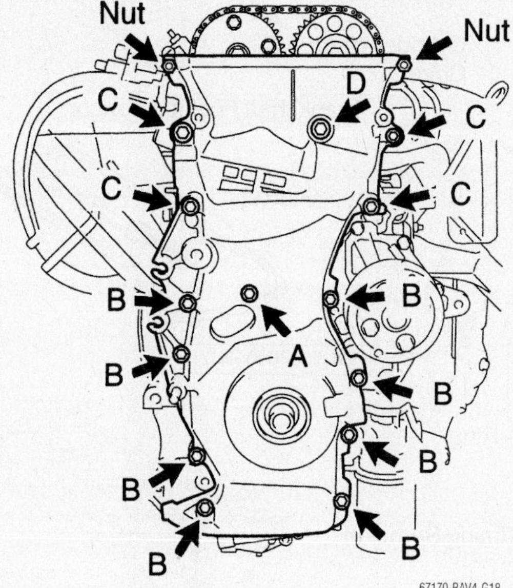

Cover bolt identification—2.4L engine

41. Install the tensioner stud bolt. Torque to 16 ft. lbs. (21 Nm).

42. Install the drive belt tensioner. Torque to 44 ft. lbs. (60 Nm).

➡The drive belt tensioner should be installed within 15 minutes of the timing cover.

43. Install the crankshaft position sensor.

44. Install the oil pan.

45. Install the chain tensioner. The plunger should be held retracted until the tensioner is installed. The gasket should be positioned as shown. Torque to 80 inch lbs. (9 Nm).

46. Turn the crankshaft counterclockwise and release the tensioner plunger. Turn the crank clockwise and check that the slipper is pushed by the plunger.

47. Turn the crank puller clockwise so that the notch in the crank pulley is aligned with the "0" mark on the timing cover. Check that all timing marks are aligned. If not, turn it 1 full turn clockwise and check again.

48. Install the cylinder head cover.

49. Install the ignition coils.

50. Install the PS pump.

51. Install the alternator.

52. Install the engine mount insulator.

53. Install the ABS actuator. Torque the brake lines to 11 ft. lbs. (15 Nm); the mounting bolts to 48 inch lbs. (5 Nm). Bleed the brakes.

54. Install the air cleaner.

55. Install the under-cover.

56. Refill the engine oil.

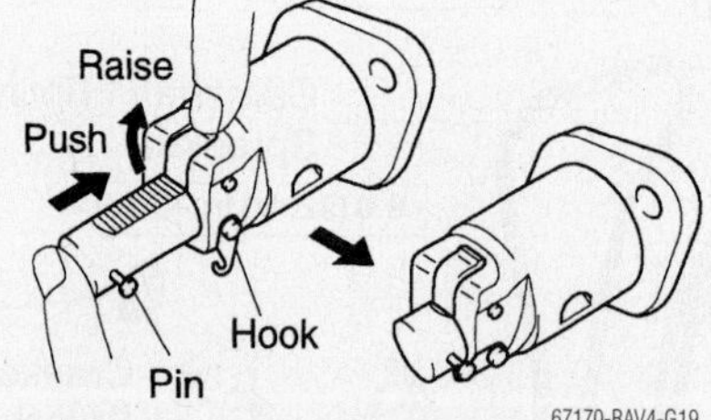

Push in the plunger—2.4L engine

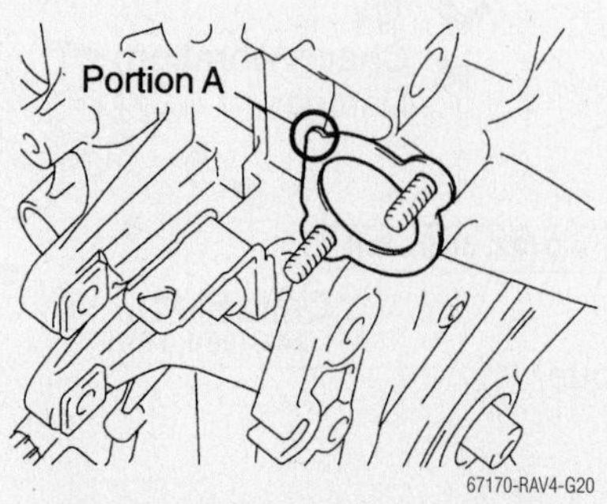

Correct gasket positioning—2.4L engine

Piston and Ring

POSITIONING

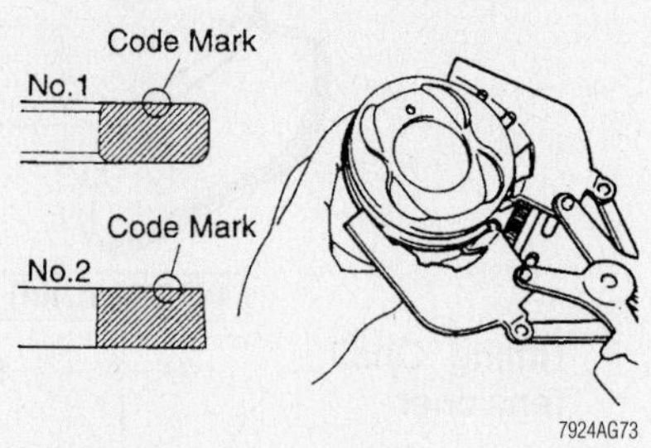

Compression ring identification mark locations–2.0L

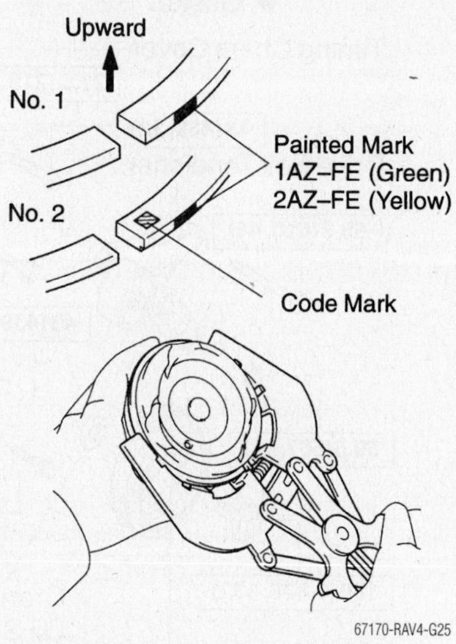

Compression ring identification marks—2.4L engine

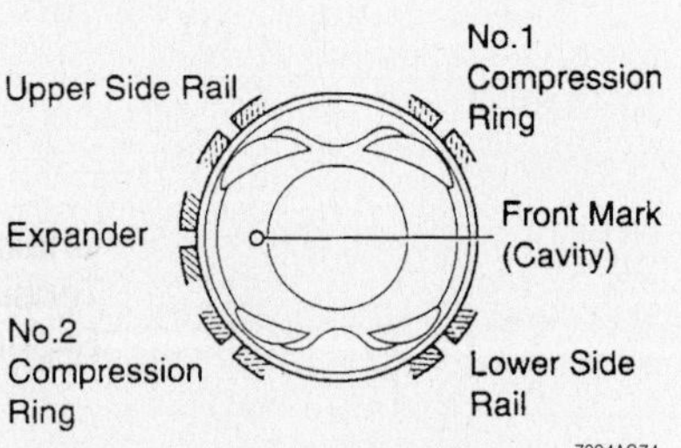

Piston ring end-gap spacing–2.0L

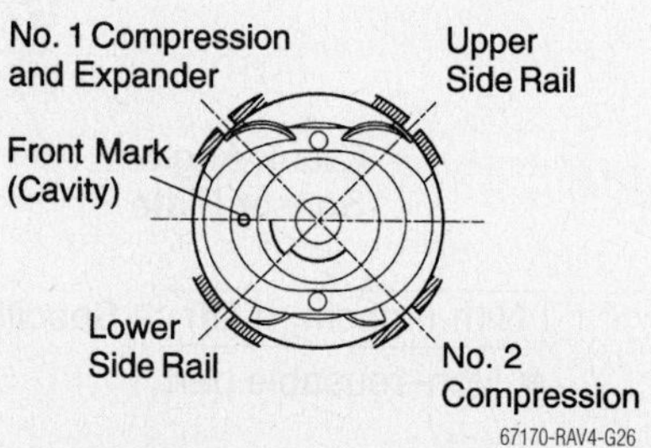

Piston ring gap spacing—2.4L engine

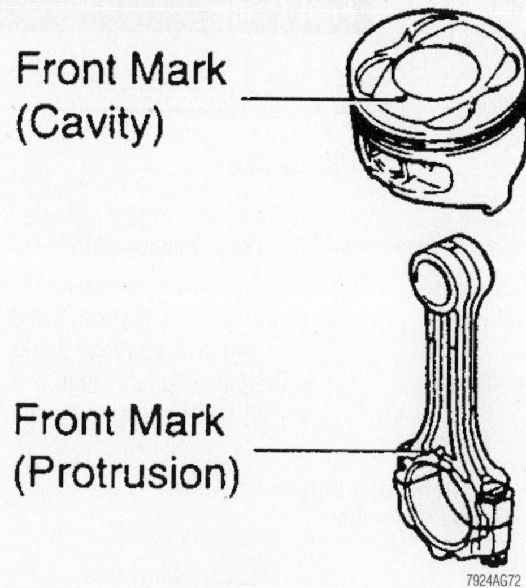

Piston-to-connecting rod assembly–2.0L

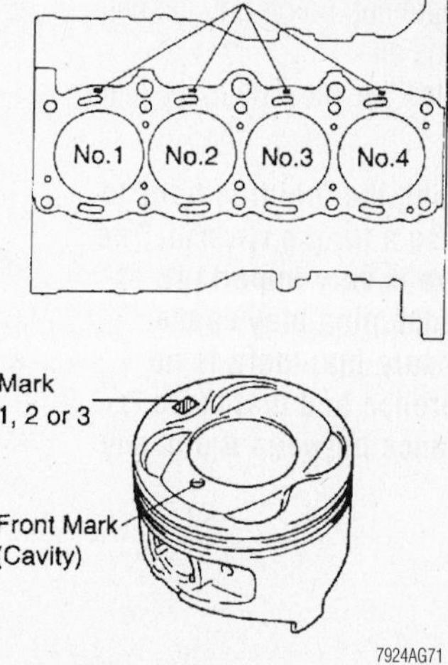

Piston-to-engine installation. Match the number on the piston crown with the number stamped on the block–2.0L

FUEL SYSTEM

Fuel System Service Precautions

Safety is the most important factor when performing not only fuel system maintenance but any type of maintenance. Failure to conduct maintenance and repairs in a safe manner may result in serious personal injury or death. Maintenance and testing of the vehicle's fuel system components can be accomplished safely and effectively by adhering to the following rules and guidelines.

- To avoid the possibility of fire and personal injury, always disconnect the negative battery cable unless the repair or test procedure requires that battery voltage be applied.
- Always relieve the fuel system pressure prior to disconnecting any fuel system component (injector, fuel rail, pressure regulator, etc.), fitting or fuel line connection. Exercise extreme caution whenever relieving fuel system pressure, to avoid exposing skin, face and eyes to fuel spray. Please be advised that fuel under pressure may penetrate the skin or any part of the body that it contacts.
- Always place a shop towel or cloth around the fitting or connection prior to loosening to absorb any excess fuel due to spillage. Ensure that all fuel spillage (should it occur) is quickly removed from engine surfaces. Ensure that all fuel soaked cloths or towels are deposited into a suitable waste container.
- Always keep a dry chemical (Class B) fire extinguisher near the work area.
- Do not allow fuel spray or fuel vapors to come into contact with a spark or open flame.
- Always use a back-up wrench when loosening and tightening fuel line connection fittings. This will prevent unnecessary stress and torsion to fuel line piping.
- Always replace worn fuel fitting O-rings with new. Do not substitute fuel hose or equivalent, where fuel pipe is installed.

Fuel System Pressure

RELIEVING

2.0L Engine

1. Before servicing the vehicle, refer to the precautions in the beginning of this section.
2. Disconnect the negative battery terminal.
3. Place a catch-pan under the joint to be disconnected. A large quantity of fuel may be released when the joint is opened.
4. Wear eye or full face protection.
5. Place a shop towel over the area and slowly loosen the joint using a wrench of the correct size. Use a back-up wrench if needed.
6. Allow the fuel left in the line to bleed off slowly before fully disconnecting the joint.
7. Plug the opened lines immediately to prevent fuel spillage or the entry of dirt.
8. Dispose of the released fuel properly.
9. After connecting fuel lines, connect the negative battery cable and start the engine.
10. Check for leaks and repair as needed.

2.4L Engine

1. Disconnect the wire at the fuel pump.
2. Start the engine and run it until it shuts off.
3. Place a catch-pan under the joint to be disconnected. A large quantity of fuel may be released when the joint is opened.
4. Disconnect the high pressure fuel line.
5. After connecting fuel lines, reconnect the wire and start the engine.
6. Check for leaks and repair as needed.

Fuel Filter

REMOVAL & INSTALLATION

2.0L Engine

1. Before servicing the vehicle, refer to the precautions in the beginning of this section.
2. Properly release fuel system pressure.
3. Remove or disconnect the following:
   - Negative battery cable
   - Fuel filter's protective shield
4. Place a pan under the delivery pipe to catch the dripping fuel and slowly loosen the union bolt or flare nut to bleed off the fuel pressure.
5. Drain the remaining fuel.
6. Remove or disconnect the following:
   - Inlet and outlet lines
   - Fuel filter

To install:

7. Coat the flare nut, union nut and bolt threads with engine oil.
8. Hand-tighten the inlet line to the fuel filter.

➡When tightening the fuel line bolts to the fuel filter, use a torque wrench. The tightening torque is very important, as under or over tightening may cause fuel leakage. Insure that there is no fuel line interference and that there is sufficient clearance between it and any other parts.

9. Install or connect the following:
   - Fuel filter. Tighten the inlet bolts to 22 ft. lbs. (30 Nm).
   - Delivery pipe using new gaskets. Tighten the union bolt to 22 ft. lbs. (30 Nm).
10. Run the engine for a few minutes and check for any fuel leaks.
11. Install the protective shield.

2.4L Engine

The filter is part of the fuel pump module and is not normally serviced.

Fuel Pump

REMOVAL & INSTALLATION

2.0L Engine

1. Before servicing the vehicle, refer to the precautions in the beginning of this section.
2. Relieve the fuel system pressure.
3. Remove or disconnect the following:
   - Negative battery cable
   - Left-hand rear seat assembly

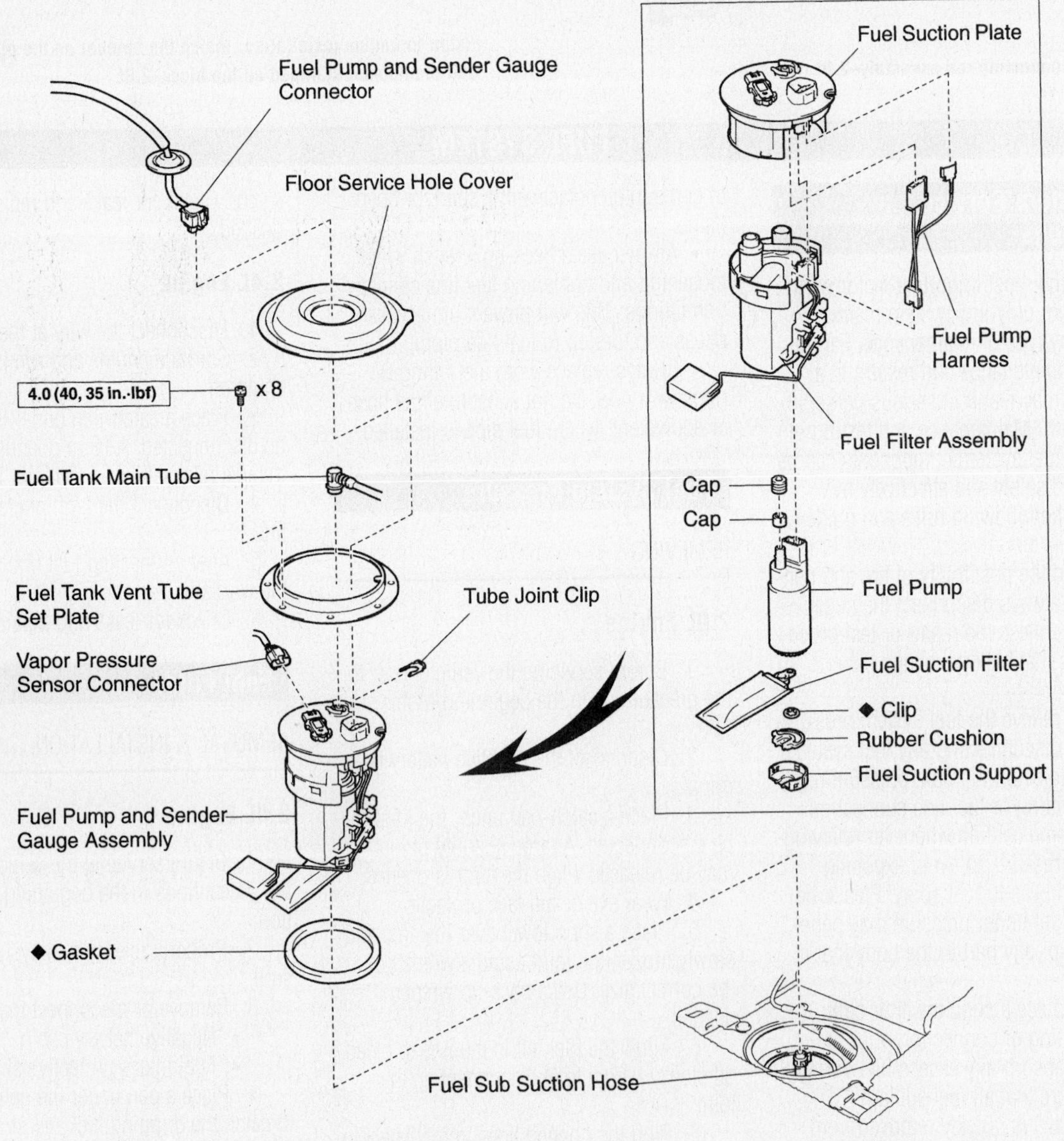

Fuel pump module—2.4L engine

- Floor service hole by pulling back the carpet; then, remove the 4 screws
- Fuel pump and sender gauge connector

➡Loosen the fuel cap to relieve any fuel pressure within the tank.

- Fuel pipe union bolt and both gaskets
- Fuel pump outlet pipe
- Return vent hose from the fuel pump
- 8 fuel pump bolts and the pump assembly from the tank

To install:

4. Install or connect the following:
   - Fuel pump to the fuel tank. Tighten the 8 bolts to 31 inch lbs. (3.5 Nm).
   - Return vent hose to the fuel pump
   - Outlet pipe to the fuel pump, using new gaskets. Tighten the union bolts to 22 ft. lbs. (29 Nm).
   - Fuel pump and sender gauge connector
   - Floor hole cover with the 4 screws
   - Carpet
   - Left rear seat assembly
   - Negative battery cable
   - Fuel cap
5. Start the vehicle and check for leaks.

2.4L Engine

1. Before servicing the vehicle, refer to the precautions in the beginning of this section.
2. Relieve the fuel system pressure.
3. Remove or disconnect the following:
   - Negative battery cable
   - Left-hand rear seat assembly
   - Floor service hole by pulling back the carpet; then, remove the 4 screws
   - Fuel pump and sender gauge connector

➡Loosen the fuel cap to relieve any fuel pressure within the tank.

- Fuel pipe
- Fuel pump outlet pipe
- Return vent hose from the fuel pump
- 8 fuel pump bolts and the pump assembly from the tank

To install:

4. Install or connect the following:
   - Fuel pump to the fuel tank. Position the unit as shown. Tighten the 8 bolts to 35 inch lbs. (4 Nm).
   - Return vent hose to the fuel pump

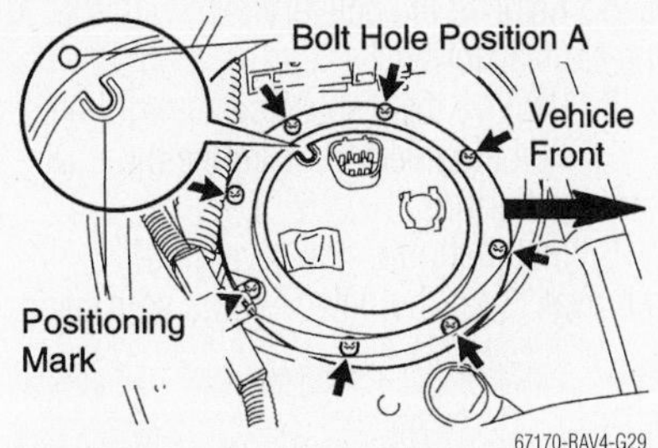

Proper fuel pump installation—2.4L engine

- Fuel pump and sender gauge connector
- Floor hole cover with the 4 screws
- Carpet
- Left rear seat assembly
- Negative battery cable
- Fuel cap

5. Start the vehicle and check for leaks.

Fuel Injector

REMOVAL & INSTALLATION

2.0L

1. Before servicing the vehicle, refer to the precautions in the beginning of this section.
2. Remove or disconnect the following:
   - Air cleaner assembly
   - Cylinder head cover
   - Throttle body from the intake manifold
   - Distributor
3. Remove or disconnect the engine wire from the intake manifold, as follows:
   - 4 injector connectors
   - Both engine wire clamps from the intake manifold wire brackets
   - Engine wire protector from the right side of the intake manifold
   - Engine wire clamp
4. Remove or disconnect the Exhaust Gas Recirculation (EGR) valve, as follows:
   - Vacuum hose from port **E** of the Vacuum Switching Valve (VSV)
   - EGR hose from the vacuum modulator
   - Loosen the EGR pipe nut from the cylinder head
   - Both nuts, EGR valve, pipe assembly and gasket
5. Disconnect the engine compartment R/B No. 2
6. Remove or disconnect the fuel inlet hose and delivery pipe, as follows:
   - Union bolt, both gaskets and the fuel inlet hose from the fuel filter outlet
   - Air assist hose from the intake manifold port
   - Air assist hose
   - Loosen both delivery pipe-to-cylinder head bolts
   - Delivery pipe from the 4 injectors
   - Delivery pipe and fuel inlet hose assembly
7. Remove or disconnect the following:
   - 4 injectors and spacers

⁂ WARNING

Be careful not to drop the injectors and spacers.

- O-rings, insulator and grommet from each injector

To install:

8. Install or connect the following:
   - New O-rings, insulator and grommet, lubricated with gasoline, to each injector
   - 4 injectors and spacers
9. Install or connect the fuel inlet hose and delivery pipe, as follows:
   - Delivery pipe and fuel inlet hose assembly
   - Delivery pipe to the 4 injectors. Tighten both delivery pipe-to-cylinder head bolts to 9 ft. lbs. (13 Nm).
   - Air assist hose
   - Air assist hose to the intake manifold port
   - Union bolt, new gaskets and the fuel inlet hose to the fuel filter outlet. Tighten the union bolt to 22 ft. lbs. (29 Nm).
10. Connect the engine compartment R/B No. 2
11. Install or connect the EGR valve, as follows:
    - New gasket, pipe assembly and EGR valve. Tighten the nut to 9 ft. lbs. (13 Nm) and the union nut to 43 ft. lbs. (59 Nm).
    - EGR hose to the vacuum modulator
    - Vacuum hose from port **E** of the VSV
12. Install or connect the engine wire to the intake manifold, as follows:
    - Engine wire clamp
    - Engine wire protector to the right side of the intake manifold
    - Both engine wire clamps to the intake manifold wire brackets
    - 4 injector connectors
13. Install or connect the following:
    - Distributor
    - Throttle body to the intake manifold
    - Cylinder head cover
    - Air cleaner assembly

2.4L Engine

1. Disconnect the PCV hose.
2. Remove the air cleaner assembly with the MAF sensor.
3. Disconnect the injector connectors.
4. Remove the 2 bolts holding the delivery pipe to the cylinder head.
5. Remove the delivery pipe with the injectors attached.
6. Remove the 2 spacers and 4 grommets.
7. Pull the injectors from the pipe. Discard the O-rings.
8. Installation is the reverse of removal. Coat the new O-rings with gasoline. Push the injectors onto the pipes and make sure they rotate freely. Position the assembly onto the head and install the 2 bolts finger tight. Make sure that the injectors still rotate freely. If not, replace the O-rings. Torque the 2 bolts to 15 ft. lbs. (20 Nm).

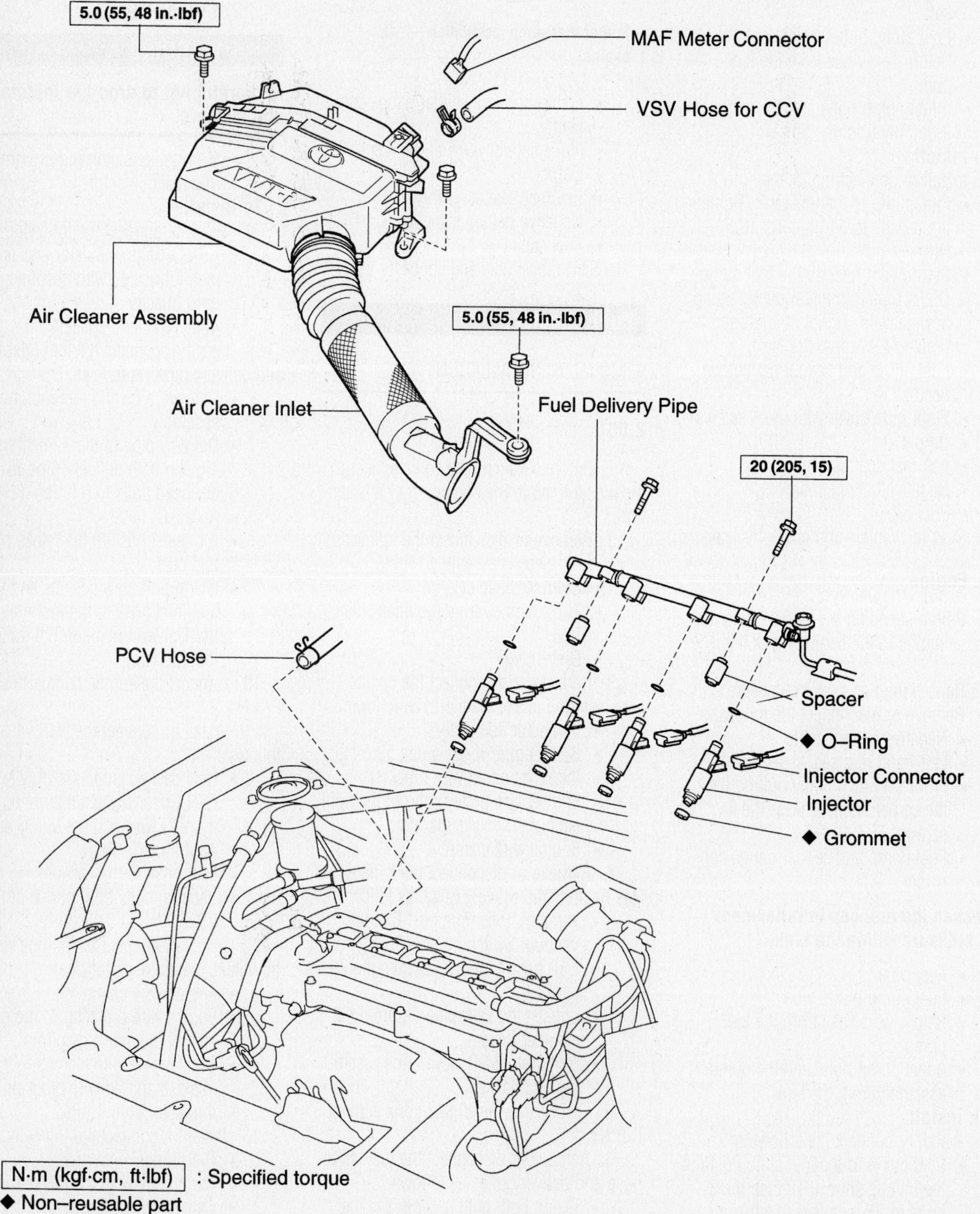

Fuel injectors and related parts—2.4L engine

DRIVE TRAIN

Manual Transmission Assembly

REMOVAL & INSTALLATION

2001–03 with 2WD

1. Before servicing the vehicle, refer to the precautions in the beginning of this section.
2. Remove or disconnect the following:
   - Negative battery cable
   - Air cleaner case assembly with hose
   - Engine coolant reservoir tank
   - Engine wire clamp set nut
   - Starter
3. Remove the clutch release cylinder, as follows:
   - Clutch line bracket-to-transaxle set bolts
   - Release cylinder and line
4. Remove or disconnect the following:
   - Ground cable from the transaxle
   - Vehicle Speed Sensor (VSS) and backup light switch connector
   - Control cable by removing the 4 clips and washers

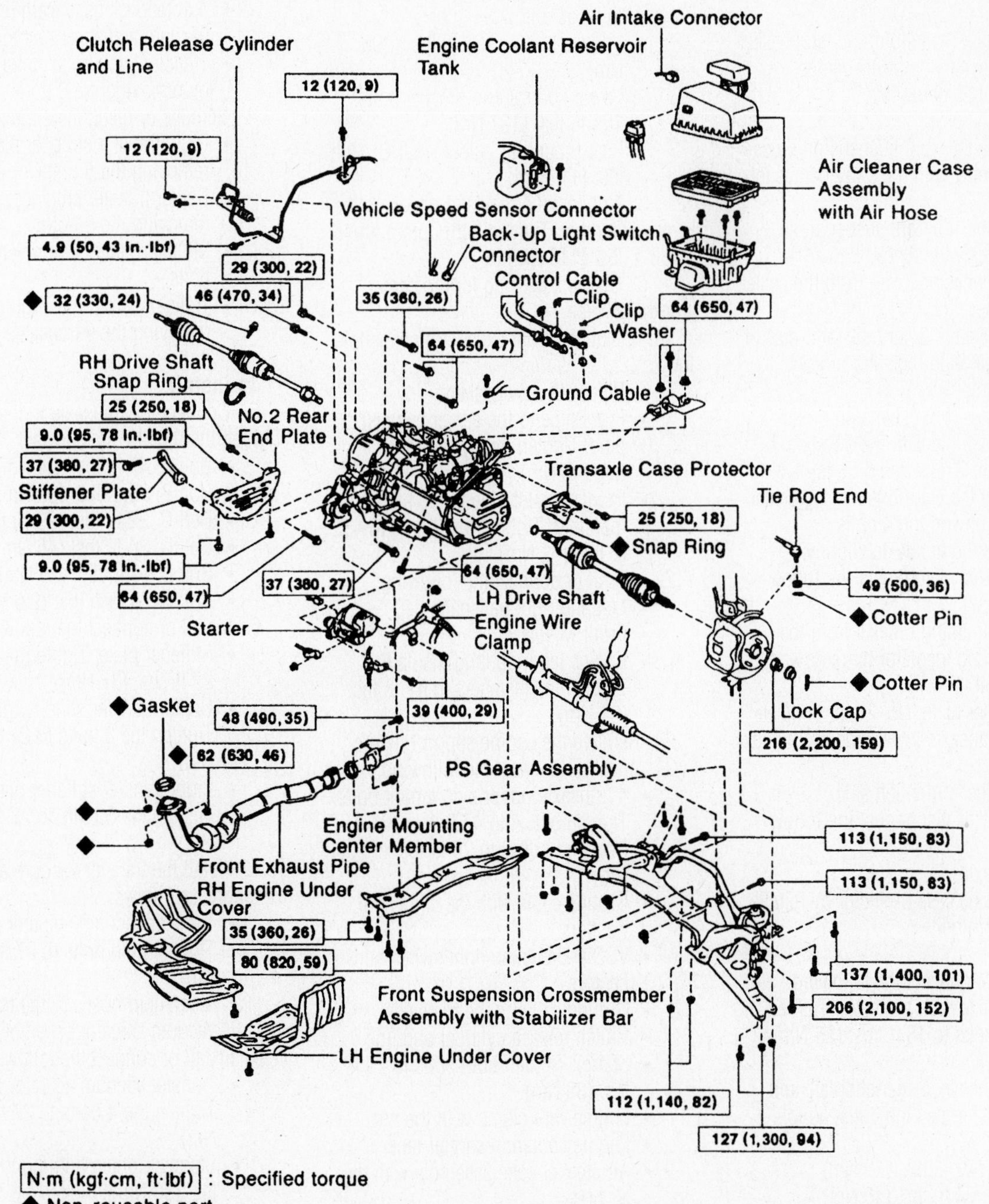

Transaxle exploded view—2001–03 2WD

- 4 upper side transaxle-to-engine bolts
- Left mount insulator

5. Install a engine support to the engine.
6. Support rack and pinion to the engine support fixture with a rope.
7. Remove or disconnect the following:
   - Front wheels
   - Left and right-hand engine under-covers
8. Drain the transaxle oil.
9. Remove the left and right halfshafts.
10. Remove the front exhaust pipe, as follows:
    - 3 exhaust manifold nuts and gasket
    - Both exhaust pipe-to-center exhaust pipe bolts
    - Exhaust pipe
11. Remove the front suspension cross-member assembly with the sway bar, as follows:
    a. Support the front suspension crossmember with a jack.
    b. Disconnect the ring from the center exhaust pipe.
    c. Remove the 2 set bolts and nuts of the power steering rack and pinion assembly.
    d. Remove the suspension cross-member assembly with the sway bar by removing the 2 nuts and 6 bolts.
12. Remove the engine mounting center-member by removing the 4 bolts.
13. Jack up the transaxle slightly.
14. Remove or disconnect the following:
    - Left mounting bracket from the mounting insulator by removing the set bolt
    - Stiffener plate, No. 2 rear endplate and transaxle lower side mounting bolt
15. Lower the engine left side
16. Remove or disconnect the following:
    - Transaxle
    - Transaxle case protector by removing both bolts

To install:

17. Install or connect the following:
    - Transaxle case protector. Tighten both bolts to 18 ft. lbs. (25 Nm).
    - Transaxle
18. Install the No. 2 rear endplate and transaxle bolts. Tighten the bolts, as follows:
    - Bolt C: 22 ft. lbs. (29 Nm)
    - Bolt D: 34 ft. lbs. (46 Nm)
    - Bolt E: 18 ft. lbs. (25 Nm)
    - Bolt F: 78 inch lbs. (9.0 Nm)
19. Install or connect the following:
    - Stiffener plate. Tighten both bolts to 27 ft. lbs. (37 Nm).
    - Engine left mounting insulator to the left mounting bracket. Tighten the bolt to 47 ft. lbs. (64 Nm).
    - Engine mount centermember. Tighten the radiator support bolts to 26 ft. lbs. (35 Nm) and the mount insulator to 59 ft. lbs. (80 Nm).
20. Install the front suspension cross-member with the sway bar, as follows:
    a. Install the sway bar and suspension crossmember. Tighten the nuts/bolts, as follows:
    - Vehicle bolt A: 152 ft. lbs. (206 Nm)
    - Lower control arm bracket bolt B: 101 ft. lbs. (137 Nm)
    - Rear mounting bracket bolt C: 82 ft. lbs. (112 Nm)

    b. Connect the rack and pinion to the crossmember. Tighten both nuts/bolts to 83 ft. lbs. (113 Nm).
    c. Connect the ring for the center exhaust pipe.
21. Install the front exhaust pipe, as follow:
    - Pipe with new gaskets
    - Front pipe to the center exhaust pipe. Tighten both bolts to 35 ft. lbs. (48 Nm).
    - Front exhaust pipe to the exhaust manifold. Tighten the 3 nuts to 46 ft. lbs. (62 Nm).
22. Install or connect the following:
    - Left and right halfshafts
    - Front wheels
    - Engine left mounting insulator. Tighten the fasteners to 47 ft. lbs. (64 Nm).
23. Remove the engine support fixture.
24. Install or connect the following:
    - 4 transaxle upper side mount bolts. Tighten bolt A to 47 ft. lbs. (64 Nm) and bolt B to 26 ft. lbs. (35 Nm).
    - Ground cable with the clips and washers
    - VSS and backup light switch connectors
    - Ground cable to the transaxle
    - Clutch release cylinder and line
    - Starter. Tighten both bolts to 29 ft. lbs. (39 Nm).
    - Engine wire clamp with the nut
    - Engine coolant reservoir tank
    - Air cleaner case assembly with the air hose
    - Negative battery cable
25. Fill the transaxle with fluid. Check all fluids.

2001–03 with 4WD

1. Before servicing the vehicle, refer to the precautions in the beginning of this section.
2. Remove or disconnect the following:
   - Transaxle/engine assembly
   - Transaxle case protector, by removing the 2 bolts
   - Starter
   - Transfer vacuum actuator bracket, by removing the 4 bolts
3. Remove the transfer vacuum actuator assembly, as follows:
   - 4 solenoid hoses from the transfer vacuum actuator assembly
   - Transfer vacuum actuator assembly, by removing the 2 bolts
4. Remove or disconnect the following:
   - Right transfer stiffener plate, by removing the 5 bolts
   - Center transfer stiffener plate, by removing the 3 bolts
   - Stiffener plate by removing the 2 bolts
   - Transaxle from the engine, by removing the 9 transaxle mount bolts

To install:

5. Connect the transaxle to the engine. Tighten the 9 bolts, as follows:
   - Bolt A: 47 ft. lbs. (64 Nm)
   - Bolt B: 26 ft. lbs. (35 Nm)
   - Bolt C: 22 ft. lbs. (29 Nm)
   - Bolt D: 34 ft. lbs. (46 Nm)
   - Bolt E: 18 ft. lbs. (25 Nm)
   - Bolt F: 78 inch lbs. (9.0 Nm)
6. Install or connect the following:
   - Stiffener plate. Tighten both bolts to 27 ft. lbs. (37 Nm).
   - Center transfer stiffener plate. Tighten the 3 bolts to 27 ft. lbs. (37 Nm).
   - Right transfer stiffener plate. Tighten the 5 bolts to 27 ft. lbs. (37 Nm).
7. Install the transfer vacuum actuator assembly, as follows:
   - Transfer vacuum actuator assembly. Tighten both bolts to 27 ft. lbs. (37 Nm).
   - 4 solenoid hoses to the transfer vacuum actuator assembly
8. Install or connect the following:
   - Transfer vacuum actuator bracket. Tighten the 4 bolts to 27 ft. lbs. (37 Nm).
   - Starter. Tighten both bolts to 29 ft. lbs. (39 Nm).
   - Transaxle case protector. Tighten both bolts to 18 ft. lbs. (25 Nm).
   - Transaxle/engine assembly

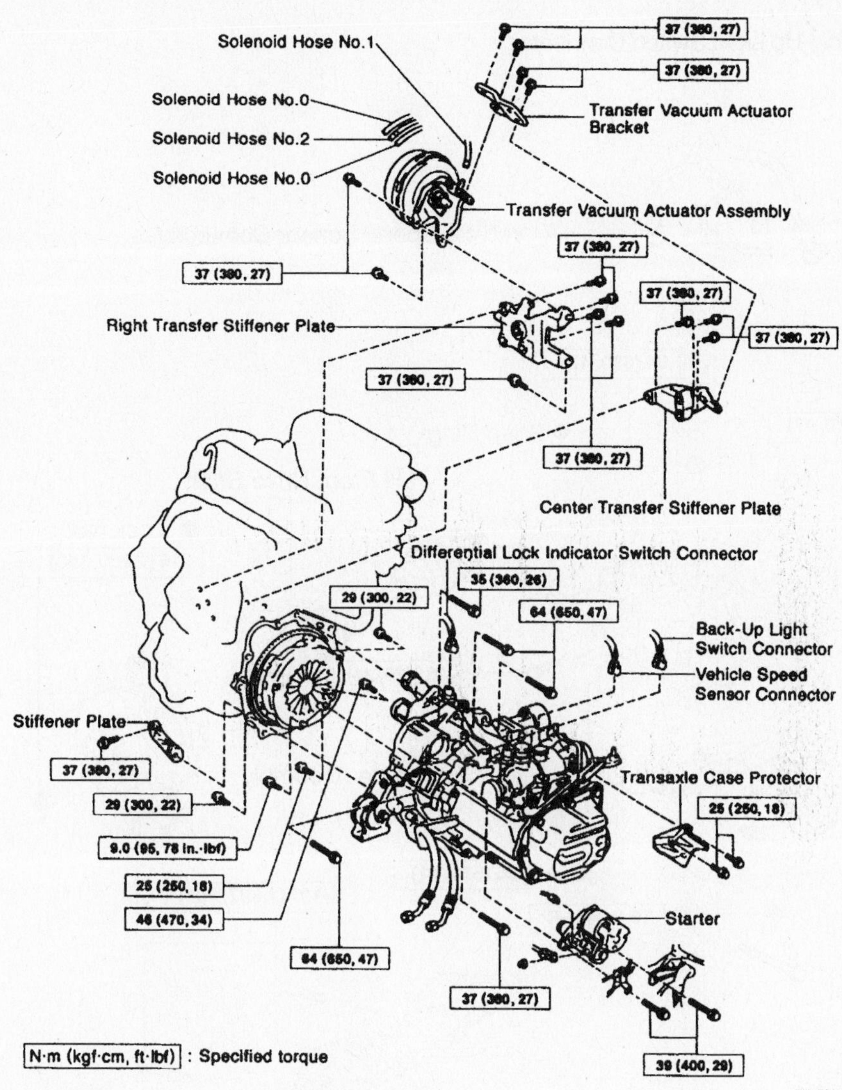

Manual transaxle exploded view—2001–03 4WD

2004 with 2WD

1. Remove the hood.
2. Remove the air cleaner assembly.
3. Remove the coolant reservoir.
4. Remove the control cables from the transmission.
5. Remove the wiring from the speed sensor and back-up light switch.
6. Remove the ground cable from the transaxle.
7. Remove the starter.
8. Remove the clutch release cylinder.
9. Remove the 5 upper transaxle-to-engine bolts.
10. Attach an engine support fixture, such as 12281-28010, or equivalent.
11. Disconnect the left engine mount from the transaxle.
12. Remove the engine under-covers.
13. Drain the transaxle.
14. Remove the front exhaust pipe.

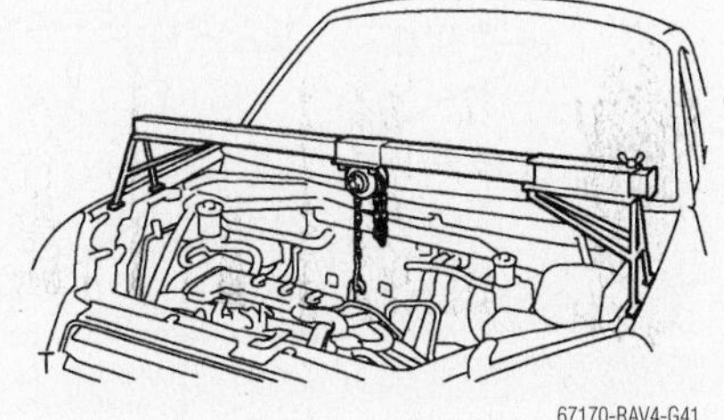

Attach an engine support fixture, such as 12281-28010, or equivalent—2004 2WD

15. Remove the halfshafts.
16. Remove the stabilizer bar.
17. Remove the power steering gear mounting bolts. Support the gear.
18. Remove the front suspension cross-member.
19. Remove the engine mount center member.
20. Using a transmission jack, slightly raise the transaxle.
21. Remove the lower transaxle-to-engine bolts and lower the transaxle from the vehicle.
22. Installation is the reverse of removal. Observe the following torques:
    - Lower transaxle-to-engine bolts (see the illustration): bolts A 34 ft. lbs. (46 Nm); bolt B 32 ft. lbs. (44 Nm)
    - Engine mount center member (see the illustration): bolt E 59 ft. lb. (80 Nm); bolts F 28 ft. lbs. (39 Nm)
    - Front suspension crossmember (see the illustration): bolt A 82 ft. lbs. (113 Nm); bolt B 115 ft. lbs. (157 Nm); bolts C & D 53 ft. lbs. (72 Nm)
    - Power steering gear set bolts: 101 ft. lbs. (137 Nm)
    - Stabilizer bar links: 32 ft. lbs. (44 Nm)
    - Exhaust pipe-to-manifold: 32 ft. lbs. (43 Nm)
    - Exhaust pipe-to-pipe: 14 ft. lbs. (19 Nm)
    - Left engine mount through-bolt: 41 ft. lbs. (56 Nm)
    - Upper transaxle-to-engine bolts (see the illustration): bolts A 47 ft. lbs. (64 Nm); bolts B 34 ft. lbs. (46 Nm)
23. Clutch release cylinder (see the

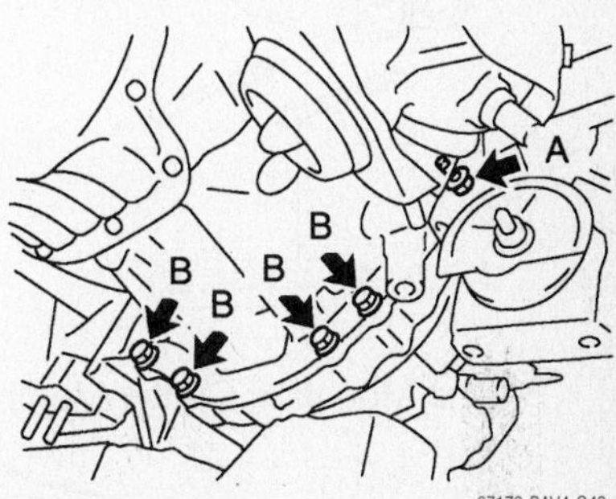

Lower transaxle-to-engine bolts—2004 2WD

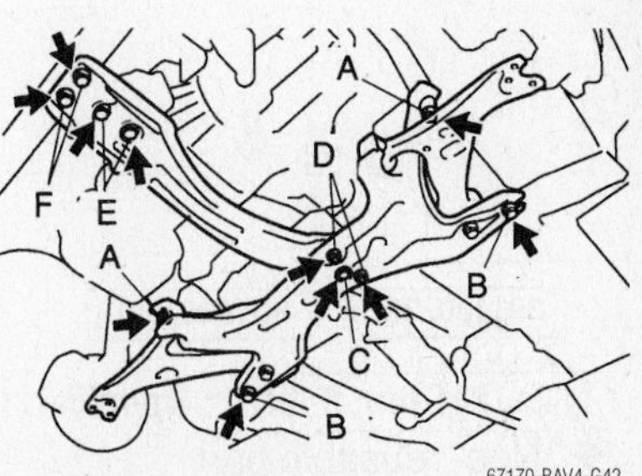

Center member and front member bolts—2004 2WD

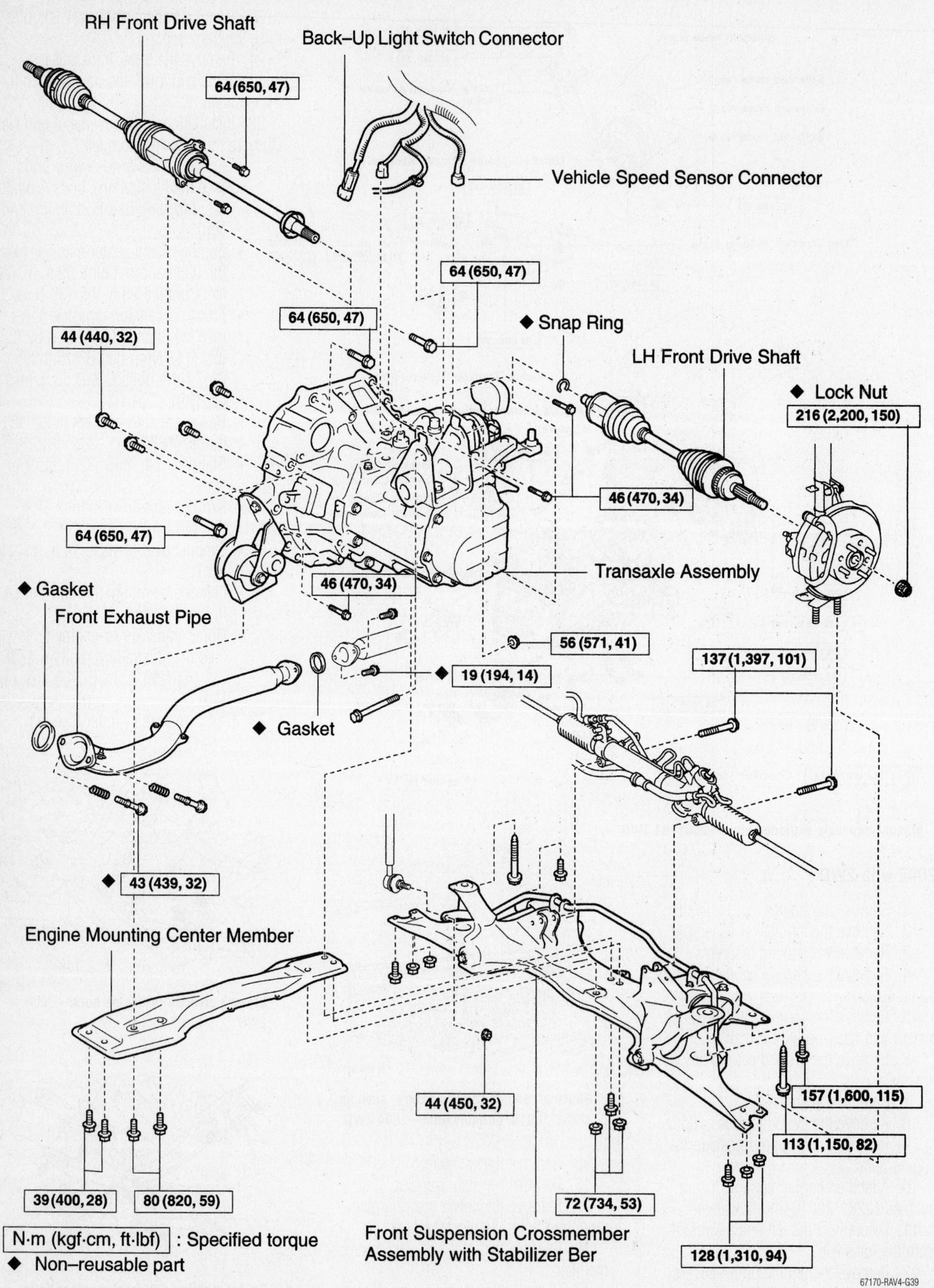

RH Front Drive Shaft
Back–Up Light Switch Connector
64 (650, 47)
Vehicle Speed Sensor Connector
64 (650, 47)
64 (650, 47)
◆ Snap Ring
44 (440, 32)
LH Front Drive Shaft
◆ Lock Nut
216 (2,200, 150)
46 (470, 34)
64 (650, 47)
Transaxle Assembly
◆ Gasket
46 (470, 34)
Front Exhaust Pipe
56 (571, 41)
137 (1,397, 101)
◆ 19 (194, 14)
◆ Gasket
◆ 43 (439, 32)
Engine Mounting Center Member
157 (1,600, 115)
44 (450, 32)
113 (1,150, 82)
39 (400, 28)
80 (820, 59)
72 (734, 53)
N·m (kgf·cm, ft·lbf) : Specified torque
◆ Non–reusable part
Front Suspension Crossmember Assembly with Stabilizer Ber
128 (1,310, 94)
67170-RAV4-G39

Manual transaxle and related parts—2004 2WD

A
B
B
A
A
67170-RAV4-G40

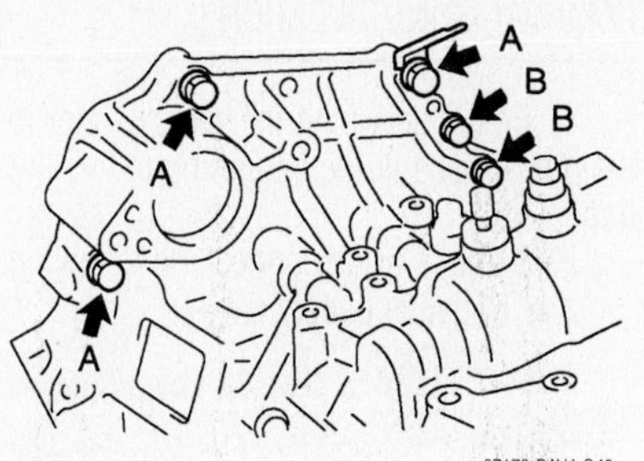

Upper transaxle-to-engine bolts—2004 2WD

Heat Insulator
A
B
C
67170-RAV4-G44

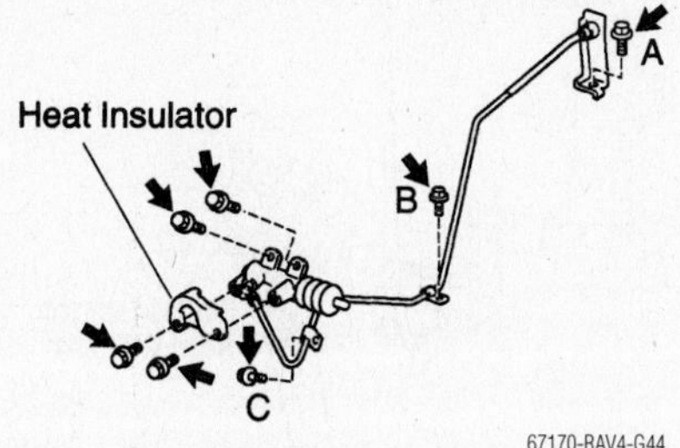

Clutch release cylinder bolts—2004 2WD

illustration): bolt A 18 ft. lbs. (25 Nm); bolt B 9 ft. lbs. (12 Nm); bolt C 44 inch lbs. (5 Nm)

- Starter: 28 ft. lbs. (37 Nm)
- Hood: 10 ft. lbs. (13 Nm)

2004 With 4WD

1. Remove the engine and transaxle as a unit.
2. Remove the transaxle case protector.
3. Remove the engine-to-transaxle wiring.

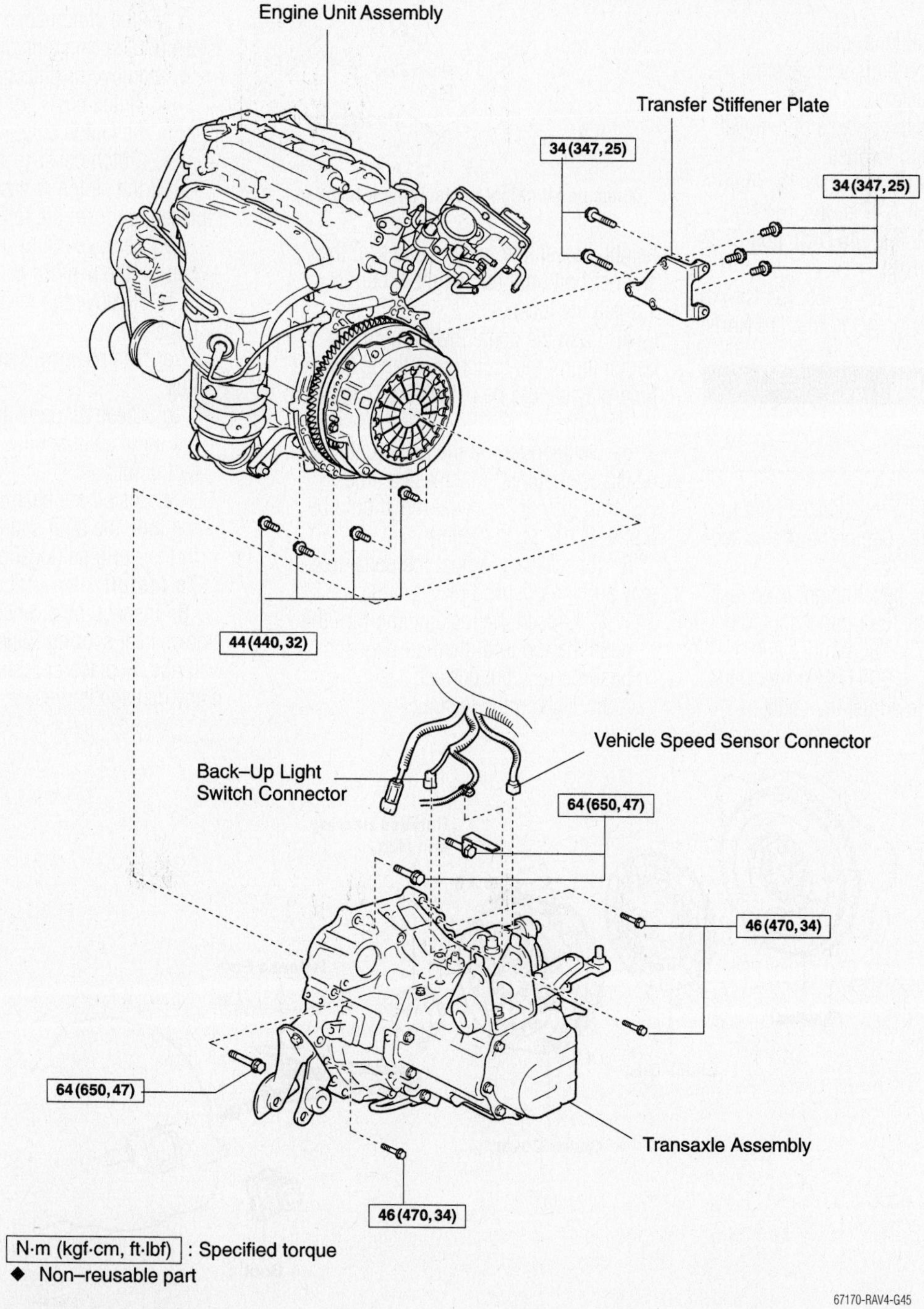

Engine Unit Assembly
Transfer Stiffener Plate
34 (347, 25)
34 (347, 25)
44 (440, 32)
Vehicle Speed Sensor Connector
Back–Up Light
Switch Connector
64 (650, 47)
46 (470, 34)
64 (650, 47)
Transaxle Assembly
46 (470, 34)
N·m (kgf·cm, ft·lbf) : Specified torque
◆ Non–reusable part
67170-RAV4-G45

Transaxle and related parts—2004 4WD

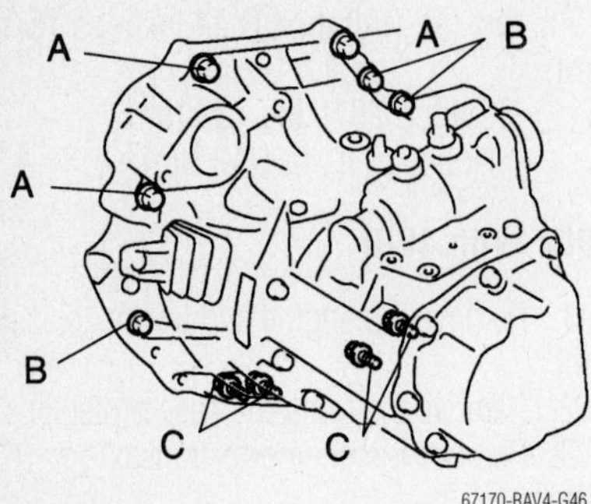

Transaxle bolt identification—2004 4WD

4. Remove the stiffener plate.

5. Remove the 10 bolts and separate the transaxle from the engine.

6. Installation is the reverse of removal. Observe the following torques:

- Transaxle-to-engine (see the illustration): bolt A 47 ft. lbs. (64 Nm); bolt B 34 ft. lbs. (46 Nm); bolt C 32 ft. lbs. (44 Nm)
- Stiffener plate: 25 ft. lbs. (34 Nm)
- Case protector: 13 ft. lbs. (18 Nm)

Clutch Assembly

ADJUSTMENTS

1. Before servicing the vehicle, refer to the precautions in the beginning of this section.

2. Check that the pedal height is correct. Pedal height from the floor panel should be: 6.889–7.283 in. (175–185mm) for 2001–03; 6.653–7.047 in. (169–179mm) for 2004.

3. If necessary to adjust the pedal

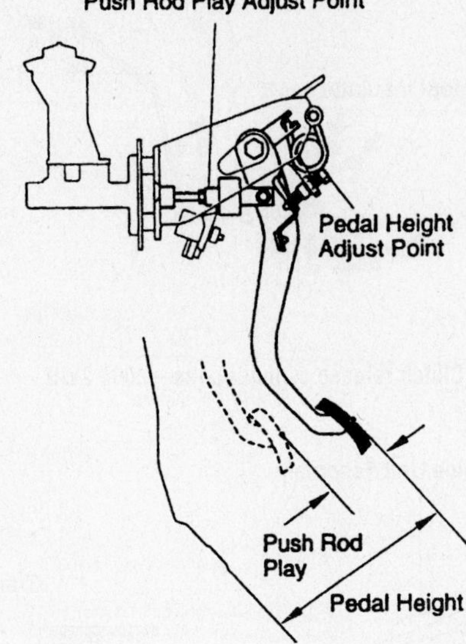

Clutch pedal height measurement location

height, loosen the locknut and turn the stopper bolt until the height is correct. Tighten the locknut.

4. Push the pedal inward until the beginning of the clutch resistance is felt. Free-play should be 0.197–0.591 in. (5–15mm).

5. Gently push on the pedal until the resistance begins to increase a little. Pushrod play at the pedal top should be 0.039–0.197 in. (1–5mm).

6. If necessary, adjust the pedal free-play and the pushrod play, as follows:

a. Loosen the locknut and turn the push the rod until the free-play and pushrod play are correct.

b. Tighten the locknut.

REMOVAL & INSTALLATION

1. Before servicing the vehicle, refer to the precautions in the beginning of this section.

2. Remove or disconnect the following:

- Negative battery cable
- Transaxle

3. Matchmark the clutch cover to the flywheel.

4. Remove the clutch pressure plate retaining bolts in small amounts and in a crisscross pattern to relieve the clutch disc spring tension.

5. At the clutch cover, loosen each bolt 1 turn until spring tension is released.

6. Remove or disconnect the following:

- Clutch cover set bolts and pull off the clutch cover with the clutch disc
- Clutch cover-to-flywheel bolts

7. If the clutch release bearing is to be replaced, perform the following:

a. Remove the bearing retaining clip(s), the bearing and hub.

b. Remove the release fork and the boot.

c. The bearing is press fitted to the hub.

d. Clean all parts and lightly grease the input shaft splines and all of the contact points.

e. Install the bearing/hub assembly, the fork, the boot and the retaining clip(s) in their original locations.

To install:

8. Inspect the flywheel surface for cracks, heat scoring (blue marks) and warpage. Replace or resurface the flywheel, if any damage is present.

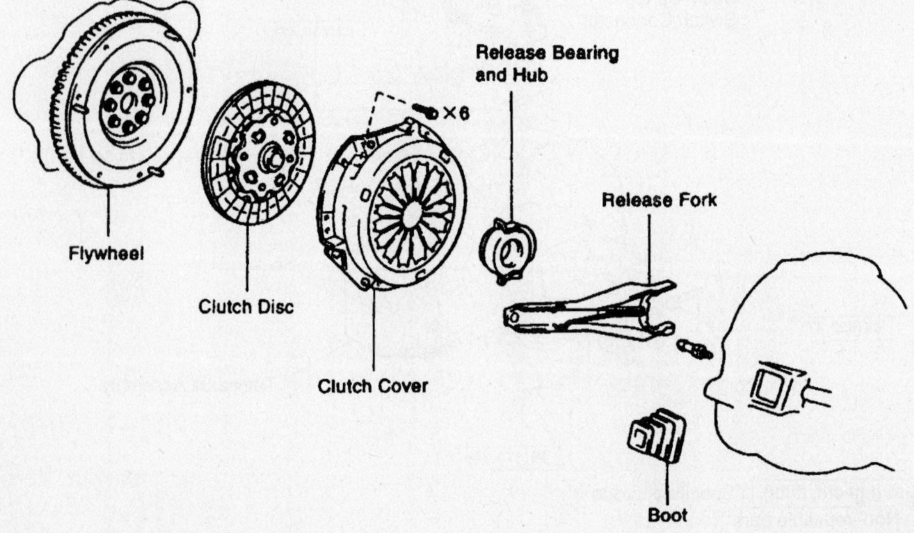

Clutch component assembly—2001–03

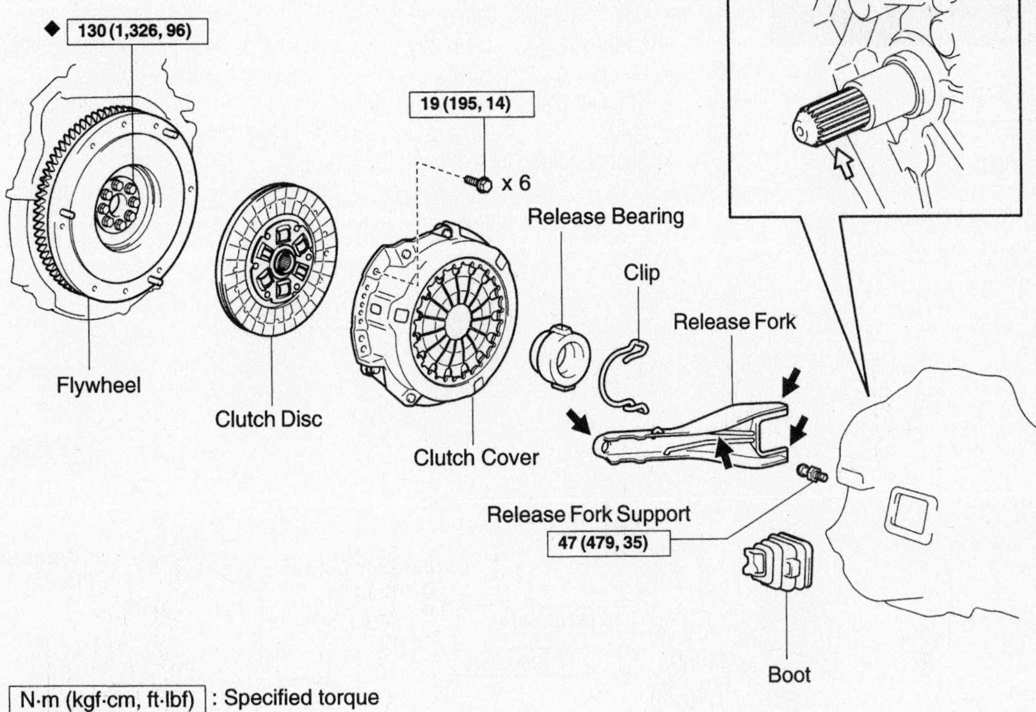

Clutch components–2004

➡Before installing any new parts, be sure they are clean. During installation, do not get grease or oil on any of the components, as this will shorten clutch life considerably.

9. Using a clutch alignment tool, position the clutch disc against the flywheel. The raised center section of the disc faces the transaxle.

10. Install or connect the following:
    - Clutch cover onto the flywheel by aligning the matchmarks
    - Clutch cover. Tighten the bolts in a crisscross pattern to 14 ft. lbs. (19 Nm).

11. Lubricate the release fork pivot and contact points, release bearing, bearing hub and input shaft spline surfaces with a suitable molybdenum disulfide lithium based or multi-purpose grease.

12. Install or connect the following:
    - Boot, release fork, hub and the bearing assemblies
    - Transaxle
    - Negative battery cable

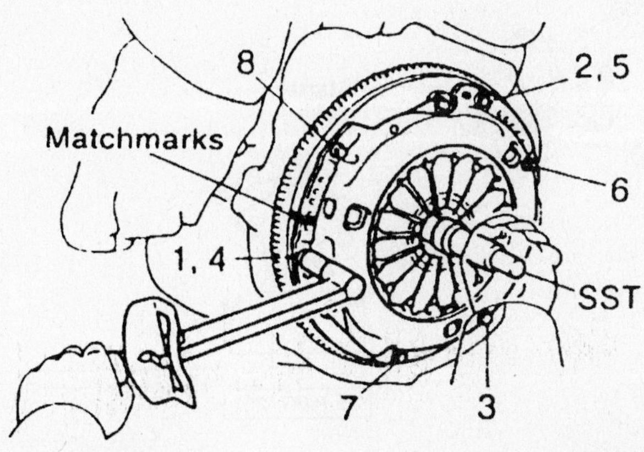

Torque sequence for the clutch cover

Hydraulic Clutch System

BLEEDING

1. Before servicing the vehicle, refer to the precautions in the beginning of this section.
2. Fill the clutch reservoir with brake fluid. Check the reservoir level frequently and add fluid as needed.
3. Connect one end of a vinyl tube to the bleeder plug on the slave cylinder and submerge the other end into a clear container half-filled with brake fluid.
4. Slowly pump the clutch pedal several times.
5. Have an assistant hold the clutch pedal down and loosen the bleeder plug until fluid and/or air starts to run out of the bleeder plug. Close the bleeder plug while the pedal is held to the floor.
6. Repeat Steps 2 and 3 until all the air bubbles are removed from the system.
7. Tighten the bleeder plug when all the air is gone.
8. Refill the master cylinder to the proper level as required.
9. Check the system for leaks.

Automatic Transmission/Transaxle Assembly

REMOVAL & INSTALLATION

2001–03 with 4WD

1. Before servicing the vehicle, refer to the precautions in the beginning of this section.

2. Remove or disconnect the following:
- Negative battery cable
- Engine/transaxle assembly
- Starter
- Stiffener plate, by removing the 3 bolts
- Rear endplate, by removing the 4 bolts
- 6 torque converter clutch mounting bolts
- Connectors and wiring harness, from the transaxle
- Center stiffener plate, by removing the 4 bolts

3. Remove the transaxle with the transfer assembly, as follows:
- 2 bolts
- 5 transaxle mounting bolts
- Transaxle from the engine

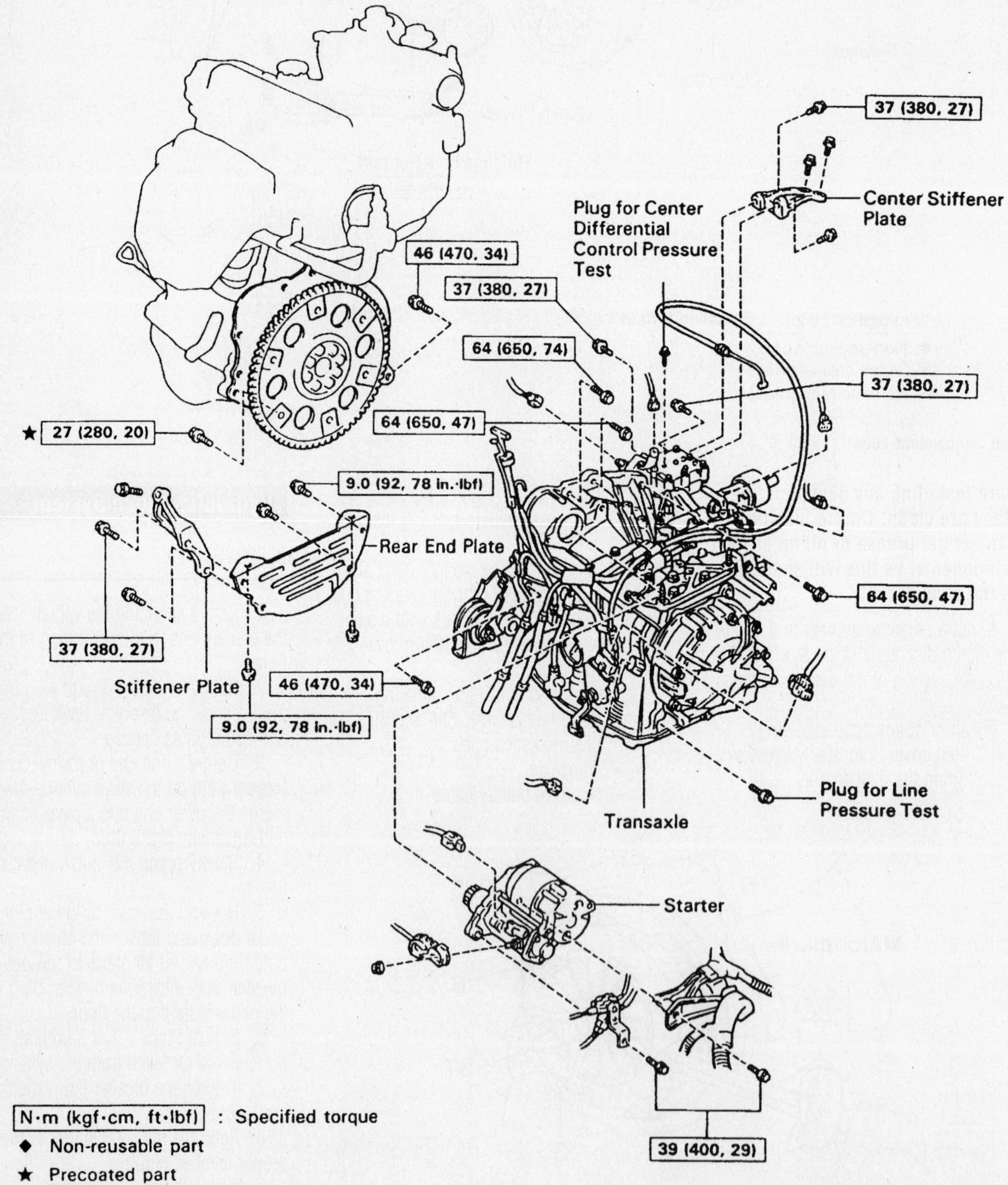

Automatic transaxle exploded view—2001–03 with 4WD

To install:

4. Install the transaxle. Tighten the 14mm head bolts to 47 ft. lbs. (64 Nm) and the 12mm head bolts to 34 ft. lbs. (46 Nm).
5. Install or connect the following:
   - Tighten both bolts to 27 ft. lbs. (37 Nm).
   - Center stiffener plate. Tighten the 4 bolts to 27 ft. lbs. (37 Nm).
   - Connectors and the wiring harness to the transaxle
   - Torque converter clutch mounting bolts. Tighten each bolt to 20 ft. lbs. (27 Nm).

➡Coat the threads of the bolts with an approved locking compound.

   - Rear endplate. Tighten the 4 bolts to 80 inch lbs. (9.0 Nm).
   - Stiffener plate. Tighten the 3 bolts to 27 ft. lbs. (37 Nm).
   - Starter. Tighten both bolts to 29 ft. lbs. (39 Nm).
   - Engine/transaxle assembly
   - Negative battery cable

2004 with 4WD

1. Remove the engine and transaxle as a unit. See "Engine Removal and Installation", above.
2. Remove the starter.

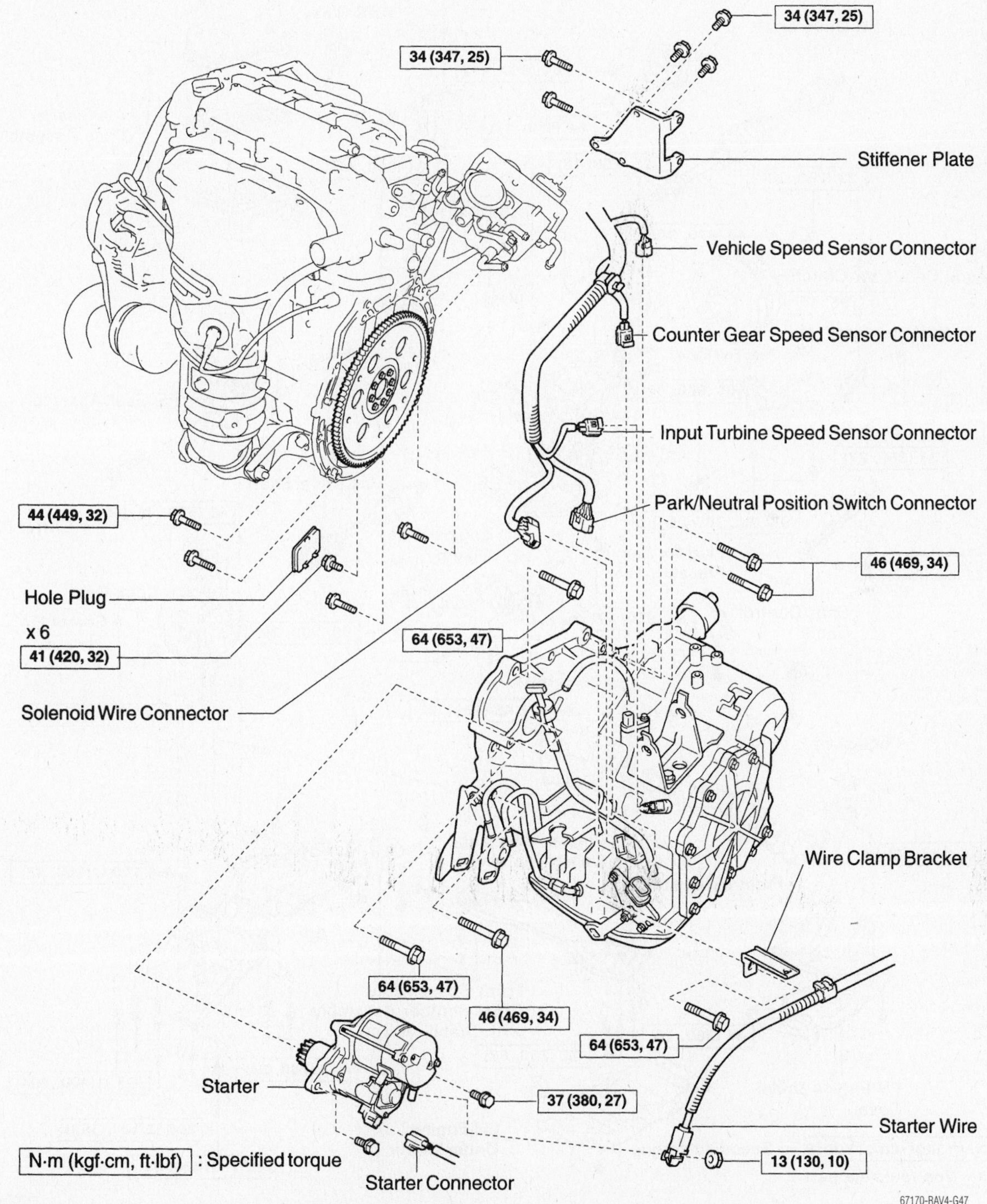

Automatic transaxle and related parts—2004 with 4WD

3. Remove the engine-to-transaxle wiring.

4. Remove the stiffener plate.

5. Remove the access plate and remove the 6 torque converter bolts.

6. Remove lower bolts and separate the transaxle from the engine.

7. Installation is the reverse of removal. Observe the following torques:

- Transaxle-to-engine lower bolts: 32 ft. lbs. (44 Nm)
- Stiffener plate: 25 ft. lbs. (34 Nm)
- Starter: 27 ft. lbs. (37 Nm)
- Torque converter bolts: 30 ft. lbs. (41 Nm). Install the green colored bolt first.

2001—03 with 2WD

1. Before servicing the vehicle, refer to the precautions in the beginning of this section.

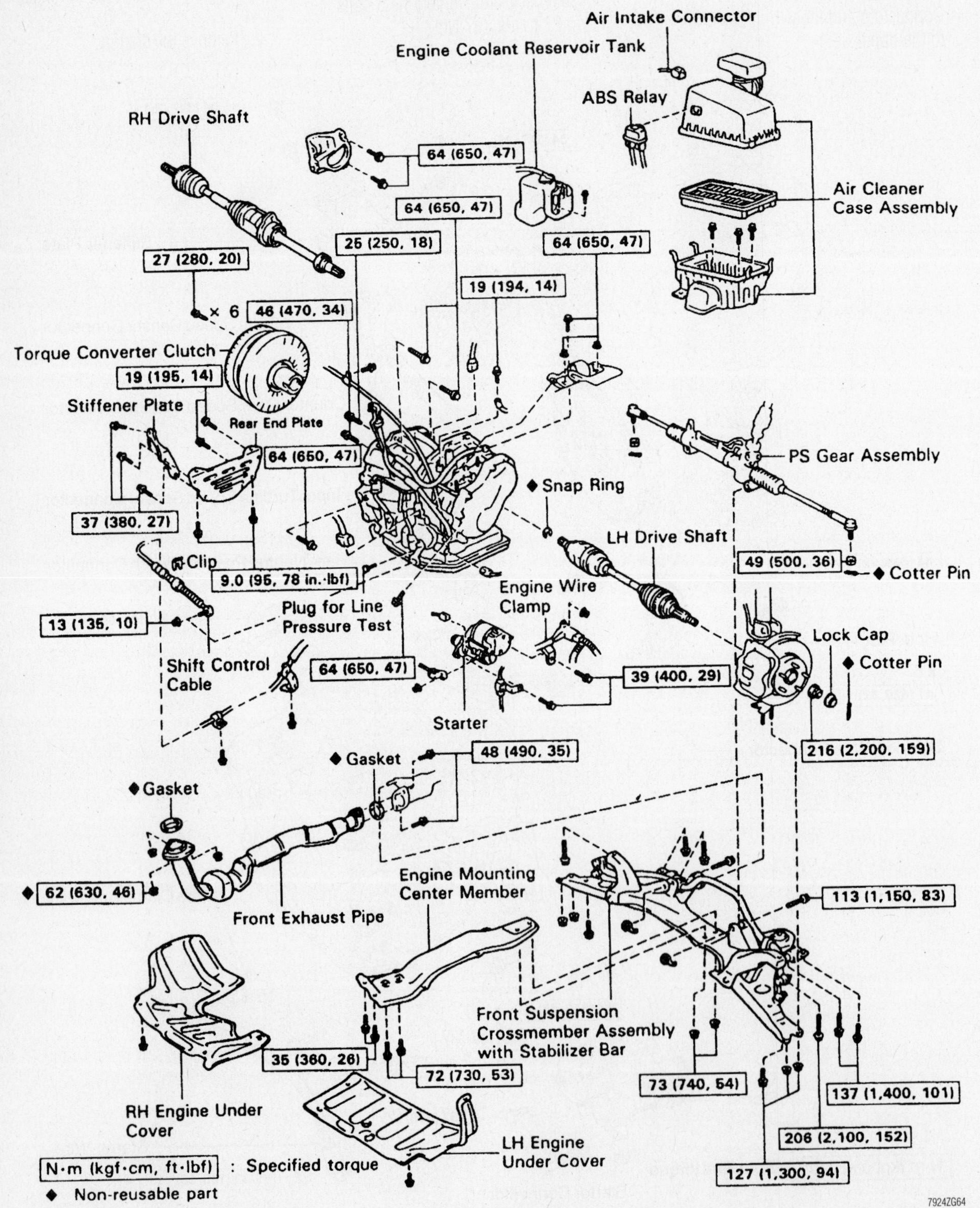

Transaxle exploded view—2001-03 with 2WD

2. Remove or disconnect the following:
   - Negative battery cable
   - Throttle cable
   - Engine coolant reservoir tank
   - Air cleaner assembly
   - Ground cable from the transaxle
   - Set nut of the engine wire clamp
3. Remove the starter, as follows:
   - Connector and nut from the starter
   - 2 bolts and the engine wire
   - Starter
4. Remove the 3 upper side transaxle mounting bolts.
5. Install an engine support fixture.
6. Remove or disconnect the following:
   - 2 bolts and 2 nuts from the left engine mount
   - Engine undercovers
7. Drain the fluid from the transaxle.
8. Remove the left and right halfshafts.
9. Remove the front exhaust pipe, as follows:
   - 2 front exhaust pipe-to-center exhaust pipe bolts and gasket
   - 3 front exhaust pipe-to-exhaust manifold nuts and gasket
   - Exhaust manifold
10. Disconnect the shift control cable from the transaxle and frame, as follows:
    - Control shaft lever nut
    - Clip and the control cable from the transaxle
    - 2 shift control cable-to-centermember bolts
    - Crossmember
11. Detach the following connectors:
    - Shift solenoid valve connector
    - Park/Neutral Position (PNP) switch connector
    - Vehicle Speed Sensor (VSS) connector
12. Remove or disconnect the following:
    - Oil cooler hoses from the transaxle
    - Rack and pinion from the crossmember by removing both nuts/bolts
13. Support the rack and pinion.
14. Support the suspension crossmember with a floor jack.
15. Remove or disconnect the following:
    - Crossmember-to-centermember fasteners
    - Crossmember with the sway bar
    - Stiffener plate by removing the 3 bolts
    - Rear endplate by removing the 4 bolts
    - 6 torque converter bolts
    - Both rear side transaxle mounting bolts
    - Transaxle

To install:

16. Install or connect the following:
    - Transaxle
    - Both rear side transaxle mounting bolts. Tighten the top bolt to 18 ft. lbs. (25 Nm) and the lower bolt to 34 ft. lbs. (46 Nm).
    - Torque converter bolts. Tighten the bolts to 20 ft. lbs. (27 Nm).

➡First install the gray bolt; then, install the 5 black bolts.

    - Rear endplate. Tighten the engine bolts to 78 inch lbs. (9.0 Nm) and the transaxle bolts to 14 ft. lbs. (19 Nm).
    - Stiffener plate. Tighten the 3 bolts to 27 ft. lbs. (37 Nm).
17. Install the front suspension crossmember and centermember with the sway bar. Tighten the nuts/bolts, as follows:
    - Bolt A: 152 ft. lbs. (206 Nm)
    - Bolt B: 101 ft. lbs. (137 Nm)
    - Bolt C: 26 ft. lbs. (35 Nm)
    - Bolt D: 53 ft. lbs. (72 Nm)
    - Nut: 54 ft. lbs. (73 Nm)
18. Install or connect the following:
    - Rack and pinion to the crossmember. Tighten the nuts to 83 ft. lbs. (113 Nm).
    - Oil cooler hoses with both clips
19. Connect the following connectors:
    - Shift solenoid valve connector
    - PNP switch connector
    - VSS connector
20. Install or connect the following:
    - Shift control cable to the transaxle. Tighten the nut to 10 ft. lbs. (13 Nm).
21. Install the front exhaust pipe, as follows:
    - Front exhaust pipe, using new gaskets
    - Front exhaust pipe to the exhaust manifold. Tighten the 3 nuts to 46 ft. lbs. (62 Nm).
    - Front exhaust pipe to the center exhaust pipe. Tighten both bolts to 35 ft. lbs. (48 Nm).
22. Install or connect the following:
    - Left and right halfshafts
    - Engine undercovers
    - Left engine mount. Tighten the both nuts/bolts to 47 ft. lbs. (64 Nm).
23. Remove the engine fixture.
24. Install or connect the following:
    - Upper side transaxle mount. Tighten the 3 bolts to 47 ft. lbs. (64 Nm).
    - Starter. Tighten both bolts to 29 ft. lbs. (39 Nm).
    - Engine wire
    - Starter wire with the nut
    - Engine wire clamp set nut
    - Ground cable to the transaxle. Tighten the bolt to 14 ft. lbs. (19 Nm).
    - Air cleaner assembly
    - Engine coolant reservoir tank
    - Throttle cable
    - Negative battery cable
25. Check all fluids.

2004 with 4WD

1. Remove the hood.
2. Remove the air cleaner assembly.
3. Remove the coolant reservoir.
4. Remove the control cables from the transmission.
5. Remove the wiring from the speed sensor and back-up light switch.
6. Remove the ground cable from the transaxle.
7. Remove the starter.
8. Remove the clutch release cylinder.
9. Remove the 3 upper transaxle-to-engine bolts.
10. Attach an engine support fixture, such as 12281-28010, or equivalent.
11. Disconnect the left engine mount from the transaxle.
12. Remove the engine under-covers.
13. Drain the transaxle.
14. Remove the front exhaust pipe.
15. Remove the halfshafts.
16. Remove the stabilizer bar.
17. Disconnect the oil cooler lines.
18. Remove the power steering gear mounting bolts. Support the gear.
19. Remove the front suspension crossmember.
20. Using a transmission jack, slightly raise the transaxle.
21. Remove the 4 lower transaxle-to-engine bolts.
22. Remove the front side mounting bolt, the 2 rear side mounting bolts and the transaxle.

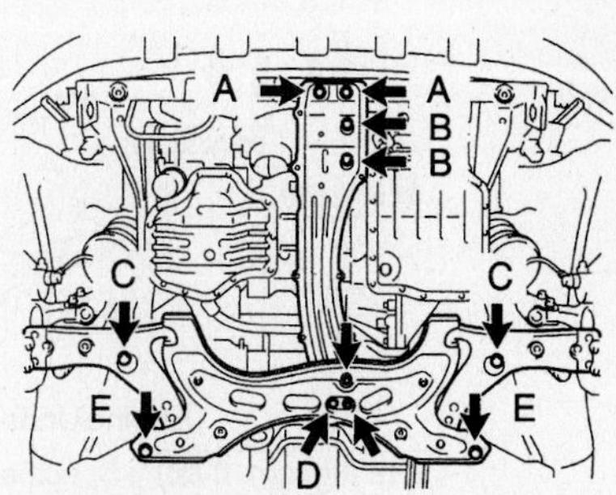

Crossmember bolt identification—2004 with 2WD

23. Installation is the reverse of removal. Observe the following torques:

- Lower transaxle-to-engine bolts: 32 ft. lbs. (44 Nm)
- Front side mounting bolt, the 2 rear side mounting bolts: 34 ft. lbs. (46 Nm)
- Torque converter bolts: 32 ft. lbs. (41 Nm). Torque the green bolt first.
- Front suspension crossmember (see the illustration): bolt A 28 ft. lbs. (39 Nm); bolt B 59 ft. lbs. (80 Nm); bolts C 80 ft. lbs. (113 Nm); bolt D 85 ft. lbs. (115 Nm); bolt E 115 ft. lbs. (157 Nm); nuts 85 ft. lbs. (115 Nm)
- Power steering gear bolts: 101 ft. lbs. (137 Nm)
- Stabilizer bar links: 32 ft. lbs. (44 Nm)

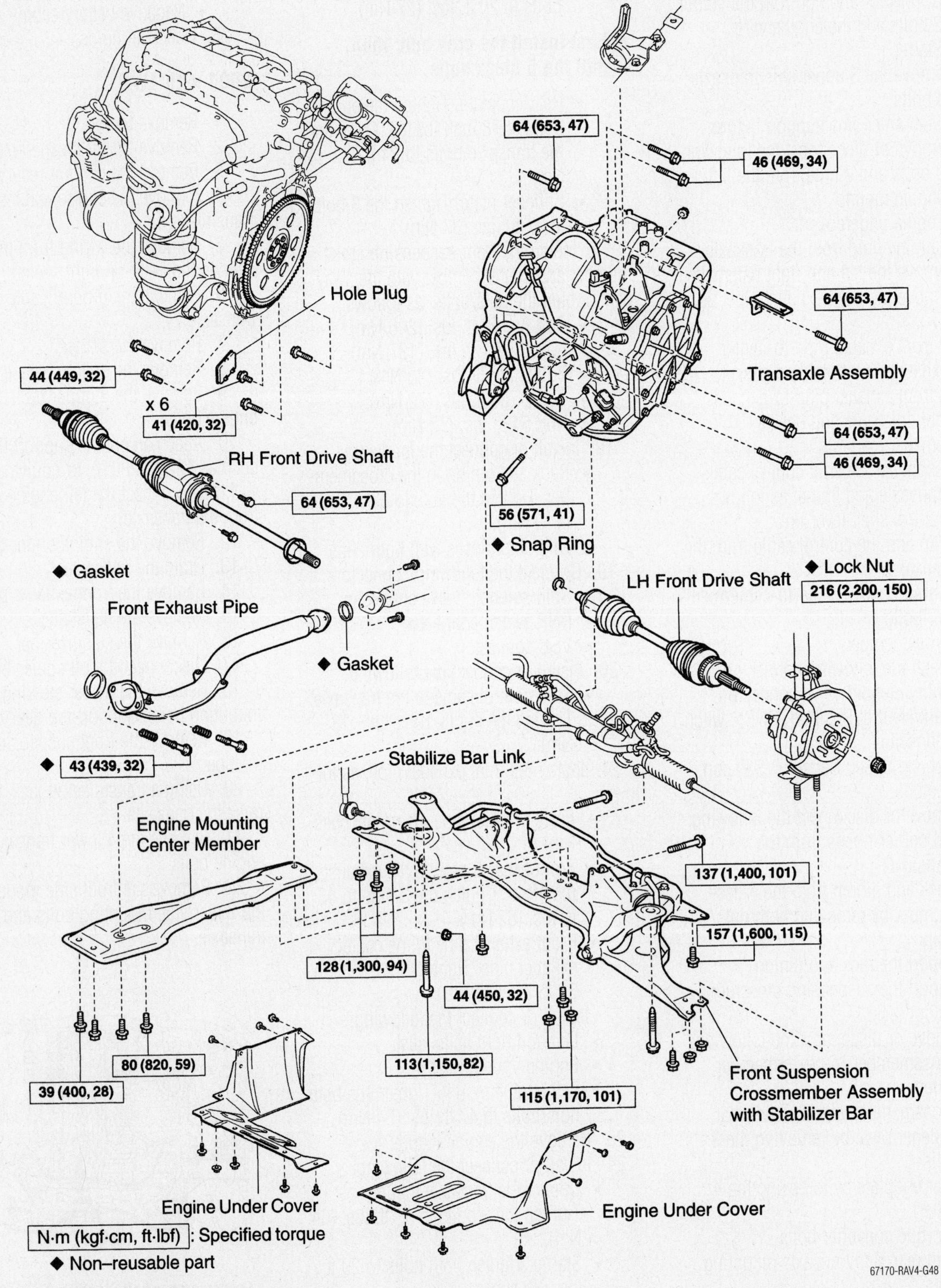

Automatic transaxle and related parts—2004 with 2WD

- Exhaust pipe-to-manifold: 32 ft. lbs. (43 Nm)
- Exhaust pipe-to-pipe: 14 ft. lbs. (19 Nm)
- Left engine mount through-bolt: 41 ft. lbs. (56 Nm)
- Upper transaxle-to-engine bolts: 47 ft. lbs. (64 Nm)
- Starter: 28 ft. lbs. (37 Nm)
- Hood: 10 ft. lbs. (13 Nm)

Transfer Case Assembly

REMOVAL & INSTALLATION

The transfer case is part of the transmission/transaxle assembly and is serviced with those units.

Halfshaft

REMOVAL & INSTALLATION

Front

2001–03

1. Before servicing the vehicle, refer to the precautions in the beginning of this section.

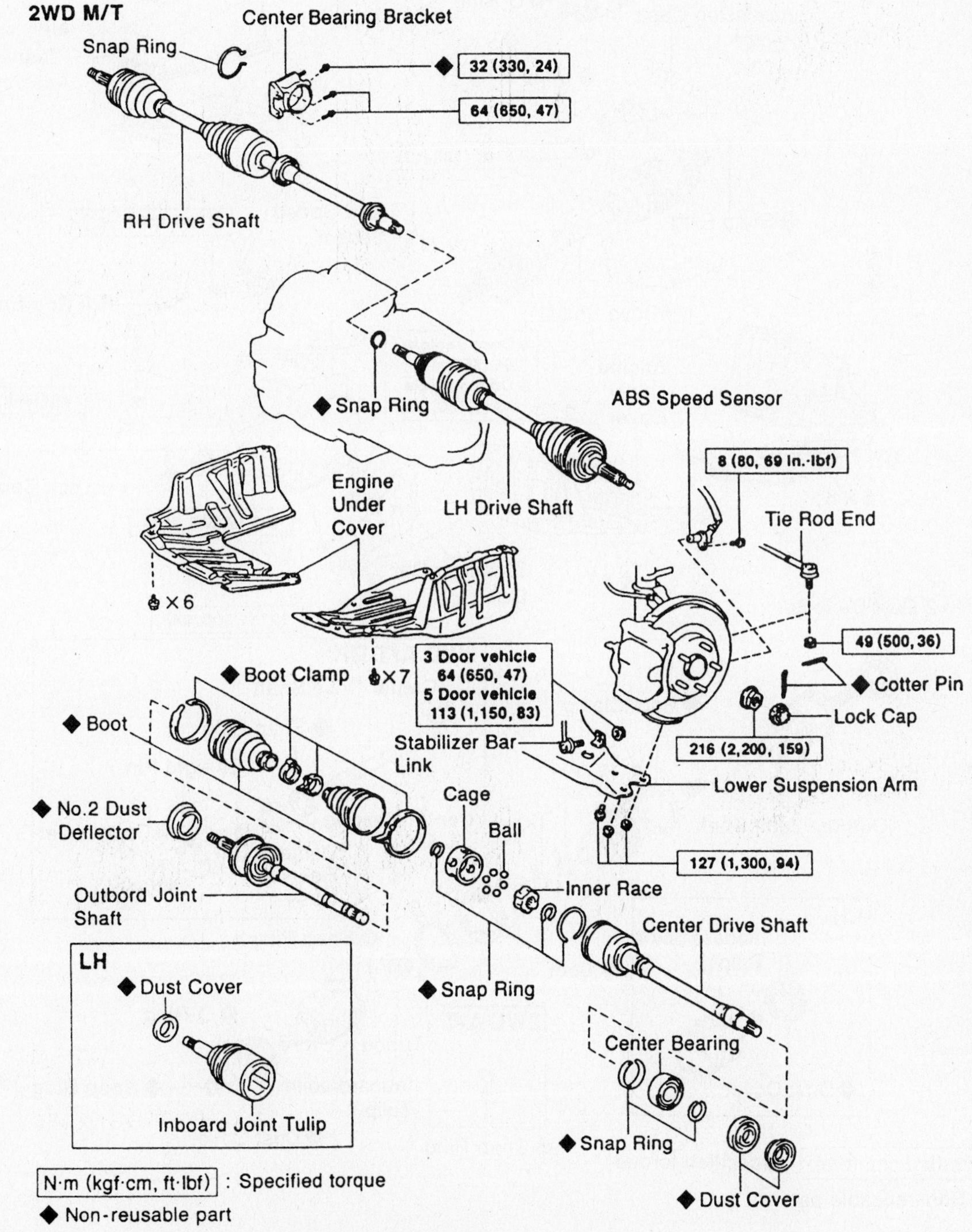

Front halfshaft exploded view (2WD with manual transmission only)—2001–03

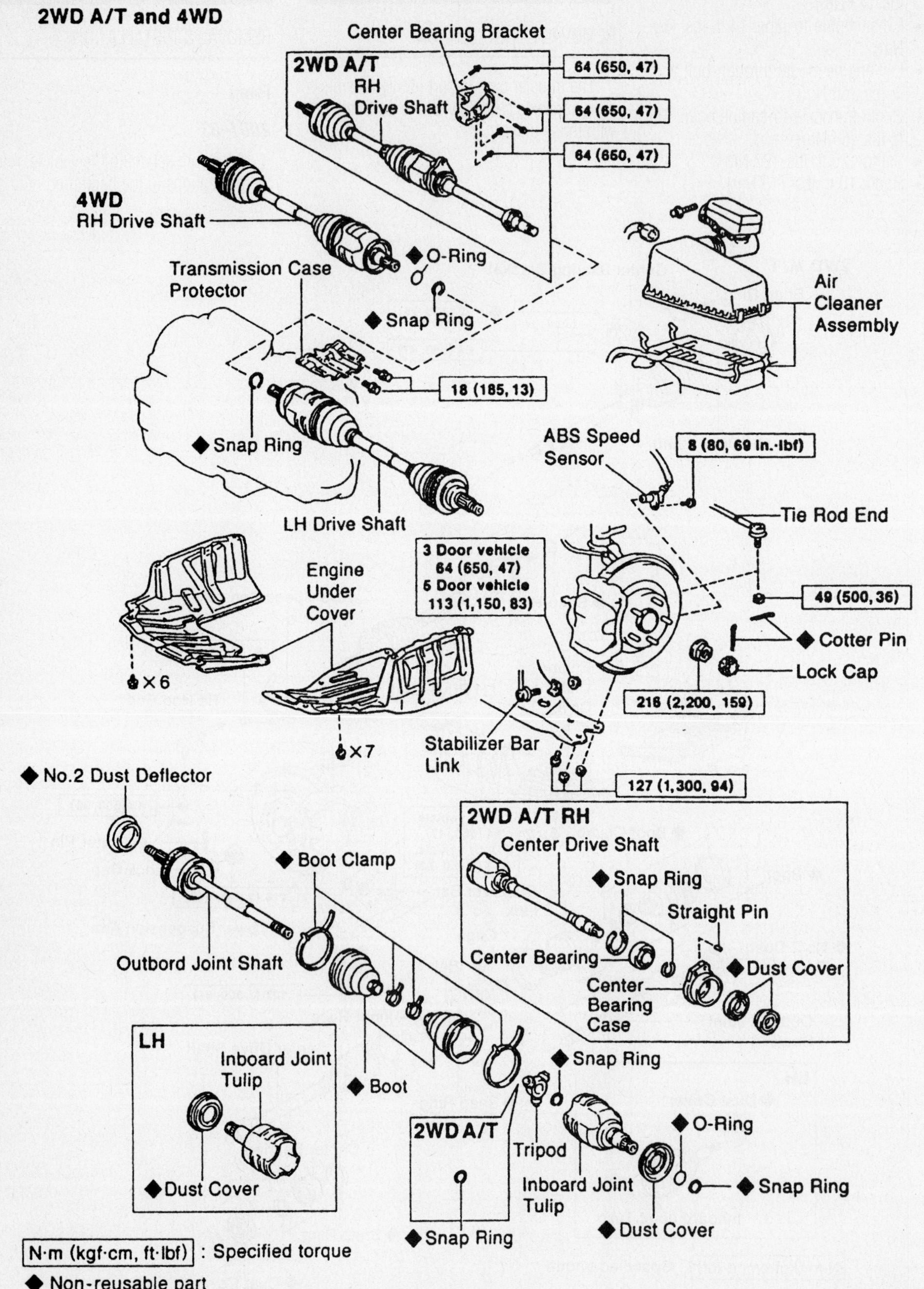

Front halfshaft exploded view (2WD with automatic transmission and 4WD)—2001–03

2. Remove or disconnect the following:
   - Negative battery cable
   - Engine undercover
3. Drain the transaxle.
4. Remove or disconnect the following:
   - Anti-lock Brake System (ABS) sensor by removing the bolt, if equipped
   - Cotter pin, lock cap and the locknut holding the halfshaft to the steering knuckle
   - Tie rod ends, from the steering knuckle
   - Sway bar link, from the lower control arm
   - Lower ball joint, from the lower control arm
   - Halfshaft from the axle hub, using a plastic hammer
5. If working on a 2WD right-hand halfshaft and the vehicle is equipped with a manual transaxle, perform the following to remove the halfshaft:
   - Snapring from the center bearing bracket, using a brass bar and hammer
   - Bolt and the center bearing bracket
   - Halfshaft with the center halfshaft
   - 2 bolts and the center bearing bracket
6. If working on a 2WD right-hand halfshaft and the vehicle is equipped with an automatic transaxle, perform the following to remove the halfshaft:
   - 2 bolts of the center bearing bracket and pull out the halfshaft together with the center bearing case and center halfshaft
   - 3 bolts and the center bearing bracket
7. If working on a 2WD left-hand, perform the following:
   - Halfshaft, using a brass bar and hammer
   - Snapring from the transaxle
8. If working of a 4WD right-hand halfshaft, perform the following:
   - Halfshaft, using a brass bar and hammer
   - Snapring from the transaxle
   - O-ring
9. If working on a 4WD left-hand side, perform the following:
   - Air cleaner
   - Transaxle case protector
   - Halfshaft, by prying it out using a hub wrench
   - Snapring

To install:

10. If working on a 4WD left-hand side, perform the following:
    - Snapring
    - Halfshaft to the transaxle
    - Transaxle case protector
    - Air cleaner
11. If working of a 4WD right-hand halfshaft, perform the following:
    - Snapring to the transaxle
    - New O-ring
    - Halfshaft to the transaxle
12. If working on a 2WD left-hand, perform the following:
    - Snapring
    - Halfshaft to the transaxle
13. If working on a 2WD right-hand halfshaft and the vehicle is equipped with an automatic transaxle, perform the following to remove the halfshaft:
    - Center bearing bracket. Tighten the 3 bolts to 47 ft. lbs. (64 Nm).
    - Halfshaft together with the center bearing case and center halfshaft. Tighten both bolts to 47 ft. lbs. (64 Nm).
14. If working on a 2WD right-hand halfshaft and the vehicle is equipped with a manual transaxle, perform the following to remove the halfshaft:
    - Center bearing bracket
    - Halfshaft with the center halfshaft
    - Center bearing bracket. Tighten the bolt to 24 ft. lbs. (32 Nm).
    - Snapring to the center bearing bracket
15. Install or connect the following:
    - Halfshaft to the axle hub
    - Lower ball joint to the lower control arm. Tighten the nuts/bolt to 94 ft. lbs. (127 Nm).
    - Sway bar link to the lower control arm. Tighten the nut to 47 ft. lbs. (64 Nm) for 3-door or to 83 ft. lbs. (113 Nm) for 5-door.
    - Tie rod end to the steering knuckle. Tighten the nut to 36 ft. lbs. (49 Nm).
    - New tie rod end cotter pin
    - Halfshaft to the axle hub. Tighten the locknut to 159 ft. lbs. (216 Nm).
    - Lock cap and cotter pin
    - ABS speed sensor with the bolt, if equipped
16. Fill the transaxle with gear oil (manual transmission) or ATF (automatic transmission).
17. Install or connect the following:
    - Engine undercover
    - Wheels
    - Negative battery cable
18. Check the ABS sensor signal.

2004

1. Remove the wheel.
2. Drain the transaxle. With 4WD, drain the transfer case.
3. Unstake and remove the locknut.
4. Unbolt the lower arm from the ball joint.
5. Remove the halfshaft from the hub, using a plastic hammer.
6. Remove the halfshaft. On 2wd vehicle, the right side shaft is retained by 2 bolts. All others are removed with a slidehammer and special tool 09520-01010, or equivalent.
7. Remove the snapring.
8. Installation is the reverse of removal. The new snapring is installed with the split facing downward. After installation, check that there is 2–3mm of axial play. Observe the following torques:
   - Right halfshaft bolts: 47 ft. lbs. (64 Nm)
   - Lower arm-to-ball joint: 94 ft. lbs. (128 Nm)
   - Hub nut: 159 ft. lbs. (216 Nm)

Rear

1. Before servicing the vehicle, refer to the precautions in the beginning of this section.
2. Remove or disconnect the following:
   - Negative battery cable
   - Rear wheels
   - Anti-lock Brake System (ABS) speed sensor from the axle assembly by removing the bolt, if equipped
   - Cotter pin, lock cap and the nut holding the halfshaft to the axle carrier
3. Place matchmarks on the halfshaft and side gear shaft.
4. Remove or disconnect the following:
   - Halfshaft from the differential side gear shaft, by removing the 4 nuts and washers
   - Halfshaft from the axle carrier, using a plastic hammer

To install:

5. Install or connect the following:
   - Halfshaft to the axle carrier
   - Halfshaft to the differential side gear shaft, by aligning the marks. Tighten the 4 nuts to 41 ft. lbs. (56 Nm).
   - Nut, lock cap and the cotter pin to hold the halfshaft to the axle carrier. Tighten the nut to 152 ft. lbs. (206 Nm) for 2001–03; 159 ft. lbs. (216 Nm) for 2004.
   - ABS sensor. Tighten the bolt to 69 inch lbs. (8 Nm).
   - Rear wheels
   - Negative battery cable
6. Check the ABS sensor signal.

2WD

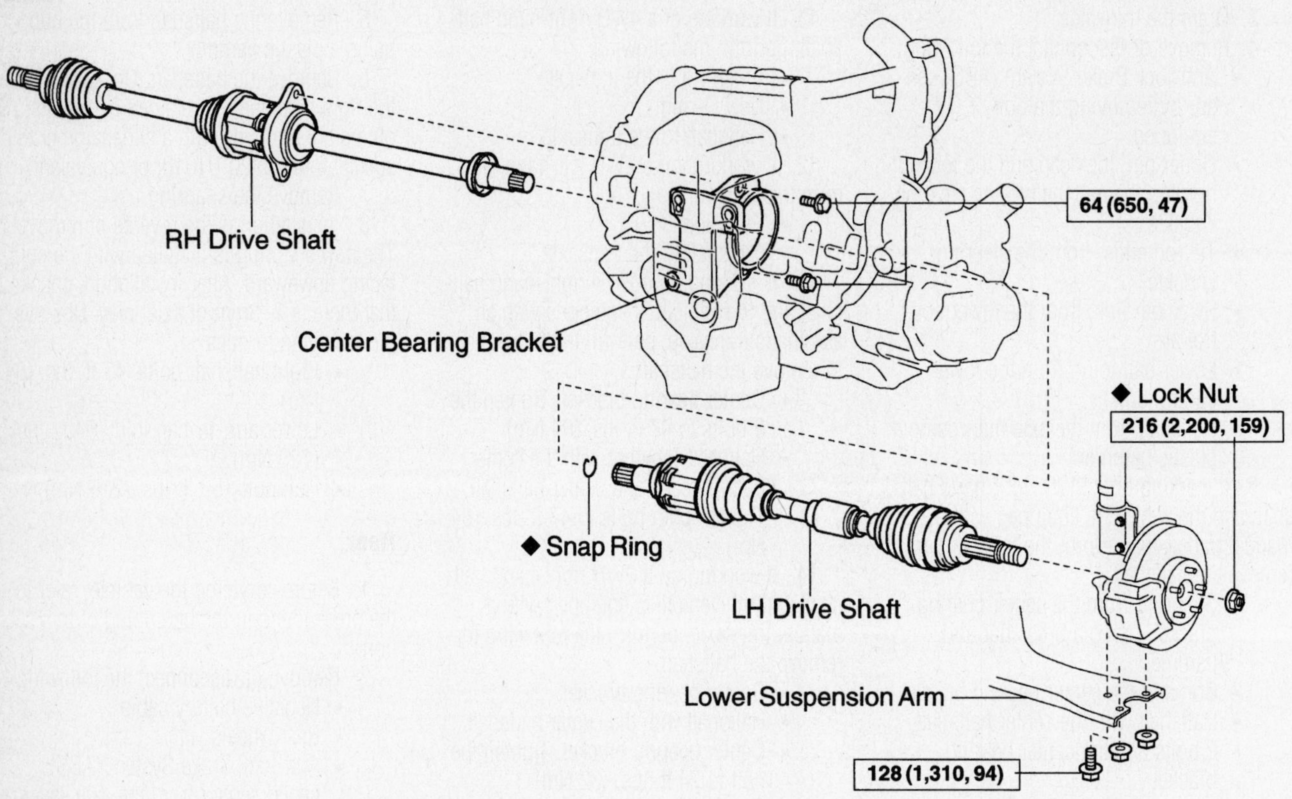

4WD

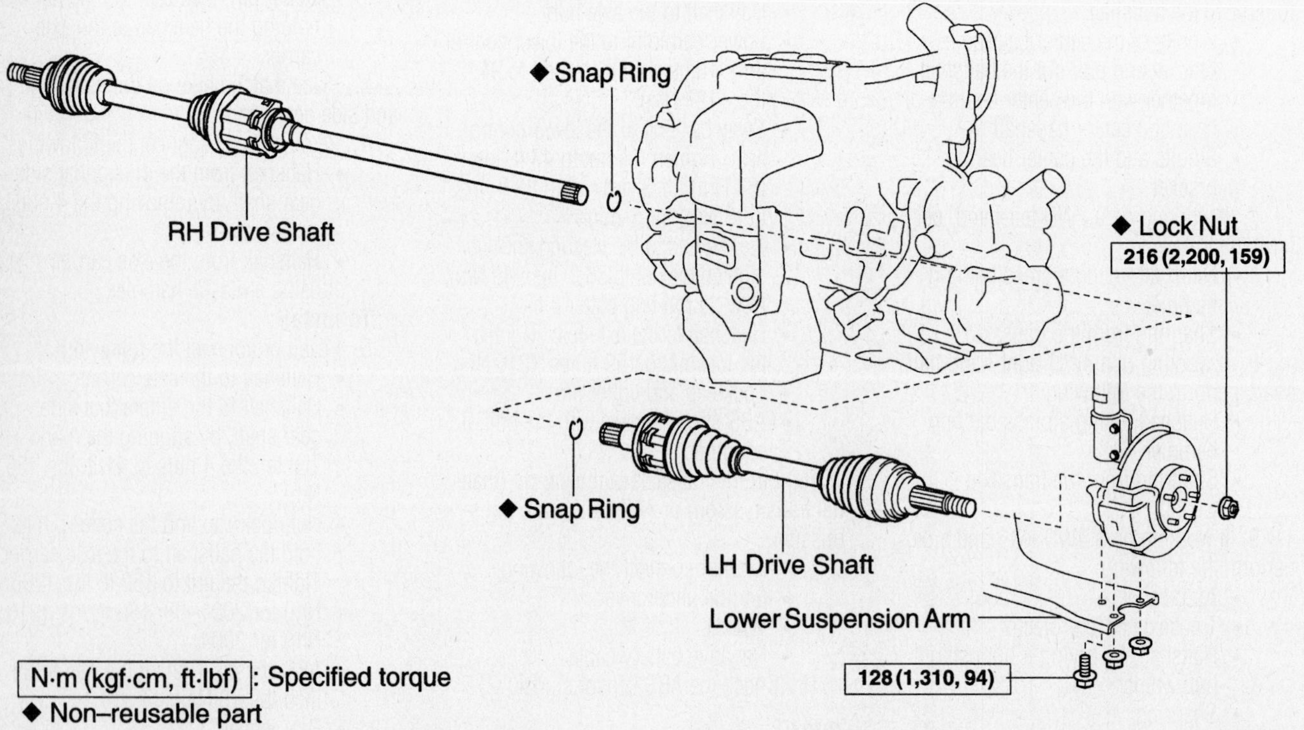

N·m (kgf·cm, ft·lbf) : Specified torque

◆ Non–reusable part

Front halfshafts—2004

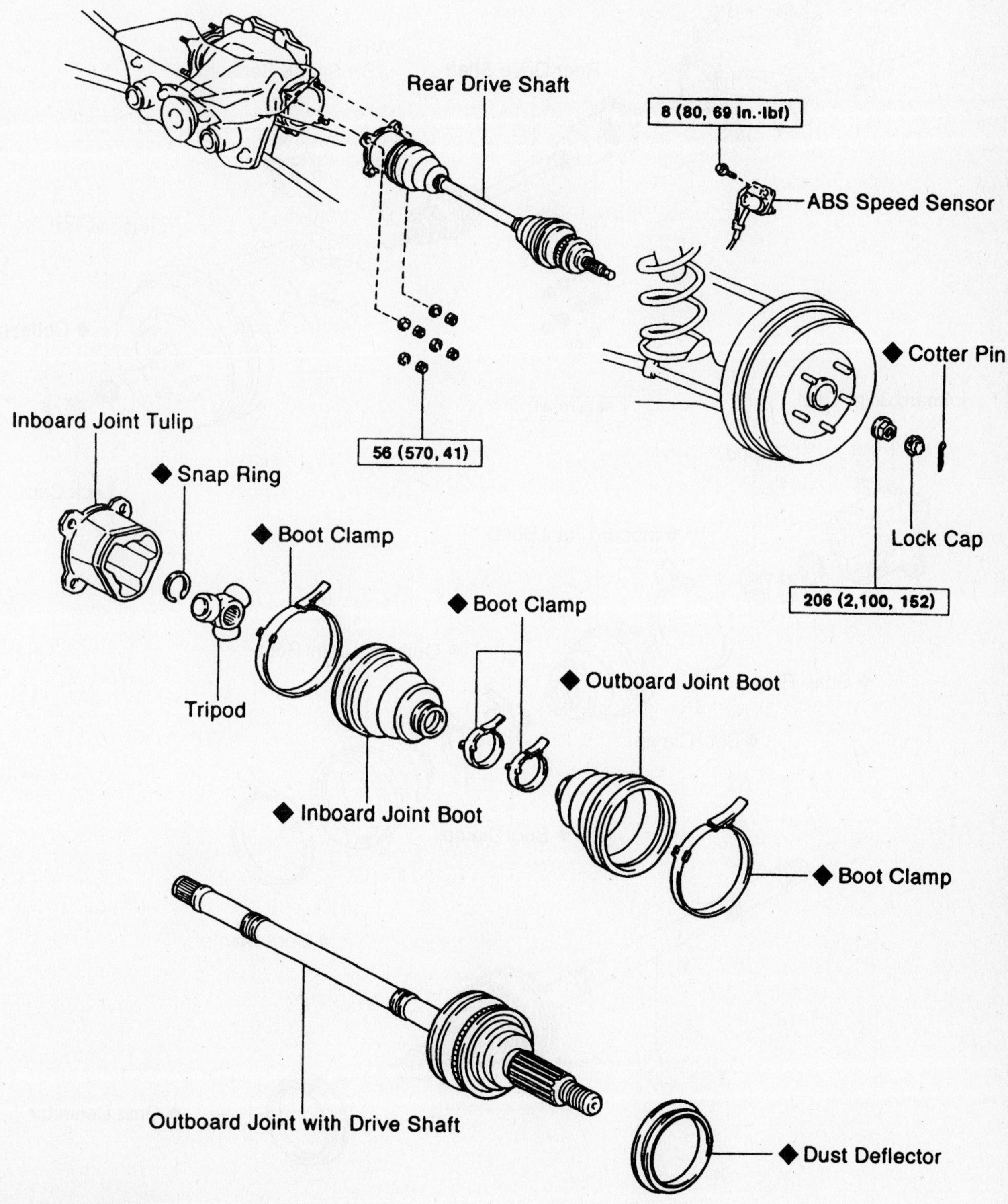

N·m (kgf·cm, ft·lbf) : Specified torque

◆ Non-reusable part

7924ZG72

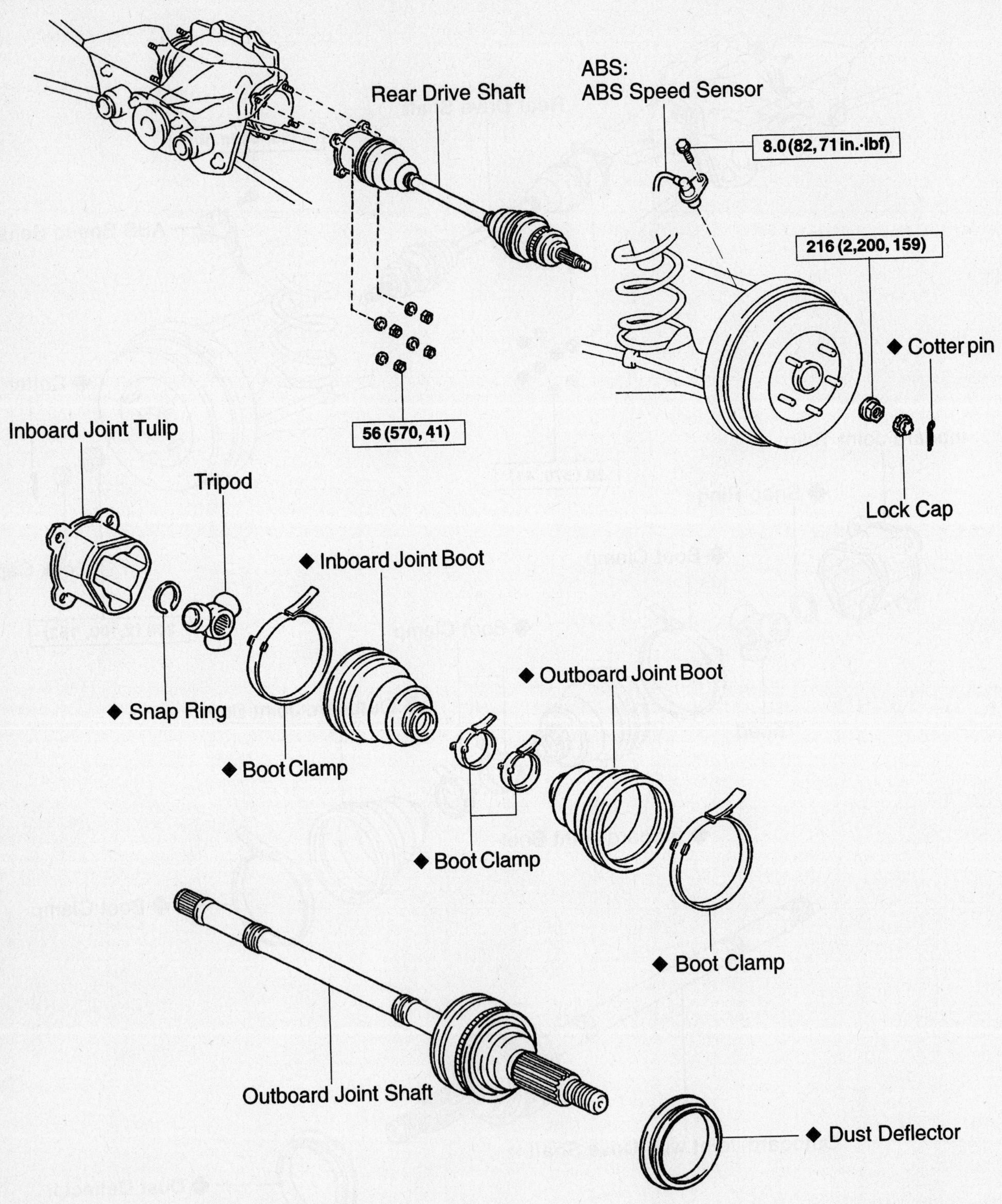

Rear halfshaft and related parts—2004

CV-Joints

OVERHAUL

Front

2WD WITH M/T

1. Before servicing the vehicle, refer to the precautions in the beginning of this section.
2. Remove the inboard and outboard joint boot clamps.
3. Disassemble the inboard joint tulip, as follows:
   - Snapring from the inboard joint tulip (center driveshaft)
   - Inboard joint tulip (center driveshaft), by matchmarking it to the shaft
4. Disassemble the inboard joint, as follows:
   - Matchmark the inner race and cage to the driveshaft
   - 6 balls and cage
   - Snapring
   - Inner race, using a brass bar and a hammer
   - Snapring
5. Remove the inboard and outboard joint boots and inboard joint clamps.

⁂ WARNING

Do not disassemble the outboard joint.

6. Remove or disconnect the following:
   - Dust cover from the center driveshaft, using a press
   - Dust cover from the inboard joint tulip, using tool 09950-00020 and a press
7. Remove the bearing, as follows:
   - Dust cover from the inboard joint tulip, using tool 09950-00020 and a press
   - Snapring
   - Bearing, using a press
   - Snapring
8. Remove the No. 2 dust deflector, using a screwdriver and a hammer.

To assemble:

9. Install a new No. 2 dust deflector, using tools 09309-36010, 09316-20011 and a press.
10. Install the bearing, as follows:
    - New snapring
    - Bearing, using a press
    - New snapring
    - Dust cover, until the clearance between the dust cover and the bearing is 0.039 in. (1.0mm)

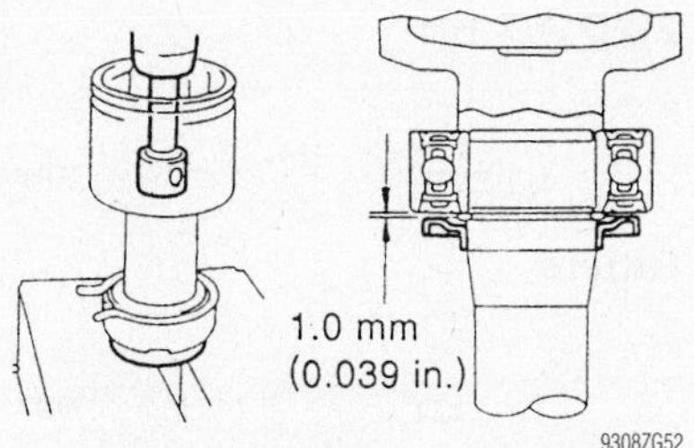

View of the bearing-to-dust cover clearance–2WD with M/T

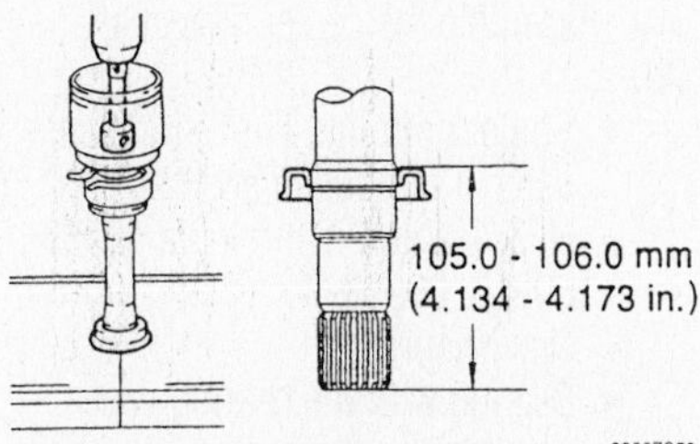

View of the dust cover-to-center drive distance–2WD with M/T

RH

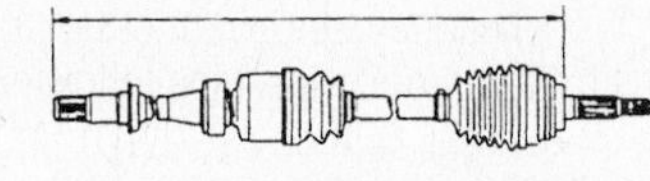

LH

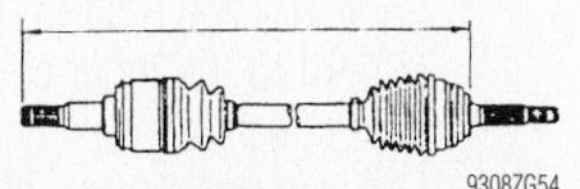

Measuring the front halfshaft lengths–2WD with M/T

11. Install or connect the following:
    - Right dust cover, until the distance from the tip of the center drive is 4.134–4.173 in. (105.0–106.0mm) to the inner edge of the dust cover
    - Left side dust cover, using a press
12. Temporarily install new outboard/inboard joint boots using new clamps, as follows:
    a. Warp tape around the driveshaft splines.
    b. Install the new outboard joint boot onto the driveshaft with both new clamps.
    c. Install the new inboard joint boot onto the driveshaft.
13. Assemble the inboard joint onto the driveshaft, as follows:
    - New snapring
    - Cage

➡The smaller diameter side must face outboard.

   - Inner race, using a brass bar and hammer by aligning the matchmarks

⁂ WARNING

Be careful not to damage the inner race.

   - New snapring
14. Install the outboard joint boot packed with grease from the boot kit.
15. Install the inboard joint tulip, as follows:
    - Cage to the inner race by aligning the matchmarks
    - 6 cage balls

➡Lubricate the balls with grease to keep them from falling.

   - Inboard joint tulip, by aligning the matchmarks
   - New snapring
   - Temporarily, install the inboard joint boot packed with grease from the kit
16. Install the boot clamps to both boots, as follows:
    - Both boots to the shaft grooves
    - Halfshaft length should be 32.988–33.382 in. (837.9–847.9mm) for the right side or 21.165–21.559 in. (537.6–547.6mm) for the left side
    - Both new clamps on the inboard joint boot
    - Crimp the new clamps using tool 09521-24010
    - Adjust the crimp clearance to 0.047–0.157 in. (1.2–4.0mm)

2WD WITH A/T

1. Before servicing the vehicle, refer to the precautions in the beginning of this section.
2. Remove the inboard and outboard joint boot clamps.
3. Disassemble the inboard joint tulip, as follows:
   - Matchmark the tri-pot, inboard joint tulip or center driveshaft to the driveshaft

⁂ WARNING

Do not use punch marks.

   - Inboard joint tulip from the driveshaft
4. Remove the inboard and outboard joint clamps.
5. Remove the tri-pot joint, as follows:
   - Snapring

- Matchmark the tri-pot joint to the driveshaft
- Tri-pot joint, using a brass bar and hammer

⁂ WARNING

Do not tap the roller.

6. Remove or disconnect the following:
   - Inboard and outboard joint boots

➡Do not disassemble the outboard joint.

   - Dust cover from the center driveshaft, using a press, for 2WD on the right side
   - Dust cover from the inboard joint tulip, using tool 09950-00020 and a press, for 2WD on the left side and 4WD
7. Disassemble the center driveshaft, as follows:
   - Snapring
   - Bearing case, using a press
   - Straight pin from the bearing case, using a pin punch and hammer
   - Dust cover, using tool 09950-00020 and a press
   - Snapring
   - Bearing, using a press
8. Remove the No. 2 dust deflector, using a screwdriver and hammer.

To assemble:

9. Install a new No. 2 dust deflector, using a press.

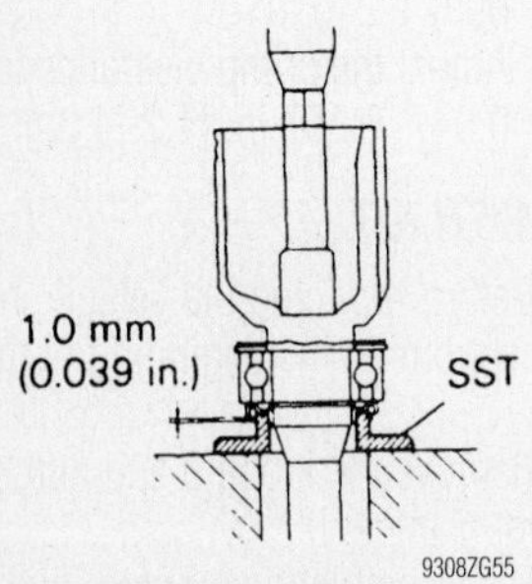

View of the bearing-to-dust cover clearance–2WD A/T and 4WD

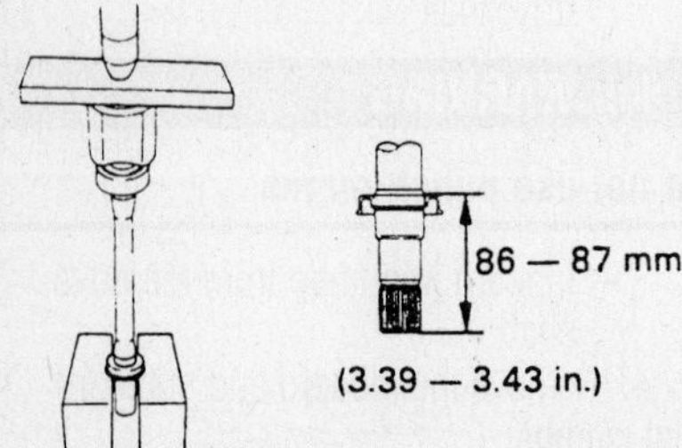

View of the dust cover-to-center drive distance–2WD A/T and 4WD

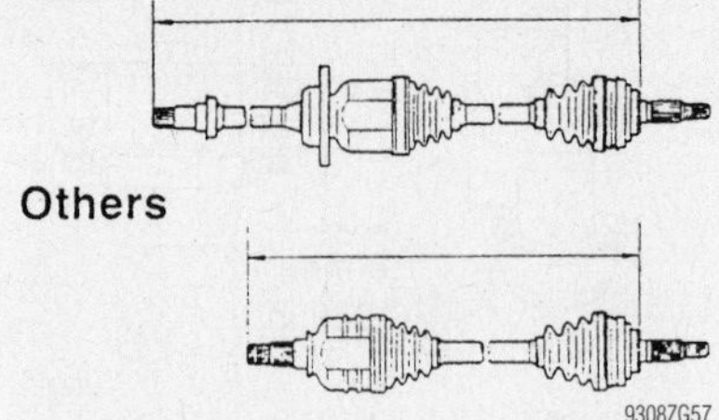

Measuring the front halfshaft lengths–2WD A/T and 4WD

10. Assemble the center driveshaft, as follows:
    - Straight pin into the bearing case, using a pin punch and hammer
    - New bearing, using tools 09959-60010, 09950-70010 and a press
    - New snapring
    - Bearing with the bearing case assembly to the center driveshaft, using tool 09710-30021 and a press
    - New snapring
    - New dust cover, until the clearance between the dust cover and the bearing is 0.039 in. (1.0mm)
11. Install or connect the following:
    - Right dust cover (2WD), until the distance from the tip of the center drive is 3.39–3.34 in. (86–87mm) to the inner edge of the dust cover
    - Left side dust cover (2WD and 4WD), using a press
12. Temporarily install new outboard/inboard joint boots using new clamps, as follows:
    a. Warp tape around the driveshaft splines.
    b. Install the new outboard joint boot onto the driveshaft.
    c. Install the new inboard joint boot onto the driveshaft.
13. Install the tri-pot joint, as follows:
    - Tri-pot joint, face the beveled side toward the outboard joint and align the matchmarks
    - Tri-pot joint onto the driveshaft, using a press

⁂ WARNING

Be careful not to tap the roller.

    - New snapring
14. Install the outboard joint boot packed with grease from the boot kit.
15. Install the inboard joint tulip, as follows:
    - Pack the inboard joint boot with grease from the boot kit
    - Inboard joint tulip, by aligning the matchmarks
    - Temporarily, install the inboard joint boot packed with grease from the kit
16. Install the boot clamps to both boots, as follows:
    - Both boots to the shaft grooves
    - Halfshaft length should be 33.055–33.449 in. (839.6–849.6mm) for the right side on 2WD with A/T, 21.397–21.791 in. (543.5–553.5mm) for the left side on 2WD with A/T, 19.929–20.323 in. (506.2–516.2mm) for the right side on 4WD or 19.803–20.197 in. (503–511mm) for the left side on 4WD
    - Both new boot clamps boot
    - Bend the band and lock it using a screwdriver

Rear

1. Before servicing the vehicle, refer to the precautions in the beginning of this section.
2. Remove the inboard and outboard joint boot clamps.
3. Disassemble the inboard joint tulip, as follows:
   - Matchmark the tri-pot, inboard joint tulip or center driveshaft to the driveshaft

⁂ WARNING

Do not use punch marks.

   - Inboard joint tulip from the driveshaft
4. Remove the tri-pot joint, as follows:
   - Snapring
   - Matchmark the tri-pot joint to the driveshaft

⁂ WARNING

Do not use punch marks.

   - Tri-pot joint, using a brass bar and hammer

⁂ WARNING

Do not tap the roller.

5. Remove or disconnect the following:
   - Inboard and outboard joint boots

➡Do not disassemble the outboard joint.

   - No. 2 dust deflector from the center driveshaft, using a screwdriver and hammer

To assemble:

6. Install a new No. 2 dust deflector, using tools 09309-36010, 09316-20011 and a press.

7. Temporarily install new outboard/inboard joint boots using new clamps, as follows:

   a. Warp tape around the driveshaft splines.

   b. Install the new outboard joint boot onto the driveshaft.

   c. Install the new inboard joint boot onto the driveshaft.

8. Install the tri-pot joint, as follows:

   - Tri-pot joint, face the beveled side toward the outboard joint and align the matchmarks
   - Tri-pot joint onto the driveshaft, using a brass bar and hammer

WARNING

Be careful not to tap the roller.

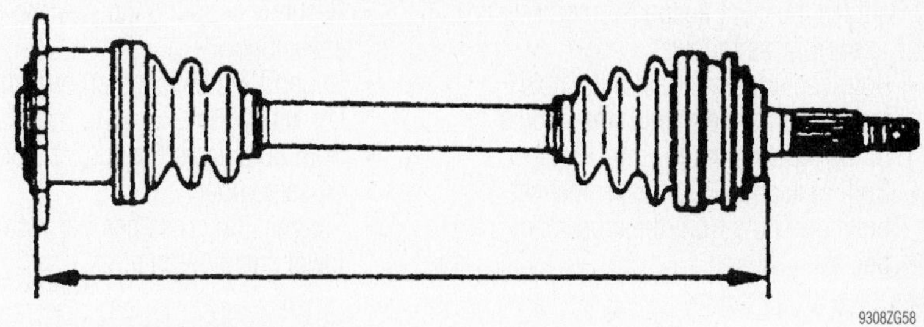

Measuring the rear halfshaft lengths

   - New snapring

9. Install the outboard joint boot packed with grease from the boot kit.

10. Install the inboard joint tulip, as follows:

   - Pack the inboard joint boot with grease from the boot kit
   - Inboard joint tulip, by aligning the matchmarks
   - Inboard joint boot packed with grease from the kit

11. Install the boot clamps to both boots, as follows:

   - Both boots to the shaft grooves
   - Halfshaft length should be 23.392–23.795 in. (594.4–604.4mm) for the right side or 21.590–21.984 (548.4–558.4mm) for the left side
   - New boot clamps boot
   - Bend the band and lock it using a screwdriver

STEERING AND SUSPENSION

Air Bag

CAUTION

Some vehicles are equipped with an air bag system. The system must be disarmed before performing service on, or around, system components, the steering column, instrument panel components, wiring and sensors. Failure to follow the safety precautions and the disarming procedure could result in accidental air bag deployment, possible injury and unnecessary system repairs.

PRECAUTIONS

Several precautions must be observed when handling the inflator module to avoid accidental deployment and possible personal injury.

- Never carry the inflator module by the wires or connector on the underside of the module.
- When carrying a live inflator module, hold securely with both hands and ensure that the bag and trim cover are pointed away.
- Place the inflator module on a bench or other surface with the bag and trim cover facing up.
- With the inflator module on the bench, never place anything on or close to the module which may be thrown in the event of an accidental deployment.

DISARMING

To avoid personal injury when working on vehicles equipped with an air bag, the negative battery cable must be disconnected and at least 90 seconds must elapse before working on the system. Failure to do so may result in deployment of the air bag.

Power Rack and Pinion Steering Gear

REMOVAL & INSTALLATION

2001–03

1. Before servicing the vehicle, refer to the precautions in the beginning of this section.

2. Disconnect the negative battery cable.

CAUTION

To avoid personal injury when working on air bag equipped vehicles, work must be started after 90 seconds or longer from the time the ignition switch is turned to the LOCK position and the negative battery terminal is disconnected. If the air bag system is disconnected with the ignition switch at the ON or ACC, diagnostic codes will be set. When removing the air bag, take care not to pull the air bag wiring harness. When carrying the wheel pad, carry it with the upper surface facing away. When storing it, keep the upper surface of the pad facing upward.

3. Turn the key to the **LOCK** position and lock the steering wheel in place.

4. Place a drain pan under the steering rack.

5. Remove or disconnect the following:

   - Front wheels
   - Right and left-hand engine undercovers
   - Right and left-hand tie rod ends from the steering knuckle
   - Front exhaust pipe
   - Sway bar with the links

6. Disconnect the No. 2 intermediate shaft from the rack and pinion, as follows:

   a. Loosen the top bolt.

   b. Remove the lower bolt holding the No. 2 intermediate shaft to the rack and pinion.

   c. Shift the No. 2 intermediate shaft and place matchmarks on the control valve shaft and the No. 2 intermediate shaft.

   d. Disconnect the No. 2 shaft from the rack and pinion.

7. Install or connect the following:

   - Pressure feed and return tubes from the rack and pinion, using a line wrench
   - Pressure feed and return tube clamps, by removing the bolt
   - Right and left lower control arms, from the steering knuckle

8. Remove the front suspension crossmember assembly, as follows:
   - Both centermember set nuts, holding the centermember to the middle of the crossmember.
   - Both rack and pinion assembly set bolts and nuts from the crossmember.
   - Securely suspend the steering gear assembly.
   - Support the suspension crossmember with a jack.
   - Both bolts from the suspension crossmember
   - Suspension crossmember with the lower suspension arms

9. Remove the rack and pinion.

To install:

10. Install or connect the following:
    - Rack and pinion

11. Install the crossmember to the vehicle, as follows:
    - Suspension crossmember with the lower control arms. Tighten

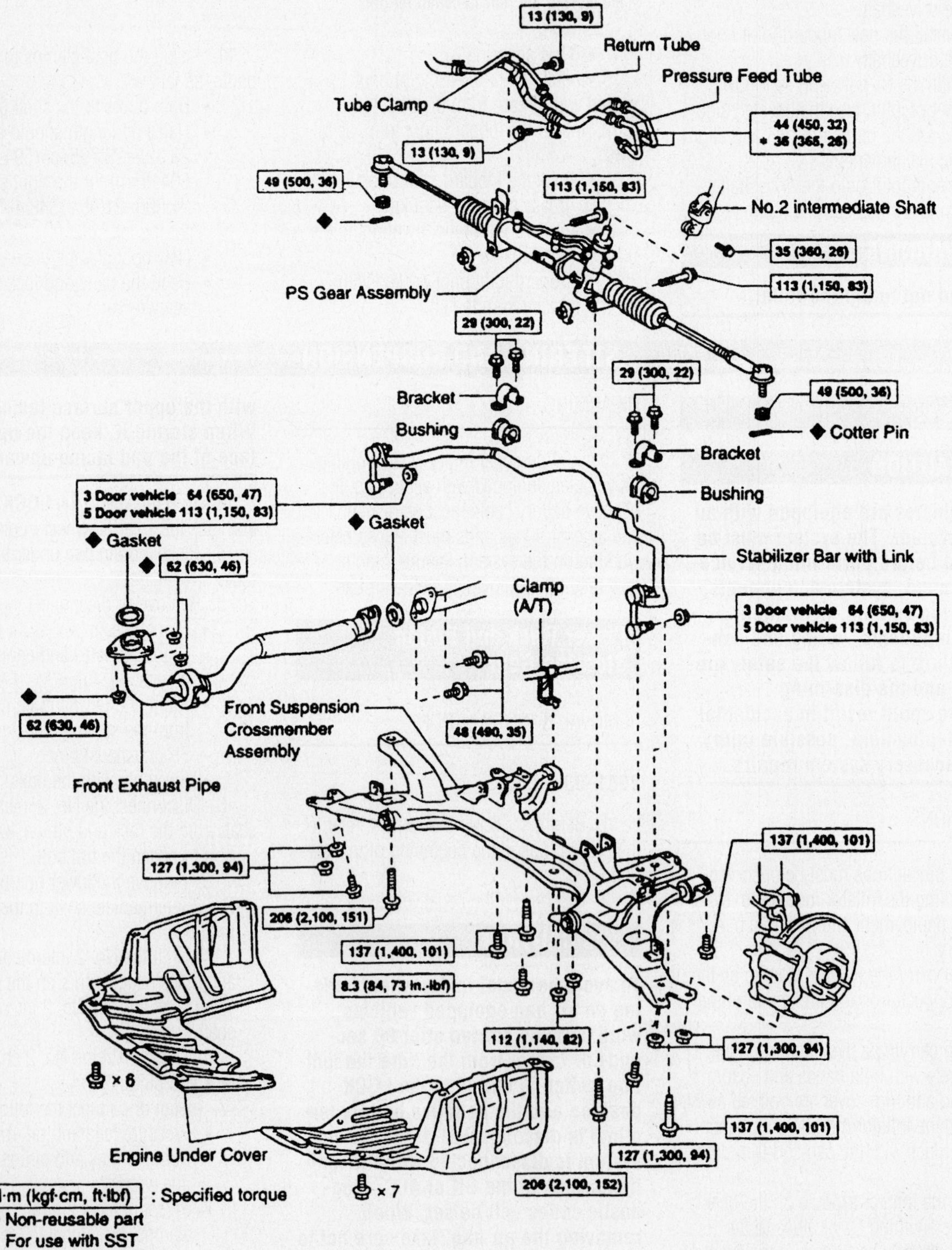

Rack and pinion exploded view—2001–03

both bolts to 152 ft. lbs. (206 Nm).

- Rack and pinion. Tighten the nuts/bolts to 83 ft. lbs. (113 Nm).
- Centermember to the crossmember. Tighten both set nuts to 82 ft. lbs. (112 Nm).

12. Install or connect the following:

- Right and left lower control arms
- Pressure feed and return tubes clamps
- Pressure feed and return tubes to the rack and pinion. Tighten the tubes to 26 ft. lbs. (36 Nm), using a torque wrench with a fulcrum length of 11.81 inches (300mm).
- Steering column No. 2 intermediate shaft to the rack and pinion. Align the marks and tighten the upper and lower pinch bolts to 26 ft. lbs. (35 Nm).
- Stabilizer bar links. Tighten the nuts to 22 ft. lbs. (29 Nm).
- Front exhaust pipe with new gaskets. Tighten the bolts to 35 ft. lbs. (48 Nm) and the nuts to 46 ft. lbs. (62 Nm).
- Right and left-hand tie rod ends to the steering knuckle. Tighten the nuts to 36 ft. lbs. (49 Nm) and install new cotter pins.
- Right and left-hand engine under-covers
- Front wheels

13. Fill the power steering unit and bleed the system. Check for leaks.

14. Check and/or adjust the front wheel alignment.

2004

1. Place the wheels in a straight-ahead position.

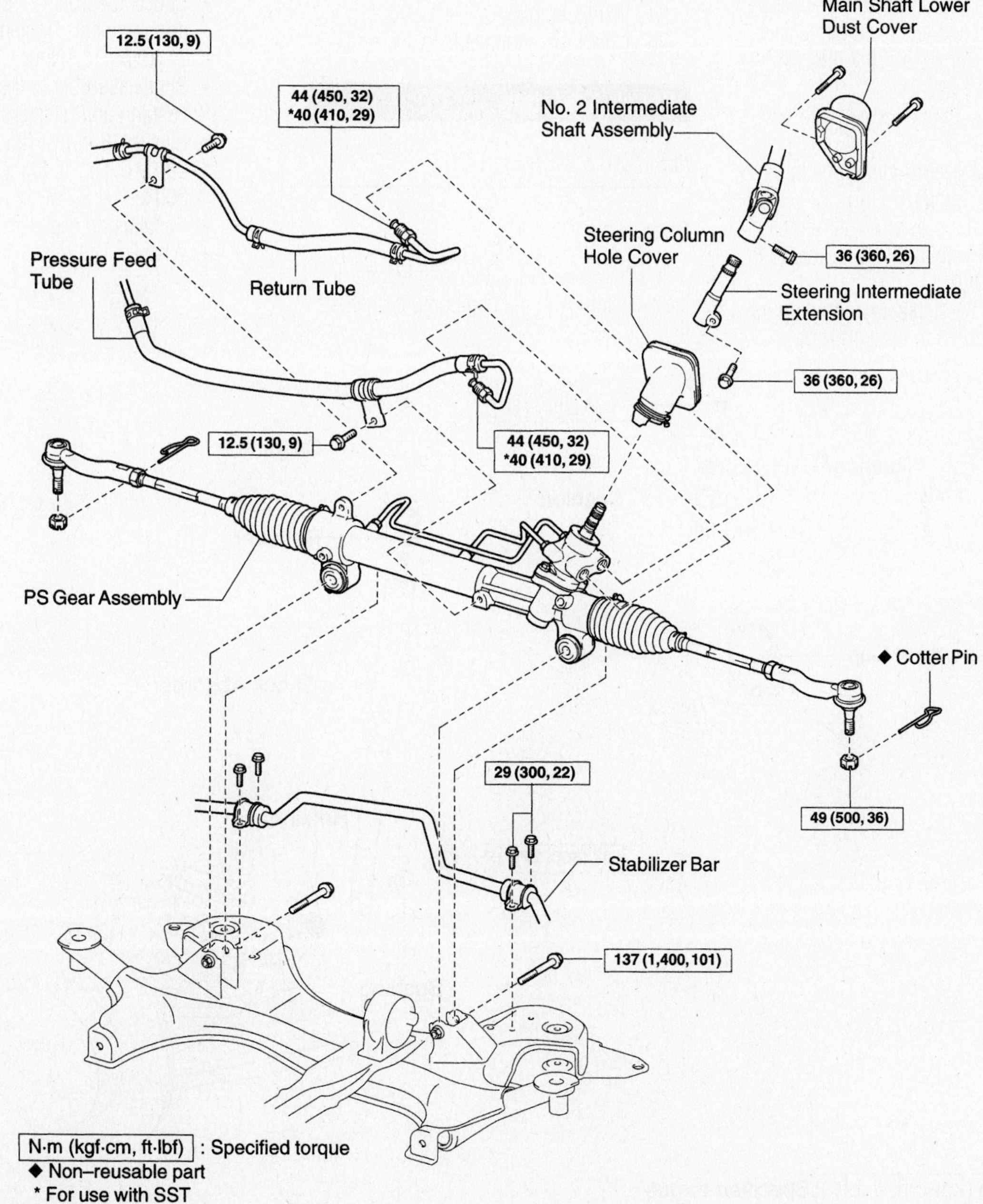

Steering rack and related parts—2004

2. Remove the steering wheel pad.
3. Remove the steering wheel.
4. Remove the intermediate shaft from the gear.
5. Remove the steering column hole cover from the dash panel.
6. Disconnect the tie rod ends.
7. Disconnect the stabilizer bar.
8. Disconnect the pressure and return lines.
9. Remove the 2 gear mounting bolts.
10. Matchmark the intermediate shaft extension and control valve shaft.
11. Remove the intermediate shaft extension bolt and remove the gear.

To install:

12. Place the gear in position. Align the matchmarks and connect the intermediate extension shaft. Torque to 26 ft. lbs. (35 Nm).
13. Install the column cover plate and clamp.
14. Install the gear mounting bolts. Torque to 101 ft. lbs. (137 Nm).
15. Connect the pressure and return lines to the gear with the 2 bolts. Torque to 9 ft. lbs. (12.5 Nm).
16. Connect the pressure and return tube fittings. Torque to 29 ft. lbs. (40 Nm).

➡Use a torque wrench with a fulcrum length of 345mm. (13.58 inches).

17. Connect the stabilizer bar. Torque to 22 ft. lbs. (29 Nm).
18. Connect the tie rods ends. Torque to 36 ft. lbs. (49 Nm).
19. Connect the intermediate shaft. Torque to 26 ft. lbs. (36 Nm).
20. Install the center spiral cable.
21. Install the power steering pump. Bleed the system.
22. Install the steering wheel. Make sure that everything is still in the straight-ahead position.
23. Torque the steering wheel nut to 37 ft. lbs. (50 Nm).
24. Install the pad.
25. Check the alignment.

Shock Absorber

REMOVAL & INSTALLATION

Rear

1. Before servicing the vehicle, refer to the precautions in the beginning of this section.
2. Remove the rear wheel.
3. Support the No. 1 control arm with a floor jack.
4. Remove or disconnect the following:
   - Suspension cap from inside the vehicle
   - Both upper shock absorber nuts, retainers and cushion
   - Shock absorber from the lower control arm by removing the bolt and 2 retainers
   - Shock absorber

To install:

5. Install or connect the following:
   - Shock absorber
   - Shock absorber to the lower control arm retainers. Tighten the bolt to 27 ft. lbs. (37 Nm).
   - Shock absorber to the chassis cushion and retainers. Tighten both nuts to 18 ft. lbs. (25 Nm) for 2001–03; 11 ft. lbs. (14.5 Nm) for 2004.
   - Suspension cap
   - Wheel

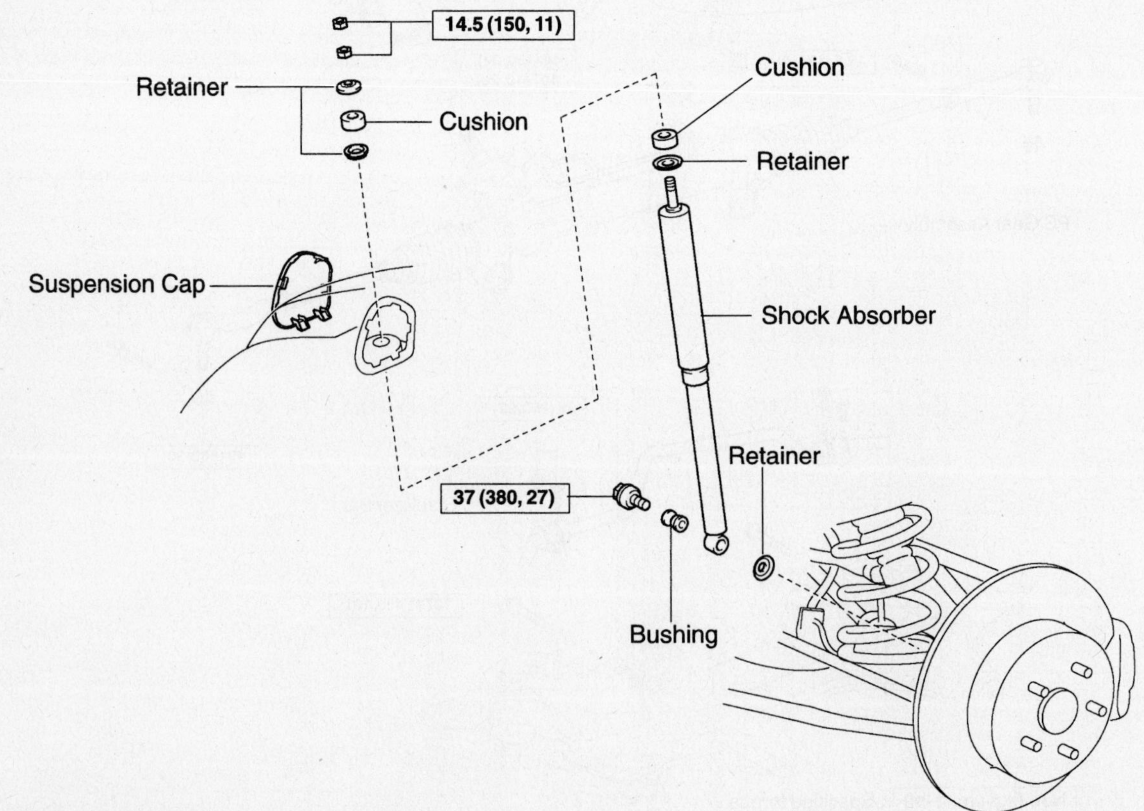

Rear shock absorber—2004 shown

Strut

REMOVAL & INSTALLATION

2001–03

1. Before servicing the vehicle, refer to the precautions in the beginning of this section.
2. Remove or disconnect the following:
   - Negative battery cable
   - Wheel

➡**Do not support the weight of the vehicle on the suspension arm.**

   - Brake hose from the strut
   - Anti-lock Brake System (ABS) electrical connection to the strut bolt, if equipped

➡**It is not necessary to disconnect the brake hose from the brake caliper.**

   - Strut from the steering knuckle
   - Suspension support bracket from the top of the strut tower
   - Strut

To install:

3. Install or connect the following:
   - Suspension support bracket to the top of the strut tower

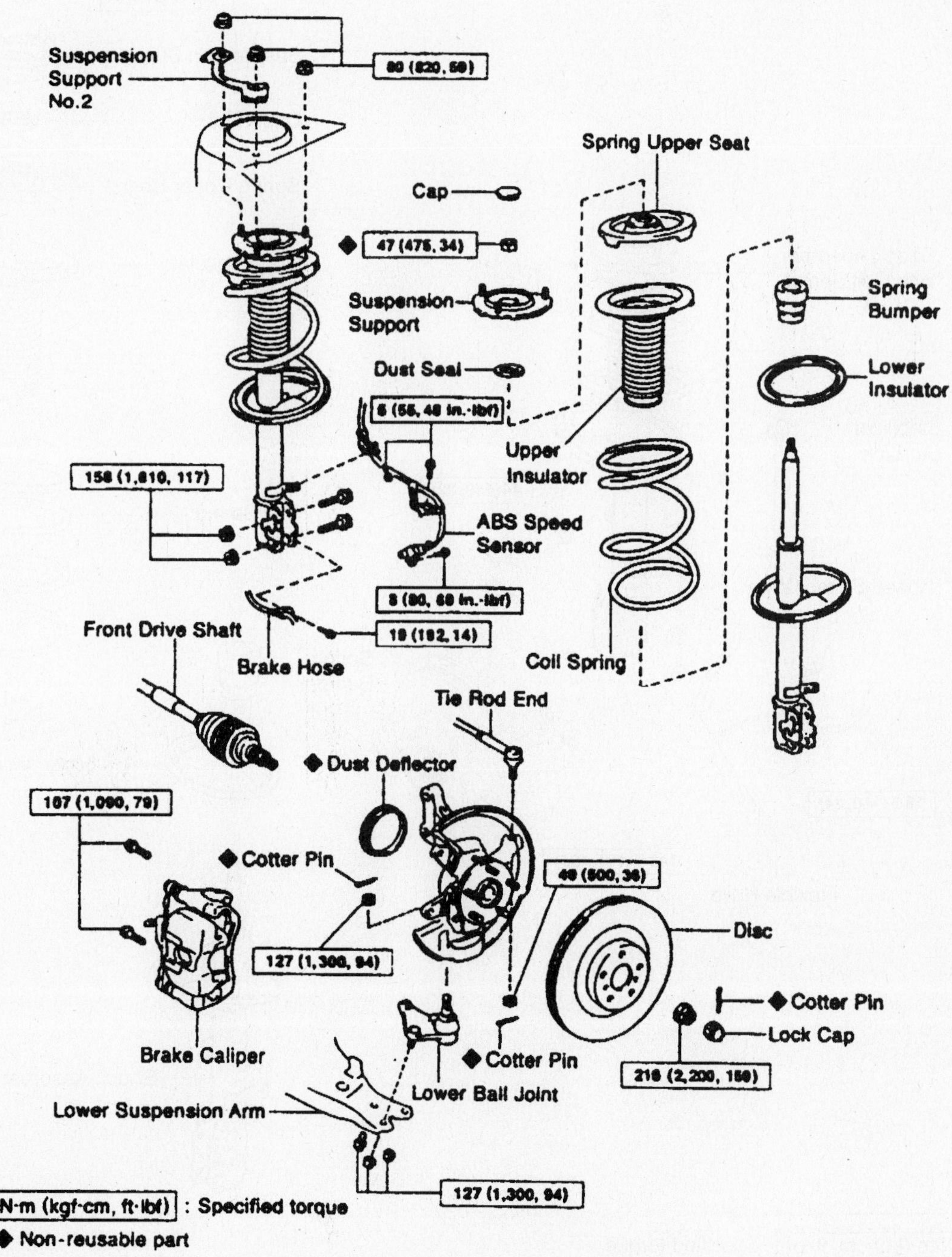

Strut assembly exploded view—2001–03

- Strut to the strut tower. Tighten the 3 nuts to 59 ft. lbs. (80 Nm).
- Steering knuckle to the strut lower bracket. Tighten the nuts to 117 ft. lbs. (158 Nm).
- ABS electrical connector to the strut. Tighten the bolt to 48 inch lbs. (5.4 Nm).
- Brake line to the strut. Tighten the bolt to 14 ft. lbs. (19 Nm).

4. If the brake lines were opened, add brake fluid and bleed the brake system.
5. Install or connect the following:
   - Wheel
   - Negative battery cable
6. Check and/or adjust the front wheel alignment.

2004

1. Remove the wheel.
2. Disconnect the stabilizer bar link from the strut.
3. Remove the brake hose and wire harness from the strut.
4. Loosen, don't remove, the 2 lower strut nuts

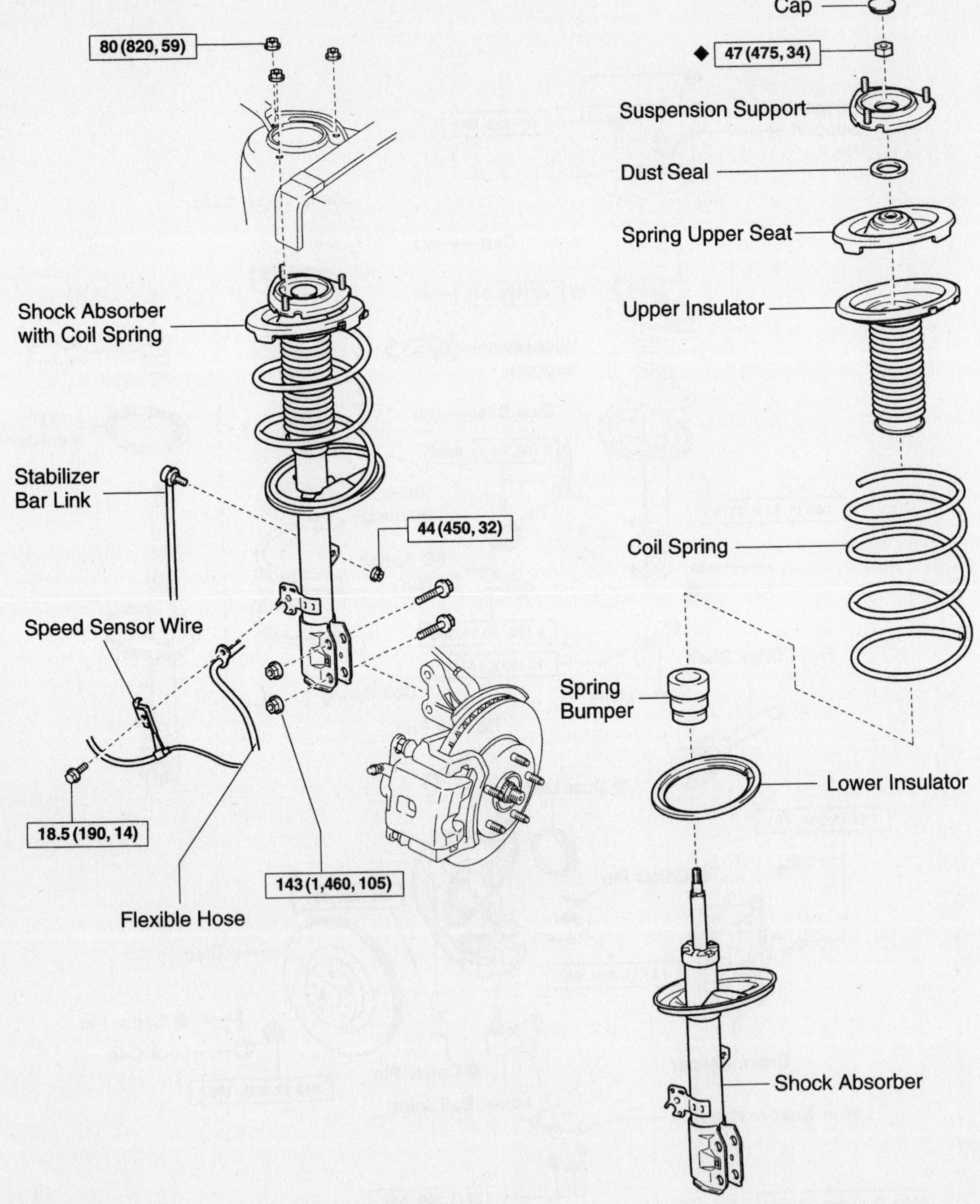

Front strut—2004

5. Remove the 3 upper end nuts.
6. Remove the 2 lower bolts and nuts.
7. Installation is the reverse of removal. Observe the following torques:
   - Upper end nuts: 59 ft. lbs. (80 Nm)
   - Lower end bolts/nuts: 105 ft. lbs. (143 Nm)
   - Stabilizer bar links: 32 ft. lbs. (44 Nm)

STRUT OVERHAUL

1. Before servicing the vehicle, refer to the precautions in the beginning of this section.
2. Remove or disconnect the following:
   - Wheel

➡If equipped, be careful not to damage the oil seal, driveshaft boot and/or speed sensor rotor when removing the steering knuckle.

   - Shock absorber (strut assembly)
3. Install a nut/bolt to the bracket at the lower portion of the strut assembly and secure it in a vise.
4. Compress the coil spring with a spring compressor.

⁂ CAUTION

The proper tools must be used for this procedure. The spring on the strut is under high pressure and can cause serious injury if not properly removed and installed.

5. Remove or disconnect the following:
   - Center retaining nut, by holding the spring seat
   - Support, dust seal, spring seat, insulator and spring from the strut assembly

To install:

6. Install the spring bumper and lower insulator to the strut assembly.
7. Compress the coil spring and fit the lower end of the spring into the spring seat gap.
8. Install or connect the following:
   - Upper insulator, spring seat, dust seal, support and spring seat. Tighten the new retaining nut to 34 ft. lbs. (47 Nm).
   - Strut
   - Wheel
9. If required, bleed the brake system and check for leaks.
10. Check and/or adjust the front wheel alignment.

Coil Spring

REMOVAL & INSTALLATION

➡For MacPherson struts, see Strut Overhaul in this section.

Rear

2001–03

1. Before servicing the vehicle, refer to the precautions in the beginning of this section.
2. Remove or disconnect the following:
   - Negative battery cable
   - Axle shaft, if equipped with 2WD
   - Halfshaft, if equipped with 4WD
   - Brake drum
   - Both brake line clamp bolts
   - Parking brake cable clamp bolt
   - Anti-lock Brake System (ABS) speed sensor and wiring harness, if equipped with ABS
   - Rear axle hub with the brake, by removing the 4 bolts
3. Support the hub securely.
4. Support the control arm with a floor jack.
5. Remove or disconnect the following:
   - Shock absorber from the control arm, by removing the bolt

➡The control arm must be supported before removing the bolt for the shock absorber. Leave the floor jack under the control arm. Later, the floor jack will be lowered to remove the coil spring.

   - Cotter pins and nuts by supporting the lower and upper suspension arms
6. Disconnect the upper and lower control arms from the control arm, using tool 09628-62011
7. Remove the coil spring and control arm, as follows:
   - Matchmark the toe adjust cam and body.
   - Coil spring and upper insulator, by loosening the bolt and lowering the control arm.
   - Bolt, toe-adjust cam, 2 attachments, nut and control arm
   - Bolt and spring bumper

To install:

8. Install or connect the following:
   - Spring bumper. Tighten the bolt to 108 inch lbs. (13 Nm).
   - Control arm, 2 attachments, toe-adjust cam, bolt and nut; do not tighten the bolt at this time
   - Spring and upper insulator
9. Raise the control arm with a floor jack.
10. Install or connect the following:
    - Upper and lower suspension arms to the control arm. Tighten the nuts to 76 ft. lbs. (103 Nm).
    - New cotter pins
    - Sock absorber to the control arm. Tighten the bolt to 27 ft. lbs. (37 Nm).
    - Rear axle hub with the brake. Tighten the 4 bolts to 59 ft. lbs. (80 Nm).
    - ABS speed sensor and wiring harness, if equipped. Tighten the ABS speed sensor to 69 inch lbs. (8 Nm) and the wiring harness to 108 inch lbs. (13 Nm).
    - Parking brake cable clamp. Tighten the bolt to 14 ft. lbs. (19 Nm).
    - Both brake line cable clamps. Tighten the bracket bolt to 13 ft. lbs. (18 Nm) and the clamp bolt to 108 inch lbs. (13 Nm).
    - Brake drum
    - Rear halfshaft, if equipped with 4WD
    - Axle shaft, if equipped with 2WD
    - Rear wheel
11. Lower the rear of the vehicle and stabilize the suspension.
12. Install or connect the following:
    - Align the matchmarks to the toe-adjust cam. Tighten the bolt to 98 ft. lbs. (132 Nm).
    - Negative battery cable
13. Check and/or adjust the wheel alignment

2004

1. Remove the wheel.
2. Remove the rear halfshaft.
3. Remove the ABS sensor wire from the lower arm and the sensor.
4. Remove the brake drum or caliper and rotor.
5. Remove the brake line from the control arm.
6. Remove the parking brake cable from the control arm.
7. Remove the rear hub.
8. Support the lower arm with a jack.
9. Remove the shock absorber.
10. Remove the stabilizer bar from the lower arm.
11. Lower the jack to remove tension from the spring.
12. Remove the 2 bolts and nuts and remove the upper arm.
13. Matchmark the camber adjusting cam and the suspension member.

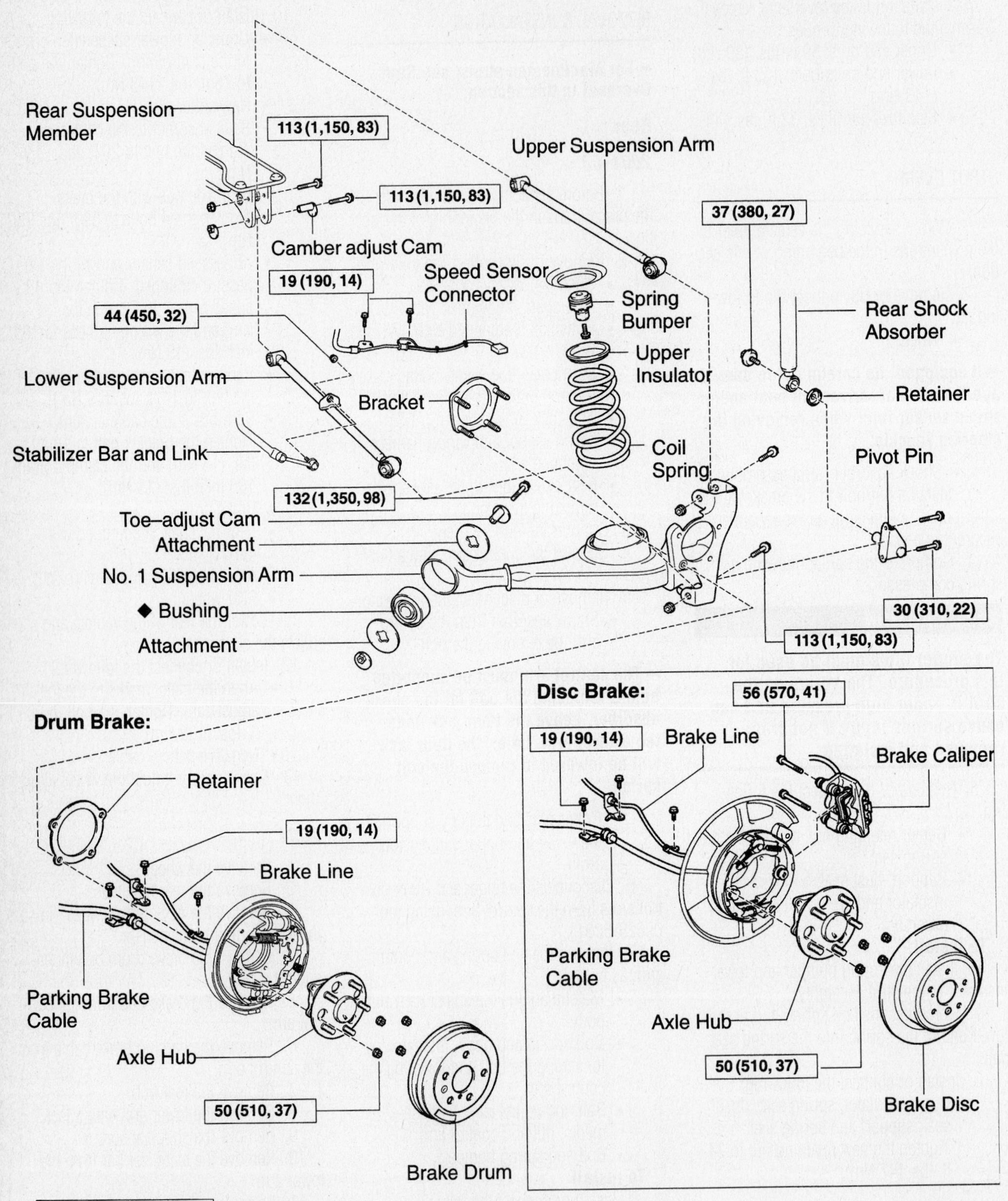

Rear Suspension Member
113 (1,150, 83)
113 (1,150, 83)
Camber adjust Cam
Upper Suspension Arm
37 (380, 27)
19 (190, 14)
Speed Sensor Connector
44 (450, 32)
Spring Bumper
Rear Shock Absorber
Upper Insulator
Lower Suspension Arm
Retainer
Bracket
Coil Spring
Stabilizer Bar and Link
Pivot Pin
132 (1,350, 98)
Toe–adjust Cam
Attachment
No. 1 Suspension Arm
30 (310, 22)
◆ Bushing
113 (1,150, 83)
Attachment
Disc Brake:
56 (570, 41)
Drum Brake:
19 (190, 14)
Brake Line
Brake Caliper
Retainer
19 (190, 14)
Brake Line
Parking Brake Cable
Parking Brake Cable
Axle Hub
Axle Hub
50 (510, 37)
50 (510, 37)
Brake Disc
Brake Drum
N·m (kgf·cm, ft·lbf) : Specified torque
◆ Non–reusable part
67170-RAV4-G59

Rear suspension—2004 2WD shown

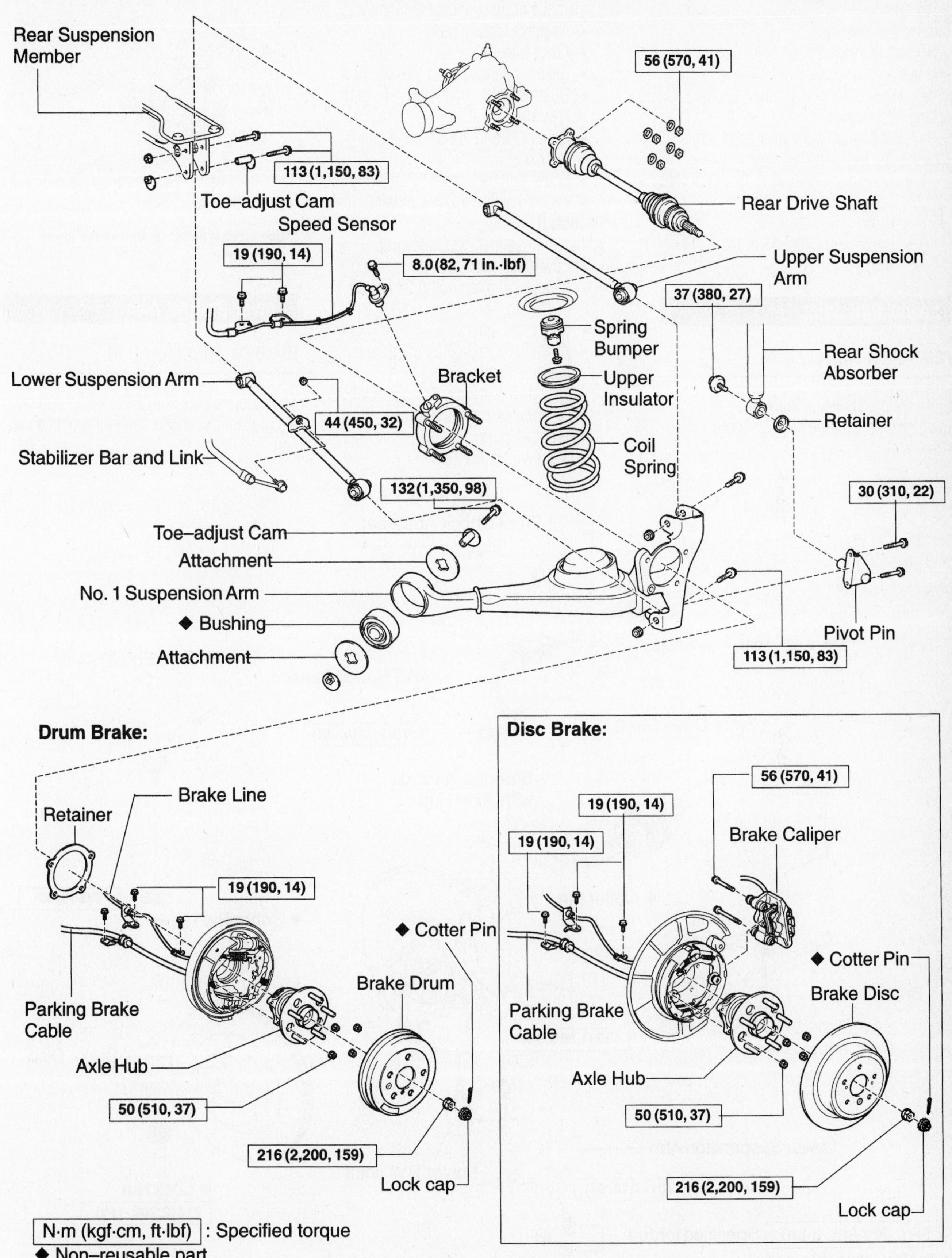
Rear Suspension Member
56 (570, 41)
113 (1,150, 83)
Toe–adjust Cam
Speed Sensor
Rear Drive Shaft
19 (190, 14)
8.0 (82, 71 in.·lbf)
Upper Suspension Arm
37 (380, 27)
Spring Bumper
Rear Shock Absorber
Lower Suspension Arm
Bracket
Upper Insulator
44 (450, 32)
Retainer
Coil Spring
Stabilizer Bar and Link
132 (1,350, 98)
30 (310, 22)
Toe–adjust Cam
Attachment
No. 1 Suspension Arm
◆ Bushing
Pivot Pin
Attachment
113 (1,150, 83)
Drum Brake:
Disc Brake:
56 (570, 41)
19 (190, 14)
Brake Line
Retainer
19 (190, 14)
Brake Caliper
19 (190, 14)
◆ Cotter Pin
◆ Cotter Pin
Brake Drum
Brake Disc
Parking Brake Cable
Parking Brake Cable
Axle Hub
Axle Hub
50 (510, 37)
50 (510, 37)
216 (2,200, 159)
Lock cap
216 (2,200, 159)
Lock cap
N·m (kgf·cm, ft·lbf) : Specified torque
◆ Non–reusable part
67170-RAV4-G60

Rear suspension—2004 4WD shown

14. Remove the 2 bolts and nuts and remove the lower arm.

15. Matchmark the toe adjusting cam and body.

16. Loosen the bolt and remove the spring and insulator.

17. Installation is the reverse of removal. Observe the following torques:

- Coil spring bolt: 98 ft. lbs. (132 Nm)
- Lower arm bolts: 83 ft. lbs. (113 Nm)
- Upper arm bolts: 83 ft. lbs. (113 Nm)

Lower Ball Joint

REMOVAL & INSTALLATION

1. Before servicing the vehicle, refer to the precautions in the beginning of this section.

2. Remove or disconnect the following:
   - Negative battery cable
   - Front wheel(s)
   - Steering knuckle with the axle hub
   - Dust deflector, by prying it from the knuckle
   - Cotter pin and nut from the ball joint stud
   - Lower ball joint from the steering knuckle, using a 2-jaw puller

To install:

3. Install or connect the following:
   - Lower ball joint onto the steering knuckle. Tighten nut to 94 ft. lbs. (127 Nm).
   - New cotter pin
   - ABS speed sensor, by aligning it the dust deflector hole
   - New dust deflector, using a driver
   - Steering knuckle and hub
   - Front wheel(s)
   - Negative battery cable

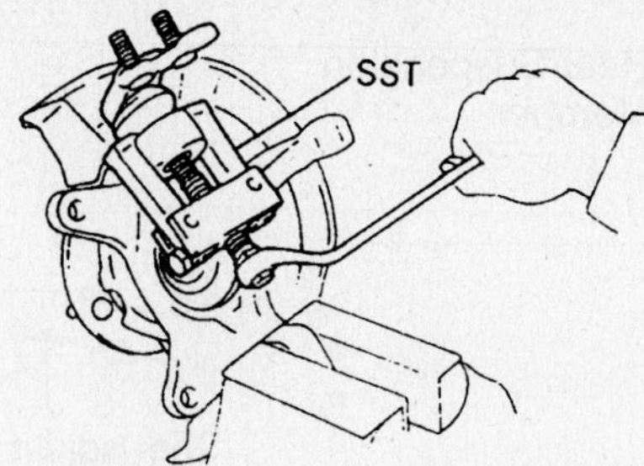

Use a 2-jaw puller to remove the lower ball joint

Lower Control Arm

REMOVAL & INSTALLATION

1. Before servicing the vehicle, refer to the precautions in the beginning of this section.

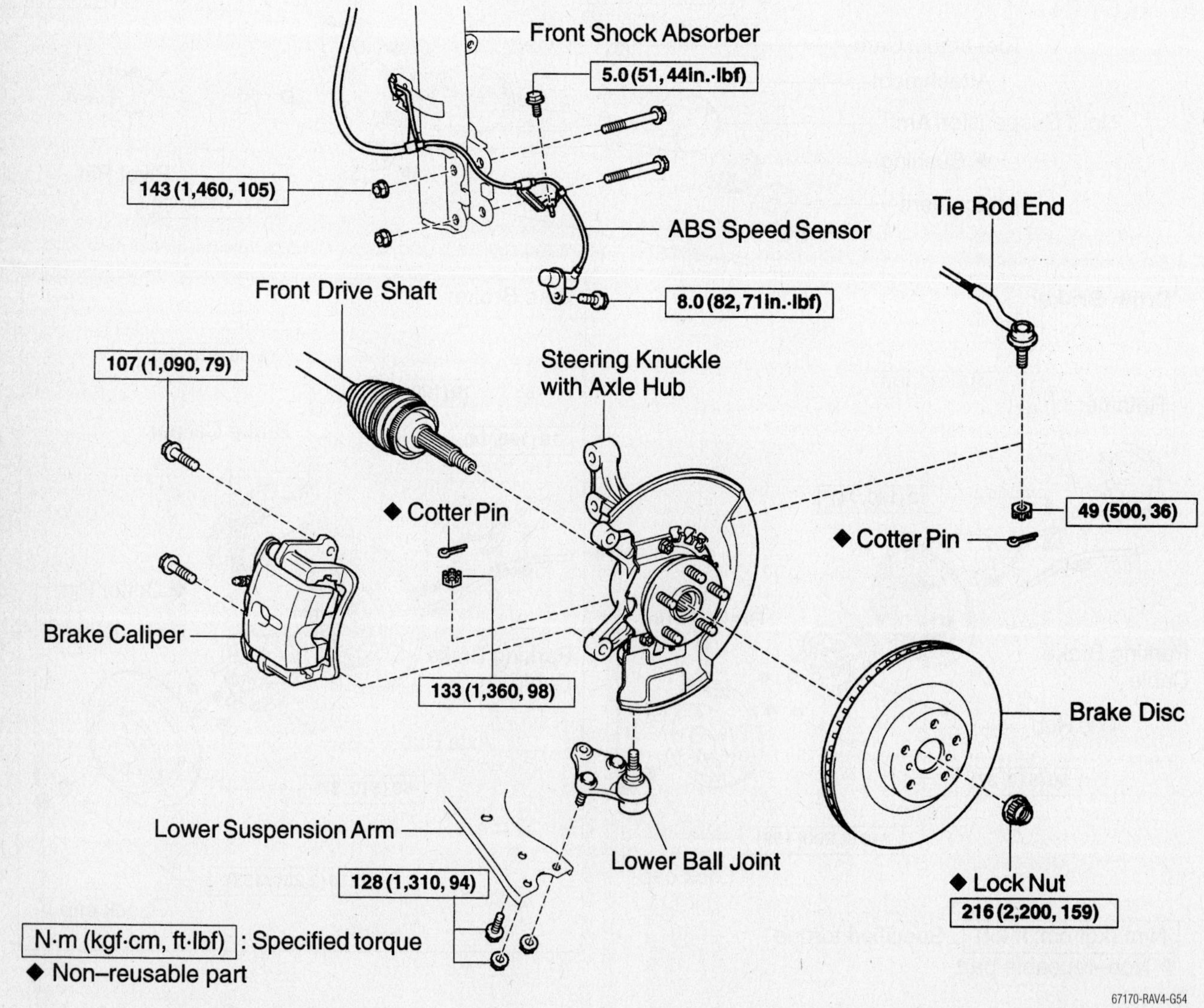

Lower ball joint removal—2004 shown

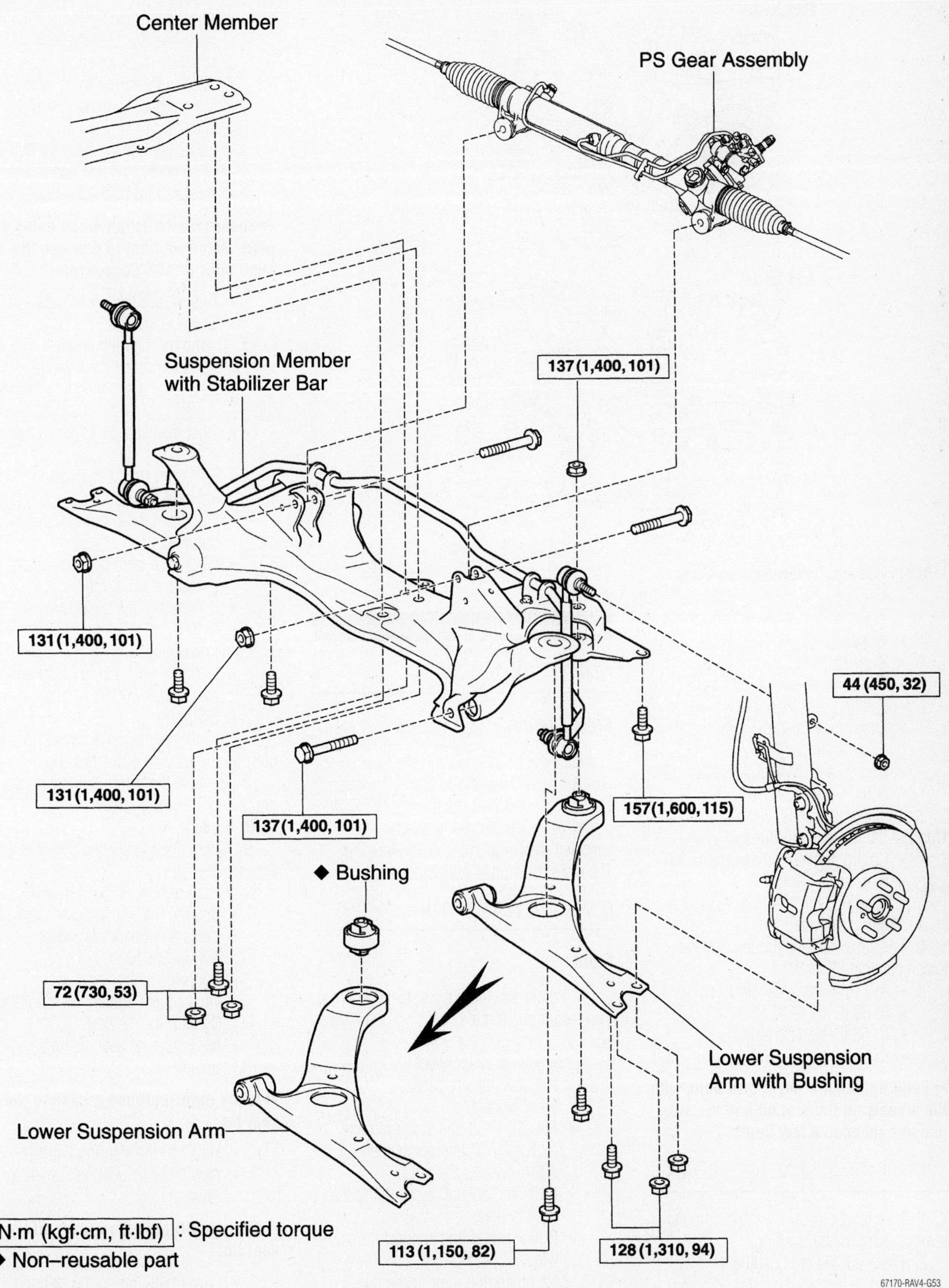

Lower control arm and related parts—2004 shown

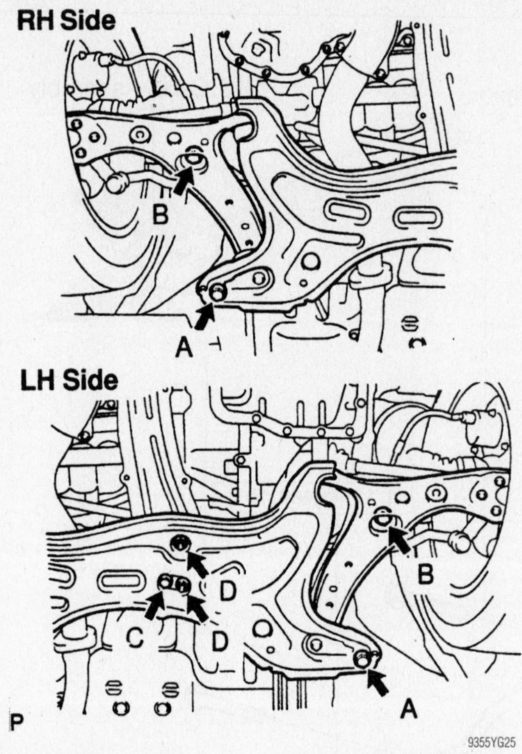

RAV4 suspension subassembly bolt torques

2. Remove or disconnect the following:
   - Wheel
   - Sway bar link
   - Control arm-to-ball joint bolts
   - Steering rack mount bolts
   - Suspension member subassembly (5 bolts and 2 nuts)
   - Control arm from subassembly (2 bolts, 1 nut)

3. Installation is the reverse of removal. Observe the following torques:
   - Control arm-to-subassembly: 101 ft. lbs. (137 Nm)
   - Steering rack mount bolts: 101 ft. lbs. (137 Nm)

4. For subassembly torques, see the accompanying illustration.
   - A: 115 ft. lbs. (157 Nm)
   - B: 82 ft. lbs. (113 Nm)
   - C: 53 ft. lbs. (72Nm)
   - D: 53 ft. lbs. (72Nm)

➡Fully tighten the lower arm bolts with the wheels on the ground and the suspension jounced a few times.

CONTROL ARM BUSHING REPLACEMENT

1. Matchmark the control arm with the triangle mark on the bushing.
2. Press out the old bushing.
3. Press in the new bushing, aligning the matchmarks.

Wheel Bearing

ADJUSTMENT

Front and Rear

Check the bearing play in the axial direction and also check the axle hub deviation. The maximum play for both checks should be 0.0020 in. (0.05mm). If greater than the specified maximum, replace the bearing. The wheel bearing is not adjustable.

REMOVAL & INSTALLATION

Front

1. Before servicing the vehicle, refer to the precautions in the beginning of this section.
2. Remove or disconnect the following:
   - Negative battery cable
   - Front wheels
   - For 2001–03, the cotter pin and lockcap from the halfshaft end. For 2004, unstake the nut.
   - Halfshaft locknut, by applying the front brakes
   - Brake caliper and support it on a wire
3. Matchmark the rotor to the hub.
4. Remove or disconnect the following:
   - Rotor
   - Anti-lock Brake System (ABS) speed sensor from the steering knuckle, if equipped
   - Loosen the strut's lower end nuts
   - Tie rod end from the steering knuckle
   - Lower control arm from the ball joint, by removing the bolt and 2 nuts
   - Halfshaft from the axle hub

➡Secure the halfshaft aside using a wire. Be careful not to damage the shaft boot or ABS sensor rotor.

   - Both strut's lower end nuts
   - Steering knuckle
5. Clamp the steering knuckle in a vise with soft jaws to protect the knuckle.
6. Remove or disconnect the following:
   - Dust deflector, by prying it from the hub
   - Ball joint from the steering knuckle
   - Hub from the knuckle, using slide hammer
   - Inner race from the hub, using press and arbor tool
   - 4 bolts and the dust cover
   - Inner oil seal, using Seal Removal tool 09308-00010
   - Outer oil seal, using Seal Removal tool 09308-00010
   - Snapring
7. Install inner race (removed from the hub) on the outside of the bearing
8. Remove the steering knuckle bearing, using a bearing driver

To install:

9. Clean bearing seating surfaces with a clean, dry rag.
10. Install or connect the following:
   - Bearing into the knuckle, using a press and Bearing Installer tool 09608-32010
   - Snapring
   - Dust cover. Tighten the 4 bolts to 74 inch lbs. (8 Nm).
   - New outer oil seal, using a seal driver

➡Apply multi-purpose grease to the oil seal lip.

   - Hub into the steering knuckle
   - New inner oil seal, using a seal driver

➡Apply multi-purpose grease to the oil seal lip.

   - Lower ball joint to the steering knuckle. Tighten the nut to 94 ft. lbs. (127 Nm).

- New cotter pin
- Dust deflector, by aligning it with the ABS speed sensor hole
- Knuckle to the lower strut and install the bolts
- Lower ball joint to the lower arm. Tighten the bolts to 94 ft. lbs. (127 Nm).
- Tie rod end to the steering knuckle. Tighten the nut to 36 ft. lbs. (49 Nm).
- Halfshaft to the hub and knuckle
- Tighten the lower strut nuts to 117 ft. lbs. (158 Nm) for 2001–03; 105 ft. lbs. (143 Nm) for 2004.
- ABS speed sensor. Tighten the bolt to 69 inch lbs. (8 Nm).

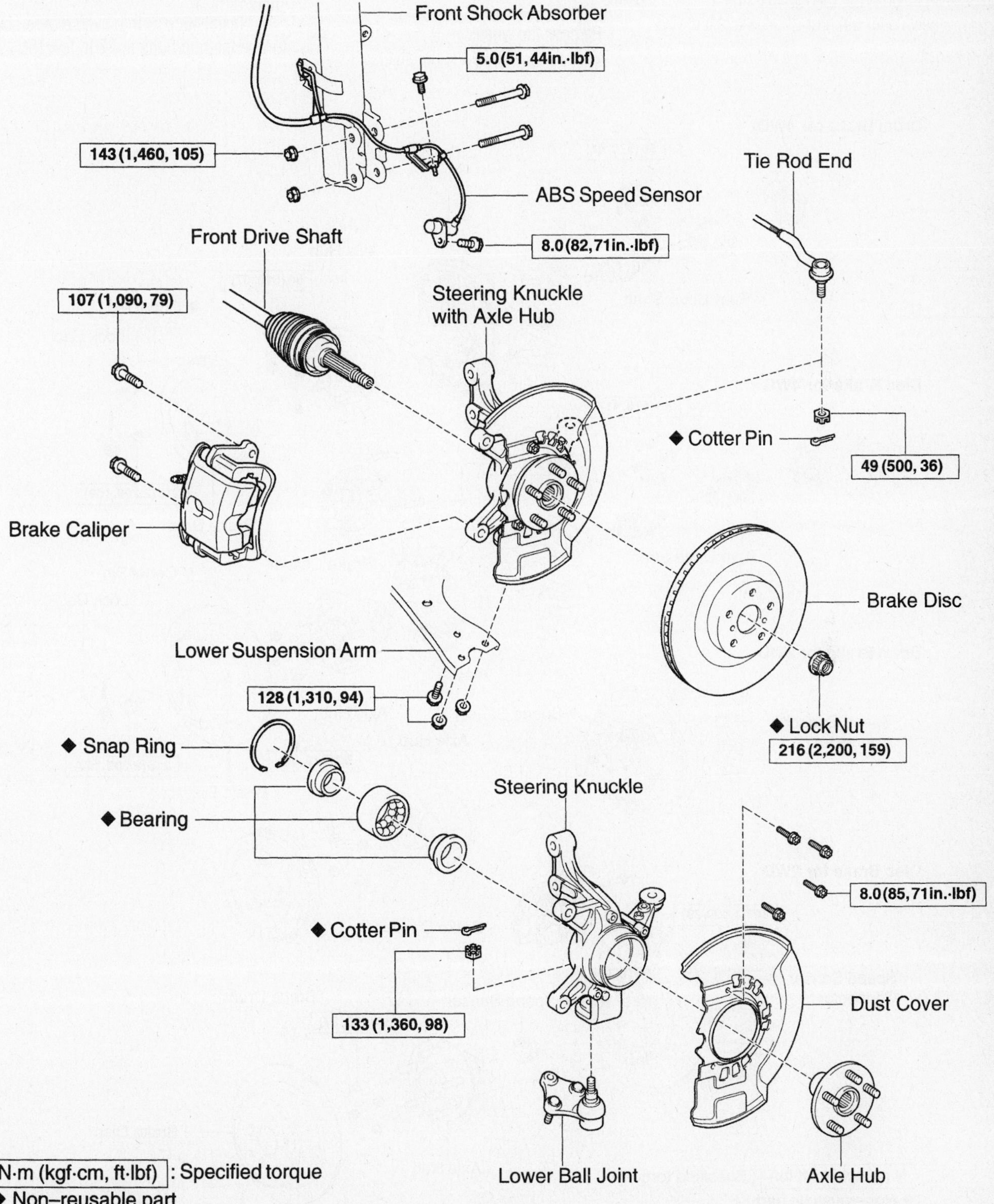

Front hub and related parts—2004 shown; others similar

- Rotor to the hub, by aligning the matchmark
- Brake caliper. Tighten the mounting bolts to 79 ft. lbs. (107 Nm).
- Axle locknut, using an assistant to apply the brakes. Tighten the nut to 159 ft. lbs. (216 Nm).
- For 2001–03, the lockcap and a new cotter pin. For 2004, stake the nut.
- Wheel
- Negative battery cable

11. Turn the wheel by hand, verify that the wheel turns without noise and without binding.
12. Check the signal from the ABS sensor.

Rear

1. Remove the wheel.
2. With 4wd, remove the halfshaft.
3. With disc brakes, remove the caliper and rotor.
4. With drum brakes, remove the brake drum.
5. With 2wd, disconnect the speed sensor.
6. Remove the 4 nuts and the hub/bearing assembly.
7. Installation is the reverse of removal. Torque the hub bolts to 37 ft. lbs. (50 Nm).

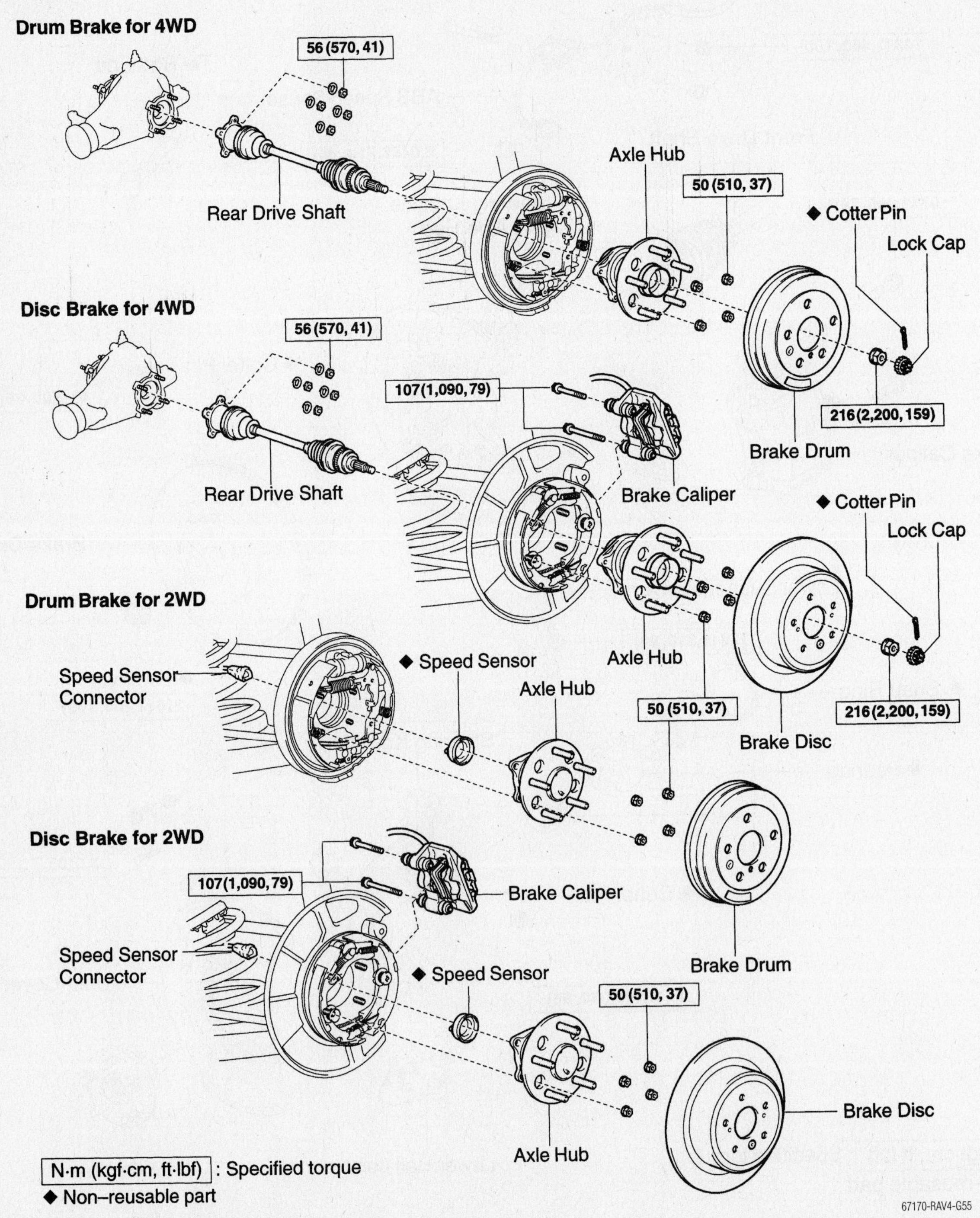

Rear hub/bearing and related parts—2004 shown; others similar

BRAKES

Front Brake Caliper

REMOVAL & INSTALLATION

1. Raise and safely support the vehicle.
2. Remove the wheel(s).
3. If the caliper is being replaced, remove the union bolt and 2 washers and remove the flexible brake hose from the caliper. Use a suitable container to catch the brake fluid as it drains out. Discard the washers.
4. Hold the sliding pin and loosen the 2 caliper mounting bolts. Remove the bolts and remove the caliper from the torque plate.
5. Remove the brake pads and brake hardware.

To install:

6. Install the brake pads and brake hardware.
7. Install the caliper to the torque plate with the 2 mounting bolts. Torque the bolts to 20 ft. lbs. (26 Nm).
8. Reconnect the flexible brake hose to the caliper with 2 new washers and the union bolt. Torque the union bolt to 22 ft. lbs. (30 Nm).
9. Refill the master cylinder with brake fluid and bleed the brake system.
10. Check for proper operation and make sure there are no leaks.

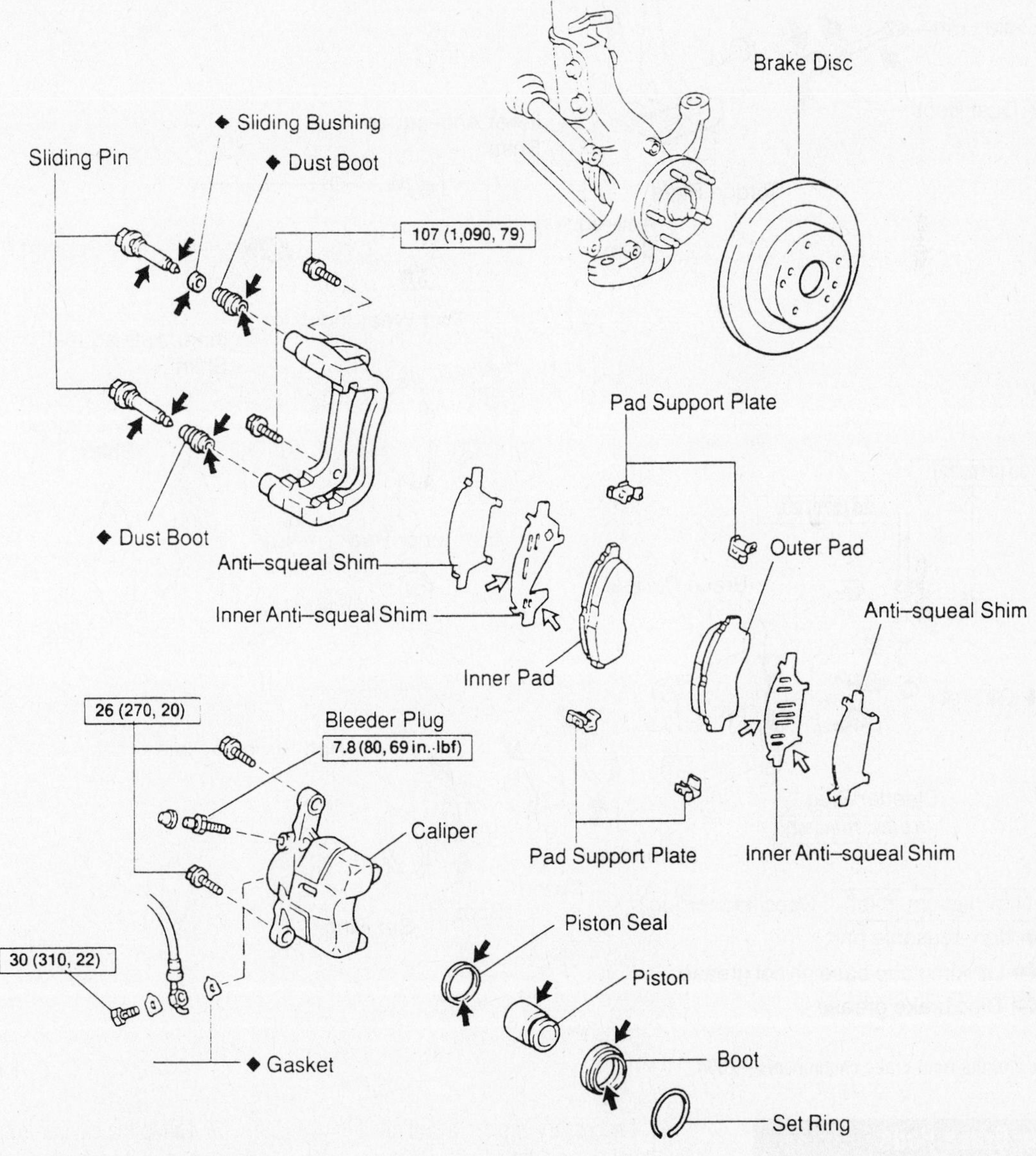

Exploded view of the front brake components—2001–03

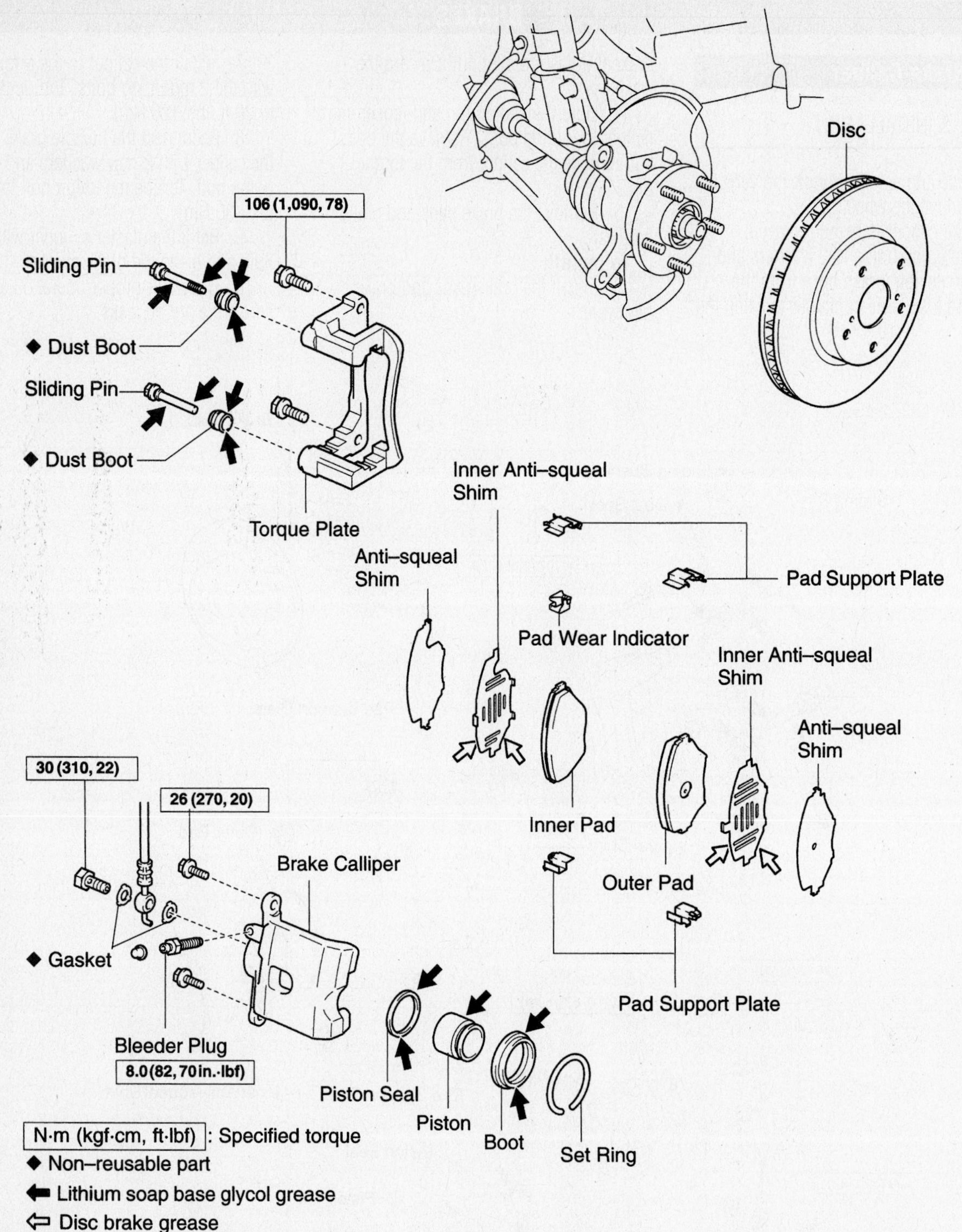

Exploded view of the front brake components—2004

Front Disc Brake Pads

REMOVAL & INSTALLATION

1. Raise and safely support the vehicle.
2. Remove the wheel(s).
3. Temporarily install 2 wheel stud nuts to hold the brake rotor in place.
4. If necessary, siphon a sufficient quantity of brake fluid from the master cylinder reservoir to prevent any brake fluid from overflowing the master cylinder when removing or installing new pads. This may be necessary, as the piston must be forced into the caliper bore to provide sufficient clearance when installing the pads.
5. Grasp the caliper from behind and carefully pull it towards you. This will start to seat the piston(s) in its bore. Using a C-clamp or other suitable tool, press the piston the remaining way into the caliper. Be careful not to cock the piston in the bore. Also, do not force the piston or the caliper and piston may be damaged.

6. Hold the sliding pin and loosen the 2 caliper mounting bolts. Remove the bolts and remove the caliper from the torque plate.

7. Secure the caliper assembly out of the way with a wire; so as not to stress the flexible hose.

8. Slide out the old brake pads along with any anti-squeal shims, springs, pad wear indicators and pad support plates. Make sure to note the position of all assorted pad hardware.

To install:

9. Check the brake disc (rotor) for thickness and run-out. Inspect the caliper and piston assembly for breaks, cracks, fluid seepage or other damage. Overhaul or replace as necessary.

10. Install the pad support plates into the torque plate.

11. Install the pad wear indicators onto the pads. Be sure the arrow on the indicator plate is pointing in the direction of rotation.

12. Install the anti-squeal shims on the outside of each pad and then install the pad assemblies into the torque plate.

13. Install the caliper to the torque plate with the 2 mounting bolts. Torque the bolts to 20 ft. lbs. (26 Nm).

14. Remove the 2 temporary wheel stud nuts and check that the rotor turns freely.

15. Reinstall the wheel(s), safely lower the vehicle, and road-test for proper brake operation.

16. Be sure to pump the brakes several times prior to moving the vehicle.

Rear Brake Caliper

REMOVAL & INSTALLATION

1. Remove the wheel.

2. If the caliper is being replaced, remove the union bolt and disconnect the brake line. If possible, plug the line to pre-

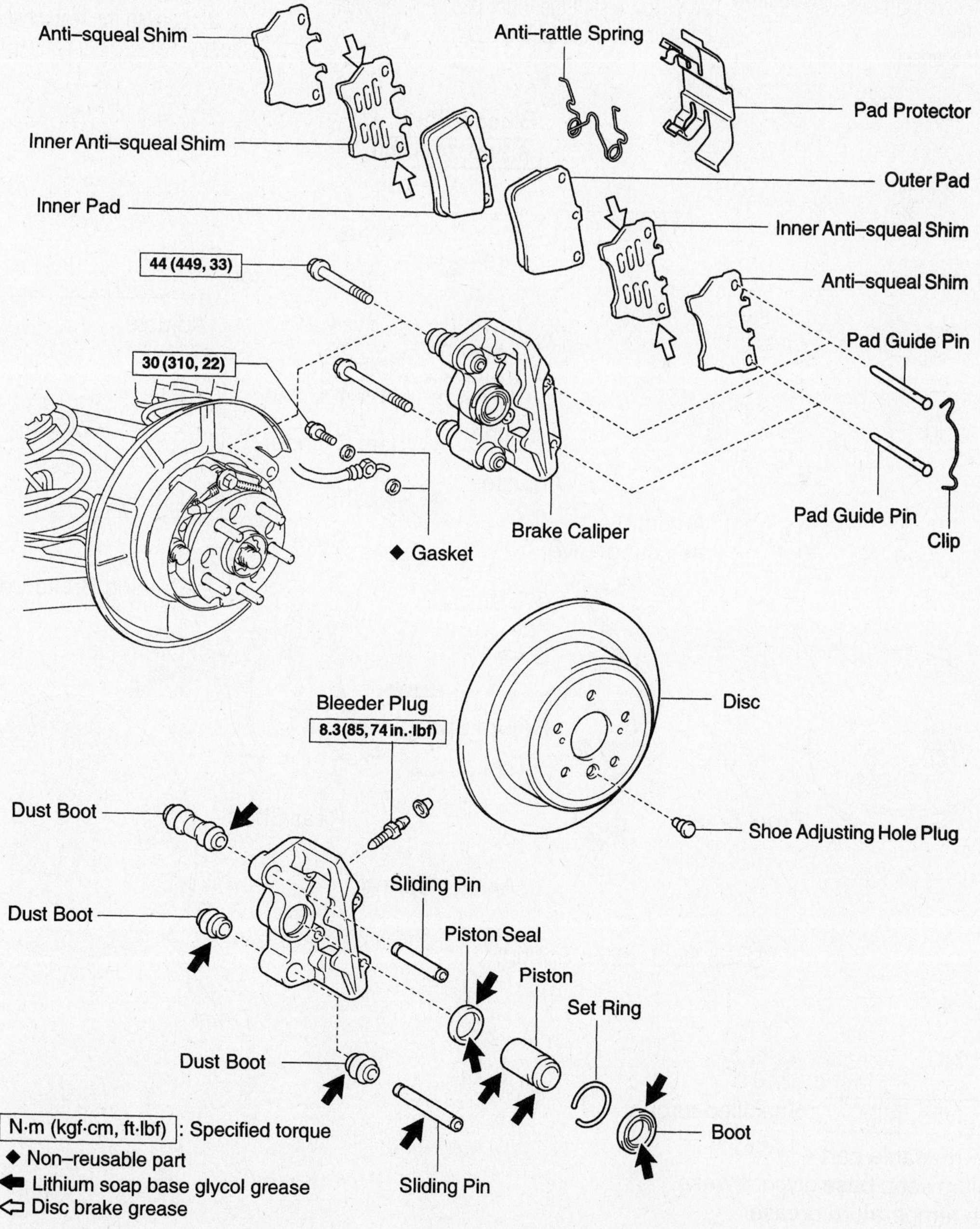

Rear disc brake components

vent fluid loss. Otherwise place a container under the line to catch the fluid. Discard the washers.

3. Remove the 2 mounting bolts and the caliper.

4. Installation is the reverse of removal. Torque the caliper bolts to 33 ft. lbs. (44 Nm). Torque the brake line union bolt to 22 ft. lbs. (30 Nm). Use new washers.

5. Bleed the brakes.

Rear Brake Pads

REMOVAL & INSTALLATION

1. Remove a small amount of fluid from the master cylinder.
2. Remove the wheel.
3. Pry off the pad protector, being careful not to bend it.
4. Remove the spring clip and the 2 pad guide pins.
5. Remove the anti-rattle spring and the pads.
6. Remove the shims from each pad.

➡When replacing worn pads, the anti-squeal shims must be replaced.

To install:

7. Apply disc brake grease to both sides of the shims.
8. Install the shims on the pads.
9. Compress the piston with a piston forcing tool.
10. Install the inner pad with the wear indicator plate facing down.
11. Install the outer pad.
12. Install the anti-rattle spring.
13. Install the guide pins.
14. Install the clip.
15. Install the protector.
16. Install the wheel.
17. Refill the master cylinder.
18. Pump the brake pedal a few times to seat the pads before moving the vehicle.

Brake Shoes

REMOVAL & INSTALLATION

1. Disconnect the negative battery cable from the battery.
2. Loosen the rear wheel lug nuts slightly. Release the parking brake.

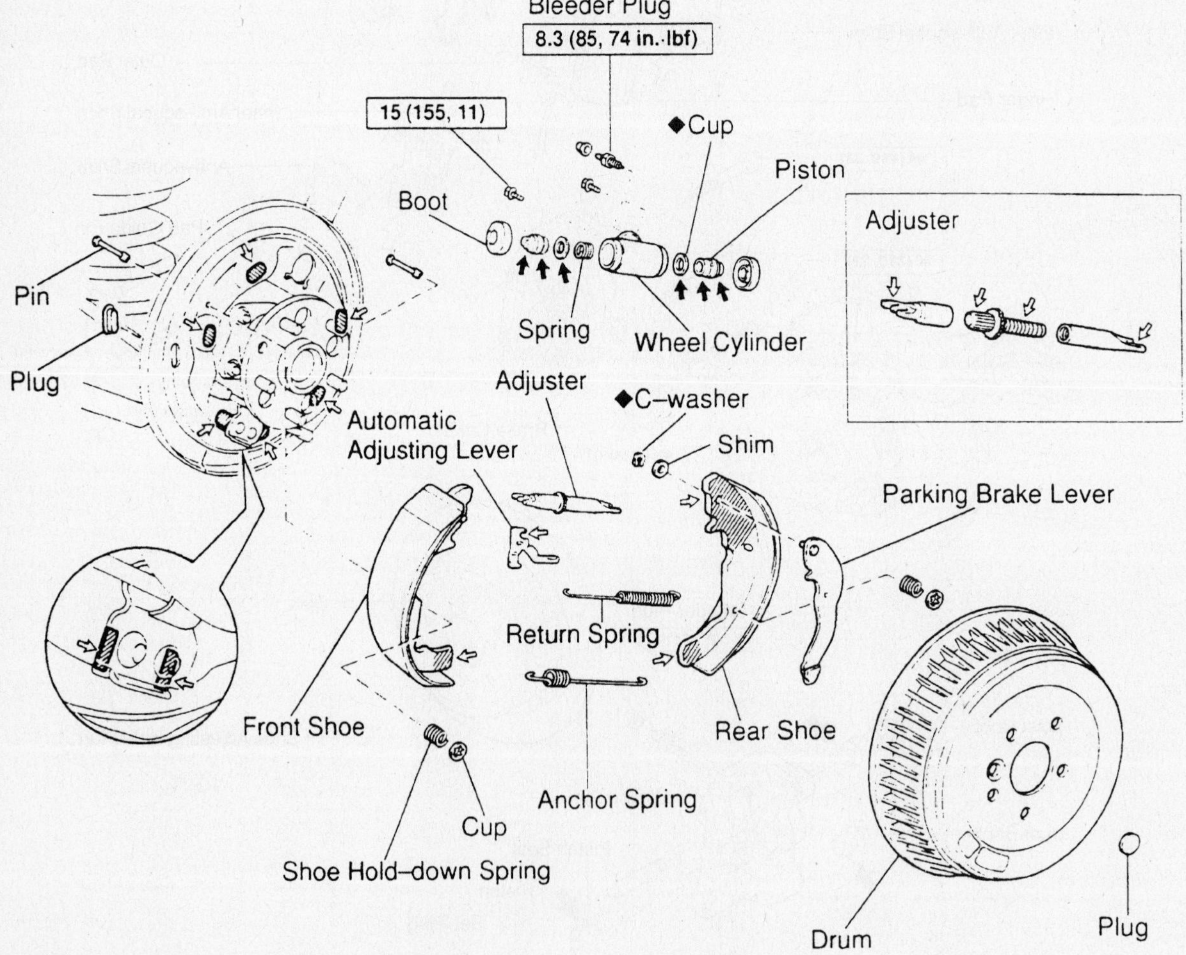

Exploded view of the rear drum brake components—2001–03

3. Block the front wheels, raise the rear of the vehicle, and safely support it with jackstands.

4. Remove the wheel lug nuts and the wheel.

5. Remove the brake drum retaining screws, if equipped. Remove the brake drum.

6. If the drum is difficult to remove, perform the following:

a. Insert the end of a bent wire (a coat hanger will do nicely) through the hole in the brake drum and hold the automatic adjusting lever away from the adjuster.

b. Reduce the brake shoe adjustment by turning the adjuster bolt with a brake tool.

c. The drum should now be loose enough to remove without much effort.

7. Carefully unhook the return spring from the leading (front) brake shoe.

8. Press the hold down spring retainer in and turn the pin on the front brake shoe.

9. Remove the hold down spring, retainers and the pin for the front brake shoe.

10. Pull out the brake shoe and unhook the anchor spring from the lower edge.

11. Remove the hold down spring from the trailing (rear) shoe. Pull the shoe out with the adjuster, automatic adjuster assembly and springs attached. Disconnect the parking brake cable. Remove the tension/return and anchor springs from the rear shoe.

12. Unhook the adjusting lever spring from the rear shoe and then remove the automatic adjuster assembly.

To install:

13. Inspect the shoes for signs of unusual wear or scoring.

14. Check the wheel cylinder for any sign of fluid seepage or frozen pistons.

15. Clean and inspect the brake backing plate and all other components. Check that the brake drum inner diameter is within specified limits. Lubricate the backing plate at the positions the brakes come in contact with the backing plate. Also lubricate the anchor plate.

16. Mount the automatic adjuster assembly onto a new rear brake shoe.

17. Connect the parking brake cable to the rear shoe and then install the automatic adjusting lever, spring and E-ring. Position the rear shoe so the lower end rides in the anchor plate and the upper end is against the boot of the wheel cylinder.

18. Install the pin and the hold down spring. Press the retainer down over the pin and rotate the pin so the crimped edge is held by the retainer.

19. Place the front brake into position and install the anchor spring between the front and rear shoes. Stretch the spring enough so the front shoe will fit as the rear did. Install the hold down spring, pin and retainer to the front brake shoe.

20. Connect the return spring to the front brake shoe.

21. Check the operation of the automatic adjuster mechanism:

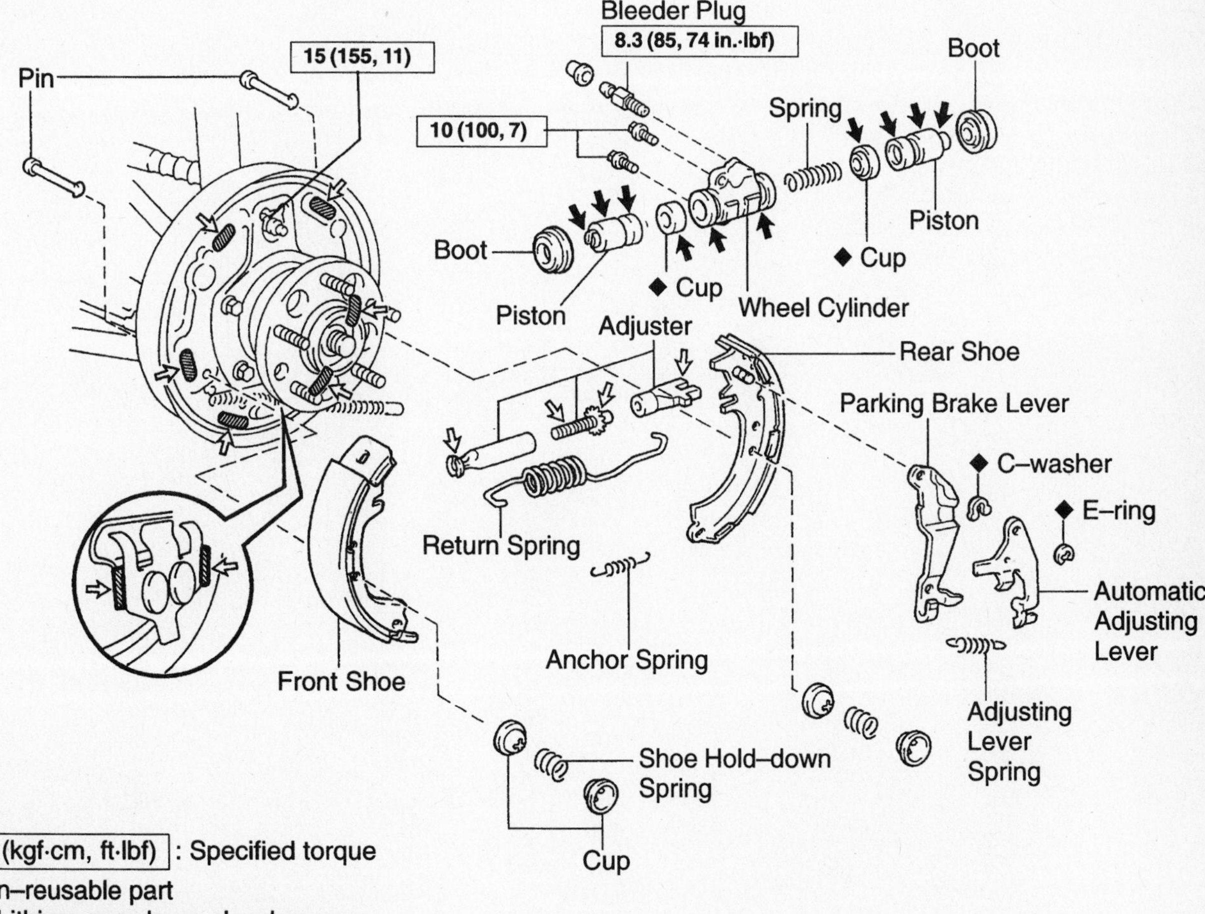

Exploded view of the rear drum brake components—2004

a. Apply the parking brake lever and verifying the adjusting bolt turns.

b. Adjust the strut to where it is the shortest possible length.

c. Install the brake drum.

d. Apply the parking brake lever until the clicking sound can no longer be heard.

22. Check the clearance between the brake shoes and drum:

a. Remove the brake drum.

b. Measure the brake drum inside diameter and diameter of the brake shoes. The difference is "Shoe-to-drum clearance" and should be approximately 0.024 inch (0.6mm). If incorrect, check the parking brake system.

➡A special brake caliper tool is required to gauge the brake drum inside diameter and shoe-to-drum clearance. However it is not required to perform brake shoe adjustment.

23. Install the brake drum.

24. Adjust the brake pedal until a slight drag is felt when the drum is spun by hand.

25. Pull the parking lever all the way up until a clicking sound can no longer be heard. Check the clearance between brake shoes and brake drum.

26. Install the rear wheels, tighten the wheel lug nuts and lower the vehicle.

27. Retighten the wheel lug nuts and pump the brake pedal a few times before moving the vehicle. Adjust the rear brakes again if necessary.

28. Check the level of brake fluid in the master cylinder, then, perform a test drive.

29. Connect the negative battery cable to the battery.

TOYOTA

Sienna

22

SPECIFICATION AND MAINTENANCE CHARTS

ENGINE AND VEHICLE IDENTIFICATION

Engine						
Code ①	Liters (cc)	Cu. In.	Cyl.	Fuel Sys.	Engine Type	Eng. Mfg.
1MZ-FE	3.0 (2995)	183	6	MFI	DOHC	Toyota
3MZ-FE	3.3 (NA)	NA	6	SFI	DOHC	Toyota

Model Year	
Code ②	Year
1	2001
2	2002
3	2003
4	2004
5	2005

MFI: Multi-port Fuel Injection

DOHC: Double Overhead Camshaft

① Stamped on the left side of the engine block

② 10th digit of the Vehicle Identification Number (VIN)

67170-SIEN-C01

GENERAL ENGINE SPECIFICATIONS

Year	Model	Engine Displacement Liters	Engine Series ID	Net Horsepower @ rpm	Net Torque @ rpm (ft. lbs.)	Bore x Stroke (in.)	Com-pression Ratio	Oil Pressure @ rpm
2001	Sienna	3.0	1MZ-FE	194@5200	209@4400	3.44x3.27	10.5:1	43-78@3000
2002	Sienna	3.0	1MZ-FE	210@5800	220@4400	3.44x3.27	10.5:1	43-78@3000
2003	Sienna	3.0	1MZ-FE	210@5800	220@4400	3.44x3.27	10.5:1	43-78@3000
2004	Sienna	3.3	3MZ-FE	230@5600	242@3600	NA	NA	36-78@3000

67170-SIEN-C02

ENGINE TUNE-UP SPECIFICATIONS

Year	Engine Displacement Liters	Engine ID	Spark Plug Gap (in.)	Ignition Timing (deg.)*	Fuel Pump (psi)	Idle Speed (rpm)	Valve Clearance Intake	Valve Clearance Exhaust
2001	3.0	1MZ-FE	0.043	10B	38-44	650-750	0.006-0.010	0.010-0.014
2002	3.0	1MZ-FE	0.043	10B	38-44	650-750	0.006-0.010	0.010-0.014
2003	3.0	1MZ-FE	0.043	10B	38-44	650-750	0.006-0.010	0.010-0.014
2004	3.3	3MZ-FE	0.039-0.043	8-12B	44-50	650-750	0.006-0.010	0.010-0.014

NOTE: The Vehicle Emission Control Information label often reflects specification changes made during production.

The label figures must be used if they differ from those in this chart.

B: Before top dead center

* With terminals TE1 and E1 connected to DLC1

67170-SIEN-C03

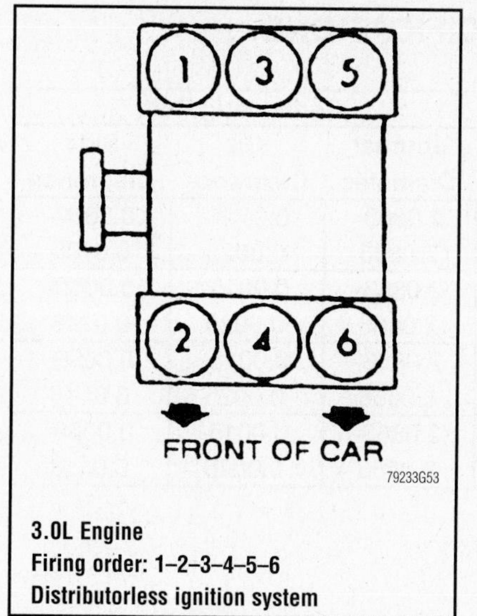

3.0L Engine
Firing order: 1–2–3–4–5–6
Distributorless ignition system

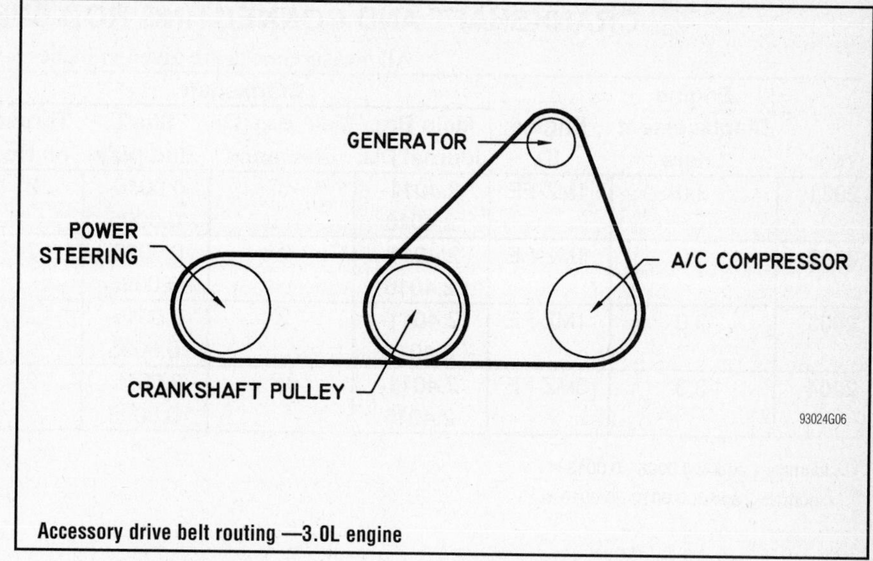

Accessory drive belt routing —3.0L engine

CAPACITIES

Year	Model	Engine Displacement Liters	Engine ID	Engine Oil with Filter (qts.)	Transmission (pts.) 5-Spd	Transmission (pts.) Auto.	Transfer Case (pts.)	Drive Axle Front (pts.)	Drive Axle Rear (pts.)	Fuel Tank (gal.)	Cooling System (qts.)
2001	Sienna	3.0	1MZ-FE	5.0	—	7.4	—	—	—	20.9	10.5
2002	Sienna	3.0	1MZ-FE	5.0	—	10.0	—	—	—	20.9	①
2003	Sienna	3.0	1MZ-FE	5.0	—	10.0	—	—	—	20.9	①
2004	Sienna	3.3	3MZ-FE	5.0	—	②	2.0	—	2.0	21.0	12.4

① w/o rear heater: 10.0
With rear heater: 11.0

② 2wd: 7.4 pts.
4wd: 7.6 pts.

67170-SIEN-C04

VALVE SPECIFICATIONS

Year	Engine Displacement Liters	Engine ID	Seat Angle (deg.)	Face Angle (deg.)	Spring Test Pressure (lbs. @ in.)	Spring Installed Height (in.)	Stem-to-Guide Clearance (in.) Intake	Stem-to-Guide Clearance (in.) Exhaust	Stem Diameter (in.) Intake	Stem Diameter (in.) Exhaust
2001	3.0	1MZ-FE	45	44.5	41.9-46.3@ 1.33	1.331	0.0010-0.0024	0.0012-0.0026	0.2154-0.2159	0.2152-0.2157
2002	3.0	1MZ-FE	45	44.5	41.9-46.3@ 1.33	1.331	0.0010-0.0024	0.0012-0.0026	0.2154-0.2159	0.2152-0.2157
2003	3.0	1MZ-FE	45	44.5	41.9-46.3@ 1.33	1.331	0.0010-0.0024	0.0012-0.0026	0.2154-0.2159	0.2152-0.2157
2004	3.3	3MZ-FE	45	40.5	41.9-46.3@ 1.437	1.331	0.0010-0.0024	0.0012-0.0026	0.2154-0.2159	0.2152 0.2156

67170-SIEN-C05

CRANKSHAFT AND CONNECTING ROD SPECIFICATIONS

All measurements are given in inches.

	Engine		Crankshaft				Connecting Rod		
Year	Displacement Liters	Engine ID	Main Brg. Journal Dia.	Main Brg. Oil Clearance	Shaft End-play	Thrust on No.	Journal Diameter	Oil Clearance	Side Clearance
2001	3.0	1MZ-FE	2.4011-2.4016	①	0.0016-0.0095	2	2.0863-2.0866	0.0015-0.0025	0.0059-0.0188
2002	3.0	1MZ-FE	2.4011-2.4016	①	0.0016-0.0095	2	2.0863-2.0866	0.0015-0.0025	0.0059-0.0118
2003	3.0	1MZ-FE	2.4011-2.4016	①	0.0016-0.0095	2	2.0863-2.0866	0.0015-0.0025	0.0059-0.0118
2004	3.3	3MZ-FE	2.4011-2.4016	①	0.0016-0.0095	2	2.0863-2.0866	0.0015-0.0026	0.0059-0.0118

① Journals 1 and 4: 0.0006 - 0.0013 in.
Journals 2 and 3: 0.0010 - 0.0018 in.

67170-SIEN-C06

PISTON AND RING SPECIFICATIONS

All measurements are given in inches.

	Engine			Ring Gap			Ring Side Clearance		
Year	Displ. Liters	Engine ID	Piston Clearance	Top Comp.	Bottom Comp.	Oil Control	Top Comp.	Bottom Comp.	Oil Control
2001	3.0	1MZ-FE	0.0033-0.0042	0.0098-0.0138	0.0138-0.0177	0.0059-0.0157	0.0008-0.0028	0.0008-0.0024	SNUG
2002	3.0	1MZ-FE	①	0.0098-0.0138	0.0138-0.0177	0.0059-0.0157	0.0008-0.0028	0.0008-0.0024	SNUG
2003	3.0	1MZ-FE	①	0.0098-0.0138	0.0138-0.0177	0.0059-0.0157	0.0008-0.0028	0.0008-0.0024	SNUG
2004	3.3	3MZ-FE	0.0013-0.0023	0.0118-0.0138	0.0197-0.0236	0.0059-0.0157	0.0012-0.0031	0.0008-0.0024	0.0012-0.0043

① AISIN piston: 0.0033-0.0042
MAHLE piston: 0.0013-0.0023

67170-SIEN-C07

TORQUE SPECIFICATIONS

All readings in ft. lbs.

Year	Engine Displacement Liters	Engine ID	Cylinder Head Bolts	Main Bearing Bolts	Rod Bearing Bolts	Crankshaft Damper Bolts	Flywheel Bolts	Manifold Intake	Manifold Exhaust	Spark Plugs	Oil Pan Drain Plug
2001	3.0	1MZ-FE	①	②	③	159	61	11	36	13	33
2002	3.0	1MZ-FE	①	②	④	159	61	11	36	13	33
2003	3.0	1MZ-FE	①	②	④	159	61	11	36	13	33
2004	3.3	3MZ-FE	①	②	④	162	61	11	36	18	33

① Step 1: 40 ft. lbs.
Step 2: Plus 90 degrees
Recessed bolt: 13 ft. lbs.

② 6-point bolts: 20 ft. lbs.
12-point bolts:
Step 1: 16 ft. lbs.
Step 2: Plus 90 degrees

③ Step 1: 29 ft. lbs.
Step 2: Plus 90 degrees

④ Step 1: 18 ft. lbs.
Step 2: +90 degrees

67170-SIEN-C08

WHEEL ALIGNMENT

Year	Model		Caster Range (+/-Deg.)	Caster Preferred Setting (Deg.)	Camber Range (+/-Deg.)	Camber Preferred Setting (Deg.)	Toe-in (in.)	Inside Wheel Angle (Deg.)
2001	Sienna	F	0.75	1.53	0.75	-0.50	0.10+/-0.08	34.32
		R	—	—	0.75	-0.92	0.09+/-0.12	—
2002	Sienna	F	0.75	1.53	0.75	-0.50	0.10+/-0.08	34.32
		R	—	—	0.75	-0.92	0.09+/-0.12	—
2003	Sienna	F	0.75	1.53	0.75	-0.50	0.10+/-0.08	34.32
		R	—	—	0.75	-0.92	0.09+/-0.12	—
2004	Sienna							
	FWD	F	0.75	2.80	0.75	-0.28	0+/-0.08	42.75
		R	—	—	0.50	-1.37	0.11+/-0.12	—
	4WD	F	0.75	2.67	0.75	-0.28	0+/-0.08	42.7
		R	—	—	0.50	-1.42	0.05+/-0.12	—

67170-SIEN-C09

TIRE, WHEEL AND BALL JOINT SPECIFICATIONS

		OEM Tires		Tire Pressures (psi)		Wheel	Ball Joint	Lugnut Torque
Year	Model	Standard	Optional	Front	Rear	Size	Inspection	(ft. lbs.)
2001	Sienna	P205/70R15 95S	P215/65R15 95S	35	35	6.5	①	76
2002	Sienna CE	P205/70R15 95S	none	35	35	6.5	①	76
	Sienna LE	P205/70R15 95S	P215/65R15 95S	35②	35②	6.5	①	76
	Sienna XLE	P215/65R15 95S	none	35②	35②	6.5	①	76
2003	Sienna CE	P205/70R15 95S	none	35	35	6.5	①	76
	Sienna LE	P205/70R15 95S	P215/65R15 95S	35②	35②	6.5	①	76
	Sienna XLE	P215/65R15 95S	none	35②	35②	6.5	①	76
2004	Sienna	P215/65R16 96T	P225/60R17 98T	35	35	NA	③	76

OEM: Original Equipment Manufacturer

NA: Information not available

PSI: Pounds Per Square Inch

① Ball joint turning torque should be 30 inch lbs.

② P215/65R15
Up to 6 passengers: 32
Trailer towing or up to vehicle capacity weight: 35

③ Ball joint turning torque should be 8.7-30 inch lbs.

67170-SIEN-C10

BRAKE SPECIFICATIONS

All measurements in inches unless noted

			Brake Disc			Brake Drum Diameter			Brake Caliper	
Year	Model		Original Thickness	Minimum Thickness	Maximum Runout	Original Inside Diameter	Maximum Machine Diameter	Minimum Lining Thickness	Bracket Bolts (ft. lbs.)	Mounting Bolts (ft. lbs.)
2001	Sienna	F	1.102	1.024	0.0020	—	—	0.039	79	25
		R	—	—	—	9.84	9.921	0.039	—	—
2002	Sienna	F	1.102	1.024	0.0020	—	—	0.039	79	25
		R	—	—	—	9.84	9.921	0.039	—	—
2003	Sienna	F	1.102	1.024	0.0020	—	—	0.039	79	25
		R	—	—	—	9.84	9.921	0.039	—	—
2004	Sienna	F	1.102	1.024	0.0020	—	—	0.039	79	25
		R	0.472	0.413	0.0039	10.00	10.08	0.039	65	25

F: Front

R: Rear

67170-SIEN-C11

SCHEDULED MAINTENANCE INTERVALS

TOYOTA—SIENNA

TO BE SERVICED	TYPE OF SERVICE	VEHICLE MILEAGE INTERVAL (x1000)																		
		5	10	15	20	25	30	35	40	45	50	55	60	65	70	75	80	85	90	95
Automatic transmission and differential fluid	S/I			✓			✓			✓			✓			✓			✓	
Ball joints and boots	S/I			✓			✓			✓			✓			✓			✓	
Brake linings, discs/drums, lines & hoses	S/I			✓			✓			✓			✓			✓			✓	
Charcoal canister	S/I												✓							
Drive belts	S/I						✓						✓						✓	
Engine coolant	R						✓						✓						✓	
Engine oil & filter	R	✓	✓	✓	✓	✓	✓	✓	✓	✓	✓	✓	✓	✓	✓	✓	✓	✓	✓	✓
Exhaust pipes & mounts	S/I			✓			✓			✓			✓			✓			✓	
Fuel lines & connections, fuel tank vapor vent system hoses, fuel tank band	S/I						✓						✓						✓	
Fuel tank cap gasket	S/I						✓						✓						✓	
Halfshaft boots & flange bolts	S/I			✓			✓			✓			✓			✓			✓	
Non-platinum spark plugs	R						✓						✓						✓	
Platinum spark plugs	R												✓							
Rack and pinion assembly	S/I			✓			✓			✓			✓			✓			✓	
Steering linkage	S/I			✓			✓			✓			✓			✓			✓	
Valves	S/I												✓							

R: Replace S/I: Service or Inspect L: Lubricate

FREQUENT OPERATION MAINTENANCE (SEVERE SERVICE)

If a vehicle is operated under any of the following conditions it is considered severe service:

- Towing a trailer or using a camper or car-top carrier.

- Repeated short trips of less than 5 miles in temperatures below freezing.

- Excessive idling or low-speed driving for long distances as in heavy commercial use, such as delivery, taxi or police cars.

- Operating on rough, muddy or salt-covered roads.

- Operating on unpaved or dusty roads.

Oil filter: service or inspect every 5000 miles or 4 months, whichever occurs first.

Brake linings and discs or drums: service or inspect every 5000 miles or 4 months, whichever occurs first.

Steering linkage: service or inspect every 5000 miles or 4 months, whichever occurs first.

Ball joints and boots: service or inspect every 5000 miles or 4 months, whichever occurs first.

Brake discs & pads (front): service or inspect every 6000 miles.

Halfshaft boots: service or inspect every 5000 miles or 4 months. Retighten the flange bolts, whichever occurs first.

Body chassis bolts and nuts: service or inspect every 5000 miles or 4 months, whichever occurs first.

Transmission and differential fluid: replace every 15,000 miles or 12 months, whichever occurs first.

Timing belt: replace every 60,000 miles or 48 months, whichever occurs first.

67170-SIEN-C12

PRECAUTIONS

Before servicing any vehicle, please be sure to read all of the following precautions, which deal with personal safety, prevention of component damage, and important points to take into consideration when servicing a motor vehicle:

- Never open, service or drain the radiator or cooling system when the engine is hot; serious burns can occur from the steam and hot coolant.
- Observe all applicable safety precautions when working around fuel. Whenever servicing the fuel system, always work in a well-ventilated area. Do not allow fuel spray or vapors to come in contact with a spark, open flame or excessive heat (a hot drop light, for example). Keep a dry chemical fire extinguisher near the work area. Always keep fuel in a container specifically designed for fuel storage; also, always properly seal fuel containers to avoid the possibility of fire or explosion. Refer to the additional fuel system precautions later in this section.
- Fuel injection systems often remain pressurized, even after the engine has been turned **OFF**. The fuel system pressure must be relieved before disconnecting any fuel lines. Failure to do so may result in fire and/or personal injury.
- Brake fluid often contains polyglycol ethers and polyglycols. Avoid contact with the eyes and wash your hands thoroughly after handling brake fluid. If you do get brake fluid in your eyes, flush your eyes with clean, running water for 15 minutes. If eye irritation persists, or if you have taken brake fluid internally, IMMEDIATELY seek medical assistance.
- The EPA warns that prolonged contact with used engine oil may cause a number of skin disorders, including cancer! You should make every effort to minimize your exposure to used engine oil. Protective gloves should be worn when changing oil. Wash your hands and any other exposed skin areas as soon as possible after exposure to used engine oil. Soap and water, or waterless hand cleaner should be used.
- All new vehicles are now equipped with an air bag system. The system must be disabled before performing service on or around system components, steering column, instrument panel components, wiring and sensors. Failure to follow safety and disabling procedures could result in accidental air bag deployment, possible personal injury and unnecessary system repairs.
- Always wear safety goggles when working with, or around, the air bag system. When carrying a non-deployed air bag, be sure the bag and trim cover are pointed away from your body. When placing a non-deployed air bag on a work surface, always face the bag and trim cover upward, away from the surface. This will reduce the motion of the module if it is accidentally deployed. Refer to the additional air bag system precautions later in this section.
- Clean, high quality brake fluid from a sealed container is essential to the safe and proper operation of the brake system. You should always buy the correct type of brake fluid for your vehicle. If the brake fluid becomes contaminated, completely flush the system with new fluid. Never reuse any brake fluid. Any brake fluid that is removed from the system should be discarded. Also, do not allow any brake fluid to come in contact with a painted surface; it will damage the paint.
- Never operate the engine without the proper amount and type of engine oil; doing so WILL result in severe engine damage.
- Timing belt maintenance is extremely important! Many models utilize an interference type, non-freewheeling engine. If the timing belt breaks, the valves in the cylinder head may strike the pistons, causing potentially serious (also time consuming and expensive) engine damage. Refer to the maintenance interval charts in the front of this manual for the recommended replacement interval for the timing belt, and to the timing belt section for belt replacement and inspection.
- Disconnecting the negative battery cable on some vehicles may interfere with the functions of the on-board computer system(s) and may require the computer to undergo a relearning process once the negative battery cable is reconnected.
- When servicing drum brakes, only disassemble and assemble one side at a time, leaving the remaining side intact for reference.

ENGINE REPAIR

Distributor

Sienna models are equipped with a distibutorless ignition system.

Alternator

REMOVAL & INSTALLATION

3.0L Engine

1. Before servicing the vehicle, refer to the precautions in the beginning of this section.
2. Remove or disconnect the following:
   - Alternator electrical connectors
   - Wiring harness from the clip
   - Pivot bolt
   - Plate washer
   - Adjusting lockbolt
   - Drive belt
   - Alternator

To install:

3. Install or connect the following:
   - Alternator
   - Drive belt. Tension the belt to 170–180 lbs. for a new belt or 95–135 lbs. for a used belt.
   - Adjusting lockbolt. Tighten the bolt to 13 ft. lbs. (18 Nm).
   - Plate washer
   - Pivot bolt. Tighten the bolt to 41 ft. lbs. (56 Nm).
   - Wiring harness from the clip
   - Alternator electrical connectors

3.3L Engine

1. Before servicing the vehicle, refer to the precautions in the beginning of this section.
2. Remove or disconnect the following:
   - Alternator electrical connectors
   - Wiring harness from the clip
   - Pivot bolt
   - Plate washer
   - Adjusting lockbolt
   - Drive belt
   - Alternator

To install:

3. Install or connect the following:
   - Alternator
   - Drive belt. Tension the belt to 170–180 lbs. for a new belt or 95–135 lbs. for a used belt.
   - Adjusting lockbolt. Tighten the bolt to 13 ft. lbs. (18 Nm).
   - Plate washer
   - Pivot bolt. Tighten the bolt to 43 ft. lbs. (58 Nm).
   - Wiring harness from the clip
   - Alternator electrical connectors

Ignition Timing

ADJUSTMENT

Ignition timing is controlled by the ECM and is not adjustable.

Engine Assembly

REMOVAL & INSTALLATION

3.0L Engine

1. Before servicing the vehicle, refer to the precautions in the beginning of this section.
2. Matchmark the hood position.
3. Remove or disconnect the following:
   - Hood
   - Wiper and blade assembly
   - Top cowl seal and panel
   - Window washer hoses from the ventilator louvers
   - Left and right ventilator louvers
   - Heater air duct
4. Properly relieve the fuel system pressure.
5. Remove or disconnect the following:
   - Both battery cables
   - Battery and tray
6. Drain the engine coolant.
7. Drain the engine oil.
8. Remove or disconnect the following:
   - Intake air cleaner and case assembly

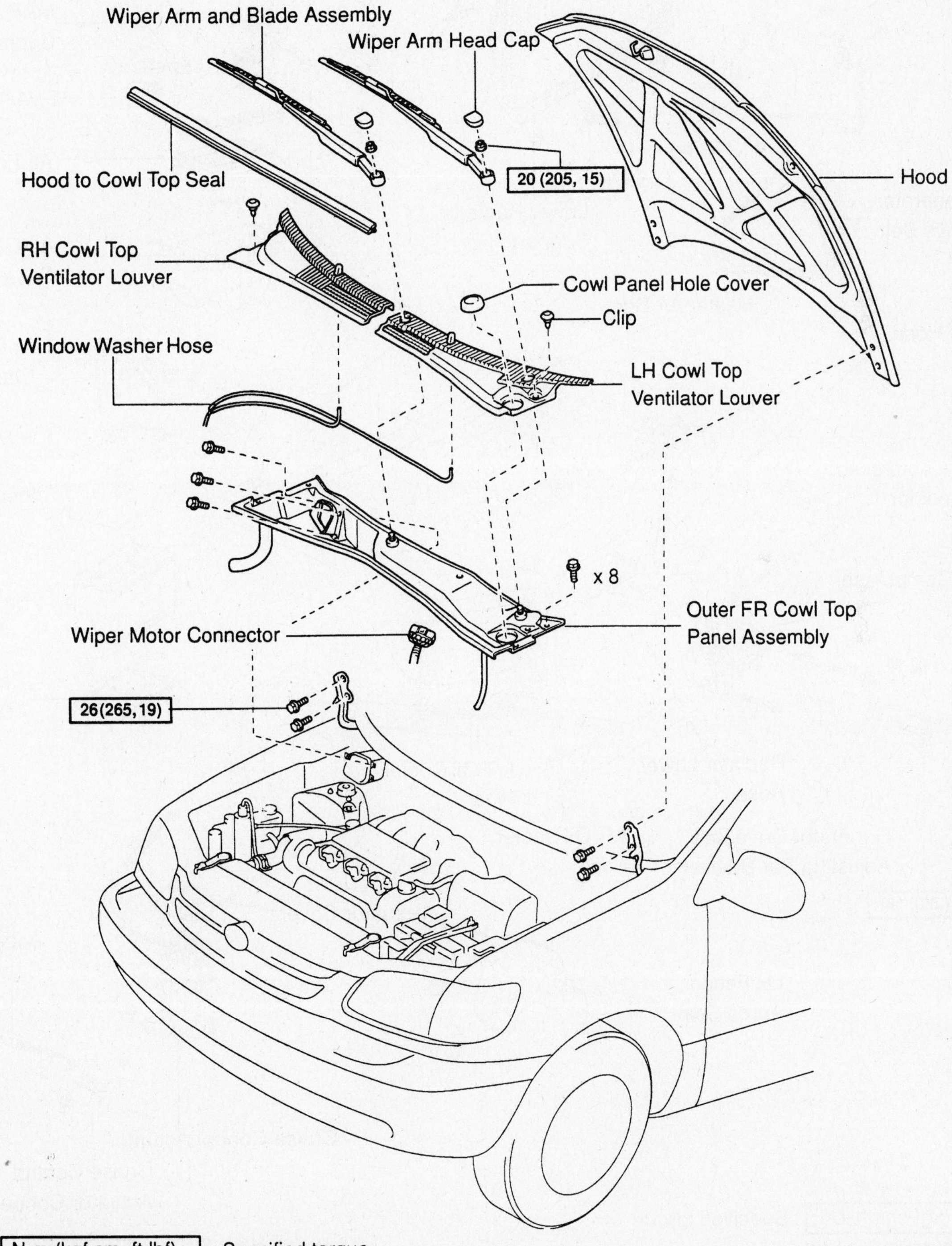

Exploded view of the top cowl and related components—2001–03

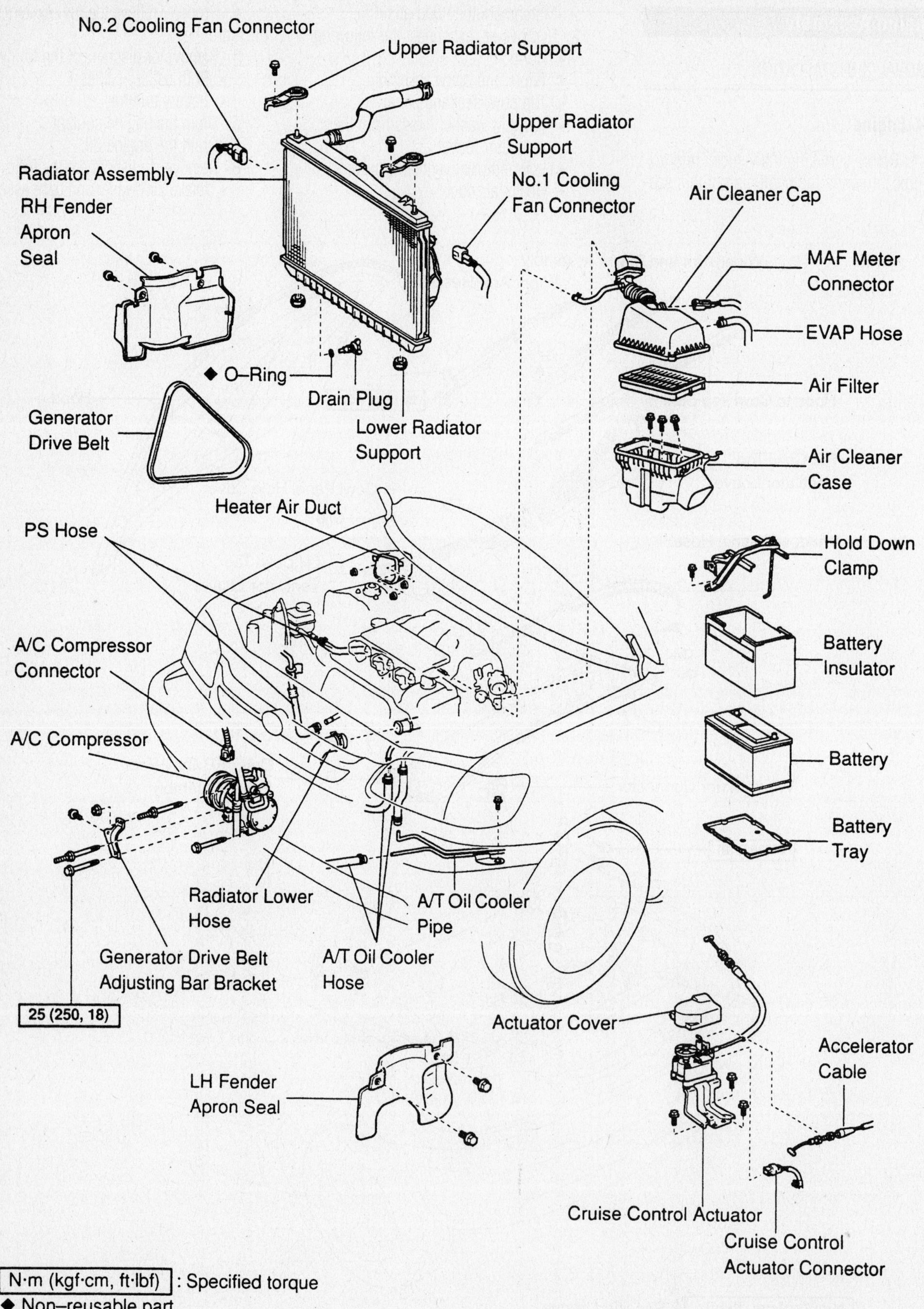
No.2 Cooling Fan Connector
Upper Radiator Support
Upper Radiator Support
Radiator Assembly
No.1 Cooling Fan Connector
Air Cleaner Cap
RH Fender Apron Seal
MAF Meter Connector
EVAP Hose
◆ O–Ring
Air Filter
Drain Plug
Generator Drive Belt
Lower Radiator Support
Air Cleaner Case
Heater Air Duct
PS Hose
Hold Down Clamp
A/C Compressor Connector
Battery Insulator
A/C Compressor
Battery
Battery Tray
Radiator Lower Hose
A/T Oil Cooler Pipe
Generator Drive Belt Adjusting Bar Bracket
A/T Oil Cooler Hose
25 (250, 18)
Actuator Cover
Accelerator Cable
LH Fender Apron Seal
Cruise Control Actuator
Cruise Control Actuator Connector
N·m (kgf·cm, ft·lbf) : Specified torque
◆ Non–reusable part
7924ZG07

Exploded view of engine pre-removal components—3.0L engine

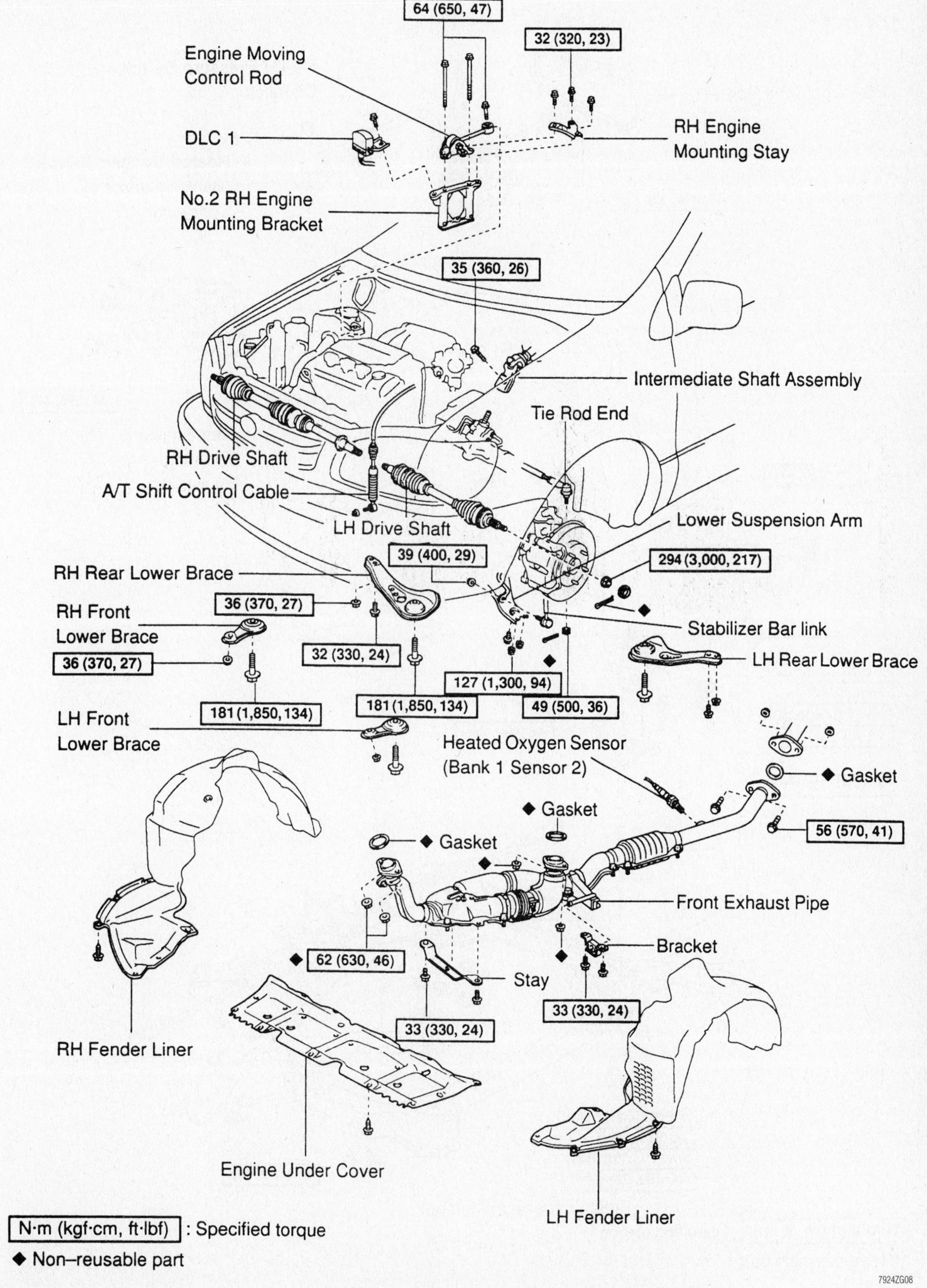

Exploded view of engine removal and installation tightening specifications of the related components—3.0L engine

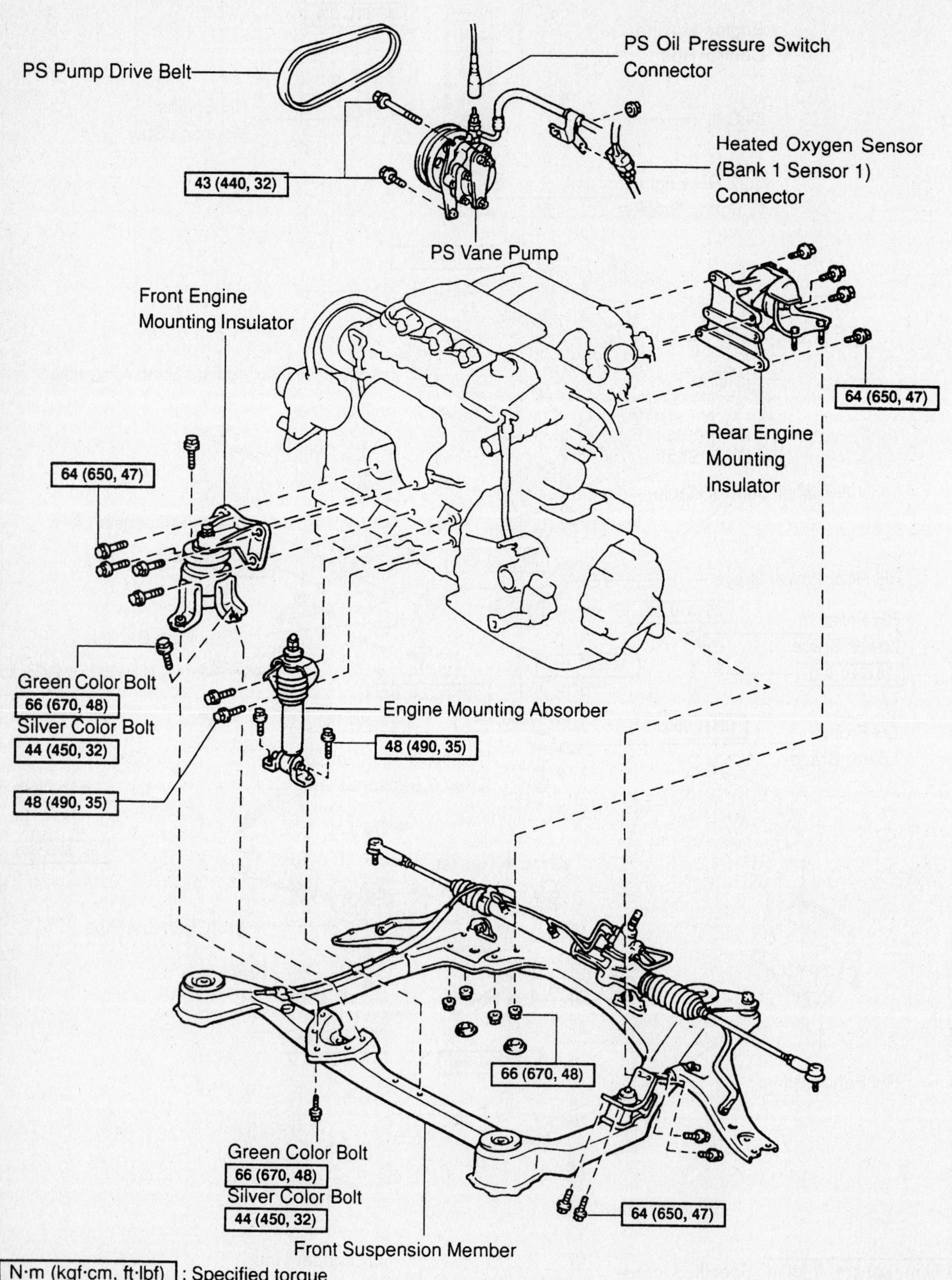

Exploded view of the suspension component removal and installation for engine removal—3.0L engine

- Cruise control actuator, if equipped
- Upper and lower radiator hoses
- Radiator
- Automatic transmission oil cooler lines
- Any connectors, hoses and sensors that would interfere with engine removal
- Engine Control Module (ECM) engine wiring harness from inside the glove box; then, pull the harness into the engine compartment
- Compressor

➡It may be necessary to remove the air conditioning compressor lines in order to remove the engine.

- Automatic transmission shifter cable from the transaxle
- Header pipes from the exhaust manifolds
- Left and right fender apron seals
- Halfshafts
- Stabilizer links and the steering intermediate shaft
- Power steering pump
- Engine undercover
- Engine hanger to the engine
- Engine sling device to the engine hangers
- Right-hand motor mount and moving control rod
- Front suspension lower braces

9. Lower the engine, transaxle and front suspension member as an assembly from the vehicle.

To install:

10. Raise the engine, transaxle and front suspension member as an assembly into the vehicle.
11. Install the front suspension lower braces, and tighten the fasteners, as follows:
    - Bolt A: 134 ft. lbs. (181 Nm)
    - Bolt B: 24 ft. lbs. (32 Nm)
    - Nut C: 27 ft. lbs. (36 Nm)
12. Install or connect the following:
    - Moving control rod. Tighten the bolts to 47 ft. lbs. (64 Nm).
    - Right-hand motor mount. Tighten the bolts to 23 ft. lbs. (32 Nm).
    - Engine sling device from the engine hangers
    - Engine undercover
    - Power steering pump hoses
    - Stabilizer links and the steering intermediate shaft
    - Halfshafts
    - Left and right fender apron seals
    - Header pipes to the exhaust manifolds
    - Automatic transmission shifter cable to the transaxle
    - Air conditioning compressor to the engine
13. Push the wiring harness into the glove box.
14. Install or connect the following:
    - ECM
    - Any connectors, hoses and sensors that were removed
    - Automatic transmission oil cooler lines
    - Upper and lower radiator hoses and fit the radiator
    - Cruise control actuator, if removed
    - Intake air cleaner and case assembly
15. Fill the engine oil to proper level.
16. Fill the engine with coolant.
17. Install or connect the following:
    - Battery tray and battery
    - Battery cables
    - Heater air duct
    - Left and right ventilator louvers
    - Window washer hoses from the ventilator louvers
    - Top cowl seal and panel
    - Wiper and blade assembly
    - Hood
    - New oil filter
18. Refill the engine with oil.
19. Refill the engine with engine coolant.
20. Install the engine undercovers.
21. Start the engine and check for leaks.

Heater Core

REMOVAL & INSTALLATION

Front Heater

1. Disconnect the negative battery cable.
2. Drain the cooling system into a clean container for reuse.
3. Disconnect the heater hoses from the heater core.
4. Remove the steering wheel by performing the following procedure:
    a. Position the front wheels facing straight-ahead.
    b. Remove the steering wheel side covers.
    c. Using a Torx® wrench, loosen the 2 screws located at each side of the steering wheel until the screw's circumference groove catches on the screw case.
    d. Pull the air bag module from the steering wheel and disconnect the electrical connector.

⁂ CAUTION

Place the air bag module in a safe place with the front side facing upward.

    e. Remove the steering wheel nut.
    f. Place alignment marks on the steering wheel and the main shaft.
    g. Using a steering wheel puller, press the steering wheel from the steering column.
5. Remove the instrument panel and reinforcement by performing the following procedure:
    a. Remove the front door scuff plates.
    b. Remove the cowl side boards.
    c. Remove the front door trim covers.
    d. Remove the front pillar garnish by disengaging the 5 clips. If equipped with a tweeter speaker, disconnect the electrical connector.
    e. Remove the steering column covers-to-steering column screws and the covers.
    f. Remove the combination switch-to-steering column screws, disconnect the electrical connector(s) and remove the combination switch.
    g. Remove the 2 hood open lever screws and the hood open lever.
    h. Remove the 2 lower finish panel bolts and disengage the panel from the 3 clips.
    i. Remove the 2 No. 1 safety pad insert bolts and the insert.
    j. Remove the 2 No. 2 finish panel bolts and disengage the panel from the 4 clips.
    k. In the left side of the glove compartment, pry out the glove box door finish plate and disconnect the air bag module connector.
    l. Remove the glove box 3 nuts and 2 screws and the glove box.
    m. Remove the center cluster finish panel by disengaging the claw (bottom center) and 4 clips (one at each corner).
    n. Remove the ashtray, the 2 ashtray receptacle box screws.
    o. Remove the 4 lower center cluster finish panel screws and disconnect the connector.
    p. Remove the clock, the No. 1 and No. 2 registers from the panel.
    q. Remove the 3 cluster finish panel screws, disengage the 8 clips and remove the panel.
    r. Remove the combination meter.
    s. Remove the radio assembly.
    t. Remove the heater control assembly.
    u. Remove 2 passenger's side air bag module bolts; then, disconnect and remove the air bag module.

✻✻ CAUTION

Place the air bag module in a safe place with the front side facing upward.

v. Remove the instrument panel-to-chassis 5 bolts and nut.

w. Remove the audio amplifier.

x. Remove the No. 1 and No. 2 braces.

y. Remove the No. 2 cowl brace.

z. Remove the instrument panel reinforcement.

6. Remove the evaporator housing by performing the following procedure:

a. Discharge and recover the air conditioning system refrigerant.

b. In the engine compartment, remove the refrigerant lines-to-cowl connector bolts; then, disconnect the lines and discard the O-rings.

c. Disconnect the electrical connector at the evaporator housing.

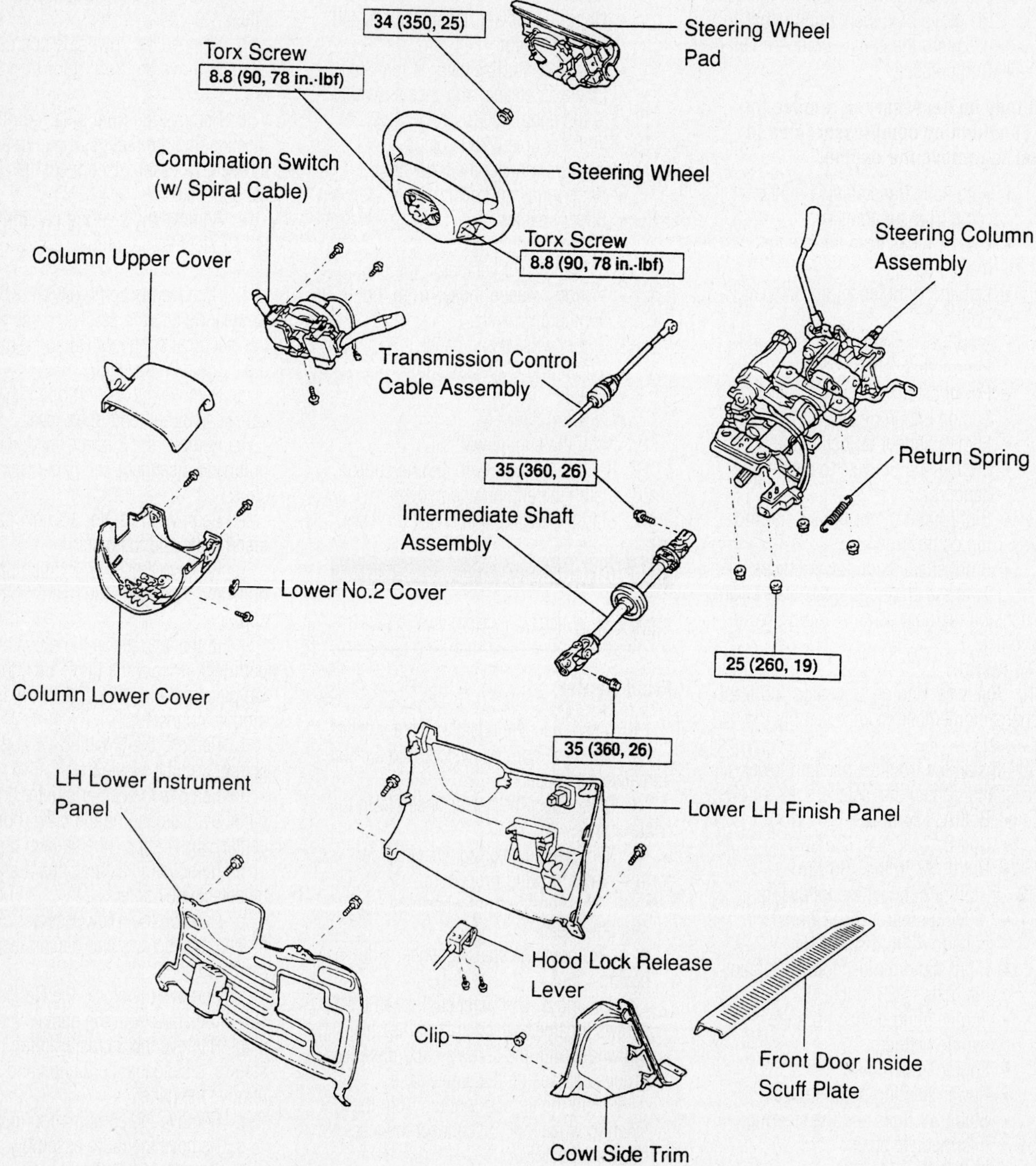

Exploded view of the steering wheel, steering column and related components

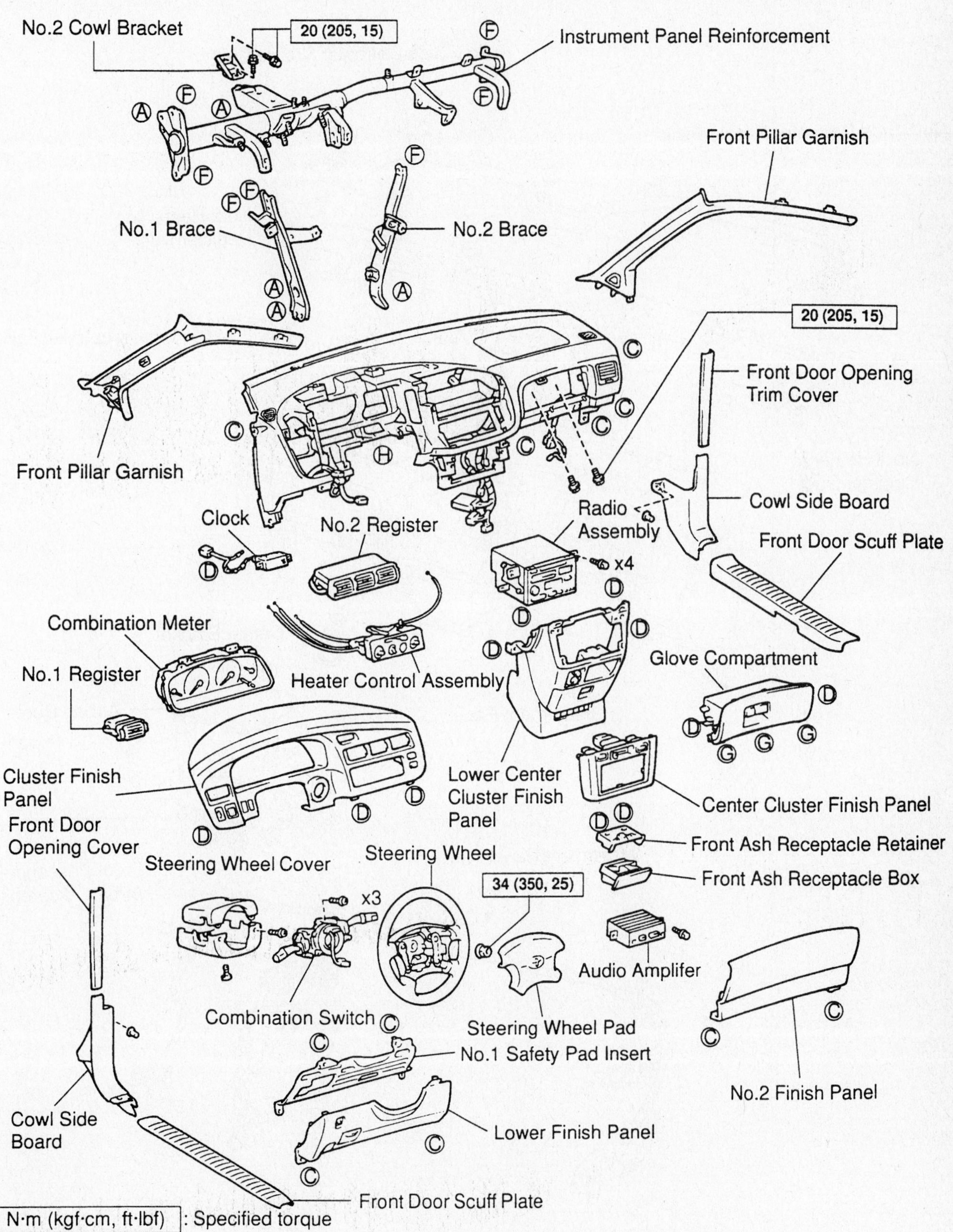

Exploded view of the instrument panel and related components

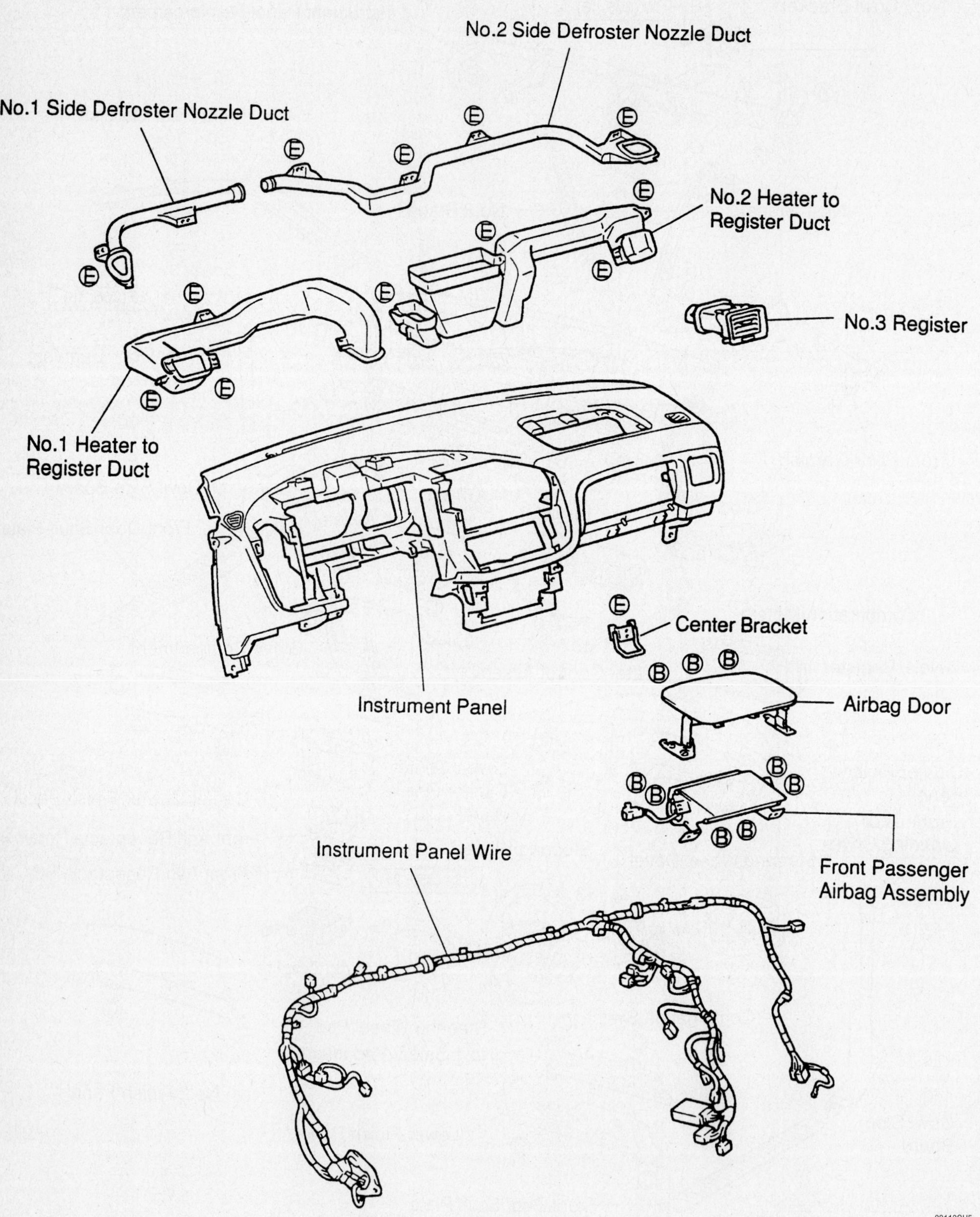

Exploded view of the ventilation system and related components

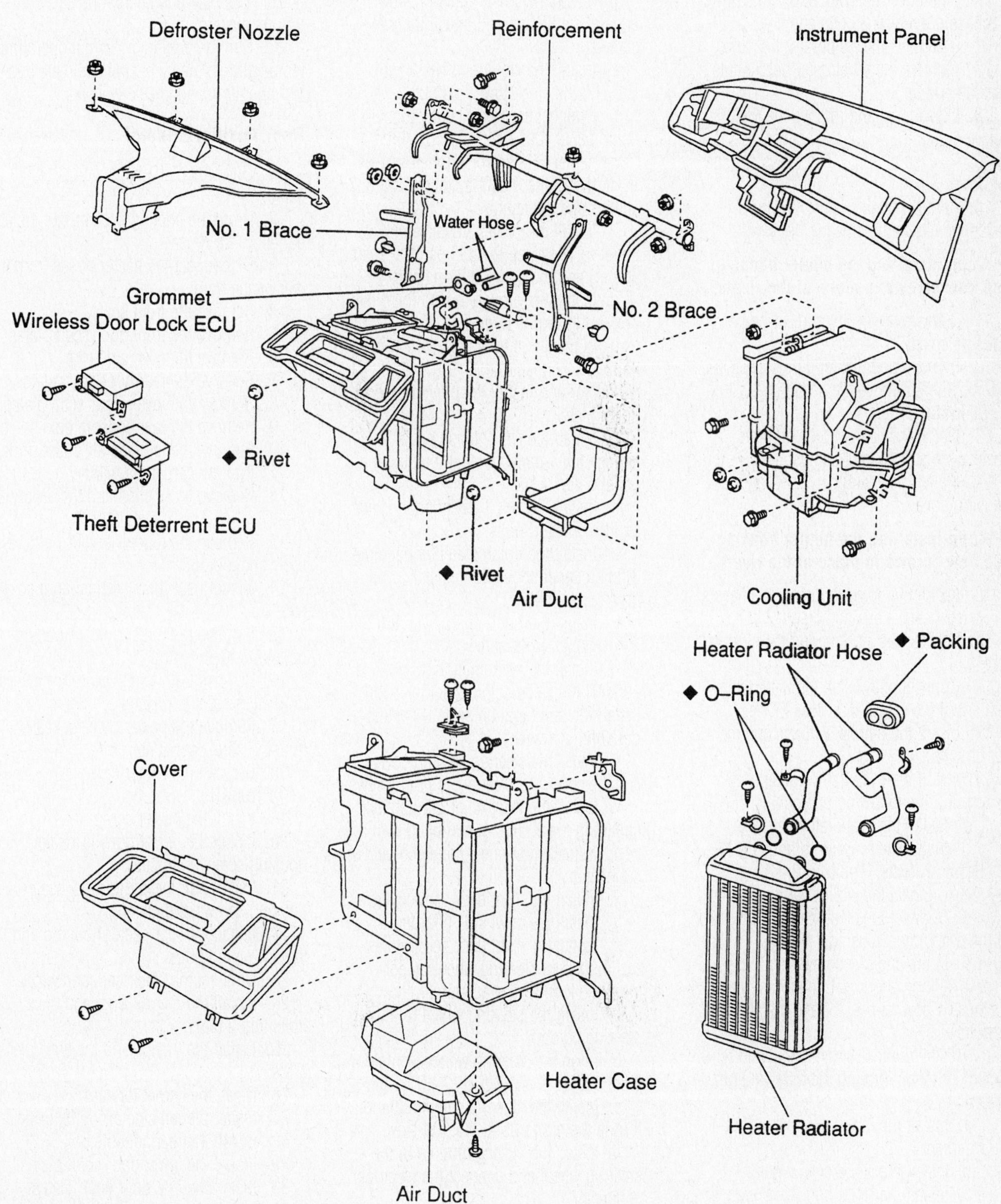

Exploded view of the heater core, heater housing, evaporator housing and related components

d. Disconnect the wiring harness clamp.

e. Remove the evaporator housing-to-chassis 2 rivets, 3 bolts and nut.

f. Remove the evaporator housing.

7. Remove the 4 defroster nozzle nuts and the nozzle.

8. Disconnect and remove the theft deterrent and the wireless door lock ECUs.

9. Release the 2 air duct claws and the air duct.

10. Remove the 2 heater housing-to-chassis rivets and the heater housing.

➡When installing the heater housing, use new screws in place of the rivets.

11. Remove the heater core-to-heater housing cover.

12. Remove both heater core screws and clamps; then, remove the heater core.

To install:

13. Install the heater core and both heater core screws and clamps.

14. Install the heater core-to-heater housing cover.

➡When installing the heater housing, use new screws in place of the rivets.

15. Install the heater housing-to-chassis and the 2 heater housing screws.

16. Release the air duct and the air duct claws.

17. Connect and install the theft deterrent and the wireless door lock ECUs.

18. Install the defroster nozzle and the 4 nozzle nuts.

19. Install the evaporator housing by performing the following procedure:

a. Install the evaporator housing.

b. Install the evaporator housing-to-chassis 2 rivets, 3 bolts and nut.

c. Connect the wiring harness clamp.

d. Connect the electrical connector at the evaporator housing.

e. In the engine compartment, use new O-rings and install the refrigerant lines-to-cowl connector and install the bolts.

20. Install the instrument panel and reinforcement by performing the following procedure:

a. Install the instrument panel reinforcement.

b. Install the No. 2 cowl brace.

c. Install the No. 1 and No. 2 braces.

d. Install the audio amplifier.

e. Install the instrument panel-to-chassis 5 bolts and nut.

f. Connect and install the air bag module and the 2 passenger's side air bag module bolts.

g. Install the heater control assembly.

h. Install the radio assembly.

i. Install the combination meter.

j. Install the cluster finish panel, engage the 8 clips and install the panel screws.

k. Install the No. 1 and No. 2 registers and the clock to the panel.

l. Connect the lower center cluster finish panel connector and install the 4 lower center cluster finish panel screws.

m. Install the 2 ashtray receptacle box screws and the ashtray.

n. Install the center cluster finish panel by engaging the 4 clips (1 at each corner) and the claw (bottom center).

o. Install the glove box and the glove box 3 nuts and 2 screws.

p. In the left side of the glove compartment, connect the air bag module connector and install the glove box door finish plate.

q. Install the No. 2 finish panel, engage the 4 panel clips and install the 3 panel bolts.

r. Install the No. 1 safety pad insert and the 2 insert bolts.

s. Install the finish panel, engage the 3 finish panel clips and install 2 lower finish panel bolts.

t. Install the hood open lever and the 2 hood open lever screws.

u. Install the combination switch, connect the electrical connector(s) and install the combination switch-to-steering column screws.

v. Install the steering column covers and the covers-to-steering column screws.

w. Install the front pillar garnish by engaging the 5 clips. If equipped with a tweeter speaker, connect the electrical connector.

x. Install the front door trim covers.

y. Install the cowl side boards.

z. Install the front door scuff plates.

21. Install the steering wheel by performing the following procedure:

a. Install the steering wheel to the steering column.

b. Align the steering wheel-to-main shaft marks.

c. Install the steering wheel nut and torque the nut to 25 ft. lbs. (34 Nm).

d. Install the air bag module to the steering wheel and connect the electrical connector.

e. Using a Torx® wrench, tighten the steering wheel screws to 78 inch lbs. (8.8 Nm).

f. Install the steering wheel side covers.

22. Connect the heater hoses to the heater core.

23. Refill the cooling system.

24. Connect the negative battery cable.

25. Evacuate and charge the air conditioning system.

26. Run the engine to normal operating temperatures; then, check the climate control operation and check for leaks.

Rear Auxiliary Heater

1. Disconnect the negative battery cable.

2. Drain the cooling system into a clean container for reuse.

3. Disconnect the heater hoses from the rear heater core.

4. Remove the front seats.

5. Remove the front door scuff plates.

6. Remove the cowl side trim.

7. Remove the rear door scuff plates.

8. Remove the lower door scuff plates.

9. Remove the rear console box.

10. Remove the left side air outlet grille.

11. Pull the carpet rearward.

12. Remove the 3 clips and the air outlet grille.

13. Remove the rear air duct 2 bolts, 2 clips and the duct.

14. Disconnect the electrical connectors.

15. Remove the 3 rear heater housing bolts and the housing.

16. Remove both heater core-to-heater housing screws and clamps.

17. Remove the heater core-to-heater housing screw and plate.

18. Remove the heater core.

To install:

19. Install the heater core.

20. Install the heater core-to-heater housing screw and plate.

21. Install both heater core-to-heater housing screws and clamps.

22. Install the rear heater housing and the 3 housing bolts.

23. Connect the electrical connectors.

24. Install the rear air duct and the 2 bolts and 2 clips.

25. Install the 3 clips and the air outlet grille.

26. Move the carpet forward.

27. Install the left side air outlet grille.

28. Install the rear console box.

29. Install the lower door scuff plates.

30. Install the rear door scuff plates.

31. Install the cowl side trim.

32. Install the front door scuff plates.

33. Install the front seats.

34. Connect the heater hoses to the rear heater core.

35. Refill the cooling system.

36. Connect the negative battery cable.

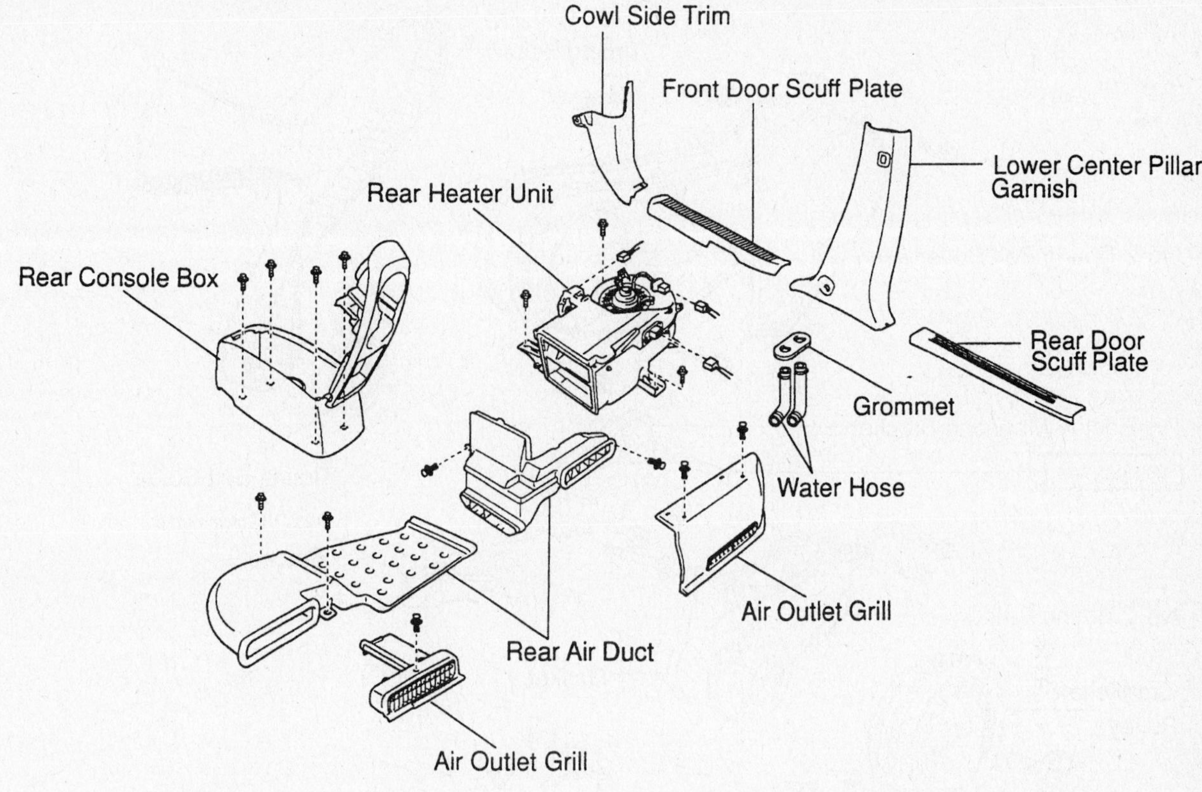

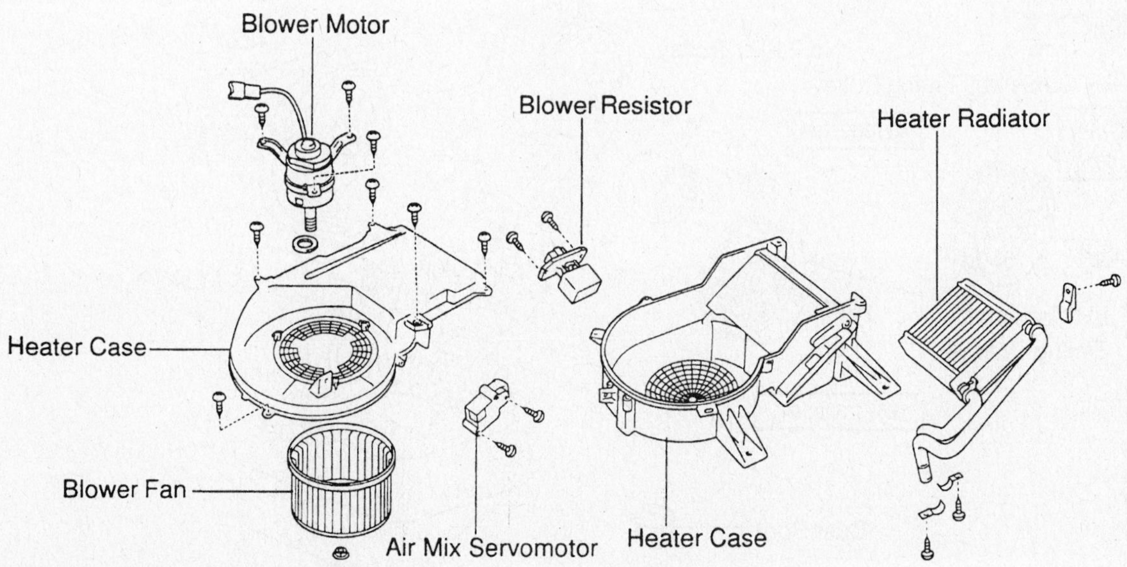

Exploded view of the rear heater core, the rear heater housing and related components

Water Pump

REMOVAL & INSTALLATION

3.0L and 3.3L Engine

1. Before servicing the vehicle, refer to the precautions in the beginning of this section.

2. Disconnect the negative battery cable.

3. Drain the engine coolant.

4. Remove or disconnect the following:
   - Wiper and blade assembly
   - Top cowl seal and panel
   - Window washer hoses, from the ventilator louvers
   - Left and right ventilator louvers
   - Heater air duct
   - Timing belt

5. Mark the left and right camshaft pulleys with a touch of paint.

6. Remove or disconnect the following:
   - Right and left camshaft pulleys bolts
   - Pulleys from the engine

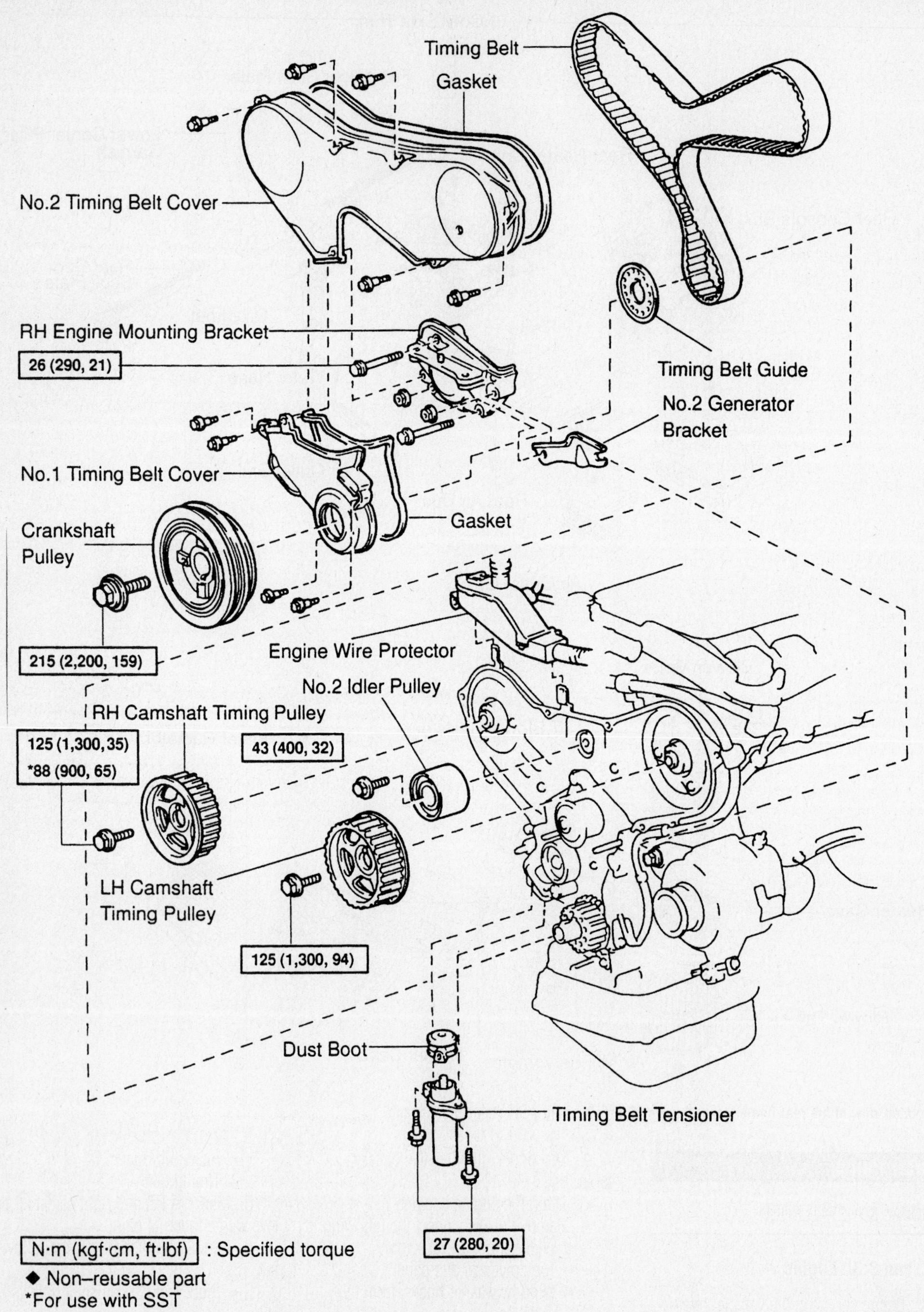

Exploded view of the components to gain access to the water pump—3.0L engine

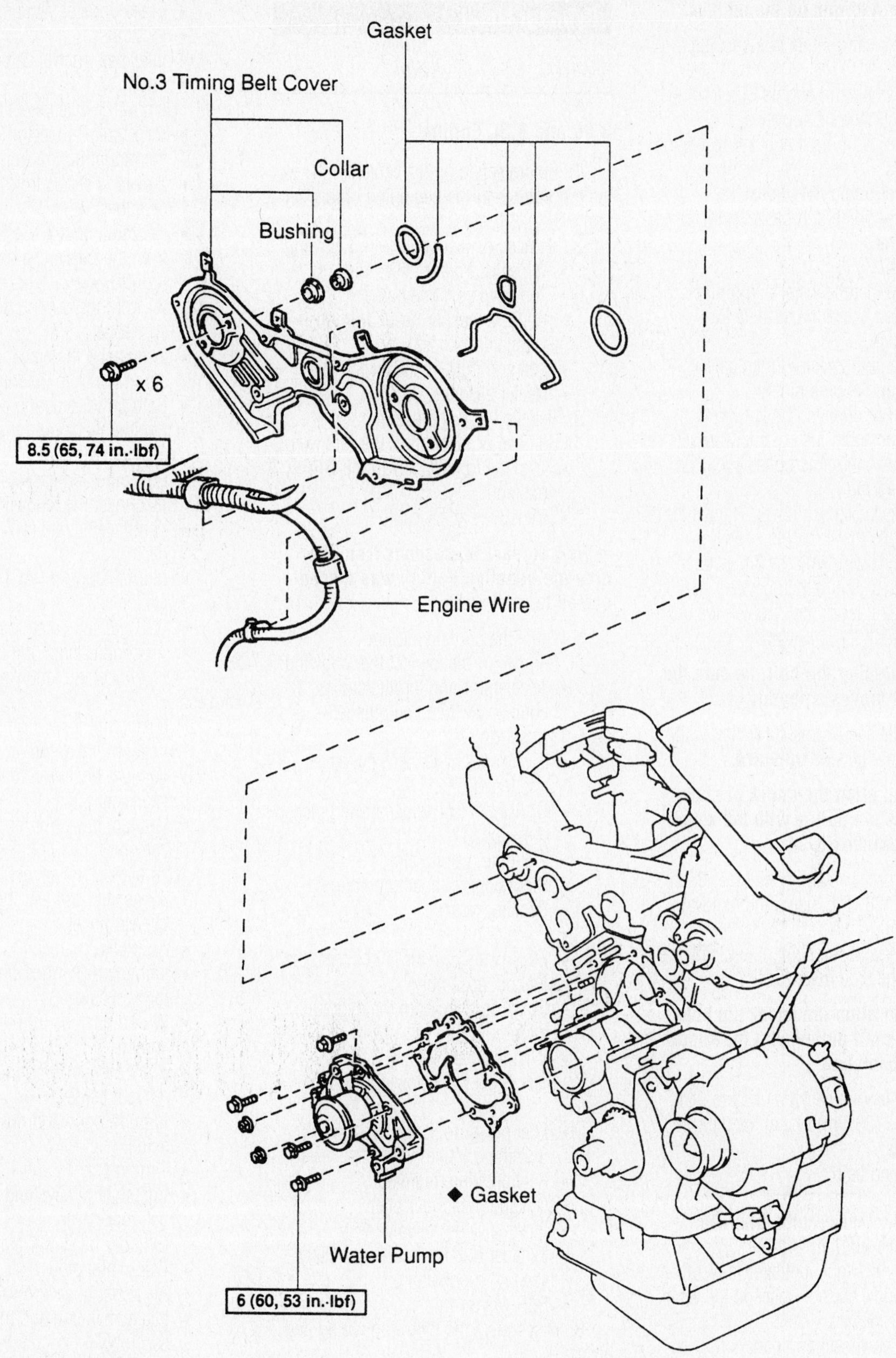

Exploded view of the water pump and related components—3.0L engine

➡**Be sure not to mix up the pulleys.**

- No. 2 idler pulley by removing the bolt
- 3 clamps and engine wire from the rear timing belt cover
- 6 No. 3 timing belt cover-to-engine bolts
- Water pump nuts/bolts
- Water pump and gasket from the engine

To install:

7. Check that the water pump turns smoothly. Also check the air hole for coolant leakage.
8. Apply liquid sealer to the gasket, water pump and engine block.
9. Install or connect the following:
   - Water pump, using a new gasket. Tighten the nuts/bolts to 53 inch lbs. (6 Nm).
   - Rear timing belt cover. Tighten the 6 bolts to 74 inch lbs. (9 Nm).
   - Engine wire with the 3 clamps to the rear timing belt cover
   - No. 2 idler pulley. Tighten the bolt to 32 ft. lbs. (43 Nm).

➡**After tightening the bolt, be sure the idler pulley moves smoothly.**

   - Right-hand camshaft pulley, with the flange side **outward**.

➡**Be sure to align the knock pin hole on the camshaft pulley with the knock pin on the camshaft.**

   - Tighten the camshaft bolt to 65 ft. lbs. (88 Nm), using the removal tools
   - Left-hand camshaft pulley, with the flange side **inward**.

➡**Be sure to align the knock pin hole on the camshaft pulley with the knock pin on the camshaft.**

   - Tighten the camshaft bolt to 94 ft. lbs. (125 Nm), using the removal tools
   - Timing belt
10. Fill the engine coolant.
11. Install or connect the following:
    - Heater air duct
    - Left and right ventilator louvers
    - Window washer hoses to the ventilator louvers
    - Top cowl seal and panel
    - Wiper and blade assembly
    - Negative battery cable
12. Start the engine.
13. Top off the engine coolant and check for leaks.

Cylinder Head

REMOVAL & INSTALLATION

3.0L and 3.3L Engine

1. Before servicing the vehicle, refer to the precautions in the beginning of this section.
2. Remove or disconnect the following:
   - Wiper and blade assembly
   - Top cowl seal and panel
   - Window washer hoses from the ventilator louvers
   - Left and right ventilator louvers
   - Heater air duct
3. Relieve the fuel pressure.
4. Remove or disconnect the following:
   - Turn the ignition key to the **OFF** position
   - Negative battery cable

➡**Wait at least 90 seconds from the time the negative battery was disconnected to start work.**

5. Drain the cooling system.
6. Remove or disconnect the following:
   - Accelerator and throttle cables, if equipped with an automatic transaxle
   - Air cleaner cover, air flow meter and the air duct
   - Cruise control actuator and bracket, if equipped
   - 2 engine ground straps
   - Right engine mounting support
   - Radiator hoses
   - 2 heater hoses
   - Fuel feed and return lines from the fuel rail assembly
   - Pressure hose from the hydraulic motor
   - V-bank cover
7. Disconnect the following vacuum hoses:
   - Fuel pressure control Vacuum Switching Valve (VSV)
   - Fuel pressure regulator
   - Cylinder head rear plate
   - Intake air control valve VSV
   - Exhaust Gas Recirculation (EGR) vacuum modulator
   - EGR valve
8. Disconnect the following wiring and hoses:
   - Intake air control valve
   - Fuel pressure regulator
   - EGR VSV
9. Remove the 2 nuts and the emission control valve set.
10. Disconnect the following hoses;
    - Brake booster vacuum hose
    - PCV hose
    - Intake air control valve vacuum hose
11. Remove or disconnect the following:
    - Data Link Connector (DLC) from the mounting bracket
    - 2 ground straps from the intake chamber
    - Hydraulic motor pressure hose from the intake chamber
    - Right Oxygen (O_2) sensor connector from the power steering pressure tube
    - 2 nuts and the power steering pressure tube from the intake chamber
    - Both power steering air hoses
    - Engine hanger and the intake chamber support
    - EGR pipe and gaskets
12. Disconnect the following wiring:
    - Throttle Position (TP) sensor connector
    - Idle Air Control (IAC) valve connector
    - EGR gas temperature connector
    - Air conditioning idle up connector
13. Disconnect the following vacuum hoses:
    - 2 vacuum hoses from the Thermal Vacuum Valve (TVV)
    - Vacuum hose from the cylinder head rear plate
    - Vacuum hose from the charcoal canister
14. Remove or disconnect the following:
    - Air assist hose and the 2 water bypass hoses
    - Air intake chamber
    - Left engine wiring harness and move it aside
    - Wiring harness from the rear of the engine
    - Right engine wiring harness and move it aside
    - Ignition coils and move them aside
    - Timing belt
    - Camshaft pulleys and the timing belt rear cover
    - Cylinder head rear plate
    - Water inlet pipe
    - Air assist hose and vacuum hose
    - Intake manifold and fuel rail assembly
    - Water outlet
    - EGR pipe from the right exhaust manifold
    - Front exhaust pipe and exhaust manifolds
    - Dipstick assembly and the power steering pump bracket

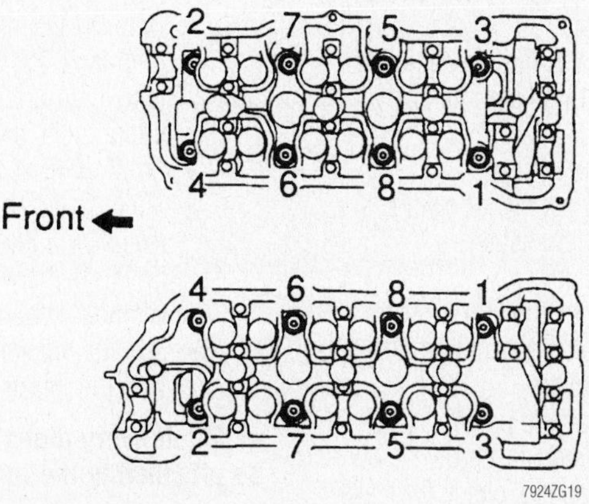

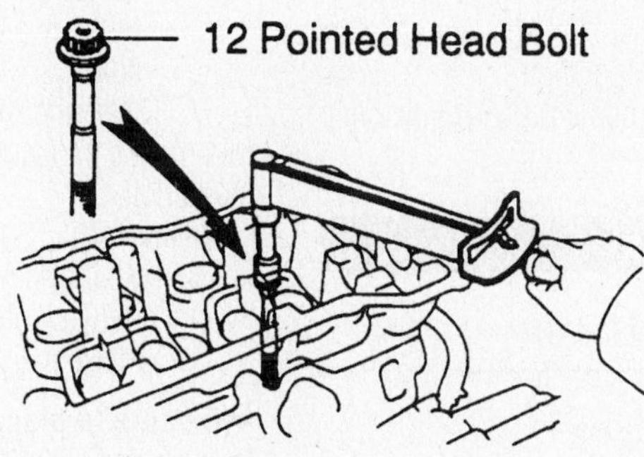

Cylinder head bolt loosening sequence

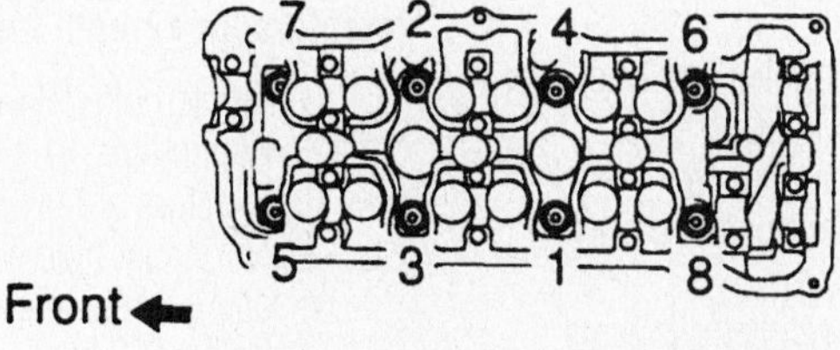

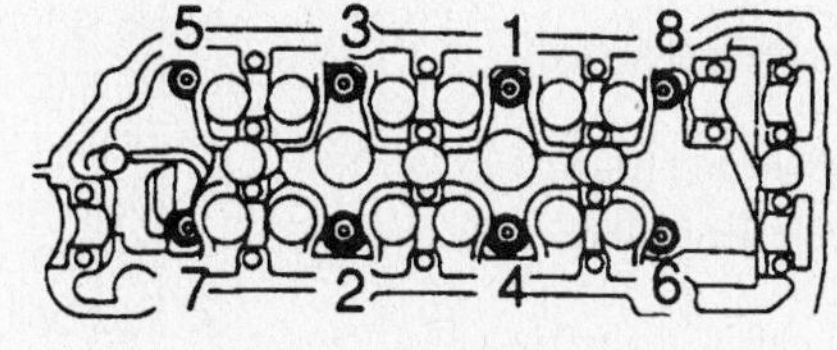

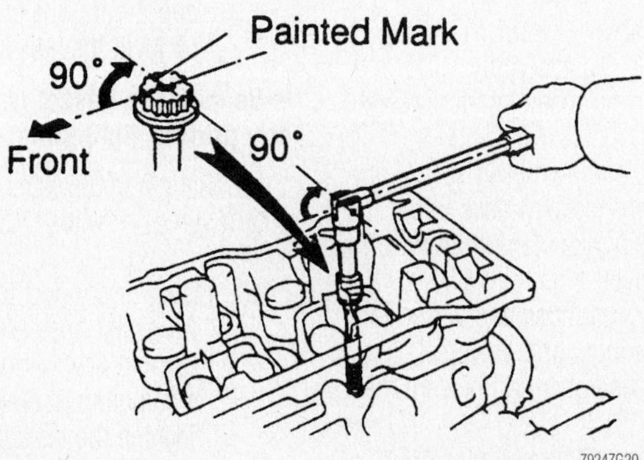

Cylinder head bolt tightening sequence—3.0L and 3.3L engines engine

- Valve covers and the Camshaft Position (CMP) sensor
- Camshafts

15. Be sure the engine is at/or near ambient temperature and remove the 2 (1 on each head) 8mm recessed hex bolts. Loosen and remove the 8 head bolts evenly, in 3 passes, in the reverse order of the installation sequence. Carefully lift the head from the engine; if necessary to pry the head loose, take great care not to damage the mating surfaces. Place the head on wood blocks in a clean work area.

➡If the cylinder head bolts are loosened out of sequence, warpage or cracking could result.

16. Remove the cylinder head gasket. With a gasket scraper, carefully remove all the old gasket material from the cylinder head and engine block surfaces.

To install:

17. Place the new cylinder head gasket onto the cylinder block.
18. Install the cylinder head, in sequence, using several steps, as follows:
    - Cylinder head onto the gasket
    - Cylinder head bolts lubricated with clean engine oil
    - Tighten the bolts in sequence in 3 steps to 40 ft. lbs. (54 Nm).

➡If any bolt does not meet the torque, replace it.

- Mark the forward edge of each bolt with paint, then tighten each bolt, in proper sequence, an additional 90 degrees.
- Check that each painted mark is now at a 90 degrees angle to the front

➡The paint mark applied to the bolt in the 9 o'clock position and should now be in the 12 o'clock position.

- Remaining 8mm bolts, lubricated with engine oil. Tighten both bolts to 13 ft. lbs. (18 Nm).

19. Install the camshafts.
20. Check and adjust the valves.
21. Apply sealant to the cylinder heads where the camshaft supports meet the cylinder heads.
22. Install or connect the following:
    - Cylinder head covers, using new gaskets
    - Dipstick and power steering pump bracket
    - Exhaust manifolds. Tighten the nuts to 36 ft. lbs. (49 Nm).
    - EGR pipe to the right exhaust manifold

- Water outlet
- Intake manifold and the fuel rail assembly. Tighten the intake manifold nuts/bolts to 11 ft. lbs. (15 Nm).
- Air assist hose and the 2 water bypass hoses
- Water inlet pipe and cylinder head rear plate
- Timing belt rear cover and camshaft pulleys
- Timing belt
- Spark plugs and ignition coils
- Right engine wiring harness
- Wiring harness to the rear of the engine
- Left engine wiring harness
- Air intake chamber
- EGR pipe, using new gaskets

23. Connect the following vacuum hoses:
    - The 2 TVV vacuum hoses
    - The vacuum hose to the rear cylinder head plate
    - Charcoal canister vacuum hose
24. Connect the following electrical wiring:
    - TP sensor connector
    - IAC valve connector
    - EGR gas temperature connector
    - Air conditioning idle up connector
25. Install or connect the following:
    - Engine hanger and the intake chamber support
    - Both power steering air hoses
    - Power steering pressure tube to the intake chamber
    - O_2 sensor connector to the pressure tube.
    - Both ground straps, to the intake chamber
    - DLC to the bracket
26. Connect the following hoses:
    - Power brake booster vacuum hose
    - PCV hose
    - IAC valve vacuum hose
27. Install or connect the following:
    - Emission control valve set and related vacuum hoses and connectors
    - V-bank cover
    - Pressure hose to the hydraulic motor
    - Fuel lines to the fuel rail assembly
    - Heater and radiator hoses
    - Right engine mounting support
    - Both engine ground straps
    - Upper front suspension brace, if removed. Tighten the nuts to 59 ft. lbs. (80 Nm).
    - Cruise control actuator and bracket
    - Air cleaner, air flow meter and air duct assembly
    - Accelerator and throttle cables, if equipped with an automatic transaxle
28. Fill the cooling system.
29. Install or connect the following:
    - Negative battery cable
    - Heater air duct
    - Left and right ventilator louvers
    - Window washer hoses from the ventilator louvers
    - Top cowl seal and panel
    - Wiper and blade assembly
30. Start the engine and check for leaks.
31. Bleed the air from the cooling system.
32. Road test the vehicle and check for unusual noise, shock, slippage, correct shift points and smooth operation.
33. Recheck the coolant and engine oil levels.

Intake Manifold

REMOVAL & INSTALLATION

3.0L and 3.3L Engine

1. Before servicing the vehicle, refer to the precautions in the beginning of this section.
2. Remove or disconnect the following:
    - Wiper and blade assembly
    - Top cowl seal and panel
    - Window washer hoses from the ventilator louvers
    - Left and right ventilator louvers
    - Heater air duct
3. Properly relieve the fuel system pressure.
4. Remove the battery and battery tray.
5. Drain and recycle the engine coolant.
6. Remove or disconnect the following:
    - Accelerator cable, on automatic transaxles
    - Throttle cable
    - Air cleaner cap assembly
    - Any wiring or hoses interfering with removal
    - Right side engine mount stay
    - Radiator and heater hoses in the way of the intake manifold removal
    - V-bank cover
    - All the vacuum hose and wiring for the emission control valve set
    - Air intake chamber and discard the gasket
    - Exhaust Gas Recirculation (EGR) pipe and discard the gaskets
    - Hydraulic motor pressure hose from the air intake chamber
    - Engine wiring harnesses from the left side, right side, rear and No. 3 timing belt cover
    - Front exhaust pipe, if necessary
    - Timing belt, camshaft timing pulleys, No. 2 idler pulley and No. 3 timing belt cover
    - Cylinder head rear plate
    - 2 bolts, nuts and plate washers with the intake manifold.

➡The delivery pipes with injectors will be attached to the manifold.

- Other fuel related components such as the No. 2 fuel pipe and pulsation damper, if needed
- Delivery pipes from the intake manifold

7. Clean and inspect the intake manifold mating surfaces. Scrape all old gasket martial off.

To install:

8. Install or connect the following:
    - Delivery pipes with injectors to the intake manifold.

➡Be sure to place 4 spacers in position on the manifold. Temporarily install 4 bolts to retain the delivery pipes to the manifold. Inspect the injectors for smooth rotation.

- Tighten the delivery pipes bolts to 84 inch lbs. (10 Nm), once the injectors are properly seated
- No. 2 fuel pipe with union bolts and gaskets. Tighten the bolts to 24 ft. lbs. (32 Nm).
- No. 1 fuel pipe with pulsation damper, using 4 new gaskets. Tighten the damper to 35 ft. lbs. (32 Nm) and the bolt to 11 ft. lbs. (15 Nm).
- Fuel pressure regulator, if removed
- Intake manifold. Tighten the 9 bolts and 2 nuts in a crisscross pattern to 11 ft. lbs. (15 Nm).

➡Be sure the gasket is in place properly prior to tightening.

9. Retighten the water outlet mounting nuts/bolts to 11 ft. lbs. (15 Nm), if loosened.
10. Install or connect the following:
    - Air assist hose and water inlet pipe, using a new O-ring, by applying a small amount of soapy water. Tighten the fastener(s) to 14 ft. lbs. (20 Nm).
    - Ground strap

- Vacuum hoses removed to the air intake chamber and vacuum tank
- Any remaining components, using new gaskets. Tighten the air intake chamber nuts/bolts to 32 ft. lbs. (43 Nm), the EGR pipe nuts to 108 inch lbs. (12 Nm) and the emission control valve set to 69 inch lbs. (8 Nm).
- Air cleaner assembly
- Heater hoses
- Battery and tray
- Throttle cable with bracket onto the throttle body
- Accelerator cable, by adjusting it, if equipped with an automatic transaxle

11. Refill the cooling system
12. Install or connect the following:
    - Negative battery cable
    - Heater air duct
    - Left and right ventilator louvers
    - Window washer hoses from the ventilator louvers
    - Top cowl seal and panel
    - Wiper and blade assembly
13. Start the engine and inspect for leaks.

Exhaust Manifold

REMOVAL & INSTALLATION

3.0L and 3.3L Engine

FRONT MANIFOLD

➡**Removing the oil filter helps gain access to a lower bolt in the front exhaust manifold.**

1. Before servicing the vehicle, refer to the precautions in the beginning of this section.
2. Remove or disconnect the following:
    - Negative battery cable
    - Engine undercovers
    - Front exhaust pipe from the exhaust manifolds, by removing the nuts

➡**Check for access to some of the manifold lower bolts, if so remove any possible.**

- Heated Oxygen (HO_2) sensor
- Exhaust manifold stay, by removing the bolt and nut
- Remaining exhaust manifold nuts; then, separate the exhaust manifold from the engine

To install:

3. Install or connect the following:
    - Exhaust manifold, using a new gasket. Uniformly, tighten the bolts to 36 ft. lbs. (49 Nm).
    - Exhaust manifold stay. Tighten the nut/bolt to 15 ft. lbs. (20 Nm).
    - Heated Oxygen (HO_2) sensor to the exhaust manifold
    - Front exhaust pipe to the exhaust manifold, using a new gasket. Tighten both nuts to 46 ft. lbs. (62 Nm).
    - Engine undercovers
    - Negative battery cable

REAR MANIFOLD

1. Before servicing the vehicle, refer to the precautions in the beginning of this section.
2. Remove or disconnect the following:
    - Negative battery cable
    - Engine undercovers
    - Front exhaust pipe from both exhaust manifolds, from below the engine
    - Exhaust Gas Recirculation (EGR) pipe from the rear exhaust manifold, by removing the 4 nuts
    - Heated Oxygen (HO_2) sensor wiring, from the right exhaust manifold
    - Exhaust manifold stay
    - 6 exhaust manifold nuts and the exhaust manifold

To install:

3. Install or connect the following:
    - Exhaust manifold to the engine, using a new gasket. Tighten the 6 nuts to 36 ft. lbs. (49 Nm).
    - Exhaust manifold stay. Tighten the nut/bolt to 15 ft. lbs. (20 Nm).
    - HO_2 sensor wiring to the exhaust manifold
    - EGR pipe to the exhaust manifold and the engine, using new gaskets. Tighten the 4 nuts to 108 inch lbs. (12 Nm).
    - Front exhaust pipe to the exhaust manifold, use a new gasket. Tighten both nuts to 46 ft. lbs. (62 Nm).
    - Engine undercovers
    - Negative battery cable

Front Crankshaft Seal

REMOVAL & INSTALLATION

3.0L and 3.3L Engine

1. Before servicing the vehicle, refer to the precautions in the beginning of this section.
2. Remove or disconnect the following:
    - Engine coolant reservoir tank and the alternator belt
    - Right front wheel and the splash shield
    - Power steering pump drive belt, by loosening both bolts
    - Both ground wire connectors
    - Right engine mounting stay
    - Engine moving control rod and the No. 2 right engine mount bracket

➡**To extract the engine bracket and control rod, raise the engine slightly.**

- No. 2 alternator bracket
- Crankshaft pulley bolt, using a pry-bar and wrench or Crankshaft Pulley Holding tool 09213-54015 and Flange Holding tool 09330-00021
- Crankshaft pulley, using a puller
- No. 1 timing belt cover

3. Remove the No. 2 timing belt cover, as follows:
    - Engine wire protector from the No. 3 (rear) timing belt cover
    - Engine wire protector clamp from the No. 3 timing belt cover
    - 5 bolts from the No. 2 timing belt cover
    - No. 2 cover

To install:

4. Install or connect the following:
    - No. 2 timing belt cover, using a new gasket

➡**Install it evenly to the part of the belt cover shaded black. After installation, press down on it so that the adhesive sticks to the belt cover firmly.**

- No. 2 timing belt cover. Tighten the 5 bolts to 74 inch lbs. (8 Nm).
- Engine wire protector clamp to the No. 3 timing belt cover
- Engine wire protector to the No. 3 timing belt cover with the bolt
- No. 3 timing belt cover, using a new gasket
- Tighten the 4 No. 1 timing belt cover bolts to 74 inch lbs. (8 Nm).
- Crankshaft pulley. Tighten the bolt to 159 ft. lbs. (215 Nm).
- No. 2 alternator bracket. Tighten the nut to 21 ft. lbs. (28 Nm). Do not tighten the pivot bolt at this time.
- No. 2 right engine mounting bracket and the moving control rod
- Right engine mount stay
- Both ground wire connectors
- Drive belts by adjusting them
- Coolant reservoir
- Right front splash shield and wheel
- Negative battery cable

5. Start the vehicle and check for any leaks.
6. Recheck the ignition timing.

Camshaft and Valve Lifters

REMOVAL & INSTALLATION

3.0L and 3.3L Engine

1. Before servicing the vehicle, refer to the precautions in the beginning of this section.
2. Remove or disconnect the following:
   - Timing belt and idler pulley
   - Camshaft timing pulleys
   - Cylinder head covers

➡The thrust clearance on both the intake and exhaust camshafts is very small; the camshafts must be kept level during removal. If the camshafts are removed without being kept level, the camshaft may be caught in the cylinder head, causing the head to break or the camshaft to seize.

3. Remove the exhaust and intake camshafts from the right side cylinder head, as follows:
   a. Turn the camshaft with a wrench until the 2 pointed marks drive and driven gears are aligned. (The right camshaft gears have 2 marks apiece; the left side camshaft gears have 1 mark each.)
   b. Secure the exhaust camshaft sub-gear to the main gear using a service bolt. A bolt 0.63–0.79 in. (16–20mm) long with a 6mm thread diameter and a 1mm pitch is recommended. When removing the exhaust camshaft be sure the sub-gear is not loaded; all the force must be eliminated.
   c. Uniformly loosen and remove the exhaust camshaft bearing cap bolts in several passes and in the proper sequence. Remove the 8 bearing cap bolts and remove the caps, keeping them in the correct order.
   d. Remove the exhaust camshaft from the engine.
   e. Uniformly loosen and remove the 10 bearing cap bolts in several passes, in the proper sequence. Remove the bearing caps, keeping them in order, remove the oil seal, then lift out the intake camshaft.
4. Remove the exhaust and intake camshafts from the left side cylinder head, as follows:
   a. Turn the camshaft with a wrench until the pointed marks on the drive and driven gears are aligned. (The right camshaft gears have 2 marks apiece; the left side camshaft gears have 1 mark each.)

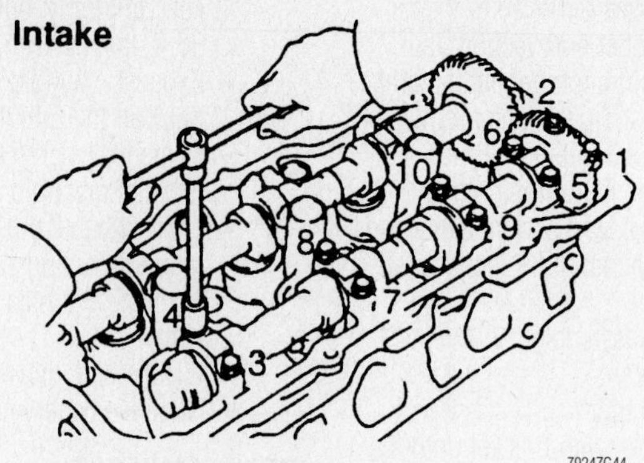

Right intake camshaft bearing cap bolt loosening sequence

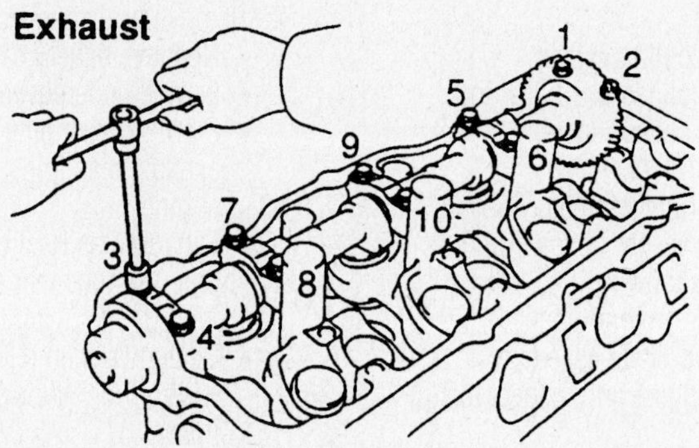

Right side exhaust camshaft bearing cap bolt loosening sequence

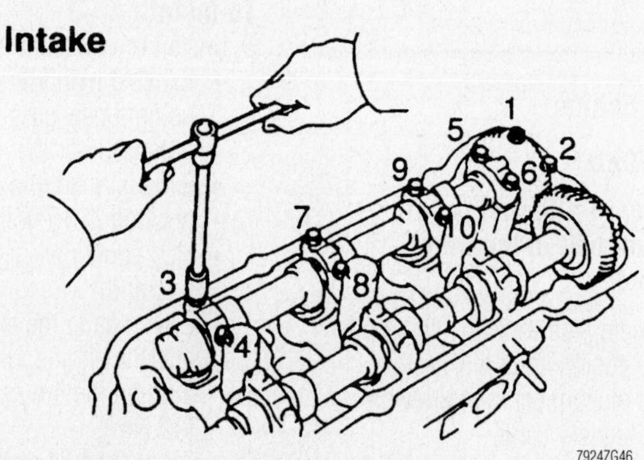

Left intake camshaft bearing cap bolt loosening sequence

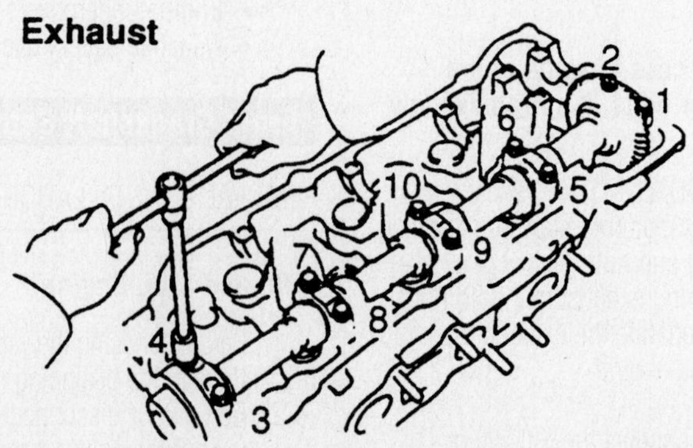

Left side exhaust camshaft bearing cap bolt loosening sequence

b. Secure the exhaust camshaft sub-gear to the main gear using a service bolt. A bolt 16–20mm long with a 6mm thread diameter and a 1mm pitch is recommended. When removing the exhaust camshaft be sure the sub-gear is not loaded; all the force must be eliminated.

c. Uniformly loosen and remove the exhaust camshaft bearing cap bolts in several passes and in the proper sequence. Remove the 8 bearing cap bolts and remove the caps. Keep the caps in the correct order.

d. Remove the exhaust camshaft from the engine.

e. Uniformly loosen and remove the 10 bearing cap bolts in several passes, in the reverse order of the installation sequence. Remove the bearing caps, keeping them in order, remove the oil seal, then lift out the intake camshaft.

5. Remove the valve lifter shims and hydraulic lifters. Identify each lifter and shim as it is removed so it can be reinstalled in the same position. If the lifters are to be reused, store them upside down in a sealed container.

To install:

6. Install the valve lifters into their original positions and install the shims. Check valve clearance and replace the shims as necessary.

7. When reinstalling, remember that the camshafts must be handled carefully and kept straight and level to avoid damage.

8. Before installing the camshafts in either cylinder head, apply multi-purpose grease to each camshaft.

9. Install the right camshafts, as follows:

a. Position the intake camshaft on the head so that the alignment marks are at a 90 degrees angle from vertical. The mark should be at the "3 o'clock" position.

b. Apply sealant to the No. 1 bearing cap.

c. Apply a light coat of clean engine oil to the bolt threads and under the bolt head. Install the bearing caps to their proper position. Tighten the bolts evenly and in several passes to 12 ft. lbs. (16 Nm) in the proper sequence.

d. Position the exhaust camshaft on the head so that the alignment marks are at a 90 degrees angle from vertical. The mark should be at the "9 o'clock" position and must align with the marks on the other gear.

e. Apply a light coat of clean engine oil to the bolt threads and under the bolt head. Install the bearing caps to their proper position. Tighten the bolts evenly

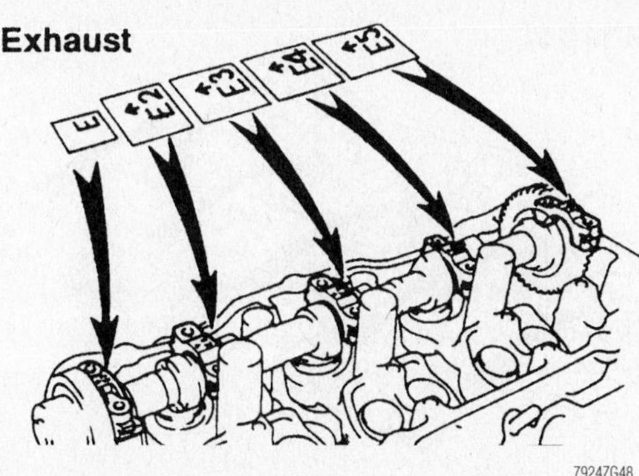

Right exhaust bearing caps must be placed in their proper locations

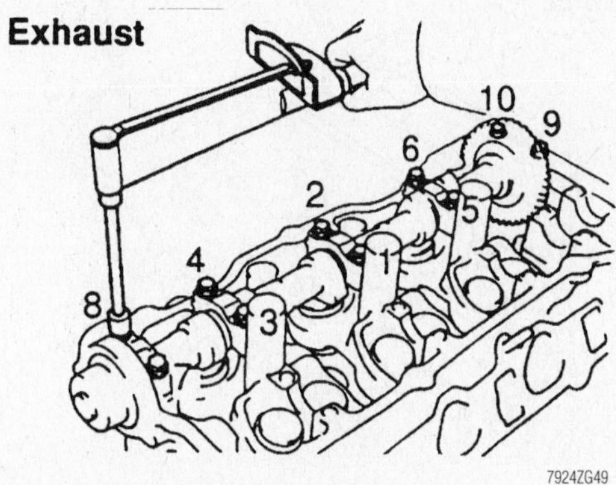

Right exhaust camshaft bearing cap bolt tightening sequence

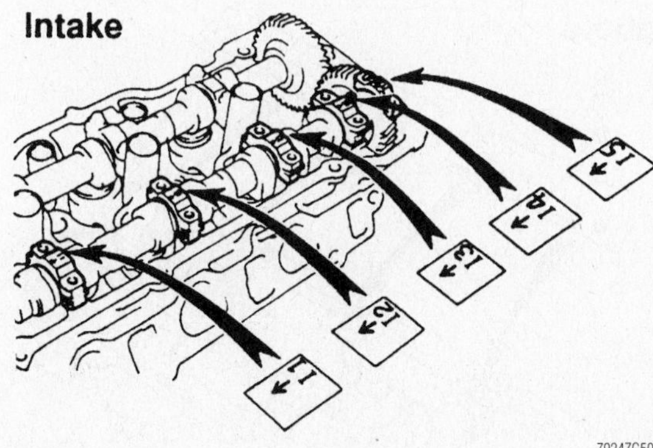

Right intake bearing caps must be placed in their proper locations

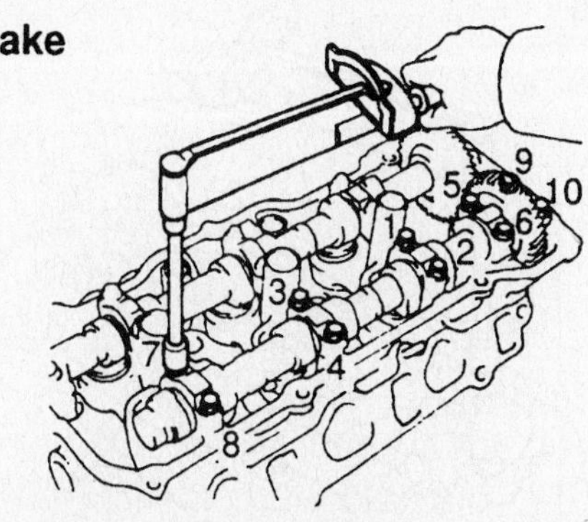

Right intake camshaft bearing cap bolt tightening sequence

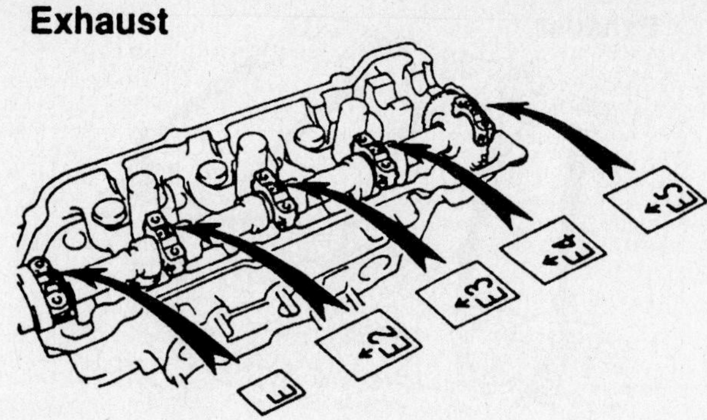

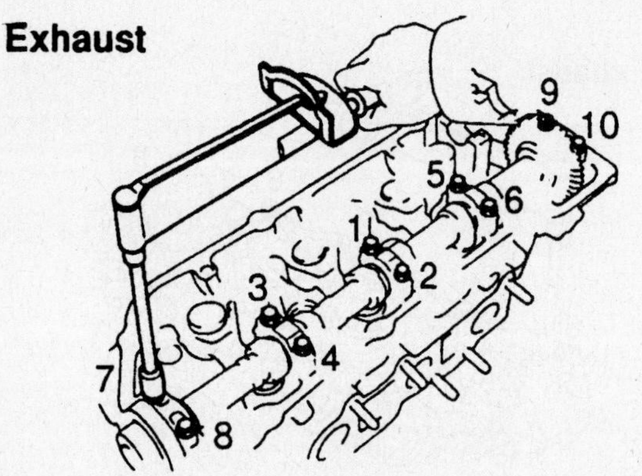

Left exhaust bearing caps locations and bolt tightening sequence

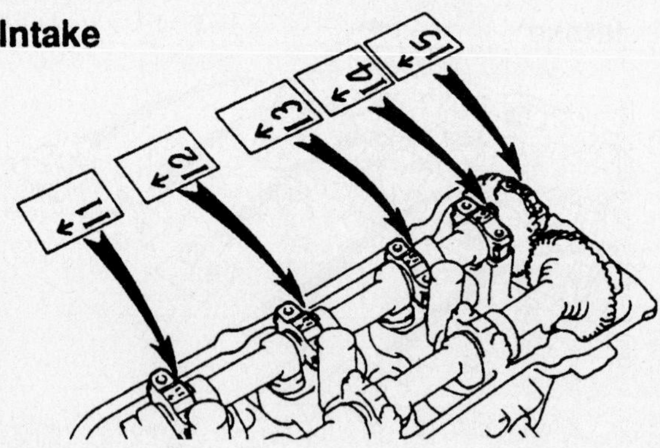

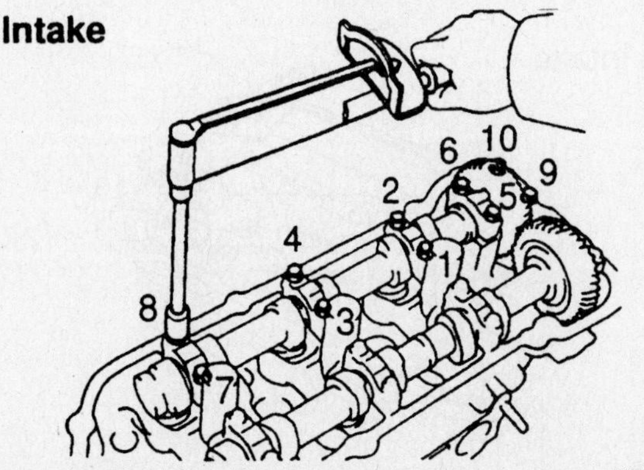

Left intake camshaft bearing cap locations and bolt tightening sequence

and in several passes to 12 ft. lbs. (16 Nm) in the proper sequence.

f. Remove the service bolt.

10. Install the left camshafts, as follows:

a. Position the intake camshaft on the head so that the alignment mark is at a 90 degrees angle from vertical. The mark should be at the "9 o'clock" position.

b. Apply sealant to the No. 1 bearing cap.

c. Apply a light coat of clean engine oil to the bolt threads and under the bolt head. Install the bearing caps to their proper position. Tighten the bolts evenly and in several passes to 12 ft. lbs. (16 Nm) in the proper sequence.

d. Position the exhaust camshaft on the head so that the alignment marks are at a 90 degrees angle from vertical. The mark should be at the "3 o'clock" position and must align with the marks on the other gear.

e. Apply a light coat of clean engine oil to the bolt threads and under the bolt head. Install the bearing caps to their proper position. Tighten the bolts evenly and in several passes to 12 ft. lbs. (16 Nm) in the proper sequence.

f. Remove the service bolt.

11. Install or connect the following:
- New camshaft oil seals, lubricated with multi-purpose grease
- No. 3 (rear) timing belt cover
- Camshaft timing gears
- Idler pulley, timing belt and covers

12. Check and adjust the valve clearance.

13. Install the cylinder head (valve) covers.

14. Start the engine. Check the ignition timing.

15. Test drive the vehicle.

16. Check all fluid levels.

Valve Lash

ADJUSTMENT

3.0L and 3.3L Engine

➡Adjust the valve clearance when the engine is cold.

1. Before servicing the vehicle, refer to the precautions in the beginning of this section.

2. Remove or disconnect the following:
- Negative battery cable. If equipped with an air bag, wait at least 90 seconds before proceeding.
- Accelerator/throttle cable from the throttle linkage

- Air cleaner cover, air flow meter and air duct assembly
- V-bank cover
- Emission control valve set
- Air intake chamber
- Engine harness from the injectors and the ignition coils
- Ignition coils and keep them in order for reassembly
- Spark plugs
- Cylinder head covers

3. Turn the crankshaft pulley and align its groove with the timing mark **0** of the No. 1 timing cover.
4. Check that the valve lifters on the No. 1 intake are loose and the No. 1 exhaust are tight. If not, turn the crankshaft 1 complete revolution (360 degrees).

➡All measurements should be written down. These recorded measurements will need to be used in conjunction with a mathematical formula to determine the thickness of the replacement shims.

5. Measure the clearance between the valve lifters and the camshaft. Record the measurements on valves No. 1 and 6 intake; No. 2 and 3 exhaust.
   a. The intake valve clearance cold is 0.006–0.010 in. (0.15–0.25mm).
   b. The exhaust valve clearance cold is 0.010–0.014 in. (0.25–0.35mm).
6. Turn the crankshaft ⅔ of a revolution (240 degrees). Record the measurements on valves No. 2 and 3 intake; No. 4 and 5 exhaust.
7. Turn the crankshaft another ⅔ of a revolution. Record the measurements on valves No. 4 and 5 intake; No. 1 and 6 exhaust.

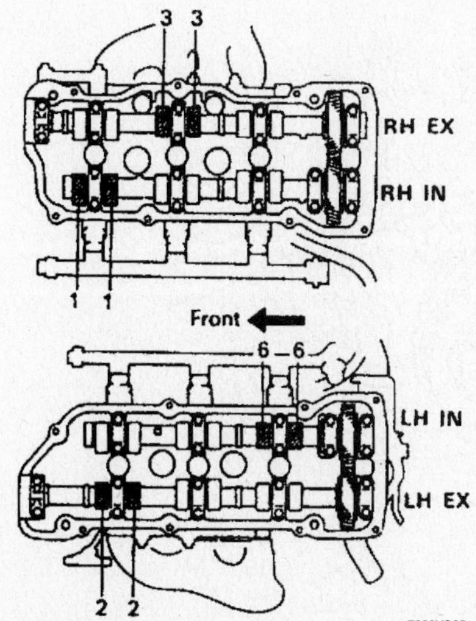

Adjust these valves during the 1st step

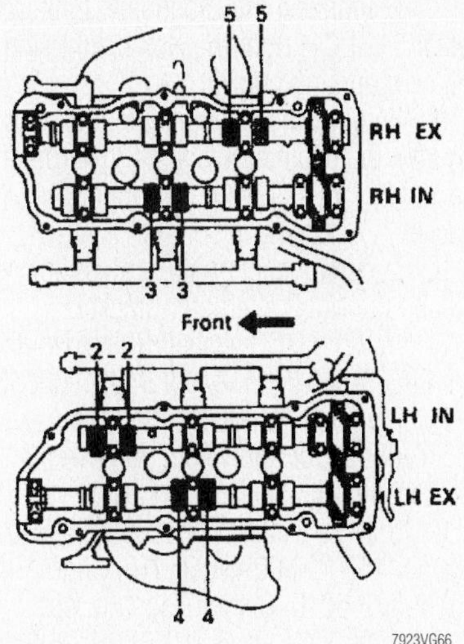

Adjust these valves during the 2nd step

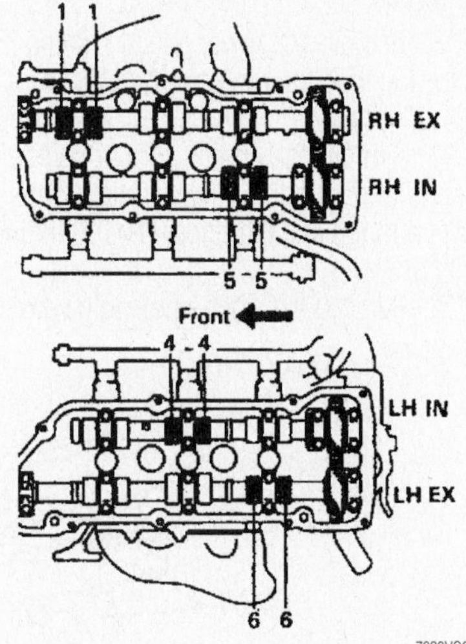

Adjust these valves during the 3rd step

8. Remove the adjusting shim by turning the crankshaft to position the cam lobe of the camshaft in the up position on the valve to be adjusted. Using a small thin flat bladed tool, turn the valve lifter so that the notches are perpendicular to the camshaft. Press down the valve lifter with tool 09248-55010 part A. Place too 09248-55010 part B between the camshaft and the valve lifter; remove part A.
9. Remove the adjusting shim with a magnet and a small screwdriver.
10. Determine the replacement adjusting shim size by either using the charts or the following formulas:
    - Intake: N = T + (A −0.008 in./0.020mm)
    - Exhaust: N = T + (A −0.012 in./0.30mm)
    - T = Thickness of removed shim
    - A = Measured valve clearance
    - N = Thickness of new shim
11. Select a new shim with a thickness as close as possible to the calculated value. Install the new replacement shim.

➡Shims are available in 17 sizes in increments of 0.0020 in. (0.050mm), from 0.0984 in. (2.500mm) to 0.1299 in. (3.300mm).

12. Recheck the valve clearance.
13. Install or connect the following:
    - Cylinder head covers
    - Spark plugs and the ignition coils
    - Engine wiring harness to the injectors and the coils
    - Intake chamber
    - Emission control valve set
    - V-bank cover
    - Air flow meter, air duct and air cleaner cover
    - Negative battery cable

Starter Motor

REMOVAL & INSTALLATION

3.0L Engine

1. Before servicing the vehicle, refer to the precautions in the beginning of this section.
2. Remove or disconnect the following:
    - Battery
    - Battery tray
3. Remove or disconnect the cruise control actuator, if equipped, as follows:
    - Actuator connector and clamp
    - 3 bolts and the actuator with the bracket
4. Remove or disconnect the following:
    - Automatic transaxle shift control cable
    - Engine wiring
    - Starter electrical connectors
    - Both bolts, shift control cable clamp and the starter

To install:

5. Install or connect the following:
    - Starter and the shift control cable clamp. Tighten the bolts to 27 ft. lbs. (37 Nm).
    - Starter electrical connectors
    - Engine wiring
    - Automatic transaxle shift control cable
6. Install or connect the following, if equipped with cruise control:
    - 3 bolts and the actuator with the bracket
    - Actuator connector and clamp
7. Install or connect the following:
    - Battery tray
    - Battery

3.3L Engine

1. Before servicing the vehicle, refer to the precautions in the beginning of this section.
2. Remove the air cleaner assembly and inlet tubes.
3. Remove the air cleaner bracket.
4. Remove the wiring from the starter.
5. Remove the 2 bolts and lower the starter.
6. Installation is the reverse of removal. Torque the starter bolts to 26 ft. lbs. (37 Nm).

Timing Belt

REMOVAL & INSTALLATION

3.0L and 3.3L Engine

1. Disconnect the negative battery cable.
2. Remove the outer front cowl top panel assembly by performing the following procedure:
   a. Remove the wiper arm/blade assemblies head caps, nuts and assemblies.
   b. Remove the head-to-cowl seal and the cowl panel hole cover.
   c. Disconnect the windshield washer clip and hose.
   d. Remove both (right and left) cowl top ventilator louvers.
   e. Disconnect the electrical connector from the windshield wiper motor.
   f. Remove the outer front cowl top panel assembly-to-cowl bolts and the panel.
3. Raise and safely support the vehicle.
4. Remove the right front wheel assembly and apron seal.
5. Remove the alternator by performing the following procedure:
   a. Loosen the pivot bolt, adjusting lockbolt and adjusting bolt, then, remove the drive belt.
   b. Disconnect the alternator's electrical connector.
   c. Remove the nut and the alternator wire.
   d. Disconnect the wiring harness from the clip.
   e. Remove the alternator-to-bracket pivot bolt, the washer, adjusting lockbolt and alternator.
6. Loosen the power steering pump's mount and adjusting bolt, then, remove the drive belt.
7. Disconnect the hose from the engine coolant reservoir.
8. Disconnect the Diagnostic Link Connector 1 (DLC1) from the No. 2 right side engine mounting bracket.
9. Remove the right side engine mounting stay, the engine moving control rod and the No. 2 right side engine mounting bracket.
10. Loosen the alternator's pivot bolt, the nut and the No. 2 alternator bracket.
11. Using the Crankshaft Pulley Holding tool 09213-54015, Bolt tool 91651-60855 and Companion Flange Holding tool 09330-00021, or equivalent, remove the crankshaft pulley bolt.
12. Using a Puller "C" Set 09950-50011 (Hanger 150 tool 09951-05010, Slide Arm tool 09952-5010, Center Bolt 100 tool 09953-05010, Center Bolt 150 tool 09953-05020 and 2 No. 2 Claw tools 09954-05020), pull the crankshaft pulley from the crankshaft.
13. Remove the lower (No. 1) timing belt cover. Remove the timing belt guide from the crankshaft.
14. Remove the engine wire protector clamps from the upper (No. 2) timing belt cover and remove the upper (No. 2) timing belt cover.
15. Remove the right side engine mounting brace

➡If reusing the timing belt, be sure that you can still read the installation marks. If not, place new installation marks on the timing belt to match the timing marks of the camshaft timing pulleys.

16. Temporarily install the crankshaft pulley bolt.
17. Set the No. 1 cylinder to Top Dead Center (TDC) of the compression stroke, as follows:
   a. Rotate the crankshaft (CLOCKWISE) to align the timing marks: dimple on the crankshaft timing sprocket with the notch on the oil pump body.
   b. Check that the timing marks on the camshaft sprockets and the rear timing belt cover are aligned; if not, rotate the crankshaft 360 degrees (1 revolution) and align the marks.
18. Remove the timing belt tensioner and the timing belt.

To install:

19. Inspect the timing belt tensioner by performing the following procedures:
   a. Inspect the seal for leakage; if leakage is suspected, replace the tensioner.
   b. Using both hands to hold the tensioner facing upward, strongly press the

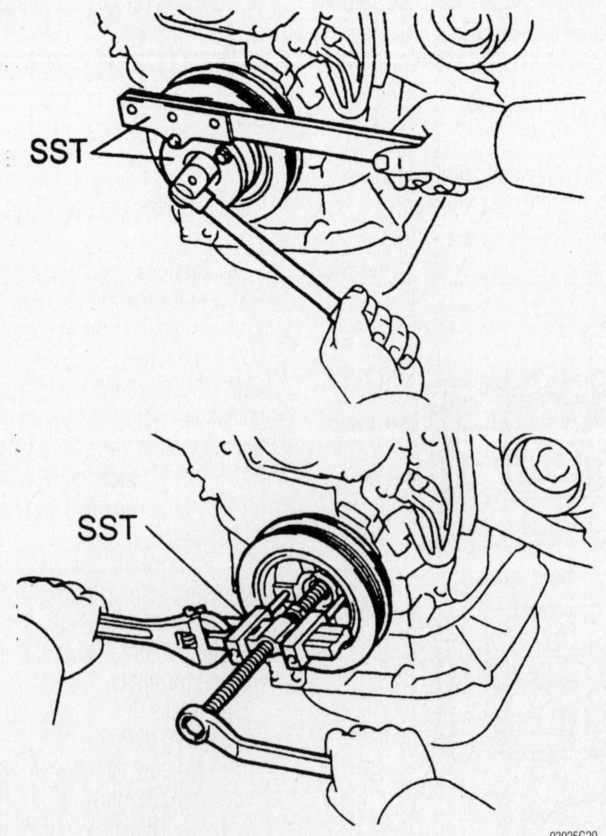

Removing/installing the crankshaft pulley

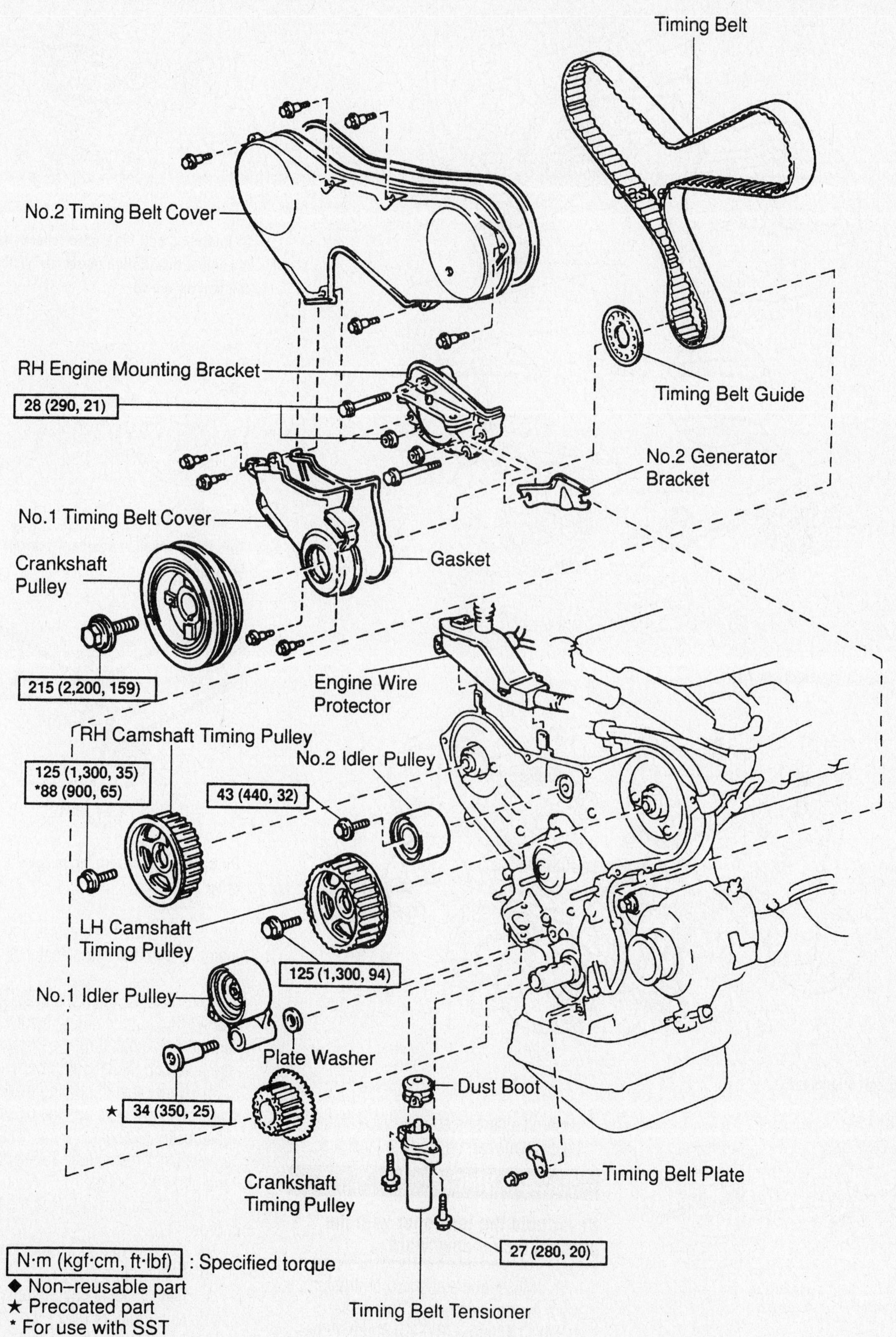

Timing Belt
No.2 Timing Belt Cover
Gasket
RH Engine Mounting Bracket
28 (290, 21)
Timing Belt Guide
No.2 Generator Bracket
No.1 Timing Belt Cover
Crankshaft Pulley
Gasket
215 (2,200, 159)
Engine Wire Protector
RH Camshaft Timing Pulley
125 (1,300, 35)
*88 (900, 65)
No.2 Idler Pulley
43 (440, 32)
LH Camshaft Timing Pulley
125 (1,300, 94)
No.1 Idler Pulley
Plate Washer
★ 34 (350, 25)
Dust Boot
Timing Belt Plate
Crankshaft Timing Pulley
27 (280, 20)
N·m (kgf·cm, ft·lbf) : Specified torque
◆ Non–reusable part
★ Precoated part
* For use with SST
Timing Belt Tensioner
93025G19

Exploded view of the timing belt assembly

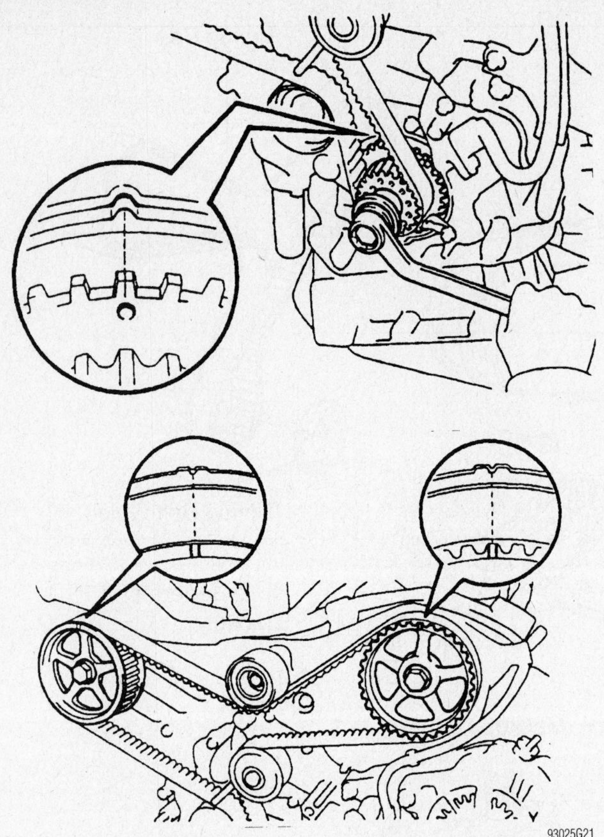

View of the timing mark locations

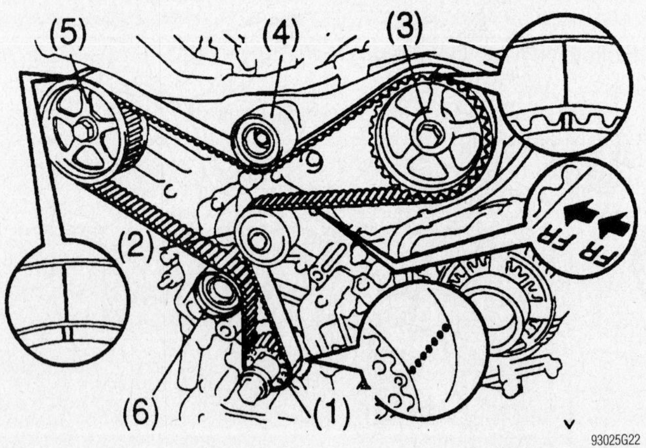

Timing belt alignment

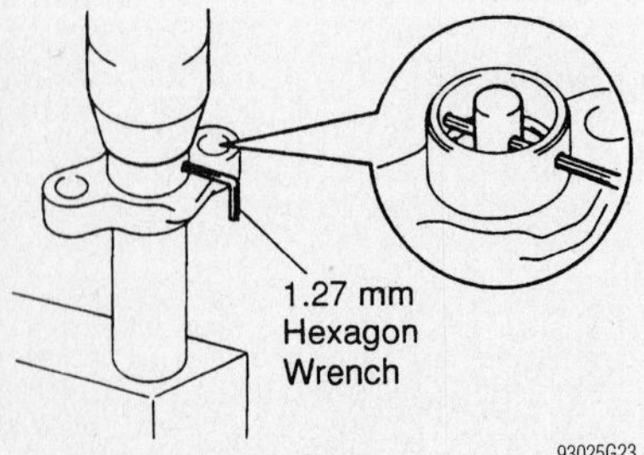

Timing belt tensioner installation preparation

pushrod against a solid surface. If the pushrod moves, replace the tensioner.

⁂ WARNING

Never hold the tensioner with the pushrod facing downward.

c. Measure the pushrod protrusion from the housing end, it should be 0.394–0.425 in. (10.0–10.8mm); if the protrusion is not as specified, replace the tensioner.

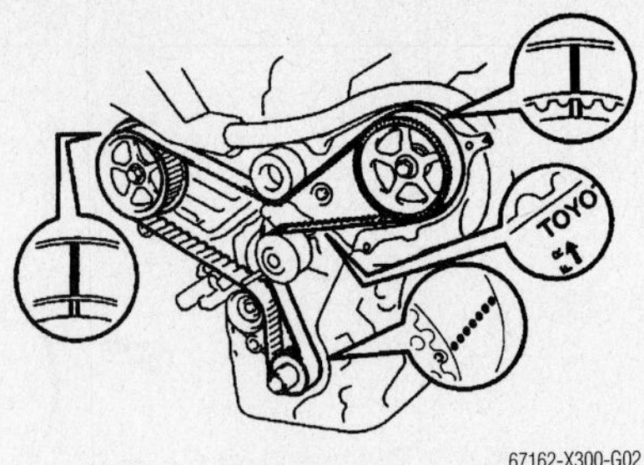

If the timing belt is re-used, check that the 3 original installation marks are visible on the belt as shown

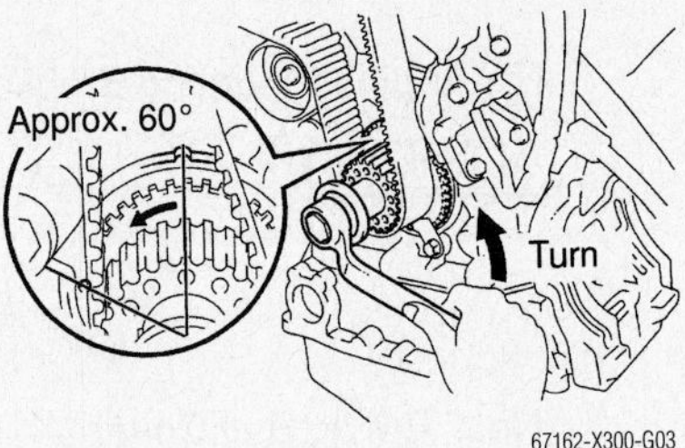

Turn the crankshaft counterclockwise by 60 degrees

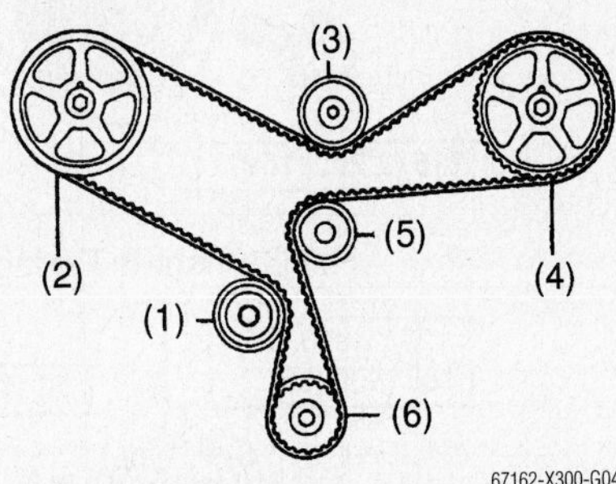

Remove the belt from the pulleys in this order

20. Set the No. 1 cylinder to Top Dead Center (TDC) of the compression stroke, as follows:
   a. Rotate the crankshaft (CLOCKWISE) to align the timing marks: dimple on the crankshaft timing sprocket with the notch on the oil pump body.
   b. Check that the timing marks on the camshaft sprockets and the rear timing belt cover are aligned; if not, rotate the crankshaft 1 revolution (360 degrees) and align the marks.
21. Install the timing belt in the following order:
   a. Crankshaft timing sprocket
   b. Water pump pulley
   c. Left camshaft timing sprocket
   d. No. 2 idler pulley
   e. Right camshaft sprocket
   f. No. 1 idler pulley
22. Using a vertical press, slowly press

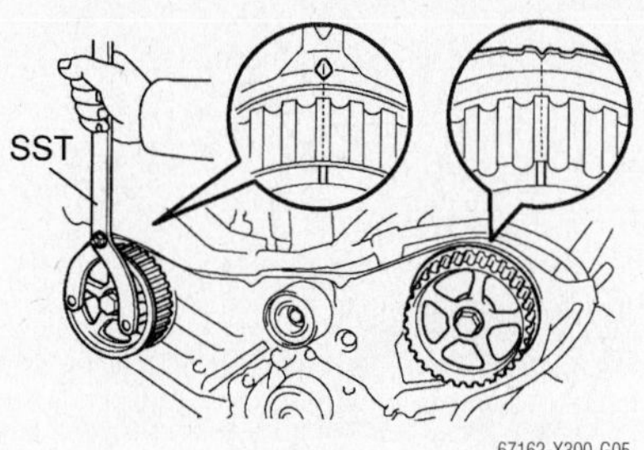

Turn the camshaft pulleys back into alignment so the marks align with the notches on the inner cover

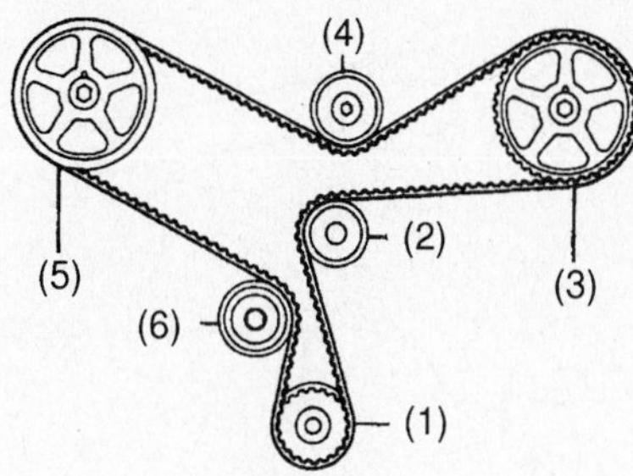

Install the belt in this order

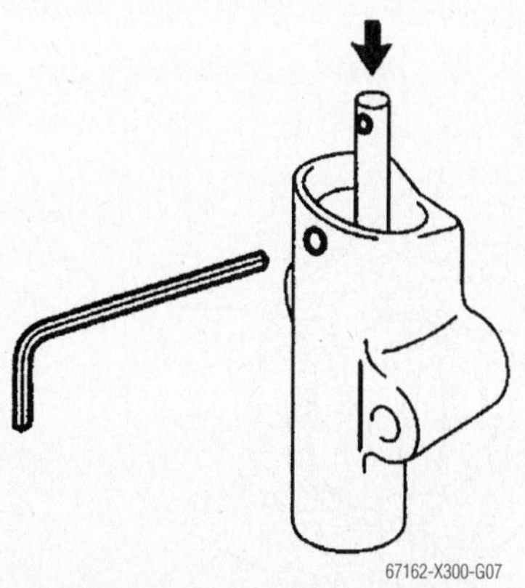

Set the tensioner in a press and collapse the plunger. Do not apply more that 2,205 lbs (9.8 kN) of force. Insert a suitable metal rod through the holes to hold the plunger in position

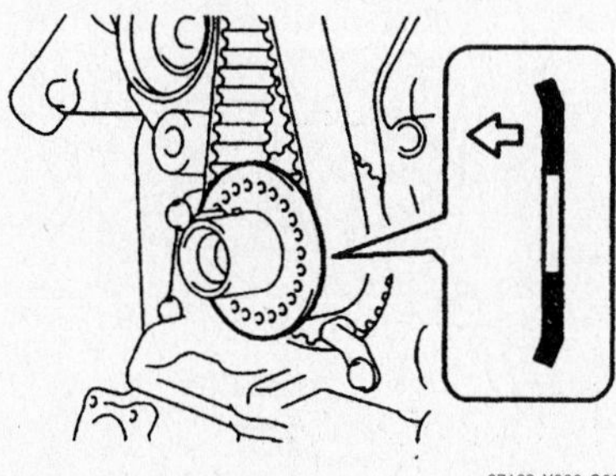

Install the timing belt guide with the cupped side facing front

the pushrod into the housing using 200–2205 lbs. (981–9807 N) until the holes align, then, install a 1.27mm Allen® wrench to secure the pushrod and release the press. Install the dust boot on the tensioner housing.

23. Install the timing belt tensioner and torque the bolts to 20 ft. lbs. (27 Nm).
24. Remove the Allen® wrench from the tensioner housing.
25. Slowly, rotate the crankshaft (CLOCKWISE) 2 complete revolutions and realign the timing marks. If the timing marks do not align, remove the timing belt and reinstall it.
26. Remove the crankshaft pulley bolt.
27. Install the right side engine mounting bracket and torque the bolts to 21 ft. lbs. (28 Nm).
28. Clean and install the upper (No. 2) timing belt cover.

➡If the gasket material on the timing belt covers is cracked, peeling or etc., replace it.

29. Install the timing belt guide on the crankshaft with the cup side facing outward.
30. Clean and install the lower (No. 1) timing belt cover.

➡If the gasket material on the timing belt covers is cracked, peeling or etc., replace it.

31. Install the crankshaft pulley.
32. Using the Crankshaft Pulley Holding tool 09213-54015, Bolt tool 91651-60855 and Companion Flange Holding tool 09330-00021, or equivalent, install the crankshaft pulley bolt and torque the bolt to 159 ft. lbs. (215 Nm).
33. To complete the installation, reverse the removal procedures.
34. Connect the negative battery cable.
35. Start the engine and check for leaks.

Oil Pan

REMOVAL & INSTALLATION

3.0L and 3.3L Engine

1. Before servicing the vehicle, refer to the precautions in the beginning of this section.
2. Remove or disconnect the following:
   - Right front wheel
   - Fender apron seal
   - Engine undercover
3. Drain the engine oil from the engine.
4. Remove or disconnect the following:
   - Front exhaust pipe
   - Front exhaust pipe bracket from the No. 1 oil pan
   - Flywheel housing undercover
   - 10 bolts and 2 nuts to the No. 2 oil pan
5. Insert the blade of the Oil Pan Seal Cutting tool 09032-00100 between the No. 1 and No. 2 oil pans. Clean the surfaces of the oil pans.
6. Remove or disconnect the following:
   - 3 oil strainer nuts and gasket
7. Remove the No. 1 oil pan, as follows:
   - 2 bolts and the flywheel housing undercover
   - 17 bolts and 2 nuts to the No. 1 oil pan

➡Make a note of the position of the each bolt. When replacing the bolts into the oil pan, place each bolt in the position from which it was removed.

   - Oil pan, by prying the portions between the cylinder block and the oil pan

➡Be careful not to damage the contact surfaces.

   - Baffle plate from the No. 1 oil pan

To install:

8. Clean all mating surfaces of the oil pans.
9. Install the baffle plate to the No. 1 oil pan and tighten to 69 inch lbs. (8 Nm).
10. Install the No. 1 oil pan, as follows:
    a. Using a non residue solvent, clean both sealing surfaces to the oil pan.
    b. Apply liquid sealant to the oil pan and engine block.
    c. Install the oil pan with the 17 bolts and 2 nuts. Uniformly tighten the bolts and nuts in several passes.
    d. Tighten the No. 1 oil pan bolts, as follows:
    - 10mm head bolt: 69 inch lbs. (8 Nm)
    - 12mm head bolt: 14 ft. lbs. (20 Nm)
    - 14mm head bolt: 27 ft. lbs. (37 Nm)

    e. Install the flywheel housing undercover with the 2 bolts. Tighten the bolts to 69 inch lbs. (8 Nm).
11. Install the oil strainer with the 3 nuts. Tighten the nuts to 69 inch lbs. (8 Nm).
12. Install the No. 2 oil pan, as follows:
    a. Using a non residue solvent, clean both sealing surfaces to the oil pan.
    b. Apply liquid sealant to the oil pan and engine block.
    c. Install the No. 2 oil pan with the 10 bolts and 2 nuts. Uniformly tighten the bolts and nuts in several passes. Tighten the bolts to 69 inch lbs. (8 Nm).
13. Install or connect the following:
    - Flywheel housing undercover

- Front exhaust pipe bracket to the No. 1 oil pan. Tighten the bolts to 15 ft. lbs. (21 Nm).

14. Install the front exhaust pipe, as follows:
    - Temporarily install the 3 new gaskets and the front exhaust pipe with the 2 bolts and 6 nuts
    - Tighten the 4 exhaust manifolds-to-front exhaust pipe nuts to 46 ft. lbs. (62 Nm).
    - Tighten the both front exhaust pipe-to-center exhaust pipe nuts/bolts to 41 ft. lbs. (56 Nm).
    - Bracket. Tighten both bolts to 14 ft. lbs. (19 Nm).
    - Support stay. Tighten both bolts to 22 ft. lbs. (29 Nm).
15. Install or connect the following:
    - Engine undercover
    - Right fender apron seal
    - Right front wheel
16. Fill the engine with oil.
17. Start the engine and check for leaks.

Oil Pump

REMOVAL & INSTALLATION

3.0L and 3.3L Engine

1. Before servicing the vehicle, refer to the precautions in the beginning of this section.
2. Remove or disconnect the following:
    - Oil pan
    - Crankshaft Position (CKP) sensor
    - 9 oil pump bolts

➡Make a note of the position of the each bolt. When replacing the bolts into the oil pump body, place each bolt in the position from which it was removed.

- Oil pump body, by prying between the oil pump and main bearing cap
- O-ring from the cylinder block
- Plug, gasket, spring and relief valve from the oil pump body
- 9 screws, pump body cover, drive and driven rotors

To install:

3. Install or connect the following:
    - Driven rotors, drive, pump body cover, using the 9 screws
    - Oil pump relief valve, spring, gasket and the plug to the oil pump body
    - New O-ring on the cylinder block
4. Using a non residue solvent, clean both sealing surfaces to the oil pump.
5. Apply liquid sealant to the oil pump and engine block.
6. Install or connect the following:
    - Oil pump

➡Be sure to engage the spline teeth of the oil pump drive gear with the large teeth of the crankshaft.

- 9 oil pump bolts. Tighten the bolts in several passes to 69 inch lbs. (8 Nm), for 10mm or to 14 ft. lbs. (20 Nm), for 12mm.
- CKP sensor. Tighten the bolt to 69 inch lbs. (8 Nm).
- Baffle plate to the No. oil pan. Tighten to 69 inch lbs. (8 Nm).
- No. 1 oil pan, oil strainer and No. 2 oil pan

7. Refill the engine with oil.
8. Start the engine and inspect for leaks.
9. Recheck the engine oil level.

Rear Main Seal

REMOVAL & INSTALLATION

3.0L and 3.3L Engine

If the rear oil seal retainer is not installed to the block, use a tapered ended screwdriver and hammer to remove the oil seal.

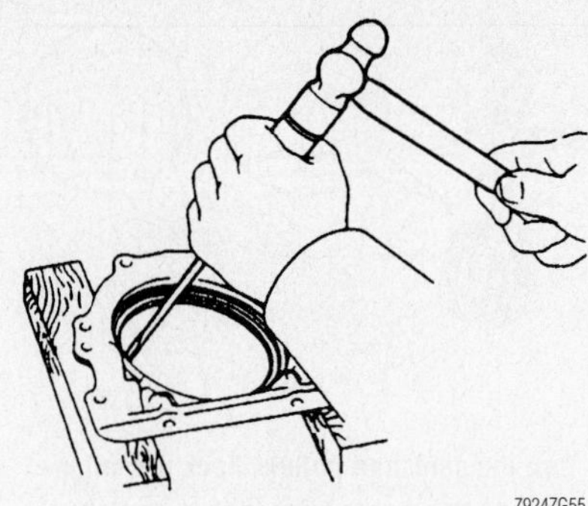

7924ZG55

Carefully tap the old seal from the retainer

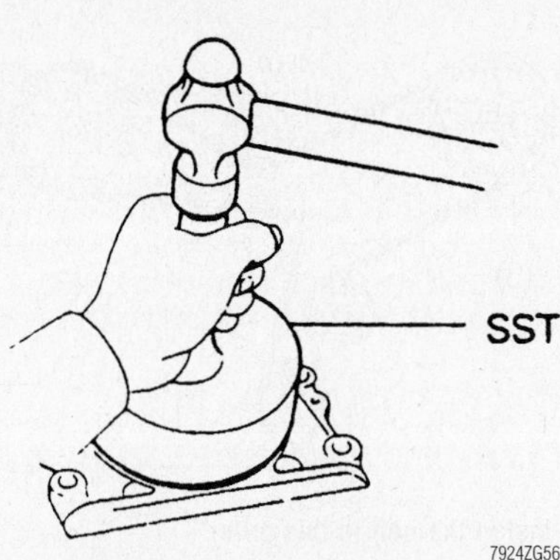

7924ZG56

Use the proper sized driver to seat the seal

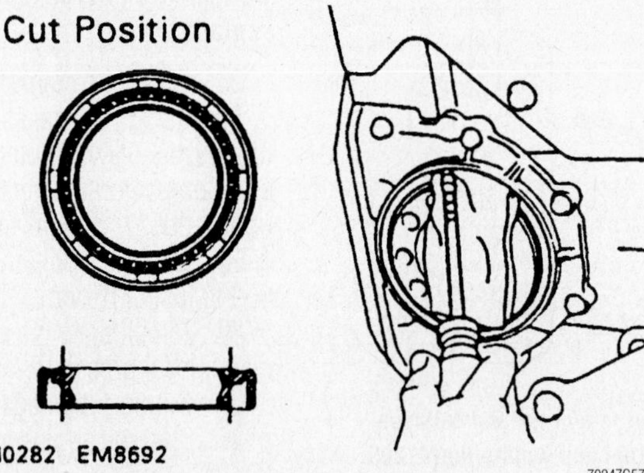

7924ZG57

Cut off the oil seal lip, then pry the seal out of the retaining plate

Apply multi-purpose grease to the new oil seal lip. Using a seal driver, tap the seal into place. Be careful not to install it slantwise.

1. Before servicing the vehicle, refer to the precautions in the beginning of this section.

If the rear oil seal retainer is installed on the cylinder block, using a knife, cut off the lip of the seal. Using a taped ended prytool, pry the old seal out of the retainer. Inspect the oil seal lip contacting surface of the crankshaft for cracks or damage. Apply multipurpose grease to the new oil seal, then tap the seal in place with a seal installer. Be careful not to install the seal slantwise.

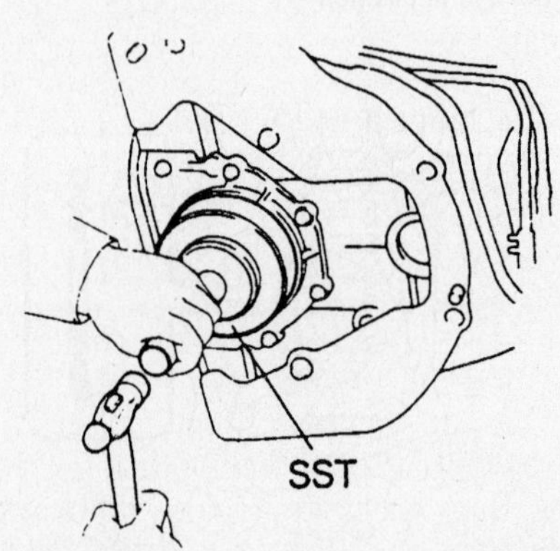

7924ZG58

Tap a new seal into place

Piston and Rings

POSITIONING

RH Piston
Front Mark (2 Cavities)
R
R
Front Mark (Mold Mark)
LH Piston
Front Mark (1 Cavity)
L
L
Front Mark (Mold Mark)
9302AG10

Piston/connecting rod-to-engine positioning

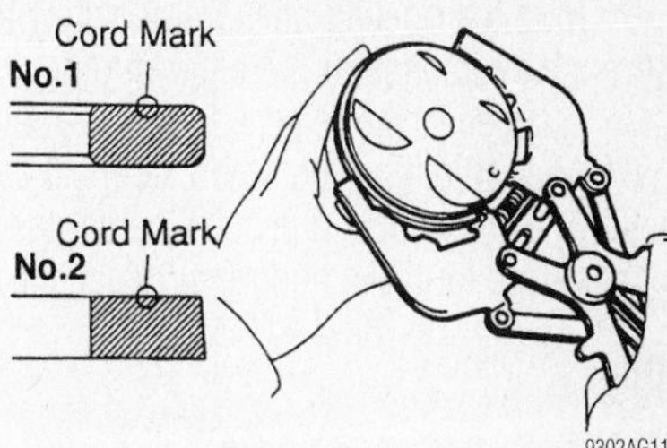

Piston ring positioning

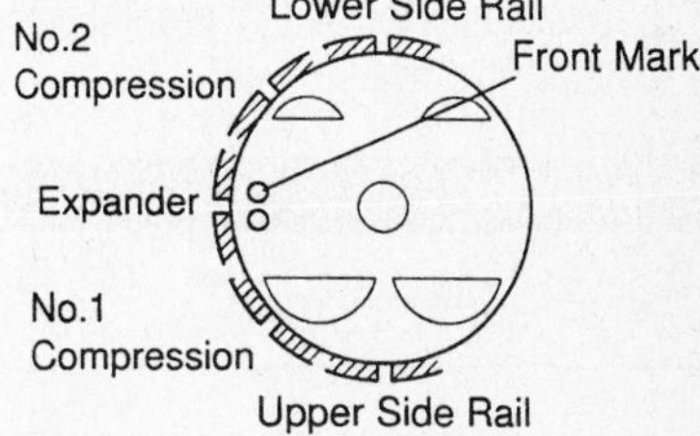

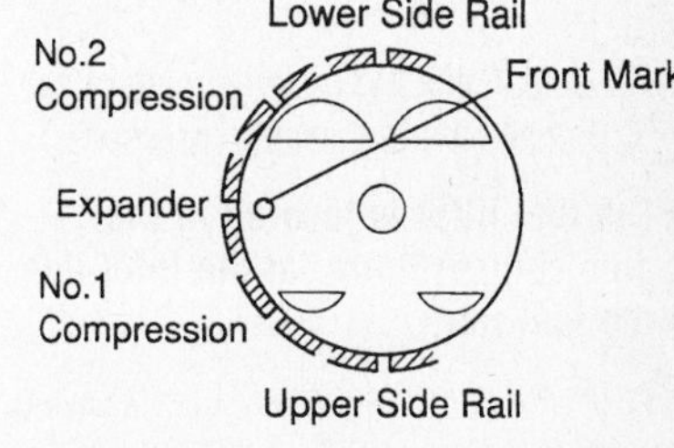

Piston ring identification

FUEL SYSTEM

Fuel System Service Precautions

Safety is the most important factor when performing not only fuel system maintenance but any type of maintenance. Failure to conduct maintenance and repairs in a safe manner may result in serious personal injury or death. Work on a vehicle's fuel system components can be accomplished safely and effectively by adhering to the following rules and guidelines.

- To avoid the possibility of fire and personal injury, always disconnect the negative battery cable unless the repair or test procedure requires that battery voltage be applied.
- Always relieve the fuel system pressure prior to disconnecting any fuel system component (injector, fuel rail, pressure regulator, etc.) fitting or fuel line connection. Exercise extreme caution whenever relieving fuel system pressure, to avoid exposing skin, face and eyes to fuel spray. Please be advised that fuel under pressure may penetrate the skin or any part of the body that it contacts.
- Always place a shop towel or cloth around the fitting or connection prior to loosening to absorb any excess fuel due to spillage. Ensure that all fuel spillage is quickly remove from engine surfaces. Ensure that all fuel-soaked cloths or towels are deposited into a flame-proof waste container with a lid.
- Always keep a dry chemical (Class B) fire extinguisher near the work area.
- Do not allow fuel spray or fuel vapors to come into contact with a light bulb, spark or open flame.
- Always use a second wrench when loosening or tightening fuel line connection fittings. This will prevent unnecessary stress and torsion to fuel piping. Always follow the proper torque specifications.
- Always replace worn fuel fitting O-rings with new ones. Do not substitute fuel hose where rigid pipe is installed.

Fuel System Pressure

RELIEVING

2001–03

1. Before servicing the vehicle, refer to the precautions in the beginning of this section.
2. Disconnect the negative battery terminal. Wait at least 90 seconds prior to working on models equipped with an airbag.
3. Place a catch-pan under the joint to be disconnected. A large quantity of fuel may be released when the joint is opened.

➡Wear eye or full-face protection.

4. Place a shop towel over the area and slowly loosen the joint using a wrench of the correct size. Use a back-up wrench if needed.
5. Allow the fuel left in the line to bleed off slowly before fully disconnecting the joint.

6. Plug the opened lines immediately to prevent fuel spillage or the entry of dirt.
7. Dispose of the released fuel properly.
8. After adjoining fuel lines, connect the negative battery cable and start the engine.
9. Check for leaks and repair as needed.

2004

1. Remove the fuel circuit opening relay from the engine compartment relay block.
2. Start the engine. After the engine stops, turn the ignition to **OFF**.
3. Check that the engine won't start.
4. Remove the fuel tank cap.
5. Disconnect the battery ground cable.
6. Install the relay.

Fuel Filter

REMOVAL & INSTALLATION

1. Before servicing the vehicle, refer to the precautions in the beginning of this section.
2. Disconnect the negative battery cable.
3. Relieve the fuel system pressure.

➡The fuel filter is located in the engine compartment, at the inlet line to the fuel rail.

4. Remove or disconnect the following:
   - Inlet and outlet lines from the filter
   - Fuel filter

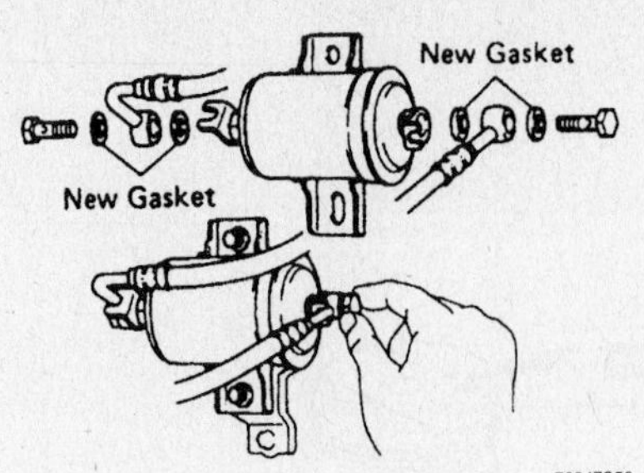

Exploded view of the fuel filter

To install:

5. Install or connect the following:
   - Fuel filter, using new O-rings. Tighten the lines to 22 ft. lbs. (29 Nm).
   - Negative battery cable
6. Start the engine and check for leaks.

Fuel Pump

REMOVAL & INSTALLATION

2001–03

1. Before servicing the vehicle, refer to the precautions in the beginning of this section.
2. Relieve the fuel pressure.
3. Disconnect the negative battery cable. Wait at least 90 seconds before proceeding on models with an airbag.
4. Drain the fuel tank.
5. Remove or disconnect the following:
   - Fuel tank
   - Access plate bolts
   - Fuel pump assembly
   - Fuel pump electrical connectors.
6. Pull the bracket from the lower side of the fuel pump.
7. Remove or disconnect the following:
   - Fuel pump from the fuel hose
   - Rubber cushion, the clip and the fuel filter from the bottom of the fuel pump.

To install:

8. Install or connect the following:
   - Fuel pump filter to the fuel pump using a new clip
   - Fuel pump to the bracket using new gaskets
   - Fuel hose to the outlet port of the fuel pump
   - Fuel pump bracket. Tighten the bolts to 26 inch lbs. (3 Nm) for 2001–02; 35 inch lbs. (4 Nm) for 2003.
   - Fuel tank
   - All electrical and fuel harness
   - Negative battery cable
9. Refill the fuel tank and check for leaks.

2004

1. Discharge the fuel system pressure.
2. Remove the charcoal canister cover.

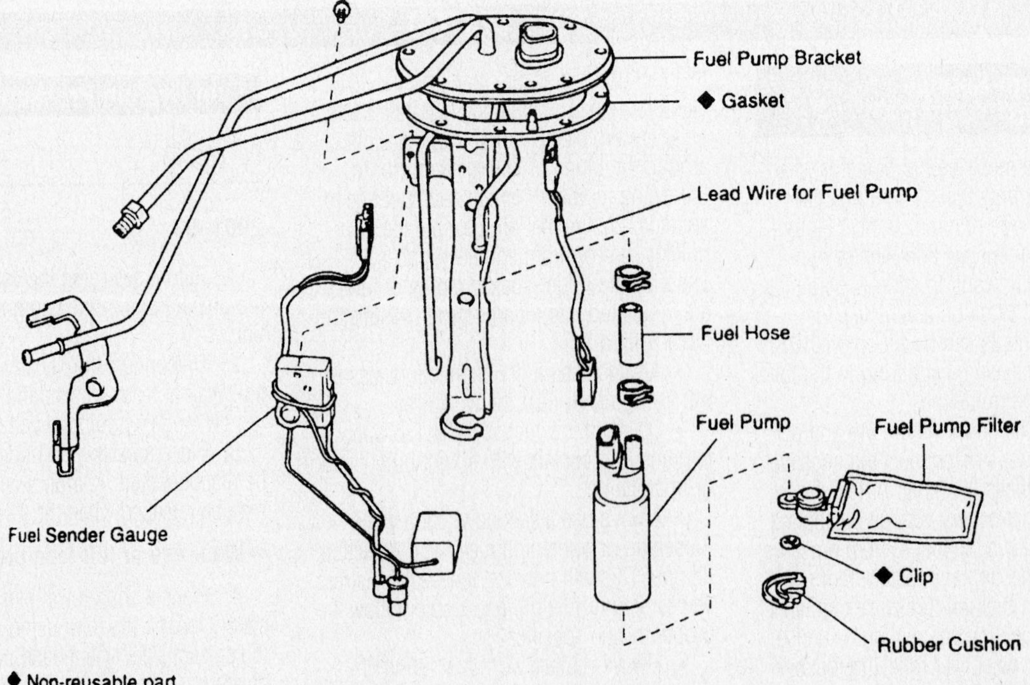

Exploded view of the fuel pump, bracket and related components—2001–02

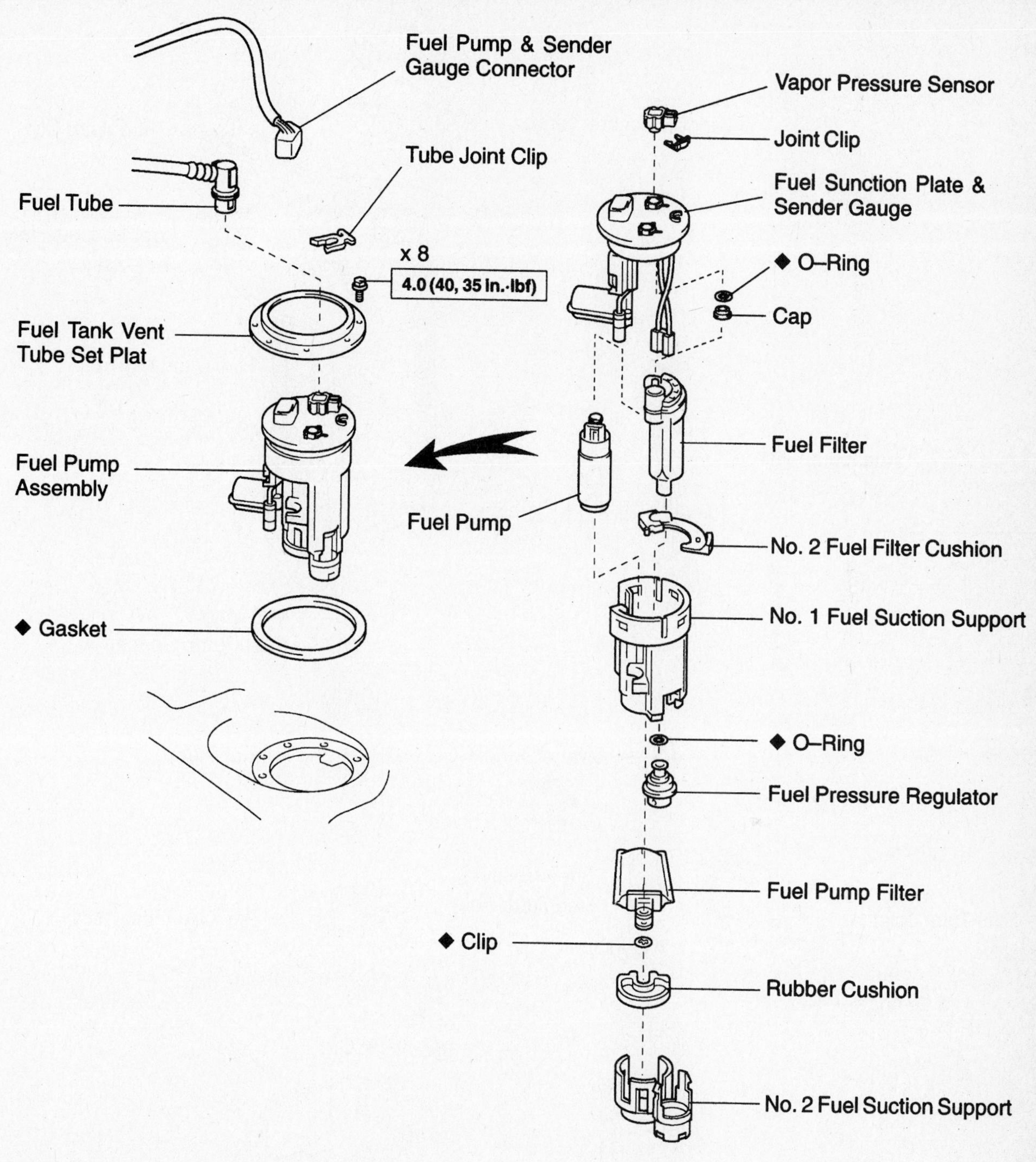

Exploded view of the fuel pump and related components—2003

3. Remove the fuel tank filler hose cover.
4. Disconnect the fuel tank vent hose.
5. Disconnect the fuel tank main tube.
6. Disconnect the filler hose.
7. Remove the wire harness clamps.
8. Place a jack under the tank, remove the bolts and the support bands.
9. Disconnect all remaining wiring and lower the tank.
10. Remove the tube joint clip and pull out the fuel main tube from the fuel pump module.
11. Remove the main tube from the tank.
12. Using tool 09808-14020, or equivalent lock ring tool, remove the lockring from the fuel pump module.
13. Remove the module from the tank.
14. Remove the joint clip and remove the pressure sensor.
15. Remove the sender assembly.
16. Wrap the tip of a small screwdriver

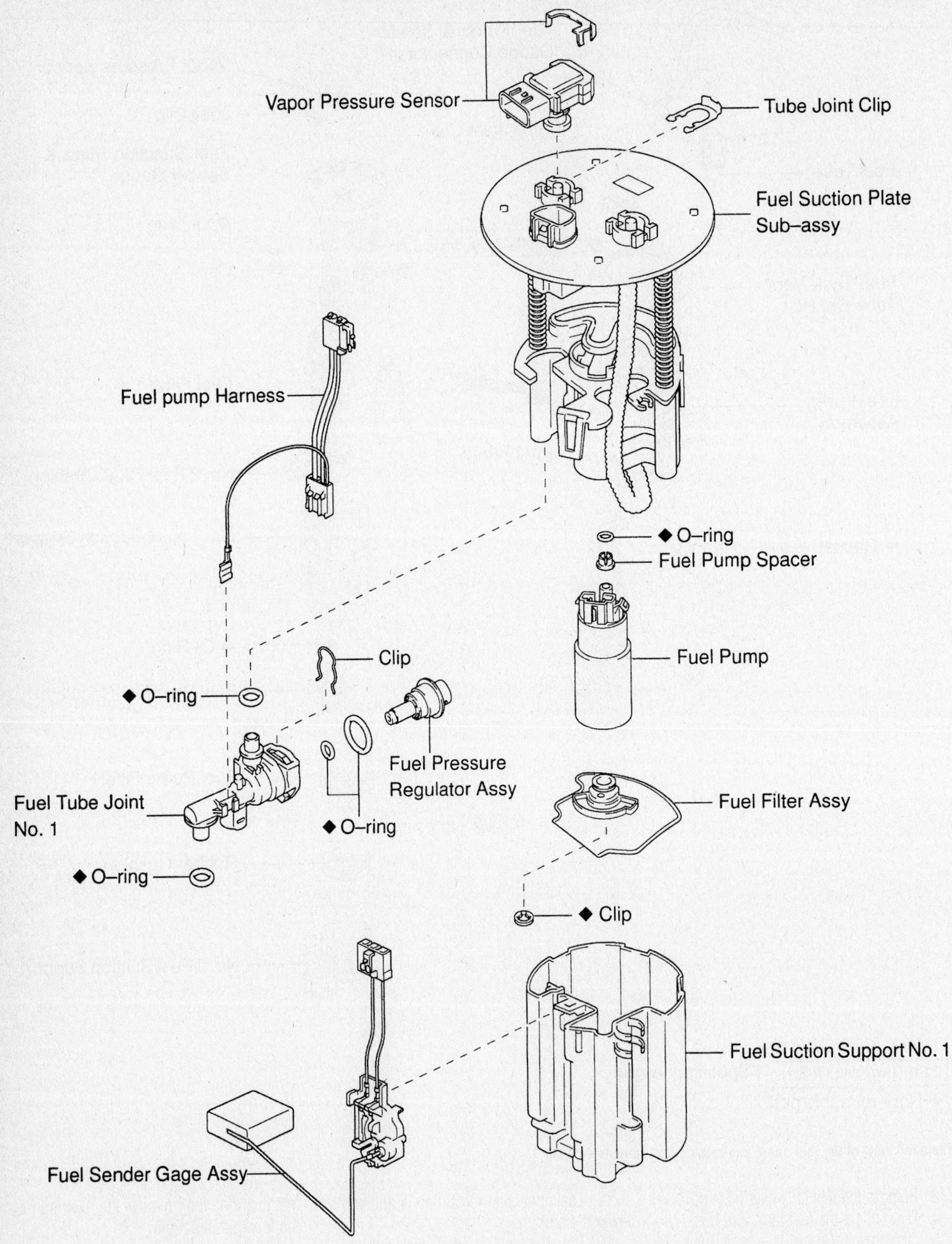

Fuel pump module exploded view—2004

with tape and disconnect the 4 snap retainers, and remove the fuel suction plate.

17. Disconnect the snap retainers, disconnect the connector, and remove the fuel pump.

18. Remove the O-ring and spacer from the pump.

19. Installation is the reverse of removal. Use a new O-ring coated with clean gasoline.

➡Prior to assembly, all new parts must be stored at room temperature for a minimum of 12 hours.

20. Make sure that the gasket groove is clean. Use a new gasket. Align the arrow on the fuel suction tube and the tank suction support.

21. Align the marks on the fuel pump module retainer and the fuel tank.

22. Position the retainer on the module and push down. Hold the module and turn the retainer by hand, one complete turn.

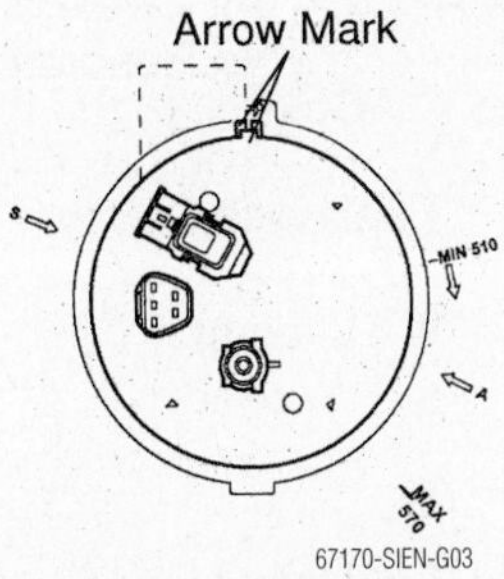

Module alignment marks—2004

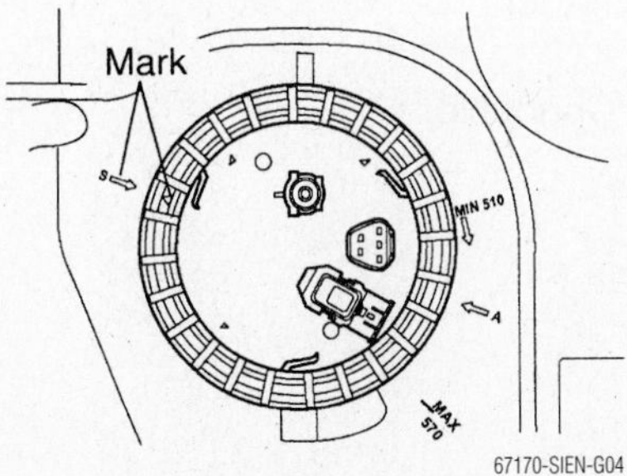

Retainer alignment marks—2004

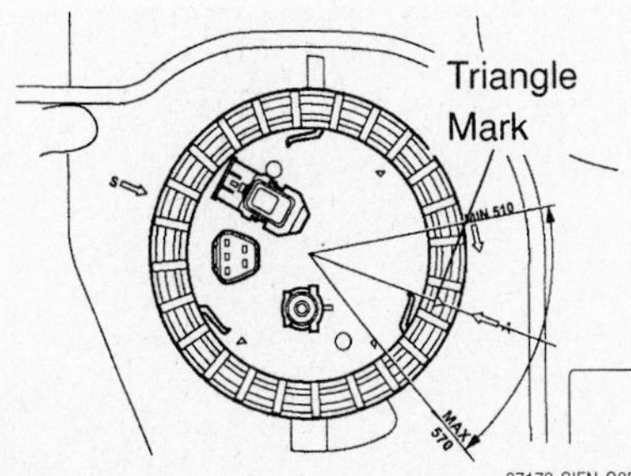

Tightening reference—2004

➡Make sure that the anti-rotation tab is in the groove during tightening. The "S" arrow on the fuel tank indicates "0" degrees position. Make sure that the retainer isn't cross-threaded.

23. Using the special tool, torque the retainer to 59–67 ft. lbs. (80–90 Nm). The triangle mark on the retainer should be about 1½ turn from the start.

Fuel Injector

REMOVAL & INSTALLATION

3.0L Engine

1. Before servicing the vehicle, refer to the precautions in the beginning of this section.
2. Remove or disconnect the following:
   - Outer front cowl top panel assembly
   - Air cleaner cap with hose
   - Negative battery cable. Work must be started approximately 90 seconds or longer after the negative battery cable has been disconnected, if equipped with an air bag.
   - Coolant
   - Accelerator and throttle cables
   - V-bank cover
   - Emission valve control set
   - No. 2 EGR pipe
   - Hydraulic motor pressure pipe from the water inlet and air inlet chamber
   - Air intake chamber assembly
   - Injector wiring
   - Air assist pipe from the bracket on the No. 1 fuel pipe
   - Air assist hoses from the intake manifold
   - Fuel return hose from the No. 1 fuel pipe
   - Fuel inlet hose for the fuel filter
   - 2 union bolts holding the No. 2 fuel pipe to the delivery pipes
   - Fuel return hose from the fuel pressure regulator
   - Union bolt for the right hand delivery pipe, 2 gaskets, 2 bolts, left hand delivery pipe together with the 3 injectors and the No. 2 fuel pipe
   - Union bolt for the delivery pipe and 2 gaskets from the No. 2 fuel pipe
   - The 3 bolts, right hand delivery pipe together with the 3 injectors and the No. 1 fuel pipe
   - The 4 spacers from the intake manifold
   - The 6 injectors from the delivery pipes
   - The two O-rings and two grommets from each injector

To install:

3. Install or connect the following:
   - 2 new grommets to each injector
   - New O-rings, with a light coat of fuel, to each injector
   - Injectors
   - The 4 spacers on the intake manifold
   - Right hand delivery pipe and the No. 1 fuel pipe together with the 3 injectors in position on the intake manifold
   - Bolt holding the right side delivery pipe, temporarily, to the intake manifold
   - Left hand delivery pipe and the No. 2 fuel pipe together with the 3 injectors in position on the intake manifold
   - Fuel return hose to the fuel pressure regulator
4. Temporarily install the 2 bolts holding the left hand delivery pipe to the intake manifold.
5. Temporarily install the No. 2 fuel pipe to the left side delivery pipe with the union bolt and 2 new gaskets.
6. Check that the injectors rotate smoothly. If they do not, Replace the O-rings.
7. Position the injector connector outward. Tighten the 4 bolts holding the delivery pipes to the intake manifold and tighten to 7 ft. lbs. (10 Nm). Tighten the bolt holding the No. 1 fuel pipe to the intake manifold to 14 ft. lbs. (20 Nm). Tighten the 2 union bolts holding the no. 2 fuel pipe to the delivery pipes to 24 ft. lbs. (32 Nm).
8. Install or connect the following:
   - Fuel inlet and return hoses. Union bolt: 22 ft. lbs. (30 Nm)
   - Fuel return hose to the No. 1 fuel pipe. Pass the fuel return hose under the heater hoses.
   - Air assist hoses to the intake manifold
   - Air assist pipe to the bracket on the No. 1 fuel pipe
   - Fuel injector wiring connectors
   - Air intake chamber assembly
   - Hydraulic motor pressure pipe to the intake chamber. Bolts: 69 inch lbs. (8 Nm)
   - No. 2 EGR pipe with new gaskets, tighten to 9 ft. lbs. (12 Nm)
   - Emission control valve set
   - V-bank cover
   - Air cleaner hose
   - Throttle and accelerator cables
   - Coolant
   - Air cleaner cap with hose
   - Outer front cowl top panel assembly
   - Negative battery cable

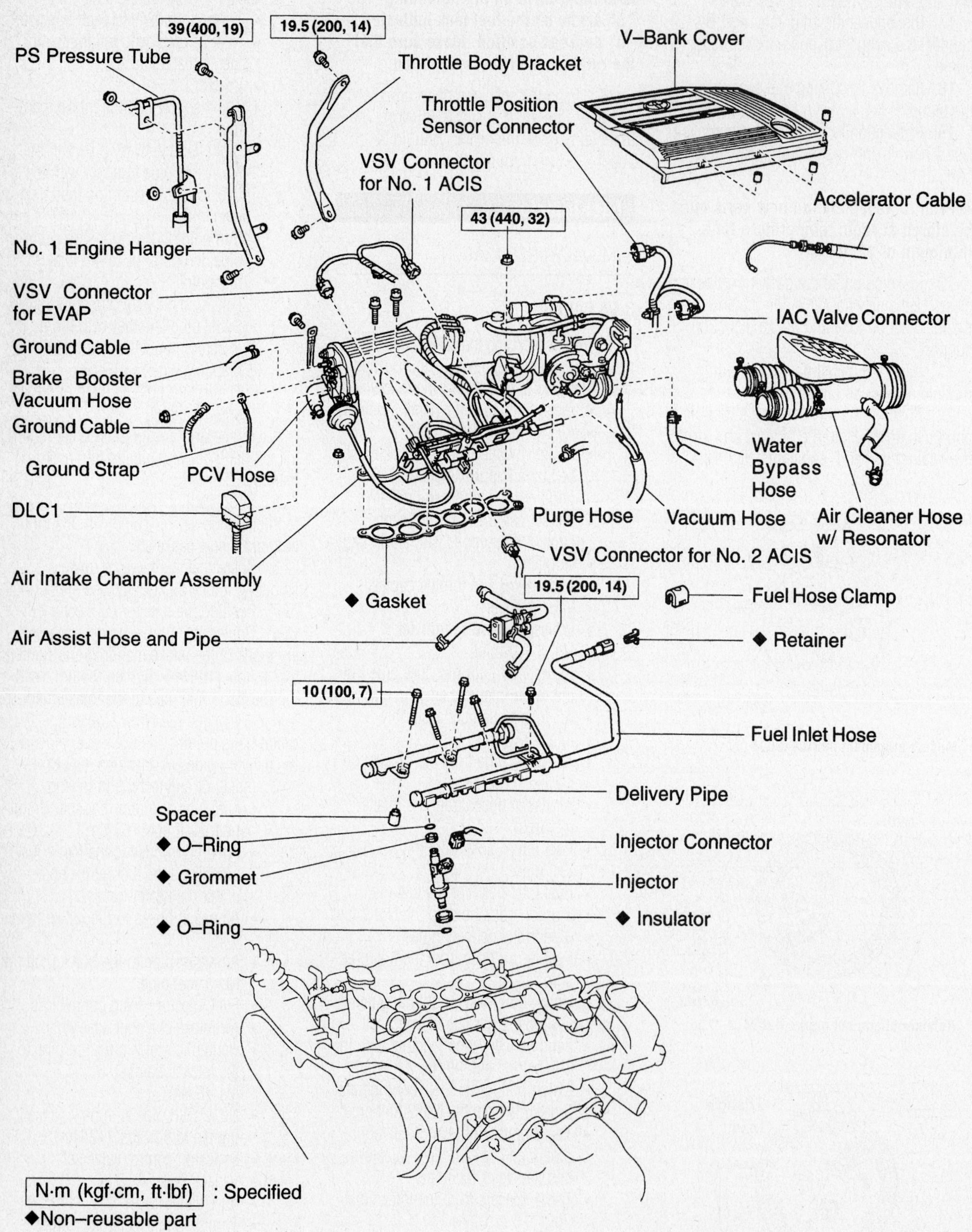

39 (400, 19)
19.5 (200, 14)
PS Pressure Tube
Throttle Body Bracket
V–Bank Cover
Throttle Position Sensor Connector
VSV Connector for No. 1 ACIS
Accelerator Cable
43 (440, 32)
No. 1 Engine Hanger
VSV Connector for EVAP
IAC Valve Connector
Ground Cable
Brake Booster Vacuum Hose
Ground Cable
Water Bypass Hose
Ground Strap
PCV Hose
DLC1
Purge Hose
Vacuum Hose
Air Cleaner Hose w/ Resonator
VSV Connector for No. 2 ACIS
Air Intake Chamber Assembly
◆ Gasket
19.5 (200, 14)
Fuel Hose Clamp
Air Assist Hose and Pipe
◆ Retainer
10 (100, 7)
Fuel Inlet Hose
Delivery Pipe
Spacer
◆ O–Ring
Injector Connector
◆ Grommet
Injector
◆ O–Ring
◆ Insulator
N·m (kgf·cm, ft·lbf) : Specified
◆Non–reusable part
67170-SIEN-G06

Fuel injectors and related parts—3.0L engine

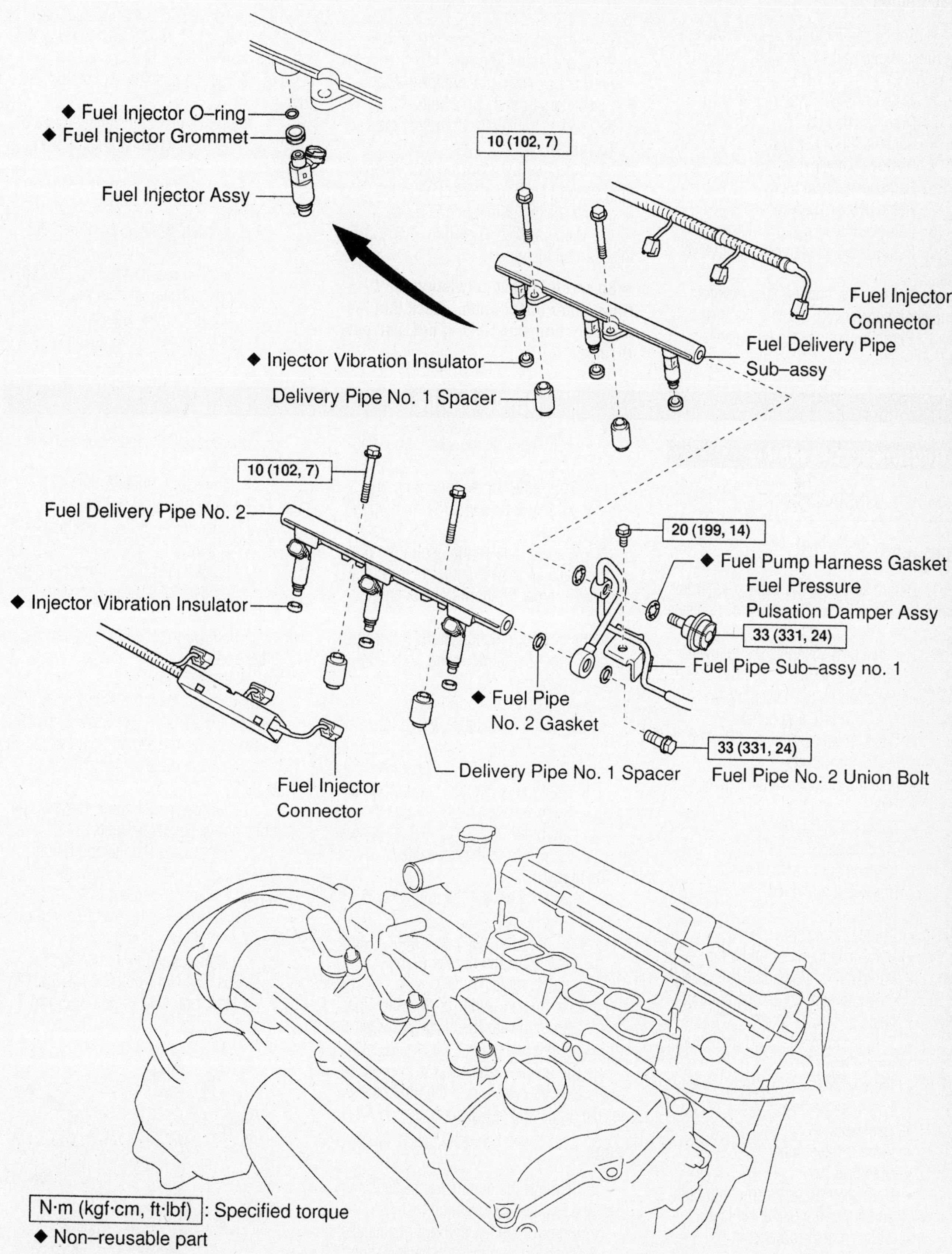

Fuel injectors and related parts—3.3L engine

3.3L Engine

1. Before servicing the vehicle, refer to the precautions in the beginning of this section.
2. Relieve the fuel system pressure.
3. Drain the coolant.
4. Remove the wiper arms.
5. Remove the wiper motor.
6. Remove the cowl tops.
7. Remove the V-bank cover.
8. Remove the air cleaner assembly.
9. Remove the emission control valve set.
10. Remove the upper intake manifold (intake air surge tank). Discard the gasket.
11. Remove the fuel pipe sub-assembly.
12. Disconnect the wiring at the injectors.
13. Remove the 4 bolts and 2 delivery pipe along with the injectors.
14. Remove the delivery pipe spacers and insulators from the manifold.
15. Pull each injector from the pipe.

To install:

16. Install new O-rings on each injector. Apply a light coating of gasoline to the O-rings and mating points on the pipes.
17. Using a twisting motion, install the injectors on the pipes.

➡Be careful to avoid twisting the O-rings. After installation, check that the injectors turn smoothly. If not, use new O-rings.

18. Install the pipes and injectors.
19. Loosely install the bolts and make sure that the injectors still turn freely. If not, replace the O-rings.
20. Torque the bolts to 84 inch lbs. (10 Nm).
21. The remainder of installation is the reverse of removal. Observe the following torques:
    - Fuel line union bolt: 24 ft. lbs. (33 Nm)
    - Pulsation damper: 24 ft. lbs. (33 Nm)
    - Fuel feed pipe: 14 ft. lbs. (20 Nm)
    - Upper intake manifold (air surge tank): 21 ft. lbs. (28 Nm)
    - Upper intake manifold stays: 14 ft. lbs. (20 Nm)

DRIVE TRAIN

Transmission Assembly

REMOVAL & INSTALLATION

2001–03

1. Before servicing the vehicle, refer to the precautions in the beginning of this section.
2. Remove or disconnect the following:
    - Hood
    - Wiper and blade assembly
    - Top cowl seal and panel
    - Window washer hoses, from the ventilator louvers
    - Left and right ventilator louvers
    - Heater air duct
    - Battery and tray
    - Throttle cable
    - Cruise control actuator with its bracket, if equipped
    - Starter
    - Shift control cable
    - Body-to-engine ground strap
    - Park/Neutral Position (PNP) switch, solenoid and ATF temperature connectors
    - 5 upper transaxle-to-engine mounting bolts
    - Front wheel
    - Engine undercover
    - Halfshafts
    - Front exhaust pipe
    - Stabilizer bar
    - Both steering gear mounting bolts and support it in the vehicle
    - Shift control cable from its bracket
    - Power steering pipe and the oil cooler clamps from the frame
    - Both left-side transaxle mounting nuts
    - Rear-side engine mounting nuts
    - Engine shock absorber mounting bolts
    - 3 front-side engine mounting bolts
3. Attach an engine sling to the engine hangers in order to support the engine weight.
4. Remove or disconnect the following:
    - Front frame mounting bolts and the frame
    - Transaxle oil cooler lines
5. Support the transaxle with a transmission jack.
    - Torque converter access cover
    - 6 torque converter mounting bolts
    - 3 lower transaxle-to-engine mounting bolts
    - Engine from the transaxle

To install:

6. Install or connect the following:
    - Transaxle
    - 3 lower transaxle-to-engine mounting bolts and tighten to the illustrated value.
    - Torque converter-to-flexplate bolts, starting with the black bolt, then the other 5.
7. The rest of installation is the reverse of the removal referring to the illustrations for the tightening specifications.

2004

1. Remove the engine/transaxle assembly. See Engine Removal and Installation".
2. Remove the left and right halfshafts.
3. Disconnect all wiring between the engine and transaxle.
4. Remove the starter.
5. Disconnect all cables between the engine and transaxle.
6. Remove the oil filler tube.
7. Remove the oil cooler tubes.
8. Remove the engine mount bracket.
9. With 2wd:
    a. Remove the torque converter cover.
    b. Remove the 6 torque converter bolts.
    c. Remove the 8 engine-to-transaxle bolts and separate the units.
10. With 4wd:
    a. Remove the 5 bolts and 1 nut and remove the transfer case stiffener plate.
    b. Remove the torque converter cover.
    c. Remove the 6 torque converter bolts.
    d. Remove the 8 engine-to-transaxle bolts and separate the units.
    e. Remove the 2 bolts and 6 nuts. Using a plastic hammer, drive the transfer case from the transaxle.
11. Installation is the reverse of removal. Observe the following torques:
    - See the illustration: bolt A 47 ft. lbs. (64 Nm); bolt B 34 ft. lbs. (46 Nm); bolt C 27 ft. lbs. (37 Nm)

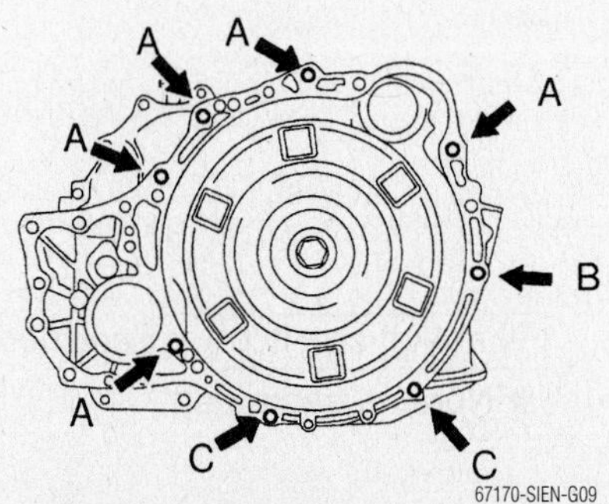

Transmission bolt identification—2004

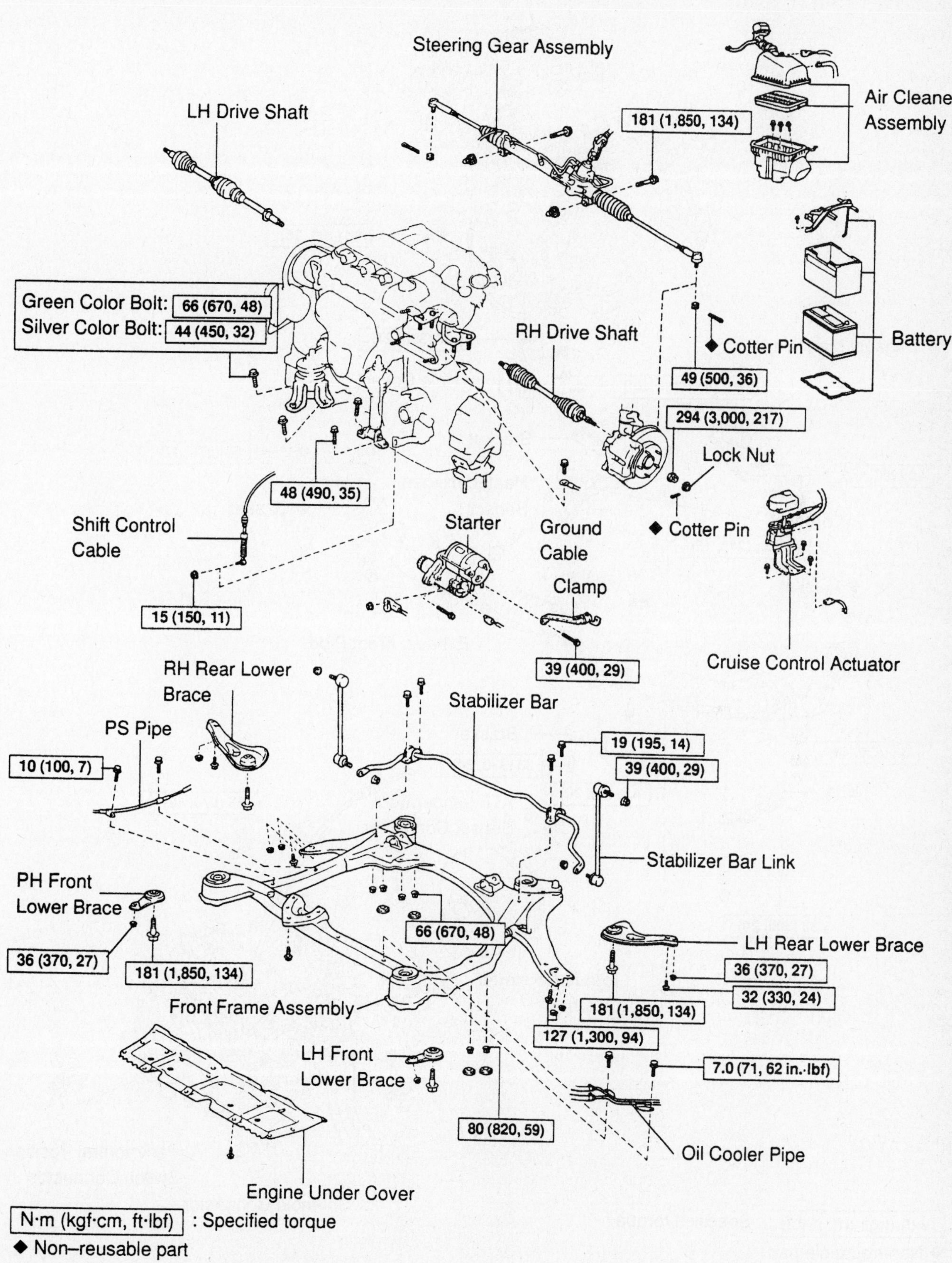

Exploded view of the transaxle removal and installation components—2001–03

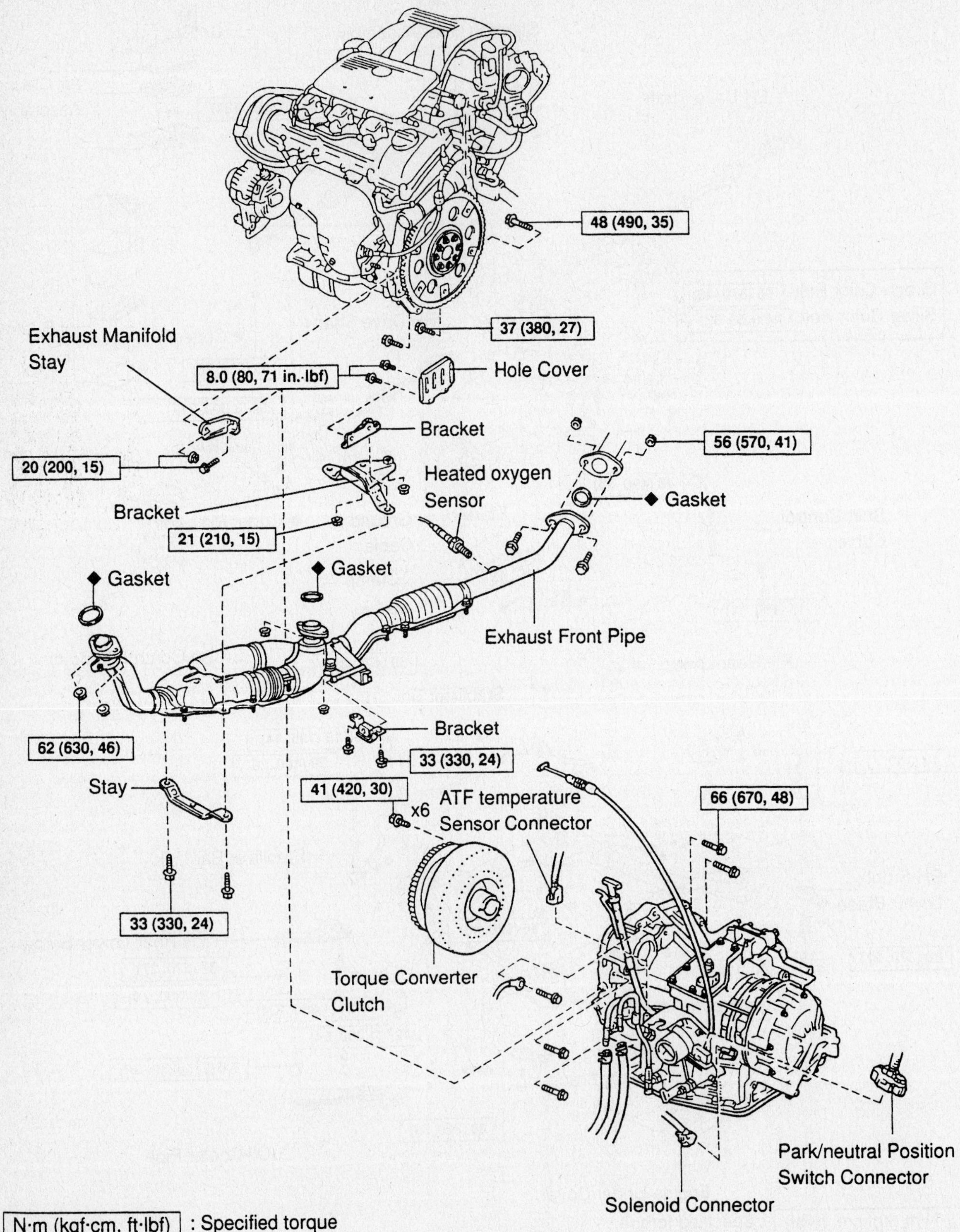

Exploded view of the transaxle removal and installation components—2001–03

- Transfer case bolts and nuts: 51 ft. lbs. (69 Nm). Use a new gasket.
- Torque converter bolts (green bolt first): 35 ft. lbs. (48 Nm)
- Stiffener plate: 25 ft. lbs. (34 Nm)
- Engine mount bracket: 47 ft. lbs. (64 Nm)

Transfer Case

REMOVAL & INSTALLATION

The transfer case is part of the transmission/transaxle assembly and is serviced with those units.

Front Halfshaft

REMOVAL & INSTALLATION

2001–03

1. Before servicing the vehicle, refer to the precautions in the beginning of this section.

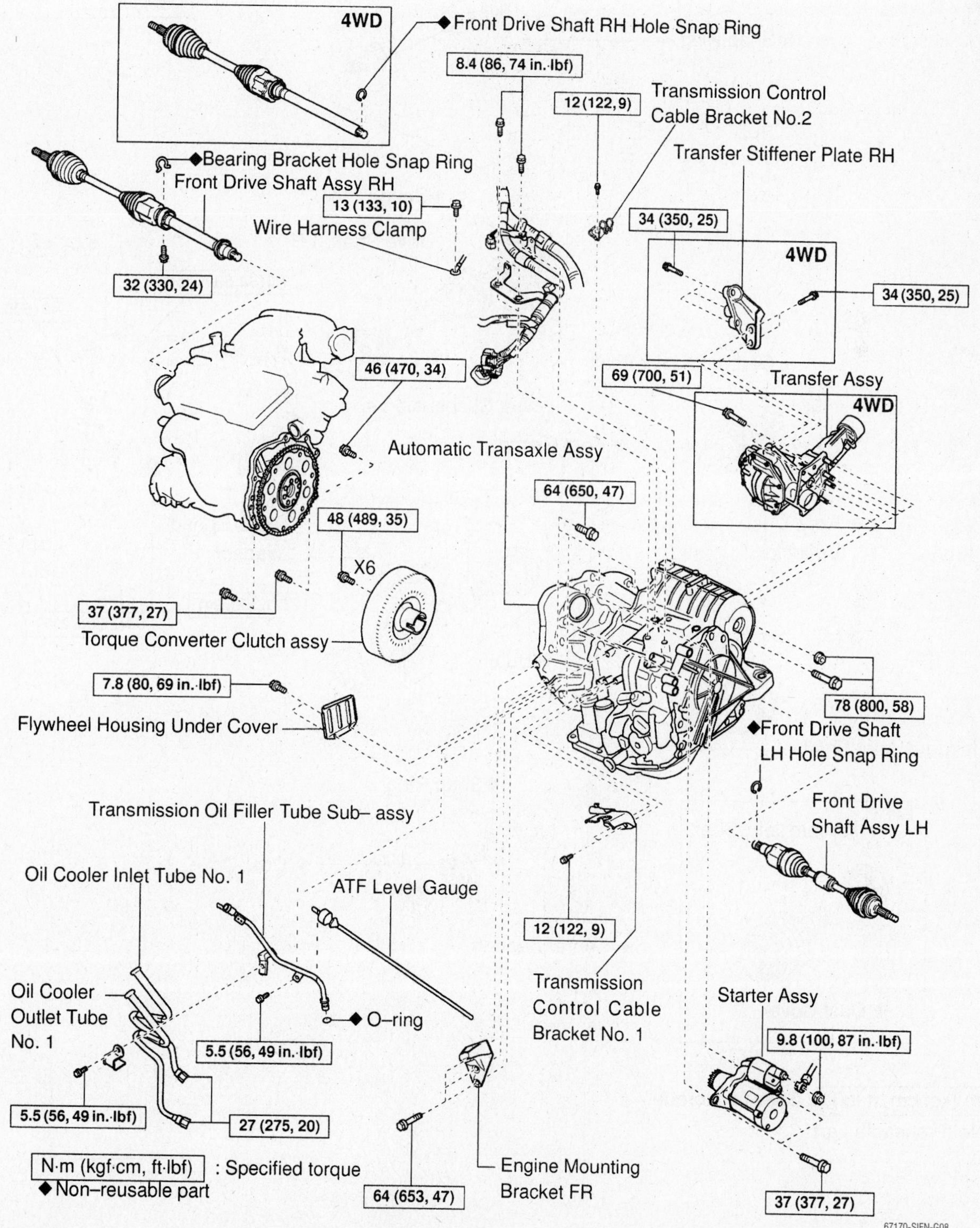

Exploded view of the transaxle removal and installation components—2004

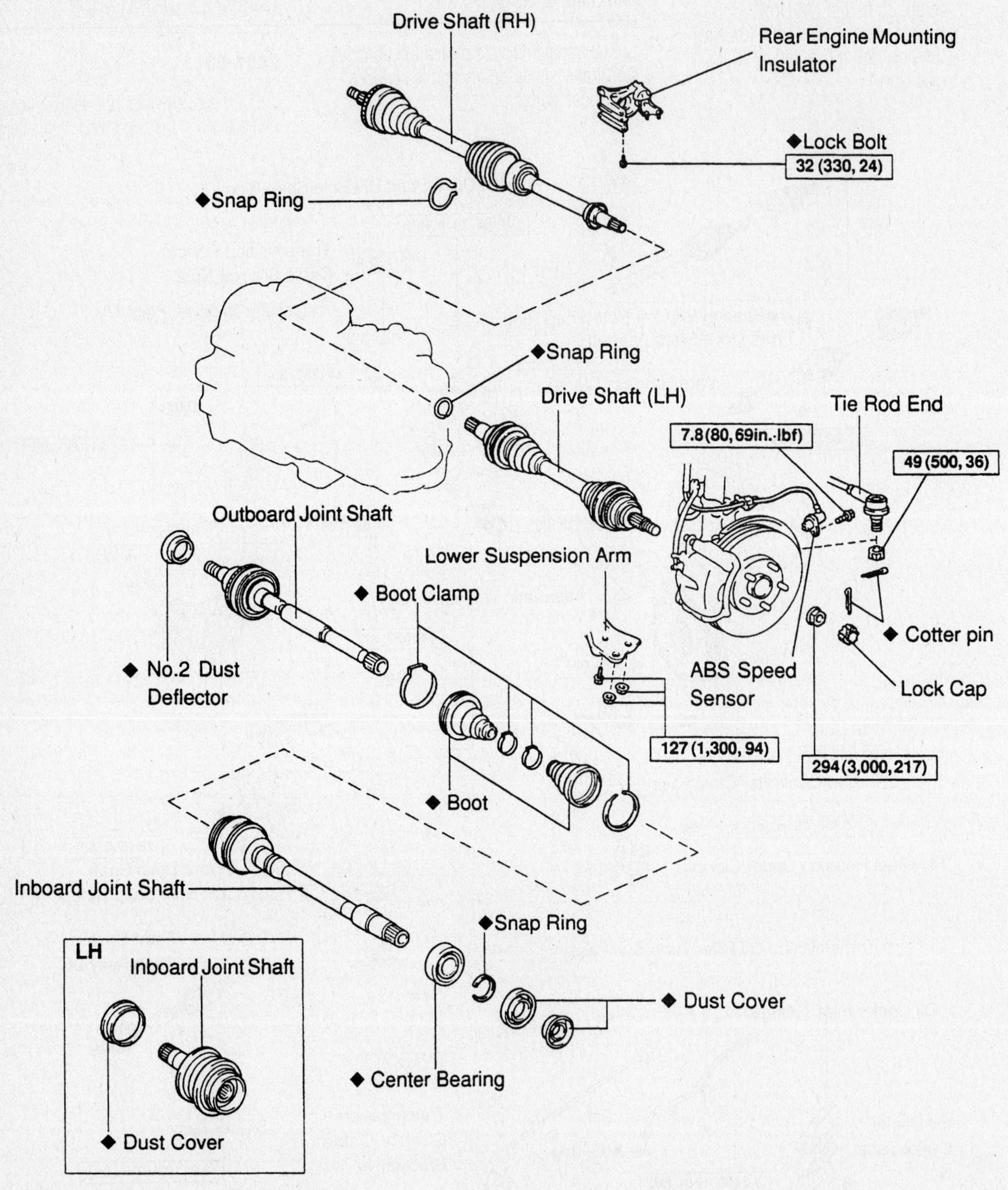
Drive Shaft (RH)
Rear Engine Mounting Insulator
◆Lock Bolt
32 (330, 24)
◆Snap Ring
◆Snap Ring
Drive Shaft (LH)
Tie Rod End
7.8(80, 69in.·lbf)
49 (500, 36)
Outboard Joint Shaft
Lower Suspension Arm
◆ Boot Clamp
◆ Cotter pin
◆ No.2 Dust Deflector
ABS Speed Sensor
Lock Cap
127 (1,300, 94)
294 (3,000, 217)
◆ Boot
Inboard Joint Shaft
◆Snap Ring
LH
Inboard Joint Shaft
◆ Dust Cover
◆ Center Bearing
◆ Dust Cover
N·m (kgf·cm, ft·lbf) : Specified torque
◆ Non–reusable part
7924ZG73

Exploded view of halfshaft—2001-03

2. Remove or disconnect the following:
   - Front wheels
   - Cotter pin and locknut cap

➡Have an assistant depress the brake pedal and loosen the bearing locknut.

   - Engine undercover
   - Fender apron seal
   - Tie rod end, from the steering knuckle
   - Steering knuckle, from the lower control arm
   - Halfshaft from the axle hub, using a plastic hammer
   - Cover the outer boot with a rag
   - Halfshaft from the transaxle, using the proper tools

To install:

3. Reverse the removal procedures to complete installation, tightening fasteners to specifications.
4. Fill the transaxle with gear oil, install the fender apron, check front end alignment and test drive.

➡If the cotter pin holes do not align, always correct by tightening the nut until the next hole aligns.

5. Install a new cotter pin.

2004

1. Drain the transaxle fluid.
2. With 4wd, drain the transfer case.

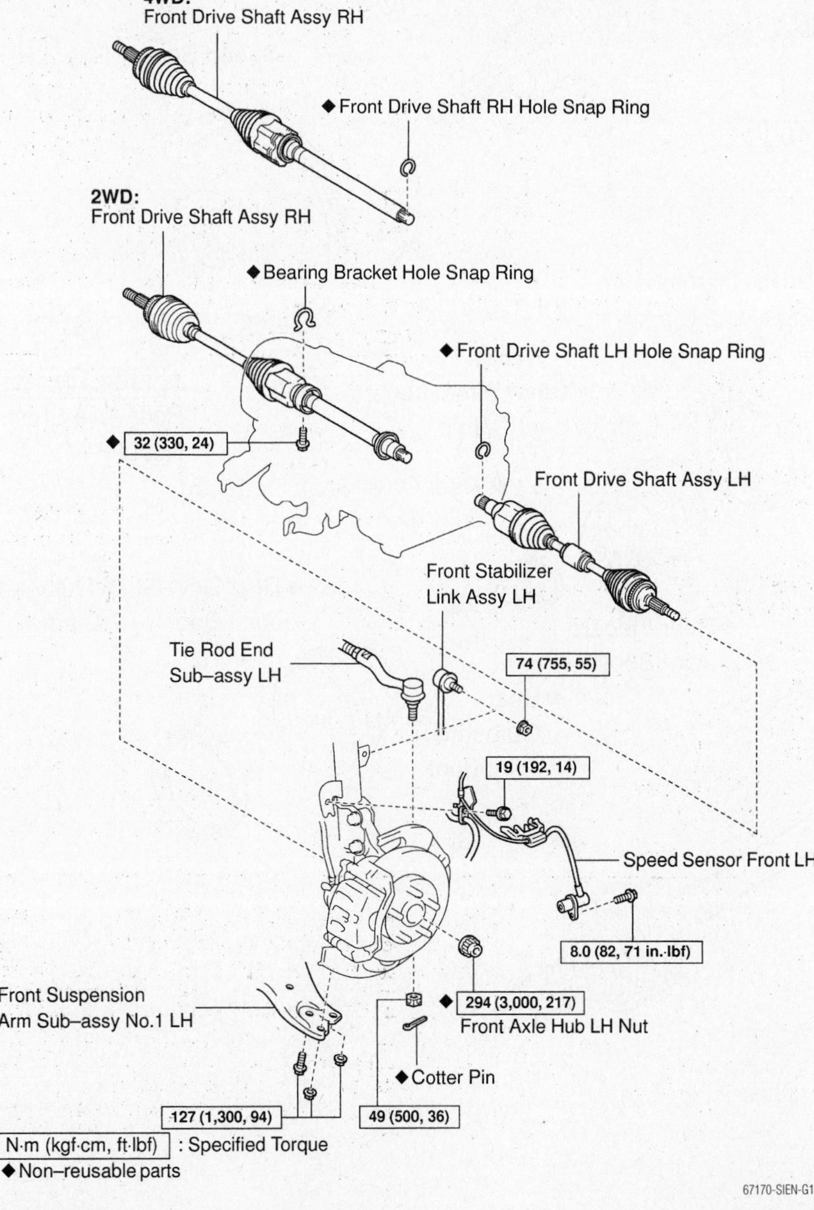

Front halfshaft exploded view—2004 models

3. Remove the wheel.
4. Unstake the hub nut, and, with the brake applied, remove the hub nut.
5. Disconnect the stabilizer link.
6. Remove the speed sensor.
7. Disconnect the tie rod end from the knuckle.
8. Disconnect the lower arm from the ball joint.
9. Using a plastic hammer, drive the halfshaft from the hub.
10. On the left side with 2wd and both sides with 4wd, using a slidehammer with adapter, pull the halfshaft from the transaxle.
11. On the right side, with 2wd, remove the halfshaft bearing bracket snapring. Remove the bolt and the halfshaft from the bearing bracket.

To install:

12. Coat the splines of the inboard end with clean ATF.
13. Drive the left shaft (2wd and both shafts 4wd) into place with a hammer and brass drift. Install the snapring with the opening downward.
14. Install the right side (2wd) shaft. Install the snapring and bolt. Torque to 24 ft. lbs. (32 Nm).
15. The remainder of installation is the reverse of removal. Observe the following torques:
    - Arm-to-ball joint: 94 ft. lbs. (127 Nm)
    - Tie rod end: 36 ft. lbs. (49 Nm). Advance the nut no more than 60 degrees to align the hole.
    - Stabilizer link: 55 ft. lbs. (74 Nm)
    - New hub nut: 217 ft. lbs. (294 Nm). Stake the nut.

Rear Halfshaft

REMOVAL & INSTALLATION

2004

1. Remove the wheel.
2. Remove the tail pipe.
3. Remove the speed sensor.
4. Unstake and remove the axle shaft nut.
5. Matchmark the halfshaft and differential side gear.
6. Remove the 4 nuts and washers and remove the shaft.
7. Installation is the reverse of removal. Torque the 4 nuts t 41 ft. lbs. (56 Nm). Torque the axle shaft nut to 159 ft. lbs. (216 Nm). Stake the nut.

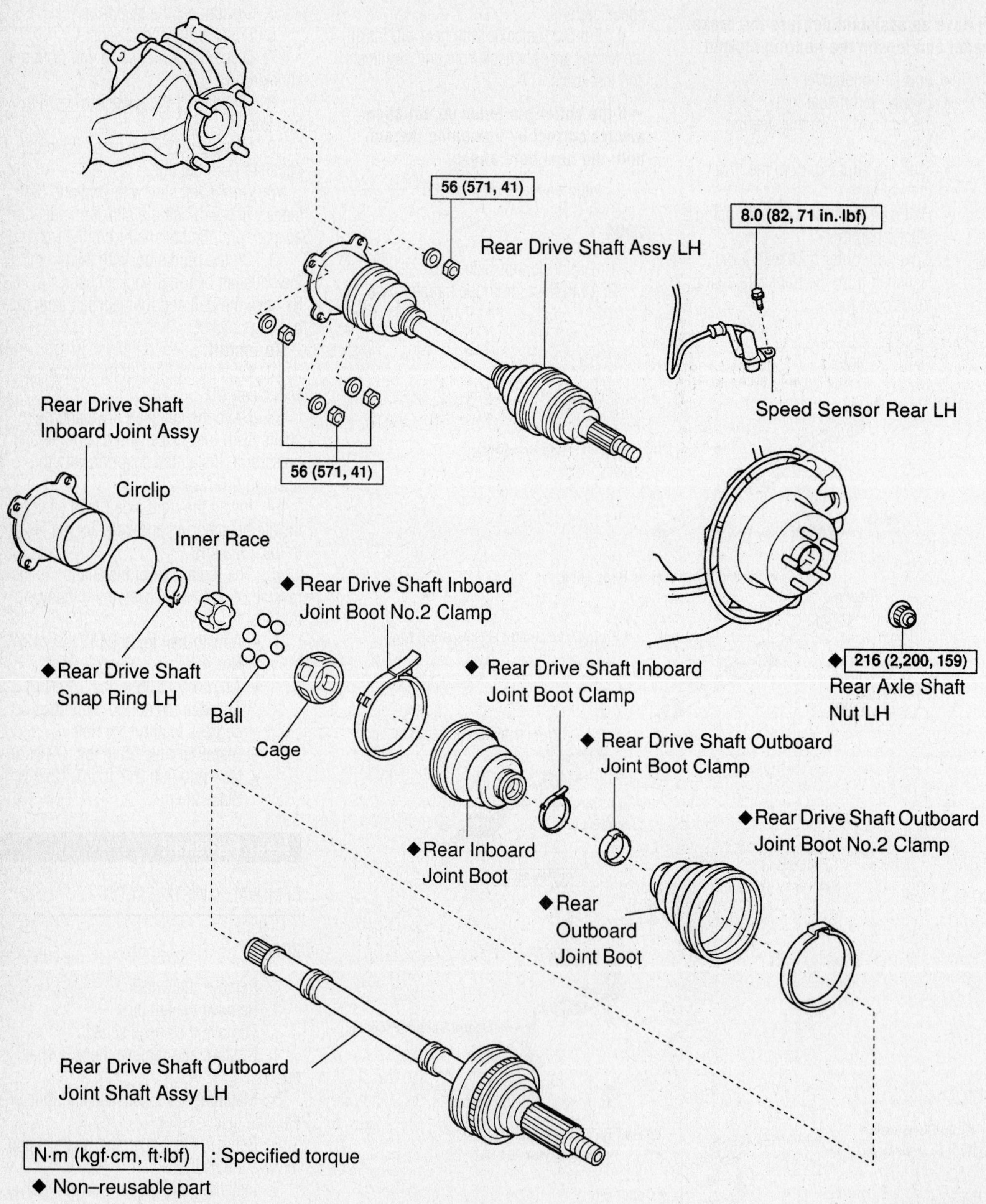

56 (571, 41)
8.0 (82, 71 in.·lbf)
Rear Drive Shaft Assy LH
Speed Sensor Rear LH
Rear Drive Shaft Inboard Joint Assy
56 (571, 41)
Circlip
Inner Race
◆ Rear Drive Shaft Inboard Joint Boot No.2 Clamp
◆ 216 (2,200, 159)
Rear Axle Shaft Nut LH
◆Rear Drive Shaft Snap Ring LH
Ball
Cage
◆ Rear Drive Shaft Inboard Joint Boot Clamp
◆ Rear Drive Shaft Outboard Joint Boot Clamp
◆Rear Drive Shaft Outboard Joint Boot No.2 Clamp
◆Rear Inboard Joint Boot
◆Rear Outboard Joint Boot
Rear Drive Shaft Outboard Joint Shaft Assy LH
N·m (kgf·cm, ft·lbf) : Specified torque
◆ Non–reusable part
67170-SIEN-G21

Rear halfshaft and related parts—2004 models

CV-Joints

OVERHAUL

2001–02

1. Before servicing the vehicle, refer to the precautions in the beginning of this section.
2. Remove or disconnect the following:
   - Halfshaft
   - Inboard joint boot clamps
3. Clean the joint before removing the boot.
4. Slide the inboard joint boot toward the outboard joint.
5. Place matchmarks on the inboard joint tulip and the shaft.
6. Remove the inboard joint tulip from the driveshaft.
7. Clamp the halfshaft in a vise.
8. Remove or disconnect the following:
   - Snapring and disassemble the tri-pot joint
   - Tri-pot joint from the halfshaft, using a brass bar and hammer

WARNING

Be careful not to punch the roller.

   - Inboard joint boot
   - Outboard joint boot clamps and boot

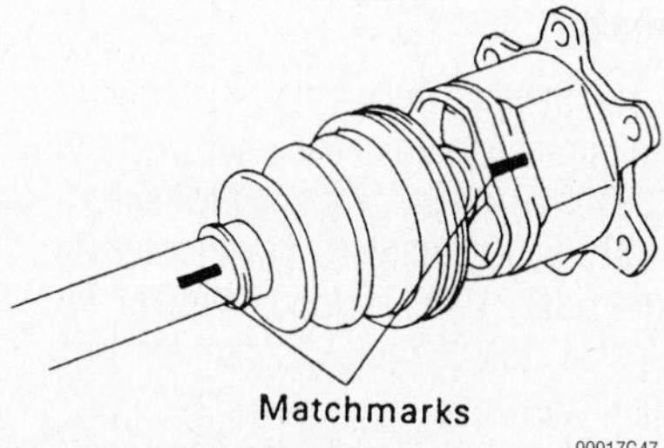

Place matchmarks on the inboard joint outer race and shaft

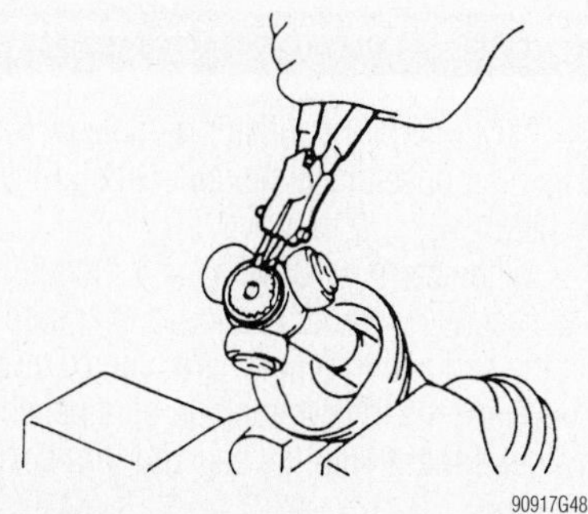

Use snapring expanders to extract the snapring from the end

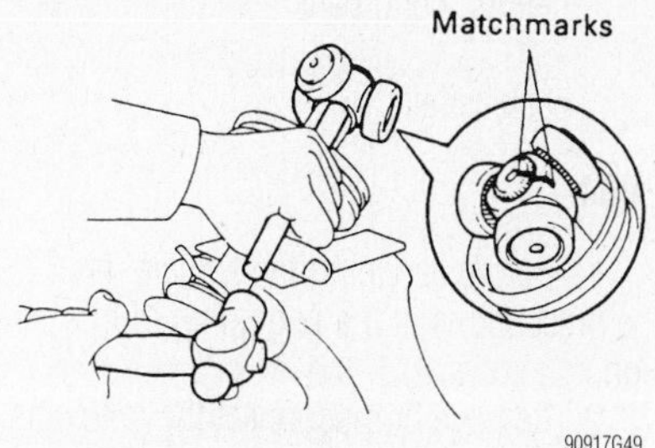

With a brass bar and hammer, tap the joint hard enough to remove

WARNING

Do not disassemble the outboard joint.

To assemble:

9. Temporarily, install the new boot and new boot clamps to the outboard joint.

➡Before installing the boot, wrap vinyl tape around the spline of the shaft to prevent damaging the boot.

10. Temporarily, install the new boot and the new boot clamps for the inboard joint to the halfshaft.
11. Assemble the tri-pot joint, as follows:
    a. Place the beveled side of the tri-pot axial spline toward the outboard joint.
    b. Align the matchmarks placed before disassembly.
    c. Using a brass bar and hammer, tap in the tri-pot joint onto the driveshaft. Do not punch the roller.
12. Install a new snapring.
13. Before assembling the boot to the outboard joint, pack the boot with grease. The capacity is 4.2–4.6 oz. (120–130 g).

➡Keep the grease off the joint connection groove of the boot. Pack in grease all over the ball and contact surface inside the joint.

14. Assemble the inboard joint to the inboard joint tulip. Pack in grease to the inboard tulip and the boot. The capacity is 7.6–7.9 oz. (215–225 g).
15. Install or connect the following:
    - Outer race, by aligning the matchmarks on the shaft
    - Inboard joint boot, without twisting it

➡Be sure the boot is on the shaft groove and inboard joint outer race groove.

16. Set the length of the shaft to 19.146–19.546 in. (486.4–496.41mm).
17. Install or connect the following:
    - Both clamps to the inboard joint boot; bend back the band and lock it
    - Halfshaft

2003

1. Before servicing the vehicle, refer to the precautions in the beginning of this section.
2. Remove or disconnect the following:
   - Halfshaft
   - Inboard joint boot clamps
3. Clean the joint before removing the boot.
4. Slide the inboard joint boot toward the outboard joint.
5. Place matchmarks on the inboard joint tulip and the shaft.
6. Remove the inboard joint tulip from the driveshaft.
7. Clamp the halfshaft in a vise.
8. Remove or disconnect the following:
   - Snapring and disassemble the tri-pot joint
   - Tri-pot joint from the halfshaft, using a brass bar and hammer

WARNING

Be careful not to punch the roller.

   - Inboard joint boot
   - Outboard joint boot clamps and boot

WARNING

Do not disassemble the outboard joint.

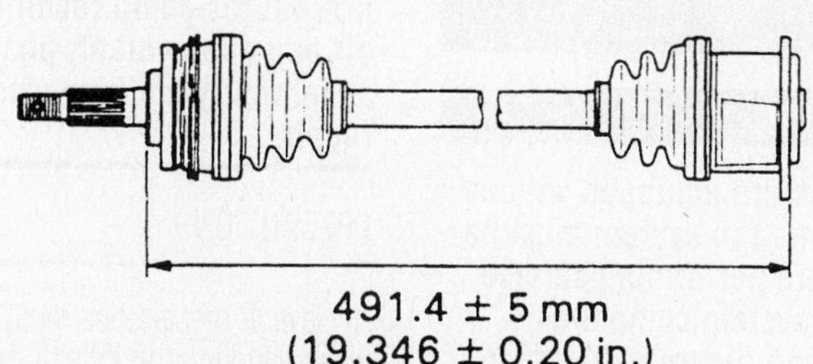

Set the length of the halfshaft at the points shown here

To assemble:

9. Temporarily, install the new boot and new boot clamps to the outboard joint.

➡Before installing the boot, wrap vinyl tape around the spline of the shaft to prevent damaging the boot.

10. Temporarily, install the new boot and the new boot clamps for the inboard joint to the halfshaft.
11. Assemble the tri-pot joint, as follows:
    a. Place the beveled side of the tri-pot axial spline toward the outboard joint.
    b. Align the matchmarks placed before disassembly.
    c. Using a brass bar and hammer, tap in the tri-pot joint onto the driveshaft. Do not punch the roller.
12. Install a new snapring.
13. Before assembling the boot to the outboard joint, pack the boot with grease. The capacity is 3.7–4.4 oz. (105–125 g).

➡Keep the grease off the joint connection groove of the boot. Pack in grease all over the ball and contact surface inside the joint.

14. Assemble the inboard joint to the inboard joint tulip. Pack in grease to the inboard tulip and the boot. The capacity is 4.2–4.6 oz. (120–130 g) for the joint side; 2.1–2.3 oz. (52.5–57.5 g) for the boot side.
15. Install or connect the following:
    - Outer race, by aligning the matchmarks on the shaft
    - Inboard joint boot, without twisting it

➡Be sure the boot is on the shaft groove and inboard joint outer race groove.

16. Set the length of the shaft (inboard end to outboard joint flange) to 23.071 in. (586mm) +/- 2mm for left side; 34.709 in. (881.6mm) +/- 2mm for the right side..
17. Install or connect the following:
    - Both clamps to the inboard joint boot; bend back the band and lock it
    - Halfshaft

2004

1. Before servicing the vehicle, refer to the precautions in the beginning of this section.
2. Remove or disconnect the following:
    - Halfshaft
    - Inboard joint boot clamps
3. Clean the joint before removing the boot.
4. Slide the inboard joint boot toward the outboard joint.
5. Place matchmarks on the inboard joint tulip and the shaft.
6. Remove the inboard joint tulip from the driveshaft.
7. Clamp the halfshaft in a vise.
8. Remove or disconnect the following:
    - Snapring and disassemble the tri-pot joint
    - Tri-pot joint from the halfshaft, using a brass bar and hammer

⁂ WARNING

Be careful not to punch the roller.

- Inboard joint boot
- Outboard joint boot clamps and boot

⁂ WARNING

Do not disassemble the outboard joint.

To assemble:

9. Temporarily, install the new boot and new boot clamps to the outboard joint.

➡Before installing the boot, wrap vinyl tape around the spline of the shaft to prevent damaging the boot.

10. Temporarily, install the new boot and the new boot clamps for the inboard joint to the halfshaft.
11. Assemble the tri-pot joint, as follows:
    a. Place the beveled side of the tri-pot axial spline toward the outboard joint.
    b. Align the matchmarks placed before disassembly.
    c. Using a brass bar and hammer, tap in the tri-pot joint onto the driveshaft. Do not punch the roller.
12. Install a new snapring.
13. Before assembling the boot to the outboard joint, pack the boot with grease. The capacity is 1.2–1.9 oz. (35–55 g) for 2wd; 1.4–2.1 oz. (40–60 g) for 4wd.

➡Keep the grease off the joint connection groove of the boot. Pack in grease all over the ball and contact surface inside the joint.

14. Install the damper on the left shaft. The distance from the outer end flange of the damper to the outer end flange of the outer joint should be 8.7 inches (221mm) +/- 2mm.
15. Assemble the inboard joint to the inboard joint tulip. Pack in grease to the inboard tulip and the boot. The capacity is 3.4–4.1 oz. (95–115 g) for 2wd; 3.9–4.6 oz. (110–130 g) for 4wd.
16. Install or connect the following:
    - Outer race, by aligning the matchmarks on the shaft
    - Inboard joint boot, without twisting it

➡Be sure the boot is on the shaft groove and inboard joint outer race groove.

17. Set the length of the shaft (inboard end to outboard joint flange) to 25.134 in. (638.4mm) +/- 5mm for 2wd left side; 37.678 in. (957mm) +/- 5mm for the 2wd right side; 21.797 in. (553.6mm) for 4wd left side; 39.065 in. (992.2mm) for 4wd right side.
18. Install or connect the following:
    - Both clamps to the inboard joint boot; bend back the band and lock it
    - Halfshaft

STEERING & SUSPENSION

Air Bag

⁂ CAUTION

Some vehicles are equipped with an air bag system. The system must be disabled before performing service on or around system components, steering column, instrument panel components, wiring and sensors. Failure to follow safety and disabling procedures could result in accidental air bag deployment, possible personal injury and unnecessary system repairs.

PRECAUTIONS

Several precautions must be observed when handling the inflator module to avoid accidental deployment and possible personal injury.

- Never carry the inflator module by the wires or connector on the underside of the module.
- When carrying a live inflator module, hold securely with both hands, and ensure that the bag and trim cover are pointed away.
- Place the inflator module on a bench or other surface with the bag and trim cover facing up.
- With the inflator module on the bench, never place anything on or close to the

module, which may be thrown in the event of an accidental deployment.

DISARMING

To avoid personal injury when working on vehicles equipped with an air bag, the negative battery cable must be disconnected and at least 90 seconds must elapse before working on the system. Failure to do so may result in deployment of the air bag.

Power Rack and Pinion Steering Gear

REMOVAL & INSTALLATION

2001–03

1. Before servicing the vehicle, refer to the precautions in the beginning of this section.

2. Remove or disconnect the following:

- Negative battery cable

➡Wait at least 90 seconds before working on the vehicle to allow the Supplemental Restraint System (SRS) system to disarm.

- Right and left side fender apron seals
- Right and left tie rod ends

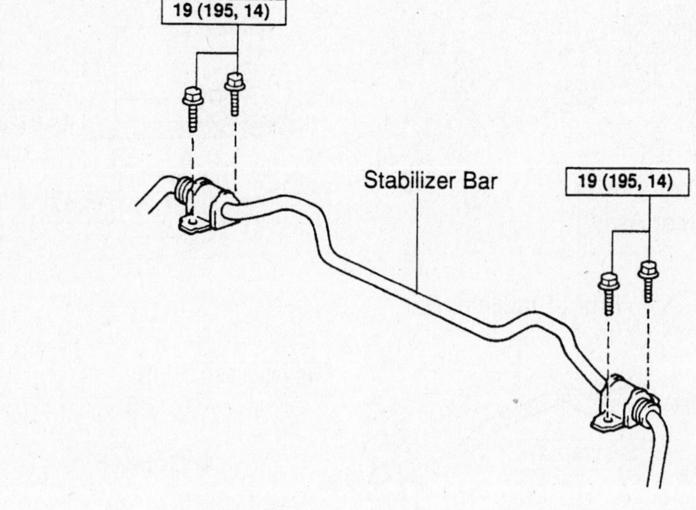

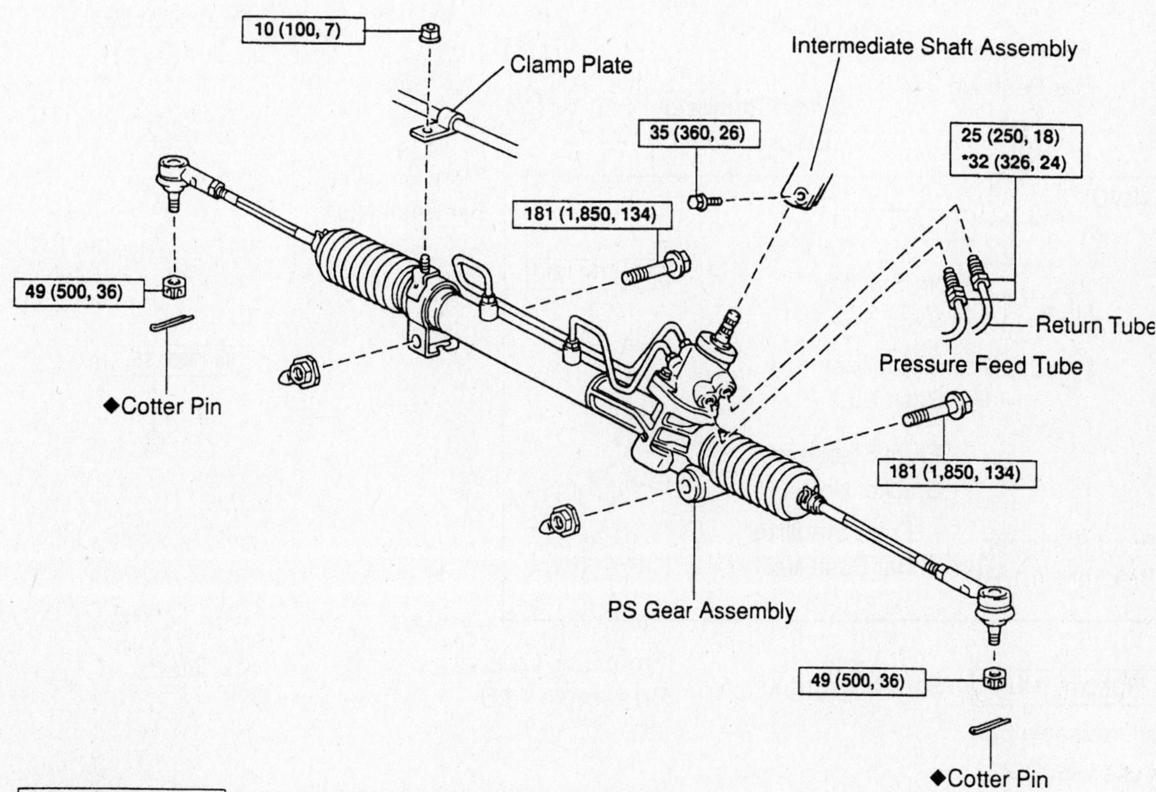

Exploded view of the power steering gear and related components—2001–03

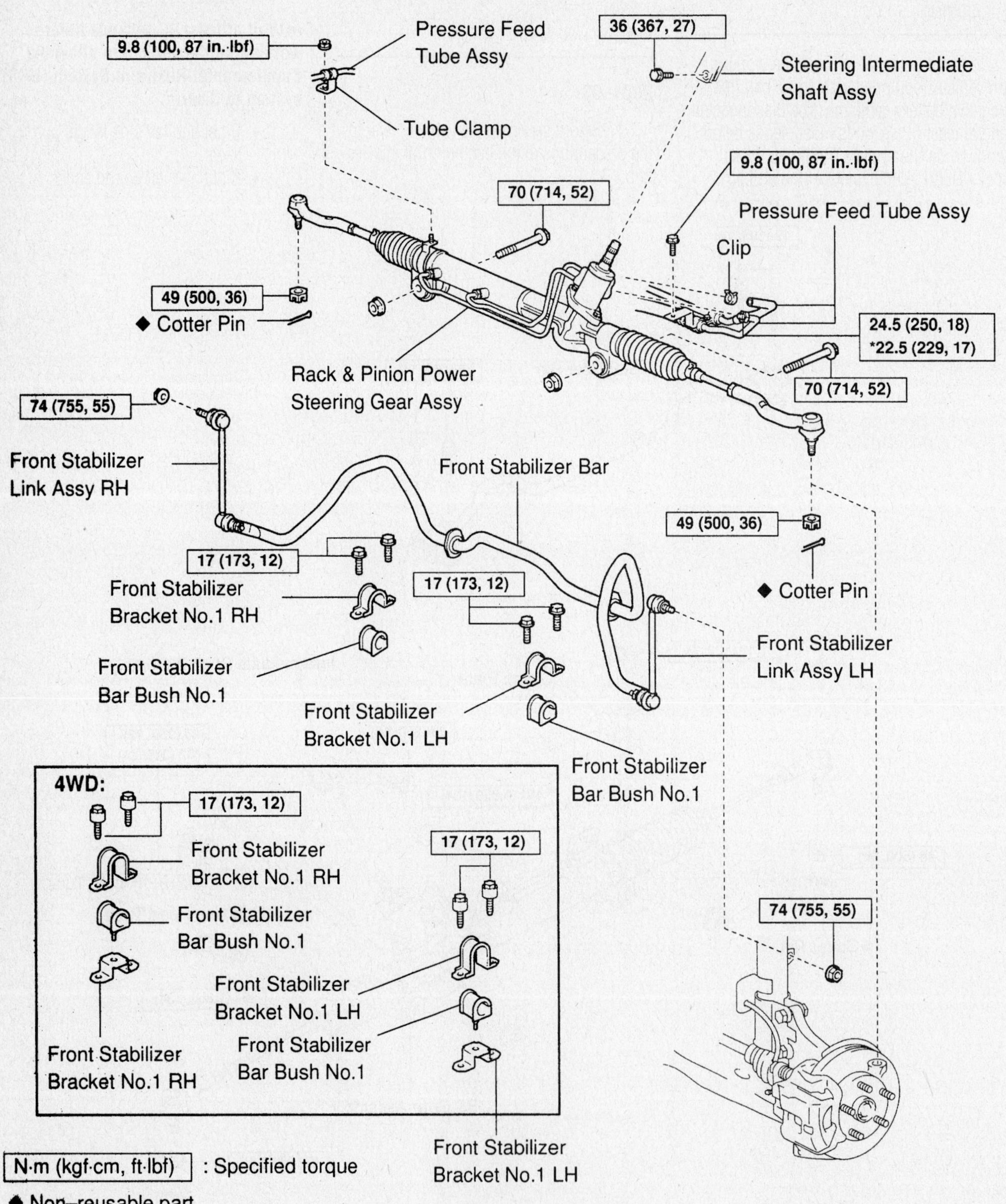

Steering gear and related parts—2004 models

3. Place matchmarks on the intermediate shaft.

4. Remove or disconnect the following:

- Pinch bolt and the intermediate shaft out from under the vehicle
- Power steering line clamp
- Pressure and feed lines
- Stabilizer bar, unbolt it but do not remove it
- Heated Oxygen (HO_2) sensor
- Both gear assembly set bolts and nuts, by lifting the stabilizer bar
- Gear assembly from the left side of the vehicle

To install:

5. Install or connect the following:

- Gear assembly to the left side of the vehicle

✱✱ WARNING

Be careful not to damage the power steering lines.

- Tighten the gear assembly set bolts and nuts to 134 ft. lbs. (181 Nm), by lifting the stabilizer bar
- HO_2 sensor
- Stabilizer bar. Tighten the bolt to 14 ft. lbs. (19 Nm) and the nut to 29 ft. lbs. (39 Nm).
- Pressure and feed return lines. Tighten them to 18 ft. lbs. (25 Nm).
- Line clamps. Tighten the nut to 84 inch lbs. (10 Nm).
- Intermediate shaft, by aligning the joint and main shaft matchmarks. Tighten to 26 ft. lbs. (35 Nm).
- Tie rod ends
- Fender apron seals. Securely tighten the bolts.

6. Remove or disconnect the following:

- Steering wheel pad
- Steering wheel

7. Position the front wheels facing straight-ahead. Do this with the front of the vehicle on jackstands.

8. Center the spiral cable.

9. Install the steering wheel at the straight-ahead position. Temporarily tighten the wheel set nut. Attach the wiring.

10. Bleed the power steering system.

11. Check the steering wheel center point. Tighten the steering nut to 26 ft. lbs. (35 Nm).

12. Check and/or adjust the front wheel alignment.

2004

1. Place the wheels in a straight-ahead position.

2. Remove the wheels.

3. Matchmark and remove the intermediate shaft from the gear.

4. Remove the steering column hole cover from the dash panel.

5. Disconnect the tie rod ends.

6. Disconnect the stabilizer bar.

7. Disconnect the pressure and return lines.

8. Remove the 2 gear mounting bolts.

9. Matchmark the intermediate shaft extension and control valve shaft.

10. Remove the intermediate shaft extension bolt and remove the gear.

To install:

11. Place the gear in position. Align the matchmarks and connect the intermediate extension shaft. Torque to 27 ft. lbs. (36 Nm).

12. Install the column cover plate and clamp.

13. Install the gear mounting bolts. Torque to 52 ft. lbs. (70 Nm).

14. Connect the stabilizer bar. Torque to 12 ft. lbs. (17 Nm).

15. Connect the tie rods ends. Torque to 36 ft. lbs. (49 Nm).

16. Connect the intermediate shaft. Torque to 26 ft. lbs. (36 Nm).

17. Check the alignment.

Strut

REMOVAL & INSTALLATION

2001–03

1. Before servicing the vehicle, refer to the precautions in the beginning of this section.

➡Do not support the weight of the vehicle on the suspension arm; the arm will deform under its weight.

2. Remove or disconnect the following:

- Wheel
- Brake hose and the Anti-lock Brake System (ABS) speed sensor wire from the strut
- Sway bar link from the strut
- Outer front cowl top panel

3. Matchmark the strut lower bracket and camber adjust cam, if equipped.

4. Remove or disconnect the following:

- Lower strut end from the steering knuckle's lower arm
- 3 upper strut mounting plate to the upper wheel arch nuts
- Strut

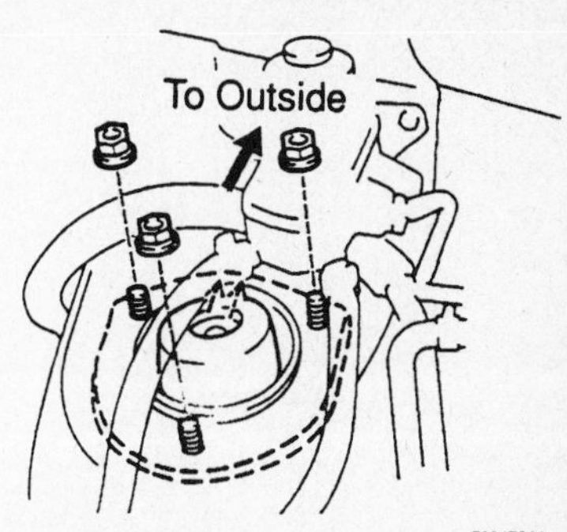

Cutaway view of the upper strut bearing position for installation—2001–03

To install:

5. Align the upper suspension support hole with the strut piston or end, so they fit properly.

6. Install or connect the following:

- Strut piston rod end to the upper suspension support. Tighten the new nut to 29–40 ft. lbs. (39–54 Nm).

✱✱ WARNING

Do not use an impact wrench to tighten the nut.

- Lubricate the suspension support bearing with multi-purpose grease. Pack the upper support space with multi-purpose grease, after installation.
- Tighten the 3 suspension support-to-wheel arch nuts to 47 ft. lbs. (64 Nm).
- Tighten the strut-to-steering knuckle arm bolts to 156 ft. lbs. (211 Nm).
- Sway bar link to the strut. Tighten the nut to 29 ft. lbs. (39 Nm).
- Outer front cowl top panel
- ABS speed sensor and the brake hose to the strut, if equipped.
- Wheel

2004

1. Remove the wheel.

2. Remove the wiper arms.

3. Remove the wiper motor.

4. Remove the top cowl.

5. Remove the stabilizer link from the strut.

6. Loosen the strut rod locknut. Don't remove it.

7. Remove the brake hose bracket from the strut.

8. Remove the lower strut bolts.

9. Remove the 3 upper strut nuts.

10. Remove the strut.

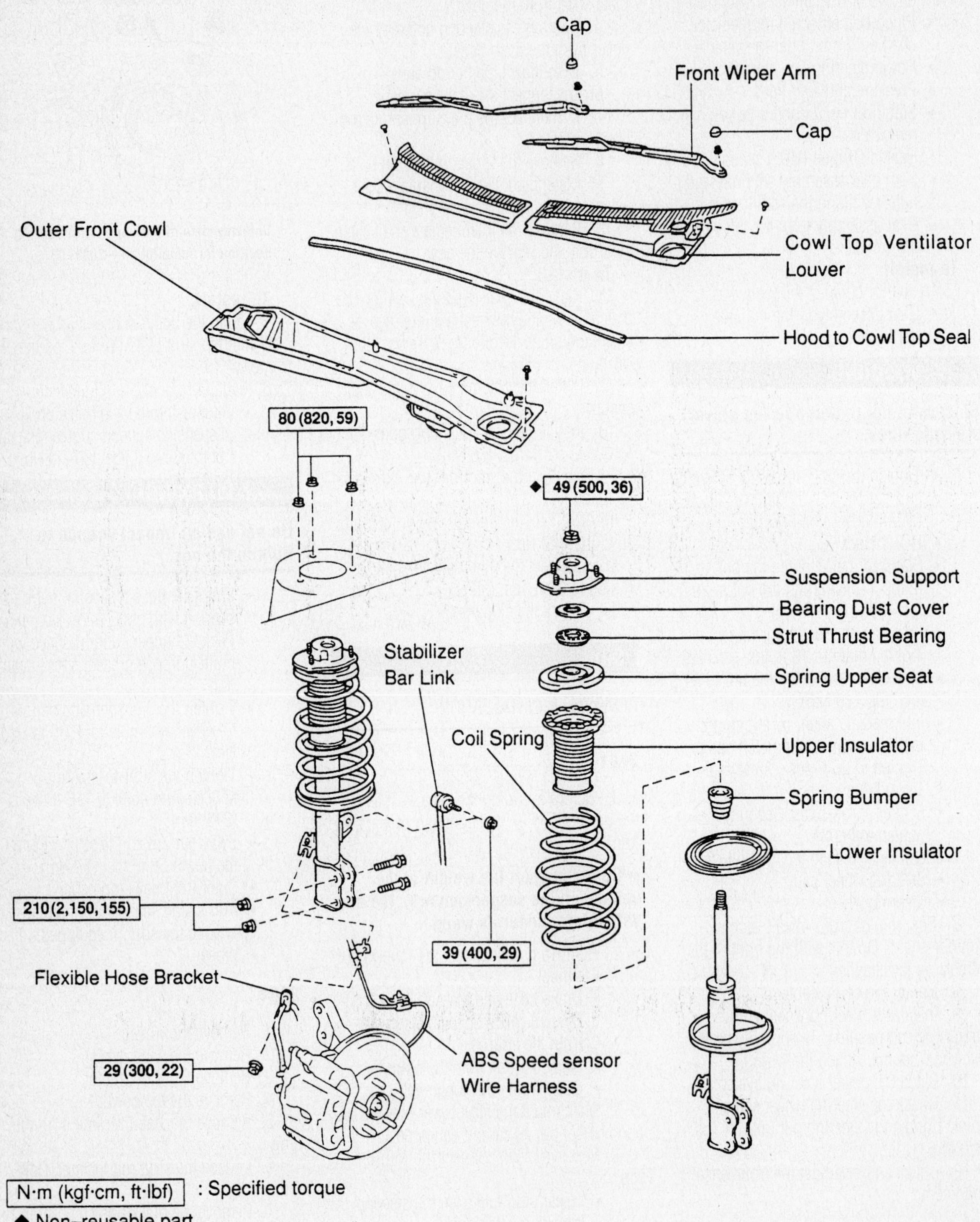

View of the front strut assembly and related components—2001–03

11. Installation is the reverse of removal. Observe the following torques:

- Upper nuts: 59 ft. lbs. (80 Nm)
- Lower nuts/bolts: 155 ft. lbs. (210 Nm)
- Strut rod locknut: 36 ft. lbs. (49 Nm)
- Stabilizer link: 55 ft. lbs. (74 Nm)

Shock Absorber

REMOVAL & INSTALLATION

2001–03

1. Before servicing the vehicle, refer to the precautions in the beginning of this section.
2. Support the axle beam with jacks.
3. Remove or disconnect the following:
   - Rear wheels
   - Interior service covers to access the upper shock mounts
   - Both nuts and retainers from the upper mount
   - Shock absorber

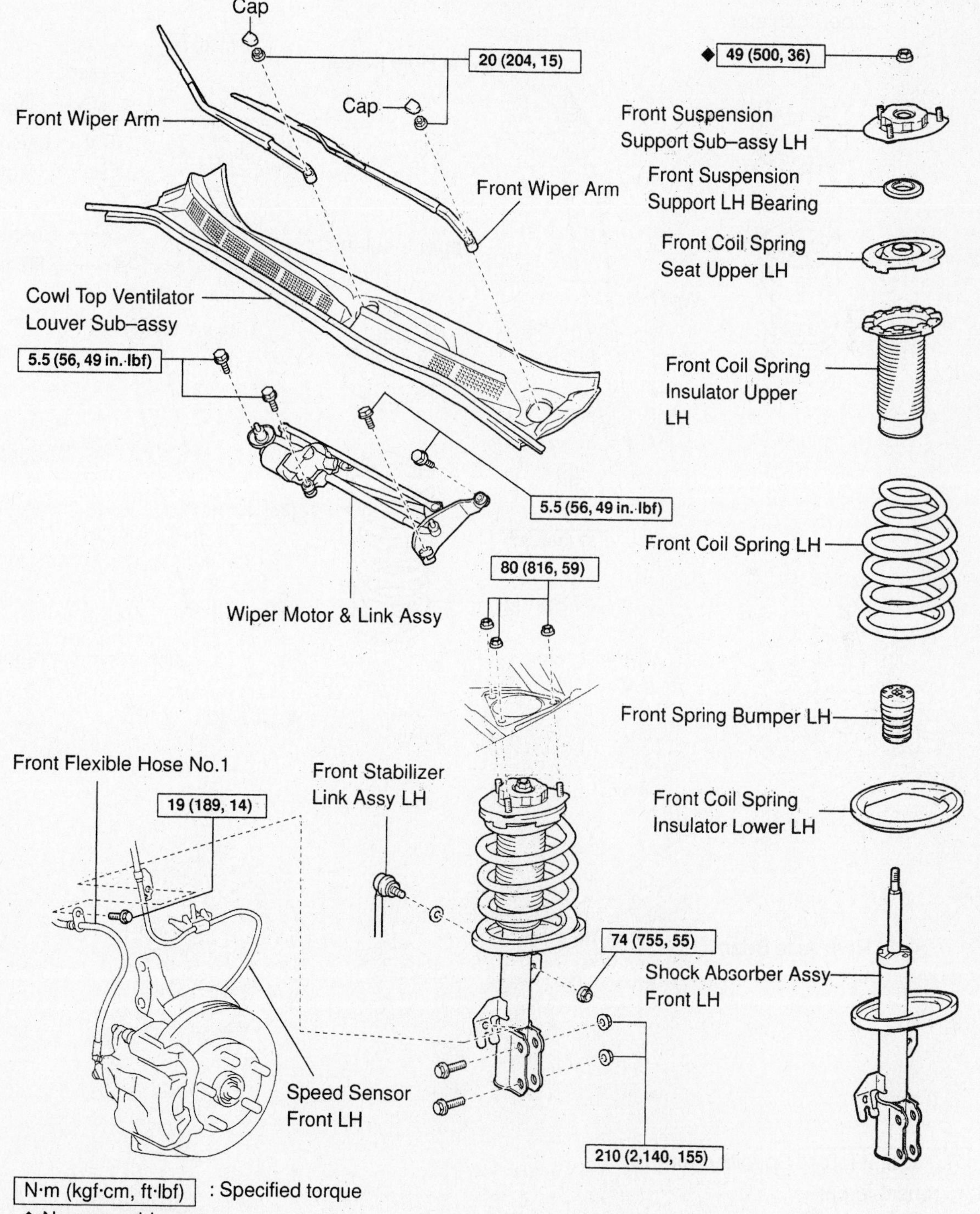

Front suspension and related components—2004

To install:

4. Install the shock absorber. Tighten the lower mounting bolt to 27 ft. lbs. (37 Nm).
5. If the upper cushion is showing signs of wear, replace it.
6. Install or connect the following:
   - Upper shock absorber. Tighten the nuts to 18 ft. lbs. (25 Nm).
   - Wheels

2004

1. Remove the wheel.
2. Remove the shock absorber cap.
3. Support the axle with a jack.
4. Remove the upper locknut, retainer and bushing.
5. Remove the lower nut and remove the shock.
6. Remove the spring bumper from the shock.

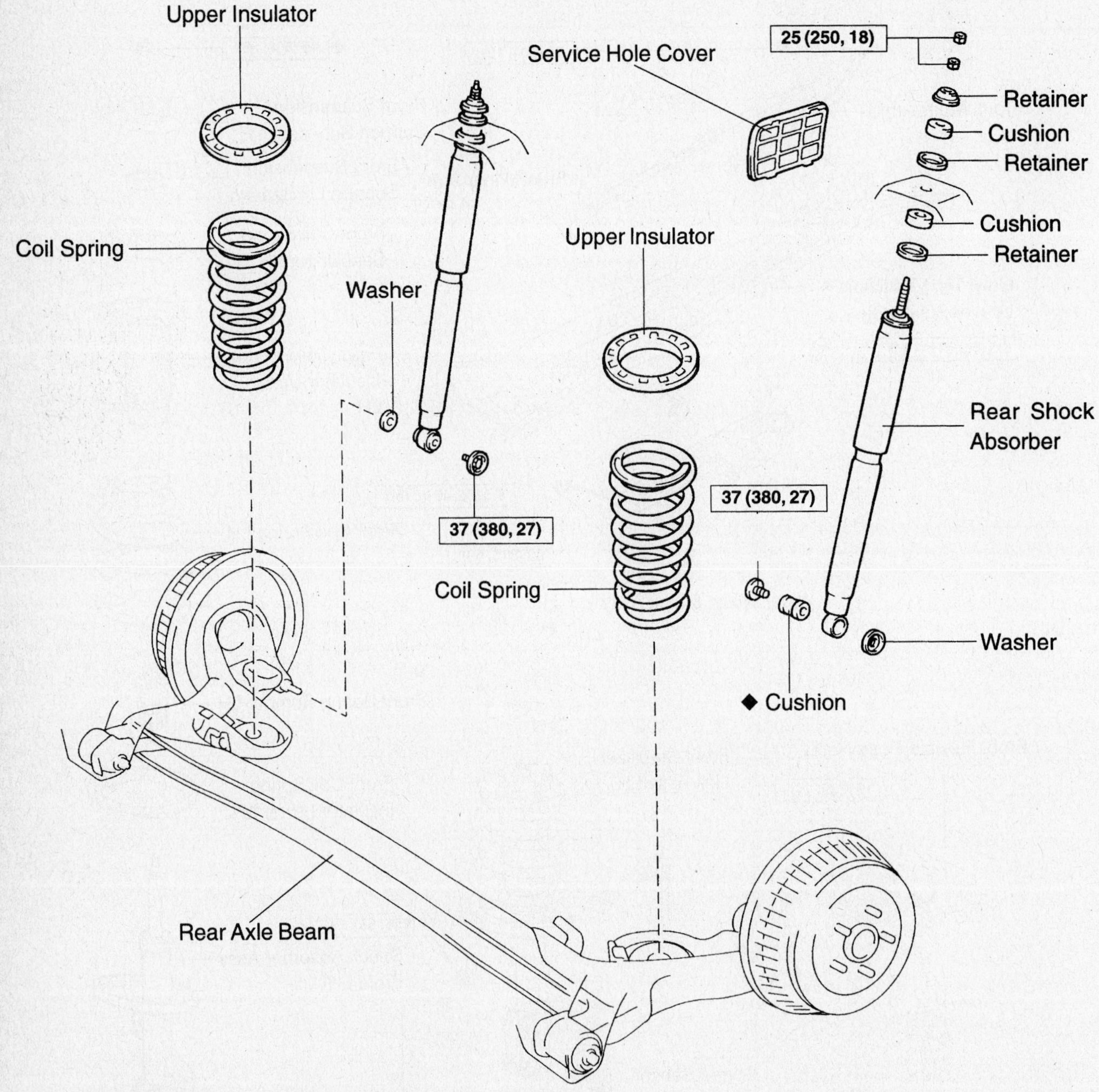

Rear suspension components—2001–03 models

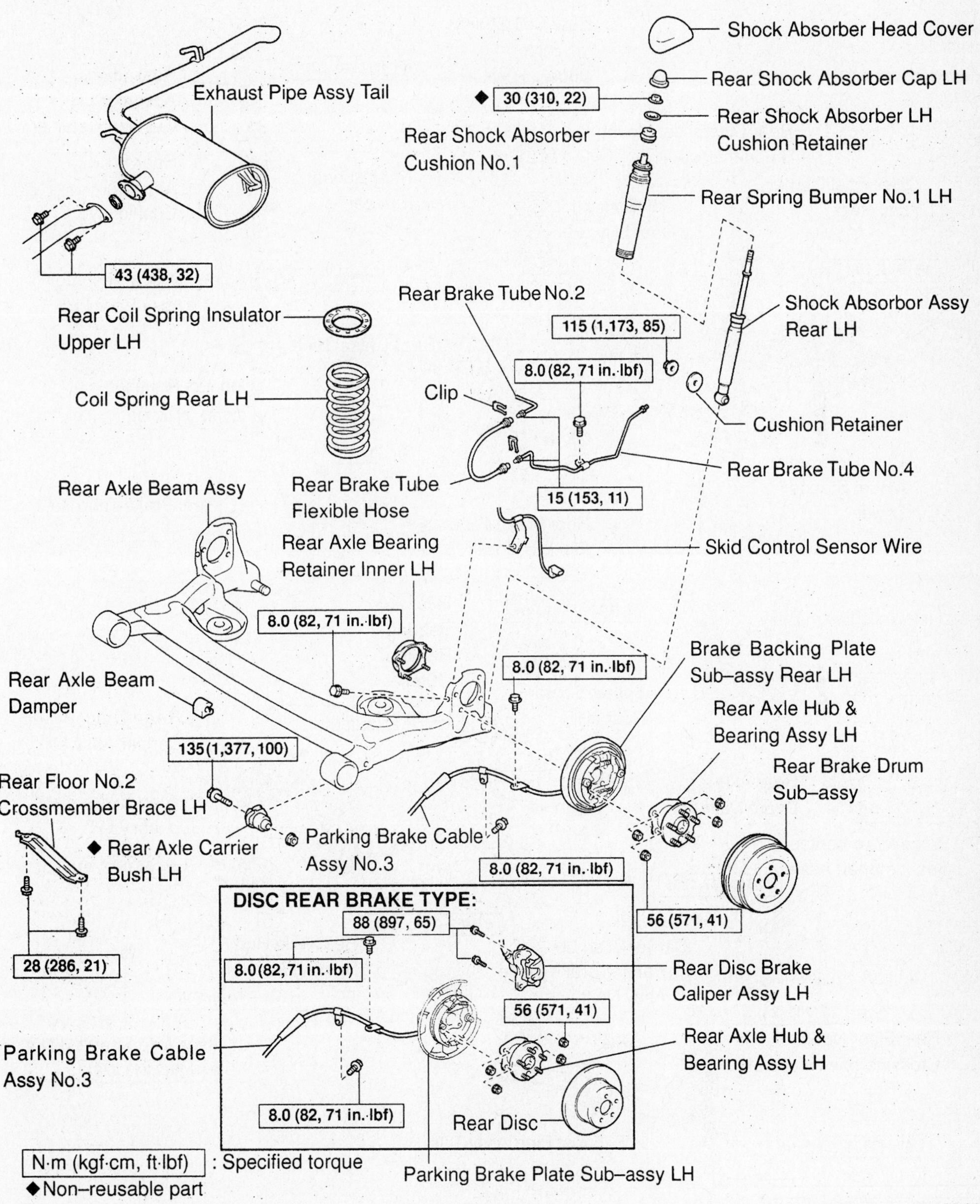
2WD DRIVE TYPE:
Shock Absorber Head Cover
Rear Shock Absorber Cap LH
◆ 30 (310, 22)
Exhaust Pipe Assy Tail
Rear Shock Absorber LH Cushion Retainer
Rear Shock Absorber Cushion No.1
Rear Spring Bumper No.1 LH
43 (438, 32)
Rear Brake Tube No.2
Shock Absorbor Assy Rear LH
Rear Coil Spring Insulator Upper LH
115 (1,173, 85)
8.0 (82, 71 in.·lbf)
Coil Spring Rear LH
Clip
Cushion Retainer
Rear Brake Tube No.4
Rear Axle Beam Assy
Rear Brake Tube Flexible Hose
15 (153, 11)
Rear Axle Bearing Retainer Inner LH
Skid Control Sensor Wire
8.0 (82, 71 in.·lbf)
Brake Backing Plate Sub–assy Rear LH
8.0 (82, 71 in.·lbf)
Rear Axle Beam Damper
Rear Axle Hub & Bearing Assy LH
135 (1,377, 100)
Rear Brake Drum Sub–assy
Rear Floor No.2 Crossmember Brace LH
◆ Rear Axle Carrier Bush LH
Parking Brake Cable Assy No.3
8.0 (82, 71 in.·lbf)
DISC REAR BRAKE TYPE:
88 (897, 65)
56 (571, 41)
28 (286, 21)
8.0 (82, 71 in.·lbf)
Rear Disc Brake Caliper Assy LH
56 (571, 41)
Rear Axle Hub & Bearing Assy LH
Parking Brake Cable Assy No.3
8.0 (82, 71 in.·lbf)
Rear Disc
N·m (kgf·cm, ft·lbf) : Specified torque
◆ Non–reusable part
Parking Brake Plate Sub–assy LH
67170-SIEN-G13

Rear suspension components—2004 2wd models

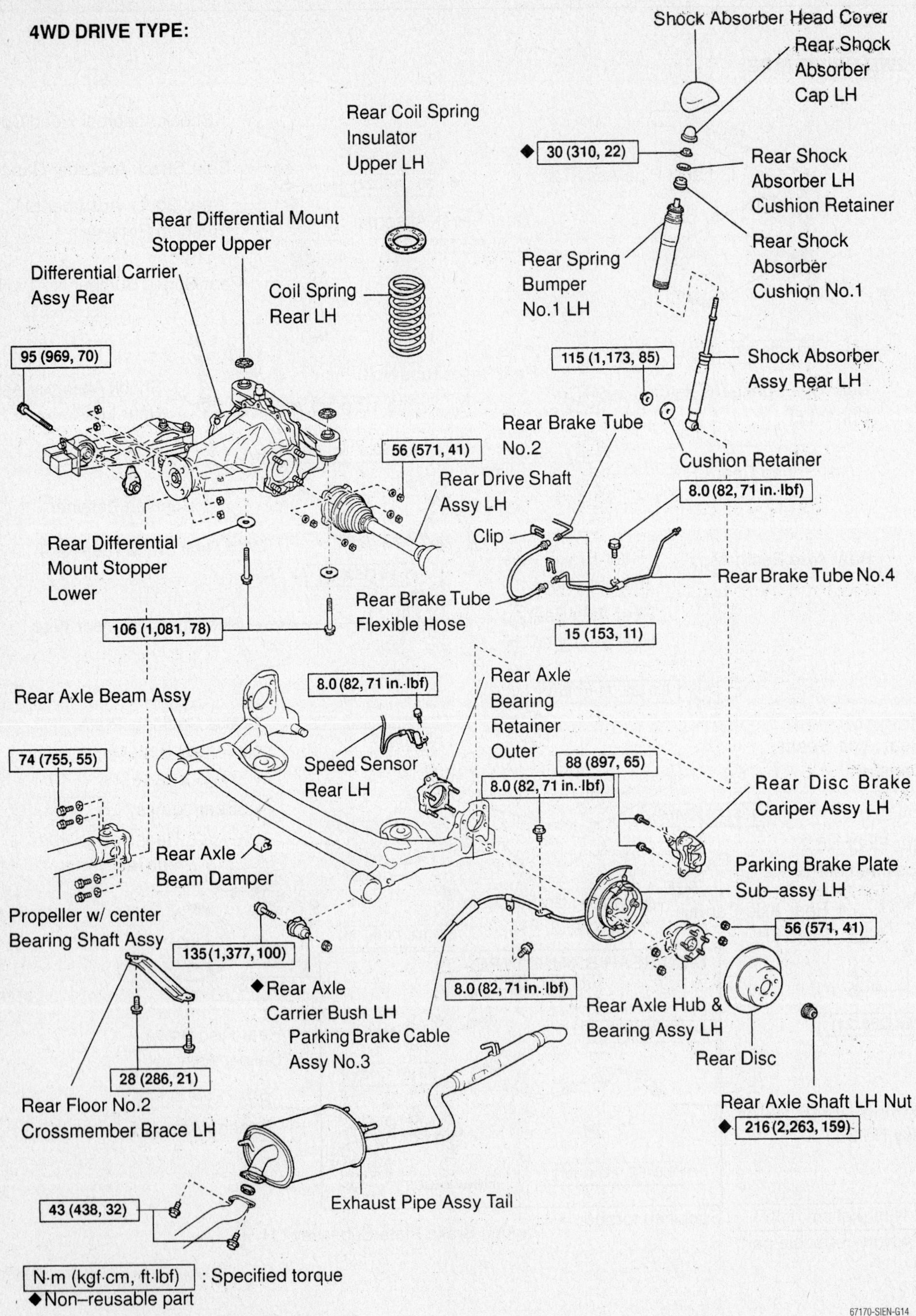
4WD DRIVE TYPE:
Shock Absorber Head Cover
Rear Shock Absorber Cap LH
Rear Coil Spring Insulator Upper LH
◆ 30 (310, 22)
Rear Shock Absorber LH Cushion Retainer
Rear Differential Mount Stopper Upper
Differential Carrier Assy Rear
Coil Spring Rear LH
Rear Spring Bumper No.1 LH
Rear Shock Absorber Cushion No.1
95 (969, 70)
115 (1,173, 85)
Shock Absorber Assy Rear LH
Rear Brake Tube No.2
56 (571, 41)
Cushion Retainer
Rear Drive Shaft Assy LH
8.0 (82, 71 in.·lbf)
Clip
Rear Differential Mount Stopper Lower
Rear Brake Tube No.4
Rear Brake Tube Flexible Hose
106 (1,081, 78)
15 (153, 11)
Rear Axle Bearing Retainer Outer
8.0 (82, 71 in.·lbf)
Rear Axle Beam Assy
Speed Sensor Rear LH
74 (755, 55)
88 (897, 65)
8.0 (82, 71 in.·lbf)
Rear Disc Brake Cariper Assy LH
Rear Axle Beam Damper
Parking Brake Plate Sub–assy LH
Propeller w/ center Bearing Shaft Assy
56 (571, 41)
135 (1,377, 100)
◆Rear Axle Carrier Bush LH
8.0 (82, 71 in.·lbf)
Rear Axle Hub & Bearing Assy LH
Parking Brake Cable Assy No.3
Rear Disc
28 (286, 21)
Rear Floor No.2 Crossmember Brace LH
Rear Axle Shaft LH Nut
◆ 216 (2,263, 159)
43 (438, 32)
Exhaust Pipe Assy Tail
N·m (kgf·cm, ft·lbf) : Specified torque
◆Non–reusable part
67170-SIEN-G14

Rear suspension components—2004 4wd models

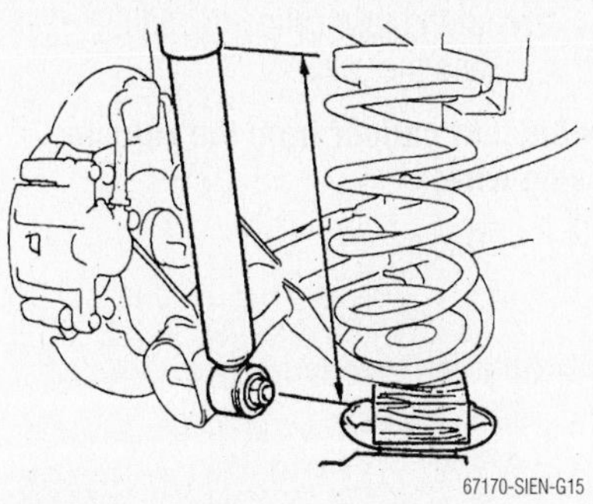

Measuring the shock absorber—2004 models

7. Installation is the reverse of removal. Torque the upper end nut to 22 ft. lbs. Install the lower end nut loosely. Raise the axle to load the shock. For 2wd models, the shock absorber length should be 9.22 in. (234mm); for 4wd, it should be 10.16 in. (258mm), then tighten the lower end nut to 85 ft. lbs. (115 Nm). If you can't reach the nut in this position, support the rear axle and place 198 lbs. (90 kg) in the trunk.

Coil Spring

REMOVAL & INSTALLATION

Front

1. Before servicing the vehicle, refer to the precautions in the beginning of this section.
2. Remove or disconnect the following:
   - Wheel

➡If equipped, be careful not to damage the oil seal, driveshaft boot and/or speed sensor rotor when removing the steering knuckle.

   - Shock absorber (strut assembly)
3. Install a nut/bolt to the bracket at the lower portion of the strut assembly and secure it in a vise.
4. Compress the coil spring with a spring compressor.

⁂ CAUTION

The proper tools must be used for this procedure. The spring on the strut is under high pressure and can cause serious injury if not properly removed and installed.

5. Remove or disconnect the following:
   - Center retaining nut, by holding the spring seat
   - Support, dust seal, spring seat, insulator and spring from the strut assembly

To install:

6. Install the spring bumper and lower insulator to the strut assembly.
7. Compress the coil spring and fit the lower end of the spring into the spring seat gap.
8. Install or connect the following:
   - Upper insulator, spring seat, dust seal, support and spring seat. Tighten the new retaining nut to 34 ft. lbs. (47 Nm) for 2001–03 models; 36 ft. lbs. (49 Nm) for 2004 models.
9. Rotate the spring seat so that the OUT mark of the spring seat faces the outside of the vehicle.
   - Strut
   - Wheel
10. If required, bleed the brake system and check for leaks.
11. Check and/or adjust the front wheel alignment.

Rear

1. Before servicing the vehicle, refer to the precautions in the beginning of this section.
2. Remove or disconnect the following:
   - Shock absorbers
   - Coil springs

To install:

3. Install or connect the following:
   - Coil springs
   - Raise the axle beam enough to apply tension on the springs
   - Shock absorbers

Lower Ball Joint

REMOVAL & INSTALLATION

2001—03

1. Before servicing the vehicle, refer to the precautions in the beginning of this section.
2. Remove or disconnect the following:
   - Wheel
   - Steering knuckle with the axle hub
   - Dust deflector, by prying it from the knuckle
   - Cotter pin and nut from the ball joint
   - Ball joint from the steering knuckle, by removing the 2 bolts
   - Lower ball joint, using a Ball Joint Separator tool 09628-62011

To install:

3. Install or connect the following:
   - Lower ball joint. Tighten the nut to 76 ft. lbs. (103 Nm) and both bolts to 94 ft. lbs. (127 Nm).
   - New cotter pin
   - Wheel

2004

1. Remove the wheel.
2. Remove the hub nut.
3. Remove the speed sensor.
4. Remove the caliper, and hang it out of the way.
5. Remove the rotor.
6. Remove the lower arm from the ball joint.
7. Remove the lower ball joint nut and cotter pin.
8. Using a puller, remove the ball joint from the knuckle.
9. Installation is the reverse of removal. Torque the ball joint stud nut to 91 ft. lbs. (123 Nm). Torque the arm-to-ball joint to 94 ft. lbs. (127 Nm).Torque the hub nut to 217 ft. lbs. (294 Nm) and stake it.

Wheel Bearings

REMOVAL & INSTALLATION

Front

1. Before servicing the vehicle, refer to the precautions in the beginning of this section.
2. Remove or disconnect the following:
   - Front wheels
   - Fender apron seal
3. Check the bearing backlash and axle hub deviation, as follows:
   a. Remove the 2 brake caliper set bolts.
   b. Hang the caliper using stiff wire on the shock absorber assembly.
   c. Remove the rotor.
   d. Place a dial indicator near the center of the axle hub and check the backlash in the bearing shaft direction.
   e. Backlash maximum should read 0.0020 inch (0.05mm). If greater than specified, replace the bearing.
   f. Using the dial indicator, check the deviation at the surface of the axle hub outside and hub bolt. Maximum is 0.0020 inch (0.05mm). If greater than specified, replace the axle hub.
4. Install the rotor and caliper assembly.
5. Remove or disconnect the following:
   - Cotter pin (discard it) and lockcap off the center hub nut
   - Driveshaft locknut, by applying the front brakes
   - Tie rod end, from the steering knuckle

- Left and right stabilizer end brackets, from the lower arms
- Both nuts and the lower arm from the ball joint
- Driveshaft from the axle hub. Secure the shaft aside using wire.

**** WARNING**

Be careful not to damage the shaft boot or Anti-lock Brake System (ABS) sensor rotor.

- Both brake caliper mounting bolts and the caliper.

➡Support caliper from the vehicle using wire.

- Brake rotor

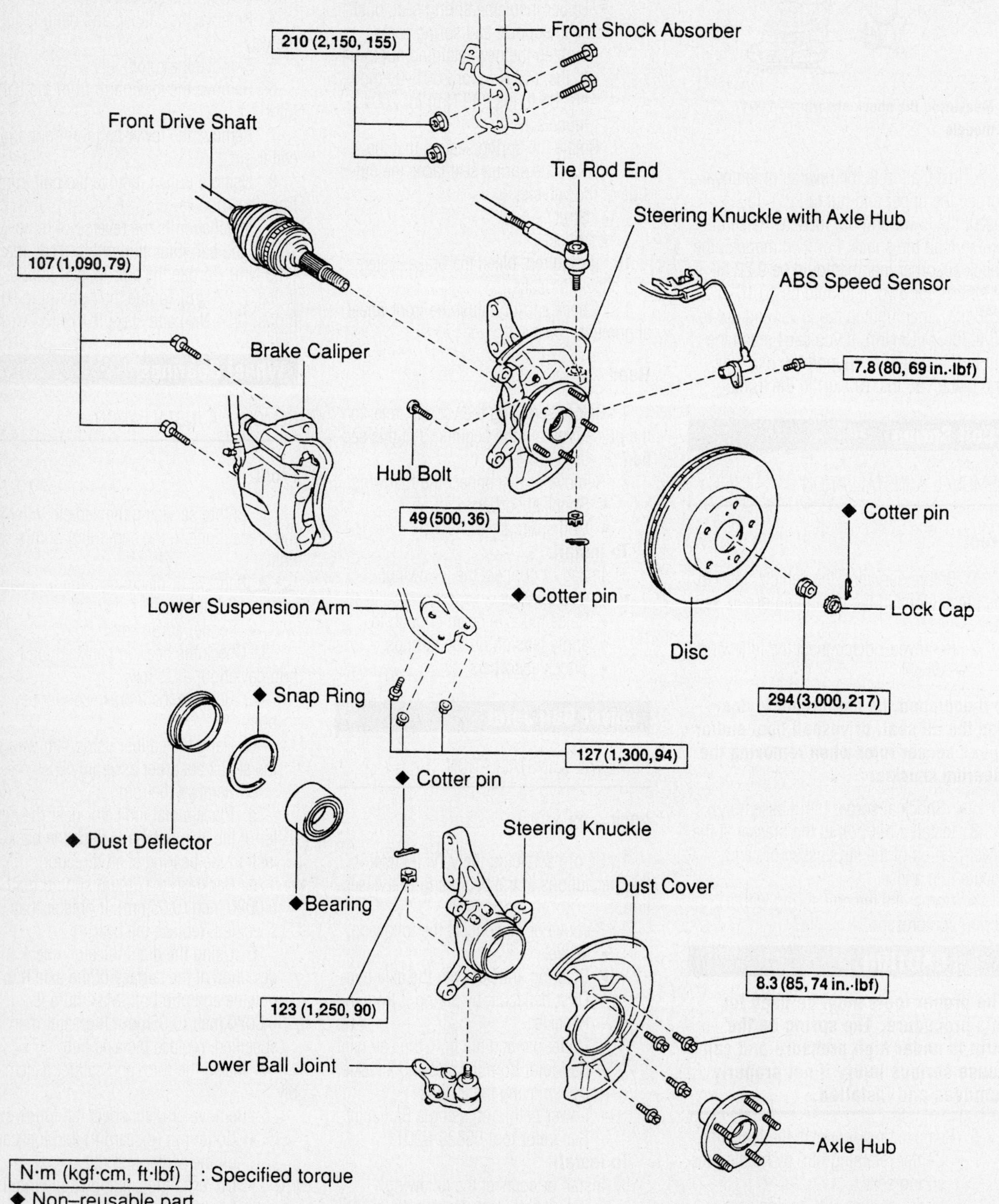

Exploded view of the front hub, bearing and steering knuckle assembly—2001–03 models

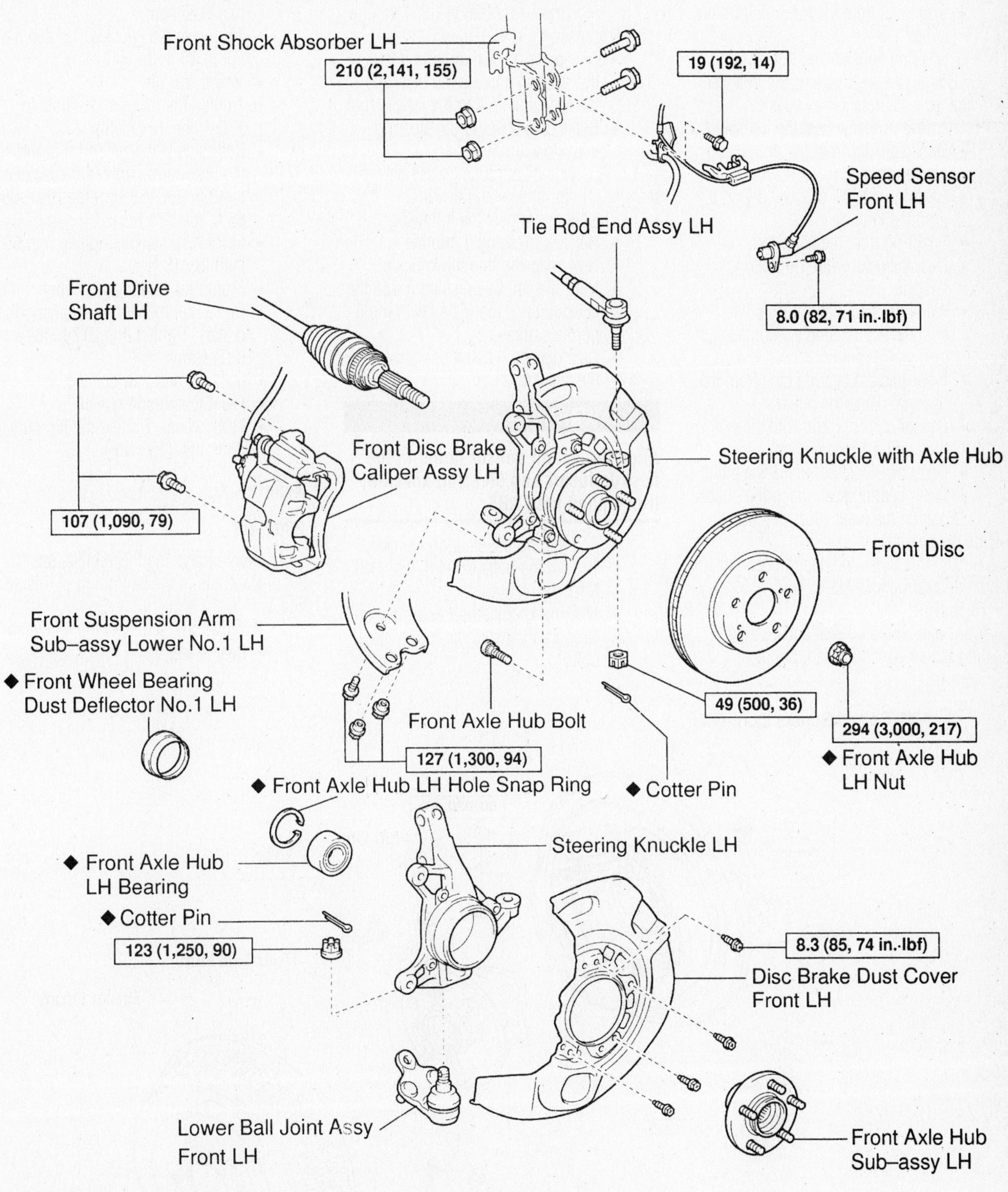

Exploded view of the front hub, bearing and steering knuckle assembly—2004 models

- Sensor from the steering knuckle, if equipped with ABS
- Both nuts from the lower end of the shock
- Steering knuckle and hub assembly

6. Clamp the steering knuckle in a vise with soft jaws to protect the knuckle.
7. Remove or disconnect the following:
   - Dust deflector from the hub, using a screwdriver
   - Bearing inner oil seal, by prying it from the knuckle
   - Snapring from the knuckle bore
   - Dust deflector from the steering knuckle
   - Axle hub, by pulling it from the dust deflector, using a 2-armed mechanical puller
   - Inner (inside) bearing race from the bearing, using the puller
   - Sensor control rotor from the axle hub, using Torx® wrench
   - Outer bearing race, using the puller
   - Outer bearing seal, using the puller
8. Position the inner (outside) race inside the bearing.
9. Using a brass rod, tap the bearing from the steering knuckle.

To install:

10. Clean all the oil seal and bearing seating surfaces with a clean, dry rag.
11. Install or connect the following:
    - Bearing into the bore, using a Bearing Driver tool 09608-32010 and a press
    - New outer oil seal, driving it into the steering knuckle, by inserting the seal side lip into the factory tool
    - Brake disc cover to the steering knuckle with the bolts
12. Apply multi-purpose grease between the oil seal lip, oil seal and bearing.
13. Install or connect the following:
    - Hub, by pressing it into the knuckle
    - New snapring into the knuckle
    - New oil seal, by pressing it into the knuckle once lubricated with multi-purpose grease
    - Dust deflector, by pressing it into the knuckle.

⁂ WARNING

Align the speed sensor holes in the dust deflector and steering knuckle, if equipped with ABS.

- Ball joint to the steering knuckle. Tighten the bolts to 94 ft. lbs. (127 Nm).
- Steering knuckle/hub assembly and temporarily install the lower shock bolts
- Lower ball joint to the lower arm. Tighten the bolt and nuts to 94 ft. lbs. (127 Nm).
- Tie rod to the knuckle. Tighten the nut to 36 ft. lbs. (49 Nm).
- New cotter pin
- Tighten the lower shock nuts to 156 ft. lbs. (211 Nm).
- Both side stabilizer end brackets to the lower arm. Tighten the fasteners to 43 ft. lbs. (58 Nm) for 2001–03; 55 ft. lbs. (74 Nm).
- Front ABS sensor. Tighten it to 69 inch lbs. (8 Nm).
- Front brake rotor and caliper.
- Driveshaft locknut, by applying the brakes. Tighten it to 217 ft. lbs. (294 Nm).
- Lockcap and new cotter pin
- Front fender apron seal
- Front wheel. Tighten the lug nuts to 76 ft. lbs. (103 Nm).

Rear

2001–03

1. Before servicing the vehicle, refer to the precautions in the beginning of this section.
2. Remove or disconnect the following:
   - Rear wheel

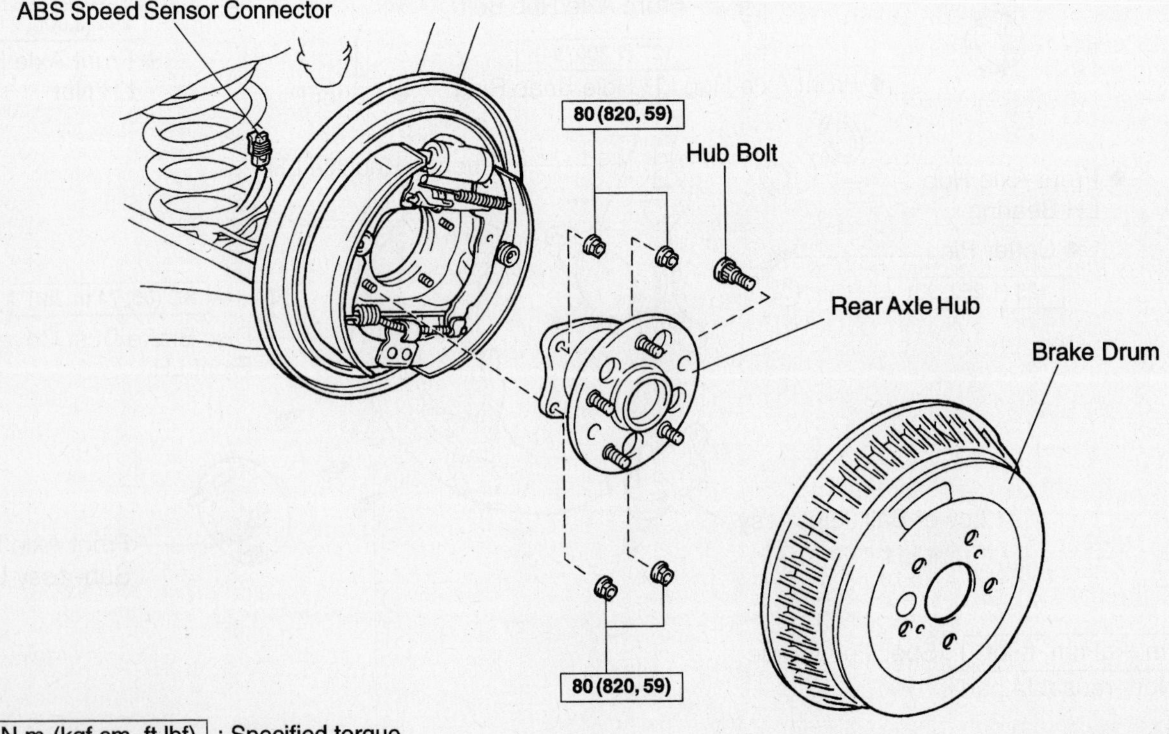

Rear hub assembly—2001–03 models

- Brake drum
- Anti-lock Brake System (ABS) speed sensor connector
- 4 rear axle hub assembly nuts
- Hub assembly

To install:

3. Install or connect the following:
   - New hub assembly. Tighten the nuts to 59 ft. lbs. (80 Nm).
   - ABS speed sensor
4. Install or connect the following:
   - Brake drum
   - Wheels
5. Test drive the vehicle.

2004 W/2WD

1. Remove the wheel.
2. Remove the brake drum (if equipped) or caliper and rotor. Hang the caliper out of the way.
   a. Place a dial indicator near the cen-

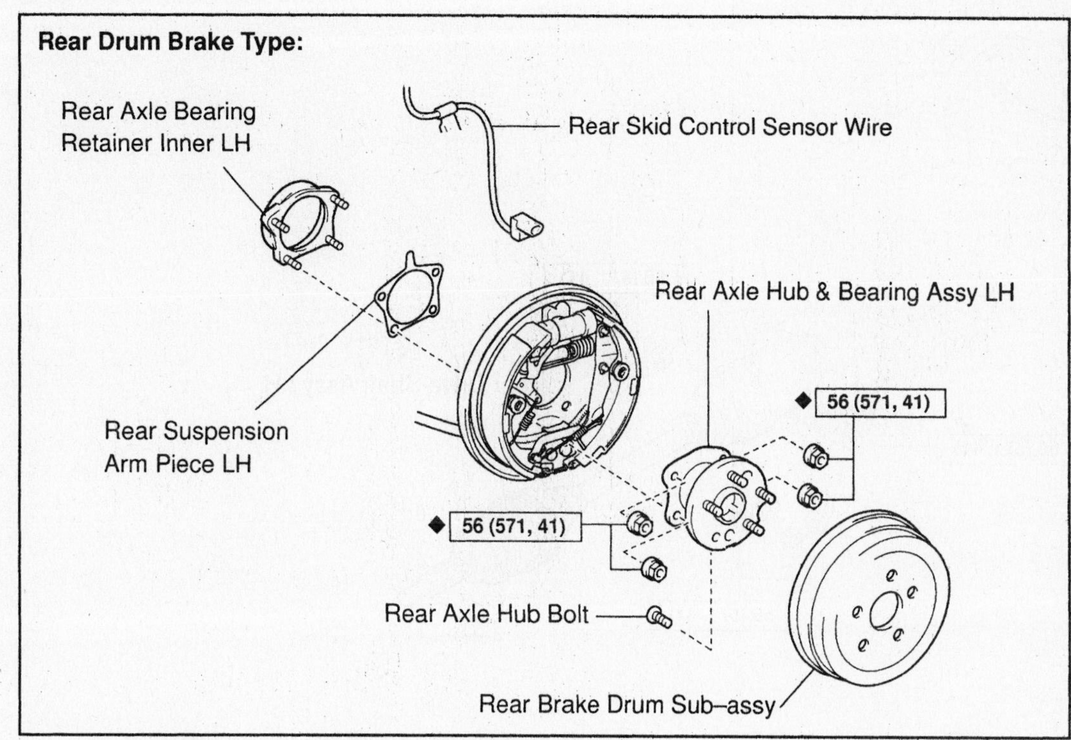

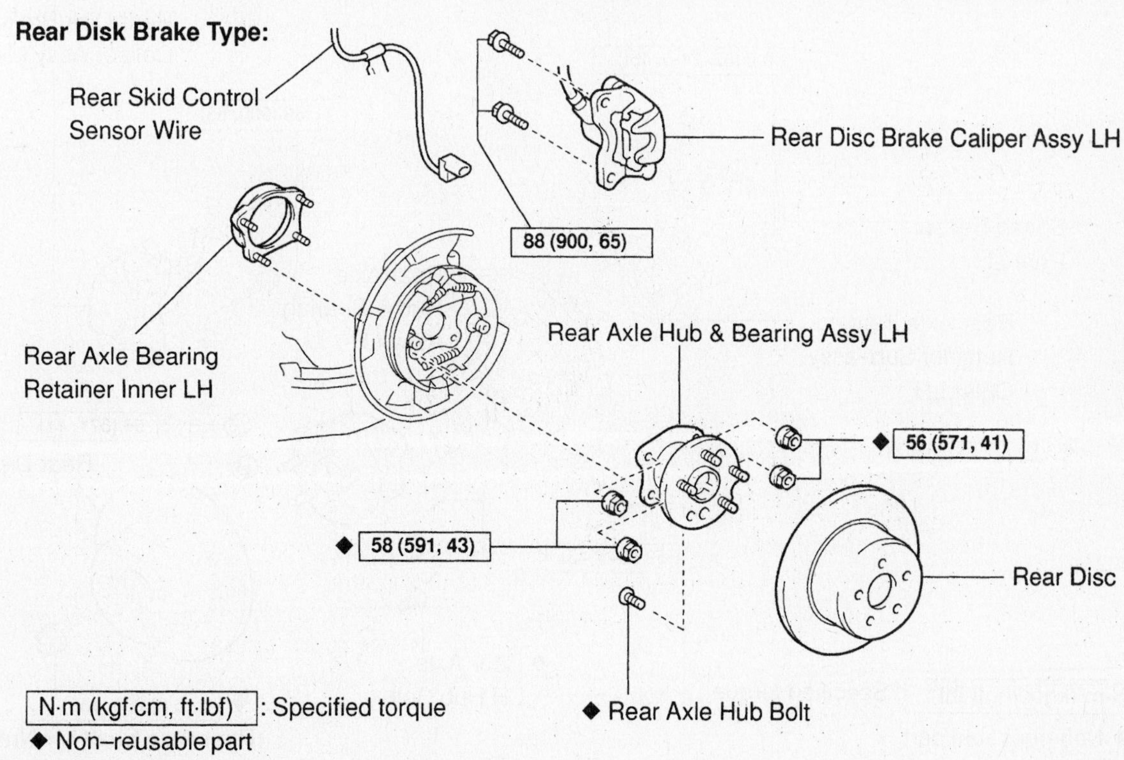

Rear hub assembly—2004 models with 2wd

ter of the axle hub and check the backlash in the bearing shaft direction.

b. Backlash maximum should read 0.0020 inch (0.05mm). If greater than specified, replace the bearing.

c. Using the dial indicator, check the deviation at the surface of the axle hub outside and hub bolt. Maximum is 0.0020 inch (0.05mm). If greater than specified, replace the axle hub.

3. Remove the ABS sensor wire.
4. Remove the 4 bolts and the hub/bearing assembly.
5. Installation is the reverse of removal. Torque the hub bolts to 41 ft. lbs. (56 Nm).

2004 W/4WD

1. Remove the wheel.
2. Unstake and remove the axle shaft nut.

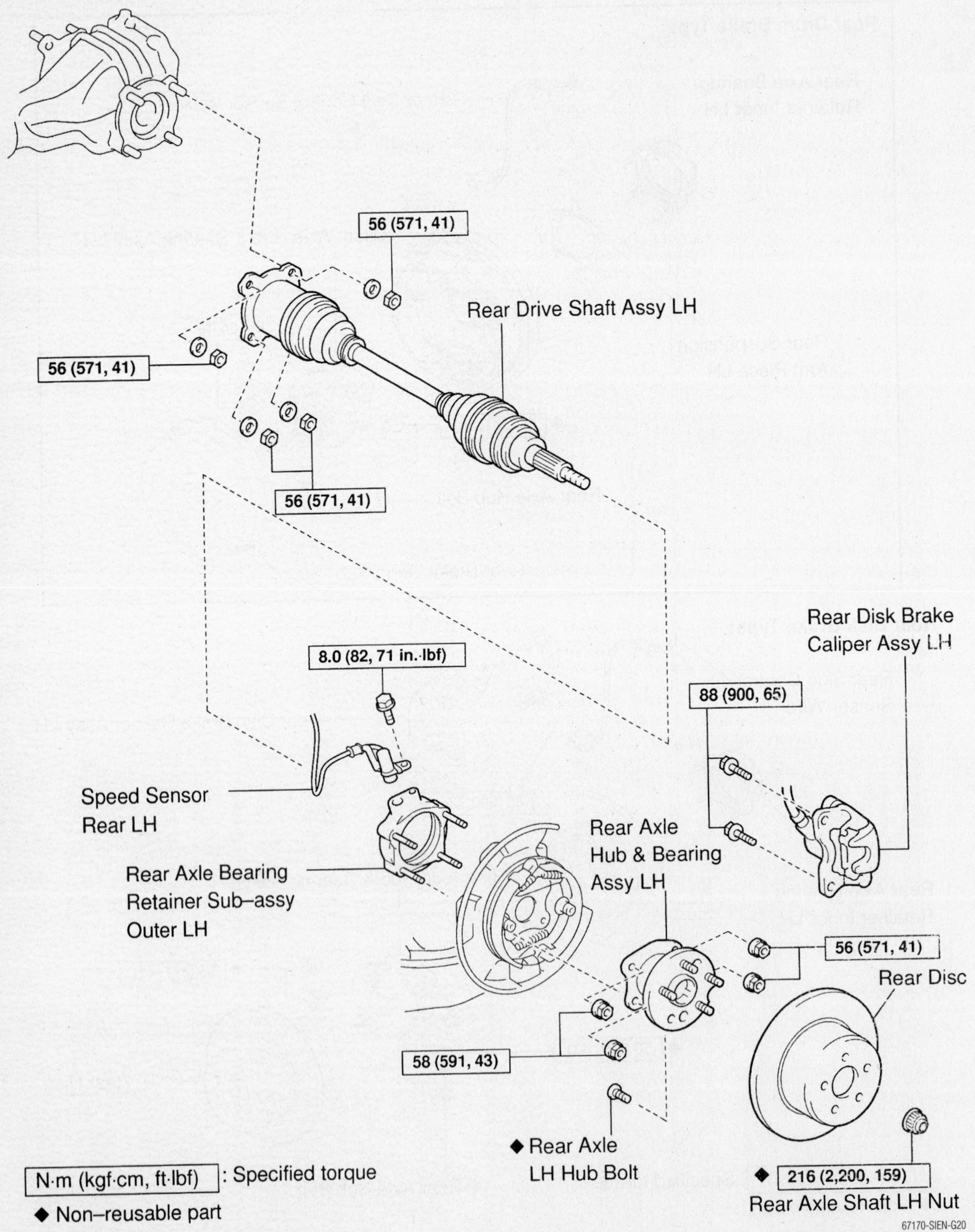

Rear hub assembly—2004 models with 4wd

3. Remove the caliper and rotor. Hang the caliper out of the way.

a. Place a dial indicator near the center of the axle hub and check the backlash in the bearing shaft direction.

b. Backlash maximum should read 0.0020 inch (0.05mm). If greater than specified, replace the bearing.

c. Using the dial indicator, check the deviation at the surface of the axle hub outside and hub bolt. Maximum is 0.0020 inch (0.05mm). If greater than specified, replace the axle hub.

4. Remove the halfshaft.

5. Remove the ABS sensor wire.

6. Remove the 4 bolts and the hub/bearing assembly.

7. Installation is the reverse of removal. Torque the hub bolts to 41 ft. lbs. (56 Nm).

BRAKES

Front Brake Caliper

REMOVAL & INSTALLATION

2001–03

1. Disconnect the negative battery cable from the battery.

2. Raise and support the vehicle safely.

3. Remove the wheels.

4. Disconnect the brake hose from the caliper by removing the union bolt and 2 gaskets. Plug the end of the hose to prevent loss of fluid.

5. Remove the bolts that attach the caliper to the torque plate.

6. Lift the bottom of the caliper up and remove the caliper assembly.

To install:

7. Grease the caliper slides and bolts with lithium grease or equivalent. Install the caliper and secure with the bolts. Torque the bolts to 25 ft. lbs. (34 Nm).

8. Reconnect the brake hose to the

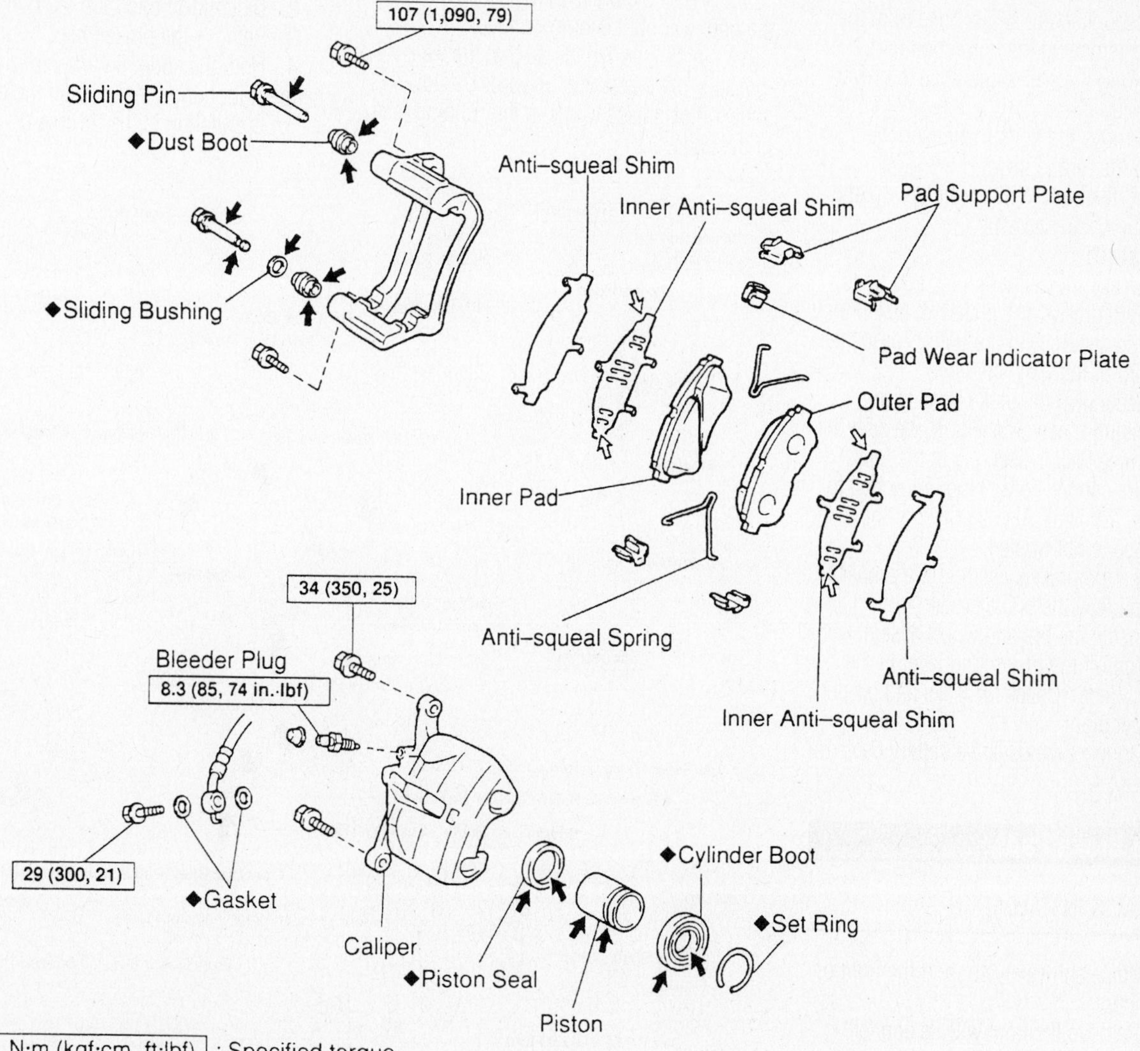

Exploded view of the front disc brake caliper assembly—2001–03 models

caliper, using 2 new washers. Make sure the flexible hose lock is securely in the lock hole of the caliper. Torque the union bolt to 21 ft. lbs. (29 Nm). Also, verify that the brake hose is not twisted.

9. Fill the brake system to the proper level and bleed the brake system.

10. Install the tire and wheel assembly.

11. Top off the brake fluid level in the master cylinder. Check for leaks and proper brake operation.

12. Connect the negative battery cable to the battery.

2004

1. Disconnect the negative battery cable from the battery.
2. Raise and support the vehicle safely.
3. Remove the wheels.
4. Disconnect the brake hose from the caliper by removing the union bolt and 2 gaskets. Plug the end of the hose to prevent loss of fluid.
5. Remove the bolts that attach the caliper to the torque plate.
6. Lift the bottom of the caliper up and remove the caliper assembly.

To install:

7. Grease the caliper slides and bolts with lithium grease or equivalent. Install the caliper and secure with the bolts. Torque the bolts to 25 ft. lbs. (34 Nm).
8. Reconnect the brake hose to the caliper, using 2 new washers. Make sure the flexible hose lock is securely in the lock hole of the caliper. Torque the union bolt to 21 ft. lbs. (29 Nm). Also, verify that the brake hose is not twisted.
9. Fill the brake system to the proper level and bleed the brake system.
10. Install the tire and wheel assembly.
11. Top off the brake fluid level in the master cylinder. Check for leaks and proper brake operation.
12. Connect the negative battery cable to the battery.

Front Brake Pads

REMOVAL & INSTALLATION

1. Raise and safely support the front of the vehicle.
2. Remove the front wheels and temporarily fasten the rotor disc with the hub nuts.
3. Hold the sliding pin on the bottom of the caliper and loosen the installation bolt.
4. Remove the lower installation bolt.
5. Lift up the caliper and suspend it securely. Do not remove the upper installation bolt.
6. Remove the following parts:
   - The 2 anti-squeal springs.
   - The 2 brake pads.
   - The 4 anti-squeal shims.
   - The 4 pad support plates.

To install:

7. Install the pad support plates.
8. Install a pad wear indicator plate to the pad. Install the anti-squeal shims and support plates to each pad.

➡It recommended that a suitable anti-squeal compound be applied to both sides of the inner anti-squeal shim.

9. Draw out a small amount of brake fluid from the brake reservoir. Press in the caliper piston with a suitable tool.
10. Press the brake piston in carefully so the boot will not become wedged.
11. Install the 2 pads so that the wear indicator plate is facing upward. Do not allow oil or grease to get in the rubbing face of the pads.
12. Lower and install the caliper. Torque the sliding main pin to 25 ft. lbs. (34 Nm).

➡When installing the sliding main pin, be careful that the plug installed in the torque plate does not come loose.

13. Install the front wheels and lower the vehicle.
14. Check the fluid level in the master cylinder and add as necessary. Be sure to pump the brake pedal a few times before road-testing the vehicle.

Rear Brake Caliper

REMOVAL & INSTALLATION

1. Remove the wheel.
2. Disconnect and plug the brake line.
3. Remove the brake hose.
4. Hold the slide pin and remove the 2 caliper mounting bolts. Lift off the caliper.
5. Installation is the reverse of removal.

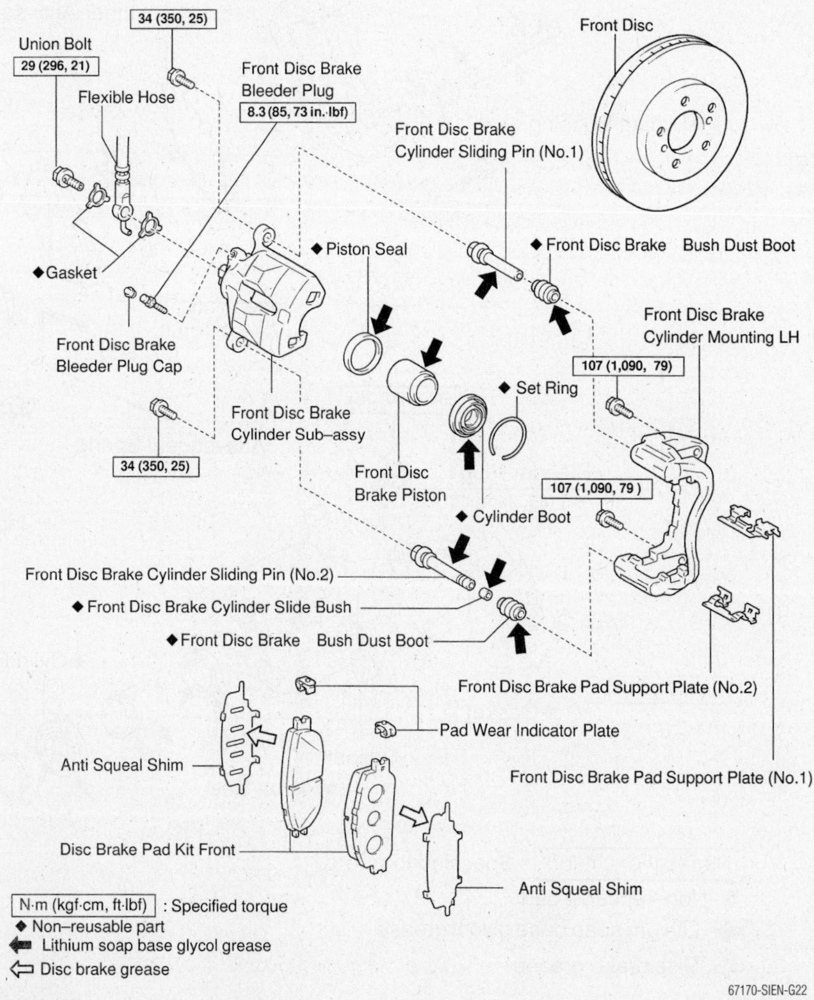

Front disc brake components—2004

Refill the system and bleed the brakes. Torque the mounting bolts to 25 ft. lbs. (34 Nm). Torque the brake hose-to-caliper to 17 ft. lbs. (23 Nm). Torque the steel brake line to 11 ft. lbs. (15 Nm).

Rear Brake Pads

REMOVAL & INSTALLATION

1. Raise and safely support the rear of the vehicle.
2. Remove the rear wheels and temporarily fasten the rotor disc with the hub nuts.
3. Hold the sliding pin on the bottom of the caliper and loosen the installation bolt.
4. Remove the lower installation bolt.
5. Lift up the caliper and suspend it securely. Do not remove the upper installation bolt.
6. Remove the following parts:
   - The 2 anti-squeal springs.
   - The 2 brake pads.
   - The 4 anti-squeal shims.
   - The 4 pad support plates.

To install:

7. Install the pad support plates.
8. Install a pad wear indicator plate to the pad. Install the anti-squeal shims and support plates to each pad.

➡It recommended that a suitable anti-squeal compound (available at your local parts house) be applied to both sides of the inner anti-squeal shim.

9. Draw out a small amount of brake fluid from the brake reservoir. Press in the caliper piston with a suitable tool.
10. Press the brake piston in carefully so the boot will not become wedged.
11. Install the 2 pads so that the wear indicator plate is facing upward. Do not allow oil or grease to get in the rubbing face of the pads.
12. Lower and install the caliper. Torque the sliding main pin to 25 ft. lbs. (34 Nm).

➡When installing the sliding main pin, be careful that the plug installed in the torque plate does not come loose.

13. Install the rear wheels and lower the vehicle.
14. Check the fluid level in the master cylinder and add as necessary. Be sure to pump the brake pedal a few times before road-testing the vehicle.

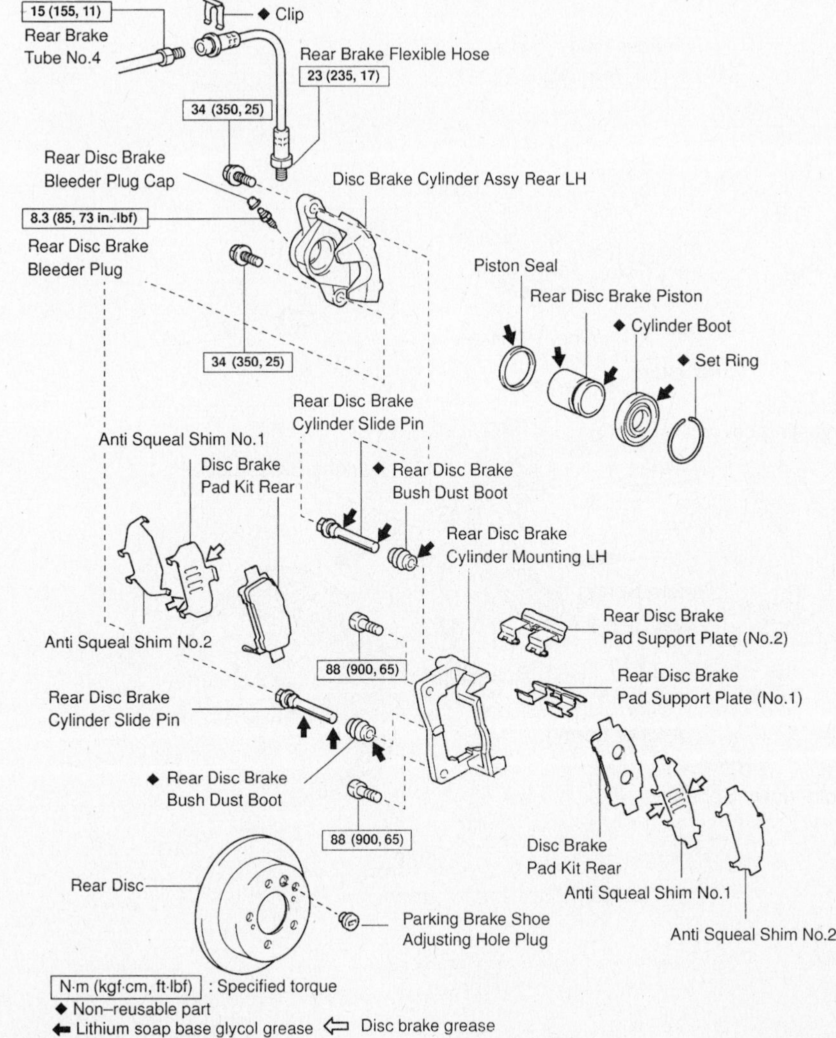

Rear disc brake components

Brake Shoes

REMOVAL & INSTALLATION

1. Disconnect the negative battery cable from the battery.
2. Loosen the rear wheel lug nuts slightly. Release the parking brake.
3. Block the front wheels, raise the rear of the vehicle, and safely support it with jackstands.
4. Remove the wheel lug nuts and the wheel.
5. Remove the brake drum retaining screws, if equipped. Remove the brake drum.
6. If the drum is difficult to remove, perform the following:
   a. Insert the end of a bent wire (a coat hanger will do nicely) through the hole in the brake drum and hold the automatic adjusting lever away from the adjuster.
   b. Reduce the brake shoe adjustment by turning the adjuster bolt with a brake tool.
   c. The drum should now be loose enough to remove without much effort.
7. Carefully unhook the return spring from the leading (front) brake shoe.
8. Press the hold down spring retainer in and turn the pin on the front brake shoe.
9. Remove the hold down spring, retainers and the pin for the front brake shoe.
10. Pull out the brake shoe and unhook the anchor spring from the lower edge.
11. Remove the hold down spring from the trailing (rear) shoe. Pull the shoe out with the adjuster, automatic adjuster assembly and springs attached. Disconnect the parking brake cable. Remove the tension/return and anchor springs from the rear shoe.
12. Unhook the adjusting lever spring from the rear shoe and then remove the automatic adjuster assembly.

To install:

13. Inspect the shoes for signs of unusual wear or scoring.

14. Check the wheel cylinder for any sign of fluid seepage or frozen pistons.

15. Clean and inspect the brake backing plate and all other components. Check that the brake drum inner diameter is within specified limits. Lubricate the backing plate at the positions the brakes come in contact with the backing plate. Also lubricate the anchor plate.

16. Mount the automatic adjuster assembly onto a new rear brake shoe.

17. Connect the parking brake cable to the rear shoe and then install the automatic adjusting lever, spring and E-ring. Position the rear shoe so the lower end rides in the anchor plate and the upper end is against the boot of the wheel cylinder.

18. Install the pin and the hold down spring. Press the retainer down over the pin and rotate the pin so the crimped edge is held by the retainer.

19. Place the front brake into position and install the anchor spring between the front and rear shoes. Stretch the spring enough so the front shoe will fit as the rear did. Install the hold down spring, pin and retainer to the front brake shoe.

20. Connect the return spring to the front brake shoe.

21. Check the operation of the automatic adjuster mechanism:

a. Apply the parking brake lever and verifying the adjusting bolt turns.

b. Adjust the strut to where it is the shortest possible length.

c. Install the brake drum.

d. Apply the parking brake lever until the clicking sound can no longer be heard.

22. Check the clearance between the brake shoes and drum:

a. Remove the brake drum.

b. Measure the brake drum inside diameter and diameter of the brake shoes. The difference is "Shoe-to-drum clearance" and should be approximately 0.024 inch (0.6mm). If incorrect, check the parking brake system.

➡A special brake caliper tool is required to gauge the brake drum inside diameter and shoe-to-drum clearance. However it is not required to perform brake shoe adjustment.

23. Install the brake drum.

24. Adjust the brake pedal until a slight drag is felt when the drum is spun by hand.

25. Pull the parking lever all the way up until a clicking sound can no longer be heard. Check the clearance between brake shoes and brake drum.

26. Install the rear wheels, tighten the wheel lug nuts and lower the vehicle.

27. Retighten the wheel lug nuts and pump the brake pedal a few times before moving the vehicle. Adjust the rear brakes again if necessary.

28. Check the level of brake fluid in the master cylinder, then perform a test drive.

29. Connect the negative battery cable to the battery.

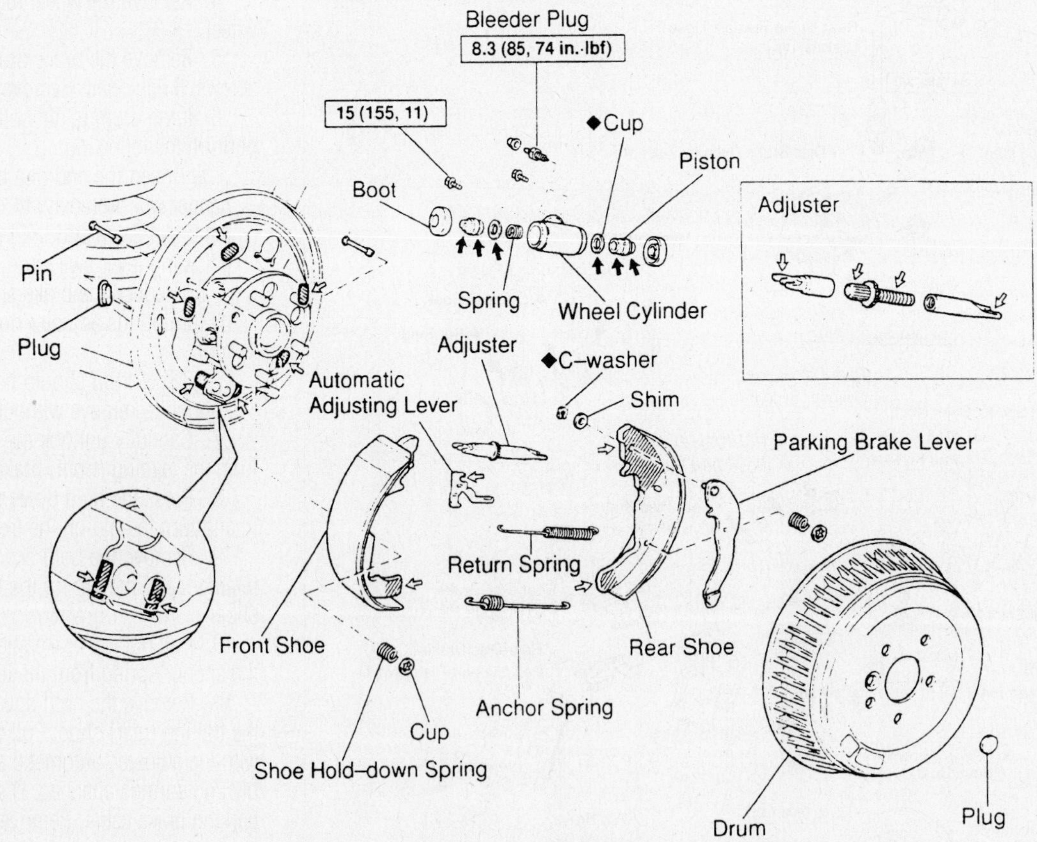

Exploded view of the rear drum brake components—Sienna

TOYOTA

Tacoma

23

SPECIFICATION AND MAINTENANCE CHARTS

ENGINE AND VEHICLE IDENTIFICATION

Engine						
Code ①	Liters (cc)	Cu. In.	Cyl.	Fuel Sys.	Engine Type	Eng. Mfg.
2RZ-FE	2.4 (2438)	149	4	MFI	DOHC	Toyota
3RZ-FE	2.7 (2693)	164	4	MFI	DOHC	Toyota
5VZ-FE	3.4 (3378)	206	6	MFI	DOHC	Toyota

Model Year	
Code ②	Year
1	2001
2	2002
3	2003
4	2004
5	2005

SFI: Sequential Fuel Injection

MFI: Multi-port Fuel Injection

DOHC: Double Overhead Camshaft

① Stamped on the left side of the engine block

② 10th digit of the Vehicle Identification Number (VIN)

67170-TACO-C01

GENERAL ENGINE SPECIFICATIONS

Year	Model	Engine Displacement Liters	Engine Series ID	Net Horsepower @ rpm	Net Torque @ rpm (ft. lbs.)	Bore x Stroke (in.)	Com-pression Ratio	Oil Pressure @ rpm
2001	Tacoma	2.4	2RZ-FE	142@5000	160@4000	3.74x3.38	9.5:1	36-71@3000
		2.7	3RZ-FE	150@4800	177@4000	3.74x3.74	9.5:1	36-71@3000
		3.4	5VZ-FE	190@4800	220@3600	3.68x3.23	9.6:1	NA
2002	Tacoma	2.4	2RZ-FE	142@5000	160@4000	3.74x3.38	9.5:1	36-71@3000
		2.7	3RZ-FE	150@4800	177@4000	3.74x3.74	9.5:1	36-71@3000
		3.4	5VZ-FE	190@4800	220@3600	3.68x3.23	9.6:1	NA
2003	Tacoma	2.4	2RZ-FE	142@5000	160@4000	3.74x3.38	9.5:1	36-71@3000
		2.7	3RZ-FE	150@4800	177@4000	3.74x3.74	9.5:1	36-71@3000
		3.4	5VZ-FE	190@4800	220@3600	3.68x3.23	9.6:1	NA
2004	Tacoma	2.4	2RZ-FE	142@5000	160@4000	3.74x3.38	9.5:1	36-71@3000
		2.7	3RZ-FE	150@4800	177@4000	3.74x3.74	9.5:1	36-71@3000
		3.4	5VZ-FE	190@4800	220@3600	3.68x3.23	9.6:1	NA

NA: Not Available

67170-TACO-C02

ENGINE TUNE-UP SPECIFICATIONS

Year	Engine Displacement Liters	Engine ID	Spark Plug Gap (in.)	Ignition Timing (deg.)*	Fuel Pump (psi)	Idle Speed (rpm) MT	Idle Speed (rpm) AT	Valve Clearance Intake	Valve Clearance Exhaust
2001	2.4	2RZ-FE	0.031	5B	38-44	650-750	—	0.006-0.010	0.010-0.014
	2.7	3RZ-FE	0.031	5B	38-44	650-750	650-750	0.006-0.010	0.010-0.014
	3.4	5VZ-FE	0.031	10B	38-44	650-750	650-750	0.006-0.010	0.010-0.014
2002	2.4	2RZ-FE	0.043	5B	38-44	650-750	—	0.006-0.010	0.010-0.014
	2.7	3RZ-FE	0.043	5B	38-44	650-750	650-750	0.006-0.010	0.010-0.014
	3.4	5VZ-FE	0.043	10B	38-44	650-750	650-750	0.006-0.010	0.010-0.014
2003	2.4	2RZ-FE	0.043	5B	38-44	650-750	—	0.006-0.010	0.010-0.014
	2.7	3RZ-FE	0.043	5B	38-44	650-750	650-750	0.006-0.010	0.010-0.014
	3.4	5VZ-FE	0.043	10B	38-44	650-750	650-750	0.006-0.010	0.010-0.014
2004	2.4	2RZ-FE	0.043	5B	38-44	650-750	—	0.006-0.010	0.010-0.014
	2.7	3RZ-FE	0.043	5B	38-44	650-750	650-750	0.006-0.010	0.010-0.014
	3.4	5VZ-FE	0.043	10B	38-44	650-750	650-750	0.006-0.010	0.010-0.014

NOTE: The Vehicle Emission Control Information label often reflects specification changes made during production.
The label figures must be used if they differ from those in this chart.

B: Before top dead center

*with Terminals TE1 and E1 connected to DLC1

67170-TACO-C03

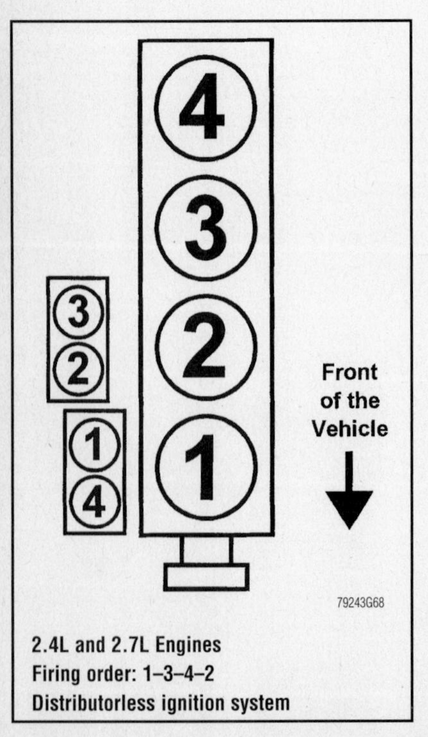

2.4L and 2.7L Engines
Firing order: 1–3–4–2
Distributorless ignition system

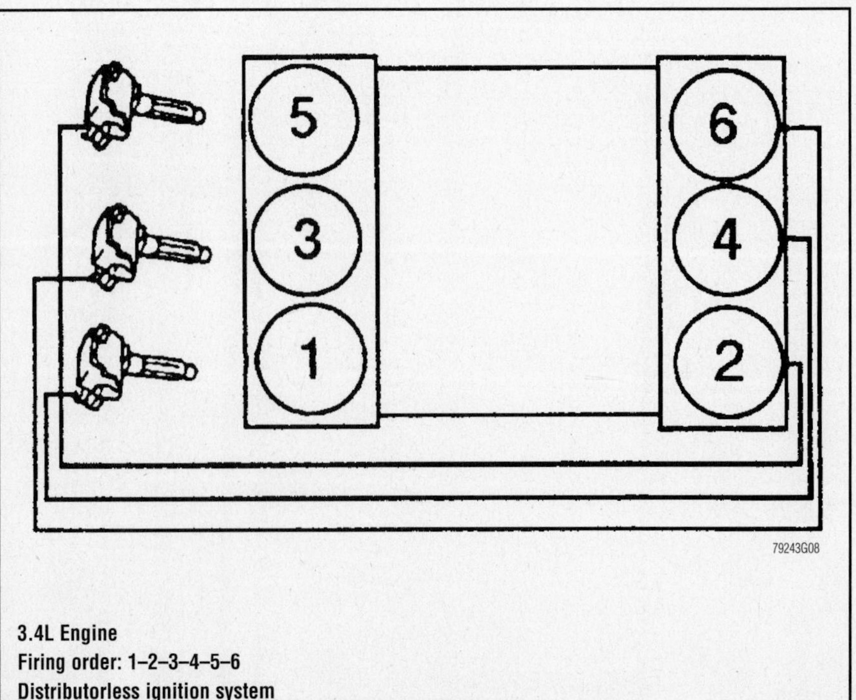

3.4L Engine
Firing order: 1–2–3–4–5–6
Distributorless ignition system

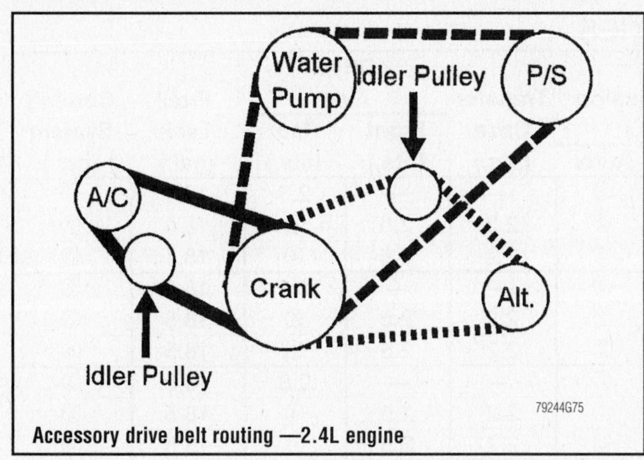

Accessory drive belt routing —2.4L engine

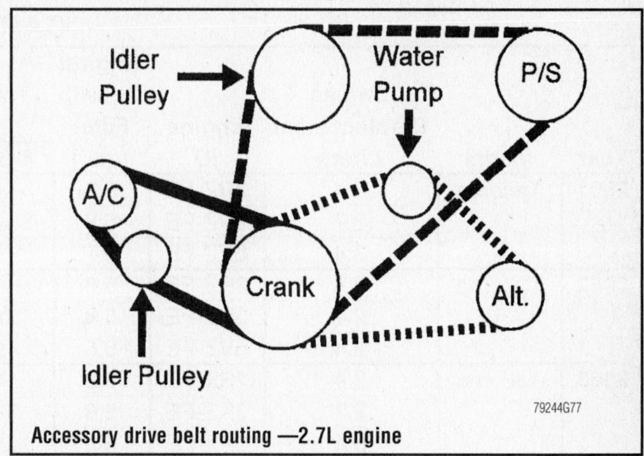

Accessory drive belt routing —2.7L engine

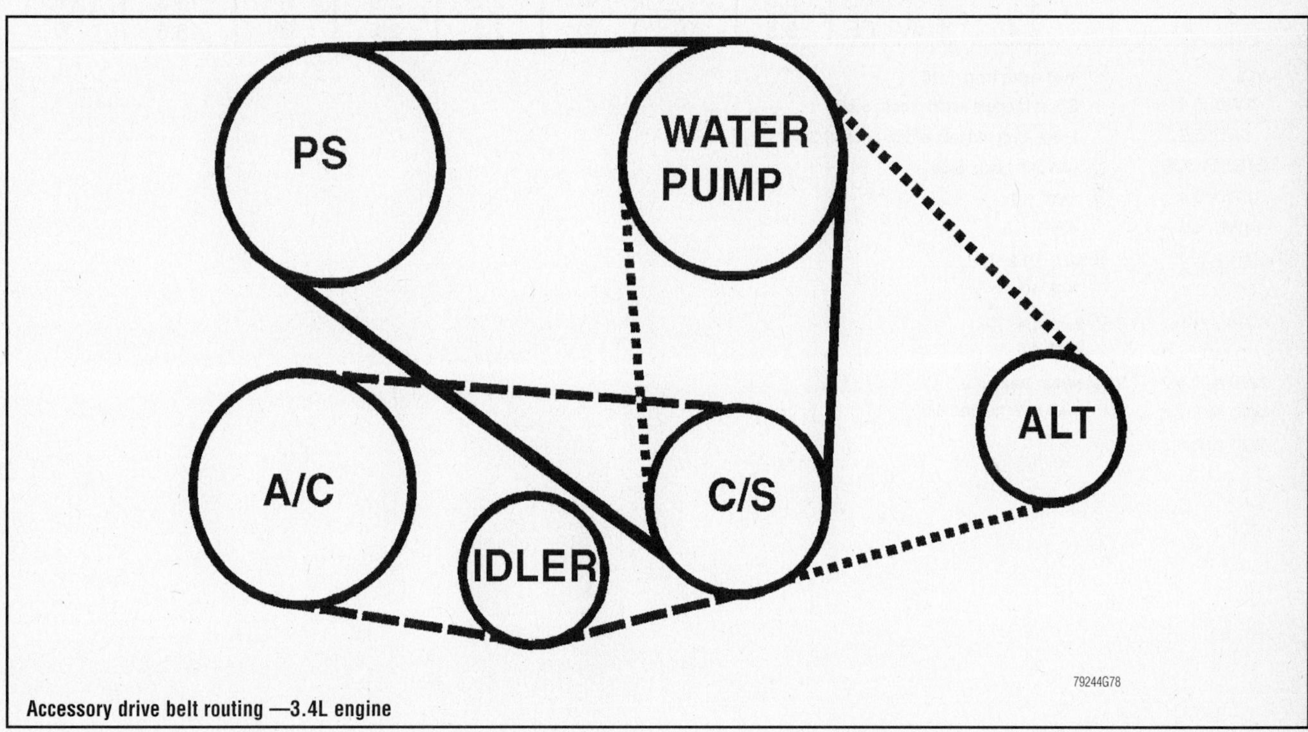

Accessory drive belt routing —3.4L engine

CAPACITIES

Year	Model	Engine Displacement Liters	Engine ID	Engine Oil with Filter (qts.)	Transmission (pts.) 5-Spd	Transmission (pts.) Auto.	Transfer Case (pts.)	Drive Axle Front (pts.)	Drive Axle Rear (pts.)	Fuel Tank (gal.)	Cooling System (qts.)
2001	Tacoma	2.4	2RZ-FE	5.8	①	②	—	—	2.8	15.1	③
		2.7	3RZ-FE	5.8	①	②	2.2	2.5	④	18.0	③
		3.4	5VZ-FE	⑥	①	②	2.2	2.5	④	18.1	⑥
2002	Tacoma	2.4	2RZ-FE	5.8	⑦	②	—	—	2.8	18.5	③
		2.7	3RZ-FE	5.8	⑦	②	2.2	2.5	④	18.5	③
		3.4	5VZ-FE	5.5	①	②	2.2	2.5	④	18.5	⑧
2003	Tacoma	2.4	2RZ-FE	5.8	⑦	②	—	—	2.8	18.5	③
		2.7	3RZ-FE	5.8	⑦	②	2.2	2.5	④	18.5	③
		3.4	5VZ-FE	5.5	①	②	2.2	2.5	④	18.5	⑧
2004	Tacoma	2.4	2RZ-FE	5.8	⑦	②	—	—	2.8	18.5	③
		2.7	3RZ-FE	5.8	⑦	②	2.2	2.5	④	18.5	③
		3.4	5VZ-FE	5.5	①	②	2.2	2.5	④	18.5	⑧

① W59:
  2WD: 5.4
  4WD: 5.2
 R150, R150F:
  2WD: 5.4
  4WD: 4.6

② A44D: 5.0
 A340E: 3.4
 A340F: 4.2

③ 2WD M/T: 8.5
 2WD A/T: 8.2
 4WD M/T: 8.8
 4WD A/T: 8.7

④ 4wd extra long: 5.16
 Short models w/diff. Lock: 5.60
 Extra long models w/diff. Lock: 6.36
 w/o diff. Lock: 5.38

⑤ 2WD: 5.7
 4WD: 5.5

⑥ M/T: 10.3
 A/T: 10.0

⑦ 2wd: 5.4
 4wd: 4.6

⑧ w/rear heater: 9.5
 w/o rear heater: 8.5

67170-TACO-C04

VALVE SPECIFICATIONS

Year	Engine Displacement Liters	Engine ID	Seat Angle (deg.)	Face Angle (deg.)	Spring Test Pressure (lbs. @ in.)	Spring Installed Height (in.)	Stem-to-Guide Clearance (in.)		Stem Diameter (in.)	
							Intake	Exhaust	Intake	Exhaust
2001	2.4	2RZ-FE	45	44.5	40.0-46.0@ 1.406	1.406	0.0010-0.0024	0.0012-0.0026	0.2350-0.2356	0.2348-0.2354
	2.7	3RZ-FE	45	44.5	40.0-46.0@ 1.406	1.406	0.0010-0.0024	0.0012-0.0026	0.2350-0.2356	0.2348-0.2354
	3.4	5VZ-FE	45	44.5	41.9-46.3@ 1.311	1.311	0.0010-0.0024	0.0012-0.0026	0.2350-0.2356	0.2348-0.2354
2002	2.4	2RZ-FE	45	44.5	40.0-46.0@ 1.406	1.406	0.0010-0.0024	0.0012-0.0026	0.2350-0.2356	0.2348-0.2354
	2.7	3RZ-FE	45	44.5	40.0-46.0@ 1.406	1.406	0.0010-0.0024	0.0012-0.0026	0.2350-0.2356	0.2348-0.2354
	3.4	5VZ-FE	45	44.5	41.9-46.3@ 1.311	1.311	0.0010-0.0024	0.0012-0.0026	0.2350-0.2356	0.2348-0.2354
2003	2.4	2RZ-FE	45	44.5	40.0-46.0@ 1.406	1.406	0.0010-0.0024	0.0012-0.0026	0.2350-0.2356	0.2348-0.2354
	2.7	3RZ-FE	45	44.5	40.0-46.0@ 1.406	1.406	0.0010-0.0024	0.0012-0.0026	0.2350-0.2356	0.2348-0.2354
	3.4	5VZ-FE	45	44.5	41.9-46.3@ 1.311	1.311	0.0010-0.0024	0.0012-0.0026	0.2350-0.2356	0.2348-0.2354
2004	2.4	2RZ-FE	45	44.5	40.0-46.0@ 1.406	1.406	0.0010-0.0024	0.0012-0.0026	0.2350-0.2356	0.2348-0.2354
	2.7	3RZ-FE	45	44.5	40.0-46.0@ 1.406	1.406	0.0010-0.0024	0.0012-0.0026	0.2350-0.2356	0.2348-0.2354
	3.4	5VZ-FE	45	44.5	41.9-46.3@ 1.311	1.311	0.0010-0.0024	0.0012-0.0026	0.2350-0.2356	0.2348-0.2354

67170-TACO-C05

CRANKSHAFT AND CONNECTING ROD SPECIFICATIONS

All measurements are given in inches.

			Crankshaft				Connecting Rod		
Year	Engine Displacement Liters	Engine ID	Main Brg. Journal Dia.	Main Brg. Oil Clearance	Shaft End-play	Thrust on No.	Journal Diameter	Oil Clearance	Side Clearance
2001	2.4	2RZ-FE	2.3617-2.3622	0.0009-0.0022	0.0008-0.0087	2	2.0861-2.0866	0.0012-0.0022	0.0063-0.0123
	2.7	3RZ-FE	2.2615-2.3620	0.0012-0.0022	0.0008-0.0087	3	2.0861-2.0866	0.0009-0.0022	0.0063-0.0123
	3.4	5VZ-FE	2.5191-2.5197	0.0008-0.0015	0.0008-0.0087	2	2.1648-2.1654	0.0009-0.0021	0.0059-0.0130
2002	2.4	2RZ-FE	2.3617-2.3622	0.0009-0.0022	0.0008-0.0087	2	2.0861-2.0866	0.0012-0.0022	0.0063-0.0123
	2.7	3RZ-FE	2.2615-2.3620	0.0012-0.0022	0.0008-0.0087	3	2.0861-2.0866	0.0009-0.0022	0.0063-0.0123
	3.4	5VZ-FE	2.5191-2.5197	0.0008-0.0015	0.0008-0.0087	2	2.1648-2.1654	0.0009-0.0021	0.0059-0.0130
2003	2.4	2RZ-FE	2.3617-2.3622	0.0009-0.0022	0.0008-0.0087	2	2.0861-2.0866	0.0012-0.0022	0.0063-0.0123
	2.7	3RZ-FE	2.2615-2.3620	0.0012-0.0022	0.0008-0.0087	3	2.0861-2.0866	0.0009-0.0022	0.0063-0.0123
	3.4	5VZ-FE	2.5191-2.5197	0.0008-0.0015	0.0008-0.0087	2	2.1648-2.1654	0.0009-0.0021	0.0059-0.0130
2004	2.4	2RZ-FE	2.3617-2.3622	0.0009-0.0022	0.0008-0.0087	2	2.0861-2.0866	0.0012-0.0022	0.0063-0.0123
	2.7	3RZ-FE	2.2615-2.3620	0.0012-0.0022	0.0008-0.0087	3	2.0861-2.0866	0.0009-0.0022	0.0063-0.0123
	3.4	5VZ-FE	2.5191-2.5197	0.0008-0.0015	0.0008-0.0087	2	2.1648-2.1654	0.0009-0.0021	0.0059-0.0130

67170-TACO-C06

PISTON AND RING SPECIFICATIONS

All measurements are given in inches.

Year	Engine Displacement Liters	Engine ID	Piston Clearance	Ring Gap Top Compression	Ring Gap Bottom Compression	Ring Gap Oil Control	Ring Side Clearance Top Compression	Ring Side Clearance Bottom Compression	Ring Side Clearance Oil Control
2001	2.4	2RZ-FE	0.0012-0.0020	0.0118-0.0169	0.0177-0.0236	0.0051-0.0150	0.0008-0.0028	0.0012-0.0028	SNUG
	2.7	3RZ-FE	0.0019-0.0028	0.0118-0.0157	0.0157-0.0194	0.0051-0.0150	0.0008-0.0028	0.0012-0.0028	SNUG
	3.4	5VZ-FE	0.0053-0.0060	0.0118-0.0197	0.0157-0.0236	0.0059-0.0217	0.0016-0.0031	0.0012-0.0028	SNUG
2002	2.4	2RZ-FE	0.0012-0.0020	0.0118-0.0169	0.0177-0.0236	0.0051-0.0150	0.0008-0.0028	0.0012-0.0028	SNUG
	2.7	3RZ-FE	0.0019-0.0028	0.0118-0.0157	0.0157-0.0194	0.0051-0.0150	0.0008-0.0028	0.0012-0.0028	SNUG
	3.4	5VZ-FE	0.0053-0.0060	0.0118-0.0197	0.0157-0.0236	0.0059-0.0217	0.0016-0.0031	0.0012-0.0028	SNUG
2003	2.4	2RZ-FE	0.0012-0.0020	0.0118-0.0169	0.0177-0.0236	0.0051-0.0150	0.0008-0.0028	0.0012-0.0028	SNUG
	2.7	3RZ-FE	0.0019-0.0028	0.0118-0.0157	0.0157-0.0194	0.0051-0.0150	0.0008-0.0028	0.0012-0.0028	SNUG
	3.4	5VZ-FE	0.0053-0.0060	0.0118-0.0197	0.0157-0.0236	0.0059-0.0217	0.0016-0.0031	0.0012-0.0028	SNUG
2004	2.4	2RZ-FE	0.0012-0.0020	0.0118-0.0169	0.0177-0.0236	0.0051-0.0150	0.0008-0.0028	0.0012-0.0028	SNUG
	2.7	3RZ-FE	0.0019-0.0028	0.0118-0.0157	0.0157-0.0194	0.0051-0.0150	0.0008-0.0028	0.0012-0.0028	SNUG
	3.4	5VZ-FE	0.0053-0.0060	0.0118-0.0197	0.0157-0.0236	0.0059-0.0217	0.0016-0.0031	0.0012-0.0028	SNUG

67170-TACO-C07

TORQUE SPECIFICATIONS

All readings in ft. lbs.

Year	Engine Displacement Liters	Engine ID	Cylinder Head Bolts	Main Bearing Bolts	Rod Bearing Bolts	Crankshaft Damper Bolts	Flywheel Bolts	Manifold Intake	Manifold Exhaust	Spark Plugs	Oil Pan Drain Plug
2001	2.4	2RZ-FE	①	②	③	193	④	22	36	14	27
	2.7	3RZ-FE	①	②	③	193	⑤	22	36	14	27
	3.4	5VZ-FE	⑥	⑦	⑧	184	63-67	13	30	13	NA
2002	2.4	2RZ-FE	①	②	③	193	④	22	36	14	27
	2.7	3RZ-FE	①	②	③	193	⑨	22	36	14	27
	3.4	5VZ-FE	⑩	⑦	⑧	184	63	13	30	13	NA
2003	2.4	2RZ-FE	①	②	③	193	④	22	36	14	27
	2.7	3RZ-FE	①	②	③	193	⑨	22	36	14	27
	3.4	5VZ-FE	⑩	⑦	⑧	184	63	13	30	13	NA
2004	2.4	2RZ-FE	①	②	③	193	④	22	36	14	27
	2.7	3RZ-FE	①	②	③	193	⑨	22	36	14	27
	3.4	5VZ-FE	⑩	⑦	⑧	184	⑪	13	30	13	NA

NA: Information not available

① Step 1: 29 ft. lbs.
Step 2: Plus 90 degrees
Step 3: Plus 90 degrees

② Step 1: 29 ft. lbs.
Step 2: Plus 90 degrees

③ Step 1: 33 ft. lbs.
Step 2: Plus 90 degrees

④ Manual transmission: 65 ft. lbs.
Automatic transmission: 54 ft. lbs.

⑤ Manual transmission: 19 ft. lbs. + 90°
Automatic transmission: 54 ft. lbs.

⑥ Step 1: 25 ft. lbs.
Step 2: Plus 90 degrees
Recessed head: 13 ft. lbs.

⑦ Step 1: 45 ft. lbs.
Step 2: Plus 90 degrees

⑧ Step 1: 18 ft. lbs.
Step 2: Plus 90 degrees

⑨ MT: 19 ft. lbs. +90 degrees
AT: 54 ft. lbs.

⑩ Step 1: 25 ft. lbs.
Step 2: Plus 90 degrees
Step 3: Plus 90 degrees
Recessed head: 13 ft. lbs.

⑪ AT: 61 ft. lbs.
MT: 63 ft. lbs.

67170-TACO-C08

WHEEL ALIGNMENT

Year	Model	Caster Range (+/-Deg.)	Caster Preferred Setting (Deg.)	Camber Range (+/-Deg.)	Camber Preferred Setting (Deg.)	Toe-in (in.)	Steering Axis Inclination (Deg.)
2001	2WD	0.75	+0.67	0.75	0	0.06+/-0.08	10.00
	4WD	0.75	0.30	0.75	+1.62	0.06+/-0.08	10.40
2002	2WD	0.75	+0.67	0.75	0	0.06+/-0.08	10.00
	4WD	0.75	0.30	0.75	+1.62	0.06+/-0.08	10.40
2003	2WD	0.75	+0.67	0.75	0	0.06+/-0.08	10.00
	4WD	0.75	0.30	0.75	+1.62	0.06+/-0.08	10.40
2004	2WD exc. PreRunner	0.75	+0.67	0.75	0	0.06+/-0.08	10.00
	4WD & PreRunner	0.75	0.30	0.75	+1.62	0.06+/-0.08	10.40

All alignment figures based on nominal ride height and standard tires

67170-TACO-C09

TIRE, WHEEL AND BALL JOINT SPECIFICATIONS

Year	Model	OEM Tires		Tire Pressures (psi)		Wheel	Ball Joint	Lugnut
		Standard	Optional	Front	Rear	Size	Inspection ①	Torque (ft. lbs.)
2001	2wd	P205/75R15	P225/75R15	Std: 29 Opt: 26	Std: 29 Opt: 29	6-JJ	4-30	83
	Double Cab	P265/70R16	None	26	26	7-JJ	4-30	83
	Prerunner	P235/55R16	None	26	29	7-JJ	4-30	83
	4wd	P225/75R15	P265/70R16	Std: 26 Opt: 26	Std: 29 Opt: 26	Std: 6-JJ Opt: 7-JJ	4-30	83
2002	2wd Reg. Cab	P205/75R15	P235/55R16	Std: 29 Opt: 29	Std: 29 Opt: 32	6-JJ Opt: 6.5J	Upper 4-30 Lower 0.8-30	83
	4wd Reg. Cab	P225/75SR15	P265/70R16	Std: 26 Opt: 26	Std: 29 Opt: 26	6-JJ Opt.: 7-JJ	Upper 6-39 Lower 1-22	83
	4wd xtracab	P225/75SR15	P265/70R16	26	29	7-JJ	Upper 6-39 Lower 1-22	83
	2wd xtracab S-Runner	P235/55R16	None	29	32	7-JJ	Upper 4-30 Lower 0.8-30	83
	2wd xtracab PreRunner	P225/75R15	P265/70R16	Std: 26 Opt: 26	Std: 29 Opt: 26	7-JJ	Upper 6-39 Lower 1-22	83
	4wd Crew Cab	P225/75SR15	P265/70R16	Std: 26 Opt: 26	Std: 29 Opt: 26	7-JJ	Upper 6-39 Lower 1-22	83
	2wd Crew Cab PreRunner	P225/75R15	P265/70R16	Std: 26 Opt: 26	Std: 29 Opt: 26	7-JJ	Upper 6-39 Lower 1-22	83
2003	2wd Reg. Cab	P205/75R15	P235/55R16	Std: 29 Opt: 29	Std: 29 Opt: 32	6-JJ Opt: 6.5J	Upper 4-30 Lower 0.8-30	83
	4wd Reg. Cab	P225/75SR15	P265/70R16	Std: 26 Opt: 26	Std: 29 Opt: 26	6-JJ Opt.: 7-JJ	Upper 6-39 Lower 1-22	83
	4wd xtracab	P225/75SR15	P265/70R16	26	29	7-JJ	Upper 6-39 Lower 1-22	83
	2wd xtracab S-Runner	P235/55R16	None	29	32	7-JJ	Upper 4-30 Lower 0.8-30	83
	2wd xtracab PreRunner	P225/75R15	P265/70R16	Std: 26 Opt: 26	Std: 29 Opt: 26	7-JJ	Upper 6-39 Lower 1-22	83
	4wd Crew Cab	P225/75SR15	P265/70R16	Std: 26 Opt: 26	Std: 29 Opt: 26	7-JJ	Upper 6-39 Lower 1-22	83
	2wd Crew Cab PreRunner	P225/75R15	P265/70R16	Std: 26 Opt: 26	Std: 29 Opt: 26	7-JJ	Upper 6-39 Lower 1-22	83
2004	Reg. Cab 2wd	P205/75R15	P235/55R16	Std: 29 Opt: 29	Std: 29 Opt: 32	std. 6-JJ opt. 6.5	②	83
	Reg. Cab 4wd	P225/75SR15	P265/70R16	Std: 26 ③	Std: 29 ③	7-JJ	②	83
	Reg. Cab PreRunner	P225/75SR15	P265/70R16	③	③	7-JJ	②	83
	Extended Cab 2wd	P205/75R15	P235/55R16	③	③	std. 6-JJ opt. 6.5	②	83
	Extended Cab 4wd	P225/75SR15	P265/70R16	③	③	7-JJ	②	83
	Extended Cab PreRunner	P225/75SR15	P265/70R16	③	③	7-JJ	②	83
	S-Runner	P235/55R16	None	③	③	6.5	②	83

OEM: Original Equipment Manufacturer

PSI: Pounds Per Square Inch

STD: Standard

OPT: Optional

① Torque required in inch lbs. to rotate ball joint when removed from the knuckle

② 2wd exc. PreRunner: Upper 4-30 inch lbs.; lower 0.8-30 inch lbs.
4wd and PreRunner: Upper 6-39 inch lbs.; lower 1-22 inch lbs.
Lower ball joint excessive play, all models: 0.020 inch

③ See placard on the vehicle

67170-TACO-C10

BRAKE SPECIFICATIONS

All measurements in inches unless noted

		Brake Disc			Brake Drum Diameter		Minimum Lining Thickness		Brake Caliper	
Year	Model	Original Thickness	Minimum Thickness	Maximum Runout	Original Inside Diameter	Maximum Machine Diameter	Front	Rear	Bracket Bolts (ft. lbs.)	Mounting Bolts (ft. lbs.)
2001	2WD	0.866	0.787	0.0028	10.00	10.08	0.039	—	80	65
	4WD	0.866	0.787	0.0028	11.61	11.69	—	0.039	—	90
2002	2WD	0.866	0.787	0.0028	10.00	10.08	0.039	—	80	65
	4WD	0.866	0.787	0.0028	11.61	11.69	—	0.039	—	90
2003	2WD	0.866	0.787	0.0028	10.00	10.08	0.039	—	80	65
	4WD	0.866	0.787	0.0028	11.61	11.69	—	0.039	—	90
2004	2WD	0.866	0.787	0.0028	10.00	10.08	0.039	—	80	65
	4WD	0.866	0.787	0.0028	11.61	11.69	—	0.039	—	90

67170-TACO-C11

SCHEDULED MAINTENANCE INTERVALS

TOYOTA—TACOMA

TO BE SERVICED	TYPE OF	VEHICLE MILEAGE INTERVAL (x1000)																		
		5	10	15	20	25	30	35	40	45	50	55	60	65	70	75	80	85	90	95
Automatic transmission and differential fluid	S/I			✓			✓			✓			✓			✓			✓	
Ball joints and boots	S/I			✓			✓			✓			✓			✓			✓	
Brake linings, discs/drums, lines & hoses	S/I			✓			✓			✓			✓			✓			✓	
Charcoal canister	S/I												✓							
Drive belts	S/I						✓						✓						✓	
Driveshaft bushing (4WD)	L						✓						✓						✓	
Engine coolant	R						✓						✓						✓	
Engine oil & filter	R	✓	✓	✓	✓	✓	✓	✓	✓	✓	✓	✓	✓	✓	✓	✓	✓	✓	✓	✓
Exhaust pipes & mounts	S/I			✓			✓			✓			✓			✓			✓	
Fuel lines & connections, fuel tank vapor vent system hoses, fuel tank band	S/I						✓						✓						✓	
Fuel tank cap gasket	S/I						✓						✓						✓	
Halfshaft boots & flange bolts	S/I			✓			✓			✓			✓			✓			✓	
Limited slip differential fluid	R						✓						✓						✓	
Manual transmission and differential fluid	S/I						✓						✓						✓	
Non-platinum spark plugs	R						✓						✓						✓	
Platinum spark plugs	R												✓							
Propeller shaft (4WD)	L			✓			✓			✓			✓			✓			✓	
Propeller shaft bolts	S/I			✓			✓			✓			✓			✓			✓	
Rack and pinion assembly	S/I			✓			✓			✓			✓			✓			✓	
Rear wheel bearing	L						✓						✓						✓	
Steering linkage	S/I			✓			✓			✓			✓			✓			✓	
Valves	S/I												✓							

R: Replace S/I: Service or Inspect L: Lubricate

FREQUENT OPERATION MAINTENANCE (SEVERE SERVICE)

If a vehicle is operated under any of the following conditions it is considered severe service:

- **Towing a trailer or using a camper or car-top carrier.**
- **Repeated short trips of less than 5 miles in temperatures below freezing.**
- **Excessive idling or low-speed driving for long distances as in heavy commercial use, such as delivery, taxi or police cars.**
- **Operating on rough, muddy or salt-covered roads.**
- **Operating on unpaved or dusty roads.**

Oil filter: service or inspect every 5000 miles or 4 months, whichever occurs first.

Brake linings and discs or drums: service or inspect every 5000 miles or 4 months, whichever occurs first.

Steering linkage: service or inspect every 5000 miles or 4 months, whichever occurs first.

Ball joints and boots: service or inspect every 5000 miles or 4 months, whichever occurs first.

Brake discs & pads (front): service or inspect every 6000 miles.

Halfshaft boots: service or inspect every 5000 miles or 4 months. Retighten the flange bolts, whichever occurs first.

Body chassis bolts and nuts: service or inspect every 5000 miles or 4 months, whichever occurs first.

Transmission and differential fluid: replace every 15,000 miles or 12 months, whichever occurs first.

Transfer case and differential fluid: replace every 15,000 miles or 12 months, whichever occurs first.

Timing belt: replace every 60,000 miles or 48 months, whichever occurs first.

67170-TACO-C12

PRECAUTIONS

Before servicing any vehicle, please be sure to read all of the following precautions, which deal with personal safety, prevention of component damage, and important points to take into consideration when servicing a motor vehicle:

- Never open, service or drain the radiator or cooling system when the engine is hot; serious burns can occur from the steam and hot coolant.
- Observe all applicable safety precautions when working around fuel. Whenever servicing the fuel system, always work in a well-ventilated area. Do not allow fuel spray or vapors to come in contact with a spark, open flame or excessive heat (a hot drop light, for example). Keep a dry chemical fire extinguisher near the work area. Always keep fuel in a container specifically designed for fuel storage; also, always properly seal fuel containers to avoid the possibility of fire or explosion. Refer to the additional fuel system precautions later in this section.
- Fuel injection systems often remain pressurized, even after the engine has been turned **OFF**. The fuel system pressure must be relieved before disconnecting any fuel lines. Failure to do so may result in fire and/or personal injury.
- Brake fluid often contains polyglycol ethers and polyglycols. Avoid contact with the eyes and wash your hands thoroughly after handling brake fluid. If you do get brake fluid in your eyes, flush your eyes with clean, running water for 15 minutes. If eye irritation persists, or if you have taken brake fluid internally, IMMEDIATELY seek medical assistance.
- The EPA warns that prolonged contact with used engine oil may cause a number of skin disorders, including cancer! You should make every effort to minimize your exposure to used engine oil. Protective gloves should be worn when changing oil. Wash your hands and any other exposed skin areas as soon as possible after exposure to used engine oil. Soap and water, or waterless hand cleaner should be used.
- All new vehicles are now equipped with an air bag system. The system must be disabled before performing service on or around system components, steering column, instrument panel components, wiring and sensors. Failure to follow safety and disabling procedures could result in accidental air bag deployment, possible personal injury and unnecessary system repairs.
- Always wear safety goggles when working with, or around, the air bag system. When carrying a non-deployed air bag, be sure the bag and trim cover are pointed away from your body. When placing a non-deployed air bag on a work surface, always face the bag and trim cover upward, away from the surface. This will reduce the motion of the module if it is accidentally deployed. Refer to the additional air bag system precautions later in this section.
- Clean, high quality brake fluid from a sealed container is essential to the safe and proper operation of the brake system. You should always buy the correct type of brake fluid for your vehicle. If the brake fluid becomes contaminated, completely flush the system with new fluid. Never reuse any brake fluid. Any brake fluid that is removed from the system should be discarded. Also, do not allow any brake fluid to come in contact with a painted surface; it will damage the paint.
- Never operate the engine without the proper amount and type of engine oil; doing so WILL result in severe engine damage.
- Timing belt maintenance is extremely important! Many models utilize an interference-type, non-freewheeling engine. If the timing belt breaks, the valves in the cylinder head may strike the pistons, causing potentially serious (also time-consuming and expensive) engine damage. Refer to the maintenance interval charts for the recommended replacement interval for the timing belt, and to the timing belt section for belt replacement and inspection.
- Disconnecting the negative battery cable on some vehicles may interfere with the functions of the on-board computer system(s) and may require the computer to undergo a relearning process once the negative battery cable is reconnected.
- When servicing drum brakes, only disassemble and assemble one side at a time, leaving the remaining side intact for reference.

ENGINE REPAIR

Distributor

All engines are equipped with distributorless ignition systems.

Alternator

REMOVAL

On some models, the alternator is mounted very low on the engine. On these models, it may be necessary to remove the gravel shield and work from beneath the vehicle in order to gain access to the alternator. Replacing the alternator while the engine is cold is recommended.

2.4L and 2.7L Engines

Remove or disconnect the following:

- Negative battery cable
- Alternator wiring

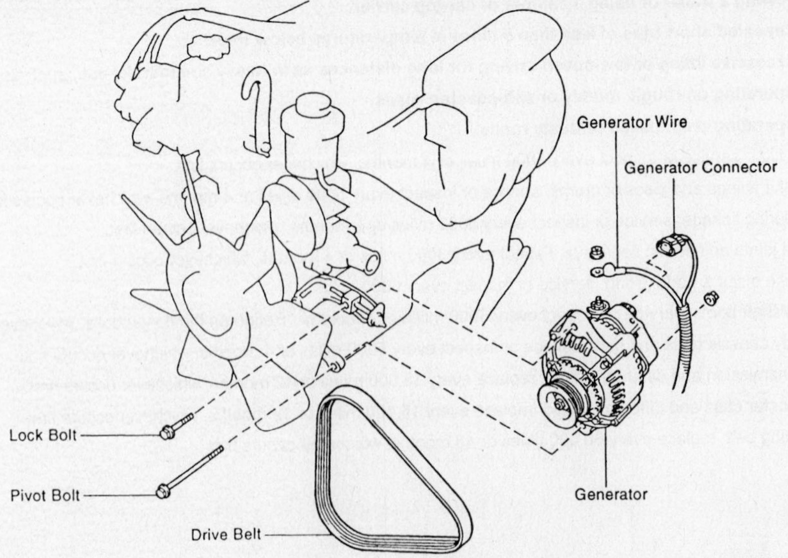

Exploded view of the alternator and drive belt—2.4L and 2.7L engines

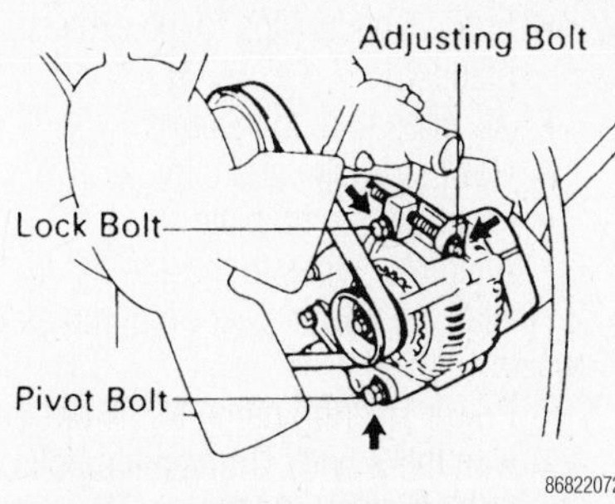

Locations of the adjusting, pivot and lock-bolts—2.4L and 2.7L engines

- Alternator lockbolt, pivot bolt, nut and adjusting bolt
- Drive belt
- Wiring harness with the clip
- Alternator

3.4L Engine

1. Remove or disconnect the following:
   - Negative battery cable
   - Alternator wiring
   - Alternator locknut, pivot bolt, nut and adjusting bolt
   - Drive belt
   - Alternator

INSTALLATION

2.4L and 2.7L Engines

Install or connect the following:

- Alternator
- Drive belt; adjust it to the proper tension. Tighten the lockbolt to 21 ft. lbs. (29 Nm) and the pivot bolt to 43 ft. lbs. (59 Nm).
- Wire harness with clip
- Alternator wiring. Tighten the nut to 7 ft. lbs. (10 Nm).
- Rubber boot over the terminal
- Wiring harness connector
- Negative battery cable

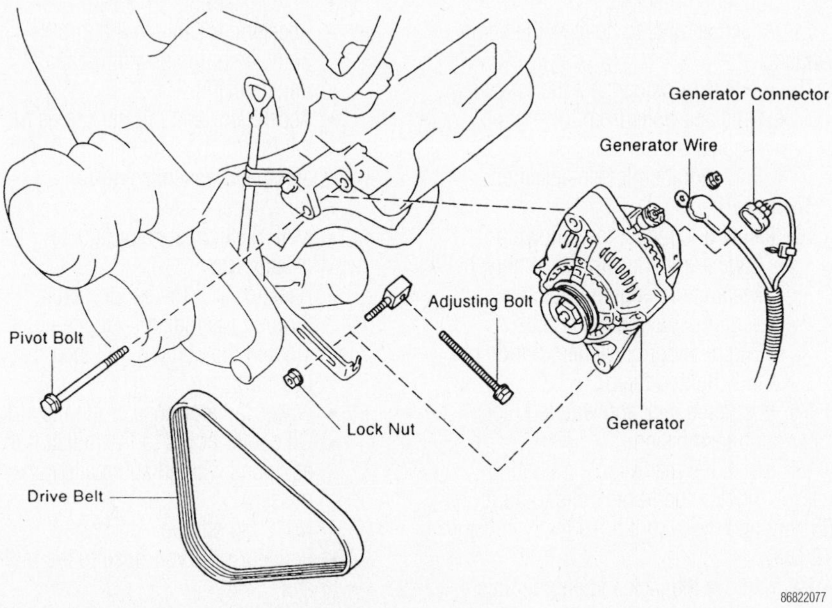

Exploded view of the alternator and drive belt—3.4L Engine

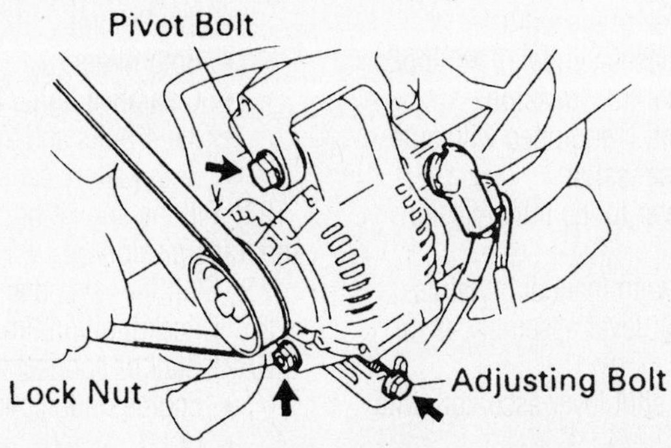

Locations of the adjusting and pivot bolts and the locknut—3.4L Engine

3.4L Engine

Install or connect the following:

- Alternator
- Drive belt. Tighten the locknut 25 ft. lbs. (33 Nm) and the pivot bolt 38 ft. lbs. (51 Nm).
- Alternator wiring
- Negative battery cable

Ignition Timing

ADJUSTMENT

All engines use a distributorless ignition system referred to as Direct Ignition System (DIS). All spark advance is permanently set by the PCM.

2.4L Engine

➡The ignition timing is not adjustable but can be checked.

1. Before servicing the vehicle, refer to the precautions in the beginning of this section.
2. Warm the engine to normal operating temperature.
3. Attach a hand-held tester to the Data Link Connector 3 (DLC3) under the dashboard on the driver's side.
4. Jumper terminals T_{E1} and E_1 of the DLC1.
5. Check the idle speed.
6. Aim the timing light at the timing indicator and check the ignition timing. Timing should be between 3–7 degrees BTDC at idle.
7. For a further check on ignition timing, disconnect the hand-held tester from the DLC3 and disconnect the jumper wire from the DLC1.
8. Point the timing light at the crankshaft pulley and read the timing. Timing should be between 7–18 degrees BTDC at idle.
9. Remove timing light from the engine.

Engine Assembly

REMOVAL & INSTALLATION

2.4L Engine

1. Before servicing the vehicle, refer to the precautions in the beginning of this section.
2. Properly relieve the fuel system pressure.
3. Turn the ignition switch **OFF**.
4. Remove or disconnect the following:

- Battery cables; negative cable first
- Hood by matchmarking the hood hinges
- Engine undercover

5. Drain the engine oil, transmission oil and cooling system.
6. Remove or disconnect the following:
   - Radiator
   - Drive belts
   - Loosen the lockbolt and adjusting bolt to the idler pulley, if equipped with power steering
   - Loosen the idler pulley nut and adjusting bolt, if equipped with air conditioning
   - Fan (with fan clutch), water pump pulley and fan shroud
   - Accelerator cable from the throttle body, if equipped with a manual transaxle
   - Accelerator and throttle cables from the throttle body, if equipped with an automatic transaxle
   - Actuator cover and cruise control cable from the actuator, if equipped with cruise control
   - Air cleaner cap, Mass Air Flow (MAF) and resonator
   - Air cleaner case
   - Intake air connector
   - Air conditioning compressor and bracket, if equipped with air conditioning
   - Alternator wires from the alternator
   - Heater hoses at the cowl panel
   - Brake booster vacuum hose
   - EVAP hose
   - Vacuum hose, if equipped with 4WD with Automatic Disconnecting Differential (ADD)
   - Both power steering hoses, if equipped with power steering
   - Fuel return hose
   - Fuel inlet hose
7. Remove the power steering pump as follows:
   - Nut and power steering pulley
   - Both bolts and the power steering pump
8. Disconnect the Engine Control Module (ECM) wiring from the ECM as follows:
   - Right front door scuff plate
   - Cowl panel side trim by removing the clip
   - 4 ECM electrical connectors.
9. Detach the engine wiring harness and connectors as follows:
   - Igniter connector
   - Ground strap from the cowl top panel
   - Both engine wiring harness clamps
   - Engine wiring harness retainer to the cowl panel nuts and pull out the engine wiring harness from the vehicle
10. Disconnect the front exhaust pipe from the exhaust manifold and catalytic converter.
11. If equipped with manual transmission, remove the shift lever assembly as follows:
    - Shift lever knob
    - 4 screws and shift lever boot
    - 6 bolts, shift lever assembly and baffle
12. Remove or disconnect the following:
    - Driveshaft
    - Speedometer cable from the transmission
    - Clutch release cylinder, if equipped with manual transmission
    - Cross-shaft, if equipped with automatic transmission
    - Wires at the starter
13. Position a jack and wooden block under the transmission and remove the rear engine mounting bracket.
14. Attach an engine hoist to the engine hangers.
15. Remove or disconnect the following:
    - Nuts and bolts from the engine mounts
    - Engine/transmission assembly

To install:

16. Install or connect the following:
    - Engine/transmission assembly, by keeping the engine level, while aligning the engine mounts
    - Engine mount fasteners but do not fully tighten them
17. Position a jack and wooden block under the transmission.
18. Install the rear engine mounting bracket. Tighten the frame bolts to 43 ft. lbs. (58 Nm) and the mount bolts to 13 ft. lbs. (18 Nm).
19. Remove the jack and engine hoist.
20. Install or connect the following:
    - Tighten the engine mounts to 28 ft. lbs. (38 Nm).
    - Starter wires to the starter
    - Clutch release cylinder, if equipped with manual transmission
    - Cross-shaft, if equipped with automatic transmission
    - Speedometer to the transmission
    - Driveshaft
21. If equipped with manual transmission, install the shift lever assembly as follows:
    - Baffle and shift lever assembly with the 6 bolts
    - Shift lever boot with the 4 screws
    - Shift lever knob
    - Front exhaust pipe to the exhaust manifold
    - All wires and connectors
    - Cowl side trim and clip
    - Front door scuff plate
    - Alternator wires to the alternator
22. Install the power steering pump as follows:
    - Power steering pump to the bracket with the 2 bolts. Tighten the bolts to 43 ft. lbs. (58 Nm).
    - Power steering pulley with the nut. Tighten the nut to 32 ft. lbs. (43 Nm).
23. Install or connect the following:
    - All hoses
    - Compressor, if equipped with air conditioning
    - Intake air connector. Tighten the 2 bolts to 13 ft. lbs. (18 Nm).
    - Water pump pulley, fan shroud, fan (with fan clutch) and alternator drive belt.
    - Drive belt, if equipped with air conditioning
    - Power steering drive belt
    - Accelerator cable to the throttle body, if equipped with manual transmission
    - Accelerator and throttle cables to the throttle body, if equipped with automatic transmission
    - Air cleaner case
    - MAF meter, resonator and air cleaner cap
    - Radiator with the support tabs through the radiator service holes. Tighten the bolts to 108 inch lbs. (13 Nm).
    - Lower radiator hose to the radiator
    - Oil cooler hoses to the radiator, if equipped with an automatic transmission
    - No. 2 fan shroud.
    - Radiator reservoir hose to the radiator
    - Upper radiator hose to the radiator
    - Air pipe
    - Radiator grille to the vehicle with the 11 clips
    - Both fillers
    - Clearance lights to the grille with the 4 bolts and 2 clips
    - Both battery cables
24. Fill the engine oil, engine coolant and transmission oil.
25. Start the engine and check for leaks.
26. Check ignition timing.
27. Install or connect the following:
    - Engine undercover
    - Hood
28. Road test the vehicle and check all fluids.

2.7L Engine

1. Before servicing the vehicle, refer to the precautions in the beginning of this section.
2. Properly relieve the fuel system pressure.
3. Turn the ignition switch **OFF**.
4. Remove or disconnect the following:
   - Battery cables; negative cable first
   - Hood
   - Engine undercover
5. Drain the engine oil, transmission oil and cooling system.
6. Remove or disconnect the following:
   - Radiator
   - Idler pulley and drive belt, if equipped with power steering
   - Idler pulley nut/bolt and drive belt, if equipped with air conditioning
   - Alternator drive belt, fan (with fan clutch), water pump pulley and fan shroud
   - Accelerator cable from the throttle body, if equipped with a manual transaxle
   - Accelerator and throttle cables from the throttle body, if equipped with a automatic transaxle
   - Actuator cover and the cruise control cable from the actuator, if equipped with cruise control
   - Air cleaner cap
   - Mass Air Flow (MAF) meter and resonator
   - Air cleaner case
   - Intake air connector
   - Air conditioning compressor and bracket, if equipped with air conditioning
   - Alternator wires
   - Heater hoses at the cowl panel
7. Disconnect the following hoses:
   - Brake booster vacuum hose
   - Evaporative Emissions (EVAP) hose
   - Vacuum hose, if equipped with 4WD with Automatic Disconnecting Differential (ADD)
   - Both power steering hoses, if equipped with power steering
   - Fuel return hose
   - Fuel inlet hose
8. Remove the power steering pump as follows:
   - Nut and power steering pulley
   - Power steering pump
9. Disconnect the Engine Control Module (ECM) wiring from the ECM as follows:
   - 4 screws to the right front door scuff plate
   - Scuff plate
   - Cowl panel side trim by removing the clip
   - 4 ECM electrical connectors
10. Detach the engine wiring harness and connectors from the vehicle as follows:
    - Igniter
    - Ground strap from the cowl top panel
    - 2 engine wiring harness clamps
    - Engine wiring harness retainer to cowl panel nut and wiring harness
11. Disconnect the front exhaust pipe from the exhaust manifold and catalytic converter.
12. If equipped with manual transmission, remove the shift lever assembly as follows:
13. Remove or disconnect the following:
    - Shift lever knob
    - 4 screws and shift lever boot
    - 6 bolts, shift lever assembly and baffle
14. Remove or disconnect the following:
    - Driveshaft
    - Speedometer cable from the transmission
    - Clutch release cylinder, if equipped with manual transmission
    - Cross-shaft, If equipped with automatic transmission
    - Wires at the starter
15. Position a jack and wooden block under the transmission.
16. Remove the rear engine mounting bracket.
17. Attach an engine hoist to the engine hangers.
18. Remove or disconnect the following:
    - Nuts and bolts from the engine mounts
    - Engine/transmission

To install:

19. Attach the engine hoist to the engine hangers.
20. Install or connect the following:
    - Engine/transmission assembly

➡**Keep the engine level, while aligning the engine mounts.**

   - Engine mount fasteners but do not fully tighten them
   - Position a jack and wooden block under the transmission
   - Rear engine mounting bracket. Tighten the frame bolts to 19 ft. lbs. (26 Nm) and the mount bolts to 13 ft. lbs. (18 Nm).
21. Remove the jack and engine hoist.
22. Install or connect the following:
    - Tighten the engine mounts to 28 ft. lbs. (38 Nm).
    - Starter wires
    - Clutch release cylinder, if equipped with manual transmission
    - Cross-shaft, if equipped with automatic transmission
    - Speedometer to the transmission
    - Driveshaft
23. If equipped with manual transmission, install the shift lever assembly as follows:
    - Baffle and shift lever assembly with the 6 bolts
    - Shift lever boot with the 4 screws
    - Shift lever knob
24. Install or connect the following:
    - Front exhaust pipe to the exhaust manifold
    - Engine wiring harness
    - Cowl side trim and clip
    - Front door scuff plate
    - Alternator wires
25. Install the power steering pump as follows:
    - Power steering pump to the bracket. Tighten the 2 bolts to 43 ft. lbs. (58 Nm).
    - Power steering pulley. Tighten the nut to 32 ft. lbs. (43 Nm).
26. Install or connect the following:
    - All hoses
    - Heater hoses at the cowl panel
    - Compressor, if equipped with air conditioning
    - Intake air connector. Tighten the 2 bolts to 13 ft. lbs. (18 Nm).
27. Install the water pump pulley, fan shroud, fan (with fan clutch) and alternator drive belt as follows:
    - Fan (with the fan clutch), water pump pulley and fan shroud in position
    - Water pump pulley but do not tighten the nuts
    - Alternator drive belt
    - Stretch the alternator belt tight. Tighten the fan nuts to 16 ft. lbs. (21 Nm).
    - Adjust the alternator drive belt
28. Install or connect the following:
    - Adjust the drive belt, if equipped with air conditioning
    - Adjust the power steering drive belt
    - Accelerator cable to the throttle body, if equipped with manual transmission
    - Accelerator and throttle cables to the throttle body, if equipped with automatic transmission
    - Air cleaner case
    - MAF meter, resonator and air cleaner cap
    - Radiator with the tabs on the supports through the radiator service

holes. Tighten the bolts to 108 inch lbs. (13 Nm).
- Lower radiator hose to the radiator
- Oil cooler hoses to the radiator, if equipped with automatic transmission
- No. 2 fan shroud
- Radiator reservoir hose to the radiator
- Upper radiator hose to the radiator
- Air pipe with the 2 bolts, if removed
- Radiator grille with the 11 clips
- 2 fillers
- Clearance lights to the grille with the 4 bolts and 2 clips
- Negative and positive cables to the battery

29. Fill the engine oil, engine coolant and transmission oil.
30. Start the engine and check for leaks.
31. Check ignition timing.
32. Install or connect the following:
    - Engine undercover
    - Hood
33. Road test the vehicle and check all fluids.

3.4L Engine

2WD

1. Before servicing the vehicle, refer to the precautions in the beginning of this section.
2. Properly relieve the fuel system pressure.
3. Remove or disconnect the following:
    - Hood
    - Battery
    - Engine under covers
4. Drain the engine coolant.
5. Drain the engine oil.
6. Remove or disconnect the following:
    - Radiator
7. Remove the power steering drive belt as follows:
    - Stretch the belt and loosen the fan pulley mounting nuts
    - Loosen the lockbolt, pivot bolt and adjusting bolt and remove the drive belt
8. Remove or disconnect the following:
    - Air conditioning drive belt by loosening the idle pulley nut and adjusting bolt, if equipped with air conditioning
    - Loosen the lockbolt, pivot bolt and adjusting bolt and the alternator drive belt
    - Fan with the fluid coupling and fan pulleys
    - Power steering pump, do not disconnect the lines from the pump
    - Compressor, if equipped with air conditioning; do not disconnect the lines from the compressor
    - Air cleaner cap
    - Mass Air Flow (MAF) meter and resonator
    - Air cleaner case and filter
9. Disconnect the following cables:
    - Actuator cable with the bracket, if equipped with cruise control
    - Accelerator cable
    - Throttle cable, if equipped with an automatic transmission
10. Disconnect the following hoses:
    - Heater hoses
    - Brake booster vacuum hose
    - Evaporative Emission (EVAP) hose
    - Fuel return hose
    - Fuel inlet hose
11. Detach the starter wire and connectors as follows:
    - Ground strap by removing the bolt
    - Positive cable from the battery
    - 3 starter wire clamps and connector
12. Detach the alternator connector and wire.
13. Detach the engine wiring harness and connectors as follows:
    - Right front door scuff plate.
    - Cowl panel side trim, by removing the clip
    - Engine Control Module (ECM)
    - 2 connectors from the cowl wire
    - Igniter
    - Ground strap
    - 6 engine wiring harness clamps
    - Engine wiring harness
14. If equipped with manual transmission, remove the shift lever assembly as follows:
    - Shift lever knob
    - 4 screws and the shift lever boot
    - Shift lever assembly and gasket, by removing the 6 bolts
15. Remove or disconnect the following:
    - Stabilizer bar
    - Driveshaft from the transmission
    - Speedometer cable
    - Front exhaust pipe
    - Clutch release cylinder, if equipped with a manual transmission
    - Cross-shaft, if equipped with an automatic transmission
16. Place a jack under the transmission.
17. Remove or disconnect the following:
    - Transmission rear mounting bracket, by removing the 8 bolts
    - Air conditioning compressor wire clamp, if equipped with air conditioning
18. If necessary, install a No. 2 engine hanger with 2 bolts. Tighten the 2 bolts to 30 ft. lbs. (40 Nm).
19. Attach the engine hoist chain to the 2 engine hangers.
20. Remove or disconnect the following:
    - 4 bolts/nuts holding the engine front mounting insulators to the frame
    - Engine/transmission assembly

To install:

21. Install or connect the following:
    - Engine
    - Engine mounts to the body mountings. Install the bolts and nuts but do not tighten at this time.
22. Remove the engine chain hoist the No. 2 engine hanger.
23. Install or connect the following:
    - Air conditioning wire, if equipped with air conditioning
    - Transmission mounting bracket. Tighten the frame bolts to 43 ft. lbs. (58 Nm) and the mounting insulator bolts to 13 ft. lbs. (18 Nm).
    - Tighten the engine mounting nuts and bolts to 28 ft. lbs. (38 Nm).
    - Cross-shaft, if equipped with an automatic transmission
    - Clutch release cylinder, if equipped with a manual transmission. Tighten the bolts to 108 inch lbs. (13 Nm).
    - Front exhaust pipe
    - Speedometer cable
    - Driveshaft
    - Stabilizer bar
24. Install the shift lever assembly, as follows:
25. Install or connect the following:
    - New gasket and shift lever assembly with the 6 bolts
    - Shift lever boot with the 4 screws
    - Shift lever knob
26. Install or connect the following:
    - All engine wiring harness, hoses and cables
    - Air cleaner case and air filter
    - MAF meter, resonator and air cleaner cap
    - Air conditioning compressor, if equipped
    - Remaining components
27. Fill the engine with oil.
28. Fill the engine and radiator with coolant.
29. Install the engine undercover.
30. Start the engine and check for leaks.

4WD

1. Before servicing the vehicle, refer to the precautions in the beginning of this section.

2. Properly relieve the fuel system pressure.
3. Remove or disconnect the following:
- Transmission
- Hood
- Battery from the vehicle
- Engine under covers
4. Drain the engine coolant.
5. Drain the engine oil.
6. Remove or disconnect the following:
- Radiator
7. Remove the power steering drive belt, as follows:
- Stretch the belt and loosen the fan pulley mounting nuts
- Loosen the lockbolt, pivot bolt and adjusting bolt
- Drive belt from the engine
8. Remove or disconnect the following:
- Air conditioning drive belt by loosening the idle pulley nut and adjusting bolt, if equipped with air conditioning
- Alternator drive belt
- Fan with the fluid coupling and fan pulleys
- Power steering pump; do not disconnect the lines from the pump
- Compressor, if equipped with air conditioning. Do not disconnect the lines from the compressor.
- Air cleaner cap, Mass Air Flow Meter (MAF) meter and resonator
- Air cleaner case and filter
9. Disconnect the following cables:
- Actuator cable with the bracket, if equipped with cruise control
- Accelerator cable
- Throttle cable, if equipped with an automatic transmission
10. Disconnect the following hoses:
- Heater hoses
- Brake booster vacuum hose
- Evaporative Emissions (EVAP) hose
- Automatic Disconnecting Differential (ADD) vacuum hose
- Fuel return hose
- Fuel inlet hose
11. Detach the starter wire and connectors as follows:
12. Remove or disconnect the following:
- Ground strap by removing the bolt
13. Disconnect the positive cable from the battery, as follows:
- 3 starter wire clamps and connector
- Automatic Disconnecting Differential (ADD) indicator switch connector
14. Detach the alternator connector and wire.
15. Detach the engine wiring harness and connectors, as follows:
- Right front door scuff plate
- Cowl panel side trim, by removing the clip
- Engine Control Module (ECM)
- 2 connectors from the cowl wire
- Igniter connector
- Ground strap
- 6 engine wiring harness clamps
- Engine wiring harness
16. Remove or disconnect the following:
- Air conditioning compressor wire clamp, if equipped with air conditioning
17. If necessary, install a No. 2 engine hanger with 2 bolts. Tighten the 2 bolts to 30 ft. lbs. (40 Nm).
18. Attach the engine hoist chain to the 2 engine hangers.
19. Remove or disconnect the following:
- 4 Engine front mounting insulators-to-frame bolts/nuts
- Engine

To install:

20. Install or connect the following:
- Engine
- Engine mounts to the body mountings. Install the bolts and nuts but do not tighten at this time.
21. Remove the engine chain hoist the No. 2 engine hanger.
22. Install or connect the following:
- Air conditioning wire with the bolt, if equipped with air conditioning
- Tighten the engine mounting nuts and bolts to 28 ft. lbs. (38 Nm).
- All engine wiring harness, hoses and cables
- Air cleaner case and air filter
- MAF meter, resonator and air cleaner cap
- Air conditioning compressor, if equipped
- Fan with the fluid coupling and fan pulleys. Tighten the nuts to 48 inch lbs. (5.4 Nm).
- Alternator drive belt
- Adjust the air conditioning drive belt, if equipped
- Power steering pump, pump pulley and the drive belt
- Radiator
23. Fill the engine with oil.
24. Fill the engine and radiator with coolant.
25. Install or connect the following:
- Hood
- Engine undercover
- Transmission
26. Start the engine and check for leaks.

Heater Core

REMOVAL & INSTALLATION

1. Disconnect the negative battery cable.

✻✻ CAUTION

After the negative battery cable has been disconnected, wait at least 1½ minutes for the air bag module to deplete its energy.

2. Drain the cooling system into a clean container for reuse.
3. Disconnect the heater hoses from the heater core.
4. Remove the steering wheel by performing the following procedure:
   a. Position the front wheels in the straight-ahead position.
   b. At both sides of the steering wheel, remove the side covers.
   c. Using a Torx® wrench, loosen the steering wheel Torx® screws until the screw's circumference ring catches on the screw case.
   d. Carefully, lift the air bag module, disconnect the electrical connector and remove the air bag.

✻✻ CAUTION

Place the air bag module in a safe location with the front facing upward.

   e. Remove the steering wheel nut.
   f. Using a steering wheel puller, press the steering wheel from the steering column.
5. Remove the instrument panel and reinforcement by performing the following procedure:
   a. Remove the steering column cover screws and the covers.
   b. Remove the combination switch-to-steering column screws and the combination switch.
   c. Remove the 2 hood lock release lever screws and the hood lock release lever.
   d. Remove the fuse box opening cover.
   e. Remove the 4 lower left side finish panel bolts, the screw and the panel.
   f. If equipped, remove the 2 rear console box bolts/screws and the rear console box.
   g. Remove the front console box.
   h. Remove the 2 upper console box mounting bracket screws and the bracket.

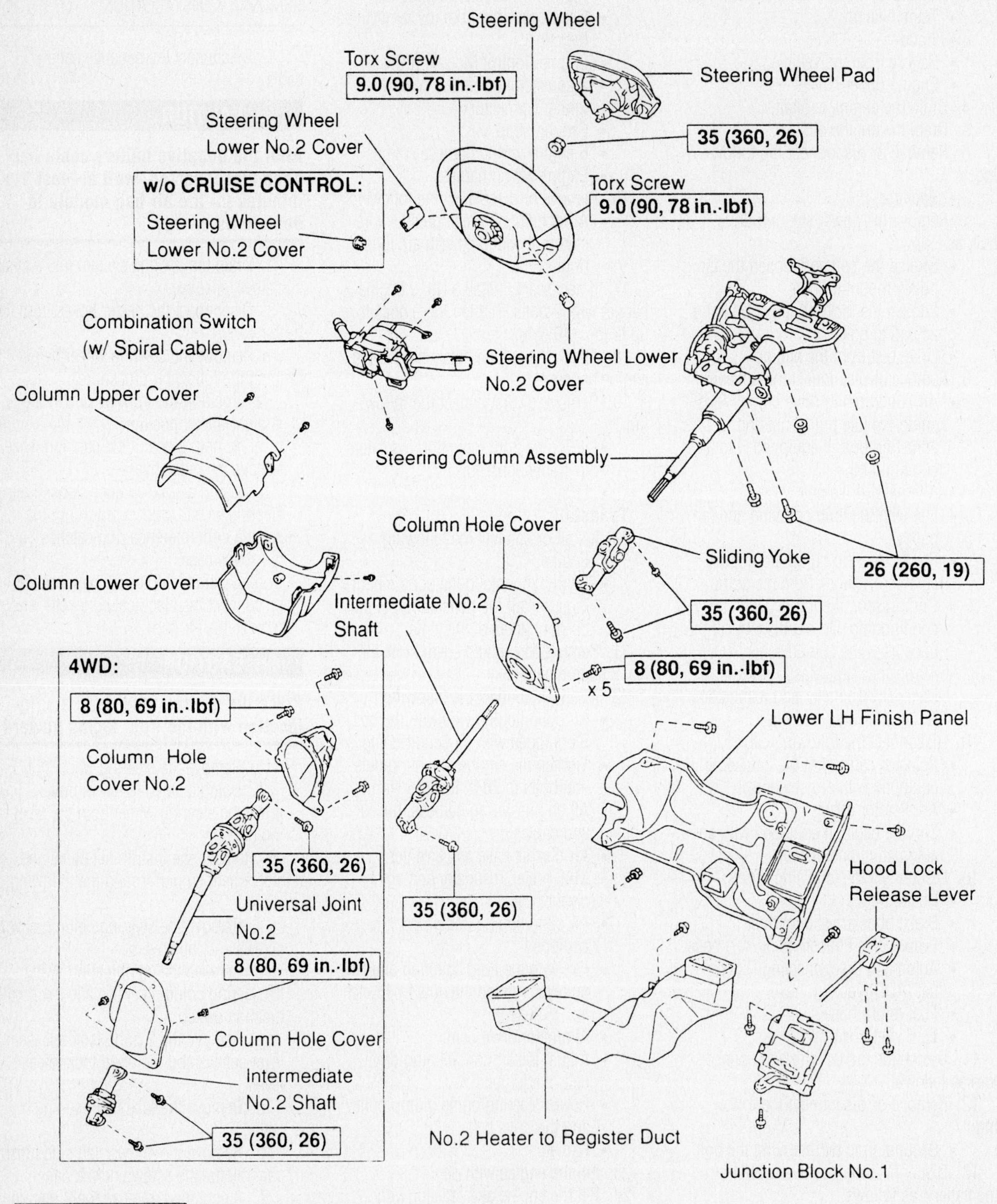

Exploded view of the steering wheel, air bag module, floor shift steering column and related components—Toyota Tacoma

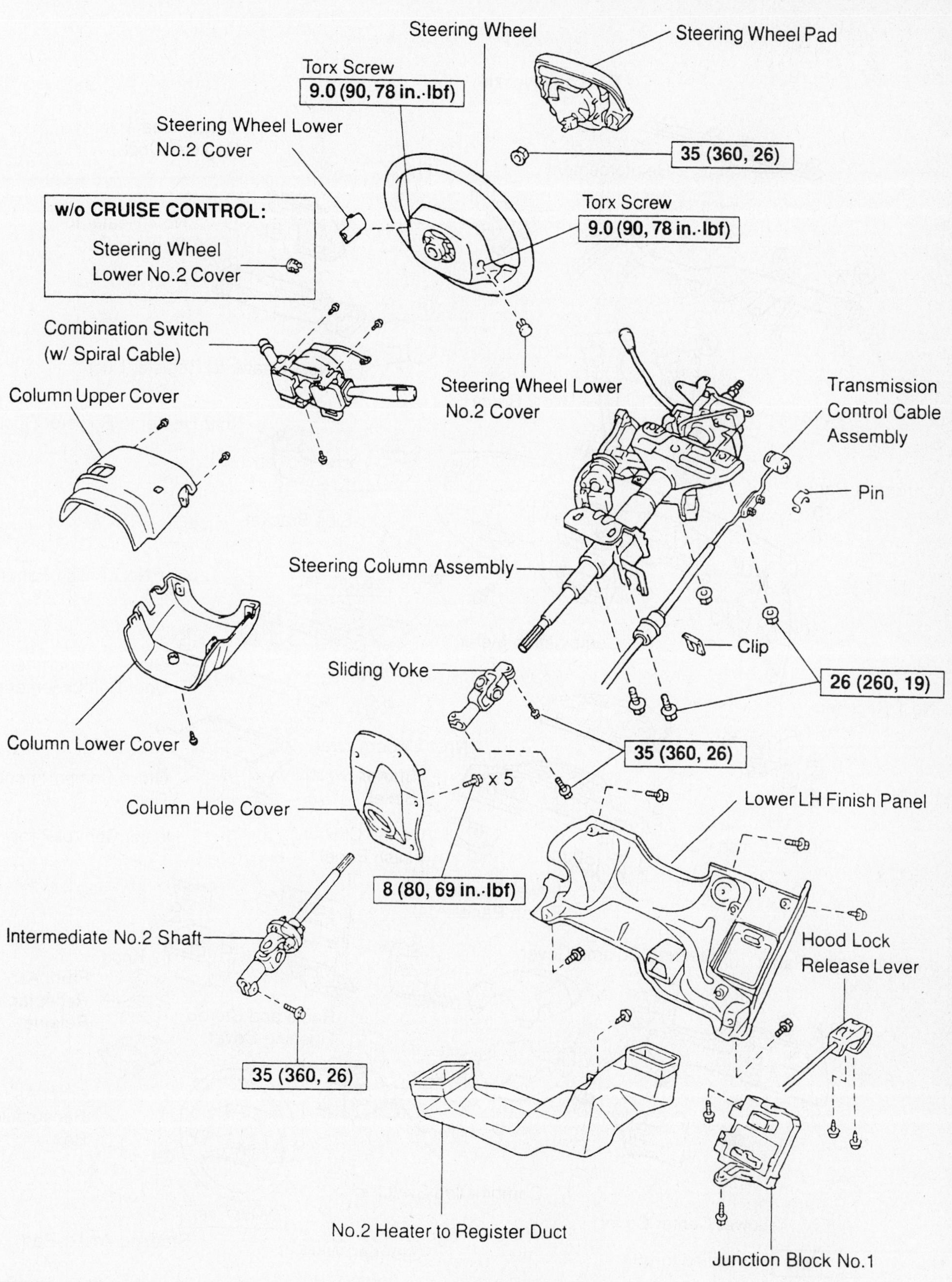

Exploded view of the steering wheel, air bag module, column shift steering column and related components

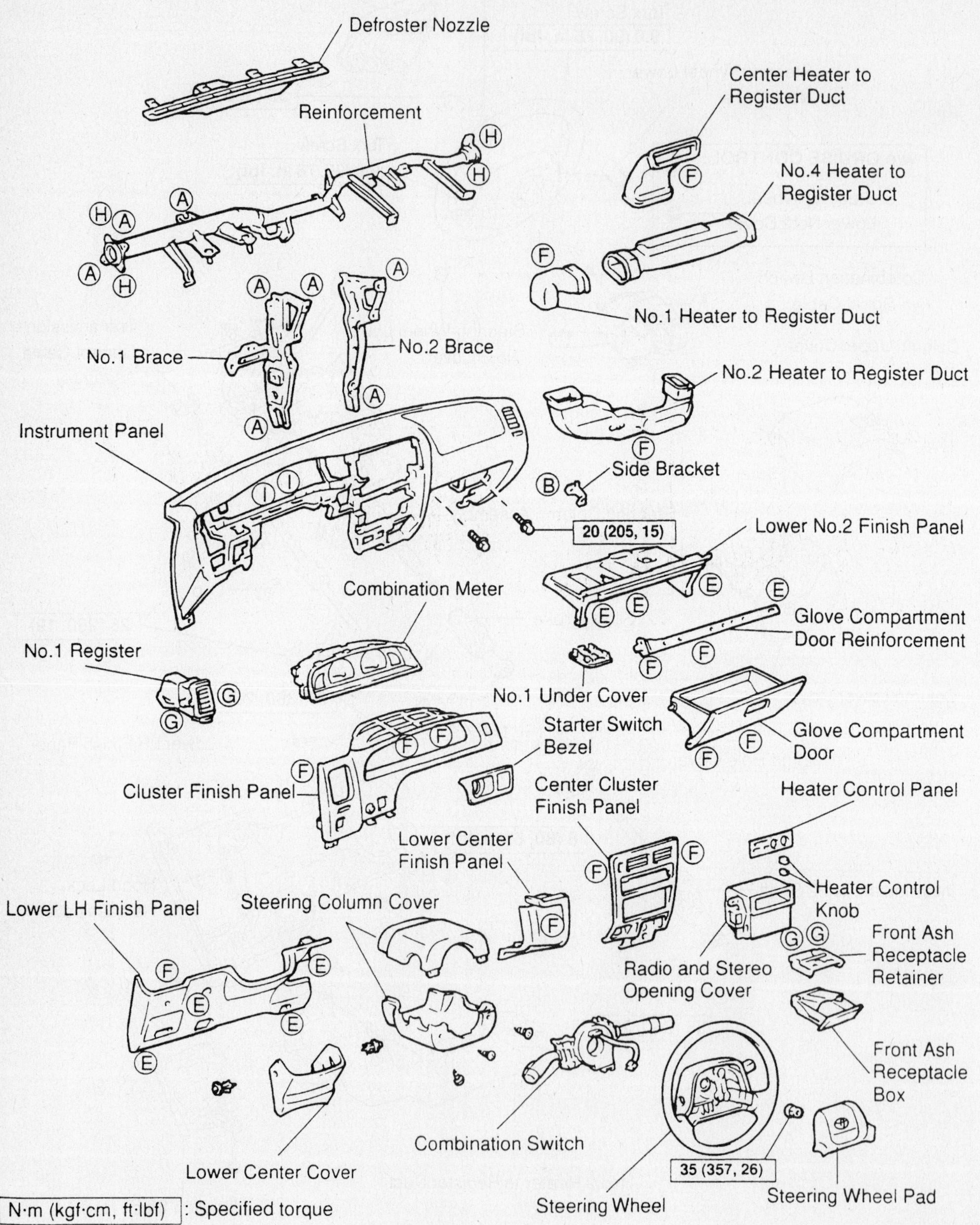

Defroster Nozzle
Reinforcement
Center Heater to Register Duct
No.4 Heater to Register Duct
No.1 Heater to Register Duct
No.1 Brace
No.2 Brace
No.2 Heater to Register Duct
Instrument Panel
Side Bracket
20 (205, 15)
Lower No.2 Finish Panel
Combination Meter
Glove Compartment Door Reinforcement
No.1 Register
No.1 Under Cover
Starter Switch Bezel
Glove Compartment Door
Cluster Finish Panel
Center Cluster Finish Panel
Heater Control Panel
Lower Center Finish Panel
Heater Control Knob
Lower LH Finish Panel
Steering Column Cover
Front Ash Receptacle Retainer
Radio and Stereo Opening Cover
Front Ash Receptacle Box
Combination Switch
Lower Center Cover
35 (357, 26)
N·m (kgf·cm, ft·lbf) : Specified torque
Steering Wheel
Steering Wheel Pad
93113GJ1

Exploded view of the instrument panel and related components

i. Remove the 2 lower center cover clips and the cover.

j. Remove the No. 2 heater-to-register duct screw and the duct.

k. Remove the heater control knobs.

l. Remove the heater control panel.

m. Remove the 2 center cluster finish panel screws, disengage the 5 clips, disconnect the electrical connectors and remove the panel.

n. Pry out the cigar lighter hole bezel.

o. Remove the front ashtray and the center cluster finish panel retainer.

p. Pry out the starter switch bezel.

q. Remove the 3 cluster finish panel screws and the panel.

r. Remove the radio and stereo opening cover.

s. Remove the combination meter and disconnect the electrical connectors.

t. Remove the 2 No. 1 register screws and the register.

u. Remove the No. 1 heater-to-register duct screw and the duct.

v. If equipped with a column shifter, disconnect the transmission control cable from the steering column.

w. Remove the steering column-to-instrument panel nuts/bolts and the lower steering column joint bolt; then, carefully, remove the steering column.

x. Remove the 2 glove compartment door screws and the door.

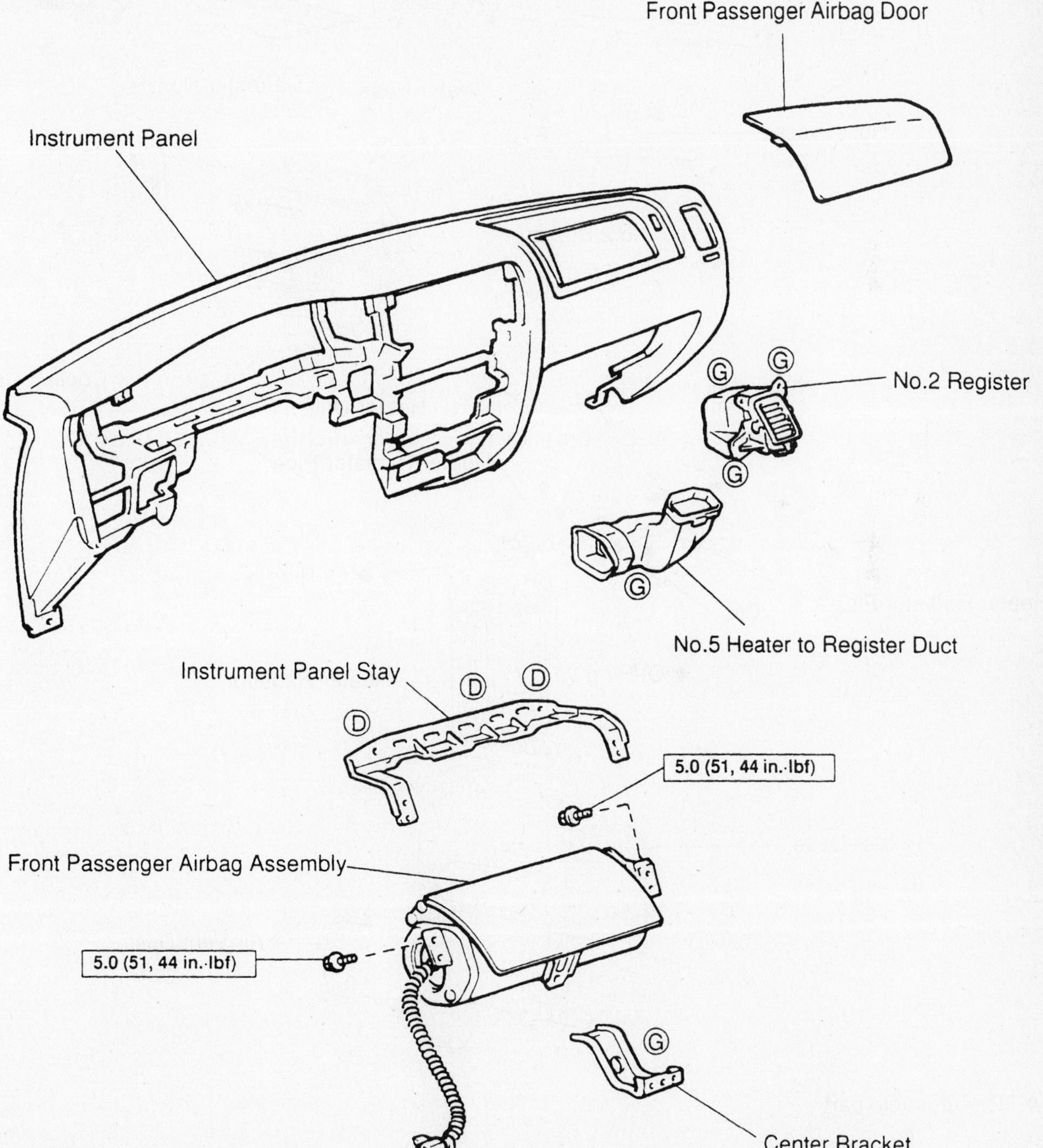

Exploded view of the instrument panel air bag module, ventilation components and brackets

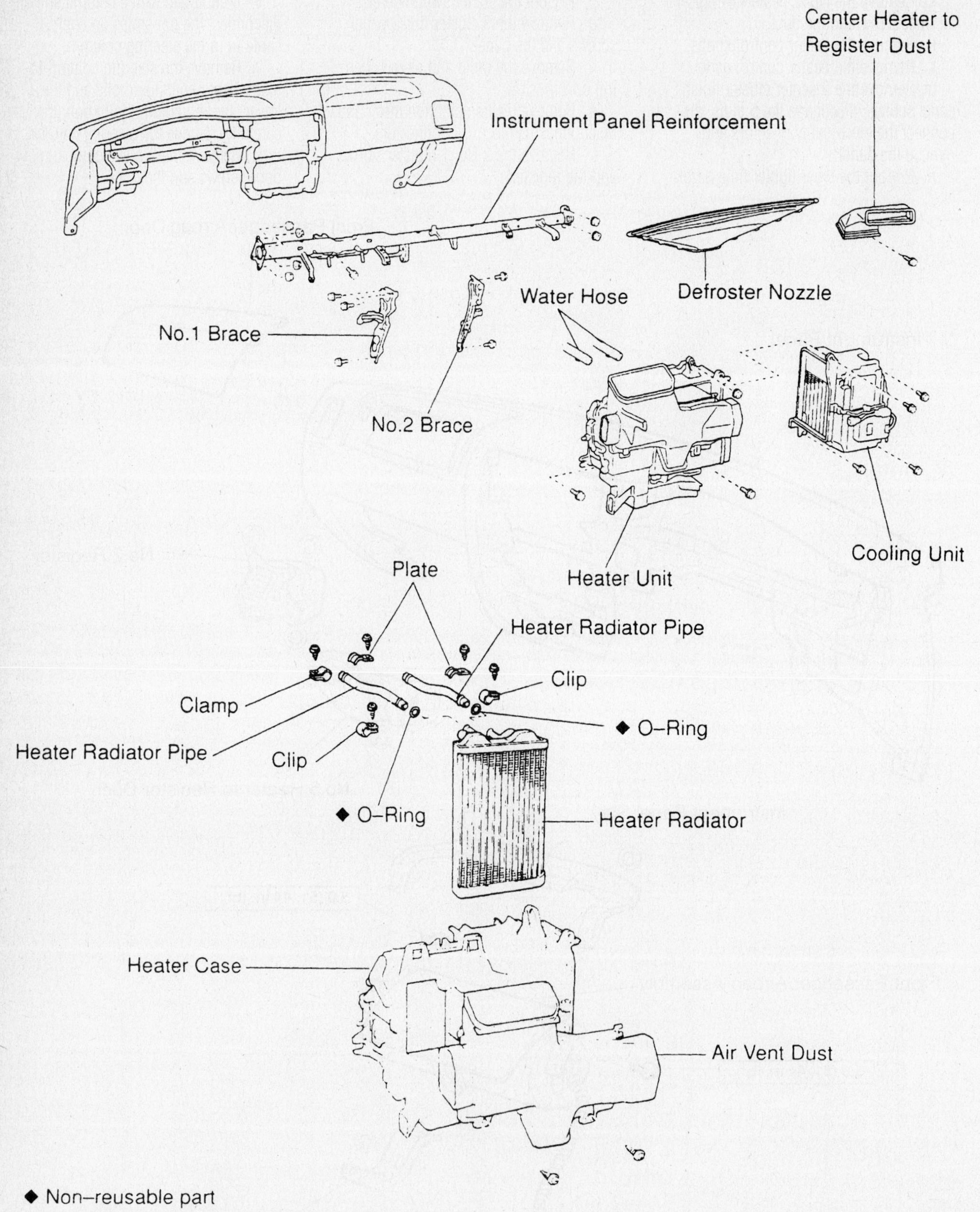

Exploded view of the front heater core, heater housing, evaporator housing and related components

y. Remove the 3 glove compartment reinforcement screws and the reinforcement.

z. Remove the No. 4 heater-to-register duct.

aa. Disconnect the No. 1 undercover and disconnect the passenger's side air bag module electrical connector.

bb. Remove the 3 lower No. 2 finish panel screws and the panel.

cc. Remove the heater control screws.

dd. Remove the lower center finish panel.

ee. Remove the instrument panel-to-chassis nuts/bolts and the instrument panel.

ff. Remove the 3 No. 1 brace bolts and the brace.

gg. Remove the 2 No. 2 brace bolts and the brace.

hh. Remove the center heater-to-register duct.

ii. Remove the defroster nozzle.

jj. Remove the reinforcement-to-chassis 3 bolts, 4 nuts and the reinforcement.

6. Remove the defroster nozzle and the heater-to-register duct.

7. Remove the evaporator housing by performing the following procedure:

a. Discharge and recover the air conditioning system refrigerant.

b. Disconnect the refrigerant lines from the evaporator core. Discard the O-rings and plug the openings to prevent contamination.

c. Remove the 2 grommets and the drain pipe grommet.

d. Disconnect the electrical connectors.

e. Remove the 3 evaporator housing-to-chassis screws, the bolt and the housing.

8. Remove the 2 heater housing-to-chassis bolts, the nut and the heater housing.

9. Remove the 3 heater core-to-heater housing screws, the 2 plates and clamp.

10. Remove the heater core from the heater housing.

To install:

11. Install the heater core to the heater housing.

12. Install the 3 heater core-to-heater housing screws, the 2 plates and clamp.

13. Install the heater housing, the nut and the 2 heater housing to-chassis bolts.

14. Install the evaporator housing by performing the following procedure:

a. Install the evaporator housing, the bolt and the 3 housing-to-chassis screws.

b. Connect the electrical connectors.

c. Install the 2 grommets and the drain pipe grommet.

d. Using new O-rings, connect the refrigerant lines to the evaporator core.

15. Install the defroster nozzle and the heater-to-register duct.

16. Install the instrument panel and reinforcement by performing the following procedure:

a. Install the reinforcement-to-chassis 3 bolts, 4 nuts and the reinforcement.

b. Install the defroster nozzle.

c. Install the center heater-to-register duct.

d. Install the No. 2 brace and the 2 brace bolts.

e. Install the No. 1 brace and the 3 brace bolts.

f. Install the instrument panel and the instrument panel-to-chassis nuts/bolts.

g. Install the lower center finish panel.

h. Install the heater control screws.

i. Install the lower No. 2 finish panel and the 3 panel screws.

j. Connect the passenger's side air bag module electrical connector and the No. 1 undercover.

k. Install the No. 4 heater-to-register duct.

l. Install the glove compartment reinforcement and the 3 reinforcement screws.

m. Install the glove compartment door and the 2 door screws.

n. Install the steering column and torque the steering column-to-instrument panel nuts/bolts to 19 ft. lbs. (26 Nm) and the lower steering column joint bolt to 26 ft. lbs. (35 Nm).

o. If equipped with a column shifter, connect the transmission control cable to the steering column.

p. Install the No. 1 heater-to-register duct and the duct screw.

q. Install the No. 1 register and the 2 register screws.

r. Install the combination meter and connect the electrical connectors.

s. Install the radio and stereo opening cover.

t. Install the cluster finish panel and the 3 panel screws.

u. Pry out the starter switch bezel.

v. Install the front ashtray and the center cluster finish panel retainer.

w. Install the cigar lighter hole bezel.

x. Install the center cluster finish panel, engage the 5 clips, connect the electrical connectors and install the 2 panel screws.

y. Install the heater control panel.

z. Install the heater control knobs.

aa. Install the No. 2 heater-to-register duct and the duct screw.

bb. Install the lower center cover and engage the 2 clips.

cc. Install the upper console box mounting bracket and the 2 bracket screws.

dd. Install the front console box.

ee. If equipped, install the rear console box and the 2 rear console box bolts/screws.

ff. Install the lower left side finish panel, the 4 bolts and the screw.

gg. Install the fuse box opening cover.

hh. Install the hood lock release lever and the 2 hood lock release lever screws.

ii. Install the combination switch-to-steering column and the combination switch screws.

jj. Install the steering column cover and the covers screws.

17. Install the steering wheel by performing the following procedure:

a. Install the steering wheel from the steering column.

b. Install the steering wheel nut and torque to 26 ft. lbs. (35 Nm).

c. Carefully, install the air bag module and connect the electrical connector.

d. Using a Torx® wrench, tighten the steering wheel screws to 78 inch lbs. (8.8 Nm).

e. At both sides of the steering wheel, install the side covers.

18. Connect the heater hoses to the heater core.

19. Refill the cooling system.

20. Connect the negative battery cable.

21. Evacuate and charge the air conditioning system refrigerant.

22. Run the engine to normal operating temperatures; then, check the climate control operation and check for leaks.

Water Pump

REMOVAL & INSTALLATION

2.4L and 2.7L Engines

1. Before servicing the vehicle, refer to the precautions in the beginning of this section.

2. Remove or disconnect the following:
   - Negative battery cable
   - Engine undercover

3. Drain the cooling system.

4. Remove or disconnect the following:
   - 2 bolts and the air pipe, for the California vehicles with 3RZ-FE engine

- Upper radiator hose from the radiator
- Oil dipstick guide, by removing the bolt
- Power steering drive belt, by loosening the lockbolt and adjusting bolt to the idler pulley, if equipped with power steering
- No. 2 fan shroud, by removing the 2 clips
- No. 1 fan shroud, by removing the 4 bolts
- Loosen the idler pulley nut and adjusting bolt and remove the air conditioning drive belt, if equipped with air conditioning

5. Remove the alternator drive belt, fan (with fan clutch), water pump pulley and the fan shroud, as follows:
   - Stretch the belt and loosen the water pump pulley mounting nuts
   - Loosen the lock, pivot and the adjusting bolts for the alternator
   - Alternator drive belt
   - 4 water pump pulley mounting nuts
   - Fan (with fan clutch) and the water pump pulley
6. Remove the water pump and discard the gasket.

To install:

7. Clean all gasket mounting surfaces.
8. Install or connect the following:
   - Apply a thin layer of liquid sealant to a new gasket
   - Place the gasket and water pump into position. Tighten the 14mm head bolts **A** to 18 ft. lbs. (25 Nm) and the 12mm head bolts to 78 inch lbs. (9 Nm).
9. Install the water pump pulley, fan shroud, fan (with fan clutch) and the alternator drive belt, as follows:
   - Fan (with the fan clutch), water pump pulley and the fan shroud in position
   - Water pump pulley mounting nuts but do not tighten the nuts at this time
   - Alternator drive belt
   - Stretch the alternator belt tight. Tighten the fan nuts to 16 ft. lbs. (21 Nm).
   - Adjust the alternator drive belt
10. Install or connect the following:
    - Adjust the drive belt, if equipped with air conditioning
    - No. 1 fan shroud, by installing the 4 bolts
    - No. 2 fan shroud, with the 2 clips
    - Adjust the power steering drive belt
    - Oil dipstick guide, with the bolt
    - Upper radiator hose to the radiator
    - Air pipe, If removed
    - Negative battery cable
11. Fill and bleed the cooling system.
12. Start the engine and check for leaks.
13. Install the engine undercover.

3.4L Engine

1. Before servicing the vehicle, refer to the precautions in the beginning of this section.
2. Remove or disconnect the following:
   - Negative battery cable
   - Engine undercover
3. Drain the engine coolant.
4. Remove the upper radiator hose.
5. Remove the power steering drive belt, as follows:
   - Stretch the belt and loosen the fan pulley mounting nuts
   - Loosen the lockbolt, pivot bolt and the adjusting bolt
   - Drive belt
6. Remove or disconnect the following:
   - Air conditioning drive belt, by loosening the idler pulley nut and adjusting bolt
   - Lockbolt, pivot bolt and the adjusting bolt
   - Alternator drive belt
   - No. 2 fan shroud, by removing the 2 clips
   - Fan with the fluid coupling and fan pulleys
   - Power steering pump and move it aside without disconnecting the lines from the pump
   - Compressor from the engine and

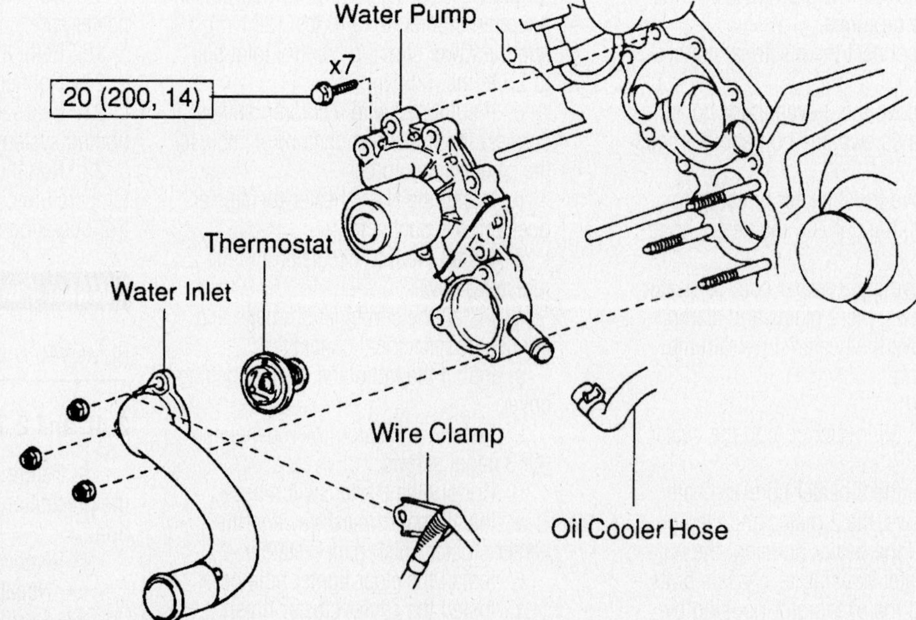

Exploded view of the water pump mounting—3.4L engine

move it aside without disconnecting the compressor lines, if equipped with air conditioning
- Air conditioning bracket, if equipped with air conditioning

7. Remove the No. 2 timing belt cover, as follows:
   - Camshaft Position (CMP) sensor connector from the No. 2 timing belt cover
   - 3 spark plug wire clamps from the No. 2 timing belt cover
   - 6 bolts and the timing belt cover
8. Remove the fan bracket, as follows:
   - Power steering adjusting strut, by removing the nut
   - Fan bracket, by removing the bolt and nut
9. Set the No. 1 cylinder to Top Dead Center (TDC) of the compression stroke, as follows:

   a. Turn the crankshaft pulley and align its groove with the timing mark **0** of the No. 1 timing belt cover.

   b. Check that the timing marks of the camshaft timing pulleys and the No. 3 timing belt cover are aligned. If not, turn the crankshaft pulley 1 revolution (360 degrees).
10. Remove the camshaft timing pulleys, as follows:

    a. Remove the timing belt tensioner by alternately loosening the 2 bolts.

    b. Using Variable Wrench Set No. 09960-10010, remove the pulley bolt, the timing pulley and the knock pin.

    c. Remove the 2 timing pulleys with the timing belt.
11. Remove or disconnect the following:
    - Thermostat
    - No. 2 oil cooler hose, from the water pump
    - Water pump, by removing the 7 bolts
12. Thoroughly clean the mating surfaces.

To install:

13. Install or connect the following:
    - Apply sealant (PN 08826-00100) to the water pump

⁂ WARNING

Parts must be assembled within 5 minutes of application. Otherwise the material must be removed and reapplied.

- Water pump. Tighten the bolts to 14 ft. lbs. (20 Nm).
- No. 2 oil cooler hose
- Thermostat
- Left camshaft timing pulley. Tighten the pulley bolt to 81 ft. lbs. (110 Nm).

14. Set the No. 1 cylinder to TDC of the compression stroke.
15. Connect the timing belt to the left camshaft timing pulley. Check that the installation mark on the timing belt is aligned with the end of the No. 1 timing belt cover, as follows:

    a. Using Variable Pin Wrench Set 09960-01000, slightly turn the left camshaft timing pulley clockwise. Align the installation mark on the timing belt with the timing mark of the camshaft timing pulley and hang the timing belt on the left camshaft timing pulley.

    b. Align the timing marks of the left camshaft pulley and the No. 3 timing belt cover.

    c. Check that the timing belt has tension between the crankshaft timing pulley and the left camshaft timing pulley.
16. Install the right camshaft timing pulley and the timing belt.
17. Set the timing belt tensioner, as follows:

    a. Using a press, slowly press in the pushrod using 220–2,205 lbs. (981–9,807 N) of force.

    b. Align the holes of the pushrod and housing, pass a 1.5mm hexagon wrench through the holes to keep the setting position of the pushrod.

    c. Release the press and install the dust boot to the tensioner.

    d. Install the timing belt tensioner and alternately tighten the bolts to 20 ft. lbs. (28 Nm).

    e. Using pliers, remove the 1.5mm hexagon wrench from the belt tensioner.
18. Check the valve timing, as follows:

    a. Slowly turn the crankshaft pulley 2 revolutions from the TDC-to-TDC; always turn the crankshaft pulley clockwise.

    b. Check that each pulley aligns with the timing marks. If the timing marks do not align, remove the timing belt and reinstall it.
19. Install or connect the following:
    - Fan bracket, with the bolt and nut
    - Remaining components
    - Negative battery cable
20. Fill with engine coolant.
21. Start the engine and check for leaks.

Cylinder Head

REMOVAL & INSTALLATION

2.4L and 2.7L Engines

1. Before servicing the vehicle, refer to the precautions in the beginning of this section.
2. Release the fuel system pressure.
3. Disconnect the negative battery cable.
4. Drain the engine coolant.
5. Remove or disconnect the following:
   - Air cleaner cap
   - Mass Air Flow (MAF) meter and resonator
   - Accelerator cable from the throttle body, if equipped with a manual transmission
   - Accelerator and throttle cables from the throttle body, if equipped with an automatic transmission
   - Cruise control cable from the actuator, if equipped with cruise control
   - Intake air connector
   - Air hose for Idle Air Control (IAC)
   - Vacuum sensing hose
   - Wire clamp for the engine wiring harness
   - Oil dipstick guide
   - Power steering belt
   - Power steering pulley, pump and bracket
   - Positive Crankcase Ventilation (PCV) hoses
   - Distributor
   - Spark plug wires from the spark plugs
   - Engine wiring harness
   - Air conditioning compressor, if equipped with air conditioning
   - Oil pressure sensor
   - Engine Coolant Temperature (ECT) sensor connector
   - ECT sender gauge connector
   - Exhaust Gas Recirculation (EGR) gas temperature sensor connector
   - Vacuum Switching Valve (VSV) connector
   - 2 vacuum hose from the VSV
   - Ground strap from the cowl top panel
   - Engine wiring harness from the air intake chamber
   - Throttle Position (TP) sensor connector
   - IAC valve connector
   - Crankshaft Position (CKP) sensor connector
   - Knock Sensor (KS) connector

- Data Link Connector 1 (DLC1) from the bracket
- Engine wiring harness clamp
- EGR pipe
- Intake chamber stay
- Air intake chamber assembly

6. Disconnect the following hoses:
   - Evaporative Emissions (EVAP) hose from the throttle body
   - Brake booster vacuum hose from the union
   - Water bypass hose from the water bypass pipe
   - Water bypass hose from the cylinder head rear cover
7. Remove or disconnect the following:
   - Injector connectors
   - Fuel inlet pipe
   - Hoses and the fuel return pipe
   - Delivery pipe and injectors
   - Intake manifold
   - Front exhaust pipe
   - Exhaust manifold and gasket
   - Water outlet
   - Cylinder head rear cover
   - Spark plugs
   - Front engine hanger
   - Engine wiring harness brackets
   - Cylinder head cover
8. Set No. 1 cylinder to Top Dead Center (TDC) of the compression stroke. The groove on the crankshaft pulley should align with the **0** mark on the timing chain cover and the timing marks (1 and 2 dots) of the camshaft gears should form a straight line in respect to the cylinder head surface. If not, turn the crankshaft 1 revolution (360 degrees).
9. Remove or disconnect the following:
   - Chain tensioner and gasket
   - Camshaft timing gear
   - Exhaust camshafts
10. Remove the intake camshaft, as follows:
    a. Uniformly, loosen and remove the bearing cap bolts in the reverse order of the tightening in several passes, in sequence.
    b. Remove the bearing caps and camshaft. Make a note of the bearing cap positions for proper installation.

➡If the camshaft is not being lifted out straight and level, reinstall the No. 3 bearing cap with the 2 bolts. Then, alternately loosen and remove the 2 bearing cap bolts with the camshaft gear pulled up.

11. Remove or disconnect the following:
    - Valve lifters and shims

➡Arrange the valve lifters and shims in correct order.

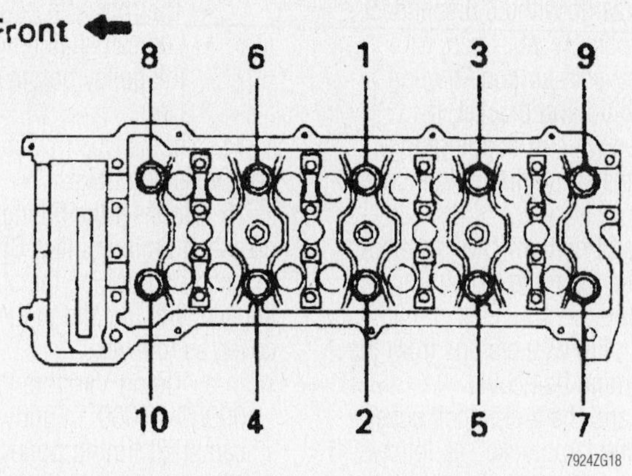

Cylinder head bolt tightening sequence—2.4L and 2.7L engines

- Cylinder head, by uniformly loosen and remove the cylinder head bolts in the reverse order of the tightening, in sequence, using several passes

To install:

12. Before installing, thoroughly clean the gasket mating surfaces and check for warpage.
13. Apply sealant (PN 08826-00080) to the 2 locations. Place a new head gasket on the block and install the cylinder head.
14. Install the cylinder head as follows:
    a. Lightly coat the cylinder head bolts with engine oil.
    b. Install the bolts and tighten, in several passes, in the sequence. Tighten all bolts to 29 ft. lbs. (39 Nm).
    c. Mark the front of the bolt with paint and retighten bolts 90 degrees in the proper sequence.
    d. Retighten an additional 90 degrees. Check that the painted mark is now facing rearward.
15. Install or connect the following:
    - Tighten the 2 front mounting bolts to 15 ft. lbs. (21 Nm).
    - Valve lifters and shims in their proper locations. Check that the valve lifter rotates smoothly by hand.
    - Intake and exhaust camshafts
16. Set No. 1 cylinder to TDC compression stroke. The groove on the crankshaft pulley should align with the **0** mark on the timing chain cover and the timing marks (1 and 2 dots) of the camshaft gears should form a straight line in respect to the cylinder head surface. If not, turn the crankshaft 1 revolution (360 degrees).
17. Install the timing gear, as follows:
    a. Place the gear over the straight pin of the intake camshaft.
    b. Hold the intake camshaft with a wrench. Install and tighten the bolt to 54 ft. lbs. (74 Nm).
    c. Hold the exhaust camshaft and install the bolt and distributor gear. Tighten the bolt to 34 ft. lbs. (46 Nm).
18. Install or connect the following:
    - Chain tensioner, using a new gasket (mark toward the front)
    - Recheck the valve timing
    - Check and adjust the valve clearance
    - Spark plugs
    - Semi-circular plug
19. Recheck the engine for proper valve timing.
20. Install or connect the following:
    - Cylinder head cover, using a new gasket
    - Engine wiring harness brackets
    - Front engine hanger. Tighten the bolts to 30 ft. lbs. (42 Nm).
    - Cylinder head rear cover. Tighten the bolts to 10 ft. lbs. (13 Nm).
    - Water outlet, using a new gasket. Tighten the bolts to 14 ft. lbs. (20 Nm).
    - Upper radiator hose
    - Exhaust manifold. Tighten the bolts to 36 ft. lbs. (49 Nm).
    - Remaining components
    - Negative battery cable
21. Fill the engine and radiator with engine coolant.
22. Start the engine and check for leaks.
23. Check the ignition timing. Road test the vehicle for proper operation.
24. Recheck all fluid levels.

3.4L Engine

1. Before servicing the vehicle, refer to the precautions in the beginning of this section.
2. Disconnect the negative battery cable.
3. Relieve the fuel system pressure.
4. Remove the engine undercover.
5. Drain the cooling system.

6. Remove or disconnect the following:
   - Front exhaust pipe
   - Air cleaner cap
   - Mass Air Flow (MAF) meter and resonator
7. Disconnect the following cables:
   - Actuator cable from the bracket, if equipped with cruise control
   - Accelerator cable
   - Throttle cable, if equipped with an automatic transmission
   - Heater hose
   - Upper radiator hose
   - Power steering drive belt
   - Air conditioning drive belt, by loosening the idle pulley nut and adjusting bolt
   - Loosen the lockbolt, pivot bolt and adjusting bolt and the alternator drive belt
   - No. 2 fan shroud by removing the 2 clips
   - Fan with the fluid coupling and fan pulleys
   - Power steering pump and move it aside without disconnecting the pump lines
   - Compressor and move it aside without disconnecting the compressor lines, if equipped with air conditioning
   - Air conditioning bracket, if equipped with air conditioning
   - Spark plug wires with the ignition coils
   - Spark plugs
   - No. 2 timing belt cover
8. Remove the fan bracket, as follows:
   a. Remove the power steering adjusting strut by removing the nut.
   b. Remove the fan bracket by removing the bolt and nut.
9. Set the No. 1 cylinder at Top Dead Center (TDC) of the compression stroke.
   a. Turn the crankshaft pulley and align its groove with the timing mark **0** on the No. 1 timing belt cover.
   b. Check that the timing marks of the camshaft timing pulleys and the No. 3 timing belt cover are aligned. If not, turn the crankshaft pulley 1 revolution (360 degrees).
10. Remove the timing belt tensioner by alternately loosening the 2 bolts.
11. Remove the camshaft timing pulleys, as follows:
   a. Using Variable Pin Wrench Set tool 09960-10010, remove the pulley bolt, the timing pulley and the knock pin.
   b. Remove the 2 timing pulleys with the timing belt.
12. Remove or disconnect the following:
   - Bolt and the No. 2 idler pulley
   - Alternator
13. Remove the Exhaust Gas Recirculation (EGR) pipe and 2 gaskets, if equipped with an EGR valve
14. Remove the intake chamber stay as follows:
   a. Remove the oil filler tube and No. 1 throttle cable clamp by removing the bolt and 2 nuts.
   b. Remove the intake chamber stay by removing the 2 bolts.
15. Remove the following connectors:
   - VSV connector for the fuel pressure control.
   - Throttle position sensor
   - IAC valve connector
   - EGR valve gas temperature sensor, if equipped
   - Vacuum Switching Valve (VSV) connector for the EGR valve, if equipped
16. Disconnect the following hoses:
   - Positive Crankcase Ventilation (PCV) hoses
   - Water bypass hoses.
   - Air assist hose from the intake air connector
   - 2 vacuum sensing hoses from the VSV
   - Evaporative Emissions (EVAP) hose
   - Air hose, from the power steering
   - Air hose from the air conditioning idle up valve, if equipped with air conditioning
17. Remove or disconnect the following:
   - 4 bolts, 2 nuts and the air intake chamber assembly
   - Intake air connector
18. Disconnect the engine wiring harness from the intake manifold, as follows:
   - Oil pressure sensor connector
   - Crankshaft position sensor connector
   - 6 injector connectors
   - Engine Coolant Temperature (ECT) sender gauge connector
   - ECT sensor connector
   - Knock (KS) sensor connector
   - Camshaft Position (CMP) sensor connector
   - 3 engine wiring harness clamps
   - 3 bolts and the engine wiring harness from the cylinder head
19. Remove or disconnect the following:
   - CMP sensor
   - No. 3 (rear) timing belt cover, by removing the 6 bolts
   - Fuel pressure regulator
   - Intake manifold assembly
   - Power steering pump bracket
   - Oil dipstick and guide

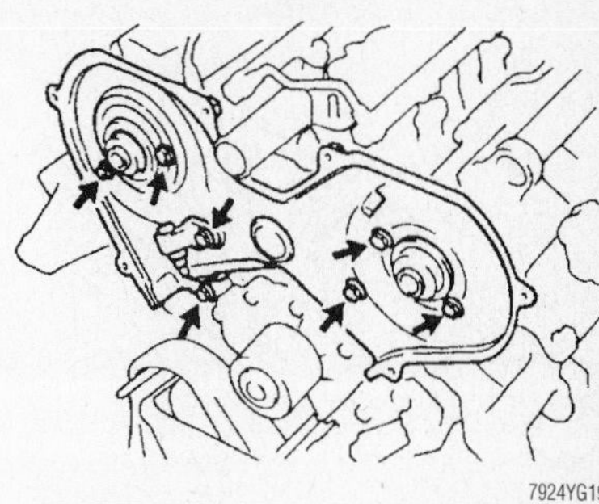

Rear timing belt cover bolt locations—3.4L engine

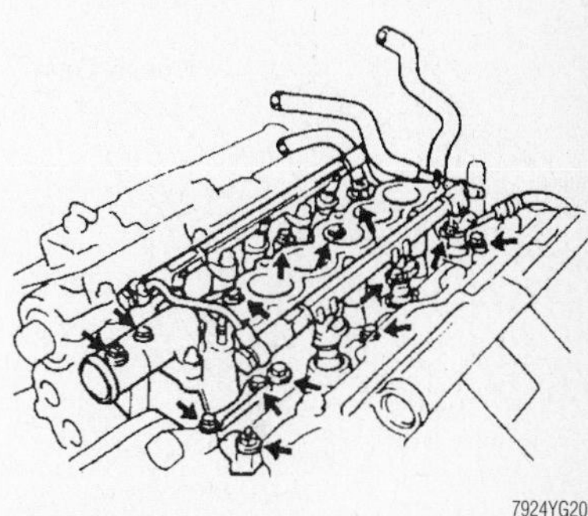

Intake manifold bolts and nuts locations—3.4L engine

   - Exhaust crossover pipe and gaskets, by removing the 6 nuts
   - Left-hand exhaust manifold, by removing the heat insulator and 6 nuts
   - Right-hand exhaust manifold, by removing the heat insulator and 6 nuts
   - 8 bolts, seal washers, cylinder head cover and gasket
   - Both cylinder head covers
   - Semi-circular plugs
   - Right exhaust camshafts
   - Right-hand intake camshaft
   - Left exhaust camshafts
   - Left-hand intake camshaft
   - Valve lifters and shims from the cylinder head; arrange the valve lifters and shims in correct order

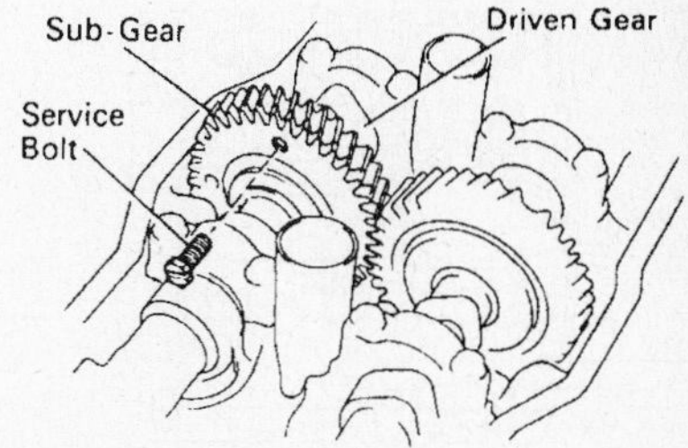

Drive gear service bolt (right side)—3.4L engine

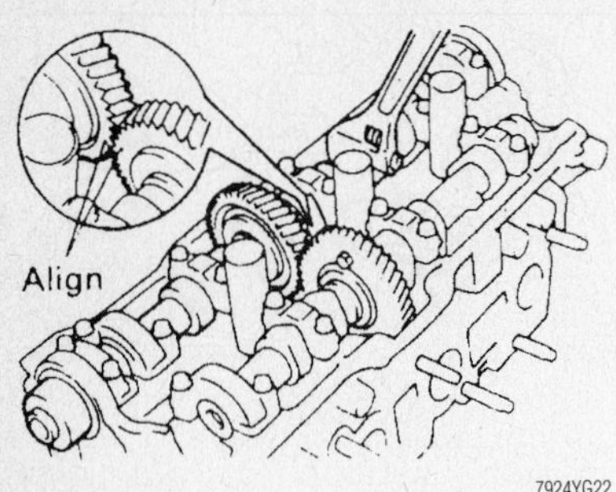

Aligning the timing mark (1 dot mark) of the left camshafts—3.4L engine

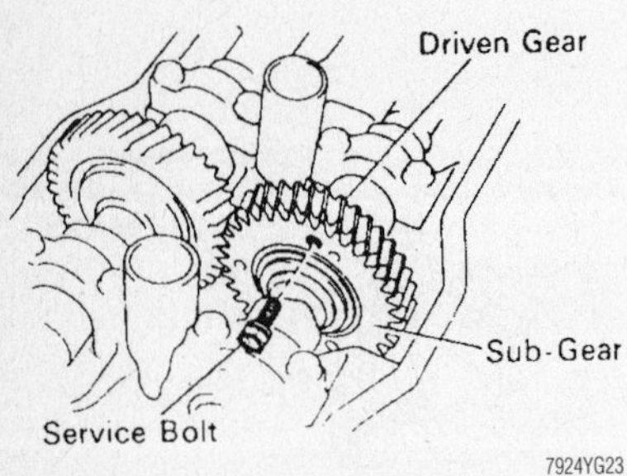

Drive gear service bolt (left side)—3.4L engine

20. Remove the cylinder heads, as follows:
  - Bolt and disconnect the ground strap
  - Cylinder head (recessed head) bolt on each cylinder head, using an 8mm hexagon wrench; then repeat for the other side
  - 8 cylinder head (12-pointed head) bolts on each cylinder head, by loosening the bolts in several

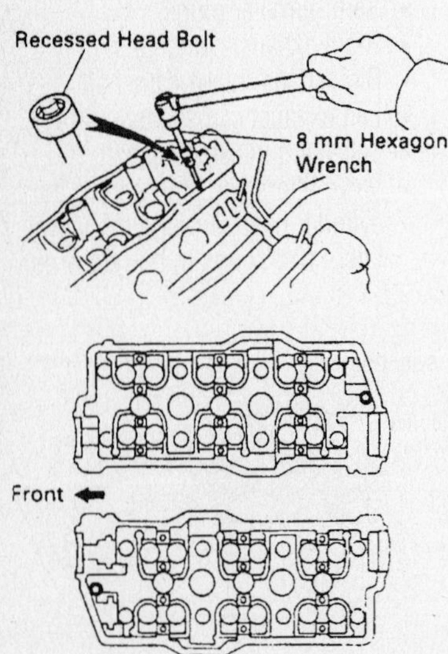

Cylinder head recessed bolts—3.4L engine

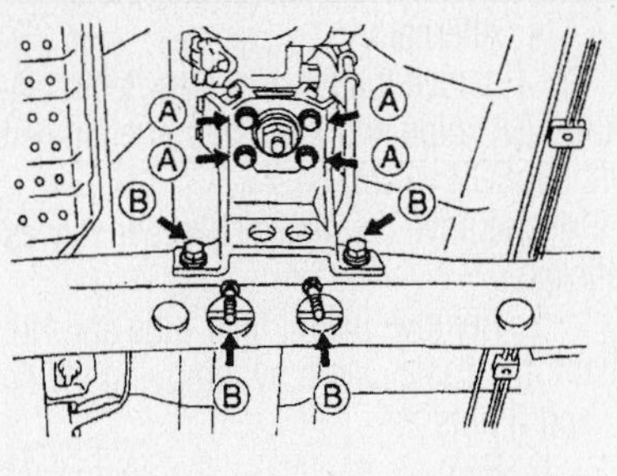

Cylinder head torque sequence—3.4L engine

passes and in the reverse order of the tightening sequence
  - 16 cylinder head bolts and plate washers.
  - Cylinder head

To install:

21. Clean all surfaces.
22. Install or connect the following:
  - New cylinder head gaskets
  - Cylinder heads
23. Apply a light coat of engine oil on the threads and under the heads of the cylinder head bolts.
24. Tighten the cylinder head bolts, in sequence, using several passes, as follows:
  - Step 1: 25 ft. lbs. (34 Nm)
  - Step 2: Mark the front of the cylinder head bolt with paint
  - Step 3: An additional 90 degrees
  - Step 4: Check that the painted mark is now at a 90 degrees angle to the front
25. Install the recessed head cylinder head bolts, as follows:
  - Apply a light coat of engine oil on the threads and under the heads of the cylinder head bolts
  - Cylinder head bolt on each cylinder head, using a 8mm hexagon wrench; then, repeat for the other side, as shown. Tighten the bolts to 13 ft. lbs. (18 Nm).
  - Bolt and the ground strap

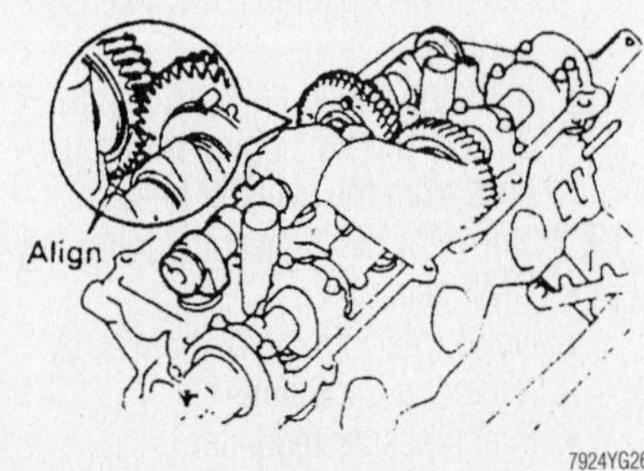

Aligning the right camshafts for installation—3.4L engine

26. Install or connect the following:
  - Valve lifters and shims

➡Check that the valve lifter rotates smoothly by hand.

  - Camshafts
  - Check and adjust the valve clearance
  - Semi-circular plugs
  - Cylinder head covers. Tighten the bolts, in several passes, to 53 inch lbs. (6 Nm).
  - Exhaust manifolds, with new gaskets. Tighten the nuts to 30 ft. lbs. (40 Nm).
  - Exhaust manifold heat insulators. Tighten the nuts to 71 inch lbs. (8 Nm).
  - Exhaust crossover pipe. Tighten the nuts to 33 ft. lbs. (45 Nm).
  - Alternator bracket. Tighten to 14 ft. lbs. (18 Nm).
  - Oil dipstick and guide, using a new O-ring
  - Power steering bracket. Tighten the fasteners to 14 ft. lbs. (18 Nm).
  - New gaskets and the intake manifold assembly. Tighten the bolts and nuts to 13 ft. lbs. (18 Nm).
  - Intake manifold stay with the 2 bolts. Tighten the bolts to 14 ft. lbs. (18 Nm).
  - Fuel inlet hose
  - Fuel pressure regulator
  - No. 3 timing belt cover. Tighten the bolts to 80 inch lbs. (9 Nm).
  - CMP sensor. Tighten to 71 inch lbs. (8 Nm).
  - Engine wiring harness
27. Install or connect the intake air connector, as follows:
  - Intake manifold. Tighten the bolts and nuts to 14 ft. lbs. (19 Nm).
  - DLC1 to the bracket on the intake manifold
  - Ground strap to the intake manifold
  - Brake booster vacuum hose to the intake air connector
  - 2 fuel return hoses
  - Engine wiring harness to the intake manifold
  - Idle up valve connector, if equipped with air conditioning
28. Install or connect the following:
  - Air intake chamber assembly. Tighten the bolts and nuts to 14 ft. lbs. (18.5 Nm).
  - Hoses
29. Attach the following connectors:
  - VSV connector for the fuel pressure control
  - TP sensor connector

- IAC valve connector
- EGR gas temperature connector, if equipped with an EGR valve
- VSV connector, if equipped with an EGR valve
- Intake chamber stay
- New gaskets and the EGR pipe. Tighten the clamp nuts to 71 inch lbs. (8 Nm) and the EGR pipe nuts to 14 ft. lbs. (18 Nm).
- Alternator but do not tighten the bolts and nuts at this time.
- No. 2 timing belt idler. Tighten the bolt to 30 ft. lbs. (40 Nm).

➡Check that the pulley bracket moves smoothly.

- Left camshaft timing pulley

30. Set the No. 1 cylinder to TDC of the compression stroke, as follows:

a. Turn the crankshaft pulley and align its groove with the timing mark **0** on the No. 1 timing belt cover.

b. Turn the camshaft, align the knock pin hole of the camshaft with the timing mark of the No. 3 timing belt cover.

c. Turn the camshaft timing pulley, align the timing marks of the camshaft timing pulley and the No. 3 timing belt cover.

31. Connect the timing belt to the left camshaft timing pulley, as follows:

a. Check that the installation mark on the timing belt is aligned with the end of the No. 1 timing belt cover.

b. Using Variable Pin Wrench Set 09960-01000, slightly turn the left camshaft timing pulley clockwise. Align the installation mark on the timing belt with the timing mark of the camshaft timing pulley and hang the timing belt on the left camshaft timing pulley.

c. Align the timing marks of the left camshaft pulley and the No. 3 timing belt cover.

d. Check that the timing belt has tension between the crankshaft timing pulley and the left camshaft timing pulley.

32. Install the right camshaft timing pulley and the timing belt, as follows:

a. Align the installation mark on the timing belt with the timing mark of the right camshaft timing pulley, and hang the timing belt on the right camshaft timing pulley with the flange side facing inward.

b. Slide the right camshaft timing pulley on the camshaft. Align the timing marks on the right camshaft timing pulley and the No. 3 timing belt cover.

c. Align the knock pin hole of the camshaft with the knock pin groove of the pulley and install the knock pin. Install the bolt and tighten to 81 ft. lbs. (110 Nm).

33. Set the timing belt tensioner as follows:

a. Using a press, slowly press in the pushrod using 220–2,205 lbs. (981–9,807 N) of force.

b. Align the holes of the pushrod and housing, pass a 1.5mm hexagon wrench through the holes to keep the setting position of the pushrod.

c. Release the press and install the dust boot to the tensioner.

34. Install the timing belt tensioner. Tighten the bolts alternately to 20 ft. lbs. (28 Nm).
35. Using pliers, remove the 1.5mm hexagon wrench from the belt tensioner.
36. Check the valve timing.
37. Install or connect the following:
    - Remaining components
    - Negative battery cable
38. Fill the radiator with engine coolant.
39. Start the engine and check for leaks.
40. Check the ignition timing.
41. Install the engine undercover.
42. Road test the vehicle.
43. Recheck all fluid levels.

Intake Manifold

REMOVAL & INSTALLATION

2.4L Engine

1. Relieve the fuel system pressure.
2. Disconnect the negative battery cable.
3. Drain the engine coolant.
4. Remove or disconnect the following:
    - Air cleaner cap
    - Mass Air Flow (MAF) meter and the resonator
    - Accelerator cable from the throttle body, if equipped with a manual transaxle
    - Accelerator and throttle cables from the throttle body, if equipped with an automatic transaxle
    - Intake air connector
    - Air conditioning idle-up valve, if equipped with air conditioning
    - No. 1 and No. 2 Positive Crankcase Ventilation (PCV) hoses
    - Spark plug wires from the spark plugs
    - Throttle body
    - Air conditioning compressor connector, if equipped with air conditioning
    - Oil pressure sensor connector
    - Engine Coolant Temperature (ECT) sensor connector
    - Exhaust Gas Recirculation (EGR) gas temperature sensor connector
    - EGR Vacuum Switching Valve (VSV) connector
5. Disconnect the engine wiring harness, as follows:
    - 2 bolts and the harness from the intake chamber
    - 5 engine harness clamps and harness
6. Remove or disconnect the following:
    - Knock (KS) sensor connector
    - Crankshaft Position (CKP) sensor connector
    - Fuel pressure control VSV connector
    - Data Link Connector 1 (DLC1) from the bracket
    - 2 engine wiring harness clamps
    - Engine wiring harness
    - Fuel injectors
    - EGR valve and vacuum modulator
    - Intake chamber stay, by removing the 2 bolts
    - Fuel return pipe, by removing the hoses and 2 bolts
7. Remove the intake chamber, as follows:
    - Vacuum hose, from the gas filter
    - Brake booster vacuum hose, from the intake chamber
    - 3 bolts, 2 nuts, air intake chamber and gasket
8. Remove the fuel inlet tube, by removing the union bolts
9. Remove the delivery pipe and injectors, as follows:
    - Vacuum hose, from the fuel pressure regulator
    - Bolts and delivery pipe together with the 4 injectors
    - 4 insulators, from the 4 spacers
    - Injectors from the delivery pipe
    - O-ring and grommet, from each injector
10. Remove or disconnect the following:
    - Intake manifold, by removing the 3 bolts and 2 nuts
    - Gasket

To install:

11. Clean the intake manifold surfaces.
12. Install or connect the following:
    - New gasket
    - Intake manifold. Tighten the bolts and nuts to 22 ft. lbs. (30 Nm).
13. Install the injectors to the delivery pipe, as follows:
    - New grommet to the injector
    - New O-ring onto the injector, by lubricating it with gasoline

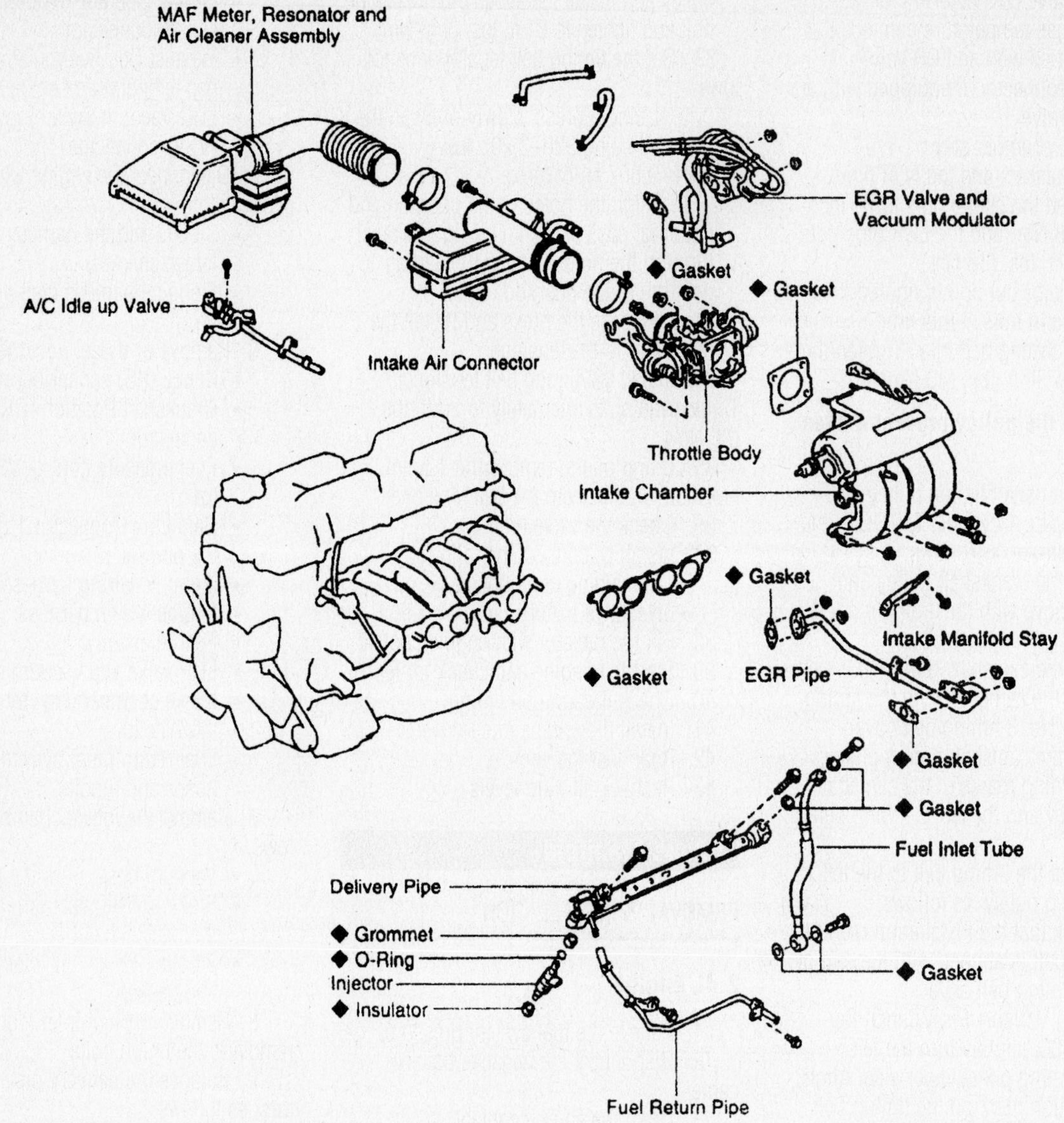

Exploded view of the intake manifold assembly—2.4L and 2.7L engines

- Injectors to the delivery pipe
- Injector with the connector facing upward
- Injectors and delivery pipe. Tighten the bolts to 15 ft. lbs. (21 Nm).

➡Check that the injectors rotate smoothly.

- Fuel tube, with new gaskets. Tighten the union bolts to 22 ft. lbs. (30 Nm).

14. Install the air intake chamber, as follows:
    - New gasket
    - Air intake chamber. Tighten the bolts and nuts to 15 ft. lbs. (21 Nm).
    - Vacuum hose, to the gas filter
    - Brake booster vacuum hose, to the intake chamber
15. Install or connect the following:
    - Fuel return pipe
    - Intake chamber stay. Tighten the bolts to 14 ft. lbs. (20 Nm).
    - EGR valve and vacuum modulator
    - Injector connectors
16. Connect the engine wiring harness to the engine, as follows:
    - Engine wiring harness to the intake manifold
    - Engine wiring harness clamps
    - DLC1 to the bracket
    - Fuel pressure control VSV connector
    - KS sensor connector
    - CKP sensor connector
    - 5 engine wiring harness clamps
    - Engine wiring harness, to the intake chamber
17. Install or connect the following:
    - EGR VSV connector
    - EGR gas temperature sensor connector
    - ECT sensor connector
    - Oil pressure sensor connector

- Compressor connector, if equipped with air conditioning
- Throttle body
- Spark plug wires, to the spark plugs
- No. 1 and No. 2 PCV hoses
- Air conditioning idle up valve, if equipped with air conditioning
- Air intake connector
- Accelerator cable, to the throttle body, if equipped with a manual transaxle
- Throttle and accelerator cables to the throttle body, if equipped with a automatic transaxle
- MAF meter, resonator and the air cleaner cap
- Negative battery cable

18. Refill the cooling system.
19. Start the engine and check for leaks.
20. Check the ignition timing. Road test the vehicle for proper operation.
21. Recheck all fluid levels.

2.7L Engine

1. Before servicing the vehicle, refer to the precautions in the beginning of this section.
2. Relieve the fuel system pressure.
3. Disconnect the negative battery cable.
4. Drain the engine coolant.
5. Remove or disconnect the following:
   - Air cleaner cap
   - Mass Air Flow (MAF) meter and resonator
   - Accelerator cable, from the throttle body, if equipped with a manual transaxle
   - Accelerator and throttle cables, from the throttle body, if equipped with an automatic transaxle
   - Intake air connector
   - Air conditioning idle-up valve, if equipped with air conditioning
   - No. 1 and No. 2 PCV hoses
   - Spark plug wires, from the spark plugs
   - Throttle body
6. Detach the following connectors:
   - Air conditioning compressor connector, if equipped with air conditioning
   - Oil pressure sensor connector
   - Engine Coolant Temperature (ECT) sensor connector
   - Exhaust Gas Recirculation (EGR) gas temperature sensor connector
   - EGR Vacuum Switching Valve (VSV) connector
7. Disconnect the engine wiring harness, as follows:
   - Engine wiring harness, from the intake chamber
   - 5 engine wiring harness clamps and engine wiring harness
   - Knock (KS) sensor connector
   - Crankshaft Position (CKP) sensor connector
   - Fuel pressure control VSV connector
   - Data Link Connector 1 (DLC1), from the bracket
   - 2 engine wiring harness clamps.
   - Engine wiring harness from the engine
8. Remove or disconnect the following:
   - Fuel injectors
   - EGR valve and vacuum modulator
   - Intake chamber stay
   - Fuel return pipe
9. Remove the intake chamber, as follows:
   - Vacuum hose, from the gas filter
   - Brake booster vacuum hose, from the intake chamber
   - Intake chamber and gasket
   - Fuel inlet tube, by removing the union bolts
10. Remove the delivery pipe and injectors, as follows:
    - Vacuum hose, from the fuel pressure regulator
    - Delivery pipe, with the 4 injectors
    - 4 insulators, from the 4 spacers
    - 4 injectors, from the delivery pipe
    - O-ring and grommet, from each injector
11. Remove or disconnect the following:
    - Intake manifold
    - Gasket

To install:

12. Clean the intake manifold surfaces.
13. Install or connect the following:
    - New gasket
    - Intake manifold. Tighten the bolts and nuts to 22 ft. lbs. (30 Nm).
14. Install the injectors to the delivery pipe, as follows:
    - New grommet onto the injector
    - New O-ring onto the injector, lubricated with gasoline
    - Injectors onto the delivery pipe, with the electrical connector upward
15. Install or connect the following:
    - Delivery pipe. Tighten the bolts to 15 ft. lbs. (21 Nm).

➡Check that the injectors rotate smoothly.

- Fuel tube, with new gaskets. Tighten the union bolts to 22 ft. lbs. (30 Nm).

16. Install the air intake chamber, as follows:
    - Air intake chamber with a new gasket. Tighten the bolts and nuts to 15 ft. lbs. (21 Nm).
    - Vacuum hose, to the gas filter
    - Brake booster vacuum hose, to the intake chamber
17. Install or connect the following:
    - Fuel return pipe
    - Intake chamber stay. Tighten the bolts to 14 ft. lbs. (20 Nm).
18. Install the EGR valve and vacuum modulator, as follows:
    - New gasket, EGR valve and vacuum modulator. Tighten the bolt to 74 inch lbs. (9 Nm) and the nuts to 14 ft. lbs. (19 Nm).
    - Vacuum hoses, to the EGR VSV
    - Water bypass hose
    - EGR pipe, using new gaskets. Tighten the bolts to 14 ft. lbs. (18 Nm), intake manifold nuts to 14 ft. lbs. (19 Nm) and the cylinder head nuts to 15 ft. lbs. (20 Nm).
    - Injector connectors
19. Connect the engine wiring harness to the engine, as follows:
    - Engine wiring harness, to the intake manifold
    - 2 engine wiring harness clamps
    - DLC1, to the bracket
    - Fuel pressure control VSV connector
    - KS sensor connector
    - CKP sensor connector
    - 5 engine wiring harness clamps.
    - Engine wiring harness, to the intake chamber
20. Attach the following connectors to the engine:
    - EGR VSV connector
    - EGR gas temperature sensor connector
    - ECT sensor connector
    - Oil pressure sensor connector
    - Compressor connector, if equipped with air conditioning
21. Install or connect the following:
    - Throttle body
    - Spark plug wires, to the spark plugs
    - No. 1 and No. 2 PCV hoses
    - Air conditioning idle up valve, if equipped with air conditioning
    - Air intake connector, by installing the 2 bolts, hose clamp and 2 air hoses
    - Accelerator cable, if equipped with a manual transaxle

- Throttle and accelerator cables, if equipped with an automatic transaxle
- MAF meter, resonator and the air cleaner cap
- Negative battery cable

22. Refill the cooling system.
23. Start the engine and check for leaks.
24. Check the ignition timing. Road test the vehicle for proper operation.
25. Recheck all fluid levels.

3.4L Engine

1. Before servicing the vehicle, refer to the precautions in the beginning of this section.
2. Disconnect the negative battery cable.
3. Relieve the fuel system pressure.
4. Drain the engine coolant.
5. Remove or disconnect the following:
   - Spark plug wires from the spark plugs
   - Air cleaner cap
   - Mass Air Flow (MAF) meter and resonator
   - Actuator cable with the bracket, if equipped with cruise control
   - Accelerator cable
   - Throttle cable, if equipped with an automatic transmission
   - Exhaust Gas Recirculation (EGR) pipe and gaskets, if equipped with an EGR valve
   - Oil filler tube and No. 1 throttle cable clamp, by removing the bolt and 2 nuts
   - Intake chamber stay, by removing the 2 bolts
   - Vacuum Switching Valve (VSV) connector, for the fuel pressure control
   - Throttle Position (TP) sensor connector
   - Idle Air Control (IAC) valve connector
   - EGR gas temperature connector, if equipped with an EGR valve
   - VSV connector for the EGR valve, if equipped with an EGR valve
   - Disconnect the Positive Crankcase Ventilation (PCV) hoses
   - Water bypass hoses
   - Air assist hose from the intake air connector
   - 2 vacuum sensing hoses from the VSV
   - Evaporative Emission (EVAP) hose
   - Air hose from the power steering
   - Air hose from the air conditioning idle up valve, if equipped with air conditioning
   - Air intake chamber assembly
   - Engine wiring harness from the intake air connector
   - 2 fuel return hoses
   - Brake booster vacuum hose, from the intake air connector
   - Ground strap, from the intake air connector
   - Data Link Connector 1 (DLC1) from the intake air connector bracket
   - Idle up valve connector, if equipped with air conditioning
   - Intake air connector
   - Upper radiator hose, from the engine
   - Oil pressure sensor connector
   - Crankshaft Position (CKP) sensor connector
   - 6 injector connectors
   - Engine Coolant Temperature (ECT) sender gauge connector
   - ECT sensor connector
   - Knock (KS) sensor connector
   - Camshaft Position (CMP) sensor connector
   - 3 engine wiring harness clamps
   - Engine wiring harness, from the cylinder head
   - Fuel pressure regulator
   - Heater hose
   - Camshaft Position sensor.
   - Fuel inlet hose
   - Intake manifold stay
   - Intake manifold assembly

To install:

6. Clean all surfaces.
7. Install or connect the following:
   - New gaskets
   - Intake manifold assembly. Tighten the bolts and nuts to 13 ft. lbs. (18 Nm).
   - Intake manifold stay. Tighten the bolts to 13 ft. lbs. (18 Nm).
   - Fuel inlet hose
   - CMP sensor. Tighten it to 71 inch lbs. (8 Nm).
   - Engine wiring harness to the cylinder head
   - 3 engine wiring harness clamps
   - Oil pressure sensor connector
   - CKP sensor connector
   - 6 injector connectors
   - ECT sender gauge connector
   - ECT sensor connector
   - KS sensor connector
   - CMP sensor connector
   - Heater hose
   - Intake manifold. Tighten the bolts and nuts to 14 ft. lbs. (18.5 Nm).
   - DLC1 to the bracket on the intake manifold
   - Ground strap to the intake manifold
   - Brake booster vacuum hose, to the intake air connector
   - 2 fuel return hoses
   - Engine wiring harness to the intake manifold
   - Idle up valve connector, if equipped with air conditioning
   - Air intake chamber assembly to the engine. Tighten the bolts and nuts to 14 ft. lbs. (18.5 Nm).
   - PCV hoses
   - Water bypass hoses
   - Air assist hose to the intake manifold
   - 2 vacuum sensing hoses to the VSV
   - EVAP hose
   - Air hose to the power steering
   - Air hose to the air conditioning idle up valve, if equipped with air conditioning
   - VSV connector for the fuel pressure control
   - TP sensor connector
   - IAC valve connector
   - EGR gas temperature connector, if equipped with an EGR valve
   - VSV connector for the EGR valve, if equipped with an EGR valve
   - Intake chamber stay. Tighten the bolts to 30 ft. lbs. (40 Nm).
   - New O-ring to the oil filler tube
   - Oil filler tube end into the tube hole in the oil pan
   - Oil filler tube and No. 1 throttle cable clamp
   - New gaskets and the EGR pipe. Tighten the clamp nuts to 71 inch lbs. (8 Nm) and the EGR pipe nuts to 14 ft. lbs. (18 Nm).
   - Fuel pressure regulator
   - 3 clamps for the spark plug wires, to the No. 2 timing belt cover
   - CMP connector to the No. 2 timing belt cover
   - Upper radiator hose
8. Fill with engine coolant.
9. Install or connect the following:
   - Spark plug wires to the spark plugs
   - Actuator cable with the bracket, if equipped with cruise control
   - Accelerator cable
   - Throttle cable, if equipped with an automatic transmission
   - Air cleaner hose
   - Negative battery cable
10. Fill the radiator with engine coolant.
11. Start the engine and check for leaks.

Exhaust Manifold

REMOVAL & INSTALLATION

2.4L and 2.7L Engines

1. Before servicing the vehicle, refer to the precautions in the beginning of this section.
2. Remove or disconnect the following:
   - Clamp from the support bracket
   - Support bracket
   - Front exhaust pipe and gaskets from the exhaust manifold
   - Heat insulator
   - Exhaust manifold and gasket

To install:

3. Install or connect the following:
   - Exhaust manifold and gasket. Tighten the nuts to 36 ft. lbs. (49 Nm).
   - Heat insulator. Tighten the bolts and nuts to 48 inch lbs. (5.5 Nm).
   - Front exhaust pipe assembly to the exhaust manifold. Tighten the nuts to 46 ft. lbs. (62 Nm).
   - Support bracket. Tighten the bolts to 29 ft. lbs. (39 Nm).
   - Clamp. Tighten the bolt to 14 ft. lbs. (19 Nm).
4. Start the engine.
5. Check for exhaust leaks.

Front exhaust pipe to exhaust manifold nut and bolt locations—2.4L and 2.7L engines

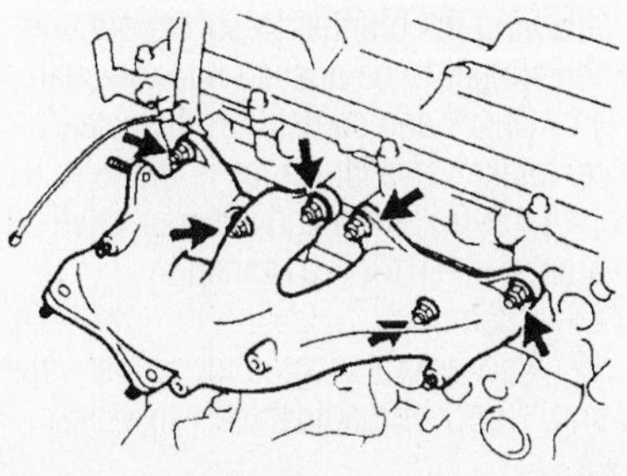

Exhaust manifold nuts—2.4L and 2.7L engines

3.4L Engine

1. Before servicing the vehicle, refer to the precautions in the beginning of this section.
2. Remove or disconnect the following:
   - Exhaust crossover pipe, from the exhaust manifold by removing the 3 nuts
   - Exhaust Gas Recirculation (EGR) pipe, from the exhaust manifold, on the left manifold equipped with an EGR valve
   - Exhaust manifold heat insulator, by removing the 3 nuts
   - Exhaust manifold

To install:

3. Install or connect the following:
   - Exhaust manifold, using a new gasket. Tighten the nuts to 30 ft. lbs. (40 Nm).
   - Exhaust heat insulator. Tighten the nuts to 71 inch lbs. (8 Nm).
   - EGR pipe to the exhaust manifold, if equipped with an EGR valve. Tighten the manifold nuts to 14 ft. lbs. (18 Nm) and the clamp nuts to 71 inch lbs. (8 Nm).
   - Crossover pipe to the exhaust manifold, using a new gasket. Tighten the nuts to 33 ft. lbs. (45 Nm).

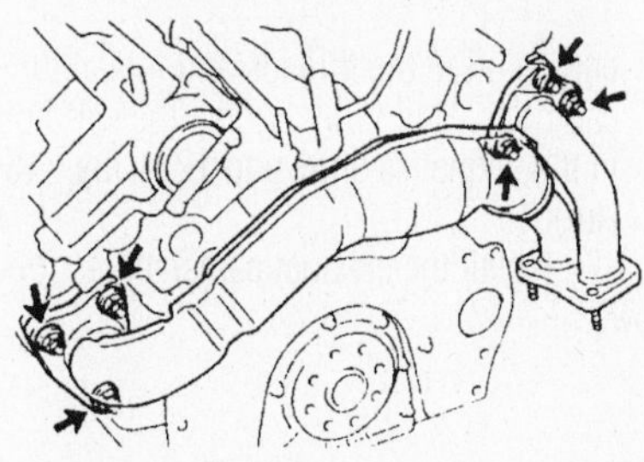

Exhaust crossover pipe mounting nut locations—3.4L engine

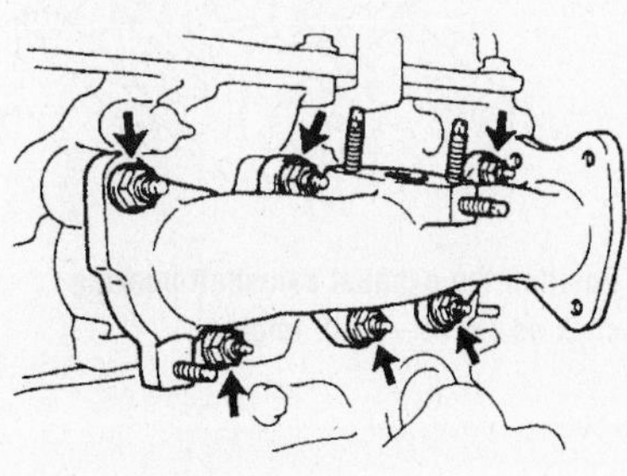

Exhaust manifold nuts—3.4L engine

Front Crankshaft Seal

REMOVAL & INSTALLATION

3.4L Engine

➡There are 2 methods to replace the oil seal, which are as follows:

OIL PUMP BODY INSTALLED

1. Before servicing the vehicle, refer to the precautions in the beginning of this section.
2. Remove or disconnect the following:
   - Negative battery cable
   - Timing belt and crankshaft pulley
   - Cut off the oil seal lip, using a knife
   - Pry out the oil seal, using a suitable tool

✱✱ WARNING

Be careful not to damage the crankshaft.

To install:

3. Install or connect the following:
   - Apply multi-purpose grease to the new oil seal lip
   - Tap in the new oil seal until its surface is flush with the oil pump case edge, using Seal Driver tool 09309-37010 and a mallet
   - Crankshaft pulley and the timing belt
   - Engine undercover, if removed
   - Negative battery cable

OIL PUMP BODY REMOVED

1. Before servicing the vehicle, refer to the precautions in the beginning of this section.
2. Carefully pry out the seal using a suitable tool.
3. Apply multi-purpose grease to the new oil seal lip.
4. Using Seal Driver tool 09309-37010, drive the new seal into place.

Camshaft and Valve Lifters

REMOVAL & INSTALLATION

2.4L Engine

1. Before servicing the vehicle, refer to the precautions in the beginning of this section.

2. Remove or disconnect the following:
   - Timing chain
   - Exhaust camshaft by bringing the service bolt hole of the driven sub-gear upwards. Turn the hexagon wrench head portion of the exhaust camshaft with a wrench.

3. Secure the exhaust camshaft sub-gear to the main gear with a service bolt. The thread diameter should be 0.23 in. (6mm) with a thread pitch of 0.04 in. (1.0mm) and a bolt length of 0.63–0.79 in. (16–20mm).

➡When removing the camshaft, be sure that the torsional spring force of the sub-gear has been eliminated by the above operation.

4. Uniformly loosen and remove the exhaust bearing cap bolts (10 of them), in several passes. Use the reverse order of the tightening sequence. Remove the 5 bearing caps and the camshaft. Do the same for the intake camshafts.

➡If the camshaft is not being lifted out straight and level, reinstall the No. 3 cap with the 2 bolts. Alternately loosen, then remove the bearing cap bolts with the camshaft pulled up. Do not pry on or force the camshaft.

5. Inspect the camshafts for excessive runout. Inspect the cam lobes and journals. The bearings are part of the cam and should be inspected for flaking or scoring. If the bearings are damaged, replace the caps and the cylinder head as a set. The camshaft journal oil and thrust clearances should be checked.

To install:

6. Install the intake camshaft, as follows:
   a. Apply multi-purpose grease to the thrust portion of the intake camshaft.
   b. Position the intake camshaft with the pin facing upward.
   c. Install the bearing caps in their proper locations. Apply a light coat of

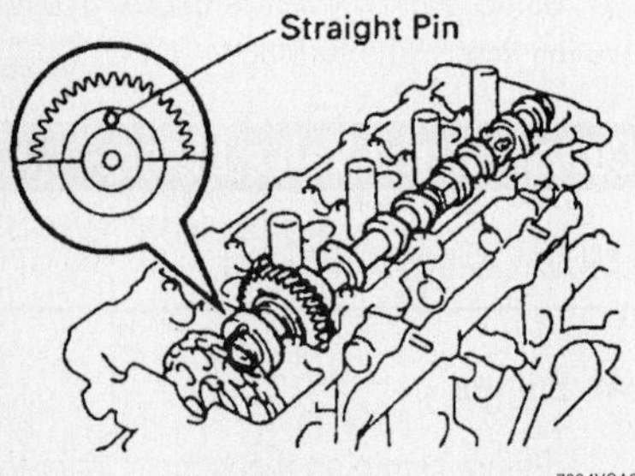

Install the camshaft with the pin facing upwards—2.4L engine

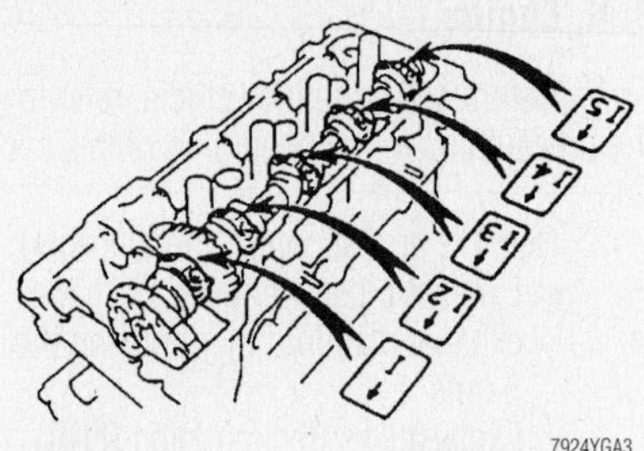

Intake camshaft bearing cap locations—2.4L engine

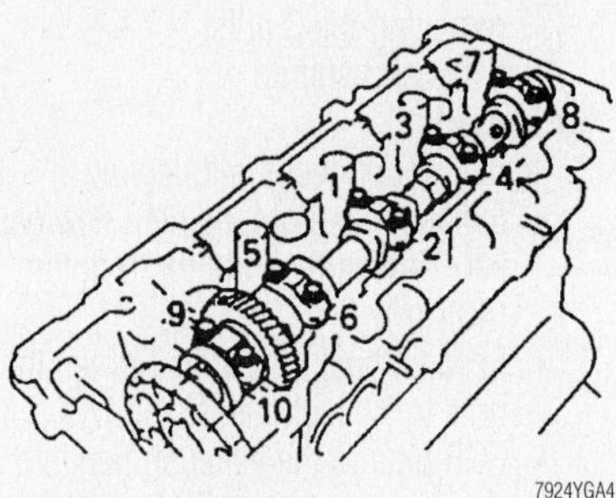

Tighten the intake bearing caps following this order—2.4L engine

engine oil to the threads and install the cap bolts. Uniformly tighten the cap bolts in the sequence shown to 12 ft. lbs. (16 Nm).

7. Install the exhaust camshaft, as follows:

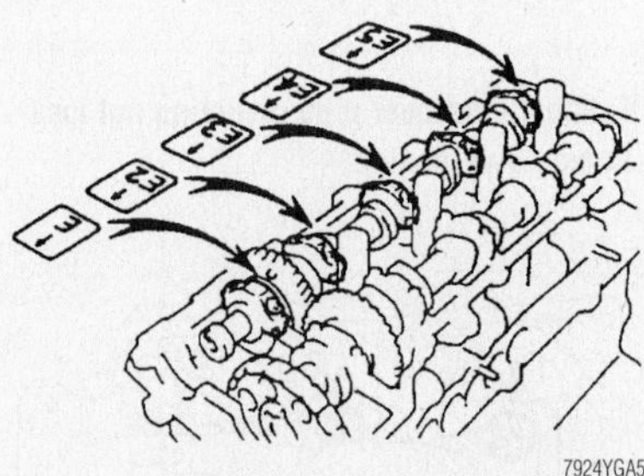

Position the exhaust camshaft bearing caps as shown—2.4L engine

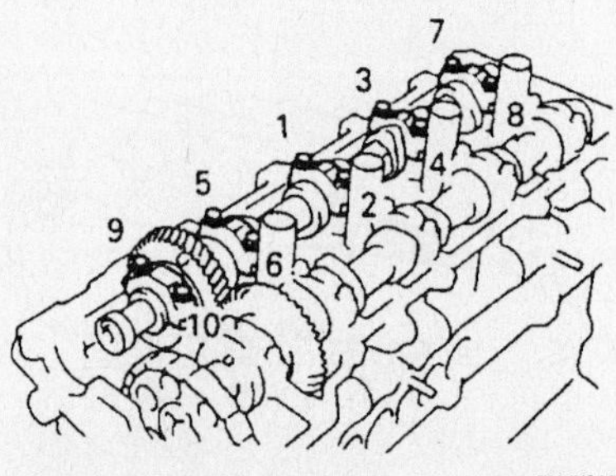

Tighten the exhaust bearing cap bolts in this order—2.4L engine

   a. Apply engine oil to the thrust portion of the intake camshaft.
   b. Engage the exhaust camshaft gear to the intake camshaft gear by matching the timing marks (1 and 2 dots) on each other.
   c. Roll down the exhaust camshaft onto the bearing journals while engaging the gears with each other. Install the bearing caps in their proper locations.
   d. Apply a light coat of engine oil to the threads and install the cap bolts. Uniformly tighten the cap bolts in the sequence shown to 12 ft. lbs. (16 Nm).
   e. Remove the service bolt from the driven sub-gear. Check that the intake and exhaust camshafts turn smoothly.

8. Set No. 1 cylinder to Top Dead Center (TDC) of the compression stroke. The crankshaft pulley groove aligns with the **0** mark on timing cover and camshaft timing marks with 1 dot and 2 dots will be in a straight line on the cylinder head surface.

9. Install the timing gear, as follows:
   a. Place the gear over the straight pin of the intake camshaft.
   b. Hold the intake camshaft with a wrench. Install and tighten the bolt to 54 ft. lbs. (74 Nm).
   c. Hold the exhaust camshaft and install the bolt and distributor gear. Tighten the bolt to 34 ft. lbs. (46 Nm).

10. Install the chain tensioner, using a new gasket (mark toward the front), as follows:
   a. Release the ratchet pawl, fully push in the plunger and apply the hook to the pin so that the plunger cannot spring out.
   b. Turn the crankshaft pulley clockwise to provide some slack for the chain on the tensioner side.
   c. Push the tensioner by hand until it touches the head installation surface, then install the 2 nuts. Tighten the nuts to 13 ft. lbs. (18 Nm). Check that the hook of the tensioner is not released.
   d. Turn the crankshaft to the left so that the hook of the chain tensioner is released from the pin of the plunger, allowing the plunger to spring out and the slipper to be pushed into the chain.

11. Check and adjust the valve clearance. Intake valve clearance is 0.006–0.010 inch (0.15–0.25mm) and exhaust valve clearance is 0.010–0.014 inch (0.25–0.35mm).

12. Recheck the engine for proper valve timing. Check and adjust the valve clearance.

13. Install the spark plugs and the semi-circular plug.

14. Recheck the engine for proper valve

timing. Install the valve cover and engine hangers. Tighten the engine hanger bolts to 30 ft. lbs. (42 Nm).

15. Reinstall all other parts from the timing chain removal. Fill any fluids, start the engine, top off the fluids.

2.7L Engine

1. Before servicing the vehicle, refer to the precautions in the beginning of this section.
2. Disconnect the negative battery cable.
3. Drain the engine coolant.
4. Remove or disconnect the following:
   - Air cleaner cap
   - Mass Air Flow (MAF) meter and the resonator
   - Accelerator cable from the throttle body, if equipped with a manual transmission
   - Accelerator and throttle cables from the throttle body, if equipped with an automatic transmission
   - Cruise control cable from the actuator, if equipped with cruise control
   - Intake air connector
   - Air hose for Idle Air Control (IAC)
   - Vacuum sensing hose
   - Wire clamp for the engine wiring harness
   - Positive Crankcase Ventilation (PCV) hoses
   - Spark plug wires from the spark plugs
   - Engine wiring harness clamps and harness
   - Air conditioning compressor connector, if equipped with air conditioning
   - Oil pressure sensor connector
   - Engine Coolant Temperature (ECT) sensor connector
   - Engine coolant temperature sender gauge connector
   - Exhaust Gas Recirculation (EGR) gas temperature sensor connector
   - Vacuum Switching Valve (VSV) connector for the EGR
   - 2 vacuum hoses from the VSV for the EGR
   - Ground strap from the cowl top panel
   - Engine wiring harness from the air intake chamber
   - Throttle Position (TP) sensor connector
   - IAC valve connector
   - Crankshaft Position (CKP) sensor connector
   - Knock (KS) sensor connector
   - Data Link Connector 1 (DLC1) from the bracket
   - Engine wiring harness clamp
   - EGR pipe
   - Intake chamber stay
   - Air intake chamber assembly.
   - Evaporative Emission (EVAP) hose from the throttle body
   - Brake booster vacuum hose from the union
   - Water bypass hose from the water bypass pipe
   - Water bypass hose from the cylinder head rear cover
   - Front engine hanger
   - Engine wiring harness brackets
   - Cylinder head cover
5. Set No. 1 cylinder to Top Dead Center (TDC) compression stroke. The groove on the crankshaft pulley should align with the **0** mark on the timing chain cover and the timing marks (1 and 2 dots) of the camshaft gears should form a straight line

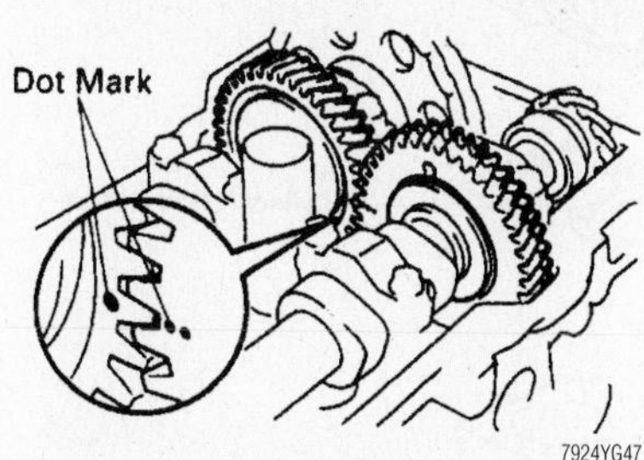

Camshafts TDC/compression timing marks. Marks with 1 and 2 dots will be in straight line on cylinder head surface—2.7L engine

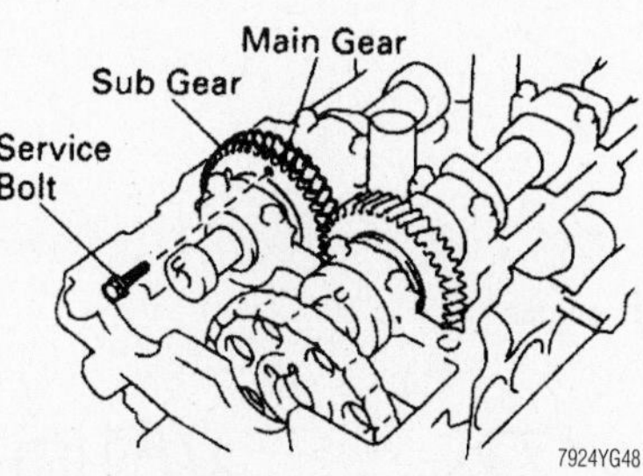

Secure the exhaust camshaft sub-gear to the main gear with a service bolt—2.7L engine

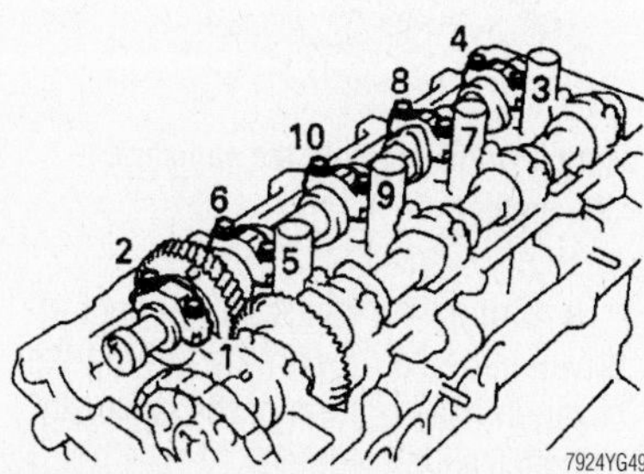

Loosen and remove the exhaust camshaft bearing cap bolts in sequence—2.7L engine

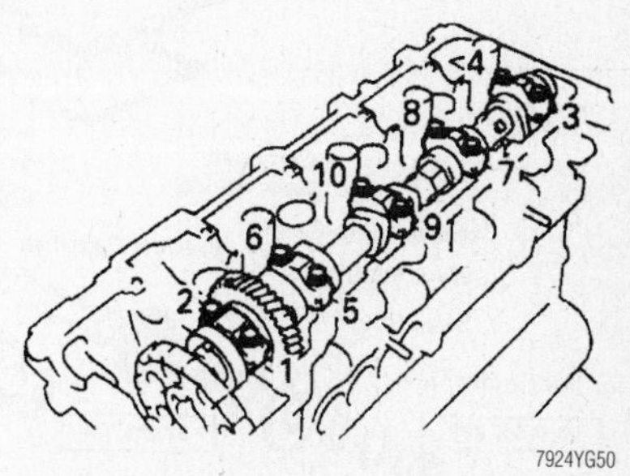

Loosen and remove the intake camshaft bearing cap bolts in sequence—2.7L engine

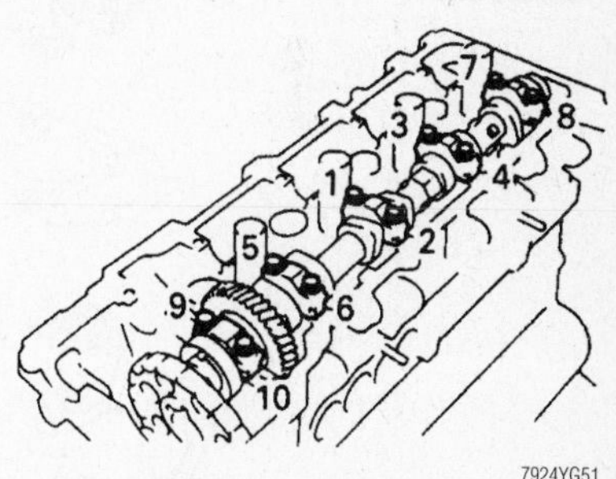

Tighten the intake camshaft bearing cap bolts in sequence—2.7L engine

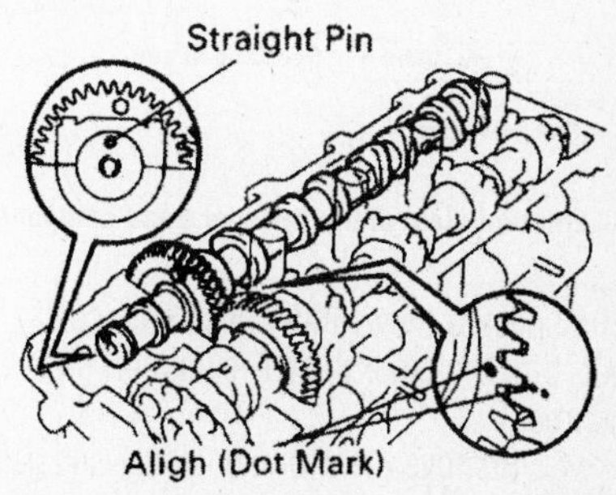

Engage both camshaft gears while matching the timing marks—2.7L engine

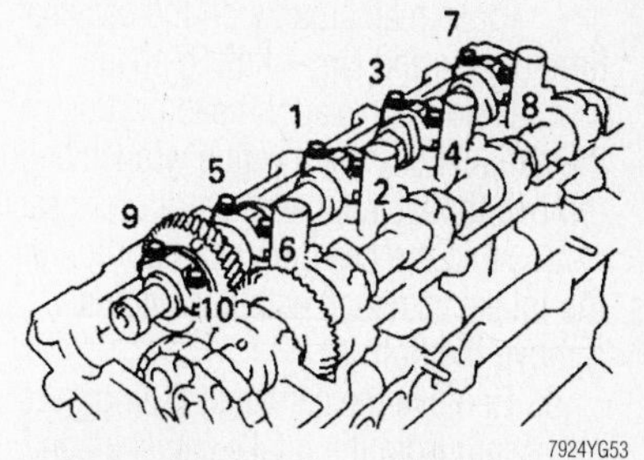

Tighten the exhaust camshaft bearing cap bolts in sequence—2.7L engine

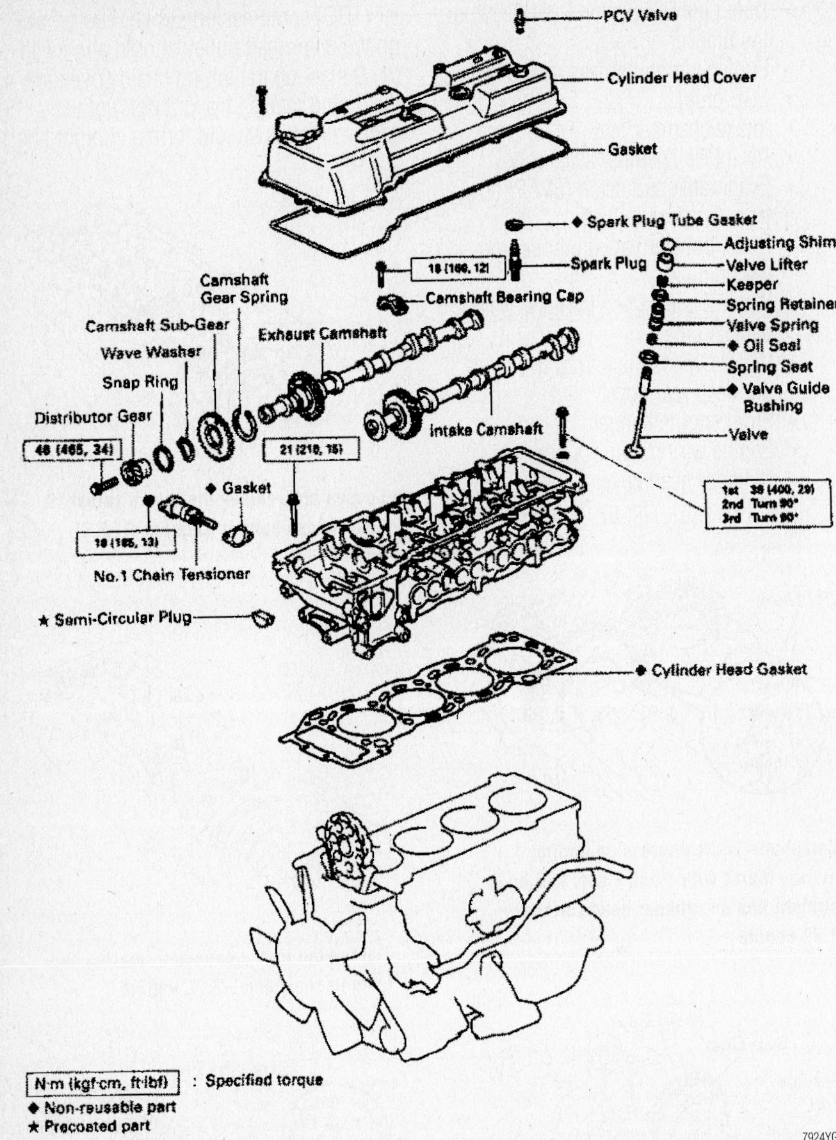

Exploded view of the cylinder head components—2.7L engine

in respect to the cylinder head surface. If not, turn the crankshaft 1 revolution (360 degrees).

6. Remove the chain tensioner and gasket.

7. Remove the camshaft timing gear as follows:

a. Remove the 2 semi-circular plugs.

b. Place matchmarks on the camshaft timing gear and No. 1 timing chain.

c. Hold the hexagon head portion of the exhaust camshaft with a wrench and remove the fastener and distributor gear.

d. Hold the hexagon head portion of the intake camshaft with a wrench and remove the bolt.

e. Remove the camshaft timing gear and chain from the intake camshaft and leave on the slipper and damper.

8. Remove exhaust camshafts:

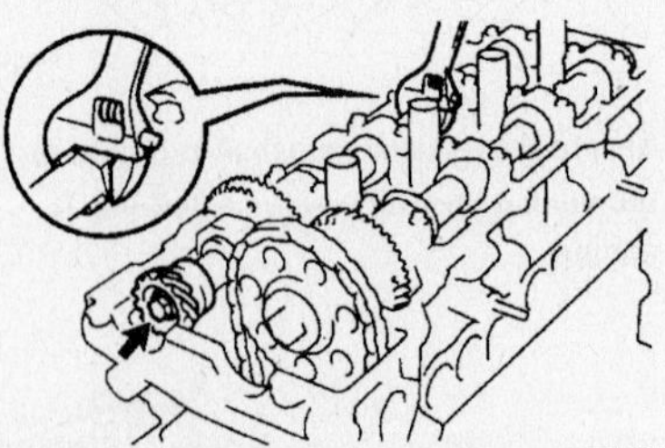

Using a wrench to hold the camshaft—2.7L engine

a. Bring the service bolt hole of the driven sub-gear upward by turning the hexagon head portion of the exhaust camshaft with a wrench.

b. Secure the exhaust camshaft sub-gear to the driven gear with a service bolt (6mm diameter, 0.63–0.79 inches in length and 1.0mm in thread pitch).

➡When removing the camshaft, be sure that the torsional spring force of the sub-gear has been eliminated by the above operation.

c. Uniformly loosen and remove the bearing cap bolts in several passes, in the sequence shown.

d. Remove the bearing caps and camshaft. Make a note of the bearing cap positions for proper installation.

9. Remove or disconnect the following:

- Intake camshaft bearing cap bolts in several passes, in the sequence shown
- Bearing caps and camshaft. Make a note of the bearing cap positions for proper installation.

➡If the camshaft is not being lifted out straight and level, reinstall the No. 3 bearing cap with the 2 bolts. Then, alternately loosen and remove the 2 bearing cap bolts with the camshaft gear pulled up.

- Valve lifters and shims from the cylinder head. Arrange the valve lifters and shims in correct order.

To install:

10. Install the valve lifters and shims in their proper locations. Check that the valve lifter rotates smoothly by hand.

11. Install the intake camshaft, as follows:

a. Apply engine oil to the thrust portion of the intake camshaft.

b. Position the intake camshaft with the knock pin facing upward.

c. Install the bearing caps in their proper locations. Apply a light coat of engine oil to the threads and install the cap bolts. Uniformly tighten the cap bolts, in sequence, to 12 ft. lbs. (16 Nm).

12. Install the exhaust camshaft, as shown:

a. Apply engine oil to the thrust portion of the intake camshaft.

b. Engage the exhaust camshaft gear to the intake camshaft gear by matching the timing marks (1 and 2 dots) on each other.

c. Roll down the exhaust camshaft onto the bearing journals while engaging the gears with each other. Install the bearing caps in their proper locations.

d. Apply a light coat of engine oil to the threads and install the cap bolts. Uniformly tighten the cap bolts in the sequence shown to 12 ft. lbs. (16 Nm).

e. Remove the service bolt from the driven sub-gear. Check that the intake and exhaust camshafts turns smoothly.

13. Set No. 1 cylinder to Top Dead Center (TDC) of the compression stroke: Crankshaft pulley groove align with **0** mark on timing cover and camshafts timing marks with 1 dot and 2 dots will be straight line on the cylinder head surface.

14. Install the timing gear. Place the gear over the straight pin of the intake camshaft.

a. Hold the intake camshaft with a wrench. Install and tighten the bolt to 54 ft. lbs. (74 Nm).

b. Hold the exhaust camshaft and install the bolt and distributor gear. Tighten the bolt to 34 ft. lbs. (46 Nm).

15. Install or connect the following:

- Chain tensioner, using a new gasket (mark toward the front)
- Recheck the engine for proper valve timing. Check and adjust the valve clearance.
- Semi-circular plug
- Recheck the engine for proper valve timing.
- Cylinder head cover with a new gasket
- Engine wiring harness brackets
- Front engine hanger. Tighten the bolts to 30 ft. lbs. (42 Nm).
- Air intake chamber assembly. Tighten the bolts to 15 ft. lbs. (20 Nm).
- Hoses
- Intake chamber stay
- Air intake chamber stay. Tighten the bolts to 15 ft. lbs. (20 Nm).
- EGR pipe. Tighten the bolts to 13 ft. lbs. (18 Nm), nut "A" to 14 ft. lbs. (19 Nm) and nut "B" to 15 ft. lbs. (20 Nm).
- Engine wiring harness
- Spark plug wires to the spark plugs
- PCV hoses
- Intake air connector. Tighten the bolts to 13 ft. lbs. (18 Nm).
- Air hose for the IAC
- Vacuum sensing hose
- Wire clamp for the engine wiring harness
- Cruise control cable to the actuator, if equipped with cruise control
- Accelerator cable to the throttle body, if equipped with a manual transmission
- Accelerator and throttle cables to the throttle body, if equipped with an automatic transmission
- Air cleaner cap, MAF meter and the resonator assembly
- Negative battery cable

16. Refill the cooling system.

17. Start the engine and check for leaks.

18. Check the ignition timing. Road test the vehicle for proper operation.

19. Recheck all fluid levels.

3.4L Engine

1. Before servicing the vehicle, refer to the precautions in the beginning of this section.

2. Remove or disconnect the following:

- Negative battery cable
- Engine undercover

3. Drain the cooling system.

4. Remove or disconnect the following:

- Air cleaner cap
- Mass Air Flow (MAF) meter and the resonator
- Actuator cable with the bracket, if equipped with cruise control
- Accelerator cable
- Throttle cable, if equipped with an automatic transmission
- Heater hose
- Upper radiator hose from the engine

5. Remove the power steering drive belt, as follows:

a. Stretch the belt and loosen the fan pulley mounting nuts.

b. Loosen the lockbolt, pivot bolt and adjusting bolt and remove the drive belt from the engine.

6. Remove or disconnect the following:

- Air conditioning drive belt, by loosening the idle pulley nut and adjusting bolt
- Loosen the alternator lockbolt, pivot bolt and adjusting bolt
- Alternator drive belt
- No. 2 fan shroud, by removing the 2 clips
- Fan with the fluid coupling and fan pulleys
- Power steering pump and move it aside without disconnecting the lines
- Compressor and move it aside without disconnecting the lines, if equipped with air conditioning
- Air conditioning bracket, if equipped with air conditioning
- Spark plug wires with the ignition coils
- Spark plugs
- Camshaft Position (CMP) sensor connector, from the No. 2 timing belt cover
- 3 spark plug wire clamps from the No. 2 timing belt cover
- 6 bolts and the timing belt cover
- Power steering adjusting strut by removing the nut
- Fan bracket by removing the bolt and nut

7. Set the No. 1 cylinder at Top Dead Center (TDC) of the compression stroke, as follows:

a. Turn the crankshaft pulley and align its groove with the timing mark **0** on the No. 1 timing belt cover.

b. Check that the timing marks of the camshaft timing pulleys and the No. 3 timing belt cover are aligned. If not, turn the crankshaft pulley 1 revolution (360 degrees).

8. Remove or disconnect the following:

- Timing belt tensioner, by alternately loosening the 2 bolts
- Pulley bolt, the timing pulley and the knock pin, using Variable Pin Wrench Set 09960-10010
- Both timing pulleys with the timing belt
- No. 2 idler pulley
- Alternator
- Positive Crankcase Ventilation PCV hoses
- Water bypass hoses
- Air assist hose from the intake air connector
- 2 vacuum sensing hoses from the Vacuum Switching Valve (VSV)
- Evaporative Emissions (EVAP) hose
- Air hose from the power steering
- Air hose from the air conditioning idle up valve, if equipped with air conditioning
- 4 bolts, 2 nuts and the air intake chamber assembly
- Intake air connector
- Camshaft Position (CMP) sensor
- No. 3 (rear) timing belt cover by removing the 6 bolts
- 8 bolts, seal washers, both cylinder head cover and gaskets
- Semi-circular plugs

9. Remove the right exhaust camshafts, as follows:

a. Bring the service bolt hole of the driven sub-gear upward by turning the hexagon head portion of the exhaust camshaft with a wrench.

b. Align the timing mark (2 dot marks) of the camshaft drive and driven gears by turning the camshaft with a wrench.

c. Secure the exhaust camshaft sub-gear to the driven gear with a service bolt (6mm diameter, 16–20mm bolt length and 1.0mm in thread pitch).

➡When removing the camshaft, be sure the torsional spring force of the sub-gear has been eliminated by the above operation.

d. Uniformly loosen and remove the bearing cap bolts in several passes, in the sequence shown.

e. Remove the bearing caps and camshaft. Make a note of the bearing cap positions for proper installation.

➡Do not pry on or attempt to force the camshaft with a tool or other object.

10. Remove the right-hand intake camshaft, as follows:

a. Uniformly loosen and remove the bearing cap bolts in several passes, in the sequence shown.

b. Remove the bearing caps, oil seal and camshaft. Make a note of the bearing cap positions for proper installation.

11. Remove the left exhaust camshafts, as follows:

a. Align the timing mark (1 dot mark) of the camshaft drive and driven gears by turning the camshaft with a wrench.

b. Secure the exhaust camshaft sub-gear to the driven gear with a service bolt (6mm diameter, 16–20mm bolt length and 1.0mm in thread pitch).

➡When removing the camshaft, be sure that the torsional spring force of the sub-gear has been eliminated by the above operation.

c. Uniformly loosen and remove the bearing cap bolts in several passes, in the sequence shown.

d. Remove the bearing caps and camshaft. Make a note of the bearing cap positions for proper installation.

12. Remove the left-hand intake camshaft, as follows:

a. Uniformly loosen and remove the bearing cap bolts in several passes, in the sequence shown.

b. Remove the bearing caps, oil seal and camshaft. Make a note of the bearing cap positions for proper installation.

13. Remove the valve lifters and shims from the cylinder head. Arrange the valve lifters and shims in correct order.

To install:

14. Clean all surfaces.

15. Install the valve lifters and shims. Check that the valve lifter rotates smoothly by hand.

16. Install the right intake camshaft, as follows:

a. Apply engine oil to the thrust portion of the intake camshaft.

b. Position the intake camshaft at 90

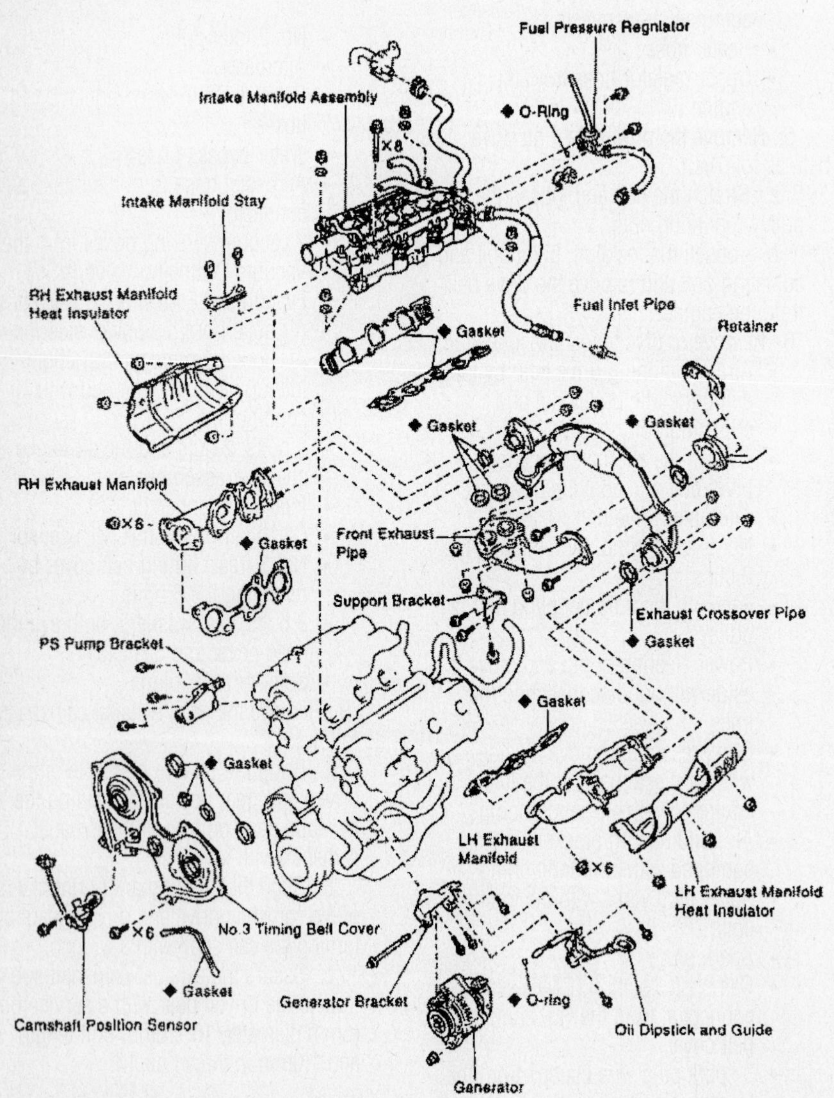

Exploded view of the cylinder head component assembly—3.4L engine

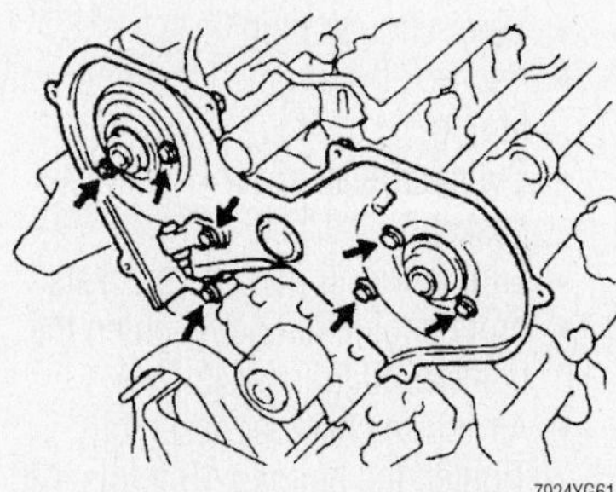

Rear timing belt cover mounting bolt locations—3.4L engine

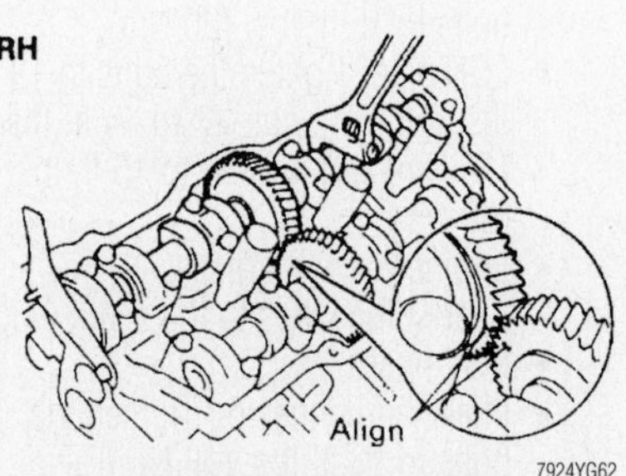

Aligning the timing marks (2 dot marks) of the right camshafts—3.4L engine

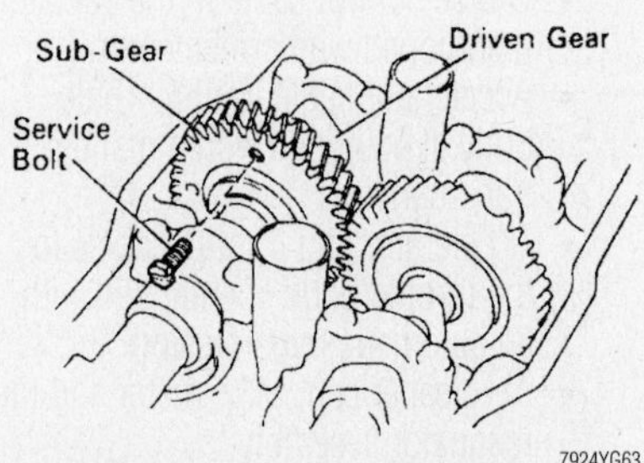

Drive gear service bolt (right side)—3.4L engine

degrees angle of the timing mark (2 dot marks) on the cylinder head.

c. Install the bearing caps in their proper locations. Apply a light coat of engine oil to the threads and install the cap bolts.

d. Apply a light coat of engine oil on the threads and under the heads of the bearing cap bolts.

e. Uniformly tighten the cap bolts in the sequence shown to 12 ft. lbs. (16 Nm).

17. Install the right exhaust camshaft, as follows:

a. Apply engine oil to the thrust portion of the intake camshaft.

b. Align the timing marks (2 dot marks) of the camshaft drive and driven gears.

c. Roll down the exhaust camshaft onto the bearing journals while engaging the gears with each other. Install the bearing caps in their proper locations.

d. Apply a light coat of engine oil to the threads and install the cap bolts.

e. Apply a light coat of engine oil on the threads and under the heads of the bearing cap bolts.

f. Uniformly tighten the cap bolts in the sequence shown to 12 ft. lbs. (16 Nm).

g. Remove the service bolt from the driven sub-gear. Check that the intake and exhaust camshafts turns smoothly.

h. Align the timing marks (2 dot marks) of the camshaft drive and driven gears by turning the camshaft with a wrench.

18. Install the left intake camshaft, as follows:

a. Apply engine oil to the thrust portion of the intake camshaft.

b. Position the intake camshaft at 90 degrees angle of the timing mark (1 dot mark) on the cylinder head.

c. Install the bearing caps in their proper locations. Apply a light coat of engine oil to the threads and install the cap bolts.

d. Apply a light coat of engine oil on the threads and under the heads of the bearing cap bolts.

e. Uniformly tighten the cap bolts in the sequence shown to 12 ft. lbs. (16 Nm).

19. Install the left exhaust camshaft, as follows:

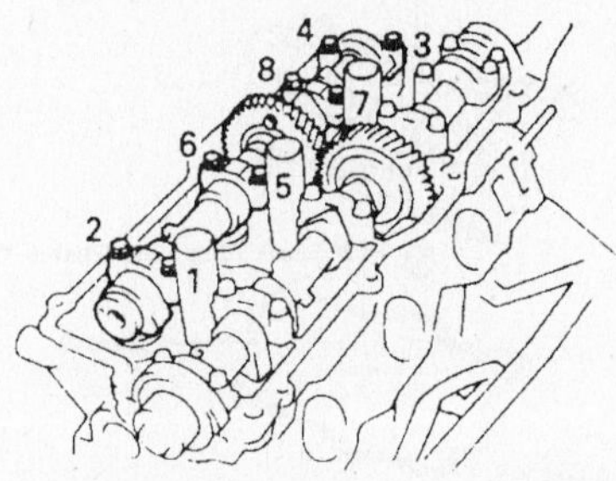

Right exhaust camshaft bolts removal sequence—3.4L engine

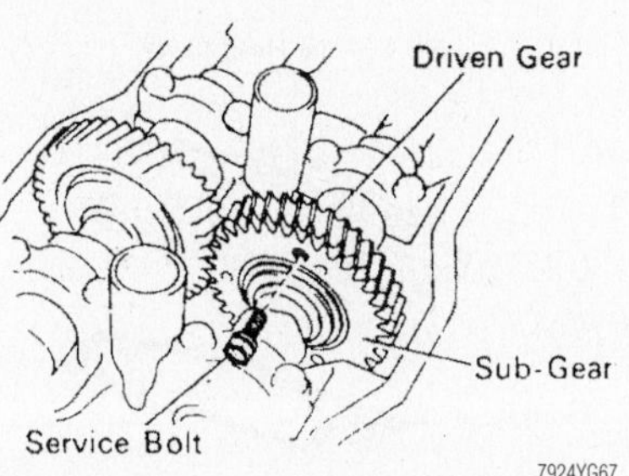

Drive gear service bolt (left side)—3.4L engine

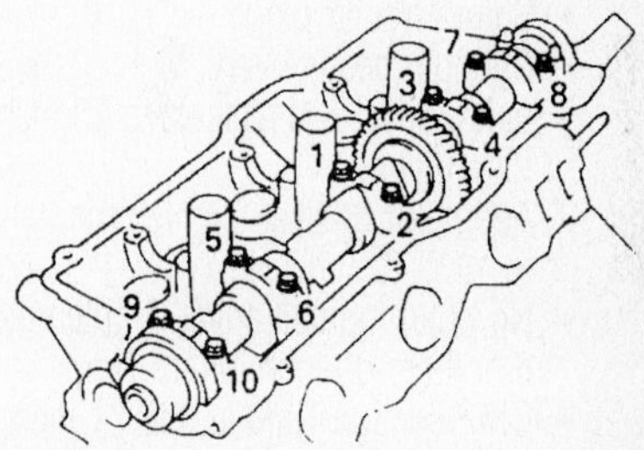

Right intake camshaft tightening sequence—3.4L engine

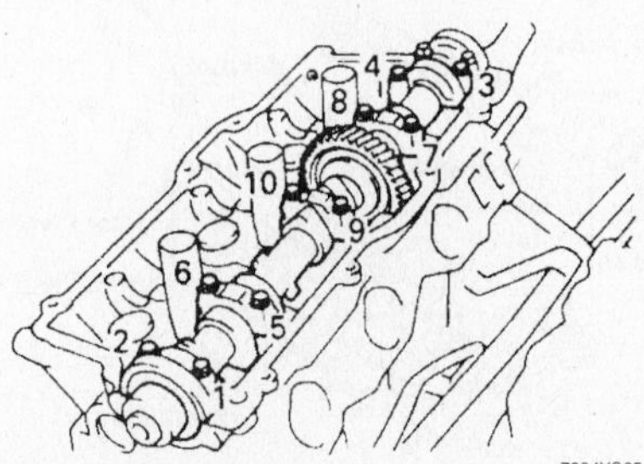

Right intake camshaft bolts removal sequence—3.4L engine

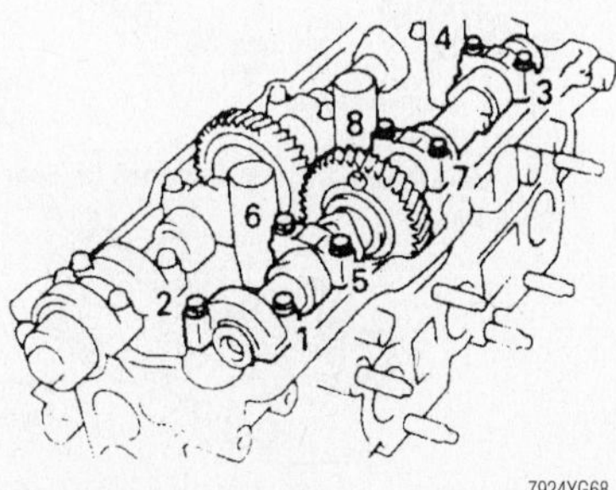

Left exhaust camshaft bolts removal sequence—3.4L engine

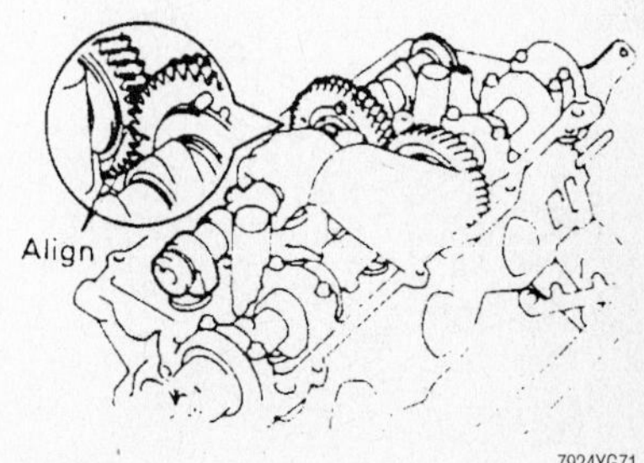

Aligning the right camshafts for installation—3.4L engine

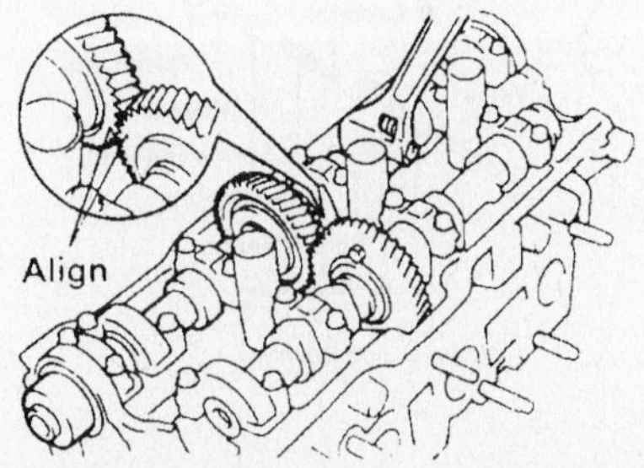

Aligning the timing mark (1 dot mark) of the left camshafts—3.4L engine

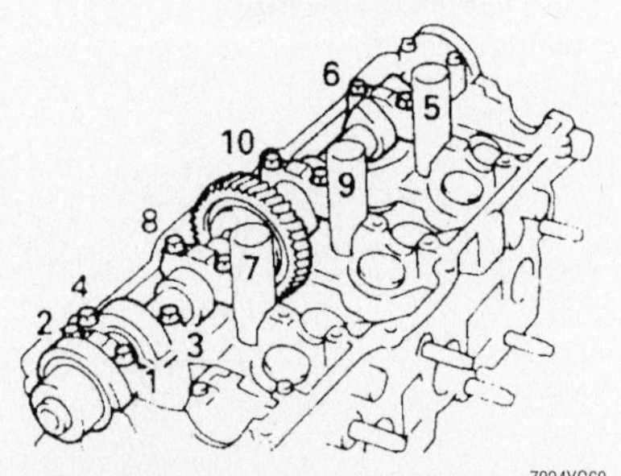

Left intake camshaft bolts removal sequence—3.4L engine

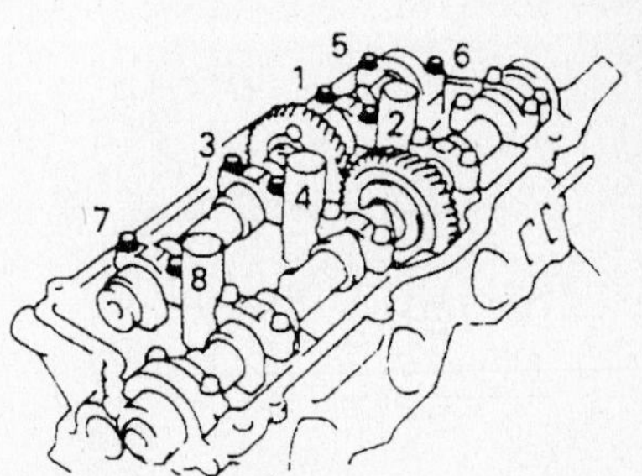

Right exhaust camshaft bolts tightening sequence—3.4L engine

a. Apply engine oil to the thrust portion of the intake camshaft.

b. Align the timing marks (1 dot mark) of the camshaft drive and driven gears.

c. Roll down the exhaust camshaft onto the bearing journals while engaging the gears with each other. Install the bearing caps in their proper locations.

d. Apply a light coat of engine oil to the threads and install the cap bolts.

e. Apply a light coat of engine oil on the threads and under the heads of the bearing cap bolts.

f. Uniformly tighten the cap bolts in the sequence shown to 12 ft. lbs. (16 Nm).

g. Remove the service bolt.

20. Check and adjust the valve clearance.

21. Install or connect the following:

- Semi-circular plugs
- Cylinder head covers. Tighten the bolts, in several passes, to 53 inch lbs. (6 Nm).
- Alternator bracket. Tighten the bolts to 14 ft. lbs. (18 Nm).
- No. 3 timing belt cover. Tighten the 6 bolts to 80 inch lbs. (9 Nm).
- CMP sensor. Tighten it to 71 inch lbs. (8 Nm).
- Intake air connector
- Hoses
- Alternator but do not tighten the bolts and nuts at this time
- No. 2 timing belt idler. Tighten the bolt to 30 ft. lbs. (40 Nm).

➡Check that the pulley bracket moves smoothly.

22. Install the left camshaft timing pulley, as follows:

a. Install the knock pin to the camshaft.

b. Align the knock pin hose of the camshaft with the knock pin groove of the timing pulley.

c. Slide the timing pulley on the camshaft with the flange side facing outward. Tighten the pulley bolt to 81 ft. lbs. (110 Nm).

23. Set the No. 1 cylinder to TDC of the compression stroke.

a. Turn the crankshaft pulley and align its groove with the timing mark **0** on the No. 1 timing belt cover.

b. Turn the camshaft, align the knock pin hole of the camshaft with the timing mark of the No. 3 timing belt cover.

c. Turn the camshaft timing pulley, align the timing marks of the camshaft timing pulley and the No. 3 timing belt cover.

24. Connect the timing belt to the left camshaft timing pulley, as follows:

a. Check that the installation mark on the timing belt is aligned with the end of the No. 1 timing belt cover.

b. Using Variable Pin Wrench Set

7924YG73

Left intake camshaft bolts tightening sequence—3.4L engine

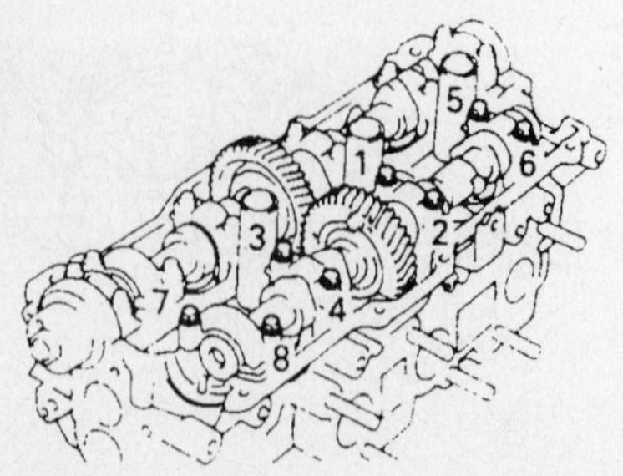

7924YG74

Left exhaust camshaft bolts tightening sequence—3.4L engine

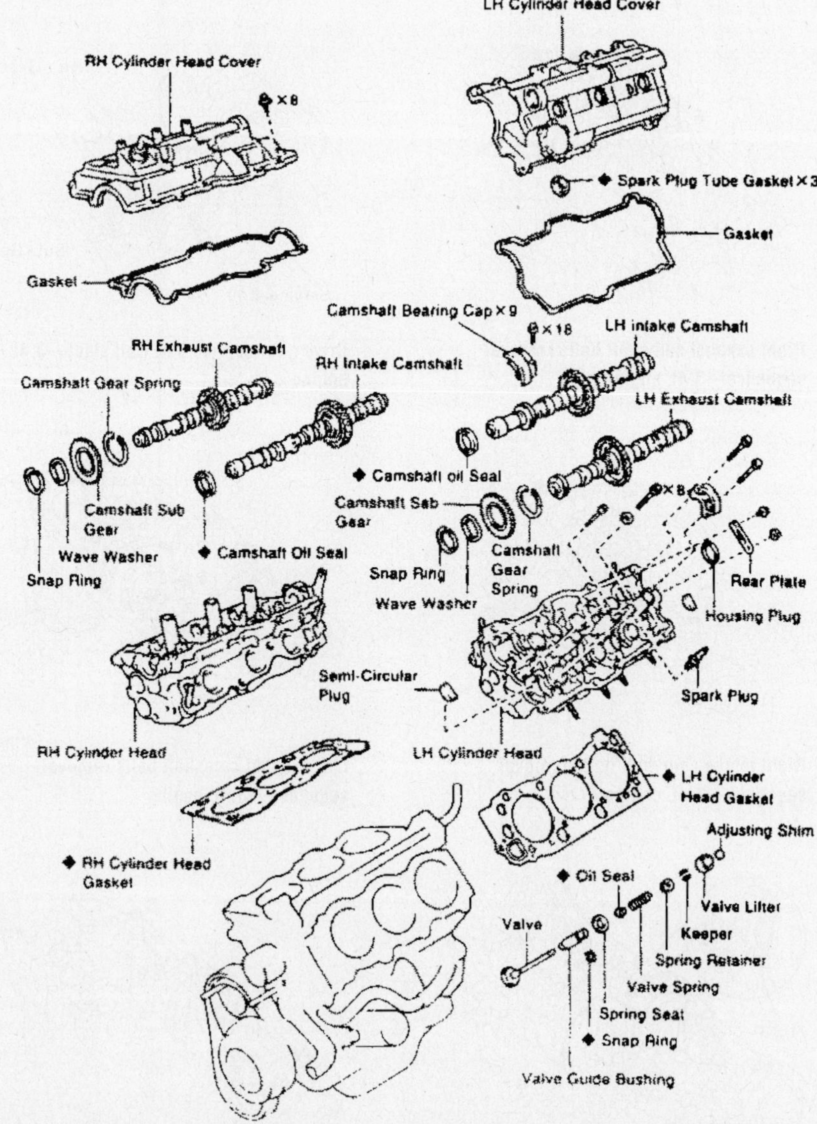

Exploded view of the cylinder head component assembly—3.4L engine

09960-01000 or equivalent, slightly turn the left camshaft timing pulley clockwise. Align the installation mark on the timing belt with the timing mark of the camshaft timing pulley, and hang the timing belt on the left camshaft timing pulley.

c. Align the timing marks of the left camshaft pulley and the No. 3 timing belt cover.

d. Check that the timing belt has tension between the crankshaft timing pulley and the left camshaft timing pulley.

25. Install the right camshaft timing pulley and the timing belt, as follows:

a. Align the installation mark on the timing belt with the timing mark of the right camshaft timing pulley, and hang the timing belt on the right camshaft timing pulley with the flange side facing inward.

b. Slide the right camshaft timing pulley on the camshaft. Align the timing marks on the right camshaft timing pulley and the No. 3 timing belt cover.

c. Align the knock pin hole of the camshaft with the knock pin groove of the pulley.

d. Install the knock pin. Tighten the bolt to 81 ft. lbs. (110 Nm).

26. Set the timing belt tensioner, as follows:

a. Using a press, slowly press in the pushrod using 220–2,205 lbs. (981–9,807 N) of force.

b. Align the holes of the pushrod and housing, pass a 1.5mm hexagon wrench through the holes to keep the setting position of the pushrod.

c. Release the press and install the dust boot to the tensioner.

27. Install the timing belt tensioner and alternately tighten the bolts to 20 ft. lbs. (28 Nm). Using pliers, remove the 1.5mm hexagon wrench from the belt tensioner.

28. Check the valve timing, as follows:

a. Slowly turn the crankshaft pulley 2 revolutions from the TDC-to-TDC. Always turn the crankshaft pulley clockwise.

b. Check that each pulley aligns with the timing marks. If the timing marks do not align, remove the timing belt and reinstall it.

29. Install or connect the following:

- Fan bracket with the bolt and nut
- Power steering adjusting strut with the nut
- No. 2 timing belt cover. Tighten the bolts to 80 inch lbs. (9 Nm).
- Negative battery cable

30. Fill the radiator with engine coolant.

31. Start the engine and check for leaks.

32. Check the ignition timing.

33. Install the engine undercover.

34. Road test the vehicle.

35. Recheck all fluid levels.

Valve Lash

ADJUSTMENT

2.4L Engine

1. Before servicing the vehicle, refer to the precautions in the beginning of this section.

2. Remove or disconnect the following:

- Negative battery cable
- Air intake connector
- Positive Crankcase Ventilation (PCV) hoses
- Spark plug wires

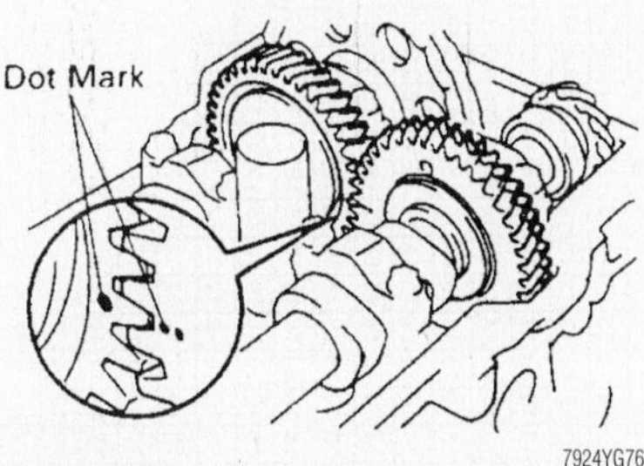

Aligning the timing marks—2.4L and 2.7L engines

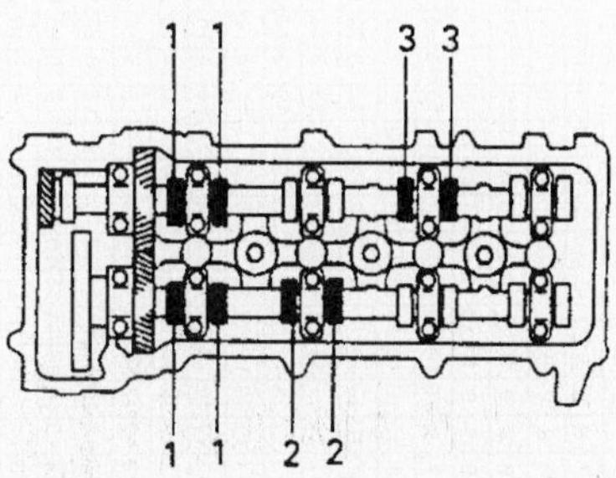

First valve adjustment—2.4L and 2.7L engines

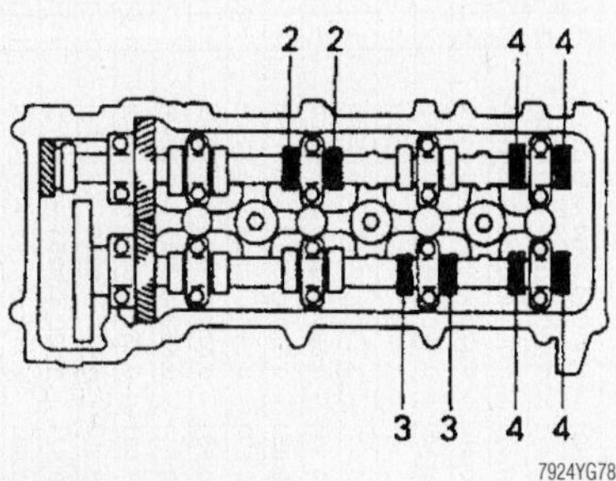

Second valve adjustment—2.4L and 2.7L engines

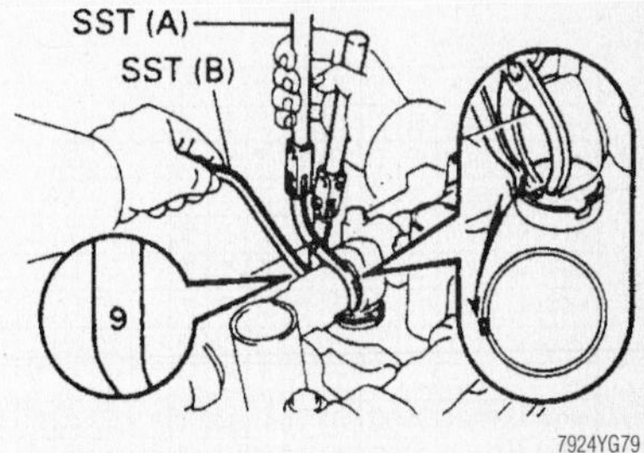

Removing adjusting shim using the special tools shown above—2.4L and 2.7L engines

- 4 clamps and the engine wiring harness
- Air conditioning compressor connector, if equipped with air conditioning
- Oil pressure sensor connector
- Engine Coolant Temperature (ECT) sensor connector
- Distributor connector
- Cylinder head cover

3. Set the No. 1 cylinder to Top Dead Center (TDC) of the compression stroke, as follows:

a. Turn the crankshaft pulley clockwise and align its groove with the **0** mark on the timing chain cover.

b. Check that the timing marks (1 and 2 dots) of the camshaft drive and driven gears are in a straight line on the cylinder head surface. If not, turn the crankshaft 1 revolution (360 degrees) and align the marks.

4. Inspect the valve clearance, as follows:

a. Measure the clearance between the valve lifter and the camshaft. Measure the 1st and 2nd intake and the 1st and 3rd exhaust valves.

b. Turn the crankshaft pulley 1 revolution (360 degrees) and align the marks as above. Measure the 3rd and 4th intake and the 2nd and 4th exhaust valves.

5. Valve clearance "cold" should be:

- Intake: 0.006–0.010 in. (0.15–0.25mm)
- Exhaust: 0.010–0.014 in. (0.25–0.35mm)

6. Adjust the valve clearance by using adjusting shims, as follows:

a. Turn the equipment driveshaft so that the cam lobe for the valve to be adjusted faces up.

b. Using SST 09248-55040, press down the valve lifter and place SST 09248-05420, between the camshaft and the valve lifter.

c. Remove SST 09248-55040.

d. Remove the adjusting shim with a

Adjusting Shim Selection Chart (Intake)

Measured clearance mm (in.) / Installed shim thickness mm (in.)	2.500 (0.0984)	2.520 (0.0992)	2.540 (0.1000)	2.550 (0.1004)	2.560 (0.1008)	2.580 (0.1016)	2.600 (0.1024)	2.620 (0.1031)	2.640 (0.1039)	2.650 (0.1043)	2.660 (0.1047)	2.680 (0.1055)	2.700 (0.1063)	2.710 (0.1067)	2.720 (0.1071)	2.730 (0.1075)	2.740 (0.1079)	2.750 (0.1083)	2.760 (0.1087)	2.770 (0.1091)	2.780 (0.1094)	2.790 (0.1098)	2.800 (0.1102)	2.810 (0.1106)	2.820 (0.1110)	2.830 (0.1114)	2.840 (0.1118)	2.850 (0.1122)	2.860 (0.1126)	2.870 (0.1130)	2.880 (0.1134)	2.890 (0.1138)	2.900 (0.1142)	2.910 (0.1146)	2.920 (0.1150)	2.930 (0.1154)	2.940 (0.1157)	2.950 (0.1161)	2.960 (0.1166)	2.970 (0.1169)	2.980 (0.1173)	2.990 (0.1177)	3.000 (0.1181)	3.010 (0.1185)	3.020 (0.1189)	3.030 (0.1193)	3.040 (0.1197)	3.050 (0.1201)	3.060 (0.1205)	3.070 (0.1209)	3.080 (0.1213)	3.090 (0.1217)	3.100 (0.1220)	3.120 (0.1228)	3.140 (0.1236)	3.150 (0.1240)	3.160 (0.1244)	3.180 (0.1252)	3.200 (0.1260)	3.220 (0.1268)	3.240 (0.1276)	3.250 (0.1280)	3.260 (0.1283)	3.280 (0.1291)	3.300 (0.1299)
0.000 – 0.030 (0.0000 – 0.0012)										1	1	1	1	2	2	2	2	2	3	3	3	3	3	4	4	4	4	4	5	5	5	5	5	6	6	6	6	6	7	7	7	7	7	8	8	8	8	8	9	9	9	9	9	10	10	10	11	11	11	12	12	12	13	13	13
0.031 – 0.050 (0.0012 – 0.0020)								1	1	1	1	1	2	2	2	2	3	3	3	3	3	4	4	4	4	4	5	5	5	5	5	6	6	6	6	6	7	7	7	7	7	8	8	8	8	8	9	9	9	9	9	10	10	10	11	11	11	11	12	12	13	13	13	13	14
0.051 – 0.070 (0.0020 – 0.0028)							1	1	1	1	1	2	2	2	3	3	3	3	3	4	4	4	4	4	5	5	5	5	5	6	6	6	6	6	7	7	7	7	7	8	8	8	8	8	9	9	9	9	9	10	10	10	10	11	11	11	11	12	12	13	13	13	13	14	14
0.071 – 0.090 (0.0028 – 0.0035)						1	1	1	1	2	2	2	3	3	3	3	3	4	4	4	4	4	5	5	5	5	5	6	6	6	6	6	7	7	7	7	7	8	8	8	8	8	9	9	9	9	9	10	10	10	10	10	11	11	11	12	12	12	13	13	13	14	14	14	15
0.091 – 0.110 (0.0036 – 0.0043)					1	1	1	1	2	2	2	3	3	3	3	4	4	4	4	4	5	5	5	5	5	6	6	6	6	6	7	7	7	7	7	8	8	8	8	8	9	9	9	9	9	10	10	10	10	10	11	11	11	11	12	12	12	13	13	13	14	14	14	15	15
0.111 – 0.130 (0.0044 – 0.0051)			1	1	1	1	1	2	2	2	3	3	3	4	4	4	4	4	5	5	5	5	5	6	6	6	6	6	7	7	7	7	7	8	8	8	8	8	9	9	9	9	9	10	10	10	10	10	11	11	11	11	11	12	12	12	13	13	13	14	14	14	15	15	15
0.131 – 0.149 (0.0052 – 0.0059)		1	1	1	1	1	2	2	3	3	3	3	4	4	4	4	5	5	5	5	5	6	6	6	6	6	7	7	7	7	7	8	8	8	8	8	9	9	9	9	9	10	10	10	10	10	11	11	11	11	11	12	12	12	13	13	13	13	14	14	15	15	15	15	16
0.150 – 0.250 (0.0059 – 0.0098)																																																																		
0.251 – 0.270 (0.0099 – 0.0106)	2	3	3	3	3	4	4	5	5	5	5	6	6	6	7	7	7	7	7	8	8	8	8	8	9	9	9	9	9	10	10	10	10	10	11	11	11	11	11	12	12	12	12	12	13	13	13	13	13	14	14	14	14	15	15	15	15	16	16	17	17	17	17	17	
0.271 – 0.290 (0.0107 – 0.0114)	3	3	3	4	4	4	5	5	5	6	6	6	7	7	7	7	7	8	8	8	8	8	9	9	9	9	9	10	10	10	10	10	11	11	11	11	11	12	12	12	12	12	13	13	13	13	13	14	14	14	14	14	15	15	15	16	16	16	17	17	17	17	17		
0.291 – 0.310 (0.0115 – 0.0122)	3	3	4	4	4	5	5	5	6	6	6	7	7	7	7	8	8	8	8	8	9	9	9	9	9	10	10	10	10	10	11	11	11	11	11	12	12	12	12	12	13	13	13	13	13	14	14	14	14	14	15	15	15	15	16	16	16	17	17	17	17				
0.311 – 0.330 (0.0122 – 0.0130)	3	4	4	4	5	5	5	6	6	6	7	7	7	8	8	8	8	8	8	9	9	9	9	10	10	10	10	10	11	11	11	11	11	12	12	12	12	12	13	13	13	13	13	14	14	14	14	14	15	15	15	15	16	16	16	16	17	17	17	17					
0.331 – 0.350 (0.0130 – 0.0138)	4	4	5	5	5	5	6	6	7	7	7	7	8	8	8	8	9	9	9	9	9	10	10	10	10	10	11	11	11	11	11	12	12	12	12	12	13	13	13	13	13	14	14	14	14	14	15	15	15	15	15	16	16	16	16	17	17	17	17						
0.351 – 0.370 (0.0138 – 0.0146)	4	5	5	5	5	6	6	7	7	7	7	8	8	8	9	9	9	9	9	10	10	10	10	10	11	11	11	11	11	12	12	12	12	12	13	13	13	13	13	14	14	14	14	14	15	15	15	15	15	16	16	16	16	17	17	17	17	17							
0.371 – 0.390 (0.0146 – 0.0154)	5	5	5	6	6	6	7	7	7	8	8	8	9	9	9	9	9	10	10	10	10	10	11	11	11	11	11	12	12	12	12	12	13	13	13	13	13	14	14	14	14	14	15	15	15	15	15	16	16	16	16	16	17	17	17	17	17								
0.391 – 0.410 (0.0154 – 0.0161)	5	5	6	6	6	7	7	7	8	8	8	9	9	9	9	10	10	10	10	10	11	11	11	11	11	12	12	12	12	12	13	13	13	13	13	14	14	14	14	14	15	15	15	15	15	16	16	16	16	16	17	17	17	17	17										
0.411 – 0.430 (0.0162 – 0.0169)	5	6	6	6	7	7	7	8	8	8	9	9	9	10	10	10	10	10	11	11	11	11	11	12	12	12	12	12	13	13	13	13	13	14	14	14	14	14	15	15	15	15	15	16	16	16	16	16	17	17	17	17	17	17											
0.431 – 0.450 (0.0170 – 0.0177)	6	6	7	7	7	7	8	8	9	9	9	9	10	10	10	10	11	11	11	11	11	12	12	12	12	12	13	13	13	13	13	14	14	14	14	14	15	15	15	15	15	16	16	16	16	16	17	17	17	17	17	17	17												
0.451 – 0.470 (0.0178 – 0.0185)	6	7	7	7	7	8	8	9	9	9	9	10	10	10	11	11	11	11	11	12	12	12	12	12	13	13	13	13	13	14	14	14	14	14	15	15	15	15	15	16	16	16	16	16	17	17	17	17	17	17	17														
0.471 – 0.490 (0.0185 – 0.0193)	7	7	7	8	8	8	9	9	9	10	10	10	11	11	11	11	11	12	12	12	12	12	13	13	13	13	13	14	14	14	14	14	15	15	15	15	15	16	16	16	16	16	17	17	17	17	17	17	17																
0.491 – 0.510 (0.0193 – 0.0201)	7	7	8	8	8	9	9	9	10	10	10	11	11	11	11	12	12	12	12	12	13	13	13	13	13	14	14	14	14	14	15	15	15	15	15	16	16	16	16	16	17	17	17	17	17	17	17																		
0.511 – 0.530 (0.0201 – 0.0209)	8	8	8	8	9	9	9	10	10	10	11	11	11	12	12	12	12	12	13	13	13	13	13	14	14	14	14	14	15	15	15	15	15	16	16	16	16	16	17	17	17	17	17	17	17																				
0.531 – 0.550 (0.0209 – 0.0217)	8	8	9	9	9	9	10	10	11	11	11	11	12	12	12	12	13	13	13	13	13	14	14	14	14	14	15	15	15	15	15	16	16	16	16	16	17	17	17	17	17	17	17																						
0.551 – 0.570 (0.0217 – 0.0224)	8	9	9	9	9	10	10	11	11	11	11	12	12	12	13	13	13	13	13	14	14	14	14	14	15	15	15	15	15	16	16	16	16	16	17	17	17	17	17	17	17																								
0.571 – 0.590 (0.0225 – 0.0232)	9	9	9	10	10	10	11	11	11	12	12	12	13	13	13	13	13	14	14	14	14	14	15	15	15	15	15	16	16	16	16	16	17	17	17	17	17	17	17																										
0.591 – 0.610 (0.0233 – 0.0240)	9	9	10	10	10	11	11	11	12	12	12	13	13	13	13	14	14	14	14	14	15	15	15	15	15	16	16	16	16	16	17	17	17	17	17	17	17																												
0.611 – 0.630 (0.0241 – 0.0248)	10	10	10	10	11	11	11	12	12	12	13	13	13	14	14	14	14	14	15	15	15	15	15	16	16	16	16	16	17	17	17	17	17	17	17																														
0.631 – 0.650 (0.0248 – 0.0256)	10	10	11	11	11	11	12	12	13	13	13	13	14	14	14	14	15	15	15	15	15	16	16	16	16	16	17	17	17	17	17	17	17																																
0.651 – 0.670 (0.0256 – 0.0264)	10	11	11	11	11	12	12	13	13	13	13	14	14	14	15	15	15	15	15	16	16	16	16	16	17	17	17	17	17	17	17																																		
0.671 – 0.690 (0.0264 – 0.0272)	11	11	11	11	12	12	13	13	13	14	14	14	15	15	15	15	15	16	16	16	16	16	17	17	17	17	17	17	17																																				
0.691 – 0.710 (0.0272 – 0.0280)	11	11	12	12	12	13	13	13	14	14	14	15	15	15	15	16	16	16	16	16	17	17	17	17	17	17	17																																						
0.711 – 0.730 (0.0280 – 0.0287)	12	12	12	12	13	13	13	14	14	14	15	15	15	16	16	16	16	16	17	17	17	17	17	17	17																																								
0.731 – 0.750 (0.0288 – 0.0295)	12	12	13	13	13	13	14	14	15	15	15	15	16	16	16	16	17	17	17	17	17	17	17																																										
0.751 – 0.770 (0.0296 – 0.0303)	12	13	13	13	13	14	14	15	15	15	15	16	16	16	16	17	17	17	17	17	17																																												
0.771 – 0.790 (0.0304 – 0.0311)	13	13	13	13	14	14	15	15	15	16	16	16	17	17	17	17	17	17	17																																														
0.791 – 0.810 (0.0311 – 0.0319)	13	13	14	14	14	15	15	15	16	16	16	17	17	17	17	17	17																																																
0.811 – 0.830 (0.0319 – 0.0327)	14	14	14	14	15	15	15	16	16	16	17	17	17	17	17																																																		
0.831 – 0.850 (0.0327 – 0.0335)	14	14	15	15	15	15	16	16	17	17	17	17	17																																																				
0.851 – 0.870 (0.0335 – 0.0343)	14	15	15	15	15	16	16	17	17	17	17	17																																																					
0.871 – 0.890 (0.0343 – 0.0350)	15	15	15	15	16	16	17	17	17	17	17																																																						
0.891 – 0.910 (0.0351 – 0.0358)	15	15	16	16	16	17	17	17	17																																																									
0.911 – 0.930 (0.0359 – 0.0366)	16	16	16	16	17	17	17	17																																																											
0.931 – 0.950 (0.0367 – 0.0374)	16	16	17	17	17	17	17																																																												
0.951 – 0.970 (0.0374 – 0.0382)	16	17	17	17	17	17																																																													
0.971 – 0.990 (0.0382 – 0.0390)	17	17	17	17																																																														
0.991 – 1.010 (0.0390 – 0.0398)	17	17	17																																																															
1.011 – 1.030 (0.0398 – 0.0406)	17	17																																																																
1.031 – 1.050 (0.0406 – 0.0413)	17																																																																	

Intake valve clearance (Cold):
0.15 – 0.25 mm (0.006 – 0.010 in.)
EXAMPLE:
The 2.800 mm (0.1102 in.) shim is installed, and the measured clearance is 0.440 mm (0.0173 in.). Replace the 2.800 mm (0.1102 in.) shim with a new No.12 shim.

New shim thickness mm (in.)

Shim No.	Thickness	Shim No.	Thickness
1	2.500 (0.0984)	10	2.950 (0.1161)
2	2.550 (0.1004)	11	3.000 (0.1181)
3	2.600 (0.1024)	12	3.050 (0.1201)
4	2.650 (0.1043)	13	3.100 (0.1220)
5	2.700 (0.1063)	14	3.150 (0.1240)
6	2.750 (0.1083)	15	3.200 (0.1260)
7	2.800 (0.1102)	16	3.250 (0.1280)
8	2.850 (0.1122)	17	3.300 (0.1299)
9	2.900 (0.1142)		

HINT:
New shims have the thickness in millimeters imprinted on the face.

67170-TACO-G01

Intake valve shim selection chart—2.4L and 2.7L engines

Adjusting Shim Selection Chart (Exhaust)

Measured clearance mm (in.) / Installed shim thickness mm (in.)	2.500 (0.0984)	2.520 (0.0992)	2.540 (0.1000)	2.550 (0.1004)	2.560 (0.1008)	2.580 (0.1016)	2.600 (0.1024)	2.620 (0.1031)	2.640 (0.1039)	2.650 (0.1043)	2.660 (0.1047)	2.680 (0.1055)	2.700 (0.1063)	2.710 (0.1067)	2.720 (0.1071)	2.730 (0.1075)	2.740 (0.1079)	2.750 (0.1083)	2.760 (0.1087)	2.770 (0.1091)	2.780 (0.1094)	2.790 (0.1098)	2.800 (0.1102)	2.810 (0.1106)	2.820 (0.1110)	2.830 (0.1114)	2.840 (0.1118)	2.850 (0.1122)	2.860 (0.1126)	2.870 (0.1130)	2.880 (0.1134)	2.890 (0.1138)	2.900 (0.1142)	2.910 (0.1146)	2.920 (0.1150)	2.930 (0.1154)	2.940 (0.1157)	2.950 (0.1161)	2.960 (0.1166)	2.970 (0.1169)	2.980 (0.1173)	2.990 (0.1177)	3.000 (0.1181)	3.010 (0.1185)	3.020 (0.1189)	3.030 (0.1193)	3.040 (0.1197)	3.050 (0.1201)	3.060 (0.1205)	3.070 (0.1209)	3.080 (0.1213)	3.090 (0.1217)	3.100 (0.1220)	3.120 (0.1228)	3.140 (0.1236)	3.150 (0.1240)	3.160 (0.1244)	3.180 (0.1252)	3.200 (0.1260)	3.220 (0.1268)	3.240 (0.1276)	3.250 (0.1280)	3.260 (0.1283)	3.280 (0.1291)	3.300 (0.1299)
0.000 – 0.030 (0.0000 – 0.0012)																		1	1	1	1	1	1	1	2	2	2	2	2	3	3	3	3	3	4	4	4	4	4	5	5	5	5	5	6	6	6	6	6	7	7	7	7	8	8	8	8	9	9	10	10	10	10	11	11
0.031 – 0.050 (0.0012 – 0.0020)															1	1	1	1	1	1	1	2	2	2	2	2	3	3	3	3	3	4	4	4	4	4	5	5	5	5	5	6	6	6	6	6	7	7	7	7	7	8	8	8	9	9	9	9	10	10	11	11	11	11	12
0.051 – 0.070 (0.0020 – 0.0028)													1	1	1	1	1	1	1	2	2	2	2	2	3	3	3	3	3	4	4	4	4	4	5	5	5	5	5	6	6	6	6	6	7	7	7	7	7	8	8	8	8	9	9	9	9	10	10	11	11	11	11	12	12
0.071 – 0.090 (0.0028 – 0.0035)												1	1	1	1	1	1	2	2	2	2	2	3	3	3	3	3	4	4	4	4	4	5	5	5	5	5	6	6	6	6	6	7	7	7	7	7	8	8	8	8	8	9	9	9	10	10	10	11	11	11	12	12	12	13
0.091 – 0.110 (0.0036 – 0.0043)											1	1	1	1	1	2	2	2	2	2	3	3	3	3	3	4	4	4	4	4	5	5	5	5	5	6	6	6	6	6	7	7	7	7	7	8	8	8	8	8	9	9	9	9	10	10	10	11	11	11	12	12	12	13	13
0.111 – 0.130 (0.0044 – 0.0051)									1	1	1	1	1	2	2	2	2	2	3	3	3	3	3	4	4	4	4	4	5	5	5	5	5	6	6	6	6	6	7	7	7	7	7	8	8	8	8	8	9	9	9	9	10	10	10	10	11	11	11	12	12	12	13	13	13
0.131 – 0.150 (0.0052 – 0.0059)								1	1	1	1	1	2	2	2	2	3	3	3	3	3	4	4	4	4	4	5	5	5	5	5	6	6	6	6	6	7	7	7	7	7	8	8	8	8	8	9	9	9	9	9	10	10	10	11	11	11	11	12	12	13	13	13	13	14
0.151 – 0.170 (0.0059 – 0.0067)							1	1	1	1	1	2	2	2	3	3	3	3	3	4	4	4	4	4	5	5	5	5	5	6	6	6	6	6	7	7	7	7	7	8	8	8	8	8	9	9	9	9	9	10	10	10	10	11	11	11	11	12	12	13	13	13	13	14	14
0.171 – 0.190 (0.0067 – 0.0075)						1	1	1	1	2	2	2	3	3	3	3	3	4	4	4	4	4	5	5	5	5	5	6	6	6	6	6	7	7	7	7	7	8	8	8	8	8	9	9	9	9	9	10	10	10	10	10	11	11	11	12	12	12	13	13	13	14	14	14	15
0.191 – 0.210 (0.0075 – 0.0083)					1	1	1	1	2	2	2	3	3	3	3	4	4	4	4	4	5	5	5	5	5	6	6	6	6	6	7	7	7	7	7	8	8	8	8	8	9	9	9	9	9	10	10	10	10	11	11	11	11	11	12	12	12	13	13	13	14	14	14	15	15
0.211 – 0.230 (0.0083 – 0.0091)			1	1	1	1	1	2	2	2	3	3	4	4	4	4	4	4	5	5	5	5	5	6	6	6	6	6	7	7	7	7	7	8	8	8	8	8	9	9	9	9	9	10	10	10	10	10	11	11	11	11	12	12	12	12	13	13	13	14	14	14	15	15	15
0.231 – 0.249 (0.0091 – 0.0098)		1	1	1	1	1	2	2	3	3	3	3	4	4	4	4	5	5	5	5	5	6	6	6	6	6	7	7	7	7	7	8	8	8	8	8	9	9	9	9	9	10	10	10	10	10	11	11	11	11	11	12	12	12	13	13	13	13	14	14	15	15	15	15	16
0.250 – 0.350 (0.0098 – 0.0138)																																																																	
0.351 – 0.370 (0.0138 – 0.0146)	2	3	3	3	3	4	4	5	5	5	5	6	6	6	7	7	7	7	7	8	8	8	8	8	9	9	9	9	9	10	10	10	10	10	11	11	11	11	11	12	12	12	12	12	13	13	13	13	13	14	14	14	14	14	15	15	15	16	16	17	17	17	17	17	
0.371 – 0.390 (0.0146 – 0.0154)	3	3	3	4	4	4	5	5	5	6	6	6	7	7	7	7	7	8	8	8	8	8	9	9	9	9	9	10	10	10	10	10	11	11	11	11	11	12	12	12	12	12	13	13	13	13	13	14	14	14	14	14	15	15	15	16	16	16	17	17	17	17	17		
0.391 – 0.410 (0.0154 – 0.0161)	3	3	4	4	4	5	5	5	6	6	6	7	7	7	7	8	8	8	8	8	9	9	9	9	9	10	10	10	10	10	11	11	11	11	11	12	12	12	12	12	13	13	13	13	13	14	14	14	14	14	15	15	15	15	16	16	16	17	17	17	17				
0.411 – 0.430 (0.0162 – 0.0169)	3	4	4	4	5	5	5	6	6	6	7	7	7	8	8	8	8	8	9	9	9	9	9	10	10	10	10	10	11	11	11	11	11	12	12	12	12	12	13	13	13	13	13	14	14	14	14	14	15	15	15	15	15	16	16	16	17	17	17	17					
0.431 – 0.450 (0.0170 – 0.0177)	4	4	5	5	5	5	6	6	7	7	7	7	8	8	8	8	9	9	9	9	9	10	10	10	10	10	11	11	11	11	11	12	12	12	12	12	13	13	13	13	13	14	14	14	14	14	15	15	15	15	15	16	16	16	17	17	17	17	17						
0.451 – 0.470 (0.0178 – 0.0185)	4	5	5	5	5	6	6	7	7	7	7	8	8	8	9	9	9	9	9	10	10	10	10	10	11	11	11	11	11	12	12	12	12	12	13	13	13	13	13	14	14	14	14	14	15	15	15	15	15	16	16	16	16	17	17	17	17	17							
0.471 – 0.490 (0.0185 – 0.0193)	5	5	5	6	6	6	7	7	7	8	8	8	9	9	9	9	9	10	10	10	10	10	11	11	11	11	11	12	12	12	12	12	13	13	13	13	13	14	14	14	14	14	15	15	15	15	15	16	16	16	16	16	17	17	17	17	17								
0.491 – 0.510 (0.0193 – 0.0201)	5	5	6	6	6	7	7	7	8	8	8	9	9	9	9	10	10	10	10	10	11	11	11	11	11	12	12	12	12	12	13	13	13	13	13	14	14	14	14	14	15	15	15	15	15	16	16	16	16	16	17	17	17	17	17										
0.511 – 0.530 (0.0201 – 0.0209)	5	6	6	6	7	7	7	8	8	8	9	9	9	10	10	10	10	10	11	11	11	11	11	12	12	12	12	12	13	13	13	13	13	14	14	14	14	14	15	15	15	15	15	16	16	16	16	16	17	17	17	17	17	17											
0.531 – 0.550 (0.0209 – 0.0217)	6	6	7	7	7	7	8	8	9	9	9	9	10	10	10	10	11	11	11	11	11	12	12	12	12	12	13	13	13	13	13	14	14	14	14	14	15	15	15	15	15	16	16	16	16	16	17	17	17	17	17	17	17												
0.551 – 0.570 (0.0217 – 0.0224)	6	7	7	7	7	8	8	9	9	9	9	10	10	10	11	11	11	11	11	12	12	12	12	12	13	13	13	13	13	14	14	14	14	14	15	15	15	15	15	16	16	16	16	16	17	17	17	17	17	17	17														
0.571 – 0.590 (0.0225 – 0.0232)	7	7	7	8	8	8	9	9	9	10	10	10	11	11	11	11	11	12	12	12	12	12	13	13	13	13	13	14	14	14	14	14	15	15	15	15	15	16	16	16	16	16	17	17	17	17	17	17	17																
0.591 – 0.610 (0.0233 – 0.0240)	7	7	8	8	8	9	9	9	10	10	10	11	11	11	11	12	12	12	12	12	13	13	13	13	13	14	14	14	14	14	15	15	15	15	15	16	16	16	16	16	17	17	17	17	17	17	17																		
0.611 – 0.630 (0.0241 – 0.0248)	7	8	8	8	9	9	9	10	10	10	11	11	11	12	12	12	12	12	13	13	13	13	13	14	14	14	14	14	15	15	15	15	15	16	16	16	16	16	17	17	17	17	17	17	17																				
0.631 – 0.650 (0.0248 – 0.0256)	8	8	9	9	9	9	10	10	11	11	11	11	12	12	12	12	13	13	13	13	13	14	14	14	14	14	15	15	15	15	15	16	16	16	16	16	17	17	17	17	17	17	17																						
0.651 – 0.670 (0.0256 – 0.0264)	8	9	9	9	9	10	10	11	11	11	11	12	12	12	13	13	13	13	13	14	14	14	14	14	15	15	15	15	15	16	16	16	16	16	17	17	17	17	17	17	17																								
0.671 – 0.690 (0.0264 – 0.0272)	9	9	9	10	10	10	11	11	11	12	12	12	13	13	13	13	13	14	14	14	14	14	15	15	15	15	15	16	16	16	16	16	17	17	17	17	17	17	17																										
0.691 – 0.710 (0.0272 – 0.0280)	9	9	10	10	10	11	11	11	12	12	12	13	13	13	13	14	14	14	14	14	15	15	15	15	15	16	16	16	16	16	17	17	17	17	17	17	17																												
0.711 – 0.730 (0.0280 – 0.0287)	9	10	10	10	11	11	11	12	12	12	13	13	13	14	14	14	14	14	15	15	15	15	15	16	16	16	16	16	17	17	17	17	17	17	17																														
0.731 – 0.750 (0.0288 – 0.0295)	10	10	11	11	11	11	12	12	13	13	13	13	14	14	14	14	15	15	15	15	15	16	16	16	16	16	17	17	17	17	17	17	17																																
0.751 – 0.770 (0.0296 – 0.0303)	10	11	11	11	11	12	12	13	13	13	13	14	14	14	15	15	15	15	15	16	16	16	16	16	17	17	17	17	17	17	17																																		
0.771 – 0.790 (0.0304 – 0.0311)	11	11	11	12	12	12	13	13	13	14	14	14	14	15	15	15	15	16	16	16	16	16	17	17	17	17	17	17	17																																				
0.791 – 0.810 (0.0311 – 0.0319)	11	11	12	12	12	13	13	13	14	14	14	15	15	15	15	16	16	16	16	16	17	17	17	17	17	17	17																																						
0.811 – 0.830 (0.0319 – 0.0327)	11	12	12	12	13	13	13	14	14	14	15	15	15	16	16	16	16	16	17	17	17	17	17	17	17																																								
0.831 – 0.850 (0.0327 – 0.0335)	12	12	13	13	13	13	14	14	15	15	15	15	16	16	16	16	17	17	17	17	17	17	17																																												
0.851 – 0.870 (0.0335 – 0.0343)	12	13	13	13	13	14	14	15	15	15	15	16	16	16	17	17	17	17	17	17	17																																														
0.871 – 0.890 (0.0343 – 0.0350)	13	13	13	14	14	14	15	15	15	16	16	16	16	17	17	17	17	17	17																																																
0.891 – 0.910 (0.0351 – 0.0358)	13	13	14	14	14	15	15	15	16	16	16	17	17	17	17	17	17																																																		
0.911 – 0.930 (0.0359 – 0.0366)	13	14	14	14	15	15	15	16	16	16	17	17	17	17	17																																																				
0.931 – 0.950 (0.0367 – 0.0374)	14	14	15	15	15	15	16	16	17	17	17	17	17																																																				
0.951 – 0.970 (0.0374 – 0.0382)	14	15	15	15	15	16	16	17	17	17	17	17																																																					
0.971 – 0.990 (0.0382 – 0.0390)	15	15	15	16	16	16	17	17	17	17	17																																																						
0.991 – 1.010 (0.0390 – 0.0398)	15	15	16	16	16	17	17	17	17																																																								
1.011 – 1.030 (0.0398 – 0.0406)	15	16	16	16	17	17	17	17																																																									
1.031 – 1.050 (0.0406 – 0.0413)	16	16	17	17	17	17	17																																																										
1.051 – 1.070 (0.0414 – 0.0421)	16	17	17	17	17	17																																																											
1.071 – 1.090 (0.0422 – 0.0429)	17	17	17	17	17																																																												
1.091 – 1.110 (0.0430 – 0.0437)	17	17	17																																																														
1.111 – 1.130 (0.0437 – 0.0445)	17	17																																																															
1.131 – 1.150 (0.0445 – 0.0413)	17																																																																

Exhaust valve clearance (Cold):
0.25 – 0.35 mm (0.010 – 0.014 in.)

EXAMPLE:
The 2.800 mm (0.1102 in.) shim is installed, and the measured clearance is 0.440 mm (0.0173 in.). Replace the 2.800 mm (0.1102 in.) shim with a new No.10 shim.

New shim thickness mm (in.)

Shim No.	Thickness	Shim No.	Thickness
1	2.500 (0.0984)	10	2.950 (0.1161)
2	2.550 (0.1004)	11	3.000 (0.1181)
3	2.600 (0.1024)	12	3.050 (0.1201)
4	2.650 (0.1043)	13	3.100 (0.1220)
5	2.700 (0.1063)	14	3.150 (0.1240)
6	2.750 (0.1083)	15	3.200 (0.1260)
7	2.800 (0.1102)	16	3.250 (0.1280)
8	2.850 (0.1122)	17	3.300 (0.1299)
9	2.900 (0.1142)		

HINT:
New shims have the thickness in millimeters imprinted on the face.

67170-TACO-G02

Exhaust valve shim selection chart—2.4L and 2.7L engines

small flat prying tool and a magnetic finger.

e. Determine the replacement adjusting shim size according to the following formula or use the adjusting shim charts.

f. Using a micrometer, measure the thickness of the removed shim. Calculate the thickness of a new shim so that the valve clearance comes within the specified value:

- T: Thickness of the removed shim
- A: Measured valve clearance
- N: Thickness of the new shim

g. Intake: N = T + (A – 0.008 in. (0.20mm))

h. Exhaust: N = T + (A – 0.012 in. (0.30mm))

i. Install a new adjusting shim. Place it on the valve lifter. Using the SST 09248-55040, press down the valve lifter.

j. Remove the SST 09248-05420.

k. Recheck the valve clearance.

7. Install or connect the following:
   - Cylinder head cover
   - Engine wiring harness and clamps
   - Distributor connector
   - ECT sensor connector
   - Oil pressure sensor connector
   - Air conditioning compressor connector, if disconnected
   - Spark plug wires
   - PCV hoses
   - Air intake connector
   - Negative battery cable
8. Check the ignition timing.

2.7L Engine

1. Before servicing the vehicle, refer to the precautions in the beginning of this section.
2. Disconnect the negative battery cable.
3. Drain the engine coolant.
4. Remove or disconnect the following:
   - Intake air connector
   - Positive Crankcase Ventilation (PCV) hoses
   - Spark plug wires
   - Engine wiring harness clamps and harness
   - Air conditioning compressor connector, if equipped with air conditioning
   - Oil pressure sensor connector
   - Engine Coolant Temperature (ECT) sensor connector
   - Distributor connector
   - Cylinder head cover
5. Set the No. 1 cylinder to Top Dead Center (TDC) of the compression stroke, as follows:

a. Turn the crankshaft pulley clockwise and align its groove with the **0** mark on the timing chain cover.

b. Check that the timing marks (1 and 2 dots) of the camshaft drive and driven gears are in a straight line on the cylinder head surface. If not, turn the crankshaft 1 revolution (360 degrees) and align the marks.

6. Inspect the valve clearance, as follows:

a. Measure the clearance between the valve lifter and the camshaft. Measure the 1st and 2nd intake and the 1st and 3rd exhaust valves.

b. Turn the crankshaft pulley 1 revolution (360 degrees) and align the marks as above. Measure the 3rd and 4th intake and the 2nd and 4th exhaust valves.

7. Valve clearance cold should be:
   - Intake: 0.006–0.010 in. (0.15–0.25mm)
   - Exhaust: 0.010–0.014 in. (0.25–0.35mm)
8. Adjust the valve clearance by using adjusting shims, as follows:

a. Turn the camshaft so the cam lobe for the valve to be adjusted faces up.

b. Using SST 09248-55040, press down the valve lifter and place SST 09248-05420, between the camshaft and the valve lifter. Remove SST 09248-55040.

c. Remove the adjusting shim with a small flat prying tool and a magnetic finger.

d. Determine the replacement adjusting shim size according to the following formula or use the adjusting shim charts.

e. Using a micrometer, measure the thickness of the removed shim. Calculate the thickness of a new shim so the valve clearance comes within the specified value.

- T: Thickness of the removed shim
- A: Measured valve clearance
- N: Thickness of the new shim

f. Intake: N = T + (A – 0.008 in. (0.20mm))

g. Exhaust: N = T + (A – 0.012 in. (0.30mm))

h. Install a new adjusting shim. Place it on the valve lifter. Using the SST 09248-55040, press down the valve lifter and remove SST 09248-05420.

i. Recheck the valve clearance.

9. Install or connect the following:
   - Cylinder head cover
   - Engine wiring harness and clamps
   - Distributor connector
   - ECT sensor connector
   - Oil pressure sensor connector
   - Air conditioning compressor connector, if disconnected
   - Spark plug wires
   - PCV hoses
   - Intake air connector
   - Negative battery cable
10. Refill with engine coolant.
11. Check the ignition timing.

3.4L Engine

1. Before servicing the vehicle, refer to the precautions in the beginning of this section.
2. Disconnect the negative battery cable.
3. Drain the engine coolant.
4. Remove or disconnect the following:
   - Air intake connector
   - Cylinder head cover
5. Set the No. 1 cylinder to Top Dead Center (TDC) of the compression stroke, as follows:

a. Turn the crankshaft pulley clockwise and align its groove with the **0** mark on the timing chain cover.

b. Check that the timing marks (1 and 2 dots) of the camshaft drive and driven gears are in a straight line on the cylinder head surface. If not, turn the crank-

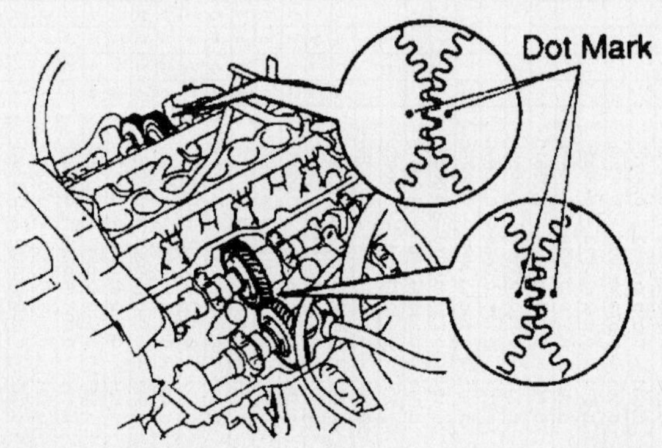

Aligning the timing marks—3.4L engine

Adjusting Shim Selection Chart (Intake)

Measured clearance mm (in.) \ Installed shim thickness mm (in.)	2.500 (0.0984)	2.520 (0.0992)	2.540 (0.1000)	2.550 (0.1004)	2.560 (0.1008)	2.580 (0.1016)	2.600 (0.1024)	2.620 (0.1031)	2.640 (0.1039)	2.650 (0.1043)	2.660 (0.1047)	2.670 (0.1051)	2.680 (0.1055)	2.690 (0.1059)	2.700 (0.1063)	2.710 (0.1067)	2.720 (0.1071)	2.730 (0.1075)	2.740 (0.1079)	2.750 (0.1083)	2.760 (0.1087)	2.770 (0.1091)	2.780 (0.1094)	2.790 (0.1098)	2.800 (0.1102)	2.810 (0.1106)	2.820 (0.1110)	2.830 (0.1114)	2.840 (0.1118)	2.850 (0.1122)	2.860 (0.1126)	2.870 (0.1130)	2.880 (0.1134)	2.890 (0.1138)	2.900 (0.1142)	2.910 (0.1146)	2.920 (0.1150)	2.930 (0.1154)	2.940 (0.1157)	2.950 (0.1161)	2.960 (0.1165)	2.970 (0.1169)	2.980 (0.1173)	2.990 (0.1177)	3.000 (0.1181)	3.010 (0.1185)	3.020 (0.1189)	3.030 (0.1193)	3.040 (0.1197)	3.050 (0.1201)	3.060 (0.1205)	3.080 (0.1213)	3.100 (0.1220)	3.120 (0.1228)	3.140 (0.1236)	3.150 (0.1240)	3.160 (0.1244)	3.180 (0.1252)	3.200 (0.1260)	3.220 (0.1268)	3.240 (0.1276)	3.250 (0.1280)	3.260 (0.1283)	3.280 (0.1291)	3.300 (0.1299)
0.000 – 0.020 (0.0000 – 0.0008)									1	1	1	1	1	1	2	2	2	2	2	3	3	3	3	3	4	4	4	4	4	5	5	5	5	5	6	6	6	6	6	7	7	7	7	7	8	8	8	8	8	9	9	9	10	10	10	11	11	11	12	12	12	13	13	13	14
0.021 – 0.040 (0.0008 – 0.0016)								1	1	1	1	1	2	2	2	2	2	3	3	3	3	3	4	4	4	4	4	5	5	5	5	5	6	6	6	6	6	7	7	7	7	7	8	8	8	8	8	9	9	9	9	10	10	10	11	11	11	12	12	12	13	13	13	14	14
0.041 – 0.060 (0.0016 – 0.0024)							1	1	1	1	2	2	2	2	2	3	3	3	3	3	4	4	4	4	4	5	5	5	5	5	6	6	6	6	6	7	7	7	7	7	8	8	8	8	8	9	9	9	9	9	10	10	10	11	11	11	12	12	12	13	13	13	14	14	14
0.061 – 0.080 (0.0024 – 0.0031)						1	1	1	2	2	2	2	2	3	3	3	3	3	4	4	4	4	4	5	5	5	5	5	6	6	6	6	6	7	7	7	7	7	8	8	8	8	8	9	9	9	9	9	10	10	10	10	11	11	12	12	12	12	13	13	14	14	14	14	15
0.081 – 0.100 (0.0032 – 0.0039)					1	1	1	2	2	2	2	3	3	3	3	3	4	4	4	4	4	5	5	5	5	5	6	6	6	6	6	7	7	7	7	7	8	8	8	8	8	9	9	9	9	9	10	10	10	10	10	11	11	12	12	12	12	13	13	14	14	14	14	15	15
0.101 – 0.120 (0.0040 – 0.0047)			1	1	1	1	2	2	2	3	3	3	3	3	4	4	4	4	4	5	5	5	5	5	6	6	6	6	6	7	7	7	7	7	8	8	8	8	8	9	9	9	9	9	10	10	10	10	10	11	11	11	12	12	12	13	13	13	14	14	14	15	15	15	16
0.121 – 0.129 (0.0048 – 0.0051)		1	1	1	1	1	2	2	3	3	3	3	3	4	4	4	4	4	5	5	5	5	5	6	6	6	6	6	7	7	7	7	7	8	8	8	8	8	9	9	9	9	9	10	10	10	10	10	11	11	11	11	12	12	13	13	13	13	14	14	15	15	15	15	16
0.130 – 0.230 (0.0051 – 0.0091)																																																																	
0.231 – 0.240 (0.0091 – 0.0094)	2	3	3	3	3	4	4	5	5	5	5	6	6	6	6	6	7	7	7	7	7	8	8	8	8	8	9	9	9	9	9	10	10	10	10	10	11	11	11	11	11	12	12	12	12	12	13	13	13	13	13	14	14	15	15	15	15	16	16	17	17	17	17	17	
0.241 – 0.260 (0.0095 – 0.0102)	2	3	3	3	4	4	4	5	5	5	6	6	6	6	6	7	7	7	7	7	8	8	8	8	8	9	9	9	9	9	10	10	10	10	10	11	11	11	11	11	12	12	12	12	12	13	13	13	13	13	14	14	14	15	15	15	16	16	16	17	17	17	17		
0.261 – 0.280 (0.0103 – 0.0110)	3	3	4	4	4	4	5	5	6	6	6	6	6	7	7	7	7	7	8	8	8	8	8	9	9	9	9	9	10	10	10	10	10	11	11	11	11	11	12	12	12	12	12	13	13	13	13	13	14	14	14	14	15	15	16	16	16	16	17	17	17	17			
0.281 – 0.300 (0.0111 – 0.0118)	3	4	4	4	4	5	5	6	6	6	6	7	7	7	7	7	8	8	8	8	8	9	9	9	9	9	10	10	10	10	10	11	11	11	11	11	12	12	12	12	12	13	13	13	13	13	14	14	14	14	14	15	15	16	16	16	16	17	17	17					
0.301 – 0.320 (0.0119 – 0.0126)	4	4	4	5	5	5	6	6	6	7	7	7	7	7	8	8	8	8	8	9	9	9	9	9	10	10	10	10	10	11	11	11	11	11	12	12	12	12	12	13	13	13	13	13	14	14	14	14	14	15	15	15	16	16	16	17	17	17	17						
0.321 – 0.340 (0.0126 – 0.0134)	4	4	5	5	5	6	6	6	7	7	7	7	8	8	8	8	8	9	9	9	9	9	10	10	10	10	10	11	11	11	11	11	12	12	12	12	12	13	13	13	13	13	14	14	14	14	14	15	15	15	15	16	16	16	17	17	17	17							
0.341 – 0.360 (0.0134 – 0.0142)	4	5	5	5	6	6	6	7	7	7	8	8	8	8	8	9	9	9	9	9	10	10	10	10	10	11	11	11	11	11	12	12	12	12	12	13	13	13	13	13	14	14	14	14	14	15	15	15	15	15	16	16	16	17	17	17	17								
0.361 – 0.380 (0.0142 – 0.0150)	5	5	6	6	6	6	7	7	8	8	8	8	8	9	9	9	9	9	10	10	10	10	10	11	11	11	11	11	12	12	12	12	12	13	13	13	13	13	14	14	14	14	14	15	15	15	15	15	16	16	16	16	17	17	17	17									
0.381 – 0.400 (0.0150 – 0.0157)	5	6	6	6	6	7	7	8	8	8	8	9	9	9	9	9	10	10	10	10	10	11	11	11	11	11	12	12	12	12	12	13	13	13	13	13	14	14	14	14	14	15	15	15	15	15	16	16	16	16	16	17	17	17											
0.401 – 0.420 (0.0158 – 0.0165)	6	6	6	7	7	7	8	8	8	9	9	9	9	9	10	10	10	10	10	11	11	11	11	11	12	12	12	12	12	13	13	13	13	13	14	14	14	14	14	15	15	15	15	15	16	16	16	16	16	17	17	17	17												
0.421 – 0.440 (0.0166 – 0.0173)	6	6	7	7	7	8	8	8	9	9	9	9	10	10	10	10	10	11	11	11	11	11	12	12	12	12	12	13	13	13	13	13	14	14	14	14	14	15	15	15	15	15	16	16	16	16	16	17	17	17	17	17													
0.441 – 0.460 (0.0174 – 0.0181)	6	7	7	7	8	8	8	9	9	9	10	10	10	10	10	11	11	11	11	11	12	12	12	12	12	13	13	13	13	13	14	14	14	14	14	15	15	15	15	15	16	16	16	16	16	17	17	17	17	17	17														
0.461 – 0.480 (0.0181 – 0.0189)	7	7	8	8	8	8	9	9	10	10	10	10	10	11	11	11	11	11	12	12	12	12	12	13	13	13	13	13	14	14	14	14	14	15	15	15	15	15	16	16	16	16	16	17	17	17	17	17	17	17															
0.481 – 0.500 (0.0189 – 0.0197)	7	8	8	8	8	9	9	10	10	10	10	11	11	11	11	11	12	12	12	12	12	13	13	13	13	13	14	14	14	14	14	15	15	15	15	15	16	16	16	16	16	17	17	17	17	17	17	17																	
0.501 – 0.520 (0.0197 – 0.0205)	8	8	8	9	9	9	10	10	10	11	11	11	11	11	12	12	12	12	12	13	13	13	13	13	14	14	14	14	14	15	15	15	15	15	16	16	16	16	16	17	17	17	17	17	17	17																			
0.521 – 0.540 (0.0205 – 0.0213)	8	8	9	9	9	10	10	10	11	11	11	11	12	12	12	12	12	13	13	13	13	13	14	14	14	14	14	15	15	15	15	15	16	16	16	16	16	17	17	17	17	17	17	17																					
0.541 – 0.560 (0.0213 – 0.0220)	8	9	9	9	10	10	10	11	11	11	12	12	12	12	12	13	13	13	13	13	14	14	14	14	14	15	15	15	15	15	16	16	16	16	16	17	17	17	17	17	17	17																							
0.561 – 0.580 (0.0221 – 0.0228)	9	9	10	10	10	10	11	11	12	12	12	12	12	13	13	13	13	13	14	14	14	14	14	15	15	15	15	15	16	16	16	16	16	17	17	17	17	17	17	17																									
0.581 – 0.600 (0.0229 – 0.0236)	9	10	10	10	10	11	11	12	12	12	12	13	13	13	13	13	14	14	14	14	14	15	15	15	15	15	16	16	16	16	16	17	17	17	17	17	17	17																											
0.601 – 0.620 (0.0237 – 0.0244)	10	10	10	11	11	11	12	12	12	13	13	13	13	13	14	14	14	14	14	15	15	15	15	15	16	16	16	16	16	17	17	17	17	17	17	17																													
0.621 – 0.640 (0.0244 – 0.0252)	10	10	11	11	11	12	12	12	13	13	13	13	14	14	14	14	14	15	15	15	15	15	16	16	16	16	16	17	17	17	17	17	17	17																															
0.641 – 0.660 (0.0252 – 0.0260)	10	11	11	11	12	12	12	13	13	13	14	14	14	14	14	15	15	15	15	15	16	16	16	16	16	17	17	17	17	17	17	17																																	
0.661 – 0.680 (0.0260 – 0.0268)	11	11	12	12	12	12	13	13	14	14	14	14	14	15	15	15	15	15	16	16	16	16	16	17	17	17	17	17	17	17																																			
0.681 – 0.700 (0.0268 – 0.0276)	11	12	12	12	12	13	13	14	14	14	14	15	15	15	15	15	16	16	16	16	16	17	17	17	17	17	17	17																																					
0.701 – 0.720 (0.0276 – 0.0283)	12	12	12	13	13	13	14	14	14	15	15	15	15	15	16	16	16	16	16	17	17	17	17	17	17	17																																							
0.721 – 0.740 (0.0284 – 0.0291)	12	12	13	13	13	14	14	14	15	15	15	15	16	16	16	16	16	17	17	17	17	17	17	17																																											
0.741 – 0.760 (0.0292 – 0.0299)	12	13	13	13	14	14	14	15	15	15	16	16	16	16	16	17	17	17	17	17	17	17																																													
0.761 – 0.780 (0.0300 – 0.0307)	13	13	14	14	14	14	15	15	16	16	16	16	16	17	17	17	17	17	17	17																																															
0.781 – 0.800 (0.0307 – 0.0315)	13	14	14	14	14	15	15	16	16	16	16	17	17	17	17	17	17	17																																																	
0.801 – 0.820 (0.0315 – 0.0323)	14	14	14	15	15	15	16	16	16	17	17	17	17	17	17	17																																																			
0.821 – 0.840 (0.0323 – 0.0331)	14	14	15	15	15	16	16	16	17	17	17	17	17	17																																																					
0.841 – 0.860 (0.0331 – 0.0339)	14	15	15	15	16	16	16	17	17	17	17	17																																																							
0.861 – 0.880 (0.0339 – 0.0346)	15	15	16	16	16	16	17	17	17	17																																																									
0.881 – 0.900 (0.0347 – 0.0354)	15	16	16	16	16	17	17	17																																																													
0.901 – 0.920 (0.0355 – 0.0362)	16	16	16	17	17	17	17																																																														
0.921 – 0.940 (0.0363 – 0.0370)	16	16	17	17	17	17																																																																
0.941 – 0.960 (0.0370 – 0.0378)	16	17	17	17	17																																																																	
0.961 – 0.980 (0.0378 – 0.0386)	17	17	17																																																																				
0.981 – 1.000 (0.0386 – 0.0394)	17	17																																																																					
1.001 – 1.020 (0.0394 – 0.0402)	17																																																																							
1.021 – 1.030 (0.0402 – 0.0406)	17																																																																							

Intake valve clearance (Cold):
0.13 – 0.23 mm (0.006 – 0.009 in.)

EXAMPLE: The 2.800 mm (0.1102 in.) shim is installed, and the measured clearance is 0.450 mm (0.0177 in.). Replace the 2.800 mm (0.1102 in.) shim with a new No.10 shim.

New shim thickness mm (in.)

Shim No.	Thickness	Shim No.	Thickness
1	2.500 (0.0984)	10	2.950 (0.1161)
2	2.550 (0.1004)	11	3.000 (0.1181)
3	2.600 (0.1024)	12	3.050 (0.1201)
4	2.650 (0.1043)	13	3.100 (0.1220)
5	2.700 (0.1063)	14	3.150 (0.1240)
6	2.750 (0.1083)	15	3.200 (0.1260)
7	2.800 (0.1102)	16	3.250 (0.1280)
8	2.850 (0.1122)	17	3.300 (0.1299)
9	2.900 (0.1142)		

HINT: New shims have the thickness in millimeters imprinted on the face.

67170-TACO-G03

Intake valve shim selection chart—3.4L engines

Adjusting Shim Selection Chart (Exhaust)

Installed shim thickness mm (in.) / Measured clearance mm (in.)	2.500 (0.0984)	2.520 (0.0992)	2.540 (0.1000)	2.550 (0.1004)	2.560 (0.1008)	2.580 (0.1016)	2.600 (0.1024)	2.620 (0.1031)	2.640 (0.1039)	2.650 (0.1043)	2.660 (0.1047)	2.670 (0.1051)	2.680 (0.1055)	2.690 (0.1059)	2.700 (0.1063)	2.710 (0.1067)	2.720 (0.1071)	2.730 (0.1075)	2.740 (0.1079)	2.750 (0.1083)	2.760 (0.1087)	2.770 (0.1091)	2.780 (0.1094)	2.790 (0.1098)	2.800 (0.1102)	2.810 (0.1106)	2.820 (0.1110)	2.830 (0.1114)	2.840 (0.1118)	2.850 (0.1122)	2.860 (0.1126)	2.870 (0.1130)	2.880 (0.1134)	2.890 (0.1138)	2.900 (0.1142)	2.910 (0.1146)	2.920 (0.1150)	2.930 (0.1154)	2.940 (0.1157)	2.950 (0.1161)	2.960 (0.1165)	2.970 (0.1169)	2.980 (0.1173)	2.990 (0.1177)	3.000 (0.1181)	3.010 (0.1185)	3.020 (0.1189)	3.030 (0.1193)	3.040 (0.1197)	3.050 (0.1201)	3.060 (0.1205)	3.080 (0.1213)	3.100 (0.1220)	3.120 (0.1228)	3.140 (0.1236)	3.150 (0.1240)	3.160 (0.1244)	3.180 (0.1252)	3.200 (0.1260)	3.220 (0.1268)	3.240 (0.1276)	3.250 (0.1280)	3.260 (0.1283)	3.280 (0.1291)	3.300 (0.1299)
0.000 – 0.020 (0.0000 – 0.0008)																						1	1	1	1	1	1	1	2	2	2	2	2	3	3	3	3	3	4	4	4	4	4	5	5	5	5	5	6	6	6	6	7	7	8	8	8	8	9	9	10	10	10	10	11
0.021 – 0.040 (0.0008 – 0.0016)																				1	1	1	1	1	1	1	2	2	2	2	2	3	3	3	3	3	4	4	4	4	4	5	5	5	5	5	6	6	6	6	6	7	7	8	8	8	8	9	9	10	10	10	10	11	11
0.041 – 0.060 (0.0016 – 0.0024)																		1	1	1	1	1	1	1	2	2	2	2	2	3	3	3	3	3	4	4	4	4	4	5	5	5	5	5	6	6	6	6	6	7	7	7	8	8	8	9	9	9	10	10	10	11	11	11	12
0.061 – 0.080 (0.0024 – 0.0031)																1	1	1	1	1	1	1	2	2	2	2	2	3	3	3	3	3	4	4	4	4	4	5	5	5	5	5	6	6	6	6	6	7	7	7	7	8	8	8	9	9	9	10	10	10	11	11	11	12	12
0.081 – 0.100 (0.0032 – 0.0039)														1	1	1	1	1	1	1	2	2	2	2	2	3	3	3	3	3	4	4	4	4	4	5	5	5	5	5	6	6	6	6	6	7	7	7	7	7	8	8	8	9	9	9	10	10	10	11	11	11	12	12	12
0.101 – 0.120 (0.0040 – 0.0047)												1	1	1	1	1	1	1	2	2	2	2	2	3	3	3	3	3	4	4	4	4	4	5	5	5	5	5	6	6	6	6	6	7	7	7	7	7	8	8	8	8	9	9	10	10	10	10	11	11	12	12	12	12	13
0.121 – 0.140 (0.0048 – 0.0055)										1	1	1	1	1	1	1	2	2	2	2	2	3	3	3	3	3	4	4	4	4	4	5	5	5	5	5	6	6	6	6	6	7	7	7	7	7	8	8	8	8	8	9	9	10	10	10	10	11	11	12	12	12	12	13	13
0.141 – 0.160 (0.0056 – 0.0063)									1	1	1	1	1	1	2	2	2	2	2	3	3	3	3	3	4	4	4	4	4	5	5	5	5	5	6	6	6	6	6	7	7	7	7	7	8	8	8	8	8	9	9	9	10	10	10	11	11	11	12	12	12	13	13	13	14
0.161 – 0.180 (0.0063 – 0.0071)								1	1	1	1	1	2	2	2	2	2	3	3	3	3	3	4	4	4	4	4	5	5	5	5	5	6	6	6	6	6	7	7	7	7	7	8	8	8	8	8	9	9	9	9	10	10	10	11	11	11	12	12	12	13	13	13	14	14
0.181 – 0.200 (0.0071 – 0.0079)							1	1	1	1	2	2	2	2	2	3	3	3	3	3	4	4	4	4	4	5	5	5	5	5	6	6	6	6	6	7	7	7	7	7	8	8	8	8	8	9	9	9	9	9	10	10	10	11	11	11	12	12	12	13	13	13	14	14	14
0.201 – 0.220 (0.0079 – 0.0087)						1	1	1	2	2	2	2	2	3	3	3	3	3	4	4	4	4	4	5	5	5	5	5	6	6	6	6	6	7	7	7	7	7	8	8	8	8	8	9	9	9	9	9	10	10	10	10	11	11	12	12	12	12	13	13	14	14	14	14	15
0.221 – 0.240 (0.0087 – 0.0094)					1	1	1	2	2	2	2	3	3	3	3	3	4	4	4	4	4	5	5	5	5	5	6	6	6	6	6	7	7	7	7	7	8	8	8	8	8	9	9	9	9	9	10	10	10	10	10	11	11	12	12	12	12	13	13	14	14	14	14	15	15
0.241 – 0.260 (0.0095 – 0.0102)			1	1	1	1	2	2	2	3	3	3	3	3	4	4	4	4	4	5	5	5	5	5	6	6	6	6	6	7	7	7	7	7	8	8	8	8	8	9	9	9	9	9	10	10	10	10	10	11	11	11	12	12	12	13	13	13	14	14	14	15	15	15	16
0.261 – 0.269 (0.0103 – 0.0106)		1	1	1	1	1	2	2	3	3	3	3	3	4	4	4	4	4	5	5	5	5	5	6	6	6	6	6	7	7	7	7	7	8	8	8	8	8	9	9	9	9	9	10	10	10	10	10	11	11	11	11	12	12	13	13	13	13	14	14	15	15	15	15	16
0.270 – 0.370 (0.0106 – 0.0146)																																																																	
0.371 – 0.380 (0.0146 – 0.0150)	2	3	3	3	3	4	4	5	5	5	5	6	6	6	6	6	7	7	7	7	7	8	8	8	8	8	9	9	9	9	9	10	10	10	10	10	11	11	11	11	11	12	12	12	12	12	13	13	13	13	13	14	14	15	15	15	15	16	16	17	17	17	17	17	
0.381 – 0.400 (0.0150 – 0.0157)	2	3	3	3	4	4	4	5	5	5	6	6	6	6	6	7	7	7	7	7	8	8	8	8	8	9	9	9	9	9	10	10	10	10	10	11	11	11	11	11	12	12	12	12	12	13	13	13	13	13	14	14	14	15	15	15	16	16	16	17	17	17	17		
0.401 – 0.420 (0.0158 – 0.0165)	3	3	4	4	4	4	5	5	6	6	6	6	6	7	7	7	7	7	8	8	8	8	8	9	9	9	9	9	10	10	10	10	10	11	11	11	11	11	12	12	12	12	12	13	13	13	13	13	14	14	14	14	15	15	16	16	16	16	17	17	17	17			
0.421 – 0.440 (0.0166 – 0.0173)	3	4	4	4	4	5	5	6	6	6	6	7	7	7	7	7	8	8	8	8	8	9	9	9	9	9	10	10	10	10	10	11	11	11	11	11	12	12	12	12	12	13	13	13	13	13	14	14	14	14	14	15	15	16	16	16	16	17	17	17					
0.441 – 0.460 (0.0174 – 0.0181)	4	4	4	5	5	5	6	6	6	7	7	7	7	7	8	8	8	8	8	9	9	9	9	9	10	10	10	10	10	11	11	11	11	11	12	12	12	12	12	13	13	13	13	13	14	14	14	14	14	15	15	15	16	16	16	17	17	17	17						
0.461 – 0.480 (0.0181 – 0.0189)	4	4	5	5	5	6	6	6	7	7	7	7	8	8	8	8	8	9	9	9	9	9	10	10	10	10	10	11	11	11	11	11	12	12	12	12	12	13	13	13	13	13	14	14	14	14	14	15	15	15	15	16	16	16	17	17	17	17							
0.481 – 0.500 (0.0189 – 0.0197)	4	5	5	5	6	6	6	7	7	7	8	8	8	8	8	9	9	9	9	9	10	10	10	10	10	11	11	11	11	11	12	12	12	12	12	13	13	13	13	13	14	14	14	14	14	15	15	15	15	15	16	16	16	17	17	17	17								
0.501 – 0.520 (0.0197 – 0.0205)	5	5	6	6	6	6	7	7	8	8	8	8	8	9	9	9	9	9	10	10	10	10	10	11	11	11	11	11	12	12	12	12	12	13	13	13	13	13	14	14	14	14	14	15	15	15	15	15	16	16	16	16	17	17	17	17									
0.521 – 0.540 (0.0205 – 0.0213)	5	6	6	6	6	7	7	8	8	8	8	9	9	9	9	9	10	10	10	10	10	11	11	11	11	11	12	12	12	12	12	13	13	13	13	13	14	14	14	14	14	15	15	15	15	15	16	16	16	16	16	17	17	17											
0.541 – 0.560 (0.0213 – 0.0220)	6	6	6	7	7	7	8	8	8	9	9	9	9	9	10	10	10	10	10	11	11	11	11	11	12	12	12	12	12	13	13	13	13	13	14	14	14	14	14	15	15	15	15	15	16	16	16	16	16	17	17	17	17												
0.561 – 0.580 (0.0221 – 0.0228)	6	6	7	7	7	8	8	8	9	9	9	9	10	10	10	10	10	11	11	11	11	11	12	12	12	12	12	13	13	13	13	13	14	14	14	14	14	15	15	15	15	15	16	16	16	16	16	17	17	17	17	17													
0.581 – 0.600 (0.0229 – 0.0236)	6	7	7	7	8	8	8	9	9	9	10	10	10	10	10	11	11	11	11	11	12	12	12	12	12	13	13	13	13	13	14	14	14	14	14	15	15	15	15	15	16	16	16	16	16	17	17	17	17	17	17														
0.601 – 0.620 (0.0237 – 0.0244)	7	7	8	8	8	8	9	9	10	10	10	10	10	11	11	11	11	11	12	12	12	12	12	13	13	13	13	13	14	14	14	14	14	15	15	15	15	15	16	16	16	16	16	17	17	17	17	17	17	17															
0.621 – 0.640 (0.0244 – 0.0252)	7	8	8	8	8	9	9	10	10	10	10	11	11	11	11	11	12	12	12	12	12	13	13	13	13	13	14	14	14	14	14	15	15	15	15	15	16	16	16	16	16	17	17	17	17	17	17	17																	
0.641 – 0.600 (0.0252 – 0.0260)	8	8	8	9	9	9	10	10	10	11	11	11	11	11	12	12	12	12	12	13	13	13	13	13	14	14	14	14	14	15	15	15	15	15	16	16	16	16	16	17	17	17	17	17	17	17																			
0.661 – 0.680 (0.0260 – 0.0268)	8	8	9	9	9	10	10	10	11	11	11	11	12	12	12	12	12	13	13	13	13	13	14	14	14	14	14	15	15	15	15	15	16	16	16	16	16	17	17	17	17	17	17	17																					
0.681 – 0.700 (0.0268 – 0.0276)	8	9	9	9	10	10	10	11	11	11	12	12	12	12	12	13	13	13	13	13	14	14	14	14	14	15	15	15	15	15	16	16	16	16	16	17	17	17	17	17	17	17																							
0.701 – 0.720 (0.0276 – 0.0283)	9	9	10	10	10	10	11	11	12	12	12	12	12	13	13	13	13	13	14	14	14	14	14	15	15	15	15	15	16	16	16	16	16	17	17	17	17	17	17	17																									
0.721 – 0.740 (0.0284 – 0.0291)	9	10	10	10	10	11	11	12	12	12	12	13	13	13	13	13	14	14	14	14	14	15	15	15	15	15	16	16	16	16	16	17	17	17	17	17	17	17																											
0.741 – 0.760 (0.0292 – 0.0299)	10	10	10	11	11	11	12	12	12	13	13	13	13	13	14	14	14	14	14	15	15	15	15	15	16	16	16	16	16	17	17	17	17	17	17	17																													
0.761 – 0.780 (0.0300 – 0.0307)	10	10	11	11	11	12	12	12	13	13	13	13	14	14	14	14	14	15	15	15	15	15	16	16	16	16	16	17	17	17	17	17	17	17																															
0.781 – 0.800 (0.0307 – 0.0315)	10	11	11	11	12	12	12	13	13	13	14	14	14	14	14	15	15	15	15	15	16	16	16	16	16	17	17	17	17	17	17	17																																	
0.801 – 0.820 (0.0315 – 0.0323)	11	11	12	12	12	12	13	13	14	14	14	14	14	15	15	15	15	15	16	16	16	16	16	17	17	17	17	17	17	17																																			
0.821 – 0.840 (0.0323 – 0.0331)	11	12	12	12	12	13	13	14	14	14	14	15	15	15	15	15	16	16	16	16	16	17	17	17	17	17	17	17																																					
0.841 – 0.860 (0.0331 – 0.0339)	12	12	12	13	13	13	14	14	14	15	15	15	15	15	16	16	16	16	16	17	17	17	17	17	17	17																																							
0.861 – 0.880 (0.0339 – 0.0346)	12	12	13	13	13	14	14	14	15	15	15	15	16	16	16	16	16	17	17	17	17	17	17	17																																									
0.881 – 0.900 (0.0347 – 0.0354)	12	13	13	13	14	14	14	15	15	15	16	16	16	16	16	17	17	17	17	17	17	17																																											
0.901 – 0.920 (0.0355 – 0.0362)	13	13	14	14	14	14	15	15	16	16	16	16	16	17	17	17	17	17	17	17																																													
0.921 – 0.940 (0.0363 – 0.0370)	13	14	14	14	14	15	15	16	16	16	16	17	17	17	17	17	17	17																																															
0.941 – 0.960 (0.0370 – 0.0378)	14	14	14	15	15	15	16	16	16	17	17	17	17	17	17	17																																																	
0.961 – 0.980 (0.0378 – 0.0386)	14	14	15	15	15	16	16	16	17	17	17	17	17	17																																																			
0.981 – 1.000 (0.0386 – 0.0394)	14	15	15	15	16	16	16	17	17	17	17	17																																																						
1.001 – 1.020 (0.0394 – 0.0402)	15	15	16	16	16	16	17	17	17	17																																																								
1.021 – 1.040 (0.0402 – 0.0409)	15	16	16	16	16	17	17	17																																																										
1.041 – 1.060 (0.0410 – 0.0417)	16	16	16	17	17	17	17																																																											
1.061 – 1.080 (0.0418 – 0.0425)	16	16	17	17	17	17																																																												
1.081 – 1.100 (0.0426 – 0.0433)	16	17	17	17	17																																																													
1.101 – 1.120 (0.0433 – 0.0441)	17	17	17																																																															
1.121 – 1.140 (0.0441 – 0.0449)	17	17																																																																
1.141 – 1.160 (0.0449 – 0.0457)	17																																																																	
1.161 – 1.170 (0.0457 – 0.0461)	17																																																																	

Exhaust valve clearance (Cold):
0.27 – 0.37 mm (0.011 – 0.014 in.)

EXAMPLE: The 2.800 mm (0.1102 in.) shim is installed, and the measured clearance is 0.450 mm
(0.0177 in.). Replace the 2.800 mm (0.1102 in.) shim with a new No.10 shim.

New shim thickness mm (in.)

Shim No.	Thickness	Shim No.	Thickness
1	2.500 (0.0984)	10	2.950 (0.1161)
2	2.550 (0.1004)	11	3.000 (0.1181)
3	2.600 (0.1024)	12	3.050 (0.1201)
4	2.650 (0.1043)	13	3.100 (0.1220)
5	2.700 (0.1063)	14	3.150 (0.1240)
6	2.750 (0.1083)	15	3.200 (0.1260)
7	2.800 (0.1102)	16	3.250 (0.1280)
8	2.850 (0.1122)	17	3.300 (0.1299)
9	2.900 (0.1142)		

HINT: New shims have the thickness in millimeters imprinted on the face.

67170-TACO-G04

Exhaust valve shim selection chart—3.4L engines

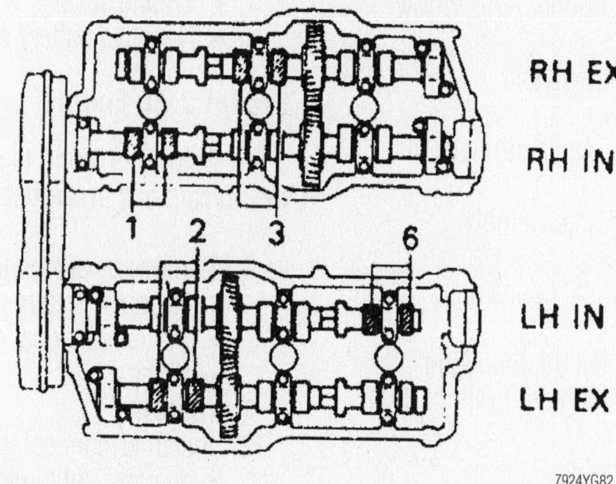

First valve adjustment—3.4L engine

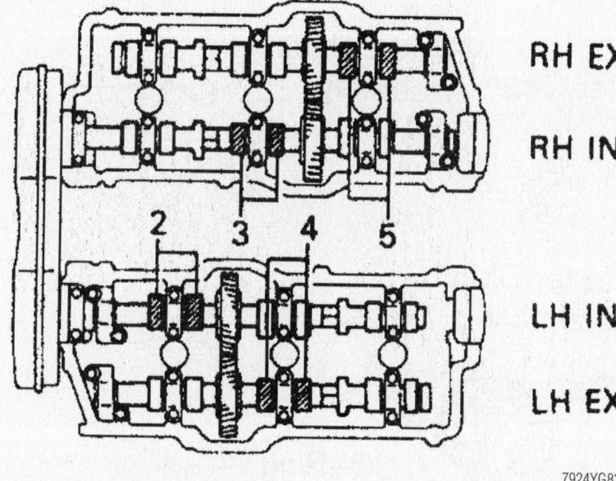

Second valve adjustment—3.4L engine

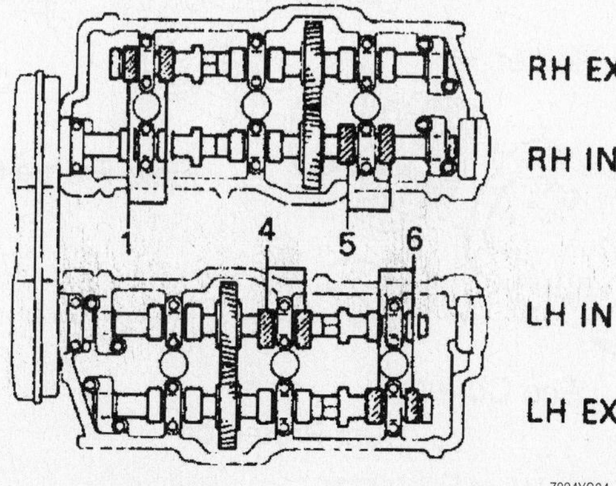

Third valve adjustment—3.4L engine

shaft 1 revolution (360 degrees) and align the marks.

6. Inspect the valve clearance, as follows:

a. Measure the clearance between the valve lifter and the camshaft. Measure the 1st intake and the 3rd exhaust valves on the right head and the 6th intake and the 2nd exhaust valves on the left head.

b. Turn the crankshaft ⅔ of a revolution (240 degrees) and adjust the 3rd intake and the 5th exhaust valves on the right head and the 2nd intake and the 4th exhaust valves on the left head.

c. Turn the crankshaft ⅔ of a revolution (240 degrees) and adjust the 5th intake and the 1st exhaust valves on the right head and the 4th intake and the 6th exhaust valves on the left head.

7. Valve clearance cold should be:

- Intake: 0.006–0.009 in. (0.13–0.23mm)

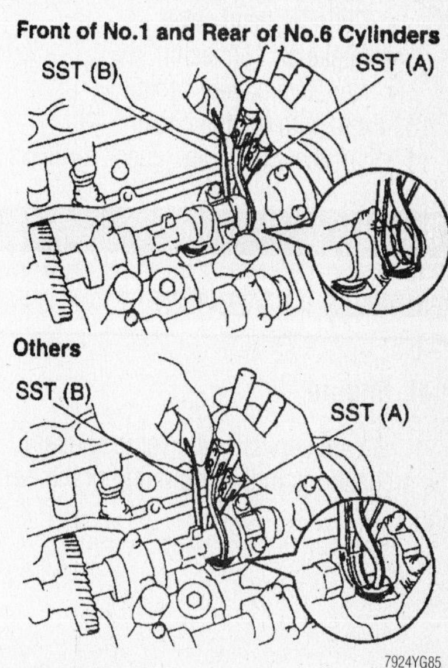

Removing the adjusting shim—3.4L engine

- Exhaust: 0.011–0.014 in. (0.27–0.37mm)

8. Adjust the valve clearance by using adjusting shims, as follows:

a. Turn the equipment camshaft so that the cam lobe for the valve to be adjusted faces up.

b. Turn the valve lifter so that the notches are perpendicular to the camshaft.

c. Using SST 09248-55040, press down the valve lifter and place SST 09248-05420, between the camshaft and the valve lifter. Remove SST 09248-55040.

d. Remove the adjusting shim with a small flat prying tool and a magnetic finger.

e. Determine the replacement adjusting shim size according to the following formula or use the adjusting shim charts.

f. Using a micrometer, measure the thickness of the removed shim. Calculate the thickness of a new shim so that the valve clearance comes within the specified value.

- T: Thickness of the removed shim
- A: Measured valve clearance
- N: Thickness of the new shim

g. Intake: N = T + (A – 0.007 in. (0.18mm))

h. Exhaust: N = T + (A – 0.013 in. (0.32mm))

i. Install a new adjusting shim. Place it on the valve lifter. Using the SST 09248-55040, press down the valve lifter and remove SST 09248-05420.

j. Recheck the valve clearance.

9. Install or connect the following:

- Cylinder head cover
- Intake air connector
- Negative battery cable

10. Refill with engine coolant.
11. Start the engine and check for leaks.

Starter

REMOVAL & INSTALLATION

2.4L Engine

1. Before servicing the vehicle, refer to the precautions in the beginning of this section.
2. Remove or disconnect the following:
   - Negative battery cable
   - Engine cover
   - Accelerator cable and intake air connector
   - Intake manifold assembly
   - Starter motor
   - Starter electrical connectors

To install:

3. Install or connect the following:
   - Starter. Tighten both bolts to 29 ft. lbs. (39 Nm).
   - Intake manifold
   - Accelerator cable and intake air connector
   - Engine cover
   - Negative battery cable

Except 2.4L Engine

1. Before servicing the vehicle, refer to the precautions in the beginning of this section.
2. Remove or disconnect the following:
   - Negative battery cable
   - Starter electrical connectors
   - Starter

To install:

3. Install or connect the following:
   - Starter. Tighten the fasteners to 29 ft. lbs. (39 Nm).

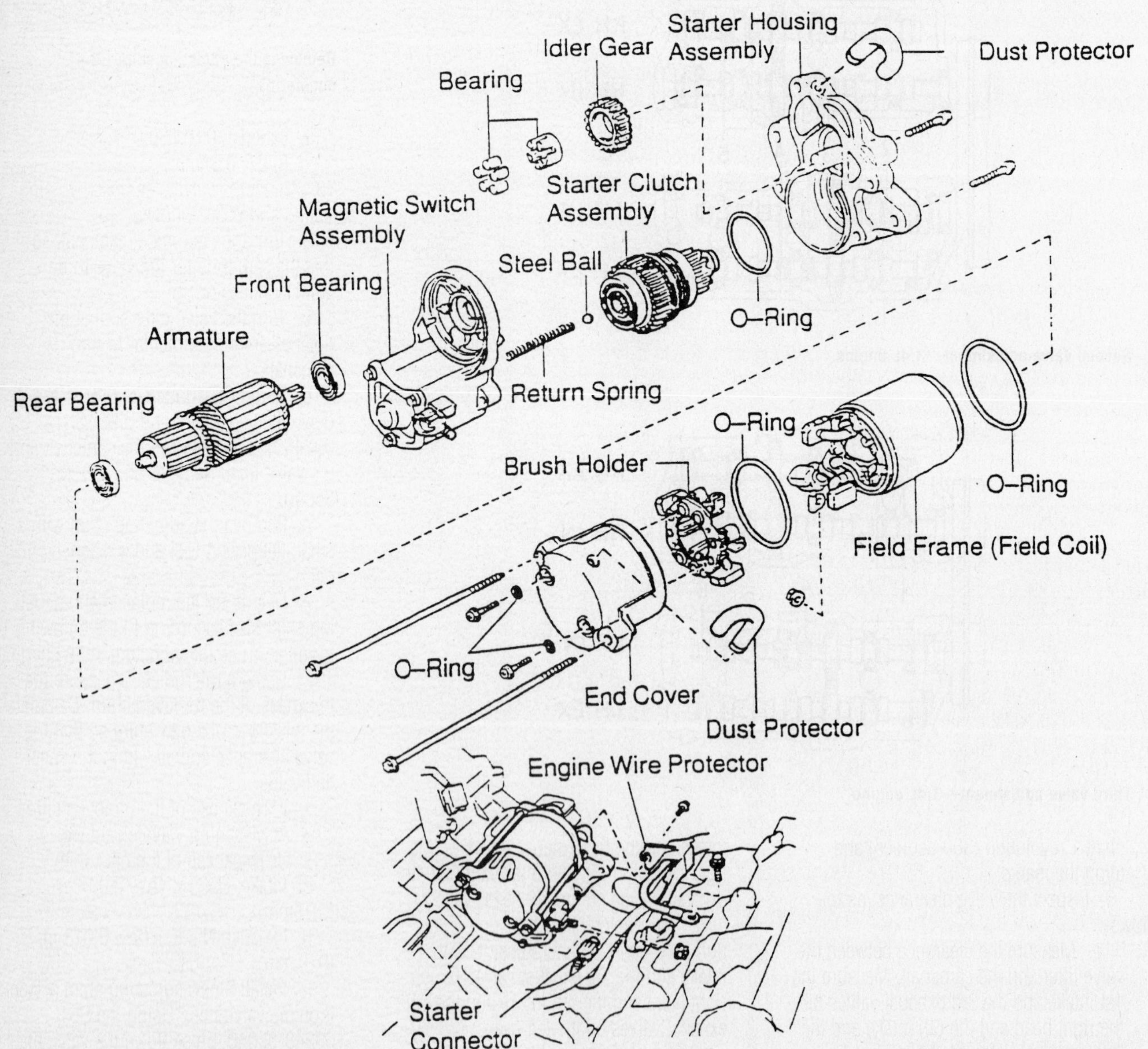

This starter motor location—2.4L engine

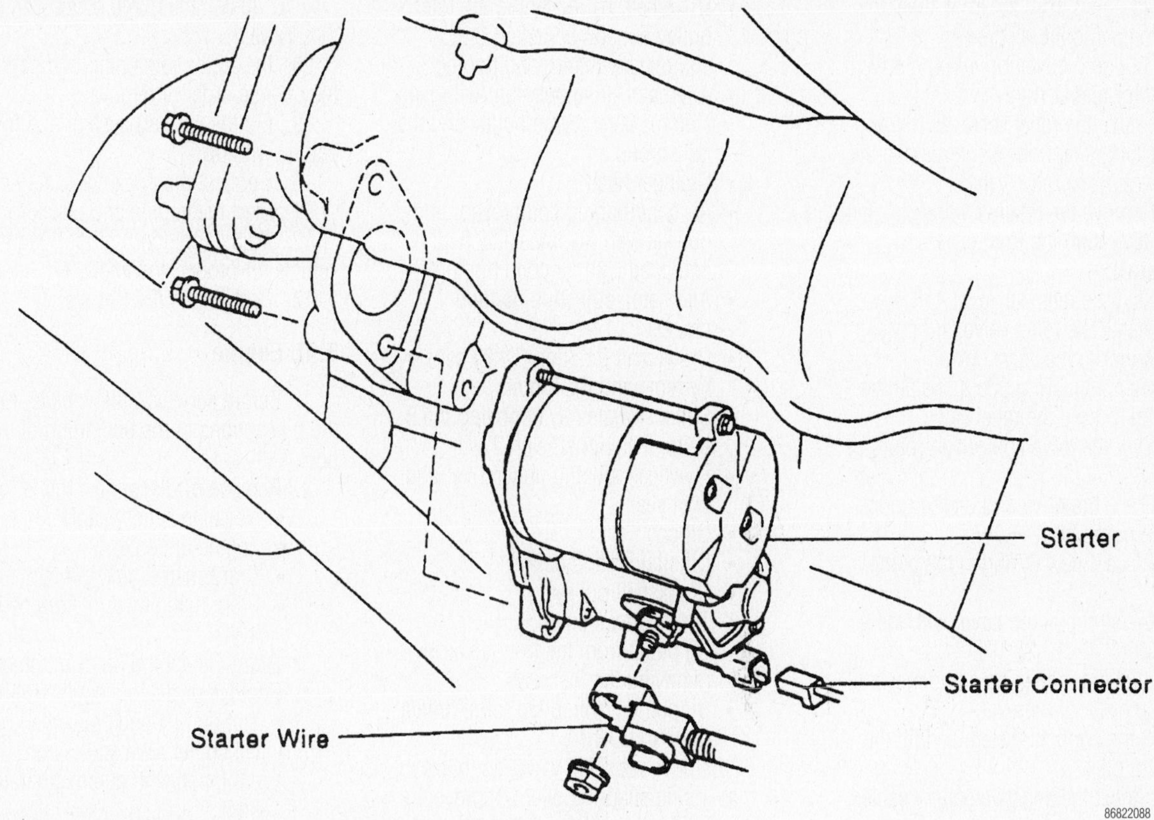

Exploded view of the starter—except 2.4L engine

- Electrical connections
- Negative battery cable

Oil Pan

REMOVAL & INSTALLATION

1. Before servicing the vehicle, refer to the precautions in the beginning of this section.
2. Disconnect the negative battery cable.
3. Drain the engine oil.
4. Remove or disconnect the following:
   - Engine undercover
   - Front differential, if equipped with 4WD
   - Oil pan, separate it from the engine using SST 09032-00100 and a brass bar

To install:

5. Apply seal packing to the oil pan.
6. Install the oil pan to the cylinder block. Tighten the nuts and bolts to 67 inch lbs. (8 Nm)

⁂ WARNING

If parts are not assembled within 5 minutes of applying time, the effectiveness of the seal packing is lost and must be removed and reapplied.

7. Install or connect the following:
   - Front differential, if removed
   - Engine undercover
   - Negative battery cable
8. Fill with engine oil.
9. Start the engine and check for leaks.

Oil Pump

REMOVAL & INSTALLATION

2.4L Engine

➡The oil pump assembly is mounted in the timing chain cover. To properly service the oil pump, the timing chain cover should be removed from the cylinder block.

1. Before servicing the vehicle, refer to the precautions in the beginning of this section.
2. Disconnect the negative battery cable.
3. Drain the oil and the cooling system.
4. Remove or disconnect the following:
   - Engine undercover
   - Front differential and halfshaft assembly, if equipped with 4WD
   - Upper radiator hose from the radiator
   - Oil dipstick guide, by removing the bolt
   - Power steering drive belt, by loosening the lockbolt and adjusting bolt, if equipped with power steering
   - No. 2 fan shroud, by removing the 2 clips
   - No. 1 fan shroud, by removing the 4 bolts
   - Drive belt, if equipped with air conditioning
   - Alternator drive belt, fan (with fan clutch), water pump pulley and the fan shroud
   - Cylinder head
   - Air conditioning compressor and bracket, if equipped with air conditioning
   - Alternator, adjusting bar and bracket
   - Crankshaft Position (CKP) sensor, by removing the 2 bolts
   - Stiffener plates by removing the 8 bolts, if equipped with 2WD
   - Flywheel housing undercover and dust seal
   - Oil pan.
   - Oil strainer and gasket
   - Crankshaft pulley
   - Timing chain cover

- Oil pump from the front cover, by removing the 9 screws
- Oil pump cover, drive rotor, driven rotor and O-ring

5. Remove the relief valve, as follows:
   a. Using snapring pliers, remove the snapring for the relief valve.
   b. Remove the retainer, spring(s) and relief valve from the front cover.

To install:

6. Install the relief valve, as follows:
   a. Install the relief valve, spring(s) and retainer to the valve cover.
   b. Using snapring pliers, install the snapring to hold the relief valve.
7. Install the drive and driven rotors as follows:
   a. Place the drive and driven rotors into the pump body.
   b. Place a new O-ring to the pump body.
   c. Install the pump cover with the 9 screws.
8. Install the remaining components in the reverse order of removal.
9. Fill the cooling system and fill the engine with oil.
10. Connect the negative battery cable.
11. Start the engine and check for leaks.
12. Adjust ignition timing. Road test the vehicle for proper operation.
13. Recheck all fluid levels.

2.7L Engine

➡The oil pump assembly is mounted in the timing chain cover. To properly service the oil pump, the timing chain cover should be removed from the cylinder block.

1. Before servicing the vehicle, refer to the precautions in the beginning of this section.
2. Disconnect the negative battery cable.
3. Drain the oil and the cooling system.
4. Remove or disconnect the following:
   - Engine undercover
   - Front differential and halfshaft assembly, if equipped with 4WD
   - 2 bolts and the air pipe, for California vehicles with 3RZ-FE engine
   - Upper radiator hose from the radiator
   - Oil dipstick guide
   - Power steering drive belt, by loosening the lockbolt and adjusting bolt, if equipped with power steering
   - No. 2 fan shroud by removing the 2 clips
   - No. 1 fan shroud by removing the 4 bolts
   - Drive belt, by loosening the idler pulley nut and adjusting bolt, if equipped with air conditioning
   - Alternator drive belt, fan (with fan clutch), water pump pulley and the fan shroud
   - Cylinder head
   - Air conditioning compressor and bracket with the lines attached, if equipped with air conditioning
   - Alternator, adjusting bar and bracket
   - Crankshaft Position (CKP) sensor by removing the 2 bolts
   - Stiffener plates by removing the 8 bolts, if equipped with 2WD
   - Flywheel housing undercover and dust seal
   - Oil pan
   - Oil strainer and gasket
   - Crankshaft pulley
   - Timing chain cover
   - Oil pump from the front cover by removing the 9 screws
   - Oil pump cover, drive rotor, driven rotor and O-ring
5. Remove the relief valve, as follows:
   a. Using snapring pliers, remove the snapring for the relief valve.
   b. Remove the retainer, spring(s) and relief valve from the front cover.

To install:

6. Install the relief valve, as follows:
   a. Install the relief valve, spring(s) and retainer to the valve cover.
   b. Using snapring pliers, install the snapring to hold the relief valve.
7. Install the drive and driven rotors, as follows:
   a. Place the drive and driven rotors into the pump body.
   b. Place a new O-ring to the pump body.
   c. Install the pump cover with the 9 screws.
8. Install the remaining components in the reverse order of removal.
9. Fill the cooling system and fill the engine with oil.
10. Connect the negative battery cable.
11. Start the engine and check for leaks.
12. Adjust ignition timing. Road test the vehicle for proper operation.
13. Recheck all fluid levels.

3.4L Engine

1. Before servicing the vehicle, refer to the precautions in the beginning of this section.
2. Remove or disconnect the following:
   - Negative battery cable
   - Engine undercover
   - Crankshaft timing pulley
   - Front differential, if equipped with 4WD
3. Drain the engine oil from the engine.
4. Remove or disconnect the following:
   - Timing belt and crankshaft gear
   - Oil cooler tube and clamp, if equipped with automatic transmission
   - Stiffener plate
   - Flywheel housing undercover and dust cover
   - Rear end cover and dust cover
   - Starter wire clamp
   - Crankshaft Position (CKP) sensor
   - Oil pan

➡Be careful not to damage the baffle plate flange.

- Oil strainer
- Oil baffle plate
- Oil pump body by removing the 8 bolts.
- O-ring from the cylinder block

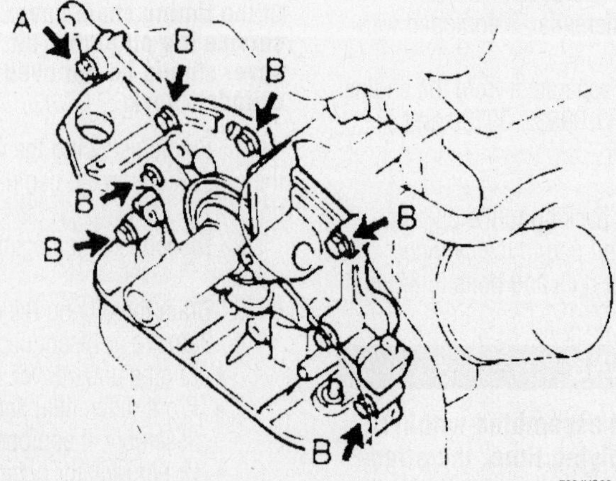

Oil pump bolt identification—3.4L engine

To install:

5. Install or connect the following:
   - Apply Seal Packing PN 08826-00080 to the oil pump
   - New O-ring into the groove of the cylinder block
   - Oil pump to the crankshaft with the spline teeth of the drive rotor engaged with the large teeth of the crankshaft. Tighten the oil pump bolts "A" 15 ft. lbs. (20 Nm) and bolts "B" 31 ft. lbs. (42 Nm)
   - CKP
   - Oil pan baffle plate
   - Oil strainer with a new gasket. Tighten the bolts to 13 ft. lbs. (18 Nm).
   - Remaining components
   - Negative battery cable
6. Fill with engine oil.
7. Start the engine and check for leaks.

Rear Main Seal

REMOVAL & INSTALLATION

Seal Retainer On Engine

1. Remove the transmission.
2. Remove the clutch cover assembly and flywheel (manual trans.) or the flexplate (automatic trans.).
3. Use a small, sharp knife to cut off the lip of the oil seal. Take great care not to score any metal with the knife.
4. Use a small prybar to pry the old seal from the retaining plate. Be careful not to damage the plate. Protect the tip of the tool with tape and pad the fulcrum point with cloth.
5. Inspect the crankshaft and seal lip contact surfaces for any sign of damage.

To install:

6. Apply a light coat of multi-purpose grease to the lip of a new oil seal. Loosely fit the seal into place by hand, making sure it is not crooked.
7. Use a seal driver such as (SST 09223–15030 and 09950–70010) of the correct size to install the seal. Tap it into place until the surface of the seal is flush with the edge of the housing.

➡Use the correct tools. Homemade substitutes may install the seal crooked, resulting in oil leaks and premature seal failure.

Seal Retainer Removed

1. Support the retainer on two thin pieces of wood.
2. Use a small prybar to pry the old seal from the retaining plate. Be careful not to damage the plate. Protect the tip of the tool with tape and pad the fulcrum point with cloth.

To install:

3. Apply a light coat of multi-purpose grease to the lip of a new oil seal. Loosely fit the seal into place by hand, making sure it is not crooked.
4. Use a seal driver such as (SST 09223–15030 and 09950–70010) of the correct size to install the seal. Tap it into place until the surface of the seal is flush with the edge of the housing.

Timing Belt

REMOVAL & INSTALLATION

3.4L Engine

1. Disconnect the negative battery cable.

**** CAUTION**

Work must be started after 90 seconds from the time the ignition switch is turned to the LOCK position and the negative battery cable is disconnected.

2. Raise and safely support the vehicle.
3. Remove the engine undercover.
4. Drain the engine coolant.

**** CAUTION**

Never open, service or drain the radiator or cooling system when hot; serious burns can occur from the steam and hot coolant. Also, when draining engine coolant, keep in mind that cats and dogs are attracted to ethylene glycol antifreeze and could drink any that is left in an uncovered container or in puddles on the ground. This will prove fatal in sufficient quantities. Always drain coolant into a sealable container. Coolant should be reused unless it is contaminated or is several years old.

5. Disconnect the upper radiator hose from the engine.
6. Remove the power steering drive belt.
7. Remove the air conditioning drive belt by loosening the idler pulley nut and the adjusting bolt.
8. Loosen the lockbolt, pivot bolt, and the adjusting bolt and the alternator drive belt.
9. Remove the No. 2 fan shroud by removing the 2 clips.
10. Remove the fan with the fluid coupling and fan pulleys.
11. Disconnect the power steering pump from the engine and set aside. Do not disconnect the lines from the pump.
12. If equipped with air conditioning, disconnect the compressor from the engine and set aside. Do not disconnect the lines from the compressor.
13. If equipped with air conditioning, disconnect the air conditioning bracket.
14. Remove the No. 2 timing belt cover, as follows:
    a. Detach the camshaft position sensor connector from the No. 2 timing belt cover.
    b. Disconnect the 3 spark plug wire clamps from the No. 2 timing belt cover.
    c. Remove the 6 bolts and remove the timing belt cover.
15. Remove the fan bracket, as follows:
    a. Remove the power steering adjusting strut by removing the nut.
    b. Remove the fan bracket by removing the bolt and nut.
16. Set the No. 1 cylinder at Top Dead Center (TDC) of the compression stroke, as follows:
    a. Turn the crankshaft pulley and align its groove with the timing mark **0** of the No. 1 timing belt cover.
    b. Check that the timing marks of the camshaft timing pulleys and the No. 3 timing belt cover are aligned. If not, turn the crankshaft pulley one revolution (360 degrees).

➡If reusing the timing belt, be sure that you can still read the installation marks. If not, place new installation marks on the timing belt to match the timing marks of the camshaft timing pulleys.

17. Remove the timing belt tensioner by alternately loosening the 2 bolts.
18. Remove the camshaft timing pulleys, as follows:
    a. Using SST 09960-10010, or equivalent, remove the pulley bolt, the timing pulley and the knock pin. Remove the 2 timing pulleys with the timing belt.
19. Remove the crankshaft pulley, as follows:
    a. Using SST 09213-54015 and 09330-00021 or their equivalents, loosen the pulley bolt.
    b. Remove the SST tool, the pulley bolt and the pulley.
20. Remove the starter wire bracket and the No. 1 timing belt cover.

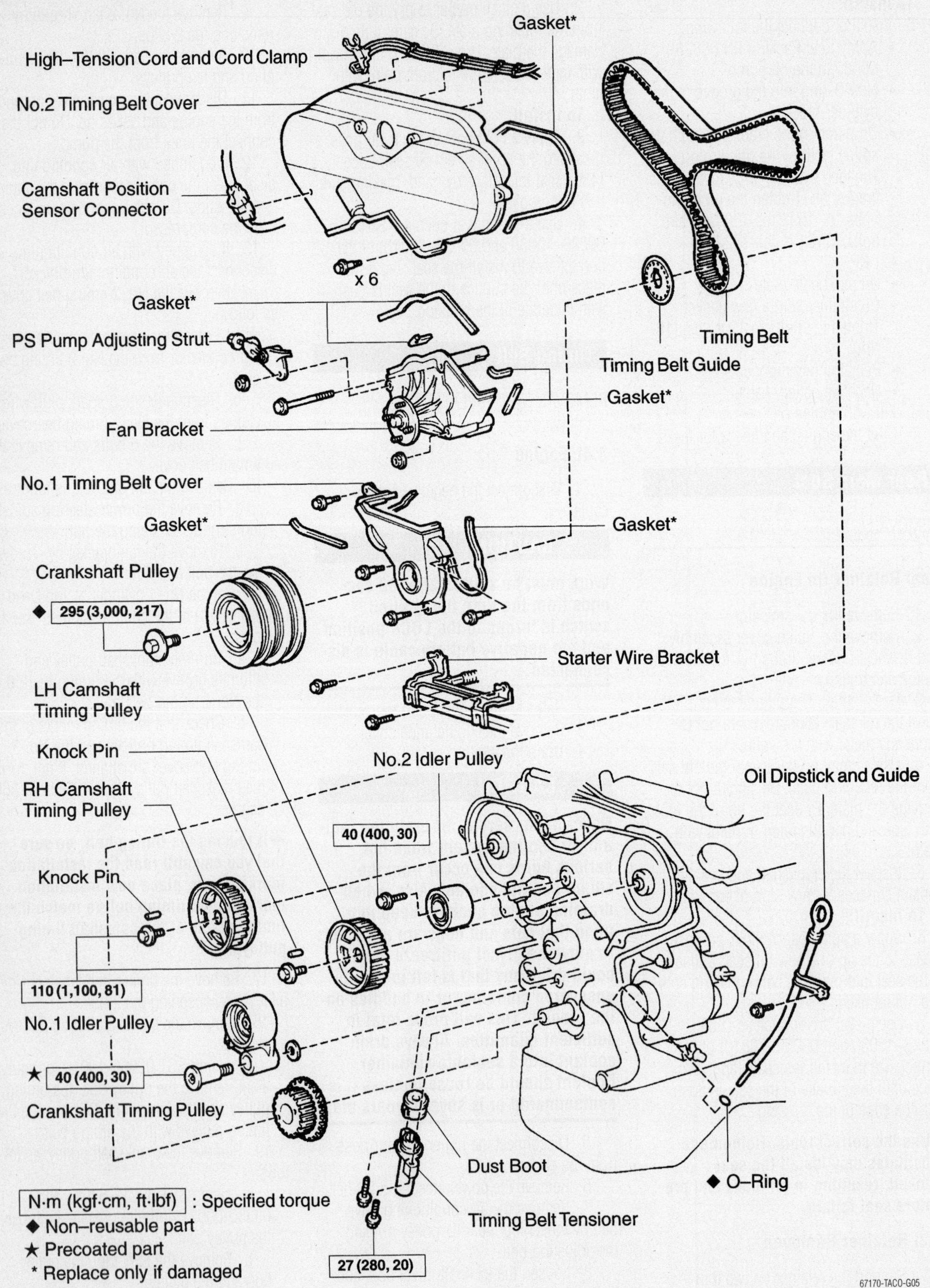

High–Tension Cord and Cord Clamp
No.2 Timing Belt Cover
Camshaft Position Sensor Connector
x 6
Gasket*
PS Pump Adjusting Strut
Fan Bracket
No.1 Timing Belt Cover
Gasket*
Crankshaft Pulley
◆ 295 (3,000, 217)
LH Camshaft Timing Pulley
Knock Pin
RH Camshaft Timing Pulley
Knock Pin
110 (1,100, 81)
No.1 Idler Pulley
★ 40 (400, 30)
Crankshaft Timing Pulley
N·m (kgf·cm, ft·lbf) : Specified torque
◆ Non–reusable part
★ Precoated part
* Replace only if damaged
Gasket*
Timing Belt
Timing Belt Guide
Gasket*
Gasket*
Starter Wire Bracket
No.2 Idler Pulley
40 (400, 30)
Oil Dipstick and Guide
Dust Boot
◆ O–Ring
Timing Belt Tensioner
27 (280, 20)
67170-TACO-G05

Timing belt and related parts—3.4L engine

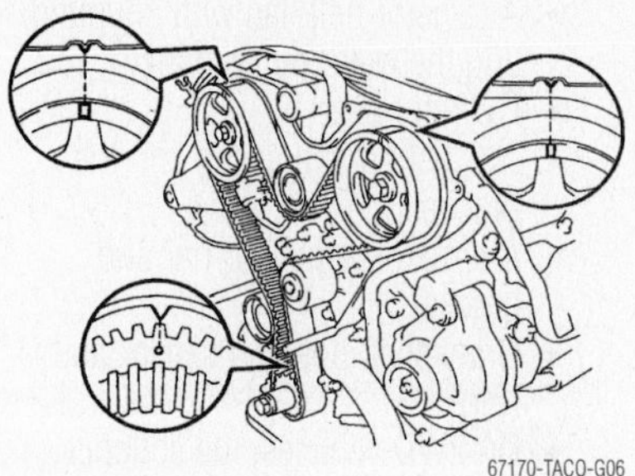

Turn the crankshaft clockwise to align the timing marks before removing the timing belt—3.4L engine

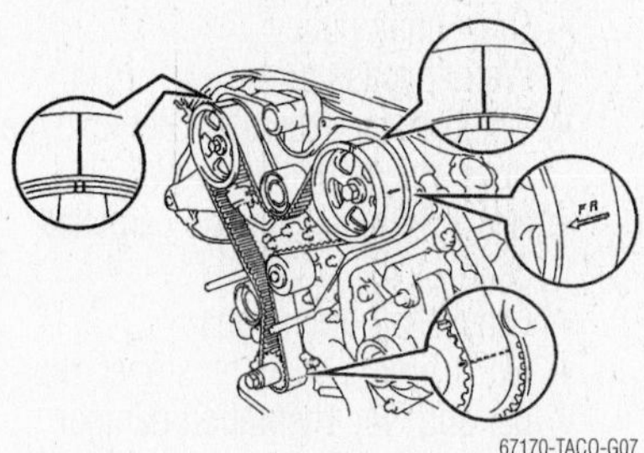

Check the installation marks on the timing belt

21. Remove the timing belt guide and remove the timing belt.

22. Remove the bolt and the No. 2 idler pulley.

23. Remove the pivot bolt, the No. 1 idler pulley and the plate washer.

24. Remove the crankshaft gear.

To install:

25. Install the crankshaft timing gear.

a. Align the timing pulley set key with the key groove of the gear.

b. Using SST 09214-60010, or equivalent, and a hammer, tap in the timing gear with the flange side facing inward.

26. Install the plate washer and the No. 1 idler pulley with the pivot bolt and tighten it to 26 ft. lbs. (35 Nm). Check that the pulley bracket moves smoothly.

27. Install the No. 2 timing belt idler with the bolt. Tighten the bolt to 30 ft. lbs. (40 Nm). Check that the pulley bracket moves smoothly.

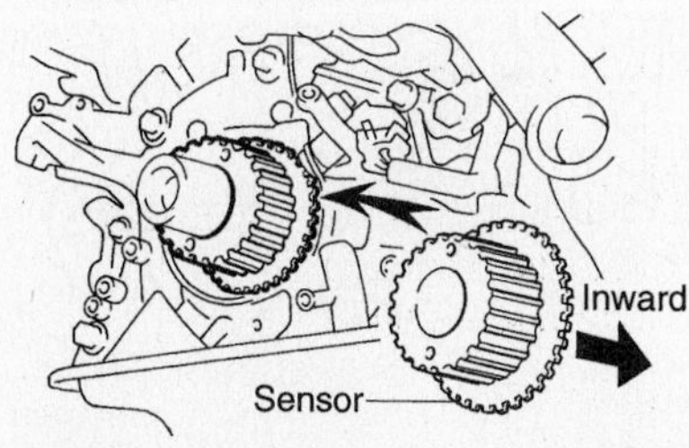

Tap in the timing gear with the flange side facing inward

28. Temporarily install the timing belt, as follows:

a. Using the crankshaft pulley bolt, turn the crankshaft and align the timing marks of the crankshaft timing pulley and the oil pump body.

b. Align the installation mark on the timing belt with the dot mark of the crankshaft timing pulley.

c. Install the timing belt on the crankshaft timing pulley, No. 1 idler pulley and the water pump pulleys.

29. Install the timing belt guide with the cup side facing outward.

30. Install the No. 1 timing belt cover and starter wire bracket. Tighten the timing belt cover bolts to 80 inch lbs. (9 Nm).

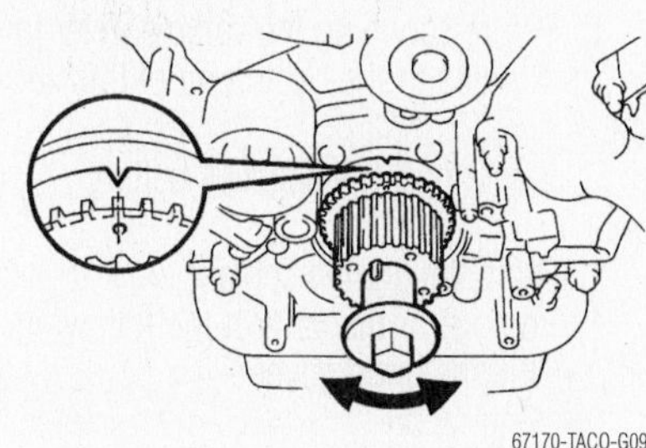

Turn the crankshaft and align the timing marks of the crankshaft timing pulley and the oil pump body, this sets the No.1 piston at TDC compression

Camshaft sprocket timing mark alignment

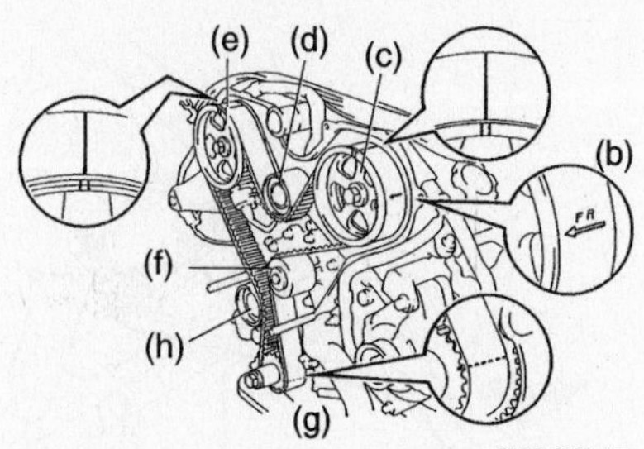

Timing belt installation sequence: b–timing belt front mark; c–left camshaft timing pulley; d–no.2 idler pulley; e–right camshaft timing pulley; f–water pump pulley; g–crankshaft; h–no.1 idler pulley

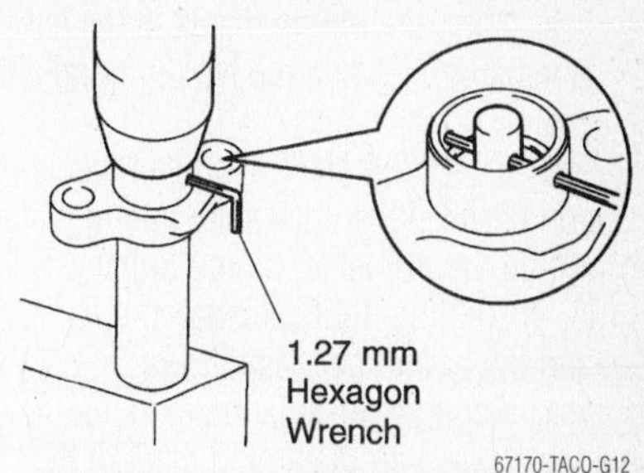

Compressing the tensioner plunger

WARNING

If any binding is felt when adjusting the timing belt tension by turning the crankshaft, STOP turning the engine, because the pistons may be hitting the valves.

31. Install the crankshaft pulley, as follows:

a. Align the pulley set key with the key groove of the crankshaft pulley.

b. Install the pulley bolt and tighten it to 184 ft. lbs. (250 Nm).

32. Install the left camshaft timing pulley.

a. Install the knock pin to the camshaft.

b. Align the knock pin hose of the camshaft with the knock pin groove of the timing pulley.

c. Slide the timing belt pulley on the camshaft with the flange side facing outward. Tighten the pulley bolt to 81 ft. lbs. (110 Nm).

33. Set the No. 1 cylinder to TDC of the compression stroke, as follows:

a. Turn the crankshaft pulley, and align its groove with the timing mark **0** of the No. 1 timing belt cover.

b. Turn the camshaft to align the knock pin hole of the camshaft with the timing mark of the No. 3 timing belt cover.

c. Turn the camshaft timing pulley, and align the timing marks of the camshaft timing pulley and the No. 3 timing belt cover.

34. Connect the timing belt to the left camshaft timing pulley, as follows:

➡Check that the installation mark on the timing belt is aligned with the end of the No. 1 timing belt cover.

a. Using SST 09960-01000, or equivalent, slightly turn the left camshaft timing pulley clockwise. Align the installation mark on the timing belt with the timing mark of the camshaft timing pulley and hang the timing belt on the left camshaft timing pulley.

b. Align the timing marks of the left camshaft pulley and the No. 3 timing belt cover.

c. Check that the timing belt has tension between the crankshaft timing pulley and the left camshaft timing pulley.

35. Install the right camshaft timing pulley and the timing belt, as follows:

a. Align the installation mark on the timing belt with the timing mark of the right camshaft timing pulley and hang the timing belt on the right camshaft timing pulley with the flange side facing inward.

b. Slide the right camshaft timing pulley on the camshaft. Align the timing marks on the right camshaft timing pulley and the No. 3 timing belt cover.

c. Align the knock pin hole of the camshaft with the knock pin groove of the pulley and install the knock pin. Install the bolt and tighten it to 81 ft. lbs. (110 Nm).

36. Set the timing belt tensioner, as follows:

a. Using a press, slowly press in the pushrod using 220–2205 lbs. (981–9807 N) of force.

b. align the holes of the pushrod and housing, pass a 1.5mm hex wrench through the holes to keep the setting position of the pushrod.

c. Release the press and install the dust boot on the tensioner.

37. Install the timing belt tensioner and alternately tighten the bolts to 20 ft. lbs. (28 Nm). Using pliers, remove the 1.5mm hex wrench from the belt tensioner.

38. Check the valve timing, as follows:

a. Slowly turn the crankshaft pulley 2 revolutions from TDC to TDC. Always turn the crankshaft pulley clockwise.

b. Check that each pulley aligns with the timing marks. If the timing marks do not align, remove the timing belt and reinstall it.

39. Install the fan bracket with the bolt and nut.

40. Install the power steering adjusting strut with the nut.

41. Install the No. 2 timing belt cover. Tighten the bolts to 80 inch lbs. (9 Nm). Install the remaining components.

42. Fill the cooling system with coolant.

43. Connect the negative battery cable.

44. Start the engine and check for leaks.

Timing Chain Cover, Seal and Timing Chain

REMOVAL & INSTALLATION

2.4L and 2.7L Engines

1. Before servicing the vehicle, refer to the precautions in the beginning of this section.
2. Disconnect the negative battery cable.
3. Drain the engine coolant.
4. Remove or disconnect the following:
   - Engine undercover
   - Engine oil
   - On 4WD vehicles, the front differential
   - Alternator belt, fan with coupling and the water pump pulley
   - Cylinder head
   - A/C belt, compressor, and the bracket
   - Alternator adjusting bar and bracket
   - Crankshaft position sensor and O-ring
   - On 2WD vehicles, the stiffener plates
   - Flywheel housing undercover and dust seal
   - Oil pan
   - Oil strainer and gasket
   - Crankshaft pulley
   - Water bypass pipe
   - Chain cover assembly. Remove the bolts shown by the arrows.
   - No. 1 timing chain and camshaft gear
   - Crankshaft timing gear
   - No. 1 timing chain tensioner slipper and No. 1 vibration damper
   - No. 1 damper
5. On the 2.4L remove the crankshaft position sensor rotor and the timing chain oil jet.
6. On the 2.7L, remove the No. 2 and No. 3 vibration dampers and the No. 2 chain tensioner as follows:

The crankshaft position sensor is located below the water pump—2.4L and 2.7L engines

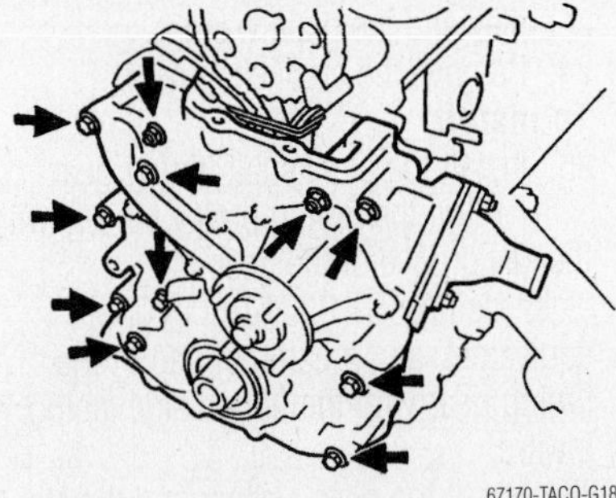

Remove the bolts securing the front cover (arrows)—2.4L and 2.7L engines

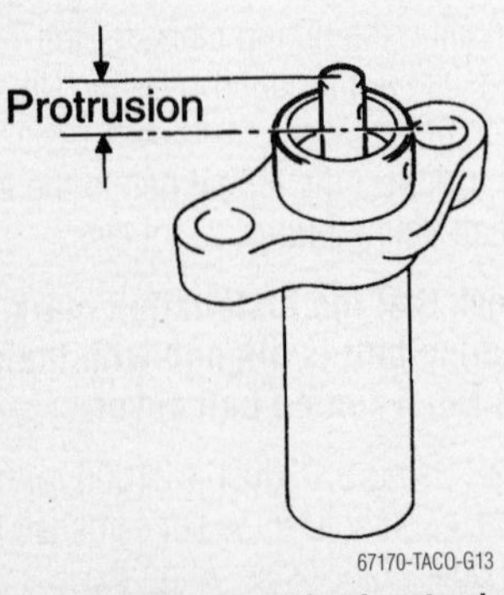

Measure the plunger pushrod protrusion as shown If the protrusion is not within 0.394–0.425 in. (10-10.8mm), replace the tensioner

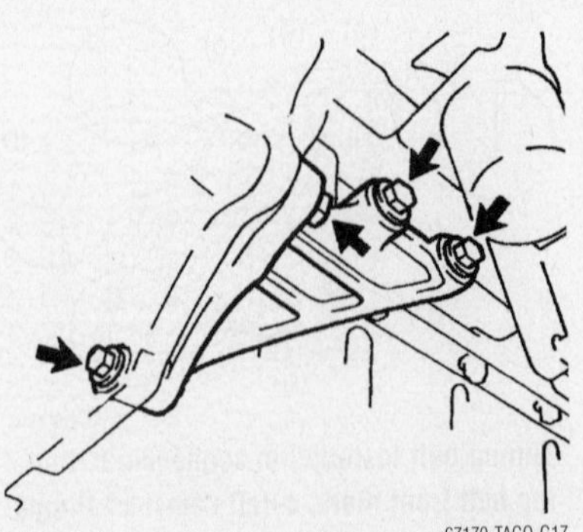

Remove the bolts securing the stiffener plates—2.4L and 2.7L engines

The oil jet is secured by a single bolt—2.4L and 2.7L engines

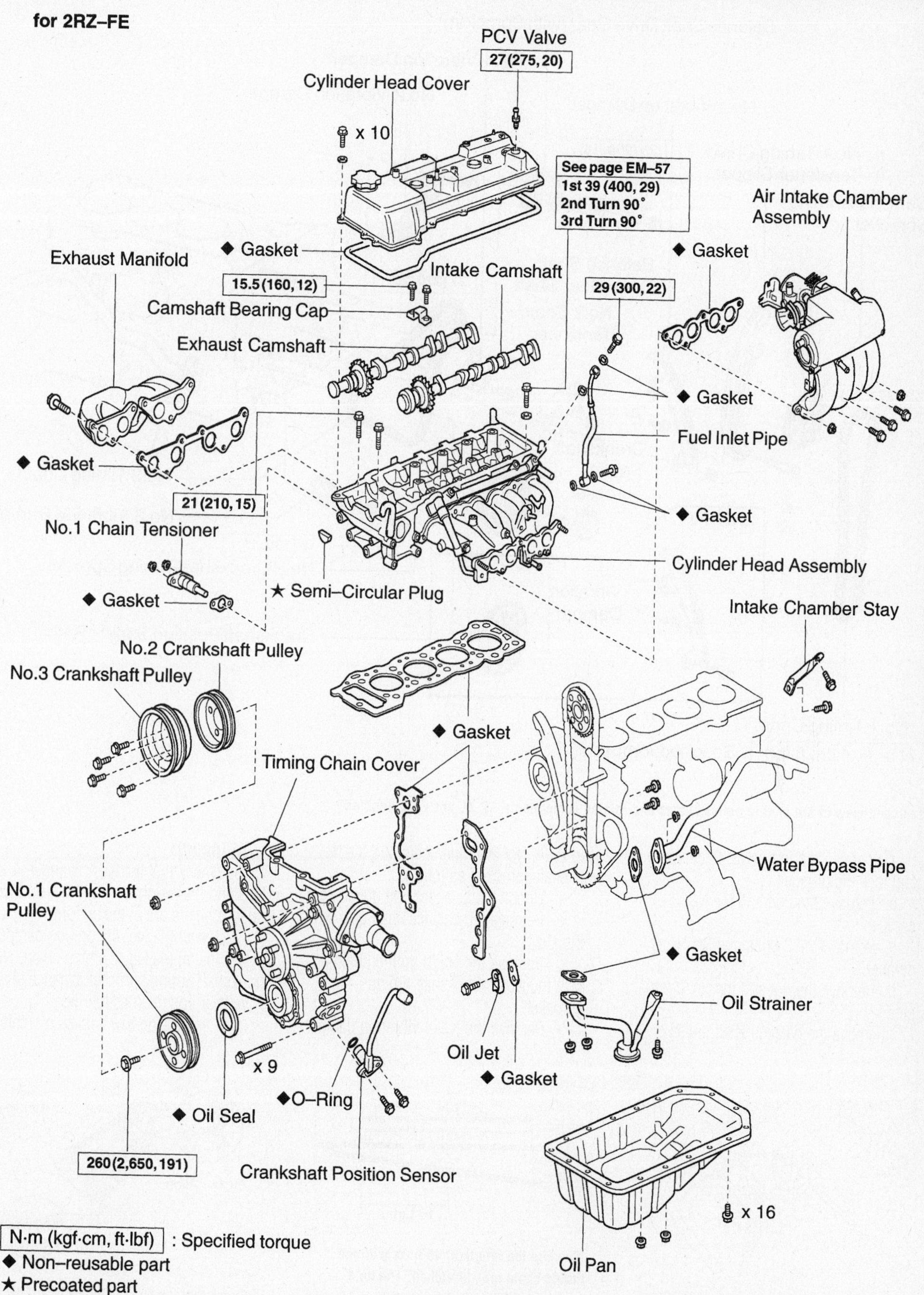
for 2RZ–FE
PCV Valve
27(275,20)
Cylinder Head Cover
x 10
See page EM–57
1st 39 (400, 29)
2nd Turn 90°
3rd Turn 90°
Air Intake Chamber Assembly
Exhaust Manifold
◆ Gasket
15.5(160,12)
Intake Camshaft
◆ Gasket
29(300,22)
Camshaft Bearing Cap
Exhaust Camshaft
◆ Gasket
Fuel Inlet Pipe
◆ Gasket
21(210,15)
◆ Gasket
No.1 Chain Tensioner
Cylinder Head Assembly
◆ Gasket
★ Semi–Circular Plug
Intake Chamber Stay
No.2 Crankshaft Pulley
No.3 Crankshaft Pulley
◆ Gasket
Timing Chain Cover
Water Bypass Pipe
No.1 Crankshaft Pulley
◆ Gasket
Oil Strainer
Oil Jet
x 9
◆ Gasket
◆O–Ring
◆ Oil Seal
260(2,650,191)
Crankshaft Position Sensor
x 16
N·m (kgf·cm, ft·lbf) : Specified torque
◆ Non–reusable part
★ Precoated part
Oil Pan
67170-TACO-G15

Front cover removal—2.4L and 2.7L engines

Balance Shaft Drive Gear (with Sprocket)
No.4 Vibration Damper
No.2 Vibration Damper
No.1 Vibration Damper
No.1 Timing Chain Tensioner Slipper
25 (250, 18)
Camshaft Timing Sprocket
Balance Shaft Drive Gear Shaft
27 (270, 20)
No.2 Chain Tensioner
27 (270, 20)
Crankshaft Timing Sprocket
No.2 Timing Chain
2RZ–FE
No.3 Vibration Damper
No.1 Vibration Damper
18 (185, 13)
No.2 Crankshaft Timing Sprocket
Crankshaft Position Sensor Rotor
No.1 Timing Chain
N·m (kgf·cm,ft·lbf) : Specified torque
67170-TACO-G14

Exploded view of the engine timing chains and related components—2.4L and 2.7L engines

a. Install a pin in the No. 2 tensioner and lock the plunger.

b. Remove the bolt and the No. 2 damper.

c. Remove the 2 bolts and the No. 3 damper.

d. Remove the nut and the No. 2 tensioner.

7. Remove the balance shaft driven gear, shaft, No. 2 timing chain and the No. 2 crankshaft sprocket, as follows:

a. Unbolt the balance shaft driven gear.

b. Remove the balance shaft gear with the shaft.

c. Remove the No. 2 timing chain with the No. 2 crankshaft timing sprocket.

- Remove the No. 4 vibration damper.

To install:

8. Check that the No.1 cylinder is at TDC and the weights of the N0.1 and No.2 balance shafts are at the bottom.

9. Install the No. 4 vibration dampener.

10. Install the No. 2 timing chain, No. 2 crankshaft timing sprocket, balance shaft drive gear and shaft as follows:

a. Install the No. 2 chain by matching

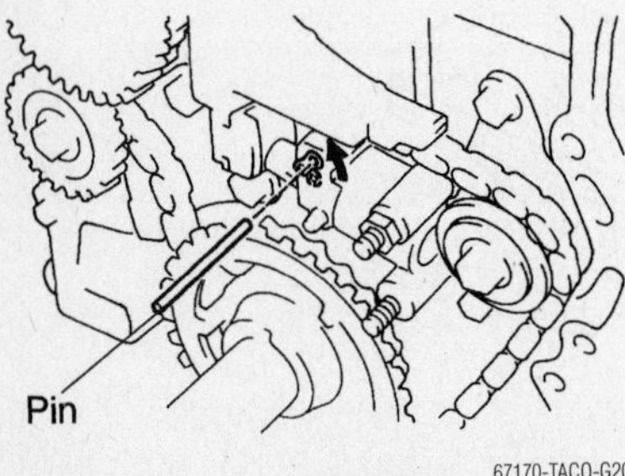

Lock the plunger with an appropriate pin—2.4L and 2.7L engines

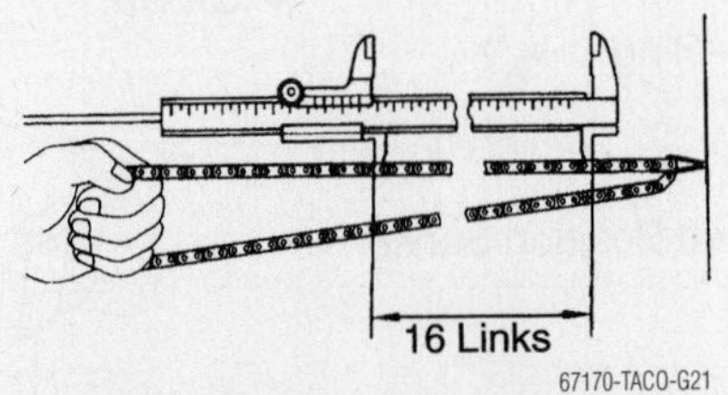

Measure the length of 16 links with the timing chain fully stretched. The no.1 chain should be 147.5mm (5.807 in.) max.; the no.2 chain should be 123.6mm (4.866 in.) max. If not, replace the chain.

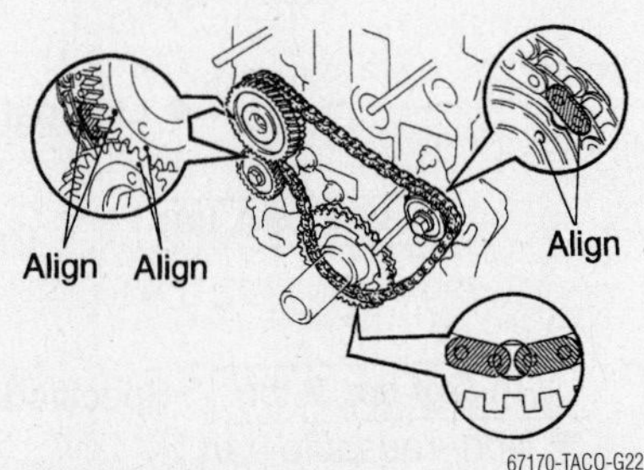

Check that the No.1 cylinder is at TDC and the weights of the No.1 and No.2 balance shafts are at the bottom

the marked links with the timing marks on the crankshaft sprocket and balance shaft timing sprocket.

b. Fit the other marked link of the No. 2 chain onto the sprocket behind the large timing mark of the balance shaft gear.

c. Insert the balance shaft gear shaft through the balance shaft drive gear so that it fits into the thrust plate hole. Align the small timing mark of the balance shaft drive gear with the timing mark of the balance shaft timing gear.

d. Install the bolt to the balance shaft gear and tighten to 18 ft. lbs. (25 Nm).

e. Check each timing mark is matched with the corresponding mark link.

11. Install the No. 2, No. 3 vibration dampers and the No. 2 chain tensioner, as follows:

➡Assemble the chain tensioner with the pin installed, then remove the pin after assembly.

a. Install the No. 2 chain tensioner with the nut, tighten to 13 ft. lbs. (18 Nm).

b. Install the No. 3 damper with the bolts, tighten to 13 ft. lbs. (18 Nm).

c. Install the No. 2 damper, tighten to 20 ft. lbs. (27 Nm).

d. Remove the pin from the No. 2 chain tensioner and free the plunger.

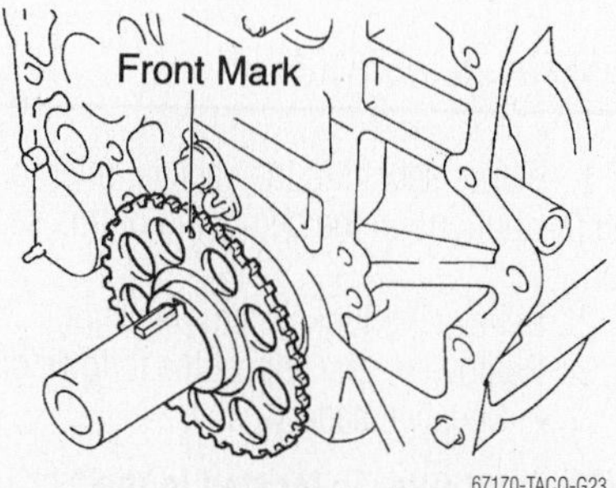

Install the crankshaft position sensor rotor

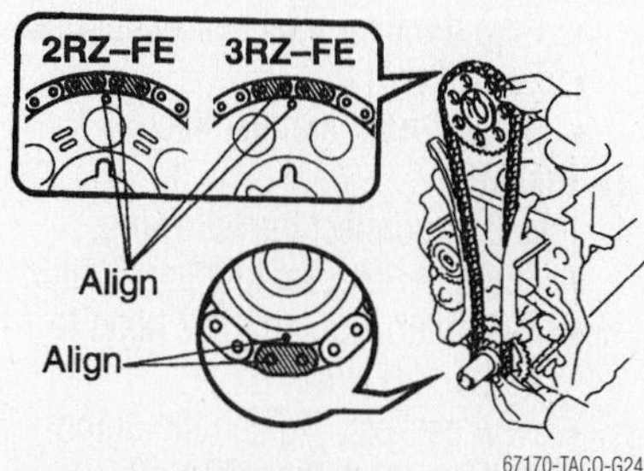

No.1 timing chain and camshaft timing sprocket installation

12. On the 2.4L, install the crankshaft position sensor rotor and the timing chain oil jet.

13. Install or connect the following:

- No. 1 timing chain tensioner slipper and the No. 1 vibration damper. Torque the No. 1 damper to 22 ft. lbs. (29 Nm); torque the slipper to 20 ft. lbs. (27 Nm). Check that the slipper moves smoothly.
- Crankshaft timing gear.
- No. 1 timing chain and camshaft timing gear.

14. Align the timing mark between the marked link of the No. 1 timing chain, and install the No. 1 timing chain to the gear.

15. Align the timing mark of the crankshaft timing gear with the mark of the No. 1 timing chain, then install the No. 1 timing chain.

16. Tie the No. 1 chain with a wire or cord, make sure it does not come loose.

17. Install or connect the following:

- Timing chain cover assembly

18. Tighten the following:

- 12mm **A** bolts—14 ft. lbs. (20 Nm)
- 12mm **B** bolts—18 ft. lbs. (25 Nm)
- 14mm bolts—32 ft. lbs. (44 Nm)
- 14mm nut—14 ft. lbs. (20 Nm)

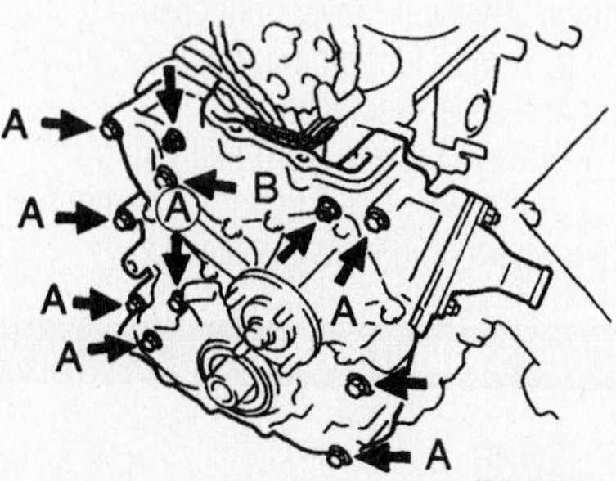

Timing cover fastener identification

19. Attach the water bypass pipe.

20. Remove the cord or wire from the chain.

21. Install or connect the following:

- Crankshaft pulley, tighten the bolt to 193 ft. lbs. (260 Nm). On A/C vehicles, install the crankshaft pulleys with bolts and tighten to 18 ft. lbs. (25 Nm).
- Oil strainer, tighten to 13 ft. lbs. (18 Nm)
- Oil pan, tighten the mounting bolts to 108 inch lbs. (13 Nm)
- Flywheel housing undercover and dust seal
- Stiffener plates on 2WD vehicles, tighten to 27 ft. lbs. (37 Nm)
- Crankshaft position sensor with a new O-ring
- Alternator, adjusting bar and bracket
- A/C compressor and bracket
- Cylinder head
- Water pump pulley and the fluid coupling with the fan
- On 4WD vehicles, the front differential and driveshaft assemblies

22. Adjust the drive belt tension.

23. Fill with engine coolant and engine oil.

24. Install the engine undercover.

25. Connect the negative battery cable.

SEAL REPLACEMENT

Cover Removed

1. Unbolt the timing chain cover assembly. Be careful to loosen only the correct bolts.

2. Pry out the seal from the cover with a flat-bladed tool.

3. It is a good idea to remove the oil pump from the timing cover and replace the O-ring.

To install:

4. Clean and inspect the timing cover area. Install new gaskets around the dowel areas and pump spline.

5. Apply multi-purpose grease to the new oil seal lip.

6. Tap the seal into place with SST 09223–50010/60010 or equivalent, and a hammer. Do this until the seal surface is flush with the cover edge.

7. Install the cover, tighten the bolts as specified for your engine.

8. If the oil pump was removed, install a new O-ring behind the pump prior to installation.

Cover Installed

1. Unbolt and remove the oil pump.

2. Using a knife, carefully cut off the oil seal lip. With a flat-bladed tool, (preferably with tape around it) pry the seal from the cover.

To install:

3. Apply multi-purpose grease to the new oil seal lip.

4. Tap the seal into place with SST 09223–50010/60011 or equivalent seal driver, and a hammer. Do this until the seal surface is flush with the cover edge.

5. Install the oil pump with a new O-ring.

Piston and Ring

POSITIONING

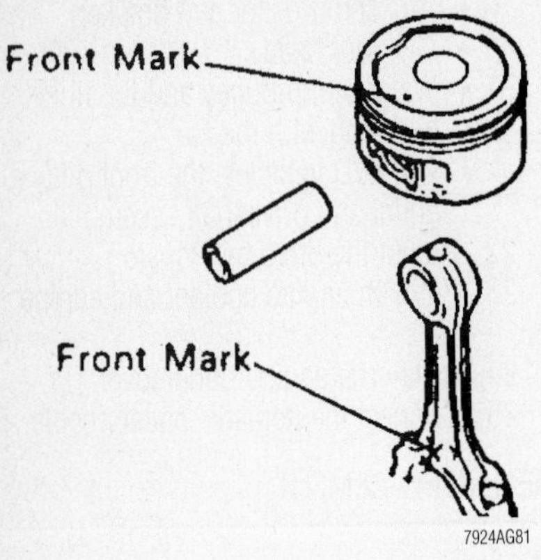

Piston to connecting rod assembly—2.4L, 2.7L and 3.4L engines

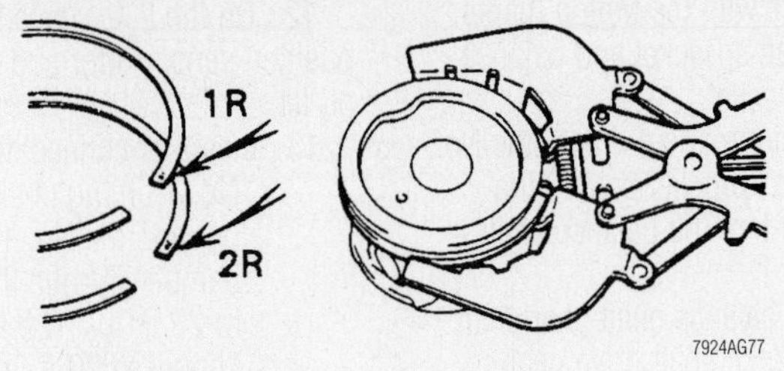

Compression ring identification mark locations—2.7L engine

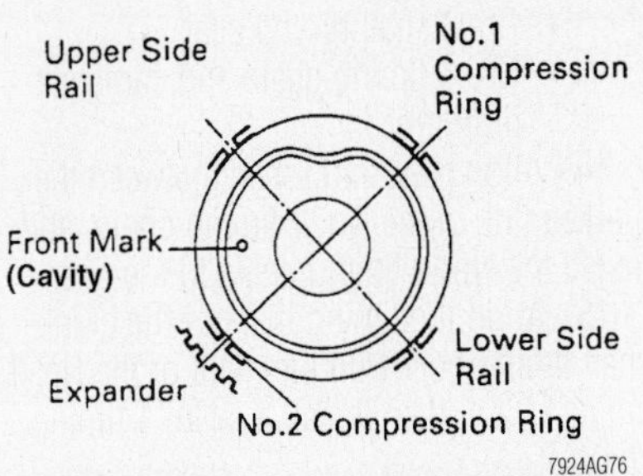

Piston ring end-gap spacing—2.4L and 2.7L engines

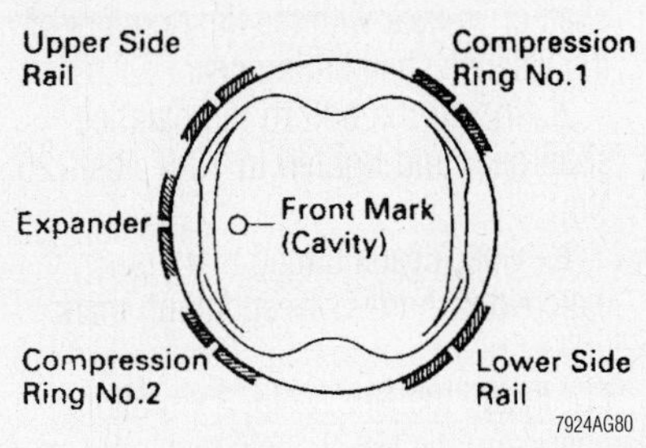

Piston ring end-gap spacing—3.4L engine

FUEL SYSTEM

Fuel System Service Precautions

Safety is the most important factor when performing not only fuel system maintenance, but any type of maintenance. Failure to conduct maintenance and repairs in a safe manner may result in serious personal injury or death. Work on a vehicle's fuel system components can be accomplished safely and effectively by adhering to the following rules and guidelines.

- To avoid the possibility of fire and personal injury, always disconnect the negative battery cable unless the repair or test procedure requires that battery voltage be applied.
- Always relieve the fuel system pressure prior to disconnecting any fuel system component (injector, fuel rail, pressure regulator, etc.) fitting or fuel line connection. Exercise extreme caution whenever relieving fuel system pressure, to avoid exposing skin, face and eyes to fuel spray. Please be advised that fuel under pressure may penetrate the skin or any part of the body that it contacts.
- Always place a shop towel or cloth around the fitting or connection prior to loosening to absorb any excess fuel due to spillage. Ensure that all fuel spillage is quickly remove from engine surfaces. Ensure that all fuel-soaked cloths or towels are deposited into a flame-proof waste container with a lid.
- Always keep a dry chemical (Class B) fire extinguisher near the work area.
- Do not allow fuel spray or fuel vapors to come into contact with a light bulb, spark or open flame.
- Always use a second wrench when loosening or tightening fuel line connection fittings. This will prevent unnecessary stress and torsion to fuel piping. Always follow the proper torque specifications.
- Always replace worn fuel fitting O-rings with new ones. Do not substitute fuel hose where rigid pipe is installed.

Fuel System Pressure

RELIEVING

1. Before servicing the vehicle, refer to the precautions in the beginning of this section.
2. Disconnect the negative battery terminal.
3. Place a catch-pan under the joint to be disconnected. A large quantity of fuel may be released when the joint is opened.
4. Wear eye or full face protection.
5. Place a shop towel over the area and slowly loosen the joint using a wrench of the correct size. Use a back-up wrench if needed.
6. Allow the fuel left in the line to bleed off slowly before fully disconnecting the joint.
7. Plug the opened lines immediately to prevent fuel spillage or the entry of dirt.
8. Dispose of the released fuel properly.
9. After connecting fuel lines, connect the negative battery cable and start the engine.
10. Check for leaks and repair as needed.

Fuel Filter

REMOVAL & INSTALLATION

1. Before servicing the vehicle, refer to the precautions in the beginning of this section.
2. Relieve the fuel system pressure.
3. Remove or disconnect the following:
   - Negative battery cable

➡The fuel filter is located in the engine compartment, at the inlet line to the fuel rail.

   - Plug the filter inlet and outlet lines
   - Fuel filter
   - Bracket from the fuel filter

To install:

4. Install or connect the following:
   - Fuel filter bracket to the fuel filter
   - Fuel filter. Tighten the 2 bolts to 14 ft. lbs. (20 Nm).
   - New gaskets. Tighten the union bolts to 22 ft. lbs. (30 Nm).
   - Negative battery cable
5. Start the engine and check for leaks.

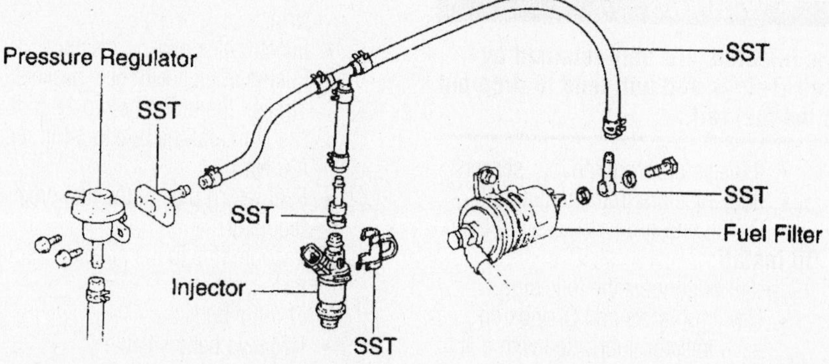

Exploded view of the fuel delivery components—2.4L and 2.7L engines

Fuel Pump

REMOVAL & INSTALLATION

1. Before servicing the vehicle, refer to the precautions in the beginning of this section.
2. Relieve the fuel pressure.
3. Disconnect the negative battery cable from the battery.
4. Drain the fuel from the fuel tank.
5. Remove or disconnect the following:
   - Fuel tank
   - Fuel pump connector from the clamp
   - Access plate bolts, then pull out the fuel pump assembly from the fuel tank
   - Gasket(s) from the pump bracket
   - Fuel pump connector
   - Bracket from the lower side of the fuel pump
   - Fuel pump from the fuel hose
   - Rubber cushion, the clip and the fuel filter at the bottom of the fuel pump

To install:

6. Install or connect the following:
   - Fuel pump filter to the fuel pump with a new clip
   - Fuel pump to the fuel pump bracket
   - Fuel hose to the outlet port of the fuel pump
   - Fuel pump connector
   - Fuel pump assembly with a new gasket(s). Tighten the bolts to 31 inch lbs. (4 Nm).
   - Fuel pump connector to the clamp
   - Fuel tank
   - All electrical and fuel connections
   - Negative battery cable
7. Refill the fuel tank and check for leaks.

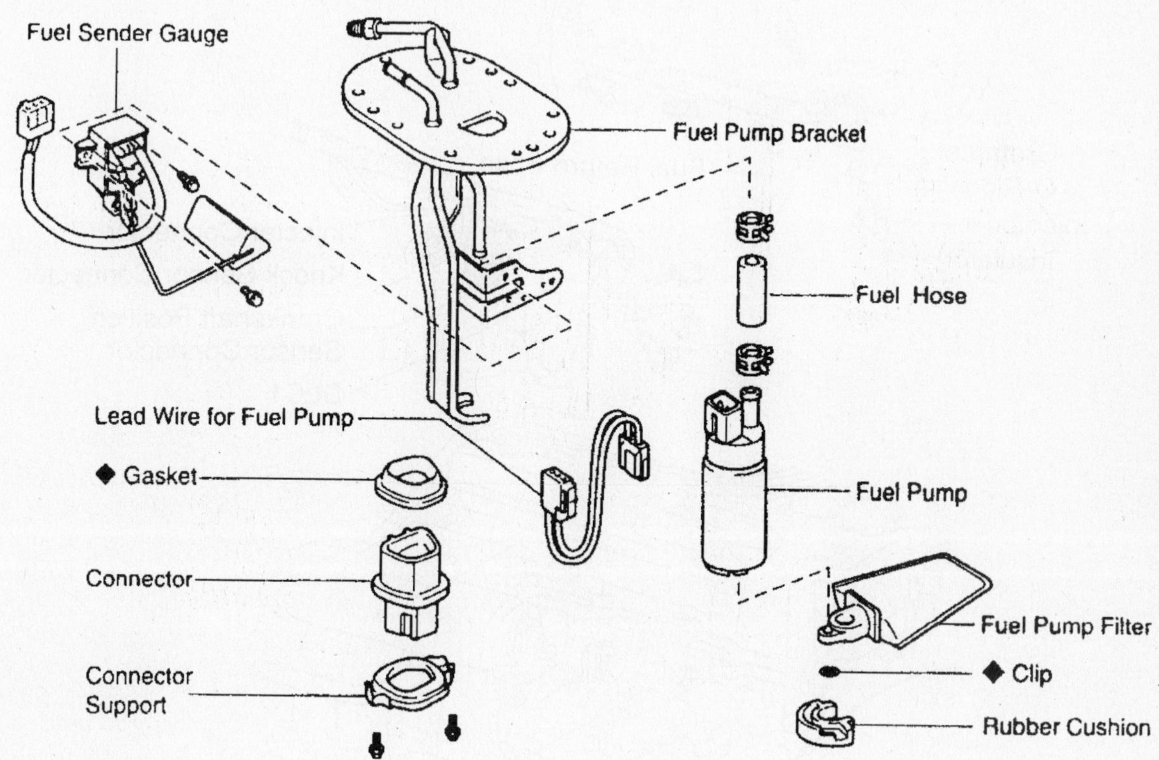

Exploded view of the fuel pump assembly and related components

Fuel Injector

REMOVAL & INSTALLATION

2.4L and 2.7L Engines

1. Before servicing the vehicle, refer to the precautions in the beginning of this section.
2. Relieve the fuel system pressure.
3. Remove or disconnect the following:
   - Throttle body
   - Fuel injector electrical connectors
   - Crankshaft Position (CKP) sensor connector
   - Knock Sensor (KS) connector
   - Data Link Connector 1 (DLC1) and wire clamp from the brackets
   - Vacuum line from the fuel pressure regulator
   - Fuel return hose from the pressure regulator
   - Union bolt and gaskets
   - Fuel inlet pipe from the fuel rail
   - Fuel rail with the injectors attached

⁂ WARNING

The injectors are only retained by their O-rings and will tend to drop out of the fuel rail.

   - 4 insulators from the four spacers
   - Fuel injectors from the fuel rail
   - O-ring and grommet, discard them

To install:

4. Install or connect the following:
   - New grommets and O-rings on each injector, lubricated with a light coat of gasoline
   - Fuel injectors, with the electrical connector facing upwards
   - New insulators and spacers on the intake manifold
5. Temporarily install the bolts holding the delivery pipe to the intake manifold.
6. Check that the injectors rotate smoothly.
7. Install or connect the following:
   - Tighten the delivery pipe-to-intake manifold bolts to 15 ft. lbs. (21 Nm).
   - Injector electrical connectors
   - Fuel inlet pipe with new gaskets. Tighten the union bolt to 22 ft. lbs. (29 Nm) and the bolt to 14 ft. lbs. (20 Nm).
   - Fuel return pipe to the fuel pressure regulator
   - Vacuum line to the pressure regulator
   - Throttle body
   - Negative battery cable

3.4L Engine

1. Before servicing the vehicle, refer to the precautions in the beginning of this section.
2. Depressurize the fuel system.
3. Remove or disconnect the following:
   - Air cleaner hose
   - Upper half of the intake manifold
   - Fuel pressure regulator

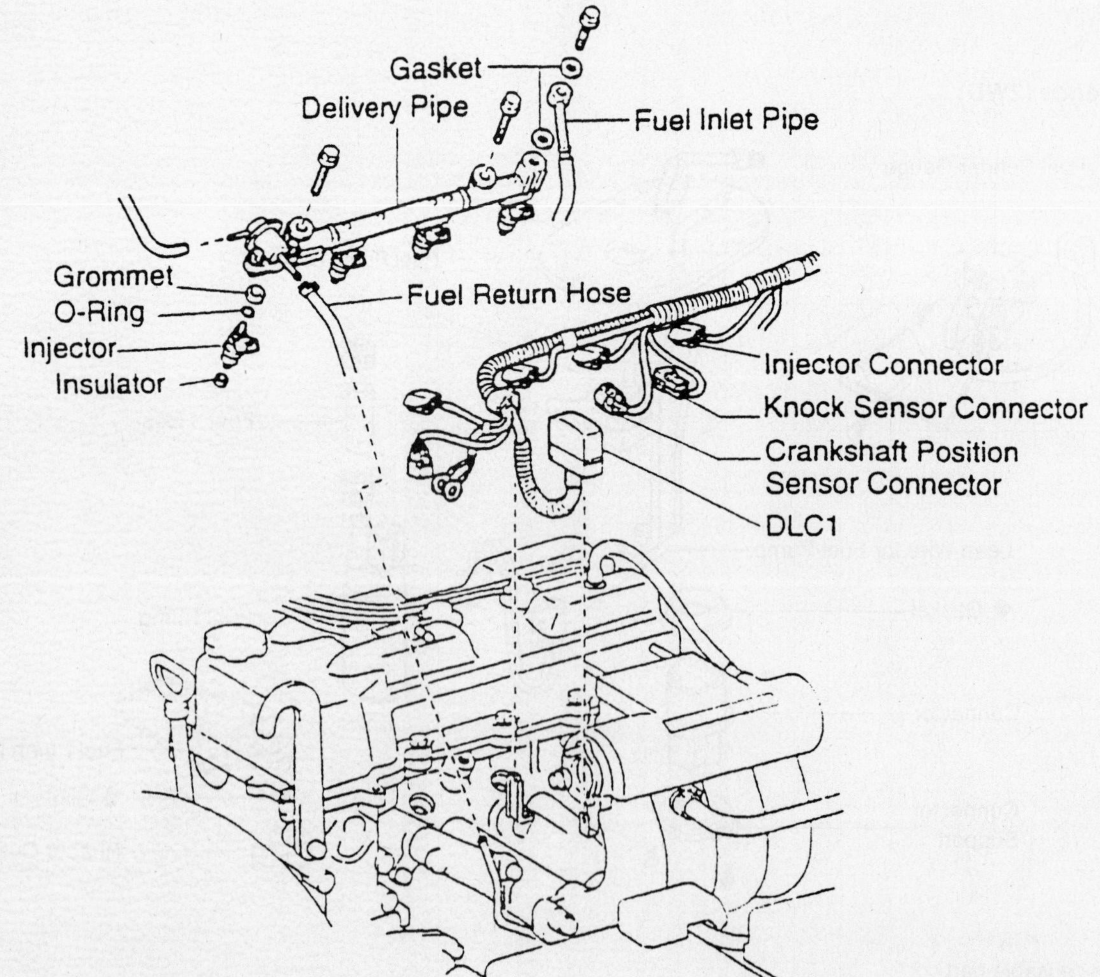

Fuel injector arrangement and related components—2.4L and 2.7L engines

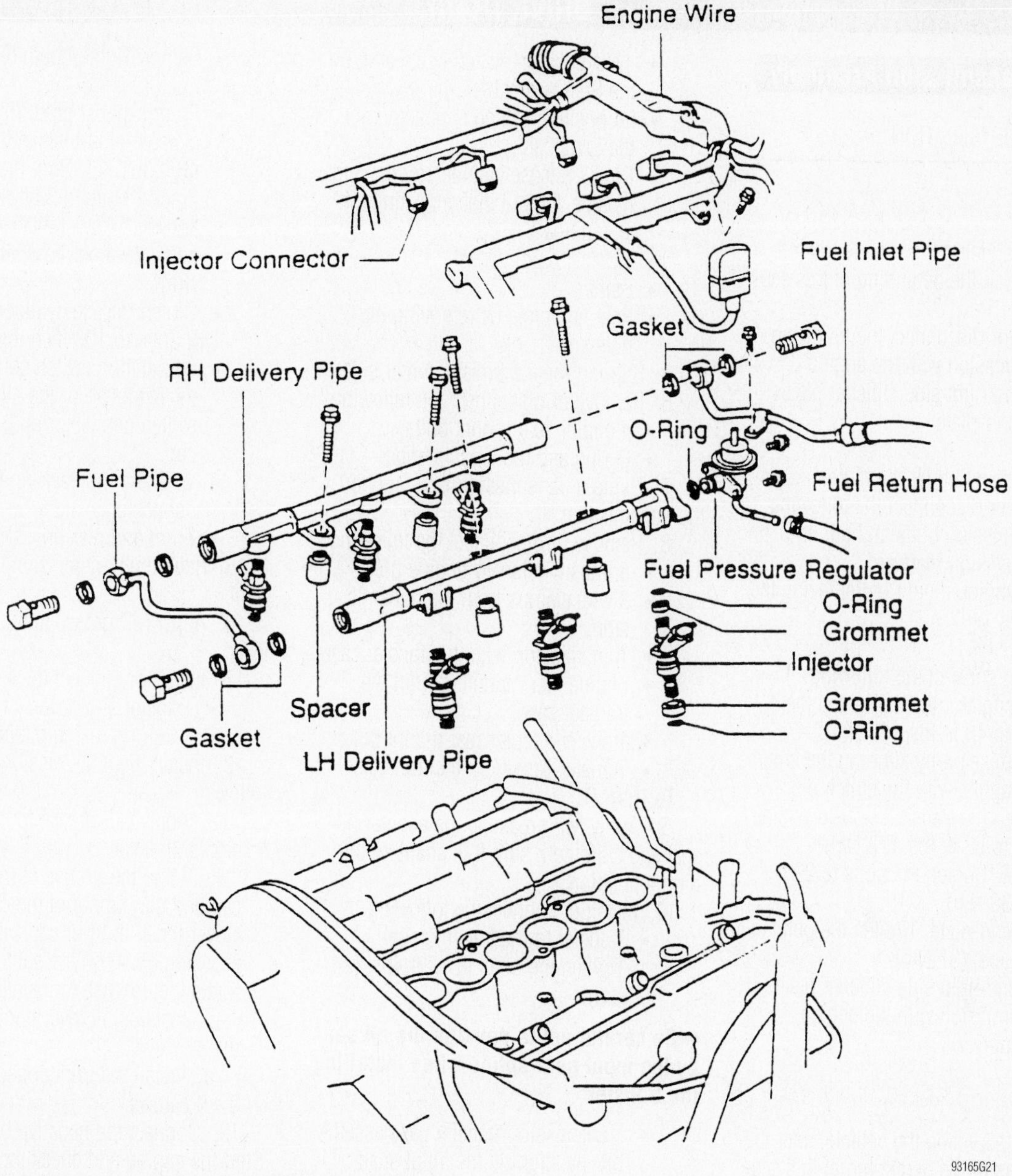

Fuel injector arrangement and related components—3.4L engine

- Fuel inlet pipe
- Fuel injector electrical connections
- Fuel rail with the injectors
- Spacers from the intake manifold
- Injectors from the delivery pipes

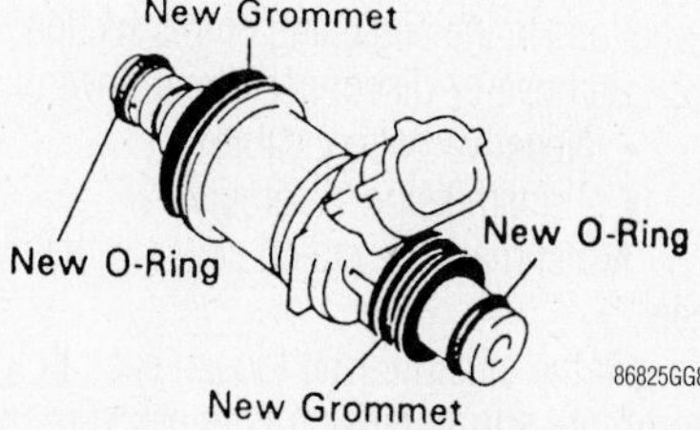

Install new O-rings and grommets on each injector—3.4L engine

- O-rings and grommets, discard them

To install:

4. Install or connect the following:
   - New grommets and O-rings on each injector, lubricated with a light coat of gasoline
   - Fuel injector with the electrical connector facing outward
   - Spacers on the intake manifold

5. Temporarily install the bolts to hold the delivery pipes to the intake manifold.

6. Check that the injectors rotate smoothly. If they do not, the O-rings have probably been installed incorrectly.

7. Install or connect the following:
   - Fuel injector electrical connectors
   - Fuel pipe with new gaskets. Tighten the bolts to 25 ft. lbs. (34 Nm) and the delivery pipes-to-intake manifold bolts to 10 ft. lbs. (13 Nm).
   - Fuel pipe union with new gaskets. Tighten the clamp bolt to 71 inch lbs. (8 Nm).
   - Fuel pressure regulator

8. Inspect the vacuum lines and connections. Look for any loose connections, sharp bends or damage.

9. Install or connect the following:
   - Air cleaner
   - Air cleaner hose

10. Start the engine and check for vacuum and fuel leaks.

DRIVE TRAIN

Manual Transmission Assembly

REMOVAL & INSTALLATION

2WD

1. Before servicing the vehicle, refer to the precautions in the beginning of this section.
2. Remove or disconnect the following:
   - Transmission with the engine
   - Left and right side stiffener plates
   - Rear end-plate
   - Starter
3. Place a stand under the transmission.
4. Remove or disconnect the following:
   - Transmission bolts and pull the transmission rearward
   - Rear engine mount by removing the 4 bolts

To install:

5. Install or connect the following:
   - Rear engine mount. Tighten the 4 bolts to 48 ft. lbs. (65 Nm).
   - Transmission by aligning the input shaft spline with the clutch disc. Tighten the transmission-to-engine bolts to 53 ft. lbs. (72 Nm).
   - Starter. Tighten the bolts to 29 ft. lbs. (39 Nm).
   - Rear end-plate. Tighten the bolts to 27 ft. lbs. (37 Nm).
   - Left and right side stiffener plates
   - Transmission with the engine assembly.

4WD

1. Before servicing the vehicle, refer to the precautions in the beginning of this section.
2. Remove or disconnect the following:
   - Negative battery cable
   - 4 screws and front console box
   - Shift lever boot retainer screws and the shift lever boot
   - Shift lever cap, by pressing downward on the shift lever cap covered with a cloth and rotating it counterclockwise to remove it
   - Transfer shift lever, using snapring pliers to pull it from the transfer case
3. Drain the transmission and the transfer oil.
4. Remove or disconnect the following:
   - Front and rear driveshafts
   - Speedometer cable and the back-up light switch connector
   - 4WD position switch connector, on the Standard cab
   - L4 position switch connector, on the Extra cab
   - Clutch release cylinder, by moving it aside without disconnecting the clutch line
   - Exhaust pipe bracket
   - Starter
   - Rear end plate by removing the nuts and 2 bolts
5. Support the transmission rear side.
6. Remove or disconnect the following:
   - 4 engine rear mount bolts
   - O-ring and the crossmember
7. Using a transmission jack, support the transmission.
8. Remove or disconnect the following:
   - 6 transmission-to-engine bolts
   - 3 wire clamps from the transmission
   - Transmission with the transfer case
   - Engine rear mounting from the transfer case
   - Transfer adapter rear mount bolts
   - Transfer case from the transmission

To install:

9. Apply MP grease to the adapter oil seal and shift the 2 shift fork shafts to the high 4 position.
10. Install or connect the following:
    - Transfer to the transmission. Tighten the bolts to 17 ft. lbs. (24 Nm).

➡Be careful not to damage the oil seal by the input gear spline when installing the transfer.

    - Transmission/transfer case assembly, by aligning the input shaft spline with the clutch disc.
11. Support the transmission with a jack.
12. Install or connect the following:
    - Tighten the engine to transmission bolts to 53 ft. lbs. (72 Nm)
    - Engine rear mount. Tighten the 4 bolts to 48 ft. lbs. (65 Nm).
13. Raise the transmission slightly with a jack.
14. Install or connect the following:
    - Crossmember. Tighten the 4 bolts to 48 ft. lbs. (65 Nm).
    - Engine rear mount. Tighten the bolts to 14 ft. lbs. (19 Nm).
    - Rear end-plate. Tighten the 4 bolts and nuts to 13 ft. lbs. (18 Nm) on R150 and R150F transmissions or to 27 ft. lbs. (37 Nm) on W59 transmissions.
    - Starter. Tighten both bolts to 29 ft. lbs. (39 Nm).
    - Front exhaust pipe. Tighten the exhaust pipe-to-manifold bolts to 46 ft. lbs. (62 Nm), the exhaust bracket bolts to 33 ft. lbs. (44 Nm) and the exhaust pipe-to-catalytic converter bolts to 35 ft. lbs. (48 Nm).
    - Clutch release cylinder. Tighten the 2 bolts to 108 inch lbs. (13 Nm).
    - L4 position switch connector on the extra cab or the 4WD position switch connector on the standard cab.
    - VSS and the back-up light switch connector.
    - Front and rear driveshafts.
15. Refill the transmission to the correct level.
16. Apply MP grease to the transfer shift lever.
17. Install or connect the following:
    - Transfer shift lever.
    - Snapring, using pliers
18. Install the transmission shift lever, as follows:
    a. Apply MP grease to the transmission shift lever.
    b. Align the groove of the shift lever cap and the pin par of the case cover. Cover the shift lever cap with a cloth. Pressing down on the shift lever cap, rotate it clockwise to install.
    c. Install shift lever boot retainer with the 4 screws.
    d. Install the front console box with the 4 screws.
19. Connect the negative battery cable. Start the engine and check for leaks.
20. Road test the vehicle for proper operation. Recheck all fluid levels.

Clutch Assembly

REMOVAL & INSTALLATION

1. Before servicing the vehicle, refer to the precautions in the beginning of this section.
2. Remove or disconnect the following:
   - Negative battery cable
   - Transmission assembly
3. Matchmark the clutch cover to the flywheel.
4. At the clutch cover, loosen each bolt 1 turn until spring tension is released.
5. Remove or disconnect the following:
   - Clutch cover set bolts and the clutch cover with the clutch disc.

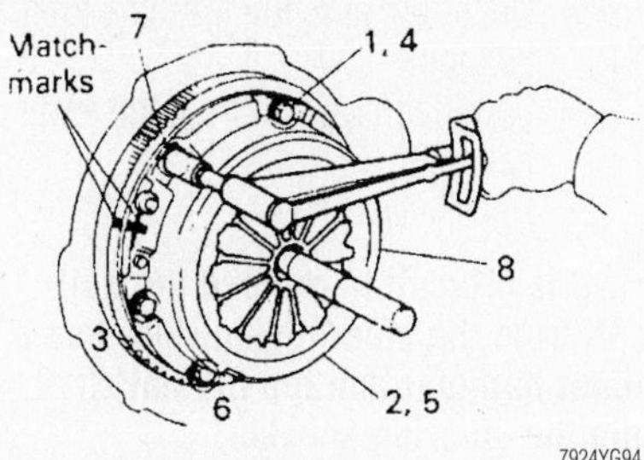

Bolt tightening sequence for the clutch cover—all engines

- Release bearing retaining clip and withdraw it
- Release fork and boot assembly

To install:

6. Install or connect the following:
   - Clutch disc onto the flywheel, using a clutch disc alignment tool
   - Clutch cover, position it onto the flywheel and if reusing the old pressure plate, align the matchmarks.
   - Clutch cover. Tighten the bolts in a crisscross pattern to 14 ft. lbs. (19 Nm).
7. Lubricate the release fork pivot and contact points, the release bearing, bearing hub and input shaft spline surfaces with a suitable molybdenum disulfide lithium based or multi-purpose grease.
8. Install or connect the following:
   - Boot, release fork, hub and the bearing assemblies
   - Transmission
   - Negative battery cable

Hydraulic Clutch System

BLEEDING

1. Before servicing the vehicle, refer to the precautions in the beginning of this section.
2. Fill the clutch reservoir with brake fluid. Check the reservoir level frequently and add fluid as needed.
3. Connect one end of a vinyl tube to the bleeder plug on the slave cylinder and submerge the other end into a clear container half-filled with brake fluid.
4. Slowly pump the clutch pedal several times.
5. Have an assistant hold the clutch pedal down and loosen the bleeder plug until fluid and/or air starts to run out of the bleeder plug. Close the bleeder plug while the pedal is held to the floor.
6. Repeat Steps 2 and 3 until all the air bubbles are removed from the system.
7. Tighten the bleeder plug when all the air is gone.
8. Refill the master cylinder to the proper level as required.
9. Check the system for leaks.

Automatic Transmission Assembly

REMOVAL & INSTALLATION

Model A340F Transmission

1. Before servicing the vehicle, refer to the precautions in the beginning of this section.
2. Remove or disconnect the following:
   - Automatic Transmission Fluid (ATF) level gauge
   - Engine undercover
3. Drain the transmission fluid.
4. Remove or disconnect the following:
   - Throttle cable
   - No. 1 fan shroud
5. Remove the transmission shift lever assembly and the transfer shift lever, as follows:
   - Rear console box
   - Front console box with the transfer shift lever knob
   - Connectors
   - Shift control rod
   - Transmission shift lever assembly
   - Snapring and pull it from the transfer case
   - Oil filler pipe, with the O-ring
   - Front and rear driveshaft
   - Exhaust pipe
   - Speedometer cable
   - No. 2 Vehicle Speed Sensor (VSS) connector
   - Solenoid connector
   - Transfer case neutral position switch connector
   - Transfer case L4 position switch connector
   - Transfer indicator switch
   - Oil cooler pipe
   - Automatic Transmission Fluid (ATF) temperature sensor connector
   - Park/Neutral Position (PNP) switch connector
   - Starter
   - 4 stabilizer bar bracket mounting bolts
6. Remove the torque converter bolts, as follows:
   - Flywheel housing undercover
   - Torque converter clutch mounting bolts, while turning the crankshaft to gain access
7. Remove the front differential rear mounting cushion, as follows:
   - Nut, using a hexagon wrench
   - Front differential by lifting it

➡Be careful not to touch the torque converter clutch housing and the front differential companion flange

   - 2 rear mount cushion bolts
8. Remove or disconnect the following:
   - Support the transmission's rear side
   - 4 engine rear mount bolts
   - 4 nuts, bolts and the crossmember, by supporting the transmission
   - Transmission

To install:

9. Install or connect the following:
   - Transmission. Tighten the engine-to-transmission bolts to 53 ft. lbs. (71 Nm).
   - Crossmember. Tighten the bolts to 48 ft. lbs. (65 Nm).
   - Engine rear mount. Tighten the bolts to 14 ft. lbs. (19 Nm).
   - Front differential rear mount cushion. Tighten the nut to 64 ft. lbs. (41 Nm).
   - Torque converter clutch mount bolt

➡Install the green colored bolt, then the 5 others. Tighten the bolts to 30 ft. lbs. (41 Nm).

   - Flywheel housing undercover. Tighten the bolts to 13 ft. lbs. (18 Nm) for 3.4L engine or to 27 ft. lbs. (37 Nm) for 2.7L engine.
   - Stabilizer bar bracket bolts. Tighten the 4 bolts to 19 ft. lbs. (25 Nm).
   - Starter. Tighten the bolts to 29 ft. lbs. (39 Nm).
   - Remaining components
   - ATF level gauge
10. Fill and check the fluid level.
11. Test drive and check for proper shifting.

A340D and A340E Transmissions

1. Before servicing the vehicle, refer to the precautions in the beginning of this section.
2. Remove or disconnect the following:
   - Transmission with the engine and place it on a stand
   - Bolts, 2 stiffener plates and rear endplate
3. Turn the crankshaft to gain access to the torque converter bolts.
4. Remove or disconnect the following:
   - Torque converter bolts
   - Starter

- Transmission-to-engine bolts
- Transmission

To install:

5. Install or connect the following:
   - Transmission to the engine. Tighten the bolts to 53 ft. lbs. (71 Nm).
   - Starter. Tighten both bolts to 29 ft. lbs. (39 Nm).
   - Torque converter. Tighten the bolts to 30 ft. lbs. (41 Nm).
   - Stiffener plate and rear endplate. Tighten the bolts to 27 ft. lbs. (37 Nm).
   - Starter wires
   - Transmission with the engine

Transfer Case Assembly

REMOVAL & INSTALLATION

1. Before servicing the vehicle, refer to the precautions in the beginning of this section.
2. Disconnect the negative battery cable.
3. Drain the transmission and the transfer case.
4. Remove or disconnect the following:
   - Transfer case with the transmission
   - Breather hose from the transfer upper cover and the transmission control retainer, if equipped with an automatic transmission
   - Rear engine mounting
   - Dynamic damper
5. Remove the driveshaft upper dust cover and the transfer from the transmissions, as follows:
   - Dust cover bolt from the bracket
   - Transfer case adapter rear mounting bolts
   - Transfer case, by pulling it straight up and away from the transmission.

⁂ WARNING

Be careful not to damage the adapter rear oil seal with the transfer input gear spline.

To install:

6. Install the transfer case and the driveshaft upper dust cover to the transmission with a new gasket, as follows:
   - Shift the 2 shift fork shafts to the high 4 position
   - Apply MP grease to the adapter oil seal
   - New gasket to the transfer adapter
   - Transfer case to the transmission.

⁂ WARNING

Take care not to damage the oil seal by the input gear spline.

   - Transfer case adapter. Tighten the rear bolts to 27 ft. lbs. (37 Nm).
   - Dust cover to the bracket. Tighten the bolt to 17 ft. lbs. (23 Nm).
7. Install or connect the following:
   - Engine rear mount. Tighten the bolts to 19 ft. lbs. (25 Nm).
   - Dynamic damper. Tighten the bolts to 27 ft. lbs. (37 Nm).
   - Breather hose, if equipped with an automatic transmission
   - Transfer case with the transmission to the engine
8. Fill the transmission and the transfer case with oil.
9. Test drive the vehicle and check the abnormal noise and smooth operation.
10. Recheck the fluid levels.

Halfshaft

REMOVAL & INSTALLATION

1. Before servicing the vehicle, refer to the precautions in the beginning of this section.
2. Drain the differential oil from the differential.
3. If not equipped with a free-wheeling hub, disconnect the halfshaft from the steering knuckle, as follows:
   - Grease cap
   - Cotter pin and lockcap, from the halfshaft
   - Locknut from the halfshaft, while having an assistant apply the brakes
4. If equipped with free-wheeling hub, remove the free wheel hub, as follows:
   - Set the control handle to FREE
   - Cover bolts and pull off the cover
   - Center bolt with washer
   - Mounting nuts and washer to the hub body
   - Cone washer, using a brass bar and hammer to tap on the bolt heads
   - Free wheel hub body and gasket
   - Snapring from the end of the halfshaft, using a snapring expander
5. Remove or disconnect the following:
   - Halfshaft from the differential, using a brass bar and hammer
   - Cotter pin and nut, from the lower ball joint
   - Lower control arm from the lower ball joint
   - Halfshaft

➡If it is difficult to remove the halfshaft from the steering knuckle, use a rubber hammer and tap the halfshaft from the steering knuckle.

   - Snapring from the inboard shaft

To install:

6. Install or connect the following:
   - New snapring to the inboard shaft
   - Halfshaft to the steering knuckle

➡Push the steering knuckle inwards and at the same time, push the halfshaft into the differential with the snapring opening facing downward. Be sure the halfshaft is fully installed to the differential by checking that it cannot be pulled out by hand.

   - Lower control arm to the lower ball joint. Tighten the nut to 112 ft. lbs. (152 Nm).
   - New cotter pin
7. If equipped with a free wheeling hub, install the hub, as follows:
   - Spacer
   - Snapring to the halfshaft, using a snapring expander
   - New gasket on the front axle hub
   - Fee wheeling hub body, with the 6 cone washers and nuts. Tighten the 6 nuts to 23 ft. lbs. (31 Nm).
   - Bolt with the washer. Tighten the bolt to 13 ft. lbs. (18 Nm).
   - Apply multi purpose grease to the inner hub splines
   - Set the control handle and clutch to the FREE position
   - New gasket on the cover
   - Cover to the hub body, with the follower pawl tabs aligned with the non-toothed portions of the hub body.
   - Tighten the cover bolts to 84 inch lbs. (10 Nm).
8. If equipped without a free wheeling hub, install the halfshaft to the steering knuckle, as follows:
   - Locknut to the halfshaft. Tighten the locknut to 174 ft. lbs. (235 Nm).
   - Lockcap and cotter pin to the halfshaft
   - Grease cap
   - Wheels
9. Fill the differential with gear oil.

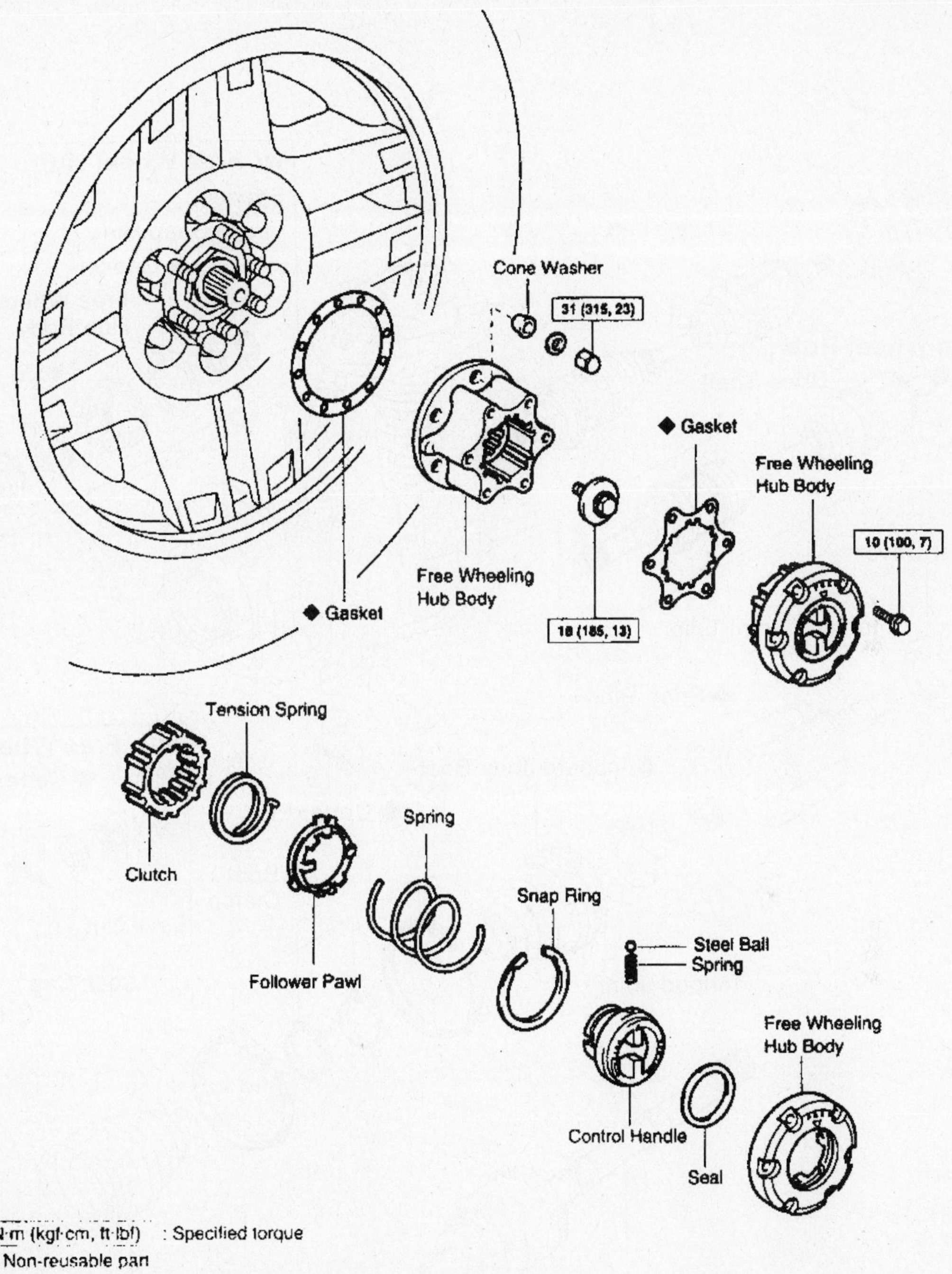

Exploded view of the free wheeling hub assembly

CV-Joints

OVERHAUL

The outboard joint is replaced with half-shaft; no overhaul is possible or necessary.

Inboard Joint

1. Before servicing the vehicle, refer to the precautions in the beginning of this section.
2. Remove the halfshaft from the vehicle.
3. Remove the large clamp from the inboard joint.
4. Remove the small clamp, using side cutters, from the inboard joint.
5. Slide the inboard joint boot toward the outboard joint.
6. Matchmark the inboard joint to the halfshaft.
7. Remove the inboard joint housing from the halfshaft.
8. Remove the snapring from the end of the halfshaft.
9. Matchmark the halfshaft to the tri-pot joint.
10. Remove the tri-pot from the half-shaft, using a brass bar and a hammer.

WARNING

Do not tap on the tri-pot joint.

11. Remove the inboard and outboard boots from the halfshaft.

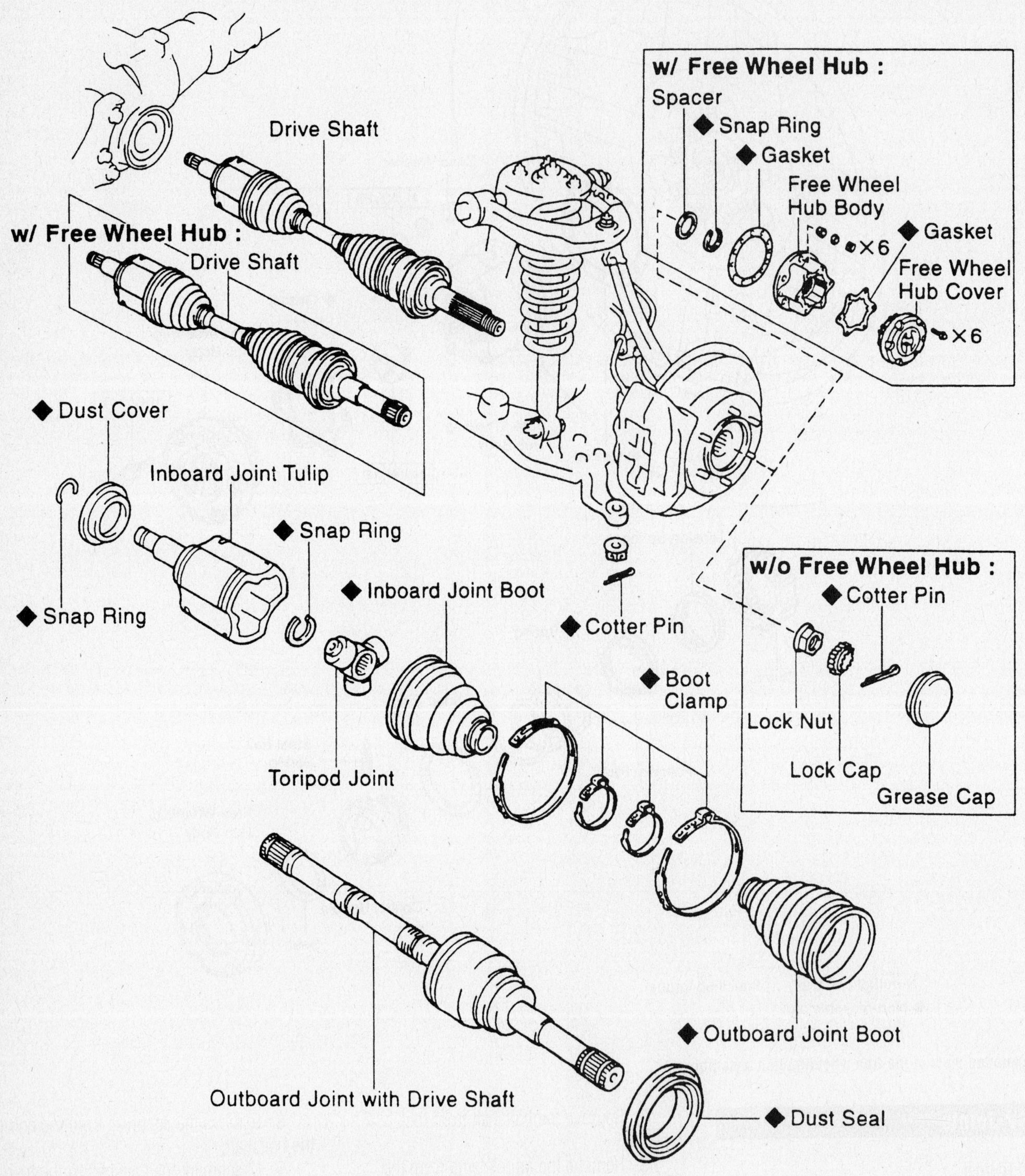

Exploded view of the halfshaft assembly—4WD

⁂ WARNING

Do not disassemble the outboard joint.

To assemble:

12. Wrap vinyl tape around the halfshaft splines to prevent damaging the boots.
13. Install the outboard and inboard boots to the halfshaft with the small end clamps.
14. Assemble the tri-pot joint to the halfshaft with the beveled side facing the outboard joint and align the matchmarks.
15. Install the tri-pot joint, using a brass bar and a hammer.

⁂ WARNING

Do not tap on the roller.

16. Install the snapring.
17. Lubricate the outboard joint with ½ of the grease supplied with the kit.
18. Assemble the boot to the outboard joint
19. Assemble the inboard joint housing to the halfshaft by aligning the matchmarks.
20. Temporarily install the boot onto the tri-pot housing.
21. Make sure the boots are positioned in the shaft grooves.
22. With the halfshaft positioned at the standard length of 17.094–17.252 in. (434.2–438.2mm), make sure that the boots are not stretched or contracted.
23. Install a new inboard joint clamp.
24. Crimp the large clamp with tool 09521-24010 so that the crimp clearance is 0.039–0.059 in. (1.0–1.5mm).
25. Install the halfshaft.

Spindle Bearings

REMOVAL, PACKING AND INSTALLATION

1. Before servicing the vehicle, refer to the precautions in the beginning of this section.
2. Remove or disconnect the following:
   - Front wheel
   - Shock absorber
   - Grease cap
   - Driveshaft
   - Cotter pin and lockcap
   - Locknut, with an assistant applying the brakes
   - Speed sensor and harness from the steering knuckle, if equipped with Anti-lock Brake System (ABS)
   - Brake line from the steering knuckle
   - Caliper and rotor
   - Lower ball joint bolts and the joint from the steering knuckle
   - Cotter pin and axle hub nut
   - Steering knuckle
   - Bearings from the steering knuckle

To install:

3. Install or connect the following:
   - Bearings to the steering knuckle
   - Steering knuckle
   - Cotter pin and axle hub nut. Tighten the nut to 80 ft. lbs. (108 Nm).
   - Lower ball joint to the steering knuckle
   - Caliper and rotor
   - Brake line to the steering knuckle
   - Speed sensor and harness to the steering knuckle, if equipped with Anti-lock Brake System (ABS)
   - Locknut, with an assistant applying the brakes. Torque the locknut to 174 ft. lbs. (235 Nm).
   - Cotter pin and lockcap
   - Driveshaft
   - Grease cap
   - Shock absorber
   - Front wheel

Axle Shaft, Bearing and Seal

REMOVAL & INSTALLATION

Front

1. Before servicing the vehicle, refer to the precautions in the beginning of this section.
2. Remove or disconnect the following:
   - Front wheel
   - Shock absorber
   - Grease cap
   - Axle shaft's cotter pin and lock cap
   - Locknut, using an assistant to apply the brakes
   - Speed sensor and harness from the steering knuckle, if equipped with Anti-lock Brake System (ABS)
   - Brake line from the steering knuckle
   - Caliper and rotor
   - Lower ball joint bolts
   - Cotter pin and loosen the axle hub nut
   - Steering knuckle
   - Axle shaft

To install:

3. Install or connect the following:
   - Axle shaft
   - Steering knuckle
   - Tighten the axle hub nut to 80 ft. lbs. (108 Nm) and the locknut to 174 ft. lbs. (235 Nm).
   - Cotter pin
   - Lower ball joint bolts
   - Caliper and rotor
   - Brake line to the steering knuckle
   - Speed sensor and harness to the steering knuckle, if equipped with Anti-lock Brake System (ABS)
   - Locknut, using an assistant to apply the brakes
   - Axle shaft's cotter pin and lock cap
   - Grease cap
   - Shock absorber
   - Front wheel

Rear

1. Before servicing the vehicle, refer to the precautions in the beginning of this section.
2. Remove or disconnect the following:
   - Rear wheel
   - Brake drum
3. Check the bearing backlash and axle shaft deviation, as follows:
   a. Using a dial indicator, check that the backlash in the bearing shaft direction. The maximum is 0.027 in. (0.7mm).
   b. If the backlash exceeds the maximum, replace the bearing.
   c. Using a dial indicator, check the deviation at the surface of the axle shaft outside the hub bolt. Maximum is 0.0039 in. (0.1mm).
   d. If the deviation exceeds the maximum, replace the axle shaft.
4. Remove or disconnect the following:
   - Anti-lock Brake System (ABS) speed sensor from the axle housing, if equipped
   - Axle shaft assembly by removing the 4 nuts from the backing plate
   - O-ring from the axle housing
   - Bearing and retainer (differential side) and ABS speed sensor rotor, if equipped
   - Snapring from the axle shaft
   - Axle shaft

➡Inspect the axle shaft and flange run-outs. The axle shaft run-out should be 0.079 in. (2.0mm) and the flange run-out should be 0.004 in. (0.1mm).

   - Outer seal
   - Bearing from the axle shaft

To install:

5. Install or connect the following:
   - Bearing from the axle shaft
   - Outer seal
   - Axle shaft
   - Snapring to the axle shaft

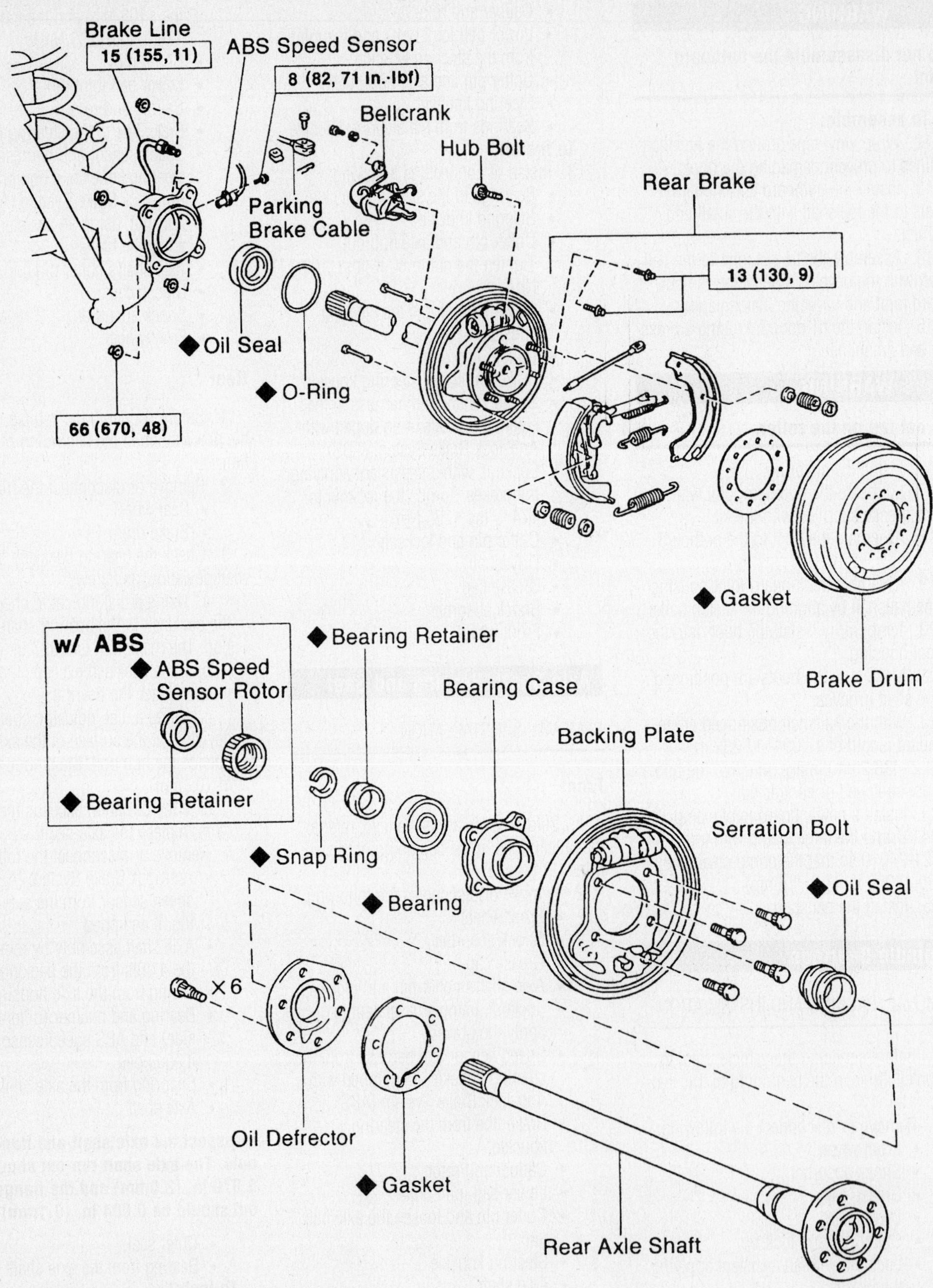

Exploded view of the rear axle shaft and components—typical

- Bearing and retainer (differential side) and ABS speed sensor rotor, if equipped
- O-ring to the axle housing
- Axle shaft assembly. Tighten the 4 backing plate nuts to 48 ft. lbs. (66 Nm).
- Anti-lock Brake System (ABS) speed sensor to the axle housing, if equipped
- Brake drum
- Rear wheel

Pinion Seal

REMOVAL & INSTALLATION

Front

1. Before servicing the vehicle, refer to the precautions in the beginning of this section.
2. Remove the engine undercover.
3. Drain the differential oil.
4. Remove or disconnect the following:
   - Front driveshaft
   - Companion flange nut, by unstaking it

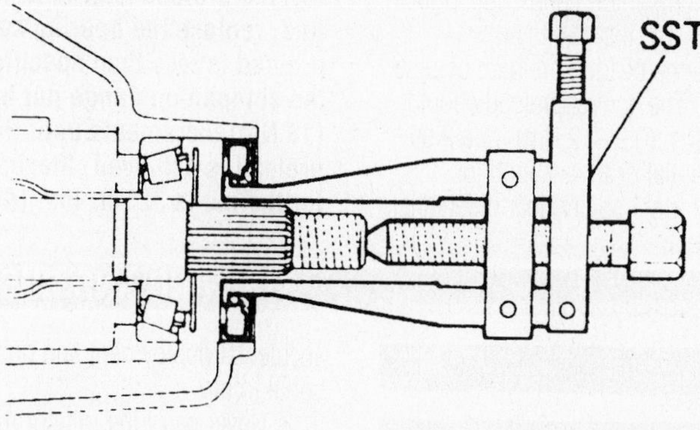

Extractor fits into the Seal Removal Tool 09308-10010

   - Companion flange
   - Pinion seal, using an extractor

To install:

5. Install a new oil seal, to a depth of 0.059 in. (1.5mm) below the lip, using a seal driver.
6. Lubricate the seal lip with multi-purpose grease.
7. Install or connect the following:
   - Companion flange, coat the threads with multi-purpose grease
   - New companion flange nut. Tighten it to 89 ft. lbs. (120 Nm).

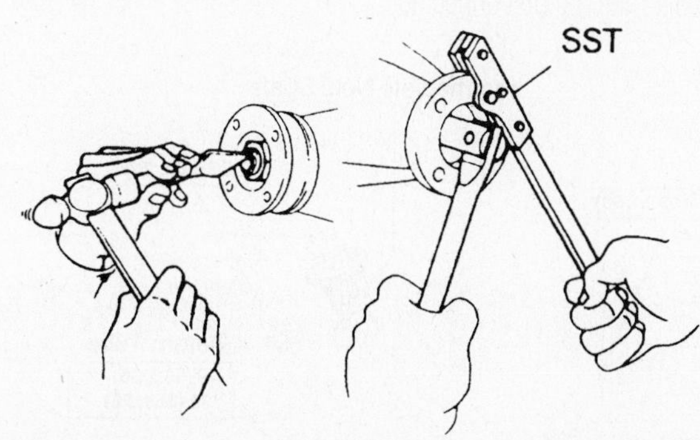

Using a chisel and hammer, loosen the staked part of the nut. Hold the flange with SST 09950-30010 or equivalent and remove the nut

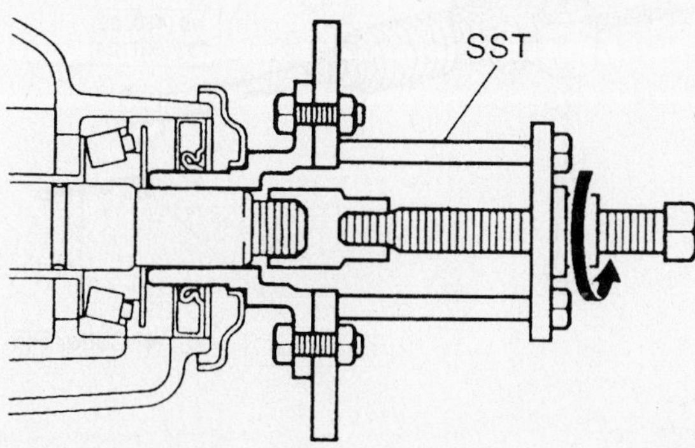

Screw-type extractor from Tool 09950-30010

8. Measure the bearing preload, using a torque wrench. The correct preload should be 5–9 inch lbs. (0.6–1.0 Nm) for a used bearing or 10–17 inch lbs. (1–2 Nm) for a new bearing.

➡If the preload is greater that specified, replace the bearing spacer. If the preload is less than specified, tighten the companion flange nut in 108 inch lbs. (13 Nm) increments until the correct preload is achieved. Maximum torque for the nut is 173 ft. lbs. (235 Nm). If the value is exceeded, the bearing spacer must be replaced; do not back off the flange nut to lower the torque or preload.

9. Install the front driveshaft by aligning the matchmarks.
10. Check the companion flange run-out; maximum allowable run-out is 0.003 in. (0.10mm).
11. Stake the pinion flange nut.
12. Refill the differential with oil.

Rear

1. Before servicing the vehicle, refer to the precautions in the beginning of this section.
2. Remove or disconnect the following:
   - Rear driveshaft by matchmarking it
   - Companion flange nut, by loosen the staked portion
   - Companion flange, using a screw-type extractor
   - Oil seal, using an extractor

To install:

3. Install a new oil seal, to a depth of 0.039 in. (1.0mm) below the lip, using a seal driver.
4. Lubricate the seal lip with multi-purpose grease.
5. Install or connect the following:
   - Companion flange, coat the threads with multi-purpose grease

• New companion flange nut. Tighten it to 109 ft. lbs. (147 Nm).

6. Measure the bearing preload, using a torque wrench. The correct preload should be 8–11 inch lbs. (0.9–1.2 Nm) for a 2 spider gear differential or to 4–7 inch lbs. (0.4–0.8 Nm) for a 4 spider gear differential.

➡If the preload is greater that specified, replace the bearing spacer. If the preload is less than specified, tighten the companion flange nut in 9 ft. lbs. (13 Nm) increments until the correct preload is achieved. Maximum torque for the nut is 325 ft. lbs. (441 Nm). If the value is exceeded, the bearing spacer must be replaced; do not back off the flange nut to lower the torque or preload.

7. Stake the pinion flange nut.
8. Install the rear driveshaft by aligning the matchmarks.

STEERING AND SUSPENSION

Air Bag

⁂ CAUTION

Some vehicles are equipped with an air bag system. The system must be disabled before performing service on or around system components, steering column, instrument panel components, wiring and sensors. Failure to follow safety and disabling procedures could result in accidental air bag deployment, possible personal injury and unnecessary system repairs.

PRECAUTIONS

Several precautions must be observed when handling the inflator module to avoid accidental deployment and possible personal injury.

- Never carry the inflator module by the wires or connector on the underside of the module.
- When carrying a live inflator module, hold securely with both hands, and ensure that the bag and trim cover are pointed away.
- Place the inflator module on a bench or other surface with the bag and trim cover facing up.
- With the inflator module on the bench, never place anything on or close to the module which may be thrown in the event of an accidental deployment.

DISARMING

To avoid personal injury when working on vehicles equipped with an air bag, the negative battery cable must be disconnected and at least 90 seconds must elapse before working on the system. Failure to do so may result in deployment of the air bag.

Power Rack and Pinion Steering Gear

REMOVAL & INSTALLATION

1. Before servicing the vehicle, refer to the precautions in the beginning of this section.
2. Remove or disconnect the following:
   - Negative battery cable
   - Right and left tie rod ends from the knuckle
   - Intermediate No. 2 shaft from the steering rack
   - Pressure feed and the return tubes, using SST 09631-22020
   - Power steering rack

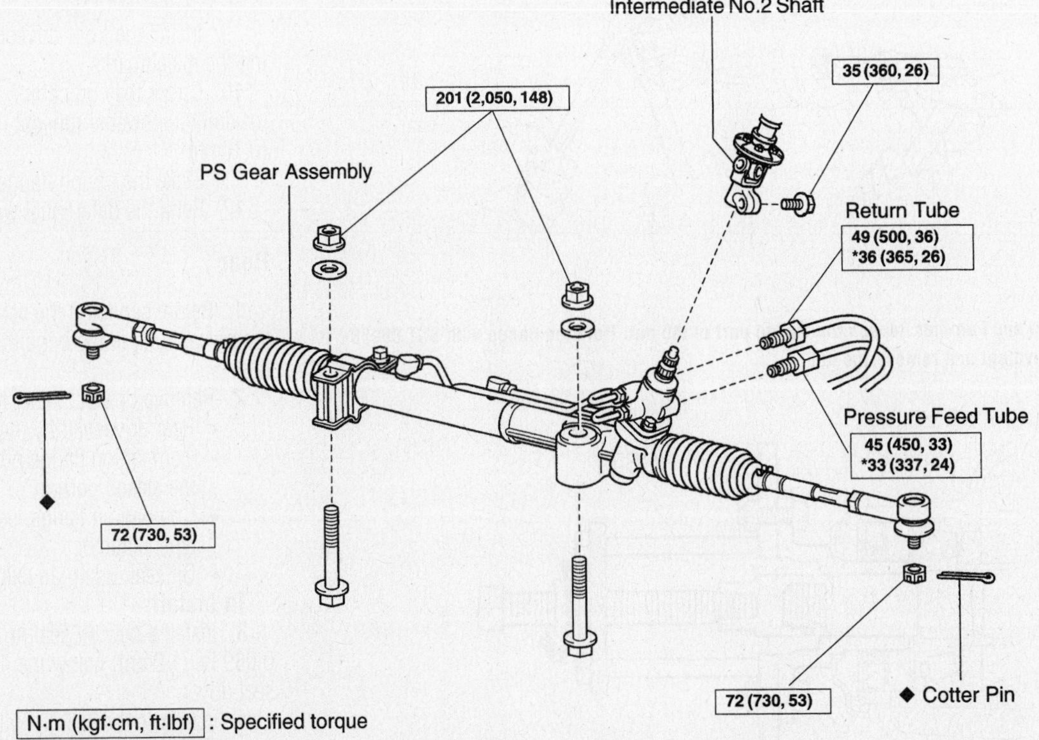

Power rack and pinion steering gear—2 wheel drive

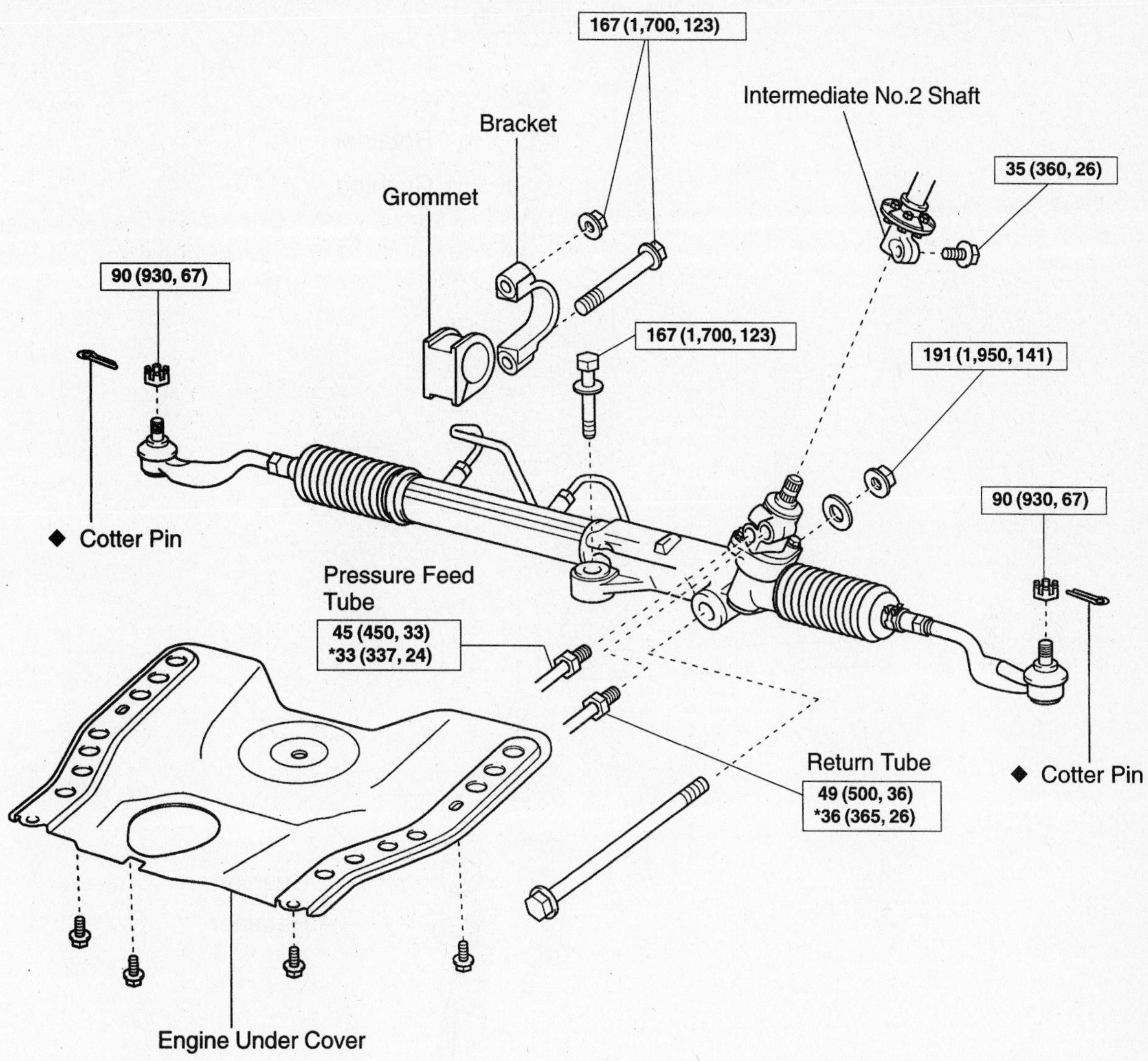

Power rack and pinion steering gear—4 wheel drive

To install:

3. Install the power steering rack.

Tighten the bolts to 148 ft. lbs. (201 Nm) for 2WD or to the following values for 4WD:

- Rack assembly bolt: 123 ft. lbs. (167 Nm)
- Rack assembly nut: 141 ft. lbs. (191 Nm)
- Bracket nut and bolt: 123 ft. lbs. (167 Nm)

4. Install or connect the following:

- New O-ring
- Pressure feed tube. Tighten it to 33 ft. lbs. (45 Nm).
- Return tube. Tighten it to 36 ft. lbs. (49 Nm) for 2WD or to 29 ft. lbs. (40 Nm) for 4WD
- Intermediate No. 2 shaft to the steering rack
- Right and left tie rod ends to the steering knuckle
- Tighten the castle nuts to specification
- New cotter pins
- Negative battery cable

5. Check the steering wheel center point.
6. Bleed the power steering system.
7. Check the front wheel alignment.

Tighten the tie rod end locknuts to 67 ft. lbs. (90 Nm)

Shock Absorbers

REMOVAL & INSTALLATION

Front

2WD

1. Before servicing the vehicle, refer to the precautions in the beginning of this section.
2. Remove or disconnect the following:
   - Front wheel
   - Shock absorber from the lower control arm

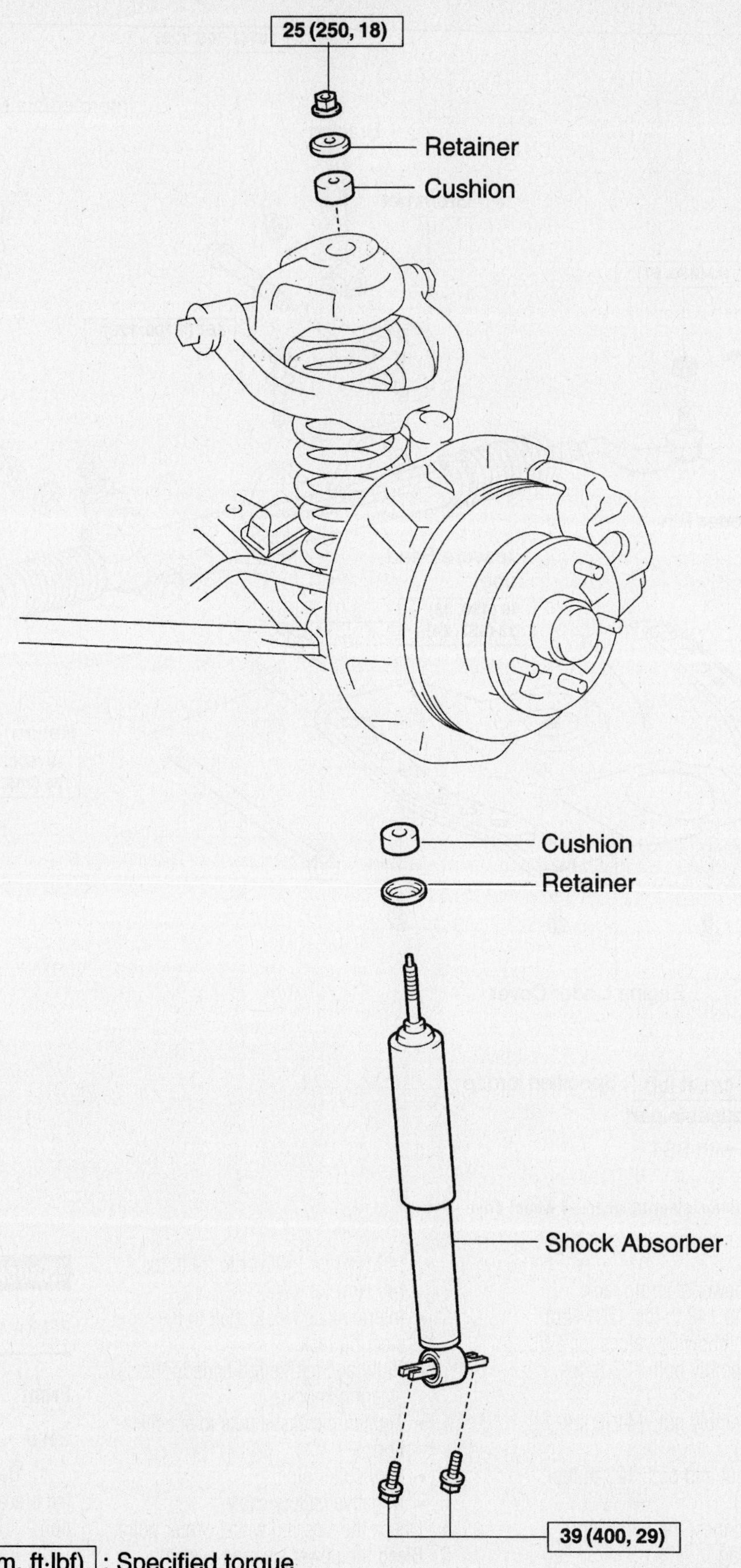

Front shock absorber—2WD

- Nut, retainers and the cushion from the top of the shock absorber
- Shock absorber
- Retainers and cushion from the shock absorber

To install:

3. Install or connect the following:
   - Retainers and cushion to the shock absorber
   - Shock absorber
   - Retainers, cushion and nut to the top of the shock absorber. Tighten the nut to 18 ft. lbs. (25 Nm).
   - Lower shock absorber-to-lower control arm. Tighten the bolts to 29 ft. lbs. (39 Nm)
   - Wheels

Rear

1. Before servicing the vehicle, refer to the precautions in the beginning of this section.
2. Remove the wheel.
3. Lower the floor jack to take tension off of the spring.
4. Remove or disconnect the following:
   - Shock absorber from the rear axle housing
   - Nut, retainers and the cushions holding the shock absorber to the frame
   - Shock absorber with the washers and bushings

To install:

5. Install the shock absorber to the frame with the washers and bushings.
6. Tighten the shock absorber-to-frame nut to the following values:
   - 2WD: 19 ft. lbs. (25 Nm)
   - 4WD and PreRunner: 53 ft. lbs. (72 Nm)
7. Connect the shock absorber to the rear axle housing. Tighten the bolt to the following specifications:
   - 2WD: 19 ft. lbs. (25 Nm)
   - 4WD and PreRunner: 53 ft. lbs. (72 Nm)
8. Install the wheels.

Struts

REMOVAL & INSTALLATION

4WD

1. Before servicing the vehicle, refer to the precautions in the beginning of this section.
2. Remove the wheel.
3. Remove the front wheel.
4. Remove the strut absorber nut and washer from the lower control arm.

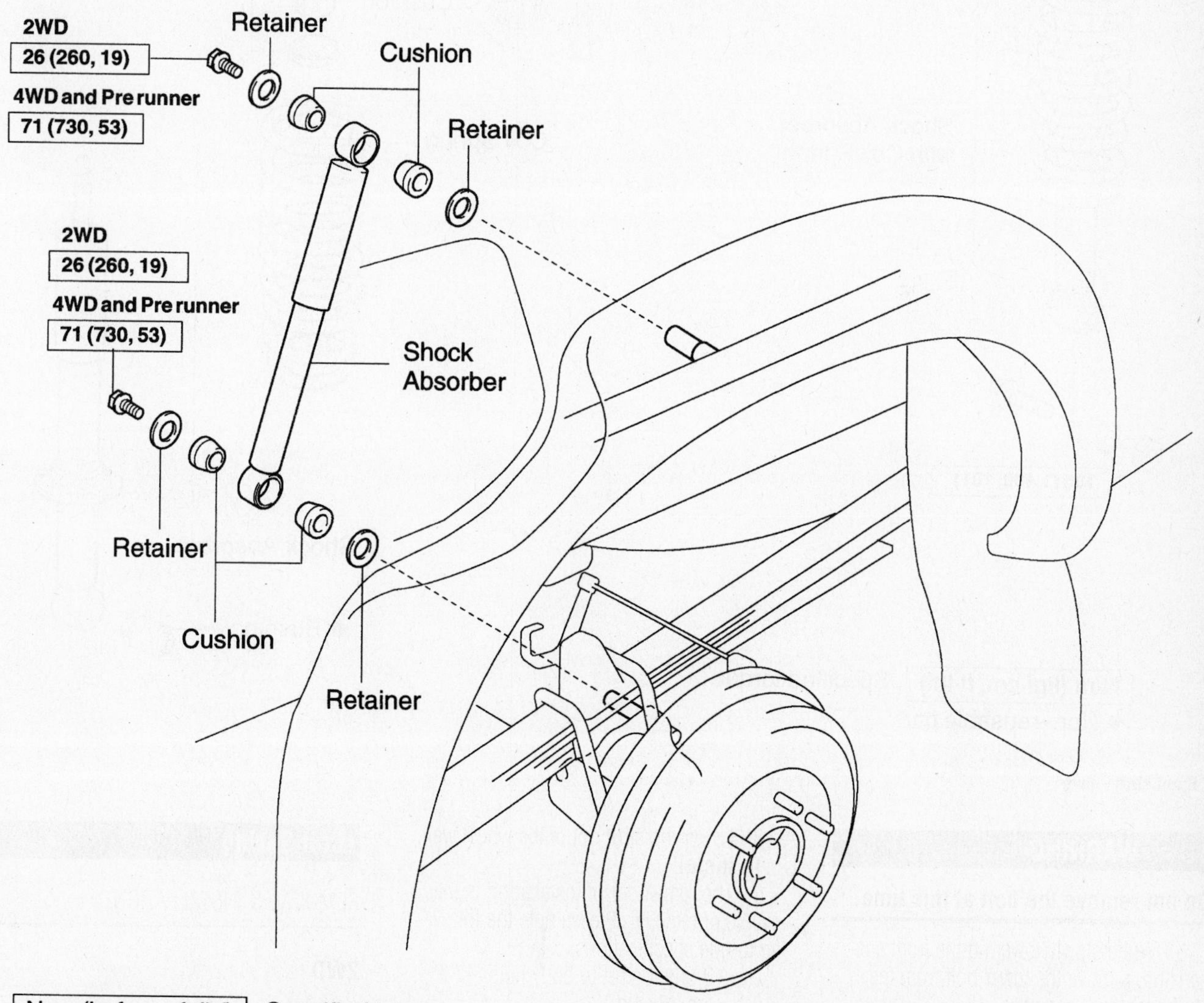

Rear shock absorber

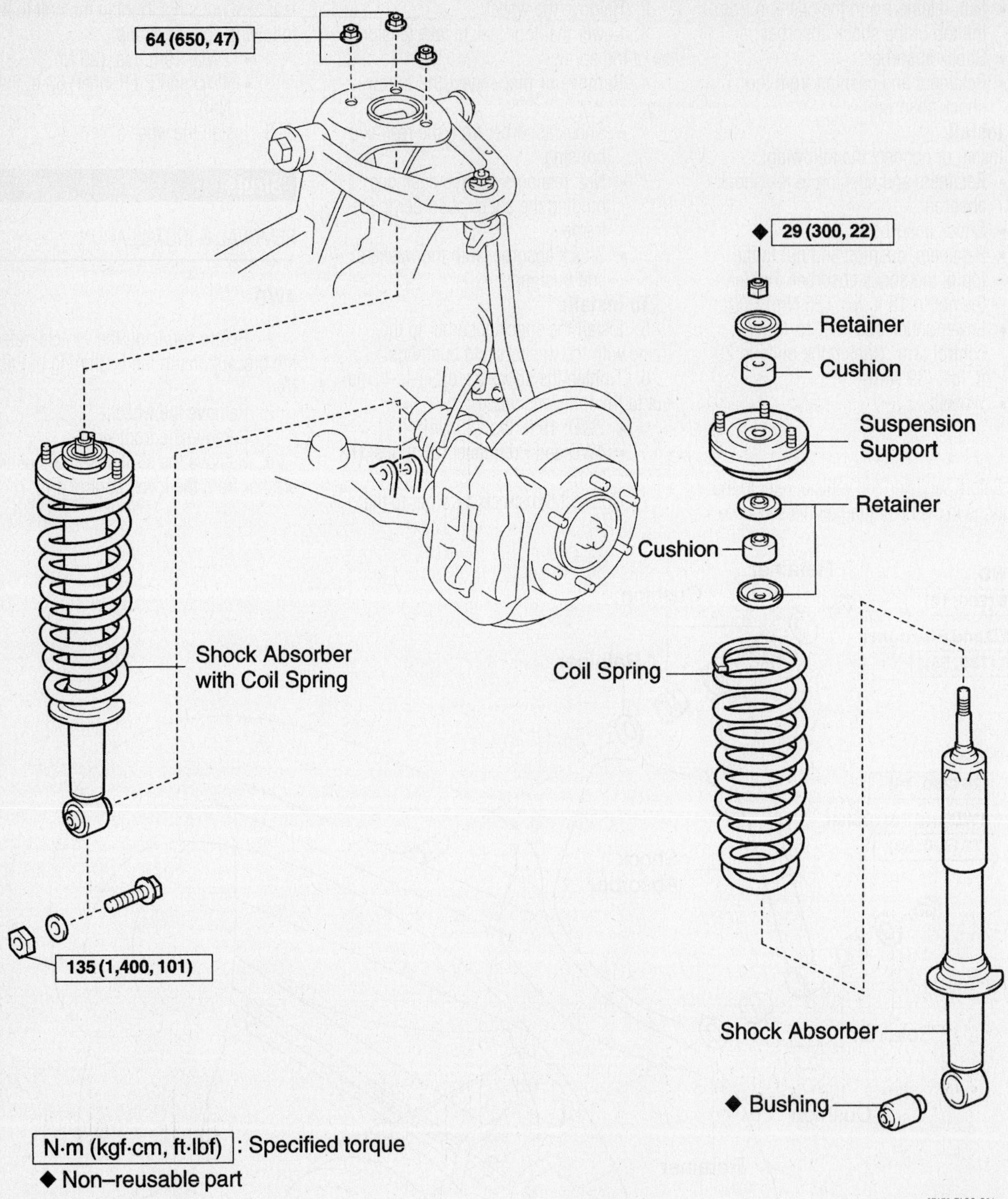

Front strut—4wd

WARNING

Do not remove the bolt at this time.

5. While slowly lowering the front suspension, remove the lower bolt from the strut absorber assembly.
6. Support the strut.
7. Remove the three nuts from the top of the strut tower.
8. Lower the strut out of the wheel well.

To install:

9. The remainder of installation is the reverse of removal. Please note the following torque specifications:
   - 3 upper strut to body nuts: 47 ft. lbs. (64 Nm)
   - Lower strut absorber mounting bolt: 101 ft. lbs. (135 Nm)

Coil Spring

REMOVAL & INSTALLATION

2WD

1. Before servicing the vehicle, refer to the precautions in the beginning of this section.

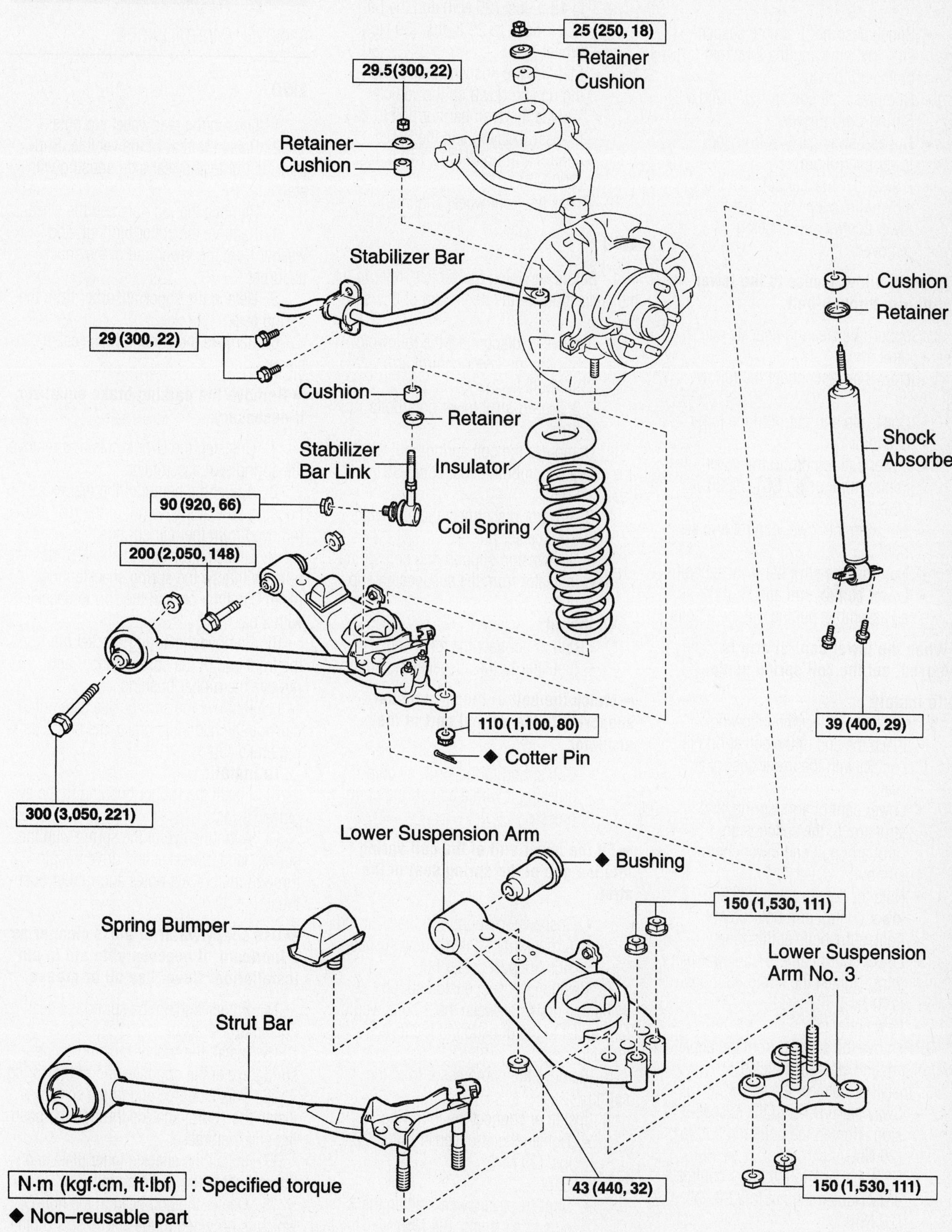
25 (250, 18)
Retainer
Cushion
29.5(300, 22)
Retainer
Cushion
Stabilizer Bar
Cushion
Retainer
29 (300, 22)
Cushion
Retainer
Stabilizer Bar Link
Insulator
Shock Absorber
90 (920, 66)
Coil Spring
200 (2,050, 148)
110 (1,100, 80)
39 (400, 29)
◆ Cotter Pin
300 (3,050, 221)
Lower Suspension Arm
◆ Bushing
150 (1,530, 111)
Spring Bumper
Lower Suspension Arm No. 3
Strut Bar
N·m (kgf·cm, ft·lbf) : Specified torque
◆ Non–reusable part
43 (440, 32)
150 (1,530, 111)
67170-TACO-G33

Coil spring and lower arm—2wd

2. Remove or disconnect the following:

- Shock absorber from the suspension, by removing the 2 bottom bolts and top nut
- Compress the coil spring, using a Spring Compressor
- Nut and sway bar link from the lower control arm
- 2 sway bar bracket bolts on the side of the suspension that the lower control arm is being removed.

➡**This will allow access to the lower control arm through-bolt.**

3. Support the steering knuckle and upper control arm.
4. Remove or disconnect the following:

- Cotter pin and nut from the lower ball joint
- Lower ball joint from the lower control arm, using SST 09628-62011
- Nut from the lower control arm set bolt
- Nut from the strut bar front set bolt
- Lower control arm and strut bar as an assembly, pulling out the 2 bolts

➡**When the lower control arm is removed, set the coil spring aside.**

To install:

5. Install or connect the following:

- Place the end of the coil spring in contact with the lower control arm seat
- Lower control arm, spring, and strut arm to the suspension.
- Strut arm bolt and lower control arm bolt
- Nuts for the strut arm bolt and lower control arm bolt; do not tighten the bolts at this time
- Lower control arm to the lower ball joint. Tighten the nut to 80 ft. lbs. (110 Nm).
- New cotter pin

6. Remove the support from the upper control arm and steering knuckle.
7. Install or connect the following:

- Sway bar bracket to the suspension. Tighten the bolts to 22 ft. lbs. (29 Nm).
- Sway bar link to the lower control arm. Tighten the nut to 22 ft. lbs. (29 Nm).

8. Making sure the coil spring is in its correct position, slowly remove the spring compressor from the coil.
9. Install or connect the following:

- Shock absorber. Tighten the top nut to 18 ft. lbs. (25 Nm) and the bottom 2 bolts to 29 ft. lbs. (39 Nm).
- Wheel
- Stabilize the suspension by pushing up and down on the vehicle
- Tighten the strut bar nut/bolt to 221 ft. lbs. (300 Nm) and the lower control arm bolt/nut to 148 ft. lbs. (200 Nm).

10. Check the front wheel alignment.

4WD

1. Before servicing the vehicle, refer to the precautions in the beginning of this section.
2. Remove or disconnect the following:

- Strut to the lower control arm nut/bolt.
- 3 strut-to-strut tower nuts/bolts
- Strut

3. Compress the coil spring until there is a clearance on both ends, using SST 09727-30030
4. Remove or disconnect the following:

- Strut center nut
- Suspension support and coil spring
- Insulator from the suspension support

To install:

5. Install or connect the following:

- Insulator to the suspension support

➡**Match the bolt of the suspension support with the cut out part of the insulator.**

- Coil spring to the strut, by compressing it with a coil spring compressor

➡**Fit the lower end of the coil spring into the gap of the spring seat of the strut.**

- Suspension support to the strut rod
- Temporarily tighten a new suspension support center nut

6. Position the suspension support so that a line drawn between the 2 bolts would be parallel to the direction of the lower bushing.
7. Remove the compressor from the spring.
8. Install or connect the following:

- Tighten the strut center nut to 22 ft. lbs. (29 Nm).
- Strut
- Strut-to-strut tower. Tighten the 3 nuts to 47 ft. lbs. (64 Nm).
- Strut-to-lower control arm. Tighten the nut/bolt to 101 ft. lbs. (135 Nm).
- Front wheels

Leaf Springs

REMOVAL & INSTALLATION

2WD

1. Loosen the rear wheel lug nuts.
2. Raise the rear of the vehicle. Support the frame and rear axle housing with stands.
3. Remove the lug nuts and the wheel.
4. Remove the cotter pin, nut, and washer from the lower end of the shock absorber.
5. Detach the shock absorber from the spring seat.
6. Remove the parking brake cable clamp.

➡**Remove the parking brake equalizer, if necessary.**

7. Unfasten the U-bolt nuts and remove the spring seat assemblies.
8. Adjust the height of the rear axle housing so that the weight of the rear axle is removed from the rear springs.
9. Unfasten the spring shackle retaining nuts. Withdraw the spring shackle inner plate. Carefully pry out the spring shackle with a bar.
10. Remove the spring bracket pin from the front end of the spring hanger and remove the rubber bushing.
11. Remove the spring. Use care not to damage the hydraulic brake line or the parking brake cable.

To install:

12. Install the rubber bushing in the eye of the spring.
13. Align the eye of the spring with the spring hanger bracket and drive the pin through the bracket holes and rubber bushings.

➡**Use soapy water or glass cleaner as a lubricant, if necessary, to aid in pin installation. Never use oil or grease.**

14. Finger-tighten the spring hanger nuts and/or bolts.
15. Install the rubber bushing in the spring eye at the opposite end of the spring.
16. Raise the free end of the spring. Install the spring shackle through the bushing and the bracket.
17. Install the shackle inner plate and finger-tighten the retaining nuts.
18. Center the bolt head in the hole which is provided in the spring seat on the axle housing.
19. Fit the U-bolts over the axle housing. Install the lower spring seat.

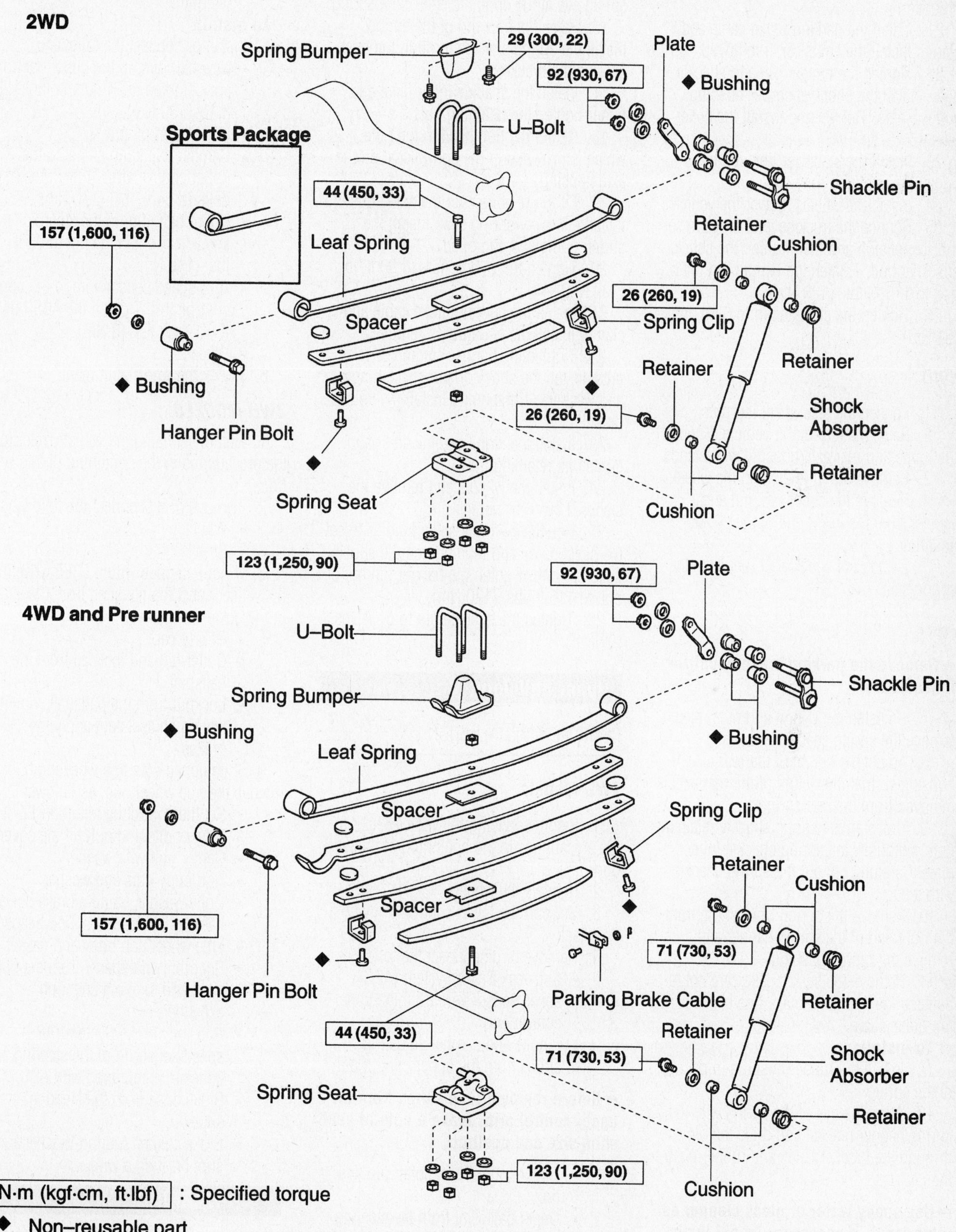

Exploded view of the rear leaf spring and related components

20. Tighten the U-bolt nuts to 90 ft. lbs. (120 Nm)

21. Install the parking brake cable and clamp. Install the equalizer, if removed.

22. Tighten the hanger pin and shackle nuts. Install the shock absorber bushings and washers. Tighten and install the cotter pins.

23. Install the stabilizer link and hand-tighten its retaining nuts.

24. Install the wheels. Lower the vehicle.

25. Bounce the truck several times to set the suspension and then tighten the shock absorber bolt. Tighten the hanger pin nut or bolt to 115 ft. lbs. (120 Nm).

26. Tighten the shackle pin to 67 ft. lbs. (91 Nm).

4WD

1. Loosen the rear wheel lug nuts.
2. Raise the rear of the vehicle. Support the frame and rear axle housing with stands.
3. Remove the lug nuts and the wheel.
4. Remove the cotter pin, nut and washer from the lower end of the shock absorber.
5. Detach the shock absorber from the spring seat.
6. Remove the parking brake cable clamp.

➡Remove the parking brake equalizer, if necessary.

7. Unfasten the U-bolt and nuts, then remove the spring seat assemblies.
8. Adjust the height of the rear axle housing so that the weight of the rear axle is removed from the rear springs.
9. Unfasten the spring shackle retaining nuts. Withdraw the spring shackle inner plate. Carefully pry out the spring shackle with a bar.
10. Remove the spring bracket pin from the front end of the spring hanger and remove the rubber bushing.
11. Remove the spring. Use care not to damage the hydraulic brake line or the parking brake cable.

To install:

12. Install the rubber bushing in the eye of the spring.
13. Align the eye of the spring with the spring hanger bracket and drive the pin through the bracket holes and rubber bushings.

➡Use soapy water or glass cleaner as a lubricant, if necessary, to aid in pin installation. Never use oil or grease.

14. Finger-tighten the spring hanger nuts and/or bolts.
15. Install the rubber bushing in the spring eye at the opposite end of the spring.
16. Raise the free end of the spring. Install the spring shackle through the bushing and the bracket.
17. Install the shackle inner plate and finger-tighten the retaining nuts.
18. Center the bolt head in the hole which is provided in the spring seat on the axle housing.
19. Fit the U-bolts over the axle housing. Install the lower spring seat. Install the spring bumper, if equipped.
20. Tighten the U-bolt nuts to 90 ft. lbs. (120 Nm)
21. Install the parking brake cable and clamp. Install the equalizer, if removed.
22. Tighten the hanger pin and shackle nuts. Install the shock absorber bushings and washers. Tighten and install the cotter pins.
23. Install the stabilizer link and hand-tighten its retaining nuts.
24. Install the wheels and remove the stands. Lower the vehicle.
25. Bounce the truck several times to set the suspension and then tighten the shock absorber bolt. Tighten the hanger pin nut or bolt to 115 ft. lbs. (120 Nm).
26. Tighten the shackle pin to 67 ft. lbs. (91 Nm).

Upper Ball Joint

REMOVAL & INSTALLATION

2WD Models

1. Before servicing the vehicle, refer to the precautions in the beginning of this section.
2. Remove the wheels.
3. Support the lower control arm with a floor jack.
4. Remove or disconnect the following:
   - Anti-lock Brake System (ABS) speed sensor wire from the upper control arm
   - 2 bolts and camber adjusting shims from the upper control arm

➡Before removing the shims from the upper control arm, make a note of each shim size and position.

- Upper control arm cotter pin and nut.
- Upper ball joint from the steering knuckle, using SST 09628-62011
- Upper control arm
- 4 upper control arm-to-upper ball joint nuts and bolts.
- Upper control arm from the upper ball joint

To install:

5. Install or connect the following:
   - New ball joint to the upper control arm. Tighten the 4 nuts/bolts to 29 ft. lbs. (39 Nm).
   - Upper control arm
   - Camber adjusting shims to the upper control arm. Tighten the 2 bolts to 94 ft. lbs. (130 Nm).
   - Upper ball joint to the steering knuckle. Tighten the nut to 80 ft. lbs. (110 Nm).
   - ABS speed sensor wire to the upper control arm. Tighten the ABS bolt to 71 inch lbs. (8 Nm).
   - Wheels
6. Check the wheel alignment.

4WD MODELS

1. Before servicing the vehicle, refer to the precautions in the beginning of this section.
2. Remove or disconnect the following:
   - Wheel
   - Strut
3. If not equipped with a FREE wheeling hub, disconnect the halfshaft from the steering knuckle, as follows:
   - Grease cap
   - Cotter pin and lockcap from the halfshaft
   - Locknut from the halfshaft, while having an assistant apply the brakes
4. If equipped with free wheeling hub, remove the free wheel hub, as follows:
   - Set the control handle to FREE
   - Cover bolts and pull off the cover
   - Center bolt with washer
   - Hub body nuts and washer
   - Cone washer, using a brass bar and hammer to tap on the bolt heads
   - Free wheel hub body and gasket
   - Snapring and spacer from the halfshaft end, using a snapring expander
   - Anti-lock Brake System (ABS) speed sensor from the steering knuckle, if equipped with ABS
   - Brake hose from the steering knuckle
   - Brake caliper support bracket and support it on a wire

✱✱ WARNING

Do not allow the caliper to hang from the brake hose.

5. Remove or disconnect the following:

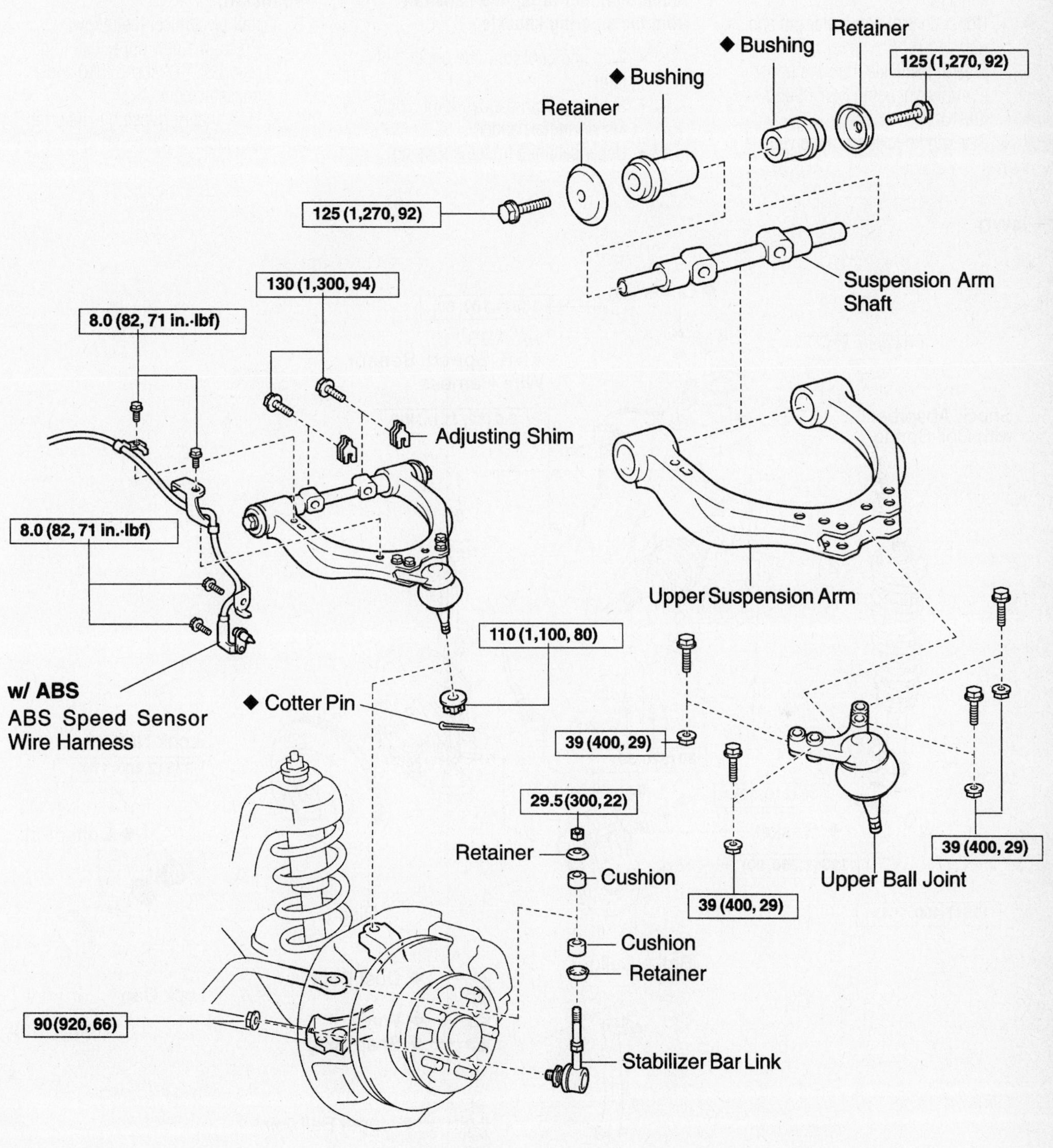
Retainer
◆ Bushing
125 (1,270, 92)
◆ Bushing
Retainer
125 (1,270, 92)
Suspension Arm Shaft
130 (1,300, 94)
8.0 (82, 71 in.·lbf)
Adjusting Shim
8.0 (82, 71 in.·lbf)
Upper Suspension Arm
110 (1,100, 80)
w/ ABS
ABS Speed Sensor Wire Harness
◆ Cotter Pin
39 (400, 29)
29.5(300,22)
Retainer
Cushion
39 (400, 29)
Upper Ball Joint
39 (400, 29)
Cushion
Retainer
90(920,66)
Stabilizer Bar Link
N·m (kgf·cm, ft·lbf) : Specified torque
◆ Non–reusable part
67170-TACO-G35

Upper control arm and related parts—2wd

- Rotor
- Lower ball joint from the steering knuckle
- Upper control arm cotter pin and nut
- Steering knuckle from the upper control arm, using SST 09950-40010
- Steering knuckle from the vehicle

➡If it is difficult to remove the halfshaft from the steering knuckle, use a rubber hammer to tap the halfshaft from the steering knuckle.

- Wire and boot from the upper ball joint
- Snapring from the ball joint, using a snapring expander
- Upper ball joint from the steering knuckle, using SST 09950-40010 (puller set) and a deep socket wrench

To install:

6. Install or connect the following:
    - Press in a new upper ball joint, using SST 09309-37010 and a socket wrench
    - New snapring, using a snapring expander

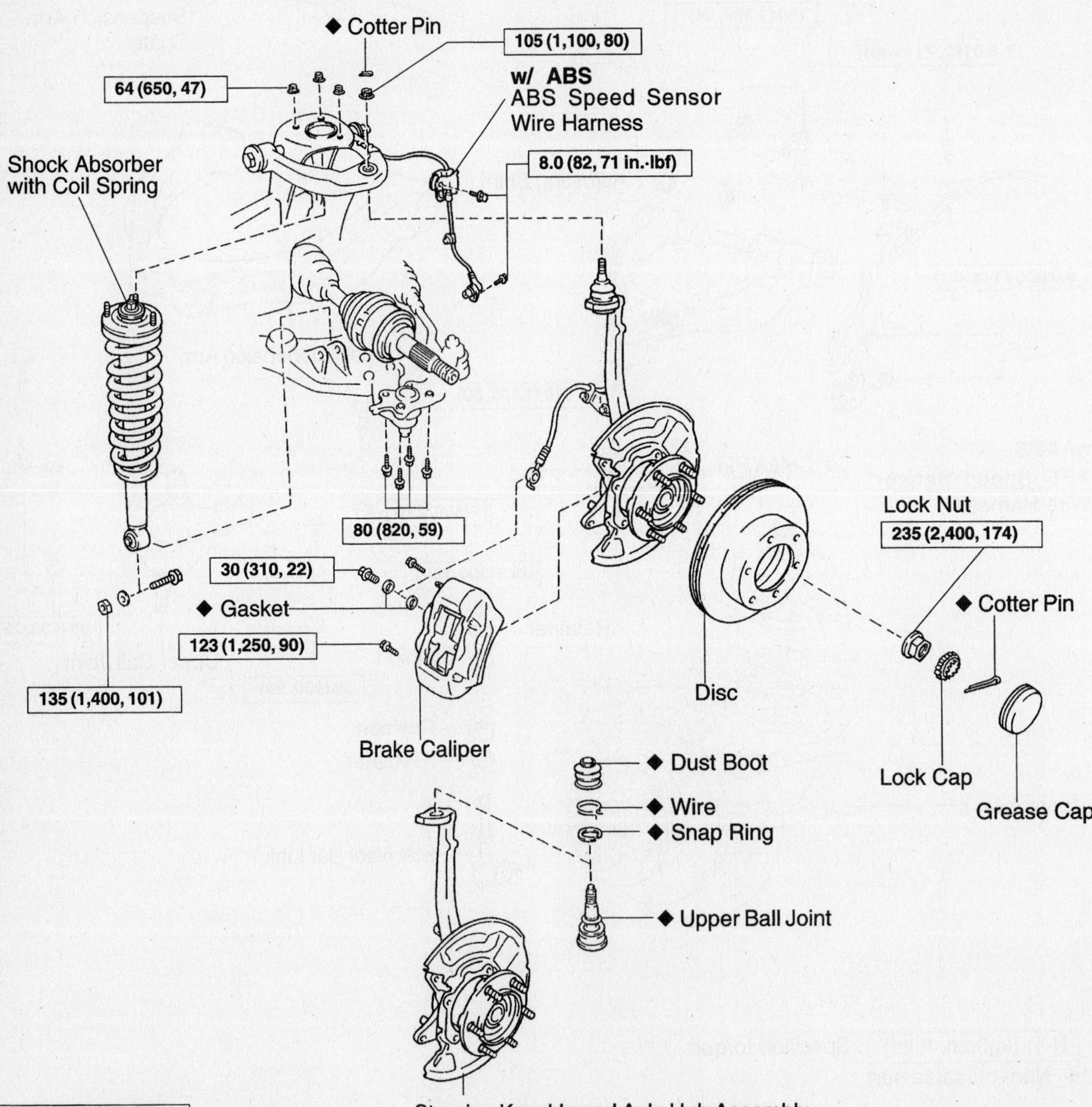

Upper ball joint and related parts—4wd, exc. PreRunner

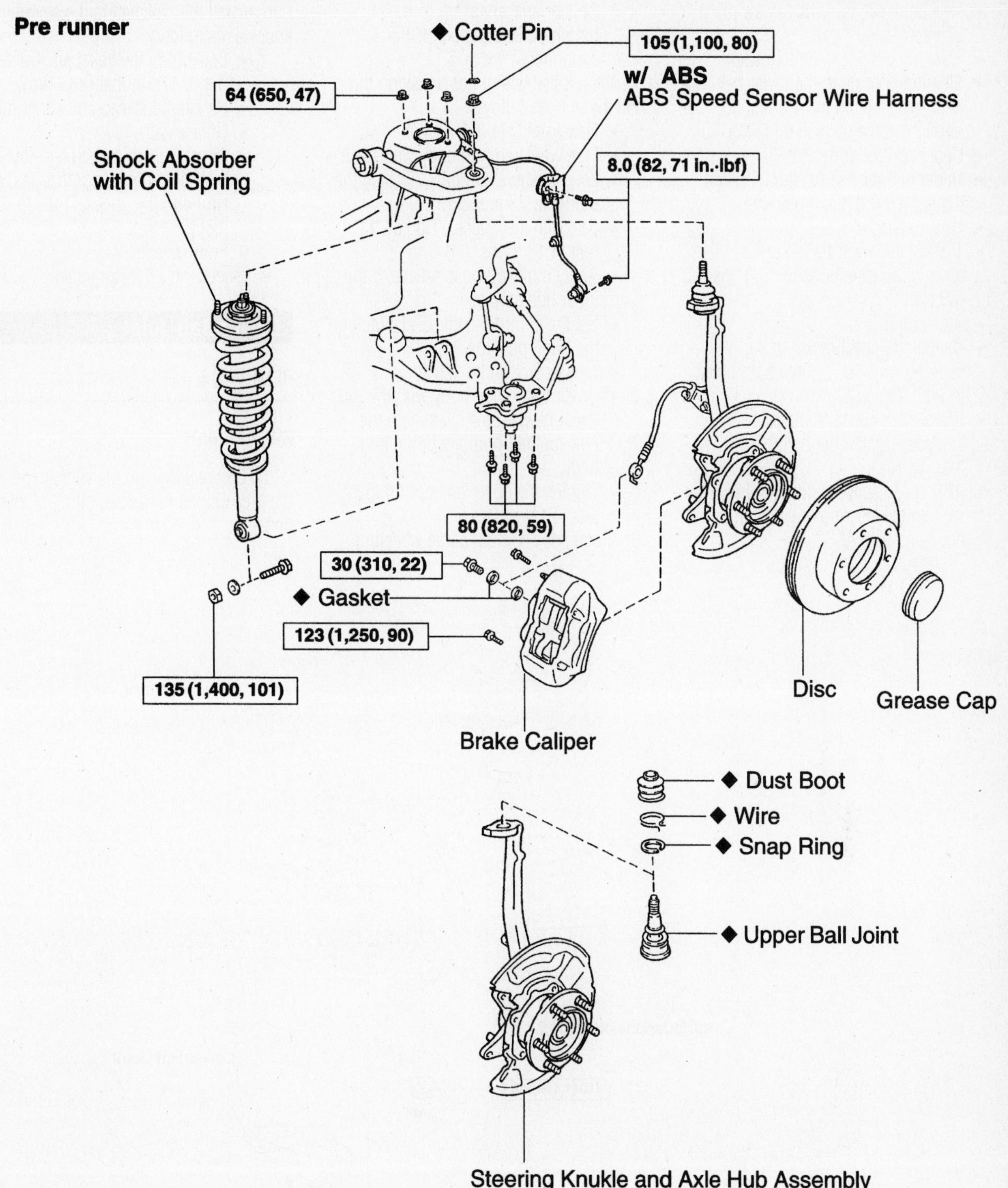

Upper ball joint and related parts—PreRunner

- New boot, secured with a new piece of wire
- Steering knuckle to the halfshaft
- Steering knuckle to the lower ball joint by installing the 4 bolts; do not tighten the bolts at this time.
- Upper control arm
- Upper ball joint to the arm. Tighten the nut to 80 ft. lbs. (105 Nm).
- New cotter pin
- Tighten the lower ball joint-to-steering knuckle bolts to 59 ft. lbs. (80 Nm).
- Brake rotor
- Caliper support bracket to the steering knuckle. Tighten both bolts to 90 ft. lbs. (123 Nm).
- Brake hose clamp to the steering knuckle. Tighten the bolt to 13 ft. lbs. (18 Nm).
- ABS speed sensor and wiring harness to the steering knuckle, if equipped with ABS
- Spacer and snapring to the halfshaft, using a snapring expander

7. If equipped with a free wheeling hub, install the hub, as follows:
   - New front axle hub gasket
   - Free wheeling hub body with the 6 cone washers and nuts. Tighten the 6 nuts to 23 ft. lbs. (31 Nm).
   - Bolt with the washer. Tighten the bolt to 13 ft. lbs. (18 Nm).
   - Apply multi-purpose grease to the inner hub splines
   - Set the control handle and clutch to the FREE position
   - New gasket on the cover
   - Cover to the hub body with the follower pawl tabs aligned with the non-toothed portions of the hub body
   - Tighten the cover bolts to 84 inch lbs. (10 Nm).
8. If equipped without a free wheeling hub, install the halfshaft to the steering knuckle, as follows:
   - Locknut to the halfshaft. Tighten the nut to 174 ft. lbs. (235 Nm).
   - Halfshaft lockcap and cotter pin
   - Grease cab
   - Strut. Tighten the strut-to-lower control arm nut to 101 ft. lbs. (135 Nm) and the upper 3 nuts to 47 ft. lbs. (64 Nm).
   - Front wheels
9. Check the wheel alignment.

Lower Ball Joint

REMOVAL & INSTALLATION

2WD Models

1. Before servicing the vehicle, refer to the precautions in the beginning of this section.

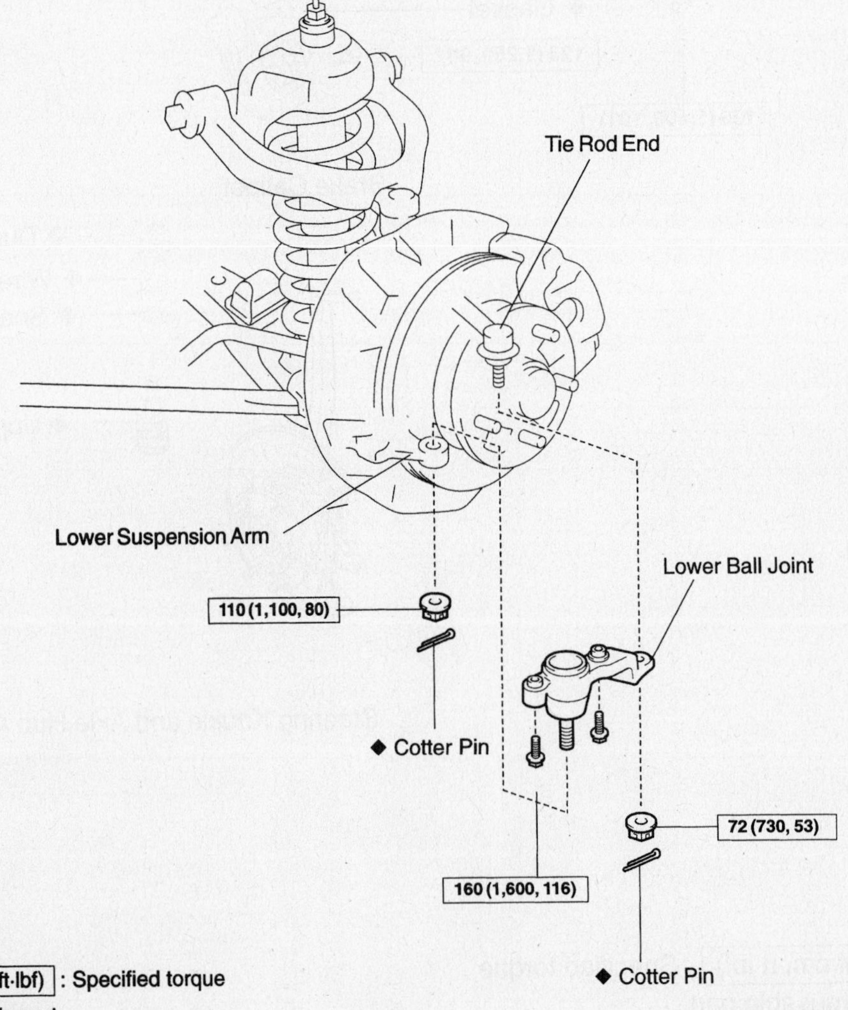

Lower ball joint installation—2wd

2. Remove the wheel.
3. Support the lower control with a floor jack.
4. Remove or disconnect the following:
   - Loosen the 2 lower ball joint set bolts
   - Cotter pin and nut from the tie rod
   - Tie rod from the ball joint bracket
   - Cotter pin and nut from the lower ball joint
   - Lower ball joint from the lower control arm
   - Both lower ball joint set bolts
   - Ball joint from the suspension

To install:

5. Install or connect the following:
   - Lower ball joint to the steering knuckle and lower control arm
   - Both lower ball joint set bolts; do not tighten the bolts at this time
   - Lower ball joint nut to hold the lower ball joint to the lower control arm. Tighten the nut to 80 ft. lbs. (110 Nm).
   - New cotter pin to the lower ball joint
   - Tie rod end to the ball joint bracket. Tighten the nut to 53 ft. lbs. (72 Nm).
   - New cotter pin to the tie rod end
   - Tighten the both lower ball joint set bolts to 116 ft. lbs. (160 Nm).
   - Wheel
6. Check the wheel alignment.

4WD Models

1. Before servicing the vehicle, refer to the precautions in the beginning of this section.
2. Remove or disconnect the following:
   - Wheel
   - Loosen the 4 lower ball joint set bolts
   - Cotter pin and nut from the tie rod
   - Tie rod from the ball joint bracket
   - Cotter pin and nut from the lower ball joint
   - Lower ball joint from the lower control arm
   - 4 lower ball joint set bolts
   - Ball joint from the suspension

To install:

3. Install or connect the following:
   - Lower ball joint to the steering knuckle and lower control arm
   - 4 lower ball joint set bolts; do not tighten the bolts at this time
   - Lower ball joint-to-lower control arm nut. Tighten the nut to 103 ft. lbs. (140 Nm).
   - New cotter pin to the lower ball joint
   - Tie rod end to the ball joint bracket. Tighten the nut to 67 ft. lbs. (90 Nm).
   - New cotter pin to the tie rod end
   - Tighten the lower ball joint set bolts to 59 ft. lbs. (80 Nm).
   - Wheel
4. Check the wheel alignment.

Upper Control Arm

REMOVAL & INSTALLATION

2WD

1. Before servicing the vehicle, refer to the precautions in the beginning of this section.
2. Remove or disconnect the following:
   - Front wheel
   - Anti-lock Brake System (ABS) speed sensor and wire harness
   - Stabilizer bar link
   - Steering knuckle from the upper bar joint
3. Loosen the 2 bolts; then, remove the front and rear alignment adjusting shims.
4. Make note of the number and thickness of the front and rear shims.

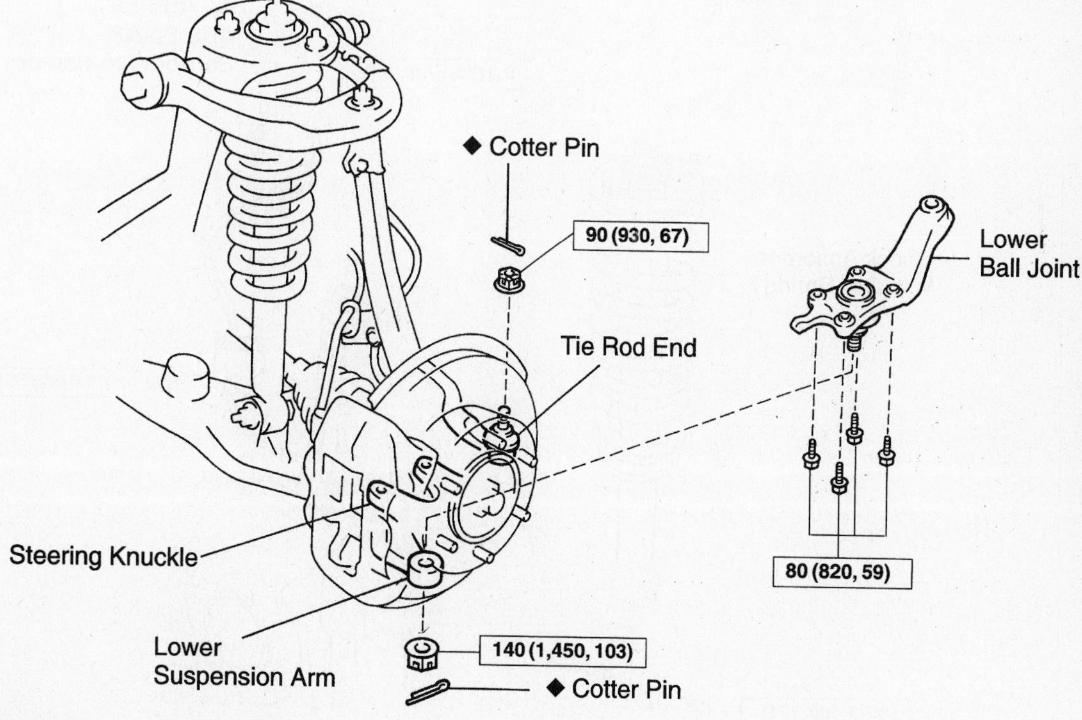

Lower ball joint installation—4wd

5. Remove or disconnect the following:
   - Upper control arm
   - Upper ball joint from the arm

To install:

6. Install or connect the following:
   - Upper ball joint to the arm. Tighten the fasteners to 29 ft. lbs. (39 Nm).

➡Do not lose the camber adjusting shims. Record the position and thickness of the camber shims so that these can be reinstalled to there original locations. Install the equal number and thickness of shims into there locations.

- Upper control arm with the shims. Tighten the mounting bolts to 94 ft. lbs. (130 Nm).
- Steering knuckle to the upper ball joint
- Stabilizer bar link
- ABS speed sensor and wire harness
- Front wheel. Tighten the lug nuts to 83 ft. lbs. (110 Nm).

4WD AND PRERUNNER

1. Before servicing the vehicle, refer to the precautions in the beginning of this section.
2. Remove or disconnect the following:
   - Front wheel
   - Shock and coil spring assembly
   - Anti-lock Brake System (ABS) speed sensor wire harness clamp
3. Upper ball joint, as follows:
   - Cotter pins and loosen the nut

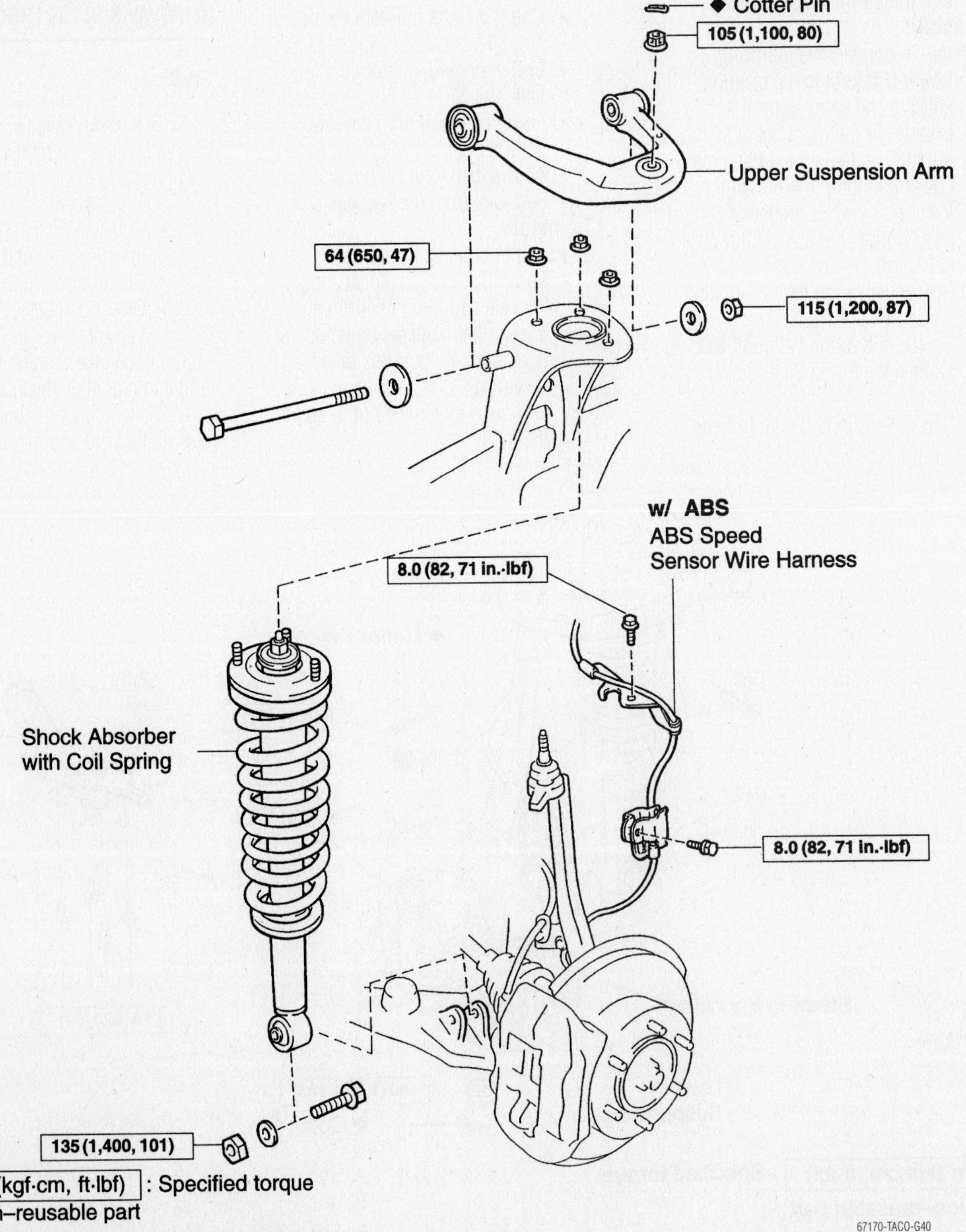

Upper control arm and related parts—4wd and PreRunner

- Upper ball joint from the control arm, using a ball joint separator
- Support the steering knuckle
- Nut

4. Detach the control arm, by removing the nut, bolt, washers and lowering the arm.

To install:

5. Install or connect the following:
   - Upper control arm with the washer, bolt and nut. Tighten the nut to 87 ft. lbs. (115 Nm).
   - Upper ball joint to the control arm. Tighten the mounting nut to 80 ft. lbs. (105 Nm).
   - New cotter pin
   - ABS speed sensor wire harness clamp. Tighten it to 71 inch lbs. (8 Nm).
   - Shock and coil spring assembly
6. Check and/or adjust the alignment.

CONTROL ARM BUSHING REPLACEMENT

4WD

1. Before servicing the vehicle, refer to the precautions in the beginning of this section.
2. Remove the upper control arm from the vehicle.
3. Pry up the bushing flange, using a chisel and a hammer.
4. Using tools 09613-26010, 09613-20060 and 09950-00020 and a shop press, remove the bushing.

To install:

5. Using tools 09223-00010, 09506-35010 and a shop press, press the new bushing into the upper control arm.
6. Install the upper control arm to the chassis.
7. Check and/or adjust the alignment.

2WD

1. Before servicing the vehicle, refer to the precautions in the beginning of this section.
2. Remove the upper control arm.
3. Cut off the outer edges of the bushing so it is flush with arm tube and the shaft.

➡Be careful not to damage the edge of the arm tubes.

4. Using a shop press with tool 09710-03031, press down the suspension arm tube until it touches tool 09710-03141.

➡Do not press the tube excessively.

5. Temporarily install a 1.8–2.0 in. (45–50mm) bolt to the arm shaft on the other side.

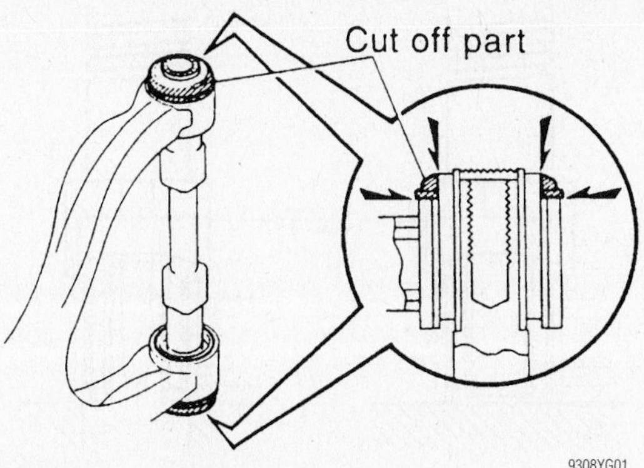

Cut the bushing flush with the upper control arm tube and shaft

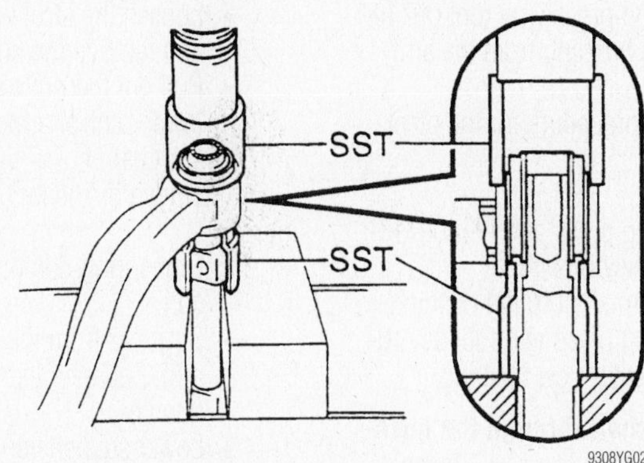

Position the upper control arm bushings with the tools

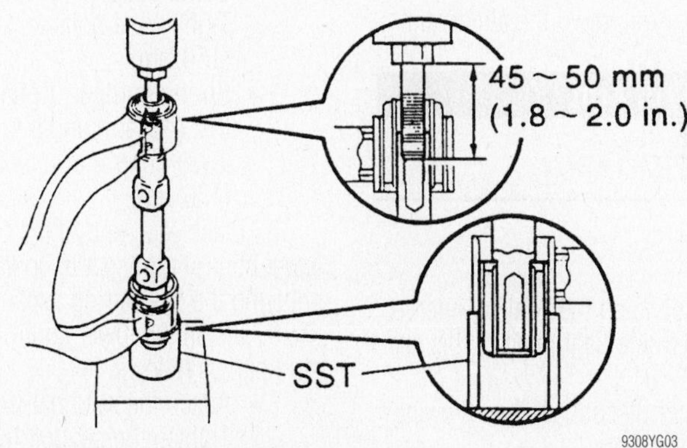

Positioning the upper control arm tube and bolt

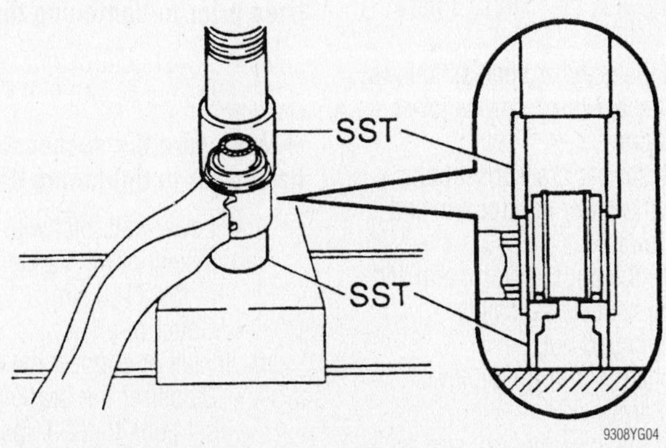

Removing the bushings from the upper control arm

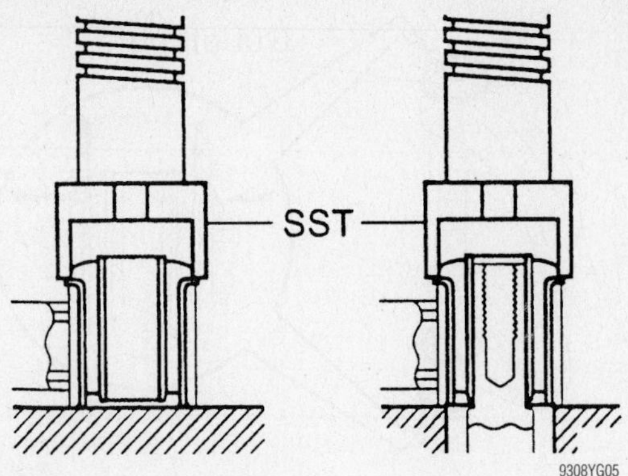

Installing the bushings to the upper control arm

6. Using a shop press with tool 09710-03141, remove the bushing from the arm shaft.

7. Repeat this procedure for the other bushing.

To install:

8. Using a shop press and tool 09710-03101, install the new bushing.

9. Place the arm shaft to the bushing.

10. Using a shop press and tool 09710-03101, install the other new bushing.

➡Pass the arm shaft through the bushing to make sure that the shaft turns freely and there is no axial play

11. Install the upper control arm.

12. Check and/or adjust the alignment.

Lower Control Arm

REMOVAL & INSTALLATION

2WD

1. Before servicing the vehicle, refer to the precautions in the beginning of this section.

2. Remove or disconnect the following:
   - Front wheel
   - Shock absorber

3. Compress the spring using a spring compressor, following the manufacturer's instructions.

4. Remove the stabilizer bar, as follows:
   - Stabilizer bar link from the lower control arm
   - 2 stabilizer bar bracket set bolts

5. Remove the lower control arm and strut bar, as follows:
   - Support the upper control arm and steering knuckle assembly
   - Cotter pin and nut
   - Lower ball joint from the lower control arm
   - Loosen the lower control arm set bolt and remove the nut
   - Loosen the strut bar front set bolt and remove the nut
   - Pull out the bolts and remove the lower control arm along with the strut bar

6. Remove or disconnect the following:
   - Coil spring compressor tool and coil
   - Strut bar from the lower control arm. Separate the nut and spring bumper.
   - Lower suspension arm No. 3

To install:

7. Install or connect the following:
   - Lower suspension arm No. 3. Tighten the bolts to 111 ft. lbs. (150 Nm).
   - Spring bumper. Tighten it to 32 ft. lbs. (43 Nm) and the strut bar-to-lower control arm to 111 ft. lbs. (150 Nm).

8. Place each end of the coil spring and lower control arm seat in contact when applying the coil spring expander.

9. Install the lower control arm and strut bar, as follows:
   - Attach the strut bar front set bolt. Tighten the set bolt to 221 ft. lbs. (300 Nm).

➡Make sure the suspension is stabilized prior to tightening the bolt.

   - Lower control arm set bolt. Tighten the nut to 148 ft. lbs. (200 Nm).

➡Make sure the suspension is stabilized prior to tightening the bolt.

   - Lower ball joint with a ball joint installer tool. Tighten the nut to 80 ft. lbs. (110 Nm).
   - Cotter pin.

10. Install or connect the following:
    - Stabilizer bar bracket. Tighten the set bolts to 22 ft. lbs. (29 Nm).
    - Stabilizer link to the lower control arm. Tighten the fasteners to 29 ft. lbs. (39 Nm).

11. Remove the spring compressing tool.

12. Install or connect the following:
    - Shock absorber
    - Wheel. Tighten the lug nuts.

13. Check and/or adjust the alignment.

4WD

1. Before servicing the vehicle, refer to the precautions in the beginning of this section.

2. Remove or disconnect the following:
   - Front wheel
   - Steering gear assembly
   - Stabilizer bar link
   - Shock absorber from lower control arm

3. Support the upper control and steering knuckle securely.

4. Remove or disconnect the following:
   - Cotter pin and nut from the lower ball joint
   - Lower ball joint from the lower control arm

5. Place matchmarks on the front and rear adjusting cams.

6. Remove or disconnect the following:
   - 2 bolts, nuts, adjusting cams and lower control arm
   - Spring bumpers, with a special tool 09922-10010

To install:

7. Install or connect the following:
   - Spring bumpers. Tighten to 17 ft. lbs. (23 Nm).
   - Lower control arm, placing it in the appropriate position with the matchmarks. Tighten the arm to 96 ft. lbs. (130 Nm).

8. Install or connect the following:
   - Lower ball joint. Tighten the nut to 103 ft. lbs. (140 Nm).
   - Shock absorber to the lower control arm
   - Stabilizer bar link
   - Steering gear assembly
   - Front wheel. Tighten the lug nuts.

9. Check and/or adjust the alignment.

CONTROL ARM BUSHING REPLACEMENT

4WD

1. Before servicing the vehicle, refer to the precautions in the beginning of this section.

2. Remove the lower control arm from the vehicle.

3. Pry up the bushing flange, using a chisel and a hammer.

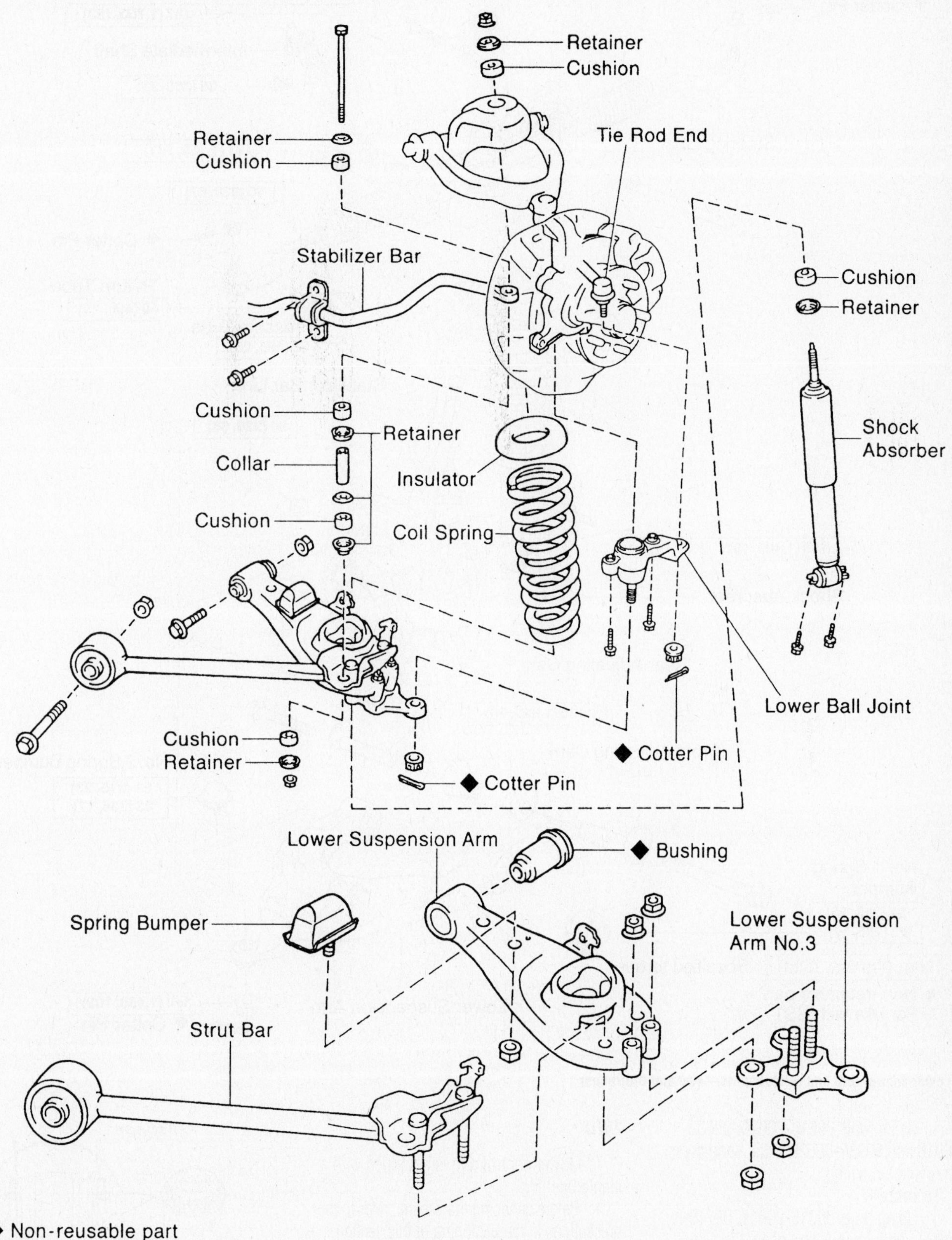

Exploded view of the lower control arm and related front suspension components—2wd

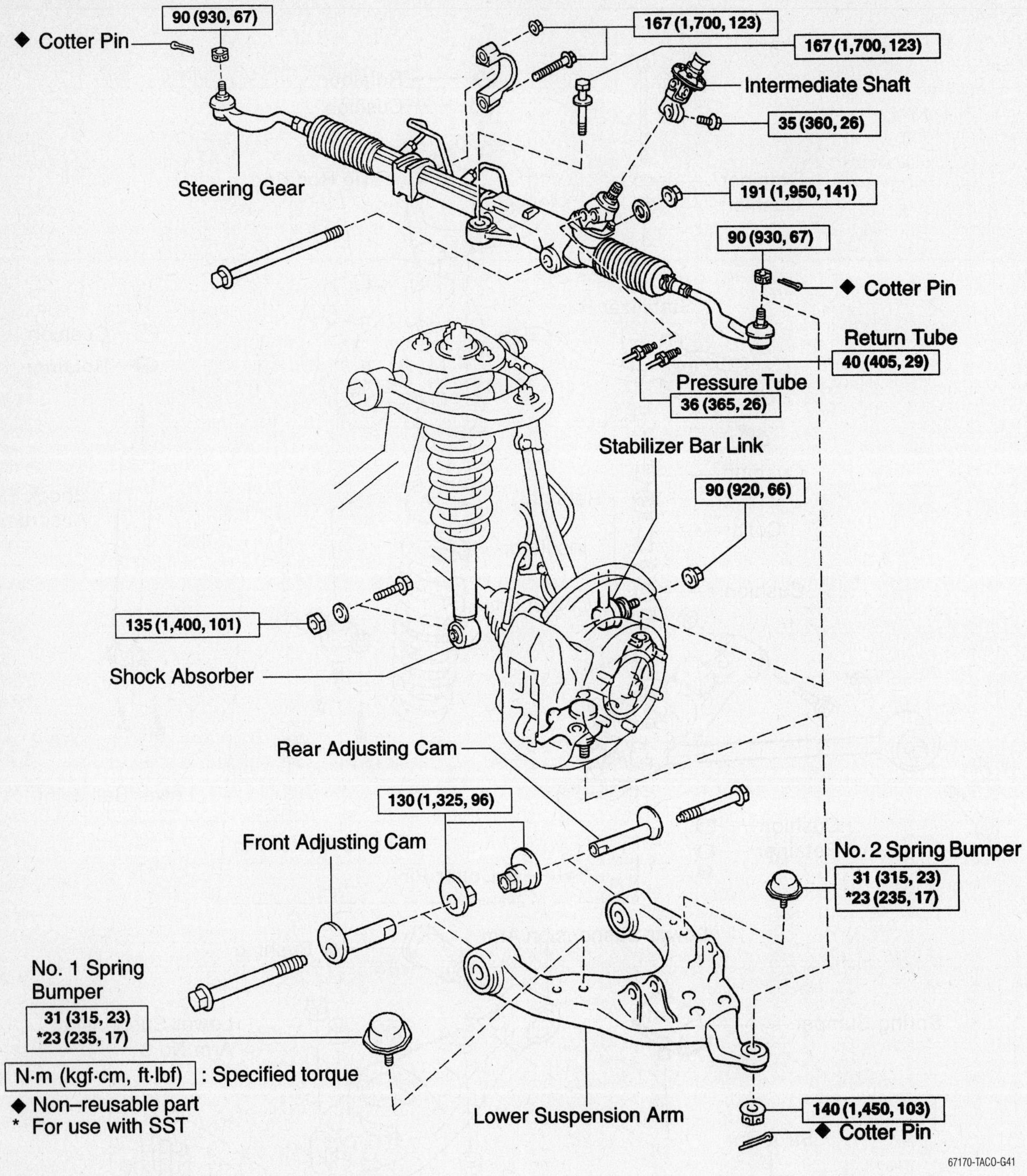

Lower control arm and related parts—4wd and PreRunner

4. Using tools 09613-26010, 09632-36010 and 09950-00020 and a shop press, remove the bushing.

To install:

5. Using tools 09316-20011, 09710-30021 and a shop press, press the new bushing into the lower control arm.
6. Install the lower control arm to the chassis.
7. Check and/or adjust the alignment.

2WD

The lower control arm is equipped with a single bushing.

1. Before servicing the vehicle, refer to the precautions in the beginning of this section.
2. Remove the lower control arm from the vehicle.
3. Cut off a portion of the bushing to expose the edge of the arm tube.

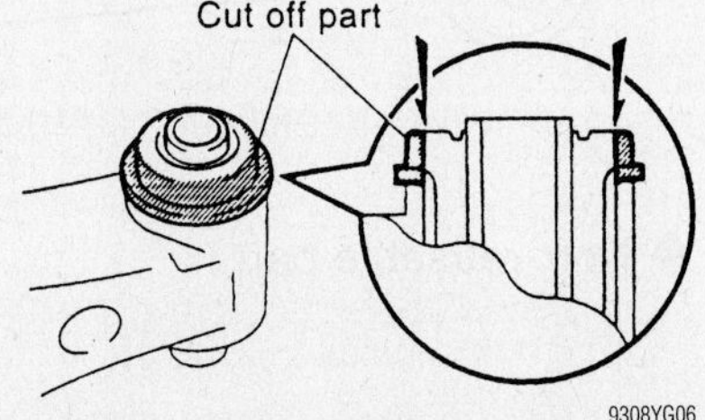

Cutting off part of the bushing—2WD

4. Position the lower control arm on a shop press with the cut side facing downward, resting on tool 09710-30021; then, press the bushing from the arm.

To install:

5. Position a new bushing onto the lower control arm.
6. Position the lower control arm on a shop press, resting on tool 09710-30021.
7. Press the bushing into the lower control arm.
8. Install the lower control arm.
9. Check and/or adjust the alignment.

Front Wheel Bearings

ADJUSTMENT

The bearings on the 4WD are not adjustable.

2WD

1. Tighten the adjusting nut to 26 ft. lbs. (35 Nm).
2. Turn the disc/hub assembly 2–3 times, from the left to the right.
3. Loosen the adjusting nut until it can be turned by hand.
4. Attach a spring tension gauge to 1 lug on the hub assembly. Pull on the gauge and measure the frictional force.

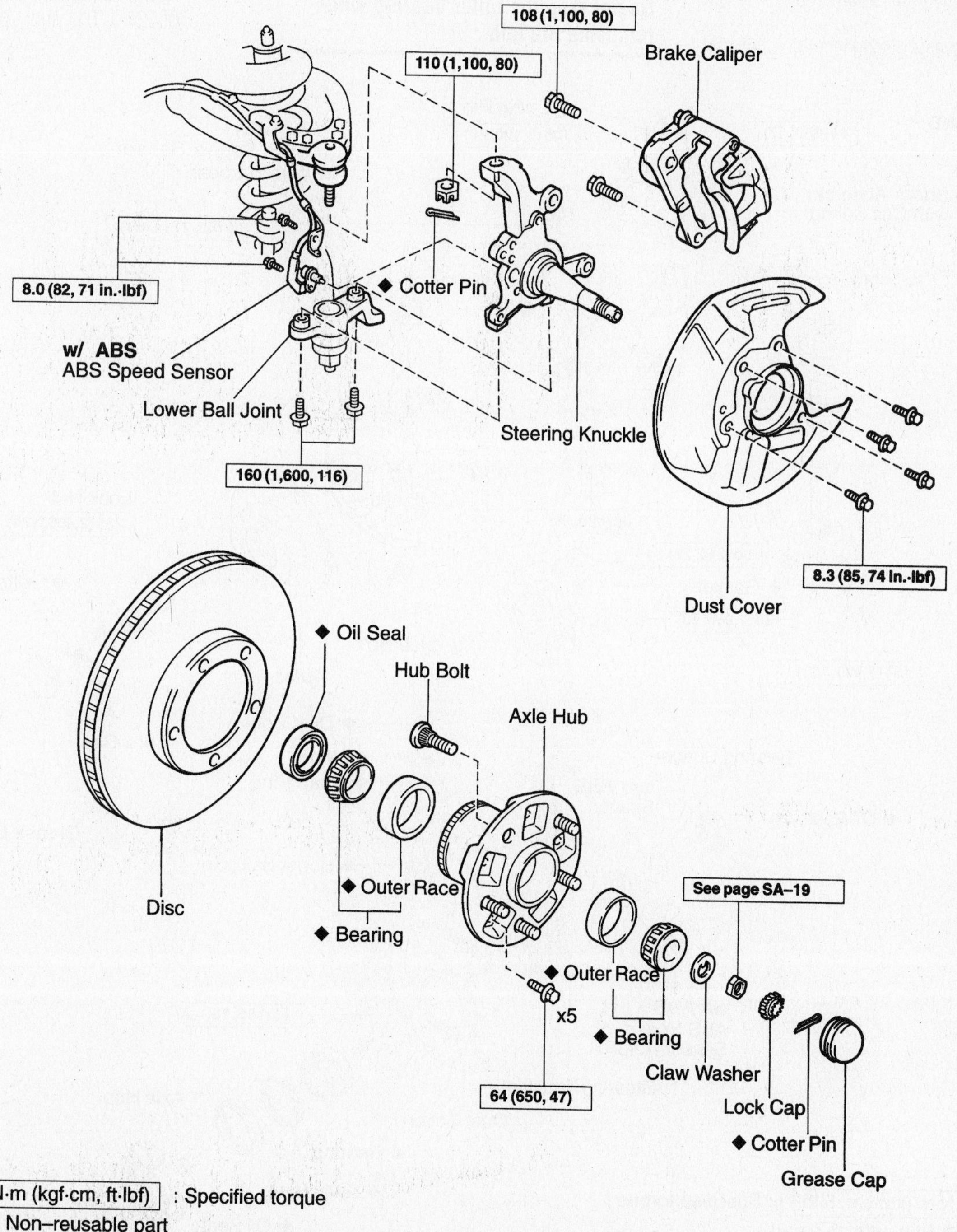

Front hub, bearing and related parts—2wd

Frictional force should be 1–3 lbs. (5.0–14.0 N).

5. Adjust the preload by tightening the nut.

6. Measure the hub axial play. The limit is 0.0020 in. (0.05mm).

REMOVAL & INSTALLATION

2WD

1. Before servicing the vehicle, refer to the precautions in the beginning of this section.

2. Remove or disconnect the following:

- Brake caliper support bracket, by removing the 2 bolts. Support the brake caliper with a piece of wire. Do not allow the caliper to hang from the brake hose.
- Cotter pin, lockcap, nut and the claw washer from the axle hub and disc
- Axle hub with the disc from the steering knuckle.

⁂ WARNING

Do not drop the outer bearing when removing the hub.

- Inner oil seal
- Inner bearing
- Bearing outer races, using SST 09527-17011, a brass bar and a hammer

3. If it is necessary to separate the hub and rotor, place matchmarks on the hub and rotor.

4. Remove the 5 bolts and remove the hub from the rotor.

To install:

5. Install or connect the following:

- New bearing races, using SST 09527-17011 and a press.

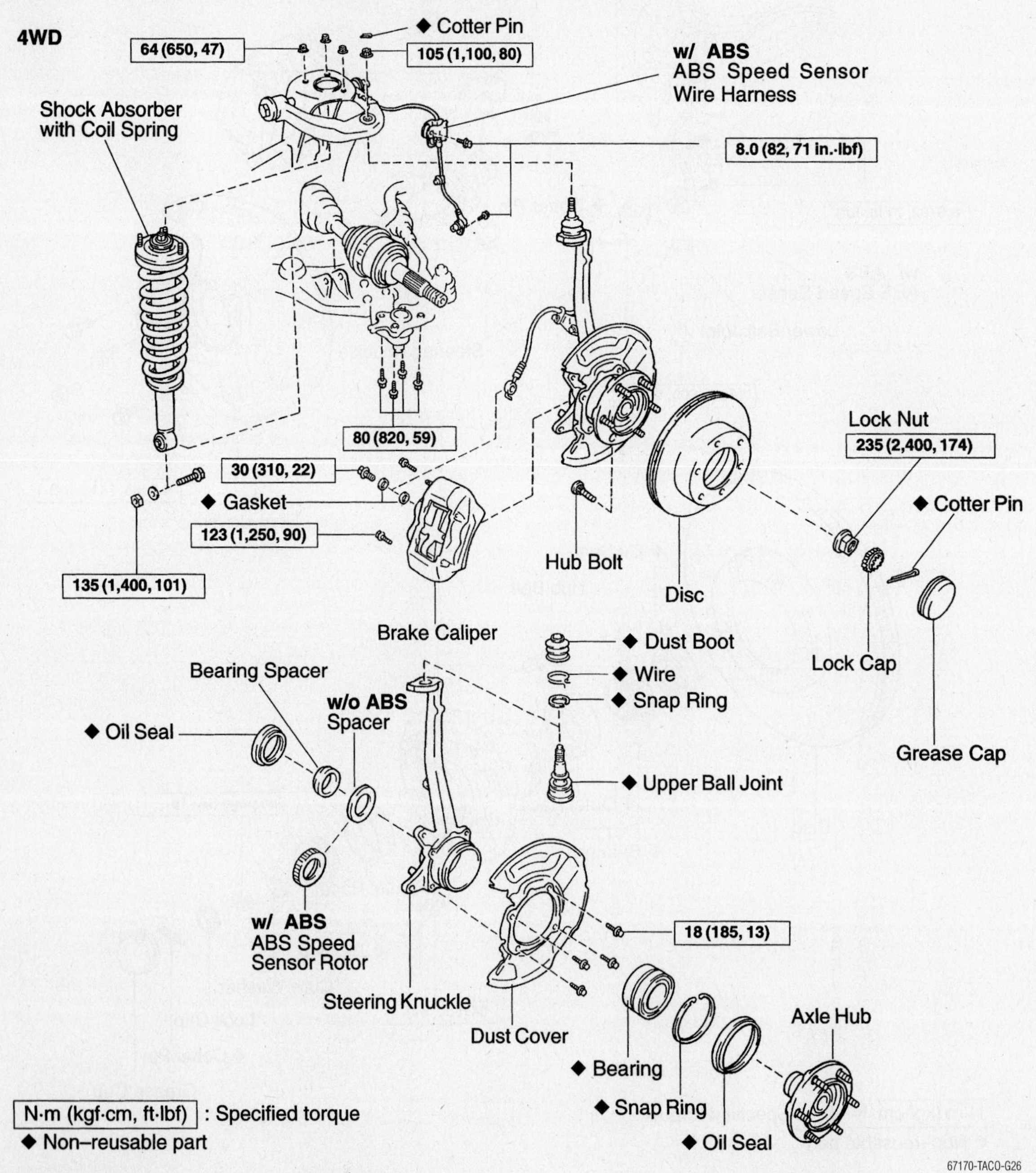

Front wheel hub—4WD components, exc. PreRunner

- Hub to the rotor. Tighten the 5 bolts to 47 ft. lbs. (64 Nm).

6. Clean all parts.

7. Repack the bearings with multi purpose grease and apply the same grease to the outer bearings.

8. Install or connect the following:

- Inner bearing and seal to the hub. Coat the inner seal with multi purpose grease.
- Outer bearing to the hub
- Hub to the steering knuckle
- Axle hub to the steering knuckle claw washer and nut

9. Adjust the bearing preload as described above.

10. Install or connect the following:

- Locknut, cotter pin and the grease cap
- Disc brake caliper. Tighten the 2 bolts to 80 ft. lbs. (108 Nm).
- Wheel

4WD

1. Before servicing the vehicle, refer to the precautions in the beginning of this section.

2. Remove the front wheel.

3. Detach the shock absorber.

4. Disconnect the driveshaft by removing the grease cap and pulling out the cotter pin and lock cap.

5. Apply the brakes to hold the axle from spinning and remove the lock nut.

6. If the vehicle is equipped with antilock brakes, detach the speed sensor and wiring harness clamp from the steering knuckle.

7. Remove or disconnect the following:

- Banjo bolt and 2 gaskets from the caliper
- Flexible brake hose from the caliper

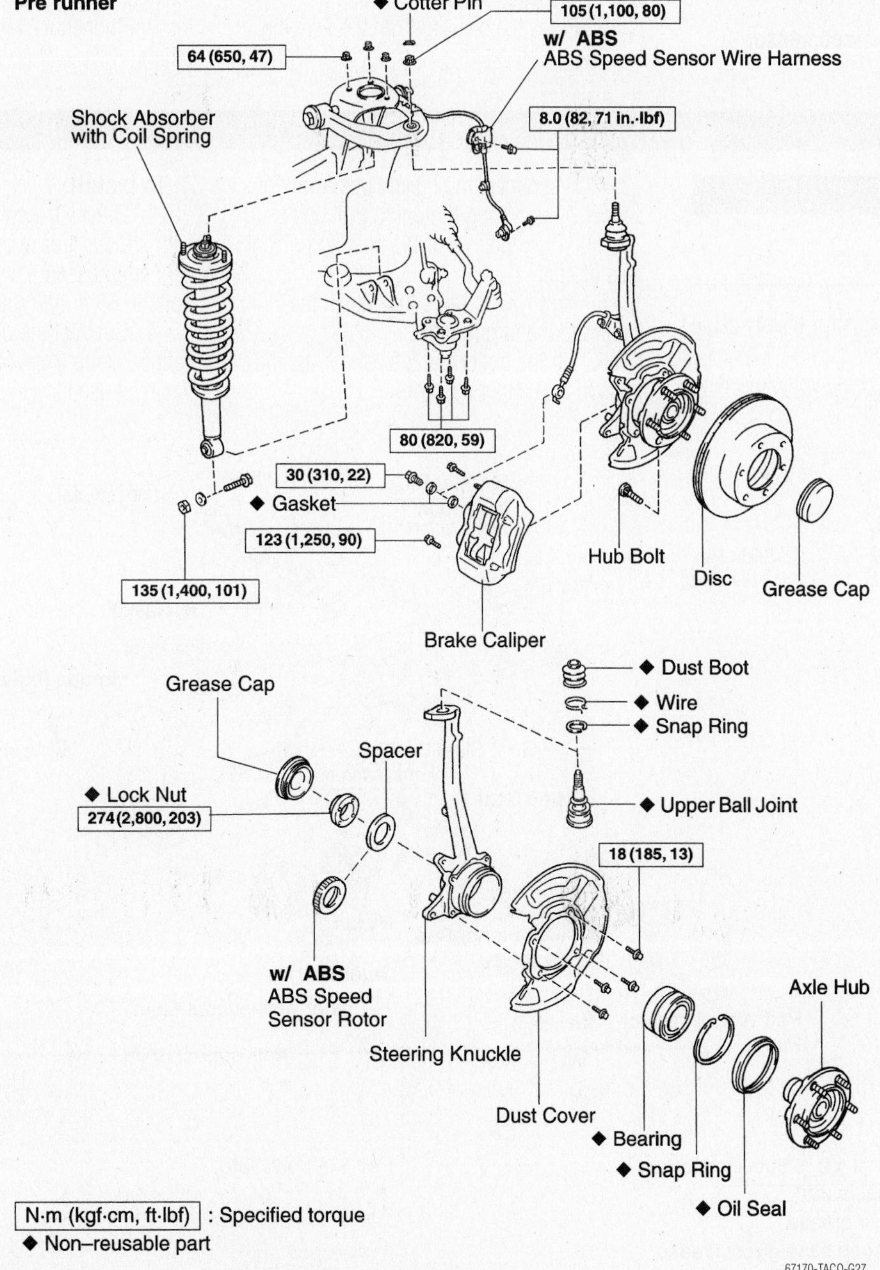

Front wheel hub—PreRunner

- Brake caliper and then the rotor
- Lower ball joint
- Steering knuckle

8. Clamp the axle hub in a soft jaw vise.

➡**Close the vise until it holds the hub bolts.**

9. Using the proper seal puller or prytool, remove the oil seal.
10. On vehicles equipped with a free wheel hub and on the Pre Runner, use a chisel and hammer to loosen the staked part of the lock nut.
11. Remove the lock nut. A special service tool may be required.
12. Remove the Antilock Brake System (ABS) speed sensor rotor/spacer.

➡**Do not scratch the speed sensor rotor.**

13. Detach the bolts to the dust shield and shift the shield towards the outside of the hub.
14. Remove the axle from the steering knuckle, a special service tool may be required.
15. Remove the dust cover from the steering knuckle.
16. On vehicles without a free wheeling hub, that are 4WD only, remove the bearing spacer and ABS speed sensor rotor spacer.
17. Remove the outside seal by prying it out with a seal puller.
18. Remove the bearing from the steering knuckle by removing the snapring with a pair of snapring pliers.
19. Press the bearing from the steering knuckle.

To install:

20. Install or connect the following:
    - New bearing
    - New oil seal
    - Axle hub to the steering knuckle. Torque the bolts to 13 ft. lbs. (18 Nm).
    - ABS speed sensor rotor
21. On vehicles that are equipped with a free wheel hub, install the bearing spacer.
22. Except Pre-Runner install a new inside oil seal.
23. On the Pre-Runner, install the grease cap.
24. The remainder of the installation procedure is the reverse of removal. New lock nut torque is 203 ft. lbs. (274 Nm) for the PreRunner; 174 ft. lbs. (235 Nm) for all others.

BRAKES

Brake Caliper

REMOVAL & INSTALLATION

1. Disconnect the negative battery cable from the battery.
2. Raise and support the vehicle safely.
3. Remove the wheels.
4. Disconnect the brake hose from the caliper by removing the union bolt and 2 gaskets. Plug the end of the hose to prevent loss of fluid.
5. Remove the bolts that attach the caliper to the torque plate.
6. Lift the bottom of the caliper up and remove the caliper assembly.

To install:

7. Grease the caliper slides and bolts with lithium grease or equivalent. Install the caliper and secure with the bolts. Torque the bolts to 65 ft. lbs. (88 Nm).
8. Connect the brake hose to the caliper, using 2 new washers. Make sure the flexible hose lock is securely in the lock

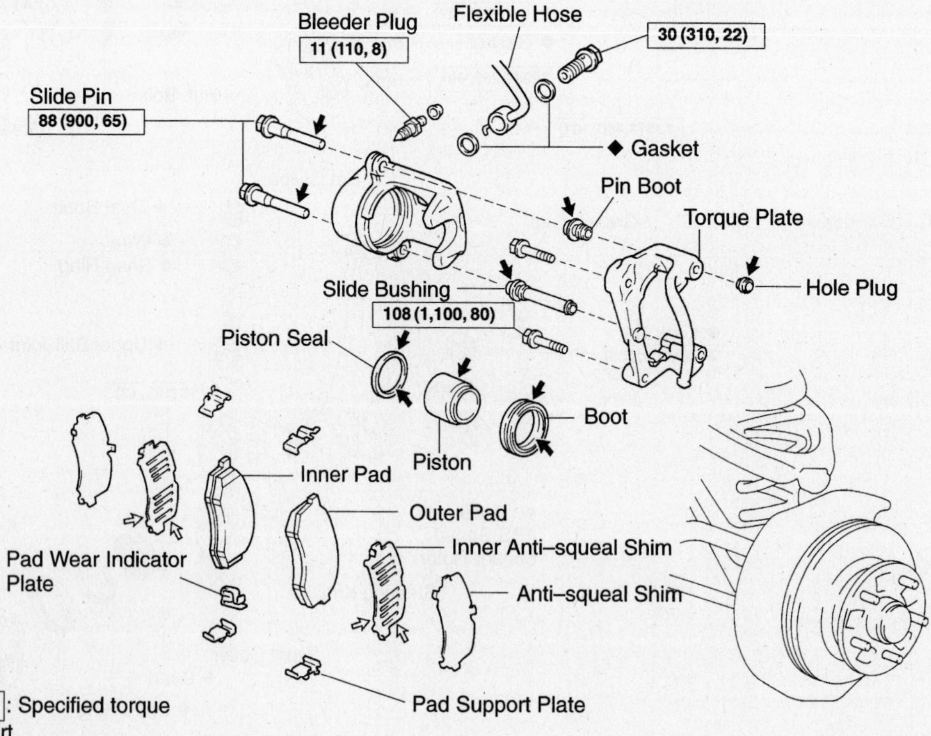

Brake caliper assembly—2WD

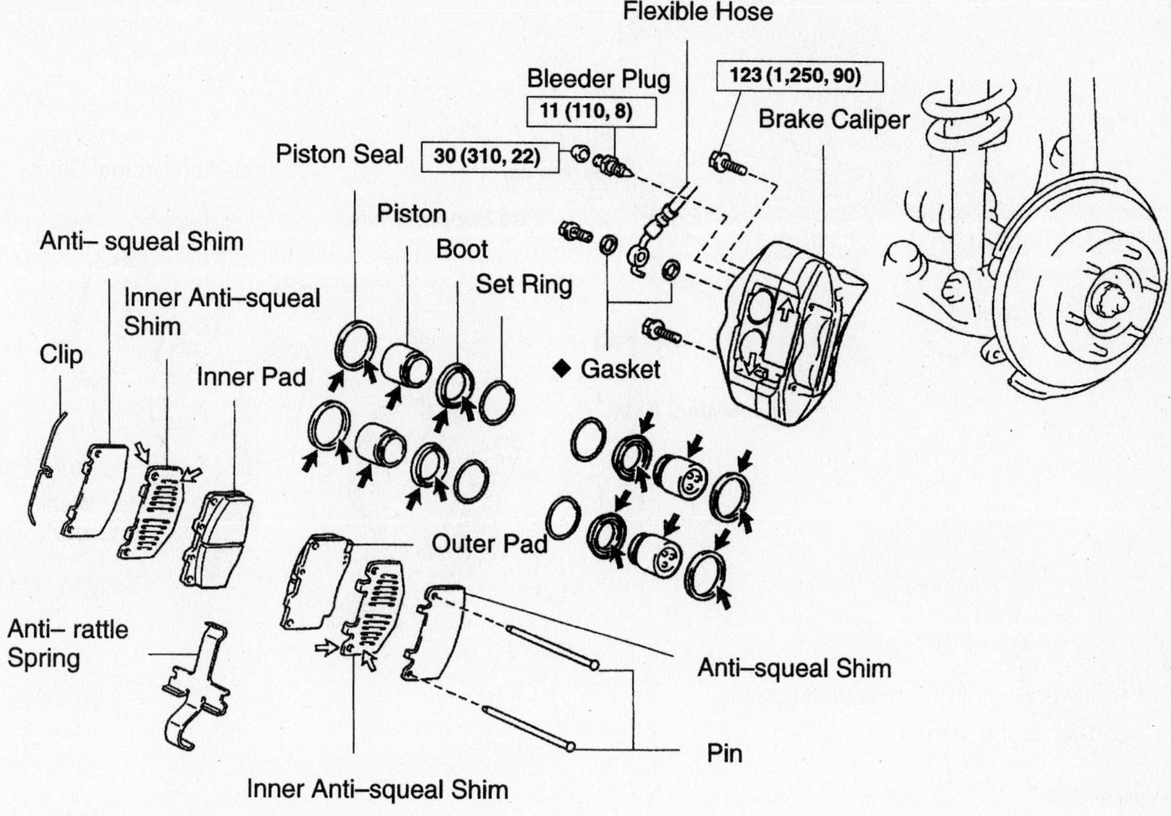

Brake caliper assembly—4WD

hole of the caliper. Torque the union bolt to 22 ft. lbs. (30 Nm).

9. Fill the brake system to the proper level and bleed the brake system.

10. Install the tire and wheel assembly.

11. Top off the brake fluid level in the master cylinder. Check for leaks and proper brake operation.

12. Connect the negative battery cable to the battery.

Disc Brake Pads

REMOVAL & INSTALLATION

2-Wheel Drive

1. Raise the vehicle and support it safely.

2. Remove the wheel and tire assembly.

3. When servicing the front pads, loosen the brake caliper upper side mounting bolt. Loosen and remove the lower side mounting bolt. Lift the cylinder and suspend it so the hose is not stretched.

4. If equipped, remove the anti-squeal spring.

5. Remove the brake pads.

To install:

6. Siphon a small amount of brake fluid from the reservoir. Press in the brake caliper piston with a hammer handle or equivalent.

7. Before installing the new pads, check the disc thickness and disc runout.

8. Install the pad support plates.

9. Install the anti-squeal shims to each pad.

➡Apply disc brake grease to both sides of the inner anti-squeal shims.

10. Install the disc pads so the wear indicator plate is facing downward.

11. If removed, install the anti-squeal springs.

12. Carefully install the brake caliper so the boot is not wedged. Torque the caliper mounting bolts, as follows:

- 2-Wheel drive w/PD60 type disc: 29 ft. lbs. (39 Nm)
- 2-wheel drive w/FS17 type disc: 65 ft. lbs. (88 Nm)

13. Install the wheel and tire assembly.

14. Check and adjust the fluid level. Apply the brake pedal several times.

15. Road-test the vehicle for proper operation.

4-Wheel Drive

1. Raise the vehicle and support it safely.

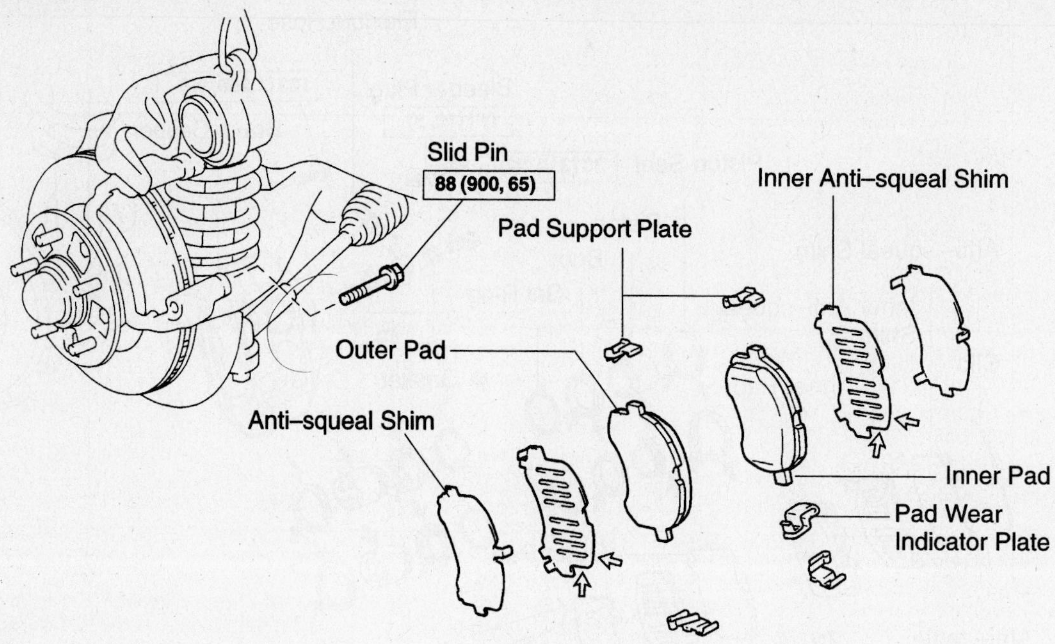
Slid Pin
88 (900, 65)
Inner Anti–squeal Shim
Pad Support Plate
Outer Pad
Anti–squeal Shim
Inner Pad
Pad Wear
Indicator Plate
N·m (kgf·cm, ft·lbf) : Specified torque
Disc brake grease
67170-TACO-G45

Brake pads—2wd

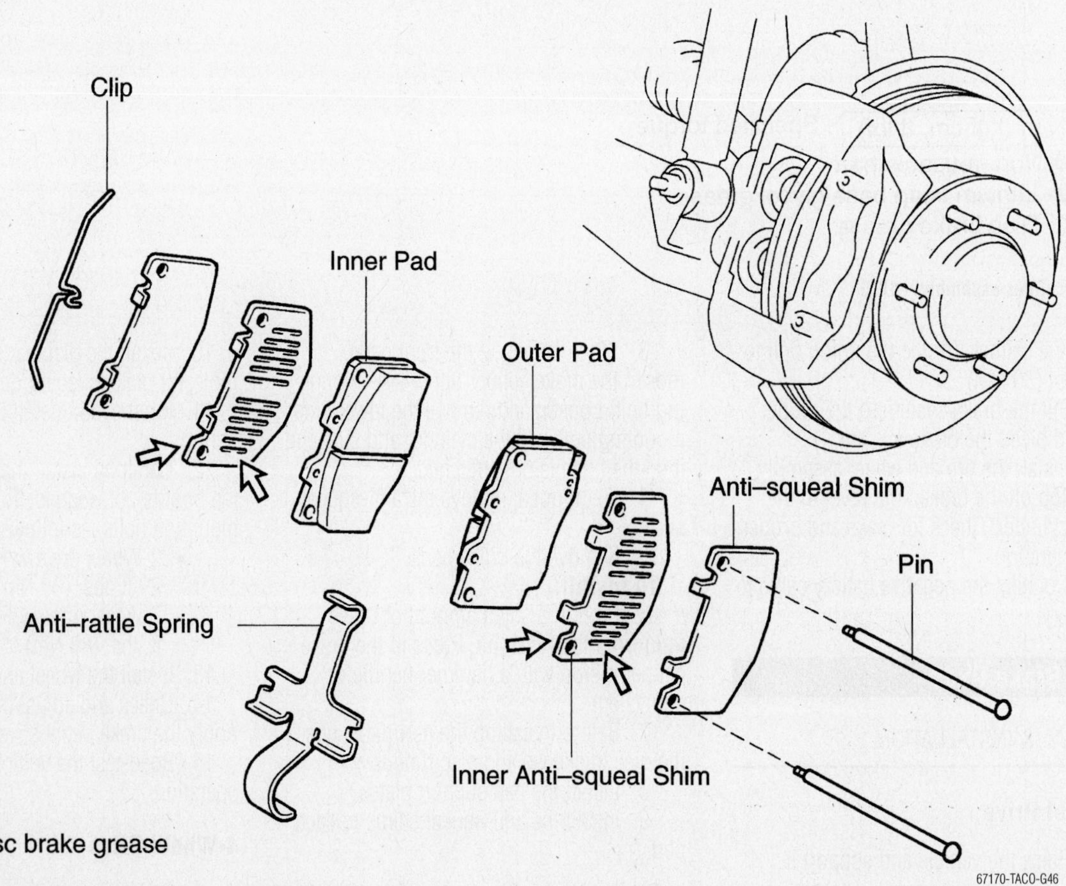
Clip
Inner Pad
Outer Pad
Anti–squeal Shim
Pin
Anti–rattle Spring
Inner Anti–squeal Shim
Disc brake grease
67170-TACO-G46

Brake pads—4wd

2. Remove the wheel and tire assembly.

3. Remove the clip, pins, and the anti-rattle spring.

4. Remove the pads and the anti-squeal shims.

5. Remove the caliper, but do not disconnect the brake hose.

To install:

6. Before installing the new pads, check the disc thickness and disc runout.

7. Siphon out a small amount of brake fluid from the reservoir.

8. Temporarily install the old inner brake pad. Press in the pistons with a C-clamp or equivalent. Remove the old inner brake pad.

9. Apply disc brake grease to both sides of the inner anti-squeal shim. Install the anti-squeal shims to the new pads.

10. Install the pads.

11. Install the anti-rattle springs and pins. Install the clip.

12. Install the caliper and the mounting bolts. Torque the mounting bolts to 90 ft. lbs. (123 Nm).

13. Install the wheel and tire assembly.

14. Check and adjust the fluid level. Apply the brake pedal several times.

15. Road-test the vehicle for proper operation.

Brake Drums

REMOVAL & INSTALLATION

1. Raise and safely support the vehicle.
2. Remove the rear wheel(s).
3. Remove the brake drum from the axle hub. If there is difficulty in removing the drum, insert a suitable tool through the hole in the rear of the backing plate, and hold the automatic adjusting lever away from the adjuster. Using another suitable tool at the same time, reduce the brake shoe adjuster by turning the adjusting wheel.

To install:

4. Install the brake drum and pull the parking brake lever all the way up until a clicking sound can no longer be heard.
5. Verify that the rear wheels will not turn. If the rear wheels turn, adjust the parking brake cable as necessary.
6. Release the parking brake and remove the brake drum. Measure the brake drum inside diameter and diameter of the brake shoes. Check that the difference between the diameters is the correct shoe clearance. Clearance is 0.024 in. (6mm).
7. If the brake shoe clearance is not correct, adjust the brake shoes until the clearance is correct.
8. Install the brake drum, replace the wheel(s), and safely lower the vehicle.
9. Road-test the vehicle for proper brake operation.

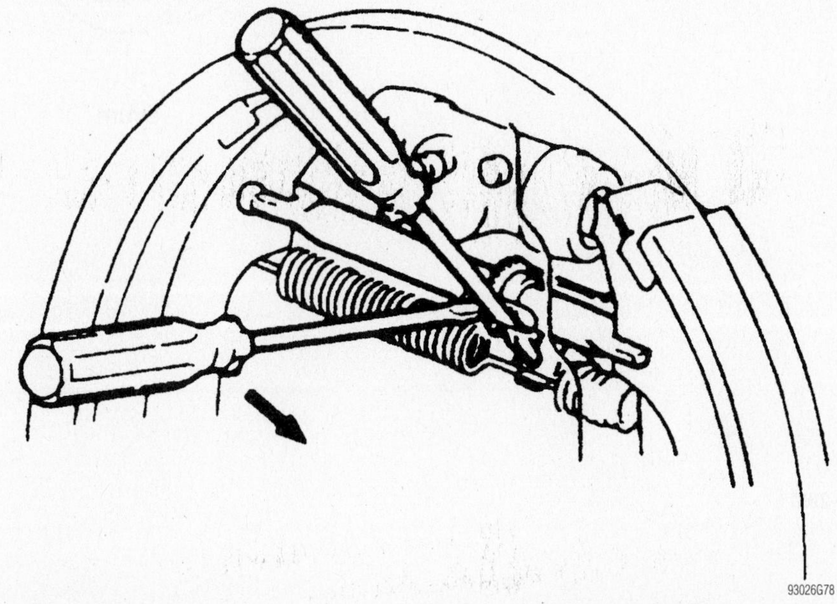

Use a brake adjusting tool (brake spoon) and a prytool to adjust the brake shoes through the adjusting hole

Brake Shoes

REMOVAL & INSTALLATION

1. Loosen the rear wheel lug nuts slightly.
2. Block the front wheels, raise the rear of the vehicle, and safely support it with jackstands.
3. Remove the wheel lug nuts and the wheel.
4. Remove the brake drum.
5. If the drum is difficult to remove, perform the following:
   a. Insert a flat prying tool through the hole in the brake drum and hold the automatic adjusting lever away from the adjuster.
   b. Reduce the brake shoe adjustment by turning the adjuster bolt with a brake tool.
   c. The drum should now be loose enough to remove without much effort.
6. Remove the rear shoe.
   a. Carefully unhook the return spring from the brake shoe.
   b. Remove the shoe hold-down spring, cups and the pin.
   c. Disconnect the anchor spring from the rear shoe and remove the rear shoe.
   d. Disconnect the anchor spring from the front shoe.
7. Remove the front shoe.
   a. Remove the shoe hold-down spring, cups and pin.
   b. Remove the return spring from the front shoe.
   c. Remove the front shoe with the adjuster.
   d. Disconnect the parking brake cable from the front shoe.

To install:

8. Inspect the shoes for signs of unusual wear or scoring.
9. Check the wheel cylinder for any sign of fluid seepage or frozen pistons.
10. Clean and inspect the brake backing plate and all other components. Check that the brake drum inner diameter is within specified limits. Lubricate the backing plate at the positions the brakes come in contact with the backing plate. Also lubricate the anchor plate.
11. Mount the automatic adjuster assembly onto a new rear brake shoe.
12. Install the front shoe.
    a. Install the parking brake cable to the front shoe.
    b. Install the front shoe with the adjuster.
    c. Install the return spring to the front shoe.
    d. Install the shoe hold-down spring, cups and pin.
13. Install the rear shoe.
    a. Install the anchor spring to the front shoe.
    b. Install the anchor spring to the rear shoe and install the rear shoe.

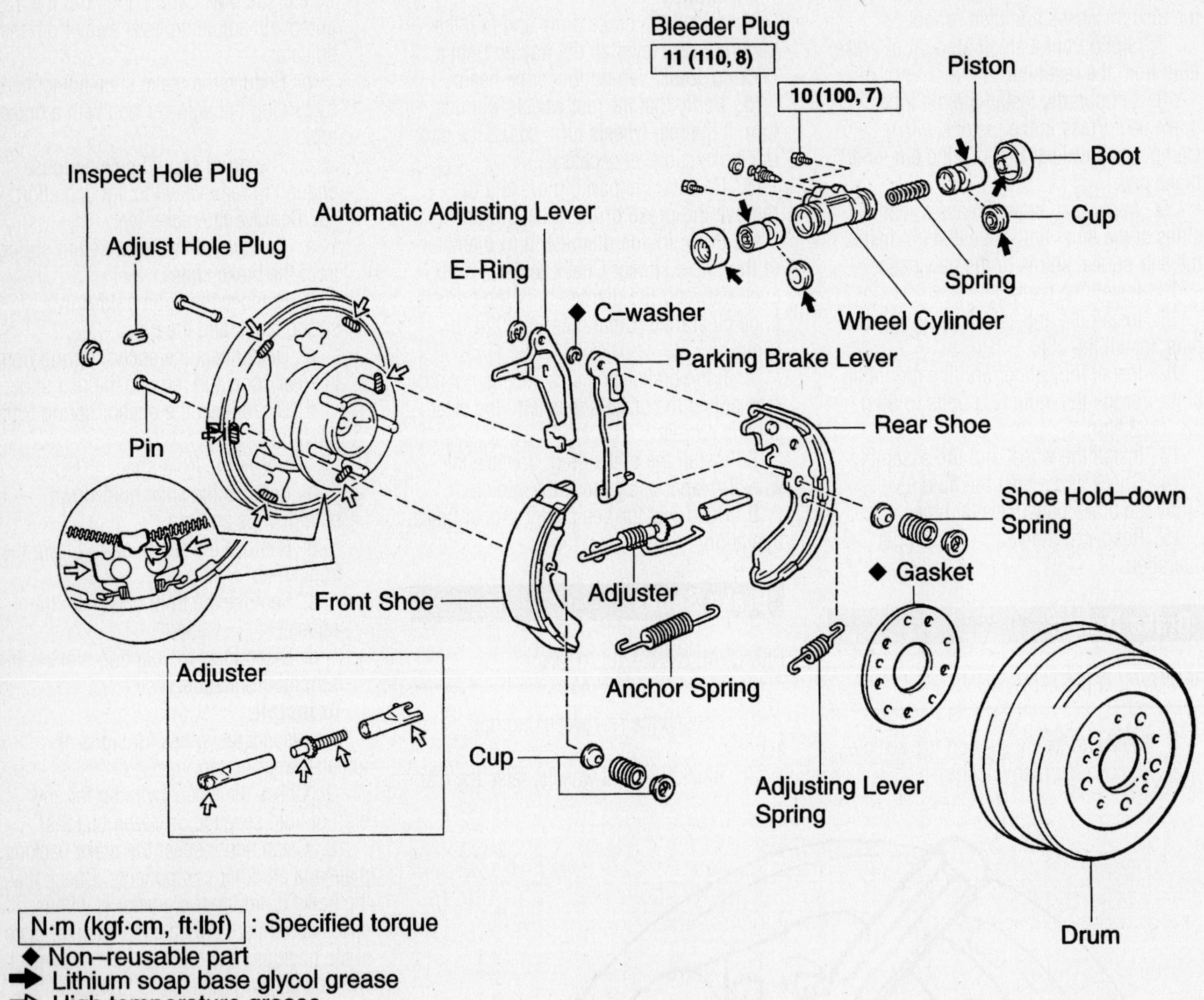

Exploded view of the rear brake drums components—2WD

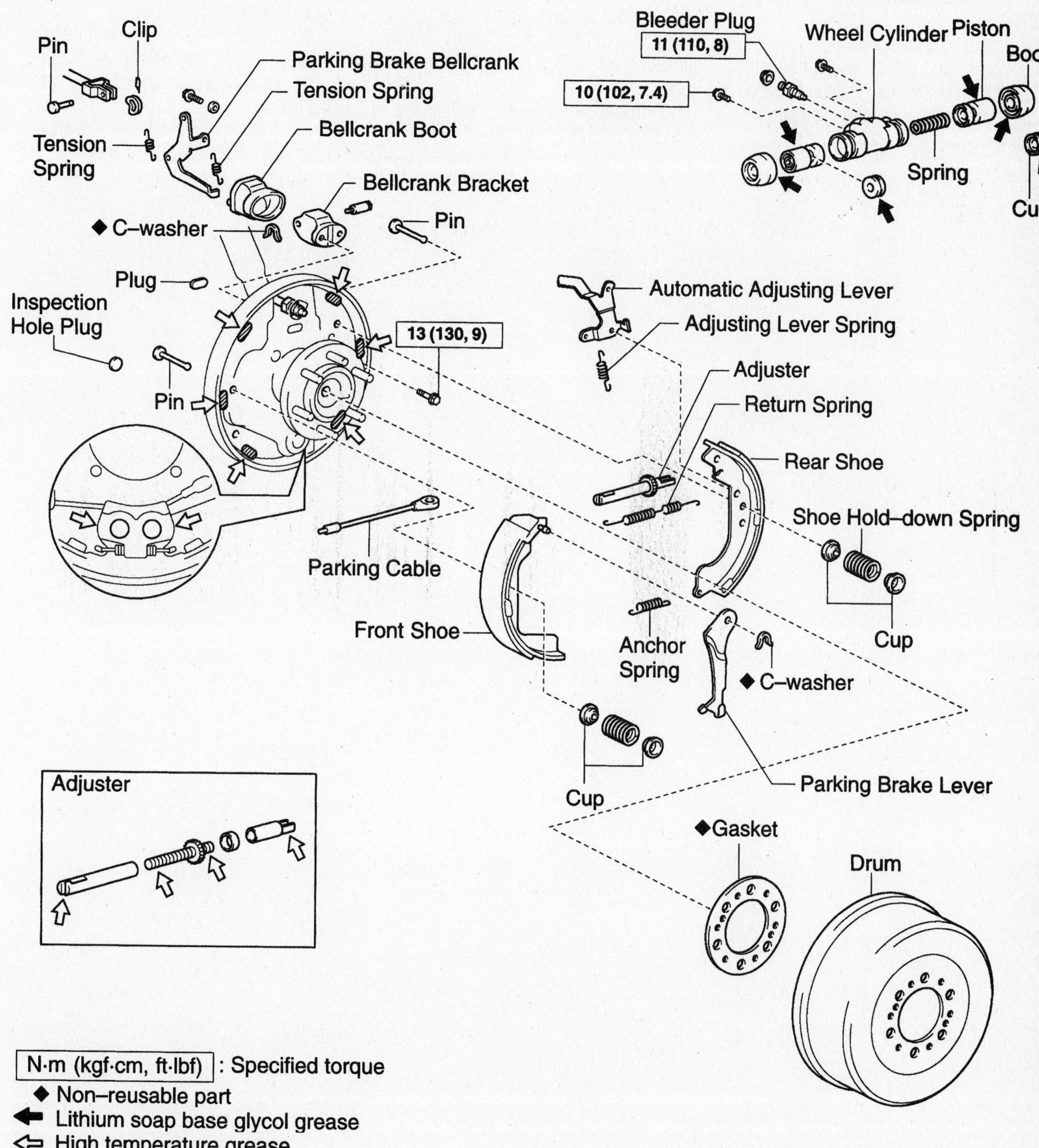
Pin
Clip
Parking Brake Bellcrank
Tension Spring
Bellcrank Boot
Tension Spring
Bellcrank Bracket
◆ C–washer
Pin
Plug
Inspection Hole Plug
Pin
13 (130, 9)
Bleeder Plug
11 (110, 8)
10 (102, 7.4)
Wheel Cylinder
Piston
Boot
Spring
Cup
Automatic Adjusting Lever
Adjusting Lever Spring
Adjuster
Return Spring
Rear Shoe
Shoe Hold–down Spring
Cup
Parking Cable
Front Shoe
Anchor Spring
◆ C–washer
Cup
Parking Brake Lever
Adjuster
◆Gasket
Drum
N·m (kgf·cm, ft·lbf) : Specified torque
◆ Non–reusable part
Lithium soap base glycol grease
High temperature grease
67170-TACO-G48

Exploded view of the rear brake drums components—4WD

c. Install the shoe hold-down spring, cups and the pin.

d. Hook the return spring to the brake shoe.

14. Install the brake drum.

15. Adjust the brake shoes until a slight drag is felt when the drum is spun by hand.

16. Remove the brake drum and check the clearance between brake shoes and brake drum. Adjust the clearance to specification.

17. Pull the parking lever all the way up until a clicking sound can no longer be heard. Verify that the drum doesn't turn. If the drum turns, adjust the parking brake cable.

18. Install the rear wheels, tighten the wheel lug nuts and lower the vehicle.

19. Retighten the wheel lug nuts and pump the brake pedal a few times before moving the vehicle. Adjust the rear brakes again if necessary.

20. Check the level of brake fluid in the master cylinder, then perform a test drive.

21. Connect the negative battery cable to the battery.

TOYOTA

Tundra

24

SPECIFICATION AND MAINTENANCE CHARTS

ENGINE AND VEHICLE IDENTIFICATION

Engine						
Code ①	Liters (cc)	Cu. In.	Cyl.	Fuel Sys.	Engine Type	Eng. Mfg.
2UZ-FE	4.7 (4664)	285	8	SFI	DOHC	Toyota
5VZ-FE	3.4 (3378)	206	6	MFI	DOHC	Toyota

Model Year	
Code ②	Year
1	2001
2	2002
3	2003
4	2004
5	2005

SFI: Sequential Fuel Injection

MFI: Multi-port Fuel Injection

DOHC: Double Overhead Camshaft

① Stamped on the left side of the engine block

② 10th digit of the Vehicle Identification Number (VIN)

67170-TUND-C01

GENERAL ENGINE SPECIFICATIONS

Year	Model	Engine Displacement Liters	Engine Series ID	Net Horsepower @ rpm	Net Torque @ rpm (ft. lbs.)	Bore x Stroke (in.)	Com-pression Ratio	Oil Pressure @ rpm
2001	Tundra	3.4	5VZ-FE	190@4800	220@3600	3.68x3.23	9.6:1	36-75@3000
		4.7	2UZ-FE	245@4800	315@3400	3.70x3.30	9.6:1	45-65@3000
2002	Tundra	3.4	5VZ-FE	190@4800	220@3600	3.68x3.23	9.6:1	36-75@3000
		4.7	2UZ-FE	245@4800	315@3400	3.70x3.30	9.6:1	45-65@3000
2003	Tundra	3.4	5VZ-FE	190@4800	220@3600	3.68x3.23	9.6:1	36-75@3000
		4.7	2UZ-FE	245@4800	315@3400	3.70x3.30	9.6:1	45-65@3000
2004	Tundra	3.4	5VZ-FE	190@4800	220@3600	3.68x3.23	9.6:1	36-75@3000
		4.7	2UZ-FE	245@4800	315@3400	3.70x3.30	9.6:1	45-65@3000

67170-TUND-C02

ENGINE TUNE-UP SPECIFICATIONS

Year	Engine Displacement Liters	Engine ID	Spark Plug Gap (in.)	Ignition Timing (deg.)*	Fuel Pump (psi)	Idle Speed (rpm)		Valve Clearance	
						MT	AT	Intake	Exhaust
2001	3.4	5VZ-FE	0.031	5B	38-44	650-750	650-750	0.006-0.010	0.010-0.014
	4.7	2UZ-FE	0.043	5B	38-44	650-750	650-750	0.006-0.009	0.011-0.014
2002	3.4	5VZ-FE	0.043	8-12	38-44	650-750	650-750	0.006-0.010	0.010-0.014
	4.7	2UZ-FE	0.043	8-12	38-44	650-750	650-750	0.006-0.009	0.011-0.014
2003	3.4	5VZ-FE	0.043	8-12	38-44	650-750	650-750	0.006-0.010	0.010-0.014
	4.7	2UZ-FE	0.043	8-12	38-44	650-750	650-750	0.006-0.009	0.011-0.014
2004	3.4	5VZ-FE	0.043	8-12	38-44	650-750	650-750	0.006-0.010	0.010-0.014
	4.7	2UZ-FE	0.043	8-12	38-44	650-750	650-750	0.006-0.009	0.011-0.014

NOTE: The Vehicle Emission Control Information label often reflects specification changes made during production. The label figures must be used if they differ from those in this chart.

B: Before top dead center

*w/terminals TC and CG connected to DLC3

67170-TUND-C03

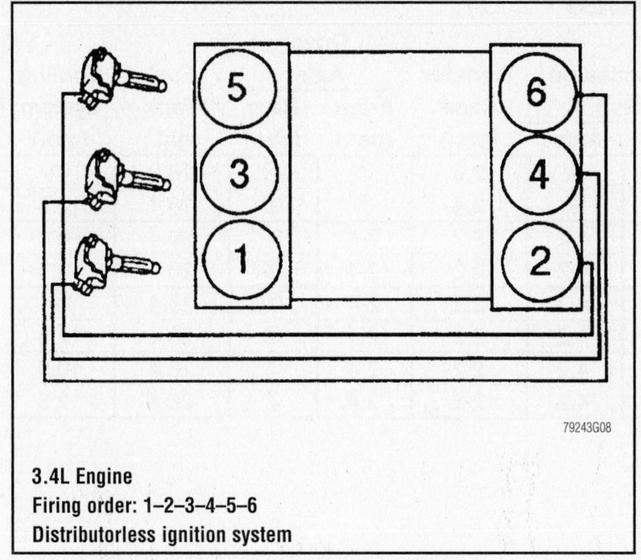

3.4L Engine
Firing order: 1–2–3–4–5–6
Distributorless ignition system

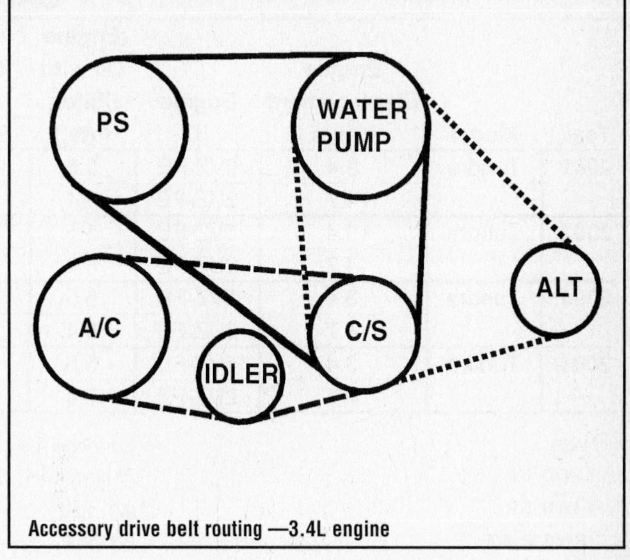

Accessory drive belt routing —3.4L engine

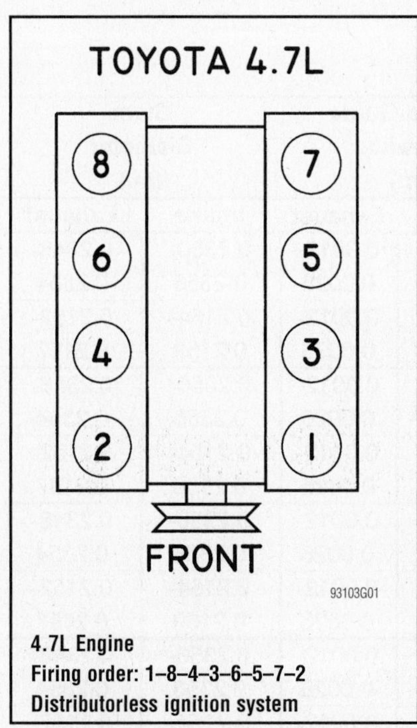

4.7L Engine
Firing order: 1–8–4–3–6–5–7–2
Distributorless ignition system

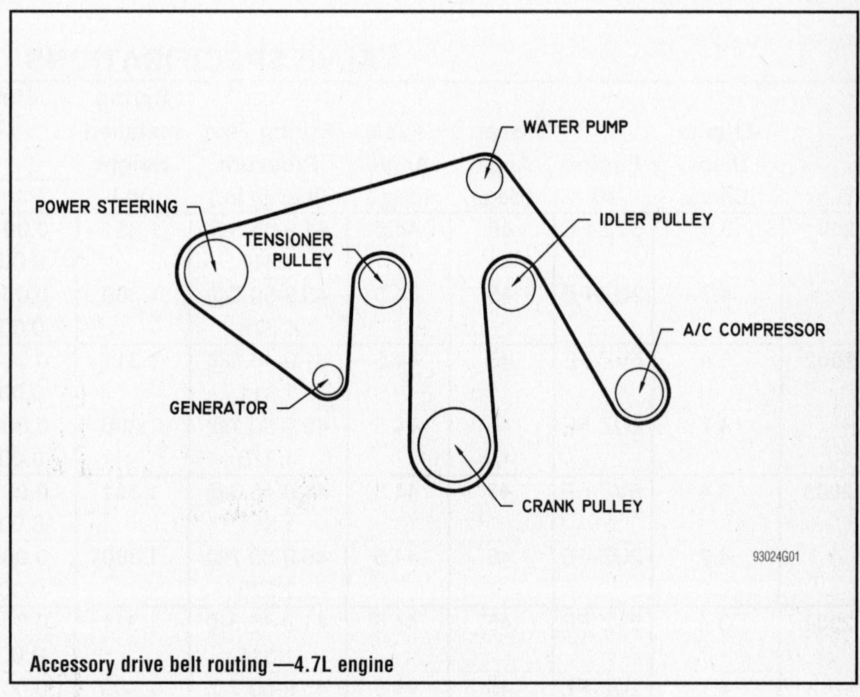

Accessory drive belt routing —4.7L engine

CAPACITIES

Year	Model	Engine Displacement Liters	Engine ID	Engine Oil with Filter (qts.)	Transmission (pts.) 5-Spd	Transmission (pts.) Auto.	Transfer Case (pts.)	Drive Axle Front (pts.)	Drive Axle Rear (pts.)	Fuel Tank (gal.)	Cooling System (qts.)
2001	Tundra	3.4	5VZ-FE	5.5	①	②	2.3	③	④	18.0	⑤
		4.7	2UZ-FE	6.4	①	②	2.4	③	④	18.1	12.3
2002	Tundra	3.4	5VZ-FE	5.5	⑥	4.2	2.2	2.4	⑦	26.4	⑤
		4.7	2UZ-FE	6.4	⑥	4.2	2.2	2.4	⑦	26.4	12.3
2003	Tundra	3.4	5VZ-FE	5.5	⑥	4.2	2.2	2.4	⑦	26.4	⑤
		4.7	2UZ-FE	6.4	⑥	4.2	2.2	2.4	⑦	26.4	12.3
2004	Tundra	3.4	5VZ-FE	5.5	⑥	4.2	2.2	2.4	⑦	26.4	10.5
		4.7	2UZ-FE	6.4	⑥	4.2	2.2	2.4	⑦	26.4	12.3

① W59:
2WD: 5.4
4WD: 5.2
R150, R150F:
2WD: 5.4
4WD: 4.6

② A43D: 5.0
A340E: 3.4
A340F: 4.2

③ Without ADD: 2.32
With ADD: 2.44

④ Extra long: 4.4
All others: 5.4

⑤ M/T: 10.3
A/T: 10.0

⑥ 2wd: 5.4
4wd: 4.6

⑦ Standard 2wd: 8.04
Standard 4wd: 7.40
LSD type 2wd: 6.66
LSD type 4wd: 6.24

67170-TUND-C04

VALVE SPECIFICATIONS

Year	Engine Displ. Liters	Engine ID	Seat Angle (deg.)	Face Angle (deg.)	Spring Test Pressure (lbs. @ in.)	Spring Installed Height (in.)	Stem-to-Guide Clearance (in.) Intake	Stem-to-Guide Clearance (in.) Exhaust	Stem Diameter (in.) Intake	Stem Diameter (in.) Exhaust
2001	3.4	5VZ-FE	45	44.5	41.9-46.3@ 1.311	1.311	0.0010-0.0024	0.0012-0.0026	0.2350-0.2356	0.2348-0.2354
	4.7	2UZ-FE	45	44.5	45.9-50.7@ 1.378	1.380	0.0010-0.0024	0.0012-0.0026	0.2154-0.2159	0.2152-0.2157
2002	3.4	5VZ-FE	45	44.5	41.9-46.3@ 1.311	1.311	0.0010-0.0024	0.0012-0.0026	0.2350-0.2356	0.2348-0.2354
	4.7	2UZ-FE	45	44.5	45.9-50.7@ 1.378	1.380	0.0010-0.0024	0.0012-0.0026	0.2154-0.2159	0.2152-0.2157
2003	3.4	5VZ-FE	45	44.5	41.9-46.3@ 1.311	1.311	0.0010-0.0024	0.0012-0.0026	0.2350-0.2356	0.2348-0.2354
	4.7	2UZ-FE	45	44.5	45.9-50.7@ 1.378	1.380	0.0010-0.0024	0.0012-0.0026	0.2154-0.2159	0.2152-0.2157
2004	3.4	5VZ-FE	45	44.5	41.9-46.3@ 1.311	1.311	0.0010-0.0024	0.0012-0.0026	0.2350-0.2356	0.2348-0.2354
	4.7	2UZ-FE	45	44.5	45.9-50.7@ 1.378	1.380	0.0010-0.0024	0.0012-0.0026	0.2154-0.2159	0.2152-0.2157

67170-TUND-C05

CRANKSHAFT AND CONNECTING ROD SPECIFICATIONS

All measurements are given in inches.

Year	Engine Displ. Liters	Engine ID	Crankshaft Main Brg. Journal Dia.	Crankshaft Main Brg. Oil Clearance	Crankshaft Shaft End-play	Crankshaft Thrust on No.	Connecting Rod Journal Diameter	Connecting Rod Oil Clearance	Connecting Rod Side Clearance
2001	3.4	5VZ-FE	2.5191-2.5197	0.0008-0.0015	0.0008-0.0087	2	2.1648-2.1654	0.0009-0.0021	0.0059-0.0130
	4.7	2UZ-FE	2.6373-2.6378	0.0016-0.0023	0.0008-0.0087	3	2.0465-2.0472	0.0011-0.0021	0.0063-0.0138
2002	3.4	5VZ-FE	2.5191-2.5197	0.0008-0.0015	0.0008-0.0087	2	2.1648-2.1654	0.0009-0.0021	0.0059-0.0130
	4.7	2UZ-FE	2.6373-2.6378	0.0016-0.0023	0.0008-0.0087	3	2.0465-2.0472	0.0011-0.0021	0.0063-0.0138
2003	3.4	5VZ-FE	2.5191-2.5197	0.0008-0.0015	0.0008-0.0087	2	2.1648-2.1654	0.0009-0.0021	0.0059-0.0130
	4.7	2UZ-FE	2.6373-2.6378	0.0016-0.0023	0.0008-0.0087	3	2.0465-2.0472	0.0011-0.0021	0.0063-0.0138
2004	3.4	5VZ-FE	2.5191-2.5197	0.0008-0.0015	0.0008-0.0087	2	2.1648-2.1654	0.0009-0.0021	0.0059-0.0130
	4.7	2UZ-FE	2.6373-2.6378	0.0016-0.0023	0.0008-0.0087	3	2.0465-2.0472	0.0011-0.0021	0.0063-0.0138

67170-TUND-C06

PISTON AND RING SPECIFICATIONS

All measurements are given in inches.

Year	Engine Displ. Liters	Engine ID	Piston Clearance	Ring Gap Top Compression	Ring Gap Bottom Compression	Ring Gap Oil Control	Ring Side Clearance Top Compression	Ring Side Clearance Bottom Compression	Ring Side Clearance Oil Control
2001	3.4	5VZ-FE	0.0053-0.0060	0.0118-0.0197	0.0157-0.0236	0.0059-0.0217	0.0016-0.0031	0.0012-0.0028	SNUG
	4.7	2UZ-FE	0.0035-0.0044	0.0118-0.0197	0.0157-0.0256	0.0051-0.0189	0.0012-0.0031	0.0012-0.0028	SNUG
2002	3.4	5VZ-FE	0.0053-0.0060	0.0118-0.0197	0.0157-0.0236	0.0059-0.0217	0.0016-0.0031	0.0012-0.0028	SNUG
	4.7	2UZ-FE	0.0035-0.0044	0.0118-0.0197	0.0157-0.0256	0.0051-0.0189	0.0012-0.0031	0.0012-0.0028	SNUG
2003	3.4	5VZ-FE	0.0053-0.0060	0.0118-0.0197	0.0157-0.0236	0.0059-0.0217	0.0016-0.0031	0.0012-0.0028	SNUG
	4.7	2UZ-FE	0.0035-0.0044	0.0118-0.0197	0.0157-0.0256	0.0051-0.0189	0.0012-0.0031	0.0012-0.0028	SNUG
2004	3.4	5VZ-FE	0.0053-0.0060	0.0118-0.0197	0.0157-0.0236	0.0059-0.0217	0.0016-0.0031	0.0012-0.0028	SNUG
	4.7	2UZ-FE	0.0035-0.0044	0.0118-0.0197	0.0157-0.0256	0.0051-0.0189	0.0012-0.0031	0.0012-0.0028	SNUG

67170-TUND-C07

TORQUE SPECIFICATIONS

All readings in ft. lbs.

Year	Engine Displacement Liters	Engine ID	Cylinder Head Bolts	Main Bearing Bolts	Rod Bearing Bolts	Crankshaft Damper Bolts	Flywheel Bolts	Manifold Intake	Manifold Exhaust	Spark Plugs	Oil Pan Drain Plug
2001	3.4	5VZ-FE	①	②	③	184	63-67	13	30	13	28
	4.7	2UZ-FE	④	⑤	③	181	⑥	13	33	13	29
2002	3.4	5VZ-FE	①	②	③	217	63	13	30	13	28
	4.7	2UZ-FE	④	⑤	③	181	⑥	13	33	13	29
2003	3.4	5VZ-FE	①	②	③	217	63	13	30	13	28
	4.7	2UZ-FE	④	⑤	③	181	⑥	13	33	13	29
2004	3.4	5VZ-FE	①	②	③	217	63	13	30	13	28
	4.7	2UZ-FE	④	⑤	③	181	⑥	13	33	13	29

① Step 1: 25 ft. lbs.
Step 2: Plus 90 degrees
Step 3: Plus 90 degrees
Recessed head: 13 ft. lbs.

② Step 1: 45 ft. lbs.
Step 2: Plus 90 degrees

③ Step 1: 18 ft. lbs.
Step 2: Plus 90 degrees

④ Step 1: 24 ft. lbs.
Step 2: Plus 180 degrees

⑤ Step 1: 20 ft. lbs.
Step 2: Plus 90 degrees

⑥ Step 1: 35 ft. lbs.
Step 2: Plus 90 degrees

67170-TUND-C08

BRAKE SPECIFICATIONS

All measurements in inches unless noted

Year	Original Thickness	Brake Disc Minimum Thickness	Maximum Runout	Brake Drum Original Inside Diameter	Brake Drum Maximum Machine Diameter	Minimum Lining Thickness Front	Minimum Lining Thickness Rear	Caliper Mounting Bolts (ft. lbs.)
2001	1.102	1.024	0.0028	11.61	11.69	0.039	0.039	90
2002	1.102	1.024	0.0028	11.61	11.69	0.039	0.039	90
2003	1.102	1.024	0.0028	11.61	11.69	0.039	0.039	90
2004	1.102	1.024	0.0028	11.61	11.69	0.039	0.039	90

67170-TUND-C09

WHEEL ALIGNMENT

		Caster		Camber			
Year	Model	Range (+/-Deg.)	Preferred Setting (Deg.)	Range (+/-Deg.)	Preferred Setting (Deg.)	Toe-in (in.)	Steering Axis Inclination (Deg.)
2001	2WD	0.75	2.05	0.75	-0.12	0.06+/-0.08	10.87
	4WD	0.75	1.07	0.75	0.33	0.07+/-0.08	10.41
2002	2WD	0.75	2.05	0.75	-0.12	0.06+/-0.08	10.87
	4WD	0.75	1.07	0.75	0.33	0.07+/-0.08	10.41
2003	2WD	0.75	2.05	0.75	-0.12	0.06+/-0.08	10.87
	4WD	0.75	1.07	0.75	0.33	0.07+/-0.08	10.41
2004	2WD	0.75	2.05	0.75	-0.12	0.06+/-0.08	10.87
	4WD	0.75	1.07	0.75	0.33	0.07+/-0.08	10.41

All alignment figures based on nominal ride height and standard tires

67170-TUND-C10

TIRE, WHEEL AND BALL JOINT SPECIFICATIONS

		OEM Tires		Tire Pressures (psi)				
Year	Model	Standard	Optional	Front	Rear	Wheel Size	Ball Joint Inspection	Lugnut Torque (ft. lbs.)
2001	Tundra	P245/70R16	P265/70R16	26	Std:35/Opt:29	NA	NS	83
2002	Tundra	P245/70R16	P265/70R16	26	Std:35/Opt:29	NA	NS	83
2003	Tundra	P245/70R16	P265/70R16	26	Std:35/Opt:29	NA	NS	83
2004	Tundra	P245/70R16	P265/70R16 P265/65R17	①	①	NA	②	83

OEM: Original Equipment Manufacturer

PSI: Pounds Per Square Inch

STD: Standard

OPT: Optional

NS: Not specified by manufacturer

NA: Not available

① See placard on vehicle

② Upper: turning torque within 6-39 inch lbs.

Lower: turning torque within 1-22 inch lbs.

67170-TUND-C11

SCHEDULED MAINTENANCE INTERVALS

TOYOTA—TUNDRA

		VEHICLE MILEAGE INTERVAL (x1000)																		
TO BE SERVICED	TYPE OF SERVICE	5	10	15	20	25	30	35	40	45	50	55	60	65	70	75	80	85	90	95
Automatic transmission and differential fluid	S/I			✓			✓			✓			✓			✓			✓	
Ball joints and boots	S/I			✓			✓			✓			✓			✓			✓	
Brake system	S/I			✓			✓			✓			✓			✓			✓	
Charcoal canister	S/I												✓							
Drive belts	S/I						✓						✓						✓	
Driveshaft bushing	L						✓						✓						✓	
Engine coolant	R						✓						✓						✓	
Engine oil & filter	R	✓	✓	✓	✓	✓	✓	✓	✓	✓	✓	✓	✓	✓	✓	✓	✓	✓	✓	✓
Exhaust system	S/I			✓			✓			✓			✓			✓			✓	
Fuel lines	S/I						✓						✓						✓	
Fuel tank cap gasket	S/I						✓						✓						✓	
Halfshaft boots & flange bolts	S/I			✓			✓			✓			✓			✓			✓	
Limited slip differential fluid	R						✓						✓						✓	
Manual transmission and differential fluid	S/I						✓						✓						✓	
Non-platinum spark plugs	R						✓						✓						✓	
Platinum spark plugs	R												✓							
Propeller shaft (4WD)	L			✓			✓			✓			✓			✓			✓	
Propeller shaft bolts	S/I			✓			✓			✓			✓			✓			✓	
Steering gear	S/I			✓			✓			✓			✓			✓			✓	
Steering linkage	S/I			✓			✓			✓			✓			✓			✓	
Valves	S/I												✓							

R: Replace S/I: Service or Inspect L: Lubricate

FREQUENT OPERATION MAINTENANCE (SEVERE SERVICE)

If a vehicle is operated under any of the following conditions it is considered severe service:

- Towing a trailer or using a camper or car-top carrier.

- Repeated short trips of less than 5 miles in temperatures below freezing.

- Excessive idling or low-speed driving for long distances as in heavy commercial use, such as delivery, taxi or police cars.

- Operating on rough, muddy or salt-covered roads.

- Operating on unpaved or dusty roads.

Oil filter: service or inspect every 5000 miles or 4 months, whichever occurs first.

Brake linings and discs or drums: service or inspect every 5000 miles or 4 months, whichever occurs first.

Steering linkage: service or inspect every 5000 miles or 4 months, whichever occurs first.

Ball joints and boots: service or inspect every 5000 miles or 4 months, whichever occurs first.

Brake discs & pads (front): service or inspect every 6000 miles.

Halfshaft boots: service or inspect every 5000 miles or 4 months. Retighten the flange bolts, whichever occurs first.

Body chassis bolts and nuts: service or inspect every 5000 miles or 4 months, whichever occurs first.

Transmission and differential fluid: replace every 15,000 miles or 12 months, whichever occurs first.

Transfer case and differential fluid: replace every 15,000 miles or 12 months, whichever occurs first.

Timing belt: replace every 60,000 miles or 48 months, whichever occurs first.

67170-TUND-C12

PRECAUTIONS

Before servicing any vehicle, please be sure to read all of the following precautions, which deal with personal safety, prevention of component damage, and important points to take into consideration when servicing a motor vehicle:

- Never open, service or drain the radiator or cooling system when the engine is hot; serious burns can occur from the steam and hot coolant.
- Observe all applicable safety precautions when working around fuel. Whenever servicing the fuel system, always work in a well-ventilated area. Do not allow fuel spray or vapors to come in contact with a spark, open flame or excessive heat (a hot drop light, for example). Keep a dry chemical fire extinguisher near the work area. Always keep fuel in a container specifically designed for fuel storage; also, always properly seal fuel containers to avoid the possibility of fire or explosion. Refer to the additional fuel system precautions later in this section.
- Fuel injection systems often remain pressurized, even after the engine has been turned **OFF**. The fuel system pressure must be relieved before disconnecting any fuel lines. Failure to do so may result in fire and/or personal injury.
- Brake fluid often contains polyglycol ethers and polyglycols. Avoid contact with the eyes and wash your hands thoroughly after handling brake fluid. If you do get brake fluid in your eyes, flush your eyes with clean, running water for 15 minutes. If eye irritation persists, or if you have taken brake fluid internally, IMMEDIATELY seek medical assistance.
- The EPA warns that prolonged contact with used engine oil may cause a number of skin disorders, including cancer! You should make every effort to minimize your exposure to used engine oil. Protective gloves should be worn when changing oil. Wash your hands and any other exposed skin areas as soon as possible after exposure to used engine oil. Soap and water, or waterless hand cleaner should be used.
- All new vehicles are now equipped with an air bag system, often referred to as a Supplemental Restraint System (SRS) or Supplemental Inflatable Restraint (SIR) system. The system must be disabled before performing service on or around system components, steering column, instrument panel components, wiring and sensors. Failure to follow safety and disabling procedures could result in accidental air bag deployment, possible personal injury and unnecessary system repairs.
- Always wear safety goggles when working with, or around, the air bag system. When carrying a non-deployed air bag, be sure the bag and trim cover are pointed away from your body. When placing a non-deployed air bag on a work surface, always face the bag and trim cover upward, away from the surface. This will reduce the motion of the module if it is accidentally deployed. Refer to the additional air bag system precautions later in this section.
- Clean, high quality brake fluid from a sealed container is essential to the safe and proper operation of the brake system. You should always buy the correct type of brake fluid for your vehicle. If the brake fluid becomes contaminated, completely flush the system with new fluid. Never reuse any brake fluid. Any brake fluid that is removed from the system should be discarded. Also, do not allow any brake fluid to come in contact with a painted surface; it will damage the paint.
- Never operate the engine without the proper amount and type of engine oil; doing so WILL result in severe engine damage.
- Timing belt maintenance is extremely important! Many models utilize an interference type, non-freewheeling engine. If the timing belt breaks, the valves in the cylinder head may strike the pistons, causing potentially serious (also time consuming and expensive) engine damage. Refer to the maintenance interval charts for the recommended replacement interval for the timing belt, and to the timing belt section for belt replacement and inspection.
- Disconnecting the negative battery cable on some vehicles may interfere with the functions of the on-board computer system(s) and may require the computer to undergo a relearning process once the negative battery cable is reconnected.
- When servicing drum brakes, only disassemble and assemble one side at a time, leaving the remaining side intact for reference.

ENGINE REPAIR

➡Disconnecting the negative battery cable on some vehicles may interfere with the functions of the on board computer system. The computer may undergo a relearning process once the negative battery cable is reconnected.

Alternator

REMOVAL

3.4L Engine

1. Remove or disconnect the following:
   - Negative battery cable
   - Alternator wiring
   - Alternator locknut, pivot bolt, nut and adjusting bolt
   - Drive belt
   - Alternator

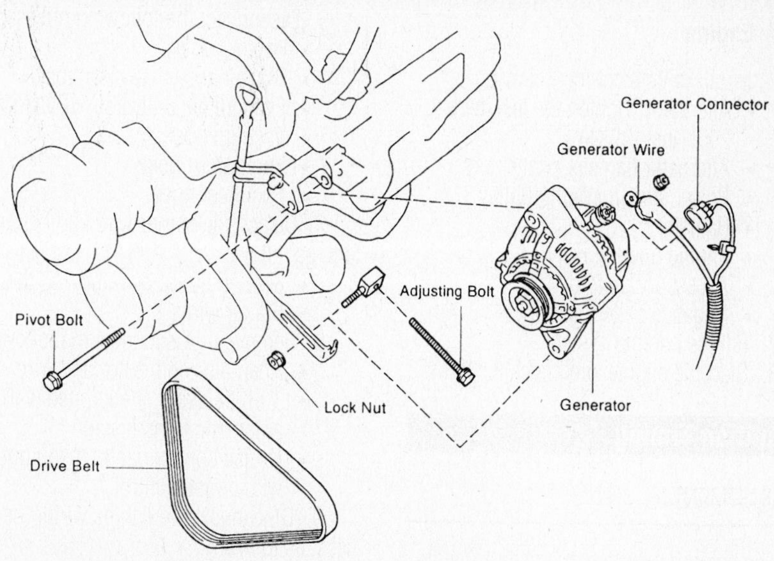

Exploded view of the alternator and drive belt—3.4L Engine

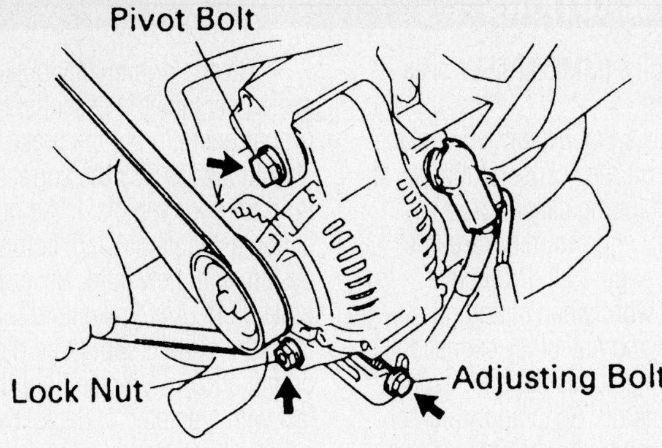

Locations of the adjusting and pivot bolts and the locknut—3.4L Engine

4.7L Engine

1. Before servicing the vehicle, refer to the precautions in the beginning of this section.
2. Drain the cooling system.
3. Remove or disconnect the following:
   - Negative battery cable
   - Accessory drive belt
   - Engine under cover
   - Radiator
   - Power steering pump pulley
   - Alternator harness connectors
   - Alternator

INSTALLATION

3.4L Engine

Install or connect the following:
- Alternator
- Drive belt. Tighten the locknut 25 ft. lbs. (33 Nm) and the pivot bolt 38 ft. lbs. (51 Nm).
- Alternator wiring
- Negative battery cable

4.7L Engine

1. Install or connect the following:
   - Alternator. Tighten the fasteners to 29 ft. lbs. (39 Nm).
   - Alternator harness connectors
   - Power steering pump pulley
   - Radiator
   - Engine under cover
   - Accessory drive belt
   - Negative battery cable
2. Fill the cooling system.
3. Start the engine and check for leaks.

Ignition Timing

ADJUSTMENT

The engines are equipped with a Distributorless Ignition System (DIS). No timing adjustment is possible.

Engine Assembly

REMOVAL & INSTALLATION

3.4L Engine

2-WHEEL DRIVE

1. Before servicing the vehicle, refer to the precautions in the beginning of this section.
2. Properly relieve the fuel system pressure.
3. Remove or disconnect the following:
   - Hood
   - Battery
   - Engine under covers
4. Drain the engine coolant.
5. Drain the engine oil.
   - Radiator
   - Fan with the fluid coupling and fan pulleys
   - Air cleaner cap
   - Air cleaner case and filter
6. Disconnect the following hoses:
   - Heater hoses
   - Brake booster vacuum hose
   - Evaporative Emissions (EVAP) hose
   - Vacuum hose
   - Fuel return hose
   - Fuel inlet hose
7. Detach the starter wire and connectors, as follows:
   - Ground strap, by removing the bolt
   - Starter wires
8. Remove or disconnect the following:
   - Alternator connector and wire
   - Throttle cable, if equipped with an automatic transmission
   - Cruise control cable, if equipped with cruise control
9. Disconnect the engine wiring harness, as follows:
   - Glove box door
   - Lower the finish No. 2 panel
   - Heater to register duct
   - 3 Engine Control Module (ECM) connectors
   - 2 cassette connectors and the 2 wire clamps from the lower finish panel
   - Engine wiring harness clamp
10. Remove or disconnect the following:
    - Igniter connector
    - Ground strap
    - 2 engine wiring harness retainer-to-cowl panel nuts and pull out the engine wiring harness
11. If equipped with a manual transmission, remove or disconnect the following:
    - Shift lever knob
    - 4 shift lever boot screws
    - 6 shift lever assembly bolts, the assembly and gasket
12. Remove or disconnect the following:
    - Driveshaft from the transmission
    - Speedometer cable

➡Do not lose the felt protector and washers.

   - Front exhaust pipe
   - Clutch release cylinder, if equipped with a manual transmission
   - Nut and the control cable
13. Place a jack under the transmission.
14. Remove or disconnect the following:
    - Transmission rear mounting bracket by removing the 8 bolts
    - Bolt and the air conditioning compressor wire clamp, if equipped with air conditioning
15. If necessary, install a No. 2 engine hanger with 2 bolts. Tighten the 2 bolts to 30 ft. lbs. (40 Nm).
16. Attach the engine hoist chain to the 2 engine hangers.
17. Remove or disconnect the following:
    - 4 engine front mounting insulators-to-frame bolts and nuts
    - Engine from the transmission

To install:

18. Install or connect the following:
    - Engine to the transmission
    - Engine mounts to the body mountings. Install the bolts and nuts but do not tighten at this time.
19. Remove the engine chain hoist the No. 2 engine hanger.
20. Install or connect the following:
    - Air conditioning wire with the bolt, if equipped with air conditioning
    - Transmission mounting bracket. Tighten the frame bolts to 43 ft. lbs. (58 Nm) and the mounting insulator bolts to 13 ft. lbs. (18 Nm).
    - Tighten the engine mounting nuts and bolts to 28 ft. lbs. (38 Nm).

- Control cable
- Clutch release cylinder, if equipped with a manual transmission. Torque the bolts to 9 ft. lbs. (12 Nm).

21. Install or connect the following:
    - Front exhaust pipe
    - Speedometer cable
    - Driveshaft
22. If equipped with a manual transmission, install or connect the following:
    - 6 shift lever assembly bolts, the assembly and gasket
    - 4 shift lever boot screws
    - Shift lever knob
23. Install or connect the following:
    - All engine wiring harness, hoses and cables
    - Fan with the fluid coupling and fan pulleys. Tighten the nuts to 48 inch lbs. (5.4 Nm).
    - Air cleaner case and air filter
    - Radiator
24. Install or connect the following hoses:
    - Fuel inlet hose
    - Fuel return hose
    - Vacuum hose
    - Evaporative Emissions (EVAP) hose
    - Brake booster vacuum hose
    - Heater hoses
25. Fill the engine with oil.
26. Fill the engine and radiator with coolant.
27. Install or connect the following:
    - Engine undercover
    - Battery
    - Hood
28. Start the engine and check for leaks.
29. Make any necessary adjustments and road test the vehicle.

4-WHEEL DRIVE

1. Before servicing the vehicle, refer to the precautions in the beginning of this section.
2. Remove or disconnect the following:
    - Transmission
    - Hood
3. Release the fuel system pressure.
4. Remove or disconnect the following:
    - Battery
    - Engine undercovers
5. Drain the engine coolant.
6. Drain the engine oil.
7. Remove or disconnect the following:
    - Radiator
    - Fan with the fluid coupling and fan pulleys
    - Air cleaner cap
    - Mass Air Flow (MAF) meter and the resonator
    - Cruise control cable, if equipped with cruise control
    - Throttle cable, if equipped with an automatic transmission
8. Disconnect the following hoses:
    - Heater hoses
    - Brake booster vacuum hose
    - Evaporative Emissions (EVAP) hose
    - Automatic Disconnecting Differential (ADD) vacuum hose
    - Vacuum hose
    - Fuel return hose
    - Fuel inlet hose
9. Detach the starter wire and connectors, as follows:
    - Ground strap, by removing the bolt
    - 3 starter wire clamps and connector
10. Detach the alternator connector and wire.
11. Disconnect the engine wiring harness, as follows:
    - Glove box door
    - Lower the finish No. 2 panel
    - 3 Engine Control Module (ECM) connectors
    - 2 cassette connectors and the 2 wire clamps from the lower finish panel
    - Igniter connector
    - Ground strap
    - Engine wiring harness clamp
12. Remove or disconnect the following:
    - 2 engine wiring harness retainer-to-cowl panel nuts and wiring harness
    - Air conditioning compressor wire clamp and compressor bracket, if equipped with air conditioning
13. If necessary, install a No. 2 engine hanger with 2 bolts. Tighten the 2 bolts to 30 ft. lbs. (40 Nm).
14. Attach the engine hoist chain to the 2 engine hangers.
15. Remove or disconnect the following:
    - 4 engine front mounting insulators-to-frame bolts and nuts
    - Engine

To install:

16. Install or connect the following:
    - Engine
    - Engine mounts-to-body mountings. Install the bolts and nuts but do not tighten at this time.
17. Remove the engine chain hoist the No. 2 engine hanger.
18. Install or connect the following:
    - Air conditioning wire with the bolt and the compressor bracket, if equipped with air conditioning
    - Tighten the engine mounting nuts and bolts to 28 ft. lbs. (38 Nm).
19. Install the engine wiring harness, as follows:
    - Engine wiring harness clamp
    - Ground strap
    - Igniter connector
    - 2 cassette connectors and the 2 wire clamps from the lower finish panel
    - 3 Engine Control Module (ECM) connectors
    - Lower the finish No. 2 panel
    - Glove box door
20. Install or connect the following:
    - Cruise control cable, if equipped with cruise control
    - Throttle cable, if equipped with an automatic transmission
21. Connect the following hoses:
    - Fuel inlet hose
    - Fuel return hose
    - Vacuum hose
    - Automatic Disconnecting Differential (ADD) vacuum hose
    - Evaporative Emissions (EVAP) hose
    - Brake booster vacuum hose
    - Heater hoses
22. Install or connect the following:
    - All wires, hoses and cables
    - Fan with the fluid coupling and fan pulleys. Tighten the nuts to 48 inch lbs. (5.4 Nm).
    - Air cleaner case and air filter
    - MAF meter, resonator and the air cleaner cap
    - Radiator
23. Fill the engine with oil.
24. Fill the engine and radiator with coolant.
25. Install or connect the following:
    - Transmission and refill it with transmission oil
    - Engine undercover
    - Battery
    - Hood
26. Start the engine, make any necessary adjustments and check for leaks.

4.7L Engine

1. Before servicing the vehicle, refer to the precautions in the beginning of this section.
2. Relieve the fuel system pressure.
3. Drain the cooling system.
4. Drain the engine oil.
5. Remove or disconnect the following:
    - Battery and tray
    - Hood
    - Engine appearance cover
    - Air intake pipe
    - Engine under covers
    - Coolant recovery tank

- Radiator hoses
- Radiator and fan shroud
- Accessory drive belt
- Cooling fan and pulley
- Powertrain Control Module (PCM) harness connectors and pass the wiring harness through the firewall
- Accelerator cable
- Power steering vacuum hoses
- Alternator harness connectors
- Heater hoses
- Engine control wiring harness and grommet at the firewall
- Ground cable connector
- Fuel lines
- Evaporative Emissions (EVAP) canister hoses
- Wire clamp at right inner fender
- Negative battery cable at the relay box and right inner fender
- Positive battery cable
- Center console
- Transmission shift lever assembly
- Transfer case shift lever and rod
- Exhaust front pipes
- Stabilizer bar
- Front and rear driveshafts
- A/C compressor
- Power steering pump

6. Attach a hoist to the engine lifting eyes.
7. Remove or disconnect the following:
   - Transfer case skid plate
   - Left and right motor mounts
   - Transmission mount crossmember
8. Attach a hoist to the engine lifting eyes and raise the powertrain out of the vehicle.

To install:

9. Lower the powertrain into the vehicle.
10. Install or connect the following:
    - Transmission mount crossmember. Tighten the bolts to 37 ft. lbs. (50 Nm) and the nuts to 55 ft. lbs. (74 Nm).
    - Transfer case skid plate
    - Left and right motor mounts. Tighten the fasteners to 22 ft. lbs. (30 Nm).
    - Power steering pump. Tighten the bolts to 13 ft. lbs. (17 Nm).
    - A/C compressor. Tighten the bolts to 36 ft. lbs. (49 Nm).
    - Front driveshaft. Tighten the fasteners to 59 ft. lbs. (80 Nm).
    - Rear driveshaft. Tighten the fasteners to 78 ft. lbs. (106 Nm).
    - Stabilizer bar. Tighten the bracket bolts to 13 ft. lbs. (18 Nm) and the link nuts to 18 ft. lbs. (25 Nm).
    - Exhaust front pipes
    - Transfer case shift lever and rod
    - Transmission shift lever assembly
    - Center console
    - Positive battery cable
    - Negative battery cable at the relay box and right inner fender
    - Wire clamp at right inner fender
    - EVAP canister hoses
    - Fuel lines
    - Ground cable connector
    - Engine control wiring harness and grommet at the firewall
    - Heater hoses
    - Alternator harness connectors
    - Power steering vacuum hoses
    - Accelerator cable
    - PCM harness connectors
    - Cooling fan and pulley
    - Accessory drive belt
    - Radiator and fan shroud
    - Radiator hoses
    - Coolant recovery tank
    - Engine under covers
    - Air intake pipe
    - Engine appearance cover
    - Hood
    - Battery and tray
11. Fill the crankcase to the correct level.
12. Fill the cooling system.
13. Start the engine and check for leaks.

Water Pump

REMOVAL & INSTALLATION

3.4L Engine

1. Before servicing the vehicle, refer to the precautions in the beginning of this section.
2. Remove or disconnect the following:
   - Negative battery cable
   - Engine undercover
3. Drain the engine coolant.
4. Remove the upper radiator hose.
5. Remove the power steering drive belt, as follows:
   - Stretch the belt and loosen the fan pulley mounting nuts
   - Loosen the lockbolt, pivot bolt and the adjusting bolt
   - Drive belt
6. Remove or disconnect the following:
   - Air conditioning drive belt, by loosening the idler pulley nut and adjusting bolt
   - Lockbolt, pivot bolt and the adjusting bolt
   - Alternator drive belt
   - No. 2 fan shroud, by removing the 2 clips
   - Fan with the fluid coupling and fan pulleys
   - Power steering pump and move it aside without disconnecting the lines from the pump
   - Compressor from the engine and move it aside without disconnecting the compressor lines, if equipped with air conditioning
   - Air conditioning bracket, if equipped with air conditioning
7. Remove the No. 2 timing belt cover, as follows:
   - Camshaft Position (CMP) sensor connector from the No. 2 timing belt cover
   - 3 spark plug wire clamps from the No. 2 timing belt cover
   - 6 bolts and the timing belt cover
8. Remove the fan bracket, as follows:
   - Power steering adjusting strut, by removing the nut
   - Fan bracket, by removing the bolt and nut
9. Set the No. 1 cylinder to Top Dead Center (TDC) of the compression stroke, as follows:
   a. Turn the crankshaft pulley and align its groove with the timing mark **0** of the No. 1 timing belt cover.
   b. Check that the timing marks of the camshaft timing pulleys and the No. 3 timing belt cover are aligned. If not, turn the crankshaft pulley 1 revolution (360 degrees).
10. Remove the camshaft timing pulleys, as follows:
    a. Remove the timing belt tensioner by alternately loosening the 2 bolts.
    b. Using Variable Wrench Set No. 09960-10010, remove the pulley bolt, the timing pulley and the knock pin.
    c. Remove the 2 timing pulleys with the timing belt.
11. Remove or disconnect the following:
    - Thermostat
    - No. 2 oil cooler hose, from the water pump
    - Water pump, by removing the 7 bolts
12. Thoroughly clean the mating surfaces.

To install:

13. Install or connect the following:
    - Apply sealant (PN 08826-00100) to the water pump

⁂ WARNING

Parts must be assembled within 5 minutes of application. Otherwise the material must be removed and reapplied.

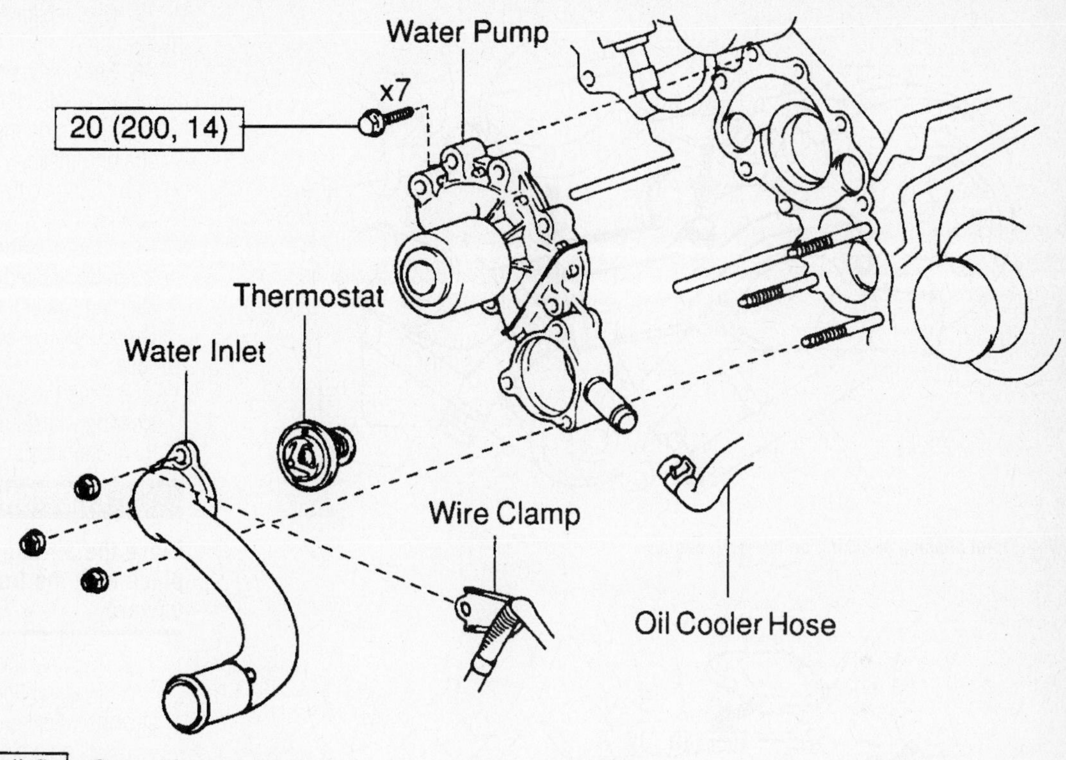

Exploded view of the water pump mounting—3.4L engine

- Water pump. Tighten the bolts to 14 ft. lbs. (20 Nm).
- No. 2 oil cooler hose
- Thermostat
- Left camshaft timing pulley. Tighten the pulley bolt to 81 ft. lbs. (110 Nm).

14. Set the No. 1 cylinder to TDC of the compression stroke.

15. Connect the timing belt to the left camshaft timing pulley. Check that the installation mark on the timing belt is aligned with the end of the No. 1 timing belt cover, as follows:

a. Using Variable Pin Wrench Set 09960-01000, slightly turn the left camshaft timing pulley clockwise. Align the installation mark on the timing belt with the timing mark of the camshaft timing pulley and hang the timing belt on the left camshaft timing pulley.

b. Align the timing marks of the left camshaft pulley and the No. 3 timing belt cover.

c. Check that the timing belt has tension between the crankshaft timing pulley and the left camshaft timing pulley.

16. Install the right camshaft timing pulley and the timing belt.

17. Set the timing belt tensioner, as follows:

a. Using a press, slowly press in the pushrod using 220–2,205 lbs. (981–9,807 N) of force.

b. Align the holes of the pushrod and housing, pass a 1.5mm hexagon wrench through the holes to keep the setting position of the pushrod.

c. Release the press and install the dust boot to the tensioner.

d. Install the timing belt tensioner and alternately tighten the bolts to 20 ft. lbs. (28 Nm).

e. Using pliers, remove the 1.5mm hexagon wrench from the belt tensioner.

18. Check the valve timing, as follows:

a. Slowly turn the crankshaft pulley 2 revolutions from the TDC-to-TDC; always turn the crankshaft pulley clockwise.

b. Check that each pulley aligns with the timing marks. If the timing marks do not align, remove the timing belt and reinstall it.

19. Install or connect the following:
- Fan bracket, with the bolt and nut
- Remaining components
- Negative battery cable

20. Fill with engine coolant.
21. Start the engine and check for leaks.

4.7L Engine

1. Before servicing the vehicle, refer to the precautions in the beginning of this section.
2. Drain the cooling system.
3. Remove or disconnect the following:
   - Negative battery cable
   - Timing belt. Refer to the Timing Belt unit repair section.
   - No. 2 idler pulley
   - Radiator hose
   - Bypass hose
   - Water inlet housing assembly
   - Water pump

To install:

4. Install or connect the following:
   - Water pump. Use a new gasket and tighten the bolts to 15 ft. lbs. (21 Nm). Tighten the stud bolt and nut to 13 ft. lbs. (18 Nm).
   - Water inlet housing assembly. Use a new O-ring and apply sealant as shown. Tighten the bolts to 13 ft. lbs. (18 Nm).
   - Bypass hose
   - Radiator hose
   - No. 2 idler pulley
   - Timing belt
   - Negative battery cable
5. Fill the cooling system.
6. Start the engine and check for leaks.

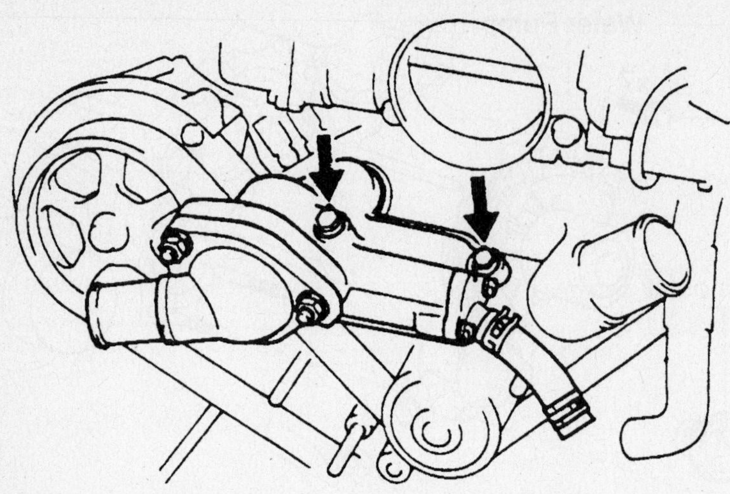

Water inlet housing attaching bolts—4.7L engine

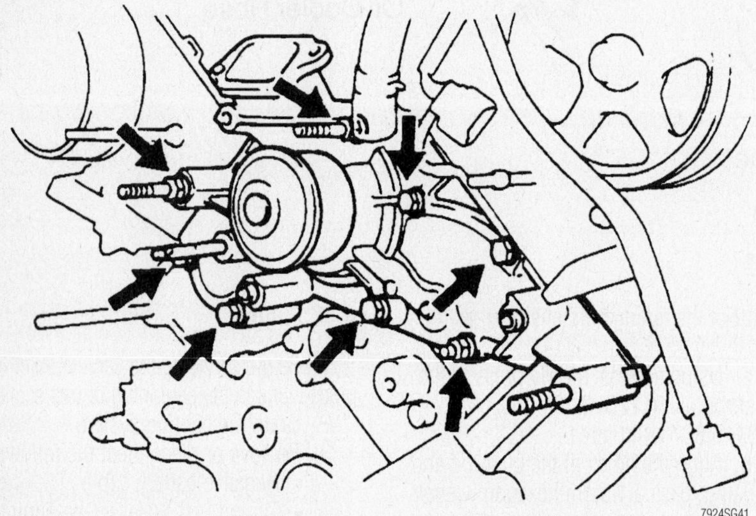

Water pump mounting bolts, stud bolts and nut locations—4.7L engine

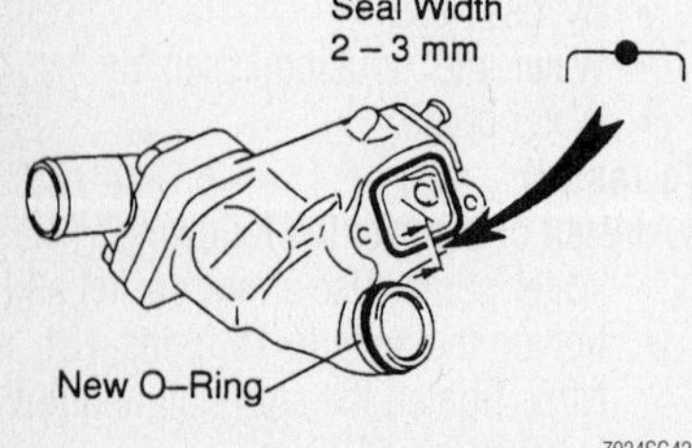

Water inlet housing sealant application—4.7L engine

Heater Core

REMOVAL & INSTALLATION

1. Disconnect the negative battery cable.
2. Drain the cooling system into a clean container for reuse.
3. Disconnect the heater hoses from the heater core.
4. Remove the steering wheel by performing the following procedure:
   a. Position the front wheels facing straight-ahead.
   b. Remove the steering wheel side covers.
   c. Using a Torx® wrench, loosen the 2 screws located at each side of the steering wheel until the screw's circumference groove catches on the screw case.
   d. Pull the air bag module from the steering wheel and disconnect the electrical connector.

CAUTION

Place the air bag module in a safe place with the front side facing upward.

   e. Remove the steering wheel nut.
   f. Place alignment marks on the steering wheel and the main shaft.
   g. Using a steering wheel puller, press the steering wheel from the steering column.
5. Remove the instrument panel and reinforcement by performing the following procedure:
   a. Remove the front door scuff plates, the cowl side trim and the front door opening trim.
   b. At the driver's side, remove the 2 assist grip plugs, the 2 screws and assist grip and the front pillar garnish.
   c. At the passenger's side, remove the 4 assist grip plugs, the 4 screws, the 2 assist grips and the front pillar garnish.

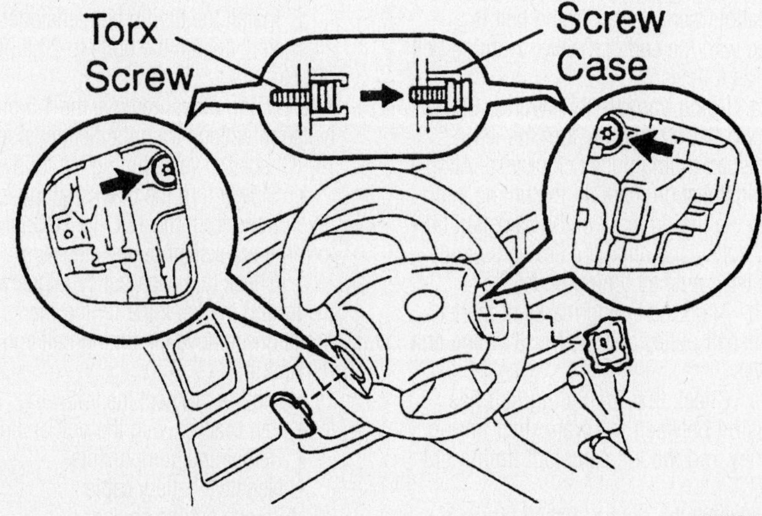

View the steering wheel's Torx® bolts

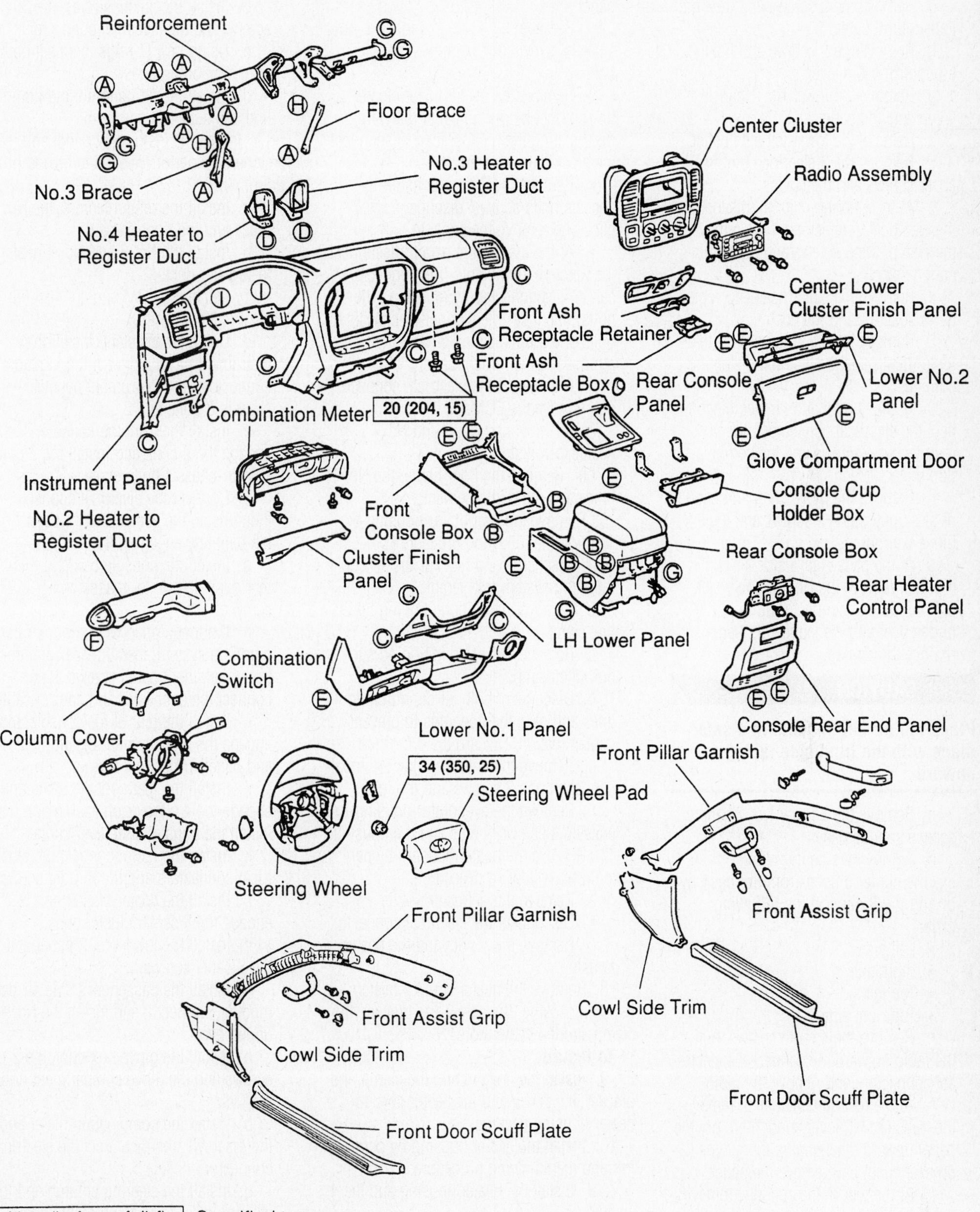

Exploded view the instrument panel and related components

d. Remove the instrument cluster finish panel.

e. Remove the 2 screws and the hood lock control cable.

f. Remove the 2 screws and the fuel lid control cable lever.

g. Remove the lower No. 1 panel screw and the panel.

h. Remove the lower left side panel.

i. Remove the 3 steering column cover screws and the covers.

j. At the steering column, disconnect the electrical connectors; then, remove the clamp, the 3 screws and the combination switch.

k. Remove the No. 2 heater-to-register duct screw and the duct.

l. Remove the steering column-to-instrument panel bolts and the steering column.

m. At the combination meter, disconnect the electrical connectors; then, remove the 4 screws and the combination meter.

n. Remove the glove compartment door stoppers, the 2 screws and the glove box door.

o. At the passenger's side air bag module, remove the No. 1 undercover, pull the air bag connector up from the undercover and disconnect it; then, remove the air bag.

⁂ CAUTION

Place the air bag module in a safe place with the front side facing upward.

p. Remove the 3 lower No. 2 panel screws and the panel.

q. Remove the center cluster; then, pry the center cluster from the dash by prying the 8 clips in the following order:

- Left side
- Right side
- Top left side
- Top right side

r. Remove the 4 radio screws, pull the radio outward, disconnect the electrical connectors and remove the radio.

s. At the rear console panel, remove the transfer shift lever knob; then, pry the panel upward disengaging the 4 clips (2 on each side) and remove the panel.

t. At the rear of the console, remove the 2 rear end panel-to-console screws; then, pry the end panel rearward disengaging the 2 clips and remove the panel.

u. If not equipped with a rear air conditioning system, disconnect the connector and control cable; then, remove the 3 rear heater control panel screws and the panel.

v. Remove the 4 rear console box-to-chassis screws/bolts and the console box.

w. Remove the center lower cluster finish panel by prying panel rearward disengaging the 5 clips; then, disconnect the electrical connector.

x. Remove the 2 front console-to-chassis bolts/screws, disengage the 2 clips and remove the console.

y. At the instrument panel, disconnect the junction connectors (the connectors can be disconnected by loosening the bolts), the instrument panel-to-chassis 8 bolts and 2 nuts. Using an assistant, remove the instrument panel.

z. Disconnect the electrical connector and remove the ECM.

aa. Remove the No. 3 and No. 4 heater-to-register ducts.

bb. Remove the floor brace, the No. 1 brace and the reinforcement.

6. Remove the evaporator housing by performing the following procedure:

a. Discharge and recover the air conditioning system refrigerant.

b. Remove the air conditioning liquid line clamp.

c. Remove the air conditioning suction line clamp.

d. Disconnect both air conditioning lines and plug the openings to prevent contamination. Discard the 4 O-rings.

e. Remove the antenna relay electrical connector, the 2 screws and the relay.

f. Remove the evaporator housing-to-chassis 4 screws/2 nuts and the housing.

7. Remove the heater housing by performing the following procedure:

a. Remove the defroster nozzle.

b. Disconnect the electrical connector.

c. Remove the 4 nuts and the heater housing.

8. Remove the heater core-to-heater housing packing, the screw, the bracket, the clamp and the heater core.

To install:

9. Install the heater core, the clamp, the bracket, the screw and the heater core-to-heater housing packing.

10. Install the heater housing by performing the following procedure:

a. Install the heater housing and the 4 nuts.

b. Connect the electrical connector.

c. Install the defroster nozzle.

11. Install the evaporator housing by performing the following procedure:

a. Install the evaporator housing and the housing-to-chassis 4 screws and 2 nuts.

b. Install the antenna relay, the 2 screws and the electrical connector.

c. Using new O-rings, connect both air conditioning lines.

d. Install the air conditioning liquid line and suction line clamp.

12. Install the instrument panel and reinforcement by performing the following procedure:

a. Install the reinforcement, the No. 1 brace and the floor brace.

b. Install the No. 3 and No. 4 heater-to-register ducts.

c. Install the ECM and connect the electrical connector.

d. Using an assistant, install the instrument panel, connect the junction connectors, the instrument panel-to-chassis 8 bolts and 2 nuts.

e. Install the front the console, engage the 2 clips and install the 2 console-to-chassis bolts/screws.

f. Connect the electrical connector; then, install the center lower cluster finish panel by engaging the 5 clips.

g. Install the console box and the 4 rear console box-to-chassis screws/bolts.

h. If not equipped with a rear air conditioning system, install rear heater control panel, the 3 panel screws; then, connect the connector and control cable.

i. Install the rear of the console and engage the 2 clips; then, install the 2 rear end panel-to-console screws.

j. Install the rear console panel and engage the 4 clips (2 on each side); then, install the transfer shift lever knob.

k. Install the radio, connect the electrical connectors and the 4 radio screws.

l. Install the center cluster and engage the 8 center cluster clips.

m. Install the lower No. 2 panel and the 3 panel screws.

n. Install the passenger's side air bag module, connect it and install the No. 1 undercover.

o. Install the glove box door, the 2 screws and the glove compartment door stoppers.

p. Install the combination meter and the 4 screws; then, connect the electrical connectors.

q. Install the steering column and the steering column-to-instrument panel bolts.

r. Install the No. 2 heater-to-register duct and the duct screw.

s. At the steering column, install the combination switch, the 3 screws and the

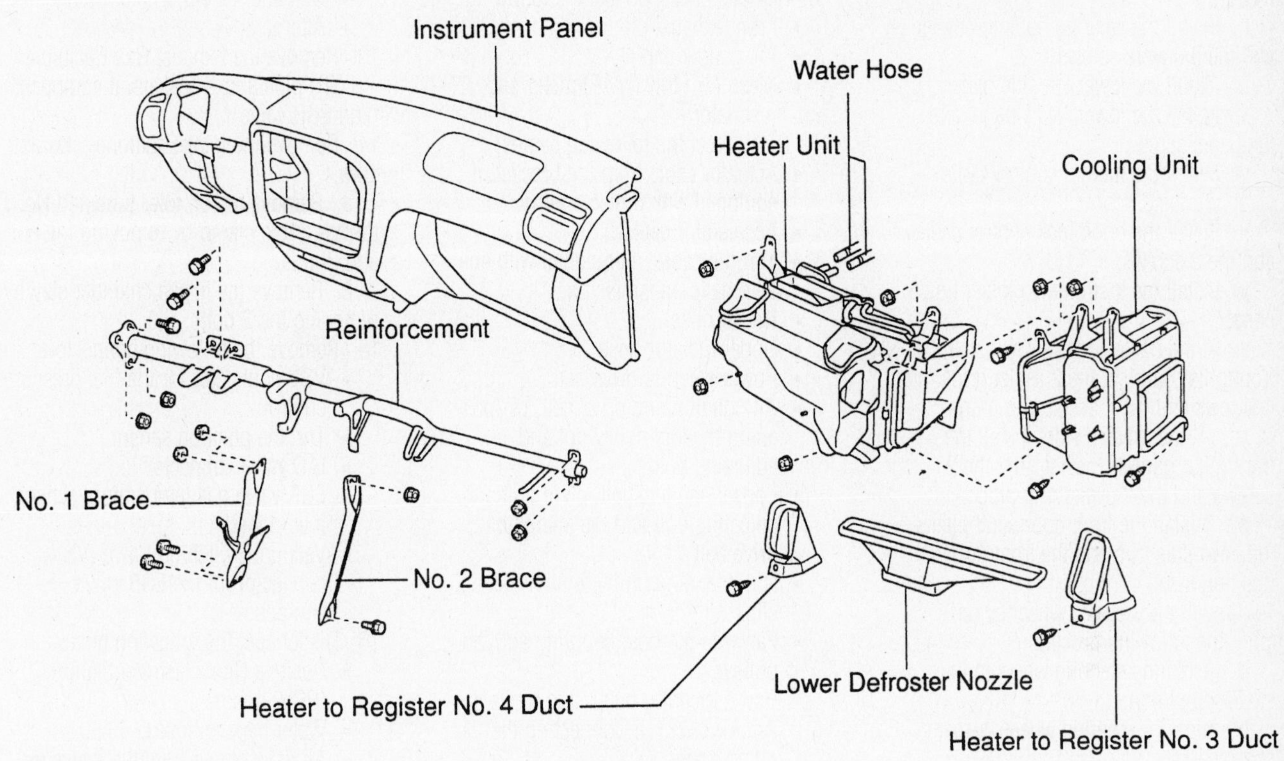

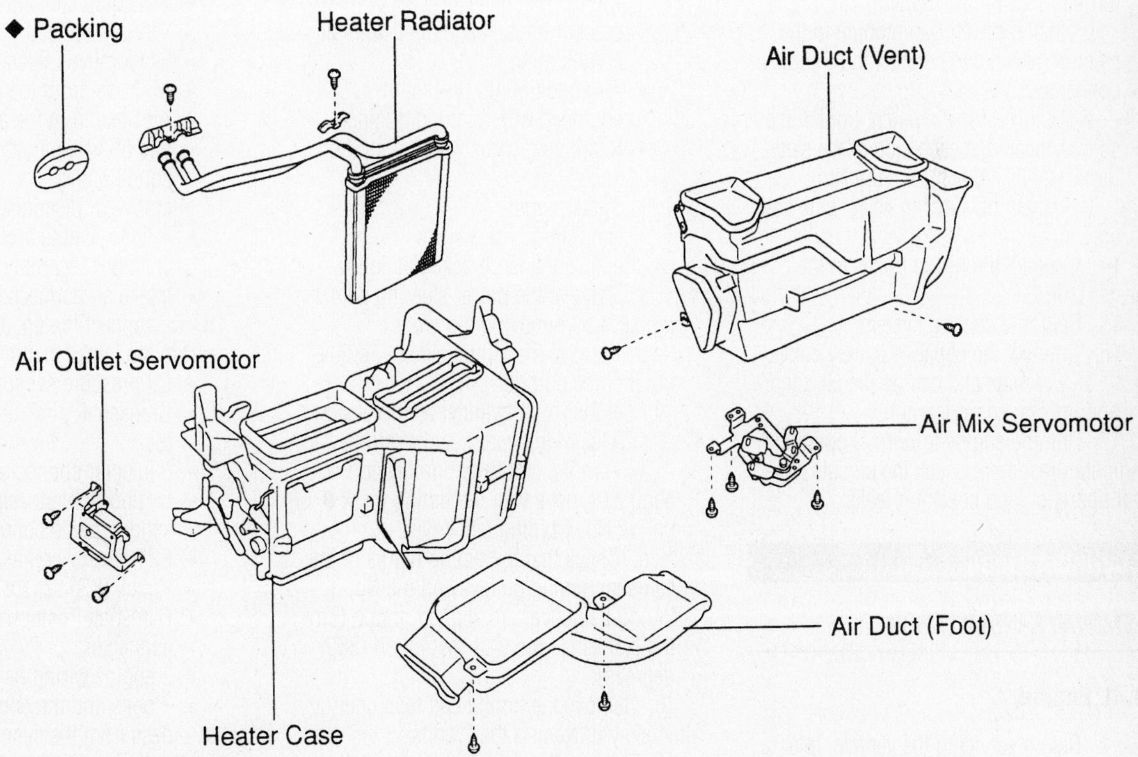

Exploded view the heater core, heater housing, evaporator housing and related components

clamp; then, connect the electrical connectors.

t. Install the steering column covers and the 3 covers screws.

u. Install the lower left side panel.

v. Install the lower No. 1 panel and the panel screw.

w. Install the fuel lid control cable lever and the 2 screws.

x. Install the hood lock control cable and the 2 screws.

y. Install the instrument cluster finish panel.

z. At the passenger's side, install the front pillar garnish, the 2 assist grips, the 4 screws and the 4 assist grip plugs.

aa. At the driver's side, install the front pillar garnish, assist grip, the 2 screws and the 2 assist grip plugs.

bb. Install the front door scuff plates, the cowl side trim and the front door opening trim.

13. Install the steering wheel by performing the following procedure:

a. Install the steering wheel to the steering column.

b. Align the steering wheel-to-main shaft marks.

c. Install the steering wheel nut and torque to 25 ft. lbs. (34 Nm).

d. Install the air bag module to the steering wheel and connect the electrical connector.

e. Using a Torx® wrench, tighten the 2 screws located at each side of the steering wheel to 78 inch lbs. (8.8 Nm).

f. Install the steering wheel side covers.

14. Connect the heater hoses to the heater core.

15. Refill the cooling system.

16. Connect the negative battery cable.

a. Evacuate and charge the air conditioning system refrigerant.

17. Run the engine to normal operating temperatures; then, check the climate control operation and check for leaks.

Cylinder Head

REMOVAL & INSTALLATION

3.4L Engine

1. Before servicing the vehicle, refer to the precautions in the beginning of this section.
2. Disconnect the negative battery cable.
3. Relieve the fuel system pressure.
4. Remove the engine undercover.
5. Drain the cooling system.
6. Remove or disconnect the following:
   - Front exhaust pipe
   - Air cleaner cap
   - Mass Air Flow (MAF) meter and resonator
7. Disconnect the following cables:
   - Actuator cable from the bracket, if equipped with cruise control
   - Accelerator cable
   - Throttle cable, if equipped with an automatic transmission
   - Heater hose
   - Upper radiator hose
   - Power steering drive belt
   - Air conditioning drive belt, by loosening the idle pulley nut and adjusting bolt
   - Loosen the lockbolt, pivot bolt and adjusting bolt and the alternator drive belt
   - No. 2 fan shroud by removing the 2 clips
   - Fan with the fluid coupling and fan pulleys
   - Power steering pump and move it aside without disconnecting the pump lines
   - Compressor and move it aside without disconnecting the compressor lines, if equipped with air conditioning
   - Air conditioning bracket, if equipped with air conditioning
   - Spark plug wires with the ignition coils
   - Spark plugs
   - No. 2 timing belt cover
8. Remove the fan bracket, as follows:

a. Remove the power steering adjusting strut by removing the nut.

b. Remove the fan bracket by removing the bolt and nut.

9. Set the No. 1 cylinder at Top Dead Center (TDC) of the compression stroke.

a. Turn the crankshaft pulley and align its groove with the timing mark **0** on the No. 1 timing belt cover.

b. Check that the timing marks of the camshaft timing pulleys and the No. 3 timing belt cover are aligned. If not, turn the crankshaft pulley 1 revolution (360 degrees).

10. Remove the timing belt tensioner by alternately loosening the 2 bolts.

11. Remove the camshaft timing pulleys, as follows:

a. Using Variable Pin Wrench Set tool 09960-10010, remove the pulley bolt, the timing pulley and the knock pin.

b. Remove the 2 timing pulleys with the timing belt.

12. Remove or disconnect the following:
    - Bolt and the No. 2 idler pulley
    - Alternator
13. Remove the Exhaust Gas Recirculation (EGR) pipe and 2 gaskets, if equipped with an EGR valve
14. Remove the intake chamber stay as follows:

a. Remove the oil filler tube and No. 1 throttle cable clamp by removing the bolt and 2 nuts.

b. Remove the intake chamber stay by removing the 2 bolts.

15. Remove the following connectors:
    - VSV connector for the fuel pressure control.
    - Throttle position sensor
    - IAC valve connector
    - EGR valve gas temperature sensor, if equipped
    - Vacuum Switching Valve (VSV) connector for the EGR valve, if equipped
16. Disconnect the following hoses:
    - Positive Crankcase Ventilation (PCV) hoses
    - Water bypass hoses.
    - Air assist hose from the intake air connector
    - 2 vacuum sensing hoses from the VSV
    - Evaporative Emissions (EVAP) hose
    - Air hose, from the power steering
    - Air hose from the air conditioning idle up valve, if equipped with air conditioning
17. Remove or disconnect the following:
    - 4 bolts, 2 nuts and the air intake chamber assembly
    - Intake air connector
18. Disconnect the engine wiring harness from the intake manifold, as follows:
    - Oil pressure sensor connector
    - Crankshaft position sensor connector
    - 6 injector connectors
    - Engine Coolant Temperature (ECT) sender gauge connector
    - ECT sensor connector
    - Knock (KS) sensor connector
    - Camshaft Position (CMP) sensor connector
    - 3 engine wiring harness clamps
    - 3 bolts and the engine wiring harness from the cylinder head
19. Remove or disconnect the following:
    - CMP sensor
    - No. 3 (rear) timing belt cover, by removing the 6 bolts
    - Fuel pressure regulator
    - Intake manifold assembly
    - Power steering pump bracket

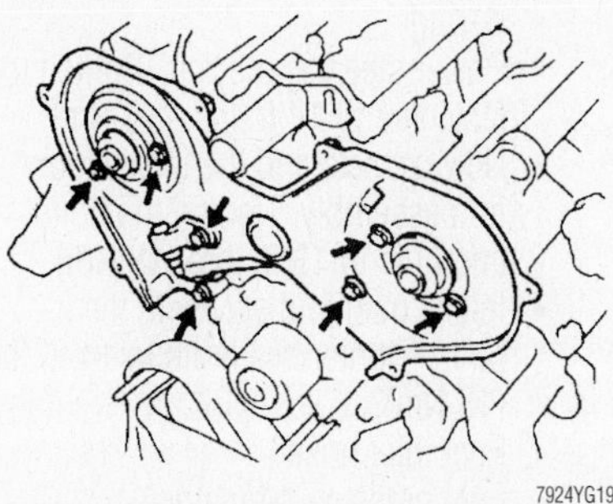

Rear timing belt cover bolt locations—3.4L engine

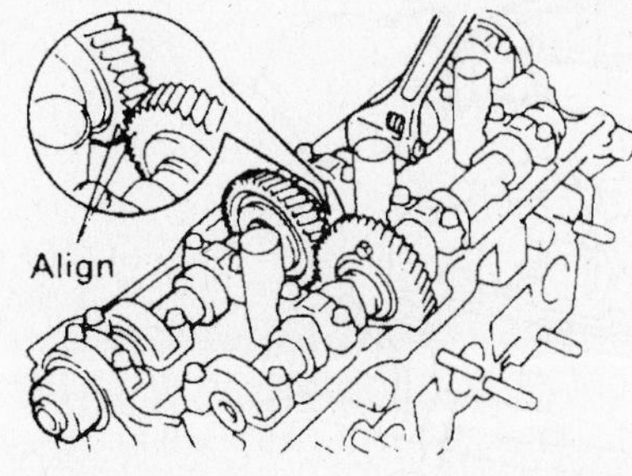

Aligning the timing mark (1 dot mark) of the left camshafts—3.4L engine

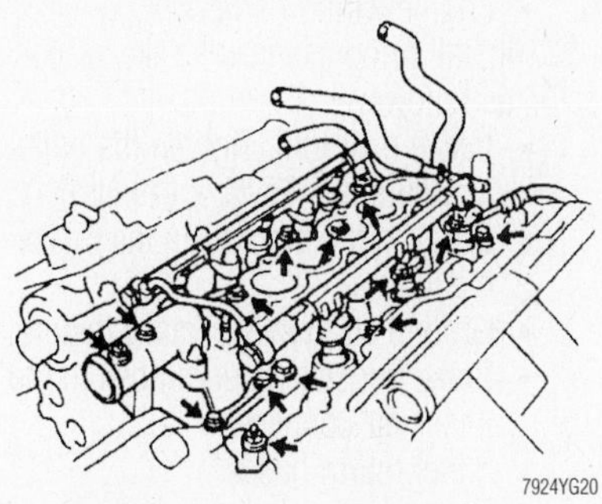

Intake manifold bolts and nuts locations—3.4L engine

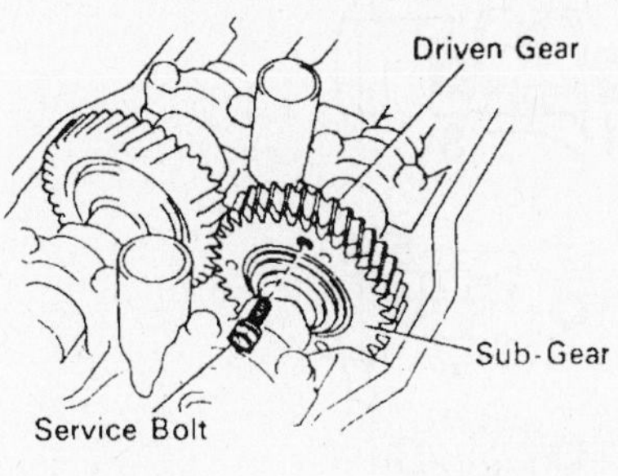

Drive gear service bolt (left side)—3.4L engine

- Oil dipstick and guide
- Exhaust crossover pipe and gaskets, by removing the 6 nuts
- Left-hand exhaust manifold, by removing the heat insulator and 6 nuts
- Right-hand exhaust manifold, by removing the heat insulator and 6 nuts
- 8 bolts, seal washers, cylinder head cover and gasket
- Both cylinder head covers
- Semi-circular plugs
- Right exhaust camshafts
- Right-hand intake camshaft
- Left exhaust camshafts

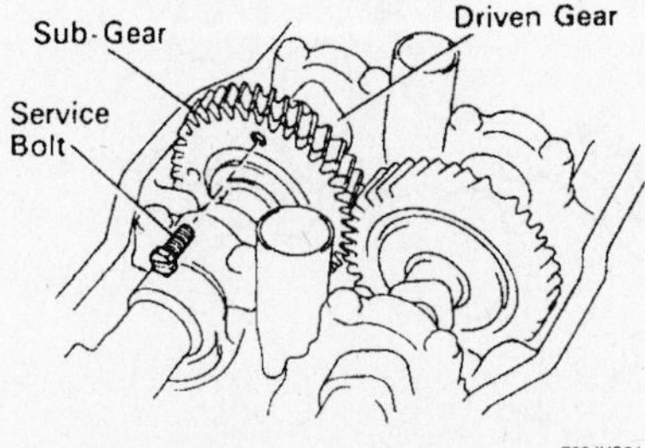

Drive gear service bolt (right side)—3.4L engine

- Left-hand intake camshaft
- Valve lifters and shims from the cylinder head; arrange the valve lifters and shims in correct order

20. Remove the cylinder heads, as follows:
    - Bolt and disconnect the ground strap
    - Cylinder head (recessed head) bolt on each cylinder head, using an 8mm hexagon wrench; then repeat for the other side
    - 8 cylinder head (12-pointed head) bolts on each cylinder head, by loosening the bolts in several passes and in the reverse order of the tightening sequence
    - 16 cylinder head bolts and plate washers.
    - Cylinder head

To install:

21. Clean all surfaces.
22. Install or connect the following:
    - New cylinder head gaskets
    - Cylinder heads
23. Apply a light coat of engine oil on the threads and under the heads of the cylinder head bolts.
24. Tighten the cylinder head bolts, in sequence, using several passes, as follows:

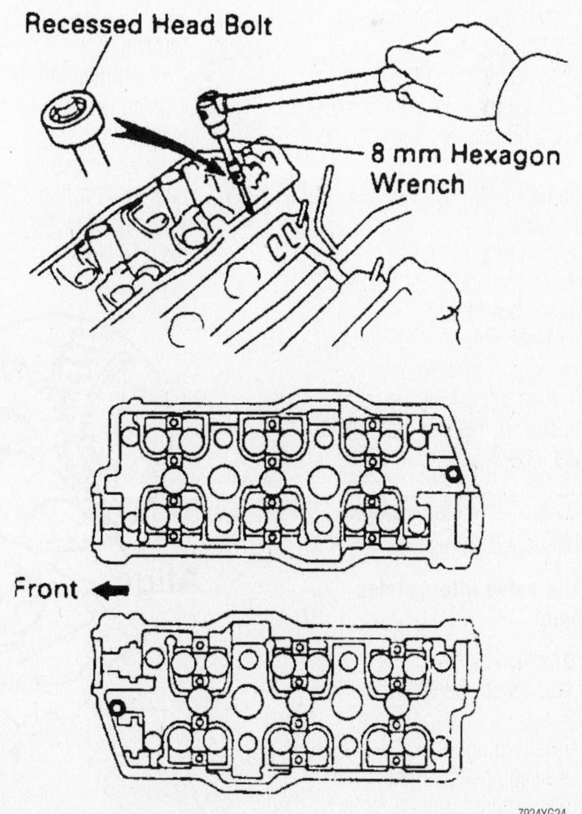

Cylinder head recessed bolts—3.4L engine

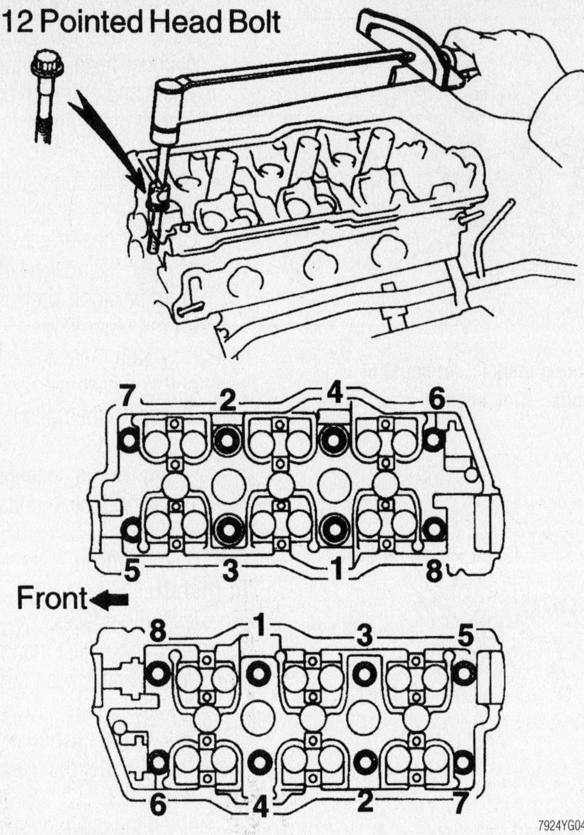

Cylinder head torque sequence—3.4L (5VZ-FE) engine

- Step 1: 25 ft. lbs. (34 Nm)
- Step 2: Mark the front of the cylinder head bolt with paint
- Step 3. An additional 90 degrees
- Step 4: Check that the painted mark is now at a 90 degrees angle to the front

25. Install the recessed head cylinder head bolts, as follows:
    - Apply a light coat of engine oil on the threads and under the heads of the cylinder head bolts
    - Cylinder head bolt on each cylinder head, using a 8mm hexagon wrench; then, repeat for the other side, as shown. Tighten the bolts to 13 ft. lbs. (18 Nm).
    - Bolt and the ground strap
26. Install or connect the following:
    - Valve lifters and shims

➡Check that the valve lifter rotates smoothly by hand.

- Camshafts
- Check and adjust the valve clearance
- Semi-circular plugs
- Cylinder head covers. Tighten the bolts, in several passes, to 53 inch lbs. (6 Nm).
- Exhaust manifolds, with new gaskets. Tighten the nuts to 30 ft. lbs. (40 Nm).
- Exhaust manifold heat insulators. Tighten the nuts to 71 inch lbs. (8 Nm).
- Exhaust crossover pipe. Tighten the nuts to 33 ft. lbs. (45 Nm).
- Alternator bracket. Tighten to 14 ft. lbs. (18 Nm).
- Oil dipstick and guide, using a new O-ring
- Power steering bracket. Tighten the fasteners to 14 ft. lbs. (18 Nm).
- New gaskets and the intake manifold assembly. Tighten the bolts and nuts to 13 ft. lbs. (18 Nm).
- Intake manifold stay with the 2 bolts. Tighten the bolts to 14 ft. lbs. (18 Nm).
- Fuel inlet hose
- Fuel pressure regulator
- No. 3 timing belt cover. Tighten the bolts to 80 inch lbs. (9 Nm).
- CMP sensor. Tighten to 71 inch lbs. (8 Nm).
- Engine wiring harness

27. Install or connect the intake air connector, as follows:
    - Intake manifold. Tighten the bolts and nuts to 14 ft. lbs. (19 Nm).
    - DLC1 to the bracket on the intake manifold
    - Ground strap to the intake manifold
    - Brake booster vacuum hose to the intake air connector
    - 2 fuel return hoses
    - Engine wiring harness to the intake manifold
    - Idle up valve connector, if equipped with air conditioning
28. Install or connect the following:
    - Air intake chamber assembly. Tighten the bolts and nuts to 14 ft. lbs. (18.5 Nm).
    - Hoses
29. Attach the following connectors:
    - VSV connector for the fuel pressure control
    - TP sensor connector
    - IAC valve connector

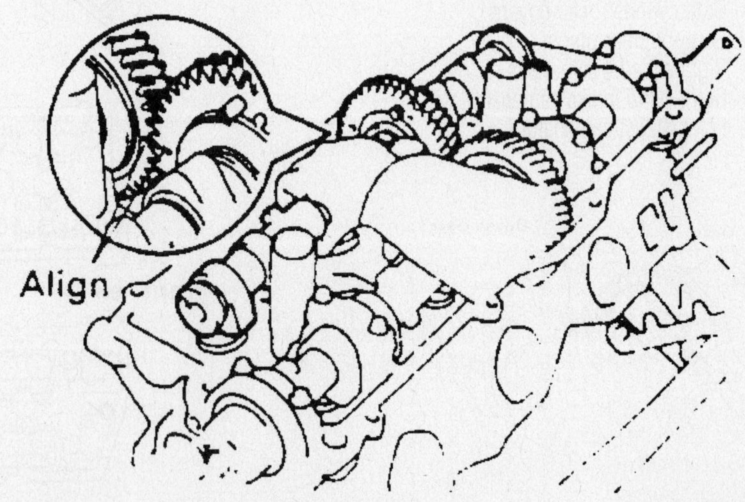

Aligning the right camshafts for installation—3.4L engine

- EGR gas temperature connector, if equipped with an EGR valve
- VSV connector, if equipped with an EGR valve
- Intake chamber stay
- New gaskets and the EGR pipe. Tighten the clamp nuts to 71 inch lbs. (8 Nm) and the EGR pipe nuts to 14 ft. lbs. (18 Nm).
- Alternator but do not tighten the bolts and nuts at this time.
- No. 2 timing belt idler. Tighten the bolt to 30 ft. lbs. (40 Nm).

➡Check that the pulley bracket moves smoothly.

- Left camshaft timing pulley

30. Set the No. 1 cylinder to TDC of the compression stroke, as follows:
    a. Turn the crankshaft pulley and align its groove with the timing mark **0** on the No. 1 timing belt cover.
    b. Turn the camshaft, align the knock pin hole of the camshaft with the timing mark of the No. 3 timing belt cover.
    c. Turn the camshaft timing pulley, align the timing marks of the camshaft timing pulley and the No. 3 timing belt cover.
31. Connect the timing belt to the left camshaft timing pulley, as follows:
    a. Check that the installation mark on the timing belt is aligned with the end of the No. 1 timing belt cover.
    b. Using Variable Pin Wrench Set 09960-01000, slightly turn the left camshaft timing pulley clockwise. Align the installation mark on the timing belt with the timing mark of the camshaft timing pulley and hang the timing belt on the left camshaft timing pulley.
    c. Align the timing marks of the left camshaft pulley and the No. 3 timing belt cover.
    d. Check that the timing belt has tension between the crankshaft timing pulley and the left camshaft timing pulley.
32. Install the right camshaft timing pulley and the timing belt, as follows:
    a. Align the installation mark on the timing belt with the timing mark of the right camshaft timing pulley, and hang the timing belt on the right camshaft timing pulley with the flange side facing inward.
    b. Slide the right camshaft timing pulley on the camshaft. Align the timing marks on the right camshaft timing pulley and the No. 3 timing belt cover.
    c. Align the knock pin hole of the camshaft with the knock pin groove of the pulley and install the knock pin. Install the bolt and tighten to 81 ft. lbs. (110 Nm).
33. Set the timing belt tensioner as follows:
    a. Using a press, slowly press in the pushrod using 220–2,205 lbs. (981–9,807 N) of force.
    b. Align the holes of the pushrod and housing, pass a 1.5mm hexagon wrench through the holes to keep the setting position of the pushrod.
    c. Release the press and install the dust boot to the tensioner.
34. Install the timing belt tensioner. Tighten the bolts alternately to 20 ft. lbs. (28 Nm).
35. Using pliers, remove the 1.5mm hexagon wrench from the belt tensioner.
36. Check the valve timing.
37. Install or connect the following:
    - Remaining components
    - Negative battery cable
38. Fill the radiator with engine coolant.
39. Start the engine and check for leaks.
40. Check the ignition timing.
41. Install the engine undercover.
42. Road test the vehicle.
43. Recheck all fluid levels.

4.7L Engine

1. Before servicing the vehicle, refer to the precautions in the beginning of this section.
2. Drain the cooling system.
3. Relieve the fuel system pressure.
4. Remove or disconnect the following:
    - Battery and tray
    - Engine appearance cover
    - Engine under covers
    - Air intake assembly
    - Accessory drive belt
    - A/C compressor and bracket
    - Cooling fan and bracket
    - Radiator
    - Idler pulley
    - Front covers
    - Timing belt. Refer to the Timing Belt unit repair section.
    - Camshaft sprockets
    - Camshaft Position (CMP) sensor
    - Power steering pump
    - Exhaust front pipes
    - Transmission dipstick tube
    - Ignition coils
    - Rear timing belt covers
    - Fuel lines

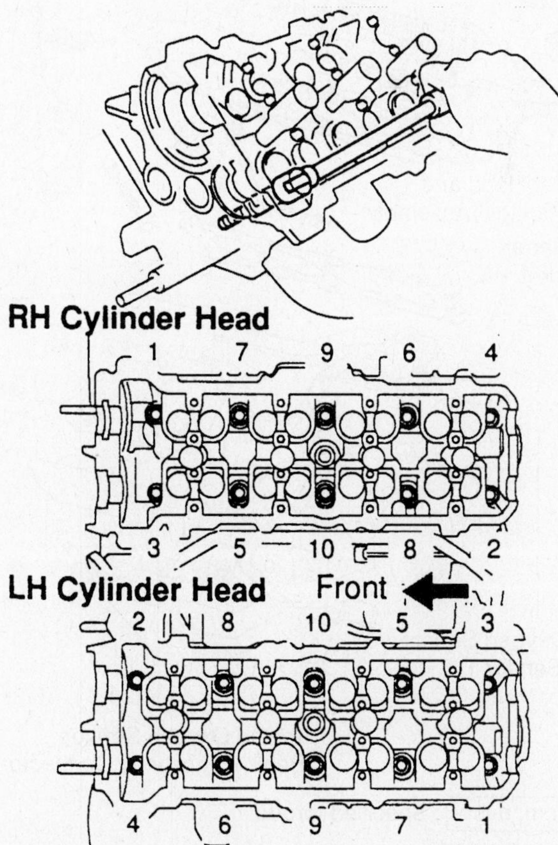

Cylinder head loosening sequence—4.7L engine

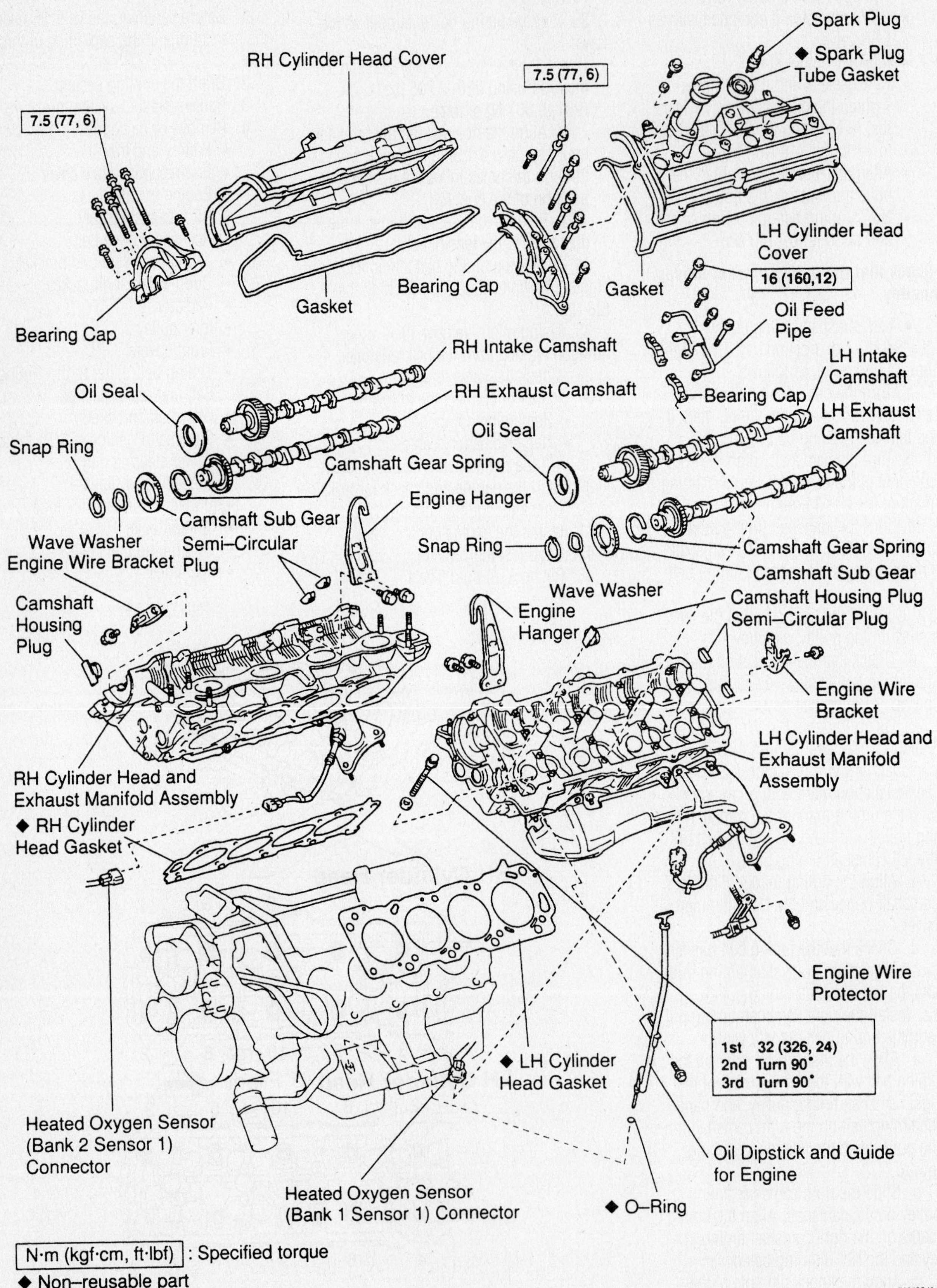

Exploded view of the cylinder head mounting—4.7L engine

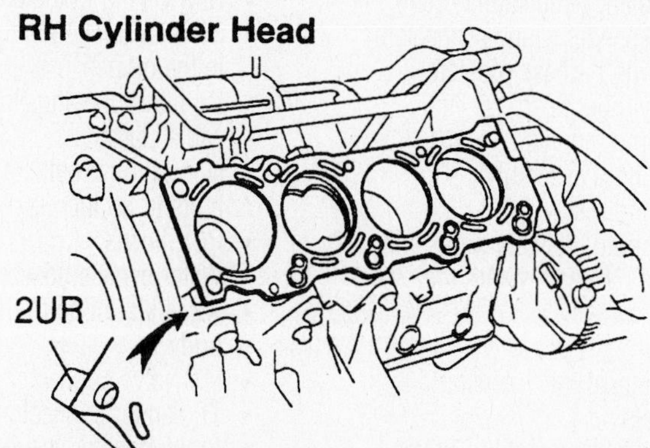

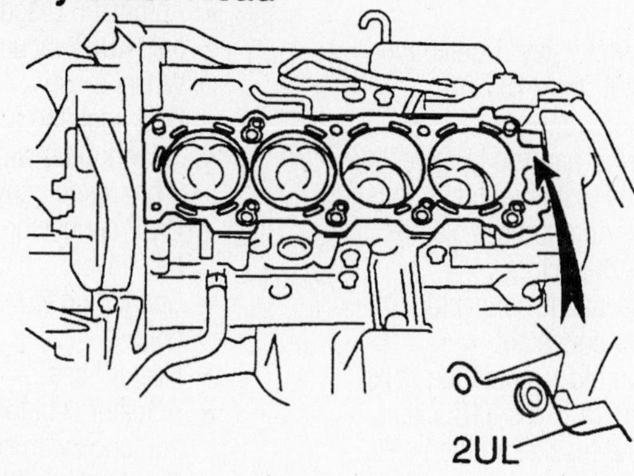

Cylinder head gasket identification—4.7L engine

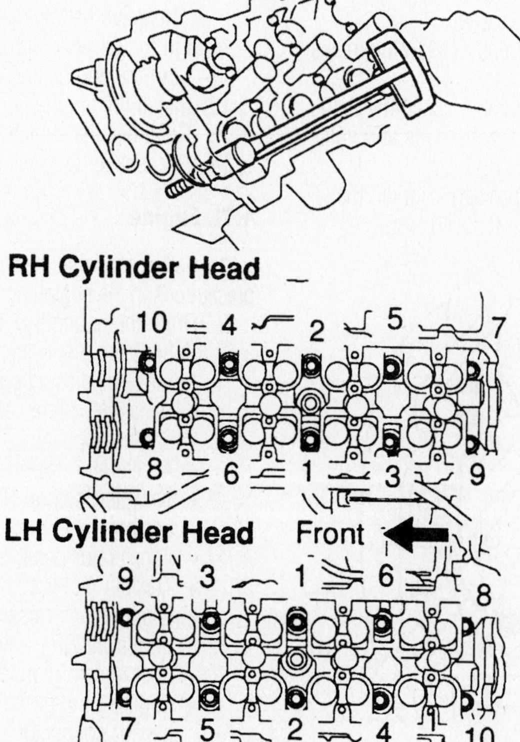

Cylinder head torque sequence—4.7L (2UZ-FE) engine

- Intake manifold
- Water inlet housing assembly
- Front and rear water bypass joints
- Engine lifting eyes
- Oil dipstick tube
- Valve covers
- Camshafts
- Cylinder heads with the exhaust manifolds attached. Loosen the bolts in the sequence shown.

To install:

5. Install the cylinder heads with new gaskets. Tighten the bolts in sequence as follows:
   a. Step 1: 24 ft. lbs. (32 Nm)
   b. Step 2: Plus 180 degrees
6. Install or connect the following:
   - Camshafts
   - Valve covers
   - Oil dipstick tube
   - Engine lifting eyes
   - Front and rear water bypass joints
   - Water inlet housing assembly
   - Intake manifold
   - Fuel lines
   - Rear timing belt covers
   - Ignition coils
   - Transmission dipstick tube
   - Exhaust front pipes
   - Power steering pump
   - CMP sensor
   - Camshaft sprockets
   - Timing belt
   - Front covers
   - Idler pulley
   - Radiator
   - Cooling fan and bracket
   - A/C compressor and bracket
   - Accessory drive belt
   - Air intake assembly
   - Engine under covers
   - Engine appearance cover
   - Battery and tray
7. Fill the cooling system.
8. Start the engine and check for leaks.

Intake Manifold

REMOVAL & INSTALLATION

3.4L Engine

1. Before servicing the vehicle, refer to the precautions in the beginning of this section.
2. Disconnect the negative battery cable.
3. Relieve the fuel system pressure.
4. Remove the engine undercover.
5. Drain the cooling system.
6. Remove or disconnect the following:

- Air cleaner cap
- Mass Air Flow (MAF) meter and the resonator
- Actuator cable with the bracket, if equipped with cruise control
- Accelerator cable
- Throttle cable, if equipped with automatic transmission

7. Disconnect the following hoses:
   - Heater hose
   - Brake booster vacuum hose
   - Evaporative Emissions (EVAP) hose
   - Automatic Disconnecting Differential (ADD) vacuum hose, for 4-Wheel drive
   - Fuel inlet and fuel return hose
8. Remove or disconnect the following:
   - Spark plug wires, with the ignition coils
   - Intake chamber stay
   - No. 2 timing belt cover
   - Air intake chamber assembly
   - Throttle Position (TP) sensor connector
   - Idle Air Control (IAC) valve connector
   - Positive Crankcase Ventilation (PCV) hoses
   - Water bypass hoses
   - Air assist hose from the throttle body
   - Intake air connector
   - Engine wiring harness
   - Fuel return hose
   - Vacuum hose, from the fuel pressure regulator
   - Ground strap, from the intake air connector
   - Data Link Connector 1 (DLC1), from the bracket
   - 6 injector connectors
   - Engine Coolant Temperature (ECT) sensor and sender gauge connectors
   - Engine wiring harness protector from the cylinder head
   - Fuel pressure regulator
   - Intake manifold assembly
   - Intake manifold stay
   - Intake manifold, delivery pipes and the injectors assembly with the gaskets

To install:

9. Install or connect the following:
   - New gaskets
   - Intake manifold assembly. Tighten the bolts and nuts to 13 ft. lbs. (18 Nm).
   - Intake manifold stay. Tighten the bolts to 14 ft. lbs. (18 Nm).
   - Fuel pressure regulator
   - Engine wiring harness to the cylinder head, by installing the 3 bolts
   - 3 engine wiring harness clamps
   - 6 injector connectors
   - ECT sender gauge connector
   - ECT sensor connector
   - Intake manifold. Tighten the bolts and nuts to 14 ft. lbs. (18.5 Nm).
   - DLC1 to the bracket on the intake manifold
   - Ground strap to the intake manifold, by installing the bolt
   - Brake booster vacuum hose, to the intake air connector
   - 2 fuel return hoses
   - Engine wiring harness to the intake manifold
   - Air intake chamber assembly to the engine. Tighten the bolts and nuts to 14 ft. lbs. (18.5 Nm).
   - Intake chamber stay. Tighten the bolts to 30 ft. lbs. (40 Nm).
   - New O-ring to the oil filler tube
   - Oil filler tube end into the tube hole in the oil pan
   - Oil filler tube and No. 1 throttle cable clamp
   - No. 2 timing belt cover. Tighten the bolts to 80 inch lbs. (9 Nm).
   - PCV hoses
   - Water bypass hoses
   - Air assist hose to the throttle body
   - IAC valve connector
   - TP sensor connector
   - Brake booster vacuum hose
   - EVAP hose
   - Automatic Disconnecting Differential (ADD) vacuum hose, for 4-Wheel drive
   - Fuel inlet and fuel return hose
   - 3 spark plug wire clamps to the No. 2 timing belt cover
   - CMP connector to the No. 2 timing belt cover
   - Spark plug wires with the ignition coils
   - Heater hose
   - Actuator cable with the bracket, if equipped with cruise control
   - Accelerator cable
   - Throttle cable, if equipped with automatic transmission
   - MAF meter, resonator and the air cleaner cap
   - Negative battery cable
10. Fill the radiator with engine coolant.
11. Start the engine and check for leaks.
12. Install the engine undercover.
13. Road test the vehicle.
14. Recheck all fluid levels.

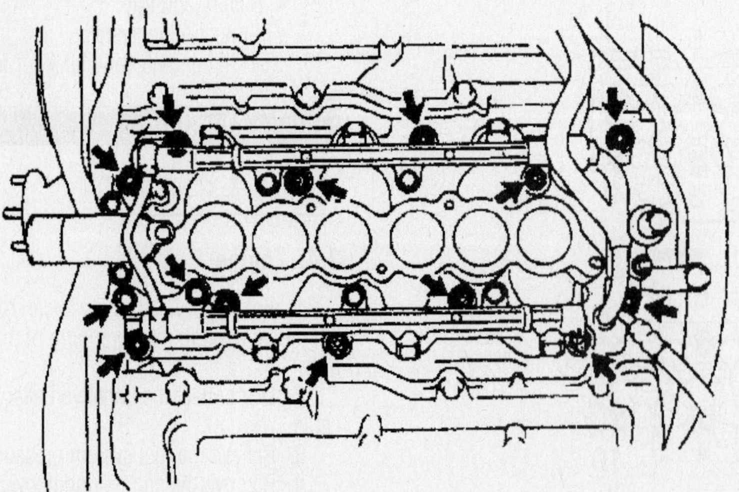

Intake manifold bolts and nuts—3.4L engine

4.7L Engine

1. Before servicing the vehicle, refer to the precautions in the beginning of this section.
2. Drain the cooling system.
3. Relieve the fuel system pressure.
4. Remove or disconnect the following:
   - Negative battery cable
   - Engine appearance cover
   - Accelerator cable
   - Throttle Position (TP) sensor connector
   - Accelerator pedal position sensor
   - Throttle motor connector
   - Evaporative Emissions (EVAP) vacuum switching valve connector
   - Fuel injector connectors
   - Engine Coolant Temperature (ECT) sensor connector
   - ETC gauge sender connector
   - Heated Oxygen (HO_2S) sensor connectors

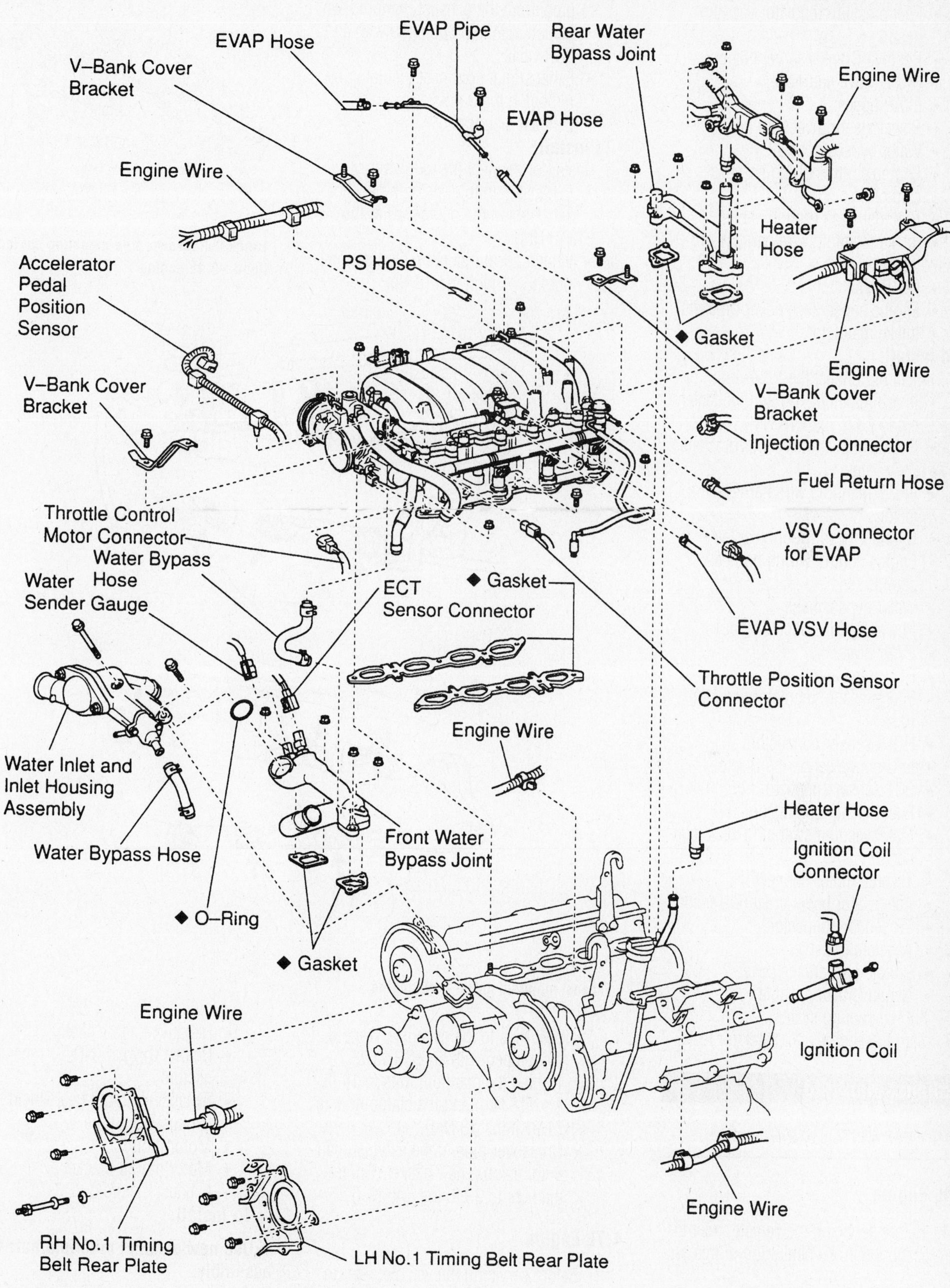
EVAP Hose
EVAP Pipe
Rear Water Bypass Joint
V–Bank Cover Bracket
Engine Wire
EVAP Hose
Engine Wire
Heater Hose
Accelerator Pedal Position Sensor
PS Hose
◆ Gasket
Engine Wire
V–Bank Cover Bracket
V–Bank Cover Bracket
Injection Connector
Fuel Return Hose
Throttle Control Motor Connector
VSV Connector for EVAP
Water Bypass Hose
Water Sender Gauge
◆ Gasket
ECT Sensor Connector
EVAP VSV Hose
Throttle Position Sensor Connector
Engine Wire
Water Inlet and Inlet Housing Assembly
Heater Hose
Water Bypass Hose
Front Water Bypass Joint
Ignition Coil Connector
◆ O–Ring
◆ Gasket
Engine Wire
Ignition Coil
Engine Wire
RH No.1 Timing Belt Rear Plate
LH No.1 Timing Belt Rear Plate
◆ Non–reusable part
7924SG50

Exploded of the intake manifold mounting—4.7L engine

- Fuel pressure regulator vacuum hose
- Positive Crankcase Ventilation (PCV) valve and hose
- EVAP hoses
- Power steering vacuum hoses
- Water bypass hose
- Engine control wiring harness clamps
- Cylinder head ground cables
- Intake manifold wire harness protector
- EVAP pipe
- Engine appearance cover brackets
- Intake manifold

To install:

5. Install or connect the following:
   - Intake manifold. Tighten the fasteners to 13 ft. lbs. (18 Nm).
   - Engine appearance cover brackets
   - EVAP pipe
   - Intake manifold wire harness protector
   - Cylinder head ground cables
   - Engine control wiring harness clamps
   - Water bypass hose
   - Power steering vacuum hoses
   - EVAP hoses
   - PCV valve and hose
   - Fuel pressure regulator vacuum hose
   - HO_2S sensor connectors
   - ETC gauge sender connector
   - ECT sensor connector
   - Fuel injector connectors
   - EVAP vacuum switching valve connector
   - Throttle motor connector
   - Accelerator pedal position sensor
   - TP sensor connector
   - Accelerator cable
   - Engine appearance cover
   - Negative battery cable
6. Fill the cooling system.
7. Start the engine and check for leaks.

Exhaust Manifold

REMOVAL & INSTALLATION

3.4L Engine

1. Before servicing the vehicle, refer to the precautions in the beginning of this section.
2. Remove or disconnect the following:
   - Exhaust crossover pipe, from the exhaust manifold by removing the 3 nuts
   - Exhaust Gas Recirculation (EGR) pipe, from the exhaust manifold, on the left manifold equipped with an EGR valve
   - Exhaust manifold heat insulator, by removing the 3 nuts
   - Exhaust manifold

To install:

3. Install or connect the following:
   - Exhaust manifold, using a new gasket. Tighten the nuts to 30 ft. lbs. (40 Nm).
   - Exhaust heat insulator. Tighten the nuts to 71 inch lbs. (8 Nm).

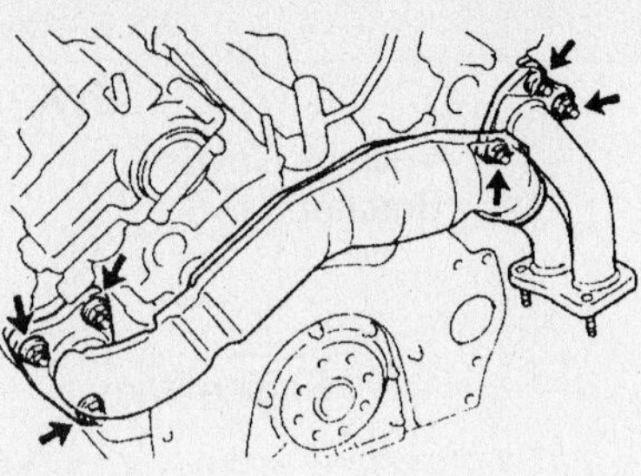

Exhaust crossover pipe mounting nut locations—3.4L engine

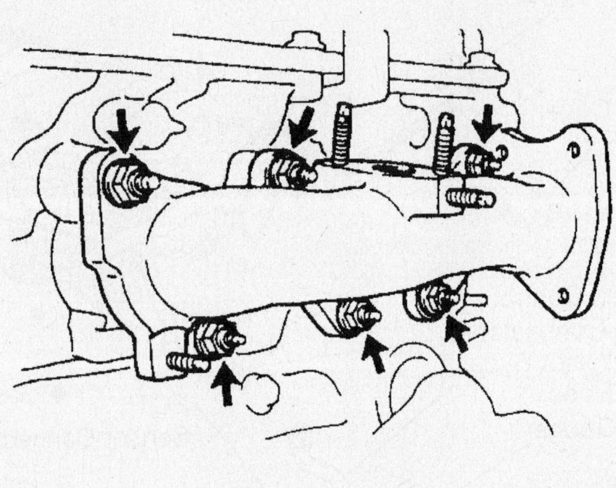

Exhaust manifold nuts—3.4L engine

   - EGR pipe to the exhaust manifold, if equipped with an EGR valve. Tighten the manifold nuts to 14 ft. lbs. (18 Nm) and the clamp nuts to 71 inch lbs. (8 Nm).
   - Crossover pipe to the exhaust manifold, using a new gasket. Tighten the nuts to 33 ft. lbs. (45 Nm).

4.7L Engine

1. Before servicing the vehicle, refer to the precautions in the beginning of this section.
2. Attach a hoist to the engine lifting eyes.
3. Remove or disconnect the following:
   - Negative battery cable
   - Heated Oxygen (HO_2S) sensor connectors
   - Exhaust manifold heat shield
   - Exhaust front pipe
   - Motor mount
   - Motor mount bracket
   - Exhaust manifold

To install:

➡Use new exhaust manifold nuts for assembly.

4. Install or connect the following:
   - Exhaust manifold. Tighten the nuts to 32 ft. lbs. (44 Nm).
   - Motor mount bracket. Tighten the bolts to 27 ft. lbs. (36 Nm).

- Motor mount. Tighten the fasteners to 22 ft. lbs. (30 Nm).
- Exhaust front pipe. Tighten the nuts to 46 ft. lbs. (62 Nm).
- Exhaust manifold heat shield
- HO_2S sensor connectors
- Negative battery cable

5. Start the engine and check for leaks.

Front Crankshaft Seal

REMOVAL & INSTALLATION

3.4L Engine

➡There are 2 methods to replace the oil seal, which are as follows:

OIL PUMP BODY INSTALLED

1. Before servicing the vehicle, refer to the precautions in the beginning of this section.
2. Remove or disconnect the following:
   - Negative battery cable
   - Timing belt and crankshaft pulley
   - Cut off the oil seal lip, using a knife
   - Pry out the oil seal, using a suitable tool

⁂ WARNING

Be careful not to damage the crankshaft.

To install:

3. Install or connect the following:
   - Apply multi-purpose grease to the new oil seal lip
   - Tap in the new oil seal until its surface is flush with the oil pump case edge, using Seal Driver tool 09309-37010 and a mallet
   - Crankshaft pulley and the timing belt
   - Engine undercover, if removed
   - Negative battery cable

OIL PUMP BODY REMOVED

1. Before servicing the vehicle, refer to the precautions in the beginning of this section.
2. Carefully pry out the seal using a suitable tool.
3. Apply multi-purpose grease to the new oil seal lip.
4. Using Seal Driver tool 09309-37010, drive the new seal into place.

4.7L Engine

1. Before servicing the vehicle, refer to the precautions in the beginning of this section.
2. Drain the cooling system.
3. Remove or disconnect the following:
   - Negative battery cable
   - Engine under cover
   - Engine appearance cover
   - Air intake assembly
   - Accessory drive belt
   - Cooling fan and pulley
   - Radiator
   - Drive belt idler pulley
   - Camshaft Position (CMP) sensor connector
   - Upper timing covers
   - Oil cooler pipe
   - Center timing cover
   - A/C compressor
   - Cooling fan bracket
   - Crankshaft pulley
   - Lower timing cover
   - Timing belt. Refer to the Timing Belt unit repair section.
   - Crankshaft timing sprocket
   - Front crankshaft seal

To install:

4. Install the oil seal so that it is flush with the oil pump housing.
5. Install or connect the following:
   - Crankshaft timing sprocket
   - Timing belt
   - Lower timing cover
   - Crankshaft pulley. Tighten the bolt to 181 ft. lbs. (245 Nm).
   - Cooling fan bracket. Tighten the 12mm bolts to 12 ft. lbs. (16 Nm) and the 14mm bolts to 24 ft. lbs. (32 Nm).
   - A/C compressor
   - Center timing cover
   - Oil cooler pipe
   - Upper timing covers
   - CMP sensor connector
   - Drive belt idler pulley. Tighten the bolt to 27 ft. lbs. (37 Nm).
   - Radiator
   - Cooling fan and pulley. Tighten the nuts to 16 ft. lbs. (21 Nm).
   - Accessory drive belt
   - Air intake assembly
   - Engine appearance cover
   - Engine under cover
   - Negative battery cable
6. Fill the cooling system.
7. Start the engine and check for leaks.

Camshaft and Valve Lifters

REMOVAL & INSTALLATION

3.4L Engine

1. Before servicing the vehicle, refer to the precautions in the beginning of this section.
2. Remove or disconnect the following:
   - Negative battery cable
   - Engine undercover
3. Drain the cooling system.
4. Remove or disconnect the following:
   - Air cleaner cap
   - Mass Air Flow (MAF) meter and the resonator
   - Actuator cable with the bracket, if equipped with cruise control
   - Accelerator cable
   - Throttle cable, if equipped with an automatic transmission
   - Heater hose
   - Upper radiator hose from the engine
5. Remove the power steering drive belt, as follows:
   a. Stretch the belt and loosen the fan pulley mounting nuts.
   b. Loosen the lockbolt, pivot bolt and adjusting bolt and remove the drive belt from the engine.
6. Remove or disconnect the following:
   - Air conditioning drive belt, by loosening the idle pulley nut and adjusting bolt
   - Loosen the alternator lockbolt, pivot bolt and adjusting bolt
   - Alternator drive belt
   - No. 2 fan shroud, by removing the 2 clips
   - Fan with the fluid coupling and fan pulleys
   - Power steering pump and move it aside without disconnecting the lines
   - Compressor and move it aside without disconnecting the lines, if equipped with air conditioning
   - Air conditioning bracket, if equipped with air conditioning
   - Spark plug wires with the ignition coils
   - Spark plugs
   - Camshaft Position (CMP) sensor connector, from the No. 2 timing belt cover
   - 3 spark plug wire clamps from the No. 2 timing belt cover
   - 6 bolts and the timing belt cover
   - Power steering adjusting strut by removing the nut
   - Fan bracket by removing the bolt and nut
7. Set the No. 1 cylinder at Top Dead Center (TDC) of the compression stroke, as follows:
   a. Turn the crankshaft pulley and align its groove with the timing mark **0** on the No. 1 timing belt cover.
   b. Check that the timing marks of the

camshaft timing pulleys and the No. 3 timing belt cover are aligned. If not, turn the crankshaft pulley 1 revolution (360 degrees).

8. Remove or disconnect the following:
- Timing belt tensioner, by alternately loosening the 2 bolts
- Pulley bolt, the timing pulley and the knock pin, using Variable Pin Wrench Set 09960-10010
- Both timing pulleys with the timing belt
- No. 2 idler pulley
- Alternator
- Positive Crankcase Ventilation PCV hoses
- Water bypass hoses
- Air assist hose from the intake air connector
- 2 vacuum sensing hoses from the Vacuum Switching Valve (VSV)
- Evaporative Emissions (EVAP) hose
- Air hose from the power steering
- Air hose from the air conditioning idle up valve, if equipped with air conditioning
- 4 bolts, 2 nuts and the air intake chamber assembly
- Intake air connector
- Camshaft Position (CMP) sensor
- No. 3 (rear) timing belt cover by removing the 6 bolts
- 8 bolts, seal washers, both cylinder head cover and gaskets
- Semi-circular plugs

9. Remove the right exhaust camshafts, as follows:

a. Bring the service bolt hole of the driven sub-gear upward by turning the hexagon head portion of the exhaust camshaft with a wrench.

b. Align the timing mark (2 dot marks) of the camshaft drive and driven gears by turning the camshaft with a wrench.

c. Secure the exhaust camshaft sub-gear to the driven gear with a service bolt (6mm diameter, 16–20mm bolt length and 1.0mm in thread pitch).

➡When removing the camshaft, be sure the torsional spring force of the sub-gear has been eliminated by the above operation.

d. Uniformly loosen and remove the bearing cap bolts in several passes, in the sequence shown.

e. Remove the bearing caps and camshaft. Make a note of the bearing cap positions for proper installation.

➡Do not pry on or attempt to force the camshaft with a tool or other object.

10. Remove the right-hand intake camshaft, as follows:

a. Uniformly loosen and remove the bearing cap bolts in several passes, in the sequence shown.

b. Remove the bearing caps, oil seal and camshaft. Make a note of the bearing cap positions for proper installation.

11. Remove the left exhaust camshafts, as follows:

a. Align the timing mark (1 dot mark) of the camshaft drive and driven gears by turning the camshaft with a wrench.

b. Secure the exhaust camshaft sub-gear to the driven gear with a service bolt (6mm diameter, 16–20mm bolt length and 1.0mm in thread pitch).

➡When removing the camshaft, be sure that the torsional spring force of the sub-gear has been eliminated by the above operation.

c. Uniformly loosen and remove the bearing cap bolts in several passes, in the sequence shown.

d. Remove the bearing caps and camshaft. Make a note of the bearing cap positions for proper installation.

12. Remove the left-hand intake camshaft, as follows:

a. Uniformly loosen and remove the bearing cap bolts in several passes, in the sequence shown.

b. Remove the bearing caps, oil seal and camshaft. Make a note of the bearing cap positions for proper installation.

13. Remove the valve lifters and shims from the cylinder head. Arrange the valve lifters and shims in correct order.

To install:

14. Clean all surfaces.

15. Install the valve lifters and shims. Check that the valve lifter rotates smoothly by hand.

16. Install the right intake camshaft, as follows:

a. Apply engine oil to the thrust portion of the intake camshaft.

b. Position the intake camshaft at 90 degrees angle of the timing mark (2 dot marks) on the cylinder head.

c. Install the bearing caps in their proper locations. Apply a light coat of engine oil to the threads and install the cap bolts.

d. Apply a light coat of engine oil on the threads and under the heads of the bearing cap bolts.

e. Uniformly tighten the cap bolts in the sequence shown to 12 ft. lbs. (16 Nm).

17. Install the right exhaust camshaft, as follows:

a. Apply engine oil to the thrust portion of the intake camshaft.

b. Align the timing marks (2 dot marks) of the camshaft drive and driven gears.

c. Roll down the exhaust camshaft onto the bearing journals while engaging the gears with each other. Install the bearing caps in their proper locations.

d. Apply a light coat of engine oil to the threads and install the cap bolts.

e. Apply a light coat of engine oil on the threads and under the heads of the bearing cap bolts.

f. Uniformly tighten the cap bolts in the sequence shown to 12 ft. lbs. (16 Nm).

g. Remove the service bolt from the driven sub-gear. Check that the intake and exhaust camshafts turns smoothly.

h. Align the timing marks (2 dot marks) of the camshaft drive and driven gears by turning the camshaft with a wrench.

18. Install the left intake camshaft, as follows:

a. Apply engine oil to the thrust portion of the intake camshaft.

b. Position the intake camshaft at 90 degrees angle of the timing mark (1 dot mark) on the cylinder head.

c. Install the bearing caps in their proper locations. Apply a light coat of engine oil to the threads and install the cap bolts.

d. Apply a light coat of engine oil on the threads and under the heads of the bearing cap bolts.

e. Uniformly tighten the cap bolts in the sequence shown to 12 ft. lbs. (16 Nm).

19. Install the left exhaust camshaft, as follows:

a. Apply engine oil to the thrust portion of the intake camshaft.

b. Align the timing marks (1 dot mark) of the camshaft drive and driven gears.

c. Roll down the exhaust camshaft onto the bearing journals while engaging the gears with each other. Install the bearing caps in their proper locations.

d. Apply a light coat of engine oil to the threads and install the cap bolts.

e. Apply a light coat of engine oil on the threads and under the heads of the bearing cap bolts.

f. Uniformly tighten the cap bolts in

the sequence shown to 12 ft. lbs. (16 Nm).

g. Remove the service bolt.

20. Check and adjust the valve clearance.
21. Install or connect the following:

- Semi-circular plugs
- Cylinder head covers. Tighten the bolts, in several passes, to 53 inch lbs. (6 Nm).
- Alternator bracket. Tighten the bolts to 14 ft. lbs. (18 Nm).
- No. 3 timing belt cover. Tighten the 6 bolts to 80 inch lbs. (9 Nm).
- CMP sensor. Tighten it to 71 inch lbs. (8 Nm).
- Intake air connector
- Hoses
- Alternator but do not tighten the bolts and nuts at this time
- No. 2 timing belt idler. Tighten the bolt to 30 ft. lbs. (40 Nm).

➡Check that the pulley bracket moves smoothly.

22. Install the left camshaft timing pulley, as follows:

a. Install the knock pin to the camshaft.

b. Align the knock pin hose of the camshaft with the knock pin groove of the timing pulley.

c. Slide the timing pulley on the camshaft with the flange side facing outward. Tighten the pulley bolt to 81 ft. lbs. (110 Nm).

23. Set the No. 1 cylinder to TDC of the compression stroke.

a. Turn the crankshaft pulley and align its groove with the timing mark **0** on the No. 1 timing belt cover.

b. Turn the camshaft, align the knock pin hole of the camshaft with the timing mark of the No. 3 timing belt cover.

c. Turn the camshaft timing pulley, align the timing marks of the camshaft timing pulley and the No. 3 timing belt cover.

24. Connect the timing belt to the left camshaft timing pulley, as follows:

a. Check that the installation mark on the timing belt is aligned with the end of the No. 1 timing belt cover.

b. Using Variable Pin Wrench Set 09960-01000 or equivalent, slightly turn the left camshaft timing pulley clockwise. Align the installation mark on the timing belt with the timing mark of the camshaft timing pulley, and hang the timing belt on the left camshaft timing pulley.

c. Align the timing marks of the left camshaft pulley and the No. 3 timing belt cover.

d. Check that the timing belt has tension between the crankshaft timing pulley and the left camshaft timing pulley.

25. Install the right camshaft timing pulley and the timing belt, as follows:

a. Align the installation mark on the timing belt with the timing mark of the right camshaft timing pulley, and hang the timing belt on the right camshaft timing pulley with the flange side facing inward.

b. Slide the right camshaft timing

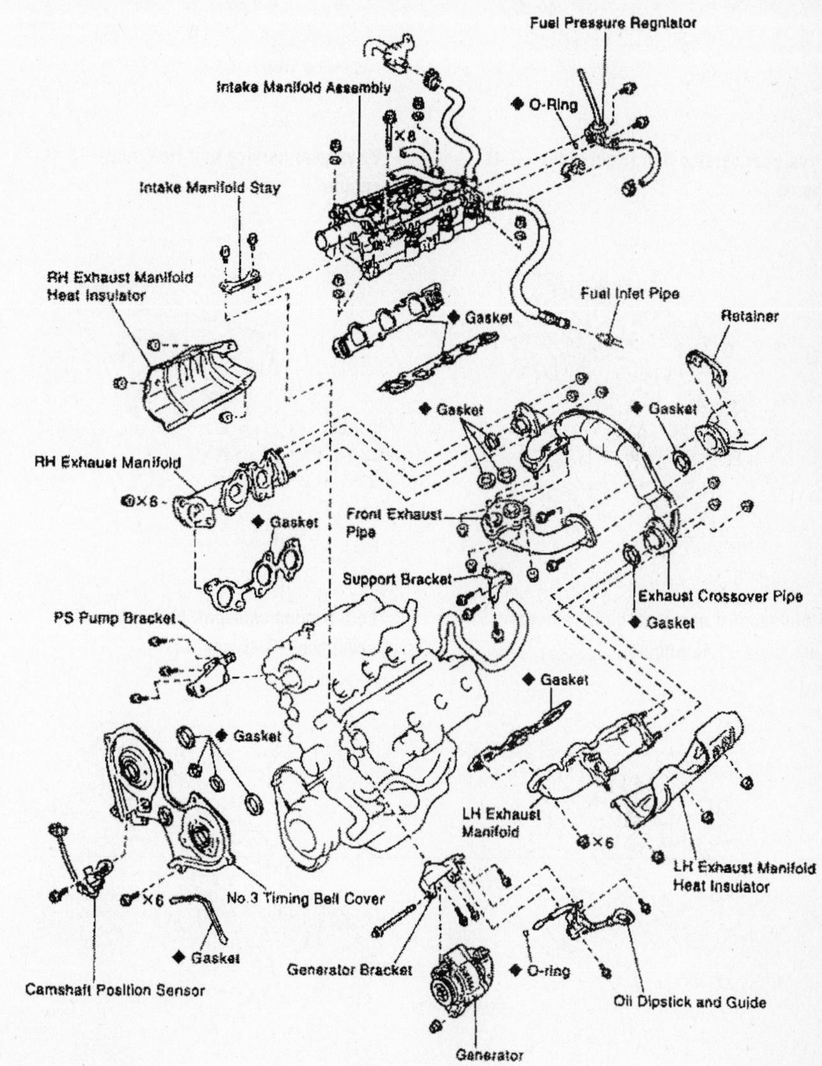

Exploded view of the cylinder head component assembly—3.4L engine

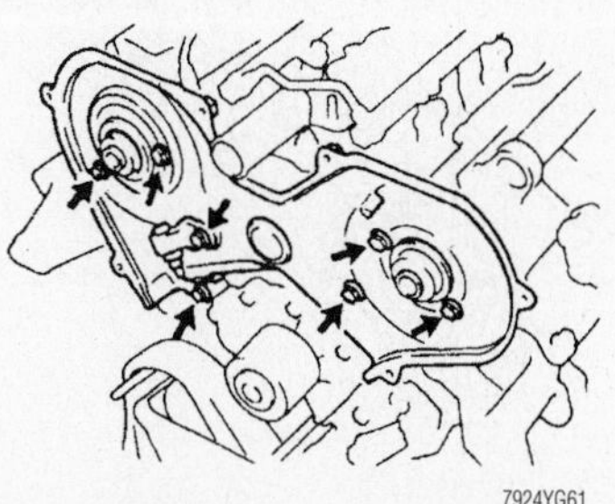

Rear timing belt cover mounting bolt locations—3.4L engine

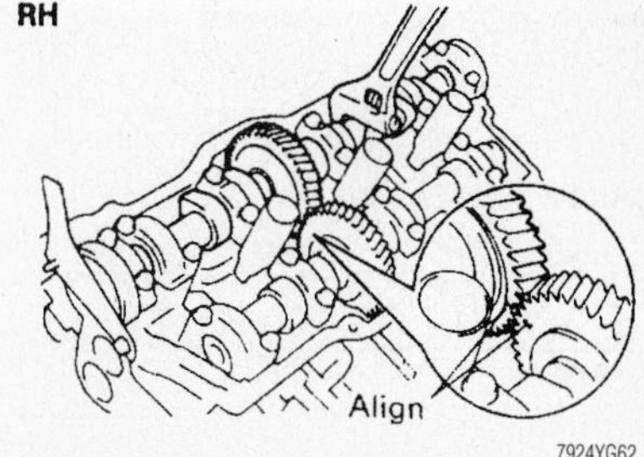

Aligning the timing marks (2 dot marks) of the right camshafts—3.4L engine

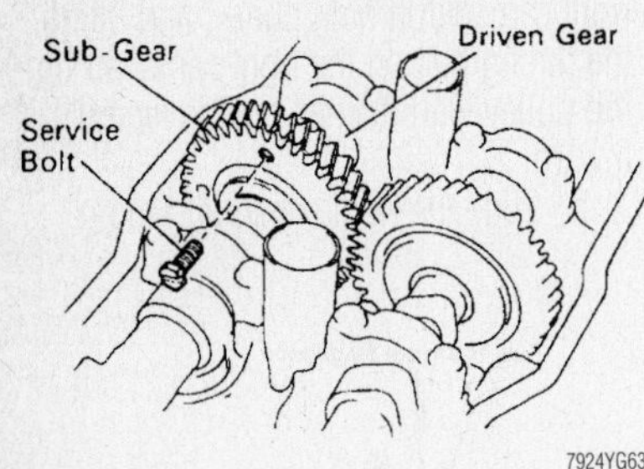

Drive gear service bolt (right side)—3.4L engine

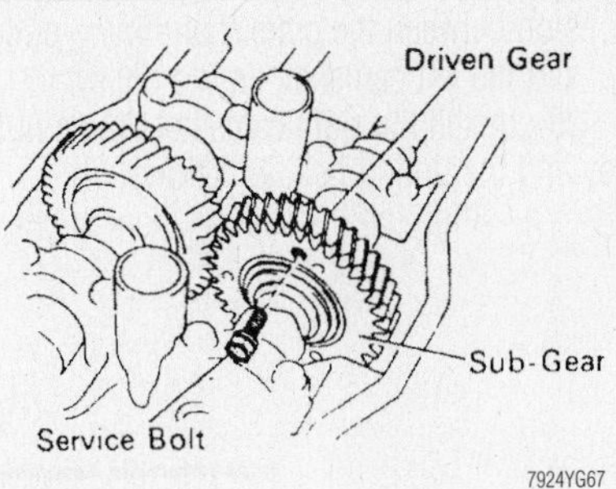

Drive gear service bolt (left side)—3.4L engine

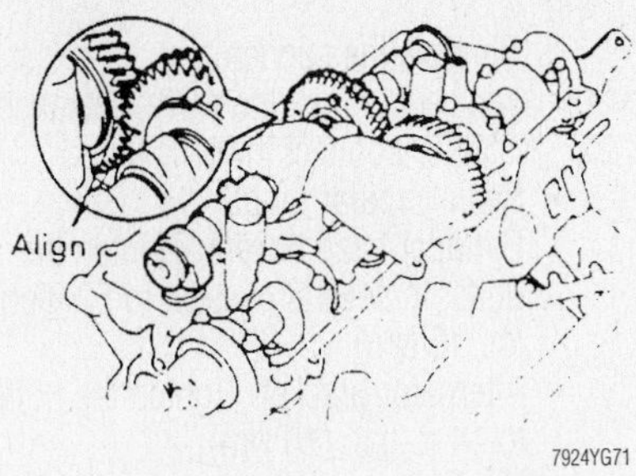

Aligning the right camshafts for installation—3.4L engine

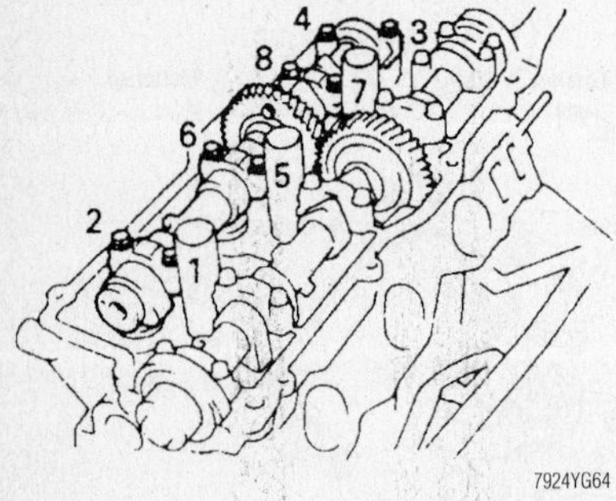

Right exhaust camshaft bolts removal sequence—3.4L engine

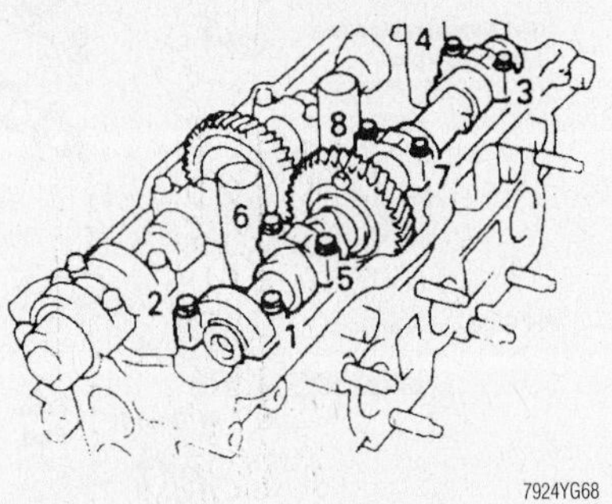

Left exhaust camshaft bolts removal sequence—3.4L engine

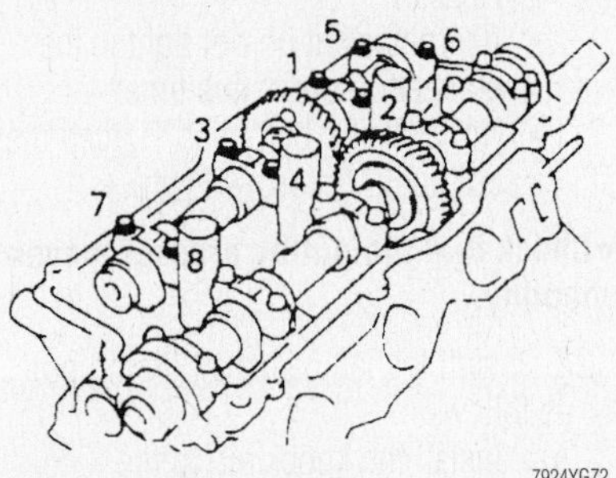

Right exhaust camshaft bolts tightening sequence—3.4L engine

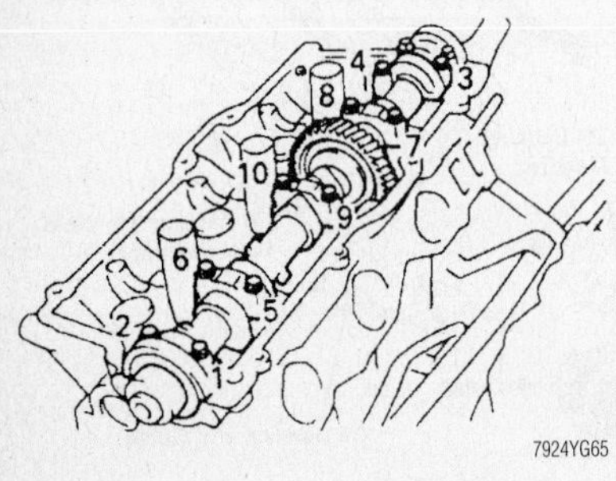

Right intake camshaft bolts removal sequence—3.4L engine

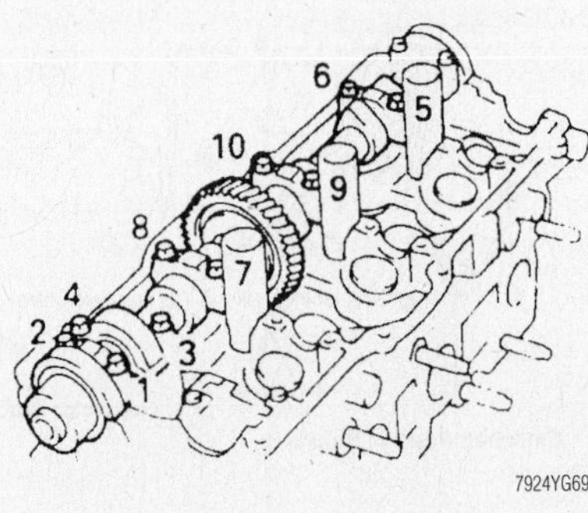

Left intake camshaft bolts removal sequence—3.4L engine

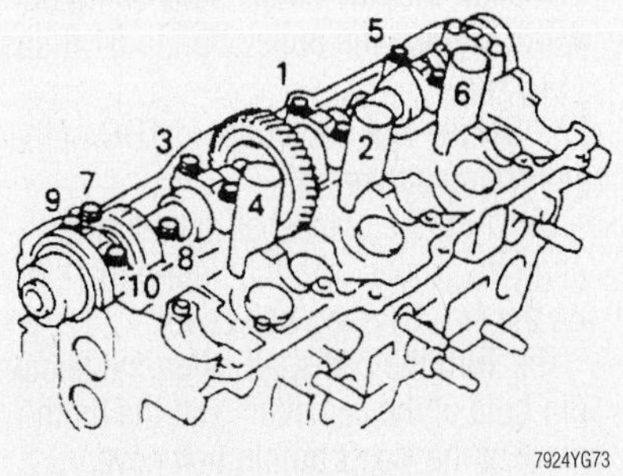

Left intake camshaft bolts tightening sequence—3.4L engine

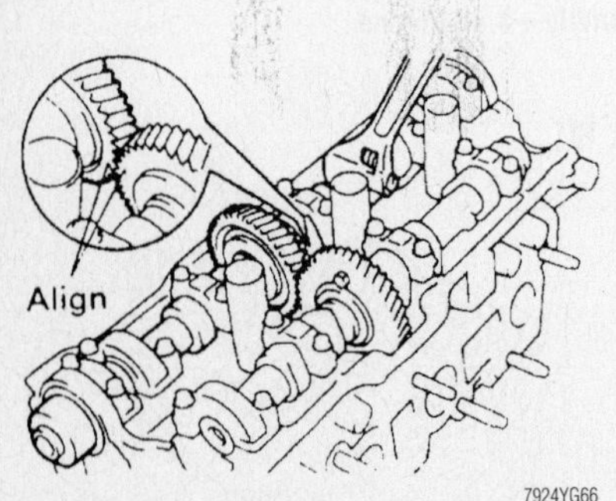

Aligning the timing mark (1 dot mark) of the left camshafts—3.4L engine

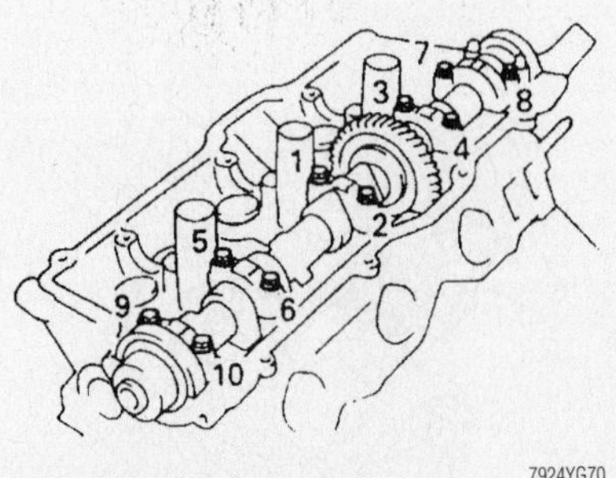

Right intake camshaft tightening sequence—3.4L engine

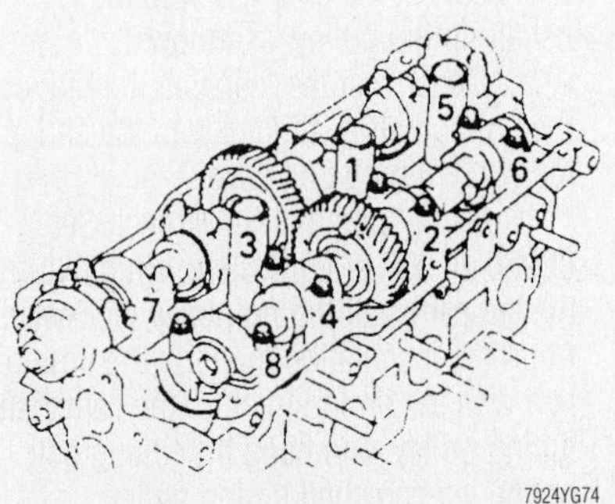

Left exhaust camshaft bolts tightening sequence—3.4L engine

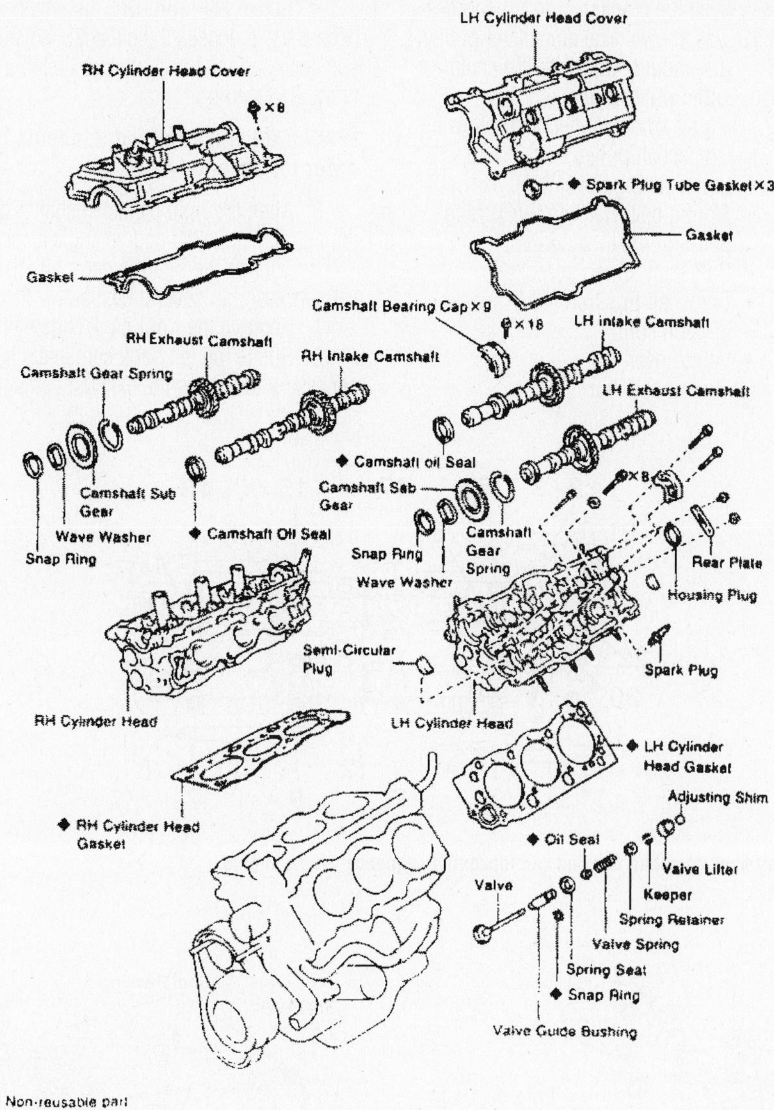

Exploded view of the cylinder head component assembly—3.4L engine

pulley on the camshaft. Align the timing marks on the right camshaft timing pulley and the No. 3 timing belt cover.

c. Align the knock pin hole of the camshaft with the knock pin groove of the pulley.

d. Install the knock pin. Tighten the bolt to 81 ft. lbs. (110 Nm).

26. Set the timing belt tensioner, as follows:

a. Using a press, slowly press in the pushrod using 220–2,205 lbs. (981–9,807 N) of force.

b. Align the holes of the pushrod and housing, pass a 1.5mm hexagon wrench through the holes to keep the setting position of the pushrod.

c. Release the press and install the dust boot to the tensioner.

27. Install the timing belt tensioner and alternately tighten the bolts to 20 ft. lbs. (28 Nm). Using pliers, remove the 1.5mm hexagon wrench from the belt tensioner.

28. Check the valve timing, as follows:

a. Slowly turn the crankshaft pulley 2 revolutions from the TDC-to-TDC. Always turn the crankshaft pulley clockwise.

b. Check that each pulley aligns with the timing marks. If the timing marks do not align, remove the timing belt and reinstall it.

29. Install or connect the following:

- Fan bracket with the bolt and nut
- Power steering adjusting strut with the nut
- No. 2 timing belt cover. Tighten the bolts to 80 inch lbs. (9 Nm).
- Negative battery cable

30. Fill the radiator with engine coolant.
31. Start the engine and check for leaks.
32. Check the ignition timing.
33. Install the engine undercover.
34. Road test the vehicle.
35. Recheck all fluid levels.

4.7L Engine

1. Before servicing the vehicle, refer to the precautions in the beginning of this section.
2. Drain the cooling system.
3. Relieve the fuel system pressure.
4. Remove or disconnect the following:

- Negative battery cable
- Engine under covers
- Engine appearance cover
- Air intake hose
- Accessory drive belt
- Cooling fan
- Radiator
- Idler pulley
- Upper and middle timing belt covers
- A/C compressor
- Cooling fan bracket
- Alternator
- Accessory drive belt tensioner

5. Set the engine to Top Dead Center (TDC) with the camshaft sprocket timing marks aligned with the rear cover timing marks.

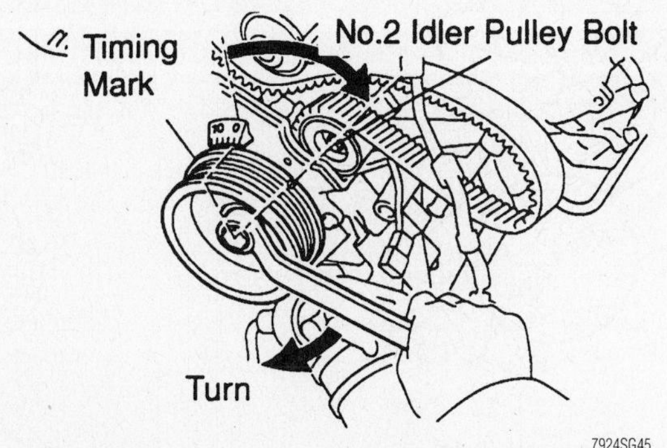

Setting the crankshaft to 50 degrees ATDC—4.7L engine

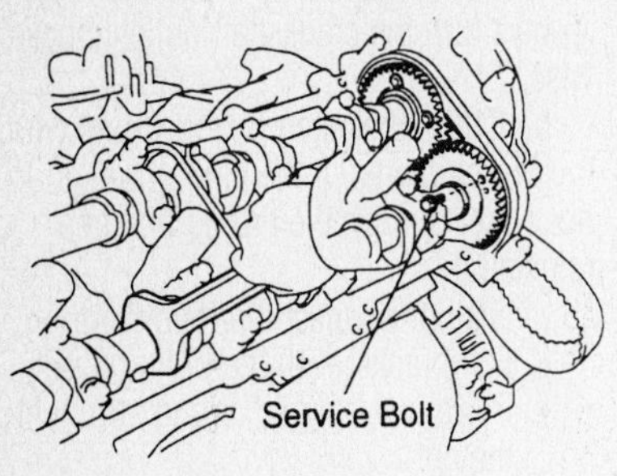

Camshaft service bolt installation—4.7L engine

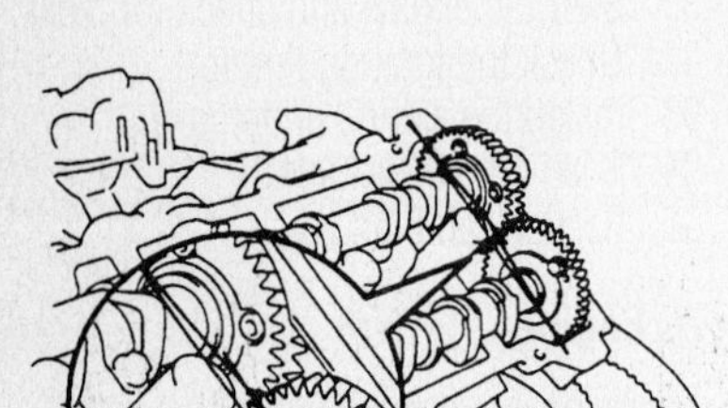

Right bank camshaft timing mark (1 dot marks) alignment—4.7L engine

6. Rotate the crankshaft to 50 degrees After TDC as shown. The crankshaft pulley timing mark should align with the center of the No. 2 idler pulley bolt.
7. Remove or disconnect the following:
   - Crankshaft pulley
   - Lower timing cover
   - Timing belt. Refer to the Timing Belt unit repair section.
   - Camshaft timing sprockets
   - Camshaft Position (CMP) sensor
   - Ignition coils
   - Valve cover
   - Timing belt rear covers

8. Rotate the right bank camshafts as necessary to access the exhaust camshaft sub-gear service bolt hole and install a 6mm x 1.0mm bolt.

➡**Keep all valvetrain components in order for assembly.**

9. Align the right bank camshaft 1 dot timing marks to a **10** degree angle as shown.
10. Loosen the bearing cap bolts in sequence and in several passes.
11. Remove the right bank camshafts.
12. Rotate the left bank camshafts as necessary to access the exhaust camshaft

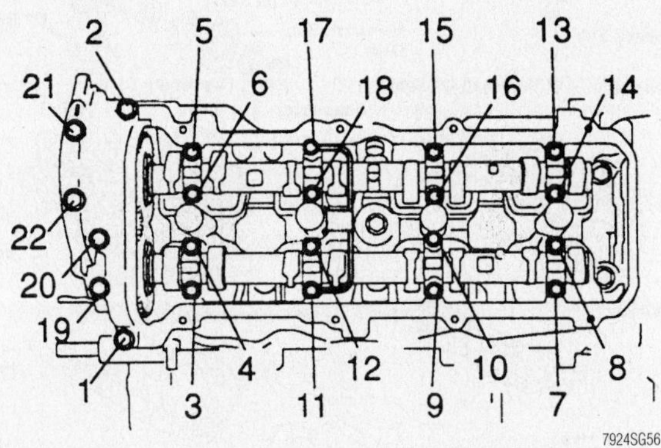

Left bank camshaft bearing cap loosening sequence—4.7L engine

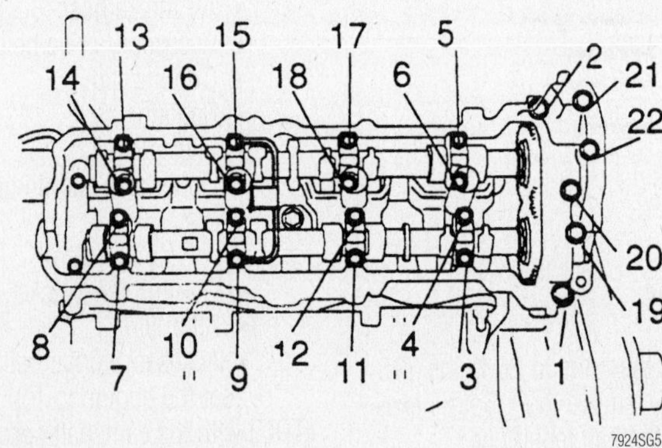

Right bank camshaft bearing cap loosening sequence—4.7L engine

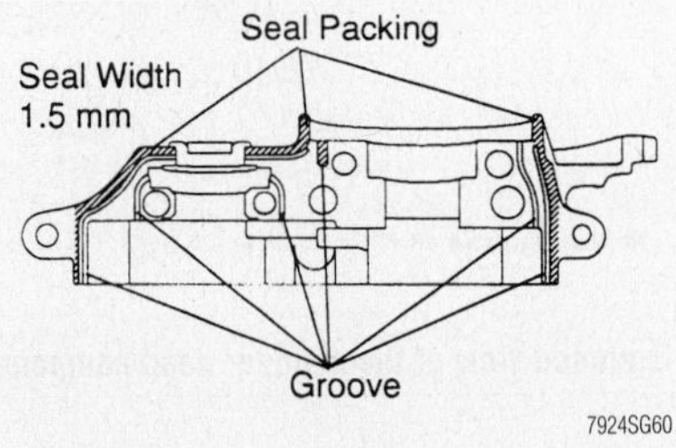

Apply a 1.5mm bead of sealant to the front bearing caps—4.7L engine

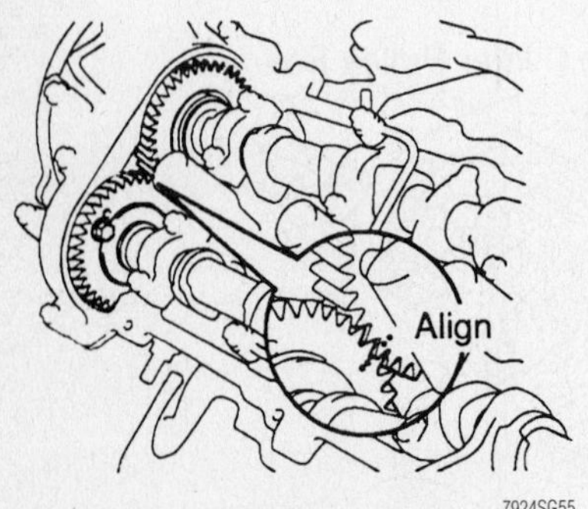

Left bank camshaft timing mark (2 dot marks) alignment—4.7L engine

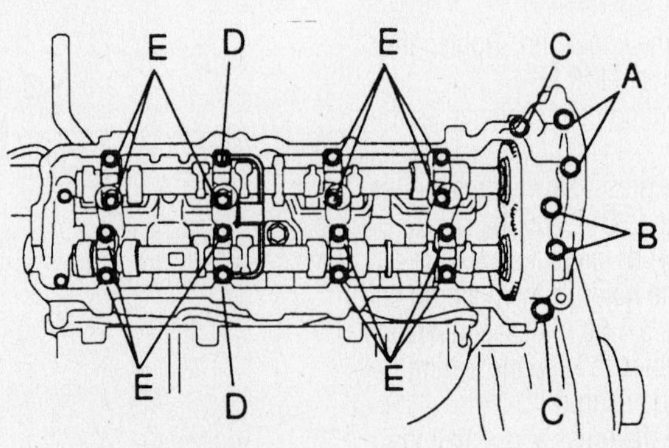

Right bank bearing cap bolt location—4.7L engine

sub-gear service bolt hole and install a 6mm x 1.0mm bolt.

13. Align the left bank camshaft 2 dot timing marks as shown.

14. Loosen the bearing cap bolts in sequence and in several passes.

15. Remove the left bank camshafts.

16. Remove the valve lifters and shims.

To install:

17. Ensure that the crankshaft is at 50 degrees After TDC.

18. Install or connect the following:

- Valve lifters and shims in their original positions
- Right bank camshafts with the 1 dot timing marks at 10 degrees
- Left bank camshafts with the 2 dot timing marks aligned
- Left and right bank camshaft bearing caps in their original positions. Apply sealant to the front bearing caps as shown.
- Camshaft oil seals

19. The bearing cap bolts vary in length and are identified as follows:

- A: 3.70 inches (94mm)
- B: 2.83 inches (72mm)
- C: 0.98 inches (25mm)
- D: 2.05 inches (52mm)
- E: 1.50 inches (38mm)

20. Bolts in positions **A**, **B** and **C** are installed dry.

21. Lubricate the threads and under the contact flange for bolts in positions **D** and **E**.

22. Install oil feed pipes and the bearing cap bolts according to position in the illustrations.

23. Tighten the camshaft bearing bolts in sequence and in several passes to the following specifications:

- Bolt C: 66 inch lbs. (7.5 Nm)
- All others: 12 ft. lbs. (16 Nm)

24. Remove the service bolts from the exhaust camshaft gears.

25. Install or connect the following:

- Timing belt rear covers
- Valve cover
- Ignition coils
- CMP sensor
- Camshaft timing sprockets. Tighten the bolts to 80 ft. lbs. (108 Nm).
- Timing belt
- Lower timing cover
- Crankshaft pulley. Tighten the bolt to 181 ft. lbs. (245 Nm).
- Accessory drive belt tensioner
- Alternator
- Cooling fan bracket
- A/C compressor
- Upper and middle timing belt covers
- Idler pulley. Tighten the bolt to 27 ft. lbs. (37 Nm).
- Radiator
- Cooling fan
- Accessory drive belt
- Air intake hose
- Engine appearance cover
- Engine under covers
- Negative battery cable

26. Fill the cooling system.

27. Start the engine and check for leaks.

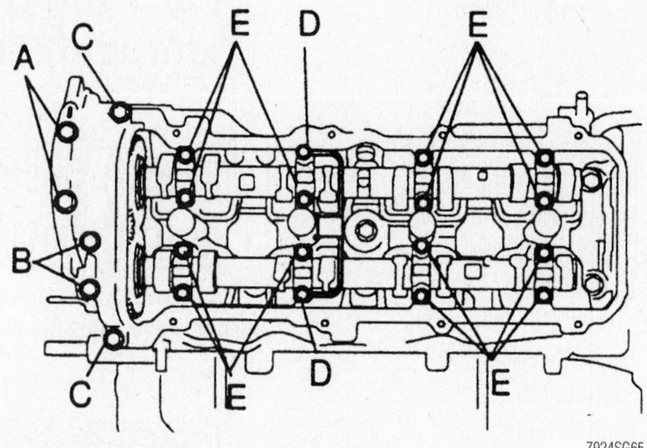

Left camshaft bearing cap bolt locations—4.7L engine

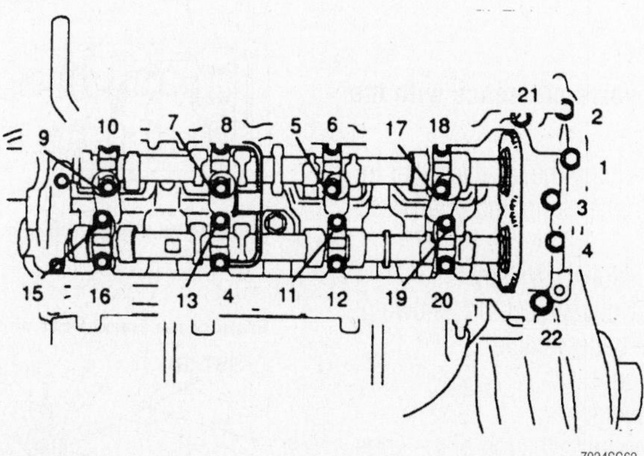

Right bank camshaft bearing cap bolt torque sequence—4.7L engine

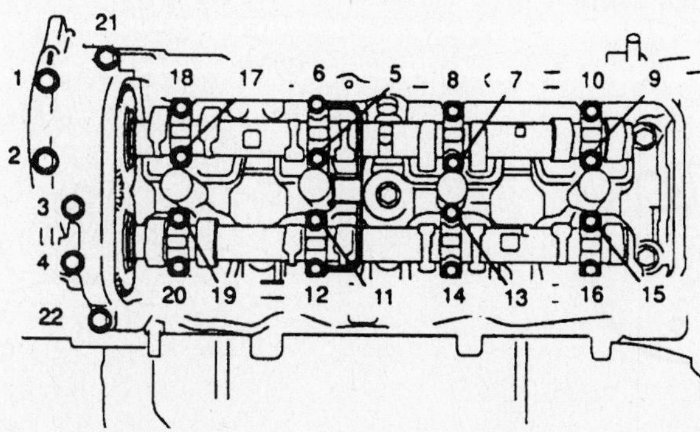

Left bank camshaft bearing cap bolt torque sequence—4.7L engine

Valve Lash

ADJUSTMENT

3.4L Engine

1. Before servicing the vehicle, refer to the precautions in the beginning of this section.
2. Disconnect the negative battery cable.
3. Drain the engine coolant.
4. Remove or disconnect the following:
   - Air intake connector
   - Cylinder head cover
5. Set the No. 1 cylinder to Top Dead

Center (TDC) of the compression stroke, as follows:

a. Turn the crankshaft pulley clockwise and align its groove with the **0** mark on the timing chain cover.

b. Check that the timing marks (1 and 2 dots) of the camshaft drive and driven gears are in a straight line on the cylinder head surface. If not, turn the crankshaft 1 revolution (360 degrees) and align the marks.

6. Inspect the valve clearance, as follows:

a. Measure the clearance between the valve lifter and the camshaft. Measure the 1st intake and the 3rd exhaust valves on the right head and the 6th intake and the 2nd exhaust valves on the left head.

b. Turn the crankshaft ⅔ of a revolution (240 degrees) and adjust the 3rd intake and the 5th exhaust valves on the right head and the 2nd intake and the 4th exhaust valves on the left head.

c. Turn the crankshaft ⅔ of a revolution (240 degrees) and adjust the 5th intake and the 1st exhaust valves on the right head and the 4th intake and the 6th exhaust valves on the left head.

7. Valve clearance cold should be:

- Intake: 0.006–0.009 in. (0.13–0.23mm)
- Exhaust: 0.011–0.014 in. (0.27–0.37mm)

8. Adjust the valve clearance by using adjusting shims, as follows:

a. Turn the equipment camshaft so that the cam lobe for the valve to be adjusted faces up.

b. Turn the valve lifter so that the notches are perpendicular to the camshaft.

c. Using SST 09248-55040, press down the valve lifter and place SST 09248-05420, between the camshaft and the valve lifter. Remove SST 09248-55040.

d. Remove the adjusting shim with a small flat prying tool and a magnetic finger.

e. Determine the replacement adjusting shim size according to the following formula or use the adjusting shim charts.

f. Using a micrometer, measure the thickness of the removed shim. Calculate the thickness of a new shim so that the valve clearance comes within the specified value.

- T: Thickness of the removed shim
- A: Measured valve clearance
- N: Thickness of the new shim

g. Intake: N = T + (A −0.007 in. (0.18mm))

h. Exhaust: N = T + (A −0.013 in. (0.32mm))

i. Install a new adjusting shim. Place it on the valve lifter. Using the SST 09248-55040, press down the valve lifter and remove SST 09248-05420.

j. Recheck the valve clearance.

9. Install or connect the following:

- Cylinder head cover
- Intake air connector
- Negative battery cable

10. Refill with engine coolant.

11. Start the engine and check for leaks.

4.7L Engine

**Measure the valve clearance with the engine cold.**

1. Before servicing the vehicle, refer to the precautions in the beginning of this section.
2. Drain the cooling system.
3. Remove or disconnect the following:

- Negative battery cable
- Ignition coils
- Valve covers

4. Set the engine to the top of the com-

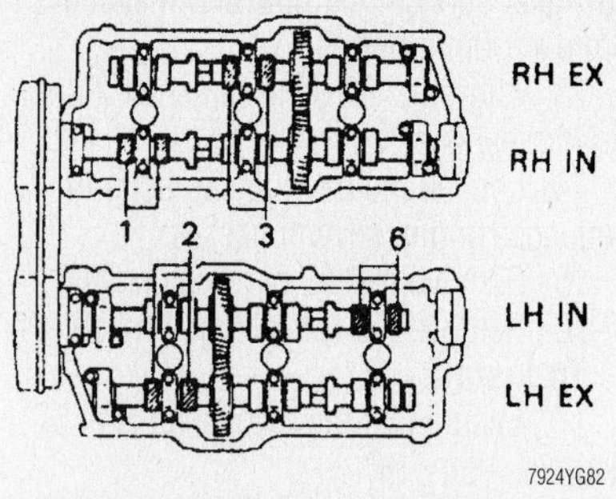

First valve adjustment—3.4L engine

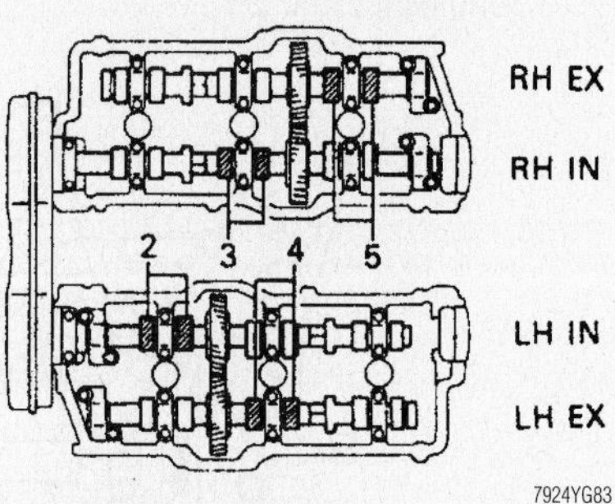

Second valve adjustment—3.4L engine

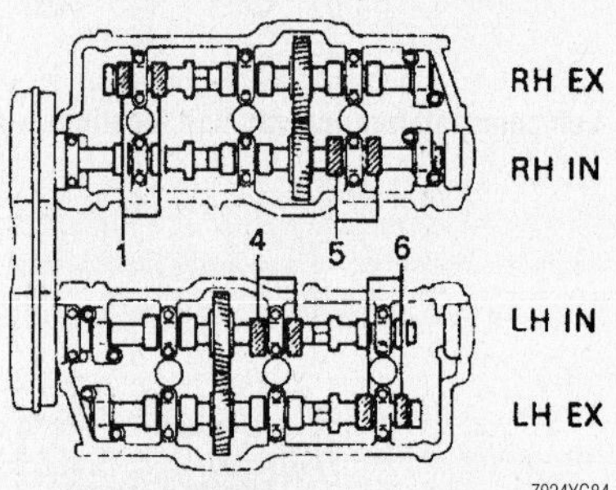

Third valve adjustment—3.4L engine

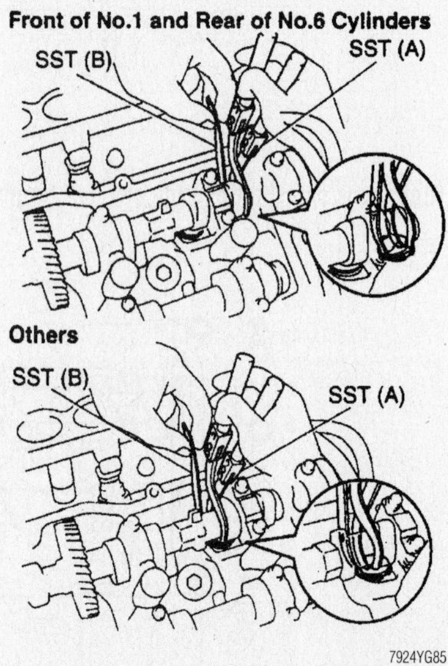

Removing the adjusting shim—3.4L engine

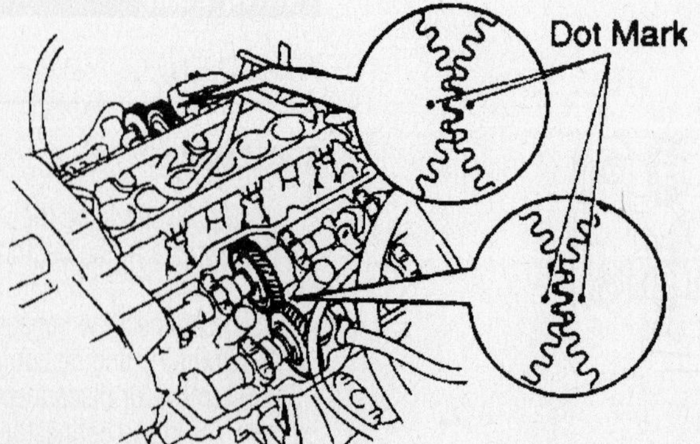

Aligning the timing marks—3.4L engine

Measured clearance mm (in.) \ Installed shim thickness mm (in.)	2.000 (0.0787)	2.020 (0.0795)	2.040 (0.0803)	2.060 (0.0811)	2.080 (0.0819)	2.100 (0.0827)	2.120 (0.0835)	2.140 (0.0843)	2.160 (0.0850)	2.180 (0.0858)	2.200 (0.0866)	2.210 (0.0870)	2.220 (0.0874)	2.230 (0.0878)	2.240 (0.0882)	2.250 (0.0886)	2.260 (0.0890)	2.270 (0.0894)	2.280 (0.0898)	2.290 (0.0902)	2.300 (0.0906)	2.310 (0.0909)	2.320 (0.0913)	2.330 (0.0917)	2.340 (0.0921)	2.350 (0.0925)	2.360 (0.0929)	2.370 (0.0933)	2.380 (0.0937)	2.390 (0.0941)	2.400 (0.0945)	2.410 (0.0949)	2.420 (0.0953)	2.430 (0.0957)	2.440 (0.0961)	2.450 (0.0965)	2.460 (0.0969)	2.470 (0.0972)	2.480 (0.0976)	2.490 (0.0980)	2.500 (0.0984)	2.510 (0.0988)	2.520 (0.0992)	2.530 (0.0996)	2.540 (0.1000)	2.550 (0.1004)	2.560 (0.1008)	2.570 (0.1012)	2.580 (0.1016)	2.590 (0.1020)	2.600 (0.1024)	2.620 (0.1031)	2.640 (0.1039)	2.660 (0.1047)	2.680 (0.1055)	2.700 (0.1063)	2.720 (0.1071)	2.740 (0.1079)	2.760 (0.1087)	2.780 (0.1094)	2.800 (0.1102)
0.000–0.030 (0.0000–0.0012)							00	00	00	00	02	02	04	04	06	06	08	08	10	10	12	12	14	14	16	16	18	18	20	20	22	22	24	24	26	26	28	28	30	30	32	32	34	34	36	36	38	38	40	40	42	44	46	48	50	52	54	56	58	60	62
0.031–0.050 (0.0012–0.0020)						00	00	00	00	02	04	06	06	08	08	10	10	12	12	14	14	16	16	18	18	20	20	22	22	24	24	26	26	28	28	30	30	32	32	34	34	36	36	38	38	40	40	42	42	44	44	46	48	50	52	54	56	58	60	62	64
0.051–0.070 (0.0020–0.0028)					00	00	00	00	02	04	06	08	08	10	10	12	12	14	14	16	16	18	18	20	20	22	22	24	24	26	26	28	28	30	30	32	32	34	34	36	36	38	38	40	40	42	42	44	44	46	46	48	50	52	54	56	58	60	62	64	66
0.071–0.090 (0.0028–0.0035)				00	00	00	00	02	04	06	08	10	10	12	12	14	14	16	16	18	18	20	20	22	22	24	24	26	26	28	28	30	30	32	32	34	34	36	36	38	38	40	40	42	42	44	44	46	46	48	48	50	52	54	56	58	60	62	64	66	68
0.091–0.110 (0.0036–0.0043)			00	00	00	00	02	04	06	08	10	12	12	14	14	16	16	18	18	20	20	22	22	24	24	26	26	28	28	30	30	32	32	34	34	36	36	38	38	40	40	42	42	44	44	46	46	48	48	50	50	52	54	56	58	60	62	64	66	68	70
0.111–0.130 (0.0044–0.0051)		00	00	00	00	02	04	06	08	10	12	14	14	16	16	18	18	20	20	22	22	24	24	26	26	28	28	30	30	32	32	34	34	36	36	38	38	40	40	42	42	44	44	46	46	48	48	50	50	52	52	54	56	58	60	62	64	66	68	70	72
0.131–0.149 (0.0052–0.0059)		00	00	00	02	04	06	08	10	12	14	16	16	18	18	20	20	22	22	24	24	26	26	28	28	30	30	32	32	34	34	36	36	38	38	40	40	42	42	44	44	46	46	48	48	50	50	52	52	54	54	56	58	60	62	64	66	68	70	72	74
0.150–0.250 (0.0059–0.0098)|																																																												
0.251–0.270 (0.0099–0.0106)	06	08	10	12	14	16	18	20	22	24	26	28	28	30	30	32	32	34	34	36	36	38	38	40	40	42	42	44	44	46	46	48	48	50	50	52	52	54	54	56	56	58	58	60	60	62	62	64	64	66	66	68	70	72	74	76	78	80	80	80	
0.271–0.290 (0.0107–0.0114)	08	10	12	14	16	18	20	22	24	26	28	30	30	32	32	34	34	36	36	38	38	40	40	42	42	44	44	46	46	48	48	50	50	52	52	54	54	56	56	58	58	60	60	62	62	64	64	66	66	68	68	70	72	74	76	78	80	80	80		
0.291–0.310 (0.0115–0.0122)	10	12	14	16	18	20	22	24	26	28	30	32	32	34	34	36	36	38	38	40	40	42	42	44	44	46	46	48	48	50	50	52	52	54	54	56	56	58	58	60	60	62	62	64	64	66	66	68	68	70	70	72	74	76	78	80	80	80			
0.311–0.330 (0.0122–0.0130)	12	14	16	18	20	22	24	26	28	30	32	34	34	36	36	38	38	40	40	42	42	44	44	46	46	48	48	50	50	52	52	54	54	56	56	58	58	60	60	62	62	64	64	66	66	68	68	70	70	72	72	74	76	78	80	80	80				
0.331–0.350 (0.0130–0.0138)	14	16	18	20	22	24	26	28	30	32	34	36	36	38	38	40	40	42	42	44	44	46	46	48	48	50	50	52	52	54	54	56	56	58	58	60	60	62	62	64	64	66	66	68	68	70	70	72	72	74	74	76	78	80	80	80					
0.351–0.370 (0.0138–0.0146)	16	18	20	22	24	26	28	30	32	34	36	38	38	40	40	42	42	44	44	46	46	48	48	50	50	52	52	54	54	56	56	58	58	60	60	62	62	64	64	66	66	68	68	70	70	72	72	74	74	76	76	78	80	80	80						
0.371–0.390 (0.0146–0.0154)	18	20	22	24	26	28	30	32	34	36	38	40	40	42	42	44	44	46	46	48	48	50	50	52	52	54	54	56	56	58	58	60	60	62	62	64	64	66	66	68	68	70	70	72	72	74	74	76	76	78	78	80	80	80							
0.391–0.410 (0.0154–0.0161)	20	22	24	26	28	30	32	34	36	38	40	42	42	44	44	46	46	48	48	50	50	52	52	54	54	56	56	58	58	60	60	62	62	64	64	66	66	68	68	70	70	72	72	74	74	76	76	78	78	80	80	80	80								
0.411–0.430 (0.0162–0.0169)	22	24	26	28	30	32	34	36	38	40	42	44	44	46	46	48	48	50	50	52	52	54	54	56	56	58	58	60	60	62	62	64	64	66	66	68	68	70	70	72	72	74	74	76	76	78	78	80	80	80	80	80									
0.431–0.450 (0.0170–0.0177)	24	26	28	30	32	34	36	38	40	42	44	46	46	48	48	50	50	52	52	54	54	56	56	58	58	60	60	62	62	64	64	66	66	68	68	70	70	72	72	74	74	76	76	78	78	80	80	80	80	80	80										
0.451–0.470 (0.0178–0.0185)	26	28	30	32	34	36	38	40	42	44	46	48	48	50	50	52	52	54	54	56	56	58	58	60	60	62	62	64	64	66	66	68	68	70	70	72	72	74	74	76	76	78	78	80	80	80	80	80	80	80											
0.471–0.490 (0.0185–0.0193)	28	30	32	34	36	38	40	42	44	46	48	50	50	52	52	54	54	56	56	58	58	60	60	62	62	64	64	66	66	68	68	70	70	72	72	74	74	76	76	78	78	80	80	80	80	80	80	80													
0.491–0.510 (0.0193–0.0201)	30	32	34	36	38	40	42	44	46	48	50	52	52	54	54	56	56	58	58	60	60	62	62	64	64	66	66	68	68	70	70	72	72	74	74	76	76	78	78	80	80	80	80	80	80	80															
0.511–0.530 (0.0201–0.0209)	32	34	36	38	40	42	44	46	48	50	52	54	54	56	56	58	58	60	60	62	62	64	64	66	66	68	68	70	70	72	72	74	74	76	76	78	78	80	80	80	80	80	80	80																	
0.531–0.550 (0.0209–0.0217)	34	36	38	40	42	44	46	48	50	52	54	56	56	58	58	60	60	62	62	64	64	66	66	68	68	70	70	72	72	74	74	76	76	78	78	80	80	80	80	80	80	80																			
0.551–0.570 (0.0217–0.0224)	36	38	40	42	44	46	48	50	52	54	56	58	58	60	60	62	62	64	64	66	66	68	68	70	70	72	72	74	74	76	76	78	78	80	80	80	80	80	80	80																					
0.571–0.590 (0.0225–0.0232)	38	40	42	44	46	48	50	52	54	56	58	60	60	62	62	64	64	66	66	68	68	70	70	72	72	74	74	76	76	78	78	80	80	80	80	80	80	80																							
0.591–0.610 (0.0233–0.0240)	40	42	44	46	48	50	52	54	56	58	60	62	62	64	64	66	66	68	68	70	70	72	72	74	74	76	76	78	78	80	80	80	80	80	80	80																									
0.611–0.630 (0.0241–0.0248)	42	44	46	48	50	52	54	56	58	60	62	64	64	66	66	68	68	70	70	72	72	74	74	76	76	78	78	80	80	80	80	80	80	80																											
0.631–0.650 (0.0248–0.0256)	44	46	48	50	52	54	56	58	60	62	64	66	66	68	68	70	70	72	72	74	74	76	76	78	78	80	80	80	80	80	80	80																													
0.651–0.670 (0.0256–0.0264)	46	48	50	52	54	56	58	60	62	64	66	68	68	70	70	72	72	74	74	76	76	78	78	80	80	80	80	80	80	80																															
0.671–0.690 (0.0264–0.0272)	48	50	52	54	56	58	60	62	64	66	68	70	70	72	72	74	74	76	76	78	78	80	80	80	80	80	80	80																																	
0.691–0.710 (0.0272–0.0280)	50	52	54	56	58	60	62	64	66	68	70	72	72	74	74	76	76	78	78	80	80	80	80	80	80	80																																			
0.711–0.730 (0.0280–0.0287)	52	54	56	58	60	62	64	66	68	70	72	74	74	76	76	78	78	80	80	80	80	80	80	80																																					
0.731–0.750 (0.0288–0.0295)	54	56	58	60	62	64	66	68	70	72	74	76	76	78	78	80	80	80	80	80	80	80																																							
0.751–0.770 (0.0296–0.0303)	56	58	60	62	64	66	68	70	72	74	76	78	78	80	80	80	80	80	80	80																																									
0.771–0.790 (0.0304–0.0311)	58	60	62	64	66	68	70	72	74	76	78	80	80	80	80	80	80	80																																											
0.791–0.810 (0.0311–0.0319)	60	62	64	66	68	70	72	74	76	78	80	80	80	80	80	80																																													
0.811–0.830 (0.0319–0.0327)	62	64	66	68	70	72	74	76	78	80	80	80	80	80																																															
0.831–0.850 (0.0327–0.0335)	64	66	68	70	72	74	76	78	80	80	80	80																																																	
0.851–0.870 (0.0335–0.0343)	66	68	70	72	74	76	78	80	80	80																																																			
0.871–0.890 (0.0343–0.0350)	68	70	72	74	76	78	80	80	80																																																					
0.891–0.910 (0.0351–0.0358)	70	72	74	76	78	80	80	80																																																						
0.911–0.930 (0.0359–0.0366)	72	74	76	78	80	80	80																																																							
0.931–0.950 (0.0367–0.0374)	74	76	78	80	80	80																																																								
0.951–0.970 (0.0374–0.0382)	76	78	80	80	80																																																									
0.971–0.990 (0.0382–0.0390)	78	80	80	80																																																										
0.991–1.010 (0.0390–0.0398)	80	80	80																																																											
1.011–1.030 (0.0398–0.0406)	80	80																																																												
1.031–1.050 (0.0406–0.0413)|80|																																																												

Intake valve clearance (Cold):
0.15 – 0.25 mm (0.006 – 0.010 in.)

EXAMPLE:
The 2.300 mm (0.0906 in.) shim is installed, and the measured clearance is 0.440 mm (0.0173 in.). Replace the 2.300 mm (0.0906 in.) shim with a No. 54 shim.

New shim thickness mm (in.)

Shim No.	Thickness	Shim No.	Thickness	Shim No.	Thickness
00	2.000 (0.0787)	28	2.280 (0.0898)	56	2.560 (0.1008)
02	2.020 (0.0795)	30	2.300 (0.0906)	58	2.580 (0.1016)
04	2.040 (0.0803)	32	2.320 (0.0913)	60	2.600 (0.1024)
06	2.060 (0.0811)	34	2.340 (0.0921)	62	2.620 (0.1031)
08	2.080 (0.0819)	36	2.360 (0.0929)	64	2.640 (0.1039)
10	2.100 (0.0827)	38	2.380 (0.0937)	66	2.660 (0.1047)
12	2.120 (0.0835)	40	2.400 (0.0945)	68	2.680 (0.1055)
14	2.140 (0.0843)	42	2.420 (0.0953)	70	2.700 (0.1063)
16	2.160 (0.0850)	44	2.440 (0.0961)	72	2.720 (0.1071)
18	2.180 (0.0858)	46	2.460 (0.0969)	74	2.740 (0.1079)
20	2.200 (0.0866)	48	2.480 (0.0976)	76	2.760 (0.1087)
22	2.220 (0.0874)	50	2.500 (0.0984)	78	2.780 (0.1094)
24	2.240 (0.0882)	52	2.520 (0.0992)	80	2.800 (0.1102)
26	2.260 (0.0890)	54	2.540 (0.1000)		

7924SG71

Intake valve clearance shim selection chart—4.7L engine

Measured clearance mm (in.) / Installed shim thickness mm (in.)	2.000 (0.0787)	2.020 (0.0795)	2.040 (0.0803)	2.060 (0.0811)	2.080 (0.0819)	2.100 (0.0827)	2.120 (0.0835)	2.140 (0.0843)	2.160 (0.0850)	2.180 (0.0858)	2.200 (0.0866)	2.210 (0.0870)	2.220 (0.0874)	2.230 (0.0878)	2.240 (0.0882)	2.250 (0.0886)	2.260 (0.0890)	2.270 (0.0894)	2.280 (0.0898)	2.290 (0.0902)	2.300 (0.0906)	2.310 (0.0909)	2.320 (0.0913)	2.330 (0.0917)	2.340 (0.0921)	2.350 (0.0925)	2.360 (0.0929)	2.370 (0.0933)	2.380 (0.0937)	2.390 (0.0941)	2.400 (0.0945)
0.000–0.030 (0.0000–0.0012)												00	00	00	00	00	00	00	00	00	02	02	04	04	06	06	08	08	10	10	12
0.031–0.050 (0.0012–0.0020)											00	00	00	00	00	00	00	02	02	04	04	06	06	08	08	10	10	12	12	14	14
0.051–0.070 (0.0020–0.0028)										00	00	00	00	00	00	02	02	04	04	06	06	08	08	10	10	12	12	14	14	16	16
0.071–0.090 (0.0028–0.0035)									00	00	00	00	00	02	02	04	04	06	06	08	08	10	10	12	12	14	14	16	16	18	18
0.091–0.110 (0.0036–0.0043)								00	00	00	00	02	02	04	04	06	06	08	08	10	10	12	12	14	14	16	16	18	18	20	20
0.111–0.130 (0.0044–0.0051)							00	00	00	00	02	04	04	06	06	08	08	10	10	12	12	14	14	16	16	18	18	20	20	22	22
0.131–0.150 (0.0052–0.0059)						00	00	00	00	02	04	06	06	08	08	10	10	12	12	14	14	16	16	18	18	20	20	22	22	24	24
0.151–0.170 (0.0059–0.0067)					00	00	00	00	02	04	06	08	08	10	10	12	12	14	14	16	16	18	18	20	20	22	22	24	24	26	26
0.171–0.190 (0.0067–0.0075)				00	00	00	00	02	04	06	08	10	10	12	12	14	14	16	16	18	18	20	20	22	22	24	24	26	26	28	28
0.191–0.210 (0.0075–0.0083)			00	00	00	00	02	04	06	08	10	12	12	14	14	16	16	18	18	20	20	22	22	24	24	26	26	28	28	30	30
0.211–0.230 (0.0083–0.0091)		00	00	00	00	02	04	06	08	10	12	14	14	16	16	18	18	20	20	22	22	24	24	26	26	28	28	30	30	32	32
0.231–0.249 (0.0091–0.0098)		00	00	00	02	04	06	08	10	12	14	16	16	18	18	20	20	22	22	24	24	26	26	28	28	30	30	32	32	34	34
0.250–0.350 (0.0098–0.0138)																															
0.351–0.370 (0.0138–0.0146)	06	08	10	12	14	16	18	20	22	24	26	28	28	30	30	32	32	34	34	36	36	38	38	40	40	42	42	44	44	46	46
0.371–0.390 (0.0146–0.0154)	08	10	12	14	16	18	20	22	24	26	28	30	30	32	32	34	34	36	36	38	38	40	40	42	42	44	44	46	46	48	48
0.391–0.410 (0.0154–0.0161)	10	12	14	16	18	20	22	24	26	28	30	32	32	34	34	36	36	38	38	40	40	42	42	44	44	46	46	48	48	50	50
0.411–0.430 (0.0162–0.0169)	12	14	16	18	20	22	24	26	28	30	32	34	34	36	36	38	38	40	40	42	42	44	44	46	46	48	48	50	50	52	52
0.431–0.450 (0.0170–0.0177)	14	16	18	20	22	24	26	28	30	32	34	36	36	38	38	40	40	42	42	44	44	46	46	48	48	50	50	52	52	54	54
0.451–0.470 (0.0178–0.0185)	16	18	20	22	24	26	28	30	32	34	36	38	38	40	40	42	42	44	44	46	46	48	48	50	50	52	52	54	54	56	56
0.471–0.490 (0.0185–0.0193)	18	20	22	24	26	28	30	32	34	36	38	40	40	42	42	44	44	46	46	48	48	50	50	52	52	54	54	56	56	58	58
0.491–0.510 (0.0193–0.0201)	20	22	24	26	28	30	32	34	36	38	40	42	42	44	44	46	46	48	48	50	50	52	52	54	54	56	56	58	58	60	60
0.511–0.530 (0.0201–0.0209)	22	24	26	28	30	32	34	36	38	40	42	44	44	46	46	48	48	50	50	52	52	54	54	56	56	58	58	60	60	62	62
0.531–0.550 (0.0209–0.0217)	24	26	28	30	32	34	36	38	40	42	44	46	46	48	48	50	50	52	52	54	54	56	56	58	58	60	60	62	62	64	64
0.551–0.570 (0.0217–0.0224)	26	28	30	32	34	36	38	40	42	44	46	48	48	50	50	52	52	54	54	56	56	58	58	60	60	62	62	64	64	66	66
0.571–0.590 (0.0225–0.0232)	28	30	32	34	36	38	40	42	44	46	48	50	50	52	52	54	54	56	56	58	58	60	60	62	62	64	64	66	66	68	68
0.591–0.610 (0.0233–0.0240)	30	32	34	36	38	40	42	44	46	48	50	52	52	54	54	56	56	58	58	60	60	62	62	64	64	66	66	68	68	70	70
0.611–0.630 (0.0241–0.0248)	32	34	36	38	40	42	44	46	48	50	52	54	54	56	56	58	58	60	60	62	62	64	64	66	66	68	68	70	70	72	72
0.631–0.650 (0.0248–0.0256)	34	36	38	40	42	44	46	48	50	52	54	56	56	58	58	60	60	62	62	64	64	66	66	68	68	70	70	72	72	74	74
0.651–0.670 (0.0256–0.0264)	36	38	40	42	44	46	48	50	52	54	56	58	58	60	60	62	62	64	64	66	66	68	68	70	70	72	72	74	74	76	76
0.671–0.690 (0.0264–0.0272)	38	40	42	44	46	48	50	52	54	56	58	60	60	62	62	64	64	66	66	68	68	70	70	72	72	74	74	76	76	78	78
0.691–0.710 (0.0272–0.0280)	40	42	44	46	48	50	52	54	56	58	60	62	62	64	64	66	66	68	68	70	70	72	72	74	74	76	76	78	78	80	80
0.711–0.730 (0.0280–0.0287)	42	44	46	48	50	52	54	56	58	60	62	64	64	66	66	68	68	70	70	72	72	74	74	76	76	78	78	80	80	80	80
0.731–0.750 (0.0288–0.0295)	44	46	48	50	52	54	56	58	60	62	64	66	66	68	68	70	70	72	72	74	74	76	76	78	78	80	80	80	80	80	80
0.751–0.770 (0.0296–0.0303)	46	48	50	52	54	56	58	60	62	64	66	68	68	70	70	72	72	74	74	76	76	78	78	80	80	80	80	80	80	80	
0.771–0.790 (0.0304–0.0311)	48	50	52	54	56	58	60	62	64	66	68	70	70	72	72	74	74	76	76	78	78	80	80	80	80	80	80	80			
0.791–0.810 (0.0311–0.0319)	50	52	54	56	58	60	62	64	66	68	70	72	72	74	74	76	76	78	78	80	80	80	80	80	80	80					
0.811–0.830 (0.0319–0.0327)	52	54	56	58	60	62	64	66	68	70	72	74	74	76	76	78	78	80	80	80	80	80	80	80							
0.831–0.850 (0.0327–0.0335)	54	56	58	60	62	64	66	68	70	72	74	76	76	78	78	80	80	80	80	80	80	80									
0.851–0.870 (0.0335–0.0343)	56	58	60	62	64	66	68	70	72	74	76	78	78	80	80	80	80	80	80	80											
0.871–0.890 (0.0343–0.0350)	58	60	62	64	66	68	70	72	74	76	78	80	80	80	80	80	80	80													
0.891–0.910 (0.0351–0.0358)	60	62	64	66	68	70	72	74	76	78	80	80	80	80	80	80															
0.911–0.930 (0.0359–0.0366)	62	64	66	68	70	72	74	76	78	80	80	80	80	80																	
0.931–0.950 (0.0367–0.0374)	64	66	68	70	72	74	76	78	80	80	80	80																			
0.951–0.970 (0.0374–0.0382)	66	68	70	72	74	76	78	80	80	80																					
0.971–0.990 (0.0382–0.0390)	68	70	72	74	76	78	80	80	80																						
0.991–1.010 (0.0390–0.0398)	70	72	74	76	78	80	80	80																							
1.011–1.030 (0.0398–0.0406)	72	74	76	78	80	80	80																								
1.031–1.050 (0.0406–0.0413)	74	76	78	80	80	80																									
1.051–1.070 (0.0414–0.0421)	76	78	80	80	80																										
1.071–1.090 (0.0422–0.0429)	78	80	80	80																											
1.091–1.110 (0.0430–0.0437)	80	80	80																												
1.111–1.130 (0.0437–0.0445)	80	80																													
1.131–1.150 (0.0445–0.0453)	80																														

Measured clearance mm (in.) / Installed shim thickness mm (in.)	2.410 (0.0949)	2.420 (0.0953)	2.430 (0.0957)	2.440 (0.0961)	2.450 (0.0965)	2.460 (0.0969)	2.470 (0.0972)	2.480 (0.0976)	2.490 (0.0980)	2.500 (0.0984)	2.510 (0.0988)	2.520 (0.0992)	2.530 (0.0996)	2.540 (0.1000)	2.550 (0.1004)	2.560 (0.1008)	2.570 (0.1012)	2.580 (0.1016)	2.590 (0.1020)	2.600 (0.1024)	2.620 (0.1031)	2.640 (0.1039)	2.660 (0.1047)	2.680 (0.1055)	2.700 (0.1063)	2.720 (0.1071)	2.740 (0.1079)	2.760 (0.1087)	2.780 (0.1094)	2.800 (0.1102)
0.000–0.030 (0.0000–0.0012)	12	14	14	16	16	18	18	20	20	22	22	24	24	26	26	28	28	30	30	32	34	36	38	40	42	44	46	48	50	52
0.031–0.050 (0.0012–0.0020)	16	16	18	18	20	20	22	22	24	24	26	26	28	28	30	30	32	32	34	34	36	38	40	42	44	46	48	50	52	54
0.051–0.070 (0.0020–0.0028)	18	18	20	20	22	22	24	24	26	26	28	28	30	30	32	32	34	34	36	36	38	40	42	44	46	48	50	52	54	56
0.071–0.090 (0.0028–0.0035)	20	20	22	22	24	24	26	26	28	28	30	30	32	32	34	34	36	36	38	38	40	42	44	46	48	50	52	54	56	58
0.091–0.110 (0.0036–0.0043)	22	22	24	24	26	26	28	28	30	30	32	32	34	34	36	36	38	38	40	40	42	44	46	48	50	52	54	56	58	60
0.111–0.130 (0.0044–0.0051)	24	24	26	26	28	28	30	30	32	32	34	34	36	36	38	38	40	40	42	42	44	46	48	50	52	54	56	58	60	62
0.131–0.150 (0.0052–0.0059)	26	26	28	28	30	30	32	32	34	34	36	36	38	38	40	40	42	42	44	44	46	48	50	52	54	56	58	60	62	64
0.151–0.170 (0.0059–0.0067)	28	28	30	30	32	32	34	34	36	36	38	38	40	40	42	42	44	44	46	46	48	50	52	54	56	58	60	62	64	66
0.171–0.190 (0.0067–0.0075)	30	30	32	32	34	34	36	36	38	38	40	40	42	42	44	44	46	46	48	48	50	52	54	56	58	60	62	64	66	68
0.191–0.210 (0.0075–0.0083)	32	32	34	34	36	36	38	38	40	40	42	42	44	44	46	46	48	48	50	50	52	54	56	58	60	62	64	66	68	70
0.211–0.230 (0.0083–0.0091)	34	34	36	36	38	38	40	40	42	42	44	44	46	46	48	48	50	50	52	52	54	56	58	60	62	64	66	68	70	72
0.231–0.249 (0.0091–0.0098)	36	36	38	38	40	40	42	42	44	44	46	46	48	48	50	50	52	52	54	54	56	58	60	62	64	66	68	70	72	74
0.250–0.350 (0.0098–0.0138)																														
0.351–0.370 (0.0138–0.0146)	48	48	50	50	52	52	54	54	56	56	58	58	60	60	62	62	64	64	66	66	68	70	72	74	76	78	80	80	80	
0.371–0.390 (0.0146–0.0154)	50	50	52	52	54	54	56	56	58	58	60	60	62	62	64	64	66	66	68	68	70	72	74	76	78	80	80	80		
0.391–0.410 (0.0154–0.0161)	52	52	54	54	56	56	58	58	60	60	62	62	64	64	66	66	68	68	70	70	72	74	76	78	80	80	80			
0.411–0.430 (0.0162–0.0169)	54	54	56	56	58	58	60	60	62	62	64	64	66	66	68	68	70	70	72	72	74	76	78	80	80	80				
0.431–0.450 (0.0170–0.0177)	56	56	58	58	60	60	62	62	64	64	66	66	68	68	70	70	72	72	74	74	76	78	80	80	80					
0.451–0.470 (0.0178–0.0185)	58	58	60	60	62	62	64	64	66	66	68	68	70	70	72	72	74	74	76	76	78	80	80	80						
0.471–0.490 (0.0185–0.0193)	60	60	62	62	64	64	66	66	68	68	70	70	72	72	74	74	76	76	78	78	80	80	80							
0.491–0.510 (0.0193–0.0201)	62	62	64	64	66	66	68	68	70	70	72	72	74	74	76	76	78	78	80	80	80	80								
0.511–0.530 (0.0201–0.0209)	64	64	66	66	68	68	70	70	72	72	74	74	76	76	78	78	80	80	80	80	80									
0.531–0.550 (0.0209–0.0217)	66	66	68	68	70	70	72	72	74	74	76	76	78	78	80	80	80	80	80	80										
0.551–0.570 (0.0217–0.0224)	68	68	70	70	72	72	74	74	76	76	78	78	80	80	80	80	80	80	80											
0.571–0.590 (0.0225–0.0232)	70	70	72	72	74	74	76	76	78	78	80	80	80	80	80	80	80													
0.591–0.610 (0.0233–0.0240)	72	72	74	74	76	76	78	78	80	80	80	80	80	80	80															
0.611–0.630 (0.0241–0.0248)	74	74	76	76	78	78	80	80	80	80	80	80	80																	
0.631–0.650 (0.0248–0.0256)	76	76	78	78	80	80	80	80	80	80	80																			
0.651–0.670 (0.0256–0.0264)	78	78	80	80	80	80	80	80	80																					
0.671–0.690 (0.0264–0.0272)	80	80	80	80	80	80	80																							
0.691–0.710 (0.0272–0.0280)	80	80	80	80	80																									
0.711–0.730 (0.0280–0.0287)	80	80	80																											
0.731–0.750 (0.0288–0.0295)	80																													
0.751–0.770 (0.0296–0.0303)																														
0.771–0.790 (0.0304–0.0311)																														
0.791–0.810 (0.0311–0.0319)																														
0.811–0.830 (0.0319–0.0327)																														
0.831–0.850 (0.0327–0.0335)																														
0.851–0.870 (0.0335–0.0343)																														
0.871–0.890 (0.0343–0.0350)																														
0.891–0.910 (0.0351–0.0358)																														
0.911–0.930 (0.0359–0.0366)																														
0.931–0.950 (0.0367–0.0374)																														
0.951–0.970 (0.0374–0.0382)																														
0.971–0.990 (0.0382–0.0390)																														
0.991–1.010 (0.0390–0.0398)																														
1.011–1.030 (0.0398–0.0406)																														
1.031–1.050 (0.0406–0.0413)																														
1.051–1.070 (0.0414–0.0421)																														
1.071–1.090 (0.0422–0.0429)																														
1.091–1.110 (0.0430–0.0437)																														
1.111–1.130 (0.0437–0.0445)																														
1.131–1.150 (0.0445–0.0453)																														

New shim thickness — mm (in.)

Shim No.	Thickness	Shim No.	Thickness	Shim No.	Thickness
00	2.000 (0.0787)	28	2.280 (0.0898)	56	2.560 (0.1008)
02	2.020 (0.0795)	30	2.300 (0.0906)	58	2.580 (0.1016)
04	2.040 (0.0803)	32	2.320 (0.0913)	60	2.600 (0.1024)
06	2.060 (0.0811)	34	2.340 (0.0921)	62	2.620 (0.1031)
08	2.080 (0.0819)	36	2.360 (0.0929)	64	2.640 (0.1039)
10	2.100 (0.0827)	38	2.380 (0.0937)	66	2.660 (0.1047)
12	2.120 (0.0835)	40	2.400 (0.0945)	68	2.680 (0.1055)
14	2.140 (0.0843)	42	2.420 (0.0953)	70	2.700 (0.1063)
16	2.160 (0.0850)	44	2.440 (0.0961)	72	2.720 (0.1071)
18	2.180 (0.0858)	46	2.460 (0.0969)	74	2.740 (0.1079)
20	2.200 (0.0866)	48	2.480 (0.0976)	76	2.760 (0.1087)
22	2.220 (0.0874)	50	2.500 (0.0984)	78	2.780 (0.1094)
24	2.240 (0.0882)	52	2.520 (0.0992)	80	2.800 (0.1102)
26	2.260 (0.0890)	54	2.540 (0.1000)		

Exhaust valve clearance (Cold):
0.25 – 0.35 mm (0.010 – 0.014 in.)

EXAMPLE:
The 2.300 mm (0.0906 in.) shim is installed, and the measured clearance is 0.440 mm (0.0173 in.). Replace the 2.300 mm (0.0906 in.) shim with a No. 44 shim.

7924SG72

Exhaust valve clearance shim selection chart—4.7L engine

pression stroke with the valves closed for the cylinder to be measured.

5. Check the valve clearance. The valve clearance specifications are as follows:

- Intake: 0.006–0.010 in. (0.15–0.25mm)
- Exhaust: 0.010–0.014 in. (0.25–0.35mm)

6. Record the measurements for each valve.

7. When all valve clearances have been measured, remove the camshafts.

8. Remove the valve shims and measure them. Note this measurement along with the clearance measurement recorded earlier.

9. Using the valve clearance and shim thickness measurements, find replacement shims in the Adjusting Shim Selection charts.

10. Install or connect the following:

- Replacement valve shims
- Camshafts
- Valve covers
- Ignition coils
- Negative battery cable

11. Fill the cooling system.

12. Start the engine and check for leaks.

Starter Motor

REMOVAL & INSTALLATION

3.4L Engine

1. Before servicing the vehicle, refer to the precautions in the beginning of this section.

2. Remove or disconnect the following:

- Negative battery cable
- Starter electrical connectors
- Starter

To install:

3. Install or connect the following:

- Starter. Tighten the fasteners to 29 ft. lbs. (39 Nm).
- Electrical connections
- Negative battery cable

4.7L Engine

1. Before servicing the vehicle, refer to the precautions in the beginning of this section.

2. Drain the cooling system.

3. Relieve the fuel system pressure.

4. Remove or disconnect the following:

- Negative battery cable
- Engine appearance cover
- Air intake tube
- Intake manifold
- Starter motor mounting bolts
- Starter wiring connectors
- Starter motor

To install:

5. Install or connect the following:

- Starter motor
- Starter wiring connectors. Tighten the cable nut to 86 inch lbs. (10 Nm).
- Starter motor mounting bolts. Tighten the bolts to 29 ft. lbs. (39 Nm).
- Intake manifold
- Air intake tube
- Engine appearance cover
- Negative battery cable

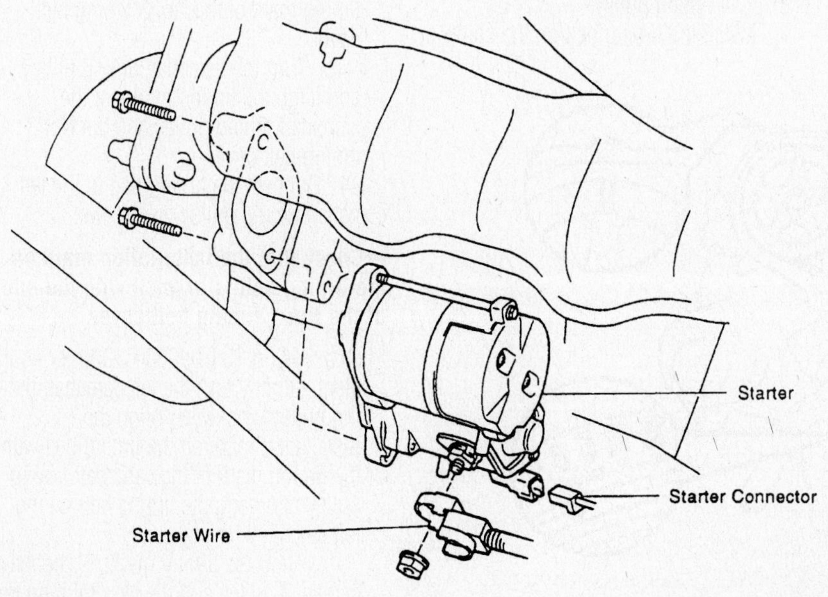

Exploded view of a common starter—3.4L engine

6. Fill the cooling system.

7. Start the engine and check for leaks.

Timing Belt

REMOVAL & INSTALLATION

3.4L Engine

1. Disconnect the negative battery cable.

✱✱ CAUTION

Work must be started after 90 seconds from the time the ignition switch is turned to the LOCK position and the negative battery cable is disconnected.

2. Raise and safely support the vehicle.

3. Remove the engine undercover.

4. Drain the engine coolant.

✱✱ CAUTION

Never open, service or drain the radiator or cooling system when hot; serious burns can occur from the steam and hot coolant. Also, when draining engine coolant, keep in mind that cats and dogs are attracted to ethylene glycol antifreeze and could drink any that is left in an uncovered container or in puddles on the ground. This will prove fatal in sufficient quantities. Always drain coolant into a sealable container. Coolant should be reused unless it is contaminated or is several years old.

5. Disconnect the upper radiator hose from the engine.

6. Remove the power steering drive belt.

7. Remove the air conditioning drive belt by loosening the idler pulley nut and the adjusting bolt.

8. Loosen the lockbolt, pivot bolt, and the adjusting bolt and the alternator drive belt.

9. Remove the No. 2 fan shroud by removing the 2 clips.

10. Remove the fan with the fluid coupling and fan pulleys.

11. Disconnect the power steering pump from the engine and set aside. Do not disconnect the lines from the pump.

12. If equipped with air conditioning, disconnect the compressor from the engine and set aside. Do not disconnect the lines from the compressor.

13. If equipped with air conditioning, disconnect the air conditioning bracket.

14. Remove the No. 2 timing belt cover, as follows:

a. Detach the camshaft position sensor connector from the No. 2 timing belt cover.

b. Disconnect the 3 spark plug wire clamps from the No. 2 timing belt cover.

c. Remove the 6 bolts and remove the timing belt cover.

15. Remove the fan bracket, as follows:

a. Remove the power steering adjusting strut by removing the nut.

b. Remove the fan bracket by removing the bolt and nut.

16. Set the No. 1 cylinder at Top Dead Center (TDC) of the compression stroke, as follows:

a. Turn the crankshaft pulley and align its groove with the timing mark **0** of the No. 1 timing belt cover.

b. Check that the timing marks of the camshaft timing pulleys and the No. 3 timing belt cover are aligned. If not, turn the crankshaft pulley one revolution (360 degrees).

➡If reusing the timing belt, be sure that you can still read the installation marks. If not, place new installation marks on the timing belt to match the timing marks of the camshaft timing pulleys.

17. Remove the timing belt tensioner by alternately loosening the 2 bolts.

18. Remove the camshaft timing pulleys, as follows:

a. Using SST 09960-10010, or equivalent, remove the pulley bolt, the timing pulley and the knock pin. Remove the 2 timing pulleys with the timing belt.

19. Remove the crankshaft pulley, as follows:

a. Using SST 09213-54015 and 09330-00021 or their equivalents, loosen the pulley bolt.

b. Remove the SST tool, the pulley bolt and the pulley.

20. Remove the starter wire bracket and the No. 1 timing belt cover.

21. Remove the timing belt guide and remove the timing belt.

22. Remove the bolt and the No. 2 idler pulley.

23. Remove the pivot bolt, the No. 1 idler pulley and the plate washer.

24. Remove the crankshaft gear.

To install:

25. Install the crankshaft timing gear.

a. Align the timing pulley set key with the key groove of the gear.

b. Using SST 09214-60010, or equivalent, and a hammer, tap in the timing gear with the flange side facing inward.

26. Install the plate washer and the No. 1 idler pulley with the pivot bolt and tighten it to 26 ft. lbs. (35 Nm). Check that the pulley bracket moves smoothly.

27. Install the No. 2 timing belt idler with the bolt. Tighten the bolt to 30 ft. lbs. (40 Nm). Check that the pulley bracket moves smoothly.

28. Temporarily install the timing belt, as follows:

a. Using the crankshaft pulley bolt, turn the crankshaft and align the timing marks of the crankshaft timing pulley and the oil pump body.

b. Align the installation mark on the timing belt with the dot mark of the crankshaft timing pulley.

c. Install the timing belt on the crankshaft timing pulley, No. 1 idler pulley and the water pump pulleys.

29. Install the timing belt guide with the cup side facing outward.

30. Install the No. 1 timing belt cover and starter wire bracket. Tighten the timing belt cover bolts to 80 inch lbs. (9 Nm).

⁂ WARNING

If any binding is felt when adjusting the timing belt tension by turning the crankshaft, STOP turning the engine, because the pistons may be hitting the valves.

31. Install the crankshaft pulley, as follows:

a. Align the pulley set key with the key groove of the crankshaft pulley.

b. Install the pulley bolt and tighten it to 184 ft. lbs. (250 Nm).

32. Install the left camshaft timing pulley.

a. Install the knock pin to the camshaft.

b. Align the knock pin hose of the camshaft with the knock pin groove of the timing pulley.

c. Slide the timing belt pulley on the camshaft with the flange side facing outward. Tighten the pulley bolt to 81 ft. lbs. (110 Nm).

33. Set the No. 1 cylinder to TDC of the compression stroke, as follows:

a. Turn the crankshaft pulley, and align its groove with the timing mark **0** of the No. 1 timing belt cover.

b. Turn the camshaft to align the knock pin hole of the camshaft with the timing mark of the No. 3 timing belt cover.

c. Turn the camshaft timing pulley, and align the timing marks of the camshaft timing pulley and the No. 3 timing belt cover.

34. Connect the timing belt to the left camshaft timing pulley, as follows:

➡Check that the installation mark on the timing belt is aligned with the end of the No. 1 timing belt cover.

a. Using SST 09960-01000, or equivalent, slightly turn the left camshaft timing pulley clockwise. Align the installation mark on the timing belt with the timing mark of the camshaft timing pulley and hang the timing belt on the left camshaft timing pulley.

b. Align the timing marks of the left camshaft pulley and the No. 3 timing belt cover.

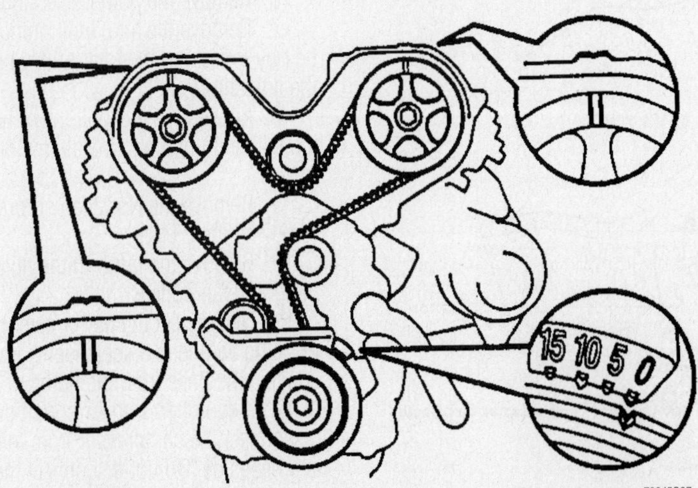

Turn the crankshaft clockwise to align the timing marks before removing the timing belt—3.4L engine

c. Check that the timing belt has tension between the crankshaft timing pulley and the left camshaft timing pulley.

35. Install the right camshaft timing pulley and the timing belt, as follows:

a. Align the installation mark on the timing belt with the timing mark of the right camshaft timing pulley and hang the timing belt on the right camshaft timing pulley with the flange side facing inward.

b. Slide the right camshaft timing pulley on the camshaft. Align the timing marks on the right camshaft timing pulley and the No. 3 timing belt cover.

c. Align the knock pin hole of the camshaft with the knock pin groove of the pulley and install the knock pin. Install the bolt and tighten it to 81 ft. lbs. (110 Nm).

36. Set the timing belt tensioner, as follows:

a. Using a press, slowly press in the pushrod using 220–2205 lbs. (981–9807 N) of force.

b. align the holes of the pushrod and housing, pass a 1.5mm hex wrench through the holes to keep the setting position of the pushrod.

c. Release the press and install the dust boot on the tensioner.

37. Install the timing belt tensioner and alternately tighten the bolts to 20 ft. lbs. (28 Nm). Using pliers, remove the 1.5mm hex wrench from the belt tensioner.

38. Check the valve timing, as follows:

a. Slowly turn the crankshaft pulley 2 revolutions from TDC to TDC. Always turn the crankshaft pulley clockwise.

b. Check that each pulley aligns with the timing marks. If the timing marks do not align, remove the timing belt and reinstall it.

39. Install the fan bracket with the bolt and nut.

40. Install the power steering adjusting strut with the nut.

41. Install the No. 2 timing belt cover. Tighten the bolts to 80 inch lbs. (9 Nm). Install the remaining components.

42. Fill the cooling system with coolant.

43. Connect the negative battery cable.

44. Start the engine and check for leaks.

4.7L Engine

1. Disconnect the negative battery cable.

2. Raise and safely support the vehicle.

3. Remove the oil pan protector and the engine under cover.

4. Drain the cooling system and store the coolant for refilling purposes.

5. Lower the vehicle and remove the battery clamp cover.

6. From the top of the engine, remove the fuel return hose, the engine cover nuts/bolts and the cover.

7. Remove the air cleaner and the intake air connector assembly.

8. Remove the cooling fan pulley by performing the following procedures:

a. Loosen the 4 fan clutch-to-fan pulley nuts.

b. Using a box-end wrench on the serpentine drive belt tensioner bolt, rotate the tensioner counterclockwise and remove the drive belt.

➡The serpentine drive belt tensioner bolt is a left-hand thread.

c. Remove the fan clutch-to-fan pulley nuts, the fan, the clutch assembly and the fan pulley.

9. Remove the radiator by performing the following procedures:

a. Disconnect the upper, lower and reservoir hoses from the radiator.

b. Disconnect and plug the automatic transmission oil cooler at the radiator. Disconnect the automatic transmission oil cooler hoses from the fan shroud clamp.

c. Remove the radiator reservoir tank.

d. Remove the fan shroud-to-radiator bolts and the shroud.

e. Remove the 2 upper radiator-to-chassis nuts.

f. Remove the middle radiator-to-chassis nut/bolts and brackets.

g. Carefully, lift the radiator from the vehicle.

10. Remove the serpentine drive belt idler pulley bolt, cover plate and pulley.

11. Remove the right side (No. 3) timing belt cover.

12. Remove the left side (No. 3) timing belt cover by performing the following procedures:

a. Disconnect the engine wire from both wire clamps.

b. Disconnect the camshaft position sensor wire from the wire clamp on the left-side (No.3) timing belt cover.

c. Disconnect the sensor connector from the connector bracket.

d. Disconnect the sensor connector.

e. Remove the wire grommet from the left-side (No. 3) timing belt cover.

f. Remove the oil cooler tube bolts and tube.

13. Remove the middle (No. 2) timing belt cover bolts and cover.

14. Remove the cooling fan bracket nuts/bolts and bracket.

➡If reusing the timing belt, make sure that there are 3 installation marks on the belt; if there are none, install them.

15. Using the Crankshaft Pulley Holding tool 09213-70010, Bolt tool 90105-08076 and Companion Flange Holding tool 09330-00021, or equivalent, loosen the crankshaft pulley bolt.

16. Position the No. 1 cylinder to approximately 50 degrees After Top Dead Center (ATDC) of the compression stroke by performing the following procedures:

a. Rotate the crankshaft pulley (CLOCKWISE) to align its groove with the timing mark "0" on the lower (No. 1) timing belt cover.

b. Check that the camshaft sprocket timing marks are aligned with the rear timing belt plate marks; if not, rotate the crankshaft 1 revolution (360 degrees).

c. Rotate the crankshaft pulley approximately 50 degrees (CLOCKWISE) and align the crankshaft pulley timing mark between the centers of the crankshaft pulley bolt and the idler pulley bolt.

✱✱ WARNING

If the timing belt is disengaged, having the crankshaft pulley in the wrong angle can cause the valve to come into contact with the piston when removing the camshaft pulley.

17. Remove the crankshaft pulley bolt.

➡If reusing the timing belt and the installation marks have disappeared, place new installation marks on the timing belt to match the camshaft timing sprocket marks.

➡To avoid meshing the timing sprocket and the timing belt, secure one with a string; then, place matchmarks on the timing belt and the right-side camshaft timing sprocket.

18. Remove the timing belt tensioner bolts and the tensioner.

19. Using the Camshaft Holding tool 09960-10010, or equivalent, slightly turn the left-side camshaft sprocket clockwise to loosen the tension spring. Then, disconnect the timing belt from the camshaft sprockets.

20. Remove the alternator by performing the following procedures:

a. Disconnect the electrical connector from the alternator.

b. Remove the rubber cap/nut and

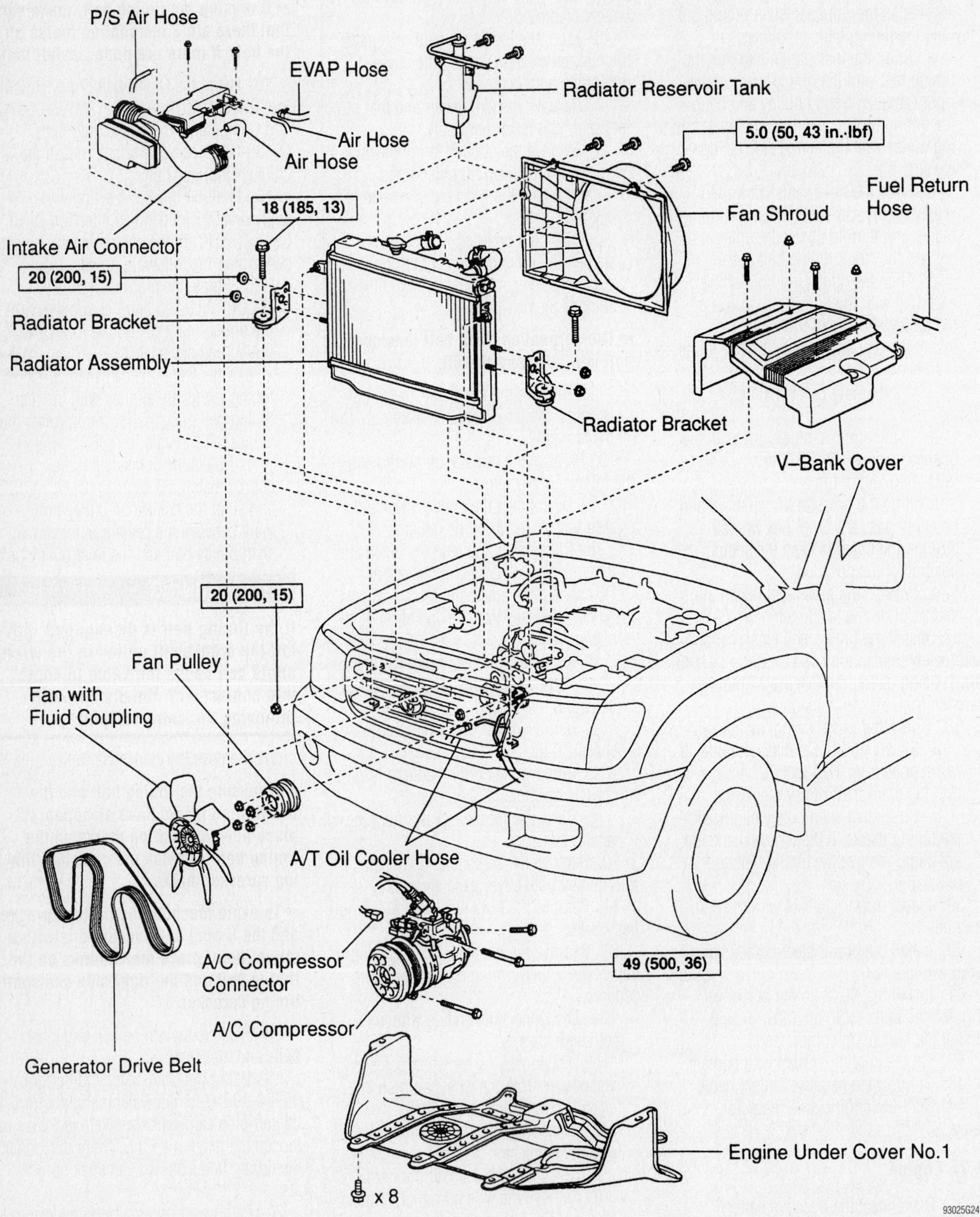

Exploded view of vehicle components for timing belt replacement—4.7L engine

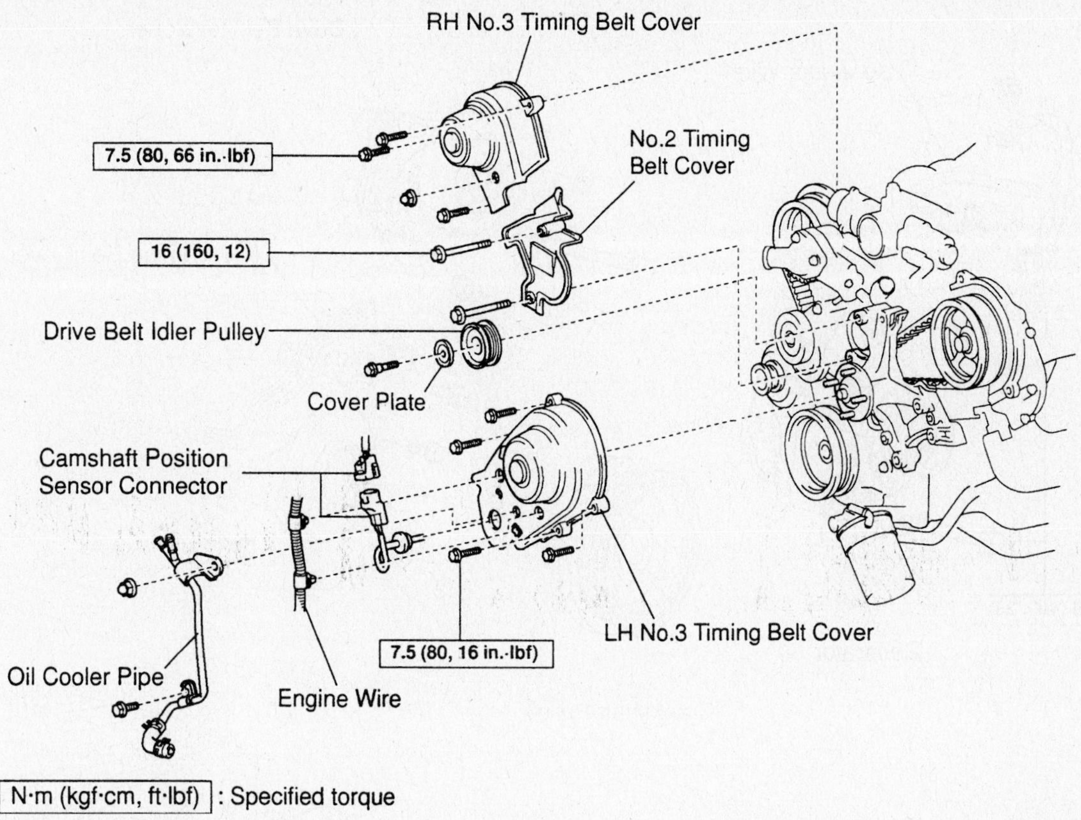

Exploded view of upper timing belt covers—4.7L engine

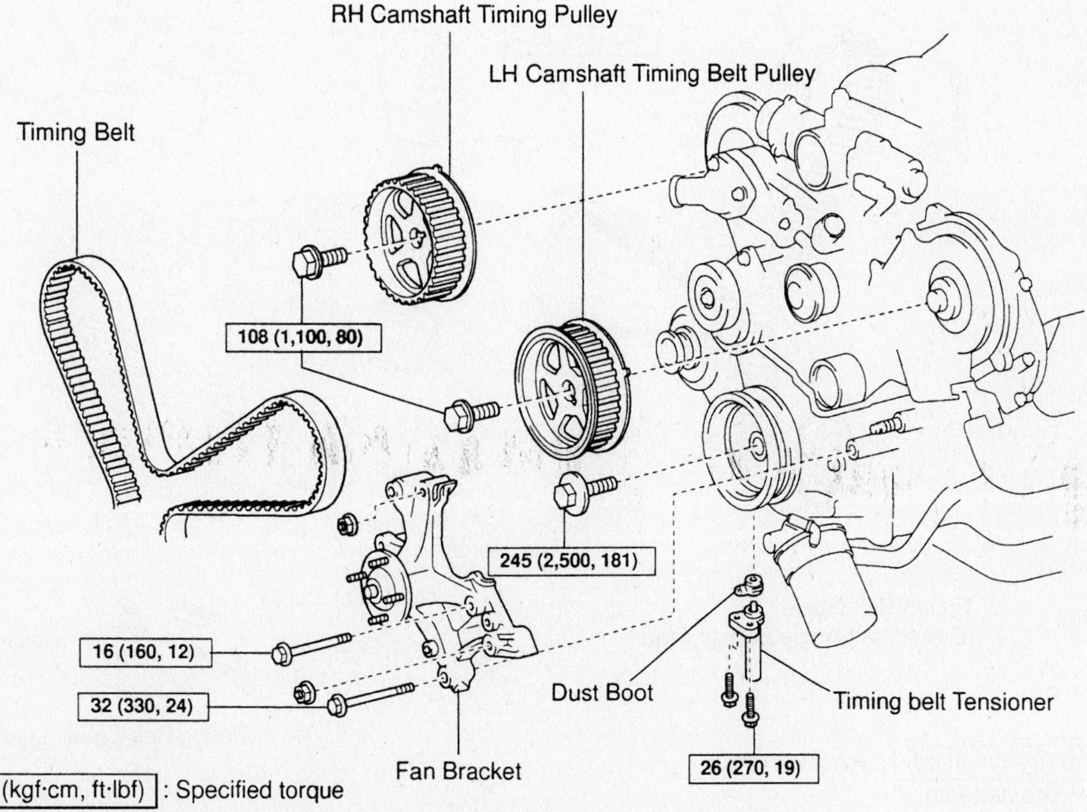

Exploded view of upper timing sprockets and components—4.7L engine

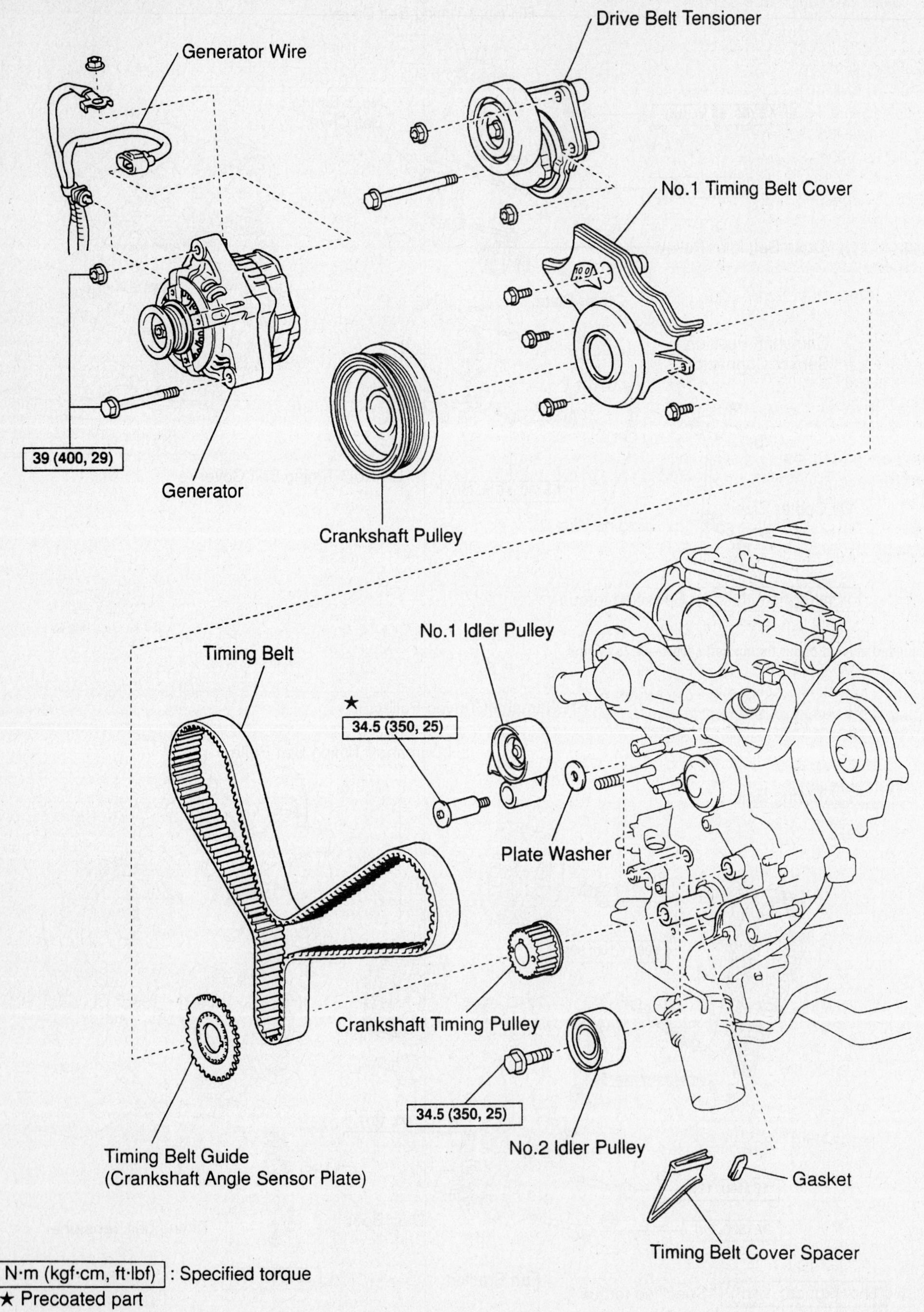

Exploded view of lower timing belt cover, sprockets and components—4.7L engine

disconnect the battery wire from the alternator.

c. Disconnect the wire clamp from the alternator cord clip.

d. Remove the alternator-to-engine nuts/bolts and the alternator.

21. Remove the serpentine drive belt tensioner nuts/bolts and the tensioner.

22. Using the Crankshaft Puller Assembly tool 09950-50012, or equivalent, press the crankshaft pulley from the crankshaft.

⁂ WARNING

DO NOT rotate the crankshaft pulley.

23. Remove the lower (No. 1) timing belt cover bolts and the cover.

24. Remove the timing belt guide, spacer and the timing belt.

To install:

➡With the timing belt removed, this is a perfect opportunity to inspect and/or replace the water pump.

25. Inspect the timing belt tensioner by performing the following procedures:

a. Inspect the seal for leakage; if leakage is suspected, replace the tensioner.

b. Using both hands to hold the tensioner facing upward, strongly press the pushrod against a solid surface. If the pushrod moves, replace the tensioner.

⁂ WARNING

Never hold the tensioner with the pushrod facing downward.

c. Measure the pushrod protrusion from the housing end, it should be 0.413–0.453 in. (10.5–11.5mm). If the protrusion is not as specified, replace the tensioner.

26. Temporarily install the timing belt by performing the following procedures:

a. Align the timing belt's installation mark with the crankshaft timing sprocket.

b. Install the timing belt on the crankshaft timing sprocket, the No. 1 idler pulley and the No. 2 idler pulley.

27. Install the gasket to the timing belt cover spacer and install the cover spacer.

28. Install the timing belt guide with the cup side facing outward.

29. Install the lower (No. 1) timing belt cover.

30. Install the crankshaft pulley by performing the following procedures:

a. Align the crankshaft pulley with the crankshaft key.

b. Using the Crankshaft Installer tool 09223-46011, or equivalent, and a ham-

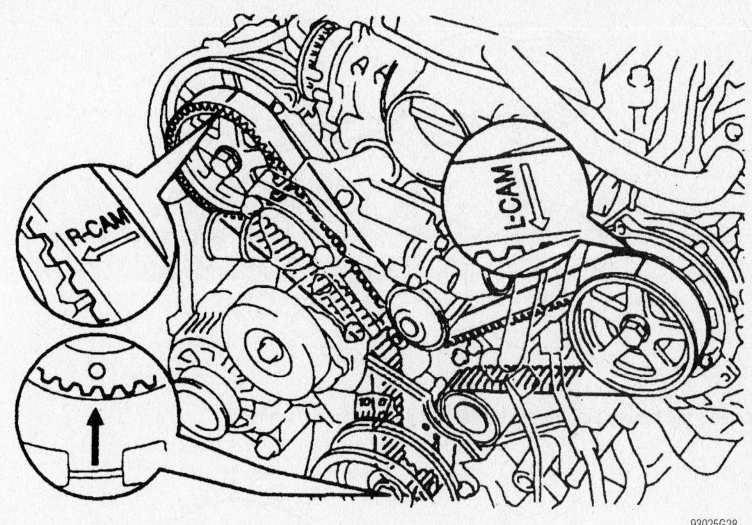

Alignment of timing belt with the timing sprockets—4.7L engine

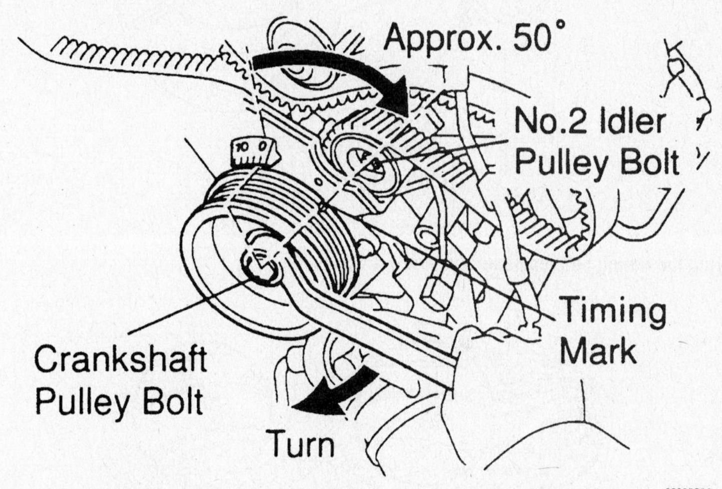

Aligning of crankshaft pulley timing mark with the center line of the crankshaft pulley bolt and the idler pulley bolt—4.7L engine

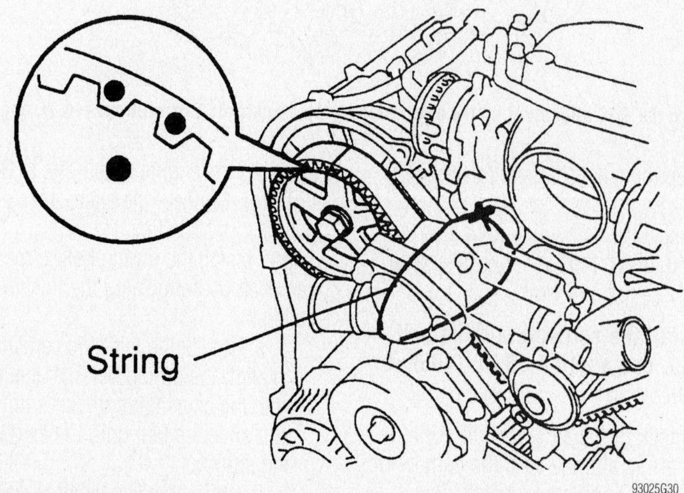

Securing the timing belt with string and matchmarking the camshaft with the timing belt—4.7L engine

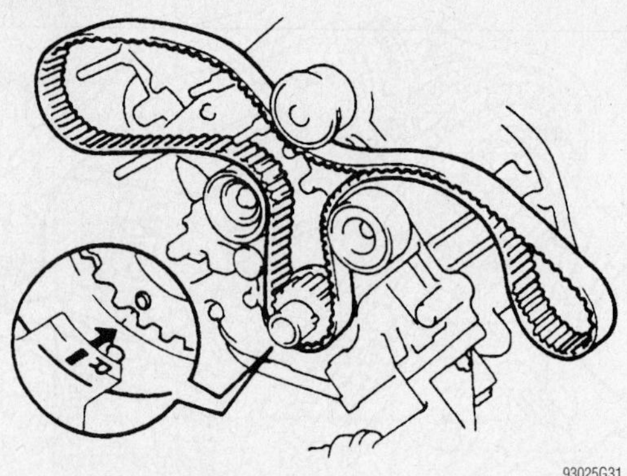

Installing the timing belt on the crankshaft sprocket—4.7L engine

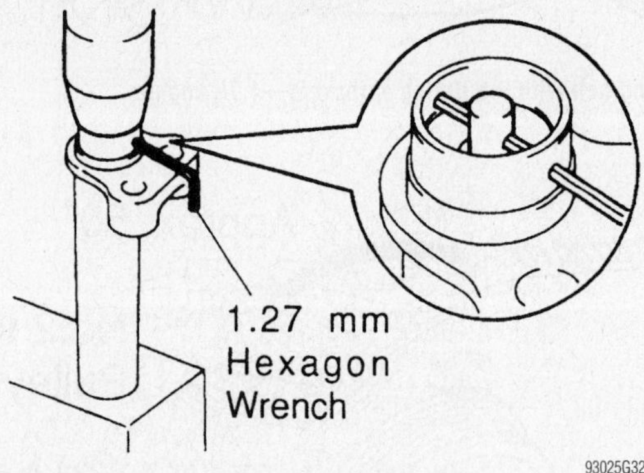

Securing the timing belt tensioner pushrod—4.7L engine

Checking the TDC alignment marks after rotating the crankshaft 2 revolutions—4.7L engine

mer, tap the crankshaft pulley into position.

31. Install the serpentine drive belt tensioner and torque the tensioner-to-engine bolts to 12 ft. lbs. (16 Nm).

➡To install the serpentine drive belt tensioner, use a bolt 4.18 in. (106mm) in length.

32. Check that the crankshaft pulley's timing mark is aligned with the centers of the idler pulley and crankshaft pulley bolts.

33. Install the alternator and torque the alternator-to-engine nuts/bolts to 29 ft. lbs. (39 Nm). Connect the alternator's electrical connectors and clip.

34. Install the timing belt to the left-side camshaft by performing the following procedures:

a. Rotate the left-side camshaft pulley to align the timing belt installation mark with the camshaft sprocket's timing mark and slide the belt onto the camshaft timing sprocket.

b. Using the Camshaft Holding tool 09960-10010, or equivalent, slightly turn the left-side camshaft sprocket counterclockwise to place tension on the timing belt between the crankshaft sprocket and the camshaft sprocket.

35. Rotate the right-side camshaft pulley to align the timing belt installation mark with the camshaft sprocket's timing mark and slide the belt onto the camshaft timing sprocket.

36. Using a vertical press, slowly press the pushrod into the housing using 200–2205 lbs. (981–9807 N) until the holes align, then, install a 1.27mm Allen® wrench to secure the pushrod and release the press. Install the dust boot on the tensioner housing.

37. Install the timing belt tensioner and torque the bolts to 19 ft. lbs. (26 Nm).

38. Using a pair of pliers, remove the Allen® wrench from the tensioner housing.

39. Check the valve timing by performing the following procedure:

a. Temporarily install the crankshaft pulley bolt.

b. Slowly, rotate the crankshaft pulley 2 revolutions (CLOCKWISE) and realign the TDC marks.

➡If the pulley/sprocket timing marks do not realign, remove the timing belt and reinstall it.

40. Using the Crankshaft Pulley Holding tool 09213-70010, Bolt tool 90105-08076 and Companion Flange Holding tool 09330-00021, or equivalent, torque the crankshaft pulley bolt to 181 ft. lbs. (245 Nm).

41. Install the cooling fan bracket and torque the 12mm (head size) bolt to 12 ft. lbs. (16 Nm) and the 14mm (head size) bolt to 24 ft. lbs. (32 Nm).

42. Install the air conditioning compressor.

43. Install the middle (No. 2) timing belt cover and torque the bolts to 12 ft. lbs. (16 Nm).

44. Install the upper right-side (No. 3) timing belt cover and torque the bolts to 66 inch lbs. (7.5 Nm).

45. Install the upper left-side (No. 3) timing belt cover by performing the following procedures:

a. Install the oil cooler tube and bolt.

b. Feed the Camshaft Position Sensor (CPS) through the left-side (No. 3) timing belt cover hole.

c. Install the left-side (No. 3) timing belt cover and torque the bolts to 66 inch lbs. (7.5 Nm).

d. Install the wire grommet to the left-side (No. 3) timing belt cover.

e. Install the sensor connector to the

connector bracket and connect the sensor connector.

f. Install the sensor wire and the engine wire to the clamps on the left-side (No. 3) timing belt cover.

46. Install the drive belt idler pulley and cover plate; then, torque the pulley bolt to 27 ft. lbs. (37 Nm).

47. To complete the installation, reverse the removal procedures.

48. Refill the cooling system and connect the negative battery cable.

Oil Pan

REMOVAL & INSTALLATION

3.4L Engine

1. Before servicing the vehicle, refer to the precautions in the beginning of this section.
2. Disconnect the negative battery cable.
3. Drain the engine oil.
4. Remove or disconnect the following:
   - Engine undercover
   - Front differential, if equipped with 4WD
   - Oil pan, separate it from the engine using SST 09032-00100 and a brass bar

To install:

5. Apply seal packing to the oil pan.
6. Install the oil pan to the cylinder block. Tighten the nuts and bolts to: 67 inch lbs. (8 Nm)

**** WARNING**

If parts are not assembled within 5 minutes of applying time, the effectiveness of the seal packing is lost and must be removed and reapplied.

7. Install or connect the following:
   - Front differential, if removed
   - Engine undercover
   - Negative battery cable
8. Fill with engine oil.
9. Start the engine and check for leaks.

4.7L Engine

1. Before servicing the vehicle, refer to the precautions in the beginning of this section.
2. Remove the engine from the vehicle and mount it on a stand.
3. Remove or disconnect the following:
   - Oil dipstick tube
   - Lower oil pan
   - Oil pan baffle
   - Upper oil pan

To install:

4. The upper oil pan bolts are different lengths and are identified as follows:
   - A: 0.79 inch (20mm) w/10mm head
   - B: 0.98 inch (25mm) w/12mm head
   - C: 2.36 inch (60mm) w/12mm head
   - D: 1.38 inch (35mm) w/10mm head
5. Apply silicone sealant to the upper oil pan as shown.
6. Install the upper oil pan and tighten the fasteners in several passes to the following specifications:
   - 10mm: 66 inch lbs. (7.5 Nm)
   - 12mm: 21 ft. lbs. (28 Nm)

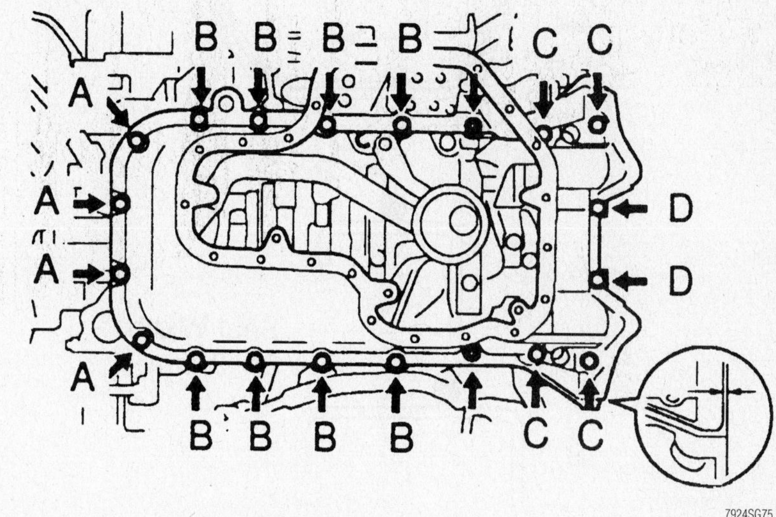

Upper oil pan bolt location—4.7L engine

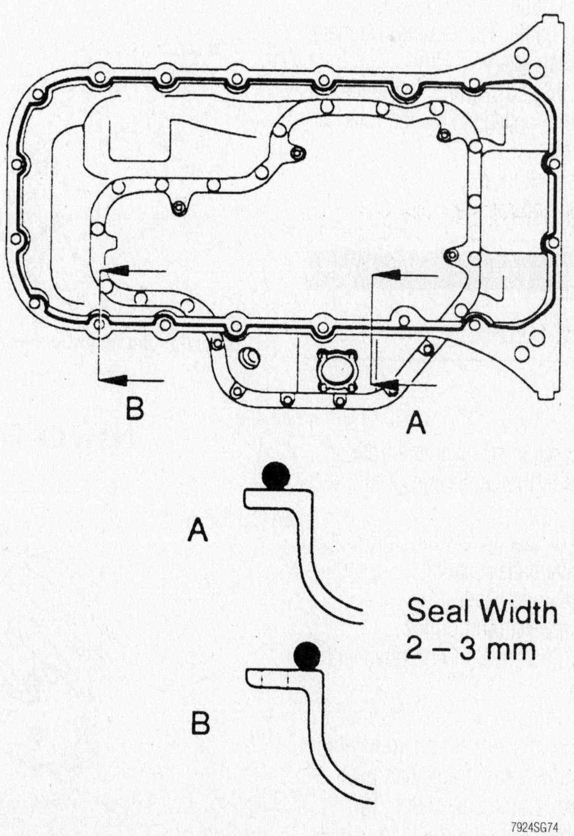

Upper oil pan sealant application—4.7L engine

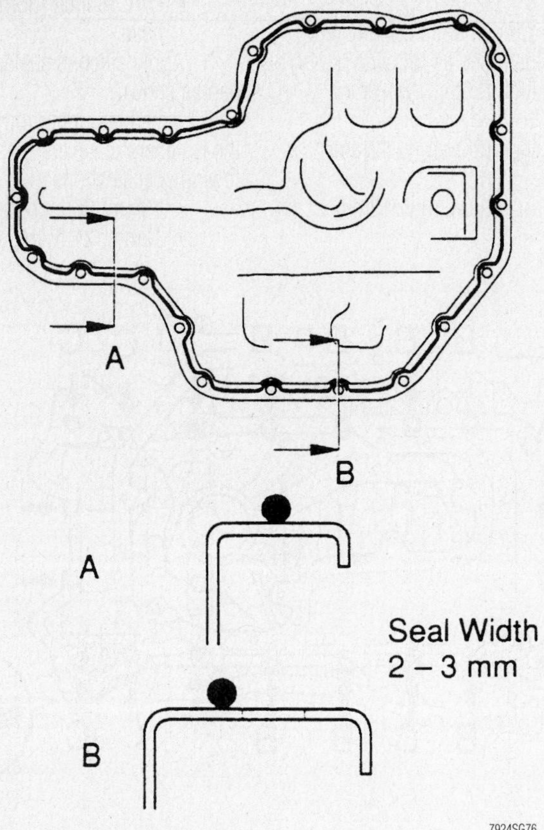

Lower oil pan sealant application—4.7L engine

7. Install or connect the following:
   - Oil pan baffle. Tighten the fasteners to 66 inch lbs. (7.5 Nm).
   - Lower oil pan. Tighten the fasteners in several passes to 66 inch lbs. (7.5 Nm).
   - Oil dipstick tube
8. Install the engine.

Oil Pump

REMOVAL & INSTALLATION

3.4L Engine

1. Before servicing the vehicle, refer to the precautions in the beginning of this section.
2. Remove or disconnect the following:
   - Negative battery cable
   - Engine undercover
   - Crankshaft timing pulley
   - Front differential, if equipped with 4WD
3. Drain the engine oil from the engine.
4. Remove or disconnect the following:
   - Timing belt and crankshaft gear
   - Oil cooler tube and clamp, if equipped with automatic transmission

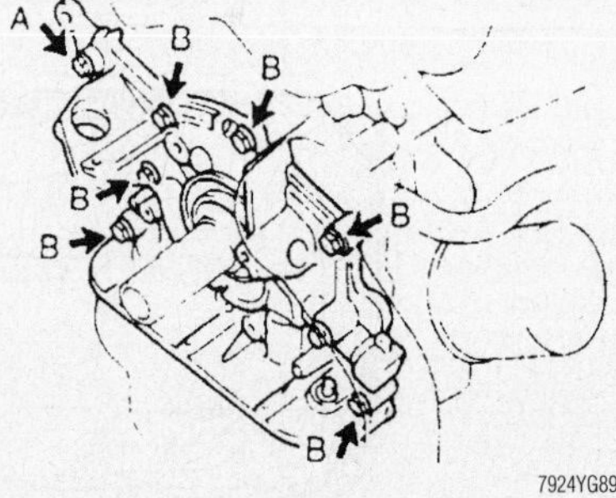

Oil pump bolt identification—3.4L engine

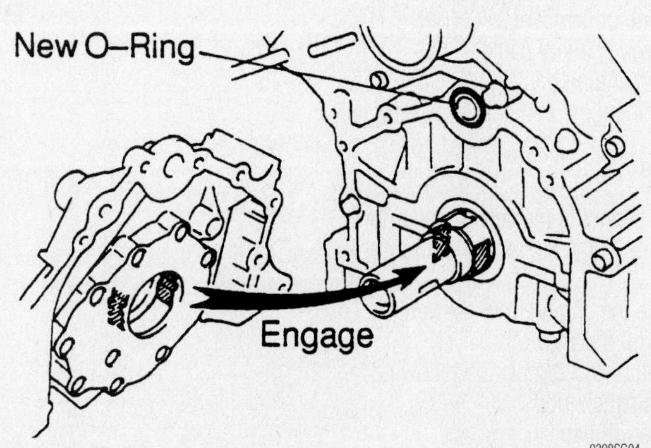

Location of the O-ring seal—4.7L engine

   - Stiffener plate
   - Flywheel housing undercover and dust cover
   - Rear end cover and dust cover
   - Starter wire clamp
   - Crankshaft Position (CKP) sensor
   - Oil pan

➡Be careful not to damage the baffle plate flange.

   - Oil strainer
   - Oil baffle plate
   - Oil pump body by removing the 8 bolts.
   - O-ring from the cylinder block

To install:

5. Install or connect the following:
   - Apply Seal Packing PN 08826-00080 to the oil pump
   - New O-ring into the groove of the cylinder block
   - Oil pump to the crankshaft with the spline teeth of the drive rotor engaged with the large teeth of the crankshaft. Tighten the oil pump bolts "A" 15 ft. lbs. (20 Nm) and bolts "B" 31 ft. lbs. (42 Nm)
   - CKP
   - Oil pan baffle plate
   - Oil strainer with a new gasket. Tighten the bolts to 13 ft. lbs. (18 Nm).
   - Remaining components
   - Negative battery cable
6. Fill with engine oil.
7. Start the engine and check for leaks.

4.7L Engine

1. Before servicing the vehicle, refer to the precautions in the beginning of this section.
2. Remove the engine from the vehicle and mount it on a stand.
3. Remove or disconnect the following:

- Front cover
- Timing belt. Refer to the Timing Belt unit repair section.
- Timing belt idler pulleys
- Crankshaft timing sprocket
- Oil dipstick tube
- Oil filter and bracket
- Crankshaft Position (CKP) sensor
- Oil pan and baffle
- Oil pump pickup tube
- Oil pump

To install:

4. The upper oil pan bolts are different lengths and are identified as follows:

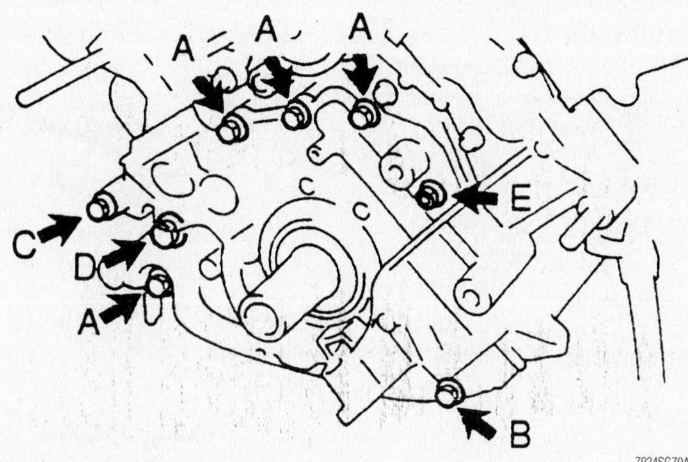

Oil pump bolt location—4.7L engine

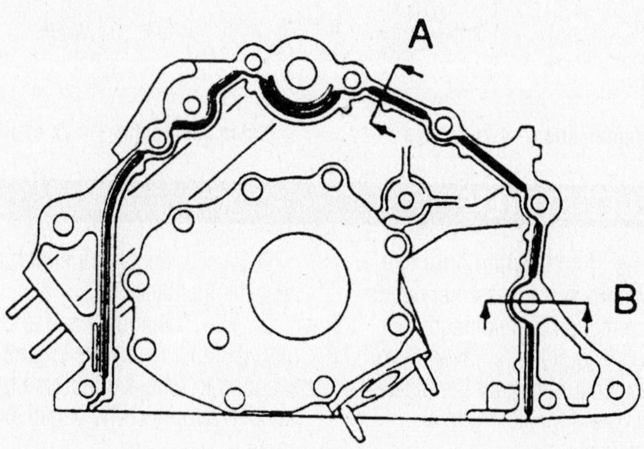

Seal Width
2 – 3 mm

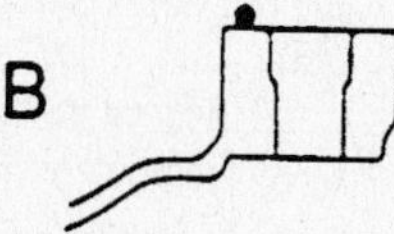

Oil pump housing sealant application—4.7L engine

- A: 1.38 inch (35mm) w/12mm head
- B: 1.97 inch (50mm) w/12mm head
- C: 4.17 inch (106mm) w/12mm head
- D: 1.57 inch (40mm) w/14mm head
- E: 1.18 inch (30mm) w/6mm hex head

5. Install a new O-ring on the engine block.
6. Apply silicone sealant to the oil pump housing as shown.
7. Install the oil pump. Tighten the bolts in several passes to the following specifications:
   - 12mm: 11 ft. lbs. (15.5 Nm)
   - 14mm: 22 ft. lbs. (30.5 Nm)
   - 6mm Hex: 11 ft. lbs. (15.5 Nm)
8. Install or connect the following:
   - Oil pump pickup tube. Tighten the bolts to 66 inch lbs. (7.5 Nm).
   - Oil pan and baffle
   - CKP sensor
   - Oil filter and bracket. Tighten the bolts to 13 ft. lbs. (18 Nm).
   - Oil dipstick tube
   - Crankshaft timing sprocket
   - Timing belt idler pulleys
   - Timing belt
   - Front cover
9. Install the engine.

Rear Main Seal

REMOVAL & INSTALLATION

1. Before servicing the vehicle, refer to the precautions in the beginning of this section.
2. Remove the transmission and flywheel/driveplate from the vehicle.
3. Cut off the rubber lip portion of the seal with a sharp knife.
4. Pry out the oil seal.

To install:

5. Install the rear main seal so that it is flush with the seal retainer housing.
6. Install or connect the following:
   - Flywheel/driveplate. Tighten the bolts to 28 ft. lbs. (38 Nm) for 3.4L engine with a manual transmission, 61 ft. lbs. (83 Nm) for 3.4L engine with an automatic transmission or 35 ft. lbs. (48 Nm) plus a 90 degree turn for 4.7L engine.
   - Transmission

Piston and Ring

POSITIONING

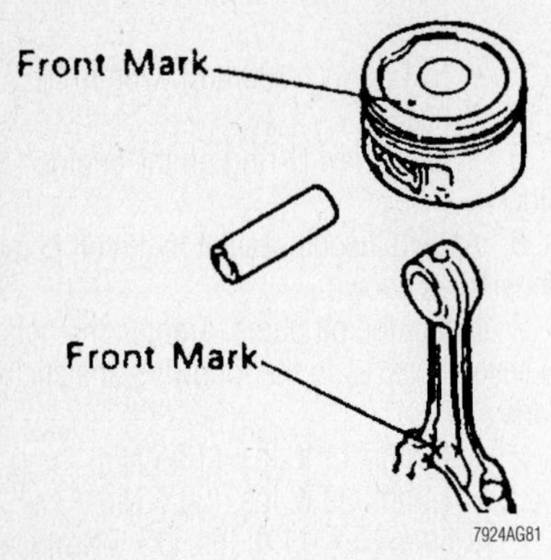

Piston to connecting rod assembly—3.4L engines

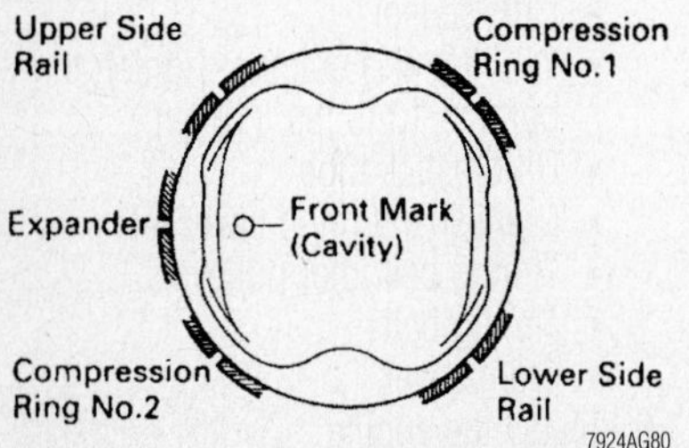

Piston ring end-gap spacing—3.4L engine

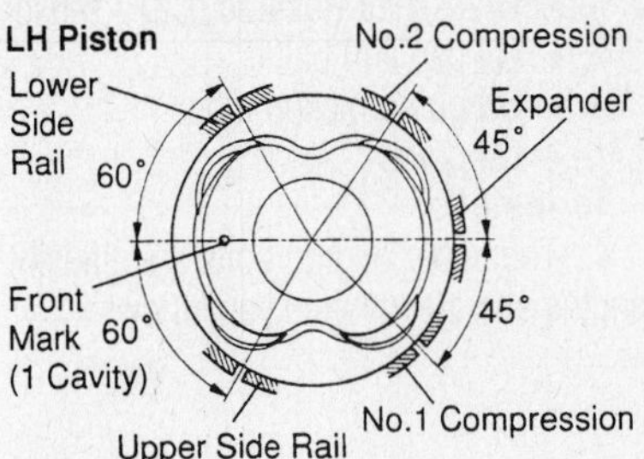

Piston ring positioning—4.7L engine

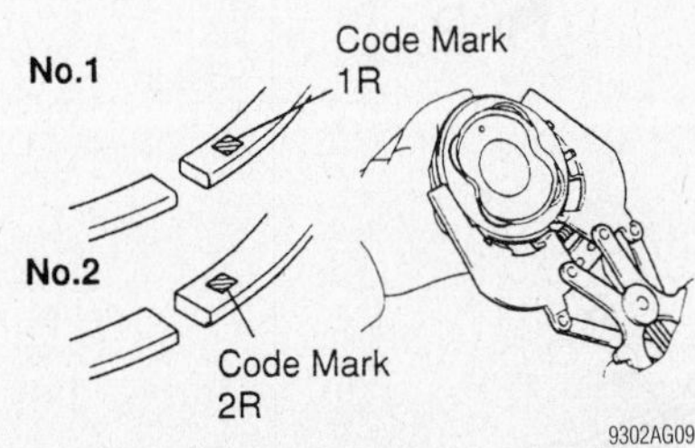

Piston ring identification—4.7L engine

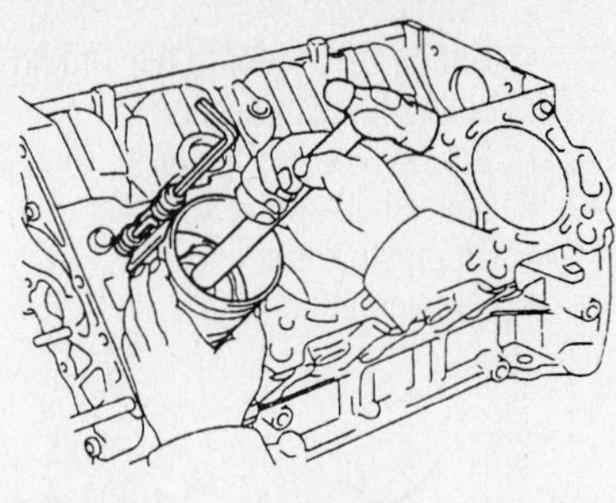

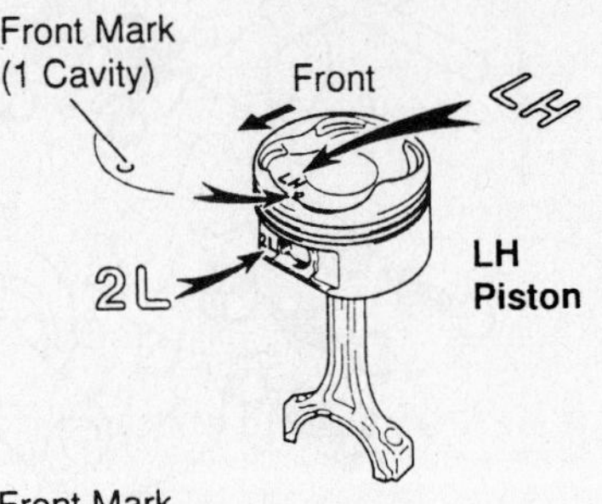

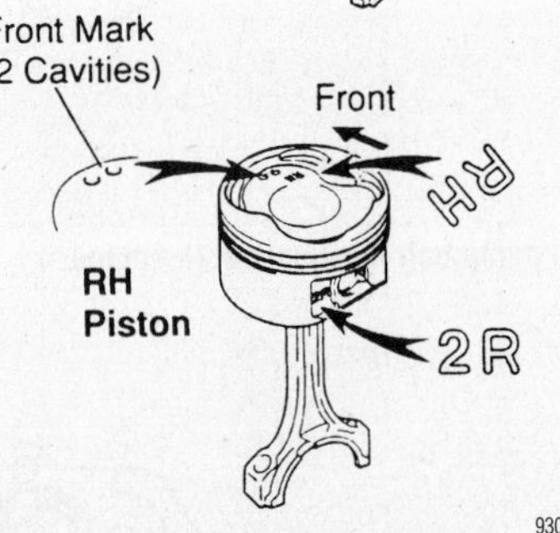

Piston positioning—4.7L engine

FUEL SYSTEM

Fuel System Service Precautions

Safety is the most important factor when performing not only fuel system maintenance but any type of maintenance. Failure to conduct maintenance and repairs in a safe manner may result in serious personal injury or death. Maintenance and testing of the vehicle's fuel system components can be accomplished safely and effectively by adhering to the following rules and guidelines.

- To avoid the possibility of fire and personal injury, always disconnect the negative battery cable unless the repair or test procedure requires that battery voltage be applied.
- Always relieve the fuel system pressure prior to disconnecting any fuel system component (injector, fuel rail, pressure regulator, etc.), fitting or fuel line connection. Exercise extreme caution whenever relieving fuel system pressure, to avoid exposing skin, face and eyes to fuel spray. Please be advised that fuel under pressure may penetrate the skin or any part of the body that it contacts.
- Always place a shop towel or cloth around the fitting or connection prior to loosening to absorb any excess fuel due to spillage. Ensure that all fuel spillage (should it occur) is quickly removed from engine surfaces. Ensure that all fuel soaked cloths or towels are deposited into a suitable waste container.
- Always keep a dry chemical (Class B) fire extinguisher near the work area.
- Do not allow fuel spray or fuel vapors to come into contact with a spark or open flame.
- Always use a back-up wrench when loosening and tightening fuel line connection fittings. This will prevent unnecessary stress and torsion to fuel line piping.
- Always replace worn fuel fitting O-rings with new. Do not substitute fuel hose or equivalent, where fuel pipe is installed.

Fuel System Pressure

RELIEVING

1. Before servicing the vehicle, refer to the precautions in the beginning of this section.
2. Disconnect the fuel pump connector near the fuel tank.
3. Start the engine and allow it to run until it stalls. Crank the engine for a few seconds to relieve additional fuel pressure.
4. Disconnect the negative battery cable.
5. When repairs are complete, connect the negative battery cable.

Fuel Filter

REMOVAL & INSTALLATION

1. Before servicing the vehicle, refer to the precautions in the beginning of this section.
2. Relieve the fuel system pressure.
3. Remove or disconnect the following:
   - Negative battery cable
   - Fuel lines
   - Fuel filter

To install:

4. Install the fuel filter.
5. Use new washers and tighten the fuel line bolts to the following specifications:

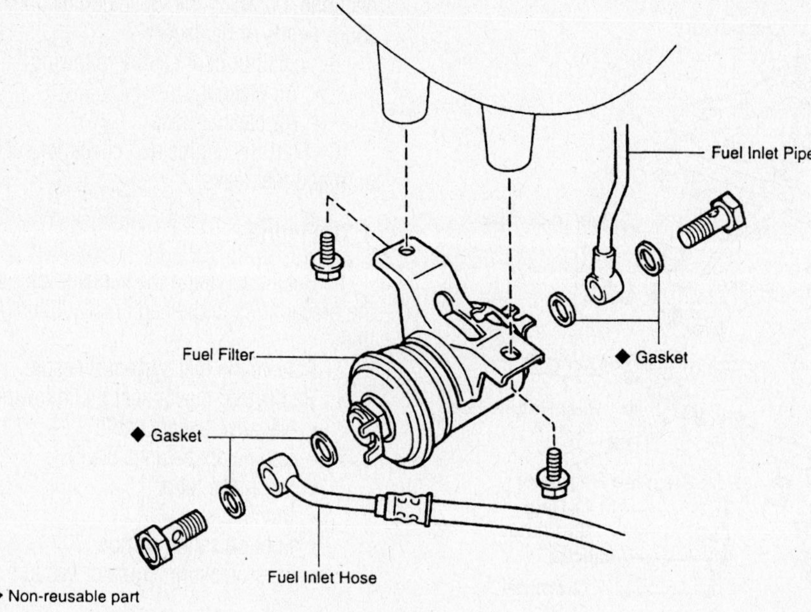

Always use new gaskets when replacing the fuel filter

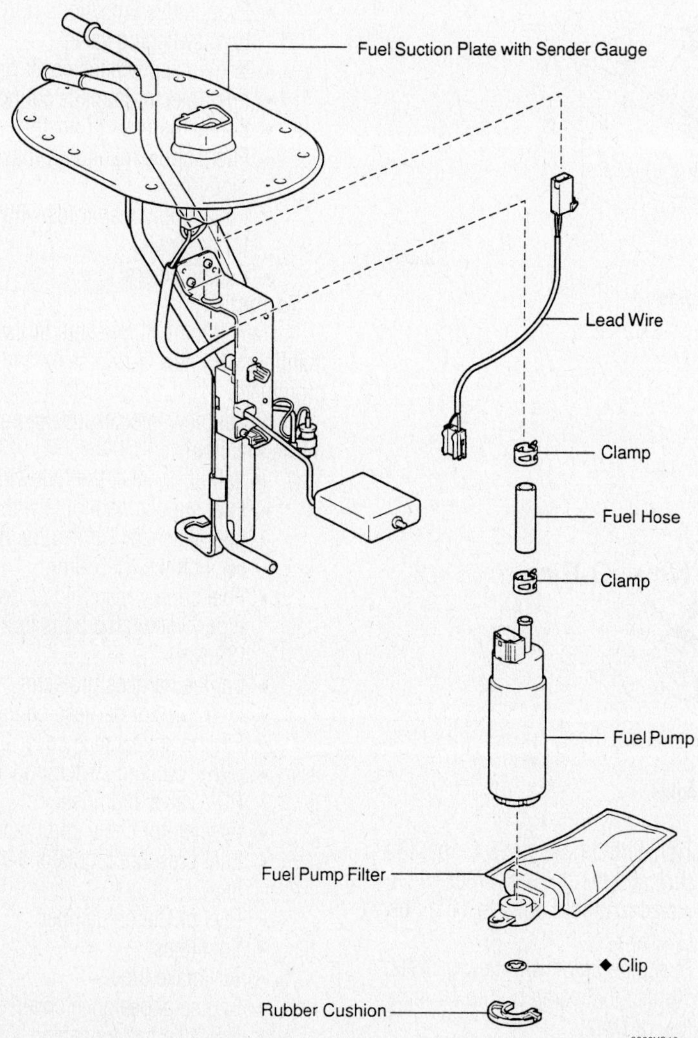

Exploded view of the fuel pump and related components

- Banjo bolt fittings: 21 ft. lbs. (29 Nm)
- Flare nut fitting: 28 ft. lbs. (38 Nm)

6. Connect the negative battery cable.
7. Start the engine and check for leaks.

Fuel Pump

REMOVAL & INSTALLATION

1. Before servicing the vehicle, refer to the precautions in the beginning of this section.
2. Relieve the fuel system pressure.
3. Remove or disconnect the following:
   - Negative battery cable
   - Fuel tank
   - Fuel pump harness connector
   - Fuel lines
   - Fuel pump module

To install:

4. Install or connect the following:
   - Fuel pump module. Tighten the bolts to 35 inch lbs. (4 Nm).
   - Fuel lines
   - Fuel pump harness connector
   - Fuel tank
   - Negative battery cable
5. Start the engine and check for leaks.

Fuel Injector

REMOVAL & INSTALLATION

3.4L Engine

1. Before servicing the vehicle, refer to the precautions in the beginning of this section.
2. Depressurize the fuel system.
3. Remove or disconnect the following:
   - Air cleaner hose
   - Upper half of the intake manifold
   - Fuel pressure regulator
   - Fuel inlet pipe
   - Fuel injector electrical connections
   - Fuel rail with the injectors
   - Spacers from the intake manifold
   - Injectors from the delivery pipes
   - O-rings and grommets, discard them

To install:

4. Install or connect the following:
   - New grommets and O-rings on each injector, lubricated with a light coat of gasoline
   - Fuel injector with the electrical connector facing outward
   - Spacers on the intake manifold

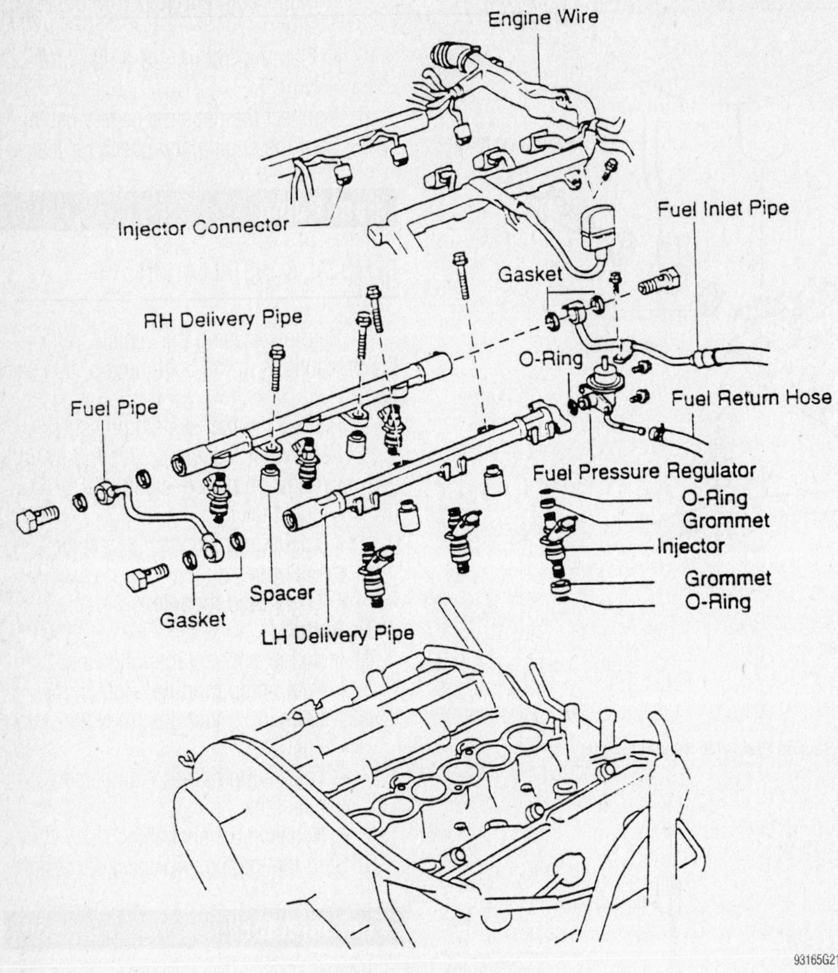

Fuel injector arrangement and related components—3.4L engine

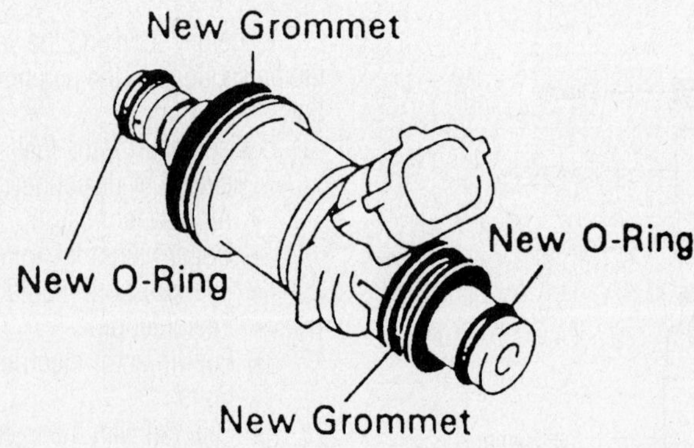

Install new O-rings and grommets on each injector—3.4L engine

5. Temporarily install the bolts to hold the delivery pipes to the intake manifold.

6. Check that the injectors rotate smoothly. If they do not, the O-rings have probably been installed incorrectly.

7. Install or connect the following:
 - Fuel injector electrical connectors
 - Fuel pipe with new gaskets. Tighten the bolts to 25 ft. lbs. (34 Nm) and the delivery pipes-to-intake manifold bolts to 10 ft. lbs. (13 Nm).
 - Fuel pipe union with new gaskets. Tighten the clamp bolt to 71 inch lbs. (8 Nm).
 - Fuel pressure regulator

8. Inspect the vacuum lines and connections. Look for any loose connections, sharp bends or damage.

9. Install or connect the following:
 - Air cleaner
 - Air cleaner hose

10. Start the engine and check for vacuum and fuel leaks.

4.7L Engine

1. Before servicing the vehicle, refer to the precautions in the beginning of this section.

2. Relieve the fuel system pressure.

3. Remove or disconnect the following:
 - Negative battery cable
 - Engine appearance cover
 - Air intake tube
 - Fuel lines
 - Fuel pulsation damper
 - Fuel pressure regulator vacuum line
 - Accelerator cable and bracket
 - Positive Crankcase Ventilation (PCV) valve and hose
 - Evaporative Emissions (EVAP) vacuum switching valve
 - Engine appearance cover brackets
 - Fuel injector harness connectors
 - Engine harness protector
 - Fuel supply manifold crossover pipe
 - Fuel supply manifolds with injectors attached
 - Fuel injectors

To install:

4. Install the fuel injectors to the supply manifold with new O-ring seals and new grommets.

5. Install new injector insulators to the intake manifold.

6. Install or connect the following:
 - Fuel supply manifolds with injectors attached. Tighten the bolts to 66 inch lbs. (7.5 Nm).
 - Fuel supply manifold crossover pipe. Tighten the bolts to 29 ft. lbs. (39 Nm).
 - Engine harness protector
 - Fuel injector harness connectors
 - Engine appearance cover brackets
 - EVAP vacuum switching valve
 - PCV valve and hose
 - Accelerator cable and bracket
 - Fuel pressure regulator vacuum line
 - Fuel pulsation damper
 - Fuel lines
 - Air intake tube
 - Engine appearance cover
 - Negative battery cable

7. Start the engine and check for leaks.

DRIVE TRAIN

Manual Transmission Assembly

REMOVAL & INSTALLATION

2WD

1. Before servicing the vehicle, refer to the precautions in the beginning of this section.
2. Remove or disconnect the following:
   - Negative battery cable
3. Drain the transmission oil.
4. Remove the shift lever assembly, as follows:
   - Shift lever knob
   - 4 screws, shift lever boot retainer and shift lever boot
   - 6 bolts, shift lever assembly and baffle
   - Turn over the dust boot
   - Shift lever cap, cover it with a cloth
   - Shift lever cap by pressing downward and rotating it counterclockwise
   - Shift lever
5. Remove or disconnect the following:
   - Driveshaft
   - Vehicle Speed Sensor (VSS) and the back-up light switch connectors
   - Oxygen (O_2) sensor connector
   - Front exhaust pipe from the exhaust manifold and catalytic converter
   - Clutch release cylinder
   - Starter wires
   - Starter
6. Position a jack and wooden block under the transmission.
7. Remove or disconnect the following:
   - Rear endplate
   - Rear engine mount bracket
   - Crossmember
8. Attach a engine hoist to the engine hangers.
9. Remove or disconnect the following:
   - Engine mounts
   - Engine/transmission assembly out of the vehicle
10. Safely support the engine/transmission assembly.
11. Remove or disconnect the following:
   - Transmission-to-engine bolts
   - Transmission mount

To install:

12. Install or connect the following:
   - Transmission
   - Transmission mount. Tighten the 4 bolts to 48 ft. lbs. (65 Nm).
   - Tighten the 6 transmission-to-engine bolts to 53 ft. lbs. (72 Nm).
   - Crossmember. Torque the bolts to 53 ft. lbs. (72 Nm).
   - Rear engine mount-to-crossmember bolts to 13 ft. lbs. (18 Nm).
   - Starter. Tighten the 2 bolts to 29 ft. lbs. (39 Nm).
   - Rear endplate. Tighten the 4 bolts to 27 ft. lbs. (37 Nm).
13. Install or connect the following:
   - Tighten the engine mounts to 28 ft. lbs. (38 Nm).
   - Starter wires
   - Clutch release cylinder
   - Driveshaft
   - Front exhaust pipe. Tighten the exhaust pipe-to-manifold bolts to 46 ft. lbs. (62 Nm) and the exhaust pipe-to-catalytic converter bolts to 35 ft. lbs. (48 Nm).
   - Remaining components
   - Negative battery cable
14. Install the shift lever assembly, as follows:
   - Shift lever
   - Shift lever cap, cover it with a cloth
   - Shift lever cap by pressing downward and rotating it clockwise
   - Turn over the dust boot
   - 6 bolts, shift lever assembly and baffle
   - 4 screws, shift lever boot retainer and shift lever boot
   - Shift lever knob
15. Fill the transmission with oil.
16. Start the engine and check for leaks.
17. Install the engine undercover.
18. Road test the vehicle and check all fluids.

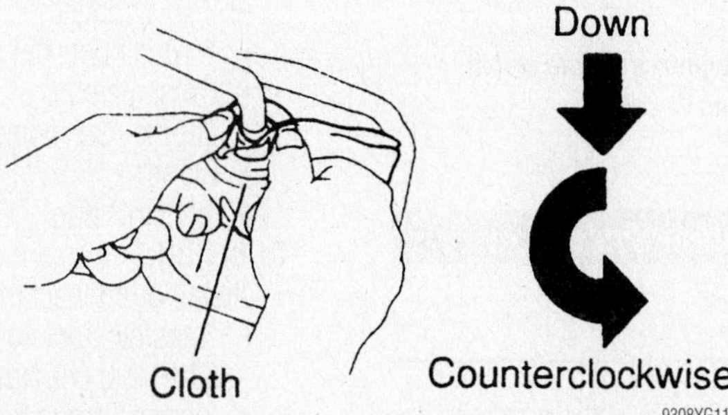

Removing the transfer shift lever

4WD

1. Before servicing the vehicle, refer to the precautions in the beginning of this section.
2. Disconnect negative battery cable.
3. Remove the transmission shift lever assembly, as follows:
   - Shift lever knob
   - 4 screws, shift lever boot retainer and shift lever boot
   - 6 bolts, shift lever assembly and baffle
   - Turn over the dust boot
   - Shift lever cap, cover it with a cloth
   - Shift lever cap by pressing downward and rotating it counterclockwise
   - Shift lever
   - Transfer shift lever, using snapring pliers to pull it from the transfer case
4. Drain the transmission and the transfer oil.
5. Remove or disconnect the following:
   - Driveshafts
   - Vehicle Speed Sensor (VSS), back-up light switch connector and the transfer indicator switch connector
   - Clutch release cylinder, move it aside without disconnecting the clutch line
   - Oxygen (O_2) sensor
   - Front exhaust pipe bracket
   - Starter
   - Rear endplate by removing the nuts and 2 bolts
6. Using a transmission jack, support the transmission.
7. Remove or disconnect the following:
   - 4 engine rear mount bolts
   - 8 bolts and the frame crossmember from the side frame
   - 6 transmission-to-engine bolts
   - 3 wire clamps from the transmission
   - Transmission with the transfer case
   - 4 engine rear mount bolts from the transfer case
   - Transfer adapter rear mount bolts

- Transfer case from the transmission

To install:

8. Apply MP grease to the adapter oil seal and shift the 2 shift fork shafts to the high 4 position.
9. Install or connect the following:
   - Transfer case to the transmission. Tighten the bolts to 17 ft. lbs. (24 Nm).

⁂ WARNING

Be careful not to damage the oil seal by the input gear spline when installing the transfer.

   - Engine rear mounting. Tighten the 4 bolts to 48 ft. lbs. (65 Nm).
   - Transmission/transfer case assembly
10. Support the transmission with a jack. Align the input shaft spline with the clutch disc and push the transmission with the transfer fully into position.
11. Install or connect the following:
   - Tighten the engine-to-transmission bolts to 53 ft. lbs. (72 Nm).
   - Crossmember. Tighten the 4 bolts to 53 ft. lbs. (72 Nm) and the 4 engine rear mount bolts to 13 ft. lbs. (18 Nm).
   - Stabilizer bar
   - Rear endplate Tighten the 2 bolts and nuts to 27 ft. lbs. (37 Nm).
   - Starter. Tighten the bolts to 29 ft. lbs. (39 Nm).
   - Front exhaust pipe. Tighten the manifold bolts to 46 ft. lbs. (62 Nm) and the converter bolts to 35 ft. lbs. (48 Nm).
   - Clutch release cylinder. Tighten the bolts to 9 ft. lbs. (12 Nm).
   - Driveshafts
   - Remaining components
12. Install the transmission shift lever assembly, as follows:
   - Apply MP grease to the shift lever
   - Shift lever
   - Shift lever cap, cover it with a cloth
   - Shift lever cap by pressing downward and rotating it clockwise
   - Turn over the dust boot
   - 6 bolts, shift lever assembly and baffle
   - 4 screws, shift lever boot retainer and shift lever boot
   - Shift lever knob
13. Connect the negative battery cable.
14. Start the engine and check for leaks.
15. Road test the vehicle for proper operation. Recheck all fluid levels.

Clutch Assembly

REMOVAL & INSTALLATION

1. Before servicing the vehicle, refer to the precautions in the beginning of this section.
2. Remove or disconnect the following:
   - Negative battery cable
   - Transmission assembly
3. Matchmark the clutch cover to the flywheel.
4. At the clutch cover, loosen each bolt 1 turn until spring tension is released.
5. Remove or disconnect the following:
   - Clutch cover set bolts and the clutch cover with the clutch disc.
   - Release bearing retaining clip and withdraw the it
   - Release fork and boot assembly

To install:

6. Install or connect the following:
   - Clutch disc onto the flywheel, using a clutch disc alignment tool
   - Clutch cover, position it onto the flywheel and if reusing the old pressure plate, align the matchmarks.
   - Clutch cover. Tighten the bolts in a crisscross pattern to 14 ft. lbs. (19 Nm).
7. Lubricate the release fork pivot and contact points, the release bearing, bearing hub and input shaft spline surfaces with a suitable molybdenum disulfide lithium based or multi-purpose grease.
8. Install or connect the following:
   - Boot, release fork, hub and the bearing assemblies
   - Transmission
   - Negative battery cable

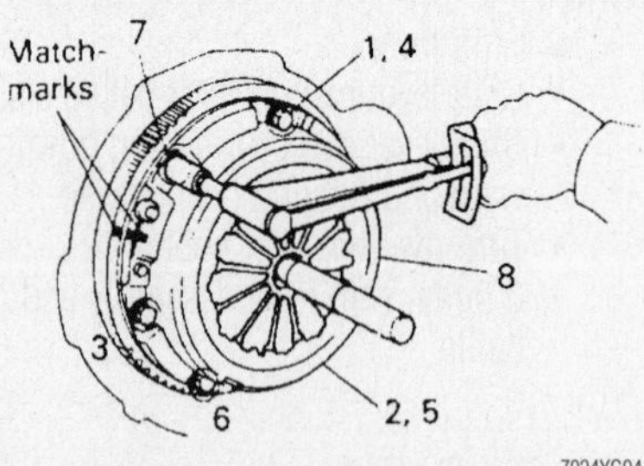

Bolt tightening sequence for the clutch cover—all engines

Hydraulic Clutch System

BLEEDING

1. Before servicing the vehicle, refer to the precautions in the beginning of this section.
2. Fill the clutch reservoir with brake fluid. Check the reservoir level frequently and add fluid as needed.
3. Connect one end of a vinyl tube to the bleeder plug on the slave cylinder and submerge the other end into a clear container half-filled with brake fluid.
4. Slowly pump the clutch pedal several times.
5. Have an assistant hold the clutch pedal down and loosen the bleeder plug until fluid and/or air starts to run out of the bleeder plug. Close the bleeder plug while the pedal is held to the floor.
6. Repeat Steps 2 and 3 until all the air bubbles are removed from the system.
7. Tighten the bleeder plug when all the air is gone.
8. Refill the master cylinder to the proper level as required.
9. Check the system for leaks.

Automatic Transmission Assembly

REMOVAL & INSTALLATION

3.4L Engine

1. Before servicing the vehicle, refer to the precautions in the beginning of this section.
2. Remove or disconnect the following:
   - Negative battery cable
   - Throttle cable
   - Transmission dipstick
   - Oil filler tube and discard the O-ring
3. Shift the transfer case into **H4** position.
4. Remove or disconnect the following:
   - Transfer case shift lever knob
   - No. 1 engine under cover
   - Front and center exhaust pipes
   - Driveshaft(s)
   - No. 1 and 2 Vehicle Speed Sensor (VSS) connectors
   - Shift control cable
   - Oil cooler pipe
   - Automatic Transmission Fluid (ATF) sensor connector
   - Park/Neutral Position (PNP) switch
   - Rear endplate
   - Torque converter bolts
   - Crossmember
   - Engine rear mounting insulator
   - Starter
   - Transmission

To install:

5. Install or connect the following:
   - Transmission. Torque the bolts to 53 ft. lbs. (71 Nm).
   - Starter. Torque the bolts to 29 ft. lbs. (39 Nm).

- Engine rear mounting insulator. Torque the bolts to 48 ft. lbs. (65 Nm).
- Crossmember. Torque the nuts to 53 ft. lbs. (72 Nm) and the bolts to 13 ft. lbs. (18 Nm)..
- Torque converter. Torque the bolts to 30 ft. lbs. (41 Nm).
- Rear endplate. Torque the bolts to 13 ft. lbs. (18 Nm).
- Park/Neutral Position (PNP) switch
- Automatic Transmission Fluid (ATF) sensor connector
- Oil cooler pipe. Torque the bolts to 25 ft. lbs. (34 Nm).
- Shift control cable. Torque the bolts to 9 ft. lbs. (12 Nm).
- No. 1 and 2 Vehicle Speed Sensor (VSS) connectors
- Driveshaft(s)
- Front and center exhaust pipes
- No. 1 engine under cover
- Transfer case shift lever knob
- Oil filler tube using a new O-ring
- Transmission dipstick
- Throttle cable
- Negative battery cable

4.7L Engine

1. Install or connect the following:
   - Transmission. Tighten the flange bolts to 53 ft. lbs. (72 Nm).
   - Transmission mount crossmember. Tighten the bolts to 37 ft. lbs. (50 Nm) and the nuts to 54 ft. lbs. (74 Nm).
   - Transmission oil cooler lines
   - Torque converter. Tighten the bolts to 35 ft. lbs. (48 Nm).
   - Motor actuator connector
   - L4 solenoid valve position switch connector
   - Center differential lock indicator switch connector
   - PNP switch connector
   - Transmission fluid temperature sensor connector
   - Solenoid harness connector
   - Overdrive clutch speed sensor connector
   - VSS sensor connectors
   - Front driveshaft. Tighten the fasteners to 59 ft. lbs. (80 Nm).
   - Rear driveshaft. Tighten the fasteners to 78 ft. lbs. (106 Nm).
   - Exhaust front pipes
   - Engine under covers
   - Transfer case shift lever and rod
   - Transmission gear select lever and rod
   - Center console
   - Transmission dipstick tube
   - Coolant recovery reservoir
   - Cooling fan and shroud
   - Air intake assembly
   - Battery and tray
2. Check the transmission and transfer case fluid levels and adjust as necessary.

Transfer Case Assembly

REMOVAL & INSTALLATION

VF2A Model

This transfer case is to be used with the 3.4L engine.

1. Before servicing the vehicle, refer to the precautions in the beginning of this section.
2. Drain the transfer case oil.
3. Shift the transfer shift lever into the **H4** position.
4. Remove or disconnect the following:
   - Transfer case shift lever knob
   - 4 screws, transfer case shift lever boot retainer and the shift lever boot
   - Snapring from the transfer case shift lever and the lever
   - Breather hose from the transfer case
   - Front and rear driveshafts
   - Dynamic damper from the transfer case
   - Crossmember from the rear of the transmission
   - 4 bolts and the engine rear mount from the transfer case adapter
   - Vehicle Speed Sensor (VSS) connector
   - Transfer Detection Switch (TDS) connector
5. Support the transfer case
6. Remove or disconnect the following:
   - 8 transfer case-to-transfer adapter bolts
   - Transfer case

To install:

7. Install or connect the following:
   - Transfer case
   - Transfer case-to-transfer adapter bolts. Torque the 8 bolts to 17 ft. lbs. (24 Nm).
   - Transfer Detection Switch (TDS) connector
   - Vehicle Speed Sensor (VSS) connector
   - Engine rear mount to the transfer case adapter. Torque the 4 bolts to 48 ft. lbs. (65 Nm).
   - Crossmember to the rear of the transmission. Torque the 4 nuts/bolts to 53 ft. lbs. (72 Nm).
   - Crossmember to the chassis. Torque the 4 bolts to 13 ft. lbs. (18 Nm).
   - Dynamic damper to the transfer case. Torque the 2 bolts to 28 ft. lbs. (38 Nm).
   - Front and rear driveshafts
   - Breather hose to the transfer case to a depth of 0.51 in. (13mm) or more
   - Snapring to the transfer case's shift lever
   - 4 screws, transfer case shift lever boot retainer and the shift lever boot
   - Transfer case shift lever knob
8. Refill the transfer case to the correct level.
9. Test drive the vehicle.

VF2BM Model

This transfer case is to be used with the 4.7L engine.

1. Before servicing the vehicle, refer to the precautions in the beginning of this section.
2. Drain the transfer case oil.
3. Turn the touch select 2–4 switch **ON**.
4. Remove or disconnect the following:
   - Breather hose from the transfer case
   - Left and right exhaust pipes
   - Front and rear driveshafts
   - Crossmember from the rear of the transmission
   - 4 bolts and the engine rear mount from the transfer case adapter
   - Vehicle Speed Sensor (VSS) connector
   - Transfer Detection Switch (TDS) connectors
   - Motor actuator connectors
5. Support the transfer case
6. Remove or disconnect the following:
   - 8 transfer case-to-transfer adapter bolts
   - Transfer case

To install:

7. Install or connect the following:
   - Transfer case
   - Transfer case-to-transfer adapter bolts. Torque the 8 bolts to 17 ft. lbs. (24 Nm).
   - Motor actuator connectors
   - Transfer Detection Switch (TDS) connector
   - Vehicle Speed Sensor (VSS) connector
   - Engine rear mount to the transfer

case adapter. Torque the 4 bolts to 48 ft. lbs. (65 Nm).

- Crossmember to the rear of the transmission. Torque the 4 nuts/bolts to 53 ft. lbs. (72 Nm).
- Crossmember to the chassis. Torque the 4 bolts to 13 ft. lbs. (18 Nm).
- Dynamic damper to the transfer case. Torque the 2 bolts to 28 ft. lbs. (38 Nm).
- Front and rear driveshafts
- Left and right exhaust pipes
- Breather hose to the transfer case to a depth of 0.51 in. (13mm) or more

8. Refill the transfer case to the correct level.
9. Test drive the vehicle.

Halfshaft

REMOVAL & INSTALLATION

1. Before servicing the vehicle, refer to the precautions in the beginning of this section.
2. Remove or disconnect the following:

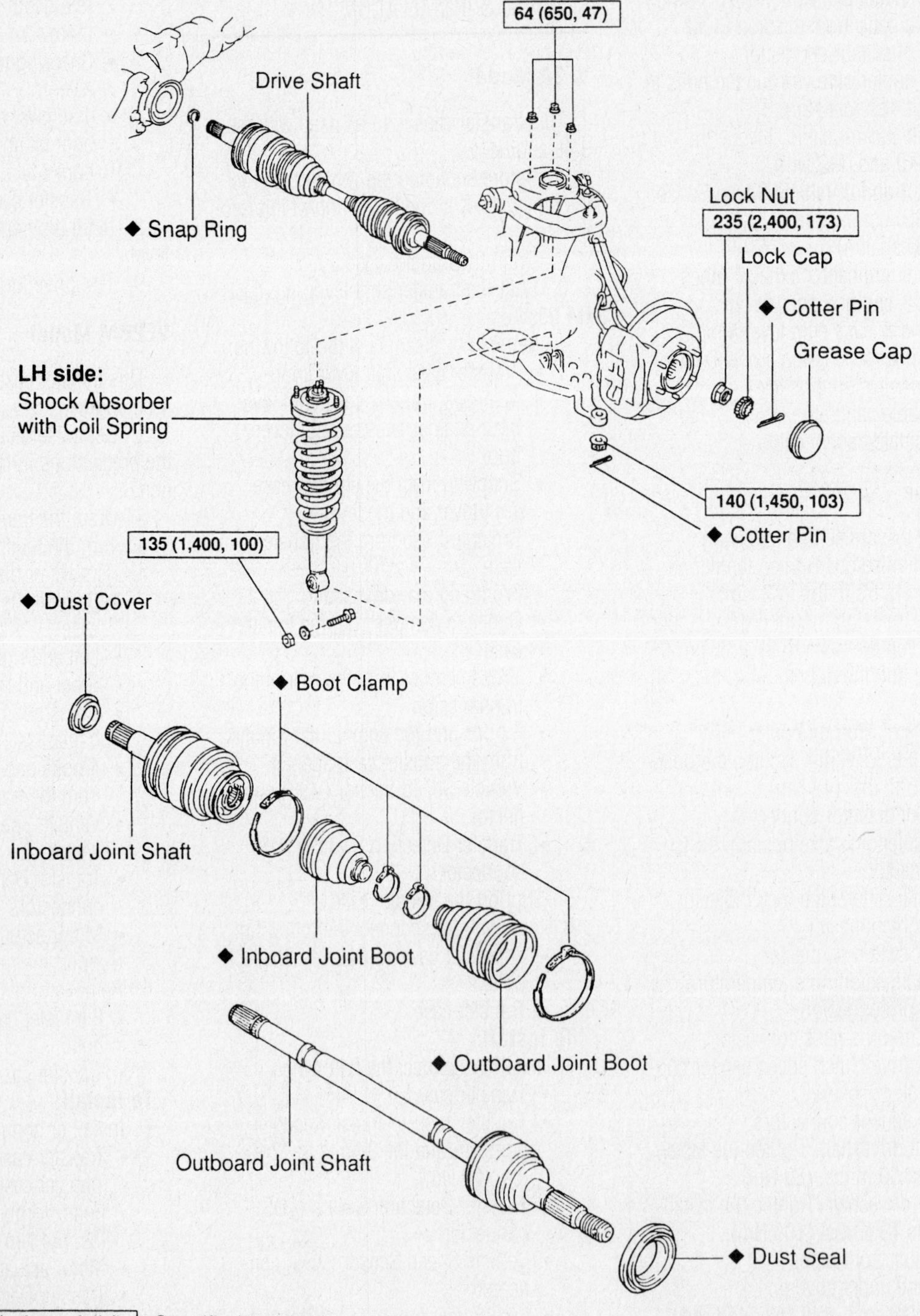

View of the halfshaft and related components

- Front wheel
- Under cover
3. Drain the differential oil.
4. Remove or disconnect the following:
   - Grease cap
   - Cotter pin and lock cap
   - Halfshaft locknut by applying the brakes
   - Lower control arm from the lower ball joint
   - Halfshaft from the steering knuckle, using a plastic hammer
   - Left strut, for the left halfshaft
   - Right halfshaft, using a brass bar and a hammer
   - Left halfshaft, using tools 09520-01010 and 09520-24010
   - Snapring from the inboard joint shaft

To install:

5. Install or connect the following:
   - New snapring, onto the inboard joint shaft with the opening facing downward
   - Halfshafts to the differential using a brass bar and a hammer
   - Halfshafts to the steering knuckles

WARNING

Be careful not to damage the oil seal, boot or dust seal.

   - Lower control arm to the lower ball joint using a new cotter pin. Torque the ball joint nut to 103 ft. lbs. (140 Nm).
   - Halfshaft locknut by applying the brakes. Torque the nut to 173 ft. lbs. (235 Nm).
   - Lock cap and a new cotter pin
   - Grease cap
6. Refill the differential with oil.
7. Install or connect the following:
   - Under cover
   - Front wheel

CV-Joints

OVERHAUL

Outer CV-joint

The outer CV-joint is serviced with the axle shaft as an assembly. The outer CV-joint boot can be serviced by removing the inner CV-joint.

Inner CV-joint

1. Before servicing the vehicle, refer to the precautions in the beginning of this section.
2. Remove or disconnect the following:
   - Halfshaft from the vehicle
   - Large boot clamps
   - Small boot clamps
3. Matchmark inboard CV-joint to the shaft
4. Remove or disconnect the following:
   - Inboard CV-joint from the shaft by expanding the snapring
   - Both CV-joint boots
   - Outer dust seal, using a shop press and tool 09950-00020
   - Outer dust cover, using a shop press and tool 09950-00020

To install:

5. Install or connect the following:
   - Outer dust cover, using a screwdriver and a hammer
   - Outer dust seal, using a screwdriver and a hammer
6. Wrap the shaft splines with tape to protect the boot from damage.
7. Install or connect the following:
   - Both CV-joint boots with clamps, temporarily
   - Inboard CV-joint to the shaft by aligning the matchmarks and expanding the snapring
8. Lubricate the outboard joint with 7.23–7.94 oz. (205–225g) grease, provided in the boot kit.
9. Lubricate the inboard joint with 6.70–7.41 oz. (190–210g) grease, provided in the boot kit.
10. Install or connect the following:
   - Both joint boots making sure the boots are in the shaft groove
   - Standard halfshaft length is 20.531–20.689 in.(521.5–525.5mm) when the shaft is not expanded or contracted
   - Large inboard boot clamp
   - All other boot clamps using tool 09521-24010. Tighten the crimping tool until the clamp clearance is 0.039–0.059 in. (1.0–1.5mm)
   - Halfshaft

Spindle Bearings

REMOVAL, PACKING AND INSTALLATION

1. Before servicing the vehicle, refer to the precautions in the beginning of this section.
2. Remove or disconnect the following:
   - Front wheel
   - Grease cap, for 2WD
   - Cotter pin and lock cap, for 4WD
   - Locknut, by applying the brakes
   - Anti-lock Brake System (ABS) speed sensor and wiring harness clamp from the steering knuckle
   - Brake line clamp from the steering knuckle

➡Be careful not to damage the brake tube.

   - Brake caliper and disc
3. Support the steering knuckle
4. Remove or disconnect the following:
   - 4 lower ball joint-to-steering knuckle bolts
   - Upper ball joint cotter pin and loosen the nut
   - Upper ball joint from the steering knuckle using tool 09950-40011
   - Steering knuckle by placing it in a soft-jawed vise
   - Inside oil seal
   - 4 bolts and shift the dust cover towards the hub side (outside)
   - Axle hub from the steering knuckle using tools 09710-30021 and 09950-40011
   - Dust cover from the steering knuckle
   - Bearing spacer and ABS speed sensor (with ABS) or spacer (without ABS)

WARNING

Be careful not to scratch the speed sensor rotor serrations

   - Outside oil seal from the steering knuckle
   - Bearing snapring from the steering knuckle
   - Bearing from the steering knuckle, using a shop press and tools 09950-60020 and 09950-70010

To install:

5. Install or connect the following:
   - Bearing to the steering knuckle, using a shop press and tools 09527-17011 and 09950-60020
   - Bearing snapring to the steering knuckle
   - New outside oil seal to the steering knuckle, using tools 09223-15030 and 09527-17011
   - Dust cover to the steering knuckle. Torque the 4 bolts to 13 ft. lbs. (18 Nm).
   - Axle hub to the steering knuckle using a shop press and tool 09649-17010
   - ABS speed sensor (with ABS) or spacer (without ABS)

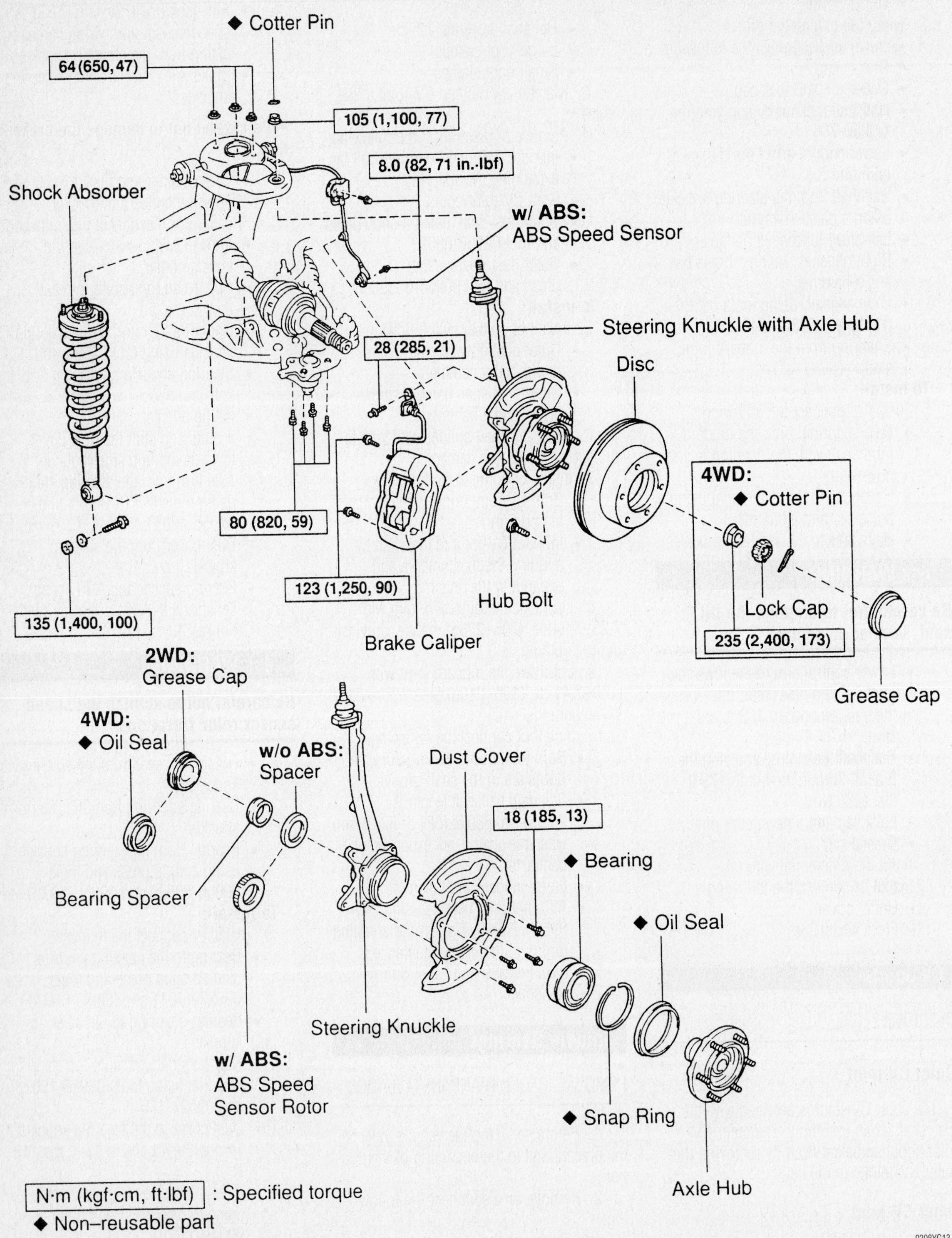

Exploded view of the front axle hub and related components

✲✲ WARNING

Be careful not to scratch the speed sensor rotor serrations

- Bearing spacer, using a shop press and tools 09950-60010 and 09950-70010
- New inside oil seal, using tool 09527-17011 and a plastic hammer

6. For 4WD, install or connect the following:
   a. Halfshaft into the axle hub and temporarily tighten the nut

✲✲ WARNING

Be careful not to damage the oil seal or boot.

   b. Steering knuckle to the upper control arm. Tighten the nut to 77 ft. lbs. (105 Nm).
   c. New cotter pin.
7. Install or connect the following:
   - Lower ball joint to the steering knuckle. Torque the 4 bolts to 59 ft. lbs. (80 Nm).
   - Strut
   - Brake disc and caliper. Torque both caliper bolts to 90 ft. lbs. (123 Nm).
   - Brake line clamp to the steering knuckle. Torque the bolt to 21 ft. lbs. (28 Nm).
   - ABS speed sensor. Torque both bolts to 7.1 ft. lbs. (8.2 Nm).
   - Halfshaft locknut. Torque the nut to 173 ft. lbs. (235 Nm), by applying the brakes.
   - Lock cap and new cotter pin
   - Grease cap, for 2WD
   - Front wheel
8. Depress the brake pedal several times.
9. Check and/or adjust the front wheel alignment.
10. Check the ABS speed sensor signal.

Axle Shaft, Bearing and Seal

REMOVAL & INSTALLATION

Rear

1. Before servicing the vehicle, refer to the precautions in the beginning of this section.
2. Remove or disconnect the following:
   - Rear wheel
   - Brake drum and gasket
3. Using a dial indicator, check the

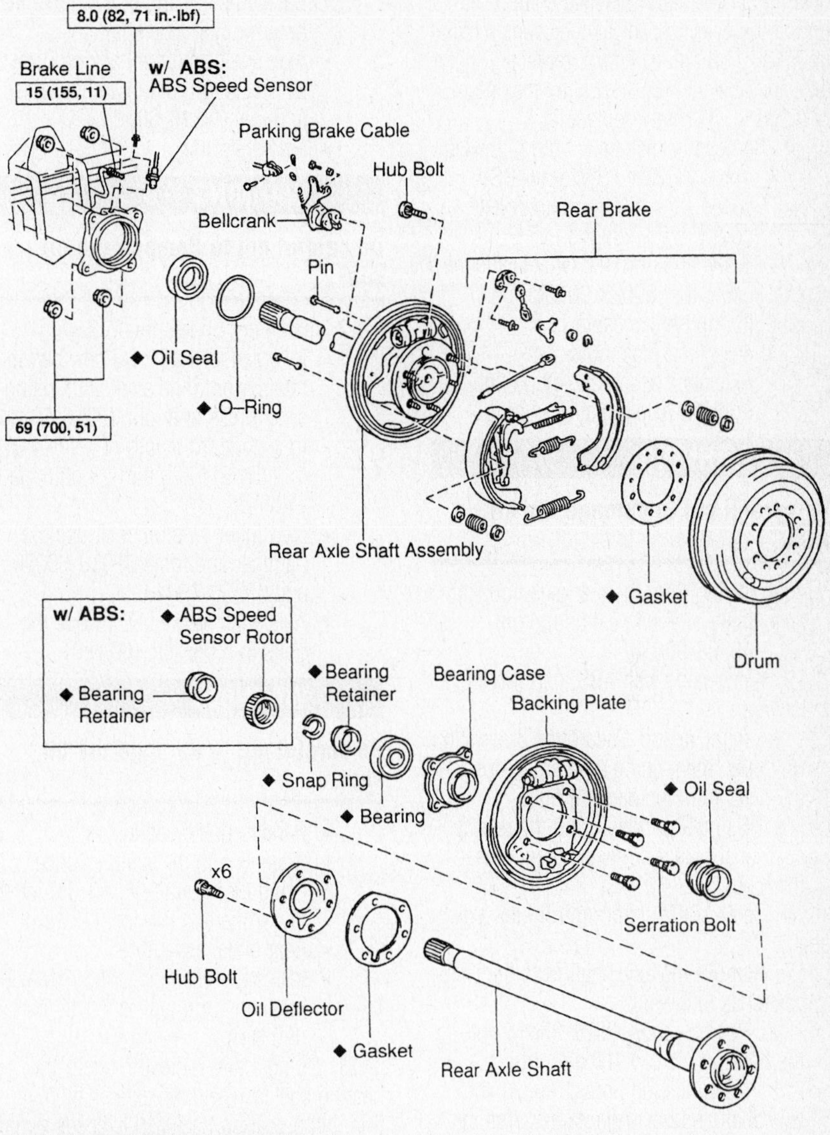

Exploded view of the rear axle

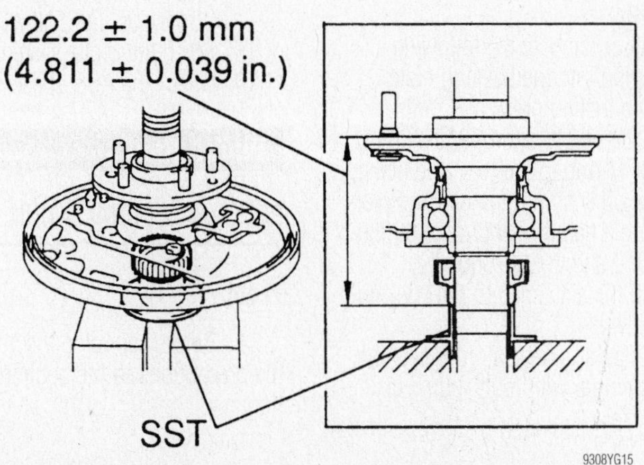

Standard length of ABS speed sensor rotor and bearing retainer—Rear axle bearing backlash and the axle shaft deviation. If the bearing backlash exceeds a maximum or 0.028 in. (0.7mm), replace it. If the axle shaft deviation exceeds the maximum of 0.004 in. (0.1mm), replace it.

4. Remove or disconnect the following:
   - Anti-lock Brake System (ABS) speed sensor from the rear axle housing, if equipped
   - Brake line from the wheel cylinder, using tool 09023-00100
   - Parking brake cable
   - 4 backing plate nuts
   - Axle shaft assembly, by pulling it from the axle housing

✱✱ WARNING

Be careful not to damage the oil seal.

   - O-ring from the rear axle housing
   - Inner side oil seal using tool 09308-00010

5. If equipped with ABS, perform the following:
   a. Remove and discard the 4 serration bolt nuts; then, using a hammer, drive the bolts from the backing plate.
   b. Using a grinder, grind the retainer and sensor rotor surfaces; then, chisel them out.
6. Remove the snapring from the axle shaft.
7. Remove the axle shaft from the backing plate, as follows:
   a. Position tool 09521-25011 onto the backing plate with the 4 nuts.
   b. Using a shop press, remove the axle shaft and bearing retainer from the backing plate.
8. Using tool 09308-00010, pull the oil seal from the backing plate.
9. Using a shop press and tools 09223-56010 and 09950-60010, press the bearing from the backing plate.

To install:

10. Install or connect the following:
    - Bearing into the backing plate, using a shop press and tools 09223-56010 and 09950-60010
    - New O-ring to the rear axle housing
    - New oil seal into the backing plate, using a hammer and tools 09950-70010 and 09950-60010
11. Install the axle shaft to the backing plate, as follows:
    - New outer side seal, lubricate the oil seal lip with multi-purpose grease
    - Backing plate and bearing retainer onto the rear axle shaft
    - Axle shaft onto the backing plate, by pressing it using a shop press and tool 09316-60011
    - New snapring

✱✱ WARNING

Be careful not to damage the oil seal.

12. Install or connect the following:
    - New sensor rotor and new bearing retainer onto the axle shaft, using a shop press and tool 09316-60011 to a standard length of 4.77–4.85 in. (121.2–123.2mm), if equipped with ABS
    - New inner side oil seal, using a hammer and tools 09950-60020 and 09950-70010
    - Axle shaft assembly. Torque the bolts to 51 ft. lbs. (69 Nm).

✱✱ WARNING

Be careful not to damage the oil seal.

    - Parking brake cable
    - Brake line to the wheel cylinder, using tool 09023-00100. Torque the brake line to 11 ft. lbs. (15 Nm).
    - Rear brake assembly
    - ABS speed sensor to the rear axle housing. Torque it to 7.1 ft. lbs. (8.0 Nm).
13. Using a dial indicator, check the bearing backlash and the axle shaft deviation. If the bearing backlash exceeds a maximum or 0.028 in. (0.7mm), replace it. If the axle shaft deviation exceeds the maximum of 0.004 in. (0.1mm), replace it.
14. Install or connect the following:
    - New gasket and brake drum
    - Rear wheel. Torque the lug nuts to 81 ft. lbs. (110 Nm).
15. Bleed the brake system.
16. Check the ABS speed sensor signal.

Pinion Seal

REMOVAL & INSTALLATION

Front

1. Before servicing the vehicle, refer to the precautions in the beginning of this section.
2. Remove the under cover.
3. Drain the differential housing oil.
4. Remove the front driveshaft.
5. Remove the companion flange, as follows:
   - Loosen the staked part of the nut, using a chisel and a hammer
   - Companion flange nut, using tool 09330-00021
   - Companion flange, using tools 09950-30011 and 09954-03010
6. Remove the oil seal and slinger, as follows:
   - Oil seal, using tool 09308-10010
   - Oil slinger

To install:

7. Install or connect the following:
   - Oil slinger
   - New oil seal, using a hammer and tool 09554-22010 to a depth of 0.153–0.189 in. (4.2–4.8mm).
8. Install the companion flange, as follows:
   - Companion flange
   - New nut, lubricated with hypoid gear oil
   - Torque the nut to 80 ft. lbs. (108 Nm), using tool 09330-00021.
9. Adjust the drive pinion preload
10. Rotate the drive pinion, using a torque wrench while tightening the flange nut to make sure the bearing preload is 10.4–16.5 inch lbs. (1.2–1.9 Nm) for a new bearing or 5.2–8.7 inch lbs. (0.6–1.0 Nm) for a used bearing. Tighten the flange nut to achieve the preload torque readings originally recorded.

✱✱ CAUTION

Never loosen the pinion nut to reduce bearing preload.

11. Install or connect the following:
    - Drive pinion nut, stake it
    - Front driveshaft. Tighten the fasteners to 54 ft. lbs. (74 Nm).
    - Under cover
12. Fill the differential with gear lubricant and check for leaks.

Rear

1. Before servicing the vehicle, refer to the precautions in the beginning of this section.
2. Drain the differential housing oil.
3. Remove the rear driveshaft.
4. Remove the companion flange, as follows:

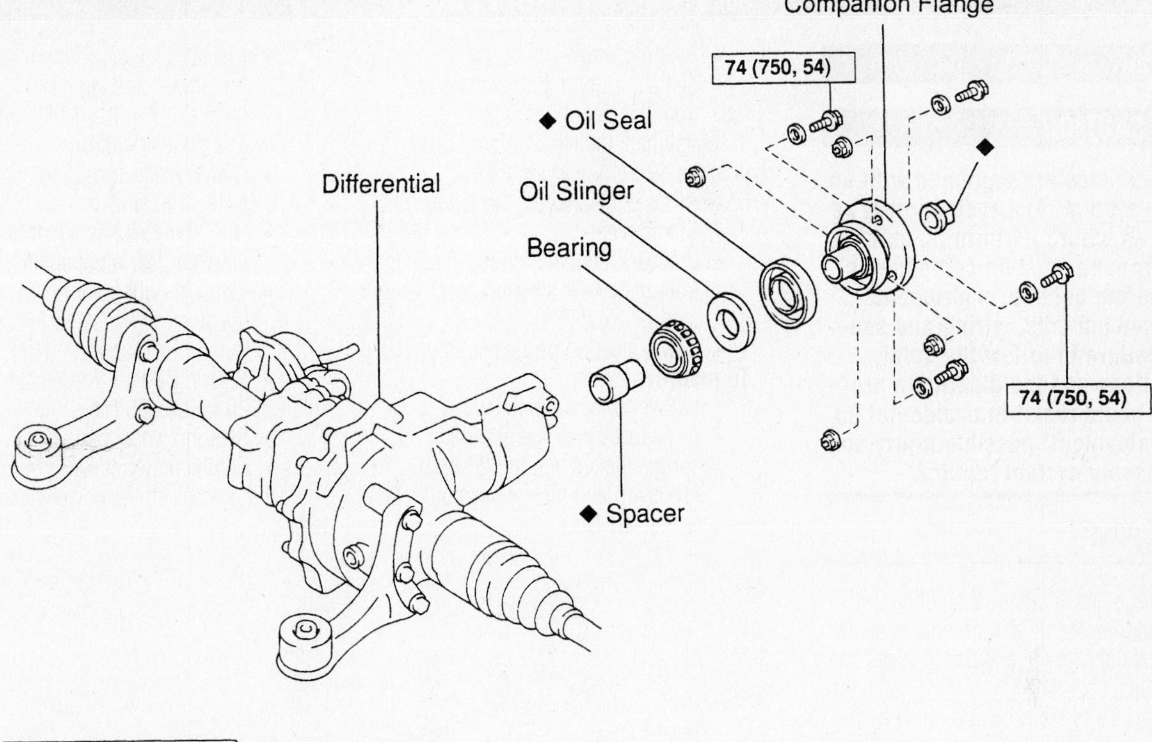

Exploded view of the front differential assembly—Rear differential assembly is similar

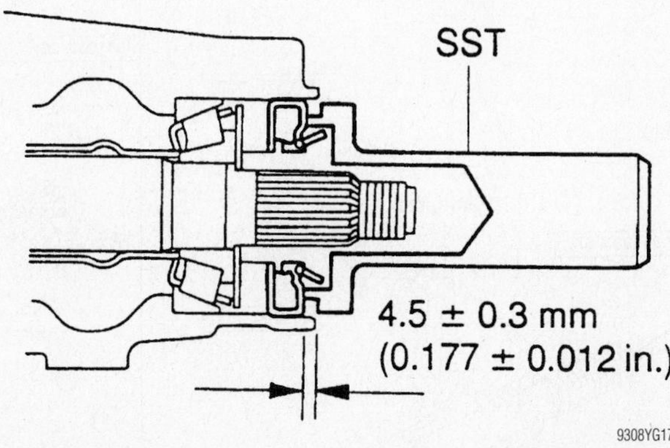

Positioning the front pinion seal in the differential housing—Rear differential assembly is similar

- Loosen the staked part of the nut, using a chisel and a hammer
- Companion flange nut, using tool 09330-00021
- Companion flange, using tools 09950-30011 and 09954-03010
- Oil seal, using tool 09308-10010

To install:

5. Install the new oil seal until it is flush with the housing, using a plastic hammer and tools 09316-12010 and 09649-17010

➡Use vinyl tape to connect both oil seal installation tools.

6. Install the companion flange, as follows:
   - Companion flange
   - New nut, lubricated with hypoid gear oil
   - Torque the nut to 109 ft. lbs. (147 Nm), using tool 09330-00021.
7. Adjust the drive pinion preload
8. Rotate the drive pinion, using a torque wrench while tightening the flange nut to make sure the bearing preload is 11.4–16.7 inch lbs. (1.3–1.9 Nm) for a new bearing or 4.3–6.9 inch lbs. (0.5–0.8 Nm) for a used bearing. Tighten the flange nut to achieve the preload torque readings originally recorded.

✻✻ CAUTION

Never loosen the pinion nut to reduce bearing preload.

9. Install or connect the following:
   - Drive pinion nut, stake it
   - Rear driveshaft. Tighten the fasteners to 54 ft. lbs. (74 Nm).
10. Refill the differential with gear lubricant and check for leaks; 3.33 qts. for 2WD or 3.12 qts. for 4WD.

STEERING AND SUSPENSION

Air Bag

✱✱ CAUTION

Some vehicles are equipped with an air bag system. The system must be disarmed before performing service on, or around, system components, the steering column, instrument panel components, wiring and sensors. Failure to follow the safety precautions and the disarming procedure could result in accidental air bag deployment, possible injury and unnecessary system repairs.

PRECAUTIONS

Several precautions must be observed when handling the inflator module to avoid accidental deployment and possible personal injury.

- Never carry the inflator module by the wires or connector on the underside of the module.
- When carrying a live inflator module, hold securely with both hands and ensure that the bag and trim cover are pointed away.
- Place the inflator module on a bench or other surface with the bag and trim cover facing up.
- With the inflator module on the bench, never place anything on or close to the module which may be thrown in the event of an accidental deployment.

DISARMING

To avoid personal injury when working on vehicles equipped with an air bag, the negative battery cable must be disconnected and at least 90 seconds must elapse before working on the system. Failure to do so may result in deployment of the air bag.

Power Rack And Pinion Steering Gear

REMOVAL & INSTALLATION

1. Before servicing the vehicle, refer to the precautions in the beginning of this section.
2. Position the front wheels in the straight-ahead position.
3. Remove or disconnect the following:
   - Engine under cover
   - Steering wheel pad
   - Steering wheel
   - Left and right outer tie-rod ends from the steering knuckles
4. Matchmark the No. 2 intermediate shaft to the steering gear input shaft.
5. Remove or disconnect the following:
   - Clamp plate
   - Pressure feed and return tubes from the power steering gear, using tool 09631-22020
   - Power steering gear assembly

To install:

6. Install or connect the following:
   - Power steering gear assembly. Torque the set bolt to 123 ft. lbs. (165 Nm) and the set nut/bolt to 96 ft. lbs. (91 Nm).
   - Pressure feed and return tubes to the power steering gear. Torque them to 27 ft. lbs. (32 Nm), using tool 09631-22020.
   - Clamp plate. Torque the bolt to 21 ft. lbs. (29 Nm).
   - No. 2 intermediate shaft to the steering gear input shaft
   - Left and right outer tie-rod ends to the steering knuckles. Torque the nuts to 67 ft. lbs. (91 Nm).
   - Steering wheel. Torque the nut to 26 ft. lbs. (35 Nm).
   - Steering wheel pad
   - Engine under cover
7. Fill and bleed the power steering system.

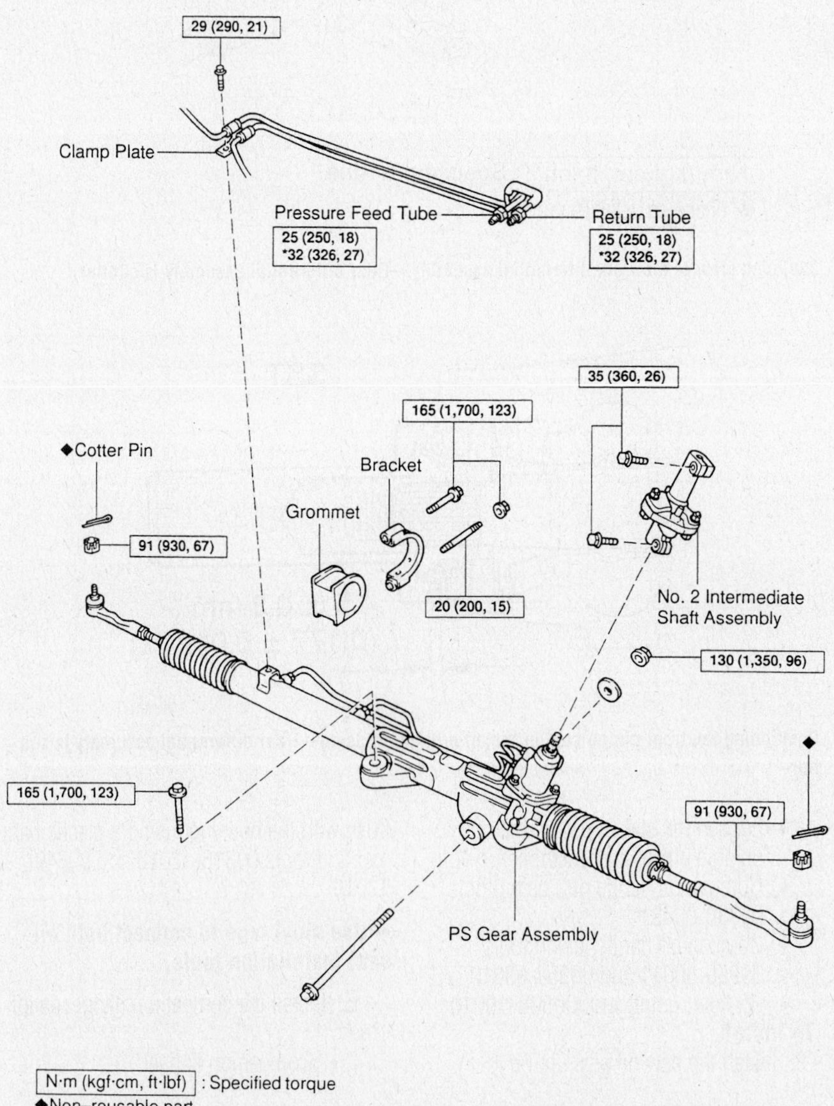

Exploded view of the power rack and pinion steering gear mounting

8. Check and/or adjust the wheel alignment, as necessary.

Strut

REMOVAL & INSTALLATION

Front

1. Before servicing the vehicle, refer to the precautions in the beginning of this section.
2. Remove or disconnect the following:
   - Front wheel
   - Strut-to-lower control arm nut/bolt and the strut
   - Strut-to-chassis 3 nuts/bolts and the strut

To install:

3. Install or connect the following:
   - Strut to the chassis. Torque the 3 nuts/bolts to 47 ft. lbs. (64 Nm).
   - Strut to the lower control arm. Torque the nut/bolt to 100 ft. lbs. (135 Nm).
   - Front wheel

Shock Absorber

REMOVAL & INSTALLATION

Rear

1. Before servicing the vehicle, refer to the precautions in the beginning of this section.
2. Remove or disconnect the following:
   - Rear wheel
   - Shock absorber

To install:

3. Install or connect the following:
   - Shock absorber. Tighten the upper nut to 15 ft. lbs. (20 Nm) and the lower nut/bolt to 64 ft. lbs. (87 Nm).
   - Rear wheel. Torque the lug nuts to 81 ft. lbs. (110 Nm).

Leaf Spring

REMOVAL & INSTALLATION

Rear

1. Before servicing the vehicle, refer to the precautions in the beginning of this section.
2. Support the vehicle at the frame.
3. Support the axle with a floor jack.
4. Remove or disconnect the following:
   - Rear wheel
   - 4 spring seat nuts and seat
   - Both leaf spring-to-chassis nuts/bolts
   - Leaf spring

To install:

5. Install or connect the following:
   - Leaf spring
   - Both leaf spring-to-chassis nuts/bolts. Torque both nuts/bolts to 125 ft. lbs. (170 Nm).
   - Spring seat. Torque the 4 nuts to 98 ft. lbs. (133 Nm).
   - Rear wheel

Upper Ball Joint

REMOVAL & INSTALLATION

1. Before servicing the vehicle, refer to the precautions in the beginning of this section.
2. Remove or disconnect the following:
   - Front wheel
   - Steering knuckle with the axle hub
   - Wire and boot
   - Snapring
   - Upper ball joint from the steering knuckle, using a deep socket wrench and tool 09050-40011

To install:

3. Install or connect the following:
   - New upper ball joint to the steering knuckle, using a deep socket and tool 09309-37010
   - New snapring
4. Using a torque wrench, inspect the upper ball joint rotation, as follows:
   a. Flip the ball joint back-and-forth 5 times.
   b. Using a torque wrench, continuously turn the nut 1 turn in 2–4 seconds.
   c. Take the reading on the 5th turn; it should be 6–39 inch lbs. (0.7–4.4 Nm). If not, replace the upper ball joint.
5. Install or connect the following:
   - New boot secured with a wire
   - Ball joint-to-knuckle: 77 ft. lbs. (105 Nm)
   - Front wheel. Torque the lug nuts to 81 ft. lbs. (110 Nm).
6. Check and/or adjust the front wheel alignment.

Lower Ball Joint

REMOVAL & INSTALLATION

1. Before servicing the vehicle, refer to the precautions in the beginning of this section.
2. Remove or disconnect the following:
   - Front wheel
   - 4 lower ball joint set bolts
   - Tie-rod end from the lower ball joint, using tool 09610-20012
   - Lower ball joint nut.
   - Lower ball joint from the lower control arm, using tool 09628-62011

To install:

3. Install or connect the following:
   - New lower ball joint to the lower control. Torque the bolts to 103 ft. lbs. (140 Nm).
   - New cotter pin
   - Tie-rod end to the lower ball joint. Torque the nut to 67 ft. lbs. (91 Nm).
   - Lower ball joint set bolts. Torque the 4 bolts to 59 ft. lbs. (80 Nm).
   - Front wheel. Torque the lug nuts to 81 ft. lbs. (110 Nm).
4. Check and/or adjust the front wheel alignment.

Upper Control Arm

REMOVAL & INSTALLATION

1. Before servicing the vehicle, refer to the precautions in the beginning of this section.
2. Remove or disconnect the following:
   - Front wheel
   - Strut
   - Wheel speed sensor harness, if equipped with Anti-lock Brake System (ABS)
3. Upper ball joint, as follows:
   - Cotter pin and loosen the nut
   - Upper ball joint from the upper control arm, using tool 09950-40011
   - Steering knuckle, support it securely
   - Upper ball joint nut
4. Remove or disconnect the following:
   - 4 clips and the fender apron seal
   - Brake/fuel line clamp nut and clamp
   - Both upper control arm-to-chassis nuts/bolts
   - Upper control arm

To install:

5. Install or connect the following:
   - Upper control arm. Torque both upper control arm-to-chassis nuts/bolts to 72 ft. lbs. (98 Nm).
   - Brake/fuel line clamp nut and clamp. Torque the clamp nut to 49 inch lbs. (5.5 Nm).

- Fender apron seal
- Upper ball joint. Torque the nut to 77 ft. lbs. (105 Nm).
- New cotter pin
- Steering knuckle
- Wheel speed sensor harness, if equipped with Anti-lock Brake System (ABS). Torque it to 71 inch lbs. (8.0 Nm).
- Strut
- Front wheel

6. Check and/or adjust the wheel alignment.

CONTROL ARM BUSHING REPLACEMENT

1. Before servicing the vehicle, refer to the precautions in the beginning of this section.
2. Remove the upper control arm from the vehicle.
3. Remove the control arm bushings, as follows:
   - Pry up the bushing flange, using a chisel and a hammer
   - Press the bushing(s) from the upper control arm, using a shop press and tools 09613-26010, 09631-20060 and 09950-00020

To install:

4. Lubricate the new control arm bushings with liquid soap.
5. Press the bushings into the control arm until the bushing flange contacts the housing edge of the control arm, using a shop press, a steel plate and tools 09631-12090 and 09710-30021
6. Install the upper control arm to the vehicle.
7. Check and/or adjust the wheel alignment.

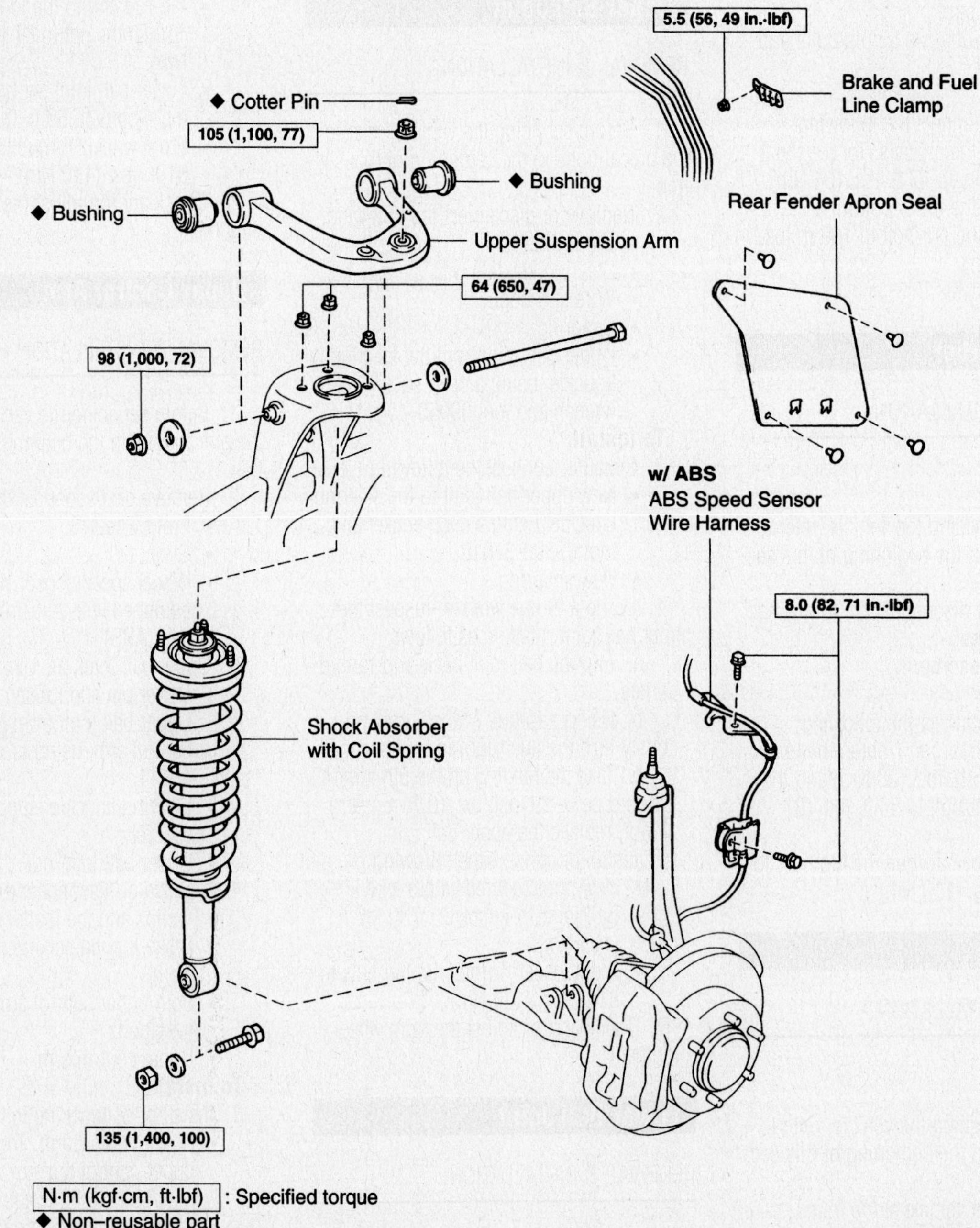

Exploded view of the front suspension and related components

Lower Control Arm

REMOVAL & INSTALLATION

1. Before servicing the vehicle, refer to the precautions in the beginning of this section.
2. Remove front wheel.
3. Disconnect the tie-rod end, as follows:
   - Cotter pin and nut
   - Tie-rod end from the lower ball joint, using tool 09610-20012
4. Remove or disconnect the following:
   - Power steering gear set bolts and nuts
   - Stabilizer bar link from the lower control arm
   - Strut from the lower control arm
5. Disconnect the lower ball joint, as follows:
   - Cotter pin and nut
   - Lower ball joint from the lower control arm
6. Matchmark both front and rear cam plates and chassis frame.
7. Remove the lower control arm while slightly shifting the power steering gear rearward.

To install:

8. Install or connect the following:
   - Lower control arm while slightly shifting the power steering gear rearward
   - Align both front and rear cam plates and chassis frame matchmarks. Torque both bolts to 96 ft. lbs. (130 Nm).
9. Connect the lower ball joint, as follows:
   - Lower ball joint to the lower control arm. Torque the nut to 103 ft. lbs. (140 Nm).
   - New cotter pin
10. Install or connect the following:
    - Strut to the lower control arm. Torque the nut/bolt to 100 ft. lbs. (135 Nm).
    - Stabilizer bar link to the lower control arm. Torque the nut to 51 ft. lbs. (69 Nm).
    - Power steering gear set bolts and nuts. Torque the set bolt and clamp nut/bolt to 122 ft. lbs. and the set nut/bolt to 96 ft. lbs. (130 Nm)
    - Tie-rod end to the lower ball joint. Torque the nut to 67 ft. lbs. (91 Nm).
    - New cotter pin
    - Front wheel. Torque lug nuts to 81 ft. lbs. (110 Nm).
11. Check and/or adjust the wheel alignment.

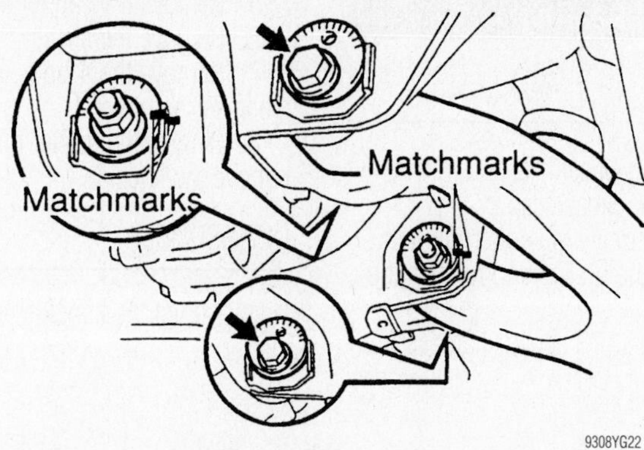

View of the lower control arm's cam plate alignment

CONTROL ARM BUSHING REPLACEMENT

1. Before servicing the vehicle, refer to the precautions in the beginning of this section.
2. Remove the lower control arm from the vehicle.
3. Remove the control arm bushings, as follows:
   - Pry up the bushing flange, using a chisel and a hammer
   - Press the bushing(s) from the upper control arm, using a shop press and tools 09613-26010, 09632-36010 and 09950-00020

To install:

4. Lubricate the new control arm bushings with liquid soap.
5. Press the No. 1 bushing into the control arm until the bushing flange contacts the housing edge of the control arm, using a shop press, a steel plate and tools 09631-12090 and 09502-12010, facing the correct direction.
6. Press the No. 2 bushing into the control arm until the bushing flange contacts the housing edge of the control arm, using a shop press, a steel plate and tools 09631-12090 and 09950-60020, facing the correct direction.
7. Install the lower control arm to the vehicle.
8. Check and/or adjust the wheel alignment.

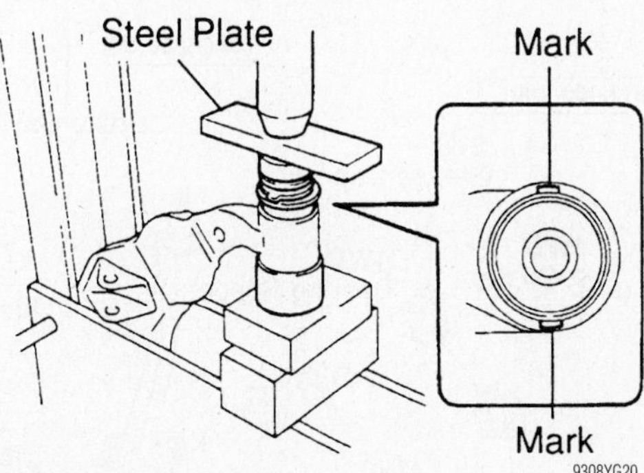

View of the No. 1 bushing's installed direction

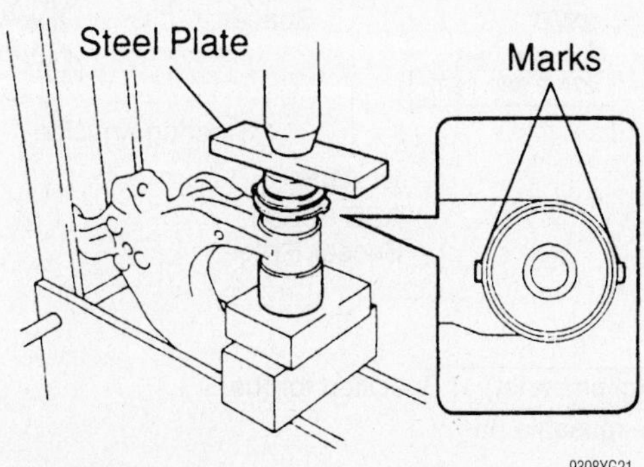

View of the No. 2 bushing's installed direction

Front Wheel Bearing

REMOVAL

1. Remove the wheel.
2. Remove the grease cap.
3. On 4wd, remove the cotter pin and lock cap. Apply the brakes and remove the locknut.
4. Remove the ABS sensor harness from the knuckle.
5. Remove the caliper and rotor.
6. Remove the strut.
7. Remove the 4 bolts and disconnect the lower ball joint.
8. Remove the cotter pin and nut and remove the knuckle.

DISASSEMBLY

1. Mount the knuckle in a soft-jawed vise.
2. With 2wd, remove the grease cap; with 4wd, remove the inner oil seal.
3. With 2wd, remove the locknut and ABS speed sensor rotor.
4. Remove the 4 bolts and shift the dust cover towards the hub.
5. Remove the hub from the knuckle with a puller.
6. Remove the outer seal.
7. Remove the snapring and press out the bearing.

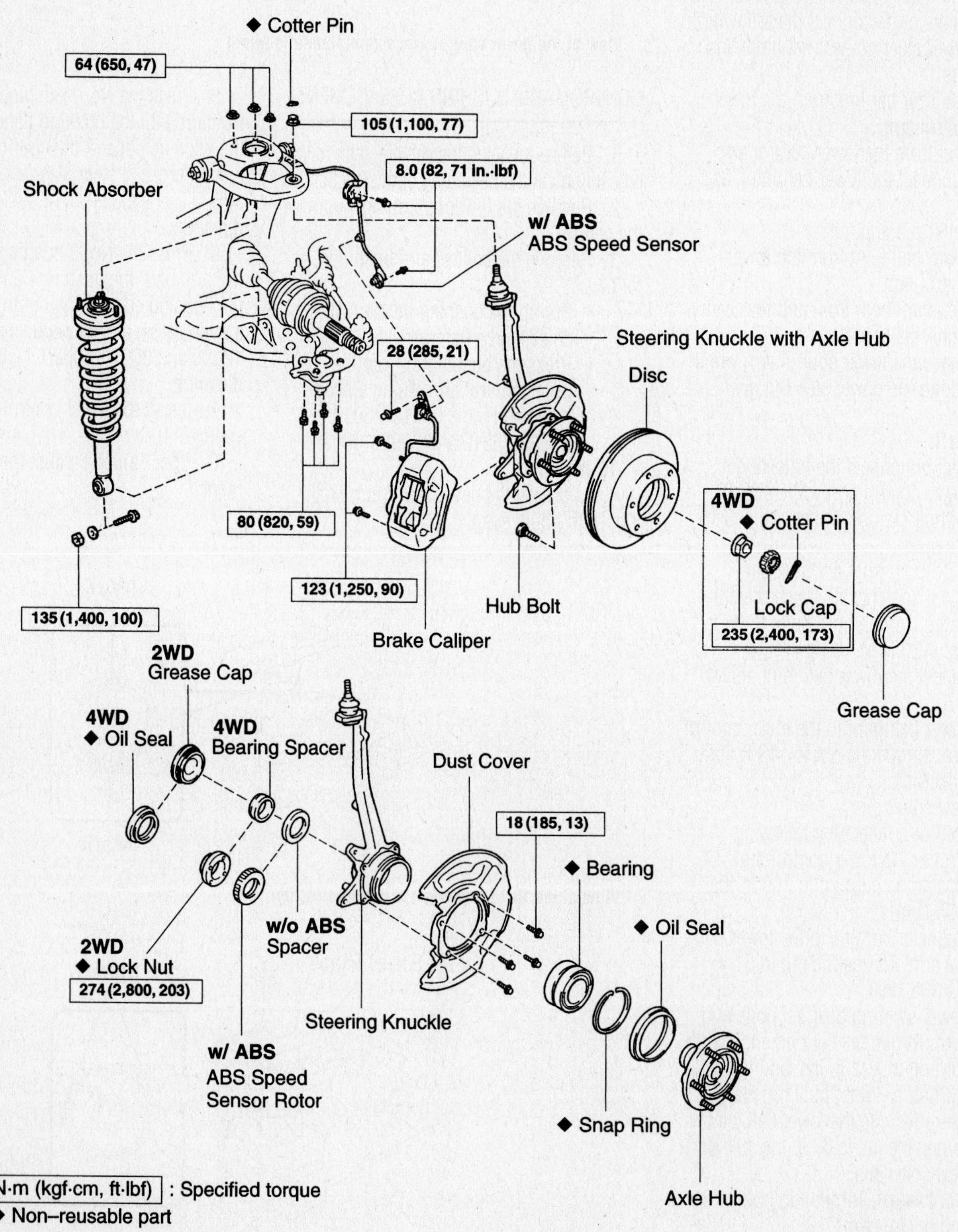

Front hub, knuckle and related components

ASSEMBLY

1. Press a new bearing into place.
2. Install a new snapring.
3. Drive a new outer seal into place. Coat the seal lip with MP grease.
4. Press the hub onto the knuckle. Torque the bolts to 13 ft. lbs. (18 Nm).
5. Install the speed sensor rotor or spacer.
6. With 2wd, install a new locknut. Torque to 203 ft. lbs. (Nm). Stake the nut.
7. With 4wd, install the bearing spacer using a driver.
8. With 2wd, install the grease cap.
9. With 4wd, install a new inner oil seal using a seal driver. Coat the seal lip with MP grease.

INSTALLATION

1. With 4wd, insert the halfshaft into the hub and temporarily tighten the nut.
2. Connect the steering knuckle to the upper arm.
3. Install the nut, torque it to 77 ft. lbs. (105 Nm) and install a new cotter pin. If the hole doesn't line up, tighten the nut up to 60 degrees more.
4. Connect the lower ball joint to the knuckle. Torque the bolts to 59 ft. lbs. (80 Nm).
5. Install the strut.
6. Install the caliper.
7. Attach the brake line clamp to the knuckle.
8. Connect the ABS wiring.
9. With 4wd, install the driveshaft locknut. Torque to 173 ft. lbs. (235 Nm). Install the lock cap and a new cotter pin. If the hole doesn't align, tighten the nut up to an additional 60 degrees.
10. Install the grease cap.
11. Install the wheel.
12. Pump the brake a few times before driving.
13. Check the alignment.

BRAKES

Brake Caliper

REMOVAL & INSTALLATION

1. Disconnect the negative battery cable from the battery.
2. Raise and support the vehicle safely.
3. Remove the wheels.
4. Disconnect the brake hose from the caliper. Plug the end of the hose to prevent loss of fluid.
5. Remove the bolts that attach the caliper to the torque plate.
6. Lift the bottom of the caliper up and remove the caliper assembly.

To install:

7. Grease the caliper slides and bolts with lithium grease or equivalent. Install the caliper and secure with the bolts. Torque the bolts to 90 ft. lbs. (123 Nm).
8. Connect the brake hose to the caliper. Torque 11 ft. lbs. (15 Nm).
9. Fill the brake system to the proper level and bleed the brake system.
10. Install the tire and wheel assembly.
11. Top off the brake fluid level in the master cylinder. Check for leaks and proper brake operation.
12. Connect the negative battery cable to the battery.

Disc Brake Pads

REMOVAL & INSTALLATION

1. Raise the vehicle and support it safely.
2. Remove the wheels.
3. Remove the clip, pins and anti-rattle spring.

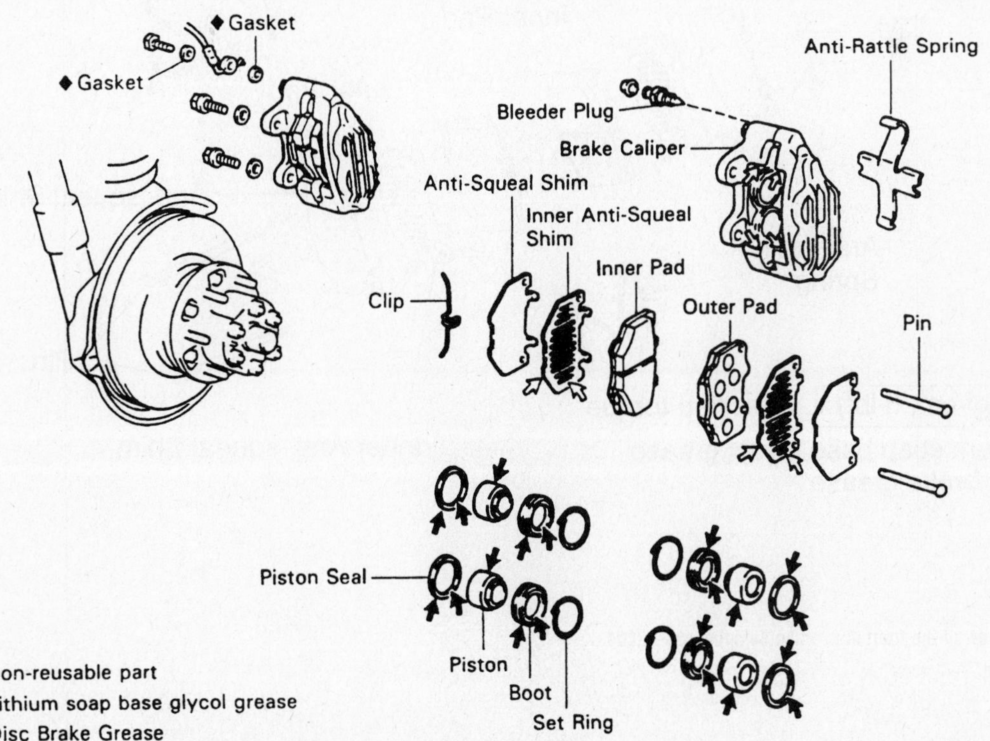

Exploded view of the front disc brake components—2001-03

4. Withdraw the pads and remove the anti-squeal shims.

To install:

5. Before installing the new pads, check the disc thickness and disc runout.
6. Siphon out a small amount of brake fluid from the reservoir.
7. Press in the pistons with a hammer handle or equivalent.
8. Apply disc brake grease to both sides of the inner anti-squeal shim. Install the anti-squeal shims to the new pads.
9. Install the pads.
10. Install the anti-rattle springs and pins. Install the clip.
11. Install the wheels.
12. Check and adjust the fluid level. Apply the brake pedal several times.
13. Road-test the vehicle for proper operation.

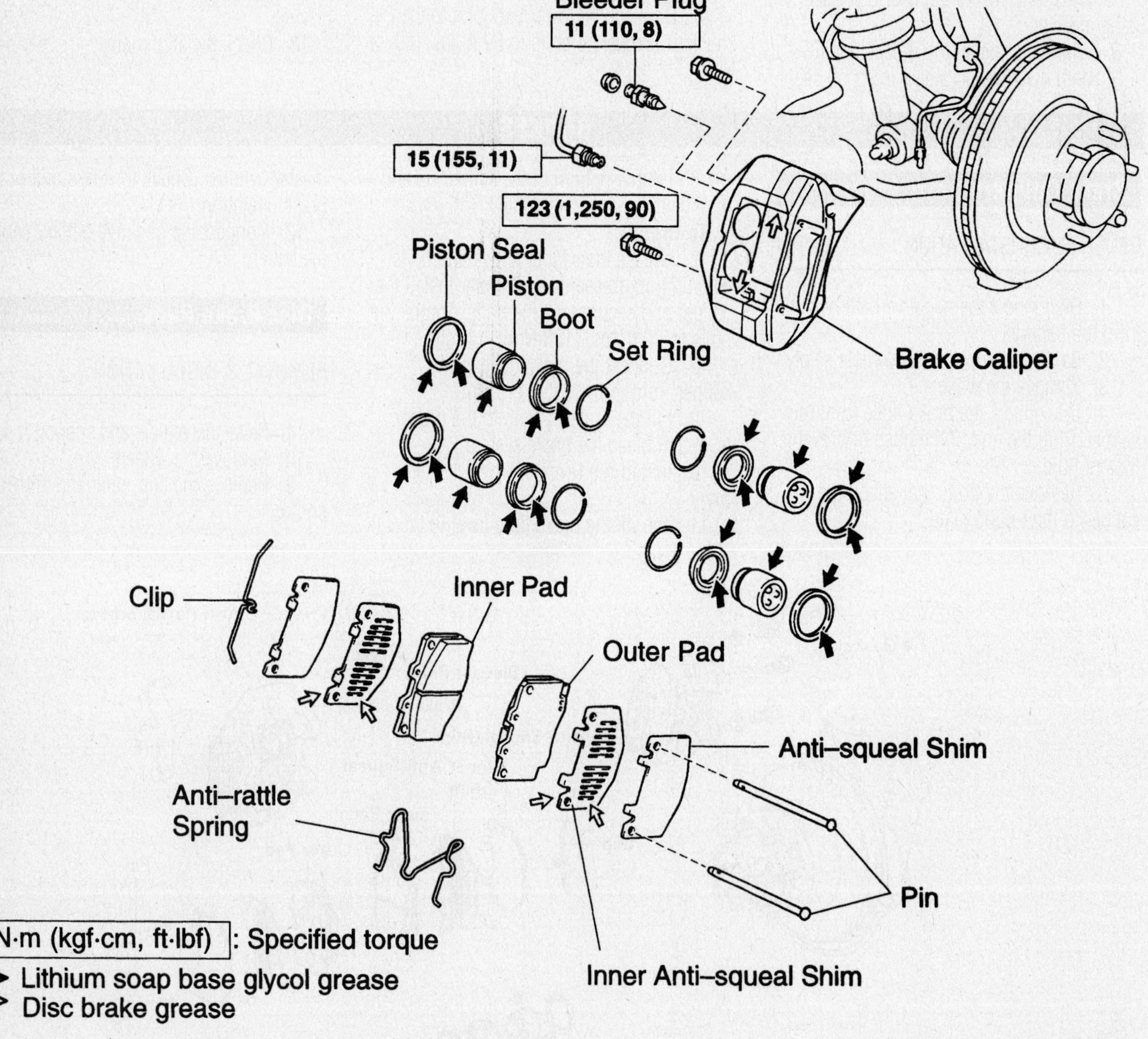

Exploded view of the front disc brake components—2004

Brake Drums

REMOVAL & INSTALLATION

1. Raise and safely support the vehicle.
2. Remove the rear wheel(s).
3. Remove the brake drum from the axle hub. If there is difficulty in removing the drum, insert a suitable tool through the hole in the rear of the backing plate, and hold the automatic adjusting lever away from the adjuster. Using another suitable tool at the same time, reduce the brake shoe adjuster by turning the adjusting wheel.

To install:

4. Install the brake drum and pull the parking brake lever all the way up until a clicking sound can no longer be heard.
5. Verify that the rear wheels will not turn. If the rear wheels turn, adjust the parking brake cable as necessary.
6. Release the parking brake and remove the brake drum. Measure the brake drum inside diameter and diameter of the brake shoes. Check that the difference between the diameters is the correct shoe clearance. Clearance is 0.024 in. (6mm).
7. If the brake shoe clearance is not correct, adjust the brake shoes until the clearance is correct.
8. Install the brake drum, replace the wheel(s), and safely lower the vehicle.
9. Road-test the vehicle for proper brake operation.

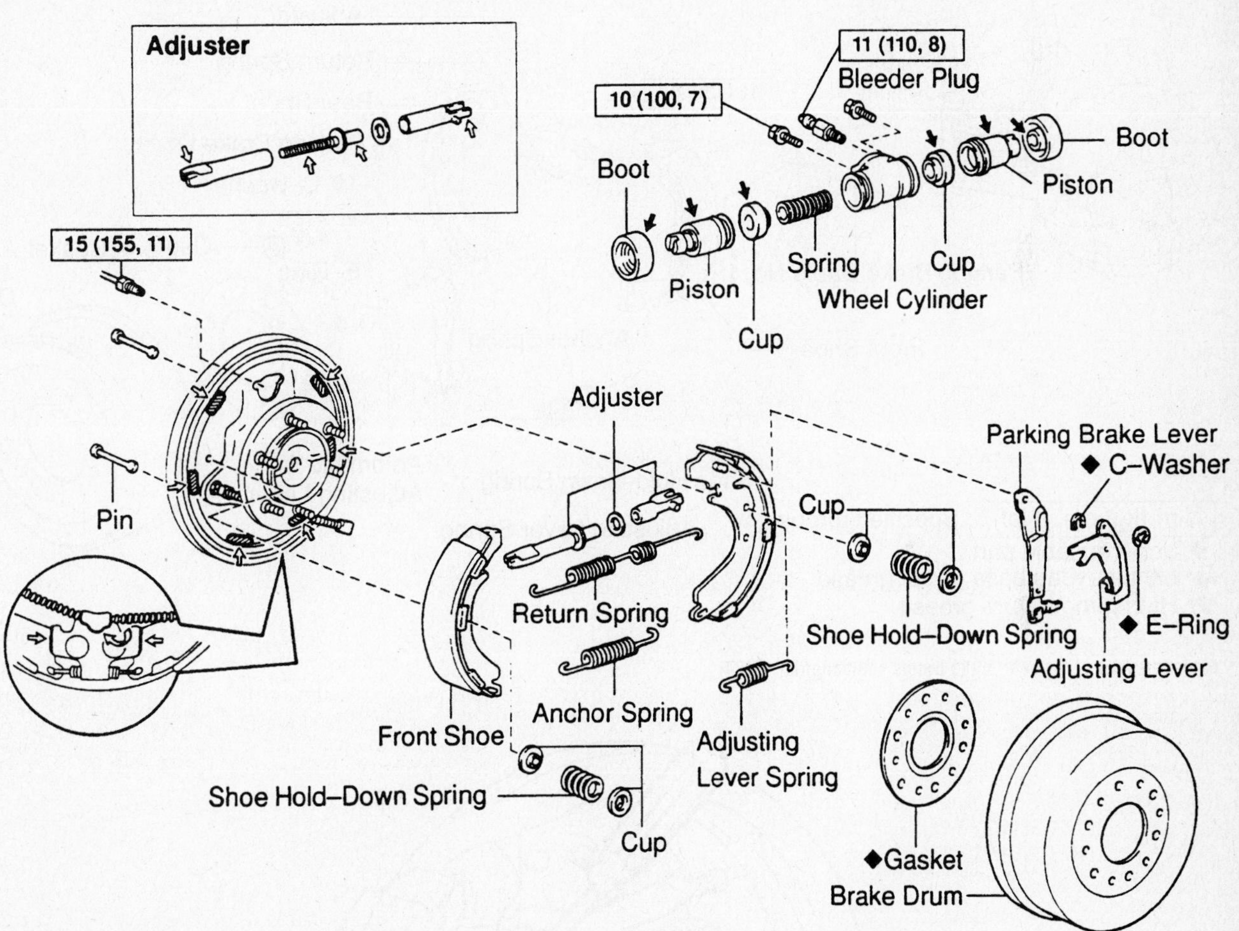

Exploded view of the rear brake drums components—2001–03

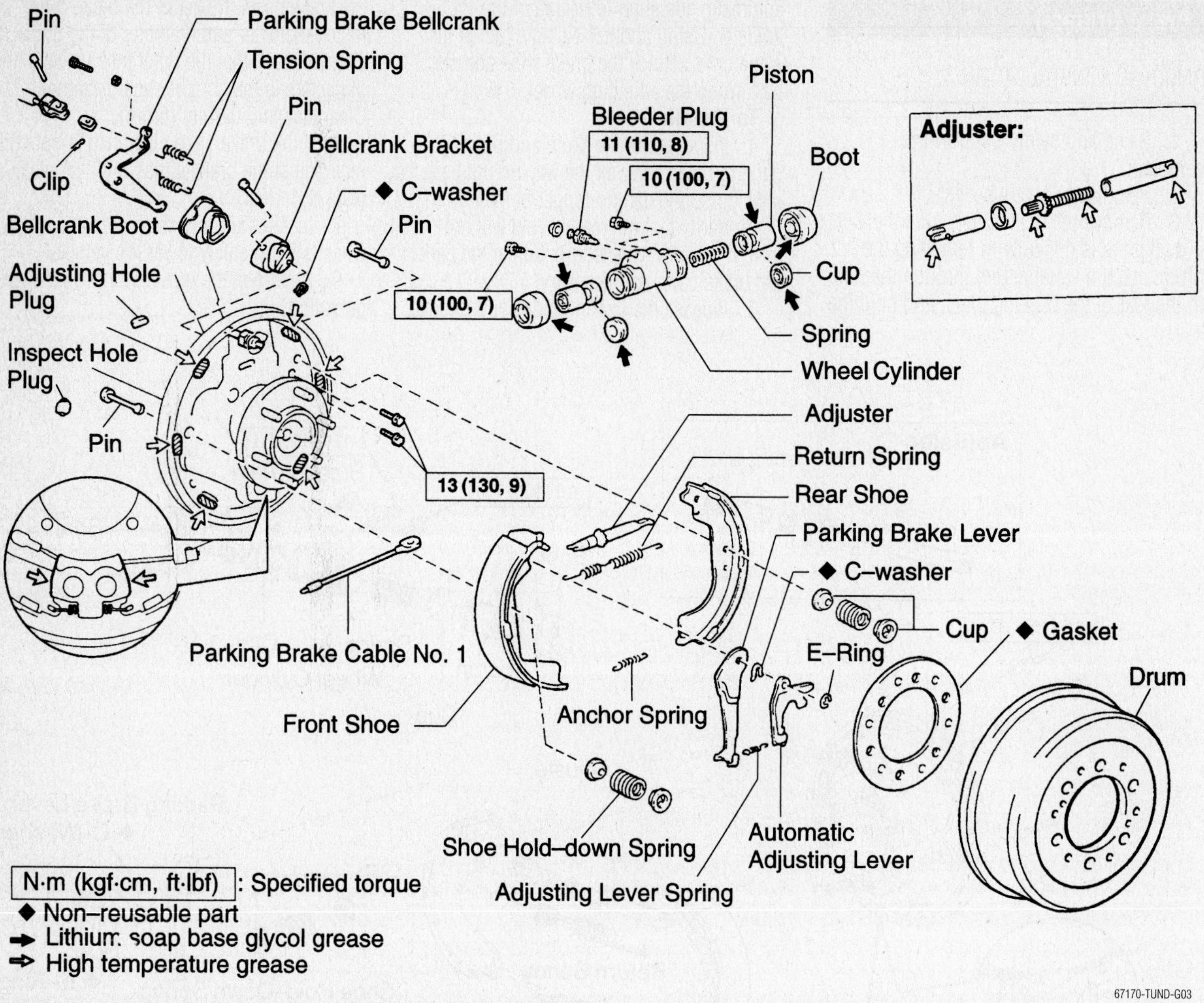

Pin
Parking Brake Bellcrank
Tension Spring
Piston
Pin
Bleeder Plug
11 (110, 8)
Bellcrank Bracket
Clip
10 (100, 7)
Boot
◆ C–washer
Adjuster:
Bellcrank Boot
Pin
Adjusting Hole Plug
Cup
10 (100, 7)
Spring
Inspect Hole Plug
Wheel Cylinder
Pin
Adjuster
13 (130, 9)
Return Spring
Rear Shoe
Parking Brake Lever
◆ C–washer
Cup
◆ Gasket
Parking Brake Cable No. 1
E–Ring
Drum
Front Shoe
Anchor Spring
Shoe Hold–down Spring
Automatic Adjusting Lever
Adjusting Lever Spring

N·m (kgf·cm, ft·lbf) : Specified torque
◆ Non–reusable part
➡ Lithium soap base glycol grease
⇨ High temperature grease

67170-TUND-G03

Exploded view of the rear brake drums components—2004

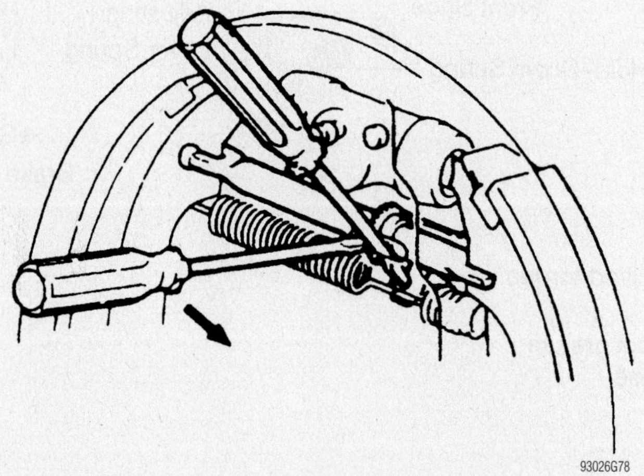

93026G78

Use a brake adjusting tool (brake spoon) and a prytool to adjust the brake shoes through the adjusting hole